HANDBOOK
— OF —
SOIL
SCIENCE

Editor-in-Chief

MALCOLM E. SUMNER

CRC PRESS

Boca Raton London New York Washington, D.C.

Library of Congress Cataloging-in-Publication Data

Handbook of soil science / Malcolm E. Sumner, editor-in-chief.
      p. cm.
    Includes bibliographical references.
    ISBN 0-8493-3136-6 (alk. paper)
    1. Soil science Handbooks, manuals, etc. I. Sumner, M. E.
(Malcolm E.), 1933–
S591.H23 1999
631.4—dc21

          99-29646
          CIP

Preface

The *Handbook of Soil Science* is a comprehensive reference work on the discipline of soil science as practiced today. It contains definitive descriptions of each major area in the discipline, including its fundamental principles, appropriate methods to measure each property, many examples of the variations in properties in different soils throughout the world, and guidelines for the interpretation of the data for various applications (agricultural, engineering, environmental).

The *Handbook of Soil Science* assembles the core of knowledge from all fields encompassed within the discipline of soil science. It is a resource rich in data which will provide professional soil scientists, agronomists, engineers, ecologists, biologists, naturalists, and students with their point of first entry into a particular aspect of soil science. The contributions serve those professionals seeking factual reference information on a particular aspect. The *Handbook* provides a thorough understanding of soil science principles and practices based on a rigorous, complete, and up-to-date treatment of the subject compiled by the leaders in each field. In general, the following critical elements are present in each subsection: description of concepts and theories, definitions, approaches, methodologies and procedures, data in tabular and figure forms, and extensive references.

The *Handbook* is organized into eight sections, covering the six traditional areas of soil science together with the interdisciplinary aspects and a final section on databases:

Soil Physics	Soil Chemistry
Soil Biology and Biochemistry	Soil Fertility and Plant Nutrition
Pedology	Soil Mineralogy
Interdisciplinary Aspects of Soil Science	Soil Databases

The subdivision of each section into a series of chapters was made by the associate editors and in some cases may appear to be somewhat arbitrary. The chapters have been arranged in such as way as to produce a thread running through each section. A complete table of contents is provided at the front of the book and gives a general outline of the scope of the covered subject material. In addition, a comprehensive subject index is provided at the back of the book.

The chapters of the *Handbook* have been written by many authors, all experts in their own fields and peer-reviewed by at least two reviewers. The eight sections have been carefully edited and integrated by the associate editors, all distinguished soil scientists in their fields. This *Handbook* is a tribute to the dedication of the authors, associate editors, the publisher and its editorial associates and my editorial assistant, Mrs. Pam Wilson and research coordinator, Mr. Gene Weeks who have worked diligently on this project. I wish to thank all the authors for their valuable contributions, the reviewers for their useful and helpful comments and criticisms, and the associate editors for all the hard work that they have put in. I also wish to recognize Ms. Marsha Baker who enticed me to undertake this task and Mr John Sulzycki who continued the management of this project. Finally, I would like to thank my wife, Priscilla, for her patience, sacrifice, understanding, support, and encouragement without which this project would not have been possible.

Malcolm E. Sumner
Editor-in-Chief

About the Editor

MALCOLM E. SUMNER (Editor-in-Chief)

Dr. Sumner is the Regents' Professor of Environmental Soil Science in the Department of Crop and Soil Sciences at the University of Georgia, Athens, GA. He received a B.Sc. Agric. degree (1954) in Chemistry and Soil Science and an M.Sc. Agric. (*cum laude*) (1957) degree in Soil Physics from the University of Natal, South Africa, and a D.Phil. degree (1961) in Soil Chemistry from the University of Oxford, UK. His *Alma Mater*, the University of Natal awarded him an Honorary Doctor of Science degree in recognition of his career contributions to Soil Science. Before coming to the University of Georgia, he was Professor and Head of the Department of Soil Science and Agrometeorology at the University of Natal, South Africa. He has spent sabbatical leaves at the Agricultural University (Wageningen, Netherlands), University of Missouri (Columbia), University of Wisconsin (Madison), University of Newcastle-upon-Tyne (United Kingdom), Vista University (South Africa), University of Adelaide (Australia) and the Commonwealth Scientific and Industrial Research Organization (Australia). He is a Fellow of both the American Society of Agronomy and the Soil Science Society of America. He holds the Agronomic Research Award and the Werner L. Nelson Award for Diagnosis of Yield Limiting Factors from the American Society of Agronomy and the Soil Science Research Award from the Soil Science Society of America. He was awarded a Doctor of Science degree (*Honoris causa*) by the University of Natal, South Africa for his contributions to Soil Science. He is the holder of many other awards and distinctions including the Sir Frederick McMasters Visiting Fellowship (Australia) and the D.W. Brooks Faculty Awards for Excellence in Research and International Agriculture (University of Georgia). He has presented addresses at more than 100 universities and research institutions on his research findings. He has served as Associate Editor for the Soil Science Society of America Journal and is a member of Editorial Boards of many other journals. His published works cover a wide range of topics including subsoil acidity, the agricultural uses of gypsum, the beneficial use of anthropogenic wastes and transport of nutrients in soils. A widely respected author, Dr. Sumner's works include *Soil Acidity* (Springer-Verlag, 1991), *Soil Crusting: Chemical and Physical Processes* (Lewis Publishers, 1992), and *Sodic Soils: Distribution, Properties, Management, and Environmental Consequences* (Oxford University Press, 1998). He has authored or co-authored over 220 refereed journal papers and has contributed chapters to over 30 books.

About the Associate Editors

ARTHUR W. WARRICK (Editor, Soil Physics)

Dr. Warrick received a B.S. in mathematics followed by M.S. and Ph.D. degrees in Soil Physics at Iowa State University. He is currently Professor of Soil Physics in the department of Soil, Water and Environmental Science at the University of Arizona, Tuscon, AZ with an Adjunct Appointment in Hydrology and Water Resources. He is a Fellow of the American Society of Agronomy, the Soil Science Society of America and the American Geophysical Union. His general areas of research are in the movement of water and solutes in soil and in the quantification of the variability of soil physical properties.

P. MING HUANG (Editor, Soil Chemistry)

Dr. Huang received his B.S.A. degree in Agricultural Chemistry from the National Chung Hsing University, Taiwan, M.Sc. Degree in Soil Science from the University of Manitoba, Canada, and the Ph.D. degree in Soil Science from the University of Wisconsin. He is currently Professor of Soil Science at the University of Sasketchewan, Saskatoon, Canada. He has served as a member of many editorial boards and is a titular member of the Commission of Fundamental Environmental Chemistry of the International Union of Pure and Applied Chemistry. He is a founding member and current Chairman of the Working Group MO "Interactions of Soil Minerals with Organic Components and Microorganisms" of the International Union of Soil Sciences. His research emphasis is on environmental soil chemistry, especially on mineral colloids and organo-mineral complexes, their reactions with nutrients and pollutants in soils and waters, and their impact on ecosystem health. Professor Huang's contributions to and impact on science and human welfare have been recognized by the award of Fellowships in the American Society of Agronomy, the Soil Science Society of America, the American Association for the Advancement of Science and the Distinguished Researcher Award from the University of Sasketchewan.

ELDOR A. PAUL (Editor, Soil Biology and Biochemistry)

Dr. Paul received B.Sc. and M.Sc. Degrees from the University of Alberta and his Ph.D. degree from the University of Minnesota. He is a Fellow of the Canadian Society of Soil Science, American Society of Agronomy, Soil Science Society of America and the American Association for the Advancement of Science and has received numerous other awards in Microbiology and Soil Science. He is a Professor in the Department of Crop and Soil Science at Michigan State University, East Lansing, MI. The textbook *Soil Microbiology and Biochemistry* published with F.E. Clark in its second edition is being translated into three other languages. His research publications and other published books reflect his broad range of research in microbial ecology, soil and organic matter dynamics in ecosystem functioning and soil carbon cycling relative to global changes and sustainable agriculture.

EUGENE J. KAMPRATH (Editor, Soil Fertility)

Dr. Kamprath received his B.S. and M.S. degrees from the University of Nebraska and his Ph.D. degree from North Carolina State University. He is a Fellow of the American Society of Agronomy and the Soil Science Society of America and is the William Neal Reynolds Distinguished Professor Emeritus of Soil Science at North Carolina State University, Raleigh, NC. He holds the Soil Science Applied Research Award and the Distinguished Service Award from the Soil Science Society of America. He served as Editor-in-Chief for the Soil Science Society of America Journal. He was awarded an Honorary Doctor of Science from the University of Nebraska. He has published widely on topics such as soil acidity and nutrient management.

LARRY P. WILDING (Editor, Pedology)

Dr. Wilding received his B.Sc and M.Sc. Agronomy degrees from South Dakota State University and a Ph.D. degree (Soil Science-Geology) from the University of Illinois. He is a Fellow of the American Society of Agronomy, the Soil Science Society of America and the American Association for the Advancement of Science. He is currently Professor of Pedology in the Department of Soil and Crop Sciences at Texas A&M University, College Station, TX. He served as President of the Soil Science Society of America, and is currently a member of the U.S. National Committee on Soil Science, National Academy of Science/National Research Council. He holds the Soil Science Research Award from the Soil Science Society of America, the Texas A&M Faculty Distinguished Achievement Award in Research, and the Texas A&M Vice-Chancellor's Awards in Excellence for Research, International Involvement, and Surface Mine Reclamation. He is holder of many other awards and distinctions for his global contributions to Soil Science and its integration with the Geosciences.

JOESPH W. STUCKI (Editor, Soil Mineralogy)

Dr. Stucki is Professor of Environmental Soil Physical Chemistry in the Department of Natural Resources and Environmental Sciences at the University of Illinois (Urbana-Champaign). He is a Fellow of the American Society of Agronomy and Fellow of the Soil Science Society of America and holds the Marion L. and Chrystie M. Jackson Soil Science Award from the Soil Science Society of America. He served as Associate Editor of the Soil Science Society of America Journal, and chaired the Soil Mineralogy Division of the Soil Science Society of America. He also received the Marion L. and Chrystie M. Jackson Mid-Career Clay Scientist Award from the Clay Minerals Society which he served as President in 1997-98.

ISAAC SHAINBERG (Editor, Interdisciplinary Aspects)

Dr. Shainberg, Fellow of the American Society of Agronomy and Fellow of the Soil Science Society of America is a Senior Research Scientist in the Institute of Soil, Water and Environmental Sciences, Volcani Center of the Agricultural Research Organization (ARO) of Israel, Bet-Dagan, Israel. Dr. Shainberg received his B.Sc. and M.Sc. degrees from the Hebrew University of Jerusalem and his Ph.D. degree from Cornell University. He has served as Deputy Director of ARO in 1975-1977 and as

Director of the Institute of Soil and Water in 1991-1994. His research interests lie in the fundamental chemistry of saline and sodic soils.

MARION F. BAUMGARDNER (Editor, Soil Databases)

Dr. Baumgardner, Fellow of the American Society of Agronomy and Fellow of the Soil Science Society of America is Professor Emeritus of Soil Science in the Department of Agronomy at Purdue University. He holds the International Soil Science Award and the Distinguished Service Award from the Soil Science Society of America. During his 39 years on the faculty at Purdue University, Baumgardner focused his research interests on the quantitative relationships among the multispectral, physical and chemical properties of soils and the application of aerospace remote sensing and spatial databases to the inventory and monitoring of soil resources at all scales from local to global. Since the mid-1980s, he has been active in research and development of the World Soils and Terrain Digital Database, known as SOTER. During the latter half of his career, his teaching has included courses in soil and water conservation, soil, water and air contamination and remote sensing of land resources.

Contributors

Robert J. Ahrens
National Resource
 Conservation Service, USDA
Lincoln, NE

Ronald G. Amundson
University of California
Berkeley, CA

D.A. Angers
Agric. and Agri-Food Canada
Ste. Foy, Canada

J.S. Angle
University of Maryland
College Park, MD

Richard W. Arnold
National Resource
 Conservation Service, USDA
Lincoln, NE

J.A. Baldock
Division of Land and Water,
 CSIRO
Glen Osmond, Australia

Stuart S. Bamforth
Tulane University
New Orleans, LA

Richmond J. Bartlett
University of Vermont
Burlington, VT

N.H. Batjes
International Soil Reference
 and Information Center
Wageningen, Netherlands

Thomas Baumgartl
Christian-Albrechts University
Kiel, Germany

Friedrich H. Beinroth
University of Puerto Rico
Mayaguez, PR

J.C. Bell
University of Minnesota
St. Paul, MN

Ellis C. Benham
National Resource
 Conservation Service, USDA
Lincoln, NE

Alfred M. Blackmer
Iowa State University
Ames, IA

Paul R. Bloom
University of Minnesota
St. Paul, MN

J.G. Bockheim
University of Wisconsin
Madison, WI

Lorenzo Borselli
Institute for Soil Genesis and
 Ecology
Firenze, Italy

J. Bouma
Agricultural University
Wageningen, Netherlands

J.M. Bradford
USDA-ARS
Weslaco, TX

Ray B. Bryant
Cornell University
Ithaca, NY

J.J. Camberato
Clemson University
Florence, SC

Oliver A. Chadwick
University of California
Santa Barbara, CA

G.J. Churchman
Division of Land and Water,
 CSIRO
Glen Osmond, Australia

David C. Coleman
University of Georgia
Athens, GA

M.E. Collins
University of Florida
Gainesville, FL

D.R. Coote
Agricultural Watersheds
  Associates
Stittsville, Canada

Clement E. Coulombe
Resource Management
  Group, Inc.
Grand Haven, MI

D.A. Crossley, Jr.
University of Georgia
Athens, GA

E.M. D'Angelo
University of Florida
Gainesville, FL

D.L. Dent
University of East Anglia
Norwich, UK

Joe B. Dixon
Texas A&M University
College Station, TX

Hari Eswaran
National Resource
  Conservation Service, USDA
Washington, DC

B.M. Evans
Pennsylvania State
  University
University Park, PA

Christine V. Evans
University of New Hampshire
Durham, NH

S.R. Evett
USDA-ARS
Bushland, TX

Stephanie Ewing
University of California
Davis, CA

D.S. Fanning
University of Maryland
College Park, MD

D.P. Franzmeier
Purdue University
West Lafayette, IN

D.W. Fryrear
Custom Products
Big Spring, TX

J.J. Germida
University of Sasketchewan
Saskatoon, Canada

Sabine Goldberg
U.S. Salinity Laboratory
Riverside, CA

Peter H. Graham
University of Minnesota
St. Paul, MN

Robert C. Graham
University of California
Riverside, CA

E.G. Gregorich
Agric. and Agri-Food Canada
Ottawa, Canada

Peter M. Groffman
Institute of Ecosystem Studies
Millbrook, NY

D.L. Grunes
USDA-ARS
Ithaca, NY

C.T. Hallmark
Texas A&M University
College Station, TX

W.G. Harris
University of Florida
Gainesville, FL

James Harsh
Washington State University
Pullman, WA

Philip A. Helmke
University of Wisconsin
Madison, WI

Paul F. Hendrix
University of Georgia
Athens, GA

Marcel R. Hoosbeek
Agricultural University
Wageningen, Netherlands

Rainer Horn
Christian-Albrechts University
Kiel, Germany

P.M. Huang
University of Sasketchewan
Saskatoon, Canada

Wayne H. Hudnall
Louisiana State University
Baton Rouge, LA

Bruce R. James
University of Maryland
College Park, MD

H.H. Jansen
Agric. and Agri-Food Canada
Ottawa, Canada

Gordon V. Johnson
Oklahoma State University
Stillwater, OK

Nestor Kämpf
University of Rio Grande do Sul
Porto Alegre, Brazil

B.D. Kay
University of Guelph
Guelph, Canada

Rami Keren
Volcani Center, ARO
Bet Dagan, Israel

J.M. Kimble
National Resource
  Conservation Service, USDA
Lincoln, NE

John Klironomos
University of Guelph
Guelph, Canada

Hannan E. LaGarry
University of Nebraska
Lincoln, NE

Loretta Landi
Dip. Scienza del Suolo e
  Nutrizione della Planta
Firenze, Italy

Inmaculata Lebron
U.S. Salinity Laboratory
Riverside, CA

Feike J. Leij
U.S. Salinity Laboratory
Riverside, CA

G.J. Levy
Volcani Center, ARO
Bet Dagan, Israel

Terry J. Logan
Ohio State University,
Columbus, OH

D.J. Lytle
National Resource
  Conservation Service, USDA
Lincoln, NE

B. MacDonald
Agric. and Agri-Food Canada
Ottawa, Canada

J.W. Massmann
University of Washington
Seattle, WA

A.B. McBratney
University of Sydney
Sydney, Australia

M.B. McBride
Cornell University
Ithaca, NY

P.A. McDaniel
University of Idaho
Moscow, ID

Robert McSorley
University of Florida
Gainesville, FL

David Mengel
Kansas State University
Manhattan, KS

D.A. Miller
Pennsylvania State
  University
University Park, PA

D.M. Miller
University of Arkansas
Lafayette, AR

W.P. Miller
University of Georgia
Athens, GA

Delbert L. Mokma
Michigan State University
East Lansing, MI

H.C. Monger
New Mexico State University
Las Cruces, NM

John J. Mortvedt
Colorado State University
Fort Collins, CO

D.J. Mulla
University of Minnesota
St. Paul, MN

Chris E. Mullins
University of Aberdeen
Aberdeen, United Kingdom

F.O. Nachtergaele
Food and Agriculture
  Organization
Rome, Italy

Paolo Nannipieri
Dip. Scienza del Suolo e
  Nutrizione della Planta
Firenze, Italy

P.N. Nelson
Division of Land and Water,
  CSIRO
Townsville, Australia

J.P. Nicot
University of Texas
Austin, TX

Egide Nizeyimana
Pennsylvania State
  University
University Park, PA

L.C. Nordt
Baylor University
Waco, TX

Jeanette M. Norton
Utah State University
Logan, UT

C.G. Olson
National Resource
 Conservation Service, USDA
Lincoln, NE

Dani Or
Utah State University
Logan, UT

W.L. Pan
Washington State University
Pullman, WA

Wm. H. Patrick, Jr.
Louisiana State University
Baton Rouge, LA

G.W. Petersen
Pennsylvania State
 University
University Park, PA

G.A. Peterson
Colorado State University
Fort Collins, CO

C.L. Ping
University of Alaska
Palmer, AK

Martin C. Rabenhorst
University of Maryland
College Park, MD

David E. Radcliffe
University of Georgia
Athens, GA

Todd C. Rasmussen
University of Georgia
Athens, GA

William R. Raun
Oklahoma State University
Stillwater, OK

K.R. Reddy
University of Florida
Gainesville, FL

George Rehm
University of Minnesota
St. Paul, MN

Paul F. Reich
Natural Resource
 Conservation Service, USDA
Washington, DC

C.J. Ritsema
Winand Staring Centre
Wageningen, Netherlands

G.P. Robertson
Michigan State University
Hickory Corners, MI

B.R. Scanlon
University of Texas
Austin, TX

Andreas C. Scheinost
University of Delaware
Newark, DE

Philip J. Schoeneberger
National Resource
 Conservation Service, USDA
Lincoln, NE

Darrell G. Schulze
Purdue University
West Lafayette, IN

A.P. Schwab
Purdue University
West Lafayette, IN

Andrew Sharpley
USDA-ARS
University Park, PA

S.D. Siciliano
University of Saskatchewan,
Saskatoon, Canada

J. Thomas Sims
University of Delaware
Newark, DE

Michael J. Singer
University of California
Davis, CA

Joseph M. Skopp
University of Nebraska
Lincoln, NE

Randal J. Southard
University of California
Davis, CA

Otto C. Spaargaren
International Soil Reference
 and Information Centre
Wageningen, Netherlands

Donald L. Sparks
University of Delaware
Newark, DE

Garrison Sposito
University of California
Berkeley, CA

D.L. Suarez
U.S. Salinity Laboratory
Riverside, CA

Malcolm E. Sumner
University of Georgia
Athens, GA

David Swanson
National Resource
 Conservation Service, USDA
Lincoln, NE

Y. Tan
Division of Land and Water,
 CSIRO
Canberra, Australia

C. Tarnocai
Agric and Agri-Food Canada
Ottawa, Canada

M.L. Thompson
Iowa State University
Ames, IA

R. Greg Thorn
University of Wyoming
Laramie, WY

Dino Torri
Institute for Soil Genesis
 and Ecology
Firenze, Italy

Goro Uehara
University of Hawaii
Honolulu, HI

H. van den Bosch
Winand Staring Centre
Wageningen, Netherlands

V.W.P. van Engelen
International Soil Reference
 and Information Centre
Wageningen, Netherlands

Martinus Th. van Genuchten
U.S. Salinity Laboratory
Riverside, CA

M.E.F. van Mensvoort
Agricultural University
Wageningen, Netherlands

P.C.J. van Vliet
University of Western Australia
Nedlands, Australia

A.L.M. van Wijk
Winand Staring Centre
Wageningen, Netherlands

Zhengping Wang
Louisiana State University
Baton Rouge, LA

Larry T. West
University of Georgia
Athens, GA

Lois M. West
Louisiana State University
Baton Rouge, LA

Robert L. Westerman
Oklahoma State University
Stillwater, OK

Larry P. Wilding
Texas A&M University
College Station, TX

S.R. Wilkinson
USDA-ARS
Watkinsville, GA

M.A. Wilson
National Resource
 Conservation Service, USDA
Lincoln, NE

Jon M. Wraith
Montana State University
Bozeman, MT

Douglas A. Wysocki
National Resource
 Conservation Service, USDA
Lincoln, NE

Introduction

Malcolm E. Sumner, Editor-in-Chief
University of Georgia

Lawrence P. Wilding
Texas A & M University

What is Soil?

We owe our existence to the extremely thin but precious skin, called soil, which covers the unweathered and partially weathered geological formations at the Earth's surface with a unique and extremely thin, fragile veneer. It is no longer rock or geological sediments but has been altered during soil formation by geological, topographical, climatic, physical, chemical and biological factors to form a living entity composed of an association of inorganic or mineral particles inextricably linked with organic matter, and perfused by gases. When wetted by vital water, the solvent and conveyor of nutrients, and major constituent of all living beings, this complex system becomes the fertile substrate from which all terrestrial life on this planet springs. This is the life-sustaining pedosphere zone which is a biologically active, porous, and structured medium that effectively integrates and dissipates mass fluxes and energy. In its pristine state, it is a self-regulating biological entity which slowly evolves as it continues to weather with time. It is analogous to a sponge, regulating and buffering the supply of nutrients and water for the growth of macro- and micro-flora and fauna and determining the partitioning of water into that which flows by surface paths to rivers and lakes and that which percolates to replenish subterranean groundwater reservoirs or aquifers.

Not only does it serve to promote and sustain life in its many forms, but it also acts as a living filter for the wastes generated by humans and animals. In this role, it cleanses, purifies and recycles water and detoxifies and renders harmless most toxins and pathogens which otherwise would irreparably contaminate and degrade the environment. Despite being contaminated by the remains of humans and animals including those from epidemics of pestilence and plague, it has succeeded in rendering most vectors harmless and is seldom, if ever, involved in the transmission of diseases. On the contrary, some of its inhabitants, the microorganisms, have yielded up great antidotes to disease and infection, the antibiotics.

Catastrophic consequences have beset prior civilizations that allowed soil and water degradation to proceed beyond levels required to sustain food production. Soil resources, either directly or indirectly, impact, undergird and transcend all of society's urban, industrial and agrarian interests. It is intimately interwoven into current local, regional, national and global policies and issues on conservation and sustainability, land use, energy, environmental quality, taxation, and food, feed and fiber production.

Having said all this, a concise definition of soil is still elusive bearing in mind the highly heterogeneous nature of the soil. Suffice it to say, at this stage, that soil is an evolving, living organic/inorganic layer at the Earth's surface in dynamic equilibrium with the atmosphere and biosphere above, and the geology below. Soil acts as an anchor and purveyor of water and nutrients for roots, as a home for a vast and still largely unidentified community of microorganisms and animals, as a sanitizer of the environment, and as a source of raw materials for construction and manufacturing. Soil is the long-term capital on which a nation builds and grows. It is the basic component of ecosystems and ecosystem management. A fundamental understanding of this elastic, porous, three-phase system (solid, liquid, gas), its components, processes and reactions, is basic to support the life of plants and animals that live in and on it. Soil serves as the foundation essential for continued human welfare, the wellspring of other renewable natural resources. In the words of Roy Simonson "Soil resources are the earthen looms that shape the lives of the people. The more completely they are understood, the better can be the fabric of life woven on these earthen looms."

Every soil consists of one to several layers called horizons, a few to hundreds of centimeters thick that reflect the physical, chemical, and biological processes which have taken place. Horizons are composed of natural aggregates called peds which are made up of associations of mineral and organic particles. Peds and particles are often separated from each other by pores that vary widely in size and shape. In addition, individual peds and particles may be coated by materials such as clay, organic matter, sesquioxides or precipitated salts. Although the internal structure of peds is not readily visible to the naked eye, the spatial arrangement of peds, particles and pores, called soil architecture, greatly influences soil behavior because the organization is frequently systematic and related to macropore distribution. At the microscale, soil architecture governs soil water/solute movement and retention, soil structure/porosity, soil strength/failure, mineral synthesis/weathering, movement of toxic and nontoxic wastes, soil/root environments, root growth/proliferation, nutrient transfers, soil erosion and oxidation/reduction reactions.

What is Soil Science?

Soil science is that spectrum of earth science that deals with soils as very slowly renewable natural resources on the earth's surface. It involves the study of soil formation, classification and mapping, the physical, chemical, biological and mineralogical properties of soil from microscopic to macroscopic scales of resolution as well as the processes and behavior of soil systems and their use and management. It is an integrative science that interlinks knowledge of the atmosphere, biosphere, lithosphere and hydrosphere. Soil science provides the tools to integrate the components of earth science systems, to understand the causes and consequences of spatial variability, and to take a more holistic approach to the dynamic processes affecting ecosystems.

Soil science focuses primarily on near-surface processes that govern the quality and distribution of soil resources relative to landform evolution (geomorphology), geochemical environment (geochemistry), and organismal habitat (ecology/biology). These pedogenic processes are interactively conditioned by lithology, climate and relief through geologic time. Soils are welded together into landscapes like chains; processes which perturb and impact higher topographic surfaces directly affect processes on adjacent lower surfaces. Pedologists and other soil scientists study the energy flows and mass fluxes, which are the dynamic driving forces of pedogenesis through and over the three-dimensional soil system. They also quantify renewal and transfer vectors for biomass production, rainfall and dusts which counter constituent losses through drainage water, lateral interflow and downslope migration or erosion products.

Soil science has its parentage in geology, chemistry, and biology, but for the past 100 years, has evolved as an independent body of knowledge with strong underpinnings to agriculture. Because of the unparalleled success that soil science has enjoyed in helping to bring ample food, fiber, feed, and fuel to the world, the development of basic soil science has come primarily as a byproduct of research in agriculture, engineering, and the environment. There is growing evidence that the complexity of these problems requires a much broader approach to the science of the soil than can be stimulated by applied research alone. Soil science is taking its place alongside basic research efforts in the biosciences, geosciences, and atmospheric sciences to provide the reservoir of fundamental knowledge needed to develop lasting solutions to the challenges of balanced use and stewardship of the Earth. Despite the knowledge accumulated over the past century, Leonardo da Vinci's statement that "We know more about the movement of the celestial bodies than about the soil underfoot" is as true today as it was then. There is need for much more investigation of our most precious resource to stave off the fate that befell our ancestors in the ancient empires of Babylon, Egypt, China, Europe and the Americas where soil erosion, salinization and siltation laid waste to their land resources. They were constrained by having only limited knowledge of the system, but we will be judged by generations to come based on the quality and stewardship of our land resources.

Modern parallels of soil and land degradation continue in both developing and industrialized countries. In addition to the problems facing the ancients which are still prevalent, chemical toxicities, irreversible land use conversion, environmental pollution and desertification are the modern plagues. Sustainability of today's culture is threatened by loss of biodiversity, population growth and land degradation leading to enhanced greenhouse gas emissions which may result in global climate change. Land degradation is a complex technical, socioeconomic and political issue without simple answers. Soil scientists can provide technical solutions but without implementation as public policy, they are likely to be futile. Currently, public attention is focusing on soil, air and water quality in efforts to maintain a clean environment. Soil science has a major role to play in this arena.

What is the Purpose of this Book?

This Handbook presents the first comprehensive reference on the discipline of soil science as practiced today. It contains definitive descriptions of each major area in the discipline, including its fundamental principles, appropriate methods to measure each property, many examples of the variations in properties in different soils throughout the world, and guidelines for the interpretation of the data for various applications (agricultural, engineering, environmental, biological, regulatory, educational, hydrological, biogeochemical). This Handbook assembles the core of knowledge from all fields encompassed within the discipline of soil science and provides a resource-rich database which will give professional soil scientists, agronomists, engineers, ecologists, biologists, naturalists, and students their point of first entry into a particular aspect of soil science. The contributions also serve those professionals seeking specific, factual reference information. The Handbook provides a thorough understanding of soil science principles and practices based on a rigorous, complete, and cutting-edge treatment of the subject compiled by leading scientists in each field. It is designed as a desk reference book.

This Handbook is divided into eight Sections dealing with Soil Physics, Soil Chemistry, Soil Biology and Biochemistry, Soil Fertility and Plant Nutrition, Pedology, Soil Mineralogy, Interdisciplinary Aspects of Soil Science and Soil Databases. The Section on Soil Physics opens with a description of the basic physical properties of soils followed by a treatment of their dynamic properties in relation to tillage and disturbance by machinery. All aspects of soil water are discussed including content/potential relationships, its movement and transfer to the atmosphere both directly and through the plant and the cotransport of solutes in moving water. The nature of soil structure and its bearing on soil behavior are explored followed by an examination of gas movement under unsaturated conditions. The role of macropores as the superhighway for bypass transport of water and solutes is considered. Finally, the heterogeneity of soils is described culminating in an evaluation of its spatial variability. After an initial description of the chemical composition of soils, the Section on Soil Chemistry elucidates the nature and dynamics of soil organic matter. The importance of the soil solution as the highway for transport and reactions in soil is then highlighted followed by an evaluation and description of the kinetics of reactions in soils. The reactions taking place when oxygen becomes limiting are then explored followed by a detailed account of soil colloidal phenomena important in predicting the behavior of the finest soil fraction. Ion exchange, sorption and precipitation reactions are quantitatively treated as a framework to evaluate the behavior of labile constituents in soils. Catalytic reactions promoted by soil and its constituents are examined followed by an account of the effects of acidity and alkalinity on soil reactions and behavior. The Section on Soil Biology and Biochemistry opens with a discussion of bacteria, fungi, cyanobacteria and algae followed by that of soil fauna comprising protozoa, nematodes, micro- and macroarthropods, enchytriads and vertebrates. The nature, function and life cycles of each are described. Then the processes mediated by these organisms including nutrient transformations are discussed. A major portion of the discussion is devoted to nitrogen transformations because of their importance in soils. The bioavailability of macro- and micronutrients as well as their interactions are discussed in the Section on Soil Fertility and Plant Nutrition prior to evaluating methods for estimating their potential availability to crops and methods of application as fertilizers. Finally, the efficiencies of nutrient and water use are discussed. Following a discussion of the pedogenic processes and models involved in soil formation, descriptions of the systems used to classify soils are given in the Section on Pedology as a framework in which to discuss the Soil Orders in Soil Taxonomy. The structure and properties of the minerals in the active clay fraction of soils are discussed in the Section on Soil Mineralogy following an introduction to mineral weathering processes. Thereafter various Interdisciplinary Aspects of Soil Science such as salinity, sodicity, hardsetting, wetland biogeochemistry, acid sulfate soils, soils and environmental quality, soil erosion by wind and water, land application of wastes, soil quality and conservation tillage are discussed. Finally, the Soil Databases available to assess worldwide soil resources are presented and discussed.

Contents

SECTION A Soil Physics

SECTION B Soil Chemistry

SECTION C   Soil Biology and Biochemistry

SECTION D   Soil Fertility and Plant Nutrition

SECTION E Pedology

SECTION F   Soil Mineralogy

SECTION G Interdisciplinary Aspects of Soil Science

SECTION H   Soil Databases

Introduction   *Marion F. Baumgardner*

The soil was deep, it absorbed and kept the water in the loamy soil, and the water that soaked into the hills fed springs and running streams everywhere
—Plato

A

Soil Physics

A. W. Warrick
University of Arizona

Soil physics includes the descriptions of the physical aspects of the soil system and transport processes. The usual setting is at or just below the soil surface, but most of the descriptions are valid to all depths and for all similar geological and extraterrestrial materials. In nature, the physical system is intertwined with the chemical and biological systems. Thus, the descriptions of the physical processes are also important in the other sections of the handbook.

Chapters 1, 2 and 7 emphasize the soil solids. Included are descriptions of the matrix as well as necessary definitions to describe both static and dynamic aspects of soils. At the start, the soil is considered as static to facilitate definitions of mass, particle size, and surface areas. Later the dynamics of tillage and temporal variations due to natural and anthropogenic actions are examined. These are related to overall effects in soil structure and mechanisms which describe how the primary particles are arranged into an overall structure of solids and pore space.

Transport processes are principally for water, energy, solutes and gas. Generally, flow laws are related to an appropriate gradient such as for soil water potential, temperature, or concentrations of solutes. The individual descriptions are related to the appropriate measurable quantities to overall mass conversion.

Soil water is the primary theme of Chapters 3 and 5, a major part of Chapters 5 and 6, and important for the other chapters as well. Measurements of soil water are of two general types, water content or water potential. Water content is related to how much is present and water potential to the energy level. Water content is appropriate to describe storage of water and water potential is needed to describe direction and magnitude of fluxes. Methods of measurement of both types (content and potential) are given. Also, soil water movement is described in the context of overall water balance and its key role in the hydrologic cycle.

Energy balance and the thermal regime are the topics of Chapters 5 and 9. Appropriate definitions, measurement techniques and the transport processes are determined. Included is a detailed description of the soil-plant-atmospheric interface, which represents a common convergence point for many of the world's problems in food production, water resources and environmental pollution.

Separate chapters are devoted to solute transport and soil-gas movement. Solute transport is of fundamental global importance in terms of environmental pollution as well as in general soil quality and fertility. Gas movement historically emphasized soil aeration relevant to a medium for plant roots and microorganisms; today, it is also a major consideration for soil remediation and global gases.

Spatial variability, Chapter 10, is treated separately in recognition of the heterogeneity of all soil properties, in fact, of all measurable natural quantities. In soil science, the development of the quantitative aspects of variation is necessary for both the description of the processes and for forecasting future events. This has been a rich area for coupling the soil physical processes with the other areas of soil science and to remote sensing.

1

Physical Properties of Primary Particles

Joseph M. Skopp
University of Nebraska

This chapter discusses the following physical properties of the primary particles: particle density, particle shape, particle size distribution, and surface area. In addition, two soil properties related to packing are also presented: bulk density and porosity. The definitions and ideas behind these properties have built into them the concept or assumption of discrete primary particles as the primary soil constituent. If organic matter or amorphous materials and cementing agents are abundant, then the importance of primary particles is reduced.

1.1 Particle Density (ρ_p)

1.1.1 Definition

Particle density represents one of the fundamental soil physical properties. One can define particle density as the mass of soil particles divided by the volume occupied by the solids (i.e., excluding voids and water). Typical values for soils range from 2.5–2.8 Mg m^{-3}, with 2.65 Mg m^{-3} being representative of many soils. Particle density provides few insights into soil physical processes. Consequently, one frequently overlooks the errors associated with particle density. However, its value enters into calculations of more useful soil properties (e.g., porosity and particle size distribution).

1.1.2 Typical Values

Each individual soil mineral has a characteristic particle density. Values for different minerals can be found in Klein and Hurlbut (1985). Quartz, a predominant soil mineral, has a value of about 2.65 Mg m^{-3} which is why this value is frequently given as representing all soils. In contrast, gypsum, biotite and hematite have values of 2.32, 2.80–3.20, and 4.80–5.30 Mg m^{-3}, respectively. The particle density of a soil is an average for the distribution of soil minerals present.

A determination of particle density may be made for individual minerals, size fractions or the soil as a whole. Organic matter removal takes place in order to reduce variation. With standard procedures and removal of organic matter, a propagated error of less than ± 0.01 Mg m^{-3} is possible. (Propagated error is the combination of all instrument errors when making a determination of a soil physical property.) Generally, determinate errors (i.e., biases) play a greater role in the analysis of particle density than indeterminate errors (i.e., random errors).

Perhaps the most interesting question in the determination of soil particle density is the role of organic matter (with a typical density of 1.0 Mg m^{-3}) in surface horizons. Most standard methods remove organic matter. Thus particle density reflects only the mineral phase. This value is the best value for use in particle size analysis, but may not be the best value for calculating porosity. Including organic matter means that changes in soil management may change the particle density.

1.1.3 Methods

Three methods of determining particle density will be examined. The most common determination of soil particle density uses a pycnometer or some variation. A pycnometer is any device which can be made to retain a reproducible or measurable volume. The soil sample is introduced into the pycnometer and the displaced volume of a fluid is determined. Water is typically the displaced fluid, but air can also be used.

The water pycnometer method generally requires the removal of organic matter prior to use. This avoids problems with trapped air and increased variability. The water pycnometer method requires good temperature control (Blake and Hartge, 1986).

An alternative air pycnometer procedure uses a gas as the displacing fluid and the ideal gas law to calculate the volume of solids. The principles of an air pycnometer are straightforward, but care must be taken to prevent temperature changes. The air pycnometer does not require the removal of organic matter which is particularly valuable when the total porosity (or total void space) or air-filled porosity is required without knowing soil bulk density or water content.

A less common method of determining particle density uses a vibrating tube (Elder, 1979) which is filled with a solution or suspension. The resonant frequency of the vibrating tube provides a very precise means of determining the density of the suspension.

1.2   Particle Shape

1.2.1   Definitions

Particle shape influences specific surface as well as particle size analysis and has a strong influence on the packing of particles and soil strength, as well as transport properties. Unfortunately, particle shape is difficult to measure and few determinations exist in the literature.

A variety of terms exist for defining particle shape. Some of the definitions require reference to a figure of an ideal particle (Fig. 1.1). Let L, B, and T represent the length, breadth, and thickness of a particle. Then Heywood (1947) defines the following:

$$\text{Flatness Ratio} = B/T$$
$$\text{Elongation Ratio} = L/B$$

Other dimensionless terms are:

$$\text{Sphericity} = \frac{\text{surface area of equivalent sphere}}{\text{actual surface area}}$$

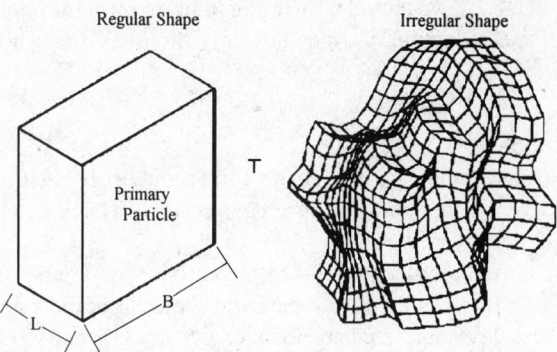

Regular Shape Irregular Shape

Fig. 1.1 Ideal individual soil particle

$$\text{Circularity} = \frac{\text{circumference of circle with same area}}{\text{actual perimeter}}$$

$$\text{Rugosity} = \frac{\text{actual perimeter}}{\text{circumference of circumscribing circle}}$$

These last two terms work best for a two-dimensional image or the projection of the particle outline onto a flat surface.

A microscopic picture of soil describes either the solids or the voids. A simple picture of the solids might be a sphere or cube while that of the voids might be a cylindrical tube or a slit between two flat surfaces. These pictures, or physical models, of the soil make it possible to deduce relations describing surface area, packing, water retention or water movement.

It is possible to apply these ideas to models that are not simple geometrical figures. Here, a characteristic length replaces the edge length of a cube and a dimensionless shape factor (C_k) describes the deviation of the shape from simple geometries (Table 1.1). These two factors are used to describe the area and volume as:

$$A = C_2 L^2 \quad \text{and} \quad V = C_3 L^3 \tag{1.1}$$

where L is the characteristic length and C_2 or C_3 is the shape factor for area or volume, respectively. For a cube, if L equals d (the length of an edge), find: $C_2 = 6$ and $C_3 = 1$. Examples of C_2 and C_3 for other

Table 1.1 Shape factors for three-dimensional physical models with L = d, where d is a diameter or edge length, and h = disc height or rod length

Geometry	Area	C_2	Volume	C_3
Sphere	πd^2	π	$\pi d^3/6$	$\pi/6$
Disc or cylinder	$(\pi dh + \pi d^2/2)$	$(\pi h/d + \pi/2)$	$\pi d^2 h/4$	$\pi h/4d$
Cube	$6d^2$	6	d^3	1
Square rod	$2d^2 + 4dh$	$2 + (4h/d)$	$d^2 h$	h/d

shapes are presented in Table 1.1. Two particles differing in size may or may not have the same shape. Typically, weathering or size reduction changes not only the total dimension but also the shape factor.

1.2.2 Methods

At least three methods exist to examine particle shape. First is direct observation under a microscope or using an image analysis system. Commercial image analysis systems exist that automatically provide shape factors or similar properties.

The second method is an indirect technique obtained from the variation of viscosity with the concentration of suspended particles. Increasing the solids concentration results in deviations from a pure liquid viscosity. These deviations are dependent on particle geometry as well as concentration of the suspension. The viscosity of the suspension (η) and the viscosity of pure solvent (η_s) usually behave as follows:

$$\frac{\eta}{\eta_s} = 1 + Kf$$ [1.2]

where f is the volume fraction of suspended material and K is an empirical constant, which is a shape parameter (Einstein, 1906). For spheres, K = 2.5 is a constant and changes with the shape of the particle. Kahn (1959) applied this technique to examine the shape of clay particles. Similar techniques (Egashira and Matsumoto, 1981) provide estimates of a/b (the ratio of major to minor axis) for montmorillonite (200–300), kaolinite (15–25) and mica (10–20).

The third method of particle shape analysis uses the scattering characteristics of light passing through a soil suspension and relies on measurements of the angles of scattered light. Instruments commercially available for particle size analysis can be used.

1.3 Particle Size Distribution

1.3.1 Definitions

Particle size distribution (PSD) is the most fundamental physical property of a soil and defines soil texture. The particle sizes present and their relative abundance sharply influence most physical properties.

Soil particle size (or effective diameter) provides the basis for a classification system. A range of diameters may be given a special designation (e.g., 2.0 mm to 1.0 mm is very coarse sand). Typically the ranges form a logarithmic scale with particles in a given size range being termed soil separates. The size boundaries vary with country or discipline. Comparisons of the names given to a size range are given elsewhere (Gee and Bauder, 1979; Sheldrik and Wang, 1993).

The phrase, equivalent diameters, is used to emphasize the role of the measurement technique in determining particle size. If identical particles are measured by different techniques, they may appear to have different diameters. It is conceivable that two soils with identical PSDs (as determined by a single method) will show differences in other physical properties resulting from distinct particle shapes.

Defining the diameter of an irregularly shaped particle is not a trivial task. A single parameter, the diameter, characterizes a smooth sphere. The symmetry of the sphere and its smoothness mean that no other information is needed to describe it. As soon as a distortion of the sphere occurs (i.e., into a jelly bean) then at least three diameters are possible. Some particle size analysis methods may orient the particle into a preferred direction (e.g., settling of a particle in a liquid). Other methods (e.g., image analysis) may observe several possible orientations.

1.3.2 Typical Distributions

Typical data for a variety of soils are presented in Table 1.2. Interpretation of particle size analysis data requires either the drawing of graphs or the calculation of summary coefficients which are discussed in Section 1.3.7 below.

Graphs of a PSD typically select the dependent variable as either the cumulative fraction up to a size or incremental fraction of soil in a size interval. The incremental fraction (F_i) is usually the mass of particles within a size interval (X_{i-1} to X_i) divided by the total mass of solids with the index i specifying the size interval. The cumulative fraction (G_i) is the sum of all fractions for particle sizes less than the X_i value.

A typical graphical expression of PSD uses the logarithm of particle diameter (Figs. 1.2 and 1.3). The shape and position of the lines provide qualitative information about soil texture. Soils frequently show a log normal distribution of particle sizes so that a plot of fraction versus the logarithm of particle diameters appears to be normally distributed.

1.3.3 Dispersion and Fractionation

Particle size mechanical analysis consists of isolating various particle sizes or size increments and then measuring the abundance of each size. Most methods accomplish this in two steps. First, the soil is dispersed, or separated into individual primary particles. Second, the dispersed sample is fractionated, or the amounts of each size interval are measured.

There are three objectives of dispersion: (1) removal of cementing agents, (2) rehydration of clays, and (3) the physical separation of individual soil particles. It is sufficient to recognize that organic matter and amorphous minerals are the primary cementing agents. When either of these is present in

Table 1.2 Particle size distribution of soil samples representing a variety of soil types from the United States. Mass percentage of total sample for the indicated size class (D.M. Hendricks, 1997, personal communication)

Soil*	Sand					Silt				Clay
	Very Coarse 2-1 mm	Coarse 1-0.5 mm	Medium 0.5-0.25 mm	Fine 0.25-0.10 mm	Very Fine 0.1-0.05 mm	Coarse 50-20 µm	Medium 20-10 µm	Fine 10-5 µm	5-2 µm	2-0 µm
	%							%		%
Anthony	18.05	13.71	17.68	12.93	8.92	7.41	2.69	2.20	1.37	15.04
Ava	0.53	0.56	1.25	0.82	0.67	12.80	21.69	15.71	9.63	30.63
Chalmers	0.74	0.62	1.67	1.38	2.52	19.42	20.18	11.18	7.44	34.89
Davidson	0.71	2.38	6.52	6.02	3.39	3.32	4.83	4.08	7.43	61.32
Fanno	8.45	4.87	2.40	9.96	9.06	5.92	4.27	4.05	5.56	46.46
Kalkaska	0.19	1.79	47.99	36.26	5.19	1.34	1.01	1.33	0.18	4.67
Mohave	15.25	11.30	12.40	8.02	5.42	30.36	5.00	1.34	0.43	10.45
Molokai	1.29	2.64	4.57	6.64	7.91	5.78	8.30	6.08	4.88	52.00
Nickolson	0.67	0.31	0.44	0.42	1.18	12.90	13.47	15.27	5.41	49.89
Wagram	7.48	20.70	32.06	21.81	5.84	2.00	1.37	1.59	3.31	3.84

*USDA Taxonomic names are: Anthony = Torrifluvents, Ava = Fragiudalf, Chalmers = Endoaquolls, Davidson = Paleudult, Fanno = Haplustalf, Kalkaska = Haplagrid, Mohave = Calciargid, Molokai = Eurostox, Nicholson = Fragiudlaf, Wagram = Paleudult

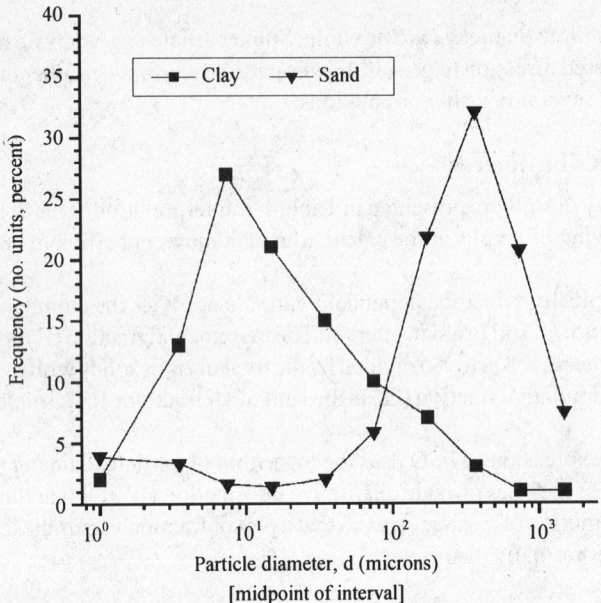

Fig. 1.2   Frequency graph of particle size distribution

large amounts (e.g., Histosols, Andisols, or Oxisols), then dispersion may be difficult or meaningless. Generally, soil dispersion is effected using a combination of chemical and mechanical means (Gee and Bauder, 1986; Loveland and Whalley, 1991; Sheldrick and Wang, 1993).

It is important to recognize that the fraction of soil that is a single size cannot be determined. What is detected is the fraction of soil within a particular particle size interval or the cumulative fraction of all particles less than a given size. The use of sieves typically determines the mass fraction (mass of particles in a size interval divided by total mass). Microscopic counting results in a number fraction (number of particles in a size interval divided by total number of particles). Photometric techniques typically determine the area fraction. Other methods result in volume or line fractions, depending on the sensing procedure. Thus, while all the methods are capable of observing PSDs, not all the methods provide results that are equal or directly comparable.

1.3.4   Sieving

The process of sieving is that of placing the particles on a pattern of holes. Small particles may fall through while the sieve retains the larger particles. Either air or water may be the fluid to support the particles as they sort on the sieve. Dry sieving has a lower practical limit of 50 μm, while wet sieving can separate somewhat smaller particle sizes. Sieve holes may be square (using a wire cloth mesh) or round, although square holes are most common. The use of sieves with square openings will not result in measurements equivalent to those using sieves with round holes (Tanner and Bourget, 1952).

The use of words such as effective or nominal diameters with sieves is in recognition of the imperfect separation that may occur. Placement of a soil sample on a sieve does not result in instantaneous separation. Several factors influence the time to achieve a fixed level of separation. These factors include: sample size, shaking intensity, particle shape, particle size and hole geometry. Since samples vary in their sieving characteristics, it is best to run a trial sample. Errors on a single set

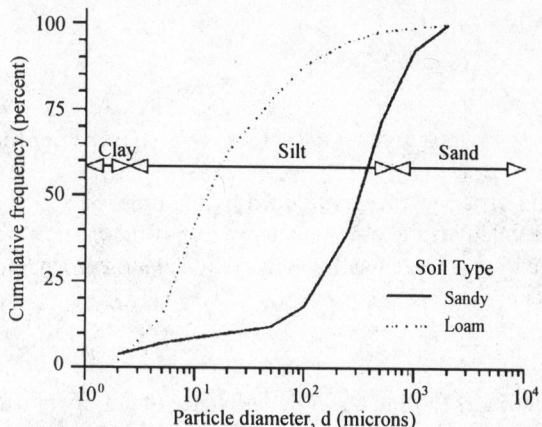

Fig. 1.3 Cumulative frequency graph of particle size distribution

of sieves typically are less than 1%, while comparisons between sieves show random errors of about 4%.

Many of the standard sieve sizes correspond to the class limits for USDA soil separates. Surprisingly, no standard sieve is available for the 50 μm cutoff between the sand and silt separates. Consequently, sieving cannot distinguish this class boundary using standard sieves.

1.3.5   Sedimentation

Below 50 μm, sieving is an inefficient and difficult procedure and for soil samples, sedimentation in water is an alternative procedure. Because a suspension of the dispersed sample settles in water, a measurement can be made of the density of particles (mass of particles in a volume of liquid) at a specified depth within the sedimentation cylinder at preselected times. Variations in the method occur as to the determination of suspension density. In all cases, Stokes' Law (Streeter amd Wylie, 1975) is central to the derivation of an equation which relates the time of settling to the size of particle sampled.

Two classic means of determining the density of a suspension exist: the hydrometer and pipet methods. In the hydrometer method, the influence of density on a floating object (the hydrometer) is observed. As density decreases due to settling out of soil particles, the hydrometer sinks. A calibration scale converts the depth of the float (i.e., hydrometer) to the suspension density (expressed as g L^{-1}). The pipet method directly removes a sample from the suspension. The concentration of solids is determined by drying the pipeted sample. Nondestructive gamma ray absorption which has the advantages of undisturbed and repeated sampling provides a commercially available alternative to both pipet and hydrometer methods (Karsten and Kotze, 1984).

The hydrometer method uses higher concentrations of soil and may be less accurate than the pipet method. However, the hydrometer method allows repeated sampling at many points on the distribution (since no sample is withdrawn). Problems in the pipet method due to convection currents near the pipet tip, effect of sample removal, and greater potential for operator error make the hydrometer method more preferable in some circumstances (Gee and Bauder, 1979).

Stokes' Law for the viscous drag on a sphere is combined with buoyant and gravitational forces to obtain a settling equation. The particle shape is also assumed to be a sphere in evaluating the other forces. Combining forces and solving for v (the particle velocity) give rise to the settling equation:

$$v = \frac{2r^2 g(\rho_p - \rho_w)}{9\eta}$$
[1.3]

where r is the equivalent radius, g is the gravitational acceleration, ρ_w is the fluid density, and η is the viscosity.

The particle velocity is the distance traveled divided by the time, or x/t. The settling equation allows a particle radius to be calculated at a particular x and t, if the particle density and solution viscosity are known. This equation is basic to all gravity sedimentation procedures but is based on a number of assumptions (Baver et al., 1972).

1.3.6 Other Methods

Alternatives exist to fractionation by settling under the influence of gravity in water. Settling can be speeded up, through the use of a centrifuge (Svedburg and Nichols, 1923). Settling can occur in air, called air elutriation (Jenson and Hansen, 1961). The settling equation for air elutriation is the same as for water except that the velocity is that of the air floating particles upward. Complete avoidance of settling may occur using microscopic methods. Image analysis in conjunction with microscopic methods can determine particle shape or geometry parameters. Other techniques (Davies, 1984; Gee and Bauder, 1986; Loveland and Whalley, 1991; Sheldrick and Wang, 1993) include conductometric (e.g., Coulter counter) and light scattering.

1.3.7 Interpretation of Results

A graph can clearly present the elements of a PSD. However, it is desirable to have a more compact means of expressing the properties of the distribution. This is the origin of soil textural classes which summarize the particle size properties in a single phrase. The phrase chosen depends on the relative abundance of sand, silt and clay, irrespective of variations within these size ranges, through the use of a textural triangle (Fig. 1.4). Typical textural triangles and the textural class names are given in Gee and Bauder (1986) and Loveland and Whalley (1991). Note that while the names of three soil separates are similar to the names of textural classes, a textural class name does not limit the sizes that may be present. In other words, a clay texture class contains sand, silt, and clay separates while the clay separate contains only clay-sized particles.

The use of only the total amount of sand, silt and clay to describe texture results in a bias which lies in the assumption that all particles within the range of the sand (or silt or clay) size are equivalent. For the clay fraction, this is a particularly misleading assumption and is partially responsible for the low correlations frequently observed when using clay in regression analysis. Two soils with the same clay (%) may differ in the amounts of fine versus coarse clay. Additional differences in the shape and mineralogy of particles can also cause variations in soil behavior.

A more interesting approach determines the moments of the size distribution (Folk, 1966) which gives a quantitative measure of the mean, spread and asymmetry of a PSD. More importantly, it increases the power of correlative studies relating texture to any other physical, chemical or biological factors of interest.

Other measures exist in the engineering literature and elsewhere to describe a PSD. One approach defines the grain diameter (D_n) at which n% passes through a sieve. Therefore, D_{40} represents that size for which 40% of the particles are smaller and 60% are larger. Various combinations of these D_n values characterize the distribution. One example is the uniformity coefficient (C_u):

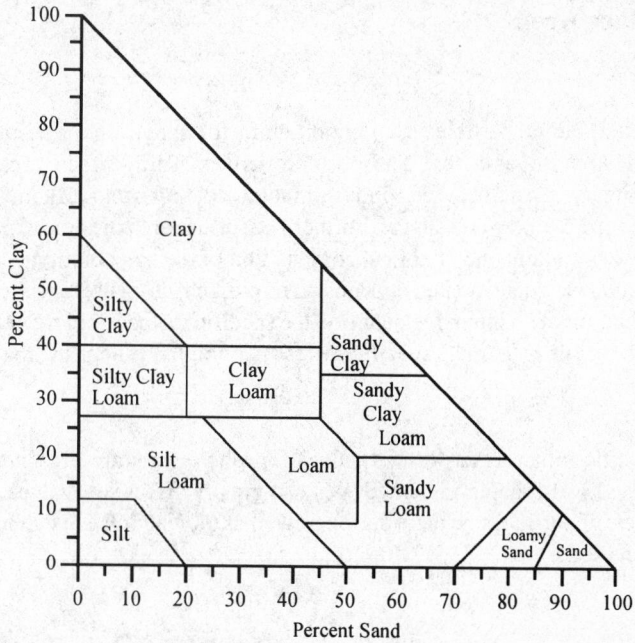

Fig. 1.4 Alternative texture triangle using clay (%) and sand (%) to determine texture class name. For example, a soil with 50% clay and 30% sand sized particles would be assigned a textural class name of clay [Reprinted from Elghamry and Elashkar, 1962. Soil Sci. Soc. Am. Proc. 26:612-613, with permission of the Soil Science Society of America].

$$C_u = \frac{D_{60}}{D_{10}}$$

[1.4]

The uniformity coefficient provides information as to how narrow the distribution is with 1 being the minimum when only a single size is present. The second ratio is the coefficient of gradation (C_g):

$$C_g = \frac{D_{30}}{D_{60}D_{10}}$$

[1.5]

Indirect descriptions of the particle size distribution are possible by using an equation or model. These models contain parameters which in turn characterize the distribution. The problem in using these models is first to determine the parameters, second, to determine the appropriateness of the model, and third, to interpret the parameters.

One model of particular interest is the power function. The use of a power function suggests an underlying fractal process. Where this applies, the exponent (δ) relates to the fractal dimension (n) as $\delta = 3 - n$, and n is between 0 and 3. Tyler and Wheatcraft (1992) apply this technique to several materials and report n values between 2 and 3. The log normal model and parameters have also been useful (Shirazi and Boersma, 1984). Campbell (1985) uses these to estimate the water holding properties of soil.

1.4 Specific Surface Area

1.4.1 Definitions

The surface area of the individual particles is an important factor in nutrient or pesticide adsorption, water absorption, soil strength and soil transport properties. The surface area of a soil has contributions from primary particles, amorphous mineral coatings and organic matter. These individual contributions may overlap and cancel. Further, the surface area of some expanding minerals may change with the water content and chemical composition of the soil solution.

Surface area is an extensive quantity (i.e., depends on how much soil is present). A more satisfying alternative is the introduction of an intensive quantity, the specific surface, which is either the surface area per mass (S_m) or per volume (S_v). The specific surface per volume changes with soil compaction.

1.4.2 Typical Values

Typical values for specific minerals as well as values for whole soils are presented in Table 1.3. Amorphous materials and soil organic matter (SOM) can greatly affect soil values. The whole soil values include the effects of particle size distribution as well as typical SOM and amorphous mineral levels for temperate zone soils.

1.4.3 Methods

Direct measurements usually rely on the adsorption of either a gas (typically N_2 or Ar under high vacuum) or a liquid (historically ethylene glycol; but more recently, ethylene glycol monoethylether [EGME] or water). Either a multi- or monomolecular film is deposited or a gas adsorption isotherm is determined. A critical point in the use of all these procedures is the means by which a multi- or monomolecular layer is obtained. Control of the adsorbed phase occurs through regulation of the gas phase. More commonly for liquid adsorption, the use of a desiccator helps to fix the total pressure and partial pressure of the adsorbed component. In the EGME method (Carter et al., 1986), a mixture of

Table 1.3 Ranges of specific surfaces of clay minerals, selected soil components and whole soils compiled from a variety of sources [Nelson and Henricks, 1943; Bower and Gschwend, 1952; Orchiston, 1955; Borggaard, 1982; Suarez and Wood, 1984]

	Specific Surface
	$m^2\ g^{-1}$
Component	
Kaolinite	15–20
Illites	80–100
Bentonite	115–260
Montmorillonite	280–500
Organic matter	560–800
Calcite	0.047
Crystalline iron oxides	116–184
Amorphous iron oxides	305–412
Alloplane and imogolite	
Soils	
Sands	< 10
Sandy loams and silt loams	5–20
Clay loams	15–40
Clay	> 25

EGME and $CaCl_2$ regulates the vapor pressure of EGME. In water absorption, a separate saturated $LiNO_3$ solution regulates the relative humidity. The $LiNO_3$ and EGME methods are convenient procedures for soils, but both are sensitive to variations in temperature. Values for specific surface vary with the method used.

Another direct method examines photomicrographs of primary particles. The classical approach determines the probability of a needle (randomly placed on the photo) either falling within a pore or intersecting the pore edge (Scheidegger, 1960). Current image analysis instrumentation allows this evaluation (through the particle size distribution) as part of many software packages. Unfortunately, most particle size distributions are not detailed enough in the smallest size range to accurately estimate specific surface.

1.4.4   Models

Models are used to describe how particle size influences specific surface and the general relation of particle geometry to specific surface. Starting with the specific surface $S_m = A/\rho_p V$, where A is the surface area of solids and V is the volume of solids, A and V are expressed in terms of the shape factors C_2 and C_3 with a characteristic dimension L:

$$S_m = \frac{C_2}{\rho_p C_3 L}$$

[1.6]

This relation states that specific surface is a function of the reciprocal of the characteristic length, and shape factors. Using Table 1.1, for spheres: $S_m = 6/\rho_p d$, while for a disc or flat plate: $S_m = [4 + 2(d/h)]/\rho_p d$. Large and small d/h ratios correspond to a flat plate and a prism or needle-like shape, respectively. A ratio of d/h = 1 is identical to the result for spheres (Fig. 1.5). The specific surface of fine clays may be one or more orders of magnitude greater than for coarse clays.

1.5   Bulk Density (ρ_b) and Porosity (ϕ)

1.5.1   Definitions

In order to quantify the state of compaction and the amount of pore space in a soil, the volume and mass of solids which are extensive quantities are related by the intensive term, bulk density (ρ_B):

$$\rho_b = \frac{(\text{mass of solids})}{(\text{volume of solids and voids})}$$

[1.7]

The voids are the pore spaces that may hold either air, water or other liquids.

A related term to density is specific gravity. Specific gravity is defined as the ratio of the density (particle, bulk, fluid or any other density) to the density of water at 4 °C and standard pressure. The specific gravity is dimensionless and the reference density of water is exactly $1.00 \, Mg \, m^{-3}$.

1.5.2   Typical Values

The bulk density is a key physical property of any porous material which changes in response to disturbance or soil management practices. Yet, there is a limit to any modification of bulk density. The particle size distribution together with packing controls the range of possible values.

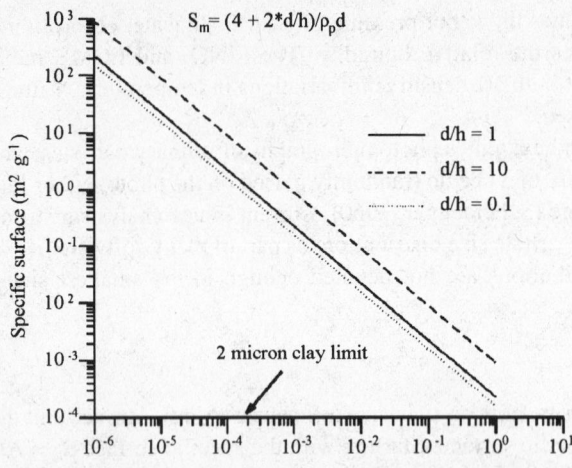

Fig. 1.5   Effect of particle diameter on specific surface for materials of a single size

Bulk density varies with the packing of the soil particles. Typically, sands pack more closely and values range between 1.4 to 1.9 Mg m^{-3}. Clays tend to bridge and cannot be packed as tightly, giving values between 0.9 to 1.4 Mg m^{-3}. Textures between sands and clay vary in their bulk density accordingly. The wide range in bulk densities for a particular texture indicates that other factors (such as SOM and compactive history) have an important influence on this property.

Field determinations of bulk density have relatively low precision (typically ± 0.05 Mg m^{-3}) which means that only about 10–20 different states of bulk density can be distinguished. An increase in measurement precision is inadequate because sampling bias and natural variability are similar in magnitude to typical measurement errors. Bulk density typically has coefficients of variation in the range of 10–40%.

Bulk density is highly dependent on soil conditions at the time of sampling. Changes in soil swelling due to changes in water content alter the bulk density. Thus, comparisons of bulk density must control or compensate for water content where swelling is significant. Other factors such as traffic patterns can also influence the bulk density. Consequently, a determination of bulk density may require the use of the above definition and the specification of the conditions at the time of sampling.

If the bulk density of a soil is fixed, then the relative amount of pore space is also fixed. To make this concept more precise, a term to describe the amount of pore space (or voids) is needed. With pore space volume rather than mass dimensions are more appropriate. Porosity, an intensive quantity, is defined as:

$$\text{Porosity} = \frac{\text{volume of voids}}{\text{volume soil}} \tag{1.8}$$

Alternately, it can be shown that this definition is equivalent to the following (if the particle density includes SOM):

$$\phi = 1 - \left(\frac{\rho_b}{\rho_p} \right) \tag{1.9}$$

This relation is the result of definitions and is not empirical. Using typical bulk densities, the total porosity of sandy soils is less than that of finer textured soils which implies, that for every value of bulk density, a given soil has only one possible value of the porosity. However, it is not true that a soil has only one possible value of bulk density.

1.5.3 Related Terms

A number of other expressions which characterize the amount of air in a soil are listed with their names, symbols and definitions:

$$\text{Air Filled Porosity} = \phi_a = \frac{\text{volume of air}}{\text{volume of soil}} = \phi - \theta_v \qquad [1.10]$$

where θ_v is the volumetric water content of a soil as defined in Section A, Chapter 2.

$$\text{Void Ratio} = \frac{\text{volume of voids}}{\text{volume of soils}} = \frac{\phi}{(1-\phi)} \qquad [1.11]$$

$$\text{Wet Density} = \rho_M = \frac{(\text{mass of solids } + \text{ water})}{\text{volume of soil}} \qquad [1.12]$$

Note that wet in the last expression refers to the inclusion of water in the numerator and not to how much water is present. It is possible to calculate the wet density of a dry soil and to convert between bulk density and wet density using the relation:

$$\rho_b = \rho_M - \theta_v \rho_w \qquad [1.13]$$

Wet density is not a preferred means of expressing the packing density of a soil.

1.5.4 Further Uses for Bulk Density

Another use for bulk density is the conversion of any gravimetric quantity (i.e., an intensive quantity that is expressed on a per mass basis) to a volumetric basis. For example, in Section 1.4, there are two kinds of specific surface: S_v and S_m. These are surface areas per quantity of volume or mass. The explicit relation between the two is

$$S_v = S_m \rho_b \qquad [1.14]$$

Some researchers have tried to relate bulk density to factors such as root penetration, soil strength and compaction (Table 1.4) generally with mixed success.

1.5.5 Methods

Typically, soil bulk density is determined by inserting a ring into the soil (Uhland, 1949). The ring is of known volume and upon extraction the soil core within the ring is dried to determine mass of solids and water present at the time of sampling. The major difficulties are first, the presence of stones or SOM (possibly alive!) and second, the compaction of the core while sampling may bias the volume. The effect of stones and compaction can be minimized by using a larger sampling ring.

Table 1.4 Approximate bulk densities that restrict root penetration [from SCS, in-house report of NSSL (1981)]

Texture	Critical bulk density for soil resistance	
	High	Low
	$(Mg\ m^{-3})$	
Sandy	1.85	1.60
Coarse-loamy	1.80	1.40
Fine-loamy	1.70	1.40
Coarse, Fine-silty	1.60	1.30
Clayey	(Depends on both clay (%) and structure)	

There will be sites where a ring technique is not feasible. For example, coarse-textured soils may not remain in the ring or it may be difficult to drive the sampling ring to the desired depth. Further, since soil compressibility depends on water content, there will be times when sampling a particular site may show bias because the soil is too wet. The presence of live or dead roots also poses a problem, particularly near the soil surface and in soils managed with reduced tillage.

An alternative field method relies on hand-excavating soil. This results in an irregularly shaped hole whose volume must be determined. The hole is then filled with a measurable volume (sand, an air filled balloon or a water filled balloon) which allows accurate calibration. Gamma ray attenuation methods have been developed for field use. Details can be found in Campbell and Henshall (1991) and Culley (1993).

Laboratory columns allow for greater precision in the direct determination of bulk density but show systematic variations with depth which depend on the packing technique used. One alternative laboratory procedure uses gamma ray attenuation (Santo and Tsuji, 1977). Other techniques that may have application include computer assisted tomography (CT) (Anderson et al., 1990) and sensing of soil dielectric properties.

1.6 References

Anderson, S.H., R.L. Peyton, and C.J. Gantzer. 1990. Evaluation of constructed and natural soil macropores using x-ray computed tomography. Geoderma 46:13-29.

Baver, L.D., W.H. Gardner, and W.R. Gardner. 1972. Soil physics. 4th Ed. John Wiley and Sons, New York, NY.

Blake, G.R. and K.H. Hartge. 1986. Particle density In A. Klute (ed.) Methods of soil analysis Part 1. Soil Science Society of America, Madison, WI.

Borggaard, O.K. 1982. The influence of iron oxides on the surface area of soil. J. Soil Sci. 33:443–449.

Bower, C.A., and F.B. Gschwend. 1952. Ethylene glycol retention by soils as a measure of surface area and interlayer swelling. Soil Sci. Soc. Am. Proc. 16:342–345.

Campbell, D.J., and J.K. Henshall. 1991. Bulk density. In K.A. Smith, and C.E. Mullins (ed.) Soil analysis physical methods. Marcel Dekker, Inc., New York, NY.

Campbell, G.S. 1985. Soil physics with basic. Elsevier. Amsterdam, Netherlands.

Carter, D.L., M.M. Mortland, and W.D. Kemper. 1986. Specific surface, p. 413–424. In A. Klute, et al. (ed.) Methods of soil analysis Part 1. Soil Science Society of America, Madison, WI.

Culley, J.L.B. 1993. Density and compressibility. In M.R. Carter (ed.) Soil sampling and methods of analysis. Lewis Publishers, Boca Raton, FL.

Davies, R. 1984. Particle size measurement: Experimental techniques. In M.E. Fayel, and L. Otten (ed.). Handbook of powder science and technology. Van Nostrand Reinhold Co., New York, NY.

Egashira, K. and J. Matsumoto. 1981. Evaluation of the axial ratio of soil clays from gray lowland soils based on viscosity measurements. Soil Sci. Plant Nutr. 27:273–279.

Einstein, A. 1906. A new determination of molecular dimensions. Annalen der Physik. 19:289–306. Translated in: Investigations on the theory of brownian movement. Dover Press, New York, NY.

Elder, J.P. 1979. Density measurements by the mechanical oscillator, p. 12–25. *In* Methods in enzymology. Academic Press, New York, NY.

Elghamry, W., and M. Elashkar. 1962. Simplified textural classification triangles. Soil Sci. Soc. Am. Proc. 26:612–613.

Folk, R.L. 1966. A review of grain-size parameters. Sedimen. 6:73–93.

Gee, G.W., and J.W. Bauder. 1979. Particle size analysis by hydrometer: A simplified method for routine textural analysis and a sensitivity test of measurement parameters. Soil Sci. Soc. Am. Proc. 43:1004–1007.

Gee, G.W., and J.W. Bauder. 1986. Particle size analysis, p. 383–411. *In* A. Klute (ed.) Methods of soil analysis, Part 1. 2nd Ed. American Society of Agronomy, Madison, WI.

Heywood, H. 1947. Symposium on particle size analysis. Trans. Inst. Chem. Eng. 22:214.

Jensen, E., and H.M. Hansen. 1961. An elutriator for particle-size fractionation in the sub-sieve range. Soil Sci. 92:94–99.

Kahn, A. 1959. Studies on the size and shape of clay particles in aqueous suspensions. Clays Clay Miner. 6:220–236.

Karsten, J.H.M., and W.A.G. Kotze. 1984. Soil particle size analysis with the gamma attenuation technique. Commun. Soil Sci. Plant Anal. 15:731–739.

Klein, C., and C.S. Hurlbut, Jr. 1985. Manual of mineralogy (after James D. Dana). 20th Ed. Wiley and Sons, New York, NY.

Loveland, P.J., and W.R. Whalley. 1991. Particle size analysis. p. 271–328. *In* K.A. Smith and C.E. Mullins (ed.). Soil analysis physical methods. Marcel Dekker, Inc., New York, NY.

Nelson, R.A., and S.B. Hendricks. 1943. Specific surface of some clay minerals, soils, and soil colloids. Soil Sci. 56:285–296.

Orchiston, H.D. 1955. Adsorption of water vapor: III. Homoionic montmorillonites at 25°C. Soil Sci. 79:71–78.

Santo, L.T., and G.Y. Tsuji. 1977. Soil bulk density and water content measurements by gamma-ray attenuation techniques. HI Agric. Expt. Stn. Tech. Bull. 98, Honolulu, HI.

Scheidegger, A.E. 1960. The physics of flow through porous media. Rev. Ed. University of Toronto Press, Toronto, Canada.

SCS. 1981. Tables for discussion of estimation of bulk density. Approximation 3. R.B. Grossman (ed.) In-house report of NSSL. Lincoln, NE.

Sheldrick, B.H., and C. Wang. 1993. Particle size distribution. p. 823. *In* M.R. Carter (ed.) Soil sampling and methods of analysis. Lewis Publishers, Boca Raton, FL.

Shirazi, M.A., and L. Boersma. 1984. A unifying quantitative analysis of soil texture. Soil Sci. Soc. Am. J. 48:142–147.

Streeter, V.L., and E.B. Wylie. 1975. Fluid mechanics. 6th Ed. McGraw-Hill Inc., New York, NY.

Suarez, D.L., and J.D. Wood. 1984. Simultaneous determination of calcite surface area and content in soils. Soil Sci. Soc. Am. J. 48:1232–1235.

Svedberg, T., and J.B. Nichols. 1923. Determination of size and distribution of size of particles by centrifugal methods. J. Amer. Chem. Soc. 45:2910–2917.

Tanner, C.B., and S.J. Bourget. 1952. Particle-shape discrimination of round and square holed sieves. Soil Sci. Soc. Am. Proc. 16:88.

Tyler, S.W., and S.W. Wheatcraft. 1992. Fractal scaling of soil particle-size distributions: Analysis and limitations. Soil Sci. Soc. Am. J. 56:362–369.

Uhland, R.E. 1949. Physical properties of soils as modified by crops and management. Soil Sci. Soc. Am. Proc. 14:361–366.

2

Dynamic Properties of Soils

Rainer Horn and Thomas Baumgartl
Christian-Albrechts University Kiel

2.1 Introduction

Soils undergo intensive changes in their physical, chemical, and biological properties during natural soil development and as a result of anthropogenic processes such as plowing, sealing, erosion by wind and water, amelioration, excavation, and reclamation of devastated land. In agriculture, soil compaction as well as erosion by wind and water are classified as the most harmful processes which not only end in a reduction of the productivity of the site but are also responsible for groundwater pollution, gas emissions and higher energy requirements to obtain a comparable yield. In forestry, normal plant and soil management, tree harvesting and clear cutting affect site-specific properties including organic matter loss and groundwater pollution and gas emissions which have the potential to cause global changes. Furthermore, soil amelioration especially by deep tillage prior to replanting often causes irreversible changes in properties and functions. These interrelationships have been described by Soane and van Ouwerkerk (1994) and quantified by Hakansson et al. (1987). Oldeman (1992) showed that about 33 million ha of arable land are already completely devastated by soil compaction in Europe alone while the total area of degraded land worldwide exceeds about 2 billion ha. Physical (soil erosion and deformation) and chemical processes are responsible for about 1.6 and 0.4 billion ha of degraded soils, respectively. Worldwide population growth will reduce the average area per person for food and fiber production from 0.27 ha today to < 0.14 ha within 40 years. Consequently, a more detailed analysis of soil and site properties is needed to manage soils in accordance with their potential properties.

With respect to physical processes and soil degradation, Voorhees and Allmaras (1998), Morras and Piccolo (1998), and Horn (1998) postulated that soil erosion can be linked to soil tillage processes, because soil in seedbeds is susceptible and easily transported by wind or water. When soils become saturated, transport as sheet or gully erosion is inititiated, often down to the plow pan. If the tilled soil dries out, transport by wind may occur resulting in severe reductions in potential site productivity. Preparation of a seedbed leads to abrupt changes in the transport of gas, water, ions, and heat between the tilled and deeper soil layers (Boone and Veen, 1994; Lipiec and Simota, 1994; Stepniewski et al., 1994). This is especially true in terms of preferential flow through structured soils. Such anthropogenic changes make the discussion of dynamics within the soil profile relevant. In addition, the interactions between soil structure, water status, and aeration of structured soils in relation to root growth and compressibility of arable land have been described by Emerson et al. (1978).

Information on dynamic soil properties including the process of structure formation is important in classifying, using and sustaining soils in accordance with their physical, mechanical, chemical or biological capacity.

2.2   Processes in Aggregate Formation

Soils containing > 15% clay tend to form structural units known as aggregates by both physical and biological processes (Hillel, 1980; McKenzie, 1988). Aggregates may vary in size from crumbs (< 2 mm) to polyhedrons or subangular blocks (0.005–0.02 m) to prisms and columns > 0.1 m (Horn et al., 1989; Becher 1992). Aggregates have either sharp rectangular edges or are defined by nonrectangular shear planes. (Hartge and Horn, 1977; Hartge and Rathe, 1983; Babel et al., 1995). Aggregated soils are always stronger than homogenized material. Physical, chemical and biological processes vary between the inter- and intraaggregate volumes. Thus, the dynamics of hydraulic, mechanical, biological and chemical processes affect soil intensity properties strongly (Goss and Reid, 1979; Tippkötter, 1988; Dexter et al., 1988; Haynes and Swift, 1990; Horn et al., 1994). Furthermore, when strength is defined mechanically and/or hydraulically, one cannot extrapolate beyond the imposed limits which must be specified.

Structural properties are always in a dynamic equilibrium and influence nearly all site properties and functions including aeration, water infiltration, capillary rise, accessibility of particle surfaces for exchange, sorption reactions and biological activity (Junkersfeld and Horn, 1997). If leaching of nutrients or organic substances alters the properties of pore walls or the outer surfaces of aggregates, chemical changes (ionic strength) have to be considered in addition to mechanical properties to adequately deal with dynamic processes in structured soils. More detailed information on aggregate formation and changes in chemical properties is presented by Horn et al. (1994), Kay (1990), and in Section A, Chapter 7.

Mechanical processes and their dynamic aspects will be described and compared in homogenized and structured soils in order to highlight the effect of soil structure, tillage and timber harvesting on properties and processes in an ecosystem. Both capacity and intensity parameters and their measurement will be discussed in relation to mechanical properties. Thereafter, the effects of structure on ion sorption and desorption, and hydraulics will be discussed and modeled.

2.3   Determination of Mechanical Parameters

Soils consist of solid, liquid and gaseous phases; chemical, physical and biological processes determine site properties and functions depending on the degree of soil development. Capacity and intensity parameters and functions will be used to either define basic material properties or to quantify material functions. The latter include the definition of well-defined limits (validity of the material properties, i.e., whether or not there are elastic or plastic changes due to mechanical loading) and the derivation of induced changes in physical, chemical and biological functions.

2.3.1   Capacity Measurements

Capacity methods quantify material properties which are constant but which differ by location in the field. The methods that can be used to compare mechanical properties can be found in Burke et al. (1986) and are summarized in Table 2.1.

Table 2.1   Methods determining soil physical properties

Method	Dimension	Soil Condition
• Atterberg Consistency test	water content (%, w/w)	homogenized soil
• Proctor test	water content (%, w/w) bulk density (g cm^{-3})	homogenized soil single aggregate
• Mean weight diameter	(-)	partly homogenized soil samples
• Wet sieving	length (cm)	single aggregates
• Percolation	water vol/vol (-)	bulk soil
• Irrigation	water vol/vol (-)	bulk soil

2.3.1.1   Soil Consistency

Atterberg limits of homogenous soil material are defined as a function of water content on a mass basis (θ_m) and are related to soil strength properties. The results are also used to predict soil workability.

Liquid Limit

The liquid limit is defined as the water content (θ_m) at a certain amount of energy applied to the soil in the Casagrande apparatus (e.g., 25 strokes). Each set of samples is homogenized, placed in a special bowl and V shape trenched from top to bottom. Thereafter, the sample is bounced up and down by the special equipment continuously and the number of strokes counted. As soon as the trench has been closed to < 1 cm, the water content (θ_m) is determined. This test is repeated for at least four water content values. The liquid limit increases with clay and organic matter contents, ionic strength, cation valency and proportion of 2:1 clay minerals in the soil.

Plastic Limit

The plastic limit is defined as the water content (θ_m) when homogenized soil samples start to crack when rolled to a diameter < 4 mm.

Plasticity Value

The difference between the water content (θ_m) at the liquid and plastic limits is the plasticity value, often used as an index of soil workability. Sensitivity to plastic deformation increases with increasing plasticity value; the smaller the value the sooner the soil can be trafficked without further soil deformation. The higher the plasticity value, the smaller the angle of internal friction for sandy soils. Many attempts have been made to correlate the plasticity value to soil strength (Kezdi, 1969; Hartge and Horn, 1992; Kretschmer, 1997). In principle, this test only gives information on mimimum strength values.

2.3.1.2   Proctor Test

The Proctor test evaluates the effects of water content (θ_m), and organic and mineral composition of homogenized soil on compactability. Optimum bulk density (ρ_{opt}) and water content (θ_{opt}) for maximum compactability are determined at a given applied energy (e.g., 3 x 25 hammer strokes of known amplitude and weight), after a series of soil tests have been performed at different water contents. The coarser the soil sample, the higher the Proctor bulk density value at a smaller water content (θ_{opt}). Sandy soils have higher values than silty or clayey soils, while the latter require a higher water content (θ_{opt}) to reach the optimum bulk density (ρ_{opt} = Proctor density). The more heterogeneous the grain size distribution, the higher the Proctor density at higher water content.

Strongly aggregated soils behave like coarser soil materials (the Proctor density is higher at a lower optimal water content). If aggregates themselves are destroyed during the test, the Proctor density gets smaller but the water content increases compared to completely homogenized material.

2.3.1.3 Mean Weight Diameter

Aggregate stability is determined by wet sieving. The soil samples are prewetted to a given pore water pressure and then sieved through a set of sieves from 8 to 2 mm diameter. The difference between the aggregate size distribution at the beginning and end of sieving under water for a given time is calculated as the mean weight diameter (MWD). This value is qualitatively related to aggregate strength and increases with increasing aggregate stability.

2.3.1.4 Penetration Resistance

Soil resistance to any kind of deformation is determined by various types of penetrometers. The most simple is a thin metal rod (< 0.5 cm diam) with a defined tip shape. Frequently, the tip angle is 30 degrees to simulate root properties or earthworm shapes. The penetrometer can be either pushed into the ground by the constant weight of a falling hammer or it can be driven by a motor at constant speed. The output readings can be either penetration depth per hammer stroke, or depth-stress depletions which have to be overcome by the penetrating body in more sophisticated models. Penetration resistance is correlated with root growth, earthworm activity, and tillage effects. When penetration resistance exceeds 2 MPa, root growth is often reduced by half, while values > 3 MPa often prevent root growth. Tillage may increase the critical stress value of a hardpan to > 3.5 MPa depending on the nature of the pore system and the type of soil structure. Because the penetrometer needle is not as flexible as a root, which can choose planes of weakness for growth, penetrometer readings quantify resistances mainly in the vertical direction. In addition, the penetrometer readings only integrate impeding effects and cannot identify the causes. Despite a voluminous literature on the effect of increasing bulk density and/or water content on penetration resistance, extrapolation to land management situations is limited. Penetrometer readings can be used to create maps of derived properties (e.g., definition of sites with a given strength irrespective of its origin). Such data can be interpreted using statistical variograms, fractal analyses or simply by stating that values are spatially different.

2.3.1.5 Other Methods

Bulk Density

In completely homogenized and/or structureless sandy soils, bulk density can be used as a first approximation of soil strength. However, because the dimensions of bulk density are mass per unit volume and not mass in a given constant volume, the interpretation is limited when dealing with structured and/or unsaturated soils. No relationship between strength and bulk density exists in aggregated soils nor is it possible to derive other properties from such data except in the case of a seedbed or a newly disturbed site.

Structural Stability in Alcohol/Water Mixtures

More detailed information on this very qualitative method is presented by Hartge and Horn (1992) and Burke et al. (1986).

2.3.2   Intensity Parameters

Soil formation including aggregate development involves changes in both physical and mechanical properties, and therefore, requires the exact definition of the limits within which properties are quantified. This is true, because *in situ* soil formation processes have to be linked to internal and external conditions (climatic, mechanical, thermal, hydrological, or chemical aspects) for a particular situation. For example, all properties such as soil strength, stress attenuation, changes in soil structure or pore distribution, water and ion fluxes, gas exchange and accessibility of exchange sites for cations are material functions (with well-defined and quantified limits). Consequently, in order to deal with dynamic soil properties, stress, strain and strength definitions are initially required to later define the limits of the material functions with respect to the application of external stresses.

2.3.2.1   Stress Theory

Definitions

Before discussing methods of field and laboratory stress measurement as well as factors influencing compaction, one needs to differentiate between several terms used to define dynamic compressive properties. These definitions have been taken from Kezdi (1969), Bradford and Gupta (1986), Hartge and Horn (1991), and Fredlund and Rahardjo (1993).

Force applied to a soil-per-unit area is defined as stress.

Stresses working along the surface will also induce stresses in the soil, which may result in a three-dimensional deformation of the soil volume or will be transmitted as a rigid body. The mechanical behavior of a soil (volume change and shear strength) can be described in terms of the soil stress state.

Strength quantifies mechanically based material functions and depends on internal parameters such as particle size distribution, type of clay mineral, nature and amount of adsorbed cations, content and type of organic matter, aggregation induced by swelling and shrinking, stabilization by roots and humic substances, bulk density, pore size distribution and pore continuity of the bulk soil and single aggregates, water content and/or water suction (Horn, 1981).

Stress State

The number of stress state variables required to define the stress state depends primarily upon the number of phases involved. The effective stress σ' can be defined as a stress variable for saturated conditions and is the difference between the total (σ) and the neutral stress (u_w) which is equal to the pore water pressure:

$$\sigma' = \sigma - u_w \tag{2.1}$$

where σ' is transmitted by solid and (u_w) by the liquid phase, respectively. In unsaturated soils stresses are transmitted by the solid, liquid and gaseous phases. Thus, Equation [2.1] becomes:

$$\sigma' = (\sigma - u_a) + X(u_a - u_w) \tag{2.2}$$

where u_a and u_w are pore air and water pressures, respectively, and X is a factor which depends on the degree of saturation. At saturation ($u_w = 0$), X = 1, while at $u_w = -10^6$ kPa, X = 0.

For sandy, less compressible and nonaggregated soils, X can be calculated as:

$$X = 0.22 + 0.78 \text{ Sr} \tag{2.3}$$

where Sr is the degree of saturation. For silty and clayey soils, the values of the parameters in Equations [2.1] and [2.2] depend on soil aggregation, pore arrangement and strength, and hydraulic properties. Thus the material function of the components in structured soils is only valid as long as the internal soil strength is not exceeded by the externally applied stresses. It changes if, for example, aggregates are destroyed during soil deformation and the structural properties reduced to those which depend on texture.

In current discussions on sustainability, soil compaction is repeatedly mentioned as one of the main threats to agriculture which should be avoided (Soane and van Ouwerkerk, 1994; Horn et al., 1995). The extent of soil deformation can be predicted by stress strain processes and by their relative proportions. In the absence of gravitational and other applied forces, stresses in three-phase soil systems are divided into three normal stresses and six shearing stresses acting on a cube; at equilibrium, the shearing forces reduce to three. Therefore, three normal stresses (σ_x, σ_y, σ_z) and three shearing stresses (τ_{xy}, τ_{xz}, τ_{yz}) must be determined to define the stress state at a point (Nichols et al., 1987; Horn et al., 1992, Harris and Bakker, 1994).

The stress state can be described completely by a symmetric matrix:

$$\begin{bmatrix} \sigma_x & \tau_{xy} & \tau_{xz} \\ \tau_{xy} & \sigma_y & \tau_{yz} \\ \tau_{xz} & \tau_{yz} & \sigma_z \end{bmatrix} \qquad [2.4]$$

For symmetric matrices, it is always possible to find a coordinate system in which the matrix becomes diagonal. In this principal axis system, the stress matrix simplifies to:

$$\begin{bmatrix} \sigma_1 & 0 & 0 \\ 0 & \sigma_2 & 0 \\ 0 & 0 & \sigma_3 \end{bmatrix} \qquad [2.5]$$

with principal stresses σ_1, σ_2 and σ_3. For a simpler characterization of the stress state two invariants of the stress matrix are often used, the mean normal stress (MNS) and the octahedral shear stress (OCTSS) (Koolen, 1994):

$$\text{MNS} = \frac{1}{3}(\sigma_1 + \sigma_2 + \sigma_3) \qquad [2.6]$$

$$\text{OCTSS} = \frac{2}{3}\sqrt{(\sigma_1 - \sigma_2)^2 + (\sigma_2 - \sigma_3)^2 + (\sigma_3 - \sigma_1)^2} \qquad [2.7]$$

Stress Propagation

Each applied stress is transmitted into the soil in three dimensions and can alter the physical, chemical and biological properties if internal mechanical strength is exceeded. The type of external force applied, time dependency and number of compaction events can either change properties to depths by divergent processes or destroy a given structure by shear forces such as kneading. The latter case may result in complete homogenization and normal shrinkage behavior (Horn 1988). Stress propagation theories are rather old and have been often modified and adapted to *in situ* situations. The fundamental theories of Boussinesque (1885) are only valid for completely elastic material, while Froehlich (1934) or Söhne (1953) (all cited in Horn, 1988) included elastoplastic properties through the introduction of concentration factor values. More complete descriptions of these models are given by Koolen and Kuipers (1985), Bailey et al. (1986, 1992), Johnson and Bailey (1990).

2.3.2.2 Strain Theory

Every change in stress state results in soil deformation. If stress exceeds the internal soil strength, the plastic (irreversible) portion increases strongly. Analogous to stress, strain can be described by normal (ε_x, ε_y, ε_z) and shear strain (ε_{xy}, ε_{xz}, ε_{yz}), being described completely by the symmetric strain matrix:

$$\begin{bmatrix} \varepsilon_x & \varepsilon_{xy} & \varepsilon_{xz} \\ \varepsilon_{xy} & \varepsilon_y & \varepsilon_{yz} \\ \varepsilon_{xz} & \varepsilon_{yz} & \varepsilon_z \end{bmatrix} \qquad [2.8]$$

In the corresponding principal axis system, the strain matrix reduces to:

$$\begin{bmatrix} \varepsilon_1 & 0 & 0 \\ 0 & \varepsilon_2 & 0 \\ 0 & 0 & \varepsilon_3 \end{bmatrix} \qquad [2.9]$$

The strain matrix completely describes the local soil deformation. For example, the volumetric strain (ε_{vol}) can be calculated as:

$$\varepsilon_{vol} = \varepsilon_1 + \varepsilon_2 + \varepsilon_3 = \varepsilon_x + \varepsilon_y + \varepsilon_z \qquad [2.10]$$

(note that the trace of a matrix is invariant under coordinate transformations). Furthermore, the strain components and their proportions depend on internal and external parameters and require the determination of all components in a three-dimensional volume.

2.3.2.3 Stress/Strain Processes

Generally, mechanical processes in soils are described by the stress/strain equation:

$$\begin{bmatrix} \sigma_x(\vec{x},t) & \tau_{xy}(\vec{x},t) & \tau_{xz}(\vec{x},t) \\ \tau_{xy}(\vec{x},t) & \sigma_y(\vec{x},t) & \tau_{yz}(\vec{x},t) \\ \tau_{xz}(\vec{x},t) & \tau_{yz}(\vec{x},t) & \sigma_z(\vec{x},t) \end{bmatrix} = f \begin{bmatrix} \varepsilon_x(\vec{x},t) & \varepsilon_{xy}(\vec{x},t) & \varepsilon_{xz}(\vec{x},t) \\ \varepsilon_{xy}(\vec{x},t) & \varepsilon_y(\vec{x},t) & \varepsilon_{yz}(\vec{x},t) \\ \varepsilon_{xz}(\vec{x},t) & \varepsilon_{yz}(\vec{x},t) & \varepsilon_z(\vec{x},t) \end{bmatrix} \qquad [2.11]$$

The function f defines soil properties. In the case of nonlinear isotropic elastic properties, the matrix notation reads as follows:

$$
\begin{bmatrix} \sigma_x & \tau_{xy} & \tau_{xz} \\ \tau_{xy} & \sigma_y & \tau_{yz} \\ \tau_{xz} & \tau_{yz} & \sigma_z \end{bmatrix} = \frac{E}{1+v} \begin{bmatrix} \varepsilon_x - \varepsilon_m & \varepsilon_{xy} & \varepsilon_{xz} \\ \varepsilon_{xy} & \varepsilon_y - \varepsilon_m & \varepsilon_{yz} \\ \varepsilon_{xz} & \varepsilon_{yz} & \varepsilon_z - \varepsilon_m \end{bmatrix} +
$$

$$
\frac{E}{3 \cdot (1 - 2 \cdot v)} \cdot \begin{bmatrix} \varepsilon_m & 0 & 0 \\ 0 & \varepsilon_m & 0 \\ 0 & 0 & \varepsilon_m \end{bmatrix}
\qquad [2.12]
$$

where   E = Young's modulus
       n = Poisson's ratio

$$
\varepsilon_m = \frac{1}{3} \cdot (\varepsilon_x + \varepsilon_y + \varepsilon_z) = \frac{1}{3} \cdot \varepsilon_{vol}
$$

2.3.2.4 Definition of Soil Deformation

Considering the soil as a continuum, soil movement is described as a translation field (d) whose local properties are usually characterized by the three components of its spatial derivative:

$$
\vec{\nabla} \circ \vec{d} = \begin{bmatrix} \dfrac{\partial}{\partial x} d_x & \dfrac{\partial}{\partial x} d_y & \dfrac{\partial}{\partial x} d_z \\ \dfrac{\partial}{\partial y} d_x & \dfrac{\partial}{\partial y} d_y & \dfrac{\partial}{\partial y} d_z \\ \dfrac{\partial}{\partial z} d_x & \dfrac{\partial}{\partial z} d_y & \dfrac{\partial}{\partial z} d_z \end{bmatrix}
\qquad [2.13]
$$

1. rotation

$$
\mathrm{rot}(\vec{d}) = \vec{\nabla} \times \vec{d} = \begin{bmatrix} \dfrac{\partial}{\partial y} d_z - \dfrac{\partial}{\partial z} d_y \\ \dfrac{\partial}{\partial z} d_x - \dfrac{\partial}{\partial x} d_z \\ \dfrac{\partial}{\partial x} d_y - \dfrac{\partial}{\partial y} d_x \end{bmatrix}
\qquad [2.14]
$$

2. compaction/decompaction = divergence = volumetric strain

$$
\varepsilon_{vol} = \mathrm{div}(\vec{d}) = \nabla \cdot \vec{d} = \mathrm{tr}(\nabla \circ \vec{d}) = \frac{\partial}{\partial x} d_x + \frac{\partial}{\partial y} d_y + \frac{\partial}{\partial z} d_z
\qquad [2.15]
$$

3. shearing

$$\varepsilon_s = \frac{1}{2}\left(\nabla \circ \vec{d} + \left(\nabla \circ \vec{d}\right)^{T}\right) - \frac{1}{3}\operatorname{div}(\vec{d}) \cdot \begin{bmatrix} 1 & 0 & 0 \\ 0 & 1 & 0 \\ 0 & 0 & 1 \end{bmatrix}$$

[2.16]

While rotation does not result in any deformation or change in soil volume (except the rotation of the principal axis system of any anisotropic material property), volumetric strain and shearing result in a deformation (Figs. 2.1 and Fig. 2.2). Thus, soil deformation is a process more complex than a volume reduction which is the same as volumetric strain expressing only one degree of freedom, whereas deformation at constant volume (shearing) summarizes 5 degrees of freedom. Although it is useful to distinguish between shear and compaction/decompaction processes as they typically exhibit very different effects, all deformations in soils are combinations of both, utilizing the full range of 6 degrees of freedom.

Compression refers to a process that describes the increase in soil mass-per-unit volume (increase in bulk density) under an externally applied load or under changes of internal pore water pressure. Examples of externally applied static or dynamic loads are vibration, rolling, trampling, etc. while

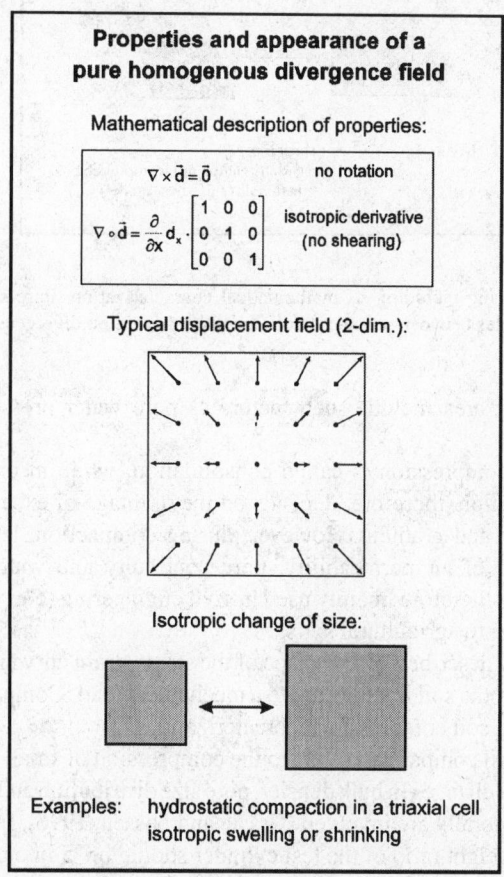

Fig. 2.1 Processes of compaction and decompaction and their mathematical characterization including some examples

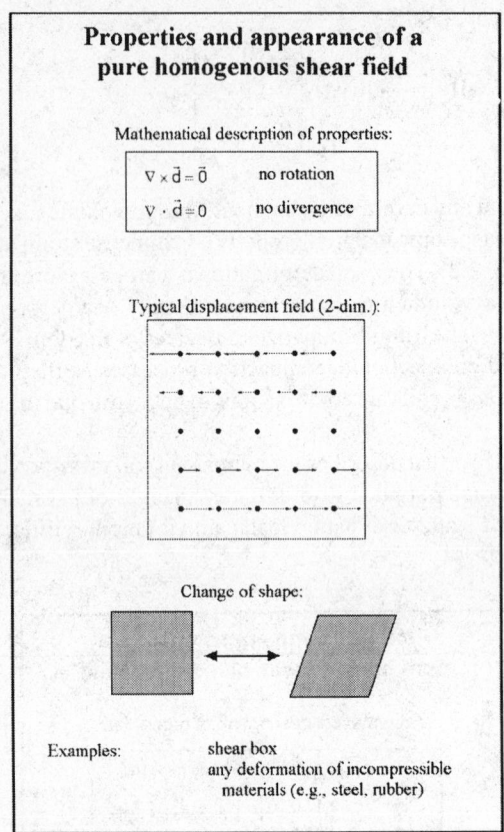

Fig. 2.2 Process of shearing including its mathematical characterization. Pure shearing results in no change in density, but changes in angles between particles. Consequently, there is no divergence and no rotation.

internal forces-per-unit area include such factors as pore water pressure or suction caused by a hydraulic gradient.

In saturated soils, compression is called consolidation, while in unsaturated soils, it is called compaction. Consolidation, therefore, depends on the drainage of excess soil water determined by hydraulic conductivity and gradient. However, during compaction, less compressible air will be expelled as a function of air permeability, pore continuity and water saturation in the profile. Consolidation tests are, therefore, mainly used in civil engineering (e.g., road construction) and have only limited application to agricultural soils.

Soil compressibility described by the shape of the stress strain curve is defined as a resistance to a volume decrease, when the soil is subjected to a mechanical load. Compaction tests are used both for laboratory and for field soil compression characterizations.

In the laboratory, soil compaction refers to the compression of small soil samples, whereas in the field three-dimensional changes in bulk density, pore size distribution and strength of an elemental soil volume to depth and laterally are involved (Gräsle and Nissen, 1996). In laboratory compaction, the optimum diameter to height ratio of the test cylinder should be 5 in order to minimize the effect of friction between cylinder wall and sample. Compaction tests to determine soil strength can be carried out on homogenized, undisturbed or bulk soil samples or single aggregate samples at different pore water pressure (i.e. water suction) (Baumgartl, 1991).

Compactability is the difference between the initial and maximum densities to which a soil can be compacted by a given amount of energy at a defined water content.

2.3.3 Stress Measurements *in situ*

Stress Distribution

Stress distribution measurements on undisturbed soil profiles under traffic with different speeds, loads and contact areas describe the type and intensity of soil strength, stress attenuation, and soil deformation. In unsaturated soils, pore water pressure at the time of loading further affects these parameters. One of the major problems in validating compaction models is sensor installation at different depths and distances from the perpendicular line. Because of the structural disturbance due to excavation and installation of the sensors, the data obtained only describe the stress patterns for homogenized or artificially mixed soil. Thus, only in nonaggregated sandy soils, can one expect to find minimal effects due to installation. In aggregated soils, sensors must be installed in undisturbed soil from the side to minimize structural disturbance so that octahedral normal and shear stresses can be obtained. The pressure sensor is a foreign body with deformation properties different than those of the soil. If the pressure sensor itself is weaker than the soil, the registered stress will underestimate the real stress at that depth. If, however, the pressure sensor stiffness exceeds that of the surrounding soil, stress concentrates on the more rigid transducer body and, therefore, the real soil stress is overestimated. The different types of stress transducers available are described in Table 2.2.

Plastic bodies in the form of pneumatic or hydraulic cylinders, balls or disks made of silicone or rubber change their volume in response to the applied stress. Prior to measurement, pressure cells must be filled with water or air and prestressed to 80 kPa which affects the stress/strain modulus of the

Table 2.2 Different types of stress state transducers (modified from Bolling, 1986)

Principle	Material	Size	Deformation	Measured Values	Calculated Values	Author
Pneumatic	rubber	cylinder	plastic	soil stiffness	-	Kögler (1933)
Hydraulic	rubber	disk	plastic	1 defined stress	-	Söhne (1951)
Hydraulic	steel	disk	elastic	1 defined stress	-	Franz (1958)
Hydraulic	rubber	sphere	plastic	mean normal stress	-	Blackwell (1978)
Hydraulic	silicon	cylinder	plastic	mean normal stress	-	Bolling (1984)
Strain gauge	silicon	sphere	plastic	mean normal stress	-	Verma (1975)
Strain gauge	steel	disk	elastic	1 defined normal	-	Cooper (1975)
Strain gauge	aluminum	disk	elastic	1 defined normal	-	Horn (1980)
Strain gauge	steel	cube	elastic	3 defined normal	-	Prange (1960)
Strain gauge	aluminum	half sphere	elastic	6 defined normal	$\sigma_{1,2,3}$ $\sigma_{xy,yz,xz}$	Nichols et al. (1987) Horn et al. (1992) Harris and Bakker (1994)
Strain gauge	aluminum	sphere	elastic	6 defined normal	$\sigma_{1,2,3}$ $\sigma_{xy,yz,xz}$	Kühner (1997)

sensor and the measured pressure value (Horn et al., 1987, 1991). Theoretically, the sensor elasticity should be the same as that of the surrounding undisturbed soil, which is almost impossible to achieve. Generally, plastic sensors are weaker than the soil and consequently, stresses will be underestimated. A plastic stress transducer measures an average normal stress. The direction of stresses cannot be identified if cylindrical or spherical transducers are used and shear stresses cannot be determined.

When rigid bodies are used as stress transducers, piezoelectric materials or strain gauges are placed on diaphragm material made of aluminum or steel. In comparison to plastic stress transducers, the stress/strain modulus of a rigid transducer cannot be matched to the surrounding soil. Peattie and Sparrow (1954) have suggested that the optimum ratio of the transducer stress/strain modulus to that of the soil should be 10.

The determination of the stress distribution in normal soils using strain gauges has been described by Horn (1981), Burger et al. (1987), van den Akker et al. (1994), Blunden et al. (1994), Ellies et al. (1995, 1996), Ellies and Horn (1996) and Horn et al. (1998) tested the applicability of such techniques for volcanic ash soils which behave in a manner different from that of normal mineral soils. Horn (1995) summarized the physical/mechanical properties of deep-tilled soils and quantified the effect of traffic on stress distribution. Kühner et al. (1994) and Kühner (1997) have quantified the stress distribution under various *in situ* conditions and Horn (1998) defined stress and strength conditions for soils under different land use systems. Blunden et al. (1992, 1994), Kirby et al. (1997), Trein (1995), Kühner (1997), van den Akker et al. (1994) describe various other aspects of stress distribution and attenuation in soils leading to soil protection strategies and engineering solutions. The stress distribution in partially deep-tilled soils (slit plow) is also described by Horn et al. (1998). Olson (1994) found that even a slight increase in moisture content or contact area always resulted in a large increase in the concentration factor value. Because measurements and calculations by Blunden et al. (1992, 1994), Kirby et al. (1997), Trein (1995), and van den Akker et al. (1994) were based on measurements of only the vertical component of applied stress, the missing data were obtained by modeling in order to predict soil stress state. However, such an approach can only verify processes which are assumed to be dominant in homogenous soil systems (seedbed). In contrast, analysis of stress distribution in unsaturated structured soils is a more complex problem with a multitude of processes operating in an interactive manner. This is particularly true for unsaturated, nonlinear, hysteretic soils whose composition changes with time or during tillage thus altering their physical and chemical properties. Consequently, for the determination of stress paths, a Stress State Transducer (SST) system (6 normal stresses on three mutually orthogonal and three nonorthogonal planes) can be used to measure all stress components at a single point. Based on continuum mechanics theory, octahedral principal and shear stresses can be partially measured and partly calculated for a cube cut from the continuum. It should be stressed that such stress propagation cannot be handled in the same way as for soil parent material (i.e., completely homogenized or stiff material as in soil mechanics).

2.3.3.1   Strain Determination

The principle involved in the determination of soil strain has been described by Koolen and Kuipers (1985). One of the earliest systems to determine the volumetric strain path *in situ* during traffic was that of Gliemeroth (1953), while those of McKibben and Green (1936), Gill and van den Berg (1967), van den Akker and Stuiver (1989), and Okhitin et al. (1991) could only be used to determine position changes of inserted particles (colored sticks, spheres, or others) relative to their original positions. The calculation of the strain path under *in situ* conditions is not possible if the adjacent soil is disturbed during installation and if one is unable to characterize sensor movement during deformation or if sensors are missing in various positions; consequently, physical properties cannot be predicted

(Erbach et al., 1991). Subsequently, Kühner et al. (1994) developed a new Deformation Transducer System (DTS), which, when installed in the undisturbed soil volume prior to traffic, records the nature of stress induced movements of soil particles. If four DTSs are installed at various positions and distances from the wheel rut, both the strain path and volumetric strain matrix can be quantified.

2.3.3.2 Stress/Strain Determination
The stress/strain apparatus described by Kühner (1997) (Fig. 2.3), consists of a SST sensor block, which is connected to a mobile measuring device in order to determine movements in the x and z directions. A data logging system with a frequency of approximately 40 observations s^{-1} is used to record stresses and sensor displacement in both directions.

2.3.4 Measurements of Soil Strength
The determination of soil strength parameters requires measurements under well-defined laboratory conditions (Table 2.3).

2.3.4.1 Uniaxial Compression Test
The uniaxial compression test is used to define the pressure at which soil begins to fail at a given water content. A vertical normal stress (σ_1) is applied to the soil sample, while the stresses on the planes mutually perpendicular to the σ_1 direction ($\sigma_2 = \sigma_3$) are zero. The uniaxial compression test is often used to determine the tensile strength of single aggregates (crushing test).

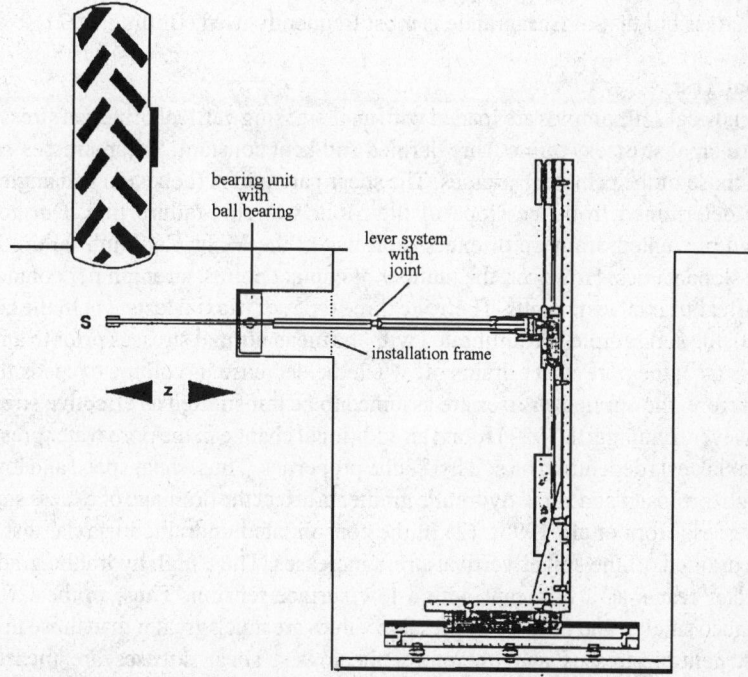

Fig. 2.3 Combined stress/strain apparatus to determine volumetric stresses and biaxial strain during traffic [Kühner, 1997]

Table 2.3 Methods of determining soil strength

Method	Dimension	Derived	Soil Condition
Uniaxial compression	pressure (Pa)		homogenized soil single aggregates structured bulk soil
Confined compression test	pressure (Pa)	precompression stress (Pa)	homogenized soil structured bulk soil
Trixial test	pressure (Pa)	cohesion (Pa) angle of internal friction (°)	homogenized soil structured bulk soil single aggregates
Direct shear test	pressure (Pa)	cohesion (Pa) angle of internal friction (°)	homogenized soil structured bulk soil single aggregates

2.3.4.2 Confined Compression Test

The soil strength relationships of undisturbed, homogenized soils and single aggregates are quantified in the confined compression test. In contrast to the uniaxial compression test, stresses in the σ_2 and σ_3 direction are undefined (rigid wall of the soil cylinder). Both the time and load dependent changes in soil deformation are measured; the slope of the virgin compression line (i.e., the compression index), and the transition from the overconsolidated to the virgin compression line (i.e., the precompression stress) can be determined. The precompression stress is defined as the stress value at the intersection of the less steep recompression curve and the virgin compression line. The latter straight line portion has a steeper slope if plotted on a semilog scale. Many methods are available to determine the precompression stress but that of Casagrande is most frequently used (Bölling, 1971).

2.3.4.3 Triaxial Test

Undisturbed cylindrical soil samples are loaded with an increasing vertical principal stress (σ_1), while the horizontal principal stresses ($\sigma_2 = \sigma_3$) are defined and kept constant. Shear stresses occur in any plane other than those of the principal stresses. The shear parameters (cohesion and angle of internal friction) can be determined from the slope of the Mohr's circles failure line. Due to aggregate deterioration, and prevented drainage of excess soil water, the Mohr Coulomb failure line is bent towards smaller slope values. However, the number of contact points, strength per contact point and pore geometry affect triaxial test results. There are three types of triaxial tests: (1) In the consolidated drained test (CD), the soil sample is equilibrated with the mean normal stresses prior to an increase in the vertical stress (σ_1); the pore water drains off when the decrease in volume exceeds the air-filled pore space. Therefore, the applied stresses are assumed to be transmitted as effective stresses via the solid phase. However, Baumgartl (1991) found an additional change in the pore water pressure during extended CD triaxial tests depending on soil hydraulic properties. Thus, shear speed and low hydraulic conductivity, high tortuosity and small hydraulic gradients affect the drainage of excess soil water and the effective stresses (Horn et al., 1995). (2) In the consolidated undrained triaxial test (CU), pore water cannot be drained off the soil as vertical stress increases. Thus, high hydraulic gradients occur and the pore water reacts as a lubricant with a low surface tension. Thus, in the CU test, shear parameters are much smaller and pore water pressure values are much greater than those in the CD test. (3) The highest neutral stresses and, therefore, the lowest shear stresses are measured in the unconsolidated undrained test (UU), where neither the effective nor the neutral stresses are equilibrated with the applied principal stresses at the beginning of the test. Thus, cohesion and angle of internal friction are strongly dependent on the compression and drainage conditions during triaxial tests. In terms of strength of agricultural soils under traffic, texture, pore water pressure (i.e., in unsaturated conditions = soil matric potential), nature of aggregates, and soil structure determine the preference for a particular test.

2.3.4.4 Direct Shear Test
In the direct shear test, the type and direction of the shear plane, which is assumed to be affected only by normal and shear stresses, are fixed. Normal stress is applied in the vertical and shear stress in the horizontal direction. As in the triaxial test, cohesion and angle of internal friction are influenced by shear speed and drainage conditions for a given soil.

2.4 Effect of Soil Structure and Dynamics on Strength and Stress/Strain Processes

2.4.1 Soil Strength
As stress is applied, soil deformation occurs at the weakest point in the soil matrix and further increases in stress result in the formation of failure zones. Therefore, the strength of the failure zone is equal to the energy required to create a new unit of surface area or to initiate a crack (Skidmore and Powers, 1982) and is called the apparent surface energy (Hadas, 1987). Consequently, soil stability is related to strength distribution in the failure zones. In principle, soil structure will be stable if the applied stress is smaller than the strength of the failure zone, i.e., if the bond strength at the points of contact exceeds the external stress.

2.4.1.1 Homogenized Soils
If only mechanical properties of homogenized soils are compared, gravelly soils are less compressible than sandy or finer textured soils (Horn, 1988) and any deviation of the particles from the spherical shape results in an increase in the shearing resistance and soil strength (Gudehus, 1981; Hartge and Horn, 1991). Ellies et al. (1995) found that the angle of internal friction for medium and coarse sands at comparable bulk density and water content but of different origin (quartz, basalt, volcanic ash and shells) ranged from 25 to 50°. Soil strength also depends on the type and content of clay and exchangable cations. At the same bulk density and pore water pressure, compressibility of homogenized soil increases with clay content and decreases with soil organic matter (SOM) content. At the same clay content, soils are more readily compressed at a lower bulk density or higher moisture content. Cohesive forces between illite, smectite and vermiculite are greater than for kaolinite which has a larger angle of internal friction because of its size and shape. Additionally, the higher the valency of the adsorbed cations, and the higher the ionic strength, the greater is the shearing resistance under a given condition (Yong and Warkentin, 1966, Mitchell, 1976).

Shearing resistance increases depending both on the nature and content of SOM. In agricultural soils with the same physical and chemical properties, the mechanical parameters (precompression stress value, angle of internal friction and cohesion) are greater when the relative contents of carbohydrates, condensed lignin subunits, bound fatty acids and aliphatic polymers are higher (Hempfling et al., 1990).

2.4.1.2 Effect of Aggregation
Mechanical strength of structured soils with comparable internal parameters depends on aggregation, actual and maximum predrying, and composition and arrangement of the pore system. For comparable grain size distribution and pore water pressure, soil strength increases with aggregation (i.e., coherent

< prismatic < blocky < crumbly, platy < subangular blocky). If the relative reduction in water-filled pores (the multiplication factor X in Equation [2.2]) is smaller than the actual decrease in pore water pressure (more negative), strength increases. Conversely, drier soils get weaker. Nevertheless, dynamic changes in mechanical properties also depend on the frequency of swell/shrink and wet/dry events and on the actual pore water pressure. Soil becomes stronger if it is redried and rewetted and if pore water pressure gradients over longer distances promote particle movement and rearrangement until entropy is minimized (Semmel et al., 1990). In addition, soil strength is promoted by two different mechanisms. The increase in strength results from either an increase in the total number of contact points between single particles (i.e., an increase in effective stress) or in shear resistance per contact point (Hartge and Horn, 1984). Thus, even if soil bulk densities are similar, strength may be quite different.

Pedogenic effects on normal mechanical strength of 3 soils are illustrated in Fig. 2.4. Luvisols derived from loess are characterized by clay illuviation, which results in decreased strength in the Al horizon but increased strength in the Bt horizon due to aggregation. Calcium precipitation in the corresponding horizon of the Mollisol also leads to a strength increase. In all three soil types, the parent material C was weakest. Processes such as annual plowing and tractor traffic create very strong plow pans and layers with precompression stress values similar to the contact pressure of a tractor tire or even higher due to lug effects (up to 300%). Strength increases due to compression by glacier movements have been detected by the determination of stress/strain behavior of undisturbed soil samples. The transition from overconsolidation stress to the virgin compression line depends to a great extent on internal parameters and applied stresses in the plow pan. In addition, strength decreases in the A horizon due to plowing and seedbed preparation can be determined until texture dependent values are reached.

Marked strength changes when crossing horizons are observed for soil profiles which have been deep-loosened or plowed (60 cm) or loosened and/or mixed with other soil material. In the case of a partial deep loosening by a trenching slit plow (Blackwell et al., 1989; Reich et al., 1985), traffic must

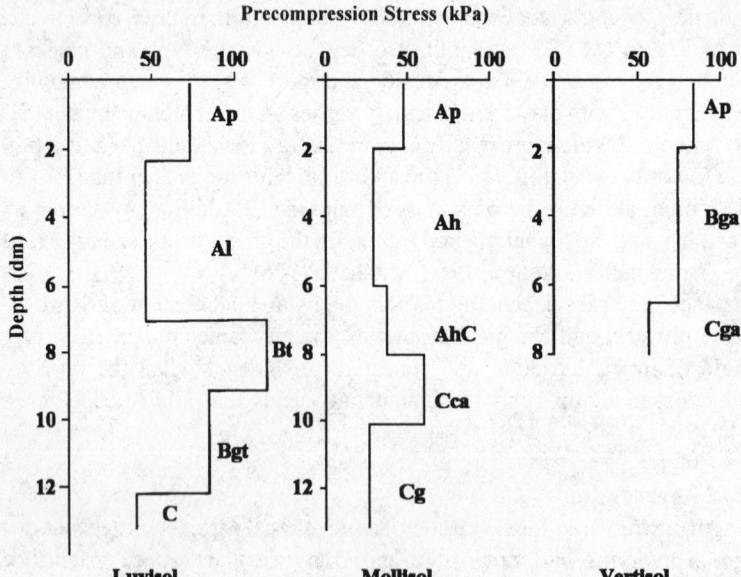

Fig. 2.4 Precompression stress values for 3 soil profiles at soil matric potential $\psi = -6$ kPa

be restricted by use of smaller machinery or controlled tracks perpendicular to the homogenized soil volume. Freezing and thawing affect soil strength, because aggregates become denser or are destroyed by ice lens formation during freezing. Aggregates may act as rigid bodies which allow the exfoliation processes to start from the outer skin because of ice lens formation. Both effects are called soil curing, but result in completely different strength values and physical properties of the bulk soil (Horn, 1985; Kay, 1990).

Soils with a pronounced vertical pore system are stronger than those with randomized or extremely horizontal pore systems because the vertical (bio-) pores are equilibrated with the major principal stress (σ_1). Thus, untilled or minimum tilled soils are stronger than conventionally tilled soils (Horn, 1986).

However, each soil deformation requires air-filled pore space and/or high hydraulic conductivity to drain off the released pore water. The smaller the hydraulic conductivity, gradient and pore continuity, the more stable will be the soil during short-term loading. In sandy soils, this effect is small, because the initial settlement equals the total strain. With increasing clay content, however, the time-dependent soil settlement (proportion of the initial to the primary and secondary settlement) is reduced. This results in an increase in the precompression stress value for short-term loading. Again, these differences are smaller in more strongly aggregated loamy and clayey soils than in sandy soils.

The development of structure always results in an increase in soil strength. Secondary large pores can only be created if the aggregates become denser and stronger so as to carry the same stresses over fewer contact points. These strength differences can also arise from changes in the angle of internal friction, cohesion, and stress dependent changes in shear strength for various applied stress ranges for single aggregates, undisturbed, and homogenized material (Fig. 2.5).

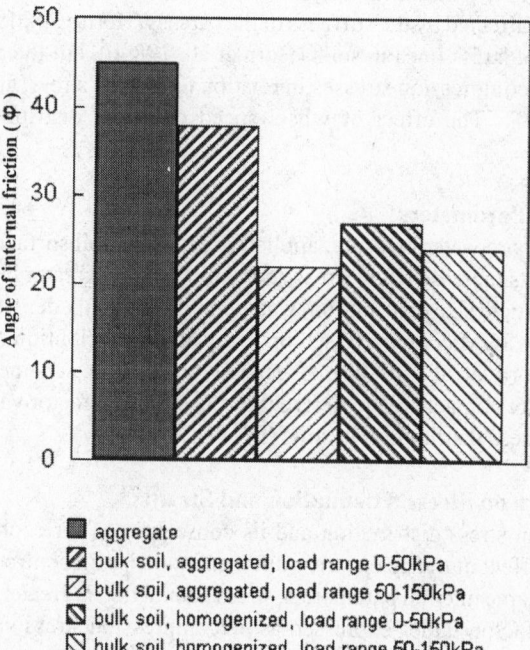

- ▓ aggregate
- ▨ bulk soil, aggregated, load range 0-50kPa
- ▨ bulk soil, aggregated, load range 50-150kPa
- ◪ bulk soil, homogenized, load range 0-50kPa
- ◪ bulk soil, homogenized, load range 50-150kPa

Fig. 2.5 Changes in shear strength at constant soil matric potential $\psi = -6$ kPa for various applied stress ranges for single aggregates, bulk soil, and homogenized material

Increased aggregation increases soil strength under comparable hydraulic conditions. Relative to the very high angle of internal friction for a single aggregate, that for bulk soil is smaller and decreases when a certain stress range is exceeded. As applied stresses increase, the angle of internal friction resembles that of homogenized material which emphasizes that each type of structure is only valid for a well-defined stress (mechanical or hydraulic) range. When this value is exceeded, only texture-dependent properties remain.

2.4.2   Stress Distribution in Soils

Any load applied at the surface is transmitted to the soil in three dimensions by the solid, liquid and gas phases. If air permeability is high enough to allow immediate deformation of the air-filled pores, soil settlement is mainly affected by fluid flow.

However, fluid flow may be delayed because changes in water content or pore water pressure depend on the hydraulic conductivity, gradient and pore continuity. Thus, the intensity and form of pressure transmission are again affected by soil strength. In the following discussion, stress distribution in both homogenized soils and in aggregated systems will be defined. Based on the theory of Boussinesq (1885) who assumed that all soils behave in a completely elastic manner, Froehlich (1934) (both cited in Horn, 1981) described a more realistic form of the equipotential lines in terms of textural-dependent concentration values. Under saturated conditions, these values should range between 3 for very strong material and 9 for soft soil, and be higher for increasing clay content. In weak soils with high concentration factor values, stresses applied are transmitted to depth but remain concentrated around the perpendicular center line. On the other hand, in strong soils with low values of the concentration factor, pressure is transmitted more horizontally in a shallower soil layer.

At the same contact pressure, stresses are transmitted deeper when the contact area is larger. Furthermore, stress distribution patterns in the soil are not only different for tire lugs and the intervening area, but are also affected by the stiffness of the carcass (Horn et al., 1987). Thus, there are no well-defined equipotential stress lines in soils (Horn et al., 1989b), but the concentration factor values must be related to precompression stresses in relation to applied stress, and contact area for a given texture (DVWK, 1995). The effect of wheel speed on stress distribution also has to be considered.

2.4.2.1   Effect of Internal Parameters

At a given bulk density and pore water pressure, applied stress at the soil surface will be transmitted deeper as silt and clay contents increase while stress attenuation will be greater in a smaller volume as the soil dries. For a given particle size distribution, water content and bulk density, stress attenuation will increase with increasing SOM content. Based on many stress distribution measurements in the field and in monoliths (Horn et al., 1989b), a general correlation scheme involving texture, precompression stress and tire contact area to derive the concentration factor value has been used to approximately predict the stress distribution in soils (DVWK, 1995).

2.4.2.2   Effect of Structure on Stress Attenuation and Strain

The effect of aggregation on stress distribution and its consequences for ecological parameters is understood well, while the effect of traffic on stress differentiation is still controversial. Under *in situ* conditions on soils with the same internal parameters, stress attenuation is greater in soils that are more aggregated. Concentration factor values expressed as precompression stress values are smaller for better aggregated and drier soils (Burger et al., 1988; DVWK, 1995). Not only the pressure but also the size and shape of the contact area affect stress distribution in unsaturated structured soils (Fig. 2.6).

If internal strength values are smaller than the external forces applied, repeated traffic results in increased soil strength. For example, if a loessial Luvisol is repeatedly traversed at constant water content, horizontal minor stresses decrease while the major vertical stress increases (Fig. 2.7). These soil strength changes increase the concentration factor values due to the more pronounced vertical stress distribution, and each loading consequently results in a smaller effective relative stress than neutral stress (i.e., negative pore water pressure). During soil loading, stress components can vary as well as the ratio between mean normal and octahedral shear stress. Shear stresses create volumetric strain by soil particle displacement and rearrangement of pores, formation of gas bubbles and increased pore water pressure resulting from reduced hydraulic conductivity. With increasing traffic frequency, measured stresses in the upper soil horizons become smaller due to a more pronounced stress attenuation, while the opposite is true for the deeper soil horizons, where even an increase in the measured stress can be detected (Horn et al., 1998). This additional increase in soil strength caused by the consecutive traffic events can be explained by the elasticity of the topsoil, which induces a plastic deformation at depth in the still weaker subsoil which irreversibly becomes compacted. Because soil amelioration to depth by ripping, deep or slit plowing only loosens compacted soil if it is dry enough to suffer brittle failure, such methods degrade those structural strengths which are dependent on

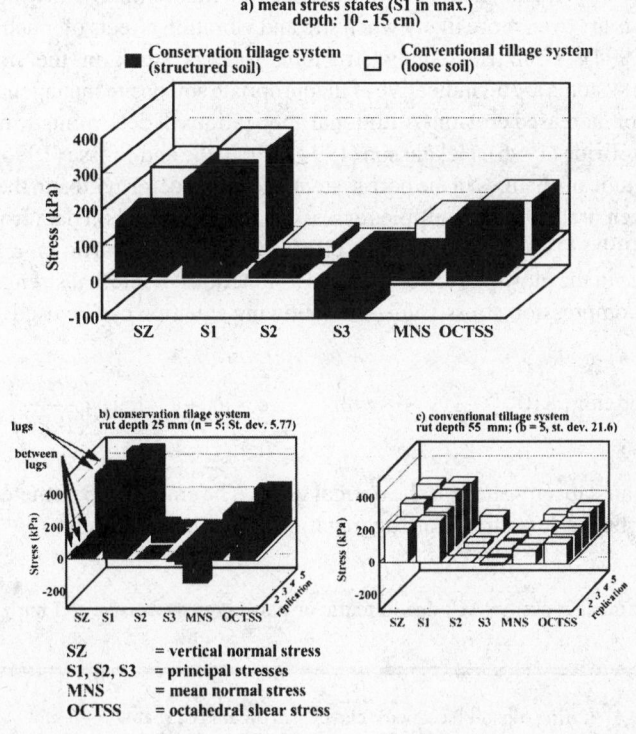

Fig. 2.6 Stress distribution under a tire with lug effects [Data from Kühner, 1997]

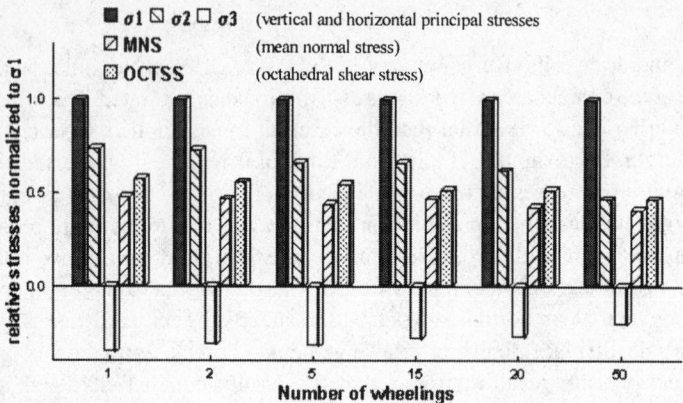

Fig. 2.7   Stress distribution in a loessial Luvisol due to repeated traffic for 1 day at constant water content

texture. Such strength declines require a complete change in tillage and agricultural practices. Such traffic experiments also provide information on soil structure deterioration due to kneading. If the soil water content is high compared to that to be drained off or the hydraulic conductivity is low, homogenization and structure deterioration can be obtained from particle displacement measurements and the change in the proportions of octahedral shear strength relative to the principal stresses during traffic. The vertical and shear components become more important, which weakens the total system. Such effects are even more likely when slip and vibration effects of machinery are considered (Kühner et al., 1994). Deep ruts, platy structure, tractors stuck in the mud and completely homogenized pore systems are all indicative of inappropriate soil or site management which will result in yield decreases or increased erosion. Additional applications of deterministic models are described by Veenhof and McBride (1996), Hakansson (1994), Mc Bride and Joosse (1996), Nissen and Gräsle (1996). The magnitude of changes in the pore system due to soil deformation in the virgin compression stress range has been predicted by multiple regression analysis based on data from more than 80 soil profiles (Nissen, 1998; DVWK, 1997). If the effect of plowing or traffic on a weak seedbed or on changes in air space in the plow pan are considered as functions of internal strength and applied stress exceeding the precompression stress value, the following equation can be used:

$$AC = b \cdot \log_{10}\left(10^{c/b} + 1\right) \qquad \text{with} \qquad c = a_0 + a_1 \cdot \log_{10}\left(\frac{\sigma_n}{1kPa}\right) \qquad [2.17]$$

in which AC is the air capacity and σ_n is the normal stress. The empirical parameters a_0, a_1 and b (Table 2.4) depend on particle size distribution, aggregation, pore water pressure.

Table 2.4   Changes in air capacity (AC) due to traffic or plowing in successive soil horizons at given initial soil physical properties

traffic, topsoil (seedbed): clayey, silty loam (Ut3) at $\psi = -6$ kPa
$a_0 = 53.73$; $a_1 = -23.38$; $b = 7$

plowing, deeper soil horizons: prism, Lt3 (very clayey loam), at $\psi = -6$ kPa
$a_0 = 29.04$; $a_1 = 15.27$; $b = 5$

2.4.3   Effect of Stress Application and Attenuation on Soil Strain

If external stress is smaller than internal soil strength, no deformation results and *vice versa*. The latter can either result in constant displacement of volume, or soil compaction (Kühner et al., 1994; Pytka et al., 1995). In order to distinguish between these processes, both the rut depth and the vertical movement of a given soil volume below the rut must be known. The extent to which soil strain occurs during traffic and the extent to which various tillage implements (conventional/conservation) deform a soil at a given pore water pressure are shown in Fig. 2.8. In the conventional tillage treatment in a loessial Luvisol, passage of a tractor (front/rear wheel) results in a pronounced vertical (up to 8 cm) and horizontal forward and backward (up to 2 cm) displacement. Under conservation tillage, these soil deformations are less because of a higher internal soil strength leading to a maximum vertical displacement of < 4 cm after 3 traffic events and a much less pronounced horizontal displacement. With increasing aggregate development, soil strength increases and aggregate deterioration is less pronounced during displacement and alteration of the pore system due to the infilling of interaggregate pores by smaller particles. Nevertheless, all stresses which are not attenuated to levels below soil strength result in volume alterations, even if the applied stresses vary for different soil types, land uses, tillage systems and environmental conditions.

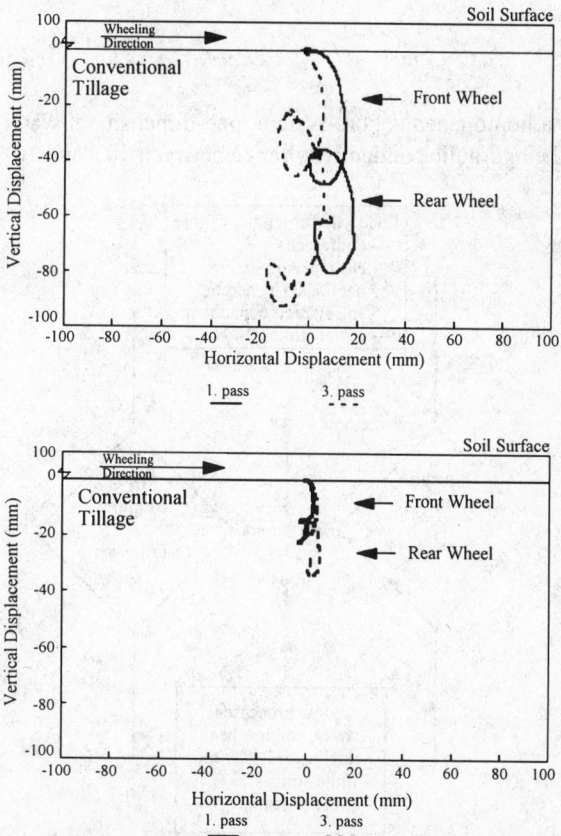

Fig. 2.8 Strain distribution in structured soils due to traffic. Particle movement is more pronounced in the conventionally tilled Luvisol while soil deformation is less intensive in the conservation tillage plot.

2.5   Further Dynamic Aspects in Soils

2.5.1   Hydraulic Aspects

2.5.1.1   Water Retention Curve

With increasing aggregation, the interaggregate pore volume becomes coarser and water retention curves start to differ from those dependent only on texture. Aggregates are normally very dense and strong, containing mostly fine with very few coarse pores and less plant available water, while their bulk densities can exceed 2 g cm^{-3} (Hartge and Horn, 1991). Thus, structure development always coincides with a change in pore system and function. As it is true for stress/strain functions which can be subdivided into a virgin compression line and an overconsolidated stress range, water retention curves can be divided into a virgin desiccation range and a rewetting or drying suction range (Junkersfeld and Horn, 1997). The effective stress equation shows that soil strength can be gained from either mechanical or pore water pressure which affect structure and pore size distribution. Richards (1992) verified these relationships (Fig. 2.9). Exceeding the degree of predryness or prestress by either mechanical or negative pore water pressure always results in a further soil deformation by normal shrinkage, rearrangement of particles, additional formation of aggregates or complete deterioration of existing pores.

2.5.1.2   Darcy's Law

Hydraulic Conductivity

Given laminar flow and a homogeneous pore system, one-dimensional water flux is described by Darcy's law. Values of the hydraulic conductivity range between 10^{-4} and 10^{-13} m s^{-1} depending on

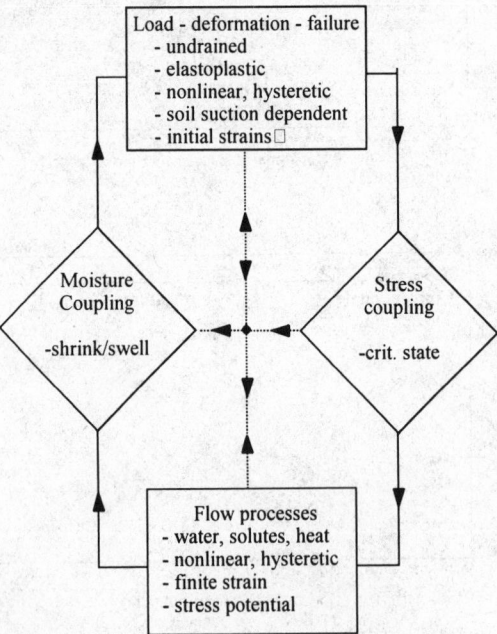

Fig. 2.9   Schematic diagram to define limits for the validity of data sets for hydraulic and mechanical parameters

water potential, texture and structure. Under saturated conditions, hydraulic conductivities range between 10^{-4} and 10^{-9} m s^{-1} in sandy or clayey soils, respectively. Since hydraulic conductivity is also affected by structure and texture, it will be higher in highly fractured or aggregated compared to compacted and dense soils. The hydraulic conductivity value (K) depends on volume and continuity of conducting pores. In structured soils with large cracks, K for the bulk soil is high but low inside single aggregates (Horn, 1990). Depending on aggregate density and pore continuity, K may be 4 orders of magnitude lower in single aggregates than in bulk soil except where the aggregate contains mostly sand in which case there may be no difference. The effects of structure on hydraulic conductivity persist under unsaturated conditions where any changes in structure directly affect hydraulic conductivity. At less negative pore water pressures, unsaturated hydraulic conductivity in single aggregates decreases with compaction of the structural elements (prisms less than polyhedral or subangular blocks) compared to fluxes in bulk soil. Only in weak aggregates differences and ranges are smaller (Gunzelmann et al., 1987). After exceeding the crossover suction at very negative pore water pressures (Hillel, 1980), hydraulic conductivity is higher in aggregates than in the bulk system. The heterogeneity of flow paths in aggregates compared to bulk soil is further enhanced, since outer surfaces of aggregates contain more clay than within where the pores are coarser (Horn, 1987). Consequently, water and/or air flow in or out of single aggregates is further reduced which can also be achieved by increasing tortuosity of the pore system at different positions in the aggregate (Glinski and Stepniewski, 1985).

Hydraulic Gradient

Generally, hydraulic gradients (m m^{-1}) vary by only half an order of magnitude, depending on pore water pressure, particle and pore size distributions (Hartge and Horn, 1977). Differences in hydraulic gradients should be large in aggregated dense soils particularly under conservation tillage, at least initially, which may result in reduced plant water uptake efficiency. Differences in root length density which can be correlated with hydraulic properties can be used as indicators of rhizosphere function (Vetterlein and Marschner, 1993).

Water Flux Density

As differences in hydraulic properties between bulk soil and single structural elements increase, fluid flow becomes more pronounced in interaggregate pores reducing the possibility of obtaining an equilibrated pore water pressure profile. Pore continuity in macropores and a more tortuous pore system in single aggregates induce preferential flow especially near saturation (Beven and Germann, 1982). Youngs and Leeds-Harrison (1990) pointed out that a pore water pressure gradient causes water to flow preferentially in macropores with little flow within aggregates when both the macro-(interaggregate) and micropores (intraaggregate) are saturated. However, when saturated macropores surround unsaturated aggregates, solutes will be transported by diffusion into or from the aggregates depending on the concentration gradient. If macropores are empty, pores inside the aggregates become isolated, severely reducing the possibility for redistribution of water and solutes between them. Booltink and Bouma (1991), Tolchel'nikov et al. (1991), and Edwards et al. (1993) showed that hydraulic properties determined for bulk soil do not account for or explain pore continuity and/or pore accessibility with respect to water flux. Methods for macropore and intraaggregate flux measurements which are given by Gunzelmann (1990), Plagge (1991). Jardine et al. (1990), Jury et al. (1991) and Kutilek and Nielsen (1994) deal with effects of soil aggregation in structured soils on changes in the hydraulic properties.

2.5.2   Chemical Aspects

Brusseau and Rao (1990) pointed out that solute transport in aggregated soils is often characterized by nonideal phenomena, usually ascribed to the presence of immobile domains within a porous medium, which influence the dynamics of physical processes.

2.5.2.1   Ion Transport

Soil ion transport rate (mmol $m^{-1} d^{-1}$) includes mass flow and diffusion for both liquid and solid phases, source/sink terms (exchange processes), ion precipitation, redox reactions and biological decay processes. Assuming laminar solute flow pattern in pores, water flow near the particle surface is retarded relative to that in the pore center because the soil solution becomes more viscous and concentrated as the particle surface is approached. Consequently, with decreasing pore size, ion diffusion in the liquid phase is reduced. With increasing tortuosity which depends on both the volume and geometry of the water pathway, the ion concentration gradient is further reduced. In unsaturated soils with a high clay content, the tortuosity is further increased by reduced thickness and discontinuity of water films, increased density of the adsorbed cation swarm, anion exclusion, and increased viscosity.

2.5.2.2   Effect of Structure on Ion Exchange and Transport

Inter- and intraaggregate porosity affects exchange site accessiblity as well as adsorption/desorption. At constant moisture content, convective ion transport (e.g., Ca and Mg) is much smaller in single aggregates than in bulk soil, especially homogenized material (Horn et al., 1989a). In well-structured soils, acidic cations are more concentrated in the soil solution from bulk soil than single aggregates where basic cations tend to accumulate (Taubner, 1993). Ion diffusion from single aggregates is lower compared to bulk soil at a given time because (1) the bulk densities of single aggregates are larger than those of bulk soil; (2) the number of accessible reaction sites depends on the ratio of outer surface to sample mass which increases with decreasing aggregate diameter (spherical aggregate); (3) flow length increases with aggregate size; and (4) the average pore size and continuity are much smaller due to a higher clay content at the outer surface of a single aggregate.

Soil solution from undisturbed soil always contains more acidic cations than homogenized material with the difference becoming more pronounced as macroporosity increases. As soil becomes less structured and/or saturated, the differences become smaller. However, even in less aggregated soils under unsaturated conditions, the values would never be the same as the equilibrium soil solution. (Hantschel et al., 1988; Kaupenjohann, 1991; Hartmann et al., 1998). This chemical disequilibrium was also verified by Hantschel and Pfirrmann (1990) under *in situ* conditions. They showed that the base status of the equilibrium soil solution from undisturbed forest soils was significantly higher than that from aggregated soil samples that essentially released Al^{3+} and H^+ during percolation. Taubner (1993) showed that the soil solution from single aggregates contains less H^+ and more basic cations than the corresponding solution from interaggregate pores of the bulk soil. These differences are more pronounced as soils become more structured and heavier textured. Exchange sites show the same heterogeneity between aggregates and the bulk soil. In general, Al^{3+} and H^+ are enriched on the outer surfaces of aggregates while, in the center the base saturation is higher. Hartmann et al. (1998) developed a model to predict the ion release under unsaturated soil conditions. The release of pore water pressure is dependent on changes in base saturation when plotted as a function of the dessication and/or flux times. Starting with a completely saturated undisturbed soil, cation release and/or accessibility of exchange sites decrease initially. Only after exceeding the crossover suction value, ion release increases due to greater hydraulic conductivity of the aggregates at that pore water pressure.

The extent to which ion diffusion is affected by the accessibility of particle surfaces inside the aggregates, and concentration gradient between soil solution and the bulk soil itself has been described by Horn and Taubner (1989). In their experiments to determine cumulative K release rates from single aggregates (polyhedral texture: loamy clay) of structured bulk and homogenized soil (< 2 mm) under saturated conditions, the release rates per unit mass were highest for homogenized material. In aggregates, however, K release from internal sites was retarded. The larger the aggregates, the smaller were the release rates. Bhadoria et al. (1991a, 1991b) furthermore quantified the impedance factor for Cl diffusion in soils as affected by bulk density and water content. An increase in bulk density from 1.38 to 1.76 Mg m^{-3} at constant gravimetric moisture content of 7% decreased the impedance value by a factor of 3, while at water contents greater than 10%, it increased linearly with increasing bulk density.

Buchter et al. (1992), Ghodrati and Jury (1990) and Gisi et al. (1996), defined the consequences of such chemical disequilibrium for ecosystem modeling approaches. Augustin (1992) also considered the microbiological differences at various positions inside the aggregate in terms of complexation and exchange processes.

2.6 Modeling Dynamic Coupled Processes

Hydraulic, thermal or gaseous transport processes in unsaturated structured soils must be mutually linked. The mathematical equations for water, gas and heat transport in soils and the effects of tillage on changes in structure are available (Jury et al., 1991). However, a schematic picture of a coupled hydraulic-elasto-plastic soil model (Gräsle et al. 1996) will be illustrated (Fig. 2.10) as one possibility for coupling processes in order to demonstrate their complexity on potential use as an environmental impact model (O'Sullivan and Simota, 1995).

2.6.1 Components of a Model to Predict Coupled Processes

The main constituents of such a model for coupled processes can be defined as follows:

2.6.1.1 Load/Deformation/Failure
This process describes the load/deformation/failure response of soil to a change in stress, load, strain, and/or displacement, with all other factors held constant. It includes (1) nonlinear elastic behavior of the material as a function of stress and its history and soil water suction (Richards, 1978); (2) changes in initial stress or strain as a result of swell/shrink behavior caused by changes in soil water suction, solute content or temperature (Richards, 1986); (3) stress/strain path dependency or hysteretic behavior in load deformation response (Richards, 1979); and (4) shear and tensile failure with dilatance or compression, and with strain softening.

2.6.1.2 Coupling Processes for a Change of Stress or Strain
During any kind of soil deformation or shearing, the changes in hydraulic properties and functions must be considered, as pointed out in the effective stress equation by Terzaghi. In addition, changes in soil stress due to swelling or shrinkage will also cause changes in soil water potential. This effect of stress on water potential is sometimes referred to as the stress potential or the stress component of field measurements of soil water suction (Richards, 1986). Load deformation analysis enables changes in soil water potential to be calculated for inclusion in subsequent water flow analyses. The displacements can also be used to calculate the new geometry and material velocities for the analysis of water flow in soils undergoing finite strain. Such analysis can be also extended to three-dimensional flow problems, which are of interest in the plow pan and in well-structured horizons below.

2.6.2   Validation of Coupled Processes by a Finite Element Model (FEM)

In order to demonstrate the capabilities of a modeling approach, Stress State Transducer (SST) data and the mechanical parameters for 3 soil horizons from a traffic experiment on a conservation tillage plot (shallow chiselling up to 8 cm, Luvisol derived from glacial till) were used to compare the measured stress and strain with those predicted by the FEM technique (Richards et al., 1997). If only the mechanical parameters of the 3 soil horizons at given rut depth and applied stress are considered, no good agreement could be obtained between the measured and calculated data with increasing depth

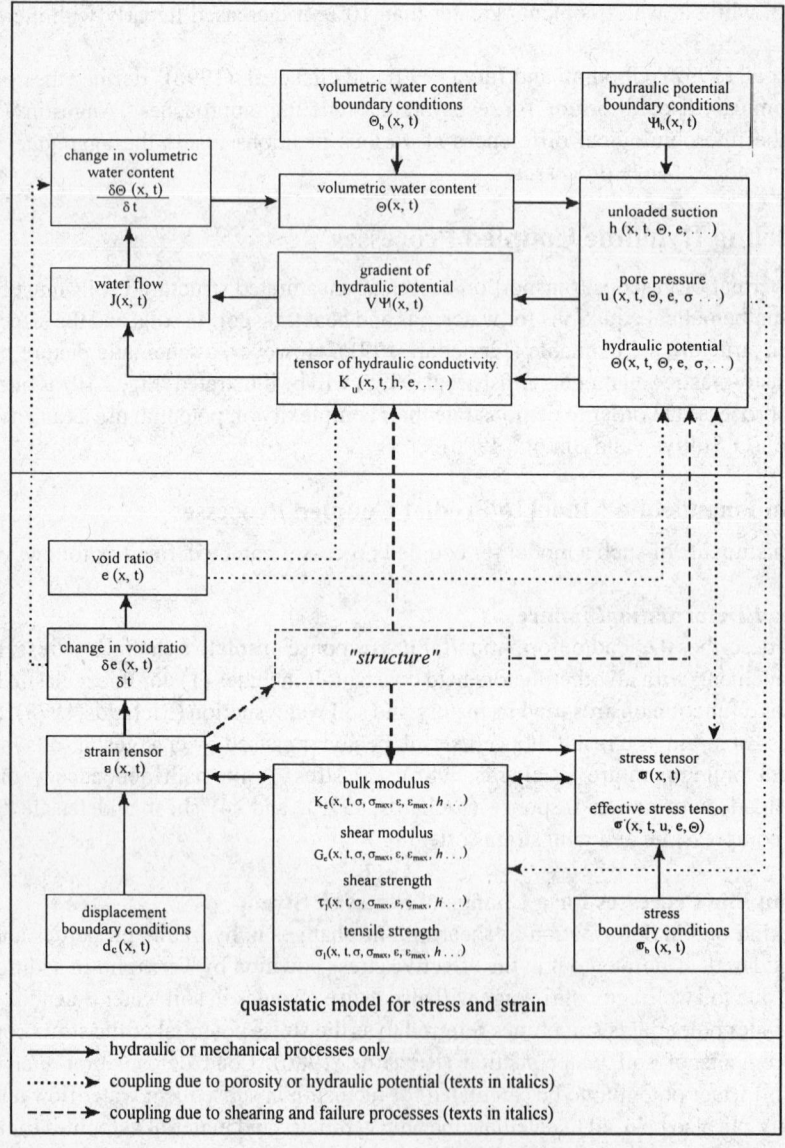

Fig. 2.10   Schematic representation of a coupled hydraulic elastoplastic model to define and to quantify coupled processes in soils

(Fig. 2.11). However, if mechanical data are included in the calculation for a depth of 20 to 40 cm where previous plowing (8 years ago) had been carried out and where the shear strength and bulk modulus data were different from those of the adjacent soil horizons, then a reasonably good agreement between the measurements and model predictions was obtained (Fig. 2.11). Corresponding measurements and modeling of coupled ion and water fluxes through unsaturated structured soils can be used to quantify aggregation effects on ion diffusion, dispersion and mass flow and to predict possible changes in pore systems and functions induced by stresses. They can be also used to estimate time-dependent effects on nutrient, heavy metal or pesticide transport through structured soils and to include the effect of drying or wetting on these processes.

2.7   Conclusions

(1) Mechanical soil properties can be determined by capacity or intensity methods, which result in relative or in actual strength values. However, the development of soil structure presents a problem because of the required homogenization during soil preparation.

(2) The determination of soil strength by intensity parameters requires the measurement of volumetric stress and strain.

(3) Structure development in arable and forest soils always results in increased soil strength. With increased aggregation, strength increases and the characteristics of the water retention curve become more pronounced.

(4) In structured soils, hydraulic properties of inter- and intraaggregate pores are different with the differences in hydraulic conductivity, plant available water and air capacity becoming more pronounced as structure development increases. In addition, the root penetrability is reduced.

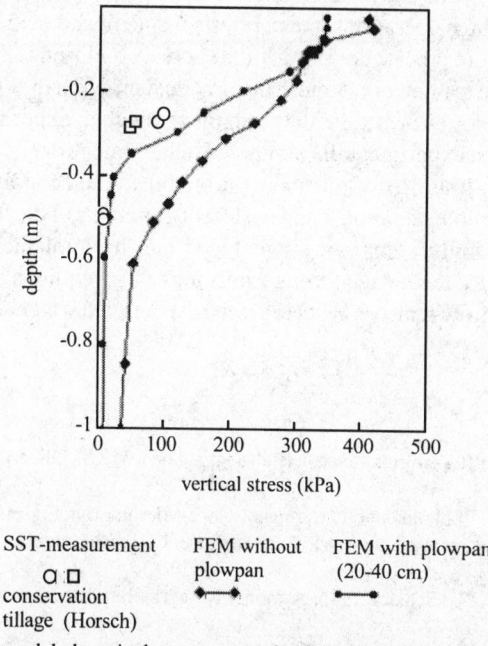

Fig. 2.11   Determined and modeled vertical stress versus depth in a Luvisol derived from glacial till under conservation tillage (type Horsch). The properties of a plowpan layer were included which had been created until 1989 and which had a thickness of 20 cm (from 20–40 cm) [Data from Kühner, 1997].

(5) The increase in strength with decreasing pore water pressure depends on the pattern of the water retention curve for single aggregates and bulk soil.

(6) With increasing degrees of aggregation, pore water pressure at the same water content in the bulk soil becomes more negative. Single aggregates are always stronger than the bulk soil or the homogenized material. This is borne out by smaller increases in pore water pressure during loading and by higher aggregate bulk density. The shear strength parameters (angle of internal friction and cohesion) are always higher for single aggregates than bulk soil for a given applied stress. As soon as the applied stress exceeds internal soil strength, the aggregate or bulk aggregated soil will become homogenized. Thus, the pattern of the Mohr Coulomb failure line resembles that of the homogenized material after exceeding this stress value.

(7) Soil strength is strongly affected by spatial differences in pore water pressure. Hydraulic properties of single aggregates and their arrangement in the bulk soil are therefore most important if soil strength and the corresponding stress distribution are to be predicted. Stress attenuation and stress distribution (intensity and spatial direction) due to loading under running wheels can vary to a great extent depending on soil and machinery parameters.

(8) Structure development always involves reduced accessibility to exchange sites for soil solutions, retarded exchange processes and stronger chemical disequilibrium. Only at the lowest entropy (at an increased accessibility), structured soils become macroscopically homogeneous. With reduced pore water potential (reduced flux rate), the exchange process becomes more complete. At higher saturations, anaerobic soil volumes in aggregates become more extensive for longer periods. Thus, soil physicochemical properties vary intensely throughout the year.

(9) Depending on the objective, either empirical or mechanistic models can be used with varying degrees of certainty and reliability. The analysis of unsaturated structured soils is a very complex problem, with many multidisciplinary processes operating in an interactive fashion. Long-term soil stability involves materials which are unsaturated, nonlinear hysteretic and the soil composition changes with time. During tillage operations, many physical and mechanical properties are altered or at least affected requiring a more precise definition of the actual conditions and the expected changes.

(10) Within this conceptual framework, almost any practical problem in soils and soft rocks, can be modeled using established theories or, where they are not applicable, experimentally based material responses. Each process can be experimentally simulated in laboratory or field tests. This can be done quickly and cheaply by simple tests carried out under the predicted field conditions of stress, moisture, density, etc. Alternatively, complex nonlinear and hysteretic functions of the load/deformation/failure process with or without the failure component included can be implemented as a linear elastic problem. The prediction of simple seepage through soils may only require a simple linear water flow process. Where contaminant flow is also considered, then the coupled water and solute flow processes must be considered.

2.8   References

Augustin, S. 1992. Mikrobielle Stofftransformationen in Bodenaggregaten, 152 S., Dissertation, Universität Göttingen, Germany.

Babel, U., H. Vogel, M. Krebs, G. Leithold, and C. Hermann. 1995. Micromorphological investigation on soil aggregates. p. 11-30. In K.H. Hartge and R. Stewart (ed.) Soil structure: Its development and function. Lewis Pubishers, Boca Raton, FL.

Bailey, A.C., C.E. Johnson, and R.L. Schafer 1986. A model for agricultural soil compaction. J. Agric. Eng. Res. 33:257–262.

Bailey, C., R.L. Raper, C.E. Johnson, and E.C. Burt 1992. An integrated approach to soil compaction modelling. p. 213–226. In Proc. Int. Agric. Eng. Conf. Uppsala, Sweden.

Baumgartl, T. 1991. Spannungsverteilung in unterschiedlich texturierten Böden und ihre Bedeutung für die Bodenstabilität. Diss. Christian-Albrechts Universität, Kiel, Germany.

Becher, H.H. 1992. Die Bedeutung der Festigkeitsverteilung in Einzelaggregaten fhr den Wasser- und Stofftransport im Boden. Z. Pflanzenernähr. Bodenk. 155:361–366.

Beven, K.J., and P. Germann. 1982. Macropores and water flow in soils. Water Resour. Res. 18:1311–1325.

Bhadoria, P.B.S., J. Kaselowsky, N. Claassen, and A. Jungk 1991a. Phosphate diffusion coefficients in soil as affected by bulk density and water content. Z. Pflanzenernähr. Bodenk. 154:53–57.

Bhadoria, P.B.S., J. Kaselowsky, N. Claassen, and A. Jungk 1991b. Impedance factor for chloride diffusion in soil as affected by bulk density and water content. Z. Pflanzenernähr. Bodenk. 154:69–72.

Blackwell, J., R. Horn, N. Jayawardane, R. White, and P. Blackwell. 1985. Vertical stress distribution under tractor wheeling in a partially deep loosened Typic Paleustalf. Soil Till. Res. 13:1-12.

Blunden, B.G., A. McLachlan, and J.M. Kirby 1992. A recording system for measuring *in situ* soil stresses due to traffic. Soil Till. Res. 25:35–42.

Blunden, B.G., R.A.S. McBride, H. Daniel, and P.S. Blackwell. 1994. Compaction of an earthy sand by rubber tracked and tyred vehicles. Aust. J. Soil Res. 32:1095–1108.

Bölling, W.M. 1971. Zusammendrükung und Scherfestigkeit von Böden. Springer Verlag, Berlin, Germany.

Bolling, L. 1986. How to predict soil compaction from agricultural tires. J. Terramech. 22:205–223.

Booltink, H.W.G., and J. Bouma 1991. Physical and morphological characterization of bypass flow in a well structured clay soil. Soil Sci. Soc. Am. J. 55:1249–1254.

Boone, F.R., and B.W. Veen. 1994. Mechanisms of crop responses to soil compaction. p. 237–264. *In* D. Soane and C. van Ouwerkerk (ed.) Soil Compaction in crop production. Elsevier Publishers, Amsterdam, Netherlands.

Bradford, J.M., and S. Gupta. 1986. Soil compressibility. p. 479–492. *In* A. Klute. (ed.) Methods of soil analysis. Part I. 2nd Ed. American Society of Agronomy, Madison, WI.

Brusseau, M.L., and P.S.C. Rao. 1990. Modeling solute transport in structured soils. A review. Geoderma 46:169–192.

Buchter, B., G. Richner, R. Schulin, and H. Flühler. 1992. Übersicht über bestehende Stofftransportmodelle zur Analyse von Mobilisierung und Auswaschungsprozessen von Schwermetallen im Boden. Schriftenreihe Umwelt 165, Bundesamt f. Umwelt, Wald und Landschaft, Bern, Switzerland.

Burger, N., M. Lebert, and R. Horn. 1987. Druckausbreitung unter fahrenden Traktoren im natürlich gelagerten Boden. Mitt. Dtsch. Bodenk. Ges. 55:135–141.

Burger, N., M. Lebert, and R. Horn. 1988. Prediction of the compressibility of arable land. Catena Suppl. 11:141–151.

Burke, W., D. Gabriels, and J. Bouma. 1986: Soil structure assessment. Balkema Publisher, Rotterdam, Netherlands.

Canarache, A. 1990. PENETR: A generalized semi-empirical model estimating soil resistance to penetration. Soil Till. Res. 16:51–70.

Croney, D., and D. Coleman. 1961. Pore pressure and suction in soil. p. 31–37. *In* Pore pressure and suction in soils. Butterworths, London, Uk.

Dexter, A.R. 1988. Advances in characterisation of soil structure. Soil Till. Res. 11:199–239.

Dexter, A.R., R. Horn, and W. Kemper. 1988. Two mechanisms of age hardening. J. Soil Sci. 39:163–175.

Drescher, J., R. Horn, and M. de Boodt. 1988. Impact of water and external forces on soil structure. Catena Suppl. 11:175.

DVWK. 1995. Merkblätter 234. Gefügestabilität ackerbaulich genutzter Mineralböden. Teil I: Mechanische Belastbarkeit. Wirtschafts– und Verlagsges. Gas und Wasser mbH, Bonn, Germany.

DVWK. 1997. Merkblätter zur Wasserwirtschaft 235. Gefügestabilität ackerbaulich genutzter Mineralböden. Teil II: Auflastabhängige Veränderung von bodenphysikalischen Kennwerten. Wirtschafts– und Verlagsges. Gas und Wasser mbH, Bonn, Germany.

Edwards, W.M., M.J. Shipitalo, and L.B. Owens. 1993. Gas, water and solute transport in soils containing macropores: a review of methodology. Geoderma 57:31–49.

Ellies, A., and R. Horn. 1996. Stress distribution in Hapludands under different use. Z. Pflanzenernähr. Bodenk. 159:113–120.

Ellies, A., R. Horn, and R. Smith. 1996. Transmision de presiones en el perfil de algunos suelos. Agro. Sur 24:4–12.

Ellies. A., R. Grez, and Y. Ramirez. 1995. Cambios en las propiedades humectantes en suelos sometidos a diferentes manejos. Turrialba 45:42–48.

Emerson, W.W., R.D. Bond, and A.R. Dexter. 1978. Modification of soil structure. John Wiley and Sons, Chichester, UK.

Erbach, D.C., G.R. Kinney, A.P. Wilcox, and A.E. Abo-Abda. 1991. Strain gauge to measure soil compaction. Trans. ASAE. 34:123–128.

Fredlund, D.G., and H. Rahardjo. 1993. Soil mechanics for unsaturated soils. John Wiley and Sons, Chichester, UK.

Ghodrati, M., and W.A. Jury. 1990. A field study using dyes to characterize preferential flow of water. Soil Sci. Soc. Am. J. 54:1558–1563.

Gill, W.R., and G.E. van den Berg. 1967. Soil dynamics in tillage and traction. USDA Agric. Man. 316.

Gisi, U., R. Schenker, R. Schulin, X.F. Stadelmann, and H. Sticher. 1996. Bodenökologie. Thieme, Stuttgart, Germany.

Gliemeroth, G. 1953. Untersuchungen über Verfestigungs- und Verlagerungsvorgänge im Ackerboden unter Rad- und Raupenfahrzeugen. Z. Acker. Pflanzenbau 96:219–234.

Glinski, J., and W. Stepniewski. 1985. Soil aeration and its role for plants. CRC Press, Boca Raton, FL.

Goss, M., and F.B. Reid. 1979. Influence of perennial ryegrass roots on aggregate stability. Agric. Res. Council Letcombe Lab. Ann. Rep. 1978:24–25.

Gräsle, W., and B. Nissen. 1996. Bestimmung der Vorbelastung bei verhinderter und zugelassener Seitendehnung. Teil I: Theoretische Grundlagen und Auswertungsverfahren. Mitt. Dtsch. Bodenkde. Ges. 80:327–330.

Gräsle, W., B.G. Richards, T. Baumgartl, and R. Horn. 1996. Interaction between soil mechanical properties of structured soils and hydraulic processes: Theoretical fundamentals of a model. p. 719–725. In E.E. Alonso and P. Delage (ed.) Unsaturated soils. Vol. II, Balkema Publishers, Paris, France.

Gudehus, G. 1981. Bodenmechanik. Enke Verlag, Stuttgart, Germany.

Gunzelmann, M. 1990. Die Quantifizierung und Simulation des Wasserhaushalts von Einzelaggregaten und strukturierten Gesamtböden unter besonderer Berücksichtigung der Wasserspannungs-/Wasserleitfähigkeitsbeziehung von Einzelaggregaten. Ph.D. Thesis, Bayreuther Bodenk. Ber., Bayreuth, Germany.

Gunzelmann, M., U. Hell, and R. Horn. 1987. Die Bestimmung der Wasserspannungs-/Wasserleitfähigkeits-Beziehung von Bodenaggregaten. Z. Pflanzenernähr. Bodenk. 150:400–402.

Hadas, A. 1987. Dependence of true surface energy of soils on air entry pore size and chemical constituents. Soil Sci. Soc. Am. J. 51:187–191.

Hakansson, I. 1994. Subsoil compaction by high axle load traffic. Soil Till. Res. 29:105–306.

Hakansson, I., W.B. Voorhees, P. Elonen, G.S.V. Raghavan, B. Lowery, A.L.M. van Wyik, K. Rasmussen, and H. Riley. 1987. Effect of high axle load traffic on subsoil compaction and crop yield in humid regions with annual freezing. Soil Till. Res. 10:259–268.

Hantschel, R., and T. Pfirrmann. 1990. Bodenkundliche Untersuchungen am Forschungsschwerpunkt Wank-Bedeutung für die Waldschäden in den Kalkalpen. Forstl. Centralbl.153:25–38.

Hantschel, R., M. Kaupenjohann, J. Gradl, R. Horn, and W. Zech. 1988. Ecologically important differences between equilibrium and percolation soil extracts. Geoderma 43:213–227.

Harris, H.D., and D.M. Bakker.. 1994. A soil stress transducer for measuring in situ soil stresses. Soil Till. Res. 29:35–48.

Hartge, K.H., and R. Horn. 1977. Spannungen und Spannungsverteilungen als Entstehungsbedingungen von Aggregaten. Mitteilgn Deutsch. Bodenkdl. Gesell. 25:23–33.

Hartge, K.H., and R. Horn. 1984. Untersuchungen zur Gültigkeit des Hooke'schen Gesetzes bei der Setzung von Böden bei wiederholter Belastung. Z. Acker- Pflanzenbau 153:200–207.

Hartge, K.H., and R. Horn. 1991. Einführung in die Bodenphysik. 2nd Ed. Enke Verlag, Stuttgart.

Hartge, K.H., and R. Horn. 1992. Bodenphysikalisches Praktikum. 3rd.Ed. Enke Verlag, Stuttgart.

Hartge, K.H., and I. Rathe. 1983. Schrumpf- und Scherrisse-Labormessungen. Geoderma 31:325–336.

Hartmann, A., W.Graesle, and R. Horn. 1998. Cation exchange processes in structured soils at various hydraulic properties. Soil Till. Res. 47:196–205.

Haynes, R.J., and R.S. Swift. 1990. Stability of soil aggregates in relation to organic constituents and soil water content. J. Soil Sci. 41:73–83.

Hempfling, A., H. Schulten, and R. Horn. 1990. Relevance of humus composition for the physical/mechanical stability of agricultural soils: A study by direct pyrolysis-mass spectrometry. J. Anal. Appl. Pyrol. 17:275–281.

Hillel, D. 1980. Fundamentals of soil physics. Academic Press, London, UK.

Horn, R. 1981. Die Bedeutung der Aggregierung von Böden für die mechanische Belastbarkeit in dem für Tritt relevanten Auflastbereich und deren Auswirkungen auf physikalische Bodenkenngrößen. Schriftenreihe FB 14 TU Berlin, Germany.

Horn, R. 1985. Die Bedeutung der Trittverdichtung durch Tiere auf physikalische Eigenschaften alpiner Böden. Z. Kulturtech. Flurber. 26:42–51.

Horn, R. 1986. Auswirkung unterschiedlicher Bodenbearbeitung auf die mechanische Belastbarkeit von Ackerböden. Z. Pflanzenernähr. Bodenk. 149:9–18.

Horn, R. 1987. The role of structure for nutrient sorptivity of soils. Z. Pflanzenernähr. Bodenk. 150:13–16.

Horn, R. 1988. Compressibility of arable land. Catena Suppl. 11:53–71.

Horn, R. 1989. Strength of structured soils due to loading: A review of the processes on macro- and microscale; European aspects. p. 9–22. *In* W.E. Larson, G.R. Blake, R.R. Allmaras, W.B. Voorhees and S. Gupta (ed.) Mechanics and related processes in structured agricultural soils. Kluwer Academic Publishers, Dordrecht, Netherlands.

Horn, R. 1990. Aggregate characterization as compared to soil bulk properties. Soil Till. Res. 17:265–289.

Horn, R. 1995. Stress distribution and recompaction in tilled and segmently disturbed subsoils under trafficking. p. 187–210. *In* Jaywardane, N.S., and B.A. Stewart (ed.) Subsoil management techniques. Lewis Publishers, Boca Raton, FL.

Horn, R. 1998. The effect of static and dynamic loading on stress distribution, soil deformation and its consequences for soil erosion. p. 233–256. *In* H.P. Blume et al. (ed.) Towards sustainable land use. Catena Verlag, Cremlingen, Germany.

Horn, R., and H. Taubner 1989. Effect of aggregation on potassium flux in a structured soil. Z. Pflanzenernähr. Bodenk. 152:99–104.

Horn, R., N. Burger, M. Lebert, and G. Badewitz. 1987. Druckfortpflanzung in Böden unter fahrenden Traktoren. Z. Kulturtech. Flurber. 28:94–102.

Horn, R., H. Taubner, and R. Hantschel. 1989a. Effect of structure on water transport, proton buffering and nutrient release. Ecol. Stud. 77:323–340.

Horn, R., M. Lebert, and N. Burger. 1989b. Vorhersage der mechanischen Belastbarkeit von Böden als Pflanzenstandort auf der Grundlage von Labor- und *in situ*-Messungen. Abschlußbericht Bayr. SMLFU, Munich, Germany.

Horn, R., T. Baumgartl, S. Kühner, M. Lebert, and R. Kayser. 1991. Zur Bedeutung des Aggregierungsgrades für die Spannungsverteilung in strukturierten Böden. Z. Pflanzenernähr. Bodenk. 154:21–26.

Horn, R., C. Johnson, H. Semmel, R. Schafer, and M. Lebert. 1992. Stress measurements in undisturbed unsaturated soils with a stress state transducer (SST)-theory and first results. J. Plant Nutr. Soil Sci. 155:269–274.

Horn, R., H. Taubner, M. Wuttke, and T. Baumgartl. 1994. Soil physical properties related to soil structure. Soil Till. Res. 35:23–36.

Horn, R., W. Stepniewski, T. Wlodarczyk, G. Walensik, and E.F.M. Eckhardt. 1994. Denitrification rate and microbial distribution within homogeneous soil aggregates. Int. Agrophys. 8:65–74.

Horn, R., T. Baumgartl, R. Kayser, and S. Baasch. 1995. Effect of aggregate strength on changes in strength and stress distribution in structured bulk soils p. 31–52. *In* Hartge, K.H. and R. Stewart (ed.) Soil structure: Its development and function. CRC Press, Boca Raton, FL.

Horn, R., W. Gräsle, and S. Kühner. 1996. Einige theoretische Überlegungen zur Spannungs- und Deformationsmessung in Böden und ihre meßtechnische Realisierung. Z. Pflanzenernähr. Bodenk. 159:137–142.

Horn, R., H. Kretschmer, T. Baumgartl, K. Bohne, A. Neupert, and A.R. Dexter 1998: Soil mechanical properties of a partly reloosened (slit plough system) and a conventionally tilled overconsolidated gleyic luvisol derived from glacial till. Int. Agrophys. 12:143-154.

Jardine, P.M., G.V. Wilson, and R.J. Luxmoore. 1990. Unsaturated solute transport through a forest soil during rain storm events. Geoderma 46:103–118.

Johnson, C.E., and A.C. Bailey. 1990. A shearing strain model for cylindrical stress states. Proc. ASAE Pap. 90–1085.

Junkersfeld, J., and R. Horn. 1997. Über die räumliche und zeitliche Variabilität scheinbar fixer Wasserhaushaltsgrößen am Beispiel von Bodenaggregaten. Z. Pflanzenernähr. Bodenk. 160:179–186.

Jury, W.A., W.R. Gardner, and W.H. Gardner. 1991. Soil physics. 5th Ed. John Wiley and Sons, New York, NY.

Kaupenjohann, M. 1991. Chemischer Bodenzustand und Nährelementversorgung immissionsbelasteter Fichtenbestände in NO-Bayern. Ph.D. Thesis. Universitat Bayreuth, Germany.

Kay, B.D. 1990. Rates of change of soil structure under different cropping systems. Adv. Soil Sci. 12:1–41.

Kezdi, A. 1969. Handbuch der Bodenmechanik. VEB Verlag Berlin, Germany.

Kirby, J.M., B.G. Blunden, and C.R. Trein. 1997. Simulating soil deformation using a critical state model. II. Soil compaction. Europ. J. Soil Sci. 48:59–70.

Kirby, M.J., and M.F. O'Sullivan. 1997. Critical state soil mechanics analysis of the constant cell volume triaxial test. Europ. J. Soil Sci. 48:71–79.

Koolen, A.J. 1994. Mechanics of soil compaction. p. 23–44. *In* Soane, B.D. and C. van Ouwerkerk (ed.) Soil compaction in crop production. Elsevier Publishers, Amsterdam, Netherlands.

Koolen, A.J., and H. Kuipers. 1985. Agricultural soil mechanics. Springer Verlag, Berlin, Germany.

Kretschmer, H. 1997. Körnung und Konsistenz. p. 60. *In* H.P. Blume, P. Felix-Henningsen, W. Fischer, H.G. Frede, R. Horn, and K. Stahr (ed.) Handbuch der Bodenkunde. Ecomed Verlag, Landsberg, Germany.

Kutilek, U., and D. Nielsen. 1994. Soil hysrology. Catena Verlag, Cremlingen, Germany.

Kühner, S. 1997. Simultane Messung von Spannungen und Bodenbewegungen bei statischen und dynamischen Belastungen zur Abschätzung der dadurch induzierten Bodenbeanspruchung. Ph.D. Thesis, Christian-Albrechts University, Kiel, Germany.

Kühner, S., T. Baumgartl, W. Gräsle, T. Way, R. Raper, and R. Horn. 1994.Three dimensional stress and strain distribution in a loamy sand due to wheeling with different slip. Proc. 13th. Int. Soil Till. Res. Org. Conf. p. 591–597.

Larson, W.E., G.R. Blake, R.R. Allmaras, W.B. Voorhees, and S. Gupta. 1989. Mechanics and related processes in structured agricultural soils. Kluwer Academic Publishers, Dordrecht, Netherlands.

Lipiec, J., and C. Simota. 1994. Role of soil and climate factors in influencing crop responses in central and eastern Europe. p. 365–390 *In* Soane, B.D. and C. van Ouwerkerk (ed.) Soil compaction in crop production. Elsevier Publishers, Amsterdam, Netherlands.

Mc Bride, R.A., and P.J. Joosse. 1996: Overconsolidation in agricultural soils: II. Pedotransfer functions for estimating preconsolidation stress. Soil Sci. Soc. Am. J. 60:373–380.

McKenzie, B. 1989. Earthworms and their tunnels in relation to soil physical properties. Ph.D. Thesis, University of Adelaide, Adelaide, Australia.

McKibben, E.G., and R.L. Green. 1936. Transport wheels for agriucultural machinery. VII: Relative effects of steel wheels and pneumatic tires on agricultural soils. Agric. Eng. 21:183–185.

Mitchell, J.K.1976: Fundamentals of soil behavior. John Wiley and Sons, New York, NY.

Morras, H., and G. Piccolo. 1998. Biological recuperation of degraded ultisols in the province of Misiones. 1345–1361. *In* H.P. Blume et al. (ed.) Towards sustainable land use. Catena Verlag, Cremlingen, Germany.

Nichols, T.A., A.C. Bailey, C.E. Johnson, and D. Grisso. 1987. A stress state transducer for soil. Trans. ASAE 30:1237–1241.

Nissen, B. 1998. Gefügestabilität ackerbaulich genutzter Mineralböden-auflastabhängige Veränderung von bodenphysikalischen Kennwerten. Ph.D. Thesis, Christian Albrechts University, Kiel, Germany.

Nissen, B., and W. Gräsle. 1996. Bestimmung der Vorbelastung bei verhinderter und zugelassener Seitendehnung. Teil II: Experimentelle Ergebnisse. Mitt. Dtsche. Bodenkde. Ges. 80:331–334.

O´Sullivan, M.F., and C. Simota. 1995. Modelling the environmental impacts of soil compaction: a review. Soil Till. Res. 35:69–84.

Okhitin, A.A., J. Lipiec, S.Tarkiewiecz, and A.V. Sudakov. 1991. Deformation of silty loam soil under a tractor tyre. Soil Till. Res. 19:187–1995.

Oldeman, L.R. 1992. Global extent of soil degradation. Proc. Symp. Soil Resil. Sustain. Land Use. Budapest, Hungary.

Olson, H.J. 1994. Calculation of subsoil compaction. Soil Till. Res. 29:105–111.

Peattie, K.R., and R.W. Sparrow. 1954. The fundamental action of earth pressure cells. J. Mech. Phys. Solids 2:141–155.

Plagge, R. 1991. Bestimmung der ungesättigten hydraulischen Leitfähigkeit im Boden. Bodenökologie und Bodengenese, Heft 3. Dissertation, TU, Berlin, Germany.

Pytka, J., R. Horn, S. Kühner, H. Semmel, and D. Blazejczak. 1995. Soil stress state determination under static load. Int. Agrophys. 9:219–227.

Reich, J., H. Unger, H. Streitenberger, C. Mäusezahl, C. Nussbaum, and P. Steinert. 1985.Verfahren und Vorrichtung zur Verbesserung verdichteter Unterböden. Agrartechnik Berlin 41:57–62.

Richards, B.G. 1978. Application of an experimentally based nonlinear constituative model to soils in the laboratory and field tests. Aust. Geomech. J. G8:20–30.

Richards, B.G. 1979. The method of analysis of the effects of volume change in unsaturated expansive clays on engineering structures. Aust. Geomech. J. G9:27–41.

Richards, B.G. 1986. The role of lateral stresses on soil water relations in swelling soils. Aust. J. Soil Res. 24:457–467.

Richards, B.G. 1992. Modelling interactive load-deformation and flow processes in soils, including unsaturated and swelling soils. p. 18–37. Proc. 6th. Aust-NZ Conf. Geomech.

Richards, B.G., T. Baumgartl, R. Horn, W. Gräsle. 1997. Modelling the effects of repeated wheel loads on soil profiles. Int. Agrophys. 11:71–87.

Semmel, H., R. Horn, U. Hell, A.R. Dexter, and E.D. Schulze. 1990. The dynamics of soil aggregate formation and the effect on soil physical properties. Soil Tech. 3:113–129.

Skidmore, E.L., and D.H. Powers. 1982. Dry soil aggregate stability: Energy based index. Soil Sci. Soc. Am. J. 46:1274–1279.

Soane, B.D., and C. van Ouwerkerk. 1994. Soil compaction in crop production. Elsevier Publishers, Amsterdam, Netherlands.

Stepniewski, W., J. Glinski, and B.C. Ball. 1994. Effects of soil compaction on soil aeration properties. p. 167–190. *In* B.D. Soane and C.van Ouwerkerk (ed.) Soil compaction in crop production. Elsevier Publishers, Amsterdam, Netherlands.

Taubner, H. 1993. Stoffdynamik unterschiedlich strukturierter immissionsbelasteter Böden-Vergleichende Untersuchungen an Gesamtboden und Aggregaten. Dissertation, Christian-Albrechts University, Kiel, Germany.

Tippkötter, R. 1988. Aspekte der Aggregierung. Habilitationsschrift. University of Hannover, Hannover, Germany.

Tolchel'nikov, Y.S., E.M. Samoylova, A.M. Grebennikov, E.A. Kondrashkin, and A.V. Mazur. 1991. Influence of fissuring on water regime of southern chernozems of Western Siberia. Sov. Soil. Sci. 11:24–32.

Trein, C.R. 1995. The mechanisms of soil compaction under wheels. Ph.D. Thesis. Silsoe College, Crainfield University, Silsoe, UK.

van den Akker, J.J.H., and H.J. Stuiver. 1989. A sensitive method to measure and visualize deformation and compaction of the subsoil with a photographed point grid. Soil Till. Res. 14:209–217.

van den Akker, J.J.H., W.B.M. Arts, A.J. Koolen, and H.J. Stuiver. 1994. Comparison of stresses, compactions and increase of penetration resistances caused by a low ground pressure tyre and a normal tyre. Soil Till. Res. 29:125–134.

Veenhof, D.W., and R.E.A. McBride. 1996. Overconsolidation in agricultural soils. I. Compression and consolidation behaviour of remoulded and structured soils. Soil Sci. Soc. Am. J. 60:362–373.

Vetterlein, D., and H. Marschner. 1993. Use of a microtensiometer technique to study hydraulic lift in a sandy soil planted with pearl millet. (*Pennisetum americanum* [L.]Leeke). Plant Soil 149:275–282.

Voorhees, W.B., and R.R. Allmaras. 1998. Long-term soil degradation in the United States by compaction from heavy agricultural machinery. p. 981–997. *In* H.P. Blume et al. (ed.) Towards sustainable land use. Catena Verlag, Cremlingen, Germany.

Yong, E., and B. Warkentin. 1966. Introduction to soil behavior. Macmillan, New York, NY.

Youngs, E.G., and P.B. Leeds-Harrison. 1990. Aspects of transport processes in aggregated soils. J. Soil Sci. 41:665–675.

3

Soil Water Content and Water Potential Relationships

Dani Or
Utah State University

Jon M. Wraith
Montana State University

3.1 Introduction

Water in soil occupies pore spaces that arise from the physical arrangement of the particulate solid phase, competitively and often concurrently with the soil gas phase (Section A, Chapters 1, 7, 8). While hidden from casual view, highly substantial volumes of water are commonly stored in soils. For example, one ha of medium textured soil 1 m deep and having field capacity water content of 20% by volume may store sufficient water to fill 4,000 200-L barrels. This reservoir serves as a substantial buffer, thus enabling consistent plant growth in areas having scattered or sporadic precipitation. Other soil organisms, many of which are beneficial, also rely heavily on the water holding characteristics of soils for their existence. On the other hand, soil water is a highly dynamic entity, exhibiting substantial variation in both time and space. This is particularly true near the soil surface, and in the presence of active plant roots. Changes in soil water content and its energy status affect many soil mechanical properties including strength, compactibility, and penetrability, and may cause changes in the bulk density of swelling soils. The liquid phase characteristics affect the soil gaseous phase and the rates of exchange between these phases, as well as other important soil properties such as the hydraulic conductivity which governs the rate of water and soluble chemical flow.

The purpose of this chapter is to introduce basic concepts related to the amount and energy state of water in soil. These concepts are prerequisite to quantify and manage soil water storage, to obtain predictions concerning rates and directions of water flow and solute transport, to utilize soils as building or foundation materials, and for many other purposes. The term soil water is used here to represent the soil liquid phase which is typically a water solution containing dissolved salts, organic substances and gases.

The water status in soils is defined by: (1) the amount of water in the soil, or soil water content (θ), and (2) the force by which water is held in the soil matrix, soil energy content or soil water potential (ψ). These soil water attributes are related to each other through a function known as the soil water characteristic (SWC).

3.2   Soil Water Content

Many agronomic, hydrologic, and geotechnical practices require knowledge of the amount of water contained in a particular soil volume. Some of the most common methods used to characterize and determine soil water content, on a mass or volume basis, in the laboratory or *in situ*, will be described.

3.2.1   Definitions

3.2.1.1   Soil Water Content on Mass Basis (Gravimetric)
Mass or gravimetric soil water content is expressed relative to the mass of oven dry soil according to:

$$\theta_m = \frac{\text{mass of water}}{\text{mass of dry soil}} = \frac{(\text{mass wet soil}) - (\text{mass oven dry soil})}{\text{mass oven dry soil}} \qquad [3.1]$$

and has units of kg kg^{-1} or other consistent mass units

3.2.1.2   Soil Water on Volume Basis
It is often desirable to express water content on a volume basis. The volumetric water content (θ_v) is defined as the volume of water per bulk volume of soil:

$$\theta_v = \frac{\text{volume of water}}{\text{bulk volume of soil}} = \frac{(\text{mass of water} / \text{density of water})}{\text{sample volume}} \qquad [3.2]$$

It also represents the depth ratio of soil water (i.e., the depth of water per unit depth of soil). The conversion between gravimetric and volumetric water contents requires knowledge of the soil dry bulk density (ρ_b) which is the ratio of oven dry soil mass to its original volume, and the density of water. The conversion formula is given by:

$$\theta_v = \theta_m \frac{\rho_b}{\rho_w} \qquad [3.3]$$

where ρ_w is the density of water (1000 kg m^{-3} at 20 C). Alternatively, a soil sample of known original volume may be processed as in the gravimetric method, and θ_v determined as: [(mass water/water density)]/sample volume.

3.2.1.3   Water Content on Relative Saturation Basis
An additional means of characterizing the soil water content is in terms of the degree of saturation:

$$\Theta = \frac{(\text{volume of water filled pore space})}{\text{total volume of soil pore space}} = \frac{\theta_v}{\theta_{vs}} \qquad [3.4]$$

where θ_{vs} is volumetric soil water content under completely water-saturated conditions. This index ranges from zero in completely dry soil to unity in a saturated soil. The degree of saturation is also commonly termed effective saturation or relative water content.

3.2.1.4 Soil Water Storage

It is often convenient to express the quantity of soil water in a specific soil depth increment in terms of soil water storage or equivalent depth of soil water (units of length). Equivalent depth of soil water, D_e (m), is calculated as $D_e = \theta_v * D$, where D is the soil depth increment (m) having volume water content θ_v. This quantity is useful to relate aboveground water dimensions of rainfall, irrigation, or evaporation (L) to belowground dimensions (θ_v, m^3 water m^{-3} soil). For example, one may wish to quantify changes in soil water content arising from addition of water by rainfall or loss of water by evaporation (Section I, Chapter 8). For suitable accuracy it is often necessary to sum the equivalent depth relationship over discrete soil depth layers having distinct water contents:

$$D_e = \sum_{i=1}^{n} \theta_{vi} D_i$$

[3.5]

where i denotes depth increments.

3.2.2 Relationships to Soil Solid and Gaseous Phases

Because the soil pore spaces are shared between soil water and soil air, a few basic interrelationships among the amount of pore space and the soil water and air contents will be addressed. These concepts are treated in greater detail in Section A, Chapter 1.

The total volume of spaces in a soil not occupied by the solid phase is termed the porosity (ϕ). The porosity indicates pore volume in the soil relative to total soil volume and its value generally lies in the range 0.3 to 0.6. Coarse textured soils tend to have less total pore space than fine-textured soils, because of the nature of intraparticle packing arising from different solid particle sizes (Section A, Chapter 1). Typical porosity values range from about 0.3 in coarse (sandy) soils to about 0.6 in fine (clayey) soils. The porosity may be quantified as $\phi = 1 - \rho_b / \rho_p$, where the ratio of soil bulk density (ρ_b) to density of the soil solids (ρ_p) characterizes the relative proportion of soil solids; thus, soil voids (pores) occupy the remaining volume. On average, volumetric, areal, and lineal porosities are assumed to be equal.

A related index is the air-filled porosity (ϕ_a) which is a complementary quantity to θ_v, and describes the relative volume of total soil volume occupied by soil air. Because ϕ may be occupied by soil water and/or air, ϕ_a may be conveniently characterized as: $\phi_a = \phi - \theta_v$. The expected ranges for ϕ_a in soils of varying texture are thus similar to those for θ_v.

3.2.3 Measurement of Soil Water Content

3.2.3.1 Thermogravimetry

This is a direct and destructive method whereby a soil sample is obtained by augering or coring into the soil; its volume need not be known. The sample is weighed at its initial wetness and then dried to remove interparticle absorbed water, but not structural water trapped within clay lattices known as crystallization water. The conventional protocol is to oven dry samples at 105°C until the soil mass becomes stable; this usually requires 24 to 48 h or more, depending on the sample size, wetness, and soil characteristics (texture, aggregation, etc.). The difference between the wet and dry weights is the mass of water held in the original soil sample (Section 3.2.1). Despite the somewhat arbitrary specification of a standard oven temperature (105°C) and the variable drying period (depending on specific conditions), the gravimetric method is considered the standard against which many indirect techniques are calibrated. The method is not without bias and/or error, however, and Gardner (1986) discussed some sources of these including the potential for water loss between sampling time and initial weighing, precision of the three weights involved (wet, dry, and tare), and the unknown amount of soil texture-dependent residual water associated with the clay fraction.

The primary advantages of gravimetry are the direct and relatively inexpensive processing of samples. The shortcomings of this method are its labor- and time-intensive nature, the time delay required for drying (although this may be shortened by use of a microwave rather than conventional oven, the methodology has not yet been standardized), and the fact that the method is destructive thereby prohibiting repetitive measurements within the same soil volume.

3.2.3.2 Neutron Scattering

This is a nondestructive but indirect method commonly used for repetitive field measurement of volumetric water content. It is based on the propensity of H nuclei to slow (thermalize) high energy fast neutrons. A typical neutron moisture meter consists of: (1) a probe containing a radioactive source that emits high energy (2-4 MeV) fast (1600 km s^{-1}) neutrons, as well as a detector of slow neutrons; (2) a scaler to electronically monitor the flux of slow neutrons; and optionally (3) a datalogger to facilitate storage and retrieval of data (Fig. 3.1). The radioactive source commonly contains a mixture of ^{241}Am and Be at 10 to 50 mCi. The ^{241}Am emits β particles which strike the Be and cause emission of fast neutrons.

When the probe is lowered into an access tube, fast neutrons are emitted radially into the soil where they collide with various atomic nuclei. Collisions with most nuclei are virtually elastic, causing only

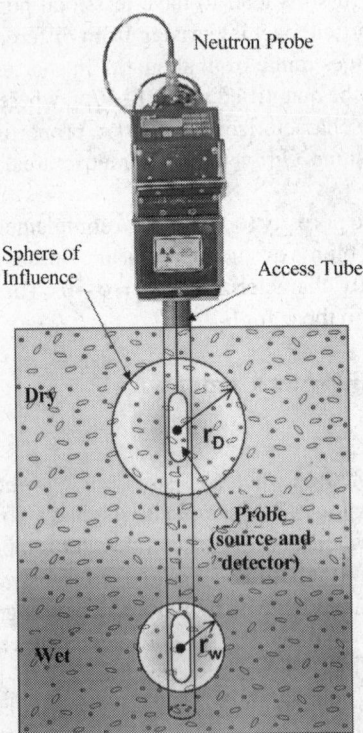

Fig. 3.1   An illustration of a neutron probe device for measuring soil water content. r_D and r_W represent different radii of measurement in dry and wet soils.

minor loss of kinetic energy by the fast neutrons. Collisions with H nuclei, which have similar mass to neutrons, cause a significant loss of kinetic energy and slow down the fast neutrons [consider a marble (neutron) colliding with a similarly sized ball bearing (H nucleus) versus a stationary bowling ball (larger atomic nucleus)]. When, as a result of repeated collisions, the speed of fast neutrons diminishes to those at ambient temperature (about 2.7 km s^{-1}), with corresponding energies of about 0.03 eV, they are called thermalized or slow neutrons. Thermalized neutrons rapidly form a cloud of nearly constant density near the probe, where the flux of the slow neutrons is measured by the detector. The average loss of the neutrons' kinetic energy, thus the relative number of slow neutrons, is therefore proportional to the amount of H nuclei in the surrounding soil. The primary source of H in soil is water; other sources of H in a given soil are assumed to be constant and are accounted for during calibration. Although several non H substances including C, Cd, Bo, Cl, and Li which may be present in trace amounts in some soils may also thermalize fast neutrons, these may generally also be effectively compensated through soil specific calibration.

Calibration of the neutron probe is thus required to account for background H sources and other local effects (soil bulk density, trace neutron attenuators), and is conveniently achieved by paired measurements of soil water content and neutron probe counts. The calibration curve (Fig. 3.2) is usually linear and relates θ_v to slow neutron counts or count ratio (CR):

$$\theta_v = a + b(CR)$$ [3.6]

where CR is the ratio of slow neutron counts at a specific location in the soil to a standard count obtained with the probe in its shield. For many soils, the calibration relation is approximately the same. Use of the count ratio rather than raw slow neutron counts compensates for the slow decay of the radioactive source over time.

The sphere of influence about the radiation source varies between about 15 cm (wet soil) to perhaps 70 cm (very dry soil), depending on how far fast neutrons must travel in order to collide with a requisite number of H nuclei. An approximate equation for the radius of influence (r, in cm) as a function of soil wetness is (IAEA, 1970):

$$r = 15\left(\theta_v\right)^{-1/3}$$ [3.7]

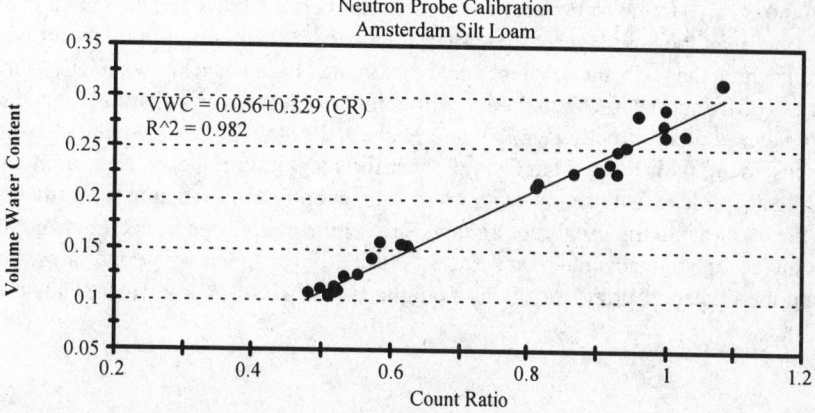

Fig. 3.2 A neutron moisture meter calibration relationship

Thus, the neutron scattering method is unsuitable for measurement near the soil surface because a portion of the neutrons may escape the soil. Advantages of this method include the ability to repeatedly measure θ_v at the same locations, averaging of the measured water content over a substantial soil volume, moderate equipment cost (generally about \$4,000 to \$6,000), and ability to measure soil water content at multiple depths and locations using the same equipment. Limitations or disadvantages include the radiation hazard and attendant licensing requirements, relatively poor (and uncertain) spatial resolution, unsuitability for near-surface measurements, and the soil-specific calibration requirement.

3.2.3.3 Electric and Dielectric Methods

Measurement methods based on changes in soil electric properties due to changes in soil water content have been used for decades (Smith-Rose, 1933; Babb, 1951), mostly in the area of exploration geophysics. Background information concerning such applications may be found in a number of sources, including Olhoeft (1985), who addressed low frequency electrical properties of porous media, Selig and Mansukhani (1975), who discussed early attempts to use resistance and capacitance techniques for soil water content measurements, and Hoekstra and Delaney (1974) concerning soil dielectric properties at very high frequencies. Presently, the most common electric approaches for soil water content measurement may be grouped according to: (1) electrical resistance techniques (Spaans and Baker, 1992), (2) capacitance methods (Dean et al., 1987), (3) time domain reflectometry (TDR) methods (Topp et al., 1980; Dalton et al., 1984), and (4) combinations of frequency domain, resonance, and capacitive techniques (von Hippel, 1954; Hilhorst and Dirksen, 1994).

3.2.3.4 Electric Resistance Methods

Changes in the electrical resistivity of soils with changes in water content (and with soluble ionic constituents) have been used to develop simple and inexpensive sensors to infer soil water status. These sensors usually consist of concentric or flat electrodes embedded in a porous matrix and connected to lead wires for measurement of electrical resistance within the sensor's porous matrix. The commonly used term gypsum block arises from early models which were in fact made of gypsum (Bouyoucos and Mick, 1940), and from the practice of saturating the matrix of many sensors made from alternative materials with gypsum to buffer local soil ionic effects. The sensor is embedded in the soil and allowed to equilibrate with the soil solution. The matric potential of water in the sensor is determined from the measured electrical resistance through previously determined calibration of the sensor itself (electrical resistance versus matric potential). Under equilibrium conditions, the sensor matric potential is equal to the soil water matric potential (below); however, the sensor water content may be different than the soil. Hence, these measurements are often used to infer soil water matric potential from which the soil water content may be estimated based on a known relationship between these quantities (Gardner, 1986). With proper calibration for a particular soil, the sensor could be used to infer soil water content directly (Kutilek and Nielsen, 1994). The main advantages of electrical resistance sensors are their low cost and simple measurement requirements. Measurements may be obtained using a simple resistance meter, or more conveniently acquired automatically using a data logger. On the other hand, the usual requirement for specific calibration of each sensor and for each soil to obtain acceptable accuracy, and lack of sensitivity under wet conditions, render this measurement method appropriate mostly as a qualitative indicator of soil water status (Spaans and Baker, 1992).

3.2.3.5 Capacitance Methods

The electrical capacitance of two electrodes inserted in the soil is dependent upon the soil dielectric constant. The bulk soil dielectric constant is dominated by the dielectric constant of soil water ($\varepsilon_w \approx 80$) because of its large magnitude relative to that of soil solids ($\varepsilon_s \approx 5$) and soil air ($\varepsilon_a = 1$). The basic relationship between the soil dielectric constant and the electrical capacitance between two parallel plates of area A and spacing d is:

$$C = \frac{A\varepsilon * \varepsilon_0}{d}$$

[3.8]

where ε^* is the relative complex dielectric constant (or permittivity) of the soil which contains both real (ε') and imaginary (ε'') components, with $\varepsilon^* = \varepsilon' - i\varepsilon''$, $i = \sqrt{-1}$, and ε_0 the dielectric permittivity of free space = 8.854×10^8 F m^{-1}. Note that the relative dielectric permittivity (ε^*) is the ratio of the complex dielectric permittivity to the permittivity of free space = $\varepsilon^*/\varepsilon_0$. In most applications, only the real part of the dielectric is considered and the subscript 'b' is commonly used to denote the relative complex dielectric constant of the bulk soil. Capacitance devices for field water content measurement are often based on annular conductor design rather than parallel plates (Kutilek and Nielsen, 1994) to facilitate depth measurements through boreholes (Dean et al., 1987).

Commercial capacitance soil water gauges often use a resonant LC circuit relating changes in resonant frequency of the circuit to changes in the soil capacitance (Dean et al., 1987; Evett and Steiner, 1995). Some gauges use a more direct capacitance bridge method to determine the unknown soil capacitance. Among the common capacitance-based measurement systems capable of soil water content profiling are the Sentry 200-AP (Troxler, NC, USA) evaluated by Evett and Steiner (1995) which uses access tubes, and a permanently installed modular bank of sensors (EnviroScan, Sentek, Australia) evaluated by Paltineanu and Starr (1997).

Advantages of capacitance methods include their lack of radiation hazard, and lower expense than transmission line approaches such as TDR (below). They share the neutron probe's variable and uncertain measurement volume, and annular air gaps around sensors that utilize access tubes can cause substantial measurement errors. Hence, buried probe designs seem to perform more reliably at present than those inserted into soil access tubes. Finally, capacitance methods share similar issues of relating the measured dielectric constant to soil water content as do other dielectric-based approaches.

3.2.3.6 Time Domain Reflectometry

Progress in devices for accurate measurement of reflected electromagnetic signals traveling in transmission cables and waveguides led to the introduction of several measurement methods for soil dielectric properties using waveguides embedded in the soil, such as time domain reflectometry (TDR) (Fig. 3.3).

TDR measures the apparent dielectric constant of the soil surrounding a waveguide, at microwave frequencies of MHz to GHz. The propagation velocity (v) of an electromagnetic wave along a transmission line (waveguide) of length L embedded in the soil is determined from the time response of the system to a pulse generated by the TDR cable tester. The propagation velocity (v = 2L/t) is a function of the soil bulk dielectric constant according to

$$\varepsilon_b = \left(\frac{c}{v}\right)^2 = \left(\frac{ct}{2L}\right)^2$$

[3.9]

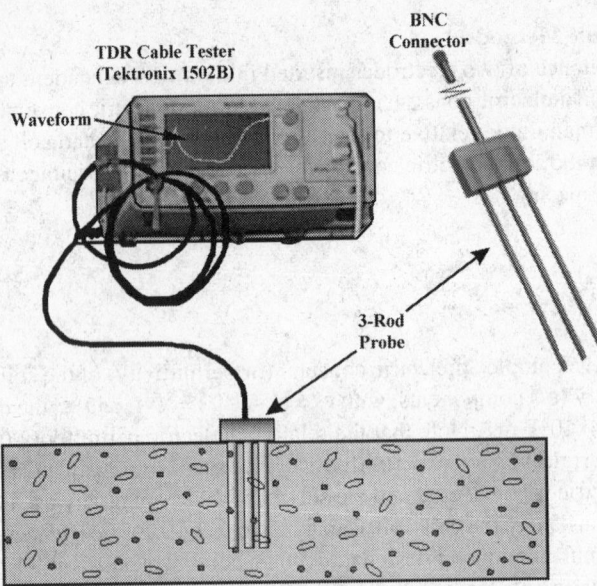

Fig. 3.3   TDR cable tester with 3-rod probe embedded vertically in surface soil layer

where c is the velocity of electromagnetic waves in vacuum (3×10^8 m s^{-1}), and t is the travel time for the pulse to traverse the length of the embedded waveguide in both directions (down and back). The soil bulk dielectric constant is governed by the dielectric constant of liquid water ($\varepsilon_w \approx 80$), as the dielectric constants of other soil constituents are much smaller, e.g., soil minerals ($\varepsilon_s \approx 3$ to 5), frozen water (ice $\varepsilon_i \approx 4$) and air ($\varepsilon_a = 1$). This large disparity of the dielectric constants makes the method relatively insensitive to soil composition and texture (other than organic matter and some clays), and thus, a good method for liquid soil water measurement. Also, because the dielectric constant of frozen water is much lower than for liquid, the method may be used in combination with a neutron probe or other technique which senses total soil water content, to separately determine the volumetric liquid and frozen water contents in frozen or partially frozen soils (Baker and Allmaras, 1990).

Two basic approaches have been used to establish the relationships between ε_b and θ_v. The first approach is empirical, whereby mathematical expressions are simply fitted to observed data without using any particular physical model. For example, Topp et al. (1980) found good agreement using a third order polynomial between ε_b and θ_v for multiple soils. The second approach uses a model of the dielectric constants and the volume fractions of each of the soil components to derive a predictive relationship between the composite (bulk) dielectric constant and soil water (i.e., a specific component). Such a physically based approach, called a dielectric mixing model, was adopted by Dobson et al. (1985), Roth et al. (1990), and others.

TDR calibration establishes the relationship between ε_b and θ_v. For this discussion, calibration is conducted in a fairly uniform soil without abrupt changes in soil water content along the waveguide. The empirical relationship for mineral soils as proposed by Topp et al. (1980):

$$\theta_v = -5.3 \times 10^{-2} + 2.92 \times 10^{-2} \varepsilon_b - 5.5 \times 10^{-4} \varepsilon_b^2 + 4.3 \times 10^{-6} \varepsilon_b^3 \qquad [3.10]$$

provides adequate description for many soils and for the water content range $\theta < 0.5$ (which covers the entire range of interest in most mineral soils), with an estimation error of about 0.013 for θ_v. However, Equation [3.10] fails to adequately describe the ε_b–θ_v relationship for water contents exceeding 0.5, and for organic rich soils, mainly because Topp's calibration was based on experimental results for mineral soils and concentrated in the range of $\theta_v < 0.5$.

Roth et al. (1990) proposed a physically based calibration model which considers dielectric mixing of the constituents and their geometric arrangement. According to this mixing model, ε_b of a three-phase system may be expressed as:

$$\varepsilon_b = \left[\theta_v \varepsilon_w^\beta + (1-\varphi)\varepsilon_s^\beta + (\varphi-\theta_v)\varepsilon_a^\beta\right]^{\frac{1}{\beta}}$$ [3.11]

where φ is soil porosity, $-1 < \beta < 1$ summarizes the geometry of the medium in relation to the axial direction of the wave guide ($\beta = 1$ for an electric field parallel to soil layering, $\beta = -1$ for a perpendicular electrical field, and $\beta = 0.5$ for an isotropic two-phase mixed medium), $1-\varphi$, θ_v and $\varphi-\theta_v$ are the volume fractions and ε_s, ε_w and ε_a are the dielectric constants of the solid, aqueous and gaseous phases, respectively. Note that these components sum to unity. Rearranging Equation [3.11] and solving for θ_v yields:

$$\theta_v = \frac{\varepsilon_b^\beta - (1-\varphi)\varepsilon_s^\beta - \varphi\varepsilon_a^\beta}{\varepsilon_w^\beta - \varepsilon_a^\beta}$$ [3.12]

which determines the relationship between ε_b as measured by TDR and θ_v. Many have used $\beta = 0.5$ which is shown by Roth et al. (1990) to produce a calibration curve very similar to the third order polynomial in Equation [3.10] for the water content range $0 < \theta_v < 0.5$. Introducing into Equation [3.12] common values for each component such as $\beta = 0.5$, $\varepsilon_w = 81$, $\varepsilon_s = 4$, and $\varepsilon_a = 1$ yields the following simplified form:

$$\theta_v = \frac{\sqrt{\varepsilon_b} - (2-\varphi)}{8}$$ [3.13]

A comparison between Equation [3.10] and a calibration curve based on Equation [3.12] with $\varphi = 0.5$ is depicted in Fig. 3.4.

The main advantages of TDR over other methods for repetitive soil water content measurement are: (1) superior accuracy to within 1 or 2% of θ_v, (2) minimal calibration requirements (in many cases soil specific calibration is not needed), (3) no radiation hazards associated with neutron probe or gamma attenuation techniques, (4) excellent spatial and temporal resolution, and (5) simple to obtain continuous soil water measurements through automation and multiplexing. Limitations of the TDR method include relatively high equipment expense, limited applicability under highly saline conditions due to signal attenuation, and the soil-specific calibration required for soils with high clay or organic matter contents. Severe attenuation of TDR waveforms in the presence of high salinity and/ or some clays having high surface area and surface charges may interfere with or even preclude water content measurements.

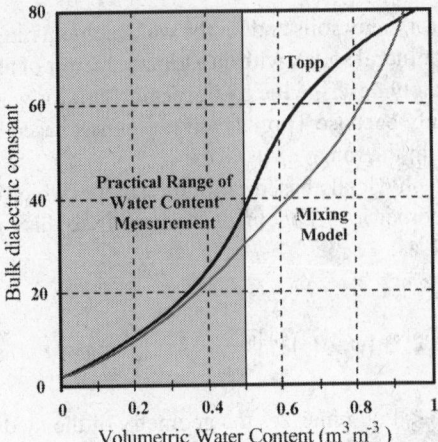

Fig. 3.4 Relationships between bulk soil dielectric constant and θ_v expressed by two commonly used TDR calibration approaches

3.2.3.7 Frequency Domain and Other Methods

Several new sensors and measurement methods are based on combinations of capacitive, reflective and frequency-shift principles, all of which are governed by the soil dielectric properties. This trend appears highly promising for the development of accurate and cost effective sensors for soil water content measurement and will likely dominate future developments in this area. An example of such a stand-alone sensor (Fig. 3.5) is the water content reflectometer (Campbell Scientific Inc., Logan, UT) which provides an indirect measurement of soil volumetric water content based on changes in soil dielectric permittivity. High speed electronic components are configured in an oscillator circuit which is connected to parallel rods acting as a waveguide. The rods are inserted in the monitored soil volume (typical rod length is 0.3 m). As soil water content changes, the resultant dielectric property causes a

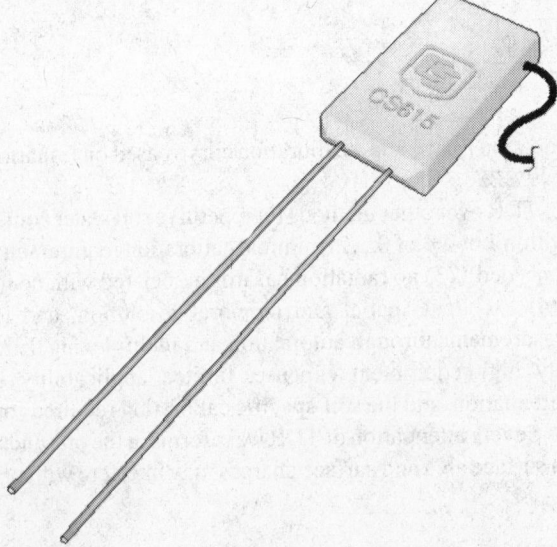

Fig. 3.5 A water content reflectometer sensor. [Source: Campbell Scientific, Inc., Logan, UT with permission]

shift in the oscillation frequency of the circuit. A calibration relationship is established between the output frequency of the circuit and θ_v. The time required for the actual measurement is less than 20 milliseconds. The method is sensitive to soil electrical conductivity and an adjustment must be made to the calibration when soil solution conductivity exceeds about 2 dS m^{-1}. Another commercially available stand-alone sensor based on frequency domain (FD) measurements is described in detail by Hilhorst and Dirksen (1994).

Other methods for soil water measurement include gamma ray attenuation techniques using dual probe apparatus for bulk density and water content, X-ray computed tomography, methods based on fiber optics, nuclear magnetic resonance (NMR), and an array of geophysical methods (including ground penetrating radar and electrical resistivity sounding). Partial dependence of porous media thermal properties on water content can form the basis for indirect inference of water content or matric potential based on measured thermal responses. A few of these are discussed in Section 3.3.4.4. Additional information on these and other methods may be found in Klute (1986), Baker (1990), Campbell and Mulla (1990), and Carter (1993).

3.2.4 Applications of Soil Water Content Information: The Water Balance

A primary use of soil water content information is for evaluation of the hydrologic water balance described by:

$$P + I = ET + D + R - \Delta W \tag{3.14}$$

where P is precipitation, I is irrigation, ET is evapotranspiration (evaporative water loss from the soil and from plants), D is drainage or deep percolation, R is surface runoff, and ΔW is change in water storage within the profile (soil water depletion). W is defined as the equivalent depth of water (De) stored in the soil profile under consideration, and $\Delta W = (W_{initial} - W_{final})$ (Fig. 3.6). These parameters are all associated with a given specific time interval. The convention used here is that inputs to the soil profile are taken as positive, and outputs negative. The concept is based on conservation of mass (water), and is similar to the familiar exercise of balancing inputs and outlays from a checking account,

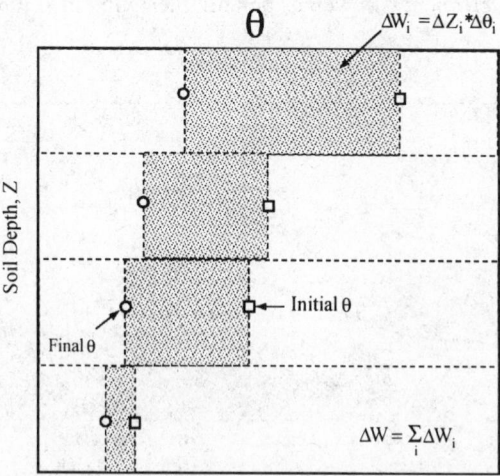

Fig. 3.6 Schematic of soil water depletion calculations for a foil profile divided into four depth increments. Total change in water storage is the sum of depletion in each layer.

for example. Under typical conditions, ΔW is fairly significant over the short term (weeks to months), but generally evens out to about zero over one to several years.

3.2.4.1 Field Capacity, Wilting Point and Plant-Available Soil Water

Observations of water content changes in the soil profile following wetting by irrigation or rainfall show that the rate of change decreases in time. In some cases, water content attains a nearly constant value within 1 to 2 days after wetting, following soil water redistribution in response to internal drainage. Field capacity is defined as the water content at which internal drainage becomes essentially negligible. The attainment of a near constant water content at field capacity (θ_{vFC}) is not always assured. It is dependent on: (1) the depth of wetting and the antecedent (initial) water content of the soil profile (for a soil which is moist at the onset of wetting and for deep wetting, the rate of redistribution is slower and the apparent value of θ_{vFC} is higher), and (2) the presence of impeding layers or a watertable which affect the rate and extent of water redistribution.

Another often misunderstood soil water content-related index is the permanent wilting point, which is defined as the water content at which plants can no longer extract soil water at a rate sufficient to meet physiological demands imposed by loss of water to the atmosphere, and thus irreversibly wilt and die. This water content (θ_{vWP}) is primarily dependent on the soil's ability to transmit water, but also to some degree on the plant's ability to withstand or mitigate drought. Though commonly taken as θ_v at -1.5 MPa (-15 bars) matric potential, there is substantial variation among plant species in their abilities (and strategies) to resist soil drought, with some plants surviving to well below this standard wilting point index. The permanent wilting point should not be confused with the phenomenon of transient wilting, which is commonly observed during the afternoon when evaporative demand is greatest. In this case plants are able to rehydrate to some extent at night.

A primary practical use of field capacity and wilting point concepts is the determination of a plant available soil water range (PASW) (Fig. 3.7). Soil water storage available for plant use is generally calculated as being between field capacity and wilting point ($\theta_{vFC} - \theta_{vWP}$), as water contents higher than θ_{vFC}, while usually plant-available, are generally not sustained for long times except under specific circumstances. Plant-available soil water storage is an important factor in the determination of irrigation amounts for a cropped field or other soil-plant system. For practical purposes, irrigation amounts in excess of field capacity are lost to deep percolation, and thus, should be avoided in the interests of water resource efficiency as well as potential leaching of soluble chemicals (Section A, Chapter 6).

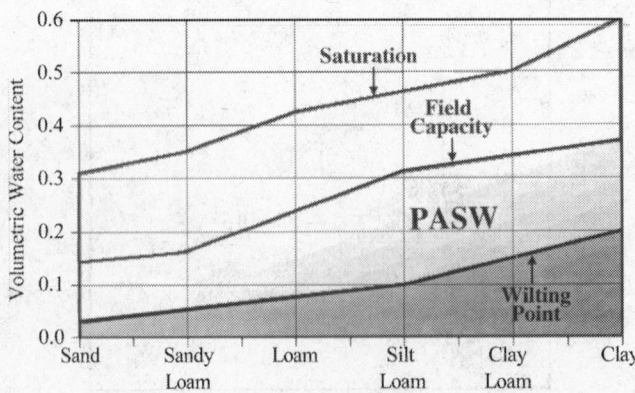

Fig. 3.7 Schematic of estimated plant-available soil water content (PASW) for a range of soil textural classes

A useful rule of thumb is to estimate θ_{FC} as $\theta_s/2$, and θ_{WP} as $\theta_{FC}/2$; in other words, a soil exhibiting this property will have lost 50% of its saturated water content at field capacity, and another 50% of the remaining water by the wilting point. Richards (1954) was probably the first to make this observation. Banin and Amiel (1970) and Dahiya et al. (1988) have also demonstrated that the assumed ratio of $1:^1/_2:^1/_4$ is a good approximation, based on measured correlations on many soils. Banin and Amiel (1970) found strong relationships between clay content or surface area (which are strongly related) and the water content at these indices.

3.2.4.2 Sources of Additional Information on Soil Water Content Indices

The information on these indices is fragmentary due to their dependency on other conditions in the soil profile (field capacity) and the plant in question (wilting point). The work of Slater and Williams (1965) provides a reasonable framework for estimating field capacity, and more recent information may be found on the Internet. A study by Waters (1980) is a good source of information regarding the wilting point. Dahiya et al. (1988) presented strong correlations among the saturation percentage, field capacity, wilting point, and available water based on data for 438 soils, with correlation coefficients in excess of 0.92. Banin and Amiel (1970) provided similar information concerning relationships among clay content, surface area, and water content at several index points. Petersen et al. (1996) also related soil surface area and other properties to their water holding capacity.

3.3 Soil Water Energy

As previously stated, water status in soils is characterized by both the amount of water present and by its energy state. Soil water is subjected to forces of variable origin and intensity, thereby acquiring different quantities and forms of energy. The two primary forms of energy of interest here are kinetic and potential. Kinetic energy is acquired by virtue of motion and is proportional to velocity squared. However, because the movement of water in soils is relatively slow (usually < 0.1 m h^{-1}), its kinetic energy is negligible. The potential energy which is defined by the position of soil water within a soil body and by internal conditions is largely responsible for determining soil water status under isothermal conditions.

Like all other matter, soil water tends to move from where the potential energy is high to where it is lower, in its pursuit of a state known as equilibrium with its surroundings. The magnitude of the driving force behind such spontaneous motion is a difference in potential energy across a distance between two points of interest. At a macroscopic scale, one can define potential energy relative to a reference state. The standard state for soil water is defined as pure and free water (no solutes and no external forces other than gravity) at a reference pressure, temperature, and elevation, and is arbitrarily given the value of zero (Bolt, 1976).

3.3.1 Total Soil Water Potential and its Components

Soil water is subject to several force fields whose combined effects result in a deviation in potential energy relative to the reference state called the total soil water potential (ψ_T) and is defined as:

> The amount of work that an infinitesimal unit quantity of water at equilibrium is capable of doing when it moves (isothermally and reversibly) to a pool of water at similar standard (reference) state (similar pressure, elevation, temperature and chemical composition).

It should be emphasized, however, that there are alternative definitions of soil water potential using concepts of chemical potential or specific free energy of the chemical species water (which is different than the soil solution termed soil water here). Some of the arguments concerning the proper definitions and their scales of application are presented by Corey and Klute (1985), Iwata et al. (1988), and Nitao and Bear (1996). Recognizing that these fundamental concepts are subject to an ongoing debate, only simple and widely accepted definitions of these quantities which are applicable at macroscopic scales, and which yield an appropriate framework for practical applications, will be presented.

The primary forces acting on soil water held within a rigid soil under isothermal conditions can be conveniently grouped as (Day et al., 1967): (1) matric forces resulting from interactions of the solid phase with the liquid and gaseous phases, (2) osmotic forces owing to differences in chemical composition of the soil solution, and (3) body forces induced by gravitational and other inertial force fields (centrifugal).

The thermodynamic approach whereby potential energy rather than forces are used is particularly useful for equilibrium and flow considerations. Equilibrium would require the vector sum of these different forces acting on a body of water in different directions to be zero; this is an extremely difficult criterion to deal with in soils. On the other hand, potential energy defined as the negative integral of the force over the path taken by an infinitesimal amount of water, when it moves from a reference location to the point under consideration, is a scalar quantity. Consequently, the total potential can be expressed as the algebraic sum of the component potentials corresponding to the different fields acting on soil water:

$$\psi_T = \psi_m + \psi_s + \psi_p + \psi_z \qquad [3.15]$$

ψ_m is the matric potential resulting from the combined effects of capillarity and adsorptive forces within the soil matrix. Dominating mechanisms for these effects include (1) adhesion of water molecules to solid surfaces due to short-range London-van der Waals forces and extension of these effects by cohesion through H bonds formed in the liquid, (2) capillarity caused by liquid-gas and liquid-solid-gas interfaces interacting within the irregular geometry of soil pores, and (3) ion hydration and binding of water in diffuse double layers. There is some disagreement regarding the definition of this component of the total potential. Some consider all contributions other than gravity and solute interactions (at a reference atmospheric pressure). Others use a tensiometer (Section 3.3.4) to measure and provide a practical definition of the matric potential in a soil volume of interest (Hanks, 1992). The value of ψ_m ranges from zero when the soil is saturated to negative numbers when the soil is dry (note that $\psi_m = 0$ mm is $> \psi_m = -1000$ mm). Because it is often more convenient or intuitive to work with positive than negative quantities, the terms matric suction or tension are commonly used. Each of these represents the absolute value of ψ_m.

ψ_s is the solute or osmotic potential determined by the presence of solutes in soil water, which lower its potential energy and its vapor pressure. The effects of ψ_s are important in the presence of: (1) appreciable amounts of solutes, and (2) a selectively permeable membrane or a diffusion barrier which transmits water more readily than salts. The effects of ψ_s are otherwise negligible when only liquid water flow is considered and no diffusion barrier exists. The two most important diffusion barriers in the soil are (1) soil-plant root interfaces (cell membranes are selectively permeable), and (2) soil water-air interfaces; thus, when water evaporates, salts are left behind. In dilute solutions the solute potential (also called the osmotic pressure) is proportional to the concentration and temperature according to:

$$\psi_s = -RTC_s \qquad [3.16]$$

where ψ_s is in kPa, R is the universal gas constant [8.314×10^{-3} kPa m^3 mol^{-1} K^{-1})], T is absolute temperature (K), and C_s is solute concentration (mol m^{-3}). A useful approximation which may be used to estimate ψ_s in kPa from the electrical conductivity of the soil solution at saturation (EC_s) in dS m^{-1} is

$$\psi_s = -36EC_s \qquad [3.17]$$

ψ_p is pressure potential defined as the hydrostatic pressure exerted by unsupported water (i.e., saturating the soil) overlying a point of interest. Using units of energy per unit weight provides a simple and practical definition of ψ_p as the vertical distance from the point of interest to the free water surface (unconfined watertable elevation). The convention used here is that ψ_p is always positive below a watertable, or zero if the point of interest is at or above the watertable. In this sense, nonzero magnitudes of ψ_p and ψ_m are mutually exclusive; either ψ_p is positive and ψ_m is zero (saturated conditions), or ψ_m is negative and ψ_p is zero (unsaturated conditions), or $\psi_p = \psi_m = 0$ at the free watertable elevation. Another definition that is used in some quarters is to combine ψ_m and ψ_p as used here into a single component that adopts negative magnitude under unsaturated conditions and positive magnitude under saturated conditions.

ψ_z is gravitational potential which is determined solely by the elevation of a point relative to some arbitrary reference point, and is equal to the work needed to raise a body against the Earth's gravitational pull from a reference level to its present position. When expressed as energy per unit weight, the gravitational potential is simply the vertical distance from a reference level to the point of interest. The numerical value of ψ_z itself is thus not important (it is defined with respect to an arbitrary reference level); what is important is the difference or gradient in ψ_z between any two points of interest. This value will not change with different reference point locations.

Soil water is at equilibrium when the net force on an infinitesimal body of water equals zero everywhere, or when the total potential is constant in the system. Though the last statement is a logical consequence of the definitions above, it is not strictly true as pointed out by Corey and Klute (1985). They argue that constant total potential is a necessary but not a sufficient condition, and for thermodynamic equilibrium to prevail, three conditions must be met simultaneously: (1) thermal equilibrium or uniform temperature, (2) mechanical equilibrium meaning no net convection producing force, and (3) chemical equilibrium meaning no net diffusional transport or chemical reaction. In most practical applications, however, the macroscopic definition of the total potential and equilibrium conditions based on it is completely adequate (Kutilek and Nielsen, 1994).

The difference in chemical and mechanical potentials between soil water and pure water at the same temperature is known as the soil water potential (ψ_w):

$$\psi_w = \psi_m + \psi_s + \psi_p \qquad [3.18]$$

Note that the gravitational component (ψ_z) is absent in this definition. Soil water potential is thus the result of inherent properties of soil water itself, and of its physical and chemical interactions with its surroundings, whereas the total potential includes the effects of gravity (an external and ubiquitous force field).

Total soil water potential and its components may be expressed in several ways depending on the definition of a unit quantity of water. Potential may be expressed as (1) energy per unit of mass, (2) energy per unit of volume, or (3) energy per unit of weight. A summary of the resulting dimensions, common symbols and units are presented in Table 3.1.

Only μ has actual units of potential; ψ has units of pressure, and h head of water. However, the above terminology (potential energy versus units of potential) is widely used in a generic sense in the soil and plant sciences. The various expressions of soil water energy status are equivalent, with:

$$\mu = \psi / \rho_w = gh \qquad\qquad\qquad\qquad\qquad\qquad\qquad\qquad\qquad [3.19]$$

where ρ_w is density of water (1000 kg m^{-3} at 20 °C) and g is gravitational acceleration (9.81 m s^{-2}).

3.3.2 Interfacial Forces and Capillarity

The matric potential is often the largest component of the total potential in partially saturated soils. To better understand the origins of this important potential which attains nonzero values only under partial saturation when all three phases (liquid, gas and solid) are present, some of the properties of water in relation to porous media which give rise to this phenomenon must be discussed.

3.3.2.1 Surface Tension

At the interface between water and solids or other fluids such as air, water molecules are exposed to different forces than are molecules within the bulk water. For example, water molecules inside the liquid are attracted by equal cohesive forces to form H bonds on all sides, whereas molecules at the air-water interface feel a net attraction into the liquid because the density of water molecules on the air side of the interface is much lower and all H bonds are towards the liquid. The result is a membrane-like water surface having a tendency to contract; thus energy is stored in the form of surface tension (as in a stretched spring). Different liquids vary in their liquid-gas (LG) surface tension (σ_{LG}) expressed as energy per unit area = force per unit length. For example, water at 20 °C: 72.7 mN m^{-1} = mJ m^{-2}; ethyl alcohol: 22 mN m^{-1} (= dynes cm^{-1} = erg cm^{-2}); and mercury: 430 mN m^{-1}.

3.3.2.2 Contact Angle

If liquid is placed in contact with a solid in the presence of a gas (three-phase system), the angle measured from the solid-liquid (SL) interface to the liquid-gas (LG) interface is the contact angle (γ). For a drop resting on a solid surface at equilibrium, the vector sum of the forces acting to spread the drop (outward) is equal to the opposing forces. This relationship is summarized by Young's equation (Adamson, 1990):

$$\sigma_{LG} \cos \gamma + \sigma_{SL} - \sigma_{GS} = 0 \qquad\qquad\qquad\qquad\qquad\qquad [3.20]$$

where σ is the respective interfacial surface tension. The equilibrium contact angle is therefore

$$\gamma = \cos^{-1}\left[\frac{\sigma_{GS} - \sigma_{SL}}{\sigma_{LG}} \right] \qquad\qquad\qquad\qquad\qquad\qquad [3.21]$$

Table 3.1 Units, dimensions, and common symbols for potential energy of soil water

Units	Symbol	Name	Dimensions	SI units	cgs units
Energy/mass	μ	Chemical potential	$L^2 t^{-2}$	J kg^{-1}	erg g^{-1}
Energy/volume	ψ	Soil water potential, suction or tension	$M (Lt^2)^{-1}$	N m^{-2} (Pa)	erg cm^{-1}
Energy/weight	h	Pressure head	L	m	cm

When liquid is attracted to the solid (adhesion) more than to other liquid molecules (cohesion), the angle is small and the solid is said to be wettable by the liquid. Conversely, when the cohesive exceeds the adhesive force, the liquid repels the solid and γ is large (Fig. 3.8). The contact angle of water on clean glass is very small, and is commonly taken as 0°. The contact angle of soil water on soil minerals is also commonly assumed $\approx$ 0°. The different contact angles of water and other liquids with soil solids, soil air, and with each other, are important contributors to the behavior of multiphase organic liquid/water/soil/air mixtures.

3.3.2.3 Curved Surfaces and Capillarity

Surface tension is associated with the phenomenon of capillarity. When the liquid-gas interface is curved rather than planar (flat), the resultant surface tension force normal to the liquid-gas interface creates a pressure difference across the interface. The pressure is greater at the concave side of the interface by an amount that is dependent on the radius of curvature and the surface tension of the fluid. For a hemispherical liquid-gas interface having radius of curvature R, the pressure difference is given by the Young-Laplace equation:

$$\Delta P = 2\sigma / R \tag{3.22}$$

Small Contact Angle
Liquid "wets" the Solid

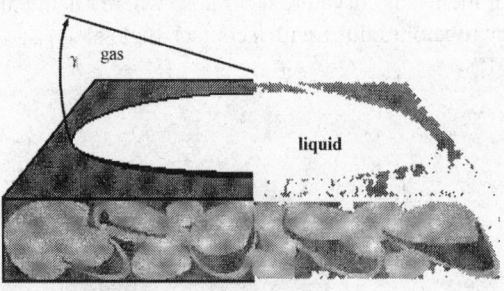

Large Contact Angle
Liquid "repels" the Solid

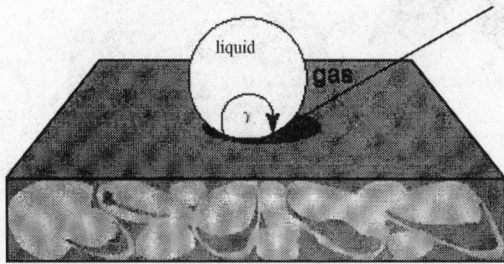

Fig. 3.8 Liquid-solid-gas contact angles

where $\Delta P = P_{liq} - P_{gas}$ when the interface curves into the gas (water droplet in air), or $\Delta P = P_{gas} - P_{liq}$ when the interface curves into the liquid (air bubble in water, water in a small glass tube). In many instances a bubble may not be spherical, or an element of liquid may be confined by irregular solid surfaces resulting in two (or more) different radii of curvature such as water in pendular rings between two spherical solid particles (Fig. 3.9). The Young-Laplace equation for this case is given by:

$$\Delta P = \sigma \left(\frac{1}{R_1} + \frac{1}{R_2} \right)$$ [3.23]

Note that (1) this equation reduces to Equation [3.22] for $R_1 = R_2$; and (2) the sign of R_i is negative for convex interfaces ($R_2 < 0$) and positive for concave interfaces ($R_1 > 0$).

3.3.3   The Capillary Model

3.3.3.1   Capillary Rise
When a small cylindrical glass tube (capillary) is dipped in free water, a meniscus is formed as a result of the contact angle between the water and the walls of the tube and from consideration of minimum surface energy. The smaller the tube the larger the degree of curvature, resulting in a larger pressure difference across the air-water (gas-liquid) interface. The pressure at the water side (P_w) will be lower than atmospheric pressure (P_0). This pressure difference will cause water to rise into the capillary tube until the upward force across the water-air interface is balanced by the weight of water in the tube (Fig. 3.10). Because the radius of meniscus curvature $R = r/\cos\gamma$ where r is the tube radius, the height that water will rise in a capillary tube of radius r with γ contact angle is

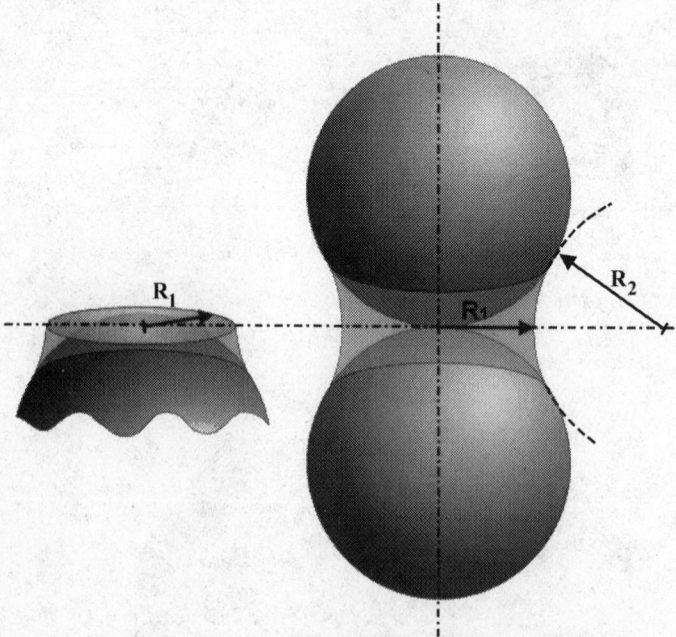

Fig. 3.9   Radii of curvature for pendular ring of water between two spherical solids

$$h = \frac{2\sigma \cos \gamma}{\rho_w gr}$$
[3.24]

Combining all the constants in Equation [3.24] (using typical values at room temperature) yields a simple and useful approximation:

$$h(m) = \frac{14.84}{r(\mu m)}$$
[3.25]

When expressed as potential energy per unit weight, $\psi_m = -h$ and thus may be related to equivalent pore radius through Equation [3.24].

3.3.3.2 Conceptual Models for Water in Soil Pore Space

The complex geometry of soil pore space creates numerous combinations of interfaces, capillaries and wedges around which films of water are formed, resulting in a variety of air-water and solid-water contact angles. Water is thus drawn into and/or held by these interstices in proportion to the resulting capillary forces. In addition, water is adsorbed onto solid surfaces with considerable force at close distances. Due to practical limitations of present measurement methods, no distinction is made between the various mechanisms affecting water in porous matrices (i.e., capillarity and surface adsorption). Common conceptual models for water retention in porous media and matric potential rely on a simplified picture of soil pore space as a bundle of capillaries. The primary conceptual steps made in such models are illustrated in Fig. 3.11. The representation of soil pores as equivalent cylindrical capillaries greatly simplifies modeling and parameterization of soil pore space. The roles of water films at very low saturation levels, and the unique contribution of surface adsorption to the matric potential are beyond the scope of this chapter. Interested readers are referred to reviews by Nitao and Bear (1996), and Parker (1986).

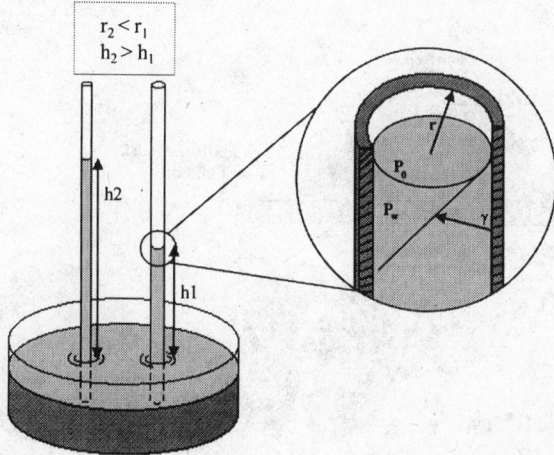

Fig. 3.10 Capillary rise in cylindrical glass tubes

3.3.4   Measurement of Soil Water Potential Components

3.3.4.1   Tensiometer for Measuring Soil Matric Potential

A tensiometer consists of a porous cup, usually made of ceramic having very fine pores, connected to a vacuum gauge through a water-filled tube (Fig. 3.12). The porous cup is placed in intimate contact with the bulk soil at the depth of measurement. When the matric potential of the soil is lower (more negative) than the equivalent pressure inside the tensiometer cup, water moves from the tensiometer along a potential energy gradient to the soil through the saturated porous cup, thereby creating suction sensed by the gauge. Water flow into the soil continues until equilibrium is reached and the suction inside the tensiometer equals the soil matric potential. When the soil is wetted, flow may occur in the reverse direction, i.e., soil water enters the tensiometer until a new equilibrium is attained. The tensiometer equation is

$$\psi_m = \psi_{gauge} + \left(z_{gauge} - z_{cup}\right) \tag{3.26}$$

The vertical distance from the gauge plane (z_{gauge}) to the cup (z_{cup}) must be added to the matric potential measured by the gauge (expressed as a negative quantity) to obtain the matric potential at the depth of the cup, when potentials are expressed per unit of weight. This accounts for the positive head at the depth of the ceramic cup exerted by the overlying tensiometer water column.

Electronic sensors called pressure transducers often replace the mechanical vacuum gauges. The transducers convert mechanical pressure into an electric signal which can be more easily and precisely measured. In practice, pressure transducers can provide more accurate readings than other gauges, and in combination with data logging equipment are able to supply continuous measurements of soil matric potential.

The standard tensiometer range is limited to suction values (absolute value of the matric potential) < 100 kPa (1 bar or 10 m head of water) at sea level, and this value decreases proportionally with elevation. Thus, other means are needed to measure or infer soil matric potential under drier conditions.

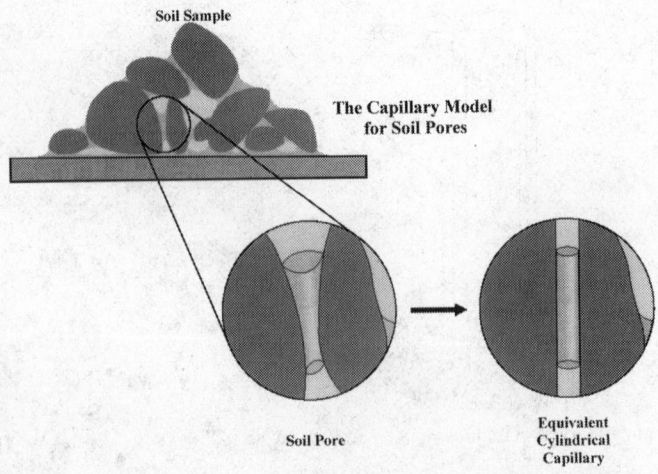

Fig. 3.11   Concept of equivalent cylindrical capillary to represent soil pore spaces

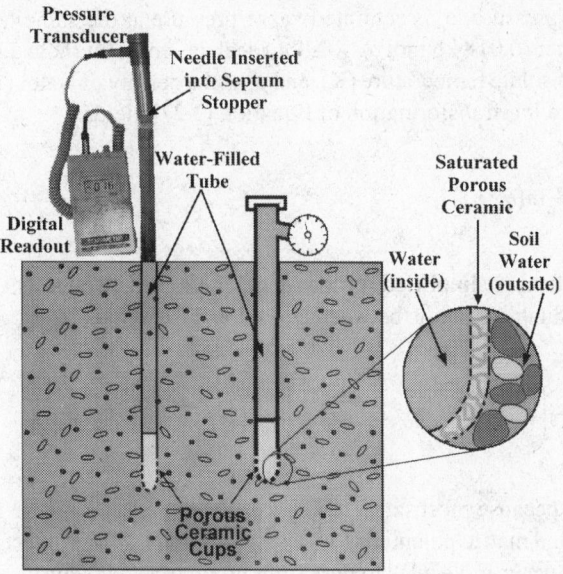

Fig. 3.12 Pressure transducer and vacuum gauge tensionmeters showing porous ceramic cup

Peck and Rabbidge (1969) described an osmotic tensiometer that relied on a confined aqueous solution, rather than pure free water as the reference state. A membrane highly impermeable to the confined solution allowed their device to cover the entire range of 0 to –1.5 MPa, unless soil solutes were excluded from the instrument by a vapor gap. Portable tensiometers that may be extended into boreholes for use at depths to several hundred meters have also been developed (Hubbell and Sisson, 1996). Achieving and maintaining adequate hydraulic contact between the porous cup and the soil at these depths was identified as an important issue, but these sensors may be highly useful in some mining, engineering, deep recharge, and hazardous waste applications.

3.3.4.2 Piezometer for Measuring Hydrostatic Pressure Potential
In a saturated soil such as below a watertable, soil water is under positive hydrostatic pressure. The pressure potential (ψ_p) equals the vertical distance from a point in the soil to the surface of the free watertable (recall that one expresses potential in terms of distance, length, or head when potential energy is expressed per unit of weight). The piezometer is a hollow tube placed in the soil to depths below the watertable. It extends to the soil surface and is open to the atmosphere. The bottom of the piezometer is perforated to allow for soil water under positive hydrostatic pressure to enter the tube. Water enters the tube and rises to a height equal to that of the free watertable. The water level within the piezometer may be determined using a variety of manual or automated measurement techniques.

3.3.4.3 Psychrometry for Measuring Water Potential
Under equilibrium conditions the soil water potential is equal to the potential of water vapor in the surrounding soil air. A psychrometer measures the relative humidity (RH) of the water vapor which is related to the water potential of the vapor (ψ_w) through the Kelvin equation:

$$RH = e / e_o = \exp^{[(M_w \psi_w)/(\rho_w RT)]}$$

[3.27]

where e is water vapor pressure, e_o is saturated vapor pressure at the same temperature, M_w is the molecular weight of water (0.018 kg mol^{-1}), R is the ideal gas constant (8.31 J K^{-1} mol^{-1} or 0.008314 kPa m^3 mol^{-1} K^{-1}), T is absolute temperature (K), and ρ_w is the density of water (1,000 kg m^{-3} at 20 °C). Rearranging and taking a log-transformation of Equation [3.27] yields

$$\psi_w = \frac{RT\rho_w}{M_w}\ln(e/e_o)$$ [3.28]

Equation [3.28] can be further simplified for the range of e/e_o near 1 often encountered in soils (the entire range of plant-available water is between $e/e_o = 1$ and 0.98):

$$\psi_w = \frac{RT\rho_w}{M_w}\left(\frac{e}{e_o}-1\right) \approx 462T\left(\frac{e}{e_o}-1\right)$$ [3.29]

for ψ_w in kPa. Note that because most salts are nonvolatile, the psychrometric measurement of ψ_w is the sum of the osmotic and matric potentials ($\psi_w = \psi_m + \psi_s$). The soil-air interface acts as a diffusion barrier in allowing only water molecules to pass from liquid to vapor state.

A psychrometer measures the difference between dry bulb and wet bulb temperatures to determine the relative humidity. The dry bulb is the temperature of the ambient air (nonevaporating surface), and the wet bulb is the temperature of an evaporating surface (generally lower than the dry bulb temperature). The relative humidity determines the rate of evaporation from the wet bulb junction, and thus the extent of temperature depression below ambient.

A thermocouple psychrometer consists of a fine wire chromel-constantan or other bimetallic thermocouple. A thermocouple is a double junction of two dissimilar metals. When the two junctions are subjected to different temperatures, they generate a voltage difference (Seebeck effect). Conversely, when an electrical current is applied through the junctions, it creates a temperature difference between the junctions by heating one while cooling the other, depending on the current's direction. For typical soil use, one junction of the thermocouple psychrometer is suspended in a thin-walled porous ceramic or stainless screen cup buried in the soil (Fig. 3.13), while another is embedded in an insulated plug to measure the ambient temperature at the same location. In psychrometric mode, the suspended thermocouple is cooled below the dew point by means of an electrical current (Peltier cooling) until pure water condenses on the junction. The cooling current then stops, and as water evaporates, it draws heat from the junction (heat of vaporization), depressing it below the temperature of the surrounding air until it attains a wet bulb temperature. The warmer and drier the surrounding air, the higher the evaporation rate and the greater the wet bulb depression. The difference in temperatures between the dry and wet bulb thermocouples is measured and used to infer the relative humidity (or relative vapor pressure) using the psychrometer equation:

$$\frac{e}{e_o} = 1 - \left[\frac{s+\gamma}{e_o}\right]\Delta T$$ [3.30]

where s is the slope of the saturation water vapor pressure curve ($s = de_o/dT$), γ is the psychrometric constant (about 0.067 kPa K^{-1} at 20 °C), and ΔT is the temperature difference (K). The slope (s), and e_o are functions of temperature only and can be approximated by closed-form expressions (Brutsaert, 1982).

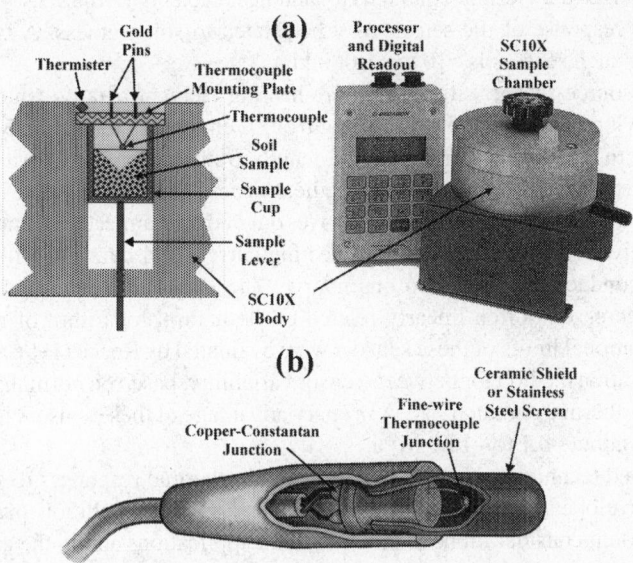

Fig. 3.13 (a) Model SC 10X sample chamber for psychrometric laboratory measurements of soil water potential [Source: Decagon Devices, Inc., Pullman, WA with permission]; and (b) a field psychrometer with porous ceramic shield [Source: Wescor, Inc., Logan UT with permission]

A typical temperature depression measurable by a good psychrometer is on the order of 0.000085 C kPa^{-1}. This means that any errors in measuring wet bulb depression can introduce large errors into psychrometric determinations. Thermal equilibrium is, therefore, a prerequisite to obtaining reliable readings, as any temperature difference between wet and dry sensors resulting from thermal gradients will be (erroneously) incorporated into the relative humidity calculation.

Summarizing, psychrometric measurements of soil water potential are based on equilibrium between liquid soil water and water vapor in the ambient soil atmosphere. The drier the soil, the fewer water molecules escape into the ambient atmosphere, resulting in lower relative humidity (lower vapor pressure). When the osmotic potential is negligible, the soil water potential measured by a psychrometer is nearly equal to the soil matric potential. In principle, soil psychrometers may be buried in the soil and left for long periods, although corrosion is a problem in some environments.

3.3.4.4   Heat Dissipation in Rigid Porous Matrix for Measuring Matric Potential

The rate of heat dissipation in a porous medium is dependent on specific heat capacity, thermal conductivity, and density. The heat capacity and thermal conductivity of a porous matrix are affected by its water content, and hence related to its matric potential. Heat dissipation sensors contain line- or point-source heating elements embedded in a rigid porous matrix with fixed pore space. The measurement is based on applying a heat pulse by passing a constant current through the heating element for a specified time, and analyzing the temperature response measured by a thermocouple fixed at a known distance from the heating source (Phene et al., 1971; Bristow et al., 1993). Sensors are individually or uniformly calibrated in terms of heat dissipation versus sensor wetness (i.e., matric potential). With the heat dissipation sensor buried in the soil, changes in soil matric potential result in a gradient between the soil and the porous matrix that induces a water flux between the two materials until a new equilibrium is established. The water flux changes the water content of the porous matrix which, in turn, changes the thermal conductivity and heat capacity of the sensor. In this manner, the measured thermal response of the sensor may be related to soil wetness. A typical useful matric potential range for such sensors is –10 to –1,000 kPa.

A similar line-source sensor with a fine-wire heating element axially centered in a cylindrical ceramic matrix having a radius of 1.5 cm and length of 3.2 cm is depicted in Fig. 3.14. A thermocouple is located adjacent to the heating element at mid-length. Both the heating wire and the thermocouple are contained in the shaft portion of a hollow needle. Because the thermocouple is located adjacent to the heating element, as the soil dries and water moves out of the ceramic, the magnitude of temperature change during a given period under constant heating current and duration will increase due to the reduced thermal conductivity of the porous matrix. The magnitude of the measured temperature increase and/or decrease is often linearly related to the natural logarithm of matric potential. The accuracy and operational limits of these sensors were evaluated by Reece (1996), who also presented an improved calibration method for between-sensor variability, based on normalizing sensor readings by oven dry sensor thermal conductivity. A primary advantage of these sensors is their wide range of applicability from about –0.1 to –12.0 MPa.

New or improved techniques based on relating sensor thermal responses to soil matric potential continue to be developed. Accuracy, repeatability, and spatial resolution of specific sensors or methods are important considerations to their potential applications and in the analysis of soil water measurements.

3.4   Soil Water Content-Energy Relationships

3.4.1   Soil Water Characteristic

A soil water characteristic (SWC) curve describes the functional relationship between soil water content (θ_m or θ_v) and matric potential under equilibrium conditions. The SWC is an important soil property related to the distribution of pore space (sizes, interconnectedness), which is strongly

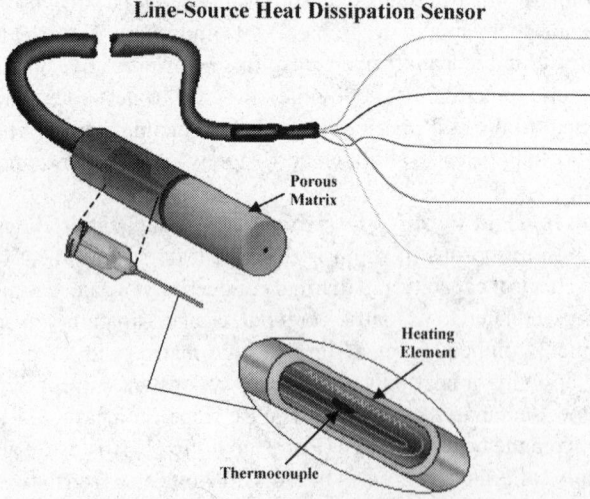

Fig. 3.14   Schematic of CSI 229 heat dissipation sensor [Source: Campbell Scientific, Inc., Logan, UT with permission]

affected by texture and structure, as well as related factors including organic matter content. The SWC is a primary hydraulic property required for modeling water flow, for irrigation management, and for many additional applications related to managing or predicting water behavior in the porous system. A SWC is a highly nonlinear function and is relatively difficult to obtain accurately. Because the matric potential extends over several orders of magnitude for the range of water contents commonly encountered in practical applications, the matric potential is often plotted on a logarithmic scale. Several SWC curves for soils of different textures demonstrating the effects on porosity (saturated water content) and on the slope of the relationships resulting from variable pore size distributions are depicted in Fig. 3.15.

3.4.2 Measurement of SWC Relationships

Several methods are available to obtain measurements needed for SWC estimation. The basic requirement is for pairs of ψ_m-θ measurements over the wetness range of interest. Among the primary experimental problems in determining a SWC are (1) the limited functional range of the tensiometer, which is often used for *in situ* measurements, (2) inaccurate θ measurements in some cases, (3) the difficulty in obtaining undisturbed samples for laboratory determinations, and (4) a slow rate of equilibrium under low matric potential (i.e., dry soils).

In situ methods are considered the most representative for determining SWCs, particularly when a wide range of ψ_m-θ values are obtained. An effective method to obtain simultaneous measurements of ψ_m and θ_v utilizes TDR probes installed in the soil at close proximity to transducer tensiometers, with the changing values of each attribute monitored through time as the soil wetness varies. Large changes in ψ_m and θ_v can be induced under highly evaporative conditions near the soil surface, or in the presence of active plant roots.

3.4.2.1 Laboratory Estimation Using Pressure Plate and Pressure Flow Cell

The pressure plate apparatus consists of a pressure chamber enclosing a water saturated porous plate, which allows water, but prevents air flow through its pores (Fig. 3.16). The porous plate is open to

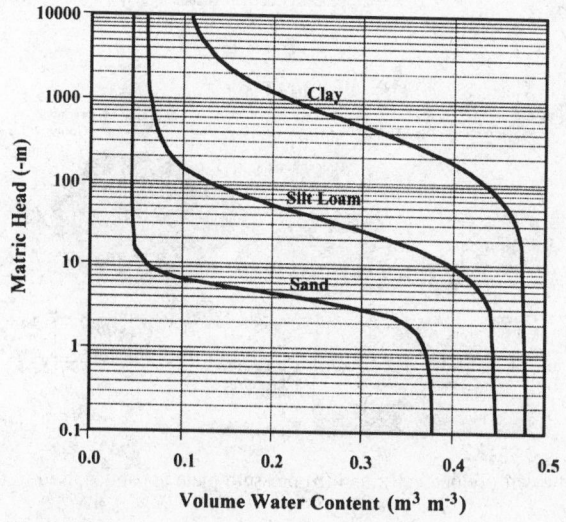

Fig. 3.15 Example soil water retention relationships for three soil textures

atmospheric pressure at the bottom, while the top surface is at the applied pressure of the chamber. Sieved soil samples (usually < 2 mm) are placed in retaining rubber rings in contact with the porous plate and left to saturate in water. After saturation is attained, the porous plate with the saturated soil samples is placed in the chamber and a known gas (commonly N_2 or air) pressure is applied to force water out of the soil and through the plate. Flow continues until equilibrium between the force exerted by the air pressure and the force by which soil water is being held by the soil (ψ_m) is reached.

Soil water retention in the low suction range of 0 to 10 m (= 1 bar = 0.1 MPa) is strongly influenced by soil structure and its natural pore size distribution. Hence, undisturbed intact soil samples (cores) are preferred over repacked samples for the wet end of the SWC. The pressure flow cell (Tempe cell) can hold intact soil samples encased in metal rings. The operation of the cell follows that of the pressure plate, except the pressure range is usually lower (0 to 10 m). The porous ceramic plates for both the pressure plate and the flow cell must be completely saturated, a process which may take a few days to achieve. Following equilibrium between soil matric potential and the applied air pressure, the soil samples are removed from the apparatus, weighed, and oven dried for gravimetric determination of water content. An estimate of the sample bulk density must be provided to convert θ_m to θ_v in the case of disturbed samples. Because of differences in pore sizes and geometry, the water content of repacked soils at a given matric potential should not be used to accurately infer θ of intact soils at the same ψ_m.

Several pressure steps may be applied to the same samples when using flow cells. The cells may be disconnected from the air pressure source and weighed to determine the change in water content from

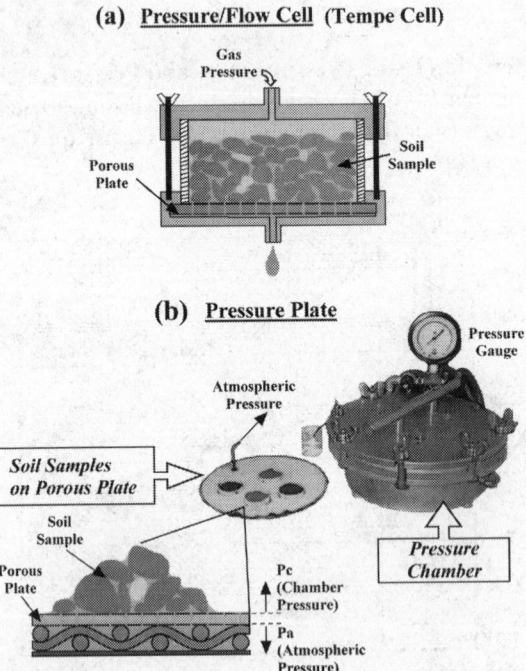

(a) Pressure/Flow Cell (Tempe Cell)

(b) Pressure Plate

Fig. 3.16   (a) Pressure flow cell (Tempe cell); and (b) pressure plate apparatus used to desaturate soil samples to desired matric potential

the previous step, then reconnected to the air pressure and a new (greater) pressure step applied. Water outflow from the cells may also be monitored and related to the change in sample water content.

3.4.2.2 Field Measurement Methods–Sensor Pairing

Despite the paramount importance of SWC determination *in situ*, suitable measurement techniques are severely lacking at present. The most common approach is to use paired sensors such as neutron moisture meter access tubes or TDR waveguides and tensiometers to determine water content and matric potential simultaneously and in the same soil volume. Single probes that combine TDR with tensiometry have also been proposed (Baumgartner et al., 1994). The limitations of most sensor pairing techniques stem from: (1) differences in the soil volumes sampled by each sensor (e.g., large volume averaging by a neutron moisture meter versus a small volume sensed by a heat dissipation sensor or a psychrometer), (2) the time required for matric potential sensors to reach equilibrium, and (3) limited ranges and deteriorating accuracy of different sensor pairs. This often results in limited overlap in SWC information measured using different techniques, and problems with measurement errors within the range of overlap.

A summary of common methods available for matric potential measurement or inference and their range of application is presented in Fig. 3.17. Note that most of the available techniques have a limited range of overlap (or do not overlap at all), and most are laboratory methods not suitable for field applications.

3.4.3 Fitting Parametric SWC Expressions to Measured Data

Measuring a SWC is laborious and time consuming. Measured $\theta-\psi$ pairs are often fragmentary, and usually constitute relatively few measurements over the wetness range of interest. For modeling and analysis purposes and for characterization and comparison between different soils and scenarios, it is therefore beneficial to represent the SWC in a continuous and parametric form. A parametric expression of a SWC model should: (1) contain as few parameters as necessary to simplify its estimation; and (2) describe the behavior of the SWC at the limits (wet and dry ends) while closely fitting the nonlinear shape of $\theta(\psi_m)$ data.

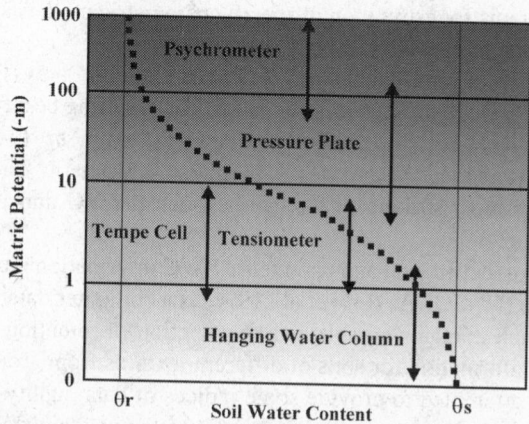

Fig. 3.17 Typical ranges of application for some common matric potential measurement or inference methods

An effective and commonly used parametric model for relating water content to the matric potential was proposed by van Genuchten (1980) and is denoted as VG:

$$\Theta = \frac{\theta - \theta_r}{\theta_s - \theta_r} = \left[\frac{1}{1 + (\alpha\psi_m)^n} \right]^m \qquad\qquad [3.31]$$

where θ_r and θ_s are the residual and saturated water contents, respectively, and α, n, and m are parameters directly dependent on the shape of the $\theta(\psi)$ curve. A considerable simplification is gained by assuming that m = 1 − 1/n. Thus the parameters required for estimation of the model are θ_r, θ_s, α, and n. θ_s is usually known and is easy to obtain experimentally with good accuracy, leaving only three unknown parameters (θ_r, α and n) to be estimated from the experimental data in many cases. Note that θ_r may be taken as $\theta_{-1.5 MPa}$, $\theta_{air dry}$, or a similar value, though it is often advantageous to use it as a fitting parameter.

Another well-known parametric model was proposed by Brooks and Corey (1964) and is denoted as BC:

$$\Theta = \frac{\theta - \theta_r}{\theta_s - \theta_r} = \left[\frac{\psi_b}{\psi_m} \right]^\lambda \qquad \psi_m > \psi_b$$
$$\Theta = 1 \qquad\qquad\qquad \psi_m \leq \psi_b \qquad\qquad\qquad [3.32]$$

where ψ_b is a parameter related to the soil matric potential at air entry ($_b$ represents bubbling pressure), and λ is related to the soil pore size distribution. Matric potentials are expressed as positive quantities in both VG and BC parametric expressions.

Estimation of VG or BC parameters from experimental data requires: (1) sufficient data points (at least 5 to 8 $\theta(\psi_m)$ pairs), and (2) a program for performing nonlinear regression. Recent versions of many computer spreadsheets provide relatively simple and effective mechanisms for performing nonlinear regression. Details of the computational steps required for fitting a SWC to experimental data using commercially available spreadsheet software are given in Wraith and Or (1998). In addition, computer programs for estimation of specific parametric models are also available such as the RETC code (van Genuchten et al., 1991).

Fitted parametric models of van Genuchten (VG) and Brooks and Corey (BC) to silt loam $\theta(\psi)$ data measured by Or et al. (1991) are presented in Fig. 3.18. The resulting best fit parameters for the VG model are $\alpha = 0.417$ m^{-1}; n = 1.75; $\theta_s = 0.513$ m^3 m^{-3}; and $\theta_r = 0.05$ m^3 m^{-3} (with r$^2 = 0.99$). For the BC model the best fit parameters are $\lambda = 0.54$; $\psi_b = 1.48$ m; $\theta_s = 0.513$ m^3 m^{-3}; and $\theta_r = 0.03$ m^3 m^{-3} (with r$^2 = 0.98$). Note that the most striking difference between the VG and the BC models is in the discontinuity at $\psi = \psi_b$ for BC.

Sources of additional experimental and parametric SWC information include (1) the Unsaturated Soil Hydraulic Database (UNSODA) (Leij et al., 1996) is a computer database compiled by the US Salinity Laboratory which contains an exhaustive collection of retention (SWC) and unsaturated hydraulic conductivity information for soils of different textures from around the world. While the authors/compilers have attempted to provide some indices of data quality or reliability, the user is advised (as always) to use their own experience and discretion in adapting others' data to their own applications; and (2) the regression studies by McCuen et al. (1981) and Rawls and Brakensiek (1989)

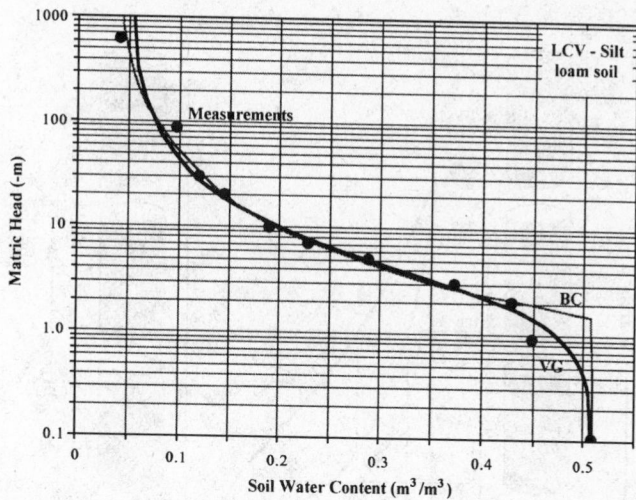

Fig. 3.18   van Genuchten (VG) and Brooks-Corey (BC) parametric models fitted to measured water retention data

provide a wealth of information on the Brooks-Corey parameter values for many soils including estimation of the hydraulic parameters based on other, often more easily available, soil properties. These estimates may be sufficiently accurate for some applications and could be used to obtain first-order approximations.

3.4.4   Hysteresis in the Soil-Water Characteristic Relation

Water content and the potential energy of soil water are not uniquely related because the amount of water present at a given matric potential is dependent on the pore size distribution and the properties of air-water-solid interfaces. A $\theta(\psi)$ relationship may be obtained by: (1) taking an initially saturated sample and applying suction or pressure to desaturate it (desorption), or by (2) gradually wetting an initially dry soil (sorption). These two pathways produce curves that in most cases are not identical; the water content in the drying curve is higher for a given matric potential than that in the wetting branch (Fig. 3.19). This is called hysteresis, defined as the phenomenon exhibited by a system in which the reaction of the system to changes is dependent upon its past reactions to change.

The hysteresis in SWC can be related to several phenomena: (1) the ink bottle effect resulting from nonuniformity in shape and sizes of interconnected pores as illustrated in Fig. 3.20a; drainage is governed by the smaller pore radius r, whereas wetting is dependent on the larger radius R; (2) different liquid-solid contact angles for advancing and receding water meniscii (Fig. 3.20b); (3) entrapped air in a newly wetted soil (e.g., pore doublet) (Dullien, 1992); and (4) swelling and shrinking of the soil under wetting and drying which may alter the porosity and pore size distribution. Based on early observations of the phenomenon by Haines (1930) as well as present day theories (Mualem, 1984; Kool and Parker, 1987), the role of individual factors remain unclear and subject to ongoing research.

3.5   Resources

A number of resources are available for readers interested in additional insight or information concerning soil water content and energy. The following suggestions are by no means inclusive, but should serve to augment and extend the discussions presented here, as well as provide a source of

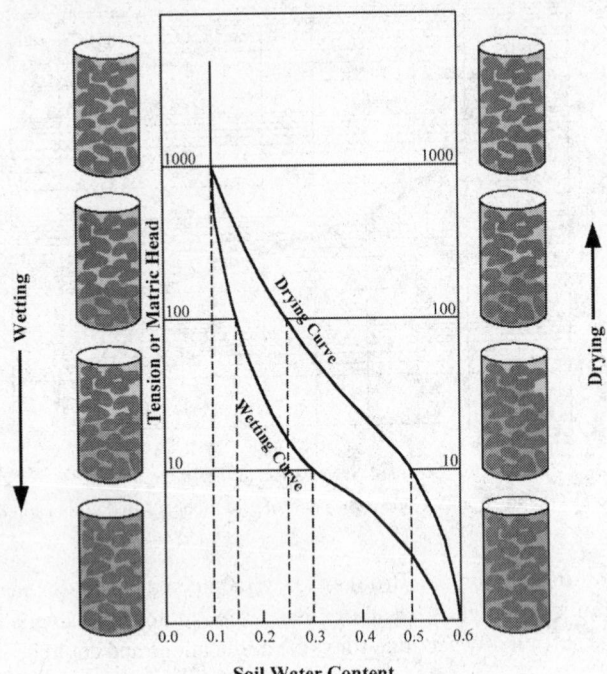

Fig. 3.19 Concept of hysteresis in soil-water characteristic relationships

additional references. We advise readers to consult the references we have cited in relevant sections of the chapter, as well as various texts, monographs, and review chapters. Soil physics and related textbooks which address soil water content and water energy include those of Childs (1969), Hanks (1992), Hillel (1998), Jury et al. (1991), Kirkham and Powers (1972), Kutilek and Nielsen (1994), and Marshall et al. (1996). Several chapters in the monograph edited by Klute (1986), as well as that edited by Carter (1993), contain valuable information. A number of excellent review papers or chapters are also available, including Baker (1990), Campbell and Mulla (1990), and Parker (1986).

Some Internet sites contain valuable information related to soil water content and energy. These include various individual or professional organization pages, discussion groups (e.g., the soil moisture group SOWACS), and others. However, Internet addresses tend to change rather frequently, and new sources may be added. This means it is generally more efficient to conduct keyword searches using one or more of the available search engines, than to rely on potentially outdated web addresses. Our own experiences in searching the World Wide Web indicate that there are presently few sites having useful or extensive soil water information. However, this may change as the web becomes more heavily utilized as a repository for soil information and soil property databases. Some government agencies are reportedly moving in this direction.

Acknowledgment
Partial funding for this work was provided by the Utah Agricultural Experimental Station (UAES) and the Montana Agricultural Experimental Station (MAES).

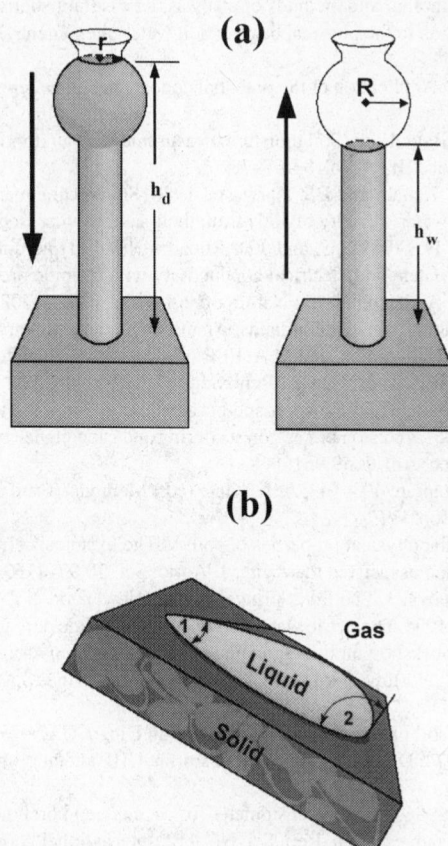

Fig. 3.20 The "ink bottle" effect (a), and the contact angle effect (b)

3.6 References

Babb, A.T.S. 1951. A radio-frequency electronic moisture meter. Analyst 76:428–433.

Baker, J.M. 1990. Measurement of soil water content. Rem. Sens. Rev. 5:263–279.

Baker, J.M., and R.R. Allmaras. 1990. System for automating and multiplexing soil moisture measurement by time-domain reflectometry. Soil Sci. Soc. Am. J. 54:1–6.

Banin A., and A. Amiel. 1970. A correlative study of the chemical and physical properties of a group of natural soils of Israel. Geoderma 3:185–198.

Baumgartner, N., G.W. Parkin, and D.E. Elrick. 1994. Soil water content and potential measured by hollow time domain reflectometry probe. Soil Sci. Soc. Am. J. 58:315–318.

Bolt, G.H. 1976. Soil physics terminology. Int. Soc. Soil Sci. Bull. 49:16–22.

Bouyoucos, G.J., and A.H. Mick. 1940. An electrical resistance method for the continous measurement of soil moisture under field conditions. MI Agric. Exp. Sta. Tech. Bull. 172:1–38.

Bristow, K. L., G.S. Campbell, and K. Calissendroff. 1993. Test of a heat-pulse probe for measuring changes in soil water content. Soil Sci. Soc. Am. J. 57:930–934.

Brooks, R.H., and A.T. Corey. 1964. Hydraulic properties of porous media. Hydrol. Pap. 3. CO State University, Fort Collins, CO.

Brutsaert, W. 1982. Evaporation into the atmosphere. D. Reidel Publishing Company. Dordrecht, Netherlands.

Campbell, G.S., and D.J. Mulla. 1990. Measurement of soil water content and potential. p. 127–142. *In* B.A. Stewart and D.R. Nielsen (ed.). Irrigation of agricultural crops. American Society of Agronomy, Madison, WI.

Carter, M.R. (ed.) 1993. Soil sampling and methods of analysis. Lewis Publishers, Boca Raton, FL.

Childs, E.C. 1969. An introduction to the physical basis of soil water phenomena. John Wiley and Sons Ltd, London, UK.

Corey, A.T., and A. Klute. 1985. Application of the potential concept to soil water equilibrium and transport. Soil Sci. Soc. Am. J. 49:3–11.

Dean, T.J., J.P. Bell, and A.J.B. Baty. 1987. Soil moisture measurement by an improved capacitance technique. Part I. Sensor design and performance. J. Hydrol. 93:67–78.

Dahiya, I.S., D.J. Dahiya, M.S. Kuhad, and P.S. Karwasra. 1988. Statistical equations for estimating field capacity, wilting point, and available water capacity of soils from their saturation presentage. J. Agric. Res. 110:515–520.

Dalton, F.N., W.N. Herkelrath, D.S. Rawlins, and J.D. Rhoades. 1984. Time-domain reflectometry: Simultaneous measurement of soil water content and electrical conductivity with a single probe. Science 224:989–990.

Day, P.R., G.H. Bolt, and D.M. Anderson. 1967. Nature of soil water. p. 193–208. In R.M. Hagan, H.R. Haise, and T.W. Edminster (ed.) Irrigation of agricultural lands. American Society of Agronomy, Madison, WI.

Dobson, M.C., F.T. Ulaby, M.T. Hallikainen, and M.A. El-Rayes. 1985. Microwave dielectric behavior of wet soil: II. Dielectric mixing models. IEEE Trans. Geosci. Rem. Sens. GE-23:35-46.

Dullien, F.A.L. 1992. Porous media: Fluid transport and pore structure. 2nd Ed., Academic Press, New York, NY.

Evett, S.R., and J.L. Steiner. 1995. Precision of neutron scattering and capacitance type soil water content gauges from field calibration. Soil Sci. Soc. Am. J. 59:961–968.

Gardner, W.H. 1986. Water content. p. 493–544. In A. Klute (ed.) Methods of soil analysis. Part 1, 2nd Ed. American Society of Agronomy, Madison, WI.

Haines, W. B. 1930. Studies in the physical properties of soil. V. The hysteresis effect in capillary properties, and the modes of moisture distribution associated therewith. J. Agric. Sci. 20:97–116.

Hanks, R.J. 1992. Applied soil physics. 2nd Ed., Springer Verlag, New York, NY.

Hilhorst, M.A., and C. Dirksen. 1994. Dielectric water content sensors: Time domain versus frequency domain. p. 23–33. In Proc. for symp. and workshop on time domain reflectometry in environmental, infrastructure, and mining applications. Sept. 7–9, 1994. Northwestern Univ., Evanston, IL Bur. Mines Spec. Publ. SP 19–94, US Department of the Interior, Washington, DC.

Hillel, D. 1998. Environmental soil physics. Academic Press, San Diego, CA.

Hoekstra, P., and A. Delaney. 1974. Dielectric properties of soils at UHF and microwave frequencies. J. Geophys. Res. 79:1699–1708.

Hubbell, J.M., and J.B. Sisson. 1996. Portable tensiometer for use in deep boreholes. Soil Sci. 161:376–381.

IAEA. 1970. Neutron moisture gauges. Tech. Rep. Ser. No. 112. International Atomic Energy Agency, Vienna, Austria.

Iwata, S., T. Tabuchi, and B.P. Warkentin. 1988. Soil water interactions. M. Dekker, New York, NY.

Jury, W.A., W.R., Gardner, and W.H. Gardner. 1991. Soil physics. John Wiley and Sons, New York, NY.

Kirkham, D., and W.L. Powers. 1972. Advanced soil physics. R.E. Krieger Publishing Company, Malabar, FL.

Klute A. (ed.). 1986. Methods of soil analysis: Physical and mineralogical methods. Part 1. 2nd Ed. Soil Science Society of America, Madison, WI.

Kool, J. B., and J. C. Parker. 1987. Development and evaluation of closed-form expressions for hysteretic soil hydraulic properties, Water Resour. Res. 23:105–114.

Kutilek, M., and D. R. Nielsen. 1994. Soil hydrology. Catena Verlag, Cremlingen-Destedt, Germany.

Leij, F.J., W.J. Alves, M.Th. van Genuchten, and J.R. Williams. 1996. The UNSODA unsaturated hydraulic database. EPA/600/R-96/095, U.S. Environmental Protection Agency, Cincinnati, OH.

Marshall, T.J., J.W. Holmes, and C.W. Rose. 1996. Soil physics. 3rd Ed. Cambridge University Press, UK.

McCuen, R.H., W.J. Rawls, and D.L. Brakensiek. 1981. Statistical analysis of the Brook-Corey and Green-Ampt parameters across soil texture. Water Resour. Res. 17:1005-1013.

Mualem, Y. 1984. A modified dependent domain theory of hysteresis. Soil Sci. 137:283-291.

Nitao, J.J., and J. Bear. 1996. Potentials and their role in transport in porous media. Water Resour. Res. 32:225-250.

Olhoeft, G.R. 1985. Low-frequency electrical properties. Geophysics 50:2492–2503.

Or, D., D. P. Groeneveld, K. Loague, and Y. Rubin. 1991. Evaluation of single and multi-parameter methods for estimating soil-water characteristic curves. Geotechnical Engineering Report No. UCB/GT/91-07, University of California, Berkeley, CA.

Paltineanu, I.C., and J.L. Starr. 1997. Real-time soil water dynamics using multisensor capacitance probes: Laboratory calibration. Soil Sci. Soc. Am. J. 61:1576–1585.

Parker, J.C. 1986. Hydrostatics of water in porous media. p. 209–296. *In* D.L. Sparks (ed.) Soil physical chemistry. CRC Press, Boca Raton, FL.

Peck, A.J., and R.M. Rabbidge. 1969. Design and performance of an osmotic tensiometer for measuring capillary potential. Soil Sci. Soc. Am. Proc. 33:196–202.

Petersen L.W., P. Moldrup, O.H. Jacobsen, and D.E. Rolston. 1996. Relations between specific surface area and soil physical and chemical properties. Soil Sci. 161:9–21.

Phene, C.J., G.J. Hoffman, and S.L. Rawlins. 1971. Measuring soil matric potential *in situ* by sensing heat dissipation within a porous body: I. Theory and sensor construction. Soil Sci. Soc. Am. Proc. 35:27–33.

Rawls, W.J., and D.L. Brakensiek. 1989. Estimation of soil water retention and hydraulic properties. p. 275–300. *In* H.J. Morel-Seytoux (ed.) Unsaturated flow in hydraulic modeling theory and practice. NATO ASI Series. Series c: Mathematical and physical science. Vol. 275.

Reece, C.F. 1996. Evaluation of line heat dissipation sensor for measuring soil matric potential. Soil Sci. Soc. Am. J. 60:1022–1028.

Richards, L.A. (ed.) 1954. Diagnosis and improvement of saline and alkali soils. USDA Handb. 60. US Government Printing Office, Washington, DC.

Roth, K., R. Schulin, H. Fluhler, and W. Attinger. 1990. Calibration of time domain reflectometry for water content measurement using composite dielectric approach. Water Resour. Res. 26:2267–2273.

Selig, E.T., and S. Mansukhani. 1975. Relationship of soil moisture to the dielectric property. ASCE Geotech. Eng. Div. GT8:755–770.

Slater, P.J., and J.B. Williams. 1965. The influence of texture on the moisture characteristics of soils. 1. A critical comparison of techniques for determining the available-water capacity and moisture characteristic curve of a soil. J. Soil Sci. 16:1–15.

Smith-Rose, R.L. 1933. The electrical properties of soils for alternating currents at radio frequencies. Proc. Roy. Soc. London 140:359.

Spaans, E.J.A., and J.M. Baker. 1992. Calibration of watermark soil moisture sensors for temperature and matric potential. Plant Soil 143:213–217.

Topp, G.C., J.L. Davis, and A.P. Annan. 1980. Electromagnetic determination of soil water content: Measurements in coaxial transmission lines. Water Resour. Res. 16:574–582.

van Genuchten, M.Th. 1980. A closed-form equation for predicting the hydraulic conductivity of unsaturated soils. Soil Sci. Soc. Am. J. 44:892–898.

van Genuchten, M.Th., F.J. Leij, and S.R. Yates. 1991. The RETC code for quantifying the hydraulic functions of unsaturated soils. EPA/600/2-91/065, US Environmental Protection Agency, Ada, OK.

von Hippel, A. R. 1954. Dielectric materials and applications. Technology Press of M.I.T., New York, NY.

Waters, P. 1980. Comparison of the ceramic and the pressure membrane to determine the 15 bar water content of soils. J. Soil Sci. 31:443–446.

Wraith J.M., and D. Or. 1998. Nonlinear parameter estimation using spreadsheet software. J. Nat. Resour. Life. Sci. Educ. 27:13–19.

4

Soil Water Movement

David E. Radcliffe and Todd C. Rasmussen
University of Georgia

4.1 Introduction

Water movement in soils occurs under both saturated and unsaturated conditions (Fig. 4.1). Saturated conditions occur below the watertable where water movement is predominately horizontal, with lesser components of flow in the vertical direction. While unsaturated conditions generally predominate above the watertable (the vadose zone), localized zones of saturation can exist especially following precipitation or irrigation. As a general rule, water movement in the unsaturated zone is vertical, but can also have large lateral components.

Saturated soils occur when soil pores are entirely filled with water. In this case, the water content (θ) is equal to the total porosity (ϕ) and the air filled porosity (θ_a) is zero. While soil pores can be assumed to be fully saturated below the watertable, even so-called saturated soils above the watertable may retain some residual entrapped air, especially near the soil surface. Here, we will assume that $\theta = \phi$ and $\theta_a = 0$ under saturated conditions, unless specified otherwise.

Water flow can be either steady or transient. While analytic solutions of the differential equations that govern soil water flow are abundant for saturated conditions, they are less available for steady flow in the unsaturated zone and almost completely lacking for transient, unsaturated flow conditions. Simplifications of the unsaturated flow equations have been developed to describe infiltration because of the importance of this transient flow process. Alternatively, numerical methods are commonly used to solve most transient flow problems in the unsaturated zone, and in saturated media where large variations in material properties make analytic solutions impossible.

The hydraulic gradient, which is the driving force behind water flow, is a vector that describes the slope of the energy distribution within the soil. The principal parameter required to predict saturated flow is the saturated hydraulic conductivity (K_s). Predicting unsaturated flow requires the unsaturated hydraulic conductivity [$K(h)$] and water retention [$\theta(h)$] functions. K_s, $\theta(h)$, and $K(h)$ are all affected by soil texture and structure. While soil texture is easily measured and not highly variable in space (in many cases), soil structure is difficult to quantify and highly variable in space and time. Although methods have been developed to measure K_s and $K(h)$, each has some disadvantage.

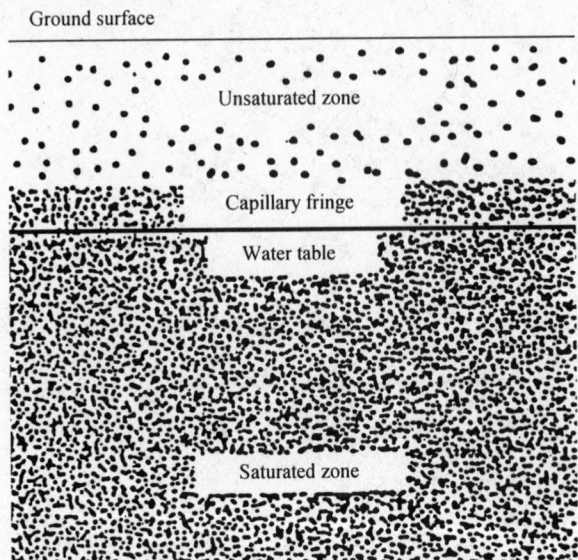

Ground surface

Unsaturated zone

Capillary fringe

Water table

Saturated zone

Fig. 4.1 Saturated and unsaturated zones [Reprinted from Fetter, 1994. Applied hydrogeology. Copyright Macmillan, New York, NY with permission]

In recent years, soil scientists have expanded their vertical scale of interest beyond the root zone to include all of the vadose zone. At the same time, there is an emphasis on expanding the horizontal scale from the traditional plot scale to the field and watershed scales. At these larger scales, soil scientists need to have a better understanding of shallow groundwater flow which is emphasized in this chapter.

This chapter introduces the conditions and equations that govern water flow, the parameters that quantify these relationships, and applications to common problems. The objective is to provide a foundation for understanding the water flow processes in soils, and to direct the reader to other resources that can clarify or extend the introductory material provided here.

4.2   Flow in Saturated Soil

The following sections examine flow under steady and transient conditions. The approach is to first examine the rule, Darcy's law, that relates the flow rate to the hydraulic gradient and then show how a conservation equation can be coupled to Darcy's law to establish the equations that govern both steady and unsteady flow. The resulting set of equations are then applied to problems in one and higher dimensions.

4.2.1   Saturated Flow Equations

4.2.1.1   Poiseuille Equation

At the microscopic scale of an individual pore, approximated as a water filled cylinder of given radius (r), the volumetric flow rate (Q) is described by Poiseuille's equation:

$$Q = \frac{\pi r^4 \rho g \Delta H}{8 \eta L}$$

[4.1]

where η is the dynamic viscosity, ρ is the density of water, g is the acceleration due to gravity, and ΔH is the difference in total head (potential on a weight basis) between two points along the cylinder separated by a distance L. Poiseuille's equation has been used to show the disproportionate effect that large diameter pores (i.e., macropores) have on transmitting water. Flow is proportional to r^4 whereas the cross-sectional area of a pore is proportional to r^2. Therefore, one large pore with the same cross-sectional area as several smaller pores will transmit considerably more water due to less viscous drag along the pore wall.

4.2.1.2 Darcy's Equation

Because it is usually not feasible to determine the size distribution and interconnectivity of pores, a macroscopic approach is used to describe flow through soils. This approach was first taken by Henry Darcy, an engineer working on sand filters used to purify the water in the city of Dijon, France. Through experimentation, he found that the volumetric flow rate per unit cross-sectional area (A) through a sand filter of a given thickness (L) was proportional to the total head gradient across the sand ($\Delta H/L$). He called the proportionality constant saturated hydraulic conductivity (K_s) (Darcy, 1856):

$$J = \frac{Q}{A} = -K_s \frac{\Delta H}{L}$$

[4.2]

where J is the fluid flux. Darcy's law can be extended to two- and three-dimensional flow by noting that the flux can be written in vector notation as:

$$\underline{J} = \left(J_x, \ J_y, \ J_z \right) = -K_s \nabla H$$

[4.3]

which implies that the total flow can be decomposed into a set of components in the two horizontal (x,y) and one vertical (z) directions. If the vertical component, J_z, is zero, then J is just (J_x, J_y). This formulation is suitable for geologic media that have uniform hydraulic conductivity. The above equation represents the hydraulic gradient as a vector:

$$\nabla H = \left(\frac{\partial H}{\partial x}, \ \frac{\partial H}{\partial y}, \ \frac{\partial H}{\partial z} \right)$$

[4.4]

An equivalent equation can be written in cylindrical and radial coordinates by noting that flow at a point is still related to the hydraulic gradient, however, the gradient is represented:

$$\nabla H = \left(\frac{\partial H}{\partial r}, \frac{\partial H}{\partial z} \right)$$

[4.5]

for axisymmetric cylindrical coordinates, where r is the radial distance and z the axial distance, and

$$\nabla H = \frac{\partial H}{\partial r}$$

[4.6]

for symmetrical spherical coordinates.

Darcy's equation takes the same form as several other important laws in science, among them, Fick's law for chemical diffusion, Ohm's law for current flow, and Fourier's law for heat flow. It clearly shows that for water to flow there must be a difference in water potential, but the rate of water flow will depend on the hydraulic gradient and the hydraulic conductivity of the soil.

Although Darcy's equation was developed empirically, it can be derived using the Navier-Stokes equation if inertial forces are assumed to be negligible compared to viscous forces (Hubbert, 1956). When inertial forces become significant, turbulent flow occurs and the Reynolds number is used to determine when the onset of turbulent flow occurs. For soil pores, turbulent flow may be expected for Reynolds numbers greater than one (Hillel, 1980a). This would occur at 20°C under gravity flow in pores with diameters larger than about 0.15 mm. Under these conditions, Darcy's equation would overestimate flux in larger diameter pores.

In a uniform soil under saturated conditions, Darcy's equation can be used to show that pressure potential varies linearly with depth (Jury et al., 1991). Thus, if water is ponded on the surface of a uniform soil column and allowed to drip from the bottom, the pressure potential should decrease linearly from the depth of ponding at the surface to zero at the bottom of the column.

4.2.1.3   Steady Flow

Coupling a conservation of mass equation that requires a mass balance at every point within the flow domain with the Darcy equation under steady flow conditions yields the Laplace equation:

$$\nabla^2 H = 0 \qquad\qquad [4.7]$$

which is equivalent to stating that there is no accumulation of water at any point within the flow domain.

4.2.1.4   Unsteady Flow

For conditions of unsteady flow, there will be an accumulation, or release, of water within the flow domain. Under these conditions, the difference between the water flowing into and away from a point within the domain must equal the accumulation, or loss, of water at the point. This is described using:

$$\nabla \cdot (K_s \nabla H) = S_s \frac{\partial H}{\partial t} \qquad\qquad [4.8]$$

where S_s is the specific storage of the medium, and t is time.

4.2.2   Saturated Flow Parameters

4.2.2.1   Saturated Hydraulic Conductivity

For saturated flow, the most important soil parameter is saturated hydraulic conductivity, which is a function of the fluid and soil properties

$$K_s = \frac{k \rho g}{\eta} \qquad\qquad [4.9]$$

where k is the intrinsic permeability of the soil. It is apparent from this equation that temperature has an effect on K_s. It is generally true that liquids other than water will have a different K_s but the same k in a given soil, if the soil structure is not affected.

Soils with low porosity, few large pores, and poor interconnectivity between pores have low values of K_s. Rawls et al. (1982) compiled values of K_s from 1,323 soils collected over 32 states (Table 4.1). Saturated hydraulic conductivities were highest in coarse-textured soils and declined in fine-textured soil, due to larger pores in the former. This can be corroborated by examining air entry potentials (h_a, the matric potential at which the largest capillary pore empties) which are more negative in the fine-textured soils, indicating that the largest capillary pore was smaller than in coarse-textured soils. Higher K_s occurred in the coarse-textured soils in spite of the generally lower porosity.

The mean values in Table 4.1 may provide unreliable values for K_s and h_a for fine-textured soils, due to the effect of structure. In Fig. 4.2, K_s for 395 soil samples from the UNSODA database (Section 4.4.3) are plotted as a function of clay content. There is great variability except at very low clay contents where there is a cluster of high K_s values. In intact soil, it is apparent that soil structure, as well as soil texture, affects water flow in all but very sandy soils.

In many cases, hydraulic conductivity is not the same in all directions. For the case where a different conductivity is observed in two (or three) different directions, this variability must be accounted for explicitly in the Darcy equation:

$$J_x = -K_x \frac{\partial H}{\partial x} \qquad J_y = -K_y \frac{\partial H}{\partial y} \qquad J_z = -K_z \frac{\partial H}{\partial z} \qquad [4.10]$$

where the three fluxes are oriented in the three directions of the coordinate system. This can be written in Einstein vector notation as:

$$\underline{J} = -\underline{\underline{K}}\nabla H \qquad [4.11]$$

where the hydraulic conductivity is no longer a constant, but is, instead:

Table 4.1 Saturated hydraulic conductivity (K_s), total porosity (ϕ), air-entry matric head (h_a), microscopic capillary length (λ_m), and macroscopic capillary length (λ_c) of soils of various textures [Adapted from Rawls et al., 1982; Brakensiek and Rawls, 1992]

Texture Class	K_s	ϕ	h_a	λ_m	λ_c
	cm h^{-1}	cm^3 cm^{-3}	cm	cm	cm
Sand	21.00	0.437	-16.0	2.83×10^{-2}	2.62
Loamy sand	6.11	0.437	-20.6	2.06×10^{-2}	3.61
Sandy loam	2.59	0.453	-30.2	9.92×10^{-3}	7.48
Sandy clay loam	0.43	0.398	-59.4	4.63×10^{-3}	16.05
Loam	1.32	0.463	-40.1	1.11×10^{-2}	6.72
Silt loam	0.68	0.501	-50.9	5.83×10^{-3}	12.74
Clay loam	0.23	0.464	-56.4	4.50×10^{-3}	16.49
Sandy clay	0.12	0.430	-79.5	3.84×10^{-3}	19.32
Silty clay loam	0.15	0.471	-70.3	3.31×10^{-3}	22.46
Silty clay	0.09	0.479	-76.5	3.02×10^{-3}	24.60
Clay	0.06	0.475	-85.6	2.77×10^{-3}	26.83

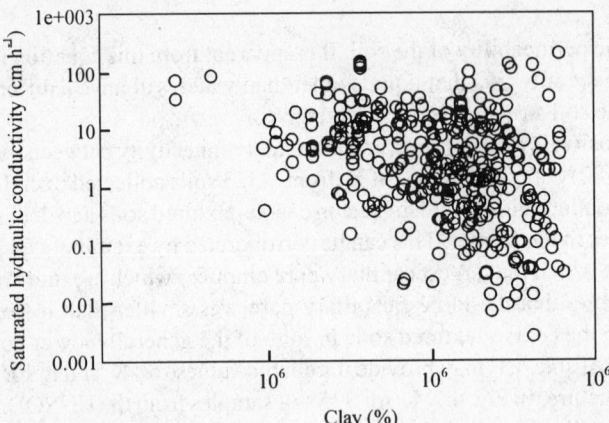

Fig. 4.2 Saturated hydraulic conductivity as a function of clay content for 395 soils from the UNSODA database

$$\underline{\underline{K}} = \begin{bmatrix} K_x & 0 & 0 \\ 0 & K_y & 0 \\ 0 & 0 & K_z \end{bmatrix}$$ [4.12]

This formulation implies that the flow direction can be different from the hydraulic gradient. The hydraulic conductivity function (K) is called a tensor, and represents the magnitude of the hydraulic conductivity as a function of direction. The example above shows K as having nonzero values along its diagonal, which only occurs when the soil layers are precisely oriented with the (x,y,z) coordinate system (e.g., for perfectly horizontal layers). For conditions when the layers are not coincident with the coordinate system, then the off-diagonal components are no longer zero, and we have:

$$\underline{\underline{K}} = \begin{bmatrix} K_{xx} & K_{xy} & K_{xz} \\ K_{yx} & K_{yy} & K_{yz} \\ K_{zx} & K_{zy} & K_{zz} \end{bmatrix}$$ [4.13]

or, equivalently:

$$J_x = -\left[K_{xx} \frac{\partial H}{\partial x} + K_{xy} \frac{\partial H}{\partial y} + K_{xz} \frac{\partial H}{\partial z} \right]$$

$$J_y = -\left[K_{yx} \frac{\partial H}{\partial x} + K_{yy} \frac{\partial H}{\partial y} + K_{yz} \frac{\partial H}{\partial z} \right]$$ [4.14]

$$J_z = -\left[K_{zx} \frac{\partial H}{\partial x} + K_{zy} \frac{\partial H}{\partial y} + K_{zz} \frac{\partial H}{\partial z} \right]$$

The fundamental importance of these expressions lies in the observation that a gradient in one direction can induce flow in a different direction. For example, when soil layers are inclined in the unsaturated zone, a downward gradient caused by gravity often induces lateral drainage. In this case, the downward gradient $(\partial H/\partial z)$ induces a lateral flux $(J_x$ or $J_y)$ due to the nonzero conductivities $(K_{xz}$ or K_{yz}, respectively).

4.2.2.2 Specific Storage Coefficient

The specific storage coefficient (S_s) is defined as the volume of water released from storage (V_w) per unit volume of soil (V) per unit decline in hydraulic head (H):

$$S_s = -\frac{1}{V}\frac{dV_w}{dH} = -\frac{d\theta}{dH}$$

[4.15]

4.2.2.3 Aquifer Transmissivity and Storativity

When flow is uniformly parallel to the soil layers, the flow-per-unit horizontal distance can be estimated using the transmissivity, $T = b\, K_s$:

$$\frac{Q}{w} = -T\frac{\Delta H}{L}$$

[4.16]

which is identical to Equation [4.2] because $A = b\, w$, where A is the cross-sectional area perpendicular to flow, b is the thickness of the flow region, and w is the width of the flow region. When the hydraulic conductivity is not uniform with depth (as is generally the case), the transmissivity is defined as the integrated hydraulic conductivity over the layer:

$$T = \int_{z_0}^{z_0+b} K_s dz \qquad \text{or} \qquad K_s = \frac{dT}{dz}$$

[4.17]

The transmissivity is important because many field situations occur where the thickness of the water conducting unit is unknown, or the hydraulic conductivity is highly variable within the unit. Measured field data can be readily used to estimate the transmissivity, but data may not exist to clearly specify the thickness of the unit. In these cases, the transmissivity serves as a parameter that describes the bulk behavior of the unit.

The aquifer storativity or storage coefficient (S) is defined as the volume of water released from storage-per-unit area of geologic medium, per-unit decline in head, or:

$$S = bS_s \qquad \text{or} \qquad S_s = \frac{dS}{dz}$$

[4.18]

Storativity is a function of whether or not the geologic unit is saturated or unsaturated, and if saturated, whether it is confined or unconfined. For unsaturated geologic units, the coefficient is related to the specific water capacity. For unconfined geologic units, the coefficient is related to the specific yield of the unit.

The storage coefficient is especially important for transient flow conditions, where water is released or added to storage. The greater the storage coefficient, the slower an energy response propagates through the system, and vice versa. Confined aquifers show rapid responses to pumping over large areas, while the influence of pumping on unconfined aquifers is more localized. Also, pulses of water moving through the unsaturated zone will be more attenuated when the specific water capacity is small.

4.2.2.4   Conductance or Leakance

The conductance, or leakance coefficient, is used for conditions of vertical flow across a low permeability layer, such as below a lake bed, across a confining layer, or under a stream. The leakance is another way of writing Darcy's equation for conditions when the thickness of the layer is unknown:

$$J = -C\, \Delta H \qquad\qquad [4.19]$$

where $C = K_s / L$ and L is the thickness of the low permeability layer. In many applications, the head drop (ΔH) across the layer is known, or can be estimated, along with the vertical flux. Because the thickness of the confining layer may not be known, one can use the leakance coefficient as a bulk parameter to describe the behavior of the system.

4.2.3   Saturated Flow Examples

4.2.3.1   Flow Perpendicular to Layers

Most soils other than Entisols and Inceptisols have well-developed horizons or layers which have different hydrologic characteristics including K_s. Above the watertable, water usually flows vertically so that flow is perpendicular to these layers. Below the watertable, flow can occur parallel to the layers (described in the following section). Steady flow through a layered soil can be described by making an analogy between Darcy's equation and Ohm's equation for current flow (Jury et al., 1991). In this case, the hydraulic resistance is the layer thickness divided by the saturated hydraulic conductivity of the layer. Because flow across N layers is analogous to electrical flow through resistances in series (all of the water must flow through each layer), the effective hydraulic resistance is the sum of the resistances in each layer. The effective saturated hydraulic conductivity is then:

$$K_{eff} = \frac{\sum\limits_{j=1}^{N} L_j}{\sum\limits_{j=1}^{N} \dfrac{L_j}{K_j}} \qquad\qquad [4.20]$$

where L_j and K_j are the thickness and saturated hydraulic conductivity of each layer. Steady vertical flow through the entire soil profile is described with Darcy's equation:

$$J = -K_{eff}\, \frac{\Delta H}{L_t} = -C\, \Delta H \qquad\qquad [4.21]$$

where potentials are measured at the top and bottom of the profile, L_t is the total thickness of the profile, and $C = K_{eff} / L_t$ is the conductance described above.

Because the flux across all layers is the same under steady flow conditions, Equation [4.23] can be used (once J is known) to determine the matric potential at each layer interface. In this manner, one can show that when water flows down through a high to a low conductivity layer (e.g., a sandy horizon over a more clayey horizon), a positive pressure occurs at the interface that is greater than would be present in a uniform soil. This "back pressure" slows flow through the high permeability layer and speeds flow through the other until there is an equivalent amount of flow through both layers. In the reverse situation (e.g., sand over a clay), the interface matric potential is less than that in a uniform soil, creating a "suction" which speeds flow through the low permeability layer and slows flow through the other.

In a layered soil, the matric potential distribution is linear with depth within a layer, but with a different rate of change in each layer. An abrupt change in slope at boundaries is usually not observed in natural soils because K_s usually varies slowly, leading to a more curvilinear change in matric potential.

For conditions where a large difference in conductance is present, negative pressures can occur at the interface that exceed the air entry potential of soil in the lower layer. This causes unsaturated conditions in the lower layer which greatly reduce flow through the profile. This often occurs when a surface seal is present. Another way to equalize flow through a layered soil with a high over a low permeability layer is for water to flow through only part of the more restrictive layer. This is a form of preferential flow called fingering. Both surface seals and fingering are discussed in Section 4.3.

4.2.3.2   Flow Parallel to Layers

For flow parallel to layers (which usually occurs below the watertable), the resistances can be considered in parallel (more water flows through high than low conductivity layers). If the soil layers are of equal thickness, then K_{eff} is the arithmetic average of the individual layer saturated hydraulic conductivities. For layers of different thickness, K_{eff} is a weighted average where the layer thicknesses are the weights. Steady horizontal flow through the profile can be described using

$$\frac{Q}{w} = K_{eff} L_t \, \nabla H = T \, \nabla H$$

[4.22]

where w is the unit width of aquifer perpendicular to flow, and T is the aquifer transmissivity.

Flow through heterogeneous aquifers is largely dominated by the highest permeability units. For an aquifer with a log-normally distributed hydraulic conductivity and a unit variance, about half of all flow occurs in only 1% of the thickness, and 90% of the flow occurs in about 16% of the unit.

4.2.3.3   Flow to a Well

Flow to a pumping well occurs when the total head in the well is lower than that in the surrounding aquifer. The hydraulic gradient induces (nearly) horizontal flow into the well casing, with some additional vertical flow due to leakage from above or below the aquifer, and from releases from storage within the aquifer. The Thiem (1906) Equation predicts the steady-state drawdown (s) in a confined aquifer with transmissivity (T) as a function of distance (r) from a well that is pumped at a rate (Q):

$$s = \frac{Q}{2\pi T} \ln\left(\frac{r}{r_0}\right)$$

[4.23]

The distance to a constant head boundary (r_0) is required to assure that the system reaches steady state. Otherwise, water levels should continue to fall over time, until a source of increased recharge is intercepted by the declining water levels. Other solutions for flow to a well including confined aquifer conditions can be found in, for example, Kruseman and deRidder (1990).

4.2.3.4 Flow to a Drain

Flow to a horizontal drain is a common concern in regions of high watertables and in arid region irrigation projects where salinity can be a problem. It can be treated as a steady two-dimensional flow problem (Hooghoudt, 1937). Radial (R_r) and horizontal (R_h) resistances to flow are calculated:

$$R_r = \frac{1}{\pi} \ln\left(\frac{0.7D}{r}\right) \qquad R_h = \frac{(L-1.4D)^2}{8DL} \qquad\qquad [4.24]$$

where D is the distance between the drains and the impermeable layer below and r is the radius of the tile drains. These resistances are used to calculate a factor (d) which corrects for radial flow near the drains:

$$d = \frac{1}{8(R_h + R_r)} \qquad\qquad [4.25]$$

The spacing (L) between drains as a function of the desired height (h_s) of the watertable above the drains (midway between the drains) is given by:

$$L^2 = \frac{8K_2 dh_s}{i} + \frac{4K_1 h_s^2}{i} \qquad\qquad [4.26]$$

where i is the infiltration rate, and K_1 and K_2 are the saturated hydraulic conductivities above and below the drains, respectively. Since L must be known to calculate R_h and d, these equations have to be solved iteratively but the sensitivity of the final value of L to the initial guess is low. In Fig. 4.3, the drain spacing required to maintain the watertable at a depth 70 cm below the soil surface is shown as a function of K_1. The drains in this example were at a depth of 100 cm. As the saturated hydraulic conductivity above the drain increased, the spacing between the drains could be increased (thereby decreasing the cost of installation) while maintaining the watertable no higher than 70 cm below the surface.

4.3 Flow in Unsaturated Soil

The unsaturated zone, also called the vadose zone, lies between the watertable and the soil surface (Fig. 4.1). In this region, the water content of the soil is less than saturation ($\theta < \phi$), many pores are air filled ($\phi_a > 0$), and pressure (matric) heads are generally negative ($h < 0$). The watertable, which separates the saturated zone from the unsaturated zone, is the surface where the water potential equals the mean atmospheric pressure.

Rising above the watertable is the capillary fringe, a zone where water is under tension, but is very near saturation. The capillary fringe is important for water flow because only a slight change in water content or total head can cause a sharp change in the watertable position. Also, lateral flow increases in this zone (Liu and Dane, 1996). The capillary fringe is also important in wetland delineation because the depth of the saturated zone extends above the watertable, as measured in a well or piezometer. The thickness of the capillary fringe depends on the water retention curve, and can be approximated by the air entry matric head (h_a). Sands have thinner capillary fringes than unstructured clays (Table 4.1). Capillary fringes are also found around zones of saturation in the unsaturated zone, such as around injection boreholes, surface seepage basins, and losing streams.

In this section, an overview of the governing equations and processes that characterize unsaturated water flow will be presented. While the subsurface should preferably be treated as a uniform porous media, the unsaturated zone possesses several features that prevent this. First, soil layers have a large effect on the behavior of water flow; even small spatial changes in soil properties can induce large variations in the direction and magnitude of water flow, especially under transient conditions. Second, soil properties vary substantially in space within a layer. Third, macropores allow water to rapidly bypass the bulk of the soil matrix.

4.3.1 Unsaturated Flow Equations

4.3.1.1 Buckingham-Darcy Equation
In the unsaturated zone, larger pores drain more readily than smaller ones, as can be noted using the capillary rise equation. Therefore, the hydraulic conductivity is much smaller under unsaturated than saturated conditions due to water moving through small pores or as films along the walls of larger pores. At very low water contents, continuous fluid paths may not exist and water may move in the vapor phase. The unsaturated hydraulic conductivity is therefore represented as a function of matric head [K(h)] or as a function of water content [K(θ)]. Buckingham (1907) modified Darcy's equation for unsaturated flow:

$$J = -K(h)\frac{\partial H}{\partial z} = -K(h)\frac{\partial h}{\partial z} - K(h)$$

[4.27]

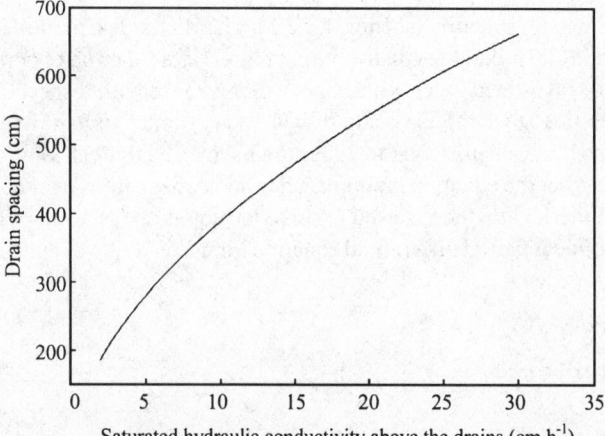

Fig. 4.3 Drain spacing as a function of hydraulic conductivity above the drains required to maintain the watertable below a depth of 70 cm with i = 0.25 cm h^{-1}

where the total head is the sum of the matric and gravitational heads, $H = h + z$. For unsaturated flow, the Laplace equation is a nonlinear differential equation because of the dependence of $K(h)$ on potential:

$$\nabla J = \nabla \cdot (-K(h)\nabla H) = -\nabla K(h) \cdot \nabla H - K(h)\nabla^2 H = 0 \qquad [4.28]$$

4.3.1.2   Richards Equation

For transient flow, the water content and matric head vary with time (t) and the Buckingham-Darcy equation must be extended. This partial differential equation was developed by Richards (1931). For one-dimensional vertical flow, the continuity equation requires that the change in volumetric water stored within a soil element be equal to the net flux into the element and any sources or sinks within the element:

$$\frac{\partial \theta}{\partial t} = -\frac{\partial J}{\partial z} + s \qquad [4.29]$$

where s is a source/sink term for water-per-unit time. When the Buckingham-Darcy equation is substituted for J:

$$\frac{\partial \theta}{\partial t} = \frac{\partial}{\partial z}\left[K(h)\left(\frac{\partial h}{\partial z} + 1\right)\right] + s \qquad [4.30]$$

which is called the *mixed form* (because it contains two dependent variables [θ and h]) of the one-dimensional Richards equation. It can also be written in terms of h (the *h form*) by using the chain rule on the left-hand side:

$$C(h)\frac{\partial h}{\partial t} = \frac{\partial}{\partial z}\left[K(h)\left(\frac{\partial h}{\partial z}\right) + 1\right] + s \qquad [4.31]$$

where $C(h)$ is the water capacity function (section 4.3.2.3). The Richards equation is also a nonlinear partial differential equation. To predict the distribution of h or θ as a function of depth for a particular time, this equation must be integrated twice with respect to z and once with respect to t, with boundary and initial conditions specific to the problem being solved. Except under very limited conditions, these integrals are not known so the Richards equation does not have an analytical solution. Therefore, for transient flow problems, either the equation is simplified in some manner (as happens with infiltration equations, below) or a numerical method is used to solve the equation (section 4.5). Only numerical solutions exist for the nonlinear partial differential equation for unsteady, two- and three-dimensional, isotropic flow:

$$\frac{\partial \theta}{\partial t} = \nabla \cdot (K(h)\nabla H) + s \qquad [4.32]$$

4.3.2 Unsaturated Flow Parameters

The most important parameters in unsaturated flow are the unsaturated hydraulic conductivity and (to a lesser extent) the water capacity functions. Both depend on pore size distribution. Values of K(h) vary by orders of magnitude with matric head, are sensitive to texture and structure, and are highly variable in space and (in some cases) time.

4.3.2.1 Unsaturated Hydraulic Conductivity

A typical K(h) curve for a hypothetical sand and an unstructured clay are shown in Fig. 4.4. The saturated hydraulic conductivity [K(0) = K_s] is higher in the sand than in the clay, as discussed in section 4.2.2.1. Conductivity drops rapidly in the sand as pores empty since there is a more narrow distribution of pore sizes compared to the clay. Conductivity declines more gradually in the clay because each decrease in potential empties only a few pores due to the wide distribution in pore size. At low matric head (water contents), the unsaturated hydraulic conductivity is greater in the clay than in the sand due to more water-filled pores and continuous water films in the clay.

The Gardner (1958) equation is commonly used to describe the unsaturated hydraulic conductivity function:

$$K(h) = K_s \exp(\alpha h)$$

[4.33]

where α is a constant. Other commonly used equations are the Brooks and Corey (1964) equation:

$$K(h) = K_s \left(\frac{h}{h_a} \right)^{-2-3\lambda}, \quad h < h_a$$

$$K(h) = K_s \quad , \quad h \geq h_a$$

[4.34]

where h_a is the air entry matric head and λ is a constant; the Campbell (1974) equation:

$$K(\theta) = K_s \left(\frac{\theta}{\theta_s} \right)^m$$

[4.35]

where m is a constant; and the Haverkamp equation (Haverkamp et al., 1977):

$$K(h) = \frac{K_s}{1 + \left(\dfrac{h}{a} \right)^N}$$

[4.36]

where a and N are constants. Another useful equation is the van Genuchten (1980) Equation [4.62] described later in section 4.4.1.2.

4.3.2.2 Capillary Length Scales

A useful way to quantify the unsaturated hydraulic conductivity function in a single parameter is in terms of the macroscopic capillary length. This is a K(h) weighted average pressure head over the

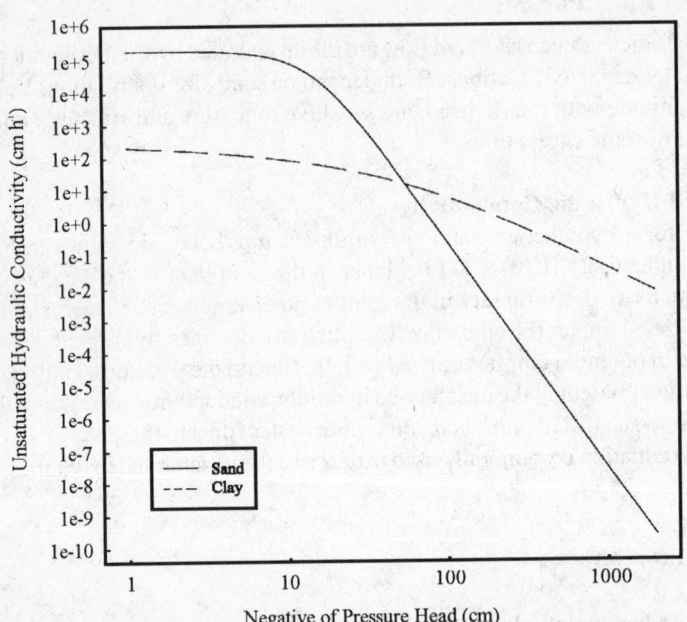

Fig. 4.4 Unsaturated hydraulic conductivity as a function of pressure head for sand and clay

range of matric heads of interest, typically the initial matric head in the soil before water is added (h_i) and the soil pressure or matric head in equilibrium with the water source (h_0) (White and Sully, 1987):

$$\lambda_c = \frac{\int_{h_i}^{h_0} K(h)dh}{K(h_0) - K(h_i)} \qquad [4.37]$$

When the Gardner Equation [4.33] is used to describe $K(h)$ and $K(h) \ll K(h_0)$, $\lambda_c = \alpha^{-1}$. Using this relationship, the effect of the shape of the $K(h)$ function on λ_c can be demonstrated. A small value of λ_c indicates that $K(h)$ rises sharply at the wet end of the curve, as it might in a structured clay with macropores (Section 4.3.6.1). Typical values for λ_c in undisturbed soils are shown in Table 4.2. The effect of structure in fine-textured soils is apparent in that an unstructured clay may have a λ_c of 25 cm, whereas a structured clay may have a λ_c of 2.8 cm. The macroscopic capillary length is also a measure of the effect of capillarity (the attraction of water to dry soil) as opposed to gravity on water movement. Water flow from a point source into a soil with a large value of λ_c (unstructured clay) will have more lateral flow than into a soil with a small value of λ_c (sand or well-structured clay).

The macroscopic capillary length can be converted to an average soil pore size called the "microscopic capillary length" (λ_m) using the capillary rise equation (White and Sully, 1987):

$$\lambda_m = \frac{\sigma}{\rho g \lambda_c} \qquad [4.38]$$

where σ is the surface tension.

Brakensiek and Rawls (1992) provided λ_c and λ_m for soils of different textures (Table 4.1). Fine-textured soils have the largest macroscopic capillary length (capillarity) and the least microscopic capillary length (average pore size).

4.3.2.3 Water Capacity Function

The water capacity function C(h) is the slope of the soil water characteristic curve (θ[h], discussed in more detail in Section A, chapter 3). It is an important parameter in numerical models of water flow (Section 4.5) and a plot of C(h) versus h is a useful way to display the pore size distribution of a soil.

4.3.3 Steady Flow

It is useful to consider several cases of steady unsaturated flow because analytical solutions to the Buckingham-Darcy equation have been developed for these special conditions. Although true steady flow may occur only rarely, transient flow usually approaches steady flow at long periods of time and the analytical solutions help to show what factors may be important in nonsteady conditions.

Steady one-dimensional flow in a uniform soil can be described by the Buckingham-Darcy Equation [4.27]. If the water content is uniform with depth, then only gravity causes flow and the flux or infiltration rate is equal to the hydraulic conductivity:

$$J_w = -K(h) \tag{4.39}$$

In this case the distribution of h with depth is a vertical line (the same at all depths). This is called *gravity flow* or *unit gradient flow*.

For steady downward flow or infiltration ($J_w = -i$) in a uniform soil from the surface to a watertable at depth L (as might occur in a soil remediation effort designed to flush contaminants from the unsaturated zone down into the capture zone of a well), the Buckingham-Darcy equation can be integrated and solved for h using the Haverkamp equation (Equation [4.36]) for K(h) (Jury et al., 1991):

$$h = a\sqrt{\frac{K_s}{i} - 1} \ \tanh\left[\frac{-\sqrt{\left(1 - \dfrac{i}{K_s}\right)\dfrac{i}{K_s}}}{a}(z + L)\right] \tag{4.40}$$

Table 4.2 Soil texture/structure categories for estimation of macroscopic capillary length (λ)

Soil texture/structure category	λ_c cm
Coarse and gravelly sands; may also include some highly structured soils with large cracks and/or macropores	2.8
Most structured soils from clays through loams; also includes unstructured medium and fine sands	8.3
Soils which are both fine textured (clayey) and unstructured	25
Compacted, structureless, clayey materials such as landfill caps and liners, lacustrine or marine sediments, etc.	100

The distribution of h versus z predicted by Equation [4.40] for a Chino clay with $K_s = 1.95$ cm day^{-1}, a = −23.8 cm, and N = 2 (Gardner and Fireman, 1958) is shown for a low (i = 0.5 cm day^{-1}) and a high (i = 1.5 cm day^{-1}) irrigation rate in Fig. 4.5. The distribution of h is now curvilinear, contrary to a uniform saturated soil (section 4.2.1.2), especially with the lower irrigation rate which causes a lower h at the surface. At the watertable, the pressure head is zero, of course. Near the surface, especially at the high irrigation rate, the curve is nearly vertical indicating that gravity flow is occurring (Equation [4.39]).

Three-dimensional water flow from a point source will approach a steady rate after a period of time. For example, steady infiltration from a shallow ponded ring on the soil surface is usually described using the Wooding (1968) equation:

$$i_s = J\big|_{z=0} = K(h_0)\left[1 + \frac{4\lambda_c}{\pi r}\right] \qquad\qquad [4.41]$$

where i_s is the steady infiltration rate, r is the radius of the ring, and $K(h_0)$ is the hydraulic conductivity corresponding to the potential of the water supply at the soil surface. Flow into the soil is considered a positive flux, in this case. It is apparent by comparing Equations [4.39] and [4.41] that the second term in Equation [4.41] accounts for the increased infiltration rate from a ring due to lateral (capillary driven) flow into dry soil. The Wooding equation assumes that the unsaturated hydraulic conductivity function can be described by the Gardner Equation [4.33].

4.3.4 Infiltration

Infiltration is a key process because it determines how much water from rainfall, irrigation, or a contaminant spill enters the soil and how much becomes runoff (or overland flow in hydrology terminology). It is also a key process in erosion in that there can be no erosion without runoff to transport and scour sediment.

During infiltration, a wetting front of higher water content moves down through the soil over time. The abruptness of the wetting front depends on the pore size distribution and shape of the K(h) function. For coarse-textured soils with a narrow pore size distribution, the wetting front will be more abrupt and in a fine-textured soil, the wetting front will be more diffuse. The wetting front is a combination of new water added by the rain and old water displaced to lower depths.

Measurements and numerical solutions have shown that the infiltration rate (i, LT^{-1}) in a uniform, initially dry, soil when rainfall does not limit infiltration, decreases with time and approaches an asymptotic minimum infiltration rate (Fig. 4.6a). Cumulative infiltration (I, L) is the area under the infiltration rate curve (Fig. 4.6b). Infiltration rate may be thought of as the Darcy flux at the soil surface with J = -i (infiltration is a special case where downward water flow is considered positive). At the soil surface, the water content increases with time so K(h) must also increase with time and cannot account for the decline in i. Initially, the wetting front is just below the soil surface and the water potential gradient at the surface is very large. As the wetting front moves deeper in the soil, the gradient at the surface decreases and this has an overriding effect on the increase in K(h) and accounts for the decrease in i with time. Eventually, the distribution of potential with depth near the surface approaches a unit gradient and at that point the infiltration rate asymptotically approaches the field-saturated hydraulic conductivity. A high antecedent water content will result in a lower initial infiltration rate due to a diminished potential gradient, but the final infiltration rate will be the same regardless of the antecedent water content.

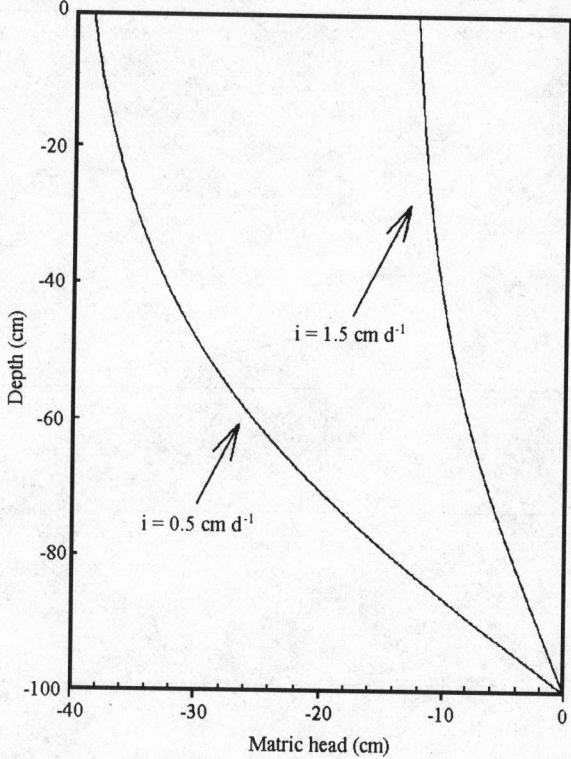

Fig. 4.5 Distribution of matric head as a function of depth for steady infiltration at a low ($i = 0.5$ cm d^{-1}) and a high ($i = 1.5$ cm d $^{-1}$) irrigation rate

Experiments have shown that this minimum final infiltration rate is less than the saturated hydraulic conductivity that would be measured on a carefully saturated, intact core in the laboratory (section 4.4.1.1). This has been attributed to the effect of entrapped air under field conditions that reduces the cross-sectional area available for water flow and the water potential gradient. Entrapped air can be divided into mobile and immobile pockets of air (Faybishenko, 1995). The immobile air resides in the fine and dead-end pores and can only be removed by dissolution. During infiltration, the mobile air is displaced from the finer into larger pores as the fine pores fill by capillarity. Thus entrapped air blocks the larger pores which have a disproportionate effect in reducing flow. It can be a matter of days for all of the mobile air to migrate to the surface (Faybishenko, 1995). Bouwer (1966) recommended using a value for field-saturated hydraulic conductivity (K_{fs}) equal to one-half of the true saturated hydraulic conductivity (K_s) for predicting the steady-state infiltration rate. In the same way, a field-saturated water content (θ_{fs}) as opposed to θ_s can be considered.

4.3.4.1   Infiltration through Crusts and Layered Soils

Another factor that can cause the infiltration rate to decrease is the formation over time of a surface seal or crust at the soil surface. A surface seal is a very thin layer (1–5 mm) at or just below the soil surface that forms due to the breakdown of soil aggregates and chemical dispersion of clay particles under raindrop impact. The clay particles fill the soil pores and create a layer with a saturated hydraulic conductivity several orders of magnitude less than the undisturbed soil (Miller and Radcliffe, 1992).

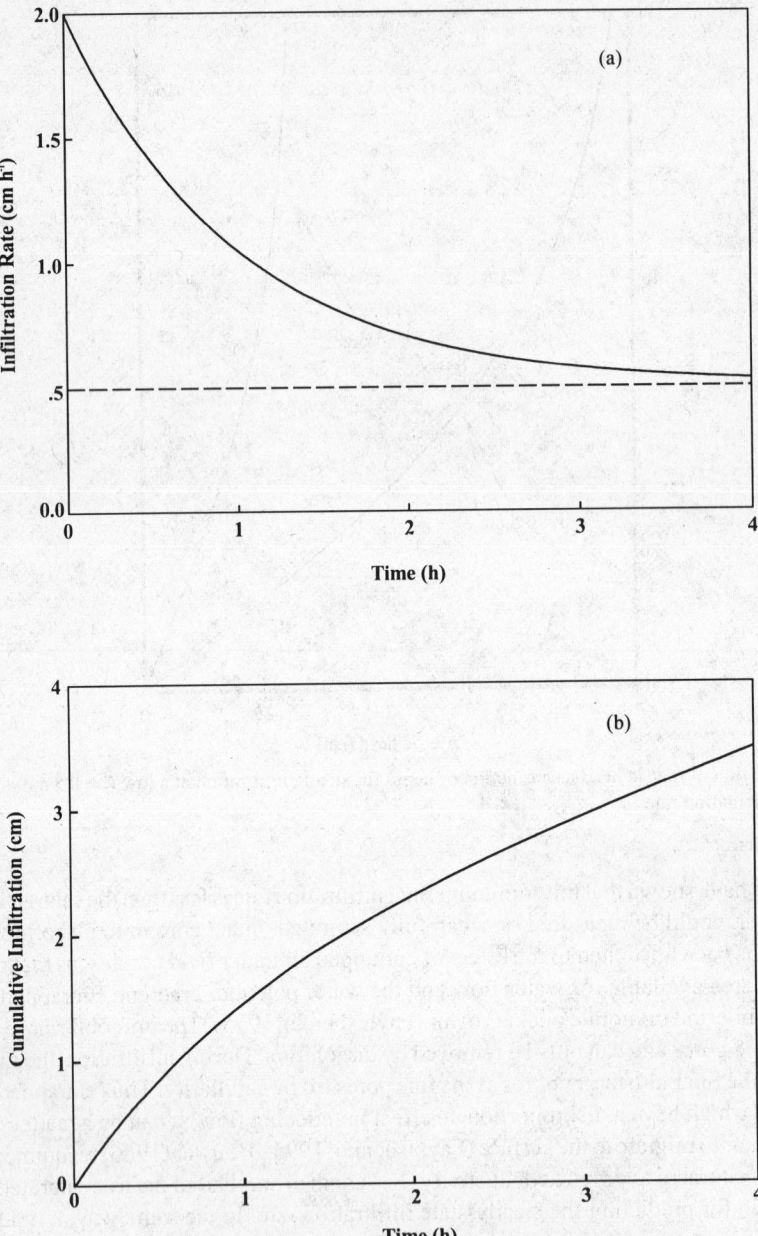

Fig. 4.6   Infiltration rate (a) and cumulative infiltration (b) as a function of time

This low conductivity layer can prevent saturation of the soil just beneath the seal due to the suction that occurs at the interface (section 4.2.3.1), further reducing the infiltration rate. Seals can be prevented by protecting the soil surface from raindrop impact with a mulch or previous crop residue, as is done in conservation tillage (Section G, chapter 12) and by preventing chemical dispersion through the use of soluble amendments (Miller and Baharuddin, 1986). Once the seal dries out, it develops a high soil strength due to the increased density of the layer and is called a crust. A crust can reduce seedling emergence, especially in dicots.

Buried clay or dry sand layers near the surface can also reduce infiltration rates. An unstructured buried clay layer will usually have a lower K_{fs} than an overlying coarse-textured layer and reduce K_{eff} (section 4.2.3.1) and i once the wetting front enters the clay layer. A buried dry sand layer under a fine-textured layer will also impede deeper movement of the wetting front and reduce i, but through a different mechanism. The water at the leading edge of the wetting front may be under several thousand cm of tension and cannot enter the smallest pores in the sand layer (which are much larger than the smallest pores in the layer above) until potential at the wetting front increases to the water entry potential for the sand. This stalls the wetting front until potentials rise to the critical level for entry. Since the pore size distribution is narrower in the sand, it is not long after water first enters the soil that the potential is high enough at the wetting front to fill the largest capillary pores in the sand. Once the sand is field saturated, it no longer impedes flow because K_{fs} is high in the sand compared to the fine-textured layer above. This is shown in Fig. 4.7 where a numerical model was used to predict water contents during infiltration from a point source into a soil with a dry sand layer at a depth of 40 cm. After 1.3 hours, the wetting front has spread laterally above the sand layer and just started to penetrate right below the source where the wetting front matric head is the greatest. Baver et al. (1972) referred to the action of a buried dry layer in temporarily impeding water flow and infiltration as a check valve. This principle is used in the design of golf greens which have a sand surface layer over a coarse gravel layer at about 50 cm. The gravel layer keeps water from frequent light irrigations in the root zone, but if there is a large rainstorm the gravel layer will fill and drain the root zone so that the green does not become waterlogged.

The effect of soil layers is to introduce bulk anisotropy within the soil profile. The hydraulic conductivity is no longer the same in all directions, hence anisotropic, and is a function of the matric head, or water content. Such moisture-dependent anisotropy causes distinctly different flow behavior as a function of ambient conditions. When the water content is such that the hydraulic conductivity of both layers is equal, the water moves vertically, but under either wetter or drier conditions there may be substantial lateral flow.

Because of the importance of the infiltration process, simplified solutions to the Richards equation have been developed to predict infiltration.

4.3.4.2 Profile Controlled Infiltration
Most of the infiltration equations have been developed for conditions when rainfall does not limit infiltration. In this case, the infiltration rate is less than the rainfall or irrigation rate and runoff occurs. Soil hydraulic properties control the infiltration rate so it is profile controlled (Hillel, 1980b), also called ponded conditions, although the depth of ponding may be negligible if surface storage is small.

Green and Ampt (1911) developed a simplified mechanistic equation for infiltration by assuming that the wetting front in a soil was a square wave or sharp front. Although this is approximately true only in coarse-textured soils, there is no error in predicting the infiltration rate as long the amount of water behind the predicted square front is equal to the amount of new water behind the true wetting front. The Green-Ampt equation for cumulative infiltration in a uniform soil, including the effect of gravity, is

$$I(t) = K(h_0)t + \Delta h\, \Delta\theta\, \ln\left(1 + \frac{I(t)}{\Delta h\, \Delta\theta}\right)$$

[4.42]

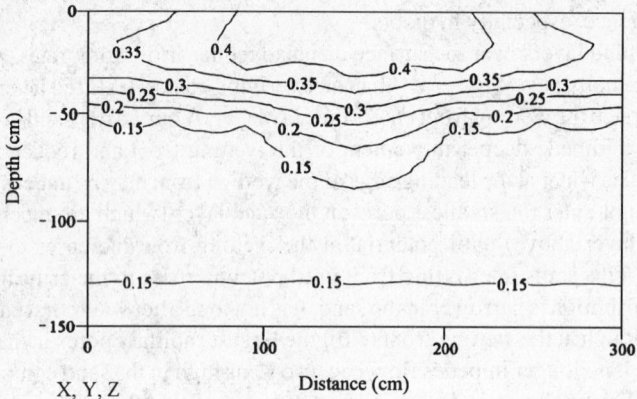

Fig. 4.7 Distribution of water after 1.3 hours of infiltration into a clay layer overlying a dry sand layer commencing at a depth of 40 cm. The hydraulic parameters were $K_s = 5$ cm day^{-1}, $\theta_s = 44$, $\theta_r = 0.15$, $h_a = -50$ cm, Brooks and Corey $\lambda = 0.6$ for the clay layer and $K_s = 100$ cm day^{-1}, $\theta_s = 0.40$, $\theta_r = 0.08$, ha $= -15$ cm, Brooks and Corey $\lambda = 1.0$ for the sand layer. The initial conditions were h $= -100$ cm. The problem was solved with VS2DT [Lapalla et al., 1987; Healy, 1990].

where $\Delta h = h_0 - h_f$, $\Delta\theta = \theta_0 - \theta_i$, θ_i is the initial water content, θ_0 is the water content extending from the wetting front to the soil surface, h_0 is the matric head corresponding to θ_0, and h_f is the matric head at the wetting front. When the rainfall rate is greater than or equal to the infiltration rate, θ_0 and $K(h_0)$ can be approximated by θ_{fs} and K_{fs}, respectively, and h_0 can be assumed to be zero. The unknowns in Equation [4.42] are then I(t) and h_f. White and Sully (1987) showed that the magnitude of the wetting front matric head can be approximated using the macroscopic capillary length λ_c (Tables 4.1 and 4.2):

$$\left|h_f\right| = \frac{\lambda_c}{2b} \qquad\qquad\qquad [4.43]$$

where b is a dimensionless factor that has a theoretical range of $^1/_2$ to $^\pi/_4$, but can be assumed to be equal to 0.55 in most cases (Warrick and Broadbridge, 1992). Hence, the only unknown in Equation [4.42] is I(t), but the equation cannot be solved directly for this variable because it appears both inside and outside the ln function. Therefore, it must be solved iteratively. Once the cumulative infiltration curve as a function of time is known, the infiltration rate can be calculated as the instantaneous slope of this curve.

The Green and Ampt Equation [4.42] can also be written as:

$$\frac{K(h_0)t}{\Delta h \Delta \theta} = \frac{\xi^2}{2} - \frac{\xi^3}{3} - \frac{\xi^4}{4} - \ldots \qquad\qquad [4.44]$$

because:

$$\ln(1 + \xi) = \xi - \frac{\xi^2}{2} + \frac{\xi^3}{3} - \ldots \qquad\qquad [4.45]$$

where $\xi = I(t) \Delta h^{-1} \Delta \theta^{-1}$. Solving for the cumulative infiltration yields an infinite series of the form (Jury et al., 1991):

$$I(t) = A_0 t^{\frac{1}{2}} + A_1 t + A_2 t^{\frac{3}{2}} + \ldots \qquad [4.46]$$

This is the same form as the Philip (1957a–e) equation for infiltration:

$$I(t) = S_0 t^{\frac{1}{2}} + A_1 t + A_2 t^{\frac{3}{2}} + \ldots \qquad [4.47]$$

where S_0 is sorptivity (L T$^{-\frac{1}{2}}$) and A_1, A_2, ... are constants that depend on soil properties, θ_0, and θ_i. Sorptivity is a measure of the capillary uptake of water and a function of the initial soil water content (θ_i) and the water content at the soil surface (θ_0). It is not surprising that it is related to the macroscopic capillary length (λ_c) (Tables 4.1 and 4.2):

$$S_0 = \frac{\sqrt{\Delta\theta \left[K(h_0) - K(h_i) \right] \lambda_d}}{b} \qquad [4.48]$$

where b is the same dimensionless factor that appears in Equation [4.43].

At early times when capillarity is much more important than gravity, cumulative infiltration can be described by discarding all but the first term in Equation [4.47]:

$$I(t) = S_0 t^{\frac{1}{2}} \qquad [4.49]$$

Rawls et al. (1990) modified the Green and Ampt Equation [4.42] for soils with a crust by multiplying $K(h_0)$ by a crust factor CF:

$$CF = \frac{K_{eff}}{K(h_0)} = \frac{SC}{1 - \dfrac{h_{SC}}{L}} \qquad [4.50]$$

where K_{eff} is the effective hydraulic conductivity of the two-layer system, h_{sc} is the subcrust matric head (just below the interface between the crust and the underlying soil), L is the depth to the Green-Ampt wetting front $[= (\theta_0 -_i \theta)/I(t)]$, and SC is a correction factor accounting for the effect of the subcrust matric head in causing desaturation, thereby reducing the hydraulic conductivity of the subcrust layer (section 4.3.4.1). They showed that both SC and h_{sc} varied with texture:

$$SC = 0.736 + 0.0019 P_s \qquad [4.51]$$

where P_s is the sand content (%) and

$$h_{SC} = -45.19 + 46.68SC \qquad [4.52]$$

Since L appears in Equation [4.50] and depends on I(t), Equations [4.42] and [4.50] must be solved iteratively.

4.3.4.3 Supply Controlled Infiltration

During the early stages of a rainfall or irrigation event before ponding occurs, it is likely that the rainfall rate will limit infiltration. In other words, the rainfall rate is less than what the potential infiltration rate would be under ponded conditions (which maximize the potential gradient). In this case, the actual infiltration rate is equal to the rainfall rate (all of the rain infiltrates) and is supply controlled (Hillel, 1980b). At a later time, called the time to ponding (t_p), the potential infiltration rate may drop below the rainfall rate (as the soil wets up and the potential gradient lessens). At that point, ponding (and runoff if surface storage is negligible) commences and rainfall no longer limits the infiltration rate (it becomes profile controlled). The crux of the problem for predicting infiltration is to determine the time to ponding and to determine how to correct infiltration rates after ponding for the fact that less water has entered the soil than would have if ponding had occurred from the beginning.

Mein and Larson (1973) used the Green-Ampt equations to solve this problem for a steady rainfall rate (r). The cumulative infiltration at the time of ponding is

$$I_p = \frac{K(h_0)\Delta h \, \Delta \theta}{r - K(h_0)} \qquad [4.53]$$

The time of ponding then can be found by dividing I_p by r. Then Equation [4.42] for ponded conditions can be used to calculate cumulative infiltration for times later than t_p, but if the actual time was used in this equation it would overpredict I(t) because it assumes that infiltration has been proceeding at the maximum (ponded) rate since the beginning of the event. Therefore, the actual time must be corrected by subtracting a time interval that adjusts for the difference between the cumulative rainfall before ponding and I(t) calculated using Equation [4.42]. This corrected time (t_c) is calculated as:

$$t_c = t + \frac{I_p - \Delta h \Delta \theta \, \ln\left(1 + \dfrac{I_p}{\Delta h \Delta \theta}\right)}{K(h_0)} - t_p \qquad [4.54]$$

Chu (1978) used the Green-Ampt equations to solve the same problem for a nonsteady rainfall rate.

4.3.4.4 Curve Number Method

The curve number method is an empirically determined rainfall-runoff relationship that provides an indirect estimate of soil infiltration. The method is based on numerous measurements of runoff for many soil types and considers soil texture, soil drainage class, antecedent moisture conditions, and vegetative cover. While not directly estimating infiltration, the method partitions the effective cumulative precipitation (P_e) over a watershed into two classes: the depth of water that quickly leaves the watershed as (surface and subsurface) runoff (Q) and that water which is abstracted, or retained, on the watershed (F) so that $P_e = Q + F$. Abstracted water can emerge later as baseflow, deep groundwater discharge, or as evapotranspiration. The empirical relationship between these variables is

$$\frac{Q}{P_e} = \frac{F}{S} = \Theta \tag{4.55}$$

where S is the water holding capacity of the soil (i.e., $F \leq S$), and Θ is the watershed average soil saturation. This function implies that the amount of runoff increases as the watershed saturation increases; the runoff is zero when no water is stored ($F \ll S$ implies that $Q = 0$), and all of the rainfall occurs as runoff when the watershed has reached its maximum capacity ($F = S$ implies that $Q = P_e$). The abstraction volume (F) can be removed to yield:

$$Q = \frac{P_e^2}{P_e + S} \tag{4.56}$$

The standard practice is to assume that P_e equals the actual precipitation, minus an initial abstraction equal to S/5. The curve number is used as a surrogate for the maximum water holding depth of the watershed:

$$CN = \frac{1000}{S + 10} \qquad \text{or} \qquad S = \frac{1000}{CN} - 10 \tag{4.57}$$

Curve numbers are tabulated and range from near zero for a dry, fully vegetated, highly permeable surface to near 100 for an impervious surface (McCuen, 1982).

Several alternative explanations can be used to explain the theoretical basis for the curve number method. A relationship between soil saturation, infiltration rate, and overland flow, can be readily hypothesized. An alternative explanation rests on the hypothesis that spatially variable soil water holding capacity rather than infiltration rate limits abstraction depths.

4.3.5 Redistribution

Water continues to move in a soil for some time after the end of a rainfall or irrigation event. Water will drain from the regions near the soil surface to deeper depths and this process is called redistribution. Changes in profile matric potentials over time are shown in Fig. 4.8 for five soils during redistribution (Bruce et al., 1985). For the three coarse-textured soils (two sands and a sandy loam), most of the change in matric potential observed over a 10-day period occurred during the first 24 hours of drainage. This was less true for the two silt loam soils. Redistribution occurs more rapidly in a coarse-textured soil because of the steeper K(h) function, compared to a fine-textured soil (Fig. 4.4). In a coarse-textured soil, K(h) is likely to be quite high in the undrained regions near the soil surface where water contents are near saturation so water drains rapidly. Once some of this water drains, however, water contents decrease, K(h) becomes much lower, and redistribution effectively stops. In a fine-textured soil, the initial drainage is slower due to the lower K(h) near saturation, but redistribution continues longer because K(h) in the drained region is not negligible.

In the absence of evapotranspiration and any further rainfall or irrigation, water will redistribute until the total potential (H) is the same at all depths and matric head (h) increases linearly with depth (at a rate of 1 cm per cm of depth). In a soil with a shallow watertable, the magnitude of the matric head at the surface will be equal to the depth of the watertable (Jury et al., 1991).

The water content of the soil near the surface once redistribution becomes negligible has been called field capacity or the drained upper limit. This has been approximated by the water content measured in the field 24–48 hours after a thorough wetting of the soil profile or by the water content corresponding to a matric potential of –33.3 kPa in a clayey soil and –10 kPa in a sandy soil. Although these approximations are not very satisfying from a theoretical point of view, the field capacity concept is a useful one in irrigation scheduling and crop modeling.

4.3.6 Preferential Flow

Preferential flow is used as a general term to describe unusually rapid or deep movement of water or solute through a fraction of the total cross-sectional area available for flow. Preferential flow includes flow through macropores, fingering, and funnel flow. Gish and Shirmohammadi (1991) recently reviewed preferential flow.

4.3.6.1 Macropore Flow

Macropores are large, continuous voids in soil and include structural, shrink-swell, and tillage fractures, old root channels, and soil fauna burrows. Suggested lower limits for macropore diameters and widths are in the 0.03–3.00 mm range (Luxmoore, 1981; Beven and Germann, 1982; White, 1985). The lower limit would include some pores that would fill by capillarity and the upper limit would exclude all capillary pores. Macropores are important because they can increase infiltration and may result in bypass flow where water and solutes move rapidly through the profile and do not interact with the soil matrix (Quisenberry and Phillips, 1976).

To a certain extent, the effect of macropores on water flow can be incorporated into conventional flow equations based on Darcy's law by careful measurement of K_{fs} and the K(h) function. If these parameters are measured in the field on a large enough sample to contain representative macropores, there is little error in using Darcy's equation (in fact, the error is that Darcy's equation will overpredict flow in pores with high Reynolds number; under prediction is usually the concern). For example, Jarvis and Messing (1995) used a tension infiltrometer (section 4.4.2.1) to measure K(h) at values of

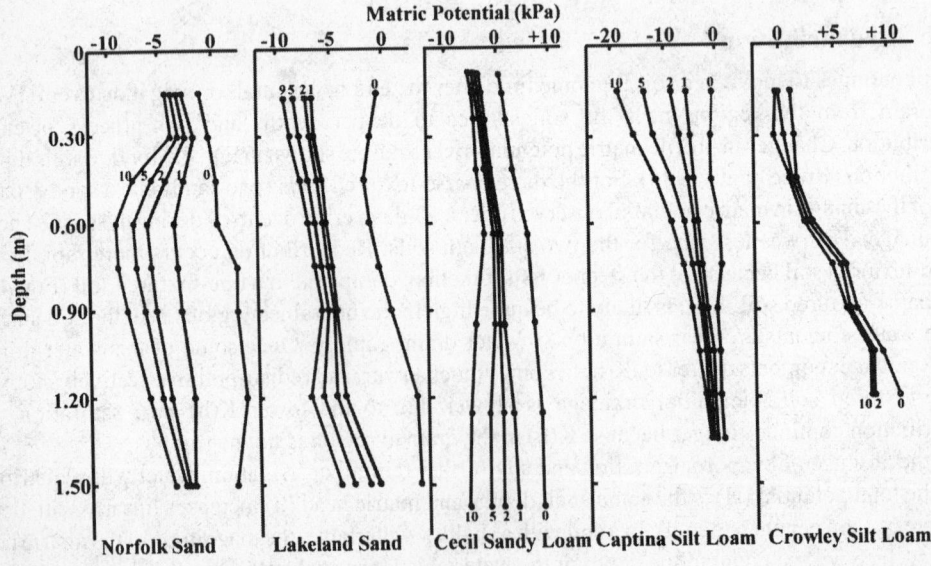

Fig. 4.8 Matric pressure potential as a function of depth for five soils during the first ten days of drainage following a thorough wetting [Bruce et al., 1980]

h between –5 and –150 mm on six Swedish soils of contrasting texture. When the data were plotted (lnK vs. h), the best fit was with two straight lines, the line near saturation being much steeper, especially in the more clayey soils (Fig. 4.9). The breakpoint occurred at h = –30 to –50 mm, which corresponds to a pore diameter near the minimum for macropores (h = –30 mm corresponds to a pore diameter of approximately 0.1 mm). The value of K(h) approached at zero pressure (an estimate of K_{fs}) was greatest for the four finer textured soils (in contrast to the data in Table 4.1). This indicated that soil structure and macropores had a greater effect on the K(h) near saturation in the fine- compared to the coarse-textured soils.

Flow in individual water-filled macropores that are cylinders or cracks can be described by a modified Poiseuille equation but the number and dimensions of macropores in a soil are usually not known (White, 1985). Also, using Poiseuille's law assumes that macropores are open ended, which is probably not the case. Beven and Germann (1981) developed a kinematic wave equation to describe water flow in macropores that allows flow down the sides of the pores that are not filled with water. It requires an estimate of the size distribution of macropores and the average distance between macropores.

Since macropores are, for the most part, noncapillary pores, it has been assumed that macropore flow cannot occur unless there is free water (incipient ponding conditions) at the soil surface. Experimentally, however, macropore flow has been observed in very dry soils at the onset of a rain when these conditions are unlikely. For example, Shipitalo and Edwards (1996) used intact soil blocks and added water equivalent to a two-year storm with a rainfall simulator. They observed more macropore flow in blocks that were initially dry than in blocks that were at a higher antecedent water content. This may be due to a hydrophobic organic soil surface that develops under dry conditions and causes free water to run across the surface and enter macropores (Miller and Wilkinson, 1979; Edwards et al., 1989).

Seyfried and Rao (1987) used dye patterns to trace water movement through intact columns and found dyed regions of the soil that were not associated with visible macropores. Gupte et al. (1996) and Shaw et al. (1997) also used a dye and found regions of dyed cross-sectional area, rather than individual dye stained macropores. Macropores were visible in some, but not all, dyed areas and there were few if any macropores in the undyed areas. Soil thin sections revealed that the dyed regions had slightly higher total porosity and more large pores than the undyed regions. This may represent spatial variability in soil development. One region of soil may develop more rapidly due to penetration by a root along a plane of weakness, which in turn leads to more intense microbial and faunal activity, more water penetration, and greater differentiation between regions. This type of flow is analogous to the two-region model of van Genuchten and Wierenga (1976) used to describe solute transport. It assumes that there are immobile regions of soil water in the interior of peds where water movement is much slower than in the mobile interped regions.

4.3.6.2 Fingering
When fine-textured soils overlie coarse-textured soils, it has been observed during infiltration that the wetting front does not uniformly penetrate the coarse layer (Hill and Parlange, 1972). Instead, the wetting front becomes unstable and separates into distinct fingers that penetrate only a portion of the coarse-textured layer. Initially, water cannot enter the coarse layer due to the low wetting front matric head created in the fine-textured layer (Hillel and Baker, 1988). As the wetting front stalls, matric head increases at the interface. Then when matric head rises to the value required to enter the smallest pores in the coarse layer, water starts to enter. Matric head continues to rise until it reaches a value where the flux through the coarse layer is equal to or greater than the flux through the top layer. Baker and Hillel (1990) called this matric head the "effective water entry" value (h_e). If the flux at this potential is

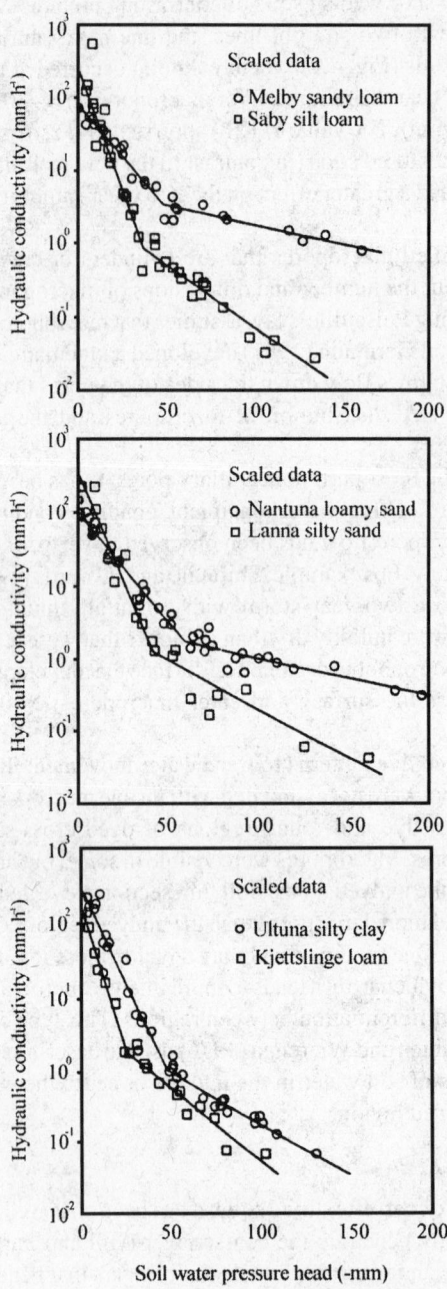

Fig. 4.9 Unsaturated hydraulic conductivity as a function of matric head in six Swedish soil of contrasting textures [Reprinted from Jarvis and Messing, 1995. Soil Sci. Soc. Am. J. 59:27-34, with permission of the Soil Science Society of America]

greater than the flux through the fine-textured layer, then the only way for flows to be equal is to confine flow through a fraction of the cross-sectional area of the coarse layer (fingering). The effective water entry matric head corresponded well with the inflection point potential on the water retention curve, which is a measure of the dominant pore size. The fraction of wetted soil in the coarse layer at steady state was equal to the flux through the top layer divided by $K(h_e)$ in the lower layer. Theoretical analyses have shown that fingering can occur under a number of conditions: (1) increasing K(h) with depth (as would occur in a fine-textured layer over a coarse-textured layer), (2) entrapped air that causes substantial air compression at the wetting front, (3) a buried hydrophobic layer, (4) an increase of water content with depth, (5) redistribution in a coarse-textured soil, and (6) continuous nonponding infiltration in a coarse-textured soil (Raats, 1973; Philip, 1975; Parlange and Hill, 1976; Diment et al., 1982; Glass et al., 1989, 1991).

4.3.6.3 Funnel Flow
Kung (1990a,b) used a dye to study water flow in sandy Wisconsin soils deposited during glacial outwash. The dye moved down uniformly from the surface until it encountered a lense of coarse sand (common in these soils) which acted as a capillary barrier. Since the lenses were not horizontal, the wetting front did not pause and wetting front matric heads never increased to the effective water entry value of the coarse sand lenses. The dye moved horizontally along the top of the lense to the downslope end where downward flow continued. Repeated encounters of the dye front with lenses resulted in funneling of the flow into a fraction of the total cross-sectional area of flow.

4.4   Measurement of Hydraulic Parameters

Accurate measurement of hydraulic parameters by both laboratory and field methods is essential to the prediction of water movement in soils. The important laboratory and field methods commonly employed in estimating hydraulic properties are discussed below with additional procedures available in Klute (1986), Smith and Mullins (1991), Topp et al. (1992), and Carter (1993). A multitude of laboratory and field methods have been developed to provide parameter inputs to predictive models because accurate measurement of hydraulic parameters is essential to the credible prediction of water movement in soils. However, uncertainties in parameter estimates arise because of spatial variation, measurement accuracy, and scale effects. For example, hydraulic conductivity typically varies over four orders of magnitude over short distances at a site. Small procedural differences in methodology can induce substantial variabilities in estimated hydraulic properties, as well. Reconciling data collected from laboratory versus field techniques often is difficult because bulk parameter estimates resulting from testing at different scales are not commensurate, as discussed in Section 4.2.

4.4.1   Laboratory Methods

Laboratory methods are used to measure K_s and K(h) on soil cores that are usually 7.6 cm in diameter, while fields, plots, or rings are used in field methods. The use of intact rather than packed cores is recommended if the parameters are intended to represent undisturbed soils, especially fine-textured soils.

4.4.1.1   Direct Methods
The constant head and falling head methods are common for measuring K_s. In the former method, a constant head of water is maintained at the top or bottom of the core and the steady water flux is recorded (Klute and Dirksen, 1986). Using Darcy's equation:

$$K_s = \frac{JL}{H_2 - H_1} \qquad\qquad [4.58]$$

where L (> 0) is the length of the core, H_1 and H_2 are total potentials at the top and bottom of the core, respectively. The flux, J, is negative in the downward direction so that K_s must always be positive. In the falling head method, a standpipe is attached to the top of the core and the head at the top is allowed to drop over time. The total head at the top of the core is measured using the standpipe at the beginning (h_1+L) and end (h_2+L) of a time interval (t). Using Darcy's equation:

$$K_s = \frac{L}{t} \ln \frac{h_1 + L}{h_2 + L} \qquad\qquad [4.59]$$

The falling head method is useful in soils with very low K_s where it is difficult to collect sufficient drainage from a core to measure J accurately within an interval of several hours. With the falling head method, h_1 can be recorded and the core left overnight, for example, before recording h_2 and there is no concern about the effect of evaporation on the drainage sample collected.

Values of K_s measured on cores in the laboratory will depend on what measures are used to remove entrapped air from the soil before making the measurement. Flushing the core with CO_2, wetting the core slowly from the bottom, and measuring K_s under conditions of an upward flux are commonly used. However, in the field some degree of entrapped air can be expected, especially near the soil surface during infiltration. Hence one can expect laboratory measurements of saturated hydraulic conductivity (K_s) to exceed field measurements (K_{fs}).

Unsaturated hydraulic conductivity can be measured on cores in a manner similar to the constant head method for K_s by fitting the core with porous plates and imposing suctions at the top and bottom of the core (Klute and Dirksen, 1986). The total potential within the core is measured at two heights using miniature tensiometers and the steady flux is recorded. The Buckingham-Darcy equation is solved for K(h) where h in this case is the average matric head within the core. Steady evaporation from the top of a long core in contact with free water at the bottom can also be used to measure K(h) by placing tensiometers at several heights, recording the steady evaporation rate, and calculating the gradient between each pair of tensiometers (Jury et al., 1991).

Steady unsaturated flow can be imposed on a long soil core by supplying water continuously at a rate less than K_{fs}. This can be done by applying a crust made of a mixture of gypsum and sand or cement and sand to the soil surface (Bouma et al., 1976, 1983). The steady water flux is measured with a constant positive head above the crust. Because of the low K_s of the crust, flow in the core is unsaturated (section 4.2.3.1). A tensiometer in the core records h and if a unit gradient is assumed then the measured flux is K(h). Successive flux measurements with different crusts, each with a different conductivity, provide measurements of K(h) over the range of matric heads achieved. Bouma et al. (1976) used this method in the field, as well.

In all laboratory methods of measuring K_s and K(h), the effect of the chemistry of the added water on dispersion must be considered. The goal is to prevent chemical dispersion so that pore size distribution and connectivity does not change during the measurement, which can be achieved by using a dilute solution with a divalent cation (Chiang et al., 1987) or native water from the site.

4.4.1.2 Indirect Method
The unsaturated hydraulic conductivity function of a soil can be estimated from its water characteristic curve and K_s by the indirect method. The characteristic curve provides information on the shape of the

$K(\theta)$ function (both depend on the pore size distribution) and the single value of K_s serves as a matching point to anchor the curve at saturation. The most common approach is based on an equation developed by Mualem (1976):

$$K(\theta) = K_s \Theta^n \left[\int_0^\theta \frac{d\theta}{h} \Big/ \int_0^{\theta_s} \frac{d\theta}{h} \right]^2$$

[4.60]

where Θ is the degree of saturation

$$\Theta = \frac{\theta - \theta_r}{\theta_s - \theta_r}$$

[4.61]

with θ_s and θ_r being the saturation and residual volumetric water contents, respectively. Mualem (1976) developed his model by considering the soil as a bundle of interconnected capillary tubes with Poiseuille's equation describing flow in each tube. The parameter n is related to the tortuosity and connectivity of the tubes and, based on measurements in 45 soils, he recommended a value of $n = \frac{1}{2}$. Some form of the relationship between and h (the water retention function) must be assumed to solve the integrals in Equation [4.60]. Most often the van Genuchten (1980) $\theta(h)$ relationship (Equation [3.31]) is used and it is assumed that the exponents in this equation are related in the manner $m = 1 - 1/n$. In this case, the integrals in Equation [4.60] can be solved analytically such that

$$K(\theta) = K_s \Theta^{\frac{1}{2}} \left[1 - \left(1 - \Theta^{\frac{1}{m}} \right)^m \right]^2$$

[4.62]

The indirect method works well for many coarse and medium textured soils, but not as well for fine-textured soils where structure has an important effect (van Genuchten and Leij, 1992). In well-structured soils, the distribution of noncapillary (macro) pores is not reflected in the water characteristic curve. In this case, the indirect method may represent the matrix $K(\theta)$ function accurately, provided a measured value of unsaturated hydraulic conductivity near saturation is used as a matching point in Equation [4.62] instead of K_s (Clothier and Smettem, 1990; Jarvis and Messing, 1995). Zurhuhl and Durner (1996) developed a bimodal form of the van Genuchten (1980) $\theta(h)$ relationship that could be used to incorporate macropores. In this case, the integrals in Equation [4.60] must be evaluated numerically.

4.4.1.3 Laboratory Inverse Method

Inverse methods measure flow under transient conditions and then use a numerical solution of the Richards Equation [4.30] to determine the parameters in an assumed form of the K(h) function. In the one-step outflow inverse method, cumulative outflow is measured from a core during an interval when the matric head at the inlet is increased substantially in a single step. Kool et al. (1985a) used a numerical model with the van Genuchten (1980) equations for water retention (Section A, chapter 3) and K(h) (Equation [4.64]) to determine the values of α, θ_r, and m that produced the best fit of the model predictions to the observed cumulative outflow. The other parameters in these equations, K_s and

θ_s, were measured independently on the core. A program called ONESTEP is available for determining the van Genuchten parameters (Kool et al., 1985b).

4.4.2 Field Methods

Several advancements in field methods of measuring K_{fs} and K(h) have occurred in recent years. These include the use of tension infiltrometers and borehole permeameters and the associated solutions to the Richards Equation [4.30], based largely on the equations developed by Philip (1957a–e).

4.4.2.1 Infiltrometers

Tension infiltrometers have become a popular method for determining unsaturated hydraulic conductivity and other hydraulic parameters in field soils. They consist of a circular porous plate or membrane which is placed on the soil surface or an excavated soil surface. Water is supplied to the plate under tension using a mariotte bottle arrangement and the rate of water entry into the soil can be measured on a graduated cylinder or with a pressure transducer (Clothier and White, 1981). In some cases, a ring is attached to the tension infiltrometer to allow ponded infiltration measurements at the same location (Perroux and White, 1988). Two approaches are common, one in which the infiltration rate is measured after it attains a steady rate, and the other in which the early nonsteady infiltration rate is measured.

Under the steady rate approach, infiltration is measured at the same location using two or more tensions, starting with the tension farthest from zero. The Wooding Equation [4.41], written in terms of volumetric flow rate (Q) instead of infiltration rate (i) where $Q = i\pi r^2$, is used to describe the steady-state flow rates (Q_1 and Q_2) at two matric heads ($h_1 < h_2$). The constant ($\alpha = 1/\lambda_c$) in the Gardner K(h) Equation [4.33] can be calculated as:

$$\alpha = \frac{1}{h_1 - h_2} \ln \frac{Q_1}{Q_2} \qquad [4.63]$$

and the field saturated hydraulic conductivity (K_{fs}) can be calculated as:

$$K_{fs} = \frac{\alpha Q_1}{r(4 + \alpha\pi r)\left(\dfrac{Q_1}{Q_2}\right)^P} \qquad [4.64]$$

where $P = h_1/(h_1 - h_2)$ (Reynolds and Elrick, 1991). Knowing K_{fs} and α, the unsaturated hydraulic conductivity at any value of h can be predicted using the Gardner Equation [4.33] on the assumption that a constant value of α enables the description of the entire K(h) function (the slope of lnK versus h is constant) which may not be the case (Jarvis and Messing, 1995). Alternatively, it can be assumed that lnK versus h is piecewise linear and that K(h) can be described by a piecewise Gardner equation (Reynolds and Elrick, 1991).

The disadvantage of the steady rate approach is that it may take several hours for flow to become constant (Warrick, 1992). For this reason, nonsteady rate approaches have been developed. These methods assume that at early times when capillarity is much more important than gravity, three-dimensional flow from a disk can be described by Philip's equation for one-dimensional infiltration into a dry soil (Equation [4.49]). In this method, S_0 is determined from a plot of I(t) versus $t^{1/2}$ during early time measurements with a tension infiltrometer. Two early time measurements of S_0 are made with different water supply potentials (h_0) and the unsaturated hydraulic conductivity is calculated at the mean of these two potentials (White and Perroux, 1989). The disadvantage of this method is that two values of S_0 are required so that measurements must be made at two locations (soil variability becomes a factor) or the soil must be allowed to dry to the antecedent water content before initiating the second measurement of S_0. Also, measurements of the antecedent water content (θ_i) and the water content of the soil in equilibrium with the supply potential of the tension infiltrometer (θ_0) are required. The latter is difficult to measure accurately when it is confined to a narrow depth.

Ring infiltrometers can be used to pond water on the soil surface and measure the infiltration rate. It is usually assumed that the Wooding Equation [4.41] describes the three-dimensional steady infiltration rate. Typical values of λ_c (Tables 4.1 and 4.2) can be used to convert steady flow rates to $K(h_0)$, which in the case of ponded infiltration is a measure of K_{fs}. Alternatively, if early time measurements of the infiltration rate are recorded, S_0 can be calculated assuming one-dimensional flow (Equation [4.49]). Assuming that the antecedent water content is low and $K(h_i)$ is negligible, Equation [4.48] may be solved for λ_c and substituted into the Wooding equation, which is solved for K_{fs} using b = 0.55:

$$K_{fs} = i_s - \frac{2.2 S_0^2}{(\theta_0 - \theta_i)\pi r}$$

[4.65]

where i_s is the steady infiltration rate and r is the radius of the ring.

Concentric double rings have been used to create one-dimensional flow in the interior ring. In this case, the steady infiltration rate is an estimate of K_{fs} in a uniform soil. Bouwer (1986) found that for the typical dimensions used in a double ring (20 cm diameter for inner ring and 30 cm diameter for outer ring), flow was not one dimensional.

4.4.2.2 Permeameters

Well permeameters are used to measure saturated hydraulic conductivity below the soil surface in the unsaturated zone. They consist of a mariotte device that maintains water at a constant level in a borehole and they allow measurement of the flow rate into the soil. The most common commercial well permeameters are the Guelph Permeameter and the Compact Constant Head Permeameter (CCHP). Steady flow into a borehole can be described by the equation:

$$i_s = K_{fs}\left[1 + \frac{H\lambda_c}{G\pi r^2} + \frac{H^2}{G\pi r^2}\right]$$

[4.66]

where H is the height of water ponded in the borehole, r is the radius of the hole, and G is a dimensionless geometric factor which depends primarily on the ratio of H/r (Elrick and Reynolds, 1992). Bosch and West (1997) fitted a polynomial equation to the data of Elrick and Reynolds (1992) to determine the value of G:

$$G = \frac{1}{2\pi}\left[A_1 + A_2\frac{H}{r} + A_3\left(\frac{H}{r}\right)^2 + A_4\left(\frac{H}{r}\right)^3\right]$$

[4.67]

The values of the coefficients $A_1...A_4$ in this polynomial depend on texture and structure (Table 4.3).

The similarity between Equation [4.66] and the Wooding Equation [4.41] is apparent. In sequence, the terms in the equation account for the effect of gravity, capillarity, and hydrostatic pressure in the borehole. There are two unknowns in this equation: K_{fs} and λ_c. The normal procedure is to measure i_s at two values of H in the same borehole and solve simultaneous equations for K_{fs} and λ_c. Since changing the level of H in the borehole necessarily changes the region of soil that is being sampled, soil heterogeneity in the form of layering or macropores can result in unrealistic and invalid (i.e., negative) K_{fs} and λ_c. As many as 30 to 80 % of measurements of K_{fs} and λ_c in structured soils may be invalid (both negative) according to Elrick and Reynolds (1992). Alternatively, a single measurement of i_s may be used and λ_c estimated using Table 4.2. Studies suggest that this method yields values of K_{fs} that are usually accurate to within a factor of 2 (Reynolds et al., 1992).

Another approach that does not require multiple measurements in the same borehole is based on the Glover solution (Zangar, 1953):

$$i_s = K_{fs} \frac{H^2}{G_G \pi r^2}$$ [4.68]

where G_G is the dimensionless geometric factor for the Glover analysis

$$G_G = \frac{\sinh^{-1}\left(\frac{H}{r}\right) - \sqrt{\left(\frac{r}{H}\right)^2 + 1} + \frac{r}{H}}{2\pi}$$ [4.69]

This approach only considers the effect of hydrostatic pressure in the borehole and ignores the effect of gravity and capillarity (Elrick and Reynolds, 1992). It can overestimate K_{fs} by an order of magnitude or more in dry, fine-textured, structureless soils (i.e., soils where capillarity is most important). On the other hand, the Glover solution can provide good estimates of K_{fs} in wet, coarse-textured, or structured soils when the ratio H/r is kept high (> 10, hydrostatic pressure dominates flow).

4.4.2.3 Instantaneous Profile and Field Inverse Methods

The instantaneous profile method is used to measure K(h) in the field during drainage under nonsteady conditions (Green et al., 1985). A soil pedon is thoroughly wetted and then allowed to drain while preventing water fluxes into or out of the soil at the surface. During the ensuing drainage period, profile distributions of water content and total head are measured periodically over the depths of interest. Values of K(h) are calculated at a given depth from an equation developed by integrating the Richards Equation [4.36] with respect to z:

Table 4.3 Coefficients for the polynomial (Equation 4.67) describing the dimensionless geometric factor G, valid for H/r < 10

Soil texture/structure	A_1	A_2	A_3	A_4
Sand	0.079	0.516	−0.048	0.002
Structured loams and clays	0.083	0.514	−0.053	0.002
Unstructured clays	0.094	0.489	−0.053	0.002

$$\frac{\partial}{\partial t}\int_0^{z_1}\theta(z,t)dz = K(h)\frac{\partial H(z,t)}{\partial z}\bigg|_{z_1} \qquad [4.70]$$

Water contents may be measured using neutron probes or TDR and gradients are usually measured with tensiometers. The range of matric head covered by this method is limited (near saturation to field capacity), but this is the range where significant unsaturated flow occurs. Many measurements of K(h) were made on soils of the southeastern United States using this method and are available in state bulletins (Cassel, 1985).

Inverse methods for field data have been developed that do not require the specific set of initial and surface boundary conditions required by the instantaneous profile (saturated profile with no surface flux). Dane and Hruska (1983) used a numerical solution to the Richards Equation [4.30], assuming that the soil profile θ(h) and K(h) could be described by the van Genuchten (1980) equations. The solution was optimized to find α and n in these equations. The method could be applied to other initial and boundary conditions, such as infiltration, but it assumed the entire profile could be described by a single set of θ(h) and K(h) functions.

4.4.2.4   Borehole Tests
Slug tests are commonly employed to estimate field hydraulic properties below the watertable because a minimum of effort is required for conducting the test. A known volume of water is either added or removed from a borehole, and the resulting water level change is recorded. The field-saturated hydraulic conductivity is estimated using (Bouwer and Rice, 1976):

$$K_{fs} = \frac{r^2}{2bt}\ \ln(R/r)\ \ln(h_o/h) \qquad [4.71]$$

where r is the radius of the borehole, b is the thickness of the permeable unit, R is the radial distance of influence of the test, h_o is the water level at the time of water injection or withdrawal, and h is the water level as a function of time (t) in the borehole. This formulation assumes horizontal flow. The radial influence of the test is an unknown, which can be approximated.

Alternatively, a steady pumping (or injection) rate can be established in one borehole, and water level changes can be observed in an observation borehole. Methods described in section 4.2 can then be used to estimate aquifer hydraulic properties.

4.4.3   Parameter Databases
There are several databases now available that contain hydraulic parameters for a large number of soils. One of these is the Unsaturated Soil Hydraulic Database (UNSODA) (Leij et al., 1996). In 1996, the database contained over 780 soils with data on water retention and saturated and unsaturated hydraulic conductivity, as well as basic soil properties such as particle size distribution, bulk density, organic matter content, etc. A large portion of the soils are sands and loamy sands. The program can be used to fit various θ(h) and K(h) functions to the measured data.

4.5   Numerical Models of Water Flow
Numerical methods of solving differential equations became feasible with the development of high speed computers, and were first applied to soil water movement in the early 1960s (Ashcroft et al., 1962; Hanks and Bowers, 1962). For many problems, it is no longer necessary to write computer code

in order to apply numerical methods to soil water flow problems. Codes are available from authors or through commercial outlets and some common codes are listed in Table 4.4. Additionally, new mathematical programming languages (e.g., MATLAB, MATHCAD, MATEMATICA) provide alternative techniques for solving nonlinear partial differential equations. It is important to note that use of numerical codes for modeling unsaturated media requires an appreciation of the difficulty in obtaining even simple solutions. As with any method, emphasis should be placed on comparing results from several different codes.

4.5.1   Finite Difference Method

Numerical approaches produce a system of simultaneous equations that must be solved in a manner similar to that used to find chemical equilibria. The finite difference and finite element methods are the most commonly used. Despite the similarity in names, the development of the sets of equations is quite different, although the equations themselves may be similar. The finite element method is more suitable for irregularly shaped flow domains but the accuracy and speed of the two approaches are nearly the same (McCord and Goodrich, 1994). The development of the finite difference equations is much more intuitive and will be discussed further.

The finite difference approach is used to develop a set of algebraic equations from the Richards equation so that it can be solved numerically (since computers cannot integrate or differentiate). The h form (Equation [4.31]) has been used most often to solve problems of transient flow in nonuniform soils, although recent articles indicate that the mixed form (Equation [4.30]) is more accurate (Celia et al., 1990; Ross, 1990). The development of the set of equations for the h form for one-dimensional vertical water movement with no source/sink term and known potentials at the boundaries will be shown.

The first step is to discretize the soil profile into N even depth intervals Δz such that $z = i \Delta z$ and $i = 0 \ldots N$ (Smith, 1985). It is not necessary that the depth intervals be even, but for simplicity we will assume even intervals. The point at which the intervals join is called a node, so there are $N + 1$ nodes in the profile. Time is also discretized by intervals of Δt such that $t = j \Delta t$. Then the derivatives in Equation [4.31] are written as discrete differences divided by the appropriate interval:

$$
\begin{aligned}
C\left(h_i^{j+\frac{1}{2}}\right)\frac{h_i^{j+1} - h_i^j}{\Delta t} &= \frac{\left[K(h)\left(\frac{\partial h}{\partial z}\right)+1\right]_{i+\frac{1}{2}} - \left[K(h)\left(\frac{\partial h}{\partial z}\right)+1\right]_{i+\frac{1}{2}}}{\Delta z} \\[2mm]
&= \frac{K\left(\frac{h_i + h_{i+1}}{2}\right)\left[\frac{h_{i+1} - h_i}{\Delta z}+1\right] - K\left(\frac{h_{i-1} + h_i}{2}\right)\left[\frac{h_i - h_{i-1}}{\Delta z}+1\right]}{\Delta z} \\[2mm]
&= \frac{K\left(\frac{h_i + h_{i+1}}{2}\right)}{(\Delta z)^2}\left[h_{i+1} - h_i + \Delta z\right] - \frac{K\left(\frac{h_{i-1} + h_i}{2}\right)}{(\Delta z)^2}\left[h_i - h_{i-1} + \Delta z\right]
\end{aligned}
\qquad [4.72]
$$

Subscripts denote depth nodes and superscripts denote time steps. At this point, the time step for evaluating the terms on the right-hand side of Equation [4.72] is unspecified. Numerical approaches consist of starting with the initial conditions, which give the values of h at all nodes at $t = 0$, then finding

Table 4.4 Examples of current numerical codes and some of their features

Model	Numerical Method	Space Dimensions	Type of Transport	References
HYSWASOR	finite element	1-D	water and solute	Dirksen et al. (1993)
LEACHM	finite difference	1-D	water and solute	Hutson and Wagenet (1992)
Dual Porosity	finite element	1-D	water and solute	Gerke and van Genuchten (1993)
SWIM	finite difference	1-D	water	Ross (1990)
WORM	finite element	1-D	water and solute	van Genuchten (1987)
VS2DT	finite difference	2-D	water and solute	Lapalla et al. (1987), Healy (1990)
VAM2D	finite element	2-D	water and solute	Huyakorn et al. (1988)
VSAFT	finite element	2-D	water and solute	Yeh et al. (1990)
SUTRA	finite element	2-D	water and solute	Voss (1984)

h at each node at the next time step ($t = \Delta t$), and repeating the process until the full time period of interest has been covered. Therefore, the crux of the problem is finding matric head at the next time step (h^{j+1}) given the value at the current time step (h^j) at each node.

If all of the terms on the right-hand side of Equation [4.72] are evaluated at the known (j) time step and $C(h_i^{j+\frac{1}{2}})$ is approximated by $C(h_i^j)$, then there is only one unknown in the equation (h_i^{j+1}) from the term on the left-hand side. The equation can be solved explicitly for h_i^{j+1} at each node (i = 1... N-1, with h_0^{j+1} and h_N^{j+1} known from the boundary conditions). This is known as the explicit finite difference method and has the disadvantage of being unstable for all but very small time steps (small time steps require more computer time). Unstable means that values of h will fluctuate sharply between time intervals and diverge from the true solution.

If all of the terms on the right-hand side of Equation [4.72] are evaluated at the unknown (j+1) time step, then there are three unknowns: h_{i-1}^{j+1}, h_i^{j+1}, and h_{i+1}^{j+1}. Collecting coefficients of these terms on the left-hand side gives:

$$a\, h_{i-1}^{j+1} + b\, h_i^{j+1} + c\, h_{i+1}^{j+1} = d \qquad [4.73]$$

and

$$a = -r\frac{K_{i-\frac{1}{2}}}{C_1}$$

$$c = -r\frac{K_{i+\frac{1}{2}}}{C_i}$$

$$b = 1 + a + c$$

$$d = h_i^f + \Delta z(a - c)$$

[4.74]

where $r = \Delta t / (\Delta z)^2$, $K_{i\pm\frac{1}{2}} = K(h_{i\pm\frac{1}{2}}^{j+1})$, $C_i = C(h_i^{j+\frac{1}{2}})$, $h_{i\pm\frac{1}{2}}^{j+1}$ is the arithmetic average of h_i^{j+1} and $h_{i\pm1}^{j+1}$, and $h_i^{j+\frac{1}{2}}$ is the arithmetic average of h_i^j and h_i^{j+1}. Equation [4.73] can be written for each node so there are N-1 equations with N-1 unknowns (h_i^{j+1}) to be solved simultaneously. This can be written in matrix notation as:

$$\underline{\underline{A}}\underline{u} = \underline{d} \qquad\qquad\qquad [4.75]$$

where A is a tridiagonal matrix with the coefficients a, b, and c along the subdiagonal, diagonal, and superdiagonal, u is a vector of the N-1 unknown values of h, and d is a vector containing the known values of h. Computer algorithms are used to invert A and solve Equation [4.75] for u at a given time step. The only remaining difficulty is that the coefficients contain $K_{i\pm\frac{1}{2}}$ and $C_{i+\frac{1}{2}}$, which require matric heads at the unknown time level in order to be evaluated. An iterative procedure is usually used at each time step wherein h_i^{j+1} is estimated, $K_{i\pm\frac{1}{2}}$ and $C_{i+\frac{1}{2}}$ are evaluated, and the set of equations is solved for an improved estimate of h_i^{j+1}. This is repeated until the difference between the two estimates is acceptably small, and then the algorithm goes to the next time step (Paniconi et al., 1991).

Specifying all of the terms on the right-hand side of Equation [75] at the j+1 time step is known as the fully implicit finite difference method. Another approach is used where the terms on the right-hand side of Equation [4.75] are evaluated at the j+½ time step (by taking an average of the right-hand side at the j and j+1 time step) and this is known as the Crank-Nicolson finite difference method (Smith, 1985). The fully implicit and Crank-Nicolson methods are stable for much larger time steps than the explicit method, but water flow problems that involve large gradients in h such as infiltration, still require very small time steps (fast computers).

4.5.2   Initial and Boundary Conditions

Numerical solution of the Richards equation requires that the initial distribution of h or θ in the profile and the conditions at the boundaries be specified, just as they must be with an analytical solution.

4.5.2.1   Initial Conditions

Transient models require the specification of conditions at the beginning of the simulation period because subsequent computations rely on previous states of the system. To avoid numerical difficulties (i.e., unstable solutions), uniform initial conditions are usually preferred. In heterogenous media, however, identifying feasible initial conditions may be difficult, in that the steady flow of water may not result in uniform initial conditions. Instead, a steady flux assumption initially leads to a distribution of potentials that result in $-K(h) = J_z$. Thus, specification of an initial flux yields the steady distribution of potentials as a function of heterogeneous unsaturated hydraulic conductivity. As a practical matter, it is best to allow sufficient time to "warm up" a model so that assumptions regarding initial conditions can be attenuated. Correct specification of initial conditions minimizes initial transients and the likelihood of model instability.

4.5.2.2   Boundary Conditions

Constant head or potentials (first kind or type 1) boundary conditions are used to prescribe known values of the boundary potential. Constant heads are prescribed for circumstances when water levels can be measured or inferred, such as using stage data from a pond, stream or well. Constant potentials are commonly prescribed along seepage faces where water is maintained at atmospheric pressure. For grid systems that do not deform with time, it is relatively easy to convert from potential to head, and vice versa.

Constant flux or gradient (second kind or type 2) boundary conditions are used to prescribe boundary values when the head or potential varies with time. In these cases, the gradient may remain fixed even when the potential is changing rapidly. For example, a prescribed flux can be specified on the upper surface due to constant infiltration, or a constant gradient can be imposed on the lower surface to represent unit drainage. Also, a zero flux or gradient can be imposed on the vertical boundaries to represent noflow conditions. Flow to a well can be modeled by specifying a constant flux at the borehole wall. Note that constant flux and constant gradient boundary conditions are not equivalent in the unsaturated zone because the hydraulic conductivity varies with potential.

Mixed potential flux (third kind or type 3) boundary conditions commonly arise for flow across a semipermeable boundary. In this case, the flux and the boundary potential are coupled, such that $J = (H_1 - H_2) / C$, where J and H_1 are the unknown boundary flux and head, respectively, C is the hydraulic impedance of the semipermeable membrane, and H_2 is a specified head exterior to the domain.

4.5.2.3 Infiltration

During a rainfall or irrigation event, the surface boundary condition must change. Initially, a constant flux into the soil equal to the rainfall rate (which can be variable) is specified (supply controlled infiltration). The matric head at the surface node is monitored and when sufficient water enters the soil to make this potential zero or positive, the boundary condition is switched to a constant potential of h equal to the depth of ponding, which is often taken as zero or some very small depth to allow for surface storage. This represents the time to ponding (t_p) and after this time the infiltration rate will be less than the rainfall rate (profile controlled infiltration) until the rain ends or the rainfall rate decreases. The calculated flux into the soil using the constant potential boundary condition is checked at each time step against the rainfall rate to ensure that the rainfall rate has not decreased below the calculated flux. If this does occur, the boundary condition is switched back to a constant flux equal to the rainfall rate.

4.5.2.4 Evapotranspiration

Evaporation also requires a change in the surface boundary condition, much like infiltration except that the flux is reversed. During the first stage of evaporation (immediately after a rainfall or irrigation event), a constant from the soil equal to the potential evaporation rate is specified. The pressure head at the soil surface node is again monitored and when sufficient water has left the soil that it drops to some critical level, usually a potential that would correspond to air dry soil, the surface boundary condition is switched to a constant matric head equal to the critical level. At this point, evaporation enters the second stage and actual soil evaporation is less than potential. Evaporation rate can vary (as it would diurnally), so the calculated flux is compared to the potential evaporation rate at each time step. When the potential evaporation rate drops below the calculated flux, the boundary condition is switched back to a constant flux equal to the potential rate. Transpiration is usually treated as a distributed sink within the root zone, rather than as a boundary condition at the soil surface.

4.6 Concluding Remarks

Soil water flow is one of the most important processes occurring in soils and is the key to predicting solute transport (Section A, Chapter 6). Although important advances have been made in understanding and predicting water flow, many challenges remain. Improvements in the methods for incorporating macropore and structure effects in water movement predictions are required (soil structure is addressed in Section A, Chapter 7). Easier, less intrusive methods for measuring saturated and unsaturated hydraulic conductivity are needed. Information on how to scale up these measurements to the field and landscape scale is needed (spatial variability is addressed in Section A, Chapter 10). Better integration of soil water and groundwater modeling is also needed.

4.7 References

Ashcroft, G., D.D. Marsh, D.D. Evans, and L. Boersma. 1962. Numerical methods for solving the diffusion equation: I. Horizontal flow in semi-infinite media. Soil Sci. Soc. Am. Proc. 26:522–525.

Baker, R.S. and D. Hillel. 1990. Laboratory tests of a theory of fingering during infiltration into layered soils. Soil Sci. Soc. Am. J. 54:20–30.

Baver, L.D., W.H. Gardner, and W.R. Gardner. 1972. Soil physics. John Wiley and Sons, Inc., New York, NY.

Beven, K.J., and P.F. Germann. 1981. Water flow in soil macropores II. A combined flow model. Soil Sci. 32:15–29.

Beven, K.J., and P.F. Germann. 1982. Macropores and water flow in soils. Water Resour. Res. 5:1311–1325.

Bosch, D.D., and L.T. West, 1997. Hydraulic conductivity variability for two sandy soils. Soil Sci. Soc. Am. J. 62:90–98.

Bouma, J., and J.H. Denning. 1972. Field measurements of unsaturated hydraulic conductivity by infiltration through gypsum crusts. Soil Sci. Soc. Am. Proc. 36:846–847.

Bouma, J., C. Gelmans, L.W. Dekker, and W.J.M. Jeurissen. 1983. Assessing the suitability of soils with macropores for subsurface liquid waste disposal. J. Environ. Qual. 12:305–311.

Bouwer, H. 1969. Infitration of water into nonuniform soil. J. Irrig. Drain. Div., Proc. ASCE IR4:451–462.

Bouwer, H. 1986. Intake rate: cylinder infiltrometer. In A. Klute (ed.) Methods of soil analysis: Part 1. Physical and mineralogical methods. 2nd Ed. Soil Science Society of America, Madison, WI.

Bouwer, H. 1966. Rapid field measurement of air entry value and hydraulic conductivity as significant parameters in flow system analysis. Water Resour. Res. 1:729–738.

Bouwer, H., and R.C. Rice. 1976. A slug test for determining hydraulic conductivity of unconfined aquifers with completely or partially penetrating wells. Water Resourc. Res. 12:423–428.

Brakensiek, D.L., and W.J. Rawls. 1992. Comment on "Fractal processes in soil water retention" by Scott W. Tyler and Stephen W. Wheatcraft. Water Resour. Res. 28:601–602.

Brooks, R.H., and A.T. Corey. 1964. Hydraulic properties of porous media. Hydrol. Pap. 3, Colorado State University, Fort Collins, CO.

Bruce, R.R., J.L. Chesness, T.C. Keisling, J.E. Pallas, Jr., D.A. Smittle, J.R. Stansell, and A.W. Thomas. 1980. Irrigation of crops in the southeastern United States. USDA-ARS, Watkinsville, GA.

Bruce, R.R., V.L. Quisenberry, H.D. Scott, and W.M. Snyder. 1985. Irrigation practice for crop culture in the southeastern United States. Adv. Irrig. 3:51–106.

Buckingham, E. 1907. Studies on the movement of soil moisture. USDA Bull. 38.

Campbell, G.S. 1974. A simple method for determining unsaturated conductivity from moisture retention data. Soil Sci. 117:311–314.

Carter, M.R. 1993. Soil sampling and methods of analysis. Lewis Publishers. Boca Raton, FL.

Cassel, D.K. 1985. Physical characteristics of soils of the southern region: Summary of in situ unsaturated hydraulic conductivity. South. Coop. Ser. Bull. 303. NC State University, Raleigh, NC.

Celia, M.A., E.T. Bouloutas, and R.L. Zarba. 1990. A general mass-conservative numerical solution for the unsaturated flow equation. Water Resour. Res. 26:1483–1496.

Chiang, S.C., D.E. Radcliffe, W.P. Miller, and K.D. Newman. 1987. Hydraulic conductivities of three southeastern soils as affected by sodium, electrolyte concentration, and pH. Soil Sci. Soc. Am. J. 51:1293–1299.

Chu, S.T. 1978. Infiltration during an unsteady rain. Water Resour. Res. 14:461–466.

Clothier, B.E., and K.R.J. Smettem. 1990. Combining laboratory and field measurements to define the hydraulic properties of soil. Soil Sci. Soc. Am. J. 54:299–304.

Clothier, B.E., and I. White, 1981. Measurement of sorptivity and soil water diffusivity in the field. Soil Sci. Soc. Am. J. 45:241–245.

Dane, J.H. and S. Hruska. 1983. In-situ determination of soil hydraulic properties during drainage. Soil Sci. Soc. Am. J. 47:619–624.

Darcy, H. 1856. Les fontaines publiques de la ville de Dijon. Dalmont, Paris, France.

Diment, G.A., K.K. Watson, and P.J. Blennerhasset. 1982. Stability anlysis of water movement in unsaturated porous materials: 1. Theoretical considerations. Water Resour. Res. 18:1248–1254.

Edwards, W.M., M.J. Shipitalo, L.B. Owens, and L.D. Norton. 1989. Water and nitrate movement in earthworm burrows within long-term no-till cornfields. J. Soil Water Cons. 44:240–243.

Elrick, D.E. and W.D. Reynolds. 1992. Infiltration from constant-head well permeameters and infiltrometers. p. 1-24. In G.C. Topp, W.D. Reynolds, and R.E. Green (ed.) Advances in measurement of soil physical properties: Bringing theory into practice. Soil Science Society of America, Inc. Madison, WI.

Faybishenko, B.A. 1995. Hydraulic behavior of quasi-saturated soils in the presence of entrapped air: laboratory experiments. Water Resour. Res. 31:2421-2435

Fetter, C.W. 1994. Applied hydrogeology. 3rd Ed. Macmillan College Publishing Co., New York, NY.

Gardner, W.R. 1958. Some steady state solutions of the unsaturated moisture flow equation with application to evaporation from a watertable. Soil Sci. 85:228–232.

Gardner, W.R., and M. Fireman. 1958. Laboratory studies of evaporation from soil columns in the presence of a watertable. Soil Sci. 85: 244–249.

Gish, T.J., and A. Shirmohammadi. 1991. Preferential flow. American Society of Agricultural Engineers, St. Joseph, MI.

Glass, R.J., J.-Y. Parlange, and T.S. Steenhuis. 1991. Immiscible displacement in porous media: Stability analysis of three-dimensional, axisymmetric disturbances with application to gravity-driven wetting front instability. Water Resour. Res. 27:1947–1956.

Glass, R.J., T.S. Steenhuis, and J.-Y. Parlange. 1989. Wetting front instability: 1. Theoretical discussion and dimensional analysis. Water Resour. Res. 25:1187–1194.

Green, R.E., L.R. Ahuja, and S.K. Chong, 1985. Hydraulic conductivity, diffusivity, and sorptivity of unsaturated soils: Field methods. p. 771–798. *In* A. Klute (ed.) Methods of soil analysis. Part 1. Soil Science Society of America, Madison, WI.

Green, W.H. and G.A. Ampt. 1911. Studies in soil physics: I. The flow of air and water through soils. J. Agric. Sci. 4:1–24.

Gupte, S.M., D.E. Radcliffe, D.H. Franklin, L.T. West, E.W. Tollner, and P.F. Hendrix. 1996. Anion transport in a Piedmont Ultisol: 2. Local-scale parameters. Soil Sci. Soc. Am. J. 60:762–770.

Hanks, R.J. and S.A. Bowers. 1962. Numerical solution of the moisture flow equation for infiltration into layered soils. Soil Sci. Soc. Proc. 26:530–534.

Haverkamp, R., M. Vauclin, J. Tovina, P.J. Wierenga, and G. Vachaud. 1977. A comparison of numerical simulation models for one-dimensional infiltration. Soil Sci. Soc. Am. Proc. 41:285–294.

Healy, R.W. 1990. Simulation of solute transport in variably saturated porous media with supplemental information on modifications to the U.S. Geological Survey's computer program VS2DT. USGS Water Resour. Invest. Rep. 90–4025.

Hill, R.E., and J.-Y. Parlange. 1972. Wetting front instability in layered soils. Soil Sci. Soc. Am. Proc. 36:697–702.

Hillel, D. 1980a. Fundamentals of soil physics. Academic Press. New York, NY.

Hillel, D. 1980b. Applications of soil physics. Academic Press. New York, NY.

Hillel, D., and R.S. Baker. 1988. A descriptive theory of fingering during infiltration into layered soils. Soil Sci. 146:51–56.

Hooghoudt, S.B. 1937. Bijdregen tot de kennis van eenige natuurkundige grootheden van de grond. Versl. Landb. Ond. 43:461–676.

Hubbert, M.K. 1956. Darcy's law and the field equations of the flow of underground fluids. Am. Inst. Min. Met. Petl. Eng. Trans. 207:222–239.

Huyakorn, P.S., J. B. Kool, and J.B. Robertson. 1989. Documentation and user's guide: VAM2SD -- Variably saturated analysis model in two dimensions. NUREG/CR-5352, HGL/89-01, Hydrogeologic, Inc. Herndon, VA.

Jarvis, N.J., and I. Messing. 1995. Near-saturated hydraulic conductivity in soils of contrasting texture measured by tension infiltrometers. Soil Sci. Soc. Am. J. 59:27–34.

Jury, W.A., W.R. Gardner, and W.H. Gardner. 1991. Soil physics. John Wiley and Sons, Inc. New York, NY.

Klute, A. 1986. Methods of soil analysis: Part 1. Physical and mineralogical methods. 2nd Ed. Soil Science Society of America, Madison, WI.

Klute, A. and C. Dirksen. 1986. Hydraulic conductivity and diffusivity: Laboratory methods. p. 687–734. *In* A. Klute (ed.) Methods of soil analysis. Part 1. Physical and mineralogical methods. 2nd Ed. Soil Science Society of America, Madison, WI.

Kool, J.B., J.C. Parker, and M.T. van Genuchten. 1985a. Determining soil hydraulic properties from one-step outflow experiments by parameter estimation: I. Theory and numerical studies. Soil Sci. Soc. Am. J. 49:1348–1354.

Kool, J.B., J.C. Parker, and M.T. van Genuchten. 1985b. ONESTEP, a non-linear parameter estimation program for evaluating soil hydrauic properties from one-step outflow experiments. VA Agric. Exp. Stn. Bull. 85–3.

Kruseman, G.P., and N.A. deRidder. 1990. Analysis and evaluation of pumping test data. 2nd Ed. International Inst. Land Reclam. Improv. Pub. 47.

Kung, K.-J.S. 1990a. Preferential flow in a sandy vadose zone: 1. Field observation. Geoderma. 46:51–58.

Kung, K.-J.S. 1990b. Preferential flow in a snady vadose zone: 2. Mechanism and implicaitons. Geoderma. 46:59–71.

Lapalla, E.G., R.W. Healy, and E.P. Weeks. 1987. Documentation of computer program VS2DT to solve the equations of fluid flow in variable saturated porous media. Water-Resources Investigations Report 83-4099. U.S. Geological Survey. Denver, CO.

Leij, F.J., W.J. Alves, and M.Th. van Genuchten. 1996. The UNSODA unsaturated soil hydraulic database. User's manual version 1.0. EPA/600/R-96/095. U.S. EPA. Cincinnati, OH.

Liu, H.H., and J.H. Dane. 1996. Two approaches to modeling unstable flow and mixing of variable density fluids in porous media. Trans. Porous Media 23:219–236.

Luxmoore, R.J. 1981. Micro-, meso-, and macroporosity of soil. Soil Sci. Soc. Am. J. 45:671.

McCord, J.T., and M.T. Goodrich. 1994. Benchmark testing and independent verification of the VS2DT computer code. Sandia National Laboratories. Albuquerque, NM.

McCuen, R.H. 1982. A guide to hydrologic analysis using SCS methods. Prentice-Hall, Inc., Englewood Cliffs, NJ.

Mein, R.G., and C.L. Larson. 1973. Modeling infiltration during a steady rain. Water Resour. Res. 9:384–394.

Miller, R.H., and J.F. Wilkinson. 1979. Nature of the organic coating on sand grains of nonwettable golf greens. Soil Sci. Soc. Am. Proc. 41:1203–1204.

Miller, W.P. , and M.K. Baharuddin. 1986. Relationship of soil dispersibility to infiltration and erosion of southeastern soils. Soil Sci. 142:235–240.

Miller, W.P., and D.E. Radcliffe. 1992. Soil crusting in the southeastern US. p. 233–266. In M.E. Sumner and B.A. Stewart (ed.) Soil crusting: Chemical and physical processes. Lewis Publishers, Boca Raton, FL.

Mualem, Y. 1976. A new model for predicting the hydraulic conductivity of unsaturated porous media. Water Resour. Res. 12:593–622.

Paniconi, C., A.A. Aldama, and E.F. Wood. 1991. Numerical evaluation of iterative and noniterative methods for the solution of the nonlinear Richards equation. Water Resour. Res. 27:1147–1163.

Parlange, J.-Y., and D.E. Hill. 1976. Theoretical analysis of wetting front instability in soils. Soil Sci. 122:236–239.

Perroux, K.M., and I. White. 1988. Designs for disc permeameters. Soil Sci. Soc. Am. J. 52:1205–1215.

Philip, J.R. 1957a. The theory of infiltration. 1. The infiltration equation and its solution. Soil Sci. 83:345–357.

Philip, J.R. 1957b. The theory of infiltration. 2. The profile at infinity. Soil Sci. 83:435–448.

Philip, J.R. 1957c. The theory of infiltration. 3. Moisture profiles and relation to experiment. Soil Sci. 84:163–178.

Philip, J.R. 1957d. The theory of infiltration. 4. Sorptivity and algebraic infiltration equations. Soil Sci. 84:257–264.

Philip, J.R. 1957e. The theory of infiltration. 5. Influence of initial moisture content. Soil Sci. 84:329–339.

Philip, J.R. 1975. Stability analysis of infiltration. Soil Sci. Soc. Am. Proc. 39:1042–1049.

Quisenberry, V.L., and R.E. Phillips. 1976. Percolation of surface-applied water in the field. Soil Sci. Soc. Am. J. 40:484–489.

Raats, P.A.C. 1973. Unstable wetting fronts in uniform and non-uniform soils. Soil Sci. Soc. Am. Proc. 37:681–685.

Rawls, W.J., D.L. Brakensiek, and K.E. Saxton. 1982. Estimation of soil water properties. Trans ASAE 25:1316–1328.

Reynolds, W.D. and D.E. Elrick. 1991. Determination of hydraulic conductivity using a tension infiltrometer. Soil Sci. Soc. Am. J. 55:633–639.

Reynolds, W.D., S.R. Vieira, and G.C. Topp. 1992. An assessment of the single-head analysis for the constant head well permeameter. Can. J. Soil Sci. 72:489–501.

Richards, L.A. 1931. Capillary conduction of liquids in porous mediums. Physics 1:318–333.

Ross, P.J. 1990. Efficient numerical methods for infiltration using Richard's equation. Water Resour. Res. 26:279–290.

Seyfried, M.S., and P.S.C. Rao. 1987. Solute transport in undisturbed columns of a aggregated tropical soil: Preferential flow effects. Soil Sci. Soc. Am. J. 51:1434–1444.

Shaw, J.N., L.T. West, C.C. Truman, and D.E. Radcliffe. 1997. Morphologic and hydraulic properties of soils with restrictive horizons in the Georgia Coastal Plain. Soil Sci. 162:875–885.

Shipitalo, M.J., and W.M. Edwards. 1996. Effects of initial water content on macropore/matrix flow and transport of surface-applied chemicals. J. Environ. Qual. 25:662–670.

Smith, G. D. 1985. Numerical solution of partial differential equations. Clarendon Press, Oxford, UK.

Smith, K.A. and C.M. Mullins. 1991. Soil analysis: Physical methods. Marcel Dekker, New York, NY.

Thiem, G. 1906. Hydrologische Methoden. J.M. Gephardt, Leipzig, Germany.

Topp, G.C. 1992. Advances in measurement of soil physical properties: Bringing theory into practice. Soil Sci. Soc. Am. Spec. Pub. 30. Soil Science Society of America, Madison, WI.

Tyler, S.W., and S.W. Wheatcraft. 1992. Reply. Water Resour. Res. 28:603–604.

van Genuchten, M.Th. 1980. A closed form equation for predicting the hydraulic conductivity of unsaturated soils. Soil Sci. Soc. Am. J. 44:892–898.

van Genuchten, M.Th. 1987. A numerical model for solute movement in and below the root zone. Res. Report. US Salinity Laboratory, Riverside, CA.

van Genuchten, M.Th., and F. J. Leij. 1992. On estimating the hydraulic properties of unsaturated soils. p. 1–14. *In* M.Th. van Genuchten, F. J. Leij, and L.J. Lund (ed.) Indirect methods for estimating the hydraulic properties of unsaturated soils. US Salinity Laboratory, Riverside, CA.

van Genuchten, M.T., and P.J. Wierenga. 1976. Mass transfer studies in sorbing porous media: I. Analytical solutions. Soil Sci. Soc. Am. J. 40:473–479.

Voss, C.I. 1984. A finite-element simulation model for saturated-unsaturated fluid-density-dependent ground-water flow with energy transport or chemically reactive single species solute transport. US Geological Survey. Denver, CO.

Warrick, A.W. 1992. Models for disc infiltrometers. Water Resour. Res. 28:1319–1327.

Warrick, A.W., and P. Broadbridge. 1992. Sorptivity and macroscopic capillary length relationships. Water Resour. Res. 28:427–431.

White, I., and K.M. Perroux. 1989. Estimation of unsaturated hydraulic conductivity from field sorptivity measurements. Soil Sci. Soc. Am. J. 53:324–329.

White, I. and M.J. Sully. 1987. Macroscopic and microscopic capillary length and time scales from field infiltration. Water Resour. Res. 23:1514–1522.

White, R.E. 1985. The influence of macropores on the transport of dissolved and suspended matter through soil. Adv. Soil Sci. 3:95–120.

Wooding, R.A. 1968. Steady infiltration from a shallow circular pond. Water Resour. Res. 4:1259–1273.

Yeh, T.-C.J., R. Srivastava, A. Guzman, and T. Harter. 1993. A numerical model for water flow and chemical transport in variably saturated porous media. Groundwater 31:634–644.

Zangar, C.N. 1953. Flow from a test hole located above groundwater level. p. 69–71. *In* Theory and problems of water percolation. Bur. Reclam. Eng. Monogr. 8, US Department of the Interior. Washington, DC.

Zurhuhl, T., and W. Durner. 1996. Modeling transient water and solute transport in a biporous soil. Water Resour. Res. 32:819–829.

5

Energy and Water Balances at Soil-Plant-Atmosphere Interfaces

S. R. Evett
USDA-ARS, Bushland, TX

5.1 Introduction

Energy fluxes at soil-atmosphere and plant-atmosphere interfaces can be summed to zero because the surfaces have no capacity for energy storage. The resulting energy balance equations may be written in terms of physical descriptions of these fluxes; and have been the basis for problem casting and solving in diverse fields of environmental and agricultural science such as estimation of evapotranspiration (ET) from plant canopies, estimation of evaporation from bare soil, rate of soil heating in spring (important for timing of seed germination), rate of residue decomposition (dependent on temperature and water content at the soil surface), and many others. The water balances at these surfaces are implicit in the energy balance equations. The soil water balance equation is different from, but linked to, the surface energy balances, a fact that has often been ignored in practical problem solving. In this chapter, energy balance will be discussed first, followed by water balance in Section 5.3.

Computer simulation has become an important tool for theoretical investigation of energy and water balances at the Earth's surface, and for prediction of important results of the mechanisms involved. This chapter will focus more on the underlying principles of energy and water balance processes, and will mention computer models only briefly. More information on computer models that include surface energy and water balance components can be found in Campbell (1985), Richter (1987), ASAE (1988), Anlauf et al. (1990), Hanks and Ritchie (1991), Pereira et al. (1995), and Peart and Curry (1998) to mention only a few.

5.2 Energy Balance Equation

The surface energy balance is

$$0 = Rn + G + LE + H \qquad [5.1]$$

where Rn is net radiation, G is soil heat flux, LE is the latent heat flux (evaporation to the atmosphere) and is the product of the evaporative flux (E) and the latent heat of vaporization (L), and H is sensible heat flux (all terms taken as positive when flux is toward the surface [W m^{-2}]). Each term may be expressed more completely as the sum of subterms that describe specific physical processes, some of which are shown in Fig. 5.1. Thus, net radiation includes the absorption and reflection of shortwave radiation (sunlight [Rsi] and the reflected portion [αRsi]), as well as the emission and reception of longwave radiation (L↑ and L↓, respectively) (Fig. 5.1). Soil heat flux involves not only diffusion of heat (G) as expressed by Fourier's law but also convective heat flux (G$_{Jw}$) as water at temperature T flows at rate J$_w$ into soil at another temperature T′. Evaporation from both the soil and from plants is an example of latent heat flux, but so also is dew formation, whether it wets the soil surface or plant canopy. Finally, sensible heat flux may occur between soil and atmosphere or between plant and atmosphere, and may be short circuited between soil and plant, for example, when sensible heat flux from the soil warms the plant. In the next few paragraphs, the fluxes and values that they may assume will be illustrated with examples from some contrasting surfaces under variable weather conditions.

Values of these energy fluxes change diurnally (Figs. 5.2–5.4) and seasonally (Figs. 5.5–5.6). Regional advection is the large scale transport of energy in the atmosphere from place to place on the Earth's surface. Regional advection events can change the energy balance greatly, as illustrated with measurements taken over irrigated wheat at Bushland, TX (35°11'N Lat; 102°06'W Long) for the 48-h period beginning on day 119, 1992 (Fig. 5.2). Total Rsi was 26.1 and 26.7 MJ m^{-2} on days 119 and 120, respectively, close to the expected maximum clear sky value of 28.6 MJ m^{-2} for this latitude and time of year. However, on day 119 strong, dry, adiabatic southwesterly winds (mean 5 m s^{-1}, mean dew point 4.1 C, mean T$_{2m}$ 20.1 C) caused H to be strongly positive, providing the extra energy needed to drive total LE to –32.8 MJ m^{-2}, even though both Rsi and Rn levels were reduced in the afternoon due to cloudiness. Total LE was much larger in actual magnitude than Rsi and Rn totals. The next day, the total LE was 39% smaller due to the absence of regional advection, even though total Rsi and Rn values were slightly higher, and G values were near zero during this period of full canopy cover when the leaf

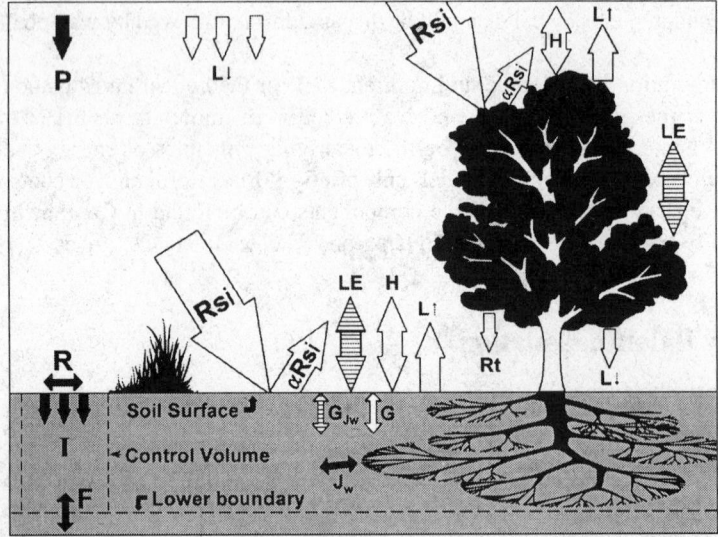

Fig. 5.1 Water and energy balance components. Water balance components are in black, energy balance in white. The shared term LE is shaded. Water balance is discussed in Section 5.2.

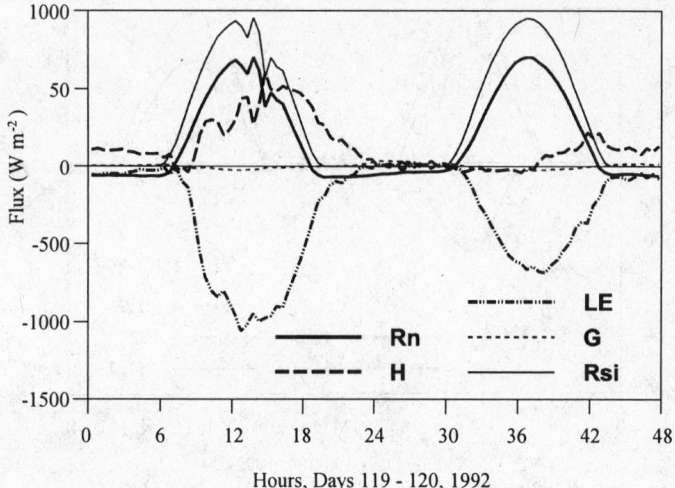

Fig. 5.2 Energy balance over irrigated winter wheat at Bushland, TX

area index (LAI. Leaf area index is defined as the single-sided surface area of leaves per unit land area) was 7. Note that net radiation was negative at night. This is indicative of strong radiational cooling of the surface, which radiates heat into the clear, low humidity night time skies common to this semiarid location at 1,170 m above mean sea level.

Over alfalfa in late summer, Rsi totals were lower (20.1 and 5.4 MJ m^{-2}, respectively, for days 254 and 255, 1997) (Fig. 5.3). On the very clear day 254, peak Rsi was 798 W m^{-2} and with regional advection occurring, LE flux was high. The 3-h period of negative H just after sunrise was due to the sun-warmed crop canopy being at higher temperature than the air. The arrival of a cool front bringing cloudy skies near midnight caused all fluxes to be much lower on day 255, with Rsi reaching only 220 W m^{-2}, and H hovering near zero for much of the day. The arrival of the cloud cover and moist air is

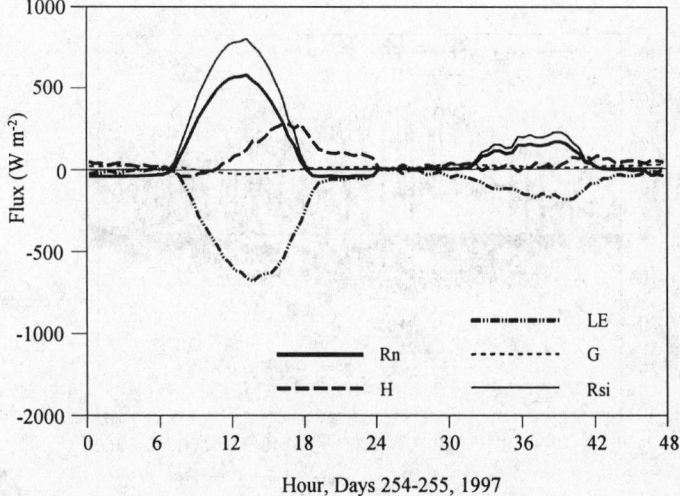

Fig. 5.3 Energy balance over irrigated alfalfa at Bushland, TX

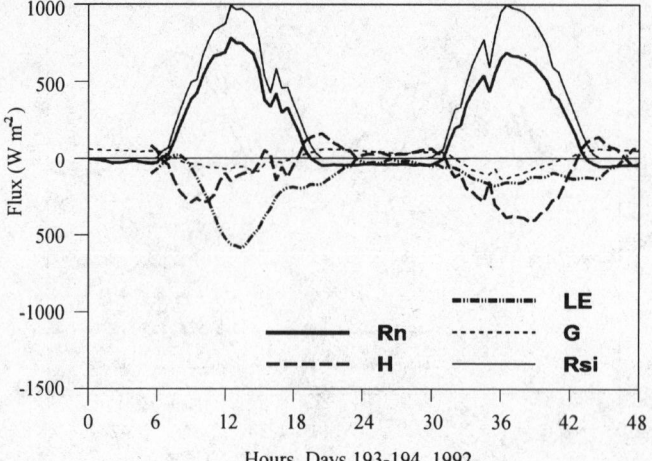

Fig. 5.4   Energy balance for bare Pullman clay loam soil after 35 mm of rain and irrigation at Bushland, TX

signaled near midnight by the abrupt change from negative values of Rn and LE to near zero values. In the case of net radiation, this is due to the increased longwave radiation from the clouds, which were warmer and had higher emissivity than the clear sky that preceded them. Latent heat flux approaches zero because the strong vapor pressure gradient from moist crop and soil to dry air is reduced by the arrival of moist air. Note that after sunset, but before midnight, latent heat flux was strong, due to continuing strong sensible heat flux, even though net radiation was negative. Again, due to full crop cover (LAI = 3), G values were low, indicating that very little energy is penetrating the soil surface.

For bare soil, G is often larger, becoming an important part of the energy balance (Fig. 5.4). After rain and irrigation totaling 35 mm over the previous two days, the soil was wet on day 193. Latent heat flux totaled -14.4 MJ m^{-2} or 6 mm of evaporation, 77% of Rn. Sensible heat flux was negative for the first few hours after sunrise because the soil was warmer than the air, which had been cooled by a night-time thunderstorm. Later in the day, H and G both approached zero, and near sunset, they became

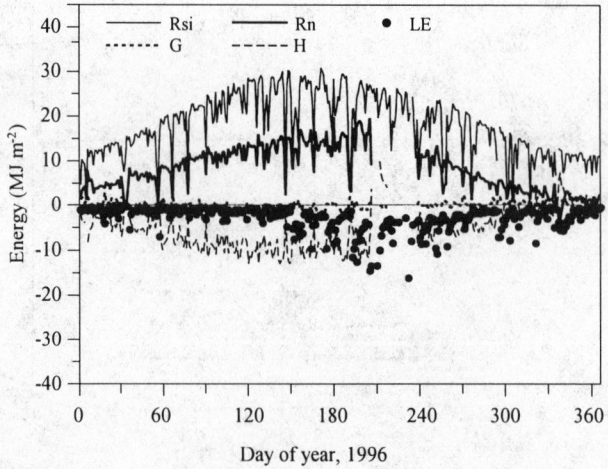

Fig. 5.5   Daily totals of energy balance terms for a fallow field (mostly bare Pullman clay loam) at Bushland, TX

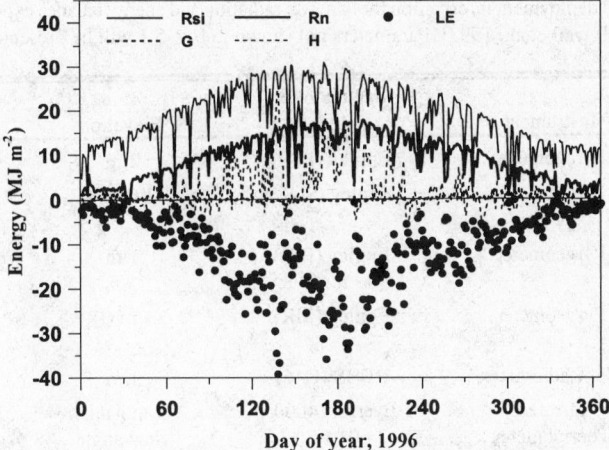

Fig. 5.6 Daily totals of energy balance terms for irrigated alfalfa at Bushland, TX

positive, supplying the energy consumed in evaporation that continued well into the night hours. Strong radiational cooling occurred on the nights of days 193 and 194 as indicated by negative values of Rn. Evaporation was probably energy limited on day 193, becoming soil limited on day 194. Latent heat flux on the second day was reduced to $-7.4\,\text{MJ m}^{-2}$, and peak daytime values were not much larger than those for G. The drying soil became warmer and contributed heat to the atmosphere during almost all daylight hours.

Seasonal variations in daily total energy flux values occur due to changes of sun angle, distance from the Earth to sun (about 3% yearly variation), seasonal weather, and surface albedo as plant and residue cover changes (Figs. 5.5 and 5.6). A curve describing clear sky solar radiation at Bushland, TX could be fitted to high points of Rsi in Figs. 5.5 and 5.6. Net radiation was similar for alfalfa and bare soil except for a rainy period beginning about day 190 when the soil was wet and dark and Rn for the fallow field was markedly larger. The big differences were in LE and H. Latent heat flux from the alfalfa was large, reaching nearly $-40\,\text{MJ m}^{-2}$ (16 mm) on day 136 during a regional advection event that allowed LE to be larger than Rsi. Sensible heat flux was positive during much of the year. Soil heat flux was small during the growing season, becoming larger as the soil cooled during the fall and winter. For the bare soil, LE values were small during the first 150 days, the latter part of a drought. Sensible heat flux was negative during this period, and remained negative after rains began until day 203. Evaporative fluxes were fairly small, rarely reaching 6 mm d^{-1} even after rains began. In contrast to alfalfa, soil heat flux for bare soil was larger and more variable throughout the year.

Methods of measurement and estimation of the energy fluxes are needed to characterize the energy balance. Examples of the instrumentation needed to measure components and subcomponents of the energy balance are given in Table 5.1.

5.2.1 Net Radiation

Net radiation is the sum of incoming and outgoing radiation:

$$\text{Rn} = \text{Rsi}(1-\alpha)- \in \sigma T^4 + L\downarrow \qquad [5.2]$$

Table 5.1 Instruments and deployment information for bare soil radiation and energy balance experiments at Bushland, TX, 1992 [Adapted from Howell et al., 1993]. Parameters not shown in Fig. 5.1 will be presented later.

Parameter	Instrument	Manufacturer[†] (Model)	Elevation	Description
R_{si}	Pyranometer	Eppley (PSP)	1 m	Solar irradiance
αR_{si}	Pyranometer	Eppley (8-48)	1 m (I[‡])	Reflected solar irradiance
$L\downarrow$	Pyrgeometer	Eppley (PIR)	1 m	Incoming long wave radiation
$L\uparrow$	Pyrgeometer	Eppley (PIR)	1 m (I)	Outgoing long wave radiation
R_n	Net Radiometer	REBS (Q*6)	1 m	Net radiation
T_s	Infrared Thermometer	Everest (4000; 60 ° fov)	1 m nadir view angle	Soil surface temperature
T_a RH	Thermistor Foil capacitor	Rotronics (HT225R)	2 m	Air temperature and relative humidity
U_2	dc generator cups	R.M. Young (12102)	2 m	Wind speed
U_d	Potentiometer vane	R.M. Young (12302)	2 m	Wind direction
T_t	Cu-Co Thermocouple	Omega (304SS)	−10 mm −40 mm	Soil temperature (4)[§]
G_{50}	Plates Thermopile	REBS (TH-1)	−50 mm	Soil heat flux (4)
θ_{v-20} θ_{v-40}	3-wire TDR probe	Dynamax TR-100/20 cm	−20 and −40 mm horizontal	Soil water content (2)
E_m	Lever-scale Load cell	Alphatron (SL50LB)	Below lysimeter box	Lysimeter mass change

[†] Manufacturers and locations are The Eppley Laboratory, Inc., Newport, RI; Radiation and Energy Balance Systems (REBS), Seattle, WA; Everest Interscience, Inc., Fullerton, CA; Rotronic Instrument Corp., Huntington, NY; R.M. Young Co., Traverse City, RI; Omega Engineering, Inc., Stamford, CT; Dynamax, Inc., Houston, TX; Alphatron, Inc., Elburn, IL.

[‡] I designates instruments that were inverted and facing the ground.

[§] Numbers in parentheses indicate replicate sensors.

where Rsi is solar irradiance at the surface, α is the albedo or surface reflectance (0 to 1), ϵ is the surface emissivity (0 to 1), σ is the Stefan-Boltzmann constant (5.67×10^{-8} W m^{-2} K^{-4}), T is surface temperature (K), and $L\downarrow$ is longwave irradiance from the sky. The sun radiates energy like a black body at about 6000 K while the Earth radiates at about 285 K. The theoretical maximum emission power spectra for these two bodies overlap very little (Fig. 5.7), a fact that leads to description of radiation from the Earth (including clouds and the atmosphere) as longwave, and radiation from the sun as shortwave. Note that the radiance of the Earth is about 4 million times lower than that of the sun (Fig. 5.7). Net radiation may be measured by a net radiometer (Fig. 5.8) or its components may be measured separately using pyranometers to measure incoming and reflected short-wave radiation, and

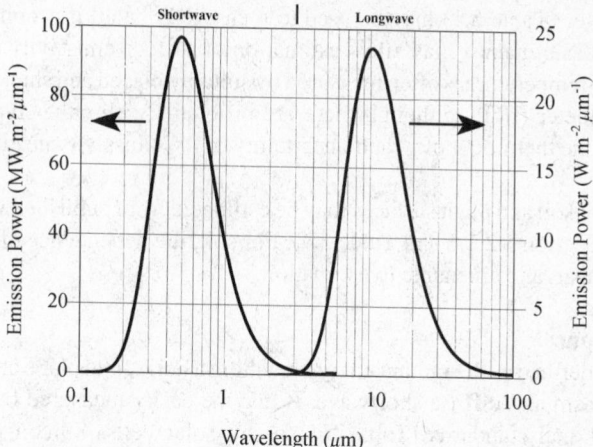

Fig. 5.7 Emission power spectra for ideal black bodies at 6000 K (left, shortwave range) and 285 K (right, longwave range)

pyrgeometers to measure incoming and outgoing long-wave radiation (first four instruments in Table 5.1). Pyranometers and pyrgeometers are thermopile devices that are equally sensitive across the spectrum.

5.2.1.1 Outgoing Long-Wave Radiation

The longwave radiance of the Earth's surface (L) is given by the Stefan-Boltzmann law for radiance from a surface at temperature T and with emissivity ϵ :

$$L\!\uparrow = \epsilon \, \sigma T^4 \tag{5.3}$$

Fig. 5.8 REBS Q*7 net radiometer

An inverted pyrgeometer (Table 5.1) may be used to measure L↑ and, if accompanied by suitable surface temperature measurements, may allow estimation of surface emissivity ($\in$) by inversion of Equation [5.3]. Surface temperature is often measured by suitably placed and shielded thermocouples, or by infrared thermometer (IRT), although there are problems with either type of measurement (radiational heating of the thermocouples, and uncertainty of the emissivity needed for accurate IRT measurements).

Values of α and $\in$ for soil and plant surfaces may be estimated from published values relating them to surface properties (Section 5.2.1.3 and Table 5.4). For soil, the dependence of α on water content is strong, but nearly linear, and amenable to estimation.

5.2.1.2 Solar Irradiance

Solar irradiance (Rsi), defined as the radiant energy reaching a horizontal plane at the Earth's surface, includes both direct beam and diffuse shortwave. It may be easily measured by pyranometer with calibration to international standards (Table 5.1) or by solar cells. Silicon photodetector solar radiation sensors are sensitive in only part of the spectrum, but are calibrated to give accurate readings in most outdoor light conditions. Silicon sensors are much cheaper than thermopile pyranometers and have found widespread use in field weather stations. Measurement of both incident (Rsi) and reflected (Rsr) shortwave allows estimation of the albedo from:

$$Rsi(1 - \alpha) = Rsi - Rsr \qquad\qquad [5.4]$$

This is done using upward and downward facing matched pyranometers (Table 5.1). Specially made albedometers are available for this purpose (Kipp & Zonen model CM-14) (Fig. 5.9).

The solar constant is the flux density of solar radiation on a plane surface perpendicular to the direction of radiation and outside the Earth's atmosphere. It is about 1,370 W m^{-2}, with a variation of about ± 3.5%, being largest in January when the sun is closest to the Earth, and smallest in July (Jones,

Fig. 5.9   Kipp and Zonen model CM-14 albedometer

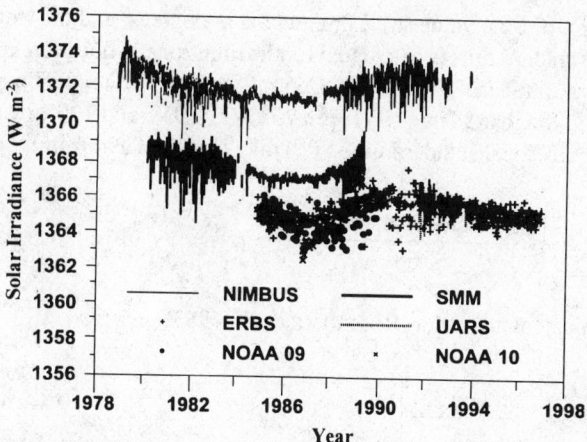

Fig. 5.10 Satellite observations of solar irradiance, Ra, outside the earth's atmosphere; corrected for earth-sun distance [Data source: NOAA, 1997]

1992). Several satellite observation platforms have recorded the value of solar irradiance over a nearly 20-year span (Fig. 5.10) and clearly show the average solar cycle of 11 years. The six sets of data shown range over about 10 W m^{-2} or ~ 0.7% of the mean value. Thus, considering the solar constant to be 1,370 W m^{-2} will introduce no more than a 1% error in calculations.

Irradiance at the Earth's surface is somewhat less, due to absorption and scattering in the atmosphere and due to sun angle effects, not often exceeding 1,000 W m^{-2}. The further the sun is from the zenith, the longer the transmission path through the atmosphere, and the more absorption and scattering occur. Also, as the sun angle above the horizon (ß) decreases (it is highest at solar noon), the radiation density on a horizontal surface decreases according to Lambert's law:

$$I = I_o \sin \beta \tag{5.5}$$

where I_o is the flux density on a surface normal to the beam. Sun angle (β) changes with time of day and year, and can be calculated from:

$$\beta = \sin^{-1}[\sin(D)\sin(L) + \cos(D)\cos(L)\cos(H)] \tag{5.6}$$

where L is latitude, D is solar declination, and H is solar time angle (all radians). Solar time angle is defined as:

$$H = \frac{(T - T_{SN})2\pi}{24} \tag{5.7}$$

where T is time (h), and T_{SN} is the time of solar noon. The time of solar noon varies with time of year and longitude according to (recall that 1° longitude = 4 min):

$$T_{SN} = 12 + \frac{4(\text{Longitude} - \text{Local Meridian})}{60} - T_{EQ} \tag{5.8}$$

where T_{EQ} is the equation of time value (h), Longitude is in degrees, and the Local Meridian is the longitude (°) for which standard time is calculated for the time zone in question. In the United States, the meridians for Eastern Standard Time (EST), Central Standard Time (CST), Mountain Standard Time (MST), and Pacific Standard Time (PST) are 75°, 90°, 105°, and 120°, respectively. Local or true solar time (T_{LS}) for any local standard time (T_{ST}) may be calculated from:

$$T_{LS} = T_{ST} - 4\frac{(\text{Longitude} - \text{Local Meridian})}{60} + T_{EQ} \qquad [5.9]$$

The declination may be calculated from (Rosenberg et al., 1983):

$$D = 0.4101\cos\left[\frac{2\pi(J-172)}{365}\right] \qquad [5.10]$$

where J is the day of the year.

List (1971) gave equation of time values to the nearest second for the 1st of each month and every 4 days after that for each month (95 values for the year). The following equation reproduces those values with a maximum error of 6 s, and can be used to estimate T_{EQ} in h for any day of the year.

$$T_{EQ} = b_0 + b_1\sin\left(\frac{J}{P_1}\right) + b_2\cos\left(\frac{J}{P_1}\right) + c_1\sin\left(\frac{J}{P_4}\right) + c_2\cos\left(\frac{J}{P_4}\right) + c_3\sin\left(\frac{2J}{P_4}\right)$$
$$+ c_4\cos\left(\frac{2J}{P_4}\right) + c_5\sin\left(\frac{3J}{P_4}\right) + c_6\cos\left(\frac{3J}{P_4}\right) + c_7\sin\left(\frac{4J}{P_4}\right) + c_8\cos\left(\frac{4J}{P_4}\right) \qquad [5.11]$$

where the coefficients b_i and c_i are given in Table 5.2, and $P_1 = 2\pi/182.5$ and $P_4 = 2\pi/365$.

Jensen et al. (1990) suggested a simpler method to calculate T_{EQ}:

$$T_{EQ} = 0.1645\sin(2b) - 0.1255(b) - 0.025\sin(b) \qquad [5.12]$$

where $b = 2\pi(J-81)/364$. The maximum error compared against List's T_{EQ} values is 88 s.

Disregarding air quality, solar irradiance is affected by latitude, time of year and day, and elevation. Latitude and time affect the sun angle (ß), and thus affect both the path length of radiation through the atmosphere (and thus absorption and scattering losses), and the flux density at the surface through

Table 5.2 Coefficients for calculating the equation of time value from Equation [5.11]

b_0	4.744×10^{-5}	c_2	9.19×10^{-3}	c_6	-1.29×10^{-3}
b_1	-0.157	c_3	-5.78×10^{-4}	c_7	-3.23×10^{-3}
b_2	-0.0508	c_4	3.61×10^{-4}	c_8	-2.1×10^{-3}
c_1	-0.122	c_5	-5.48×10^{-3}		

Equation [5.5]. Elevation affects the path length. Methods for calculating extraterrestrial (Rsa) and clear sky solar irradiance at the surface (Rso) are given by McCullough and Porter (1971), Campbell (1977, Chapter 5), Jensen et al. (1990, Appendix B), and Jones (1992, Appendix 7). Calculation of Rsa depends on latitude and time of day. Once Rsa is calculated Rso may be estimated from considerations of adsorption and scattering in the atmosphere, which depend mainly on the path length through the atmosphere and its density. Thus latitude, time of day and elevation are factors in estimating Rso from Rsa. The value of Rso is an important quantity against which to check measured Rsi; and it can be used in estimates of Rn, either to replace Rsi in Equation [5.2], or using regression relationships of Rn = f(Rso) (Jensen et al., 1990, Appendix B). Duffie and Beckman (1991) presented the following method of calculating Rsa (MJ m^{-2} h^{-1}) for any period [P (h)]:

$$Rsa = \left[\frac{24(60)}{2\pi}\right] G_{SC} d_r \{\cos(L)\cos(D)[\sin(\omega_2) - \sin(\omega_1)] + (\omega_1 - \omega_2)\sin(L)\sin(D)\}$$

[5.13]

where G_{SC} is the solar constant (0.08202 MJ m^{-2} min^{-1}), d_r is the relative Earth-sun distance, and ω_1 and ω_2 are the solar time angles at the beginning and end of the period, respectively (all angles in radians). The term $24(60)/(2\pi)$ is the inverse angle of rotation per minute. The relative Earth-sun distance is given by:

$$d_r = 1 + 0.033\cos\left(\frac{2\pi J}{365}\right)$$

[5.14]

where J is the day of year. The factors ω_1 and ω_2 are the solar time angles at the beginning and end of the period in question

$$\omega_1 = \omega - \frac{\pi}{(24/P)}$$

[5.15]

$$\omega_2 = \omega + \frac{\pi}{(24/P)}$$

[5.16]

where ω is the solar time angle at the center of the period (radians), and P is the length of the period in h.

The sunset time angle (angle from noon to sunset) is given by:

$$\omega_s = \cos^{-1}[-\tan(L)\tan(D)]$$

[5.17]

from which it is clear that day length [T_D (h)] is

$$T_D = \frac{24\omega_s}{\pi}$$

[5.18]

Equation [5.13] can be rewritten for total daily Rsa as

$$Rsa = \left[\frac{24(60)}{\pi}\right] G_{SC} d_r \left[\cos(L)\cos(D)\sin(\omega_s) + \omega_s \sin(L)\sin(D)\right] \qquad [5.19]$$

For example, on day 119 at latitude 35° 11' N, longitude 102° 6' W, Rsa calculated using Equation [5.13] on a half hourly basis was 38.097 MJ m^{-2} compared with 38.100 MJ m^{-2} calculated with Equation [5.19].

Jensen et al. (1990) recommended estimating daily total clear sky solar irradiance as:

$$Rso = 0.75 Rsa \qquad [5.20]$$

Somewhat in agreement with this, Monteith and Unsworth (1990) stated that direct beam radiation rarely exceeded 1,030 W m^{-2}, about 75% of the solar constant.

Jones (1992) and Monteith and Unsworth (1990) suggest:

$$Rsi = Rsi_{max} \sin\left(\frac{\pi t}{N}\right) \qquad [5.21]$$

for instantaneous values of Rsi on clear days, where Rsi_{max} is the maximum instantaneous irradiance occurring at solar noon, t is time after sunrise (h), and N is daylength (h).

It is more common to know daily total Rsi. Collares-Pereira and Rabl (1979) gave the ratio of hourly irradiance, [Rsi,h] to daily irradiance, [Rsi,d] as

$$\frac{Rsi,h}{Rsi,d} = \left(\frac{\pi}{24}\right)(a + b\cos\omega)\frac{(\cos\omega - \cos\omega_S)}{(\sin\omega_S - \omega_S \cos\omega_S)} \qquad [5.22a]$$

where

$$a = 0.409 + 0.5016 \sin\left(\omega_S - \frac{\pi}{3}\right) \qquad [5.22b]$$

and

$$b = 0.6609 + 0.4767 \sin\left(\omega_S - \frac{\pi}{3}\right) \qquad [5.22c]$$

Equation [5.22] performed well when applied to data from Bushland, TX (Fig. 5.11).

More complex methods of estimating Rso account for attenuation of direct beam radiation using Beer's law, coupled with Lambert's law to calculate irradiance on a horizontal surface, plus an accounting of diffuse irradiance (List, 1971; Rosenberg et al., 1983; Jones, 1990). Beer's law describes the intensity (I) of radiation after passing a distance (x) through a medium in terms of an extinction coefficient (k), and the initial intensity (Ia) as:

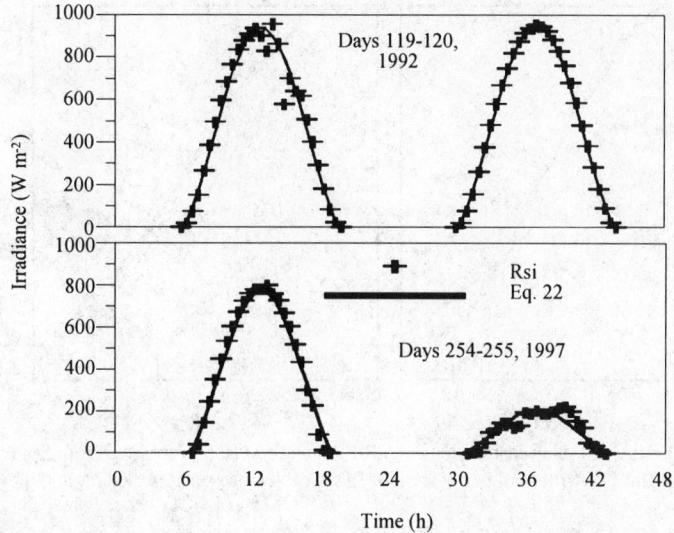

Fig. 5.11 Solar irradiance measured at Bushland, TX in 1992 and 1997 on clear and cloudy days; and Equation [5.22] half-hourly predictions

$$I = Ia\ e^{kx} \qquad [5.23]$$

For solar radiation, the distance is expressed in terms of air mass number (m) as (List, 1971):

$$m = \sec\left(\frac{\pi}{2} - \beta\right) \qquad [5.24]$$

The air mass is referenced to the length of the path when the sun is directly overhead. For ß < 0.175 radians (10°), the measured air mass number is less than that given by Equation [5.24] due to refraction and reflection at these low angles. List (1971) gives corrections, and notes also that for pressures (p) less than standard sea level pressure (p_0), m should be corrected by m = m(p/p_0). Rewriting Equation [5.23]:

$$Io = Ia\ e^{k\sec\left(\frac{\pi}{2} - \beta\right)} \qquad [5.25]$$

where Io is direct beam radiation at the Earth's surface. Monteith and Unsworth (1990) give a range of values of k for England as 0.07 for very clean air to 0.6 for very polluted air.

Assuming that both direct (Io) and diffuse (Id) radiation are known, the total irradiance at the surface is:

$$Rsi = Io(\sin\beta) + Id \qquad [5.26]$$

Diffuse radiation is quite difficult to estimate because it is so dependent on cloud cover, and aerosol concentration in the air. Yet, summarizing several data sets, Spitters et al. (1986) found that the

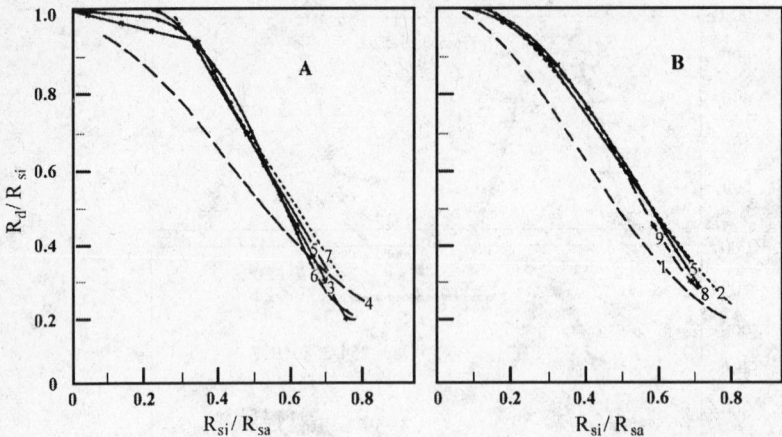

Fig. 5.12 Daily (A) and hourly (B) relationships between Rd/Rsi and Rsi/Rsa [Reprinted from Spitters et al., 1986. Agric. Forest Meteorol. 38:217-229 with kind permission of Elsevier Science - NL, Amsterdam, Netherlands]

proportion of Rd to Rsi is a function of the ratio of Rsi to Rsa (Fig. 5.12) described for daily total Rsi by:

$$Rd,d = 1, \qquad\qquad Rsi,d/Rsa,d < 0.07 \qquad\qquad [5.27a]$$

$$Rd,d/Rsi,d = 1-2.3(Rsi,d/Rsa,d-0.07)^2, \qquad 0.07 \le Rsi,d/Rsa,d < 0.35 \qquad [5.27b]$$

$$Rd,d/Rsi,d = 1.33-1.46(Rsi,d/Rsa,d), \qquad 0.35 \le Rsi,d/Rsa,d < 0.75 \qquad [5.27c]$$

$$Rd,d/Rsi,d = 0.23(Rsi,d/Rsa,d), \qquad 0.75 \le Rsi,d/Rsa,d \qquad [5.27d]$$

and for hourly values by:

$$Rd,h = 1, \qquad\qquad Rsi,h/Rsa,h \le 0.22 \qquad\qquad [5.28a]$$

$$Rd,h/Rsi,h = 1-6.4(Rsi,h/Rsa,h-0.22)^2, \qquad 0.22 < Rsi,h/Rsa,h \le 0.35 \qquad [5.28b]$$

$$Rd,h/Rsi,h = 1.47-1.66(Rsi,h/Rsa,h), \qquad 0.35 < Rsi,h/Rsa,h \le K \qquad [5.28c]$$

$$Rd,h/Rsi,h = R, \qquad\qquad K < Rsi,h/Rsa,h \qquad\qquad [5.28d]$$

where

$$R = 0.847-1.61\sin\beta + 1.04\sin^2\beta \qquad\qquad [5.29]$$

and

$$K = (1.47-R)/1.66 \qquad\qquad [5.30]$$

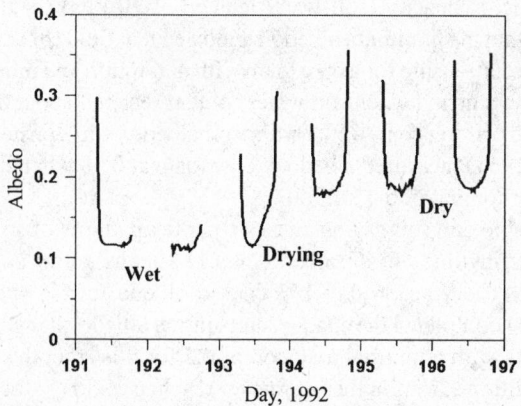

Fig. 5.13 Albedo for smooth, bare Pullman clay loam at Bushland, TX when wet and dry

5.2.1.3 Surface Albedo and Emissivity

Because Rsi provides most of the energy that is partitioned at the Earth's surface, albedo plays a major role in the energy balance. The mean albedo of the Earth is 0.36 ± 0.06 (Weast, 1982). But albedo varies diurnally (Fig. 5.13), with higher albedo corresponding to lower sun angle (see also bare soil data of Monteith and Szeicz [1961], Idso et al. [1974], and Aase and Idso [1975]). Soil and plant surfaces are often considered optically rough, but in some cases, specular (mirror like) rather than diffuse reflection may occur. Some plant leaves are shiny and reflect specularly when the angle of incident radiation is low. Wet soil surfaces may also reflect specularly. These mechanisms lead to higher albedo when sun angle is low. The albedo of plant stands is also lower at midday because more sunlight penetrates deeply within the canopy and is trapped by multiple reflections. Wilting and other physiological changes during the day may also contribute to changes in albedo.

Soil albedo decreases as water content increases. Bowers and Hanks (1965) found the relationship to be curvilinear, as did Skidmore et al. (1975). Idso and Reginato (1974) found that bare soil albedo changed linearly with the water content of the surface 2 mm of soil (smooth clay loam) (Fig. 5.14). For thicker layers the relationship was curvilinear. The maximum albedo (0.3) occurred for air dry soil, but the minimum albedo (0.14) occurred at about 0.23 m^3 m^{-3} water content, well before the soil was

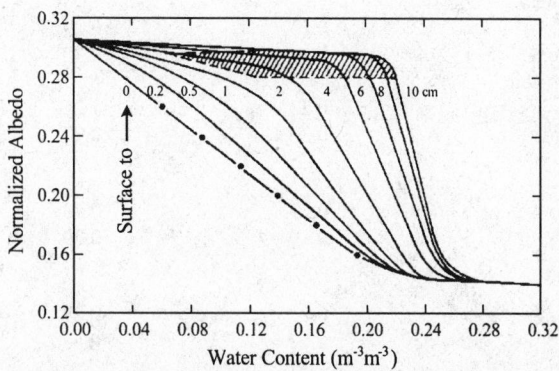

Fig. 5.14 Albedo, normalized according to sun angle, vs. soil water content for different surface layer thicknesses of Avondale clay loam at Phoenix, Arizona. Data for shaded area are uncertain. [Reprinted from Idso et al. 1974. Soil Sci. Soc. Am. Proc. 38:831-837 with permission of the Soil Science Society of America]

saturated. This represents field capacity (soil water tension of 30 kPa) for this soil, and Idso and Reginato (1974) postulated that the minimum albedo would occur at field capacity for all soils. Kondo et al. (1992) found a similar relationship for a bare loam with maximum and minimum albedos of 0.24 and 0.13 with the minimum occurring when soil water content reached about 0.22 $m^3 m^{-3}$. Idso et al. (1974,1975) showed that the difference in wet and dry soil albedos was constant despite time of day and day of year. Monteith (1961) measured albedo of clay loam at 0.18 when dry, decreasing to 0.11 when at field capacity water content of 0.35 $m^3 m^{-3}$.

The interaction of sun angle and soil drying causes complex patterns of soil albedo change over time. Fig. 5.13 illustrates low daytime wet soil albedos of 0.11 after irrigation and rain on days 191 and 192, 1992. Rapid soil surface drying on day 193 caused albedo to rise sharply during the day. Additional drying on day 194 completed the change, and diurnal albedo changes on days 195 and 196 reflect only sun angle effects with a minimum albedo of 0.2 for this smooth soil surface. The same surface in a roughened condition earlier in the year never reached midday albedo values higher than 0.13.

Other than water content, major determinants of soil albedo are color, texture, organic matter content, and surface roughness. Dvoracek and Hannabas (1990) presented a model of albedo dependence on sun angle, surface roughness, and color:

$$\alpha = p^{(c \, SIN \, \beta + 1)} \qquad\qquad [5.31]$$

where p was a color coefficient, c was a roughness coefficient, and ß is solar angle. They demonstrated good fits with measured data (Table 5.3). Albedo values modeled using p and c values from Table 5.3 for wheat and cotton (day of year 192, latitude 41°N) appear realistic (Fig. 5.15). However, the physical meaning of the p and c coefficients is not well understood.

Daily mean albedos may be calculated as the ratio of daily total reflected shortwave energy to daily total Rsi. Using data from Figs. 5.5 and 5.6, daily mean albedos for fallow (soybean residue) and

Table 5.3   Color (p) and roughness (c) coefficients for Equation [5.31] [Modified from Dvoracek and Hannabas, 1990. Proc. 3rd Nat. Irrig. Symp., Phoenix, AZ with permission of the American Society of Agricultural Engineers]

Surface and condition	Color coefficient p	Roughness coefficient c	Mean r^2
Lakes and ponds, clear water			
waves, none	0.13	0.29	0.82
waves, ripples up to 2.5 cm	0.16	0.70	0.74
waves, larger than 2.5 cm with			
occasional whitecaps	0.23	1.25	0.83
waves, frequent whitecaps	0.30	2.00	0.85
Lakes and ponds,			
green water, ripples up to 2.5 cm	0.22	0.70	0.90
muddy water, no waves	0.19	0.29	0.76
Cotton			
winds, calm to 4.5 m s^{-1}	0.27	0.27	0.80
winds, over 4.5 m s^{-1}	0.27	0.43	0.88
Wheat			
winds, calm to 4.5 m s^{-1}	0.31	0.92	0.85
winds, over 4.5 m s^{-1}	0.37	1.30	0.85

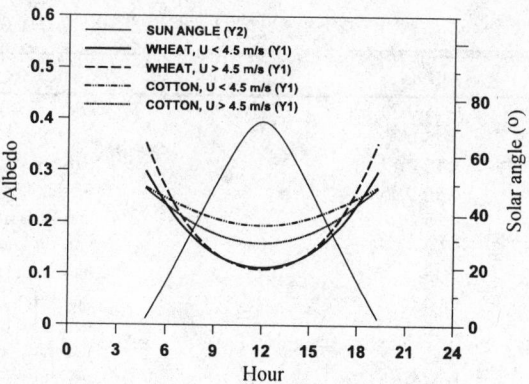

Fig. 5.15   Albedo values for wheat and cotton modeled with Equation [5.31]

alfalfa differ by about 0.10 when the soil is very dry (Fig. 5.16). The gradual decline in fallow albedo in early 1996 may be due to decomposition of the soybean residue. Albedo for the alfalfa field declined at each cutting to nearly that of the fallow field, which was initially rougher than the soil under the alfalfa. But, during heavy rains in the latter part of the year, the fallow soil surface was slaked and smoothed and its albedo increased to near that of the alfalfa. Thus, after the 4th cut, the alfalfa field albedo was lower than that of the fallow field for a brief time, probably because the alfalfa was irrigated and the fallow field had dried out again. Peaks of albedo exceeding 0.8 were due to snow early and late in the year.

In contrast to soil, albedo of closed canopies (well watered) is relatively constant (Table 5.4). Albedo values for many plant covers may be found in Gates (1980). For surfaces with plants, the amount of radiation reaching the soil surface (Rt) (Fig. 5.1) depends on the leaf area index (LAI) and the canopy structure. Numerical models have been developed that take into account leaf orientation and distribution in the canopy to calculate absorption of radiation at different levels in the canopy (Goudriaan, 1977; Chen, 1984). Lascano et al. (1987) used Chen's model to calculate polynomials representing the dependence of albedo on LAI, as well as the dependence of the view factor (proportion of sky visible from the soil) on LAI, and incorporated these into their ENergy and WATer BALance model (ENWATBAL). Monteith and Unsworth (1990) present equations describing the

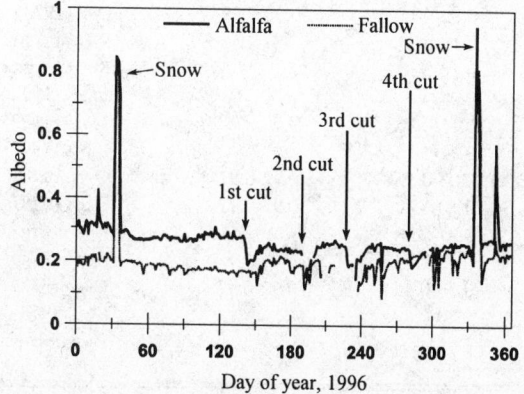

Fig. 5.16   Daily mean albedos for irrigated alfalfa, and fallow after soybean on Pullman clay loam at Bushland, TX

Table 5.4   Some albedo and emissivity values for various soil and plant surfaces

Surface	Albedo	Emissivity	Source
soils, dark, wet to light, dry	0.05-0.50	0.90-0.98	Oke (1978)
dry sandy soil	0.25-0.45		Rosenberg et al. (1983)
bare dark soil	0.16-0.17		Rosenberg et al. (1983)
dry clay soil	0.20-0.35		Rosenberg et al. (1983)
quartz sand	0.35		van Wijk and Scholte Ubing (1963)
sand, wet	0.09	0.98	van Wijk and Scholte Ubing (1963)
sand, dry	0.18	0.95	van Wijk and Scholte Ubing (1963)
dark clay, wet	0.02-0.08	0.97	van Wijk and Scholte Ubing (1963)
dark clay, dry	0.16	0.95	van Wijk and Scholte Ubing (1963)
fields, bare	0.12-0.25		van Wijk and Scholte Ubing (1963)
fields, wet, plowed	0.05-0.14		van Wijk and Scholte Ubing (1963)
dry salt cover	0.50		Rosenberg et al. (1983)
snow, fresh	0.80-0.95		Rosenberg et al. (1983)
snow, old	0.42-0.70		Oke (1978)
snow, fresh	0.95	0.99	Oke (1978)
snow, old	0.40	0.82	van Wijk and Scholte Ubing (1963)
snow, fresh	0.80-0.85		van Wijk and Scholte Ubing (1963)
snow, compressed	0.70		van Wijk and Scholte Ubing (1963)
snow, melting	0.30-0.65		Oke (1978)
grass, long (1 m)	0.16	0.90	Oke (1978)
short (0.02 m)	0.26	0.95	van Wijk and Scholte Ubing (1963)
grass, green	0.16-0.27	0.96-0.98	van Wijk and Scholte Ubing (1963)
grass, dried	0.16-0.19		van Wijk and Scholte Ubing (1963)
prairie, wet	0.22		van Wijk and Scholte Ubing (1963)
prairie, dry	0.32		van Wijk and Scholte Ubing (1963)
stubble fields	0.15-0.17		van Wijk and Scholte Ubing (1963)
grain crops	0.10-0.25		Jensen et al. (1990)
green field crops full cover, LAI>3	0.20-0.25		Jensen et al. (1990)
leaves of common farm crops		0.94-0.98	Rosenberg et al. (1983)
most field crops	0.18-0.30		Monteith and Unsworth (1990)
field crops, latitude 22-52°	0.22-0.26	0.94-0.99	Monteith and Unsworth (1990)
field crops, latitude 7-22°	0.15-0.21	0.94-0.99	Rosenberg et al. (1983)
deciduous forest	0.15-0.20	0.96[†]	Oke (1978)
decid. forest, bare	0.15	0.97	Oke (1978)
leaved	0.20	0.98	Rosenberg et al. (1983)
coniferous forest	0.10-0.15	0.97[1]	Oke (1978)
coniferous forest	0.05-0.15	0.98-0.99	Rosenberg et al. (1983)
vineyard	0.18-0.19		Rosenberg et al. (1983)
mangrove swamp	0.15		Jones (1992)
grass	0.24		Jones (1992)
crops	0.15-0.26		Jones (1992)
forest	0.12-0.18		Oke (1978)
water, high sun	0.03-0.10	0.92-0.97	Oke (1978)
water, low sun	0.10-1.00	0.92-0.97	Rosenberg et al. (1983)
sea, calm	0.07-0.08		Rosenberg et al. (1983)
sea, windy	0.12-0.14		Oke (1978)
ice, sea	0.30-0.45	0.92-0.97	Oke (1978)
ice, glacier	0.20-0.40		Rosenberg et al. (1983)
ice, lake, clear	0.10		Rosenberg et al. (1983)
ice, lake, w/ snow	0.46		Rosenberg et al. (1983)

[†] van Wijk and Scholte Ubing (1963)

albedo of a deep canopy with a spherical distribution of leaves for sun angles higher than 25°. More discussion of these concepts can be found in Russell et al. (1989). For field studies, one can either measure albedo, or directly measure the components of net radiation, or use a net radiometer (Table 5.1). The transmitted radiation can be measured below the canopy with tube solarimeters.

5.2.1.4   Incoming Long-Wave Radiation

Methods of estimating long-wave irradiance from the sky ($L\downarrow$) usually take the form:

$$L\downarrow = \epsilon \sigma \ (T_a + 273.16)^4$$ [5.32]

where T_a (C) is air temperature at the reference measurement level (often 2 m), and the emissivity may be calculated from the vapor pressure of water in air at reference level (e_a) or e_a and T_a. The vapor pressure is

$$e_a = RH(e_s)$$ [5.33]

where RH is the relative humidity of the air and e_s is the saturation vapor pressure (kPa) at T_a (C) given by (Murray, 1967):

$$e_s = 0.61078 \ \exp\left(\frac{17.269T_a}{237.3 + T_a}\right)$$ [5.34]

If the dew point temperature, rather than the RH, is known then:

$$e_a = 0.61078 \ \exp\left(\frac{17.269T_{dew}}{237.3 + T_{dew}}\right)$$ [5.35]

Hatfield et al. (1983) compared several methods for estimating ϵ and concluded that methods using only air temperature performed less well than those that used vapor pressure or both vapor pressure and air temperature. Among the best methods was Idso's (1981) equation:

$$\epsilon_a = 0.70 + 5.95 \times 10^{-4} e_a \ \exp\left[\frac{1500}{T_a + 273.1}\right]$$ [5.36]

where e_a is in kPa. Idso (1981) showed fairly conclusively that ϵ is a function of both e_a and T_a.

Howell et al. (1993) measured $L\downarrow$ (Table 5.1) and calculated ϵ by inverting Equation [5.32]. Applying Equation [5.36] as well as Brunt's (1932) equation

$$\epsilon_a = 0.52 + 0.206e_a^{0.5}$$ [5.37]

and Brutsaert's (1982) equation

$$\epsilon_a = 0.767e_a^{1/7}$$ [5.38]

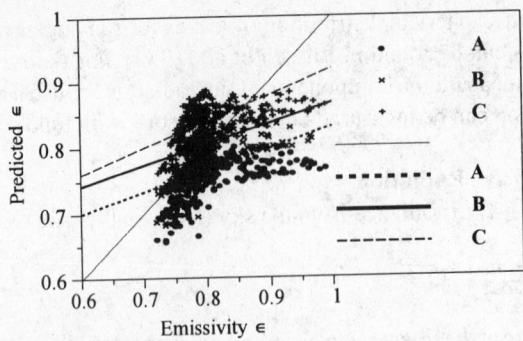

Fig. 5.17 Comparison of predictions with measured emissivity for two periods in 1992 at Bushland, TX. Points plotted at extreme right were associated with nighttime and overcast conditions. A = Eq. [5.37], B = Eq. [5.38], C = Eq. [5.36]. Lines are for linear regressions shown in Table 5.5

to their data shows that all three equations gave good predictions for clear sky conditions but probably underestimated ϵ for cloudy and night time conditions (Fig. 5.17). For regressions of predicted versus measured ϵ, the Idso (1981) equation gave a slightly higher correlation coefficient and a slope closer to unity (Table 5.5). Under heavy clouds, sky emissivity approaches unity, and none of these models predicts this well.

Despite the difficulty of estimating sky emissivity well, uncertainty in the value of $L\downarrow$ usually causes little difficulty in estimating net radiation because $L\downarrow$ is very often a small component of the energy balance.

5.2.1.5 Comparison of Net Radiation Estimates with Measured Values

It has become commonplace to have data from field weather stations that include Rsi, and air temperature (T_{az}), wind speed (U_z), and relative humdity (RH_z) measured at some reference height (z) (often 2 m). Measurement of Rn is still not common, probably due to several factors including additional expense, fragility of the plastic domes used on some models of net radiometers, and problems with calibration. Net radiometer calibration changes with time, and experience shows that even new radiometers may not agree within 10%. If a net radiometer is used, it is prudent, as with all instruments, to check measured Rn values against estimated ones. Methods presented in previous sections can be used to estimate Rn, but simpler methods exist that are adequate for most cases. Jensen et al. (1990) compared four methods of estimating Rn, including those of Wright and Jensen (1972), Doorenbos and Pruitt (1977), a combination of Brutsaert (1975) and Weiss (1982), and Wright (1982), against values measured at Copenhagen, Denmark, and Davis, CA. The Wright (1982) method was best overall, but underestimated Rn in the peak month at Copenhagen by 9%. The Wright and

Table 5.5 Regressions of predicted emissivity (ϵ_p) versus measured values (ϵ) for data from day 133 through 140 and 192 through 197, 1992 at Bushland, TX

Method	Regression Equation	r^2	SE
Brunt, Equation [5.37]	$\epsilon_p = 0.505 + 0.325\epsilon$	0.33	0.024
Brutsaert, Equation [5.38]	$\epsilon_p = 0.556 + 0.311\epsilon$	0.32	0.024
Idso, Equation [5.36]	$\epsilon_p = 0.522 + 0.398\epsilon$	0.37	0.027

Jensen (1972) method was almost as good. These methods all assume that surface temperature is not measured, so that only air temperature is used in the calculations.

Jensen et al. (1990) calculated net long-wave radiation (Rnl) as:

$$Rnl = -\left[a\left(\frac{Rsi}{Rso}\right) + b\right]\left(a_1 + b_1 e_d^{0.5}\right)\sigma T_{az}^4 \qquad [5.39]$$

where e_d is the saturation vapor pressure of water in air at dew point temperature (kPa); and the term $(a_1 + b_1 e_d^{0.5})$ is a net emittance ($\in '$) of the surface. The net emittance attempts to compensate for the fact that surface temperature is not measured, the assumption being that T_{az} can substitute reasonably well for both sky and surface temperature. The coefficients a, b, a_1, and b_1 are climate specific, a and b being cloudiness factors. Some values are presented by Jensen et al. (1990, Table 3.3).

Many weather stations report only daily totals of solar radiation, and maximum and minimum of air temperature (T_x and T_n), respectively (K). If this is the case, the term σT^4 can be estimated as:

$$\sigma T^4 \cong \frac{\sigma\left(T_x^4 - T_n^4\right)}{2} \qquad [5.40]$$

If mean dew point temperature is not available, it may be estimated as equal to T_n in humid areas.

Allen et al. (1994a,b) presented slightly modified versions of the methods presented by Jensen et al. (1990) in a proposed FAO standard for reference evapotranspiration estimation. As an example, estimates of daily total net radiation were made for Bushland, TX using the following equations from Allen et al. (1994b):

$$Rn = (1-\alpha)Rs - \left[a_c\left(\frac{Rsi}{Rso}\right) + b_c\right]\left(a_1 + b_1 e_d^{0.5}\right)\sigma\left(\frac{T_m^4 + T_n^4}{2}\right) \qquad [5.41]$$

where the cloud factors were $a_c = 1.35$ and $b_c = -0.35$, the emissivity factors were $a_1 = 0.35$ and $b_1 = -0.14$, the albedo was $\alpha = 0.23$, Rsi was measured, e_d was calculated from mean dew point temperature, and Rso was calculated from:

$$Rso = (0.75 + 0.00002\ ELEV)Rsa \qquad [5.42]$$

where Rsa is from Equation [5.19], and ELEV is elevation (m) above mean sea level. This is similar to Equation [5.20] but with a correction increasing Rso for higher elevation sites. The mean daily saturated vapor pressure at dew point temperature was estimated from mean daily dew point temperature (T_d):

$$e_d = 0.611\ \exp\left[\frac{17.27\ T_d}{T_d + 237.3}\right] \qquad [5.43]$$

Additional estimates were calculated from half-hourly measured values of Rsi, T_a, and T_d using equations given by Allen et al. (1994b) equivalent to Equations [5.7, 5.8, 5.10, 5.12, 5.13, 5.14, 5.15, and 5.16] to estimate half-hourly Rsa, and Equation [5.41] to estimate half-hourly Rso. Equation [5.42] was applied to half-hourly dew point temperatures to estimate half-hourly e_d values. Equation [5.40] was written for half-hourly values of air temperature (T_a) as:

$$Rn = (1-\alpha)Rs - \left[a_c\left(\frac{Rsi}{Rso}\right) + b_c\right]\left(a_1 + b_1 e_d^{0.5}\right)\sigma\left(T_a^4\right) \qquad [5.44]$$

where the ratio of Rsi to Rso was set to 0.7 for night-time estimates of Rn.

Comparison of daily Rn estimates, calculated using half-hourly data means, with measurements made with a REBS Q*5 (Seattle, WA) net radiometer over irrigated grass shows excellent agreement for alfalfa (Fig. 5.18) and grass (Fig. 5.19) at Bushland, TX. But, there was a consistent bias for Rn estimated from daily means, with underestimation of Rn at high measured values, and overestimation at low measured values (Fig. 5.20). The bias evident when daily means and maximum/minimum temperatures were used is probably tied to both poor estimates of vapor pressure from the max/min temperature data, and the inadequacy of Equation [5.39].

Estimates of half-hourly net radiation for alfalfa at Bushland, Texas using half-hourly data and these methods also gave good results (Fig. 5.21). Allen et al. (1994a, b) give detailed methods for estimating Rn when measurements are missing for Rsi and/or e_d.

5.2.2 Latent Heat Flux Measurement

Latent heat flux is the product of the evaporative flux (E) (kg s^{-1} m^{-2}), and the latent heat of evaporation (L) (2.44 x 10^6 J kg^{-1} at 25 C). The value of L is temperature dependent, but is well described (in J kg^{-1} x 10^6) by:

$$L = 2.501 - 2.370 \times 10^{-3} T \qquad \left(r^2 = 0.99995\right) \qquad [5.45]$$

where T is in C. Methods of measurement of E include weighing lysimeter (including microlysimeters), and other mass balance techniques that rely on measurements of change in soil water storage (ΔS) as well as eddy correlation and Bowen ratio measurements. Because ΔS is a component

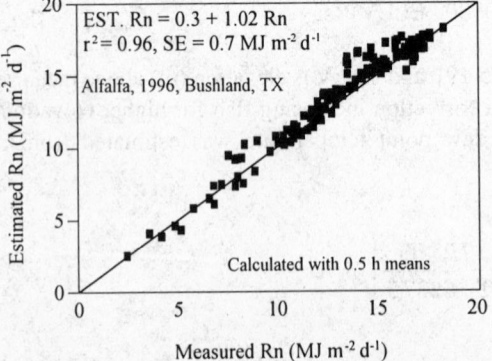

Fig. 5.18   Net radiation estimated with methods from Allen et al. (1994a,b) compared to measurements over sprinkler irrigated alfalfa in 1996 at Bushland, TX

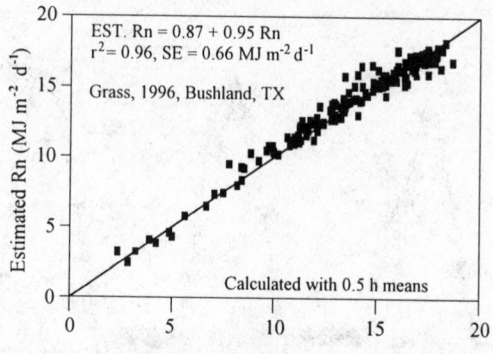

Fig. 5.19 Daily net radiation, estimated with methods from Allen et al. (1994a,b) using half-hourly data, compared to measurements with a REBS Q*5 net radiometer over drip irrigated grass in 1996 at Bushland, TX

of the soil water balance, and lysimetry is a key tool for investigations of soil water balance, discussion of lysimetric techniques will be deferred to Section 5.3.

5.2.2.1 Boundary Layers

Evaporative fluxes move between plant, or soil surfaces and the air by both diffusion and convection. Diffusive processes prevail in the laminar sublayer close (mm's) to these surfaces. In this layer, air movement is parallel to the surface and little mixing occurs. Vapor flux across the laminar sublayer is well described by a Fickian diffusion law relating flux rate to vapor pressure gradient factored by a conductance term. But in the turbulent layer beyond the laminar layer, the flux is mostly convective in nature so that water vapor is moved in parcels of air that are moved and mixed into the atmosphere in turbulent flow. These moving parcels of air are often referred to as eddies, similar to eddies seen in a stream. Usually, the eddies are not visible, but in foggy, smoky or dusty air they may be apparent. Certainly, anyone who has felt the buffeting of the wind can attest to the force of eddies and the turbulence of the air stream in which they occur. As wind speeds increase, the depth of the laminar sublayer decreases. Surface roughness enhances this process, resulting in thinner laminar sublayers. Because the resistance to vapor transport across the laminar air stream is much larger than the

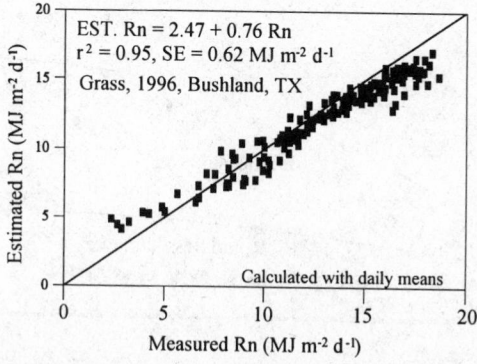

Fig. 5.20 Daily net radiation, estimated with methods from Allen et al. (1994a,b) using daily means and temperature maxima and minima, compared to measurements with a REBS Q*5 net radiometer over drip irrigated grass in 1996 at Bushland, TX

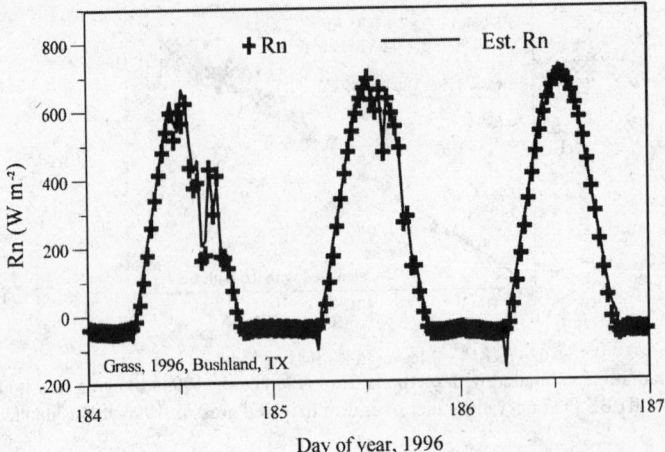

Fig. 5.21 Half-hourly measured net radiation compared with values estimated with methods of Allen et al. (1994a,b) using half-hourly data for drip irrigated grass at Bushland, TX

resistance across a turbulent air stream of similar dimensions, increasing roughness and wind speed both tend to enhance vapor transport. If the air is still, then eddies due to turbulent flow do not exist, but eddies due to free convection may well be present. Free convection occurs when an air parcel is warmer (or colder) than the surrounding air, and thus moves upward (or downward) because it is lighter (or heavier). These buoyancy effects can predominate at very low wind speeds when the surface is considerably warmer than the air. As opposed to free convection, transport in eddies due to wind is called forced convection.

A full discussion of the fluid mechanics of laminar and turbulent flow, Fickian diffusion, and forced and free convection is well beyond the scope of this chapter. Discussions relevant to soil and plant surfaces are presented by Rosenberg et al. (1983), Monteith and Unsworth (1990), and Jones (1992). Here the discussion will concentrate on some results and methods of measurement. These methods are valid within the constant flux layer (Fig. 5.22) which is a layer of moving air that develops from the point at which the air stream first reaches a surface of a given condition, for example, the wheat field shown in Fig. 5.22. As the air moves over the field, it mixes, equilibrating with the new surface condition, and forming a layer of gradually increasing thickness (δ) within which the flux of heat and

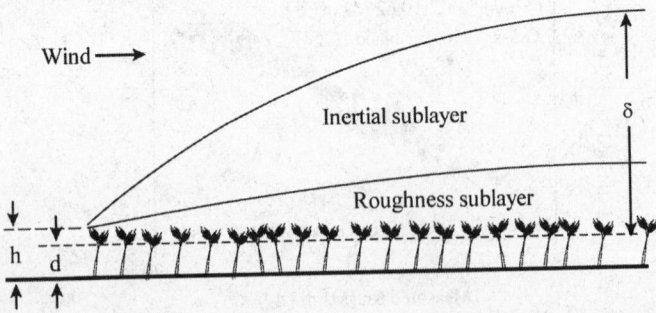

Fig. 5.22 Schematic of sublayers of the surface boundary layer over a wheat crop. The height, h, of the crop and the depth, δ, of the constant flux layer are noted. The height d is the zero plane displacement height, which is the height to which a logarithmic wind profile, measured above the crop, would extrapolate to zero wind speed.

vapor is constant with height. This is the fully adjusted or equilibrium layer. Within this layer is a layer, extending from the roughness elements (wheat plants in this schema) upward, within which air flow is more turbulent due to the influence of the roughness elements. This is called the roughness sublayer (Monteith and Unsworth, 1990). For any measurement of air temperature, humidity or wind speed, the fetch is the distance upwind from the point of observation to the edge of the new surface. The ratio of the fetch to the value of δ is dependent on the roughness of the surface, the stability of the air, and wind speed. For many crop surfaces, it may be as small as 20:1 or as large as 200:1. For smooth surfaces such as bare soil, the ratio may well be larger than 200:1. Measurements should be made in the constant flux layer but above the roughness sublayer.

5.2.2.2 Eddy Correlation Measurements

The observation of turbulent flow and the concept of eddies lead to the eddy correlation method of latent heat flux measurement. The main idea here is that if eddies with a vertical velocity component upward are correlated with humidities on average higher than the humidities correlated with downward moving eddies, then the net flux of water vapor is upward. In this method, very fast response sensors are used to measure the vertical wind speed and humidity simultaneously at a rate of, for example, 20 Hz. This gives a direct measure of the flux at the measurement height (but see fetch requirements below) according to (Rosenberg et al., 1983):

$$E = \left(\frac{M_w}{M_a P} \right) \rho_a \overline{w' e_a'}$$ [5.46]

where the overscore indicates time averages of vertical wind speed (w') and vapor pressure (e_a'), the primes indicate instantaneous deviations from the mean, P is atmospheric pressure (Pa), ρ_a is air density, and M_w and M_a are the molecular weights of water and air. The rate of data acquisition must be faster for measurements nearer the surface. Monteith and Unsworth (1992) state that eddy sizes increase with surface roughness and wind speed, and with height above the surface, and they suggest 1 kHz rates may be needed near a smooth surface, while 10 Hz or slower may be adequate at several m above a forest. Because the measurements should take place within the fully adjusted boundary layer, simply increasing sensor height will not eliminate the need for fast sensor response. Eddy correlation methods are difficult to carry out due to the data handling and sensor requirements. Data processing requirements are large, but modern data logging and computing equipment are capable of handling these. Commercial systems including data processing software are now available, although expensive (Campbell Scientific Inc., Logan, UT and The Institute of Ecology and Resource Management at the University of Edinburgh, Scotland). The sonic anemometer is the wind sensor of choice for eddy correlation work due to its fast response and sensitivity. At this time, a single axis unit costs about $2,500, and a three-dimensional sonic anemometer costs about $8,000. Suitable vapor pressure sensors include the krypton hygrometer and infrared gas analyzer (IRGA) available currently in the $5,000 to $10,000 range.

Eddy correlation measurements may be made for sensible heat flux as well (Section 5.2.4), and if both E and H are measured by eddy correlation, the performance of the system may be checked (Houser et al., 1998) by rearranging Equation [5.1] to:

$$LE + H = -Rn - G$$ [5.47]

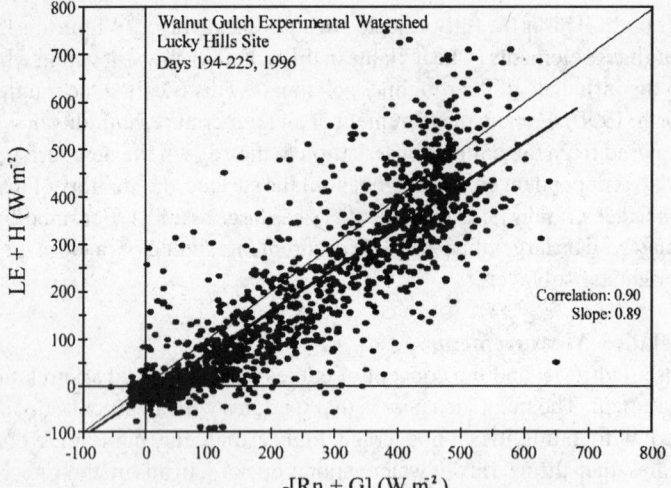

Fig. 5.23 Check of eddy correlation system LE and H values against measured Rn and G values [Adapted from Houser, 1998 with permission]

and measuring Rn and G (Fig. 5.23). Fast response thermocouples for measuring air temperature are used in eddy correlation systems for measuring H. Because these are very much less expensive than fast response vapor pressure sensors, it is sometimes sensible to measure Rn and G, and H by eddy correlation, and find LE as the residual:

$$LE = -Rn - G - H \qquad [5.48]$$

Comparisons of eddy correlation and Bowen ratio systems are found in Houser et al. (1998) and Dugas et al. (1991). Some specifics of eddy correlation system design are given in Unland et al. (1996) and Moncrieff et al. (1997).

5.2.2.3 Bowen Ratio Measurement

The Bowen ratio is the ratio of sensible to latent heat flux (ß = H/LE). Introducing this into Equation [5.1] and rearranging gives the Bowen ratio method for estimating LE:

$$LE = \frac{-(Rn+G)}{\beta+1} \qquad [5.49]$$

In the constant flux layer, it is possible to measure temperature and vapor pressure differences at two heights, z_1 and z_2, and evaluate ß from a finite difference form:

$$\beta = \frac{\dfrac{K_H \rho_a c_p \left(T_{z_2} - T_{z_1}\right)}{z_2 - z_1}}{\dfrac{K_V \left(\dfrac{\epsilon_m}{P}\right) \rho_a L \left(e_{z_2} - e_{z_1}\right)}{z_2 - z_1}} \approx \frac{c_p \left(T_{z_2} - T_{z_1}\right)}{\left(\dfrac{\epsilon_m}{P}\right) L \left(e_{z_2} - e_{z_1}\right)} = \gamma \frac{\left(T_{z_2} - T_{z_1}\right)}{\left(e_{z_2} - e_{z_1}\right)} \qquad [5.50]$$

where the second and third entities assume equivalency of the exchange coefficients for sensible heat flux (K_H) and latent heat flux (K_V), $\in_m$ is M_w/M_a, and $\gamma = c_pP/(\in_m L)$ is the psychrometric constant. Commonly, values of T and e are half-hour or hourly means. Because the sensor response time does not have to be very short, Bowen ratio equipment is much less expensive than that for eddy correlation, with complete systems available for under $10,000. Systems are available from Radiation and Energy Balance Systems (REBS), Seattle, WA, Campbell Scientific, Inc., Logan, UT, and others.

Because slight differences in instrument calibration may lead to large errors, it is advisable to switch instruments between the measurement heights. The moving arm system popularized by REBS is one way to do this. Bowen ratio measurements are usually valid only during daylight hours. At night, the sum of Rn and G approaches zero causing Equation [5.49] to become imprecise. For periods just after sunrise and before sunset, the gradients of T and e may become small at the same time that Rn becomes small, leading to instability in Equation [5.49] and imprecision in the estimate of LE. Under advective conditions, Bowen ratio systems tend to underestimate LE when regional sensible heat advection occurs (Blad and Rosen̶̶̶̶̶, 1974; Todd, 1998b), probably because $K_H/K_V > 1$ under the stable conditions that prevail then (Verma et al., 1978). Four Bowen ratio systems were compared by Dugas et al. (1991) who discuss the merits of different designs. Three eddy correlation systems agreed well with each other but LE measurements from them were consistently lower than those from the four Bowen ratio systems.

5.2.2.4 Fetch Requirements

Both eddy correlation and Bowen ratio methods are sensitive to upwind conditions. The LE and H values from these methods represent an areal mean for a certain upwind area, often called the footprint. Both methods require considerable upwind fetch, often running to hundreds of m, of surface that is essentially the same as where the measurement is made, if the measurement is to be representative of that surface. Also, the longer the same-surface fetch is, the deeper the fully adjusted layer, and the higher the instruments can be placed above the surface. Issues of instrument height and fetch are discussed by Savage et al. (1995,1996) who recommended placing the sonic anemometer no closer than 0.5 m above a short grass cover. Because eddies are smaller nearer the surface, placement of the sonic anemometer too near the surface may lead to eddies being smaller than the measurement window of the anemometer. Fetch requirements may be stated as a ratio of fetch distance to instrument height. Heilman et al. (1989) studied fetch requirements for Bowen ratio systems, and concluded that a fetch ratio of 20:1 was adequate for many measurements, down from the 100:1 ratios reported earlier. Fetch requirements increase as measurement height (z_m) increases. This poses some additional problems for Bowen ratio systems because these incorporate two sensors and the sensors must be separated enough that the vapor pressure and temperature gradients between them are large enough to be accurately sensed. The rougher the surface, the smaller the gradients. For many surfaces and common instrument resolutions, these facts lead to separation distances on the order of 1 m. The lower measurement should be above the roughness sublayer, typically at least 0.5 m above a crop (more for a very rough surface such as a forest), so the upper measurement may well be nearly 2 m above the crop surface. This could easily lead to a fetch requirement of 100 m. Analysis of relative flux and cumulative relative flux for an alfalfa field under moderately stable conditions using the methods of Schuepp et al. (1990) leads to rather large fetch requirements (Todd, 1998a) (Fig. 5.24). For unstable conditions, mixing is enhanced and the boundary layer becomes adjusted more quickly over a new surface so that fetch requirements are lessened. Fetch requirements are more severe for the Bowen ratio than for eddy correlation measurements (Schmid, 1997).

Because of the direct way in which fluxes are measured in eddy correlation schemes, this method is sometimes stated to be the only true measure of latent (or sensible) heat flux. However,

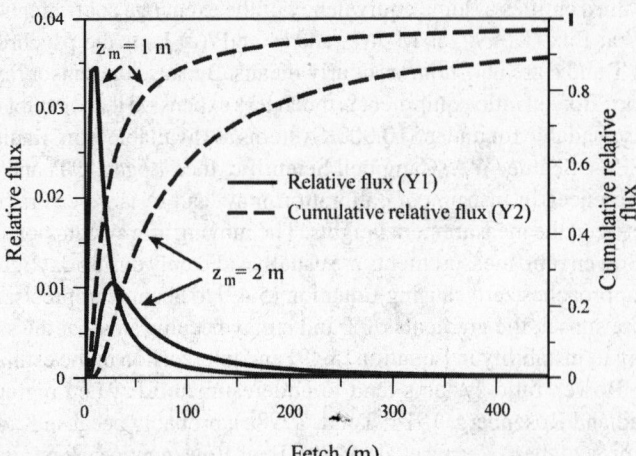

Fig. 5.24 Relative and cumulative relative flux of an alfalfa field for measurement heights (z_m) of 1 and 2 m, moderately stable thermal conditions, and canopy height of 0.5 m. Cumulative relative flux reaches 0.8 at 65 m for $z_m = 1$ m, and at 225 m for $z_m = 2$ m.

consideration of fetch requirements leads to a conclusion that both eddy correlation and Bowen ratio measurements are true only for a constantly changing footprint area upwind of the measurement location. The footprint area and the true flux are poorly defined because the location and size of the footprint change with wind direction and speed. There is strength in this kind of areal averaging, because it reduces noise due to the spatial variability of evaporation. But the measurement cannot be said to be true for any specific location. Indeed, as wind direction changes the measurement area may change completely. By contrast, the soil water balance methods of estimating E, discussed in Section 5.3, provide measures for specific locations. In the case of weighing lysimeters, these are, in fact, direct measurements of E, specific to a well-defined location, for all times during which precipitation and runoff are not occurring (neglecting the negligible change in plant mass over short periods).

5.2.2.5 Penman-Monteith Estimates of Latent Heat Flux
Since Penman (1948) published his famous equation describing evaporation from wet surfaces based on the surface energy balance, there have been developments, additions and refinements of the theory too numerous to mention. Notable examples are the van Bavel (1966) formulation, which includes a surface roughness length term (z_0), and the Penman-Monteith (PM) formula (Monteith, 1965), which includes aerodynamic and surface resistances. The van Bavel equation tends to overestimate in windy conditions and is very sensitive to the value of z_0 (Rosenberg, 1969). Howell et al. (1994) compared several ET equations for well-watered, full-cover winter wheat and sorghum and found that the PM formula performed best. Because it is widely used in agricultural and environmental research, and because it has been presented by ASCE (Jensen et al., 1990) and FAO (Allen et al., 1994a,b) as a method of computing estimates of reference crop water use, the Penman-Monteith equation will be discussed:

$$LE = -\frac{\Delta(R_n + G) + \rho_a c_p (e_s - e_a)/r_a}{\Delta + \gamma(1 + r_s / r_a)} \qquad [5.51]$$

where LE is latent heat flux, R_n is net radiation, and G is soil heat flux (all in MJ m^{-2} s^{-1}), Δ is the slope of the saturation vapor pressure-temperature curve (kPa C^{-1}), ρ_a is air density (kg m^{-3}), c_p is the specific heat of air (kJ kg^{-1} C^{-1}), e_a is vapor pressure of the air at reference measurement height z, and e_s is the saturated vapor pressure at a dew point temperature equal to the air temperature at z (kPa), (e_s-e_a) is the vapor pressure deficit (VPD), r_a is the aerodynamic resistance (s/m), r_s is the surface (canopy) resistance (s/m), and γ is the psychrometric constant (kPa C^{-1}). Penman's equation and those derived from it were developed as a means of eliminating canopy temperature from energy balance considerations. Besides measurements of Rn and G, the user must know the vapor pressure of the air (e_a), and air temperature (from which e_s may be calculated) at reference measurement height (z) (often 2 m). The values of r_a and r_s may be difficult to obtain. The surface or canopy resistance is known for only a few crops and is dependent on plant height, leaf area, irradiance, and water status of the plants.

Jensen et al. (1990) and Allen et al. (1994a,b) presented methods of calculating E for well-watered, full-cover grass and alfalfa. The following example, drawn from recent studies at Bushland, TX, employs those methods. Aerodynamic resistance was estimated for neutral atmospheric conditions from:

$$r_a = \frac{\ln\left(\dfrac{z_m - d}{z_{0m}}\right)\ln\left(\dfrac{z_H - d}{z_{0H}}\right)}{k^2 U_z} \quad [5.52]$$

where z_m (m) is the measurement height for wind speed (U_z) (m/s), z_H (m) is measurement height for air temperature and relative humidity, k is 0.41, z_{0m} and z_{0H} are the roughness length parameters for momentum (wind) and sensible heat transport, and d is the zero plane displacement height. The value of r_a calculated from Equation [5.51] will be too high for highly unstable conditions and too low for very stable conditions. Stability corrections should be made to Equation [5.51] for those conditions (see Monteith and Unsworth, 1990, p. 234 for some examples), but were not made for this example.

Surface resistance was calculated from:

$$r_s = \frac{r_1}{(0.5\ \text{LAI})} \quad [5.53]$$

where r_1 is the stomatal resistance taken as 100 s/m, and the leaf area index (LAI) was taken as:

$$\text{LAI} = 5.5 + 1.5\ln(h_c) \quad [5.54]$$

where the crop height (h_c) was taken as 0.12 cm for grass, and 0.5 m for alfalfa.

The zero plane displacement height (d) was calculated as:

$$d = \frac{2}{3}h_c \quad [5.55]$$

The roughness length for momentum (z_{0m}) was calculated as:

$$z_{0m} = 0.123\ h_c \qquad [5.56]$$

and the roughness length for sensible heat transport was

$$z_{0H} = 0.1\ z_{0m} \qquad [5.57]$$

Net radiation was calculated as shown in Section 5.2.1.5. All calculations were on a half-hourly basis. For well watered mixed fescue grass in 1996, the Penman-Monteith (PM) equation underestimated ET, as measured by a weighing lysimeter, at ET rates > 4 mm day⁻¹ (Fig. 5.25), even though Rn and G were well estimated. The underestimation of ET was due to systematic error in the surface and/or aerodynamic resistances. For well-watered, full-cover alfalfa in 1996, the PM estimates of ET were close to values measured with a weighing lysimeter (Fig. 5.26). Because Rn and G were well estimated, it is presumed that r_a and r_s were predicted well also. Examination of diurnal dynamics showed that the PM method was capable of closely reproducing those dynamics.

Although important as a research model, the PM method is not much used for direct prediction of LE due to the difficulty of knowing r_a and r_s. However, it is commonly used to predict a theoretical reference evapotranspiration (ET_r) for use in irrigation scheduling (Allen et al., 1994a,b). In this application, crop water use or ET is predicted from daily values of ET_r and a dimensionless crop coefficient (K_c) which is dependent on the crop variety and time since planting or growing degree days:

$$ET = K_c ET_r \qquad [5.58]$$

The crop coefficients are determined from experiments that measure daily crop water use and ET_r and compute:

$$K_c = \frac{ET}{ET_r} \qquad [5.59]$$

Many details on this methodology are found in Jensen et al. (1990).

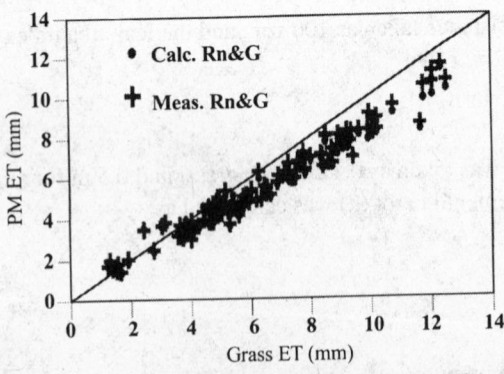

Fig. 5.25 Daily Penman-Monteith estimates of ET using both measured and estimated values of Rn and G were not significantly different from each other for well-watered, full-cover mixed fescue grown at Bushland, TX in 1996. Both PM ET values were less than values measured by a weighing lysimeter for values above 4 mm d⁻¹.

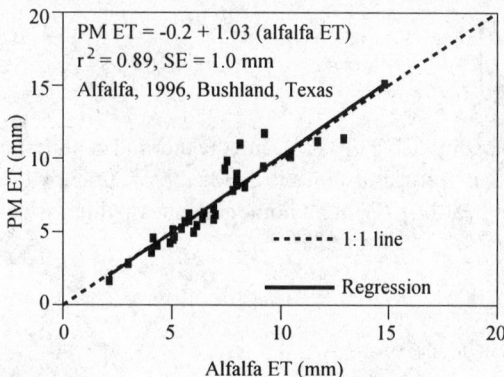

Fig. 5.26 Daily Penman-Monteith estimates of ET using estimated Rn and G for well-watered, full-cover (LAI > 3) alfalfa at Bushland, TX in 1996

5.2.2.6 Bare Soil Evaporation Estimates

Fox (1968) and later Ben-Asher et al. (1983) and Evett et al. (1994) described an LE prediction method based on subtracting the energy balance equations (Equation [5.1]) written for a dry and a drying soil. Because LE is zero for a dry soil, this gives an expression for LE from the drying soil in terms of the other energy balance terms. The method requires a column of dry soil embedded in the field of drying soil, and measurements of the surface temperatures of the dry soil and of the drying field soil. The surface temperature difference between the dry and drying soils explains most of E, but prediction accuracy is only moderately good ($r^2 = 0.82$ for daily predictions) (Evett et al., 1994). Evett et al. (1994) showed that the aerodynamic resistance over the dry soil surface was reduced, probably by buoyancy of air heated over the relatively hot surface, and that the resistance was relatively independent of wind speed. They also showed that consideration of the soil albedo change with drying could improve the E estimates. Although the method shows promise, it does not provide an estimate of surface soil water content that would be needed to calculate the albedo change.

When the soil is wet, the evaporative flux can be estimated using the Penman or Penman-Monteith equations with surface resistance set to an appropriate low value (Howell et al., 1993). This wet period is the energy limited stage of evaporation. As the soil dries, E becomes limited by soil properties. van Bavel and Hillel (1976) addressed this using a finite difference model of soil water and heat flux that later was developed into the CONSERVB model of evaporation from bare soil. This model described one-dimensional soil water movement with Darcy's law, including the dependence of hydraulic conductivity (K) (m s^{-1}), on soil water potential (h) (m), and the soil water retention function [$\theta_v(h)$]. The surface energy balance was solved implicitly for surface temperature (T), resulting in calculated values of E, H, Rn, and G at each time step. The value of E was used as the upper boundary condition for soil water flux at the next time step. The elements of CONSERVB were included in the ENWATBAL model by Lascano et al. (1987), and the latter model was upgraded to model albedo changes dependent on surface soil water content by Evett and Lascano (1993). Although CONSERVB was not validated against directly measured E, the 1993 version of ENWATBAL was shown to more accurately predict E than either the Penman or Penman-Monteith equations (Howell et al., 1993).

5.2.3 Soil Heat Flux

Briefly, heat conduction in one dimension is described by a diffusion equation:

$$C \frac{\partial T}{\partial t} = \lambda \frac{\partial}{\partial z} \left[\frac{\partial T}{\partial z} \right] \qquad\qquad [5.60]$$

where the volumetric heat capacity (C) $(J\,m^{-3}\,K^{-1})$, and the thermal conductivity (λ) $(J\,s^{-1}\,m^{-1}\,K^{-1})$, are assumed constant in space, and vertical distance is denoted by z, time by t, and temperature by T.

The one-dimensional soil heat flux (G) for a homogeneous medium is described by:

$$G = -\lambda \frac{\partial T}{\partial z} \qquad\qquad [5.61]$$

The thermal conductivity is a single-valued function of water content and is related to the thermal diffusivity (D_T) $(m^2\,s^{-1})$ by:

$$\lambda = D_T C \qquad\qquad [5.62]$$

where the volumetric heat capacity (C) $(J\,m^{-3}\,K^{-1})$ can be calculated with reasonable accuracy from the volumetric water content (θ_v) $(m^3\,m^{-3})$, and the soil bulk density (ρ_b) $(Mg\,m^{-3})$ by:

$$C = \frac{2.0 \times 10^6 \rho_b}{2.65} + 4.2 \times 10^6 \theta_v + 2.5 \times 10^6 f_o \qquad\qquad [5.63]$$

for a soil with a volume fraction (f_o) of organic matter (Hillel, 1980).

Table 5.6 lists thermal conductivities at wet and dry points for several soils. For coarse soils, the thermal conductivity versus water content curve is S-shaped (e.g., Campbell et al., 1994), with a rapid rise at water contents corresponding to about 33 kPa soil water tension (about field capacity). For fine soils, the relationship is more linear, and the thermal conductivity between dry and wet conditions in Table 5.6 can be linearly interpolated from the values given, with reasonably small errors. But for water contents below the dry value, the thermal conductivity should be taken as the value corresponding to the dry state.

de Vries (1963) developed a method of estimating soil thermal conductivity from soil texture, bulk density, and water content. The method, while including most important soil properties affecting conductivity, is limited in that it requires knowledge of parameters called shape factors that describe how the soil particles are packed together. The shape factors are specific to a given soil and perhaps pedon and must be measured. They are, in effect, fitting parameters (Kimball et al., 1976). de Vries' method tends to overestimate thermal conductivity at water contents above about 0.15 (Asrar and Kanemasu, 1983; Evett, 1994). Campbell et al. (1994) developed modifications of de Vries' theory that allowed them to match measured values well. They showed that, as temperature increased, the thermal conductivity versus water content curve assumed a pronounced S-shape for the 8 soils in their study, with the curve deviating from monotonicity at temperatures above 50 °C.

Horton et al. (1983) developed a measurement method for D_T based on harmonic analysis. The method entailed fitting a Fourier series to the diurnal soil temperature measured at 1-h intervals at 0.01 m depth followed by the prediction of temperatures at a depth (z) (0.1 m), based on the Fourier series solution to the one-dimensional heat flux problem using an assumed value of D_T. The value of D_T was changed in an iterative fashion until the best fit between predicted and measured temperatures at z was obtained. The best fit was considered to occur when a minimum in the sum of squared differences

Table 5.6   Thermal conductivity (λ) of some soil materials

Soil	Dry θ_v	λ W m^{-1} K^{-1}	Wet θ_v	λ W m^{-1} K^{-1}	ρ_b Mg m^{-3}	Source
Fairbanks sand	0.003	0.33	0.18	2.08	1.71	1
quartz sand	0.00	0.25	0.40	2.51	1.51	1
sand	0.02	0.9	0.38	2.25	1.60	2
sand	0.00	0.27	0.38	1.77	1.64	3
sand	0.003	0.32	0.38	2.84	1.66	4
gravelly coarse sand (pumice)	0.02	0.13	0.40	0.52	0.76	5
medium and coarse gravel (pumice)	0.01	0.09	0.43	0.39	0.44	5
loamy sand	0.01	0.25	0.40	1.59	1.69	6
loam	0.01	0.20	0.60	1.05	1.18	6
Avondale loam	0.08	0.46	0.23	0.88	1.35-1.45	7
Avondale loam	0.03	0.31	0.30	1.20	1.40	9
silt loam	0.09	0.40	0.50	1.0	1.25	2
Yolo silt loam	0.14	0.49	0.34	1.13	1.25	8
Muir silty clay loam	0.03	0.30	0.30	0.90	1.25	9
silty clay loam	0.01	0.20	0.59	1.09	1.16	6
Pullman silty clay loam	0.07	0.16	0.29	0.89	1.3	10
Healy clay	0.04	0.30	0.30	0.91	1.34	1
Fairbanks peat	0.03	0.06	0.61	0.37	0.34	1
forest litter	0.02	0.10	0.55	0.40	0.21	2

1: de Vries (1963); 2: Riha et al. (1980); 3: Watts et al. (1990); 4: Howell and Tolk (1990); 5: Cochran et al. (1967); 6: Sepaskhah and Boersma (1979); 7: Kimball et al. (1976); 8: Wierenga et al. (1969); 9: Asrar and Kanemasu (1983); 10: Evett (1994).

between predicted and measured temperatures was found (i.e., minimum sum of squared error, SSE). Poor fits with this and earlier methods are often due to the fact that field soils usually exhibit increasing water content with depth and changing water content with time while the method assumes a homogeneous soil. Costello and Braud (1989) used the same Fourier series solution and a nonlinear regression method, with diffusivity as a parameter to be fitted, for fitting the solution to temperatures measured at depths of 0.025, 0.15 and 0.3 m.

Neither Horton et al. (1983) nor Costello and Braud (1989) addressed the dependency of diffusivity on water content or differences in water content between the different depths. Other papers have dealt with thermal diffusivity in nonuniform soils but did not result in functional relationships between thermal properties and water content, probably due to a paucity of depth dependent soil water content data (Nasser and Horton, 1989, 1990). Soil water content often changes quickly with depth, time, and horizontal distance. Moreover, diffusivity is not a single-valued function of soil water content and so is difficult to directly use in modeling. The ability of time domain reflectometry (TDR) to measure water contents in layers as thin as 0.02 m (Baker and Lascano, 1989; Alsanabani, 1991) provided the basis for design of a system that simultaneously measures water contents and temperatures at several depths. Evett (1994) used measurements of soil temperature at several depths (e.g., 2, 4, 6, 8... cm), coupled with TDR measurements of soil water content at the same depths, to find a relationship between thermal conductivity and water content in a field soil. He used the minimum SSE method of Horton et al. (1983) to find the thermal diffusivity for each soil layer between vertically adjacent measurements of water content and temperature. The water content for this layer was used to calculate C and thus λ corresponding to that water content. A function of λ versus θ_v was developed by regression analysis on the λ and θ_v data (Fig. 5.27). Because both C and λ were known for each layer, this method also gave the soil heat flux.

Single-probe heat pulse methods have been developed to measure thermal diffusivity, and a dual-probe heat pulse method (Campbell et al., 1991) can measure the thermal diffusivity (D_T) as well as λ and C (Kluitenberg et al., 1995). Noborio et al. (1996) demonstrated a modified trifilar (three-rod) TDR probe that measured θ_v by TDR, and λ by the dual-probe heat pulse method. Their measured λ compared well to values calculated from de Vries (1963) theory.

Soil heat flux is commonly measured using heat flux plates (Table 5.1). These are thermopiles that measure the temperature gradient across the plate, and knowing the conductivity of the plate, allow calculation of the heat flux from Equation [5.61]. Heat flux plates are impermeable and block water movement. Because of this, the plates should be installed at a minimum of 5 cm below the soil surface so that the soil above the plate does not dry out or wet up appreciably more than the surrounding soil. Typical installation depths are 5 cm or 10 cm. Even at these shallow depths, the heat flux is greatly reduced from its value at the soil surface, and corrections must be applied to compute surface heat flux. The most common correction involves measuring the temperature and water content of the soil at midlayer depths (z_j) in N layers (j = 1 to N) between the plate and surface, and applying the combination equation over some time period (P) defined by beginning and ending times t_i and t_{i+1}:

$$G = G_z + \frac{\sum_{j=1}^{N}\left(T_{zj\_i+1} - T_{zj\_i}\right)\Delta z_j C_{zj}}{\left(t_{i+1} - t_i\right)} \qquad [5.64]$$

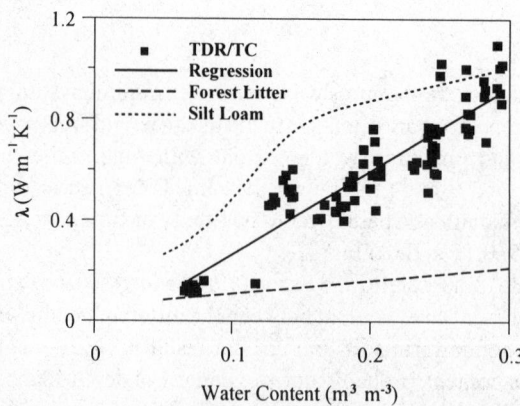

Fig. 5.27 Thermal conductivity of Pullman silty clay loam determined from TDR probe and thermocouple arrays compared to functions from Campbell (1985) for forest litter and silt loam

where G is the surface heat flux during P, G_z is the flux at depth z, T_{zj} are temperatures at the N depths (z_j) at times (t_i and t_{i+1}), Δz_j is the depth of the layer with midpoint z_j, and where the volumetric heat capacities (C_{zj}) at depths z_j are calculated from Equation [5.63] rewritten as:

$$C_{zj} = \frac{2.0 \times 10^6 \rho_{bzj}}{2.65} + 4.2 \times 10^6 \theta_{vzj} + 2.5 \times 10^6 f_{ozj} \qquad [5.65]$$

where θ_{vzj}, ρ_{bzj}, and f_{ozj} are the water contents, soil bulk densities, and volume fractions of organic matter, respectively, at depths z_j. The estimate of G is not much changed by the exact form of the combination equation as shown by data from Bushland, Texas for four forms of Equation [5.64] (Fig. 5.28). For situations where water content and temperature change rapidly with depth, or bulk density or f_o change rapidly with depth, the multiple layer approach will work better.

The four methods of combining temperature and water content data to correct heat flux for bare soil data collected at Bushland, TX in 1992 (Fig. 5.28) used the following measurements. Temperatures were measured at 2- and 4-cm depths (2 replicates) with thermocouples (T_2 and T_4); at the surface with a single infrared thermometer (T_0), and as a mean temperature of the surface to 5-cm depth soil layer using thermocouples wired in parallel and buried (4 replicates) at 1- and 4-cm depths (T_{1-4}). Water contents were measured by TDR probes (2 replicates) inserted horizontally at 2- and 4-cm depths (θ_2 and θ_4). Soil heat flux at 5-cm depth (G_5) was measured with heat flux plates (4 replicates). For all methods, the product of soil bulk density and heat capacity of soil solids was set to 1.125 MJ m^{-3}. For the first method, the surface heat flux (G_0) was

$$G_0 = G_5 + \frac{(1.125 + 4.2\theta_w) \times 10^6 (0.05)(T_{w+1} - T_w)}{1800} \qquad [5.66a]$$

where 1800 was the period in s, the weighted water content for the surface to 5 cm depth layer (θ_w) was

$$\theta_w = \frac{3\theta_2}{5} + \frac{2\theta_4}{5} \qquad [5.66b]$$

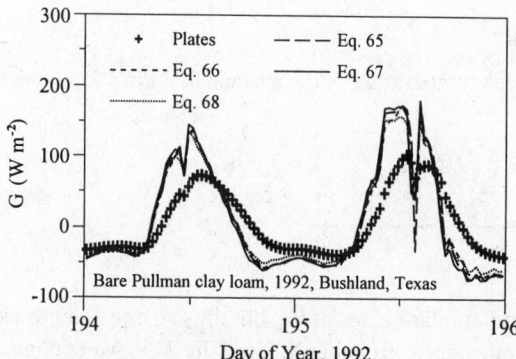

Fig. 5.28 Four methods of correcting heat flux, measured with plates at 5-cm depth, to surface heat flux

the weighted temperature for the surface to 5 cm depth layer (T_w) was

$$T_w = \frac{3T_2}{5} + \frac{2T_4}{5}$$ [5.66c]

and T_{w+1} was the same, but was for the previous measurement.

For the second method, θ_w and the series wired thermocouple temperature were used:

$$G_0 = G_5 + \frac{(1.125 + 4.2\theta_w) \times 10^6 (0.05)(T_{1\_4+1} - T_{1\_4})}{1800}$$ [5.67]

For the third method, θ_w was used:

$$G_0 = G_5 + \frac{(1.125 + 4.2\theta_w) \times 10^6 (0.05)(T_{024+1} - T_{024})}{1800}$$ [5.68a]

but the depth weighted mean (T_{024}) of infrared thermometer temperature and those measured at 2 and 4 cm were used.

$$T_{024} = \frac{\left[\dfrac{(T_0 + T_2)/2 + T_0}{2}\right]}{5} + \frac{2T_2}{5} + \frac{2T_4}{5}$$ [5.68b]

For the fourth method, a modified layer approach was used:

$$G_0 = G_5 + \frac{(1.125 + 4.2\theta_2) \times 10^6 (0.05)(T_{02+1} - T_{02})}{1800}$$
$$+ \frac{(1.125 + 4.2\theta_4) \times 10^6 (0.05)(T_{4+1} - T_4)}{1800}$$ [5.69a]

where the depth weighted mean temperature in the surface to 3 cm deep layer (T_{02}) was

$$T_{02} = \frac{\left[\dfrac{(T_2 + T_0)/2 + T_0}{2}\right]}{3} + \frac{2T_2}{3}$$ [5.69b]

All of these methods produced similar values of G_0, but those using a depth weighted water content tended to overestimate extreme values, probably because the 2 cm water content was lower than that at 4 cm. The weighted mean approach for both water content and temperature (Equation [5.68]), with surface temperature included, produced generally the largest diurnal swing in G_0. Methods that did not include the surface temperature, but used the weighted mean approach for both water content and temperature (Equations [5.66] and 5.67]), produced intermediate results. The layer approach (Equation [5.69]) produced the smallest diurnal swing in G_0, despite using the surface temperature, and is probably the most accurate approach. All methods corrected both the amplitude and the phase of the diurnal cycle of G_0 appropriately.

Convective heat flux can play an important role in soil heating or cooling. This is the heat transported by moving air or water, the latter denoted in Fig. 5.1 by G_{Jw} for heat transported by infiltrating water. Because of the low heat capacity of air, the convective heat flux due to air movement is usually small, but convective heat flux due to infiltration of water can be much larger than that due to diffusion on a diurnal basis. For example, irrigation with 5 cm of water at 15 °C on a soil at 25 °C with an initial water content of 0.1 $m^3 m^{-3}$ and a bulk density of 1.48 kg m^{-1} would immediately lower the temperature of the 11.6 cm deep wetted layer to 20 °C (assuming negligible heat of wetting, the soil brought to saturation, and a heat capacity of 1.54 MJ $m^{-3} K^{-1}$). The heat of wetting is usually not large enough to be important in heat balance calculations. It can be large for clays with large surface area if they are extremely dry, ranging from 40 J g^{-1} for kaolinitic clays to 125 J g^{-1} for allophanic clays (Iwata et al., 1988). But, it decreases quickly as the initial water content of the soil increases, and is not likely to be important for the normal range of field water contents.

5.2.4   Sensible Heat Flux (H)

Sensible heat flux is the transfer of heat away from or to the surface by conduction or convection. Because air is not a very good conductor of heat, most sensible heat flux is by convection (movement) of air. This occurs in eddies of different scales depending on the turbulence of the atmosphere near the surface. Turbulence is influenced by the aerodynamic roughness of the surface, the wind speed, and the temperature differential between the surface and the air. Perhaps the most common method of evaluating sensible heat flux is to measure the other terms in Equation [5.1] as accurately as possible and then set H equal to the residual:

$$H = -Rn - G - LE \qquad [5.70]$$

Of course, this approach lumps all the errors in the other terms into H. More importantly, it does not allow for a check on the accuracy of the energy balance. By definition, if H is defined by Equation [5.69], then Equation [5.1] will sum to zero. Only an independent measure of H can provide a check sum for Equation [5.1].

As noted in Section 5.2.2.2, eddy correlation is a direct method of measuring H:

$$H = \overline{\rho_a} \, c_p \, \overline{w'T'} \qquad [5.71]$$

where the overbars denote short time averages of air density (ρ_a), vertical wind speed (w'), and air temperature (T'), measured at some height within the constant flux layer.

The Bowen ratio method can be applied to sensible heat flux as well as to latent heat flux as outlined in Section 5.2.2.3. For sensible heat flux, the Bowen ratio is (Rosenberg et al., 1983, p. 256):

$$H = \frac{-(Rn + G)}{\left(1 + \dfrac{1}{\beta}\right)} \qquad [5.72]$$

The considerations of fetch, measurement height, equipment, etc. mentioned in Section 5.2.2 for Bowen ratio and eddy correlation measurements apply as well to sensible heat flux measurements made with these methods.

Though obviously a dynamic and complex process, sensible heat flux (H) (W m^{-2}) is sometimes estimated using a straightforward resistance equation:

$$H = \frac{\rho c_p (T_z - T_0)}{r_{aH}}$$ [5.73]

where ρ is the density of air ($\rho = 1.291 - 0.00418 T_a$, with less than 0.005 kg m^{-3} error in the -5 to 40 °C range, T_a in °C), c_p is the heat capacity of air (1.013×10^3 J kg^{-1} K^{-1}), T_z is the air temperature at measurement height, r_{aH} is the aerodynamic resistance to sensible heat flux (s m^{-1}), and T_0 is the temperature of the surface. [For vegetation, the surface for aerodynamic resistance is the height at which the logarithmic wind speed profile, established by measurements of wind speed above the surface, extrapolates to zero. This height is $d + z_{om}$ and is often well below the top of the canopy, typically at 2/3 to 3/4 h. Measurements of surface temperature (with, for instance, an infrared thermometer) may not be the mean temperature at the same height as the aerodynamic surface, thus causing some problems with r_{aH} estimation. Also, the roughness length for momentum (z_{om}) may be different from that for sensible heat (z_{oH})].

A general form for r_a is

$$r_a = \frac{1}{k^2 u_z} \left\{ \ln\left(\frac{[z-d]}{z_0} \right) \right\}^2$$ [5.74]

where k is the von Kármán constant = 0.41, z_0 is the roughness length (m), z is the reference measurement height (m), u_z is the wind speed (m s^{-1}) at that height, and d is the zero plane displacement height (m). Equation [5.74] only holds for neutral stability conditions. Unstable conditions occur when the temperature (and thus air density) gradient from the surface upward is such that there is warm air rising through the atmosphere. Stable conditions prevail when the air is much cooler and denser near the surface, thus inhibiting turbulent mixing. Neutral conditions obtain when neither stable nor unstable conditons do.

For bare soil, Kreith and Sellers (1975) simplified Equation [5.74] to:

$$r_a = \frac{1}{k^2 u_z} \left[\ln\left(\frac{z}{z_0} \right) \right]^2$$ [5.75]

where u_z is the wind speed (m s^{-1}) at the reference height (z) (m). They found a value of $z_0 = 0.003$ m worked well for smooth bare soil.

For non neutral conditions, a variety of stability corrections have been proposed (Rosenberg et al., 1983, p. 140-144; Monteith and Unsworth, 1990, p. 234-238). Because many models of the soil-plant-atmosphere continuum use Equation [5.73] to model H, it is important to note that, while stability corrections can improve model predictions of H and surface temperature, the stability corrections are implicit in terms of H. This leads to a requirement for iterative solution of sensible heat flux at each time step in these models.

Knowledge of appropriate values for d and z_0 in the above equations can be hard to obtain. Campbell (1977) suggests estimating these from plant height (h) as:

$$d = 0.64h \qquad [5.76]$$

for densely planted agricultural crops, and

$$z_{0m} = 0.13h \qquad [5.77]$$

for the roughness length for momentum for the same condition. Campbell (1977) gives the roughness length parameters for sensible heat (z_{0H}), and vapor transport (z_{0v}) as:

$$z_{0H} = z_{0v} = 0.2z_{0m} \qquad [5.78]$$

Note that Equation [5.78] differs from Equation [5.57] where Jensen et al. (1990) used $z_{0H} = 0.1 \, z_{0m}$. For coniferous forest, Jones (1992) gives:

$$d = 0.78h \qquad [5.79]$$

and

$$z_{0m} = 0.075h \qquad [5.80]$$

for these parameters. As wind speeds increase, many plants change form and height, with resulting decrease in h, d, and z_{0m}. It is unlikely that the relationships given in Equations [5.76–5.80] hold true for high wind speeds.

5.3 Water Balance Equation

Water balance is written for a control volume of unit surface area, and with a vertical dimension that extends from the soil surface to a lower boundary that is usually assigned a depth at or below the bottom of the root zone (Fig. 5.1)

$$0 = \Delta S - P + R - F - E \qquad [5.81]$$

where ΔS is the change in soil water storage in the profile, P is precipitation or irrigation, R is the sum of runoff and runon, F is flux across the lower boundary of the profile, and E is water lost to the atmosphere through evaporation from the soil or plant or gained by dew formation. The value of P is always positive or zero, but values of ΔS, R, F, and E may have either sign. By convention, R is taken to be positive when there is more runoff than runon. Also conventionally, E is often taken to be positive when flux is out of the control volume. Here, in order to be compatible with the energy balance equation, a break with convention will be made and E will be taken as positive toward the surface of the soil. The equation is often rearranged to provide values of E when suitable measurements or estimates of the other terms are available, but it can and has been used to estimate runoff, soil water available for plants, and deep percolation losses (flux downward out of the profile). Here, F is taken as positive when flux is upward across the lower boundary into the control volume. The term F is used rather than P for deep percolation, both to avoid confusion with precipitation, and to avoid the common misconception that flux is only downward when P is used to indicate deep percolation.

Usually, the values of terms in Equation [5.81] are given as equivalent depths of water-per-unit area (e.g., mm m^{-2}). In the case of E, the units of kg m^{-2} may be conveniently converted to mm by dividing by 1 kg m^{-2} mm^{-1}, with little loss of accuracy because the density of water (kg L^{-1}) is not quite unity (1 L = 1,000 cm^3 = volume of a right rectangular prism 1 m, 1 m, and 1mm). The change in storage (ΔS) is often determined by measuring soil water content changes by methods that give volumetric water content (θ_v) (m^3 m^{-3}). Multiplying the water content by the depth of the layer gives the depth of water stored. In the United States, the term evapotranspiration (ET) is used to represent the sum of evaporative fluxes from the soil and plants. By convention ET is taken as positive for fluxes from plant or soil surface to the atmosphere. Thus, ET = –E/(1 kg m^{-2} mm^{-1}) and the water balance may be rearranged as:

$$\Delta S = P - R + F - ET \qquad\qquad [5.82]$$

This provides a use for the ET term for those who prefer to say evaporation rather than evapotranspiration. Examination of Equation [5.82] will satisfy the reader that soil water storage increases with precipitation, decreases if runoff from precipitation occurs, decreases with increasing ET, and increases with flux upward into the control volume.

5.3.1   Measuring ΔS and ET

Probably the most accurate method of measuring ΔS is the weighing lysimeter (Wright, 1991). Although large weighing lysimeters involve considerable expense, they can give very precise measurements (0.05 mm = 0.05 kg m^{-2}) (Howell et al., 1995). An excellent review of the use of weighing lysimeters is given by Howell et al. (1991). Careful design, installation, and operation will overcome any of the serious problems reported with some lysimeters including disturbance of the soil profile (less with monolithic lysimeters), interruption of deep percolation and horizontal flow components, uneven management of lysimeter compared to field soil (Grebet and Cuenca, 1991), and other sources of bias (Ritchie et al., 1996). Other drawbacks include heat flux distortions caused by highly conductive steel walls (Black et al., 1968; Dugas and Bland, 1991, but minimal for large lysimeters), and high cost, e.g., $65,000 (Lourence and Moore, 1991) and $80,000 (Marek et al., 1988).

Schneider et al. (1996) described simplified monolithic weighing lysimeters (Fig. 5.29) that were considerably less expensive than, and nearly as accurate as, the monolithic weighing lysimeters described by Marek et al. (1988) (Fig. 5.30). These two designs represent contrasts in mode of operation and each presents some advantages and disadvantages. The design in Fig. 5.30 allows access to all sides and the bottom of the lysimeter for installation or repair of sensors and weighing or drainage systems. The Campbell Scientific, Inc. CR7 data logger that handles all measurements is installed in the underground chamber and typically is subject to only a small diurnal temperature swing of 1 °C, reducing temperature-induced errors in low level measurements such as load cell transducer bridges and thermocouples. Other equipment installed in the chamber includes a system for TDR measurements of soil water content and concurrent measurements of soil temperature, and an automatic vacuum drainage system that continuously monitors drainage rate. The drainage tanks are suspended by load cells from the bottom of the lysimeter tank, allowing measurement of tank mass

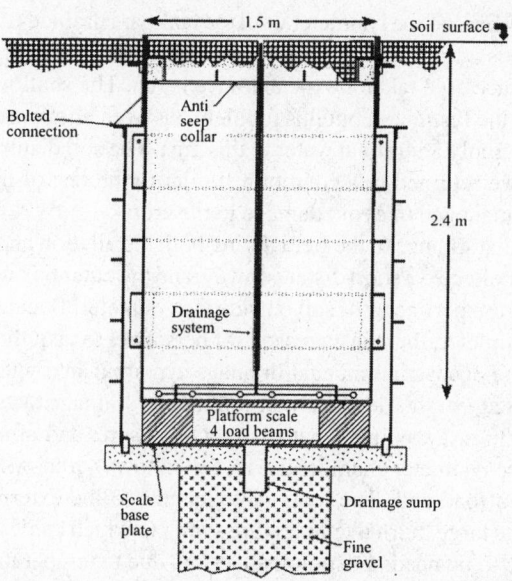

Fig. 5.29 Cross-sectional view of the simplified weighing lysimeter installed for grass reference ET measurements at Bushland, TX (Schneider, 1998a)

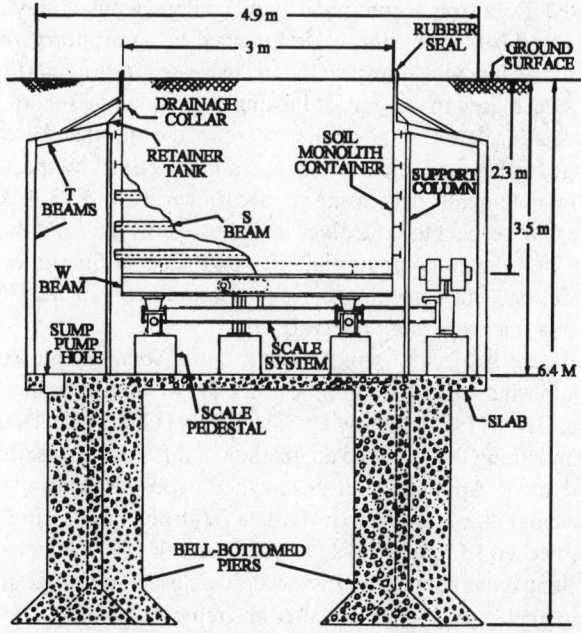

Fig. 5.30 Cross-sectional view of one of the four large weighing lysimeters at Bushland, TX (Schneider, 1998b)

change without changing the mass of the lysimeter until the tanks are drained (manual but infrequent). The main disadvantages of this design are the shallow soil depth over the ceiling around the periphery of the chamber, and the surface area taken by the entrance hatch. The shallow soil depth can cause uneven plant growth next to the lysimeter, but this problem has been eliminated with the installation of a drip irrigation system to apply additional water to this area. The soil disturbed to install the outer chamber wall appears to have returned to a condition similar to the rest of the field. Access to the lysimeter must be carefully managed to avoid damage to the crop.

A design that minimizes disturbance to the field during both installation and operation is shown in Fig. 5.29. The monolith was collected a short distance away, and the outer box was installed in a square hole that disturbed only a 15 cm perimeter of soil outside the lysimeter. Because there is no access to the sides or bottom of the lysimeter, there is no reason for personnel to visit the lysimeter area except for crop management and the occasional manual drainage accomplished with a vacuum pump and collection bottle. A disadvantage of this design is that continuous drainage rates are not available. The CR7 data logger is located 30 m away in a weather-tight enclosure, and all cables are buried. The location of the CR7 inside the lysimeter chamber in Fig. 5.30 allows a four-wire bridge to be used for reading the weighing system load cell. The long cable lengths to the external CR7 used with the lysimeter in Fig. 5.29, and the large diurnal temperature swing to which cables and CR7 are exposed, both cause a six-wire bridge to be needed to eliminate errors due to temperature-induced resistance changes when reading the platform scale load cells. Measurement precision with the lever beam scale in Fig. 5.30 is 0.05 mm while that with the platform scale in Fig. 5.29 is 0.1 mm.

Weighing lysimeters measure mass change over a given time (ΔM). If mass is measured in kg, then dividing the mass change by the surface area in m^2 of the lysimeter will give the change in water storage (ΔS) of the lysimeter as an equivalent depth of water in mm, with only slight inaccuracy due to the density of water not being quite equal to 1 kg L^{-1}. If only daily ET values are needed, then ΔS is computed from the 24-h change in lysimeter mass, usually midnight to midnight. Some averaging of readings around midnight may be needed to smooth out noise. By adding any precipitation or drainage, the daily ET is computed. Data from a continuously weighing lysimeter may be presented as a time sequence of mass (or depth of water storage) referenced to an arbitrary zero (Fig. 5.31). Often irrigation or precipitation events will show as obvious increases in storage (Fig. 5.31), and drainage events will show up as decreases in storage. Adjusting the sequential record of storage amount by adding the rainfall or irrigation depth, or the drainage depth at the time that these occurred, will remove these changes in storage, and is equivalent to the operations defined by the +P and +F in Equation [5.82], resulting in the monotonically decreasing storage shown in Fig. 5.32. Taking the first derivative of the adjusted storage with respect to time gives the adjusted ΔS rate, and thus the ET rate if R and F are zero (Fig. 5.33). In order to compute ET rates on the same time interval as lysimeter mass measurements are made, one must have concurrent measurements of irrigation, precipitation and drainage on the same or a finer recording interval.

Weighing lysimeters are subject to wind loading, more so when the soil surface is bare, as evidenced by Fig. 5.34. In windy regions, it may be necessary to smooth the data to remove noise when calculating the ET rate. Gorry (1990), following Savitsky and Golay (1964), described a method for general least squares smoothing that allowed application of different levels of smoothing to both raw data and their first derivative. Application of this method to post-processing of data is preferable to real time smoothing that may eliminate detail in the data. With post-processing, one can apply only the amount of smoothing needed to reduce noise to acceptable levels. A computer program to apply Savitsky-Golay smoothing is available (http://www.cprl.ars.usda.gov/programs/).

Microlysimeters are small enough to be installed and removed by hand for weighing daily or more often. They can give good precision but are sensitive to spatial variability. Lascano and Hatfield

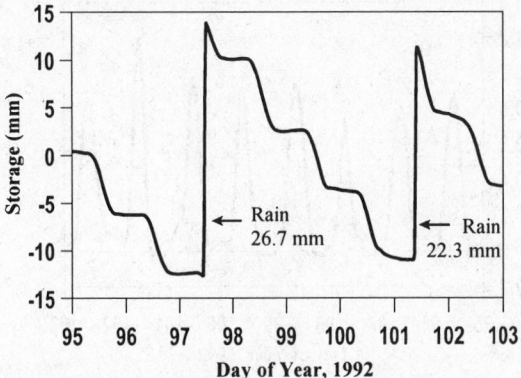

Fig. 5.31   Unadjusted weighing lysimeter storage for winter wheat at Bushland, TX

(1992) showed that 182 microlysimeters were required to measure field average E with precision of 0.1 mm d^{-1} at a 90% confidence level when their soil was wet, but only 39 when dry. This was due to the greater variability of E for wet soil. For a precision of 0.5 mm d^{-1}, only 7 microlysimeters were required for any soil wetness. To avoid heat conduction to and from the surface, microlysimeter walls should be made of low thermal conductivity materials such as plastic, and to avoid trapping heat at the bottom of the microlysimeter, the bottom end cap should be made of a thermally conductive material such as metal (Evett et al., 1995). Plastic pipe makes good microlysimeters. Typical dimensions are 7.6 or 10 cm in diameter, and from 10 cm to 40 cm high. Beveling the bottom end eases insertion into the soil. Typical practice is to insert the microlysimeter vertically until its top is level with the soil surface, then dig it out, or rotate it to shear the soil at the bottom, and pull it out. After capping the bottom with a water tight seal, it is weighed before reinsertion into the original or a new hole; sometimes lined with a material (e.g., plastic sheet or bag) to prevent soil sticking to the microlysimeter. After a period of time, the microlysimeter is reweighed and the difference in initial and final weights is the evaporative loss. Short microlysimeters should be replaced daily, as the water supply is soon used up to the point that the soil inside the lysimeter is no longer at the same water content as the soil outside. In a study of spatial variability of evaporation from bare soil, Evett et al. (1995) used 30-cm-high microlysimeters to avoid daily replacement so that the spatial relationship

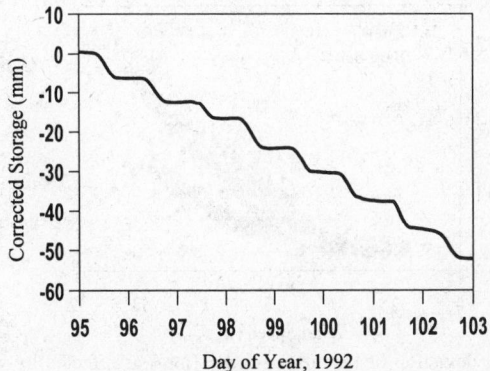

Fig. 5.32   Lysimeter storage from Fig. 5.31 adjusted by subtracting precipitation amounts

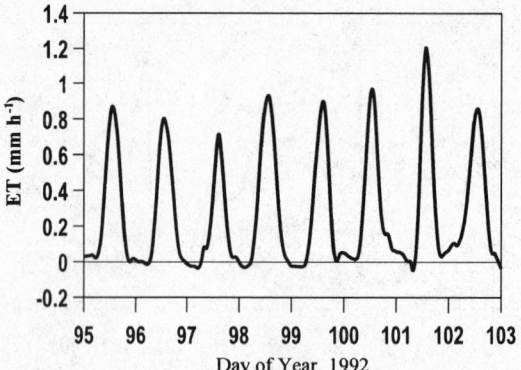

Fig. 5.33 Evapotranspiration rate calculated by taking the negative of the first derivative of adjusted storage from Fig. 5.32. The negative ET rates shown for some nights are caused by dew formation.

would not be changed. They showed that, for their clay loam soil, the 30-cm height was adequate for 9 days. If plant roots are present, microlysimeters should be replaced daily to lessen errors associated with root water uptake that occurs elsewhere in the field but not in the microlysimeters.

Alternatives to weighing lysimetry include soil water measurement methods for assessing ΔS for a soil profile of given depth over a given time. In this case, the soil volume of interest is unbounded below the surface and F is, strictly speaking, uncontrolled. Measurements of soil water content can give the change in soil water stored in a profile of a given depth with good accuracy, and can give good values for E if water flux across the bottom of the profile is known or can be closely estimated. Baker and Spaans (1994) described a microlysimeter with TDR probe installed vertically to measure the water content. Comparison of E calculated from the change in storage closely matched the E measured by weighing the microlysimeter. Young et al. (1997) showed that a single 800-mm long probe installed vertically from the surface could account for 96% of ET from weighing lysimeters irrigated on a 6-d interval, but the standard error for the probe was about 4 times larger than that for the lysimeter (0.46 and 0.07 mm, respectively). In a container study with a sorghum plant, Wraith and Baker (1991) showed that a TDR system could measure ET with high resolution and provide measurements of change in storage on a 15-min interval that compared very well to those measured by an electronic scale.

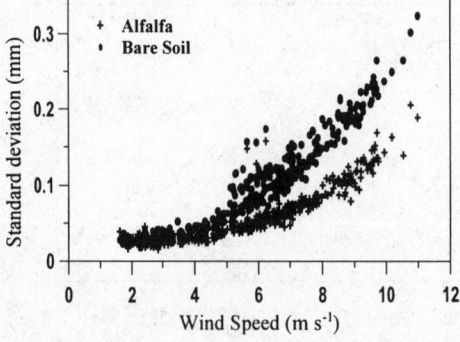

Fig. 5.34 Half-hourly standard deviation of lysimeter storage (mm) as affected by wind speed over contrasting surfaces for days 97-105 at Bushland, TX in 1994

Evett et al. (1993) showed that change in storage in the upper 35 cm of the profile under winter wheat could be accurately tracked with horizontally placed TDR probes, with an average of 88% of daily ΔS occurring in the upper 30 cm. But, E estimates were incorrect (compared to a weighing lysimeter) when flux across the 30-cm boundary occurred. However, combination of the TDR system with neutron probe measurements of deeper soil moisture allowed measurement of E to within 0.7 mm of lysimeter measured E over a 16-d period, five times better than the accuracy achieved using only neutron probe measurements.

The soil water storage (referenced to arbitrary zero) as measured for winter wheat by weighing lysimeter and two TDR arrays is shown in Fig. 5.35. Each TDR array consisted of 7 probes inserted horizontally into the side of a pit and the pit backfilled after wheat planting. Probe depths were 2, 4, 6, 10, 15, 20, and 30 cm, and the probes were read every half hour. Rains on days 101, 104, and 106 can be seen as increases in the storage amount. Changes in storage as measured by the two systems were nearly identical in the 7-d period shown (Fig. 5.36); and ET amounts were closely similar (Fig. 5.37).

Water balance measurement intervals commonly range between hours and weeks and are usually no smaller than the required period of LE measurement. Measurement of each variable in Equation [5.81] presents its own unique problems. These include measurement errors in determination of lysimeter mass or ΔS, and errors in P and R measurement. Problems of P and R measurement are essentially identical for either weighing lysimetry or soil profile water content methods, because the surface area of the control volume can be defined for both methods with a watertight border, often consisting of a sheet metal square or rectangle pressed into or partially buried in the soil surface. When the soil volume is unbounded below the surface, as in the soil profile water content method, there are additional errors due to uncontrolled horizontal flow components and deep percolation that are difficult to measure or estimate. Nevertheless, the profile water balance technique is applicable in many situations for which lysimetry is inappropriate or impossible and is, in addition, much less expensive. In many cases, the horizontal flow components may be assumed to sum to zero, and deep percolation may be nil if the soil profile water content measurements are made to sufficient depth (Wright, 1990).

When neutron scattering alone is used to measure soil water content, the soil water balance method is suitable for periods of several days or more if closure (F = 0) at the bottom of the measured soil profile can be obtained (Wright, 1990). Neutron scattering (NS) is the most common water content

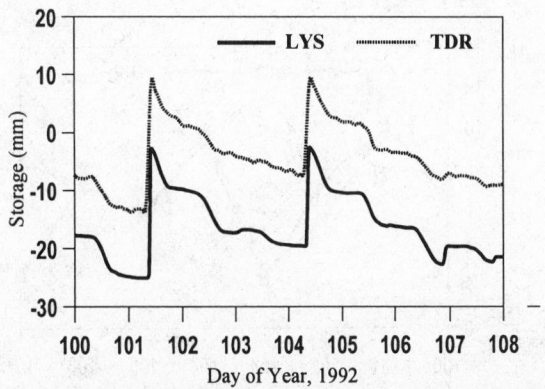

Fig. 5.35   Soil water storage in the upper 35 cm of the soil profile as measured by two TDR probe arrays compared to storage in a 2.4-m-deep weighing lysimeter. Zero reference is arbitrary. Winter wheat, Bushland, TX, 1992.

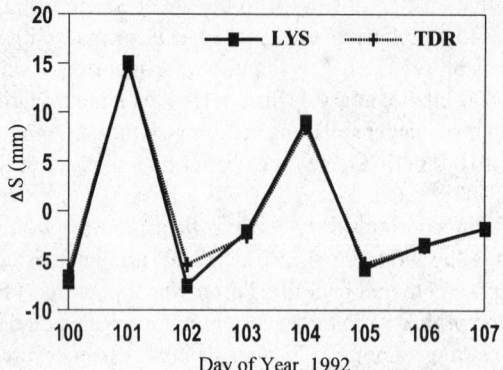

Fig. 5.36   Daily change in storage for the 35-cm and 2.4-m profiles from Fig. 5.35

measurement technique used (Cuenca, 1988; Wright, 1990), but due to the small changes in water content associated with daily ET and the limited precision of NS near the soil surface, this water balance method has usually been restricted to measurement of ET over several day periods (Carrijo and Cuenca, 1992). Evett et al. (1993) showed that TDR measurements of soil water content near the surface could be coupled with deeper water content measurements by NS to close the water balance considerably more precisely than NS alone, opening up the prospect for daily ET measurements by this method.

5.3.2   Estimating Flux Across the Lower Boundary

One of the great advantages of lysimeters is that they control the soil water flux (F) into and out of the control volume. To date, a reliable soil water flux meter has not been developed, so F must be estimated if it is not controlled. If water flux across the lower boundary of the control volume is vertical, it may sometimes be estimated by measurements (preferably multiple) of soil water potential (h) at different depths separated by distance (Δz) and knowledge of the dependence of hydraulic conductivity (K) (m s^{-1}), on soil water potential [K(h) curve]. The potential difference (Δh) coupled with the unit hydraulic gradient for vertical flux gives the hydraulic head difference (ΔH) driving soil water flux. Averaging the measurements allows estimation of the mean hydraulic conductivity for the

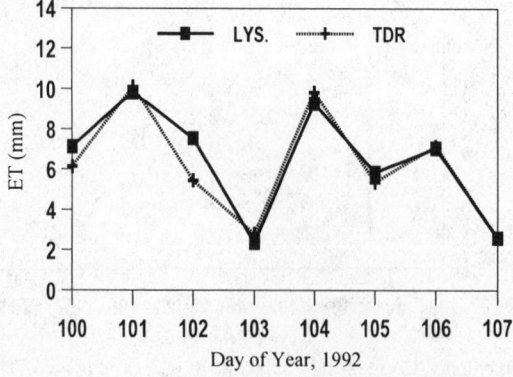

Fig. 5.37   Daily evapotranspiration as calculated from the TDR and weighing lysimeter data shown in Figs. 5.35 and 5.36

soil layer between the measurements from the K(h) curve, and thus estimation of the soil water flux (J_w) (m s^{-1}), from a finite difference form of Darcy's law:

$$J_w = -K\left(\frac{\Delta H}{\Delta z}\right)$$ [5.83]

Soil water potential may be measured by tensiometer or other means described in Section A, Chapter 3, where methods of measuring or estimating the K(h) curve may also be found (van Genuchten et al., 1991). For fluxes across boundaries too deep for the installation of tensiometers, the soil water content may be measured at two or more depths by neutron scattering and the soil water potential inferred by inverting the θ(h) relationship, which may be estimated or measured (van Genuchten et al., 1991). Due to the hysteresis of the θ(h) relationship, there is more room for error when basing J_w estimates on θ measurements. But, for many cases, the soil water potential will be in the range where hysteresis is not a large source of error (drier soils), and hydraulic conductivity is not large either. Thus, both the value of and the error in J_w may be small enough for practical use.

5.3.3   Precipitation and Runoff

An in-depth discussion of precipitation and runoff measurement and modeling is beyond the scope of this chapter. A classic and still valuable reference on field hydrologic measurements is Brakensiek et al. (1979). A more up-to-date and extensive reference is the *ASCE Hydrology Handbook,* 2nd Edition (ASCE, 1996). Flow measurement in channels is detailed in Bos et al. (1983), while Haan et al. (1982) include useful material on stochastic modeling, precipitation and snowmelt modeling, runoff modeling, etc. and list some 75 hydrologic models available at that time. For soil water balance measurements, runoff is often controlled with plot borders or edging driven into the ground or included as the aboveground extension of a lysimeter. Steel borders driven into the soil to a depth of 20 cm will suffice in many situations. Sixteen-gauge galvanized steel in rolls 30 cm wide is useful for this, and can be reinforced by rolling over one edge. If runoff must be measured, this can be done with flumes such as the H-flume and recording station shown in Fig. 5.38.

Precipitation varies so much from location to location that it is rarely useful to attempt to estimate it. Measurement methods include standard US Weather Bureau rain gauges read manually, various tipping bucket rain gauges, heated gauges to capture snow fall (e.g., Qualimetrics model 6021, Sacramento, CA), snow depth stations, etc. If possible, a rain gauge should be surrounded by a wind shield to avoid catch loss associated with wind flow over the gauge (Fig. 5.39). A standard for the capture area or throat of a rain gauge is that it should be 20 cm in diameter because smaller throats lead to more variability in amount captured. Various designs of tipping bucket rain gauges have become standard equipment on field weather stations. These are capable of providing precipitation data needed to solve the soil water balance for short intervals. Two problems are sometimes associated with tipping bucket type gauges. First, most of these devices count the tips using a Hall effect sensor for detecting the magnetic field of a magnet attached to the tipping bucket, and the sensing system is sometimes susceptible to interference from sources of electromagnetic noise such as vehicle ignition systems. Second, tipping bucket gauges do not keep up with very high rainfall rates. At Bushland, TX, tipping bucket errors of 10–15% have been observed for totals of rainfall from high intensity convective thunderstorms compared to amounts collected in standard rain gauges and sensed by weighing lysimeters. If accuracy is very important, then a tipping bucket gauge should be supplemented with a standard gauge that captures and stores all the rainfall. For solving the soil water

Fig. 5.38 H-flume and recorder (in white box) for measuring runoff rate from graded bench terrace at Bushland, TX. Note dike diverting flow from uphill flume

Fig. 5.39 Wind shield installed around a heated tipping bucket rain gauge

balance, experience shows that the rain gauge(s) should be placed directly adjacent to the location of ΔS measurement. Separation of even 100 m can lead to large errors due to the spatial variability of precipitation.

For studies and operations at scales larger than small plot size, there are now precipitation estimates from Doppler radar-based systems that offer calibrated rainfall data on a 24-h basis (Fig. 5.40) (see also Legates et al., 1996; and Vieux and Farajalla, 1996). Although Fig. 5.40 shows large grid sizes and only 16 levels of precipitation, grid sizes of 4 km on a side, with 256 levels of rainfall, are available. Data for these maps are generated by the WSR-88D radar system, usually known as the NEXRAD weather radar system, in widespread use in the United States. The Center for Computational Geosciences at the University of Oklahoma has developed a radar-based precipitation interface (RPI) for the radar data to generate the maps. Radar data from two or more stations are used and calibrated against rain gauge measurements available from, for example, the Oklahoma MESONET system of weather stations.

Acknowledgments

Much of the data presented in this chapter was collected by the Evapotranspiration Research Team at the USDA-ARS Conservation and Production Research Laboratory at Bushland, TX, of which the author is a member. Other past and present team members are (alphabetically): Karen S. Copeland, Donald A. Dusek, Terry A. Howell, Arland D. Schneider, Jean L. Steiner, Rick W. Todd, and Judy A. Tolk. Many technicians supported the data collection efforts including C. Keith Brock, Jim L. Cresap, and Brice B. Ruthardt, among many others. Some of the data shown here have not yet been published and may appear at a later date in a peer-reviewed journal. This work was supported in part by USAID under the subproject title Water Requirements and Management for Maize under Drip and Sprinkler Irrigation, a part of the Agricultural Technology Utilization Project, Egypt. "Everything can be taken from a person except the freedom to choose one's attitude in any situation," a favorite quote of Ruby O. Crosby.

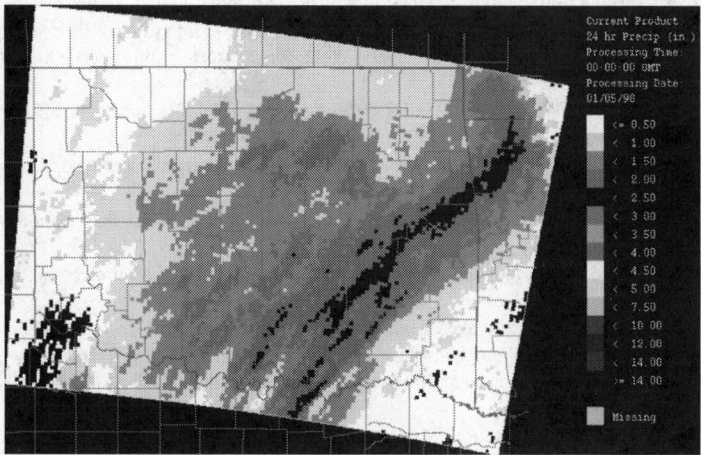

Fig. 5.40 Map of 24-h rainfall accumulation over Oklahoma [From NEXRAD radar data processed by the RPI. Image downloaded from http://ccgwww.ou.edu/rip_images and converted to gray scale with permission.]

5.4 References

Aase, J.K., and S.B. Idso. 1975. Solar radiation interactions with mixed prairie rangeland in natural and denuded conditions. Arch. Met. Geoph. Biokl. Ser. B 23:255–264.

Allen, R.G., M. Smith, A. Perrier, and L.S. Pereira. 1994a. An update for the definition of reference evapotranspiration. ICID Bull. 43:1–34.

Allen, R.G., M. Smith, A. Perrier, and L.S. Pereira. 1994b. An update for the calculation of reference evapotranspiration. ICID Bull. 43:35–92.

Alsanabani, M.M. 1991. Soil water determination by time domain reflectometry: Sampling domain and geometry. Ph.D. Thesis. University of Arizona, Tucson, AZ.

Anlauf, R., K. Ch. Kersebaum, L.Y. Ping, A. Nuske-Schüler, J. Richter, G. Springob, K.M. Syring, and J. Utermann. 1990. Models for processes in the soil-Programs and exercises. Catena Verlag, Cremlingen, Germany.

ASAE. 1988. Modeling agricultural, forest, and rangeland hydrology. American Society of Agricultural Engineers, St. Joseph, MI.

ASCE. 1996. Hydrology handbook. 2nd Ed. American Society of Civil Engineers, New York, NY.

Asrar, G., and E.T. Kanemasu. 1983. Estimating thermal diffusivity near the soil surface using Laplace Transform: Uniform initial conditions. Soil Sci. Soc. Am. J. 47:397–401.

Baker, J.M., and R.J. Lascano. 1989. The spatial sensitivity of time-domain reflectometry. Soil Sci. 147:378–384.

Baker, J.M., and E.J.A. Spaans. 1994. Measuring water exchange between soil and atmosphere with TDR-microlysimetry. Soil Sci. 158:22–30.

Ben-Asher, J., A.D. Matthias, and A.W. Warrick. 1983. Assessment of evaporation from bare soil by infrared thermometry. Soil Sci. Soc. Am. J. 47:185–191.

Black, T.A., G.W. Thurtell, and C.B. Tanner. 1968. Hydraulic load cell lysimeter, construction, calibration, and tests. Soil Sci. Soc. Am. Proc. 32:623–629.

Blad, B.L., and N.J. Rosenberg. 1974. Lysimetric calibration of the Bowen ratio energy balance method for evapotranspiration estimation in the central Great Plains. J. Appl. Meteorol. 13:227–236.

Bos, M.G., J.A. Replogle, and A.J. Clemmens. 1983. Flow measuring and regulating flumes. I.L.R.I., Wageningen, Netherlands.

Bowers, S.A., and R.J. Hanks. 1965. Reflection of radiant energy from soils. Soil Sci. 100:130–138.

Brakensiek, D.L., H.B. Osborn, and W.J. Rawls. 1979. Field manual for research in agricultural hydrology. USDA Agric. Handb. 224. US Government Printing Office, Washington, DC.

Brunt, D. 1932. Notes on radiation in the atmosphere. Quart. J. Roy. Met. Soc. 58:389–418.

Brutsaert, W. 1975. The roughness length for water vapor, sensible heat and other scalars. J. Atm. Sci. 32:2028–2031.

Brutsaert, W. 1982. Evaporation into the atmosphere. D. Reidel Publishing Co., Boston, MA.

Campbell, G.S. 1977. An introduction to environmental biophysics. Springer-Verlag, New York, NY.

Campbell, G.S. 1985. Soil physics with BASIC-Transport models for soil-plant systems. Elsevier, New York, NY.

Campbell, G.S., C. Calissendorff, and J.H. Williams. 1991. Probe for measuring soil specific heat using a heat-pulse method. Soil Sci. Soc. Am. J. 55:291–293.

Campbell, G.S., J.D. Jungbauer, Jr., W.R. Bidlake, and R.D. Hungerford. 1994. Predicting the effect of temperature on soil thermal conductivity. Soil Sci. 158:307–313.

Carrijo, O.A., and R.H. Cuenca. 1992. Precision of evapotranspiration measurements using neutron probe. J. Irrig. and Drain. Engrg. ASCE 118:943–953.

Chen, J. 1984. Mathematical analysis and simulation of crop micrometeorology. PhD Dissertation. Agricultural University, Wageningen, Netherlands.

Cochran, P.H., L. Boersma, and C.T. Youngberg. 1967. Thermal conductivity of a pumice soil. Soil Sci. Soc. Am. Proc. 31:454–459.

Collares-Pereira, M., and A. Rabl. 1979. The average distribution of solar radiation-Correlations between diffuse and hemispherical and between daily and hourly insolation values. Solar Energy 22:155–164.

Costello, T.A., and H.J. Braud, Jr. 1989. Thermal diffusivity of soil by nonlinear regression analysis of soil temperature data. Trans. ASAE 32:1281–1286.

Cuenca, R.H. 1988. Model for evaptranspiration using neutron probe data. J. Irrig. Drain. Eng., ASCE 114:644–663.

de Vries, D.A. 1963. Thermal properties of soils. In W.R. van Wijk (ed.) Physics of plant environment. North-Holland Publishers, Amsterdam, Netherlands.

Doorenbos, J., and W.O. Pruitt. 1977. Guidelines for predicting crop water requirements. FAO Irrig. Drain. Pap. 24. Food and Agriculture Organization, Rome, Italy.

Duffie, J.A., and W.A. Beckman. 1991. Solar engineering of thermal processes. 2nd Ed. John Wiley and Sons, New York, NY.

Dugas, W.A., and W.L. Bland. 1991. Springtime soil temperatures in lysimeters in Central Texas. Soil Sci. 152:87–91.

Dugas, W.A., L.J. Fritchen, L.W. Gay, A.A. Held, A.D. Matthias, D.C. Reicosky, P. Steduto, and J.L. Steiner. 1991. Bowen ratio, eddy correlation, and portable chamber measurements of sensible and latent heat flux over irrigated spring wheat. Agric. For. Meteor. 56:1–20.

Dvoracek, M.J., and B. Hannabas. 1990. Prediction of albedo for use in evapotranspiration and irrigation scheduling. p. 692-699. *In* Visions of the future. American Society of Agricultural Engineers, St. Joseph, MI .

Evett, S.R. 1994. TDR-Temperature arrays for analysis of field soil thermal properties. p. 320–327. *In* Proc. Symp. Time Dom. Reflect. Environ., Infrastr. Mining Appl. Northwestern University, Evanston, IL.

Evett, S.R., T.A. Howell, J.L. Steiner, and J.L. Cresap. 1993. Evapotranspiration by soil water balance using TDR and neutron scattering. p. 914–921. *In* R.G. Allen and C.M.U. Neale (ed.) Management of irrigation and drainage systems, Integrated perspectives. American Society of Civil Engineers, New York, NY.

Evett, S.R., and R.J. Lascano. 1993. ENWATBAL: A mechanistic evapotranspiration model written in compiled BASIC. Agron. J. 85:763–772.

Evett, S.R., A.D. Matthias, and A.W. Warrick. 1994. Energy balance model of spatially variable evaporation from bare soil. Soil Sci. Soc. Amer. J. 58:1604–1611.

Evett, S.R., A.W. Warrick, and A.D. Matthias. 1995. Wall material and capping effects on microlysimeter performance. Soil Sci. Soc. Amer. J. 59:329–336.

Fox, M.J. 1968. A technique to determine evaporation from dry stream beds. J. Appl. Meteorol. 7:697–701.

Gates, D.M. 1980. Biophysical ecology. Springer-Verlag, New York, NY.

Gorry, P.A. 1990. General least-squares smoothing and differentiation by the convolution (Savitsky-Golay) method. Anal. Chem. 62:570–573.

Goudriaan, J. 1977. Crop micrometeorology: A simulation study. PUDOC, Wageningen, Netherlands.

Grebet, R., and R.H. Cuenca. 1991. History of lysimeter design and effects of environmental disturbances. p. 10–19. *In* R.G. Allen et al. (ed.). Lysimeters for evapotranspiration and environmental measurements. American Society of Civil Engineers, New York, NY.

Hanks, J., and J.T. Ritchie. 1991. Modeling plant and soil systems. Soil Science Society of America, Madison, WI.

Haan, C.T., H.P. Johnson, and D.L. Brakensiek. 1982. Hydrologic modeling of small watersheds. American Society of Agricultural Engineers, St. Joseph, MI.

Hatfield, J.L., R.J. Reginato, and S.B. Idso. 1983. Comparison of long-wave radiation calculation methods over the United States. Water Resour. Res. 19:285–288.

Heilman, J.L., C.L. Brittin, and C.M.U. Neale. 1989. Fetch requirements for Bowen ratio measurements of latent and sensible heat fluxes. Agric. For. Meteor. 44:261–273.

Hillel, D. 1980. Fundamentals of soil physics. Academic Press, San Diego, CA.

Horton, R., P.J. Wierenga, and D.R. Nielsen. 1983. Evaluation of methods for determining the apparent thermal diffusivity of soil near the surface. Soil Sci. Soc. Am. J. 47:25–32.

Houser, P.R. 1998. Personal communication from Paul R. Houser, Hydrological Sciences Branch, Data Assimilation Office, NASA Goddard Space Flight Center, Houston, TX.

Houser, P.R., C. Harlow, W.J. Shuttleworth, T.O. Keefer, W.E. Emmerich, and D.C. Goodrich. 1998. Evaluation of multiple flux measurement techniques using water balance information at a semi-arid site. p. 84–87. *In* Proc. Spec. Symp. Hydrol. American Meteorological Society, Phoenix, AZ.

Howell, T.A., A.D. Schneider, D.A. Dusek, T.H. Marek, and J.L. Steiner. 1995. Calibration and scale performance of Bushland weighing lysimeters. Trans. ASAE 38:1019–1024

Howell, T.A., A.D. Schneider, and M.E. Jensen. 1991. History of lysimeter design and use for evapotranspiration measurements. p. 1–9. *In* R.G. Allen et al. (ed.). Lysimeters for evapotranspiration and environmental measurements. American Society of Civil Engineers, New York, NY.

Howell, T.A., J.L. Steiner, S.R. Evett, A.D. Schneider, K.S. Copeland, D.A. Dusek, and A. Tunick. 1993. Radiation balance and soil water evaporation of bare Pullman clay loam soil. p. 922–929. *In* R.G. Allen and C.M.U. Neale (ed.) Management of irrigation and drainage systems, Integrated perspectives. American Society of Civil Engineers, New York, NY.

Howell, T.A., J.L. Steiner, A.D. Schneider, S.R. Evett, and J.A. Tolk. 1994. Evapotranspiration of irrigated winter wheat, sorghum and corn. ASAE Pap. 94–2081.

Howell, T.A., and J.A. Tolk. 1990. Calibration of soil heat flux transducers. Theor. Appl. Climatol. 42:263–272.

Idso, S.B. 1981. A set of equations for full spectrum and 8- to 14-μm and 10.5- to 12.5-μm thermal radiation from cloudless skies. Water Resour. Res. 17295–304.

Idso, S.B., and R.J. Reginato. 1974. Assessing soil-water status via albedo measurement. Hydrol. Water Resour. Ariz. Southwest 4:41–54.

Idso, S.B., R.D. Jackson, R.J. Reginato, B.A. Kimball, and F.S. Nakayama. 1975. The dependence of bare soil albedo on soil water content. J. Appl. Meteorol. 14:109–113.

Idso, S.B., R.J. Reginato, R.D. Jackson, B.A. Kimball, and F.S. Nakayama. 1974. The three stages of drying in a field soil. Soil Sci. Soc. Amer. Proc. 38:831–837.

Iwata, S., T. Tabuchi, and B.P. Warkentin. 1988. Soil-water interactions: Mechanisms and applications. Marcel Dekker, Inc., New York, NY.

Jensen, M.E., R.D. Burman, and R.G. Allen. 1990. Evapotranspiration and irrigation water requirements. ASCE Man. Rep. Engin. Pract. 70, American Society of Civil Engineers, New York, NY.

Jones, H.G. 1992. Plants and microclimate: A quantitative approach to environmental plant physiology. 2nd Ed. The Cambridge University Press, Cambridge, UK.

Kimball, B.A., R.D. Jackson, R.J. Reginato, F.S. Nakayama, and S.B. Idso. 1976. Comparison of field measured and calculated soil heat fluxes. Soil Sci. Soc. Am. J. 40:18–24.

Kluitenberg, G.J., K.L. Bristow, and B.S. Das. 1995. Error analysis of heat pulse method for measuring soil heat capacity, diffusivity, and conductivity. Soil Sci. Soc. Am. J. 59:719–726.

Kondo, J., N. Saigusa, and T. Sato. 1992. A model and experimental study of evaporation from bare-soil surfaces. J. Applied Meteor. 31:304–312.

Kreith, F., and W.D. Sellers. 1975. General principles of natural evaporation. p. 207–227. In D.A. de Vries and N.H. Afgan (ed.) Heat and mass transfer in the biosphere. Part 1. John Wiley and Sons, New York, NY.

Lascano, R.J., C.H.M. van Bavel, J.L. Hatfield, and D.R. Upchurch. 1987. Energy and water balance of a sparse crop: simulated and measured soil and crop evaporation. Soil Sci. Soc. Am. J. 51:1113–1121.

Lascano, R.J., and J.L. Hatfield. 1992. Spatial variability of evaporation along two transects of a bare soil. Soil Sci. Soc. Am. J. 56:341–346.

Legates, D.R., K.R. Nixon, T.D. Stockdale, and G. Quelch. 1996. Soil water management using a water resource decision support system and calibrated WSR-88D precipitation estimates. Proc. AWRA Symp. GIS Water Resour. American Water Resources Associates, Fort Lauderdale, FL.

List, R.J. 1971. Smithsonian meteorological tables. Smithsonian Institution Press, Washington, DC.

Lourence, F., and R. Moore. 1991. Prefabricated weighing lysimeter for remote research stations. p. 423–439. In R.G. Allen, T.A. Howell, W.O. Pruitt, I.A. Walter, and M.E. Jensen (ed.) Lysimeters for evapotranspiration and environmental measurements. Proc. Int. Symp. Lysimetry. American Society of Civil Engineers, New York, NY.

Marek, T.H., A.D. Schneider, T.A. Howell, and L.L. Ebeling. 1988. Design and construction of large weighing monolithic lysimeters. Trans. ASAE 31:477–484.

McCullough, E.C., and W.P. Porter. 1971. Computing clear day solar radiation spectra for the terrestrial ecological environment. Ecol. 52:1008–1015.

Moncrieff, J.B., J.M. Massheder, H.A.R. De Bruin, J. Elbers, T. Friborg, B. Heusinkveld, P. Kabat, S. Scott, H. Soegaard, and A. Verhoef. 1997. A system to measure surface fluxes of momentum, sensible heat, water vapour and carbon dioxide. J. Hydrol. 189:589–611.

Monteith, J.L. 1961. The reflection of short-wave radiation by vegetation. Quart. J. Roy. Meteorol. Soc. 85(366):386–392.

Monteith, J.L. 1965. Evaporation and the environment. p. 205–234. In The state and movement of water in living organisms. Academic Press, New York, NY.

Monteith, J.L., and G. Szeicz. 1961. The radiation balance of bare soil and vegetation. Quart. J. Roy. Meteorol. Soc. S7 (372):159–170.

Monteith, J.L., and M.H. Unsworth. 1990. Principles of environmental physics. 2nd Ed. Edward Arnold, London, UK.

Murray, F.W. 1967. On the computation of saturation vapor pressure. J. Applied Meteor. 6:203–204.

Nassar, I.N., and R. Horton. 1989. Determination of the apparent thermal diffusivity of a nonuniform soil. Soil Sci. 147:238–244.

Nassar, I.N., and R. Horton. 1990. Determination of soil apparent thermal diffusivity from multiharmonic temperature analysis for nonuniform soils. Soil Sci. 149:125–130.

NOAA. 1997. Data files UARS96.PLT, ERBS.PLT, NOAA09.PLT, NOAA10.PLT, NIMBUS.PLT, and SMM.PLT. Accessed January 14, 1998. From ftp://ftp.ngdc.noaa.gov/STP/SOLAR_DATA/SOLAR_IRRADIANCE/.

Noborio, K., K.J. McInnes, and J.L. Heilman. 1996. Measurements of soil water content, heat capacity, and thermal conductivity with a single TDR probe. Soil Sci. 161:22–28.

Oke, T.R. 1978. Boundary layer climates. Methuen, New York, NY.

Peart, R.M., and R.B. Curry. 1998. Agricultural systems modeling and simulation. Marcel Dekker, Inc., New York, NY.

Penman, H.L. 1948. Natural evapotranspiration from open water, bare soil and grass. Proc. Roy. Soc. London Ser. A. 193:120–145.

Pereira, L.S., B.J. van den Broek, P. Kabat, and R.G. Allen. 1995. Crop-water-simulation models in practice. Wageningen Pers, Wageningen, Netherlands.

Richter, J. 1987. The soil as a reactor: Modeling processes in the soil. Catena Verlag, Cremlingen, Germany.

Riha, S.J., K.J. McKinnes, S.W. Childs, and G.S. Campbell. 1980. A finite element calculation for determining thermal conductivity. Soil Sci. Soc. Am. J. 44:1323–1325.

Ritchie, J.T., T.A. Howell, W.S. Meyer, and J.L. Wright. 1996. Sources of biased errors in evaluating evapotranspiration equations. p. 147–157. *In* C.R. Camp, E.J. Sadler, and R.E. Yoder (ed.) Proc. Int. Conf. Evapotran. Irrig. Sched. American Society of Agricultural Engineers, San Antonio, TX.

Rosenberg, N.J. 1969. Seasonal patterns of evapotranspiration by irrigated alfalfa in the Central Great Plains. Agron. J. 61:879–886.

Rosenberg, N.J., B.L. Blad, and S.B. Verma. 1983. Microclimate, the biological environment. John Wiley and Sons, New York, NY.

Russell, G., B. Marshall, and P.G. Jarvis. 1989. Plant canopies: Their growth, form and function. Cambridge University Press, Cambridge, UK.

Savage, M.J., K.J. McInnes, and J.L. Heilman. 1995. Placement height of eddy correlation sensors above a short turfgrass surface. Agric. For. Meteor. 74:195–204.

Savage, M.J., K.J. McInnes, and J.L. Heilman. 1996. The "footprints" of eddy correlation sensible heat flux density, and other micrometeorological measurements. S. Afr. J. Sci. 92:137–142.

Savitsky, A., and M.J.E. Golay. 1964. Smoothing and differentiation of data by simplified least squares. Anal. Chem. 36:1627–1639.

Schmid, H.P. 1997. Experimental design for flux measurements: matching scales of observations and fluxes. Agric. For. Meteorol. 87:179–200.

Schneider, A.D. 1998a. Personal communication. Drawing of cross-section of simplified weighing lysimeter installed at Bushland, TX.

Schneider, A.D. 1998b. Personal communication. Drawing of cross-section of large weighing lysimeter at Bushland, TX.

Schneider, A.D., T.A. Howell, T.A. Moustafa, S.R. Evett, and W.S. Abou-Zeid. 1996. A simplified weighing lysimeter for developing countries. p. 289–294 *In* C.R. Camp, E.J. Sadler, and R.E. Yoder (ed.) Proc. Int. Conf. Evapotran. Irrig. Sched. San Antonio, TX.

Schuepp, P.J., M.Y. LeClerc, J.I. MacPherson, and R.L. Desjardins. 1990. Footprint prediction of scalar fluxes from analytical solutions of the diffusion equation. Boundary-Layer Meteorol. 50:355–373.

Sepaskhah, A.R., and L. Boersma. 1979. Thermal conductivity of soils as a function of temperature and water content. Soil Sci. Soc. Am. J. 43: 439–444.

Skidmore, E.L., J.D. Dickerson, and H. Schimmelpfennig. 1975. Evaluating surface-soil water content by measuring reflection. Soil Sci. Soc. Am. J. 39:238–242.

Spitters, C.J.T., H.A.J.M. Toussaint, and J. Goudriaan. 1986. Separating the diffuse and direct component of global radiation and its implications for modeling canopy photosynthesis. Part I. Components of incoming radiation. Agric. For. Meteorol. 38:217–229.

Todd, R.M. 1998a. Personal communication, analysis of footprint of Bowen ratio system over alfalfa field at Bushland, TX.

Todd, R.M. 1998b. Personal communication, Bowen ratio results for period between 3rd and 4th cuttings of alfalfa at Bushland, TX.

Unland, H.E., P.R. Houser, W.J. Shuttleworth, and Z-L. Zang. 1996. Surface flux measurement and modeling at a semi-arid Sonoran desert site. Agric. For. Meteorol. 82:119–153.

van Bavel, C.H.M. 1966. Potential evaporation: The combination concept and its experimental verification. Water Resour. Res. 2:455–467.

van Bavel, C.H.M., and D.I. Hillel. 1976. Calculating potential and actual evaporation from a bare soil surface by simulation of concurrent flow of water and heat. Agric. Meteorol. 17:453–476.

van Genuchten, M. Th., F.J. Leij, and S.R. Yates. 1991. The RETC code for quantifying the hydraulic functions of unsaturated soils. EPA/600/2-91/065. R.S. Kerr Environ. Res. Lab., US Environmental Protection Agency, Ada, OK.

van Wijk, W.R., and D.W. Scholte Ubing. 1963. Radiation. p. 62–101. *In* W.R. van Wijk (ed.) Physics of plant environment. North-Holland Publishing Co., Amsterdam, Netherlands.

Verma, S.B., N.J. Rosenberg, and B.L. Blad. 1978. Turbulent exchange coefficients for sensible heat and water vapor under advective conditions. J. Appl. Meteorol. 17:330–338.

Vieux, B.E., and N.S. Farajalla. 1996. Temporal and spatial aggregation of NEXRAD rainfall estimates on distributed storm runoff simulation. Intl. 3rd Conf. GIS Environ. Model. National Center for Geographic Information and Analysis, Santa Barbara, CA.

Watts, D.B., E.T. Kanemasu, and C.B. Tanner. 1990. Modified heat-meter method for determining soil heat flux. Agric. For. Meteor. 49:311–330.

Weast, R.C. 1982. Handbook of chemistry and physics. CRC Press, Boca Raton, FL.

Weiss, A. 1982. An experimental study of net radiation, its components and prediction. Agron. J. 74:871–874.

Wierenga, P.J., D.R. Nielson, and R.J. Hagan. 1969. Thermal properties of a soil, based upon field and laboratory measurements. Soil Sci. Soc. Am. J. 44: 354–360.

Wraith, J.M., and J.M. Baker. 1991. High-resolution measurement of root water uptake using automated time-domain reflectometry. Soil Sci. Soc. Am. J. 55:928–932.

Wright, J.L. 1982. New evapotranspiration crop coefficients. J. Irrig. and Drain. Div. ASCE. 108(IR1):57–74.

Wright, J.L. 1990. Comparison of ET measured with neutron moisture meters and weighing lysimeters. p. 202–209. *In* Proc. Natl. Conf. Irrig. Drain. American Society of Civil Engineers, New York, NY.

Wright, J.L. 1991. Using weighing lysimeters to develop evapotranspiration crop coefficients. *In* R.G. Allen, et al. (ed.) Lysimeters for evapotranspiration and environmental measurements. Proc. Int. Symp. Lysimetry. American Society of Civil Engineers, New York, NY.

Wright, J.L., and M.E. Jensen. 1972. Peak water requirements of crops in Southern Idaho. J. Irrig. Drain. Div. ASCE 96(IR1):193–201.

Young, M.H., P.J. Wierenga, and C.F. Mancino. 1997. Monitoring near-surface soil water storage in turfgrass using time domain reflectometry and weighing lysimetry. Soil Sci. Soc. Am. J. 61:1138–1146.

6

Solute Transport

Feike J. Leij
U.S. Salinity Laboratory, USDA-ARS, Riverside, CA

Martinus Th. van Genuchten
U.S. Salinity Laboratory, USDA-ARS, Riverside, CA

6.1 Introduction

Soil scientists and agricultural engineers have traditionally been interested in the behavior and effectiveness of agricultural chemicals (fertilizers, pesticides) applied to soils for enhancing crop growth, as well as in the effect of salts and other dissolved substances in the soil profile on plant growth. More recently, concern for the quality of the vadose zone and possible contamination of groundwater has provided a major impetus for studying solute transport in soils.

The movement and fate of solutes in the subsurface is affected by a large number of physical, chemical and microbiological processes requiring a broad array of mathematical and physical sciences to study and describe solute transport. A range of experimental and mathematical procedures may be employed to quantify transport in soils. Transport of a dissolved substance (solute) depends on the magnitude and direction of the solvent (water) flux; considerable experimental and numerical effort may be needed to determine the transient flow regime in unsaturated soils. Furthermore, the determination of solute concentrations is not always straightforward, particularly if the solute is involved in partitioning between different phases or subject to transformations.

A vast body of work on solute transport can be found in the soil science literature. An equally vast amount of pertinent studies on solute transport in porous media has been published by civil and environmental engineers, geophysicists and geochemists, physical chemists and others. The scope of this chapter permits only an introductory treatment of the subject. First, the standard transport mechanisms pertaining to the fundamental advection-dispersion equation (ADE) will be introduced in Section 6.2. This equation, also known as the convection-dispersion equation, is most often used to model solute transport in porous media. The movement of a solute that undergoes adsorption by the soil requires modifications of the ADE, particularly if several solute species are present that may participate in a number of different reactions. Section 6.3 is devoted to analytical and numerical methods for quantifying solute concentrations as a function of time and space. The traditional advection-dispersion concept is not always adequate to describe solute transport in field soils. Section 6.4 describes the stream tube model as an example of an alternative transport model that may be better suited to model transport in real world situations.

6.2   The Advection-Dispersion Equation

Consider the transport of a chemical species in a three-phase soil-air-water system, and assume that the chemical species (the solute) is completely miscible with water (the solvent). At the macroscopic level and for one-dimensional flow, the mass balance equation for a solute species subject to arbitrary reactions is given as:

$$\frac{\partial \theta C}{\partial t} = -\frac{\partial J_s}{\partial x} + \theta R_s \qquad \qquad [6.1]$$

where θ is the volumetric water content ($L^3 L^{-3}$), C is the solute concentration expressed as solute mass-per-solvent volume ($M L^{-3}$), t is time (T), x is position (L), J_s is the solute flux expressed in solute mass-per-cross-sectional area of soil-per-unit time ($M L^{-2} T^{-1}$), and R_s denotes arbitrary solute sinks (< 0) or sources (> 0) ($M L^{-3} T^{-1}$). Similar equations may be derived for multidimensional flow and transport. The solute flux is usually distinguished in an advective and a dispersive component according to:

$$J_S = J_w C + J_D \qquad \qquad [6.2]$$

where J_w is a vector quantifying the water flux (LT^{-1}), namely, the Darcian velocity expressed as volume of water-per-cross-sectional area of soil-per-unit time, and J_D is a solute flux to quantify transport caused by a gradient in the solute concentration ($M L^{-2} T^{-1}$), also per unit area of soil.

6.2.1   Transport   Mechanisms

The movement of a solute with flowing water, described by the solute flux ($J_w C$), is referred to as advection or convection. Because dissolved substances move in a passive fashion, advective transport can be readily quantified when the solvent flux (J_w) is known. The water flux is generally a function of time and position. However, for transport in laboratory soil columns, J_w is often constant while for field studies, approximations may sometimes be made to facilitate a simpler steady-state one-dimensional flow description.

   Even if the macroscopic water flux is known or can be measured, the velocity at smaller pore scales is not easily determined. Variations in the microscopic velocity will lead to unequal solute movement in the direction of flow. This movement is quantified by means of the dispersive solute flux. If, during steady water flow, the solute concentration of the solution at the inlet of a water saturated soil column is changed abruptly at time $t = 0$, the observed breakthrough of a solute at the column outlet at times $t > 0$ will not exhibit an equivalent abrupt change (Nielsen and Biggar, 1961). The solute concentration will change more gradually with time as a result of (hydrodynamic) dispersion, which quantifies the effects of both mechanical dispersion and diffusion. Molecular diffusion and mechanical dispersion will be discussed first for free solutions and then for soil solutions.

6.2.1.1   Diffusion

Molecular or ionic diffusion is an important mechanism for solute transport in soils in directions where there is little or no water flow. A net transfer of molecules of a solute species usually occurs from regions with higher to lower concentrations as the result of diffusion as described by Fick's first law. For a free or bulk solution, the one-dimensional mass flux [J_{dif} ($M L^2 T^{-1}$)] due to molecular diffusion is given by:

$$J_{dif} = -D_o \frac{\partial C}{\partial x}$$ [6.3]

where D_o is the coefficient of molecular diffusion for a free or bulk solution ($L^2 T^{-1}$). Many publications exists that provide data on D_o (Kemper, 1986; Lide, 1995).

The experimentally observed proportionality between J_{dif} and the concentration gradient can be described at the molecular or ionic level with a balance of forces. The driving force for particle movement from higher to lower concentrations is the gradient in the chemical potential. For mixtures with ideal behavior, the chemical potential can be expressed in terms of its mole fraction. For solutions with nonideal mixing behavior, the activity coefficient of the solute needs to be determined. Ionic activity coefficients can be estimated with the extended Debye-Hückel equation or the Davies extension for solutions up to 0.5 M (Section B, Chapter 10). Activity coefficients for a greater concentration range up to 16 M can be estimated with the Pitzer virial equations (Pitzer, 1979; Harvie and Weare, 1980). Codiffusion or counterdiffusion occurs in systems with multiple ion species. Diffusion rates for individual species as predicted solely by Fick's model would violate the electroneutrality principle. The ionic diffusion flux consists then of a term for ordinary Fickian diffusion and a term accounting for electric transference of ions. The corresponding diffusion coefficient is related to the ionic mobility using the Nernst-Planck equation (Helfferich, 1962).

To characterize diffusion in soils, the diffusivity in a free solution is typically adjusted with terms accounting for a reduced solution phase (a smaller cross-sectional area available for diffusion), and an increased path length. A general treatment of diffusion in soils can be found in Olsen and Kemper (1968) and Nye (1979). The macroscopic diffusive flux per unit area of soil can be written as:

$$J_{dif} = -\theta D_{dif} \frac{\partial C}{\partial X}$$ [6.4]

where D_{dif} is the coefficient of molecular or ionic diffusion for the liquid phase of the soil. The diffusion coefficients for the soil liquid and a free liquid are related by (Epstein, 1989):

$$D_{dif} = \frac{D_o}{\left(L_{dif} / L\right)^2} \equiv \frac{D_o}{\tau^2} \equiv D_o \tau_o$$ [6.5]

where L_{dif} and L are the actual and the shortest path lengths for diffusion (L), $\tau = L_{dif}/L$ is known as the tortuosity, and τ^2 as the tortuosity factor, while $\tau_a = (L/L_{dif})^2$ is often designated as an apparent tortuosity factor. The tortuosity L_{dif}/L appears twice in Equation [6.5] to account for changes in the concentration gradient along the streamline and the travel distance as compared to a higher concentration difference and a shorter travel distance along a straight path with length L in a free solution (Olsen and Kemper, 1968). It should be noted that the terms tortuosity and tortuosity factor have not been used consistently in the literature. Furthermore, some authors include the water content in their definition of tortuosity (Dykhuizen and Casey, 1989) or solute adsorption (retardation) in the expression for either τ or D_{dif} (Nye, 1979; Robin et al., 1987).

For unsaturated conditions, it is convenient to quantify the dependency of the diffusion coefficient on water content. Assuming that the tortuosity affects diffusion in the liquid phase in the same way as in the gaseous phase, the tortuosity term previously derived for gaseous diffusion in soils by Millington (1959) and Millington and Quirk (1961), can be adapted to describe aqueous diffusion in variably saturated soils. The following expressions then result:

$$D_{dif} = \frac{\theta^{10/3}}{\varepsilon^2} D_o \quad , \quad D_{dif} = \frac{\theta^2}{\varepsilon^{2/3}} D_o \qquad\qquad [6.6]$$

where ε is the soil porosity ($L^3 L^{-3}$). An equivalent of the first expression, using a volumetric air content instead of θ, has been used frequently to describe gaseous diffusion in soils although Jin and Jury (1996) reported that the lesser known second version provided a better description.

Diffusion coefficients in soil systems are usually determined by mathematically analyzing solute concentration profiles in the soil as a function of time or position. Van Rees et al. (1991) measured diffusivities by allowing diffusion from (1) a spiked solution into a soil having a zero or low initial concentration, (2) a spiked soil into a solution, and (3) a spiked soil into the soil. In the first two procedures, the concentration of the solution is observed as a function of time. A mathematical solution is then fitted to the observation to determine the diffusion coefficient.

The third procedure of diffusion from a soil with a higher to a lower concentration has been widely applied (Kemper, 1986; Oscarson et al., 1992). Two blocks of soil with different concentrations are brought together at time $t = 0$. After sufficient time has elapsed for solute diffusion to occur from the block with the higher to the lower concentration, the joined soil blocks are sectioned. The solute concentration of each section is determined, for example, by using extraction, centrifugation, and chemical analysis of the supernatant liquid. This approach yields a concentration profile versus distance from which the diffusion coefficient may be estimated using an appropriate analytical solution of the governing solute diffusion equation. Consider the diffusion equation,

$$\frac{\partial C}{\partial t} = D_{dif} \frac{\partial^2 C}{\partial x^2} \qquad\qquad [6.7]$$

subject to the initial condition,

$$C(x,0) = \begin{cases} C_o & -\infty < x < 0 \\ C_i & 0 < x < \infty \end{cases} \qquad\qquad [6.8]$$

and the approximate boundary conditions,

$$C(-\infty,t) = C_o \quad , \quad C(\infty,t) = C_i \qquad\qquad [6.9]$$

The solution for this problem is given by (Crank, 1975):

$$C(x,t) = C_i + 0.5(C_o - C_i)\,\mathrm{erfc}\!\left(x/\sqrt{4D_{dif}t}\right) \qquad\qquad [6.10]$$

where erfc is the complementary error function (Gautschi, 1964). The distributions of the solute concentration as a function of distance for different times after joining two soil blocks with concentrations $C_i = 0$ and C_o, and assuming $D_{dif} = 1$ cm^2 d^{-1}, are presented in Fig. 6.1.

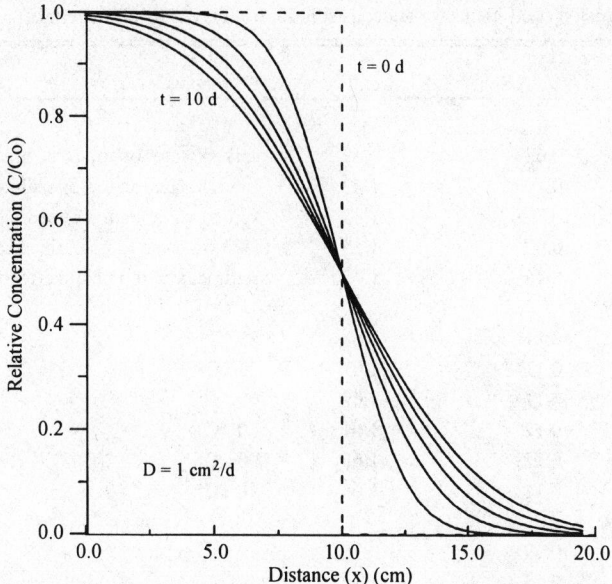

Fig. 6.1 Concentration profiles resulting from diffusion (D_{dif} = 1 d⁻¹) at different times after two soil columns with $C = 0$ and $C = C_o$ were joined at $t = 0$

Typical values for diffusion coefficients in clays and soils with accompanying τ_a are provided in Table 6.1. Additional soil diffusivity data are given by, among others, Hamaker (1972) and Nye (1979).

6.2.1.2 Dispersion

Local variations in water flow in a porous medium will lead to mechanical dispersion. Several mechanisms that are commonly used to contribute to mechanical dispersion are illustrated with hypothetical tracer particles in Fig. 6.2. Dispersion may occur because of (1) the development of a velocity profile within an individual pore such that the highest velocity occurs in the center of the pore, and presumably little or no flow at the pore walls; (2) different mean flow velocities in pores of different sizes; (3) the mean water flow direction in the porous medium being different from the actual streamlines within individual pores, which differ in shape, size, and orientation; and (4) solute particles converging to or diverging from the same pore. All of these processes contribute to increased spreading, in which initially steep concentration fronts become smoother during movement along the main flow direction.

The effects of dispersion can be illustrated with a hypothetical laboratory experiment in which water and a dissolved tracer are applied to an initially tracer-free, uniformly packed soil column of length L (Fig. 6.3). The column is subjected to steady-state water flow with a uniform water content. As more of the tracer is added, the initially very sharp concentration front near the soil surface becomes spread out because of dispersion. Eventually, a smooth and sigmoidally shaped effluent curve can be monitored at the column exit as shown in Fig. 6.3d. In the absence of dispersion, the front of a perfectly inert tracer will travel as a square wave through the column (a process often called piston flow) to reach the bottom of the column at time $t = L/v$, where v is the average pore water or interstitial velocity. This velocity is the ratio of the Darcian water flux density (J_w), and the volumetric water content (θ). Notice that v is given per unit area of fluid whereas J_w is defined per unit area of (bulk) soil.

Table 6.1 Selected diffusion coefficients for aqueous solutions in clay and soil materials

D_* cm^2 day^{-1}	τ_a -	ρ_b g cm^{-3}	Comments
van Rees et al. (1991)[†]			
1.46	0.64	1.42	spiked water on top of sediment ($\theta = 0.42$)
1.47	0.70	1.42	spiked water on top of sediment ($\theta = 0.42$)
1.66	0.79	1.35	spiked water on top of sediment ($\theta = 0.45$)
1.46	0.69	1.35	lake water on top of spiked sediment ($\theta = 0.45$)
0.97	0.46	1.42	sediment on top of spiked sediment ($\theta = 0.42$)
Robin et al. (1987)[‡]			
0.19	0.11	1.63	25 °C
0.20	0.11	1.61	25 °C
0.36	0.11	1.62	60 °C
0.40	0.12	1.61	60 °C
0.54	0.11	1.63	90 °C
0.56	0.11	1.64	90 °C
Oscarson et al. (1992)[§]			
0.33	0.19	0.90	
0.27	0.15	1.12	
0.17	0.10	1.31	
0.08	0.05	1.50	

[†] ^{3}H$_2$O tracer in litoral sediment
[‡] ^{36}Cl tracer in bentonite-sand mixture using a spiked ($C_o \approx 0.27$ M) and unspiked soil plug for different temperatures
[§] ^{125}I tracer in compacted bentonite using spiked and unspiked clay plugs

For piston flow, the tracer reaches the column exit exactly after one pore volume of tracer solution has been injected (or collected at the column exit). Pore volume is defined as the amount of water stored in that column.

The degree of spreading is usually related to the solute travel time, although some constraints do exist on the amount of spreading. Dispersion, as sketched in Fig. 6.2a, is limited because of transverse molecular diffusion which causes solutes to move from the center of a tube to areas near the pore walls, or *vice versa*, in response to local concentration gradients. Such transverse diffusion counteracts spreading caused by variations in the longitudinal flow velocity. Dispersion is also limited since capillaries in a porous medium generally are not independent cylindrical tubes, but branch and join or rejoin each other at distances characteristic of the pore or particle size distribution of the medium. This branching and rejoining promotes lateral mixing of solutes from different pores as sketched in Fig. 6.2d.

The macroscopic solute flux due to mechanical dispersion is often conveniently described by Fick's first law of diffusion, despite the conceptual differences between the diffusion and dispersion mechanisms (Scheidegger, 1974). A mathematical foundation for the Fickian description of mechanical dispersion is provided in a classical paper by Taylor (1953). He considered a circular tube, with tube radius r_o (L), filled with water flowing according to a parabolic velocity profile, with v_o being the maximum velocity at the axis (L T^{-1}) and a mean velocity $<v> = v_o/2$ over a cross-section of the tube. Taylor (1953) obtained the following expression for the coefficient of mechanical dispersion:

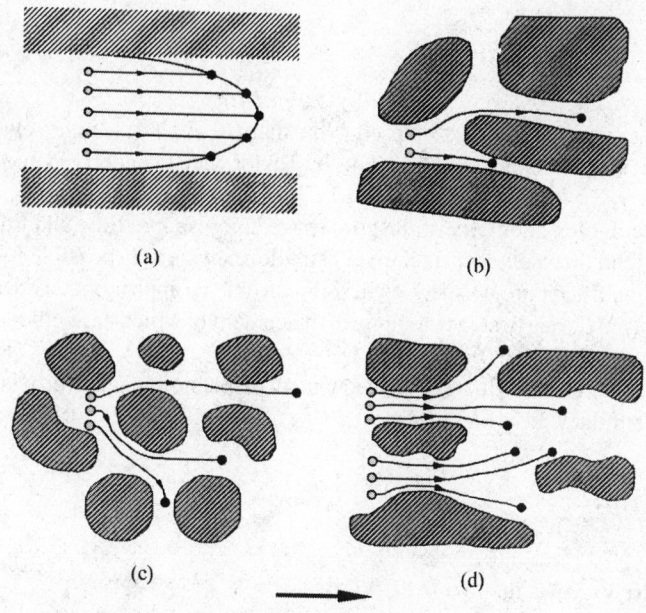

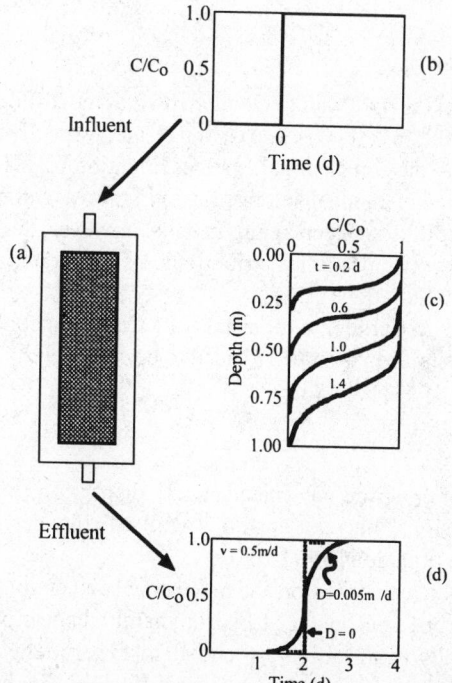

Fig. 6.2 Schematic concepts contributing to mechanical dispersion

Fig. 6.3 Hypothetical laboratory tracer experiment: (a) column of soil, (b) influent curve, (c) concentration distributions inside the column, and (d) breakthrough curves with and without dispersion [Modified after van Genuchten, 1988. p. 360-362. *In* S.P. Parker (ed.) McGraw-Hill Yearbook of Science and Technology, McGraw-Hill, New York]

$$D_{dis} = \frac{r_o^2 v_o^2}{192 D_o} = \frac{r_0^2 <v>^2}{48 D_o}$$ [6.11]

which is valid after sufficient time has elapsed. Note that D_{dis} ($L^2 T^{-1}$) is inversely proportional to the coefficient of molecular diffusion, D_o. The results by Taylor were later derived by Aris (1956) for more general conditions.

Because of the complex geometry of the pore space, microscopic flow and transport processes in soils do not easily lend themselves to a relatively simple analysis as is possible for solute transport in a well-defined, water-filled pore. Dispersion in soils can only be approximately described as a Fickian process, particularly at the early stage of solute displacement in which case other models may need to be employed (Jury and Roth, 1990).

The one-dimensional solute flux due to mechanical dispersion in a uniform isotropic soil may be approximated in a similar way as Fick's law:

$$J_{dis} = -\theta D_{dis} \frac{\partial C}{\partial x}$$ [6.12]

where J_{dis} is the dispersive solute flux ($M L^{-2} T^{-1}$).

The above one-dimensional geometry may be too simplistic for many transport problems. Three-dimensional dispersion is quantified with a dispersion tensor. The components of the symmetric dispersion tensor for an isotropic soil are given as (Bear and Verruijt, 1987):

$$D_{ij} = \delta_{ij} \alpha_T |\vec{v}| + (\alpha_L + \alpha_T) v_i v_j / |\vec{v}|$$ [6.13]

where $|\vec{v}|$ denotes the magnitude of the pore water velocity with v_i as the ith component (LT^{-1}), δ_{ij} is the Kronecker delta ($\delta_{ij} = 1$, if $i = j$ and $\delta_{ij} = 0$, if $i \neq j$), and α_L and α_T are, respectively, the longitudinal and transverse dispersivity (L). For a one-dimensional system, Equation [6.13] reduces to $D_{dis} = \alpha_L |v|$. Mechanical dispersion, as sketched in an idealized fashion in Fig. 6.2a, can be reversed by changing the flow direction to make a smooth front steep again. In soils, however, dispersion is not reversible since mixing erases antecedent concentration distributions, as illustrated in Fig. 6.2d. Absolute values are, therefore, used for v in Equation [6.13].

In the case of uniform water flow parallel to the x-axis of a Cartesian coordinate system, only the following three main components of Equation [6.13] need to be considered:

$$D_{xx} = \alpha_L v \quad , \quad D_{yy} = \alpha_T v \quad , \quad D_{zz} = \alpha_T v$$ [6.14]

where D_{xx} is the coefficient of longitudinal (mechanical) dispersion, and D_{yy} and D_{zz} are the coefficients of transverse dispersion. This relationship is similar to that derived by Taylor since $<v>$ is inversely proportional to r_o^2 in Equation [6.11].

The macroscopic similarity between diffusion and mechanical dispersion has led to the practice of describing both processes with one coefficient of hydrodynamic dispersion ($D = D_{dis} + D_{dif}$). This practice is consistent with results from laboratory and field experiments which do not permit a distinction between mechanical dispersion and molecular diffusion. The hydrodynamic dispersive flux (J_D) (Equation [6.2]) consists of contributions from molecular diffusion (Equation [6.4]) and mechanical dispersion (Equation [6.12]):

$$J_D = J_{dif} + J_{dis} \quad [6.15]$$

Since diffusion is independent of flow, the contribution of diffusion to hydrodynamic dispersion diminishes if the soil water flow rate increases. Hydrodynamic dispersion is often simply referred to as dispersion, as will be done in the remainder of this chapter.

Dispersion coefficients may be determined by fitting a mathematical solution to observed concentration (Toride et al., 1995). Additional procedures to determine dispersion coefficients are given by Fried and Combarnous (1971) and van Genuchten and Wierenga (1986). Values of the longitudinal dispersivity (α_L) for laboratory experiments typically vary between 0.1 and 10 cm with six to 20 times smaller values for α_T (Klotz et al., 1980). Dispersivities for field soils are generally much higher. Gelhar et al. (1992) reviewed published field-scale dispersivities determined in aquifer materials that are typically one or two orders of magnitudes larger, even more so for relatively large experimental scales.

6.2.2 Advection-Dispersion Equation

The expressions for the advective and dispersive solute fluxes can be substituted in mass balance Equation [6.1]. The one-dimensional advection-dispersion equation for solute transport in a homogeneous soil becomes:

$$\frac{\partial \theta C}{\partial t} = -\frac{\partial}{\partial x}\left(J_w C - \theta D \frac{\partial C}{\partial x} \right) + \theta R_s \quad [6.16]$$

In the case where the water content is invariant with time and space, the ADE may be simplified to ($v = J_w/\theta$):

$$\frac{\partial C}{\partial t} = D \frac{\partial^2 C}{\partial x^2} - v \frac{\partial C}{\partial x} + R_s \quad [6.17]$$

This is a second order linear partial differential equation. Similar to the diffusion equation, the ADE is classified as a parabolic differential equation. To complete the mathematical formulation of the transport, several concentration types and mathematical conditions will be reviewed in Section 6.3.1.

A variety of solute source or sink terms may be substituted for R_s. The most common source/sink term is due to adsorption/desorption and ion exchange stemming from chemical and physical interactions between the solute and the soil solid phase. Many other processes such as radioactive decay, aerobic and anaerobic transformations, volatilization, photolysis, precipitation/dissolution, reduction/oxidation, and complexation may also affect the solute concentration. A further refinement of the transport model is necessary in the case of nonuniform interactions between the solute and the soil, or if there is adsorption on moving particles and colloids. In the following, only interactions at the solid-liquid interface will be considered.

6.2.3 Adsorption

Dissolved substances in the liquid phase can interact with several soil constituents such as primary minerals, oxides, and inorganic or organic colloids. Dissolved ions in the soil solution counterbalance the surface charge of soil particles caused by isomorphous substitution of one element for another in the crystal lattice of clay minerals, by the presence of hydronium or hydroxyl ions at the solid surface, or other mechanisms. The net surface charge of an assemblage of soil particles produces an electric field that affects the distribution of cations and anions within water films surrounding the soil particles. The mechanisms and characteristics of reactions in solid-solution-solute systems are further discussed in Section B, Chapter 3.

Adsorption of solute (adsorbate) by the soil (adsorbent) is an important phenomenon affecting the fate and movement of solutes. The ADE for one dimensional transport of an adsorbed solute may be written as:

$$\frac{\partial C}{\partial t} + \frac{\rho_b}{\theta}\frac{\partial S}{\partial t} = D\frac{\partial^2 C}{\partial x^2} - v\frac{\partial C}{\partial x} \qquad [6.18]$$

where S is the adsorbed concentration, defined as mass of solute per mass of dry soil (M M^{-1}). The above equation can be expressed in terms of one dependent variable by assuming a suitable relationship between the adsorbed and liquid concentrations. This is typically done with a simple adsorption isotherm to quantify the adsorbed concentration as a function of the liquid concentration at a constant temperature. In addition to temperature, the adsorption isotherm is generally also affected by the solution composition, total concentration, the pH of the bulk solution, and sometimes the method used for measuring the isotherm. A mathematically pertinent distinction is often made between linear and nonlinear adsorption. Although most adsorption isotherms are nonlinear, the adsorption process may often be assumed linear for low solute concentrations or narrow concentration ranges.

6.2.3.1   Linear Adsorption
Consider the general case of nonequilibrium adsorption, where a change in C is accompanied by a delayed change in S. The adsorption rate can be described assuming first order kinetics:

$$\frac{\partial S}{\partial t} = kh(C, S) \qquad [6.19]$$

where k is a rate parameter (T^{-1}) and h is a function to quantify how far the adsorption or desorption process is removed from equilibrium. A single-valued isotherm for equilibrium adsorption $\Gamma(C)$ as in Equation [6.24], is used to define $h(C,S)$ according to:

$$h(C, S) = \Gamma(C) - S \qquad [6.20]$$

For equilibrium adsorption $k \to \infty$, and hence $h(C,S) \to 0$, which implies that $S = \Gamma(C)$. For a linear adsorption isotherm, the relation between Γ and C can simply be given as:

$$\Gamma = K_d C \qquad [6.21]$$

where K_d is a partition coefficient, often referred to as the distribution coefficient, expressed in volume of solvent per mass of soil (L^3 M^{-1}). For $\Gamma = S$, substitution of Equation [6.21] into Equation [6.18] leads to the following ADE commonly used to describe transport of a solute that undergoes linear equilibrium exchange:

$$R\frac{\partial C}{\partial t} = D\frac{\partial^2 C}{\partial x^2} - v\frac{\partial C}{\partial x}$$

[6.22]

in which the retardation factor R is given by

$$R = 1 + \frac{\rho_b}{\theta}K_d$$

[6.23]

with ρ_b as soil bulk density. The advective and dispersive fluxes are reduced by a factor R as a result of adsorption. The movement of the solute is said to be retarded with respect to the average solvent movement. If there is no interaction between the solute and the soil ($K_d = 0$), the value for R is equal to unity. The value for R can be readily calculated from K_d as obtained from chemical analyses of the solution and adsorbed phases. Alternatively, R can be estimated from solute displacement studies on laboratory soil columns (Fig. 6.3). A mathematical solution may then be used to estimate R from observed concentrations with nonlinear optimization programs. The change in the amount of solute in the soil column should be equal to the net solute flux into the column; the following mass balance can hence be formulated to estimate R:

$$v[g(t) - C_e] = R\int_0^L [C(x,t) - f(x)]dx$$

[6.24]

where C_e is the effluent concentration and $f(x)$ and $g(t)$ are the initial and influent concentrations.

The effects of linear adsorption on solute transport in a homogeneous soil profile are shown in Fig. 6.4. Analytically predicted solution and adsorbed concentrations are plotted four days after the start of a one-day application of influent with a unit solute concentration (units may be selected arbitrarily) to an initially solute-free soil profile subject to steady saturated water flow. Other parameters for this example are $J_w = 10\,\text{cm}\,\text{d}^{-1}$, $\theta = 0.40\,\text{cm}^3\,\text{cm}^{-3}$, and $D = 62.5\,\text{cm}^2\,\text{d}^{-1}$. The pore water velocity ($v = J_w/\theta$) is hence 25 cm d^{-1} and $\alpha_L = 2.5$ cm. Solute distributions are plotted (Fig. 6.4) for three values of the retardation factor, R. When R is increased from 1.0 to 2.0, the apparent solute velocity (v/R) is reduced

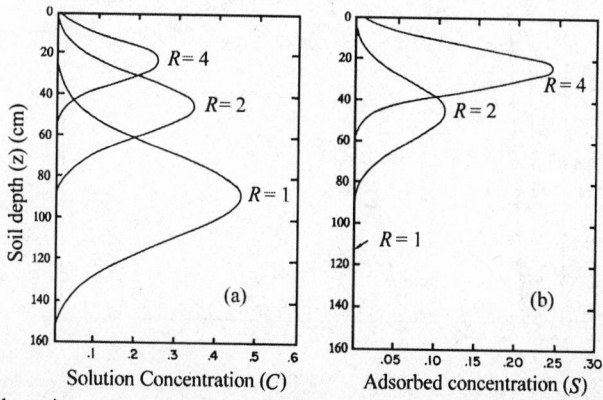

Fig. 6.4 Effect of adsorption, as accounted for by the retardation factor R, on solution (C) and adsorbed (S) concentration distributions in a homogeneous soil profile

by one-half (Fig. 6.4a), causing a shallower penetration of the solute pulse into the profile. At the same time, the area under the curve in Fig. 6.4a is also reduced by one-half. When $R = 4$, the apparent solute velocity and the area under the curve are again reduced by half. Distributions for the adsorbed concentration (S) which increases from zero (no adsorption) when $R = 1$ to a maximum when $R = 4$ are similar (Fig. 6.4b). Assuming a soil bulk density (ρ_b) of 1.25 g cm^{-3} and the same water content as before ($\theta = 0.40 \, \text{cm}^3 \, \text{cm}^{-3}$), one may calculate, using Equation [6.23], that the distribution coefficient K_d = 0, 0.32 and 0.96 cm^3 g^{-1} for $R = 1$, 2, and 4, respectively.

Anion exclusion occurs when negatively charged surfaces of clays and ionizable organic matter are present; anions are repelled from such surfaces and accumulate in the center of pores. Because water flow velocities are zero at pore walls and maximum in the center of pores (Fig. 6.2a), the average anion movement will be faster than the average water movement. Many displacement experiments also suggest faster anion than water movement simply because the apparent displacement volume is smaller for anions than water. The quantity $(1-R)$ is the relative anion exclusion volume. The exclusion volume-per-unit mass of soil can also be estimated as:

$$V_{ex} = \int (1 - c/C_o) dV \qquad [6.25]$$

where V_{ex} is the exclusion volume (L^3 M^{-1}), c is the local concentration of the anion (M L^{-3}) and C_o its bulk concentration (M L^{-3}), and V is the entire volume encompassing the liquid phase. Instead of using $R < 1$, anion transport may be modeled with a model, with $R = 1$, which restricts the accessible liquid volume (Krupp et al., 1972).

Anions are also adsorbed by the soil through surface complexation and adsorption onto positively charged areas of the solid matrix (Section B, Chapter 7). If the effect of adsorption exceeds exclusion, the anion will be retarded. The retardation factor should be viewed as an effective parameter since it quantifies a variety of adsorption and exclusion processes to which the solute (anion) is subjected.

Breakthrough curves typical for the transport of an excluded anion (Cl$^-$) a nonreactive solute (tritiated water, ^{3}H$_2$O), and an adsorbed cation (Ca^{2+}) are presented in Fig. 6.5. The first two tracers pertain to transport through 30 cm columns containing disturbed Glendale clay loam soil (P. J.

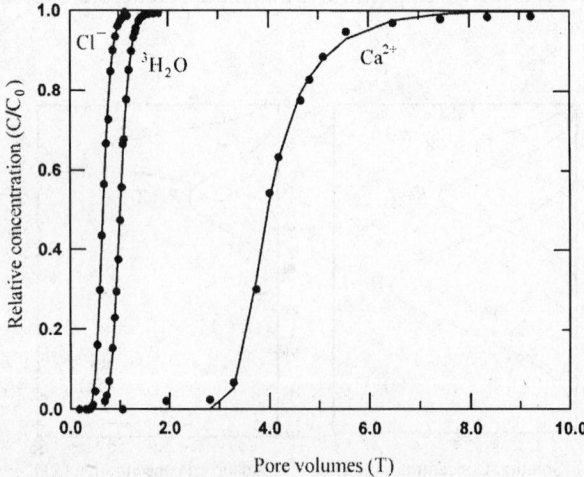

Fig. 6.5 Observed and fitted ADE breakthrough curves for three tracers typifying the transport of anions (Cl$^-$), a (nearly) nonreactive solute (^{3}H$_2$O), and an adsorbing solute (Ca^{2+})

Wierenga, personal communication; van Genuchten and Cleary, 1982), while the Ca^{2+} data are for transport through a 30-cm long column containing a Troup loam and a Savannah fine loam (Leij and Dane, 1989). Analysis of the three breakthrough curves in terms of the ADE, using inverse procedures (Parker and van Genuchten, 1984b), yielded R values of 0.681, 1.027, and 4.120 for Cl^-, ^3H_2O, and Ca^{2+}. Hence, the Cl^- curve was strongly affected by anion exclusion, while ^3H_2O transport was subject to very minor adsorption/exchange.

6.2.3.2 Nonlinear Adsorption

In many cases adsorption, and hence, the retardation factor, cannot be described using a simple K_d approach. For nonlinear equilibrium adsorption, R is given as:

$$R(C) = 1 + \frac{\rho_b}{\theta} \frac{\partial \Gamma}{\partial C}$$

[6.26]

Two common nonlinear adsorption isotherms are the Langmuir and Freundlich equation

$$\Gamma = \frac{k_1 C}{1 + k_2 C} \qquad \text{Langmuir}$$

[6.27]

$$\Gamma = k_3 C^n \qquad \text{Freundlich}$$

[6.28]

where k_1, k_2, k_3, and n are empirical constants. Many other equations for adsorption exist, including some that account for differences between adsorption and desorption isotherms (van Genuchten and Sudicky, 1999).

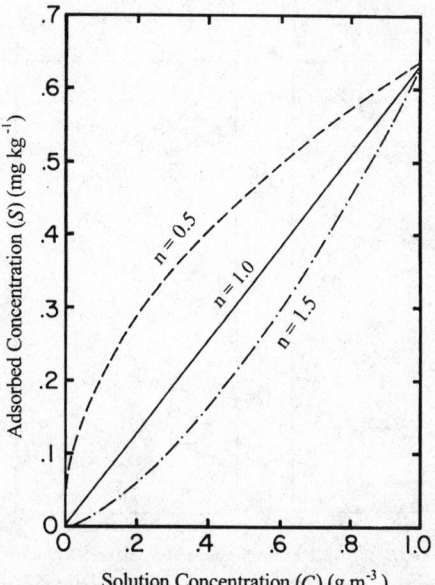

Fig. 6.6 Freundlich equilibrium plots for $k_3 = 0.64$ and three values of the exponent n

The Freundlich isotherm will be used in the following to illustrate the effects of nonlinear equilibrium adsorption on solute transport. In order to keep the calculations simple, the value of k_3 in Equation [6.28] is taken to be 0.64. Three different values of the exponent n are used, viz., 0.5, 1.0, and 1.5, to demonstrate favorable, linear, and unfavorable adsorption isotherms (Fig. 6.6). Calculated distributions of the solution (C) and adsorbed (S) concentrations versus soil depth (z) eight days after application of a 4-day long solute pulse to the soil surface are shown in Fig. 6.7. The same pore water velocity is used as for the example illustrated in Fig. 6.5, but with a smaller dispersion coefficient of D = 25 cm^2 d^{-1} (α_L = 1 cm). Notice that, as in Fig. 6.5, the solution concentration distribution for $n = 1$ (linear adsorption) has a nearly symmetrical shape versus depth. The other two n values yield nonsymmetric profiles.

If $n = 0.5$, a very sharp concentration front develops, while the curve near the soil surface becomes more dispersed. The sharp front can be explained by considering the retardation factor (R) for nonlinear adsorption (Equation [6.26]), which for $n = 0.5$, $\rho_b = 1.25$ g cm^{-3}, $\theta = 0.40$, and $k_3 = 0.64$ leads to $R = 1 + 1/C$. This shows that R increases rapidly when C decreases with the extreme $R \rightarrow \infty$, if $C = 0$. Consequently, the apparent solute velocity $v_a = v/R$ is very small at the lower liquid concentrations, but increases at higher values. Of course, higher concentrations cannot move faster than lower concentrations; front sharpening will lead to a steep solute front. This front never becomes a step function because the large concentration gradient across the front will create a large diffusion/ dispersion flux. When $n \ll 1$, an estimate of the front can be obtained from the average slope of the isotherm between the initial and the maximum concentration. Because in the present example these are zero and approximately one, the average slope of the isotherm is exactly the same as the linear distribution coefficient ($d\Gamma/dC = 0.64$). Substituting this value into Equation [6.26] yields $R = 3$. Hence, the apparent solute velocity (v_a) equals 25/3 or 8.33 cm d^{-1}, and the solute front after 8 days is located at a depth of about 67 cm (Fig. 6.7). Transport of favorably adsorbed solutes is frequently modeled with traveling wave solutions (van der Zee, 1990; Simon et al., 1997).

A reverse scenario occurs if $n > 1$ (unfavorable exchange). Adsorption at the lower concentrations is now relatively small and, as displayed in Fig. 6.7, the toe of the front moves through the profile at a

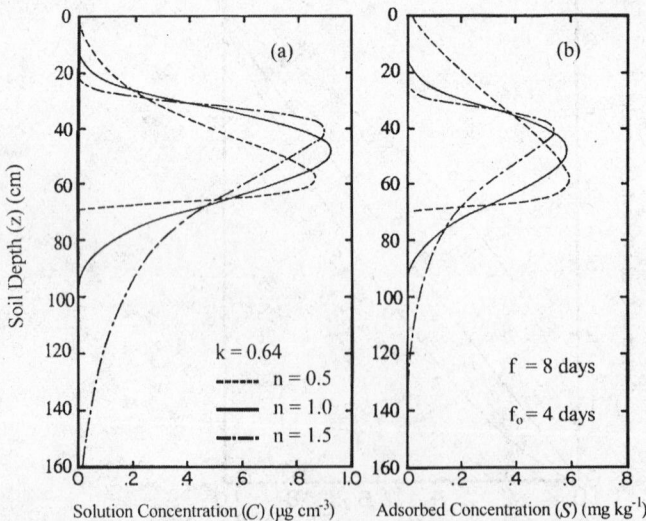

Fig. 6.7 Effect of nonlinear adsorption on solution (C) and adsorbed (S) concentrations in a deep homogeneous soil profile during steady-state flow. The distributions were obtained for the three isotherms shown in Fig. 6.6.

velocity nearly equal to that of an inert solute. Adsorption at the higher concentrations, on the other hand, is much more extensive, resulting in a lower apparent solute velocity in the higher range of concentrations. As a result, the concentration front becomes increasingly dispersed over time. Ignoring dispersion, the velocity of the solute front ($v_a = v/R$) at any given value of C is given by:

$$v_a = \frac{v}{1 + 3\sqrt{C}}$$

[6.29]

while the depth of the solute front can be approximated by:

$$z(C, t) = z(C, 0) + v_a t$$

[6.30]

where $z(C, 0)$ is the location of a solute concentration with value C at $t = 0$.

The above discussion pertained to adsorption of a single ion species. Cation exchange processes in transport studies involve at least two species. The simplest case arises when two cations of the same valency and total concentration such as Ca^{2+} and Mg^{2+} are considered. The resulting exchange process is then approximately linear for relatively small changes in the composition of the soil solution. Exchange between Na^+ and Ca^{2+}, on the other hand, is considerably more nonlinear. Equations that quantify the exchange reaction have been proposed by Gapon, Kerr, Vanselow, Eriksson, and others (Section B, Chapter 7).

6.2.4 Nonequilibrium Transport

Solute breakthrough curves for aggregated soils will exhibit asymmetrical distributions or nonsigmoidal concentration fronts. The concept behind physical nonequilibrium models is that differences between regions of the liquid phase lead to mostly lateral gradients in the solute concentration resulting in a diffusive type of solute transfer process. Depending upon the exact pore structure of the medium, asymmetry is sometimes enhanced by desaturation when the relative fraction of water residing in the marginally continuous immobile region increases.

Since most of the sorption sites are only accessible after diffusion through the immobile region of the liquid phase, a corresponding delay in adsorption will occur. The delayed adsorption can also be explained with a kinetic description of the adsorption process. Both cases may be described with chemical nonequilibrium models, which distinguish between sites with equilibrium and kinetic sorption.

Bi-continuum or dual-porosity nonequilibrium models are the most widely used. Only two concentrations need to be considered and the equilibrium ADE (Equation [6.18]) can be readily modified for this purpose. The same dimensionless mathematical formulation can be used for physical and chemical nonequilibrium models. If necessary, the ADE can be modified to incorporate additional nonequilibrium processes and continua.

6.2.4.1 Physical Nonequilibrium

Consider one-dimensional solute movement in an isotropic soil with uniform flow and transport properties during steady flow, and assume that the solute is subject to linear retardation, that is, equilibrium sorption can be described with a linear exchange isotherm. The physical nonequilibrium approach is based on a partitioning of the liquid phase into a mobile or flowing region and an immobile or stagnant region. Solute movement in the mobile region occurs by both advection and dispersion, whereas solute exchange between the two regions occurs by first order diffusion (Coats and Smith, 1964). Following van Genuchten and Wierenga (1976), the governing equations for the two region model are

$$\left(\theta_m + f\rho_b K_d\right)\frac{\partial C_m}{\partial t} = \theta_m D_m \frac{\partial^2 C_m}{\partial x^2} - \theta_m v_m \frac{\partial C_m}{\partial x} - \alpha\left(C_m - C_{im}\right)$$ [6.31]

$$\left[\theta_m + (1-f)\rho_b K_d\right]\frac{\partial C_{im}}{\partial t} = \alpha\left(C_m - C_{im}\right)$$ [6.32]

where f represents the fraction of sorption sites in equilibrium with the fluid of the mobile region, α is first order mass transfer coefficient (T^{-1}), and the subscripts m and im, respectively, refer to the mobile and immobile liquid regions (with $\theta = \theta_m + \theta_{im}$), while ρ_b and K_d are the soil bulk density and distribution coefficient for linear sorption. Transport Equation [6.31] follows directly from addition of a source/sink term (R_s) to Equation [6.18].

Anion exclusion can be viewed as a particular example of physical nonequilibrium, the exclusion volume roughly corresponds to the immobile region (Krupp et al., 1972). The physical nonequilibrium concept may, therefore, be adapted to describe transport of excluded anions (van Genuchten, 1981) instead of using a retardation factor of less than one.

6.2.4.2   Chemical Nonequilibrium

Sorption of solute, especially for organic chemicals, has often been described with a combined equilibrium and kinetic sorption expression so as to better simulate transport in soils with a wide variety of soil constituents (clay minerals, organic matter, and oxides). The lack of an instantaneous equilibrium for the sorption process is sometimes referred to as chemical nonequilibrium. This terminology is somewhat misleading since the rate of adsorption or exchange is usually determined mostly by physical phenomena such as diffusion through the liquid film around soil particles and inside the aggregates (Boyd et al., 1947; Sparks, 1989).

The simplest and by far most popular approach distinguishes between type-1 sites, with instantaneous adsorption, and type-2 sites, where adsorption obeys a kinetic rate law (Selim et al., 1976). In the case of first-order kinetics, the general adsorption rates can be given with a model similar to Equations [6.19] and [6.20] as:

$$\frac{\partial S_1}{\partial t} = \alpha_1\left[\Gamma_1(C) - S_1\right]$$ [6.33]

$$\frac{\partial S_2}{\partial t} = \alpha_2\left[\Gamma_2(C) - S_2\right]$$ [6.34]

where α is again a rate constant (T^{-1}), S is the actual adsorbed concentration (M M^{-1}), Γ is the final adsorbed concentration at equilibrium as prescribed by the adsorption isotherm, the subscripts 1 and 2 refer to the type of adsorption site, and $\Gamma_1 + \Gamma_2 = \Gamma$. Because type-1 sites are always at equilibrium, $S_1 = \Gamma_1$ and Equation [6.33] can further be ignored. The transport equation becomes:

$$\frac{\partial C}{\partial t} + \frac{\rho_b}{\theta}\frac{\partial \Gamma}{\partial t} + \frac{\alpha_2 \rho_b}{\theta}\left(\Gamma_2 - S_2\right) = D\frac{\partial^2 C}{\partial x^2} - v\frac{\partial C}{\partial x} \tag{6.35}$$

If the fraction of exchange sites that is at equilibrium (type-1) equals f, and if equilibrium adsorption is governed by the same linear isotherm for both types 1 and 2 ($\Gamma_1 = \Gamma_2$) then:

$$\Gamma_1 + \Gamma_2 = fK_dC + (1-f)K_dC \tag{6.36}$$

Of course, nonlinear equilibrium isotherms may also be used in nonequilibrium transport models. The complete transport problem can now be written as:

$$\left(1 + \frac{\rho_b fK_d}{\theta}\right)\frac{\partial C}{\partial t} = D\frac{\partial^2 C}{\partial x^2} - v\frac{\partial C}{\partial x} - \frac{\alpha\rho_b}{\theta}\left[(1-f)K_dC - S_2\right] \tag{6.37}$$

$$\frac{\partial S_2}{\partial t} = \alpha\left[(1-f)K_dC - S_2\right] \tag{6.38}$$

where the subscript for α has been dropped. This two-site chemical nonequilibrium model reduces to a one-site kinetic nonequilibrium model by setting $f = 0$. The two-site chemical nonequilibrium model was applied successfully to describe solute breakthrough curves by Selim et al. (1976), van Genuchten (1981), and Nkedi-Kizza et al. (1983), among others.

6.2.4.3 General Nonequilibrium Formulation

The two-site and the two-region nonequilibrium models can be cast in the same (dimensionless) model according to Nkedi-Kizza et al. (1984):

$$\beta R\frac{\partial C_1}{\partial T} = \frac{1}{P}\frac{\partial^2 C_1}{\partial X^2} - \frac{\partial C_1}{\partial X} + \omega(C_2 - C_1) \tag{6.39}$$

$$(1-\beta)R\frac{\partial C_2}{\partial T} = \omega(C_1 - C_2) \tag{6.40}$$

where β is a partition coefficient, R is a retardation factor, C_1 and C_2 are dimensionless equilibrium and nonequilibrium concentrations, T is time, X is distance, P is the Peclet number, ω is a mass transfer coefficient, and the subscripts 1 and 2 refer to the equilibrium and nonequilibrium phases, respectively. The common dimensionless parameters are defined using an arbitrary characteristic concentration (C_o) and length (L):

$$T = vt/L \quad , \quad X = x/L \quad , \quad P = vL/D \quad , \quad R = 1 + \rho_b K_d/\theta \tag{6.41}$$

For the physical nonequilibrium model, the remaining dimensionless parameters are

$$\beta = \frac{\theta_m + f\rho_b K_d}{\theta + \rho_b K_d} \quad , \quad \omega = \frac{\alpha L}{\theta v} \quad , \quad C_1 = \frac{C_m}{C_o} \quad , \quad C_2 = \frac{C_{im}}{C_o} \qquad [6.42]$$

whereas for the chemical nonequilibrium model:

$$\beta = \frac{\theta_m + f\rho_b K_d}{\theta + \rho K_d} \quad , \quad \omega = \frac{\alpha(1-\beta)RL}{v} \quad , \quad C_1 = \frac{C}{C_o} \quad , \quad C_2 = \frac{S_2}{(1-f)K_d C_o} \qquad [6.43]$$

In the chemical engineering literature, $\alpha L/v$ is known as the Damköhler number; it quantifies the rate of the reaction or exchange relative to advective transport.

6.3 Solutions of the Advection-Dispersion Equation

The research and management of solute behavior in soils almost invariably require that the temporal and spatial solute distribution be known. Solute distributions as a function of time and/or space can be estimated with a variety of analytical and numerical solutions of the ADE, some of which will be briefly reviewed in the following.

6.3.1 Basic Concepts

A complete mathematical formulation of the transport problem requires that the pertinent dependent variable or concentration type is used and that the proper auxiliary conditions are specified.

6.3.1.1 Concentration Types

Concentration is conventionally defined as the amount of solute-per-unit volume of the liquid. Since microscopic concentrations are based on a relatively small scale, a concentration at a larger scale needs to be introduced to allow use of the ADE which is based on larger macroscopic variables and parameters. For this purpose, a macroscopic resident or volume-averaged concentration (C_R) is defined as:

$$C_R = \frac{1}{\Delta V} \iiint c \, dV \qquad [6.44]$$

where c is the variable local-scale (microscopic) concentration ($M\,L^{-3}$) in a volume element (V) and V is some representative elementary volume (Bear and Verruijt, 1987).

A different concentration type may be encountered at soil boundaries. In many solute displacement experiments, the concentration is determined from effluent samples as the ratio of the solute flux (J_s) and water flux (J_w) densities:

$$C_F = J_s / J_w \qquad [6.45]$$

where C_F is the flux averaged concentration. This concentration represents the mass of solute-per-unit volume of fluid passing through a soil cross-section during an elementary time interval (Kreft and Zuber, 1978). For a one-dimensional solute flux consisting of an advective and a dispersive component, the flux-averaged concentration can be derived from the resident concentration according to the transformation:

$$C_F = C_R - \frac{D}{v}\frac{\partial C_R}{\partial x}$$

[6.46]

The resident concentration may be determined from the flux averaged concentration using (van Genuchten et al., 1984):

$$C_R(x,t) = \frac{v}{D}\exp\left(\frac{vx}{D}\right)\int\limits_x^\infty \exp\left(-\frac{v\xi}{D}\right)C_F(\xi,t)d\xi$$

[6.47]

Additional transformations between flux and resident type concentrations are given by Parker and van Genuchten (1984a).

The difference between C_R and C_F is usually small, except when the second term on the right-hand side of Equation [6.46] is relatively large. It should be noted that a distinction between flux and resident type can be made for both the application and the detection of solutes (Kreft and Zuber, 1978, 1986). In soil science, a flux type application mode is often implicitly assumed (Parker and van Genuchten, 1984a). Flux-averaged concentrations are typically used when it is not possible to determine or specify a reliable value for the (resident) concentration. Resident concentrations are used for solute detection with, for example, time domain reflectometry, and to specify most initial conditions. Flux-averaged concentrations, on the other hand, are used for effluent samples, and to specify the influent concentration in most boundary value problems. Unless stated otherwise, it is assumed that solute concentrations are of the resident type.

Averaged concentrations can also be defined in terms of the observation scale, the latter exceeding the macroscopic scale associated with using the ADE. A time-averaged concentration (C_T), is obtained by averaging over a time interval (Δt) about a discrete time (t_o) (Fischer et al., 1979):

$$C_T(x,t_o) = \frac{1}{\Delta t}\int\limits_{t_o-\Delta t/2}^{t_o+\Delta t/2} C(x,t)dt$$

[6.48]

where C is a continuous solution of the solute concentration, which can be obtained by solving the ADE. This type of concentration occurs if solute breakthrough curves are measured using, for example, fraction collectors or gamma ray attenuation. Similarly, a one-dimensional spatial average can be defined as:

$$C_L(x_o,t) = \frac{1}{\Delta x}\int\limits_{x_0-\Delta x/2}^{x_0+\Delta x/2} C(x,t)dx$$

[6.49]

This concentration may be used to describe experimental results obtained for samples with centroid (x_o) and length (Δx). This situation occurs, for example, when the measured concentration of a large core sample is to be modeled as a point value (Leij and Toride, 1995).

6.3.1.2 Boundary and Initial Conditions

Initial and boundary conditions need to be specified in order to obtain a meaningful solution of the ADE. For a finite or semi-infinite soil, the initial condition can be formulated as:

$$C(x,0) = f(x) \qquad x \geq 0 \tag{6.50}$$

where $f(x)$ is an arbitrary function. Initial concentrations are almost invariably of the resident type.

The selection of the most appropriate boundary conditions for a transport problem is a somewhat esoteric topic that has received considerable attention in the literature. This is partly due to a lack of detailed experimental information for evaluating and applying boundary conditions, and inherent shortcomings of the transport equation itself at boundaries.

Many transport problems involve the application to the soil of a solute, whose influent concentration may be described by a function $g(t)$. The application method may be pumping, ponding, or sprinkling. Two different types of inlet conditions are used, which assume either continuity in solute concentration or solute flux density. Simultaneous use of both conditions is seldom possible. It is generally more desirable to ensure mass conservation in the whole system than a continuous concentration at the inlet. The solute fluxes at the inlet boundary are, therefore, equated to obtain the following third or flux type inlet condition:

$$\left(vC - D\frac{\partial D}{\partial x} \right)_{x=0^+} = vg(t) \tag{6.51}$$

where 0^+ indicates a position just inside the soil. It is assumed that there is no dispersion outside the soil. The alternative condition requires the concentration to be continuous across the interface at all times. At smaller scales, such continuity will likely exist. However, at the scale of the ADE, it appears difficult to maintain a constant concentration at the interface, particularly during the initial stages of solute displacement for low influent fluxes and high dispersive fluxes in the soil. Mathematically, the first or concentration type condition is expressed as:

$$C(0,t) = g(t) \qquad t > 0 \tag{6.52}$$

The outlet condition can be defined as a zero gradient at a finite or infinite distance from the inlet. The infinite outlet condition,

$$\frac{\partial C}{\partial x}(\infty, t) = 0 \tag{6.53}$$

is more convenient for mathematical solutions than the finite condition,

$$\frac{\partial C}{\partial x}(L^-, t) = 0 \tag{6.54}$$

The use of Equation [6.53] implies that there is a semi-infinite fictitious soil layer beyond $x = L$, with identical properties as the actual soil. Such a layer does not affect the movement of the solute in the actual soil upstream of the exit boundary. Since the transport at the outlet cannot be precisely described, the intuitive contradiction of an infinite mathematical condition to describe a finite physical system is often more acceptable than using Equation [6.54], which precludes dispersion inside the soil near the outlet.

The formulation of the boundary and inlet conditions should account for the injection and detection modes in order to arrive at a mathematically consistent formulation of the problem with the same concentration type as independent variable. Only differences in detection mode for finite and semi-infinite systems will here be explored. The ADE in terms of the (usual) resident concentration is given as:

$$\frac{\partial C_R}{\partial t} = D\frac{\partial^2 C_R}{\partial x^2} - v\frac{\partial C_R}{\partial x} \qquad [6.55]$$

subject to a uniform initial condition, a third-type inlet condition, and a finite or infinite outlet condition:

$$C_R(x,0) = C_i \qquad [6.56]$$

$$\left(vC_R - D\frac{\partial C_R}{\partial x}\right)_{x=0^+} = vg(t) \qquad [6.57]$$

$$\frac{\partial C_R}{\partial x}(\infty, t) = 0 \quad \text{or} \quad \frac{\partial C_R}{\partial x}(L, t) = 0 \qquad [6.58]$$

This problem can be written in terms of a flux-averaged concentration using Equation [6.48] according to (Parker and van Genuchten, 1984b):

$$\frac{\partial C_F}{\partial t} = D\frac{\partial^2 C_F}{\partial x^2} - v\frac{\partial C_F}{\partial x} \qquad [6.59]$$

subject to

$$C_F(x,0) = C_i \qquad [6.60]$$

$$C_F(0,t) = g(t) \qquad [6.61]$$

$$\frac{\partial C_F}{\partial x}(\infty, t) = 0 \quad \text{or} \quad \frac{\partial C_F}{\partial x}(L,t) = -\frac{D}{v}\frac{\partial^2 C_R}{\partial x^2}(L,t) \qquad [6.62]$$

Notice that the mathematical problem for the flux mode involves a simpler first-type inlet condition with mass being conserved, unlike the use of a first-type condition for a resident concentration. The solution for C_R for a semi-infinite system involving a first-type inlet condition is the same as the solution for C_F that conserves mass. As shown by Toride et al. (1994), the transformation is less convenient for nonuniform initial conditions or finite systems. Solutions for C_F are then more easily obtained by transforming C_R according to Equation [6.46].

Differences between the preferred third type solution for C_R and its first type solution are usually small except for low values of the dimensionless time $[\varsigma = v^2t/(RD)]$ (van Genuchten and Parker, 1984). The relative mass balance error in a semi-infinite soil profile if a first type rather than a third-type inlet condition is used, is given in Fig. 6.8 (van Genuchten and Parker, 1984). The error pertains to the transport of a tracer solution of concentration (C_o) into an initially solute-free semi-infinite soil profile. Especially for small ς, a substantial error may occur. Unless stated otherwise, a resident concentration is used in conjunction with a third-type inlet condition.

6.3.2 Analytical Solutions

Analytical solutions can formally be obtained only for linear transport problems. It would appear that analytical solutions are not very useful for transport in field soils where there is (1) spatial and temporal variability of flow and transport parameters, (2) transient flow, especially for unsaturated soils, and (3) nonuniformity in the boundary and initial conditions. However, analytical solutions can still be quite valuable. A nonlinear transport problem may be linearized through a suitable transformation to obtain a problem for which an analytical solution is available. Also, analytical solutions provide quick estimates of solute behavior over large temporal and spatial scales while they may offer insight into the underlying transport processes. Moreover, there is usually a lack of input parameters for field problems, which diminishes the advantage of numerical over analytical model results. Analytical solutions are also routinely used to evaluate the performance of numerical schemes. Finally, the mathematical and physical conditions tend to be well defined for laboratory settings and an analytical solution can often be used, especially to estimate transport parameters by fitting analytical solutions to experimental data (Parker and van Genuchten, 1984; van Genuchten and Parker, 1987).

6.3.2.1 Variable Transformation

One straightforward way to obtain an analytical solution is to transform the ADE to an equation for which a solution already exists. As an example, consider transport in an infinite system given by:

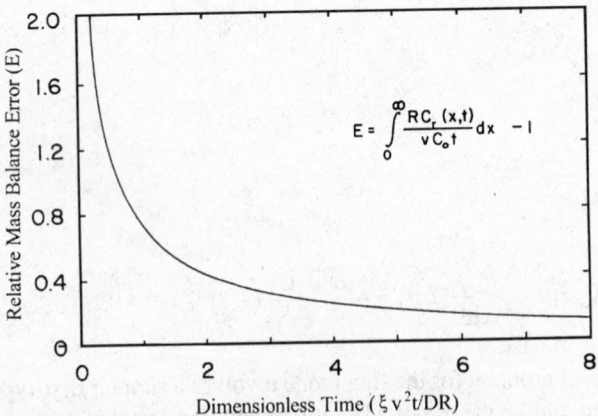

$$E = \int_0^\infty \frac{RC_r(x,t)}{vC_o t}dx \ - 1$$

Fig. 6.8 Plot of the relative mass balance error versus dimensionless time for a semi-infinite profile when a first-type rather than a third-type inlet condition is used [After van Genuchten and Parker, 1984]

$$R\frac{\partial C}{\partial t} = D\frac{\partial^2 C}{\partial x^2} - v\frac{\partial C}{\partial x}$$

[6.63]

$$C(x,0) = \begin{cases} C_o & x < 0 \\ 0 & x > 0 \end{cases}$$

[6.64]

The new coordinates

$$\xi = x - vt$$
$$\tau = t$$

[6.65]

transform the ADE into a heat or solute diffusion problem given by:

$$R\frac{\partial C}{\partial \tau} = D\frac{\partial^2 C}{\partial \xi^2}$$

[6.66]

$$C(\xi,0) = \begin{cases} C_0 & \xi < 0 \\ 0 & \xi > 0 \end{cases}$$

[6.67]

The solution for this problem can be readily found in the literature on diffusion problems (Carslaw and Jaeger, 1959; Crank, 1975):

$$C(x,t) = \frac{C_o}{2}\,\mathrm{erfc}\!\left(\frac{Rx - vt}{\sqrt{4RDt}}\right)$$

[6.68]

Other transformations to the diffusion problem have been employed as well (Brenner, 1962; Selim and Mansell, 1976; Zwillinger, 1989). Transformation of time to a time-integrated flow variable sometimes also allows one to derive an analytical solution of the nonlinear ADE for transient flow (Wierenga, 1977; Parker and van Genuchten, 1984b; Huang and van Genuchten, 1995).

6.3.2.2 Laplace Transformation

The ADE is commonly solved directly with the method of Laplace transforms. The solution procedure will be illustrated here for an initially solute-free semi-infinite soil with a constant solute flux (vC_o) or concentration (C_o) at the inlet boundary. The mathematical problem consists of solving the ADE given by Equation [6.63] subject to:

$$C - \delta\frac{D}{v}\frac{\partial D}{\partial x} = C_o \qquad \delta = \begin{cases} 0 & \text{first type} \\ 1 & \text{third type} \end{cases}$$

[6.69]

$$\frac{\partial C}{\partial x}(\infty, t) = 0 \tag{6.70}$$

with δ as a coefficient depending on the type of inlet condition. The Laplace transform ($\mathscr{L}$) of the solute concentration with respect to time is defined as (Spiegel, 1965):

$$\overline{C}(x, s) = \mathscr{L}[C(x,t)] = \int_0^\infty C(x,t)\exp(-st)dt \tag{6.71}$$

where s is the (complex) transformation variable (T^{-1}). This transformation changes the transport equation from a partial to an ordinary differential equation:

$$\frac{d_2\overline{C}}{dx^2} - \frac{v}{D}\frac{d\overline{C}}{dx} - \frac{sR}{D}\overline{C} = 0 \tag{6.72}$$

subject to

$$\overline{C} - \delta\frac{D}{v}\frac{d\overline{C}}{dx} = \frac{C_o}{s} \tag{6.73}$$

$$\frac{d\overline{C}}{dx}(\infty, s) = 0 \tag{6.74}$$

where the bar denotes a transformed variable. The following solution for the concentration in the Laplace domain is obtained with help of the inlet condition:

$$\overline{C}(x, s) = \frac{v}{v - \delta\lambda^- D}\frac{C_o}{s}\exp(\lambda^- x) \quad , \quad \lambda^- = \frac{v}{2D} - \left[\left(\frac{v}{2D}\right)^2 + \frac{SR}{D}\right]^{1/2} \tag{6.75}$$

Inversion of this solution may be done with a table of Laplace transforms, by applying the inversion theorem, or by using a numerical inversion program. It should be noted that the solution for a finite outlet condition is also possible with the Laplace transform, although a bit more cumbersome (Brenner, 1962; Leij and van Genuchten, 1995).

6.3.2.3 Equilibrium Transport

Van Genuchten and Alves (1982) provided a compendium of available analytical one-dimensional solutions for a variety of mathematical conditions and physical processes. Four common analytical solutions for a zero initial condition involving a first- or third-type inlet condition and an infinite or finite outlet condition are listed in Table 6.2. The solutions may be expressed in terms of the dimensionless variables (P), (T), and (X) (Equation [6.41]). Typically, L is equal to the position of the outlet (the column length) for a finite system, whereas for a semi-infinite soil system, L can be assigned to any arbitrary length.

Fig. 6.9 contains solute profiles, $[C/C_o(X)]$ according to the solutions listed in Table 6.2 using $R = 1$ and two different values for P and T for a first- (A1) or a third-type (A2) inlet condition assuming an infinite outlet condition, or a first- (A3) or third-type (A4) inlet condition in case of finite outlet condition. The predicted profile for a first-type condition (A1, A3) for the lower Peclet number ($P = 5$) lies considerably above the line predicted for a third-type condition (A2, A4). The effect of the outlet condition is initially minor, but when the solute front reaches the outlet (L), a clear difference between a finite and an infinite outlet condition can be observed for both a first (A1, A3) and a third-type (A2, A4) inlet condition. The simpler solution for a semi-infinite system can, in many cases, be used to approximate the solution for a finite condition; van Genuchten and Alves (1982) formulated the empirical restriction:

$$X < 0.9 - 8/P \qquad [6.76]$$

on the position for which such an approximation is reasonable. For smaller times ($T \ll 1$), when the solute has not reached the outlet, the finite and infinite outlet condition obviously lead to a similar solution.

Table 6.2 Analytical solutions of the ADE for different boundary conditions after van Genuchten and Alves (1982)

Case	Inlet Condition	Exit Condition	Analytical Solution $C(x,t)$
A1	$C(0,t) = C_o$	$\dfrac{\partial C}{\partial x}(\infty, t) = 0$	$\dfrac{1}{2}\mathrm{erfc}\left(\dfrac{Rx - vt}{(4RDt)^{1/2}}\right) + \dfrac{1}{2}\exp\left(\dfrac{vx}{D}\right)\mathrm{erfc}\left(\dfrac{Rx + vt}{(4RDt)^{1/2}}\right)$
A2	$\left(vC - D\dfrac{\partial C}{\partial x}\right)_{x=0} = vC_o$	$\dfrac{\partial C}{\partial x}(\infty, t) = 0$	$\dfrac{1}{2}\mathrm{erfc}\left(\dfrac{Rx - vt}{(4RDt)^{1/2}}\right) - \dfrac{1}{2}\left(1 + \dfrac{vx}{D} + \dfrac{v^2 t}{DR}\right)\mathrm{erfc}\left(\dfrac{Rx + vt}{(4RDt)^{1/2}}\right)$ $+ \left(\dfrac{v^2 t}{\pi RD}\right)^{1/2}\exp\left(\dfrac{(Rx - vt)^2}{4RDt}\right)$
A3	$C(0,t) = C_o$	$\dfrac{\partial C}{\partial x}(L, t) = 0$	$1 - \displaystyle\sum_{m=1}^{\infty} \dfrac{2\beta_m \sin\left(\dfrac{\beta_m x}{L}\right)\exp\left(\dfrac{vx}{2D} - \dfrac{v^2 t}{4DR} - \dfrac{\beta_m^2 Dt}{L^2 R}\right)}{\beta_m^2 + \left(\dfrac{vL}{2D}\right)^2 + \dfrac{vL}{2D}}$ $\beta_m \cot(\beta_m) + \dfrac{vL}{2D} = 0$
A4	$\left(vC - D\dfrac{\partial C}{\partial x}\right)_{x=0} = vC_o$	$\dfrac{\partial C}{\partial x}(L, t) = 0$	$1 - \displaystyle\sum_{m=1}^{\infty} \dfrac{\dfrac{2vL}{D}\beta_m\left[\beta_m \cos\left(\dfrac{\beta_m x}{L}\right) + \dfrac{vL}{2D}\sin\left(\dfrac{\beta_m x}{L}\right)\right]\exp\left(\dfrac{vx}{2D} - \dfrac{v^2 t}{4DR} - \dfrac{\beta_m^2 Dt}{L^2 R}\right)}{\left[\beta_m^2 + \left(\dfrac{vL}{2D}\right)^2 + \dfrac{vL}{2D}\right]\left[\beta_m^2 + \left(\dfrac{vL}{2D}\right)^2\right]}$ $\beta_m \cot(\beta_m) - \dfrac{\beta_m^2 D}{vL} + \dfrac{vL}{2D} = 0$

For a third-type inlet condition, the concentration at $X = 0$ just inside the soil is not equal to the influent concentration, even at time $T = 1$. Although the jump in concentration is physically odd, mass conservation is ensured. For the higher Peclet number ($P = 20$), deviations between a first- and a third-type inlet condition are significantly reduced. This is in accordance with Fig. 6.8, which shows a smaller error for increased v^2t/RD.

Large differences in the predicted concentration may occur if the solute front approaches the outlet at $X = 1$ or $x = L$. Calculated concentrations according to solutions A1, A2, A3, and A4 versus the Peclet number at the outlet for $T = 1$ are illustrated in Fig. 6.10. The greatest difference occurs for small Peclet numbers, namely, when hydrodynamic dispersion is relatively important. The nature of hydrodynamic dispersion suggests that C/C_o should be approximately 0.5 for $X = T = R = 1$, the average of the zero initial concentration and the influent concentration (C_o). Because of the effect of the boundary conditions, this only happens when the Peclet number exceeds 10 or more, depending on the type of solution. For the first-type inlet condition (A1, A3), C/C_o exceeds 0.5 at low Peclet numbers since a considerable amount of solute is forced to diffuse into the column to establish a constant inlet concentration.

Differences between calculated solute breakthrough curves because of boundary conditions are further depicted in Fig. 6.11 for three different Peclet numbers. Notice that for $P = 60$, the curves are almost indistinguishable, considering the margin of error of most solute displacement experiments. The results in Fig. 6.11 show that the choice of inlet and outlet conditions for determining parameters from breakthrough experiments becomes less important when P exceeds about 30.

Finally, for displacement experiments involving finite columns, it may be of interest to quantify the amount of solute that can be stored in the liquid and sorbed phases of the soil (Equation [6.24]). When, beginning at $t = 0$, a solution with concentration (C_o) is applied to a soil column, holdup can be defined as (Nauman and Buffham, 1983):

$$H = \frac{v}{L}\int_0^\infty \frac{C_o - C(L,t)}{C_o}dt \qquad [6.77]$$

This amounts to the integration of the complementary solute concentration versus dimensionless time, namely, the area above the breakthrough curves in Fig. 6.11. Van Genuchten and Parker (1984)

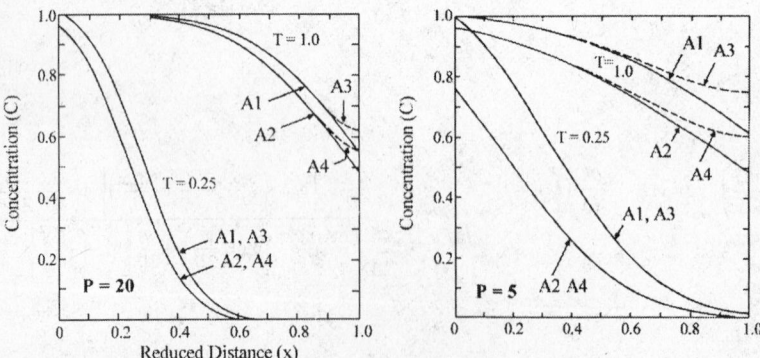

Fig. 6.9 Plot of the concentration (C/C_o) as a function of distance (X) calculated for four different combinations of boundary conditions according to the solutions in Table 6.2 at two different times for a Peclet number of 5 and 20 [After van Genuchten and Alves, 1982]

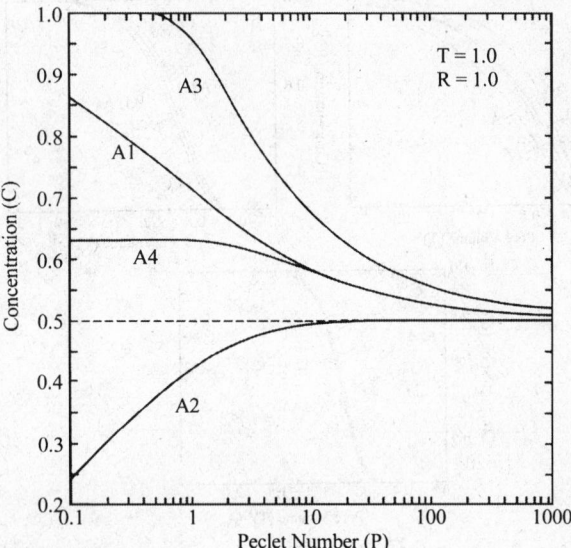

Fig. 6.10 Solute concentration predicted with solutions A1, A2, A3, and A4 as a function of the Peclet number at $T = 1$ and $X = 1$ [After van Genuchten and Alves, 1982]

showed that $H = R$ for solutions A1 and A4, $H = R[1+(1/P)]$ for solution A2 and $H = R\{1 - (1/P)[1 - \exp(-P)]\}$ for solution A3. In case of anion exclusion, the relative exclusion volume equals $1-R$ and the column holdup will be less than one.

6.3.2.4 Nonequilibrium Transport

Analytical solutions for one-dimensional bimodal nonequilibrium transport have been presented by, among others, Lindstrom and Narasimhan (1973), Lindstrom and Stone (1974), Lassey (1988), and Toride et al. (1993). The boundary value problem involving solute application with a constant concentration may be specified by Equations [6.39], [6.40], and [6.53] subject to the following conditions:

$$C_1(X,0) = C_2(X,0) = 0 \tag{6.78}$$

$$\left(C_1 - \frac{1}{P}\frac{\partial C_1}{\partial X}\right)_{X=0^+} = 1 \tag{6.79}$$

$$\frac{\partial C_1}{\partial X}(\infty, T) = 0 \tag{6.80}$$

Solutions for the equilibrium and nonequilibrium concentrations are

$$C_1 = \int_0^T J(a,b)G(X,\tau)d\tau \tag{6.81}$$

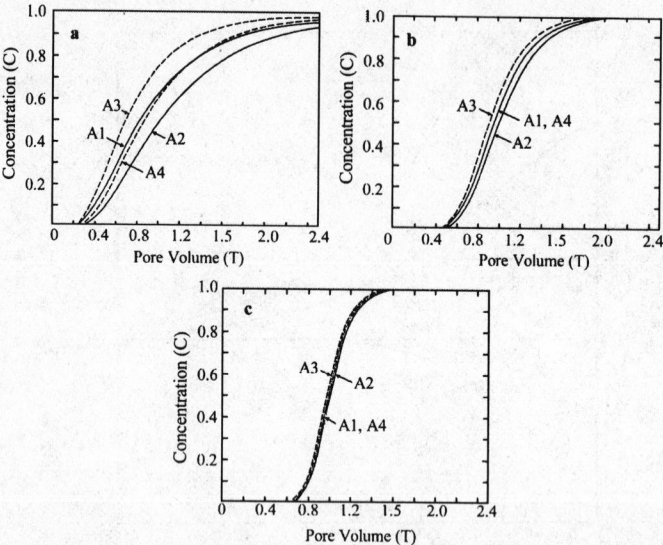

Fig. 6.11 Solute breakthrough curves predicted with the four analytical solutions of Table 6.2 for three different Peclet numbers [After van Genuchten and Alves, 1982]

$$C_2 = \int_0^T [1 - J(b,a)]G(X,\tau)d\tau \qquad\qquad [6.82]$$

where the auxiliary (equilibrium) function $G(X,\tau)$ for resident concentrations is defined as:

$$G(X,\tau) = \sqrt{\frac{P}{\pi\beta R\tau}}\exp\left(-\frac{(\beta RX-\tau)^2}{4\beta R\tau/P}\right) - \frac{p}{2\beta R}\,\mathrm{erfc}\left[\frac{\beta RX+\tau}{\sqrt{4\beta R\tau/P}}\right] \qquad\qquad [6.83]$$

and for flux-averaged concentrations by:

$$G(X,\tau) = \sqrt{\frac{\beta RPX^2}{4\pi\tau^3}}\exp\left(-\frac{(\beta RX-\tau)^2}{4\beta R\tau/P}\right) \qquad\qquad [6.84]$$

Furthermore, J denotes Goldstein's J function (Goldstein, 1953), which is defined as:

$$J(a,b) = 1 - \exp(-b)\int_0^a \exp(-x)I_0\left(2\sqrt{ab}\right)d\xi \qquad\qquad [6.85]$$

with I_0 as the zero order modified Bessel function. The variables a and b are given by

$$a = \frac{\omega\tau}{\beta R} \quad , \quad b = \frac{\omega(T-\tau)}{(1-\beta)R} \qquad\qquad [6.86]$$

The above solution for a flux-averaged concentration was used to describe breakthrough data for the pesticide 2,4,5-T (2,4,5-trichlorophenoxyacetic acid) as observed from a 30-cm long soil column containing aggregated (< 6 mm) Glendale clay loam (van Genuchten and Parker, 1987). The nonequilibrium model with three adjustable parameters (P, β, ω) provided an excellent description of the data (Fig. 6.12b). The retardation factor (R) was estimated independently, using the distribution coefficient obtained from batch experiments, according to Equation [6.23]. A one parameter ADE fit (using P as adjustable parameter) did not yield a good description of the data (Fig. 6.12a) while a similar two parameter (P, R) fit, which is not shown, gave results that were only marginally different from those shown in Fig. 6.12a.

As pointed out by van Genuchten and Dalton (1986), the main disadvantage of the first order physical nonequilibrium approach is the obscure dependency of the transfer coefficient (α or ω) on the actual diffusion process in the aggregate, particularly the value for the diffusion coefficient and the aggregate geometry. For well-defined structured or aggregated porous media (media for which the size and geometry of all aggregates are known), the diffusion process inside the aggregate can be modeled, which allows a more detailed description of the concentration inside the aggregate. Analytical solutions are available for several aggregate geometries. The simplified immobile concentration, which is used in the general nonequilibrium formulation, can always be obtained by averaging the more detailed solution over the aggregate volume.

6.3.2.5 Time Moments

Moments are frequently used to characterize statistical distributions such as those of solute particles (concentrations) versus time or positions. Analytical expressions for lower order moments are

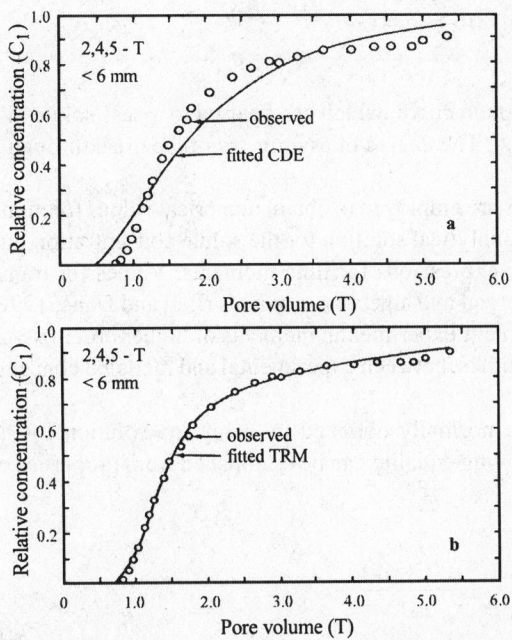

Fig. 6.12 Observed and fitted effluent curves for 2,4,5-T movement through Glendale clay loam. The fitted curves were based on: (a) the equilibrium ADE and (b) the nonequilibrium ADE.

sometimes derived for deterministic transport models, especially when a direct analytical solution may be difficult to obtain. Moment analysis is more widely employed in stochastic than deterministic transport models since the uncertainty in both model parameters and predicted results is most conveniently quantified with moments. Only time moments will here be considered.

The pth (time) moment of the breakthrough curve, as obtained, say, from effluent samples collected from a soil column (with length $x = L$) to which a solute pulse is applied during steady water flow, is defined as:

$$m_p(L) = \int_0^\infty t^p C(L,t)dt \qquad (p = 0,1,2,\ldots) \tag{6.87}$$

The zero moment is proportional to the total solute mass, the first moment quantifies the mean displacement, the second moment is indicative of the variance (dispersion), whereas the third moment quantifies the asymmetry or skewness of the breakthrough curve. Normalized moments (μ_p) are obtained as follows:

$$\mu_p = \frac{m_p}{m_0} \tag{6.88}$$

The mean breakthrough time is given by μ_1. Central moments are defined with respect to this mean according to:

$$\mu_p'(L) = \frac{1}{m_0} \int_0^\infty (t - \mu_1)^p C(L,t)dt \tag{6.89}$$

The variance of a breakthrough curve, which can be used to assess solute dispersion, is given by the second central moment (μ_2'). The degree of asymmetry of the breakthrough curve is indicated by its skewness ($\mu_3'/(\mu_2')^{3/2}$).

The previous definitions are employed to obtain numerical values for moments from experimental results. Substitution of an analytical solution for the solute concentration into the definitions allows the derivation of algebraic expressions for time moments. Values for transport parameters can be obtained by equating numerical and algebraic moments (Leij and Dane, 1992; Jacobsen et al., 1992). This procedure is not reliable if experimental moments of higher order ($p > 2$) are needed since even small deviations, at larger times, between experimental and modeled concentrations will greatly bias such moments.

Algebraic moments are normally obtained by using the solution for the concentration in the Laplace domain. The following equality can be established from properties of the Laplace transform (Spiegel, 1965):

$$m_p(x) = (-1)^p \lim_{s \to 0} \frac{d^p \overline{C}(x,s)}{ds^p} \tag{6.90}$$

where, as before, $\overline{C}(x,s)$ is the concentration in the Laplace domain, and s is the (complex) transformation variable. Expressions for moments can hence be obtained by differentiating the solution in the Laplace domain and letting the Laplace variable go to zero (Aris, 1958; van der Laan, 1958). This task is conveniently handled by mathematical software.

Time moments will now be considered for three different transport problems. First, the mathematical problem for physical nonequilibrium transport can be written as:

$$\theta_m R \frac{\partial C_m}{\partial t} + \theta_{im} R \frac{\partial C_{im}}{\partial t} = \theta_m D_m \frac{\partial^2 C_m}{\partial x^2} - \theta v \frac{\partial C_m}{\partial x}$$

[6.91]

$$\theta_{im} R \frac{\partial C_{im}}{\partial t} = \alpha (C_m - C_{im})$$

[6.92]

The conditions for instantaneous solute application to a soil are

$$C(x,0) = 0 \qquad 0 \le x < \infty$$

[6.93]

$$C(0,t) = \frac{m_0}{v} \delta(t)$$

[6.94]

$$\frac{\partial C}{\partial x}(\infty, t) = 0$$

[6.95]

where m_0 is the solute mass that is applied per unit area of soil solution at $t = 0$, and $\delta(t)$ is the Dirac delta function (T^{-1}). The first-type inlet condition is used to describe flux-averaged concentrations such as effluent samples from column displacement experiments. Second, the equilibrium problem is defined by the same set of equations by setting $\theta_m = \theta$, $\theta_{im} = 0$, and $\alpha \to 0$. Third, the chemical nonequilibrium transport equations are as follows:

$$\frac{\partial C}{\partial t} + \frac{\rho_b}{\theta} \frac{\partial S}{\partial t} = D \frac{\partial^2 C}{\partial x^2} - v \frac{\partial C}{\partial x}$$

[6.96]

$$\frac{\partial S}{\partial t} = \alpha (K_d C - S)$$

[6.97]

These equations are also subject to boundary and initial conditions in Equations [6.93] through [6.95].

Formulas for the mean breakthrough time (μ_l) and the variance (μ_2') of the breakthrough curve predicted for these three models are presented in Table 6.3. The expressions suggest that nonequilibrium conditions do not affect the mean travel time but they do increase solute spreading. Since only the solution in the Laplace domain is needed, moment analysis is particularly useful for more complex transport problems to study the general behavior of the breakthrough curve.

6.3.3 Numerical Solutions

The solution of many practical transport problems requires the use of numerical methods because of changes in water saturation (as the result of irrigation, evaporation, and drainage), spatial and temporal variability of soil properties, or complicated boundary and initial conditions. Numerical methods are based on a discretization of the spatial and temporal solution domain, and subsequent calculation of the concentration at discrete nodes in the domain. This approach is in contrast with analytical methods, which offer a continuous description of the concentration. In some cases, a combination of analytical and numerical techniques may be employed (Sudicky, 1989; Moridis and Reddell, 1991; Li et al., 1992).

6.3.3.1 Introduction

The accuracy of the numerical results depends on the input parameters, the approximation of the governing partial differential equation, the discretization, and implementation of the numerical solution in a computer code solving the simulated problem. Numerous texts exist on the numerical modeling of flow and transport in porous media (Pinder and Gray, 1977; Huyakorn and Pinder, 1983; Campbell, 1985; van der Heijde et al., 1985; Istok, 1989).

The many numerical methods for solving the ADE can be classified into three groups (Neuman, 1984): (1) Eulerian, (2) Lagrangian, and (3) mixed Lagrangian-Eulerian. In the Eulerian approach, the transport equation is discretized by the method of finite differences or finite elements using a fixed mesh. For the Lagrangian approach, the mesh deforms along with the flow while it is stable in a moving coordinate system. A two-step procedure is followed for a mixed approach. First, advective transport is solved using a Lagrangian approach and concentrations are obtained from particle trajectories. Subsequently, all other processes including sinks and sources are modeled with an Eulerian approach using finite elements, finite differences, etc.

The method of finite differences (Bresler and Hanks, 1969; Bresler, 1973) and the Galerkin method of finite elements (Gray and Pinder, 1976; van Genuchten, 1978) belong to the first group as do the previously mentioned combination of analytical and numerical techniques. The finite difference and finite element methods were the first numerical methods used for solute transport problems and, in spite of their problems discussed below, are still the most often utilized methods. Numerical experiments have shown that both methods give very good results if significant dispersion exists as quantified with the Peclet number. If advection is dominant, however, numerical oscillations may

Table 6.3 Mean breakthrough time (μ_1) and variance (μ_2') for the equilibrium and nonequilibrium solution of the ADE at a distance x from the inlet as a result of a Dirac delta input described with a first-type inlet condition (i.e., flux-averaged concentration).

ADE model	Mean breakthrough time, μ_1	Variance μ_2'
Equilibrium	$\dfrac{Rx}{v}$	$\dfrac{2DR^2x}{v^3}$
Nonequilibrium		
Two-Region	$\dfrac{Rx}{v}$	$\dfrac{2\theta_m D_m R^2 x}{\theta v^3} + \dfrac{2\theta(1-\beta)^2 R^2 x}{\alpha v}$
Two-Site	$\dfrac{Rx}{v}$	$\dfrac{2DR^2x}{v^3} + \dfrac{2(1-\beta)Rx}{\alpha v}$

occur for both methods and small spatial increments should be used. It may not always be possible to decrease the spatial step size due to the associated increase in computations and a variety of approaches have been developed to overcome the oscillations (Chaudhari, 1971; van Genuchten and Gray, 1978; Donea, 1991).

Lagrangian solution methods will result in very few numerical oscillations (Varoglu and Finn, 1982). However, Lagrangian methods, which are based on the method of characteristics, suffer from inherent diffusion and do not conserve mass. They are difficult to implement for two- and three-dimensional problems. Instabilities resulting from inadequate spatial discretization may occur during longer simulations due to deformation of the stream function, especially if the solute is subject to sorption, precipitation, and other reactions.

The mixed approach has been applied by several authors (Konikov and Bredehoeft, 1978; Molz et al., 1986; Yeh, 1990). In view of the different behavior of the diffusive (parabolic) and advective (hyperbolic) terms of the ADE, the problem is decomposed into an advection and a diffusion problem. Advective transport is solved with the Lagrangian approach while all other terms are solved with the Eulerian approach. The trajectories of flowing particles are obtained using continuous forward particle tracking (to follow a set of particles as they move through the flow domain), single-step reverse particle tracking (the initial position of particles arriving at nodal points was calculated for each time step), and a combination of both approaches. Ahlstrom et al. (1977), among others, used the attractive random walk model to describe the movement of individual solute particles by viewing the travel distance for a particular time step as the sum of a deterministic and a stochastic velocity component.

In the following, only a brief introduction to the finite difference and finite element methods to solve transport problems is provided. Both methods encompass a wide variety of more specific numerical approaches. As a rule of thumb, the finite difference method is attractive because of its simplicity and the availability, at least early on, of handbooks and computer programs simulating flow and transport in porous media. On the other hand, the finite element method has proven to be more suitable for problems involving irregular geometries of the flow and transport problem, such as situations involving flow to drainage pipes and along sloping soil surfaces.

6.3.3.2   Finite Difference Methods

For one-dimensional transient solute transport the dependent variable $[C(x,t)]$ can be discretized according to

$$C(x,t) = C(i\Delta x, j\Delta t) = C_{ij} \qquad (i = 0,1,2,....,n; j = 0,1,2,....,m) \qquad [6.98]$$

Consider the simulation of the one-dimensional ADE for steady flow of a conservative tracer as given by Equation [6.63] with $R = 1$. Temporal and spatial derivatives are approximated with Taylor series expansions. Assume that the concentrations are known at the current time $(t = j\Delta t)$ and that the objective is to calculate the concentration distribution $\{C[i\Delta x,(j+1)t]\}$ at the next time. A forward-in-time finite difference scheme where the unknown concentration is given explicitly in terms of known concentrations, is written as

$$\frac{C_{i,j+1} - C_{i,j}}{\Delta t} \approx D \frac{C_{i+1,j} - 2C_{i,j} + C_{i-1,j}}{(\Delta x)^2} - v \frac{C_{i+1,j} - C_{i-1,j}}{2\Delta x} \qquad [6.99]$$

On the other hand, a backward-in-time or implicit scheme is given by:

$$\frac{C_{i,j+1} - C_{i,j}}{\Delta t} \approx D \frac{C_{i+1,j+1} - 2C_{i,j+1} + C_{i-1,j+1}}{(\Delta x)^2} - v \frac{C_{i+1,j+1} - C_{i-1,j+1}}{2\Delta x}$$

[6.100]

which contains several concentrations at the next time level. The problem is solved implicitly after combining the difference schemes of all nodes, and subsequently using a matrix equation solver. Finally, a weighted scheme can be defined as:

$$\frac{C_{i,j+1} - C_{i,j}}{\Delta t} \approx D \frac{\omega\left(C_{i+1,j+1} - 2C_{i,j+1} + C_{i-1,j+1}\right) + (1-\omega)\left(C_{i+1,j} - 2C_{i,j} + C_{i-1,j}\right)}{(\Delta x)^2}$$

$$-v \frac{\omega\left(C_{i+1,j+1} - C_{i-1,j+1}\right) + (1-\omega)\left(C_{i+1,j} - C_{i-1,j}\right)}{2\Delta x}$$

[6.101]

in which ω is a weighting constant between 0 and 1. The scheme is said to be fully explicit for $\omega = 0$ and fully implicit for $\omega = 1$, while a Crank-Nicolson central-in-time scheme is derived if $\omega = 0.5$. For transient flow, the scheme becomes more complicated; the velocity (v) and water content (θ) are usually obtained by solving the Richards equation prior to solving the transport problem.

Errors associated with the discretization and the solution procedure can be evaluated by comparison to analytical results, provided that the problem can be sufficiently idealized (linearized) to permit the use of such solutions. This is true for other numerical solutions as well. For a convergent scheme, the difference between the numerical and analytical solutions should decrease if smaller space and time steps are used in the numerical solution; the difference should become zero if $\Delta x \to 0$ and $\Delta t \to 0$, barring round off and computational errors.

An even more important question relates to the stability of the finite difference approximation, namely, the degree to which the numerical solution is affected by errors that occur during the simulation. Such errors can usually not be eliminated completely; they depend on the implemented discretization, the values of the input parameters, and the type of numerical scheme used for approximation of the governing transport equation. Errors are damped during the course of a simulation when a stable scheme is used, while unstable schemes allow such errors to grow unboundedly.

Implicit systems are unconditionally stable, but their results may not be as amenable to changes in grid sizes as is the case with explicit systems. The Crank-Nicolson method offers an attractive compromise of being unconditionally stable and having a truncation error of order $O[(\Delta x)^2 + (\Delta t)^2]$. These properties imply, among other things, that a variable time step can be used independently of the spatial step to effectively balance the needs of an accurate approximation and a limited number of computations. However, oscillations near the concentration front may develop even for unconditionally stable methods due to the hyperbolic (convection) term in the solute transport equation. To avoid these oscillations, stability criteria in terms of the grid Peclet (P or Pe) and Courant (Cr) numbers are frequently formulated. Huyakorn and Pinder (1983) provided the following conservative guidelines for one-dimensional transport of a nonreactive solute:

$$Pe = v\Delta x / D < 2$$

[6.102]

$$Cr = v\Delta t / \Delta x < 1 \qquad\qquad [6.103]$$

For transport of a reactive solute, the retardation factor R must be included in the denominator of the Courant number.

6.3.3.3 Finite Element Methods

The application of the finite element methods for modeling solute transport in soils involves several steps, which will be briefly reviewed in a qualitative manner. The specifics of the method can be found in, among others, Huyakorn and Pinder (1983) and Istok (1989). The (spatial) solution domain should first be discretized. The finite element mesh, consisting of nodal points marking the elements, is usually tailored to the problem at hand. The selection of mesh size is a somewhat subjective process that considers the required degree of accuracy, the geometry of the problem, the ease of mesh generation, and the mathematical complexity associated with the use of a particular mesh. The shape of the finite elements is determined by the dimensionality and the geometry of the transport problem. One-dimensional elements consist of lines between nodal points along the coordinate where (one-dimensional) transport occurs. Examples of two-dimensional elements are triangles, rectangles, and parallelograms, while many different three-dimensional elements can be constructed.

A second step in implementing the finite element method is the description of the transport problem for each element. Generally, the method of weighted residuals is used to arrive at an integral formulation for the governing transport equation. An approximate or trial solution for the concentration in each element is formulated as the weighted sum of the unknown concentration at the nodes of that element. A wide variety of so-called basis functions can be selected to assign weights to the nodes; the functions depend on the type of element being selected. The approximation for the element concentration will not be exact; substitution of the element concentration in the ADE yields an expression for this error (residual). The objective is to minimize the residual over the entire domain with some kind of weighting procedure for the elements. The method of weighted residuals forces the weighted average of the residuals at the nodes to be zero and gives an expression for the residual of each element. After applying Green's theorem for each element, the residual can be written in matrix form.

The third step is the assembly of all element matrices into a global matrix system that contains the nodal concentrations and its temporal derivative as unknown variables.

The last step involves the solution of this global matrix equation. Since there are, in general, no obvious advantages for also using a finite element approximation for the temporal derivative, the latter is normally dealt with using the (simpler) finite difference method. The time step may be constant or variable. As noted earlier, the explicit or forward scheme ($\omega = 0$) is conditionally stable whereas the implicit or backward ($\omega = 1$) and the centered or Crank-Nicolson ($\omega = 0.5$) schemes are uncon-ditionally stable. The nodal concentration can then be solved using standard numerical procedures.

6.3.3.4 Application

Numerical oscillations and dispersion are illustrated for several numerical schemes using results by Huang et al. (1997). The numerical solution of the ADE is particularly difficult for relatively sharp concentration fronts with advection-dominated transport characterized by small dispersivities. As mentioned earlier, undesired oscillations can often be prevented with judicious space and time discretizations. The Peclet number increases when advection dominates dispersion; the potentially adverse effect on the numerical solution can be compensated by selecting a smaller grid size. Numerical oscillations can be virtually eliminated if the local Peclet number is always less than 2.

However, acceptable results may still be obtained for local Peclet numbers as high as 10 (Huyakorn and Pinder, 1983). The time discretization is based on a second dimensionless number, the Courant number (Equation [6.103]).

Consider one-dimensional, steady-state flow in a soil column with $v = 4$ cm d^{-1}. The column was initially free of any solute, a nonreactive solute ($C_o = 1$) was subsequently applied for 20 days. For negligible molecular diffusion and a longitudinal dispersivity ($\alpha_L = 0.02$ cm), the grid Peclet for a spatial discretization ($\Delta x = 2$ cm) is given by:

$$Pe = \frac{v\Delta x}{D} = \frac{\Delta x}{\alpha_L} = 100 \qquad\qquad [6.104]$$

This grid Peclet number of 100 indicates solute transport that is dominated by advection. In all calculations, the Courant number was less than 1, which is the stability condition for Eulerian methods.

The concentration predicted with four different finite element methods and with the analytical solution is shown in Fig. 6.13. The first numerical method is based on the central Crank-Nicolson scheme for time with weighting constant ($\omega = 0.5$). Second, the analytical solution is shown as a solid line. Third, an implicit scheme ($\omega = 1.0$) is used. The last two numerical methods implement upstream weighting (Huyakorn and Pinder, 1983) in the central and the implicit schemes, respectively. The results obtained with the Crank-Nicolson method have significant numerical oscillations (both overshoot and undershoot) (Fig. 6.13). This is not surprising given the large value for Pe. The oscillations are reduced by using an implicit scheme. Upstream weighting leads to very similar results for the central and implicit schemes. The oscillations are virtually eliminated but the solute profile exhibits more numerical dispersion; there is a greater discrepancy with the steep analytical profile than for the regular implicit scheme.

6.4   Stream Tube Models

Considerable errors may be made when applying deterministic methods to the field since model parameters are actually stochastic due to spatial and temporal variability, measurement errors, and different averaging scales. Usually, it is not possible or important to obtain discrete values for parameters in deterministic models of field-scale transport. Instead, transport properties and model results are described with statistical functions. Stochastic modeling is no substitute for data collection or model development, but merely a method to deal with uncertainty of model parameters and complexity of flow and transport processes. The scale at which solute movement is observed or modeled is important. The averaging process associated with larger field-scale descriptions tends to filter out the variability at smaller scales.

The stochastic modeling of actual field problems is seldom possible. The stream tube model provides a possible exception (Dagan and Bresler, 1979; Amoozegar-Fard et al., 1982; Rubin and Or, 1993). The field is conceptualized as a system of parallel tubes illustrated in Fig. 6.14. A process-based model is used to describe a one-dimensional, autonomous transport in each tube as a function of time and depth. Transport parameters are either deterministic or stochastic. The problem may be solved analytically at the scale of the tube and the field. The stream tube model is suitable for inverse procedures to estimate transport and statistical properties (Toride et al., 1995).

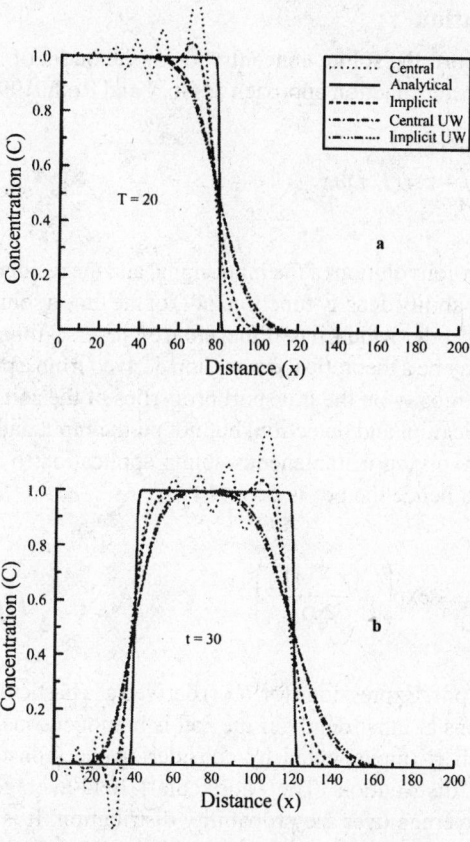

Fig. 6.13 Concentration profiles predicted with different numerical schemes for (a) solute infiltration ($t = 30$ d) and (b) solute leaching ($t = 30$ d)

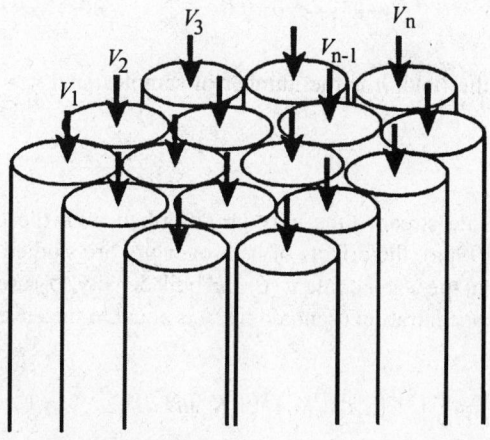

Fig. 6.14 Schematic of the stream tube model

6.4.1 Model Formulation

For one-dimensional transport, the solute concentration at the outlet of a stream tube of length (L) may be written with the transfer function approach as (Jury and Roth, 1990):

$$C(L,t) = \int_0^t C(0, t - \tau) f(L, \tau) d\tau \qquad [6.105]$$

The outlet concentration is a convolution of the input signal and the residence time distribution $f(L,\tau)$. The latter is, in effect, a probability density function (pdf) of the time a solute particle resides in the soil between $x = 0$ and L. The pdf, which has dimension of inverse time, can be determined from experimental results or it may be a theoretical expression derived from a process-based model such as the ADE. The pdf depends mostly on the transport properties of the soil, it depends to some degree on the mode of solute application and detection, but not on the input and initial concentrations. For the equilibrium problem involving instantaneous solute application to a solute-free soil, the flux-averaged concentration and hence the pdf is given by:

$$f(x,t) = \frac{x}{\sqrt{4\pi D t^3}} \exp\left(-\frac{(x - vt)^2}{4Dt}\right) \qquad [6.106]$$

Note that this is a Gaussian pdf. Expressions for $f(x,t)$ derived at a particular depth can be readily used for predicting concentrations at other depths if the soil is homogeneous.

The stochastic approach is implemented by considering the (constant) column parameters as realizations of a random distribution. The (horizontal) field-averaged solute concentration is considered the ensemble average over the probability distribution. It is assumed that each random parameter obeys a distribution function that is independent of location and that the ensemble of possible concentrations may be estimated from a sufficient number of samples taken at different locations. The average concentration across the field is then identical to the ensemble average:

$$\langle C(x,t) \rangle = \frac{1}{A} \int_A C(x,t) dA = \lim_{n \to \infty} \frac{1}{n} \sum_{i=1}^n C_i(x,t) \qquad [6.107]$$

where A denotes the area of the field, n is the number of samples, and $< >$ indicates an ensemble average.

6.4.2 Application

Solute transport in a local-scale stream tube may be described with the one-dimensional ADE. Following Toride and Leij (1996a), the effects of heterogeneity are studied using pairs of random parameters (v and K_d). Note that the water content (θ) and bulk density (ρ_b) are the same for all stream tubes. The field-scale mean concentration (denoted by ^) is equal to the ensemble average:

$$\hat{C}(x,t) = \langle C(x,t) \rangle = \int_0^\infty \int_0^\infty C(x,t; v, K_d) f(v, K_d) dv \, dK_d \qquad [6.108]$$

where the bivariate lognormal joint probability density function $f(v, K_d)$ is given by:

$$f(v,K_d) = \frac{1}{2\pi\sigma_v\sigma_{Kd}vK_d\sqrt{1-\rho_{vKd}^2}}\ \exp\left(-\frac{Y_v^2 - 2\rho_{vKd}Y_vY_{Kd} + Y_{Kd}^2}{2(1-\rho_{vKd}^2)}\right)$$ [6.109]

with

$$\rho_{vKd} = \langle Y_vY_{Kd}\rangle = \int_0^\infty \int_0^\infty Y_vY_{Kd}f(v,K_d)dv\ dK_d$$ [6.110]

$$Y_v = \frac{\ln(v)-\mu_v}{\sigma_v}, \qquad Y_{Kd} = \frac{\ln(K_d)-\mu_{Kd}}{\sigma_{Kd}}$$ [6.111]

where μ and σ are the mean and standard deviation of the log transformed variable, and ρ_{vKd} is the correlation coefficient between Y_v and Y_{Kd} (K_d tends to increase with v for positive ρ_{vKd} while K_d decreases with v for negative ρ_{vKd}). Ensemble averages of v and K_d are given by (Aitcheson and Brown, 1963)

$$\langle v\rangle = \exp\left(\mu_v + \frac{1}{2}\sigma_v^2\right), \qquad \langle K_d\rangle = \exp\left(\mu_{Kd} + \frac{1}{2}\sigma_{Kd}^2\right)$$ [6.112]

with coefficients of variation (CV):

$$CV(v) = \sqrt{\exp(\sigma_v^2)-1}, \qquad CV(K_d) = \sqrt{\exp(\sigma_{Kd}^2)-1}$$ [6.113]

Based upon the detection mode of the local-scale concentration, three types of field-scale concentrations can be defined: (1) the ensemble average of the flux-averaged concentration ($<C_F>$); (2) the field-scale resident concentration ($\hat{C}_R$), which is equal to the ensemble average of the resident concentration ($<C_R>$); and (3) the field-scale flux-averaged concentration ($\hat{C}_F$) which is defined as $<vC_F>/<v>$. The second type of concentration is obtained from averaging values of the resident concentration at a particular depth across the field. The third type is defined as the ratio of ensemble solute and water fluxes in a similar manner as for deterministic transport. It should be noted that $<vC_F>/<v> \neq <C_F>$ because v is a stochastic variable.

The use of the stream tube model for field-scale transport is illustrated in Fig. 6.15. The local-scale concentration only depends on the particular realizations of the two stochastic parameters (v and K_d) after the independent variables (t and x) have been specified. The solution for the ADE at $x = 100$ cm and $t = 5$ day as a function of v and K_d is shown in Fig. 6.15a and the bivariate lognormal pdf for v and K_d in Fig. 6.15b; the distribution is skewed with respect to v since σ_v is fairly high, the smaller σ_{Kd} results in a more symmetric distribution for K_d. The negative ρ_{vKd} results in an increasing v with a decreasing K_d, and *vice versa*. The expected concentration is shown in Fig. 6.15c, which is obtained by weighting the local concentration (Fig. 6.15a) by multiplying it with the joint pdf (Fig. 6.16b). The peak in Fig. 6.15c suggests that stream tubes with approximately $v = 25$ cm d^{-1} and $K_d = 1$ cm^3 g^{-1} contribute the most to solute breakthrough when $x = 100$ cm and $t = 5$ days. The volume of the distribution in Fig. 6.15c corresponds to the ensemble average ($<C_R>$).

Variations in the local-scale concentration between stream tubes, at a particular depth and time, can be characterized by its variance. The variance across the horizontal plane is given by (Bresler and Dagan, 1981; Toride and Leij, 1996b):

$$Var\big[C(x,t)\big] = \int_0^\infty \int_0^\infty \big[C(x,t)-\langle C(x,t)\rangle\big]^2 f(v,K_d)dv\,dK_d = \langle C^2(x,t)\rangle - \langle C(x,t)\rangle^2 \qquad [6.114]$$

For a deterministic distribution coefficient (K_d) Equation [6.108] reduces to:

$$\langle C(x,t)\rangle = \int_0^\infty C(x,t;v)f(v)dv \qquad\qquad\qquad\qquad\qquad [6.115]$$

where the lognormal pdf for the single stochastic variable (v) is given by:

$$f(v) = \frac{1}{\sqrt{2\pi}\sigma_v v}\exp\left(-\frac{\big[\ln(v)-\mu_v\big]^2}{2\sigma_v^2}\right) \qquad\qquad\qquad [6.116]$$

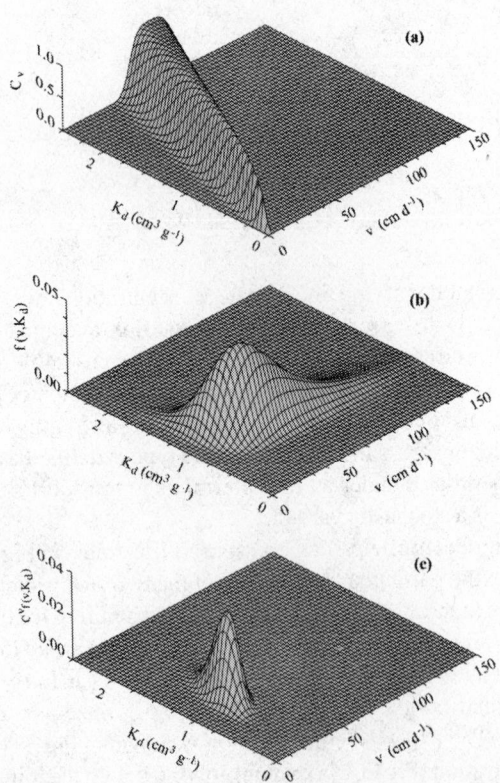

Fig. 6.15 Illustration of the stream tube model for field-scale transport for a 2-d solute application with $D = 20$ cm^2 d^{-1} and $\rho_b/\theta = 4$ g cm^{-3}: (a) the local-scale concentration, C_R, as a function of v and K_d at $x = 100$ cm and $t = 5$ days; (b) a bivariate lognormal pdf for $\rho_{vKd} = 0.5$, $<v> = 50$ cm d^{-1}, $\sigma_v = 0.2$, $<K_d> = 1$ cm^3 g^{-1}, and $\sigma_{Kd} = 0.2$; and (c) the expected concentration, C^R, at $x = 100$ cm and $t = 5$ days.

The mean (Fig. 6.16a) ($\hat{C}_R = <C_R>$) and the variance (Fig. 6.16b) are shown according to Equation [6.114], as a function of depth for three values of σ_v at $t = 3$ day for a nonreactive solute. More solute spreading occurs in the $\hat{C}_R$-profile when σ_v increases (Figure 6.16a). The variation in the local-scale C_R also increases with σ_v, indicating a more heterogeneous solute distribution in the horizontal plane (Fig. 6.16b). Because flow and transport become more heterogeneous as σ_v increases, more observations are needed to estimate $\hat{C}_R$ for $\sigma_v = 0.5$ than for $\sigma_v = 0.1$. The double peak in the variance profiles of Fig. 6.16b is due to the relative minimum for this pulse application at around $x = 30$ cm, where the highest concentration occurs.

For reactive solutes, the variability in the distribution coefficient (K_d) must also be considered. The field-scale resident concentration ($\hat{C}_R$) at $t = 5$ day is plotted versus depth in Fig. 6.17 for perfect and no correlation between v and K_d. The figure shows that the negative correlation between v and K_d causes additional spreading. Such a negative correlation seems plausible since coarse-textured soils generally have a relatively high v, due to their higher hydraulic conductivity, and a small K_d, due to their lower exchange capacity. The opposite may hold for fine-textured soils. Robin et al. (1991) reported a weak negative correlation between K_d and the saturated hydraulic conductivity for a sandy aquifer.

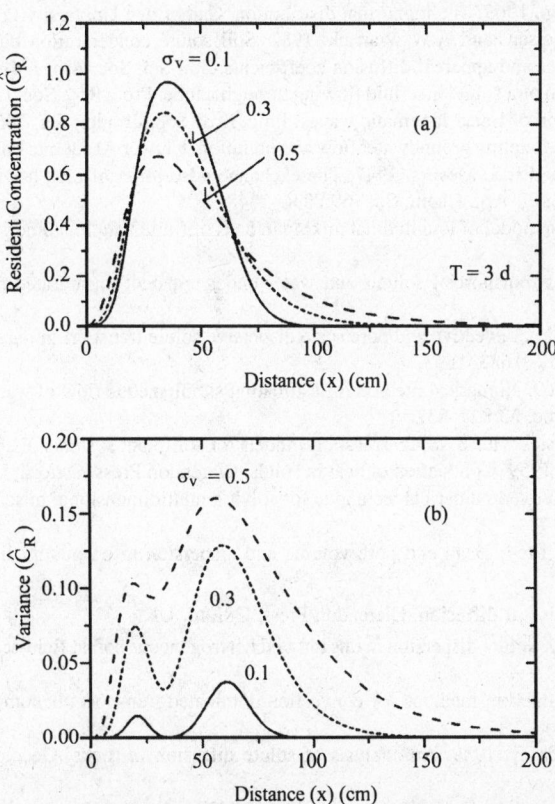

Fig. 6.16 The effect of variability in the pore-water velocity, v, on (a) the field scale resident concentration profile, and (b) the distribution of the variance for $\hat{C}_R$ in the horizontal plane; for a 2-d application of a nonreactive solute ($R = 1$) assuming $<v> = 20$ cm d^{-1} and $D = 20$ cm^2 d^{-1}

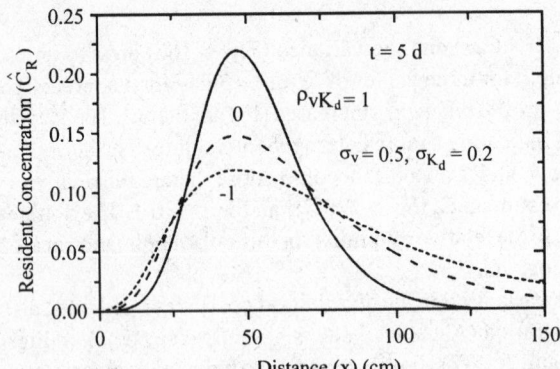

Fig. 6.17   The field-scale resident concentration profile, $\hat{C}_R(x)$, for three different correlations between v and K_d with $<v> = 50$ cm d^{-1}, $D = 20$ cm^{2}d^{-1}, $<K_d> = 1$ g^{-1}cm^3, $\sigma_{Kd} = 0.2$, $<R> = 5$, and $\rho_b/\theta = 4$ g cm^{-3}

6.5   References

Ahlstrom, S.W., H.P. Foote, R.C. Arnett, C.R. Cole, and R.J. Serne. 1977. Multicomponent mass transport model theory and numerical implementation. Battelle Pacific Northwest Laboratory Report BNWL-2127. Richland, WA.

Aitcheson, J., and J.A.C. Brown. 1963. The lognormal distribution. Cambridge University Press. London, UK.

Amoozegar-Fard, A., D.R. Nielsen, and A.W. Warrick. 1982. Soil solute concentration distribution for spatially varying pore water velocities and apparent diffusion coefficients. Soil Sci. Soc. Am. J. 46:3–9.

Aris, R. 1956. On the dispersion of a solute in a fluid flowing through a tube. Proc. Roy. Soc. London Ser. 235:67–77.

Aris, R. 1958. On the dispersion of linear kinematic waves. Proc. Roy. Soc. London Ser. 245:268–277.

Bear, J., and A. Verruijt. 1987. Modeling groundwater flow and pollution. Kluwer Academic Publishers, Norwell, MA.

Boyd, G.E., A.W. Adamson, and L.S. Meyers. 1947. The exchange adsorption of ions from aqueous solutions by organic zeolites. II. Kinetics. J. Am. Chem. Soc. 69:2836–2848.

Brenner, H. 1962. The diffusion model of longitudinal mixing in beds of finite length. Numerical values. Chem. Eng. Sci. 17:229–243.

Bresler, E. 1973. Simultaneous transport of solutes and water under transient unsaturated flow conditions. Water Resour. Res. 9:975–986.

Bresler, E., and G. Dagan. 1981. Convective and pore scale dispersive solute transport in unsaturated heterogeneous fields. Water Resour. Res. 17: 1683–1693.

Bresler, E., and R.J. Hanks. 1969. Numerical method for estimating simultaneous flow of water and salt in saturated soils. Soil Sci. Soc. Am. Proc. 33:827–832.

Campbell, G.S. 1985. Soil physics with BASIC: Transport models for soil-plant systems. Elsevier, New York, NY.

Carslaw, H.S., and J.C. Jaeger. 1959. Conduction of heat in solids. Clarendon Press, Oxford, UK.

Chaudhari, N.M. 1971. An improved numerical technique for solving multidimensional miscible displacement. Soc. Pet. Eng. J. 11:277–284.

Coats, K.H., and B.D. Smith. 1964. Dead end pore volume and dispersion in porous media. Soc. Petrol. Eng. J. 4:73–84.

Crank, J. 1975. The mathematics of diffusion. Clarendon Press, Oxford, UK.

Dagan, G., and E. Bresler. 1979. Solute dispersion in unsaturated heterogeneous soil at field-scale. I. Theory. Soil Sci. Soc. Amer. J. 43:461–467.

Donea, J. 1991. Generalized Galerkin methods for convection dominated transport phenomena. Appl. Mech. Rev. 44:205–214.

Dykhuizen, R.C., and W.H. Casey. 1989. An analysis of solute diffusion in rocks. Geochim. Cosmochim. Acta 53:2797–2805.

Epstein, N. 1989. On tortuosity and the tortuosity factor in flow and diffusion through porous media. Chem. Eng. Sci. 44:777–780.

Fischer, H.B., E. List, R.C.Y. Koh, J. Imberger, and N.H. Brooks. 1979. Mixing in inland and coastal waters. Academic Press, New York, NY.

Fried, J.J., and M.A. Combarnous. 1971. Dispersion in porous media. Adv. Hydrosci. 9:169–282.

Gautschi, W. 1964. Error function and Fresnel integrals. p. 295–329. *In* M. Abramowitz and I.A. Stegun handbook of mathematical functions. Appl. Math. Ser. 55. National Bureau of Standards, Washington, DC.

Gelhar, L.W., C. Welthy, and K.R. Rehfeldt. 1992. A critical review of data on field-scale dispersion in aquifers. Water Resour. Res. 28:1955–1974.

Goldstein, S. 1953. On the mathematics of exchange processes in fixed columns. I. Mathematical solutions and asymptotic expansions. Proc. Roy. Soc. London Ser. A. 219:151–185.

Gray, W.G., and G.F. Pinder. 1976. An analysis of the numerical solution of the transport equation. Water Resour. Res. 12:547–555.

Hamaker, J.W. 1972. Diffusion and volatilization, p. 341–397. *In* C.A.I. Goring and J.W. Hamaker (ed.) Organic chemicals in the soil environment. Marcel Dekker, Inc., New York, NY.

Harvie, C.E., and J.H. Weare. 1980. The prediction of mineral solubilities in natural waters: The Na-K-Mg-Ca-Cl-SO_4-H_2O system from zero to high concentration at 25 C. Geochim. Cosmochim. Acta 44:981–997.

Helfferich, F. 1962. Ion exchange. McGraw-Hill. New York, NY.

Huang, K., and M. Th. van Genuchten. 1995. An analytical solution for predicting solute transport during ponded infiltration. Soil Sci. 159:217–223.

Huang, K., J. Šimunek, and M. Th. van Genuchten. 1997. A third-order numerical scheme with upwind weighting for modeling solute transport. Int. J. Numer. Meth. Eng. 40:1623–1637.

Huyakorn, P.S., and G.F. Pinder. 1983. Computational methods in subsurface flow. Academic Press, New York, NY.

Istok, J. 1989. Groundwater modeling by the finite element method. American Geophysical Union, Washington, DC.

Jacobsen, O.H., F.J. Leij, and M. Th. van Genuchten. 1992. Parameter determination for chloride and tritium transport in undisturbed lysimeters during steady flow. Nordic Hydrol. 23:89–104.

Jin, Y., and W.A. Jury. 1996. Characterizing the dependence of gas diffusion coefficient on soil properties. Soil Sci. Soc. Am. J. 60:66–71.

Jury, W.A., and K. Roth. 1990. Transfer functions and solute movement through soil. Theory and applications. Birkhäuser Verlag, Basel, Switzerland.

Kemper, W.D. 1986. Solute Diffusivity. p. 1007–1024. *In* A. Klute (ed.) Methods of soil analysis. I. Physical and mineralogical methods. Soil Science Society of America, Madison, WI.

Klotz, D., K.P. Seiler, H. Moser, and F. Neumaier. 1980. Dispersivity and velocity relationship from laboratory and field experiments. J. Hydrol. 45:169–184.

Konikov, L.F.,and J.D. Bredehoeft. 1978. Computer model of two-dimensional solute transport and dispersion in groundwater. *In* Techniques of water resources investigation, Book 7. US Geological Survey, Reston, VA.

Kreft, A., and A. Zuber. 1978. On the physical meaning of the dispersion equations and its solutions for different initial and boundary conditions. Chem. Eng. Sci. 33:1471–1480.

Kreft, A., and A. Zuber. 1986. Comments on "Flux-averaged and volume-averaged concentrations in continuum approaches to solute transport" by J. C. Parker and M. Th. van Genuchten. Water Resour. Res. 22:1157–1158.

Krupp, H.K., J.W. Biggar, and D.R. Nielsen. 1972. Relative flow rates of salt and water in soil. Soil Sci. Soc. Amer. Proc. 36:412–417.

Lassey, K.R. 1988. Unidimensional solute transport incorporating equilibrium and rate-limited isotherms with first-order loss. 1. Model conceptualizations and analytic solutions. Water Resour. Res. 3:343–350.

Leij, F.J., and J.H. Dane. 1992. Moment method applied to solute transport with binary and ternary exchange. Soil Sci. Soc. Am. J. 56:667–674.

Leij, F.J., and N. Toride. 1995. Discrete time- and length-averaged solutions of the advection-dispersion equation. Water Resour. Res. 31:1713–1724.

Leij, F.J., and M. Th. van Genuchten. 1995. Approximate analytical solutions for transport in two-layer porous media. Trans. Porous Media 18:65–85.

Li, S-G, F. Ruan, and D. McLaughlin. 1992. A space-time accurate method for solving solute transport problems. Water Resour. Res. 28:2297–2306.

Lide, D.R. 1995. Handbook of chemistry and physics. CRC Press, Boca Raton, FL.

Lindstrom, F.T., and M.N.L. Narasimhan. 1973. Mathematical theory of a kinetic model for dispersion of previously distributed chemicals in a sorbing porous medium. SIAM J. Appl. Math. 24:469–510.

Lindstrom, F.T., and W.M. Stone. 1974. On the start up or initial phase of linear mass transport of chemicals in a water saturated sorbing porous medium. SIAM J. Appl. Math. 26:578–591.

Millington, R.J. 1959. Gas diffusion of porous solids. Science 130:100–102.

Millington, R.J., and J.P. Quirk. 1961. Permeability of porous solids. Trans. Farad. Soc. 57:1200–1206.

Molz, F.J., M.A. Widdowson, and L.D. Benefield. 1986. Simulation of microbial growth dynamics coupled to nutrient and oxygen transport in porous media. Water Resour. Res. 22:1207–1216.

Moridis, G.J., and D.L. Reddell. 1991. The Laplace transform finite difference method for simulation of flow through porous media. Water Resour. Res. 27:1873–1884.

Nauman, E.B., and B.A. Buffham. 1983. Mixing in continuous flow systems. John Wiley and Sons, New York, NY.

Neuman, S.P. 1984. Adaptive Eulerian-Lagrangian finite element method for advection-dispersion. Int. J. Numer. Meth. Eng. 20:321–337.

Nielsen, D.R., and J.W. Biggar. 1961. Miscible displacement in soils. I. Experimental information. Soil Sci. Soc. Amer. Proc. 25:1–5.

Nkedi-Kizza, P., J.W. Biggar, M. Th. van Genuchten, P.J. Wierenga, H.M. Selim, J.M. Davidson, and D.R. Nielsen. 1983. Modeling tritium and chloride 36 transport through an aggregated oxisol. Water Resour. Res. 19:691–700.

Nkedi-Kizza, P., J.W. Biggar, H.M. Selim, M. Th. van Genuchten, P.J. Wierenga, J.M. Davidson, and D.R. Nielsen. 1984. On the equivalence of two conceptual models for describing ion exchange during transport through an aggregated Oxisol. Water Resour. Res. 20:1123–1130.

Nye, P.H. 1979. Diffusion of ions and uncharged solutes in soils and soil clays. Adv. Agron. 31: 225–272.

Olsen, S.R., and W.D. Kemper. 1968. Movement of nutrients to plant roots. Adv. Agron. 20:91–151.

Oscarson, D.W., H.B. Hume, N.G. Sawatsky, and S.C.H. Cheung. 1992. Diffusion of iodide in compacted bentonite. Soil Sci. Soc. Am. J. 56:1400–1406.

Parker, J.C., and M. Th. van Genuchten. 1984a. Flux-averaged concentrations in continuum approaches to solute transport. Water Resour. Res. 20:866–872.

Parker, J.C., and M. Th. van Genuchten. 1984b. Determining transport parameters from laboratory and field tracer experiments. VA Agric. Exp. Stat. Bull. 84–3.

Pinder, G.F., and W.G. Gray. 1977. Finite elements in surface and subsurface hydrology. Academic Press, New York, NY.

Pitzer, K.S. 1979. Activity coefficients in electrolyte solutions. CRC Press, Boca Raton, FL.

Robin, M.J.L., R.W. Gillham, and D.W. Oscarson. 1987. Diffusion of strontium and chloride in compacted clay-based materials. Soil Sci. Soc. Am. J. 51:1102–1108.

Robin, M.J.L., E.A. Sudicky, R.W. Gillham, and R.G. Kachanoski. 1991. Spatial variability of Strontium distribution coefficients and their correlation with hydraulic conductivity in the Canadian Forces Base Borden aquifer. Water Resour. Res. 27: 2619– 2632.

Rubin, Y., and D. Or. 1993. Stochastic modeling of unsaturated flow in heterogeneous soils with water uptake by plant roots: The parallel columns model. Water Resour. Res. 29:619–631.

Scheidegger, A.E. 1974. The physics of flow through porous media. University of Toronto Press, Toronto, Canada.

Selim, H.M., J.M. Davidson, and R.S. Mansell. 1976. Evaluation of a two-site adsorption-desorption model for describing solute transport in soil. p. 444–448. In Proc. Computer Simulation Conference, American Institute of Chemical Engineering, Washington, DC.

Selim, H.M., and R.S. Mansell. 1976. Analytical solution of the equation for transport of reactive solutes through soils. Water Resour. Res. 12:528–532.

Simon, W., P. Reichert, and C. Hinz. 1997. Properties of exact and approximate traveling wave solutions for transport with nonlinear and nonequilibrium sorption. Water Resour Res. 33:1139–1147.

Sparks, D.L. 1989. Kinetics of soil chemical processes. Academic Press, San Diego, CA.

Spiegel, M.R. 1965. Theory and problems of Laplace transforms. Schaum's Outline Series, McGraw-Hill, New York, NY.

Sudicky, E.A. 1989. The Laplace transform Galerkin technique: A time-continuous finite element theory and application to mass transport in groundwater. Water Resour. Res. 25:1833–1846.

Taylor, G.I. 1953. Dispersion of soluble matter in solvent flowing through a tube. Proc. Roy. Soc. London Ser. 219:186–203.

Toride, N., and F.J. Leij. 1996a. Convective-dispersive stream tube nodel for field-scale solute transport: I. Moment analysis. Soil Sci. Soc. Am. J. 60:342–352.

Toride, N., and F.J. Leij. 1996b. Convective-dispersive stream tube nodel for field-scale solute transport: II. Examples and calibration. Soil Sci. Soc. Am. J. 60:352–361.

Toride, N., F.J. Leij, and M. Th. van Genuchten. 1993. A comprehensive set of analytical solutions for nonequilibrium solute transport with first-order decay and zero-order production. Water Resour. Res. 29:2167–2182.

Toride, N., F.J. Leij, and M. Th. van Genuchten. 1994. Flux-averaged concentrations for transport in soils having nonuniform initial solute distributions. Soil Sci. Soc. Am. J. 57:1406–1409.

Toride, N., F.J. Leij, and M. Th. van Genuchten. 1995. The CXTFIT code for estimating transport parameters from laboratory or field tracer experiments. Version 2.0. US Salinity Lab. Res. Rep. 137, Riverside, CA.

van der Heijde, P., Y. Bachmat, J. Bredehoeft, B. Andrews, D. Holtz, and S. Sebastian. 1985. Groundwater management: The use of numerical models. 2nd Ed. American Geophysical Union, Wahington, DC.

van der Laan, E.T. 1958. Notes on the diffusion-type model for the longitudinal mixing in flow. Chem. Eng. Sci. 7:187–191.

van der Zee, S.E.A.T.M. 1990. Analytical traveling wave solutions for transport with nonlinear and nonequilibrium adsorption. Water Resour. Res. 26:2563–2578.

van Genuchten, M. Th. 1978. Mass transport in saturated-unsaturated media. One-dimensional solutions. Princeton University Research Rep. 78-WR-11.

van Genuchten, M. Th. 1981. Non-equilibrium transport parameters from miscible displacement experiments. US Salinity Lab. Res. Rep. 119, Riverside, CA.

van Genuchten, M. Th. 1988. Solute transport. p. 360-362. *In* S.P. Parker (ed.) McGraw-Hill Yearbook of Science and Technology, McGraw-Hill Book Co., New York, NY.

van Genuchten, M. Th., and W.J. Alves. 1982. Analytical solutions of the one-dimensional convective-dispersive solute transport equation. USDA Tech. Bull. 1661.

van Genuchten, M. Th., and R.W. Cleary. 1982. Movement of solutes in soils: Computer-simulated and laboratory results. p. 349-386. *In* G.H. Bolt (ed.). Soil Chemistry. B. Physico-chemical models. Elsevier. Amsterdam, Netherlands.

van Genuchten, M. Th., and F.N. Dalton. 1986. Models for simulating salt movement in aggregated field soils. Geoderma 38:165–183.

van Genuchten, M. Th., and W.G. Gray. 1978. Analysis of some dispersion corrected numerical schemes for solution of the transport equation. Int. J. Num. Meth. Eng. 12:387–404.

van Genuchten, M. Th., and J.C. Parker. 1987. Parameter estimation for various contaminant transport models, p. 273–295. *In* C.A. Brebbia and G.A. Keramidas (ed.) Reliability and robustness of engineering software. Elsevier, New York, NY.

van Genuchten, M. Th., and J.C. Parker. 1984. Boundary conditions for displacement experiments through short laboratory columns. Soil Sci. Soc. Amer. J. 48:703–708.

van Genuchten, M. Th., and E.A. Sudicky. 1999. Recent advances in vadose zone flow and transport modeling. p. 155–193. *In* M.B. Parlange and J.W. Hopmans (ed.) Vadose zone hydrology: Cutting across disciplines. Oxford University Press, New York, NY.

van Genuchten, M. Th., D.H. Tang, and R. Guennelon. 1984. Some exact solutions for solute transport through soils containing large cylindrical macropores. Water Resour. Res., 20:335–346.

van Genuchten, M. Th., and P.J. Wierenga. 1976. Mass transfer studies in sorbing porous media. I. Analytical solutions. Soil Sci. Soc. Amer. J. 40:473–480.

van Genuchten, M. Th., and P.J. Wierenga. 1986. Solute dispersion coefficients and retardation factors. *In* A. Klute (ed.) Methods of soil analysis. I. Physical and mineralogical methods. Soil Science Society of America, Madison, WI.

van Rees, K.C.J., E.A. Sudicky, P.S.C. Rao, and K.R. Reddy. 1991. Evaluation of laboratory techniques for measuring diffusion coefficients in sediments. Env. Sci. Tech. 25:1605–1611.

Varoglu, E., and W.D.L. Finn. 1982. Utilization of the method of characteristics to solve accurately two-dimensional transport problems by finite elements. Int. J. Num. Fluids 2:173–184.

Wierenga, P.J. 1977. Solute distribution profiles computed with steady-state and transient water movement. Soil Sci. Soc. Amer. J. 41:1050–1055.

Yeh, G.T. 1990. A Lagrangian-eulerian method with zoomable hidden fine-mesh approach to solving advection-dispersion equations. Water Resour. Res. 26:1133–1144.

Yeh, G.T., and V.S. Tripathi. 1989. A critical evaluation of recent developments in hydrogeochemical transport models of reactive multichemical components. Water Resour. Res. 25:93–108.

Yeh, G.T., and V.S. Tripathi. 1991. A model for simulating transport of reactive multispecies components: Model development and demonstration. Water Resour. Res. 27:3075–3094.

Zwillinger, D. 1989. Handbook of differential equations. Academic Press, San Diego, CA.

7

Soil Structure

B. D. Kay
University of Guelph

D. A. Angers
Agriculture and Agri-Food Canada

Soil structure has a major influence on the ability of soil to support plant growth, cycle C and nutrients, receive, store and transmit water, and to resist soil erosion and the dispersal of chemicals of anthropogenic origin. Particular attention must be paid to soil structure in managed ecosystems where human activities can cause both short- and long-term changes that may have positive or detrimental impacts on the functions that soil fulfills.

The importance of soil structure was recognized by researchers more than 150 years ago and the large volume of research on its nature has been summarized in a number of comprehensive reviews (Harris et al., 1966; Oades, 1984; Dexter, 1988; Kay, 1990; Horn et al., 1994). This paper builds on previous reviews and provides an overview to assist readers in interpreting data on soil structure in relation to land use. Emphasis will be placed on soils containing less than 40 to 60% clay from the temperate regions of the world where the relations between soil structure and land use have been studied most extensively. Although the nature of these relations changes with depth in the profile, attention will be primarily directed to the A horizon where the effects of land use practices on soil structure are most pronounced. Because research on soil structure has, until recently, been carried out primarily in the context of agriculture, much of the following discussion is necessarily focused on this aspect.

7.1   Characteristics, Significance, and Measurement of Soil Structure

Field experience and decades of research have given rise to a multitude of descriptions of soil structure. The term tilth is often used to describe the quality of soil structure for plant growth and is a popular term that predates modern agriculture. The term soil tilth embodies an integration of many of the characteristics of soil structure and reflects the practical experience of generations of people who have worked the soil and the understanding that they have of conditions that lead to greatest productivity and ease of management. A soil considered to possess good tilth is one that readily fractures whether the stress arises from tillage, emerging seedlings or growing roots and provides an optimal environment for the growth of plants and microorganisms. Consequently, Gupta (1986) has attempted to define tilth using an index incorporating several variables. Tilth has also been defined using single characteristics such as those related to compaction (Scott Blair, 1937) and tensile failure (Utomo and Dexter, 1981a). Characteristics of soil structure that influence different functions of soil (including plant growth) will be considered in this chapter and, although tilth will not be considered in detail, several of these characteristics relate to the concept of tilth.

7.1.1   Form

The term structural form applies to a group of characteristics that describe the heterogeneous arrangement of void and solid space that exists in soil at any given time. An assessment of soil structure at a site normally involves a visual assessment of structural form that is complemented by quantitative analyses of samples collected from the site. Visual analyses include a description of the morphology of soil at the surface and the variation in morphology with depth. Most national soil classification systems include procedures and terminology for describing and classifying soil structure based on morphology (Soil Survey Staff, 1975; Hodgson, 1978; Canada Expert Committee on Soil Survey, 1987). For example, the Canadian system classifies soil structure in terms of grade or distinctness (structureless, weak, moderate, strong), class or size (very fine, fine, medium, coarse, very coarse) and type (single grain, amorphous or massive, angular blocky, subangular blocky, granular, platy, prismatic, columnar). Grades and types of soil structure are defined on the basis of visual evidence of the extent of development of peds and their shape (peds are secondary structural units made up of primary particles that are distinguished from adjacent structures on the basis of failure zones and are formed by natural processes, in contrast with clods which are formed artificially). Grades of soil structure are distinguished as follows by the Canada Expert Committee on Soil Survey (1987): (1) *weak:* peds are barely observable in place; (2) *moderate:* peds are moderately evident; on disturbance, soil will break down into a mixture of many distinct entire peds, some broken peds, and little disaggregated material; and (3) *strong:* peds are quite evident in undisturbed soil; on disturbance, peds retain their identity with some broken peds, and little disaggregated material. The types of structure are diagrammatically illustrated in Fig. 7.1 and details on class or size of the different types are given in Table 7.1

Although morphological descriptions are largely qualitative, an experienced pedologist can make some quantitative predictions from these analyses (McDonald and Julian, 1966; McKeague et. al.,1982; Wang et. al., 1985). Visual analyses during a site examination may also include an examination of the distribution of roots, soil color, or the infiltration of precipitation and, together with observations of morphology, can provide evidence of the influence of structural form on functions that the soil is fulfilling. Visual analyses are normally complemented by quantitative measurements of soil physical properties. A number of structural characteristics have been used to characterize the soil matrix.

7.1.1.1   Pores

Total Porosity and Bulk Density
Total porosity is seldom measured directly but is normally calculated from the bulk density and the particle density. Total porosity is not very sensitive to the variation in particle density that is normally encountered in the field, and consequently, discussions on total porosity are often presented using bulk density as the measured variable. Details on the measurement of these characteristics are given in Section A, Chapter 1.

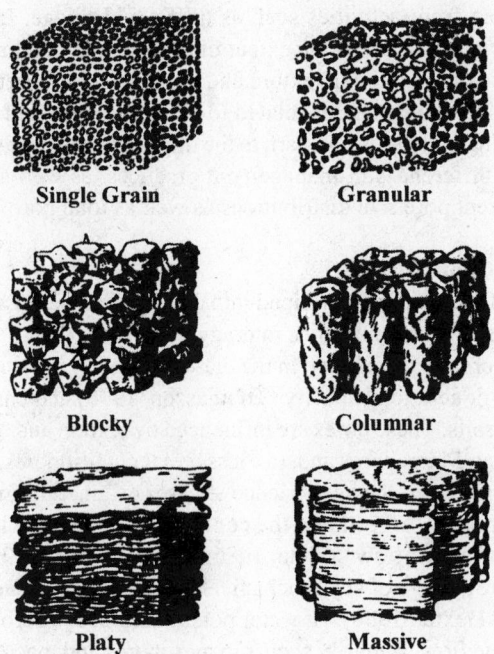

Single Grain   **Granular**

Blocky   **Columnar**

Platy   **Massive**

Fig. 7.1 Diagrammatical representation of types of structure

Table 7.1 Details on types of structure and their size [Canada Expert Committee on Soil Survey, 1987]

Type	v. fine	fine	medium	coarse	v. coarse
	----------------------------Class----------------------------				
	----------------------------mm----------------------------				
Single grain: loose incoherent mass of individual particles as in sands	---	---	---	---	---
Amorphous (massive): coherent mass showing no evidence of any distinct arrangement of soil particles	---	---	---	---	---
Angular blocky: faces of peds rectangular and flattened, vertices sharply angular	< 5	< 10	10-20	20-50	> 50
Subangular blocky: faces of peds subangular, vertices mostly oblique or subrounded	< 5	< 10	10-20	20-50	> 50
Granular: spheroidal peds characterized by rounded vertices	< 1	< 2	2-5	5-10	---
Platy: horizontal planes more or less developed	< 10	< 2	2-5	> 5	---
Prismatic: vertical faces of peds well defined and edges sharp	< 10	< 20	20-50	50-100	> 100
Columnar: vertical edges near top of columns are not sharp (columns may be flat or round topped, or irregular)		< 20	20-50	50-100	> 100

Bulk density is strongly influenced by texture and organic C content and, for a given soil, reflects the impact of stresses arising from activities such as traffic and tillage. Interpreting data on bulk density from different soils with respect to the extent of compaction that has occurred is, therefore, complicated by the influence of variation in texture and organic C. Interpreting data on bulk density from different soils with respect to processes related to the growth of plants or water flow, has also proven to be difficult. This difficulty is due, in part, to the importance of pore size distribution on these processes and the fact that differences in management practices, as well as texture and organic C contents, can result in different pore size distributions as well as total porosity.

Pore Size Distribution and Continuity

Pores of different size are often arbitrarily grouped into different classes (Table 7.2). Pores > 30 µm include biopores, shrinkage cracks, and other interaggregate pores, and the largest pores within aggregates and peds. This porosity is included in the class referred to as structural porosity (Stengel, 1979; Derdour et al., 1993) or aeration capacity (Thomasson, 1978) and can represent as much as a third of the total porosity of soils. These pores are influenced by texture and organic C content and are very sensitive to management. Macropores and, to a lesser extent mesopores, are the least stable of all pore size classes, and collapse when they experience stresses originating from a range of sources.

The volume fraction of pores > 30 µm and the continuity or connectivity of these pores have a major influence on water and solute flow (White, 1985; Blackwell et al., 1990b; Ahuja et al., 1993), aeration (Thomasson, 1978), a range of soil mechanical characteristics (Carter, 1990b), and on root development (Jakobsen and Dexter, 1988). At water potentials close to zero (most of these pores are water filled), water and solute flow primarily occurs in mesopores and macropores and the rest of the pore system is bypassed. These pores drain quickly and, as air is redistributed in the soil, aeration is enhanced. Macroporosity and mesoporosity represent failure zones of low strength and therefore a reduction in the volume fraction of these pore classes results in an increase in tensile strength and resistance to penetration.

Pores with equivalent diameters of 0.1 to 30 µm are often referred to as storage pores and include the volume fraction of pores defining the water that is potentially available to plants (Veihmeyer and Hendrickson, 1927). These pores also provide a habitat for microorganisms and smaller soil fauna. Micropores and ultramicropores are strongly influenced by texture and organic C content (Ratliff et. al.,1983; da Silva and Kay, 1997a) but are not strongly influenced by increases in bulk density that can arise from traffic or other stresses (da Silva and Kay, 1997a).

Pores < 0.1 µm are least influenced by management of any of the pore classes. Although these pores remain water filled for a much larger proportion of time than the other pore classes, little of the water in these pores is available to plants and the rate of water flow through them is very slow. These pores remain largely inactive biologically because they are not penetrated by roots or by most microorganisms and are, therefore, only accessible to molecular sized byproducts of biological

Table 7.2 Pore size classification [After Soil Science Society of America, 1997]

Class	Class limits equivalent diam.
	(µm)
Macropores	> 75
Mesopores	30-75
Micropores	5-30
Ultramicropores	0.1-5
Cryptopores	< 0.1

activity. The relative inaccessibility of these pores to microorganisms may also give rise to one of their most important functions, namely the physical protection of organic C.

Pore size distribution can be estimated from the total porosity and measurements during the progressive removal of water or the intrusion of Hg (Danielson and Sutherland, 1982; Carter and Ball, 1993). Details of the shape, tortuosity and connectivity of macro- and mesopores can be obtained from an examination of thin sections that often utilizes image analyses techniques (Moran et al. 1989; Crawford et al.,1995) or from measurements of gas movement (Ball and Smith, 1991). The nature of macropores has also been studied using computed tomography (Warner et al., 1989).

7.1.1.2 Failure Zones

The application of mechanical stress (such as that associated with tillage) to undisturbed bulk soil results in failure of the soil matrix into peds, broken peds and disaggregated material. Failure occurs through zones made up of elemental volumes in which the bonding between the elementary particles is small and, therefore, failure zones include pores as well as elemental volumes containing less cementing materials than the surrounding matrix. Failure zones, therefore, reflect the characteristics of both the void and the solid space.

Failure zones can be characterized in terms of their average strength (or strength at a given percentile of failure in a frequency distribution) and the distribution of the magnitude of these strengths. Zones with a given strength will fracture when the force that is applied exceeds this value. The spatial distribution of failure zones of different strengths leads to the creation of aggregates with characteristic size distributions. The application of increasing energy causes fragmentation to occur along zones of increasing strength and closer proximity thereby resulting in smaller aggregates. The increase in the tensile strength of aggregates with decreasing size of aggregates has been used to quantify the friability of soils (Utomo and Dexter, 1981b).

Characterizing the strength of failure zones and their distribution involves the application of stress and measurement of the nature of failure that has occurred. The strength of failure zones has been determined from measurements of the maximum tensile stress required to cause failure of the matrix (Braunack et al., 1979), and measurements of aggregate size distributions after the sample has fallen through a given distance (a drop-shatter test) (Marshall and Quirk, 1950; Ingles, 1965).

7.1.1.3 Peds, Aggregates, and Clods

Morphological descriptions of soil structure include reference to the extent of development and shape of peds (Table 7.1). Peds are considered to be naturally formed (in contrast to formation by anthropogenic means) and the extent of ped development relates to the strength of failure zones; peds that are very visible (structure is strongly developed), are separated by failure zones of very low strength. The term aggregate may be used interchangeably with ped where the type of soil structure is granular. A granular structure is often found in topsoils that have been continuously under grass for many years (McKeague et al., 1987). The term aggregate is also used for structural units that result from fragmentation of the soil matrix through the application of mechanical energy. The energy may be applied in the field through tillage or seedbed preparation and in the laboratory through crushing or sieving. Aggregates > 10 cm are commonly referred to as clods. The distinction between aggregates and peds becomes greater as more energy is applied to the bulk soil. Most studies on soil structural units in the laboratory have involved the application of energy and, therefore, the term aggregate will be used in the remainder of this chapter in relation to these structural units. It is important to note, however, that the characteristics of aggregates reflect, in part, the amount of energy that has been applied during the fragmentation process.

Aggregates may be characterized by their size, shape and surface roughness, although aggregate size has the greatest practical importance. Aggregate size is estimated by distributing the sample across sieves with openings of different size and weighing the amount of soil on each sieve (Kemper and Rosenau, 1986; White, 1993). Any soil sample will contain aggregates of different size and the soil can be characterized in terms of the distribution of sizes of aggregates and the properties of aggregates in the different size fractions. The distribution of weights of aggregates among the different size fractions (dry aggregate size distribution), has been described using parameters related to the center of the distribution such as mean weight diameter and to the shape of the distribution. An assessment of different parameters used for this purpose is given by Perfect et al. (1993). Aggregates < 250 μm are often referred to as microaggregates, in contrast to larger aggregates which are referred to as macroaggregates. This distinction is based on the hypotheses that large aggregates are made up of small aggregates (a hierarchical structure), that the building blocks of the large aggregates of most soils are aggregates that are < 250 μm, and that the cementing materials within the microaggregates are different from those that bind the microaggregates into macroaggregates (Tisdall and Oades, 1982). Individual primary soil particles that are released from the matrix when stress is applied represent disaggregated material (Table 7.1) and may include gravel, sand, silt and clay sized material. Part of the clay, and to a lesser extent the silt sized material, may be readily suspended in water and are often measured as a dispersible fraction. Material that becomes dispersed simply as a consequence of wetting aggregates to saturation is referred to as spontaneously dispersible material, whereas, material that only becomes dispersed after energy is applied to the aggregates is referred to as mechanically dispersible material.

The distribution of aggregate sizes in the seedbed is related to type of tillage, number of passes and soil water content. When energy input is kept constant, dry aggregate size distribution is influenced by soil properties and water content at the time of tillage, and reflects the influence of these factors on the characteristics of the failure zones in the soils. While the effects of tillage or cropping practices on aggregate size distribution can disappear quickly after tillage due to crop growth and climatic factors, these effects can have a significant impact on the germination, emergence and early growth of seedlings. Measurements of aggregate size distributions are most relevant to the germination and early growth of plants on soils that are tilled, structurally stable, and are not compacted by traffic. The measurements will have less relevance to later growth, or to early growth on untilled soils, or tilled soils that are unstable or compacted by traffic.

The size distribution of aggregates in the seedbed can also influence the flux of water and solutes. Distributions dominated by large aggregates will have a higher proportion of macropores (Dexter et al., 1982), and consequently under saturated or near saturated conditions, will have higher hydraulic conductivities with larger fractions of water and solutes bypassing the rest of the pore network. These pores will exist between aggregates. Little information on the distribution of pore sizes within aggregates is provided directly by aggregate size distribution (Wu and Vomocil, 1992), although the development of fractal theory to describe the organization of solid and void space in a manner that is scale independent offers the potential for predicting transport characteristics from information on solid space (Crawford, 1994).

7.1.2 Stability
The term structural stability is used to describe the ability of soil to retain its arrangement of solid and void space when exposed to different stresses. Stability characteristics are specific for a characteristic of structural form and the type of stress applied. Stresses may arise from processes as diverse as tillage, traffic and wetting or drying. The response of both the void and the solid space to the

application of these stresses reflects, in part, the strengths of the failure zones and their spatial distribution. Extensive research has been directed to understanding the stability of aggregates when they are exposed to stresses arising from wetting under different conditions and then exposed to mechanical stresses arising from sieving, shaking in water, or exposed to ultrasonic energy. Consequently, the term structural stability is often considered to be synonymous with aggregate stability. There are, however, important characteristics that reflect the ability of a soil to retain its arrangement of solid and void space when other stresses are applied and conventional measurements of aggregate stability may not provide meaningful measures of these properties. For instance, the compressibility index is a measure of the ability of the total porosity to withstand compressive stresses. This parameter is particularly important where changes in agricultural practices are leading to greater traffic.

7.1.2.1 Pores

Pores experience stress arising from tillage, traffic, wetting or drying, and root growth. Compaction of soil by vehicular traffic or other means results in a decrease in total pore space. The resistance of pores to compression can be measured using compression cells of various types (Bradford and Gupta, 1986). Compression indices can be calculated and related to soil properties and management (Larson et al., 1980; Angers et al., 1987). Although macropores and mesopores tend to be lost first during compaction, as discussed earlier, the stability of macropores may also relate to the orientation of the pores in relation to the applied stress. The stability of biopores is greatest if the stress is applied parallel to the axis of the pores (Blackwell et al., 1990a).

Pore space may experience stress during wetting if irrigation or precipitation result in rapid wetting (Mitchell et al., 1995). The stresses arise from differential swelling of clay minerals and from air entrapment subsequently causing increased air pressure. The stability of the pore system under these circumstances can be characterized by comparing the permeability of the soil to water with the permeability to a nonreactive fluid such as air (Reeve, 1953) or by determining changes in pore size distribution arising from fast and slow wetting rates (Collis-George and Figueroa, 1984).

In many soils the pore space also varies with water content. The loss of porosity upon drying can be characterized by the shrinkage curve. Smaller slopes of the shrinkage curve under no till versus tilled soils (McGarry and Smith, 1988) or under cropped versus bare soils (Mitchell and van Genuchten, 1992) were attributed to greater pore stability in the case of no-till and cropped soils, respectively.

7.1.2.2 Aggregates

Aggregates may experience stresses related to tillage, traffic, wetting and mechanical abrasion by flowing water. The ability of aggregates to resist stresses from wetting followed by mechanically shaking in water is referred to as wet aggregate stability. Although this parameter was originally used to characterize the erodibility of soil (Yoder, 1936), it has been increasingly used to study the cohesion of aggregates and the dynamics of changes in the nature of bonding between particles. The stability of aggregates is strongly dependent on the rate of wetting. Aggregate stability declines as the rate of wetting increases and, as for pores, the decline has been attributed to increased stresses related to entrapment of air within the aggregates and differential swelling of clays (Panabokke and Quirk, 1957; Concaret, 1967; Emerson, 1977).

Measurements of the loss of stability due to rapid rates of wetting are most relevant to aggregates at the soil surface in environments where the soil surface is very dry and rapid wetting occurs from intense rains or irrigation. In environments where wetting occurs more slowly (surface aggregates that have high water contents or experience less intense rainfall events, or aggregates beneath the surface), measurements of stability in which air entrapment and differential swelling are minimized are more relevant. Soil samples are often not dried if measurements involve slow wetting and, in these cases, stability is very strongly influenced by the water content of the sample prior to saturation (Section 7.2.4.2).

Measurements of stability in water can be restricted to macroaggregates, microaggregates, dispersible material or can include a wide range of size classes. The measurements will be strongly influenced by size class of the aggregates selected for study, their initial water content, conditions under which wetting occurs, and size classes that are measured after application of the stress. Early methods include those of Yoder (1936) and Hénin et al. (1958). Details on current methodologies for measuring wet aggregate stability are summarized in Kemper and Rosenau (1986) and in Angers and Mehuys (1993).

The stability of aggregates to stresses representative of those encountered during tillage have also been measured (Watts et al., 1996a,b), although this characteristic has been explored in much less detail than wet aggregate stability. Watts and coworkers used a falling weight to impose different specific amounts of mechanical energy on aggregates and measured mechanically dispersible clay as the stability parameter.

7.1.3   Resiliency

The term structural resiliency describes the ability of soil to recover its structural form through natural processes when the stresses are reduced or removed. Recovery relates, in the first instance, to changes in pore characteristics and the distribution of strengths of failure zones. Resiliency arises from a number of processes such as freezing/thawing, wetting/drying and biological activity (e.g., root development and activity of soil fauna). Soils differ in their ability to recover structural form when stresses are reduced or removed. Soils that are most resilient are known as self-mulching soils (Wenke and Grant, 1994). The structural form of such soils are not particularly sensitive to mechanical processes. For instance, if much of the macroporosity of the upper few cm are destroyed through tillage when the soil is wet, a desirable structure can be recreated by wetting and drying. The creation of failure zones by wetting events has been referred to as tilth mellowing by Utomo and Dexter (1981a). Resiliency is also exhibited by the partial recovery (rebound or relaxation) of porosity after removal of a compressive stress (Guérif, 1979; Stone and Larson, 1980; McBride and Watson, 1990). The maximum change or recovery that is possible by the different mechanisms, and the rate at which this change can occur, are particularly important characteristics of soils where management practices are leading to the application of increasing stress (Kay et al., 1994a).

7.1.4   Vulnerability

The term structural vulnerability reflects the combined characteristics of stability and resiliency (Fig. 7.2). Soils which are least stable under a given stress and which do not recover when the stress is reduced or removed (or which recover very slowly) are most vulnerable to stress, whether the origin of the stress is natural or anthropogenic. Conversely, soils which have a great resistance to stress and are very resilient are the least vulnerable.

7.1.5   Spatial and Temporal Variability and Significance of Scale

The size of pores, the distance between failure zones of similar strength, and the size of peds or aggregates can range across several orders of magnitude. Consequently, stability and resiliency can also be considered at different scales. At any given scale, these characteristics are spatially variable

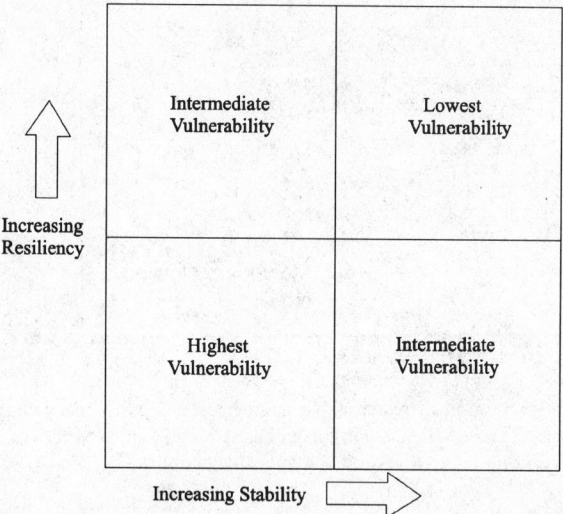

Fig. 7.2 Diagrammatic representation of the relation between stability, resiliency and vulnerability [Kay, 1998. Reprinted with permission from Lal et al. 1998. Soil processes and the carbon cycle. CRC Press, Boca Raton, FL]

and Dexter (1988) has argued that this spatial variability or heterogeneity is, itself, a fundamental characteristic of soil structure. Sampling strategies should, therefore, be designed to provide data on the spatial variability in structural characteristics.

Structural characteristics also vary temporally and changes can occur at scales ranging from hours to decades. The extent of temporal variation is also dependent on the spatial scale of interest. For instance, macroporosity is very dynamic and can change over short time periods (e.g., as a consequence of tillage) whereas pores < 0.1 μm are much less dynamic but may change over several years or decades (as a consequence of changes in organic C contents arising from management practices). Sampling strategies should therefore also take into account the scale of temporal variation in the structural characteristics being measured. Data presented in Fig. 7.3 illustrate the temporal variation in the proportion of water-stable aggregates during a four-year period in a silty clay soil submitted to two contrasting tillage systems. Short-term variations (within a growing season) represent the immediate effects of climatic events and tillage practices whereas long-term trends are a result of cumulative effects of both these factors. For instance, the drop in aggregation observed in months 9 to 13 coincides with conditions of dry soils during which aggregates were subject to greater slaking. Despite these short-range variations, the difference in the level of stable aggregation between the two tillage treatments kept increasing with time reflecting the cumulative effects of less disturbance and greater accumulation of C at the surface of the no till soil.

7.2   Soil Factors Influencing Structure

The dominant soil characteristics influencing structure are texture and clay mineralogy, organic matter, inorganic noncrystalline materials, composition of the pore fluid and adsorbed or exchangeable solutes, plants and soil organisms, and depth in the profile. Few, if any, of these factors function in isolation. In addition, the magnitude of their influence on soil structure may vary with factors unrelated to soils such as climate, and land use or management practices.

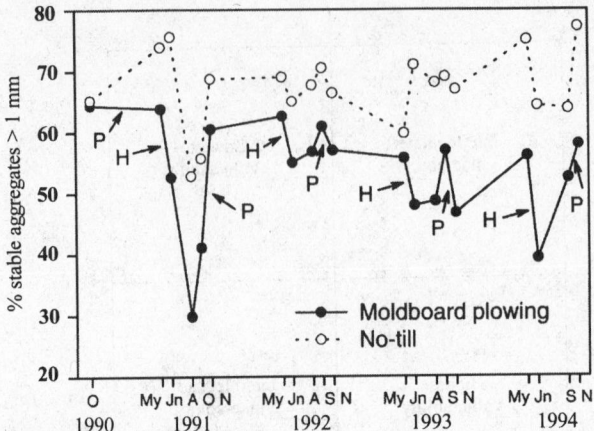

Fig. 7.3 Temporal variation in the proportion of water stable aggregates > 1 mm under two contrasting tillage systems in silty clay soil (Humic Gleysol, Humic Haplaquept) under barley production in eastern Canada. P = Plowing event, H̄ = Disk harrowing event [J. Laford, D.A. Angers and D. Pageau, unpublished]

7.2.1   Texture and Mineralogy

Soil texture has a major influence on the form, stability and resiliency of soil structure as well as the response of soil structure to weather, biological factors and management. The nature of the influence of texture on structural characteristics is related to the soil matrix. For instance, sands may be viewed, in the simplest case, as being made up of single grains and all characteristics of structural form are determined by the distribution of grain sizes, and modification introduced by tillage or traffic. Clay or silt sized materials that may be present, exist as coatings on the sand grains or fill the interstices between the sand grains. The structure does not shrink or swell and is not very responsive to freezing. Organic materials, together with fine clays and other amorphous and crystalline inorganic materials, provide what little cementation exists between the grains. The resulting aggregates have low stability. The single grain structure is, however, relatively stable under compressive stresses. The structural form possesses limited resiliency and has, therefore, low to intermediate vulnerability (Fig. 7.2). As the clay content increases, the characteristics of the soil matrix (including both structural form and stability) are increasingly dominated by the characteristics of the clay (including mineralogy and exchangeable ions), the nature and quantities of cementing materials and the composition of the pore fluid. Rengasamy et al. (1984) has suggested that the physical properties of soils change with increasing clay content until the clay content reaches about 30%; at greater clay contents, the behavior of the soil depends on the type of clay, and cannot be predicted from clay content alone.

Structural resiliency, related to weather, biological processes, and management play an important role and the magnitude of the impacts of these processes are determined by the characteristics of the matrix. For instance, the formation of pores and other failure zones by biological processes may be more important in medium textured soils or low activity clay soils that undergo limited shrinking and swelling than is the case for finer textured soils that are particularly responsive to wetting and drying (Oades, 1993). In the latter soils, biological processes may complement abiotic processes which exert the dominant control over soil structure.

7.2.1.1   Form

The effect of texture on pores, failure zones and aggregation can be influenced by mineralogy and depth in the profile and, therefore, it is difficult to identify broad generalizations. For instance,

Manrique and Jones (1991) found, using data largely from the United States, that bulk density was positively correlated with clay content in the B and C horizons but was not correlated with clay in the A horizon. Difficulties in relating bulk density to texture in the A horizon are created by the strong influence of tillage, traffic, and organic C content on bulk density and variation in their influence with texture. When the data used by Manrique and Jones (1991) were grouped by soil order, the bulk density was positively related to clay content in Aridisols, Entisols, Inceptisols, and Mollisols, negatively related in Ultisols and Vertisols, and not related in Oxisols and Spodosols. Differences may relate to mineralogy with clay content having a negative relation with bulk density as the amount of swelling clay minerals increase.

Few studies have related pore size distribution to texture or mineralogy. The contribution of shrinkage cracks to macroporosity would be expected to be positively related to the amounts of swelling clay minerals present, with the magnitude of the contribution relating to exchangeable ions and the composition of the pore fluid. The influence of texture and mineralogy on pores 0.2–30 μm equivalent diameter in nonswelling soils can be estimated from the water release curve. The influence of texture on these curves has been described by regression equations (pedotransfer functions) although the descriptions have not always been consistent (Kay et al., 1997). Inconsistent relations between texture and the water release curve may be due, in part, to variations in particle size distribution not reflected in broad textural characteristics (clay, silt or sand content) which can alter the water release curve (Schjønning, 1992). Inconsistent relations between texture and the water release curve may also arise from variation in mineralogy. Few studies have systematically examined the role of mineralogy on pore size distribution. Williams et al. (1983) found that montmorillonite, Fe oxide, vermiculite, and quartz introduced greater variation in the water release curve of Australian soils than illite, kaolinite, halloysite, or randomly interstratified materials. The influence of texture and mineralogy on aggregates reflects the influence of these variables on failure zones and their response to stress (stability). Aggregate size distributions that are dominated by large aggregates and clods are more common in finer textured soils.

7.2.1.2 Stability

The influence of the clay fraction on stability varies with the nature of the stress applied. Larson et al. (1980) found that the compression index, a measure of the susceptibility to compressive stress, increased approximately linearly with clay content up to about 33% and thereafter remained approximately constant. Soils dominated by kaolinite or Fe oxides were less compactible (lower compression index) than soils with predominantly 2:1 type clays. On the other hand, the stability of aggregates to wetting and mechanical abrasion during sieving usually increases with increasing clay content, but the relation varies with depth in the profile. For instance, measurements of the stability of vapor wetted aggregates from soils across the western United States and Canada indicated that aggregate stability increased curvilinearly as the clay content increased to 90% (Kemper and Koch, 1966). Similar results have been obtained from other regions of the world and over narrower ranges of texture (le Bissonnais, 1988; Elustondo et al., 1990; Rasiah et al., 1992). The influence of mineralogy is most apparent at the scale of dispersible mineral material where the strength of bonds between platelets, quasicrystals and domains are controlled by mineralogy, exchangeable ions and composition of the pore fluid (Rengasamy et al., 1993; Rengasamy and Sumner, 1998) (section 7.2.4).

7.2.1.3 Resiliency and Vulnerability

The resiliency of soils related to weather is more strongly influenced by the clay fraction than that due to biological factors. The degree of reversibility of soil compaction by development of microcracks caused by wetting and drying increases with shrink/swell potential (Barzegar et al., 1995). Wenke and

Grant (1994) have noted that soils that are self-mulching have clay contents > 35%. Large changes in the strength of failure zones can also be caused by freezing and thawing events. The morphology of the ice lenses are strongly influenced by texture as well as mineralogy (Czeratzki and Frese, 1958). Although many of the pores associated with these lens are unstable and collapse on melting (Kay et al., 1985), their spatial distribution corresponds with the distribution of failure zones and associated aggregates (Pawluk, 1988). Resiliency arising from rebound or relaxation following compaction also varies with mineralogy, being larger in soils containing expanding than nonexpanding clays (Stone and Larson, 1980).

7.2.2 Organic Matter

The most general characteristic of soil organic matter (SOM) is the total organic C content, and the impact of SOM on soil structure is often described using this parameter. Structural form and stability generally improve with increasing SOM contents; however, exceptions to this generalization exist because SOM is not homogeneous in composition or location. SOM represents materials of plant, animal and microbial origin that are in various stages of decomposition and are associated with the mineral fraction with different degrees of intimacy. In most recent conceptual and predictive models, SOM is defined as being composed of three pools which vary in chemical composition, physical location and decomposition kinetics representing a continuum along the decomposition process. The forms and location of organic C associated with these various pools are first briefly reviewed as an introduction to this section; the reader is referred to Section B, Chapter 2 and Section C, Chapter 4, for more comprehensive discussions of SOM and decomposition processes, respectively.

7.2.2.1   Forms and Location of Organic Matter and Association with Minerals

Particulate Organic Matter

Plant and animal materials enter the soil through a range of different processes and the nature of the space that is occupied by these materials and their degree of contact with the soil vary with the material and the process. For instance, plant shoots that have been incorporated into the soil by tillage will be located in macropores and the contact area between the crop residue and the soil will be less than that of plant roots and root hairs which will be located in smaller pores and be in more extensive contact with the soil. Plant and animal materials are attacked by soil flora and fauna at the outset of the decomposition process. Organic C that is relatively free, or has least association with the mineral fraction, is the lightest and coarsest fraction (the density of this fraction may be < 1.6 g cm^{-3} and its size often defined as > 50 μm) and is often referred to as particulate organic matter (POM). Golchin et al. (1994) distinguished between POM that was free and that which was occluded (requiring ultrasonic dispersion for separation). They found the two fractions to be present in roughly similar proportions in the light fraction of five cultivated soils and to make up as much as 38% of the total organic C. Although these proportions vary with soil type and cropping history (Cambardella and Elliott, 1993; Besnard et al., 1996; Gregorich et al., 1997), it is clear that a large proportion of the POM is in close contact with the mineral particles and is often occluded within the soil matrix (Figs. 7.4 and 7.5).

Microbial Biomass and Metabolites

The presence of readily mineralizable C will result in growth in the microbial biomass and their predators and the production of extracellular materials including polysaccharides and other compounds (Section C, Chapter 4). Electron micrographs show that the microorganisms (Fig. 7.5) and extracellular material (Foster, 1988) can become intimately associated with the mineral material at this stage. Part of the microbial biomass may grow into pores adjacent to the organic residue or may be squeezed into this space along with the extracellular material. Dense mucilage may impregnate the soil fabric as much as 50 μm from the plant residue (Foster, 1988). The polysaccharides may exist as fibrous or granular materials and become more closely associated with the mineral surface as drying occurs (Foster, 1988). The matrix of decomposing plant residues, microorganisms, extracellular materials and mineral material has been proposed by Golchin et al. (1994) to coincide with material in the 1.8 to 2.0 g cm^{-3} density fraction. Its chemical composition changes during decomposition. As decomposition of plant residues proceeds, less C is quantitatively and qualitatively available to the decomposers and microbial activity decreases. Chemical analyses suggest that after consumption of the more labile portions of the organic debris (proteins and carbohydrates), the more resistant plant structural materials (which have low O-alkyl, high alkyl and aromatic content) are concentrated as occluded particles within soil aggregates. Electron micrographs obtained by Waters and Oades (1991) show an abundance of void spaces within the aggregates that would fall in the range between 0.1 and 30 μm diameter. Some of these pores are associated with cellular debris while others represent voids within aggregates. These pores may be completely closed, have one entrance (bottle shaped pores) or may be tubes with two or more entrances. As mineralization approaches advanced stages, the site of the decomposing plant material may become a failure zone of low strength.

Recalcitrant Materials
Chemical and biological transformations of the organic materials ultimately lead to their stabilization in the soil (Section B, Chapter 2; Section C, Chapter 4). At this stage, the organic material is intimately associated with the mineral phase and would appear in the dense fraction of the soil (> 2.0 g cm^{-3}). This fraction accounted for 25 to 50% of the organic C in the soils considered by Golchin et al. (1994). Analyses of this fraction using ^{13}C CP/MAS NMR spectroscopy indicated that about half of its C was O-alkyl or carbohydrate C . There was no microscopic evidence of particulate material with a cellular

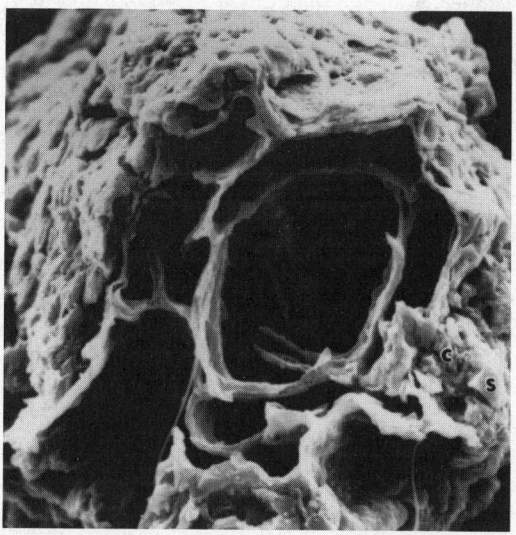

———— 0.01 mm

Fig. 7.4 Particulate organic matter (POM) encrusted with mineral soil particles. Remnants of conducting vessels are coated by clay and silt particles. Conventional SEM. White bar is 10 μm, c = clay particles, s = silt particles [P. Puget and C. Chenu, unpublished].

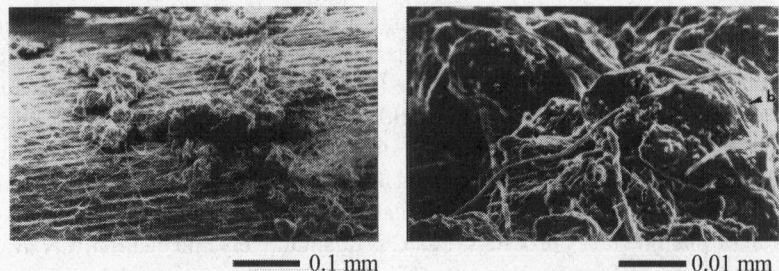

■■■■ 0.1 mm   ■■■ 0.01 mm

Fig. 7.5   Aggregate development on decomposing wheat straw residues. Straw was incubated for 3 months in a silty soil. Fungi enmesh the soil aggregates within mycelial strands and the newly stabilized aggregates are also colonized by bacteria. Low temperature SEM. Bar is 100 μm (a) and 10 μm (b), b = bacteria, f = fungi [D.A. Angers, C. Chenu, and S. Recous, unpublished].

structure in that fraction, and spectroscopic analyses indicated that the composition was relatively uniform irrespective of the composition of the original C source (Golchin et al.,1994) suggesting that the material is largely of microbial origin. If so, its recalcitrance must be partially due to physical protection from attack by microorganisms as a consequence of being adsorbed or located in pores that are too small to be accessible. A more complete discussion on the characteristics of the stabilized SOM is presented in Section B, Chapter 2.

7.2.2.2   Effects of Organic Matter on Pores and their Stability

Pores

Organic matter influences total porosity and pore size distribution. The total porosity that is measured at any given time in nonswelling soils has generally been found to increase (and bulk density to decrease) with increasing organic C content (Anderson et al., 1990; Lal et al., 1994; Schjønning et al., 1994). The magnitude of the increase varies with the soil order (Manrique and Jones, 1991). The dynamic nature of pores with an equivalent diameter > 30 μm may account for the poor correlations between SOM content and macro- or mesoporosity (Thomasson and Carter 1992) or inconsistent correlations (Douglas et al., 1986; Kay et al., 1997). The influence of organic C on these pore size classes would be most obvious where there is least variation in texture, mineralogy, pore fluid composition, climate and management conditions and variability in these factors may also contribute to the difficulty in generalizing across different studies.

The role of organic C on 0.1–30 μm pores has frequently been overlooked or underestimated. Data from site-specific studies from various parts of the world (Karlen et al., 1994; Schjønning et al., 1994; Emerson, 1995) have shown an increase in the volume of 0.2–30 μm pores ranging from 1–10 mL g^{-1} organic C. However, Anderson et al. (1990) found that, although increases in organic C due to the annual additions of manure to a montmorillonitic soil for 100 years increased the volume fraction of pores > 25 μm, the volume fraction of 5–25 μm pores decreased, suggesting that the latter pore class may be less responsive to increases in organic C in soils that are dominated by swelling clay minerals. The increase in the volume fraction of 0.1–30 μm pores with organic C may vary with texture, although there are inconsistencies in the conclusions that have been drawn by several researchers (Emerson, 1995; Hudson 1994; Kay et al., 1997) and further research is required. The increase in the volume fraction of 0.1–30 μm pores with organic C may be attributed to several mechanisms. Although none have been rigorously assessed, possible mechanisms include: pores in POM, sites of plant or animal residue from which most of the C has been mineralized but which are encased in mineral materials enriched in microbial decomposition products, and pores within a matrix of extracellular polysaccharides (Kay, 1998). The volume fraction of small pores also tends to increase with increasing organic C. Calculations based on data from Emerson (1995), Schjønning et al. (1994) and Karlen et al. (1994) show that the increase in the volume of pores < 0.2 μm ranged from 0–3 mL g^{-1} organic C.

Stability of Pores

Where the process of creating the pore does not substantially increase the organic C content of the pore wall, the stability of the pore wall would be expected to be equivalent to the stability of the surrounding matrix which is often inadequate to allow the pore to withstand the effective stress created by drying, rewetting or swelling pressures, or stresses from overburden or traffic. For instance, the organic C content of the pore walls is not increased during tillage and many of the pores created by this process are unstable (Derdour et al., 1993). Gusli et al. (1994) showed that the structural collapse of beds of aggregates on wetting and draining was caused by the development of failure zones on wetting followed by consolidation on draining and that the extent of collapse was greater for soils having lower organic C contents. Macropores created by shrinkage or by earthworms may also be unstable in some situations. For instance, Mitchell et al. (1995) found that these pores failed to provide preferential flow paths within 10 minutes of the onset of flood irrigation of a swelling soil with low organic C content (0.9% C). Pores created by the tap roots of alfalfa (*Medicago sativa* L.) were, however, much more stable, presumably due to increased C content of the pore wall and consequent enhanced stability.

Traffic can cause a loss of porosity and a decrease in pore continuity, the extent of which is determined by the magnitude of the stresses imposed (contact pressure, the number of passes) and the compactibility of soils. Increasing organic C content diminishes the impact of stress on total porosity (Soane, 1975; Angers and Simard, 1986) and on pore continuity (Ball and Robertson, 1994). The influence of organic C on the stability of pores under compaction may also vary with texture and form of the organic C (Soane, 1990). Angers (1990) found increasing organic C only resulted in reduced compactibility of soils at clay content > 35%. Ball et al. (1996) reported a strong negative correlation between maximum bulk density after compaction and NaOH extractable carbohydrates. Fresh and decomposing plant residues may have variable effects on the stability of pores under compaction (Guérif, 1979; Gupta et al., 1987; Rawitz et al., 1994).

7.2.2.3 Effects of Organic Matter on Failure Zones and Strength

The influence of organic matter on failure zones is dependent on the nature of the failure zones, the forms of the organic C, and its spatial distribution and care must be taken into account in interpreting these relations. An increase in porosity, particularly macro- and mesoporosity, would be expected to result in a decrease in strength. An increase in organic C in soil with the same water content can, however, also cause a decrease in water potential (because of the effect on the pore size distribution) with a concomitant increase in effective stress and strength. Kay et al. (1997) used pedotransfer functions to assess the influence of increasing organic C contents on the resistance to penetration of soils with a range of clay contents that were used for corn production under two different tillage treatments and found that (1) an increase in the organic C content caused a decrease in soil resistance when the water potential was constant; and (2) the magnitude of the decrease increased with clay content and with declining potential.

Increases in organic C content that result in increasing levels of polysaccharides would also be expected to increase the cementation between mineral particles (Chenu and Guérif, 1991). However, the impact of increasing polysaccharide content on failure zones may vary with the nature of the soil and the spatial distribution of the organic C. The strength of some soils is strongly influenced by dispersible clay and silt material which functions as cementing material as soils dry. Under these conditions, increasing organic C content (and in particular, the fraction closely associated with the mineral phase) reduces the amounts of spontaneously dispersible mineral material resulting in reduced strength on drying. This mechanism appears to be particularly important in hardsetting soils (Young and Mullins, 1991; Chan, 1995). However, an increase in the strength of bonds between mineral material may also result in an overall increase in strength, if the cementation occurs uniformly throughout failure zones. Measurements on disturbed soils (44% clay, 1.5–2.3% C) packed to bulk densities of 1.25 g cm^{-3} have indicated (Davies, 1985) that shear strength increased with organic C content when measurements were made at similar potentials. Emerson et al. (1994) have inferred a similar trend from measurements of penetration resistance noting that high levels of polysaccharide gels could lead to compacted parts of the mineral matrix becoming so strongly bonded together that, at a water potential of 0.01 MPa, the penetration resistance would be sufficiently high to limit root penetration. Fine roots and fungal hyphae enmesh the soil matrix, and therefore, management practices that increase the amount of this form of organic C can result in an increase in the strength of failure zones. Additional details on the impact of plant roots and fungi on the strength of soils are given in Sections 7.2.5 and 7.2.6.

7.2.2.4   Effects of Organic Matter on Aggregates and their Stability

The role of SOM on aggregates reflects the cumulative effects of the method of incorporation, the dynamics of decomposition (including the transformation of different forms of organic C, their spatial location and their degree of association with the mineral materials), and the influence of organic C on pore characteristics and failure zones (and therefore aggregate stability). Several studies have found a close relationship between the level of aggregation and SOM (some are reviewed by Tisdall and Oades, 1982), and numerous others have shown that incorporation of organic residues usually results in the formation of water-stable aggregates (Lynch and Bragg, 1985). Organic matter and its various fractions can contribute to both the formation and the stabilization of soil aggregates but, under some circumstances, specific fractions of organic matter can also destabilize aggregates and increase the dispersibility of clay and silt sized materials.

Increasing total organic C content normally results in an increase in the size and stability of aggregates, irrespective of the origin of the stress and the trends are most obvious in soils with widely varying SOM contents. The response of soils to wetting with mechanical abrasion will be used to illustrate this generalization. In a comprehensive study of soils from western United States and Canada (< 0.6–10% C), Kemper and Koch (1966) found that the stability of vacuum saturated soils increased with log C with different relations for subsurface layers, and surface layers under sod and cultivation, respectively. The stability decreased most rapidly at organic C contents below 1.2–1.5%. Greenland et al. (1975) found a critical level of 2% C below which soils from England and Wales were very liable to structural deterioration, especially in the absence of $CaCO_3$. Albrecht et al. (1992) also suggested that for a vertisol and a ferrisol, the aggregates were stable if they contained a minimum of 2% organic C.

The physical mechanisms by which organic matter affects aggregate stability are complex. The stability of soil aggregates is controlled by two opposing factors: the development of stresses within the aggregate pore space during wetting due to the compression of the entrapped air, and the strength of the interparticle bonds (Yoder, 1936; Hénin, 1938 cited by Concaret, 1967). SOM influences stability by reducing the rate of wetting and increasing the resistance to stresses generated during wetting (Monnier, 1965; Quirk and Murray, 1991; Rasiah and Kay, 1995; Caron et al., 1996b).

As is the case for other properties that are strongly dependent on the characteristics of failure zones, the water stability of aggregates from a soil can exhibit large changes due to cropping treatments before changes in the total organic C content are observed (Baldock and Kay, 1987). Changes in aggregate stability have been, therefore, attributed to changes in the amounts of various organic fractions such as POM (including fine roots and fungal hyphae), polysaccharides and lipids. These materials are considered to be labile and represent only a fraction of the total C content. Their amounts in the soil at any given time are determined by the rates of plant C input and the mineralization of this C and the microbial byproducts. It is the macroaggregate fraction (> 250 μm) which is mostly influenced by the labile SOM fractions. This generalization is consistent with the model which suggests that for most soils, the nature of the SOM responsible for providing stability varies with the size of the aggregates; water-stable macroaggregates are stabilized by transient and relatively undecomposed organic binding agents while microaggregates would be stabilized by more processed SOM (Tisdall and Oades, 1982). Incubation and isotopic studies have confirmed that the turnover and nature of SOM varies with aggregate size and that macroaggregates are enriched in young and labile SOM (Elliott, 1986; Puget et al., 1995).

The role of POM in stabilizing aggregates can be through several mechanisms. First, as suggested by Tisdall and Oades (1982), fungal hyphae and fine roots can directly bind soil particles (Sections 7.2.5 and 7.2.6). Second, POM serves as substrate for microbial activity which can produce microbial bonding material (Golchin et al., 1994; Cambardella and Elliott, 1993; Jastrow, 1996; Besnard et al., 1996). Tiessen and Stewart (1988) and Oades and Waters (1991) have observed that many aggregates have cores of plant debris, leading to the proposal that encrustation of plant fragments by mineral particles was a mechanism of aggregate formation. POM entering the soil is rapidly colonized by microbes (Fig. 7.5) resulting in byproducts that have strong adhesive properties, causing mineral particles to adhere to them. The dynamics of aggregation and decomposition are linked. Plant residues and POM can be assigned different aggregating potentials depending on their intrinsic decomposability and stage of decomposition, namely, on their ability to support microbial growth (Monnier, 1965; Golchin et al., 1994). Finally, it has been proposed that organic matter can enhance aggregate stability by obstructing soil pores, thereby reducing the rate of water entry and the extent of slaking (Caron et al., 1996). It is conceivable that POM could play this role.

Because polysaccharides are strongly adsorbed on mineral materials, they are particularly effective, at the interparticle level, in strengthening failure zones (Chenu, 1989; Chenu and Guérif, 1991) accounting for the frequent significant correlations that have been observed between aggregate stability and polysaccharide or readily extractable carbohydrate content (Chaney and Swift, 1984; Haynes and Swift, 1990; Roberson et al., 1991; Angers et al., 1993). Further supporting evidence comes from observations showing (1) an improvement in aggregate stability following addition of microbial polysaccharides (Lynch and Bragg, 1985) and (2) disruption of aggregates by periodate treatment (Greenland et al., 1962; Clapp and Emerson, 1965; Cheshire et al., 1983). Using ultrathin sections and specific staining, Foster (1981) observed that polysaccharides were widely distributed in soils and suggested that their locations in very small pores and as clay coatings explained their inaccessibility to microbial degradation and their role in stabilizing clay into aggregates. Because polysaccharides are readily mineralized, their effects on the strength of failure zones will be transient (Tisdall and Oades, 1982) unless the residue source is continually renewed or the polysaccharides are physically protected from attack by microorganisms and extracellular enzymes. Guckert et al. (1975) proposed that following labile organic matter additions, polysaccharides were involved in the early stage of aggregate formation and stabilization and that longer lasting effects involved more humified substances.

Decomposition of organic materials yields increasing proportions of long chain aliphatic compounds (lipids) (Section C, Chapter 4,) which may contribute to biologically induced changes in aggregate stability (Dinel et al., 1991, 1992). Capriel et al. (1990) and Monreal et al. (1995) found high correlations between the quantity of lipidic compounds and the stability of field soils under various management systems. Giovannini et al. (1983) extracted a hydrophobic fraction with benzene and found that this treatment reduced the water stability of aggregates from a water repellent soil. The rate of water entry into an aggregate largely determines its resistance to slaking. Lipidic compounds are hydrophobic and would, therefore, reduce the rate of water entry thus increasing water stability (Capriel et al., 1990). The effects of the long-chain fatty acids was also attributed to polyvalent cation bridging (Dinel et al., 1992).

A significant proportion of the SOM is composed of organic molecules with chemical structures which do not allow them to be classified as biomolecules (Section B, Chapter 2,). These so-called humic substances have strong adhesive properties due to their charge characteristics. Correlations have been found between aggregate stability and the content of humic material (Chaney and Swift, 1984), and their addition to the soil can increase aggregate stability (Piccolo and Mbagwu, 1994). Extraction of humic compounds complexed with Fe and Al with either pyrophosphate (Baldock and Kay, 1987) or acetylacetone (Giovannini and Sequi, 1976) has confirmed the role of humic substances in aggregate stability. Monnier (1965) suggested that humic substances, as the ultimate organic matter transformation products, had smaller but longer lasting effect on aggregate stability than early fermentation products such as polysaccharides. Further, Chaney and Swift (1986a,b) concluded that humic substances provided long lasting stability to microaggregates whereas microbial polysaccharides were more involved in the short-term binding of these microaggregates into macroaggregates, which is in accordance with the Tisdall and Oades (1982) conceptual model.

Although it is clear that specific fractions of SOM have different roles in aggregate formation and stabilization, it is also doubtless that under field conditions most if not all fractions would be involved but at different degrees and at different scales depending on prevailing soil, climatic and cropping conditions. This may explain why in several field studies, correlation analysis shows that most fractions studied are correlated with one another and with aggregate stability (Capriel et al., 1990; Haynes and Francis, 1993; Angers et al., 1993).

Specific fractions of SOM may also, in some circumstances, destabilize aggregates, enhance dispersion, or stabilize materials in suspension that have become dispersed. The effects appear to be most pronounced in kaolinitic soils with low SOM content (Emerson, 1983). The fractions that are responsible are not well defined although soluble organic anions have been shown to cause dispersion (Shanmuganathan and Oades, 1983; Visser and Caillier, 1988). Shanmuganathan and Oades (1983) found the extent of dispersion by soluble organic anions was determined by the magnitude of the change in net surface charge arising from adsorption of the anions. Organic materials exuded by the growing roots of some plants (particularly maize) have also been reported to exhibit dispersive characteristics (Reid and Goss, 1981) although Pojasok and Kay (1990), using exudates extracted from growing maize plants, were unable to demonstrate this behavior.

7.2.2.5 Effects of Organic Matter on Resiliency and Vulnerability

The role, if any, that SOM plays in self-mulching is not well understood although it is probable that it alters the rate of wetting and drying. The rebound of soils after removal of a compressive stress increases with increasing SOM (Guérif, 1979), but the magnitude of recovery is inhibited by high initial bulk densities (McBride and Watson, 1990). Although the strength of failure zones appears to be particularly sensitive to wetting/drying or freezing/thawing events, there is less evidence that pore characteristics are as responsive and very little is known about the role of SOM on changes in either failure zones or pore characteristics. There is a dearth of information on the effects of SOM on soil structural vulnerability. Although the vulnerability of soils probably decreases with increasing C content (since increasing organic C generally increases stability and is unlikely to decrease resiliency), research is needed to confirm this speculation. The degree of vulnerability of soil structure has an important bearing on the long-term impact of land use practices on processes controlling crop productivity and environmental quality. Consequently, the role of organic C on structural vulnerability needs to be much better understood.

7.2.3 Inorganic Sesquioxides and Amorphous Materials

7.2.3.1 Inorganic Cementing Materials

Sesquioxides which increase bonding between mineral particles have the greatest influence on structural characteristics by increasing the strength of failure zones. These materials exist as coatings on the surface of mineral particles and are amorphous or poorly crystallized. Examples include Fe and Al oxides, precipitated aluminosilicates or combinations of these materials that may include SOM. Because of their variable charge and pH dependent solubility, the extent of cementation can be modified by pH. In addition, their effectiveness may be related to the mineralogy of the matrix, exchangeable ions, pore fluid composition and water content. The interaction of Al and Fe oxides with clay minerals and their effect on soil physical properties have been reviewed by Goldberg (1989).

7.2.3.2 Effects of Inorganic Materials on Failure Zones, Pores, Aggregates, and their Stability

The role of cementing materials has been studied most extensively in soils of high strength, with less emphasis on their influence on pore and aggregate characteristics. The strength of hardsetting soils and fragipans has been correlated positively with extractable Si and negatively with Al (Chartres and Fitzgerald 1990; Franzmeier et al., 1996; Norfleet and Karathanasis 1996) (Section G, Chapter 4). Hardsetting soils have low wet and high dry strength. Franzmeier et al. (1996) have hypothesized that hardsetting horizons are formed when silica bonds with Fe oxide minerals on clay surfaces and the bond is strengthened on drying. The influence of amorphous materials on the properties of soils of lower strength is more ambiguous.

Amorphous inorganic materials have dominantly a variable charge nature and can alter the surface charge characteristics of clay minerals when adsorbed, or existing as a surface film. Changes in surface charge characteristics result in changes in particle-particle interaction. For instance, addition of Fe(III) polycations to clay suspensions from a red brown earth resulted in clay flocculation, increased microaggregation and volume of 40–100 μm pores, and decreased bulk density and modulus of rupture (Shanmuganathan and Oades, 1982). Under field conditions, however, many of the positive charges on the oxides may be balanced by SOM functional groups and the role of the inorganic oxides may be linked to that of SOM. For instance, Kemper and Koch (1966) found that the stability of aggregates from 519 soils from the western United States and Canada was positively correlated with Fe_2O_3 but not with Al_2O_3. In a review of inorganic oxide-clay mineral interactions, Emerson (1983) concluded that such effects could be mainly due to the strong interaction between Fe and SOM. Bartoli et al. (1988) have attempted to assess the relative importance of SOM and poorly ordered Fe and Al hydroxides and concluded that no single constituent is likely to adequately describe aggregate stability. In soils with < 6% C and small quantities of metal hydroxides, Bartoli et al. (1988) found evidence that soil polysaccharides were most involved in the aggregation process. The contribution of metal oxides to aggregate stability in soils with low levels of organic C might be expected to increase as the metal oxide content increases. In a study of the aggregation of B horizon of Alfisols and Inceptisols, Barberis et al. (1991) concluded that Fe oxides did enhance aggregation and that oxides in the amorphous form were more effective than those in the crystalline form (Al oxides were not found to have a major effect on stability). In soils with > 6% C and richer in noncrystalline materials, Bartoli et al. (1988) found that the cementing materials were mainly poorly ordered Fe and Al hydrous oxides associated with Ca or Al higher molecular weight organic materials. These analyses suggest that there is a complex relation between noncrystalline inorganic cementing materials and particle-particle interaction and that the relation is strongly influenced by a number of other characteristics of the soil environment. The sensitivity of failure zones and aggregate stability to variation in surface charge characteristics diminishes with increasing crystallinity of the sesquioxides. For instance, Oxisol aggregates rich in gibbsite are extremely stable and nonelectrostatic bonds are much more important than electrostatic ones (Bartoli et al., 1992). Stability increases with SOM content, but the increases have been attributed to lower porosities and greater hydrophobicity (Bartoli et al., 1992).

7.2.4 Pore Fluid

7.2.4.1 Solute Composition, Exchangeable Ions and their Relation to Structural Form and Stability

Pore fluid composition and the exchangeable ion suite influence the interaction between mineral particles, the extent of swelling and, ultimately, the dispersibility of the particles. Changes in the ease of dispersion of particles, particularly at the scale of clay sized particles, result in changes in strength characteristics and the stability of pores and aggregates. The attraction or bonding between mineral particles is sensitive to soil pH, exchangeable Na and/or Mg percentage (ESP, EMgP), and electrical conductivity (EC) of the pore fluid. The influence of one factor, for example, composition of the pore fluid, can be influenced by other factors, such as pH. In addition, the impact of these factors on the dispersibility of clay is strongly influenced by clay mineralogy. The cumulative effect of these factors on clay dispersibility is reflected in the difference in osmotic pressure between the diffuse electrical double layer and the pore fluid; as the difference in the osmotic pressure increases clays become more dispersible (Rengasamy et al., 1993; Rengasamy and Sumner, 1998). The manifestation of differences in osmotic pressure on the dispersibility of clay is also influenced by organic and inorganic cementing materials, which can restrain particles from swelling. Among common clay minerals, only smectites with a high ESP show extensive swelling (Churchman et al., 1993); the extent of swelling decreases as the EC (or osmotic pressure) of the pore fluid increases. Illites with a high ESP often remain dispersed in solutions of high EC, partly because the shape of the particles prevents strong cohesion. Increasing proportions of Ca as exchangeable or dissolved species (and to a lesser degree K [Chen et al., 1983]) reduce dispersion. The dispersion of kaolinites varies with pH since a significant part of their charge is variable. The sensitivity of clay sized materials in soils to dispersion with changes in pH may also be related to the effect of pH on the charge of metal oxides; as pH increases the oxides become more negatively charged and clay minerals which may be coated by these oxides become more dispersible in the presence of Na. Exchangeable Mg enhances the dispersion arising from exchangeable Na, but apparently to a greater extent in illitic soils than in smectitic soils. Clay dispersion is discussed in detail by Emerson (1983), Rengasamy and Sumner (1998) and in Section B, Chapter 6.

As the clay becomes more dispersible, aggregate stability declines and the formation of crusts become more common. Pore stability decreases, dispersed clay lodges in pore necks, and the hydraulic conductivity is reduced. In addition, increased dispersible clay leads to dramatic increases in strength of most soils on drying (Fig. 7.6). The effects of dispersion on soil physical properties are reviewed in more detail by So and Aylmore (1993).

7.2.4.2   Water Content/Potential and Relation to Structural Form and Stability

Soil water content varies continuously under field conditions. Water loss through drainage or evapotranspiration results in decreasing potentials and can cause shrinkage. The decrease in potential increases the effective stress (Bishop and Blight, 1963; Groenevelt and Kay, 1981) with a concomitant increase in strength. In addition, the effectiveness of bonds between cementing materials and mineral particles may increase with decreasing water content. Changes in the effectiveness of cementing materials with decreasing water content can arise from changes in orientation or position or from adsorption/precipitation. Dispersed clay can be reoriented or accumulate in the menisci between larger mineral particles as soils dry (Brewer and Haldane, 1957; Horn and Dexter, 1989). Adsorption, precipitation or crystallization of solutes occurs when drying concentrates solutes in the remaining pore fluid (Gifford and Thran, 1974; Kemper et al., 1989). Changes in the effectiveness of organic cementing materials may also occur (Caron et al., 1992b). These effects may be only slowly reversible and increases in water content can result in decreases in the cohesion of aggregates and increases in the dispersibility of clay that may continue for several days (Caron and Kay, 1992; Caron et al., 1992c).

   The dependence of shrinkage, effective stress and cementation on soil water content cause many other structural characteristics to vary with water content. Measurements of aggregate stability made under conditions in which air entrapment and differential swelling are minimized have shown that the stability of a soil declines with increasing water content (Reid and Goss, 1982; Gollany et al., 1991). The loss in stability applies at the scale of macroaggregates (> 250 μm) as well as at the scale of material < 2 μm (Perfect et al., 1990a; Kay and Dexter, 1990). In cases where air entrapment and differential swelling are present, slaking can occur and water content may have a different effect on

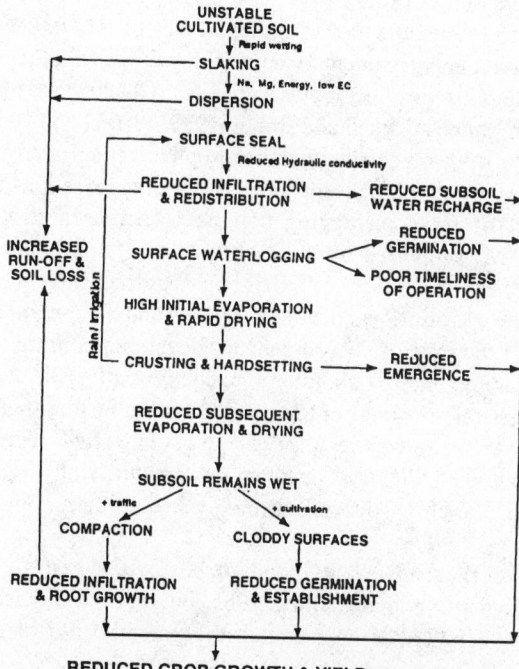

Fig. 7.6   Diagrammatic representation of processes arising from increased dispersion of soil clays [Reprinted from So and Aylmore, 1993. Aust. J. Soil Res. 31:761-777, with permission]

stability (Concaret, 1967). In such cases, stability initially increases with water content and then decreases. The relationship can vary with soil type and cropping system (Kemper et al., 1985; le Bissonnais, 1988; Angers, 1992). Threshold water contents can be used to identify the lower limit of the range in water contents in which a structural characteristic is sensitive to changes in water content (Emerson 1983; Watts et al., 1996 a,b). For instance, after imposing different levels of mechanical energy on aggregates at different water contents, Watts et al. (1996a) found that the dispersibility of clay in aggregates was a function of both energy input and water content. Below a certain threshold water content, approximating the plastic limit, even relatively large energy inputs had little effect on clay dispersion. As the water content increased above the plastic limit, the soil became increasingly sensitive to mechanical disruption.

7.2.5 Soil Microorganisms

Evidence for the involvement of microorganisms in soil aggregate formation and stabilization comes from early work on the effects of adding organic materials to soil. Organic matter additions have little or no effect unless microorganisms are present (Lynch and Bragg, 1985). The role of plant residues in providing a nucleus for aggregate formation was discussed earlier (Section 7.2.2). There are several mechanisms by which microorganisms are involved in stabilizing soil structure: first, by directly providing mechanical binding between soil particles; second, by producing cementing materials during the decomposition of organic materials; and last, microorganisms also serve as a substrate for further microbial growth.

Fungi are believed to be the most efficient group of soil microorganisms in terms of soil aggregation (Lynch and Bragg, 1985). Soil fungal hyphae ramify throughout the soil and bind soil particles together (Dorioz et al., 1993). The degree of soil aggregation has been correlated with hyphal length and biomass indicators (Tisdall et al., 1997; Chantigny et al., 1997). Fungi are believed to be involved in both the formation and stabilization of aggregates (Degens et al., 1996; Tisdall et al., 1997). In addition to providing physical enmeshment by networking in the soil, thereby contributing to the formation of aggregates, fungi also produce polysaccharides (Chenu, 1989; Beare et al., 1997) and other proteic and lipidic compounds (Lynch and Bragg, 1985; Wright and Upadhyaya, 1996) which promote aggregate stability. Both mycorrhizal and saprophytic fungi were found to be effective. Mycorrhizae are believed to be partially responsible for the aggregation of soil particles in the rhizosphere and the effect of roots on aggregation has often been associated with vesicular arbuscular mycorrhiza (VAM) supported by the root systems of some plant species (Tisdall and Oades, 1979; Thomas et al., 1986; Jastrow, 1987). This role was also clearly demonstrated in studies of the revegetation of unstable maritime sand dunes (Sutton and Sheppard, 1976; Forster, 1990). Saprophytic fungi, as part of the suite of microorganisms developing during residue decomposition, are also involved in soil aggregation. Laboratory incubations of soils with or without organic amendments have shown the role of fungi in the initial increase in soil aggregation and aggregate stabilization (Molope et al., 1987; Metzger et al., 1987). As pointed by Degens (1997), the growth of saprophytic fungi is restricted to the soil surrounding the decaying organic residue and their aggregating efficiency will, therefore, depend on the even distribution and fragmentation of the residues in the soil.

Other filamentous microorganisms were found to be involved in soil aggregation such as streptomycetes and microalgae (Lynch and Bragg, 1985) but their mechanisms of action have been much less studied. Metting (1987) has investigated the potential use of a microalgae as a soil conditioner.

Although bacteria are often associated with mineral particles, because of their size they are not likely involved in the direct binding of soil particles to form aggregates (Dorioz et al., 1993). It is rather the product of their activity which is the active binding mechanism. Several compounds are produced by bacteria and microorganisms in general during decomposition of organic residues. Extracellular polysaccharides having binding capacities can be produced in large quantities by soil bacteria (Lynch and Bragg, 1985) and their role in stabilizing soil aggregates was discussed earlier as well as those of long-chain aliphatics or lipids which are also produced during residue decomposition (Section 7.2.2.4).

7.2.6 Plants
Root development is strongly influenced by soil structure. In return, plants also influence the form, stability and resiliency of the structure.

7.2.6.1 Effects on Structural Form
Growing in existing pores or through the soil matrix, roots create compressive and shear stresses which result in the creation of new pores and the enlargement of existing ones. Root influence on pores varies with the stage of growth and decay. Infiltration rates can be reduced by actively growing roots which fill pores. Transport efficiency in root channels is enhanced as the root decays and tissue remnants and associated microflora remain as pore coatings on channel walls (Barley, 1954). For example, alfalfa (*Medicago sativa* L.) is characterized by a large diameter, long and straight tap root and has been reported to create macropores that improve water and solute flow properties (Meek et al., 1989 and 1990; Mitchell et al.,1995; Caron et al., 1996a) and reduce subsoil penetration resistance (Radcliffe et al., 1986). While probably less noticeable than the effect on macropores, the effect of plant roots on smaller pores may be very important to other soil structural properties. Radial and axial pressures exerted by the growing roots will affect the small pores (Guidi et al., 1985; Bruand et al., 1996) by compressing the soil in the vicinity of the roots and decreasing pore size in that immediate zone.

Depending on their clay content and type, some soils show potential to shrink and swell with variations in water content. During shrinking, the loss of water is associated with a loss of volume and the development of cracks in planes of weakness. The type of vegetation present has a marked influence on the soil cracking pattern since the plants cause shrinkage of the soil when extracting water for their use. Grevers and de Jong (1990) found differences due to grass species in macropore structure of a swelling clay soil. They attributed the structural differences to differences in water uptake and thus desiccation of the soil. The greater the plant biomass production, the greater were the area and length of macropores. Since plant distribution varies in space because of associated cultural practices, the cracking pattern will also vary in relation to plants. For row crops, water is used first at the row position and the cracking pattern will develop at the interrow position (Fox, 1964; Chan and Hodgson, 1984). Wetting and drying cycles also influence aggregate formation and fragmentation (Horn and Dexter, 1989), and plant growth as well as precipitation or watering will influence the magnitude, frequency and effects of these cycles on aggregation (Caron et al., 1992a; Materechera et al., 1992). Materechera et al. (1992) attributed the production of small aggregates associated with root growth to soil cracking caused by tensile stress induced by heterogeneous water uptake by the plants.

7.2.6.2 Effects on Structural Stability
The root systems of many plant species form a dense network in soils leading to soil stabilization and reinforcement as found on streambanks (Kleinfelder et al., 1992). Soil shear resistance was found to be greatly improved by the root systems of various plants (Waldron and Dakessian, 1982). On a smaller scale, plant roots and root hairs can also directly enmesh and stabilize soil aggregates of millimeter size (Tisdall and Oades, 1982). Visual (Fig.7.7) and microscopic observations (Foster and Rovira, 1976;

Tisdall and Oades, 1982; Dormaar and Foster, 1991) clearly show that aggregates are formed and stabilized in the immediate vicinity of plant roots. Field and greenhouse studies have demonstrated that growing plants induce the rapid formation and stabilization of soil aggregates (Tisdall and Oades, 1979; Reid and Goss, 1981; Angers and Mehuys, 1988; Stone and Buttery, 1989). Statistical correlations have been found between root length or mass and soil aggregation (Dufey et al., 1986; Miller and Jastrow, 1990). Although fine roots can form a dense network which can probably entangle or enmesh soil particles and form aggregates, indirect effects such as associated microbial activity or the release of binding material have most often been invoked to explain the apparent relationship between fine roots and aggregate stability.

Soil structural stability is influenced by soil water content and its variations with time (Section 7.2.4.2). Water uptake by the plant and the consequent decrease in soil moisture will usually result in increased soil strength (Horn and Dexter, 1989; Lafond et al., 1992). At the aggregate level, soil cohesion is greatly enhanced by decreasing water content and the dispersion of clay size material decreases accordingly (Caron and Kay, 1992). The drying of soil by roots may also act synergistically with the aggregate binding material produced in the rhizosphere and increase soil structural stability. As will be discussed later, organic materials released by the roots and microbial population of the rhizosphere can be efficient in binding soil particles. The drying that occurs in the zone of mucilage production contributes to the efficiency of the binding agents through increased sorption of the binding material onto colloid surfaces (Reid and Goss, 1981; Caron et al., 1992b). Plant roots can promote soil aggregation by releasing material, which can directly stabilize soil particles, or by favoring microbial activity in the rhizosphere which, in turn, will affect soil structure. Morel et al. (1991) provided evidence that intact mucilage released by maize root tips significantly increased soil aggregate stability independent of any microbial activity since it occurred immediately after the incorporation of the exudates in the soil. Pojasok and Kay (1990) found that the release of nutrient ions and C in the exudates of bromegrass and corn roots increased aggregate stability.

The rhizosphere presents a very high level of microbial activity induced by root exudation and mucilage, root sloughing and favorable aeration and water conditions (Bowen and Rovira, 1991). The rhizosphere microbial community is also very diverse. The presence of mycorrhiza in the rhizosphere of many plants is noticeable and their role in soil aggregation has been discussed earlier (Section 7.2.5). Few studies have investigated the contribution of other specific rhizosphere microorganisms to soil aggregation (Gouzou et al., 1993). Microbial exocellular polysaccharides are found in the

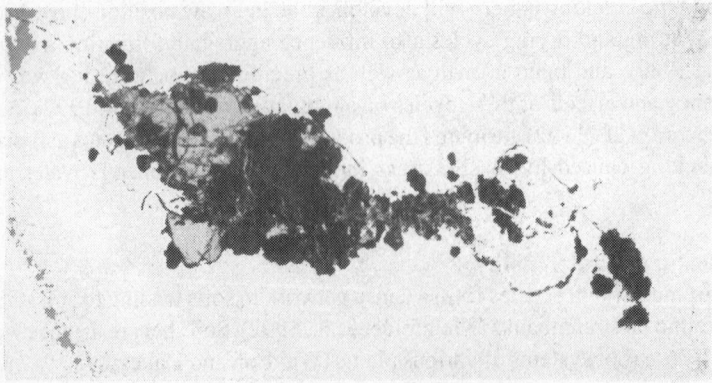

Fig. 7.7   Aggregate formation around timothy roots [Reprinted from Angers and Caron, 1998. Biogeochem. 42:55-72, with kind permission from Kluwer Academic Publishers]

rhizosphere of plants (Bowen and Rovira, 1991) and could act as cementing materials, but their effect cannot be distinguished easily from that of plant mucilages. Rhizobial polysaccharides have also been shown to be efficient in promoting soil aggregation (Clapp et al., 1962). Much remains to be determined about the mechanisms of aggregate formation and stabilization in the plant rhizosphere, and the respective contributions of roots and specific rhizosphere microorganisms is still unclear. Moreover, both biological and physical processes such as drying contribute to aggregate formation and stabilization in the immediate vicinity of the roots through complex interactions.

Aside from the immediate and short-term effects of roots on soil structure which have already been described, plant roots and litter also have a longer term influence through their contribution to bulk SOM. A large proportion of the C fixed by plants is distributed below ground. Consequently, in many ecosystems, plant roots constitute the most important source of SOM, and so have a predominant effect on biologically induced changes in soil structure. Moreover, in many ecosystems, a large part of the aboveground plant is returned to the soil as litter or crop residue, which also constitutes an important C source potentially affecting soil structure.

7.2.7 Soil Fauna

Soil is the habitat of a large number of animals whose biomass and work may well exceed those outside it. Through their activity, animals affect soil chemical, biological and physical properties in various ways (Hole, 1981), but their effects on soil structure are mostly through their movement in the soil and their feeding/excreting activities. Because earthworms have been by far the most widely studied, the following discussion will be limited largely to their effects on soil structure.

7.2.7.1 Effects on Porosity

Soil fauna form channels as they move in soil contributing greatly to the formation and maintenance of porosity. Among the different groups of soil animals, earthworms, termites and ants are believed to contribute to the formation of pore whereas smaller animals like microarthropods are confined to preexisting burrows because of their small size (Lee and Foster, 1991). The pores formed by earthworms are often large (macropores) and cylindrical. When connected to the soil surface, they contribute to water infiltration (Ehlers, 1975) and gas exchange. Water infiltration rate is well correlated with earthworm population (Bouché and Al-Addan, 1997). Pores formed by soil animals also improve root growth (Edwards and Lofty, 1978). Anecic species, which feed at the soil surface, create vertical pores whereas endogenic species feed within the soil and provide more or less horizontal voids, which can be connected with vertical voids. Biopores formed by earthworms are believed to be relatively stable, although this may not be the case in swelling soils in which lateral pressure can be high (Mitchell et al., 1995). Dexter (1978) showed that the soil around burrows was not compacted and concluded that earthworms contributed more to the formation of pores by ingesting soil than by compressing it. Pores are coated with a thin lining called the drilosphere (Bouché, 1975, cited by Lee and Foster, 1991) enriched in various components and oriented clays, which presumably provide some stability to the walls.

Termites and ants live in colonies, and therefore, their effect on soil porosity is concentrated in discrete areas (Lee and Foster, 1991). Although their population may be very high they affect, at a given time, only a small proportion of the soil volume. More work is required to determine the effects of termites on soil structure.

7.2.7.2 Aggregate Formation and Stabilization

In the case of earthworms, ingestion of organic and fine-textured inorganic materials results in intimate mixing and the excretion of casts which have structural characteristics vastly different from the bulk soil. Aggregates created in the presence of earthworms are generally more stable than those created in their absence (Blanchart, 1992) and the size of the formed aggregates depends on the size of the earthworms (Blanchart et al., 1997). However, earthworm casts can be unstable while they are fresh and wet (Shipitalo and Protz, 1988; Marinissen and Dexter, 1990). Preexisting bonds between soil particles prior to ingestion are probably disrupted in the gut of the animal where large amounts of watery mucus are produced and the soil is mechanically dispersed. At excretion, casts are very moist, slurry-like and relatively unstable but will gain stability with drying and aging (Shipitalo and Protz, 1988). Microbial activity and, in particular, fungi (Marinissen and Dexter, 1990) and the presence of coarse organic fragments (Shipitalo and Protz, 1988) also stabilize earthworm casts. Marinissen et al. (1996) concluded that orientation of clay particles following passage in the worm also leads to closer contact resulting in the stabilization of the casts.

As emphasized by Lee and Foster (1991), little quantitative information is available on effects of termites or ants on aggregate formation and stabilization. A termite species (*Thoracotermes macrothorax*) was shown by Garnier-Sillam et al. (1988) to have a positive influence on structural stability which was attributed to the enrichment of the termitary soil with organic matter and cations.

7.3   Other Factors Influencing Soil Structure

7.3.1   Climate

Climate controls the temporal variation in the water content and temperature of soils and has direct effects on a range of physical and biological processes linked to soil structure. Combeau (1965) illustrated how temperature and soil moisture controlled the seasonal variation in structural stability for both tropical and temperate soils. The impact of temporal variation in water content on structural characteristics is strongly influenced by the rate at which the water content changes, and the swelling characteristics of the soils. Slow changes in water content arising from low intensity rainfall events, drainage and evapotranspiration can cause changes in the pore characteristics in both swelling and nonswelling soils. In the case of swelling soils, the changes are a result of swelling or shrinkage and crack formation, whereas in nonswelling soils, changes are restricted to the seedbed where decreasing water content and a concomitant increase in effective stress causes a loss of interaggregate pore space and a progressive consolidation of the seedbed.

The seasonal variation in structural characteristics that are strongly influenced by failure zones (tensile strength and aggregate stability) is often much larger than changes caused by different cropping practices and this variation can be related to wetting events preceding the time of sampling (Kay et al., 1994b). The magnitude of the variation in water content and its rate of change decrease rapidly with depth in the profile. It is, therefore, not surprising that major differences in the strength of failure zones can be found with differences in depth within the A horizon as small as 5 cm (Kay et al., 1994b).

Changes in the liquid water content of surface soils, as a consequence of ground freezing, are common in many soils. The pore water may freeze *in situ*, and where the soil is saturated, the increased volume can create stresses that may be expected to result in failure zones. A much more common feature, particularly on medium and fine-textured soils, is the migration of water in response to gradients in water potential at the freezing front and the subsequent accumulation and freezing of water in the form of lamella or ice lenses just behind the freezing front (Fig. 7.8). The pores that are created under these conditions do not appear to be stable and most are lost as the soil consolidates during the thaw period (Kay et al., 1985). The sites of ice lenses must, however, represent zones of very low strength and undoubtedly contribute to the loss in stability (Willis, 1955; Bullock et al., 1988) and strength (Utomo and Dexter, 1981b; Voorhees, 1983; Douglas et al., 1986) of large aggregates that are often observed as a consequence of freezing. Zones between ice lenses are desiccated and the rearrangement and flocculation of clay sized particles (Rowell and Dillon, 1972; Richardson, 1976) in these areas may account for the increased stability of small aggregates that is often observed after freezing (Perfect et al., 1990b,c).

7.3.2 Management

Land is managed for a range of uses and each use may involve a suite of management practices. Management of soil, water and plants influences soil structure by controlling the form and amount of C entering the soil and its spatial distribution in the soil. These practices also influence the populations of soil macro- and microorganisms and the mineralization rate of C. The nature of soil cover provided by plants or crop residue influences soil structure through its influence on raindrop impact, soil water content, the rate of wetting and the depth of freezing. The nature of the root system of the crops selected will influence the depth of water extraction, and therefore, the depth to which shrinkage may occur. In addition, macropores can be created by tillage and destroyed by traffic.

There are an infinite number of combinations of management options used in crop production under various soil and climatic conditions. Consequently, the number of case studies reported in the literature is large. The following discussion is an attempt to illustrate how selected management combinations affect soil structure under selected soil and climatic conditions. Detailed reviews on management effects on soil structure, with particular reference to aggregation and organic matter, are available for various soil and climatic conditions (Haynes and Beare, 1996; Dalal and Bridge, 1996; Feller et al., 1996; Angers and Carter, 1996).

7.3.2.1 Effects of Crops

The specific effects of plants on both soil structural form and stability and the mechanisms involved were discussed in Section 7.2.6. This section will review some site-specific studies, mostly from Eastern Canada, to illustrate the effects of various crop sequences on soil aggregation and aggregate stability. For a discussion on the role of plants on porosity, the reader is referred to Section 7.2.6.

Fig. 7.8 Macromorphology of ice enriched frozen silt loam [Reprinted from Kay et al., 1985. Soil Sci. Soc. Am. J. 49:973-978, with permission of the Soil Science Society of America]

Beneficial effects of perennial forages on soil aggregation are well-known. Several studies have shown that water-stable aggregation increases rapidly when arable or degraded lands are put into continuous perennial forages. Maximum stability is often achieved after 3–5 yr. (Low, 1955; Perfect et al., 1990c; Angers, 1992). Choice of perennial forage crop species and cultivar will also influence the extent of aggregate formation and stabilization. Alfalfa was as efficient as bromegrass but slightly less than timothy (*Phleum pratense* L.) and reed canarygrass (*Phalaris arundinacea* L.) in improving the water-stable aggregation of two fine-textured soils (Chantigny et al., 1997). Carter et al. (1994) showed that the potential for increasing the stability of a loamy soil from the eastern seaboard of Canada varied with grass species but also with cultivar. Management of perennial forages should also affect aggregation through factors which influence crop C production. However, Perfect et al. (1990c) did not observe any effect of a cutting regime of red clover (*Trifolium pratense* L.) or N fertilization of bromegrass on aggregation. Experimental results on the effects of short-term rotations and cover crops on soil aggregation have been inconclusive. In some cases cover crops and rotation with a legume or a grass legume mixture significantly improved soil macroaggregation (Webber, 1965; Raimbault and Vyn, 1991) but in others little or no effects were observed (MacRae and Mehuys, 1987; Carter and Kunelius, 1993). The effects of short-term rotations and cover crops on water-stable aggregation are highly dependent on crop species and, in particular, the amount of residues that each crop of the rotation returns to the soil in the form of either roots or aboveground residues as well as the tillage and fertilizer practices, and the soil water uptake pattern associated with the plant grown in each phase of the rotation.

7.3.2.2   Organic Amendments and Fertilization

There is a multitude of organic materials applied to agricultural soils. The incorporation of organic amendments such as cattle manure, compost, and industrial and municipal organic wastes usually results in a decrease in bulk density and a consequent increase in soil porosity (Hardan and Al-Ani, 1978; Pagliai et al., 1981; Churchman and Tate, 1986; N'Dayegamiye and Angers, 1990). Pagliai et al. (1981) found that the different pore size fractions were all increased by the addition of municipal sludge. The beneficial effects of farmyard manure are also observed on soil aggregation under various soil and climatic conditions (Mazurak et al., 1977; Ketcheson and Beauchamp, 1978; Sommerfeldt and Chang, 1985; Dormaar et al., 1988; N'Dayegamyie and Angers, 1990). The application of other organic materials such as compost, sewage sludge and lignocellulosic materials is also probably beneficial to soil aggregation, although they vary widely in composition and decomposability and their efficiency to promote the aggregation. Their aggregating efficiency has been related to the decomposability of the material (Martin and Waksman, 1940; Monnier, 1965; Tisdall et al., 1978).

Commercially produced organic polymers have been used to stabilize structure thereby reducing seal formation, increasing infiltration, and diminishing runoff and erosion. Their effectiveness varies with polymer characteristics (type and amount of charge, configuration and molecular weight), soil properties (clay content and mineralogy, EC, and pH), methods of application (spraying on soil surface or applying in irrigation water) and amounts applied (Stewart, 1975; Seygold, 1994). The cost of commercially produced polymers and the need for repeated applications continue to constrain their widespread use in agriculture.

Mineral fertilizers, in particular N, can have both positive and negative effects on soil aggregation. In the short term, N fertilization can accelerate mineralization of organic binding agents (Acton et al., 1963), whereas in the long term, N fertilization increases C content by improving crop yield and crop residue inputs to the soil and potentially results in increased aggregation (Campbell et al., 1993). Kay (1990) suggested that the potential improvement in soil aggregation associated with increased production of roots due to fertilization may not be fully realized if fertilization also causes increased rates of mineralization of the binding material.

7.3.2.3   Tillage Practices and Controlled Traffic

The impact of tillage on soil structural form will depend on the type of equipment used, initial structural form, soil water content at time of tillage and frequency of tillage. In the short term, tillage usually results in a decrease in bulk density and, therefore, increases total porosity in the tilled soil layer. Large pores are usually created which favor fluid transmission and root growth in the surface soil (Ehlers, 1977; Carter, 1988; Klute, 1982). However, tillage can disrupt the continuity of the pore system created by root and faunal activity in the surface soil and between this layer and deeper nontilled horizons (Ball, 1981; Goss et al., 1984). It may also create compacted zones at the bottom of the plowed layer (Bowen, 1981).

In the long term (> 5 yr), the effects of the absence of tillage on total porosity compared to plowed soils may be either positive, negative or absent (Voorhees and Lindstrom, 1984; Heard et al., 1988). More importantly, no-till soils usually show greater macropore continuity which results in higher hydraulic conductivity (Heard et al., 1988; Edwards et al., 1988). van Lanen et al. (1992) observed a better structure, as assessed by greater porosity and hydraulic conductivity, in a permanent grassland soil than in an young arable soil after 8 yr due to intensive tillage in the latter.

As discussed earlier, compaction by vehicular traffic or animals can be very detrimental to soil structure as it reduces soil porosity and increases soil strength, and ultimately can reduce crop yield and increase the risk of water erosion. Although damage caused by compaction can, in some cases, be alleviated by subsoiling or by natural processes of wetting/drying, it is generally agreed that prevention by controlling traffic or maintaining SOM levels is more efficient. Controlled traffic involves traversing the same tracks in the field, thus restricting structural damage to zones of traffic (Unger, 1996).

Tillage practices also have a strong influence on soil aggregation and structural stability. Several studies have shown that surface soils (0–10 cm) under no till contain larger and more stable aggregates than their tilled counterparts (Carter, 1992; Beare et al., 1994a) due to the combined effects of crop residue accumulation at the surface and the absence of mechanical disturbance in no-till systems. Mechanical stress would result in both a direct breakup of the aggregates and oxidation of SOM which stabilizes the aggregates. Labile C fractions which are involved in stabilizing aggregates such as polysaccharides and fungal hyphae are generally present in lower concentrations in tilled than in no-till soils (Beare et., 1994b; Angers et al., 1993). The spatial distribution of incorporated crop residue, which may influence C dynamics and subsequently aggregate stability, varies with the type of tillage (Allmaras et al., 1996).

7.3.2.4   Managing Water

Management of soil water influences soil structure primarily by altering the resistance of structural form to deformation when stress is applied. Soil water can be considered from the perspective of both quantity and quality and is primarily managed through irrigation and drainage, and to a lesser extent through management of runoff and evaporation. Farmers have supplemented rainfall with irrigation to meet the needs of growing crops for centuries. There is ample historical evidence of the need to prevent the accumulation of Na in irrigated soils and the associated decline in stability, infiltration and drainage and increased runoff and hardsetting at the soil surface (So and Aylmore, 1993).

Drainage is used to remove excess water and, in rainfed environments, the improved aeration and timeliness of operations can result in changes in both soil and crop management practices. The effects of drainage on soil structure arise from changes in organic C dynamics (including the quantity and quality of C added as well as the rates of mineralization of soil C and added C), and changes in the probability of degradation by tillage and traffic (related to changes in stability and the magnitude of stress applied). Little research has related changes in soil structure to changes in drainage. Hundal et al. (1976) found that drainage increased alfalfa and timothy yields on a clay soil resulting in greater hydraulic conductivity, lower strength and decreased bulk density after 16 yr. The tillage/traffic and yield response accounted for the improvements in structure with improved drainage. Although yields of corn increased with improved drainage in soils of similar texture (Fausey and Lal, 1989), plant available water was lower on the drained treatments (Lal and Fausey, 1993). These changes may have been less influenced by the production of corn (for 5 yr.) than by the decline in soil organic C content with improved drainage (Fausey and Lal, 1992).

7.4   Interpreting Data on Soil Structure

Soil structure has a major influence on the functions of soils in ecosystems to support plant growth, cycle carbon and nutrients, receive, store and transmit water, and resist soil erosion and the dispersal of chemicals of anthropogenic origin. A case has been made that particular attention must be paid to soil structure in managed ecosystems where human activities can cause both short- and long-term changes that may have positive or detrimental impacts on these functions. Consequently, emphasis will be placed on interpreting data on soil structure in relation to functions that soils fulfill in managed ecosystems, particularly agroecosystems.

There have been two major practical objectives for doing research on soil structure in agroecosystems during the past 50 yr. The first one emerged from observations that the ability of soil to fulfill different functions was related to its resistance to various types of stresses (initially those related to raindrop impact and rapid wetting, wind and later those stresses caused by tillage and traffic). This led to a large body of work on various aspects of soil aggregation and structural stability (particularly aggregate stability). The second purpose was to relate its various characteristics to seedling development, plant growth and productivity in an attempt to better understand factors controlling yield. More recently, interest related to the protection of the environment has stimulated research in two additional areas: the role of soil structure in C and nutrient cycling, and in controlling water and solute flow (particularly with respect to macropore or bypass flow). The degree to which data on soil structure can be interpreted with respect to the different functions of soil is therefore related to their evolutionary stage. Emphasis here will be placed on the relation of structure to plant growth, C and nutrient cycling, and soil erosion while the aspects of soil structure dealing with the ability of soil to receive, store and transmit water and to resist the dispersal of chemicals are covered in Section A, Chapters 4 and 6.

7.4.1   Plant Growth

7.4.1.1   Germination, Emergence and Growth of Seedlings
Structural conditions in the seedbed that promote rapid germination, uniform emergence and rapid growth of seedlings are desirable. Where seedbeds are created by tillage or seeding operations, the principal characteristic of soil structure that can be varied is aggregate size distribution which determines the availability of O_2 and water under a given climatic condition and the resistance offered to emerging shoots and roots. Ideal conditions for a seedbed are produced by stable aggregates not <0.5–1.0 mm and not >5–6 mm diameter (Russell, 1973, Schneider and Gupta, 1985). Qualitative ratings have been developed for aggregate size distributions by assigning different weighting values to

aggregate size fractions (MacRae and Mehuys, 1987; Braunack and Dexter, 1989). Although the relative values of the weightings that have been used by investigators differ, the greatest weighting is applied to 1–2 mm aggregates and lowest to > 5 mm and < 0.2 mm aggregates. While these criteria represent useful broad generalizations, the optimum aggregate size distribution will also vary with both crop and climate. For instance, finer seedbeds may be required when crops with smaller seeds are planted (Hadas and Russo, 1974) or where precipitation is low (Braunack and Dexter, 1989). Measurements of aggregate size distributions are most relevant to the germination and early growth of plants on soils that are tilled, structurally stable, and are not compacted by traffic. The measurements will have less relevance to later growth, or to early growth on untilled soils or tilled soils with characteristics that do not persist.

The persistence of the seedbed characteristics during the relatively short period of germination, emergence and early growth is related to the structural stability and the amount of energy that may be applied. During this period, the surface of the seedbed is particularly vulnerable to stresses arising from rainfall that can break up aggregates and disperse clay and fine silt if the structural stability is low. Subsequent drying results in the formation of a crust. A crust reduces infiltration rates and the availability of O_2 but has the most profound impact on the emergence of seedlings. When the mechanical strength of the crust approaches or exceeds the force that can be exerted by the growing seedling, emergence becomes nonuniform or even prevented. Although there are few generalized measures of soil structure to interpret the susceptibility of soils to form crusts of high strength, the mechanical strength of the crust that arises from a storm of a given intensity must be related to the susceptibility to disaggregate and disperse (le Bissonnais, 1996) and to the susceptibility of the resulting matrix to achieve high strength on drying.

7.4.1.2   Plant Growth, Development and Yield

Subsequent to emergence, the growth and development of plants and final crop yield continue to be influenced by structural characteristics that control aeration, the availability of water and the resistance offered to penetration of the soil by the growing roots. The structure will reflect the coalescence of the seedbed, traffic-related compaction and soil and environmental characteristics. The extent to which soil structure influences water and O_2 supply and the mechanical impedance offered to root development strongly depends on water content and evaporative demand. Under rainfed conditions, this means that the impact of a given structure on plant growth varies with climate.

After tillage, the coalescence of the structure of the seedbed will be controlled, in part by the pattern of rainfall and subsequent evapotranspiration (and associated stresses) and structural stability. Macropores will be progressively lost (Kwaad and Mucher, 1994), relative compaction will increase (Carter, 1990), and hydraulic conductivity (particularly near saturation) will decline (Mapa et al., 1986). The rates of change in these characteristics may vary with the initial conditions of the seedbed (including those related to position relative to the row [Cassel, 1983]) and are not readily predictable since the stresses related to wetting and drying will vary from year to year depending on climatic conditions (Carter, 1990a). In addition, there have been few attempts to define quantitative relations between stability characteristics, stress and the rate of coalescence of the seedbed.

Traffic and associated compaction also lead to changes in the structure of the seedbed. The processes and mechanics of soil compaction are described in Section A, Chapter 2. As mentioned earlier, attempts to relate soil aggregate characteristics to susceptibility to compaction, which is related to the preconsolidation state of the soil and its water content, have been relatively unsuccessful (Angers et al., 1987, Angers, 1990). Soil structural characteristics more likely to be related to compactibility are, therefore, initial bulk density and soil water content which determine soil strength. The degree of compaction of soils, whether arising from traffic or from coalescence, is best described by the relative compaction (RC), namely, the observed bulk density divided by the maximum bulk density under a standard compaction treatment (Häkansson, 1990; Carter, 1990a), when soils with different inherent characteristics (texture and organic C contents) are compared. The RC tends to normalize bulk density with respect to variation in inherent characteristics and differences in RC primarily reflect changes in macro- and mesoporosity (Carter, 1990a). Studies of the relation between RC and relative yields (observed yield divided by the maximum yield) have shown similar relations in Scandinavia (Häkansson, 1990) and eastern Canada (Carter, 1990a); relative yields of cereals vary curvilinearly with RC and are maximum at RC values between 0.77 and 0.84. The relation between RC and relative yield (including the optimum values of RC) must reflect an integration of functional relations between (1) compaction and the structural characteristics controlling the availability of O_2 and water to the plant as well as resistance of the soil to penetration by roots, (2) compaction and the temporal variation in water content, and (3) plant response to stresses arising from detrimental soil structure/water content conditions. The degree to which these relations may vary with soils, climate and crop species (or perhaps even varieties) is not well understood. Although RC appears to be a particularly useful characteristic to quantify the extent of compaction of different soils arising from wetting and drying or traffic, interpretations with respect to the yield of different crops under different climates should be made with caution.

The availability of O_2 to plant roots is determined by O_2 diffusion rates, which decreases as an increasing proportion of the pore space becomes water filled, and approaches zero when there are no continuous air-filled pores remaining. This limiting condition is strongly influenced by the volume fraction and continuity of macro- and mesopores. It is, therefore, not surprising that aeration is spatially variable between these pores and within aggregates (Zausig et al., 1993). While direct measurements of O_2 diffusion rates are the preferred way to assess the influence of soil structure on aeration since the measurements reflect the influence of both pore size distribution and continuity (Hodgson and MacLeod, 1989a), such measurements often cannot be made and aeration must be inferred from measurements of other structural characteristics, most often those related to pore size distribution. One such characteristic, the volume fraction of air-filled pores, is about $0.10 \, m^3 \, m^{-3}$ (Xu et al., 1992) although limiting values of 0.05 and 0.145 $m^3 \, m^{-3}$ have been reported (Ehlers et al., 1995; Hodgson and MacLeod, 1989b) when the diffusion rates approach zero. Other investigators have found no relation between diffusion rates and air-filled porosity (Zausig et al., 1993). Inconsistent interpretations of the influence of air-filled porosity on O_2 diffusion may reflect variation in the contribution of pore continuity to the measurements. The air-filled porosity, as a measure of aeration, has also been linked to cessation of rapid drainage (Thomasson, 1978). Under the soil conditions in England and Wales, Thomasson (1978) found that the water potential of mineral topsoils approached a value of –0.005 MPa as rapid drainage ceased (this potential corresponds to pores with an equivalent diameter of > 60 μm being air-filled) and he referred to the air-filled porosity at this potential as aeration capacity. Soils with an aeration capacity > 0.15, 0.10–0.15, 0.05–0.10 and < 0.05 $m^3 \, m^{-3}$ were considered to have aeration characteristics that were very good, good, moderate and poor, respectively. He also noted that the aeration capacity requirement was dependent on climate and should be increased in wetter parts of the country. Aeration conditions that are limiting for plant growth will also vary with plant species and with the demands for O_2 by the soil microbial population.

Pore characteristics also determine the amount of water that is potentially available to plants and the rate at which water is transported to the growing roots. Thomasson (1978) defined potentially available water as that held between potentials of –0.005 and –1.5 MPa (corresponding to water held in pores with equivalent diameters of 60 to 0.2 μm) and considered values of 0.20, 0.15–0.20, 0.10–0.15 and < 0.10 $m^3 \, m^{-3}$ to be very good, good, moderate and poor, respectively. Once again, it was recognized that the qualitative ratings were climate dependent and the water contents should be adjusted downward in wetter regions.

Different sized pores dominate the influence of pores on aeration and on potentially available water and, therefore, there is value in trying to represent the effects of pore size distribution on both aeration and potentially available water. Thomasson (1978) applied qualitative assessment to both aeration and available water (Fig. 7.9) to develop a classification of the structural quality of topsoils. Although the numerical values in such a representation may be climate dependent, there is obvious value in combining the two structural characteristics for the purpose of interpreting soil structure information in relation to plant growth. Perhaps the greatest limitation in this approach is that there is no direct provision for the influence of structure on the resistance of soil to penetration by roots. Root growth can occur in macropores which have dimensions similar to those of the root, but most growth involves deformation of the existing pore structure to accommodate the root. Root growth decreases as the resistance offered to penetration increases. Soil resistance to penetration has been measured with metal probes (penetrometers) with diameters ranging from 150 μm to about 1 cm. Neither the size, shape nor friction characteristics of these probes simulate roots very well but the penetration resistance that is measured is well correlated with root length and elongation rate in soils with few macropores. However, in soils with an abundance of macropores, measurements of penetration resistance may be less relevant (Stypa et al., 1987). Although the functional dependence of root growth on penetration resistance depends on a number of factors including species (Materechera et al., 1991), root growth is often found to approach minimum values when the soil resistance is about 2 MPa (Taylor et al., 1966). Resistance to penetration increases with bulk density and decreasing water content.

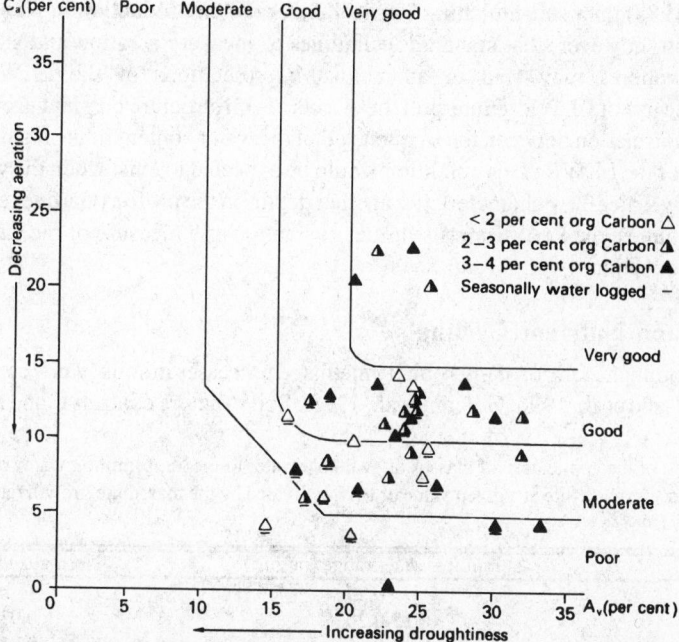

Fig. 7.9 A classification of structural quality of topsoils based on aeration (C_a) and available water (A_v) [Reprinted from Thomasson, 1978. J. Soil Sci. 29:38-46, with permission from Blackwell Science Ltd]

Aeration, available water and soil strength have been incorporated into a single characteristic referred to as the least (or non) limiting water range (LLWR) (Letey, 1985). The LLWR is defined by water contents at which aeration, water potential and mechanical impedance reach values that are critical or limiting to plant growth. The upper limit of this range is defined by the water content at field capacity or the water content at which aeration becomes limiting, whichever is smaller. The lower limit is defined by the water content at the permanent wilting point or the water content at which penetration resistance becomes limiting, whichever is higher. The LLWR integrates many of the characteristics of pores into a single parameter and does so in a way that is directly related to plant growth. Using limiting values for aeration and penetration resistance of 0.10 $m^3 m^{-3}$ and 2.0 MPa, respectively, and the potential at field capacity and permanent wilting point of 0.01 and 1.5 MPa, respectively, da Silva and Kay (1997b) calculated the LLWR for different soils under humid temperate conditions in southwestern Ontario, Canada and found that the growth of corn plants decreased linearly with increasing frequency of the soil water content falling outside of the LLWR. Under similar climatic conditions, the frequency of this occurrence would be expected to increase as the LLWR gets smaller in soils that drain freely, and was found to follow a logistic type function (da Silva and Kay, 1997c). These data could be used to define the quality of soil structure for plant growth using classes of the LLWR based on classes for available water given by Thomasson (1978) (Table 7.3). Studies in Australian orchards (Emerson et al., 1994), using the same limiting values for aeration, available water and resistance to penetration as used by da Silva and Kay (1987b), showed that small values of the LLWR coincided with a paucity of peach roots.

The value of using the LLWR to interpret the effects of soil structure on plant growth rather than using one of aeration capacity, potentially available water or soil resistance to penetration is illustrated in a survey of eight Canadian soils (Topp et al., 1994). Over 90% of the horizons tested developed a penetrometer resistance > 2 MPa at water potentials higher (less negative) than -1.5 MPa and nearly 50% of the horizons had aeration limitations at field capacity. Preliminary studies (McKenzie et al., 1988) on a self-mulching Vertisol under cotton production in New South Wales, Australia, suggested, however, that standard techniques to measure aeration and strength on soils with vertical macropores may lead to anomalous interpretations of the LLWR. Additional complications of the use of LLWR to interpret the effects of soil structure on plant growth arise when there is little or no correlation between the frequency that the water content falls outside of the LLWR and the magnitude of the LLWR. This condition would be expected to exist when the water content is largely controlled by water flow characteristics at other depths in the profile (Gardner et al., 1984). On the basis of these studies, the LLWR merits further evaluation as a measure of the structural quality of soils for crop production.

7.4.2 Carbon and Nutrient Cycling

Soil structure is a dominant control of microbially mediated processes in soils (van Veen and Kuikman, 1990; Juma, 1993; Ladd et al., 1996; Golchin et al., 1998). Soil structure controls C and nutrient cycling

Table 7.3 Illustration of the application of classes of available water to the least limiting water range (LLWR) and normalized growth rates (growth rate at a given value of LLWR divided by the maximum growth rate) calculated from da Silva and Kay (1997b)

Structural Quality	Least limiting water range ($m^3 m^{-3}$)	Normalized growth rate
very good	> 0.20	> 0.97
good	0.15-0.20	0.93-0.97
moderate	0.10-0.15	0.82-0.93
poor	< 0.10	< 0.83

through its influence on habitable pore space and soil water content (Killham et al., 1993). Soil structure is characterized by its high spatial heterogeneity even on a very small scale, which contributes to heterogeneity in microbial activities. As pointed out by van Veen and Kuikman (1990), large amounts of readily decomposable compounds can be found in the vicinity of starving microbial populations, suggesting that organic matter may be rendered inaccessible to decomposers through chemical and physical mechanisms. The latter is often referred to as physical protection.

In many soils, a large proportion of the pores are too small to be accessible to microbes (Elliott et al., 1980; van Veen and Kuikman, 1990) and the SOM located therein may be physically protected from decomposition. Assessment of critical pore neck sizes necessary to constrain specific biological activities is difficult considering the wide diversity of organisms involved in the processes of C and nutrient turnover in soils (Ladd et al., 1996). A few attempts have been made to measure the influence of soil pore characteristics on specific microbial activities. Killham et al. (1993) observed that glucose turned over faster when incorporated into larger pores (6–30 μm) than into smaller pores (< 6 μm). Compaction, which reduces total pore volume and increases the percentage of small pores less accessible to decomposers, may retard the decomposition of labile C compounds (van der Linden et al., 1989 cited by van Veen and Kuikman, 1990) and may reduce N mineralization from plant residues (Breland and Hansen, 1996).

The biological formation and stabilization of soil aggregates also provide stability to the organic compounds located inside aggregates. Indirect evidence for this feedback mechanism is found in experiments showing slower turnover of organic molecules in aggregated than in nonaggregated soils (Adu and Oades, 1978). Encrustation of plant residues with minerals provides protection from decomposition (Golchin et al., 1994). Within aggregates, POM had a slower apparent decomposition rate than free POM (Besnard et al., 1996; Gregorich et al., 1997) due to physical protection by the aggregates. The formation of stable aggregates was held partly responsible for the accrual of stable C observed in restored prairie soils or under conservation management (Jastrow, 1996; Beare et al., 1997; Angers and Chenu, 1998). Disruption of soil aggregates usually leads to a flush in C and N mineralization which has been attributed to the release or increased accessibility of physically protected SOM (Rovira and Greacen, 1957; Hassink, 1992; Beare et al., 1994a). Further discussion on mechanisms providing recalcitrancy to SOM in soils can be found in the reviews previously cited and in Section B, Chapter 2. Although soil structure undoubtedly influences carbon and nutrient cycling, the heterogeneous nature of soil structure and the complexity of the relationships between organic compounds and the mineral material do not yet allow definitive interpretations and generalizations in the form of models.

7.4.3 Erosion

Very early on erosion by both wind and water was recognized to be largely dependent on the degree and stability of soil aggregation. Erosion by water is determined by a number of factors related to soil and climatic conditions, and land management (Section G, Chapter 7). Soil erodibility, which defines the intrinsic susceptibility of soils to erosion, is largely a function of soil aggregate stability, which is in turn influenced by intrinsic soil properties. As mentioned earlier, aggregate stability was originally used to characterize erodibility (Yoder, 1936). Increasing the aggregate stability reduces the detachability of the particles and reduces the susceptibility to surface crusting, which are two factors favoring erosion. Due to the difficulty of making field erodibility measurements on a large scale, predictive tools based on easily measured soil properties are necessary. However, assessment of aggregate stability to predict water erodibility has not always been successful. This is due to the complexity of the processes involved and the numerous interactions among factors. Among those, the nature and rate of wetting of aggregates which should simulate natural processes are probably the most significant. As discussed earlier, methods involving rapid wetting are most relevant to simulate conditions of environments involving intense rains or quick irrigation, whereas slow wetting is more relevant to subsurface soil or in areas with less intense rainfall. le Bissonnais and Arrouays (1997) illustrated the close relationship that exists between the proportion of water-stable aggregates, soil organic matter and the susceptibility to crust formation and erodibility of French loamy soils. Their results emphasized the importance of wetting procedures in assessing erodibility from aggregate stability measurements. Their best relationship was with stability measurements involving slow wetting. They proposed that rapid wetting would be more appropriate in discriminating between soils of different stability.

Wind erodability is directly influenced by the size, shape and density of the structural units, and by their mechanical stability (Chepil and Woodruff, 1963) (Section G, Chapter 8). Predictions of wind erodibility from soil structural properties have been attempted. The Wind Erosion Equation relates proportion of aggregates > 0.84 µm and wind erodibility (Woodruff and Siddoway, 1965). Black and Chanasyk (1989) showed that wind erosion was difficult to predict from intrinsic soil properties because of the great variability in the amount of dry aggregates > 0.84 µm due to management and in particular tillage. Using a different dry soil aggregate stability test for wind erodibility based on the resistance of clods to mechanical stress, Skidmore and Layton (1992) could predict stability from clay or water content.

Although there is no doubt that water and wind erodibility are directly related to soil structural form and stability, our ability to predict and model erodibility from soil structural information is still limited.

7.5   References

Acton, C.J., D.A. Rennie, and E.A. Paul. 1963. Dynamics of soil aggregation. Can. J. Soil Sci. 43:201–209.

Adu, J.R., and J.M. Oades. 1978. Utilization of organic materials in soil aggregates by bacteria and fungi. Soil Biol. Biochem. 10:117–122.

Ahuja, L.R., D.G. Decoursey, B.B. Barnes, and K. Rojas. 1993. Characteristics of macropore transport studied with the ARS Root Zone Water Quality Model. Trans. ASAE 36:369–380.

Albrecht, A., L. Rangon, and P. Barret. 1992. Effets de la matière organique sur la stabilité structurale et la détachabilité d'un vertisol et d'un ferrisol (Martinique). Cah. ORSTOM, sér. Pédol. 27:121–133.

Allmaras, R.R., S.M. Copeland, P.J. Copeland, and M. Oussible. 1996. Spatial relations between oat residue and ceramic spheres when incorporated sequentially by tillage. Soil Sci. Soc. Am. J. 60:1209–1216.

Anderson, S.H., C.J. Gantzer, and J.R. Brown. 1990. Soil physical properties after 100 years of continuous cultivation. J. Soil Water Cons. 45:117–121.

Angers, D.A. 1990. Compression of agricultural soils from Québec. Soil Till. Res. 18:357–365.

Angers, D.A. 1992. Changes in soil aggregation and organic C under corn and alfalfa. Soil Sci. Soc. Am. J. 56:1244–1249.

Angers, D.A., and R.R. Simard. 1986. Relation entre la teneur en matière organique et la masse volumique apparente du sol. Can. J. Soil Sci. 66:743–746.

Angers, D.A., and G.R. Mehuys. 1988. Effects of cropping on macroaggregation of a marine clay soil. Can. J. Soil Sci. 68:723–732.

Angers, D.A., and G.R. Mehuys. 1993. Aggregate stability to water p. 651–657. In M.R. Carter (ed.) Manual on soil sampling and methods of analysis. CRC Press, Boca Raton, FL.

Angers, D.A., and M.R. Carter. 1996. Aggregation and organic matter storage in cool, humid agricultural soils. p. 193–211. In M. R. Carter and B. A. Stewart (ed.) Structure and organic matter storage in agricultural soils. CRC Press, Boca Raton, FL.

Angers, D.A., and C. Chenu. 1998. Dynamics of soil aggregation and C sequestration. p. 199–206. In R. Lal et al. (ed.) Soil processes and the carbon cycle. CRC Press, Boca Raton, FL.

Angers, D.A., and J. Caron. 1998. Plant-induced changes in soil structure: processes and feedbacks. Biogeochem. 42:55-72.

Angers, D.A., B.D. Kay, and P.H. Groenevelt. 1987. Compaction characteristics of a soil cropped to corn and bromegrass. Soil Sci. Soc. Am. J. 51:779-783.

Angers, D.A., N. Samson, and A. Légère. 1993. Early changes in water-stable aggregation induced by rotation and tillage in a soil under barley production. Can. J. Soil Sci. 73:51-59.

Baldock, J.A., and B.D. Kay. 1987. Influence of cropping history and chemical treatments on the water-stable aggregation of a silt loam soil. Can. J. Soil Sci. 67:501-511.

Ball, B.C. 1981. Pore characteristics of soils from two cultivation experiments as shown by gas diffusivities and permeabilities and air-filled porosities. J. Soil Sci. 32:483-498.

Ball, B.C., and E.A.G. Robertson. 1994. Effects of uniaxial compaction on aeration and structure of ploughed or direct drilled soils. Soil Till. Res. 31:135-149.

Ball, B.C., and K.A. Smith. 1991. Gas movement. p. 511-549. *In* K.E. Smith and C.E. Mullins (ed.). Soil analysis. Marcel Dekker Inc., New York, NY.

Ball, B.C., M.V. Cheshire, E.A.G. Robertson, and E.A. Hunter. 1996. Carbohydrate composition in relation to structural stability, compactibility and plasticity of two soils in a long-term experiment. Soil Till. Res. 39:143-160.

Barberis, E., F.A. Marsan, V. Boero, and E. Arduino. 1991. Aggregation of soil particles by iron oxides in various size fractions of soil B horizons. J. Soil Sci. 42:535-42.

Barley, K.P. 1954. Effects of root growth and decay on the permeability of a synthetic sandy loam. Soil Sci. 78:205-211.

Bartoli, F., R. Philippy, and G. Burtin. 1988. Aggregation in soils with small amounts of swelling clays: Aggregate stability. J. Soil Sci. 39:593-616.

Bartoli, F., G. Burtin, and J. Guérif. 1992. Influence of organic matter on aggregation in Oxisols rich in gibbsite or in geothite. II. Clay dispersion, aggregate strength and water stability. Geoderma 54:259-274.

Barzegar, A.R., P. Rengasamy, and J.M. Oades. 1995. Effects of clay type and rate of wetting on the mellowing of compacted soils. Geoderma 68:39-49.

Beare, M.H., P.F. Hendrix, and D.C. Coleman. 1994a. Water-stable aggregates and organic matter fractions in conventional- and no-tillage soils. Soil Sci. Soc. Am. J. 58:777-786.

Beare, M.H., M.L. Cabrera, P.F. Hendrix, and D.C. Coleman. 1994b. Aggregate-protected and unprotected organic matter pools in conventional and no-tillage soils. Soil Sci. Soc. Am. J. 58:787-795.

Beare, M.H., S. Hu, D.C. Coleman, and P.F. Hendrix. 1997. Influences of mycelial fungi on soil aggregation and organic matter storage in conventional and no-tillage soils. Appl. Soil Ecol. 5:211-219.

Besnard, E., C. Chenu, J. Balesdent, P. Puget, and D. Arrouays. 1996. Fate of particulate organic matter in soil aggregates during cultivation. Europ. J. Soil Sci. 47:495-503.

Bishop, A., and G.E. Blight. 1963. Some aspects of effective stress in saturated and partly saturated soils. Geotech. 13:177-197.

Black, J.M.W., and D.S. Chanasyk. 1989. The wind erodability of some Alberta soils after seeding: Aggregation in relation to field parameters. Can. J. Soil Sci. 69:835-847.

Blackwell, P.S., T.W. Green, and W.K. Mason. 1990a. Responses of biopore channels from roots to compression by vertical stresses. Soil Sci. Soc. Am. J. 54:1088-91.

Blackwell, P.S., A.J. Ringrose-Voase, N.S. Jayawardane, K.A. Olsson, D.C. McKenzie, and W.K. Mason. 1990b. The use of air-filled porosity and intrinsic permeability to air to characterize structure of macropore space and saturated hydraulic conductivity of clay soils. J. Soil Sci. 41:215-228.

Blanchart, E. 1992. Restoration by earthworms (Megascolecidae) of the macroaggregate structure of a destructed savanna soil under field conditions. Soil Biol. Biochem. 24:1587-1594.

Blanchart, E., P. Lavelle, E. Braudeau, Y. le Bissonnais, and C. Valentin. 1997. Regulation of soil structure by geophagous earthworm activities in humid savannas of Côte d'Ivoire. Soil Biol. Biochem. 29:431-439.

Bouché, M.B. and F. Al-Addan. 1997. Earthworms, water infiltration and soil stability: some new assessments. Soil Biol. Biochem. 29:441-452.

Bowen, G.D., and A.D. Rovira. 1991. The rhizosphere. p. 641-669. *In* Y. Waisel et al. (ed.) Plant roots. The hidden half. Marcel Dekker, Inc., New York, NY.

Bowen, H.D. 1981. Alleviating mechanical impedance. p. 21-53. *In* G.F. Arkin and H.M. Taylor (ed.). Modifying the root environment to reduce stress. American Society of Agricultural Engineers, St. Joseph, MI.

Bradford, J.M., and S.C. Gupta. 1986. Compressibility. p. 479–492. *In* A.L. Page (ed.) Methods of soil analysis. Part 1. Physical and mineralogical methods. Soil Science Society of America, Madison WI.

Braunack, M.V., and A.R. Dexter. 1989. Soil aggregation in the seedbed: a review. 2. Effect of aggregate sizes on plant growth. Soil Till. Res. 14:281–298.

Braunack, M.V., J.S. Hewitt, and A.R. Dexter. 1979. Brittle fracture of soil aggregates and the compaction of aggregate beds. J. Soil Sci. 30:653–67.

Breland, T.A., and S. Hansen. 1996. Nitrogen mineralization and microbial biomass as affected by soil compaction. Soil Biol. Biochem. 28:655–663.

Brewer, R., and A.D. Haldane. 1957. Preliminary experiments in the development of clay orientation in soils. Soil Sci. 84:301–9.

Bruand, A., I. Cousin, B. Nicoullaud, O. Duval, and J.C. Bégon. 1996. Backscatter electron scanning images of soil porosity for analyzing soil compaction around roots. Soil Sci. Soc. Am. J. 60:895–901.

Bullock, M.S., W.D. Kemper, and S.D. Nelson. 1988. Soil cohesion as affected by freezing, water content, time and tillage. Soil Sci. Soc. Am. J. 52:770–776.

Cambardella, C.A., and E.T. Elliott. 1993. Carbon and nitrogen in aggregates from cultivated and native grassland soils. Soil Sci. Soc. Am. J. 57:1071–1076.

Campbell, C.A., D. Curtin, S. Brandt, and R.P. Zentner. 1993. Soil aggregation as influenced by cultural practices in Saskatchewan: II. Brown and dark brown Chernozemic soils. Can. J. Soil Sci. 73:597–612.

Canada Expert Committee on Soil Survey. 1987. The Canadian system of soil classification. Agriculture Canada Publ. 1646. Canadian Government Publication Centre, Ottawa, Canada.

Capriel, P., T. Beck, H. Borchert, and P. Harter. 1990. Relationship between soil aliphatic fraction extracted with supercritical hexane, soil microbial biomass, and soil aggregate stability. Soil Sci. Soc. Am. J. 54:415–420.

Caron, J., and B.D. Kay. 1992. Rate of response of structural stability to a change in water content: Influence of cropping history. Soil Till. Res. 25:167–185.

Caron, J., B.D. Kay, and E. Perfect. 1992a. Short term decrease in soil structural stability following bromegrass establishment on a clay loam soil. Plant Soil. 145:121–130.

Caron, J., B.D. Kay, and J.A. Stone. 1992b. Improvement of structural stability of a clay loam with drying. Soil Sci. Soc. Am. J. 56:1583–1590.

Caron, J., B.D. Kay, J.A. Stone, and R.G. Kachanoski. 1992c. Modeling temporal changes in structural stability of a clay loam soil. Soil Sci. Soc. Am. J. 56:1597–1604.

Caron, J., O. Banton, D.A. Angers, and J.P. Villeneneuve. 1996a. Preferential bromide transport through a clay loam under alfalfa and corn. Geoderma 69:175–191.

Caron, J., C.R. Espindola, and D.A. Angers. 1996b. Soil structural stability during rapid wetting: Influence of land use on some aggregate properties. Soil Sci. Soc. Am. J. 60:901–908.

Carter, M.R. 1988. Temporal variability of soil macroporosity in a fine sandy loam under molboard ploughing and direct drilling. Soil Till. Res. 12:37–51.

Carter, M.R. 1990a. Relative measures of soil bulk density to characterize compaction in tillage studies on fine sandy loams. Can. J. Soil Sci. 70:425–433.

Carter, M.R. 1990b. Relationship of strength properties to bulk density and macroporosity in cultivated loamy sand to loam soils. Soil Till. Res. 15:257–268.

Carter, M.R. 1992. Influence of reduced tillage systems on organic matter, microbial biomass, macro-aggregate distribution and structural stability of the surface soil in a humid climate. Soil Till. Res. 23:361–372.

Carter, M.R., and B.C. Ball. 1993. Soil porosity. p. 581–588. *In* M. R. Carter (ed.). Manual on soil sampling and methods of analysis. CRC Press, Boca Raton, FL.

Carter, M.R., and H.T. Kunelius. 1993. Effect of undersowing barley with annual ryegrasses or red clover on soil structure in a barley-soybean rotation. Agric. Ecosys. Environ. 43:245–254.

Carter, M.R., D.A. Angers, and H.T. Kunelius. 1994. Soil structural form and stability, and organic matter under cool-season perennial grasses. Soil Sci. Soc. Am. J. 58:1194–1199.

Cassel, D.K. 1983. Spatial and temporal variability of soil physical properties following tillage of Norfolk loamy sand. Soil Sci. Soc. Am. J. 47:196–201.

Chan, K.Y. 1995. Strength characteristics of a potentially hardsetting soil under pasture and conventional tillage in the semi-arid region of Australia. Soil Till. Res. 34:105–113

Chan, K.Y., and A.S. Hodgson. 1984. Moisture regimes of a cracking clay used for cotton production. Rev. Rural Sci. 5:176–180.

Chaney, K., and R.S. Swift. 1984. The influence of organic matter on aggregate stability in some British soils. J. Soil Sci. 35:223–230.

Chaney, K., and R.S. Swift. 1986a. Studies on aggregate stability. I. Re-formation of soil aggregates. J. Soil Sci. 37:329–335

Chaney, K., and R.S. Swift. 1986b. Studies on aggregate stability. II. The effect of humic substances on the stability of re-formed soil aggregates. J. Soil Sci. 37:337–343.

Chantigny, M.H., D.A. Angers, D. Prévost, L.P. Vézina, and F.P. Chalifour. 1997. Soil aggregation, and fungal and bacterial biomass under annual and perennial cropping systems. Soil Sci. Soc. Am. J. 61:262–267.

Chartres, C.J., and J.D. Fitzgerald. 1990. Properties of siliceous cements in some Australian soils and saprolites. p.199–205. *In* L.A. Douglas (ed.) Soil micromorphology. Elsevier Science Publishers, Amsterdam, Netherlands.

Chen, Y., A. Banin, and A. Borochovitch. 1983. Effect of potassium on soil structure in relation to hydraulic conductivity. Geoderma 30:135–47.

Chenu, C. 1989. Influence of a fungal polysaccharide, scleroglucan, on clay microstructures. Soil Biol. Biochem. 21:299–305.

Chenu, C., and J. Guérif. 1991. Mechanical strength of clay minerals as influenced by an adsorbed polysaccharide. Soil Sci. Soc. Am. J. 55:1076–1080.

Chepil, W.S., and N.P. Woodruff. 1963. The physics of wind erosion and its control. Adv. Agron. 15:211–302.

Cheshire, M.V., G.P. Sparling, and C.M. Mundie. 1983. Effect of periodate treatment of soil on carbohydrate constituents and soil aggregation. J. Soil Sci. 34:105–112.

Churchman, G.J., and K.R. Tate. 1986. Effect of slaughterhouse effluent and water irrigation upon aggregation in seasonally dry New Zealand soil under pasture. Aust. J. Soil Res. 24:505–516.

Churchman, G.J., J.O. Skjemstad, and J.M. Oades. 1993. Influence of clay minerals and organic matter on effects of sodicity on soils. Aust. J. Soil Res. 31:779–800.

Clapp, C.E., and W.W. Emerson. 1965. The effect of periodate oxidation on the strength of soil crumbs. Soil Sci. Soc. Am. Proc. 29:127–134.

Clapp, C.E., R.J. Davis, and S.H. Waugaman. 1962. The effect of rhizobial polysaccharides on aggregate stability. Soil Sci. Soc. Am. Proc. 26:466–469.

Collis-George, N., and B.S. Figueroa. 1984. The use of high energy moisture characteristic to assess soil stability. Aust. J. Soil Res. 22:349–56.

Combeau, A. 1965. Variations saisonnières de la stabilité structurale du sol en région tempérée. Comparaison avec la zone tropicale. Cah. ORSTOM Sér. Pédol. 3:123–135.

Concaret, J. 1967. Etude des mécanismes de la destruction des agrégats de terre au contact de solutions aqueuses. Ann. Agron. 18:99–144.

Crawford, J.W. 1994. The relationship between structure and the hydraulic conductivity of soil. Europ. J. Soil Sci. 45:493–502.

Crawford, J.W., N. Matsui, and I.M. Young. 1995. The relation between the moisture release curve and the structure of soil. Europ. J. Soil Sci. 46: 369–375.

Czeratzki, W., and H. Frese. 1958. Importance of water in formation of soil structure. Hwy. Res. Bd. Spec. Rep. 40:200–211. Highway Research Board, Washington, DC.

da Silva, A., and B.D. Kay. 1997a. Estimating the least limiting water range of soils from properties and management. Soil Sci. Soc. Amer. J. 61:877–883.

da Silva, A., and B.D. Kay. 1997b. The sensitivity of shoot growth of corn to the least limiting water range of soils. Plant Soil 184:323–329.

da Silva, A., and B.D. Kay. 1997c. Effect of soil water content variation on the least limiting water range. Soil Sci. Soc. Am. J. 61:994–888.

Dalal, R.C., and B.J. Bridge. 1996. Aggregation and organic matter storage in sub-humid and semi-arid soils. p. 263–307. *In* M.R. Carter and B.A. Stewart (ed.) Structure and organic matter storage in agricultural soils. CRC Press, Boca Raton, FL.

Danielson, R.E., and L.P. Sutherland. 1982. Porosity. p. 443–461. *In* A. Klute (ed.) Methods of soil analysis. Part 1. Physical and mineralogical methods. Soil Science Society of America, Madison, WI.

Davies, P. 1985. Influence of organic matter content, moisture status and time after reworking on soil shear strength. J. Soil Sci. 36:299–306.

Degens, B.P. 1997. Macro-aggregation of soils by biological bonding and binding mechanisms and the factors affecting these: a review. Aust. J. Soil Res. 35:431–459.

Degens, B.P., G.P. Sparling, and L.K. Abbott. 1996. Increasing the length of hyphae in a sandy soil increases the amount of water-stable aggregates. Appl. Soil Ecol. 3:149–159.

Derdour, H., D.A. Angers, and M.R. Laverdière. 1993. Caractérisation de l'space poral d'un sol argileux : Effets des ses constituants et du travail du sol. Can. J. Soil Sci. 73:299–307.

Dexter, A.R. 1978. Tunnelling of soil by earthworms. Soil Biol. Biochem. 10:447–449.

Dexter, A.R. 1988. Advances in the characterization of soil structure. Soil Till. Res. 11:199–238.

Dexter, A.R., D. Hein, and J.S. Hewitt. 1982. Macro-structure of the surface layer of a self-mulching clay in relation to cereal stubble management. Soil Till. Res. 2:251–64.

Dinel, H., P.E.M. Lévesque, and G.R. Mehuys. 1991. Effects of beeswax, a naturally occurring source of long-chain aliphatic compounds on the aggregate stability of lacustrine silty clay. Soil Sci. 151:228–239.

Dinel, H., P.E.M. Lévesque, P. Jambu, and D. Righi. 1992. Microbial activity and long-chain aliphatics in the formation of stable soil aggregates. Soil Sci. Soc. Am. J. 56:1455–1463.

Dorioz, J.M., M. Robert, and C. Chenu. 1993. The role of roots, fungi and bacteria on clay particle organization. An experimental approach. Geoderma 56:179–194.

Dormaar, J.F., and R.C. Foster. 1991. Nascent aggregates in the rhizosphere of perennial ryegrass (*Lolium perenne* L.). Can. J. Soil Sci. 71:465–474.

Dormaar, J.F., C.W. Lindwall, and G.C. Kosub. 1988. Effectiveness of manure and commercial fertilizer in restoring productivity of an artificially eroded dark brown chernozemic soil under dryland conditions. Can. J. Soil Sci. 68:669–679.

Douglas, J.T. 1986. Effects of season and management on the vane shear strength of a clay topsoil. J. Soil Sci. 37:669–679.

Douglas, J.T., M.G. Jarvis, K.R. Howse, and M.J. Goss. 1986. Structure of a silty soil in relation to management. J. Soil Sci. 37:137–51.

Dufey, J.E., H. Halen, and R. Frankart. 1986. Evolution de la stabilité structurale du sol sous l'influence des racines de trèfle (*Trifolium pratense* L.) et de ray-grass (*Lolium multiflorum* Lmk.). Observations pendant et après culture. Agronomie 6:811–817.

Edwards, C.A., and J.R. Lofty. 1978. The influence of arthropods and earthworms upon root growth of direct drilled cereals. J. Appl. Ecol. 15:789–795.

Edwards, W.M., L.D. Norton, and C.E. Redmond. 1988. Characterizing macropores that affect infiltration into nontilled soil. Soil Sci. Soc. Am. J. 52:483–487.

Ehlers, W. 1975. Observation on earthworm channels and infiltration on tilled and untilled loess soils. Soil Sci. 119:242–249.

Ehlers, W. 1977. Measurement and calculation of hydraulic conductivity in horizons of tilled and untilled loess-derived soil, Germany. Geoderma 19:293–306.

Ehlers, W., O. Wendroth, and F. de Mol. 1995. Characterizing pore organization by soil physical parameters. p. 257–75. *In* K.H. Hartge, and B.A. Stewart (ed.). Soil structure: Its development and function. Lewis Publishers, Boca Raton, FL.

Elliott, E.T. 1986. Aggregate structure and carbon, nitrogen and phosphorus in native and cultivated soils. Soil Sci. Soc. Am. J. 50:627–633.

Elliott, E.T., R.V. Anderson, D.C. Coleman, and C.V. Cole. 1980. Habitable pore space and microbial trophic interactions. Oikos 35:327–335.

Elustondo, J., D.A Angers, M.R. Laverdière, and A. N'Dayegamiye. 1990. Étude comparative de l'agrégation et de la matière organique associée aux fractions granulométriques de sept sols sous culture de maïs ou en prairie. Can. J. Soil Sci. 70:395–402.

Emerson, W.W. 1977. Physical properties and structure. p. 78–104. *In* J.S. Russel and E.L. Greacen (ed.) Soil factors in crop production in a semi-arid environment. Queensland University Press, Brisbane, Australia.

Emerson, W.W. 1983. Inter-particle bonding. p. 477–98. *In* Soils: An Australian viewpoint. Academic Press, London, UK.

Emerson, W.W., R.C. Foster, J.M. Tisdall, and D. Weissmann. 1994. Carbon content and bulk density of an irrigated Natrixeralf in relation to tree root growth and orchard management. Aust. J. Soil Res. 32:939–951.

Emerson, W.W. 1995. Water retention, organic C and soil texture. Aust. J. Soil Res. 33:241–251.

Fausey, N.R., and R. Lal. 1989. Drainage-tillage effects on Crosby-Kokomo soil association in Ohio. Soil Tech. 2:359–70.

Fausey, N.R., and R. Lal. 1992. Drainage-tillage effects on a Crosby-Kokomo soil association in Ohio, 3. Organic matter content and chemical properties. Soil Tech. 5:1–12.

Feller, C., A. Albrecht, and D. Tessier. 1996. Aggregation and organic matter storage in kaolinitc and smectitic tropical soils. p. 309–359. *In* M.R. Carter and B.A. Stewart (ed.) Structure and organic matter storage in agricultural soils. CRC Press, Boca Raton, FL.

Forster, S.M. 1990. The role of microorganisms in aggregate formation and soil stabilization: types of aggregation. Arid Soil Res. Rehab. 4:85–98.

Foster, R.C. 1981. Polysaccharides in soil fabrics. Science 214:665–667.

Foster, R.C. 1988. Microenvironments of soil microorganisms. Biol. Fert. Soils 6:189–203.

Foster, R.C., and A.D. Rovira. 1976. Ultrastructure of wheat rhizosphere. New Phytol. 76:343–352.

Fox, W.E. 1964. Cracking characteristics and field capacity in a swelling soil. Soil Sci. 98:413.

Franzmeier, D.P., C.J. Chartres, and J.T. Wood. 1996. Hardsetting soils in southeast Australia: Landscape and profile processes. Soil Sci. Soc. Am. J. 60:1178–87.

Gardner, E.A., R.J. Shaw, G.D. Smith, and K.J. Coughlan. 1984. Plant available water capacity: Concept, measurement and prediction. p.164–175. *In* J.W. McGarity, E.H. Hoult and H.B. So (ed.). The properties and utilization of cracking clay soils. University of New England, Armidale, Australia.

Garnier-Sillam, E., F. Toutain, and J. Renoux. 1988. Comparaison de l'influence de deux termitières (humivore et champignonniste) sur la stabilité structurale des sols forestiers tropicaux. Pedopedologia. 32:89–97.

Gifford, R.O., and D.F. Thran. 1974. Bonding mechanisms for soil crusts: Part II. Strength of silica cementation. AZ Agric. Res. Sta. Tech. Bull. 214:28–32.

Giovannini, G., and P. Sequi. 1976. Iron and aluminum as cementing substances of soil aggregates: Changes in stability of soil aggregates following extraction of iron and aluminum by acetylacetone in a non-polar solvent. J. Soil Sci. 27:148–153.

Giovannini, S. Lucchesi, and S. Cervelli. 1983. Water-repellent substances and aggregate stability in hydrophobic soil. Soil Sci. 135:110–113.

Golchin, A., J.M. Oades, J.O. Skjemstad, and P. Clarke. 1994. Soil structure and carbon cycling. Aust. J. Soil Res. 32:1043–1068.

Golchin, A., J.A. Baldock, and J.M. Oades. 1998. A model linking organic matter decomposition, chemistry and aggregate dynamics. p.245–266. *In* R. Lal et al. (ed.) Soil processes and the carbon cycle. CRC Press, Boca Raton, FL.

Goldberg, S. 1989. Interaction of aluminum and iron oxides and clay minerals and their effect on soil physical properties: A review. Commun. Soil Sci. Plant Anal. 20:1181–207.

Gollany, H.T., T.E. Schumacher, P.D. Evanson, M.J. Lindstrom, and G.D. Lemme. 1991. Aggregate stability of an eroded and desurfaced Typic Haplustoll. Soil Sci. Soc. Am. J. 55:811–816.

Goss, M.J., W. Ehlers, F.R. Boone, I. White, and K.R. Howse. 1984. Effects of soil management practice on soil physical conditions affecting root growth. J. Agric. Eng. Res. 30:131–140.

Gouzou, L., G. Burtin, R. Philippy, F. Bartoli, and T. Heulin. 1993. Effect of inoculation with Bacillus polymyxa on soil aggregation in the wheat rhizosphere: preliminary examination. Geoderma 56:479–491.

Greenland, D.J., G.R. Lindstrom, and J.P. Quirk. 1962. Organic materials which stabilize natural soil aggregates. Soil Sci. Soc. Am. Proc. 26:366–371.

Greenland, D.J., D. Rimmer, and D. Payne. 1975. Determination of the structural stability class of English and Welsh soils using a water coherence test. Soil Sci. 26:294–303.

Gregorich, E.G., C.F. Drury, B.H. Ellert, and B.C. Liang. 1997. Fertilization effects on physically protected light fraction organic matter. Soil Sci. Soc. Am. J. 61:482–484.

Grevers, M.C.J., and E. de Jong. 1990. The characterization of soil macroporosity of a clay soil under ten grasses using image analysis. Can. J. Soil Sci. 70:93–103.

Groenevelt, P.H., and B.D. Kay. 1981. On pressure distribution and effective stress in unsaturated soils. Can. J. Soil Sci. 61:431–443.

Guckert, A., T. Chone, and F. Jacquin. 1975. Microflore et stabilité structurale du sol. Rev. Ecol. Biol. Sol. 12:211–223.

Guérif, J. 1979. Rôle de la matière organique sur le comportement d'un sol au compactage. II. Matières organiques libres et liées. Ann. Agron. 30:469–480.

Gupta, R.P. 1986. Criteria for physical rating of soils in relation to crop production. Proc. XIII Intl. Cong. Soil Sci. Hamburg. 2:69–70.

Gupta, S.C., E.C. Schneider, W.E. Larson, and A. Hadas. 1987. Influence of corn residue on compression and compaction behavior of soils. Soil Sci. Soc. Am. J. 51:207–212.

Gupta, V.V.S.R., and J.J. Germida. 1988. Distribution of microbial biomass and its activity in different soil aggregate size classes as affected by cultivation. Soil Biol. Biochem. 20:777–786.

Guidi, G., G. Poggio, and G. Petruzelli. 1985. The porosity of soil aggregates from bulk soil and soil adhering to roots. Plant Soil 87:311–314.

Gusli, S., A. Cass, D.A. MacLeod, and, P.S. Blackwell. 1994. Structural collapse and strength of some Australian soils in relation to hardsetting: 1. Structural collapse on wetting and draining. Europ. J. Soil Sci. 45:15–21.

Hadas, A., and D. Russo. 1974. Water uptake by seeds as affected by water stress, capillary conductivity, and seed-soil contact. II Analysis of experimental data. Agron. J. 66:647–652.

Häkansson, I. 1990. A method for characterizing the state of compactness of the plough layer. Soil Till. Res. 16:105–120.

Hardan, A., and A.N. Al-Ani. 1978. Improvement of soil structure by using date and sugar beet waste products. p. 305–308. In W.W. Emerson, R.D. Bond and A.R. Dexter (ed.). Modification of soil structure. A Wiley-Interscience Publication, Brisbane, Australia.

Harris, R.F., G. Chesters, and O.N. Allen. 1966. Dynamics of soil aggregation. Adv. Agron. 18:107–169.

Hassink, J. 1992. Effects of soil texture and structure on C and N mineralization in grassland soils. Biol. Fertil. Soils 14:126–134.

Haynes, R.J., and G.S. Francis. 1993. Changes in microbial biomass C, soil carbohydrates and aggregate stability induced by growth of selected crop and forage species under field conditions. J. Soil Sci. 44:665–675.

Haynes, R.J., and R.S. Swift. 1990. Stability of soil aggregates in relation to organic constituents and soil water content. J. Soil Sci. 41:73–83.

Haynes, R.J., and M.H. Beare. 1996. Aggregation and organic matter storage in meso-thermal, humid soils. p. 213–262. In M.R. Carter and B.A. Stewart (ed.). Structure and organic matter storage in agricultural soils. CRC Press, Boca Raton, FL.

Haynes, R.J., R.S. Swift, and R.C. Stephen. 1991. Influence of mixed cropping rotations (pasture-arable) on organic matter content, water-stable aggregation and clod porosity in a group of soils. Soil Till. Res. 19:77–87.

Heard, J.R., E.J. Kladivko, and J.V. Mannering. 1988. Soil macroporosity, hydraulic conductivity and air permeability of silty soils under long-term conservation tillage in Indiana. Soil Till. Res. 11:1–18.

Hénin, S., G. Monnier, and A. Combeau. 1958. Méthode pour l'étude de la stabilité struturale des sols. Ann. Agron. 9:73–92.

Hodgson, A.S., and D.A. MacLeod. 1989a. Oxygen flux, air-filled porosity and bulk density as indices of vertisol structure. Soil Sci. Soc. Am. J. 53:540–543.

Hodgson, A.S., and D.A. MacLeod. 1989b. Use of oxygen flux density to estimate critical air-filled porosity of a vertisol. Soil Sci. Soc. Am. J. 53:355–61.

Hodgson, J.M. 1978. Soil sampling and soil description. Clarendon Press, Oxford, UK.

Hole, F.D. 1981. Effects of animals on soil. Geoderma. 25:75–112.

Horn, R., and A.R. Dexter. 1989. Dynamics of soil aggregation in a desert loess. Soil Till. Res. 13:253–266.

Horn, R., H. Taubner, M. Wuttke, and T. Baumgartl. 1994. Soil physical properties related to soil structure. Soil Till. Res. 30:187–216.

Hudson, B.D. 1994. Soil organic matter and available water capacity. J. Soil Water Conser. 49:189–194.

Hundal, S.S., G.O. Schwab, and G.S. Taylor. 1976. Drainage system effects on physical properties of a lakebed clay soil. Soil Sci. Soc. Am. J. 40:300–305.

Ingles, O.G., 1965. The shatter test as an index of the strength of soil aggregates. p. 284–302. In C.J. Osborn (ed.) Proc. Tewksbury Symp. Fract. Faculty of Engineering, University of Melbourne, Melbourne, Australia.

Jakobsen, B., and A.R. Dexter. 1988. Influence of biopores on root growth, water uptake and grain yield of wheat. Predictions from a computer model. Biol. Fert. Soils 6:315–321.

Jastrow, J.D. 1987. Changes in soil aggregation associated with tallgrass prairie restoration. Amer. J. Bot. 74:1656–1664.

Jastrow, J.D. 1996. Soil aggregate formation and the accrual of particulate and mineral-associated organic matter. Soil Biol. Biochem. 45:665–676.

Juma, N.G. 1993. Interrelationships between soil structure/texture, soil biota/soil organic matter and crop production. Geoderma 57:3–30.

Karlen, D.L., N.C. Wollenhaupt, D.C. Erbach, E.C. Berry, J.B. Swan, N.S. Eash, and J.L. Jordahl. 1994. Crop residue effects on soil quality following 10-years of no-till corn. Soil Till. Res. 31:149–167.

Kay, B.D. 1990. Rates of change of soil structure under different cropping systems. Adv. Soil Sci. 12:1–52.

Kay, B.D. 1998. Soil structure and organic C: A review. p. 169–197. *In* R. Lal et al. (ed.) Soil processes and the carbon cycle. CRC Press, Boca Raton, FL.

Kay, B.D., and A.R. Dexter. 1990. Influence of aggregate diameter, surface area and antecedent water content on the dispersibility of clay. Can. J. Soil Sci. 70:655–671.

Kay, B.D., C.D. Grant, and P.H. Groenevelt. 1985. Significance of ground freezing on soil bulk density under zero tillage. Soil Sci. Soc. Am. J. 49: 973–978.

Kay, B.D., V. Rasiah, and E. Perfect. 1994a. Structural aspects of soil resiliency. p. 449–468. *In* D.J. Greenland and I. Szabolcs (ed.) Soil resilience and sustainable land use. CAB International, London, UK.

Kay, B.D., A.R. Dexter, V. Rasiah, and C.D. Grant. 1994b. Weather, cropping practices and sampling depth effects on tensile strength and aggregate stability. Soil Till. Res. 32: 135–148.

Kay, B.D., A.P. da Silva, and J.A. Baldock. 1997. Sensitivity of soil structure to changes in organic C content: predictions using pedotransfer functions. Can. J. Soil Sci. 77:655–667.

Kemper W.D., and E.J. Koch. 1966. Aggregate stability of soils from the western portions of the United States and Canada. USDA Tech. Bull. 1355.

Kemper, W.D., and R.C. Rosenau. 1986. Aggregate stability and size distribution. p. 425–442. *In* A.L. Page (ed.) Methods of soil analysis. Part 1. Physical and mineralogical methods. Soil Science Society of America, Madison, WI.

Kemper, W.D., R. Rosenau, and S. Nelson. 1985. Gas displacement and aggregate stability of soils. Soil Sci. Soc. Am. J. 49:25–28.

Kemper, W.D., M.S. Bullock, and A.R. Dexter. 1989. Soil cohesion changes. p. 81–95. *In* W.E. Larson, G.R. Blake, R.R. Allmaras, W.B. Voorhees, and S.C. Gupta (ed.). Mechanics and related processes in structured agricultural soils. Kluwer Academic Publishers, Boston, MA.

Ketcheson, J.W., and E.G. Beauchamp. 1978. Effects of corn stover, manure, and nitrogen on soil properties and crop yield. Agron. J. 70:792–797.

Killham, K., M. Amato, and J.N. Ladd. 1993. Effect of substrate location in soil and soil pore-water regime on carbon turnover. Soil Biol. Biochem. 25:57–62.

Kleinfelder, D., S. Swanson, G. Norris, and W. Clary. 1992. Unconfined compressive strength of some streambank soils with herbaceous roots. Soil Sci. Soc. Am. J. 56:1920–1925.

Klute, A. 1982. Tillage effects on the hydraulic properties of soil: A review. p. 29–43. *In* P. Unger and D.M. van Doren (ed.) Predicting tillage effects on soil physical properties and processes. Soil Science Society of America, Madison, WI.

Kwaad, F.J.P.M., and H.J. Mucher. 1994. Degradation of soil structure by welding – A micromorphological study. Catena 23:253–68.

Ladd, J.N., R.C. Foster, and J.M. Oades. 1996. Soil structure and biological activity. p. 23–78. *In* G. Stotzky and J.M. Bollag (ed.) Soil biochemistry. Marcel Dekker Inc., New York, NY.

Lafond, J., D.A. Angers, and M.R. Laverdière. 1992. Compression characteristics of a clay soil as influenced by crops and sampling dates. Soil Till. Res. 22:233–241.

Lal, R., and N.R. Fausey. 1993. Drainage and tillage effects on a Crosby-Kokomo soil association in Ohio. 4. Soil physical properties. Soil Tech. 6:123–35.

Lal, R., A.A. Mahboubi, and N.R. Fausey. 1994. Long-term tillage and rotation effects on properties of a central Ohio soil. Soil Sci. Soc. Amer. J. 57:517–522.

Larson, W.E., S.C. Gupta, and R.A. Useche. 1980. Compression of soils from eight soil orders. Soil Sci. Soc. Am. J. 44:450–457.

le Bissonnais, Y. 1988. Comportement d'agrégats terreux soumis à l'action de l'eau : Analyse des mécanismes de désagrégation. Agronomie 8:915–924.

le Bissonnais, Y. 1996. Aggregate stability and assessment of soil crustability and erodibility: I. Theory and methodology. Europ. J. Soil Sci. 47:425–437.

le Bissonnais, Y., and D. Arrouays. 1997. Aggregate stability and assessment of soil crustabiltiy and erodibility: II. Application to humic loamy soils with various organic C contents. Europ. J. Soil Sci. 48:39–48.

Lee, K.E., and R.C. Foster. 1991. Soil fauna and soil structure. Aust. J. Soil Res. 29:745–775.

Letey, J. 1985. Relationship between physical properties and crop productions. Adv. Soil Sci. 1:277–294.

Low, A.J. 1955. Improvements in the structural state of soil under leys. J. Soil Sci. 6:177–199.

Lynch, J.M., and E. Bragg. 1985. Microorganisms and soil aggregate stability. Adv. Soil Sci. 2:133–171.

MacRae, R.J., and G.R. Mehuys. 1987. Effects of green manuring in rotation with corn on the physical properties of two Quebec soils. Biol. Agric. Hortic. 4:257–270.

Manrique, L.A., and C.A. Jones. 1991. Bulk density of soils in relation to soil physical and chemical properties. Soil Sci. Soc. Am. J. 55:476–481.

Mapa, R.B., R.E. Green, and L. Santo. 1986. Temporal variability of soil hydraulic properties and wetting and drying subsequent to tillage. Soil Sci. Soc. Am. J. 50:1133–38.

Marinissen, J.C.Y., and A.R. Dexter. 1990. Mechanisms of stabilization in earthworm casts and artificial casts. Biol. Fertil. Soils. 9:163–167.

Marinissen, J.C.Y., E. Nijhuis, and N. van Breemen. 1996. Clay dispersibility in moist earthworm casts of different soils. Appl. Soil Ecol. 4:83–92.

Marshall, T.J., and J.P. Quirk. 1950. Stability of structural aggregates of dry soil. Aust. J. Agric. Sci. 1:266–275.

Martin, J.P., and S.A. Waksman. 1940. Influence of microorganisms on soil aggregation and erosion. Soil Sci. 50:29–47.

Materechera, S.A., A.R. Dexter, and A.M. Alston. 1991. Penetration of very strong soils by seedling roots different plant species. Plant Soil 135:31–41.

Materechera, S.A., A.R. Dexter, and A.M. Alston. 1992. Formation of aggregates by plant roots in homogenised soils. Plant Soil 142:69–79.

Mazurak, A.P., L. Chesnin, and A. Amin Thijeel. 1977. Effects of beef cattle manure on water-stability of soil aggregates. Soil Sci. Soc. Am. J. 41:613–615.

McBride, R.A., and G.C. Watson. 1990. An investigation of the re-expansion of unsaturated, structured soils during cyclic static loading. Soil Till. Res. 17:241–53.

McDonald, D.C., and R. Julian. 1966. Quantitative estimation of soil total porosity and macro-porosity as part of the pedological description. NZ J. Agric. Res. 8:927–946

McGarry, D., and K.J. Smith. 1988. Indices of residual shrinkage to quantify the comparative effects of zero and mechanical tillage on a Vertisol. Aust. J. Soil Res. 26:543–548.

McKeague, J.A., C. Wang, and G.C. Topp. 1982. Estimating saturated hydraulic conductivity from soil morphology. Soil Sci. Soc. Am. J. 46:1239–1244.

McKeague, J.A., C.A. Fox, J.A. Stone, and R. Protz. 1987. Effects of cropping system on structure of Brookston clay loam in long-term experimental plots at Woodslee, Ontario. Can. J. Soil Sci. 67:571–84.

McKenzie, D.C., P.J. Hulme, T.S. Abbott, D.A. MacLeod, and A. Cass. 1988. Vertisol structure dynamics following an irrigation of cotton, as influenced by prior rotation crops. p. 33–37. In K.J. Coughlan and P.N. Truong (ed.). Effects of management practices on soil physical properties. Proc. Natl. Worksh., Toowoomba, Queensland Department Primary Industries, Brisbane, Australia.

Meek, B.D., E.A. Rechel, L.M. Carter, and W.R. DeTar. 1989. Changes in infiltration under alfalfa as influences by time and wheel traffic. Soil Sci. Soc. Am. J. 53:238–241.

Meek, B.D., W.R. De Tar, D. Rolph, E.R. Rechel, and L.M. Carter. 1990. Infiltration rate as affected by an alfalfa and no-till cotton cropping system. Soil Sci. Soc. Am. J. 54:505–508.

Metting, B. 1987. Dynamics of wet and dry aggregate stability from a three-year microalgal soil conditioning experiment in the field. Soil Sci. 143:139–143.

Metzger, L., D. Levanon, and U. Mingelgrin. 1987. The effect of sewage sludge on soil structural stability: microbiological aspects. Soil Sci. Soc. Am. J. 51:346–351.

Miller, R.M., and J.D. Jastrow. 1990. Hierarchy of root and mycorrhizal fungal interactions with soil aggregation. Soil Biol. Biochem. 5:579–584.

Mitchell, A.R., and M.T. van Genuchten. 1992. Shrinkage of bare and cultivated soil. Soil Sci. Soc. Am. J. 56:1036–1042.

Mitchell, A.R., T.R. Ellsworth, and B.D. Meek. 1995. Effect of root systems on preferential flow in swelling soil. Commun. Soil Sci. Plant Anal. 26:2655–2666.

Molope, M.B., I.C. Grieve, and E.R. Page. 1987. Contributions of fungi and bacteria to aggregate stability of cultivated soils. J. Soil Sci. 38:71–77.

Monnier, G. 1965. Action des matières organiques sur la stabilité structurale des sols. Ann. Agron. 16:327–400.

Monreal, C.M., M. Schnitzer, H.-R. Schulten, C.A. Campbell, and D.W. Anderson. 1995. Soil organic structures in macro- and microaggregates of a cultivated brown chernozem. Soil Biol. Biochem. 27:845–853.

Moran, C.J., A.B. McBratney, and A.J. Koppi. 1989. A rapid method for analysis of soil macropore structure. I. Specimen preparation and digital binary image production. Soil Sci. Soc. Am. J. 53:921–28.

Morel, J.L., L. Habib, S. Plantureux, and A. Guckert. 1991. Influence of maize root mucilage on soil aggregate stability. Plant Soil 136:111–119.

N'Dayegamiye, A., and D.A. Angers. 1990. Effets de l'apport prolongé de fumier de bovins sur quelques propriétés physiques et biologiques d'un loam limoneux Neubois sous culture de maïs. Can. J. Soil Sci. 70:259–262.

Norfleet, M.L., and A.D. Karathanasis. 1996. Some physical and chemical factors contributing to fragipan strength in Kentucky soils. Geoderma 71:289–301.

Oades, J.M. 1984. Soil organic matter and structural stability mechanisms and implications for management. Plant Soil 76:319–337.

Oades, J.M. 1993. The role of biology in the formation, stabilization and degradation of soil structure. Geoderma 56:377–400.

Oades, J.M., and A.G. Waters. 1991. Aggregate hierarchy in soils. Aust. J. Soil Res. 29:815–828.

Pagliai, M., G. Guidi, M. La Marca, M. Giachetti, and G. Lucanante. 1981. Effects of sewage sludges and compost on porosity and aggregation. J. Envir. Qual. 10:556–561.

Panabokke, C.R., and J.P. Quirk. 1957. Effect of initial water content on the stability of soil aggregates in water. Soil Sci. 83:185–189

Pawluk, S. 1988. Freeze-thaw effects on granular structure reorganization for soil materials of varying texture and moisture content. Can. J. Soil Sci. 68:485–94.

Perfect, E., B.D. Kay, W.K.P. van Loon, R.W. Sheard, and T. Pojasok. 1990a. Factors influencing soil structural stability within a growing season. Soil Sci. Soc. Am. J. 54:173–179.

Perfect, E., W.K.P. van Loon, B.D. Kay, and P.H. Groenevelt. 1990b. Influence of ice segregation and solutes on soil structural stability. Can. J. Soil Sci. 70:571–581.

Perfect, E., B.D. Kay, W.K.P. van Loon, R.W. Sheard, and T. Pojasok. 1990c. Rates of change in soil structural stability under forages and corn. Soil Sci. Soc. Am. J. 54:179–186.

Perfect, E., B.D. Kay, J.A. Ferguson, A.P. da Silva, and K.A. Denholm. 1993. Comparison of functions for characterizing the dry aggregate size distribution of tilled soil. Soil Till. Res. 28:123–139.

Piccolo, A., and J.S.C. Mbagwu. 1994. Humic substances and surfactants effects on the stability of two tropical soils. Soil Sci. Soc. Am. J. 58:950–955.

Pojasok, T., and B.D. Kay. 1990. Effect of root exudates from corn and bromegrass on soil structural stability. Can. J. Soil Sci. 70:351–362.

Puget, P., C. Chenu, and J. Balesdent. 1995. Total and young organic matter distributions in aggregates of silty cultivated soils. Europ. J. Soil Sci. 46:449–459.

Quirk, J.P., and R.S. Murray. 1991. Towards a model for soil structural behavior. Aust. J. Soil Res. 29:828–867.

Radcliffe, D.E., R.L. Clark, and M.E. Sumner. 1986. Effect of gypsum and deep-rooting perennials on subsoil mechanical impedance. Soil Sci. Soc. Am. J. 50:1566–1570.

Raimbault, B.A., and T.J. Vyn. 1991. Crop rotation and tillage effects on corn growth and soil structural stability. Agron. J. 83:979–985.

Rasiah, V., and B.D. Kay. 1995. Characterizing rate of wetting: Impact on structural destabilization. Soil Sci. 160:176–182.

Rasiah, V., B.D. Kay, and T. Martin. 1992. Variation of structural stability with water content: Influence of selected soil properties. Soil Sci. Soc. Am. J. 56:1604–1609.

Ratliff, L., J.T. Ritchie, and D.K. Cassel. 1983. Field-measured limits of soil water availability as related to laboratory-measured properties. Soil Sci. Soc. Am. J. 47:770–775.

Rawitz, E., A. Hadas, H. Etkin, and M. Margolin. 1994. Short-term variations of soil physical properties as a function of the amounts and C/N ratio of decomposing cotton residues. II. Soil compressibility, water retention and hydraulic conductivity. Soil Till. Res. 32:199–212.

Reeve, R.C. 1953. A method for determining the stability of soil structure based upon air and water permeability measurements. Soil Sci. Soc. Am. Proc. 17: 324–29.

Reid, J.B., and M.G. Goss. 1981. Effect of living roots of different plant species on the aggregate stability of two arable soils. J. Soil Sci. 32:521–541.

Reid, J.B., and M.G. Goss. 1982. Interactions between soil drying due to plant water use and decreases in aggregate stability caused by maize roots. J. Soil Sci. 33:47–53.

Rengasamy, P., and M.E. Sumner. 1998. *In* M.E. Sumner and R. Naidu (ed.) Sodic soils. Oxford University Press, New York, NY.

Rengasamy, P., R.S.B. Greene, and G.W. Ford. 1984. The role of clay fraction in the particle arrangement and stability of soil aggregates: A review. Clay Res. 3:53–67.

Rengasamy, P., R. Beech, T.A. Naidu, K.Y. Chan, and C. Chartres. 1993. Rupture strength as related to dispersive potential in Australian soils. Catena Suppl. 24:65–75.

Richardson, S.J. 1976. Effect of artificial weathering cycles on the structural stability of a dispersed silt soil. J. Soil Sci. 27: 287–294.

Roberson, E.B., S. Sarig, and M.K. Firestone. 1991. Cover crop management of polysaccharide-mediated aggregation in an orchard soil. Soil Sci. Soc. Am. J. 55:734–739.

Rovira, A.D., and E.L. Greacen. 1957. The effect of aggregate disruption on enzyme activity of microorganisms in the soil. Aust. J. Agric. Res. 8:659–673.

Rowell, D.L., and P.J. Dillon. 1972. Migration and aggregation of Na and Ca clays by the freezing of dispersed and flocculated suspensions. J. Soil Sci. 23: 442–447.

Russell, E.W. 1973. Soil conditions and plant growth. Longman Group, London, UK.

Schjønning, P. 1992. Size distribution of dispersed and aggregated particles and of soil pores in 12 Danish soils. Acta Agric. Scand. Sec. B - Soil Plant Sci. 42:26–33.

Schjønning, P., B.T. Christensen, and B. Carstensen. 1994. Physical and chemical properties of a sandy loam 'receiving animal manure, mineral fertilizer or no fertilizer for 90 years. Europ. J. Soil Sci. 45:257–68.

Schneider, E.C. and S.C. Gupta. 1985. Corn emergence as influenced by soil temperature, matric potential, and aggregate size distribution. Soil Sci. Soc. Am. J. 49:415–422.

Scott Blair, G.W. 1937. Compressibility curves as a quantitative measure of soil tilth. J. Agric. Sci. 27: 541–56.

Seygold, C.A. 1994. Polyacrylamide review: Soil conditioning and environmental fate. Commun. Soil Sci. Plant Anal. 25:2171–2185.

Shanmuganathan, R.T., and J.M. Oades. 1982. Modification of soil physical properties by manipulating the net surface charge on colloids through addition of Fe (III) polycations. J. Soil Sci. 33: 451–65.

Shanmuganathan, R.T., and J.M. Oades. 1983. Influence of anions on dispersion and physical properties of the A horizon of a red-brown earth. Geoderma 29:257–277.

Shipitalo, M.J., and R. Protz. 1988. Factors influencing the dispersibility of clay in worm casts. Soil Sci. Soc. Am. J. 52:764–769.

Skidmore, E.L., and J.B. Layton. 1992. Dry-aggregate stability as influenced by selected soil properties. Soil Sci. Soc. Am. J. 56:557–561.

So, H.B., and L.A.G. Aylmore. 1993. How do sodic soils behave? The effects of sodicity on soil physical behavior. Aust. J. Soil Res. 31:761–777.

Soane, B.D. 1975. Studies on some physical properties in relation to cultivation and traffic. p. 160–183. *In* Soil physical conditions and crop production. MN Agric. Food Fish. Tech. Bull. 29.

Soane, B.D. 1990. The role of organic matter in soil compactibility: A review of some practical aspects. Soil Till. Res. 16:179–201.

Soil Science Society of America. 1997. Glossary of soil science terms. Soil Science Society of America, Madison. WI.

Soil Survey Staff. 1975. Soil Taxonomy. USDA Handb. 436. US Government Printing Office, Washington DC.

Sommerfeldt, T.G., and C. Chang. 1985. Changes in soil properties under annual applications of feedlot manure and different tillage practices. Soil Sci. Soc. Am. J. 49:983–987.

Stengel, P. 1979. Utilisation de l'analyse des systèmes de porosité pour la caractérisation de l'etat physique du sol *in situ*. Ann. Agron. 30:27–51.

Stewart, B.A. 1975. Soil conditioners. Soils Sci. Soc. Am. Spec. Pub. 7. Soil Science Society of America, Madison, WI.

Stone J.A., and W.E. Larson. 1980. Rebound of five unidimensionally compressed unsaturated granular soils. Soil Sci. Soc. Am. J. 44:819–822.

Stone, J.A. and B.R. Buttery. 1989. Nine forages and the aggregation of a clay loam soil. Can. J. Soil Sci. 69:165–169.

Stypa, M., A. Nunez-Barrios, D.A. Barry, and M.H. Miller. 1987. Effects of subsoil bulk density, nutrient availability and soil moisture on corn root growth in the field. Can. J. Soil Sci. 67: 293–308.

Sutton, J.C., and B.R. Sheppard. 1976. Aggregation of sand dune soil by endomycorrhizal fungi. Can. J. Bot. 54:326–333.

Taylor, H.M., G.M. Roberson, and J.J. Parker Jr. 1966. Soil strength-root penetration relations for medium to coarse-textured soil materials. Soil Sci. 102:18–22.

Thomas, R.S., S. Dakessian, R.N. Ames, M.S. Brown, and G.J. Bethlenfalvay. 1986. Aggregation of a silty loam soil by mycorrhizal onion roots. Soil Sci. Soc. Am. J. 50:1494–1499.

Thomasson, A.J. 1978. Towards an objective classification of soil structure. J. Soil Sci. 29:38–46.

Thomasson, A.J., and A.D. Carter. 1992. Current and future uses of the UK soil water retention data set. p. 355–359. *In* M. Th. van Genuchten, F. J. Leij, and L. J. Lund (ed.) Proc. Int. Worksh. Indir. Meth. Estim. Hydr. Prop. Unsat. Soils. USDA/ARS/University of California, Riverside, CA.

Tiessen, H., and J.W.B. Stewart. 1988. Light and electron microscopy of stained microaggregates: the role of organic matter and microbes in soil aggregation. Biogeochem. 5:312–322.

Tisdall, J.M., and J.M. Oades. 1979. Stabilization of soil aggregates by the root systems of ryegrass. Aust. J. Soil Res. 17:429–441.

Tisdall, J.M., and J.M. Oades. 1982. Organic matter and water-stable aggregates. J. Soil Sci. 33:141–163.

Tisdall, J.M., B. Cockroft, and N.C. Uren. 1978. The stability of soil aggregates as affected by organic materials, microbial activity and physical disruption. Aust. J. Soil Res. 16:9–17.

Tisdall, J.M., S.E. Smith, and P. Rengasamy. 1997. Aggregation of soil by fungal hyphae. Aust. J. Soil Res. 35:55–60.

Topp, G.C., Y.T. Galganov, K.C. Wires, and J.L.B. Culley. 1994. Non-limiting water range (NLWR): An approach for assessing soil structure. Soil Qual. Eval. Prog. Tech. Rep. 2. Centre for Land and Biological Resources Research, Agriculture Canada, Ottawa, Canada.

Unger, P.W. 1996. Soil bulk density, penetration resistance, and hydraulic conductivity under controlled traffic conditions. Soil Till. Res. 37:67–75.

Utomo, W.H., and A.R. Dexter. 1981a. Tilth mellowing. J. Soil Sci. 32:187–201

Utomo, W.H., and A.R. Dexter. 1981b. Soil friability. J. Soil Sci. 32:203–213

van Lanen, H.A.J., G.J. Reinds, O.H. Boersma, and J. Bouma. 1992. Impact of soil management systems on soil structure and physical properties in a clay loam soil, and the simulated effects on water deficits, soil aeration and workability. Soil Till. Res. 23:203–20.

van Veen, J.A., and P.J. Kuikman. 1990. Soil structural aspects of decomposition of organic matter by micro-organisms. Biogeochem. 11:213–233.

Veihmeyer, F. J., and A.H. Hendrickson. 1927. Soil moisture conditions in relation to plant growth. Plant Physiol. 2:71–82.

Visser, S.A., and M. Caillier. 1988. Observations on the dispersion and aggregation of clays by humic substances, I. Dispersive effects of humic acids. Geoderma 42:331–337.

Voorhees, W.B. 1983. Relative effectiveness of tillage and natural forces in alleviating wheel-induced soil compaction. Soil Sci. Soc. Am. J. 47:129–133.

Voorhees, W.B., and M.J. Lindstrom. 1984. Long-term effects of tillage method on soil tilth independent of wheel traffic compaction. Soil Sci. Soc. Am. J. 48:152–156.

Waldron, L.J., and S. Dakessian. 1982. Effect of grass, legume, and tree roots on soil shearing resistance. Soil Sci. Soc. Am. J. 46:894–899.

Wang, C., J.A. McKeague, and G.C. Topp. 1985. Comparison of estimated and measured horizontal Ksat values. Can. J. Soil Sci. 65:707–715.

Warner, G.S., J.L. Neiber, I.D. Moore, and R.A. Giese. 1989. Characterizing macropores in soil by computed tomography. Soil Sci. Soc. Am. J. 53:653–660.

Waters, A.G., and J.M. Oades. 1991. Organic matter in water-stable aggregates. p. 163–174. *In* W.S. Wilson (ed.) Advances in soil organic matter research. The impact on agriculture and the environment. Royal Society of Chemistry, Cambridge, UK.

Watts, C.W., A.R. Dexter, E. Dumitru, and J. Arvidsson. 1996a. An assessment of the vulnerability of soil structure to destabilization during tillage. 1. A laboratory test. Soil Till. Res. 37:161–74.

Watts, C.W., A.R. Dexter, and D.J. Longstaff. 1996b. An assessment of the vulnerability of soil structure to destabilization during tillage. 2. Field trials. Soil Till. Res. 37:175–90.

Webber, L.R. 1965. Soil polysaccharides and aggregation in crop sequences. Soil Sci. Soc. Am. Proc. 29:39–42.

Wenke, J.F., and C.D. Grant. 1994. The indexing of self-mulching behavior in soils. Aust. J. Soil Res. 32:201–211

White, R.E. 1985. The influence of macropores on the transport of dissolved and suspended matter through soil. Adv. Soil Sci. 3:94–120.

White, W.M. 1993. Dry aggregate distribution. p. 659–662. *In* M.R. Carter (ed.). Manual on soil sampling and methods of analysis. CRC Press, Boca Raton, FL.

Williams, J., R.E. Prebble, W.T. Williams, and C.T. Hignett. 1983. The influence of texture, structure and clay mineralogy on the soil moisture characteristic. Aust. J. Soil Res. 21:15–32.

Willis, W. 1955. Freezing and thawing, and wetting and drying in soils treated with organic chemicals. Soil Sci. Soc. Amer. Proc. 19:263–267.

Woodruff, N.P., and F.H. Siddoway. 1965. A wind erosion equation. Soil Sci. Soc. Am. Proc. 29:602–608.

Wright, S.F, and A. Upadhyaya. 1996. Extraction of an abundant and unusual protein from soil and comparison with hyphal protein of arbuscular mycorrhizal fungi. Soil Sci. 161:575–586.

Wu, L., and J.A. Vomocil. 1992. Predicting the soil water characteristic from the aggregate-size distribution. p. 139–45. *In* M.Th. van Genuchten, F.J. Leij, and L.J. Lund (ed). Proc. Int. Worksh. Indir. Meth. Estim. Hydr. Prop. Unsat. Soils. USDA/ARS/University of California, Riverside, CA.

Xu, X., J.L. Nieber, and S.C. Gupta. 1992. Compaction effect on the gas diffusion coefficient in soils. Soil Sci. Soc. Am. J. 56:1743–50.

Yoder, R.E. 1936. A direct method of aggregate analysis of soils and a study of the physical nature of erosion losses. J. Am. Soc. Agron. 28:337–351.

Young, I., and C.E. Mullins. 1991. Water-suspensible solids and structural stability. Soil Till. Res. 19: 89–94.

Zausig, J., W. Stepniewski, and R. Horn. 1993. Oxygen concentration and redox potential gradients in unsaturated model soil aggregates. Soil Sci. Soc. Am. J. 57:908–16.

8

Soil Gas Movement in Unsaturated Systems

B.R. Scanlon
University of Texas at Austin

J.P. Nicot
University of Texas at Austin

J.M. Massmann
University of Washington

8.1 General Concepts Related to Gas Movement

An understanding of gas transport in unsaturated media is important for evaluation of soil aeration or movement of O_2 from the atmosphere to the soil. Soil aeration is critical for plant root growth because roots generally cannot get enough O_2 from leaves. Evaluation of gas movement is also important for estimating transport of volatile and semivolatile organic compounds from contaminated sites through the unsaturated zone to the groundwater. The use of soil venting, or soil vapor extraction, as a technique for remediating contaminated sites has resulted in increased interest in gas transport in the unsaturated zone (Rathfelder et al., 1995). Migration of gases from landfills, such as methane formed by decomposition of organic material, is important in many areas (Moore et al., 1982; Thibodeaux et al., 1982). Soil gas composition has also been used as a tool for mineral and petroleum exploration and for mapping organic contaminant plumes. An understanding of gas transport is important for evaluating movement of volatile radionuclides, such as ^3H, ^{14}C, and Rd from radioactive waste disposal facilities. The adverse health effects of radon and its decay products have led to evaluation of transport in soils and into buildings (Nazaroff, 1992). A thorough understanding of gas transport is required to evaluate these issues.

Gas in the unsaturated zone is generally moist air, but has higher CO_2 concentrations than atmospheric air because of plant root respiration and microbial degradation of organic compounds. Oxygen concentrations are generally inversely related to CO_2 concentrations because processes producing CO_2 generally deplete O_2 levels. Contaminated sites may have gas compositions that differ markedly from atmospheric air, depending on the type of contaminants.

This chapter will focus primarily on processes of gas transport in unsaturated media. The following issues will be evaluated:

(1) How does gas move through the unsaturated zone?
(2) How does one evaluate single gas and multicomponent gas transport?
(3) How does one measure or estimate the various parameters required to quantify gas flow?
(4) How does one numerically simulate gas flow?

Although chemical reactions are discussed in another part of this book, the model equations developed in this chapter need to be incorporated into them. The reader may find the discussion of water flow in unsaturated media in Section A, Chapters 3 and 4 helpful in understanding many of the concepts in this chapter.

8.1.1 Gas Content

Unsaturated media consist generally of at least three phases: solid, liquid, and gas. In some cases, a separate nonaqueous liquid phase may exist if the system is contaminated by organic compounds. In most cases the pore space is only partly filled with gas. The volumetric gas content (θ_G) is defined as

$$\theta_G = \frac{V_G}{V_T} \qquad\qquad [8.1]$$

where V_G (L^3) is the volume of the gas and V_T (L^3) is the total volume of the sample. This definition is similar to that used for volumetric water content in unsaturated-zone hydrology. In many cases, the volumetric gas content is referred to as the gas porosity. The saturation with respect to the gas phase (S_G) is

$$S_G = \frac{V_G}{V_v} \qquad\qquad [8.2]$$

where V_v (L^3) is the volume of voids or pores. Saturation values range from 0 to 1. Volumetric gas content and gas saturation are related as follows:

$$\theta_G = \phi S_G \qquad\qquad [8.3]$$

where ϕ is porosity (V_p/V_T). If only two fluids, gas and water, are in the system, the volumetric gas and water contents sum to the porosity. Therefore, volumetric gas content can be calculated if the volumetric water content and porosity are known. Volumetric water content can be measured using procedures described in Gardner (1986).

In unsaturated systems, water is the wetting and gas is the nonwetting phase. Therefore, water wets the solids and is in direct contact with them whereas gas is generally separated from the solid phase by the water phase. Water fills the smaller pores, whereas gas is restricted to the larger pores.

8.1.2 Differences between Gas and Water Properties

Gas and water properties differ greatly and can be compared as follows: (1) gas density is much lower than liquid density. The density of air varies with composition but ranges generally from 1 to 1.5 kg m^{-3}, whereas the density of water is close to 1,000 kg m^{-3} (Table 8.1); (2) although water is generally

Table 8.1 Variations in density and viscosity with temperature for water, water vapor, and air. Moist air corresponds to a vapor pressure of 4.6 mm at 0 °C to 261 mm at 100 °C.

T (°C)	[1]Density water (kg m^{-3})	[2]Density water vapor (kg m^{-3})	[3]Density moist air (kg m^{-3})	[4]Viscosity water (Pa s)	[5]Viscosity water vapor (Pa s)	[6]Density dry air (Pa s)
0	1000	0.005	1.290	1.79E-03	8.03E-06	1.73E-05
10	1000	0.009	1.242	1.31E-03	8.44E-06	1.77E-05
20	998	0.017	1.194	1.00E-03	8.85E-06	1.81E-05
30	996	0.030	1.146	7.97E-04	9.25E-06	1.85E-05
40	992	0.051	1.097	6.53E-04	9.66E-06	1.89E-05
50	988	0.083	1.043	5.48E-04	1.01E-05	1.93E-05
60	983	0.130	0.981	4.70E-04	1.05E-05	1.97E-05
70	978	0.198	0.909	4.10E-04	1.09E-05	2.01E-05
80	971	0.293	0.823	3.62E-04	1.13E-05	2.05E-05
90	965	0.423	0.718	3.24E-04	1.17E-05	2.09E-05
100	959	0.598	0.588	2.93E-04	1.21E-05	2.13E-05

[1,3,4]Weast, 1986; [2]Childs and Malstaff (1982) in Hampton (1989); [5,6]White and Oostrom (1996); Vapor pressure of moist air ranges from 4.6 mm at 0 °C to 761 mm at 100 °C.

considered incompressible, the incompressibility assumption is not always valid for gas flow. Gas density depends on gas pressure, which results in a nonlinear flow equation; (3) at solid surfaces, water is assumed to have zero velocity whereas gas velocities are generally non zero, and such velocities result in slip flow or the Klinkenberg effect (Fig. 8.1); (4) because the dynamic viscosity of air is ~ 50 times lower than that of water, significant air flow can occur at much smaller pressure gradients. Gas viscosity increases with temperature, whereas water viscosity decreases with temperature; (5) air conductivities are generally an order of magnitude less than water or hydraulic conductivities in the same material because of density and viscosity differences between the fluids; (6) because gas molecular diffusion coefficients are about four orders of magnitude greater than those of water, gas diffusive fluxes are generally much greater than those of water.

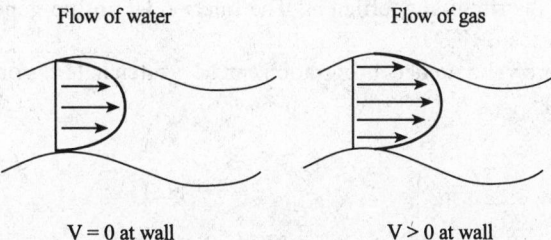

Fig. 8.1 Schematic demonstrating zero velocities at pore walls of liquid flow and nonzero velocities of gas flow

8.1.3   Terminology

Gas refers to a phase that may be single or multicomponent. Gas flux can be expressed in a variety of ways, such as volume ($L^3 L^{-2} t^{-1}$), mass ($M L^{-2} t^{-1}$), molar ($mol\, L^{-2} t^{-1}$), molecular (number of molecules $L^{-2} t^{-1}$) fluxes. If a gas is incompressible (density constant), then volume and mass are conservative whereas if a gas is compressible only mass is conservative. The mean mass flux is weighted by the mass of each molecule. In the mean molar or molecular flux all molecules behave identically and contribute equally to the mean. The distinction between mass and molar flux will be clarified in Section 8.1.5.4. Parameters important in describing gas transport mechanisms include mean free path (λ), which is the average distance a gas molecule travels before colliding with another gas molecule; pore size (λ_p), which is equivalent to the average distance between soil particles in dry porous media; and particle size (r_p). Equimolar component gases are components that have the same molecular weight, such as N_2 and CO, whereas nonequimolar component gases have different molecular weights. The terms *concentration* ($M L^{-3}$) and *partial pressure* ($M L^{-1} t^{-2}$) refer to individual gas components. The partial pressure of gas component i is equal to the mole fraction times the total pressure and can be related to concentration through the ideal gas law for ideal gases. Advective flux refers to bulk gas phase and occurs in response to unbalanced mechanical driving forces acting on the phase as a whole.

8.1.4   Mass Transfer versus Mass Transport

Most of this chapter deals with transport of gas and associated chemicals through the unsaturated zone. Mass transfer refers to transfer of mass or partitioning of mass among gas, liquid, and solid phases. Partitioning of chemicals into other phases retards their transport in the gas phase and can be described by

$$C_T = \rho_b C_{ad} + \theta_l C_l + \theta_G C_G \qquad [8.4]$$

where C_T is the total mass concentration ($M L^{-3}$ soil); C_G, C_l, and C_{ad} are the mass concentrations in the gas and liquid phases and adsorbed on the solid phase; and ρ_b is the bulk density ($M L^{-3}$). The relationship between gas and water mass concentrations can be described by Henry's law:

$$C_G = K_H C_l \qquad [8.5]$$

where K_H is the dimensionless Henry's law constant. Many types of isotherms describe the adsorption onto the solid phase, the simplest being the linear adsorption isotherm:

$$C_{ad} = K_d C_l \qquad [8.6]$$

where K_d ($L^3 M^{-1}$) is the distribution coefficient. The linear relationship generally applies to low polarity compounds (Brusseau, 1994).

If linear relationships are valid, total concentration can be written in terms of gas concentration as follows (Jury et al., 1991):

$$C_T = \left(\frac{\rho_b K_d}{K_H} + \frac{\theta_l}{K_H} + \theta_G \right) C_G = B_G C_G \qquad [8.7]$$

where B_G is the bulk gas phase partition coefficient (Charbeneau and Daniel, 1992). In some studies this partitioning behavior is used to quantify the amount of a liquid phase in the system. For example, transport of gas tracers that partition into water or nonaqueous phase liquids is compared with transport of those that do not partition (conservative) into that phase to determine the amount of water or organic compound in the system (Jin et al., 1995).

8.1.5 Mechanisms of Gas Transport

Many readers may be more familiar with transport mechanisms in the liquid than in the gas phase. Primary mechanisms of solute transport in the liquid phase include advection (movement with the bulk fluid) and hydrodynamic dispersion (mechanical dispersion and molecular diffusion) (Freeze and Cherry, 1986). Mechanical dispersion, resulting from variations in fluid velocity at the pore scale, is the product of dispersivity and advective velocity. Transport in the gas phase may also be described by advection and dispersion. Although some studies have found mechanical dispersion or velocity dependent dispersion to be important for chemical transport in the gas phase (Rolston et al., 1969; Auer et al., 1996), in most cases mechanical dispersion is ignored because gas velocities are generally too small and the effects of diffusion are generally much greater than dispersion in the gas phase. Molecular diffusion coefficients are approximately four orders of magnitude greater in the gas than in the liquid phase.

Diffusive transport in the liquid phase is described by molecular diffusion. Traditionally, diffusive transport in the gas phase has also been described by molecular diffusion. Diffusion in the gas phase may, however, be much more complicated and may include Knudsen, molecular, and nonequimolar diffusion. Surface diffusion of adsorbed gases, generally not significant, is not discussed in this chapter. Pressure diffusion is generally negligible at depths of less than 100 m, which include most unsaturated sections (Amali and Rolston, 1993). Although temperature gradients in unsaturated media are generally too low to result in significant diffusion except at the land surface, thermal diffusion is important for water vapor transport (Section 8.6.3).

8.1.5.1 Advective Flux

If a total pressure gradient exists in a soil as a result of external forces such as atmospheric pumping (Section 8.2), gases will flow from points of higher to those of lower pressure. It has been shown that relatively small gradients in total pressure can result in advective gas fluxes that are much larger than diffusive gas fluxes (Alzaydi and Moore, 1978; Thorstenson and Pollock, 1989; Massmann and Farrier, 1992). The driving force for flow is the total pressure gradient, and the resistance to flow is caused by viscosity of the gas. Other terms for advective flux are pressure driven or viscous flux. Under a total pressure gradient, advection is dominant when the mean free path of the gas molecules (λ) is much less than the pore radius (λ_p) and the particle radius (r_p) ($\lambda \ll \lambda_p$ and $\lambda \ll r_p$), resulting in intermolecular collisions being dominant relative to collisions between gas molecules and the pore walls (Cunningham and Williams, 1980). Because the mean free path is inversely proportional to the mean pressure, low mean free paths relative to pore size may occur in dry, coarse grained media and/or under high mean pressure. A pressure gradient is required to maintain advective flux because of the velocity reduction near the pore walls caused by gas molecules rebounding from collisions with the pore walls. In unsaturated media the pore walls will generally consist of the gas liquid interface. Advective flux is also termed nonsegregative or nonseparative because bulk flow as a result of a pressure gradient does not segregate the gas into individual components. The viscous flux of gas component i is proportional to its mole fraction (x_i) in the mixture:

$$N_i^V = x_i N^V \tag{8.8}$$

where N_i^V is the molar viscous flux (mol L^{-2} t^{-1}) of gas i and N^V is the total molar viscous flux. As the mean pressure and/or as the pore size decrease as a result of decreasing grain size or gas saturation (increasing water saturation), a continuum results, from advective or viscous flux to viscous slip flux (Klinkenberg effect) to Knudsen diffusive flux. Viscous slip flux occurs when the mean free path of the gas molecules becomes approximately the same as the pore radius and results from nonzero velocity at the pore wall. Stonestrom and Rubin (1989) and Detty (1992) found that errors resulting from ignoring viscous slip flux in dry, coarse grained porous media (sands) were less than 7%.

8.1.5.2 Knudsen Diffusive Flux

Knudsen flux occurs when the gas mean free path is much greater than the pore radius ($\lambda >> \lambda p$) Knudsen number ($K_n = \lambda/\lambda_p$) = 10 corresponding to the Knudsen regime (Alzaydi, 1975; Abu-El-Sha'r, 1993). Molecule-wall collisions dominate over molecule-molecule collisions. The term *free molecule flux* is also used to describe Knudsen flux because rebounding molecules do not collide with other gas molecules. The Knudsen diffusive flux depends on the molecular weights and temperatures of gases and the radius of the pores. It is not influenced by the presence of other species of gas and is described as follows:

$$N_i^k = -D_i^k \nabla C_i \tag{8.9}$$

where N_i^k is the Knudsen molar flux of component i (mol L^{-2} t^{-1}), D_i^k is the effective Knudsen diffusivity (L^2 t^{-1}), and C_i is the molar concentration of gas i (mol L^{-3}) (Cunningham and Williams, 1980). D_i^k is defined as

$$D_i^k = Q_p \left(\frac{RT}{M_i} \right)^{0.5} \tag{8.10}$$

where Q_p is the Knudsen radius (L), R is the ideal gas constant (M L^2 t^{-2} T^{-1} mol^{-1}), T is temperature (T), and M_i is the molecular weight of gas i (M mol^{-1}).

8.1.5.3 Molecular Diffusive Flux

Molecular diffusion is the only type of transport mechanism that occurs under isothermal, isobaric conditions when equimolar pairs of gases (e.g., N_2, CO) counterdiffuse in pores whose size is much greater than the mean free path of the gas molecules. In this case molecule-molecule collisions dominate over molecule-wall collisions. The molecular diffusive flux depends on gas molecular weights and temperatures in the pore space and is unaffected by the physical nature of the pore walls. Because molecular diffusion results in segregation of the different component gases, it is termed *segregative*. The molar diffusive flux of component i resulting from molecular diffusion $\left(J_{iM}^m \right)$, in a binary gas mixture under isothermal, isobaric conditions is proportional to the concentration gradient and is described by Mason and Malinauskas (1983):

$$J_{iM}^m = -D_{ij}^e \nabla n_{iM} \tag{8.11}$$

where D_{ij}^e is the effective molecular diffusion coefficient ($L^2\,t^{-1}$) and n_{iM} is the molar concentration of gas i (mol L^{-3}) (Abu-El-Sha'r, 1993). The effective diffusion coefficient in porous media is calculated from the free air diffusion coefficient as follows:

$$D_{ij}^e = \tau\theta_G D_{ij} \qquad [8.12]$$

where τ is the tortuosity, θ_G is the gas content, and D_{ij} is the free air diffusion coefficient ($L^2\,t^{-1}$). Volumetric gas content (θ_G) accounts for the reduced cross-sectional area in porous media relative to free air, and tortuosity (τ) accounts for increased path length. A variety of equations have been developed to calculate tortuosity. The most commonly used equations are those of Penman (1940a,b) and Millington and Quirk (1961) (Table 8.2). The effective diffusion coefficient decreases with increased water content in unsaturated systems; the rate of decrease is low at low water contents because gas is in the large and water in the small pore spaces. The effective diffusion coefficient decreases sharply as soils become saturated with water because the large pores become occluded.

8.1.5.4 Bulk Diffusive Flux

Bulk diffusion includes molecular and nonequimolar diffusion. Nonequimolar diffusion occurs when gas components have different molecular weights. According to the kinetic theory of gases, gas molecules have the same kinetic energy in an isothermal, isobaric system; therefore, lighter gas molecules have higher velocities than heavier gas molecules. In a binary mixture of nonequimolar gases, the more rapid diffusion of the lighter gas molecules results in a pressure gradient. Such pressure gradients should be distinguished from external pressure gradients that result in advective flux. The flux resulting from the buildup of pressure is diffusive and is called the *nonequimolar flux* or the *diffusive slip flux* (Cunningham and Williams, 1980). Because the nonequimolar flux does not result in separation of gas components (nonsegregative), the molar bulk diffusive flux includes the segregative molecular diffusive flux and the nonsegregative nonequimolar diffusive flux:

$$N_i^D = J_{iM}^m + x_i \sum_{j=1}^{v} N_j^D \qquad [8.13]$$

Table 8.2 Models for estimating tortuosity, θ_G is the gas content, and n is the porosity; a more complete listing is provided in Abu-El-Sha'r (1993).

Tortuosity (τ) times θ_g	Reference	Comments
$0.66\,\phi$	Penman (1940a, b)	experimental
$(\theta_G)^{3/2}$	Marshall (1958)	pseudo theoretical
$(\theta_G)^{4/3}$	Millington (1959)	pseudo theoretical
$\theta_G^{10/3}/\phi^2$	Millington and Quirk (1961)	theoretical
$(1-S_w)^2 *(\phi-\theta)^{2x}$	Millington and Shearer (1970)	theoretical
$\theta_G^{7/3}$	Lai et al. (1976)	experimental
$0.435\,\phi$	Abu-El-Sha'r and Abriola (1997)	experimental

where N_i^D is the bulk molar diffusive flux of component i (mol $L^{-2} t^{-1}$), J_{iM}^m is the molar diffusive flux of component i resulting from molecular diffusion, x_i is the mole fraction of component i, v is the number of gas components, and the second term on the right $x_i \sum_{j=1}^{v} N_j^D$ is the nonequimolar flux.

It is important to distinguish between diffusive gas flux with respect to average velocity of the gas molecules (molar velocity) and that with respect to velocity of the center of mass of the gas (mass velocity) because the two are not necessarily the same (Cunningham and Williams, 1980). In nonequimolar gas mixtures, because the lighter gas molecules diffuse more rapidly, the molar gas flux moves in the direction in which the lighter gas molecules are moving; whereas the center of mass of the gas moves in the direction in which the heavier molecules diffuse. The molar and mass fluxes can, therefore, move in opposite directions.

8.1.6   Gas Transport Models

Darcy's law is used to model advective or viscous gas transport, and traditionally Fick's law has been used to model molecular diffusion. The Stefan Maxwell equations and the dusty gas model can also be used to simulate diffusive gas flux. If there is a total pressure gradient in a system, then the diffusive fluxes are calculated relative to the bulk molar flux. The total molar flux is calculated by adding the molar viscous or advective flux to the bulk molar diffusive flux:

$$N_i^T = N_i^V + N_i^D \tag{8.14}$$

where N_i^T is the total gas molar flux, N_i^V is the molar viscous flux, and N_i^D is the molar diffusive flux of component i. For analysis of transient gas flux, the constitutive flux equations are incorporated into the conservation of mass. If the mass density of the fluid and the diffusion coefficients are assumed constant, then the transport equation can be written as:

$$\frac{\partial (Cx_i)}{\partial t} + \nabla \cdot \left(N_i^T \right) = 0 \tag{8.15}$$

where C is the molar concentration of the gas (mol L^{-3}) and x_i is the mole fraction of component i (Massmann and Farrier, 1992). This equation is strictly valid for relatively dilute gas mixtures or for vapors with molecular weights close to the average molecular weight of air (Bird et al., 1960).

8.1.6.1   Darcy's Law

Darcy's law is used to describe advective gas transport. The simplest form of Darcy's law is as follows:

$$J_G = -\frac{k_G}{\mu_G} \nabla P \tag{8.16}$$

where J_G is the volumetric flux ($L^3 L^{-2} t^{-1}$), k_G is the permeability (L^2), μ_G is the dynamic gas viscosity (M $L^{-1} t^{-1}$), and ∇P is the applied pressure gradient. The ideal gas law can be used to convert the volumetric gas flux (J_G) to the molar gas flux (N^V):

$$PV = nRT \qquad \rho_G = \frac{nM}{V} = \frac{PM}{RT} \tag{8.17}$$

where n is the number of moles of the gas, M is the molar mass of the gas, R is the ideal gas constant $(M \, L^2 \, t^{-2} \, T^{-1} \, mol^{-1})$, and T is temperature (K). The molar viscous or advective flux is

$$N^V = -\frac{P}{RT} \frac{k_G}{\mu_G} \nabla P \tag{8.18}$$

The validity of Darcy's law can be evaluated by plotting the gas flux for a single component against the pressure gradient. The relationship should be linear with a zero intercept if Darcy's law is valid.

8.1.6.2 Fick's Law
Fick's law is an empirical expression originally developed to describe molecular diffusion of solutes in the liquid phase. Fick's law is generally used to describe molecular diffusion of gas i in gas j (Bird et al., 1960; Jaynes and Rogowski, 1983):

$$J_{iM}^m = -D_{ij} C \nabla x_i \tag{8.19}$$

where C is the total molar concentration (mol $L^{-2} \, t^{-1}$, constant in an isothermal, isobaric system) and x_i is the mole fraction of gas i, and $D_{ij} = D_{ji}$ (Bird et al., 1960). Fick's law is strictly applicable to molecular diffusion of equimolar gases in isothermal, isobaric systems. Fick's law excludes the effects of Knudsen diffusion and nonequimolar diffusion. Fick's law can predict the flux of only one component. The adequacy of Fick's law will be discussed after the other models have been described.

8.1.6.3 Dusty Gas Model
In contrast to Fick's law, which is empirical, the dusty gas model is based on the full Chapman Enskog kinetic theory of gases. Application of the kinetic theory of gases is only possible by considering the dust particles (porous medium) as giant molecules that constitute another component in the gas phase. The multicomponent equations, which are based on the dusty gas model originally proposed by Mason et al. (1967), are presented by Satterfield and Cadle (1968), Cunningham and Williams (1980), Mason and Malinauskas (1983), and Thorstenson and Pollock (1989). The equations for the dusty gas model of binary and multicomponent gas diffusion are (Cunningham and Williams, 1980):

$$\frac{x_1 N_2^D - x_2 N_1^D}{D_{12}^e} - \frac{N_1^D}{D_1^k} = \frac{\nabla P_1}{RT}$$

$$\sum_{\substack{i=1 \\ i \neq j}}^{v} \frac{x_i N_j^D - x_j N_i^D}{D_{ij}^e} - \frac{N_i^D}{D_i^k} = \frac{\nabla P_i}{RT} \tag{8.20}$$

$$\underset{\substack{\text{molecular} \\ \text{diffusion}}}{} \quad \underset{\substack{\text{Knudsen} \\ \text{diffusion}}}{} \quad \underset{\substack{P \text{ gradient} \\ \text{term}}}{}$$

where x_i is the mole fraction of component i, N_i^D is the bulk molar diffusive flux of gas component i (mol $L^{-2} \, t^{-1}$), D_{ij}^e is the effective molecular diffusivity ($L^2 \, t^{-1}$), D_i^k is the effective Knudsen diffusivity ($L^2 \, t^{-1}$), and P_i is the partial pressure of component i (M $L^{-1} t^{-2}$). Although one generally writes equations in terms of a flux of a component being proportional to a gradient, multicomponent gas equations are simplest when written in terms of a gradient of gas component i being proportional to the fluxes of all other components (Mason and Malinauskas, 1983). Derivation of the above form of these equations for a binary gas mixture is shown in Section 8.7. Knudsen diffusion is important when the second term in the dusty gas model is larger than the first term. This requires that $D_i^k << D_{ij}^e$ which appears counterintuitive (Thorstenson and Pollock, 1989), but which can be understood if one considers the reciprocal of D_i^k as a resistance. Knudsen diffusion is thus important when $1/D_i^k$ is large and D_i^k is small. Although there is a gradient in the natural log of temperature in the dusty gas model (Thorstenson and Pollock, 1989), gradients in temperature in the subsurface are generally small and gradients in the natural log of temperature are generally negligible. It has, therefore, been omitted in Equation [8.20]. The primary assumptions of the dusty gas model are that the dust particles are spherical and that there are no external forces on the gas. The dusty gas model can predict the flux of all components in a gas mixture. The system of equations can be solved analytically (Jackson, 1977) or numerically (Massmann and Farrier, 1992). Many analyses of the dusty gas model assume steady-state flow (Thorstenson and Pollock, 1989; Abu-El-Sha'r, 1993).

When both partial and total pressure gradients are important for gas flux, the combined effects of advection and diffusion need to be considered. Knudsen diffusion, which is proportional to the total mole fraction, provides the link between advection and diffusion. Because the dusty gas model is the only model that includes Knudsen diffusion, it is the only model that can theoretically link advection and diffusion through the Knudsen diffusion term. The dusty gas model for combined advection and diffusion is as follows (Thorstenson and Pollock, 1989, Equation [8.40]):

$$\sum_{\substack{j=1 \\ i \neq 1}}^{v} \frac{x_i N_j^T - x_j N_i^T}{D_{ij}^e} - \frac{N_i^T}{D_i^k} = \frac{P \nabla x_i}{RT} + \left(1 + \frac{kP}{D_i^k \mu_G}\right) \frac{x_i \nabla P}{RT} \qquad [8.21]$$

where x_i is the mole fraction of component i, N_i^T is the total molar gas flux of component i relative to a fixed coordinate system (mol L^{-2}, t^{-1}), D_{ij}^e is the effective molecular diffusion coefficient of component i in j ($L^2 t^{-1}$), D_{ij}^k is the effective Knudsen diffusion coefficient of component i ($L^2 t^{-1}$), P is pressure ($M L^{-1} t^{-2}$), R is the ideal gas constant, T is absolute temperature, k is permeability (L^2), and μ_G is gas viscosity ($M L^{-1} t^{-1}$).

8.1.6.4   Stefan Maxwell Equation

The Stefan Maxwell equations, which can predict the fluxes of all but one gas component in a multicomponent gas mixture, can be obtained from the dusty gas model by assuming negligible Knudsen diffusion (no molecule particle collisions):

$$\sum_{\substack{j=1 \\ i \neq 1}}^{v} \frac{x_i N_j^D - x_j N_i^D}{D_{ij}^e} = \frac{\nabla P_i}{RT} \qquad [8.22]$$

8.2   Transport of a Homogeneous Gas

Transport of a homogeneous gas in dry, coarse grained media can be described by advection. Because such a gas can be considered as a single component, the only type of diffusion possible is Knudsen diffusion. Single component gases in dry, coarse grained media are dominated by advective or viscous flux because Knudsen diffusion is negligible in such systems. This analysis of gas transport is appropriate when gas velocities are high. The single fluid, nonreactive, noncompositional approximation is appropriate only when one is interested in the bulk flow of a homogenous gas.

Natural advective gas transport can occur in response to barometric pressure fluctuations, wind effects, watertable fluctuations, density effects, and can also be induced by injection or extraction, as in soil vapor extraction systems. Barometric pressure fluctuations consist of (1) diurnal fluctuations due to thermal and gravitational effects, which are on the order of a few millibars, and (2) longer term fluctuations that result from regional scale weather patterns, which are on the order of tens of millibars within a few hours of when a high or low pressure front moves through. The penetration depth of barometric pressure fluctuations increases with the thickness of the unsaturated zone and with the permeability of the medium. Because highs are balanced by lows, the net transport of the gas may be negligible, except in fractured media, where contaminants may migrate large distances (Nilson et al., 1991). Smaller scale fluctuations, such as gusts and lulls related to wind, may be important in fractured media (Weeks, 1993). Watertable fluctuations, resulting in changes in the gas volume, can produce advective flow; however, advective fluxes as a result of watertable fluctuations are considered important only if the rate of rise or decline of the watertable is rapid, the permeability of the material is high, and the watertable is shallow.

8.2.1   Governing Equations

Darcy's law, which was developed for water flow in saturated media, can also be applied to single phase, gas flow. Darcy's law is an empirical expression and the general form is

$$J_G = -\frac{k_G}{\mu_G}\left(\nabla P + \rho_G g \nabla z\right)$$ [8.23]

where J_G is the volumetric flux density ($L^3 L^{-2} t^{-1}$), k_G is gas permeability (L^2), ρ_G is gas density ($M L^{-3}$), g is gravitational acceleration ($L t^{-2}$), μ_G is gas viscosity ($M L^{-1} t^{-1}$), P is pressure ($M L^{-1} t^{-2}$), and z is elevation (L). The negative sign in Equation [8.23] is required because flow occurs in the direction of decreasing pressure. The first term in parentheses is the driving force due to pressure and the second term is the driving force due to gravity. Generally gas density is a function of pressure and composition, and the above form of Darcy's law is the only valid form in these cases. The pressure gradient provides the main driving force for advective gas transport, and the resistance to flow results from the gas viscosity. Small pressure gradients can result in substantial advective gas fluxes because the resistance to flow is small. The small flow resistance is attributed to the negligible viscosities of gases relative to those of liquids and to the gas being present in the largest pores. If the components of a gas all have the same molecular mass, then gas density is independent of composition and is only a function of pressure.

Under static equilibrium and isothermal conditions gas does not move. If the only forces operating on the gas are pressure and gravity, then these two forces must be balanced. The condition of zero motion results when

$$\nabla P = -\rho_G g \nabla z$$ [8.24a]

$$\frac{dP}{dz} = -\rho_G g$$ [8.24b]

If gas density varies because of variable composition, the above equation cannot be simplified further. If the gas is barotropic (i.e., $\rho_G = \rho_G(P)$ only), however, the ideal gas law (Equation [8.17]) can be used to relate density and pressure when pressure variations are close to atmospheric. Equation [8.24a] can be solved through integration by inserting the ideal gas law:

$$\frac{RT}{gM}\int_{P_0}^{P}\frac{dP}{P} = -\int_{z_0}^{z}dz; \qquad \frac{RT}{gM}\ln\left(\frac{P}{P_0}\right) = -z$$

$$P = P_0\exp\left(-\frac{Mg}{RT}z\right) \tag{8.25}$$

where P is assumed to be equal to P_0 at $z = 0$. If $z = 1$ m, $T = 290$ K, R is 8.314 J mol^{-1} K^{-1}, M is 28.96 g mol^{-1} (air), and g is 9.8 m s^{-2}, then $P = 0.99988$ P_0. Gas pressure, therefore, changes by 0.012% over a height of 1 m. Because the gas pressure changes are so small, pressure can generally be considered independent of height.

If gas density is only a function of pressure (compressible gas), Equation [8.23] can be manipulated to obtain:

$$q_G = -\frac{k_G\rho_G}{\mu_G}\left(\frac{\nabla P}{\rho_G} + g\nabla z\right) = \frac{k_G\rho_G}{\mu_G}\left(\nabla\int_{P_0}^{P}\frac{dP}{\rho_G} + g\nabla z\right) \tag{8.26}$$

If gas density is independent of pressure, then the gas is incompressible:

$$\nabla\int_{P_0}^{P}\frac{dP}{\rho_G} = \nabla\left(\frac{P}{\rho_G}\right) \tag{8.27}$$

$$q_G = -\frac{k_G\rho_G g}{\mu_G}\left(\nabla\left(\frac{P}{\rho_G g}\right) + \nabla z\right) = -\frac{k_G\rho_G g}{\mu_G}(\nabla h + \nabla z) = -\frac{k_G\rho_G g}{\mu_G}(\nabla H) \tag{8.28}$$

where h is pressure head (L) and H is total head (L, pressure + gravitational head). Because gas densities are generally very low (density of air is approximately three orders of magnitude less than water density), the gravitational term in Equation [8.23] ($\rho_G g\nabla z$) is generally $< 1\%$ of the pressure term and is ignored in most cases, as shown in Equation [8.16]. For transient gas flow, Darcy's law (ignoring gravity, Equation [8.16]), substituted into the conservation of mass equation, results in:

$$\frac{\partial(\rho_G\theta_G)}{\partial t} = -\nabla\cdot\left(-\frac{\rho_G k_G}{\mu_G}\nabla P\right) \tag{8.29}$$

Equation [8.29] is nonlinear because gas density depends on gas pressure and can be linearized in terms of P^2 by assuming ideal gas behavior (Equation [8.17]), which results in (Section 8.7):

$$\frac{\partial P^2}{\partial t} \approx \frac{k_G P_0}{\phi \mu_G} \nabla^2 P^2 \approx \alpha \nabla^2 P^2 \qquad\qquad [8.30]$$

where $\dfrac{k_G P_0}{\phi \mu_G}$ is the pneumatic diffusivity (α).

If pressure variations are small, another linear approximation to Equation [8.29] can be obtained:

$$\frac{\partial P}{\partial t} = \frac{k_G P_0}{\phi \mu_G} \nabla^2 P \qquad\qquad [8.31]$$

Derivations of Equations [8.30] and [8.31] are given in Section 8.7.

8.2.2 Gas Permeability

Gas permeability describes the ability of the unsaturated zone to conduct gas. Permeability (k) should be simply a function of the porous medium if the fluid does not react with the solid. Gas permeability (k_G; L^2) is related to gas conductivity (K_G) as follows:

$$K_G = k_G \frac{\rho_G g}{\mu_G} \qquad\qquad [8.32]$$

Some groups use the term *gas permeability* for K_G, which is called *gas conductivity*, and use the term *intrinsic permeability* for k_G, which is called *gas permeability*. Equation [8.32] shows that gas conductivity increases with gas density and decreases with gas viscosity.

The above material describes single phase gas flow. In unsaturated media, the pore space generally contains at least two fluids, gas and liquid. Darcy's law is applied to each fluid in the system, which assumes that there is no interaction between the fluids (Dullien, 1979). Because the cross-sectional area available for flow of each fluid is less than if the system were saturated with a single fluid, the permeability with respect to that fluid is also less. The gas permeability decreases as the gas saturation decreases. The relative permeability (k_{rG}) is a function of the gas saturation (S_G) and is defined as the permeability of the unsaturated medium at a particular gas saturation ($k_G(S_G)$) divided by the permeability at 100% saturation (k_G):

$$k_{rG}(S_G) = k_G(S_G)/k_G \qquad\qquad [8.33]$$

Relative permeability varies with (1) fluid saturation, (2) whether the fluid is wetting or nonwetting, and (3) whether the system is wetting or drying (hysteresis).

The relative permeability of the gas phase is greater than that of the liquid phase at low to moderate liquid saturations because the gas generally occurs in larger pores (Demond and Roberts, 1987). Relative permeabilities do not sum to one. This has been attributed to flow pathways traversed by two fluid phases being more tortuous than those traversed by a single phase (Scheidegger, 1974) or to pores with static menisci that cannot result in flow (Demond and Roberts, 1987). Zero relative permeabilities correspond to nonzero fluid saturations (Fig. 8.2). For example, in systems initially water saturated that begin to drain, gas will not begin to flow until a minimum gas saturation has been reached, the residual gas saturation (S_{rG}; Fig. 8.2). The gas zero relative permeability region corresponds to trapped gas and disruption of gas connectivity caused by water blockages (Stonestrom and Rubin, 1989). Liquid-phase permeability also exhibits a zero-permeability region that corresponds to residual liquid or water saturation (Section A, Chapter 3). In natural systems, the zone of residual water saturation corresponds to a zone of constant gas relative permeability because gas content does not increase. The dashed line in Fig. 8.2 in this zone corresponds with further increases in gas relative permeability (k_{rG}), which is related to increased gas content if the soil is oven dried and the water content is reduced to zero. A similar effect occurs with the water relative permeability (k_{rw}), where the dashed line shows increased water relative permeability corresponding to vacuum saturation or saturation of the sample under back pressure. Gas permeability is hysteretic at low water content, which indicates that gas permeability is not a unique function of gas saturation but depends on saturation history; i.e., whether the system is drying or wetting. At a given saturation, gas permeability is generally greater for wetting than for drying (Stonestrom and Rubin, 1989).

Various expressions have been developed to relate relative permeability to gas saturation (Table 8.3). Many of the expressions were developed to estimate relative permeabilities with respect to water; the corresponding expressions for gas relative permeability were obtained by replacing the effective liquid saturation (S_e) by $1-S_e$, where S_e is

$$S_e = \frac{S - S_r}{1 - S_r} \qquad [8.34]$$

where S_r is the residual water saturation (Brooks and Corey, 1964).

Darcy's law can be written as

$$J_G = -\frac{k_{rG}(S_G)k_G}{\mu_G}(\nabla P) = -\frac{k_G(S_G)}{\mu_G}\nabla P \qquad [8.35]$$

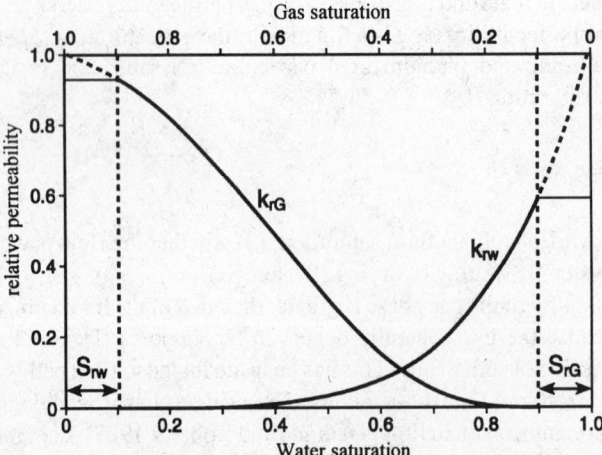

Fig. 8.2   Schematic relative permeabilities with respect to gas and water saturation

The velocity of the gas particles (V_G) $(L\ t^{-1})$ can be calculated from the volumetric flux density by dividing by the volumetric gas content:

$$V_G = J_G / \theta_G \qquad [8.36]$$

8.2.3 Deviations from Darcy's Law

Non-Darcian effects such as viscous slip flow and inertial flow can occur under many circumstances. As mentioned earlier, gas flow differs from liquid flow in that the velocity of the gas molecules is nonzero at the pore walls (Fig. 8.1) and is called *slip velocity* (Dullien, 1979). Slip flow, or the Klinkenberg effect, results in underestimation of gas flux by Darcy's law. Because of slip flow, permeability depends on pressure and the flow equation is nonlinear (Collins, 1961). Klinkenberg (1941) evaluated the relationship between slip-enhanced, or apparent, permeability k_G (L^2) and the permeability at infinite pressure k (L^2) when the gas behaves like a liquid:

$$k_G = k\left[1 + \frac{4c\lambda}{\overline{\lambda}_p}\right] \qquad [8.37]$$

where c is a constant characteristic of the porous medium, λ is the mean free path of the gas at the mean pressure, and $\overline{\lambda}_p$ is the mean pore radius (Klinkenberg, 1941). Because the mean free path is inversely proportional to the mean pressure in the system, the equation can be modified to

$$k_G = k\left(1 + \frac{b_i}{P}\right) \qquad [8.38]$$

Table 8.3 Expressions relating relative gas permeability and saturation. S_e is the effective saturation with respect to water; $S_e = \frac{S_w - S_{rw}}{1 - S_{rw}}$, S_{rw} is the residual wetting phase saturation (water in water/air system), λ is the bubbling pressure, and m and n are fitting parameters

Brooks and Corey (1964)	$k_{rG} = \left(1 - S_e\right)^2\left(1 - S_e^{\frac{2+\lambda}{\lambda}}\right)$	
Corey (1954)	$k_{rG} = \left(1 - S_e\right)^2\left(1 - S_e^2\right)$	
Falta et al. (1989)	$k_{rG} = \left(1 - S_e\right)^3$	
van Genuchten (1980); Mualem (1976)	$k_{rG} = \left(1 - S_e^{0.5}\right)\left\{1 - \left[1 - \left(1 - S_e\right)^{1/m}\right]^m\right\}^2$	$m = 1 - 1/n$ $0 < m < 1$
van Genuchten (1980); Burdine (1953)	$k_{rG} = \left(1 - S_e\right)^2\left\{1 - \left[1 - \left(1 - S_e\right)^{1/m}\right]^m\right\}$	$m = 1 - 2/n$ $0 < m < 1; n > 2$

where $\overline{P}$ is the mean pressure and b_i (M L^{-1} t^{-2}) is a constant (slip parameter) that depends on the porous medium and the gas i used in the measurement (Klinkenberg, 1941). These two equations show that gas slippage, or slip-enhanced permeability, is enhanced when the mean pressure is low. As the mean pressure increases, gas permeability decreases and approaches liquid permeability. Detty (1992) found that the slip parameter is not only a function of the reciprocal mean pressure as indicated by Klinkenberg (1941) but also a function of the pressure gradient and saturation. Slip correction factors measured in unconsolidated sands are generally fairly low (1 to 6%) (Stonestrom and Rubin, 1989; Detty, 1992) and resembled those measured in consolidated sands (Estes and Fulton, 1956).

Non Darcian behavior can also occur at high flow velocities as a result of inertial effects. According to Darcy's law, the flux is linearly proportional to the pressure gradient. At high flow velocities, the relationship becomes nonlinear and inertial effects result in fluxes lower than those predicted by Darcy's law. Forcheimer (1901) modified Darcy's law for high velocities:

$$\nabla P = \frac{\mu_G J_G}{k_G} + \beta \rho_G J_G^2 \qquad [8.39]$$

where P is pressure, μ_G is gas viscosity, J_G is volumetric gas flux, and β is the inertial flow factor (L^{-1}) (Detty, 1992). The second order term results from kinetic energy losses from high velocity flows. There is a spectrum from viscous (linear-laminar or Darcy flow) to inertial flow (fully turbulent) with a visco-inertial regime between. Laminar flow is not restricted to the region where Darcy's law is valid but extends into the visco-inertial regime where nonlinear-laminar flow occurs ($q \sim \nabla p^n$) (Detty, 1992). Detty (1992) indicated that deviations from viscous flow can be significant. The flow rates and pressure gradients that result in visco-inertial flow vary with water saturation.

8.3 Multicomponent Gas Transport

In isobaric systems, gas transport occurs by diffusion, whereas in nonisobaric systems gas transport occurs by advection (Darcy's law) and diffusion. A variety of models are available to describe multicomponent gas diffusion. Traditionally, gas diffusion has been described by Fick's law. As noted earlier, Fick's law is valid strictly for isothermal, isobaric, and equimolar countercurrent diffusion of a binary gas mixture. Theoretical analysis by Jaynes and Rogowski (1983) shows that the diffusion coefficient of Fick's law is only a function of the porous medium under certain conditions: equimolar countercurrent diffusion in a binary gas mixture, diffusion of a dilute component gas in a multicomponent gas mixture, and diffusion in a ternary system when one gas component is stagnant (zero flux, no sources or sinks, no reactions). Studies by Leffelaar (1987) indicate that if binary gas diffusion coefficients differ by more than a factor of 2, then Fick's law is invalid. Amali and Rolston (1993) also added that the total mole fraction of the diffusing gas components needs to be considered and should be low in order for Fick's law to be applicable.

If the flux of each gas component depends on the flux of the other gas components, then Fick's law no longer applies. In some cases, the calculated Fick's law flux is not only different in magnitude relative to multicomponent molecular diffusive flux, but opposite in direction also, as a result of molecular diffusion against a concentration gradient (Thorstenson and Pollock, 1989). Because Fick's law predicts the diffusion of only one component, variations in concentration of other components are attributed to other processes. Baehr and Bruell (1990) showed that physical displacement of naturally occurring gases such as O_2 by organic vapors and evaporative advective fluxes can be incorrectly attributed to aerobic microbial degradation or other sink terms when Fick's law is used. Unlike Fick's law, which can only consider binary gas mixtures, the Stefan-Maxwell equations or the dusty gas model can evaluate multicomponent gas transport. For multicomponent analysis, the transport of one component depends upon the transport of all other components that are present in the gas mixture. It is impossible to separate fully the effects of diffusion from those of advection in multicomponent gas systems (Thorstenson and Pollock, 1989). As noted earlier, Stefan Maxwell equations are valid strictly for isobaric systems where Knudsen diffusion is negligible.

The validity of the single component advection-diffusion (Fick's law) equation for simulating gas phase transport depends on the pore sizes and permeability of the porous media, the relative concentrations of the gas components that are involved, and their molecular weights, viscosities, and pressure gradients. Massmann and Farrier (1992) compared fluxes using the single-component advection-diffusion equation with those calculated using the multicomponent Stefan Maxwell equation and dusty gas model. The comparisons were developed for transport conditions similar to what might be observed in volatile organic compounds (VOCs) in the near surface vadose zone. For total pressure gradients ranging from 100 to 1,000 Pa m^{-1} (1 to 10 mbar m^{-1}), the single component advection-diffusion (Fick's law) equation significantly overestimates fluxes for toluene and trichloroethylene (TCE) in soils having permeabilities on the order of 10^{-16} to 10^{-17} m^2 (Fig. 8.3a). These permeabilities might correspond to unweathered clays, glacial tills, or very fine silts (Table 8.4). The Stefan Maxwell equations also overestimate fluxes in this permeability range because Knudsen diffusion is not included in Fick's law or in the Stefan Maxwell equations (Fig. 8.3b). The flux predicted by the single component advection-diffusion equation is within a factor of 2 of that predicted by the dusty gas model in materials having permeabilities greater than 10^{-14} m^2. This permeability corresponds to a relatively dry silty-sand material (Table 8.4). The Stefan Maxwell equation underestimates fluxes in high permeability materials for moderate total pressure gradients (0.1 to 1 mbar m^{-1}; 10 to 100 Pa m^{-1}) (Fig. 8.3b). Massmann and Farrier (1992) also illustrated how multicomponent equations may result in situations in which diffusion can occur in a direction opposite that of the partial pressure gradients. In multicomponent equations, the diffusive flux of one component depends on the diffusive fluxes of all other components in the system. For example, a large partial pressure gradient for TCE can cause the diffusive flux of N$_2$ and O$_2$ to occur in a direction opposite to their partial pressure gradients. In general, single component equations will be more valid for dilute gases having molecular weights similar to those of other species in the gas mixture.

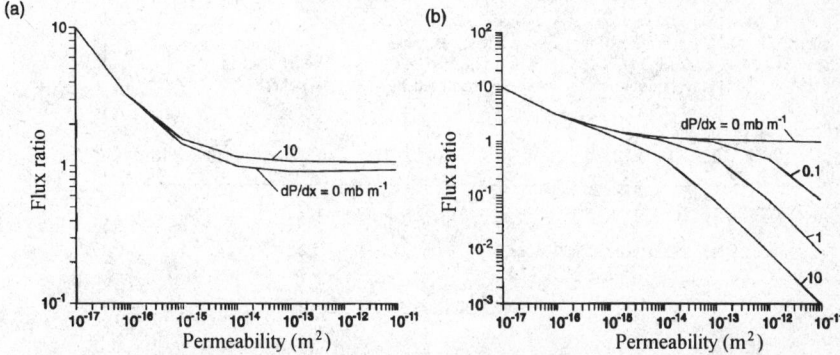

Fig. 8.3 (a) Ratios of TCE fluxes calculated using the single component equation (Fick's law for diffusion) divided by fluxes calculated using the dusty gas model; (b) ratios of TCE fluxes calculated using the Stefan-Maxwell equation divided by fluxes calculated using the dusty gas model [Reprinted from Massmann and Farrier, 1992. Water Resour. Res. 28:777-791, with permission]

Abriola et al. (1992) evaluated advective and diffusive fluxes in a system comprising two components. They reported that under an applied total pressure gradient of 0.05 mbar m^{-1} (5 Pa m^{-1}), the single and multicomponent models give virtually indistinguishable results for transport predictions in a sandy soil. For transport in a low permeability material, such as a clay under a total pressure gradient of 0.05 mbar m^{-1}, diffusive fluxes dominate and single and multicomponent equations give different transport predictions. Multicomponent gas experiments that evaluated transport of methane and TCE in air were examined by Abu-El-Sha'r (1993). The system was evaluated as either a binary (considering N_2 and O_2 in air as a single component with methane or TCE as the other component) or a ternary system. Results of the experiments showed that model predictions based on Fick's law, the Stefan Maxwell equations, and the dusty gas model were indistinguishable in sand samples, whereas the dusty gas model provided slightly better predictions in kaolinite. All three models can, therefore, predict diffusional gas fluxes under isobaric conditions when transport is dominated by molecular diffusion. However, under pressure gradients, deviations of measured and predicted values of gas fluxes increased with increasing total pressure gradients. All three models underestimated gas fluxes under nonisobaric conditions. Deviations from measured fluxes were lowest for the dusty gas model, increased for the Stefan Maxwell equations, and were greatest for Fick's law. Theoretical analysis by Thorstenson and Pollock (1989) indicates that multicomponent analysis using the dusty gas model is required for evaluation of stagnant gases such as N_2 or Ar in the air. Analysis of transient gas flux on the basis of column experiments using benzene and TCE shows that Fick's law and the Stefan Maxwell equations give similar results in the transient phase of the experiment, but that Fick's law underestimates measured fluxes by as much as 10% in the steady-state phase. The Stefan-Maxwell equations gave much better results (Amali et al., 1996).

These studies were used to develop Table 8.5, which generally outlines which models are most applicable under which pressure, permeability, and concentration conditions. For low permeability material the dusty gas model is required because it incorporates Knudsen diffusion. For high permeability material under isobaric conditions, Fick's law can be used if the diffusing gas component has a low concentration, whereas the Stefan Maxwell equation is required if concentration is high

Table 8.4 Typical values of permeabaility for different types of sediments and rocks [Terzaghi and Peck, 1968. Soil mechanics in engineering practice. John Wiley, New York, with permission]

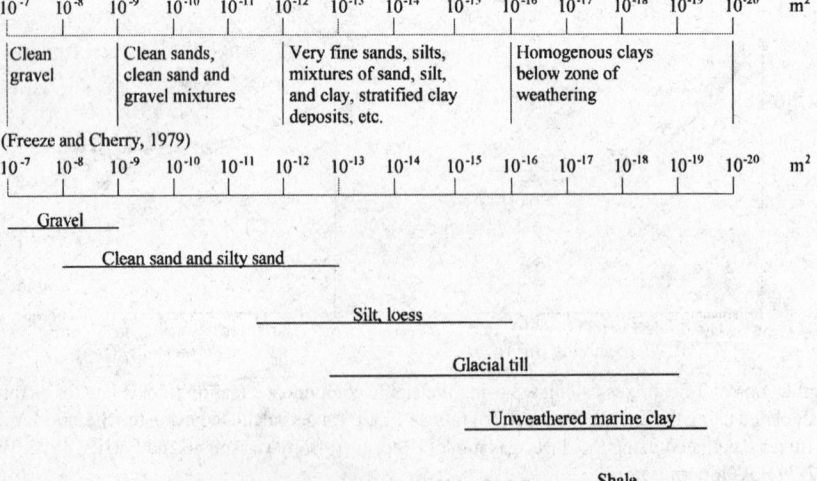

because of the interdependence of flux. The Stefan Maxwell equations or the dusty gas model can be used when Fick's law applies. Similarly, the dusty gas model can also be used when the Stefan Maxwell equations are valid. Under nonisobaric conditions, Darcy's law is combined with a diffusion model. The single component advection diffusion (Fick's law) model can be used in low concentration situations in high permeability material whereas the dusty gas model is required in high concentration situations. The Stefan Maxwell equations generally do not apply under nonisobaric conditions (Massmann and Farrier, 1992; pressure gradients > 10 Pa m^{-1}, Fig. 8.3b).

8.3.1 Density Driven Advection

Density driven advective transport, a specific type of multicomponent gas transport, is important for high molecular weight compounds that have high vapor densities, such as dense volatile organic contaminants (Falta et al., 1989; Mendoza and Frind, 1990; Mendoza and McAlary, 1990). Important factors controlling density driven advective flow include saturated vapor density, molecular mass of the chemical, and gas phase permeability. Vapor densities are maximized when the air mixture is saturated by the vapor phase of the chemical. Such high saturations are restricted to areas close to the free phase. Numerical simulations indicate that density driven advective flow in homogeneous media is important at the high permeabilities (> 10^{-11} m^2) typical of sands and gravels (Falta et al., 1989). The presence of fractures can greatly enhance density driven advective flow.

An order of magnitude estimate of the advective gas flux resulting from density variations may be obtained from (Falta et al., 1989):

$$J_G = -\frac{k_G g}{\mu_G} \nabla(\rho - \rho_0)$$

[8.40]

Equation [8.40] describes steady-state downward flux of fluid of density ρ through a stagnant fluid of density ρ_o in material of permeability k_G and ignores diffusion and phase partitioning. An estimate of the density potential of the gas mixture (ρ) with respect to air (ρ_0) can be obtained from relative vapor density (ρ_{rv}) (Mendoza and Frind, 1990):

$$\rho_{rv} = \frac{\rho}{\rho_0}\frac{x_c M_c + (1 - x_c)M_a}{M_a}$$

[8.41]

Table 8.5 Most appropriate model for describing gas flux under different pressure, permeability, and concentration conditions. Pressure gradient refers to the existence of an external pressure gradient, concentration refers to concentration of the diffusing gas component, dusty gas model* includes Darcy's law to described advective gas flow.

Pressure Gradient	Permeability	Low Concentration	High Concentration
Isobaric	low	dusty gas model	dusty gas model
Isobaric	high	Fick's law	Stefan Maxwell
Nonisobaric	low	dusty gas model*	dusty gas model*
Nonisobaric	high	advection diffusion	dusty gas model

where x_c is the fractional molar concentration of the compound, and M_c and M_a are the molecular weights of the compound and air, respectively. Relative vapor density of the source depends on the vapor pressure and molecular weight of the compound (Mendoza and Frind, 1990). Density driven advection occurs generally in areas contaminated by high molecular weight volatile organic contaminants that have high vapor densities.

8.4   Methods

8.4.1   Estimated Parameters

Because theoretically permeability should be independent of fluid used, gas and liquid permeabilities should be the same. In fine grained media or under small mean pressures, gas permeability is greater than liquid permeability because of gas slippage (Klinkenberg effect). Permeability can be corrected for gas slippage using Equation [8.38]. If gas slippage is negligible, as in coarse media having high mean pressures, gas conductivity can be estimated from hydraulic conductivity by:

$$K_G = K_w \left(\frac{\rho_G}{\rho_w} \right) \left(\frac{\mu_w}{\mu_G} \right) \qquad\qquad [8.42]$$

Air conductivities are about one-tenth of hydraulic conductivities because the air density is approximately three orders of magnitude less than water density and air viscosity is approximately 50 times less than water viscosity (Table 8.1). The estimated gas conductivities (Equation [8.42]) assume that pores are completely filled with gas. Typical values of permeability for different sediment textures are found in Freeze and Cherry (1986) and Terzaghi and Peck (1968) (Table 8.4), and permeability varyies over 13 orders of magnitude from clay to gravel.

A rough estimate of gas permeability in gas saturated systems can be obtained from particle size data:

$$k = C\left(D_{10}\right)^2 \text{ (Hazen, 1911)} \qquad\qquad [8.43]$$

$$k = 1,250\left(D_{15}\right)^2 \text{ (Massmann, 1989)} \qquad\qquad [8.44]$$

where C is a dimensionless shape factor and D_{10} and D_{15} correspond to the grain diameters at which 10 or 15% of the particles by weight are finer. Other estimates of exponents include 1.65 and 1.85 instead of 2 in the Hazen formula (Shepherd, 1989).

The Klinkenberg b parameter can be estimated according to the following empirical equation developed by Heid et al. (1950) for air-dry consolidated media (standard correction for the Klinkenberg effect by the American Petroleum Industry):

$$b_{air} = \left(3.98 \times 10^{-5}\right)k_\infty^{-0.39} \qquad\qquad [8.45]$$

where b is in atmospheres and k_∞ is in cm². The Klinkenberg b parameter for any gas (b_i) can be estimated from that for air (b_{air}), developed by Heid et al. (1950) according to the following (Thorstenson and Pollock, 1989):

$$b_i = (\mu_i / \mu_{air})(M_{air} / M_i)^{1/2} b_{air} \qquad [8.46]$$

The Knudsen diffusion coefficient is related to the Klinkenberg effect because both are related to the ratio of the mean free path of the gas molecules to the pore radius. Thorstenson and Pollock (1989) showed how the Knudsen diffusion coefficient (D_i^k) can be estimated from the true permeability and the Klinkenberg b parameter for gas i:

$$D_i^k = kb_i / \mu_i \qquad [8.47]$$

Gas conductivities and permeabilities vary with gas content. Estimates of gas conductivity at different gas saturations are provided by equations listed in Table 8.3.

8.4.2 Laboratory Techniques

A variety of laboratory techniques are available for measuring parameters related to gas movement, such as gas pressure and permeability. Gas pressure can be measured with U-tube manometers containing different fluids, such as water or mercury. Such manometers generally measure differential pressure because one end of the tube is inserted into soil or rock and the other is exposed to the atmosphere. Manometers can be inclined to increase sensitivity. Pressure transducers are used widely to measure absolute or differential gas pressure. The operational range of these instruments varies, and their precision is generally a percentage of the full scale measurement range. Manufacturers include Setra (Acton, Massachusetts) and Microswitch Honeywell (Freeport, Illinois). Because transducers are subject to drift, they have to be calibrated periodically. Data can be recorded automatically in a data logger.

Gas permeability can be measured in the laboratory on undisturbed or repacked cores. Repacking should be done only on samples having low clay content. Because repacking changes structure, gas transport parameters are affected. The more structured the soil, the bigger the potential change. Clay soils tend to be more structured. Permeability measurements include determination of flow rate of each phase under an applied pressure gradient and measurement of saturation in unsaturated systems (Scheidegger, 1974). Various methods used to measure air and water permeabilities include those in which both phases move and are measured at the same time and those in which the permeability of one fluid phase is measured while the other phase remains stationary. Steady-state methods that hold the wetting phase stationary are used most widely (Corey, 1986). Tempe cells (Soilmoisture Equipment Corp., Santa Barbara, California) used for measuring water retention functions can be adapted as permeameters and include a sample holder with ceramic end plates. Alternatively, if sample shrinkage is expected, flexible wall permeameters can be used to minimize air flow along the annulus between the sample and the holder. Sharp et al. (1994) described an electronic minipermeameter for measuring gas permeability in the laboratory. Gas is injected at a constant pressure and the steady-state flow is measured by electronic mass flow transducers. Permeabilities can be measured over a wide range $(10^{-15}$ to 10^{-8} m^2). The Hassler method (Hassler, 1944) controls the capillary pressure at both ends of a soil core by means of capillary barriers and measures air and water relative permeabilities at the same time under a pressure gradient. The capillary barriers separate wetting and nonwetting phases. The Hassler method was used by Stonestrom (1987) and by Springer (1993). Other procedures for laboratory measurement of gas permeability were described in Springer et al. (1995) and Detty (1992). Darcy's law is used to estimate gas permeability:

$$k_G(S_G) = -(J_G\mu_G)/(dP/dz) \tag{8.48}$$

Additional equipment required for gas permeability measurements includes a pressure transducer or manometer, a flow sensor such as a soap film bubble meter, and a temperature sensor such as a thermistor. Gas permeabilities are measured at different gas saturations or water contents. If the sample is initially saturated with water, a minimum pressure must be reached (air entry pressure) before a continuous gas phase is achieved. Gas permeability increases as water content decreases.

The Klinkenberg b parameter can be estimated by plotting gas permeabilities measured at different mean pressures ($\overline{P}$) as a function of the reciprocal mean pressure at which the tests were performed (Equation [8.38]). Rearranging Equation [8.38] results in:

$$k_G = k(1 + b_i/\overline{P}) = k + kb_i(1/\overline{P}) \tag{8.49}$$

Therefore, the slope is the product of k times b_i and the intercept is the true permeability.

Gas diffusivities can be measured in open, semi-open, and closed systems (Abu-El-Sha'r, 1993). An open system involves component gases flowing past the edges of the soil. Pressure gradients and absolute pressures can be controlled by regulating the flow rate of the component gases. A semi-open system is generally termed a Stefan tube that consists of the diffusing substance in liquid form at the base and either free air (if measuring free air diffusivity) or the porous medium (if measuring effective diffusivity of the porous medium) at the top. Closed systems generally consist of two chambers connected by a capillary or chamber filled with the porous medium (Glauz and Rolston, 1989). This system can be used in the study of noxious gases. Semi-open systems are generally used in hydrology to measure effective binary diffusion coefficients. Fick's law is generally used to analyze these experiments. Although most studies in the past used nonequimolar pairs of gases, none of these studies considered nonequimolar diffusion. Experiments conducted by Abu-El-Sha'r (1993) were the first to use an equimolar pair of gases (N_2 and CO) to determine effective binary molecular diffusion coefficients. Details of various procedures for measuring molecular diffusion coefficients are outlined in Rolston (1986).

Single gas experiments are used to measure the Knudsen diffusion coefficient by applying the dusty gas model (Abu-El-Sha'r, 1993):

$$N_i^T = -\left(\frac{D_i^k}{RT} + \frac{\overline{P}k}{\mu_G RT}\right)\nabla P \tag{8.50}$$

Rearranging Equation [8.50] results in

$$N_i^T = -\left(\frac{D_i^k \mu_G}{\overline{P}} + k\right)\frac{\overline{P}}{\mu_G RT}\nabla P \tag{8.51}$$

Plotting $\dfrac{N_i^T LRT}{\Delta P}$ versus $\overline{P}$ should result in a straight line with an intercept of D_i^k and a slope of k/μ_G, where L is the length of the column in the experiment (Abu-El-Sha'r, 1993).

8.4.3   Field Techniques

8.4.3.1   Estimation of Gas Permeability for Advective Gas Flow

Advective transport of gases depends on gas permeability and pressure gradient. Gas permeability can be estimated from (1) analysis of atmospheric pumping data, (2) pneumatic tests, or (3) measurements by air minipermeameters. Comparison of gas permeabilities from laboratory and field indicates that field derived estimates of gas permeabilities generally exceed laboratory derived estimates by as much as orders of magnitude (Weeks, 1978; Edwards, 1994). These differences in permeability are attributed primarily to the increase in scale from laboratory to field measurements and inclusion of macropores, fractures, and heterogeneities in field measurements. Field permeability measurements in low permeability media include the effects of viscous slip and Knudsen diffusion.

8.4.3.2   Analysis of Atmospheric Pumping Data

Comparison of temporal variations in gas pressure, monitored at different depths in the unsaturated zone, with atmospheric pressure fluctuations at the surface can be used to determine minimum vertical air permeability between land surface and monitoring depth (Weeks, 1978; Nilson et al., 1991). Differential pressure transducers are used to monitor gas pressures in the unsaturated zone. Gas ports generally consist of screened intervals in boreholes of varying diameter. Flexible tubing (Cu or nylon) connects the gas port at depth with a differential pressure transducer at the surface. One port of the transducer is left open to the atmosphere. Atmospheric pressure is monitored at the surface by a barometer.

Data analysis consists of expressing the variations in atmospheric pressure as time harmonic functions. The attenuation of surface waves at different depths in the unsaturated zone provides information on how well or poorly unsaturated sections are connected to the surface. The accuracy of the results increases with the amplitude of the surface signals. Pressure fluctuations resulting from irregular weather variations change by as much as 20 to 30 mbar (2,000 to 3,000 Pa) during a 24-hr period (Massmann and Farrier, 1992).

If the surface pressure (upper boundary condition) is assumed to vary harmonically with time and the watertable or a low permeability layer acts as a no-flow boundary, an analytical solution can be derived (Carslaw and Jaeger, 1959 in Nilson et al., 1991). Pneumatic diffusivity can be estimated graphically by means of the amplitude ratio. The ratio of the amplitude at a certain depth z compared with the amplitude at the surface can be obtained graphically or by time series analysis (Rojstaczer and Tunks, 1995). Air permeability is estimated from the pneumatic diffusivity by dividing by the volumetric air content.

8.4.3.3   Pneumatic Tests

Pneumatic tests are also used widely to evaluate gas permeability in the unsaturated zone. In these tests, air is either injected into or extracted from a well and pressure is monitored in gas ports installed at different depths in surrounding monitoring wells (Fig. 8.4). A reversible air pump is used to inject or extract air. Most analyses of pneumatic tests assume that the gas content (θ_G) is constant over time; i.e., that no redistribution of water occurs during the test. To evaluate this assumption, injection or extraction tests should be conducted at several different rates. If results from the different rates are similar, the assumption of constant gas content is valid. The tests can be conducted in horizontal or vertical wells.

A variety of techniques are available for analyzing pneumatic tests. The initial transient phase of the test or the steady-state portion of the test can be analyzed. If transient data are available, volumetric gas content can also be estimated. Analysis of pneumatic tests resembles the inverse problem in well hydraulics, where permeabilities are estimated from pressure data. Solutions for estimating gas permeability differ in terms of the boundary conditions that are assumed at the ground surface (such as unconfined, leaky confined, and confined) and the method of solution. The lower boundary is generally assumed to be the watertable or an impermeable layer. All solutions assume radial flow to a vertical well.

As discussed previously, the gas flow equation is nonlinear because of the pressure dependence of the density, viscosity, and permeability (Klinkenberg effect). In many cases the pressure dependence of the density is approximated by ideal gas behavior (Equation [8.17]). Under low to moderate pressures and pressure gradients typical of unsaturated media, pressure dependence of the viscosity can be neglected. In most analyses, the Klinkenberg effect is also neglected (Massmann, 1989; Baehr and Hult, 1991). If pressure variations are assumed to be small, the transient gas flow Equation [8.52] can be written in a form similar to the groundwater flow Equation [8.53]:

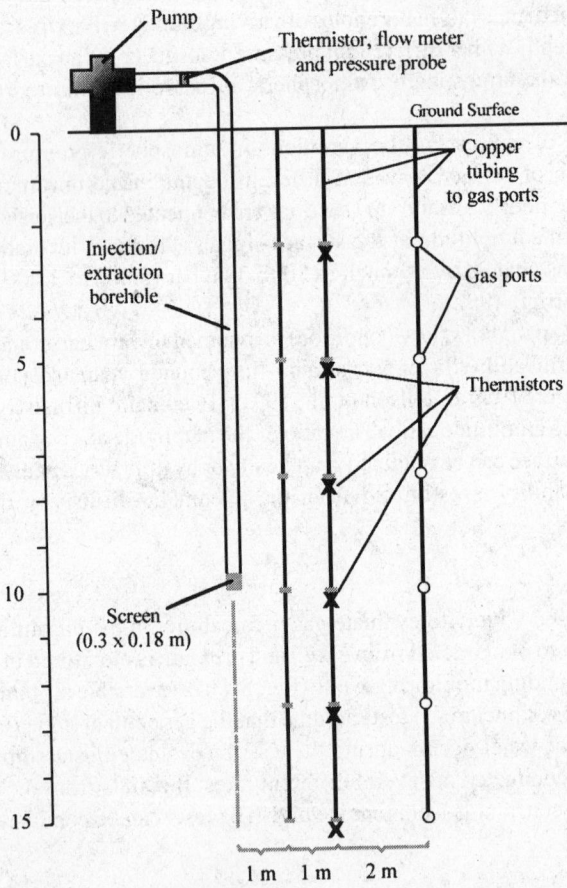

Fig. 8.4   Schematic design for a field pneumatic test

$$\frac{\theta_G \mu_G}{P_0} \frac{\partial P}{\partial t} = \nabla(k_G \nabla P)$$

[8.52]

$$S_s \frac{\partial h_w}{\partial t} = \nabla(K_w \nabla h_w)$$

[8.53]

where S_s is specific storage, h_w is hydraulic head (L), and K_w is hydraulic conductivity (L t^{-1}).

A summary of various techniques for evaluating field-scale pneumatic tests is provided in Table 8.6. Massmann (1989) used many techniques developed for groundwater hydraulics to analyze transient gas tests and used a modified Theis solution for systems with no leakage from the ground surface and a modified Hantush solution for systems with leakage. The computer software AQTESOLV (Duffield and Rumbaugh, 1989) was used to estimate the parameters according to the Marquadt nonlinear least squares technique. Johnson et al. (1990a) developed an analytical model to evaluate transient, 1-D, radial flow in a homogeneous, isotropic flow field. McWhorter (1990) developed an analytical model to analyze transient flow in a 1-D, radially symmetrical flow field that includes the nonlinear effects of compressible flow. He found that the nonlinear effects resulting from the pressure dependence of permeability were negligible for pressures within ±20% of P_0.

The transient phase of gas tests is generally short (approximately seconds to hours) (Edwards, 1994), and it is sometimes difficult to collect reliable data. Many studies analyze the steady-state portion of the pneumatic test. Baehr and Hult (1991) developed analytical solutions for steady-state, 2-D, axisymmetric gas flow to a partially screened well for an open system and a leaky confined system. Analysis of the leaky confined system required a Hantush type leakage term. A computer code (AIR2D) is available that includes these analytical solutions (Joss and Baehr, 1997). Falta (1995) developed analytical solutions for steady-state and transient gas pressure and stream function fields using parallel horizontal injection and extraction wells where the ground surface is open to the atmosphere. Shan et al. (1992) also developed an analytical solution for steady-state flow in homogeneous and anisotropic media that includes the effects of leakage. A constant pressure upper boundary is used that assumes that the system is completely open. Horizontal and vertical gas

Table 8.6 Various approaches for evaluating field scale pneumatic tests. * includes horizontal wells, all other studies do not include horizontal wells.

Approach	Transient or steady state	No. of dimensions	Layered or homogeous	Anisotropic or isotropic	Data
Massman (1989)	transient	1-D	layered	anisotropic	data
Johnson et al. (1990a,b)	transient	1-D	homogeous	isotropic	data
McWhorter (1990)	transient	1-D	homogeous	isotropic	data
Baehr and Hult (1991)	steady state	2-D	layered	anisotropic	data
Shan et al. (1992)	steady state	2-D	homogeous	anisotropic	no data
Croise and Kaleris (1992)	steady state	2-D	layered	anisotropic	data
Massmann and Madden (1994)	transient	2-D	layered	isotropic	data*

permeabilities are estimated using type curves. Kaluarachchi (1995) developed an analytical solution for 2-D, axisymmetric flow with anisotropic gas permeability that includes the Klinkenberg effect. Errors resulting from neglecting the Klinkenberg effect are highest in low permeability materials and near the well. Edwards (1994) used a numerical model of steady-state, radial and vertical, anisotropic, heterogeneous compressible flow with an optimization routine to estimate gas permeabilities. All tests reached steady state before 5 s (Edwards, 1994). Results of this analysis indicate that vertical heterogeneities were significant. Variations in gas permeabilities with depth were, therefore, attributed to increases in water content with depth. Test data can also be analyzed by means of groundwater flow models such as MODFLOW with preprocessing (Joss and Baehr, 1995), which is discussed in the Section 8.5.

8.4.3.4   Air Minipermeameters

A minipermeameter is a device for measuring gas flow that is used to determine permeability in the field at a localized scale. Measurements are made rapidly and are nondestructive. A mechanically based field minipermeameter was described by Goggin et al. (1988). Compressed N_2 is injected at a constant pressure through a tip pressed against the measurement surface. The steady-state flow rate is measured by a series of rotameters, and the gas injection pressure is measured at the tip seal. The flow rate at a particular injection pressure is calibrated against measurements on core plugs of known permeability.

8.4.3.5   Evaluation of Diffusive Transport Parameters

Knudsen diffusion coefficients are measured in the laboratory using single gas experiments (Section 8.4.2) while the effective molecular diffusion coefficient can be determined from field experiments (Raney, 1949; McIntyre and Philip, 1964; Lai et al., 1976; Rolston, 1986). The simplest method for measuring gas diffusivity in near surface soils involves inserting a tube into the soil surface and supplying gas from a chamber over the inserted tube (McIntyre and Philip, 1964; Rolston, 1986). Gas diffusivity can be calculated by measuring initial gas concentration in the soil and chamber and gas concentrations in the chamber at different times. An independent measurement of volumetric gas content in the soil is also required. If gas samplers are installed at different depths in the soil and the gas concentration gradient is calculated, gas flux can be estimated by using information on the diffusivity and gas concentration gradient in conjunction with Fick's law. Kreamer et al. (1988) used atmospheric fluorocarbons to determine field tortuosities. A permeation device was used that slowly released fluorocarbon gases, the concentrations of which were monitored for several days. Nicot (1995) used instantaneous release of a tracer and continuous monitoring of gas concentration to determine field tortuosities. The data resulting from the tracer tests can be evaluated by analytical or numerical methods.

Carslaw and Jaeger (1959) presented a derivation of an analytic solution for diffusion in an isotropic, homogeneous, and infinitely porous medium. The partial differential equation reduces to

$$\frac{\partial C}{\partial t} = D_e \frac{\partial^2 C}{\partial r^2} + D_e \frac{2}{r} \frac{\partial C}{\partial r} \qquad\qquad [8.54]$$

where C is the concentration (M L^{-3}), D_e is the effective diffusion coefficient, and r is the radial distance (L). The solution to the above equation, based on instantaneous release of mass M at a point source, is

$$C = \frac{M}{8(\pi D_e t)^{3/2}} \exp\left(-\frac{r^2}{4D_e t}\right)$$ [8.55]

where M is the molecular mass (M). The inverse problem is solved by estimating effective diffusion coefficients from the measured concentration data, a procedure similar to the one described for permeability estimation.

8.5 Applied Numerical Modeling

8.5.1 Single Phase Flow

A summary of the various types of codes available for simulating gas flow is provided in Table 8.7. Most numerical simulations of water flow in unsaturated media have generally ignored the gas phase by assuming that the gas is at atmospheric pressure. The equation that is solved is the Richards equation, which describes a single phase (liquid), single component (water) system. The Richards approximation is generally valid for most cases of unsaturated flow (Section A, Chapters 3 and 4).

A variety of numerical models have been developed to simulate gas flow in unsaturated systems. An important consideration in choosing a code for evaluating gas transport is the assumptions of each code. Groundwater flow models can be used to simulate gas flow in cases where the vapor behaves as an ideal gas and where pressure fluctuations are small and gas content is constant (no water redistribution) (Massmann, 1989). Such assumptions are generally valid for vapor extraction remediation systems. The code most widely used to simulate groundwater flow is MODFLOW (McDonald and Harbaugh, 1988). Joss and Baehr (1995) developed a sequence of computer codes (AIR3D) adapted from MODFLOW to simulate gas flow in the unsaturated zone. The codes can be used to simulate 3-D air flow in a heterogeneous, anisotropic system including induced air flow in dry wells or trenches. Pre- and postprocessors are included. AIR3D can also be used to simulate natural advective air transport in response to barometric pressure fluctuations in shallow, unsaturated systems when gravity and temperature gradients can be neglected. AIR3D transforms the air flow Equation

Table 8.7 Summary of types of codes available to simulate gas flow

Code	Dim.	No. Phases	Components	Energy Balance	Porous/Fractured Systems
AIR3D	3	1	gas	no	porous
BREATH	1	2	water	yes	porous
SPLaSHWaTr	1	2	water	yes	porous
UNSAT-H	1	2	water	yes	porous
Princeton Code	2	2	water and air	no	porous
FEHM	3	2	water and air	yes	porous/fractured
TOUGH	3	2	water and air	yes	porous/fractured
STOMP	3	3	water, air, NAPL	yes	porous/fractured

[8.52] into a form similar to the groundwater flow Equation [8.53] that is solved by MODFLOW. Air compressibility is approximated by the ideal gas law. The simulations can be conducted (1) in a calibration mode to evaluate parameters such as permeability from pneumatic tests or (2) in a predictive mode.

8.5.2   Two-Phase Flow (Water and Air)

Codes have also been developed to simulate two-phase (liquid and gas), two component (water and air) systems. In situations where the air phase retards infiltration of the water phase, a two-phase code is required (Touma and Vauclin, 1986). Two-phase codes are also required to simulate transport of volatile organic compounds. Various approaches have been developed to simulate two-phase flow. In the petroleum industry, Buckley and Leverett (1942) developed an approach that excluded the effects of capillary pressure and gravity. The basic theory of Buckley and Leverett (1942) was extended by Morel-Seytoux (1973) and Vauclin (1989). If the fluids are considered incompressible, the 1-D pressure equation reduces to (Celia and Binning, 1992):

$$q_w + q_a = constant \qquad\qquad\qquad [8.56]$$

where q_w is the water flux and q_a is the air flux. Only one equation, therefore, has to be solved. The fractional flow equation was solved by Morel-Seytoux and Billica (1985) using a finite difference approximation. Because for 2-D flow both the pressure and saturation equations must be solved, the problem no longer reduces to one equation.

Celia and Binning (1992) developed a finite element, 1-D code that considers dynamic coupling of the gas and water phases. The code is inherently mass conservative because the mixed formulation is used: temporal differentiation in terms of water content and spatial differentiation in terms of pressure head. The code was used to simulate laboratory two-phase flow experiments conducted by Touma and Vauclin (1986). The results of these simulations were compared with those achieved by means of a finite element, two-phase flow code developed by Kaluarachchi and Parker (1989) that uses the traditional h-based formulation and a finite difference code that was based on fractional flow theory used by Morel-Seytoux and Billica (1985). The simulations were used to evaluate the conditions under which water flow is significantly altered by air flow. Infiltration experiments in a bounded column that resulted in ponded conditions showed significantly altered water flow because the air could not readily escape. An important insight gained from these simulations was that the reduction in water velocity resulted from a reduction in hydraulic conductivity with increased air content, rather than from a buildup of air pressure ahead of the wetting front (Celia and Binning, 1992).

More general codes have been developed by the National Laboratories to simulate multiphase flow and transport. TOUGH (Transport of Unsaturated Groundwater and Heat) is a 3-D code that simulates nonisothermal flow and transport of two fluid phases (liquid and gas) and two components (water and air) (Pruess, 1987). Subsequent upgrades include ATOUGH, VTOUGH, CTOUGH, ITOUGH (Finsterle and Pruess, 1995), and TOUGH2 (Pruess, 1991). A separate module has been developed for simulating transport of volatile contaminants (T2VOC). FEHM (Finite Element transport of Heat and Mass) is a 3-D, nonisothermal code that simulates two-phase flow and transport of multiple components (Zyvoloski et al., 1996). Both TOUGH and FEHM are used for simulation of flow and transport in variably saturated fractured systems at Yucca Mountain, Nevada, the proposed high level radioactive waste disposal site. STOMP (Subsurface Transport Over Multiple Phases) simulates nonisothermal flow and transport in porous media (White and Oostrom, 1996). STOMP was developed specifically to evaluate remediation of sites contaminated by organics and includes a separate NAPL phase. Both TOUGH and STOMP use an integrated finite difference method to solve mass- and energy-balance equations.

There has been considerable interest in simulating flow in fractured systems because the proposed high level nuclear waste facility will be located in fractured tuff. Mathematical models for simulating flow in fractured media can be subdivided into continuum and discrete fracture models (Rosenberg et al., 1994; NRC, 1996). In addition, models can be deterministic or stochastic. Continuum models can be further subdivided into equivalent continuum, dual porosity, and dual permeability models (Rosenberg et al., 1994). In equivalent continuum models the system is described by a single equivalent porous medium with average hydraulic properties. This approach is considered valid if transport between fractures and matrix is rapid relative to transport within fractures (Pruess et al., 1990). Dual porosity models consider the fractures and the matrix as two interacting media with interaction between the fractures and the matrix restricted to a local scale (Pruess and Narasimhan, 1985). There is no continuous flow in the matrix. Dual porosity models are generally used to simulate transient flow and transport in saturated zones. Dual permeability models allow continuous flow in either the fractures or the matrix and an interaction term is used to describe coupling between the two media. The dual permeability formulation is generally used to simulate transient flow and transport in unsaturated systems. Dual porosity and dual permeability models treat fractures as high permeability porous media (Rosenberg et al., 1994). Discrete fracture network models assume that the matrix is not involved in flow and transport and that fluid flow can be predicted from information on the geometry of the fractures and the permeability of individual fractures (NRC, 1996). Stochastic simulation is generally used to produce several realizations of the fracture system.

8.6 Applications of Gas Transport Theory

8.6.1 Soil Vapor Extraction

A detailed review of various issues related to soil vapor extraction is provided in Rathfelder et al. (1995). Soil vapor extraction has become the most common innovative technology for treating subsurface soils contaminated by volatile and semivolatile organic compounds (USEPA, 1992). This popularity is partially due to its low cost relative to other available technologies, especially when contamination occurs relatively deep below the ground surface. Vapor extraction systems are also attractive because mitigation is completed *in situ*, reducing the exposure of chemical contaminants to onsite workers and the offsite public. Vapor extraction also offers flexibility in terms of installation and operation. This flexibility allows systems to be installed at locations crowded by existing structures, roadways, and other facilities and to be readily adjusted during the course of remediation to improve mass removal efficiency.

Vapor extraction involves two major processes: mass transfer and mass transport. Mass transfer is the movement from one phase to another at a particular location. Volatile and semivolatile compounds will enter the vapor phase by desorption from the soil particles through volatilization from the soil water and by evaporation from nonaqueous phase liquids (NAPLs), such as gasoline or liquid solvents. The rate at which this mass transfer occurs depends on subsurface temperature, humidity, and pressure; the properties of the contaminant, including vapor pressure and solubility; and the sorptive properties of the soil.

The second major process involved in vapor extraction is mass transport, the movement of vapor from one location to another. This transport, which is primarily due to advection, is caused by pressure gradients that are developed by using extraction wells or trenches. In some instances, mass transport is enhanced through passive or active injection wells and trenches and through low permeability soil covers and cutoff walls. The rate at which mass transport occurs is a function of the distribution of soil permeabilities, soil moisture content, and pressure gradients induced through the extraction and injection systems. In highly heterogeneous systems, flow may be concentrated in high permeability layers, with flow bypassing low permeability material.

The applicability of vapor extraction at a site depends on the volatility of the contaminants and the ability to generate advective gas flow through the subsurface (Hutzler et al., 1989, Johnson et al., 1995). High volatility contaminants and uniformly high permeability soils are optimal for successful vapor extraction.

The time required to clean a particular site by means of vapor extraction depends on the amount and distribution of contaminants in the subsurface and the rates at which mass transfer and mass transport occur. It is often assumed in vapor extraction applications that the rate of mass transfer is faster than mass transport, so that local equilibrium conditions exist in terms of vapor concentrations (Baehr et al., 1989; Hayden et al., 1994). Under conditions of local equilibrium, the uncertainty in cleanup time is due primarily to uncertainties in the amount and distribution of subsurface contaminants and in the amount and distribution of air flow induced by the extraction system. The distribution of both contaminants and air flow are controlled, at least partially, by the spatial distribution of soil permeability near the spill or leak. In other cases, laboratory and field experiments indicate that mass transfer from pore water to the gas phase is rate limited (Gierke et al., 1992; McClellan and Gillham, 1992). The rate limitation has been attributed to intraaggregrate or intraparticle diffusion or bypassing of low permeability zones (Gierke et al., 1992; Travis and McGinnis, 1992).

Vapor extraction systems are most often used to treat soils that have been contaminated from leaks or spills of NAPLs that occur near the ground surface. The NAPL that is released from the spill migrates through the unsaturated soil as a separate phase. As the NAPL moves through the soil, portions are trapped within the pores by capillary forces. The amount of NAPL that is trapped is termed *residual saturation*. This residual saturation depends upon the volume and rate of the NAPL release, characteristics of the soil, and properties of the contaminant. In typical situations, between 5 and 50% of the pore space may be filled by this residual saturation (Mercer and Cohen, 1990). The spatial distribution of the residual NAPL will depend upon heterogeneities in the soil column. In general, lower permeability soils tend to trap more of the NAPL than higher permeability materials (Pfannkuch, 1983; Hoag and Marley, 1986; Schwille et al., 1988). After the contaminant has entered the subsurface, it will become partitioned among the NAPL, solid, aqueous, and vapor phases. In relatively fresh spills, the major fraction of the contaminant will remain in the NAPL form.

Operation of a vapor extraction system at the spill location will induce air flow through subsurface soils. The spatial distribution of air flow in the subsurface will depend on the distribution of the soil permeability: the flow will tend to occur through higher permeability channels. As the air flows through the subsurface, it will become saturated with contaminant vapors that evaporate from the residual NAPL. After the contaminant has volatilized, it will migrate in the vapor phase to the extraction wells.

The design and operation of vapor extraction systems are described in a variety of texts, including Hutzler et al. (1989), Baehr et al. (1989), Johnson et al. (1990a,b), Pedersen and Curtis (1991), and the USEPA (1991). The simplest design consists of a single vapor extraction well. In most cases, several extraction wells are used. The radius of influence of the extraction wells is often used to design the optimal distance between extraction wells (Pedersen and Curtis, 1991). Horizontal wells are used when contamination is restricted to the shallow subsurface (Connor, 1988; Hutzler et al., 1989; Pedersen and Curtis, 1991). Air injection and passive vents may be used to increase gas flow. Many studies have shown that surface seals can greatly increase the radius of influence of the extraction well and increase the efficiency of vapor extraction systems (Rathfelder et al., 1991; 1995). Modeling of soil vapor extraction systems was reviewed by Rathfelder et al. (1995) and described in references including Massmann (1989), Gierke et al. (1992), Benson et al. (1993), and Falta et al. (1993).

8.6.2 Radon Transport

Radon transport has created considerable interest because of the radiological health hazard (lung cancer) associated with the decay products of radon. Radium radioactively decays to radon, and the greatest doses of radiation result from ^{222}Rn. Many of the issues related to radon transport were reviewed in Nazaroff (1992). Recent evaluations of contaminated sites have focused on risk assessment and suggest that one of the most critical exposures to radon and volatile organic contaminants is from movement of soil gas into buildings. Much research is, therefore, currently being conducted on gas movement from soil into buildings.

The typical range of radium content in surficial sediments in the United States is 10 to 100 Bq kg^{-1} (Bq, becquerel is the SI unit for activity, which corresponds to the number of atoms needed to yield a radioactive decay rate of one per second) (Nazaroff, 1992). Much higher concentrations of radon occur near uranium mines and mill tailings. Radon is also emitted from low level radioactive waste disposal sites. The half-life of ^{222}Rn is 3.8 days. Radon partitions among solid, liquid, and gas phases. Diffusive transport is sufficient to account for radon migration from unsaturated material into the atmosphere. Radon fluxes estimated on the basis of diffusion are consistent with experimental measurement (Nazaroff, 1992). The diffusion coefficient of ^{222}Rn in air is 1.2×10^{-5} m^2 s^{-1} (Hirst and Harrison, 1939). The effective diffusion coefficient, including the effect of reduced cross-sectional area and increased path length (tortuosity), is about a factor of four smaller than the free air diffusion coefficient in fairly dry soils (Nazaroff, 1992). Effective diffusivities in saturated systems are about four orders of magnitude less than those in dry soils (Tanner, 1964). This difference has been used in the design of engineered barriers for uranium mill tailings to minimize upward movement of radon into the atmosphere. Thick clay layers (0.6 to 2 m) are used as radon barriers because they retain much more water than does sand and thus reduce radon diffusion. Desiccation of clay, however, results in cracking and development of preferred pathways for gas transport. Typical diffusive fluxes for radon estimated by Nazaroff (1992) average 0.022 Bq m^{-2} s^{-1}, which is the same average as diffusive fluxes in Australian soils estimated by Schery et al. (1989).

Advection is generally considered more important than diffusion for radon transport into buildings because concrete slabs provide a barrier to diffusion and ventilation causes a pressure differential between indoor and outdoor air that drives advective gas transport. Advective gas transport, described by Darcy's law, varies with permeability and pressure gradient. Because permeability generally decreases with grain size, Nazaroff and Sextro (1989) suggested that diffusion is dominant in low permeability materials and advection dominant in high permeability materials, the cutoff being ~ 10^{-11} m^2. Gas permeability is generally anisotropic because sediment layering results in higher permeabilities horizontally than vertically. Openings in the substructures of buildings provide pathways for advective radon transport into buildings. The average indoor ^{222}Rn concentration in buildings is 50 Bq m^{-3}; in soil, ~ 30 Bq m^{-3} (Nazaroff, 1992). Pressure differentials across buildings may result from ventilation caused by open windows and doors, fans, heating, and air conditioning. Heating generally increases radon entry into buildings; air conditioning reduces it.

The construction of building substructures can have important implications for advective gas transport. Concrete slabs are commonly underlain by gravel layers. Penetrations at floor wall joints and perimeter drain systems allow gas transport. Temperature differences, barometric pressure fluctuations, wind, and rain cause pressure differentials (Nazaroff, 1992).

Diffusive fluxes can account for about 10% of mean radon concentration in single family dwellings (Nero et al., 1986). Nazaroff (1992) presented several different paradigms for radon entry into dwellings. In the base case, transport of radon from the soil into buildings is attributed primarily to advective air flow driven by weather and ventilation. An alternative paradigm would be soil with strongly varying permeability, in which case diffusion would be important for transporting gas from high to low permeability regions, such as gravel seams or cracked clays. Another alternative is a high permeability material at the interface between the building and the soil, such as a gravel layer or a thin air gap. In this case, molecular diffusion from the native low permeability soil to the high permeability gravel is important, and advection alone would result in much slower transport of radon.

8.6.3   Water Vapor Diffusion

Analysis of water vapor transport in natural systems can be done by approximating soil gas as a two component system containing water vapor and air. Diffusion of water vapor is important in near-surface sediments where evaporation is occurring and in arid systems where water contents are extremely low. The simple theory of water vapor diffusion assumes that water vapor behaves like other gases. The flux of water vapor (water in the gas phase) is

$$q_G^w = -\theta_G \tau D_G^w \nabla \rho_v \qquad [8.57]$$

where D_G^w is the binary diffusion coefficient for water vapor and ρ_v is the water vapor density. The Kelvin equation can be used to express water vapor density in terms of temperature and matric potential head:

$$\rho_v = \rho_v^0 \cdot RH = \rho_v^0 \exp \frac{h_m V_w}{\rho_w RT} \qquad [8.58]$$

where ρ_v^0 is saturated vapor density, RH is relative humidity, h_m is matric potential (Pa), V_w is the molar volume of water (1.8×10^{-5} $m^3 mol^{-1}$), R is the gas constant, and T is Kelvin temperature. Laboratory experiments showed that water vapor fluxes estimated by Equation [8.58] underestimated measured water vapor fluxes by as much as one order of magnitude. Philip and de Vries (1957) attributed the discrepancy between laboratory measured and estimated water vapor fluxes to (1) liquid island enhancements and (2) increased temperature gradients in the air phase. Liquid island enhancement refers to the fact that when the liquid phase is discontinuous, liquid islands act as short circuits for thermal vapor diffusion. Water vapor condenses on one side of liquid water and evaporates on the other side. This results in an increased cross-sectional area for diffusion from the volumetric gas content to the sum of the volumetric gas and liquid contents (porosity) when the water phase is discontinuous. Temperature gradients in the gas phase are much higher than the average temperature gradients in porous media because thermal conductivity in the gas phase is much lower than that in the liquid and solid phases:

$$q_G^w = -\theta_G \tau_G D_G^w \left.\frac{\partial \rho_G^w}{\partial \psi}\right|_T \nabla\psi - f\frac{\nabla T_G}{\nabla T}\tau_G D_G^w \left.\frac{\partial \rho_v}{\partial T}\right|_\psi \nabla T \qquad [8.59]$$

where ∇T_G is the average temperature gradient in the gas phase, ∇T is the average bulk temperature gradient in the system, and f is a correction factor for liquid island enhancement and is equal to the porosity when the liquid phase is discontinuous (Milly and Eagleson, 1980). Thermal vapor flux, resulting from variations in saturated vapor pressure with temperature, is generally considered much more important than isothermal vapor flux. A temperature difference of 1 C at 20 C results in a greater difference in vapor density (1.04×10^{-3} kg m^{-3}) than does a 1.5 MPa difference in matric potentials from -0.01 to -1.5 MPa (0.17×10^{-3} kg m^{-3}) (Hanks, 1992, p. 95). The temperature dependence of saturated vapor pressure is given in Table 8.1. Temperature gradients are generally high in near surface sediments, and nonisothermal vapor diffusion is important.

One-dimensional liquid and vapor transport in arid and semi-arid systems has been simulated by two-phase (liquid and gas), single component (water), nonisothermal codes (BREATH, SPLaSHWaTr, UNSAT-H) (Stothoff, 1995; Milly, 1982; Fayer and Jones, 1990). Advective gas transport is not included. In many applications of these codes, liquid and vapor flow is simulated in response to atmospheric forcing functions. Mass and energy equations are solved sequentially, and the resultant tridiagonal equations are readily solved by the efficient Thomas algorithm. These simulators solve the continuity equation for water:

$$\frac{\partial}{\partial t}\left(\rho_l^w \theta_l + \rho_G^v \theta_G\right) = -\nabla\cdot\left(\rho_l^w q_l + \rho_G^v q_v\right) \qquad [8.60]$$

where ρ_l^w is the density of water in the liquid phase and ρ_G^v is the density of water vapor in the gas phase. Darcy's law is used to solve the liquid water flux, and the theory of Philip and de Vries (1957) is used to solve the vapor diffusive flux. The energy balance equation is also solved in these codes. Numerical simulations of liquid and vapor flux provide insights on flow processes in the unsaturated zone. Simulations in response to 1 yr of atmospheric forcing in the Chihuahuan Desert of Texas showed net downward water flux in response to seasonally varying temperature gradients that was consistent with isotopic tracer data (Scanlon and Milly, 1994). Simulations at Yucca Mountain, Nevada, evaluated the impact of hydraulic properties on infiltration (Stothoff, 1997). For low permeabilities in this system, vapor transport was dominant. The simulations demonstrated the importance of the alluvial cover on fractured bedrock in decreasing downward water flux. The UNSAT-H code was used to evaluate the performance of engineered barriers, including a capillary barrier at the Hanford site (Fayer et al., 1992). Hysteresis was found to be important in reproducing drainage measured through the capillary barrier.

8.6.4 Preferential Flow

Preferential flow refers to nonuniform movement of a fluid. In most cases, preferential flow has been examined with respect to water flow. Preferential flow can occur in macropores that are noncapillary size pores, such as desiccation cracks in clays, worm holes and root tubules in soils, and fractures in rocks. Because macropores drain rapidly they are generally dry, and preferential flow of gases should, therefore, be much greater than that of liquids.

There has been much interest in fractured systems in recent years because the proposed high level radioactive waste disposal facility is located in fractured tuff. Fractures allow rapid transport of contaminants in unsaturated systems. The cubic law is used to describe fracture flow. Fracture flow is proportional to the fracture aperture cubed for a given head gradient. Laminar flow between smooth parallel plates is given by:

$$Q = Aq_G = (bw)\frac{b^2}{12\mu_G}\frac{\partial P}{\partial z} = \frac{b^3 w}{12\mu_G}\frac{\partial P}{\partial z} \qquad [8.61]$$

where Q is the volumetric flow rate ($L^3\,t^{-1}$), b is aperture opening (L), w is the width of the fracture perpendicular to the direction of flow (L), bw is the cross-sectional area (L^2), and P is the hydraulic head (Pa) (Snow, 1968). Fractures are important because they account for most of the permeability in the system, whereas the matrix is important because it accounts for most of the porosity. Nilson et al. (1991) evaluated contaminant gas transport in fractured permeable systems. For example, a system that has 1 mm wide fractures (δ_f) separated by 1 m slabs of matrix (δ_m) with a matrix porosity of 0.10 (ϕ_m) will have a ratio of capacitance volumes (V) as follows (Nilson et al., 1991):

$$\frac{V_m}{V_f} = \frac{\phi_m \delta_m}{\delta_f} \sim 100 \qquad [8.62]$$

Most mass flow occurs through the fractures as shown by the following (Nilson et al., 1991):

$$\frac{Q_f}{Q_m} = \frac{A_f q_f}{A_m q_m} \sim \frac{\delta_f}{\delta_m}\frac{(\delta_f^2/12)}{k_m} \sim \frac{1}{12}\frac{\delta_f^3}{\delta_m k_m} \sim 10^5 \qquad [8.63]$$

where Q is the mass flow, q is the mass flux density, and k_m is the gas permeability of the matrix (10^{-15} m^2). Even with high matrix permeability ($\sim 10^{-12}$ m^2) and narrow fractures (10^{-4} m), about 99% of the flow will occur in the fractures.

In fractured, permeable media, advective fluxes resulting from barometric pressure fluctuations may be orders of magnitude greater than diffusive fluxes and could result in upward movement of contaminated gases into the atmosphere (Nilson et al., 1991). Theoretical analyses by Nilson et al. (1991) showed that in homogeneous media, a differential barometric pressure fluctuation of 7% ($\Delta P/P_0$) would result in the upward movement of an interface between contaminated and pure gas of 20 m if the interface were 300 m deep. In contrast, in fractured systems the volume expansion of the gas is the same as that in porous media; however, gas expansion occurs primarily in the fractures and the interface moves as much as two orders of magnitude higher in the system. The conceptual model developed for upward movement of contaminants involves upward gas movement in the fractures as a result of barometric pressure fluctuations and lateral diffusion from the fractures into the surrounding matrix, which holds the contaminants at that level until the next barometric pressure fluctuation. This is a racheting mechanism for transporting gases upward. A gas tracer experiment conducted by Nilson et al. (1992) confirmed the results of earlier theoretical analyses. These experiments showed upward movement of gas tracers in a period of months from a spherical cavity (depth ~300 m) created by underground nuclear tests at the Nevada Test site. This rapid upward movement of gases was attributed to the effects of barometric pumping in fractured tuff.

Weeks (1987; 1993) also conducted detailed studies of flow in fractured tuff at Yucca Mountain. In areas of steep topography such as at Yucca Mountain, temperature and density driven topographic effects result in continuous exhalation of air through open boreholes at the mountain crest in the winter, as cold dry air from the flanks of the mountain replaces warm moist air within the rock/borehole system (Weeks, 1987). Wind also results in air discharge from the boreholes that is about 60% of that resulting from temperature induced density differences (Weeks, 1993). Open boreholes greatly enhance the advective air flow at this site; numerical simulations indicate that water fluxes resulting from advective air flow under natural conditions (0.04 mm yr^{-1}) are five orders of magnitude less than those found in the borehole (Kipp, 1987) and similar in magnitude to estimated vapor fluxes as a result of the geothermal gradient (0.025 to 0.05 mm yr^{-1}; Montazer et al., 1985). These processes could cause drying of fractured rock uplands and could expedite the release of gases from underground repositories to the atmosphere (Weeks, 1993).

8.7   Derivation of Equations

Derivation of Equation [8.30]:

$$\frac{\partial \rho_G \phi}{\partial t} = -\nabla \cdot \left(-\frac{\rho_G k_G}{\mu_G} \nabla P \right)$$

[8.64; 8.29]

Assuming ϕ is constant and writing gas density in terms of pressure (Equation [8.17]):

$$\phi \frac{\partial \frac{PM}{RT}}{\partial t} = \nabla \cdot \left(\frac{\frac{PM}{RT} k_G}{\mu_G} \nabla P \right)$$

[8.65]

The term M/RT is constant and can be canceled:

$$\frac{M}{RT} \phi \frac{\partial P}{\partial t} = \frac{M}{RT} \nabla \cdot \left(\frac{P k_G}{\mu_G} \nabla P \right)$$

[8.66]

Assuming permeability and viscosity are constant:

$$\phi 2P \frac{\partial P}{\partial t} = \frac{k_G}{\mu_G} 2P \nabla \cdot (P \nabla P)$$

[8.67]

$$\frac{\partial P^2}{\partial t} \approx \frac{k_G P_0}{\phi \mu_G} \nabla^2 P^2$$

[8.68; 8.30]

Equation [8.68] is Equation [8.30] in Section 8.2.2. The derivation assumes single-phase flow. Similar expressions can be developed for two-phase flow (air and water) by using the volumetric gas content (θ_G) instead of porosity.

Equation [8.31] is derived by assuming that the pressure fluctuations are small, an assumption satisfied in many cases in the unsaturated zone:

$$P = P_0 + \Delta P$$
$$P^2 = P_0^2 + 2\Delta P P_0 + (\Delta P)^2$$

[8.69]

The first term on the right of Equation [8.69] is a constant; therefore, its derivative is zero. The third term is negligible and is neglected. Equation [8.30] can be converted to Equation [8.31] using the following:

$$\frac{\partial P^2}{\partial t} = 2P_0 \frac{\partial P}{\partial t}$$
$$\nabla^2 P^2 = 2P_0 \nabla^2 P$$

[8.70]

$$\frac{\partial P}{\partial t} = \alpha \nabla^2 P$$

[8.71; 8.31]

Derivation of the form of equations (Section 8.1.6.3) for a binary gas mixture is shown in the following. Diffusive transport of nonequimolar gases in an isobaric system mixture is described by (Cunningham and Williams, 1980):

$$N_i^D = J_{iM}^m + N_i^N$$

[8.72]

The molar diffusive flux of component i (N_i^D) results from the molecular diffusive flux $\left(J_{iM}^m\right)$ and the nonequimolar diffusive flux (N_i^N) of gas component i. Because the nonequimolar diffusive flux is nonsegregative, the contribution of each species is proportional to its mole fraction x_i and Equation [8.72] can be reformulated:

$$N_i^D = J_{iM}^m + x_i N^N$$

[8.73]

$$\sum_{i=1}^{v} N_i^D = \sum_{i=1}^{v} j_{iM}^m + \left(\sum_{i=1}^{v} x_i\right) N^N$$

[8.74]

Because $\sum_{i=1}^{v} x_i = 1$, it can be proved that $\sum_{i=1}^{v} J_{iM}^m = 0$; Equation [8.74] can be rewritten as

$$\sum_{i=1}^{v} N_i^D = N^N$$

[8.75]

If Equation [8.73] is applied to the two components A and B of an isobaric, binary mixture, Equation [8.73] becomes:

$$N_A^D = -D_{AB}\nabla n_{AM} + x_A\left(N_A^D + N_B^D\right) \qquad [8.76]$$

$$N_B^D = -D_{BA}\nabla n_{BM} + x_B\left(N_A^D + N_B^D\right) \qquad [8.77]$$

where n_{AM} and n_{BM} are the number of moles of gases A and B. Because the system is isobaric, the total flux is zero (otherwise a pressure gradient would result), and Equations [8.77] and [8.78] reduce to Fick's law applied to the total diffusive flux. This is the only case when Fick's law is strictly valid. In deriving Equations [8.76] and [8.77], the segregative diffusive flux is assumed to follow a gradient law:

$$J_{AM}^m = -D_{AB}\nabla n_{AM} \qquad [8.78]$$

The following two equations result:

$$N_A^D = -D_{AB}\nabla n_{AM} + x_A\left(N_A^D + N_B^D\right) \qquad [8.79]$$

$$N_B^D = -D_{BA}\nabla n_{BM} + x_B\left(N_A^D + N_B^D\right) \qquad [8.80]$$

Multiplying Equation [8.79] with x_B and Equation [8.80] with x_A results in:

$$x_B N_A^D = -x_B D_{AB}\nabla n_{AM} + x_A x_B\left(N_A^D + N_B^D\right) \qquad [8.81]$$

$$x_A N_B^D = -x_A D_{BA}\nabla n_{BM} + x_A x_B\left(N_A^D + N_B^D\right) \qquad [8.82]$$

Subtracting Equation [8.82] from Equation [8.81] results in:

$$x_B N_A^D - x_A N_B^D = -D_{AB}\left(x_B \nabla n_{AM} - x_A \nabla n_{BM}\right) \qquad [8.83]$$

Because $n_{AM} + n_{BM} = n = $ constant and $x_A + x_B = 1$, Equation [8.83] can be rewritten as:

$$x_B N_A^D - x_A N_B^D = -D_{AB}\left((1 - x_A)\nabla n_{AM} - x_A\left(\nabla n_{BM}\right)\right)$$
$$x_B N_A^D - x_A N_B^D = -D_{AB}\left(\nabla n_{AM} - x_A \nabla\left(n_{AM} + n_{BM}\right)\right) \qquad [8.84]$$
$$x_B N_A^D - x_A N_B^D = -D_{AB}\nabla n_{AM}$$

According to the ideal gas law $P_A = \dfrac{n_{AM}}{RT}$; therefore,

$$\frac{x_B N_A^D - x_A N_B^D}{D_{AB}} = -\frac{\nabla P_A}{RT}$$ [8.85]

8.8 References

Abriola, L.M., C.S. Fen, and H.W. Reeves. 1992. Numerical simulation of unsteady organic vapor transport in porous media using the dusty gas model. Proc. Int. Assoc. Hydrol. Conf. Subsurf. Contamin. Immisc. Fluids. pp. 195–202. A.A. Balkema, Rotterdam, Netherlands.

Abu-El-Sha'r, W.Y. 1993. Experimental assessment of multicomponent gas transport mechanisms in subsurface systems. Ph.D. Dissertation, University of Michigan, Ann Arbor, MI.

Abu-El-Sha'r, W.Y., and L.M. Abriola. 1997. Experimental assessment of gas transport mechanisms in natural porous media: Parameter evaluation. Water Resour. Res. 33:505–516.

Alzaydi, A.A. 1975. Flow of gases through porous media. Ohio State University, Columbus, OH.

Alzaydi, A.A., and C.A. Moore. 1978. Combined pressure and diffusional transition region flow of gases in porous media. AICHE J. 24:35–43.

Amali, S., and D.E. Rolston. 1993.Theoretical investigation of multicomponent volatile organic vapor diffusion: steady-state fluxes. J. Environ. Qual. 22:825–831.

Amali, S., D.E. Rolston, and T. Yamaguchi. 1996. Transient multicomponent gas-phase transport of volatile organic chemicals in porous media. J. Environ. Qual. 25:1041–1047.

Auer, L.H., N.D. Rosenberg, K.H. Birdsell, and E.M. Whitney. 1996. The effects of barometric pumping on contaminant transport. J. Contam. Hydrol. 24:145–166.

Baehr, A.L., and C.J. Bruell. 1990. Application of the Stefan-Maxwell equations to determine limitations of Fick's law when modeling organic vapor transport in sand columns. Water Resour. Res. 26:1155–1163.

Baehr, A.L., G.E. Hoag, and M.C. Marley. 1989. Removing volatile contaminants from the unsaturated zone by inducing advective air-phase transport. J. Contam. Hydrol. 4:1–26.

Baehr, A.L., and M.F. Hult. 1991. Evaluation of unsaturated zone air permeability through pneumatic tests. Water Resour. Res. 27:2605–2617.

Benson, D.A., D. Huntley, and P.C. Johnson. 1993. Modeling vapor extraction and general transport in the presence of NAPL mixtures and nonideal conditions. Ground Water 31:437–445.

Bird, R.B., W.E. Stewart, and E.N. Lightfoot. 1960. Transport phenomena. John Wiley, New York, NY.

Brooks, R.H., and A.T. Corey. 1964. Hydraulic properties of porous media. Colorado State University Hydrol. Pap. 3.

Brusseau, M L. 1994. Transport of reactive contaminants in heterogeneous porous media. Rev. Geophys. 32:85–313.

Buckley, S. E., and M.C. Leverett. 1942. Mechanisms of fluid displacement in sands. Trans. Am. Inst. Mining Metallur. Eng. 146:107–116.

Burdine, N.T. 1953. Relative permeability calculations from pore size distribution data. Petrol. Trans. AIME, 198:71–78.

Carslaw, H.S., and J.C. Jaeger. 1959. The conduction of heat in solids. Oxford University Press, London.

Celia, M.A., and P. Binning. 1992. A mass conservative numerical solution for two-phase flow in porous media with application to unsaturated flow. Water Resour. Res. 28:2819–2828.

Charbeneau, R.J., and D.E. Daniel. 1992. Contaminant transport in unsaturated flow. p. 15.1–15.54. In D.R. Maidment (ed.) Handbook of hydrology. McGraw-Hill, Inc., New York, NY.

Childs, S.W., and G. Malstaff. 1982. Heat and mass transfer in nsaturated porous media. Pacific Northwest Laboratory, Richland, WA.

Collins, R E. 1961. Flow of fluids through porous media. van Nostrand-Reinhold, Princeton, NJ.

Connor, J.R. 1988. Case study of soil venting. Pollut. Eng. 20:74–78.

Corey, A.T. 1954. The interrelation between gas and oil relative permeabilities. Produc. Month. 19:38–44.

Corey, A.T. 1986. Air permeability. p. 1121–1136. In A. Klute (ed.) Methods of soil analysis. American Society of Agronomy, Madison, WI.

Croise, J., and V. Kaleris. 1992. Field measurements and numerical simulations of pressure drop during air stripping in the vadose zone. In K.U. Weger (ed.) Subsurface contamination by immiscible fluids. A.A. Balkema, Rotterdam, Netherlands.

Cunningham, R.E., and R.J.J. Williams. 1980. Diffusion of gases and porous media. Plenum, New York, NY.

Demond, A.H., and P.V. Roberts. 1987. An examination of relative permeability relations for two-phase flow in porous media. Water Resour. Bull. 23:617–628.

Detty, T.E. 1992. Determination of air and water relative permeability relationships for selected unconsolidated porous materials. Ph.D. Dissertation, University of Arizona, Tucson, AZ.

Duffield, G.M., and J.O. Rumbaugh. 1989. AQTESOLV aquifer test solver documentation; version 1.00, Geraghty and Miller Modeling Group, Reston, VA.

Dullien, F.A.L. 1979. Porous media: Fluid transport and pore structure. Academic Press, New York, NY.

Edwards, K.B. 1994. Air permeability from pneumatic tests in oxidized till. J. Environ. Engin. 120:329–346.

Estes, R.K., and P.F. Fulton. 1956. Gas slippage and permeability measurements. Trans. Am. Inst. Min. Metall. Pet. Eng. 207:338–342.

Falta, R.W. 1995. Analytical solutions for gas flow due to gas injection and extraction from horizontal wells. Ground Water 33:235–246.

Falta, R.W., I. Javandel, K. Pruess, and P.A. Witherspoon. 1989. Density-driven flow of gas in the unsaturated zone due to the evaporation of volatile organic compounds. Water Resour. Res. 25:2159–2169.

Falta, R.W., K. Pruess, and D.A. Chesnut. 1993. Modeling advective contaminant transport during soil vapor extraction. Ground Water 31:1011–1020.

Fayer, M.J., and T.L. Jones. 1990. UNSAT-H Version 2.0: Unsaturated soil water and heat flow model. Pacific Northwest Laboratory, Richland, WA.

Fayer, M.J., M.L. Rockhold, and M.D. Campbell. 1992. Hydrologic modeling of protective barriers: comparison of field data and simulation results. Soil Sci. Soc. Am. J. 56:690–700,.

Finsterle, S., and K. Pruess. 1995. Solving the estimation-identification problem in two-phase flow modeling. Water Resour. Res. 31:913–924.

Forcheimer, P. 1901. Wasserbewegung durch boden. Z. Des Verin. Deutch Ing. 49:1736–1749.

Freeze, R.A., and J.A. Cherry. 1986. Groundwater. Prentice Hall, New Jersey.

Gardner, W. H. 1986. Water content. p. 493–545. *In* A. Klute (ed.) Methods of soil analysis. Part 1. Physical and mineralogical methods. American Society of Agronomy, Madison, WI.

Gierke, J.S., N.J. Hutzler, and D.B. McKenzie. 1992. Vapor transport in unsaturated soil columns: Implications for vapor extraction. Water Resour. Res. 28:323–335.

Glauz, R.D., and D.E. Rolston. 1989. Optimal design of two-chamber, gas diffusion cells. Soil Sci. Soc. Am. J. 53:1619–1624.

Goggin, D.J., R.L. Thrasher, and L.W. Lake. 1988. A theoretical and experimental analysis of minipermeameter response including gas slippage and high velocity flow effects. *In situ.* 12:79–116.

Hampton, D. 1989. Coupled heat and fluid flow in saturated-unsaturated compressible porous media. Colorado State University, Fort Collins, CO.

Hanks, R.J. 1992. Applied soil physics. Springer-Verlag, New York, NY.

Hassler, G.L. 1944. Method and apparatus for permeability measurements. Patent Number 2,345,935 April 2, 1944. US Patent Office, Washington, DC.

Hayden, N.J., T.C. Voice, M.D. Annable, and R.B. Wallace. 1994. Change in gasoline constituent mass transfer during soil venting. J. Environ. Eng. 120:1598–1614.

Hazen, A. 1911. Discussion: dams on sand foundations. Trans. Am. Soc. Civ. Eng. 73:199.

Heid, J.G., J.J. McMahon, R.F. Nielsen, and S.T. Yuster. 1950. Study of the permeability of rocks to homogeneous fluids. A.P.I. Drilling and Production Practice, A.P.I., Alexandria, VA.

Hirst, W., and G.E. Harrison. 1939. The diffusion of radon gas mixtures. Proc. Roy. Soc. London, Ser. A 169:573–586.

Hoag, G.E., and M.C. Marley. 1986. Gasoline residual saturation in unsaturated uniform aquifer materials. J. Environ. Eng. 112:586–604.

Hutzler, N.J., B.E. Murphy, and J.S. Gierke. 1989. State of technology review: Soil vapor extraction systems. Final Report USEPA, Hazardous Waste Engineering Research Laboratory, USEPA, Washington, DC.

Jackson, R. 1977. Transport in porous catalysts. Elsevier Science Publishing Co., New York, NY.

Jaynes, D.B., and A.S. Rogowski. 1983. Applicability of Fick's law to gas diffusion. Soil Sci. Soc. Am. J. 47:425–430.

Jin, M., M. Delshad, B. Dwarakanath, D.C. McKinney, G.A. Pope, K. Sepehrnoori, C. Tilburg, and R.E. Jackson. 1995. Partitioning tracer test for detection, estimation, and remediation performance assessment of subsurface nonaqueous phase liquids. Water Resour. Res. 31:1201–1211.

Johnson, P.C., A.L. Baehr, R.A. Brown, R.E. Hinchee, and G.E. Hoag. 1995. Innovative site remediation technology: Vol. 8. Vacuum vapor extraction. American Academy of Environmental Engineers, Annapolis, MD.

Johnson, P.C., C.C. Stanley, M.W. Kemblowski, D.L. Byers, and J.D. Colthart. 1990b. A practical approach to the design, operation, and monitoring of *in situ* soil-venting systems. Ground Water Monit. Rev. 10:159–178.

Johnson, P.C., M.W. Kemblowski, and J.D. Colthart. 1990a. Quantitative analysis for the cleanup of hydrocarbon-contaminated soils by *in situ* soil venting. Ground Water 28:413–429.

Joss, C.J., and A.L. Baehr. 1997. AIR2D—A computer code to simulate two-dimensional, radially symmetric airflow in the unsaturated zone. US Geol. Surv. Open File Rep. 97-588.

Joss, C.J., and A L. Baehr. 1995. Documentation of AIR3D, an adaptation of the ground water flow code MODFLOW to simulate three-dimensional air flow in the unsaturated zone. US Geol. Surv. Open File Rep. 94-533.

Jury, W.A., W.R. Gardner, and W.H. Gardner. 1991. Soil physics. John Wiley and Sons, Inc., New York, NY.

Kaluarachchi, J.J. 1995. Analytical solution to two-dimensional axisymmetric gas flow with Klinkenberg effect. J. Envir. Eng. 121:417–420.

Kaluarachchi, J.J., and J.C. Parker. 1989. An efficient finite element method for modeling multiphase flow. Water Resour. Res. 25:43–54.

Kipp, K.L.J. 1987. Effect of topography on gas flow in unsaturated fractured rock: Numerical simulation. p. 171–176. *In* D.D. Evans and T.J. Nicholson (ed.) Flow and transport through unsaturated fractured rock. American Geophysical Union, Washington, DC.

Klinkenberg, L.J. 1941. The permeability of porous media to liquids and gases. A.P.I. Drilling and Production Practice, A.P.I., Alexandria, VA.

Kreamer, D.K., E.P. Weeks, and G.M. Thompson. 1988. A field technique to measure the tortuosity and sorption-affected porosity for gaseous diffusion of materials in the unsaturated zone with experimental results from near Barnwell, South Carolina. Water Resour. Res. 24:331–341.

Lai, S.H., J.M. Tiedje, and A.E. Erickson. 1976. *In situ* measurement of gas diffusion coefficient in soils. Soil Sci. Soc. Am. J. 40:3–6.

Leffelaar, P.A. 1987. Dynamic simulation of multinary diffusion problems related to soil. Soil Sci. 143:79–91.

Marshall, T.J. 1958. A relation between permeability and size distribution of pores. J. Soil Sci. 9:1–8.

Mason, E.A., and A.P. Malinauskas. 1983. Gas transport in porous media: The Dusty-Gas Model. Chem. Eng. Monogr. 17. Elsevier, New York, NY.

Mason, E.A., A.P. Malinauskas, and R.B. Evans. 1967. Flow and diffusion of gases in porous media. J. Chem. Phys. 46:3199–3126.

Massmann, J.W. 1989. Applying groundwater flow models in vapor extraction system design. J. Env. Eng. 115:129–149.

Massmann, J.W., and D.F. Farrier. 1992. Effects of atmospheric pressures on gas transport in the vadose zone. Water Resour. Res. 28:777–791.

Massmann, J.W., and M. Madden. 1994. Estimating air conductivity and porosity from vadose-zone pumping tests. J. Env. Eng. 120:313–328.

McClellan, R.D., and R.W. Gillham. 1992. Vapour extraction of trichloroethylene under controlled conditions at the Borden site. p. 89–96. *In* Subsurface contamination by immiscible fluids. A. A. Balkema, Rotterdam, Netherlands.

McDonald, M. G., and A.W. Harbaugh. 1988. A modular three-dimensional finite-difference ground-water flow model. *In* Techniques of water resources investigations. US Geological Survey, Denver, CO.

McIntyre, D.S., and J.R. Philip. 1964. A field method for measurement of gas diffusion into soils. Aust. J. Soil Res. 2:133–145.

McWhorter, D.B. 1990. Unsteady radial flow of gas in the vadose zone. J. Contam. Hydrol. 5:297–314.

Mendoza, C.A., and E.O. Frind. 1990. Advective-dispersive transport of dense organic vapors in the unsaturated zone: Sensitivity analysis. Water Resour. Res. 26:388–398.

Mendoza, C.A., and T.A. McAlary. 1990. Modeling of ground-water contamination caused by organic solvent vapors. Ground Water 28:199–206.

Mercer, J.W., and R.M. Cohen. 1990. A review of immiscible fluids in the subsurface: properties, models, characterization, and remediation. J. Contam. Hydrol. 6:107–163.

Millington, R.J. 1959. Gas diffusion in porous media. Science 130:100–102.

Millington, R.J., and J.P. Quirk. 1961. Permeability of porous solids. Farad. Soc. Trans. 57:1200–1207.

Millington, R.J., and R.C. Shearer. 1970. Diffusion in aggregated porous media. Soil Sci. 3:372–378.

Milly, P.C.D. 1982. Moisture and heat transport in hysteretic, inhomogeneous porpous media: a matric head-based formulation and a numerical model. Water Resour. Res. 18:489–498.

Milly, P.C.D., and P.S. Eagleson. 1980. The coupled transport of water and heat in a vertical soil column under atmospheric excitation. Ralph M. Parsons Laboratory, Massachusetts Institute of Technology, Cambridge, MA.

Montazer, P., E.P. Weeks, F. Thamir, S.N. Yard, and P.B. Hofrichter. 1985. Monitoring the vadose zone in fractured tuff, Yucca Mountain, Nevada. Proc. Charact. Monit Vadose (Unsaturated) Zone, p. 439–469. National Water Well Association, Denver, CO.

Moore, C., I.S. Rai, and J. Lynch. 1982. Computer design of landfill methane migration control. J. Env. Eng. Div. Am. Soc. Civ. Eng. 108:89–107..

Morel-Seytoux, H.J. 1973. Two-phase flow in porous media. Adv. Hydrosci. 9:119–202.

Morel-Seytoux, H.J., and J.A. Billica. 1985. A two-phase numerical model for prediction of infiltration: Application to a semi-infinite column. Water Resour. Res. 21:607–615.

Mualem, Y. 1976. A new model for predicting the hydraulic conductivity of unsaturated porous media. Water Resour. Res. 12:513–521.

Muskat, M. 1946. Flow through porous media. McGraw Hill, New York, NY.

Nazaroff, W.W. 1992. Radon transport from soil to air. Rev. Geophys. 30:137–160..

Nazaroff, W.W., and R.G. Sextro. 1989. Technique for measuring the indoor ^{222}Rn source potential of soil. Environ. Sci. Technol. 23:451–458.

Nero, A.V., A.J. Gadgil, W.W. Nazaroff, and K.L. Revzan. 1986. Distribution of airborne radon–222 concentrations in U.S. homes. Science 234:992–997.

Nicot, J.P. 1995. Characterization of gas transport in a playa subsurface, Pantex Site, Amarillo, Texas. M.S. Thesis. University of Texas, Austin, TX.

Nilson, R.H., W.B. McKinnis, P.L. Lagus, J.R. Hearst, N.R. Burkhard, and C.F. Smith. 1992. Field measurements of tracer gas transport induced by barometric pumping. Proc. Third Int. Conf. High Level Radioact. Waste Manag. p. 710–716. American Nuclear Society, LaGrange Park, IL..

Nilson, R.H., E.W. Peterson, K.H. Lie, N.R. Burkard, and J.R. Hearst. 1991. Atmospheric pumping: a mechanism causing vertical transport of contaminated gases through fractured permeable media. J. Geophys. Res. 96(B13):21933–21948.

NRC. 1996. Rock fractures and fluid flow contemporary understanding and applications. National Academy Press, Washington, DC.

Pederson, T.A., and J.T. Curtis. 1991. Soil vapor extraction technology—Reference handbook. Office of Research and Development, USEPA, Washington, DC.

Penman, H.L. 1940a. Gas and vapor movements in the soil. I. The diffusion of vapors through porous solids. J. Agr. Sci. 30:437–462.

Penman, H.L. 1940b. Gas and vapor movements in the soil. II. The diffusion of carbon dioxide through porous solids. J. Agr. Sci. 30:570–581.

Pfannkuch, H. 1983. Hydrocarbon spills, their retention in the subsurface and propagation into shallow aquifers. Office of Water Resources Technology, USEPA, Washington, DC.

Philip, J.R., and D.A. de Vries. 1957. Moisture movement in porous materials under temperature gradients. Trans. AGU 38:222–232.

Pruess, K. 1987. TOUGH User's Guide. NUREG/CR-4645, Nuclear Regulatory Commission, Washington, DC.

Pruess, K. 1991. TOUGH2—A general-purpose numerical simulator for multiphase fluid and heat flow. Lawrence Berkeley Laboratory, Berkeley, CA.

Pruess, K., and T.N. Narasimhan. 1985. A practical method for modeling fluid and heat flow in fractured porous media. Soc. Petrol. Engin. J. 25:14–26.

Pruess, K., J.S.Y. Wang, and Y.W. Tsang. 1990. On thermohydrologic conditions near high-level nuclear wastes emplaced in partially saturated fractured tuff, I. Simulation studies with explicit consideration of fracture effects. Water Resour. Res. 26:1235–1248.

Raney, W. A. 1949. Field measurement of oxygen diffusion through soil. Soil Sci. Soc. Am. Proc. 14:61–65.

Rathfelder, K., W.W.-G. Yeh, and D. Mackay. 1991. Mathematical simulation of soil vapor extraction systems: model development and numerical examples. J. Contam. Hydrol. 8:263–297.

Rathfelder, K., J.R. Lang, and L.M. Abriola. 1995. Soil vapor extraction and bioventing: Applications, limitations, and future research directions. Rev. Geophys. 33:067–1081.

Rojstaczer, S., and J.P. Tunks. 1995. Field-based determination of air diffusivity using soil-air and atmospheric pressure time series. Water Resour. Res. 12:3337–3343.

Rolston, D.E. 1986. Gas diffusivity. p. 1089–1102. *In* A. Klute (ed.) Methods of soil analysis. American Society of Agronomy, Madison, WI.

Rolston, D.E., D. Kirkham, and D.R. Nielsen. 1969. Miscible displacement of gases through soils columns. Soil Sci. Soc. Am. Proc. 33:488–492.

Rosenberg, N.D., W.E. Soll, and G.A. Zyvoloski. 1994. Microscale and macroscale modeling of flow in unsaturated fractured rock: Chapman Conf. Aqu. Phase Multiphase Transp. Fractured Rock, American Geophysical Union, Burlington, VT.

Satterfield, C.N., and P J. Cadle. 1968. Diffusion and flow in commercial catalysts at pressure levels about atmospheric. J. Ind. Eng. Chem. Fund. 7:202 .

Scanlon, B.R., and P.C.D. Milly. 1994. Water and heat fluxes in desert soils. 2. Numerical simulations. Water Resour. Res. 30:721–733.

Scheidegger, A.E. 1974. The physics of flow through porous media. 3rd Ed. University of Toronto Press, Toronto, Canada.

Schery, S.D., S. Whittlestone, K.P. Hart, and S.E. Hill. 1989.The flux of radon and thoron from Australian soils. J. Geophys. Res. 94:8567–8576.

Schwille, F., W. Bertsch, R. Linke, W. Reif, and S. Zauter. 1988. Dense chlorinated solvents in porous and fractured media: Model experiments. Lewis Publishers, Chelsea, MI.

Shan, C., R.W. Falta, and I. Javandel. 1992. Analytical solutions for steady state gas flow to a soil vapor extraction well. Water Resour. Res. 28:1105–1120.

Sharp, J.M., L. Fu, P. Cortez, and E. Wheeler. 1994. An electronic minipermeameter for use in the field and laboratory. Ground Water 32:41–46.

Shepherd, R G. 1989. Correlations of permeability and grain size. Ground Water 27:633–638.

Snow, D.T. 1968. Rock fracture spacings, openings, and porosities. J. Soil Mech. 94:73–91.

Springer, D.S. 1993. Determining the air permeability of porous materials as a function of a variable water content under controlled laboratory conditions. University of California, Santa Barbara, CA.

Springer, D.S., S.J. Cullen, and L.G. Everett. 1995. Laboratory studies on air permeability. p. 217–247. *In* L.G. Wilson, L.G. Everett and S.J. Cullen (ed.) Handbook of vadose zone characterization and monitoring., Lewis Publishers, Boca Raton, FL.

Stonestrom, D.A. 1987. Co-determination and comparisons of hysteresis-affected, parametric functions of unsaturated flow: Water-content dependence of matric pressure, air trapping, and fluid permeabilities in a mon-swelling soil. Stanford University, Palo Alto, CA.

Stonestrom, D.A., and J. Rubin. 1989. Air permeability and trapped-air content in two soils. Water Resour. Res. 25:1959–1969.

Stothoff, S.A. 1995. BREATH Version 1.1 - coupled flow and energy transport in porous media, Simulator description and user guide. US Nuclear Regulatory Commission, Washington, DC.

Stothoff, S.A. 1997. Sensitivity of long-term bare soil infiltration simulations to hydraulic properties in an arid environment. Water Resour. Res. 33:547–558.

Tanner, A.B. 1964. Radon migration in the ground: A review. p. 161–190. *In* J.A.S. Adams and W.M. Lowder (ed.) Natural radiation environment. Chicago Press,Chicago, IL.

Terzaghi, K., and R.B. Peck. 1968. Soil mechanics in engineering practice. John Wiley and Sons, New York, NY.

Thibodeaux, L.J., C. Springer, and L.M. Riley. 1982. Models and mechanisms for the vapor phase emission of hazardous chemicals from landfills. J. Hazard. Mater. 7:63–74.

Thorstenson, D.C., and D.W. Pollock. 1989. Gas transport in unsaturated porous media, the adequacy of Fick's law. Rev. Geophys. 27:61–78.

Touma, J., and M. Vauclin. 1986. Experimental and numerical analysis of two-phase infiltration in a partially saturated soil. Transp. Porous Media 1:22–55.

Travis, C.C., and J. M.McGinnis. 1992. Vapor extraction of organics from subsurface soils - Is it effective? Env. Sci. Technol. 26:1885–1887.

USEPA. 1991. Soil vapor extraction technology reference handbook. US Environmental Protection Agency, Washington, DC.

USEPA. 1992. A technology assessment of soil vapor extraction and air sparging. Office of Research and Development, US Environmental Protection Agency, Washington, DC.

van Genuchten, M.T. 1980. A closed-form equation for predicting the hydraulic conductivity of unsaturated soils. Soil Sci. Soc. Am. J. 44:892–898.

Vauclin, M. 1989. Flow of water and air in soils: Theoretical and experimental aspects. p. 53–91. *In* H.J. Morel-Seytoux (ed.) Unsaturated flow in hydrologic modelling, theory and practice. Kluwer Academic, Dordrecht, Netherlands.

Weast, R.C. 1986. CRC handbook of chemistry and physics. CRC Press, Boca Raton, FL.

Weeks, E.P. 1993. Does the wind blow through Yucca Mountain. p. 45–53. *In* D.D. Evans and T.J. Nicholson (ed.) Flow and transport through unsaturated fractured rock related to high-level radioactive waste disposal. NUREG CP-0040, US Nuclear Regulatory Commission, Washington, DC.

Weeks, E.P. 1987. Effect of topography on gas flow in unsaturated fractured rock: concepts and observations. p. 165–170. *In* D.D. Evans and T.J. Nicholson (ed.) Flow and transport through unsaturated fractured rock. American Geophysical Union, Washington, DC.

Weeks, E.P. 1978. Field determination of vertical permeability to air in the unsaturated zone. US Geol. Sur. Prof. Pap. US Geological Survey, Denver, CO.

White, M.D., and M. Oostrom. 1996. STOMP: Subsurface transport over multiple phases theory guide. Pacific Northwest National Laboratory, Richland, WA.

Zyvoloski, G.A., B.A. Robinson, Z. Dash, and L.L. Trease. 1996. Models and methods summary for the FEHM application. Los Alamos National Laboratory, Los Alamos, NM.

9

Soil Spatial Variability

D. J. Mulla
University of Minnesota

A. B. McBratney
University of Sydney

9.1 Variability in Soil Properties from Soil Classification

A key feature of soil is the variation with depth in soil properties. Soil is formed as a result of the influences of climate, plants, and time acting on geologic parent material in different landscape positions. Soils are uniquely different from geologic parent material such as loess, glacial till, or sedimentary rock because soils develop horizonation, in which each horizon has a distinct set of characteristic and diagnostic soil properties. Horizons may differ in organic matter content, color, structure, texture, pH, base saturation, cation exchange capacity, bulk density, water holding capacity, as well as many other soil physical and chemical properties.

Variation in soil properties also occurs across the landscape and in response to regional variations in climate and parent material. Broad differences in soil types at the soil order level are identified based on horizon characteristics, such as thickness and organic C content of the A horizon, base saturation of the B horizon, or the presence of a pale colored, extensively leached subsurface E horizon. Differences in soil horizonation have been extensively described and classified, giving rise to the 12 major soil orders.

Changes in soil properties at the order level of classification generally, but not always, take place over relatively large distances, often across a significant climatic and/or orographic gradient. Within fields and across shorter distances, soil properties also vary significantly, even across locations that involve only one soil order. This variation has also been extensively described and classified, giving rise to the soil series level of classification. In the United States, the variability in soil properties at the series mapping unit level is depicted using County Soil Series maps. Although a given field or farm may be mapped as having from 5 to 10 or more soil mapping units, it is rare for these units to represent more than a few soil orders.

The variability in soil properties at the soil series level is often caused by small changes in topography that affect the transport and storage of water across and within the soil profile. Hillslopes are typically divided into 5 landscape positions which are closely related to patterns in water transport and storage, which strongly influence soil development. These landscape positions are the summit, shoulder, backslope, footslope, and toeslope positions. The soil variation across a hillslope is usually referred to as a catena, which is an association of soil mapping units that are closely linked to hillslope position.

One of the most common soil catenas in the midwestern United States is the Clarion-Nicollet-Webster catena. These are all silt loam textured soils with a thick organic C rich A horizon (mollic epipedon) formed from calcareous glacial till parent material. The Clarion soil is typically well drained and occurs at the summit and shoulder positions. The Nicollet is moderately to somewhat poorly to well drained and occurs at the backslope and footslope positions. The Webster is a poorly drained soil which occurs at the toeslope position. Thus, variations in soil properties at the scale of a hillslope are often controlled by landscape position and water flow behavior.

9.1.1   Variability within Soil Mapping Units

It would be wrong to give the impression that soil variability at the field/scale can be completely described by soil mapping units because there is considerable variation in soil properties that are not accounted for by soil mapping unit classification. Two types of variation can be identified. (1) There is soil variabilty within soil mapping units that is caused by mapping and classification error (Burrough, 1991). The scale of most soil series maps in the United States is 1:24,000. At this scale, up to 40% of the region within a soil mapping unit can consist of dissimilar inclusions, which are soils that are not classified the same as the dominant soil in the mapping unit. Errors in mapping may occur anywhere within the soil mapping unit, whether near the boundary with other mapping units or in the center of the mapping unit. Often, these errors appear as lighter or darker colored regions within a mapping unit when viewed in an aerial photograph of bare soil (Thompson and Robert, 1995). (2) There is variation in soil properties within soil mapping units due to the effects of human management activities. For example, the application of animal manure to soil results in significant increases of soil nutrient levels and soil electrical conductivity relative to unmanured soil. Since the soil survey does not classify soils based on nutrient content or electrical conductivity, these human-induced variations would not be diagnostic criteria used in soil classification. Alternatively, the deposition of wind-blown ash from an industrial site may bring with it trace metals that contaminate soil in an isolated area. A large variety of other human activities can alter or disturb the natural characteristics of soils, including landfill disposal, industrial discharge of solvents, mining and logging operations, tillage and drainage of agricultural fields, application of fertilizers and pesticides, rotation of crops, irrigation practices, and installation of septic tanks.

Thus, it is evident that soil maps published in the County Soil Surveys are not sufficient alone for describing the detailed patterns in variation of soil properties that occur across hillslopes, within fields or parcels of land, and across regions. A significant amount of research has been conducted on methods for sampling, characterizing, and representing soil variability at a finer resolution than is given by soil mapping units (Di et al.,1989; Rogowski and Wolf, 1994).

9.2   Classical Measures of Variability

Soil variability can be estimated using either continuous or discrete sampling. In continuous sampling, measurements of a selected soil property are obtained at all locations in a field. This is usually accomplished through analysis of a satellite or aerial image, or through ground collection of soil data by noninvasive remote sensing techniques. An example of the latter is sensing of soil electrical conductivity with the Geonics EM-38 electromagnetic induction sensor. Continuous sampling results in data for all locations of a field, with no need for interpolation between measurements or emphasis on soil sampling design strategies.

Discrete sampling is usually accomplished by collecting soil samples at predetermined locations and depths using destructive sampling techniques. Unlike continuous sampling, only a subset of the sample population is observed in discrete sampling. If enough samples have been collected, the characteristics of the sample population can be inferred using statistical techniques.

9.2.1 Classical Measures of Central Tendency and Spread

Two key characteristics of sample populations are measures of central tendency and measures of data variation (Snedecor and Cochran, 1980; Upchurch and Edmonds, 1991). Measures of central tendency include the mean, median, and mode. The mean ($\bar{z}$) is the arithmetic sample average from:

$$\bar{z} = \frac{1}{N} \sum_{i=1}^{N} z_i \qquad [9.1]$$

where N is the number of sampled locations, and z_i is the measured value at the ith location. At the median, half the sampled population has a value greater than the median, and half has a value less than the median. When the data for a sampled population are ranked by value, the mode is the value that occurs with the greatest frequency.

The spread of sample values around the mean is an important measure of variability in the sample population and is computed as the standard deviation (s), which is the square root of the sample variance (s^2). The variance is estimated from:

$$s^2 = \frac{1}{N} \sum_{i=1}^{N} \left[z_i - \bar{z} \right]^2 \qquad [9.2]$$

where N is the number of measured values z_i. For a normal frequency distribution, approximately 95% of the population will have a value of the mean plus or minus 2 standard deviations.

Other measures of spreading about the mean include the range, quartile range, and coefficient of variation. As the value of the range, standard deviation, and coefficient of variation increases, so, too, does population variability. The range is the difference between the maximum and minimum values, while the quartile range is the difference between the first and third quartile values. The coefficient of variation (CV) gives a normalized measure of spreading about the mean, and is estimated using:

$$CV = \frac{s}{\bar{z}} \times 100\% \qquad [9.3]$$

Properties with larger CV values are more variable than those with smaller CV values.

Wilding (1985) has described a classification scheme for identifying the extent of variability for soil properties based on their CV values in which CV values of 0–15, 16–35, and > 36 indicate little, moderate and high variability, respectively. Typical ranges of CV values from published studies of soil properties (Jury, 1986; Jury et al., 1987; Beven et al., 1993; Wollehaupt et al., 1997) are given in Table 9.1. Properties such as soil pH and porosity are among the least variable, while those pertaining to water or solute transport are among the most variable.

In classical statistics, a sample population is characterized by frequency distributions which are parameterized using the mean and standard deviation. A frequency distribution is the number of observations occurring in several class intervals between the minimum and maximum values of the

Table 9.1   Ranges in values for the coefficient of variation (CV) of selected soil and crop properties

Property	CV (%)	Magnitude of Variability
pH	2-15	Low
Porosity	7-11	Low
Bulk Density	3-26	Low to Moderate
Crop Yield	8-29	Low to Moderate
%Sand	3-37	Low to Moderate
0.01 MPa Water Content	4-20	Low to Moderate
Pesticide Adsorption Coeff.	12-31	Moderate
Organic Matter Content	21-41	Moderate to High
1.5 MPa Water Content	14-45	Moderate to High
%Clay	16-53	Moderate to High
Soil Nitrate-N	28-58	Moderate to High
Soil Water Infiltration Rate	23-97	Moderate to High
Soil Available Potassium	39-157	High
Soil Available Phosphorus	39-157	High
Soil Electrical Conductivity	91-263	High
Saturated Hydraulic Conductivity	48-352	High
Solute Saturated Velocity	78-194	High
Solute Dispersion Coeff.	79-178	High
Solute Dispersivity	78-539	High

ample population. Frequency distributions may either be normally or non-normally distributed, being symmetrical and asymmetrical about the mean, respectively. In a normal frequency distribution, the mean, median, and mode are all equal whereas in a non-normal distribution, they are not. One of the most common non-normal distributions is the log normal distribution in which the median is larger than the mode but smaller than the mean. For log normal frequency distributions, the mean is heavily influenced by a few large measurement values; hence, the median is a more accurate measure of central tendency than the mean.

Asymmetrical frequency distributions can often be transformed to normal distributions by taking the natural logarithm of the data. Examples of properties where a log transform produced normally distributed data include studies of soil hydraulic properties (Vieira et al., 1988; Wierenga et al., 1991; Russo and Bouton, 1992), soil nutrient content (Wade et al., 1996), and trace metal soil concentrations (Markus and McBratney, 1996). The mean (m) and variance (s) of the distribution for the log transformed data can be used to estimate the arithmetic mean of the untransformed data using the expression (Webster and Oliver, 1990):

$$\bar{z} = \exp\!\left(m + 0.5s^2\right) \tag{9.4}$$

The spread of the normal frequency distribution about the mean is an important measure of variability in the sample population. As an example, consider the frequency distributions for two artificial data sets generated with mean values of 50, and a standard deviation of either 10 (Fig. 9.1A) or 5 (Fig. 9.1B). A sampled population with more variability (Fig. 9.1A) will have a wider spread in the frequency distribution than a population with less variability (Fig. 9.1B). The frequency distribution for a sampled population with a larger standard deviation appears relatively flat and spread out. In contrast, a sampled population with little variability will have a narrow spread in the frequency distribution, and most of the sampled values will be close to the mean and median.

The frequency distribution characterizes the mean and spread of sample values about the mean. Comparison of two frequency distributions for the same soil property (for instance, soil moisture content) allows comparison of the mean values for each sampled population, as well as comparisons about the relative amount of sample variation observed. Comparisons of the mean values for the normal frequency distributions from two samples of the same property are often achieved using t-tests (Snedecor and Cochran, 1980; Upchurch and Edmonds, 1991).

9.3 Geostatistics

The frequency distribution is somewhat limited in its ability to describe variability of a sampled population because it does not provide any information about the spatial correlation between soil samples at a given location. Most often, soil properties do not occur across the landscape in a random fashion. Typically, the values for a soil property from samples taken at close spacings will be similar or spatially correlated (Oliver, 1987). Soil properties from large sample spacings will typically be dissimilar and spatially uncorrelated.

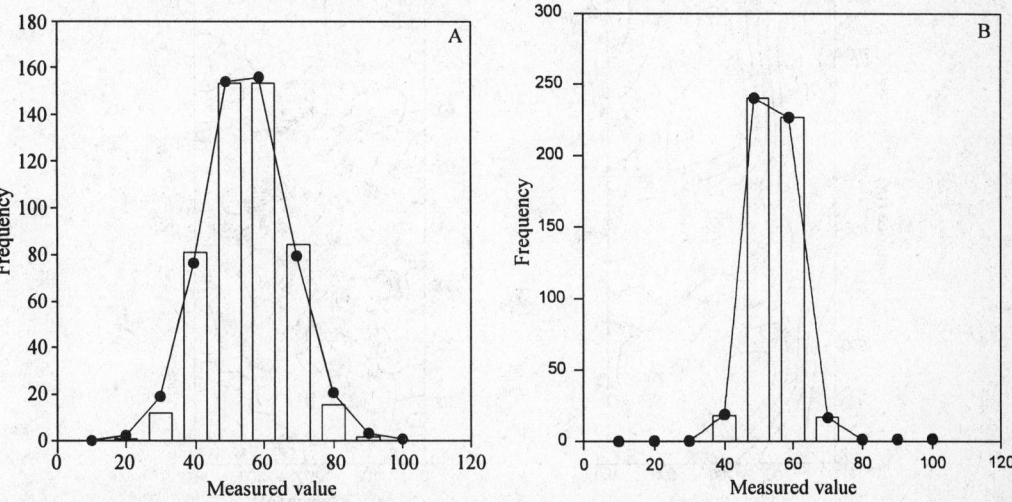

Fig. 9.1 (A) Normal frequency plot (●) and sample frequency distribution (bars) for a population with a mean of 50 and a standard deviation of 10; (B) normal frequency plot (●) and sample frequency distribution (bars) for a population with a mean of 50 and a standard deviation of 5

The important differences between spatially correlated and randomly distributed data can be illustrated using the following example. Suppose soil surface samples are taken from a large field, and the organic C content of each sample is measured. One would expect samples collected close to one another to have similar organic C contents, while samples taken farther apart would be less similar Fig. 9.2A shows the measured patterns in surface organic C content for a study site in Minnesota from 234 samples collected on a regular grid at spacings of 30 x 45 m. For comparison, a random artificial data set was generated having the same mean and standard deviation as the measured organic C contents (Fig. 9.2B). Comparison of Figs. 9.2A and 9.2B shows the great differences in appearance of the spatial patterns for organic C content when spatial correlation is absent. Notice that the randomly distributed values for organic C content appear much patchier than the spatially correlated values, which show larger spatial groupings where either large or small values for organic C occur. This is an illustration of spatial dependence, which manifests itself as a greater similarity between closely spaced than widely spaced samples.

Thus, in addition to knowing what the mean and standard deviation of a sample population are, one is often interested in knowing the spatial correlation structure of the population. The spatial structure of a population can be estimated using approaches developed in geostatistics (Journel and Huijbregts, 1978; Isaaks and Srivastava, 1989; Cressie, 1991). Geostatistics is a branch of applied statistics that quantifies the spatial dependence and spatial structure of a measured property and, in turn, uses that spatial structure to predict values of the property at unsampled locations. These two steps typically involve spatial modeling (variography) and spatial interpolation (kriging).

In the jargon of geostatistics, a regionalized variable is a property such as soil moisture or hydraulic conductivity which can be sampled. This regionalized variable has a value at every point within the region to be studied, and the collection of regionalized variables at all sampled points is known as a random function. One realization of this random function is a regionalized variable $Z(x)$, having data values $z(x_i)$, or simply z_i, at sampling locations x_i. There can be many realizations of the regionalized

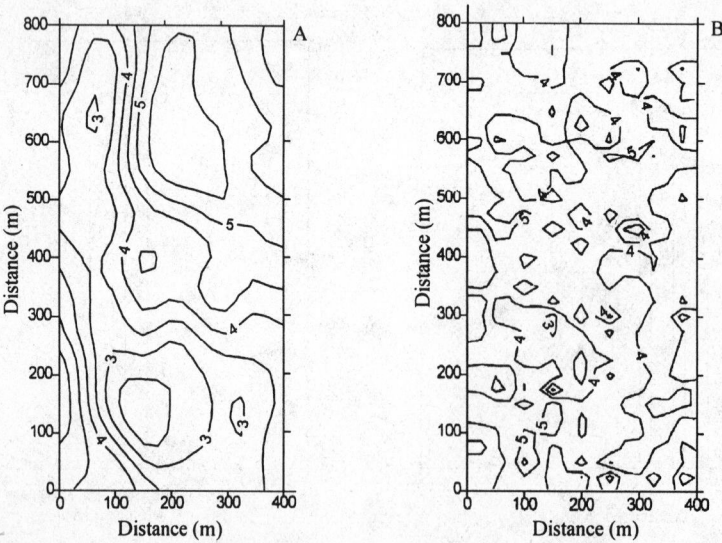

Fig. 9.2   (A) Spatially correlated patterns in measured soil surface organic carbon content (%) from a field in southern Minnesota; (B) Random distribution of soil surface organic carbon contents (%) having the same mean and standard deviation as the data presented in Fig. 9.2A

variable if the region is sampled many different times, with each realization being a subsample of the sample population.

9.3.1 Variography

Spatial dependence can be quantified and modeled using the semivariogram (Burgess and Webster, 1980a). The semivariogram $\gamma(h)$ is estimated using the equation:

$$\hat{\gamma}(h) = \frac{1}{2n(h)} \sum_{i=1}^{n(h)} \left[z_i - z_{i+h} \right]^2 \tag{9.5}$$

where h is the separation distance between locations x_i and x_{i+h}, z_i and z_{i+h} are the measured values for the regionalized variable at locations x_i or x_{i+h}, and $n(h)$ is the number of pairs at any separation distance h.

The semivariogram is theoretically related to the autocorrelation function [r(h)] by the expression (Burgess and Webster, 1980a; Vauclin et al., 1982):

$$\gamma(h) = s^2 \left[1 - \rho(h) \right] \tag{9.6}$$

where s^2 is the population variance. This relationship applies best to data which have no trend, meaning that the population mean and variance are constant over the sampled area (Rossi et al., 1992). Based on this relationship, the semivariogram should theoretically be equal to the population variance when the separation distance is very large. At very small separation distances, the semivariogram approaches a value of zero. Thus, the semivariogram is a quantitative measure of how the variance between sampled points is reduced as separation distance decreases.

Often, the separation distances are normalized by dividing separation distance by the smallest separation distance between locations in the sampling design. The resulting normalized separation distance is termed a lag or a lag separation distance (h) in geostatistical jargon. Lags are integers taking on values 1, 2, 3, ... etc., whereas separation distances (for example, 30, 60, 90, ..., etc.) are the actual averaged distances between pairs.

If pairwise squared differences are averaged without regard to direction, the semivariogram is considered to be isotropic and omnidirectional. When pairs of a given separation distance are considered as a function of direction, the semivariogram is considered to be directional. There are two types of anisotropy (Webster and Oliver, 1990): zonal where the sill varies with direction, and geometric where the range varies with direction. Anisotropic forms of the semivariogram can also be estimated using explicitly two-dimensional equations (Mulla, 1988a; Webster and Oliver, 1990), rather than simply computing a collection of one-dimensional semivariograms in various directions, and modeling their direction dependence (Nash et al., 1988).

9.3.1.1 Binning

A key consideration in the estimation of semivariograms is deciding how to group pairs when calculating the squared differences at varying separation distances. In geostatistical jargon, this is a decision about bin distance, or simply binning procedures. The term bin refers to a group of pairs having a narrow range of separation distances. For example, consider a situation where samples are collected on a regular grid at spacings of 30 x 50 m, and an omnidirectional semivariogram is to be estimated. The separation distances between pairs of points are 30, 50, 58.3, 60, 78.1, 90, 100,..., etc.

It is inefficient to estimate the semivariogram at each of these separation distances because there is likely to be little gain in useful information by having separate values for the semivariogram at closely spaced separation distances (e.g., 58.3 and 60). Therefore, one can average values for the semi-variogram in bins which represent a narrow range of separation distances, i.e., 55–60 inclusive.

To illustrate binning procedures, various ways of grouping pairs by separation distance from the previous example will be investigated. As a first step, one might estimate the semivariogram using the bins 48.7, 71.7, 104.4, ..., etc. Note that this involves averaging the data from pairs at separation distances of 30, 50, and 58.3 m, and averaging data from pairs separated by 60, 78.1, and 90 m. Much of the information about variability at short separation distances is lost using this approach. A second approach could involve bins with pairs separated by an average distance of 30, 56.6, 82.1, 108.3, ..., etc. Clearly, there is more information about short-range variability in this approach, but still considerable averaging at larger separation distances. A third approach which produces even less averaging of pairs with dissimilar separation distances is to use bins of 30, 55.4, 60, 78.1, 90, 102.9, ..., etc. Clearly this approach provides more detail about spatial variability at separation distances representative of the actual sampling pattern but fewer data points are available in each bin.

The previous example on binning of pairs shows that the best choice of bins on a regular grid is one that best represents the actual separation distances between sampled points with a minimum of averaging, especially at short separation distances. The only reason to use an approach that involves more spatial averaging is when there is an insufficient number of pairs (e.g., less than 30 to 50 pairs) at a given separation distance. In this case, binning provides a way to increase the number of sample pairs upon which a semivariogram is estimated.

Some general guidelines for binning procedures are as follows. For omnidirectional semivariograms, use bins that include the closest separation distances in the sample design so that the maximum amount of information is obtained about the semivariogram at small separation distances. Avoid making bins so narrow that the number of pairs in any bin is less than 30. For directional semivariograms, it is necessary to have bins for both separation distance and direction (Webster and Oliver, 1990). As an example, if semivariograms are to be estimated in the four principal directions, the directional bins could be $0° \pm 22.5°$, $90° \pm 22.5°$, $180° \pm 22.5°$, and $270° \pm 22.5°$. Narrower directional bins could also be selected, but wider ones should probably be avoided, since they would lead to a loss of information about directional effects.

9.3.1.2 Modeling the Semivariogram

The semivariogram can be modeled using any of several authorized models (Journel and Huijbregts, 1978; Oliver, 1987; Isaaks and Srivastava, 1989) that are commonly fitted to the semivariogram data using nonlinear least squares with either uniform weighting of points or weighting that favors data at small over large separation distances (Cressie, 1985). The most common models are linear, spherical, and exponential models. The linear model (Fig. 9.3A) is given by:

$$\gamma(h) = C_0 + mh \qquad h < a \qquad\qquad\qquad\qquad\qquad\qquad\text{[9.7a]}$$

and

$$\gamma(h) = C_0 + C_1 \qquad for\ h \geq a \qquad\qquad\qquad\qquad\qquad\qquad\text{[9.7b]}$$

The spherical model (Fig. 9.3B) is given by:

$$\gamma(h) = C_0 + 0.5C_1 \left[\left(\frac{3h}{a} \right) - \left(\frac{h}{a} \right)^3 \right] \quad for\ h < a \tag{9.8a}$$

and

$$\gamma(h) = C_0 + C_1 \quad for\ h \geq a \tag{9.8b}$$

The exponential model (Fig. 9.3C) is given by:

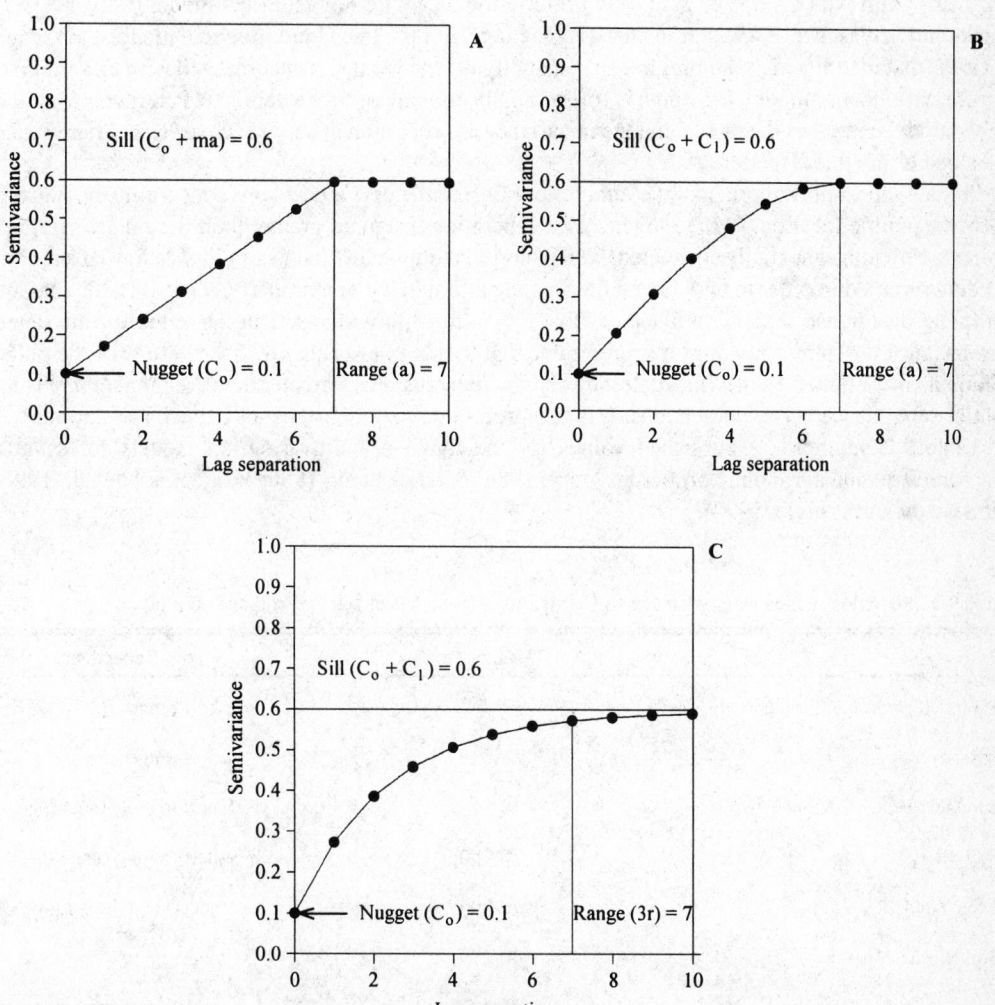

Fig. 9.3 (A) Linear semivariogram model with a range of 7, a sill of 0.6, and a nugget of 0.1; (B) Spherical semivariogram model with a range of 7, a sill of 0.6, and a nugget of 0.1; (C) Exponential semivariogram model with a range of 7, a sill of 0.6, and a nugget of 0.1

$$\gamma(h) = C_0 + C_1\left[1 - \exp\left(\frac{-h}{r}\right)\right]$$ [9.9]

The semivariogram model and its parameters provide a quantitative expression of spatial structure for the measured property. In these models, the fitted parameters have the following definitions (Journel and Huijbregts, 1978; Oliver, 1987; Oliver and Webster, 1991). The nugget parameter (C_0) is a measure of the amount of variance due to errors in sampling, measurement, and other unexplained sources of variance. The sum of parameters C_0 and C_1 is the sill, theoretically equal to the variance of the sampled population at large separation distances if the data have no trend. If the nugget parameter is about equal to the sample variance, it is an indication that the sampled property has very little spatial structure or varies randomly. In this case, the semivariogram would be best described using a linear semivariogram with a slope of zero, often referred to as a pure nugget effect model (Oliver, 1987; Isaaks and Srivastava, 1989). Parameter a is the range. For the linear and spherical models, the range (a) is the distance at which samples become spatially independent and uncorrelated with one another. For the exponential model, the range is operationally equivalent to the value $3r$. Parameter m of the linear model expresses the rate of change in variance as separation distance increases, in other words, the slope of the linear model.

Probably the most important of the semivariogram parameters for decisions concerning the spacing between sample locations is the range (a). At separation distances greater than the range, sampled points are no longer spatially correlated, which has great implications for sampling design. If a region is being sampled in order to understand the spatial pattern of a given property, it is advisable that the sampling design use separation distances that are, at most, no greater than the value for the range parameter of the semivariogram. It is preferable that the sample spacing be from ¼ to ½ of the range (Flatman and Yfantis, 1984). If samples are separated by distances greater than the range, there is no spatial dependence between locations. It is inappropriate to use geostatistics in such a situation.

Table 9.2 summarizes published values for the range of semivariogram models for several measured soil and agronomic properties (Jury, 1986; Warrick et al., 1986; Wollenhaupt et al., 1997; McBratney and Pringle, 1997).

Table 9.2   Variation in the range parameter for semivariogram models of selected soil and crop properties

Property	Range (m)	Spatial Dependence
Saturated Hydraulic Conductivity	1-34	short range
%Sand	5-40	short range
Saturated Water Content	14-76	short to moderate range
Soil pH	20-260	short to long range
Crop Yield	70-700	moderate to long range
Soil Nitrate-N	40-275	moderate to long range
Soil Available Potassium	75-428	moderate to long range
Soil Available Phosphorus	68-260	moderate to long range
Organic Matter Content	112-250	long range

When fitting data to semivariogram models, there are a few important guidelines to follow. First, the number of sample data values used in estimating the semivariogram must not be < 30 (Journel and Huijbregts, 1978), while some authors recommend at least 100 to 200 sampling locations for accurate estimation of the semivariogram (Oliver and Webster, 1991). For regularly spaced sample designs, the maximum number of pairs in the semivariogram calculation at the smallest separation distance is always equal to N-1, where N is the number of sample data points. Thus, it follows that the number of sample pairs used in the estimation of the semivariogram at any separation distance should not be < 30. Second, the semivariogram model should not be fitted to semivariogram data at separation distances greater than ½ of the largest dimension of the study area (Burrough, 1991). This is because the pairs of measured data at large separation distances are representative of the variance structure at the edges of the field, not the majority of the samples.

9.3.1.3 Influential Observations

When there are influential observations in the measured data, estimating the semivariogram can be problematic leading to significant increases in the nugget and sill values for the semivariogram. In the worst case, a few outliers can obscure all spatial structure in the semivariogram and make it appear as if the measured data are spatially independent (Rossi et al., 1992).

The following example shows the effect of a few outliers on semivariogram modeling. A study site in Minnesota was sampled at 234 locations and analyzed for organic C content. Spatial structure from the resulting data set obeyed a spherical semivariogram (Equation [9.8] and Fig. 9.4A) with a nugget of ~ 0.12, a range of ~ 260 m, and a sill of ~ 1.4. For illustration, 5 outliers were inserted into the data set and the semivariogram was recomputed. The resulting semivariogram was significantly affected by the 5 outliers (Fig. 9.4B), with the main effects being an increase in the nugget to ~ 0.7, and a decrease in the range to ~ 210 m. The overall effect of the outliers was to increase the perceived short-range variability, and to decrease the range of spatial dependence.

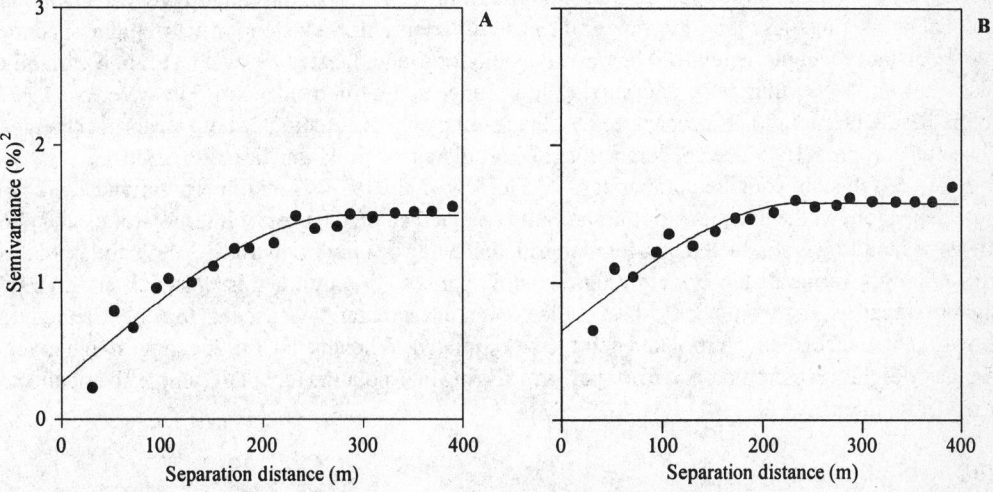

Fig. 9.4 (A) Spherical semivariogram model for measured soil surface organic carbon contents displayed in Fig. 9.2A; (B) spherical semivariogram model for measured soil surface organic carbon contents from Fig. 9.2A with 5 added influential observations

Influential observations can be removed using several methods. One of the best is to remove influential observations when computing the semivariogram using the method of Hawkins (1980). Another good approach is to estimate the semivariogram using a robust resistant rather than the classical semivariogram estimator (Cressie and Hawkins, 1980; Birrell et al., 1996).

9.3.2 Interpolation by Kriging

Kriging is a general term describing a geostatistical approach for interpolation at unsampled locations. There are several types of univariate kriging methods including punctual (Burgess and Webster, 1980a), indicator (Journel, 1986), disjunctive (Yates et al., 1986a,b), universal (Webster and Burgess, 1980), and block kriging (Burgess and Webster, 1980b). A multivariate form of kriging is known as cokriging (Vauclin et al., 1983). Conditional simulation has also been applied to kriging (Hoeksma and Kitanidis, 1985; Warrick et al., 1986; Gutjahr, 1991). The methods used for punctual kriging, cokriging, block kriging, and indicator kriging will be described in this chapter.

The science of kriging was developed for the mining industry by Matheron (1963), and first applied in the mining industry by Krige (1966) and applied to soil science by Burgess and Webster (1980a,b) and Webster and Burgess (1980) in Europe and in the United States by Vauclin et al. (1982, 1983) and Vieira et al. (1983).

No other linear interpolation method provides less bias in predictions than kriging which is known as a best linear unbiased predictor (BLUP) (Burgess and Webster, 1980a). This is because the interpolated or kriged values are computed from equations that minimize the variance of the estimated value. In fact, when interpolating at a location where a measurement exists, kriging will always generate a value equal to the measured value. For this reason, kriging is often loosely described as being a type of linear regression in which the regression line always passes through every one of the measured data points.

Several studies have compared kriging and classical methods for interpolation such as inverse distance weighting and cubic splines (Dubrule, 1984; Laslett et al., 1987). As a general rule of thumb, kriging methods are equivalent or superior to classical methods when the data to be interpolated have well-developed spatial structure, have a semivariogram without a significant nugget effect, and are sampled at spacings less than the range of the semivariogram in clusters or at irregular spacings. Inverse distance weighting tends to be more suitable for use with data having short-range variability (Cooke et al., 1993) than with data having long-range spatial dependence (Gotway et al., 1996). Spline-based interpolation is perhaps better than kriging only in situations where an abrupt change in measured soil property values occurs across a short distance (Voltz and Webster, 1990).

A key conclusion from the comparisons by Gotway et al. (1996) is that interpolation accuracy is more dependent on the adequacy of the sampling design than on the type of interpolator used. When intensive sampling is conducted on a regular grid, there may be only small differences in interpolation with kriging, inverse distance weighting, or cubic splines. In view of this, Warrick et al. (1988) suggested that kriging would be the best choice for an interpolator, because it is the only method that allows the variance of an interpolated point to be estimated. Whelan et al. (1996) suggest that inverse distance weighting interpolators will outperform kriging interpolators for small sample sizes collected at moderate intensity.

9.3.2.1 Stationarity

The kriging interpolator uses information from the semivariogram model about spatial structure of the measured property. In order to use kriging methods, one very important condition must be met which is the intrinsic hypothesis (Warrick et al., 1986). It states that the mean of the measured property is stationary and the semivariogram at any separation distance is finite and they do not depend on location. This means that the expectation of the squared differences between values depends only on separation distance and not on location. If these conditions are not met, the measured property is said to be nonstationary, and kriging is not appropriate without removing the causes of nonstationarity in the data.

Nonstationarity can result from three types of problems. The first is a long-range systematic change in the mean with location, often referred to as trend. This type of effect can be removed by fitting a regression model to the data and removing the long-range trend. Alternatively, mean or median polish can be used to remove trend (Cressie, 1991). The latter method works best when the trend is aligned with the rows or columns of the sample design. The second is a short-range stochastic change in the mean with location, often referred to as drift. This is much more problematic, and the primary method for removing drift is to use universal kriging. The third type of nonstationarity involves a change in variance with location. Universal kriging can also be used to remove this effect.

9.3.2.2 Punctual Kriging

Punctual kriging is essentially a linear interpolator, which sums up weighted values for measurements at locations neighboring the unsampled location. The expression used in kriging is

$$z_0 = \sum_{i=1}^{N} \lambda_i z_i \qquad\qquad [9.10]$$

where z_0 is the interpolated value, N is the number of points neighboring the interpolated location, z_i are the measured values at neighboring locations, and λ_i is the kriging weighting factor for each of the neighboring measured values. The weighting factors must sum to unity so that the expected value of the interpolated points minus the measured points is zero. To further illustrate this concept, if interpolation occurs at a location where a measured value exists, the weighting factor is unity for the measured point at the location to be interpolated, and the weighting factors for all neighboring locations are zero.

9.3.2.3 Neighborhood Search Strategy

Identifying the N locations that neighbor the unsampled location is an important step in the kriging process. There are a few important guidelines concerning the search for neighboring locations around an unsampled location. First, for omnidirectional kriging, the search radius should be set so that at least 6–8 locations, but not more than 16 to 24 locations, are identified as neighboring locations (Wollenhaupt et al., 1997). With fewer than 6-8 locations, the matrix solver may not be able to converge to a solution. The search radius should not exceed the range of the semivariogram, and is typically less than ½ the range. For systematic grid sample designs, the search radius typically describes a circle which extends to and includes the closest 4 neighbors as well as the next 4 neighbors along a diagonal from the unsampled location. If the measured data are not regularly spaced, or are clustered, it may be difficult to use a fixed search radius. Some unsampled locations may have too few nearest neighbors in order to include at least 8 neighbors. In this case, it is best to use a flexible search radius so that each unsampled location has approximately the same number of neighbors. Second, the neighbors should be evenly distributed in all directions outward from the unsampled location, if possible. This avoids low weighting factors that result from data redundancies. A quadrant or octant search strategy can be used to find neighbors that are located in the 4 or 8 principal directions away from the unsampled location (Webster and Oliver, 1990). Third, using a large search radius may give rise to a large number of neighbors and many small or negative weights. In extreme cases, this can lead to a negative value at the unsampled location which can be avoided by placing an upper limit on either the search radius or the maximum number of neighbors.

For randomly oriented or clustered sample designs, the search radius should be set to a value slightly larger than the average spacing between data points (Wollenhaupt et al., 1997). If anisotropic kriging is to be used, the search neighborhood is usually elliptical in shape. The longest axis of the search ellipse corresponds to the direction in which the range of the semivariogram is greatest, while the shortest axis corresponds to the direction in which the range is smallest.

9.3.2.4 Estimation Variance

The kriging technique has one major advantage over all other interpolation methods. Only with kriging is it possible to estimate the variance of the interpolated values. The estimation variance depends upon only two factors. The first is the spatial structure for the measured property as shown in the semivariogram, and the second is the weighting factors for neighbors of the interpolated location. These weighting factors depend, in turn, upon the arrangement of sampled locations around the unsampled location. The estimation variance is calculated using the expression:

$$\sigma^2(z_o) = \mu + \sum_{i=1}^{N} \lambda_i \gamma_{io} \qquad [9.11]$$

where σ^2 is the estimation variance, μ is the Lagrangian multiplier, and γ_{io} is the semivariance at the separation distance between the locations for measured (z_i) and interpolated values (z_o).

If the sample population can be assumed to follow a normal distribution, the kriging estimation variance can be used to estimate the confidence interval for prediction at the unsampled location (McBratney and Pringle, 1997). The 95% confidence interval about the mean is estimated from the expression given by Birrell et al. (1996):

$$\left(\bar{z} \pm 1.96 \frac{\sigma_o}{\sqrt{N}} \right) \qquad [9.12]$$

where σ_o is the maximum estimation variance obtained at all unsampled locations.

As shown by Burrough (1991), the estimation variance is affected primarily by two factors. The first of these is the form and shape of the semivariogram. As the nugget effect increases, so, too, does the estimation variance. In many cases, the nugget effect can be minimized by sampling at close spacings and by compositing or bulking samples. The optimum number of bulked samples depends upon the CV for the measured property and the shape of its semivariogram. A cost-effective and optimum number of composite samples may be anywhere from 4 to 16 samples (Webster and Burgess, 1984; Burrough, 1991; Oliver et al., 1997). The second factor is the number and spatial arrangement of neighboring locations. As separation distance increases and the number of neighboring locations decreases, the estimation variance at an unsampled location increases. Generally, the arrangement of sampled neighbors which minimizes estimation variance is along a regular square grid or an equilateral triangle (Webster and Burgess, 1984; Burrough, 1991).

9.3.2.5 Solving for Weighting Factors

The key step in kriging is to estimate the n weighting factors for locations that neighbor the unsampled location where interpolation is to occur. This is accomplished by solving a set of $N+1$ simultaneous equations with $N+1$ unknowns (the N weighting factors and the undetermined Lagrangian multiplier). These equations can be written in matrix notation using:

$$[F][L] = [B] \tag{9.13}$$

where $[F]$ is an $N+1$ by $N+1$ matrix of semivariogram values between measured locations, $[L]$ is an $N+1$ by 1 matrix of weighting factors and the Lagrangian multiplier, and $[B]$ is an $N+1$ by 1 matrix of semivariogram values between the interpolated location and its neighboring locations. The full mathematical details for each matrix are given below:

$$[F] = \begin{bmatrix} 0 & \gamma_{12} & \gamma_{13} & \cdots & \gamma_{1n} & 1 \\ \gamma_{21} & 0 & \gamma_{23} & \cdots & \gamma_{2n} & 1 \\ \vdots & & & & \vdots & 1 \\ \gamma_{n1} & \gamma_{n2} & \gamma_{n3} & \cdots & 0 & 1 \\ 1 & 1 & 1 & \cdots & 1 & 0 \end{bmatrix} \tag{9.14}$$

$$[L] = \begin{bmatrix} \lambda_1 \\ \lambda_2 \\ \vdots \\ \lambda_n \\ \mu \end{bmatrix} \tag{9.15}$$

$$[B] = \begin{bmatrix} \gamma_{o1} \\ \gamma_{o2} \\ \vdots \\ \gamma_{on} \\ 1 \end{bmatrix} \tag{9.16}$$

9.3.3 Example: Spatial Variability of Alachlor Sorption

Soil surface samples were collected from a field in south central Minnesota at 35 locations (Fig. 9.5). The Freundlich adsorption partition coefficients (K_f) were obtained from experiments conducted in the laboratory on each sample. A subset of these K_f values (shown with circles in Fig. 9.5) are given in Table 9.3 to illustrate the estimation of the semivariogram, and for an example of interpolation by kriging. The interpolated value is to be estimated at location o, which has the x,y coordinates 150,550.

9.3.3.1 Semivariogram Example

The semivariogram for this subset of alachlor adsorption data is estimated from Equation [9.5] as follows. First, one estimates the value of the semivariogram at the closest separation distance (also known as the first lag distance) which is 30.5 m. As is evident from Table 9.3 and Fig. 9.5 for this subset of the data, there are four pairs of data separated by 30.5 m.

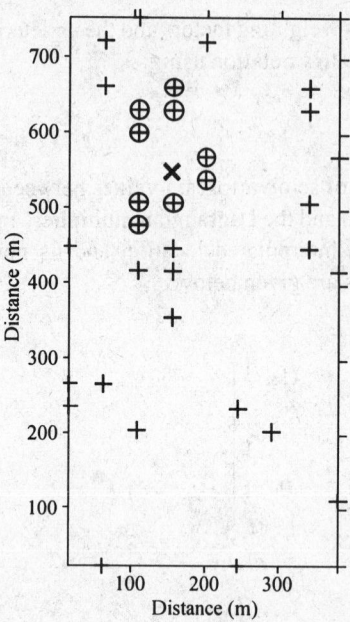

Fig. 9.5 Sample locations of alachlor Freundlich adsorption partition coefficient measurements (+) from a field in southern Minnesota. Sample locations denoted with circles are further described in Table 9.3. Location denoted with the symbol X is an unsampled location.

$$\gamma(h = 30.5) = 0.5\left(\frac{1}{4}\right)\left[(7.6 - 6.4)^2 + (5.9 - 5.3)^2 + (5.5 - 5.9)^2 + (6.8 - 9.9)^2\right]$$

$$= \left(\frac{1}{8}\right)(11.57) = 1.45 \tag{9.17}$$

The semivariogram for the second lag is estimated from all pairs of measurements separated by either 54.9 m or 45.7 m. These two distances correspond to the distance between sample locations 1 and 5 or 2 and 5, respectively. There are a total of 6 pairs of data with these separation distances, for an average separation distance of 51.9 m. The semivariogram at the second lag ($h = 51.9$ m) is given by:

$$\gamma(h = 51.9) = 0.5\left(\frac{1}{6}\right)\left[(12.2 - 6.4)^2 + (12.2 - 7.6)^2 + (5.9 - 5.3)^2 + (5.9 - 5.9)^2\right] +$$

$$(5.5 - 5.9)^2 + (9.9 - 12.2)^2\right] = \left(\frac{1}{12}\right)(60.61) = 5.05 \tag{9.18}$$

The semivariogram for the full set of alachlor K_f values is shown in Fig. 9.6 and was best described using a linear model (Equation [9.7]). Notice that the semivariogram values for the first and second lag distances in the full model are similar to those estimated using the partial dataset.

Table 9.3 Location and value for alachlor Freundlich adsorption partition coefficients (K_f) from a field in southern Minnesota

Location No.	X Distance (m)	Y Distance (m)	K_f Value (mL g^{-1})
1	106.7	478.5	6.4
2	106.7	509.0	7.6
3	106.7	600.5	5.3
4	106.7	630.9	5.9
5	152.4	509.0	12.2
6	152.4	630.9	5.9
7	152.4	661.4	5.5
8	198.1	539.5	9.9
9	198.1	570.0	6.8
0	150.0	550.0	unsampled

9.3.3.2 Kriging Example

For the interpolation of measured K_f values at the unsampled location (150, 550), all measured values within a search radius of 100 m were considered. From Fig. 9.5, it is evident that there are 8 (not 9) neighbors within a distance of 100 m from the unsampled location. Using the linear semivariogram model shown in Fig. 9.6, the semivariance values for the [F] and [B] matrices of Equations [9.14] and [9.16] were estimated. These values are shown in Equation [9.19] in the first and last matrix. The lower left portion of the [F] matrix is left blank for convenience; but since the matrix is symmetrical, the values omitted are the same as those across the diagonal at an equivalent position. After inversion of the [F] matrix, values were determined for the weighting factors and Lagrangian multiplier. These solutions are shown in the second matrix (the [L] matrix).

$$
\begin{bmatrix}
0 & 1.6 & 6.4 & 8.0 & 2.9 & 8.3 & 5.7 & 6.8 & 1 \\
 & 0 & 4.8 & 6.4 & 2.4 & 6.8 & 5.0 & 5.7 & 1 \\
 & & 0 & 1.6 & 5.3 & 2.9 & 5.7 & 5.0 & 1 \\
 & & & 0 & 6.8 & 2.4 & 6.8 & 5.7 & 1 \\
 & & & & 0 & 6.4 & 2.9 & 4.0 & 1 \\
 & & & & & 0 & 5.3 & 4.0 & 1 \\
 & & & & & & 0 & 1.6 & 1 \\
 & & & & & & & 0 & 1 \\
1 & 1 & 1 & 1 & 1 & 1 & 1 & 1 & 0
\end{bmatrix}
\times
\begin{bmatrix}
-0.08 \\
0.19 \\
0.22 \\
-0.04 \\
0.33 \\
0.06 \\
0.14 \\
0.18 \\
-0.49
\end{bmatrix}
=
\begin{bmatrix}
4.37 \\
3.12 \\
3.48 \\
4.80 \\
2.15 \\
4.23 \\
2.57 \\
2.72 \\
1
\end{bmatrix}
\qquad [9.19]
$$

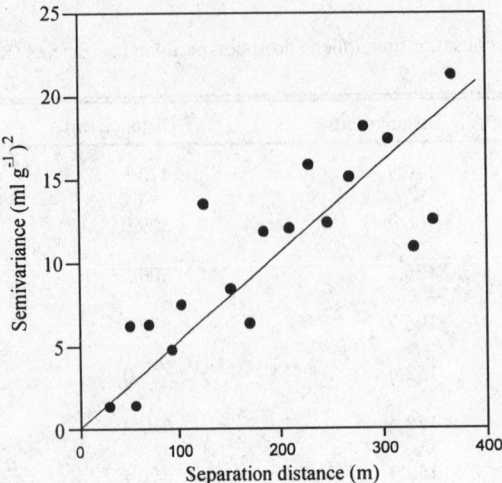

Fig. 9.6 Linear semivariogram model for alachlor Freundlich adsorption partition coefficients from a field in southern Minnesota

Note that the sum of the weighting factors in the [L] matrix is one, as required. Also note that the heaviest weight (0.33) is for location 5, with moderately large weights for locations 2 (0.19), 3 (0.22), 8 (0.14), and 9 (0.18). This makes sense, because location 5 is closest to the unsampled location.

Note that in spite of being farther from the unsampled location, the weights for locations 2 and 3 are each larger than the weights for locations 8 and 9. This is because locations 8 and 9 are very close to one another, and so their values are somewhat redundant. Rather than give each location a large weighting factor, the kriging approach accounts for the redundancy in closely spaced values. The combined weights for locations 8 and 9 is 0.32, about the same as the weighting factor for location 5. Because measured values at locations 2 and 3 are in different directions from the unsampled location, there is little redundancy in the measured values, and the weighting factors are not corrected for redundancy.

Note also that neighboring locations 1, 4, and 6 have very small or even negative weighting factors for two reasons. The first is that the latter neighboring locations are farther from the unsampled location than the other neighbors and so their spatial correlation with the unsampled location is smaller. Second, the low weighting results from a screening effect. The path from each of the low weighted neighbors to the unsampled location is nearly blocked by a closer neighbor. This makes the information contained in the screened out neighbors redundant, so the weighting factors are adjusted downward.

Now that the weighting factors have been obtained, one can use them to estimate the alachlor K_f value at the unsampled location. The interpolated value (8.85 mL g^{-1}) is obtained using Equation [9.10] as follows:

$$8.85 = (-0.08)6.4 + (0.19)7.6 + (0.22)5.3 + (-0.04)5.9 +$$
$$(0.33)12.2 + (0.06)5.9 + (0.14)9.9 + (0.18)6.8$$

[9.20]

Note that the interpolated value (8.85) is higher than the average value of 7.5 for the 8 neighboring locations. In averaging, each of the measured values is assigned an equal weighting factor, which is 1/8 in this example. In kriging, the weighting factors depend upon the distance between the measured value and the unsampled location, as well as upon the arrangement of the sampled location relative to all other sampled locations.

The weighting factors can also be used to obtain the estimation variance for the unsampled location. The estimation variance [2.12 (mL g^{-1})^2] is obtained as follows:

$$2.12 = (-0.49) + (-0.08)4.4 + (0.19)3.1 + (0.22)3.5 + (-0.04)4.8 +$$
$$(0.33)2.1 + (0.06)4.2 + (0.14)2.6 + (0.18)2.7$$

[9.21]

Of course, this is a very simplified example of estimating the semivariogram and using kriging for interpolation. In reality, a computer program would use the procedures described above at hundreds of unsampled locations, rather than just one.

A primary objective of kriging is to produce a set of interpolated values from which spatial patterns of the measured property can be mapped. To illustrate this, the alachlor K_f values were interpolated throughout the Minnesota study site using all 35 measured values and kriging on a regular grid at spacings of 25 x 25 m. The resulting map for K_f values (Fig. 9.7A) shows a region of relatively large sorption centered near location (250,200). In addition, a region with relatively weak sorption is centered near location (200,650). This map, produced using 35 measured K_f values shows the usefulness of information produced by kriging. One could use the map as an aid to precision application of the alachlor herbicide by applying lower rates of herbicide in the regions where sorption is relatively weak, and higher rates in the regions where sorption is relatively strong. Or one could use the maps to estimate spatial patterns in leaching and runoff losses of the herbicide using a fate and transport model.

The estimation variance is also useful in this example, because it shows the uncertainty in interpolated values at various locations for the study site. In this case, there are a few regions (175, 125) and (400,300) which are undersampled as shown by larger estimation variances (Fig. 9.7B). The uncertainty of interpolated values in regions with larger estimation variances is greater than that in regions with smaller estimation variances. Additional samples could be collected in regions with large estimation variances to improve the accuracy of prediction.

9.3.4 Block Kriging

As the previous section shows, punctual kriging is a method for estimating the value of a property at an unsampled point location. There are times when it is more appropriate to estimate the average value of a property in a block or small region around an unsampled point location. For example, data from the Landsat Thematic Mapper satellite are typically available at a spatial resolution of 30 m. If data are missing from some sections of the image due to interference from clouds, it would be useful to be able to estimate the average value for the missing data in the unsampled blocks. Alternatively, one may wish to estimate the average value of a soil property for mapping units of the soil survey, or the effective permeability in a particular stratum of the subsurface geology (Journel et al., 1986). The geostatistical approach for estimating the average value of a property in an unsampled block is called block kriging (Burgess and Webster, 1980b).

Block kriging is analogous to punctual kriging in many ways. The average value of a block, $z_o(B)$ is estimated from a linear weighting of measured values inside and outside of the block using the expression (Webster and Oliver, 1991):

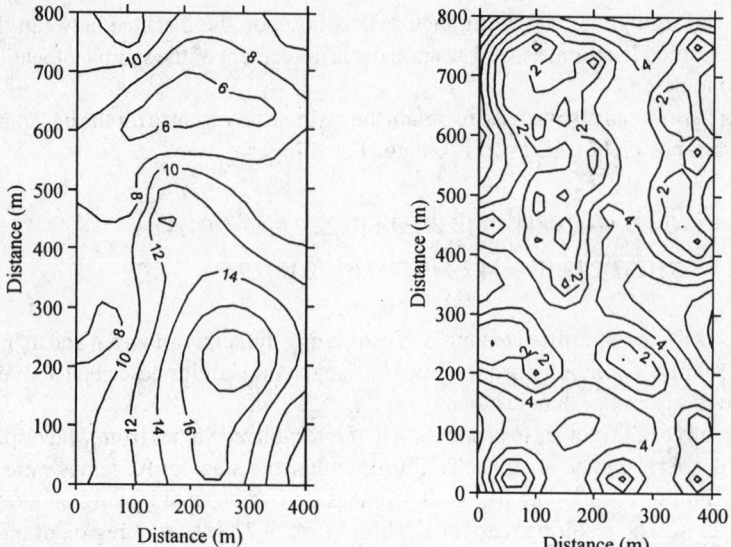

Fig. 9.7 (A) Kriged alachlor Freundlich adsorption partition coefficients for a field in southern Minnesota; (B) estimation variances from kriging of alachlor Freundlich adsorption partition coefficients for a field in southern Minnesota

$$z_o(B) = \sum_{i=1}^{N} \lambda_i z_i \qquad\qquad [9.22]$$

where the weighting factors sum to unity. The weighting factors are estimated by solving an equation similar to Equation [9.13], except that the semivariograms in this equation are not estimated between points, but between points and blocks, or within a block.

The average semivariogram between a sampled point and the block is estimated using:

$$\overline{\gamma_{iB}} = \frac{1}{B} \int_B \gamma(x_i, x)dx \qquad\qquad [9.23]$$

where B is the area of the block, x_i is the measured point and x is a point inside the block. Similarly, the average semivariogram within the block is estimated using:

$$\overline{\gamma_{BB}} = \frac{1}{B^2} \int_B \int_B \gamma(x, x')dxdx' \qquad\qquad [9.24]$$

where both x and x' are points within the block. The integrations required to estimate semivariograms for block kriging can be obtained using analytical functions (Clark, 1979; Webster and Burgess, 1984) based on the type of semivariogram and the geometry of the block, or by direct numerical integration.

Finally, the estimation variance of a block $\sigma^2(B)$ is obtained using the expression:

$$\sigma^2(B) = \sum_{i=1}^{N} \lambda_i \overline{\gamma_{iB}} + \mu - \overline{\gamma_{BB}} \qquad\qquad [9.25]$$

where μ is the Lagrangian multiplier. One of the advantages of block kriging is that the estimation variances are typically much smaller than those obtained from punctual kriging.

For normally distributed data, there is no need to estimate block-averaged semivariograms using analytical functions in order to use block kriging. A simpler approach is to use punctual kriging with a fine resolution grid spacing, and then to average the interpolated values within blocks of a specified size and geometry. The average value of a block is equivalent to the value obtained by averaging all of the point kriged values generated on a dense grid within the block (Isaaks and Srivastava, 1989).

9.3.5   Cokriging

Cokriging is an interpolation technique that uses information about the spatial patterns of two different, but spatially correlated properties to interpolate only one of the properties (Vauclin et al., 1983). Typically, cokriging is used to map the property that is more difficult or expensive to measure (z_2) based on its spatial dependence on a property that is easier or less expensive to measure (z_1). An example of this is the interpolation by cokriging of sparsely sampled soil test P levels using intensively sampled soil organic C content values (Bhatti et al., 1991; Mulla, 1997). Another example might be the interpolation by cokriging of sparsely sampled soil moisture content levels (Yates and Warrick, 1987; Mulla, 1988b; Stein et al., 1988).

Cokriging requires that the semivariogram models [$\gamma_1(h)$, $\gamma_2(h)$] be estimated for both of the measured properties. In addition, cokriging requires estimation of the cross-semivariogram model [$\gamma_{12}(h)$] describing spatial dependence between the two measured properties. This cannot be done unless measurements of the more intensively sampled property are available at each of the locations where the sparsely sampled property is measured.

The cross-semivariogram is estimated using the following expression:

$$\gamma_{12}(h) = \frac{1}{2n(h)} \sum_{i=1}^{n(h)} \left[z_{1i} - z_{1,i+h} \right] \left[z_{2i} - z_{2,i+h} \right]$$

[9.26]

where z_{1i} and $z_{1,i+h}$ are the measured values for property 1 at locations i and $i+h$, respectively, and z_{2i} and $z_{2,i+h}$ are the measured values for property 2 at locations i and $i+h$, respectively. Zhang et al. (1992) have proposed a method for estimating pseudo cross-semivariograms using data that are sampled at nearly the same, but not identical, locations.

Interpolation with cokriging involves an approach that is similar to that for kriging, with terms for weighting factors of both measured properties. Interpolated values for the second property (z_{20}, the sparsely measured property) are obtained using the expression:

$$z_{20} = \sum_{i=1}^{N_1} \lambda_{1i} z_{1i} + \sum_{i=1}^{N_1} \lambda_{2i} z_{2i}$$

[9.27]

where λ_{1i} and λ_{2j} are the weighting factors for property 1 at location i and property 2 at location j, respectively, and z_{1i} and z_{2j} are the measured values for property 1 at location i and property 2 at location j, respectively. Thus, the interpolated value is simply a linear combination of the measured values for both properties at locations that neighbor the unsampled location.

As in kriging, the cokriging prediction is a best linear unbiased predictor. This is ensured by requiring that the weighting factors for property 1 sum to zero, and the weighting factors for property 2 sum to unity.

The cokriging technique provides an estimate for the variance of the cokriging prediction at all unsampled locations. At locations where a measured value exists for the cokriged property, the estimation variance is zero, and the interpolated value equals the measured value. The expression for the cokriging estimation variance (σ_o^2) is given by:

$$\sigma_o^2 = \mu_2 + \sum_{i=1}^{N_1} \lambda_{1i} \gamma_{12,io} + \sum_{j=1}^{N_2} \lambda_{2j} \gamma_{22,jo} \qquad [9.28]$$

where μ_2 is the Lagrangian multiplier for property 2, λ_{1i} is the weighting factor for the ith location of property 1, λ_{2j} is the weighting factor for the jth location of property 2, $\gamma_{12,io}$ is the cross-semivariance at a separation distance between the ith location and the unsampled location, and $\gamma_{22,jo}$ is the semivariance for property 2 at a separation distance between the jth location and the unsampled location.

9.3.6 Indicator Kriging

Punctual kriging, block kriging, and cokriging are all useful for estimating the value of a property at an unsampled location. The arithmetic average of all interpolated values is an estimate of the mean. Often times, rather than estimating the mean, it is necessary to estimate the proportion of values at a site which is above or below some critical value. For example, the extent of cleanup and remediation required at a site contaminated with trace metals depends upon the proportion of the site where the soil concentrations of trace metals exceed statutory limits. The kriging methods discussed up to this point are unsuitable for making such estimates.

Indicator kriging is an approach for estimating the proportion of values that fall within specified class intervals or the proportion of values that are below a threshold level (Journel, 1986; Isaaks and Srivastava, 1989). The basic approaches in indicator kriging are (1) to transform measured values into indicator variables, (2) to estimate the semivariogram for the indicator transformed values, (3) to use simple kriging to interpolate the indicator transformed values across the study site, and (4) to compute the average of the kriged indicator values. This average is the proportion of values that is less than the threshold.

The indicator transform (I_i) is simply defined as follows (Rossi et al., 1992):

$$I_i = \begin{array}{lll} 1 & if & z_i \leq k \\ 0 & if & z_i > k \end{array} \qquad [9.29]$$

where k is the specified threshold cutoff value. Indicator values can be averaged over the site using a series of different threshold cutoff values if there is some uncertainty about the most appropriate value to use for the threshold. For each threshold value specified, a new semivariogram model should be estimated. However, it has often been found that the semivariogram estimated using the population median as a threshold is similar to the semivariograms estimated using other threshold values (Isaaks and Srivastava, 1989).

9.4 Sampling Design

The variability and spatial structure of a sample population are of great importance in developing a valid soil sampling design. Soil sampling design is concerned with developing a statistically rigorous and unbiased strategy for collecting soil samples (Brown, 1993; Carter, 1993). The main considerations in sampling design include determining the optimal number of samples to collect, and determining the spatial arrangement of the samples to be collected (Wollenhaupt et al., 1997). A rigorous sampling design ensures that sampling points are representative of the region studied (ASTM, 1997a), meaning that the mean of the sampled points is a very good estimate for the mean of the population. It also minimizes the bias or systematic error caused by human judgement. Sample designs may also be developed to provide quality control and quality assurance, involving replicate measurements and split samples (ASTM, 1997b). Finally, the sampling design should provide the basis for accurate identification of spatial patterns in the measured property.

The optimal number of samples to collect depends upon the variability of the sample population, the desired level of accuracy in estimating the population mean, the confidence interval desired for estimation of the population mean, and considerations for cost of sampling and sample analysis, and availability of labor or equipment. When samples are expected to obey a normal distribution and are uncorrelated, the following formula (Snedecor and Cochran, 1980; Warrick et al., 1986; Burrough, 1991; ASTM, 1997a) is often used to estimate the number (N) of samples required to achieve a desired precision in estimating the mean value for the property studied:

$$N = \frac{\left(t^2 s^2\right)}{d^2}$$

[9.30]

where t is the tabulated value of student's t for a two sided confidence interval at a given probability level, s is a preliminary estimate for the standard deviation of the population, and d is the deviation desired between the population and measured means. In theory, since N is unknown, the proper number of degrees of freedom for estimating t in Equation [9.30] is also unknown. Skopp et al. (1995) have provided an iterative approach for successively estimating an initial value for N, its corresponding degrees of freedom (t) and the true value of N.

In practice, one often assumes that the sample size (N) is large enough that the degrees of freedom can be obtained for a large sample size. Consider an example in which one computes the number of samples required to estimate the mean within a 95% confidence interval. The number of samples required is $(1.96)^2[(s^2/d^2]$. For this example, the number of samples needed to estimate the mean for the two sample populations whose frequency distributions were plotted in Figs. 9.1A and B, will be computed. The two populations had means of 50 and standard deviations of 10 and 5, respectively. If the desired deviation cannot exceed $d = 1$, one would have to collect 384 and 96 samples to characterize the means of the first and second populations, respectively. If the desired deviation cannot exceed $d = 3$, one would have to collect 43 and 11 samples, respectively. Finally, if the desired deviation is $d = 10$, only 4 and 1 sample would be required. Clearly, as the desired deviation from the mean decreases, the number of samples required increases dramatically, although the rate of increase depends on the standard deviation of the population.

For sample populations in which spatial correlation is expected, the formula for estimating the number of samples is modified to take spatial correlation into account. Generally, when spatial correlation is present, the amount of information in any sample is diminished, and a greater number of samples are required to estimate the mean. This increase in the number of samples needed can be estimated by computing the equivalent number of independent observations giving the same population variance. In general, the number of equivalent independent observations is much less than the number of correlated samples. The following formula shows how to estimate the equivalent coefficient (r).

$$N' = N\left[1 + 2\{\rho/(1-\rho)\}\{1-(1/N)\} - 2\rho/(1-\rho)2(1-\rho N-1)/N\right] \qquad [9.31]$$

For spatially correlated sample populations, the value of N' is always smaller than N. Thus, the true number of samples needed to estimate the mean is N^2/N'. Another consequence of correlation is that the confidence intervals computed for independent samples are too narrow for correlated samples.

Other considerations for soil sampling in addition to the number of samples include sampling depth, sampling time, sample volume or support, and compositing (Wollenhaupt et al., 1997). The proper sampling depth is dependent on many factors, including the type of property measured, the type of equipment used to collect samples, the type and depth of tillage (Kitchen et al., 1990), the application of broadcast or banded fertilizer (Mahler, 1990; Tyler and Howard, 1991), and the condition of the soil (wet, dry, compacted, etc.). In some cases, it is desirable to sample at depths corresponding to distinct soil horizons. When soil horizons are very thick, this may not be practical. The proper timing of sample collection is particularly important when measuring a temporally variable soil property such as soil NO_3-N (Lockman and Molloy, 1984), soil moisture content (Bertuzzi et al., 1994), or soil hydraulic conductivity.

The proper sample support or volume depends to some extent on the small scale variability of the property (Johnson et al., 1990; Lame and Defize, 1993; Starr et al., 1995). For instance, undisturbed cores used to estimate saturated soil hydraulic conductivity should be large enough to represent all of the structural heterogeneities (especially macropores) that are likely to affect infiltration. A rigorous approach for determining the optimum sample support from the shape of the semivariogram was developed by Zhang et al. (1990). van Wesenbeeck and Kachanoski (1991) used the range of the semivariogram for solute travel times to estimate the optimum support size for studies of spatial variability in solute leaching. Kamgar et al. (1993) determined the optimum size of field plots and the optimum number of neutron probe access tubes in each plot using semivariograms in a study of water storage; a theoretical relationship between plot size and sample variance developed by Smith (1938) was used. Gilbert and Doctor (1985) developed an approach for determining optimum sample volume with consideration for not only the variance-volume relationship, but also the cost of sampling and analysis at different sample volumes.

Composite disturbed samples are often collected from a small region and mixed when undesirable small-scale spatial patterns would obscure broad-scale spatial patterns. The optimum number of composite samples collected depends on the variance of the measured property, and its spatial structure (Webster and Burgess, 1984; Giesler and Lundstrom, 1993). Typically, a series of from 4 to 8 subsamples are collected and composited from an area having a radius no larger than 5 m (Oliver et al., 1997).

To avoid bias in sampling, a statistically rigorous arrangement of sample locations is needed. Several strategies for determining sample locations have been developed, including judgement, simple random, stratified random, systematic, stratified systematic unaligned, targeted, adaptive, and geostatistical sampling. Each of these approaches will be described below.

9.4.1 Judgemental Sampling

This is an approach used when there is little interest in statistical rigor. The objective of the sampling program may be to evaluate soil properties at several locations where some type of problem is thought to exist. This problem could, for example, be low crop yield, soil contamination, poor drainage, or severe erosion. The decisions about where to sample in this approach are often based on visual evidence of a change in soil properties, which could include changes in color, soil moisture, vegetation height and color, or salinity.

Judgemental sampling is not statistically rigorous. Sample numbers are often not large enough for accurate estimation of the population mean. Because the choice of sampling locations is based on personal experience, judgement, and opinion, there is almost always bias in the sample design. For example, the objective of judgemental sampling may be to diagnose the reason for low crop yields at several locations in the field. Samples could be collected only from those areas with low yields, and analyzed for soil nutrient availability or permeability. The samples collected are not representative of the whole field.

Even more problematic, there is judgement involved in determining what constitutes a low yielding portion of the field. Two people may have very different criteria for deciding what portions of the field are low yielding, and hence, their sampling programs may be quite different. If one of the two people makes a living from selling fertilizer, his or her judgement about where to sample may be biased by a desire to sell more fertilizer. Thus, judgement sampling is subject to human bias.

9.4.2   Simple Random Sampling

Bias can be avoided by using statistically rigorous methods to identify sampling locations. Random selection of sampling locations avoids bias. In this approach, a random number generator is used to select locations for sampling (Fig. 9.8A). All locations have an equal probability of being selected, and successive selections for sampling locations are completely independent of the previous selections.

It may not be easy to actually sample from the locations selected by the random number generator. This is because there may be features in the field, such as a tree, a pond, a fence, or a building, which were not accounted for when selecting sample locations. In such cases, the desired sampling location may be replaced by another nearby one which is selected by randomly selecting a direction and distance to the new point. Unless sophisticated position locator equipment is available (GPS), it may also prove difficult to find the randomly selected location with any great degree of accuracy. Finally, the random sampling design often has an uneven or clustered appearance, and large areas of the field may be unsampled. This leads to a large degree of uncertainty about the soil properties in the unsampled regions.

9.4.3   Stratified Random Sampling

Stratified random sampling is a refinement of simple random sampling in which the area is divided into cells or strata that can be regularly or irregularly shaped. The most common types of strata include grid cells that are square in shape, or irregular cells that correspond to soil mapping units. If square cells are chosen, there is some judgement involved in specifying the dimension of the cells, which can lead to bias.

Sampling locations within each cell are selected at random, using a random number generator (Fig. 9.8B). This reduces, but may not eliminate, the unevenness and clustering that often occur in simple random sampling. The number of sampling locations within each cell is preset based upon an evaluation of the total number of samples needed to accurately define the population mean.

Stratification of the study area by soil type may be most useful when the soil variation studied is strongly linked to soil mapping units. This may be especially true when studying a soil morphological property such as depth of the A horizon. When studying soil properties that are strongly influenced by soil management, this approach for stratifying sample locations may not lead to improvements in understanding sample variation. An example of this might be the study of soil P availability, in which the major source of variability is due to previous management effects rather than soil morphology and mapping unit delineations.

9.4.4   Systematic Sampling

To avoid clustered sampling and difficulties associated with finding randomly spaced locations in the field, many sampling designs use systematic sampling (Wollenhaupt et al., 1994). In this approach, sample locations occur either at the center of regularly shaped grid cells or at the intersection of the grid lines. When samples are located at the center of cells, the resulting values are thought to be representative of the soil within the grid cell (Fig. 9.8C). When they are located at the intersections of grid lines, all of the values along the edge of the grid cell can be averaged to obtain a representative value for the grid cell.

Grid cells can be of varying size and shape. The most common shapes are square, rectangular, hexagonal, and triangular grid cells. Triangular grid cells are widely considered to be more efficient than the other methods. Rectangular grid cells may be used when there is some reason to believe that the spatial variability in the sampled property exhibits anisotropy due to topographic, tillage, or other types of influences. The square grid cell is probably the most widely used approach in systematic sampling, and is often simply referred to as grid sampling.

One reason for the popularity of the square grid cell sampling design is the ease in finding or surveying sampling locations. On flat fields, two parallel rows of sampling locations can be surveyed on adjacent transects. Then, starting at the midpoint of these two rows, two more columns can be surveyed in a direction perpendicular to the first two rows. Stakes or flags can be driven at each sampling location along the four transects. All other sampling locations can then be found by sighting toward the two rows and the two columns until the two flags at the right sampling distance in those transects are both perfectly lined up with your location. Alternatively, global positioning satellites, distance measuring devices, and dead reckoning can be used to find sampling locations.

The major disadvantage of systematic sampling is that the sample rows may be perfectly aligned with soil or management features that vary systematically (Wollenhaupt et al., 1997). For instance, sample rows may be alternately aligned with irrigation furrows and then hills, or with tillage rows and then wheel tracks. In either case, the analysis of spatial patterns in the direction of the columns might show periodicity. To avoid this bias, random sampling may be done within cells (Fig. 9.8D).

9.4.5   Stratified, Systematic, Unaligned Sampling

The bias introduced by the systematic sampling strategy can be overcome by reducing the alignment of sampling locations in the columns and rows. Stratified systematic unaligned sampling involves stratifying the study area into regular sized cells which may be square, rectangular, triangular, or hexagonal. Each cell is further subdivided into many smaller cells of the same shape. The approach for choosing sampling locations within each larger cell is described by Wollenhaupt et al. (1997). The stratified, systematic, unaligned sampling strategy combines the best features of the systematic and stratified random sampling designs. Finding sample locations in the field is relatively easy, yet the random element of the design overcomes problems associated with regular features caused by row spacing and tillage.

9.4.6   Targeted or Directed Sampling

There are two major types of variation in soil properties, namely, those that can be seen and those that cannot. An example of the former is the spatial variation in soil color as a result of patterns in soil

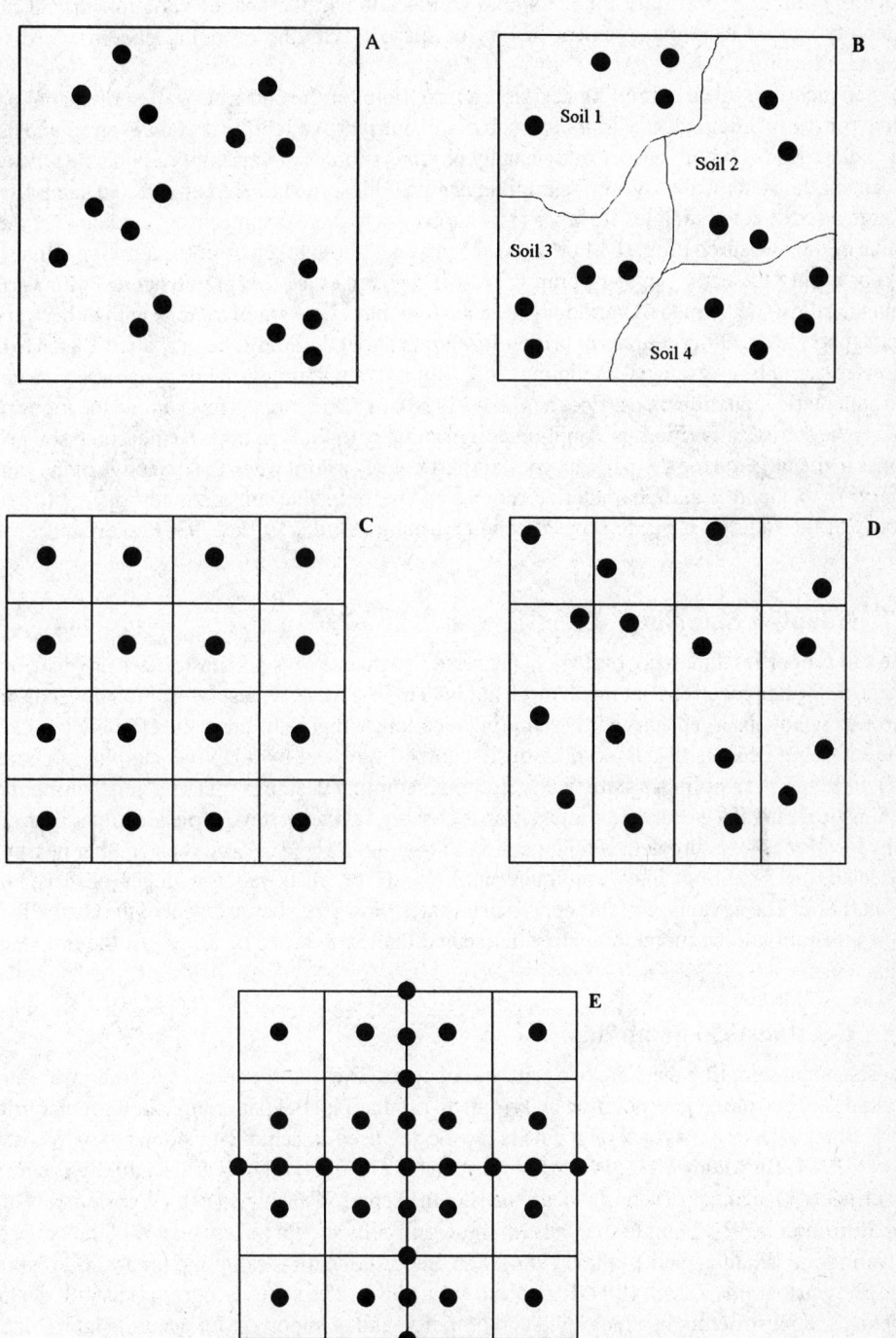

Fig. 9.8   (A) Example of a simple random sampling design; (B) example of a stratified random sampling design; (C) example of a systematic sampling design; (D) example of a sampling design for random sampling within cells; (E) example of a geostatistical sampling design

surface organic C or moisture contents, and of the latter is the spatial variation in soil nutrient availability caused by a management history of spatially varying cropping patterns, fertilizer, or manure applications.

A statistically rigorous sampling design is appropriate for the characterization of spatial patterns arising from unseen sources of variability such as soil nutrient availability. If, however, spatial patterns in the field arise from both unseen and visually obvious sources of variability, it may be advisable to supplement the statistically rigorous sampling design with some targeted or directed samples.

Targeted or directed samples are taken at locations where there is some visual evidence for a change in value of the measured property. For instance, an aerial photo taken prior to sampling (Bhatti et al., 1991) or during the early stages of crop growth (Ferguson et al., 1996) may show color variations within the study site. A map of variations in crop yield may show small regions with either very poor or very good yields. Other sources of prior information about the study site may also be used to design a directed sampling strategy, including soil, digital elevation, ground penetrating radar, and electromagnetic induction maps (Pocknee et al., 1996). In these cases, a few sample locations may be added to the statistically rigorous sampling design strategy to find out how the measured soil property changes in the targeted region. This approach is particularly useful when the objective of the sampling program is to identify and characterize regions of the field that are distinctly different from the majority of the field, as is needed for Precision Farming (Mulla, 1991, 1993; Francis and Schepers, 1997).

9.4.7 Adaptive Sampling

Some soil properties and many biological properties such as weeds and pests often are distributed in clusters. Sampling locations are much more likely to have a weed or pest infestation if weeds or pests were or were not observed at a nearby sampling location. If this is the case, the efficiency for any one of the sampling designs discussed previously can be improved by adaptive sampling (Thompson, 1997). In adaptive sampling, a statistically rigorous sampling design is first selected. During the field sampling program, if a weed or pest infestation is observed at one of the sample locations, then another nearby location is also sampled. If that location also shows a weed or pest, then another nearby point is selected for sampling. This continues until weeds or pests are no longer observed in the neighborhood. The advantage of this approach is that it allows the spatial extent of the cluster for weed or pest infestation to be more competely delineated than in the case of using just the initial sample design.

9.4.8 Geostatistical Sampling

In geostatistical sampling there are two primary concerns. The first concern is systematic sampling for the purposes of accurate interpolation by kriging to produce spatial pattern maps. A regular grid with square, triangular, or hexagonal elements is most often used to achieve this objective (Webster and Burgess, 1984; Burrough, 1991; Wollenhaupt et al., 1997). For a regular grid sampling program, the most efficient placement of sample locations is in the center of each grid cell (Webster and Burgess, 1984; Burrough, 1991). Sample spacings for these grid cells should be less than ½ of the range for the semivariogram (Flatman and Yfantis, 1984). The second concern is sampling for accurate estimation of the semivariogram (Russo, 1984; Russo and Jury, 1988). The semivariogram is useful as a tool for modeling spatial structure in a measured soil property, and is important for accurate interpolation by kriging. A typical approach for accurate estimation of the semivariogram is to supplement the systematic sampling grid with one or more transects consisting of closely spaced sample locations (Fig. 9.8E). This helps define the shape of the semivariogram at small separation distances.

One of the dilemmas in designing geostatistically based sampling schemes is that the design depends upon the semivariogram, yet the semivariogram is often unknown until the study site is sampled. There are two approaches for solving this dilemma. The first is to use pre-existing information about the range of the semivariogram for the soil property of interest, whether this information is from study site, a nearby site, or a site in the same region. Once an estimate for the range is obtained, it can be used to set the approximate sample spacings in the sampling design of interest. The second approach is used if no pre-existing information about the semivariogram can be obtained. In this case, it is necessary to conduct a preliminary sampling survey of the study site along several transects (Flatman and Yfantis, 1984).

If the semivariogram is known, it is a powerful tool which can be used to evaluate various sampling strategies before any samples are collected (McBratney et al., 1981; Rouhani, 1985; Burrough, 1991). The first type of evaluation is the determination of optimum spacing for sample locations (Rouhani, 1985; Warrick et al., 1986; Oliver and Webster, 1987; Burrough, 1991; Gutjahr, 1991). This is achieved by computing kriging estimation variances for a wide range of sample spacings and arrangements. The estimation variances can be computed even when no measurements are available at sample locations because the estimation variance depends only upon the semivariogram and separation distances between potential sampled and unsampled locations. The second type of evaluation is determination of the optimum number of composite or bulked samples to collect at each sample location (Webster and Burgess, 1984; Oliver et al., 1997). Significant reductions in estimation error are possible by mixing samples taken from small blocks around a sampled location and analyzing the composite sample.

9.5   References

ASTM. 1997a. Standard guide for general planning of waste sampling. p. 460–469. *In* J. Azara, N.C. Baldini, E. Barszcewski, L. Bernhardt, E.L. Gutman, J.G. Kramer, C.M. Leinweber, V.A. Mayer, P.A. McGee, T.J. Sandler, and R.F. Wilhelm (ed.) Standards on environmental sampling. 2nd Ed. American Society for Testing Materials, West Conshohocken, PA.

ASTM. 1997b. Standard guide for sampling strategies for heterogeneous wastes. p. 521–537. *In* J. Azara, N.C. Baldini, E. Barszcewski, L. Bernhardt, E.L. Gutman, J.G. Kramer, C.M. Leinweber, V.A. Mayer, P.A. McGee, T.J. Sandler, and R.F. Wilhelm (ed.) Standards on environmental sampling. 2nd Ed. American Society for Testing Materials, West Conshohocken, PA.

Bertuzzi, P., L. Bruckler, D. Bay, and A. Chanzy. 1994. Sampling strategies for soil water content to estimate evapotranspiration. Irrig. Sci. 14:105–115.

Beven, K.J., D.E. Henderson, and A.D. Reeves. 1993. Dispersion parameters for undisturbed partially saturated soil. J. Hydrol. 143:19–43.

Bhatti, A.U., D.J. Mulla, and B.E. Frazier. 1991. Estimation of soil properties and wheat yields on complex eroded hills using geostatistics and thematic mapper images. Remote Sens. Environ. 37:181–191.

Birrell, S.J., K.A. Sudduth, and N.R. Kitchen. 1996. Nutrient mapping implications of short-range variability. p. 207–216. *In* P.C. Robert, R.H. Rust, and W.E. Larson (ed.) Precision agriculture. American Society of Agronomy, Madison, WI.

Brown, A.J. 1993. A review of soil sampling for chemical analysis. Aust. J. Exp. Agric. 33:983–1006.

Burgess, T.M. and R. Webster. 1980a. Optimal interpolation and isarithmic mapping of soil properties. I. The semivariogram and punctual kriging. J. Soil Sci. 31:315–331.

Burgess, T. M. and R. Webster. 1980b. Optimal interpolation and isarithmic mapping of soil properties. II. Block kriging. J. Soil Sci. 31:333–341.

Burrough, P.A. 1991. Sampling designs for quantifying map unit composition. p. 89–125. *In* M.J. Mausbach and L.P. Wilding (ed.) Spatial variabilities in soils and landforms. Soil Sci. Soc. Am. Spec. Pub. 28. Soil Science Society of America, Madison, WI.

Carter, M.R. 1993. Soil sampling and methods of analysis. Lewis Publishers, Boca Raton, FL.

Clark, I. 1979. Practical geostatistics. Applied Science Publishers, London, UK.

Cooke, R.A., S. Mostaghimi, and J.B Campbell. 1993. Assessment of methods for interpolating steady-state infiltrability. Trans. ASAE 36:1333–1341.

Cressie, N. 1985. Fitting variogram models by weighted least squares. Math. Geol. 17:563–586.

Cressie, N. 1991. Statistics for spatial data. Wiley Interscience. New York, NY.

Cressie, N. and D.M. Hawkins. 1980. Robust estimation of the variogram. J. Inter. Assoc. Math. Geol. 12:115–125.

Di, H.J., B.B. Trangmar, and R.A. Kemp. 1989. Use of geostatistics in designing sampling strategies for soil survey. Soil Sci. Soc. Am. J. 53:1163–1167.

Dubrule, O. 1984. Comparing splines and kriging. Computers Geosci. 10:327–338.

Ferguson, R.B., C.A. Gotway, G.W. Hergert, and T.A. Peterson. 1996. Soil sampling for site-specific nitrogen management. p. 13–22. In P.C. Robert, R.H. Rust, and W.E. Larson (ed.) Precision agriculture. American Society of Agronomy, Madison, WI.

Flatman, G.T. and A.A. Yfantis. 1984. Geostatistical strategy for soil sampling: The survey and the census. Environ. Monit. Assess. 4:335:349.

Francis, D.D., and J.S. Schepers. 1997. Selective soil sampling for site-specific nutrient management. p. 119–126. In J.V. Stafford (ed.) Precision agriculture '97. Vol. 1: Spatial variability in soil and crop. BIOS Scientific Publishers Ltd., Oxford, UK.

Giesler, R., and U. Lundstrom. 1993. Soil solution chemistry: Effects of bulking soil samples. Soil Sci. Soc. Am. J. 57:1283–1288.

Gilbert, R.O., and P.D. Doctor. 1985. Determining the number and size of soil aliquots for assessing particulate contaminant concentrations. J. Environ. Qual. 14:286–292.

Gotway, C.A., R.B. Ferguson, G.W. Hergert, and T.A. Peterson. 1996. Comparison of kriging and inverse-distance methods for mapping soil parameters. Soil Sci. Soc. Am. J. 60:1237–1247.

Gutjahr, A. 1991. Geostatistics for sampling designs and analysis. ACS Symp. Ser. 465:48–90.

Hawkins, D.M. 1980. Identification of outliers. Chapman and Hall. London, UK.

Hoeksma, R.J., and P.K. Kitanidis. 1985. Comparison of Gaussian conditional mean and kriging estimation in the geostatistical solution of the inverse problem. Water Resour. Res. 21:825–836.

Isaaks, E.H., and R.M. Srivastava. 1989. An introduction to applied geostatistics. Oxford University Press, New York, NY.

Johnson, C.E., A.H. Johnson, and T.G. Huntington. 1990. Sample size requirements for the determination of changes in soil nutrient pools. Soil Sci. 150:637–644.

Journel, A.G. 1986. Constrained interpolation and qualitative information-The soft kriging approach. Math. Geol. 18:269–286.

Journel, A.G. and C.J. Huijbregts. 1978. Mining geostatistics. Academic Press, London, UK.

Journel, A.G., C. Deutsch, and A.J. Desbarats. 1986. Power averaging for block effective permeability. Soc. Petrol. Engin. Pap. SPE 15128.

Jury, W.A. 1986. Spatial variability of soil properties. p.245–269. In S.C. Hern and S.M. Melancon (ed.) Vadose zone modeling of organic pollutants. Lewis Publishers, Chelsea, MI.

Jury, W.A., D. Russo, G. Sposito, and H. Elabd. 1987. The spatial variability of water and solute transport properties in unsaturated soil: I. Analysis of property variation and spatial structure with statistical models. Hilgardia 55:1–32.

Kamgar, A., J.W. Hopmans, W.W. Wallender, and O. Wendroth. 1993. Plot size and sample number for neutron probe measurements in small field trials. Soil Sci. 156:213–224.

Kitchen, N.R., J.L. Havlin, and D.G. Westfall. 1990. Soil sampling under no-till banded phosphorus. Soil Sci. Soc. Am. J. 54:1661–1665.

Krige, D.G. 1966. Two dimensional weighted moving average trend surfaces for ore-evaluation. J. S. Afr. Inst. Min. Metall. 66:13–38.

Laslett, G.M., A.B. McBratney, P.J. Pahl, and M.F. Hutchinson. 1987. Comparison of several spatial prediction methods for soil pH. J. Soil Sci. 38:325–341.

Lame, F.P.J., and P.R. Defize. 1993. Sampling of contaminated soil: Sampling error in relation to sample size and segregation. Environ. Sci. Technol. 27:2035–2044.

Lockman, R.B., and M.G. Molloy. 1984. Seasonal variations in soil test results. Commun. Soil Sci. Plant Anal. 15:741–757.

Mahler, R.L. 1990. Soil sampling fields that have received banded fertilizer applications. Commun. Soil Sci. Plant Anal. 21: 1793–1802.

Markus, J.A., and A.B. McBratney. 1996. An urban soil study: Heavy metals in Glebe, Australia. Aust. J. Soil Res. 34:453–465.

Matheron, G. 1963. Principles of geostatistics. Econ. Geol. 58:146–1266.

McBratney, A.B., R. Webster, and T.M. Burgess. 1981. The design of optimal sampling schemes for local estimation and mapping of regionalized variables. I. Theory and method. Comp. Geosci. 7:331–334.

McBratney A.B., and M.J. Pringle. 1997. Spatial variability in soil: Implications for precision agriculture. p. 3–31. *In* J.V. Stafford (ed.) Precision agriculture '97. Volume 1: Spatial variability in soil and crop. BIOS Scientific Publishers Ltd., Oxford, UK.

Mulla, D.J. 1988a. Using geostatistics and spectral analysis to study spatial patterns in the topography of southeastern Washington State, U.S.A. Earth Surf. Proc. Land. 13:389–405.

Mulla, D.J. 1988b. Estimating spatial patterns in water content, matric suction, and hydraulic conductivity. Soil Sci. Soc. Am. J. 52:1547–1553.

Mulla, D.J. 1991. Using geostatistics and GIS to manage spatial patterns in soil fertility. p. 336–345. *In* G. Kranzler (ed.) Automated agriculture for the 21st century. American Society of Agricultural Engineers, St. Joseph, MI.

Mulla, D.J. 1993. Mapping and managing spatial patterns in soil fertility and crop yield. p. 15–26. *In* P.C. Robert, R.H. Rust, and W.E. Larson (ed.) Site specific crop management. American Society of Agronomy, Madison, WI.

Mulla, D.J. 1997. Geostatistics, remote sensing and precision farming. p. 100–114. *In* J.V. Lake, G.R. Bock, and J.A. Goode (ed.) Precision agriculture: Spatial and temporal variability of environmental quality. John Wiley and Sons, Chichester, UK.

Nash, M.H., L.A. Daugherty, A. Gutjahr, P.J. Wierenga, and S.A. Nance. 1988. Horizontal and vertical kriging of soil properties along a transect in Southern New Mexico. Soil Sci. Soc. Am. J. 52:1086–1090.

Oliver, M.A. 1987. Geostatistics and its application to soil science. Soil Use Manag. 3:8–20.

Oliver, M.A. and R. Webster. 1987. The elucidation of soil pattern in the Wyre Forest of the West Midlands, England. II. Spatial distribution. J. Soil Sci. 38:293–307.

Oliver, M.A., and R. Webster. 1991. How geostatistics can help you. Soil Use Manag. 7:206–217.

Oliver, M.A., Z. Frogbrook, R. Webster, and C.J. Dawson. 1997. A rational strategy for determining the number of cores for bulked sampling of soil. p. 155–162. *In* J.V. Stafford (ed.) Precision agriculture '97. Volume 1: Spatial variability in soil and crop. BIOS Scientific Publishers Ltd., Oxford, UK.

Pocknee, S., B.C. Boydell, H.M. Green, D.J. Waters, and C.K. Kvien. 1996. Directed soil sampling. p. 159–168. *In* P.C. Robert, R.H. Rust, and W.E. Larson (ed.) Precision agriculture. American Society of Agronomy, Madison, WI.

Rogowski, A.S., and J.K. Wolf. 1994. Incorporating variability into soil map unit delineations. Soil Sci. Soc. Am. J. 58:163–174.

Rossi, R.E., D.J. Mulla, A.G. Journel, and E.H. Franz. 1992. Geostatistical tools for modeling and interpreting ecological spatial dependence. Ecol. Mono. 62:277–314.

Rouhani, S. 1985. Variance reduction analysis. Water Resour. Res. 21:837–846.

Russo, D. 1984. Design of an optimal sampling network for estimating the variogram. Soil Sci. Soc. Am. J. 48:708–716.

Russo, D., and M. Bouton. 1992. Statistical analysis of spatial variability in unsaturated flow parameters. Water Resour. Res. 28:1911–1925.

Russo, D., and W.A. Jury. 1988. Effect of the sampling network on estimates of the covariance function of stationary fields. Soil Sci. Soc. Am. J. 52:1228–1234.

Skopp, J., S.D. Kachman, and G.W. Hergert. 1995. Comparison of procedures for estimating sample numbers. Commun. Soil Sci. Plant Anal. 26:2559–2568.

Smith, H.F. 1938. An empirical law describing heterogeneity in the yields of agricultural crops. J. Agri. Sci. 28:1–23.

Snedecor, G.W. and W.G. Cochran. 1980. Statistical methods. 7th Ed. Iowa State University Press, Ames, IA.

Starr, J.L., T.B. Parkin, and J.J. Meisinger. 1995. Influence of sample size on chemical and physical soil measurements. Soil Sci. Soc. Am. J. 59:713–719.

Stein, A., W. van Dooremolen, J. Bouma, and A.K. Bregt. 1988. Cokriging point data on moisture deficit. Soil Sci. Soc. Am. J. 52:1418–1423.

Thompson, S.K. 1997. Spatial sampling. p. 161–168. *In* J.V. Lake, G.R. Bock, and J.A. Goode (ed.). Precision agriculture: Spatial and temporal variability of environmental quality. John Wiley and Sons, Chichester, UK.

Thompson, W.H., and P.C. Robert. 1995. Evaluation of mapping strategies for variable rate applications. p. 303–323. *In* P.C. Robert, R.H. Rust, and W.E. Larson (ed.) Site-specific management for agricultural systems. American Society of Agronomy, Madison, WI.

Tyler, D.D., and D.D. Howard. 1991. Soil sampling patterns for assessing no-tillage fertilization techniques. J. Fert. Issues 8:52–56.

Upchurch, D.R., and W.J. Edmonds. 1991. Statistical procedures for specific objectives. p. 49–71. *In* M.J. Mausbach and L.P. Wilding (ed.) Spatial variabilities of soils and landforms. Soil Sci. Soc. Am. Spec. Publ. 28. Soil Science Society of America, Madison, WI.

van Wesenbeeck, I.J., and R.G. Kachanoski. 1991. Spatial scale dependence of *in situ* solute transport. Soil Sci. Soc. Am. J. 55:3–7.

Vauclin, M., S.R. Vieira, R. Bernard, and J.L. Hatfield. 1982. Spatial variability of surface temperature along two transects of a bare soil. Water Resour. Res. 18:1677–1686.

Vauclin, M., S.R. Vieira, G. Vachaud, and D.R. Nielsen. 1983. The use of cokriging with limited field soil observations. Soil Sci. Soc. Am. J. 47:175–284.

Vieira, S.R., J.L. Hatfield, D.R. Nielsen, and J.W. Biggar. 1983. Geostatistical theory and application to variability of some agronomical properties. Hilgardia 51:1–75.

Vieira, S.R., W.D. Reynolds, and G.C. Topp. 1988. Spatial variability of hydraulic properties in a highly structured clay soil. p. 471–483. *In* P.J. Wierenga and D. Bachelet (ed.) International conference and workshop on the validation of flow and transport models for the unsaturated zone. NM State University, Las Cruces, NM.

Voltz, M., and R. Webster. 1990. A comparison of kriging, cubic splines and classification for predicting soil properties from sample information. J. Soil Sci. 41:473–490.

Wade, S.D., I.D.L. Foster, and S.M.J. Baban. 1996. The spatial variability of soil nitrates in arable land and pasture landscapes: Implications for the development of geographical information system models of nitrate leaching. Soil Use Manag. 12:95–101.

Warrick, A.W., D.E. Myers, and D.R. Nielsen. 1986. Geostatistical methods applied to soil science. p. 53–82. *In* A. Klute (ed.) Methods of soil analysis. Part 1. 2nd Ed. Soil Science Society of America, Madison, WI.

Warrick, A.W., R. Zhang, M.K. El-Harris, and D.E. Myers. 1988. Direct comparisons between kriging and other interpolators. p. 505–510. *In* P.J. Wierenga and D. Bachelet (ed.) International Conference and workshop on the validation of flow and transport models for the unsaturated zone. NM State University, Las Cruces, NM.

Webster, R., and T. M. Burgess. 1980. Optimal interpolation and isarithmic mapping of soil properties. III. Changing drift and universal kriging. J. Soil Sci. 31:505–524.

Webster, R., and T.M. Burgess. 1984. Sampling and bulking strategies for estimating soil properties in small regions. J. Soil Sci. 35:127–140.

Webster, R., and M.A. Oliver. 1990. Statistical methods in soil and land resource survey. Oxford University Press, New York, NY.

Whelan, B.M., A.B. McBratney, and R.A. Viscarra Rossel. 1996. Spatial prediction for precision agriculture. p. 331–342. *In* P.C. Robert, R.H. Rust, and W.E. Larson (ed.) Precision agriculture. American Society of Agronomy, Madison, WI.

Wierenga, P.J., R.G. Hills, and D.B. Hudson. 1991. The Las Cruces trench site: Characterization, experimental results, and one-dimensional flow predictions. Water Resour. Res. 27:2695–2705.

Wilding, L. P. 1985. Spatial variability: its documentation, accomodation and implication to soil surveys. p. 166–194. *In* D. R. Nielsen and J. Bouma (ed.) Soil spatial variability. Pudoc, Wageningen, Netherlands.

Wollenhaupt, N.C., D.J. Mulla, and C.A. Gotway Crawford. 1997. Soil sampling and interpolation techniques for mapping spatial variability of soil properties. p. 19–53. *In* F.J. Pierce and E.J. Sadler (ed.) The state of site specific management for agriculture. American Society of Agronomy, Madison, WI.

Wollenhaupt, N.C., R.P. Wolkowski, and M.K. Clayton. 1994. Mapping soil test phosphorus and potassium for variable-rate fertilizer application. J. Prod. Agric. 7:441–448.

Yates, S.R., A.W. Warrick, and D.E. Myers. 1986a. Disjunctive kriging. 1. Overview of estimation and conditional probability. Water Resour. Res. 22:615–621.

Yates, S.R., A.W. Warrick, and D.E. Myers. 1986b. Disjunctive kriging. 2. Examples. Water Resour. Res. 22: 623–630.

Yates, S.R., and A.W. Warrick. 1987. Estimating soil water content using cokriging. Soil Sci. Soc. Am. J. 51:23–30.

Zhang, R., A.W. Warrick, and D.E. Myers. 1990. Variance as a function of sample support size. Math. Geol. 22:107–121.

Zhang, R., D.E. Myers, and A.W. Warrick. 1992. Estimation of the spacial distribution of soil chemicals using pseudo-cross-variograms. Soil Sci. Soc. Am. J. 56:1444–1452.

We know more about the movement of the celestial
bodies than about the soil underfoot
—Leonardo da Vinci

Soil Chemistry

P. M. Huang
University of Saskatchewan

Soil is a dynamic system in which the soil solution is the medium of physical, chemical, and biological processes in soil environments. Soil solution is in dynamic equilibrium with minerals, organic matter, microorganisms, and soil atmosphere. Thus, it is the bottle neck of transformations and transport of vital and detrimental molecules and ions in the ecosystem.

Chemistry of the soil covers chemical reactions and processes of the soil pertaining to plant and animal growth and human development environments. Soil chemical processes are fundamental to the evolution of geoderma, the biosphere, and the human environment. Chemistry of the soil thus plays a vital role in the development of natural resources, the protection of the environment, and the sustainability of ecosystem health. Therefore, fundamental understanding of soil chemical reactions and processes at the atomic, molecular, and microscopic levels is essential to developing innovative resource management strategies and to understanding and regulating the behavior of the terrestrial ecosystem at regional and global scales.

The significance of soil chemistry in the scientific community has recently become more prominent. The National Science Foundation is working on several initiatives on environmental research pertaining to new opportunities for soil chemists. On a global scale, the International Union of Pure and Applied Chemistry (IUPAC), which is the highest international organization of pure and applied chemistry, has recently restructured its organization and created a new division, namely, the Division of Chemistry and the Environment. There are several commissions in this new division. The Commission of Fundamental Environmental Chemistry, and the Commissions of Soil and Water Chemistry are particularly noteworthy. These recent developments are very encouraging signs indicating that the future prospects for soil chemistry are indeed bright.

Therefore, it is time to look back, to integrate the existing knowledge of soil chemistry, and to identify future directions in this very important and exciting area of science. There are ten chapters in this section which address the issues on the fundamentals of soil chemical reactions and processes and their impact on the terrestrial system. This section covers chemical composition of soils, nature of soil organic matter, chemistry of soil solution, kinetics and mechanisms of important soil chemical reactions, redox reactions, soil colloidal behavior, ion exchange phenomena, chemisorption and precipitation reactions, abiotic catalysis in transformations of metals, metalloids, other inorganics, and natural and anthropogenic organic compounds, and soil acidity and alkalinity. This section should be useful to researchers, educators consultants, personnel in government regulatory agencies, and students in environmental, ecotoxicological, and soil sciences.

I am grateful to the authors who have contributed chapters to this section. Appreciation is extended to the external referees who have provided invaluable critical inputs to maintain the quality of this section.

1

The Chemical Composition of Soils

Philip A. Helmke
University of Wisconsin-Madison

1.1 Introduction

Soils are created on land surfaces by weathering processes that include chemical, geological, hydrological, and on the planet Earth, biological phenomena. The continual influence of percolating water and the biota produce a vertical stratification (a soil profile) of the concentration of elements and the chemical compounds in which they occur. The solid phase of soils is composed of two broad classifications of compounds: inorganic (mineral) compounds and organic matter. The Earth's crust contributes the chemical elements found in the mineral component while the dominant elements in the organic matter are derived from CO_2 and water contributed from the atmosphere through their utilization by the biota. The proportions between the mineral and organic components in soils can vary widely. Coarse-textured soils can contain < 1% organic matter, while organic soils can contain up to 95% organic matter.

Variations in the chemical composition of soils result mainly from variability in the ratio of the mineral and organic matter contents in soils and from variability of the types of compounds comprising the mineral fraction. While the concentration of elements in the mineral and organic components varies from soil to soil, the variations in the concentrations of elements in the mineral fraction are usually much greater than those found for soil organic matter (SOM). Part of the variation in the concentration of elements in the mineral fraction is inherited from variations in the composition of the soil parent material, granite versus basalt for example. Additional differences in soil composition result from the flux of matter and energy into or from soils over geologic time because soils are open biogeochemical systems. Some of the most important fluxes of elements in soils that affect their chemical composition are the loss of cations by leaching, the loss or gain of clay and silt sized particles by eolian and fluvial transport processes, and for elements active in biological process, their enrichment at the soil surface.

It is useful to classify elements as either major elements or trace elements when discussing element concentrations in soils and rocks. The distinction between the two groups is not always exact, but major elements are characterized by being present in sufficient concentrations that they form discrete phases whose properties affect the properties of the soils or rocks which they comprise. Trace elements are often found at such low concentrations that they do not form discrete phases or the phases are present at such inconsequential abundances that they do not contribute to the bulk properties of the system. Only the eight elements O, Si, Al, Fe, Ca, Mg, Na, and K occur at concentrations exceeding 1% in the Earth's crust, and they are always classified as major elements. The same elements plus C are the most abundant elements in soils. Hydrogen would be added to the list if the water content of soil is considered. All of the other elements together comprise a few percent or less of the Earth's crust and soil.

The concentrations of elements in soils are usually based on the mass of the soil, even though the solid phases (mineral and organic) make up only about one-half of the soil volume. The other half of the soil volume is occupied by the soil solution and air. The proportions between the soil solution and soil air vary widely with changes in the moisture content of soils. The density of soil organic matter is about 0.5 g cm^{-3} and density of soil mineral matter is about 2.7 g cm^{-3}. The bulk density of soil, therefore, varies with the ratio of the mineral and organic contents and with the porosity.

All soils contain some of all of the naturally occurring chemical elements. The answer to the question "Is a given element present in a soil?" is always "yes." In an investigation of the composition of soil, the pertinent question for any given element is "What is its concentration?" or "What is its chemical form or speciation?" Even though the concentrations of elements in soils are affected by many factors, the chemical properties of the elements restrict the range of their potential concentrations to a finite set of important compounds in soils. The elements are distributed in slightly more than 2000 minerals that are found in the Earth's crust, but only a few dozen minerals comprise the bulk of most crustal rocks and soils.

It may at first seem surprising that the concentrations of elements and their relative proportions in modern soils on Earth reflect the relative abundances of the elements when they were first formed by processes of nucleosynthesis in ancient stars. The predominating abundance of elements with low atomic numbers in soils, with the notable exception of Li, Be, and B, and the trace occurrence of all of the heavy elements in soils, can be best appreciated with a brief understanding of the processes responsible for the formation of the elements. The nuclear properties of the various isotopes of each element and the processes of nucleosynthesis restrict the potential range of element concentrations in soils and rocks by their influence on the original production of elements in stars.

1.2   Origin and Abundance of Elements

1.2.1   Elements and Isotopes

The 10 most abundant elements in most dry soils are O, Si, Al, Fe, Ca, K, Na, Mg, Ti, and C. The list for igneous crustal rocks is similar except that C is replaced by P. The reason that these elements are the most abundant in soils and rocks becomes very clear when one considers the nuclear processes responsible for the formation of the elements and their subsequent fractionation by chemical and physical processes.

Over 99% of the total atoms in the universe are H atoms (Ross and Aller, 1976). The same is true for our solar system because most of its mass resides in the Sun, which is predominantly H and He (Fig. 1.1). From the study of distant stars and galaxies, it appears that the composition of the solar system is fairly typical of the present universe. The results of these observations also suggest that H and He were the original constituents of the universe, and that the other elements have been produced from H since its beginning (Burbridge et al., 1957; Selbin, 1973; Trimble, 1975; Fowler, 1984). The same nuclear reactions that convert mass into energy in stars also produce the elements. All elements heavier than H are produced in stars. Exceptions may be Li, Be, and B as discussed later. Stars slightly larger than our Sun end their nuclear lives with a supernova event, which seeds interstellar space with the elements produced by the star.

Nuclei are composed of positively charged protons and uncharged neutrons. These elementary particles, which have approximately the same mass, are collectively called nucleons. The nuclear

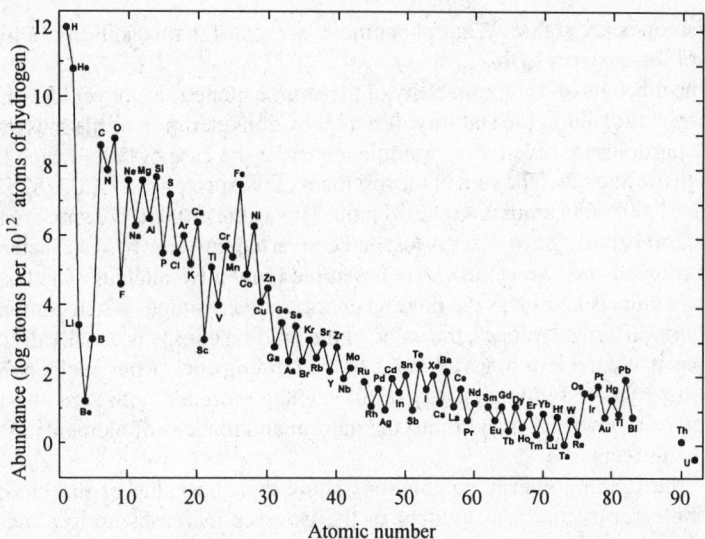

Fig. 1.1 Abundances of the elements in the solar system showing the relative number of atoms on a logarithmic scale normalized to 10^{12} atoms of H [Data from Ross and Aller, 1976]

charge, proportional to the number of protons, is balanced in a neutral atom by an equal number of electrons. The number of protons is the atomic number (Z) of an element. The number of electrons in the atom of an element controls its chemical properties and its position in the periodic table. The mass of an atom depends predominantly on the total number of nucleons. Each different nuclide can be specified by giving its atomic number Z and mass number A. The mass number is the sum of the number of protons and neutrons.

With the invention of the mass spectrometer, it became possible to measure the masses of individual atoms, and it was found that many elements consist of a mixture of atoms with different masses. These are known as isotopes. Isotopes have the same number of protons but different numbers of neutrons, thus giving different mass numbers. For example, magnesium, with Z = 12, has three stable isotopes, with A = 24, 25, and 26, denoted as ^{24}Mg, ^{25}Mg and ^{26}Mg. The isotopes of magnesium are present in very nearly the same proportions, 79.99, 10.00 and 11.01%, respectively, in all samples of matter from our solar system. The average mass number of natural Mg is, therefore, 24.320, which is very close to its relative atomic mass [24.305 atomic mass units (amu)]. Relative atomic masses are measured against ^{12}C, assuming that its mass is exactly 12.00000000 amu.

The chemical properties of different isotopes of an element are very similar, and almost indistinguishable for many purposes. However, the difference in mass does result in small differences in the rates of some chemical and physical reactions. However, the different number of neutrons in the isotopes of a single element can result in very different nuclear properties of each isotope of the element. For some elements, natural samples from different sources do not have quite the same isotopic composition and this gives rise to small variations in relative atomic masses. The origin of these deviations often contains interesting biogeochemical or cosmogonic information.

1.2.2 Binding Energy of Nuclei

Electrical charges of the same sign repel each other, with the force of repulsion increasing as the inverse square of the distance between the charged particles decreases. The protons in a nucleus of an atom are in very close proximity, and they must experience tremendous repulsive forces, yet the nuclei of many isotopes are stable. What phenomena account for this stability? Physicists are still searching for all of the answers to this problem.

Even though the mechanism for the stability of the atomic nucleus is not yet known, one can easily calculate the energy involved in the stability of nuclei by consideration of the masses, or the energy equivalent, of the particles involved. For example, consider the case of ^4He, which has two protons and two neutrons in the nucleus. The sum of the rest mass of two protons (2 x 1.0078252 amu) plus that of two neutrons (2 x 1.0086654 amu) is 4.032981 amu. This value is 0.030377 amu greater than the rest mass of ^4He (4.002604 amu). Some mass is lost and converted into energy via Einstein's equation (E = mc^2) when two protons and two neutrons are assembled into a He nucleus. The loss in mass when converted to energy units is known as the binding energy of the isotope. When comparing the values of binding energy for different isotopes, the value of the binding energy is usually divided by the total number of nucleons in the nucleus to give a value of the binding energy per nucleon for each isotope. Consideration of the relative binding energy of the nuclei combined with some basic principles of nuclear physics gives a powerful insight into the natural abundances of elements in soils, our solar system, and for the universe.

The results of the binding energy calculations show that the value of the binding energy per nucleon becomes greater as the mass number of the isotopes increases up to a mass of about 56, which is the mass number of Fe. After mass number 56, the binding energy per nucleon decreases with increasing mass number. Therefore, fusion of any two light nuclei in a stellar environment to produce a product nucleus having a mass number ≤ 56 will be an exothermic reaction because the product nucleus will have a greater binding energy than that of the reactant nuclei. Formation of nuclei with mass numbers > 56 by fusion of two nuclei always will be an endothermic reaction, and thus such reactions are not energetically favored.

1.2.3   Synthesis of Elements up to Iron

Elements up to Fe are produced by fusion processes starting with protons from H. These reactions are exothermic because the binding energy per nucleon increases as a function of mass number, reaching a maximum at Fe. Such fusion events usually occur by two body collisions and may involve a renewable cycle of isotopes, such as the C cycle. Fusion reactions that produce elements heavier than Fe have a low probability of occurring because such reactions are endothermic and because of the high coulombic repulsion between the large nuclei needed to make heavy elements. First-generation stars formed initially after the big bang of the early universe are composed originally of H. They can efficiently synthesize elements only up to and including Fe. Stars slightly larger than our Sun, which is a very average star, undergo a supernova event when their fuel is exhausted and insufficient heat is generated to prevent gravitational collapse. The resulting implosion and rebound from the shock wave disperses a portion of the star's mass into interstellar space.

1.2.4   Synthesis of Elements from Cobalt through Bismuth

Incorporation of some of the material dispersed by the supernova of a first-generation star into the material forming a new star, a second-generation star, allows a new set of nuclear reactions to occur simultaneously with the fusion of light nuclei. The new nuclear reactions produce a small, but steady flux of free neutrons. Some of these neutrons are captured by Fe nuclei to produce heavier Fe isotopes that are radioactive. Their decay by electron emission (β^-) decay produces a stable isotope of Co (^{60}Co). Capture of an additional neutron by ^{60}Co and subsequent decay of the resulting radioactive isotope produces Ni. This sequence of successive neutron capture followed by B^- decay produces elements up through Bi. About 100 to 1000 years are needed for each neutron capture step. The process is known as the slow or "s" process. This process is stopped after Bi because the half-lives of all of the isotopes for the next several elements are short compared the time needed for each neutron capture step [^{209}Po (103 y), ^{211}At (7.2 h), ^{211}Rn (16 h), ^{222}Fr (22 m), and ^{227}Ac (21.2 y)].

1.2.5 Synthesis of Heavy Elements

The heaviest elements (U and Th) are radioactive but have isotopes with such long half-lives (e.g., ^{238}U with a half-life of 4.5 x 10^9 y) that a fraction has survived to the present time. For the practical purposes of discussing element concentrations, the long lived isotopes of Th and U can be considered stable. Their production requires an intense flux of free neutrons so that several successive neutron capture-ß decay events occur in a time span shorter than the half-life of isotopes of the elements between Bi and Th. Such conditions are found for a short time during the supernova event of a second-generation star. This is the rapid or "r" process. This process bridges the gap in element synthesis imposed by the group of elements with short half-lives in the slow process.

The fact that the elements U and Th are found in our planet Earth and in other material throughout our solar system implies that our solar system, including the Sun, is a third generation or later solar system. Studies of the isotopic anomalies in Pb produced by the radioactive decay of U and Th show that the age of the Earth, Moon, Mars, meteorites, and presumably our solar system, is 4.48 x 10^9 yr. The discovery that some elements in meteorites have isotopic abundance anomalies resulting from the decay of extinct isotopes (short-lived radioisotopes produced by nucleosynthesis) implies that the cloud of gas and dust that formed meteorites, and presumably the planets and other bodies in our solar system, condensed into discrete solid phases within a few hundred thousand years after an ancient supernova event that gave our solar system some of its present elements.

1.2.6 Relation between Nuclear Properties and Element Abundances

The light elements are many orders of magnitude more abundant than are the heavy elements in our solar system (Fig. 1.1). The abundance of the elements declines exponentially as the atomic numbers of the elements increase with the notable exception of Li, Be, B and Fe. The abundance of Fe is about two orders of magnitude greater than that predicted by interpolation of the abundance of its nearest element neighbors. The unusually high abundance of Fe is a consequence of its isotopes having the maximum values of binding energy per nucleon. The fusion of light nuclei produces isotopes that tend to accumulate at Fe during the normal fusion reactions in stars.

The abundance of Li, B, and Be are 10 to 100 million times less abundant than predicted by interpolation of the abundance of their nearest element neighbors. This startling anomaly to an otherwise regular pattern of element abundances is because the binding energy per nucleon for ^{4}He is unusually high, much higher than that for Li, Be, and B. The result is that Li, Be, and B make good nuclear fuels and they are rapidly consumed in stellar reactions, or they are less likely to be formed in the first place. Present thought is that most of the Li, Be, and B in the universe is formed by infrequent reactions of ions in interstellar space. The low abundance of B in soils is thus a result of nucleosynthesis, and not a consequence of geochemical fractionation processes, although soil weathering processes may still further reduce the concentration of B in soils to levels that induce deficiency symptoms in plants.

The pattern of element abundances shown in Fig. 1.1 also exhibits a pronounced alternation effect. With very few exceptions, elements with an even atomic number are more abundant than those with an odd atomic number. This phenomenon is called the even-odd effect. Neutrons and protons have quantum mechanical spin properties analogous to those of electrons. Proton-proton and neutron-neutron spin paired configurations are energetically favored states. Such spin pairing of nucleons has significant effects on the stability of nuclei and the resulting number of stable isotopes per element. Elements with an even atomic number have an even number of protons. The spin pairing of all of the protons in the isotopes of an even atomic numbered element results in a slightly greater binding energy per nucleon of all of the isotopes of that element. Elements with even atomic numbers, therefore, have a greater number of stable isotopes than do elements with odd atomic numbers. A greater number of stable isotopes results in greater elemental abundance.

Protons and neutrons also occur in shells as predicted by their quantum mechanical properties. The so-called magic numbers represent filled shells of protons or neutrons. Filled shells of protons or neutrons have a lower energy configuration than partially filled shells. Half-filled shells also have a lower energy configuration than partially filled shells but not as low as filled shells. The filled shells contain 2, 8, 20, 28, 50, 82, and 126 protons or neutrons. The extraordinary high value of the binding energy of the ^{4}He nucleus compared to those of Li, Be, and B arises because the ^{4}He nucleus has a filled shell of protons and a filled shell of neutrons. Oxygen, Ca, Ni, Sn, and Pb are slightly more abundant than their nearby element neighbors (Fig. 1.1) because the nuclei of their isotopes contain a full shell of protons, which results in a larger number of stable isotopes. Silicon is slightly more abundant because the nuclei of its isotopes contain a half-filled shell of protons.

A good example of the interplay of these factors is the relative abundance of K, Ca, and Sc with atomic numbers 19, 20, and 21, respectively. Calcium, with an even atomic number and a filled shell of 20 protons, would be expected to have the greatest number of stable isotopes and be the most abundant element. This is the case. Calcium has seven stable isotopes. Calcium also has two isotopes that have a filled shell of neutrons (^{40}Ca and ^{48}Ca). In contrast, Sc with an odd atomic number has only one stable isotope, while K has two stable isotopes and one (^{40}K) that is radioactive, but which is considered stable because of its long half-life (1.28×10^9 y). The relative abundance of these elements in the universe and in terrestrial soils and rocks can then be accurately predicted to be Ca > K > Sc.

1.3   General Abundances of Elements in the Earth's Crust and Soils

The abundance of silicate minerals in the Earth suggests that the composition of the Earth is greatly different from the composition of the solar system shown in Fig. 1.1. The atypical composition of the Earth is a result of chemical fractionation. The Earth's continental crust and soils are highly fractionated artifacts of the overall inventory of elements in the universe. However, even with the numerous opportunities for intense chemical fractionation of the nonvolatile elements during the primordial accretion of the Earth, the segregation of terrestrial material into core, mantle, and crust, and the weathering of crustal rocks to soil, the abundance pattern of elements in soils clearly retains many of the features imposed initially by nucleosynthesis.

The concentrations of selected elements in the continental crust and soils as a function of atomic number are shown in Fig. 1.2. Comparison of the element abundance patterns shown in Figs. 1.1 and 1.2 shows that although the major geochemical events that have led to the formation of the continental crust have severely fractionated several groups of elements, the two patterns exhibit remarkable similarity. The major differences are the severe depletion of the noble gases (He, Ne, Ar, Kr, and Xe) and the siderophilic (Fe-loving) elements (Co, Ni, Ru, Rh, Pd, Re, Os, Ir, Pt, and Au) from the Earth's crust.

It is proposed that the noble gases were not retained during the initial condensation and accretion of the primordial Earth because they do not readily form nonvolatile compounds. Current models of the formation of the solar system hypothesize that the planets formed from a gas and dust cloud or nebula which surrounded the Sun as it formed. The inner planets were either too hot or too small to retain an atmosphere while there was sufficient gas around to be picked up.

In the case of the siderophilic elements, they were scavenged from the silicate phase of the early Earth by the formation and segregation of the Fe core of the Earth. The chalcophiles (S, Se, Te, As, Sb) were responsive to segregation into the FeS phase of the outer core and are partially depleted in the crust. The remaining elements, the lithophiles, occur in relative abundances that are very similar to the solar abundances (compare Figs. 1.1 and 1.2). None of the isotopes of the elements Tc, Pm, Po, At, Rn,

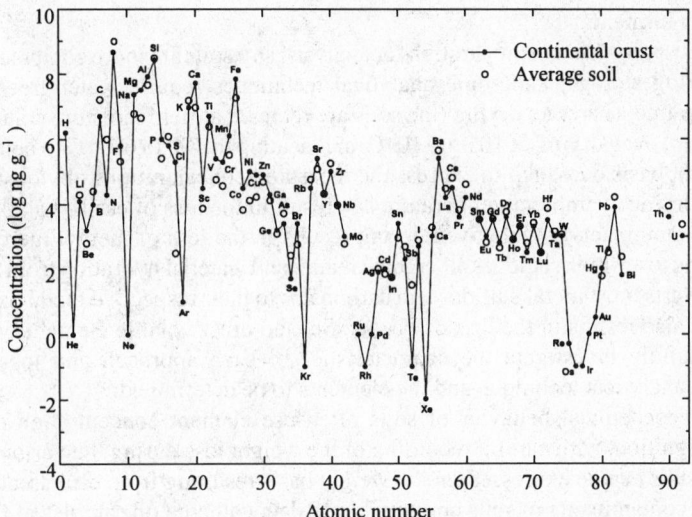

Fig. 1.2 Concentrations of elements in soils (open circles) are compared to the concentrations of elements in the continental crust (closed circles connected by a line). Soil data are from Table 1.2. Most values are the means for the Conterminous United States and Wisconsin plus values from other columns to complete the list of elements [Data for the continental crust from Mason and Moore (1982), Taylor and McLennan (1985) and Ranov and Yaroshevsky (1969)].

Fr, Ra, and Ac are stable and they occur only at ultratrace concentrations (generally $< 10^{-15}$ g g^{-1}) in soils and rocks as a result of the radioactive decay and natural fission of U and Th.

The concentrations of most elements in soil are very similar to those in the Earth's crust (Fig. 1.2). The concentrations of B, C, and N are higher in soils. This likely results from the accumulation of SOM. The concentrations of Na, Mg, Ca, and the first row transition elements (Sc–Zn) are lower in soils by as much as a factor of 10 and that of Cl is lower by a factor of a 1,000. With the exception of Na and Cl, these elements are relatively enriched in the minerals olivine and pyroxene, which are among the first minerals to undergo weathering when exposed to surface environmental conditions. These elements are partially lost from soils by leaching during weathering processes. Tellerium appears to be enriched in soil, but this may be an artifact of the low number of determinations of this element in soils. The concentrations of most of the heavier trace elements are nearly identical in soil and crust, indicating that these elements are strongly conserved during soil formation processes.

1.4   Concentrations of Elements in Soils

1.4.1   Techniques for Elemental Analysis of Soils

1.4.1.1   Soil Analysis
Determining the concentrations of elements in soils is not a trivial endeavor. Detailed procedures for determining the concentrations of most elements in soils by the common analytical techniques are given in Sparks et al. (1996). Soils are a mixture of inorganic and organic compounds that have refractory and volatile properties. Some of these compounds resist dissolution by simple treatments with strong acids or oxidizing agents. Special care must be exercised to ensure that soil samples are completely dissolved in preparation for analytical procedures that require aqueous samples.

1.4.1.2   Sample Treatment

Samples of soil are usually dried in preparation for analysis. Dry soils are more compositionally stable than wet soils during storage, and some analytical techniques require water free samples. The common drying treatments are: no drying (the soils are retained at field moisture content), drying at ambient temperature, oven drying at 105 to 110 °C, and heating to 400 to 600 °C to determine loss on ignition. The weight basis on which to express the measured concentrations obviously differs with these treatments. Drying at ambient temperature can result in the loss of easily volatile compounds (e.g., petroleum contaminants), while oven drying results in the loss of methyl mercury and other compounds. Heating to ignition removes all organic matter and water of hydration from minerals. This treatment also converts most metal sulfides and carbonates to their respective metal oxides with loss of CO_2 and SO_2. It also results in the loss of Pb, As, Se, and other volatile elements. The analyst is encouraged to carefully investigate the characteristics of each approach and to select the one appropriate for the analytical technique and the elements to be determined.

Studies of the geochemical behavior of soils often use element concentration data based on samples heated to ignition, with careful recording of the weight loss during the various drying steps so that the final results can be expressed on the weight basis resulting from each treatment if this is needed. Studies of contaminants in soils commonly use data gathered on samples at field moisture, ambient temperature dried, or oven dried at 105 °C, depending on the properties of the target elements or compounds and the requirements of the analytical technique. The mass of the soil after heating to ignition is the only treatment that gives a reproducible weight basis across all soil types.

1.4.1.3   Sample Dissolution

Many analytical techniques require that the sample be in solution. This can be a challenge with soil samples because the inorganic and organic materials in soils require different treatments for their dissolution. The easiest method is dissolution of the sample after heating to ignition because all of the organic matter is already removed. Grinding the samples to pass a 100-mesh plastic sieve helps ensure complete dissolution. A sequential treatment of the soil sample with HF and HNO_3 is often effective. A closed vessel, such as a Teflon® (du Pont de Nemours & Co) container, enclosed in a metal bomb is needed to prevent loss of Si as SiF_4 if this element will be measured. Excess fluoride ion attacks the glass nebulizer used in several common spectroscopic techniques, but HF resistant nebulizers are available from most manufacturers. Fusion of soil samples with lithium borate or other fluxes is very effective in destroying zircon and other minerals resistant to dissolution in strong acids.

Soil samples containing SOM require sequential treatments with a strong oxidizing agent and HF. An initial treatment with $HClO_4$ or fuming HNO_3 destroys soil organic matter. These are potentially dangerous operations and appropriate safety precautions are required. The inorganic residue is then treated with HF and HNO_3 until everything is dissolved.

1.4.1.4   Historical Analytical Techniques

The concentrations of elements in soils and rocks were traditionally determined by gravimetric and titration techniques, spectrophotometry, and spark source emission spectroscopy. The wet chemical techniques produce accurate results under the care of a skilled analyst, but they are labor intensive and time consuming. Spark source emission spectroscopy is a multi-element instrumental technique, but it is limited to semiquantitative measurements. These analytical techniques have been superseded by instrumental techniques capable of producing highly accurate results with excellent precision and low levels of detection. Most new techniques also have multi-element capability. Skilled analysts are needed to produce the accurate results these techniques are capable of yielding. Standard samples of soils and rocks that have been carefully calibrated by multiple analyses in many different laboratories selected for their analytical competence are available to serve as checks on the quality of element concentration measurements.

1.4.1.5 Optical Spectroscopy Techniques

Inductively coupled plasma-optical emission spectroscopy (ICP-OES) and atomic absorption spectrophotometry (AAS) and the graphite furnace method of sample introduction for AAS known as GFAAS are the most common analytical techniques now used for determining element concentrations in soils. Measurement of the characteristic light emitted by elements excited in the plasma torch is the basis for ICP-OES. The basis of AAS is the measurement of the absorption of each element's characteristic light, generated by a hollow cathode lamp, by the neutral analyte atoms in an air-acetylene flame. The graphite furnace variation uses an electrically heated C tube to vaporize the sample, and it is capable of analyzing sample volumes as small as 20 μL. The sensitivity for most elements ranges from μg to ng mL^{-1} for these techniques. The samples must be in solution for both techniques. Most ICP-OES instruments determine many elements simultaneously in a single sample, while AAS determines elements sequentially.

1.4.1.6 Mass Spectroscopy Techniques

The technique of inductively coupled plasma-mass spectroscopy (ICP-MS) couples a quadrupole mass spectrometer to an inductively coupled plasma sampling device to determine many elements simultaneously. Individual ions of elements in the sample are created in the plasma torch. The cations are then separated by mass in the mass spectrometer and counted by the detector on the mass spectrometer. The cost of analyses by ICP-MS is several times greater than that by ICP-OES, but ICP-MS is becoming common for geochemical analyses. Any element that forms cations in the plasma state can be determined with extreme sensitivity by ICP-MS, down to 0.01 ng mL^{-1} or less. The ratio of the stable isotopes can also be measured to a precision of a few parts per thousand. The technique can produce concentration values of very high precision if stable isotope dilution techniques are used. Sample volumes can be as small as 50 μL if a direct injection nebulizer is used. A laser ablation sampling device enables direct analysis of solid samples, but this has found little use for the analysis of soils. The preferred matrix is dilute HNO_3. High concentrations of Cl^- must be avoided because it produces complex cations that have masses similar to important analyte elements. The carrier gas in the plasma is Ar, and this prevents the measurement of K and Fe because K has mass spectral overlaps with Ar, and Fe and ArO have the same mass number.

1.4.1.7 X-Ray Fluorescence

X-ray fluorescence (XRF) techniques are limited to the measurement of major and minor elements in soils as they have a detection limit of several tens to hundred μg g^{-1}. The technique is based on the measurement of characteristic X-rays emitted by elements in the sample when it is irradiated with white X-rays (X-rays of continuous energy distribution). The detection limit becomes better as the atomic number of the element increases. Measurement of elements with atomic numbers lower than that of Na require special techniques. The greatest accuracy is achieved when the samples are diluted and fused with a lithium borate flux to form a glass disc with uniform composition. This requires that soil samples be heated to at least 600 ºC prior to fusion to remove organic matter, water, and carbonates. When available, this is the technique of choice for the rapid, inexpensive, and accurate determination of the major elements in soils and rocks.

1.4.1.8 Neutron Activation Analysis

Analysis of samples by neutron activation (NAA) requires access to a research nuclear reactor and a gamma ray spectrometer coupled to a high resolution germanium detector. The technique is based on the measurement of the characteristic gamma rays emitted by the radioactive isotopes created by capture of thermal neutrons in the reactor by the atoms in the sample. The most common technique is instrumental neutron activation analysis (INAA), which implies that the samples are radioassayed after neutron irradiation without any post irradiation chemical separations. About 25 elements in soils and rocks can be easily determined by INAA. Many additional elements can be determined if the irradiated sample is dissolved in the presence of a few mg of carrier for each element to be measured. Separation and purification of each element or group of elements remove elements with isotopes emitting potentially interfering gamma rays and decreasing the background in the gamma ray spectrum, thus greatly improving the limit of detection and the number of elements that can be measured.

No sample preparation is needed except grinding and sieving to produce a homogenous or representative sample. The potential effects of contamination are thus minimized and no sample blanks are needed. The sample size can range from a few mg to several g. The limit of detection ranges from ng to $\mu g\ g^{-1}$, depending on the element determined. The precision is limited by counting statistics and ranges from 1% upward. The cost of analysis is between that of ICP-OES and ICP-MS. It is the only common technique that can measure O in soils and rocks, although the very short half-life of the O isotope requires automated sample handling equipment. When the facilities are available, it is the technique of choice for fast and accurate measurements of Na, K, and Cl because of its sensitivity and freedom of sample contamination by these common elements. It also excels in the determination of the alkali metals, several of the first row transition elements, and the rare earth elements.

1.4.1.9   Carbon, Nitrogen, and Sulfur Analyzers

Carbon and N cannot be easily determined by any of the techniques discussed above. Sulfur can be determined by ICP-OES if the instrument has a vacuum pathway in the spectrometer. These three elements are obviously very important in soils. Carbon, N, and S (CNS) analyzers are available that can easily measure these elements in soil samples with great sensitivity and precision. These instruments oxidize the SOM and the concentrations of the oxides of C, N, and S in the resulting gas are determined by measuring the absorption of light from a laser. An alternative is to use a quadrupole mass spectrometer as the detector, which provides greater sensitivity and the ability to determine the isotopic composition of each of the elements.

1.4.2   Standard Reference Soils

The difficulty in obtaining accurate analyses of soils makes it imperative that analysts use certified reference materials to validate their analytical techniques. Such samples for soils have been available since the late 1970s, but the perusal of the literature required to collect the data given in this chapter shows that very few analysts utilize this important resource. The most comprehensive set of reference soil materials for total element analysis is available from the Canada Centre for Mineral and Energy Technology (Steger, 1980). The selected values of element concentrations that are used in the author's laboratory for these soils are given in Table 1.1, although the reader is cautioned that some of these values differ from the certified values. A compilation of the published values for these reference samples in 1983 showed that the values of concentrations for most elements in uniform samples ranged over a factor of two to three (Gladney and Burns, 1985). It is clear that many analytical laboratories could greatly improve the quality of their experimental results.

1.4.3   Variation of Element Concentrations Across Soil Orders

Soils within a soil order or soil series are generally not characterized by unique elemental compositions. The concentration of elements in soils has a set of characteristic values only within a few soil orders and then usually only for a small subset of the elements, such as Fe in Oxisols and C in Histosols. This is unlike the classification of rocks, where mineralogical and elemental compositions are combined with characteristics of texture and fabric to delineate the gradational nature of all rock series. It is, therefore, impossible to give exact representative concentrations of elements in a soil order.

Table 1.1 Mean concentrations of elements ($\mu g\ g^{-1}$) and one standard deviation from the mean from multilaboratory determination of elements in four standard soils from Canada (Gladney and Burns, 1985; Koons and Helmke, 1978). See Steger (1980) for recommended values and confidence intervals for selected elements when these samples are used for interlaboratory comparisons

Element	SO - 1 Sample from the C horizon of an Entisols (35 - 75 cm)		SO - 2 Sample from the B horizon of a Spodosols (10 - 30 cm)		SO - 3 Sample from calcareous glacial-till parent material of an Alfisols		SO - 4 Sample from the A horizon of a Mollisols	
	Mean	Std dev	Mean	Std dev	Mean	Std dev	Mean	Std dev.
Li	48	3	8.4	1.9	9	3	19.6	1.3
Be	1.4	0.8	1.7	0.4	0.86	0.14	1.5	0.2
B	19.3	1.5	2.3	0.4	24	6	42.9	1.6
C	2,700	400	48,000	1,200	66,300	1,500	44,200	1,600
N	390	110	2,100	200	140	70	3,800	200
F	800	150	480	50	320	40	300	10
Na	20,100	700	18,600	500	7,500	300	10,000	300
Mg	23,200	800	5,400	500	50,700	1,600	5,500	400
Al	92,700	2,000	80,200	2500	30,600	1,400	54,600	1,700
Si	257,200	2,900	249,700	3,100	157,200	1,400	319,600	2,800
P	662	30	3,000	230	455	50	880	70
S	108	14	410	80	167		595	110
Cl	150		84		210		30	
K	26,300	600	24,400	500	11,700	600	17,200	500
Ca	17,600	900	19,300	1,600	150,000	9,000	11,000	800
Sc	18.0	0.8	11.4	0.4	5.2	0.3	8.5	0.5
Ti	5,160	290	8,530	280	2,030	230	3,310	280
V	136	8	62	11	36	6	84	3
Cr	160	15	16	2	26	3	61	6
Mn	890	35	716	36	535	33	595	35
Fe	59,900	2,200	55,700	2,500	15,500	800	23,700	900
Co	31	5	8.2	2.5	5.6	1.2	9.9	1.0
Ni	94	10	10	6	14	4	25	4
Cu	61	5	6.6	1.8	16	2	21.0	2.1
Zn	146	5	124	5	52	3	94	3
Ga	28	5	25	5	10	4	13	4
Ge	1		1.5		1		1.5	
As	2.03	0.26	1.21	0.11	2.6	0.2	7.1	0.5
Se	0.088	0.007	0.36	0.06	0.040	0.015	0.44	0.08
Br	1.4	1	15	1	5.2		5.2	
Rb	145	11	79	8	40	2	72	7
Sr	317	34	334	49	225	32	172	16
Y	23.8	1.6	38.0	1.7	16	2	22	2
Nb	12		27		6		13	
Mo	2.1	1.2	2.6	1.4	2.0	1.2	1.5	0.7
Ag	<2		<2		<2		<2	
Cd	0.13	0.04	0.13	0.04	0.12	0.03	0.34	0.07
In	0.1		0.3		0.1		0.1	
Sn	2.6	0.6	2.6	0.5	0.97	0.15	3.2	0.5
Sb	0.25	0.05	0.11	0.02	0.310	0.013	0.70	0.02
I	12		17		1		3	
Cs	5.10	0.13	0.42	0.03	1.12	0.11	2.9	0.2
Ba	890	80	980	90	295	50	730	70

Table 1.1 (cont.)

Element	SO - 1 Sample from the C horizon of an Entisols (35 - 75 cm)		SO - 2 Sample from the B horizon of a Spodosols (10 - 30 cm)		SO - 3 Sample from calcareous glacial-till parent material of an Alfisols		SO - 4 Sample from the A horizon of a Mollisols	
	Mean	Std dev	Mean	Std dev	Mean	Std dev	Mean	Std dev
Zr	81	12	790	70	161	21	278	29
La	54.2	1.2	46.1	0.8	17.5	1.2	28.6	1.6
Ce	105	8	114	3	34.3	1.6	53.2	0.9
Pr	13		14		5		8	
Nd	45	3	57	2	16.7		26	2
Sm	8.1	0.2	11.2	0.9	3.5	0.2	4.7	0.2
Eu	1.65	0.15	3.3	0.2	0.79	0.13	0.99	0.08
Gd	6.2	0.3	11.3	0.8	3.1	0.1	4.03	0.06
Tb	0.89	0.07	1.6	0.2	0.49	0.01	0.60	0.01
Dy	4.8		9.2	1.6	2.8		3.8	
Ho	1		1.8		0.5		1	
Er	2.0		4.4		1.8		2.2	
Tm	0.4		0.67		0.3		0.4	
Yb	2.3	0.2	3.6	0.4	1.78		2.31	0.02
Lu	0.32	0.02	0.50	0.02	0.21	0.02	0.39	0.02
Hf	2.5	0.3	17.4	1.3	4.3	0.2	8.1	0.3
Ta	0.7	0.1	1.15	0.13	0.43	0.06	0.62	0.11
W	0.7		0.4		0.6		1.0	
Hg	0.021	0.003	0.082	0.012	0.016	0.005	29	4
Tl	0.9		0.7		0.3		0.4	
Pb	21	8	20	6	13	4	14	5
Bi	0.37		0.07		0.065		0.15	
Th	12.3	0.4	4.1	0.3	4.00	0.13	8.5	0.7
U	1.70	0.04	0.964	0.025	1.08	0.19	2.4	0.2

The composition of the soil parent material is only one factor affecting the composition of a soil, the others are climate, vegetation, topography, and time. Thus, soils with properties parallel enough to be classified within a single order can develop on substrates of varying parent material and elemental composition. The current system for classifying soils accentuates their morphology with less emphasis on their genesis and soil forming factors. The factors of climate, vegetation, topography, and time do not introduce sufficiently reliable patterns of element fractionation to use element concentrations as a useful criterion for classifying soils. Pedologists, therefore, do not generally generate or collect data on the elemental composition of soils. This has resulted in a scarcity of elemental concentration data for soils. Indeed, much of the data on the chemical composition of soils has been collected by scientists outside of the soil science discipline.

1.4.4 Concentrations of Elements in Soils by Geographic Region

Representative concentrations of elements in soils over several geographic regions are given in Table 1.2. It has been observed that element concentrations in natural materials are distributed in a log normal pattern. Therefore, the geometric mean gives a better estimate of the most probable value in large data sets. The arithmetic mean is used for smaller data sets or in cases where the soils are closely related. The mean values of concentration of most elements in soils vary less than a factor of two to three across geographic regions and many of the mean values are almost identical, such as those for

Table 1.2 Typical concentrations of elements (μg g⁻¹) in soils from various geographic regions

Element	World Soils[a] Median	World Soils[a] Range	Scotland[b] Arithmetic Mean	China[c] Geometric Mean	Tibet[d] Geometric Mean	Conterminous United States[e] Geometric Mean	Conterminous United States[e] Range	Wisconsin–United States[f] Arithmetic Mean	Wisconsin–United States[f] Range
Li	30	7–200	51.2	29	38	20	<5–140	17	5–31
Be	6	0.1–40	2.72	1.8	2.6	0.63	<1–15		
B	10	2–100	16.0	39	61	26	<20–300	50	3–123
C	20,000			12,000		16,000	600–370,000		
N	1,000	200–2,500							
F	200	20–300	410	440	525	210	<10–3,700		
Na	6300	750–7,500		7,000	11,000	5,900	<500–100,000	6,400	220–10,600
Mg	5,000	600–6,000		6,400	6,400	4,400	50–>100,000	3,500	310–8,800
Al	71,000	10,000–300,000		64,000	62,000	47,000	5000–>100,000	35,500	6,100–56,400
Si	333,000	250,000–350,000				310,000	1,700–440,000		
P	650	30–900	1525			260	40–6,800	350	110–1,330
S	700		769			1,200	<800–48,000	180	9–640
Cl	100		310						
K	14,000	400–30,000		18,000	19,000	15,000	50–63,000	16,700	8,900–24,900
Ca	13,700	7,000–500,000		7,100	12,000	9,200	100–320,000	4,900	190–15,100
Sc	7	10–25	19.0	10.5	9.3	7.5	<5–50	5.5	0.6–12
Ti	5,000	1,000–10,000	6400	5,800	3,200	2,400	70–20,000		
V	100	20–500	194	2.2	73	58	<7–500	42	5–90
Cr	100	5–3000	97.5	54	70	37	1–2000		
Mn	850	100–4,000	1030	432	566	330	<2–7000	620	66–1,720
Fe	38,000	7,000–550,000		28,000	20,000	18,000	100–>100,000	19,000	2,700–40,500
Co	8	1–40	16.5	11	11	6.7	<0.3–70	7.7	0.6–20
Ni	40	10–1,000	25.4	23	29	13	<5–700	26	2–82
Cu	20	2–100	20	20	20	17	<1–700	23	7–113
Zn	50	10–300	93.8	67	71	48	<5–2,900		
Ga	30	0.4–300	31.1	16	18	13	<5–70		
Ge	1	1–50	3.12	1.7	1.8	1.2	<0.1–2.5		
As	6	0.1–40	8.37	9.2	17	5.2	<0.1–97	5.7	1.1–26
Se	0.2	0.01–2	0.17	0.22	0.14	0.26	<0.1–4.3	1.4	0.2–4.5
Br	5	1–10	13.3	3.4	3.6	0.56	<0.5–11	3.7	1.0–11
Rb	100	20–600	91.0	107	134	58	<20–210	62	6.6–100
Sr	300	50–1,000	206	120	135	120	<5–3000		
Y	50	25–250	33.2	22	20	21	<10–200		
Zr	300	60–2,000	604	237	229	180	<20–2,000		
Nb			77.7			9.3	<10–100		
Mo	2	0.5–5	1.84	1.2	1.1	0.59	<3–15		
Ag	0.1	0.01–5	0.41	0.11	0.10				
Cd	0.06	0.01–0.7	0.77	0.074	0.080				

Table 1.2 (Cont.)

Element	World Soils[a] Median	World Soils[a] Range	Scotland[b] Arithmetic Mean	China[c] Geometric Mean	Tibet[d] Geometric Mean	Conterminous United States[e] Geometric Mean	Conterminous United States[e] Range	Wisconsin–United States[f] Arithmetic Mean	Wisconsin–United States[f] Range
In				0.060	0.049				
Sn	10	2–200	3.82	2.3	3.1	0.89	<0.1–10		
Sb			0.64	1.1	1.4	0.48	<1–8.8		
Te				0.021	0.037				
I	5		1.32	2.3	1.3	0.75	<0.5–9.6		
Cs	6	0.3–25	3.63	7.2	17			1.9	0.3–4.5
Ba	500	100–3,000	605	450	370	440	10–5000	420	47–650
La	30	1–5,000	34.7	37	40	30	<30–200	21	5.4–60
Ce	50		67.9	65	67	63	<150–300	39	9.2–76
Pr			6.46	6.7	6.2				
Nd			29.3	25	26	40	<70–300	21	6–80
Sm			5.8	4.9	5.0			3.3	0.67–6.7
Eu			1.23	0.98	0.9			0.64	0.1–1.4
Gd			3.49	4.4	4.4				
Tb			0.84	0.58	0.50			0.44	0.07–0.9
Dy			5.66	3.9	3.2				
Ho			0.80	0.83	0.83				
Er			2.08	2.4	2.2				
Tm			0.62	0.35	0.34				
Yb			4.43	2.3	2.1	2.6	<1–50	2.0	0.3–4.5
Lu			0.51	0.35	0.31			0.31	0.05–0.64
Hf	6		14.4	7.3	6.6			7.3	1.1–12
Ta				1.1	1.1			1.4	0.1–18
W			1.13	2.2	2.8				
Hg	0.03	0.01–0.3		0.040	0.021	0.058	<0.01–4.6		
Tl	0.1		0.31	0.58	0.66				
Pb	10	2–100	28.0	24	28	16	<10–700		
Bi			0.23	0.32	0.49				
Th	5	0.1–12	11.2	13	16	8.6	2.2–31	7.0	1.3–12.8
U	1	0.9–9	4.0	2.8	3.2	2.3	0.29–11	2.0	0.4–5.2

[a] Concentrations of elements in oven-dried soils summarized from data published in 1964 and earlier (Bowen, 1966)

[b] Concentrations of elements in 12 air-dried soils representing Spodosols and Inceptisols (Ure et al., 1979)

[c,d] Concentrations of elements in soils and other regoliths from 3893 locations in mainland China and 202 locations in Tibet. Soils were sampled at three depths at locations away from roads and mining activities (Chen et al., 1991)

[e] Concentrations of elements in oven-dried soils and other regoliths from 1,318 locations in the United States. Samples were collected from a depth of 20 cm. Locations supporting native vegetation were emphasized (Schacklette and Boerngen, 1984)

[f] Concentrations of elements in oven-dried soils from 33 locations in Wisconsin representing the top 60 cm of Mollisols, Alfisols, Spodosols, and Inceptisols (Boerth, 1997)

Cu. Some of the differences in the mean concentrations may result from the different depths of sampling for the various studies. Some of these data sets also include data on nonsoil regoliths.

1.4.5 Concentrations of Selected Trace Elements in Surface Soils

The awareness that potentially toxic elements can accumulate in plants from soils has led to intensive research efforts on the effects of element contamination of soils on the concentrations of such elements into the food chain. The mean concentrations of selected trace elements in surface soils from the United States and England and Wales are compared in Table 1.3. Both data sets, which are based on the analyses of several thousand soils, appear to be of high quality. The data set for the United States was obtained on soils from sample locations that were thought to be unaffected by obvious sources of element contamination. The data set for England and Wales was obtained by sampling on a national grid pattern and included some areas with natural Pb-Zn mineralization. These data sets are useful guidelines for typical concentrations of these elements in topsoil.

1.4.6 Concentrations of Selected Elements in Soil Orders

The geometric mean concentrations of selected trace elements in the United States topsoils as a function of soil order are given in Table 1.4. The concentrations of Ni, Cu, Zn, Cd, and Pb are similar for most of these soil orders. The Ultisols are the most leached of these soil orders and they have the lowest concentrations of these elements, in agreement with the earlier observation that the first row transition elements are depleted in soils compared to their concentrations in crustal rocks. The concentration of Cu is highest in the Histosols, which is expected because of the strong association of Cu with soil humic substances. The concentrations of Ni and Zn are highest in the Vertisols. This may be a consequence of the high content of the clay mineral montmorillonite in Vertisols. Montmorillonite has a high cation exchange capacity (CEC) plus the ability to accommodate divalent cations of various ionic radii within its structure. Both of these properties would tend to retain Ni and Zn during soil forming processes.

Table 1.3 Concentrations (μg g^{-1}) of selected trace elements in 3045 samples of air-dried agricultural soils from the United States [from Holmgren et al. (1993)] and in 5692 samples of topsoil (0–15 cm) from England and Wales [from McGrath and Loveland (1992)]

| Element | United States Agricultural Soils | | England and Wales Topsoils | |
	Geometric Mean	Range	Geometric Mean	Range
Cr			41.2	0.2–838
Mn			3,740	41–62,700
Co			10.6	0.2–322
Ni	17	0.7–269	24.5	0.8–440
Cu	18	<0.6–495	23.1	1.2–1,510
Zn	43	<3.0–264	97.1	5.0–3,650
Cd	0.18	<0.01–2.0	0.8	<0.2–41
Ba			141	11–2,970
Pb	11	7.5–135	74	3.0–16,300

Table 1.4 Geometric mean concentrations (µg g⁻¹) of selected trace elements in 3045 samples of United States agricultural soils by taxonomic soil order [Adapted from Holmgren et al., 1993. J. Environ. Qual. 22:335–348, with permission of the American Society of Agronomy]

Element	Ultisols	Alfisols	Spodosols	Mollisols	Vertisols	Aridisols	Inceptisols	Entisols	Histosols
Ni	7.4	12.6	22	22.8	75.9	24.3	25.6	21	11.3
Cu	6.2	10.9	48.3	19.1	48.5	25	28.4	21.1	193
Zn	13.8	31.3	44.1	54.4	93.1	70.1	69.4	65.5	62.6
Cd	0.049	0.112	0.2	0.227	0.239	0.304	0.223	0.246	0.622
Pb	8	9.6	10	10.7	17.1	10.6	15.2	10	12.5

The geometric mean concentrations of selected trace elements as a function of soil order in topsoils from mainland China are given in Table 1.5. There is little noteworthy difference in the mean concentrations of elements across orders except that the highest concentrations of F, Cr, V, Co, Ni, Zn, Cu, Cd, Hg, and Pb occur in the Lithosols. The slight extent of weathering in the Lithosols would provide little opportunity for these elements to be leached.

The concentrations of selected elements as a function of soil order on a world scale are given in Table 1.6. The mean concentrations are similar across all orders, with the exception of some elements in the Histosols. These data were compiled from published sources. They represent the results of many different analytical laboratories and soils of the same order developed from different parent materials. They would be expected to be less uniform than the data in Tables 1.3 and 1.4, which had the advantage of a more restricted geographic range and a single analytical laboratory.

Table 1.5 Geometric mean concentrations (µg g⁻¹) in soils from mainland China by soil order [Adapted from Chen et al., 1991. Soil Air Water Pollut. 57–58:699–705, with kind permission from Kluwer Academic Publishers]

Soil Order	No	F	Cr	V	Mn	Co	Ni	Cu	Zn	As	Se	Cd	Hg	Pb
Lithosols	205	589	76.4	114	577	16.8	35.9	26.5	94.6	16.2	0.05	0.24	0.084	29.8
Cold High-land Soils	196	490	69.6	78.5	619	12.0	31.4	23.8	78.9	16.9	0.19	0.098	0.021	26.0
Mollisols	240	378	46.2	67.7	563	11.3	21.9	10.0	60.0	9.00	0.14	0.070	0.022	18.3
Aridisols	108	394	48.0	69.0	582	12.0	23.6	21.7	59.3	8.63	0.11	0.087	0.011	18.6
Inceptisols	508	446	56.4	82.1	613	13.8	27.6	21.8	71.4	10.1	0.18	0.080	0.009	20.0
Alfisols	186	366	46.8	69.0	1130	11.2	17.7	15.1	75.3	4.95	0.16	0.090	0.055	19.8
Ultisols	796	483	49.8	86.2	323	9.50	20.6	17.8	71.9	9.50	0.38	0.055	0.077	26.8
Oxisols	262	281	34.2	57.0	197	5.55	10.2	10.9	33.6	5.70	0.33	0.030	0.040	26.0
Vertisols	223	447	53.2	77.5	586	11.8	23.3	19.6	64.4	8.60	0.20	0.070	0.030	24.2
Entisols	265	517	63.0	82.9	604	11.4	28.9	22.2	69.7	9.40	0.15	0.090	0.032	21.4

Table 1.6 World means and ranges of concentrations ($\mu g\ g^{-1}$) of selected trace elements in surface soils from some soil orders [Adapted from Kabata-Pendias and Pendias, 1992. Reprinted with permission from Trace elements in soils and plants. Copyright CRC Press, Boca Raton, FL]

Element	Spodosols Arithmetic Mean	Range	Inceptisols Arithmetic Mean	Range	Mollisols Rendolls Arithmetic Mean	Range	Mollisols Ustosols - Borolls Arithmetic Mean	Range	Histosols Arithmetic Mean	Range
Li	22	<5–72	46	1.4–130	56	6–105	53	9–175	1.3	0.01–3.2
B	22	1–134	40	<1–128	40	1–210	45	11–92	25	4–100
F	130	10–1,100	385	<10–800	360	<10–840	530	10–1,200	220	10–335
Sc	5	0.8–30	8	2.4–20	8	<5–15	10	<5–20	---	---
Ti	26,000	200–17,000	3,300	500–24,000	4,800	400–10,000	3,500	700–7,000	2,300	80–6,700
V	67	10–260	76	15–330	115	10–500	78	20–150	18	6.3–150
Cr	47	1.4–530	51	4–1,100	83	5–500	77	11–195	12	1–100
Mn	270	7–2,000	525	45–9,200	445	50–7,750	480	100–3,900	465	7–2,200
Co	5.5	0.1–65	10	3–58	12	1–70	7.5	0.5–50	4.5	0.2–49
Ni	13	1–110	26	3–110	34	2–450	25	6–61	12	0.2–119
Cu	13	1–70	23	4–100	23	6.8–70	24	6.5–140	16	1–113
Zn	45	3.5–220	60	9–360	100	10–570	65	20–770	50	5–250
As	4.4	<0.1–30	8.4	1.3–27	---	---	8.5	1.9–23	9.3	<0.1–67
Se	0.25	0.005–1.3	0.34	0.02–1.9	0.38	0.1–1.4	0.33	0.1–1.2	0.37	0.1–1.5
Sr	87	5–1,000	210	15–1,000	195	15–1,000	145	70–500	100	5–300
Mo	1.3	0.17–3.7	2.8	0.1–7.2	1.5	0.3–7.4	2	0.4–6.9	1.5	0.3–3.2
Cd	0.37	0.01–2.7	0.45	0.08–1.6	0.62	0.38–0.84	0.44	0.18–0.71	0.78	0.19–2.2
I	2.3	<0.1–10	1.7	0.3–8.3	3.4	0.3–9.5	2.4	0.3–11	4	1–10
Ba	330	20–1,500	520	19–1,500	520	150–1,500	520	100–1,000	175	10–700
Hg	0.05	0.008–0.7	0.1	0.01–1.1	0.05	0.01–0.5	0.1	0.02–0.53	0.26	0.04–1.1
Pb	22	2.3–70	28	1.5–70	26	10–50	23	8–70	44	1.5–176

1.4.7 Concentrations of Elements in Selected Soil Orders as a Function of Depth or Horizon

Soil horizons develop during the soil forming process as a result of the additions, removals, transfers, and transformation of compounds and elements as a function of soil parent material, climate, vegetation, topography, and time. These processes result in some stratification of element concentrations in soils. The effects of these processes on the concentrations of the major elements are much better documented than for the trace elements. In general, the stratification of element concentrations during horizon development is too variable to provide distinctive criteria at the level of soil orders.

The effects of profile development on element concentrations in Mollisols may be relatively small as shown by the data in Table 1.7. The largest change is some surface depletion of Ca and Mg compared to their concentrations at depth. The high concentration of Ca in the deeper horizons results from the presence of $CaCO_3$ in the parent material. While this is common in many Mollisols, other Mollisols develop under regimes with less rainfall and they show Ca accumulation in the upper horizons.

A well-developed Spodosol profile is a striking example of soil genesis as the horizons are sharply delineated. The leaching of Fe from the bleached eluvial horizon and its accumulation in the reddish brownish illuvial horizon result in the distinctive ash gray horizon of Spodosols developed in cool, temperate, humid regions, as shown by the Fe concentrations in these horizons in Table 1.8. In contrast, intense leaching over a long period of time results in the accumulation of Fe and Al compounds and other insoluble oxides and the depletion of Ca, Mg, and Si. Both features are characteristic of Oxisols (Table 1.9).

The concentrations of most elements are much lower in Histosols than in mineral soils. The profiles of Histosols exhibit some unique features of element concentrations. The surface layers of Histosols often contain the highest concentrations of elements that are active in biological processes or are

Table 1.7 Major element concentrations ($\mu g \; g^{-1}$ of oven dried soil) in a Mollisols by depth [Adapted from Byers et al., 1935]

Element	A_1 0 - 23 cm	B_1 23 - 43 cm	B_2 43 - 84 cm	C 84 - 152 cm
Na	8,460	8,610	6,230	6,600
Mg	5,550	4,340	12,100	14,300
Al	60,300	59,900	53,900	55,000
Si	324,000	344,000	250,000	269,000
P	370	22	392	327
S	560	440	280	240
K	14,900	14,600	11,800	13,400
Ca	11,500	10,900	95,900	75,200
Ti	2,940	2,940	2,760	2,880
Mn	1,240	930	1,160	1,400
Fe	26,900	27,700	25,900	27,500

Table 1.8 Major element concentrations (µg g^{-1} soil) in a Spodosols by depth [Adapted from McKeague et al., 1983]

| | E | Bh | Bir$_1$ | Bir$_2$ | Bir$_3$ | BC | C |
Element	0-10 cm	10-20 cm	20-35 cm	35-42 cm	42-56 cm	56-70 cm	70-130 cm
Na	20,000	17,000	19,000	22,000	23,000	25,000	25,000
Mg	4,000	4,000	5,000	6,300	6,200	6,300	6,300
Al	63,000	63,000	85,000	90,000	90,000	87,000	87,000
Si	322,000	234,000	260,000	269,000	279,000	288,000	289,000
K	29,000	21,000	25,000	24,000	27,000	30,000	29,000
Ca	18,000	18,000	21,000	25,000	28,000	28,000	28,000
Ti	11,000	6,700	6,500	7,400	8,000	8,200	8,500
Fe	36,000	62,000	47,000	54,000	52,000	52,000	53,000

strongly bound to soil humic substances. This is illustrated by the data in Table 1.10, which shows higher concentrations of Mg, Ca, Co, Ni, Zn Se, Mo, Cd, and Pb in the surface layers compared to their concentrations at depth. Elements that are not bioactive or are not associated with humic substances, such as Sc, Ti, and Zr, occur at higher concentrations at depth because the ratio of mineral to organic compounds in Histosols often increases with depth.

1.4.8 Effect of Quartz on Element Concentrations in Soils

The mineral composition of soils affects the concentrations of elements in soils. The greatest effect results from variations in the content of quartz (SiO_2) in soils that are not highly leached (Oxisols and Ultisols). Histisols are also exempt. This occurs because the structure of quartz precludes substitution of other elements into its structure. The concentrations of all elements except Si and O are much lower in quartz than in soils. The greatest concentrations of quartz in soils occur in the sand sized fraction and smaller amounts occur in the silt sized fraction. The amount of quartz in the clay sized (< 2 µm) fraction is generally small. Quartz particles reduced to clay sized dimensions have sufficient excess surface free energy that their solubility is greatly increased and their longevity on the geologic time scale is short. The result is that fine-textured soils tend to have low concentrations of quartz, and thus, the highest concentrations of all elements except Si. Soils with high concentrations of quartz are generally coarse-textured and tend to have the lowest concentrations of elements. This phenomenon contributes to the high concentrations of trace elements in Vertisols.

Table 1.9 Major element concentrations (µg g^{-1} of oven dried soil) in an Oxisols by depth [Adapted from Bennett, 1926]

| | | | Depth | | |
Element	0–66 cm	66–102 cm	102–396 cm	396–488 cm	488 + cm
Na	3,640	2,220	3,560	2,900	2,670
Mg	1,990	2,890	3,860	3,620	10,700
Al	97,700	58,900	65,400	77,600	10,600
Si	15,300	10,500	8,600	7,250	196,000
P	65	trace	trace	trace	trace
K	500	660	170	420	660
Ca	860	trace	71	1,100	10,700
Ti	4,800	1,600	4,800	4,800	300
Mn	3,250	2,170	2,940	3,640	930
Fe	441,000	487,000	497,000	476,000	54,800

Table 1.10   Element concentrations ($\mu g\ g^{-1}$ air-dried weight) in a Histosols by depth [Reprinted from Chattopadhyay and Jervis, 1974. Anal. Chem. 46:1630–1639. Copyright American Chemical Society with permission]

				------------Soil depth in cm------------				
Element	Surface -1	Surface - 2	0 - 7.5	7.5 - 15.0	15.0 - 22.5	22.5 - 30.0	30.0 - 37.5	37.5 - 45.0
Na	6,000	5,950	5,000	4,320	4,170	3,860	3,700	3,610
Mg	780	765	640	420	400	525	546	583
Cl	72.2	68.5	45.1	32.6	26.0	38.7	11.6	8.42
Ca	20,000	20,000	14,000	11,500	12,300	10,550	11,400	12,200
Sc	3.22	3.41	3.65	3.06	4.31	5.26	5.52	5.88
Ti	101	105	122	145	160	168	173	185
V	10.5	11.6	15.2	21.4	26.1	30.0	31.4	32.3
Cr	14.3	12.1	18.5	30.2	35.4	45.8	28.6	26.7
Mn	1,350	1,275	840	625	715	1,120	1,200	1,430
Fe	28,000	28,350	21,200	18,000	19,750	22,400	24,000	27,700
Co	8.52	8.65	8.24	7.66	5.48	7.32	4.37	5.28
Ni	8.11	7.85	6.21	6.64	5.23	7.55	4.81	3.57
Zn	165	152	148	128	94	65	48	42
As	1.72	1.61	1.85	0.96	0.78	0.85	0.50	0.42
Se	1.10	1.43	1.22	0.81	0.62	1.05	0.91	0.53
Sr	80.5	82.8	65.6	52.3	41.4	56.7	75.0	89.3
Zr	200	200	278	323	405	467	500	676
Mo	2.12	2.00	1.32	1.05	0.84	1.18	1.25	1.47
Ag	0.92	0.85	0.68	0.52	0.40	N. D.[a]	N. D.	N. D.
Cd	1.45	1.52	1.71	0.97	0.85	1.15	1.32	0.67
In	2.60	2.12	2.95	1.56	1.12	1.86	0.93	0.56
Sn	14.0	13.5	12.2	12.5	10.1	12.3	13.1	14.2
Sb	2.11	1.85	1.73	0.66	0.58	0.93	0.75	0.70
Te	6.25	6.10	4.33	2.17	1.2	N. D.	N. D.	N. D.
Cs	4.82	4.91	2.22	1.95	1.83	2.65	1.78	2.26
Ba	285	270	252	293	300	328	295	330
Hg	1.45	2.50	1.18	1.62	1.80	1.20	1.30	N. D.
Tl	0.22	0.20	0.17	N. D.	N. D.	0.18	N. D.	N. D.
Pb	32.2	30.0	20.1	18.5	17.8	15.2	12.4	11.6
Bi	1.52	1.46	1.33	N. D.	1.0	N. D.	N. D.	1.21

[a] N. D., not detected

Much of the variation in element concentrations within soils from a limited geographic region is a result of the phenomena described above (Helmke et al., 1977; Boerth, 1997). Separation and analysis of the clay sized fraction show that this component of soils has very uniform element concentrations over hundreds km^2. This feature is very useful in determining the areal extent of element contamination of soils and sediments by multi-element analysis of the clay sized fraction.

1.5   Composition of the Soil Solution

1.5.1   Sampling the Soil Solution

The soil solution represents the aqueous phase of the soil at or below field moisture capacity. The soil solution is differentiated from soil water, which is the moisture that fills the pores between soil particles at above field moisture capacity. The soil solution is considered to be the medium where soil chemical reactions occur (Section B, Chapter 3). The simplicity of the foregoing definition belies the difficulty in studying the soil solution in practice. This occurs because no techniques are commonly available to study the chemical components in the soil solution *in situ*, and because the composition of the soil solution varies with moisture content. The soil solution is, therefore, operationally defined by the particular method used to obtain samples.

The common methods to sample the soil solution are displacement by a fluid immiscible with water and direct extraction by vacuum, applied pressure, or centrifugation. Displacement methods use high density organic liquids (such as trichlorotrifluoroethane). They are mixed with the soil and centrifuged so that the soil solution is displaced and floats to the top of the centrifuge tube. Displacement methods give a high yield of soil solution and are applicable to field moist soils. The soil solution can also be separated from the soil by withdrawing it through a filter by application of a vacuum or high pressure, or by centrifuging the soil at high speed.

1.5.2 Concentrations of Solutes in Soil Solutions

Only five cations (Na^+, K^+, Mg^{2+}, Ca^{2+}, and NH_4^+) and four anions (NO_3^-, HCO_3^-, Cl^-, and SO_4^{2-}) are the dominant solutes in almost all soil solutions. These species commonly occur at mM or lower concentrations in most soils. All of the other elements occur at μM or lower concentrations. Some examples of the concentrations of the dominant solutes in several soil orders are given in Table 1.11. These values should not be considered typical values for the various soil orders as the composition of the soil solution varies with moisture content and by the method of sample collection. The concentration of HCO_3^- varies with the partial pressure of CO_2 (P_{CO2}) in the soil air. Values of P_{CO2} in soil range from 0.006–0.015 atm for well-drained surface soils supporting abundant plant life, up to 0.2 atm under frozen soil horizons.

Table 1.11 Typical concentrations of major dissolved components in soil solutions collected from diverse soil orders by depth

Soil	pH	Na	K	Mg	Ca	NH$_4$	Al	Si	NO$_3$	HCO$_3$	Cl	SO$_4$
Alfisols[a]	6.2	0.79	0.67	0.25	0.75	0.24	0.062	0.36	1.34	0.14	0.23	0.5
Alfisols[d]	5.3	0.38	0.62	0.14	1.15		0.013	0.844	1.78		0.56	0.47
Oxisols[b]												
0-10 cm	5	0.86	0.79	0.73	0.4	5.22			0.53	1.62	3.06	1.53
20-30 cm	5.8	0.3	0.18	0.09	0.01	1.52			0.05	0.12	0.45	0.23
90-120 cm	4.8	0.28	0.01	0.02	<0.01	0.03			<0.01	<0.01	0.25	0.05
Inceptisols[b]												
0-10 cm	7.9	0.07	0.3	0.44	0.64	0.03			0.01	2.12	0.22	0.11
60-90 cm	6.6	0.14	0.05	0.05	0.05	0.01			0.01	0.14	0.04	0.02
Ultisols[c]												
15-30 cm	4.7	0.17	0.12	0.07	0.11		<0.0004	0.14	0.31		0.28	0.11
45-60 cm	4.9	0.04	0.12	0.06	0.15		0.0004	0.08	0.3		0.66	0.05
75-90 cm	4.6	0.21	0.64	0.07	0.11		0.011	0.1	0.3		0.54	0.09
Spodosols[d]	4.3	0.13	0.21	0.08	0.35		0.025	0.142	0.02		0.82	0.31
Mollisols[e]												
Saturation	7.65	1.4	0.3	0.85	3.2	0.4			4.4	3.4	1.9	0.85
10 kPa moisture	7.52	2.3	0.3	1.1	5.55	0.2			8.8	3.8	2.3	1.4

(header span: -------mmol L^{-1}------- over Na through SO$_4$)

[a] Average of 8 topsoils from New Zealand. Solution collected by centrifugal displacement from May to September (Edmeades et al., 1985)

[b] Solution collected by centrifugal displacement from Australian soils (Gillman and Bell, 1978)

[c] Solution collected by centrifugal displacement from Georgia, USA, soils at 7.5 kPa soil-water potential (Gillman and Sumner, 1987)

[d] Solution collected by displacement with inert flurocarbon liquid from topsoils from the United Kingdom at field moisture content (Kinniburgh and Miles, 1983)

[e] Solution collected by pressure plate extraction from California, USA, soils at saturation and 10 kPa soil-water potential (Eaton et al., 1960)

1.6   References

Bennett, H.H. 1926. Some comparisons of the properties of humid-tropical and humid-temperate American soils, with special reference to indicated relations between chemical composition and physical properties. Soil Sci. 21:349–358.

Boerth, T.J. 1997. Indigenous element concentrations in Wisconsin soils. M.S. Thesis. University of Wisconsin, Madison, WI.

Bowen, H.J.M. 1966. Trace elements in biochemistry. Academic Press, London, UK.

Burbidge, E.M., G.R. Burbidge, W.A. Fowler, and F. Hoyle. 1957. Synthesis of the elements in stars. Rev. Modern Phys. 29:547–552.

Byers, H.G., L.T. Alexander, and R.S. Holmes. 1935. The composition and constitution of the colloids of certain of the great groups of soils. USDA Tech. Bull. 484.

Chattopadhyay, A., and R.E. Jervis. 1974. Multielement determination in market-garden soils by instrumental photon activation analysis. Anal. Chem. 46:1630–1639.

Chen, J., F. Wei, C. Zheng, Y. Wu, and D.C. Adriano. 1991. Background concentrations of elements in soils of China. Water Air Soil Pollution 57–58:699–705.

Eaton, F.M., R.B. Harding, and T.J. Ganje. 1960. Soil solution extractions at tenth-bar moisture percentages. Soil Sci. 90:253–258.

Edmeades, D.C., D.M. Wheeler, and O.E. Clinton. 1985. The chemical composition and ionic strength of soils solutions from New Zealand topsoils. Aust. J. Soil Res. 23:151–165.

Fowler, W.A. 1984. Experimental and theoretical nuclear astrophysics: the quest for the origin of the elements. Rev. Modern Phys. 56:149–159.

Gillman, G.P. and M.E. Sumner. 1987. Surface charge characterization and soil solution composition of four soils from the southern piedmont in Georgia. Soil Sci. Soc. Am. J. 51:589–594.

Gillman, G.P. and L.C. Bell. 1978. Soil solution studies on weathered soils from tropical North Queensland. Aust. J. Soil Res. 16:67–77.

Gladney, E.S., and C.E. Burns. 1985. 1983 Compilation of elemental concentration data for samples SO-1 to SO-4. Geostand. Newsl. 9:38–43.

Helmke, P.A., R.D. Koons, P.J. Schomberg, and I.K. Iskandar. 1977. Determination of trace element contamination of sediments by multielement analysis of clay-size fraction. Environ. Sci. Tech. 11:984–989.

Holmgren, C.G.S., M.W. Meyer, R.L. Chaney, and R.B.J. Daniels. 1993. Cadmium, lead, zinc, copper, and nickel in agricultural soils of the United States of America. J. Environ. Qual. 22:335–348.

Kabata-Pendias, A. and H. Pendias. 1992. Trace elements in soils and plants. 2nd Ed. CRC Press, Boca Raton, FL.

Kinniburgh, D.G. and D.L. Miles. 1983. Extraction and chemical analysis of interstitial water from soils and rocks. Environ. Sci. Tech. 17:362–368.

Koons, R.D., and P.A. Helmke. 1978. Neutron activation analysis of standard soils,. Soil Sci. Soc. Am. J. 42:237–240.

Mason, B., and C.B. Moore. 1982. Principles of geochemistry. 4th Ed. John Wiley and Sons, New York, NY.

McGrath, S.P. and P.J. Loveland. 1992. The soil geochemical atlas of England and Wales. Blackie Academic and Professional, Glasgow, UK.

McKeague, J.A., F. DeConinck, and D.P. Franzmeier. 1983. Spodosols. p. 217–252. In L.P. Wilding, N.E. Smeck and G.G. Hall (ed.) Pedogenesis and soil taxonomy. Elsevier, Amsterdam, Netherlands.

Ranov, A.B., and A.A. Yaroshevsky. 1969. Chemical composition of the Earth's crust. Am. Geophys. Union Monogr. 13:37–41.

Ross, J.E., and L.H. Aller. 1976. The chemical composition of the sun. Science 191:1223–1226.

Schacklette, H.T., and J.G. Boerngen. 1984. Element concentrations in soils and other surficial materials of the conterminous United States. US Geol. Surv. Prof. Pap. 1270.

Selbin, J. 1973. The origin of the chemical elements. J. Chem. Edu. 50:380–387.

Sparks, D.L., A.L. Page, P.A. Helmke, R.H. Loeppert, P.N. Soltanpour, M.A. Tabatabai, C.T. Johnston, and M.E. Sumner. 1996. Chemical methods of soil analysis. Soil Science Society of America, Madison, WI.

Steger, H.F. 1980. Certified reference materials. CANMET Report M38-13/80-6E, Energy, Mines, and Resources Canada, Ottawa, Canada.

Taylor, S.R., and S. McLennan. 1985. The continental crust: Its composition and evolution. Blackwells Scientific Publications, Palo Alto, CA.

Trimble, V. 1975. The origin and abundance of the chemical elements. Rev. Modern Phys. 47:877–898.

Ure, A.M., J.R. Bacon, M.L. Berrow, and J.J. Watt. 1979. The total trace element content of some Scottish soils by spark source mass spectrometry. Geoderma 22:1–23.

2

Soil Organic Matter

J.A. Baldock
Division of Land and Water, CSIRO, Glen Osmond, Australia

P.N. Nelson
Division of Land and Water, CSIRO, Townsville, Australia

2.1   Introduction and Definitions

Research pertaining to the organic fraction of soils can be traced back more than 200 yr. Achard (1786) isolated a dark amorphous precipitate upon acidification of an alkaline extract from peat. The effect of organic matter on soil N fertility (Liebig, 1840), and studies on the use of animal manures for maintaining soil fertility (Lawes, 1861) and the influence of soil and tree species on the development of humus form (Muller, 1887) demonstrated the importance of organic matter in soil processes. The advancement of organic chemical methodologies and confirmation of the presence of various chemical structures in soil organic matter (SOM) lead to theories that SOM was composed of a heterogeneous mixture of dominantly colloidal organic substances containing acidic functional groups and N. More recently, the polyphenol theory was proposed in which quinone structures of lignin and microbial origin polymerize in the presence of N containing groups (amino acids, peptides, and proteins) to produce nitrogenous polymers (Flaig et al., 1975). The early research pertaining to SOM has been reviewed by Stevenson (1994). While alkaline extraction of SOM is still practiced, modern analytical techniques, including solid-state ^{13}C nuclear magnetic resonance spectroscopy (^{13}C NMR), infrared spectroscopy (IR), and pyrolysis gas chromatography/mass spectroscopy (Py-GCMS) allow selective probing of the chemistry of SOM within samples of whole soil, and avoid the problems of incomplete extraction, lack of biological significance, and artifact synthesis often ascribed to alkaline extraction procedures. The combination of these techniques with novel approaches capable of identifying biologically important fractions of SOM has significantly advanced our knowledge of the organic fraction of soils and its dynamics over the last 20 years.

Despite such a long history of research and new methodological and technological advancements, many questions related to the genesis and chemical composition of organic materials in soils and their impacts on soil fertility, pedogenesis, and soil physical and chemical properties persist today. Many excellent texts and review papers have been written on the topic of SOM; some of the more recent of these are given in Table 2.1.

An examination of terms used to describe SOM and its components in the literature revealed a lack of precise definitions of what SOM and its various fractions represent. Such a problem exists because

Table 2.1   List of texts and review articles pertaining to the study of soil organic matter released since 1985. Individual review articles within texts have not been identified separately [modified from Hedges and Oades, 1997]

Texts

Humic substances in soil, sediment and water (Aiken et al., 1985)

Soil organic matter and biological activity (Vaughan and Malcolm, 1985)

Interactions of soil minerals with natural organics and microbes (Huang and Schnitzer, 1986)

Cycles in soil: carbon, nitrogen, phosphorus, sulfur, micronutrients (Stevenson, 1986)

Chemistry of soil organic matter (Kumada, 1987)

Soil organic matter (Tate, 1987)

Humic substances and their role in the environment (Frimmel and Christman, 1988)

Dynamics of soil organic matter in tropical ecosystems (Coleman et al., 1989)

Humic substances II: In search of structure (Hayes et al., 1989)

Soil microbiology and biochemistry (Paul and Clark, 1989)

Soil Biology and Biochemistry, Vol. 6 (Bollag and Stotzky, 1990)

Humic substances in soil and crop science (MacCarthy et al., 1990a)

Advances in soil organic matter research: The impact on agriculture and the environment (Wilson, 1991)

Soil Biology and Biochemistry, Vol. 7 - Vol. 9 (Stotzky and Bollag, 1992, 1993, 1996)

Soil organic matter dynamics and sustainability of tropical agriculture (Mulongoy and Merckx, 1993)

Environmental organic chemistry (Schwartzenbach et al., 1993)

Humic substances in the global environment (Senesi and Miano, 1994)

Humus chemistry: genesis, composition, reactions (Stevenson, 1994)

Soil organic matter management for sustainable agriculture (Lefroy et al., 1995)

Humic substances of soils and general theory of humification (Orlov, 1995)

Carbon forms and functions in forest soils (McFee and Kelly, 1995)

The role of nonliving organic matter in the earth's carbon cycle (Zepp and Sonntag, 1995)

Driven by nature: plant litter quality and decomposition (Cadisch and Giller, 1997)

Soil organic matter in temperate ecosystems (Paul et al., 1997)

Review Articles

The retention of organic matter in soils (Oades, 1988)

An introduction to organic matter in mineral soils (Oades, 1989)

Soil organic matter - the next 75 years (Schnitzer, 1991)

Physical fractionation of soil organic matter in primary particle size and density separates (Christensen, 1992)

A hierarchical model for decomposition in terrestrial ecosystems - application to soils of the humid tropics (Lavelle et al., 1993)

Organic matter in tropical soils - current conditions, concerns and prospects for conservation (Ross, 1993)

Modelling food webs and nutrient cycling in agro-ecosystems (Deruiter et al., 1994)

Towards a minimum data set to assess soil organic matter quality in agricultural soils (Gregorich et al., 1994)

The chemical composition of soil organic matter in classical humic compound fractions and in bulk samples - a review (Beyer, 1996)

Carbon in primary and secondary organo-mineral complexes (Christensen, 1996)

Applications of NMR to soil organic matter analysis - history and prospects (Preston, 1996)

Characterization of humic and soil particles by analytical pyrolysis and computer modeling (Schulten and Leinweber, 1996)

Life after death - lignin-humic relationships reexamined (Shevchenko and Bailey, 1996)

Stabilization and destabilization of soil organic matter - mechanisms and controls (Sollins et al., 1996)

of the heterogeneity of organic material found in soil in terms of its source, chemical and physical composition, diversity of function, and dynamic, ever changing character. The term SOM has been used to encompass all organic materials found in soil (Stevenson, 1994), excluding charcoal (Oades, 1988), or excluding nondecayed plant and animal tissues, their partial decomposition products, and the living soil biomass (MacCarthy et al., 1990b). As suggested by MacCarthy et al. (1990b), it is most important that readers establish how particular authors apply the various terms to fully understand and assess the implications of comments made. However, it is also important that a set of definitions for SOM and its components is derived and applied consistently. The definitions of SOM and its components to be utilized in this chapter (Table 2.2) have been derived from several sources (Oades,

Table 2.2 Definitions of soil organic matter and its components

Component	Definition
Soil organic matter (SOM)	The sum of all natural and thermally altered biologically derived organic material found in the soil or on the soil surface irrespective of its source, whether it is living or dead, or stage of decomposition, but excluding the aboveground portion of living plants.
Living Components Phytomass	Living tissues of plant origin. Standing plant components which are dead (e.g., standing dead trees) are also considered as phytomass.
Microbial Biomass	Organic matter associated with cells of living soil microorganisms.
Faunal Biomass	Organic matter associated with living soil fauna.
Nonliving Components Particulate organic matter	Organic fragments with a recognizable cellular structure derived from any source but usually dominated by plant derived materials.
Litter	Organic materials devoid of mineral residues located on the soil surface.
Macroorganic matter	Fragments of organic matter > 20 μm or > 50 μm (i.e., greater than the lower size limit of the sand fraction) contained within the mineral soil matrix and typically isolated by sieving a dispersed soil.
Light Fraction	Organic materials isolated from mineral soils by flotation of dispersed soil suspensions on water or heavy liquids of densities 1.5 - 2.0 Mg m^{-3}.
Dissolved organic matter	Water soluble organic compounds found in the soil solution which are < 0.45 μm by definition. Typically this faction consists of simple compounds of biological origin (e.g., metabolites of microbial and plant processes) including sugars, amino acids, low molecular weight organic acids (e.g., citrate, malate, etc.) but may also include large molecules.
Humus	Organic materials remaining in the soil after removal of macroorganic matter and dissolved organic matter.
Non-humic biomolecules	Identifiable organic structures which can be placed into discrete categories of biopolymers including polysaccharides and sugars, proteins and amino acids, fats, waxes and other lipids, and lignin.
Humic substances	Organic molecules with chemical structures which do not allow them to be placed into the category of non-humic biomolecules.
Humic acid	Organic materials which are soluble in alkaline solution but precipitate on acidification of the alkaline extracts.
Fulvic acid	Organic materials which are soluble in alkaline solution and remain soluble on acidification of the alkaline extracts.
Humin	Organic materials which are insoluble in alkaline solution.
Inert organic matter	Highly carbonized organic materials including charcoal, charred plant materials, graphite and coal with long turnover times.

1988; MacCarthy et al., 1990b; Stevenson, 1994) and are put forward in an effort to reduce existing variations in the use of the terms.

2.2   Functions of Soil Organic Matter

The organic fraction of soils often accounts for a small but variable proportion of total soil mass. Organic C concentrations ranging from < 5 g C kg^{-1} soil for desert loams (Aridisols) to > 130 g C kg^{-1} soil for alpine humus soils (Histosols and Mollisols) were presented for the 0–10 cm layer of the great soil groups of Australia (Spain et al., 1983). Sombroek et al. (1993) presented topsoil and subsoil organic C content values ranging from 0.5–3 g C kg^{-1} for Yermosols up to 310–555 g C kg^{-1} for Histosols using 400 soil profiles grouped according to FAO soil units. Contents of organic C on the order of 441–549 g C kg^{-1} and 440–550 g C kg^{-1} were measured for Canadian peats (Preston et al., 1989b) and German forest litter layers (Zech et al., 1992), respectively.

Despite its often minor contribution to the total mass of mineral soils, the organic fraction can exert a profound influence on soil properties, ecosystem functioning, and the magnitude of various obligatory ecosystem processes (Table 2.3). The soil properties which organic matter influences have been classified into three broad groups: biological, chemical, and physical properties. It should be noted that strong interactions and interdependencies exist between these groups. For example, the ability of organic matter to chelate multivalent cations can affect its potential to stabilize soil structure and also its biodegradability (Juste and Delas, 1970; Juste et al., 1975; Gaiffe et al., 1984; Muneer and Oades, 1989a,b,c). In addition, the effects of organic matter on soil properties often involve interactions with the soil mineral fraction, and variations in soil properties noted across different soils may not be solely a consequence of qualitative or quantitative variations in the soil organic component.

2.2.1   Biochemical Functions

2.2.1.1   Reservoir of Metabolic Energy

The most fundamental function of the organic fraction of soils is the provision of metabolic energy which drives soil biological processes and the direct and indirect effects which this has on other soil properties and processes. Photosynthesis fixes CO_2 into glucose which is then converted into a wide range of organic compounds (e.g., cellulose, hemicellulose, lignin, lipids, proteins, etc.) by various enzymatic processes. The C fixed into such compounds is deposited in or on the soil during plant growth and as the plant or a portion of its tissues senesce, thereby providing C substrates for soil macro- and microdecomposer organisms. As the organic C is utilized by decomposer organisms, it is either assimilated into body tissues, released as metabolic products, or respired as CO_2. Oades (1989) presented an estimate of the C flow through a fertile grassland where primary production via photosynthesis was 10 Mg C ha^{-1} yr^{-1}. Of the 3 Mg C ha^{-1} yr^{-1} that was added to the soil organic C fraction, the soil fauna were estimated to utilize 0.3–0.45 Mg C ha^{-1} yr^{-1}, while the soil microbial biomass was estimated to utilize 2.4 Mg C ha^{-1} yr^{-1}. The majority of organic matter processing is thought to be completed by the soil microbial biomass. However, other activities of the soil fauna enhance the ability of soil microbial decomposers to utilize organic residues added to soil. These include (1) fragmentation of plant debris, which enhances the surface area per unit weight of plant residue available to microbial attack, and (2) distributing organic materials throughout the soil matrix, which provides an avenue for greater contact between decomposer microorganisms and substrates.

2.2.1.2   Source of Macronutrients

A result of the utilization of organic materials by soil organisms is a conversion of macronutrients (N, P, and S) within organic chemical structures into inorganic forms, which are either immobilized and

Table 2.3 Properties and functions of organic matter in soil

Property	Function
Biological Properties	
Reservoir of metabolic energy	Organic matter provides the metabolic energy which drives soil biological processes.
Source of macronutrients	The mineralization of soil organic matter can significantly influence (positively or negatively) the size of the plant available macronutrient (N, P, and S) pools.
Ecosystem resilience	The build up of significant pools of organic matter and associated nutrients can enhance the ability of an ecosystem to recover after imposed natural or anthropogenic perturbations.
Stimulation and inhibition of enzyme activities and plant and microbial growth	The activity of enzymes found in soils and the growth of plants and microorganisms can be stimulated or inhibited by the presence of soil humic materials.
Physical Properties	
Stabilization of soil structure	Through the formation of bonds with the reactive surfaces of soil mineral particles, organic matter is capable of binding individual particles and aggregations of soil particles into water-stable aggregates at scales ranging from $< 2\mu m$ for organic molecules through to mm for plant roots and fungal hyphae.
Water retention	Organic matter can directly affect water retention because of its ability to absorb up to 20 times its mass of water and indirectly through its impact on soil structure and pore geometry.
Low solubility	Ensures that the bulk of the organic materials added to the soil are retained and not leached out of the soil profile
Color	The dark color which soil organic matter imparts on a soil may alter soil thermal properties.
Chemical Properties	
Cation exchange capacity	The high charge characteristics of soil organic matter enhance retention of cations (e.g., Al^{3+}, Fe^{3+}, Ca^{2+}, Mg^{2+}, NH_4^+, and transition metal micronutrients).
Buffering capacity and pH effects	In slightly acidic to alkaline soils, organic matter can act as a buffer and aids in the maintenance of acceptable soil pH conditions.
Chelation of metals	Stable complexes formed with metals and trace elements enhance the dissolution of soil minerals, reduce losses of soil micronutrients, reduce the potential toxicity of metals, and enhance the availability of phosphorus.
Interactions with xenobiotics	Organic matter can alter the biodegradability, activity and persistence of pesticides in soils.

used in the synthesis of new tissues within soil organisms or mineralized and released into the soil mineral nutrient pool. With the exception of intensively managed soil receiving significant inputs of macronutrients in the form of fertilizers, organic matter provides the largest pool of macronutrients in the soil. McGill and Cole (1981) proposed that the mineralization of C, N, P, and S followed a dichotomous system involving both biological and biochemical mineralization. Biological mineralization is driven by the need of decomposer organisms for C as an energy source and accounts for the mineralization of N and C bonded S. Biochemical mineralization refers to the release of phosphate and sulfate from the P and S ester pool via enzymatic hydrolysis outside of the cell membrane. As a result and in contrast to organic N, organic P and S accumulation and mineralization in soils can occur independently of C and N dynamics. This leads to the potential for large variations in C:N:P:S ratios in SOM. An average C/N/P/S ratio of 107:7.7:1:1 was presented by Stevenson (1986) for SOM. In the next several paragraphs, the contents and chemical structures of soil organic N, P, and S and ratios of organic C/N, C/P, and C/S will be discussed to provide an indication of the organic forms and potential stores of macronutrients within SOM. More detailed presentations of the forms and availability of organic N, P and S, can be found in Section C, Chapters 3 and 4.

Nitrogen

The soil N pool is dominated by N found in organic structures. In soils with significant contents of K^+-containing clay minerals (e.g., illite) capable of fixing NH_4^+, approximately 90% of the soil N is contained in organic structures, 8% exists as fixed NH_4^+, and 1–3% can be found in the inorganic plant available pool (NO_3^- and NH_4^+). In soils with little capacity to fix NH_4^+ in clay minerals, the proportion of organic N is > 97% and the inorganic fraction is 1–3%. On a global scale, Söderlund and Svensson (1976) estimated that the organic N fraction of soils accounted for 95% of the total soil N pool, which is equivalent to the average value presented by Bremner (1967). The C/N ratio of SOM depends on the C/N ratio of the vegetational inputs and the degree to which they are decomposed. Organic materials that have cycled through the decomposer biomass generally have C/N ratios of 12–16, whereas undecomposed fragments of plant litter and organic materials in peat deposits, where decomposition is hindered by anaerobic conditions, can have much higher C/N ratios.

Soil organic N has been traditionally divided into the following five fractions based on a variety of acid hydrolysis procedures: (1) acid insoluble N, (2) ammonia N recovered after hydrolysis, (3) amino acid N, (4) amino sugar N, and (5) hydrolyzable unidentified N. Data summarized by Stevenson (1994) for 11 studies where acid hydrolysis procedures were applied to different soil types showed that there was as much variation in the contents of each form of N within similar soils as between different soil types. The proportions of each form of organic N were 7–44% acid insoluble N, 9–37% ammonia N, 13–50% amino acid N, 1–14% amino sugar N, and 4–40% hydrolyzable unidentified N. Although methodological differences may account for a portion of the large variations noted in the composition of soil organic N, it is evident that approximately 50% of the total soil N cannot be identified by acid hydrolysis procedures (acid insoluble N + hydrolyzable unidentified N).

Initial attempts to identify the chemical composition of unidentifiable organic N utilized gel filtration followed by acetylation and gas chromatography/mass spectroscopy (GC/MS) (Schnitzer, 1985; Schnitzer and Spiteller, 1986). Schulten et al. (1995, 1997) utilized Curie point pyrolysis-GC/MS with N selective detection of the pyrolysis products. These studies suggested that heterocyclic N compounds represented an important component of unidentified soil organic N (Schulten et al., 1997 presented examples of the chemical structure of the heterocyclic N compounds). The formation of heterocyclic N compounds via nonbiological fixation of $^{15}NH_3$ by humic substances (IHSS Suwannee River fulvic acid and peat and leonardite humic acids) and by reacting ^{15}N labeled analine with humic materials was noted by Thorn and Makita (1992) and by Thorn et al. (1996), respectively. In contrast to these results, studies utilizing solid-state ^{15}N nuclear magnetic resonance (NMR) spectroscopy have failed to observe substantial contributions from heterocyclic N, and spectra tend to be dominated by signals arising from amides and terminal amino groups (Knicker and Ludemann, 1995; Knicker et al., 1995; Clinton et al., 1995). Further effort is required to address these inconsistencies, and to quantitatively characterize the composition of the fraction of N, which cannot be identified by conventional acid hydrolysis procedures.

Phosphorus

The composition and cycling of soil organic P have been recently reviewed by Stevenson (1986, 1994), and Sanyal and DeDatta (1991). As a result of potential adsorption and inorganic precipitation reactions capable of reducing the availability of P in soils, mineralization of organic P is important to soil fertility (Tiessen et al., 1984; Beck and Sanchez, 1994). The relative importance of organic P as a nutrient source tends to be greater on highly weathered soils (Duxbury et al., 1989). The principal organic P-containing compounds in soils and their approximate proportions include inositol phosphates (2–50%), phospholipids (1–5%), nucleic acids (0.2–2.5%), trace amounts of phospho-proteins, and metabolic phosphates (Stevenson, 1994). Soil organic P accounts for a variable proportion of the total soil P. Halstead and McKercher (1975) and Uriyo and Kessaba (1975) presented soil organic P values ranging from 4–1400 μg g^{-1} soil which accounted for 3–90% of the total soil P. Uriyo and Kessaba (1975) derived the relationship between organic P and organic C given in Equation [2.1] which produces an organic C/P ratio of 115 and is consistent with the average value of 117 proposed by Stevenson (1994). However, large variations in the organic C/P ratio (61–526) have been noted for Finnish soils (Kiala, 1963).

$$\text{Organic C (mg g}^{-1}\text{ soil)} = 4.9 + 0.059 \text{ Organic P (}\mu\text{g g}^{-1}\text{ soil) (R}^2 = 0.49) \qquad [2.1]$$

Sulfur

Reviews of the cycling and chemical composition of soil organic S include Stevenson (1986, 1994), Germida et al. (1992), and Nguyen and Goh (1994). Sulfur-containing organic compounds found in soils are generally grouped into two pools: compounds in which the S can be reduced to H_2S by hydroiodic acid (HI), and compounds in which the S is directly bound to C. The HI reducible fraction consists mainly of ester sulfates (C-O-S bonds) and some ester sulfamates (C-N-S bonds). The C bonded S fraction contains amino acid S (C-S bonds) or sulfonates (C-SO$_3^-$ bonds). The ester sulfates and sulfamates are typically associated with aliphatic side chains of soil organic compounds (Bettany et al., 1979), while the C bonded S is incorporated along with C and N into the core of soil organic compounds and is generally less biologically accessible (McGill and Cole, 1981; Stewart and Cole, 1989). Organic S typically accounts for > 90% of the total S found in non-saline and non-tidal soils (Nguyen and Goh, 1994; Stevenson, 1994).

2.2.1.3 Ecosystem Resilience

The resilience of an ecosystem is defined as its capacity to return to its initial state after being subjected to some form of disturbance or stress. The important role played by SOM in determining the resilience of an ecosystem can be exemplified by a comparison of the contents of chemical energy and nutrients stored within the soil organic fractions in several ecosystems. In temperate grasslands, high organic matter levels are built up in soils as a result of large below-ground additions of photosynthate, limited leaching, and slow decomposition rates. Storage of C in such ecosystems is greater in the soil than in the vegetation (Szabolcs, 1994). The large store of chemical energy and nutrients contained in the soil organic fraction offers resistance to the loss of soil fertility induced by natural or by agricultural disturbance. Temperate grassland soils, typically Mollisols, have remained agriculturally productive with limited inputs for many years, despite the mining of energy and nutrient reserves contained within the soil organic fraction (Janzen, 1987; Tiessen et al. 1994b). Such systems can be considered resilient, at least initially, but one must question for how long such systems can be sustained. This was exemplified by Tiessen et al. (1983), who showed that rates of organic P mineralization in a grassland soil were in excess of crop requirements over the first 60 years of agricultural production, but that subsequent to 60 years, only the less labile, low energy providing forms of organic matter remained, and organic P mineralization rates decreased below crop demands. In temperate forests, SOM contents are less than those of temperate grasslands and more C and nutrients are stored in aboveground vegetation than in the readily available soil organic materials (Szabolcs, 1994). As a result, the impact of a natural disturbance such as fire can significantly deplete ecosystem stores of energy and nutrients, and ecosystem recoveries (resilience) are slow due to low residual contents of SOM and associated nutrients. Where temperate forests are cleared and agricultural production is initiated, SOM and nutrient losses must be minimized; however, production systems which increase SOM and nutrient reserves (e.g., crop rotations including legume pastures) can lead to highly productive and sustainable agriculture. In tropical forest ecosystems, the storage of energy and nutrients in vegetation dominates, and the rapid utilization of plant residues by decomposer organisms and cycling of nutrients maintain ecosystem stability. This, when coupled with the low stores of energy and nutrients in organic matter of tropical soils, indicates a reduced importance of SOM in ecosystem resilience (Anderson, 1995). A comparison of a temperate grassland Mollisol with a tropical Oxisol (Tiessen et al. 1994b) demonstrated the important contribution of organic matter to the resilience of the grassland soil and its reduced significance in the tropical soil.

2.2.1.4   Stimulation and Inhibition of Enzyme Activities and Plant and Microbial Growth

Research pertaining to the impacts of SOM on plants, microorganisms and enzyme activities has typically utilized humic substances (e.g., humic and fulvic acids) as surrogates for SOM. The influence of humic and fulvic acids, tannins, and melanins on the activity of various enzymes was summarized by Ladd and Butler (1975), Müller-Wegener (1988), and Gianfreda and Bollag (1996). Based on earlier studies, Ladd and Butler (1975) concluded that the effect of humic acids on the activity of proteolytic enzymes varied and that the mechanism of humic acid-enzyme interaction involved primarily the carboxyl groups of humic acids. Inhibition of nonproteolytic enzyme activities by humic acids have also been demonstrated (Sarkar and Bollag, 1987). Müller-Wegener (1988) indicated that possible humic acid-enzyme interactions which could impact on enzyme activity included (1) a direct interaction of the humic acid with the enzyme resulting in a modification of enzyme structure or changes in the functioning of active sites, (2) interference in the equilibrium of the enzyme reaction via the humic substances acting as analog substrates, and/or (3) a reduction in the availability of cations, which often act as cofactors required for enzyme catalysis or structural stabilization of the protein molecule by fixation on the humic acid molecule.

The effect of soil humic substances on plant and microbial growth involves the absorption or adsorption of the humic species and their impacts on biochemical properties at cell walls, cell membranes, and/or in the cytoplasm. Information on the impacts of humic materials in field studies is scarce and often confounded with other impacts of humic materials on soil properties (e.g., CEC, nutrient status, etc.). The effects of humic materials on plant growth were reviewed by Chen and Aviad (1990). Favorable effects on plant growth included (1) increased uptake of water and germination rate of seeds, (2) enhanced growth of shoots and roots as assessed by measurements of length and fresh and/or dry mass, and (3) increased root elongation, number of lateral roots, and root initiation. These effects result from increased permeability of cell membranes, increased chlorophyll content and rates of photosynthesis and respiration, enhanced protein synthesis resulting from a stimulation of ribonucleic acid synthesis, and enhanced enzyme activity (Vaughan and Malcolm, 1985). The influence of humic substances on the growth of microorganisms (bacteria and fungi) also involves a penetration and alteration of cell membranes. Addition of humic substances at concentrations ≤ 30 mg L^{-1} to a nutrient solution increased growth rates in microbial cultures (Visser, 1995), and *in vitro* growth and activity of nitrifying bacteria have been increased by the addition of humic acid (Valdrighti et al., 1996; Vallini et al., 1997).

2.2.2 Physical Functions

2.2.2.1 Stabilization of Soil Aggregates

Organic matter is considered important to the maintenance of the structural stability of a wide range of soil types including Mollisols, Alfisols, Ultisols and Inceptisols. Its importance tends to be less in Oxisols and Andisols, where hydrous oxides play an important stabilizing role, and in self-mulching soils (e.g., some Vertisols) that contain clays with a high shrink/swell potential. In soils where organic matter is an important agent binding mineral particles together, a hierarchical arrangement of soil aggregates exists in which aggregates break down in a stepwise manner as the magnitude of an applied disruptive force increases (Tisdall and Oades, 1982; Oades and Waters, 1991; Oades, 1993). Golchin et al. (1998) and others have proposed the existence of three levels of aggregation: (1) the binding together of clay plates into packets < 20 μm, (2) the binding of clay packets into stable microaggregates (20–250 μm), and (3) the binding of stable microaggregates into macroaggregates (> 250 μm).

The importance and nature of the organic materials associated with each level of aggregation vary. At the scale of packets of clays, aggregation is primarily dictated by soil mineralogical and chemical properties important in controlling the extent of dispersion, and is often a function of pedological processes. The binding together of clay packets to form microaggregates occurs via a range of mechanisms. The dominant mechanism is proposed to involve polysaccharide-based glues (mucilages or mucigels) produced by plant roots and soil microorganisms (Ladd et al., 1996). Emerson et al. (1986) presented transmission electron micrographs showing mucilage located between packets of clay plates. Small microaggregates (< 53 μm) held together by humified organic matter and biologically processed materials are bound together around a particulate organic core (Oades, 1984; Elliott, 1986; Beare et al., 1994b; Golchin et al., 1994b) to produce larger microaggregates and small macroaggregates < 2,000 μm. Macroaggregates > 2,000 μm are stabilized by the presence of roots, fungal hyphae and larger fragments of plant residues, which interconnect soil aggregates via bonding to aggregate surfaces, penetration into or through aggregates, and/or physical enmeshment (Tisdall and Oades, 1982; Churchman and Foster, 1994; Foster, 1994). Additional information pertaining to the involvement of organic matter in stabilizing soil structure is presented in Section A, Chapter 7.

2.2.2.2 Water Retention

Organic materials can influence soil water retention directly and indirectly. SOM can absorb and hold substantial quantities of water, up to twenty times its mass (Stevenson, 1994). This direct effect, however, depends on the morphological structure of the organic materials and will not impart any beneficial effect to the soil unless it serves to enhance the ability of soil to hold water at potentials within the plant available range. Organic matter in the form of surface residues can also influence water retention directly by reducing evaporation and increasing the infiltration of water.

The indirect effect of organic matter on water retention arises from its impact on soil aggregation and pore size distribution, and thus on the plant available water holding capacity (AWHC) of soil (the difference between volumetric water content at field capacity and at permanent wilting point). This effect is best exemplified by the inclusion of soil organic C content as a significant parameter in pedotransfer functions which predict pore size distribution (Vereecken et al., 1989; da Silva and Kay, 1997; Kay et al., 1997). Equation [2.2] presents the pedotransfer function derived by da Silva and Kay

(1997) to describe the relationship between volumetric water content, θ_v (m³ m⁻³), and matric potential; ψ (MPa), clay content, CL (%); organic C content, OC (%); and bulk density, BD (Mg m⁻³). Using this equation Kay et al. (1997) calculated predicted changes in AWHC for soils ranging in clay content from 7 to 35% when organic C content was increased by 0.01 kg kg⁻¹. Increases in AWHC of 0.039 and 0.020 (m³ m⁻³) were obtained for the soils with 7 and 35% clay, respectively, at a relative bulk density of 0.75. Application of the same equations to a data set acquired by Wegner et al. (1989) for 80 South Australian red brown earths (Alfisols) showed that the increase in AWHC induced by increasing organic C content by 0.01 kg kg⁻¹ soil could be expressed by Equation [2.3]. These results indicate that the presence of additional organic matter enhances AWHC of soils; however, the magnitude of the increase decreases with increasing clay content.

$$\theta_\omega = \varepsilon^{(4.15 + 0.68 \ln CL + 0.42 \ln OC + 0.27 \ln BD)} \psi_m^{(-0.54 + 0.11 \ln CL + 0.02 \ln OC + 0.10 \ln BD)} \quad (R^2 = 0.94) \qquad [2.2]$$

$$\text{Change in AWHC} = -0.0012 \, (\%\text{clay}) + 0.055 \quad (R^2 = 0.82) \qquad [2.3]$$

2.2.2.3   Soil Thermal Properties
The typical dark color of SOM contributes to the dark color of surface mineral soils and can enhance soil warming and promote biological processes related to temperature in cooler climates (e.g., plant growth and mineralization of C and nutrients contained in SOM). However, the presence of litter layers or organic horizons can insulate a soil against fluctuations in air temperature and solar heating. On several Canadian forest soils subject to cold winters and cool springs, average soil temperatures and the growth of fertilized seedlings were greater where the litter layers were removed compared to where they were left intact (Burgess et al., 1995a,b). Similar effects have been observed in a comparison of cropping systems that leave different amounts of crop residue on the surface of mineral soil (Fortin, 1993).

2.2.3   Chemical Functions

2.2.3.1   Cation Exchange Capacity
Organic matter contributes 25–90% of the cation exchange capacity (CEC) of surface layers of mineral soils and practically all of the CEC of peats and forest litter and humus layers (Stevenson, 1994). The contribution is greatest for soils with low clay content or where the clay fraction is dominated by minerals with a low charge density, such as kaolinite, and is lowest for soils with high contents of highly charged minerals such as vermiculite or smectite. In sandy soils, organic matter plays a critical role in providing CEC. As a general rule of thumb, each weight percentage of organic matter in soils contributes approximately 3 $cmol_c$ kg⁻¹ soil to the CEC of neutral permanent charge soils (McBride, 1994) and approximately 1 $cmol_c$ kg⁻¹ soil to the CEC of variable charge soils (Oades, 1989). The CEC of organic matter is derived principally from carboxyl functional groups but also from phenol, enol and imide groups, and is, therefore, dependent on soil pH. However, because of potential organo-mineral interactions in soils and the ability of organic matter to complex cations, the contribution of organic matter to soil CEC values is often much less than could be expected based on its carboxyl content. For example, the protonation of carboxylate groups, interaction with positively charged sites on inorganic colloids, and complexation of Al^{3+} and Fe^{3+} can reduce the ability of organic matter to contribute to the CEC of acidic soils (McBride, 1994). Measurements of the CEC of SOM have yielded values ranging from 60–300 $cmol_c$ kg⁻¹ (Leinweber et al., 1993; Stevenson, 1994).

2.2.3.2 Buffering Capacity and Soil pH

The presence of weakly acidic chemical functional groups on soil organic molecules, that can act as conjugate acid/base pairs, makes SOM an effective buffer. The diversity in chemical composition of the functional groups (e.g., carboxylic, phenolic, acidic alcoholic, amine, amide and others) provides organic matter with the ability to act as a buffer over a wide range of soil pH. James and Riha (1986) reported buffer capacities of 18–36 and 1.5–3.5 $cmol_c$ kg^{-1} (pH unit)$^{-1}$ for the organic and mineral horizons of forest soils, respectively. Starr et al. (1996) obtained a good correlation between acid buffer capacity and organic matter content for 29 organic and 87 mineral soil horizons (E, B, and C horizons) exhibiting buffering capacities of 9.8–40.8 and 0.1–5.2 $cmol_c$ kg^{-1} (pH unit)$^{-1}$, respectively. For 59 agricultural soil samples taken from the 0–15 cm layer of cultivated fields, Curtin et al. (1996) noted that titratable acidity could be described by Equation [2.4] in which the terms OC and Clay represent the soil organic C and clay contents expressed in units of kg kg^{-1} soil and ΔpH is the reference pH (e.g., 8) minus the initial pH. Assuming that the organic C content of SOM is 58%, Equation [2.4], indicates that the buffering capacity offered by organic matter was approximately 34 $cmol_c$ kg^{-1} (pH unit)$^{-1}$ and is an order of magnitude greater than that offered by clay [34 versus 3 $cmol_c$ kg^{-1} (pH unit)$^{-1}$]. The average clay/organic C ratio for the soils studied by Curtin et al. (1996) was 7.9/ 1, indicating that even though most soils contained much more clay than organic C, organic C accounted for about two-thirds of the soil buffering capacity.

$$\text{Titratable acidity to pH 8} = 0.02 + 59 \text{ OC } \Delta pH + 3.0 \text{ Clay } \Delta pH \quad R^2 = 0.95 \quad [2.4]$$

Addition of organic matter to soil may result in increases or decreases in soil pH, depending on the influence that the addition has on the balance of the various processes that consume and release protons. A detailed presentation of these soil processes and their ability to release or consume protons is given by van Breeman et al. (1983). Factors which need to be considered include the chemical nature of the soil and that of the organic materials added as well as environmental properties including water content and extent of leaching. The net effect of adding organic matter to acidic soils is generally an increase in pH values (e.g., Yan et al., 1996; Pocknee and Sumner, 1997) with the main processes leading to the increase being (1) a decomplexation of metal cations, (2) mineralization of organic N, and (3) denitrification. Pocknee and Sumner (1997) found that on the acid Cecil soil, the extent of the increase in pH was controlled by the N content to basic cation content ratio. The decarboxylation of organic acids has also been shown to increase the pH of acid soils (Yan et al., 1996). Under alkaline soil conditions, however, these processes would be ineffective and would contribute to a reduction in soil pH as a result of their influence on soil CO_2 concentrations. The addition of organic matter to alkaline soils tends to acidify them especially under waterlogged and leaching conditions (Nelson and Oades, 1998). The main processes involved in the acidification of alkaline soils on addition of organic materials include (1) mineralization of organic S and P, (2) mineralization followed by nitrification of N, (3) leaching of the mineralized and nitrified organic N, (4) dissociation of organic ligands, and (5) dissociation of CO_2 during decomposition.

2.2.3.3 Complexation of Inorganic Cations

The presence of various functional groups on soil organic materials provides the capacity for interaction with inorganic cations. The possible interactions can take the form of simple cation exchange reactions, such as those between negatively charged carboxyl groups and monovalent cations, or more complex interactions where coordinate linkages with organic ligands are formed, such as occur between amino acids and Cu^{2+}. The mechanisms involved in the complexation of inorganic cations by soil organic materials are discussed in detail in Section B, Chapter 8. The influence that the complexation of inorganic cations by soil organic materials has on soil properties and processes includes the following: (1) increased availability of insoluble mineral P through complexation of Fe^{3+} and Al^{3+} in acid soil and Ca^{2+} in calcareous soil, competition for P adsorption sites, and displacement of adsorbed P (Stevenson, 1994; Cajuste et al., 1996), (2) the release of plant nutrients through the weathering of rocks and soil parent materials by the removal of structural cations from silicate minerals (Tan, 1986; Robert and Berthelin, 1986), (3) enhanced availability of trace elements in the upper portion of the soil profile as a result of upward translocation by plant roots and subsequent deposition on the soil surface and complexation during residue decomposition (Stevenson, 1994), (4) facilitated adsorption of organic materials to soil minerals which aids in the generation and/ or stabilization of soil structure (Oades, 1984; Emerson et al., 1986), (5) buffering of excessive concentrations of otherwise toxic levels of metal cations (e.g., Al^{3+}, Cd^{2+}, and Pb^{2+}) (Anderson, 1995), and (6) pedogenic translocation of metal cations to deeper soil horizons (McKeague et al., 1986) and the formation of minerals (Huang and Violante, 1986).

2.3   Quantifying Soil Organic Matter Contents

Soil organic matter contents are difficult to measure directly. Most methods measure organic C content, and multiply the resultant values by conversion factors ranging from 1.72 to 2.0 to obtain organic matter contents. The value of the conversion factor used is determined by the C content of SOM, which is assumed to range from 50 to 58%. Recent studies have suggested that the C content of SOM is variable and that no single factor is appropriate for all soils. As a result, researchers are encouraged to determine and report organic C contents.

Methods currently used to estimate soil organic C contents have been reviewed recently by Nelson and Sommers (1996) and include dry and wet combustion methods for total C and dichromate wet oxidation methods for organic C. Total C analyses involve the complete conversion of all C in the soil to CO_2 and quantification of the evolved CO_2 by various means (e.g., infrared detection, increased mass of an Ascarite trap, or others). Corrections for the presence of inorganic C must be performed when using total C methods if carbonates are present in the soil. Where organic C contents are measured using dichromate methods (e.g., Walkely and Black, 1934; and subsequent modified methods), an excess of dichromate is added. The unused dichromate is titrated with $Fe(NH_4)_2(SO_4)_2 \cdot 6H_2O$ to determine the amount of dichromate used during the reaction and thus the amount of organic C present. A basic assumption of the dichromate procedure is that soil organic C has a valance of 0. In addition, dichromate oxidation methods are subject to interferences from Cl^-, Fe(II) or Fe(III) oxides, and Mn. The presence of significant amounts of Cl^- and Fe^{2+} will lead to an overestimation, while the presence of MnO_2 will result in an underestimation of organic C levels. The extent of heating used in the analysis is also important. Where the heat of dilution or minimal external heating is used, complete oxidation of soil organic compounds does not occur and correction factors must be applied to the results of the oxidation procedure. Correction factors ranging from 1.0 to 2.86 have been noted in the literature and an average value of 1.3 is recommended if an experimentally derived value is not available (Nelson and Sommers, 1996). Dichromate methods that incorporate extensive heating (Tinsley, 1950) do not require a correction factor as oxidation of the soil organic C is considered complete. Heating has also been implicated in the ability of dichromate oxidation procedures to recover the organic C contained in carbonized materials (charcoal, charred plant materials, etc.); however, the impact of heating is variable and Nelson and Sommers (1996) suggest that recovery is highly dependent on the characteristics of the carbonized materials. Recent results obtained by J. Skjemstad (Personal communication) show that the particle size of charcoal is an important factor determining its susceptibility to oxidation by dichromate, with increased detection as particle size decreased. Of the methodologies currently available, a dry combustion automated analyzer which measures CO_2 evolution with an infrared detector is the preferred methodology for determining soil organic C, provided accurate estimates of inorganic C contents can be obtained where required.

2.4   Factors Determining Soil Organic C Levels

The amount of organic C contained in a particular soil is a function of the balance between the rate of deposition of plant residues in or on soil and the rate of mineralization of the residue C by soil biota. Losses by erosion or leaching may be significant in some cases. With the exceptions of peatland and wetland soils (Histosols), which have been estimated to accumulate 0.1–0.3 Pg C yr^{-1} globally (Post et al., 1990), organic C levels in soil do not increase indefinitely but rather tend to equilibrium values dictated by the soil forming factors of climate, biota (vegetation and soil organisms), parent material, time, and topography. During the initial phases of soil development, a lack of available nutrients places a ceiling on the amount of organic C which can be fixed by photosynthetic organisms. The low rates of C addition and low nutrient status are not capable of supporting large populations of decomposer organisms. Thus, soil organic C accumulates slowly, aided by mechanisms capable of biologically stabilizing organic C against processes of decomposition. The mechanisms through which soil organic C can be biologically stabilized depend on the composition of the soil mineral phase and the chemical structure of the organic residues added to the soil. Thus, each soil could potentially offer a different protective capacity. With continued soil development, organic C content and the activity of decomposer organisms increase to a point where a continued supply of nutrients in a plant available form is reached. At this point, the rate of organic C synthesis and deposition is much greater than mineralization, and organic C accumulates at an exponential rate. With increasing organic C content, the proportion of potential sites of biological stabilization which remains vacant decreases, and an increasing proportion of added organic C cannot be biologically stabilized. As a result, the increase in organic C per unit time proceeds through an inflection point and then begins to decrease. Once the capacity for biological stabilization offered by soil mineral components is approached, the rate of mineralization of organic materials tends toward the rate of deposition of fresh organic residues and soil organic C levels approach an equilibrium value.

It is important to note that biological protection rarely equates to a permanent and complete removal of organic C from the decomposing pool, but rather to a reduction in its rate of decomposition when compared to similar materials existing in an unprotected state. As the older protected C is slowly mineralized, its position in the biologically protected pool is replaced with younger modern organic C, but the proportion of such modern organic C in the protected pool remains low. It is possible that some forms of soil organic C may be rendered unavailable to decomposition for prolonged periods (potentially on a geologic time scale) through mechanisms such as entrapment between layers of clay plates (Theng et al., 1986) or burning to create charred plant residues and/or charcoal. Pieces of charcoal selectively removed from soil and unassociated with soil minerals have been shown to have radiocarbon ages equivalent to or greater than soil humin fractions (Pressenda et al., 1996). Typically, the proportion of the total soil organic C held in such more highly stable forms would be minor. However, in recent soil fractionation/ultraviolet oxidation studies, J. Skjemstad (Personal communication) has estimated that charcoal C can account for 0–40% of the total organic C found in Australian soils.

The progression of soil organic C content with soil development and the magnitude of the equilibrium level of soil organic C will depend on interactions which occur between the factors of soil formation. Where cold and water saturated soil conditions persist, decomposition is confined to slow anaerobic processes and organic C contents may continue to increase leading to the formation of Histosols. In sandy soils, the extent of biological protection offered by the soil mineral component will be much different than that offered by clay rich soils and large differences in organic C content of the two textural classes of soil develop.

An independent evaluation of the influence of any single soil-forming factor on soil organic C content is difficult because of the requirement that all other factors remain constant. Variations in the soil-forming factors experienced on a landscape scale and the interdependence of these factors contribute to the large variability noted for soil organic C contents, even within localized areas. Computer simulation models of soil organic C dynamics (e.g., Parton et al., 1987) can be used to provide valuable information pertaining to the interaction of soil-forming factors on soil organic C levels and thus, ecosystem functioning, but field data are still required to validate predictions (Burke et al., 1989).

An additional factor which must be considered in an examination of factors influencing soil organic C contents in disturbed soils (those used for agriculture and forestry) is management. Management can induce rapid and drastic changes to equilibrium contents of soil organic C attained under natural undisturbed conditions and completely override the influence of soil-forming factors. Combining this observation with early results presented by Jenny (1930) indicates that the relative importance of the soil-forming factors on SOM content can be viewed as: management > climate > biota (vegetation and soil organisms) > topography = parent material > time.

2.4.1   Climate

Climate impacts on soil organic C content primarily through the effects of temperature, moisture, and solar radiation on the array and growth rate of plant species, and on the rate of soil organic C mineralization. Post et al. (1982) found that amounts of soil organic C were positively correlated with precipitation and, at a given level of precipitation, negatively correlated with temperature. In the Great Plains of North America, precipitation controls net primary productivity and temperature controls rates of soil organic C mineralization (Parton et al., 1987; Sala et al., 1988; Burke et al., 1989; USDA-SCS, 1994). Ladd et al. (1985) compared the mean loss of ^{14}C-labeled plant residues from four soils in South Australia with those obtained by Jenkinson and Ayanaba (1977) for soils in England and Nigeria and observed a doubling of the rate of substrate C mineralization for an 8–9 °C increase in mean annual temperature. An influence of temperature on decomposition can also be inferred from ^{14}C signatures of soil organic C, which showed a latitudinal gradient in the mean residence time of soil organic C (Bird et al., 1996).

The observed trend of decreasing soil organic C content with increasing temperature implies that the relative temperature sensitivity of decomposition is greater than that of net primary productivity. Because of the strong interactions between temperature, water availability and substrate quantity, it is difficult to assess the temperature dependence of decomposition without confounding effects. In a compilation of data extracted from controlled incubation studies where water limitations were avoided and a common substrate was used at all temperatures, Kirschbaum (1995) showed that the Q_{10} value of C mineralization from soil was greater than that for net primary productivity developed by Lieth (1973), especially at temperatures < 15 °C. Increases in temperature, particularly when starting from temperatures < 15 °C, will enhance decomposition more than net primary productivity.

Climate has also been shown to affect the chemical structure of soil organic C. Using pyrolysis-gas chromatography to characterize the chemical structure of soil organic C in a climosequence of nine New Zealand soils, Bracewell et al. (1976) observed significant correlations between changes in the intensity of peaks in the chromatograms and mean annual precipitation and temperature. By including both temperature and precipitation in a regression analysis, the resultant regression line explained

90% of the variation in chromatogram peak intensities. Amelung et al. (1997) used ten grassland samples originating from different climatic zones of the North American Great Plains to investigate the impacts of mean annual temperature and precipitation on the chemical structure of soil organic C using a combination of chemical methods and ^{13}C NMR. Mean annual precipitation was capable of accounting for only 10% of the variation in alkaline CuO oxidizable lignin. Higher precipitation tended to favor an accumulation of polysaccharide C; however, at a given mean annual precipitation, polysaccharide C tended to decrease with increasing temperature. Amelung et al. (1997) suggested that the increased content of polysaccharide C in more humid conditions may have resulted from (1) a positive feedback mechanism in which increased plant production enhanced microbial activity and soil structural conditions, thereby offering the potential for stabilizing microbial polysaccharides within aggregates, and/or (2) an enhanced activity of earthworms increased polysaccharide content relative to the surrounding soils (Guggenberger et al., 1995b) and offered organic C some physical protection against mineralization (Lavelle and Martin, 1992). Accompanying the decrease in polysaccharide C noted with increasing temperature, Amelung et al. (1997) noted an increase in aliphatic C content. Accumulation of alkyl C at high temperature may be explained by (1) enhanced mineralization of carbohydrates and selective preservation of plant or microbially derived alkyl structures by adsorption onto clay particles (Baldock et al., 1989, 1992), and/or (2) higher inputs of plant-derived alkyl C in plant residues due to the presence of thicker cuticles on plants growing in warmer climates.

2.4.2 Soil Mineral Parent Materials and Products of Pedogenesis

The mineral phase of soils can exert a strong influence on soil organic C contents as a result of mechanisms capable of stabilizing organic materials against biological attack. As noted above in Section 2.4, each soil has a given capacity to protect soil organic C dictated by the following soil characteristics: (1) the chemical nature of soil minerals, (2) the presence of multivalent cations and their ability to form complexes with organic molecules in soils, (3) the adsorptive capacity of soil minerals for organic materials as governed by particle size and surface area, (4) physical protection mechanisms which restrict access of organic materials to biological attack. As with other aspects of soil organic C dynamics addressed so far in this review, strong interactions can exist between these characteristics (e.g., the presence and type of multivalent cations will undoubtedly be related to the chemical nature of the minerals present).

2.4.2.1 Chemical Nature of the Soil Mineral Fraction

An analysis of different soil types indicates that soils with high contents of active $CaCO_3$ and amorphous Al and Fe tend to have higher organic C contents (Sombroek et al., 1993). In a study of the influence of soil properties on soil organic C genesis, Duchaufour (1976) suggested that the presence of $CaCO_3$ in a Rendzina could stabilize fresh and humified organic materials. Thin carbonate coatings visible under stereoscan examination, and a precipitation of organic molecules induced by Ca^{2+} complexation were implicated in the stabilization of fresh and humified organic residues, respectively, and helped to explain the observed impedance of mineralization. Stabilization of organic C in high base status soils with less reactive or low contents of $CaCO_3$ results predominantly from the formation of Ca organic linkages. In such soils, the initial decomposition of plant residues is rapid, but the subsequent utilization of initial decomposition products is slow leading to higher soil organic C contents, lower C/N ratios and longer retention times. Soils with high base status typically have higher clay contents, are more fertile, and have greater annual vegetative inputs than similar low base status soils. Establishment of causative relationships between base status and organic C contents must, therefore, be examined carefully because of the potential confounding effects of increased vegetative inputs and stabilization mechanisms involving clay minerals.

Soils derived from volcanic ash (Andisols) are typically characterized by large accumulations of organic C, high C/N ratios, and high allophane contents. The formation of Al organic complexes is considered to be important to the biological protection of organic C in Andisols. Boudot et al. (1986, 1988) obtained a significant correlation between the amount of native C mineralized from 10 French highland soils and the contents of amorphous Al and allophane, without observing significant correlations with clay content, exchangeable Al^{3+}, or crystalline Fe oxides. Decreased organic C mineralization rates from ^{14}C-labeled organic substrates in allophanic and nonallophanic soils amended with allophane, relative to that noted in unamended nonallophanic soils, also demonstrated a protective effect of allophanic material on soil organic C (Zunino et al., 1982; Boudot et al., 1988; Boudot et al., 1989). Zunino et al. (1982) demonstrated that the influence of allophane on mineralization of C from an organic substrate varied with the chemical structure of the substrate. The presence of amorphous Fe compounds and Fe^{3+} cations has been shown to have a similar effect to that of allophane and Al^{3+} cations on the mineralization of C from organic materials; however, the magnitude of the protective effect was reduced (Boudot et al., 1989).

2.4.2.2   Impacts of Multivalent Cations

The presence of multivalent cations in soil has important implications on the behavior of clays and organic materials, and the biological availability of organic C. When saturated with multivalent cations, clays remain flocculated, which reduces exposure of organic materials adsorbed onto their surfaces (Section 2.4.2.3) and macromolecular organic materials bearing functional groups become more condensed, and thus, less susceptible to biological attack. The dominant multivalent cations present in soils include Ca^{2+} and Mg^{2+} in neutral and alkaline soils and hydroxypolycations of Fe^{3+} and Al^{3+} in acidic, ferrallitic and andic soils. A stabilizing effect of Ca^{2+}, relative to Na^+, on organic C mineralization was effectively demonstrated by Sokoloff (1938), where the extent of mineralization and solubility of organic C in two soils was reduced by addition of Ca^{2+} salts and enhanced by addition of Na^+ salts. Other studies have also shown a decreased solubility of organic C in the presence of Ca^{2+} (Muneer and Oades 1989c) and reduced mineralization of native organic materials and organic substrates on addition of Ca^{2+} in incubation studies (Linhares, 1977; Muneer and Oades, 1989a,b). In such studies, the question remained as to whether the effect of Ca^{2+} addition on mineralizable C resulted from an indirect effect on colloidal dispersibility or from a direct effect of Ca^{2+} complexation on the biodegradability of organic molecules.

A direct effect of multivalent cation complexation on biodegradability in soil was demonstrated by the following results: (1)   a reduced oxygen absorption on incubation of humic acids saturated with Ca^{2+}, Al^{3+}, or Fe^{3+} in the same soil, relative to that noted for Na^+-saturated humic acids (Juste and Delas, 1970; Juste et al., 1975), (2) an increased stability of Al^{3+} and Fe^{3+} forms of plant and microbial polysaccharides (Martin et al., 1966, 1972), and (3) a three-fold increase in the amount of C mineralized from an organic soil after replacing Ca^{2+} cations with K^+ during a 25-week incubation (Gaiffe et al., 1984). Indirect evidence for the involvement of cations in the accumulation of organic C in soils can also be obtained through a comparison of the organic C contents of a variety of soil types. Using data derived from Spain et al. (1983) for the organic C contents of 29 Australian great soil groups, Oades (1988) showed that, excluding soils subject to waterlogging, there was a positive correlation between organic C contents and either high base status or the presence of substantial contents of Al and Fe oxides. Of interest was the comparison of siliceous and calcareous sands which have little or no clay, but indicate an increased soil organic C content in the presence of Ca^{2+}-containing mineral fractions (< 0.5–1.5% versus 1.5–> 4% organic C, respectively.)

2.4.2.3   Adsorption of Organic Materials onto Mineral Surfaces

Clay particles provide a reactive surface onto which organic materials can be adsorbed and it is generally accepted that such adsorption reactions provide a mechanism of stabilizing soil organic C against microbial attack. Correlations between soil organic C and clay contents have been observed (Schimel et al., 1985a,b; Spain, 1990; Feller et al., 1991) and the various interactions between soil clays and organic materials have been summarized by Oades (1989). Such interactions are principally defined by the chemical nature of organic materials (functional group content, molecular size, etc.) and the type of clay mineral (kaolinite, illite, smectite, etc.). Numerous studies utilizing isotopically labeled organic substrates have shown a positive relationship between the contents of residual substrate C and soil clay content (Amato and Ladd, 1992).

In a field experiment where ^{14}C-labeled plant residues were added to four cultivated soils varying in clay content (5–42%) but having similar clay mineralogy, climatic conditions, and no other organic inputs, the amounts of residual ^{14}C and total organic C in the topsoil (0–10 cm) remaining after 8 years of decomposition were nearly proportional to soil clay content (Ladd et al., 1985). Saggar et al. (1996) completed a similar study in which the decomposition of ^{14}C-labeled ryegrass was monitored over six years in four soils having variable clay content (16–60%) and clay mineralogy. The mean residence time of the ^{14}C-labeled ryegrass was not related to clay content but rather to surface area as measured by adsorption of *p*-nitrophenol. The increase in mean residence time with increasing soil surface area suggested that the protective capacity of the soils toward transformed metabolites derived from plant residues was principally controlled by adsorption onto soil surfaces. Since the data presented by Ladd et al. (1985) were derived from soils with a similar clay mineralogy, soil surface areas would have been well correlated with clay content. The importance of available surface area was also suggested by the results of Sørensen (1972, 1975) where the addition of high surface area montmorillonite to a soil/sand mixture stabilized microbial metabolites, but addition of low surface area kaolinite had little influence. These results suggest that the potential protective capacity of soil clay minerals is more a function of the surface area available for adsorption of organic C than the actual amount of clay.

2.4.2.4   Physical Protection Within Soil Matrix Offered by Soil Architecture

The architecture or structural condition of a soil can exert significant control over processes of biological decomposition by limiting the accessibility of soil organic C to decomposer microorganisms and of microorganisms to their faunal predators. This limitation results from the ability of clays to encapsulate organic materials (Tisdall and Oades, 1982), the burial of organic C within aggregates (Golchin et al., 1994b, 1997b), and the entrapment of organic C within small pores (Elliott and Coleman, 1988). As outlined by van Veen and Kuikman (1990) and Hassink (1992), evidence of the importance of these processes in the protection of organic C in soils can be inferred from the following observations: (1) a faster turnover rate of organic substrates in liquid microbial cultures relative to that of similar substrates in mineral soils, (2) an enhanced mineralization of C and N when soils are disrupted prior to incubation, and (3) a more rapid mineralization of organic C and plant residues in sandy soils than clay soils.

A continuum of pore sizes exist in soils, starting with large macropores (> 20 µm) and decreasing to micropores (< 0.1µm). Kilbertus (1980) suggested that bacteria can only enter pores > 3 µm, which suggests that a significant proportion of the soil pore space may not be accessible to microbial decomposers. Organic materials adsorbed onto clay particles contained in pores < 3 µm would only be decomposed as a result of diffusion of extracellular enzymes released by microorganisms. With increasing soil clay content, the proportion of the total soil pore space contained in micropores increases, and the potential for stabilization due to the exclusion of soil microorganisms increases. This concept of exclusion can be extended to the predation of microorganisms by soil fauna. Van der

Linden et al. (1989) suggested that Protozoa and nematodes are excluded from pores $< 5\ \mu m$ and $< 30\ \mu m$, respectively. Killham et al. (1993) showed that although placing glucose into pores $< 6\ \mu m$ or $< 30\ \mu m$ did not impact on the rate of glucose decomposition, the turnover of glucose C incorporated into the microbial biomass was slower where glucose was only added to pores $< 6\ \mu m$.

The ability of clay particles to adsorb organic materials can also contribute to a biological stabilization of soil organic C through an encapsulation of organic residues in soils and the formation of stable aggregates. Encapsulation of particulate organic residues in soils not only places a physical barrier between decomposer organisms indigenous to soils and substrates, but can also can limit the movement of water and O_2 to sites of potentially active decomposition. A similar situation develops within soil aggregates. Relative to the larger pores between aggregates, the smaller pores within aggregates are more likely to remain filled with water during drying events, and therefore, restrict O_2 movement into the aggregate. The presence of organic cores in aggregates (Beare et al., 1994a,b; Golchin et al. 1994, 1997a,b) will serve to increase this effect by enhancing O_2 consumption within the aggregate. It has been found that anaerobic conditions can exist in the core of moist aggregates even under well-aerated conditions (Sexstone et al., 1985). In native grasslands, Amelung and Zech (1996) demonstrated that the exterior 0.5 mm of > 2 mm diameter peds contained less organic C and had a higher C/N ratio, less lignin, and more microbial derived saccharides than ped interiors. The organic C associated with ped surfaces, therefore, appeared to turn over more rapidly, and exhibit a greater degree of decomposition than that contained within peds.

2.4.3   Biota: Vegetation and Soil Organisms

2.4.3.1   Vegetative Inputs: Variations across and within Ecosystems
Vegetation can influence soil organic C levels as a result of the amount, placement and biodegradability (chemical recalcitrance) of plant residues returned to the soil. The greatest effects of vegetation on soil organic C contents are confined to the A horizon. Concentrations of organic C detected below the A horizon result from pedogenic processes, which occur over much longer time scales than the lifetime of current vegetation. Volkoff and Cerri (1988) showed that for Brazilian soil profiles, current vegetative cover was only in direct equilibrium with topsoil (A horizon) organic C, while that in subsoils was largely unaffected by the nature of vegetative cover. Once the organic C moves to depth (e.g., argillic or spodic horizons), it becomes less accessible to decomposer organisms, as exemplified by the increased radiocarbon ages of soil with depth (Pressenda et al., 1996).

Scharpenseel et al. (1992) provided estimates of the amount of organic C contained in the vegetation, soil, and annual litterfall associated with various ecosystems (Fig. 2.1). Across the tropical, temperate, and boreal forests, a continuous decrease in the amount of plant biomass and litter C is noted, with little change in the amount of organic C stored in soils. The decrease in the ratio of plant biomass C: soil organic C was associated with an increase in turnover time from 18 to 60 yr. Presumably, most of this variation was related to the effect of temperature on litterfall decomposition; however, significant changes in litterfall quality and morphology are also evident. On the basis of the information reviewed in Section 2.4.1, the amount of residue returned to the soil under similar types of vegetation appears to be a function of climatic factors, principally the amount of precipitation; however, this will depend on the nature of the factor most limiting plant growth. Where ample water is available, the amount of residues returned to the soil may be a function of some other factor such as nutrient supply. Where climatic and soil factors are constant, residue placement may become important. A comparison of the amounts of organic C contained in the plant biomass and soils of temperate grassland and forest ecosystems reveals that despite a much smaller amount of plant biomass in the grassland, annual litter C inputs and soil organic C contents were approximately twice

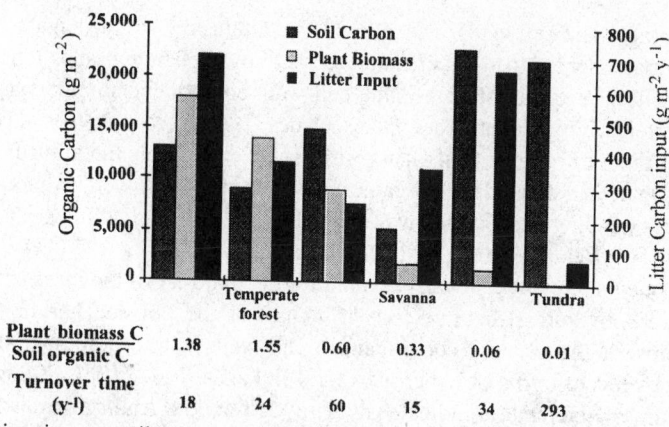

Fig. 2.1 Variations in mean soil organic C contents, plant biomass C contents and rate of litter deposition in various ecosystems [Drawn from data of Scharpenseel et al., 1992]

that of the forests (Fig. 2.1). The occurrence of deep organic rich mineral horizons in temperate grassland soils (e.g., Mollisols), in comparison to the concentration of organic materials in litter layers in boreal forest soil (e.g., Spodosols), is an example of the influence which vegetation can have on soil organic C content and distribution within the soil profile. The apparent larger input of belowground residues in grassland soils, compared to forest soils, places organic C in close proximity to the soil mineral components, thereby enhancing the potential for biological stabilization via the mechanisms discussed in Section 2.4.2. Turnover times in grassland soils are greater than in forest soils (34 versus 24 yr, respectively).

The fate of surface deposited residues depends on the activity of soil microorganisms and fauna and their ability to mix these residues into the surface mineral horizons. In well-drained soils with high Ca status, the activity of earthworms and other soil fauna is high, leading to a mixing of organic residues through processes of particle size diminution, ingestion and casting, and bioturbation. Under such conditions, a mull type humus layer is formed and litter layers do not develop. Plant residues and their decomposition products are intimately mixed with soil mineral particles, which facilitates potential biological stabilization through organomineral interactions (Section 2.4.2). Soils low in Ca do not support as active soil faunal populations and plant residues tend to accumulate on the soil surface forming organic rich, mor-type humus layers. Within mor-type humus little potential exists for biological stabilization other than that due to the chemical recalcitrance of highly decomposed residues. The intermediate form of humus is referred to as a moder.

2.4.3.2 Composition of Plant Materials: The Parent Material for Soil Organic C

Plant materials can be viewed as the parent material for soil organic C in much the same manner as one views primary minerals as the parent materials of soil mineral components. Plant materials are altered by soil fauna and microorganisms, predominantly after deposition in or on the soil, resulting in changes in the original chemical structure and in the synthesis of new compounds, just as some soil minerals dissolve and others precipitate during pedogenesis. An understanding of the chemical nature of plant materials is, therefore, important to studies of soil organic C genesis and composition.

Plant materials consist of a range of different compounds varying in concentration across plant species, plant components (e.g., conducting, supporting or photosynthetic tissues), growth stages, and space (distribution in the landscape). Plant cells can be divided up into three components: the cytoplasm, cell membranes and cell walls. The cytoplasm contains the simple sugars, organic acids, amino acids, and enzymes essential to maintain metabolic activity. Cell membranes consist of globular proteins embedded within a lipid bilayer. Plant cell-wall components include hemicelluloses, celluloses, lignins, proteins, cuticular and root waxes. Oades (1989) presented the following average contents for the major types of organic C in plant residues: (1) extractable materials including water extractables (simple sugars, amino acids and organic acids) and organic solvent extractables (free and bound alkyl molecules including fats, oils and waxes) – 200 g kg^{-1}, (2) hemicelluloses – 200 g kg^{-1}, (3) celluloses – 300 g kg^{-1}, (4) lignins – 200 g kg^{-1}, and (5) proteins – 60 g kg^{-1}.

The organic components of plant cell walls account for the majority of the mass of plant residues deposited in soils. Carbohydrate structures consist mainly of the polysaccharides cellulose and hemicellulose. Cellulose is the primary component of cell walls with a dominant structure of D-glucopyranose residues linked into a polymer via β-1,4 linkages (Fig. 2.2). Cellulose can exist in either a crystalline or amorphous state as indicated by X-ray diffraction (Atalla and van der Hart, 1984) and solid-state ^{13}C nuclear magnetic resonance (van der Hart and Atalla, 1984). The crystalline state is more highly resistant to microbial and enzymatic degradation than the amorphous form (Ljungdahl and Eriksson, 1985). Hemicellulose is defined as the polysaccharides extractable in alkali solution. The hemicelluloses exist as linear and branched polymers of D-xylose, L-arabinose, D-mannose, D-glucose, D-galactose, and D-glucuronic acid monomers, which may be acetylated or methylated. Most hemicelluloses are composed of 2 to 6 of these monomers linked together primarily via a β–1,4 linkage backbone as shown in Fig. 2.2 for pectin, a glucuronic acid polymer.

Lignin represents the second most abundant organic compound in plant residues, and accounts for approximately 5% of the mass of grasses and up to 30% of the mass of hardwood forest species (Haider, 1992). The basic building block of lignin, coumaryl alcohol, can be substituted with zero,

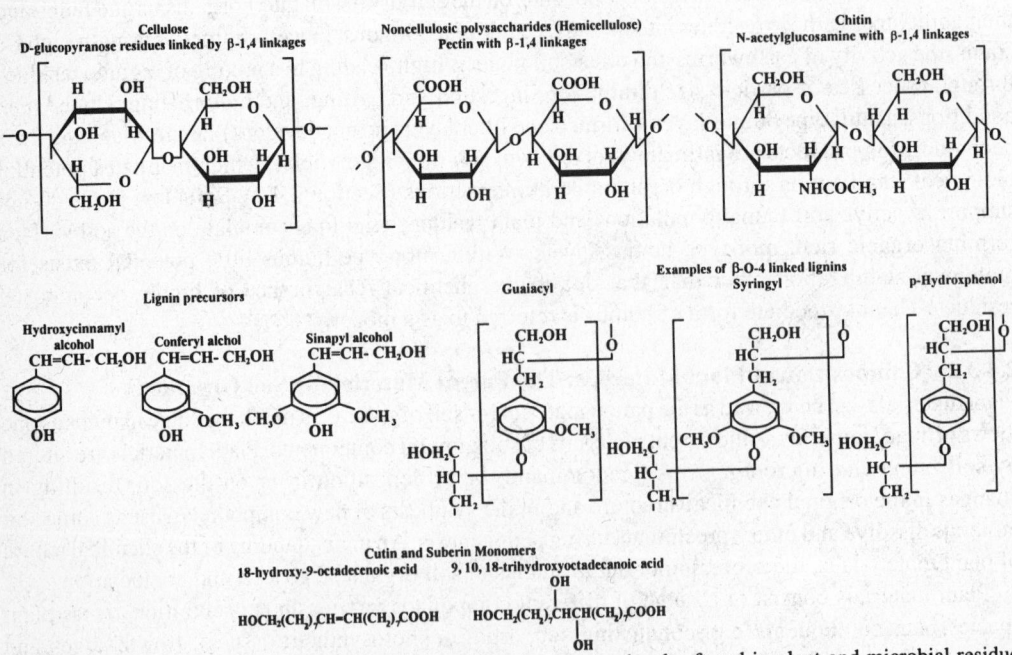

Fig. 2.2 Representative chemical structures of the organic macromolecules found in plant and microbial residues entering the soil.

one, or two methoxyl groups at the C-3 and C-5 positions on the benzene ring to produce the *p*-hydroxyphenol, guaiacyl and syringyl lignin monomeric units, respectively (Fig. 2.2). The units are then linked together by more than 12 possible interunit linkages based on C-O or C-C bonds (McDougall et al., 1993). The major interunit linkage, accounting for about 60% of the linkages, is the β-O-4 linkage depicted in Fig. 2.2 for the three lignin monomeric units. The nature of the lignin molecule changes with plant type: softwoods (gymnosperms) are dominated by guaiacyl-based lignin, hardwoods (angiosperms) contain a mixture of guaiacyl- and syringyl-based lignin, and grasses are dominated by syringyl lignin. Results presented by Hedges et al. (1985) suggested that such changes in lignin composition can affect its biodegradability, with syringyl lignin being more susceptible to decomposition than guaiacyl lignin.

The protein and water soluble components of plant residues, unless stabilized against biological attack, provide a readily decomposable substrate capable of supplying the chemical energy and nutrients required to drive soil biological processes. Enzymatic cleavage of the peptide linkages to form amino acids and mineralization of amino acid N to form NH_4^+ provide sources of N for soil biological processes, and the abiotic chemical processes, to be discussed subsequently.

Alkyl components of plant materials include free and bound lipids, polyesters, and nonsaponifiable alkyl C-dominated biopolymers. Free and bound lipids represent a heterogeneous group of neutral and polar molecules which are classified together based on their solubility in organic solvents (Tegelaar et al., 1989). The neutral component consists of triacylglycerols and waxes, which serve to protect external plant surfaces and to store energy. The polar component is dominated by the esterified fatty acids found in cell membranes. Insoluble polyesters derived from hydroxy fatty acids are found in cutin in plant cuticles and in suberin in roots. Cutin and suberin are composed of various long-chain (C_{16} and C_{18}) substituted fatty acids. The main substituent group is hydroxyl with lesser amounts of epoxy, ketone and carboxyl groups also present (Holloway, 1982). Two examples of cutin and suberin monomers are presented in Fig. 2.2 and Tegelaar et al. (1989) have presented figures showing additional monomeric chemical structures and a proposed model of the structure of intact cutin and suberin. Plant cuticles and roots have also been shown to contain nonsaponifiable aliphatic biopolymers which have been labeled cutan and suberan (Nip et al., 1986a,b, 1987). Cutan and suberan are considered similar to cutin and suberin with the exception that they are highly crosslinked by nonester bonds.

In order to assess the influence of plant residue composition on decomposition and mineralization, it is essential to remove other confounding effects such as climatic, soil and biological parameters. In field studies, this can be accomplished by examining decomposition of all residues of interest at a single site. Vedrova (1997) assessed the impact of forest species on litter decomposition rates by placing litter collected from each species on small plots located within a single unforested site. Mean rates of mineralization measured for cedar, pine, larch, spruce, aspen, and birch litter over ~ 2 years were 1.93, 1.57, 1.85, 2.20, 2.56, and 2.57 mg C g^{-1} litter C d^{-1}, respectively. A limitation of such studies is demonstrated, however, by the work of Elliott et al. (1992), in which the decomposition of four different forest litters (mixed hardwood, red pine, beech, and hemlock) was examined in each of the oril four forest types. The rates of decomposition were principally a function of litter type, with mixed hardwood litter decomposing the fastest and hemlock litter the slowest. However, with the exception of the mixed hardwood litter, decomposition rates of the individual litter types were highest when they were placed in the forest type from which they were derived (i.e., decomposition of the hemlock litter was greatest in the hemlock forest). This interaction between litter type and forest type suggests that decomposition pathways in any given ecosystem may be tailored to the type of litter deposited. Thus, the results of decomposition studies where litters are removed from their ecosystem of origin, or where the community structure of the decomposer organisms is altered, may not accurately reflect the relative effects of residue composition on decomposibility.

2.4.3.3 Relative Impacts of Soil Fauna and Microorganisms

The requirement of soil organisms for chemical energy and nutrients drives processes of heterotrophic decomposition in soils, which account for the major pathways through which soil organic C is mineralized. Abiotic chemical oxidation is unlikely to account for > 20% of total C mineralization (Moorehead and Reynolds, 1989) and more often accounts for < 5% (Lavelle et al., 1993). Microorganisms are the major contributors to soil respiration and are responsible for 80–95% of the mineralization of C. Hassink et al. (1994) calculated that the contribution of the fauna to C mineralization in two sandy and two loam grassland soils ranged from 5–13% of the total C mineralization. The pattern of C mineralization by the soil fauna through time, noted by Hassink et al. (1994), differed from that of total C mineralization, suggesting that the activity of the soil fauna did not contribute substantially to the differences in total C mineralization observed between the soils. Hassink et al. (1993) concluded that soil protozoa and nematodes did not significantly influence soil C mineralization despite a positive response of bacterivorous nematodes on the amount of N mineralized. Several other studies have shown that soil fauna enhanced nutrient mineralization, and had both positive and negative effects on soil organic C mineralization (Griffiths, 1994; Kajak, 1995; Alphei et al., 1996). In a study including protozoa, nematodes and earthworms, Alphei et al. (1996) noted that none of the fauna studied significantly affected basal respiration.

The role of soil fauna in decomposition processes should not be based only on their direct contribution to C mineralization. Soil fauna also act to reduce the particle size of litter, distribute it within the soil, transport otherwise immobile microorganisms to new sites within the soil matrix, and prime microorganism activities by the production of easily metabolizable substrates (e.g., earthworm intestinal mucus). In so doing, soil fauna generally enhance microbial activity and rates of decomposition. Soil conditions which limit (e.g., water saturation and the development of anaerobic conditions) or enhance (e.g., tillage or installation of drains in imperfectly drained soils) the activity of soil microorganisms or fauna will also impact significantly on organic C mineralization rates and thus, alter soil organic C levels.

2.4.3.4 Composition of the Microbial Community

The population of decomposer microorganisms in soil is extensive; densities up to 1010 bacteria and several km of fungal hyphae per gram of soil have been measured in a wide range of soils (Lavelle et al., 1993). As a result of the diversity of decomposer organisms, the existence of interactions between specific types of organic residue and species of decomposer organisms can have pronounced effects on the chemical structure and biological availability of residual organic materials. The decomposition of woody materials provides an excellent example of how the species composition of the decomposer population can influence the chemical nature of decomposition products. Laboratory incubations of Eucryphia cordifolia wood with a brown rot fungus (unidentified species) and a white rot fungus (Ganoderma australe) showed a more selective utilization of carbohydrate C by the brown rot fungus and a delignification by the white rot fungus (Martínez et al. 1991). Using the same white rot fungus in a solid-state fermentation procedure with beech wood, Martínez et al. (1991) noted little change in the chemical composition of the wood, despite a 36% mass loss. Barrassa et al. (1992) obtained similar results in an ultrastructural study. Selective delignification of *Laurelia philippiana* wood by the white rot fungus *Phlevia chrysocrea* was noted, but decomposition of the same wood by *Ganoderma australe* resulted in increased lignin contents. The selective degradation of carbohydrates by brown rot fungi appears to occur independently of the fungal or wood species involved. However, the presence of a selective or nonselective degradation process for white rot fungi appears to depend on interactions between the species of fungus and wood. Under anaerobic conditions, the activity of obligate aerobes such as wood degrading fungi is limited and bacterial decomposition processes dominate. In examinations of buried woods, it has been found that decomposition processes invariably result in a preferential utilization of carbohydrates and a concentration of lignin (Bates and Hatcher, 1989; Bates et al., 1991). Such data indicate that changes in species composition of the decomposer community can significantly alter the decomposition processes and thus, rates of accumulation or loss of organic C from soils. At present, insufficient data exist to extend the results obtained for woody residues more generally to nonwoody organic materials, but the governing principle would be expected to be similar.

2.4.3.5 Relationship Between Organic Residue Composition and Chemical Recalcitrance

All organic C in soils can serve as a substrate. In addition to the potential mechanisms of biological stabilization of organic materials offered by the soil mineral fraction, chemical structure can also impart a degree of chemical recalcitrance. Rates of decomposition of known organic substances in soils were reviewed by Paul and van Veen (1978). Although variations in decomposition rates for any single substrate were evident as a result of differences in soils and incubation conditions, simple organic molecules and monomeric compounds decomposed most rapidly. Oades (1989) showed that the extent and rate of mineralization of C for a series of polysaccharides (glucose, dextran, cellulose) and a fungal polysaccharide) decreased with increasing molecular complexity and branching. Similar results were obtained by Martin and Haider (1975) for the mineralization of C from specifically ^{14}C-labeled benzoic and caffeic acid monomers and polymers. C mineralization was most extensive from carboxylic acid groups, less extensive from the aromatic ring C of the monomers, and least extensive from the polymeric aromatic ring C. Of the polymeric materials contained in plant residues, lignin and other polyphenolic C and aliphatic C appear to be the most recalcitrant, but as discussed in the previous section, the stability of lignin C will be related to the species composition of the decomposer community.

Many studies have demonstrated a relationship between decomposition and plant residue characteristics thought to be indicative of residue quality (Edmons and Thomas, 1995; Hobbie, 1996; Cortez et al., 1996; Ågren and Bosatta, 1996). Included in these residue characteristics are N concentration, C:N ratios, lignin and/or polyphenol concentration, lignin:N ratios, and acid soluble carbohydrates. In a laboratory incubation experiment, examining the impact of temperature on decomposition of six species of tundra vegetation, patterns of decomposition were better correlated with substrate C composition than nutrient; mass losses were positively correlated with acid-soluble carbohydrate content and negatively correlated with lignin content (Hobbie, 1996). In a separate incubation study, mass loss from four types of one-year-old mediterranean ratio of the litter (Cortez et al., 1996; Ågren and Bosatta, 1996). Included in these residue characteristics are N concentration, C:N ratios, lignin and/or polyphenol concentration, lignin/N ratios and acid soluble carbohydrates. In a laboratory incubation experiment, examining the impact of temperature on decomposition of six species of tundra vegetation, patterns of decomposition were better correlated with substrate C composition than nutrient; mass loss from four types of one year old Mediterranean deciduous leaf litter as well correlated with the %N, C:N ratio, lignin concentration and lignin:N ratio of the litter (Cortex et al., 1996). Ågren and Bosatta (1996) found that the proportions of extractable, acid soluble, and acid insoluble C obtained from a conventional chemical fractionation could be used to assess the quality of forest litter, particularly when the acid insoluble fraction did not dominate. During the decompisition of plant residues, significant changes in chemical composition of residual C are evident

(Baldock et al., 1997). In response to such changes, Berg and Staaf (1980) proposed a model of litter decay in which decomposition was controlled initially by N content, but subsequently, by lignin concentration. This was supported by the results of Edmonds and Thomas (1995), which showed that organic C mineralization rates from green needles of western hemlock and pacific silver fir were initally similar, but became more a function of litter chemistry (e.g., lignin:N ratio) as decomposition progressed.

2.4.4 Topography

Topography exerts its major control over soil organic C contents through a modification of climate and soil textural factors, and through its impacts on the redistribution of water within a landscape. Soils in downslope positions are often wetter, have warmer average temperatures, and have finer textures than soils in upslope positions or at the top of knolls. Burke et al., (1995) examined the extent to which soil organic C varied at a landscape scale at two sites differing in soil texture, but having similar climatic characteristics. Burke et al. (1995) noted increased organic C contents (and clay and silt contents) in downslope positions relative to the summits at both sites. Such a finding has been attributed to the downslope movement of organic C and organic rich clay (Reiners, 1983). However, additional gradients in available water along slopes, especially in water limited systems, influence plant production (Peterson et al., 1988) with greater biomass inputs and a greater potential biological stabilization of organic C via higher clay contents at the base of slopes. Where excessive water exists, drainage of depressions in the landscape can be restricted, leading to the development of anaerobic conditions and preservation of organic C relative to the better drained higher landscape elements during wetter times of the year.

2.4.5 Land Management Practices

Paustian et al. (1997) have comprehensively reviewed the influence of agricultural management practices on soil organic C levels. The influence of forestry management practices has been reviewed by Johnson (1992). The most dramatic influence of agricultural practices occurs when soils are first brought into production. Typically, soil organic C levels decrease for the first few years after cultivation and then stabilize at a new equilibrium level which is dictated principally by the ability of the soil to stabilize organic C and amount, quality, and distribution of plant residues inputs. For example, 28–59% of the soil C was lost following 30 to 43 yr of cropping at 11 sites within the North American prairies (Haas et al., 1957). The following characteristics of cereal production systems, in comparison to those of native grasslands, help to explain the observed losses of soil organic C induced by cultivation: (1) 80% lower allocation of organic C to soils (Buyanovsky et al., 1987), (2) reduced below ground allocation of photosynthate (Anderson and Coleman, 1985), (3) enhanced aggregate disruption and exposure of physically protected organic C due to cultivation (see Section 2.4.2.4), and (4) enhanced rates of decomposition of available organic C substrates due to more favorable abiotic conditions (e.g., aeration, temperature and water content).

The intensity with which soil is cultivated can impact both the total amount of soil organic C and its distribution with soil depth. No-till systems tend to concentrate residue inputs at the soil surface and generally enhance soil organic C and N contents in soil surface layers. Therefore, to accurately evaluate the influence of tillage practices on soil organic C stores, it is important to sample at least to the depth of tillage, and preferably, deeper. Paustian et al. (1997) presented data from a number of long-term field trails indicating that soil organic C retention is typcally enhanced under no-till relative to more intensive conventional tillage systems. The selection of annual crops and the inclusion of annual and perennial pastures or mechanical fallows in rotation with annual crops can significantly impact on soil organic C levels. In a long-term crop rotation trial located at the Waite Institute in Australia, soil organic C contents have increased under permanent pasture and declined to varying degrees under cropping systems, with the extent of the delcine being related to the intensity of soil cultivation (Tisdall and Oades, 1982). Field trials set up to examine the impact of fertilizer additions on soil organic C content have revealed that the addition of N fertilizers typically enhances soil organic C contents. Explanatory mechanisms suggest that N fertilizer additions result in a greater return of plant residues to soils, a reduction in decomposition rates due to enhanced soil drying (Andrén, 1987), a promotion of soil acidification (Thurston et al., 1976), a repression of lignilytic enzymes, and a formation of recalcitrant humic materials through a reaction of amino acides with humic precursors (Fog, 1988).

In tropical systems, the establishment of pastures after clearing of forests is becoming increasingly widespread (Sombroek et al., 1993). Pasture establishment immediately after deforestation, using species with high proportions of below-ground biomass, may increase soil organic C contents as demonstrated in the Brazilian Amazon (Serrâo et al., 1979), Latin America (Ligel, 1992) and East Africa (Boonman, 1993). If pastures are installed after one or more cycles of shifting cultivation, productivity is low and soil organic C levels remain low (Sombroek et al., 1993).

2.5 Chemical Structure of Soil Organic Matter

2.5.1 Characterizing the Chemical Structure of Soil Organic C

2.5.1.1 Wet Chemical Methods of Extraction and Characterization

Classical approaches to chemical characterization of soil organic C involved the use of water- and organic solvent-based chemical extractants, and various degradative procedures considered selective in their attack on specific molecular structures. After the removal of macroorganic materials via sieving or flotation and extraction of dissolved organic C (DOC), the remaining organic C was partitioned into humic substances and nonhumic biomoleculres. The distinction between these two forms of organic C is based on structural chemistry (Table 2.2). Several of the potential methodologies capable of being used to quantify and examine the composition of different classes of nonhumic biomolecules are indicated in Fig. 2.3. Although not indicated in Fig. 2.3, such selective methodologies can be applied equally to the macroorganic and DOC fractions and to the humic fractions to selectively remove nonhumic biomolecule contamination.

In the classical extraction scheme (Fig. 2.3), humic substances are differentiated simply on the basis of their extractability in alkali solution at pH values ranging from 10–13, and subsequent solubility on acidification of the alkali extract of pH 2. The unextracted alkali insoluble fraction is referred to as humin. The material which remains soluble in the acidified alkaline extract is the fulvic acid fraction while that which precipitates is the humic acid fraction. All three humic fractions are not discrete compounds, with each fraction containing a multitude of different chemical structures which can be further fractionated and purified. Each fraction typically contains nonhumic biomolecules as a result of the nonspecific nature of the alkaline extraction procedure. Purification of the humic and fulvic acid fractions, by removal of the nonhumic biomolecules, results in the isolation of humic and fulvic acids.

The use of alkaline reagents to extract soil organic fractions prior to the application of characterization procedures has been criticized of late, especially with the development of techniques which allow the chemical composition of soil organic C to be assessed *in situ*. Prior to the development of such spectroscopic techniques, extraction of soil organic materials from the mineral components was a prerequisite to their selective characterization. The main criticisms put forward include (1) the

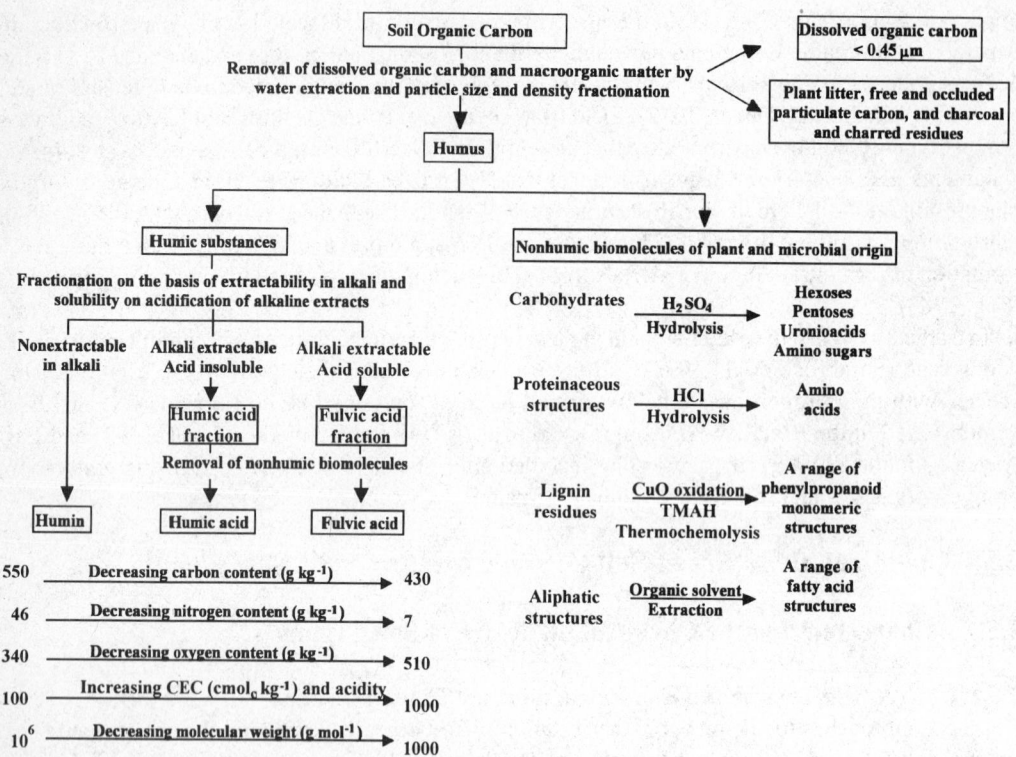

Fig. 2.3   Fractionation of soil organic matter based on chemical and physical properties [Modified from Oades, 1989]

questionable ability of the alkaline extractable organic matter to represent the composition of the non-extracted and whole soil organic fractions, (2) the apparent lack of a relationship between the biological functioning of organic C in soils and its extractability in alkaline reagents based on C and N isotopic tracer studies (Oades, 1995), (3) differences in the chemical characteristics displayed by the extracted organic molecules when compared to those of the same materials existing in soils in an adsorbed state (e.g., conformation, cation binding capacity, hydrophobicity, etc.), and (4) the creation of artifacts during the extraction procedure.

Of greatest concern is the creation of artifacts during the alkaline extraction procedure. A recent study by Zuman and Rupp (1995) used differential pulse polarography (DPP) and spectrophotometry to examine the impact of alkaline solutions on the cleavage of lignin. The rate of lignin cleavage and release of soluble aromatic aldehydes were shown to follow first order kinetics, with the magnitude of the rate constant describing lignin cleavage increasing exponentially over the pH range of 7.5–10.5. Negligible release was noted at pH values < 7.5. A similar effect of solution pH was noted for the rate constant associated with the release of aromatic aldehydes directly from a sandy loam. The maximum release of aromatic aldehydes occurred within 1 or 2 h for the sandy loam and the lignin, respectively. Therefore, under the recommended conditions of alkaline extraction of humic materials from soils, namely, the exposure of soil to a solution of pH 13 for 4 h (Swift, 1996), a significant release of aromatic aldehydes, and potentially of many other aromatic species from lignin may occur. Acidification of alkaline solutions containing aromatic fragments of lignin results in an acid-catalyzed polymerization, and the formation of polymers having chemical structures and molecular masses different from those of lignin. The release of aromatic aldehydes from humic acids extracted from the sandy loam used by Zuman and Rupp (1995) was found to differ from that of the whole soil and lignin in two ways: (1) release of aromatic aldehydes was initiated at pH 3; and (2) the kinetics of release were much faster (maximum release was attained in 10–20 min). The presence of two different mechanisms of release was suggested from the polarographic data. The similar patterns of aromatic aldehyde release from the whole soil and lignin and the contrasting pattern noted for humic acids suggested that the predominant aromatic components found in the whole soil were derived from unaltered lignin and/ or partially degraded lignin fragments, and not from humic acids. This observation, taken in conjunction with the potential acid catalyzed polymerization of lignin fragments extracted by alkaline reagents, suggests that at least a portion, and potentially a significant component, of the humic acid fraction of soil organic materials is not representative of materials naturally present in soils, but is an artifact of the extraction procedure, especially when high pH extractants (pH > 10) are used. In addition, the cleaved aromatic species which do not polymerize on acidification of the alkaline extract end up in the fulvic acid fraction. Therefore, some of the organic species that end up in the fulvic acid fraction may also be a result of the alkaline extraction procedure employed.

Swift (1996) suggested that the problem of artifact production can be reduced by using mild extraction reagents (e.g., neutral sodium pyrophosphate); however, the pH of the extractant would have to be decreased to < 7.5. In doing so, the efficiency of organic C extraction would decrease to the point where the extracted organic C would no longer be an adequate representation of the entire soil organic C fraction. It is strongly suggested that the use of alkaline extractants be avoided, and that a combination of modern spectroscopic techniques and/or wet chemical degradative procedures known to be selective in their action (molecular techniques) be utilized to characterize soil organic C.

2.5.1.2 Spectroscopic Methods of Characterization

Modern spectroscopic techniques that can be used to probe the chemical structure of soil organic C *in situ* include solid-state ^{13}C MNR, analytical pyrolysis, and Fourier transform infrared spectroscopy (FTIR). Selected review and research articles addressing the details of the techniques, recent developments, and their application to the study of organic materials in soils are presented in Table 2.4. Modern spectroscopic methods are complementary and, where possible, efforts should be made to utilize combinations of these methods to confirm and enhance data pertaining to the chemical structure of soil organic materials.

Solid-State ^{13}C Nuclear Magnetic Resonance Spectroscopy (^{13}C NMR)

Only the application of solid-state ^{13}C NMR to soils will be examined in this chapter; however, other forms of NMR including solid-state ^{15}N NMR and solution state ^{13}C, ^{15}N, and ^{31}P NMR have potential applications to the study of organic materials in soils (see Table 2.4 for indicative studies). The greatest positive aspect of the application of solid-state ^{13}C NMR to soils and soil fractions is its ability to characterize the chemical structure of soil organic C *in situ* and nondestructively. A typical spectrum acquired by applying the conventional solid-state cross-polarization magic angle spinning (CPMAS) ^{13}C NMR pulse sequence to soil humus is presented in Fig. 2.4a (Skjemstad et al., 1997). Organic C found in different chemical environments can be differentiated on the basis of chemical shift (the x axis in Fig. 2.4a expressed in units of ppm of the applied magnetic field). A review of the solid-state ^{13}C NMR chemical shift values of organic C found in a wide variety of known chemical structures has been presented by Duncan (1987). Solid-state ^{13}C NMR spectra acquired for most mineral soils with < 50 g C kg^{-1} soil are typically divided into four regions of chemical shift because of the large heterogeneous mixture of different types of organic C: alkyl C (0–45 ppm), O-alkyl C (45–110 ppm), aromatic C (110–165 ppm), and carbonyl C (165–220 ppm). The labels given to each chemical shift region are only indicative of the dominant form of organic C present, and there will undoubtedly be a range of

Table 2.4 List of review and research articles pertaining to the application of and recent developments in modern spectroscopic techniques to the characterization of the chemistry of soil organic materials

Title
Nuclear Magnetic Resonance Spectroscopy (NMR)
Applications of NMR to soil organic matter analysis - history and prospects (Preston, 1996)
N.M.R. techniques and applications in geochemistry and soil science (Wilson, 1987)
Characterisation of soil organic matter by solid-state ^{13}C NMR spectroscopy (Skjemstad et al., 1997)
Nuclear magnetic resonance in agriculture (Pfeffer and Gerasimowicz, 1989)
^{13}C NMR studies of soil organic matter in whole soils: I. Quantitation possibilities (Kinesch et al., 1995)
Pyrolysis-gas chromatography-mass spectrometry (PyGCMS)
Characterization of humic and soil particles by analytical pyrolysis and computer modeling (Schulten and Leinweber, 1996)
Thermal degradation of humic substances relevant to structural studies (Bracewell et al., 1989)
Analytical pyrolysis of humic substances and soils: geochemical, agricultural and ecological consequences (Schulten, 1993)
Pyrolysis and soft ionization mass spectrometry of aquatic/terrestrial humic substances and soils (Schulten, 1987)
Infrared spectroscopy (IR)
Characterization of humic acids, composts, and peat by diffuse reflectance Fourier transform infrared spectroscopy (Niemeyer et al., 1992)
Vibrational, electronic, and high energy spectroscopic methods for characterising humic substances (Bloom and Leenheer, 1989)

structures within each region. Spectra acquired for mineral soils with > 50 g C kg^{-1}, peats, forest litter, and other materials with high C contents such as particulate soil organic fractions can often be divided into narrower chemical shift regions indicative of more discrete types of C, because of higher signal to noise ratios.

The relative contribution of each type of C can be estimated by expressing the integral of signal intensity under a given peak or across a given chemical shift region as a proportion of the total integrated area of the entire spectrum. Such an approach to assessing the distribution of chemical structures within a sample should be considered as a semiquantitative analysis unless detailed NMR experiments are performed which examine the relative rates of signal generation (T_{CH}) and decay ($T_1\rho H$) and the rates of relaxation ($T_1 H$) for each type of carbon present in the sample (Wilson, 1987; Baldock et al., 1989; Pfeffer and Gerasimowicz, 1989; Kinesch et al., 1995). Even with the completion of such detailed experiments, however, nonquantitative results can be obtained when paramagnetic materials are present or where ^{13}C nuclei are separated from ^{1}H nuclei. The presence of paramagnetic species (metal cations with unpaired electrons or organic free radicals) can drastically reduce the spin-lattice relaxation time ($T_1 H$) such that signals from C in close proximity to such species are not observed. Of principal concern in the application of solid-state NMR to soil is the ratio of organic C to Fe in the samples (Arshad et al., 1988; Skjemstad et al., 1994). Arshad et al. (1988) suggested that, at organic C:Fe ratios > 1, adequate spectra can be obtained, but at ratios < 1, steps must be taken to remove the Fe prior to acquisition of NMR spectra. Skjemstad et al. (1994) showed that organic C:Fe ratios > 1 do not necessarily indicate that reasonable spectra can be accumulated if samples have high magnetic susceptibilities, and that the relative visibility (RV) (integrated spectral area g^{-1} organic C in the sample) of soil organic C could be estimated by Equation [2.5], where OC/Fe is the ratio of the gravimetric percentages of organic C to that of Fe in the sample. RIF is the remaining inorganic

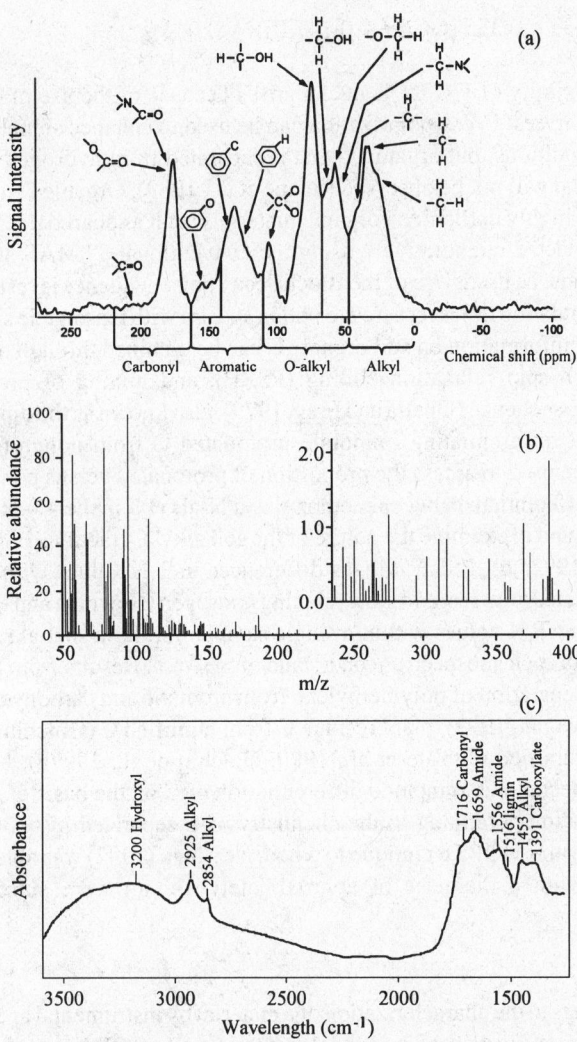

Fig. 2.4 Examples of data obtained using modern spectroscopic techniques to study the chemical structure of soil organic matter. a) Solid-state ^{13}C NMR spectrum of soil humus showing four general chemical shift regions and typical chemical shift assignments (modified from Skjemstad et al., 1997). b) Pyrolysis-field ionization mass spectrum of a humic acid (Schulten, 1987). c) Diffuse reflectance Fourier transform infrared spectrum acquired for the 0-2.5 cm layer of a mineral soil (52 g C kg^{-1}) by subtraction of the spectrum obtained for a sample heated at 350 $^{\circ}$C from that of the unheated whole soil.

fraction expressed as a gravimetric percentage of the total mass of the sample after HF treatment, and χ is the mass magnetic susceptibility (μm^3 kg^{-1}). Repeated pretreatment of soils with a 2% HF solution reduced Fe concentrations and concentrated organic C through the dissolution of a portion of the inorganic matrix, thus enhancing the NMR visibility of the organic C in soil samples considerably. The use of such pretreatments must be examined carefully, since the HF pretreatment may remove significant quantities of organic C.

$$RV = 5.33 + 0.375 \, (OC/Fe - 0.147 \, RIF + 0.043 \, \chi^{-1}) \hspace{3cm} [2.5]$$

The issue of the proximity of ^{13}C to ^{1}H nuclei arises because of the use of cross-polarization in solid-state ^{13}C NMR analyses. Cross-polarization can be used to enhance signal intensity by a factor of 4 under optimum conditions, but organic C that is separated from hydrogen by > 5 bond lengths cannot cross-polarize and will not be observed (Snape et al., 1989). Organic C in soils which contain appreciable contents of highly carbonized organic materials such as charcoal, charred plant material and coal will not be detected quantitatively using the conventional CPMAS analysis, and a Bloch decay pulse sequence must be used. Use of the Bloch decay pulse sequence is restricted because of the exceedingly long accumulation times required to obtain spectra with reasonable signal-to-noise ratios.

Additional chemical information on soil organic C can be obtained through the use of interrupted decoupling (ID), proton spin relaxation editing (PSRE), and mixing of proton spins (MOPS), techniques. The ID pulse sequence (Opella and Frey, 1979), also known as the dipolar dephasing pulse sequence, is capable of differentiating immobile protonated C from nonprotonated C or mobile protonated C. It has been used to assess the proportion of protonated versus nonprotonated aromatic C (Hatcher, 1987), to differentiate between methoxyl and N-alkyl C in the 45–60 ppm chemical shift region (Hatcher, 1987), and to examine the nature of the soil alkyl C fraction (Baldock et al., 1990a,b; Kögel-Knabner et al., 1992a,b). PSRE utilizes differences in T_1H values within a solid sample to derive subspectra associated with fast and slow relaxing hydrogen (Newman and Hemmingson, 1990). The presence of different T_1H values within a single sample implies a spatial separation of organic materials associated with each subspectrum on a scale of > 30 nm. Results from PSRE analyses have demonstrated a spatial separation of polymethylene from aromatic and carbohydrate C in HF-treated humin (Preston and Newman, 1992), plant residue C from humified C (Golchin et al., 1997a,b), and charcoal-like C from humified C (Tate et al., 1990; Golchin et al., 1997b). MOPS represents an extension of the PSRE technique to include differentiation of C on the basis of both T_1H and $T_{1\rho}H$, which provides information pertaining to the chemistry of C separated by > 30 nm and 2–30 nm, respectively. In applying the MOPS technique to wood, Newman (1992) was able to demonstrate that cellulose microfibrils with a diameter of approximately 14 nm were surrounded by a lignin hemicellulose mixture.

Analytical Pyrolysis

Analytical pyrolysis refers to the characterization of a material by instrumental analysis of its pyrolysis products and encompasses a group of methodologies including offline pyrolysis, pyrolysis-mass spectrometry, derivatization pyrolysis-mass spectrometry, and pyrolysis-gas chromatography/mass spectrometry. A detailed description of these techniques is presented by Schulten and Leinweber (1996).

Application of thermal energy to organic materials results in a cleavage of the weaker bonds in the organic molecules present, and a release of a range of reaction products. The thermal energy can be applied using two techniques: quasi-instantaneous heating (Curie point pyrolysis) or a controlled temperature programmed heating. Temperature programmed methods have been criticized because of the potential to form secondary reaction products unrelated to the organic materials present in the sample. Curie point pyrolysis methods are rapid, sensitive, highly reproducible and can be applied to small samples.

In offline pyrolysis, the loss of C and/or N from a sample during pyrolysis is measured with no attempt to characterize the chemical nature of the pyrolysis products. Using a set of 64 soil samples with organic C contents ranging from 1 to 460 g kg^{-1}, Schulten and Leinweber (1996) showed that the amount of volatilization was proportional to the soil organic C content. It was also shown that the proportions of organic N which volatilized were often greater than those of organic C. In pyrolysis-mass spectrometry (Py-MS), the pyrolysis reaction products pass directly into a mass spectrometer that is capable of separating the products on the basis of m/z values (mass/charge ratios). Several methods of ionization are available to promote the movement of reaction products through the mass spectrometer: field ionization (FI), chemical ionization (CI), fast atom bombardment (FAB), and laser ionization (LI). Irrespective of the method of ionization utilized, distinctive patterns of pyrolysis products, often referred to as fingerprints, are produced (Fig. 2.4b). Signals located at various m/z values within these fingerprints can be considered diagnostic for particular types of organic molecules. For example, based on the work of Schulten et al. (1987), Schnitzer (1990) tentatively identified several of the signals in Fig. 2.4b as follows: polysaccharides (m/z values of 60, 82, 96, 110, and 126), heterocyclic N-containing structures (m/z values of 79, 92, and 117), and lignin (m/z values of 212, 302, 344). After normalization of the m/z peak intensities, a qualitative assessment of changes in the chemical nature of the organic C in samples of whole soils or soil fractions can be determined. For example, differences in the chemical composition of organic C contained in particle size fractions isolated from two mineral soils were presented by Baldock et al. (1991). Lists of other applications of Py-MS and the appropriate references are presented by Schnitzer (1990) and Stevenson (1994).

A cautionary note regarding the comparison of different of m/z signal intensities is required. First, the intensities observed at any single m/z value may result from multiple pyrolysate fragments if they have the same mass:charge ratio. Second, the volatilization of different types of pyrolysis fragments varies (e.g., in Py-FIMS volatilization decreases with increasing polarity of the fragments). As a result, differences in signal intensities at various m/z values within any one soil sample do not necessarily correlate with contents of the parent molecules present in the original sample. It is appropriate, however, to utilize variations in a given m/z signal intensity to infer difference in structural chemistry across a series of soil samples run under a constant set of analytical conditions.

Various on- and offline derivatization processes have been developed to reduce differences in the volatilization of the various pyrolysis products. Derivatization Py-MS techniques utilizing tetramethylammonium hydroxide (TMAH) convert hydroxyl and carboxyl groups into their equivalent methyl ethers and methyl esters, respectively. Pyrolysis methylation with TMAH enabled detection of aliphatic $C_2–C_{39}$ monocarboxylic acid methyl esters, $C_4–C_{30}$ dicarboxylic acid dimethyl esters, benzenecarboxylic acid methyl esters and a range of other methylated compounds (Schulten and Leinweber, 1996) The incorporation of a gas chromatograph between the pyrolysis chamber and the mass spectrometer can further aid in the separation of similar pyrolysis fragments prior to detection and analysis by the mass spectrometer.

Fourier Transform Infrared Spectroscopy (FTIR)
In its application to studies of the chemical structure of soil organic C, the absorption of infrared radiation of different frequencies can be used to determine the type of the atoms to which C is bound as well as the nature of the bond. Detailed identification of chemical structure is possible for pure simple organic molecules, but the dominance of signals derived from mineral components limits the application of FTIR spectroscopy to the study of soil organic C in mineral soils. Even after extraction of organic materials from soils, the complex and heterogeneous nature of the organic C results in spectra exhibiting absorption bands spanning wide ranges of frequency with few well-resolved peaks. However, comparisons of spectra obtained before and after various chemical degradation or derivation treatments (e.g., carbohydrate hydrolysis and methylation or acetylation) can yield useful information. The use of difference spectra (spectrum of an untreated sample minus that of a pretreated sample) can provide insight into chemical structures. An example of the difference spectrum associated with the removal of organic C from a soil is given in Fig 2.4c and can be used to provide

an approximation of the chemical nature of the organic materials removed in the pretreatment, provided that a significant alteration of the soil mineral components has not occurred.

Infrared spectra contain a vast amount of information pertaining to chemical structure. The problem experienced to date has been how to get at the information and utilize it in a meaningful manner. The use of multivariate statistics such as partial least squares (PLS) allows those portions of an FTIR spectrum that are correlated to other measurements (e.g., total sugar concentration as identified by H_2SO_4 hydrolysis and subsequent colorimetric determination) to be identified across a range of soil samples. Once such relationships have been identified, the content of the required parameter of interest can be estimated with a known degree of precision using FTIR in conjunction with the predictive equations. The rapid nature of modern infrared analyses (several min./sample) when utilized in this manner, would facilitate the acquisition of data without the need for utilizing laborious routine procedures. Skjemstad et al. (1997) have presented the results of such an evaluation in which a strong correlation ($r^2 = 0.977$) was found between aromatic C content as measured by solid-state CPMAS ^{13}C NMR and that generated using PLS procedures in conjunction with FTIR. Given that the typical time required to acquire a solid-state ^{13}C NMR spectrum for a mineral soil is 6–24 h, an enormous potential exists to utilize FTIR data to facilitate an assessment of the chemical nature of soil organic C, provided that enough ^{13}C NMR data exists to complete an appropriate calibration procedure.

2.5.2 Dissolved Organic Carbon

Dissolved organic C (DOC) is a small but important and dynamic fraction of soil organic C. Its importance relates largely to its mobility, both within the soil and from the soil into groundwater or surface water bodies. From a biological point of view, it provides a mobile source of energy and the nutrients associated with it can make an important contribution to nutrient availability and cycling. From a chemical point of view it behaves as a reactive component of the soil solution, and facilitates transport of other materials. A detailed review of information on DOC in soils was recently completed by Herbert and Bertsch (1995).

The term dissolved refers to materials in solution that do not settle out under the influence of gravity. The smaller and more polar an organic molecule, the more likely that it will stay in aqueous solution. Whether or not a particular organic molecule is dissolved depends on the water content of the soil, and the nature of the surfaces and other solutes. At the larger, less polar end of the range, the boundaries between dissolved, colloidal and particulate organic matter are not definite. The definitions are usually made operationally. For example, DOC is usually defined as organic C that has passed through a particular suction cup or filter or that occurs in the supernatant after centrifuging a soil suspension at a given relative centrifugal force for a given period. These operational parameters should be clearly stated in discussions of DOC in soil, and it is important to note that processes such as adsorption onto or clogging of filters or suction cups may significantly influence the nature and amount of material obtained (Grossman and Udluft, 1991). In this chapter, DOC has been defined as that which passes through a 0.45 μm filter.

2.5.2.1 Isolation and Measurement

Before measurement or characterization, DOC must be extracted from the soil. The means of extraction determine the amount and nature of the material extracted (Herbert and Bertsch, 1995). While extraction at or near the water content of interest is desirable, such a procedure may result in very small volumes being collected. At high soil water potential (more negative and drier), extraction becomes impractical, but at low soil water potential (less negative and near saturation), sufficient volumes are often easily obtained. In the field, DOC has been extracted by the use of tension or zero-tension porous cups or plates, or by wick lysimeters (Grossman and Udluft, 1991). In the laboratory, DOC can be extracted by forcing solution out of the soil through a porous membrane using vacuum or centrifugal force or by leaching repacked columns or undisturbed cores (Dunnivant et al., 1992). For convenience, many studies have used extracts made by shaking soil with water or dilute salt solutions, followed by separation of the solution by centrifugation or filtration. Material extracted by alkaline solutions has also been used to represent DOC. However, this practice is likely to be inappropriate, because materials dissolved by alkali differ from those found dissolved in soil solution (Novak and Bertsch, 1991).

Once a soil extract has been obtained, DOC contents are estimated by oxidizing the dissolved organic C to CO_2, and then measuring the amount of CO_2 produced or oxidant consumed. A common technique that appears satisfactory in most cases is oxidation by persulfate and UV radiation followed by detection of the CO_2 evolved by IR absorbance or other techniques. Techniques that measure the amount of oxidant consumed face similar problems as those discussed in Section 2.3 for the analysis of soil organic C contents. The yellow to brown color of DOC has resulted in the use of absorbance in the visible or UV range to quantify DOC concentrations. Absorbance is simple and convenient and readily lends itself to continual monitoring and logging. However, it should be kept in mind that it is not a direct measure of concentration, as different organic materials differ widely in their absorption properties.

2.5.2.2 Gains, Losses and Amounts

DOC enters the soil as leachate from vegetation and litter, and is generated within the soil through the processes of excretion from organisms and desorption or dissolution from the solid phase. It is lost through biological uptake, extracellular mineralization, sorption, precipitation and as leachate into surface water courses or groundwater. The concentration of DOC is governed by all these processes, and typically ranges from 5–50 mg L^{-1} in surface and litter horizons, and from 0.5–5 mg L^{-1} or less in B and C horizons (McDowell and Likens, 1988).

Plant roots and microorganisms excrete a large variety of soluble organic materials. For example, in anaerobic soils, organic acids (mostly low molecular weight fatty acids) are major byproducts of decomposition. Uptake of organic molecules by roots has been documented, but this removal mechanism is probably of minor importance.

Microbial uptake, decomposition, and mineralization are the major mechanisms determining the concentration and nature of DOC. The amount of DOC in soil has been correlated with soil respiration and denitrification potential (Burford and Bremner, 1975; Davidson et al., 1987). From this, it is sometimes inferred that DOC is the active fraction of SOM, or that it is the pool turning over most rapidly. However, equating DOC with the active pool is only a very rough approximation. Although many organic substrates (e.g., cellulose) are solubilized by enzymes prior to mineralization, the flux of readily mineralized substrates through the DOC pool is not necessarily reflected in the size of the pool, because the concentration of readily available materials such as monosaccharides, simple organic acids, or amino acids is kept very low due to their rapid assimilation or mineralization. Nevertheless, the concentration of readily mineralizable DOC may be temporarily higher in some situations, such as where roots are active or where fresh organic residues have been added (Nelson et al., 1994). The proportion of DOC that is mineralized in short-term incubation studies generally varies between 3 and 40% (Boyer and Groffman, 1996; Nelson et al., 1994), and varies with the soil depth from which it was extracted (Nelson et al., 1994). DOC, as a whole, may not be any more easily mineralized than particulate or adsorbed fractions. Some DOC appears to be biologically recalcitrant (Zsolnay and Steindl, 1991), with the reasons not being well understood. While physical entrainment may protect particulate organic matter from microbial attack, it is doubtful as a stabilization mechanism for DOC which can diffuse towards organisms (Adu and Oades, 1978).

The processes of adsorption/desorption and dissolution/precipitation are governed by the nature and concentration of the DOC and other solutes, pH, and the nature of the solid phase materials. Sorption occurs on many soil materials through a variety of mechanisms (Oades, 1989). It is greatest at low pH (Jardine et al., 1989), and large or hydrophobic molecules are preferentially adsorbed compared to small or hydrophilic molecules (Dunnivant et al., 1992; Kaiser et al., 1996). In soils with high surface area or high clay content, concentrations of DOC are kept low by adsorption of organic molecules onto mineral surfaces (Nelson et al., 1993). Where multivalent cations, such as Fe or Al in acid soils and Ca in neutral to alkaline soils, are abundant DOC concentrations are kept low because organic complexes of multivalent cations do not ionize readily and are of relatively low solubility. Ong and Bisque (1968) showed that the flocculating effect of cations on DOC was proportional to the sixth power of their valence. Multivalent cations may also form bridges between dissolved organic molecules and mineral particles, thereby taking the organic molecules out of solution (Nelson and Oades, 1998). In water-saturated sediments, DOC concentrations can be high due to dissolution of Fe and Al complexes (Thurman, 1985), and this is probably also the case in waterlogged soils. In soils with low surface area or in soils with a high proportion of monovalent cations, a higher proportion of organic C tends to be dissolved (Nelson and Oades, 1998). Seasonal factors such as temperature and moisture and management factors such as acidification, tillage and application of fertilizer, lime, or manure all influence the amount and nature of DOC in soils (Herbert and Bertsch, 1995). In a leaching environment, DOC that is not retained in the soil by adsorption or precipitation mechanisms is lost to groundwater or surface water (Nelson et al., 1993), where it has a significant influence on water chemistry. Preferential flow through macropores can greatly enhance transport through the soil and loss to water bodies (Jardine et al., 1990).

2.5.2.3 Chemical Structure

The chemical structure of soil DOC is complex and varied. Molecular weight ranges from a few hundred to hundreds of thousands (Homann and Grigal, 1992). Conventional wet chemical molecular techniques can typically identify about 50% of the structures in DOC; the rest remains unidentifiable. Easily identified structures include carbohydrates, hydrocarbons, polyphenolic compounds, and amino, aliphatic and aromatic acids (Stevenson, 1994). Complex material that is not easily identified is commonly referred to as humic substances, most of which can be classified as fulvic acid because it remains dissolved at low pH.

Due to the complexity of DOC, samples are frequently fractionated, either as a characterization or preparative technique for other characterization methods. One of the most widely used fractionation techniques involves sorption on nonionic and ion exchange resins (Herbert and Bertsch, 1995). These and related techniques yield fractions termed hydrophilic and hydrophobic acids, bases and neutrals. DOC has also been fractionated on the basis of molecular size, using gel permeation chromatography or ultrafiltration (Herbert and Bertsch, 1995). Materials in DOC fractions are still complex, and it is not always clear to what extent materials are lost, gained or altered during the fractionation process.

Important characteristics of DOC include its acidity, chemical structure, and sorption behavior. Titratable acidity, largely due to low molecular weight organic acids, varies from 6 to 15 mol_c kg^{-1} C (Herbert et al., 1993). Examination by ^{13}C NMR spectroscopy has shown that DOC consists of carbohydrate, aromatic, aliphatic and carbonyl structures (Novak and Bertsch, 1991). In litter and surface horizons, DOC contains a high proportion of aliphatic material and hydrophobic acids, whereas, in deeper horizons, it is less aliphatic and dominated by hydrophilic acids (Easthouse et al., 1992).

2.5.3   Particulate Organic Matter

Particulate organic C (POC) is defined as the C found in fragments or organic debris with a recognizable cellular structure. POC may be derived from any source, but is usually dominated by pieces of plant structures. In the following discussion, POC found in Histosols and on the mineral soil surface (litter) will be dealt with separately from that found within the mineral matrix.

2.5.3.1   POC in Histosols and Forest Litter Layers

Histosols and forest litter layers present a simple system in which to characterize the chemistry of POC and the chemical changes associated with POC decomposition. The absence of mineral particles facilitates a selective chemical characterization of POC, and reduces the biological stabilization of labile organic C via adsorption onto mineral surfaces or physical entrapment within aggregations of mineral particles. In these highly organic materials, decomposition and chemical transformation processes are controlled by the composition of the plant residues and the changes imparted by the decomposer community.

The chemical composition of organic C contained in Histosols has been examined by numerous researchers using ^{13}C NMR. Using rubbed fiber (see Lévesque et al., 1980 for a definition) as an indicator of extent of decomposition, positive correlations were obtained with the contents of alkyl and aromatic C and negative correlations were obtained with the contents of O-alkyl C (Hammond et al., 1985). Cultivating Histosol surface layers induced an increase in alkyl C and reductions in O-alkyl and aromatic C, relative to the uncultivated Histosol (Preston et al., 1987). The reduction in aromatic C presumably resulted from the presence of a more oxidizing environment in the cultivated soil, and an enhanced activity of aerobic lignin degrading fungi. Preston et al. (1989b) and Nordén et al. (1992) examined the chemical changes associated with decomposition of peat by performing ^{13}C NMR analyses on particle size fractions. Nordén et al. (1992) observed a progressive loss of identifiable plant structures and an accumulation of alkyl C and loss of O-alkyl C as particle size decreased. Similar results were obtained by Preston et al. (1989b), but a loss of aromatic C was also noted, and the magnitude of the chemical changes observed across the particle size fractions increased as the overall extent of decomposition exhibited by the Histosols increased.

Assessments of the chemical changes induced by decomposition in forest soils typically include a characterization of the different litter layers encountered in progressing from the fresh litter located at the top of the forest floor through to the well-humified materials located at the top of the upper mineral soil horizon. ^{13}C NMR data collected for a range of German forest soils exhibiting examples of the mull, moder, and mor humus forms showed an increase in alkyl C and a decrease in O-alkyl C contents with depth for all soils (Zech et al., 1992). Changes in aromatic C content were less consistent, but in general, they tended to decrease with increasing extent of decomposition. Baldock and Hatcher (unpublished data, see Baldock et al., 1997) observed a similar pattern of alkyl and O-alkyl C contents in litter layers found under red pine and tamarack plantations, but aromatic C content tended to increase with extent of decomposition. Variations in the aromatic C content with increasing extent of decomposition content are suspected to arise from differences in the activity of lignin degrading fungi. An assessment of the possible mechanisms accounting for an accumulation of alkyl C in forest litter layers is presented in Section 2.5.4.2.

In studies where depth profiles are used to assess the chemical changes associated with decomposition, it is assumed that the composition of the original plant residues deposited on the soil surface has not changed as the litter layers have developed. Where significant differences in the vegetative source of organic C exist, variations in the chemical composition of litter and humified organic C are possible. Krosshavn et al. (1992) observed differences in the chemical composition of organic C found in peats and forest litters derived from different types of vegetation. For samples exhibiting a similar degree of decomposition and organic C content, the greatest differences in chemical composition were noted between peat and forest organic C; however, differences between forest types were also noted. Thus, when characterizing the changes in chemical composition induced by decomposition in different deposits of terrestrial organic C, care must be taken to ensure that differences in vegetative background are not confounded in the analysis.

2.5.3.2   Particulate Organic Materials Found Within the Mineral Soil Matrix

POC found within mineral soils is usually dominated by plant-derived materials, but can also contain fungal hyphae, spores, pollen grains, and faunal skeletons, as demonstrated by microscopic examinations (Waters and Oades, 1991; Oades and Waters, 1991). This fraction of soil organic C serves as a source of both energy and nutrients for soil organisms, and as a source of nutrients for plants. POC fills an intermediate position between fresh undecomposed plant materials and the more decomposed humus fraction. POC has been separated from the mineral matrix by two main methods: (1) sieving of dispersed soil samples to yield a macroorganic fraction, and (2) collection of dispersed soil materials which float on a heavy liquid to yield a light fraction (Gregorich and Janzen, 1996). A discussion of the methodologies used to separate the macroorganic and light fractions from soil is given in Gregorich and Ellert (1993). Although both POC fractions are dominated by plant residues, the nature of the materials collected in each fraction may differ. In fractions collected on the basis of particle size, humified organic materials bound strongly to large inorganic particles and organic debris coated with mineral particles will be retained on the sieves and included in the macroorganic fraction. In the light fractions, such organo-mineral complexes may have a density higher than that of the fractionating solution and will not be included in the light fraction.

A prerequisite for the collection of either type of POC fraction is a dispersion of the soil, preferably using physical processes which minimize alterations to the chemical composition and particle size distribution of organic fragments. Ultrasonic dispersion is now utilized widely to accomplish this task; despite the potential for a fragmentation of the organic particles and a redistribution of organic materials within the soil with this method (Elliott and Cambardella, 1991; Cambardella and Elliott, 1993). The amount of organic C which accumulates in the macroorganic and light fractions is a function of the intensity of dispersion and the nature of the dispersing medium. Golchin et al. (1994a) observed a loss of light fraction C (< 1.6 Mg m^{-3}) with increasing duration of ultrasonic dispersion, when the dispersion was accomplished in water. The procedure which maximized the extraction of light fraction C involved an initial removal of a free light fraction by gentle end-over-end shaking of soil in the heavy liquid, followed by an extraction of the POC occluded within mineral particle associations by using a 300 s ultrasonic treatment in the heavy liquid.

The lower size limit of the macroorganic fraction varies between investigations, but usually corresponds to the lower size limit of the sand fraction, 20 μm for the ISSS and 50 μm for the American particle size classification systems. Solid-state ^{13}C NMR was used by Oades et al. (1987, 1988) and Baldock et al. (1992) to determine the chemical composition of the organic materials associated with various particle size fractions after ultrasonic dispersion and fractionation by sieving and sedimentation. A density separation was applied to the macroorganic fractions to concentrate the organic C and allow acquisition of acceptable ^{13}C NMR spectra. The chemical composition of the largest macroorganic particles (250–2000 μm or 53–2000 μm) resembled that of plant materials. The reduction of macroorganic particle size to 20–53 μm, induced via an increase in the extent of decomposition, was associated with a loss of O-alkyl C and an accumulation of aromatic and alkyl C. Similar results were obtained by Skjemstad et al. (1993) for four additional Australian soils. The changes in chemical composition noted by Baldock et al. (1992) were consistent with a preferential utilization of carbohydrate structures by soil microorganisms and a preservation of lignin and aliphatic structures during decomposition. In progressing into the finer fractions, where the dominant form of organic C would be microbial metabolic products and humified materials adsorbed onto mineral surfaces, aromatic C contents decreased and alkyl C contents increased significantly. The increased alkyl C content could not be explained by a selective preservation of plant-derived alkyl C alone, and it was postulated that a contribution was made from microbial alkyl C synthesized as a metabolic product of decomposition processes. These observations led to the development of a simplified model which describes the oxidative decomposition of POC fragments in mineral soils (Baldock et al., 1992). The model is consistent with a biopolymer degradation model proposed by Hatcher and Spiker (1988) for the genesis of humic substances.

The separation of light fractions from soils utilizes the difference in particle density between organic and mineral particles. Organic particles generally have a density of ≤ 1.0 Mg m^{-3} while that of mineral particles is > 2.0 Mg m^{-3}. Most studies, where light fractions have been isolated, have used heavy liquid densities ranging from $1.5 - 2.0$ Mg m^{-3}; however, liquids with densities as low as 1.0 Mg m^{-3} have been used for the isolation of light fractions from soils having a low mineral particle density, such as Andisols (Golchin et al., 1997b). Studies on soil aggregation have suggested that many soil aggregates have cores of organic particles (Waters and Oades, 1991) and that macroaggregates form around particles of plant residues (Beare et al., 1994a; Buyanovsky et al., 1994; Golchin et al., 1998). As a result, two forms of light fraction POC exist in soils: (1) free POC without significant association with mineral particles, and (2) occluded POC strongly associated with mineral particles or buried within soil aggregates. The extent to which a soil is disrupted will, therefore, dictate the amount of free and occluded light fraction released and measured, particularly in highly aggregated fine-textured soils. The chemical composition of free and occluded light fractions was examined by ^{13}C NMR for five different soil types (Golchin et al., 1994a). Relative to the free light fractions, which had a chemical composition similar to that of the plant materials deposited on the soil, the occluded light fraction always showed decreased contents of O-alkyl C and increased contents of alkyl C. A significant increase in aromatic C was observed for two of the five soils when progressing from the free to occluded light fractions and only minor changes in the carbonyl C region were detected. This result indicated that the free and occluded POM in soils are chemically distinct, and that the occluded fraction exhibits a chemical structure consistent with a more highly decomposed POC fraction. This work was extended to examine the nature of the organic materials associated with the occluded light fraction in more detail by fractionating it into four density classes: < 1.6, $1.6–1.8$, $1.8–2.0$, and > 2.0 Mg m^{-3} (Golchin et al., 1994b). Again, different chemical compositions were observed for the different classes of occluded organic C. Progressing from the $1.8–2.0$ to $1.6–1.8$ to < 1.6 Mg m^{-3} occluded fractions, chemical changes were again consistent with an enhanced extent of decomposition. Microscopic examination of these occluded fractions indicated that the < 1.6 and $1.6–1.8$ Mg m^{-3} fractions existed as fragments of plant debris exhibiting characteristics associated with extensive decomposition (e.g., the presence of lignin coils and large pores). The $1.8–2.0$ Mg m^{-3} fraction was found to exist as only slightly decomposed plant fragments strongly bound to and buried within a mineral matrix. The chemical composition of organic C contained in the > 2.0 Mg m^{-3} fraction did not resemble that of plant fragments, and was considered representative of microbial residues and humified organic materials of plant origin adsorbed onto mineral surfaces. These results were summarized by Golchin et al. (1994b) in a model linking the chemical composition of the light fraction POC to decomposition and aggregation. The model was supported by subsequent work examining the organic C turnover rates for the various free and occluded POC light fractions (Golchin et al., 1995) and then was extended by Golchin et al. (1998)

In addition to the contributions of decomposed plant residues to soil POC fractions, evidence of significant contributions to POC fractions from charcoal or charred residues in some soils has been presented (Skjemstad et al., 1996b). After treating soil silt fractions (< 53 μm) with an ultraviolet photo-oxidation process capable of oxidizing most of the biologically derived organic components found in soils (for a description of the method, see Skjemstad et al., 1993), the residual organic materials of three soils showed a particulate morphology and a chemical composition consistent with charcoal. The large broad aromatic C peak observed in ^{13}C NMR spectra acquired for the ultraviolet photo-oxidized and HF-treated samples was similar to that found for chars produced by heating cellulose to > 400 °C (Shafizadeh, 1984), and for thermally degraded cork (Pascoal Neto et al., 1995). The significant enhancements noted in the aromatic region of ^{13}C NMR spectra acquired using a Bloch decay pulse sequence, relative to the standard cross-polarization pulse sequence, were also consistent with the highly condensed/nonprotonated aromatic C typically found in charcoal and charred plant residues. Skjemstad et al. (1996) calculated charcoal C contents of 3.2, 3.4, and 8.3 mg charcoal C g^{-1} soil which accounted for up to approximately 30% of the total organic C found in the soils. Golchin et al. (1997a,b) examined the nature of the POM fraction in a series of Andisols, in which the length of time since annual burning of grassland had ceased, differed. UV photo-oxidation was unsuccessful in identifying a residual charcoal component within the POC fractions because of the presence of inorganic cementing agents which are not removed by the photo-oxidation procedure. However, data acquired from the PSRE and Bloch decay NMR techniques indicated that, relative to the site where annual burning had not been practiced, annual burning of grasses led to the presence of a charcoal-like fraction as evidenced by the following characteristics: (1) nonlignin highly aromatic C separated physically from undecomposed plant residue like C and (2) the presence of condensed and proton deficient aromatic C (Golchin, 1997a). Subsequent density fractionation of these Andisols, followed by application of PSRE and Bloch decay NMR experiments (Golchin et al., 1997b), showed that charcoal was distributed throughout the 1.0–2.0 Mg m^{-3} fractions. However, the charcoal accounted for a greater proportion of the organic C in the < 1.6 and 1.6–1.8 Mg m^{-3} occluded POC fractions of all sites, even those without a recent history of burning.

The presence of significant quantities of nonlabile charcoal in soils will have significant implications for studies of POC composition and dynamics. The aromatic portion of soil fractions, after accounting for lignin and lignin like structures (e.g., tannins), is typically considered to be derived from humic substances. Using estimates of lignin and tannin contents based on ^{13}C NMR data, Skjemstad et al. (1996) demonstrated that much, if not all, of the aromatic C remaining, after accounting for lignin and tannin structures, could be accounted for by their estimates of charcoal C, and that extracted humic acid fractions had chemical structures consistent with charcoal. Models of POC dynamics typically utilize total organic C measurements to quantify the size of POC pools. The presence of even small quantities of inert charcoal C in POC fractions could significantly alter estimates of turnover times and radiocarbon ages associated with the fraction of soil POC not derived from charcoal or charred plant materials.

2.5.4   Soil Humus

Soil humus refers to the amorphous organic materials remaining in soils after extraction of the water-soluble fraction and exclusion of particulate organic materials (Fig. 2.3). Humus consists of a mixture of humic substances and nonhumic biomolecules. In alkaline and neutral soils, humus dominates the soil organic fraction because of the rapid decomposition of plant residues by soil fauna and microorganisms. In acidic soils, plant fragments make a larger contribution to the soil organic fraction; however, humus still represents an important fraction and can contribute significantly to soil processes (e.g., podzolization and mineral dissolution). Although recent evidence questions the importance and validity of the classical humic fractions (Section 2.5.1.1), it is not yet possible to simply dispel their importance in soils and a discussion of their chemical properties is required.

2.5.4.1 Humic Substances

The component fractions of soil humic substances, humin, humic acid and fulvic acid, exhibit a continuum of several chemical characteristics (Fig. 2.3). Since the differentiation of the three humic fractions is based simply on their solubility in alkaline and acid solutions, fractionation procedures are method dependent. Quantitative, reproducible fractionation can only be achieved by strict adherence to a given methodology. Changes in methodology (e.g., the type and concentration of solute in the extractant) can significantly alter the partitioning of soil humus C into these fractions and the chemical characteristics observed for each fraction (Hayes et al., 1975). A standard extraction methodology exists (Swift, 1996) and well-characterized reference and standard humic substances are available from the International Humic Substances Society.

The ability to extract and de-ash soil humic and fulvic acids facilitated the acquisition of detailed data pertaining to their chemical composition and physical properties. Much less information has been collected for soil humin fractions; however, the introduction of spectroscopic methods capable of *in situ* analysis has enhanced our understanding of the composition of soil humin (Saiz-Jimenez et al., 1979; Preston and Newman, 1992). Average chemical formulae for humic and fulvic acids as suggested by Steelink (1985) are $C_{10}H_{12}O_5N$ and $C_{12}H_{12}O_9N$, respectively. Humic acids contain more C and N but less O than fulvic acids. Stevenson (1994) and Swift (1996) have provided detailed compilations of the chemical methods used to quantify the content of functional groups, the degradative methods used to identify structural components, and the physical methods used to measure properties such as molecular weight and particle size. Data collected by Schnitzer (1977) on the functional group content of humic and fulvic acids extracted from soils found in different climatic zones showed significant variations. Solid-state ^{13}C NMR analyses of humic acids and fulvic acid fractions from several soil types also revealed structural differences (Malcolm, 1990).

Detailed reviews of the chemical composition and structure and physical properties of humic substances can be found in Oades (1989), Hayes et al. (1989), Stevenson (1994), and Beyer (1996). On the basis of the information presented in these reviews, the chemical and physical properties of soil humic substances can be summarized as follows: (1) aromatic rings are a significant component and multiple substitution with carboxyl, hydroxyl, carbonyl and alkyl groups exists; (2) significant quantities of C1-C20 alkyl C chains either unsubstituted or substituted with O-containing functional groups are present, with smaller chain lengths predominating; (3) aromatic and alkyl groups are bound together principally by C-C bonds and ether linkages to form the backbone structure of the humic molecules; (4) the aromatic alkyl backbone structure is random and is not characterized by a regular sequence of aromatic and alkyl groups; (5) simple and polymeric proteinaceous and carbohydrate groups may be bound to the aromatic alkyl backbone or physically associated with humic surfaces; and (6) molecular weights vary from 10^3 to in excess of 10^6 and the molecules exist in random coils, which are more tightly cross-linked and coiled in the center. An attempt to combine these characteristics has led to the generation of chemical models representative of an average structure for humic materials such as that presented by Schulten and Schnitzer (1993) for humic acid.

The range in both type and content of function groups, the presence of numerous chemically different monomeric components, and the inclusion of biopolymer-like components in humic structures, suggest that a variety of different mechanisms may be involved in their genesis. The mechanisms of humic substance genesis have been reviewed (Hatcher and Spiker, 1988; Hedges, 1988; Stevenson, 1994; Shevchenko and Bailey, 1996), and there is general consensus that humic substances result from the transformation of organic plant residues. The proposed mechanisms of genesis can be placed within two contrasting categories: (1) partial biotic biopolymer degradation, where the integrity of the biopolymer is not destroyed, and the modified biopolymer forms the humic substance backbone, and (2) abiotic condensation polymerization, in which simple products of

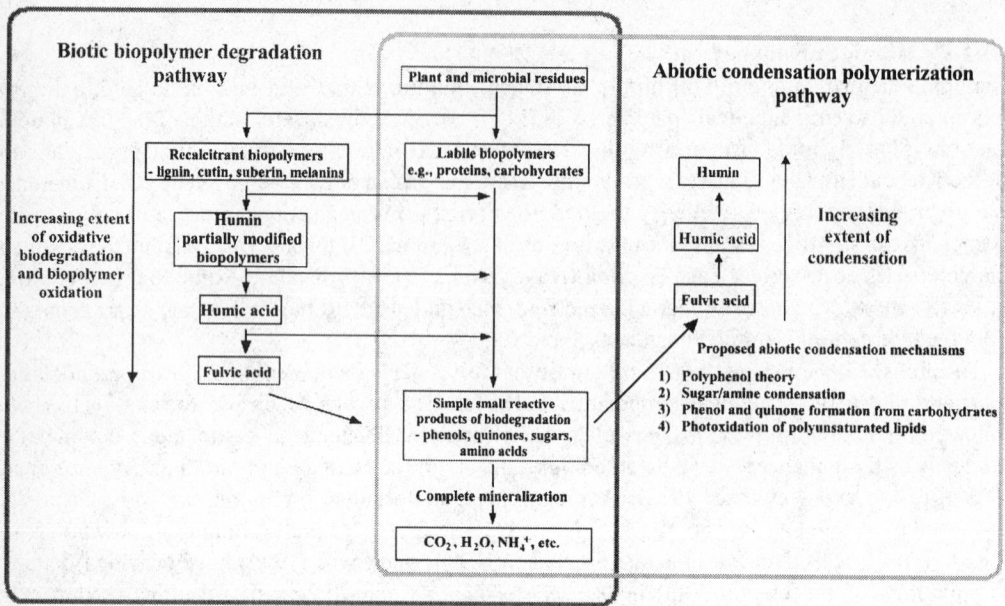

Fig. 2.5 Schematic representation of the mechanisms postulated to be involved in the genesis of soil humic substances [Based on models presented by Hedges, 1988 and Hatcher and Spiker, 1988]

biopolymer degradation repolymerize to form humic substances (Fig. 2.5). It is important to note, as suggested by Hedges (1988), that the two mechanisms of humic substance genesis are not mutually exclusive, since biopolymer degradation is prerequisite for the abiotic condensation pathway. The term abiotic is used to refer to the later steps in the pathway in which chemical reactions between simple compounds released by biotic processes condense to form the humic substances.

In the proposed biotic biopolymer degradation pathway, recalcitrant macromolecules enter the humin fraction in a partially altered state. ^{13}C NMR data have indicated the presence of lignin-like humin structures in anaerobic peats, whereas the humin fraction of aerobic soils lacks the methoxyl, aryl ether, and phenolic groups typical of lignin (Hatcher et al., 1985). Significant quantities of aliphatic C presumed to originate from plant cuticles (Nip et al., 1986b), microbial sources (Hatcher et al. 1983; Baldock et al., 1989), or the cleavage of lignin ring structures (Flaig, 1966) are also present in the soil humin fraction. As the extent of decomposition of humin molecules increases, small organic fragments are released and the content of O-containing functional groups (carbonyl and hydroxyl) increases on the residual molecule. Such increases in functional group content result in an increased extractability of the residual molecule in alkaline solution, indicative of a transition from humin to humic acid. Although similarities in chemical structure between humin and humic acids are noted, humic acids typically contain less methoxyl and more carboxyl C than humin (Hatcher et al., 1985), and pyrolysis studies indicate that the monomeric species of residual lignin structures in humic acids exhibit a more advanced stage of decomposition than those found in humin (Saiz-Jiminez et al., 1979). Further oxidative degradation fragments the humic acid molecules, and increases the content of O-bearing functional groups to produce fulvic acids and simple monomeric species. Ertel and Hedges (1984) have demonstrated that fulvic acids also contain lignin-derived structures, and that the extent of alteration was greater than that noted for humic acids. To account for the N contained in humic substances, a reaction mechanism has been proposed whereby amino groups react with modified lignin (Stevenson, 1994).

In the abiotic condensation polymerization pathway, molecular fragments released during the decomposition of precursor molecules and/or molecules released as metabolic byproducts from microorganisms polymerize via chemical reactions. As the molecular weight of the synthesized polymer increases, a progression from fulvic acid to humic acid to humin occurs. Four abiotic mechanisms have been proposed and schematic representations of the various chemical reactions are presented by Hedges (1988) and Stevenson (1994): (1) the polyphenol theory, (2) the melanoidin model (a sugar/amine condensation reaction), (3) phenol and quinone formation from carbohydrates, and (4) photo-oxidation of polyunsaturated lipids. In the polyphenol theory, monomeric phenolic species (mono-, di-, and trihydroxy phenols) are produced by the enzymatic degradation of lignin (Kirk, 1984) or synthesized by various soil microorganisms (Martin and Haider, 1971). These monomeric phenolic species are capable of forming a quinone structure in the presence of O_2 or polyphenoloxidase enzymes, which spontaneously polymerize with each other or with amines or ammonia to produce polymeric compounds. Laboratory experiments have shown quinones generated from *o*- and *p*-polyphenols condense with each other or with amino acids to form synthetic polymers with chemical structures similar to those of natural humic substances (Ertel and Hedges, 1983). In the melanoidin model, sugars and amines react to form an N-substituted glucosamine which can dehydrate, rearrange, or condense to form simple fragments (e.g., glyceraldehyde) and structurally complex brown nitrogenous polymers. Dark colored polymers can also form from carbohydrates at a similar rate in the absence of nitrogenous compounds (Popoff and Theander, 1976; Hedges, 1978). Auto-oxidative crosslinking reactions between polyunsaturated fatty acids, induced by photolysis, were postulated to be a mechanism of humic substance formation in marine environments (Harvey and Boran, 1985). Such reactions may help to explain the presence and persistence of highly crosslinked alkyl C in terrestrial systems (Kögel-Knabner et al., 1992a,b).

2.5.4.2 Nonhumic Biopolymers

Carbohydrates

The chemical structure, properties, and importance of carbohydrates in soil have been reviewed by Cheshire (1979), Oades (1989), and Stevenson (1994). Carbohydrates account for the largest fraction of soil organic C found in nonhumic biopolymers (100–250 g kg^{-1} soil organic C). They exhibit a range in molecular size from simple monosaccharides to oligosaccharides containing several monosaccharide units, to polysaccharides of high molecular weight which contain large numbers of monosaccharide units. Plants contribute directly to the soil carbohydrate fraction through the deposition of simple sugars, hemicellulose and cellulose in particulate residues and various simple sugars and polysaccharides in root exudates and mucilaginous materials. Various products of the metabolic activity of soil organisms also contribute to the soil carbohydrate fraction. These include extracellular mucilages encapsulating bacteria and fungi, cell wall structural polysaccharides (e.g., chitin) (Fig. 2.2), and intracellular polysaccharides. Relative to many other forms of soil organic C, soil carbohydrates are vulnerable to decomposition unless: (1) decomposition processes are limited by environmental conditions; (2) the carbohydrates are buried within a matrix of biochemically recalcitrant material (e.g., cellulose and/or hemicellulose buried within a lignin matrix in plant materials); or (3) the carbohydrates are biologically stabilized by adsorption onto soil mineral particles or by their physical separation from decomposer microorganisms.

In peats, forest litter layers and composts, reduced rates of decomposition and/or the preponderance of plant materials lead to a carbohydrate fraction dominated by structures of plant origin, whereas in mineral soils the carbohydrate fraction is typically dominated by microbially derived materials. Although the carbohydrate structures presented for plant materials in Fig. 2.2 are derived from single monosaccharides, the large variety of monosaccharides found in soil and different glycosidic linkages (e.g., α, β1-4 and β1-6) gives rise to a potential for large variations in chemical structure. However, both Oades (1989) and Stevenson (1994) indicated that within any single polysaccharide molecule, it is unlikely that more than three or four different component monosaccharides can be identified.

Hydrolysis procedures result in the degradation of plant polysaccharide structures and the release of various classes of monosaccharides including neutral sugars, amino sugars, acidic sugars, methylated sugars, and sugar alcohols. The neutral sugars consist of hexoses (glucose, galactose and mannose), pentoses (arabinose and xylose), and deoxyhexoses (rhamnose and fucose) with the hexoses, particularly glucose, generally accounting for the largest component (Cheshire, 1979). Excluding glucose, polysaccharides synthesized by microorganisms are dominated by galactose, manose, rhamnose, and fucose, whereas plant polysaccharides contain appreciable contents of arabinose and xylose (Cheshire, 1979). This has led to the use of the galactose+mannose/ xylose+arabinose (g+m/x+a) and rhamnose+fucose/arabinose+xylose (r+f/a+x) ratios as indices of the contribution made to the soil carbohydrate fraction by plants and microorganisms (Oades, 1984; Murayama, 1994). Values of < 0.5 for the (g+m/x+a) ratio and < 0.01 for the (r+f/a+x) ratio are indicative of carbohydrates of plant origin, while respective values of > 2.0 and > 0.40 are indicative of microbially derived carbohydrates. Guggenberger et al. (1994) used these two ratios to examine the origin of organic C associated with different particle size fractions. The changes in the two ratios across several land management treatments indicated an increased proportion of plant-derived carbohydrate with increasing particle size.

Amino sugars account for 20–60 g kg^{-1} of soil organic C. They are generally assumed to be of microbial origin. In excess of 25 different amino sugars are known to exist as products of microbial metabolism (Sharon, 1965). Monomeric amino sugars can be identified by chromatographic analysis of HCl hydrolysates. The most prevalent of the amino sugars is D-glucosamine, the monomeric component of chitin, the N-acetylglucosamine structural polymer found in fungal mycelia. Significant quantities of D-galactosamine can also be found in soils, and changes in the ratio of D-glucosamine/ D-galactosamine may be indicative of the composition of the microbial decomposer community (Stevenson, 1994). The higher ratios noted by Sowden (1959) for acidic soils are consistent with the enhanced relative importance of fungi in acidic environments. Other amino sugars identified in soils include muramic acid and D-mannosamine. The dominant forms of the acidic sugars, uronic acids, are glucuronic and galacturonic acids. Measurement of the contents of these acidic sugars in soil hydrolysates is difficult because of their rapid decarboxylation which may result in losses in excess of 50%. Using a carbazole method of analysis, acidic sugars have been estimated to account for approximately 1–5% of the soil organic C; however, Greenland and Oades (1975) consider this to be a minimum value. The methylated sugars 2-O-methyl-L-rhamnose and 4-O-methyl-D-galactose and the sugar alcohols mannitol and inositol have also been identified in the hydrolysates of soils and peats (Cheshire, 1979)

The O-alkyl or (O-alkyl + di-O-alkyl) region of solid-state ^{13}C NMR spectra is often used to provide an estimate of the content of polysaccharide C in soils. Numerous studies have reported higher carbohydrate C contents based on ^{13}C NMR analysis than could be accounted for by wet chemical hydrolysis followed by quantification of monosaccharide contents of hydrolysates (Ogner, 1985; Oades et al., 1987; Preston et al., 1989a,b). For example, Oades et al. (1987) observed that in soil particle size fractions > 20 μm the content of O-alkyl C as measured by ^{13}C NMR was approximately twice that of carbohydrate C determined by acid hydrolysis and quantification of neutral, acidic, and amino sugars. Preston et al. (1989a) found that O-alkyl C accounted for 40–50% of the signal intensity of particle size fractions of peats, a significantly greater proportion of the total organic C than was accounted for by wet chemical analysis of total carbohydrate C. Cheshire et al. (1992) proposed that the over estimation of carbohydrate C contents by ^{13}C NMR analyses resulted from the presence of secondary or pseudo polysaccharide structures which exhibited O-alkyl resonances in ^{13}C NMR, well-defined C-O vibration absorption bands in IR, and a CH_2O elemental composition. The pseudo polysaccharides differed from the normal polysaccharides in that they did not contain hydrolyzable monomeric sugars and did not yield dehydrated sugar derivatives on pyrolysis. Several organic structures proposed to account for the pseudo polysaccharide signals included highly degraded polysaccharides and melanoidins; however, any structure containing C-O bonds, including the three C atoms in lignin propanoid chains, could give rise to NMR signals within the chemical shift region typically ascribed to carbohydrate C. The three oxygenated propanoid C atoms of lignin appear over the chemical shift range of 60–90 ppm in NMR spectra (Hatcher, 1987). The use of the terms pseudo or secondary polysaccharides, therefore, appears misleading since what was really being compared was the relationship between the quantity of organic C found in C-O structures derived from many possible sources versus carbohydrate C.

Alkyl Compounds

Initial applications of solid-state ^{13}C NMR and analytical pyrolysis to the study of soil organic C and humic substances showed that significant amounts of alkyl C were present in these materials, contrary to the previous traditional beliefs of a dominance of aromatic C. Numerous solid-state ^{13}C NMR studies have shown significant increases in alkyl C content as the extent of decomposition in peats, forest litter horizons, composts and mineral soils increases (Baldock et al., 1997). The majority of research addressing the nature of soil alkyl C and the structural changes associated with decomposition processes has been carried out on forest soils. In such ecosystems, the high organic C contents of litter layers and Ah horizons, and the reduced potential for organic-mineral interactions facilitates analysis and interpretation. Extension of these results to agricultural mineral soils, especially to high clay content soils with large adsorptive capacities, should be made with caution.

Alkyl C generally accounts for 15–20% of the organic C contained in fresh litter horizons and 30–40% for the more humified litter horizons and mineral Ah horizons (Kögel-Knabner et al., 1992b). Organic materials derived from a variety of compounds contained in plant residues can contribute to the fraction of alkyl C. Soil microorganisms also synthesize alkyl structures from nonaliphatic substrate (Baldock et al., 1989; Golchin et al., 1996). Alkyl C in soil organic C consists of the following compounds (Kögel-Knabner et al., 1992a): (1) solvent extractable free and bound lipids (fatty acids, and waxes derived from plants and soil microorganisms), (2) insoluble polyesters and nonpolyesters contained in the plant cuticles and walls of cork cells in roots and bark, and (3) macromolecules synthesized by soil microorganisms.

Solvent extractable free and bound lipids account for approximately 30% of the ^{13}C NMR signal intensity in the alkyl C region observed for forest soils (Kögel-Knabner et al., 1988; Ziegler, 1989; Ziegler and Zech, 1989). The chemical nature of free and bound lipids has been discussed by Stevenson (1994). They include waxes, organic acids, steroids and terpenoids, carotenoids, porphyrins, glycerides and phospholipids. The nature of petroleum ether extractable free lipids and their relationships to the original litter materials have been examined by Almendros et al. (1996). These materials are readily metabolized by soil microorganisms in forest soils and are not considered a major contributor to the accumulation of alkyl C in humified structures (Tegelaar et al., 1989; Ziegler, 1989).

Cutin and suberin (insoluble polyesters) can also be readily decomposed by those bacteria and fungi that produce cutinase. Mammals effectively degrade cutin so that contributions to the soil via faecal deposition are minimal (Tegelaar et al., 1989). Riederer et al. (1993) measured changes in the content of cutin-derived monomers in *Fagus sylvatica* litter decomposed for 446 d. Over the first 100 d, the content of cutin monomer C, released by saponification of the residual litter, decreased proportionately faster than bulk organic C, but then decreased at a rate similar to bulk organic C over the 100–446 d period. Similar results were obtained by Ziegler and Zech (1990) and Kögel-Knabner et al. (1992b) indicating that a selective preservation of cutin structures was not responsible for the increased proportion of alkyl C noted with increasing extent of decomposition. Further, decreased concentrations of cutin and suberin were noted in progressing from litter to well-humified organic and mineral Ah horizons when concentrations were expressed per unit mass of total organic C (Kögel-Knabner et al., 1989; Riederer et al., 1993). A selective preservation of intact cutin or suberin is, therefore, not likely to contribute to the increased proportion of alkyl C observed with increasing extent of decomposition; however, microbially or chemically transformed cutin and suberin may accumulate.

Nonsaponifiable cutan and suberan (insoluble nonpolyesters) appear to be resistant to biological and chemical degradation (Tegelaar et al., 1989) and although present at low concentrations in plant residues, concentration by selective preservation could represent a mechanism for the accumulation of alkyl C in soils. Riederer and Schönherr (1988) noted that the nonpolyester fraction of *Clivia miniata* Reg. cuticle increased with ontogenetic development through a continuous transformation of polyester cutin into nonpolyester cutan. Such a reaction may be occurring via a photo-oxidative crosslinking mechanism, similar to that proposed by Harvey and Boran (1985) for the genesis of marine humic substances. Kögel-Knabner et al. (1992a,b) used a combination of selective degradation techniques and ID ^{13}C NMR to distinguish between mobile (little crosslinking) and rigid (highly crosslinked) forms of nonsaponifiable alkyl C in forest soils. Fresh forest litters contained 55–60% of their C in rigid structures and 40–45% in mobile structures. With the increasing extent of decomposition the mobile fraction was lost and the Oh and Ah nonsaponifiable alkyl C was composed almost exclusively of rigid structures. Possible mechanisms that may account for such a conversion during decomposition processes include (1) a selective preservation of rigid structures present in the initial plant residues, (2) the production of rigid metabolic products of microbial origin synthesized during decomposition, and/or (3) the conversion of mobile alkyl C to rigid alkyl C, with enhanced cross-linking being induced by chemical, physical or biological processes. In addition, rigidity as assessed by ID ^{13}C NMR may be influenced by adsorption onto mineral surfaces; however, this remains to be determined. ID ^{13}C NMR spectra acquired, for a bulk culture of intact soil microorganisms and their nonsaponifiable alkyl residues, indicated the presence of both mobile and rigid structures (Kögel-Knabner et al. 1992a). In a more detailed application of ID ^{13}C NMR to characterizing untreated intact cultures of soil microorganisms, Baldock et al. (1990b) found that 13% of fungal and 66% of bacterial alkyl C exemplified a rigid structure.

On the basis of the results obtained from forest soils, the increase in alkyl C during decomposition appears to result from an accumulation of nonsaponifiable rigid structures which may originate from selective preservation of plant and microbial residues or be produced *in situ* via biological and/or chemical transformations of existing alkyl structures. In soils with appreciable mineral surface areas, the potential exists for other forms of alkyl C (extractable free and bound lipids and cutin and suberin polyesters) to contribute through a biological stabilization imparted by adsorption onto mineral surfaces. Supporting evidence for the importance of adsorption reactions with mineral surfaces is provided by the data of Oades et al. (1987) and Baldock et al. (1992). They found significant accumulations of alkyl C in the fine particle size fractions, with a large highly reactive surface area, for a range of mineral soils including an Alfisol, two Mollisols and two Oxisols.

Lignin

The chemical composition of lignin and its residues in soil is a function of its initial chemical structure and the extent of modification by soil organisms. Lignin is generally more resistant to biological decomposition than the other major biopolymers found in plant residues, because of its chemical structure (Fig. 2.2). The most efficient lignin degrading organisms are the group of aerobic filamentous basidiomycetes collectively referred to as white rot fungi. These organisms fragment the lignin polymer at irregular positions within both the side-chain and aromatic ring structures by the synthesis and excretion of enzymes which catalyse free radical formation (Haider, 1992; Kirk, 1987). Brown rot basidiomycetes do not fragment the lignin polymer extensively, but are capable of demethylating methoxyl groups on the guaiacyl or syringyl units to produce *o*-hydroquinonoid structures which can be easily oxidized to quinones. The quinone structures can then induce condensation reactions as discussed for the polyphenol theory. For plant litter, it is generally assumed that basidiomycetes, similar to the white rot fungi, aid in lignin decomposition after other organisms have removed the more labile components. Bacteria play a smaller role in lignin degradation, but as noted for the brown rot fungi, they can modify the nature of functional groups attached to the lignin polymer. Organisms do not gain energy or metabolites from lignin degradation (Haider, 1994), but benefit through an exposure of labile cellulose and hemicellulose buried within the lignin/polysaccharide matrix. The role of various organisms and the mechanisms of lignin degradation have been discussed recently by Shevchenko and Bailey (1996) and Hammel (1997).

The chemical state of lignin in plant residues and soils has been examined by spectroscopic and chemical degradative methods. Hatcher (1987) presented solid-state ^{13}C NMR spectra of isolated natural lignin. On the basis of this work and others, the signals from each of the various types of C contained in lignin polymers can be assigned as follows: 56 ppm = methoxyl C, 60–90 ppm = propyl side-chain O-alkyl C, 105–145 ppm = C- and H-substituted aromatic C, and 145–160 ppm = phenolic C. The chemical shift values and form of the phenolic peak can be used to provide information on the type of lignin monomers present. The aromatic C-O associated with the methoxyl group of guaiacyl units appears at 148 ppm while that of syringyl units appears at 153 ppm. Thus, softwood lignin, dominated by guaiacyl monomers would exhibit a phenolic peak at 148 ppm, hardwood lignin containing both guaiacyl and syringyl monomers exhibits two peaks at 148 and 153 ppm, and grass lignin dominated by syringyl monomers exhibits a peak at 153 ppm. On average, of the 10–11 C atoms contained in the two major lignin monomers (guaiacyl and syringyl), 2–3 are phenolic, 3–4 are C- and H-substituted aromatic, 3 are O-alkyl, and 1–2 are methoxyl. The actual values depend on the distribution of the two major monomeric species (guaiacyl and syringyl) (Fig. 2.2) and the nature of the bonds between monomeric units. Significant deviations from these values would be indicative of how the average lignin molecule was altered by decomposition processes.

In soils and peats, lignin is just one component of the decomposing residues and several signals from other types of C overlap and mask changes in the lignin structure. The propane side-chain C signals are completely hidden by the C-O C of carbohydrates, and amine C overlaps the methoxyl signals. For the methoxyl signal, however, the ID ^{13}C NMR pulse sequence can be used to differentiate between the content of methoxyl and amine C. The ID pulse sequence can also be used to determine the average amount of protonated versus nonprotonated aromatic C. Readers are referred to Hatcher (1987) for an excellent example of the application of ID to the study of lignin. Decreases in the proportions of phenolic and methoxyl C are typical of the results obtained when solid-state ^{13}C NMR has been used to characterize the chemistry of organic C in soil fractions of decreasing particle size

(Oades et al., 1987; Baldock et al., 1992; Guggenberger et al., 1995a). Such observations are consistent with the observed patterns of decomposition noted for bacteria and brown rot fungi decomposing plant residues mentioned above. Nevertheless, a degree of uncertainty exists with solid-state ^{13}C NMR data because the analysis does not indicate conclusively from which structures the various types of C were derived.

Lignin degradation is most effectively examined by procedures that quantitatively break the linkages between component monomers and measure the amount of each type of monomer. Two such methods are an alkaline CuO oxidation procedure (Hedges and Ertel, 1982) and a tetramethylammonium hydroxide (TMAH) thermochemolysis procedure (Hatcher et al., 1995). In both procedures the lignin polymer is fragmented into its component monomers, which are separated and quantified using reverse phase high performance liquid chromatography (Kögel and Bochter, 1985), gas chromatography (Hedges and Ertel, 1982; Baldock et al., 1997) or gas chromatography-mass spectrometry (Hatcher et al., 1995).

The CuO-oxidation procedure has been shown to release 25–90% of the lignin C in the form of simple phenolic monomeric structures, depending on the types of lignin monomers involved (Hedges and Ertel, 1982; Ertel and Hedges, 1984). A comparison of the lignin contents between samples can be made by summing the contents of the guaiacyl, syringyl and p-hydroxyphenol derivatized reaction products. The ratio of the quantity of acidic to aldehydic forms of each monomer (Ac/Al) can give an indication of the state of decomposition or structural alteration of each monomeric unit within the lignin polymer, or of the entire polymer (Ertel and Hedges, 1984; Ertel et al., 1984; Moran et al. 1991). The ratio of lignin-derived dimers/monomers present in CuO oxidation products of sedimentary lignins has also been used as a measure of the extent of fungal-induced degradation (Goni et al., 1993).

Several recent examples of the application of the CuO oxidation methodology to the study of lignin in various forms of soil organic C include work competed by deMontigny et al. (1993), Guggenberger et al. (1994a) and Amelung and Zech (1996). deMontigny et al. (1993) studied the extent of lignin alteration in nonwoody and woody forest horizons. In progressing from the surface fibric to the more humified humic layer of the nonwoody horizons, little change in the total amount of phenolic C released was noted (31–33.5 g kg^{-1} organic C); however, the guaiacyl Ac/Al ratio increased significantly from 0.43 to 1.01, suggesting that, although the lignin content did not change significantly, it became more structurally modified with increasing depth, and presumably forest floor residence time. The CuO oxidation products of lignin contained in particle size fractions (sand, silt and clay fractions) in mineral soil collected from four different ecosystems showed a decrease in lignin content and an increase in the extent of lignin alteration with decreasing particle size (Guggenberger et al., 1994). Amelung and Zech (1996) selectively removed the external 0.5 mm of soil peds and used CuO oxidation to examine the amount and state of decomposition of lignin in soil from the ped surface and ped interior. The sum of the lignin-derived phenolic compounds was significantly less for the ped surface than ped interior soil, suggesting an enhanced extent of lignin decomposition at ped surfaces. The increased Ac/Al ratios for guaiacyl and syringyl monomers and the selective loss of the more easily decomposable syringyl monomers in ped surface soil supported the hypothesis of enhanced oxidative decomposition at ped surfaces relative to ped interiors.

The TMAH thermochemolysis procedure is analogous to the CuO oxidation process except that the derivatives are methylated and ready for direct injection into a GC or GC/MS system. Hatcher et al. (1995) compared the results of the TMAH and CuO oxidation methods when performed on the samples analysed by deMontigny et al. (1993). The distribution of products obtained with both analyses was similar, as evidenced by a linear correlation between the Ac/Al ratios obtained using the two methods. However, the TMAH method provided a more sensitive indicator of the extent of decomposition because of the larger range of values obtained. The TMAH method also resulted in a greater preservation of monomer side-chain structures than the CuO oxidation. It is, therefore, important in the TMAH analysis to include all the various derivatives containing methoxylated side chains, when estimating the amount of C contained in lignin structures from the products of the thermochemolysis reaction. The enhanced preservation of the side-chain structures may, with further research, be utilized to provide additional structural information pertaining to the lignin polymer and its alteration through decomposition. Additional studies that have utilized the TMAH procedure to examine the amount and extent of decomposition of lignin have thus far concentrated on relatively pure natural organic materials (Baldock et al., 1997; Nanny , 1997. Personnal communication) and its extension to mineral soils or mineral soil fractions has not yet occurred. Baldock et al. (1997) used the TMAH procedure in conjunction with solid-state ^{13}C NMR to examine the chemical changes associated with the decomposition of three species of wood exhibiting different stages of decomposition. The two analytical techniques were complementary, but the TMAH results showed that significant alterations to the structure of lignin could occur before significant changes in the NMR spectral intensities became apparent.

2.6 Conclusions

The diverse chemical nature of soil organic C contributes to the important role which it plays in defining the magnitude of a variety of soil properties and processes, and has made its selective chemical characterization challenging. The continual development of new technologies and new procedures or applications of existing technologies capable of studying the properties and chemical characteristics of soil organic C *in situ*, will undoubtedly further advance our understanding of this important soil component. Several research areas which offer significant potential are (1) further development and use of methods which can selectively characterize the chemical composition of SOM *in situ* and thus avoid the use of chemical extractants, (2) definition of relationships between the species composition of the decomposer community and the resultant changes in chemical structure of plant residues and soil organic materials, (3) identification and quantification of biologically significant pools of C and the importance of these pools in defining the magnitude of soil properties and processes, (4) quantification of the capacity of soils to protect organic materials from mineralization through identification and measurement of the role of soil mineral components and soil architecture, and (5) renewed application of wet chemical molecular techniques to the study and quantification of the contents of specific classes of biomolecules found in soil organic fractions, particularly in association with quantitative applications of spectroscopic techniques (solid-state ^{13}C NMR). In particular, work is required to address the problems associated with identifying the pseudo-polysaccharide and the unidentified organic N fractions.

Acknowledgments
Useful suggestions and improvements to the manuscript made by Jeff Ladd, Ron Smernick, Jan Skjemstad and the editors are gratefully acknowledged.

2.7 References

Achard, F.K. 1786. Chemische utersunchungen des torts. Crell's Chem. Ann. 2:391–403.

Adams, J.M., H. Faure, L. Faure-Denard, J.M. McGlade, and F.I. Woodward. 1990. Increases in terrestrial carbon storage from the last glacial maximum to the present. Nature 348:711–714.

Adu, J.K., and J.M. Oades. 1978. Physical factors influencing decomposition of organic materials in soil aggregates. Soil Biol. Biochem. 10:109–115.

Ågren, G.I., and E. Bosatta. 1996. Quality: a bridge between theory and experiment in soil organic matter studies. Oikos 76:522–528.

Aiken, G.R., D.M. McKnight, R.L. Wershaw, and P. McCarthy. 1985. Humic substances in soil, sediment and water. John Wiley and Sons, New York, NY.

Almendros, G., J. Sanz, and F. Velasco. 1996. Signatures of lipid assemblages in soils under continental Mediterranean forests. Europ. J. Soil Sci. 47:183–196.

Alphei., J., M. Bonkowski, and S. Scheu. 1996. Protozoa, nematoda and lumbricidae in the rhizosphere of *Hordelumus europaeus* (Poaceae): faunal interactions, response of microorganisms and effects on plant growth. Oecologia 106:111–126.

Amato, M.A., and J.N. Ladd. 1992. Decomposition of ^{14}C-labelled glucose and legume material in soils: properties influencing the accumulation of organic residue C and microbial biomass C. Soil Biol. Biochem. 24:455–464.

Amelung, W., and W. Zech. 1996. Organic species in ped surface and core fractions along a climosequence in the prairie, North America. Geoderma 74:193–206.

Amelung, W., K.W. Flach, and W. Zech. 1997. Climatic effects on soil organic matter composition in the Great Plains. Soil Sci. Soc. Am. J. 61:115–123.

Anderson, D.W. 1995. The role of nonliving organic matter in soils. p. 81–92. *In* R.G. Zepp and Ch. Sonntag (ed.) The role of nonliving organic matter in the Earth's carbon cycle. John Wiley and Sons, Chichester, UK.

Anderson, D.W., and D.C. Coleman. 1985. The dynamics of organic matter in grassland soils. J. Soil Water Conserv. 40:211–216.

Andrén, O. 1987. Decomposition in the field of shoots and roots of barley, lucerne and meadow fescue. Swed. J. Agric. Res. 17:113–122.

Arshad, M.A., J.A. Ripmeester, and M. Schnitzer. 1988. Attempts to improve solid-state ^{13}C NMR spectra of whole mineral soils. Can. J. Soil Sci. 68:593–602.

Atalla, R.H., and D.L. van der Hart. 1984. Native cellulose: a composite of two distinct crystalline forms. Science 222:283–285.

Baldock, J.A., G.J. Currie, and J.M. Oades 1991. Organic matter as seen by solid-state ^{13}C NMR and pyrolysis tandem mass spectrometry. p. 45–60. *In* W.S. Wilson (ed.) Advances in soil organic matter research: The impact on agriculture and the environment. Royal Society of Chemistry, London, UK.

Baldock, J.A., J.M. Oades, P.N. Nelson, T.M. Skene, A. Golchin, and P. Clarke. 1997. Assessing the extent of decomposition of natural organic materials using solid-state ^{13}C NMR spectroscopy. Aust. J. Soil Res. 35:1061–1083.

Baldock, J.A., J.M. Oades, A.M. Vassallo, and M.A. Wilson. 1989. Incorporation of uniformly labelled ^{13}C-glucose into the organic fraction of a soil. Carbon balance and CP/MAS ^{13}C NMR measurements. Aust. J. Soil Res. 27:725–746.

Baldock, J.A., J.M. Oades, A.M. Vassallo, and M.A. Wilson. 1990a. Solid-state CP/MAS ^{13}C NMR analysis of particle size and density fractions of a soil incubated with uniformly labelled ^{13}C-glucose. Aust. J. Soil Res. 28:193–212.

Baldock, J.A., J.M. Oades, A.M. Vassallo, and M.A. Wilson.. 1990b. Solid-state CP/MAS ^{13}C NMR analysis of bacterial and fungal cultures isolated from a soil incubated with glucose. Aust. J. Soil Res. 28:213–225.

Baldock, J.A., J.M. Oades, A.G. Waters, X. Peng, A.M. Vassallo, and M.A. Wilson. 1992. Aspects of the chemical structure of soil organic materials as revealed by solid-state ^{13}C NMR spectroscopy. Biogeochem. 16:1–42.

Baldock, J.A., T. Sewell, and P.G. Hatcher. 1997. Decomposition induced changes in the chemical structure of fallen red pine, white spruce and tamarack logs. p. 75–83. *In* G. Cadisch, and K.E. Giller (ed.) Driven by nature: Plant litter quality and decomposition. CAB International, Wallingford, UK.

Balesdent, J., and B. Guillet. 1982. Les datations par le ^{14}C des matiéres organiques des sols. Sci. Sol. 2:93–112.

Ball-Caelho, B., H. Salcedo, H. Tiessen, and J.W.B. Stewart. 1993. Short- and long-term phosphorus dynamics in a fertilized Ultisol under sugarcane. Soil Sci. Soc. Am. J. 57:1027–1034.

Barassa, J.M., A.E. Gonzĉlez, and A.T. Martinez. 1992. Ultrastructural aspects of fungal delignification of Chilean wood by *Ganoderma australe* and *Phelbia chrysocrea*. A study of natural and *in vitro* degradation. Holzforschung 46:1–8.

Bates, A.L., and P.G. Hatcher. 1989. Solid-state ^{13}C NMR studies of a large fossil gymnosperm from the Yallourn Open Cut, Latrobe Valley, Australia. Org. Geochem. 14:609–17.

Bates, A.L., P.G. Hatcher, H.E. Lerch, C.B. Cecil, S.G. Neuzil, and M. Supardi. 1991. Studies of a peatified angiosperm log cross section from Indonesia by nuclear magnetic resonance spectroscopy and analytical pyrolysis. Org. Geochem. 17:37–45.

Beare, M.H., M.L. Caberra, P.F. Hendrix, and D.C. Coleman. 1994b. Aggregate-protected and unprotected organic matter pools in conventional- and no-tillage soils. Soil Sci. Soc. Am. J. 58:787–795.

Beare, M.H., P.F. Hendrix, and D.C. Coleman. 1994a. Water-stable aggregates and organic matter fractions in conventional and no-tillage soils. Soil Sci. Soc. Am. J. 58:777–786.

Beck, M.A., and P.A. Sanchez. 1994. Soil phosphorus fraction dynamics during 18 years of cultivation on a Typic Paleudult. Soil Sci. Soc. Am. J. 58:1424–1431.

Berg, B., and H. Staaf. 1980. Decomposition rate and chemical changes of Scots pine needle litter. II. Influence of chemical composition. Ecol. Bull. 32:363–372.

Berner, R.A., and A.J. Lasaga. 1990. Simulation des geochemishchen Kreislaufs. Spektrum der Wissenschaft 5:56. (as cited by Scharpenseel and Becker-Heidmann, 1990)

Bettany, J.R., J.W.B. Stewart, and S. Saggar. 1979. The nature and forms of sulphur in organic matter fractions of soils selected along an environmental gradient. Soil Sci. Soc. Am. J. 43:981–985.

Beyer, L. 1996. The chemical composition of soil organic matter in classical humic compound fractions and in bulk samples - a review. Z. Pflanzenernähr. Bodenk. 159:527–539.

Bird, M.I., A.R. Chivas, and J. Head. 1996. A latitudinal gradient in carbon turnover times in forest soils. Nature 381:143–146.

Bloom, P.R., and J.A. Leenheer. 1989. Vibrational, electronic, and high-energy spectroscopic methods for characterising humic substances. p. 409–446. *In* M.H.B. Hayes, P. MacCarthy, R. Malcolm, and R.S. Swift (ed.) Humic substances. II. In search of structure. Wiley-Interscience, Chichester, UK.

Bolin, B. 1983. The carbon cycle. p. 41–45. *In* B. Bolin and R.B. Cook (ed.) The major biogeochemical cycles and their interactions. John Wiley and Sons, Chichester, UK.

Bollag, J.-M., and G. Stotzky. 1990. Soil biochemistry. Vol. 6. Marcel Dekker, New York, NY.

Boonman, J.G. 1993. East Africa's grasses and fodders: Their ecology and husbandry. *In* Tasks for vegetation science 29. Kluwer Academic Publishers, Dordrecht, Netherlands.

Boudot, J.P., A. Bel Hadi Brahim, and T. Chone. 1986. Carbon mineralization in andosols and aluminium-rich highland soils. Soil Biol. Biochem. 18:457–461.

Boudot, J.P., A. Bel Hadi Brahim, and T. Chone. 1988. Dependence of carbon and nitrogen mineralization rates upon amorphous metallic constituents and allophanes in highland soils. Geoderma 42:245–260.

Boudot, J.P., A. Bel Hadi Brahim, R. Steiman, and F. Seigle-Murandi. 1989. Biodegradation of synthetic organo-metallic complexes of iron and aluminum with selected metal to carbon ratios. Soil Biol. Biochem. 21:961–966.

Boyer, J.N., and P.M. Groffman. 1996. Bioavailability of water extractable organic carbon fractions in forest and agricultural soil profiles. Soil Biol. Biochem. 28:783–790.

Bracewell, J.M., K. Haider, S.R. Larter, and H.-R. Schulten. 1989. Thermal degradation of humic substances relevant to structural studies. p. 181–222. *In* M.B.H. Hayes, P. MacCarthy, R.L. Malcolm, and R.S. Swift (ed.) Humic substances. II. In search of structure. John Wiley and Sons, Chichester, UK.

Bracewell, J.M., G.W. Robertson, and K.R. Tate. 1976. Pyrolysis-gas chromatography studies on a climosequence of soils in tussock grassland, New Zealand. Geoderma 15:209–215.

Bremner, J.M. 1967. The nitrogenous constituents of soil organic matter and their role in soil fertility. Pontif. Acad. Sci. Scripta Varia 32:143–193.

Burford, J.R., and J.M. Bremner. 1975. Relationships between the denitrification capacities of soils and total, water-soluble and readily decomposable soil organic matter. Soil Biol. Biochem. 7:389–394.

Burgess, D., J.A. Baldock, S. Wetzel, and D.G. Brand. 1995a. Scarification, fertilization and herbicide treatment effects on planted conifers and soil fertility. Plant Soil 168–169:513–522.

Burgess, D., J.A. Baldock, S. Wetzel, and D.G. Brand. 1995b. Scarification, fertilization and herbicide treatment effects on seedling growth and quality and soil fertility. p. 95–101. *In* Innovative silviculture systems in boreal forests. Natural Resources Canada/Canadian Forest Service, Edmonton, Canada.

Burke, I.C., E.T. Elliott, and C.V. Cole. 1995. Influence of macroclimate, landscape position, and management on soil organic matter in agroecosystems. Ecol. Appl. 5:124–131.

Burke, I.C., C.M. Yonker, W.J. Parton, C.V. Cole, K. Flach, and D.S. Schimel. 1989. Texture, climate and cultivation effects on soil organic matter in U.S. grassland soils. Soil Sci. Soc. Am. J. 53:800–805.

Buyanovsky, G.A., C.L. Kucera, and G.H. Wagner. 1987. Comparative analysis of carbon dynamics in native and cultivated ecosystems. Ecol. 68:2023–2031.

Buyanovsky, G.A., M. Aslam, and G.H. Wagner. 1994. Carbon turnover in physical fractions. Soil Sci. Soc. Am. J. 58:1167–1173.

Cadisch, G., and K.E. Giller. 1997. Driven by nature: plant litter quality and decomposition. CAB International, Wallingford, UK.

Cajuste, L.J., R.J. Laird, L. Cajuste, and B.G. Cuevas. 1996. Citrate and oxalate influence on phosphate, aluminum, and iron in tropical soils. Commun. Soil Sci. Plant Anal. 27:1377–1386.

Cambardella, C.A., and E.T. Elliott. 1993. Methods for physical separation and characterisation of soil organic matter fractions. Geoderma 56:449–457.

Chen, Y., and T. Aviad. 1990. Effects of humic substances on plant growth. p. 161–186. *In* P. McCarthy, C.E. Clapp, R.L. Malcolm, and P.R. Bloom (ed.) Humic substances in soil and crop sciences: selected readings. Soil Science Society of America, Madison, WI.

Cheshire, M.V. 1979. Nature and origin of carbohydrates in soils. Academic Press, London, UK.

Cheshire, M.V., J.D. Russell, A.R. Fraser, J.M. Bracewell, B.W. Robertson, L.M. Benzing-Purdie, C.I. Ratcliffe, J.A. Ripmeester, and B.A. Goodman. 1992. Nature of soil carbohydrate and its association with soil humic substances. J. Soil Sci. 43:359–373.

Christensen, B.T. 1992. Physical fractionation of soil and organic matter in primary particle size and density separates. Adv. Soil Sci. 20:1–90.

Christensen, B.T. 1996. Carbon in primary and secondary organo-mineral complexes. Adv. Soil Sci. 24:97–165.

Christman, R.F., and T. Gjessing. 1981. Aquatic and terrestrial humic materials. Ann Arbor Science, Ann Arbor, MI.

Churchman, G.J., and R.C. Foster. 1994. The role of clay minerals in the maintenance of soil structure. Trans. 15th Cong. of Int. Soil Sci. Soc. 8a:17–34.

Clinton, P.W., R.H. Newman, and R.B. Allen. 1995. Immobilization of ^{15}N in forest litter studies by ^{15}N CPMAS NMR spectroscopy. Europ. J. Soil Sci. 46:551–556.

Coleman, D.C., J.M. Oades, and G. Uehara. 1989. Dynamics of soil organic matter in tropical ecosystems. Niftal, Honolulu, HI.

Cortez, J., J.M. Demard, P. Bottner, and L.J. Monrozier. 1996. Decomposition of Mediterranean leaf litters - a microcosm experiment investigating relationships between decomposition rates and litter quality. Soil Biol. Biochem. 28:443–452.

Curtin, D., C.A. Campbell, and D. Messer. 1996. Prediction of titratable acidity and soil sensitivity to pH change. J. Environ. Qual. 25:1280–1284.

da Silva, A.P., and B.D. Kay. 1997. Estimating the least limiting water range of soils from properties and management. Soil Sci. Soc. Am. J. 61:877–883.

Davidson, E.A., L.F. Galloway, and M.K. Strand. 1987. Assessing available carbon: comparison of techniques across selected forest soils. Commun. Soil Sci. Plant Anal. 18:45–64.

deMontigny, L.E., C.M. Preston, P.G. Hatcher, and I. Kögel-Knabner. 1993. Comparison of humus fraetions from two ecosystem phases on northern Vancouver Island using ^{13}C CPMAS NMR spectroscopy and CuO oxidation. Can. J. Soil Sci. 73:9–25.

Deruiter, P.C., A.M. Neutel, and J.C. Moore. 1994. Modelling food webs and nutrient cycling in agro-ecosystems. Trends Ecol. Evol. 9:378–383.

Duchaufour, P. 1976. Dynamics of organic matter in soils of temperate regions: its action on pedogenesis. Geoderma 15:31–40.

Duncan, T.M. 1987. ^{13}C chemical shielding in solids. J. Phys. Chem. Ref. Data 16:125–151.

Dunnivant, F.M., P.M. Jardine, D.L. Taylor and J.F. McCarthy. 1992. Transport of naturally occurring dissolved organic carbon in laboratory columns containing aquifer material. Soil Sci. Soc. Am. J. 56:437–444.

Duxbury, J.M., M.S. Smith, and J.W. Doran. 1989. Soil organic matter as a source and sink of plant nutrients. p. 33–67. In D.C. Coleman, J.M. Oades, and G. Uehara (ed.) Dynamics of soil organic matter in tropical ecosystems. University of Hawaii Press, Honolulu, HI.

Dyke, G.V. 1993. John Lawes of Rothamsted: Pioneer of science farming and industry. Hoos Press, Harpenden, UK.

Easthouse, K.B., J. Mulder, N. Christophersen, and H.M. Seip. 1992. Dissolved organic carbon fractions in soil and stream water during variable hydrological conditions at Birkenes, southern Norway. Water Resour. Res. 28:1585–1596.

Edmonds, R.L., and T.B. Thomas. 1995. Decomposition and nutrient release from green needles of western hemlock and pacific silver fir in an old-growth temperate rain forest, Olympic national park, Washington. Can. J. For. Res. 25:1049–1057.

Elliot, E.T. 1986. Aggregate structure and carbon, nitrogen, phosphorus in native and cultivated soils. Soil Sci. Soc. Am. J. 50:627–633.

Elliot, E.T., and D.C. Coleman. 1988. Let the soil do the work for us. Ecol. Bull. 39:23–32.

Elliott, E.T., and C.A. Cambardella. 1991. Physical separation of soil organic matter. Agric. Ecosyst. Environ. 34:407–419.

Elliott, W.M., N.B. Elliott, and R.L. Wyman. 1992. Relative effect of litter and forest type on rate of decomposition. Am. Midl. Nat. 129:87–95.

Emerson, W.W., R.C. Foster, and J.M. Oades. 1986. Organo-mineral complexes in relation to soil aggregation and structure. p. 521–528. In P.M. Huang and M. Schnitzer (ed.) Interactions of soil minerals with natural organics and microbes. Soil Sci. Soc. Am. Spec. Publ. 17, Soil Science Society of America, Madison, WI.

Ertel, J.R., and J.I. Hedges. 1983. Bulk chemical and spectroscopic properties of marine and terrestrial humic acids, melanoidins and catechol-based synthetic polymers. p. 143–163. In R.F. Christman and E.T. Gjessing (ed.) Aquatic and terrestrial humic materials. Ann Arbor Science, Ann Arbor, MI.

Ertel, J.R., and J.I. Hedges. 1984. The lignin component of humic substances: distributions among soil and sedimentary humic, fulvic, and base-insoluble fractions. Geochim. Cosmochim. Acta 48:2065–2074.

Ertel, J.R., J.I. Hedges, and E.M. Perdue. 1984. Lignin signature of aquatic humic substances. Science 223:485–487.

Feller, C., E. Fritsch, R. Poss, and C. Valentin. 1991. Effect of the texture on the storage and dynamics of organic matter in some low activity clay soils (West Africa, particularly). Cah. ORSTOM, Ser. Pedol. XXVI:25–36.

Flaig, W. 1966. The chemistry of humic substances. p. 103–127. *In* The use of isotopes in soil organic matter studies, Report of FAO/IAEA Technical Meeting. Pergamon Press, New York, NY.

Flaig, W., H. Beutelspacher, and E. Rietz. 1975. Chemical composition and physical properties of humic substances. p. 1–211. *In* J.E. Gieseking (ed.) Soil components: Vol. I. Organic components. Springer-Verlag, New York, NY.

Fog, K. 1988. The effect of added nitrogen on the rate of decomposition of organic matter. Biol. Rev. 63:433–462.

Fortin, M.C. 1993. Soil temperature, soil water, and no-till corn development following in row residue removal. Agron. J. 85:571–576.

Foster, R.C. 1994. Microorganisms and soil aggregates. p. 144–155. *In* C.E. Pankurst, B.M. Doube, V.V.S.R. Gupta, and P.R. Grace (ed.) Soil biota. Management in sustainable farming systems. CSIRO, East Melbourne, Australia.

Frimmel, F.H., and R.F. Christman. 1988. Humic substances and their role in the environment. Wiley, New York, NY.

Gaiffe M., B. Duquet, H. Tavant, Y. Tavant, and S. Brukert. 1984. Stabilité biologique et comportement physique d'un complex argilo-humique placé dans différentes condition de saturation en calcium ou en potassium. Plant Soil 77:271–284.

Germida, J.J., M. Wainwright, and V.V.S.R. Gupta. 1992. Biochemistry of sulfur cycling in soil. p.1–53. *In* G. Stotzky and J.-M. Bolag (ed.) Soil biochemistry. Vol. 7. Marcel Dekker, New York, NY.

Gianfreda, L., and J.-M. Bollag. 1996. Influence of natural and anthropogenic factors on enzyme activity in soil. p. 123–193. *In* G. Stotzky and J.-M. Bollag (ed.) Soil biochemistry. Vol. 9. Marcel Dekker, New York, NY.

Golchin, A., J.A. Baldock, and J.M. Oades. 1998. A model linking organic matter decomposition, chemistry and aggregate dynamics. p. 245–266. *In* R.Lal, J.M. Kimble, R.F. Follett and B.A. Stewart (ed.) Soil processes and the carbon cycle. CRC Press, Inc., Boca Raton, FL.

Golchin, A., J.A. Baldock, P. Clarke, T. Higashi, and J.M. Oades. 1997b. The effects of vegetation and burning on the chemical composition of soil organic matter in a volcanic ash soil. II. Density fractions. Geoderma 76:175–192.

Golchin, A., P. Clarke, J.A. Baldock, T. Higashi, J.O. Skjemstad, and J.M. Oades. 1997a. The effects of vegetation and burning on the chemical composition of soil organic matter in a volcanic ash soil. I. Whole soil and humic fraction. Geoderma 76:155–174.

Golchin, A., P. Clarke, and J.M. Oades. 1996. The heterogeneous nature of microbial products as shown by solid-state ^{13}C CP/MAS NMR spectroscopy. Biogeochem. 34:71–87.

Golchin, A., J.M. Oades, J.O. Skjemstad, and P. Clarke. 1994a. Study of free and occluded particulate OM in soils by solid-state ^{13}C CP/MAS NMR spectroscopy and scanning electron microscopy. Aust. J. Soil Res. 32:285–309.

Golchin, A., J.M. Oades, J.O. Skjemstad, and P. Clarke. 1994b. Soil structure and carbon cycling. Aust. J. Soil Res. 32:1043–1068.

Golchin, A., J.M. Oades, J.O. Skjemstad, and P. Clarke, P. 1995. Structural and dynamic properties of soil organic matter as reflected by ^{13}C natural abundance, pyrolysis-mass spectrometry and solid-state ^{13}C NMR spectroscopy in density fractions of an Oxisol under forest and pasture. Aust. J. Soil Res. 33:59–76.

Goni, M.A., B. Nelson, R.A. Blanchette, and J.I. Hedges. 1993. Fungal degradation of wood lignins: geochemical perspectives from CuO-derived phenolic dimers and monomers. Geochim. Cosmochim. Acta 57:3985–4002.

Greenland, D.J., and J.M. Oades. 1975. Saccharides. p. 213–261. *In* J.E. Gieseking (ed.) Soil components: Vol. 1. Organic components. Springer-Verlag, New York, NY.

Gregorich, E.G., and B.H. Ellert. 1993. Light fraction and macroorganic matter in mineral soils. p. 397–407. *In* M.R. Carter (ed.) Soil sampling and methods of analysis. Lewis Publishers, Boca Raton, FL.

Gregorich, E.G., and H.H. Janzen. 1996. Storage of soil carbon in the light fraction and macroorganic matter. p. 167–190. *In* M.R. Carter and B.A. Stewart (ed.) Structure and organic matter storage in agricultural soils. CRC Press, Boca Raton, FL.

Gregorich, E.G., M.R. Carter, D.A. Angers, C.M. Monreal, and B.H. Ellert. 1994. Towards a minimum data set to assess soil organic matter quality in agricultural soils. Can. J. Soil Sci. 74:367–385.

Griffiths, B.S. 1994. Microbial-feeding nematodes and Protozoa in soil: their effects on microbial activity and nitrogen mineralization in decomposition hotspots and the rhizosphere. Plant Soil 164:25–33.

Grossman, J., and P. Udluft. 1991. The extraction of soil water by the suction cup method: A review. J. Soil Sci. 42:83–93.

Guggenberger, G., B.T. Christensen, and W. Zech. 1994. Land-use effects on the composition of organic matter in particle-size separates of soil. I. Lignin and carbohydrate signature. Europ. J. Soil Sci. 45:449–458.

Guggenberger, G., W. Zech, and R. Thomas. 1995b. Lignin and carbohydrate alteration in particle size separates of an Oxisol under tropical pastures following native savanna. Soil Biol. Biochem. 27:1629–1638.

Guggenberger, G., W. Zech, L. Haumaier, and B.T. Christensen. 1995a. Land-use effects on the composition of organic matter in particle-size separates of soils: II. CPMAS and solution ^{13}C NMR analysis. Europ. J. Soil Sci. 46:147–158.

Haas, H.J., C.E. Evans, and E.F. Miles. 1957. Nitrogen and carbon changes in great plains soils as influenced by cropping and soil treatments. USDA Tech. Bull. 1164.

Haider, K. 1994. Advances in the basic research of the biochemistry of humic substances. p. 91–107. *In* N. Senesi and T.M. Miano (ed.) Humic substances in the global environment and implications on human health. Proceedings of the 6th Meeting of the International Humic Substances Society, Monopoli, Italy, Sept. 20–25, 1992.

Haider, K. 1992. Problems related to the humification processes in soils of temperate climates. p. 55–94. *In* G. Stotzky and J.-M. Bollag (ed.) Soil biochemistry. Vol. 7. Marcel Dekker, New York, NY.

Halsted, R.L., and R.B. McKercher. 1975. Biochemistry and cycling of phosphorus. p. 31–63. *In* E.A. Paul and A.D. McLaren (ed.) Soil Biochemistry. Vol. 4. Marcel Dekker, New York, NY.

Hammel, K.E. 1997. Fungal degradation of lignin. p. 33–45. *In* G. Cadisch, and K.E. Giller (ed.) Driven by nature: Plant litter quality and decomposition. CAB International, Wallingford, UK.

Hammond, T.E., D.G. Cory, W.M. Ritchey, and H. Morita. 1985. High resolution solid-state ^{13}C NMR of Canadian peats. Fuel 64:1687–1695.

Harvey, G.R., and D.A. Boran. 1985. Geochemistry of humic substances in seawater. p. 233–247. *In* G.R. Aiken, D.M. McKnight, R.L. Wershaw, and P. MacCarthy (ed.) Humic substances in soil, sediment and water. John Wiley and Sons, New York, NY.

Hassink, J. 1992. Effects of soil texture and structure on carbon and nitrogen mineralization in grassland soils. Biol. Fert. Soils 14:126–134.

Hassink, J., L.A. Bouwman, K.B. Zwart, and L. Brussaard. 1993. Relationships between habitable pore space, soil biota and mineralization rates in grassland soils. Soil Biol. Biochem. 25:47–55.

Hassink, J., A.M. Neutel, and P.C. De Ruiter. 1994. C and N mineralization in sandy and loamy grassland soils: the role of microbes and microfauna. Soil Biol. Biochem. 26:1565–1571.

Hatcher, P.G. 1987. Chemical structural studies of natural lignin by dipolar dephasing solid-state ^{13}C nuclear magnetic resonance. Org. Geochem. 11:31–39.

Hatcher, P.G., and E.C. Spiker. 1988. Selective degradation of plant biomolecules. p. 59–74. *In* F.H. Frimmel and R.F. Christman (ed.) Humic substances and their role in the environment. John Wiley and Sons, New York, NY.

Hatcher, P.G., I.A. Breger, L.W. Dennis and G.E. Maciel. 1983. Solid-state ^{13}C-NMR of sedimentary humic substances: new revelations on their chemical composition in aquatic and terrestrial humic materials. p. 37–81. *In* R.F. Christman and E.T. Gjessing (ed.) Aquatic and terrestrial humic materials. Ann Arbor Science, Ann Arbor, MI.

Hatcher, P.G., I.A. Breger, G.E. Maciel and N.M. Szeverenyi. 1985. Geochemistry of humin. p. 275–302. *In* G.R. Aiken, D.M. McKnight, R.L. Wershaw, and P. MacCarthy (ed.) Humic substances in soil, sediment and water. John Wiley and Sons, New York, NY.

Hatcher, P.G., N.A. Nanny, R.D. Minard, S.D. Dible, and D.M. Carson. 1995. Comparison of two thermochemolytic methods for the analysis of lignin in decomposing wood: the CuO oxidation methods and the method of thermochemolysis with tetramethylammonium hydroxide (TMAH). Org. Geochem. 23:881–888.

Hatcher, P.G., E.C. Spiker, N.M Szeverenyi, and G.E. Maciel. 1983. Selective preservation and origin of petroleum-forming aquatic kerogen. Nature 305:498–501.

Hayes, M.H.B., P. MacCarthy, R. Malcolm, and R.S. Swift. 1989. Humic substances II. In search of structure. Wiley-Interscience, Chichester, UK.

Hayes, M.H.B., R.S. Swift, R.E. Wardle, and J.K. Brown. 1975. Humic materials from an organic soil: a comparison of extractants and of properties of extracts. Geoderma 13:231–245.

Heal, O.W., J.M. Anderson, and M.J. Swift. 1997. Plant litter quality and decomposition: an historical overview. p. 3–30. *In* G. Cadisch and K.E. Giller (ed.) Driven by nature: Plant litter quality and decomposition. CAB International, Wallingford, UK.

Hedges, J.I. 1978. The formation and clay mineral reactions of melanoidins. Geochim. Cosmochim. Acta 42:69–76.

Hedges, J.I. 1988. Polymerization of humic substances in natural environments. p. 45–58. *In* F.H. Frimmel and R.F. Christman (ed.) Humic substances and their role in the environment. John Wiley and Sons, New York, NY.

Hedges, J.I., and J.R. Ertel. 1982. Characterization of lignin by gas capillary chromatography of cupric oxide oxidation products. Anal. Chem. 54:174–178.

Hedges, J.I., and J.M. Oades. 1997. Comparative organic geochemistries of soils and sediments. Org. Geochem. 27:319–361.

Hedges, J.I., G.L. Cowie, J.R. Ertel, R.J. Barbour, and P.G. Hatcher. 1985. Degradation of carbohydrates and lignins in buried woods. Geochim. Cosmochim. Acta 49:701–711.

Herbert, B.E., and P.M. Bertsch. 1995. Characterization of dissolved and colloidal organic matter in soil solutions: A review. p. 63–88. *In* J.M. Kelley, and W.W. McFee (ed.) Carbon forms and functions in forest soils. Soil Science Society of America, Madison, WI.

Herbert, B.E., P.M. Bertsch, and J.M. Novak. 1993. Pyrene sorption to water-soluble organic carbon. Environ. Sci. Tech. 27:398–403.

Hetier, J.M., B. Guillet, R. Brousse, G. Delibrias, and R.C. Maury. 1983. ^{14}C dating of buried soils in the volcanic Chaine des Puys (France). Bull. Volcanol. 46:193–201.

Hobbie, S.E. 1996. Temperature and plant species control over litter decomposition in Alaskan tundra. Ecol. Monogr. 66:503–522.

Holloway, P.J. 1982. The chemical constitution of plant cutin. p. 45–85. *In* D.F. Cutler, K.L. Alvin, and C.E. Price (ed.) The plant cuticle. Academic Press, London, UK.

Homann, P.S., and D.F. Grigal. 1992. Molecular weight distribution of soluble organics from laboratory manipulated surface soils. Soil Sci. Soc. Am. J. 56:1305–1310.

Huang, P.M., and M. Schnitzer. 1986. Interactions of soil minerals with natural organics and microbes. Soil Science Society of America, Madison, WI.

Huang, P.M., and A. Violante. 1986. Influence of organic acids on crystallization and surface properties of precipitation products of aluminum. p. 160–221. *In* P.M. Huang and M. Schnitzer (ed.) Interactions of soil minerals with natural organics and microbes. Soil Science Society of America, Madison, WI.

James, B.R., and S.J. Riha. 1986. pH buffering in forest soil organic horizons: Relevance to acid precipitation. J. Environ. Qual. 15:229–234.

Janzen, H.H. 1987. Soil organic matter characteristics after long-term cropping to various spring wheat rotations. Can. J. Soil Sci. 67:845–856.

Jardine, P.M., N.L. Weber, and J.F. McCarthy. 1989. Mechanisms of dissolved organic carbon adsorption on soil. Soil Sci. Soc. Am. J. 53:1378–1385.

Jardine, P.M., G.V. Wilson, J.F. McCarthy, R.L. Luxmoore, D.L. Taylor, and L.W. Zelazny. 1990. Hydrogeochemical processes controlling the transport of dissolved organic carbon through a forested hillslope. J. Contam. Hydrol. 6:3–19.

Jenkinson, D.S., and A. Ayanaba. 1977. Decomposition of carbon-14 labelled plant material under tropical conditions. Soil Sci. Soc. Am. J. 41:912–915.

Jenkinson, D.S., and J.H. Rayner. 1977. The turnover of soil organic matter in some of the Rothamsted classical experiments. Soil Sci. 123:298–305.

Jenny, H. 1930. A study on the influence of climate upon the nitrogen and organic matter content of the soil. MO Agric. Exp. Stn. Bull. 152.

Johnson, D.W. 1992. The effects of forest management on soil carbon storage. Water Air Soil Pollut. 64:83–120.

Juste, C., and J. Delas. 1970. Comparison par une méthode repirométrique, des solubilités bioloiques d'un humate de calcium et d'un humate de sodium. C.R. Acad. Sci. Paris Serie D 270:1127–1129.

Juste, C., J. Delas and M. Langon. 1975. Comparison de la stabilités biologique de différents humates metalliques. C.R. Acad. Sci. Paris Serie D 281:1685–1688.

Kaiser, K., G. Guggenberger, and W. Zech. 1996. Sorption of DOM and DOM fractions to forest soils. Geoderma 74:281–303.

Kajak, A. 1995. The role of soil predators in decomposition processes. Europ. J. Soil Ento. 92:573–580.

Kaliz, P.J., and E.L. Stone. 1980. Cation exchange capacity of acid forest humus layers. Soil Sci. Soc. Am. J. 44:407–413.

Kay, B.D., A.P. da Silva and J.A. Baldock. 1997. Sensitivity of the structure of different soils to changes in organic carbon content: predictions using pedotransfer functions. Can. J. Soil Sci. 77:655–667.

Kiala, A. 1963. Organic phosphorus in Finnish soils. Soil Sci. 95:38–44.

Kilbertus, G. 1980. Microhabitats in soil aggregates. Their relationship with bacterial biomass and the size of procaryotes present. Rev. Ecol. Biol. Sol 17:543–557.

Killham, K., M. Amato, and J.N. Ladd. 1993. Effect of substrate location in soil and soil pore-water regime on carbon turnover. Soil Biol. Biochem. 25:57–62.

Kinesch, P., D.S. Powlson, and E.W. Randall. 1995. ^{13}C NMR studies of soil organic matter in whole soils: I. Quantitation possibilities. Europ. J. Soil Sci. 46:125–138.

Kirk, T.K. 1984. Degradation of lignin. Micrbiol. Ser. 13:399–437.

Kirk, T.K. 1987. Lignin-degrading enzymes. Philos. Trans. Royal Soc. London. A. Math. Phys. Sci. 321:461–474.

Kirschbaum, M.U.F. 1995. The temperature dependence of soil organic matter decomposition, and the effect of global warming on soil organic C storage. Soil Biol. Biochem. 27:753–760.

Knicker, H., G. Almendros, F.J. Gonzalezvila, H.D. Ludemann, and F. Martin. 1995. C-13 and N-15 NMR analysis of some fungal melanins in comparison with soil organic matter. Org. Geochem. 23:1023–1028.

Knicker, H., and H.D. Ludemann. 1995. N-15 and C-13 CPMAS and solution NMR studies of N-15 enriched plant material during 600 days of microbial degradation. Org. Geochem. 23:329–341.

Kögel, I., and R. Bochter. 1985. Characterization of lignin in forest humus layers by high-performance liquid chromatography of cupric oxide oxidation products. Soil Biol. Biochem. 17:637–640.

Kögel-Knabner, I., F. Ziegler, M. Riederer, and W. Zech. 1989. Distribution and decomposition pattern of cutin and suberin in forest soils. Z. Pflanzenernähr. Bodenk. 152:409–413.

Kögel-Knabner, I., J.W. de Leeuw, and P.G. Hatcher. 1992a. Nature and distribution of alkyl carbon in forest soil profiles: implications for the origin and humification of aliphatic biomacromolecules. Sci. Total Environ. 117/118:175–185.

Kögel-Knabner, I., P.G. Hatcher, E.W. Tegelaar, and J.W. de Leeuw. 1992b. Aliphatic components of forest soil organic matter as determined by solid-state ^{13}C NMR and analytical pyrolysis. Sci. Total Environ. 113:89–106.

Kögel-Knabner, I., W. Zech, and P.G. Hatcher. 1988. Chemical composition of the organic matter in forest soils. III. The humus layer. Z. Pflanzenernähr. Bodenk. 151:331–340.

Krosshavn, M.K., T.E. Southon, and E. Steinnes, E. 1992. The influence of vegetational origin and degree of humification of organic soils on their chemical composition, determined by solid-state ^{13}C NMR. J. Soil Sci. 43:485–93.

Kumada, K. 1987. Chemistry of soil organic matter. Elsevier, Tokyo, Japan.

Ladd, J.N., M. Amato, and J.M. Oades. 1985. Decomposition of plant materials in Australian soils. III. Residual organic and microbial biomass C and N from isotope-labelled legume materials and soil organic matter, decomposing under field conditions. Aust. J. Soil Res. 23:603–611.

Ladd, J.N., and J.H.A. Butler. 1975. Humus-enzyme systems and synthetic organic polymer-enzyme analogs. p. 143–194. In E.A. Paul and A.D. McLaren (ed.) Soil biochemistry. Vol. 4. Marcel Dekker, New York, NY.

Ladd, J.N., R.C. Foster, P. Nannipieri, and J.M. Oades. 1996. Soil structure and biological activity. p. 23–78. In G. Stotzky and J.-M. Bollag (ed.) Soil biochemistry. Vol. 9. Marcel Dekker, New York, NY.

Lavelle, P., and A. Martin. 1992. Small-scale and large-scale effects of endogenic earthworms on soil organic matter dynamics in soils of the humid tropics. Soil Biol. Biochem. 24:1491–1498.

Lavelle, P., E. Blanchart, A. Martin, S. Martin, A. Spain, F. Toutain, I. Barois, and R. Schaefer. 1993. A hierarchical model for decomposition in terrestrial ecosystems: Application to soils of the humid tropics. Biotropica 25:130–150.

Lawes, J. 1861. On the application of different manures to different crops and on their proper distribution on the farm. Private publication cited by Dyke (1993) (as per Heal et al., 1997).

Lefroy, R.D.B., G.J. Blair and E.T. Craswell. 1995. Soil organic matter management for sustainable agriculture. Australian Centre for International Agricultural Research, Canberra, Australia.

Leinweber, P., G. Reuter, and K. Brozio. 1993. Cation exchange capacity of organo-mineral particle size fractions in soils from long-term experiments. J. Soil Sci. 44:111–119.

Lévesque, M., H. Morita, M. Schnitzer, and S.P. Mathur. 1980. The physical, chemical and morphological features of some Quebec and Ontario peats. Agric. Can. Pub. 1155.

Liebig, J. von. 1840. Die Chemie in ihrer Anwendung auf Agrikultur and Physiologie Braunschweig. Vieweg. As cited by Hedges and Oades (1997).

Lieth, H. 1973. Primary production: terrestrial ecosystems. Hum. Ecol. 1:303–332.

Ligel, L.H. 1992. An overview of carbon sequestration in soils of Latin America. In F.H. Beinroth (ed.) Organic carbon sequestration in the soils of Puerto Rico. Agronomy and Soils, University of Puerto Rico, Mayaguez, PR.

Linhares, M. 1977. Contribution de l'ion calcium à la stabilisation biologique de la matière organique des sols. Thèse Doc. Spéc., Université de Bordeaux III.

Ljungdahl, L.G., and K.-E. Eriksson. 1985. Ecology of microbial cellulose degradation. p. 237–299. K.C. Marshall (ed.) Advances in microbial ecology, Vol. 8, Plenum, New York, NY.

MacCarthy, P., R.L. Malcolm, C.E. Clapp, and P.R. Bloom. 1990a. Humic substances in crop and soil sciences: selected readings. American Society of Agronomy/Soil Science Society of America, Madison, WI.

MacCarthy, P., R.L. Malcolm, C.E. Clapp, and P.R. Bloom. 1990b. An introduction to soil humic substances. p. 1–12. In P. MacCarthy, R.L. Malcolm, C.E. Clapp, and P.R. Bloom. (ed.) Humic substances in crop and soil sciences: Selected readings. Soil Science Society of America, Madison, WI.

Malcolm, R.L. 1990. Variations between humic substances isolated from soils, stream waters, and ground waters as revealed by ^{13}C-NMR spectroscopy. p. 13–35. In P. MacCarthy, R.L. Malcolm, C.E. Clapp, and P.R. Bloom. (ed.) Humic substances in crop and soil sciences: Selected readings. Soil Science Society of America, Madison, WI.

Martin, J.P., and K. Haider. 1971. Microbial activity in relation to soil humus formation. Soil Sci. 47:54–63.

Martin, J.P., and K. Haider. 1975. Decomposition of specifically labelled benzoic acid and cinnamic acid derivatives in soil. Soil Sci. Soc. Am. Proc. 39:657–662.

Martin, J.P., J.O. Ervin and S.J. Richards. 1972. Decomposition and binding action in soil of some mannose containing microbial polysaccharides and their Fe, Al, Cu, and Zn complexes. Soil Sci. 113:322–327.

Martin, J.P., J.O. Ervin, and R.A. Shepherd. 1966. Decomposition of the iron, aluminium, zinc, and copper salts or complexes of some microbial and plant polysaccharides in soils. Soil Sci. Soc. Am. Proc. 30:196–200.

Martínez, A.T., A.E. González, M. Valmaseda, B.E. Dale, M.J. Lambregts, and J.F. Haw. 1991. Solid-state NMR studies of lignin and plant polysaccharide degradation by fungi. Holzforschung 45:49–54.

McBride, M.B. 1994. Environmental chemistry of soils. Oxford University Press, Oxford, UK.

McDougall, G.J., I.M. Morrison, D. Stewart, J.D.B. Weyers, and J.R. Hillman. 1993. Plant fibres: botany, chemistry and processing for industrial use. J. Sci. Food Agric. 62:1–20.

McDowell, W.H., and G.E. Likens. 1988. Origin, composition and flux of dissolved organic carbon in the Hubbard Brook Valley. Ecol. Monogr. 58:177–195.

McFee, W.W., and J.M. Kelly. 1995. Carbon forms and functions in forest soils. Soil Science Society of America, Madision, WI.

McGill, W.B., and C.V. Cole. 1981. Comparative aspects of cycling of organic C, N, S and P through soil organic matter. Geoderma 26:267–286.

McKeague, J.A., M.V. Cheshire, F. Andreux, and J. Berthelin. 1986. Organo-mineral complexes in relation to pedogenesis. p. 549–592. *In* P.M. Huang and M. Schnitzer (ed.) Interactions of soil minerals with natural organics and microbes. Soil Science Society of America, Madison, WI.

Moorehead, D.L., and J.F. Reynolds. 1989. The contribution of abiotic processes to buried litter decomposition in northern Chihuahuan desert. Oecologia 79:133–135.

Moran, M.A., L.R. Pomeroy, E.S. Sheppard, L.P. Atkinson, and R.E. Hodson. 1991. Distribution of terrestrially derived dissolved organic matter in the southeastern US continental shelf. Limnol. Oceanogr. 36:1134–1149.

Muller, P.E. 1887. Studien über die natürlichen Humusformen und deren Einwirkungen auf Vegetation und Boden. Springer, Berlin, Germany.

Müller-Wegener, U. 1988. Interactions of humic substances with biota. p. 179–192. *In* F.H. Frimmel and R.F. Christman (ed.) Humic substances and their role in the environment. John Wiley and Sons, New York, NY.

Mulongoy, K., and R. Merckx. 1993. Soil organic matter dynamics and sustainability of tropical agriculture. John Wiley and Sons, Chichester, UK.

Muneer, M., and J.M. Oades. 1989a. The role of Ca-organic interactions in soil aggregate stability. 1. Laboratory studies with ^{14}C-glucose, $CaCO_3$ and $CaSO_4.2H_2O$. Aust. J. Soil Res. 27:389–399.

Muneer, M., and J.M. Oades. 1989b. The role of Ca-organic interactions in soil aggregate stability. Field studies with ^{14}C-labelled straw, $CaCO_3$ and $CaSO_4.2H_2O$. Aust. J. Soil Res. 27:401–409.

Muneer, M., and J.M. Oades. 1989c. The role of Ca-organic interactions in soil aggregate stability. 3. Mechanisms and models. Aust. J. Soil Res. 27:411–423.

Murayama, S. 1984. Changes in the monosaccharide composition during the decomposition of straws under field conditions. Soil Sci. Plant Nutr. 30:367–381.

Nelson, D.W., and L.E. Sommers. 1996. Total carbon, organic carbon and organic matter. p. 961–1010. *In* D.L. Sparks et al. (ed.) Methods of soil analysis. Part 3. Chemical methods. Soil Science Society of America, Madison, WI.

Nelson, P.N., and J.M. Oades. 1998. Organic matter, sodicity and soil structure. p. 67–91. *In* M.E. Sumner, and R. Naidu (ed.) Sodic soils: Distribution, processes, management and environmental consequences. Oxford University Press, New York, NY.

Nelson, P.N., J.A. Baldock, and J.M. Oades. 1993. Concentration and composition of dissolved organic carbon in streams in relation to catchment soil properties. Biogeochem. 19:27–50.

Nelson, P.N., M.-C. Dictor, and G. Soulas. 1994. Availability of organic carbon in soluble and particle-size fractions from a soil profile. Soil Biol. Biochem. 26:1549–1555.

Newman, R.H. 1992. Nuclear magnetic resonance study of the spatial relationships between chemical components in wood cell walls. Holzforschung 46:205–210.

Newman, R.H., and J.A. Hemmingson. 1990. Determination of the degree of cellulose crystallinity in wood by carbon-13 nuclear magnetic resonance spectroscopy. Holzforschung 44:351–355.

Nguyen, M.L., and K.M. Goh. 1994. Sulphur cycling and its implications on sulphur fertilizer requirements of grazed grassland ecosystems. Agric. Ecosys. Environ. 49:173–206.

Niemeyer, J., Y. Chen, and J.-M. Bollag. 1992. Characterization of humic acids, composts, and peat by diffuse reflectance Fourier transform infrared spectroscopy. Soil Sci. Soc. Am. J. 56:135–140.

Nip, M., J.W. de Leeuw, P.J. Holloway, J.P.T. Jensen, J.C.M. Sprenkels, M. de Pooter, and J.J.M. Sleeckx. 1987. Comparison of flash pyrolysis differential scanning calorimetry, ^{13}C-NMR and IR spectroscopy in the analysis of a highly aliphatic biopolymer from plant cuticles. J. Anal. Appl. Pyrol. 11:287–295.

Nip, M., E.W. Tegelaar, H. Brinkhuis, J.W. de Leeuw, P.A. Schenck, and P.J. Holloway. 1986b. Analysis of modern and fossil plant cuticles by Curie point Py-GC and Curie point Py-GC-MS: recognition of a new highly aliphatic and resistant biopolymer. Org. Geochem. 10:769–778.

Nip, M., E.W. Tegelaar, J.W. de Leeuw, P.A. Schenck, and P.J. Holloway. 1986a. A new non-saponifiable highly aliphatic and resistant biopolymer in plant cuticles: evidence from pyrolysis and ^{13}C-NMR analysis of present-day and fossil plants. Naturwissenschaften 73:579–585.

Norden, B., E. Bohlin, M. Nilsson, A. Albano, and C. Ršckner. 1992. Characterization of particle size fractions of peat. An integrated biological, chemical and spectroscopic approach. Soil Sci. 153:382–96.

Novak, J.M., and P.M. Bertsch. 1991. The influence of topography on the nature of humic substances in soil organic matter at a site in the Atlantic Coastal Plain of South Carolina. Biogeochem. 15:111–126.

Oades, J.M. 1984. Soil organic matter and structural stability: mechanisms and implications for management. Plant Soil 76:319–337.

Oades, J.M. 1984. Soil organic matter and water-stable aggregates in soils. Plant Soil 76:319–337.

Oades, J.M. 1988. The retention of organic matter in soils. Biogeochem. 5:35–70.

Oades, J.M. 1989. An introduction to organic matter in mineral soils. p. 89–159. In J.B. Dixon and S.B. Weed (ed.) Minerals in soil environments, 2nd Ed. Soil Science Society of America, Madison, WI.

Oades, J.M. 1993. The role of biology in the formation, stabilisation and degradation of soil structure. Geoderma 56:377–400.

Oades, J.M. 1995. Organic matter: chemical and physical fractions. p. 135–139. In R.D.B. Lefroy, G.J. Blair, and E.T. Craswell (ed.) Soil organic matter management for sustainable agriculture. Australian Centre for International Agricultural Research, Canberra, Australia.

Oades, J.M., and A.G. Waters. 1991. Aggregate hierarchy in soils. Aust. J. Soil Res. 29:815–828.

Oades, J.M., A.M. Vassallo, A.G. Waters, and M.A. Wilson, M.A. 1987. Characterization of organic matter in particle size and density fractions from a red-brown earth by solid-state ^{13}C NMR. Aust. J. Soil Res. 25:71–82.

Oades, J.M., A.G. Waters, A.M. Vassallo, M.A. Wilson, and G.P. Jones. 1988. Influence of management on the composition of organic matter in a red-brown earth as shown by ^{13}C nuclear magnetic resonance. Aust. J. Soil Res. 26:289–299.

Ogner, G. 1985. A comparison of four different raw humus types in Norway using chemical degradations and CPMAS ^{13}C NMR spectroscopy. Geoderma 35:343–353.

Ong, H.L., and R.E. Bisque. 1968. Coagulation of humic colloids by metal ions. Soil Sci. 106:220–224.

Opella, S.J., and M.H. Frey. 1979. Selection of nonprotonated carbon resonances in solid-state nuclear magnetic resonance. J. Am. Chem. Soc. 101:5854–5856.

Orlov, D.S. 1995. Humic substances of soils and general theory of humification. Russian translations series 111. A.A. Balkema Publishers, Brookfield, VT.

Parton, W.J., D.C. Schimel, C.V. Cole, and D.S. Ojima. 1987. Analysis of factors controlling soil organic matter levels in Great Plains grasslands. Soil Sci. Soc. Am. J. 51:1173–1179.

Pascoal Neto, C., J. Rocha, A. Gil, N. Cordeiro, A.P. Esculcas, S. Rocha, I. Delgadillo, J.D. Pedrosa de Jesus, and A.J. Ferrer Correia. 1995. ^{13}C solid-state nuclear magnetic resonance and Fourier transform infrared studies of the thermal decomposition of cork. Solid State Nucl. Magn. Res. 4:143–151.

Paul, E.A., and F.E. Clark. 1989. Soil microbiology and biochemistry. Academic Press, San Diego, CA.

Paul, E.A., and H. van Veen. 1978. The use of tracers to determine the dynamic nature of organic matter. Trans. 11th. Int. Cong. Soil Sci. 3:61–102.

Paul, E.A., K. Paustian, E.T. Elliott, and C.V. Cole. 1997. Soil organic matter in temperate agroecosystems. Long-term experiments in North America. CRC Press, Boca Raton, FL.

Paustian, K., H.P. Collins, and E.A. Paul. 1997. Management controls on soil carbon. p. 15–49. In E.A. Paul, E.T. Elliott, K. Paustian, and C.V. Cole (ed.) Soil organic matter in temperate agroecosystems. Long-term experiments in North America. CRC Press, Boca Raton, FL.

Pearson, R.W., and R.W. Smith. 1939. Organic phosphorus in seven Iowa soil profiles: distribution and amounts as compared to organic carbon and nitrogen. Soil Sci. Soc. Am. Proc. 4:162–167.

Peterson, G.A., D.G. Westfall, C.W. Wood, and S. Ross. 1988. Crop and soil management in dryland agroecosystems. CO State Univ. Tech. Bull. LTB88-6.

Pfeffer, P.E. and W.V. Gerasimowicz. (ed.) 1989. Nuclear magnetic resonance in agriculture. CRC Press, Boca Raton, FL.

Pocknee, S., and M.E. Sumner. 1997. Cation and nitrogen contents of organic matter determine its soil liming potential. Soil Sci. Soc. Am. J. 61:86–92.

Popoff, T., and O. Theander. 1976. Formation of aromatic compounds from carbohydrates. IV. Chromones from reaction of hexuronic acids in slightly acidic, aqueous solution. Acta Chem. Scand. B30:705–710.

Post, W.M., W.R. Emmanuel, P.J. Zinke, and A.G. Stangenberger. 1982. Soil carbon pools and world life zones. Nature 298:156–159.

Post, W.M., T.H. Peng, W.R. Emmanuel, A.W. King, V.H. Dale, and D.L. De Angelis. 1990. The global carbon cycle. Am. Sci. 78:310–326.

Pressenda, L.C.R., R. Aravena, A.J. Melfi, E.C.C. Telles, R. Boulet, E.P.E. Vanencia, and M. Tomazello. 1996. The use of carbon isotopes (C-13, C-14) in soil to evaluate vegetation changes during the Holocene in central Brazil. Radiocarbon 38:191–201.

Preston, C.M., S.-E. Shipitalo, R.L. Dudley, C.A. Fyfe, S.P. Mathur, and M. Levesque. 1987. Comparison of ^{13}C CPMAS NMR and chemical techniques for measuring the degree of decomposition in virgin and cultivated peat profiles. Can. J. Soil Sci. 67:187–98.

Preston, C.M. 1996. Applications of NMR to soil organic matter analysis: History and prospects. Soil Sci. 161:144–166.

Preston, C.M., and R.H. Newman. 1992. Demonstration of the spatial heterogeneity in the organic matter of de-ashed humin samples by solid-state ^{13}C CPMAS NMR. Can. J. Soil Sci. 72:13–19.

Preston, C.M., D.E. Axelson, M. Lévesque, S.P. Mathur, H. Dinel, and R.L. Dudley. 1989a. Carbon-13 NMR and chemical characterization of particle-size separates of peats differing in degree of decomposition. Org. Geochem. 14:393–403.

Preston, C.M., M. Schnitzer, and J.A. Ripmeester. 1989b. A spectroscopic and chemical investigation on the de-ashing of a humin. Soil Sci. Soc. Am. J. 53:1442–1447.

Reiners, W.A. 1983. Transport processes in the biogeochemical cycles of carbon, nitrogen, phosphorus, and sulfur. p. 143–176. *In* B. Bolin and R.B. Cook (ed.) The major biogeochemical cycles and their interactions. Scope 21. John Wiley and Sons, New York, NY.

Riederer, M., and J. Schönherr. 1988. Development of plant cuticles: fine structure and cutin composition of *Clivia miniata* Reg. leaves. Planta 174:127–138.

Riederer, M., K. Matzke, F. Ziegler, and I. Kögel-Knabner. 1993. Occurrence, distribution and fate of the lipid plant biopolymers cutin and suberin in temperate forest soil. Org. Geochem. 20:1063–1076.

Robert, M., and J. Berthelin. 1986. Role of biological and biochemical factors in soil mineral weathering. p. 455–495. *In* P.M. Huang and M. Schnitzer (ed.) Interactions of soil minerals with natural organics and microbes. Soil Science Society of America, Madison, WI.

Ross, S.M. 1993. Organic matter in tropical soils - current conditions, concerns and prospects for conservation. Prog. Phys. Geogr. 17:265–305.

Saggar, S., A. Parshotam, G.P. Sparling, C.W. Feltham, and P.B.S. Hart. 1996. ^{14}C-labelled ryegrass turnover and residence times in soils varying in clay content and mineralogy. Soil Biol. Biochem. 28:1677–1686.

Saiz-Jimenez, C., K. Haider, and H.L.C. Meuzelaar. 1979. Comparisons of soil organic matter and its fractions by pyrolysis mass-spectroscopy. Geoderma 22:25–37.

Sala, O.E., W.J. Parton, L.A. Joyce, and W.K. Lauenroth. 1988. Primary production of the central grassland region of the United States. Ecol. 69:40–45.

Sanyal, S.K., and S.K. DeDatta. 1991. Chemistry of phosphorus transformations in soil. Adv. Soil Sci. 16:1–119.

Sarkar, J.M., and J.-M. Bollag. 1987. Inhibitory effect of humic and fulvic acids on oxidoreductases as measured by the coupling of 2,4-dichlorophenol to humic substances. Sci. Total Environ. 62:367–377.

Scharpenseel, H.W., H.U. Neue, and A. St. Singer. 1992. Biotransformations in different climatic belts; source-sink relationships. p. 91–105. *In* J. Kubat (ed.) Humus, its structure and role in agriculture and environment. Elsevier Science Publishers, Amsterdam, Netherlands.

Schimel, D.S., D.C. Coleman, and K.A. Horton. 1985a. Soil organic matter dynamics in paired rangeland and cropland toposequences in North Dakota. Geoderma 36:201–214.

Schimel, D.S., M.A. Stillwell, and R.G. Woodmansee. 1985b. Biogeochemistry of C, N, and P in a soil catena of the short grass steppe. Ecol. 66:276–282.

Schnitzer, M. 1977. Recent findings on the caharacterization of humic substances extracted from soils from widely differing climatic zones. Proc. IAEA Symp. on Soil Orgic Matter. FAO, Rome Italy.

Schnitzer, M. 1985. Nature of nitrogen in humic substances. p. 303–325. *In* G.R. Aiken, D.M. McKnight, R.L. Wershaw, and P. MacCarthy (ed.) Humic substances in soil, sediment, and water. Wiley and Sons, Chichester, UK.

Schnitzer, M. 1990. Selected methods for the characterisation of soil humic substances. p. 65–89. *In* P. MacCarthy, R.L. Malcolm, C.E. Clapp, and P.R. Bloom. (ed.) Humic substances in crop and soil sciences: selected readings. Soil Science Society of America, Madison, WI.

Schnitzer, M. 1991. Soil organic matter: The next 75 years. Soil Sci. 151:41–58.

Schnitzer, M., and M. Spiteller. 1986. The chemistry of the "unknown" soil nitrogen. Trans. 13th Int. Congr. Soil Sci., 3:473–474.

Schulten, H.-R. 1987. Pyrolysis and soft ionization mass spectrometry of aquatic/terrestrial humic substances and soils. J. Anal. Appl. Pyrol. 12:49–186.

Schulten, H.-R. 1993. Analytical pyrolysis of humic substances and soils: geochemical, agricultural and ecological consequences. J. Anal. Appl. Pyrol. 25:97–122.

Schulten, H.-R., G. Abbt-Braun, and F.H. Frimmel. 1987. Time-resolved pyrolysis field ionization mass spectrometry of humic material isolated from freshwater. Environ. Sci. Technol. 21:349–357.

Schulten, H.-R., and R. Hempfling. 1992. Influence of agriculture and management on humus composition and dynamics: classical and modern analytical techniques. Plant Soil 142:259–271.

Schulten, H.-R., and P. Leinweber. 1996. Characterization of humic and soil particles by analytical pyrolysis and computer modeling. J. Anal. Appl. Pyrol. 38:1–53.

Schulten, H.-R., and M. Schnitzer. 1993. A state of the art structural concept for humic substances. Naturwissenschaften 80:29–30.

Schulten, H.-R., C. Sorge, and M. Schnitzer. 1995. Structural studies on soil nitrogen by Curie-point pyrolysis-gas chromatography/mass spectrometry with nitrogen selective detection. Biol. Fertil. Soils 20:174–184.

Schulten, H.-R., C. Sorge-Lewin and M. Schnitzer. 1997. Structure of "unknown" soil nitrogen investigated by analytical pyrolysis. Biol. Fertil. Soils 24:249–254.

Schwartzenbach, R.P., P.M. Gschwend, and D.M. Imboden. 1993. Environmental organic chemistry. John Wiley and Sons, New York, NY.

Sensi, N., and T.M. Miano. 1994. Humic substances in the global environment and implications on human health. Elsevier, Amsterdam, Netherlands.

Serrâo, E.A.S., I.C. Falesi, J.B. da Vega, and J.R. Teizeira Neto. 1979. Productivity of cultivated pastures on low fertility soils of the Amazon region of Brazil. p. 195–225. In P.A. Sanchez and L.W. Tergas (ed.) Pasture production on acid soils of the tropics. CIAT, Cali, Columbia.

Sexstone, A.J., N.P. Revsbech, T.B. Parkin, and J.M. Tiedje. 1985. Direct measurements of oxygen profiles and denitrification rates in soil aggregates. Soil Sci. Soc. Am. J. 49:45–651.

Shafizadeh, F. 1984. The chemistry of pyrolysis and combustion. p. 489–529. In R. M. Rowell (ed.) The chemistry of solid wood. American Chemical Society, Washington, DC.

Sharon, N. 1965. Distribution of amino sugars in microorganisms, plants, and invertebrates. p. 1–45. In R.W. Jeanloz and E.A. Balazs (ed.) The amino sugars. Vol. IIA. Academic Press, New York, NY.

Shevchenko, S.M., and G.W. Bailey. 1996. Life after death: lignin-humic relationships reexamined. Crit. Rev. Environ. Sci. Technol. 26:95–153.

Skjemstad, J.O., P. Clarke, A. Golchin, and J.M. Oades. 1997. Characterisation of soil organic matter by solid-state ^{13}C NMR spectroscopy. p. 253–271. In G. Cadisch, and K.E. Giller (ed.) Driven by nature: Plant litter quality and decomposition. CAB International, Wallingford, UK.

Skjemstad, J.O., P. Clarke, J.A. Taylor, J.M. Oades, and R.H. Newman. 1994. The removal of magnetic materials from surface soils. A solid-state ^{13}C CP/MAS NMR study. Aust. J. Soil. Res. 32:1215–1229.

Skjemstad, J.O., P. Clarke, J.A. Taylor, J.M. Oades, and S.G. McClure. 1996. The chemistry and nature of protected carbon in soil. Aust. J. Soil Res. 34,:251–71.

Skjemstad, J.O., L.J. Janik, M.J. Head, and S.G. McClure. 1993. High energy ultraviolet photo-oxidation: a novel technique for studying physically protected organic matter in clay- and silt-sized aggregates. J. Soil Sci. 44:485–499.

Snape, C.E., D.E. Axelson, R.W. Botto, J.J. Delpuech, P. Tekely, B.C. Gerstein, M. Pruski, G.E. Maciel, and M.A. Wilson. 1989. Quantitative reliability of aromaticity and related measurements on coals by ^{13}C NMR. A debate. Fuel 68:547–560.

Söderlund, R. and B.H. Svensson. 1976. The global nitrogen cycle. Ecol. Bull. 22: 23–73.

Sokoloff, V.P. 1938. Effect of neutral salts of sodium and calcium on carbon and nitrogen in soils. J. Agric. Res. 57:201–216.

Sollins, P., S.P. Sline, R. Verhoeven, D. Sachs, and G. Spycher. 1987. Patterns of log decay in old-growth Douglas-fir forests. Can. J. For. Res. 17:1585–1595.

Sollins, P., P. Homann, and B.A. Caldwell. 1996. Stabilization and destabilization of soil organic matter - mechanisms and controls. Geoderma 74:65–105.

Sombroek, W.G., F.O. Nachtergaele, and A. Hebel. 1993. Amounts, dynamics and sequestering of carbon in tropical and subtropical soils. Ambio 22:417–426.

Sørensen, L.H. 1972. Stabilization of newly formed amino acid metabolites in soil by clay minerals. Soil Sci. 114:5–11.

Sørensen, L.H. 1975. The influence of clay on the rate of decay of amino acid metabolites synthesized in soils during decomposition of cellulose. Soil Biol. Biochem. 7:171–177.

Sowden, F.J. 1959. Investigations on the amounts of hexosamines found in various soils and methods for their determination. Soil Sci. 88:138–143.

Spain, A. 1990. Influence of environmental conditions and some soil chemical properties on the carbon and nitrogen contents of some Australian rainforest soils. Aust. J. Soil Res. 28:825–839.

Spain, A.V., R.F. Isbell, and M.E. Probert. 1983. Soil organic matter. p. 551–563. *In* Soils, an Australian viewpoint. CSIRO, Melboune, Australia / Academic Press, London, UK.

Starr, M., C.J. Westman, and J. Ala-Reini. 1996. The acid buffer capacity of some Finnish forest soils: Results of acid addition laboratory experiments. Water Air Soil Pollut. 89:147–157.

Steelink, C. 1985. Elemental characteristics of humic substances. p. 457–476. *In* G.R. Aiken, D.M. McKnight, R.L. Wershaw, and P. MacCarthy (ed.) Humic substances in soil, sediment and water. John Wiley and Sons, New York, NY.

Stevenson, F.J. 1986. Cycles of soil: Carbon, nitrogen, phosphorus, sulfur, micronutrients. John Wiley and Sons, New York, NY.

Stevenson, F.J. 1994. Humus chemistry. Genesis, composition, reactions. 2nd Ed. John Wiley and Sons, New York, NY.

Stevenson, F.J. 1982. Origin and distribution of nitrogen in soils. p. 1–42. *In* F.J. Stevenson (ed.) Nitrogen in agricultural soils. American Society of Agronomy, Madison, WI, USA.

Stewart, J.W.B., and C.V. Cole. 1989. Influence of elemental interactions and pedogenic processes in organic matter dynamics. Plant Soil 115:199–209.

Stotzky, G., and J.-M. Bollag. 1992. Soil biochemistry. Vol. 7. Marcel Dekker, New York, NY.

Stotzky, G., and J.-M. Bollag. 1993. Soil biochemistry. Vol. 8. Marcel Dekker, New York, NY.

Stotzky, G., and J.-M. Bollag. 1996. Soil biochemistry. Vol. 9. Marcel Dekker, New York, NY.

Swift, R.S. 1996. Organic matter characterisation. p. 1011–1069. *In* D.L. Sparks et al. (ed.) Methods of soil analysis. Part 3. Chemical methods. Soil Science Society of America, Madison, WI.

Szabolcs, I. 1994. The concept of soil resilience. p. 33–39. *In* D.J. Greenland, and I. Szabolcs (ed.) Soil resilience and sustainable land use. CAB International, Wallingford, UK.

Tan, K.H. 1986. Degradation of soil minerals by organic acids. p. 1–27. *In* P.M. Huang and M. Schnitzer (ed.) Interactions of soil minerals with natural organics and microbes. Soil Science Society of America, Madison, WI.

Tate, K.R., K. Yamanoto, G.J. Churchman, R. Meinhold, and R.H. Newman. 1990. Relationships between the type and carbon chemistry of humic acids from some New Zealand and Japanese soils. Soil Sci. Plant Nutrit. 36:611–621.

Tate, R.L. III. 1987. Soil organic matter. Wiley, New York, NY.

Tegelaar, E.W., J.W. de Leeuw, and C. Saiz-Jiminez. 1989. Possible origin of aliphatic moieties in humic substances. Sci. Total Envrion. 81/82:1–17.

Theng, B.K.G., G.J. Churchman, and R.H. Newman. 1986. The occurrence of interlayer clay-organic complexes in two New Zealand soils. Soil Sci. 142:262–266.

Thorn, K.A., and M.A. Makita. 1992. Ammonium fixation by humic substances: A nitrogen-15 and carbon-13 NMR study. Sci. Total Environ. 113:67–87.

Thorn, K. A., P.J. Pettigrew, and W.S. Goldenberg. 1996. Covalent binding of aniline to humic substances. 2. N-15 NMR studies of nucleophilic addition reactions. Environ. Sci. Tech. 30:2764–2775.

Thurman, E.M. 1985. Organic geochemistry of natural waters. Martinus Nijhoff/ Dr W Junk Publishers, Dordrecht, Netherlands.

Thurston, J.M., E.D. Williams, and A.E. Johnston. 1976. Modern developments in an experiment on permanent grassland started in 1856: effects of fertilizers and lime on botanical composition and crop and soil analyses. Ann. Agron. 27:1043–1082.

Tiessen, H., E. Cuevas, and P. Chacon. 1994. The role of soil organic matter in sustaining soil fertility. Nature 371:783–785.

Tiessen, H., J.W.B. Stewart, and D.W. Anderson. 1994. Determinants of resilience in soil nutrient dynamics. p. 157–170. *In* D.J. Greenland, and I. Szabolcs (ed.) Soil resilience and sustainable land use. CAB International, Wallingford, UK.

Tiessen, H., J.W.B. Stewart, and J.O. Moir. 1983. Changes in organic and inorganic phosphorus composition of two grassland soils and their particle size fractions during 60-70 years of cultivation. J. Soil Sci. 34:815–823.

Tiessen, H., J.W.B. Stewart, and J.O. Moir. 1984. Pathways of phosphorus transformations in soils of differing pedogenesis. Soil Sci. Soc. Am. J. 48:853–858.

Tinsley, J. 1950. Determination of organic carbon in soils by dichromate mixtures. Trans. 4th Int. Congr. Soil Sci. 1:161–169.

Tisdall, J.M., and J.M. Oades. 1982. Organic matter and water-stable aggregates in soils. J. Soil Sci. 33:141–163.

Uriyo, A.P. and A. Kessaba. 1975. Amounts and distribution of organic phosphorus in some profiles in Tanzania. Geoderma 13:201–210.

USDA-SCS. 1994. Soils of the United States 391. CD-ROM. USDA National Soil Survey Laboratory, Lincoln, NE.

Valdrighti, M.M., A. Pera, M. Agnolucci, S. Frassinetti, D. Junardi, and G. Vallini. 1996. Effects of compost-derived humic acids on vegetable biomass production and microbial growth within a plant (*Cichorium intybus*)-soil system: A comparative study. Agric. Ecosyst. Environ. 58:133–144.

Vallini, G., A. Pera, M. Agnolucci, and M.M. Valdrighti. 1997. Humic acids stimulate growth and activity of *in vitro* tested axenic cultures of soil autotrophic nitrifying bacteria. Biol. Fertil. Soils 24:243–248.

van Breeman, N., J. Mulder, and C.T. Driscoll. 1983. Acidification and alkalinization of soils. Plant Soil 75:283–308.

van der Hart, D.L., and R.H. Atalla. 1984. Studies of microstructure in native celluloses using solid-state ^{13}C NMR. Macromol. 17:1465–1472.

van der Linden, A.M.A., L.J.J. Jeurisson, J.A., Van Veen, and B. Schippers. 1989. Turnover of soil microbial biomass as influenced by soil compaction. p. 25–36. *In* J. Attansen and K. Henriksen (ed.) Nitrogen in organic wastes applied to soil. Academic Press, London, UK.

van Veen, J.A., and P.J. Kuikman. 1990. Soil structural aspects of decomposition of organic matter by micro-organisms. Biogeochem. 11:213–233.

Vaughan, D. R.E. Malcolm, and B.G. Ord. 1985. Influence of humic substances on biochemical processes in plants. p. 77–108. *In* D. Vaughan and R.E. Malcolm (ed.) Soil organic matter and biological activity. Martinus Nijhoff/Dr. W. Junk Publishers, Dordrecht, Netherlands.

Vaughan, D., and R.E. Malcolm. 1985. Soil organic matter and biological activity. Junk, Boston, MA.

Vedrova, E.F. 1997. Organic matter decomposition in forest litters. Eurasian Soil Sci. 30:181–188.

Vereecken, H., J. Maes, J. Feyen, and P. Darus. 1989. Estimating the soil moisture retention characteristic from texture, bulk density, and carbon content. Soil Sci. 148:389–403.

Visser, S.A. 1995. Physiological action of humic substances on microbial cells. Soil Biol. Biochem. 17:457–462.

Volkoff, B., and C.C. Cerri. 1988. L'humus des sols du Brésil-Nature et relations avec l'environnement. Cah. ORSTOM. Sér. Pédol. 24:83–95.

Walkely, A., and I.A. Black. 1934. An examination of the Degtjareff method for determining soil organic matter and a proposed modification of the chromic acid titration method. Soil Sci. 37:29–38.

Waters, A.G. and J.M. Oades. 1991. Organic matter in water-stable aggregates. p. 163–174. *In* W.S. Wilson (ed.) Advances in soil organic matter research. The impact on agriculture and the environment. Royal Society of Chemistry, Cambridge, UK.

Wegner, P.F., C.J. McDowall, and A.B. Frensham. 1989. Monitoring farming systems. Effects of cropping practices. Limitations to wheat yields. Dep. Agric. S. Aust. Tech. Pap. 23.

Wilson, M.A. 1987. N.M.R. techniques and applications in geochemistry and soil chemistry. Pergamon Press, Oxford, UK.

Wilson, W.S. 1991. Advances in organic matter research. Redwood, Melksham, UK

Yan, F., S. Schubert, and K. Mengel. 1996. Soil pH increase due to biological decarboxylation of organic anions. Soil Biol. Biochem. 28:617–624.

Zech, W., F. Ziegler, I. Kögel-Knabner, and L. Haumaier. 1992. Humic substances distribution and transformation in forest soil. Sci. Total Environ. 117/118:155–174.

Zepp, R.G., and Ch. Sonntag. 1995. The role of nonliving organic matter in the Earth's carbon cycle. John Wiley and Sons, Chichester, UK.

Ziegler, F. 1989. Changes of lipid content and lipid composition in forest humus layers derived from Norway spruce. Soil Biol. Biochem. 21:237–243.

Ziegler, F., and W. Zech. 1989. Distribution pattern of total lipids and lipid fractions in forest humus. Z. Pflanzenernähr. Bodenk. 152:287–290.

Ziegler, F., and W. Zech. 1990. Decomposition of beech litter cutin under laboratory conditions. Z. Pflanzenernähr. Bodenk. 153:373–374.

Zsolnay, A., and H. Steindl. 1991. Geovariability and biodegradability of the water-extractable organic material in an agricultural soil. Soil Biol. Biochem. 23:1077–1082.

Zuman, P., and E.B. Rupp. 1995. Polarography in the investigation of alkaline cleavage of lignin, soil organic matter and humic acids. Chemia Analityczna 40:549–563.

Zunino, H., F. Borie, S. Aguilera, J.P. Martin, and K. Haider. 1982. Decomposition of ^{14}C-labeled glucose, plant and microbial products and phenols in volcanic ash-derived soils of Chile. Soil Biol. Biochem. 14:37–43.

3

The Soil Solution

A.P. Schwab
Purdue University

3.1 Basic Concepts

The soil solution is the center of chemical and biological activities in the soil. Many critical functions slow significantly in the absence of water. For example, soil organisms become dormant or die, mineral transformations become imperceptibly slow, and soil chemical weathering and formation processes become greatly impeded. Addition of moisture to previously dry systems reinitiates chemical reactions and revives dormant organisms.

3.1.1 Definitions

The simplest definition of the soil solution is the aqueous liquid phase associated with the soil. Although this definition is technically correct, it is not meaningful from many physical, chemical, and biological perspectives, because a significant fraction of the soil solution seemingly does not participate in some important phenomena. Many examples exist in different subdisciplines of soil science that illustrate this point. In soil physics, the concepts of mobile and immobile water have been introduced to explain why water often is observed to move more rapidly through soil than would be predicted, assuming that all water participates in the transport process (Section A, Chapter 4). Higher plants and microorganisms undergo serious moisture deficit even when measurable water still exists in the soil. The composition and kinetics of reaction of water are significantly different very near solid surfaces in the soil than in the bulk solution. Therefore, a more functional definition of soil solution is needed than simply the aqueous liquid phase. However, the definition of soil solution may be dependent upon the application and, in some instances, the methodology used to obtain this solution from the soil.

From the viewpoint of a soil chemist, the soil solution can be defined as "the aqueous liquid phase in soil whose composition is influenced by flows of matter and energy between it and its surroundings and by the gravitational field of the Earth" (Sposito, 1989). By this definition, the soil solution is an open system that exchanges energy and other constituents with the surrounding water, air, and biological entities. Also, defining it as a phase implies that the soil solution has uniform bulk properties (such as composition and temperature) and can be isolated from the soil. The requirement of uniformity can be met only on small scales of time and space because of the dynamic and spatially variable nature of soils.

Viewing the soil solution on a molecular level would reveal that it is not a distinct entity, but rather a part of the "continuum of phases exhibiting indistinct interfaces at the molecular level. Solutes in the aqueous phase may be associated with bound water at the surfaces of soil colloids, free water percolating through soil macropores, water in the free space of plant roots, or immobile water in soil micropores" (Wolt, 1994, p. 13).

3.1.2   Composition

Because soil solutions are highly variable over space and time, their composition can be discussed only in general terms. Trends in relative concentrations of soluble inorganic and organic constituents (Section B, Chapter 1) are similar for most soils, but natural and anthropogenic factors can have radical influences on some or all components.

3.1.2.1   Inorganic Constituents

The inorganic constituents most commonly found in the soil solution are given in Table 3.1. Predominant cations are usually Ca^{2+}, Mg^{2+}, and K^+ (in that order), and a large number of minor cations occur, including Na^+, Fe^{2+}, Fe^{3+}, Cu^{2+}, Cu^+, and Zn^{2+}. The most prevalent anions are HCO_3^-, Cl^-, and SO_4^{2-}. Soils that become flooded are influenced strongly by reducing conditions and microbial activities, and Fe^{2+}, HS^-, and HCO_3^- can take on greater importance. In contaminated environments, the composition of the soil solution will change to reflect the most soluble components of the contaminants.

The sum of soluble cations and anions usually is less than 10^{-2} mol L^{-1}. Among the cations, Ca^{2+} will comprise at least half of this total; for the anions, the division among HCO_3^-, Cl^-, and SO_4^{2-} will be dependent upon soil pH and composition of the soil solids. Naturally saline soils (Section G, Chapter 1) will contain elevated concentrations of Ca^{2+}, Mg^{2+}, Cl^-, SO_4^{2-}, and sometimes Na^+, depending upon pH and the source of the salts. Most soil solutions will contain $Si(OH)_4^o$ as the main inorganic constituent of Si.

Concentrations of inorganic constituents in the soil solution are controlled by many factors, including pH, redox potential, and solid phase composition. These relationships will be addressed later in this chapter and elsewhere (Section B, Chapters 5–8).

3.1.2.2   Organic Constituents

The composition of soluble organic compounds in the soil solution reflects the organic components of the solid phase (Section B, Chapter 2) but is not as complex. Soluble organics also originate from organisms in the soil, including exudates from plant roots and soil microbes. Because these compounds are degraded by microbial activity, their concentrations in the soil solution can be more dynamic and more variable than those of the inorganic constituents.

Table 3.1   Major inorganic components found in soil solutions. Concentration ranges are estimates and could change depending upon specific environments. The components listed are not necessarily the dominant solution species.

Category	Major Components (10^{-4} to 10^{-2} mol L^{-1})	Minor Components (10^{-6} to 10^{-4} mol L^{-1})	Others*
Cations	Ca^{2+}, Mg^{2+}, Na^+, K^+	Fe^{2+}, Mn^{2+}, Zn^{2+}, Cu^{2+}, NH_4^+, Al^{3+}	Cr^{3+}, Ni^{2+}, Cd^{2+}, Pb^{2+}
Anions	HCO_3^-, Cl^-, SO_4^{2-}	$H_2PO_4^-$, F^-, HS^-	CrO_4^{2-}, $HMoO_4^+$
Neutral	$Si(OH)_4^o$	$B(OH)_3^o$	

*Components normally found in concentrations $<10^{-6}$ mol L^{-1} unless in a contaminated environment.

As with the inorganics, the organic constituents can be positively or negatively charged, or neutral. The chemical reactions of these compounds will be influenced by their electronic charge, but hydrophobic and hydrophylic tendencies are critical as well. For example, positively charged compounds that are water soluble can be adsorbed by cation exchange sites, negatively charged organics can form strong complexes with iron or aluminum oxides, and uncharged compounds can adsorb onto hydrophobic sites of soil organic matter (SOM).

The low molecular weight carboxylic acids typify soluble organics (Table 3.2). They are exuded continually by plant roots and soil microorganisms and are degraded readily, having a half-life of hours. In carboxylic acids which contain the COOH functional group, the proton usually is dissociated in soil solutions, and the resulting negatively charged ligand often can act as a complexing agent for metals. Formic and tartaric acids are exuded by grass roots, acetic acid is generated under anaerobic conditions, oxalic acid often is associated with ectomycorrhizae, and citric acid is excreted by fungi and plant roots (Sposito, 1989). Typical concentrations for these acids range from 10^{-5} to 10^{-3} mol L^{-1}. Stevenson (1967) published an extensive review of these and other organic constituents in the soil solution.

Organic pollutants include those compounds that have been introduced artificially into the soil, whether by design or accident. Most organic pesticides are applied at very low concentrations (0.1–10 kg ha^{-1}). Organics that have been spilled or accidentally leaked onto the soil can be present at many orders of magnitude higher. The resulting concentration in the soil solution will be dependent upon the total concentration in the soil, water solubility, strength of adsorption, volatility, and degradability of the compound in question. For example, the triazine herbicide, atrazine, normally is applied at a rate of approximately 2 kg ha^{-1}, has a water solubility of 35 mg L^{-1}, is adsorbed weakly by the soil, has a half-life of approximately 30 d, and has limited volatility. In contrast, glyphosate is applied at about the same rate as atrazine but is 1000 times more soluble, very strongly adsorbed, and not volatile. Immediately after application, atrazine concentrations in the soil solution can exceed 5 mg L^{-1} and represent a serious environmental threat to surface and groundwater. In contrast, aqueous glyphosate concentrations are nearly always <1 μg L^{-1} and seldom are detected in drinking water. McBride (1994) provides an excellent overview of the chemical behavior of pesticides in soil. Organic solvents and petroleum hydrocarbons tend to have very low water solubilities (< 1 μg L^{-1}), but their toxicities are higher, and even trace concentrations in the soil solution can be environmentally significant.

3.2 Sampling the Soil Solution

The definition of the soil solution given in Section 3.1.1 is idealized and serves as a point from which the soil solution can be conceptualized. At the experimental level, the soil solution is defined by the

Table 3.2 Major organic components found in soil solutions. Concentration ranges are estimates and could change depending upon specific environments. The components listed are not necessarily the dominant solution species.

Source	Major Components (10^{-5} to 10^{-3} mol L^{-1})	Minor Components (<10^{-5} mol L^{-1})
Natural	Carboxylic acids, amino acids, simple sugars	carbohydrates, phenolics, proteins, alcohols, sulfhydryls
Anthropogenic	(see note below)*	herbicides, fungicides, insecticides, PCBs, PAHs, petroleum hydrocarbons, surfactants, solvents

* Most organic contaminants are present in low concentrations except in the cases of spills, leaks, or accidental releases.

method used to separate the aqueous phase from the rest of the soil. As discussed below, sampling methodology has a profound influence on the composition of the soluble constituents. Thus, consistent with the Heisenberg uncertainty principle for the study of subatomic particles, one cannot study the soil solution without altering it.

For both laboratory and field studies, one of the major problems associated with obtaining an unaltered soil solution is that the moisture content of field soils can range from air dry to saturated over a very short period of time and influences the chemical and microbiological dynamics in the aqueous phase. Most techniques for obtaining samples of the soil solution function poorly when the moisture content is below saturation, and very few function at all if moisture tensions are ≤ 33 kPa (1/3 bar). The method (and moisture content) chosen for obtaining a sample of the soil solution must minimize the impact on its composition, although the act of sampling necessarily has changed it.

An example of the influence of moisture content on soil chemical properties is the often observed change in pH at different soil:solution ratios (Fig. 3.1). Soil pH was measured for two soils with moisture contents ranging from saturation to a ratio of 10:1 water:soil (on a volume/mass basis). As the amount of water increased, the pH increased significantly. The magnitude of change was similar to that observed by Mubarak and Olsen (1976). These observations lead to the following inescapable conclusion:

"Consideration of soil at field moisture contents is necessitated by the inability to predict consistently the effects of variation in soil to water ratios across broad ranges of soil:solution composition; neither variation in total electrolyte concentration nor the activity ratios of specific ion components of the soil solution can be adequately resolved when water to soil ratios vary from field moisture contents to ratios > 1. This is the main limitation to the use of water extracts as models of soil solution." (Wolt, 1994).

3.2.1   Laboratory Methods

Many laboratory methods have been developed for sampling the soil solution; only a few will be summarized here. The methods can be categorized broadly as aqueous extracts, column displacement, and pressure extraction. Each technique has advantages and limitations. Proper sampling and handling techniques for the samples also must be considered. For example, the simple act of air drying the soil

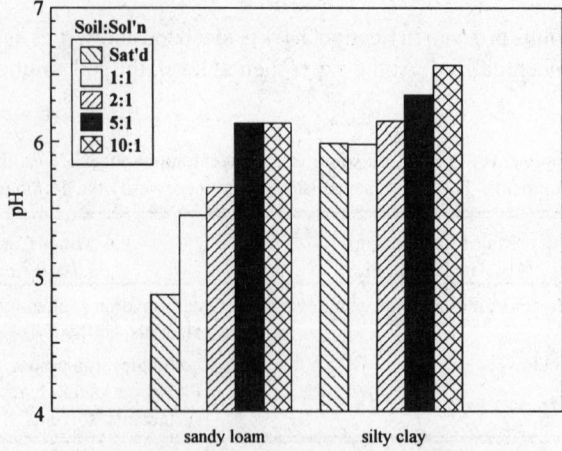

Fig. 3.1   The impact of soil:solution ratio on the measured pH of two soils [Schwab, 1992]

can have a profound effect on its chemical and microbiological properties (Bartlett and James, 1980; West et al., 1992).

The steps to obtaining aqueous extracts include adding water to the soil to the point of saturation or beyond, equilibrating, and removing solution. Equilibration times and separation techniques will depend upon the method used and desired application. If a saturated paste is prepared (U.S. Salinity Laboratory Staff, 1954) for the assessment of soil salinity, the paste is equilibrated for 16 h followed by vacuum removal of the soil solution. If equilibrium with soil solid phases is desired, equilibrations of weeks or years may be required (Kittrick, 1977; Schwab, 1989). The advantages of these techniques include ease of preparation, simple separation of soil and solution, and the ability to control many experimental parameters such as aeration and shaking. Limitations are unrealistic moisture contents, abrasion of soil surfaces during shaking, and uncertain impacts of wide soil:solution ratios on solution composition. This method has been applied with success in many instances (Lindsay, 1979).

Column displacement consists of forcing a fraction of the solution from the soil by leaching with an aqueous solution (miscible displacement) or a water insoluble organic solvent (immiscible displacement). These techniques have a long history, perhaps beginning with Thompson (1850) and Way (1850) leaching $(NH_4)_2SO_4$ solutions through columns of soil and finding that the NH_4^+ had replaced Ca^{2+}. In 1866, Schloesling used miscible displacement to obtain a sample of the soil solution; this method is quite similar to modern miscible displacement techniques (Adams, 1974). All column displacement procedures involve packing moist soil into columns to a desired bulk density, sealing the surface by mechanically dispersing the clays, and leaching with the displacing liquid. The soil solution is collected in fractions until the displacing liquid appears in the leachate. This system can be modified to include pressure from the top (Ross and Bartlett, 1990) or vacuum applied to the bottom of the column (Wolt and Graveel, 1986).

The advantages of column displacement methods are obtaining soil solution from a soil at field moisture conditions, no requirement for grinding or shaking, and maintaining the dissolved gas content (except in the vacuum modification). Disadvantages include generally short incubation times to prevent unusual microbial growth or changes in moisture content, a lack of knowledge of the fraction of the true soil solution that is obtained because of chemical and physical heterogeneity of the columns and a large fraction of the water-filled pore space being occupied by the diffuse double layer, and (in the case of immiscible displacement) an unknown influence of organic solvents on the displacement of ions.

Pressure extraction is the use of positive pressure, vacuum (negative pressure), or force applied by a centrifuge to remove the soil solution. For soils with moisture contents below saturation, the centrifuge method is more efficient than the positive pressure or vacuum methods. The efficiency of the latter two methods can be increased by employing a displacing solution, similar to some of the techniques described immediately above. Positive pressure methods generally refer to the use of a pressure membrane (Richards, 1941). A cylinder equipped with small pore filters on the inlet and outlet is packed with the moist soil. Pressure of approximately 2 MPa (20 bar) is applied to the inlet, and the solution is collected from the outlet. The centrifugation method similarly requires specialized equipment. The apparatus usually consists of a centrifuge tube to contain the moist soil, a permeable frit, filter, or ceramic beneath the soil, and a small volume at the bottom of the apparatus for collection of the extruded solution (Elkhatib et al., 1986). Large samples (up to 1 kg) usually are centrifuged at low speeds, but smaller samples can be subjected to much greater speeds and result in a greater fraction of the soil water being collected. An immiscible liquid also can be placed on the top of a small sample prior to centrifugation, resulting in a modification of immiscible displacement.

The pressure membrane approach has not been used as widely as the centrifuge method because of lower yields, requirement for large samples, and specialty apparatus. The high speed centrifuge method is an excellent choice because it uses equipment that is normally available in a soil chemistry laboratory, yields a sample very quickly (< 1 h), and is amenable to a large number of samples. The disadvantages are similar to those of other displacement type methods, namely, short incubation times, an unknown fraction of the true soil solution that is sampled, uncertain impact of organic solvents (if used), and potential changes in the dissolved gas composition.

3.2.2   Field Methods

As is often the case in soil science, field methods of sampling the soil solution are more challenging than laboratory methods, particularly when one is interested in obtaining samples that are truly representative. Field methods cover a wide range of configurations, including block or monolithic, zero-tension or pan, and porous cup vacuum lysimeters.

Monolithic lysimeters are large blocks of soil (undisturbed or refilled) contained in a structure that has some means of collecting leachates. The apparatus is labor intensive and expensive to construct, but generally is placed in a typical field setting to allow growth of vegetation while a variety of soil parameters are measured. The leachate collection system can be free drainage (i.e., zero-tension or air entry potential) or can use a vacuum system to impose a moderate moisture tension. Soil in contact with a free drainage collection system must be very near saturation to allow water to move from the column to the collection area. The thickness of the near saturated zone will be dependent upon the texture of the soil, and the saturated condition will impact the chemistry, microbiology, and physics of the soil. Addition of a vacuum system, although adding to the complexity and cost of construction, will overcome the saturation problem. Unfortunately, the vacuum will at least partially degas the collected leachates.

Water flow through large lysimeters will affect directly the chemical composition of the leachates. In refilled versions, the original structure is destroyed, and, in the absence of plants, preferential flow within the soil is eliminated. However, this is unrepresentative of natural conditions. Block type, undisturbed lysimeters preserve much of the original soil structure, and water movement will be similar to the original field condition. In both block and refilled lysimeters, wall flow (movement of water in the spaces between the lysimeter walls and the soil) can be an important form of preferential transport of water and solutes (Till and McCabe, 1976). Wall flow can be minimized through careful design and construction, and its relative impact is lessened as the ratio of surface area of the side walls to the total volume of soil increases.

Zero-tension lysimeters include any device installed in the soil that collects water by free drainage of the soil above, including pans, troughs, funnels, and plates. These lysimeters usually are installed from a trench by excavating a slot into the soil (Fig. 3.2). Leachate is collected in the lysimeter, and the sample is removed by means of a hose or tube leading to the trench or to the soil surface. This design is useful for monitoring the solutes that move with water. The assumption is made that the soil disturbance from digging the trench and installing the pan will have little impact on water flow and solute composition. However, water will not flow into the pan until the soil is nearly saturated. Water can accumulate above the interface between the soil and lysimeter and cause incoming water to move laterally away from the lysimeter. Modifications to this procedure such as putting the lysimeter under tension can partly overcome this problem (Cole et al., 1961).

Porous cup vacuum lysimeters (Fig. 3.3) are used widely in sampling the soil solution. Their design is extremely simple, and they are installed easily using standard soil sampling equipment. Entire assemblies are available commercially or can be built readily. A length of PVC pipe is fitted with and glued to a ceramic cup. A rubber stopper with two air lines is placed in the opposite end of the pipe. The lysimeter is buried in the soil with the ceramic cup facing down and the air lines extending to the soil surface. One air line is used for drawing a vacuum, either with a hand pump or a portable electric

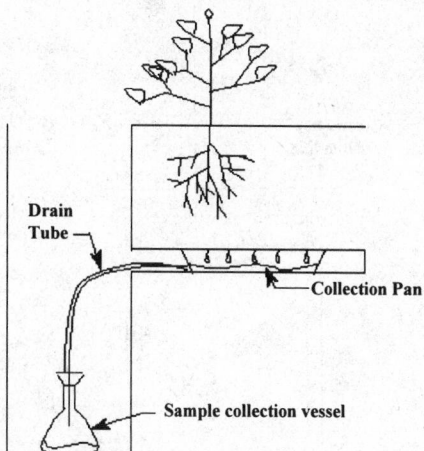

Fig. 3.2 General set up for installation of a zero-tension lysimeter. The collection vessel and drain tube are located in an excavated trench cut into the soil along the side of the sample area.

pump. The lines are sealed, and the lysimeter is allowed to draw the soil solution into the ceramic cup. (A vacuum may be applied continuously or intermittently.) The second air line extends to the bottom of the cup and is used for retrieving the collected sample either by drawing a vacuum on the retrieval line (marked B in Fig. 3.3) or applying pressure on the other line (A).

Porous cup lysimeters have been criticized as field samplers of the soil solution. The ceramic cup has the potential to retain analytes or to contaminate the sample (Wood, 1973), particularly for inorganic species. Likewise, the PVC tube can retain organic compounds of interest. The area of soil that is sampled is not known (Warrick and Amoozegar-Fard, 1977) but will be influenced by the vacuum applied (Morrison and Lowry 1990), the texture of the soil, the method of installation, and soil moisture content. Unlike the other field methods described above, these samplers will operate only when the operator engages them. Unless the porous cup is in saturated soil, the soil solution will not flow into it until vacuum is applied.

Each field method has positive and negative aspects. The monolithic lysimeters are, in essence, a field laboratory with all parameters either controllable or measurable. However, they are difficult and expensive to construct and maintain, and wall effects can be dominant. Zero-tension lysimeters are less expensive but require significant disturbance to the soil adjacent to the site, and water flow patterns may be altered immediately above the interface between soil and lysimeter. Porous-cup vacuum lysimeters are the least expensive of all field methods but collect the smallest fraction of the soil solution. In all cases, the soil solution will be defined by the specifics of the method used.

3.3   Thermodynamics of the Soil Solution

In discussing the thermodynamics of aqueous systems, strong distinctions are made between chemical equilibrium and nonequilibrium (kinetic) systems. Overlap exists between the statics and dynamics of solutions, on both a theoretical and a practical basis. In this section, only equilibrium thermodynamics will be addressed. Kinetics are addressed in Section B, Chapter 4. The assumption of equilibrium may be applied safely to only a fraction of the reactions occurring in soils; many reactions do not achieve equilibrium before new perturbations are imposed. Despite serious practical limitations, equilibrium models are important because "they are simpler in that they require less information, but they are nevertheless powerful when applied within their proper limits" (Stumm and Morgan, 1996).

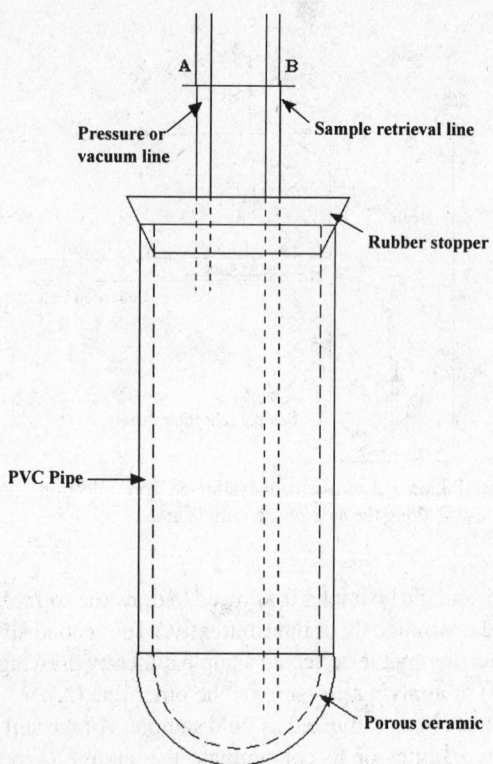

Fig. 3.3 Design of a typical porous-cup vacuum lysimeter

Soil solutions are open systems in which energy and matter are exchanged readily with the surrounding environment. Experimentalists are often more familiar with closed systems in which such exchanges do not occur, and rigorous mass and energy balances can be applied. Thus, the study of soils in natural settings can require some adjustments in design, execution, and interpretation. Open systems will be assumed in all derivations discussed below unless otherwise noted.

Chemical thermodynamics of aqueous solutions have been developed to various degrees of depth and detail, depending upon the application. Stumm and Morgan (1996) provided an excellent, in-depth discussion of equilibrium thermodynamics for aqueous systems. Wolt (1994) provided a discussion of similar depth but with the ultimate application to soil solutions. Sposito (1981) dedicated an entire textbook to the subject of the thermodynamics of soil solutions. In this chapter, important thermodynamic laws and equations will be stated and discussed; they will not be derived.

3.3.1 Fundamentals, Units, and Variables

As mentioned previously, true equilibrium is achieved in only a fraction of the reactions that occur in soil solutions but is a powerful tool when properly applied. Some of the reasons for investigating the application of equilibrium thermodynamics to systems that are frequently not in equilibrium include determining whether some or all of the components of the solution are in equilibrium; comparing the measured system with systems in equilibrium; quantifying the energy of disequilibrium (i.e., the energy input necessary to achieve equilibrium); and calculating the effects of temperature on equilibria. The equations necessary to obtain these goals are presented in following sections.

   Important variables and functions and associated units are given in Table 3.3. Nearly all thermodynamic equations relative to soil solutions are derived from the four principles of thermodynamics and the First and Second Laws of Thermodynamics (Stumm and Morgan, 1996). The four principles of thermodynamics establish an absolute temperature scale, define the internal energy of a system, and describe the relationship between entropy and temperature.

3.3.2   First and Second Laws of Thermodynamics

The First Law of Thermodynamics for equilibrium systems of fixed compositions,

$$dE = dq - dw \tag{3.1}$$

represents changes in internal energy (E) as affected by heat transferred to the system (q) and work done by the system (w). For a reversible process, the Second Law is given by:

$$dS_{sys} = \frac{dq}{T} \tag{3.2}$$

and relates the temperature and heat transferred to a system with entropy (S). Other forms of this equation exist for the system plus its surroundings as well as irreversible changes in a system.

   From the basic (not differentiated) equations $H = E + PV$ and $G = E + PV - TS$, one can obtain the important relationship, $G = H - TS$, where G is the Gibbs free energy, H is enthalpy, P is pressure and V is volume. For a finite state change at constant pressure and temperature:

$$\Delta G = \Delta H - T\Delta S \tag{3.3}$$

This equation, a restatement of the Second Law of Thermodynamics in terms of state functions of the system, is used in the application of thermodynamics to numerical solutions of chemical problems.

3.3.3   Associated Thermodynamic Relationships

The development of theoretical thermodynamics can be quite detailed, but its utility can be realized only if expressed in measurable variables. For example, enthalpy change (dH) in Table 3.3 can be described as $dH = dq + VdP$; if $dP = 0$ (constant pressure process), then $dH = dq_p$. This expression is particularly useful in determining temperature effects, because constant pressure heat capacity (C_p) is measurable and is

Table 3.3   Variables, thermodynamic functions, and equations of state

Variable	Units	Function	Equation of State
Temperature (T)	degrees K	Enthalpy (H)	$dH = TdS + VdP$
Entropy (S)	J deg^{-1}	Helmholtz free energy (A)	$dA = -SdT - PdV$
Pressure (P)	kPa	Gibbs free energy (G)	$dG = -SdT + VdP$
Volume (V)	L	Internal energy (E)	$dE = TdS - PdV$
Chemical potential (μ)	J mol^{-1}		
Quantity	mol		

$$C_P = \left(\frac{dq}{dT}\right)_P \qquad\qquad [3.4]$$

In the absence of external work, $q_p = \Delta H$.

Under conditions of constant pressure and temperature, the state of a system is characterized by dG. For irreverisible changes in a system,

$$dG - VdP + SdT < 0 \qquad\qquad [3.5]$$

and for reversible changes,

$$dG - VdP + SdT = 0 \qquad\qquad [3.6]$$

The above equations are applicable only for systems of constant chemical composition, $dn_i = 0$. When the composition of only species i is allowed to change,

$$dG = \left(\frac{\partial G}{\partial T}\right)_{P,n_i} dT + \left(\frac{\partial G}{\partial P}\right)_{T,n_i} dP + \sum_i \left(\frac{\partial G}{\partial n}\right)_{P,T,n_j} dn_i \qquad\qquad [3.7]$$

3.3.4   Chemical Potential and Free Energy

The chemical potential (μ) of a species (i) is defined as:

$$\mu_i = \left(\frac{\partial G}{\partial n_i}\right)_{P,T,n_j} \qquad\qquad [3.8]$$

Employing the relationships $(\partial G/\partial T)_{P,nj} dT = -S$ and $(\partial G/\partial P)_{T,nj} dP = V$, yields the equation:

$$dG = SdT + VdP + \sum_i \mu_i dn_i \qquad\qquad [3.9]$$

and, for a single phase system at constant temperature and pressure,

$$dG = \sum_i \mu_i dn_i \qquad\qquad [3.10]$$

For a multiphase system, dG for the entire system is obtained by summing $\mu_i dn_i$ over all phases. When equilibrium is established for all reactions and phases, $dG = 0$.

These basic equations can be manipulated further to yield equations for chemical potentials of components in various phases. The resulting relationships depend upon assumptions made during the derivation. Equations for gases and electrolytes are summarized in Table 3.4. As an example derivation for an ideal gas, consider the partial free energy of this gas with respect to P at constant T and n: $(\partial G/\partial P)_{T,n} = V$ and, thus, $(\partial G/\partial P)_{T,n} = (nRT)/P$. If the second equation is rearranged and integrated from G^o (standard free energy) to G and from P^o (standard pressure) to P, the result is

$$G - G^O = nRT \ln\left(\frac{P}{P^O}\right) \tag{3.11}$$

Differentiation with respect to n and recalling the definition for chemical potential, gives:

$$\mu_i = \mu_i^o + RT \ln\left(\frac{P_i}{P^O}\right) \tag{3.12}$$

where μ_i^o is the chemical potential for the ideal gas in its standard state (e.g., as defined by the conditions given in Table 3.4) at $T = 298.15$ K and $P = 101.33$ kPa (1 atm). Similar derivations for real gases, condensed phases that obey Raoult's Law, and condensed phases that obey Henry's Law are summarized succinctly by Wolt (1994). Sposito (1994) discussed the standard states for phases and elements relevant to the study of soil solutions.

3.3.5 Chemical Potential and Activities in Ideal Solutions

Perhaps the most important expressions for chemical potential in dealing with soil solutions are those for solutions of electrolytes. A classical case is that of NaCl dissolved in water. Sodium chloride is a strong electrolyte, indicating that it is dissociated fully leaving no NaCl complexes in the aqueous solution. The chemical potential of this solution is the sum of the chemical potentials for the Cl^- and Na^+ species:

$$\mu_{NaCl} = \mu_{Na^+} + \mu_{Cl^-} \tag{3.13}$$

Table 3.4 Expressions for the chemical potential of components in gas and condensed phases as influenced by the assumptions of the behavior of the component [Modified from Wolt, 1994)

Phase	Chemical Potential	Coefficients	Standard State
Ideal gas	$\mu_i = \mu_i^o + RT \ln(P_i/P^o)$		298.15 K, 101.33 kPa pure, ideal gas $P_i = 101.33$ kPa
Nonideal gas	$\mu_i = \mu_i^o + RT \ln(P_i\lambda_i/P^o)$	$\lambda_i \equiv$ fugacity coefficient	298.15 K, 101.33 kPa pure, non-ideal gas $P_i = 101.33$ kPa
Condensed phase; follows Henry's Law	$\mu_i = \mu_i^o + RT \ln(C_i/C^o)$	$C_i = P/K_{hi}$	pure condensed phase component, $C_i = 1$
Condensed phase; does not follow Henry's Law	$\mu_i = \mu_i^o + RT \ln(C_i\lambda_i/C^o)$	$\lambda_i \equiv$ Henry's Law coefficient	pure condensed phase component, $C_i = 1$ and $\lambda_i = 1$
Condensed phase; follows Raoult's Law	$\mu_i = \mu_i^o + RT \ln\chi_i$	$\chi_i = P_i/P^o_i$	pure condensed phase component, $\chi_i = 1$
Condensed phase; does not follow Raoult's Law	$\mu_i = \mu_i^o + RT \ln(\chi_i\lambda_i)$	$\lambda_i \equiv$ Raoult's Law coefficient	pure condensed phase component, $\chi_i = 1$ and $\lambda_i = 1$
Solute in ideal, dilute solution	$\mu_i = \mu_i^o + RT \ln C_i$		pure condensed phase, $C_i = 1$
Solute in non-ideal solution	$\mu_i = \mu_i^o + RT \ln C_i\gamma_i$	$\gamma_i =$ single ion activity coefficient	pure condensed phase, $C_i = 1$ and $\lambda_i = 1$
Solute in non-ideal solution	$\mu_\pm = \mu_\pm^o + RT \ln C_\pm\gamma_\pm$	$\gamma_\pm =$ mean activity coefficient	pure condensed phase, $C_\pm = 1$ and $\lambda_\pm = 1$

In an ideal, dilute solution, the equations for the chemical potentials of the individual ions with concentrations C_{na+} and C_{Cl-} would be

$$\mu_{Na^+} = \mu^o_{Na^+} + RT \ln\left(C_{Na^+}\right) \tag{3.14}$$

$$\mu_{Cl^-} = \mu^o_{Cl^-} + RT \ln\left(C_{Cl^-}\right) \tag{3.15}$$

The standard potential for an aqueous solution of NaCl ($\mu^o_{NaCl(aq)}$) is defined as:

$$\mu^o_{NaCl(aq)} = \mu^o_{Na^+} + \mu^o_{Cl^-} \tag{3.16}$$

and the total chemical potential is:

$$\mu_{NaCl(aq)} = \mu^o_{NaCl(aq)} + RT \ln\left(C_{Na^+} C_{Cl^-}\right) \tag{3.17}$$

If the concentration of the salt is defined as C_{NaCl}, and realizing that $C_{NaCl} = C_{Na+} = C_{Cl-}$, then:

$$\mu_{NaCl(aq)} = \mu^0_{NaCl(aq)} + RT \ln\left(C_{NaCl}\right)^2 \tag{3.18}$$

3.3.6   Activities and Activity Coefficients: Non-ideal Solutions

For non-ideal solutions, the concept of activity and activity coefficient is introduced. A non-ideal solution of NaCl would have a chemical potential of

$$\mu_{NaCl(aq)} = \mu^o_{NaCl(aq)} + RT \ln\left(a_{Cl^-} a_{Na^+}\right) \tag{3.19}$$

or

$$\mu_{NaCl(aq)} = \mu^o_{NaCl(aq)} + RT \ln\left(m_{Cl^-} \gamma_{Cl^-} m_{Na^+} \gamma_{Na^+}\right) \tag{3.20}$$

in which a_i is the activity of component i, and γ_i is the activity coefficient of component i. The product $a_{Na+} a_{Cl-}$ is designated by a_{NaCl}, which is the activity of the aqueous solute, and its mean activity is $a_{\pm} = (a_{Na+} a_{Cl-})^{1/2}$, which can be determined experimentally. Similarly, the mean activity coefficient, $\gamma_{\pm}$ is equal to $(\gamma_{Na+} \gamma_{Cl-})^{1/2}$ and $\gamma_{Na+} \gamma_{Cl-}$ can be determined experimentally by measuring water vapor pressures over NaCl solutions of varying concentrations.

3.3.7   Equilibrium Constants

The Gibbs free energy of a reaction can be related to the system composition through the expressions for chemical potential

$$\mu_i = \mu_i^o + RT\ln(a_i)$$ [3.21]

and ΔG

$$\Delta G = \sum_i v_i \mu_i$$ [3.22]

in which v_i is the stoichiometric coefficient of a component i in the reaction. Combining these equations gives:

$$\Delta G = \sum_i v_i \mu_i^o + RT\sum_i v_i \ln(a_i)$$ [3.23]

or

$$\Delta G = \Delta G^O + RT\ln\prod_i (a_i)^{v_i}$$ [3.24]

ΔG^o is the standard Gibbs free energy change of the reaction, and Π_i is the quotient of concentrations of products over reactants. Consider the reaction,

$$aA + bB \rightleftharpoons cC + dD$$ [3.25]

in which a and b are the stoichiometric coefficients of reactants A and B, and c and d are the stoichiometric coefficients of products C and D. The expression for Π_i would be

$$\left[\frac{a_C^c a_D^d}{a_A^a a_B^b}\right]$$ [3.26]

The expression often is given the symbol Q. Thus,

$$\Delta G = \Delta G^O + RT\ln Q$$ [3.27]

At equilibrium, $\Delta G = 0$, $Q = K$ (the equilibrium constant), and

$$\Delta G^O = -RT\ln K$$ [3.28]

Both K and Q are written in terms of activities (a_i), but activities of individual species are not measurable. However, both can be written in terms of concentrations by using the relationship between activity and concentration: $a_i = C_i\gamma_i$:

$$K = \left[\frac{a_C^c a_D^d}{a_A^a a_B^b}\right] = \left[\frac{(C_C\gamma_C)^c (C_D\gamma_D)^d}{(C_A\gamma_A)^a (C_B\gamma_B)^b}\right]$$ [3.29]

3.3.8 Effects of Temperature and Pressure

The general expression for the effect of temperature on the free energy of reaction is

$$\left(\frac{\partial\left(\Delta G^{o}/T\right)}{\partial T}\right)_{P}=-\frac{\Delta H^{o}}{T^{2}} \tag{3.30}$$

and the van't Hoff equation for the corresponding equilibrium constant is

$$\frac{d\ln K}{dT}=-\frac{\Delta H^{o}}{RT^{2}} \tag{3.31}$$

The temperature dependence of enthalpy of reaction is

$$H_{2}-H_{1}=\int_{T_{1}}^{T_{2}}C_{P}dT \tag{3.32}$$

If ΔH^{o} is independent of temperature,

$$\ln\frac{K_{2}}{K_{1}}=\frac{\Delta H^{o}}{R}\left(\frac{1}{T_{1}}-\frac{1}{T_{2}}\right) \tag{3.33}$$

or when the heat capacity (ΔC_{p}^{o}) is independent of temperature,

$$\ln\frac{K_{2}}{K_{1}}=\frac{\Delta H^{o}}{R}\left(\frac{1}{T_{1}}-\frac{1}{T_{2}}\right)+\frac{\Delta C_{P}^{o}}{R}\left(\frac{T_{1}}{T_{2}}-1-\ln\frac{T_{1}}{T_{2}}\right) \tag{3.34}$$

or

$$\ln K=B-\left(\frac{\Delta H^{o}}{RT}\right)+\frac{\Delta C_{p}^{o}}{R}\ln T \tag{3.35}$$

where B and ΔH^{o} are constants (Stumm and Morgan, 1996). When ΔC_{p}^{o} is a function of temperature, the form of the final equation will reflect the temperature-dependent expression for the heat capacity.

The effects of pressure on free energy and equilibrium constants are handled in a fashion similar to temperature (Stumm and Morgan, 1996). The general expression, when ΔV^{o} is independent of pressure, is

$$\ln\frac{K_{P}}{K_{1}}=-\frac{\Delta V^{o}(P-1)}{RT} \tag{3.36}$$

With specific reference to aqueous solutions,

$$\mu_i = \mu_i^o + RT \ln \gamma_i C_i \qquad [3.37]$$

$$\left(\frac{\partial \ln K}{\partial P}\right)_{T,C} = -\frac{\Delta V^O}{RT} \qquad [3.38]$$

$$\left(\frac{\partial \ln K}{\partial P}\right)_{T,C} = \frac{\overline{V}_i - \overline{V}_i^{o}}{RT} \qquad [3.39]$$

where $\overline{V}$ is the partial molar volume and $\overline{V}_i^{o}$ is the standard partial molar volume of species i.

3.3.9   Single Ion Activity Coefficients

The mean activity coefficient of a salt in solution, $\gamma_\pm$, is measurable by experimental methods. However, the activity coefficient of a single ion, such as γ_{Na+} or γ_{CL}, is not measurable and must be estimated by theoretical models. Such a model was provided by the Debye-Hückel theory, which combined thermodynamic and electrostatic expressions to describe the interaction between charged species in solution. In its first configuration, the theory was based upon the assumption that the ions act as point charges. The resulting equation, the Debye-Hückel Limiting Law, was

$$\log \gamma_i = -AZ_i^2 I^{0.5} \qquad [3.40]$$

in which Z_i is the charge of the ion, I is the ionic strength of the solution ($I = 0.5 \, \Sigma C_i Z_i^2$), and A is related to the dielectric constant for water and has a value of 0.509 at 298.15 K and 101.33 kPa. Activity coefficients calculated from this equation begin to deviate from measurements when $I > 0.005$. The Debye-Hückel theory was extended to greater ionic strengths ($I = 0.1$ mol L^{-1}) by adding terms that account for the spatial interaction of the ions:

$$\log \gamma_i = \frac{-AZ_i^2 I^{0.5}}{1 + \beta a_i^o I^{0.5}} \qquad [3.41]$$

in which β is a constant characteristic of the solvent, equal to 0.328×10^8 at 298.15 K and 101.33 kPa, and a_i^o is a parameter of effective diameter of the ion ranging from (2.5 to 9) x 10^{-8} for some individual ions in aqueous solutions and must be obtained from a compilation of values (Kielland, 1937). The Davies equation is a further modification of the Debye-Hückel and is applicable when $I < 0.5$ mol L^{-1}:

$$\log \gamma_i = -AZ_i^2 \left(\frac{I^{0.5}}{1 + I^{0.5}} - 0.2I \right) \qquad [3.42]$$

The advantages of the Davies over other equations are the applicability to higher ionic strengths and the elimination of using β values.

3.3.10 Electrochemical Potential

The standard electrochemical potential of an oxidation/reduction reaction, E^o, is the potential of the reaction relative to the oxidation of $H_2(g)$ to H^+ in aqueous solution. The balanced chemical reaction of such a cell involving the reduction of Cu^{2+} to the metal would be

$$Cu^{2+}(aq) + H_2(g) \rightleftharpoons 2H^+ + Cu(s) \tag{3.43}$$

The half-cell reactions are written conveniently as follows, with the implicit understanding that the hydrogen half-cell is always present

$$Cu^{2+}(aq) + 2e^- \rightleftharpoons Cu(s) \tag{3.44}$$

For any given reaction, the Nernst equation can be derived:

$$E_H = E_H^o - \frac{2.303RT}{nF} \log \frac{\Pi_i \{ox\}^{ni}}{\Pi_i \{red\}^{nj}} \tag{3.45}$$

in which E_H is the measured potential, E_H^o is the standard potential of the cell, R is the gas constant, T is temperature in degree K, n is the number of electrons involved in the reaction, F is the Faraday constant, and Π designates the product of either the reactants or products in the equation.

For any oxidation/reduction reaction, the relationship between the electrode potential and pe, negative logarithm of the electron activity, $-\log(e^-)$, can be derived

$$pe = \frac{F}{2.303RT} E \tag{3.46}$$

or $pe = E_H/59.2$ when determined at 298.15 K and where E_H is in mV. As with H^+, aqueous solutions do not contain free electrons, and the concentration of solvated electrons is vanishingly small. Nevertheless, pe is a very convenient parameter in manipulating equilibrium equations and plotting data.

3.3.11 Heat of Hydration of Ions

One of the most important reactions for ions in solution is hydration, i.e., the electrostatic interaction between the polar water molecules and the charged ion. When an ion is released into aqueous solution, heat is released as water molecules form a somewhat ordered structure around the ion. The water that surrounds the ion tends to insulate the charged species from other ions in solution. In infinitely dilute solutions, this solvation effect completely isolates the ions from interacting with each other.

The heat released during solvation of an ion by water is the heat of hydration, $\Delta H_{hydration}$. The strength of the water-ion interaction increases with increasing valence because higher charged ions have the capacity to react with more water molecules. Within a group of ions of the same valence, $\Delta H_{hydration}$ decreases (becomes more negative) linearly with decreasing crystallographic radius (Fig. 3.4). Thus, a small monovalent ion such as Li^+ releases more heat upon hydration than the much larger Cs^+ ion. A direct result of this is that Li^+ also has a greater hydrated radius than Cs^+ which is partly responsible for some of the differences in strength of retention of these ions by cation exchange sites in soil (Section B, Chapter 7).

3.4 Interactions of Gases with the Soil Solution

Chemical reactions between gases and the liquid aqueous phase are important not only in the soil solution but in biological systems, surface water, groundwater, and the atmosphere. For example, acid rain is a long recognized problem that results from the combustion of fossil fuels. The oxidation of C, S, and N in the fuels generates several gaseous oxides including CO_2, NO_2, NO, SO_2, and SO_3. When these oxides dissolve in water, they generate acids: H_2CO_3, HNO_3, HNO_2, H_2SO_4, and H_2SO_3. The extent to which the gaseous oxides dissolve in the water can be described by Henry's Law (Section 3.3.4) (Table 3.4).

3.4.1 Henry's Law

This law describes the equilibrium distribution of a species between the gas phase and the aqueous phase. The expression is based upon the thermodynamic parameter, chemical potential, and requires the determination of a partitioning coefficient for each gas. The thermodynamic expression is

$$f_A = Ka_A \qquad\qquad [3.47]$$

in which K is a constant, f_A is the fugacity of the gas, and a_A is the activity of the species in the aqueous phase. The transition between the above thermodynamic equation and a usable expression with

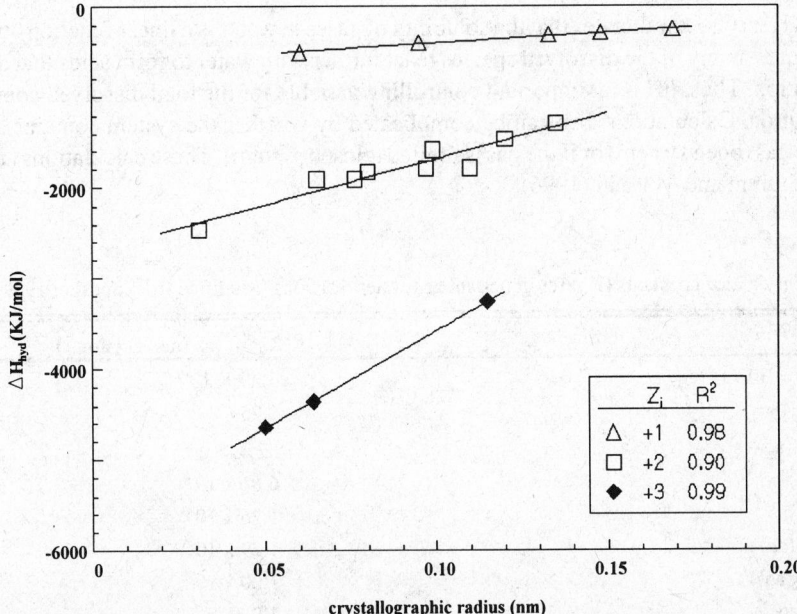

Fig. 3.4 Relationship between crystallographic radius and heat of hydration of monovalent, divalent, and trivalent ions. The R^2 values correspond to linear regression [Data from Bohn et al., 1985].

measurable terms is made simpler if one first assumes the condition of dilute solutions. Then, Henry's Law can be written in one of two ways, either of which is correct. In the first expression, the partitioning coefficient (H) is dimensionless:

$$\frac{\left[A(aq)\right]}{\left[A(g)\right]} = H \tag{3.48}$$

The units of concentration for the gaseous and aqueous species must be the same (e.g., mol L^{-1}). In the second form of Henry's Law, the partitioning coefficient is not dimensionless:

$$\frac{\left[A(aq)\right]}{P_A} = K_H \tag{3.49}$$

If the units of concentration for the aqueous component are mol L^{-1}, and the partial pressure is in atmospheres, then K_H must have units of mol L^{-1} atm^{-1}. The conversion between H and K_H is

$$K_H = \frac{H}{RT} \tag{3.50}$$

Table 3.5 contains a compilation of K_H values for important gases in soils and other settings. Of particular note in this table is that the values generally range from 10^{-2} to 10^{-4} except for $HNO_3(g)$, $SO_2(g)$, $NH_3(g)$, and $H_2S(g)$.

Although Henry's Law dictates that the solubility of gases in water is a linear function only of their partial pressures, many of the dissolved species react further with water to form acids that are subject to deprotonation. Thus, pH is an important controlling variable for the total dissolved component in aqueous solution. Calculations also can be complicated by whether the system contains an infinite source of the gas (open system) or if the gas is limited (closed system). These calculations are handled in detail by Stumm and Morgan (1996).

Table 3.5 Henry's Law constants (K_H) for important gas-water reactions [Modified from Sposito, 1989]

Reaction	$K_{H, 298.15 K}$ (mol L^{-1} atm^{-1})
$CO_2 + H_2O(l) = H_2CO_3{}^*(aq)$	3.39×10^{-2}
$SO_2(g) + H_2O(l) = SO_2 \cdot H_2O(aq)$	1.25
$NH_3(g) = NH_3(aq)$	57
$N_2(g) = N_2(aq)$	6.61×10^{-4}
$O_2(g) = O_2(aq)$	1.26×10^{-3}
$CH_4(g) = CH_4(aq)$	1.29×10^{-3}
$NO_2(g) = NO_2(aq)$	1.00×10^{-2}
$NO(g) = NO(aq)$	1.9×10^{-3}
$N_2O(g) = N_2O(aq)$	2.57×10^{-2}
$HNO_3(g) = H^+(aq) + NO_3{}^-(aq)$	3.46×10^6
$H_2S(g) = H_2S(aq)$	1.05×10^{-1}

3.4.2 Volatile Organic Compounds

Many organic compounds are subject to loss from solid, liquid, or aqueous phases through volatilization. As with inorganic gases, the tendency for volatile organic compounds to partition between the aqueous phase and the atmosphere can be described by Henry's Law. Constants for Henry's Law can be calculated by measuring aqueous and gaseous phase concentrations in equilibrium systems. The K_H values in Table 3.6 cover a range similar to that for the inorganic gases (Table 3.5).

3.4.3 Rates of Dissolution of Gases in Water

Quantification of equilibrium distributions of volatile compounds between the atmosphere and aqueous phase is more powerful when accompanied by knowledge of rates of reactions. Transfer of a gas across the interface between the liquid and gas can be approximated by a diffusion model, assuming that the bulk phases are well mixed up to two diffusion films, one in the liquid phase and one in the gas phase (Stumm and Morgan, 1996). The diffusion films are assumed to be within a very small distance of the interface.

The flux across the films of thickness z is expressed by Fick's Law:

$$F = -D\frac{dc}{dz} \qquad [3.51]$$

The units for flux, F, will depend upon the units of concentration (c), diffusion coefficient (D), and thickness; assuming concentrations in mol m^{-3}, D in m^2 s^{-1}, and distance in m, then F will have units of mol m^{-2} s^{-1}. A steady state will have been achieved when the flux across the two films is equal. The thickness of the air diffusion film is assumed to be z_{air}, the thickness of the water diffusion film is z_{water}, the concentrations in the air and water are c_{air} and c_{water}; the concentration in the air-water interface is c_{a-w}; and, the concentration in the water-air interface is c_{w-a}. Therefore, at steady state

$$F = -\frac{D_w}{z_w}\left(c_{w-a} - c_w\right) = -\frac{D_a}{z_a}\left(c_a - c_{a-w}\right) \qquad [3.52]$$

Table 3.6 Water solubilities, vapor pressures, and Henry's Law constants for selected organic compounds [Modified from Stumm and Morgan, 1996]

Compound	Water solubility (mol L^{-1})	Vapor pressure, P_A (atm)	K_H (mol L^{-1} atm^{-1})
Hexane	7.0×10^{-4}	0.25	2.8×10^{-3}
n-Octane	5.8×10^{-6}	1.8×10^{-2}	3.1×10^{-4}
Dieldrin	5.8×10^{-7}	6.6×10^{-9}	88
Lindane	2.6×10^{-5}	8.3×10^{-8}	313
Naphthalene	2.6×10^{-4}	1.0×10^{-4}	2.6
Benzene	2.3×10^{-2}	0.12	0.19
Toluene	5.6×10^{-3}	3.7×10^{-2}	0.15
Biphenyl	4.9×10^{-5}	7.5×10^{-5}	0.65
Dimethyl sulfide	0.35	0.63	0.56

Assuming that the transfer across the interfaces is much faster than any ensuing chemical reactions that may occur, Henry's Law can be applied to the concentrations in the interface regions (using the dimensionless constant from Section 3.4.1)

$$H = \frac{C_{w-a}}{C_{a-w}} = K_H RT \qquad\qquad [3.53]$$

Substituting into the steady-state equation,

$$F = \frac{D_w}{z_w}\left(c_w - c_{w-a}\right) = \frac{D_a}{z_a}\left(\frac{c_{w-a}}{H} - c_a\right) \qquad\qquad [3.54]$$

Using data from Lovelock et al. (1976) in Stumm and Morgan (1996), one can calculate that the flux of freon across the water-gas interface in marine environments is 1.5×10^{-9} g cm^{-2} yr^{-1}. Extended derivations and similar calculations can be made for other chemical systems after making allowances for chemical reactions. For example, the steady-state flux of $CO_2(g)$ from a lake into the atmosphere is 6×10^{-9} mol cm^{-2} s^{-1} at pH 6.7, total alkalinity of 3×10^{-3} mol L^{-1}, and 298.15 K.

3.5   Acid/Base Reactions in the Soil Solution

Acid and base reactions are the most fundamental and often the most important in soil solutions. The weathering of primary minerals often generates alkaline conditions, and many natural and anthropogenic activities can generate acidic conditions. The rates and extents of many chemical reactions are dependent upon soil solution pH including biological activity, mineral dissolution, partitioning of some gases, and bioavailability of critical nutrient and pollutant elements. This section will address some of the important acid/base concepts and discuss the numerical handling of related equilibria.

3.5.1   Fundamentals of Acid/Base Chemistry in Soil Solution

The central component of acid/base reactions is the proton or hydrogen ion, H$^+$. This ion actually does not exist as H$^+$ in aqueous solutions but is hydrated to form H_3O^+, $H_7O_3^+$, $H_9O_4^+$, etc. For the sake of simplicity in representing the equilibria involving the proton, the symbol H$^+$ will be used.

According to the Brønsted-Lowry concept, an acid is any substance that donates a proton to another substance. Similarly, a Brønsted-Lowry base is any substance that accepts a proton from another substance. According to these definitions, hundreds of reactions can occur in soil solutions that are examples of acids and bases. For example, the bicarbonate ion dissociates readily to form carbonate and H$^+$

$$HCO_3^- \rightleftharpoons CO_3^{2-} + H^+ \qquad\qquad [3.55]$$

In this case, bicarbonate is acting as an acid. Bicarbonate also can act as a base, accepting a proton to form carbonic acid

$$HCO_3^- + H^+ \rightleftharpoons H_2CO_3^\circ \qquad\qquad [3.56]$$

Although the discussions of Brønsted-Lowry theory for acids and bases presented here will be limited to aqueous systems, the theory is applicable to all solvents in which protons can be exchanged, including ammonia, sulfuric acid, and ethanol.

Metal ions in aqueous solutions are solvated readily and, thus, exist as hydrates rather than bare ions. The number of water molecules surrounding a cation in the first hydration layer will be dependent upon the ionic radius of the cation and, to a lesser extent, its charge. The water in the hydration sphere tends to act as a weak acid, donating a proton to the solution and (in essence) contributing a hydroxyl ion to the metal.

$$\left[Fe(H_2O)_6\right]^{3+} \rightleftharpoons \left[Fe(H_2O)_5 OH\right]^{2+} + H^+ \qquad [3.57]$$

The acidity of the water associated with the hydrolysis reaction increases with increasing valence and decreasing ionic radius of the central cation.

Another acid/base concept was formulated by G.N. Lewis (Lewis and Randall, 1923). A Lewis acid accepts a pair of electrons from a Lewis base. All Brønsted-Lowry acids and bases are also Lewis acids and bases, but the Lewis definition encompasses more reactions. In the neutralization reaction of H^+ with OH^- to give water, a lone pair of electrons on the hydroxyl is donated to hydrogen; thus, this reaction fits both acid/base definitions. When an orthophosphate ion reacts with Fe(III) in the structure of goethite, the Fe(III) accepts an electron pair from the oxygen on the phosphate group to form the bond. This fits the Lewis definition of an acid/base reaction but is clearly not a Brønsted-Lowry acid/base pair.

3.5.2 Calculations for Acid/Base Equilibria

Graphical representations of acid/base equilibria can take several forms, but the underlying calculations are built on a single theoretical foundation. Whether the system is simple or complex, the approach is the same, although organizing and executing the computations can be challenging for large systems. The steps include identifying the participating components, assembling the pertinent equations, identifying the master variables, and solving the equations in terms of the master variables. A simple system (aqueous solution of carbonate) will be used as an example. More complex systems can be handled in the same fashion and will be developed later in this chapter.

In this system, the species in solution would be OH^-, H^+, $H_2CO_3^\circ$, HCO_3^-, and CO_3^{2-} with $CO_2(g)$ in the gas phase. The defining reactions and equilibrium constants are given in Table 3.7. The equilibrium constants (K°) are given at zero ionic strength, 298.15 K, and 101.33 kPa (1 atm) and are taken from Lindsay (1979). Defining the equilibrium constants at zero ionic strength assumes that all species are given in terms of activities. In Fig. 3.5, activities of the three inorganic carbon solution species in equilibrium with 0.0003 atm $CO_2(g)$ are plotted as functions of pH. The predominant solution species below pH 6.33 is $H_2CO_3^\circ$ at an activity of $10^{-4.98}$. Between pH 6.33 and 10.36, the dominant carbonate species is HCO_3^-, and CO_3^{2-} is present at the greatest activities above pH 10.36. However, the total carbonate in solution approaches molar quantities near pH 10 in equilibrium with $CO_2(g)$ at this pressure, concentrations rarely seen in natural solutions. Thus, the kinetics of $CO_2(g)$ dissolution probably predominate over equilibrium predictions at these high pH values in open systems.

Information similar to that in Fig. 3.5 can be provided without the overlying assumption of an open system or predicting equilibrium activities. Figure 3.6 is a mole fraction distribution diagram depicting the relative concentrations of the three carbonate species as a function of pH. The mole fraction is

Table 3.7 Controlling equations for equilibria involving $CO_2(g)$ and $H_2O(l)$ [Lindsay, 1979]

Equation	$\log K^o$
$CO_2(g) + H_2O(l) \rightleftharpoons H_2CO_3{}^o(aq)$	−1.46
$H_2CO_3{}^0 \rightleftharpoons H^+ + HCO_3^-$	−6.36
$H_2CO_3^- \rightleftharpoons H^+ + HCO_3^{2-}$	−10.33

defined as the ratio of the concentrations of a given species to the sum of the concentrations of all the carbonate species.

The equilibrium expressions for these species in Table 3.4 are substituted appropriately and solved across pH (Lindsay, 1979, p. 81). The resulting diagram (assuming that $\gamma_i = 1$) is applicable to either open or closed systems and requires equilibrium only among the solution species not between the aqueous and gaseous phases. As in Fig. 3.5, Fig. 3.6 illustrates that bicarbonate ion is dominant between pH 6.36 and 10.33, carbonic acid is dominant below pH 6.36, and carbonate above pH 10.33.

3.5.3 Buffering Capacity of Soil Solutions

Because of the many acids and bases present in the soil and dissolved in the soil solution, the acid/base chemistry of soils is highly complicated. One of the results of this complex system is that the pH of soils and, to a smaller extent, soil solutions does not change greatly in response to inputs of acid or base. The ability of a system to resist pH changes is called the buffering capacity and can be determined experimentally by titrating the solution and measuring the pH at each increment of acid or base addition. A simple, qualitative illustration of the buffering power of carbonate in water is illustrated in Fig. 3.7, in which aqueous solutions are titrated theoretically with base from pH 7 to 11. In the first

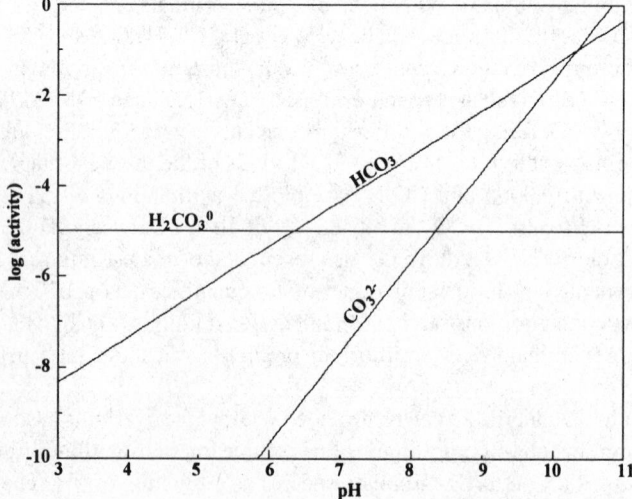

Fig. 3.5 Activities of carbonate species are affected by pH assuming equilibrium with $P_{CO2} = 0.0003$ atm.

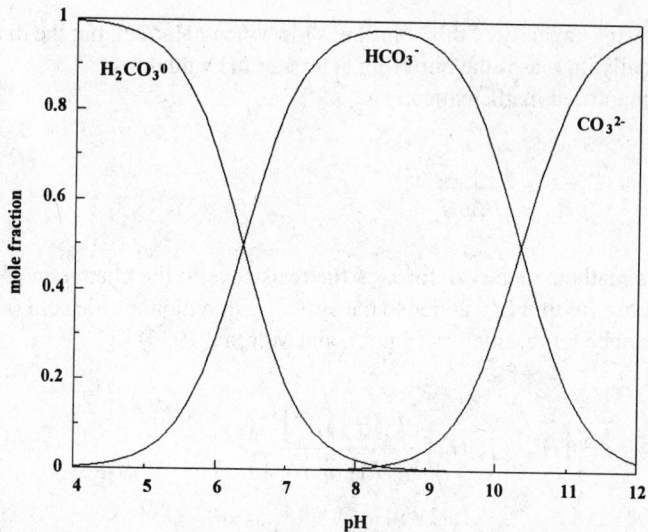

Fig. 3.6 Mole fraction distribution of carbonate species as a function of pH. For all calculations, activity coefficients are assumed to be unity ($\gamma_i = 1.0$).

case, only pure water is titrated, and the solution shows no capacity to resist pH change; with each increment of base, a large pH change is noted. In the second case, a solution containing 0.001 M total carbonate is adjusted initially to pH 7 and titrated with NaOH to pH 11. In this theoretical titration, activity coefficients are ignored ($\gamma_i = 1$), and the volume is assumed to remain constant. The carbonate solution has a significant resistance to pH change (buffering capacity) compared to the pure solution. The third case in Fig. 3.7 is an open system in equilibrium with atmospheric $CO_2(g)$ assuming 0.0003

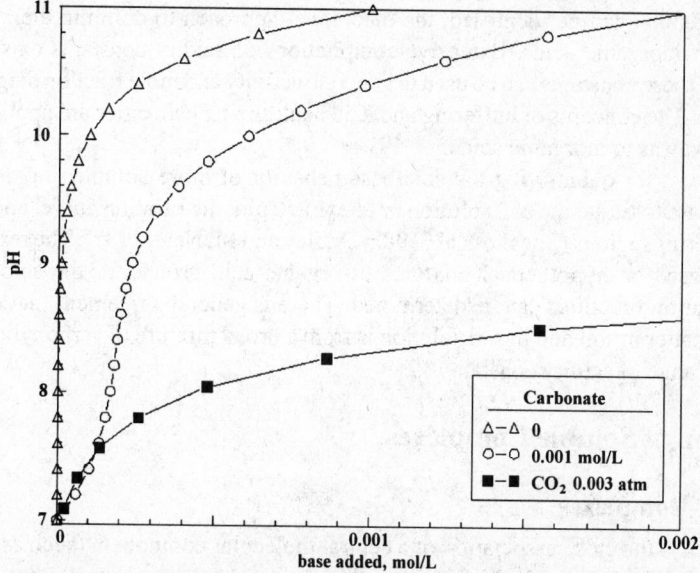

Fig. 3.7 Theoretical titration curve of an aqueous solution containing pure water (Δ), 0.001M total carbonate in a closed system (o), or an open system in equilibrium with atmospheric $CO_2(g)$ at 30 Pa[0.0003 atm] (■)

atm (30 Pa). The buffering capacity of this solution is low when pH < 7.5, but the dissolution of CO_2 into the solution radically increases the buffering at greater pH values.

Buffering can be quantified mathematically as

$$buffering = \frac{\Delta C_{base}}{\Delta pH} = -\frac{\Delta C_{acid}}{\Delta pH}$$ [3.58]

Thus, the buffering is mathematically defined as the resistance to the change in pH induced by an increment of acid or base (in mol L^{-1}) added to the system. For a monoprotic acid (HA) dissolved in water, the buffering can be represented as (Stumm and Morgan, 1996)

$$buffering = 2.3\left([H^+]+[OH^-]+\frac{[HA][A^-]}{[HA]+[HA^-]}\right)$$ [3.59]

This equation defines the buffering of the system at any point in the titration of the monoprotic acid, HA. Similar expressions can be derived for mixtures of acids or for polyprotic acids.

3.5.4 Soluble Organic Acids

The chemistry of soluble organic acids in soil solutions has not been studied as much as that of inorganic acids, although extensive reviews exist (Stevenson, 1967). The relative lack of information is not because of the lack of importance of soluble organic acids in soils but is a reflection of the difficulty in quantitatively characterizing the organic components in solution and the dynamic nature of soil organic molecules. Many simple organic acids have been identified in solution ranging from formic and acetic acids to more complex aromatic acids such as catechin (McKeague et al., 1986). The aliphatic carboxylic acids are degraded very rapidly, but the aromatic acids tend to be more persistent.

For organic acids that can be identified, the theoretical approach to defining their chemistry is identical to that for inorganic acids. Extensive compilations of acidity constants exist (Smith and Martell, 1976), and these constants can be used to generate activity and mole fraction diagrams similar to Figs. 3.5 and 3.6. The concepts of buffering and acid neutralizing capacities are applied to organic acids in the same way as to inorganic acids.

Methods also exist for quantifying the acid/base behavior of more complex organic mixtures. Direct approaches are to titrate the soil solution or to extract, plot the titration curve, and identify the characteristic buffering regions (Sposito et al., 1982; Dudley and McNeal, 1987). The resulting curves can be modeled based on hypothetical mixtures of organic acid groups, or the acid neutralizing capacity and formation functions can be determined. There is general agreement that the acidity of complex organic matter in soil and in soil solution is from a broad mixture of carboxylic and benzoic acids with a wide range of acidity constants.

3.6 Formation of Soluble Complexes

3.6.1 Types of Complexes

A solution complex is the close association of a central molecular component (such as a cation) and other atoms or molecules. Very often, metals or other positively charged species act as the central components that attract neutral or negatively charged ligands. In the hydrolysis of Fe^{3+} to form

$Fe(H_2O)_5OH^{2+}$, the Fe^{3+} cation acts as the central unit, and water and hydroxyl act as the ligands. This is a special case of complex formation called a solvation complex. Two other categories of complexes can be formed depending upon the strength of bonding between the central unit and the ligand. If the interaction between the cation and ligand is strong enough for the ligand to displace the solvation sphere, the resulting association is termed an inner sphere complex. Typical examples of inner sphere complexes are AlF_n^{3-n} and Fe(III)citrate. If the attraction between the cation and ligand is not strong enough to displace the hydration layer, then it is termed an outer sphere complex or an ion pair. An inner sphere complex can form if the heat evolved during the association exceeds the energy needed to displace the hydration sphere. If the heat evolved is less than the energy necessary to displace the water, then the hydration sheath remains intact, and an outer sphere complex results.

A ligand that occupies more than one coordination site in the complex is referred to as multidentate. For example, two of the oxygens in oxalate can bond simultaneously with a central Fe(III) ion. Such a complex is called a metal chelate, and the process is chelation. The multidentate nature of the interaction can significantly increase the strength of the bond, and the resulting formation constant is much higher than those of ordinary complexes.

3.6.2 Hard and Soft Acid-Base Rules

Other useful concepts in predicting the associations between cations and ligands are the hard and soft acid-base rules (Pearson, 1963). Hard acids and bases are those species that tend to have smaller radii and are not readily deformable (Table 3.8). Soft acids and bases are often larger and more polarizable (have a more easily deformed electron sheath). Hard acids tend to bond preferentially with hard bases, and soft acids with soft bases. Thus, Ca^{2+} would tend to bond with phosphate and carbonate rather than chloride and sulfide.

3.6.3 Rates of Formation of Solution Complexes

The rates of formation of solution complexes often are rapid, establishing equilibrium very quickly. However, certain reactions proceed very slowly. For example, the reaction in Equation [3.60]

$$Al^{3+} + F^- \rightleftharpoons AlF^{2+} \tag{3.60}$$

requires nearly 20 minutes to proceed half way to completion. This is in contrast to the formation of $MnSO_4^o$, which requires 10^{-9} s.

Hydrolysis reactions generally are quite rapid for monovalent and divalent cations but can proceed slowly for the higher charged cations. The rate of these reactions can be approximated by examining the rate of water exchange from a hydrated cation, $Me(H_2O)_x^{y+}$ (Stumm and Morgan, 1996).

Table 3.8 Hard and soft acids and bases [Pearson, 1963]

	Hard	Intermediate	Soft
Bases	F^-, CO_3^{2-}, OH^-, CH_3COO^-, PO_4^{3-}, SO_4^{2-}, NH_3, $R\text{-}NH_2$, H_2O, R-OH, NO_3^-	SO_3^{2-}, NO_2^-, $C_6H_5NH_2$	S^{2-}, CN^-, Cl^-, I^-, R-SH
Acids	H^+, Li^+, Na^+, K^+, Mg^{2+}, Ca^{2+}, Sr^{2+}, Al^{3+}, La^{3+}, Si^{4+}, Zr^{4+}, Th^{4+}, Cr^{3+}, Mn^{3+}, Fe^{3+}	Mn^{2+}, Fe^{2+}, Ni^{2+}, Cu^{2+}, Zn^{2+}, Pb^{2+}, Bi^{3+}	Ag^+, Cu^+, Cd^{2+}, Hg^{2+}, Sn^{2+}

$$Me(H_2O)_x^{y+} + H_2O^* \rightleftharpoons Me(H_2O)_x^{y+} \cdot H_2O^* \qquad [3.61]$$

$$Me(H_2O)_x^{y+} + H_2O^* \rightleftharpoons Me(H_2O)_{x-1}(H_2O^*)^{y+} + H_2O \qquad [3.62]$$

in which H_2O^* is the water being exchanged into the solvation complex. The rate of this reaction is

$$\frac{d\left[Me(H_2O)_{x-1}(H_2O^*)^{y+}\right]}{dt} = k_{-w}\left[Me(H_2O)_x^{y+} \cdot H_2O^*\right] \qquad [3.63]$$

After further manipulation and assuming $k_{-1} \gg k_{-w}$ and steady-state conditions

$$\frac{d\left[Me(H_2O)_{x-1}(H_2O^*)^{y+}\right]}{dt} = k_{-w}K_{OS}\left[Me(H_2O)_x^{y+}\right]\left[H_2O^*\right] \qquad [3.64]$$

where K_{os} is the equilibrium constant for outer sphere complex formation. Thus, the rate constant for the exchange of water in the hydration shell is estimated by $k_{-w}K_{os}$.

Stumm and Morgan (1996) compiled values of k_{-w} for water exchange reactions for several metals. For Pb^{2+}, Hg^{2+}, Cu^{2+}, Ca^{2+}, Cd^{2+}, La^{3+}, Zn^{2+}, Mn^{2+}, Co^{2+}, $Fe(OH)_2^+$, and $Fe(OH)_4^-$, $k_{-w} \geq 10^6\ s^{-1}$. Values of k_{-w} for other metals were 2×10^2 for Fe^{3+}, 1.0 for Al^{3+}, and $5 \times 10^{-7}\ s^{-1}$ for Cr^{3+}. Rate constants for ligands other than water can be treated the same as those for water by merely substituting the new ligand for the H_2O^*. Also, sequential reactions can be "coupled" and solved. Exact, analytical solutions to these models can be obtained (Sposito, 1994), but experimental observations typically are used to help establish simplifying assumptions.

The formation of the solution complex $CdCl^+$ in the presence of $10^{-3}\ mol\ L^{-1}$ Cl can be used to illustrate the application of these equations.

$$Cd(H_2O)_x^{2+} + Cl^- \rightleftharpoons Cd(H_2O)_{x-1}Cl^+ + H_2O \qquad [3.65]$$

The rate of reaction can be given by

$$\frac{d\left[Cd(H_2O)_{x-1}Cl^+\right]}{dt} = K_{OS}k_{-w}\left[Cd(H_2O)_x^{2+}\right]\left[Cl^-\right] = -\frac{d\left[Cd(H_2O)_x^{2+}\right]}{dt} \qquad [3.66]$$

where $k_{-w} = 2 \times 10^6\ (s^{-1})$, $K_{OS} = 10^{1.98}\ (mol\ L^{-1})^{-1}$, and $(k_{-w})(K_{OS}) = 1.91 \times 10^8\ L\ mol^{-1}s^{-1}$. Substituting these values into the above equation,

$$-\frac{d\left[Cd(H_2O)_x^{2+}\right]}{dt} = (1.91 \times 10^8)(10^{-3})\left[Cd(H_2O)_x^{2+}\right] = 1.91 \times 10^5\left[Cd(H_2O)_x^{2+}\right] \qquad [3.67]$$

For a first order reaction such as this, the half-life of the reaction (that is, the time required in seconds for the reaction to proceed halfway to completion) is equal to $(\ln 2)/k$. Thus, the half-life for this reaction is $0.693/1.91 \times 10^5$ or 3.6×10^{-6} s. Sposito (1989) tabulated the half-lives for several solution reactions and found the values ranging from 10^{-9} s for the formation of $MnSO_4^\circ$ to 10^3 s for the formation of AlF^{2+}.

The relationships presented in this section demonstrate that most reactions in the soil solution are rapid enough to assume that equilibrium can be achieved during the course of of a typical experiment. Even the slowest example given above, the formation of AlF^{2+} in 10^3 s, will be nearly complete in less than 1 h. However, the rate of certain oxidation/reduction reactions can be very slow, particularly reactions that involve oxyanions (such as arsenate or chromate) or are not microbially catalyzed. These reactions can take days to reach completion, if at all. Therefore, knowledge of equilibrium predictions alone can be of limited utility if the kinetics of reaction are unfavorable.

3.7 Application of Thermodynamic and Equilibrium Concepts to Soil Solutions

The chemical composition and dynamics of the soil solution are reflections of all the processes that depend upon the aqueous phase: biological activity, mineral dissolution/precipitation, adsorption/desorption, physical transport, and anthropogenic inputs. Wolt (1994) described the soil solution as "a window to chemically reacting soil systems where the intensity and distribution of chemicals in the soil aqueous phase represent the integration of multiple physical, chemical, and biological processes occurring concurrently within the soil environment." The information provided by the chemical composition is enhanced further by knowledge of time trends and the application of kinetic and thermodynamic theories.

3.7.1 Determination of Speciation and Single Ion Activity

The determination of free ion activities (or concentrations) for soil solution components provides a powerful interpretative tool for chemical reactions in soils. A limited number of ion activities can be determined directly (such as the H^+ activity using a glass electrode), but the rest must be approximated through calculations.

The steps in calculating ion activities and the distribution of solution species include obtaining and analyzing the soil solution, identifying the important solution species that can form, assembling relevant equilibrium constants and reactions for the formation of these species, obtaining an analytical or numerical solution for the series of equations, and calculating activity coefficients. All the steps in the calculations can be handled by any of the many geochemical models currently available (discussed later); however, understanding the chemical concepts behind the models and how the activities are determined will be helpful in properly executing the computer models and obtaining the best possible data. The following sections provide a brief review of the important chemical reactions and activity calculations.

3.7.1.1 Hydrolysis, Complexation, and Oxidation/Reduction Reactions

To help illustrate the steps involved in activity calculations, soluble Fe will be used as an example. The first step is to determine the important reactions in solution, and this can be done with the aid of some of the compilations of thermodynamic data for aqueous systems, such as Garrels and Christ (1965), Lindsay (1979), Sadiq and Lindsay (1979), or Smith and Martell (1976). Many of the metals undergo hydrolysis reactions, and Fe(III) and Fe(II) have several important hydrolysis species as indicated in Table 3.9. The distribution of these species as a function of pH (Fig. 3.8) shows that the predominant species will be strongly dependent upon pH with mole fractions ranging from 0.0 to over 0.9. Examination of Table 3.9 indicates that the respective complexes listed must be considered if $(Cl^-) >$

10^{-2}, $(F^-) > 10^{-7}$, $(SO_4^{2-}) > 10^{-5}$. The constants in Table 3.9 and their corresponding equations can be used to solve for ionic activities.

3.7.1.2 Solution Composition Example

Consider the calculation of Fe^{2+} and Fe^{3+} activities in a soil solution with the following composition: 0.2 µmol L^{-1} total Fe, 20 mmol L^{-1} total Cl, 10 µmol L^{-1} total F, 1 mmol L^{-1} total SO_4, ionic strength 0.01, pH 6.5, and a redox potential of 600 mV.

The solution of this problem is fairly complex and requires a few assumptions to make it solvable with the present information. The first assumption is that the ligands will not form complexes with any cations in solution other than Fe^{2+} and Fe^{3+}. This is a poor assumption, because each ligand forms a number of complexes with other cations, particularly Al^{3+}, Ca^{2+}, and Mg^{2+}. The second assumption is that the only species of Fe to be considered are those found in Table 3.9.

The approach given here is to identify all potential variables, establish the same number of independent equations as the number of variables, and solve the equations simultaneously. The system variables are the activities of H^+, OH^-, e^-, Fe^{3+}, Fe^{2+}, Cl^-, SO_4^{2-}, F^-, $FeOH^{2+}$, $Fe(OH)_2^+$, $Fe(OH)_3^\circ$, $Fe(OH)_4^-$, $FeOH^+$, $Fe(OH)_2^\circ$, $FeCl^{2+}$, FeF^{2+}, $FeSO_4^+$, and $FeSO_4^\circ$, for a total of 18 variables. The relevant equations include 11 complexation equations from Table 3.9 (all equations except for the formation of $FeHPO_4^+$), K_w (water dissociation equation), pH = 6.5, Eh = 600 mV (pe = 10.1), and the four mass balance equations for the components:

$$\Sigma Fe = \left[Fe^{3+}\right] + \left[Fe^{2+}\right] + \left[FeOH^{2+}\right] + \left[Fe(OH)_2^+\right] + \left[Fe(OH)_3^\circ\right] + \left[Fe(OH)_4^-\right] +$$
$$\left[Fe(OH)^+\right] + \left[Fe(OH)_2^\circ\right] + \left[FeCl^{2+}\right] + \left[FeF^{2+}\right] + \left[FeSO_4^+\right] + \left[FeSO_2^\circ\right] \tag{3.68}$$

$$\Sigma SO_4^{2-} = \left[SO_4^{2-}\right] + \left[FeSO_4^+\right] + \left[FeSO_4^\circ\right] \tag{3.69}$$

Table 3.9 Reactions and equilibrium constants for selected solution complexes of iron

Reaction	log K^o
Fe(III) Hydrolysis	
$Fe^{3+} + H_2O = FeOH^{2+} + H^+$	−2.19
$Fe^{3+} + 2H_2O = Fe(OH)_2^+ + 2H^+$	−5.69
$Fe^{3+} + 3H_2O = Fe(OH)_3^\circ + 3H^+$	−13.09
$Fe^{3+} + 4H_2O = Fe(OH)_4^- + 4H^+$	−21.59
Fe(II) Hydrolysis	
$Fe^{2+} + H_2O = FeOH^+ + H^+$	−6.74
$Fe^{2+} + 2H_2O = Fe(OH)_2^\circ + 2H^+$	−16.04
Redox	
$Fe^{3+} + e^- = Fe^{2+}$	13.04
Complexation	
$Fe^{3+} + Cl^- = FeCl^{2+}$	1.48
$Fe^{3+} + F^- = FeF^{2+}$	6.00
$Fe^{3+} + SO_4^{2-} = FeSO_4^+$	4.15
$Fe^{3+} + H_2PO_4^- = FeHPO_4^+ + H^+$	3.71
$Fe^{2+} + SO_4^{2-} = FeSO_4^\circ$	2.20

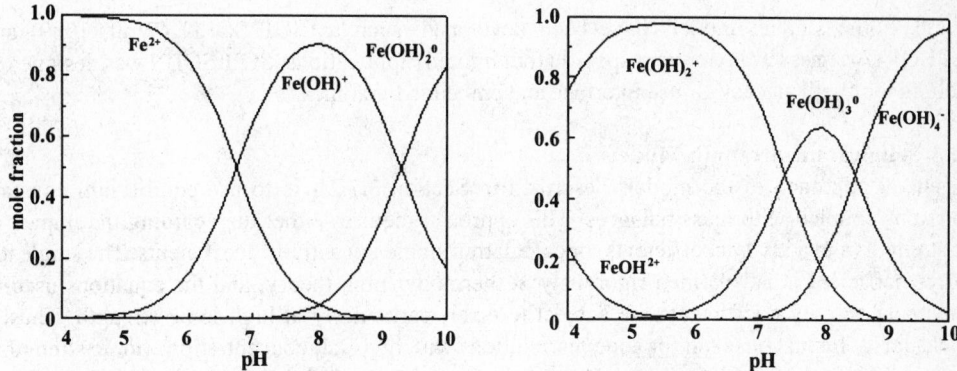

Fig. 3.8 Distribution of Fe (III) and Fe(II) hydrolysis species in solution as a function of pH

$$\sum Cl = \left[Cl^-\right] + \left[FeCl^{2+}\right] \tag{3.70}$$

$$\sum F = \left[F^-\right] + \left[FeF^{2+}\right] \tag{3.71}$$

In the mass balance expressions, brackets [x] represent concentrations of species x. Activity coefficients must be calculated for each species to tie together the mass balance equations with the formation equations. Thus, there are a total of 18 equations for the 18 variables. Many methods exist for solving this system of equations including back substitution and using a matrix. These systems can be solved by hand, but solutions are reached rapidly using computers.

3.7.2 Geochemical Models

Over the past 2 or 3 decades, many scientists and research groups have recognized the need for computer programs to solve chemical equilibrium systems, and several computer models have been established. Models vary widely in their construction and composition as well as their intended use (Baham, 1984; Melchior and Bassett, 1990). Most of the models are based upon single ion activities, single ion activity coefficients, and formation constants, Pitzer's equations for applications at high ionic strength, or a free energy minimization approach.

3.7.2.1 Formation Constant Approach

In Section 3.7.1.2, an example was given outlining the formation constant approach to determine single ion activities in soil solutions. Several series of published models take this same approach. The WATEQ series (Truesdall and Jones, 1974; Ball et al., 1979) was developed by the U.S. Geological Survey to compute single ion activities and predict the fate of critical elements in geochemical environments. Emphasis was placed on obtaining a well-documented thermodynamic database with the best available equilibrium constants; a reaction would be kept out of the database if the constant was estimated or suspected to be of poor quality. The mathematical solution method was back substitution and required recompilation of the programming code if new reactions were added or if constants were changed. The MINEQL series (Westall et al., 1976) was developed using the mathematical approach of REDEQL (Morel and Morgan, 1972) to rapidly solve equilibrium problems and provide information about trends in observed data. The mathematical solution method is rapid and efficient, but the thermodynamic database was not reviewed rigorously. The model SOILCHEM (Sposito and Coves, 1988) was an extension of the MINEQL model with emphasis on expanding the database. Some smaller models have been developed, such as SOILSOLN (Wolt, 1989) and CALPHOS (Adams, 1971), for very specific (but limited) applications. SOILSOLN was designed as a teaching tool with an easy-to-use interface and embedded graphics.

3.7.2.2   High Ionic Strength Models

The general approach in the models described in Section 3.7.2.1 is to use equilibrium constant expressions coupled with mass balances. This approach requires either disregarding the impact of ionic strength (and activity coefficients) or calculating single ion activity coefficients. The single ion activity coefficient is not defined rigorously in thermodynamic theory, and the equations used to approximate activity coefficients are subject to error, particularly at high ionic strength. Thus, a special approach must be taken for aqueous solutions with high salt concentrations (ionic strength > 0.5). One such approach is to use the Pitzer equations (Pitzer, 1979) for calculation of single ion activity coefficients, and at least two published models currently use this approach. Felmy and Weare (1995) discussed their model and experimental results for simple systems with high ionic strengths. The agreement between theory and experimental determinations was promising. The C-SALT model (Smith et al., 1995) also uses the Pitzer equations. The model has been tested against experimental data.

The Pitzer equations require a relatively large number of parameters that are specific to the ions and systems under consideration. The number of required parameters increases rapidly as the systems become more complex. The Pitzer, Davies, and Debye-Hückel equations all calculate single ion activity coefficients. Unfortunately, these activity coefficients cannot be defined unambiguously, because they are dependent upon a clear, thermodynamic definition of single ion concentrations. As stated by Sposito (1994) when discussing activity coefficients for a metal ion, γ_M, and a ligand, γ_L: "For γ_M and γ_L to have chemical significance, the species molalities, m_M and m_L, must have a well-defined operational meaning. Thus, the single ion activity coefficient has no meaning apart from the set of operational procedures used to define ionic species and to determine their concentrations in an aqueous solution." Although the determination of single ion activities can provide useful information, single ion activity coefficients cannot be determined experimentally and do not have a thermodynamic foundation.

3.7.2.3   Solving Example Problem with MINTEQ

The geochemical model, MINTEQ (Allison et al., 1990), was developed by merging the mathematics and computer code from MINEQL with the database from the WATEQ series. Although the MINTEQ database is not as extensive as SOILCHEM's, all values are documented fully as to their source and reason for selection.

The hypothetical analytical data for the example problem in Section 3.7.1.2 was placed in a standard MINTEQ input file, and the program executed. The resulting activities (Table 3.10) show that the hydrolysis species of Fe(III) and Fe(II) are much more important than the chloro, fluoro, and sulfate species. This is because these ligands react only with the unhydrolyzed cations, and, in the case of Fe(III), the activity of Fe^{3+} is very low. In the case of Fe(II), the ligand concentrations simply are not high enough for the respective complexes to become the predominant solution species.

Comparisons of the Fe(III) and Fe(II) species also are worth noting. In Table 3.9, the constant for the equilibrium between Fe^{3+} and Fe^{2+} suggests that Fe^{2+} will be present at higher activities than Fe^{3+} when the pe < 13 (770mV). In this case, (Fe^{2+}) is nearly 1000 times greater than (Fe^{3+}), but the Fe(III) hydrolysis species are by far the predominant Fe solution species. At pH 6.5, the oxidation/reduction potential would have to be pe < 5.5 (325 mV) before the Fe(II) species would be present at a greater activity than the predominant Fe(III) species.

3.7.3   Oxidation/Reduction Reactions

Chemical reactions involving oxidation and reduction are important in all biological systems, and soils are no exception. Localized areas can be found in all soils that are highly oxidized or highly reduced. Areas of reduction generally are associated with small microsites that are saturated with water and have restricted $O_2(g)$ diffusion. Microbial activity within these sites rapidly depletes the available O_2, and the redox potential of the soil solution decreases.

The first step in handling oxidation/reduction equilibria in soil solution is defining the limits of stability of aqueous systems. If a solution is subject to a high oxidizing potential, then the water spontaneously can be degraded into oxygen:

$$H^+ + e^- + 0.25 O_2(g) \rightleftharpoons 0.5 H_2 O(l) \qquad [3.72]$$

with an equilibrium constant expression:

$$K^o = \frac{(H_2 O)^{0.5}}{(H^+)(e^-)(O_2(g))^{0.25}} = 10^{20.78} \qquad [3.73]$$

Substituting, rearranging, and solving for *pe* in terms of pH gives

$$pe = 20.78 + 0.25 \log(O_2) - pH \qquad [3.74]$$

The pe at which water spontaneously decomposes to oxygen gas would correspond to $O_2(g)$ pressure of 1 atm, or *pe* = 20.78 - pH. Using the combined parameter, *pe*+pH, the upper oxidizing limit of the stability field of water would be

$$pe + pH = 20.78 \qquad [3.75]$$

A strongly reducing potential can generate H_2 gas

Table 3.10 Activities of selected aqueous species calculated using MINTEQ for the example solution composition given in Section 3.7.1.2

Species	Log (activity)	Species	Log (activity)
Fe^{3+}	−14.10	Fe^{2+}	−11.20
$Fe(OH)^{2+}$	−9.79	$FeOH^+$	−11.44
$Fe(OH)_2^+$	−6.79	$Fe(OH)_2^o$	−14.24
$Fe(OH)_3^o$	−7.69	$FeSO_4^o$	−12.19
$Fe(OH)_4^-$	−9.69	F^-	−5.05
FeF^{2+}	−13.15	SO_4^{2-}	−3.18
FeF_2^+	−14.99	Cl^{--}	−1.74
FeF_3^o	−17.54		
$FeSO_4^+$	−13.13		
$FeCl^{2+}$	4.37		

$$H^+ + e^- \leftrightarrow 0.5H_2(g) \tag{3.76}$$

Again solving for pe,

$$pe = -0.5 \log\left(H_2(g)\right) - pH \tag{3.77}$$

At 1 atm $H_2(g)$, this reduces to $pe = -pH$. Converting to the combined parameter, $pe + pH$ gives

$$pe + pH = 0 \tag{3.78}$$

In a plot of pe versus pH, the region between the lines $pe = -pH$ and $pe = 20.78 - pH$ represents the stability field for water. Water is thermodynamically unstable at potentials outside this region. Another important aspect of these equations is the equilibrium partial pressures of $H_2(g)$ and $O_2(g)$ at typical redox potentials in soil solutions. For example, highly reduced soils seldom have redox potentials below $pe + pH = 4$, corresponding to $P_{H2} = 10^{-8}$ atm. Measured redox potentials in fully oxidized soils seldom exceed $pe + pH = 18$, and equilibrium at this potential would require $10^{-11.23}$ atm $O_2(g)$. Assuming that equilibrium is established rapidly with these gases, then the partial pressures of H_2 and O_2 are expected to be quite small in all situations.

This same approach can be used to investigate many redox reactions in soil solutions. The Fe(III)/Fe(II) system was discussed in a prior section, and many other redox active species exist in soil solutions including Cu^{2+}, Mn^{2+}, and CrO_4^{2-}. Chromium is an environmentally important metal that has an interesting redox behavior. The Cr(III) species is considered environmentally less hazardous than the Cr(VI), which is closely regulated. Published equilibrium constants will be used to determine the redox potential at pH 7 at which the predominant Cr(III) species $[Cr(OH)_2^+]$ converts to the predominant Cr(VI) species $[CrO_4^{2-}]$. The following reaction and constant were obtained from the MINTEQ database:

$$CrO_4^{2-} + 6H^+ + 3e^- \rightleftharpoons Cr(OH)_2^+ + 2H_2O \qquad \log K^o = 67.38 \tag{3.79}$$

The expression for the equilibrium constant, converted to logarithms, would be

$$\log\left[\frac{\left(Cr(OH)_2^+\right)}{CrO_4^{2-}}\right] + 6pH + 3pe = 67.38 \tag{3.80}$$

The redox potential at which Cr(III) converts to Cr(VI) occurs when the predominant species of each oxidation state of Cr has equal activity, or $(Cr(OH)_2^+) = (CrO_4^{2-})$. The ratio of these activities would equal unity, the logarithm of which is zero. Therefore, after dividing by three,

$$2pH + pe = 22.46 \tag{3.81}$$

or, after converting to $pe + pH$,

$$pe + pH = 22.46 - pH \tag{3.82}$$

and, at pH = 7,

$$pe + pH = 15.46 \tag{3.83}$$

This is a moderately reducing condition for soil solutions, roughly the same redox potential at which denitrification occurs (Lindsay, 1979).

3.7.4 Successful Applications of Geochemical Modeling to Soil Solutions

Application of equilibrium thermodynamics to soil solutions will have several requirements for success, and the definition of success will depend upon individual interpretations. The kinetics of some reactions in soil solutions were discussed in Section 3.6.3; most reactions were found to proceed quickly, but others are quite slow. Therefore, care must be taken to apply equilibrium thermodynamics only to systems that have had long enough equilibration periods. The method of sample collection can dictate whether or not enough time is available for equilibrium. For example, miscible displacement methods often have equilibration periods of less than 24 hours, which is not long enough for some solid phase equilibria or redox reactions involving oxyanions to be completed.

The success of applying a model to a chemical system will increase as more information is available. For example, many models do not contain adequate data for the formation of polymeric aluminum species, Al^{3+}-phosphate solution complexes, or a means to approximate the impact of soluble organic matter on metal speciation. In many cases, so little is known about these phenomena that the magnitudes of their impacts cannot be approximated to any degree of certainty.

Most applications of geochemical models to natural systems have not been validated. Fortunately, some published examples exist in which models have been validated by laboratory tests followed by extending the model to similar systems in natural settings. Felmy and Weare (1995) incorporated Pitzer's equations into an ion interaction model, the purpose of which was to allow chemical speciation in high ionic strength solutions. The Davies equation for activity coefficients was found to inadequately predict $\gamma_\pm$ for many salts at ionic strengths as low as 0.2 for 1:1 electrolytes and 0.05 for 2:1 electrolytes (Fig. 3.9). The parameters for the ion interaction model were obtained experimentally and applied to $Na_2B_4O_7$-Na_2SO_4-H_2O and $Na_2B_4O_7$-Na_2CO_3-H_2O systems. Concentrations of the Na salt were adjusted from 0 to 3 molal, and $Na_2B_4O_7$ molalities were determined experimentally. The agreement between modeling predictions and experimental results was excellent ($\pm10\%$) (Fig. 3.10), even with very high borate concentrations and ionic strengths as high as 14.

The solubility of manganese phosphates in soils was hypothesized by Lindsay (1979) as a possible activity controlling mechanism for both PO_4^{3-} and Mn^{3+}. Boyle and Lindsay (1985) pursued this by synthesizing various Mn-phosphates and measuring their solubilities in the laboratory. The trivalent Mn solid, $MnPO_4 \cdot H_2O$, was found to be the least soluble. In a follow-up study, they measured the ion activity products (IAPs) for this compound in several alkaline soils (Boyle and Lindsay, 1986) and found good agreement between the IAP and the previously determined solubility product. Schwab (1989) measured the IAPs in acid soils and also found excellent agreement with the solubility product, particularly for high P soils that had not been fertilized for 20 yr (Fig. 3.11). In soil that had been fertilized recently, the IAPs were consistently higher by one order of magnitude, indicating the presence of a poorly crystalline solid.

3.7.5 Limitations to Applying Geochemical Models to Soil Solutions

Four major limitations exist to applying equilibrium geochemical models to soil solutions: the dynamic (nonequilibrium) nature of soils, poorly defined equilibria, limitations in the analytical

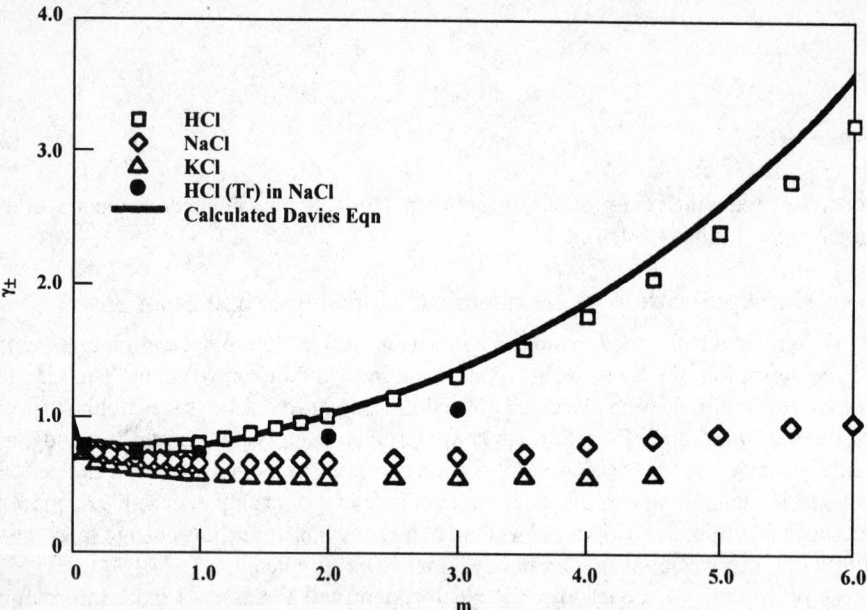

Fig. 3.9 Changes in experimentally determined mean activity coefficients ($\gamma\pm$) for various electrolytes compared to those generated by the Davies equation. Measured values are symbols, predicted values are lines [Reprinted from Felmy and Weare, 1995. Soil Sci. Soc. Am. Spec. Pub. 42 with permission of the Soil Science Society of America].

chemistry, and lack of knowledge of how to accurately handle soluble organic compounds. The ever-changing nature of soils leads to nonequilibrium conditions. Although equilibrium within the soil solution can be attained quickly, solid and gas phases often establish equilibria at a slow rate, if at all. Unfortunately, these systems have a very strong impact on the composition of the soil solution. Nonequilibrium can be overcome only by the passage of time. The degree of nonequilibrium and the rate at which a steady state is being approached can be discerned by sampling the soil solution over time and monitoring the progress of the chemical reactions. Suarez (1995) discussed the merits of a nonequilibrium approach to modeling the carbonate system.

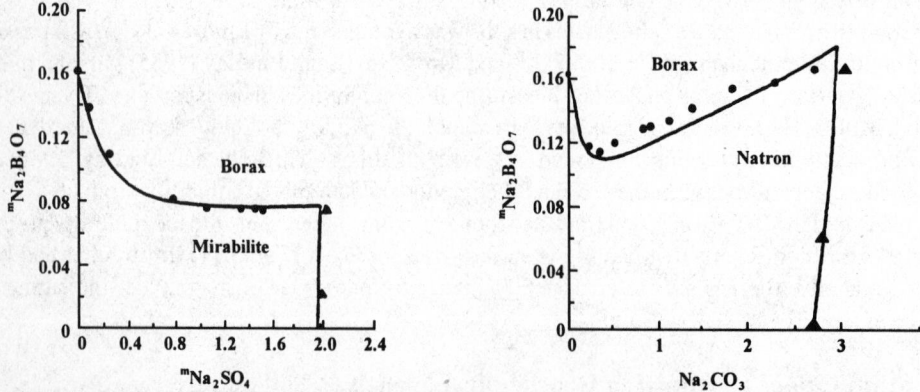

Fig. 3.10 Comparison of ion interaction predictions (lines) to experimentally determined measurements for $Na_2B_4O_7$-Na_2SO_4-H_2O and $Na_2B_4O_7$-Na_2CO_3-H_2O systems [Felmy and Weare, 1995. Reprinted with permission from Loeppert et al. (ed.) Soil Sci. Soc. Am. Spec. Pub. 42]

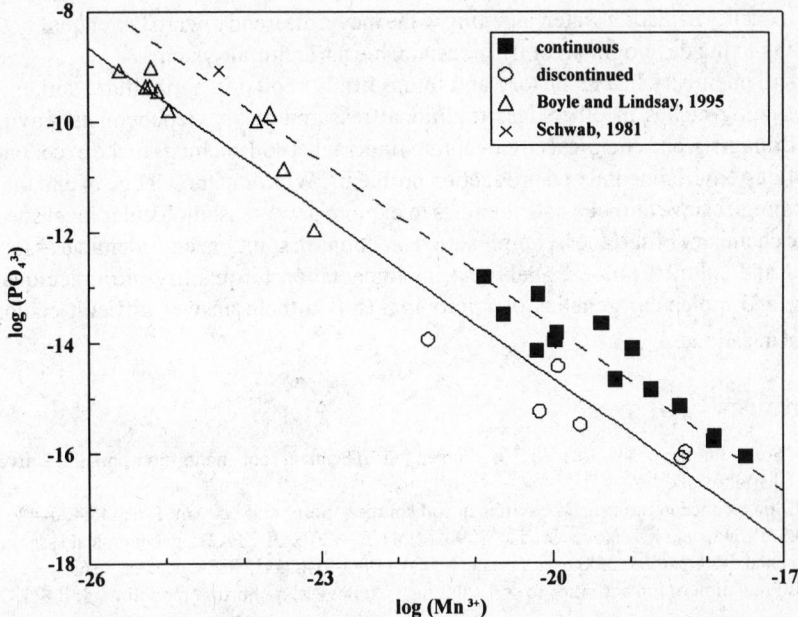

Fig. 3.11 Maganese phosphate solubilities for several soils. The solid line represents the solubility of $MnPO_4$ 1.5 H_2O as detemined by Boyle and Lindsay (1985), and the dashed line represents an amorphous analog. "Continuous" refers to annual application of P fertilizer since 1946; "discontinued" refers to the plots in which P applications ceased in 1965.

The problem of poorly defined equilibria arises when information about the system is lacking or the database accompanying the geochemical model does not contain all the equilibrium constants to fully characterize important reactions. If many data are lacking, then a speciation approach may not be possible. However, estimating experimental parameters or equilibrium constants may be an acceptable alternative. Limitations in analytical chemistry (quantification limits, capability to analyze certain species) must be dealt with in the same way.

As discussed in previous sections, soluble organic compounds are among the most important components in the soil solution, and we have the least information about them. These compounds can be very complex and contain a wide range of functional groups and molecular chemical environments, making approximation of complexation constants difficult. Thus, a decision must be made to either ignore the effects of soluble organics completely because their impact cannot be quantified exactly or to apply estimates of the constants even though serious errors can be generated.

3.8   Current Status and Future Research Directions

This review of the chemistry of the soil solution has been given a distinct theoretical flavor and has not focused on some of the excellent, recent advances in other aspects of the science. Topics not covered include colloidal chemistry (both inorganic and organic), spectroscopic approaches to identifying important complexes, salinity, xenobiotics, and heavy metals in the soil solution. Fortunately, most of these topics are covered in other chapters in Section B.

Study of the soil solution has its origins in agriculture, and a great portion of soil chemical research still is agriculturally oriented and will be for years to come. Many challenges still exist in which soil chemists will play a pivotal role, including managing soil acidity, growing crops in arid regions with limited or poor quality irrigation water, adapting to the inevitible trend toward lower quality fertilizers, and contributing to the development of truly sustainable agricultural systems.

Although soil chemistry has its history and future firmly rooted in agriculture, soil chemists are increasingly finding niches in other, less traditional research areas. Geochemists, environmental engineers, hydrologists, and chemical engineers are finding that soil chemists make excellent research partners in solving environmental and production problems. Working knowledge of and the ability to use new technologies have allowed soil chemists to explore new areas: molecular level spectroscopy to quantify the chemistry of surfaces, complexation reactions in solution, and chemical associations in both the solid and solution phases, analytical instrumentation to quantify ultratrace quantities of contaminants, and molecular genetics of microorganisms to help answer difficult ecological and environmental questions.

3.9   References

Adams, F. 1974. Soil Solution. p. 441–481. *In* E.W. Carson (ed.) The plant root and its environment. University Press of Virginia, Charlottesville, VA.

Adams, F. 1971. Ionic concentrations and activities in soil solution. Soil Sci. Soc. Am. Proc. 35:420–426.

Allison, J.D., D.S. Brown, and K.J. Novo-Gradac. 1990. MINTEQA2/PRODEFA2, a geochemical assessment model for environmental systems: Ver. 3.00 user's manual. EPA-600/3-91-021. USEPA, Athens, GA.

Baham, J. 1984. Prediction of ion activities in soil solutions: Computer equilibrium modeling. Soil Sci. Soc. Am. J. 48:525–531.

Ball, J.W., E.A. Jenne, and D.K. Nordstrom. 1979. WATEQ - a computerized chemical model for trace and major element speciation and mineral equilibria of natural waters. *In* E.A. Jenne (ed.) Chemical modeling in aqueous systems. Am. Chem Soc. Symp. Series N. 93. American Chemical Society, Washington, DC.

Bartlett, R., and B. James. 1980. Studying dried, stored soil samples: Some pitfalls. Soil Sci. Soc. Am. J. 44:721–724.

Bohn, H., B. McNeal, and G. O'Connor. 1985. Soil chemistry. Wiley Interscience, New York, NY.

Boyle, F.W., and W.L. Lindsay. 1985. Preparation, X-ray diffraction pattern, and solubility product of manganese(III) phosphate hydrate. Soil Sci. Soc. Am. J. 49:758–768.

Boyle, F.W., and W.L. Lindsay. 1986. Manganese phosphate equilibrium relationships in soils. Soil Sci. Soc. Am. J. 50:588–593.

Cole, D.W., S.P. Gessel, and E.E. Held. 1961. Tension lysimeter studies of ion and moisture movement in glacial till and coral atoll soils. Soil Sci. Soc. Am. Proc. 25:321–325.

Dudley, L.M., and B.L. McNeal. 1987. A model for electrostatic interaction among charged sites of water-soluble organic polyions: I. Description and sensitivity. Soil Sci. 143:329–340.

Elkhatib, E.A., O.L. Bennett, V.C. Baligar, and R.J. Wright. 1986. A centrifugation method for obtaining soil solution using an immiscible liquid. Soil Sci. Soc. Am. J. 50:297–299.

Felmy, A.R., and J.H. Weare. 1995. The development and application of aqueous thermodynamic models: the specific ion-interaction approach. *In* R. Loeppert et al. (ed.) Chemical equilibrium and reaction models. Soil Sci. Soc. Am. Spec. Pub. 42. Soil Science Society of America, Madison, WI.

Garrels, R.M., and C.L. Christ. 1965. Solutions, minerals, and equilibria. Freeman, Cooper, and Company, San Francisco, CA.

Kielland, J. 1937. Individual activity coefficients of ions in aqueous solutions. J. Am. Chem. Soc. 59:1675–1678.

Kittrick, J.A. 1977. Mineral equilibria and the soil system. p. 1–25. *In* J.B. Dixon and S.B. Weed (ed.) Minerals in soil environments. Soil Science Society of America, Madison, WI.

Lewis, G.N., and M. Randall. 1923. Thermodynamics and the free energy of chemical substances. McGraw-Hill, New York, NY.

Lindsay, W.L. 1979. Chemical equilibria in soils. Wiley Interscience, New York, NY.

McBride, M.B. 1994. Environmental chemistry of soils. Oxford University Press, New York, NY.

McKeague, J.A., M.V. Cheshire, F. Andreux, and J. Berthelin. 1986. Organo-mineral complexes in relation to pedogensis. p. 549–592. *In* P.M. Huang and M. Schnitzer (ed.) Interactions of soil minerals with natural organics and microbes. Soil Science Society of America, Madison, WI.

Melchior, D.C., and R.L. Bassett. 1990. Chemical modeling of aqueous systems II. ACS Symp. Ser. 416. American Chemical Society, Washington, DC.

Morel, F.M.M., and J.J. Morgan. 1972. A numerical method for computing equilibrium in aqueous chemical systems. Environ. Sci. Technol. 6:58–67.

Morrison, R.D., and B. Lowry. 1990. Effect of cup properties, sampler geometry and vacuum on the sampling rate of porous cup samplers. Soil Sci. 149:308–316.

Mubarek, A., and R.A. Olsen. 1976. Immiscible displacement of the soil solution by centrifugation. Soil Sci. Soc. Am. J. 40:329–341.

Pearson, R.G. 1963. Hard and soft acids and bases. J. Amer. Chem. Soc. 85:3533.

Pitzer, K.S. 1979. Theory: ion interaction approach. *In* R. M. Pytkowicz (ed.) Activity coefficients in electrolyte solutions. CRC Press, Boca Raton, FL.

Richards, L.A. 1941. A pressure-membrane extraction apparatus for soil solution. Soil Sci. 51:377–386.

Ross, D.S., and R.J. Bartlett. 1990. Effects of extraction methods and sample storage on properties of solutions obtained from forested Spodosols. J. Environ. Qual. 19:108–113.

Sadiq, M., and W.L. Lindsay. 1979. Selection of standard free energies of formation for use in soil chemistry. CO State Univ. Exp. Sta. Tech. Bull. 134.

Schwab, A.P. 1992. Chemical and physical characterization of soils. Conf. Proc. Hazardous Substance Res. Center pp. 326–344.

Schwab, A.P. 1989. Manganese-phosphate soubility relationships in an acid soil. Soil Sci. Soc. Am. J. 53:1654–1660.

Smith, G.R., K.K. Tanji, R.G. Burau, and J.J. Jurinak. 1995. C-Salt, a chemical equilibrium model for multicomponent solutions. *In* R. Loeppert et al. (ed.) Chemical equilibrium and reaction models. Soil Sci. Soc. Am. Spec. Pub. 42, Soil Science Society of America, Madison, WI.

Smith, R.M., and A.E. Martell. 1976. Critical stability constants. Plenum Press, New York, NY.

Sposito, G. 1994. Chemical equilibria and kinetics in soils. Oxford University Press, New York, NY.

Sposito, G. 1989. The chemistry of soils. Oxford University Press, New York, NY.

Sposito, G. 1981. The thermodynamics of soil solutions. Oxford University Press, New York, NY.

Sposito, G., and J. Coves. 1988. SOILCHEM: a computer program for the calculation of chemical equilibria in soil solutions and other natural water systems. Kearney Foundation of Soil Science, University of California, Riverside, CA.

Sposito, G., K.M. Holtzclaw, C.S. LeVesque, and C.T. Johnston. 1982. Trace metal chemistry in arid-zone field soils amended with sewage sludge: II. Comparative study of fulvic acid fraction. Soil Sci. Soc. Am. J. 46:265–270.

Stevenson, F. J. 1967. Organic acids in soil. p.119–146. *In* A.D. McLaren and G.H. Peterson (ed.) Soil biochemistry. Wiley Interscience, New York, NY.

Stumm, W., and J.J. Morgan. 1996. Aquatic chemistry. Chemical equilibria and rates in natural waters. Wiley Interscience, New York, NY.

Suarez, D.L. 1995. Carbonate chemistry in computer programs and application to soil chemistry. *In* R. Loeppert et al. (ed.) Chemical equilibrium and reaction models. Soil Sci. Soc. Am. Spec. Pub. 42, Soil Science Society of America, Madison, WI.

Thompson, H.S. 1850. On the absorbent power of soils. Royal Agric. Soc. Eng. J. 11:68–74.

Till, A.R., and T.P. McCabe. 1976. Sulfur leaching and lysimeter characterization. Soil Sci. 121:44–47.

Truesdall, A.H., and B.F. Jones. 1974. WATEQ, a computer program for calculating chemical equilibria of natural waters. J. Res. U.S. Geol. Surv. 2:223.

US Salinity Laboratory Staff. 1954. Diagnosis and improvement of saline and alkali soils. USDA Handb. 60. US Government Printing Office, Washington, DC.

Warrick, A.W. and A. Amoozegar-Fard. 1977. Soil water regimes near porous cup water samplers. Water Resour. Res. 13:203–207.

Way, J.T. 1850. On the power of soils to absorb manure. Royal Agric. Soc. Eng. J. 11:313–379.

West, A.W., G.P. Sparling, C.W. Feltham, and J. Reynolds. 1992. Microbial activity and survival in soils dried at different rates. Aust. J. Soil Res. 30:209–222.

Westall, J.C., J.L. Zachary, and F.M.M. Morel. 1976. MINEQL - a computer program for the calculation of chemical equilibrium composition of aqueous systems. Tech. Note 18, Department of Civil Engineering, Massachusetts Institute Technology, Cambridge, MA.

Wolt, J.D. 1994. Soil solution chemistry. Applications to environmental science and agriculture. John Wiley and Sons, New York, NY.

Wolt, J.D. 1989. SOILSOLN: A program for teaching equilibria modeling of soil solution composition. J. Agron. Ed. 18:40–42.

Wolt, J.D., and J.G. Graveel. 1986. A rapid routine method for obtaining soil solution using vacuum displacement. Soil Sci. Soc. Am. J. 50:602–605.

Wood, W.G. 1973. A technique using porous cups for water sampling at any depth in the unsaturated zone. Water Resour. Res. 9:468–488.

4

Kinetics and Mechanisms of Soil Chemical Reactions

Donald L. Sparks
University of Delaware

4.1 Introduction

Since its inception in the mid 1850s, soil chemistry has focused on the macroscopic, equilibrium aspects of soil chemical reactions and processes. From these studies, much was learned about important soil chemical processes including sorption, desorption, precipitation, complexation, dissolution and oxidation/reduction. However, such investigations do not convey information on reaction rates or reaction mechanisms. In the past two decades, as concerns and interests about soil and water quality have increased, soil and environmental chemists, environmental and chemical engineers and geochemists have increasingly realized that reactions in subsurface environments are time dependent. Thus, to accurately predict the fate, mobility, speciation, and bioavailability of environmentally important plant nutrients, trace elements, radionuclides, and organic chemicals in soils, one must understand the kinetics and mechanisms of the reactions.

While major progress has been made in better understanding the kinetics of soil chemical processes, much uncertainty remains. In part, this is due to the complex, heterogeneous nature of natural materials such as soils. However, with development of kinetic techniques that can be used to measure a wide range of time scales, time-dependent models that can describe both chemical reaction and mass transfer processes, and the employment of state-of-the-art *in situ* spectroscopic and microscopic surface techniques in combination with rate studies, major advances are being made in understanding the kinetics and mechanisms of soil chemical reactions. Arguably, this will be a major leitmotif in soil chemistry research for decades to come.

In this review, the application of chemical kinetics to heterogeneous systems such as soils and soil components (clay minerals, organic matter, and humic substances), with emphasis on sorption/release processes will be discussed. A critical review of kinetic models that can be used to describe reaction rates on heterogeneous surfaces will be covered. The review will also present discussions on the rates of important soil chemical reactions and processes including inorganic and organic sorption/desorption, dissolution and redox. For additional details on these topics and other aspects of kinetics of soil chemical and geochemical processes, the reader should consult a number of recent books and monographs (Sparks, 1989; Sparks and Suarez, 1991; Stumm, 1992; Schwarzenbach et al., 1993; Sposito, 1994; Sparks, 1995).

4.2   Time Scales of Soil Chemical Processes

A variety of chemical reactions occur in soils and often in combination with one another. Reaction time scales can vary from microseconds and milliseconds for many ion association and some ion exchange and sorption reactions to years for many mineral solution and mineral crystallization phenomena and for some sorption/release reactions (Fig. 4.1). Ion association reactions include ion pairing, inner and outersphere complexation, and chelation in solution. Gas-water reactions involve gaseous exchange across the air-liquid interface. Ion exchange reactions occur when cations and anions are adsorbed (outersphere complexation) and desorbed from soil surfaces by electrostatic attractive forces. Ion exchange reactions are reversible and stoichiometric. Sorption reactions can involve adsorption processes including partitioning, outersphere and innersphere complexation, and multinuclear complexation (e.g., surface precipitation). Mineral-solution reactions include precipitation/dissolution of minerals, and coprecipitation reactions in which small constituents become a part of mineral structures (Sparks, 1989; Amacher, 1991).

The type of soil component can drastically affect the reaction rate. For example, sorption reactions are often more rapid on clay minerals such as kaolinite and smectites than on vermiculitic and micaceous minerals. This is in large part due to the availability of sites for sorption. For example, kaolinite has readily available planar external sites and smectites have primarily internal sites that are also quite available for retention of sorbates. Thus, sorption reactions on these soil constituents are often quite rapid, even occurring on time scales of seconds and milliseconds (Sparks, 1989).

On the other hand, vermiculite and micas have multiple sites for retention of metals and organics, including planar, edge, and interlayer sites, with some of the latter sites being partially to totally collapsed. Consequently, sorption and desorption reactions on these sites can be slow, tortuous, and mass transfer controlled. Often, an apparent equilibrium may not be reached even after several days or weeks. Thus, with vermiculite and mica, sorption can involve two to three different reaction rates: high rates on external sites, intermediate rates on edge sites, and low rates on interlayer sites (Jardine and Sparks, 1984; Comans and Hockley, 1992).

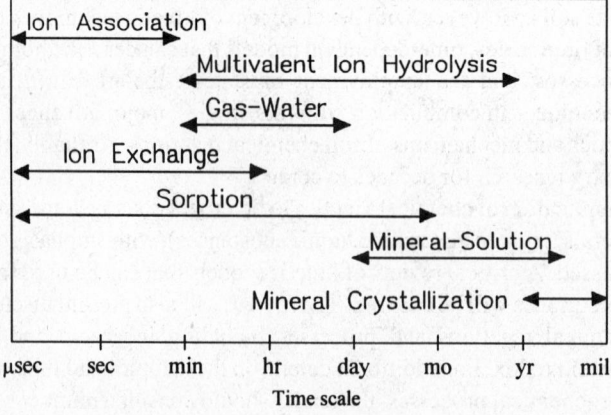

Fig. 4.1   Time ranges required to attain equilibrium by different types of reactions in soil environments [Reprinted from Amacher, 1991. Soil Sci. Soc. Am. Spec Pub. 27 with permission of the Soil Science Society of America]

Metal sorption reactions on oxides, hydroxides, and humic substances depend on the type of surface and metal being studied, but the chemical reaction rate appears to be rapid. For example, chemical reaction rates of metals and oxyanions on goethite occurred on millisecond time scales (Zhang and Sparks, 1989, 1990 a,b; Grossl et al., 1994; Grossl et al., 1997). Half-times for divalent Pb, Cu, and Zn sorption on peat ranged from 5 to 15 s (Bunzl et al., 1976). A number of studies have shown that heavy metal sorption on oxides (Bruemmer et al., 1988; Ainsworth et al., 1994; Scheidegger et al., 1996a, 1998) and clay minerals (Lövgren et al., 1990) increase with longer residence times. The mechanism for these lower reaction rates is not well understood, but has been ascribed to diffusion phenomena, sites of lower reactivity, and surface nucleation/precipitation (Scheidegger and Sparks, 1996b; Sparks, 1998, 1999). Recent findings on slow metal retention rates and mechanisms at the mineral/water interface will be discussed later.

Sorption/desorption of metals and organic chemicals on soils is often very slow which has been attributed to diffusion into micropores of inorganic minerals and into humic substances, retention on sites of varying reactivity, and surface nucleation/precipitation (Scheidegger and Sparks, 1996b; Sparks, 1997, 1999). These reactions will be discussed in more detail later.

4.3 Application of Chemical Kinetics to Heterogeneous Surfaces

The study of chemical kinetics, even in homogeneous systems, is complex and often arduous. When one attempts to study the kinetics of reactions in heterogeneous systems such as soils, sediments and even soil components such as clay minerals, hydrous oxides, and humic substances, the difficulties are greatly magnified. This is largely due to the complexity of soils which are made up of a mixture of inorganic and organic components. These components often interact with each other and display different types of sites with various reactivities for inorganic and organic sorptives. Moreover, the variety of particle sizes and porosities in soils and sediments further adds to their heterogeneity. In most cases, both chemical kinetics and multiple transport processes are occurring simultaneously. Thus, the determination of chemical kinetics, which can be defined as "the investigation of rates of chemical reactions and of the molecular processes by which reactions occur where transport is not limiting" (Gardiner, 1969) is extremely difficult, if not impossible, in heterogeneous systems. In these systems, one is studying kinetics, which is a generic term referring to time dependent or nonequilibrium processes. Thus, apparent and non-mechanistic rate laws and rate parameters are determined (Skopp, 1986; Sparks, 1989).

4.3.1 Rate Limiting Steps

Both transport and chemical reaction processes can affect the reaction rates in the subsurface environment. Transport processes include (Aharoni and Sparks, 1991): (1) transport in the solution phase; (2) transport across a liquid film at the particle/liquid interface (film diffusion [FD]); (3) transport in liquid filled macropores (> 2 nm), all of which are nonactivated diffusion processes and occur in mobile regions; (4) particle diffusion (PD) processes which include diffusion of sorbate occluded in micropores (< 2 nm) (pore diffusion) and along pore wall surfaces (surface diffusion); and (5) diffusion processes in the bulk of the solid, all of which are activated diffusion processes (Fig. 4.2). Pore and surface diffusion within the immediate region can be referred to as interaggregate (interparticle) diffusion while that in the solid is intraaggregate diffusion. The actual chemical reaction (CR) at the surface, for example, adsorption, is usually instantaneous. The slowest of the CR and transport processes is rate limiting.

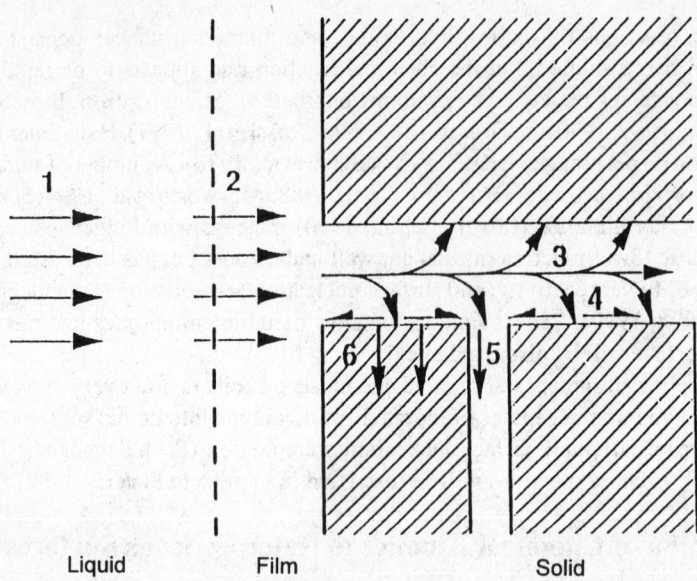

Fig. 4.2 Transport processes in solid-liquid soil reactions–nonactivated processes: (1) transport in the soil solution, (2) transport across a liquid film at the solid-liquid interface, (3) transport in a liquid-filled macropore; activated processes, (4) diffusion of a sorbate at the surface of the solid, (5) diffusion of a sorbate occluded in a micropore, (6) diffusion in the bulk of the solid [Reprinted from Aharoni and Sparks, 1991. Soil Sci. Soc. Am. Spec Pub. 27 with permission of the Soil Science Society of America]

4.3.2 Rate Laws

There are two important reasons for investigating the rates of soil chemical processes (Sparks, 1989,1995,1998, 1999): (1) to determine how rapidly reactions attain equilibrium, and (2) to infer information on reaction mechanisms. One of the most important aspects of chemical kinetics is the establishment of a rate equation or law. By definition, a rate law is a differential equation. For the following reaction (Bunnett, 1986; Sparks, 1989, 1995, 1998, 1999):

$$aA + bB \rightarrow yY + zZ \tag{4.1}$$

the rate is proportional to some power of the concentrations of reactants A and B and/or other species (C, D, etc.) in the system and a, b, y, and z are stoichiometric coefficients and are assumed to be equal to one in the discussion that follows on rate laws. The power to which the concentration is raised may equal zero (i.e., the rate is independent of that concentration), even for reactant A or B. Rates are expressed as a decrease in reactant concentration or an increase in product concentration per unit time. Thus, the rate of conversion of reactant A above, which has a concentration [A] at any time t, is ($-d[A]/(dt)$) while the rate with regard to product Y having a concentration [Y] at time t is ($d[Y]/(dt)$).

The rate expression for Equation [4.1] is therefore,

$$\left| d[Y]/dt \right| = \left| -d[A]/dt \right| = k[A]^{\alpha}[B]^{\beta} \tag{4.2}$$

where k is the rate constant and α and β are the orders of the reaction with respect to reactants A and B, respectively, and can be referred to as a partial orders for the total reaction. These orders are experimentally determined and not necessarily integral numbers. The sum of all the partial orders (α, β) is the overall order (n) of the total reaction and may be expressed as:

$$n = \alpha + \beta + ... \qquad [4.3]$$

Once the values of α, β, etc., are determined experimentally, the rate law is defined. Reaction order provides only information about the manner in which rate depends on concentration. Order does not mean the same as molecularity which concerns the number of reactant particles (atoms, molecules, free radicals, or ions) entering into an elementary reaction. One can define an elementary reaction as one in which no reaction intermediates have been detected, or need to be postulated to describe the chemical reaction on a molecular scale. An elementary reaction is assumed to occur in a single step and to pass through a single transition state (Bunnett, 1986).

To demonstrate that a reaction is elementary, one can use experimental conditions that are different from those employed in determining the reaction rate law. For example, if one conducted kinetic studies using a flow technique with set steady-state flow rates, one could see if reaction rate and rate constants changed with flow rate. If they did, one would not be determining mechanistic rate laws (see definition below).

Rate laws serve three purposes: (1) they assist one in predicting the reaction rate; (2) mechanisms can be proposed; and (3) reaction orders can be ascertained. There are four types of rate laws that can be determined for soil chemical processes (Skopp, 1986): mechanistic, apparent, transport with apparent, and transport with mechanistic. Mechanistic rate laws assume that only chemical kinetics are operational and transport phenomena are not occurring. Consequently, it is difficult to determine mechanistic rate laws for most soil chemical systems due to the heterogeneity of the system caused by different particle sizes, porosities, and types of retention sites. There is evidence that with some kinetic studies, using chemical relaxation techniques (Sparks, 1989; Sparks and Zhang, 1991) and pure systems (e.g., clay minerals, oxides), mechanistic rate laws are determined or closely approximated since the agreement between equilibrium constants calculated from both kinetic and equilibrium studies are comparable (Hachiya et al., 1984; Tang and Sparks, 1993).

Since natural materials are heterogeneous and transport processes often affect the reaction rate, apparent rate laws are usually determined for such systems. Apparent rate laws include both chemical kinetics and transport-controlled processes. Thus, soil structure, stirring, mixing, and flow rate all would affect the kinetics. Transport with apparent rate laws emphasize transport limited phenomena. One often assumes first order or zero order reactions (see discussion below on reaction order). In determining transport with mechanistic rate laws, one attempts to describe simultaneously transport-controlled and chemical kinetics phenomena. One is thus trying to explain accurately both the chemistry and physics of the system.

4.3.3 Determination of Reaction Order and Rate Constants/Coefficients

There are three basic ways to determine rate laws and rate constants/coefficients (Bunnett, 1986; Skopp, 1986; Sparks, 1989, 1995, 1998, 1999): (1) initial rates, (2) directly using integrated equations and graphing the data, and (3) using nonlinear least square analysis.

Let us assume the following elementary reaction between species A, B, and Y,

$$A + B \underset{k_{-1}}{\overset{k_1}{\rightleftharpoons}} Y \tag{4.4}$$

A forward reaction rate law can be written as

$$d[A]/dt = -k_1[A][B] \tag{4.5}$$

where k_1 is the forward rate constant and α and β (see Equation [4.2]) are each assumed to be 1.

The reverse reaction rate law for Equation [4.4] is

$$d[A]/dt = +k_{-1}[Y] \tag{4.6}$$

where k_{-1} is the reverse rate constant.

Equations [4.5] and [4.6] are only applicable far from equilibrium where back or reverse reactions are insignificant. If both forward and reverse reactions are occurring, Equations [4.5] and [4.6] must be combined such that:

$$d[A]/dt = -k_1[A][B] + k_{-1}[Y] \tag{4.7}$$

Equation [4.7] applies the principle that the net reaction rate is the difference between the sum of all reverse reaction rates and the sum of all forward reaction rates.

One way to ensure that back reactions are not important is to measure initial rates. The initial rate is the limit of the reaction rate as time reaches zero. With an initial rate method, one plots the concentration of a reactant or product over a short reaction time period during which the concentrations of the reactants change so little that the instantaneous rate is hardly affected. Thus, by measuring initial rates, one could assume that only the forward reaction in Equation [4.4] predominates. This would simplify the rate law to that given in Equation [4.5] which as written would be a second order reaction, first order in reactant A and first order in reactant B. Equation [4.5], under these conditions, would represent a second order irreversible elementary reaction. To measure initial rates, one must have available a technique that can measure rapid reactions such as a chemical relaxation method and an accurate analytical detection system to determine product concentrations.

Integrated rate equations can also be used to determine rate constants/coefficients. If one assumes that reactant B in Equation [4.5] is in large excess of reactant A, which is an example of the method of isolation to analyze kinetic data, and $Y_o = 0$, where Y_o is the initial concentration of product Y, Equation [4.5] can be simplified to:

$$d[A]/dt = -k_1'[A] \tag{4.8}$$

where $k_1' = k_1[B]$

The first order dependence of $[A]$ can be evaluated using the integrated form of Equation [4.8] using the initial conditions at $t = 0$, $A = A_o$,

$$\log[A]_t = \log[A]_o - \frac{k_1' t}{2.303} \tag{4.9}$$

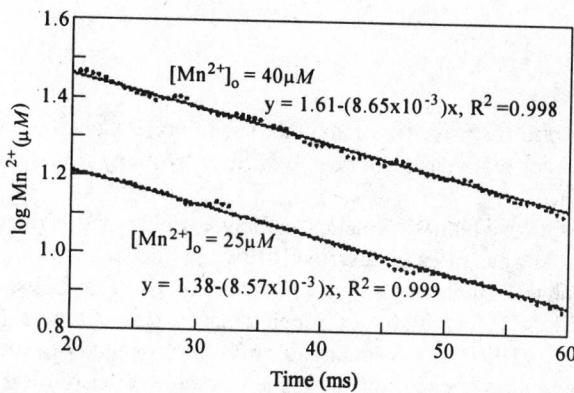

Fig. 4.3 Initial reaction rates depicting the first order dependence of Mn^{2+} sorption as a function of time for initial Mn^{2+} concentrations ($[Mn^{2+}]_0$) 25 and 40 μM [Reprinted from Fendorf et al., 1993. Soil Sci. Soc. Am. J. 57:57-62 with permission of the Soil Science Society of America]

The half-time ($t_{1/2}$) for the above reaction is equal to $0.693/k'_1$ and is the time required for half of reactant A to be consumed.

If a reaction is first order a plot of log $[A]_t$ versus t should result in a straight line with a slope $= k'_1/2.303$ and an intercept of log $[A]_o$. An example of first order plots for Mn^{2+} sorption on δ-MnO_2 at two initial Mn^{2+} concentrations, $[Mn^{2+}]_o$ (25 and 40 μM), is shown in Fig. 4.3. One sees that the plots are linear at both concentrations which would indicate that the sorption process is first order. The $[Mn^{2+}]_o$ values, obtained from the intercepts in Fig. 4.3, were 24 and 41 μM, which are in good agreement with the two $[Mn^{2+}]_o$ values and the rate constants were 3.73×10^{-3} and 3.75×10^{-3} s^{-1}, respectively. The fact that the rate constants do not significantly change with concentration is a good indication that the reaction in Equation [4.8] is first order under the imposed experimental conditions.

It is dangerous to conclude that a particular reaction order is correct, based simply on the conformity of data to an integrated equation. As illustrated above, multiple initial concentrations that vary considerably should be employed to see if the rate is independent of concentration. One should also test multiple integrated equations. It may also be useful to show that reaction rate is not affected by a species whose concentration does not change considerably during an experiment, such as substances not consumed or present in large excess (Bunnett, 1986; Sparks, 1989, 1991, 1995, 1998, 1999).

Least squares analysis can also be used to determine rate constants/coefficients. With this method, one fits the best straight line to a set of points that are linearly related as $y = mx + b$ where y is the ordinate and x is the abscissa datum point, respectively. The slope (m) and the intercept (b) can be calculated by least squares analysis.

When kinetic data are plotted, curvature may be observed due to an incorrect assumption of reaction order. If first-order kinetics is assumed and the reaction is really second order, downward curvature results. If second order kinetics is assumed but the reaction is first order, upward curvature is observed. Curvature can also be due to fractional, third, higher, or mixed reaction order. Nonattainment of equilibrium often results in downward curvature. Temperature changes during the study can also cause curvature; thus, it is important that temperature be accurately controlled during a kinetic experiment.

4.4   Kinetic Models

4.4.1   Ordered Models

First order kinetic models often describe reactions at the soil mineral/water interface. Both single first order and multiple first order reactions have been described by many investigators (Sparks, 1989, 1991; Sparks et al., 1993).

It is not uncommon to observe biphasic kinetics, namely, a rapid reaction rate followed by a much slower reaction rate. Such data can often be described by two first order reactions. Some investigators have interpreted such biphasic kinetics to suggest reactions on two types of sites such as external, readily accessible sites (Slope 1) and internal, difficult to access sites (Slope 2) (Jardine and Sparks, 1984; Comans and Hockley, 1992) or molecular sites of differing reactivity such as high reactivity innersphere complex sites and low reactivity outersphere complex sites (Grossl et al., 1994, 1997).

However, it is unsound to conclude anything about mechanisms based solely on multiple rate constants that are calculated from multiple slopes of kinetic plots. There are other ways to definitively ascertain reaction mechanisms such as calculating energies of activation, elucidating rate limiting steps through stopped flow and interruption approaches, using independent or direct methods to determine mechanisms such as spectroscopic and microscopic techniques, and employing blocking agents that are specific for certain reaction sites. An example of the latter approach is found in the research of Jardine and Sparks (1984) who studied K-Ca exchange on a Delaware soil at three temperatures and observed two apparent simultaneous first order reactions at 283 and 298 K (Fig. 4.4). They hypothesized that the first, more rapid reaction was predominantly due to adsorption on external planar sites of the organic matter and kaolinite in the soil. The slower reaction was ascribed to vermiculitic clay sites that promoted slow pore and surface diffusion. These hypotheses were seemingly validated by using a large organic polymer, cetyltrimethylammonium bromide (CTAB), which because of its size, is sterically hindered from internal sites. Thus, CTAB should only block

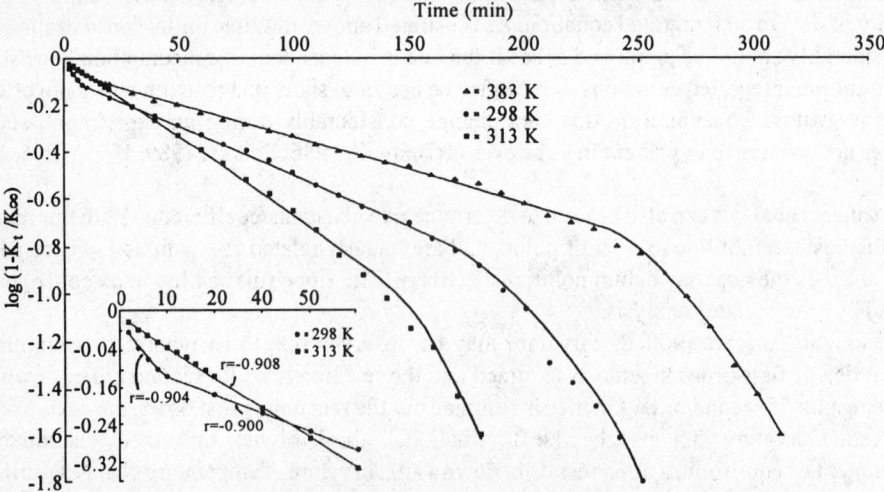

Fig. 4.4   First-order kinetics for potassium adsorption at three temperatures on Evesboro soil with inset showing the initial 50 min of the first-order plots at 298 and 313 K [Reprinted from Jardine and Sparks, 1984. Soil Sci. Soc. Am. J. 48:39-45 with permission of the Soil Science Society of America]

Table 4.1   Linear forms of kinetic equations commonly used in environmental soil chemistry[†]

Zero-order[‡]

$$[A]_t = [A]_o - k_1't$$

First order

$$\log[A]_t = \log[A]_o - \frac{k_1't}{2.303}{}^c$$

Second order[§]

$$\frac{1}{[A]_t} = \frac{1}{[A]_o} + kt$$

Elovich

$$q = (1/\beta)\ln(\alpha\beta) + (1/\beta)\ln t$$

Parabolic Diffusion

$$\left(\frac{1}{t}\right)\left(\frac{Q_t}{Q_\infty}\right) = \frac{4}{\pi^{1/2}}\left(\frac{D}{r^2}\right)^{1/2}\frac{1}{t^{1/2}} - \frac{D}{r^2}$$

Power Function

$$\ln q_t = \ln k + v \ln t$$

[†] From Sparks (1995); terms in equations are defined in the text of the chapter.
[‡] Describing the reaction $A \rightarrow Y$.
[§] ln x = 2.303 log x, is the conversion from natural logarithms (ln) to base 10 logarithms (log).

external planar sites. When CTAB was applied to the soil, the first slope was eliminated, while the second slope was still present, suggesting multireactive sites.

While first order models have been used widely to describe the kinetics of chemical reactions on natural materials, a number of other simple kinetic models also have been employed. These include various ordered equations such as zero, second, and fractional order, and Elovich, power function or fractional power, and parabolic diffusion models. A brief discussion of some of these will be given; the final forms of the equations are given in Table 4.1. For more complete details and applications of these models, one may consult Sparks (1989, 1995, 1998, 1999).

4.4.2   Elovich Equation

The Elovich equation was originally developed to describe the kinetics of heterogeneous chemisorption of gases on solid surfaces (Low, 1960). It seems to describe a number of reaction mechanisms including bulk and surface diffusion and activation and deactivation of catalytic surfaces.

In soil chemistry, the Elovich equation has been used to describe the kinetics of sorption and desorption of various inorganic materials on soils (Sparks, 1989, 1995, 1998, 1999). It can be expressed as (Chien and Clayton, 1980):

$$q = (1/\beta)\ln(\alpha\beta) + (1/\beta)\ln t \tag{4.10}$$

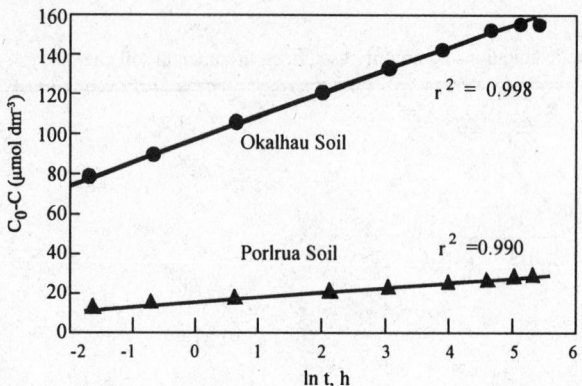

Fig. 4.5 Plot of Elovich equation for phosphate sorption on two soils where C_0 is the initial phosphorus concentration in the soil solution at time t. The quantity $(C_0\text{-}C)$ can be equated to q, the amount sorbed at time t [Reprinted from Chien and Clayton, 1980. Soil Sci. Soc. Am. J. 44:265-268 with permission of the Soil Science Society of America].

where q is the amount of sorbate per unit mass of sorbent at time t and α and ß are constants during any one experiment. A plot of q versus ln t should give a linear relationship if the Elovich equation is applicable with a slope of $(1/\beta)$ and an intercept of $(1/\beta)\ln(\alpha\beta)$. An application of Equation [4.10] to P sorption on soils is shown in Fig. 4.5.

Some investigators have used the α and β parameters from the Elovich equation to estimate reaction rates. For example, it has been suggested that a decrease in β and/or an increase in α would increase reaction rate. However, this is questionable. The slope of plots using Equation [4.10] changes with the concentration of the sorptive and with the solution-to-soil ratio (Sharpley, 1983). Therefore, the slopes are not always characteristic of the soil, but may depend on various experimental conditions.

Some researchers also have suggested that breaks or multiple linear segments in Elovich plots could indicate a changeover from one type of binding site to another (Atkinson et al., 1970). However, such mechanistic suggestions may not be correct (Sparks, 1989, 1995).

4.4.3 Parabolic Diffusion Equation

The parabolic diffusion equation is often used to suggest that diffusion-controlled phenomena are rate limiting. It was originally derived from radial diffusion in a cylinder where the ion concentration on the surface is constant, and initially, the ion concentration within the cylinder is uniform. It is also assumed that ion diffusion through the upper and lower faces of the cylinder is negligible. Following Crank (1976), the parabolic diffusion equation, as applied to soils can be expressed as

$$(Q_t / Q_\infty) = \frac{4}{\pi^{1/2}}\left(\frac{Dt}{r^2}\right)^{1/2} - \frac{Dt}{r^2} - \frac{1}{3\pi^{1/2}}\left(\frac{Dt}{r^2}\right)^{3/2}$$

[4.11]

where r is the radius of the cylinder, Q_t is the quantity of diffusing substance that has left the cylinder at time t, Q is the corresponding quantity after infinite time, and D is an apparent diffusion coefficient.

For the relatively short times in most experiments, the third and subsequent terms may be ignored, and thus

$$\frac{Q_t}{Q_\infty} = \frac{4}{\pi^{1/2}} \left(\frac{Dt}{r^2} \right)^{1/2} - \frac{Dt}{r^2}$$

or   [4.12]

$$\frac{1}{t} \left(\frac{Q_t}{Q_\infty} \right) = \frac{4}{\pi^{1/2}} \left(\frac{Dt}{r^2} \right)^{1/2} \frac{1}{t^{1/2}} - \frac{Dt}{r^2}$$

and thus, a plot of $\dfrac{Q_t / Q_\infty}{t}$ versus $1/t^{1/2}$ should give a straight line with a slope of

$$\frac{4}{\pi^{1/2}} \left(\frac{D}{r^2} \right)^{1/2}$$

and intercept $(-D/r^2)$. Thus, if r is known, D may be calculated from both the slope and intercept.

The parabolic diffusion equation has successfully described metal reactions on soils and soil constituents (Chute and Quirk, 1967; Jardine and Sparks, 1984; Krishnamurti and Huang, 1992; Krishnamurti et al., 1997), feldspar weathering (Wollast, 1967), and pesticide reactions (Weber and Gould, 1966).

4.4.4 Fractional Power or Power Function Equation

This equation can be expressed as

$$q = kt^v \qquad\qquad\qquad [4.13]$$

where q is amount of sorbate per unit mass of sorbent at time t, k and v are constants and v is positive and < 1. Equation [4.13] is empirical, except for the case where $v = 0.5$, when it is similar to the parabolic diffusion equation.

Equation [4.13] and various modified forms have been used by a number of researchers to describe the kinetics of reactions in natural materials (Kuo and Lotse, 1974; Havlin and Westfall, 1985).

4.4.5 *Z(t)* and Diffusion Models

In a number of studies, several simple kinetic models have described rate data well, based on correlation coefficients and standard errors of the estimate (Chien and Clayton, 1980; Onken and Matheson, 1982; Sparks and Jardine, 1984; Allen et al, 1995). Despite this, there is often an inconsistent relation between the equation that gives the best fit and the physicochemical and mineralogical properties of the sorbent(s) being studied. Another problem with some of the kinetic models is that they are empirical and no meaningful rate parameters can be obtained.

Aharoni and Ungarish (1976) and Aharoni (1984) noted that some simple kinetic models are approximations of more general expressions within certain limited time ranges. They suggested a generalized empirical equation by examining the applicability of power function, Elovich, and first order equations to experimental data. By writing these as the explicit functions of the reciprocal of the rate (Z) which is $(dq/dt)^{-1}$, one can show that a plot of Z versus t should be convex if the power function

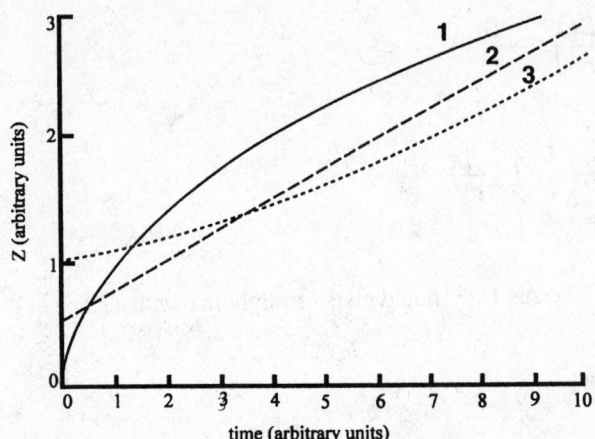

Fig. 4.6 Plot of Z vs. time implied by (1) power function model, (2) Elovich model, and (3) first order model. The equations for the models were differentiated and expressed as explicit functions of the reciprocal of the rate, Z [Reprinted from Aharoni and Sparks, 1991. Soil Sci. Soc. Am. Spec Pub. 27 with permission of the Soil Science Society of America].

equation is operational (1 in Fig. 4.6); linear, if the Elovich equation is appropriate (2 in Fig. 4.6); and concave, if the first order equation is appropriate (3 in Fig. 4.6). However, such plots for soil systems (Fig. 4.7) are usually S shaped, convex at small, concave at large, and linear at some intermediate t value suggesting that the reaction rate can best be described by the power function, first order and Elovich equations in these ranges of t, respectively. Thus, the S-shaped curve indicates that the above equations may be applicable, each over some limited time range.

One of the reasons a particular kinetic model appears to be applicable may be that the study is conducted during the time range when the model is most appropriate. While sorption, for example, decreases over many orders of magnitude before equilibrium is approached, with most methods and

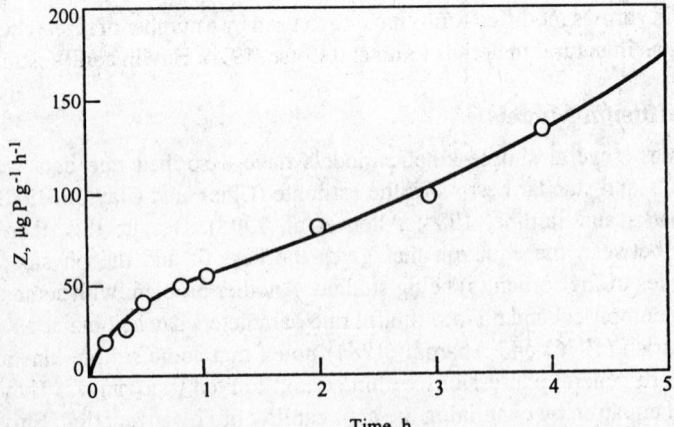

Fig. 4.7 Sorption of phosphate by a Typic Dystrochrept soil plotted as Z vs. time. The circles represent the experimental data of Polyzopoulos et al. (1986). The solid line is a curve calculated according to a homogeneous diffusion model [Reprinted from Aharoni and Sparks, 1991. Soil Sci. Soc. Am. Spec Pub. 27 with permission of the Soil Science Society of America].

experiments, only a portion of the entire reaction is measured, and over this time range, the assumptions associated with a simple kinetic model (power function, Elovich, and first order) are valid. Aharoni and Suzin (1982a,b) showed that the S-shaped curves could be well described using homogeneous and heterogeneous diffusion models. In homogeneous diffusion situations, the final and initial portions of the S-shaped curves (conforming to the power function and first order equations, respectively) predominated, whereas in instances where the heterogeneous diffusion model was operational, the linear portion of the S-shaped curve that conformed to the Elovich equation predominated. Derivations of homogeneous and heterogeneous diffusion models can be found in Aharoni and Sparks (1991).

4.4.6 Implications of Diffusion Models

The finding that slower reactions at the soil mineral/liquid interface can be described by diffusional models indicates that the kinetics of chemical processes cannot be considered separately from physically limited transport phenomena. Thus, such a combination of processes cannot be treated using first or other order chemical kinetics equations. When one states that a reaction between the molecular species A and B is first order with respect to A, one assumes that the molecules of A have equal chances of participating in the reaction, and therefore, the rate is proportional to the concentration (C_A). This reasoning can be extended to a reaction between a sorbing surface and a sorptive. In this case, C_A refers to the number of reactive sites per unit area, which corresponds to the number of unoccupied sites per unit area $(1-\theta_A)$. However, by using first order kinetics (or other order kinetics), one tacitly assumes that all the surface sites are potential reactants at any time, and have an opportunity of participating in the sorption process. If one assumes that there are sites that cannot be reached directly from the fluid phase, but can be reached after the sorbate has undergone sorption and desorption at other sites, one cannot separate chemical kinetics from diffusion limited kinetics. The overall kinetic process obeys a diffusion equation since diffusion is the rate limiting process. However, the diffusion coefficient, which reflects the rate at which the sorbate jumps from one site to another, is determined by the rate of the chemical reactions by which the sorbent-sorbate bonds are created and destroyed. Additionally, the activation energy for diffusion is equivalent to the activation energy of the chemical reaction.

4.4.7 Multiple Site Models

Based on the previous discussion, it is evident that simple chemical kinetics models such as ordered reaction, power function and Elovich models may not be appropriate to describe reactions in heterogeneous systems such as soils, sediments, and soil components. In these systems where there is a range of particle sizes and multiple retention sites, both chemical kinetics and transport phenomena are occurring simultaneously, and a fast reaction is often followed by a slower reaction(s). In such systems, nonequilibrium models that describe both chemical and physical nonequilibrium and that consider multiple components and sites are more appropriate. Physical nonequilibrium is ascribed to some rate limiting transport mechanism such as FD or PD, while chemical nonequilibrium is due to a rate limiting mechanism at the particle surface (CR). Nonequilibrium models include two-site, multiple site, radial diffusion (pore diffusion), surface diffusion, and multiprocess models (Table 4.2). Emphasis here will be placed on the use of these models to describe sorption phenomena.

The term sites can have a number of meanings (Brusseau and Rao, 1989): (1) specific, molecular scale reaction sites; (2) sites of differing degrees of accessibility (external, internal); (3) sites of differing sorbent type (organic matter and inorganic mineral surfaces); and (4) sites with different sorption mechanisms. With chemical nonequilibrium sorption processes, the sorbate may undergo two or more types of sorption reactions, one of which is rate limiting. For example, a metal cation may sorb to organic matter by one mechanism and to mineral surfaces by another mechanism, with one of the mechanisms being time dependent.

4.4.7.1   Chemical Nonequilibrium Models

Chemical nonequilibrium models describe time-dependent reactions at sorbent surfaces. The one-site model is a first order approach that assumes that the reaction rate is limited by only one process or mechanism on a single class of sorbing sites and that all sites are of the time-dependent type. In many cases, this model appears to describe soil chemical reactions quite well. However, often it does not. This model would seem not appropriate for most heterogeneous systems since multiple sorption sites exist.

The two-site (two compartment, two box) or bicontinuum model has been widely used to describe chemical nonequilibrium (Leenheer and Ahlrichs, 1971; Hamaker and Thompson, 1972; Karickhoff, 1980; McCall and Agin, 1985; Karickhoff and Morris, 1985; Jardine et al., 1992) and physical nonequilibrium (Nkedi-Kizza et al., 1984; Lee et al., 1988; van Genuchten and Wagenet, 1989) (Table 4.2). This model assumes that there are two reactions occurring, one that is fast and reaches equilibrium quickly and a slower reaction that can continue for long time periods. The reactions can occur either in series or in parallel (Brusseau and Rao, 1989).

In describing chemical nonequilibrium with the two-site model, two types of sorbent sites are assumed. One site involves an instantaneous equilibrium reaction and the other, the time-dependent reaction. The former is described by an equilibrium isotherm equation while a first order equation is usually employed for the latter.

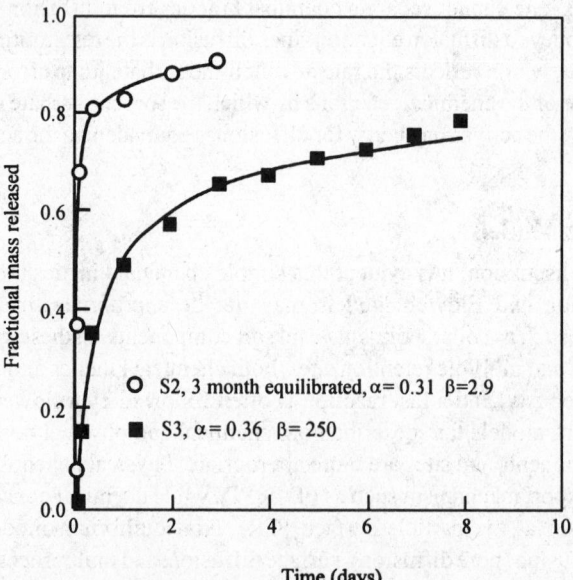

Fig. 4.8 Mass fractional release of naphthalene from two soils, S2 and S3 (S2 is a freshly contaminated soil, reacted with naphthalene for 3 months and S3 is a field [aged] contaminated soil) fitted with a multisite continuum compartment (Γ) model [Reprinted with permission from Connaughton et al., 1993. Environ. Sci. Technol. 27:2397-2403. Copyright American Chemical Society]

With the two-site model, there are two adjustable or fitting parameters, the fraction of sites at local equilibrium (X_1) and the rate constant (k). A distribution (K_d) or partition coefficient (K_p) is determined independently from a sorption/desorption isotherm.

To account for the multiple sites that may exist in heterogeneous systems, Connaughton et al. (1993) developed a multisite compartment (continuum) model (Γ) that incorporates a continuum of sites or compartments with a distribution of rate coefficients that can be described by a gamma density function. A fraction of the sorbed mass in each compartment is at equilibrium with a desorption rate coefficient or distribution coefficient for each compartment or site (Table 4.2). The multisite model has two fitting parameters α, a shape parameter, and $1/\beta$, which is a scale parameter that determines the mean standard deviation of the rate coefficients. Fig. (4.8) shows application of the Γ model to desorption of naphthalene from contaminated soils. The entire desorption process was described well with this model.

4.4.7.2 Physical Nonequilibrium Models

A number of models can be used to describe physical nonequilibrium reactions. Since transport processes in the mobile phase are not usually rate limiting, physical nonequilibrium models focus on diffusion in the immobile phase or interaggregate/diffusion processes such as pore and/or surface diffusion. The transport between mobile and immobile regions is accounted for in physical nonequilibrium models in three ways (Brusseau and Rao, 1989): (1) explicitly with Fick's law to describe the physical mechanism of diffusive transfer, (2) explicitly by using an empirical first order mass transfer expression to approximate solute transfer, and (3) implicitly by using an effective or lumped dispersion coefficient that includes the effects of sink/source differences and hydrodynamic dispersion and axial diffusion.

A pore diffusion model (Table 4.2) has been used by a number of investigators to study sorption processes using batch systems (Wu and Gschwend, 1986; Steinberg et al., 1987; Ball and Roberts, 1991; Harmon et al., 1992; Pignatello et al., 1993). Wu and Gschwend (1986) successfully used the pore diffusion model to describe chlorobenzene congener sorption/desorption on soils and sediments. Fig. 4.9 shows experimental and model fits for tetrachlorobenzene and pentachlorobenzene sorption on soils. The sole fitting parameter in this model is the effective diffusion coefficient (D_e) which may be estimated *a priori* from chemical and colloidal properties. However, this estimation is only valid if the sorbent material has a narrow particle size distribution so that an accurate, average particle size can be defined. Moreover, in the pore diffusion model, an average representative D_e is assumed, which means there is a continuum in properties across an entire pore size spectrum. This is not a valid assumption for micropores in which there are higher adsorption energies of sorbates causing increased sorption. The increased sorption reduces diffusive transport rates and results in nonlinear isotherms for sorbents with pores less than several sorbate diameters in size. Other factors can cause reduced transport rates in micropores including steric hindrance, which increases as the pore size approaches the solute size and greatly increased surface area to pore volume ratios (which occurs as pore size decreases).

Another problem with the pore diffusion model is that sorption and desorption kinetics may have been measured over a narrow concentration range. This is a problem since a sorption/desorption mechanism in micropores at one concentration may be insignificant at another concentration.

Fuller et al. (1993) used a pore space diffusion model (Table 4.2) to describe arsenate adsorption on ferrihydrite that included a subset of sites whereby sorption was at equilibrium. A Freundlich model was used to describe sorption on these sites. Diffusion into the particle was described by Fick's second

Table 4.2 Comparison of sorption kinetic models[†,‡]

Conceptual model	Fitting parameter(s)	Model limitations
one-site model $$\overset{k_d}{S \to C}$$	k_d	cannot describe biphasic sorption/desorption
two-site model $$\overset{X_1 K_p k_d}{S_1 \to C \leftarrow S_2}$$	k_d, $K_p^{\S}$, X_1	may not describe the "bleeding" or slow, reversible, nonequilibrium desorption for residual sorbed compounds (Karickhoff, 1980)
radial diffusion penetration retardation (pore diffusion) model (Wu and Gschwend, 1986) $$\overset{K_p \quad D_{eff}}{S \leftrightarrow C'' \to C}$$	$D_{eff}^{\P} = f(n,t)D_m n/(1-n)\rho_s K_p$	cannot describe instantaneous uptake without additional correction factor; did not describe kinetic data for times greater than 10^3 min (Wu and Gschwend, 1986)
dual-resistance surface diffusion model (Miller and Pedit, 1992) $$\overset{D_s \quad k_b}{S' \to C'_s \to C}$$	D_s, k_b	model calibrated with sorption data predicted more desorption than occurred in the desorption experiments (Miller and Pedit, 1992)
multisite continuum compartment model (Connaughton et al., 1993) $$F(t) = 1 - \frac{M(t)}{M} = 1 - \left(\frac{\beta}{\beta + t}\right)^{\alpha}$$	α, β	assumption of homogeneous, spherical particles and diffusion only in aqueous phase
pore space diffusion model (Fuller et al., 1993) $$\left(\epsilon + \frac{S_a}{n} K_s C(r)^{(1-1/n)}\right)\frac{\partial C(r)}{\partial t} =$$ $$D_e\left(\frac{\partial^2 C(r)}{\partial r^2} + \frac{2\partial C(r)}{r\partial r}\right)$$	D_e, ϵ, K_s, $1/n$, F_{eq}	
multiple particle class pore diffusion model (Pedit and Miller, 1995) $$\left(\frac{\theta_p^i + \rho_a^i}{\partial C_{p_i}^i}\right)\frac{\partial q_r^i(r,t)}{(r,t)} \frac{\partial C_p^i(r,t)}{\partial t} =$$ $$\frac{\theta_p^i D_p^i}{r^2}\frac{\partial}{\partial r}\left(\frac{r^2 \partial C_p^i(r,t)}{\partial r}\right)$$ $$-\theta_p^i \lambda_p^i C_p^i(r,t) - \rho_a^i \lambda_r^i q_r^i(r,t)$$	θ_p^i, ρ_a^i, D_p^i, λ_p^i, λ_r^i	multiple fitting parameters; variations in sorption equilibrium and rates that might occur within a particle class or an individual particle grain are not addressed.

[†] Partially adapted from Connaughton et al. (1993).

[‡] Abbreviations used are as follows: S, concentration of the bulk sorbed contaminant (g g^{-1}); C, concentration of the bulk aqueous-phase contaminant (g mL^{-1}); k_d, first-order desorption rate coefficient (min^{-1}); S_2, concentration of the sorbed contaminant that is rate limited (g g^{-1}); S_1, concentration of the contaminant that is in equilibrium with the bulk aqueous concentration (g g^{-1}); X_1, fraction of the bulk sorbed contaminant that is in equilibrium with the aqueous concentration; K_p, sorption equilibrium partition coefficient (mL g^{-1}); D_{eff}, effective diffusivity of sorbate molecules or ions in the particles (cm^2 s^{-1}); S', concentration of contaminant in immobile bound state (mol g^{-1}); C', concentration of contaminant free in the pore fluid (mol cm^{-3}); n, porosity of the sorbent (cm^3 of fluid cm^{-3}); D_m, pore fluid diffusivity of the sorbate (cm^2 s^{-1}); ρ_s, specific gravity of the sorbent (g cm^{-3}); f (n,t), pore geometry factor; k_b, boundary layer mass transfer coefficient (m s^{-1}); r, radius of the spherical solid particle, assumed constant (m); ρ, macroscopic particle density of the solid phase (g m^{-3}); C'_s, solution-phase solute concentration corresponding to an equilibrium with the solid-phase solute concentration at the exterior of the particle (g L^{-1}); D_s, surface diffusion coefficient (m s^{-1}). § K_p can be determined independently. ¶ K_p, D_m, and ρ, can be determined independently; F(t), fraction of mass released through time t; M(t), mass remaining after time t; M, total initial mass; β, scale parameter necessary for determination of mean and standard deviation of k_s; α, shape parameter; ϵ, internal porosity of sorbent; C(r), concentration of sorptive in the aqueous phase in the pore fluid at radial distance r; S_a is the surface of sorbent per unit volume of solid; 1/n, the adsorption isotherm slope; K_s, adsorption isotherm intercept; D_e, effective diffusion coefficient; a, radius of the aggregate; F_{eq}, equilibrium fraction of adsorption sites; θ_p^i, intraparticle porosity of particle class I; ρ_a^i, apparent particle density of particle class I; r, radial distance; C_p^i (r, t), intraparticle fluid-phase solute concentration of the particle class I; D_p^i, pure diffusion coefficient for particle class I; λ_p^i, intraparticle fluid-phase first-order reaction rate coefficient for particle class I; λ_r^i, intraparticle solid-phase first-order reaction rate coefficient for particle class I, q_r^i (r, t), intraparticle solid-phase solute concentration of particle class I.

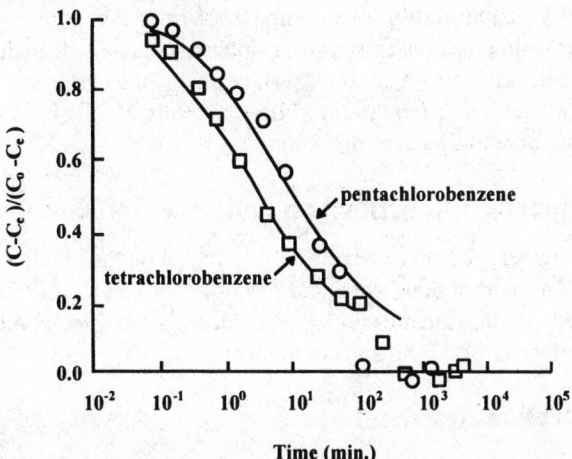

Fig. 4.9 Experimental and model fitting results for pentachlorobenzene and tetrachlorobenzene sorption on Iowa soils where C is the dissolved concentration of organic chemical in the bulk solution, C_0 is the initial concentration and C_e is the equilibrium concentration. The points represent experimental data and the solid lines represent fit of the data to the radial diffusion (pore diffusion) model [Reprinted with permission from Wu and Gschwend, 1986. Environ. Sci. Technol. 20:717-725. Copyright American Chemical Society].

law of diffusion; homogeneous, spherical aggregates, and diffusion only in the aqueous phase were assumed. Fig. 4.10 shows the fit of the model when sorption at all sites was controlled by intra-aggregate diffusion. The fit was better when sites that had attained sorption equilibrium were included based on the assumption that there was an initial rapid sorption on external surface sites before intra-aggregate diffusion.

Pedit and Miller (1995) have developed a general multiple particle class pore diffusion model that accounts for differences in physical and sorptive properties for each particle class (Table 4.2). The model includes both instantaneous equilibrium sorption and time-dependent pore diffusion for each

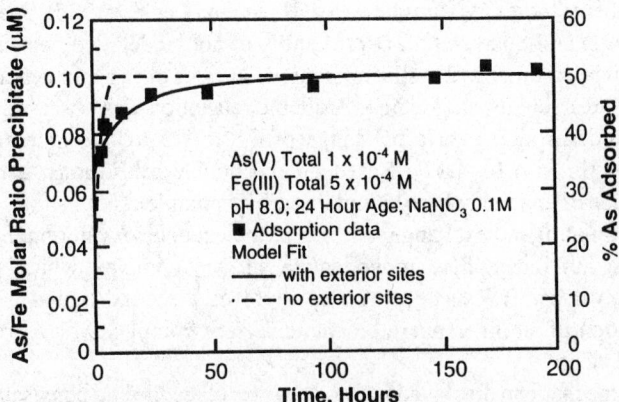

Fig. 4.10 Comparison of pore space diffusion model fits of As (V) sorption with experimental data (dashed curve represents sorption where all surface sites are diffusion limited and the solid curve represents sorption on equilibrium sites plus diffusion limited sites) [Reprinted from Fuller et al., 1993. Surface chemistry of ferrihydrite. 2. Geochim. Cosmochim. Acta 57:2271-2282, with kind permission of Elsevier Science, Amsterdam, Netherlands]

particle class. The pore diffusion portion of the model assumes that solute transfer between the intraparticle fluid and the solid phases is fast *vis-à-vis* interparticle pore diffusion processes.

Surface diffusion models, assuming a constant surface diffusion coefficient, have been used by a number of researchers (Weber and Miller, 1988; Miller and Pedit, 1992). The dual resistance model (Table 4.2) combines both pore and surface diffusion.

4.5 Kinetics of Important Reactions on Soils and Soil Components

In the past two decades, numerous studies have been conducted on the kinetics of metal, oxyanion, radionuclide, plant nutrient, and organic chemical reactions on natural materials. In this section, emphasis will be placed on the kinetics of sorption/desorption, precipitation/dissolution, and oxidation/reduction reactions on soils and soil components.

4.5.1 Sorption/Desorption Reaction Rates

4.5.1.1 Heavy Metals and Oxyanions

Chemical reactions of heavy metals on soil components are rapid, occurring on a millisecond time scale. For such rapid reactions, chemical techniques such as pressure jump (p-jump) relaxation must be employed (Hayes and Leckie, 1986; Sparks, 1989; Sparks and Zhang, 1991; Grossl and Sparks, 1995; Sparks et al., 1996).

The use of p-jump relaxation to measure the kinetics of ion sorption/desorption on metal oxide surfaces was pioneered by several Japanese chemists. Their research includes some of the following sorption/desorption kinetic studies: divalent metal ion (Hachiya et al., 1984), phosphate (Mikami et al., 1983a), chromate (Mikami et al., 1983b), and uranyl (Mikami et al., 1983c) sorption reactions on γ-Al_2O_3. Hayes and Leckie (1986) were the first to use p-jump relaxation to study sorption/desorption kinetics of a metal ion contaminant (Pb^{2+}) on goethite (α-FeOOH). Other successive studies monitored the rapid sorption/desorption kinetics of molybdate (Zhang and Sparks, 1989), sulfate (Zhang and Sparks, 1990a), selenate and selenite (Zhang and Sparks, 1990b), Cu^{2+} (Grossl et al., 1994), and arsenate and chromate (Grossl et al., 1997) on goethite. Additional studies have investigated borate sorption/desorption kinetics on pyrophyllite (Keren et al., 1994) and on γ-Al_2O_3 (Toner and Sparks, 1995).

Details of many of these studies are summarized in Hayes and Leckie (1986), Sparks (1989, 1995), Sparks and Zhang (1991) and Sparks et al. (1996) and will not be detailed here. A recent study of Grossl et al. (1997) will be summarized to illustrate rapid CR rates of two environmentally important oxyanions (chromate and arsenate) on goethite. A double relaxation was observed for both arsenate and chromate sorption/desorption over a pH range of 6.5 to 7.5 for arsenate and 5.5 to 6.5 for chromate, respectively (Figs. 4.11–4.12). Based on the double relaxations, a two-step process, resulting in the formation of an innersphere bidentate surface complex (Fig. 4.13) was proposed. The first step involves an initial ligand exchange reaction of the aqueous oxyanion (H_2AsO^- or $HCrO_4^-$) with goethite, forming an innersphere monodentate surface complex which produces signals associated with fast τ values. The succeeding step involves a second ligand exchange reaction, resulting in the formation of an innersphere bidentate surface complex which produces the signal associated with slow τ values.

To determine if the mechanism displayed in Fig. 4.13 was plausible and consistent with the kinetic data, the following linearized rate equations relating reciprocal relaxation time values (τ^{-1}) to the concentrations of reactive species were used:

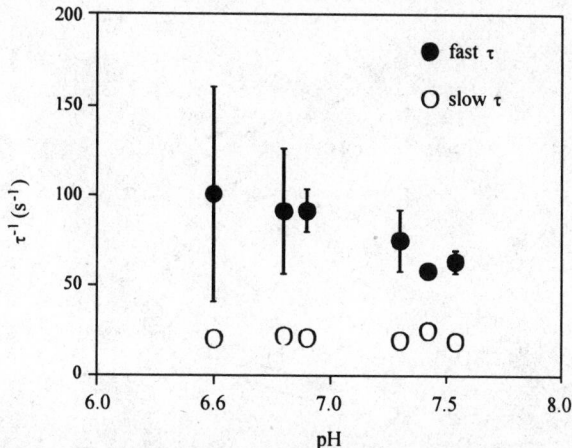

Fig. 4.11　τ^{-1} values determined from p-jump experiments for arsenate adsorption/desorption on goethite, as a function of pH [Reprinted with permission from Grossl et al. (1997). Environ. Sci. Technol. 31:321-326. Copyright American Chemical Society]

$$\tau^{-1}_{fast} + \tau^{-1}_{slow} = k_1 \quad ([XOH]+[ion\ species]) + k_{-1} + k_2 + k_{-2} \qquad [4.14]$$

$$\tau^{-1}_{fast} \cdot \tau^{-1}_{slow} = k_1[k_2 + k_{-2}] \quad ([XOH]+[ion\ species]) + k_{-1}k_{-2} \qquad [4.15]$$

where the ion species are $H_2AsO_4^-$ or $HCrO_4^-$. The derivation of these equations was obtained from Bernasconi (1976) and is based on the two-step reaction system ($A + B \leftrightarrow C \leftrightarrow D$). If the mechanism portrayed in Fig. 4.13 is accurate, then a plot of $\tau^1_{fast} + \tau^{-1}_{slow}$ and $\tau^1_{fast} \cdot \tau^{-1}_{slow}$ as a function of the concentration term ($[XOH] + [ion\ species]$) should be linear. Plots of Equations [4.14] and [4.15] were linear for both arsenate and chromate suggesting that the proposed mechanism was plausible (Figs. 4.14–4.15).

From the plots in Figs. 4.14–4.15, forward and reverse rate constants were obtained for the sorption and desorption reactions of both the monodentate and bidentate steps: where k_1 = slope (Fig. 4.14);

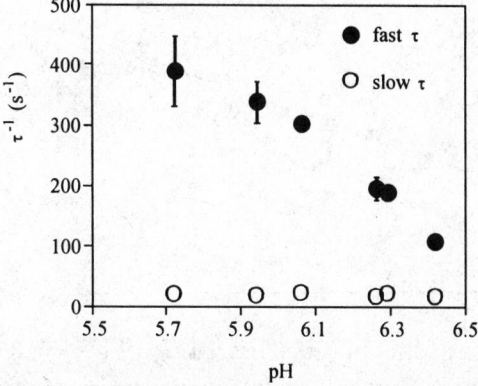

Fig. 4.12　τ^{-1} values determined from p-jump experiments for chromate adsorption/desorption on goethite, as a function of pH [Reprinted with permission from Grossl et al. (1997). Environ. Sci. Technol. 31:321-326. Copyright American Chemical Society]

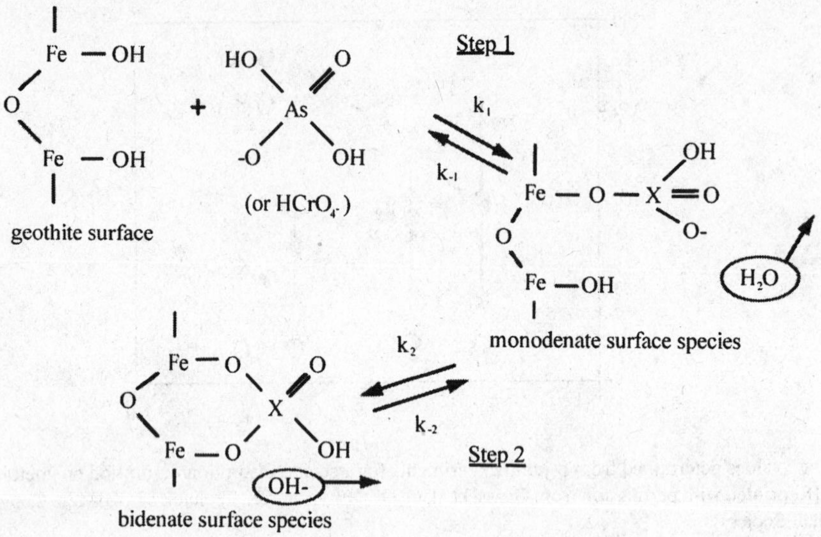

Fe —OH   HO   O   Step 1

O   +   As   k₁

Fe —OH   -O   OH   k₋₁   Fe — O — X ═ O

(or HCrO₄⁻)   O

geothite surface   Fe —OH   H₂O

                                               monodenate surface species

Fe — O   O

O   X   k₂

Fe — O   OH   k₋₂   Step 2

OH-

bidenate surface species

Fig. 4.13 Proposed mechanism for oxyanion adsorption/desorption on goethite. The X represents either As(V) or Cr(VI) [Reprinted with permission from Grossl et al., 1997. Environ. Sci. Technol. 31:321-326. Copyright American Chemical Society].

k_{-1} = intercept (Fig. 4.14) – slope (Fig. 4.15)/slope (Fig. 4.14); k_2 = intercept (Fig. 4.14) $- k_{-1} - k_{-2}$; and k_{-2} = intercept (Fig. 4.15)/k_{-1}. The calculated rate constants for both chromate and arsenate adsorption/ desorption on goethite are listed in Table 4.3.

   Overall, the forward rate constants associated with the formation of the innersphere oxyanion/ goethite surface complexes were more rapid than the reverse rate constants. Therefore, the rate limiting steps were the reverse reactions. The equilibrium constants listed in Table 4.3 were calculated

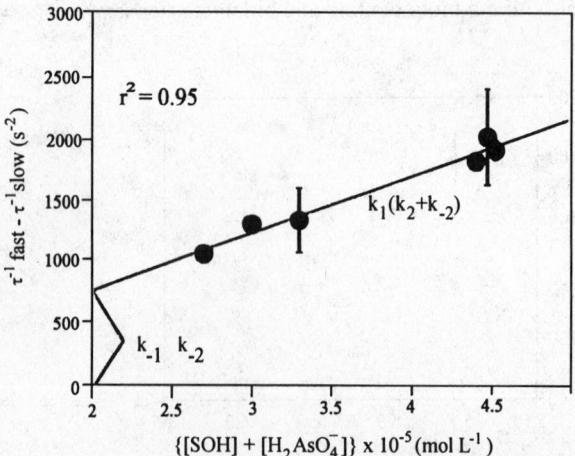

Fig. 4.14 Evaluation of the linearized rate Equation [4.14] for the mechanism displayed in Fig. 4.13 for arsenate [Reprinted with permission from Grossl et al. (1997). Environ. Sci. Technol. 31:321-326. Copyright American Chemical Society]

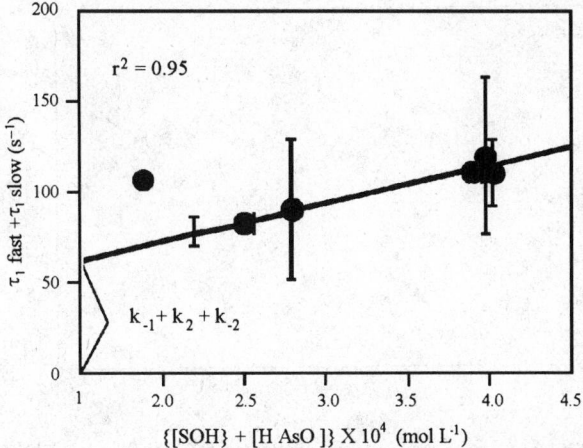

Fig. 4.15 Evaluation of the linearized rate Equation [4.15] for the mechanism displayed in Fig. 4.13 for arsenate [Reprinted with permission from Grossl et al. (1997). Environ. Sci. Technol. 31:321-326. Copyright American Chemical Society]

using the rate constants for each reaction step in the proposed mechanism (Fig. 4.13) from the following relationship:

$$K_{eq} = k_1 / k_{-1} \qquad [4.16]$$

The calculated equilibrium constant for step 1 for arsenate was $10^{5.35}$ and for step 2 was $10^{0.26}$, while the calculated K_{eq} for step 1 for chromate was $10^{3.7}$ and for step 2 was $10^{-0.4}$. The sorption of both oxyanions and subsequent formation of innersphere surface complexes are thermodynamically favorable, with the exception of the equilibrium constant for the second step associated with chromate sorption (slightly less than 1). Thus, the monodentate chromate/goethite surface complex is slightly favored over the bidendate surface complex. This is in agreement with spectroscopic data obtained from X-ray absorption fine structure (XAFS) analyses (Fendorf et al., 1997) which indicate a mixture of both monodentate and bidentate arsenate, and chromate surface complexes; but at low surface coverage, a greater proportion of chromate is associated with the monodentate complex than the

Table 4.3 Calculated rate constants for chromate and arsenate adsorption/desorption on goethite [Reprinted with permission from Grossl et al. (1997). Environ. Sci. Technol. 31:321-326. Copyright American Chemical Society]

	Step I	Step II
Arsenate	$k_1 = 10^{6.3}$ L mol^{-1} s^{-1}	$k_2 = 15$ s^{-1}
	$k_{-1} = 8$ s^{-1}	$k_{-2} = 8$ s^{-1}
	$K_{eq} = 10^{5.35}$ L mol^{-1} s^{-1}	$K_{eq} = 10^{0.26}$
Chromate	$k_1 = 10^{5.8}$ L mol^{-1} s^{-1}	$k_2 = 16$ s^{-1}
	$k_{-1} = 129$ s^{-1}	$k_{-2} = 38$ s^{-1}
	$K_{eq} = 10^{3.7}$ L mol^{-1} s^{-1}	$K_{eq} = 10^{-0.4}$

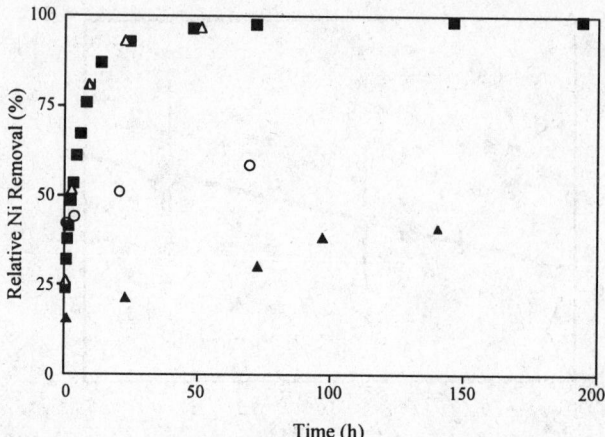

Fig. 4.16   Kinetics of Ni sorption [%] on pyrophyllite (■), kaolinite (△), gibbsite (▲) and montmorillonite (o) from a 3 mM Ni solution at pH = 7.5 and an ionic strength $I = 0.1M$ (NaNO$_3$). The last sample of each experiment was collected and analyzed by XAFS [Reprinted from Scheidegger et al.,1997. J. Colloid Interf. Sci. 186:118-128, with permission]

bidentate complex. The results from both kinetic and XAFS experiments suggest that arsenate is more likely to form an innersphere surface complex with goethite than chromate.

While the initial sorption of heavy metals is rapid, with the CR step occurring on millisecond time scales, further sorption is usually quite slow (Fig. 4.16) occurring over time scales of days and longer. This slow sorption has been ascribed to several mechanisms including interparticle or intraparticle diffusion in pores and solids, sites of low reactivity, and surface precipitation/nucleation (Sparks, 1998; Strawn and Sparks, 1997, unpublished data). With heterogeneous soils and even soil components, there may be a continuum between the three sorption mechanisms, for example, between adsorption and surface precipitation/nucleation. While it has generally been assumed that adsorption in comparison to surface precipitation/nucleation is much more rapid, recent studies (Scheidegger et al., 1998) that will be discussed later, have shown that surface precipitation/nucleation processes can occur on time scales as short as 15 min which indicates that sorption and nucleation processes can

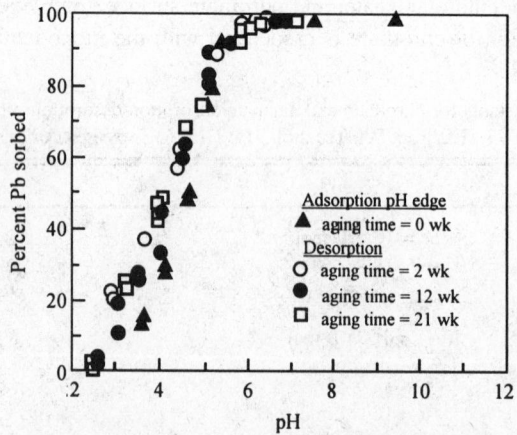

Fig. 4.17   Fractional sorption-desorption of Pb^{2+} to hydrousFe(III) as a function of pH and HFO-Pb^{2+} aging time [reprinted from Ainsworth et al., 1994. Soil Sci. Soc. Am. J. 58:1615-1623 with permission of the Soil Science Society of America]

occur simultaneously. However, in some cases, depending on reaction conditions and the metal involved, a particular sorption mechanism can dominate.

Obviously, an important factor affecting the degree of slow sorption/desorption of metals (and for that matter also of organic chemicals} is the time period the sorbate has been in contact with the sorbent (residence time). Bruemmer et al. (1988) studied Ni^{2+}, Zn^{2+}, and Cd^{2+} sorption on goethite, a porous Fe oxide that has defect structures in which metals can be incorporated to satisfy charge imbalances. Bruemmer et al. (1988) found, at pH 6, that as reaction time increased from 2 hr to 42 days (at 293 K), adsorbed Ni^{2+} increased from 12 to 70% of total adsorption, and total Zn^{2+} and Cd^{2+} adsorption over this time increased by 33 and 21%, respectively. The kinetic reactions could be well described using a Fickian diffusion model. Metal uptake was hypothesized to occur by a three-step mechanism: (1) adsorption of metals on external surfaces, (2) solid-state diffusion of metals from external to internal sites, and (3) metal binding and fixation at positions inside the goethite particle.

Ainsworth et al. (1994) studied the adsorption/desorption of Co^{2+}, Cd^{2+}, and Pb^{2b+} on hydrous Fe(III) as a function of aging and metal oxide residence time. Oxide aging did not cause hysteresis of metal cation sorption/desorption. Aging the oxide with the metal cations resulted in hysteresis with Cd^{2+} and Co^{2+} but little with Pb^{2+}. With Pb^{2+} between pH 3 and 5.5 there was slight hysteresis over a 21-week aging process (hysteresis varied from < 2% difference between sorption and desorption to ~10%). At pH 2.5, Pb^{2+} desorption was complete within a 16-h period and was not affected by aging time (Fig. 4.17). However, with Cd^{2+} and Co^{2+}, extensive hysteresis was observed over a 16-week aging period and the hysteresis increased with aging time (Figs. 4.18–4.19). After 16 weeks of aging, 20% of the Cd^{2+} and 53% of the Co^{2+} was not desorbed, and even at pH 2.5, hysteresis was observed. The extent of reversibility with aging for Co^{2+}, Cd^{2+}, and Pb^{2+} was inversely proportional to the ionic radius of the ions, namely, $Co^{2+} < Cd^{2+} < Pb^{2+}$. Ainsworth et al. (1994) attributed the hysteresis to Co and Cd incorporation into a recrystallizing solid (probably goethite) by isomorphic substitution and not to micropore diffusion.

Fuller et al. (1993) combined kinetic sorption and desorption experiments with spectroscopic observations (Waychunas et al., 1993) to study As sorption on ferrihydrite. Using XAFS spectroscopy, they found that As was sorbed predominantly as innersphere bidentate complexes, regardless of whether the As was adsorbed after the mineralization of the ferrihydrite, or it was present during precipitation. No As surface precipitates were observed. Slow As sorption and desorption were

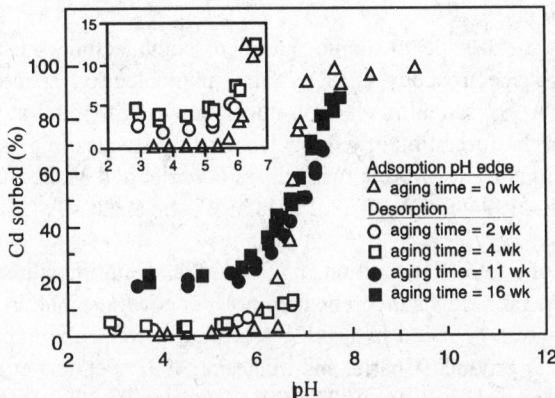

Fig. 4.18 Fractional sorption-desorption of Cd^{2+} to hydrous Fe-oxide (HFO) as a function of pH and HFO-Cd^{2+} aging time; insert shows adsorption-desorption of Cd^{2+} to HFO at 2- and 4-wk aging times [Reprinted from Ainsworth et al., 1994. Soil Sci. Soc. Am. J. 58:1615-1623 with permission of the Soil Science Society of America]

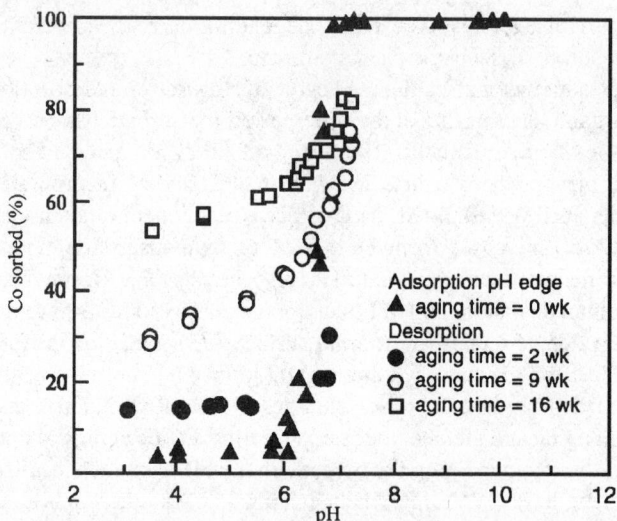

Fig. 4.19 Fractional adsorption of Co^{2+} to hydrous Fe-oxide (HFO) as a function of pH and HFO-Co^{2+} aging time [Reprinted from Ainsworth et al., 1994. Soil Sci. Soc. Am. J. 58:1615-1623 with permission of the Soil Science Society of America]

explained as slow diffusion of the As to or from interior surface complexation sites that exist within disordered aggregates of crystallites. The kinetic reactions could be described using a Fickian diffusion model.

Slow metal sorption/desorption has also been ascribed to conversion of the metal sorbate from a high to a low energy state (Kuo and Mikkelsen, 1980; Padmanabham, 1983; Schultz et al., 1987; Backes et al., 1995). Lehmann and Harter (1984) measured the kinetics of chelate-promoted Cu^{2+} release from a soil to assess the strength of the bond formed. Sorption/desorption was biphasic, which was attributed to high and low energy bonding sites. With increased residence time from 30 min to 24 h, Lehmann and Harter (1984) speculated that there was a transition of Cu from low to higher energy sites (as evaluated by release kinetics). Incubations for up to four days showed a continued uptake of Cu and a decrease in the fraction released within the first 3 min, which was referred to as the low energy sorbed fraction.

Recent studies using surface spectroscopic and microscopic techniques such as XAFS, electron paramagnetic resonance spectroscopy (EPR), X-ray photoelectron spectroscopy (XPS), auger electron spectroscopy (AES), scanning electron microscopy (SEM) and atomic force microscopy (AFM) have shown that the formation of polynuclear surface species (e.g., surface precipitates) on natural materials is an important sorption mechanism (Charlet and Manceau, 1993; Fendorf et al., 1994; Junta and Hochella, 1994; O'Day et al., 1994a,b; Wersin et al., 1994; Scheidegger et al., 1996, 1997, 1998; Towle et al., 1997).

Surface precipitates of Co, Cr(III), Cu, and Ni on oxides and aluminosilicates have been observed at metal surface loadings far below a theoretical monolayer coverage, and in a pH range well below the pH where the formation of metal hydroxide precipitates would be expected according to the thermodynamic solubility product (Charlet and Manceau, 1992; Fendorf et al., 1994; O'Day et al., 1994a,b, 1996; Scheidegger et al., 1996, 1997, 1998; Towle et al., 1997; Xia et al., 1997).

Three different types of polynuclear surface species have been proposed. Formation or sorption of polymers (dimers, trimers, etc.) on the surface as (1) a solid solution or coprecipitate that involves co-ions dissolved from the adsorbent, (2) a precipitate formed on the surface composed of ions from the bulk solution, or (3) their hydrolysis products (Farley et al., 1985; Sposito, 1986; Brown, 1990; Chisholm-Brause et al., 1990; Scheidegger et al., 1996b).

Recent studies have shown that the formation of polynuclear surface species could be a significant cause of slow metal sorption on soil surfaces (Scheidegger et al., 1997, 1998). Such polynuclear surface species do not seem to occur with larger metals such as Pb^{2+} at surface loadings where such species have been observed with smaller metals such as Co^{2+}, Cu^{2+} and Ni^{2+} (Strawn and Sparks, unpublished data). This appears to be related to the mismatch in size between Pb^{2+} (1.19Å), and Al^{3+} (0.535Å) that is contained in the structure of the clay minerals and Al oxides. The Pb^{2+} ion is too large to fit into the mineral structure, while ions such as Ni^{2+} (0.69Å) and Co^{2+} (0.65Å) can.

Initial research with clay mineral systems demonstrated the formation of surface precipitates using XAFS spectroscopy, but the polynuclear structure was not strictly identified (O'Day et al., 1994a,b). Thus, the exact mechanism for surface precipitate formation remained unknown. Recent research in our laboratory suggests that during sorption of metals such as Ni, dissolution of the clay mineral or Al oxide surface leads to precipitation of mixed Ni/Al hydroxide surface species at the mineral/water interface (de la Caillerie et al., 1995; Scheidegger et al., 1996a, 1997, 1998). The precipitates are structurally similar to the mineral takovite (Scheidegger et al., 1997) or synthetic layered double hydroxides (de la Caillerie et al., 1995).

In general, the formation of surface precipitates has been considered a slow phenomenon. However, Scheidegger et al. (1998) have demonstrated that the rate of surface precipitate formation can be quite rapid. The appearance of surface precipitates during sorption of Ni to pyrophyllite at pH 7.5 occurred over time scales less than one hour. Similar results were observed for Ni sorption to kaolinite. However, the kinetics of Ni sorption onto gibbsite, and subsequent surface precipitate formation, were slower than for pyrophyllite (Figs. 4.16 and 4.20). Time resolved XAFS studies

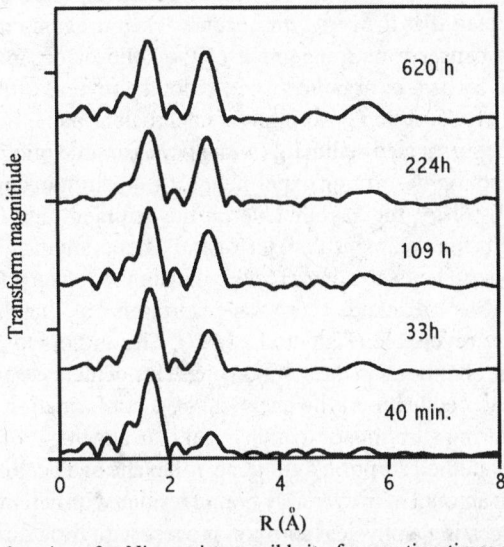

Fig. 4.20 Radial structure functions for Ni sorption to gibbsite for reaction times up to 620 h demonstrating the appearance and growth of second shell contributions due to surface precipitation and growth of a hydrotalcite-like phase [Reprinted from Scheidegger et al., 1998. The kinetics of mixed Ni-Al hydroxide formation on clays and aluminum oxide. Geochim. Cosmochim. Acta 62:2233-2245 with kind permission of Elsevier Science, Amsterdam, Netherlands]

demonstrated continued growth of the surface precipitate during Ni uptake as the structure and Ni:Al stoichiometry of the sorption complex approached that of a takovite-like phase (Scheidegger et al., 1998).

4.5.1.2   Organic Contaminants

Numerous studies on the kinetics of organic chemical sorption/desorption reactions on soils and soil components have shown that sorption/desorption is characterized by a rapid, reversible stage followed by a much slower, nonreversible stage (Karickhoff et al., 1979; DiToro and Horzempa, 1982; Karickhoff and Morris, 1985; Kan et al., 1997), or biphasic kinetics. The rapid phase has been ascribed to retention of the organic chemical in a labile form that is easily desorbed. However, the much slower reaction phase involves the entrapment of the chemical in a nonlabile form that is difficult to desorb. The labile form of the chemical is available for microbial attack, while the nonlabile portion is resistant to biodegradation.

This slower sorption/desorption reaction has been ascribed to intraparticle and interparticle diffusion of the chemical into organic matter and inorganic soil components (Wu and Gschwend, 1986; Steinberg et al., 1987; Ball and Roberts, 1991). Weber and Huang (1996) and Werth and Reinhard (1997) theorize that the slow intraparticle diffusion into SOM can be ascribed to the soft amorphous humic materials and the hard condensed microcrystalline materials. Cornelissen et al. (1997), who studied the temperature dependence of slow adsorption and desorption kinetics of some chlorobenzenes, polychlorinated biphenyls (PCBs) and polycyclic aromatic hydrocarbons (PAHs) in laboratory and field contaminated sediments, obtained activation enthalpies of 60–70 kJ mol^{-1} which are in the range for diffusion in polymers. These values are much higher than those for pore diffusion (20–40 kJ mol^{-1}) suggesting that intraorganic matter diffusion may be a more important mechanism for slow organic chemical sorption than interparticle pore diffusion.

However, some investigators have recently questioned the hypothesis that diffusion processes are responsible for the irreversible sorption of some organic chemicals (Kan et al., 1997; Chang et al., 1997). Kan et al. (1997) noted that irreversibility occurs when organic chemicals sorb on organic materials and the sorption causes a rearrangement of the solid or organic C matrix. Subsequent desorption from the altered sorbent or organic matrix is not the reverse of the sorption process. Kan et al. (1997) studied repeated adsorption/desorption of naphthalene and 2,2',5,5'-tetrachlorobiphenyl on a sediment and an antase surrogated sediment (with no measurable microporosity) amended with a surfactant. In repeated sorption/desorption experiments, the maximum concentrations that resisted desorption (q^{irr}_{max}) were 10 μg g^{-1} for naphthalene on the sediment and 0.36 μg g^{-1} for 2,2',5,5'-tetrachlorobiphenyl on both the sediment and surrogate. The amount of organic chemical in the irreversibly sorbed component increased linearly with the number of sorption steps until q^{irr}_{max} was reached. After q^{irr}_{max} was reached, sorption/desorption of naphthalene and 2,2',5,5'-tetrachlorobiphenyl became reversible (Kan et al., 1997). The authors hypothesized that since the concentration of the organic chemicals in the solution phase is not increased with desorption time, the irreversibility may be due to occlusion of the chemical by a conformation change in the organic C associated with the solid during sorption, or to a physical rearrangement of the organic C phase.

Chang et al. (1997) who studied sorption of toluene, n-hexane and acetone on pressed humic acid disks found an insignificant amount of irreversibly bound residue with activation energies in the range of 42.3–65.8 kJ mol^{-1}, suggesting a physical sorption process with little diffusion.

An example of the biphasic kinetics that is observed for many organic chemical reactions in soils/sediments is shown in Fig. 4.21. In this study, 55% of the labile polychlorinated biphenyl (PCB) was desorbed from sediments in a 24-h period, while little of the remaining 45% nonlabile fraction was

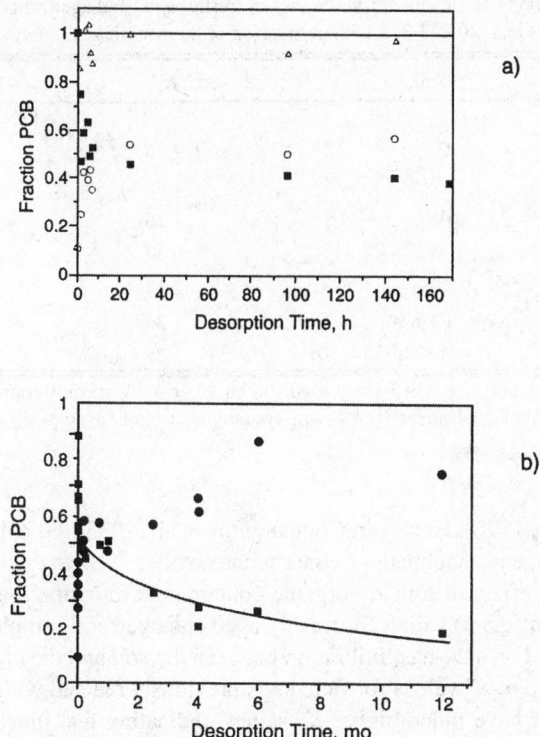

Fig. 4.21 (a) Short-term polychlorinated biphenyl (PCB) desorption in hours (h) from Hudson River sediment con-taminated with 25 mg kg^{-1} PCB. Distribution of the PCB between the sediment (■) and XAD-4 resin (o) is shown, as well as the overall mass balance (Δ). The resin acts as a sink to retain the PCB that is desorbed (Carroll et al., 1994). (b) Long-term PCB desorption in months (mo) from Hudson River sediment contaminated with 25 mg kg^{-1} PCB. Distribution of the PCB between the sediment (■) and XAD-4 resin (•) is shown. The line represents a nonlinear regression of the data by a two-site model [Reprinted with permission from Carroll et al., 1994. Environ. Sci. Technol. 28:253-258. Copyright American Chemical Society]

desorbed in 170 h (Fig. 4.21a). Over another 1-yr period, ~ 50% of the remaining nonlabile fraction desorbed (Fig. 4.21b).

Pavlostathis and Mathavan (1992) observed a biphasic desorption process for field soils contaminated with trichloroethylene (TCE), tetrachloroethylene (PCE), toluene (TOL) and xylene (XYL). A fast desorption reaction occurred in 24 h followed by a much slower desorption reaction beyond 24 h. In 24 h, 9–29%, 14–48%, 9–40%, and 4–37% of the TCE, PCE, TOL, and XYL, respectively, were released. However, the apparent irreversibility or hysteresis may be an artifact caused by not reaching a true sorption equilibrium. For example, DiVincenzo and Sparks (1997), studying pentachlorophenol sorption/desorption on a soil, found that if desorption was initiated after an apparent sorption equilibrium (i.e., slow sorption was measured) was reached, hysteresis or irreversibility was significantly reduced.

A number of studies have also shown that with aging the nonlabile portion of the organic chemical in the soil/sediment becomes more resistant to release (McCall and Agin, 1985; Steinberg et al., 1987; Pignatello and Huang, 1991; Pavlostathis and Mathavan, 1992; Scribner et al., 1992; Alexander,

Table 4.4 Sorption distribution coefficients for herbicides in freshly aged and aged soils [Adapted from Pignatello and Huang, 1991. J. Environ. Qual. 20:222-228 with permission of the American Society of Agronomy]

Herbicide	Soil	K_d[‡]	K_{app}[§]
		L kg^{-1}	
Metolachlor	CVa	2.96	39
	CVb	1.46	27
	W1	1.28	49
	W2	0.77	33
Atrazine	CVa	2.17	28
	CVb	1.32	29
	W3	1.75	4

[‡] Sorption distribution coefficient (L kg^{-1}) of freshly aged soil based on a 24-hr equilibration period.
[§] Apparent sorption distribution coefficient (L kg^{-1}) in contaminated soil (aged soil) determined using a 24-hr equilibration period.

1995; Loehr and Webster, 1996). However, Connaughton et al. (1993) did not observe the nonlabile fraction increasing with age for naphthalene contaminated soils.

One way to gauge the effect of time on organic contaminant retention in soils is to compare K_d (sorption distribution coefficient) values for freshly aged and aged soil samples. In most studies, K_d values are measured based on a 24-h equilibration between the soil and the organic chemical. When these values are compared to K_d values for field soils previously reacted with the organic chemical (aged samples), the latter have much higher K_d values, indicating that much more of the organic chemical is in a sorbed state. For example, Pignatello and Huang (1991) measured K_d values in freshly aged (K_d) and "aged" soils (K_{app}, apparent sorption distribution coefficient) reacted with atrazine and metolachlor, two widely used herbicides. The aged soils had been treated with the herbicides 15–62 months before sampling. The K_{app} values ranged from 2.3–42 times higher than the K_d values (Table 4.4).

Scribner et al. (1992) studying simazine (a widely used triazine herbicide for broadleaf and grass control in crops) desorption and bioavailability in aged soils found that K_{app} values were 15 times higher than K_d values. Scribner et al. (1992) also showed that 48% of the simazine added to the freshly aged soils was biodegradable over a 34-day incubation period while none of the simazine in the aged soil was biodegraded.

One of the implications of these results is that while many transport and degradation models for organic contaminants in soils and waters assume that the sorption process is an equilibrium process, the above studies clearly show that kinetic reactions must be considered when making predictions about the mobility and fate of organic chemicals. Moreover, calculation of K_d values based on a 24-h equilibration period, which are commonly used in fate and risk assessment models, can be inaccurate since 24-h K_d values often overestimate the amount of organic chemical in the solution phase.

The finding that many organic chemicals are quite persistent in the soil environment has both good and bad features. The beneficial aspect is that the organic chemicals are less mobile and may not be readily transported in groundwater supplies. The negative aspect is that their persistence and inaccessibility to microbes may make decontamination more difficult, particularly if *in situ* remediation techniques such as biodegradation are employed.

4.5.2   Kinetics of Mineral Dissolution

4.5.2.1   Rate Limiting Steps

Dissolution of minerals involves several steps (Stumm and Wollast, 1990): (1) mass transfer of dissolved reactants from the bulk solution to the mineral surface, (2) adsorption of solutes, (3) interlattice transfer of reaction species, (4) surface chemical reactions, (5) removal of reactants from the surface, and (6) mass transfer of products into the bulk solution. Under field conditions mineral dissolution is slow and mass transfer of reactants or products in the aqueous phase (Steps 1 and 6) is not rate limiting. Thus, the rate limiting steps are either transport of reactants and products in the solid phase (Step 3) or surface CRs (Step 4) and removal of reactants from the surface (Step 5).

Transport controlled dissolution reactions or those controlled by mass transfer or diffusion can be described using a parabolic rate law given below (Stumm and Wollast, 1990):

$$r = \frac{dC}{dt} = kt^{-1/2} \qquad\qquad\qquad\qquad [4.17]$$

where r is the reaction rate, C is the concentration in solution, t is time, and k is the reaction rate constant. Integrating, C increases with $t^{1/2}$,

$$C = C_0 + 2kt^{1/2} \qquad\qquad\qquad\qquad [4.18]$$

where C_0 is the initial concentration in solution.

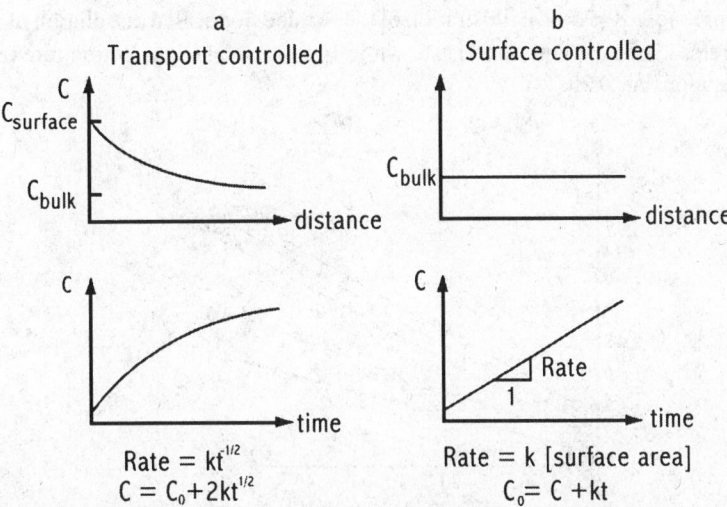

Fig. 4.22   Transport versus surface-controlled dissolution. Schematic representation of concentration in solution, C, as a function of distance from the surface of the dissolving mineral. In the lower part of the figure, the change in concentration is given as a function of time [Reprinted from Stumm, 1992. Chemistry of the solid-water interface. with permission of John Wiley & Sons, Inc.].

If the surface reactions are slow compared to the transport reactions, dissolution is surface controlled, which is the case for most dissolution reactions of silicates and oxides. In surface-controlled reactions, the concentrations of solutes next to the surface are equal to the bulk solution concentrations and the dissolution kinetics are zero order if steady-state conditions are operational on the surface. Thus, the dissolution rate (r) is

$$r = \frac{dC}{dt} = kA \qquad\qquad\qquad [4.19]$$

and r is proportional to the mineral's surface area, A. Thus, for a surface-controlled reaction the relationship between t and C should be linear. Fig. 4.22 compares transport- and surface-controlled dissolution mechanisms.

Intense arguments have ensued over the years concerning the mechanism for mineral dissolution. Those that supported a transport-controlled mechanism believed that a leached layer formed as mineral dissolution proceeded and that subsequent dissolution took place by diffusion through the leached layer (Wollast, 1967; Petrovic et al., 1976). Advocates of this theory found that dissolution was described by the parabolic rate law (Equation [4.18]). However, the apparent transport-controlled kinetics may be an artifact caused by dissolution of hyperfine particles formed on the mineral surfaces after grinding that are highly reactive sites, or by use of batch methods that cause reaction products to accumulate causing precipitation of secondary minerals. These experimental artifacts can cause incongruent reactions and pseudoparabolic kinetics. Studies employing XPS and nuclear resonance profiling (Schott and Petit, 1987; Casey et al., 1989) have demonstrated that, although some incongruency may occur in the initial dissolution process, which may be diffusion controlled, the overall reaction is surface controlled. Energies of activation from 60–86 kJ mol^{-1} have been observed for dissolution of oxides and silicates, further suggesting surface-controlled dissolution (Lasaga, 1984; Jordan and Rammensee, 1996). An illustration of the surface-controlled dissolution of γ-Al$_2$O$_3$ resulting in a linear release of Al^{3+} with time is shown in Fig. 4.23. The dissolution rate (r) can be obtained from the slope of Fig. 4.23.

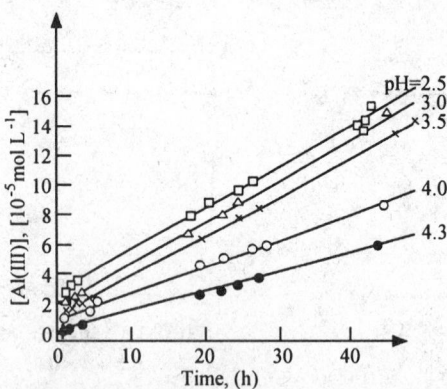

Fig. 4.23 Linear dissolution kinetics observed for the dissolution of γ-Al$_2$O$_3$. Representative of processes whose rates are controlled by a surface reaction and not by transport [Reprinted from Furrer and Stumm, 1986. The coordination chemistry of weathering. Geochim Cosmochim. Acta 50:1847-1860 with kind permission of Elsevier Science, Amsterdam, Netherlands]

Scanning force microscopy (SFM) which has also been used increasingly as an *in situ* technique for imaging mineral surfaces immersed in aqueous solution during the course of dissolution (Hellman et al., 1992; Hillner et al., 1992a,b; Johnsson et al., 1992; Bosbach and Rammensee, 1994; Maurice et al., 1995) permits a direct measure of surface-controlled dissolution rates by providing 3-dimensional data on changes in microtopography. *In situ* SFM has the unique ability to detect separate processes, such as dissolution and secondary phase formation, occurring simultaneously on a mineral surface (Maurice, 1997).

4.5.2.2 Surface-Controlled Dissolution Mechanisms

Dissolution of oxide minerals through a surface-controlled reaction by ligand and proton promoted processes has been described by Furrer and Stumm (1986), Zinder et al. (1986), and Stumm and Furrer (1987) using a surface coordination approach. The important reactants in these processes are H_2O, H^+, OH^-, ligands, and reductants and oxidants. The reaction mechanism occurs in two steps (Stumm and Wollast, 1990):

$$\text{Surface sites } + \text{ Reactants } (H^+,\ OH^-,\ \text{ligands}) \xrightarrow{\text{fast}} \text{Surface species} \qquad [4.20]$$

$$\text{Surface species} \xrightarrow[\text{detachment of metal(M)}]{\text{slow}} M(aq) \qquad [4.21]$$

Thus, the attachment of the reactants to the surface sites is fast and detachment of metal species from the surface into solution is slow and rate limiting.

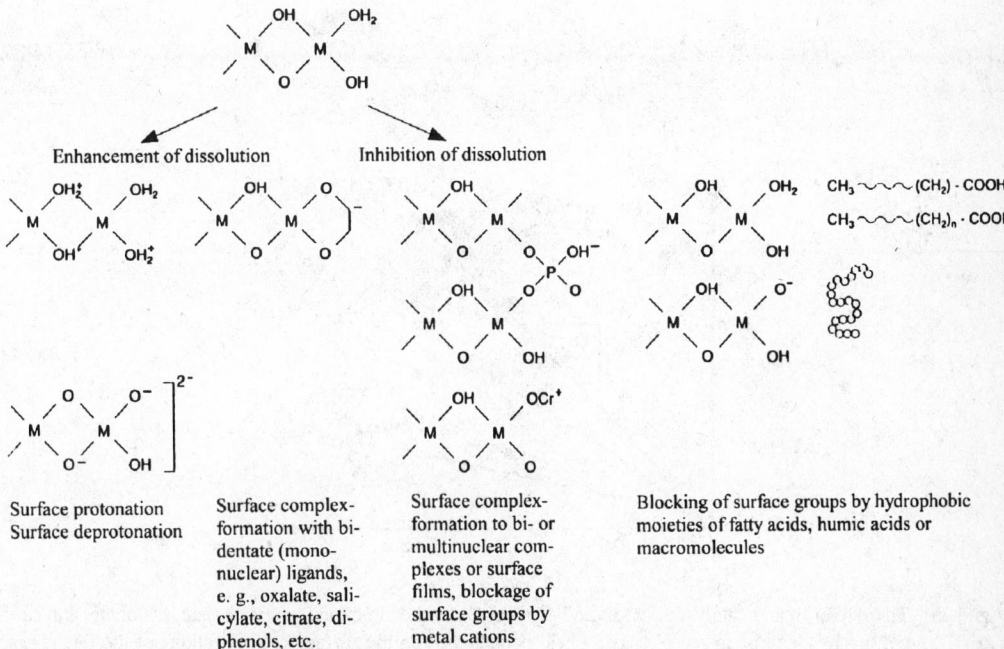

Fig. 4.24 The dependence of surface reactivity and of kinetic mechanisms on the coordinative environment of the surface groups [Reprinted from Stumm and Wollast, 1990. Rev. Geophys. 28:53-69, with permission]

4.5.2.3   Ligand Promoted Dissolution

Fig. 4.24 shows how the surface chemistry of the mineral affects dissolution. One sees that surface protonation of the surface ligand increases dissolution by polarizing interatomic bonds close to the central surface ions that promote the release of a cation surface group into solution. Hydroxyls that bind to surface groups at higher pHs can ease the release of an anionic surface group into the solution phase.

Ligands that form surface complexes by ligand exchange with a surface hydroxyl add negative charge to the Lewis acid center coordination sphere, and lower the Lewis acid acidity. This polarizes the M-O bonds causing detachment of the metal cation into the solution phase. Thus, innersphere surface complexation plays an important role in mineral dissolution. Ligands such as oxalate, salicylate, F^-, EDTA, and NTA increase dissolution, but others, such as SO_4^{2-}, CrO_4^{2-} and benzoate, inhibit dissolution. Phosphate, arsenate, and selenite enhance dissolution at low pH and dissolution is inhibited at pH > 7 (Bondietti et al., 1993).

The reason for these differences may be that bidentate species that are mononuclear promote dissolution while binuclear bidentate species inhibit dissolution. With binuclear bidentate complexes, more energy may be needed to remove two central atoms from the crystal structure. With phosphate and arsenate, at low pH, mononuclear species are formed while at higher pH (~ pH 7), binuclear or trinuclear surface complexes form. Mononuclear bidentate complexes are formed with oxalate while binuclear bidentate complexes form with CrO_4^{2-}. Additionally, the electron donor properties of CrO_4^{2-} and oxalate are also different. With CrO_4^{2-}, a high redox potential is maintained at the oxide surface which restricts reductive dissolution (Stumm and Wollast, 1990; Stumm, 1992). Dissolution can also be inhibited by cations such as VO^{2+}, Cr(III) and Al(III) that block surface functional groups (Bondietti et al., 1993).

One can express the rate of the ligand promoted dissolution (R_L) as:

$$R_L = k_L'(\equiv ML) = k_L'C_L^s$$ [4.22]

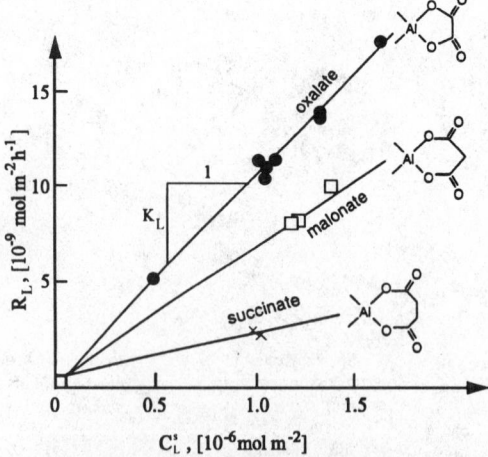

Fig. 4.25   The rate of ligand catalyzed dissolution of $\gamma\text{-}Al_2O_3$ by the aliphatic ligands oxalate, malonate, and succinate, R_L (nmol m^{-2} h^{-1}), can be interpreted as a linear dependence on the surface concentrations of the ligand complexes, C_L^s. In each case the individual values for C_L^s were determined experimentally [Reprinted from Furrer and Stumm, 1986. The coordination chemistry of weathering. Geochim. Cosmochim. Acta 50:1847-1860 with kind permission of Elsevier Science, Amsterdam, Netherlands].

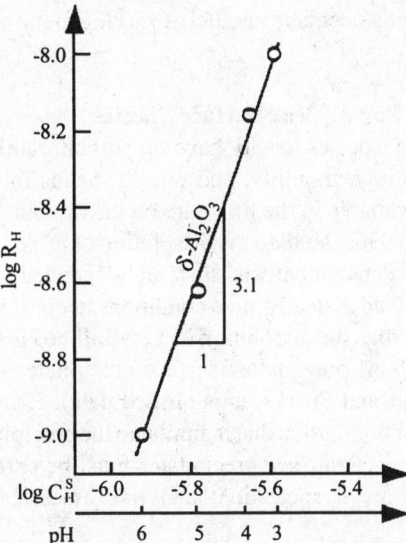

Fig. 4.26 The dependence of the rate of proton promoted dissolution of γ-Al_2O_3, R_H (mol m^{-2} h^{-1}), on the surface concentration of the proton complexes, C^s_H (mol m^{-2}). [Reprinted from Furrer and Stumm, 1986. The Coordination chemistry of weathering. Geochim Cosmochim. Acta 57:2271-2282 with kind permission of Elsevier Science, Amsterdam, Netherlands]

where k'_L is the rate constant for ligand promoted dissolution (t^{-1}), $\equiv ML$ is the metal-ligand complex, and C^s_L is the surface concentration of the ligand (mol m^{-2}). Equation [4.22] adequately describes ligand promoted dissolution of γ-Al_2O_3 (Fig. 4.25).

4.5.2.4 Proton Promoted Dissolution

Under acid conditions, protons can promote mineral dissolution by binding to surface oxide ions, causing bonds to weaken which is followed by detachment of metal species into solution. The proton promoted dissolution rate (R_H) can be expressed as (Stumm and Wollast, 1990):

$$R_H = k'_H \left(\equiv MOH_2^+ \right)^j = k'_H \left(C^s_H \right)^j \qquad [4.23]$$

where k'_H is the rate constant for proton promoted dissolution, $\equiv MOH_2^+$ is the metal-proton complex, C^s_H is the concentration of the surface adsorbed proton complex (mol^{-2}) and j corresponds to the oxidation state of the central metal ion in the oxide structure (i.e., $j = 3$ for Al(III) and Fe(III) in simple cases). If dissolution occurs by only one mechanism, j is an integer. Fig. 4.26 shows an application of Equation [4.23] for the proton promoted dissolution of γ-Al_2O_3.

4.5.2.5 Overall Dissolution Mechanisms

The rate of mineral dissolution, which is the sum of the ligand, proton, and deprotonation promoted (or bonding of OH$^-$ ligands) dissociation [$R_{OH} = k'_{OH} \left(C^s_{OH} \right)^i$] rates along with the pH independent portion of the dissolution rate $\left(k'_{H_2O} \right)$ which is due to hydration, can be expressed as (Stumm and Wollast, 1990):

$$R = k'_L \left(C^s_L \right) + k'_H \left(C^s_H \right)^j + k'_{OH} \left(C^s_{OH} \right)^i + k'_{H_2O} \qquad [4.24]$$

Equation [4.24] is valid if dissolution occurs in parallel at varying metal centers (Furrer and Stumm, 1986).

4.5.2.6   Dissolution Kinetics of Polynuclear Surface Species

While polynuclear metal surface species could have important ramifications with respect to environmental quality (bioavailability, mobility, and fate of metals in soils and waters) through dissolution, little information is available in the literature on the dissolution kinetics of polynuclear species. Scheidegger and Sparks (1996a) studied the dissolution of mixed Ni-Al polynuclear surface species from pyrophyllite. Nickel detachment was slow and depended strongly on the pH and the experimental method (Fig. 4.27). Under steady-state conditions, a constant Ni detachment rate was observed, which was much slower than the dissolution of a crystalline $Ni(OH)_2$ reference compound.

The structures of the mixed Ni-Al polynuclear surface precipitates are also not changed after extensive dissolution (Scheidegger and Sparks, unpublished data). EXAFS analysis shows that a tacovite-like structure remains after dissolution that is similar to the precipitate before commencement of dissolution. It is obvious that metal surface precipitates must be carefully considered in metal surface complexation modeling, metal speciation, and risk assessments for the migration of contaminants in polluted sites.

4.5.3   Redox Kinetics

It is well-known that Mn (III/IV), Fe(III), Co(III), and Pb(IV) oxides/hydroxides are thermodynamically stable in oxygenated systems at neutral pH. However, under anoxic conditions, reductive dissolution of oxides/hydroxides by reducing agents occurs as shown below for MnOOH and MnO_2 (Stone, 1991):

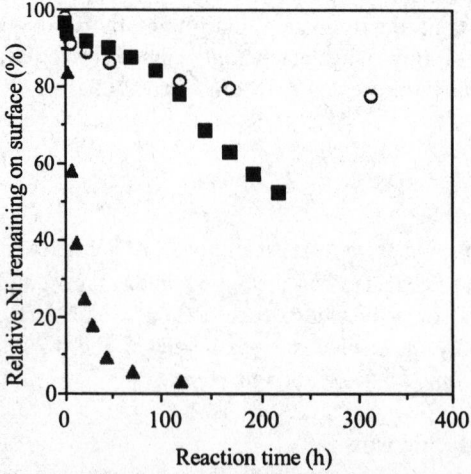

Fig. 4.27   Kinetics of Ni detachment from surface precipitates at pH = 4. Relative Ni remaining on the surface (%) is shown for the *conventional method* (O) and the *replenishment method* (■) as a function of the reaction time. 98% of the initial Ni was sorbed in the beginning of the detachment experiment. The dissolution of an equivalent amount of crystalline $Ni(OH)_2$ (in mol) at pH = 4 is given for comparison (▲) [Reprinted from Scheidegger and Sparks (1996a). Chem. Geol. 132:157-164, with permission].

$$Mn(III)OOH(s) + 3H^+ + e^- = Mn^{2+} + 2H_2O \qquad E^0 = +1.50V \qquad [4.25]$$

$$Mn(IV)O_2(s) + 4H^+ + 2e^- = Mn^{2+} + 2H_2O \qquad E^0 = +1.23V \qquad [4.26]$$

Changes in the oxidation state of the metals associated with the oxides above can greatly affect their solubility and mobility in soil and aqueous environments. The reductants can be either inorganic or organic.

There are a number of natural and xenobiotic organic functional groups that are good reducers of oxides and hydroxides. These include carboxyl, carbonyl, phenolic, and alcoholic functional groups of SOM. Microorganisms in soils and sediments are also examples of organic reductants. Stone (1987a) showed that oxalate and pyruvate, two microbial metabolites, could reduce and dissolve Mn(III/IV) oxide particles. Inorganic reductants include As(III), Cr(III), and Pu(III).

4.5.3.1 Mechanisms for Reductive Dissolution of Metal Oxides/Hydroxides

The reductive dissolution of metal oxides/hydroxides appears to occur in the following sequential steps (Stone, 1986; 1991): (1) diffusion of the reductant molecules to the oxide surface, (2) a surface chemical reaction, and (3) release of reaction products and diffusion away from the oxide surface. Steps (1) and (3) are transport steps. The rate-controlling step in reductive dissolution of oxides appears to be surface chemical reaction control. Reductive dissolution can be described by both inner- and outersphere complex mechanisms that involve (1) precursor complex formation, (2) electron transfer, and (3) breakdown of the successor complex (Fig. 4.28). Innersphere and outersphere precursor complex formation are adsorption reactions that increase the density of reductant molecules at the oxide surface which promotes electron transfer (Stone, 1991). In the innersphere mechanism, the reductant enters the inner coordination sphere by ligand exchange and bonds directly to the metal center prior to electron transfer. With the outersphere complex, the inner coordination sphere is left intact and electron transfer is enhanced by an outersphere precursor complex (Stone, 1986). Kinetic studies have shown that high rates of reductive dissolution are favored by high rates of precursor complex formation (i.e., large k_1 and low k_{-1} values), high electron transfer rates (i.e., large k_2), and high rates of product release (i.e., high k_3) (Fig. 4.28).

Specifically adsorbed cations and anions may reduce reductive dissolution rates by blocking oxide surface sites or by causing release of Mn(II) into solution. Stone and Morgan (1984a) showed that PO_4^{3-} inhibited the reductive dissolution of Mn(III/IV) oxides by hydroquinone. Addition of 10^{-2} M

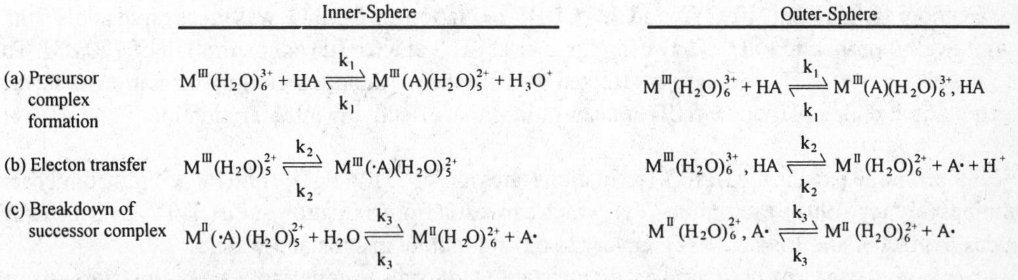

Fig. 4.28 Reduction of $M(H_2O)_6^{3+}$ by phenol (HA) in homogeneous solution [From Stone, 1986. ACS Symp. Ser 323:446-461 with permission of the American Chemical Society]

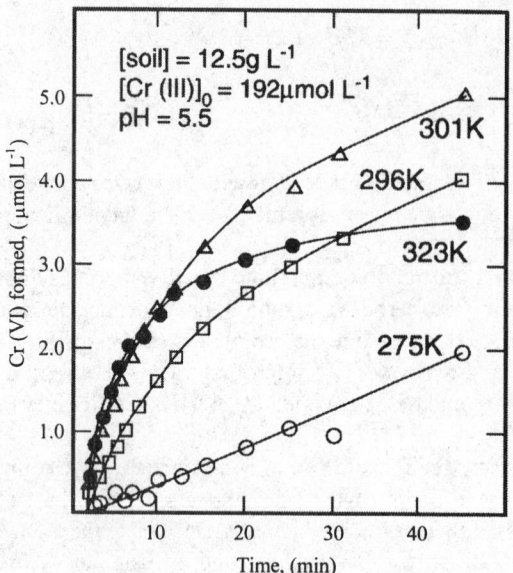

Fig. 4.29   Effect of temperature on the kinetics of Cr(III) oxidation in moist Hagerstown silt loam soil [From Amacher and Baker, 1982]

PO_4^{3-} at pH 7.68 caused the dissolution rate to be only 25% of the rate when PO_4^{3-} was not present. Phosphate had a greater effect than Ca^{2+}.

4.5.3.2   Oxidation of Pollutants

As mentioned earlier, Mn oxides can oxidize a number of environmentally important ions that can be toxic to humans and animals. Chromium and plutonium are similar in their chemical behavior in aqueous settings (Rai and Serne, 1977; Bartlett and James, 1979). They can exist in multiple oxidation states and as both cationic and anionic species. Chromium (III) is quite stable and innocuous, and occurs as Cr^{3+} and its hydrolysis products or as CrO_2^{-}. Chromium (III) can be oxidized to Cr(VI) by Mn(III/IV) oxides (Bartlett and James, 1979; Fendorf and Zasoski, 1992). Chromium(VI) is mobile in the soil environment and is a suspected carcinogen. It occurs as the dichromate ($Cr_2O_7^{2-}$) or chromate ($HCrO_4^{-}$ and CrO_4^{2-}) anions (Huang, 1991).

Fig. 4.29 shows the oxidation kinetics of Cr(III) to Cr(VI) in a soil. Most of the oxidation occurred during the first hour. At higher temperatures, there was a rapid oxidation rate, followed by a slower rate. Fendorf and Zasoski (1992) found that Cr(III) oxidation on δ-MnO_2 was more rapid at pH 5 than 3 with overall production of Cr(VI) being greater at pH 3 at a Cr(III) concentration of 770 μM. The rate and extent of Cr(III) oxidation is affected by a number of factors including formation of surface precipitates at higher pHs and Cr(III) concentrations that effectively inhibit oxidation (Fendorf et al., 1992).

Plutonium can exist in the III to VI oxidation states as Pu^{3+}, Pu^{4+}, PuO_2^{+}, and PuO_2^{2+} in strongly acid solutions (Huang, 1991). Plutonium (VI), which can result from oxidation of Pu (III/IV) by Mn(III/IV) oxides (Amacher and Baker, 1982), is very toxic and mobile in soils and waters.

Arsenic (As) can exist in several oxidation states and forms in soils and waters. In waters, As can exist in the $+5, +3, 0$, and -3 oxidation states. Arsenite, As(III), and arsine (AsH_3, where the oxidation state of As is – 3) are much more toxic to humans than arsenate, As(V). Manganese (III/IV) oxides can oxidize As(III) to As(V) as shown below where As(III) as $HAsO_2$ is added to MnO_2 to produce As(V) as H_3AsO_4 (Oscarson et al., 1983):

$$HAsO_2 + MnO_2 = (MnO_2) \cdot HAsO_2 \qquad [4.27]$$

$$(MnO_2) \cdot HAsO_2 + H_2O = H_3AsO_4 + MnO \qquad [4.28]$$

$$H_3AsO_4 = H_3AsO_4^- + H^+ \qquad [4.29]$$

$$H_2AsO_4^- = HAsO_4^{2-} + H^+ \qquad [4.30]$$

$$(MnO_2)HAsO_2 + 2H^+ = H_3AsO_4 + Mn^{2-} \qquad [4.31]$$

Equation [4.28] involves the formation of an adsorbed layer. Oxygen transfer occurs and $HAsO_2$ is oxidized to H_3AsO_4 (Equation [4.28]). At pH ≤ 7, the predominant As(III) species is arsenious acid ($HAsO_2$), but the oxidation product, H_3AsO_4, will dissociate and form the same quantities of $H_2AsO_4^-$ and $HAsO_4^{2-}$ with little H_3AsO_4 present at equilibrium (Eqs. 4.29 and 4.30). Each mole of As(III) oxidized releases about 1.5 mol H^+. The H^+ produced after H_3AsO_4 dissociation reacts with the adsorbed $HAsO_2$ on MnO_2, forming H_3AsO_4, and leads to the reduction and dissolution of Mn(IV) (Equation [4.31]). Thus, every mole of As(II) that is oxidized to As(V) results in 1 mol of Mn(IV) in the solid phase being reduced to Mn(II) and partially dissolved in solution (Oscarson et al., 1981).

Oscarson et al. (1980) studied the oxidation of As(III) to As(V) in sediments from five lakes in Saskatchewan, Canada. Oxidation of As(III) to As(V) occurred within 48 hr. In general, > 90% of the added As was sorbed on the sediments within 72 hr. Scott and Morgan (1995) studied the oxidation of As(III) by synthetic birnessite. The depletion of As(III) was rapid with 50% of the initial As(III) removed from solution in 10 minutes and after 90 minutes the As(III) concentration was below the detection limit. Arsenic (V) was released into solution as fast as As(III) was depleted and the total

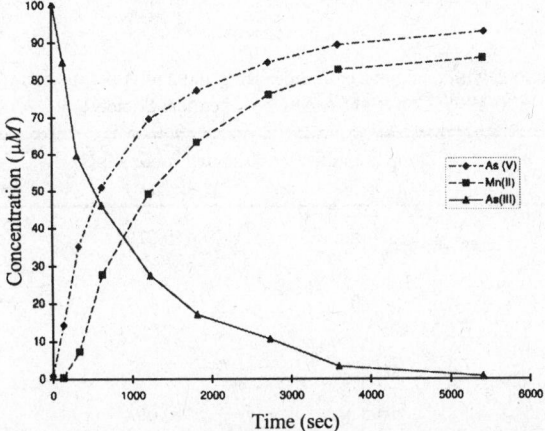

Fig. 4.30 Experimental behavior of aqueous As(III), As(V) and Mn(II) following 99.6 μM As(III) addition to a 0.21 g L^{-1} of γ-MnO_2 particle suspension at pH 4, 25 $^\circ$C, and 0.1 M NaClO$_4$ [Reprinted with permission from Scott and Morgan, 1995. Environ. Sci. Technol. 29:1898-1905. Copyright American Chemical Society]

concentration of aqueous As was about constant over the duration of the experiment (Fig. 4.30). Scott and Morgan (1995) concluded that the process of electron transfer and release of As(V) were fast compared to the sorption of As(III), the rate limiting step.

Manganese oxides and hydroxides (e.g., Mn_3O_4 and MnOOH) may also catalyze the oxidation of other trace metals such as Co^{2+}, Co^{3+}, Cu^{2+}, Ni^{2+}, Ni^{3+}, and Pb^{2+} by disproportionation to Mn^{2+} and MnO_2 (Hem, 1978). The disproportionation results in vacancies in the Mn oxide structure. Since the Mn^{2+} and Mn^{3+} in the oxides have similar physical sizes as Co^{2+}, Co^{3+}, Cu^{2+}, Ni^{2+}, Ni^{3+}, and Pb^{2+}, these metals can occupy the vacancies in the Mn oxide and become part of the structure. With disproportionation or with other redox processes involving the Mn oxides, the solubility of the metals can be affected. For example, if during the disproportionation process Co_3O_4, the oxidized form of the metal forms from Co^{2+}, the reaction can be expressed as (Hem, 1978):

$$2Mn_3O_4(s) + 3Co^{2+} + 4H^+ \rightarrow MnO_2(s) + Co_3O_4(s) + 5Mn^{2+} + 2H_2O \qquad [4.32]$$

and the equilibrium constant ($K°$) is (Hem, 1978)

$$\left(Mn^{2+}\right)^5 / \left(Co^{2+}\right)^3 \left(H^+\right)^4 = 10^{18.73} \qquad [4.33]$$

Thus, the oxidation of Co(II) to Co(III) reduces its solubility and mobility in the environment. Using XPS analyses (Murray and Dillard, 1979), this reaction has been shown to occur. More recent evidence for heterogeneous redox reactions of trace metals is discussed in Section B, chapter 9. Scott and Morgan (1996) studied the oxidation of Se(IV) by synthetic birnessite. Se(IV) was oxidized to Se(VI) with Se(VI) first appearing in the aqueous suspension after 12 h and was produced at a constant rate over the duration of the experiment (28 days). Scott and Morgan (1996) suggested the following oxidation mechanism: (1) birnessite directly oxidized Se(IV) through a surface complex mechanism; (2) the rate limiting step in the production of Se(VI) was the electron transfer step involving a transfer of two electrons from the anion to the metal ion, breaking of two Mn-O bonds, and addition of an O

Table 4.5 Inorganic redox reactions with manganese dioxides [Reprinted with permission from Scott and Morgan, 1995. Environ. Sci. Technol. 29:1898-1905. Copyright American Chemical Society]

System	Time to oxidize 50%	Driving force at pH 4 $\Delta e°$ (V)[‡]	Source
δ-MnO_2: As(III) → As(V) pH 4, 25 °C, 14 $m^2 L^{-1}$	10 min	+0.529	Scott and Morgan (1995)
δ-MnO_2: Se(IV) → Se(VI)			
pH 4, 35 °C, 14 $m^2 L^{-1}$	10 days		
pH 4, 25 °C, 28 $m^2 L^{-1}$	16 days	+0.092	
pH 4, 25 °C, 14 $m^2 L^{-1}$	30 days		
β-MnO_2: Cr(III) → Cr(VI) pH 4, 25 °C, 71 $m^2 L^{-1}$	95 days	+0.011	Eary and Rai (1987)

‡ The activity ratio for each oxidant/reductant pair is taken as unity.

from water to Se(VI); and (3) the reaction products Se(VI) and Mn(II) were released from the surface by different steps.

Scott and Morgan (1996) compared their results to those of Eary and Rai (1987) who studied Cr(III) oxidation by pyrolusite (ß-MnO_2) between pH 3.0 and 4.7 and Scott and Morgan (1995) who studied As(III) oxidation by birnessite (δ-MnO_2) at pH values between 4.0 and 8.2 (Table 4.5). The Cr(III) redox transformation on pyrolusite was slowest which Scott and Morgan (1996) attributed to unfavorable adsorption on both a positively charged surface and aqueous species and the small thermodynamic driving force. Also, the transfer of three electrons from Cr(III) to Mn(IV) requires the involvement of more than one Mn(IV) atom per Cr(III) atom.

Manganese oxides also appear to play an important role in ligand facilitated metal transport. Using soil columns that consisted of fractured saprolite coated with amorphous Fe- and Mn-oxides, Jardine et al. (1993) studied the transport of Co(II) $EDTA^{2-}$, a mixture of Co(II) $EDTA^{2-}$ and Co(III) $EDTA^{-}$ and Sr $EDTA^{2-}$. The Mn-oxides oxidized Co(II)-$EDTA^{2-}$ into Co(III)-$EDTA^{-}$, a very stable complex (log K value of 41.4, Xue and Traina, 1996). The formation of this complex resulted in enhanced transport of Co.

Xue and Traina (1996) found that an aerobic goethite suspension catalyzed oxidation of Co(II) $EDTA^{2-}$ to Co(III)-$EDTA^{-}$ by dissolved O_2. The kinetics were described using a pseudo-first-order rate constant, k_1 of 0.0078 ± 0.0002 h^{-1} at pH 5 and a goethite concentration of 3.09 g L^{-1}.

A number of investigators have studied the reductive dissolution of Mn oxides by organic pollutants such as hydroquinone (Stone and Morgan, 1984a), substituted phenols (Stone, 1987b), and other organic compounds (Stone and Morgan, 1984b). With substituted phenols, the rate of dissolution was proportional to substituted phenol concentration and the rate increased as pH decreased (Stone, 1987b). Phenols containing alkyl, alkoxy, or other electron donating substituents were more slowly degraded; p-nitrophenol reacted slowly with Mn(III/IV) oxides. The increased rate of reductive dissolution at lower pH may be due to more protonation reactions that enhance the creation of precursor complexes or increases in the protonation level of the surface precursor complexes that increase electron transfer rates (Stone, 1987b). Further discussions on this topic can be found in Section B, Chapter 9.

4.6   Conclusions

Research on the kinetics and mechanisms of soil chemical reactions will be a common theme in soil and environmental sciences for decades to come. This research emphasis is in large part due to the need to more accurately understand and predict the long-term fate and transport of contaminants in the subsurface environment. Without such data, economically sound decisions about soil remediation cannot be made and risk assessment models are incomplete and most probably inaccurate.

While some very fine and informative research on rates and mechanisms of soil chemical reactions/ processes has been conducted in the past few decades, there are many gaps that need to be filled. The following research is needed: (1) more accurate kinetic models that describe reactions on multireactive, heterogeneous particle surfaces; (2) long-term sorption and particularly, desorption rate studies; (3) a better understanding of residence time effects on plant nutrient, radionuclide, metal, and organic retention/release mechanisms on soils and other natural materials; (4) an increased knowledge of nucleation/precipitation and dissolution reaction rate phenomena at the mineral/water interface and their effect on nutrient/contaminant mobility and bioavailability in the soil environment; (5) more studies on the kinetics and mechanisms of redox processes in soils, particularly the role that soil components such as Mn oxides have on oxidation/reduction of inorganic and organic pollutants; and, (6) increased use of time resolved, *in situ* spectroscopic and microscopic techniques to confirm reaction mechanisms.

4.7   References

Aharoni, C. 1984. Kinetics of adsorption: The S shaped Z(t) plot. Adsorpt. Sci. Technol. 1:1–29.

Aharoni, C., and D.L. Sparks. 1991. Kinetics of soil chemical reactions: A theoretical treatment. p. 1–18. *In* D.L. Sparks and D.L. Suarez (ed.) Rates of soil chemical processes. Soil Sci. Soc. Am. Spec. Pub. 27, Soil Science Society of America, Madison, WI.

Aharoni, C., and Y. Suzin. 1982a. Application of the Elovich equation to the kinetics of occlusion: Part 1. Homogenous microporosity. J. Chem. Soc. Farad. Trans. 178:2313–2320.

Aharoni, C., and Y. Suzin. 1982b. Application of the Elovich equation to the kinetics of occlusion: Part 3. Heterogenous microporosity. J. Chem. Soc. Farad. Trans. 178:2329–2336.

Aharoni, C., and M. Ungarish. 1976. Kinetics of activated chemisorption. Part 1. The non-Elovichian part of the isotherm. J. Chem. Soc. Farad. Trans. 172:400–408.

Ainsworth, C.C., J.L. Pilou, P.L. Gassman, and W.G.V.D. Sluys. 1994. Cobalt, cadmium, and lead sorption to hydrous iron oxide: Residence time effect. Soil Sci. Soc. Am. J. 58:1615–1623.

Alexander, M. 1995. How toxic are toxic chemicals in soils? Environ. Sci. Technol. 29:2713–2717.

Allen, E.R., D.W. Ming, L.R. Hossner, and D.L. Henninger. 1995. Modeling transport kinetics in clinoptilolite-phosphate rock systems. Soil Sci. Soc. Am. J. 59:248–255.

Amacher, M.C. 1991. Methods of obtaining and analyzing kinetic data. p.19–59. *In* D.L. Sparks and D.L. Suarez (ed.) Rates of soil chemical processes. Soil Sci. Soc. Am. Spec. Pub. 27, Soil Science Society of America, Madison, WI.

Amacher, M.C., and D.E. Baker. 1982. Redox reactions involving chromium, plutonium, and manganese in soils. Pennsylvannia State University, University Park, PA.

Atkinson, R.J., F.J. Hingston, A.M. Posner, and J.P. Quirk. 1970. Elovich equation for the kinetics of isotope exchange reactions at solid-liquid interfaces. Nature 226:148–149.

Backes, C.A., R.G. McLaren, A.W. Rate, and R.S. Swift. 1995. Kinetics of cadmium and cobalt desorption from iron and manganese oxides. Soil Sci. Soc. Am. J. 59:778–785.

Ball, W.P., and P.V. Roberts. 1991. Long-term sorption of halogenated organic chemicals by aquifer material: 1. Equilibrium. Environ. Sci. Technol. 25: 1223–1236.

Bartlett, R.J., and B.R. James. 1979. Behavior of chromium in soils. III. Oxidation. J. Environ. Qual. 8:31–35.

Bernasconi, C.F. 1976. Relaxation kinetics. Academic Press, New York, NY.

Bondietti, G., J. Sinniger, and W. Stumm. 1993. The reactivity of Fe(III) (hydr)oxides: Effects of ligands in inhibiting the dissolution. Colloid. Surf. A 79:157–167.

Bosbach, D., and W. Rammensee. 1994. *In situ* investigation of growth and dissolution on the (010) surface of gypsum by scanning force microscopy. Geochim. Cosmochim. Acta. 58:843–849.

Brown Jr., G.E. 1990. Spectroscopic studies of chemisorption reaction mechanisms at oxide-water interfaces. p. 309–353. *In* M.F. Hochella and A.F. White (ed.) Mineral-water interface geochemistry. Mineralogical Society of America, Washington, DC.

Bruemmer, G.W., J. Gerth, and K.G. Tiller. 1988. Reaction kinetics of the adsorption and desorption of nickel, zinc and cadmium by goethite: I. Adsorption and diffusion of metals. J. Soil Sci. 39:37–52.

Brusseau, M.L., and P.S.C. Rao. 1989. Sorption nonideality during organic contaminant transport in porous media. CRC Crit. Rev. Environ. Control 19:33–99.

Bunnett, J.F. 1986. Kinetics in solution. p.171–250. *In* C.F. Bernasconi (ed.) Investigations of rates and mechanisms of reactions. John Wiley and Sons, New York, NY.

Bunzl, K., W. Schmidt, and B. Sansoni. 1976. Kinetics of ion exchange in soil organic matter. IV. Adsorption and desorption of Pb^{2+}, Cu^{2+}, Zn^{2+}, and Ca^{2+} by peat. J. Soil Sci. 27:32–41.

Carroll, K.M., M.R. Harkness, A.A. Bracco, and R.B. Balcarcel. 1994. Application of a permeant/polymer diffusional model to the desorption of polychlorinated biphenyls from Hudson River sediments. Environ. Sci. Technol. 28:253–258.

Casey, W.H., H.R. Westrich, G.W. Arnold, and J.F. Banfield. 1989. The surface chemistry of dissolving labradorite feldspar. Geochim. Cosmochim. Acta 53:821–832.

Chang, M., S. Wu, and C. Chen. 1997. Diffusion of volatile organic compounds in pressed humic acid disks. Environ. Sci. Technol. 31:2307–2312.

Charlet, L., and A. Manceau. 1992. X-ray absorption spectroscopic study of the sorption of Cr(III) at the oxide-water interface. II: Adsorption, coprecipitation and surface precipitation on ferric hydrous oxides. J. Colloid Interf. Sci. 148:443–458.

Charlet, L., and A. Manceau. 1993. Structure, formation, and reactivity of hydrous oxide particles: Insights from X-ray absorption spectroscopy. p. 117–164. *In* J. Buffle and H.P. van Leeuwen (ed.) Environmental particles. Lewis Publishers, Boca Raton, FL.

Chien, S.H., and W.R. Clayton. 1980. Application of Elovich equation to the kinetics of phosphate release and sorption in soils. Soil Sci. Soc. Am. J. 44:265–268.

Chisholm-Brause, C.J., P.A. O'Day, G.E. Brown, Jr., and G.A. Parks. 1990. Evidence for multinuclear metal-ion complexes at solid/water interfaces from X-ray absorption spectroscopy. Nature 348:528–530.

Chute, J.H., and J.P. Quirk. 1967. Diffusion of potassium from mica-like materials. Nature 213:1156–1157.

Comans, R.N.J., and D.E. Hockley. 1992. Kinetics of cesium sorption on illite. Geochim. Cosmochim. Acta 56:1157–1164.

Connaughton, D.F., J.R. Stedinger, L.W. Lion, and M.L. Shuler. 1993. Description of time-varying desorption kinetics: Release of naphthalene from contaminated soils. Environ. Sci. Technol. 27:2397–2403.

Cornelissen, G., P.C.M. VanNoort, J.R. Parsons, and H.A.J. Govers. 1997. Temperature dependence of slow adsorption and desorption kinetics of organic compounds in sediments. Environ. Sci. Technol. 31:454–460.

Crank, J. 1976. The mathematics of diffusion. Oxford University Press (Clarendon), London, UK.

de la Caillerie, J.B., M. Kermarec, and O. Clause. 1995. Impregnation of γ-alumina with Ni (II) and Co (II) ions at neutral pH: Hydrotalcite-type coprecipitate formation and characterization. J. Am. Chem. Soc. 117:11471–11481.

DiToro, D.M., and L.M. Horzempa. 1982. Reversible and resistant components of PCB adsorption-desorption: Isotherms. Environ. Sci. Technol. 16:594–602.

DiVincenzo, J.P., and D.L. Sparks. 1997. Slow sorption kinetics of pentachlorophenol on soil: Concentration effects. Environ. Sci. Technol. 31:977–983.

Eary, L.E., and D. Rai. 1987. Kinetics of chromium(III) oxidation to chromium (VI) by reaction with manganese dioxide. Environ. Sci. Technol. 21:1187–1193.

Farley, K.J., D.A. Dzombak, and F.M.M. Morel. 1985. A surface precipitation model for the sorption of cations on metal oxides. J. Colloid Interf. Sci. 106:226–242.

Fendorf, S.E., M.J. Eick, P.R. Grossl, and D.L. Sparks. 1997. Arsenate and chromate retention mechanisms on goethite. 1. Surface structure. Environ. Sci. Technol. 31:315–320.

Fendorf, S.E., M. Fendorf, D.L. Sparks, and R. Gronsky. 1992. Inhibitory mechanisms of Cr (III) oxidation by δ-MnO_2. J. Colloid Interf. Sci. 153:37–54.

Fendorf, S.E., G.M. Lamble, M.G. Stapleton, M.J. Kelley, and D.L. Sparks. 1994. Mechanisms of chromium (III) sorption on silica: 1. Cr(III) surface structure derived by extended X-ray absorption fine structure spectroscopy. Environ. Sci. Technol. 28:284–289.

Fendorf, S.E., D.L. Sparks, J. A. Franz, and D.M. Camaioni. 1993. Electron paramagnetic resonance stopped-flow kinetic study of manganese (II) sorption-desorption on birnessite. Soil Sci. Soc. Am. J. 57:57–62.

Fendorf, S.E., and R.J. Zasoski. 1992. Chromium (III) oxidation by γ-MnO_2: I. Characterization. Environ. Sci. Technol. 26:79–85.

Fuller, C.C., J.A. Davis, and G.A. Waychunas. 1993. Surface chemistry of ferrihydride: Part 2. Kinetics of arsenate adsorption and coprecipitation. Geochim. Cosmochim. Acta 57:2271–2282.

Furrer, G., and W. Stumm. 1986. The coordination chemistry of weathering. I. Dissolution kinetics of γ-Al_2O_3 and BeO. Geochim. Cosmochim. Acta 50:1847–1860.

Gardiner, W.C., Jr. 1969. Rates and mechanisms of chemical reactions. Benjamin, New York, NY.

Grossl, P.R., M.J. Eick, D.L. Sparks, S. Goldberg, and C.C. Ainsworth. 1997. Arsenate and chromate retention mechanisms on goethite. 2. Kinetic evaluation using a pressure-jump relaxation technique. Environ. Sci. Technol. 31:321–326.

Grossl, P.R., and D.L. Sparks. 1995. Evaluation of contaminant ion adsorption/desorption on goethite using pressure-jump relaxation kinetics. Geoderma 67:87–101.

Grossl, P.R., D.L. Sparks, and C.C. Ainsworth. 1994. Rapid kinetics of Cu (II) adsorption/desorption on goethite. Environ. Sci. Technol. 28:1422–1429.

Hachiya, K., M. Sasaki, I. Ikeda, N. Mikami, and T. Yasunaga. 1984. Static and kinetic studies of adsorption-desorption of metal ions on a γ-Al_2O_3 surface. 2. Kinetic studies by means of pressure-jump technique. J. Phys. Chem. 88:27–31.

Hamaker, J.W., and J.M. Thompson. 1972. Adsorption. p. 39–151. *In* C.A.I. Goring and J.W. Hamaker (ed.) Organic chemicals in the environment, Marcel Dekker, New York, NY.

Harmon, T.C., L. Semprini, and P.V. Roberts. 1992. Simulating solute transport using laboratory-based sorption parameters. J. Environ. Eng. 118:666–689.

Havlin, J.L., and D.G. Westfall. 1985. Potassium release kinetics and plant response in calcareous soils. Soil Sci. Soc. Am. J. 49:366–370.

Hayes, K.F., and J.O. Leckie. 1986. Mechanism of lead ion adsorption at the goethite-water interface. ACS Symp. Ser. 323:114–141.

Hellman, R., B. Drake, and K. Kjoller. 1992. Using atomic force microscopy to study the structure, topography and dissolution of albite surfaces. p. 149–152. In Y.K. Kharaka and A.S. Maest (ed.) Water-rock interaction VIIA. A. Balkema, Rotterdam, Netherlands.

Hem, J.D. 1978. Redox processes at the surface of manganese oxide and their effects on aqueous metal ions. Chem. Geol. 21:199–218.

Hillner, P.E., A.J. Gratz, S. Manne, and P.K. Hansma. 1992a. Atomic-scale imaging of calcite growth and dissolution in real-time. Geol. 20:359–362.

Hillner, P.E., S. Manne, A.J. Gratz, and P.K. Hansma. 1992b. AFM images of dissolution and growth on a calcite crystal. Ultramicrosc. 44:1387–1393.

Huang, P.M. 1991. Kinetics of redox reactions on manganese oxides and its impact on environmental quality. p. 191–230. In D.L. Sparks and D.L. Suarez (ed.) Rates of soil chemical processes, Soil Science Society of America, Madison, WI.

Jardine, P.M., F.M. Dunnivant, H.M. Selim, and J.F. McCarthy. 1992. Comparison of models for describing the transport of dissolved organic carbon in aquifer columns. Soil Sci. Soc. Am. J. 56:393–401.

Jardine, P.M., G.K. Jacobs, and J.D. O'Dell. 1993. Unsaturated transport processes in undisturbed heterogenous porous media: II. Co-contaminants. Soil Sci. Soc. Am. J. 57:954–962.

Jardine, P.M., and D.L. Sparks. 1984. Potassium-calcium exchange in a multireactive soil system. I. Kinetics. Soil Sci. Soc. Am. J. 48:39–45.

Johnsson, P.A., M.F. Hochella, Jr., G.A. Parks, A.E. Blum, and G. Sposito. 1992. Direct observation of muscovite basal-plane dissolution and secondary phase formation: An XPS, LEED, and SFM study. p. 159–162. In Y.K. Kharaka and A.S. Maest (ed.) Water-rock interaction VIIA. A. Balkema, Rotterdam, Netherlands.

Jordan, G., and W. Rammensee. 1996. Dissolution rates and activation energy for dissolution of brucite (001): A new method based on the microtopography of crystal surfaces. Geochim. Cosmochim. Acta 60:5055–5062.

Junta, J.L., and M.F. Hochella, Jr. 1994. Manganese (II) oxidation at mineral surfaces: A microscopic and spectroscopic study. Geochim. Cosmochim. Acta 58:4985–4999.

Kan, A.T., G. Fu, M.A. Hunter, and M.B. Tomson. 1997. Irreversible adsorption of napthalene and tetrachlorobiphenyl to Lula and surrogate sediments. Environ. Sci. Technol. 31:2176–2185.

Karickhoff, S.W. 1980. Sorption kinetics of hydrophobic pollutants in natural sediments. p. 193–205. In R.A. Baker (ed.) Contaminants and sediments. 2. Ann Arbor Science, Ann Arbor, MI.

Karickhoff, S.W., D.S. Brown, and T.A. Scott. 1979. Sorption of hydrophobic pollutants on natural sediments. Water Res. 13:241–248.

Karickhoff, S.W., and K.R. Morris. 1985. Sorption dynamics of hydrophobic pollutants in sediment suspensions. Environ. Toxicol. Chem. 4:469–479.

Keren, R., P.R. Grossl, and D.L. Sparks. 1994. Equilibrium and kinetics of borate adsorption-desorption on pyrophyllite in aqueous suspensions. Soil Sci. Soc. Am. J. 58:1116–1122.

Krishnamurti, G.S.R., and P.M. Huang. 1992. Dynamics of potassium chloride induced manganese release in different soil orders. Soil Sci. Soc. Am. J. 56:1115–1123.

Krishnamurti, G.S.R., G. Cieslinski, P.M. Huang, and K.C.J. Van Rees. 1997. Kinetics of cadmium release from soils as influenced by organic acids: Implication in cadmium availability. J. Environ. Qual. 26:271–277.

Kuo, S., and E.G. Lotse. 1974. Kinetics of phosphate adsorption and desorption by lake sediments. Soil Sci. Soc. Am. Proc. 38:50–54.

Kuo, S., and D.S. Mikkelsen. 1980. Kinetics of zinc desorption from soils. Plant Soil 56:355–364.

Lasaga, A.C. 1984. Chemical kinetics of water-rock interactions. J. Geophys. Res. B 6:4009–4025.

Lee, L.S., P.S.C. Rao, M.L. Brusseau, and R.A. Ogwada. 1988. Nonequilibrium sorption of organic contaminants during flow through columns of aquifer materials. Environ. Toxicol. Chem. 7:779–793.

Leenheer, J.A., and J.L. Ahlrichs. 1971. A kinetic and equilibrium study of the adsorption of carbaryl and parathion upon soil organic matter surfaces. Soil Sci. Soc. Am. Proc. 35:700–704.

Lehmann, R.G., and R.D. Harter. 1984. Assessment of copper-soil bond strength by desorption kinetics. Soil Sci. Soc. Am. J. 48:769–772.

Loehr, R.C., and M.T. Webster. 1996. Behavior of fresh vs. aged chemicals in soils. J. Soil Contam. 5:361–383.

Lövgren, L., S. Sjöberg, and P.W. Schindler. 1990. Acid/base reactions and Al(III) complexation at the surface of goethite. Geochim. Cosmochim. Acta. 54:1301–1306.

Low, M.J.D. 1960. Kinetics of chemisorption of gases on solids. Chem. Rev. 60: 267–312.

Maurice, P.A. 1997. Scanning probe microscopy of mineral surfaces. *In* P.M. Huang, N. Senesi and J. Buffle (ed.) Structure and surface reactions of soil particles. Vol. 4. John Wiley and Sons, New York, NY.

Maurice, P.A., M.F. Hochella, Jr., G.A. Parks, G. Sposito, and U. Schwertmann. 1995. Evolution of hematite surface microtopography upon dissolution by simple organic acids. Clays Clay Miner. 43:29–38.

McCall, P.J., and G.L. Agin. 1985. Desorption kinetics of picloram as affected by residence time in the soil. Environ. Toxicol. Chem. 4:37–44.

Mikami, N., M. Sasaki, K. Hachiya, R.D. Ikeda, and T. Yasunaga. 1983a. Kinetics of the adsorption of PO_4 on the γ-Al_2O_3 surface using the pressure-jump technique. J. Phys. Chem. 87:1454–1458.

Mikami, N., M. Sasaki, T. Kikuchi, and T. Yasunaga. 1983b. Kinetics of the adsorption-desorption of chromate on γ-Al_2O_3 surfaces using the pressure-jump technique. J. Phys. Chem. 87:5245–5248.

Mikami, N., M. Sasaki, K. Hachiya, and T. Yasunaga. 1983c. Kinetic study of the adsorption-desorption of the uranyl ion on a γ-Al_2O_3 surface using the pressure-jump technique. J. Phys. Chem. 87:5478–5481.

Miller, C.T., and J. Pedit. 1992. Use of a reactive surface-diffusion model to describe apparent sorption-desorption hysteresis and abiotic degradation of lindane in a subsurface material. Environ. Sci. Technol. 26:1417–1427.

Murray, J.W., and J.G. Dillard. 1979. The oxidation of cobalt (II) adsorbed on manganese dioxide. Geochim. Cosmochim. Acta 43:781–787.

Nkedi-Kizza, P., J.W. Biggar, H.M. Selim, M.T. van Genuchten, P.J. Wierenga, J.M. Davidson, and D.R. Nielsen. 1984. On the equivalence of two conceptual models for describing ion exchange during transport through an aggregated Oxisol. Water Resour. Res. 20:1123–1130.

O'Day, P.A., G.A. Parks, and G.E. Brown, Jr. 1994a. Molecular structure and binding sites of cobalt(II) surface complexes on kaolinite from X-ray absorption spectroscopy. Clays Clay Miner. 42:337–355.

O'Day, P.A., G.E. Brown, Jr., and G.A. Parks. 1994b. X-ray absorption spectroscopy of cobalt (II) multinuclear surface complexes and surface precipitates on kaolinite. J. Coll. Interf. Sci. 165:269–289.

O'Day, P.A., C.J. Chisholm-Brause, S.N. Towle, G.A. Parks, and G.E. Brown, Jr. 1996. X-ray absorption spectroscopy of Co(II) sorption complexes on quartz (α-SiO_2) and rutile (TiO_2). Geochim. Cosmochim. Acta. 60:2515–2532.

Onken, A.B., and R.L. Matheson. 1982. Dissolution rate of EDTA-extractable phosphate from soils. Soil Sci. Soc. Am. J. 46:276–279.

Oscarson, D.W., P.M. Huang, C. Defosse, and A. Herbillon. 1981. The oxidative power of Mn (IV) and Fe (III) oxides with respect to As (III) in the terrestrial and aquatic environment. Nature 291:50–51.

Oscarson, D.W., P.M. Huang, and U.T. Hammer. 1983. Oxidation and sorption of arsenite by manganese dioxide as influenced by surface coatings of iron and aluminum oxides and calcium carbonate. Water Air Soil Pollut. 20:233–244.

Oscarson, D.W., P.M. Huang, and W.K. Liaw. 1980. The oxidation of arsenite by aquatic sediments. J. Environ. Qual. 9:700–703.

Padmanabham, M. 1983. Adsorption-desorption behavior of copper (II) at the goethite-solution interface. Aust. J. Soil Res. 21:309–320.

Pavlostathis, S.G., and G.N. Mathavan. 1992. Desorption kinetics of selected volatile organic compounds from field contaminated soils. Environ. Sci. Technol. 26:532–538.

Pedit, J.A., and C.T. Miller. 1995. Heterogenous sorption processes in subsurface systems. 2. Diffusion modeling approaches. Environ. Sci. Technol. 29:1766–1772.

Petrovic, R., R.A. Berner, and M.B. Goldhaber. 1976. Rate control in dissolution of alkali feldspars. I. Study of residual feldspar grains by X-ray photoelectron spectroscopy. Geochim. Cosmochim. Acta 40:537–548.

Pignatello, J.J., F.J. Ferrandino, and L.Q. Huang. 1993. Elution of aged and freshly added herbicides from a soil. Environ. Sci. Technol. 27:1563–1571.

Pignatello, J.J., and L.Q. Huang. 1991. Sorptive reversibility of atrazine and metolachlor residues in field soil samples. J. Environ. Qual. 20:222–228.

Polyzopoulos, N.A., V.Z. Keramidas, and A. Pavlatou. 1986. On the limitations of the simplified Elovich equation in describing the kinetics of phosphate sorption and release from soils. J. Soil Sci. 37:81–87.

Rai, E., and R.J. Serne. 1977. Plutonium activities in soil solutions and the stability and formation of selected plutonium minerals. J. Environ. Qual. 6:89–95.

Scheidegger, A.M., G.M. Lamble, and D.L. Sparks. 1996. Investigation of Ni sorption on pyrophyllite: An XAFS study. Environ. Sci. Technol. 30:548–554.

Scheidegger, A.M., and D.L. Sparks. 1996a. Kinetics of the formation and the dissolution of nickel surface precipitates on pyrophyllite. Chem. Geol. 132:157–164.

Scheidegger, A.M., and D.L. Sparks. 1996b. A critical assessment of sorption-desorption mechanisms at the soil mineral/water interface. Soil Sci. 161:813–831.

Scheidegger, A.M., G.M. Lamble, and D.L. Sparks. 1997. Spectroscopic evidence for the formation of mixed-cation hydroxide phases upon metal sorption on clays and aluminum oxides. J. Colloid Interf. Sci. 186:118–128.

Scheidegger, A.M., D.G. Strawn, G.M. Lamble, and D.L. Sparks. 1998. The kinetics of mixed Ni-Al hydroxide formation on clays and aluminum oxide: A time-resolved XAFS study. Geochim. Cosmochim. Acta.62:2233–2245.

Schott, J., and J.C. Petit. 1987. New evidence for the mechanisms of dissolution of silicate minerals. p. 293–312. *In* W. Stumm (ed.) Aquatic surface chemistry, John Wiley and Sons, New York, NY.

Schultz, M.F., M.M. Benjamin, and J.F. Ferguson. 1987. Adsorption and desorption of metals on ferrihydrite: Reversibility of the reaction and sorption properties of the regenerated solid. Environ. Sci. Technol. 21:863–869.

Schwarzenbach, R.T., P.M. Gschwend, and D.M. Boden. 1993. Environmental organic chemistry. John Wiley and Sons, New York , NY.

Scott, M.J., and J.J. Morgan. 1995. Reactions at oxide surfaces. 1. Oxidation of As (III) by synthetic birnessite. Environ. Sci. Technol. 29:1898–1905.

Scott, M.J., and J.J. Morgan. 1996. Reactions at oxide surfaces. 2. Oxidation of Se(IV) by synthetic birnessite. Environ. Sci. Technol. 30:1990–1996.

Scribner, S.L., T.R. Benzing, S. Sun, and S.A. Boyd. 1992. Desorption and bioavailability of aged simazine residues in soil from a continuous corn field. J. Environ. Qual. 21:115–120.

Sharpley, A.N. 1983. Effect of soil properties on the kinetics of phosphorus desorption. Soil Sci. Soc. Am. J. 47:462–467.

Skopp, J. 1986. Analysis of time dependent chemical processes in soils. J. Environ. Qual. 15:205–213.

Sparks, D.L. 1989. Kinetics of soil chemical processes. Academic Press, San Diego, CA.

Sparks, D.L. 1991. Chemical kinetics and mass transfer processes in soils and soil constituents. p. 585–637. *In* J. Bear and M.Y. Corapcioglu (ed.) Transport processes in porous media. Kluwer Academic Publishers, Dordecht, Netherlands.

Sparks, D.L. 1995. Environmental soil chemistry. Academic Press, San Diego, CA.

Sparks, D.L. 1998. Kinetics of sorption/release processes on natural surfaces. *In* P.M. Huang, N. Senesi, and J. Buffle (ed.) Structure and surface reaction of soil particles. Vol. 4. John Wiley and Sons, New York, NY.

Sparks, D.L. 1999. Kinetics of soil chemical phenomena: Future directions. p. 81–102. *In* P.M. Huang, D.L. Sparks, and S.A. Boyd (ed). Future prospects for soil chemistry. Soil Sci. Soc. Am. Spec. Pub., Soil Science Society of America, Madison, WI.

Sparks, D.L., and P.M. Jardine. 1984. Comparison of kinetic equations to describe K-Ca exchange in pure and in mixed systems. Soil Sci. 138:115–122.

Sparks, D.L., and D.L. Suarez (eds). 1991. Rates of soil chemical processes. Soil Sci. Soc. Am. Spec. Pub. 27, Soil Science Society of America, Madison, WI.

Sparks, D.L., and P.C. Zhang. 1991. Relaxation methods for studying kinetics of soil chemical phenomena. p. 61–94. *In* D.L. Sparks, and D.L. Suarez (eds.) Rates of soil chemical processes. Soil Sci. Soc. Am. Spec. Pub. 27, Soil Science Society of America, Madison, WI.

Sparks, D.L., S.E. Fendorf, P.C. Zhang, and L. Tang. 1993. Kinetics and mechanisms of environmentally important reactions on soil colloidal surfaces. p. 141–168. *In* D. Petruzzelli and F.G. Helfferich (ed.) Migration and fate of pollutants in soils and subsoils. Springer-Verlag, Berlin, Germany.

Sparks, D.L., S.E. Fendorf, C.V. Toner, IV, and T.H. Carski. 1996. Kinetic methods and measurements. p. 1275–1307. *In* D.L. Sparks (ed.) Methods of soil analysis. Part 3. Chemical methods. Soil Science Society of America, Madison, WI.

Sposito, G. 1986. Distinguishing adsorption from surface precipitation. ACS Symp. Ser. 323:217–228.

Sposito, G. 1994. Chemical equilibria and kinetics in soils. John Wiley and Sons, New York, NY.

Steinberg, S.M., J.J. Pignatello, and B.L. Sawhney. 1987. Persistence of 1,2 dibromoethane in soils: Entrapment in intra particle micropores. Environ. Sci. Technol. 21:1201–1208.

Stone, A.T. 1986. Adsorption of organic reductants and subsequent electron transfer on metal oxide surfaces. ACS Symp. Ser. 323:446–461.

Stone, A.T. 1987a. Microbial metabolites and the reductive dissolution of manganese oxides: Oxalate and pyruvate. Geochim. Cosmochim. Acta 51:919–925.

Stone, A.T. 1987b. Reductive dissolution of manganese (III/IV) oxides by substituted phenols. Environ. Sci. Technol. 21:979–988.

Stone, A.T. 1991. Oxidation and hydolysis of ionizable organic pollutants at hydrous metal oxide surfaces. p. 231–254. *In* D.L. Sparks and D.L. Suarez (ed.) Rates of soil chemical processes. Soil Sci. Soc. Am. Spec. Pub. 27, Soil Science Society of America, Madison, WI.

Stone, A.T., and J.J. Morgan. 1984a. Reduction and dissolution of manganese (III) and manganese (IV) oxides by organics. 1. Reaction with hydroquinone. Environ. Sci. Technol. 18:450–456.

Stone, A.T., and J.J. Morgan. 1984b. Reduction and dissolution of manganese (III) and manganese (IV) oxides by organics. 2. Survey of the reactivity of organics. Environ. Sci. Technol. 18:617–624.

Stumm, W. 1992. Chemistry of the solid-water interface. John Wiley and Sons, New York, NY.

Stumm, W., and G. Furrer. 1987. The dissolution of oxides and aluminum silicates: Examples of surface-coordination-controlled kinetics. p. 197–219. *In* W. Stumm (ed.) Aquatic surface chemistry, John Wiley and Sons, New York, NY.

Stumm, W., and R. Wollast. 1990. Coordination chemistry of weathering. Kinetics of the surface-controlled dissolution of oxide minerals. Rev. Geophys. 28:53–69.

Takahashi, M.T., and R.A. Alberty. 1969. The pressure-jump methods. p. 31–55. *In* K. Kustin (ed.) Methods in enzymology 16. Academic Press, New York, NY.

Tang, L., and D.L. Sparks. 1993. Cation exchange kinetics on montmorillonite using pressure-jump relaxation. Soil Sci. Soc. Am. J. 57:42–46.

Toner IV, C.V., and D.L. Sparks. 1995. Chemical relaxation and double layer model analysis of boron adsorption on alumina. Soil Sci. Soc. Am. J. 59:395–404.

Towle, S.N., J.R. Bargar, G.E. Brown, Jr., and G.A. Parks. 1997. Surface precipitation of Co(II) (aq) on Al_2O_3. J. Colloid Interf. Sci. 187:62–82.

van Genuchten, M.T., and R.J. Wagenet. 1989. Two-site/two-region models for pesticide transport and degradation: Theoretical development and analytical solutions. Soil Sci. Soc. Am. J. 53:1303–1310.

Waychunas, G.A., B.A. Rea, C.C. Fuller, and J.A. Davis. 1993. Surface chemistry of ferrihydrite: Part 1. EXAFS studies of the geometry of coprecipitated and adsorbed arsenate. Geochim. Cosmochim. Acta 57:2251–2269.

Weber, W.J., Jr., and J.P. Gould. 1966. Sorption of organic pesticides from aqueous solution. Adv. Chem. Ser. 60:280–305.

Weber, W.J., Jr., and C.T. Miller. 1988. Modeling the sorption of hydrophobic contaminants by aquifer materials. 1. Rates and equilibria. Water Res. 22:457–464.

Weber, W.J., Jr., and W. Huang. 1996. A distributed reactivity model for sorption by soils and sediments. 4. Intraparticle heterogeneity and phase-distribution relationships under nonequilibrium conditions. Environ. Sci. Technol. 30:881–888.

Wersin, P., M.F. Hochella, Jr., P. Persson, G. Redden, J.O. Leckie, and D.W. Harris. 1994. Interaction between aqueous uranium (VI) and sulfide minerals: Spectroscopic evidence for sorption and reduction. Geochim. Cosmochim. Acta 58:2829–2843.

Werth, C.J., and M. Reinhard. 1997. Effects of temperature on trichloroethylene desorption from silica gel and natural sediments. 2. Kinetics. Environ. Sci. Technol. 31:697–703.

Wollast, R. 1967. Kinetics of the alteration of K-feldspar in buffered solutions at low temperature. Geochim. Cosmochim. Acta 31:635–648.

Wu, S., and P.M. Gschwend. 1986. Sorption kinetics of hydrophobic organic compounds to natural sediments and soils. Environ. Sci. Technol. 20:717–725.

Xia, K., A. Mehadi, R.W. Taylor, and W.F. Bleam. 1997. X-ray absorption and electron paramagnetic resonance studies of Cu(II) sorbed to silica: Surface-induced precipitation at low surface coverages. J. Coll. Interf. Sci. 185:252–257.

Xu, J., and P.M. Huang. 1995. Zinc adsorption-desorption on short-range ordered iron oxide as influenced by citric acid during its formation. Geoderma 64:232–356.

Xue, Y., and S.J. Traina. 1996. Oxidation kinetics of Co (II) - EDTA in aqueous and semi-aqueous goethite suspensions. Environ. Sci. Technol. 30:1975–1981.

Zhang, P.C., and D.L. Sparks. 1989. Kinetics and mechanisms of molybdate adsorption/desorption at the goethite/water interface using pressure-jump relaxation. Soil Sci. Soc. Am. J. 53:1028–1034.

Zhang, P.C., and D.L. Sparks. 1990a. Kinetics and mechanisms of sulfate adsorption/desorption on goethite using pressure-jump relaxation. Soil Sci. Soc. Am. J. 54:1266–1273.

Zhang, P.C., and D.L. Sparks. 1990b. Kinetics of selenate and selenite adsorption/desorption at the goethite/water interface. Environ. Sci. Technol. 24:1848–1856.

Zinder, B., G. Furrer, and W. Stumm. 1986. The coordination chemistry of weathering. II. Dissolution of Fe(III) oxides. Geochim. Cosmochim. Acta 50:1861–1869.

5

Redox Phenomena

Bruce R. James
University of Maryland

Richmond J. Bartlett
University of Vermont

5.1 Concepts, Principles, and Theories

5.1.1 Nature of the Electron and Proton in Soil Solutions

Electrons are subatomic particles with wave like properties that have defied exact characterization since their discovery in 1897, despite their central role in chemical reactions (Castellan, 1983). Electrons are considered fundamental subatomic particles, but recent findings suggest that they may possess a substructure of lepto quarks, that electron orbitals do not exist, and that the role of electrons in predicting periodicity of the elements is only approximate (Scerri, 1997). In addition, new frontiers are expanding to explain electron transfer processes at interfaces (Tributsch and Pohlmann, 1998). Studying and understanding electron transfer processes in soils, therefore, is challenging and must be based on sources of information from relatively simple aqueous systems and natural waters; as well as on a large base of empirical studies using soils.

To understand and characterize the nature of "electron activity" and oxidation-reduction reactions in heterogeneous, multi-phase soils and soil solutions, an appreciation of the characteristics of electrons and closely-allied protons is needed (James, 1989; Bartlett, 1998). An examination of the complementary nature of electrons and protons affirms the importance of hydrogen ion and electron activities as master variables in soils (Sillén, 1967).

The H atom, composed of one proton and one electron may be visualized and modeled as a spherical puff of cotton candy with a radius of approximately 10 cm and a proton nucleus with a diameter 0.005 mm: essentially an invisible fleck of unspun sugar in the center! The remaining volume of the atom is occupied by the electron: the density of the spun sugar represents the probability of finding the electron in any one location, and it becomes increasingly thinner (less likely) with distance away from the positively-charged proton. (Castellan, 1983). The radius of the H atom (0.3 Å), therefore, is approximately 20,000 times that of the proton ($\sim 1.5 \times 10^{-5}$ Å). The proton also may be visualized as the size of a 0.1 μm colloidal clay particle, compared to a 2000 μm sand grain in a soil. In contrast to the large proportion of the volume of the H atom occupied by the negatively charged, wave-like electron, the electron has only neglible mass equal to approximately 550 μg mol^{-1}, 1/1836 of the mass of the H atom (10^6 μg mol^{-1}).

The tiny, heavy proton persists in hydrated form in aqueous media as H_3O^+, the hydroxonium or hydronium ion. The H atom can only form the H^+ ion when its compounds are dissolved in media that solvate protons, so the H^+ cannot exist in solid phases (Cotton and Wilkinson, 1980). The solvation enthalpy of -1091 kJ mol^{-1} provides the energy for bond rupture. Protons migrate rapidly between water molecules, and an individual H_3O^+ ion has a lifetime of approximately 10^{-13} sec. Its concentration and activity can, therefore, be measured and understood in terms of the hydration of a cation in solution. The single ion activity coefficient can be calculated based on a knowledge of ionic strength (I), temperature, effective diameter and solvent characteristics, and H^+ activity can be calculated or measured (Westcott, 1978; Bates, 1981).

The extremely large "charge-to-size ratio" of the electron prevents it from persisting in free form in aqueous systems, as does the solvated H^+. The ephemeral "hydrated electron" is a powerful reducing agent with a potential of -2.7 V relative to the H^+/H_2 reference potential of 0.0 V, and has a half-life of < 1 msec. It reacts rapidly with second order rate constants of 10^8 to 10^{10} M^{-1} sec^{-1}, near the diffusion-controlled limit (Sullivan et al., 1976).

In soil chemical calculations and theory, one considers the electron as a "species," designated "e$^-$," with neglible mass and thermodynamically as a ligand, reactant, and product (Sposito, 1981). The electron is not ionic, but it is "negatively charged" as the carrier of negative electricity (Thompson, 1923). Its "activity" is conceptually analogous to that of H^+, but its concentration in "mol L^{-1}" is undefined. All these caveats about the electron require that one understands that electron activity in soils and natural waters should be regarded as related strictly to energy functions. Such functions can be described simply as "the ability to do work", "electrochemical potential," and more colloquially, "electron pressure."

Generically, for a reduction half-reaction: ox + ne$^-$ = red, the electron activity, (e$^-$), equals $[(red)/K(ox)]^{1/n}$, in which ox is the oxidized species, red the reduced species, and K is the equilibrium constant. Therefore, electron activity is a function of the ratio of reduced-to-oxidized species activities and of the equilibrium constant (or Gibbs free energy for the reaction, since $\Delta G_r^o = -RT\ln K$, where R is the gas law constant and T is the absolute temperature). A "favorable reaction" with a large positive value for K (or large negative ΔG_r^o) will therefore be associated with low electron activity. That is, the reaction will proceed at relatively low electron pressures.

The concept of H^+ activity and pH also can be described in terms of thermodynamic work (dw) defined as the product of an intensive property and extensive variation, such as P(dV), where P is pressure and dV is change in volume (Stumm and Morgan, 1996). Similarly, electron activity may be defined as E (de) where E is potential and de is change in charge of the system. Proton and electron activities also may defined as chemical work, μ(dn), where μ is chemical potential and dn is change in moles (Sposito, 1981).

Viewing the sibling concepts of "proton activity" and "electron activity" in soils, they must not be considered twins. Recognition must be given to similarities and differences in the formulation of conceptual and operational definitions for these key variables, and such comparisons are based on the differences in the nature of the proton and electron, as described above. In both cases, however, thermodynamic activity is defined as the ratio of e$^-$ or H^+ activity in a given system relative to its activity under standard state conditions. In the case of the e$^-$, the standard state is at $(H^+) = 1M$, a partial pressure of $H_2 = 1$ atm, and a temperature of 298K, characteristics of the standard hydrogen electrode (SHE) with 0.00 V under these conditions (Lindsay, 1979; Stumm and Morgan, 1996; Compton and Sanders, 1996). For H^+ activity, the reference state is 1 M or pH 0 (Bates, 1981).

The familiar concept of buffer capacity for pH is defined as the change in acid or base added to effect a one unit change in pH (Stumm and Morgan, 1996). Similarly, the capacity factor in redox is referred to as poise and is defined as the change in added equivalents of reductant or oxidant to bring about a one unit change in pe (or Eh change of 59 mV, as discussed in Section 5.1.2). Poise in soils and natural waters has been less intensively studied than pH buffering, but it is a central concept governed by reductant and oxidant activities and microbial processes.

The choice of oxidant or reductant to add during the titration obviously will affect the resulting calculated poise. Heron et al. (1994) titrated aquifer sediments with the reducing agent, Ti^{3+} EDTA, to determine the poise associated with oxidants, such as Fe(III)- and Mn(III,IV)-(hydr)oxides. Barcelona and Holm (1991) titrated with Cr^{2+} and $Cr_2O_7^{2-}$ to determine poise associated with oxidation and reduction, respectively.

Redox intensity measurements and estimates of poise as the redox capacity of soils and sediments are also distinguished from those only associated with H^+ in that microbial catalysis controls the rates of many of the oxidation and reduction reactions governing soil redox status (Russell, 1973). Also, the adaptation and growth of higher plants and algae in soils affect redox intensity and capacity. Selected examples of such microbial processes governing redox in soils include Fe(III) and SO_4^{2-} reduction in sediments (Coleman et al., 1993), denitrification in soil microenvironments (Højberg et al., 1994), N_2-fixation in flooded soils (Buresh et al., 1980), methane oxidation in flooded rice soils (Wang et al., 1997), the reactions of H_2 in anoxic groundwater (Lovley et al., 1994), and the formation of oxidized root channels (rhizospheres) in soils (Mendelssohn et al., 1995).

The poise and pe of soils also differ from natural waters in that the solids-to-solution ratio is high, and subsequently, redox reactions across colloid-solution interfaces are centrally important. Relatively few studies have been reported on redox reactions across soil colloid-water interfaces, but this area is critically important for predicting valence state and potential mobility of nutrients and pollutants in soil environments. Examples include reductive dissolution of Fe(III) and Mn(III,IV)(hydr)oxides and redox reactions on such surfaces or with constituent ions (Bartlett and James, 1979; Stone and Morgan, 1987; Suter at al., 1991; Stumm and Sulzberger, 1992; Postma and Jakobsen, 1996; White and Peterson, 1996).

5.1.2 Thermodynamic Relationships for Electron Activity in Soils

Since electrons are transferred from reductants (e^- donors or reducing agents) to oxidants (e^- acceptors or oxidizing agents) and do not exist free in soil solution, reduction reactions must be coupled to oxidation reactions to describe complete oxidation-reduction processes (redox reactions). By convention, reduction half-reactions can be written and thermodynamic relationships derived from them, despite the fact that they do not occur in isolation. The log K values (where K is the equilibrium constant for the half-reaction) for the coupled reactions may be compared to predict the likelihood of a reaction occurring spontaneously as written.

For example, development of reduction half-reactions to predict whether or not gaseous H_2S could reduce colloidal FeOOH across a sulfidic-aerobic interface in a wetland soil may be completed in the following manner:

(1) Write the oxidized and reduced species on the left and right sides of the equation, respectively, with the appropriate number of electrons per mole of oxidant written on the left side:

$$FeOOH + e^- = Fe^{2+} \tag{5.1}$$

$$SO_4^{2-} + 8e^- = H_2S \tag{5.2}$$

The number of electrons needed for reduction is calculated from the oxidation numbers of the elements in the oxidized and reduced species (Vincent, 1985). In this example, the oxidation numbers are 3+ and 6+ for Fe and S in FeOOH and SO_4^{2-}, respectively.

(2)   Add H_2O to the equation (usually to the right side) to balance the moles of O:

$$FeOOH + e^- = Fe^{2+} + 2H_2O \qquad\qquad [5.3]$$
$$SO_4^{2-} + 8e^- = H_2S + 4H_2O \qquad\qquad [5.4]$$

(3)   Then add H^+ to balance moles of H, usually on the left side of the equation:

$$FeOOH + e^- + 3H^+ = Fe^{2+} + 2H_2O \qquad\qquad [5.5]$$
$$SO_4^{2-} + 8e^- + 10H^+ = H_2S + 4H_2O \qquad\qquad [5.6]$$

(4)   Check charge and mass balances for the equations.

These equations can now be used to develop expressions for electron activity, based on equilibrium expressions and free energy of formation (ΔG_f^o) data.

The principles for doing this can be understood by starting with a generic reduction half-reaction:

$$Aox + Be^- + CH^+ = Dred + EH_2O \qquad\qquad [5.7]$$

where A, B, C, D, and E are stoichiometric coefficients. The expression for the equilibrium constant (K) is:

$$K = [(red)^D (H_2O)^E] / [(ox)^A (e^-)^B (H^+)^C] \qquad\qquad [5.8]$$

where ( ) denotes activity and (H_2O) has a value of 1 by convention.

Taking the log of both sides of the equation,

$$\log K = \log [(red)^D/(ox)^A] + \log [1/(e^-)^B] + \log [1/(H^+)^C] \qquad\qquad [5.9]$$

and,

$$\log K = D\log (red) - A\log (ox) + Bpe + CpH \qquad\qquad [5.10]$$

where pe and pH are defined as –log electron activity and H^+ activity, respectively.

The p notation means power and denotes the exponent for the H^+ activity (pH 4, 10^{-4} mol H^+ kg^{-1} water). Therefore, (H^+) = 10^{-pH}. In contrast, the exponent for e^- activity is not 10^{-pe}, since the electron activity is not defined in terms of mol e^- L^{-1}. Both pe and pH are analogous, though, if viewed as related to the ability of e^- and H^+ to do thermodynamic work (section 5.1.1).

Further rearrangement of Equation [5.10] yields a pe–pH relationship of the following form:

$$[(1/B)\log K - (D/B)\log(red) + (A/B)\log(ox)] - (C/B)pH = pe \qquad\qquad [5.11]$$

which gives a straight line with slope C/B and the intercept in square brackets that is a function of log K for the half-reaction and the activities of the oxidized and reduced species.

For a one electron transfer (B = 1) coupled with one proton consumption (C = 1), and when D = A and (red) = (ox),

$$pe + pH = \log K \qquad [5.12]$$

and when pH = 0,

$$pe = \log K \qquad [5.13]$$

Knowledge of pe and pH is pertinent to describing the equilibrium condition of a soil as defined by the master variables, pe and pH, and their sum (Lindsay, 1979). The concepts of electron activity and H^+ ion activity are closely coupled and cannot be separated in assessing the oxidation/reduction status of a soil system. The pe + pH parameter has been described metaphorically by Bartlett (1998) as a seesaw with a hungry baby bird on one side (the electron-deficient player) and earthworms on the other (the electron-rich source). The parent bird carries the earthworms to the baby, representing electron transfer and causing a shift in the postion of the seesaw, quantified by the pe-pH balance. As worms are transferred, the pe increases and the pH decreases, counteracting and interacting controls on the redox equilibrium. As the worms are consumed by the baby, the seesaw shifts back again, representing a dynamic, metastable system governed by the flow of energy and the cycling of nutrients, characteristics of ecosystems. The position of the seesaw governed by pe and pH reflects the equilibrium condition of a soil, and the values of pe and pH are intensity measures of redox status. The change in measured pe and pH per mole of electrons plus H^+ (H atoms in principle) transferred reflects the buffer capacity or poise of the system, $\Delta(pe + pH)/\Delta(mol\ H\ transferred)$

To relate the concept of log K to the Gibbs free energy change (ΔG_r°) for a given half-reaction in the soil, the following expressions are pertinent:

$$\Delta G_r^\circ = - RT \ln K \qquad [5.14]$$

where ΔG_r° is Gibbs free energy of reaction under standard conditions (298 K and 1 atm), R is the universal gas law constant (0.001987 kcal mol^{-1} K^{-1} or 0.008314 kJ mol^{-1} K^{-1}), and T is absolute temperature (298 K). Converting to $\log_{10}$ (ln K = 2.303 log K) yields:

$$\Delta G_r^\circ /{-1.364} = \log K \text{ (based on kcal)} = \Delta G_r^\circ /{-5.71} \text{ (based on kJ)} \qquad [5.15]$$

Log K may, therefore, be estimated simply from knowledge of free energies of formation of H_2O, red, and ox, since those of H^+ and e^- are zero, by convention. To relate log K directly to pe as defined by Equation [5.11] for one electron transfers, log K values must be divided by B, mol of e^- consumed in the reaction. Therefore, pe at a given pH and other defined conditions is log K, and the parameter pe + pH couples the master variables in defining the equilibrium redox state of soil-water system.

Log K values also are related to thermodynamic electrochemical potentials (Eh) according to the following expressions (Compton and Sander, 1996):

$$\Delta G_r^0 = -nF\ Eh \qquad [5.16]$$

where n is the number of e^- transferred mol^{-1}, F is the Faraday constant (23.1 kcal V^{-1} equiv.$^{-1}$ or 96,500 coulombs mol^{-1}: 1.60 x 10^{-19} coulomb/e^- x 6.02 x 10^{23} e^-/mol).

Since both Equations [5.16] and [5.14] are expressions for ΔG_r°,

$$-nF\,Eh = -RT \ln K \tag{5.17}$$

and, when T = 298 K,

$$Eh = \frac{RT\,2.303 \log K}{nF} = 0.059 \log K \tag{5.18}$$

If n = 1 and (red) = (ox), pe + pH = log K (Equation [5.12]), and

$$Eh(V) = 0.059\,pe \tag{5.19}$$

or,

$$Eh/0.059 = pe \tag{5.20}$$

Therefore, calculating and interpreting Eh values rigorously and linking soil measurements to thermodynamics requires a knowledge of pH of the soil/water system.

Applying these principles to the H_2S-FeOOH problem above (Equations [5.5] and [5.6]), the following pe-pH expressions can be derived:

For FeOOH-Fe^{2+}:

$$pe = \log K_{Fe} - \log (Fe^{2+}) - 3\,pH \tag{5.21}$$

where activity of FeOOH is assumed to be 1, an assumption that may or may not be valid for redox processes.

For SO_4^{2-}-H_2S:

$$pe = (^{1}/_{8})\log K_S - (^{1}/_{8}) \log [P_{H_2S}/(SO_4^{2-})] - (^{5}/_{4})pH \tag{5.22}$$

where K_{Fe} and K_S are the equilibrium constants for the FeOOH and SO_4^{2-} reduction half-reactions, and P_{H_2S} is the partial pressure of H_2S. Calculating log K values, and substituting into Equations [5.21] and [5.22] yield

$$pe = 13.0 - \log (Fe^{2+}) - 3pH \tag{5.23}$$

and

$$pe = 5.21 - (^{1}/_{8}) \log [P_{H_2S}/(SO_4^{2-})] - (^{5}/_{4})pH \tag{5.24}$$

At defined activities for the ions and H_2S, pe may be calculated, and if pe for Fe reduction > pe for SO_4^{2-} reduction, then FeOOH will be reduced and H_2S oxidized at equilibrium, and at the hypothetical soil interface. The same result will be obtained if after subtracting the log K value for S from that for Fe, a positive answer is obtained.

The United States Environmental Protection Agency has developed MINTEQ (Mineral Thermal Equilibria), a DOS-based computer program that calculates the distribution of myriad species based on minimizing the Gibbs free energy of a system of multiple chemical equilibria. Different from earlier such programs, such as GEOCHEM, it incorporates algorithms for redox reactions and allows the estimation of pe, pH, and reductant and oxidant activities in conjunction with dissolution and exchange equilibria. The use of MINTEQ also permits an easy way of conducting a sensitivity analysis for estimated pe, pH, or ion activities when just one value for a given parameter is varied (Allison and Brown, 1995).

5.1.3 Kinetic Derivation of Thermodynamic Parameters for Redox

This thermodynamic derivation of log K and its relationship to pe and pH of soils is based on free energy of formation data for oxidants and reductants, and it is not related to reaction mechanisms or rates. An alternate procedure to obtain log K values, in theory, is the kinetic approach suggested by Sparks (1985) for cation exchange and by Harter and Smith (1981) for adsorption processes. This approach has been little used in redox soil chemistry, probably due to difficulties in obtaining rate coefficients for many electron transfer processes in soils and due to the irreversibility of most redox reactions. The application of such approaches should, however, be appropriate for certain reversible redox reactions in soils, especially in situations where metastability and lack of chemical equilibrium prevail, or when accurate free energy of formation data are unavailable for oxidants and reductants.

In principle, this is the method for calculating kinetic parameters: the equilibrium constant (K) for redox equilibria can be estimated from the ratio of forward and reverse rate coefficients for a given reversible reaction in soils:

$$K = \frac{k_f}{k_r} \tag{5.25}$$

where k_f and k_r are the rate constants for forward and reverse reactions, respectively.

The standard Gibbs free energy for the reaction can then be calculated by substituting ln K into Equation [5.14]. Preparation of Arrhenius plots of the rate coefficients as a function of 1/T permits estimation of activation energies for the forward and reverse reactions, and standard enthalpies for the reactions can be calculated as:

$$\Delta H^o = E_f^* - E_r^* \tag{5.26}$$

where ΔH^o is enthalpy for the redox reaction, and ΔE_f^* and ΔE_r^* are the activation energies for the forward and reverse reactions. With a knowledge of ΔG^o and ΔH^o, standard entropy of the reaction can be calculated as:

$$\Delta S^o = (\Delta H^o - \Delta G^o)/T \tag{5.27}$$

Since the e^- and H^+ are key reactants and products in the thermodynamic sense, a refined knowledge of mechanisms for particular redox reactions is needed because of our current limitations in understanding how similar or different e^- and H^+ reactions are in soils and natural waters. A knowledge of thermodynamic stability for a redox system does not necessarily predict kinetic lability, a concept directly pertinent to reactivity of different types of complexes (Cotton and Wilkinson, 1980). The kinetic lability of redox processes in soils has not been compared systematically with predictions of stability based on thermodynamic data, but linear free energy relationships (LFERs) are useful plots of observed rate constants vs. equilibrium constants or Eh values (Stumm and Morgan, 1996; Bürge and Hug, 1997). Such information could affirm the reliability of using thermodynamic data to predict bioavailability of nutrients and pollutants, and to estimate true reactivity of electron donors and acceptors in nonequilibrium, kinetically governed soil chemical environments.

Given the constraining limitation of irreversibility of most redox reactions in soils for using experimental data to calculate values for thermodynamic parameters, the use of kinetic knowledge on soil redox processes is less widely used than are pe (and Eh) and pH for assessing equilibrium conditions of soils. Empirical studies in relatively pure, aqueous systems and complex soil suspensions have, however, provided pertinent information on the kinetics of redox processes and their relation to thermodynamic parameters and reaction pathways (Stumm, 1992; Stumm and Morgan, 1996; Hug et al., 1997; Typrin, 1998).

5.2   Methods and Procedures

5.2.1   Uses of pe-pH Thermodynamic Information

Values of log K derived from Gibbs free energy of formation data or kinetic evaluations of redox reactions provide tools for predicting if a reduction half-reaction coupled with an oxidation half-reaction will allow the spontaneous transfer of electrons from reductant to oxidant. Since soils are only metastable and highly heterogeneous in nature, such predictive capability is necessary to formulate hypotheses for many processes that may occur in the field, even if they require catalysis or other coupled reactions to occur at ambient temperatures and pressures of soil-water-plant systems.

A listing of reduction half-reactions for species of N, O, Mn, Fe, S, C, various pollutants sensitive to redox conditions in soils, and several reactions pertinent to the analysis or characterization of redox conditions is presented in Table 5.1. Within groups, the half-reactions are arranged in descending order of log K values, calculated as described above. These values are pe values at pH = 0 when activities of oxidant and reductant are 1 and may be considered standard, reference pe values for the reactions. The pe values listed at pH 5 and 7 are calculated to represent typical activities of ions and partial pressures of gases in soil environments.

Higher log K or pe values indicate greater "ease of reduction" of an oxidant (left side of equation) to its reduced form than do lower values. This means that for predictive purposes, an oxidant in a particular reduction half-reaction is able to oxidize the reductant in another half-reaction with a lower pe, at a specified pH. For example, oxides of Mn (III, IV) would be expected to oxidize $Cr(OH)_3$ to Cr(VI) at pH 5 since the range of pe values for reduction of Mn (12.8-16.7) is greater than that for Cr(VI) reduction (10.9). This has been demonstrated to occur in most field-moist soils in the pH range of 4-7 containing oxides of Mn(III, IV) (Bartlett and James, 1979; James and Bartlett, 1983). Conversely, these oxides would not be expected to oxidize N_2 to N_2O (pe at pH 5 = 22.9).

Even though the log K or pe for one reduction half-reaction is less than that for a second half-reaction, the reduced form in the first reaction may still be oxidized by the oxidized species in the second half-reaction (the reverse of the above concept). For example, the pe at pH 5 for reduction of CO_2 to $C_6H_{12}O_6$ is -5.9 and that for reduction of O_2 to H_2O is 15.6. Based on this difference, one would predict that reduction of O_2 to H_2O would be coupled to oxidation of $C_6H_{12}O_6$ to CO_2, and coupling oxidation of H_2O to O_2 with reduction of CO_2 to $C_6H_{12}O_6$ would not be thermodynamically possible. In fact, both respiration (the predicted reaction) and photosynthesis (the second, "impossible" reaction) occur together, and the balance of the two is responsible for the existence of the aerobic lifestyle and the persistence of soil organic matter (SOM). Photosynthesis (represented simply as $CO_2 \longrightarrow C_6H_{12}O_6$) is made possible by a complex series of coupled reactions that make an overall thermodynamically-impossible reaction occur rapidly in sunlight.

Table 5.1 Selected reduction half-reactions pertinent to soil, natural waters, plant growth, and microbial systems

	log K[a]	pe[b]	
		pH 5	pH 7
Nitrogen Species			
$^1/_2N_2O + e^- + H^+ = ^1/_2N_2 + ^1/_2H_2O$	29.8	22.9	20.9
$NO + e^- + H^+ = ^1/_2N_2O + ^1/_2H_2O$	26.8	19.8	17.8
$^1/_2NO_2^- + e^- + ^3/_2H^+ = ^1/_4N_2O + ^3/_4H_2O$	23.6	15.1	12.1
$^1/_5NO_3^- + e^- + ^6/_5H^+ = ^1/_{10}N_2 + ^3/_5H_2O$	21.1	14.3	11.9
$NO_2^- + e^- + 2H^+ = NO + H_2O$	19.8	9.8	5.8
$^1/_4NO_3^- + e^- + ^5/_4H^+ = ^1/_8N_2O + ^5/_8H_2O$	18.9	12.1	9.6
$^1/_6NO_2^- + e^- + ^4/_3H^+ = ^1/_6NH_4^+ + ^1/_3H_2O$	15.1	8.4	5.7
$^1/_8NO_3^- + e^- + ^5/_4H^+ = ^1/_8NH_4^+ + ^3/_8H_2O$	14.9	8.6	6.1
$^1/_2NO_3^- + e^- + H^+ = ^1/_2NO_2^- + ^1/_2H_2O$	14.1	9.1	7.1
$^1/_6NO_3^- + e^- + ^7/_6H^+ = ^1/_6NH_2OH + ^1/_3H_2O$	11.3	5.4	3.1
$^1/_6N_2 + e^- + ^4/_3H^+ = ^1/_3NH_4^+$	4.6	-0.7	-3.3
Oxygen Species			
$O_3^{•-} + e^- + 2H^+ = H_2O + O_2$	--[c]	--	30.5[d]
$^1/_2O_3 + e^- + H^+ = ^1/_2O_2 + ^1/_2H_2O$	35.1	28.4	26.4
$OH^• + e^- = OH^-$	33.6	33.6	33.6
$O_2^- + e^- + 2H^+ = H_2O_2$	32.6	22.6	18.6
$^1/_2H_2O_2 + e^- + H^+ = H_2O$	30.0	23.0	21.0
$HO_2^• + e^- + H^+ = H_2O_2$	--	--	17.9[d]
$O_3 + e^- = O_3^{•-}$	--	--	15.1[d]
$^1/_4O_2 + e^- + H^+ = ^1/_2H_2O$	20.8	15.6	13.6
$^1/_2O_2 + e^- + H^+ = ^1/_2H_2O_2$	11.6	8.2	6.2
$^3O_2 + e^- = O_2^{•-}$ (triplet)	-9.5	-6.2	--6.2
$^1O_2 + e^- = O_2^{•-}$ (singlet)	--	--	14.1[d]
Sulfur Species			
$^1/_8SO_4^{2-} + e^- + ^5/_4H^+ = ^1/_8H_2S + ^1/_2H_2O$	5.2	-1.0	-3.5
$^1/_6SO_4^{2-} + e^- + ^4/_3H^+ = ^1/_6S^0 + ^2/_3H_2O$	5.3	-1.4	-4.0
$^1/_2SO_4^{2-} + e^- + 2H^+ = ^1/_2SO_2 + H_2O$	2.9	-7.1	-11.1
Iron and Manganese Compounds			
$^1/_2Mn_3O_4 + e^- + 4H^+ = ^3/_2Mn^{2+} + 2H_2O$	30.7	16.7	8.7
$^1/_2Mn_2O_3 + e^- + 3H^+ = Mn^{2+} + ^3/_2H_2O$	25.7	14.7	8.7
$Mn^{3+} + e^- = Mn^{2+}$	25.5	25.5	25.5
$\gamma\text{-}MnOOH + e^- + 3H^+ = Mn^{2+} + 2H_2O$	25.4	14.4	8.4
$0.62MnO_{1.8} + e^- + 2.2H^+ = 0.62Mn^{2+} + 1.1H_2O$	22.1	13.4	8.9
$^1/_2Fe_3(OH)_8 + e^- + 4H^+ = ^3/_2Fe^{2+} + 4H_2O$	21.9	7.9	-0.1
$^1/_2MnO_2 + e^- + 2H^+ = ^1/_2Mn^{2+} + H_2O$	20.8	12.8	8.8
$[Mn^{3+}(PO_4)_2]^{3-} + e^- = [Mn^{2+}(PO_4)_2]^{4-}$	20.7	20.7	20.7
$Fe(OH)_2^+ + e^- + 2H^+ = Fe^{2+} + 2H_2O$	20.2	10.2	6.2

Table 5.1 (cont.)

	log K^a	pe^b	
		pH 5	pH 7
$\frac{1}{2}Fe_3O_4 + e^- + 4H^+ = \frac{3}{2}Fe^{2+} + 2H_2O$	17.8	3.9	−4.1
$MnO_2 + e^- + 4H^+ = Mn^{3+} + 2H_2O$	16.5	0.54	−7.5
$Fe(OH)_3 + e^- + 3H^+ = Fe^{2+} + 3H_2O$	15.8	4.8	−1.2
$FeOH^{2+} + e^- + H^+ = Fe^{2+} + H_2O$	15.2	10.2	8.2
$\frac{1}{2}Fe_2O_3 + e^- + 3H^+ = Fe^{2+} + \frac{3}{2}H_2O$	13.4	2.4	−3.6
$FeOOH + e^- + 3H^+ = Fe^{2+} + 2H_2O$	13.0	2.0	−4.0
$Fe^{3+} + e^- = Fe^{2+}$ (phenanthroline)	18.0	$--^c$	--
$Fe^{3+} + e^- = Fe^{2+}$ (H_2O only)	13.0	13.0	13.0
$Fe^{3+} + e^- = Fe^{2+}$ (acetate)	--	5.8	--
$Fe^{3+} + e^- = Fe^{2+}$ (malonate)	--	4.4 (pH 4)	--
$Fe^{3+} + e^- = Fe^{2+}$ (salicylate)	--	4.4 (pH 4)	--
$Fe^{3+} + e^- = Fe^{2+}$ (hemoglobin)	--	--	2.4
$Fe^{3+} + e^- = Fe^{2+}$ (EDTA)	--	--	2.0^d
$Fe^{3+} + e^- = Fe^{2+}$ cyt b_3 (plants)	--	--	0.68
$Fe^{3+} + e^- = Fe^{2+}$ (oxalate)	--	--	0.034
$Fe^{3+} + e^- = Fe^{2+}$ (pyrophosphate)	-2.4	--	--
$Fe^{3+} + e^- = Fe^{2+}$ (peroxidase)	--	--	−4.6
$Fe^{3+} + e^- = Fe^{2+}$ (ferredoxin); spinach	--	--	−7.3
$Fe^{3+} + e^- = Fe^{2+}$ (ferritin)	--	--	$−3.2^d$
$\frac{1}{3}KFe_3(SO_4)_2(OH)_6 + e^- + 2H^+ = Fe^{2+} + 2H_2O + \frac{2}{3}SO_4^{2-} + \frac{1}{3}K^+$	8.9	6.9	2.9
$[Fe(CN)_6]^{3-} + e^- = [Fe(CN)_4]^{4-}$	--	--	6.1
Carbon Species			
$CH_3CH_2{}^\bullet + e^- + H^+ = CH_3CH_3$	--	--	32.2^d
$RO^\bullet + e^- + H^+ = ROH$ (aliphatic)	--	--	27.1^d
$CO_3{}^{\bullet-} + e^- + H^+ = CO_3^{2-}$	--	--	25.1^d
catechol-$O^\bullet + e^- + H^+ =$ catechol	--	--	9.0^d
benzosemiquinone + $e^- + H^+ =$ hydroquinone	--	--	7.8^d
$\frac{1}{2}$ o-quinone + $e^- + H^+ = \frac{1}{2}$ diphenol	--	--	5.9
$\frac{1}{2}$ p-quinone + $e^- + H^+ = \frac{1}{2}$ hydroquinone	--	--	4.7
$\frac{1}{2}CH_3OH + e^- + H^+ = \frac{1}{2}CH_4 + \frac{1}{2}H_2O$	9.9	4.9	2.9
$\frac{1}{12}C_6H_{12}O_6 + e^- + H^+ = \frac{1}{4}C_2H_5OH + \frac{1}{4}H_2O$	4.4	0.1	−1.9
Pyruvate + $e^- + H^+ =$ Lactate	--	--	−3.1
$\frac{1}{8}CO_2 + e^- + H^+ = \frac{1}{8}CH_4 + \frac{1}{4}H_2O$	2.9	−2.1	−4.1
$\frac{1}{2}CH_2O + e^- + H^+ = \frac{1}{2}CH_3OH$	2.1	−2.9	−4.9
$\frac{1}{2}HCOOH + e^- + H^+ = \frac{1}{2}CH_2O + \frac{1}{2}H_2O$	1.5	−3.5	−5.5
$\frac{1}{4}CO_2 + e^- + H^+ = \frac{1}{24}C_6H_{12}O_6 + \frac{1}{4}H_2O$	−0.21	−5.9	−7.9
$\frac{1}{2}$ dehydroascorbate + $e^- + H^+ = \frac{1}{2}$ ascorbic acid	1.0	−3.5	−5.5

Table 5.1 (cont.)

	log K[a]	pe[b]	
		pH 5	pH 7
$^1/_4CO_2 + e^- + H^+ = ^1/_4CH_2O + ^1/_4H_2O$	−1.2	−6.1	−8.1
$^1/_2CO_2 + e^- + H^+ = ^1/_2HCOOH$	−1.9	−6.7	−8.7
$CO_2 + e^- = CO_2^{\bullet-}$	--	--	−30.5
Pollutant/nutrient Group			
$Co^{3+} + e^- = Co^{2+}$	30.6	30.6	30.6
$^1/_2NiO_2 + e^- + 2H^+ = ^1/_2Ni^{2+} + H_2O$	29.8	21.8	17.8
$PuO_2^+ + e^- = PuO_2$	26.0	22.0	22.0
$^1/_2PbO_2 + e^- + 2H^+ = ^1/_2Pb^{2+} + H_2O$	24.8	16.8	12.8
$PuO_2 + e^- + 4H^+ = Pu^{3+} + 2H_2O$	9.9	−6.1	−14.1
$^1/_3HCrO_4^- + e^- + ^4/_3H^+ = ^1/_3Cr(OH)_3 + ^1/_3H_2O$	18.9	10.9	8.2
$^1/_2AsO_4^{3-} + e^- + 2H^+ = ^1/_2AsO_2^- + H_2O$	16.5	6.5	2.5
$Hg^{2+} + e^- = ^1/_2Hg_2^{2+}$	15.4	13.4	13.4
$^1/_2MoO_4^{2-} + e^- + 2H^+ = ^1/_2MoO_2 + H_2O$	15.0	3.0	−1.0
$^1/_2SeO_4^{2-} + e^- + H^+ = ^1/_2SeO_3^{2-} + ^1/_2H_2O$	14.9	9.9	7.9
$^1/_4SeO_3^{2-} + e^- + ^3/_2H^+ = ^1/_4Se + ^3/_4H_2O$	14.8	6.3	3.3
$^1/_6SeO_3^{2-} + e^- + ^4/_3H^+ = ^1/_6H_2Se + ^1/_2H_2O$	7.6	1.0	−1.7
$^1/_2VO_2^+ + e^- + ^1/_2H_3O^+ = ^1/_2V(OH)_3$	6.9	2.4	1.4
$Cu^{2+} + e^- = Cu^+$	2.6	2.6	2.6
$PuO_2 + e^- + 3H^+ = PuOH^{2+} + H_2O$	2.9	−8.1	−14.1
Paraquat $+ e^- = $ paraquat$^{\bullet-}$	--	--	−7.6
Analytical Couples			
$CeO_2 + e^- + 4H^+ = Ce^{3+} + 2H_2O$	47.6	31.6	23.6
$^1/_2ClO^- + e^- + H^+ = ^1/_2Cl^- + ^1/_2H_2O$	29.0	24.0	22.0
$HClO + e^- + H^+ = ^1/_2Cl_2 + H_2O$	27.6	20.6	18.6
$^1/_2Cl_2 + e^- = Cl^-$	25.0	25.0	25.0
$^1/_6IO_3^- + e^- + H^+ = ^1/_6I^- + ^1/_2H_2O$	18.6	13.6	11.6
$^1/_2Pt(OH)_2 + e^- + H^+ = ^1/_2Pt + H_2O$	16.6	11.6	9.6
$^1/_2I_2 + e^- = I^-$	9.1	11.1	11.1
$^1/_2Hg_2Cl_2 + e^- = Hg + Cl^-$	4.5	3.9	3.9
$e^- + H^+ = ^1/_2H_2$	0	−5	−7
$^1/_2PtS + e^- + H^+ = ^1/_2Pt + ^1/_2H_2S$	−5.0	−10.0	−12.0

[a] Calculated for reaction as written according to Eq. 14. Free energy of formation data were taken from Lindsay (1979) as a primary source, and when not available from that source, from Garrels and Christ (1965) and Loach (1976).

[b] Calculated using tabulated log K values, (red) and (ox) = 10^{-4} soluble ions and molecules, activities of solid phases = 1, and partial pressures for gases that are pertinent to soils: 10^{-4} atm for trace gases, 0.21 atm for O_2, 0.78 for N_2, and 0.00032 atm for CO_2.

[c] values not listed by Loach (1976) or Larson (1997)

[d] values from Larson (1997)

Similarly, a reaction predicted to be thermodynamically possible may not occur under natural conditions. The pe for NO_3^- reduction to N_2 at pH 5 (14.3) is greater than that for $HCrO_4^-$ reduction to $Cr(OH)_3$ (10.9), but this NO_3^- oxidation of $Cr(III)$ has not been demonstrated in soils or plants, probably because the reduction of NO_3^- requires enzymatic catalysis to lower the energy of activation at such a high pH.

The order of log K values for reduction-half reactions also has been used to predict the sequence of reduction reactions carried out by respiring soil microorganisms following saturation of a soil (Ponnamperuma, 1972). The descending order of preference (pe) for the electron acceptors at pH 7 (proportional to free energy derived from the reduction) is O_2/H_2O (13.6), NO_3^-/N_2 (11.9), MnO_2/Mn^{2+} (8.8), $Fe(OH)_3/Fe^{2+}$ (–1.2), SO_4^{2-}/H_2S (–3.5), and CO_2/CH_4 (–4.1). Heterotrophic bacteria are using organic compounds as the electron donors (energy source) in their respiration to produce CO_2 or organic acids (pe range at pH 7 of –8.7 to –3.1), so most of the organic compounds can be used throughout the reduction sequence following depletion of atmospheric O_2.

The predicted pe values at a given pH are determined by chosen activities of reductant and oxidant, and by the values of ΔG_f^o used to calculate log K values. As shown in Table 5.2, the sensitivity of calculated pe due to variation or error in log K and activities varies considerably among reduction half-reactions. Predicted pe values for O_2, NO_3^-, CO_2, and SO_4^{2-} reduction changed less than 0.5 units in response to decreasing activity from 10^{-4} to 10^{-6}. In contrast, an error of only 10 kcal mol^{-1} in the ΔG_f^o resulted in changes in the calculated pe values of 3.6, 1.5, 0.3, and 0.9, respectively, for the above half-reactions. This result indicates that wide variation and error in estimates of activities will have a smaller effect on pe than will errors in free energy of formation data, especially for O_2 and NO_3^- reduction reactions.

In contrast to these half-reactions involving the reduction or oxidation of gases, those involving oxides and oxyhydroxides of Mn and Fe are subject to greater error (1.0 to 3.0 units) in estimation of pe due to variation in activity of Fe^{2+} or Mn^{2+} between 10^{-4} and 10^{-6} M. Fewer electrons (one or two) are accepted per mole of these oxides than are accepted by the oxyanions (five to eight). This difference has the algebraic effect of making the calculated pe more sensitive, the smaller the number of electrons transferred.

Similar to the gas group, pe values for the oxide group showed considerable change due to error in G_f^o, ranging from 3.7 to 7.4 pe units. Differences between MnOOH, Mn_3O_4, and MnO_2 also were large, and similar differences were found for the Fe oxides. This indicates that correct identification of Mn and Fe oxide mineralogy significantly affects pe-pH relationships and predicted energy changes associated with these reduction reactions in soils at certain pH values.

These observations of the sensitivity of predicted pe values to variations in activities and errors in G_f^o indicate that exact pe values for the reduction of a particular oxidant are difficult to obtain, and ranges of values may be more accurate. The use of such ranges also recognizes the heterogeneity of soil minerals, gaseous composition, and ion activity in space and time.

Values for log K for reduction half-reactions are especially sensitive to changes in activity of oxidant or reductant if O atoms are transferred from the oxidant to H_2O, as in the reduction of MnOOH to Mn^{2+}, in contrast to the reduction of Mn^{3+} to Mn^{2+}. The formation of H_2O is favorable because it increases entropy of aqueous systems (Cotton and Wilkinson, 1980) and thereby tends to lower calculated ΔG_r^o. Water molecules formed as a product of a reduction half-reaction are balanced by H^+ on the left side of the equation, resulting in a larger ratio of H^+-to-e^- consumed. This increases the slope of the pe-pH relationship, resulting in a greater error in the estimation of pe due to error in pH measurement.

5.2.2   Use of pe-pH Diagrams

Individual pe-pH relationships can be defined by the equation for a straight line (Equation [5.11]) in which log K and the activities of the oxidant and reductant determine the y intercept, and the ratio of H⁺-to-e⁻ consumed determines the slope. When such lines are plotted together, they predict whether or not the oxidation-reduction reaction can occur spontaneously, that is, with $\Delta G_r^o < 0$. Effects of pH on each member of the redox couple are clearly shown, so that the interactions of pe and pH as master variables are delineated.

5.2.2.1   Oxygen Species

While the pe for reduction of O_2 to H_2O ranges from 20.8 at pH 0 to 13.6 at pH 7 (Fig. 5.1 and Table 5.1), the intermediates associated with one electron transfers show a wide fluctuation in their oxidizing power, a property of O_2 that is pertinent to understanding the transition in soils from aerobic to anaerobic conditions. Anaerobic respiratory enzymes are produced when Po_2 reaches approximately 1% of atmospheric levels (0.2 kPa or 0.0021 atm). The data in Table 5.2 also indicate that the pe for reduction of O_2 is relatively insensitive to Po_2 in this range.

Ozone (O_3) and the hydroxyl free radical (OH•) are the most powerful oxidants among the O species (Table 5.1), and the latter may be formed during stepwise, four electron reduction of O_2 to O_2^-, H_2O_2, and H_2O (Fridovich, 1978). The high and low positions of the pe-pH lines for superoxide (O_2^-) reduction to H_2O_2 and for superoxide oxidation to O_2 indicate that both a powerful oxidizing and reducing agent is formed in the first step of reduction of O_2. The enzyme superoxide dismutase scavenges O_2^- in living cells using O_2 as the terminal electron acceptor (respiring aerobically), but relatively little is known about its reactivity in biological and chemical processes in soils that may be pertinent to our understanding of the formation of highly reduced components such as SOM and highly oxidized species such as NO_3^- that coexist in soil at chemical quasi-equilibrium.

The wide range of reduction potentials for O_2 and its partially reduced intermediates, coupled with biological processes controlling the partial pressure of this gas in soil solution, create conditions under which the O_2/H_2O may rarely attain equilibrium. Therefore, only metastable conditions and a slow approach to chemical equilibrium characterize O_2 behavior, thereby making thermodynamic predictions difficult for aerobic soils (Bartlett, 1981).

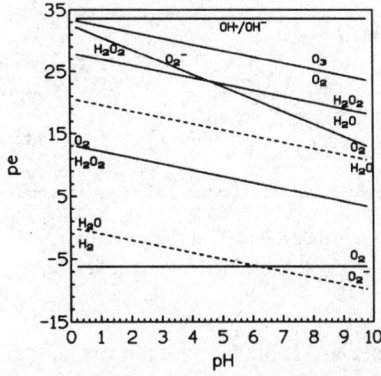

Fig. 5.1   pe-pH diagram for O_2 reduction reactions, including partially reduced intermediates: superoxide (O_2^-), hydroxyl free radical (OH•), and hydrogen peroxide (H_2O_2). Reduction of O_3 to O_2 is also included for comparison with O_2 reduction reactions. Activity of oxygen-containing ionic or molecular species is 10^{-4}, except Po_2 is 21 kPa.

Table 5.2 Sensitivity of calculated pe at pH 0 to variation in ΔG_f^o and activities of reductant or oxidant in selected reduction half-reactions pertinent to soils

Couple	– log Activity		ΔG_f^o		pe	Δpe^a
	Ox	Red	Ox	Red		
O_2/H_2O	0.68	0	0	–56.70	20.61	
	2.68	0	0	–56.70	20.11	–0.50
	0.68	0	0	–66.70	24.20	3.59
NO_3^-/N_2	4	0.11	–26.43	0	20.30	
	6	0.11	–26.43	0	19.90	–0.40
	4	0.11	–36.43	0	18.80	–1.50
Mn_3O_4/Mn^{2+}	0	4	–306.20	–54.40	36.70	
	0	6	–306.20	–54.4	39.70	3.00
	0	4	–316.20	–54.40	33.00	–3.70
MnO_2/Mn^{2+}	0	4	–111.10	–54.40	22.80	
	0	6	–111.10	–54.40	23.80	1.00
	0	4	–121.10	–54.40	19.10	–3.70
$MnOOH/Mn^{2+}$	0	4	–133.10	–54.40	29.40	
	0	6	–133.10	–54.40	31.40	2.00
	0	4	–143.10	–54.40	22.10	–7.30
$CO_2/C_6H_{12}O_6$	3.5	4	–94.26	–218.58	–0.91	
	3.5	6	–94.26	–218.58	–0.83	0.08
	3.5	4	–94.26	–228.58	–0.60	–0.31
$Fe(OH)_3/Fe^{2+}$	0	4	–170.40	–21.80	19.80	
	0	6	–170.40	–21.80	21.80	2.00
	0	4	–180.40	–21.80	12.40	–7.40
Fe_2O_3/Fe^{2+}	0	4	–177.10	–21.80	17.40	
	0	6	–177.10	–21.80	19.40	2.00
	0	4	–187.10	–21.80	13.70	–3.70
SO_4^{2-}/H_2S	4	4	–177.95	–8.02	5.20	
	6	4	–177.95	–8.02	5.50	0.30
	4	4	–167.95	–8.02	6.10	0.90

[a]change in calculated pe resulting from change in activity of Ox or Red (column 1 or 2) or resulting from use of ΔG_f^o 10 kcal mol^{-1} different from the published value (first row, column 3 or 4)

5.2.2.2 Nitrogen Species

Most reduction reactions of N species (Table 5.1) are not reversible, are biologically mediated, and therefore, are not well defined by thermodynamic pe-pH relationships. The series of half-reactions composing the process of denitrification, though, is instructive in that it identifies the wide range in pe for reduction of each of the intermediates believed to form in the sequence of electron acceptors used by microbes:

$$NO_3^- \rightarrow NO_2^- \rightarrow NO \rightarrow N_2O \rightarrow N_2$$

Step (1) (2) (3) (4)

Step 1 of the sequence occurs at pe values less than those for reduction of O_2 to H_2O, while those for steps 2, 3, and 4 are increasingly higher (Fig. 5.2). The overall reduction of NO_3^- to N_2 is almost identical to that for the O_2/H_2O couple. The pe overlap for the O_2 and NO_3^- reduction intermediates indicates that denitrification and aerobic respiration may occur at the same time under certain conditions when organic C is used as the electron donor. They may not be mutually exclusive, as predicted from log K values for the overall reactions, O_2 to H_2O and NO_3^- to N_2.

5.2.2.3 Manganese Oxide Species

Manganese exists in soils in the II+, III+, and IV+ valence states, and the latter two are most stable as oxides or oxyhydroxides. Trivalent Mn may exist as Mn^{3+}, especially if stabilized by ligands, such as pyrophosphate or citrate. The pe-pH relationships of Fig. 5.3 predict that different valences of Mn in Mn_3O_4, $MnOOH$, and MnO_2 affect the pe at which Mn^{2+} would be expected to form at pH values < 7, but they are all similar at pH values near 7. The Mn^{3+}/Mn^{2+} line indicates that at pH 4, Mn^{3+} is a powerful oxidant similar to O_2^- and Mn_3O_4 (Fig. 5.3) if in equilibrium with Mn^{2+}. At pH ≈ 6.5, Mn^{3+} in equilibrium with MnO_2 is a powerful reductant similar to H_2 and O_2^-. This predicted reducing ability of Mn^{3+} may be pertinent to anaerobic soils that are exposed to O_2, and in which Mn^{2+} is oxidizing to form Mn(III, IV) oxides via Mn^{3+}. In oxidized soils containing MnO_2, flooding and the process of becoming reduced may produce Mn^{3+} that is a powerful reducing agent. The trivalent Mn species may be an ephemeral intermediate in such processes at redox interfaces, such as in the rhizosphere of plant roots or between soil water and groundwater. As the metal analog of superoxide in its oxidizing/ reducing power and as a free radical, Mn^{3+} is appropriately referred to as the super manganese ion.

Since many Mn(III, IV) oxides are non-stoichiometric and no compound with the exact composition of MnO_2 is known (Arndt, 1981), predictions of their redox properties as a function of mineralogy or valence in heterogeneous soils may be hard to formulate. Despite the uncertainty of thermodynamic predictions for the redox behavior of Mn, the chemistry of this element is pertinent to a number of processes governing speciation and valence state of trace elements and pollutants found in soils. The pe-pH data indicate that oxides of Mn may oxidize Pu(III) to Pu(IV), V(III) to V(V), As(III) to As(V), Se(IV) to Se(VI), N(III) to N(V), and Cr(III) to Cr(VI) since the pe for each of these couples

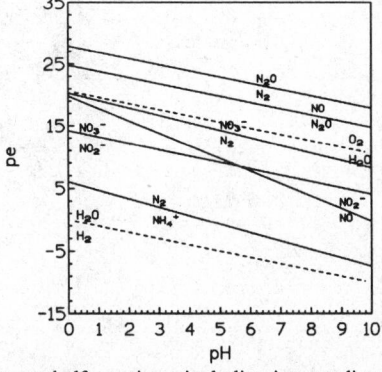

Fig. 5.2 pe-pH diagram for nitrogen half-reactions, including intermediates formed or consumed in denitrification and dinitrogen fixation. Partial pressures of gaseous intermediates are 0.01 kPa, except N_2 is 78 kPa; activity for NO_3^-, NO_2^-, and NH_4^+ is 10^{-4}

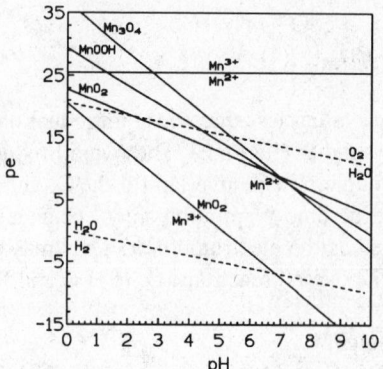

Fig. 5.3   pe-pH diagram for Mn oxides, trivalent Mn ions, and superoxide. Ion activities are 10^{-4} and H_2O_2 concentrations are 10^{-4} M, and that for Po_2 is 21 kPa.

falls below that for Mn oxides (Fig. 5.4 and Table 5.1). The oxidations of Pu(III), As(III), Se(IV), N(III), and Cr(III) all have been demonstrated to occur in soils containing Mn(III,IV)(hydr)oxides or by synthetic Mn(III,IV)(hydr)oxides (Amacher and Baker, 1981; Bartlett and James, 1979; Bartlett, 1981; Blaylock and James, 1994, Moore et al., 1990).

The instability of Mn^{3+} and its ability to dismutate, similar to H_2O_2 and O_2^-, mean that kinetic constraints and very low steady state concentrations in soil solution may be particularly important in understanding the redox behavior of Mn in soils undergoing transitions between anaerobic and aerobic conditions. The kinetic lability of these species is poorly understood and new knowledge could contribute significantly to predictions of bioavailability and toxicity of numerous plant nutrients and pollutants in a range of soil types from rice paddies and wetlands to well-drained agricultural and forest soils.

5.2.2.4   Iron Species
While predictions of redox behavior of Fe(II) and Fe(III) species indicate that they fall below most Mn oxide species (lower pe values and less free energy released per equivalent upon reduction), intermediate hydrolysis products, such as $Fe(OH)_2^+$, theoretically can oxidize Cr(III) to Cr(VI) at pH

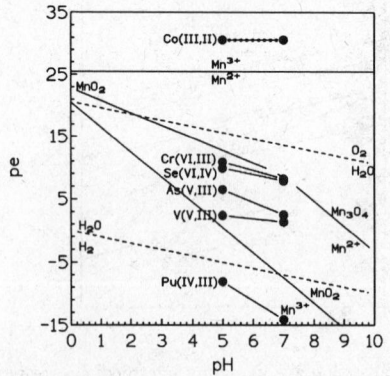

Fig. 5.4   pe-pH diagram for Mn^{3+}, MnO_2, and Mn_3O_4; as compared with reduction values between pH 5 and 7 for Co, Cr, Se, As, V, and Pu. Activity for ionic species is 10^{-4}.

values < 4 (Fig. 5.5 and Table 5.1). In addition, thermodynamically more stable complexation of Fe^{3+} by organic and inorganic ligands such as OH^-, relative to complexation of Fe^{2+}, lowers the pe at which Fe^{3+} is reduced to Fe^{2+}. These Fe(II)-Fe(III)-ligand reduction potentials are similar to those of Fe(III) oxides in the pH range 5 to 7 (Fig. 5.5). This phenomenon suggests that Fe(II)/Fe(III)-ligand systems create Fe(II) species that are more powerful reductants that hexaquo Fe^{2+} if the the the Fe(III)-ligand complex is more thermodynamically stable than the Fe(II)-ligand complex. Conversely, if the complex with Fe(II) is more stable than that with Fe(III) (e.g., with phenanthroline), the Fe(II) becomes a less powerful reductant than hexaquo Fe^{2+}. This may explain the ability of a Fe(II,III) system to act as a cofactor in enzymes involved in redox processes, such as peroxidases and superoxide dismutases. These enzymes reduce or dismutate H_2O_2 and O_2^-. The application of such concepts to abiotic redox processes in soils remains a key area for future research.

5.2.2.5 Carbon and Sulfur Species

Reduced forms of C and S are normally viewed as reductants in soils, either in chemical or biological processes. Thermodynamic predictions support this idea for carbohydrates produced in photosynthesis, CH_4 from methanogenesis, and H_2S from reduction of SO_4^{2-} (Fig. 5.6 and Table 5.1). The reduction reactions of *o*- and *p*-quinone suggest that these compounds may be reduced at higher pe values than are CO_2 and SO_4^{2-}. The low position of these lines, however, coincides with the MnO_2/Mn^{3+} couple at pH 7, suggesting that Mn^{3+} may act as a reducing agent for certain organic species in near neutral soils. Coupling of reduction of the organic with oxidation of Mn may result in formation of free radical species. This is pertinent to understanding the formation and persistence of SOM in high pH soils that may contain reactive forms of Mn oxides.

Reactions of H_2S and H_2Se are predicted to be similar with respect to SO_4^{2-} and SeO_3^{2-} formation (Fig. 5.6). While SeO_4^{2-} and SO_4^{2-} are similar chemically, the oxidation of SeO_3^{2-} to SeO_4^{2-} is predicted to occur at higher pe than that of H_2S-to-SO_4^{2-} (Fig. 5.6 and Table 5.1). Blaylock and James (1994) observed that Mn oxides in soils or in pure form will oxidize SeO_3^{2-} to SeO_4^{2-}, as predicted by thermodynamics (Fig. 5.4). They also observed that adding "reducing" phenolic acids, such as gallic and ascorbic acids, actually enhanced this oxidation reaction of SeO_3^{2-}. They hypothesized that partial reduction of MnO_2 in soils converted the Mn(IV) oxide into a Mn(III) oxide or ion that is a more powerful oxidant for SeO_3^{2-} than is Mn(IV) oxide. Such a hypothesis is supported by the relative oxidizing power of MnO_2, MnOOH, and Mn_3O_4, where the latter two oxides contain Mn(III) (Fig. 5.3).

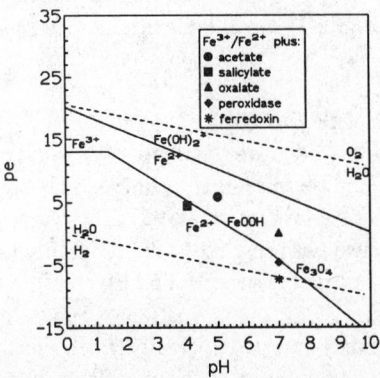

Fig. 5.5 pe-pH diagram for Fe(III) oxides and dihydroxy species, Fe^{3+}/Fe^{2+} in the presence or absence of five organic complexing ligands, and the $HCrO_4^-/Cr(OH)_3$ redox couple. Activity of ionic species is 10^{-4}.

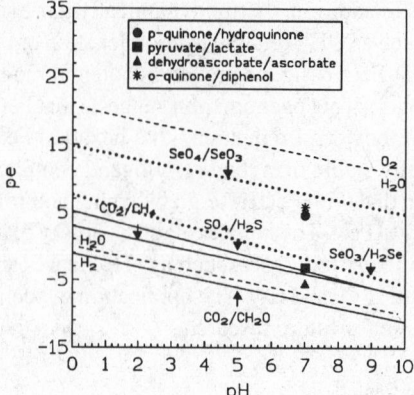

Fig. 5.6 pe-pH diagram for S, C, and Se species. Ion activity is 10^{-4} and molecular concentrations are 10^{-4} M, and P_{CO_2} is 0.032 kPa.

As discussed above in Section 5.2.1, the sensitivity of pe predictions (Table 5.2) can be represented graphically in pe-pH diagrams (Figs. 5.7-5.12). Such responses in predicted pe values due to perturbations or variation in free energy data and mineralogy indicate inherent weaknesses and points of departure for further research on the use of thermodynamic data to assess redox phenomena in soils.

5.2.3 Measurement of Soil Redox and Acid-Base Status

The most common method for quantifying electron activity of soils and natural waters is to measure the potential difference between a Pt indicator electrode and a calomel or Ag/AgCl reference electrode, both connected to a voltmeter or pH meter (Pearsall and Mortimer, 1939; Patrick and DeLaune, 1972; Rowell, 1981; Bricker, 1982; Patrick et al., 1996). In this method, the Pt electrode is presumed to be inert and to not react chemically while coming into equilibrium with electroactive species in soil solution and on soil colloids (Compton and Sanders, 1996). Generally this potentiometric measurement is unreliable for accurate assessments of redox status, especially of aerobic soils (Whitfield, 1974; Bartlett, 1981; Grenthe et al., 1992; Hostettler, 1992; Grundl, 1994). Other methods that employ analysis of soil solution analytes indicative of redox status, along with thermodynamic half-reactions, as discussed above, may prove more reliable for calculating pe ranges for aerobic and anaerobic soil systems (Peiffer et al., 1992; Kludze et al., 1994; Lovley et al., 1994; Stumm and Morgan, 1996; Typrin, 1998).

5.2.3.1 Construction and Use of Pt Electrodes

Platinum and suitable reference electrodes are relatively easy and inexpensive to construct (Mueller et al., 1985; Farrel et al., 1991), but measurement technique may significantly alter measured voltages (Bartlett, 1981; Bricker, 1982; Matia et al., 1991). These researchers have described several aspects of electrode use and misuse with respect to the reliability of recorded voltages for natural systems. Comparisons have been made between H_2 and Eh measurements for redox status in a contaminated aquifer (Chapelle et al., 1996); quantification of H_2 was more reliable than Eh measurements for identifying anoxic redox process, expecially when considered with respect to electron acceptor availability. The limits and limitations of Eh measurements have been described for natural systems, especially with respect to how long the electrodes may be left in place in the soil or water

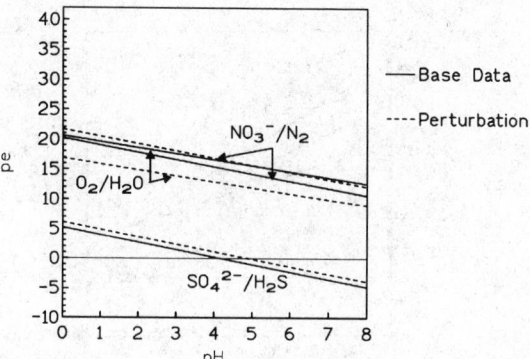

Fig. 5.7 pe-pH diagram showing the effect of a variation in the Gibbs free energy of formation of 10 kcal mol^{-1} on predicted pe: oxygen, nitrate, and sulfate as oxidants

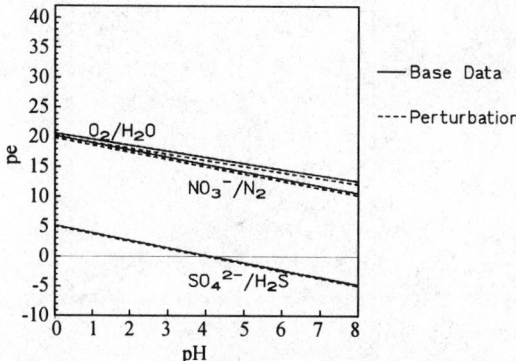

Fig. 5.8 pe-pH diagram showing the effect of a variation in activity values (10^{-4} to 10^{-6} for nitrate and sulfate; P_{O_2} from 0.21 to 0.0021 atm for oxygen predicted pe: oxygen, nitrate, and sulfate as oxidants

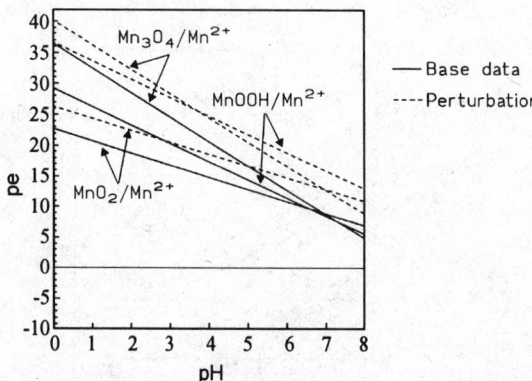

Fig. 5.9 pe-pH diagram showing the effect of a variation in the Gibbs free energy of formation of 10 kcal mol^{-1} on predicted pe: manganese (III, IV)(hydr)oxides as oxidants

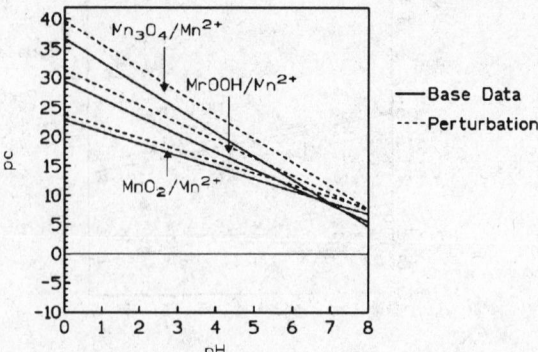

Fig. 5.10   pe-pH diagram showing the effect of a variation in activity values (10^{-4} to 10^{-6}) on predicted pe: manganese(III, IV) (hydr)oxides as oxidants

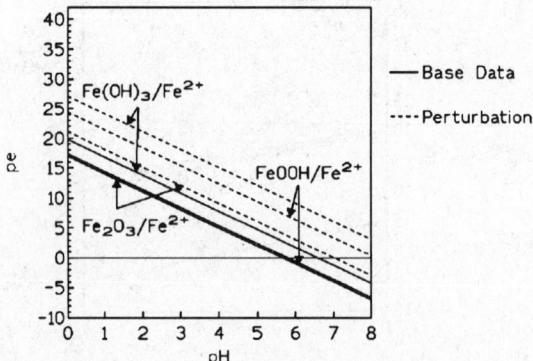

Fig. 5.11   pe-pH diagram showing the effect of a variation in the Gibbs free energy of formation of 10 kcal mol^{-1} on predicted pe: iron (III)(hydr)oxides as oxidants

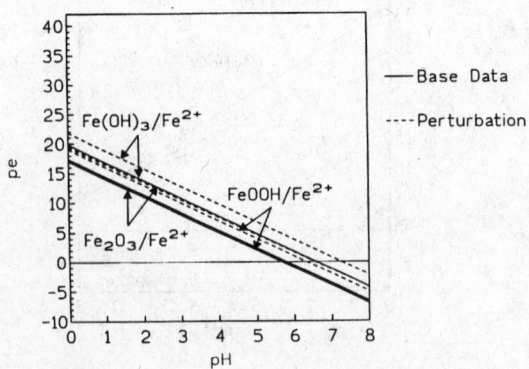

Fig. 5.12   pe-pH diagram showing the effect of a variation in activity values (10^{-4} to 10^{-6}) on predicted pe: iron(III)(hydr)oxides as oxidants

(Mansfeldt, 1993; Norrström, 1994) and with respect to interpreting the Eh values obtained (Baas-Becking, 1960; Whitfield, 1974; Lindberg and Runnells, 1984; Yu, 1992). Platinum electrode systems also have been incorporated into potential-controlling systems for long-term laboratory studies (Patrick, 1966; Petrie et al., 1998).

5.2.3.2 Inadequacies of Pt Electrode Potentials

Assessing "electron activity" in soils relates strictly to an evaluation of the ability of the electron to be transferred, or do thermodynamic work, and not of its concentration in soil solution, as can be defined for H^+. Because of the nature of the electron and its differences from H^+, a number of caveats must be described and recognized when evaluating Pt electrode potentials.

Dissolved Oxygen Status

While a stable potential can be obtained for a Pt-reference electrode pair immersed in an oxygenated soil suspension, this potential is unreliable as a measure of dissolved oxygen status (Bricker, 1982; Stumm and Morgan, 1996). The Pt surface may react with O_2 to form $Pt(OH)_2$ that develops a potential with elemental Pt with a pe of 9.6 at pH 7 (Table 5.1). In addition, the measurement may not be that of the O_2-H_2O couple, but may be responding to O_2 reduction intermediates, such as H_2O_2 and O_2^- (Bricker, 1982). In addition, predicted pe values are relatively insensitive to changes in dissolved O_2 between 0.21 and 0.0021 atm (Table 5.2; Fig. 5.8), the range of O_2 partial pressures in which aerobic respiration occurs (Russell, 1973). For these reasons, Pt electrode potentials cannot be used reliably as a measure of redox status for aerobic soils, but empirical values for pe may be obtained for comparison purposes (Bartlett, 1981; 1998).

While more faith is placed in measurements of soil pH, it also should be considered an empirical measurement because of uncertainty about the form of the H^+ ion in colloidal environments and about the behavior of the glass electrode in such systems. For these reasons, both pe and pH measured with electrodes in soils may be very uncertain for accurate descriptions of the redox status of soil environments containing air-filled pores.

Irreversibility of Redox Couples

Many of the important redox processes involving C, H, N, O, and S (the "light" elements, relative to the "heavy" metals) are irreversible in the thermodynamic sense, and nonelectroactive gases and molecules may be consumed or formed. As a result, potentials generated by redox couples for these elements are difficult to obtain and interpret using a Pt electrode. In addition, many of these reactions do not reach true chemical equilibrium, and activities measured in soil solution may be kinetically constrained (Liu and Narasimhan, 1989). Since soil redox status is often set by "microbial potentials," consuming or producing compounds or ions containing one or more of these elements may render Pt electrode measurements inaccurate.

Mixed Potentials

Measured redox potentials in soils normally are governed by more than one redox couple ("mixed potentials"), and these couples usually are not at true chemical equilibrium. Therefore, thermodynamic interpretations of measured Pt electrode potentials in soil suspensions need to consider several couples, and analyses of soil solution concentrations of redox active species may be indicative of the couples controlling the potential at the Pt surface (Typrin, 1998). For example, a Fe^{2+}-Fe^{3+} couple at activities greater than 10^{-5} generates sufficient anodic and cathodic currents to obtain a measurable voltage for the system, and this may coexist with the relatively non-electroactive couple, O_2/H_2O in heterogeneous soils.

The establishment of a measurable balance between the cathodic and anodic current at the Pt electrode at the point of zero applied voltage requires activities of $> 10^{-5}$, a condition that may not exist for many redox active species of interest in soils. Activities $< 10^{-6}$ may be common, especially for certain plant nutrients and soil pollutants. For this reason, application of measured Pt electrode potentials to predicting soil composition may not be possible.

Coupling of pH and pe

Based on the complementary nature of pH and pe (Stumm and Morgan, 1996) summing "pe + pH" to describe log K for soils is possible theoretically, as described above. Due to the fact that the Pt electrode responds to pH (almost in Nerntsian fashion) as well as to electron activity, pH should always be measured and reported with pe. The negative slopes of the pe-pH (Figs. 5.1-5.6) relationships of many of the reduction half-reactions also indicate that the energy change associated with a particular reduction reaction decreases with increasing pH. Therefore, an Eh measurement cannot be used to predict the presence of a particular redox couple unless pH is known. Since higher pe values at lower pH's correspond to larger releases of free energy, reduction reactions are expected to be more likely to occur at lower pH; that is, such systems are ones in which "reduction is favored." In contrast, loss of electrons from reductants of a particular couple is favored at higher pH, or the system is more "prone to oxidation."

5.2.4   Proposed Alternative Strategies for More Accurate Measurement of Soil Redox Status

5.2.4.1   Using Electrochemical Relations in Reverse

Given the uncertainty associated with measured Pt electrode potentials in soils to quantify Eh or pe, actual measurements of reductant and oxidant activities, along with a reliable pH measurement, may be a better approach (Stumm and Morgan, 1996; Typrin, 1998). The activities are substituted into appropriate half-reactions relating pe and pH, and pe is thereby obtained by calculation.

As shown in Table 5.2, the pe predicted by such a technique will be inaccurate to different degrees for different half-reactions. For example, an error of two log units for H_2S or O_2 partial pressures will only produce errors of 0.3 to 0.5 pe units. In contrast, similar errors in measurement of Mn^{2+} or Fe^{2+} will result in pe errors of 1.0 to 3.0 pe units. Since Mn and Fe are relatively easy to measure accurately by atomic absorption or colorimetric methods, and mineralogy of associated oxides can be made with infrared or X-ray techniques, assessing pe for Mn and Fe oxide-dominated systems could be reliable. Detailed research is needed to prove this hypothesis. In contrast, dissolved gases are harder to measure accurately, and qualitative estimates may be sufficient to obtain accurate evaluations of pe.

If calculated pe values obtained with this "reverse electrochemical technique" are equal for two different reduction half-reactions, then chemical equilibrium may be assumed to exist. If the pe values are unequal, then disequilibrium and a metastable, kinetically limited soil system probably exists. This latter condition is common in soils due to spatial heterogeneity of soil solution, oxide mineralogy, organic matter reactivity, and microbial controls on key reactions. New thinking and hands-on research are needed to provide new ideas for evaluation of the "electron activity" for soils by this combination of analytical and thermodynamic approaches. A key to its accuracy is knowing what redox couples are contributing to the electron activity, assumed to exist as a quantifiable parameter at chemical equilibrium.

5.2.4.2 Redox Ranges for Empirical pe Values

These limitations to assessing soil pe based on Pt electrode and reverse electrochemical methods indicate that our sense of accuracy for soil redox status must be modified. If we surrender in our efforts to conceptualize and operationally define soil redox status, we will have lost a challenging scientific crusade. This means that while new ideas are being developed, we accept a lack of knowledge of pe values more accurate than ranges bracketed by whole numbers.

Liu and Narasimhan (1989) have described "redox zones" in which a range in Eh or pe defines an electron activity condition. The oxygen-nitrogen range is defined by Eh values of +250 to +100 mV, the iron range is +100 to 0.0 mV, the sulfate range is 0.0 to –200 mV, and that for methane-hydrogen is defined at < –200 mV. Sposito (1989) proposed "oxic" soils as those with pe > 7, "suboxic" ones in the range of pe between +2 and +7. and "anoxic" soils with pe < +2; all at pH 7. These ranges correspond roughly to redox control by oxygen-nitrogen, manganese-iron, and sulfur couples.

Berner (1981) proposed categories for redox named "oxic, post-oxic, sulfidic, and methanic" controlled by transformations of oxygen/nitrogen, iron, sulfur, and methane-hydrogen, respectively. James (1989) has proposed these names be assigned to ranges in EMpe (Bartlett, 1981; 1998) of +7 to +13, +2 to +7, –2 to +2, and –6 to –2 (at pH 7). The appropriateness of these categories and names will require further evaluation of new operational definitions for the concept of "redox status" in soils.

Bartlett and James (1995) proposed a new system for categorizing soil redox status using chemical field tests such as tetramethylbenzidine for Mn status, Cr oxidation-reduction reactions, Fe speciation, sulfide levels, and pH. They proposed the following categories: superoxic, manoxic, suboxic, redoxic, anoxic, and sulfidic which relate to electron lability in heterogeneous soil systems under field conditions. Bartlett (1998) has refined this system of redox classification and applied it to interfacial processes in soils pertinent to wetlands, hydric soils, and other soil-water systems containing contrasting zones of redox status.

5.3 Other Sources on Redox Phenomena and Databases

There are numerous general reviews and examples of redox measurements, processes, applications, and data in soils and natural waters. Interested readers are referred to the following sources: Bartlett (1998), Bartlett and James (1993), Blough and Zepp (1995), Buxton et al. (1988), Compton and Sanders (1996), Helz et al. (1994), Lindsay (1979), Neta et al. (1988), Sawyer (1991), Schwarzenbach et al. (1993), Stumm and Morgan (1996), Wardman (1989).

Remarkable, almost prescient insight is provided in several references on electron activity and its measurement by the Pt electrode in the four decades following the discovery of the electron (Gillespie, 1920; Willis, 1932; Pearsall and Mortimer, 1939). These researchers linked thermodynamic theory, biological processes in soils, and applications of redox chemistry to societal needs of their times, an accomplishment that can serve as a guide for future studies on the elusive electron in soil environments.

5.4 References

Allison, J.D., and D.S. Brown. 1995. MINTEQA2/PRODEGA2—A geochemical speciation model and interactive processor. p. 241–252. *In* R.H. Loeppert et al. (ed.) Chemical equilibria and reaction models. Soil Sci. Soc. Am. Spec. Pub., Madison, WI.

Amacher, M.C., and D.E. Baker. 1982. Redox reactions involving chromium, plutonium, and manganese in soils. Final Rep. Div. Energy Tech., US Dep. Energy DOE/DP/0415-1, Washington, DC.

Arndt, D., 1981. Manganese compounds as oxidizing agents in organic chemistry. Open Court Publishing Co., La Salle, IL.

Baas-Becking, L.G.M., I.R. Kaplan, and D. Moore. 1960. Limits of the natural environment in terms of pH and oxidation-reduction potentials. J. Geol. 68:243–284.

Barcelona, M.J., and T.R. Holm. 1991. Oxidation-reduction capacities of aquifer solids. Environ. Sci. Technol. 25:1565–1572.

Bartlett, R.J., 1981. Oxidation-reduction status of aerobic soils. p.77–102 *In* D. Baker (ed.) Chemistry in soil environments. Soil Science Society of America, Madison, WI.

Bartlett, R.J. 1998. Characterizing soil redox behavior. pp. 371–397. *In* D.L. Sparks (ed.) Soil physical chemistry. 2nd Ed. CRC Press, Boca Raton, FL.

Bartlett, R.J., and B. James. 1979. Behavior of chromium in soils. III. Oxidation. J. Environ. Qual. 8:31–35.

Bartlett, R.J. and B.R. James. 1993. Redox chemistry of soils. Adv. Agron. 50:151–208.

Bartlett, R.J., and B.R. James. 1995. System for categorizing soil redox status by chemical field testing. Geoderma 68:211–218.

Bates, R.G. 1981. The modern meaning of pH. Crit. Rev. Anal. Chem. 10:247–278.

Berner, R.A., 1981. A new geochemical classification of sedimentary environments. J. Sediment. Petrol. 51:359–365.

Blaylock, M.J., and B.R. James. 1994. Redox transformations and plant uptake of selenium resulting from root-soil interactions. Plant Soil 158:1–12.

Blough, N.V., and R.G. Zepp. 1995. Reactive oxygen species in natural waters. p. 280–332. *In* C.S. Foote et al. (ed.) Active oxygen in chemistry. Chapman and Hall, New York.

Bricker, O.P., 1982. Redox measurement: Its measurement and importance in water systems. Water Analysis, Vol. 1. Academic Press, Orlando, FL.

Buresh, R.J., M.E. Casselman, and W.H. Patrick, Jr. 1980. Nitrogen fixation in flooded soil systems, A review. Adv. Agron. 33:149–192.

Bürge, I., and S. Hug. 1997. Kinetics and pH dependence of chromium(VI) reduction by iron(II). Environ. Sci. Technol. 31:1426–1432.

Buxton, G.V., C.L. Greenstock, W.P. Helman, and A.B. Ross. 1988. Critical review of rate constants for reactions of hydrated electrons, hydrogen atoms, and hydroxyl radicals (OH/ OH⁻) in aqueous solution. J. Phys. Chem. Ref. Data. 17:513–886.

Castellan, G.W., 1983. Physical chemistry. 3rd Ed. Addison-Wesley Publishing Co., Reading, MA.

Chapelle, F.H., S.K. Haack, P. Adriaens, M.A. Henry, and P.M. Bradley. 1996. Comparison of Eh and H_2 measurements for delineating redox processes in a contaminated aquifer. Environ. Sci. Technol. 30:3565–3569.

Coleman, M.L., D.B. Hedrick, D.R. Lovley, D.C. White, and K. Pyle. 1993. Reduction of Fe(III) in sediments by sulphate-reducing bacteria. Nature 361:436–438.

Compton, R.G., and G.H.W. Sanders. 1996. Electrode potentials. Oxford Science Publications, Oxford, UK.

Cotton, F.A., and G. Wilkinson. 1980. Advanced inorganic chemistry. John Wiley and Sons, New York, NY.

Farrell, R.E., G.D.W. Swerhone, and C. van Kessel. 1991. Construction and evaluation of a reference electrode assembly for use in monitoring *in situ* soil redox potentials. Comm. Soil Sci. Plant Anal. 22:1059–1068.

Fridovich, I., 1978. The biology of oxygen radicals. Science 201:875–880.

Garrels, R.M., and C.L. Christ. 1965. Solutions, minerals, and equilibria. Freeman, Cooper, and Co., San Francisco, CA.

Gillespie, L.J. 1920. Reduction potentials of bacterial cultures and of water-logged soils. Soil Sci. 9:199–216.

Grenthe, I., W. Stumm, M. Laaksuharju, A.-C. Nilsson, and P. Wikberg. 1992. Redox potentials and redox reactions in deep groundwater systems. Chem. Geol. 98:131–150.

Grundl, T. 1994. A review of the current understanding of redox capacity in natural, disequilibrium systems. Chemosphere 28:613–626.

Harter, R.D., and G. Smith. 1981. Langmuir equation and alternate methods for studying "adsorption" reactions in soils. p. 167–182. *In* D. Baker (ed.) Chemistry in soil environments. Soil Science Society of America, Madison, WI.

Helz, G.R., R.G. Zepp, and D.G. Crosby (ed.) 1994. Aquatic and surface photochemistry. Lewis Publishers, Boca Raton, FL.

Heron, G., T.H. Christensen, and J.C. Tjell. 1994. Oxidation capacity of aquifer sediments. Environ. Sci. Technol. 28:153–158.

Højberg, O., N.P. Revsbech, and J.M. Tiedje. 1994. Denitrification in soil aggregates analyzed with microsensors for nitrous oxide and oxygen. Soil Sci. Soc. Am. J. 58:1691–1698.

Hostettler, J.D. 1992. The physical basis of Eh related to Eh measurements in natural waters. Preprint extended abstract, American Chemical Society Meetings, San Francisco, CA.

Hug, S.J., B.R. James, and H-U. Laubscher. 1997. Iron(III) catalyzed photochemical reduction of chromium(VI) by oxalate and citrate in aqueous solutions. Environ. Sci. Technol. 31:160–170.

James, B.R., 1989. Electron activity in soils: A key master variable. Agron. Abstr. p. 201.

James, B.R., and R.J. Bartlett. 1983. Behavior of chromium in soils. VI. Interactions between oxidation-reduction and organic complexation. J. Environ. Qual. 12:173–176.

Kludze, H.K., R.D. DeLaune, and W.H. Patrick, Jr. 1994. A colorimetric method for assaying dissolved oxygen loss from container-grown rice roots. Soil Sci. Soc. Am. J. 86:483–487.

Larson, R.A. 1997. Naturally occurring antioxidants. Lewis Publishers, Boca Raton, FL.

Lindberg, R.D., and D.D. Runnells. 1984. Ground water redox reactions: An analysis of equilibrium state applied to Eh measurements and geochemical modeling. Science 225:925–927.

Lindsay, W.L., 1979. Chemical equilibria in soils. Wiley-Interscience, New York, NY.

Liu, C.W., and T.N. Narasimhan. 1989. Redox-controlled multiple-species reactive chemical transport. 1. Model development. Water Resour. Res. 25:869–882.

Loach, P.A., 1976. Oxidation-reduction potentials, absorbency bands, and molar absorbency of compounds used in biochemical studies. p. 122–130. *In* G. Fasman (ed.) Handbook of biochemistry and molecular biology. Chemical Rubber Co., Cleveland, OH..

Lovley, D.R., F.H. Chappelle, and J.C. Woodward. 1994. Use of dissolved H_2 concentrations to determine distribution of microbially catalyzed redox reactions in anoxic groundwater. Environ. Sci. Technol. 28:1205–1210.

Mansfeldt, T. 1993. Redoxpotentialmessungen mit dauerhaft installierten Platinelektroden unter reduzierenden Bedingungen. Z. Pflanzenernähr. Bodenk. 156:287–292.

Matia, L., G. Rauret, and R. Rubio. 1991. Redox potential measurement in natural waters. Fres. J. Anal. Chem. 339:455–462.

Mendelssohn, I.A., B.A. Kleiss, and J.S. Wakeley. 1995. Factors controlling the formation of oxidized root channels: A review. Wetlands 15:37–46.

Mueller, S.C., L.H. Stolzy, and G.W. Fick. 1985. Constructing and screening platinum microelectrodes for measuring soil redox potential. Soil Sci. 139:558–560.

Moore, J.N., J.R. Walker, and T.H. Hayes. 1990. Reaction scheme for the oxidation of As(III) to As(V) by birnessite. Clays Clay Min. 38:549–555.

Neta, P., R.E. Huie, and A.B. Ross. 1988. Rate constants for reactions of inorganic radicals in aqueous solution. J. Phys. Chem. Ref. Data. 17:1027–1284.

Norrström, A.C. 1994. Field-measured redox potentials in soils at the groundwater-surface-water interface. Eur. J. Soil Sci. 45:31–36.

Oscarson, D.W., P.M. Huang, C. Defosse, and A. herbillon. 1981. The oxidative power of Mn(IV) and Fe(III) oxides with respect to As(III) in the terrestrial and aquatic environment. Nature 291:50–51.

Patrick, W.H., Jr. 1966. Apparatus for controlling the oxidation-reduction potential of waterlogged soils. Nature 212:1278–1279.

Patrick, W.H., Jr., and R.D. DeLaune. 1972. Characterization of the oxidized and reduced zones in flooded soils. Soil Sci. Soc. Am. Proc. 36:573–576.

Patrick, W.H., R.P. Gambrell, and S.P. Faulkner. 1996. Redox measurements of soils. pp. 1255–1273. *In* D.L. Sparks (ed.) Methods of soil analysis. Part 3. Soil Science Society of America, Madison, WI.

Pearsall, W.H., and C.H. Mortimer. 1939. Oxidation-reduction potentials in waterlogged soils, natural waters, and muds. J. Ecol. 27:483–501.

Peiffer, S., O. Klemm, K. Pecher, and R. Hollerung. 1992. Redox measurements in aqueous solutions: A theoretical approach to data interpretation, based on electrode kinetics. J. Contam. Hydrol. 10:1–18.

Petrie, R.A., P.R. Grossl, and R.C. Sims. 1998. Controlled environment potentiostat to study solid-aqueous systems. Soil Sci. Soc. Am. J. 62:379–382.

Ponnamperuma, F.N., 1972. The chemistry of submerged soils. Adv. Agron. 24:29–96.

Postma, D.. and R. Jakobsen. 1996. Redox zonation: Equilibrium constraints on the Fe(III)/SO_4-reduction interface. Geochim. Cosmochim. Acta 60:3169–3175.

Rowell, D.L., 1981. Oxidation and reduction. p. 401–461. *In* D.J. Greenland and M.H.B. Hayes (ed.) The chemistry of soil processes. John Wiley and Sons, Cleveland, OH.

Russell, E.W., 1973. Soil conditions and plant growth. 10th Ed. Longman, London, UK.

Sawyer, D.T. 1991. Oxygen chemistry. Oxford University Press, New York, NY.

Scerri, E.R. 1997. The periodic table and the electron. Am. Scientist 85:546–553.

Schwarzenbach, R.P., P.M. Gschwend, and D.M. Imboden. 1993. Environmental organic chemistry. John Wiley and Sons, New York, NY.

Sillén, L.G., 1967.*In* Equilibrium concepts in natural water systems. Adv. Chem. Ser. no. 67. American Chemical Society, Washington, DC.

Sparks, D.L., 1985. Kinetics of ionic reactions in clay minerals and soils. Adv. Agron. 38:231–265.

Sposito, G., 1981. The thermodynamics of soil solutions. Oxford University Press, New York, NY.

Sposito, G., 1989. The chemistry of soils. Oxford University Press, New York, NY.

Stone., A. T., and J.J. Morgan. 1987. Reductive dissolution of metal oxides. p. 221–254. *In* W. Stumm (ed.) Aquatic surface chemistry. John Wiley, New York.

Stumm, W., 1992. Chemistry of the solid-water interface. Wiley-Interscience, New York, NY.

Stumm, W., and J.J. Morgan. 1996. Aquatic chemistry. 3rd Ed. Wiley-Interscience, New York, NY.

Stumm, W. and B. Sulzberger. 1992. The cycling of iron in natural environments: Considerations based on laboratory studies of hetergeneous redox processes. Geochim. Cosmochim. Acta 56:3233–3258.

Sullivan, J.C., S. Gordan, D. Cohen, W. Mulac, and K.H. Schmidt. 1976. Pulse radiolysis studies of uranium (VI), neptunium (VI), neptunium (V), and plutonium (VI) in aqueous perchlorate media. J. Phys. Chem. 8:1684–1686.

Suter, D., S. Banwart, and W. Stumm. 1991. Dissolution of hydrous iron(III) oxides by reductive mechanism. Langmuir 7:809–813.

Thompson, J.J., 1923. The electron in chemistry. Franklin Institute Press, Philadelphia, PA.

Tributsch, H., and L. Pohlmann. 1998. Electron transfer: Classical approaches and new frontiers. Science 279:1891–1895.

Typrin, L.R. 1998. Using a thermodynamically based approach to assess the redox status and predict the valence state of chromium in simple, aqueous systems and chromium-enriched soils. M.S. thesis, Univ. of Maryland, College Park, MD.

Vincent, A., 1985. Oxidation and reduction in inorganic and analytical chemistry. John Wiley and Sons, Chichester, UK.

Wardman, P. 1989. Reduction potentials of one-electron couples involving free radicals in aqueous solution. J. Phys. Chem. Ref. Data. 18:1637–1756.

Wang, Z.P., D. Zeng, and W.H. Patrick, Jr. 1997. Characteristics of methane oxidation in a flooded rice profile. Nutrient Cycling Agroecosystems 49:97–103.

Westcott, C.C., 1978. pH measurements. Academic Press, New York, NY.

White, A.F., and M.L. Peterson. 1996. Reduction of aqueous transition metal species on the surfaces of Fe(II)-containing oxides. Geochim. Cosmochim. Acta 60:3799–3814.

Whitfield, M. 1974. Thermodynamic limitations on the use of the platinum electrode in Eh measurements. Limnol. Oceanogr. 19:857–865.

Willis, L.G. 1932. Oxidation-reduction potentials and the hydrogen ion concentration of a soil. J. Agric. Res. 45:571–575.

Yu, T.R. 1992. Electrochemical techniques for characterizing soil chemical properties. Adv. Agron. 48:205–250.

6

Soil Colloidal Behavior

Sabine Goldberg, Inmaculata Lebron, and D.L. Suarez
USDA-ARS, U.S. Salinity Laboratory

6.1 Nature of Soil Colloids

6.1.1 Significance of Colloidal Phenomena

The importance of colloids in soil science has been appreciated for many years. However, recent understanding that organic and inorganic contaminants are often transported via colloidal particles has increased interest in colloid science. Essentially, all chemicals and individual species are to some extent reactive with soils, including species such as Cl^-, which undergo repulsion from negatively charged surfaces. With few exceptions, soil chemistry is primarily the chemistry of colloids and surfaces. The primary importance of colloids in soil science stems from their surface reactivity and charge characteristics. The overwhelming majority of surface area and electrostatic charge in a soil resides in the less than 1 μm size fraction, with particles with radii between 20 and 1,000 nm constituting the major part of the soil surface area (Borkovec et al., 1993).

Characterizations of size, shape, surface area, surface charge density and changes in surface charge are required for understanding the processes of adsorption, flocculation, dispersion, and transport in soils and the resultant changes in soil hydraulic properties as well as chemical migration. Since the major part of the surface area is in the colloidal fraction of the soil, almost all surface-controlled processes including adsorption reactions, nucleation and precipitation/dissolution involve colloids. Colloids are reactive not only because of their total surface area but because of enhanced reactivity related to rough surfaces and highly energetic sites, as well as the effects of electrostatic charge. Colloid charge is associated with substitution of lower charge cations for those of higher charge in the mineral lattice (which results in a net permanent charge) as well as surface charge associated with broken bonds. The charge associated with broken bonds is characterized as variable charge in as much as the solution influences the surface speciation (Section B, Chapter 7). In addition to these chemical processes, colloids are mobile in soils, and thus affect not only the chemical transport of otherwise immobile chemicals, but also exert a strong influence on soil hydraulic properties.

6.1.2 Types of Soil Colloids

Colloidal particles are defined as having an equivalent spherical radius smaller than 1 μm (van Olphen, 1977). A homogeneous dispersion of colloidal particles in a liquid is called a colloidal dispersion. If the particles are large and settle rapidly, the dispersion is called a suspension. A colloidal dispersion is defined as a system where particles of colloidal dimensions are dispersed in a continuous phase of a different composition (van Olphen, 1977).

6.1.2.1   Oxides

Oxides including hydroxides and oxyhydroxides are ubiquitous constituents of soils occurring as both discrete particles and as coatings on other soil surfaces. Oxide minerals that are commonly found in the soil clay fraction are discussed in Section F, Chapter 3.

Hydroxylation of oxide minerals can either be structural and/or by chemisorption of water in an aqueous medium (Schwertmann and Taylor, 1977). Edge hydroxyl groups on oxide and clay minerals represent the most abundant and reactive surface functional groups in soils (Sposito, 1984). Any one type of oxide mineral contains various groups of surface hydroxyls that are distinguishable by crystal plane location and/or extent of coordination to the cations of the bulk structure. However, as a simplification it is often assumed that each oxide mineral has a single set of homogeneous functional groups. Surface hydroxyl groups on most oxide minerals are amphoteric, exhibiting positive charge at low pH and negative charge at high pH. For this reason oxide minerals are often referred to as variable charge soil minerals. Table 6.1 provides densities of surface hydroxyl groups for some common oxide minerals in soils.

Boehmite and gibbsite are the only crystalline Al oxides common in soils. Aluminum oxides are the products of intense weathering of aluminosilicate minerals and are most abundant in tropical soils. Gibbsite can also be found in volcanic ash soils of humid regions (Brown et al., 1978). Noncrystalline Al oxides, which have similar structure and chemical characteristics but smaller particle size than crystalline varieties, often dominate the chemical reactions with anions in soils (Hsu, 1977). Aluminum oxides play an important role in ion adsorption, stabilization of aggregates, and flocculation of soil particles.

Iron oxides are found in most soils, providing soil horizons with their red, yellow, and brown colors (Brown et al., 1978). Goethite, which is the most common Fe oxide in temperate, subtropical, and tropical soils, is usually thermodynamically the most stable (Schwertmann and Taylor, 1977). Soil goethites are usually fine grained and contain appreciable substituted Al. Lepidocrocite is a minor constituent of waterlogged temperate soils undergoing alternating oxidizing and reducing conditions while hematite is a common soil mineral that can be inherited from parent materials or formed pedogenically in warm regions (Brown et al., 1978). Two magnetic Fe minerals, magnetite and maghemite, occur in soils; the former being inherited from parent rock, while the latter is formed

Table 6.1   Densities of surface hydroxyl groups on oxide minerals [Adapted from Davis and Kent, 1990]

Solid	Site density range (sites nm^{-2})
Gibbsite	2-12
Goethite	2.6-16.8
Hematite	5-22
Ferrihydrite	1.1-10.1
MnO_2	6.2
TiO_2	2-12
Amorphous SiO_2	4.5-12

pedogenically in highly weathered soils (Brown et al., 1978). Ferrihydrites are poorly crystalline, have indefinite composition, and occur as very small particles with high surface area (Schwertmann and Taylor, 1977). Ilmenite is an uncommon mineral usually inherited from igneous or metamorphic parent rocks (Brown et al., 1978). Iron oxides play an important role in ion adsorption and aggregation and cementation of soil particles.

Manganese oxides which supply Mn for plants occur widely in soils as minor constituents, mainly as dark coatings on particle surfaces. Manganese oxides are chemically complex, existing as a continuous range of compositions between MnO and MnO_2 (Brown et al., 1978). Birnessite, vernadite, lithiophorite, and hollandite are the most common crystalline manganese minerals in soils (McKenzie, 1989). Birnessite occurs in both acid and alkaline soils, while lithiophorite occurs mainly in neutral to acid soils (Brown et al., 1978). Manganese oxides exhibit a strong adsorption capacity for metal cations, especially Cu, due to their variable charge, small particle size, and large surface area (McKenzie, 1989).

Rutile and anatase, the common Ti oxides in soils, are the high and low temperature forms, respectively, the former occurring in igneous and metamorphic rocks and the latter as an alteration product of Ti-containing minerals such as ilmenite, much less abundant than rutile (Hutton, 1977; Brown et al., 1978). Titanium oxides are present in both the coarse and fine fractions of soils and are very insoluble (Hutton, 1977).

Quartz is not only the most abundant Si oxide but also the most abundant mineral in most soils. Most quartz is found predominantly in the sand, silt, and coarse clay fraction of soils (Wilding et al., 1977). Silicon oxides are generally considered inert having a small surface area and little surface charge.

6.1.2.2   Clay Minerals

The clay fraction of most soils is dominated by various layer silicate clay minerals. Layer silicate clay minerals are classified as 1:1 where each layer consists of one tetrahedral silica sheet and one octahedral alumina sheet, 2:1 where each layer consists of one octahedral sheet sandwiched between two tetrahedral sheets, or 2:1:1 where a metal hydroxide sheet is sandwiched between the 2:1 layers. Layer silicate minerals common in soils are discussed in Section F, Chapter 2. A discussion of silicate structures is provided by Schulze (1989).

Layer silicate clay minerals are characterized by isomorphic substitution of lower valence cations in either or both the tetrahedral and octahedral sheets. This excess of negative charge is balanced by other cations, either inside the crystal or on the external surface (McBride, 1994). Layer charge is charge that is balanced outside of the structural unit and determines the strength and type of bonding occurring between the basal planes. Charge arising from isomorphic substitution is called permanent charge because it is independent of solution pH. Layer silicate clay minerals also possess variable charge located at the broken edges of clay particles. At the edges of the octahedral sheet, hydroxyl ions attached to Al cations are called aluminol groups. Similar to the hydroxyl groups on oxide minerals, aluminol groups are amphoteric. At the edges of the tetrahedral sheet hydroxyl groups attached to Si cations are called silanol groups. Silanol groups do not undergo protonation, but dissociate and become negatively charged at high pH. Adsorbed cations on clay minerals balance both variable and permanent charge. Table 6.2 provides charge characteristics and cation exchange capacities (CEC) for some common clay minerals in soils.

Kaolinite is one of the most widespread clay minerals in soils, being most abundant in soils of warm moist climates (Dixon, 1977); while halloysite is formed through acid weathering and in soils of volcanic origin. The halloysite structure is the same as the kaolinite structure but contains a sheet of water molecules between the layers. Both have low colloidal activity, surface area, and CEC and anion

Table 6.2 Charge characteristics and cation exchange capacities of clay minerals [Bohn et al., 1985; McBride, 1994]

Solid	Charge per unit half cell		Cation exchange capacity (cmol$_c$ kg^{-1})
	Tetrahedral	Octahedral	
Kaolinite	0	0	1-10
Smectite			80-120
Montmorillonite	0	−0.33	
Beidellite	−0.5	0	
Vermiculite	−0.85	+0.23	120-150
Mica			20-40
Muscovite	−0.89	−0.05	
Chlorite			10-40

exchange capacities (AEC) (Dixon, 1977) which are predominantly variable in nature (McBride, 1994).

Micas are abundant in soils, occurring as primary minerals inherited from soil parent materials. Micas strongly retain interlayer K ions, rendering them nonexchangeable and reducing the CEC of these minerals. Through weathering, micas release K and provide an important natural source of this plant nutrient (Fanning and Keramidas, 1977). Illite is a secondary mineral that is less crystalline, contains less K, and more water than muscovite mica (McBride, 1994). Micas and illites are nonswelling minerals.

Smectites constitute an important group of 2:1 clay minerals. Members that are important in soils include montmorillonite, beidellite, and nontronite. Smectites are most significant in moderately weathered soils and have high colloidal activity and surface area. Smectites are responsible for a large part of the CEC and the majority of the shrink/swell properties of smectitic soils (Borchardt, 1977).

Vermiculite is an important clay mineral in soil that is formed as an alteration product of muscovite and biotite micas (Douglas, 1977). Vermiculite, which is widely distributed and has a wide particle size range, contains hydrated Mg cations that can be readily exchanged by K and NH$_4$ ions, resulting in collapse of the clay layers and fixation of these nutrient ions. Vermiculites have high CEC and surface area but exhibit limited swelling.

Chlorites are 2:1:1 layer silicates that occur extensively in soils. The hydroxide interlayer sheet is usually dominated either by brucite [Mg(OH)$_2$] or by gibbsite [Al(OH)$_3$]. This interlayer sheet restricts swelling, decreases effective surface area and effective cation exchange capacity (ECEC) (McBride, 1994). Chlorites are nonswelling silicates.

6.1.2.3 Organic Matter

Soil organic matter (SOM) refers to the mixture of products resulting from microbial and chemical transformations of organic residues and is discussed in Section B, Chapter 2. Humus is a complex and microbially resistant mixture of amorphous and colloidal substances modified from original tissues or synthesized by soil microorganisms. Humic substances are subdivided into humic acid, fulvic acid, and humin using a separation scheme based on solubility in strong acid and base (McBride, 1994). The structure and composition of humus are complex and incompletely known. The structure contains a variety of reactive functional groups including carboxyl R–COOH, phenol C_6H_5OH, alcohol R–CH_2OH, enol R–CH=CH–OH, ketone R–CO–R′, quinone O=C_6H_4=O, ether R–CH_2–O–CH_2–R′, and amino R–NH_2 (Stevenson, 1982). Humus is amorphous and highly colloidal; its surface area, ion adsorption and CEC are greater than those of layer silicate clay minerals (McBride, 1994). The presence of humus usually promotes aggregation of soil particles.

6.1.3 Properties of Soil Colloids

6.1.3.1 Particle Size and Shape

Colloids in natural systems are characterized by a continuous particle size distribution of extreme complexity and diversity. Organisms, organic macromolecules, clays, oxides and combinations of any of them constitute the colloidal fraction in soils. The distribution of shapes, densities, surface chemical properties, and chemical composition may vary widely with size. Some fractions of the size spectrum may be living, and all particulates are subject to diverse physical, chemical and biological processes that can alter size distribution, shape, or chemical composition (Kavanaugh and Leckie, 1980).

Colloids are dynamic particles; the distribution of particle sizes in natural systems is the result of a number of processes, which either bring the particles together (coagulation mechanisms) or disrupt existing aggregates (Filella and Buffle, 1993; Buffle and Leppard, 1995a,b). Sequential gravimetric analysis following filtration or centrifugation is the classical method for measuring particle size distributions in colloidal systems. However, this technique does not provide a reliable size distribution for particles less than 1 μm. Techniques such as electron microscopy and light scattering methods are better suited for submicron particle sizing. High resolution transmission electron microscopy (HRTEM) and photon correlation spectroscopy (PCS) are promising techniques for this size range, provided that artifacts are minimized (Filella and Buffle, 1993).

Minimization of sample handling and processing is recommended when analyzing colloidal systems. Unfortunately, no *in situ* technique that allows direct measurement of the particle size distribution has yet been developed. Furthermore, most of the methods commercially available do not allow quantification of small particles in the presence of a high proportion of larger particles, a condition frequently found in soils. A combination of fractionation techniques and other methods of analysis is advised due to the unstable nature of the system under study (Filella and Buffle, 1993).

Measurement of colloid particle size distribution is challenging because of the heterogeneous characteristics of particulates. Shape, density, refractive index, and other physical properties are usually nonuniform throughout the six to seven orders of magnitude size range of the particulate fraction. No single measurement technique is able to measure the particle size distribution over this wide size range (Kavanaugh et al., 1980). The interpretation of these measurements is further complicated because changes in the size distribution occur due to particle aggregation or breakup during sampling or sample preparation.

Measurement Methodology

There are many methods to determine the particle size distribution of a soil sample. These methods can be divided into direct and indirect methods. Direct methods are those that provide information by direct observations. In this category, all the microscopic and ultramicroscopic techniques are included. Indirect methods, which are more extensively used, determine the size of the particles in dispersions. Certain bulk physical properties of the suspensions are measured and the average particle dimensions are computed on the basis of theoretical relations between particle dimensions and physical properties. The following physical properties of suspensions have been used for the evaluation of particle size and shape: sedimentation velocity, viscosity, decay of optical birefringence induced by an orientation of the particles in an electric field, dynamic light scattering, intensity of transmitted light, and low angle X-ray scattering, among others (van Olphen, 1977; Hunter, 1993). All of these techniques have intrinsic assumptions and limitations. It is not unusual to obtain different particle size distributions for the same sample using different techniques.

Electron Microscopy

The electron microscope is the only instrument capable of measuring the size of particles in the colloidal range (1 nm to 1 μm). Both the transmission and scanning electron microscopes require sample evacuation and in most cases coating of the particulates with a conducting material, typically carbon or gold. Thus sample preparation can alter the original particle size distribution, unless freeze drying techniques are used. From the weight concentration of the suspension and the number of particles in a given volume, the average weight of a single particle can be calculated (van Olphen, 1977). New techniques, such as image analysis, facilitate the quantification of particle sizes with electron microscopy. These methods are not widely used for routine particle size analysis because of the high cost of this equipment. Additionally, detection limits are poor, because small sample volumes are scanned at any given magnification. Despite this limitation, microscopy remains the essential technique for determination of particulate shape factors.

Sedimentation. The steady-state velocity of a particle in suspension, settling in the gravitational field, is given by:

$$M_e g = f\left(\frac{dx}{dt}\right) \qquad [6.1]$$

where M_e is the effective particle mass, f is the frictional coefficient, dx/dt is the particle velocity, and g is the gravitational acceleration. Steady-state velocities are normally expressed in terms of the sedimentation coefficient (S). Thus, for Equation [6.1],

$$S = \frac{1}{g}\frac{dx}{dt} = \frac{M_e}{f} \qquad [6.2]$$

The sedimentation coefficient is the ratio between the particle effective mass and the frictional coefficient. The relationship between sedimentation data and particle size is a function of the particle shape and density and the density and viscosity of the medium. The results from a sedimentation experiment can be given as a distribution of S. If the particle density (ρ) and the density of the medium (ρ_0) are known, S can be expressed as:

$$S = \frac{V(\rho - \rho_0)}{f} \qquad [6.3]$$

where V is the particle volume. Assuming that the particles are spherical with the frictional coefficient given by Stokes' Law as $f = 3\pi_0 \eta d_p$, the sedimentation coefficient for a particle of diameter d_p suspended in a medium with viscosity (η), then becomes:

$$S = \frac{V(\rho - \rho_0)d_p^2}{18\eta} \qquad [6.4]$$

from which the particle diameter may be calculated.

For particles other than spheres, similar expressions can be obtained using frictional coefficients appropriate to the particle dimensions and geometry. A more detailed theoretical treatment is provided in Jackson (1979) and Bunville (1984).

A particle falling through an infinite fluid will eventually fall at a terminal constant velocity determined by the size of the particle and the resistance offered by the fluid. In a centrifugal field, the terminal velocity is not constant, because the centrifugal field is a function of the centrifugal radius. Measurement of the terminal velocity is necessary in order to calculate the particle size. The relationship between the movement of the particle and the movement of fluid around that particle may be reduced to the Stokes equation. For fluid moving past a particle of diameter (d_p), the ratio of the inertial transfer is described by the dimensionless parameter, the Reynolds number (R_E):

$$R_E = \frac{\rho_0 u d_p}{\eta} \qquad [6.5]$$

where u is the velocity. If $R_E \leq 0.2$, the fluid conditions are described as streamlined or laminar, and the drag on the particle is due mainly to viscous force within the fluid. Particles with high densities or large particle diameters may be moving with velocities that exceed $R_E = 0.2$, and in this situation, are likely to enter the region of turbulent flow, where velocities are more difficult to calculate.

In a centrifugal field, Stokes' equation has the form:

$$u = \frac{(\rho - \rho_0)\omega^2 d_p^2}{18\eta} = \frac{\ln(r/S_0)}{t} \qquad [6.6]$$

where ω is the rotational velocity (rad s^{-1}) and t is the time (s) required for a particle of diameter d_p to move from its starting point radius (S_0) to the analytical radius (r).

Application of the Stokes equation requires certain assumptions that are not always achieved. All these assumptions are critical to the measurement of the size of the sedimenting particles. The first assumption is that the particles are spherical, smooth, and rigid. Since this is almost never valid, the diameter calculated is an equivalent or Stokes diameter (d_{st}). It is assumed that the particle terminal velocity is reached instantly, although calculations show that a finite but small time period is actually required before this condition is reached. The particle is assumed to be moving without interference or interaction from other particles in the system. This assumption is only true at high dilutions (< 1%) that ensure considerable separation between the particles. Also, it is assumed that inertial effects are not present and that the fluid only exhibits Newtonian flow properties. Since water is generally the dilution medium in soils and colloidal particles are < 1 μm, these assumptions are usually valid.

A more detailed analysis of the methodology as well as a description of different centrifugation methods are provided in Bunville (1984), Groves (1984), Koehler et al. (1987), Holsworth et al. (1987), and Coll and Oppenheimer (1987).

Electrical Properties

The increase in the resistance across a small aperture produced by a nonconducting particle in a conducting medium is called the Coulter effect. The magnitude of this increase in resistance (ΔR) for a spherical particle of diameter (d_p) suspended in an aperture of diameter (D_a) is

$$\Delta R = \frac{8P_f d_p}{3\pi D_a^4}\left[1 + \frac{4}{5}\left(\frac{d_p}{D_a}\right)^2 + \frac{24}{35}\left(\frac{d_p}{D_a}\right) + \ldots\right]$$ [6.7]

where P_f is the resistivity of the conducting medium. This resistance pulse (ΔR) results in a voltage pulse ($i\Delta R$) for a sphere of diameter d_p, where i is the current across the aperture. The resulting voltage pulses are counted and scaled using a multichannel analyzer.

Instruments utilizing the resistive pulse technique require calibration using standards of particles with known diameter to assign a particle size to each of the thresholds. This procedure takes into account the dimensions and electrical characteristics of the aperture and the conducting medium.

The advantage of the resistive pulse technique is that no other property of the particle, such as refractive index or specific gravity, is required for the interpretation of the data in terms of a particle size distribution. Developments in instrumentation for particle size analysis using the resistive pulse technique allow the analysis of particle sizes < 1 µm (Bunville, 1984).

Photon Correlation Spectroscopy

Photon correlation spectroscopy (PCS) or dynamic light scattering measures the fluctuation in scattered intensity of a laser beam over small time intervals when it passes through a small volume of particles under Brownian motion. The polarized intensity (I_{vv}) time correlation function is given by the expression (Pecora, 1983; Schurtenberger and Newman, 1993):

$$\langle I_{vv}(0)I_{vv}(t)\rangle = A = B\exp\left[-2q^2 D_t \tau\right]$$ [6.8]

where A and B are constants, τ is the delay time, D_t is the translational diffusion coefficient, and q is the length of the scattering vector, which is related to the wavelength of the incident light (λ) and the scattering angle (θ) by the expression:

$$q = 4\frac{\pi}{\lambda}\sin\left(\frac{\theta}{2}\right)$$ [6.9]

Assuming that the particles are spheres, this translational diffusion constant is related to the radius (r) of the particle according to the Stokes-Einstein relationship at infinite dilution:

$$D_t = k_B \frac{T}{6\pi\eta r}$$ [6.10]

where k_B is the Boltzmann constant and T is the absolute temperature (Pecora, 1983).

When the particles have a regular shape other than spherical, the depolarized component is used to study the particle rotational diffusion coefficient. The rotational and translational diffusion coefficients obtained in conjunction with theoretical, hydrodynamic relationships contain information on the particle dimensions.

For a nonspherical particle larger than the incident wavelength, light is scattered from different parts of the same particle producing interferences which are dependent on the angle of the scattered intensity and characteristic of a particular particle shape. The time dependence of fluctuations in orientation, shape, or size is important and cannot be observed unless the fluctuation exhibits a sufficiently large amplitude, compared to q^{-1}, for the different states (Pecora, 1983).

When the particles are not spheres or have no unique shape the theoretical model is complex. Despite limitations, PCS has been applied to measure particle size of colloids in natural systems (Lebron et al., 1993; Filella et al., 1997).

Field-Flow Fractionation
Field/flow fractionation (FFF) is a group of separation techniques capable of fractionating and characterizing the particle size distribution of colloids in the range 0.01 to 1 μm. Like chromatography, FFF is an elution methodology in which constituents are differentially retained, and thus separated in a flow channel (Beckett et al., 1997). With this technique, a colloidal sample is introduced into a stream of liquid and subjected to a field (such as gravitational, centrifugal, third crossflow, thermal gradient, electrical, or magnetic) acting perpendicular to the stream direction (Beckett and Hart, 1993). According to theory, the rate at which particles are displaced downstream, measured as emergence times, can be related exactly to particle properties such as mass, size, and density. However, since different kinds of particles move at different velocities in this system, broad particle populations are sorted into graded size (or mass) distributions along the length of the flow channel. Observation of the shape of the emerging distribution, combined with theory, yields particle size distribution curves.

If applied as indicated above, FFF provides highly detailed size distribution curves and is a highly flexible technique which can be adapted to different particle types in almost any suspending medium. Giddings et al. (1987) provide a detailed explanation of this technique.

Applications
Several instruments perform automated particle size analysis based on gravimetric sedimentation but differing in the method used to monitor the particle concentration. The more commonly used methods are X-ray and white light. At the beginning of the analysis the sample cell is positioned relative to a stationary detector at a depth sufficient to include the largest particles expected in the sample. The cell is then moved downward through the X-ray beam at a programmed rate. The time and depth in the suspension define a given particle diameter. The attenuation of the beam monitors the concentration of dispersed phase having this diameter. The results of the analysis are obtained as a graph of the cumulative mass percent as a function of the equivalent spherical diameter (Fig. 6.1).

The analytical application of PCS for size distribution measurements of natural colloidal suspensions provides valuable information about the system and it is one of the few nondestructive techniques that yields an estimate of the size distribution of suspended submicron colloids with a minimum of sample processing.

Precise information about colloidal size and shape is important because submicron size colloids often act as vehicles that govern the transport and fate of adsorbed pollutants (hydrophobic organic compounds, toxic trace metals and radionuclides). Submicron size colloids have been insufficiently studied because methods for their isolation, detection, and characterization have been inadequate except for a few examples (Kretzschmar et al., 1993; Kaplan et al., 1993). Little is known about

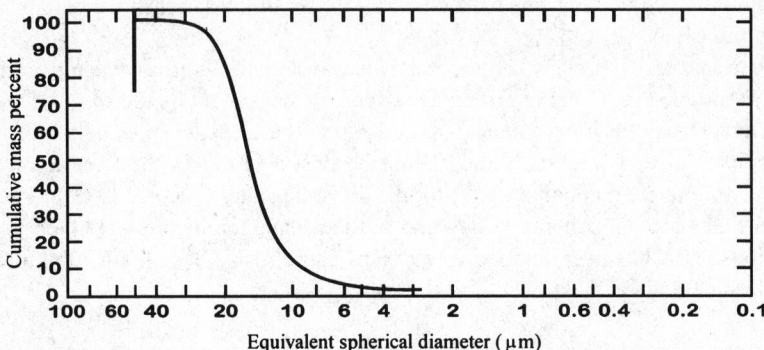

Fig. 6.1   Particle size distribution of glass spheres obtained using the Sedigraph 5000-D [Reprinted by permission of John Wiley & Sons, Inc. from Bunville, 1984. H.G. Barth (ed.) Modern methods of particle size analysis]

colloidal materials in natural environments, in particular, the chemical nature of colloids present and their structure, size and shape distributions (Filella et al., 1997).

The interpretation of many properties of clay suspensions has been traditionally based on a model of separated clay plates with well-developed electrical double layers. However, single plates of montmorillonite tend to assemble in an organized manner as quasicrystals. A group of aligned montmorillonite quasicrystals or illite crystals is described as a clay domain (Quirk and Aylmore, 1971). Schramm and Kwak (1982), using light transmission and viscosity measurements, found that quasicrystal size increased in the order of Li < Na < K < Mg < Ca. The average number of plates per quasicrystal relative to Li-montmorillonite varied from 1.5 for Na- to 6.1 for Ca-montmorillonite. Using PCS particle size analysis, Lebron et al. (1993) demonstrated the presence of illite domains in suspensions of sodium adsorption ratio [(SAR = $Na^+/(Ca^{2+}+Mg^{2+})^{0.5}$ where the cation concentration is expressed in mmol L^{-1})] < 15. Illite registered a decrease in the mean of the two dimensions of the domains (b and c crystal axis) when the SAR increased, while montmorillonite quasicrystal size decreased only in the c dimension. The b dimension of montmorillonite remained constant in the range of SAR studied. These differences are indicative of a different particle arrangement for illite domains than for montmorillonite quasicrystals (Fig. 6.2).

6.1.3.2   Surface Area

Surface area must be regarded as a relative term, in as much as it is scale dependent, as well as often dependent on the chemical and physical conditions in a system. Determinations of surface area range

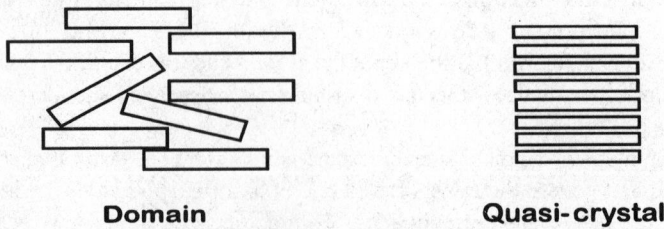

Domain   **Quasi-crystal**

Fig. 6.2   Representation of mica domains and montmorillonite quasi-crystals based on photon correlation spectroscopy measurements [Reprinted from Lebron et al., 1993. Clays Clay Min. 41:380-388, with permission]

from particle size calculations assuming smooth surfaces and simplified geometry, generally termed geometric surface area, to possible molecular level calculations based on the distances between surface ions in a mineral structure. Because the measurement is scale and system dependent, there is no best way to measure surface area. Determination as to which measurement system to utilize should consider the scale and chemical conditions required by the application. Kinetic reactions which are diffusion controlled should likely consider geometric surface area while surface-controlled reactions (such as some adsorption and some dissolution/precipitation reactions) should consider surface area at the scale of the reacting molecule. In most instances, surface area is related to surface reactivity, either adsorption or surface-controlled kinetic processes.

Measurement Methodology

The results of surface area determinations must be interpreted within the context of the size and orientation of the adsorbate as well as the attractive forces between the surface and the adsorbate. This distinction is particularly important for clays such as smectites, which can be considered to have internal as well as external surface area. Internal surface area is representative of the surface area of the interlayers. Inert gases such as N_2 are not able to enter the interlayer positions, and thus, measure only external surface area of clay particles. In contrast, polar molecules such as ethylene glycol and water are able to cause expansion of the layers and penetrate into interlayer positions. Use of such molecules results in measurement of internal and external surface area. Since water is the solvent in environmental systems, these total surface area measurements are appropriate for adsorption studies.

To characterize adsorption methods of surface area determination, it is useful to distinguish between chemical and physical adsorption. Physical adsorption is characterized by: (1) low heats of adsorption without structural changes at the surface, (2) fully reversible and rapid reactions since no activation energy is required, (3) surface coverage of the entire surface rather than specific sites, (4) little or no adsorption at elevated temperatures, and (5) potential coverage by more than one layer of adsorbate (Lowell, 1979). Chemisorption is characterized by high heats of adsorption, localization of adsorption at specific surface sites, and irreversible reaction. The term specific surface area has been used to denote the surface area of a material expressed on a mass basis ($m^2 \ kg^{-1}$).

Gas Adsorption Using the BET Equation

The BET equation is commonly used in conjunction with physical gas adsorption to measure surface area. The BET equation is named after Brunauer, Emmett, and Teller (1938), who extended the Langmuir theory for monolayer gas adsorption to multilayer adsorption. The Langmuir equation (Langmuir, 1918) is given by:

$$P / V = 1 / kV_m + P / V_m \qquad [6.11]$$

where P is the pressure, V is the volume of gas adsorbed per kg of adsorbent at that pressure, k is a constant, and V_m is the volume of gas adsorbed per kg of adsorbent at monolayer surface coverage. The surface area is obtained by determination of $1/ V_m$, which is the slope of the P/ V versus P plot. The specific surface area is then equal to $1/ V_m$ multiplied by the cross-sectional area of the adsorbate and the number of molecules in volume V_m.

The BET relation assumes that there is a dynamic equilibrium between the molecules in the various layers such that the number of molecules in each layer remains constant although different sites may or may not be occupied at any given time. Use of the equation enables calculation of the number of molecules in a monolayer despite the fact that complete monolayer coverage may not have occurred. The BET equation is written as (Lowell, 1979):

$$\frac{1}{W[P_o/P]} = \frac{1}{W_mC} + \frac{C-1}{W_mC}\frac{P}{P_o}$$ [6.12]

where P is the adsorbate gas pressure, P_o is the adsorbate pressure at saturation for the temperature of the experiment, W is the weight adsorbed in the monolayer, W_m is the weight adsorbed in the complete monolayer, and C is the BET constant. Multipoint BET plots are created by plotting $1/(W(P_o/P-1))$ on the y axis and P/P_o on the x axis. The value of W_m is calculated from the slope and intercept. The specific surface area is determined by dividing the total surface area by the sample weight. The region of P/P_0 between 0.05 and 0.35 is usually linear and within the region of pressures corresponding to sufficient adsorption to complete monolayer coverage, and thus, best suited for determination of W_m (Lowell, 1979).

Often the BET surface area can be determined from a single pressure measurement without much loss of accuracy. For relatively high values of C, the intercept value is small relative to the slope and can be approximated by zero. The BET equation is thus reduced to (Lowell, 1979):

$$W_m = W\left(1 - \frac{P}{P_o}\right)$$ [6.13]

Soil and mineral surface areas are most commonly measured by N_2 sorption using the BET equation. The calculation is made using the N_2 cross-sectional area of 0.162 nm^2 (Gregg and Sing, 1982). Often, this area is referred to as the effective or occupied area. Alternatively for surface area < 1,000 m^2 kg^{-1}, the use of Kr is recommended.

Most commonly, the BET method consists of adsorption of N_2 at a fixed partial pressure P in a He-N_2 mixture and measurement of the desorbed N_2 in a pure He gas stream using gas chromatography. Alternative methods include measurement of the mass of N_2 sorbed. In this method, the sample is evacuated to high vacuum, heated, then cooled to liquid N_2 temperature and weighed. Quantities of N_2 are then added to the system and a series of weighings is made at various pressures. In this instance, the N_2 partial pressures are equal to the total pressure in the system.

Organic Molecules

Ethylene Glycol. Ethylene glycol was utilized by Dyal and Hendricks (1950) for determination of the total surface area of clays. The method consists of adding excess of ethylene glycol to soil or clays and allowing the excess to evaporate under vacuum. It is assumed that when the rate of weight loss of the sample decreases only a monolayer of ethylene glycol remains. Dyal and Hendricks (1950) calibrated the method assuming a bentonite surface area of 810,000 m^2 kg^{-1} and calculated that 0.31 mg of adsorbed ethylene glycol corresponded to each m^2 of surface area. The method was modified by Bower and Goertzen (1959) by using $CaCl_2$-monoglycolate to maintain an ethylene glycol vapor pressure just below that of the saturation vapor pressure. In this method the sample and the liquid are placed in separate open vessels in an evacuated system and the sample is weighed until it is in equilibrium with the vapor pressure of the ethylene glycol.

Ethylene Glycol Monoethyl Ether. Ethylene glycol monoethyl ether (EGME) has replaced ethylene glycol as the polar solvent of choice for determination of surface area. Since EGME has a higher vapor pressure than ethylene glycol, it requires a shorter reaction time to equilibrate the sample (Carter et al., 1986). A solvate of EGME and $CaCl_2$ is used in the evacuated chamber to lower the vapor pressure of EGME to just below the saturation pressure. Open vessels of EGME/$CaCl_2$ and soil are placed in the chamber and the soil is periodically weighed until no further weight gain is observed. It is assumed that the EGME surface coverage is 5.2×10^{-19} m^2 per molecule and that 0.286 mg adsorbed corresponds to one m^2 of surface area (Carter et al., 1986). The method is limited in that the EGME affinity for cations results in greater than monolayer coverage at those sites, the assumption that EGME covers all surfaces cannot be properly evaluated, and the large size of the molecule may prevent coverage in small surface voids.

Methylene Blue. Methylene blue, an organic cation, is reacted at various concentrations with soil suspensions (typically with organic material removed) under pH buffered conditions. Measurement of methylene blue concentration before and after reaction with soil is made using spectrophotometry at a wavelength of 665 nm. The adsorbed concentration of methylene blue is used to calculate surface area assuming an area of 1.3 nm^2 per methylene blue molecule (Pham and Brindley, 1970). This surface area corresponds to the molecule attaching parallel to its long axis. This method can only be used in hydrated systems. Alternatively, Borkovec et al. (1993) obtained relatively good agreement between methylene blue and BET surface areas for 4 soil samples assuming a methylene blue surface area of 0.247 nm^2, corresponding to a molecular attachment of methylene blue perpendicular to its long axis. Orientation of the organic molecules may be related to surface site characteristics as well as concentration of adsorbing molecules. These differences illustrate one of the disadvantages of using charged molecules for surface area determination; their use is not recommended.

Scanning Electron Microscopy Image Analysis

Transmission (TEM) as well as scanning electron microscopy (SEM) can be used for determination of geometric surface area using a variety of methods, including calculation of planar surface area and assumptions regarding geometry to calculate total external surface area. With this method, only edge roughness can be measured. The SEM method, which offers the possibility of measuring the external surface roughness, consists of collecting two images at different sample tilt angles and constructing a 3-dimensional representation of the surface. Surface area is calculated within a grid of fixed lines by summation of the planar surfaces in the grid. The ratio of the calculated surface to the area within the grid gives a measure of surface roughness. Measurements can also be made at different scales, providing information about the size distribution of the surface features. Since these measurements are typically made at the micrometer scale and measure external surface area, it is not surprising that the values are intermediate between geometric surface areas based on particle size and BET values. Determination of specific surface area requires conversion of particle surface area to a mass basis using the particle density. The assumption that particle density is equal to density of specimen samples of the mineral is reasonable and introduces relatively minor errors in comparison to other assumptions made in the calculation. The advent of large computer storage devices and suitable software makes this method the most useful of the geometric methods.

Small Angle X-ray Scattering

In this application, $K\alpha$ radiation from a conventional X-ray tube is scattered by freeze-dried samples in glass capillary tubes. The background corrected X-ray scattering intensity is plotted against the scattering vector (q) to obtain the apparent surface fractal dimension (D_s) using the relation (Borkovec et al., 1993):

$$I_q = Aq^{D_s-5} + B \qquad [6.14]$$

where I_q is the scattering intensity, A is a proportionality constant, and B is the background correction. Values of D_s obtained for soils using this method are in relatively good agreement with values calculated from gas adsorption surface area (a) for soil particle radii (r) using the relation (Borkovec, 1993):

$$a = C\lambda^{2-D_s} r^{D_s-3} \qquad [6.15]$$

where λ is the size of the probing molecules and C is the proportionality constant.

Negative Adsorption

Negative adsorption refers to a deficiency in the concentration of an ion in solution adjacent to a solid surface relative to the concentration of the ion in the bulk solution. The deficit is caused by electrostatic repulsion between ions and surfaces of similar charge. Since most charged surfaces in soils are negatively charged, negative adsorption usually relates to solution anion concentrations and negative surfaces. For a 1:1 electrolyte, the diffuse double layer model produces the following approximation (Sposito, 1984):

$$d_e \approx \frac{2}{\sqrt{\beta c}} - \delta \qquad [6.16]$$

where d_e is the exclusion distance, c is the concentration of the bulk electrolyte, β is a constant equal to 1.08×10^{16} m.mol^{-1}, and δ is the distance between two planes. The exclusion volume and area (S_e) are related by:

$$V_e = S_e d_e \qquad [6.17]$$

where V_e is the exclusion volume (the hypothetical volume from which the ion is completely excluded). Substituting into Equation [16] yields:

$$V_e = \frac{2S_e}{\sqrt{\beta c}} - \delta S_e \qquad [6.18]$$

The exclusion volume is calculated by first removing the bulk solution, determining the remaining liquid volume, displacing the liquid and analyzing for the bulk solution and extract concentration. The exclusion volume is then equal to the mass deficit in the extract divided by the concentration in the bulk solution. A plot of V_e versus $c^{-0.5}$ should yield a straight line whose slope is proportional to the exclusion specific surface area as shown by inspection of Equation [6.18].

This method was first described by Schofield (1949) who used it to determine montmorillonite surface area from reactions with NaCl, NaNO$_3$, and Na$_2$SO$_4$ solutions. Subsequent work by Edwards et al. (1965a,b) demonstrated that the specific area determined with this method varied with cation selected. Calculated values for illites ranged in values close to the BET values with Li to 0 with Cs. In contrast, for montmorillonite, LiCl exclusion values were ten times greater than BET, while Cs values were comparable to those obtained by BET.

Applications

Surface area measurements are required for a variety of calculations. In most soils the bulk soil surface area is dominated by the surface area of the clay minerals. Surface charge density, which requires measurement of both surface area and particle charge, has been related to cation exchange selectivity. As expected, increasing surface charge density favors adsorption of the higher valence cation in heterovalent exchange (Maes and Cremers, 1977). Determination of specific surface area is required for chemical studies on many different minerals. In this case, bulk surface area is not appropriate. Controlled laboratory studies are often performed using addition of quantities of a well-characterized mineral having a known surface area. This method is used to study a variety of chemical reactions, such as kinetics of calcite, gypsum, and dolomite dissolution, surface area of calcite for prediction of phosphate adsorption, and addition of various Fe or Mn oxides for study of adsorption and redox processes. Among the various applications of the EGME method, Ross (1978) related shrink/swell properties of soils to surface area and Supak et al. (1978) related the specific surface area of clays to the adsorption of the pesticide aldicarb.

6.1.3.3 Surface Charge

The total net surface charge on a particle (σ_p) is

$$\sigma_p = \sigma_s + \sigma_H + \sigma_{is} = \sigma_{os} - \sigma_d \qquad [6.19]$$

where σ_s is the permanent structural charge, σ_H is the proton surface charge resulting from the specific adsorption of protons and hydroxyl ions, σ_{is} is the inner-sphere complex charge resulting from specific ion adsorption, σ_{os} is the outer-sphere complex charge resulting from nonspecific adsorption, and σ_d is the dissociated charge. This definition is similar to the one provided by Sposito (1984) with the exception that total net surface charge results from isomorphic substitution and is generated by specifically adsorbing ions (Hunter, 1981). Inner- and outer-sphere surface complexes differ in containing no water (inner) and at least one water molecule (outer) between the adsorbing ion and the surface functional group (Sposito, 1984). Examples of surface functional groups are reactive surface hydroxyl groups on oxide minerals, aluminol and silanol groups on clay minerals, and carboxyl and phenol groups on SOM.

Measurement Methodology

The total net surface charge can be measured directly using electrokinetic experiments. The point of zero charge (p.z.c.) of a particle is the solution pH value where total net particle charge is zero. The p.z.c. can be measured directly using electrokinetic experiments or indirectly from potentiometric titrations under certain experimental conditions (Sposito, 1984).

Electrokinetic phenomena are processes where a relative velocity exists between two parts of the electrical double layer (Hiemenz, 1977). In this motion, a thin layer of liquid remains with the solid and a shear plane is located between the solid and liquid phases at some distance from the solid surface (van Olphen, 1977). The electric double layer potential at the shear plane is called the zeta potential (ζ). The assumption that ζ is equal to or very close to the diffuse double layer potential (ψ_d)

is supported indirectly by a large body of data on a variety of surfaces (Hunter, 1981, 1989). The principal electrokinetic phenomena that measure ζ potential are discussed below.

Electrophoresis

Electrophoresis measures the movement of a suspended charged particle in response to an applied electric field. This movement is called electrophoretic mobility (u_E) and is given by the Smoluchowski equation (Hunter, 1989):

$$u_E = \frac{\varepsilon\zeta}{\eta}$$ [6.20]

where ε is the relative permittivity and η is the viscosity. The Smoluchowski equation applies when the particle dimensions are much greater than the double layer thickness (Hunter, 1987). The complete formula relating ζ and u_E is derived by the theoretical evaluation of the electric force on the charged particle (f_1), the hydrodynamic frictional force on the particle by the liquid (f_2), the electrophoretic retardation force (f_3), a frictional force resulting from the movement of water with the counter ions, and the relaxation force (f_4), caused by distortion of the double layer around the particle (van Olphen, 1977). Additional considerations arise for nonspherical particles and those carrying two double layers such as clays. For these reasons, the Smoluchowski equation is, in general, only approximate and it is advisable to report electrophoresis results as electrophoretic mobility rather than attempt to convert to ζ potential (van Olphen, 1977).

Electrophoresis is the most common method of determining ζ potential. For colloidal systems the most important technique is microelectrophoresis where the movement of individual particles is followed directly by microscopy (Hunter, 1981). Microelectrophoresis is only applicable at very low particle concentrations. Electrophoresis can also be studied using laser Doppler velocimetry and photon correlation spectroscopy. The mass transport mobility apparatus measures electrophoretic mobility from the mass of colloids transported to a suitable electrode compartment (Hunter, 1981). This apparatus can be used at much higher particle concentrations than microelectrophoresis.

Electroosmosis

Electroosmosis measures the movement of the liquid adjacent to a flat, charged surface in response to an electric field applied parallel to the surface. This movement is called electroosmotic velocity (v_{eo}) and is also obtained from the Smoluchowski equation (Hunter, 1987):

$$v_{eo} = \frac{-\varepsilon\zeta}{\eta}$$ [6.21]

where E is the electric field strength. While it is possible to measure electroosmotic velocity directly using microscopy, it is more common to measure the volume of liquid transported per unit time (Hunter, 1981):

$$\frac{V}{i} = \frac{\varepsilon\zeta}{\eta\lambda_0}$$ [6.22]

where V is volume, i is the electric current and λ_0 is the electrical conductivity. The material whose ζ potential is being measured is formed into a porous plug and the transport of liquid across a tube in response to an electric field may be obtained by measuring the movement of an air bubble in the capillary providing the return path (Adamson, 1976). It is also possible to measure the electroosmotic flow by applying a counter pressure until the flow is exactly compensated (Hunter, 1981).

Streaming Potential

The streaming potential is an electric potential difference generated when liquid adjacent to a charged surface is set in motion by an applied pressure gradient (Hunter, 1987). The streaming potential (Φ_{st}) is also governed by the Smoluchowski equation and given by (Sposito, 1984):

$$\Phi_{st} = \frac{\varepsilon\zeta}{\lambda_0\eta}\Delta P \qquad\qquad [6.23]$$

where ΔP is the applied pressure difference. The streaming potential can be measured in similar fashion as the electroosmotic velocity. Liquid is forced under pressure through a porous plug and Φ_{st} is measured by electrodes in the solution on either end (Adamson, 1976).

Sedimentation Potential

When particles having charged surfaces sediment in a liquid under the force of gravity, a plane of shear is developed. As the particles settle, the interfacial charge is separated since a portion inside the shear plane moves with the particle and the remainder is left behind (Sposito, 1984). An electric potential difference arises from the separation of charge called the sedimentation potential. The gradient for the sedimentation potential ($d\Phi_{sed}$) is given by (Sposito, 1984):

$$\frac{d\Phi_{sed}}{dz} = \frac{\varepsilon\zeta}{\lambda_0\eta}n\Delta\rho g \qquad\qquad [6.24]$$

where n is the number of particles per unit volume, $\Delta\rho$ is the difference in mass density between the particles and the liquid phase, g is gravitational acceleration and z is distance. The potential difference is measured by inserting reversible electrode probes at two different heights in the column of settling particles (Hunter, 1981). For low particle concentration the sedimentation potential is also governed by the Smoluchowski equation (Hunter, 1981).

Potentiometric Titration

Potentiometric titration measures the surface density of proton surface charge (σ_H) defined as (Sposito, 1984):

$$\sigma_H = \frac{F}{A}(q_H - q_{OH}) \qquad\qquad [6.25]$$

where F is the Faraday constant, A is specific surface area, q_H is the complexed proton charge (moles), and q_{OH} is the complexed hydroxyl charge (moles) per unit mass of solid. Titration data consist of pH readings obtained while known amounts of acid or base are added to a solid suspension. A net titration curve is obtained by subtracting a calibration curve obtained by titrating the equivalent supernatant solution. The values of $(q_H - q_{OH})$ are given by (Sposito, 1984):

$$q_H - q_{OH} = \frac{C_A - C_B - [H^+] + [OH^-]}{C_s}$$ [6.26]

where C_A is the molar concentration of acid added, C_B is the molar concentration of base added, $[H^+]$ is the molar proton concentration and $[OH^-]$ is the molar hydroxyl concentration obtained from pH measurement, and C_s is the particle concentration. In order for Equation [6.26] to be valid, added protons and hydroxyl ions must only react with variable charge surface reactive groups. Usually, other reactions occur which are also pH dependent such as soluble complex formation, dissolution of solid phases, or complexation with surfaces whose charge is not variable (Parker, 1979). Thus without these corrections no surface chemical significance can be provided by Equation [6.26]. A detailed

Table 6.3 Representative points of zero charge for various minerals

Solid	p.z.c.
	Electrophoresis
Goethite	8.8
Hematite	8.5
Magnetite	6.9
Amorphous iron oxide	8.0
Gibbsite	9.8
Bayerite	9.2
Boehmite	9.4
Pseudoboehmite	9.2
Amorphous aluminum oxide	9.3
δ-MnO$_2$	2.3
Rutile	4.8
Anatase	5.9
Kaolinite	2.9
	Streaming Potential
γ-Al$_2$O$_3$	9.1
α-Al$_2$O$_3$	9.2
	Titration
Goethite	8.7
Hematite	8.6
Magnetite	6.9
Gibbsite	9.8
Boehmite	8.5
Pseudoboehmite	9.3
Amorphous aluminum oxide	9.5
δ-MnO$_2$	3.6
SiO$_2$	3.0
Rutile	5.8
Anatase	6.0
Kaolinite	2.9

description of the use of potentiometric titration to determine surface charge is provided by Huang (1981).

Applications

One of the most important applications of electrokinetic experiments and potentiometric titrations is the determination of the p.z.c. A characteristic of variable charge minerals is the p.z.c. obtained in the presence of an inert electrolyte. This p.z.c. is determined electrokinetically as the pH value where the ζ potential is zero or indirectly from the point of zero net proton charge (p.z.n.p.c) or from the point of zero salt effect (p.z.s.e.) obtained potentiometrically. The p.z.n.p.c and the p.z.s.e. are discussed in Section B, Chapter 7. The p.z.n.p.c and the p.z.s.e. are equivalent to the p.z.c. in the absence of surface complex formation. Table 6.3 provides characteristic values of p.z.c. obtained using electrokinetic experiments and potentiometric titrations for a variety of variable charge minerals.

Electrokinetic experiments and potentiometric titrations can be used to infer adsorption mechanisms for adsorbing ions on surfaces. Adsorption of ions that form inner-sphere surface complexes is characterized by shifts in the p.z.c of the particles and reversals of their electrophoretic mobility with increasing ion concentration (Hunter, 1981). Adsorption of ions that form outer-sphere surface complexes does not produce p.z.c. shifts since they are assumed to lie outside the shear plane. Fig. 6.3 presents the shifts in p.z.c. and charge reversals observed for gibbsite upon the specific adsorption of increasing amounts of molybdate. These results are indirect evidence for inner-sphere surface complexation of this ion.

Net particle surface charge is a primary factor in dispersion of clay minerals. Zeta potential as a measure of particle surface charge was also related to percentage of dispersible clay (Chorom and Rengasamy, 1995). Fig. 6.4 indicates the relationship between ζ potential measured using electrophoresis, and dispersible clay for Na saturated kaolinite, montmorillonite, and illite.

The plane interface technique determines the electroosmotic velocity at large plane interfaces and can be used to determine the ζ potential of two different surfaces at the same time under the same conditions (Nishimura et al., 1992). These authors used the plane interface technique to simultaneously study silica plates and muscovite mica basal planes. The ζ potential values for silica presented in Fig. 6.5 indicate that the asymmetric silica-mica cell provides results comparable to those of the symmetrical silica-silica cell.

Zeta potentials of muscovite mica have also been determined using a flat plate streaming potential apparatus (Scales et al., 1990). Zeta potential of clays becomes less negative with increasing

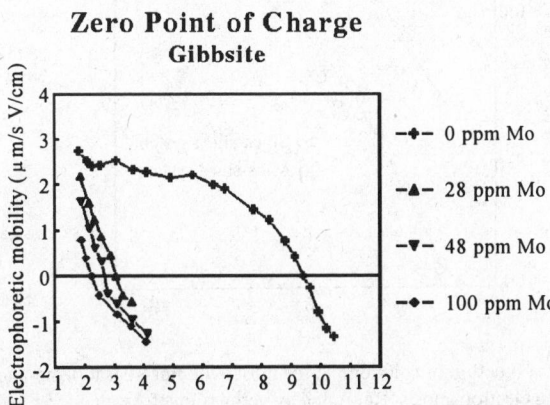

Fig. 6.3   Shifts in p.z.c. and charge reversal of gibbsite in the presence of molybdate [From Goldberg et al., 1996]

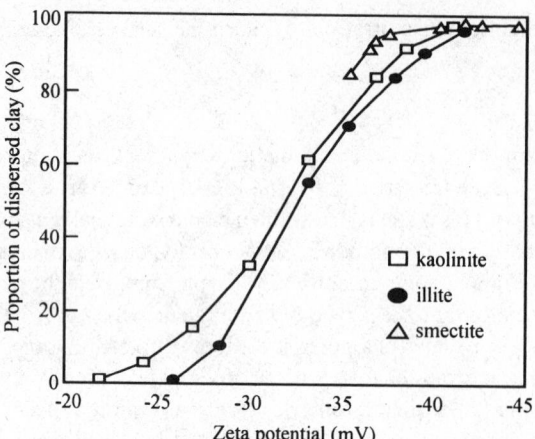

Fig. 6.4 Relation between zeta potential and dispersible clay of Na-clay minerals obtained using electrophoresis [Reprinted by permission of Blackwell Science Ltd, Oxford, UK from Chorom and Rengasamy, 1995. Europ. J. Soil Sci. 46:657-665]

electrolyte concentration as a result of double layer compression (Fig. 6.6). The resulting charge reduction causes clay flocculation.

A vastly different application of electrokinetic experiments is the application to dewatering and decontamination of soils and clays. For example, electroosmosis has been used for removing organic contaminants from kaolinite clays (Shapiro and Probstein, 1993). These authors present an application for *in situ* hazardous waste remediation of soil.

A potentiometric titration method has been developed to account for changes in solubility of the solid with changes in pH (Schulthess and Sparks, 1986). This is a batch method where the reference for each sample is the supernatant specific to that sample backtitrated to pH 7. This method is considered to account for all sources of proton consumption (Schulthess and Sparks, 1986).

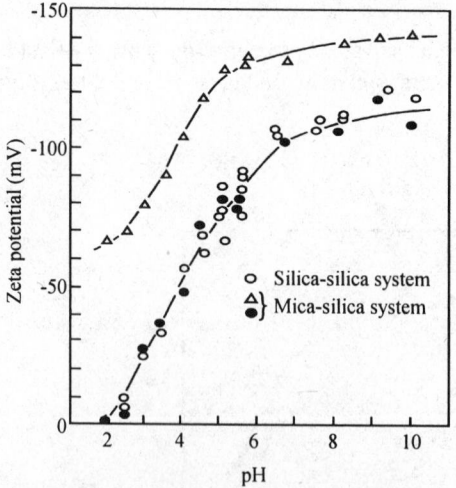

Fig. 6.5 Zeta potentials as a function of solution pH for muscovite mica basal plane-and silica plate-aqueous solution interfaces obtained using electroosmosis [Reprinted by permission of Academic Press, Inc. from Nishimura et al., 1992. J. Colloid Interf. Sci. 152:359-367]

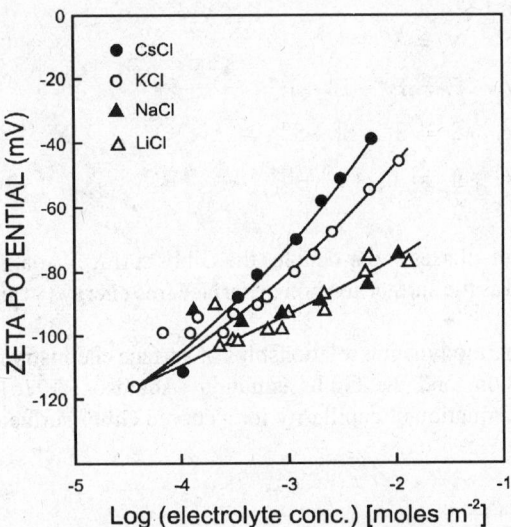

Fig. 6.6 Relation between zeta potential and electrolyte concentration for various 1:1 electrolytes obtained from streaming potential [Reprinted from Scales et al., 1990. Langmuir 6:582-589. Copyright American Chemical Society]

6.1.4 Thermodynamics of Colloid Surfaces

To develop the thermodynamic treatment of the surface region, a few definitions are useful. The interfacial region is a space between two adjoining phases (gas-liquid, gas-solid, liquid-liquid, liquid-solid, solid-solid) which is characterized by inhomogeneity in its properties. The Gibbs surface is a mathematical dividing surface, without volume, drawn parallel to the boundaries of the interfacial region, which is used to define the volumes of the two adjoining bulk phases. A schematic of the interfacial region and the Gibbs surface is presented in Fig. 6.7. The actual values for the system as a whole will differ from the sum of the values for the bulk phases by an excess or deficiency due to the Gibbs surface (Adamson, 1976). The following relations hold for the variables of state:

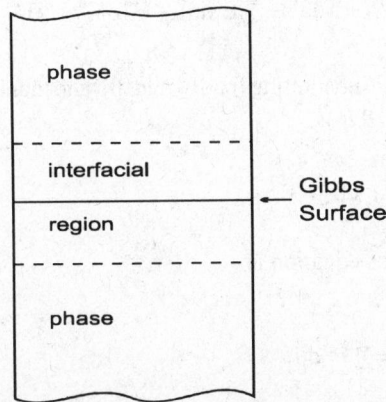

Fig. 6.7 Representation of the interface region and the Gibbs surface

$$\text{Volume:} \quad V = V^\alpha - V^\beta$$

$$\text{Internal Energy:} \quad E = E^\alpha + E^\beta + E^\sigma$$

$$\text{Entropy:} \quad S = S^\alpha + S^\beta + S^\sigma \qquad [6.27]$$

$$\text{Moles:} \quad n_i = n_i^\alpha + n_i^\beta + n_i^\sigma$$

where α and β denote the bulk phases and σ denotes the Gibbs surface (Adamson, 1976). Additional variables of state are defined as the surface tension or surface free energy (γ) and the area of the Gibbs surface (A).

The three fundamental thermodynamic relationships of surface chemistry are the Young-Laplace equation, the Kelvin equation, and the Gibbs equation (Adamson, 1976). The Young-Laplace equation is the fundamental equation of capillarity for a curved Gibbs surface:

$$P^\beta - P^\alpha = \gamma \left(\frac{1}{r_1} + \frac{1}{r_2} \right) \qquad [6.28]$$

where ($P^\beta - P^\alpha$) is the capillary pressure and r_1 and r_2 are the radii of curvature.

The Kelvin equation gives the effect of surface curvature on the molar free energy of a substance. The free energy of a substance can be related to its vapor pressure assuming the vapor to be ideal (Adamson, 1976). The Kelvin equation is

$$\ln\left(\frac{P}{P^0} \right) = \frac{\gamma V}{RT} \left(\frac{1}{r_1} + \frac{1}{r_2} \right) \qquad [6.29]$$

where P^0 is the normal vapor pressure of the liquid, P is the vapor pressure observed over the curved surface, and R is the molar gas constant, and T is temperature.

For a small, reversible change dE in the energy of the system (Adamson, 1976):

$$dE = dE^\alpha + dE^\beta + dE^\sigma = TdS^\alpha + \sum \mu_i dn_i^\alpha - P^\alpha dV^\alpha + TdS^\beta$$

$$+ \sum \mu_i dn_i^\beta - P^\beta dV^\beta + TdS^\sigma + \sum \mu_i dn_i^\sigma - P^\sigma dV^\sigma + \gamma dA \qquad [6.30]$$

where μ_i is chemical potential. Substituting for dE^α and dE^β and manipulating the equation for dE^σ lead to the expression (Adamson, 1976):

$$S^\alpha dT = Ad\gamma + \sum n_i^\sigma d\mu_i = 0 \qquad [6.31]$$

At constant T and A, the Gibbs equation is

$$-d\gamma = \sum \frac{n_i^\sigma}{A} d\mu_i = \sum \Gamma_i^\sigma d\mu_i \qquad [6.32]$$

where Γ_i^σ is a surface excess concentration per unit area defined as $\Gamma_i^\sigma = n_i^\sigma/A$. The Gibbs equation can be applied to liquid-liquid and liquid-vapor interfaces where the surface tension can be measured to calculate the surface concentration of the adsorbed species causing the surface tension change. Similarly, if the surface concentration can be measured directly but the surface tension cannot, the Gibbs equation can be used to calculate the lowering of γ from the measured adsorption in solid-gas and solid-liquid systems (Hunter, 1987).

6.2 Interparticle Forces

6.2.1 Electrical Double Layer

Double layer theory describes the distribution of ionic concentrations near electrostatically charged particles. The charge on the colloidal particles is due to isomorphic substitution in the particle lattices or arises from the preferential adsorption of one ionic species from the solution phase (Babcock, 1963). Such charge requires the presence of a layer of ions of opposite charge. The double layer consists of an excess of ions of opposite sign and a deficiency of ions of the same sign that are electrostatically repelled by the particle. Double layer theory assumes that the surface of the colloidal particles is represented by an infinite flat surface having continuous and uniform electrostatic charge density immersed in an electrolyte with a uniform dielectric constant (Babcock, 1963). All electrolyte ions are assumed to be point charges. The electrical potential, ion charge, and ion distributions can be calculated from the Poisson-Boltzmann equation:

$$\frac{d^2\psi}{dx^2} = -\frac{1}{\varepsilon_0 D} \sum cFz_i \exp(-z_i F\psi(x)/RT)$$

[6.33]

where $\psi(x)$ is the inner potential at a distance x from the surface, e_0 is the permittivity of free space, D is the dielectric constant of water, c is the concentration, R is the gas constant, T is the temperature (K), F is the Faraday constant, and z_i is the valence of the charged species (Sposito, 1984). Figs. 6.8 and 6.9 present the ion distribution and the electric potential distribution, respectively, in the double

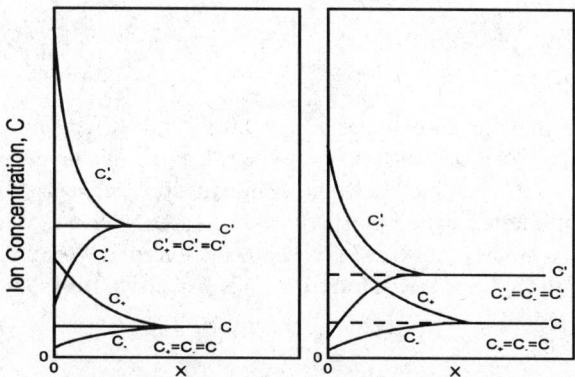

Fig 6.8 Distribution of ions in the electric double layer at two electrolyte concentrations (c' > c). Left: constant potential surface. Right: constant surface charge [Adapted from van Olphen, 1977. An introduction to clay colloid chemistry. Copyright John Wiley & Sons, New York with permission]

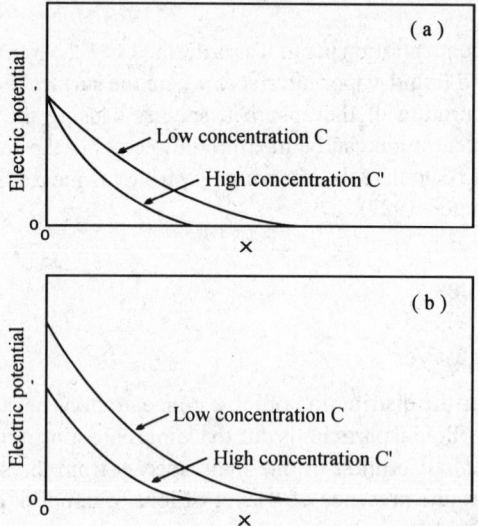

Fig. 6.9 Electrical potential distribution in the double layer at two electrolyte concentrations. Top: constant potential surface. Bottom: constant surface charge [Adapted from van Olphen, 1977. An introduction to clay colloid chemistry. Copyright John Wiley and Sons, New York with permssion]

layer at two electrolyte concentrations. The extent of the double layer is given by the distance $(1/\kappa)$ in units of m:

$$\frac{1}{\kappa} = \sqrt{\frac{\varepsilon_0 RT}{2000 F^2 I}} \qquad\qquad\qquad [6.34]$$

where I is the ionic strength $(=\frac{1}{2}\Sigma\, c_i\, z_i^2)$. Important findings of diffuse double layer theory are (1) the excess cations near the negative surface neutralize more of the charge than the anion deficit; (2) the electric potential decreases as the electrolyte concentration increases; (3) the double layer distance $(1/\kappa)$ decreases as the electrolyte concentration increases. Some important limitations of double layer theory are that it applies only to infinitely dilute suspensions and to low surface charge densities (Babcock, 1963).

6.2.2 Attractive Force

The attractive force acting on colloidal particles is called the van der Waals force and acts to bring particles closer together. The basis of the attractive force is that the fluctuating dipole of one atom polarizes another and the two atoms attract each other. This attraction between atom pairs is additive, and therefore, the energy of interaction between particles decreases much more slowly with distance than that between individual atoms (Quirk, 1994). The interaction energy per unit area between two opposing planar solids for the van der Waals force (Φ_{vdW}) is (Israelachvili, 1992):

$$\Phi_{vdW} = -\frac{A_H}{12\pi d^2} \qquad\qquad\qquad [6.35]$$

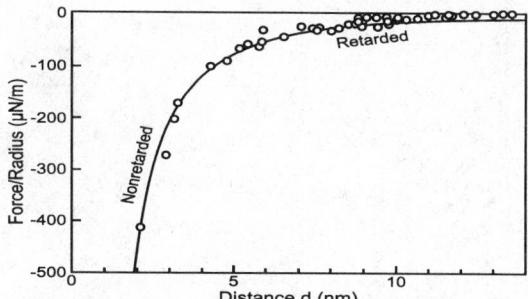

Fig. 6.10 van der Waals forces between two mica surfaces in aqueous electrolyte solutions. The measured Hamaker constant is A = 2.2x10^{-20} J. Retarded van der Waals forces are observed above 5 nm [Reprinted from Israelachvili, 1992. Intermolecular and surfaces forces. Academic Press Ltd, London, UK with permission].

where A_H is the Hamaker constant and d is the distance separating the solid surfaces. At distances > 5 nm, the correlations between the induced dipole distributions weaken and the interaction energy per unit area corresponds to the retarded van der Waals force (Φ_{RvdW}):

$$\Phi_{RvdW} = -\frac{B_{rH}}{3d^3}$$

[6.36]

where B_{rH} is the retarded Hamaker constant. Fig. 6.10 shows values of the van der Waals force obtained experimentally between two mica surfaces.

6.2.3 Repulsive Force

The electrostatic force results from the charge on the colloidal particles and acts to repel them. A force operates on charged surfaces as a result of their interacting double layers. This force is repulsive if the charges on the particles are the same. The repulsion described in terms of interaction energy per unit area (Φ_R) is given by the force times the distance through which it operates (Hiemenz, 1977):

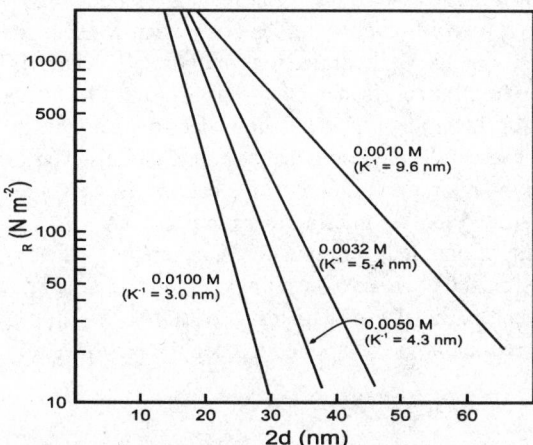

Fig. 6.11 Repulsive force between two plates for different concentrations of a 1:1 electrolyte where K is the inverse double layer distance [Reprinted by permission of Marcel Dekker, Inc. from Hiemenz, 1977. Principles of colloid and surface chemistry]

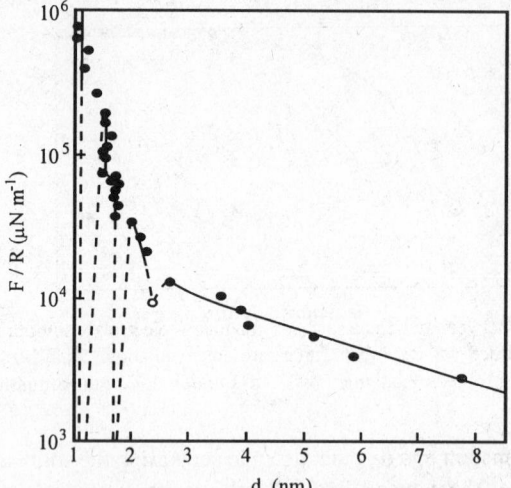

Fig. 6.12 Forces between two mica surfaces in 7 mmol L^{-1} NaCl electrolyte. F is the total force scaled by the radius (R) of the curved surfaces. At distance > 1.5 nm, the forces are described by double layer theory. Hydration forces are observed at distances < 1.5 nm [Reprinted with permission from Ducker and Pashley, 1992. Langmuir 8:109-112. Copyright American Chemical Society].

$$\Phi_R = -\frac{64a^2}{z} cRT \exp(-z\kappa d)$$ [6.37]

where a = tanh($z\psi_0/4RT$), z is the charge on the electrolyte ions, c is the concentration of the electrolyte ions, κ is the inverse double layer distance, and d is ½ of the surface separation.. This equation is valid only when the surface separation 2d $\gg$ 1/κ and a $\approx$ 1. The electrostatic force between two plates for different electrolyte concentrations is presented in Fig. 6.11.

When surfaces are brought closer and closer together or, when water is the medium (hydration force), an additional repulsive force (solvation force) becomes important. Solvation forces are short range and oscillatory and arise whenever liquid molecules are induced to order between surfaces. Between colloid surfaces, repulsive hydration forces arise when water molecules strongly bind to hydrogen bonding surface groups such as hydrated ions or hydroxyl groups (Israelachvili, 1992). The effective range of hydration forces in clays and silicates is 3–5 nm. The interaction of mica surfaces in dilute solution obeys double layer theory, but at higher electrolyte concentration, a hydration force develops due to the energy needed to dehydrate surface bound cations. The strength of the hydration force which increases with the hydration number of the cation $Mg^{2+} > Ca^{2+} > Li^+ \approx Na^+ > K^+ > Cs^+$ (Israelachvili, 1992) is illustrated for two mica surfaces in Fig. 6.12.

Clay swelling is the result of double layer repulsion between the surfaces of individual particles. Under confining conditions, a fluid pressure or swelling pressure is created that is a direct measure of the balance of forces between particles (van Olphen, 1977). The swelling pressure is obtained by measuring the confining force that must be applied to keep the clay layers at a given distance. The swelling pressure (Π) is

$$\Pi = P_{vdW} + P_R$$ [6.38]

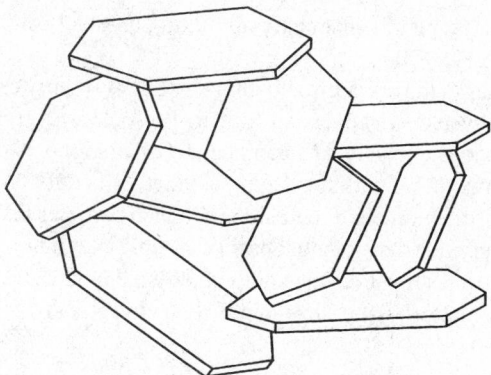

Fig. 6.13   Representation of card house structure [Reprinted by permission of Oxford University Press from Hunter, 1987. Foundations of colloid science Vol. 1]

where P_{vdw} is the van der Waals force and P_R is the electrostatic force (Greathouse et al., 1994). At short distances, hydration forces become significant in the swelling pressure. At greater distances, measured swelling pressures are of similar magnitude to calculated double layer repulsions (van Olphen, 1977).

Interparticle forces can be measured experimentally using the surface force apparatus (SFA), total internal reflectance microscopy, and the atomic force microscope (AFM) (Israelachvili, 1992). The SFA has been used to measure attractive van der Waals forces, repulsive double layer forces, and repulsive hydration forces in aqueous solutions at the 10^{-10} m level of resolution (Israelachvili, 1992). Because of its smooth surface and ease of handling, mica has been the primary solid used in SFA studies. Total internal reflection microscopy has been used to study forces between a surface and an individual colloidal particle. Atomic force microscopy has been used to measure both short- and long-range forces. Interactions between charged mica surfaces have been investigated using a combination of SFA and AFM experiments. Results from both methods agree with theoretical predictions (Kékicheff et al., 1993).

6.3   Colloidal Stability

6.3.1   Flocculation and Dispersion

The stability of colloidal suspensions is a balance between repulsive and attractive forces acting among the suspended particles. If net repulsive forces predominate, particles do not coagulate and they remain dispersed. When the attractive forces are dominant, interacting particles coagulate and the resulting floccules settle more rapidly from the suspension than the smaller dispersed particles. This theory is referred to as DLVO after Derjaguin, Landau, Verwey, and Overbeek who were largely responsible for its development (Hunter, 1987).

Flocculation is a thermodynamically favorable process; however, the kinetics of coagulation determine the stability of colloidal suspensions. Generally, all colloidal suspensions will spontaneously flocculate given sufficient time, but potential energy barriers retard the rate of flocculation. These barriers are analogous to activation energies considered in chemical kinetics (Hiemenz, 1977). Particles in a primary minimum are adhesive and are not readily separated. In contrast the dispersion/

flocculation transition is the result of a secondary minimum, involves card house type structures (Fig. 6.13), and is readily reversible (Quirk, 1994).

There are many examples in the literature showing that DLVO theory can account for the observed kinetic behavior of colloid systems (Napper and Hunter, 1974). However, this theory must be applied with caution since it treats ions exclusively as point charges and ignores their chemical properties.

The potential barrier preventing particles from coagulating is defined by the stability ratio (W), which is the fraction of the total number of collisions between particles that result in coagulation. The rate of coagulation in the absence of a potential barrier or rapid coagulation (R_f) is limited only by the rate of diffusion of the particles towards one another. When the particles have a potential barrier to overcome, the rate of coagulation (R_s) is slow and is related to R_f by:

$$R_s = \frac{R_f}{W}$$
[6.39]

Rapid coagulation occurs when no significant repulsive forces act between particles and van der Waals or long-range coulombic attractions predominate. The quantification of the coagulation rate under these conditions was examined by von Smoluchowski (1916, 1917) and is discussed by Overbeek (1952).

Slow coagulation occurs over distances of the order of 1-100 nm when the approaching particles experience a barrier as their double layers overlap. Diffusion over this distance results from many individual Brownian events, some of which bring the particles closer together and some of which take them further apart. Since the rates of rapid and slow coagulation are directly proportional to the number of particles diffusing in the direction of a central particle (J), it follows from Equation [6.39] that the stability ratio is given by:

$$W = \frac{R_f}{R_s} = \frac{J_f}{J_s} = 2r\int_{2r}^{\infty} \exp\left(\frac{E_T}{K_B T}\right)\frac{dr}{d^2}$$
[6.40]

where r is the particle radius and $d \cong 2r$. Verwey and Overbeek (1948) showed that W was determined almost entirely by the value of the total potential energy (E_T) at the maximum (Fig. 6.14). In Fig. 6.14, E_R and E_A are the potential energies due to repulsive and attractive forces. A more complete analysis of the kinetics of colloid flocculation is presented in Hunter (1987).

Gravity removes suspended particles by sedimentation while inducing differential sedimentation coagulation, thereby decreasing Brownian coagulation rates. Ultimately, sedimentation limits the time of a flocculation series test. The test should be long enough to detect relative changes in suspended particle numbers, but not so long that all dispersed particles settle from a stable suspension (Hesterberg and Page, 1990a).

The reverse of flocculation is called dispersion. Ideally, the amount of energy required to separate two particles coagulated into a potential energy minimum is approximately equal to the difference between the interaction energy at the minimum and that at the adjacent maximum (van Olphen, 1977). Under constant chemical conditions, separation of particles could presumably be induced by an input of kinetic energy (e.g., thermal or mechanical shear). Alternatively, changing the chemical conditions of the bulk solution surrounding coagulated particles could provide chemical and/or electrochemical energy to produce dispersion.

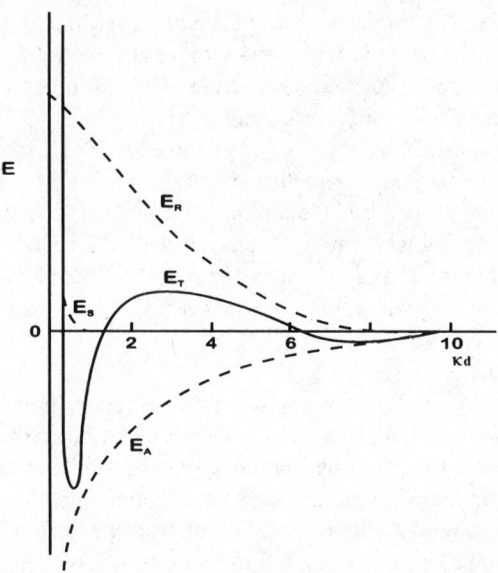

Fig. 6.14 Total potential energy of interaction of two colloidal particles: $E_T = E_S + E_R + E_A$, where E_S is the potential energy of repulsion due to the solvent layers, E_R is the potential energy due to the repulsive forces, and E_A is the potential energy due to the attractive forces, κD is particle separation [Reprinted by permission of Oxford University Press from Hunter, 1987. Foundations of colloid science Vol. 1]

6.3.1.1 Modes of Particle Association

Edge-Edge, Edge-Face, Face-Face

Clay crystals have a net negative surface charge as a consequence of isomorphic substitutions of electropositive elements for elements with a lower valence. This negative charge generates an ionic reorganization in the solution medium that has been described above as the double layer. Clay particles also have edge surfaces with atomic structure different from the faces. At the edge of the platelet, the tetrahedral sheet of Si and the octahedral layer of Al exhibit broken bonds which, in turn, generate another electric double layer.

Double layer theory assumes that the surfaces of the clay minerals are of semi-infinite spatial extent and show no edge effects, a simplification that is not always satisfactory. Clay mineral particles have finite dimensions. Below the p.z.c., edge surfaces carry a positive charge due to specific adsorption of protons. Using the Poisson-Boltzmann equation, Secor and Radke (1985) calculated the effect of edge-face corners on the electrical potential distribution around an idealized, symmetrical montmorillonite disk. Spillover of the negative electrical potential from the particle faces into the edge region can result in a negative potential everywhere around the particle.

The electrical potential at the edge surface strongly depends on the electrolyte concentration and the ratio of the face to edge charge density. The extent of spillover also is a weak function of the particle shape (Secor and Radke, 1985), which implies that attraction between positively charged edges and negatively charged faces of phyllosilicate mineral particles will depend on the extent of the edge protonation, electrolyte concentration, and shape of the particle. A phyllosilicate structure that is collapsed in the c-dimension, such as mica, should be able to acquire a larger edge surface charge density than a layer silicate like smectite where structural expansion increases the distance between edge surface aluminol groups (Hesterberg, 1988).

When there is a reduction in the thickness of the double layer, the particles can associate among themselves in three different ways: face-face, face-edge, or edge-edge. Face-face association is also called parallel aggregation and does not produce floccules while the other two associations do produce three-dimensional structures called card houses (Fig. 6.13).

In concentrated suspensions of clay, the edge-edge and edge-face associations form a continuum, with chains of particles in the card house structures mentioned above. The rigidity of the gel depends on the number and strength of the bonds in the continuum structure. Some attempts to characterize the gel structure using freeze drying techniques have been made (Norrish and Rausell-Colom, 1961). When the water is eliminated from the suspension, the volume of the system does not change and the final product is a dry clay structure with some strength that has been called aerogel (van Olphen, 1977).

Domains and Quasicrystals

Some colloidal systems may, under certain circumstances, show a reversible clustering among particles. The earliest reported example is the Fe hydroxide sol described by Cotton and Mouton (1907). The term tactoid was first used by Freundlich (1932) and was more precisely defined by Overbeek (1952) as the association of particles at a certain distance affected by electrolyte concentration and pH. Quirk and Aylmore (1971) proposed the term quasicrystal to describe the regions of parallel alignment of individual aluminosilicate lamellae in montmorillonite and the term domain to describe the regions of parallel alignment of crystals for illite which has been adopted in this chapter.

The distribution model of adsorbed ions in mixed mono- and divalent systems for smectite quasicrystals was described by Shainberg and Otoh (1968) and Bar-On et al. (1970). According to this theory, called the ion demixing model, when Na is added to Ca saturated montmorillonite, most of the adsorbed Na will concentrate on the external surfaces of the quasicrystals until 10% of the adsorbed Ca has been replaced by Na. Initially, the size and shape of the particles are not altered by the addition of adsorbed Na. As further Na is added to the system, Na penetrates into the quasicrystals and brings about disintegration of the clay packets (Bar-On et al., 1970).

Table 6.4 Critical coagulation concentrations of phyllosilicates under various conditions [Adapted from Hesterberg, 1988]

Mineral	CCC (mol$_c$ L^{-1})	Background electrolyte	pH	Solids concentration (g kg^{-1})
Kaolinite-4	2-40	NaNO$_3$	4-10	0.025
Kaolinite-9	8, 30	NaHCO$_3$, Na$_2$CO$_3$	8.3, 9.5	0.6-0.9
Kaolinite (Georgia)	5, 245, 75	NaCl, NaHCO$_3$, Na$_2$CO$_3$	7, 8.3, 9.5	0.6-0.9
Kaolinite (KGa-1)	< 0.19-54.6	NaCl	5.8-9.1	0.67
Kaolinite-4	0.1-0.3	Ca(NO$_3$)$_2$	4-10	0.025
Montmorillonite-23	1-10	NaNO$_3$	3.8-10	0.25
Montmorillonite-23	20, 48, 68	NaCl, NaHCO$_3$, Na$_2$CO$_3$	7, 8.3, 9.5	0.6-0.9
Montmorillonite-27	14, 47,17	NaCl, NaHCO$_3$, Na$_2$CO$_3$	7, 8.3, 9.5	0.6-0.9
Montmorillonite (SAz-1)	14-28	NaCl	6.4-9.4	0.67
Montmorillonite (SAz-1)	1.09, 1.56	CaCl$_2$	6.1, 7.6	0.67
Montmorillonite (SAz-1)	0.93, 2.02, 0.88	MgCl$_2$	6.1, 8.4, 9.0	0.67
Vermiculite	38, 58, 30	NaCl, NaHCO$_3$, Na$_2$CO$_3$	7, 8.3, 9.5	0.6-0.9
Illite-36	9, 185, 95	NaCl, NaHCO$_3$, Na$_2$CO$_3$	7, 8.3, 9.5	0.6-0.9
Illite (Grundy)	7.24	NaCl	~ 6	1
Illite (Grundy)	0.2	CaCl$_2$	~ 6	1

Lebron and Suarez (1992a) used electrophoretic mobility experiments to show that the demixing model can be applied to micaceous clays. The electrophoretic mobility of micaceous domains increased when the sodium adsorption ratio (SAR) increased from 5 to approximately 10.

6.3.1.2 Critical Coagulation Concentration

The critical coagulation concentration (CCC) sometimes called the critical flocculation concentration (CFC) is the minimum concentration of indifferent electrolyte that induces rapid coagulation and is strongly dependent on counterion valence (Table 6.4). This observation is known as the Schulze-Hardy rule. An estimate of the CCC (mol L^{-1}) is given by (Hunter, 1987):

$$CCC = \frac{0.107 \, \epsilon^3 \left(K_B T\right)^5 Z^4}{N_A A_H^2 (ze)^6} \qquad [6.41]$$

where $Z = \tanh(ze\psi_0/4k_B T)$, N_A is Avogadro's number, A_H is the Hamaker constant, ϵ is the relative dielectric constant, k_B is the Boltzman constant, and e is the proton charge.

The variation of CCC with counterion valence depends approximately on the inverse sixth power of z. At 25°C in water, using the experimental observation that coagulation usually occurs between low potential surfaces, the result is

$$CCC \propto \frac{\psi_0^4}{z^2} \qquad [6.42]$$

where now the CCC depends on the inverse of the square of the valence. However, the CCC calculated with Equation [6.42] agrees well with experimental results if $\psi_0 \propto 1/z$. When calculating the CCC in soils, one must consider that soils consist of mixtures of permanent and variable charge minerals. The net surface charge, and consequently, the electrical potential around the mixture of particles are dependent upon variables such as pH, specifically adsorbed ions, ionic strength, and mineralogy. A more detailed evaluation of the effect of these variables on the stability of colloids is presented below.

6.3.2 Factors Affecting Colloidal Stability

6.3.2.1 Solution Composition

The magnitude of the repulsion barrier is determined by the nature of the material adsorbed on the particle surface. In the case of a charged colloid, repulsion depends on the magnitude of the surface charge and on the extent of the electrical double layer which, in turn, depends on the total electrolyte concentration. It is necessary here to distinguish between the concentration of the potential determining ions and that of other ions that have no direct interaction with the surface. If the surface potential of the particles is determined by the concentration of potential determining ions, the magnitude of this potential is not affected by the addition of an indifferent electrolyte. For this type of double layer, when salt concentration increases, the double layer thickness decreases, the surface charge of the particles increases, and the surface potential remains constant (Fig. 6.9a).

If the surface charge of the particle is determined by interior crystal imperfections, the surface charge does not change with increasing electrolyte concentration. The diffuse double layer compresses, but in this case, the surface potential decreases with increasing electrolyte concentration (Fig. 6.9b) (van Olphen, 1977).

6.3.2.2 Exchange Complex Composition

As explained above, clay particles are surrounded by cations as a consequence of the net negative electrical charge on the surface. Cations bonded to the surface can be exchanged for other cations in solution. Consequently, the cations on the exchange complex are dependent on the solution composition.

There is an equilibrium between the cations on the exchange complex and the cations in the solution. Not all the cations are adsorbed with the same affinity. Cations with larger hydrated radii are less strongly adsorbed than those with smaller radii. In solutions with equal initial concentrations of different cations, the amounts of Ca and Mg adsorbed are several times greater than the amount of Na adsorbed. In general, polyvalent cations are adsorbed more strongly than monovalent cations and are not easily displaced by other cations. The order of adsorption strength is Al > Ca > Mg > H > K > Na (Douchaufour, 1970).

The flocculation/dispersion behavior of soil colloids depends on salt concentration, exchange complex cation, cation valence, and dominant clay mineralogy. In general, divalent cations are more effective in flocculating colloids than monovalent cations. For example, Quirk and Schofield (1955) found that the flocculating power of $CaCl_2$ is 50 to 100 times higher than NaCl. For monovalent cations, there is also a difference, with KCl showing greater flocculation power than NaCl (Pashley, 1981).

6.3.2.3   pH

pH is an important determinant of the electrical potential of the clay surface. Changes in pH affect the edge charge on clays and the surface charge of variable charge minerals such as Fe and Al oxides. There is considerable variability depending on structural composition and degree of crystallinity, but Fe and Al oxides generally undergo a surface charge reversal around pH 7 to 9 (positively charged below that pH and negatively charged above). This is also the region in which kaolinite exhibits its edge charge reversal as evidenced by Cl^- adsorption studies (Schofield and Samson, 1954). Soil colloids consist of a mixture of minerals, each with a different p.z.c. At low pH, edge to face bonding, as well as bonding of positive Fe and Al oxides to negative clay surfaces is expected to occur (van Olphen, 1977; Kretzschmar et al., 1993, 1997). This type of bonding should hinder dispersion and should thus result in flocculation. With increasing pH, as the p.z.c. is approached, edge to face clay bonding decreases and Fe and Al hydrous oxide bonding to clays is also expected to decrease (Suarez et al., 1984). In variable charge systems, flocculation is at a maximum when the soil is at its p.z.c.

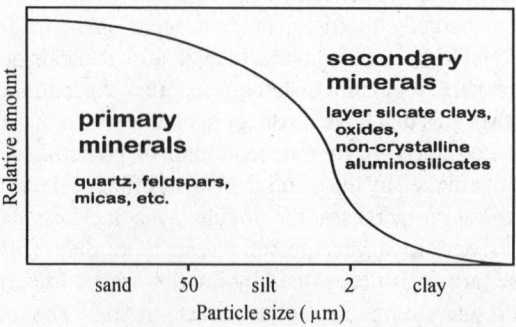

Fig. 6.15 Typical abundance of primary and secondary minerals in different size fractions of the soil [Reprinted from McBride, 1994. Environmental chemistry of soils. Copyright Oxford University Press, Inc. used by permission of Oxford University Press]

Critical coagulation concentration increased at high SAR values with increasing pH for three soil clays whose clay mineralogy was dominantly kaolinite, montmorillonite, or illite (Goldberg and Forster, 1990). Hesterberg and Page (1990b) found also an increase in CCC for a Na and a K illite with increasing pH. Similar results were found for illite and three micaceous soil clays when SAR > 15 (Lebron and Suarez, 1992a,b). The electrophoretic mobility of these materials increased when the SAR was higher than 20 and the pH was above the p.z.c. No pH effect was observed at SAR < 15 for either mobility or CCC.

6.3.2.4 Mineralogy

The colloidal fraction of a soil consists primarily of secondary minerals (Fig. 6.15). Layer silicate minerals differ in chemical composition and charge characteristics leading to different physico-chemical behavior (Tables 6.3 and 6.4). The stoichiometry of each mineral varies due to isomorphic substitutions in the crystal lattice during the formation or evolution of the mineral structure.

The siloxane (Si-O-Si) surfaces of 2:1 layer silicates without structural charge are hydrophobic. Therefore, the surface oxygens coordinated to Si shows little tendency to H bond with water molecules. Smectites, such as montmorillonite, with structural charge located mainly in the octahedral sheet form weak H bonds with water as a result of the delocalization of some structural charge into the surface oxygens (Farmer, 1978). Smectites, such as beidellite, with a high proportion of charge in the tetrahedral sheet form stronger H bonds. The siloxane surfaces of vermiculite are the most hydrophilic of the 2:1 layer silicate clays because they possess a large tetrahedral charge partially distributed onto surface oxygens (Farmer, 1978). Tetrahedral charge is much more localized on fewer surface oxygens than octahedral charge, explaining the stronger H bonding of adsorbed water on vermiculite (McBride, 1989). An extensive description of the structural characteristics of soil minerals can be found in Dixon and Weed (1989).

An example of how these mineral differences affect the flocculation/dispersion behavior of soil colloids is shown in Table 6.4. An overview of the data reveals that reported CCC values of kaolinite, montmorillonite, vermiculite, and illite are quite variable within and between these mineral groups.

Many of the differences in Table 6.4 are due to differences in methodology in the determination of CCC and/or differences in the stoichiometry of the silicate minerals. However, these factors do not account for all of the variability. Lebron and Suarez (1992a) found substantial differences in CCC within samples from the same soil type. Differences in content of organic matter and other minerals can drastically change the behavior of soil colloids. Consequently, general guidelines for reclamation of agricultural land or the use of amendments to maintain colloids in a flocculated state must be implemented with caution because soils usually require higher concentrations than the corresponding pure clay minerals to maintain a flocculated condition.

Table 6.5 Spatial extensions of electrical double layers and polymers of different molecular weight [Reprinted by permission of Oxford University Press from Hunter, 1987. Foundations of colloid science. Vol. 1]

1:1 Electrolyte concentration (mol L^{-1})	Double layer thickness, $1/\kappa$ (nm)	Polymer molecular weight	Spatial extension (nm)
10^{-5}	100	1,000,000	60
10^{-4}	30	100,000	20
10^{-3}	10	10,000	6
10^{-2}	3	1000	2
10^{-1}	1		

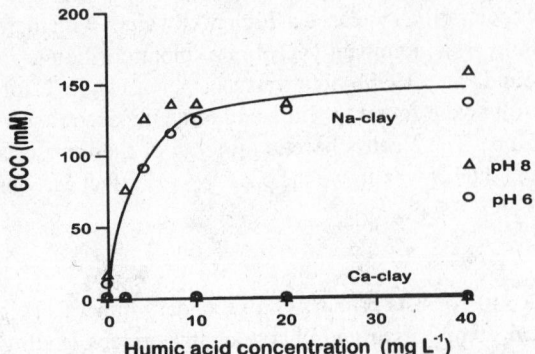

Fig. 6.16   Effect of humic acid concentration on the critical coagulation concentration (CCC) of a Na-clay and a Ca-clay [Used by permission of Oxford University Press, Inc. from Nelson and Oades, 1998. M.E. Sumner and R. Naidu (ed.) Sodic soils: Distribution, properties, management and environmental consequences. Copyright Oxford University Press, Inc.]

6.3.2.5   Organic Matter

Organic matter constitutes a small portion of the soil mass (0.5-10%) but is intimately associated with inorganic particles and plays an important role in the improvement of soil structure (Nelson and Oades, 1998). Aeration, water-holding capacity, and permeability increase when the SOM content increases in a soil. However, organic matter can promote dispersion of the soil particles. Organic coatings under certain conditions maintain a dispersed state for soil colloids in suspension through a combination of electrostatic and steric mechanisms (Stevenson, 1982). Table 6.5 shows the spatial extensions for non-ionic polymer molecules of different molecular weights. Like electrical double layers, macromolecules of at least a few thousand molecular weight also extend in space over distances comparable to, or greater than, the van der Waals attraction. When dissolved organic matter (DOM) adsorbs onto mineral surfaces, the humic substances behave as polyelectrolytes generating hydrophobic and hydrophilic surfaces (Hunter, 1987).

Goldberg and Forster (1990) and Kretzschmar et al. (1993) found that the removal of SOM enhanced soil flocculation (decreased the CCC). Similarly, addition of small amounts of organic material substantially increased dispersion of Na-saturated soil or clay in the order humic acid > soil polysaccharide $\geq$ anionic polysaccharide (Gu and Doner, 1993; Kretzschmar et al., 1997). Using smectite, kaolinite, and three soils whose clay fractions were dominated by one of these minerals, Frenkel et al. (1992) showed that the CCC values of Na-soils were much higher, and much more affected by organic matter than those of Ca soils. Tarchitzky et al. (1993) showed similar comparisons between Na and Ca montmorillonite suspensions with varying additions of humic and fulvic acids (Fig. 6.16). The effect of organic matter on stability of soil colloids is a function of its size. Large organic materials such as polysaccharides and hyphae act to bind colloid particles together. Small organic molecules such as fulvic acid and organic acids increase dispersion of soil colloids through their effect on particle charge. A more detailed review of organic matter chemistry and its effect on soils is provided in Stevenson (1982) and Nelson and Oades (1998).

6.3.3   Measurement of Colloidal Stability

6.3.3.1   Flocculation Series Test

Critical coagulation concentrations are commonly determined using the flocculation series test. The experiment can be performed by taking a series of test tubes containing the same concentration of the colloid and adding varying amounts of the coagulating electrolyte. The tubes are then shaken quickly and allowed to stand for a given time. They are then briefly reshaken in order to remix the contents and again allowed to stand. If the electrolyte range has been properly chosen, the CCC is defined as the concentration above which the settling material leaves behind it a perfectly clear supernatant solution. Below the CCC, the supernatant retains some of the uncoagulated colloid. The concentration of colloidal particles, at which the flocculation test must be performed, should not exceed 10 g L^{-1}, thus avoiding interferences from different particle interactions such as gel formation.

6.3.3.2 Dispersion Indices

There are several methods (qualitative, semiquantitative, or quantitative) to determine the dispersion or flocculation status of soil colloids. Qualitative analyses are those based on direct observation of small particles when the soil is immersed in water. This observation can be made with the optical microscope or with the naked eye, as is the case for the test of Emerson (1967). Methods based on turbidity of a suspension of dispersed soil can be considered semiquantitative when comparative measurements are made. Normally, a standard curve is constructed using known amounts of dispersed clay. The soil under evaluation is assigned a dispersion value by comparison with the standard. The most commonly used quantitative method to determine soil flocculation state is the dispersion index which is the ratio of the amount of colloid in a water suspension after shaking to that when the soil has been treated with a dispersant. This procedure is recommended by the Soil Conservation Service (Sherard et al., 1977). Different variations regarding particle size, dispersing agent, and manipulation of samples are found in the methods proposed by Dong et al. (1983) and El-Swaify et al. (1970), among others.

6.3.3.3 Bingham Yield Stress

Rheology is the study of the flow and deformation of colloidal systems under the influence of mechanical forces (van Olphen, 1977). A Newtonian liquid when confined between two parallel plates moves at a constant shear rate ($\dot{\gamma}$) which is proportional to the applied shear stress (τ):

$$\tau = \eta\dot{\gamma} \tag{6.43}$$

Non-Newtonian fluids obey different relations between shear stress and shear rate. Plastic flow is flow that occurs only above a certain finite stress (τ_0) called the yield stress. Ideal plastic flow exhibits a

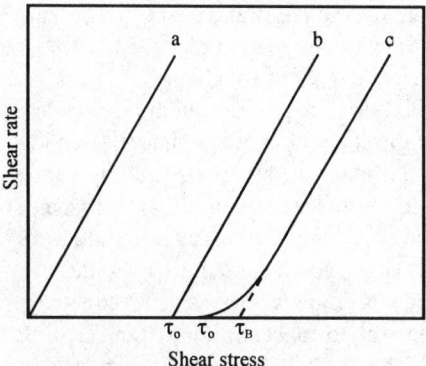

Fig. 6.17 Shear rate versus stress relationships. Curve a represents Newtonian flow, curve b represents ideal plastic flow, and curve c represents Bingham flow [Adapted from van Olphen, 1977]

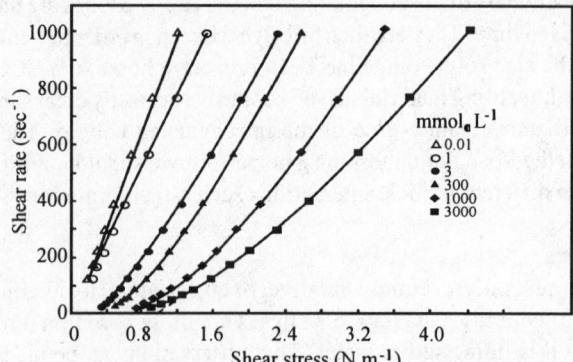

Fig. 6.18 Shear rate versus shear stress for Na-kaolinite. The open symbols represent dispersed systems and ideal plastic flow. The closed symbols represent coagulated systems and Bingham flow [Adapted from Yong et al., 1979]

linear relationship between shear rate and shear stress over all rates of shear. Many colloidal dispersions exhibit Bingham flow which is characterized by the equation:

$$\tau = \tau_B - \eta_\Delta \dot{\gamma} \qquad\qquad [6.44]$$

where τ_B is the Bingham yield stress found by extrapolating Equation [6.43] to zero shear rate and η_Δ is the differential or plastic viscosity. The differential viscosity is the derivative of shear stress with respect to shear rate at a given shear rate (van Olphen, 1977). Shear rate versus shear stress relationships for Newtonian, ideal plastic, and Bingham flow are presented in Fig. 6.17.

Bingham yield stress is a measure of the degree of coagulation of a colloidal suspension and the mode of particle interaction. Bingham yield stress is a function of both the number of particle-particle linkages in the coagulated structure and the energy required to break these linkages (Rand and Melton, 1977). A stable clay dispersion exhibits ideal plastic flow, while flocculated suspensions exhibit Bingham flow as can be seen in Fig. 6.18 for kaolinite. Differential viscosity can also be used to assess the extent of particle flocculation, but is a much less sensitive parameter than Bingham yield stress (Heath and Tadros, 1983).

Rheological studies have been carried out on kaolinite (Flegman et al., 1969; Rand and Melton, 1977; Yong et al., 1979; Diz and Rand, 1989), illite (Yong et al., 1979; Ohtsubo et al., 1991; Hesterberg and Page, 1993), montmorillonite (Rand et al., 1980; Heath and Tadros, 1983; Tombácz et al., 1989; Keren, 1988, 1989a), clay mixtures (Yong et al., 1979; Keren, 1989b, 1991), and soil clays (Zhao et al., 1991). A series of curves of Bingham yield stress as a function of pH obtained at successively increasing electrolyte concentration should coincide at one point. This point is characteristic of the pH value of the edge p.z.c. of the mineral (Rand and Melton, 1977). Edge p.z.c. values have been determined in this manner for various kaolinites and range in value from pH 5.6 to 8.8 (Diz and Rand, 1989). Kaolinite particles occur in edge-face associations below the edge p.z.c. and in edge-edge associations around the edge p.z.c. (Rand and Melton, 1977).

Montmorillonite exhibits no edge-face associations over the pH range 4 to 11; coagulation produced by electrolyte additions is initially the result of edge-edge interactions with face-face interactions occurring at high electrolyte concentrations (Rand et al., 1980). No edge p.z.c. could be determined on montmorillonite. This is likely because the edge area is small and attraction between edges and faces is small compared to repulsion between faces (Rand et al., 1980). In distilled water Ca montmorillonite exhibited Newtonian flow. With increasing ESP, differential viscosity, deviation from Newtonian flow and Bingham yield stress of montmorillonite all increased (Keren, 1988). These increases are likely due to the increased number of particles in solution as tactoids break down. Differential viscosity and Bingham yield stress decreased with increasing electrostatic charge density of smectites (Keren, 1989a), which is attributed to the reduced swelling of higher charge density smectites. Kaolinite exhibited Newtonian flow at all ESPs. The introduction of even 5% montmorillonite into the kaolinite systems resulted in deviations from Newtonian flow and increased differential viscosity (Keren, 1989b). Bingham yield stress and deviations from Newtonian flow of kaolinite-montmorillonite mixtures also increased with ESP (Keren, 1991).

6.4 Colloid Transport

Colloid transport is important in predicting movement of organic chemicals, biological entities such as viruses and bacteria, inorganic species such as heavy metals, and clay movement which affects soil hydraulic properties. The colloids important to transport in soils include clay minerals, Fe, Mn, and Al oxyhydroxides, humic substances, bacteria, and viruses. Understanding colloid transport requires not only knowledge of the chemical and biological processes and reactions but also physical principles of filtration and deposition in porous media. In this section, both movement of the entity of interest as well as transport of a chemical after attachment to a colloid will be discussed.

In addition to the chemical factors affecting colloid stability, transport of colloids is related to soil physical factors such as pore size distribution and continuity. Coarse-textured soils have a larger distribution of pore sizes (with larger pores) than fine-textured soils; suggesting greater potential for colloid movement. This is not generalizable since some fine-textured materials experience cracking and formation of large macropores.

6.4.1 Principles of Colloid Transport

Various factors, both physical and chemical, affect the transport of colloids. Colloid transport can be represented by the physical processes of molecular diffusion (Brownian movement), liquid flow, and gravitational forces. Molecular diffusion, the random motion of particles caused by thermal effects is related to temperature and viscosity. The diffusion coefficient D is described by:

$$D = \frac{kt}{3\pi\eta d_p}$$

[6.45]

where k is the Boltzmann constant, t is time, η is viscosity and d_p is particle diameter.
. Velocity differences within the fluid cause velocity differences in particle transport. These differences can result in interparticle contact. Gravitational forces within the moving fluid cause differential velocities among the particles, as related to their differences in density and size.

Filtration theory has been developed to describe the processes of aggregation and deposition, which can be readily adopted for colloid transport in soils. Aggregation is the term used to describe the attachment of particles which come in contact while moving. Deposition or coagulation is the process whereby moving particles are attached to stationary particles (O'Melia and Tiller, 1993). The rate of aggregation can be represented by:

$$\frac{dn}{dt} = -k_a n^2$$

[6.46]

where n is the concentration of the particles in the fluid and k_a is the rate constant which is a function of the physical and chemical properties of the system. Deposition can be described by the relation

$$\frac{dn}{dL} = -k_a n \qquad [6.47]$$

where L is distance in the flow path (O'Melia and Tiller, 1993). For a monodisperse suspension flowing into a clean filter bed, the rate of particle deposition can be described by:

$$\frac{dn}{dL} = -\frac{3}{2}\alpha_d \eta(p,c)\frac{(1-\theta)}{d_c}n \qquad [6.48]$$

where α_d is the dimensionless sticking coefficient (determined experimentally), $\eta(p,c)$ is the dimensionless collision frequency number for contact between the suspended particles and the filtering medium particles, θ is the porosity, and d_c is the diameter of the filtering media particles (O'Melia and Tiller, 1993). The collision frequency function is taken as the sum of the collisions from molecular diffusion, liquid flow, and gravitational processes.

Numerical treatment of colloid transport has ranged from relatively simple models to complex ones. Harvey and Garbadian (1991) developed the following simple model incorporating reversible and irreversible adsorption, as well as dispersion and advection at constant water content and velocity:

$$\theta\frac{\partial c}{\partial t} + \rho_b\frac{\partial s}{\partial t} = D\theta\frac{\partial^2 c}{\partial x^2} - v\theta\left(\frac{\partial c}{\partial x} + k_p c\right) \qquad [6.49]$$

where θ is the porosity, c is the concentration of the contaminant in solution, s is the concentration of reversibly adsorbed contaminant, D is the dispersion coefficient, v is the water velocity, t is time ρ_b is bulk density, x is the spatial coordinate and k_p is the irreversible adsorption constant.

Corapcioglu and Choi (1996) have recently developed a detailed model for description of colloid transport for unsaturated porous media. In this model, four phases are considered: air, water, solid, and colloids. The model considers both capture/release of the contaminant on the air-water interface as well as on the solid. Previous studies established that the air-water interface can be important in retention of colloids in soils (Wan and Wilson, 1994). Ibaraki and Sudicky (1995) describe a two-dimensional numerical model which includes colloid transport in fractured porous media.

6.4.2 Colloid Mediated Contaminant Transport

Numerous studies have established that organic and inorganic chemicals which are highly sorbed onto soil nonetheless are transported in the subsurface. Significant concentrations of colloids have been found in groundwaters. The major transport mechanism for colloids in groundwaters is considered to be convective diffusion (O'Melia, 1990), while convective transport is the major process for soils. Colloid mediated transport is, of course, dependent on the sorption processes. These processes are generally classified into two groups: equilibrium or instantaneous reactions, and slow or kinetically controlled reactions.

Many contaminants applied to soils were initially considered immobile, and thus, presenting little danger to groundwaters. Subsequently, it has been realized that migration of tightly held contaminants has occurred via colloid transport. Among these are metals, radionuclides, nonpolar organic contaminants, biomolecules such as viruses and bacteria, and macromolecular dissolved organic carbon. Transport of the contaminants should consider partitioning between the aqueous and solid phases and transport via dissolved and colloid movement. Treatment of colloid movement in soils must also consider that an important component of transport is by preferential flow, such as through macropores and cracks in the soil. Some of the general issues concerning contaminant migration by colloid facilitated transport are discussed in McCarthy and Zachara (1989). McDowell-Boyer et al. (1986) also provide a review of studies on contaminant transport.

Kaplan et al. (1993) examined the relation between colloid transport and flow rate for various soils packed into lysimeters. Increased flow rates increased colloid transport for all soils but the extent of increase varied greatly. Mobile colloids generally contained greater concentrations of organic C than did the bulk soil. This may explain the high negative charge of the colloids and their resultant stability in the aqueous phase.

Viruses range in size from 20 to 200 nm (Bitton, 1975). The surface charge of a virus is generated by the protonation/deprotonation of amino acid functional groups, The major charged functional entities are carboxyl, primary amine, and phenolic hydroxyl groups with the reactions given below (Taylor, 1981):

$$
\begin{aligned}
-COOH &\rightleftharpoons -COO^- + H^+ \\
-NR_2H^+ &\rightleftharpoons -NR_2 + H^+ \\
Ph-OH &\rightleftharpoons Ph-O^- + H^+
\end{aligned}
\qquad [6.50]
$$

These surfaces can be treated in a manner similar to variable charge mineral surfaces, discussed earlier. The particle charge can be determined using electrophoresis. Based on size (both of the bacteria and the size definition of a colloid) not all bacteria are strictly colloids, but they are nonetheless included in discussions of colloid transport since some of them are true colloids.

Among the studies on bacteria transport are those of Harvey and Garbadian (1991) who used Equation [6.49] in combination with the colloid filtration model to determine k_p as follows:

$$
k_p = \frac{3}{2}\frac{(1-\theta)}{d}\alpha\eta_e \qquad [6.51]
$$

where α is the collision efficiency factor, d is the diameter of the porous media grains, and η_e is the single collector efficiency. Yates and Yates (1991) provide further details on microbial transport modeling.

Radionuclide transport has been examined by Ryan et al. (1998), among others, who described Pu mobilization and subsurface transport under simulated rainfall conditions and observed the importance of macropore flow. Solution composition is an important factor affecting colloid-mediated transport in soils. Elevated pH and low ionic strength enhanced Ni transport related to colloid mobility despite the fact that elevated pH increased Ni sorption (Roy and Dzombak, 1997). Similarly, Grolimund et al. (1996) demonstrated that Pb^{2+} transport was dominantly associated with colloid mobility, the colloids being mobilized by a decrease in the ionic strength of the infiltrating water. Conditions of low ionic strength, high exchangeable Na and elevated pH all contribute to dispersion of clays and subsequent colloid transport (Kaplan et al., 1996). Prediction of clay colloid mobility is thus related to the CCC value, allowing application of extensive data on the chemistry of dispersion and flocculation to colloid transport.

6.4.3   Effect of Colloid Transport on Hydraulic Conductivity

Reductions in hydraulic conductivity caused by colloid transport can be divided into two groups: formation of crusts and migration and deposition within the media. Crust formation results when the colloids are not able to enter the soil medium and form a compact layer above the soil. The description of this process and the resultant flow under saturated conditions can be treated using the filtration technology developed in the engineering literature. This process is called surface filtration (Herzig et al.,1970). Alternatively, the particles may enter the medium, flow with the water in a suspension, and be deposited within the medium. This process may result from either mechanical filtration of large colloids or physicochemical processes of attraction/repulsion of small colloids. Numerical description of the process under saturated flow, termed deep bed filtration (Herzig et al., 1970) is often made with a three-parameter model. The filter coefficient and the filtration efficiency parameters control the rate of deposition of the particles and the extent of removal of the particles from the suspension, respectively, while the flow restriction parameter is used to simulate the reduction in permeability (Ibaraki and Sudicky, 1995).

Research in soil science related to the effects of colloid transport on hydraulic conductivity has focused primarily on description of the chemical process affecting clay movement, rather than on mathematical expressions for prediction of flow rates. Conditions of low ionic strength, high exchangeable Na and elevated pH were related to both dispersion of clays (Goldberg and Forster, 1990; Miller et al., 1990; and Suarez et al., 1984) and subsequent migration and reduction in soil hydraulic conductivity (Suarez et al., 1984). In highly weathered acid soils where positively charged colloids are important, Seaman et al. (1995) determined that addition of $CaCl_2$ initially enhanced colloid mobility likely due to release of exchangeable Al and a decrease in pH.

6.4.4   Effect of Colloid Transport on Soil Formation

Soil profile development is often affected by colloid movement. Among these processes are the formation of impermeable clay layers in the subsurface and movement of Fe and Al, most probably as organic metal complexes. Moderately to strongly developed soils are generally characterized by depleted clay contents in the A horizon and larger amounts of clay generally in the upper portion of the B horizon. A substantial portion of the argillic B horizon is related to migration of clays from the upper portion of the profile and subsequent deposition (Birkeland, 1974). Flocculation in the lower part of the profile is enhanced by increased electrolyte concentration relative to the surface horizons. Clay colloid migration is evident by the presence of clay films over ped surfaces and inside voids and the presence of oriented clay particles. This process is also observed in Aridisols in which low organic matter and elevated exchangeable Na enhance colloid transport. Clay deposition is enhanced by increasing electrolyte concentration and removal of water by evapotranspiration. In general, formation of argillic horizons requires that the soil be at least partially dry during some part of the season.

Formation of Fe- and Al-rich layers by translocation (podzolization) is probably caused by transport of metal-fulvic acid complexes, rather than movement of the dissolved metals. The spodic B horizon marks the location at which these chelates either flocculate due to increases in electrolyte concentration or by decomposition of the organic matter (Birkeland, 1974).

6.5   References

Adamson, A.W. 1976. Physical chemistry of surfaces. 3rd ed. John Wiley and Sons, New York, NY.

Babcock, K.L. 1963. Theory of chemical properties of soil colloidal systems at equilibrium. Hilgardia 34:417–542.

Bar-On, P., I. Shainberg, and I. Michaeli. 1970. Electrophoretic mobility of montmorillonite particles saturated with Na/Ca ions. J. Coll. Interf. Sci. 33:471–472.

Beckett, R., and B.T. Hart. 1993. Use of field-flow fractionation techniques to characterize aquatic particles, colloids and macromolecules. p. 165–205. *In* J. Buffle and H.P. van Leeuwen (ed). Environmental Particles, Vol. 2. Lewis Publishers, Boca Raton, FL.

Beckett, R., D. Murphy, S. Tadjiki, D.J. Chittleborough, and J.C. Giddings. 1997. Determination of thickness, aspect ratio and size distribution for platey particles using sedimentation field-flow fractionation and electron microscopy. Coll. Surf. A 120:17–26.

Birkeland, P.W. 1974. Pedology, weathering, and geomorphological research. Oxford University Press, New York, NY.

Bitton, G. 1975. Adsorption of viruses onto surfaces in soil and water. Water Res. 9:473–484.

Bohn, H.L., B.L. McNeal, and G.A. O'Connor. 1985. Soil chemistry. John Wiley and Sons, New York, NY.

Borchardt, G.A. 1977. Montmorillonite and other smectite minerals. p. 293–330. *In* J.B. Dixon and S.B. Weed (ed.) Minerals in soil environments, Soil Science Society of America, Madison, WI.

Borkovec, M., Q. Wu, G. Degovics, P. Laggner, and H. Sticher. 1993. Surface area and size distribution of soil particles. Coll. Surf. A 73:65–76.

Bower, C.A., and J.O. Goertzen. 1959. Surface area of soils by an equilibrium ethylene glycol method. Soil Sci. 87:289–292.

Brown, G., A.C.D. Newman, J.H. Rayner, and A.H. Weir. 1978. The structure and chemistry of soil clay minerals. p. 29–178. *In* D.G. Greenland and M.H.B. Hayes (ed.) The chemistry of soil constituents, John Wiley and Sons, New York, NY.

Brunauer, S., P.H. Emmett, and E. Teller. 1938. Adsorption of gases in multimolecular layers. J. Am. Chem. Soc. 60:309–319.

Buffle, J., and G.G. Leppard. 1995a. Characterization of aquatic colloids and macromolecules. 1. Structure and behavior of colloidal materials. Environ. Sci. Technol. 29:2169–2175.

Buffle, J., and G.G. Leppard. 1995b. Characterization of aquatic colloids and macromolecules. 2. Key role of physical structures on analytical results. Environ. Sci. Technol. 29:2176–2184.

Bunville, L.G. 1984. Commercial instrumentation for particle size analysis. p. 1–42. *In* H.G. Barth (ed.) Modern methods of particle size analysis. John Wiley and Sons, New York, NY.

Carter, D.L., M.M. Mortland, and W.D. Kemper. 1986. Specific surface. p. 413–423. *In* A. Klute (ed.) Methods of soil analysis. Part 1. Physical and mineralogical methods. 2nd. Ed. Soil Science Society of America, Madison, WI.

Chorom, M., and P. Rengasamy. 1995. Dispersion and zeta potential of pure clays as related to net particle charge under varying pH, electrolyte concentration and cation type. Europ. J. Soil Sci. 46:657–665.

Coll, H., and L.E. Oppenheimer. 1987. Improved techniques in disc centrifugation. p. 202–214. *In* T. Provder (ed.) Particle size distribution. Assessment and characterization. American Chemical Society, Washington, DC.

Corapcioglu, M.Y., and H. Choi. 1996. Modeling colloid transport in unsaturated porous media and validation with laboratory column data. Water Resour. Res. 32:3437–3449.

Cotton, A., and H. Mouton. 1907. Magneto-optical properties of colloids and heterogeneous liquids. Ann. Chim. Phys. 11:145–203, 289–339.

Davis, J.A., and J.D. Hem. 1989. The surface chemistry of aluminum oxides and hydroxides. p. 185–219. *In* G. Sposito (ed.) The environmental chemistry of aluminum. CRC Press, Boca Raton, FL.

Davis, J.A., and D.B. Kent. 1990. Surface complexation modeling in aqueous geochemistry. Rev. Mineral. 23:177–260.

Dixon, J.B. 1977. Kaolinite and serpentine group minerals. p. 357–403. *In* J.B. Dixon and S.B. Weed (ed.) Minerals in soil environments. Soil Science Society of America, Madison, WI.

Dixon, J.B., and S.B. Weed. 1989. Minerals in soil environments. Soil Science Society of America, Madison, WI.

Diz, H.M.M., and B. Rand. 1989. The variable nature of the isoelectric point of the edge surface of kaolinite. British Ceram. Trans. J. 88:162–166.

Dong, A., G. Chesters, and G.V. Simsiman. 1983. Soil dispersibility. Soil Sci. 136:208–212.

Douchafour, P. 1970. Precis de pedologie. Masson et Cie, Paris, France.

Douglas, L.A. 1977. Vermiculites. p. 259–292. *In* J.B. Dixon and S.B. Weed (ed.) Minerals in soil environments. Soil Science Society of America, Madison, WI.

Ducker, W.A., and R.M. Pashley. 1992. Forces between mica surfaces in the presence of rod-shaped divalent counterions. Langmuir 8:109–112.

Dyal, R.S., and S.B. Hendricks. 1950. Total surface of clays in polar liquids as a characteristic index. Soil Sci. 69:421–432.

Dzombak, D.A., and F.M.M. Morel. 1990. Surface complexation modeling: Hydrous ferric oxide. John Wiley and Sons, New York, NY.

Edwards, D.G., A.M. Posner, and J.P. Quirk. 1965a. Repulsion of chloride ions by negatively charged clay surfaces. Part 1. Monovalent cation Fithian illites. Trans. Farad. Soc. 61:2808–2815.

Edwards, D.G., A.M. Posner, and J.P. Quirk. 1965b. Repulsion of chloride ions by negatively charged clay surfaces. Part 2. Monovalent montmorillonites. Trans. Farad. Soc. 61:2816–2819.

El-Swaify, S.A., S. Ahmed, and L.D. Swindale. 1970. Effects of adsorbed cations on physical properties of tropical red and tropical black earths. II. Liquid limit, degree of dispersion, and moisture retention. J. Soil Sci. 21:188–198.

Emerson, W.W. 1967. A classification of soil aggregates based on their coherence in water. Aust. J. Soil Res. 15:255–262.

Fanning, D.S., and V.Z. Keramidas. 1977. Micas. p. 195–258. *In* J.B. Dixon and S.B. Weed (ed.) Minerals in soil environments. Soil Science Society of America, Madison, WI.

Farmer, V.C. 1978. Water on particle surfaces. p. 405–448. *In* D.J. Greenland and M.H.B. Hayes (ed.) The chemistry of soil constituents. John Wiley and Sons, New York, NY.

Filella, M., and J. Buffle. 1993. Factors controlling the stability of submicron colloids in natural waters. Coll. Surf. A 73:255–273.

Filella, M., J. Zhang, M.E. Newman, and J. Buffle. 1997. Analytical applications of photoncorrelation spectroscopy for size distribution measurements of natural colloidal suspensions: capabilities and limitations. Coll. Surf. A. 120:27–46.

Flegman, A.W., J.W. Goodwin, and R.H. Ottewill. 1969. Rheological studies on kaolinite suspensions. Proc. British Ceram. Soc. 13:31–45.

Frenkel, H., G.J. Levy, and M.V. Fey. 1992. Clay dispersion and hydraulic conductivity of clay-sand mixtures as affected by the addition of various anions. Clays Clay Miner. 40:515–521.

Freundlich, H. 1932. Kapillarchemie II. Eine Darstellung der Chemie der Kolloide und verwandter Gebiete. 4. Akademische Verlagsgesellschaft. Leipzig, Germany.

Giddings, J.C., K.D. Cadwell, and H.K. Jones. 1987. Measuring particle size distribution of simple and complex colloids using sedimentation field-flow fractionation. p. 215–230. *In* T. Provder (ed.) Particle size distribution. Assessment and characterization. American Chemical Society, Washington, DC.

Goldberg, S., and H.S. Forster. 1990. Flocculation of reference clays and arid-zone soil clays. Soil Sci. Soc. Am. J. 54:714–718.

Goldberg, S., H.S. Forster, and C.L. Godfrey. 1996. Molybdenum adsorption on oxides, clay minerals, and soils. Soil Sci. Soc. Am. J. 60:425–432.

Greathouse, J.A., S.E. Feller, and D.A. McQuarrie. 1994. The modified Gouy-Chapman theory: Comparisons between electrical double layer models of clay swelling. Langmuir 10:2125–2130.

Gregg, S.J., and K.S. W. Sing. 1982. Adsorption, surface area and porosity. Academic Press, New York, NY.

Grolimund, D., M. Borkovec, K. Bartmettler, and H. Sticher. 1996. Colloid-facilitated transport of strongly sorbing contaminants in natural porous media: A laboratory column study. Environ. Sci. Technol. 30:3118–3123.

Groves, M.J. 1984. The application of particle characterization methods to submicron dispersion and emulsions. p. 43–91. *In* H.G. Barth (ed.) Modern methods of particle size analysis. John Wiley and Sons, New York, NY.

Gu, B., and H.E. Doner. 1993. Dispersion and aggregation of soils as influenced by organic and inorganic polymers. Soil Sci. Soc. Am. J. 57:709–716.

Harvey, R.W., and S.P. Garbadian. 1991. Use of colloid filtration theory in modeling movement of bacteria through a contaminated sandy aquifer. Environ. Sci. Technol. 25:178–185.

Hayes, M.H.B., and R.S. Swift. 1978. The chemistry of soil organic colloids. p. 179–320. *In* D.G. Greenland and M.H.B. Hayes (eds.) The chemistry of soil constituents. John Wiley and Sons, New York, NY.

Heath, D., and Th.F. Tadros. 1983. Influence of pH, electrolyte and poly(vinyl alcohol) addition on the rheological characteristics of aqueous dispersions of sodium montmorillonite. J. Coll. Interf. Sci. 93:307–319.

Herzig, J.P., D.M. Leclerc, and P. Le Groff. 1970. Flow of suspensions through porous media: Applications to deep filtration. Ind. Eng. Chem. 62:8–35.

Hesterberg, D.L. 1988. Critical coagulation concentrations and rheological properties of illite. Ph.D. Thesis. University of California, Riverside CA.

Hesterberg, D., and A.L. Page. 1990a. Flocculation series test yielding time-invariant critical coagulation concentrations of sodium illite. Soil Sci. Soc. Am. J. 54:729–735.

Hesterberg, D., and A.L. Page. 1990b. Critical coagulation concentration of sodium and potassium illite as affected by pH. Soil Sci. Soc. Am. J. 54:735–739.

Hesterberg, D., and A.L. Page. 1993. Rheology of sodium and potassium illite suspensions in relation to colloidal stability. Soil Sci. Soc. Am. J. 57:697–704.

Hiemenz, P.C. 1977. Principles of colloid and surface chemistry. Marcel Dekker, Inc., New York, NY.

Holsworth, R.M., T. Provder, and J.J. Stansbrey. 1987. External-gradient-formation method for disc centrifuge photosedimentometric particle size distribution analysis. p. 191–201. *In* T. Provder (ed.) Particle size distribution. Assessment and characterization. American Chemical Society, Washington, DC.

Hsu, P.H. 1977. Aluminum hydroxides and oxyhydroxides. p. 99–143. *In* J.B. Dixon and S.B. Weed (ed.) Minerals in soil environments. Soil Science Society of America, Madison, WI.

Huang, C.P. 1981. The surface acidity of hydrous solids. p. 183–217. *In* M.A. Anderson and A.J. Rubin (ed.) Adsorption of inorganics at solid-liquid interfaces. Ann Arbor Science, Ann Arbor, MI.

Hunter, R.J. 1981. Zeta potential in colloid science. Academic Press, London, UK.

Hunter, R.J. 1987. Foundations of colloid science. Vol. 1. Oxford University Press, New York, NY.

Hunter, R.J. 1989. Foundations of colloid science. Vol. 2. Oxford University Press, New York, NY.

Hunter, R.J. 1993. Introduction to modern colloid science. Oxford Science Publications, Oxford, UK.

Hutton, J.T. 1977. Titanium and zirconium minerals. p. 673–688. *In* J.B. Dixon and S.B. Weed (ed.) Minerals in soil environments. Soil Science Society of America, Madison, WI.

Ibaraki, M., and E.A. Sudicky. 1995. Colloid-facilitated contaminant transport in discretely fractured porous media. I. Numerical formulation and sensitivity analysis. Water Resour. Res. 31:2945–2960.

Israelachvili, J.N. 1992. Intermolecular and surface forces. 2nd Ed. Academic Press, London, UK.

Jackson, M.L. 1979. Soil chemical analysis: Advanced course. 2nd Ed. Published by the author, Madison, WI.

James, R.O., and G.A. Parks. 1982. Characterization of aqueous colloids by their electrical double-layer and intrinsic surface chemical properties. Surf. Coll. Sci. 12:119–126.

Kaplan, D.I., P.M. Bertsch, D.C. Adriano, and W.P. Miller. 1993. Soil-borne colloids as influenced by water flow and organic carbon. Environ. Sci. Technol. 27:1193–1200.

Kaplan, D.I., M.E. Sumner, P.M. Bertsch, and D.C. Adriano. 1996. Chemical conditions conducive to the release of mobile colloids from Ultisol profiles. Soil Sci. Soc. Am. J. 60:269–274.

Kavanaugh, M.C., and J.O. Leckie. 1980. Particulates in water. American Chemical Society, Washington, DC.

Kavanaugh, M.C., C.H.Tate, A.R. Trussell, R. R. Trussell, and G. Treweek. 1980. Use of particle size distribution measurements for selection and control of solid/liquid separation processes. p. 305–328. *In* M.C. Kavanaugh and J.O. Leckie (ed.) Particulates in water. American Chemical Society, Washington, DC.

Kékicheff, P., S. Marcelja, T.J. Senden, and V.E. Shubin. 1993. Charge reversal seen in electrical double layer interaction of surfaces immersed in 2:1 calcium electrolyte. J. Chem. Phys. 99:6098–6113.

Keren, R. 1988. Rheology of aqueous suspension of sodium/calcium montmorillonite. Soil Sci. Soc. Am. J. 52:924–928.

Keren, R. 1989a. Effect of clay charge density and adsorbed ions on the rheology of montmorillonite suspension. Soil Sci. Soc. Am. J. 53:25–29.

Keren, R. 1989b. Rheology of mixed kaolinite-montmorillonite suspensions. Soil Sci. Soc. Am. J. 53:725–730.

Keren, R. 1991. Adsorbed sodium fraction's effect on rheology of montmorillonite-kaolinite suspensions. Soil Sci. Soc. Am. J. 55:376–379.

Koehler, M.E., R.A. Zandler, T. Gill, T. Provder, and T.F. Niemann. 1987. An improved disc centrifuge photosedimentometer and data system for particle size distribution analysis. p. 180–190. *In* T. Provder (ed.) Particle size distribution. Assessment and characterization. American Chemical Society, Washington, DC.

Kretzschmar, R., W.P. Robarge, and S.B. Weed. 1993. Flocculation of kaolinitic soil clays: Effects of humic substances and iron oxides. Soil Sci. Soc. Am. J. 57:1277–1283.

Kretzschmar, R, D. Hesterberg, and H. Sticher. 1997. Effects of adsorbed humic acid on surface charge and flocculation of kaolinite. Soil Sci. Soc. Am. J. 61:101–108.

Langmuir, I. 1918. The adsorption of gases on plane surfaces of glass, mica, and platinum. J. Am. Chem. Soc. 40:1361–1402.

Lebron I., and D.L. Suarez. 1992a. Electrophoretic mobility of illite and micaceous soil clays. Soil Sci. Soc. Am. J. 56:1106–1115.

Lebron I., and D.L. Suarez. 1992b. Variations in soil stability within and among soil types. Soil Sci. Soc. Am. J. 56:1412–1421.

Lebron, I., D.L. Suarez, C. Amrhein, and J.E. Strong. 1993. Size of mica domains and distribution of the adsorbed Na-Ca ions. Clays Clay Miner. 41:380–388.

Ledin, A., S. Karlsson, A. Duker, and B. Allard. 1993. Applicability of photon correlation spectroscopy for measurement of concentration and size distribution of colloids in natural waters. Anal. Chem. Acta 281:421–428.

Lowell, S. 1979. Introduction to powder surface area. John Wiley and Sons, New York, NY.

Maes, A., and A. Cremers. 1977. Charge density effects in ion exchange. Part 1. Heterovalent exchange equilibria. J. Chem. Soc. Faraday Trans. I 73:1807–1814.

McBride, M.B. 1989. Surface chemistry of soil minerals. p. 35–88. In J.B. Dixon and S.B. Weed (ed.) Minerals in soil environments. 2nd Ed., Soil Science Society of America, Madison, WI.

McBride, M.B. 1994. Environmental chemistry of soils. Oxford University Press, New York, NY.

McCarthy, J.E., and J.M. Zachara. 1989. Subsurface transport of contaminants. Environ. Sci. Technol. 23:496–502.

McDowell-Boyer, L.M., J.R. Hunt, and N. Sitar. 1986. Particle transport through porous media. Water Resour. Res. 22:1901–1921.

McKenzie, R.M. 1989. Manganese oxides and hydroxides. p. 439–465. In J.B. Dixon and S.B. Weed (ed.) Minerals in soil environments. 2nd Ed. Soil Science Society of America, Madison, WI.

Miller, W.P., H. Frenkel, and K.D. Newman. 1990. Flocculation concentration and sodium/calcium exchange of kaolinitic soil clays. Soil Sci. Soc. Am. J. 54:346–351.

Napper, D.H., and R.J. Hunter. 1974. Hydrosols. p.161–213. In M. Kerker (ed.) M.T.P. Int. Review of Science Phys. Chem. Series. Butterworths, London, UK.

Nelson, P.N., and J.M. Oades. 1998. Organic matter, sodicity and soil structure. p. 67–84. In Sodic Soils. M.E. Sumner and R. Naidu (ed.) Oxford University Press, New York, NY.

Nishimura, S., H. Tateyama, K. Tsunematsu, and K. Jinnai. 1992. Zeta potential measurement of muscovite mica basal plane-aqueous solution interface by means of plane interface technique. J. Coll. Interf. Sci. 152:359–367.

Ohtsubo, M., A. Yoshimura, S.-I. Wada, and R.N. Yong. 1991. Particle interaction and rheology of illite-iron oxide complexes. Clays Clay Miner. 39:347–354.

O'Melia, C.R. 1990. Kinetics of colloid chemical processes in aquatic systems. p. 447–422. In W. Stumm (ed.) Aquatic chemical kinetics. Reaction rates of processes in natural waters. Wiley Interscience, New York, NY.

O'Melia, C.R, and C.L. Tiller. 1993. Physicochemical aggregation and deposition in aquatic environments. p. 353–385. In J. Buffle and H. van Leeuwen (ed.) Environmental particles. Vol. 2. Lewis Publishers, Boca Raton, FL.

Norrish, K., and J.A. Rausell-Colom. 1961. Low-angle X-ray diffraction studies of the swelling of montmorillonite and vermiculite. Clays Clay Min. 10:123–149.

Overbeek, J.Th.G. 1952. Stability of hydrophobic colloids and emulsions. Coll. Sci. 1:302–341.

Parker, J.C., L.W. Zelazny, S. Sampath, and W.G. Harris. 1979. A critical evaluation of the extension of zero point of charge (ZPC) theory to soil systems. Soil Sci. Soc. Am. J. 43:668–674.

Pashley, R.M. 1981. DLVO and hydration forces between mica surfaces in Li^+, Na^+, K^+, and Cs^+ electrolyte solutions: A correlation of double-layer and hydration forces with surface cation. J. Coll. Interf. Sci. 83:531–546.

Pecora, R. 1983. Quasi-elastic light scattering of macromolecules and particles in solution and suspension. p. 3–30. In B.E. Dahneke (ed.) Measurement of suspended particles by quasi-elastic light scattering. John Wiley and Sons, New York, NY.

Pham, P.T., and G.W. Brindley. 1970. Methylene blue adsorption by clay minerals: Determination of surface areas and cation exchange capacities. Clays Clay Miner. 18:203–212.

Quirk, J.P. 1994. Interparticle forces: A basis for the interpretation of soil physical behavior. Adv. Agron. 53:121–183.

Quirk, J.P., and L.A.G. Aylmore. 1971. Domain and quasi-crystalline regions in clay systems. Soil Sci. Soc. Am. Proc. 35:652–654.

Quirk, J.P., and R.K. Schofield. 1955. The effect of electrolyte concentration on soil permeability. J. Soil Sci. 6:163–178.

Rand, B., and I.E. Melton. 1977. Particle interactions in aqueous kaolinite suspensions. I. Effect of pH and electrolyte upon the mode of particle interaction in homoionic sodium kaolinite suspensions. J. Coll. Interf. Sci. 60:308–320.

Rand, B., E. Pekenc, J.W. Goodwin, and R.W. Smith. 1980. Investigation into the existence of edge-face coagulated structures in Na-montmorillonite suspensions. J. Chem. Soc. Farad. I. 76:225–235.

Ross, G. 1978. Relationships of specific surface area and clay content to shrink-swell potential of soils having different clay mineralogical compositions. Can. J. Soil Sci. 58:159–166.

Roy, S.B., and D.A. Dzombak. 1997. Chemical factors influencing colloid-facilitated transport of contaminants in porous media. Environ. Sci. Technol. 37:656–664.

Ryan, J.N., T.H. lllangasekare, M.I. Litaor, and R. Shannon. 1998. Particle and plutonium mobilization in macroporous soils during rainfall simulations. Environ. Sci. Technol. 32:476–482.

Scales, P.J., F. Grieser, and T.W. Healy. 1990. Electrokinetics of the muscovite mica-aqueous solution interface. Langmuir 6:582–589.

Schofield, R. K. 1949. Calculation of surface areas of clays from measurements of negative adsorption. Trans. British Ceramic Soc. 48:207–213.

Schofield, R.K., and H.R. Samson. 1954. Flocculation of kaolinite due to the attraction of oppositely charged crystal faces. Disc. Farad. Soc. 18:135–145.

Schramm L.L., and J.C.T. Kwak. 1982. Influence of exchangeable cation composition on the size and shape of montmorillonite particles in dilute suspensions. Clays Clay Miner. 30:40–48.

Schulthess, C.P., and D.L. Sparks. 1986. Backtitration technique for proton isotherm modeling of oxide surfaces. Soil Sci. Soc. Am. J. 50:1406–1411.

Schulze, D.G. 1989. An introduction to soil mineralogy. p. 1–34. *In* J.B. Dixon and S.B. Weed (ed.). Minerals in soil environments. 2nd Ed. Soil Science Society of America, Madison, WI.

Schurtenberger, P., and M.E. Newman. 1993. Characterization of biological and environmental particles using static and dynamic light scattering. p. 37–115. *In* J. Buffle and H.P. van Leeuwen (ed.). Environmental particles, Vol. 2. Lewis Publishers, Boca Raton, FL.

Schwertmann, U., and R.M. Taylor. 1977. Iron oxides. p. 145–180. *In* J.B. Dixon and S.B. Weed (ed.) Minerals in soil environments. Soil Science Society of America, Madison, WI.

Seaman, J.C., P.M. Bertsch, and W.P. Miller. 1995. Chemical controls on colloid generation and transport in a sandy aquifer. Environ. Sci. Technol. 29:1808–1815.

Secor, R.B., and C.J. Radke. 1985. Spillover of the diffuse double layer on montmorillonite particles. J. Coll. Interf. Sci. 103:237–244.

Shainberg, I., and H. Otoh. 1968. Size and shape of montmorillonite particles saturated with Na/Ca ions (inferred from viscosity and optical measurements). Israel J. Chem. 6:251–259.

Shapiro, A.P., and R.F. Probstein. 1993. Removal of contaminants from saturated clay by electroosmosis. Environ. Sci. technol. 27:283–291.

Sherard, J.L., L.P. Dunningan, and R.S. Decker. 1977. Identification and nature of dispersive soils. J. Geotech. Eng., Am. Soc. Chem. Eng. 4:287–301.

Sposito, G. 1984. The surface chemistry of soils. Oxford University Press, New York, NY.

Stevenson, F.J. 1982. Humus chemistry. Genesis, composition, reactions. John Wiley and Sons, New York, NY.

Stumm, W., and J.J. Morgan. 1996. Aquatic chemistry. Chemical equilibria and rates in natural waters. John Wiley and Sons, New York, NY.

Suarez, D.L., J.D. Rhoades, R. Lavado, and C.M. Grieve. 1984. Effect of pH on saturated hydraulic conductivity and soil dispersion. Soil Sci. Soc. Am. J. 48:50–55.

Supak, J.R., A.R. Swoboda, and J.B. Dixon. 1978. Adsorption of aldicarb by clays and soil organo-clay complexes. Soil Sci. Soc. Am. J. 42:244–248.

Tarchitzky, J., Y. Chen, and A. Banin. 1993. Humic substances and pH effects on sodium and calcium-montmorillonite flocculation and dispersion. Soil Sci. Soc. Am J. 57:367–372.

Taylor, D. 1981. Interpretation of the adsorption of viruses by clays from their electrokinetic properties. p. 595–612. *In* J.U. Cooper (ed.) Chemistry in water reuse. Ann Arbor Science, Ann Arbor, MI.

Tombácz, E., J. Balázs, J. Lakatos, and F. Szántó. 1989. Influence of the exchangeable cations on stability and rheological properties of montmorillonite suspensions. Colloid Polym. Sci. 267:1016–1025.

van Olphen, H. 1977. An introduction to clay colloid chemistry. 2nd Ed. John Wiley and Sons, New York, NY.

Verwey, E.J.W., and J.Th.G. Overbeek. 1948. Theory of stability of lyophobic colloids. Elsevier, Amsterdam, Netherlands.

von Smoluchowski, M. 1916. Three discourses on diffusion, Brownian movements, and the coagulation of colloid particles. Physik. Z.17:557–571; 585–599.

von Smoluchowski, M. 1917. Mathmetical theory of the kinetics of the coagulation of colloidal solutions. Z. Physik. Chem. 92:129–168.

Wan, J., and J.L. Wilson. 1994. Colloid transport in unsaturated porous media. Water Resour. Res. 30:857–864.

Wilding, L.P., N.E. Smeck, and L.R. Drees. 1977. Silica in soils: Quartz, cristobalite, tridymite, and opal. p. 471–552. *In* J.B. Dixon and S.B. Weed (ed.) Minerals in soil environments. Soil Science Society of America, Madison, WI.

Yates, M.V. and S.R. Yates. 1991. Modeling microbial transport in the subsurface: A mathematical discussion. p. 48–76. *In* C.J. Hurst (ed.). Modeling the environmental fate of microorganisms. American Society for Microbiology, Washington, DC.

Yong, R.N., A.J. Sadh, H.P. Ludwig, and M.A. Jorgensen. 1979. Interparticle action and rheology of dispersive clays. J. Geotech. Eng. Div. 105:1193-1209.

Zhao, H., P.F. Low, and J.M. Bradford. 1991. Effects of pH and electrolyte concentration on particle interaction in three homoionic sodium soil clay suspensions. Soil Sci. 151:196–207.

7

Ion Exchange Phenomena

Garrison Sposito
University of California at Berkeley

7.1 Origin of Surface Charge

The surfaces of the solid particles in soils, irrespective of their chemical composition, can develop electrical charge in two principal ways: either permanently, from isomorphic substitutions of component ions in the bulk structure of a solid, or conditionally, from the reactions of surface functional groups on a solid with adsorptive ions in aqueous solution. [A surface functional group is a chemically reactive molecular unit bound into the structure of a solid adsorbent at its periphery, such that the reactive portion of the functional group is exposed to an aqueous solution contacting the adsorbent (Section B, Chapter 8)]. Surface functional groups occur on both organic and inorganic adsorbents (e.g., surface hydroxyls occur on both humus and metal oxide minerals). After reaction with an adsorptive ion in aqueous solution (which then becomes an adsorbate), they can form adsorption complexes, which are defined as immobilized molecular entities comprising the adsorbate and the surface functional group to which it is bound (Everett, 1972). A further classification of adsorption complexes can then be made as either inner or outer-sphere surface complexes (Sposito, 1981a). An inner-sphere surface complex has no water molecule interposed between the surface functional group and the small ion or molecule it binds, whereas an outer-sphere surface complex has at least one such interposed water molecule. Outer-sphere surface complexes thus comprise solvated adsorbate ions or molecules. These fundamental concepts and their accepted definitions are discussed further by Everett (1972) and Sposito (1992a).

Surface complexes involving metal cations are illustrated in Fig. 7.1 for a 2:1 layer type clay mineral, such as montmorillonite (Section B, Chapter 8). Similar illustrations of bivalent cations bound in surface complexes on metal oxides are given by Bargar et al. (1996, 1997). (For a description of the molecular structures and surface characteristics of clay minerals and metal oxides, see Section F, Chapters 2 and 3.) In the examples shown in Fig. 7.1, the surface functional group is the siloxane ditrigonal cavity formed by a ring of oxygen ions in six corner-sharing silica tetrahedra in the layer structure. This cavity has a diameter of about 0.26 nm and can mediate negative charge resulting from isomorphic cation substitutions in the clay layer (e.g., Al^{3+} for Si^{4+} in tetrahedral coordination, or Mg^{2+} for Al^{3+} in octahedral coordination). When charged, this cavity then can form inner-sphere or outer-sphere surface complexes with aqueous cations, as shown in Fig. 7.1.

MODES OF CATION ADSORPTION
BY 2:1 LAYER TYPE CLAY MINERALS

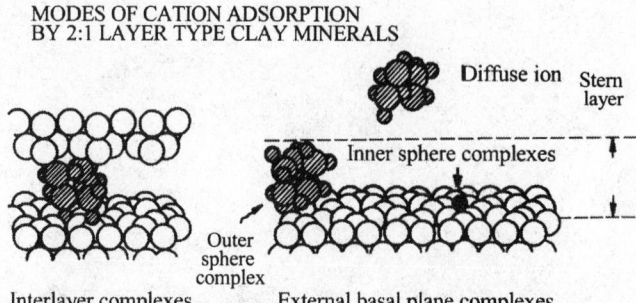

Fig. 7.1   Adsorbed metal cation species on a 2:1 layer type clay mineral [From Sposito, 1992a. Reprinted from J. Buffle and H.P. van Leeuwen (ed.) Environmental particles, with permission of Lewis Publishers, Chelsea, MI]

Ions in surface complexes are to be distinguished from those adsorbed in the diffuse layer (Fig. 7.1), because the former species remain immobilized on an adsorbent surface over molecular time scales that are long when compared, e.g., with the 4–10 ps required for a diffusive step by a solvated ion in aqueous solution (Ohtaki and Radnai, 1993). For example, the well-known outer-sphere surface complex formed by bivalent metal cations in the interlayer of montmorillonite (cf. the left side of Fig. 7.1) is immobile on the molecular time scale of about 100 ps that is probed by electron spin resonance spectroscopy (Sposito and Prost, 1982). This clear separation of residence time scales will not always be as sharp for outer-sphere surface complexes versus diffuse layer ions. Moreover, if the diffuse layer ions reside in very close proximity to the adsorbent surface [i.e., counterion condensation occurs (Sposito, 1992b)], then it becomes more difficult to contrast their behavior with that of ions in outer-sphere surface complexes, since both of these species are solvated ions.

The three types of surface species (inner-sphere complex, outer-sphere complex, and diffuse layer) represent three modes of adsorption of aqueous ions that contribute to the formation of the electrochemical double layer on charged particle surfaces (Everett, 1972). No inference of special planes containing adsorbed ions is required by these chemical speciation concepts, nor is any detailed molecular structure implied, other than the notions of surface complexes and dissociated ions. It is sometimes convenient to group all surface complexes into a Stern layer (Everett, 1972) in order to distinguish them from diffuse layer ions (Fig. 7.1). This purely geometric distinction among surface species, however, does not mean that diffuse layer ions necessarily approach an adsorbent surface less closely than do the Stern layer ions (Sposito, 1992b).

With these basic notions of the structure of the electrochemical double layer in mind, it is possible to define the five components of surface charge in soils (Sposito, 1992a). These five components are summarized in Table 7.1.

The net structural surface charge density (σ_o in Coul m^{-2}) is created by isomorphic substitutions in both primary and secondary minerals, but they produce significant surface charge only in the 2:1 layer type clay minerals. For these minerals, $\sigma_o < 0$ invariably because of cation substitutions (Section F, Chapter 2). The net proton surface charge density (σ_H in Coul m^{-2}) is proportional to the difference between the moles of protons and the moles of hydroxide ions complexed by surface functional groups:

$$\sigma_H = F \frac{(q_H - q_{OH})}{a_s}$$

[7.1]

Table 7.1 Components of surface charge

Symbol	Name	Source
σ_0	net structural surface charge density	isomorphic substitutions
σ_H	net proton surface charge density	H^+/OH^- surface complexation
σ_{IS}	inner-sphere complex surface charge density	inner-sphere surface complexes of ions
σ_{OS}	outer-sphere complex surface charge density	outer-sphere surface somplexes of ions
σ_d	diffuse layer surface charge density	diffuse layer adsorption of ions

where q_i is the specific adsorbed ion charge ($mol_c\ kg^{-1}$) of ion i complexed by surface groups (i.e., q_i is the product of the absolute value of the valence of ion i and Γ_i the part of its specific surface excess [Γ_i, in $mol\ kg^{-1}$] that can be attributed to adsorption complexes), F is the Faraday constant, and a_s is specific surface area ($m^2\ kg^{-1}$). Conceptually, diffuse layer protons are not included in the definition of σ_H. The most important surface functional groups that complex protons are the hydroxyl groups found on soil humus, metal oxides, and 1:1 layer type clay minerals (e.g., kaolinite). The values of σ_H can be negative, zero, or positive, depending on pH, ionic strength, etc. Note that q_{OH} is quantitatively the same as the quantity of protons dissociated from acidic surface functional groups ($mol_c\ kg^{-1}$).

The specific surface excess that appears implicitly in Equation [7.1] is defined conceptually in terms of two gedanken laboratory operations (Sposito, 1989): (1) reaction of a soil with an aqueous solution of known composition at fixed temperature and applied pressure for a prescribed period of time, and (2) chemical analysis of the reacted soil, the soil solution, or both, to determine their composition. The reaction in step 1 can take place either in a closed system, with the reactant solution and soil particles mixed uniformly (batch reactor), or in an open system, with the reactant solution in uniform motion relative to a column of soil (flow through reactor). The reaction time should be long enough to permit a detectable accumulation of the adsorbate, but short enough to avoid competing side reactions, such as redox, precipitation, or dissolution reactions (Section B, Chapters 4 and 8). In batch reactors, the chemical analysis in step 2 is usually carried out after separation of the soil from the reacted solution by centrifugal or gravitational force. In this separation, some reacted solution will always be entrained with the soil to form a slurry. In flow-through reactors, the composition of effluent solution is analyzed to determine the changes in its composition caused by adsorption.

The moles of chemical species i adsorbed per kilogram of dry soil after contacting an aqueous solution, termed the specific surface excess Γ_i, is calculated with the equation (Everett, 1972):

$$\Gamma_i = n_i - M_w m_i \qquad [7.2]$$

where n_i is the total amount of species i ($mol\ kg^{-1}$) of dry soil in the soil slurry (batch reactor) or in the wet soil column (flow-through reactor); M_w is the gravimetric water content of the slurry or soil column (kg water kg^{-1} dry soil); and m_i is the molality (mol kg^{-1} water) of species i in the supernatant solution (batch reactor) or in the effluent solution (flow-through reactor). No direct reference to an adsorbing surface appears in Equation [7.2], but a simple calculation will show that Γ_i is identically zero if the adsorbing species i is water. Therefore, strictly speaking, Γ_i refers to an interface at which there is no net adsorption of water. The value of Γ_i in Equation [7.2] can be positive, zero, or negative.

Consider, for example, an acidic soil that has been reacted in a batch process with KNO_3 solution. After a 24-h reaction, the soil and aqueous solution are separated by centrifugation. The soil slurry is found to contain 2.58 mmol K kg^{-1} and to have a gravimetric water content of 0.45 kg kg^{-1}. The supernatant solution contains K at a molality of 3.96 mmol kg^{-1}. According to Equation [7.2],

$$\Gamma_K = 2.58 - (0.45)(3.96) = +0.8 \text{ mmol kg}^{-1} \qquad [7.3]$$

is the positive specific surface excess of K in the soil. Suppose also that the molality of NO_3 in the supernatant solution is 20 mmol kg^{-1} and the soil slurry contains 2.8 mmol NO_3 kg^{-1}. Then

$$\Gamma_{NO_3} = 2.8 - (0.45)(20) = -6.2 \text{ mmol kg}^{-1} \qquad [7.4]$$

is the negative specific surface excess of NO_3 in the soil. In both examples, Γ_i is the net excess of species i (mol kg^{-1} dry soil), as referenced to an aqueous solution containing M_w (kg) of water and species i at the molality m_i. This net excess is attributed to the presence of adsorbing soil particles.

Besides σ_H, particle surfaces can have an inner-sphere complex surface charge density (σ_{IS}) and an outer-sphere complex surface charge density (σ_{OS}) (Coul m^{-2}). Contributing to σ_{IS} is the net total charge of the ions, other than H^+ or OH^-, which are bound into inner-sphere surface complexes. Similarly, σ_{OS} receives contributions from the net total charge of the ions, other than H^+ or OH^-, that are bound into outer-sphere surface complexes. [These two components of surface charge density do not include H^+ or OH^- because they appear in σ_H, reflecting the traditional emphasis given to these two latter ions as components of surface reactive solid phases and aqueous solutions. Other adsorptive ions (e.g., Ca^{2+}, K^+) that form surface complexes are less ubiquitous in mineral particle structures than are H^- or OH^-.] It is useful in applications to group these four-component surface charge densities into an intrinsic surface charge density (Sposito, 1992a),

$$\sigma_{in} = \sigma_o + \sigma_H \qquad [7.5]$$

and a Stern layer surface charge density (Everett, 1972),

$$\sigma_S = \sigma_{IS} + \sigma_{OS} \qquad [7.6]$$

The intrinsic surface charge density reflects particle charge developed from isomorphic substitutions and surface complex formation involving H^+ or OH^-. The Stern layer surface charge density reflects particle charge developed from counterions other than H^+ or OH^- immobilized on an adsorbent surface. The net total particle surface charge density then can be defined as (Sposito, 1992a):

$$\sigma_p \equiv \sigma_{in} + \sigma_S = \sigma_o + \sigma_H + \sigma_{IS} + \sigma_{OS} \qquad [7.7]$$

Laboratory methods for the measurement of the components of surface charge, including σ_{in} and σ_p, have been discussed by Sposito (1992a) and, more recently, by Chorover and Sposito (1995), Zelazny et al. (1996), and Schroth and Sposito (1997) (Section B, Chapter 6).

The net total particle surface charge density (σ_p) must be balanced, when it is nonzero, by another kind of surface charge. This balancing charge arises from ions in the diffuse layer which, although they move about freely in aqueous solution, remain in a swarm near enough to particle surfaces to create an effective surface charge density (σ_d) that balances σ_p. On the molecular scale, this effective surface charge density can be apportioned to diffuse layer ions according to the equation (Everett, 1972; Zelazny et al., 1996):

$$\sigma_d = \frac{F}{Ma_s} \Sigma_i Z_i \int_V \left[c_i(x) - c_i(\infty) \right] dV \qquad [7.8]$$

where Z_i is the valence of diffuse layer ion i, $c_i(x)$ is its concentration at point x in aqueous solution, and $c_i(\infty)$ is its concentration in bulk solution, far enough from any particle surface to avoid being counted as adsorbed in a diffuse layer. The integral in Equation [7.8] is over the entire volume V of aqueous solution contacting a mass M of soil adsorbent whose specific surface area is a_s. Thus, Equation [7.8] represents the excess surface charge created by the ions in aqueous solution; if $c_i(x) = c_i(\infty)$ uniformly, there is no contribution of ion i to σ_d. Equation [7.8] applies to all ions in solution, including H^+ and OH^-, and defines the diffuse layer surface charge density (Coul m^{-2}) that is required to balance σ_p in order to maintain electrical neutrality (Everett, 1972):

$$\sigma_p + \sigma_d = 0 \qquad [7.9]$$

Equation [7.9], which expresses the balance of surface charge, can be applied both to an individual soil particle and to an entire soil. It serves as a general conservation law for the characterization of soil particle surface charge.

A useful alternative form of Equation [7.9] is obtained by defining the net adsorbed ion charge,

$$F \frac{(q_+ - q_-)}{a_s} \equiv \sigma_{IS} + \sigma_{OS} + \sigma_d \qquad [7.10]$$

where q is adsorbed cation (+) or anion (−) charge (mol$_c$ kg^{-1}), irrespective of surface speciation. The methods of measurement of $(q_+ - q_-) \equiv \Delta q$ are discussed by Sposito (1992a) and Zelazny et al. (1996). Chorover and Sposito (1995) have proposed Equation [7.9] as a check on independent measurements of Δq, σ_H, and σ_o. Given Equations [7.7], [7.9], and [7.10], a graph of $F\Delta q/a_s$ versus σ_H must have a slope equal to -1 and both x- and y-intercepts equal to σ_o (see Fig. 7.2). Chorover plots of this type have also been presented by Polubesova et al. (1995) and Schroth and Sposito (1997). One notes in passing that the ratio F/a_s can be deleted from Equations [7.1], [7.8], and [7.10], to leave all surface charge components in the convenient units of moles of charge per kilogram of soil adsorbent, without affecting Equations [7.5], [7.6], [7.7], or [7.9].

7.2 Points of Zero Charge

In soil chemistry parlance, points of zero charge are pH values at which one or more of the surface charge components in Equation [7.7] vanishes at fixed temperature, applied pressure, and aqueous solution composition. A standard nomenclature for points of zero charge has only recently been established (Table 7.2); previously the terminology has been highly erratic. For example, the point of zero net proton charge (p.z.n.p.c.) has often been called the zero point of charge and, in much of the extant surface chemistry literature concerning natural particles, the point of zero salt effect (p.z.s.e.) has been termed the point of zero charge, as has been the point of zero net charge (p.z.n.c.).

Table 7.2 Points of zero charge

Symbol	Name	Definition
p.z.c.	point of zero charge	$\sigma_p = 0$
p.z.n.p.c.	point of zero net proton charge	$\sigma_H = 0$
p.z.n.c.	point of zero net charge	$\sigma_{in} = 0$
p.z.s.e.	point of zero salt effect	$\partial\sigma_H/\partial I = 0$

Irrespective of this unfortunate variability in terminology, agreement does exist on the great importance of points of zero charge to understanding soil surface chemistry. It should be borne in mind, however, that pH is not the only chemical master variable which can lead to a vanishing component of the surface charge density (Hunter, 1993; Sposito, 1994).

The point of zero charge (p.z.c.) is the pH value at which the net total particle charge vanishes: $\sigma_p = 0$ (Everett, 1972). Thus, by Equation [7.9], at the p.z.c., there is no net particle surface charge to be neutralized by ions in the diffuse layer and all adsorbed ions are immobilized in surface complexes. In principle, the p.z.c. can be measured by ascertaining the pH value at which soil particles do not respond to an applied electric field, or, more generally, by determining the pH value at which perfect charge balance exists in an aqueous solution in which the particles are suspended. The first method actually determines the pH value at which the particle electrophoretic mobility vanishes (Section B, Chapter 6), and is termed the isoelectric point (i.e.p.) (Everett, 1972). Equality between i.e.p. and p.z.c. requires that no part of the diffuse layer be carried along with the particle when it moves steadily in response to a uniform, constant electric field (Hunter, 1993). If this condition is not met, then the i.e.p. will correspond instead to the vanishing of the poorly defined sum, σ_p, plus a portion of the diffuse layer charge, σ_d. Complicating matters further is the fact that particles for which $\sigma_p = 0$ because of mutually canceling patches of surface charge (i.e., a heterogeneous surface charge distribution, but with zero overall net charge) can still exhibit a nonzero electrophoretic mobility, unless they are perfectly spherical (Fair and Anderson, 1989). In this case, the i.e.p. would not correspond to $\sigma_p = 0$ and, therefore, would differ from the p.z.c.

The p.z.n.p.c. is the pH value at which the net proton charge vanishes: $\sigma_H = 0$. A thermodynamic stability property of σ_H is that it either decreases (or remains unchanged) as the pH increases (i.e., $\Delta\sigma_H/\Delta pH \leq 0$). This stability criterion applies regardless of the composition or ionic strength of an aqueous solution, and independently of the nature of the soil particles interacting with it.

The p.z.n.c. is the pH value at which the intrinsic charge vanishes; thus, $\sigma_{in} = 0$, the same as requiring the net adsorbed ion charge to vanish,

$$F\frac{(q_+ - q_-)}{a_s} = \sigma_{IS} + \sigma_{OS} + \sigma_d = -\sigma_{in} \qquad [7.11]$$

at the p.z.n.c. Diffuse layer charge can exist at the p.z.n.c., whereas it cannot exist at the p.z.c. It is common practice to utilize index ions, like K^+ and Cl^-, in the measurement of the p.z.n.c. (Zelazny et al., 1996). Evidently, the value of the p.z.n.c. will depend on the choice of index ions, although experience shows that this dependence is often small if these ions are chosen from the group: Li^+, Na^+, K^+. The p.z.s.e., although frequently measured in soil surface chemistry (Zelazny et al., 1996), is not strictly a point of zero charge, in that it is defined by the invariance of σ_H under changes of ionic strength (I) instead of the vanishing of a surface charge component. Thus, the relationship of the p.z.s.e. to particle surface charge is indirect, and it is necessary to appeal to a specific model of the particle/aqueous solution interface in order to interpret p.z.s.e. values (Lyklema, 1984; Hunter, 1993).

A set of four general statements about the points of zero charge can be proved using only the law of conservation of surface charge (Equation [7.9]) and the thermodynamic stability criterion ($\partial \sigma_H / \partial pH < 0$). These statements, the p.z.c. Theorems, do not require knowledge of the molecular details of chemical speciation at the soil particle/aqueous solution interface, and so may be applied to validate surface speciation models, such as modified Gouy-Chapman theory or site binding models (as discussed in Section B, Chapter 8), or to examine experimental surface speciation data for internal consistency. Proofs of the p.z.c. Theorems follow.

7.2.1 Theorem 1

Let q_+ be the specific adsorbed cation charge and let q_- be the specific adsorbed anion charge. If ($\partial \sigma_H / \partial pH$) < 0, then:

$$\sigma_0 = -(q_+ - q_-) \text{ at pH} = \text{p.z.n.p.c.} \tag{7.12}$$

and

$$\sigma_0 \overset{>}{\underset{<}{=}} 0 \text{ if p.z.n.c.} \overset{>}{\underset{<}{=}} \text{p.z.n.p.c.} \tag{7.13}$$

where σ_0 is expressed in $mol_c \, kg^{-1}$ (as is σ_H).

Proof. Given the units of σ_0 and σ_H, Equations [7.7], [7.9], and [7.10] combine to yield the surface charge balance expression:

$$\sigma_0 + \sigma_H + \Delta q = 0 \tag{7.14}$$

Equation [7.12] follows from setting $\sigma_H = 0$ (Table 7.2) in this expression. Equation [7.13] is derived after noting that

$$\sigma_H(pH = \text{p.z.n.p.c.}) \overset{>}{\underset{<}{=}} 0 \text{ if p.z.n.c.} \overset{>}{\underset{<}{=}} \text{p.z.n.p.c.} \tag{7.15}$$

since σ_H (pH = p.z.n.p.c.) ≡ 0 and $\partial \sigma_H / \partial pH < 0$. Because $\sigma_0 = -\sigma_H$ at the p.z.n.c. (Table 7.2), Equation [7.13] then follows. QED

If $|\sigma_H/\sigma_0| \ll 1$ (permanent charge soil), Equation [7.12] applies at any pH value, not just at the p.z.n.p.c.. Therefore, for soils whose adsorption reactions mainly involve 2:1 layer type clay minerals, σ_0 dominates σ_{in} and can be measured directly with a suitable choice of adsorbing index ions (Sumner and Miller, 1996). If σ_H cannot be neglected, then σ_0 can be inferred from the intercepts of a Chorover plot, as described in Fig. 7.2 (dotted lines). Equation [7.13] shows that p.z.n.p.c. = p.z.n.c. when $\sigma_0 = 0$ (i.e., for oxides and organic matter). Thus the p.z.n.c. can be used to determine absolute values of σ_H when it is measured by proton titration (Sposito, 1992a; Schroth and Sposito, 1997). More directly, graphs of Δq versus pH can be applied to determine p.z.n.p.c. (and, therefore, the pH value when σ_H

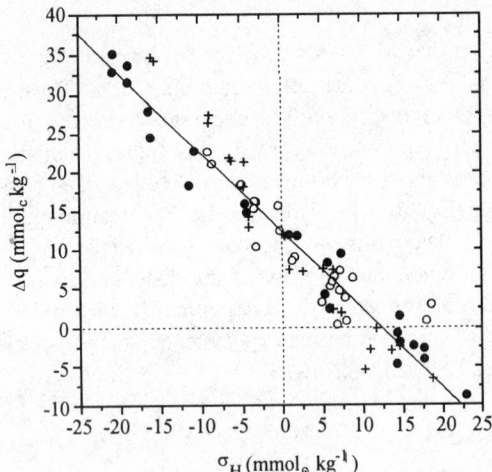

Fig. 7.2 "Chorover plot" (Chorover and Sposito, 1995) of the net adsorbed ion charge against the net proton surface charge density for a Brazilian Oxisol (Manaus soil). The combined data, for ion strengths of 0.001 (○), 0.005 (+), and 0.01 (•) mol kg^{-1}, can be fit to the regression equation (solid line):

$$\Delta q = -1.01(\pm 0.07)\sigma_H + 12.5(\pm 0.8) \qquad r^2 = 0.92 ***$$

where Δq and σ_H are in mmol$_c$ kg^{-1}. Charge balance is confirmed by the values of the slope and both intercepts ($\sigma_0 =$ $= 12.5 \pm 0.4$ mmol$_c$ kg^{-1} from direct measurement).

truly equals zero) after noting the pH value at which Equation [7.12] holds (Chorover and Sposito, 1995). This approach requires prior determination of σ_0 (Zelazny et al., 1996).

7.2.2   Theorem 2

The p.z.c. will equal the p.z.n.c. if and only if $(\sigma_{IS} + \sigma_{OS}) = 0$ when pH = p.z.n.c.

Proof. Equations [7.5], [7.6], and [7.9] combine to yield the surface charge balance expression:

$$\sigma_{in} + \sigma_s + \sigma_d = 0 \qquad\qquad\qquad\qquad [7.16]$$

Theorem 2 is then demonstrated by noting (by Equation [7.9]) that $\sigma_d = 0$ at the p.z.c.(sufficiency) and that $\sigma_{in} = 0$ at the p.z.n.c. (necessity). QED

Note that Theorem 2 is identically true if the diffuse layer provides the only mechanism of ion adsorption. If adsorption involves only outer-sphere surface complexes and the diffuse layer ions, Theorem 2 will still follow if there is no contribution to the net particle surface charge density when pH = p.z.n.c. from ions in the outer-sphere surface complexes. Electrolytes for which Theorem 2 is true either under this latter circumstance or identically (i.e., diffuse layer adsorption only) are termed indifferent electrolytes (Hunter, 1993).

7.2.3   Theorem 3a

If $(q_+ - q_-)$ decreases (resp. increases), then the p.z.n.p.c. increases (resp. decreases).

Proof. Given Equation [7.14], a shift in the net adsorbed ion charge by any adsorption mechanism is subject to the charge balance constraint:

$$\delta\sigma_H + \delta\Delta q = 0 \qquad\qquad\qquad [7.17]$$

at any pH value, since σ_0 cannot be changed by a shift in ion adsorption (i.e., through adjustment of aqueous solution composition or temperature). If Equation [7.17] is applied at the p.z.n.p.c. ($\sigma_H = 0$) and the stability condition ($\partial\sigma_H/\partial pH < 0$) is invoked, Theorem 3a follows. QED

7.2.4   Theorem 3b

If ($\sigma_{IS} + \sigma_{OS}$) decreases (resp. increases), then the p.z.c. decreases (resp. increases).

Proof. Given Equation [7.16] and the definition of σ_{in} (Equation [7.5]), the charge balance constraint applicable at the p.z.c. is

$$\sigma_H = -(\sigma_0 + \sigma_s) \qquad (pH = p.z.c.) \qquad\qquad [7.18]$$

The pH value at which Equation [7.18] holds must increase (decrease) as σ_s increases (decreases), since $\partial\sigma_H/\partial pH < 0$. QED

Theorem 3a shows that (positive) adsorption of cations decreases the p.z.n.p.c., whereas (positive) adsorption of anions increases the p.z.n.p.c. (negative adsorption of anions evidently would decrease the p.z.n.p.c.). Theorem 3b shows that the surface complexation of cations increases the p.z.c., whereas the surface complexation of anions decreases the p.z.c. Note that neither of these shifts in a point of zero charge indicates specific adsorption, defined on the molecular level as inner-sphere surface complexation (Sposito, 1989). Moreover, shifts in the p.z.n.p.c. cannot be used to identify the mechanism of adsorption (because any one of them can cause a shift), and shifts in the p.z.c. do not distinguish specific adsorption from outer-sphere surface complexation.

7.2.5   Theorem 4

If $\partial(q_+ - q_-)/I = 0$ at the p.z.n.c. (I = ionic strength), then p.z.s.e. = p.z.n.c. If $\sigma_0 = 0$ as well, then p.z.s.e. = p.z.n.p.c. also, by Theorem 1. If ($\sigma_{IS} + \sigma_{OS}$) = 0 as well, then p.z.s.e. = p.z.c. also, by Theorem 2.

Proof. Given Equation [7.14] and the independence of σ_0 from changes in aqueous solution properties, a charge balance constraint applicable at any pH value is

$$\frac{\partial\sigma_H}{\partial I} + \frac{\partial\Delta q}{\partial I} = 0 \qquad\qquad\qquad [7.19]$$

Theorem 4 follows from applying Equation [7.19] at the p.z.n.c., noting the definition of the p.z.s.e., then adding the conditions that make either Theorem 1 or Theorem 2 apply. QED

Theorem 4 addresses the relationship between p.z.s.e. and true points of zero charge. If the net adsorbed ion charge is invariant under changes in ionic strength at the p.z.n.c., then p.z.s.e. = p.z.n.c. If there is no structural particle charge (e.g., oxides and humus), then p.z.s.e. = p.z.n.p.c. as well. Equality between p.z.s.e. and p.z.c. requires the further condition that electrical neutrality is satisfied among the surface complexes alone, or that only a diffuse layer of adsorbed ions exists. Experimental verification of the invariance of the net adsorbed ion charge under changes in ionic strength is essential to all of these relationships, making clear the point that there is no vanishing surface charge component necessarily to be associated with a crossover point that has been determined through measurements of σ_H carried out at different ionic strengths.

Methods of determining points of zero charge in soils have been reviewed by Sposito (1992a) and Zelazny et al. (1996), who also provide tabulations of representative values for specimen minerals. As a general rule, the p.z.n.c. values for silica, humus, clay minerals, and Mn oxides are well below pH 4, whereas those for Al and Fe oxides and for calcite are well above pH 7. Schroth and Sposito (1997) recently have reported the first measurements of the points of zero charge for a specimen sample of Georgia kaolinite that were made with explicit verification of surface charge balance, as described in Equation [7.14]. They found a p.z.n.c. value of 3.5 and a p.z.n.p.c. value of 5.0. These results indicate the presence of negative structural charge in the kaolinite sample (Equation [7.13]), which they attributed to 2:1 layer type clay mineral inclusions.

7.3   Ion Exchange Phenomenology

Readily exchangeable ions can be replaced easily by leaching with an index electrolyte solution of prescribed composition, concentration, and pH value. Despite the empirical nature of this concept, there is a consensus that ions which are usually adsorbed specifically (e.g., Pb^{2+} or HPO_4^{2-}, Section B, Chapter 8) are not to be counted among those readily exchangeable ions in soils. Thus, experimental methods to determine readily exchangeable adsorbed ions must avoid extracting specifically adsorbed ions. From this point of view, only fully solvated ions adsorbed on soil particles are readily exchangeable ions, with the definition of readily exchangeable based, therefore, on the diffuse ion swarm and outer-sphere surface complex mechanisms of adsorption (Fig. 7.1).

The ion exchange capacity of a soil is the number of moles of adsorbed ion charge that can be desorbed from unit mass of soil, under given conditions of temperature, pressure, soil solution composition, and soil solution mass ratio. In most applications, an ion exchange capacity refers to the maximum adsorption (positive surface excess) of readily exchangeable ions that adsorb on soil particle surfaces solely in outer-sphere complexes and the diffuse layer.

The measurement of an ion exchange capacity involves the replacement of native, readily exchangeable ions by an index cation or anion whose surface excess then is determined following the principles discussed above. Detailed laboratory procedures for this measurement are described by Sumner and Miller (1996) and Zelazny et al. (1996). For soils in which the readily exchangeable cations are typically monovalent or bivalent (e.g., Aridisols), the index ions can be Ca^{2+} and Cl^-, whereas for soils also bearing trivalent readily exchangeable cations (e.g., Spodosols), Ba^{2+} or K^+ is the index cation of choice. Often NH_4^+ has been used as an "index" cation. However, this cation forms inner-sphere surface complexes with 2:1 layer type clay minerals, and can even dislodge cations from easily weathered primary soil minerals. The use of NH_4^+ to measure the soil cation exchange capacity thus has significant potential for inaccuracy (Sumner and Miller, 1996). With regard to anion exchange capacity, the use of Cl^- as the index anion is widespread because of its nonspecific adsorption characteristics (Zelazny et al., 1996).

Controversy exists over the surface chemical interpretation of ion exchange capacities, particularly those measured for cations. If a high concentration of the index cation is used in a solution at high pH (e.g., pH $\geq$ 8.2), the measured surface excess of the index cation can approximate the maximum (absolute value of) intrinsic surface charge of a soil. This parameter is then equivalent to the largest value expected for the adsorbed cation charge. On the other hand, if the pH value, or some other chemical property of the solution containing the index electrolyte, is such that the maximum intrinsic surface charge (σ_{in}) is not neutralized by the adsorption of the index ions, then the measured surface excess of the latter will simply reflect the chemical conditions chosen. This latter situation is illustrated for Li^+ and Cl^- adsorption under varying pH in the Oxisols investigated by Chorover and Sposito (1995). Thus both the maximum and the less than maximum varieties of cation exchange measurement are useful in soil chemistry. The maximum intrinsic surface charge measurement indicates the potential capacity of a soil for adsorbing cations or anions, whereas a less than maximum intrinsic surface charge measurement indicates the actual capacity of a soil for adsorbing index ions under given experimental conditions.

These conceptual issues are formalized in the surface charge balance condition expressed by Equations [7.11] and [7.16]:

$$q_+ - q_- = \sigma_{os} + \sigma_d = -\sigma_{in} \qquad\qquad [7.20]$$

where q_+ and q_- refer only to readily exchangeable index ions (mol$_c$ kg^{-1}). Equation [7.20] is valid only if conditions are such that the index ions have fully neutralized the structural and net proton surface charge densities (Equation [7.5]). Otherwise, the second equality would not be appropriate. A more subtle aspect of Equation [7.20] occurs if either q_+ or q_- is negative (cation or anion exclusion; Section B, Chapter 6). Given that σ_0 and σ_H have been neutralized fully by the adsorption of index ions, no error occurs in the use of Equation [7.20] to measure σ_{in} in the presence of ion exclusion, so long as both q_+ and q_- are measured and then subtracted as prescribed by charge balance. An error will occur, however, if only one of the two adsorbed ion charge components is measured and equated presumptively to $-\sigma_{in}$ (Amrhein and Suarez, 1990). Returning to the example in Equations [7.3] and [7.4], one can see that equating $|\sigma_{in}|$ to q_K alone (because σ_H has been assumed to be negligible) would lead to a significant underestimation of $|\sigma_{in}|$ ($q_K = 0.8$ mmol$_c$ kg^{-1}, whereas $|\sigma_{in}| = 7.0$ mmol$_c$ kg^{-1}).

Table 7.3 lists representative cation exchange capacity (CEC) values based primarily on measurements made for United States soils using as the index cation NH_4^+ in a solution at pH 7 (Sposito, 1989). Large variability of the CEC within each soil order exists, but the very low values for Ultisols and Oxisols, and the high values for Histosols and Vertisols are significant trends. Detailed studies of the CEC show that it is correlated positively with both the content of soil organic matter and with soil pH. Anion exchange capacity is significant only for Oxisols, Spodosols, Andisols, and Ultisols approaching the value of 0.01 mol$_c$ kg^{-1} (Parfitt, 1980; Ji, 1997).

The composition of readily exchangeable ions in a soil can be determined by chemical analysis after reaction with an index ion, as outlined by Sumner and Miller (1996). In alkaline soils, the readily exchangeable cations are Ca^{2+}, Mg^{2+}, Na^+, and K^+, decreasing in their contribution in the order shown. In acidic soils, the most abundant readily exchangeable metal cation is Al^{3+} (including $AlOH^{2+}$ and $Al(OH)_2^+$), followed by Ca^{2+} and Mg^{2+}. Once the surface excess of each readily exchangeable ion in a soil has been measured under varying conditions of soil solution composition, an exchange isotherm can be constructed. An exchange isotherm is analogous to an adsorption isotherm (Section B, Chapter

Table 7.3 Representative cation exchange capacities (CEC) of surface soils

Soil Order	CEC mol$_c$ kg^{-1}	Soil Order	CEC mol$_c$ kg^{-1}
Alfisols	0.12	Mollisols	0.22
Aridisols	0.16	Oxisols	0.05
Entisols	0.13	Spodosols	0.11
Histosols	1.40	Ultisols	0.06
Inceptisols	0.19	Vertisols	0.37

8), except that the variables plotted are charge fractions instead of surface excesses and solution concentrations. The charge fraction of an adsorbed ion is

$$E_i = \frac{q_i}{Q} \qquad\qquad [7.21]$$

where q_i is the specific adsorbed charge of ion i, and Q is either the CEC or the AEC of a soil. The charge fraction of an ion in aqueous solution is

$$\tilde{E}_i = \frac{|Z_i| m_i}{\tilde{Q}} \qquad\qquad [7.22]$$

where Z_i is the valence and m_i is the molality (or other concentration variable) of ion i and

$$\tilde{Q} = \Sigma_k \ |Z_k| m_k \qquad\qquad [7.23]$$

with the sum extending over all ions of the same valence sign (i.e., over all cations or all anions) as ion i in Equation [7.22]. An exchange isotherm then is defined as a graph of E_i against $\tilde{E}_i$ at fixed temperature and applied pressure.

Exchange isotherms for Na $\rightarrow$ Mg and Ca $\rightarrow$ Mg exchange on Altamont soil (a Vertisol) are illustrated in Fig. 7.3 (Fletcher et al., 1984) at 298 K and 1 atm pressure, for $\tilde{Q}$=0.05 $mol_c \ kg^{-1}$. The maximum range of the charge fractions in each graph is from 0 to 1. In Fig. 7.3, only the pairs of cations indicated explicitly were investigated, so the isotherms refer to binary exchange systems. For the open circle data in Fig. 7.4, however, adsorbed Na$^+$ was present during a Ca $\rightarrow$ Mg exchange experiment on Domino soil (an Aridisol), and the resulting isotherm refers to a ternary exchange system (Sposito et al., 1986). In natural soils, ternary, quaternary, or even higher order exchange systems are common. The binary exchange reaction is convenient for laboratory study, under the critical assumption that naturally occurring, higher order exchange systems can be understood in terms of binary exchange reactions. That this assumption may be true is indicated in Fig. 7.4, which shows Ca $\rightarrow$ Mg exchange reactions both in the absence and in the presence of adsorbed Na$^+$. The charge fractions of Mg^{2+} plotted for an exchangeable sodium percentage (ESP) of 23% (Section G, Chapter 2) are based on Equations

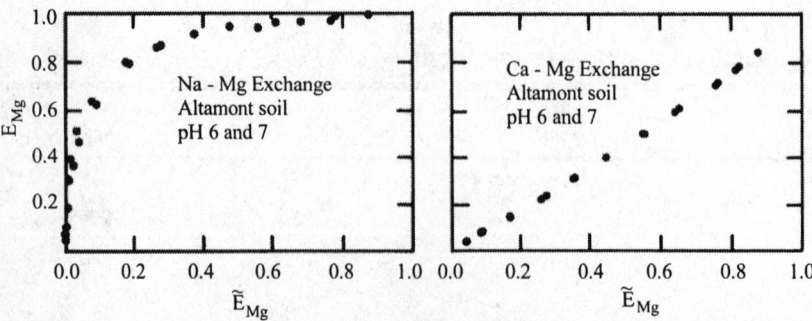

Fig. 7.3 Exchange isotherms for the (silt + clay) fraction of a Vertisol (Altamont soil) with data at pH 6 and 7 combined (Fletcher et al., 1984)[Reprinted from Sposito, 1994. Chemical equilibria and kinetics in soils. Oxford University Press. Used by permission of Oxford University Press, Inc.]

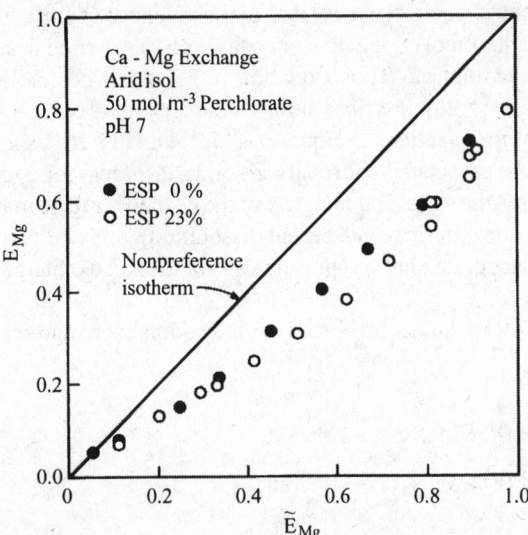

Fig. 7.4 Homovalent exchange isotherms for an Aridisol (Domino soil) at pH 7 with Exchangeable Sodium Percentage (ESP) maintained at 23% (Sposito et al., 1986). The solid line represents the thermodynamic nonpreference isotherm. [Reprinted from Sposito, 1989. The chemistry of soils. Oxford University Press. Used by permission of Oxford University Press, Inc.]

[7.21] and [7.22], but with Q and limited to contributions from Mg^{2+} and Ca^{2+} so as to allow direct comparison with the binary exchange data (ESP = 0%). The closeness of the two exchange isotherms in Fig. 7.4 suggests that Ca $\rightarrow$ Mg exchange on the soil was largely independent of the presence of adsorbed Na^+ in the ESP range investigated.

The straight line drawn in Fig. 7.4 is the thermodynamic nonpreference exchange isotherm (bivalent-bivalent). For bivalent-bivalent exchange, and for any exchange reaction involving ions of the same valence (homovalent exchange), the thermodynamic nonpreference isotherm is represented mathematically by the equation:

$$E_i = \tilde{E}_i \qquad [7.24]$$

which plots as a straight line making a 45° angle with both coordinate axes. According to the data in Fig. 7.4, the Ca $\rightarrow$ Mg exchange isotherms lie below the nonpreference isotherm, indicating a preference for Ca^{2+} over Mg^{2+}. The cause of this selectivity can be determined only after detailed chemical investigation of the soil adsorbent (Section B, Chapter 8).

7.4 Ion Exchange Models

Stoichiometric, binary, ion exchange reactions can be expressed by the chemical equations:

$$bAX_a(s) + aB^{b+}(aq) \Leftrightarrow aBX_b(s) + bA^{a+}(aq) \qquad [7.25]$$

$$dCY_c(s) + cD^{d-}(aq) \Leftrightarrow cDY_d(s) + dC^{c-}(aq) \qquad [7.26]$$

where a, b, c, d are stoichiometric coefficients related to the valences of the cations A^{a+}, B^{b+}, or the anions, C^{c-}, D^{d-}, X or Y represents 1 mol of negative or positive charge carried by a soil exchanger, and s = solid phase, aq = aqueous solution. Thus Equation [7.25] describes cation exchange on the exchanger X^-, whereas Equation [7.26] describes anion exchange on the exchanger Y^+. Fundamental to a chemical interpretation of the reactions in Equations [7.25] and [7.26] as applied to soils is the assumption that X^- and Y^+ can be associated with charged solids, insofar as ion exchange is concerned. This assumption, which requires the structural integrity of the exchanger to be maintained throughout an ion exchange reaction (i.e., no significant adsorbent dissolution), must be verified experimentally in each application. If it is not accurate, then a representation of the soil exchanger as merely 1 mol of surface charge is not possible.

An example of Equation [7.25] can be developed for the cation exchange reactions illustrated in Fig. 7.3:

$$2NaX(s) + Mg^{2+}(aq) \Leftrightarrow MgX_2(s) + 2Na^+(aq) \tag{7.27}$$

$$CaX_2(s) + Mg^{2+}(aq) \Leftrightarrow MgX_2(s) + Ca^{2+}(aq) \tag{7.28}$$

where $X^-(s)$ represents 1 mol of negative surface charge borne by the Altamont soil (Vertisol). (Note that a common factor of 2 has been deleted from all terms in Equation [7.28].) Similarly, on the Manaus soil (Oxisol) investigated by Chorover and Sposito (1995), $Cl \rightarrow NO_3$ exchange exemplifies Equation [7.26]:

$$ClY(s) + NO_3^-(aq) \Leftrightarrow NO_3Y(s) + Cl^-(aq) \tag{7.29}$$

where $Y^+(s)$ represents 1 mol of positive surface charge borne by this kaolinitic Oxisol. Equations [7.27]–[7.29] are meaningful chemically as long as the dissolution of montmorillonite in the Altamont soil, and the dissolution of kaolinite in the Manaus soil, are negligible processes on the time scale over which the ion exchange reactions occur.

A conditional equilibrium constant (K_c) (Section B, Chapter 3) can be formulated to describe the reaction in Equation [7.25], following conventional chemical thermodynamics methods (Sparks, 1995):

$$K_c = \frac{x_B^a \left(A^{a+}\right)^b}{x_A^b \left(B^{b+}\right)^a} \tag{7.30}$$

where

$$x_i = \frac{\Gamma_i}{\Gamma_A + \Gamma_B} \qquad (i = A \text{ or } B) \tag{7.31}$$

x_i is the mole fraction of adsorbed ion i (i = (A or B), Γ_i is defined in Equation [7.2], and (A^{a+}) is the thermodynamic activity of the aqueous ionic species A^{a+} (aq), etc. Methods for determining this latter parameter are discussed by Sposito (1994) and by Mattigod and Zachara (1996). The particular combination of variables on the right side of Equation [7.30] usually will not remain constant as the exchanger and aqueous solution composition of the soil system to which it refers changes (at fixed temperature and applied pressure). Nonetheless, it is often a good first approximation [especially for homovalent exchange reactions (a = b) and unibivalent exchange reactions (a = 1, b = 2)] to model K_c as a composition independent constant (K_V). The resulting equilibrium constraint,

$$K_V = \frac{x_B^a \left(A^{a+}\right)^b}{x_A^b \left(B^{b+}\right)^a} \qquad [7.32]$$

is termed the Vanselow model (Sposito, 1981b).

For the case of homovalent cation exchange, it is straightforward to show that there is no loss in generality from setting a = b = 1 in Equation [7.32] to derive, by rearrangement of the expression, the exchange isotherm equation:

$$E_B = \frac{K_V \tilde{E}_B}{1 + \left(K_V - 1\right)\tilde{E}_B} \qquad [7.33]$$

where E_B and $\tilde{E}_B$ are the charge fractions defined in Equations [7.21] and [7.22], respectively. (In deriving Equation [7.33], the assumption is made that the ratio of activities of aqueous cations A and B is equal to the ratio of their molalities.) By convention, the exchange process described by Equation [7.25] is termed selective for cation B if $K_V > 1$ in Equation [7.32]; selective for cation A if $K_V < 1$ in Equation [7.32]; and to show no thermodynamic preference, if $K_V = 1$ in Equation [7.32]. Thus, the homovalent exchange isotherms illustrated in Figs. 7.3 and 7.4 can be interpreted with Equation [7.33] using $K_V = 1$ and $K_V < 1$, respectively.

For the case of unibivalent exchange, a rearrangement of Equation [7.32] yields (Sposito et al., 1981):

$$E_B = 1 - \left\{ 1 + \frac{2K_V}{S\tilde{Q}} \cdot \left[\frac{1}{\left(1 - \tilde{E}_B\right)^2} - \frac{1}{\left(1 - \tilde{E}_B\right)} \right] \right\}^{-1/2} \qquad [7.34]$$

where $\tilde{Q}$ is defined in Equation [7.23] and $S \equiv \gamma_A^2 / \gamma_B$ is the ratio of single ion activity coefficients (γ) for the aqueous cations A and B (Sposito, 1984; Sparks, 1995) (Section B, Chapter 3.) Equation [7.34] can be applied to describe Na → B exchange reactions on soils with 2:1 layer type clay minerals rather well, where B is a bivalent cation, such as Ca^{2+} or Cu^{2+} (Oster and Sposito, 1980; Sposito et al., 1981; Amhrein and Suarez, 1991; Zhang and Sparks, 1996). The determination of cation exchange selectivity is more subtle than for Equation [7.33], however, because of the aqueous solution factor $\tilde{Q}$ (and, to some extent, S) in Equation [7.34]. The thermodynamic nonpreference exchange isotherm still corresponds to $K_V = 1$, but the resulting curve will not usually be a linear relationship between E_B and $\tilde{E}$ unless $\tilde{Q} \approx 1$ molal [i.e., a rather concentrated aqueous solution (Sposito, 1994)]. It is, therefore, necessary to compare the observed exchange isotherm to Equation [7.34] with the appropriate values of $\tilde{Q}$ and S inserted, along with $K_V = 1$, in order to ascertain exchange selectivity. If the observed isotherm lies above the (curvilinear) nonpreference isotherm, then $K_V > 1$ and the cation exchange process described by Equation [7.25] is "selective for cation B". Sumner and Miller (1996) have described methods for measuring K_V and applying the nonpreference isotherm in order to interpret unibivalent cation exchange isotherms. Their approach is illustrated in Fig. 7.5 for Na $\rightarrow$ Ca and Na $\rightarrow$ Mg exchange on the Aridisol described in Fig. 7.4 (Sposito et al., 1986). The curve drawn in the figure represents Equation [7.34] with $\tilde{Q} = 0.05$ mol kg^{-1} and S = 1.5. Evidently the soil exchanger shows preference for both Ca^{2+} and Mg^{2+} relative to Na^+ under the given experimental conditions. Note that the difference between the two exchange isotherms shown in Fig. 7.5 implies the preference for Ca^{2+} relative to Mg^{2+} that was deduced from Fig. 7.4.

Implicit in the Vanselow model is the assumption that the cations adsorbed by the soil exchanger mix ideally, in that their occupation of negatively charged surface sites depends only on their valence, and not on their size, or any detail of their electronic structure that would lead to stereochemical effects or to a nonrandom occupation of charged surface sites (Sposito, 1989). It is this assumption of ideal mixing of the exchangeable cations that leads to the use of mole fractions as the exchanger composition variables in Equation [7.32] (Sposito, 1990, 1994). In the more complex situation wherein ideal mixing does not occur, a popular choice for replacing the mole fractions in Equation [7.32] is the Rothmund-Kornfeld model (Sposito, 1981b, 1989):

$$K_{RK} = \frac{\left[E_B^{b/\beta}\right]^a \left(A^{a+}\right)^b}{\left[E_A^{a/\beta}\right]^b \left(B^{b+}\right)^a} \qquad [7.35]$$

where ß $(0 < ß \le 1)$ is an adjustable parameter. The composition variable $E_i^{Z_i/\beta}$ thus replaces the mole fraction x_i for a cation whose valence is Z_i and whose charge fraction on the exchanger is E_i. Equation [7.35] can be derived from a cation exchange kinetics model in which the rate coefficients for adsorption or desorption of the cations are each assumed to take on a continuum of values that is

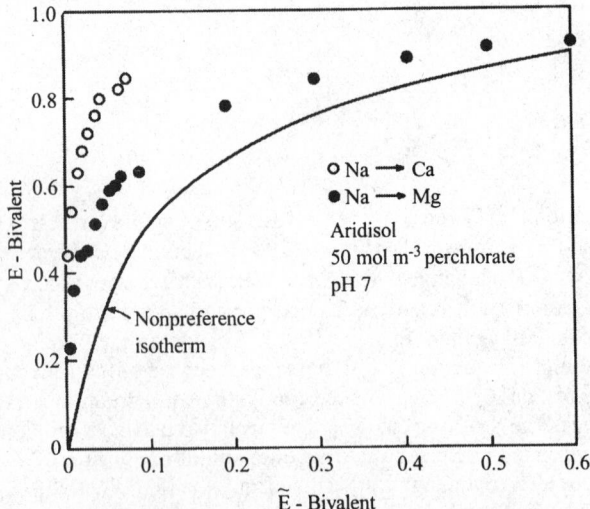

Fig. 7.5 Unibivalent exchange isotherms for an Aridisol (Domino soil) at pH 7 (Sposito et al., 1986). The solid curve represents the thermodynamic nonpreference isotherm [Reprinted from Sposito, 1989. The chemistry of soils. Oxford University Press. Used by permission of Oxford University Press, Inc.].

described well by a two-parameter gamma probability distribution (Sposito, 1994). The parameter ß in Equation [7.35] then can be interpreted as a quantitative measure of the breadth of the (gamma) distribution of adsorption-desorption rate coefficients, with small values of ß implying a very broad distribution (Sposito, 1994). The physical picture is one of surface site heterogeneity in a soil, leading to a continuum of adsorption/desorption rate coefficients with a corresponding distribution of site affinities for the exchanging cations. In the limit of a very narrow distribution of site affinities, ß↑ 1.0 and Equation [7.35] reduces to the Gapon model (Sposito, 1989):

$$K_G \equiv \left(K_{RK}\right)^{1/ab} = \frac{E_B\left(A^{a+}\right)^{1/a}}{E_A\left(B^{b+}\right)^{1/b}} \qquad [7.36]$$

which has met with limited success in applications to Na → B (B primarily Ca) exchange reactions (Oster and Sposito, 1980; Amrhein and Suarez, 1991). Bond (1995) has demonstrated the application of Equation [7.35] to Na → K exchange reactions on soils of mixed clay mineralogy, for which a distribution of site affinities would seem reasonable.

7.5 Ion Exchange Kinetics and Thermodynamics

A variety of data suggests that rates of ion exchange processes often are transport controlled, not reaction controlled (Sparks and Suarez, 1991). This is likely because the ions involved form solvated adsorbate species, the extreme case being two mobile exchanging ions that merely replace one another in the diffuse layer. Thus, readily exchangeable ions most probably engage in reactions whose rates are transport controlled, whereas specifically adsorbed ions likely participate in reactions that are surface controlled (Sposito, 1994).

Even with the selection of diffusion as the rate limiting process for most ion exchange reactions, there remains a need to choose between film diffusion and intraparticle diffusion. This can be done experimentally by the interruption test (Bunzl, 1974). Once an ion exchange reaction has been initiated, the exchanger particles are separated physically from the aqueous solution phase, and then reimmersed in it after waiting a short time interval. If film diffusion is the rate limiting step, no significant effect of this interruption on the ion exchange kinetics should be observed. If intraparticle diffusion is the rate limiting step, the concentration gradient driving the exchange process should relax to near zero during the interruption, and the rate of ion exchange should increase after the exchanger particles are reimmersed in the aqueous solution phase and a steep gradient is re-established. In general, exchanger particles with large specific surface areas should favor the film diffusion mechanism, whereas those with significant microporosity should favor the intraparticle diffusion mechanism (Petruzelli et al., 1991).

A rate law for ion exchange controlled by film diffusion can be developed after considering the Fickian rate laws (Petruzelli et al., 1991):

$$j_i = \left(D/\delta\right)\left([i]_{bulk} - [i]_{surf}\right) \qquad (i = 1,2,\ldots) \qquad [7.37]$$

where j_i is the rate at which ion i arrives at an adsorbent surface per unit area of the latter (mol m^{-2} s^{-1}), $[i]_{bulk}$ is its concentration in the bulk aqueous solution, $[i]_{surf}$ is its concentration at the interface of the liquid boundary layer (Nernst film) and the solid adsorbent surface, D_i is its diffusion coefficient, and δ is the thickness of the Nernst film through which the ion must diffuse from the (well-stirred) bulk aqueous solution phase to the adsorbent surface. The rate laws in Equation [7.37] are subject to the constraint of charge balance:

$$\sum_i \frac{dq_i}{dt} = 0 \qquad [7.38]$$

where q_i is the specific adsorbed charge of ion i and the sum is over all exchanging ions. The rate (dq_i/dt) is proportional to the adsorptive flux (j_i) in Equation [7.37]:

$$\frac{dq_i}{dt} = Ma_s |Z_i| j_i \qquad [7.39]$$

where M is the mass of exchanger particles and a_s is their specific surface area. Equations [7.37]–[7.39] provide the basis for a complete mathematical description of the kinetics of cation exchange governed by film diffusion.

A useful model expression that can be derived with the coupled Equations [7.37]–[7.39] is the Bunzl rate law (Bunzl, 1974). For a binary ion exchange system, this rate law takes the form (Sposito, 1994):

$$\frac{dE_1}{dt} = \left(\frac{D_1 D_2 Ma_s}{\delta Q} \right) \frac{\left(\alpha_{21} |Z_1| [1]_{bulk} E_2 - |Z_2| [2]_{bulk} E_1 \right)}{\alpha_{21} D_2 E_2 + D_1 E_1} \qquad [7.40]$$

where E_i (i = 1, 2) is a charge fraction in the exchanger, Q is either the CEC or the AEC (Equation [7.21]), and α_{21} is the exchange separation factor:

$$\alpha_{ij} \equiv \Gamma_j [i]_{surf} / \Gamma_i [j]_{surf} \qquad [7.41]$$

for two ions, i and j. Equation [7.40] describes the rate of change of the charge fraction of adsorbate ion 1 during an exchange reaction with ion 2. The prefactor on the right side reflects the film diffusion mechanism, whereas the numerator it multiplies compares the forward and backward rates of the exchange reaction. At equilibrium this numerator vanishes and

$$\alpha_{21}\big|_{\frac{dE_1}{dt} \to 0} = |Z_2| [2]_{bulk}^{eq} E_1^{eq} / |1|_{bulk}^{eq} E_2^{eq} = \Gamma_1^{eq} [2]_{bulk}^{eq} / \Gamma_2^{eq} [1]_{bulk}^{eq} \equiv \alpha_{21}^{eq} \qquad [7.42]$$

Unlike the value of α_{21} in Equation [7.40], the equilibrium value of α_{21} in Equation [7.42] depends only on the equilibrium bulk concentrations of ions 1 and 2 in aqueous solution. Aside from the appearance of α_{21}, however, the Bunzl rate law itself displays only bulk ion concentrations, regardless of the extent of reaction.

The chemical significance of the Bunzl rate law can be appreciated by examining the roles played by the parameters it contains. Its principal dependence on the film diffusion mechanism, for example, is epitomized in the film diffusion rate constant,

$$k_{idiff} \equiv D_i Ma_s / \delta V \qquad [7.43]$$

to which the prefactor in Equation [7.40] reduces (with $i = 1$) after multiplication of both sides of the equation by Q/V, in order to express the rate formally in concentration units, and then division by D_2 factored from the numerator to leave the ratio D_1/D_2 as the coefficient of E_1 in its second term. The first-order rate coefficient k_{idiff} increases as either the diffusion coefficient of ion i or the exchanger surface area per unit volume of suspension (Ma_s/V) increases, and it decreases with increasing film thickness.

The numerator that multiplies k_{idiff} in Equation [7.40] has the appearance of a rate law for a cation exchange reaction based on reaction control. With α_{21} interpreted as the ratio of an adsorption rate coefficient to a desorption rate coefficient for ion 1 (Sparks and Suarez, 1991), the numerator has the same mathematical form as a conventional difference between forward and backward rate laws for an exchanging ion, based on Equation [7.25] or [7.26]. The Bunzl rate law also is similar in mathematical form to a mixed transport-reaction control rate law for adsorption/desorption processes, as presented by Sposito (1994). These two comparisons demonstrate the pitfalls in attempting to attribute a particular reaction mechanism to a particular mathematical form of a rate law. Despite the separation factor in Equation [7.40], the observed rate of change in the charge fraction of ion 1 with time is, in fact, determined solely by the relative rates of diffusion of ions 1 and 2 through a boundary layer around the exchanger, as represented in Equations [7.37]–[7.39]. The rate of the chemical exchange reaction at the adsorbent surface is, in fact, assumed to be so large that, on the diffusion time scale, it has no influence on ion exchange kinetics. Applications of Equation [7.40] have been discussed by Bunzl and Schimmack (1991) and Sposito (1994).

Irrespective of the kinetics or mechanisms of ion exchange, a thermodynamic equilibrium constant can be formulated by conventional methods (Sposito, 1989; Sparks, 1995) for the reactions in Equations [7.25] and [7.26]. Taking cation exchange as an example, the exchange equilibrium constant is

$$K_{ex} = \frac{f_B^a x_B^a \left(A^{a+}\right)^b}{f_A^b x_A^b \left(B^{b+}\right)^a} = \frac{f_B^a}{f_A^b} K_c \qquad [7.44]$$

where f_A and f_B are rational activity coefficients for the cations A^{a+} and B^{b+} on the soil exchanger and K_c is the conditional equilibrium constant in Equation [7.30]. The left side of Equation [7.44] can depend only on temperature and applied pressure, whereas K_c on the right side of Equation [7.44] depends on composition as well.

The causes of the composition dependence of K_c are the interactions between adsorbed species, exchanger heterogeneity, and often, changes in soil particle arrangement associated with colloidal phenomena that occur as the exchanger composition varies. If mole fractions alone were adequate to reflect the effects of composition changes on K_{ex}, the exchanger mixture of $AX_a(s)$ and $BX_b(s)$ would be termed ideal according to thermodynamic convention. Real mixtures are not usually ideal, so rational activity coefficients must be introduced to correct the mole fractions in K_c for this effect and thereby maintain a constant value of K_{ex}. For example, if the Rothmund-Kornfeld model is applicable, it follows from Equations [7.35] and [7.44] that, for a homovalent cation exchange, $f_i = x_i^{[(1-\beta)/\beta]}$ ($i = A$, B; $0 \le x_i \le 1$), which is dependent on exchanger composition as represented by the mole fraction of ion i (Bond, 1995).

If the conditional equilibrium constant has been measured as a function of exchanger composition (Sumner and Miller, 1996), standard thermodynamic methods, described in a lucid manner by Sparks (1995), can be applied to show that

$$\ln K_{ex} = \int_0^1 \ln K_c dE_B \qquad [7.45]$$

It follows from Equations [7.44] and [7.45] that the rational activity coefficients must always satisfy the integral condition:

$$\int_0^1 \ln\left(f_A^b / f_B^a\right) dE_B \equiv 0 \qquad [7.46]$$

Equation [7.46] can be applied to examine either calculated values or model expressions for K_c or the activity coefficients to ensure consistency with chemical thermodynamics.

Another kind of experimental consistency check can be made if K_{ex} is determined for three or more binary exchange reactions in an N-ary exchange system (N = 3, 4, ...). Consider, for example, a ternary system with the three following cation exchange reactions:

$$bAX_a(s) + aB^{b+}(aq) = aBX_b(s) + bA^{a+}(aq) \qquad [7.47a]$$
$$eBX_b(s) + bE^{e+}(aq) = bEX_e(s) + eB^{b+}(aq) \qquad [7.47b]$$
$$eAX_a(s) + aE^{e+}(aq) = aEX_e(s) + eA^{a+}(aq) \qquad [7.47c]$$

Since these three reactions are coupled, only two can be independent, and the dependence of the remaining reaction on the other two is reflected in a constraint on the three equilibrium constants:

$$K_{ex}^{(3)b} = K_{ex}^{(2)a} K_{ex}^{(1)e} \qquad [7.48a]$$

which is verified readily by introducing Equation [7.44] for Reactions [7.47a]–[7.47c] into Equation [7.48a]. A less cumbersome notation results by indexing the three cations, A^{a+}, B^{b+}, and E^{e+} with the numbers 1, 2, and 3, respectively, such that Equation [7.44] becomes:

$$K_{ijc} \equiv K_{ij}\left(f_i^{Z_j} / f_j^{Z_i}\right) = 1/K_{jic} \qquad (i,j = 1,2,3) \qquad [7.49]$$

where Z_i, Z_j are valences and K_{ij} replaces K_{ex}. Strictly speaking, both K_{ijc} and the rational activity coefficients should also carry a designation for the total number of exchanging cations (e.g., T for ternary) to emphasize their composition dependence. This designation is not necessary for the exchange equilibrium constant K_{ij} because it is independent of composition. Equation [7.48a] can then be expressed in the equivalent logarithmic form:

$$Z_1 \ln K_{23} + Z_2 \ln K_{31} + Z_3 \ln K_{12} = Z_1 \ln K_{23c} + Z_2 \ln K_{31c} + Z_3 \ln K_{12c} = 0 \qquad [7.48b]$$

where the second equality can be verified by direct substitution of Equation [7.49]. Note that the subscripts in the three terms in Equation [7.48b] involve only the three cyclic permutations of the indices {123}: 123, 312, 231.

An expression having the simplicity and generality of the Davies equation (Sposito, 1984) for the activity coefficients of aqueous ions has not been discovered yet for the rational activity coefficients of exchangeable cations. If the conditional exchange constant has been determined as a function of composition, however, chemical thermodynamic methods can be applied to derive equations for the rational activity coefficients as functions of composition (Chu and Sposito, 1981). The result for a ternary exchange system is, in differential form (extension to an N-ary system is direct):

$$Z_i Z_j d \ln f_k^T = -Z_i (1 - E_k) d \ln K_{jkc}^T - Z_k E_i d \ln K_{ijc}^T \qquad [7.50]$$

where {ijk} is any cyclic permutation of {123}. Equation (50) can be integrated along a path in the three-dimensional Euclidean space defined by {E_1, E_2, E_3} between the points ($E_i = 0, E_j = 0, E_k = 1$) and ($E_i, E_j, E_k$) to calculate the value of $\ln f_k^T$ at the latter point in exchanger composition space. (Recall that, in any N-ary system, $\ln f_k^T \equiv 0$ when the exchanger comprises only the cation k.) This integration is facilitated if the natural logarithms of and in Equation [7.50] can be fit to polynomial expressions involving the charge fractions E_1, E_2, and E_3 (Sposito, 1994). It is well to observe, in this regard, that the quantitation of the exchanger composition in the case of a ternary or N-ary ion exchange system can impose significant experimental data requirements.

Because any K_{ij} is independent of exchanger composition, it can always be calculated with data on K_{ijc} for a binary exchange system. Generalizations of Equation [7.45] exist for an N-ary system (Sposito, 1994), but they are not necessary, or even desirable, in applications because of the formidable data set needed for their evaluation. In a thermodynamic context, K_{ij} is independent of whatever path is taken in exchanger composition space in order to perform integrals containing $\ln K_{ijc}$. The simplest path to take is the one-dimensional path exemplified in Equation [7.45], which imposes the smallest experimental data requirement. Unfortunately, activity coefficients are not, in principle, independent of composition and, therefore, they must be evaluated with data on the N-ary system to which they refer.

It is important to emphasize that N-ary ion exchange relationships like Equation [7.50], which do not enjoy composition independent status, must be examined case by case to determine whether they can be built up from binary exchange data. For example, there is no reason to expect a conditional equilibrium constant K_{ijc} to remain invariant under changes in the composition of an ion exchange system from binary to ternary. As another example, one can pose the question as to whether or not binary exchange data for rational activity coefficients can be used, in general, to predict those in an N-ary system. The answer to this question, provided by chemical thermodynamics, is no (Chu and Sposito, 1981). The fundamental difficulty is that the rational activity coefficients for, say, a ternary exchange system cannot be expressed uniquely in terms of those for a binary exchange system for mathematical reasons (Sposito, 1994). In some cases, a unique relationship may exist because of the specific nature of the exchange system, but this cannot be assumed arbitrarily (Sposito, 1994; Bond and Verburg, 1997).

Acknowledgments
Preparation of this chapter was supported in part by NSF Grant EAR 9505629. Gratitude is expressed to Ms. Angela Zabel for excellent preparation of the typescript.

7.6 References

Amhrein, C., and D.L. Suarez. 1990. Procedure for determining sodium-calcium selectivity in calcareous and gypsiferous soils. Soil Sci. Soc. Am. J. 54:999–1007.

Amhrein, C., and D.L. Suarez. 1991. Sodium-calcium exchange with anion exclusion and weathering corrections. Soil Sci. Soc. Am. J. 55:698–706.

Bargar, J.R., S.N. Towle, G.E. Brown, Jr., and G.A. Parks. 1996. Outer-sphere Pb(II) adsorbed at specific surface sites on a single crystal α-alumina. Geochim. Cosmochim. Acta 60:3541–3547.

Bargar, J.R., G.E. Brown, Jr., and G.A. Parks. 1997. Surface complexation of Pb(II) at oxide-water interfaces: II. XAFS and bond-valence determination of mononuclear Pb(II) sorption products and surface functional groups on iron oxides. Geochim. Cosmochim. Acta 61:2639–2652.

Bond, W.J. 1995. On the Rothmund-Kornfeld description of cation exchange. Soil Sci. Soc. Am. J. 59:436–443.

Bond, W.J., and K. Verburg. 1997. Comparison of methods for predicting ternary exchange from binary isotherms. Soil Sci. Soc. Am. J. 61:444–454.

Bunzl, K. 1974. Kinetics of ion exchange in soil organic matter. III. Differential ion exchange reactions of Pb²⁺ in humic acid and peat. J. Soil Sci. 25:517–532.

Bunzl, K., and W. Schimmack. 1991. Kinetics of ion sorption on humic substances. p. 119–134. *In* D.L. Sparks and D.L. Suarez (ed.) Rates of soil chemical processes. Soil Science Society of America, Madison, WI.

Chorover, J., and G. Sposito. 1995. Surface charge characteristics of kaolinitic tropical soils. Geochim. Cosmochim. Acta 59:875–884.

Chu, S.-Y., and G. Sposito. 1981. The thermodynamics of ternary cation exchange systems and the subregular model. Soil Sci. Soc. Am. J. 45:1084–1089.

Everett, D.H. 1972. Definitions, terminology and symbols in colloid and surface chemistry. Pure Applied Chem. 31:578–638.

Fair, M.C., and J.L. Anderson. 1989. Electrophoresis of nonuniformly charged ellipsoidal particles. J. Coll. Interf. Sci. 127:388–400.

Fletcher, P.F., G. Sposito, and C.S. LeVesque. 1984. Sodium-calcium-magnesium exchange reactions on a montmorillonite soil: I. Binary Exchange reactions. Soil Sci. Soc. Am. J. 48:1016–1021.

Hunter, R.J. 1993. Introduction to modern colloid science. Oxford University Press, New York, NY.

Ji, G.L. 1997. Electrostatic adsorption of anions. p. 112–139. *In* T.R. Yu (ed.) Chemistry of variable charge soils. Oxford University Press, New York, NY.

Lyklema, J. 1984. Points of zero charge in the presence of specific adsorption. J. Coll. Interf. Sci. 99:109–117.

Mattigod, S.V., and J.M. Zachara. 1996. Equilibrium modeling in soil chemistry. p. 1309–1358. *In* D.L. Sparks (ed.) Methods of soil analysis. Part 3. Chemical methods. Soil Science Society of America, Madison, WI.

Ohtaki, H., and T. Radnai. 1993. Structure and dynamics of hydrated ions. Chem. Rev. 93:1157–1204.

Oster, J.D., and G. Sposito. 1980. The Gapon coefficient and the exchangeable sodium percentage-sodium adsorption ratio relation. Soil Sci. Soc. Am. J. 44:258–260.

Parfitt, R.L. 1980. Chemical properties of variable charge soils. p. 167–194. *In* B.K.G. Theng (ed.) Soils with variable charge. New Zealand Society of Soil Science, Lower Hutt, NZ.

Petruzelli, D., F.G. Helfferich, and L. Liberti. 1991. Ion-exchange kinetics on reactive polymers and inorganic soil constituents. p. 95–118. *In* D.L. Sparks and D.L. Suarez (ed.) Rates of soil chemical processes. Soil Science Society of America, Madison, WI.

Polubesova, T.A., J. Chorover, and G. Sposito. 1995. Surface charge characteristics of podzolized soil. Soil Sci. Soc. Am. J. 59:772–777.

Schroth, B.K., and G. Sposito. 1997. Surface charge properties of kaolinite. Clays Clay Miner. 45:85–91.

Sparks, D.L. 1995. Environmental soil chemistry. Academic Press, San Diego, CA.

Sparks, D.L., and D.L. Suarez. (ed.). 1991. Rates of soil chemical processes. Soil Science Society of America, Madison, WI.

Sposito, G. 1981a. The operational definition of the zero point of charge in soils. Soil Sci. Soc. Am. J. 45:292–297.

Sposito, G. 1981b. Cation exchange in soils: An historical and theoretical perspective. p. 13–30. *In* Chemistry in the soil environment. Soil Science Society of America, Madison, WI.

Sposito, G. 1984. The future of an illusion: Ion activities in soil solutions. Soil Sci. Soc. Am. J. 48:531–536.

Sposito, G. 1989. The chemistry of soils. Oxford University Press, New York, NY.

Sposito, G. 1990. Molecular models of ion adsorption on mineral surfaces. p. 261–279. *In* M.F. Hochella and A.F. White (ed.) Mineral-water interface geochemistry. Mineralogical Society of America, Washington, DC.

Sposito, G. 1992a. Characterization of particle surface charge. p. 291–314. *In* J. Buffle and H.P. van Leeuwen (ed.) Environmental particles. Vol. 1. Lewis Publishers, Chelsea, MI.

Sposito, G. 1992b. The diffuse-ion swarm near smectite particles suspended in 1:1 electrolyte solutions: Modified Gouy-Chapman theory and quasicrystal formation. p. 128–155. *In* N. Güven and R. Pollastro (ed.) Clay-water interface and its rheological implications. The Clay Minerals Society, Boulder, CO.

Sposito, G. 1994. Chemical equilibria and kinetics in soils. Oxford University Press, New York, NY.

Sposito, G., and R. Prost. 1982. Structure of water adsorbed on smectites. Chem. Rev. 82:553–573.

Sposito, G., K.M. Holtzclaw, C.T. Johnston, and C.S. LeVesque-Madore. 1981. Thermodynamics of sodium-copper exchange in Wyoming bentonite at 298 K. Soil Sci. Soc. Am. J. 45:1079–1084.

Sposito, G., C.S. LeVesque, and D. Hesterberg. 1986. Calcium-magnesium exchange on illite in the presence of adsorbed sodium. Soil Sci. Soc. Am. J. 50:905–909.

Sumner, M.E., and W.P. Miller. 1996. Cation exchange capacity and exchange coefficients. p. 1201–1229. *In* D.L. Sparks (ed.) Methods of soil analysis. Part 3. Chemical methods. Soil Science Society of America, Madison, WI.

Zelazny, L.W., L. He, and A. Vanwormhoudt. 1996. Charge analysis of soils and anion exchange. p. 1231–1253. *In* D.L. Sparks (ed.) Methods of soil analysis. Part 3. Chemical methods. Soil Science Society of America, Madison, WI.

Zhang, Z.Z., and D.L. Sparks. 1996. Sodium-copper exchange on Wyoming montmorillonite in chloride, perchlorate, nitrate, and sulfate solutions. Soil Sci. Soc. Am. J. 60:1750–1757.

8

Chemisorption and Precipitation Reactions

M.B. McBride
Cornell University

8.1 Introduction

Soils have a remarkable, yet finite, ability to remove ions and molecules from water by sorption reactions. Sorption is defined broadly here as the transfer of ions and molecules from the solution phase to the solid phase. In Section B, Chapter 7, it was shown that those elements which take the form of cations are retained in soils by cation exchange on clays and humus. However, more selective and less reversible sorption reactions such as complexation with organic functional groups and bonding on variable charge minerals (e.g., oxides, allophane) can also retain and even immobilize metal cations. Elements in anionic form in solution are commonly more mobile than metal cations (phosphate is an obvious exception). Nevertheless, they can be retained in soils, primarily by selective bonding (chemisorption) processes at variable charge mineral surfaces and layer silicate particle edges. These types of cation and anion adsorption are collectively referred to as specific adsorption to distinguish them from ion exchange. In addition, chemical precipitation may participate in the removal of ions from water and is treated in this chapter as a part of sorption.

This chapter will discuss the important mechanisms by which ions, whether present naturally in the soil or introduced by pollution, are sorbed by specific adsorption and precipitation processes that are inherently less reversible than ion exchange. The chapter is intended to provide a fundamental basis for understanding inorganic cation and anion adsorption and precipitation, the reactions which limit solubility of many of the elements.

8.2 Surface Functional Groups

8.2.1 Mineral Surfaces

Noncrystalline aluminosilicates (allophanes), oxides and hydroxides of Fe, Al and Mn, and even the edges of layer silicate clays, provide surface sites in soils for the chemisorption of transition and heavy metals. All of these minerals present a similar type of adsorptive site to the soil solution: a valence-unsatisfied (terminal) OH^- or H_2O ligand bound to a metal ion (usually Fe^{3+}, Al^{3+}, or $Mn^{3+,4+}$). For example, on Fe oxides, a trace metal (M) may bind according to the reaction:

$$> Fe-OH]^{-1/2} + M(H_2O)_6^{n+} \rightarrow Fe-O-M(H_2O)_5]^{(n+3/2)^+} + H_2O^+ \qquad [8.1]$$

This reaction is distinguished from cation exchange by at least 4 features: (1) release of as many as n H^+ ions for each M^{n+} cation adsorbed; (2) a high degree of specificity shown by particular minerals for particular trace metals; (3) tendency toward irreversibility, or at least a desorption rate that is orders of magnitude slower than the adsorption; and (4) change in the measured surface charge toward a more positive value. This last feature implies that the adsorbed metal and its charge become part of the mineral surface, thereby shifting the point of zero charge (p.z.c.) to higher pH.

Like metal cations, anions chemisorb on soil minerals which possess reactive surface hydroxy groups. The most important minerals in this regard are noncrystalline aluminosilicates (allophanes), oxides and hydroxides of Fe, Al and Mn, and layer silicate clays (edge sites only). Again, it is the valence-unsatisfied OH^- or H_2O ligands bound to surface metal ions (usually Fe, Al, or Mn) that are the sites of chemisorption. In general terms, the surface reaction can be written:

$$> S-OH + A^{n-} = \; > S-A^{(n-1)-} + OH^- \qquad\qquad K_{A1} \qquad\qquad [8.2]$$

or, for the binuclear reaction :

$$2 > S-OH + A^{n-} = \quad \begin{array}{c} >S \\ \diagdown \\ \diagup \\ >S \end{array} A^{(n-1)-} + 2OH^- \qquad\qquad K_{A2} \qquad\qquad [8.3]$$

where A^{n-} is an anion of charge $-n$, and $> S-OH$ is a reactive surface hydroxyl group. Reactions [8.2] and [8.3] are termed ligand exchange reactions because the anion displaces OH^- or H_2O from coordination positions of a metal ion at the surface. These reactions are favored by low pH, as is evident from the release of OH^- into solution. Low pH causes surface OH^- groups to accept protons, and since H_2O is an easier ligand to displace from metal bonding sites than OH^-; this facilitates the ligand exchange. Therefore, from the point of view of kinetics as well as equilibrium thermodynamics, low pH promotes anion adsorption.

Anions may also adsorb on positively charged mineral surfaces by anion exchange, a process involving nonspecific electrostatic forces. Ligand exchange is distinguished from anion exchange based on the following characteristics: (1) release of up to 2 OH^- into solution for each anion adsorbed; (2) a high degree of specificity shown toward particular anions; (3) a tendency to be nonreversible, or at least for desorption to be much slower than adsorption; and (4) a change in the measured surface charge to a more negative value.

The rules that describe reactivity of minerals toward anions are much the same as those outlined for cation adsorption. Specifically, the most reactive groups are valence unsatisfied surface OH^- ions, found at layer silicate edges, on oxides of Fe, Al and Mn and on noncrystalline aluminosilicates. Reactive groups are most numerous on noncrystalline minerals such as ferrihydrite, allophane, and imogolite.

The above features are indicative of strong chemical interactions with surfaces, or chemisorption, in which a covalent or short-range electrostatic bond forms between the molecule and the surface. Weak adsorption, on the other hand, is characteristic of physical adsorption, in which the bonding interaction is not very energetic (typically < 10 kcal mol^{-1} adsorbate). Physical adsorption is more typical of nonpolar and weakly polar organic molecules, and will not be discussed further. The contrasting features of chemisorption and physical adsorption are listed in Table 8.1.

Some minerals are much more reactive than others on a weight basis. Reactivity is limited by the type and number of terminal (valence-unsatisfied) groups at the surface. For example, gibbsite has no such groups on its (001) crystal faces because each OH group is coordinated to two Al ions, and is valence satisfied according to Pauling's rules (Fig. 8.1). Since (001) surfaces constitute most of the surface area of the plate-like gibbsite crystals, this mineral, even in the microcrystalline form, chemisorbs very small quantities of metals such as Cu^{2+}. Bonding probably occurs at edges that possess H_2O or OH$^-$ groups coordinated to a single Al^{3+} ion. In contrast, noncrystalline oxides and allophanes possess larger numbers of valence-unsatisfied groups because of their structural disorder, and can chemisorb more trace metals.

8.2.2 Organic Surfaces

With the possible exception of some noncrystalline minerals with very high surface areas, soil organic matter (SOM) has the greatest capacity and strength of bonding with most trace metals of any soil component. It is, therefore, common to observe statistically significant correlations between solubility and SOM content for metals such as Cu, Hg and Cd (Boekhold and van der Zee, 1992; Lee et al.,1996; Tichy et al.,1997; Yin et al.,1997; McBride et al., 1997). In the field, trace metals are commonly associated with organic rich soils or horizons within soils (Boekhold and van der Zee, 1992; Tichy et al., 1997). A regression analysis of total metals content of United States soils versus soil properties (Holmgren et al., 1993) reveals that several trace metals (Cd, Cu, Zn, Pb, Ni) are statistically correlated with SOM content. The strongest correlation ($p < 0.01$) was for total soil Cd (Cd_T):

$$Cd_T = 0.10 + 0.0094 \text{ SOM} \qquad r = 0.51 \qquad [8.4]$$

where SOM is in units of g kg^{-1} soil. The relationship of Cd content to SOM is presumably due to its bioaccumulation and retention against long-term leaching, although soils which had received heavy applications of metal contaminated animal manures or sewage sludges could have contributed to the

Table 8.1 Characteristics of physical and chemical adsorption

Property	Physical	Chemical
Heat of adsorption	< 10 kcal mol^{-1}	> 20 kcal mol^{-1}
Temperature range of adsorption	Only below boiling point of adsorbate	Both low and high temperature
Slope of adsorption isotherm	Greater at higher adsorbate concentration	Less at higher adsorbate concentration
Dependence on properties of adsorbent	Relatively little	Great
Dependence on properties of adsorbate	Great	Great
Activation energy for adsorption	Low or none	May be high
Number of layers of adsorbed molecules	Multiple (at most)	Single (at most)

correlation. Also, organic soils frequently contain higher concentrations of Cd (and several other trace metals) than mineral soils in the same region.

Soil organic matter contains many different types of functional groups that act as Lewis bases in metal bonding reactions. They include carboxylic, phenolic, amine, carbonyl, and sulfhydryl groups, with the first two being the most abundant. The sulfhydryl (and to some degree, amine) groups have the greatest ability to form strong covalent bonds with soft metals such as Hg^{2+}, Cd^{2+}, Tl^+, and Ag^+. Given this diversity in functional groups associated with SOM, a very wide range of bonding strengths for metals can be expected, depending on the level of metal addition (Bresnahan et al., 1978; Waller and Pickering, 1993). Large metal additions force bonding to occur at the predominant functional groups (carboxyl). Reported bonding strengths for trace metals such as Cu are low in these cases (Bresnahan et al., 1978), but should not be taken to suggest that similarly weak bonding would occur in field soils at realistic metal loadings. Recent research suggests that some metals, such as Hg^{2+} (Skyllberg et al., 1996) and Cd^{2+} (Mench et al., 1994) are bonded to S-containing ligands in the soil, reactions that could form very strong complexes.

8.3 Surface Complexes

8.3.1 Mineral Surfaces

In chemisorption, the adsorbing cation or anion bonds directly by an inner sphere mechanism to atoms at the surface. Consequently, the properties of the surface, and therefore, the nature of the metal constituting the adsorption site, should dictate to some extent the tendency for adsorption. For example, if oxides of different metals are compared, the valence and coordination number of the constituent metal is correlated with reactivity. Comparing silica to alumina, >Si-OH groups are found to be more acidic than > Al-OH groups. The fact that Si has a valence of +4, a charge shared more or less equally by four coordinated O atoms, means that a surface OH group (terminal) on silica feels a charge of $+^4/_4 = +1$ from Si. By the same reasoning, OH (terminal) on alumina feels $+^3/_6 = +^1/_2$ charge from Al^{3+}. This means that electron density is drawn more strongly away from the OH of silica, thereby creating a stronger acid on the silica surface. Furthermore, the dissociated -Si-O⁻ group is a weak Lewis base because of this polarization effect, and chemisorption of trace metals is not very favorable on silica.

In general, then, the ratio of metal valence to coordination number, referred to hereafter as shared charge, seems to be a useful parameter in predicting surface acidity and reactivity of oxides, and the basic concept outlined here for silica and alumina can be applied to other metal oxides. The

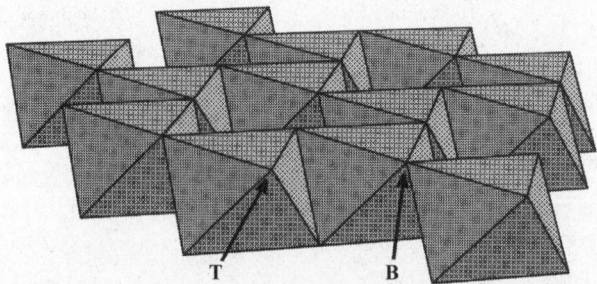

Fig. 8.1   Polyhedral depiction of a section of the planar (001) surface of gibbsite, indicating positions of bridging (B) and terminal (T) OH⁻ groups in edge $Al(OH)_6$ octahedra

Table 8.2　Acid dissociation constants of oxide surface OH groups [Adapted from Schindler and Stumm, 1987]

Group	pK$_1$*	pK$_2$**	Valence/Coordination No.
>Si-OH	< 2	6 - 7	4/4 = 1
>Ti-OH	3 - 4	7 - 9	4/6 = 0.67
>Fe-OH	6.5	9	3/6 = 0.5
>Al-OH	5 - 7.5	8 - 10	3/6 = 0.5

* $K_1 = \{S\text{-}OH\}[H^+]/\{S\text{-}OH^{2+}\}$
** $K_2 = \{S\text{-}O^-\}[H^+]/\{S\text{-}OH\}$.
 S is the metal ion of the particular surface group.

approximate surface acid dissociation constants for several oxides, compared to their ratios of valence to coordination number, are listed in Table 8.2.

Extending the argument above, which is based on simple electrostatics, the tendency of O atoms at surface sites to chemisorb metals should vary depending on the valence and coordination number of the metal constituting the adsorbent; that is, the adsorbing groups listed toward the bottom of Table 8.2 should chemisorb metals more effectively than those toward the top. For groups with very similar values of valence/coordination number, say Al-OH and Fe(III)-OH, the electronegativity of the constituent metal also has to be considered in explaining differences in trace metal adsorption behavior.

For oxyanion adsorption on minerals, the above shared charge concept is reversed, because in this case, the Lewis base is the adsorbate and the Lewis acid is the adsorbent (surface). Therefore, the shared charge of the O ligand of the anion is used as a predictor of tendency to chemisorb on a given mineral surface (Section 8.9). The tendency of oxyanions to adsorb on a given mineral, goethite for example, increases as the positive charge formally accepted by the O ligand from the central atom in the oxyanion is reduced (McBride, 1994).

8.3.2　Organic Surfaces

Organic matter in soils is very effective in chemisorbing metal cations, particularly polyvalent ones. Although the metal bonding reaction can be written as an ion exchange process between H$^+$ and the metal at acidic functional groups or ligands (L):

$$M^{x+} + >LH_y = >L-M^{(x-y)+} + yH^+ \tag{8.5}$$

the high degree of selectivity shown by SOM for certain metals signifies that some metals coordinate directly (form inner sphere complexes) with the functional groups. Strong ionic and covalent bonds are formed in these complexes.

The nature of the organic ligands involved in coordination bonding with the metals determines the strength and reversibility of the metal sorption. Soil organic matter contains numerous ligands capable of forming coordination bonds with metals, including carboxylic, phenolic, amine, carbonyl, and sulfhydryl groups. These ligands can be classified as Lewis bases according to their hardness, with the order of hardness being O > N > S, where S is in the reduced state (valence of –2). Calcium, for example, is a hard acid, preferentially complexing with O containing groups such as carboxylate. Conversely, Cd^{2+}, an ion with the same charge and radius as Ca^{2+}, is a soft acid, and preferentially complexes with S^{2-} containing groups such as sulfhydryl. If Cd^{2+} adsorption behavior in soils is now considered, it is found that low levels of the metal are adsorbed with very high preference over Ca^{2+}, but at high levels, preference over Ca^{2+} diminishes to nearly equal affinity. It seems that the soil contains a limited number of soft bases (ligands), and once these complex with Cd^{2+}, the remaining bases (presumably the harder carboxylic and phenolic groups) fail to show selectivity for Cd^{2+} over Ca^{2+}. A similar tendency has been noted for other trace metals, and in the case of Cu^{2+}, there is electron spin resonance (ESR) spectroscopic evidence that more covalent bonds are formed with humus at low than high metal loadings (McBride, 1989). This result could indicate preferential complexation of Cu^{2+} with amine, polyphenolic, or other ligands softer than the prevalent carboxylic acid groups. Because of the limited capacity in SOM for highly selective bonding, an important principle can be stated: the degree of preference for one metal over another on SOM reduces with increasing adsorption of the preferred metal.

A further contribution to stability in metal organic matter complexes is the entropy created by the complexation reaction. This is only a consideration if the metal bonds to two or more functional groups, for example:

$$
M^{X+} + 2 > LH = \begin{matrix} > L \\ \diagdown \\ \diagup \\ > L \end{matrix} M^{(x-2)+} + 2H^+
\tag{8.6}
$$

There is a net release of one mole of ions into solution by this reaction, because two moles of protons are released for each mole of metal ions complexed. The greater degrees of rotational and translational freedom associated with this release contribute a positive entropy term, and hence, a more negative free energy for Reaction [8.6]. Consequently, the reaction is likely to be spontaneous in the direction written.

8.3.3 Ternary Complexes

Soil solutions contain numerous metal cations and anions simultaneously, each of which may be able to chemisorb on surfaces. For example, because the same type of reactive Fe or Al hydroxyl group on soil oxides is thought to be responsible for both cation and anion adsorption on these surfaces, there can be competition for these sites both among and between groups of cations and anions. Consequently, anions which are particularly active in chemisorption, such as phosphate and sulfate, should suppress metal cation adsorption on variable charge surfaces. What is sometimes found, however, is that metal ion adsorption is enhanced by the presence of certain anions, and *vice versa* (McBride, 1989). This appears to be a process in which adsorption of anions and cations is not competitive, and is sometimes greater in combination than separate. The explanation of this behavior is based on ternary complex formation at variable charge mineral surfaces, schematically described as follows:

$$
> S-OH + M + A = > S-O-M-A + H^+
\tag{8.7}
$$
$$
> S-OH + M + A = > S-A-M-A + OH^-
\tag{8.8}
$$

Reaction [8.7] forms a Type A ternary complex in which the metal ion (M^{x+}) links the surface to the anion (A^{n-}). Reaction [8.8] forms a Type B ternary complex in which the anion forms a bridge between the metal ion and the surface. Type A complexes seem to be the more common of the two.

Examples of observed ternary complexes include Cu^{2+}, Pb^{2+}, Cd^{2+} and Zn^{2+} with PO_4^{3-} on Fe and Al hydroxide surfaces, in which the presence of PO_4 in solution promotes trace metal adsorption and lowers metal solubility (McBride, 1989). In soils, the presence of exchangeable Ca^{2+} lowers the solubility of PO_4, perhaps because of Ca phosphate ternary complexes formed on minerals. Many Type A complexes of trace metals with chelating organic ligands are known. Ternary complexes are also believed to form on SOM as multivalent cations such as Al^{3+} and Fe^{3+} bond simultaneously to the functional groups of humus and to anions such as PO_4. This affords a way for SOM to adsorb certain anions to a limited extent.

Ternary complexes seem only to form between multivalent cations (particularly transition and heavy metals) and anions (or uncharged molecules) possessing at least two metal coordinating ligand positions. Consequently, organic ligands with chelating tendencies such as oxalate, bipyridine, glycine, and ethylenediamine form particularly stable Type A complexes. Organic ligands with three or more coordinating positions may not form ternary complexes, at least, when present in excess of the metal, since the metal cation must dissociate from the surface in order to maximize bonding with the ligand. The metal chelating anion (EDTA) tends to solubilize chemisorbed metals, bringing them into solution as metal EDTA complexes. In fact, if any anion is able to form a soluble complex with a metal cation, that anion competes with adsorbing surfaces to diminish metal ion adsorption. Whether or not an anion or ligand enhances metal cation adsorption by ternary complex formation, or suppresses it by competition with the surface may depend on the anion/metal ratio. A large mole excess of the anion generally suppresses metal cation adsorption, while molar parity with the metal favors adsorption by ternary complex formation. The direct consequence of ternary complex formation in soils is likely to be that solubilities of numerous anions and trace metal cations are lowered below that expected from either chemisorption or precipitation.

8.3.4 Identification of Surface Complexes by Spectroscopy

The identification of metal complexes, both inner- and outer-sphere, at metal oxide, silicate clay and SOM surfaces, has been made by spectroscopic techniques such as electron spin resonance (McBride, 1989; McBride, 1990) and Extended X-ray Absorption Fine Structure Spectroscopy (EXAFS) (Brown et al., 1989; Manceau et al., 1994; Scheidegger and Sparks, 1996). Even ternary complexes such as oxide-Cu-glycine can be identified by electron spin resonance (ESR) due to the pertubation of the ligand field of surface-bound Cu upon the replacement of inner sphere H_2O by glycine (McBride, 1985).

Outer-sphere metal complexes are distinguished by the observation that metals bound to surfaces in this way retain their inner sphere of hydration (at least) and undergo rapid thermal rotation at room temperature. Rotational motion averages the magnetic parameters of the ESR spectrum to produce an isotropic signal. In contrast, the spectrum of a metal in an inner sphere complex at a surface is anisotropic, referred to as the rigid limit spectrum. On oxides of Fe, Al and Mn, as well as on SOM, most heavy metals and oxyanions studied to date form inner sphere complexes, as identified by ESR and EXAFS. Still, some relatively weakly adsorbing species, such as Mn^{2+} on humus (McBride, 1978a), and selenate on Fe oxides (Hayes et al., 1987), may form outer-sphere complexes. The latter observation has been contradicted by a later EXAFS study, which reported that selenate, like the more strongly adsorbing selenite and arsenate, forms a binuclear bidentate (inner sphere) complex on Fe oxide surfaces (Manceau and Charlet, 1994). Infrared spectroscopic studies have reported inner sphere complexes of other weakly adsorbing anions such as sulfate and nitrate on oxides, but it is likely that in these studies the drying of the samples prior to analysis shifted the anions from outer sphere to inner sphere positions by the removal of water.

8.4 Precipitation and Coprecipitation of Inorganic Ions

8.4.1 Concepts of Precipitation

For many of the more abundant elements in soils, such as Al, Fe, Si, Mn, Ca, Mg (and perhaps P and S), precipitation of mineral forms is common and may control their solubility. For most of the trace metals, however, precipitation is less likely than chemisorption because of the low soil concentrations of these metals. Only when soils become heavily polluted does the metal solubility reach a level that supports precipitation of pure mineral phases. The ensuing discussion is relevant if a high level of a trace metal has accumulated in the soil, either from pollution or from natural geochemical processes.

As the concentration of an ion in a solution is increased, precipitation of a new solid phase will not occur until the solubility product of that phase has been exceeded. That is, some degree of supersaturation is required because crystal nuclei can only be formed after an energy barrier has been overcome. The status of the solution phase with respect to precipitation can be quantified by comparing the ion activity product (IAP) to the solubility product of the crystalline solid (K_{so}). The ratio (IAP/K_{so}) defines undersaturation ($IAP/K_{so} < 1$), saturation ($IAP/K_{so} = 1$), and supersaturation ($IAP/K_{so} > 1$) of the solution with respect to a particular solid phase. Because of the higher K_{so} of smaller crystallites and nuclei, precipitation can only begin in homogeneous solutions if supersaturation is exceeded by a large margin ($IAP/K_{so} > 100$). Extreme supersaturation leads to rapid formation of crystal nuclei and produces many very small crystallites or even noncrystalline solids. Minimal supersaturation, on the other hand, results in exceedingly slow nuclei formation, so that crystal growth occurs at these few nuclei only and a highly crystalline product with large crystals, if any, results.

In soil solutions, heterogeneous is much more likely than homogeneous nucleation because mineral and organic surfaces are present that can catalyze the nucleation step of crystallization. The energy barrier to nucleation is reduced or removed by these surfaces, especially in cases where there is crystallographic similarity between the surface and the precipitating phase. This reduces the extent of supersaturation necessary for precipitation to be initiated. For example, solutions supersaturated with respect to gibbsite do not always form a precipitate, but the presence of smectite promotes gibbsite precipitation. Similarly, $CaCO_3$ in soils seems to promote the heterogeneous nucleation of $CdCO_3$, thereby preventing solutions in Cd^{2+}-contaminated calcareous soils from becoming supersaturated (McBride, 1989). Nevertheless, precipitation reactions are often much slower than chemisorption reactions in soils, so that time-dependent sorption of metals and other ions is often a characteristic of precipitation.

8.4.2 Likely Precipitates in Soils

The solubility products of some of the least soluble minerals, which are most likely to precipitate in the chemical environment of a soil thereby setting an upper limit on trace and heavy metal solubility, are presented in Table 8.3. For example, Cd or Pb solubility is most likely to be limited by carbonate or sulfide precipitation, depending on the redox potential of the soil. The solubility of Hg may also be controlled by sulfide in reducing soils, while the relatively insoluble oxide can form in nonacidic aerobic soils. Unfortunately, the hydrolysis of Hg^{2+} produces $Hg(OH)_2^0$, a moderately soluble molecule that maintains a hazardous concentration of Hg in solution even though the concentration of the free Hg^{2+} cation is extremely low in the neutral pH range (Baes and Mesmer, 1986). This example illustrates the pitfalls of using solubility products alone to assess potential mobility and toxicity without also considering speciation of the dissolved metal.

Table 8.3 Solubility products of metal carbonates, oxides and sulfides [Data from Stumm and Morgan,1996, and Baes and Mesmer, 1986]

Carbonates: $K_{SO} = (M^{2+})(CO_3^{2-})$

	Pb	Cd	Fe	Mn	Zn	Ca
$-\log K_{SO}$	13.1	11.7	10.7	10.4	10.2	8.42

Oxides and Hydroxides: $K_{SO} = (M^{n+})(OH^-)^n$

	Fe^{3+}	Al^{3+}	Hg^{2+}	Cu^{2+}	Zn^{2+}	Pb^{2+}	Fe^{2+}	Cd^{2+}	Mn^{2+}	Mg^{2+}
$-\log K_{SO}$	39	31.2	25.4	20.3	16.9	15.3	15.2	14.4	12.8	11.2

Sulfides: $K_{SO} = (M^{2+})(S^{2-})$

	Hg	Cu	Pb	Cd	Zn	Fe	Mn
$-\log K_{SO}$	52.1	36.1	27.5	27.0	24.7	18.1	13.5

Other precipitates that can sometimes form in soils include silicates, phosphates, and sulfates. In isolated instances, these solids may limit the solubility of certain trace metals. In the case of Pb and Zn, phosphate concentrations in well-fertilized soils may be high enough to favor metal phosphate precipitates over oxides or carbonates. Although phosphate minerals, notably pyromorphite, should limit the activity of Pb^{2+} to below 10^{-9} at pH 7 according to published solubility products (Lindsay, 1979), a substantially higher Pb^{2+} activity in pure systems containing Pb and phosphate has been measured (Sauvé et al., 1998) suggesting that early published solubility products for pyromorphite and related Pb phosphates are too low. Nevertheless, apatite (Ca phosphate) is a rather effective sink for trace metals, particularly Pb, even at low pH (Chen et al., 1997; Xu et al., 1994), providing phosphate for the formation of pyromorphite. Addition of soluble phosphate to Pb-contaminated soils is an effective way to reduce Pb solubility and bioavailability in severely contaminated soils (Berti and Cunningham, 1997). Still, it does not appear that separate Pb phosphate phases comprise a significant fraction of the total Pb in heavily contaminated soils; rather, dispersal of the heavy metal over colloidal mineral and organic surfaces is the more likely condition (Cotter-Howells, 1996).

8.5 Adsorption Versus Precipitation

8.5.1 The Adsorption-Precipitation Continuum

Chemisorption reactions in soils, which are two-dimensional surface processes, can rarely be separated experimentally from the three-dimensional nucleation and precipitation reactions. It is perhaps best to view the removal of adsorbate ions from solution, termed generically as sorption, as a continuous process that ranges from chemisorption at the low to precipitation at the high end of solubility. Unless a new solid phase can be detected, the onset of precipitation and termination of chemisorption during sorption is usually not recognized by experimentalists.

The metal-surface bonding reaction tends to be favored by the same conditions that favor hydrolysis and precipitation reactions of metals in solution, namely, metals with high charge, small radius, and softness (polarizability), and high pH. This is not coincidence as there are parallels between bonding of a metal ion with a surface O^{2-} or OH^-, and coordination of the same metal ion with OH^- in solution. The parallels are made apparent by writing the chemisorption and hydrolysis reaction for the metal (M^{n+}) as:

$$> S-OH \; + \; M^{n+} \; = \; > S-O-M^{(n-1)+} + H^+ \qquad\qquad [8.9]$$
$$\textit{chemisorption}$$

$$H-OH \; + \; M^n \; = \; H-O-M^{(n-1)+} + H^+ \qquad\qquad [8.10]$$
$$\textit{hydrolysis}$$

It should not be surprising that the following orders of preference for adsorption on hematite: Pb > Cu > Zn > Co > Ni > Mn; goethite: Cu > Pb > Zn > Co > Ni > Mn; calcite: Cd > Zn > Mn > Co > Ni > Ba, Sr bear some relationship to the solubility of these metals as pure hydroxides and carbonates (Table 8.3).

The tendency for easily hydrolyzed metals to appear to be strongly adsorbed on soil minerals may be caused in part by the manner in which adsorption is traditionally measured. Typically, the pH is adjusted over a range, and the amount of metal removed from solution is calculated from the change in solution concentration. As the pH is raised from an acid value, chemisorption is initially favored, but adsorption sites may become saturated, and metal ions then cluster into metal oxide or hydroxide nuclei at the surface. Ultimately, precipitation of the metal oxide or hydroxide ensues. This sequence of processes usually appears as a sorption continuum. Because the onset of metal chemisorption and the beginning of metal hydroxide precipitation are not often separated by a wide margin of pH, the chemisorption-nucleation-precipitation sequence is rarely resolved into discrete processes by clear discontinuities in the sorption curves.

Since precipitation ultimately removes all of the strongly hydrolyzing metals from solution as the pH is raised, isotherms of the type depicted in Fig. 8.2 for metal adsorption on Fe oxides necessarily

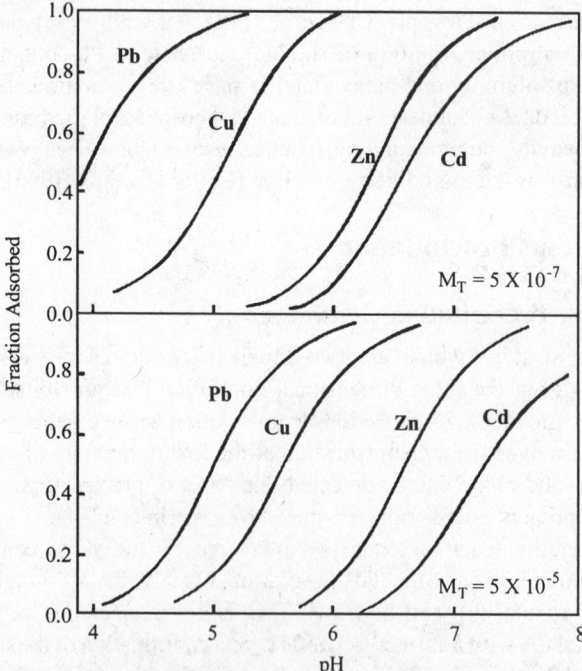

Fig. 8.2 Adsorption edges for Cd^{2+}, Cu^{2+}, Pb^{2+} and Zn^{2+} on amorphous iron hydroxide at two metal/iron hydroxide ratios. [Isotherms redrawn from Benjamin and Leckie, 1981]

reach 100 % sorption at high pH regardless of the nature and quantity of adsorbing surfaces. It is not possible to tell from the isotherm how much of the reported sorption is chemisorption and how much is precipitation; this depends on the relative quantities of reactive surface groups and adsorptive metals. Recent results with La sorption indicate that different mineral surfaces promote metal hydrolysis and metal hydroxide precipitation over different pH ranges, depending on the fundamental charge properties (p.z.c.) of the minerals (Fendorf and Fendorf, 1996). Thus, generalizations about the relative degree of chemisorption and precipitation on different surfaces may be impossible.

8.5.2 Metal Cluster Formation

Metal hydroxy polymer formation and nucleation on mineral surfaces occur when the solution phase is still undersaturated with respect to homogeneous precipitation of the metal hydroxide or oxide. As studies of Cr(III) sorption on silica and goethite have shown, sorption can be dominated by the formation of metal hydroxy clusters at isolated locations on the surface, even when the adsorbate loading is well below monolayer coverage (Fendorf et al., 1996). Isolated cluster formation is more favorable on silica, a rather weak adsorbent for Cr(III), than on goethite, where the higher density of strong bonding sites appears to favor a more even surface distribution of clusters. Segregation into metal hydroxy clusters at surfaces is probably symptomatic of relatively weak metal-surface interaction or a low number or density of reactive surface sites.

Cluster formation appears to be a common, perhaps dominant, sorption process for strongly hydrolyzing metal cations on various oxide and hydroxide minerals (McBride, 1982; Bleam and McBride, 1985,1986). Similar processes may also occur on silicate clays (Scheidegger et al., 1996), although clays with high CECs may adsorb metals predominantly by ion exchange except at rather high pH. During sorption, the slow formation of mixed surface phases involving both the adsorbing metal and metals from the mineral (e.g., Al) may occur, gradually lowering the solubility and extractability of the sorbed metal. This process will be discussed later in the Section 8.11.

Sorption, then, is described by several processes, beginning with chemisorption of metals or oxyanions at discrete sites, the most stable form of surface retention. At some higher level of sorption, polymerization into clusters is promoted by the surface, perhaps at the chemisorption sites. Finally, the clusters grow or coalesce to the point that a separate precipitated phase can be identified, and the influence of the surface in controlling solubility is overcome. A schematic picture of this proposed sequence is shown in Fig. 8.3.

8.5.3 Solubility Diagrams

The nucleation and clustering processes observed at surfaces produce IAP values in solution that can not be described by standard solubility products for pure precipitated phases. In fact, the solubility product approach to ion solubility is usually not successful for trace elements such as Cu and Zn unless the soil is grossly contaminated with the element in question, generally being more useful for elements with moderate to high total concentration in soils (Fe, Al, Ca, P, etc.). At low to moderate concentrations, trace metal solubility is usually much lower than that expected from the solubility product of likely precipitates due to solubility control by surface adsorption and other processes. For example, when Zn^{2+} is added to a soil that has been adjusted to different pH values, Zn^{2+} solubility levels in the soil solutions (Fig. 8.4) reveal undersaturation with respect to the solubility lines of all known pure solid precipitates of Zn^{2+} that could reasonably be expected to form in the soil. Possible explanations for the low solubility in the soil include (1) the chance that solid phases not previously considered might control Zn^{2+} solubility at an even lower level; (2) the possibility that chemisorption sites are not saturated, or that cation exchange is involved, so that adsorption limits Zn^{2+} solubility at

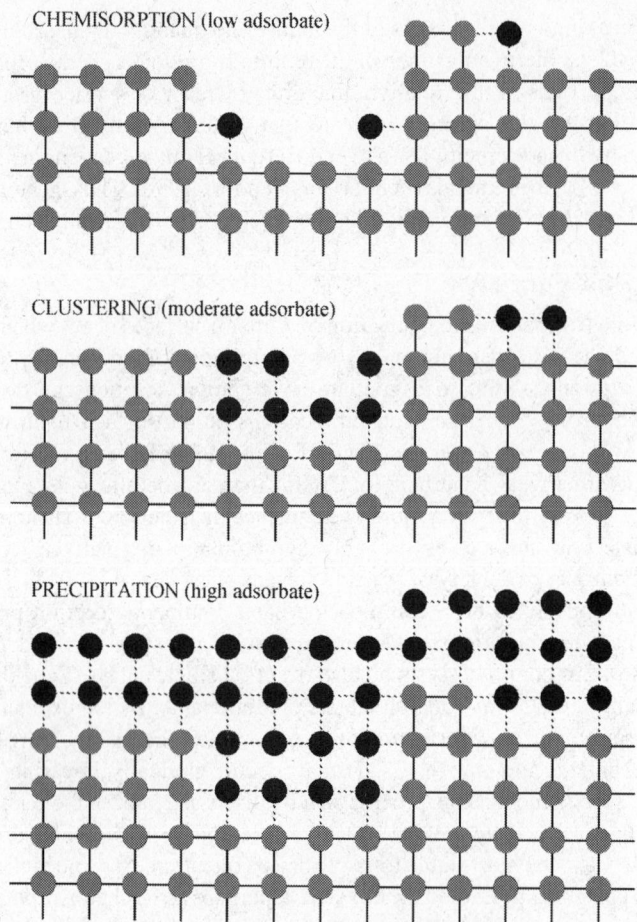

Fig. 8.3 Schematic view of the progression from metal chemisorption at discrete surface sites (IAP « K_{SO}), through clustering (IAP < K_{SO}), to surface precipitation (IAP = K_{SO}). Structural metal ions of adsorbent are shaded, while adsorbate ions are shown in black. Structural anions are not shown for simplicity.

a lower level than precipitation; and (3) the existence of mixed solid phases (coprecipitates) in which the trace ion has lower solubility than in pure Zn phases.

The first explanation can probably be ruled out despite thermodynamic data indicating that certain spinel type structures such as $ZnFe_2O_4$ and Zn_2SiO_4 are less soluble (more stable) than any of the pure solid phases represented in this figure (Lindsay, 1979). There is no evidence that these minerals can crystallize at the pressure-temperature regime typical for soils. They are sometimes found in soils as residual minerals, evidently having formed under metamorphic conditions. The second explanation may apply in most situations. Cation exchange and specific adsorption reactions of trace levels of metals can lower metal solubility below that of the pure solid phases. The formation of coprecipitates can also lower the solubility of trace elements, and will be discussed later.

8.6 Adsorption Isotherms

8.6.1 Isotherm Classification by Shape

Adsorption data are most commonly represented by an adsorption isotherm, which is a plot of the quantity of ion (or molecule) adsorbed by a solid as a function of the concentration of that ion in the bathing solution phase that is at equilibrium with the solid. The shape of this isotherm line suggests (but does not confirm) information about the adsorbate-adsorbent (ion surface) interaction; to this end, isotherms have been classified into 4 types, illustrated in Fig. 8.5:

(1) The **L-type** (Langmuir) isotherm reflects a relatively high affinity between the adsorbate and adsorbent, and is usually indicative of chemisorption.

(2) The **S-type** isotherm suggests cooperative adsorption, which operates if adsorbate-adsorbate is stronger than the adsorbate-adsorbent interaction. This condition favors the clustering of adsorbate molecules at the surface because they bond more strongly with one another than with the surface.

(3) The **C-type** (constant partitioning) isotherm, which suggests a constant relative affinity of the adsorbate molecules for the adsorbent, is usually observed only at the low range of adsorption. Deviation from the linear isotherm is likely at high adsorption levels. Nevertheless, because many nonpolar organic compounds of interest in soils are adsorbed at quite low concentrations, the linear C-type isotherm is often a reasonable description of reality.

(4) The **H-type** isotherm, indicative of very strong adsorbate-adsorbent interaction (i.e., chemisorption), is really an extreme case of the L-type. This isotherm is not often encountered with organic molecules because few of them form strong ionic or covalent bonds with soil colloids.

More complex isotherm shapes are possible, but these can be generally recognized as hybrids of the four main types listed here. Only the H-and L-types are typical for chemisorption, with the S-and C-types being more indicative of physical adsorption (Table 8.1). It must be remembered that isotherm features can never prove the adsorption mechanism involved; they can only point to reasonable

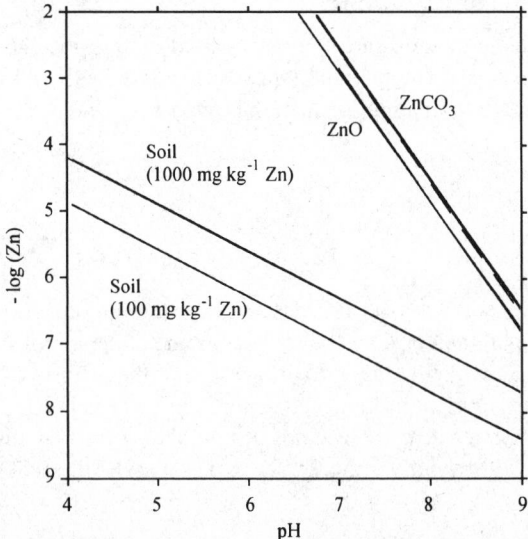

Fig. 8.4 Solubility of Zn [$-\log(Zn^{2+})$] as a function of pH for long-contaminated soil at two levels of total Zn. Solubility lines of pure Zn mineral phases are shown, assuming atmospheric CO_2 pressure.

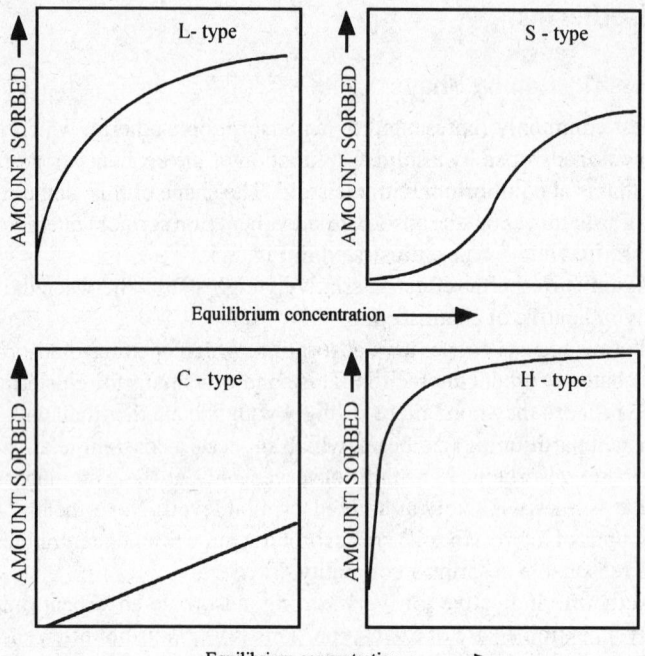

Fig. 8.5 Classification of adsorption isotherms by shape [Reprinted with permission from McBride. 1994. Environmental chemistry of soils. Copyright Oxford University Press, Inc.]

mechanisms that must ultimately be confirmed by more direct methods such as molecular spectroscopy.

8.6.2 Adsorption Isotherm Equations

The Langmuir and Freundlich equations are commonly used to describe chemisorption isotherms, both producing the general shape of L-type (or H-type) isotherms, with the slope decreasing at higher adsorption levels. The form of the Langmuir equation is (Sposito, 1989):

$$q_i = b\left(\frac{Kc_i}{(1 + Kc_i)}\right) \qquad\qquad [8.11]$$

where the quantity (q_i) of an adsorbate (i) adsorbed is related to the equilibrium solution concentration of the adsorbate (c_i), by the parameters K and b. The magnitude of the initial slope (steepness) of the isotherm is determined by K, which is therefore a measure of the affinity of the adsorbate for the surface. The value of b is the upper limit for q_i and hence, b represents the maximum adsorption of i determined by the number of reactive surface adsorption sites. One can calculate b and K from adsorption data (Sposito, 1989) by converting Equation [8.11] to the linear form:

$$\frac{q_i}{c_i} = bK - Kq_i \qquad\qquad [8.12]$$

and plotting q_i / c_i (the distribution coefficient) against q_i. If the Langmuir equation is applicable, the data should fall on a straight line with slope of $-K$ and x intercept of bK.

The Freundlich equation has the form:

$$q_i = ac_i^n \qquad [8.13]$$

where a and n are adjustable positive valued parameters, but n can range only between 0 and 1 ($n = 1$ would produce the linear C-type isotherm). The parameters are estimated by plotting $\log q_i$ against $\log c_i$, with the resulting straight line having a y intercept of log a and slope of n. The Freundlich equation will fit data generated from the Langmuir equation. The sum of two Langmuir equations will fit any L-shaped isotherm (Sposito, 1989), including the Freundlich equation. Because of this nonunique fitting of functions to a given set of data, mechanistic information about surface bonding cannot be inferred from isotherm shape. Nevertheless, the fitting parameters in the equation have often been correlated statistically to properties of soils when sorption data have been analyzed.

Converted to the logarithmic form, the Freundlich Equation [8.13] becomes:

$$\log q_i = \log a + n \log c_i \qquad [8.14]$$

Considering the adsorption of metals by soils and soil materials, q_i is equated to the total adsorbed metal concentration (M_T) (mg kg^{-1}), and c_i is equated to the dissolved metal concentration (M_s) (mg L^{-1}) in the soil solution at equilibrium with the solid. Defining log a as a constant (C), the equation becomes:

$$\log M_T = C + n \log M_s \qquad [8.15]$$

The form of this equation has been used to relate the amount of Cu adsorbed on humus and Fe hydroxide (FeOOH) to the dissolved concentration of free Cu^{2+} ions. These relationships are

$$\text{FeOOH:} \quad \log Cu_T = 0.34 + 1.44\,pH + 0.93 \log Cu_s \qquad [8.16]$$

$$\text{humus:} \quad \log Cu_T = 2.86 + 1.16\,pH + 0.86 \log Cu_s \qquad [8.17]$$

and show that humus controls Cu^{2+} solubility at a substantially lower value than the oxide, especially at lower pH (McBride, unpublished data). A comparison of the forms of Equations [8.15], [8.16] and [8.17] indicates that pH must be held constant (incorporated into the constant, C) in order for the Freundlich equation to apply. In these simple systems, the value of the parameter n is about 0.9.

In more complex systems such as soils, a similar form of the equation has been tested (McBride et al., 1997). In Cu-contaminated soils, the best fit equation for free Cu^{2+} solubility is

$$\log Cu_T = (0.66 + 0.70\,pH + \log SOM) + 0.51 \log Cu_s \qquad [8.18]$$

where the bracketed terms can be equated to the constant C if soil pH and SOM content (g kg^{-1}) do not vary. In Cd contaminated soils, a similar equation:

$$\log Cd_T = (4.60 + 0.52\,pH + 0.47 \log SOM) + 1.04 \log Cd_s \qquad [8.19]$$

with dissolved Cd (Cd_S) $(mol\ L^{-1})$ simplifies approximately to:

$$\log\left(\frac{Cd_T}{Cd_S}\right) = \log K_D = f(pH,\ SOM) \qquad [8.20]$$

where K_D is defined as the distribution coefficient of Cd between the soil solids and solution and f symbolizes a mathematical function that is sensitive to soil pH and SOM content. Since K_D is a measure of potential leachability in soils, Equation [8.20] indicates that, as SOM and pH increase, the leachability of Cd is decreased (K_D increases). Several other studies have confirmed that higher SOM contents and pH reduce the relative solubility of Cd, influencing the magnitude of K_D in Equation [8.20] or the bracketed constant in Equation [8.19] (Boekhold and van der Zee, 1992; Lee et al., 1996).

8.7   Adsorption Kinetics and Equilibria

8.7.1   Adsorption/Desorption Reaction Reversibility

Because metal ion release from bonding sites controls metal mobility (and therefore, plant availability) in soils, the rate of desorption is at least as important to the understanding of metal behavior as the rate of adsorption. Specific adsorption, and especially desorption, of metal cations on soil clays and oxides is slower than cation exchange. Trace metal cations adsorbed on these clays gradually lose much of their initial lability (as measured by diminishing self-exchange rates) over a period of days (Swift and McLaren, 1991). Consequently, metal adsorption on soils and the mineral components of soils is considered to be partially nonreversible, that is, less metal desorbs at each particular pH than expected (assuming that the forward adsorption curve represents equilibrium).

Generally, sorption of metals seems to be more nearly reversible at low than at high pH. This may arise if the monodentate complexation Reaction [8.1] gives way to a bidentate reaction at higher pH. The latter reaction, involving two metal-surface bonds, could have a very slow rate in the reverse direction (desorption). Studies of heavy metal bonding on pure oxides have indicated that the adsorption reaction step is fast (msec) and probably diffusion controlled ($K_f \approx 2 \times 10^5\ mol^{-1}\ L\ sec^{-1}$), whereas the desorption reaction step has a rate constant that may be as much as three orders of magnitude slower (McBride, 1989; Grossl and Sparks, 1995; Scheidegger and Sparks, 1996). While Pb^{2+} shows quite reversible adsorption on goethite, Cu^{2+} does not (Padmanabham, 1983; Gunneriusson et al., 1994), suggesting that smaller ions such as Cu^{2+}, Ni^{2+} and Mg^{2+} are more likely than larger ions, including Ca^{2+}, Mn^{2+} and Pb^{2+}, to enter into structural positions at or near the surface (McBride, 1978b).

Some anion adsorption processes seem relatively reversible. For example, selenite and borate reversibly adsorb on reactive soil minerals (Swift and McLaren, 1991). Chromate and selenite adsorption is reversible to change in pH (Zachara et al., 1989; Balistrieri and Chao, 1987). A sizable fraction of phosphate adsorbed by soils is rapidly converted to nonlabile forms, with desorption being slow. Even so, a labile fraction remains that is capable of rapid exchange with dissolved phosphate. In the case of phosphate, both chemisorption and precipitation may be occurring, with the latter reaction perhaps accounting for the nonlabile fraction. For arsenate and chromate adsorbed on goethite, EXAFS spectroscopy has revealed that the favored surface complexes are probably monodentate (Fendorf et al., 1997), although bidentate-binuclear complexes have also been reported for these anions (at higher surface loadings) as well as selenate, selenite and phosphate (Manceau and Charlet,

1994). Monodentate bonding would be consistent with rapidly reversible adsorption, since only a single bond would have to be broken to allow desorption. Since it appears that bidentate complexes may be less favored energetically than monodentate complexes, it may be that bond strain inhibits the bidentate geometry, or possibly the strongest Lewis acid sites do not occur at adjacent surface coordination positions.

8.7.2 Metal Adsorption and Exchange Rates on Organic Matter

Sorption reactions of trace metals on humic acids are rapid, following the order (Jin et al., 1996): $Cr(III) > Pb(II) > Cu(II) > Ag(I) > Cd(II) = Co(II) = Li(I)$. Generally, those metal cations that bond most strongly to SOM (e.g., Pb^{2+}, Cu^{2+}) tend also to be the most rapidly adsorbed (Bunzl et al., 1976). This is a likely result of the tendency for chemical reactions which are furthest from equilibrium to be the fastest. Conversely, these strongly bonding metals are the most slowly desorbed, with dissociation from surfaces being several orders of magnitude slower than adsorption. Sluggish desorption is a characteristic of the inner sphere complexes formed by strongly adsorbed cations. A large activation energy is needed to break the ligand-metal bond of these nonlabile complexes. The degree of lability of metal-surface complexes is a key predictor of how mobile and available the metal will be in soils.

Lability of metal-surface complexes can be described by the rate at which bound metal ions exchange with metal ions of the same element in solution, according to the reaction:

$$M^*(H_2O)_n^{x+} + M–L = M(H_2O)_n^{x+} + M^*–L \qquad [8.21]$$

The asterisk indicates that the metal ions initially in solution have somehow been labelled (perhaps by a radioactive or stable isotope) in order to experimentally distinguish them from the metal ions complexed with the surface groups (L). How quickly Reaction [8.21] mixes the labeled with the unlabeled ions depends on the rate at which the hydrated metals undergo ligand exchange with the surface group. The actual exchange process involves the displacement of a water ligand by the ligand L at a metal coordination site. An estimate of the time required for this exchange is made from the known water-water exchange rate of the metal cation of interest, that is, for the reaction:

$$M(H_2O)_n^{x+} + H_2O^* = M(H_2O^*)(H_2O)_{n-1}^{x+} + H_2O \qquad [8.22]$$

For any particular metal, the ligand exchange rates of Reactions [8.21] and [8.22] are likely to be correlated, as reaction rates are determined more by the radius, charge and electronic properties of the metal cation than the identity of the ligand. Metal cations of low charge and large size (low ionic potential) generally have the fastest ligand exchange because the M-L and M-OH$_2$ bonds are weakest for such cations. Consequently, as a group, the alkali metals and most of the alkaline earth metals have rapid exchange. Transition metals, however, do not follow this simple trend because of d-orbital effects on bonding. For example, $Cr(H_2O)_6^{3+}$ has an extremely slow water exchange rate (half-time > 10^5 sec). The $Cu(H_2O)^{2+}$ ion, in contrast, has an unusually rapid exchange rate for a small divalent cation, attributed to distortion in its octahedral geometry. The very low bioavailability of Cr^{3+} in soils is perhaps attributable in part to the kinetic stability of Cr^{3+} humus complexes. Jin et al. (1996) found that Cr^{3+} preadsorbed on humic acid markedly diminished the subsequent adsorption of the other trace metals.

Measurements of dissociation rates and lability of metal complexes with SOM indicate that most metals (Pb^{2+}, Cd^{2+}, Cu^{2+} and Fe^{3+}) when complexed to solid organic matter (humus), have low lability (Waller and Pickering,1993; McLaren and Crawford,1974; Florence,1986; Sedlacek et al.,1987).

Dissolved humic and fulvic acid complexes of metals such as Cu^{2+}, Ni^{2+}, and Co^{2+}, in contrast, appear to be largely labile (as defined by release of the metal to a competing strong ligand) (Lavigne et al.,1987; Cabannis,1990; Swift and McLaren, 1991; Rate et al.,1992). However, lability is always operationally defined and sensitive to the technique used to measure it. Lability is particularly sensitive to pH and metal/organic matter ratio, decreasing as pH is raised and as the metal/organic ratio is decreased.

Metals with fast water exchange rates such as Pb^{2+} and Cd^{2+} dissociate from model organic ligands much more quickly than metals with slow water exchange rates such as Al^{3+} and Cu^{2+} (Hering and Morel, 1990). Consequently, complexes of the latter metals should not be as toxic as the free metals. In soils, slow ligand exchange rates for potentially phytotoxic metals could provide some protection for plant roots. In contrast, labile complexes such as $CdCl^+$ and low molecular weight Cd organic acids could be nearly as available and toxic as the Cd^{2+} ion to soil biota (Smolders and McLaughlin, 1996). Generally, low molecular weight complexes of most trace metals, including simple organic acid complexes, amino acid complexes and inorganic ion pairs (with Cl^-, CO_3^{2-}, etc.) are electrochemically labile and likely to be more available to biota than high molecular weight humic and fulvic acid complexes (Florence, 1986).

8.7.3   Relating Reaction Reversibility to Activation Energy

Adsorption may or may not require a significant activation energy (E_a^*) depending upon whether a strong bond must be broken to allow the adsorbing ion to coordinate to the surface. A metal cation may have to lose one or more H_2O molecules of its hydration shell, for example. In contrast, assuming that the adsorption reaction is energetically favorable (exothermic), desorption always requires an activation energy (E_d^*) to at least overcome the adsorption energy. Consequently, many chemisorption reactions have a much higher activation energy in the reverse than the forward direction. The Arrhenius equation then predicts that the adsorption reaction should be faster than the desorption reaction. The commonly observed nonreversibility in metal adsorption may be the result of the long time period that would be required for desorption to be complete. This would be particularly noticeable in the case of bidentate metal bonding, presumed to be an energetically very favorable reaction. The large activation energy for desorption would derive from the need to break two metal-surface bonds. However, there are other possible explanations for slow desorption reactions, and these will be discussed later in Section 8.12.

8.8   Sorption of Cations

8.8.1   Bonding Preference for Metals

Electronegativity is an important factor in determining which of the trace metals chemisorb on minerals with the highest preference. The more electronegative metals should form the strongest covalent bonds with O atoms on any particular mineral surface. For some commonly studied divalent metals, the expected order of bonding preference on the basis of electronegativity would be: Cu > Ni > Co > Pb > Cd > Zn > Mg > Sr. On the other hand, on the basis of electrostatics, the strongest bond should be formed by the metal with the greatest charge-to-radius ratio. This would produce a different order of preference for the same metals: Ni > Mg > Cu > Co > Zn > Cd > Sr > Pb and would also predict that trivalent trace metals such as Cr^{3+} and Fe^{3+} would chemisorb in preference to all of the divalent metals listed above.

Manganese oxides show particularly high selectivity for Cu^{2+}, Ni^{2+}, Co^{2+} and Pb^{2+} (McKenzie, 1980), perhaps indicating an important contribution from covalent bonding to adsorption. On the other hand, Fe and Al oxides as well as silica adsorb Pb^{2+} and Cu^{2+} most strongly of the divalent metals listed above, suggesting that neither a purely electrostatic or covalent model of surface bonding is adequate. Because Pb^{2+} and Cu^{2+} happen to be the two most easily hydrolyzed of the listed metals, it might be suggested that adsorption and hydrolysis are correlated in some way. Perhaps the metal adsorbs in a hydrolyzed form, for example, by the reaction:

$$> S - OH + M^{n+} = > S - O - M - OH^{(n-2)+} + 2H^+ \qquad [8.23]$$

where S is the metal of the adsorbing surface.

In SOM, metals of smaller radius generally tend to form the stronger complexes. This produces the well-known Irving-Williams series of complexing strengths for the divalent metal ions: Ba < Sr < Ca < Mg < Mn < Fe < Co < Ni < Cu > Zn. The Irving-Williams series places Cu at the apex of complex stability for first row transition elements. At the same time, the metal softness factor cannot be ignored, as ionic radius fails to explain tendencies such as the ability of Hg^{2+} and Pb^{2+} to complex strongly with soil organics. The metal softness factor measures ability to form covalent bonds with organic bases, and is related to electronegativity. In fact, it appears that electronegativity gives a fair indication of the tendency of metals to complex with SOM (Table 8.4). Metals with high electronegativities draw electrons in Lewis bases toward them with the greatest energy, favoring softer bases. Because metals such as Cu^{2+} and Ni^{2+} are more electronegative (are softer acids) than Mg^{2+}, Ca^{2+} or Mn^{2+}, they have a greater tendency to complex with amine (or other less hard) ligands in humus.

A typical affinity sequence of metals for SOM at pH 5 is given in Table 8.4. The metals listed first tend to form inner sphere complexes with SOM, while those toward the end of the list are inclined to retain a hydration shell and remain exchangeable.

8.8.2 Factors Influencing Selectivity and Adsorption Efficiency

Universally consistent rules of metal selectivity for bonding to minerals and SOM cannot be given, as selectivity depends on a number of factors beyond the properties of the metals themselves, including (1) the chemical nature of the reactive surface group(s) (e.g., type of functional groups in organic matter that complex the metal); (2) the level of adsorption (adsorbate/adsorbent ratio); (3) the pH at which adsorption is measured (some metals compete more effectively with H^+ for bonding on functional groups than others); (4) the ionic strength of the solution in which adsorption is measured (determines intensity of competition by other cations for the bonding sites); and (5) the presence of soluble ligands that could complex the free metal.

All of these variables can shift metal adsorption isotherms. Thus, the pH adsorption edges of heavy metals on Fe oxides shift toward higher pH with increasing adsorbate (metal) or decreasing adsorbent (oxide) concentration, as shown in Fig. 8.2. Low adsorption densities of metals lead to stronger intrinsic bonding than high densities (Benjamin and Leckie, 1981).

The presence of dissolved organic matter (DOM) can also modify adsorption, producing S-shaped adsorption isotherms rather than the expected L-type (Langmuir) isotherms for metals added to soils

Table 8.4 Approximate order of affinity of divalent metal ions for soil organic matter related to Pauling electronegativity [McBride, 1989]

Affinity Sequence	Hg >	Cu >	Ni >	Pb >	Co >	Ca* >	Zn >	Cd >	Mn >	Mg
Electronegativity	2.0	2.0	1.91	1.87	1.88	1.00	1.65	1.69	1.55	1.31

* The affinity of humus for the essential macronutrients, Ca^{2+} and K^+, is (fortunately for plants and animals) higher than electronegativity would predict.

or humus (Neal and Sposito, 1986; Yin et al., 1997). In effect, these ligands inhibit the adsorption of metals added at low concentrations, and their effect on the isotherm is only overcome at higher metal concentrations. Halide anions in solution can markedly suppress the adsorption of metals such as Cd^{2+} and Hg^{2+} on minerals and soils by a similar mechanism (Hahne and Kroontje, 1973; Gunneriusson et al., 1994, Lumsdon et al., 1995).

Competition from monovalent metals in background electrolytes has relatively little effect on adsorption of heavy metals such as Cd^{2+}, although the presence of Ca^{2+} does suppress adsorption on Fe oxide, shifting the Cd adsorption edge to higher pH (Cowan et al., 1991). Other alkaline earth elements have little or no effect. Competition is observed between heavy metals simultaneously adsorbed on oxides, but there is to some degree a discrimination in the competition, which suggests the presence of several types of surface sites with different degrees of susceptibility to this process (Benjamin and Leckie, 1981; Zasoski and Burau, 1988).

The importance of metal/adsorbent ratio has been confirmed experimentally in Zn^{2+} and Cd^{2+} goethite systems, where reducing the quantity of goethite while increasing the less reactive adsorbent, kaolinite, to maintain the same total quantity of adsorbent, shifts the S-shaped adsorption curves of trace quantities of Zn^{2+} and Cd^{2+} to higher pH (Tiller et al., 1984). The same general tendency is observed for trace metal adsorption in soils, but the explanation may lie partly in the heterogeneous nature of soil surfaces and the fact that the first occupied adsorption sites have a higher bonding strength with the metal. Furthermore, clays and soils possess an ability to adsorb trace levels of metals at much lower pH than expected from the isotherms for pure oxide systems. There is some indication that this low pH adsorption is due to cation exchange on permanent charge clays or SOM, even though most experiments that attempt to measure specific adsorption employ a background electrolyte such as 10^{-2} M $CaCl_2$ to competitively inhibit exchange adsorption of the trace metal cation. However, only a very small fraction of exchange sites need be occupied to adsorb a significant fraction of low concentration of trace metal in solution, so that excess Ca^{2+} in solution may not fully inactivate ion exchange as a mechanism of trace metal removal from solution. In any event, trace metal sorption in soil clays containing layer silicates tends to begin at lower pH and increase more gradually with pH than the S-shaped curves of Fig. 8.2 would suggest.

The critical importance of metal (adsorbate) to soil (adsorbent) ratio can be stated in a basic rule: the preference of adsorbing surfaces for trace metal cations (relative to the prevalent cation in soil solution, usually Ca^{2+}) diminishes with increasing adsorption level. Preference or affinity is measured by a selectivity or distribution coefficient, (K_D) (Section 8.6.2). The reduction of this selectivity with increased adsorption is observed for metal adsorption on both soils and pure minerals (Hendrickson and Corey, 1981; Benjamin and Leckie,1981).

In severely metal polluted soils, hydrolysis and precipitation can remove hydrolysis-prone metals from solution as the pH approaches neutrality, so that experimental sorption curves, which include both chemisorption and precipitation, tend to be more abrupt than the adsorption edges shown in Fig. 8.2. For example, as the pH of the $Cu/Al(OH)_3$ system is adjusted upward, $Cu(OH)_2$ can precipitate if insufficient adsorption has occurred to keep the $(Cu^{2+})(OH^-)^2$ activity product below its solubility product. In the absence of adsorption, 10^{-5}, 10^{-4}, and 10^{-3} M Cu^{2+} would begin to be removed from solution as $Cu(OH)_2$ at pH 6.8, 6.3, and 5.8, respectively. This means that, in contrast to metal adsorption curves, metal hydroxide and oxide precipitation curves shift to lower pH as the total metal in the system increases.

Each trace metal has its own characteristic S-shaped sorption curve, as exemplified by the metal sorption curves of Fig. 8.2. In general, as was pointed out earlier, tendency toward sorption seems to be correlated with ease of metal hydrolysis. Consequently, the sorption curve for a strongly hydrolyzing metal such as Cu^{2+} is centered at lower pH than that for a more weakly hydrolyzing metal such as Cd^{2+}.

8.9   Sorption of Anions

8.9.1   Anion Bonding Preference

Anions adsorb to the oxide, noncrystalline aluminosilicate (allophane) and silicate mineral fraction of soils. However, certain anions are able to bond to SOM as well. Borate $[B(OH)_4^-]$ is notable in this regard, bonding according to a reaction of the type:

$$
\begin{array}{c}
>C\text{-}OH \\
\\
\quad + \ B(OH)_4^- \ = \\
\\
>C\text{-}OH
\end{array}
\qquad
\begin{array}{c}
>C\text{-}O \qquad OH^- \\
\diagdown \ \diagup \\
B \qquad + \ 2H_2O \\
\diagup \ \diagdown \\
>C\text{-}O \qquad OH
\end{array}
\qquad\qquad [8.24]
$$

where the involved organic group can be aliphatic or aromatic. It is not clear whether other anions which appear to adsorb on humus, such as arsenate and arsenite, do so by this mechanism or by bonding to inorganic impurities in the humus (Thanabalasingam and Pickering, 1986). Some anions may bond indirectly to organic groups through a bridging metal ion such as Al^{3+} or Fe^{3+}. Because most anions are not adsorbed on humus, anion bonding at mineral surfaces accounts for most of the anion retention in soils. It is this latter process which will be discussed in detail in this section.

The reasons for selectivity shown by particular anions for mineral surfaces are not very well understood. Many elements of environmental concern, Cr, As, and Se, for example, occur as

Table 8.5   Chemical characteristics of important anions in soil chemistry*

Oxyanion	Formula	"Shared charge"	Electronegativity
Borate	$B(OH)_4^-$	$3/4 = 0.75$	2.04
Silicate	SiO_4^{4-}	$4/4 = 1.0$	1.90
Hydroxyl	OH^-	$1/1 = 1.0$	2.2
Phosphate	PO_4^{3-}	$5/4 = 1.25$	2.19
Arsenate	AsO_4^{3-}	$5/4 = 1.25$	-
Selenite	SeO_3^{2-}	$4/3 = 1.33$	-
Carbonate	CO_3^{2-}	$4/3 = 1.33$	2.55
Molybdate	MoO_4^{2-}	$6/4 = 1.5$	2.35
Chromate	CrO_4^{2-}	$6/4 = 1.5$	-
Sulfate	SO_4^{2-}	$6/4 = 1.5$	2.58
Selenate	SeO_4^{2-}	$6/4 = 1.5$	-
Nitrate	NO_3^-	$5/3 = 1.67$	3.04
Perchlorate	ClO_4^-	$7/4 = 1.75$	3.16
Halide			
Fluoride	F^-	-	3.98
Chloride	Cl^-	-	3.16
Bromide	Br^-	-	2.96
Iodide	I^-	-	2.66

* The shared charge and the electronegativity (Pauling) are properties of the oxygen atom (Lewis base) and central atom of the oxyanion, respectively. For the halide anions, shared charge is not a meaningful concept.

oxyanions in soils, some adsorbing much more strongly than others on oxides and other variable charge minerals. Some anions of interest in soil chemistry are listed in Table 8.5, ranked according to a measure of the positive charge that the oxyanion's central atom shares formally with each bonded O atom. This shared charge is determined by dividing the valence of the central atom by the number of bonded O atoms. Assuming that it is the deprotonated O atoms of oxyanions which actually bond with metals such as Fe^{3+} and Al^{3+} on mineral surfaces, then the smaller the shared charge, the greater the effective negative charge residing on each O atom, and the stronger is the metal-oxyanion ionic bond. The concept seems to work fairly well because the ranking of anions in Table 8.5 according to shared charge follows approximately the order of anion affinity for oxides and aluminosilicates. In cases of oxyanions with comparable values of shared charge, further chemical factors such as electronegativity have to be considered in order to gauge covalency contributions to the strength of the oxyanion-surface bond. For example, chromate bonds more strongly than selenate on Fe hydroxide at any particular pH even though these two oxyanions have the same shared charge (Benjamin, 1983).

With the exception of F^-, the halide anions listed in Table 8.5 bond by outer sphere electrostatic attraction, and are therefore adsorbed only if variable charge mineral surfaces are positively charged. This means that all halides but F^- rank low on the scale of selectivity of minerals for anions. The strong F^- surface bond is explained by the high electronegativity and the small charge to radius ratio of F^-, ensuring an energetic anion surface association of the inner sphere type.

In summary, those anions which rank high in Table 8.5 are chemisorbed by ligand exchange (inner sphere) on soil minerals, while those which rank lower tend to be adsorbed by nonspecific anion exchange (outer sphere). Anions of high rank readily displace those of lower rank by a competitive process of ligand exchange, unless adsorption sites are available in excess. For example, phosphate is more effective than molybdate in displacing arsenate from bonding sites on oxides (Manning and Goldberg, 1996), and sulfate suppresses chromate adsorption by soil minerals (Zachara et al., 1989). The inhibition of phosphate adsorption on Fe oxide was observed to follow the order (Ryden et al., 1987): arsenate > selenite > silicate > molybdate, with arsenate being the most effective in competing with phosphate. Competition with selenite adsorption at pH 7 on goethite followed the order (Balistrieri and Chao, 1987): phosphate > silicate > citrate > molybdate > (bi)carbonate > oxalate > fluoride > sulfate. Several of the oxyanions, including phosphate, arsenate and selenite, may chemisorb on oxides by a binuclear bridging mechanism:

$$\begin{array}{c} \text{>S-OH} \\ \\ \text{+ } MO_y^{n-} \text{ =} \\ \\ \text{>S-OH} \end{array} \qquad \begin{array}{c} \text{>S-O} \\ \diagdown \\ \quad\quad MO_{y-2}^{(n-2)-} \text{ + } 2OH^- \\ \diagup \\ \text{>S-O} \end{array}$$

[8.25]

Because of the energy of such an association, nonreversibility in adsorption/desorption behavior is expected. The association is further stabilized by entropy, since desorption requires the improbable event that two bonds be simultaneously broken.

One problem with the shared-charge rule is classifying those oxyanions which form weak acids, such as silicate, arsenite and borate. These anions are expected to form intrinsically strong bonds with minerals such as oxides because of their low shared charge (Table 8.5). However, they also dissociate very little at normal soil pH, and therefore, adsorb weakly except at relatively high pH. In effect, the shared charge is increased by the bonded proton, suppressing the tendency of weak acid oxyanions to adsorb at low pH. Silicate in solution takes the form of uncharged monosilicic acid $[Si(OH)_4]$ unless the pH is quite high. Consequently, Si adsorption on oxides increases as the pH is raised from 3 to about 9, a result of the dissociation of the weakly acidic $Si(OH)_4$ to silicate anions at high pH. Similarly, B in solution takes the form of the neutral $B(OH)_3^\circ$ molecule, a very weak acid. Boron adsorbs to a greater extent at pH 9 than at lower pH, as the higher pH favors conversion to the borate anion, $B(OH)_4^-$. In contrast, most of the oxyanions listed in Table 8.5, being anions of strong acids (H_3PO_4, H_2SO_4, HNO_3), form negatively charged species in solution even at low pH, and their dissociation is sufficiently easy that it does not impede chemisorption. These anions, therefore, adsorb most effectively at low pH, as the general anion adsorption Equations [8.2] and [8.3] predict should be the case. As shown in Fig. 8.6, most anions such as fluoride and antimony adsorb most favorably on minerals and soils at low pH, but the weak acid oxyanions, such as arsenite and borate, adsorb most strongly above pH 7.

Given that the effect of pH on adsorption can be different for different anions, generalizations about the preference of mineral surfaces for one anion over another must be qualified. The following rules result: (1) oxyanions of weak acids chemisorb optimally at moderate to high pH; (2) oxyanions of strong acids chemisorb optimally at low pH; (3) oxyanion adsorption has a tendency to maximize at a pH near the pK_a value of the protonated form of the adsorbing anion; and (4) anion adsorption at high pH is disfavored by competition with OH^- and carbonate, and by surface charge. The last rule describes the influence that carbonate and hydroxy anions have in suppressing anion adsorption. Dissolved organic matter in soils also reduces oxyanion adsorption, since carboxylic acids and other functional groups compete for the same bonding sites.

8.9.2 Factors Influencing Selectivity and Adsorption

As with metals, selectivity of soils and soil minerals for anions depends on a number of factors besides the properties of the anions themselves, including (1) the chemical nature of the reactive surface group(s); (2) the level of adsorption (adsorbate/adsorbent ratio); (3) the pH at which adsorption is measured; (4) the ionic strength of the solution in which adsorption is measured (determines intensity of competition by other anions for the bonding sites); and (5) the presence of soluble ligands and oxyanions that could compete with the anion. The adsorbent/adsorbate ratio affects anion adsorption,

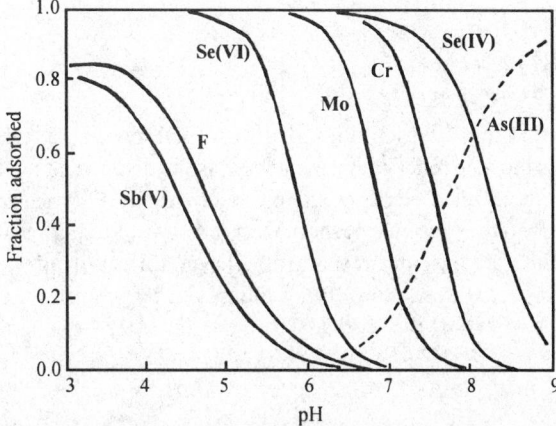

Fig. 8.6 Predicted adsorption of several anionic elements on hydrous ferric oxides, based on bonding constants summarized in Dzombak and Morel (1990) [Reproduced with permission from Dr. L.J. Evans]

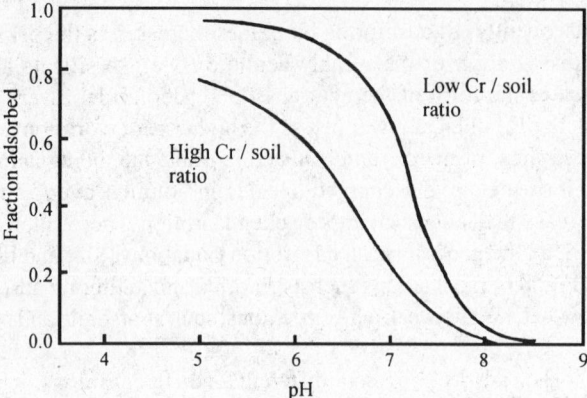

Fig. 8.7 Adsorption edge for chromate (CrO_4^{2-}) on oxidic subsoils at two different chromate/soil ratios. [Drawn from data of Zachara et al.,1989]

with the adsorption edges being shifted to higher pH at higher solid/anion ratios (Balistrieri and Chao, 1990,1987; Zachara et al., 1989) (Fig. 8.7).

8.10   Molecular Adsorption Models

8.10.1   Mineral Surfaces

Surface functional groups serve as Lewis bases toward metal cations. The metals complex with one, or possibly two, of the deprotonated hydroxyl groups on oxides as follows :

$$> S-OH \ + \ M^{n+} \ = \ > S-OM^{(n-1)+} + H^+ \qquad\qquad [8.26]$$

$$2 > S-OH \ + \ M^{n+} \ = \ > (S-O)_2 M^{(n-2)+} + 2H^+ \qquad\qquad [8.27]$$

An average binding constant (K_M) can be defined by:

$$K_M = \frac{(S-OM)_T (H^+)^x}{(S-OH)_T (M)} \cdot CF \qquad\qquad [8.28]$$

where ($S-OM_T$) represents the sum of the activities of all adsorbed species, ($S-OH_T$) is the activity of all protonated surface groups that can react with a metal, (M) is the dissolved metal concentration and x is the average number of protons released per metal ion adsorbed. The CF term corrects for nonideal behavior, suggested to be due to the presence of more than one type of bonding site at the surface, or alternatively, electrostatic interactions that result from the modification of surface charge upon specific adsorption of cations. However, simplifying the model by ignoring the CF term does not necessarily reduce the effectiveness of the model (Cowan et al., 1991), giving the nonelectrostatic model equation (NEM):

$$K_M = \frac{(S-OM)_T (H^+)^x}{(S-OH)_T (M)}$$ [8.29]

As long as metal adsorption density is low (less than a few percent of the reactive surface groups are bonded to metals), the value of K_M in Equation [8.29] remains relatively constant (Benjamin and Leckie, 1981). At high metal loadings, a substantial fraction of adsorption sites is occupied, and the average binding constant gets smaller. This is not predicted from the model, and suggests that several kinds of bonding sites exist on the surface, with the metals preferentially adsorbing on those sites where the binding strength is greatest.

If θ_M is defined as the fraction of surface sites occupied by the metal, then $1-\theta_M$ is the fraction of unoccupied sites, and it is apparent that:

$$\frac{(S-OM)_T}{(S-OH)_T} = \frac{\theta_M}{1-\theta_M}$$ [8.30]

Rearrangement of Equation [8.29] then gives:

$$(M) = \frac{(H^+)^x}{K_M} \frac{\theta_M}{(1-\theta_M)}$$ [8.31]

and it is apparent that Equation [8.31] is a Langmuir function if the term $(H^+)^x/K_M$ can be assumed to be constant. Thus, at low surface adsorption and constant pH, metal adsorption on oxides should obey the Langmuir equation.

Dzombak and Morel (1990) have summarized existing data for metal adsorption experiments on Fe oxides, and their best estimates of K_M are reported in Table 8.6. The wide range of bonding affinities is shown, ranging from Hg^{2+} (strongest) to Ag^+ (weakest). However, these K_M values probably

Table 8.6 Best estimate $\log K_M$ values for the metal adsorption reaction, $>Fe-OH^o + M = >Fe-O-M + x\ H^+$, at strong sites, where $x = 1$ for all metals except Cr^{3+} ($x = 2$) and Ca^{2+}, Sr^{2+}, Ba^{2+} ($x = 0$) [Data from Dzombak and Morel, 1990]

Metal	$\log K_M$	Assumed Surface Complex
Pb(II)	4.65	$>Fe-O-Pb^+$
Zn(II)	0.99	$>Fe-O-Zn^+$
Cd(II)	0.47	$>Fe-O-Cd^+$
Hg(II)	7.76	$>Fe-O-Hg^+$
Cu(II)	2.89	$>FeO-Cu^+$
Ag(I)	−1.72	$>Fe-O-Ag^0$
Ni(II)	0.37	$>Fe-O-Ni^+$
Co(II)	−0.46	$>Fe-O-Co^+$
Cr(III)	2.06	$>Fe-O-CrOH^+$
Ca(II)	4.97	$>Fe-OH-Ca^{2+}$
Sr(II)	5.01	$>Fe-OH-Sr^{2+}$
Ba(II)	5.46	$>Fe-OH-Ba^{2+}$

Table 8.7 Best estimate log K_A values for the predominant anion adsorption reactions, $>Fe-OH^\circ + A^{n-} + x\,H^+ = >Fe-H_{x-1}A^{(n-x)-} + H_2O$, where possible values of $x = 1,2,...n$ [Data from Dzombak and Morel, 1990]

Anion A	Value of x	log K_A	Assumed surface complex
PO_4^{3-}	3	31.3	$>FeH_2PO_4^0$
PO_4^{3-}	2	25.4	$>FeHPO_4^-$
PO_4^{3-}	1	17.7	$>FePO_4^{2-}$
AsO_4^{3-}	3	29.3	$>FeH_2AsO_4^0$
AsO_4^{3-}	2	23.5	$>FeHAsO_4^-$
SO_4^{2-}	1	7.78	$>FeSO_4^-$
SeO_4^{2-}	1	7.73	$>FeSeO_4^-$
SeO_3^{2-}	1	12.7	$>FeSeO_3^-$
CrO_4^{2-}	1	10.9	$>FeCrO_4^-$

overestimate affinity for oxide surfaces in real soils, because of the competitive effects of Ca^{2+}, Al^{3+} and other cations for surface sites as well as the ability of DOM to remove metals from surfaces.

The model for chemisorption of an anion (A) is conceptually similar to that for metal cations. Thus, the equation defining the bonding constant (K_A) that corresponds to Equation [8.29] is

$$K_A = \frac{(S-A)_T (OH^-)^y}{(S-OH)_T (A)}$$
[8.32]

where $(S-A)_T$ represents the sum of the activities of all adsorbed species, $(S-OH)_T$ is the activity of all protonated surface groups that can be displaced by the anion, (A) is the concentration of the anion in solution, and y is the average number of OH^- ions released for each anion adsorbed. Since both monodentate and bidentate surface complexes are possible, as many as two OH^- could be displaced by one anion, although the value of y is generally much less than 2. Reported values for anion bonding constants are summarized by Dzombak and Morel (1990), and presented in Table 8.7.

As with metal adsorption, the equation defining the bonding constant can be rearranged to give a Langmuir function:

$$(A) = \frac{(OH^-)^y}{K_A} \frac{\theta_A}{1-\theta_A}$$
[8.33]

where θ_A is defined as the fraction of surface sites occupied by the anion. Thus, at low surface adsorption (where K_A should hold constant) and constant pH, anion adsorption on oxides should obey the Langmuir equation (McBride, 1994). Strongly retained anions such as phosphate tend to adsorb on soil minerals according to this function. The success of the Langmuir equation for soils means that phosphate solubility is often controlled by the degree of saturation of sorption sites, and as the adsorption capacity of a soil is approached, solubility increases nonlinearly. Consequently, soils near their saturation limit for adsorption are much more likely to allow anions to leach.

In most soil systems, adsorption of anions such as phosphate shows more complex behavior than that implied by this model because the adsorption capacity seems to increase over time. Such a time-dependent phenomenon may be the consequence of slow precipitation reactions superimposed on the more rapid chemisorption (van Riemsdijk and van der Zee, 1991; van der Zee and van Riemsdijk, 1991).

8.10.2 Organic Surfaces

Metal adsorption on SOM involves proton displacement from acidic functional groups (ligands) such as phenolic, carboxylic, sulfhydryl and other ligands. The stability of such a metal-ligand complex (ML) is defined by the equation:

$$K_{ML} = \frac{(ML_n)}{(M^{x+})(L^{y-})^n}$$ [8.34]

where the round brackets denote activities. Although K_{ML} is ideally a true thermodynamic stability constant which would remain invariable for different pH and ionic strength conditions, K_{ML} cannot be expected to be a true constant for SOM-metal complexes for several reasons. First, for solid soil organics, ML_n is a surface complex rather than a dissolved molecule, and activity is not straighforwardly defined for such a species. Secondly, the metal bonding functional groups are heterogeneous, with a wide range of Lewis basicity and chelating ability. Thus, the metal bonding constant is always larger for small additions of a trace metal to SOM than for large additions. It is important to recognize that many K_{ML} values for metal-organic complexes reported in the literature are for high loadings of metals (approaching the capacity of the SOM to adsorb metals), and therefore, suggest much weaker bonding of metals by SOM than actually occurs in most situations. These low metal K_{ML} values predict, for example, that DOM is unlikely to complex Cd^{2+} significantly, yet there is evidence to suggest that concentrations of dissolved Cd in soils can be increased by soluble organics (Neal and Sposito, 1986; McBride et al., 1997). At low Cd and Zn concentrations in solution, conditional stability constants for these metals with DOM can be comparable to those for Cu (Aguirre-Gomez, 1995).

Assuming $n = 1$ in Equation [8.34], some values for the K_{ML} constant for natural DOM are (Lövgren and Sjöberg,1989; Lövgren et al.,1986) $K_{AlL} = 4.29$, $K_{CdL} = 2.89$, $K_{CuL} = 3.75$, and $K_{HgL} = 10.8$. These bonding constants predict the pH dependent complexation shown in Fig. 8.8, and reveal the high-affinity of Hg^{2+}, as well as the relatively low affinity of Cd^{2+}, for SOM.

Although the effect of pH on metal complexation is often of primary interest in soil studies, the stability constant defined by Equation 8.34 reveals no apparent involvement of pH. Nevertheless, pH affects M^{x+} and L^{y-} activity through various side reactions that are not explicitly accounted for in Equation [8.34]. By controlling the activities of these two ions, the free metal and the uncomplexed but fully dissociated ligand, pH alters the degree of metal complexation while maintaining accordance with Equation 8.34. Equation 8.34 is commonly modified slightly in actual usage, replacing the dissociated ligand activity by the more easily estimated total dissolved ligand concentration ($[L]_T$). This modified equation defines a conditional stability constant (β_{ML}) where it is now clear that (β_{ML}) is strongly pH dependent.

Empirically, metal bonding to SOM can be modelled by reactions analogous to those used for metal chemisorption on mineral surfaces (Equations [8.26]–[8.31]). Consequently, the Langmuir adsorption function (Equation [8.31]) should describe the control of metal ion activity by SOM reasonably well. This form of equation has been shown to explain the relationship between (Al^{3+}) and pH in acid forest soils (Wesselink et al., 1996):

$$\left(Al^{3+}\right) = K_{AlL}\left(H^{+}\right)^{x}\left(Al_{org}\,/\,OC\right) \tag{8.35}$$

where K_{AlL} is the average Al complexation constant and OC is the organic C content. The form of this equation is analogous to Equation [8.31], since the ratio of organically bound Al [Al_{org}] to soil organic C content is equivalent to the metal adsorption density term [$\theta_M\,/(1-\theta_M)$] at least in soils where the dominant adsorbent for metals is SOM. A function of similar form may also predict reasonably well the activity of a trace metal, such as Cu^{2+}, Cd^{2+}, and Zn^{2+}, although organic C content is not always a significant predictor in the equation (McBride et al., 1997).

For many heavy metals, particularly those with a strong tendency to hydrolyze, pH strongly controls solubility, for example, by the reaction:

$$M^{x} + xOH^{-} = M(OH)_{x} \tag{8.36}$$

The solubility product of the metal hydroxide (K_{SO}) then determines the free metal activity:

$$K_{SO} = \left(M^{x+}\right)\left(OH^{-}\right)^{x} \tag{8.37}$$

Furthermore, most complexing ligands on SOM are weak or very weak acids, and must dissociate in order to form the free ligand:

$$LH_{y} = L^{y-} + yH^{+} \tag{8.38}$$

The dissociation constant for this reaction is given by:

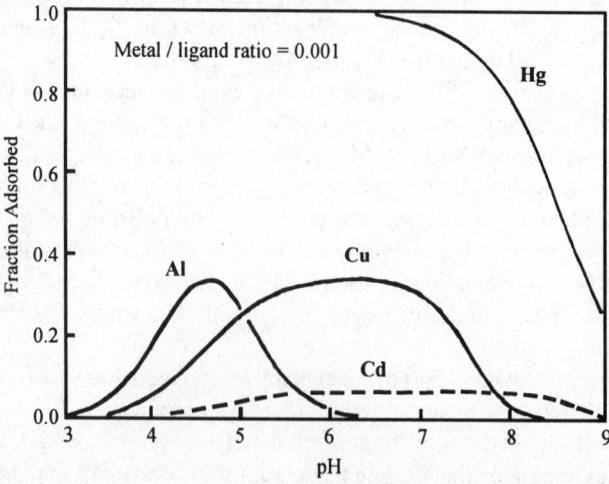

Fig. 8.8 Predicted fraction of Al^{3+}, Cu^{2+}, Cd^{2+} and Hg^{2+} complexed by natural organic matter as a function of pH. Based on bonding constants given by Lövgren et al.(1986) and Lövgren and Sjöberg(1989) [Reproduced with permission from Dr. L.J. Evans]

$$K_{DISS} = \frac{(L^{y-})(H^+)^y}{(LH_y)}$$ [8.39]

It becomes evident, therefore, that at extreme pH, complexation of metals by SOM would diminish because of metal hydroxide precipitation (high pH) and association of the ligands with H^+ (low pH). Optimal complexation of metals with SOM must then occur at some intermediate pH (Lumsdon et al., 1995). This is illustrated in Fig. 8.8 for trace metal adsorption on SOM. For most metals, adsorption is expected to decrease above pH 7-8 and below pH 5, the result of metal hydroxide/oxide precipitation at high pH, and proton competition for ligand binding sites at low pH. However, some metals that do not form highly insoluble hydroxides or carbonates, such as Ag and Hg, could remain complexed with SOM even under very alkaline conditions.

8.11 Coprecipitation of Trace Elements in Solid Solutions

8.11.1 Theory of Solid Solutions

The theory of solid solutions predicts that the solubility of an ion can be lowered in a mixed ionic compound relative to the solubility of the pure compound. Consider, for example, a trace metal cation (B) isomorphously substituted into a solid composed of metal cations (A) and anions (Y). The chemical formula $[A_{1-x}B_xY]$ is variable because x can range from 0.0 to 1.0 if AY and BY (the solid solution end members) form a continuous series. Unlike ionic compounds of fixed composition, solid solutions do not have constant solubility products. Instead, equations for both components (AY and BY) must be specified:

$$(A)(Y) = K_{so}{}^{AY} a_{AY}$$ [8.40]

$$(B)(Y) = K_{so}{}^{BY} a_{BY}$$ [8.41]

Here, parentheses symbolize the activities of ions in solution, while a_{AY} and a_{BY} are the activities of the components AY and BY in the solid solution. $K_{so}{}^{AY}$ and $K_{so}{}^{BY}$ are the solubility products of pure AY and BY. Since the trace metal (B) when incorporated at low levels into the solid results in a low value of a_{BY}, then Equation [8.41] predicts an effective solubility product $[(B)(Y)]$ that is lower than that of pure BY. Consequently, solid solution formation is an effective means of lowering the solubility of a trace element.

At equilibrium, the distribution of A and B ions between the aqueous (aq) and solid (s) phase is represented by the reaction:

$$AY(s) + B(aq) = A(aq) + BY(s)$$ [8.42]

If the solid phase is homogeneous, the reaction can be described by:

$$a_{BY} / a_{AY} = D(B) / (A)$$ [8.43]

where D is the distribution coefficient that specifies the relative degree to which cations A and B are incorporated into the solid. It is easily shown, from Equations [8.40] and [8.43], that D equals the ratio of solubility products of the pure phases (K_{so}^{AY}/K_{so}^{BY}).

To form an ideal solid solution of formula $A_{1-x}B_xY$, the two cations A and B must be closely matched such that one cation can substitute for the other with no change in energy of the solid structure. The solid solution is then said to have zero heat of mixing. On the other hand, the theoretical entropy of mixing in this ideal solution is not generally zero, but is dependent on the mole fractions of A and B, according to the equation:

$$S = 2.303R \left[x \log x + (1-x) \log(1-x) \right] \qquad [8.44]$$

It can be shown from this equation that the activities of the components in ideal solid solutions are simply given by:

$$a_{AY} = 1 - x \qquad a_{BY} = x \qquad [8.45]$$

That is, activities of AY and BY can be equated to mole fractions of AY and BY in the solid. Symbolizing these mole fractions as X_{AY} and X_{BY}, Equation [8.43] becomes:

$$X_{BY} / X_{AY} = D(B)/(A) \qquad [8.46]$$

For dilute aqueous solutions, (B) and (A) in this equation can be replaced by [B] and [A], the mole concentrations in solution, giving:

$$X_{BY} / X_{AY} = D[B]/[A] \qquad [8.47]$$

If D is very large or small, this means that a solid solution precipitated from solution cannot be homogeneous because one of the two metals, A or B, is selectively scavenged from solution as the solid precipitates. This more preferred ion ends up occluded in the center of the growing crystallites, while the less preferred ion concentrates in the outer layers.

This segregating mechanism could limit metal solubility and bioavailability in soils, providing a means of literally burying toxic or essential metals in a form inaccessible to desorption. Furthermore, the solubility of a toxic or essential metal could be lowered well below that predicted from the solubility of the pure solid phase. That is, if the trace metal of interest is symbolized by B, the solubility of B in equilibrium with the solid solution would be lower than that in equilibrium with pure BY. This fact is demonstrated by combining Equations [8.41] and [8.45] to give:

$$(B)(Y) = x K_{so}^{BY} \qquad [8.48]$$

Consequently, a trace metal substituted at low levels ($x \ll 1$) into the structure produces an effective solubility product [(B)(Y)] that can be orders of magnitude less than the K_{so} of the pure solid (BY).

8.11.2   Evidence for Solid Solutions

Solid solution formation may play a role during Cd^{2+} and Mn^{2+} sorption in suspensions of calcite (McBride, 1979,1980; Papadopoulos and Rowell, 1988). Calcite adsorbs these cations initially by a fast reaction which seems to involve exchange of Ca^{2+} by the trace metal at the surface:

$$Cd^{2+} + CaCO_3(s) = Ca^{2+} + CdCO_3(s) \qquad\qquad [8.49]$$

Generally, in reactions between two metal carbonates as exemplified by Reaction [8.49], the metal cation of the least soluble carbonate (in this case, Cd^{2+}) is preferentially adsorbed at the carbonate surface. The reaction can be viewed as a chemisorption process rather than precipitation because: (1) it proceeds to a degree determined by the calcite surface area; and (2) it occurs even when the suspension is undersaturated with respect to solid $CdCO_3$.

Subsequent slow removal of Cd^{2+} from solution is observed, and may be the result of recrystallization into a thin calcite surface layer to form a Cd^{2+}/Ca^{2+} solid solution at the surface. This would explain the observed reduction in extractability and exchangeability of the Cd^{2+} over a period of time (Davis et al., 1987). Because calcite is in a dynamic state of dissolution and recrystallization in suspension, even when at equilibrium, a solid solution can form at the surface fairly quickly. If it is assumed that the $Ca_{1-x}Cd_xCO_3$ solid solution is ideal, then the distribution coefficient equals the quotient of solubility products of the pure carbonate solids:

$$\mathbf{D} = K_{SO}^{CaCO_3} / K_{SO}^{CdCO_3} = 10^{-8.47} / 10^{-11.3} = 680 \qquad\qquad [8.50]$$

This large $\mathbf{D}$ value signifies a strong tendency for the recrystallizing calcite to incorporate Cd^{2+}.

In soils, the prevalence and importance of solid solution formation in controlling cation and anion solubility has not yet been determined. Certain minerals readily incorporate only those metal ions with radii equal to or less than the radius of the structural metal ion. For calcite, this means that Mn^{2+}, Cd^{2+} and Fe^{2+} readily enter the calcite structure upon precipitation while smaller ions such as Cu^{2+} and Zn^{2+} probably do not (Angus et al., 1979; Pingitore et al., 1988; Konigsberger et al., 1991; Rock et al., 1994). Some of the coprecipitated metals appear to be concentrated within trace phases in the carbonate, not distributed evenly in the structure. Ideal solid solutions may be unusual in minerals. Besides metal cations, anions such as selenate and borate can substitute for CO_3^{2-} in calcite (Reeder et al., 1994).

Oxides of Fe and Al form coprecipitates with trace metals, and it is likely that the most stable of these occur for metals with radii and charge like those of the Fe^{3+} or Al^{3+} ions that they replace in the structure. Thus, Cr^{3+} appears to substitute readily in Fe oxides (Sass and Rai, 1987) and other metals with relatively unstable +3 oxidation states, such as Co^{3+}, Mn^{3+}, and V^{3+} (Schwertmann and Pfab, 1994), may be stabilized in the oxide lattice. In contrast, the coprecipitation of divalent metals such as Cu^{2+}, Zn^{2+}, Cd^{2+}, and Pb^{2+} into oxides distorts the structure and produces a less crystalline solid (McBride, 1978b; Gerth, 1990). These metals appear to reside largely with the least crystalline fraction of the oxide, probably at or near surfaces (Nalovic et al., 1975; McBride, 1989; Bibak et al., 1995). They may be segregated into deposits within the solid, rather than evenly distributed (Ford and Bertsch, 1996), particularly if the metal induces strong distortion in the oxide lattice. Those metals with radii allowing facile substitution into octahedral sites, such as, Ni^{2+}, Cu^{2+} and Zn^{2+}, appear to be retained more nonreversibly in the oxide than metals with larger radii, such as Mn^{2+}, Cd^{2+}, and Pb^{2+} (McBride, 1978b; Schultz et al., 1987; Ainsworth et al., 1994). Thus, exclusion or expulsion of Pb^{2+} and Cd^{2+} from the coprecipitate has been observed to occur as Fe oxides recrystallize during aging (Ford and Bertsch, 1996; Martinez and McBride, 1998). Although Pb^{2+} chemisorbs readily on $Al(OH)_3$, it does not readily substitute during coprecipitation (Keizer and Bruggenwert, 1991, Lothenbach et al., 1997). The rules which determine selectivity in chemisorption at a solid surface do not necessarily apply to substitution within the solid. One might expect that heavy metals coprecipitated into Fe oxides would have access to a greater number of binding sites than those adsorbed on Fe oxides, but coprecipitation often does not offer any advantage over adsorption in terms of achieving a lower measured metal solubility (Crawford et al., 1993; Benjamin and Leckie, 1981; Martinez and McBride, 1998), again suggesting that, for some metals, a disproportionate fraction of coprecipitated trace metals occupies surface or near surface positions.

Small divalent cations, including Zn^{2+}, Cu^{2+} and Mg^{2+}, form solid solutions with Al hydroxide polymers and particles, while larger cations like Pb^{2+}, Cd^{2+}, Mn^{2+}, and Ca^{2+} may not (McBride,1978b; Keizer and Bruggenwert 1991; Lothenbach et al.,1997). In fact, solid solutions with the general formula $\left[M^{2+}_{1-x} Al_x (OH)_2 \right]^{x+} A^{n-}_{x/n}$ where A^{n-} is an anion, and M^{2+} has an ionic radius that favors octahedral coordination, are well-known and referred to as hydrotalcites, layered structures possessing the unusual property of permanent positive charge that is manifested as a high anion exchange capacity. These solids form at high pH, and are effective in sequestering Mg^{2+} and heavy metals cations with radii similar to that of Mg^{2+}. At the same time, contaminant anions such as borate and arsenate can replace for OH^- in the hydrotalcite (Reardon and Della Valle, 1997). The drawback of this apparently versatile solid as a scavenger of toxic elements is the fact that it is formed under alkaline conditions and is unlikely to persist under the acidifying conditions of leaching in soils of wet climates.

Calcium phosphate minerals form solid solutions with both Cd^{2+} and Pb^{2+}, metals that fit well into the Ca^{2+} site of hydroxyapatite and related minerals (Driessens, 1986). Thus, heavy metals added in solution to hydroxyapatite form pure metal phosphates or coprecipitates with the mineral (Xu et al., 1994; Chen et al., 1997). In severely Pb-contaminated soils, a Pb/Ca phosphate solid solution has been identified (Cotter-Howells, 1996). Calcium hydrosilicates also appear to possess sites conducive to the immobilization of Cd^{2+}, Pb^{2+}, Zn^{2+}, and even the chromate anion (Bobrowski et al., 1997), allowing toxic trace elements to be stabilized in soils and water by treatment with alkaline cement products. Calcium/magnesium silicates, which neoform from these alkaline silica rich products, contain structural sites that readily accept cations that favor octahedral (e.g., Cu^{2+}, Ni^{2+}) and higher coordination number (e.g., Cd^{2+}, Pb^{2+}).

The importance of coprecipitation in controlling metal solubility and selectivity in soils under natural conditions is not known at this time. Coprecipitates of oxides are often formed in the laboratory under conditions atypical of soil environments, particularly very alkaline conditions and/or high temperature (Schultz et al., 1987; Bibak et al., 1995). The potential for solid solution formation in soils is low for minerals with small solubility products, such as Al oxides and aluminosilicates, because spontaneous dissolution and recrystallization are very slow in these minerals at near neutral pH and 20 C. Without recrystallization, trace metals cannot be incorporated into the mineral structures. On the other hand, phosphate and carbonate minerals have high enough solubility that the time scale of sorption allows solid dissolution and reprecipitation with the sorbed trace metal (Xu et al., 1994; Chen et al., 1997). Movement of metal ions into intact mineral crystals by solid diffusion is not possible on the time scale of adsorption experiments; ionic diffusion into crystalline solids is negligibly slow at all but extremely high temperatures. Nevertheless, metals could diffuse into imperfect solids along interstices, pores or other structural defects.

The opportunity for coprecipitation and solid solution formation is higher with Fe and Mn oxides than Al oxides or aluminosilicates. The reason for this is the higher solubility of the former two minerals under anaerobic conditions. Soil reduction generates the soluble ions, Fe^{2+} and Mn^{2+}, which then reoxidize again to form the insoluble oxides once the soil is reaerated. The possibility exists for coprecipitation of trace metals during these cycles of alternating reduction and oxidation. Furthermore, the fresh precipitates may also be effective adsorbates for trace metals (Iu et al., 1981).

8.12 Aging Effects During Sorption

Certain time-dependent features of metal and anion sorption on pure minerals, organic matter and soil are consistent with slow reaction processes that may be redistributing or otherwise altering the adsorbed form and accessibility of these adsorbates. These include a sorption capacity that is ill-defined and increasing with time, and decreasing reversibility of sorption with time that coincides with a decreasing lability of the sorbed ion (Davis et al., 1987; Schultz et al., 1987; Bruemmer et al., 1988; Willett et al., 1988; Swift and McLaren, 1991; Ainsworth et al., 1994; Onken and Adriano, 1997). The effect of time in mitigating the phytotoxic and zootoxic effects of metal salts added to soils has been demonstrated repeatedly. This aging effect is more pronounced for metals that adsorb strongly on soil solids (e.g., Cu and Pb) than for those that do not (e.g., Zn and Cd). Consequently, Cd and Zn have more pronounced residual effects in soils that persist for decades following metal contamination, revealed by high solubility and ready uptake into crops (McBride, 1995). The following are several possible mechanisms which could account for the aging effect.

Solid-State Diffusion. Although this has often been proposed, measurement of a very high activation energy for solid diffusion of metal ions in silicate minerals (~ 200 kJ mol^{-1}) indicates an exceedingly slow process at ambient temperature (Morioka, 1980). Adsorbate diffusion into microcrystalline minerals probably occurs along interstices of particle aggregates (Willett et al., 1988), not into the lattice structure.

High Activation Energy for Complex Formation. It is possible that the energy required to form strong bonds between the adsorbate and the surface is high, as might be the case for weak acid sites on SOM or oxides which require protons to be displaced for adsorption to occur. Since adsorption on minerals and organic matter is very rapid, this explanation can only be consistent with observation if the ions were to first bond at sites with low activation energies, and then gradually form stronger bonds at sites with high activation energy. This suggestion is supported by the fact that ligands in organic matter that form the strongest bonds with trace metals also have the slowest dissociation reactions. Conversely, the predominant carboxylate ligands of organic matter display rapid proton and metal dissociation/exchange (Hering and Morel, 1990). Thus, the activation energy explanation for aging

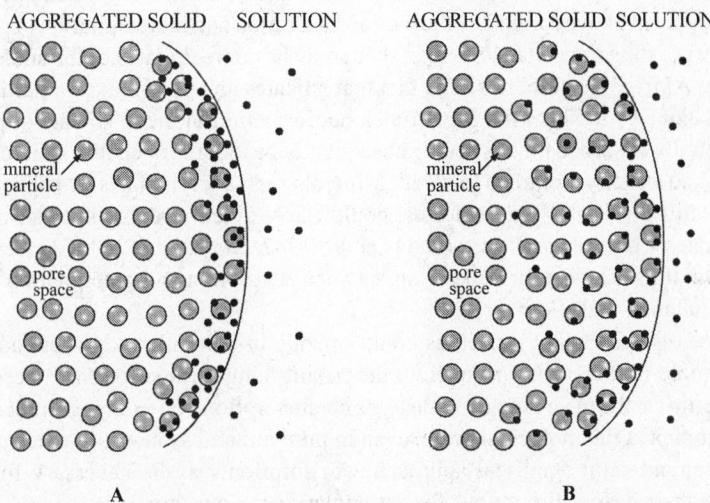

Fig. 8.9 An idealized view of the gradual process of migration of an adsorbate into an aggregate of particles by interparticle diffusion

effects in soils would argue that introduced metals would initially complex predominantly at the weakest complexation sites (carboxylates) of SOM for two reasons: (1) carboxylate groups predominate and are, therefore, the most probable to be encountered by chance; and (2) carboxylate groups rapidly release H^+ or alkaline earth metal cations to allow complexation of the trace metal. This means, however, that a metastable state is reached initially, and there should be a gradual shift of the complexed metals from carboxylate to other ligands forming stronger complexes. Hence, the activity of the metal ion should decrease gradually toward some limiting value.

Surface and Micropore Diffusion. The introduction of a locally high concentration of adsorbate into an aggregated and microporous solid, whether it be a pure mineral, organic matter or soil, necessarily generates a temporary, large concentration and activity gradient. In homogeneous solution, such gradients dissipate rapidly by diffusion. However, within the matrix of an aggregate, pictured schematically in Fig. 8.9, the gradient is likely to persist because, at any instant in time, most of the adsorbate ions or molecules are adsorbed and cannot diffuse. Local differences in activity can, therefore, persist for long periods, buffered by the spatially variable adsorbed fraction. Consequently, although immediate adsorption of most of the added metals or anions occurs, subsequent redistribution toward physically less accessible (probably) interior sites (micropores) lowers the activity (toxicity) and extractability of the adsorbate at the soil aggregate-soil solution interface. The redistribution of adsorbate is driven by the entropy of mixing (Equation [8.44]), which is maximized when random thermal motion eventually spreads the adsorbate evenly over all accessible adsorption sites. This model provides an explanation for the slow reaction of metals, phosphate and other chemicals with aggregated, porous solids. It predicts that reversibility of adsorption should diminish with time, as the free energy of the system only reaches a minimum when the adsorbate is evenly distributed over all adsorption sites. It also suggests that attempts to reverse adsorption processes by lowering the adsorbate concentration in solution cannot be wholly successful, because adsorbate will continue to migrate further into aggregates (as well as out) along a concentration gradient irrespective of changes in concentration outside the aggregate.

Neoformation of Precipitates during Adsorption. A final explanation for aging effects is found in recent metal adsorption experiments on silicate minerals and oxides in which EXAFS spectroscopy detected new polymeric phases at the surfaces (Charlet and Manceau, 1994; Scheidegger et al., 1997). These phases appear to have formed by the partial dissolution and reprecipitation of components of the minerals, incorporating the metals. Some of them may be coprecipitates of the adsorbate metals with metals such as Al from the minerals. The fact that silicates and phosphates in particular may have a high enough solubility to allow the fairly quick neoformation of metal silicate or phosphate phases means that initially adsorbed metals could subsequently be incorporated into solid phases where their extractability and toxicity would be lowered. Minerals such as carbonates and phosphates, with even higher recrystallization rates than silicates, could fairly quickly redistribute adsorbed metals into mixed carbonate or phosphate phases (Davis et al., 1987; Ma et al., 1993; Chen et al., 1997). Such processes again illustrate that the separation between adsorption and precipitation reactions may be more conceptual than real.

All or some of these aging processes could operate in particular adsorbate-adsorbent systems, preventing simple models from adequately describing long-term sorption. Processes imagined as simple and distinct at the macroscopic scale (e.g., chemisorption, ion exchange, precipitation) are now seen to be complex at the molecular scale, even in pure mineral systems. In materials as complex as soils, empirical and semi-empirical approaches to sorption may be necessary for the foreseeable future, despite their serious limitations for extrapolation to new situations.

8.13 References

Aguirre-Gomez, A. 1995. Electrochemical studies of cadmium, copper, lead and zinc complexation in synthetic and soil solutions. Ph.D. Thesis, Cornell University, Ithaca, NY.

Ainsworth, C.C., J.L. Pilon, P.L. Gassman, and W.G. van der Sluys. 1994. Cobalt, cadmium and lead sorption to hydrous iron oxide: residence time effect. Soil Sci. Soc. Am. J. 58:1615–1623.

Angus, J.G., J.B. Raynor, and M. Robson. 1979. Reliability of experimental partition coefficients in carbonate systems: Evidence for inhomogeneous distribution of impurity cations. Chem. Geol. 27:181–205.

Baes, C.F., and R.E. Mesmer. 1986. The hydrolysis of cations. John Wiley and Sons, New York, NY.

Balistrieri, L.S., and T.T. Chao. 1987. Selenium adsorption by goethite. Soil Sci. Soc. Am. J. 51:1145–1151.

Balistrieri, L.S., and T.T. Chao. 1990. Adsorption of selenium by amorphous iron oxyhydroxide and manganese dioxide. Geochim. Cosmochim. Acta 54:739–751.

Benjamin, M.M. 1983. Adsorption and surface precipitation of metals on amorphous iron oxyhydroxide. Environ. Sci. Technol. 17:686–692.

Benjamin, M.M., and J.O. Leckie. 1981. Multiple-site adsorption of Cd, Cu, Zn and Pb on amorphous iron oxyhydroxide. J. Coll. Interf. Sci. 79:209–221.

Berti, W.R., and S.D. Cunningham. 1997. In-place inactivation of Pb in Pb-contaminated soils. Environ. Sci. Technol. 31:1359–1364.

Bibak, A., J. Gerth, and O.K. Borggaard. 1995. Retention of cobalt by pure and foreign-element associated goethites. Clays Clay Min. 43:141–149.

Bleam, W.F., and M.B. McBride. 1985. Cluster formation versus isolated-site adsorption. A study of Mn(II) and Mg(II) adsorption on boehmite and goethite. J. Coll. Interf. Sci. 103:124–132.

Bleam, W.F., and M.B. McBride. 1986. The chemistry of adsorbed Cu(II) and Mn(II) in aqueous titanium dioxide suspensions. J. Coll. Interf. Sci. 110:335–346.

Bobrowski, A., M. Gawlicki, and J. Malolepszy. 1997. Analytical evaluation of immobilization of heavy metals in cement matrices. Environ. Sci. Technol. 31:745–749.

Boekhold, A.E., and S.E.A.T.M. van der Zee. 1992. Significance of soil chemical heterogeneity for spatial behavior of cadmium in field soils. Soil Sci. Soc. Am. J. 56:747–754.

Bresnahan, W.T., C.L. Grant, and J.H. Weber. 1978. Stability constants for the complexation of copper(II) ions with water and soil fulvic acids measured by an ion-selective electrode. Anal. Chem. 50:1675–1679.

Brown, G.E., G.A. Parks, and C.J. Chisholm-Brause. 1989. *In situ* X-ray absorption spectroscopic studies of ions at oxide-water interfaces. Chimia 43:248–256.

Bruemmer, G.W., J. Gerth, and K.G. Tiller. 1988. Reaction kinetics of the adsorption and desorption of nickel, zinc and cadmium by goethite. I. Adsorption and diffusion of metals. J. Soil Sci. 39:37–52.

Bunzl, K., W. Schmidt, and B. Sansoni. 1976. Kinetics of ion exchange in soil organic matter. IV. Adsorption and desorption of Pb^{2+}, Cu^{2+}, Cd^{2+}, Zn^{2+} and Ca^{2+} by peat. Soil Sci. 27:32–41.

Cabannis, S.E. 1990. pH and ionic strength effects on nickel-fulvic acid dissociation kinetics. Environ. Sci. Technol. 24:583–588.

Charlet, L,. and A. Manceau. 1994. Evidence for the neoformation of clays upon sorption of Co(II) and Ni(II) on silicates. Geochim. Cosmochim. Acta 58:2577–2582.

Chen, X., J.V. Wright, J.L. Conca, and L.M. Peurrung. 1997. Effects of pH on heavy metal sorption on mineral apatite. Environ. Sci. Technol. 31:624–631.

Cotter-Howells, J. 1996. Lead phosphate formation in soils. Environ. Pollut. 93:9–16.

Cowan, C.E., J.M. Zachara, and C.T. Resch. 1991. Cadmium adsorption on iron oxides in the presence of alkaline-earth elements. Environ. Sci. Technol. 25:437–446.

Crawford, R.J., I.H. Harding, and D.E. Mainwaring. 1993. Adsorption and coprecipitation of single heavy metal ions onto the hydrated oxides of iron and chromium. Langmuir 9:3050–3056.

Davis, J.A., C.C. Fuller, and A.D. Cook. 1987. A model for trace metal sorption processes at the calcite surface: adsorption of Cd^{2+} and subsequent solid solution formation. Geochim. Cosmochim. Acta 51:1477–1490.

Driessens, F.C.M. 1986. Ionic solid solutions in contact with aqueous solutions. p. 524–560. *In* J.A. Davis and K.F. Hayes (ed.) Geochemical processes at mineral surfaces. ACS Symp. Ser. 323, American Chemical Society, Washington, DC.

Dzombak, D.A., and F.M.M. Morel. 1990. Surface complexation modeling. Hydrous ferric oxides. John Wiley and Sons, New York, NY.

Fendorf, S., M.J. Eick, P. Grossl, and D.L. Sparks. 1997. Arsenate and chromate retention mechanisms on goethite.1. Surface structure. Environ. Sci. Technol. 31:315–320.

Fendorf, S., and M. Fendorf. 1996. Sorption mechanisms of lanthanum on oxide minerals. Clays Clay Min. 44:220–227.

Fendorf, S.E., G. Li, and M.E. Gunter. 1996. Micromorphologies and stabilities of chromium(III) surface precipitates elucidated by scanning force microscopy. Soil Sci. Soc. Am. J. 60:99–106.

Florence, T.M. 1986. Electrochemical approaches to trace element speciation in waters. A review. Analyst 111:489–505.

Ford, R.G., and P.M. Bertsch. 1996. Changes in divalent metal partitioning during iron (hydr)oxide aging. p. 62. Abstracts of the Clay Minerals Society Meeting, Gatlinburg, TN.

Gerth, J. 1990. Unit-cell dimensions of pure and trace metal-associated goethites. Geochim. Cosmochim. Acta 54:363–371.

Grossl, P.R., and D.L. Sparks. 1995. Evaluation of contaminant ion adsorption/desorption on goethite using pressure-jump relaxation kinetics. Geoderma 67:87–101.

Gunneriusson, L., L. Lövgren, and S. Sjöberg. 1994. Complexation of Pb(II) at the goethite (γ-FeOOH)/water interface: The influence of chloride. Geochim. Cosmochim. Acta 58:4973–4983.

Hahne, H.C.H., and W. Kroontje. 1973. Significance of pH and chloride concentration on behavior of heavy metal pollutants: Mercury(II), cadmium(II), zinc(II) and lead(II). J. Environ. Qual. 2:444–450.

Hayes, K.F., A.L. Roe, G.E. Brown, K.O. Hodgson, J.O. Leckie, and G.A. Parks. 1987. *In situ* X-ray absorption study of surface complexes: selenium oxyanions on γ-FeOOH. Science 238:783–786.

Hendrickson, L.L., and R.B. Corey. 1981. Effect of equilibrium metal concentrations on apparent selectivity coefficients of soil complexes. Soil Sci. 131:163–171.

Hering, J.G., and F.M.M. Morel. 1990. The kinetics of trace metal complexation: implications for metal reactivity in natural waters. p.145–171. *In* W. Stumm (ed.) Aquatic chemical kinetics. John Wiley and Sons, New York, NY.

Holmgren, G.G.S., M.W. Meyer, R.L. Chaney, and R.B. Daniels. 1993. Cadmium, lead, zinc, copper and nickel in agricultural soils of the United States of America. J. Environ. Qual. 22:335–348.

Iu, K.L., I.D. Pulford, and H.J. Duncan. 1981. Influence of waterlogging and lime or organic matter additions on the distribution of trace metals in an acid soil. II. Zinc and copper. Plant Soil 59:327–333.

Jin, X., G.W. Bailey, Y.S. Yu, and A.T. Lynch. 1996. Kinetics of single and multiple metal ion sorption processes on humic substances. Soil Sci. 161:509–520.

Keizer, P., and M.G.M. Bruggenwert. 1991. Adsorption of heavy metals by clay-aluminum hydroxide complexes. p. 177–203. *In* G.H. Bolt, M.F. DeBoodt, M.H.B. Hayes and M.B. McBride (ed.) Interactions at the soil colloid-soil solution interface. Kluwer Academic Publishers, Dordrecht, Netherlands.

Konigsberger, E., R. Hausner, and H. Gamsjager. 1991. Solid-solute phase equilibria in aqueous solution. IV: The system $CdCO_3$-$CaCO_3$-CO_2-H_2O. Geochim. Cosmochim. Acta 55:3505–3514.

Lavigne, J.A., C.H. Langford, and M.K.S. Mak. 1987. Kinetic study of speciation of nickel(II) bound to a fulvic acid. Anal. Chem. 59:2616–2620.

Lee, S.-Z., H.E. Allen, C.P. Huang, D.L. Sparks, P.F. Sanders, and W.J.G.M. Peijnenburg. 1996. Predicting soil-water partition coefficients for cadmium. Environ. Sci. Technol. 30:3418–3424.

Lindsay, W.L. 1979. Chemical equilibria in soils. John Wiley and Sons, New York, NY.

Lothenbach, B., G. Furrer, and R. Schulin. 1997. Immobilization of heavy metals by polynuclear aluminum and montmorillonite compounds. Environ. Sci. Technol. 31:1452–1462.

Lövgren, L., T. Hedlund, L.-O.Öhman, and S. Sjöberg 1986. Equilibrium approaches to natural water systems. 6. Acid-base properties of a concentrated bog-water and its complexation reactions with aluminium(III). Water Res. 21:1401–1407.

Lövgren, L,. and S. Sjöberg 1989. Equilibrium approaches to natural water systems. 7. Complexation reactions of copper(II), cadmium(II) and mercury(II) with dissolved organic matter in a concentrated bog-water. Water Res. 23:327–332.

Lumsdon, D.G., L.J. Evans, and K.A. Bolton. 1995. The influence of pH and chloride on the retention of cadmium, lead, mercury and zinc by soils. J. Soil Contam. 4:137–150.

Ma, Q.Y., S.J. Traina, and T.J. Logan. 1993. *In situ* lead immobilization by apatite. Environ. Sci. Technol. 27:1803–1810.

Manceau, A., and L. Charlet. 1994. The mechanism of selenate adsorption on goethite and hydrous ferric oxide. J. Coll. Interf. Sci. 168:87–93.

Manceau, A.L., L. Charlet, M.C. Boisset, B. Didier, and L. Spadini. 1994. Sorption and speciation of heavy metals on hydrous Fe and Mn oxides. Appl. Clay Sci. 7:201–223.

Manning, B.A., and S. Goldberg. 1996. Modeling competitive adsorption of arsenate with phosphate and molybdate on oxide minerals. Soil Sci. Soc. Am. J. 60:121–131.

Martinez, C.E., and M.B. McBride. 1998. Solubility of Cd^{2+}, Cu^{2+}, Pb^{2+}, and Zn^{2+} in aged coprecipitates with amorphous iron hydroxides. Environ. Sci. Technol. 32:743–748.

McBride, M.B. 1978a. Transition metal bonding in humic acid. Soil Sci. 126:200–209.

McBride, M.B. 1978b. Retention of Cu^{2+}, Ca^{2+}, Mg^{2+}, and Mn^{2+} by amorphous alumina. Soil Sci. Soc. Am. J. 42:27–31.

McBride, M.B. 1979. Chemisorption and precipitation of Mn^{2+} at $CaCO_3$ surfaces. Soil Sci. Soc. Am. J. 43:693–698.

McBride, M.B. 1980. Chemisorption of Cd^{2+} on calcite surfaces. Soil Sci. Soc. Am. J. 44:26–28.

McBride, M.B. 1982. Cu^{2+} adsorption characteristics of aluminum hydroxide and oxyhydroxides. Clays Clay Min. 30:21–28.

McBride, M.B. 1985. Influence of glycine on Cu^{2+} adsorption by microcrystalline gibbsite and boehmite. Clays Clay Min. 33:397–402.

McBride, M.B. 1989. Reactions controlling heavy metal solubility in soils. Adv. Soil Sci. 10:1–56.

McBride, M.B. 1990. Electron spin resonance spectroscopy. p. 233–281. *In* D.L. Perry (ed.) Instrumental surface analysis of geologic materials. VCH Publishers, New York, NY.

McBride, M.B. 1994. Environmental chemistry of soils. Oxford University Press, New York, NY.

McBride, M.B. 1995. Toxic metal accumulation from agricultural use of sludge: Are USEPA regulations protective? J. Environ. Qual. 24:5–18.

McBride, M.B., S. Sauvé, and W. Hendershot. 1997. Solubility control of Cu, Zn, Cd and Pb in contaminated soils. Europ. J. Soil Sci. 48:337–346.

McKenzie, R.M. 1980. The adsorption of lead and other heavy metals on oxides of manganese and iron. Aust. J. Soil Res. 18:61–73.

McLaren, R.G., and D.V. Crawford. 1974. Studies on soil copper. III. Isotopically exchangeable copper in soils. J. Soil Sci. 25:111–119.

Mench, M.J., E. Martin, and P. Solda. 1994. After effects of metals derived from a highly metal-polluted sludge on maize (*Zea mays* L.). Water Air Soil Poll. 75:277–291.

Morioka, M. 1980. Cation diffusion in olivine. I. Cobalt and magnesium. Geochim. Cosmochim. Acta 44:759–762.

Nalovic, L., G. Pedro, and C. Janot. 1975. Demonstration by Mossbauer spectroscopy of the role played by transitional trace elements in the crystallogenesis of iron hydroxides(III). p. 601–610. *In* S.W. Bailey (ed.) Proc. Int. Clay Conf., Mexico City, Mexico.

Neal, R.H., and G. Sposito.1986. Effects of soluble organic matter and sewage sludge amendments on cadmium sorption by soils at low cadmium concentrations. Soil Sci. 142:164–172.

Onken, B.M., and D.C. Adriano. 1997. Arsenic availability in soil with time under saturated and subsaturated conditions. Soil Sci. Soc. Am. J. 61:746–752.

Padmanabham, M. 1983. Comparative study of the adsorption-desorption behaviour of copper(II), zinc(II), cobalt(II) and lead(II) at the goethite-solution interface. Aust. J. Soil Res. 21:515–525.

Papadopoulos, P., and D.L. Rowell. 1988. The reactions of cadmium with calcium carbonate surfaces. J. Soil Sci. 39:23–36.

Pingitore, N.E., M.P. Eastman, M. Sandidge, K. Oden, and B. Freiha. 1988. The coprecipitation of manganese(II) with calcite: an experimental study. Mar. Chem. 25:107–120.

Rate, A.W., R.G. McLaren, and R.S. Swift. 1992. Evaluation of a log-normal distribution first-order kinetic model for copper(II)-humic acid complex dissociation. Environ. Sci. Technol. 26:2477–2483.

Reardon, E.J., and S. Della Valle. 1997. Anion sequestering by the formation of anionic clays: lime treatment of fly ash slurries. Environ. Sci. Technol. 31:1218–1223.

Reeder, R.J., G.W. Lamble, J.F. Lee, and W.J. Staudt. 1994. Mechanism of SeO_4^{2-} substitution in calcite: An XAFS study. Geochim. Cosmochim. Acta 58:5639–5646.

Rock, P.A., W.H. Casey, M.K. McBeath, and E.M. Walling. 1994. A new method for determining Gibbs energies of formation of metal-carbonate solid solutions: 1. The $Cs_xCd_{1-x}CO_3(s)$ system at 298 K and 1 bar. Geochim. Cosmochim. Acta 58:4281–4291.

Ryden, J.C., J.K. Syers, and R. W. Tillman. 1987. Inorganic anion sorption and interactions with phosphate sorption by hydrous ferric oxide gel. J. Soil Sci. 38:211–217.

Sass, B.M., and D. Rai. 1987. Solubility of amorphous chromium (III)- iron (III) hydrooxide solid solutions. Inorg. Chem. 26:2228–2232.

Sauvé, S., M. McBride, and W. Hendershot. 1998. Lead phosphate solubility in water and soil suspensions. Environ. Sci. Technol. 32:388–393.

Scheidegger, A.M., M. Fendorf, and D. L. Sparks. 1996. mechanisms of nickel sorption on pyrophyllite:macroscopic and microscopic approaches. Soil Sci. Soc. Am. J. 60:1763–1772.

Scheidegger, A.M., G.M. Lamble, and D.L. Sparks. 1997. Spectroscopic evidence for the formation of mixed-cation hydroxide phases upon metal sorption on clays and aluminum oxides. J. Coll. Inter. Sci. 186:118–128.

Scheidegger, A. M., and D.L. Sparks. 1996. A critical assessment of sorption-desorption mechanisms at the soil mineral/water interface. Soil Sci. 161:813–831.

Schwertmann, U., and G. Pfab. 1994. Structural vanadium in synthetic goethite. Geochim. Cosmochim. Acta 58:4349–4352.

Schindler, P.W., and W. Stumm. 1987. The surface chemistry of oxides, hydroxides, and oxide minerals. p. 83–110. *In* W. Stumm (ed.) Aquatic surface chemistry. John Wiley and Sons, New York, NY.

Schultz, M. F., M. M. Benjamin, and J.F. Ferguson. 1987. Adsorption and desorption of metals on ferrihydrite: Reversibility of the reaction and sorption properties of the regenerated solid. Environ. Sci. Technol. 21:863–869.

Sedlacek, J., E. Gjessing, and J.P. Rambaek. 1987. Isotope exchange between inorganic iron and iron naturally complexed by aquatic humus. Sci. Total Environ. 62:275–279.

Skyllberg, U., P. R. Bloom, E. A. Nater, K. Xia, and W. Bleam, 1996. Binding of mercury(II) in soil organic matter. Agron. Abs. 88:212.

Smolders, E., and M. J. McLaughlin. 1996. Chloride increases cadmium uptake in Swiss Chard in a resin-buffered nutrient solution. Soil Sci. Soc. Am. J. 60:1443-1447.

Sposito, G. 1989. The chemistry of soils. Oxford University Press, New York, NY.

Stumm, W., and J. J. Morgan. 1996. Aquatic chemistry. 3nd Ed. Wiley and Sons, New York, NY.

Swift, R. S., and R. G. McLaren. 1991. Micronutrient adsorption by soils and soil colloids. p. 257–292. *In* G.H. Bolt, M.F. DeBoodt, M.H.B. Hayes, and M.B. McBride (ed.) Interactions at the soil colloid-soil solution interface. Kluwer Academic Publishers, Dordrecht, Netherlands.

Thanabalasingam, P., and W.F. Pickering. 1986. Arsenic sorption by humic acids. Environ. Poll. (Series B) 12:233–246.

Tichý, R., V. Nýdl, S. Kužel, and L. Kolář. 1997. Increased cadmium availability to crops on a sewage-sludge amended soil. Water Air Soil Poll. 94:361–372.

Tiller, K.G., J. Gerth, and G. Brummer. 1984. The relative affinities of Cd, Ni, and Zn for different soil clay fractions and goethite. Geoderma 34:17–35.

van der Zee, S.E.A.T.M., and W.H. van Riemsdijk. 1991. Model for the reaction kinetics of phosphate with oxides and soil. p. 205–240. *In* G.H. Bolt, M.F. DeBoodt, M.H.B. Hayes, and M.B. McBride (ed.) Interations at the soil colloid-soil solution interface. Kluwer Academic Publishers, Dordrecht, Netherlands.

van Riemskijk, W.H., and S.E.A.T.M. van der Zee. 1991. Comparison of models for adsorption, solid solution and surface precipitation. p. 241–256. *In* G.H. Bolt, M.F. DeBoodt, M.H.B. Hayes, and M.B. McBride (ed.) Interations at the soil colloid-soil solution interface. Kluwer Academic Publishers, Dordrecht, Netherlands.

Waller, P.A., and W.F. Pickering. 1993. The effect of pH on the lability of lead and cadmium sorbed on humic acid particles. Chem. Spec. Bioavail. 5:11–22.

Wesselink, L.G., N. van Breemen, J. Mulder, and P.H. Janssen. 1996. A simple model of soil oganic matter complexation to predict the solubility of aluminium in acid forest soils. Europ. J. Soil Sci. 47:373–384.

Willett, I.R., C.J. Chartres, and T.T. Nguyen. 1988. Migration of phosphate into aggregated particles of ferrihydrite. J. Soil Sci. 39:275–282.

Xu, T., F.W. Schwartz, and S.J. Traina. 1994. Sorption of Zn^{2+} and Cd^{2+} on hydroxyapatite surfaces. Environ. Sci. Technol. 28:1472–480.

Yin, Y., H.E. Allen, C.P. Huang, and P.F. Sanders. 1997. Adsorption/desorption isotherms of Hg(II) by soil. Soil Sci. 162:35–45.

Zachara, J.M., C.C. Ainsworth, C.E. Cowan, and C.T. Resch. 1989. Adsorption of chromate by subsurface soil horizons. Soil Sci. Soc. Am. J. 53:418–428.

Zasoski, R.J., and R.G. Burau. 1988. Sorption and sorptive interaction of cadmium and zinc on hydrous manganese oxide. Soil Sci. Soc. Am. J. 52:81–87.

9

Abiotic Catalysis

P.M. Huang
University of Saskatchewan

9.1 Introduction

Abiotic catalytic reactions in soils and associated environments are more common than had previously been thought. Clay minerals, metal oxides, and dissolved metals often demonstrate the ability to catalyze the transformations of natural and anthropogenic organic compounds, metals, metalloids and other inorganics (Bartlett, 1986; Mortland, 1986; Wang et al., 1986; Huang, 1991a, 1995; McBride, 1994; Stone and Torrents, 1995; Smolen and Stone, 1998).

Clay minerals play a vital role in oxidative polymerization of polyphenols and the subsequent transformations to humic substances (Kumada and Kato, 1970; Wang et al., 1971; Wang and Li, 1977; Wang et al., 1980). Iron oxides (Scheffer et al., 1959) and especially Mn oxides (Shindo and Huang, 1982, 1984a,b; Kung and McBride, 1988; Wang and Huang, 1992) are most reactive in mediating the transformation of phenolic compounds. These mineral colloids catalyze the ring cleavage of polyphenols, the deamination, decarboxylation, and dealkylation of amino acids, and polycondensation of phenolic compounds and amino acids (Wang and Huang, 1987, 1992, 1997). Further, other organic compounds such as aromatic amines (Furukawa and Brindley, 1973; McBride, 1979) and organic acids (Jauregui and Reisenauer, 1982; Stone and Morgan, 1984b) can be oxidatively degraded.

Synthetic organic compounds such as pesticides and plasticizers introduced into the environment can be transformed through Brønsted and Lewis acidity and hydrolysis (Theng, 1974, 1979; Cheng, 1991; Stone and Torrents, 1995; Smolen and Stone, 1998). Heterogeneous systems such as soils and sediments contain a series of solid surfaces and dissolved constituents that can catalyze the transformation reactions. These species can alter transformation pathways and rates through phase partitioning, general acid, base, and metal catalysis.

Besides oxidation of various organics, Mn oxides and Fe-containing minerals have the ability to catalyze the transformations of metals, metalloids, and/or other elements. Manganese oxides are effective catalysts in promoting many reactions such as transformation of Cr(III) to Cr(VI) (Bartlett and James, 1979; Amacher and Baker, 1982), As(III) to As(V) (Oscarson et al., 1981a, 1983a), Fe(II) to Fe(III) (Krishnamurti and Huang, 1987), Pu(III) to Pu(IV) (Cleveland, 1970; Amacher and Baker, 1982), the auto-oxidation of Mn(II) (Ross and Bartlett, 1981), and oxidation of nitrite to nitrate (Bartlett, 1981). Heterogeneous oxidation/reduction reactions involving electron transfer between transition metals and Fe-bearing minerals have been demonstrated (Wehrli and Stumm, 1989; Ilton and Veblen, 1994; Peterson et al., 1996; White and Peterson, 1996). Therefore, abiotic catalytic reactions merit close attention in understanding chemical processes in soils and the impact on environmental quality and ecosystem health.

9.2 Fundamentals of Catalysis

9.2.1 Definition of Catalysis

The process of changing the rate of a chemical reaction by use of a catalyst is termed catalysis, coined by Berzelius in 1836 to describe some enhanced chemical reactions (Williams, 1965). The accepted definition of a catalyst, due to Oswald, is that it is a substance that changes the speed of a chemical reaction without itself undergoing any permanent chemical change. Since a reactant or a product may also be a catalyst, Bell (1941) suggests the definition, "A catalyst is a substance which appears in the rate expression to a power higher than that to which it appears in the stoichiometric equation." Actually many substances classified as catalysts are destroyed either as a result of the process that gives them their catalytic activity or because of subsequent combination with the products (Moore and Pearson, 1981). From a practical point of view, a catalyst is a substance that changes the rate of a desired reaction, regardless of the fate of the catalyst itself. Although catalysts are frequently defined as materials which accelerate chemical reactions without themselves undergoing change, as the manager of any plant using a catalytic process knows, this is too optimistic a definition: the properties of all real catalysts do change with use (Twigg, 1989).

An important criterion of a catalyst is that it changes the mechanism of the parent reaction (Moore and Pearson, 1981). Without this change in mechanism, the observed change in rate could not occur. Since catalysts increase the rate of reaction, the mechanism must change to one that is easier for the system to follow, involving, in general, a lower energy barrier. Therefore, the catalyst provides an alternative pathway by which the reaction comes to equilibrium, although it does not alter the position of the equilibrium (Daintith, 1990). The catalyst itself takes part in the reaction. In certain circumstances, very small quantities of catalyst can speed up very large reactions. Some catalysts are also highly specific in the type of reaction they catalyze, particularly in biochemical reactions.

9.2.2 Homogeneous and Heterogeneous Catalysis

The process of changing the rate of a chemical reaction by use of a catalyst that has the same phase as the reactant is homogeneous catalysis (e.g., dissolved metals in catalyzing organic reactions or enzymes in biochemical reactions) (Daintith, 1990). The process that is driven by a catalyst that has a phase different from the reactant is heterogeneous catalysis (e.g., metal oxides in catalyzing organic and inorganic reactions). In heterogeneous catalysis, the catalyzed reaction steps take place very close to the solid surface. These steps may be between molecules adsorbed on the catalyst surface or extensive reaction can take place involving the topmost atomic layers of the catalysts (Twigg, 1989). The sequence of stages in the catalysis of a reaction by a heterogeneous catalyst is shown in Table 9.1. Any of these stages, if slow, may limit the overall rate of a catalytic reaction. Distinctions are often drawn between catalysts which are film-diffusion controlled (i.e., limited by stages 1 and/or 7), pore-diffusion controlled (i.e., limited by stages 2 and/or 6), and reaction controlled (i.e., limited by stages 3, 4 and/or 5). There is a complex interaction between the relative importance of these different stages and the resulting catalytic effect on organic and inorganic reactions.

Both homogeneous and heterogeneous catalytic reactions are significant in soil and environmental sciences (Bartlett, 1986; Huang, 1990, 1991a; Stumm, 1992; Stone and Torrents, 1995; Smolen and

Table 9.1   Sequence of stages in the catalysis of a reaction by a heterogeneous catalyst [Modified from Twigg, 1989. Catalyst handbook. Copyright Wolfe Publishing Ltd with permission]*

1.	Transport of reactants through the liquid or gas phase to the exterior of the catalyst.
2.	Transport of reactants through the pore system of the catalyst to a catalytically active site.
3.	Adsorption of reactants at the catalytically active site.
4.	Chemical reactions between reactants at the catalytically active site (frequently several steps).
5.	Desorption of products from the catalytically active site.
6.	Transport of products through the catalyst pore system from the catalytically active site to the exterior of the catalyst
7.	Transport of products into the liquid or gas phase from the exterior of the catalyst.

*Several different catalytically active sites may be involved. Adsorption, possibly followed by reaction, may occur at one site, followed by transport of an intermediate product to a different site for further reactions.

Stone, 1998). Advances in surface science and catalysis made during the last 40 years are presented in a treatise edited by Hightower et al. (1996).

9.2.3   Proton and Electron Transfer Catalysis

Brønsted acid-base catalysis is effective because proton transfers are generally rapid compared to the making and breaking of other chemical bonds. Therefore, reactions involving proton transfer in a typical acid or base catalysis are rapid compared to similar reactions of comparable free energy. The difference in rates of chemical reactions is related to the steric hindrance involved. Steric hindrance, that is, the repulsion of nonbonded atoms in the activated complex, is a most important factor in determining activation energies (Moore and Pearson, 1981). Since the proton lacks the filled inner electron shells usually responsible for the repulsion and is not surrounded by other groups, it is quite free from steric effects. Proton transfers involving oxygen-hydrogen bonds are generally rapid but they are not instantaneous. For instance, in the ionization of water, the activation energy is at least as great as 57 kJ mol^{-1} (the heat of the reverse reaction) and the entropy of activation is negative; therefore, the rate constant (5×10^{-7} M s^{-1}) is small (Moore and Pearson, 1981). This example demonstrates that an unfavorable equilibrium constant must necessarily make a reaction slow, even if other factors are quite favorable. Given a favorable equilibrium, proton transfers involving O and N bonds to H are almost always very fast, approaching diffusion control in many cases. Exceptions can occur if the proton is in a well-shielded position (Kresge, 1975).

Catalysis by proton transfer is by far the most common in homogeneous reactions. For those reactions that are subject to proton transfer catalysis, an expected relationship exists between the strength of the acid or base, as determined by its ionization constant, and its efficiency as a catalyst, determined by the observed rate constant. This relationship is best shown by the Brønsted catalysis law (Brønsted, 1928),

$$k_a = C_A K_a^\alpha \qquad k_b = C_B K_b^\beta \tag{9.1}$$

where k_a and k_b are the rate constants (also termed the catalytic constants) for acid and base catalytic reactions, respectively; K_a and K_b are the acid and base ionization constants; and C_A, C_B, α and β are constants characteristic of the reaction, the solvent, and the temperature. Normally α and β are positive and have values between 0 and 1. In the Brønsted equation, a low value of α and β signifies a low sensitivity of the catalytic constant to the strength of the catalyzing acid or base, and *vice versa*. Proton transfer catalysis is of significance in soils and associated environments (Theng, 1974, 1979; Cheng, 1991; Nannipieri and Gianfreda, 1998) as discussed below.

In acid catalysis of organic molecules, placing a proton on negatively charged molecules reduces their negative charge, and thus, facilitates the transfer of electrons (Steinberger and Westheimer, 1949, 1951). In agreement with this explanation, a number of multiply charged cations act as catalysts in the transformation of organic compounds (Stone and Torrents, 1995). Presumably a metal-organic complex is formed that reduces the negative charge and increases the electron transfer. The catalytic efficiency of a metal ion depends both on its positive charge and on its ability to form a stable complex of the chelate type. The metal ions may be considered to be acting as generalized acids. However, metal ions have some significant advantages over the proton. They can have greater charges, which lead to greater polarization of the reactant molecules. Unlike the proton, metal ions can be stabilized by other ligands and, thus, can exist in neutral or even basic solutions. The high coordination numbers of metal ions permit the binding of a substrate at more than one site. This advantage helps to make metal ions very efficient catalysts for the hydrolysis of many organic compounds (Kroll, 1952). Further, the metal ion has the ability to simultaneously bind both a substrate and a reagent. This can have the effect of a template, in which the two reactants are assembled prior to combination (Basolo and Pearson, 1968).

Many metal ions, especially of the transition series, have several stable oxidation states which enables them to act as catalysts in certain redox reactions. Transition metals are the best catalysts, in most cases, for catalyzing reactions that are slow for symmetry reasons (Pearson, 1976). In addition to the slow three body reactions, a second class of slow reactions is that forbidden by orbital symmetry (Moore and Pearson, 1981). Even when the reaction is thermodynamically favorable, a large energy barrier can exist. Such reactions are prime candidates for catalysis. The reason why transition metals are the best catalysts is chiefly because of their partly filled d orbitals which have symmetry properties that are different from those of s and p orbitals.

Because of special properties of metal ions, particularly transition metal ions, they can catalyze a wide variety of organic and inorganic reactions in soils and related ecosystems (Siegel, 1976; Bartlett, 1986; Mortland, 1986; Huang, 1990, 1991a; McBride, 1994; Stone and Torrents, 1995; Smolen and Stone, 1998).

9.3   Abiotic Catalysis of Natural and Anthropogenic Organic Compounds

9.3.1   Oxidative Transformations of Phenolic and Other Organic Compounds

The oxidative polymerization of polyphenols can be accelerated enzymatically and nonenzymatically (Haider et al., 1975; Suflita and Bollag, 1981; Huang, 1990). Soil minerals play an important role in catalyzing the abiotic polymerization of phenolic compounds and the subsequent formation of humic substances (Wang et al., 1986; Huang, 1991a, 1995; Pal et al., 1994).

Manganese oxides (birnessite, cryptomelane, and pyrolusite), which occur in soils and sediments (McKenzie, 1989), are very reactive in promoting the polymerization of phenolic compounds (Shindo and Huang, 1982). They can act as Lewis acids which accept electrons from diphenols, leading to their oxidative polymerization. The rate determining step in the formation of humic acids (HAs) from phenols is apparently the formation of a semiquinone radical involving a single electron transfer reaction (Schnitzer, 1982). Semiquinones couple with each other to form a stable HA polymer. The coupling of free radicals requires little activation energy, in contrast to electron transfer reactions (Chang and Allan, 1971). Therefore, coupling of semiquinones rather than the formation of quinones should be kinetically the preferred reaction pathway in the transformation of polyphenols to humic macromolecules.

   The transformation of pyrogallol (1,2,3-trihydroxybenzene) to humic macromolecules is strongly
promoted by birnessite (δ-MnO$_2$) (Wang and Huang, 1992) with total yields 10.5-fold higher than
those formed in the pyrogallol system alone. The infrared (IR) spectra of the pyrogallol-derived HA
(Wang and Huang, 1992) resemble those of natural humic substances (Schnitzer, 1977). During the
catalytic transformation of pyrogallol to HA, birnessite also promotes the abiotic generation of CO$_2$
through its ability to cleave the ring structure of pyrogallol under ambient temperature and pressure in
a system free of microbial activity (Wang and Huang, 1992). The abiotic ring cleavage of polyphenols
by soil inorganic components may partially account for the findings of the high aliphaticity of natural
humic substances (Wilson and Goh, 1977; Hatcher et al., 1981).
   Recent data (Lee and Huang, 1995) show that the abiotic release of CO$_2$ in the birnessite-
polyphenol and polyphenol systems increases with light intensity, a consequence of a photo-
fragmentation of polyphenolics catalyzed by birnessite. These findings also imply that the pathways
of C turnover in the photic zones of soils and aquatic environments may differ from those in their
subsurface and submerged layers.
   Polyhydroxyphenolic acids with *p*- and *o*-OH groups are rapidly oxidized by Mn oxides (Pohlman
and McMoll, 1989) (Table 9.2) to polymeric humic products in both soil and Mn suspensions. On the
other hand, *m*-polyhydroxyphenolic acids are not readily oxidized. Birnessite has also been shown to
oxidize dihydroxybenzene (McBride, 1989a,b); hausmannite has the ability to oxidize hydroquinone
(Kung and McBride, 1988).
   The catalytic ability of Fe oxides in the rapid oxidative polymerization of polyphenols (Scheffer
et al., 1959) increases in the following order: ferrihydrite > goethite > maghemite > lepidocrocite >
hematite. Ferrihydrite, which is short-range-ordered Fe oxide with high surface area, is most reactive
in catalyzing the oxidative polymerization reaction. Besides the nature of Fe oxides, the catalytic
ability of Fe is related to the structure and functionality of phenolic compounds (Shindo and Huang,
1984a).

Table 9.2   Kinetic constants for polyhydroxyphenolic acid oxidation by soil and manganese oxide suspension [From
Pohlman and McColl, 1989. Soil Sci. Soc. Am. J. 53:686-690 with permission of the Soil Science Society of America]*

	Rate constants (L mol^{-1} sec^{-1})	
Compound	Challenge A horizon	MnO$_2$
2,3-dihydroxybenzoic acid	ND[‡]	0.03
2,5-dihydroxybenzoic acid	0.06	0.04
2,6-dihydroxybenzoic acid	0.00[§]	ND
3,4-dihydroxybenzoic acid	0.10	0.04
3,5-dihydroxybenzoic acid	0.00	0.00
Gallic acid	0.25	0.05
Syringic acid	ND	0.01
Vanillic acid	ND	0.00

*Oxidations were run using 10 g L^{-1} soil and 2.5 x 10^{-4} *M* phenolic acid or 0.19 g L^{-1} MnO$_2$ and 5.0 x 10^{-4} *M*
phenolic acid at pH 4.5 and 30 °C
[‡]ND = not determined
[§]No oxidation of phenolic acid within 120 min of reaction

The oxidative polymerization of polyphenols is substantially influenced by the catalysis of Al hydroxides (Wang et al., 1983). The Al^{3+} ions apparently promote delocalization of electrons from phenolic oxygen atoms into the π orbital system through displacement of protons from the phenolic groups catalyzing the oxidation of polyphenols.

Soluble silicic acid in aqueous solution and precipitated short-range ordered silica can catalyze the oxidation of polyphenols (Ziechmann, 1959). The surface of ground quartz has a disturbed layer which is short-range ordered in nature (Iler, 1979). Similar disturbed surface layers are present on quartz grains in soils (Ribault, 1971). Oxidative polymerization of polyphenols may be catalyzed by the disturbed surface of quartz in soils.

Among metal oxides, Mn(IV) oxides are most reactive in the abiotic oxidation of polyphenols in the systems free of microbial activity (Shindo and Huang, 1982, 1984a). Shindo and Huang (1992) compared the catalytic effects of Mn oxide and tyrosinase on the oxidative polymerization of diphenols over the pH common in soil environments. Manganese oxide influences the oxidative polymerization of hydroquinone and resorcinol to a larger extent than does tyrosinase, whereas the reverse is true for catechol. The yields of HAs are significantly influenced by the kind of catalyst and polyphenol. In the Mn oxide system, the yield of HAs is in the order: hydroquinone > catechol > resorcinol. In the tyrosinase system, catechol produces the highest yield of HA, followed by hydroquinone and resorcinol. These findings indicate that the relative catalytic effects of Mn(IV) oxides and enzymes such as tyrosinase would vary with the type of polyphenols in soils.

Phenolic acids have been shown to be oxidized rapidly in the presence of MnO_2 to form a number of soluble products (Lehmann and Cheng, 1988). Mass spectrometric data show that some of the soluble products of the reaction have somewhat higher molecular weights than the parent compounds. However, the soluble products of the reaction of ferulic acid and MnO_2 do not contain any ferulic acid-derived polymers. By contrast, the products of enzymatic polymerization range from dimers to hexamers (Liu et al., 1981; Bollag et al., 1982). The oxidized products of ferulic acids are apparently rapidly sorbed on the surfaces of MnO_2. Recently, FTIR data (Naidja et al., 1998) clearly show that, in the birnessite-catechol system, some of the reaction products are adsorbed on the mineral surface with the spectrum of the reaction products resembling that of synthetic HA (Niemeyer et al., 1992) obtained from catechol by oxidation with sodium periodate (Hannien et al., 1987) rather than that of humic substances obtained from catechol by oxidation with horseradish peroxidase (Schnitzer et al., 1984). Humic acids formed by mineral catalysis have a better defined chemical structure than those formed by enzymatic oxidative polymerization. Further, mass spectrometry data (Naidja et al., 1998) indicate that in the abiotic catalysis by birnessite, adsorption of reaction products on the mineral surface limits the extent of polymerization, favoring the formation of components with lower degrees of aromatic ring condensation and lower molecular weights compared with those generated in the presence of tyrosinase.

Besides metal oxides, clay size layer silicates have the ability to catalyze the oxidative transformation. Before the pioneering work on the catalytic role of clay size layer silicates in oxidative polymerization of phenolic compounds and the subsequent formation of humic substances (Kumada and Kato, 1970; Wang and Li, 1977; Filip et al., 1977), the conversion of many aromatic amines into their color derivatives by clay minerals and clays had been investigated (Faust, 1940; Hauser and Leggett, 1940). Solomon et al. (1968) reported that, except for talc, a large number of representative minerals produce a blue color of varying intensity when brought in contact with a saturated solution of benzidine hydrochloride. The active sites for the oxidation of benzidine are located on the crystal edges and on transition metal atoms in the higher oxidation state that occupy octahedral sites in the silicate layers. Thompson and Moll (1973) measured the oxidative power of smectites by oxidation of hydroquinone to p-benezoquinone in a clay slurry. Oxidation occurs in the presence of O_2 (air), but not of N_2 unless Fe^{3+} or Cu^{2+} are the exchangeable cations. Adsorbed O_2 molecules or radicals on the clay surface are apparently partially responsible for the oxidation.

Pinnavaia et al. (1974) reported aromatic radical cation formation on the intracrystal surfaces of transition metal-saturated layer silicates. Aromatic molecules, when reacted under very moderate conditions with Cu(II) or Fe(II) ions, may donate electrons to the metal cations, leading to the formation of polymers (Mortland and Halloran, 1976). Montmorillonite, vermiculite, illite, and kaolinite accelerate the formation of HAs to varying degrees (Table 9.3). The promoting effect of 2:1 layer silicates is higher than that of 1:1 layer silicates because of the larger specific surface and lattice imperfections which favor the adsorption of O_2 molecules or radicals.

One of the well-identified precursors (Flaig et al., 1975; Hayes and Swift, 1978) for the formation of humic substances, hydroquinone, can be transformed in aqueous solution at near neutral pH (6.5) to humic macromolecules and deposited in the interlayers of nontronite saturated with Ca which is the most common and most abundant exchangeable cation in soils and sediments (Wang and Huang, 1986). Most of the interlayer humic macromolecules are highly resistant to alkali extraction and may, thus, be humin type materials. Therefore, besides Al interlayering of clays (Barnhisel and Bertsch, 1989), the formation of humic substance interlayers in 2:1 expansible layer silicates, through polymerization of phenol monomers and the associated reactions in soils and sediments, deserves close attention.

The ability of nontronite to promote the oxidation of polyphenols is related to the structure and functionality of the polyphenols, and part of the reaction process may proceed as shown below (Wang and Huang, 1994):

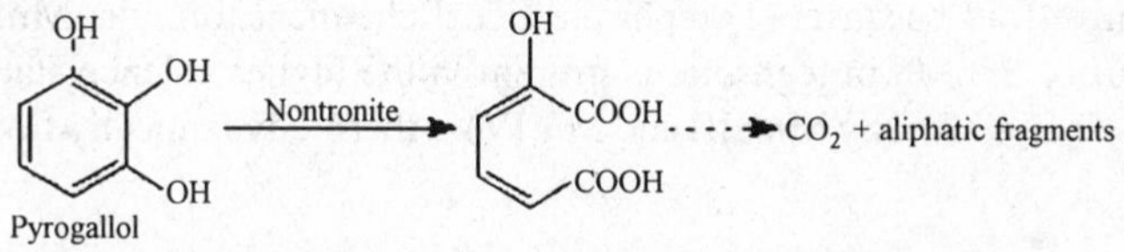

$$[9.2]$$

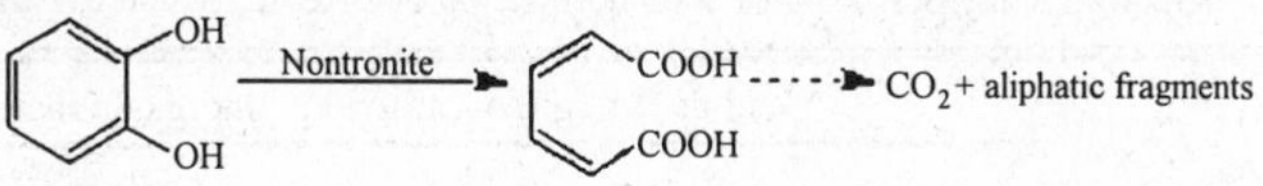

$$[9.3]$$

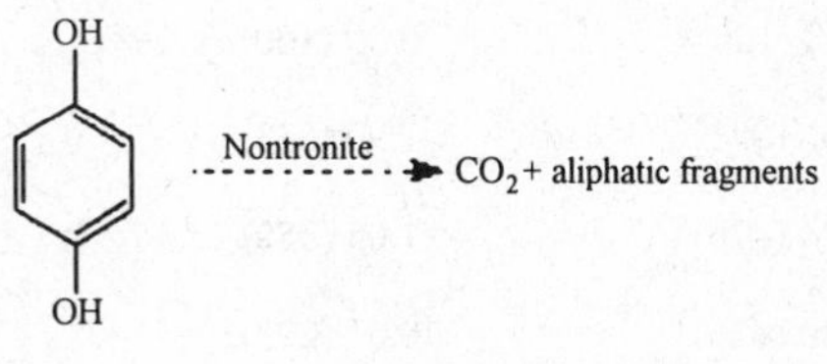

$$[9.4]$$

Catechol with two *o*-OHs is evidently more easily cleaved than hydroquinone with two *p*-OHs while pyrogallol with three hydroxyls in adjacent positions is even more easily cleaved than catechol. The resultant carboxyl group-containing intermediates are further oxidized to form CO_2 and aliphatic fragments. In the reaction systems, intermediate products and aliphatic fragments may form polycondensates. The structure and functionality of polyphenols, thus, have an important role in influencing the extent of catalytic transformations by nontronite.

Table 9.3 Effects of clay minerals on the synthesis of humic acids (HA) at an initial pH of 5.5 at the end of 7 days [From Shindo and Huang, 1985. Appl. Clay Sci. 1:71-81 with permission of Elsevier Science, Amsterdam]

| System | Yield of HA (g HA-carbon kg^{-1} inorganic material)* | | |
	S fraction[‡]	P fraction**	Total
Control[§]	0.68 (100)[¶]	0.30 (100)	0.98 (100)
Montmorillonite	1.25 (184)	0.31 (103)	1.56 (159)
Vermiculite	0.98 (144)	0.31 (103)	1.29 (132)
Illite	0.77 (113)	0.32 (107)	1.09 (111)
Kaolinite	0.68 (100)	0.31 (103)	0.99 (101)

*1 ml of 0.02 M KMnO$_4$ consumed was calculated as corresponding to 0.45 mg carbon
[‡]Soluble fraction
**Precipitated fraction
[§]In the absence of inorganic material
[¶]The index of the yield of HA in the control system is assigned 100 as the basis for comparison.

Primary minerals which are commonly present in soil environments (Dixon and Weed, 1989) can catalyze the oxidative polymerization of polyphenols (Table 9.4). The degree of acceleration of the oxidative polymerization of hydroquinone is greatest in the tephroite system which increases the total HA yields more than nine-fold because: (1) tephroite (ideal chemical formula, MnSiO$_4$) is a Mn-bearing silicate, (2) part of the Mn in tephroite is present in the higher valence states, and (3) the oxidation of diphenols [C$_6$H$_4$(OH)$_2$] by Mn(III) and Mn(IV) is thermodynamically favorable (Weast, 1978).

Table 9.4 Effects of primary minerals on the synthesis of humic acids (HA) at an initial pH of 5.5 at the end of 7 days [From Shindo and Huang, 1985. Appl. Clay Sci. 1:71-81 with permission of Elsevier Science, Amsterdam]

| System | Yield of HA (g HA-carbon kg^{-1} inorganic material)* | | |
	S fraction[‡]	P fraction**	Total
Control[§]	0.68 (100)[¶]	0.30 (100)	0.98 (100)
Tephroite	6.60 (971)	1.90 (633)	8.50 (867)
Hornblende	3.24 (476)	1.05 (350)	4.29 (438)
Augite	2.90 (426)	0.67 (223)	3.57 (364)
Biotite	2.03 (299)	0.55 (183)	2.58 (263)
Quartz	1.15 (169)	0.32 (107)	1.47 (150)
Microcline	0.98 (144)	0.32 (107)	1.30 (133)

*1 ml of 0.02 M KMnO$_4$ consumed was calculated as corresponding to 0.45 mg carbon
[‡]Soluble fraction
**Precipitated fraction
[§]In the absence of inorganic material
[¶]The index of the yield of HA in the control system is assigned 100 as the basis for comparison.

The hydroquinone-derived polymers with molecular weights of approximately 3,500 and higher formed in the presence of the tephroite system (Shindo and Huang, 1985b) have similar IR absorption bands to those of humic substances (Schnitzer, 1978). The surface features of these polymers (Fig. 9.1) are similar to those of soil HA and FA (Stevenson and Schnitzer, 1982) with the smallest discrete particles being spheroids with diameters of 0.1 to 0.2 µm (Fig. 9.1a) and some aggregation of individual spheroids (Fig. 9.1b,c). Small aggregates (Fig. 9.1c) resemble moss while the large aggregates are nodule like (1–5 µm diam.) and doughnut like (6–8 µm diam.) (Fig. 9.1a, b). The polymers do not appear to be associated with the surfaces of tephroite particles (Fig. 9.1d). Consequently, the role of primary minerals in the oxidative polymerization of polyphenols and the subsequent formation of humic substances in soils should not be overlooked (Shindo and Huang, 1985b).

Short-range ordered (SRO) aluminosilicates such as allophane are known to act as catalysts in the oxidative degradation of polyphenols (Kyuma and Kawaguchi, 1964; Kumada and Kato, 1970), but the role of other SRO aluminosilicates remains obscure. On the other hand, the formation of SRO aluminosilicates is significantly affected by inorganic ligands, low-molecular-weight organic acids, humic substances, metallic cations, and expansible layer silicates (Huang, 1991b) resulting in the formation of ill-defined aluminosilicate complexes and hydroxyaluminosilicate-intercalated layer silicates. The catalytic ability of these SRO mineral colloids in the transformation of polyphenols and other organic compounds in soils has yet to be investigated.

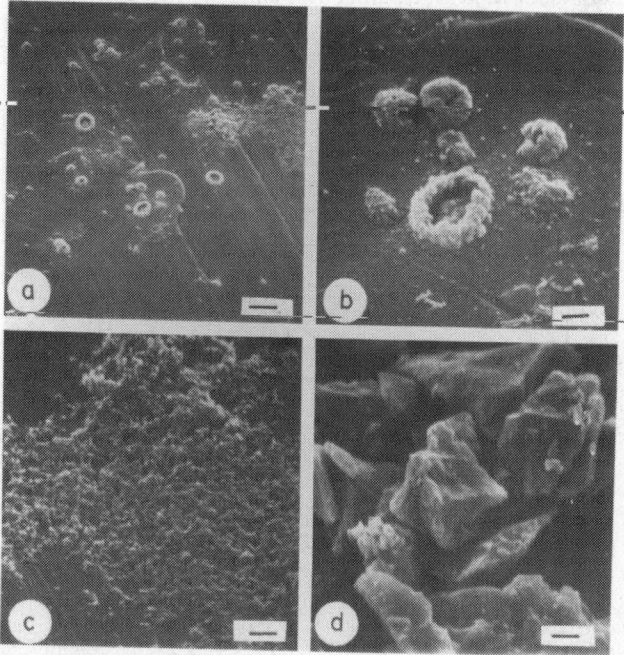

Fig. 9.1   SEM micrographs of hydroquinone polymers in the supernatant and mineral particles settled in the tephroite system at the ratio of mineral to hydroquinone solution of 0.01 at the initial pH of 6.0 at the end of 7 d. a, b, and c: hydroquinone polymers; d: tephroite particles after reaction with hydroquinone. Bar in Fig. 9.1a = 10 μm; bars in Fig. 9.1b to d = 2 μm [Reprinted from Shindo and Huang, 1985. Soil Sci. 139:505-511. Copyright Williams & Wilkins, Baltimore, MD with permission]

In addition to polyphenolic compounds, many other organics including salicylic, pyruvic, oxalic, and malic acids are degraded by mineral colloids such as Mn oxides by electron transfer reactions (Jauregui and Reisenauer, 1982; Stone and Morgan, 1984a,b). Even monophenolic compounds, particularly those containing electron-donating substituent groups on the aromatic ring, can be oxidatively transformed by Mn oxides (Lehmann et al., 1987; Stone, 1987).

Birnessite is also an efficient solid-state catalyst of the breakdown of organic pollutants such as 2,4-D and ethyl ether which are adsorbed on birnessite and rapidly oxidized (Cheney et al., 1996), both producing CO_2 as a major product (Fig. 9.2), but by somewhat different mechanisms. In the case of 2,4-D only, methanol extractable Mn^{2+} is detected. Thus, the solid degradation of organochlorine herbicides can occur at significant rates in the presence of birnessite, a common component of soils. Possible contributions of solid-state degradation to herbicide breakdown by abiotic catalysis should be considered in modeling herbicide breakdown or when designing experiments for soil remediation.

9.3.2 Polycondensation of Phenolic Compounds and Amino Acids

Birnessite efficiently catalyzes the polycondensation of phenolic compounds and amino acids such as in hydroquinone-glycine systems in the common pH range (4–8) of soils with the formation of NH_3-N and nitrogenous polymers (Table 9.5). The proposed processes are as follows: (1) the Mn oxide acts as a Lewis acid by accepting electrons from hydroquinone, which is thus oxidized and subsequently polymerized; (2) the products of the reaction of glycine with hydroquinone polymerize to form nitrogenous polymers, thereby incorporating glycine into the polymers during the oxidative polymerization of hydroquinone; and (3) a partial deamination of glycine occurs during process (2).

Birnessite also catalyzes the polycondensation of glycine and pyrogallol and deamination and decarboxylation of glycine especially in the presence of pyrogallol (Wang and Huang, 1987). The formation of HA and FA is not evident in the presence of glycine alone. In a N_2 atmosphere, CO_2 release in the birnessite-glycine-pyrogallol system is drastically reduced. Birnessite also greatly enhances the formation of N-containing humic polymers. Most of the released NH_3 can be attributed

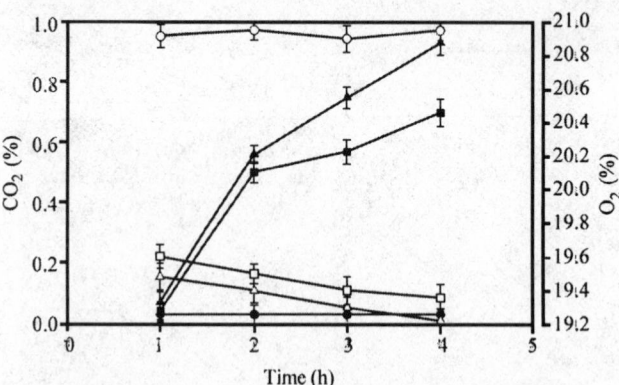

Fig. 9.2 Gas chromatographic measurements of CO_2 and oxygen in the headspace of reaction vials containing birnessite and 2,4-D at 1, 2, 3 and 4 h, respectively. The values indicate the mole percent of CO_2 and oxygen in the headspace gas. ●, CO_2 from air control; ■, CO_2 from ethyl ether control; ▲, CO_2 from 0.3 μmol 2,4-D plus ethyl ether sample; O, oxygen from air control; □, oxygen from the ethyl ether control; △, oxygen from 0.3 μmol 2,4-D plus ethyl ether. The error bars indicate the standard error for two measurements [Reprinted from Cheney et al., 1996. Coll. Surf. A. Phy. Eng. Asp. 107:131-140 with permission of Elsevier Science, Oxford].

Table 9.5 Formation of NH_3-N and polymer-N (in mg kg^{-1}) in the hydroquinone-glycine system both in the absence and presence of Mn oxide at different pH values at the end of 24 h [From Shindo and Huang, 1984. Nature 308:57-58 with permission of Macmillan, London]

	Initial pH of 4.0			Initial pH of 6.0				Initial pH of 7.3			
	Final pH	NH_3-N	Polymer-N	Sum-N	Final pH	NH_3-N	Polymer-N	Sum-N	Final pH	NH_3-N	Polymer-N	Sum-N
Hydroquinone glycine	4.1	4.4	3.0	7.4	6.1	6.0	2.0	8.0	6.9	11.7	3.4	15.1
Mn oxide hydroquinone glycine	4.1	30.5	12.6	43.1	7.6	30.2	40.1	70.3	7.9	34.1	41.2	75.3

20 mg of Mn(IV) oxide (birnessite) was suspended in 20 mL of 0.1 M sodium acetate solution at different pH values (adjusted with dilute acetic acid) containing 200 μmol of hydroquinone and 1,000 μmol of glycine. The flasks (125 mL) containing the suspensions were placed in an oscillating water bath at 25°C. The control solution (hydroquinone and glycine) without Mn oxide was also placed in the same conditions. At the end of the reaction period of 24 h, the reaction solutions were centrifuged at 38,000 g for 20 min. The supernatant obtained was used in the determination of NH_3-N formed. The supernatant was also dialysed against water until the dialysed water was colorless. The dialysis tube (~3,500 molecular weight cutoff, from Spectra Medical Industries, Inc., Los Angeles, CA) was used. Dialysis eliminated unreacted glycine and hydroquinone, lower molecular weight polymers and acetate. The materials retained in the dialysis tube were used for the determination of polymer-N. NH_3-N is the ammonia-N formed, which was determined by steam distillation using MgO. Polymer-N is nitrogen that is present in the fraction with a molecular weight of ~3,500 and higher, which was determined by Kjeldahl method. Sum-N = NH_3-N + Polymer-N.

to the deamination of glycine by pyrogallol-derived free radicals as catalyzed by birnessite but an appreciable amount can also be directly derived from the deamination of glycine by birnessite.

Recently, ^{14}C labeling of glycine was used to measure the extent of incorporation of carboxyl C and alkyl C into the polycondensates of pyrogallol and glycine (Wang and Huang, 1997). Birnessite promotes the incorporation of carboxyl and especially alkyl C of glycine into the polycondensates formed in the presence of pyrogallol. The results indicate that the formation of the C-C bond is more prevalent than that of the C-O bond in the polycondensation of glycine and pyrogallol catalyzed by birnessite.

Besides birnessite, clay-size layer silicates also have the ability to catalyze the polycondensation of phenolic compounds and amino acids. Wang et al. (1985) show the catalytic effect of Ca-illite on the formation of N-containing HAs in the systems containing phenolic compounds and amino acids at neutral pH. The yields and N contents of the synthesized HAs depend on the nature of amino acids (Table 9.6).

Nontronite in which structural Fe(III) acts as an electron acceptor analogous to Mn(IV) in MnO_2 (Solomon, 1968; Theng, 1974) also catalyzes the polycondensation of glycine and pyrogallol (Wang and Huang, 1991). The IR and ESR spectra of the HA and FA formed, which show the presence of a stabilized organic free radical (semiquinone), are similar to those of natural soil HA and FA (Schnitzer, 1977; Senesi and Schnitzer, 1977; Schnitzer and Levesque, 1979; Hatcher et al., 1980; Schnitzer and Ghosh, 1982). The formation of semiquinone free radicals appears to proceed through electron transfer from pyrogallol to Fe(III) or other variable-valence transition metal ions on the edges or in the nontronite structure (Solomon, 1968). Electrons diffuse or tunnel to octahedral sites from layer edges or basal surfaces (Tennakoon et al., 1974). The strong oxidation power of chemisorbed O_2 on silicates (Solomon and Hawthorne, 1983) such as nontronite seems to cause the ring cleavage of pyrogallol and the decarboxylation and deamination of glycine (Wang and Huang, 1991).

Table 9.6 Yields and analytical results of the model humic acids (MHA) formed by various amino acids with the solution of phenolic compounds in the suspension method [From Wang et al., 1985. Soil Sci. 140:3-10. Copyright Williams & Wilkins, Baltimore, MD]

Kinds of amino acids	MHA yield	Recovery of reactants as*	Catalytic effect[‡]	C content of MBA	N content of MHA		mmole of amino acid supplying N to MHA	S content of MHA		mmole of amino acid supplying S to MHA
	mg	%		%	%	mg		%	mg	
Blank	a 119.1§	12.0	2.7	54.9	0	0	0	--	--	--
	b 44.7	4.5		52.2	0	0	0	--	--	--
Glycine**	a 156.6	13.7	3.9	52.8	1.7	2.6	0.19	--	--	--
	b 39.4	3.5		52.5	0	0	0	--	--	--
Leucine	a 139.2	11.3	3.5	52.0	1.7	2.4	0.17	--	--	--
	b 40.0	3.2		54.4	0	0	0	--	--	--
Cystein	a 228.8	18.8	1.8	48.1	1.8	4.1	0.29	4.0	9.2	0.29
	b 126.9	10.4		48.9	1.6	2.0	0.14	3.5	4.4	0.14
Methionine	a 172.9	13.6	1.8	48.0	1.1	1.9	0.14	1.7	2.9	0.09
	b 93.6	7.6		49.5	1.1	1.0	0.07	1.5	1.4	0.04
Proline	a 113.1	9.4	1.2	50.3	0.8	0.9	0.06	--	--	--
	b 93.8	7.8		53.6	0.4	0.4	0.03	--	--	--
Phenylalanine	a 256.8	19.8	2.9	54.8	2.2	5.6	0.40	--	--	--
	b 89.4	6.9		57.3	0.3	0.3	0.05	--	--	--
Tryptophan	a 543.1	39.7	1.8	55.1	4.9	26.6	0.95	--	--	--
	b 307.6	22.4		54.8	4.1	12.6	0.45	--	--	--
Lysine	a 272.4	20.5	3.6	51.2	3.9	10.6	0.38	--	--	--
	b 76.4	5.8		52.5	1.8	1.4	0.05	--	--	--
Histidine	a 284.3	22.2	3.1	51.6	5.8	16.5	0.39	--	--	--
	b 92.4	7.2		51.3	5.2	4.8	0.11	--	--	--
Arginine	a 325.2	24.8	3.1	49.8	4.5	14.6	0.26	--	--	--
	b 104.1	7.9		53.0	3.0	3.1	0.06	--	--	--
Tyrosine	a 41.2	25.3	1.9	54.0	3.4	1.4	0.10	--	--	--
	b 22.0	13.8		54.3	1.0	0.2	0.01	--	--	--

*Recovery = (Yield/Total reactant) x 100.
[‡]Catalytic effect = (Yield of a/Yield of b).
§a, 200 mg Ca-illite used as the catalyst; b, no catalyst.
**Amount of each amino acid added was 1.8 mmol 100 mL^{-1}. Due to its low solubility, only 0.12 mmol tyrosine was added.

Semiquinone radicals, aliphatic fragments, and glycine apparently undergo polycondensation to form humic polymers.

Because polycondensation of phenolic compounds and amino acids and the subsequent formation of humic polymers is catalyzed by clay-size layer silicates and metal oxides, further work is warranted on these new abiotic pathways of C turnover and N transformations.

9.3.3 Surface Brønsted and Lewis Acidity and Hydrolysis of Organic Compounds

Mineral colloids may detoxify adsorbed pesticides by catalyzing their decomposition through the ability of mineral colloids to behave as Brønsted acids and donate protons or to act as Lewis acids and accept electron pairs. Brønsted acidity derives essentially from the dissociation of water molecules coordinated to surface bound cations, resulting in the formation of active protons. This acidity is, thus, strongly influenced by the hydration status and polarizing power of surface bound and structural cations on mineral colloids (Mortland, 1970, 1986). Low water contents and highly polarizing cations (small, high charge) promote Brønsted acidity. Lewis acidity arises from constituent ions such as Al and Fe ions exposed at the edges of mineral colloids (Sposito, 1984; McBride, 1994).

Many uncharged organic molecules that require extremely acid conditions to accept a H^+ ion in solution can be protonated on the surfaces of mineral colloids (McBride, 1989c, 1994; Huang, 1990). The Brønsted acid strength and the degree of protonation are related to the electronegativity and polarizing power of the exchangeable and structural metal cations in the following order (McBride, 1994): $H^+ > Al^{3+}$, $Fe^{3+} > Mg^{2+} > Ca^{2+} > Na^+ > K^+$. As clay surfaces become drier, the Brønsted acidity increases and protons are concentrated in a smaller volume of water resulting in more extreme surface acidity. Even very weak bases, i.e., poor proton acceptors can be protonated on the surfaces of such mineral colloids.

The degradation of certain pesticides is catalyzed as the result of their adsorption on mineral colloids which was first observed by formulation chemists who used clays as carriers and diluents (Fowker et al., 1960). Surface-catalyzed degradation of pesticides on clays was later demonstrated for several organophosphate and *s*-triazine pesticides (Brown and White, 1969; Saltzman et al., 1974; Mingelgrin et al., 1977) and has been attributed to the surface acidity of clay minerals (Brown and White, 1969). Mingelgrin and Saltzman (1979) demonstrated that the nature of the clay, its saturating cation, and hydration status determine the rate and mechanism of degradation of parathion. The adsorption-catalyzed degradation of parathion is a hydrolysis that proceeds either directly or through a rearrangement step.

Magnesium-saturated montmorillonite converts the amino form of 3-aminotriazole to the imino form (Russell et al., 1968a) as shown below:

$$[9.5]$$

Triazine compounds can also be protonated on dry mineral colloids. Armstrong et al. (1967) showed that the addition of sterilized soil to atrazine solutions increased the hydrolysis rate of atrazine tenfold. Soil catalytic degradation of atrazine through hydrolysis is an important pathway. Mineral surface acidity also catalyzes hydrolysis of the chloro-*s*-triazine herbicides to the non-phytotoxic 2-hydroxy-*s*-triazines (Russell et al., 1968b), and may be of significance in a wide range of other organic compounds (McBride, 1994).

Besides the Brønsted acid, the Lewis acid properties of metals appear to be of significance in mineral-catalyzed hydrolysis reactions. The oxides of Fe and Al in water, and especially in the dry state, have some catalytic function in organic hydrolysis reactions, at least for those that are known to be hydroxyl ion-catalyzed (Hoffmann, 1990). Those hydrolyzable organics that have a suitable structure to chelate with the surface metal cations are, in general, the most susceptible to mineral catalyzed degradation.

Dissolved metals and metal-containing surfaces play an important role in catalyzing the hydrolysis of organic pollutants. The catalytic effectiveness of a metal ion depends on its ability to complex reactant molecules (e.g., ester linkage, leaving group, and attacking nucleophile) and shift electron density and conformation in ways favorable to reaction (Hoffmann, 1980). Metals vary in their complex formation constants, which reflect differences in metal-ligand bond strengths and solvation forces. The following rules can explain the general features: (1) complex formation constants generally increase as the charge to radius ratio of the metal ion is increased, because the electrostatic contribution to bond formation is increased; (2) polarizable metals and ligands gain additional complex stability through covalent bond formation; and (3) as complex formation constants increase, competition for the metal among available ligands (including OH$^-$ and other inorganic constituents) becomes more pronounced (Stone and Torrents, 1995). Trivalent metals such as Al and Fe(III) and tetravalent metals such as Ti(IV) are classic hard metals which form strong complexes with classic hard ligands (Pearson, 1966). Organic compounds possessing O and N donor atoms are classified as hard ligands, whereas those possessing S donor atoms are classified as soft ligands. Oxygen donor organic ligands form strong complexes with metals such as Al, Fe(III) and Ti(IV). However, they compete with hydroxo(OH$^-$) and oxo(O^{2-}) species that limit metal solubility near neutral pH values. Although Ca^{2+} and Mg^{2+} are also classified as "hard" metals, their s and p orbitals form bonds that are primarily ionic in nature (Stone and Torrents, 1995).

Complex formation constants of metals with a ligand primarily vary with the ionic radius of metals, although increases in polarizability with increasing atomic number are also important (Houghton, 1979). Even if differences in metal ligand complex formation constants are taken into consideration, differences in electronic structure cause some metals to emerge as particularly effective catalysts.

Metals can form a series of metal hydroxo complexes (Bases and Mesmer, 1986). Although metal hydroxo complexes are less reactive nucleophiles than hydroxide ions, their concentrations can be substantially higher, especially in the neutral to acidic pH range. Because the pK$_1$ values for strong Lewis acid metals are low, they generate hydroxo species across a wider pH range. The nucleophilites of various metal hydroxo complexes are governed by pK$_1$ and the polarizability of the metal (Stone and Torrents, 1995).

The reactivity of hydrolyzable organic pollutants arises from the presence of electrophilic (electron deficient) sites within the molecules (Stone and Torrents, 1995). The S$_N$2 mechanism (neucleophilic substitution) involves attack of the electrophilic sites by OH$^-$ or H$_2$O, generation of a higher coordination number intermediate, subsequent elimination of the leaving group, and the formation of a hydrolysis product:

$$\text{[9.6]}$$

In the case of the S$_N$1 mechanism (nucleophilic substitution monomolecular), the reaction proceeds with the loss of the leaving group to generate a lower coordination number intermediate and then followed by generation of the hydrolysis product by nucleophilic addition as shown below:

$$\text{[9.7]}$$

It is generally accepted that metal ions can catalyze hydrolysis in a way similar to acid catalysis. Metal ions and protons coordinate to the pollutants so that electron density is shifted away from the site of nucleophilic attack to facilitate the reaction. Because protons have an extremely high charge density and great polarizing power, metal catalysis is insignificant in acidic solutions. However, metal ions can readily coordinate two or more ligand donor sites on a molecule and can greatly outnumber protons in neutral and alkaline conditions (Plastourgou and Hoffmann, 1984). A list of metal catalysis mechanisms are given in Table 9.7.

The hydrolysis of compounds possessing good leaving groups is limited by the rate of nucleophilic attack. These compounds are susceptible to types 1, 3, 4, 6 and possibly 5 of metal catalysis. The hydrolysis of compounds possessing poor leaving groups are limited by the breakdown of the tetragonal intermediate; therefore, types 2, 7 and possibly 5 are important. A number of pesticides and other pollutants are in the intermediate region, where metal catalysis may shift the rate limiting step from nucleophilic attack to breakdown of the tetrahedral intermediate or *vice versa*.

Many organic compounds are susceptible to metal ion catalysis. These include carboxylic acid esters, amides, anilides, phosphate-containing esters, and other hydrolyzable compounds (Stone and Torrents, 1995). In the mid 1950s, the catalytic ability of metals in hydrolysis of thionophosphate pesticides such as parathion and EPN (thionobenzene phosphoric acid $\underline{O}$-ethyl-$\underline{O}$-p-nitrophenyl ester) was recognized (Ketelaar et al., 1956). Rate enhancements arising from the addition of 1.0×10^{-4} M Cu^{2+} are as high as 20-fold for parathion at pH 8.5 and 48-fold for EPN at pH 8.2. Coordination of the S donor group by Cu^{2+} activates the P center to nucleophilic attack (type 1 catalysis). Chlorpyrifos, a phosphothionate pesticide, contains a pyridyl N capable of forming a six-membered chelate ring with the Cu^{2+} ion (Blanchet and St. George, 1982).

Metal ion catalysis arises from the increased electrophilicity of the P atoms in the chelate ring (type 1 catalysis) (Blanchet and St. George, 1982) and from possible leaving group effects (type 2 catalysis). Copper (II) is by far the most reactive catalyst among a number of metals [including Mg(II) and Al(III)] which have been observed to catalyze Chlorpyrifos hydrolysis. However, Cu^{2+} concentration would have to exceed 10^{-5} M, which seldom occurs in soil solutions and other natural waters, in order for the metal-catalyzed pathway to surpass the rate of the uncatalyzed hydrolysis pathway (Mill and Mabey, 1988).

Many metals, which are potentially catalytic, form low-solubility inorganic solids within the pH domain of soils and associated environments. Therefore, surface-bound metals must be accessible to reaction with solute species in order for a metal-catalyzed effect to be observed. Since the mid 1950s,

Table 9.7 Mechanisms of metal catalysis [From Stone and Torrents, 1995. P.M. Huang et al. Environmental impact of soil component interactions. Vol. 1. Copyright CRC Press, Boca Raton, FL with permission]

Type 1	The metal coordinates the electrophile, shifting the electronic distribution in the molecule in a way that enhances its reactivity.
Type 2	The metal coordinates to the leaving group, increasing its leaving ability.
Type 3	The metal acts as a center for simultaneous attachment of both the electrophile and attacking nucleophile (template effect).
Type 4	The metal coordinates the nucleophile and induces deprotonation (which increases the reactivity of the nucleophile).
Type 5	Coordination of the substrate with the metal induces confirmation changes that facilitate reaction.
Type 6	Coordination of the substrate with the metal makes the molecule more positive, lessening unfavorable electrostatic interaction with the nucleophile.
Type 7	Coordination with the metal blocks inhibitory reverse reaction paths, such as (1) loss of OH^- from a tetrahedral intermediate instead of loss of the leaving group X^-, or (2) nucleophilic attack by X^-.

research has established that metal oxide/hydroxide precipitates act as hydrolysis catalysts for phosphate esters (Butcher and Westheimer, 1955; Wilkins, 1991). Much has been subsequently learned about the chemical properties and reactivities of naturally occurring solids such as oxides, carbonates, sulfides and aluminosilicates (Stone and Torrents, 1995; Smolen and Stone, 1998).

Complex formation equilibria among dissolved and surface-bound metal species are similar in many aspects. Surface complex formation constants for the metal ions and organic ligands increase in magnitude as the formation constants of analogous complexes in solution are increased (Kummert and Stumm, 1980; Stumm et al., 1980; Davis and Hayes, 1986). Attempts have recently been made to compare reactivities of dissolved and surface bound metal species (Wehrli, 1990; Wehrli et al., 1990). To compare the catalytic reactivity of dissolved and surface-bound metals, any unusual or unique characteristic of the mineral/water interface must be investigated. The high Lewis and Brønsted acidity of unoccupied coordinative sites is responsible for the well-known hydrolytic reactivity of dehydrated and partially dehydrated mineral surfaces (Theng, 1982; Voudrias and Reinhard, 1986). Although the catalysis by surface-bound metals is influenced by hydration, dramatic effects of solid surfaces on hydrolysis rates in aqueous suspensions have been demonstrated (Sanchez-Camazano and Sanchez-Martin, 1983).

The catalysis by surface-bound metals is observed when all participating reactants are adsorbed to a significant extent, and when rate constants for the reactions at the mineral-water surface are comparable to or exceed rate constants for the reaction in homogeneous solution. Although adsorption phenomena have accounted for catalysis, relatively little is known about the conformation and stoichiometry of adsorbed species. This hampers our understanding of surface catalysis on a fundamental level. However, based on the existing literature, two generalities can be made: (1) auxiliary donor groups that facilitate metal catalysis by metal ions in solution should also facilitate surface catalysis, and (2) phenomena of only secondary importance in reactions of dissolved complexes, for example, electrostatic and hydrophobic interactions may play a much greater role in surface catalysis (Stone and Torrents, 1995). The nature and potential significance of surface catalysis are summarized below.

The hydrolysis of phenyl picolinate (PHP) is catalyzed both by surface bound and homogeneous solution metal ions (type 1 catalysis) (Fife and Przystas, 1985; Torrents and Stone, 1991). Appropriate metal ions chelate the heterocyclic N and carbonyl O of PHP and increase the partial positive charge at the carbonyl C, thus, facilitating nucleophilic attack and increasing its susceptibility to hydrolysis. Suspensions containing FeOOH or TiO_2 dramatically accelerate PHP hydrolysis rate, while in particle-free solution and suspensions containing Al_2O_3 or SiO_2, hydrolysis is negligible (Fig. 9.3). Similar results are obtained with methyl picolinate (MEP) (Stone and Torrents, 1995). Catalysis arises from reaction at the oxide surface and not from release of soluble metals, since removal of particles by filtration causes an immediate halt to any catalytic activity.

The susceptibility of hydrolyzable compounds to metal catalysis depends on the nature of the auxiliary donor group. The pyridyl N of PHP and MEP encourages chelate formation with Ti(IV) and Fe(III) in preference to less polarizable Al. Therefore, TiO_2 and FeOOH are effective hydrolysis catalysts and Al_2O_3 is not. In contrast, O donor groups, such as the phenolic group in phenyl salicylate and the O heteroatom of methyl furanoate, are harder and thus, more favorable to Al chelate formation (Stone and Torrents, 1995). Hydrolysis of PHP in the presence of TiO_2 is virtually pH independent and in the presence of FeOOH exhibits only a slight pH dependence. In particle-free solution, PHP exhibits classic base-catalyzed hydrolysis behavior; the reaction rate of the hydrolysis increases by an order of magnitude for every unit increase in pH. Activation of the ester linkage through chelate formation with

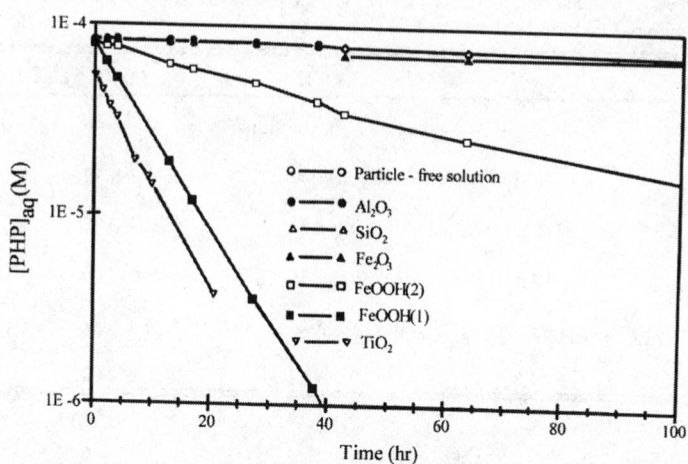

Fig. 9.3 Effect of various metal oxides on the loss of phenyl picolinate (PHP) from solution via hydrolysis. All suspensions contained 10 g L^{-1} oxide, 1.0 mM acetate buffer (pH 5.0), and 50 mM NaCl [Reprinted from Torrents and Stone, 1991. Environ. Sci. Technol. 25:143-149 with permission of the American Chemical Society]

a surface-bonded metal ion is apparently strong enough to react with the weak neucleophile H$_2$O, which overshows the pH-dependent reaction with OH$^-$.

Positively charged surfaces such as Al$_2$O$_3$ (below pH 8.6) and TiO$_2$ (below pH 6.4) accelerate the hydrolysis of monophenyl terephthalate (MPT) by an order of magnitude or more (Stone and Torrents, 1995). Hydrolysis of MPT catalyzed by Al$_2$O$_3$ and TiO$_2$ exhibits elements of both type 3 and type 6 catalysis. Positive oxide surfaces serve as the template to accumulate both anionic reactants, OH$^-$ and MPT ions, increasing their encounter frequency. Type 6 catalysis (electrostatic interactions favoring encounter between like-charge species) is more important for surface chemical reactions than for solution reactions due to the additive electrostatic effect arising from neighboring charged groups. Type 3 catalysis (simultaneous attachment of substrate and nucleophile) is also typical of surface chemical reactions.

Many pesticides are susceptible to surface catalyzed hydrolysis (Torrents, 1992). For example, catalysis of Chlorpyrifos hydrolysis by TiO$_2$, FeOOH, and Al$_2$O$_3$ apparently involves chelate formation between the surface-bound metal ion, the thionate S, and pyridinyl N.

A surface-catalyzed reaction following type 4 catalysis has been postulated for the degradation of the fungicide oxycarboxin stored in borosilicate glass (Stanton, 1987). Because of low hydroxide ion concentration in neutral and acidic pH, hydrolytic ring cleavage is not observed in the absence of a catalyst. The glass exerts its catalytic effect by providing sufficiently strong nucleophiles (surface bound OH$^-$ and O$_2^{2-}$) for attacking the oxathiin ring.

Quartz and aluminosilicate surfaces present in soils and associated environments could also catalyze oxycarboxin hydrolysis. It has recently been reported that metal oxides catalyze the hydrolysis of *p*-nitrophenyl acetate by this mechanism (Hoffmann, 1990). However, in surface-catalyzed hydrolysis reactions, the nature and relative significance of type 4 catalysis is poorly understood and, thus, merits closer attention.

Table 9.8 Oxidation of As(III) and sorption of As by Mn(IV) oxide [From Oscarson et al., 1981. Nature 291:50-51 with permission of Macmillan, London]

As(III) or As(V) added	As(III)	As(V)	Mn	Final pH
μg mL^{-1}	--------------------------μg mL^{-1} in solution------------------------			
100 As(III)	ND*	83.5 ± 1.4[‡]	0.41 ± 0.12	7.1
300 As(III)	63.2 ± 7.0	186 ± 5	8.08 ± 0.36	7.1
500 As(III)	213 ± 4	205 ± 1	6.06 ± 0.60	7.3
1000 As(III)	665 ± 5	216 ± 4	4.16 ± 0.86	7.5
300 As(V)	ND	298 ± 1	0.06 ± 0.02	7.5

*ND = not detectable
[‡]Mean ± SD; n = 3

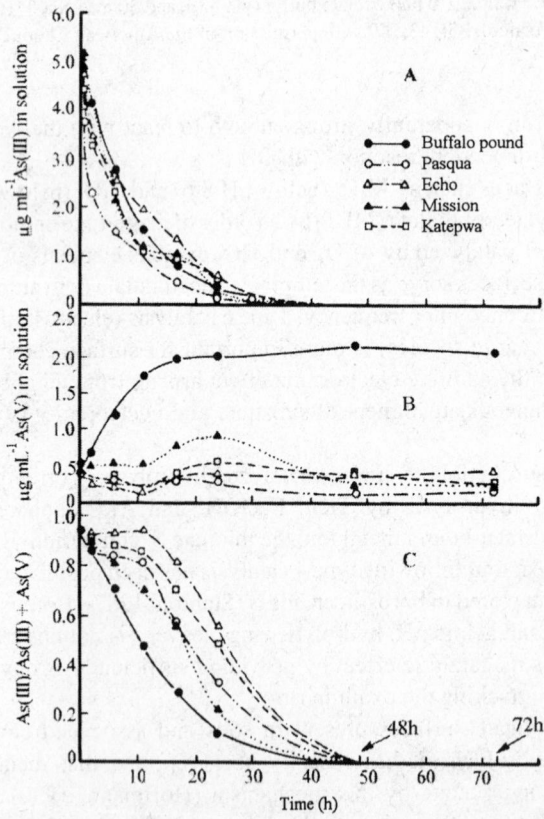

Fig. 9.4 Oxidation of As(III) to As(V) and the sorption of As by lake sediments as a function of time. (A) concentration of As(III) in solution, (B) concentration of As(V) in solution, and (C) As(III)/As(III) + As(V). Ten μg/mL of As(II) were added initially. During the reaction period, the pH of the As-sediment suspension ranged from 8.0 to 8.2 for the Buffalo Pound sediment and from 7.3 to 7.6 for the other four sediments [Reprinted from Oscarson et al., 1980. J. Environ. Qual. 9:700-703 with permission of the American Society of Agronomy].

Table 9.9 Specific surface and point of zero charge (p.z.c.) of the Mn dioxides and rate constants and energies of activation for the depletion of As(III) by the Mn dioxides [From Oscarson et al., 1983. Soil Sci. Soc. Am. J. 47:644-648 with permission of the Soil Science Society of America]

Mineral	Specific surface	p.z.c.	Rate constant x 10^{-3}			Energies of activation
			278 K	298 K	318 K	
	$m^2\,g^{-1}$			h^{-1}		$kJ\,mol^{-1}$
Birnessite	277 ± 5*	2.3 ± 0.1	126 ± 13	267 ± 6	533 ± 38	26.0 ± 0.2
Cryptomelane	346 ± 4	2.8 ± 0.1	54 ± 10	189 ± 8	318 ± 22	32.3 ± 6.7
Pyrolusite	8 ± 1	6.4 ± 0.3	0.12 ± 0.02	0.44 ± 0.03	0.58 ± 0.05	29.0 ± 9.8

*Mean ± SE

9.4 Abiotic Catalysis in the Transformation of Metals, Metalloids, and Other Inorganics

9.4.1 Transformations of Metals and Metalloids

9.4.1.1 Arsenic

Arsenic bioavailability and toxicology depend on its chemical state (Huang and Fujii, 1996) with As(III) being much more toxic than As(V). Birnessite is a very effective oxidant of As(III) as shown in Table 9.8. In the control experiment, no detectable As(III) is oxidized in the absence of Mn(IV) oxide.

Although As(V) is a thermodynamically stable species in oxygenated water at common pH values (Penrose, 1974), the kinetics of oxidation of As(III) with O_2 is very slow at near neutral pH values (Kolthoff, 1921). Lake sediments from Saskatchewan, Canada, can oxidize As(III) (700 µg As) to As(V) within 48 h (Fig. 9.4). The oxidation of As(III) is not detectable within 72 h in the absence of sediment. The oxidation of As(III) to As(V) is not affected by flushing N_2 or air through the sediment suspensions, nor does the addition of $HgCl_2$ to the system eliminate the conversion of As(III) to As(V). This indicates that the oxidation of As(III) to As(V) is an abiotic process.

When the lake sediments are treated with hydroxylamine hydrochloride or sodium acetate, which are effective extractants for Mn, the oxidation of As(III) to As(V) by the treated sediments is greatly decreased relative to the untreated sediments (Oscarson et al., 1981b). Hydroxylamine hydrochloride treatment also removes Fe oxide; however, the evidence obtained from colorimetry and X-ray photoelectron spectroscopy shows that a redox reaction between Fe oxide and As(III) does not occur within 72 h, indicating that the redox reaction between As(III) and Fe(III) is kinetically slow (Oscarson et al., 1981a). This supports the evidence that Mn oxide is the primary component responsible for catalyzing the conversion of As(III) to As(V). The rate constant increases with increasing temperature from 278 to 298 K; the heat of activation for the process varies from 13.8 to 35.6 kJ mol^{-1}, indicating that the depletion of As(III) is predominantly a diffusion controlled process (Oscarson et al., 1981c).

The ability of Mn oxides to deplete As(III) from solution (oxidation plus sorption) varies with their crystallinity and specific surface (Oscarson et al., 1983a). The depletion of As(III) by Mn oxides follows first-order kinetics. Pyrolusite is highly ordered and has a low specific surface; conversely, birnessite and cryptomelane are poorly crystalline and have relatively high specific surfaces (Table

9.9). Because birnessite and cryptomelane have relatively high specific surface areas due to their porous nature (Huang, 1991a), their rate constants for the depletion of As(III) are much higher than those for pyrolusite (Table 9.9). On the other hand, the rate constants for the depletion of As(III) by birnessite are significantly greater than those for cryptomelane despite its higher specific surface. Birnessite has a greater negative charge density than cryptomelane at pH 7 (McKenzie, 1981). Because As(V) is also negatively charged at pH 7, the repulsive interaction energy would be greater between birnessite than cryptomelane and As(V) which is why birnessite does not sorb a detectable amount of As(V) relative to cryptomelane (Oscarson et al., 1983a). Differences in point of zero charge (p.z.c.) of birnessite and cryptomelane and their ability to sorb As(V) explain the greater As(III) depletion by birnessite than by cryptomelane even though cryptomelane has the greater surface area. Little As(V) is sorbed from solution by birnessite upon oxidation of As(III) and less total As is sorbed by birnessite than by cryptomelane; the electron accepting sites on the surface of birnessite are, thus, blocked to a lesser extent than those on the surface of cryptomelane. Consequently, the rate of depletion of As(III) from solution is greater for birnessite than for cryptomelane.

The surfaces of many soil Mn oxides are coated with various chemical species (McKenzie, 1989). The rate constants for the depletion (oxidation plus sorption) of As(III) are generally substantially smaller for the Mn oxides with higher levels of coatings of Fe and Al oxides and $CaCO_3$ than those for the untreated MnO_2 or MnO_2 with lower levels of coatings (Oscarson et al., 1983b). The electron

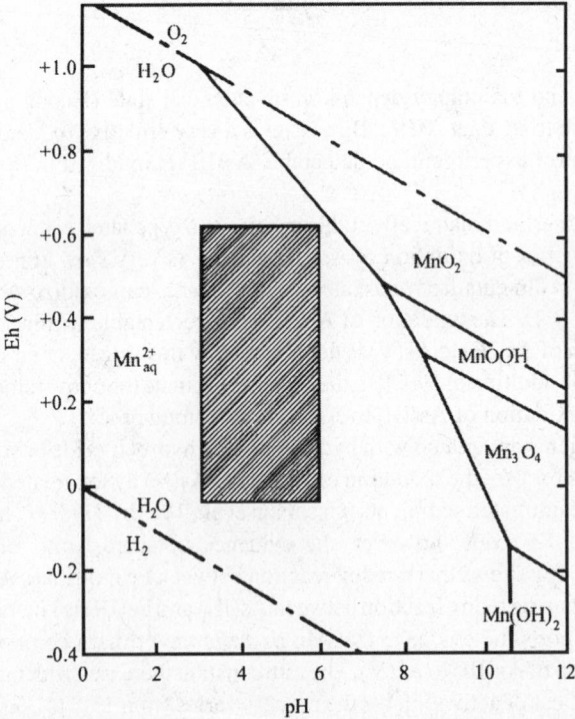

Fig. 9.5 Stability relations of different species of Mn at 25 °C, 0.101 MPa, and $a_{Mn2+}=10^{-6}$ (Bricker, 1965). The Eh and pH ranges of the systems during the formation of iron oxide precipitates in the presence of Mn oxides in the present study are shown as shaded region [Reprinted from Krishnamurti and Huang, 1988. Clays Clay Miner. 36:467-475 with permission of the Clay Minerals Society, Boulder, CO].

accepting sites on MnO_2 are partially masked by the oxides and $CaCO_3$. The corroborating evidence shows that Fe and Al oxides and $CaCO_3$ do not oxidize As(III) to As(V) (Oscarson et al., 1981a, 1983b). The fraction of the depletion of As(III) by the oxidation of As(III) to As(V) on the coated MnO_2 is due solely to MnO_2.

The mechanism of conversion of As(III) to As(V) catalyzed by Mn oxides has been treated in depth (Oscarson et al., 1981a, 1983b; Huang, 1991a). The conversion of As(III) to As(V) by uncoated and coated Mn oxides has important implications for the transport, fate, and toxicity of As in terrestrial and associated environments. In some environments that have been contaminated with As(III), the addition of reactive Mn oxides may alleviate the toxicity of As(III) through their catalytic reactions.

9.4.1.2 Iron

Manganese oxides, which have different structural and surface properties, vary in their ability to catalyze the conversion of Fe(II) to Fe(III) and the subsequent formation of Fe oxides and oxyhydroxides (Krishnamurti and Huang, 1987, 1988). The standard electrode potential (E°) of the redox pairs Fe^{2+}-MnO_2 and Fe^{2+}-Mn_3O_4 can be described by the following equations (Bricker, 1965):

$$2Fe^{2+} + MnO_2 + 4H^+ = Mn^{2+} + 2Fe^{3+} + 2H_2O \qquad E° = +0.438V \qquad [9.8]$$

$$2Fe^{2+} + Mn_3O_4 + 8H^+ = 3Mn^{2+} + 2Fe^{3+} + 4H_2O \qquad E° = +1.04V \qquad [9.9]$$

These E° values indicate that oxidation of Fe^{2+} by Mn oxides is thermodynamically feasible. The Eh–pH diagram (Fig. 9.5) indicates the feasibility of the conversion of Mn^{4+} and Mn^{3+} to Mn^{2+} in the Mn oxides in the Eh–pH ranges of the formation of Fe oxides and oxyhydroxides. Further, the ESR spectra of the filtrates after the reaction of Fe^{2+} with Mn oxides show the presence of a significant amount of Mn^{2+}, demonstrating the reduction of Mn(IV) and Mn(III) oxides to Mn^{2+} in the presence of Fe^{2+} in solution. Simultaneously, the oxidation of Fe^{2+} to Fe^{3+} by Mn oxides leads to subsequent hydrolysis of Fe^{3+} to form a series of the precipitation products including lepidocrocite, åkaganeite, feroxyhyte and magnetite. Because various Mn oxides differ in their ability to catalyze the formation of Fe oxides and oxyhydroxides, their role in the transformation of Fe deserves attention.

9.4.1.3 Manganese

Although oxidation of Mn in soils by atmospheric O_2 is a thermodynamic possibility through the pe and pH range common in well-aerated soils, Mn is not readily oxidized in solutions unless the pH is raised above pH 8.5 (Ross and Bartlett, 1981). A catalytic mechanism of some sort is apparently necessary. Oxidation of Mn in soil environments is generally assumed to be microbially mediated.

Oxidation of Mn(II) by soils has been shown to be proportional to the level of existing reactive Mn oxides (Ross and Bartlett, 1981). Arrhenius plots of rates of the oxidation between 1 and 35 °C show that the oxidation of Mn(II) by soils has nonbiological characteristics. Generally, oxidation of Mn(II) is initially rapid, with no lag period and may be related to the mechanism proposed for the accumulation of Co and Zn on Mn oxide surfaces (Loganathan et al., 1977). The adsorption on the oxide surfaces is specific and not a simple function of surface charge. Above pH 5 to 6, this adsorption reverses the surface charge of the Mn oxides from negative to positive, as measured by electrophoretic mobility. A positively charged surface should result in a higher OH⁻ concentration near the surface which could enable oxidation of adsorbed Mn(II) by atmospheric O_2. The mechanism of specific adsorption is not clear, but the preferential adsorption of transition metals by Mn oxides is well documented (Jenne, 1977). The oxidation of Mn(II) by Mn oxides is theorized to be autocatalytic, involving specific adsorption of Mn(II) on existing Mn oxide surfaces (Ross and Bartlett, 1981).

9.4.1.4   Trace Metals

In addition to oxidation of metalloids such as As(III) and metals such as Fe(II) and Mn(II), Mn oxides and oxyhydroxides can catalyze the oxidation of trace metals by disproportionation of Mn^{2+} and MnO_2. The disproportionation facilitates electron transfer processes that can either greatly decrease or increase the equilibrium solubility of certain metals (Hem, 1978).

Chromium and Pu are similar in chemical behavior in aqueous environments (Rai and Serne, 1977; Bartlett and James, 1979). Both elements can exist in multiple oxidation states and as cationic and anionic species in aqueous systems. Chromium occurs in the II, III and VI oxidation states in water. Chromium (II) is unstable. Chromium (III) has broad stability, exists as the cation Cr^{3+} and its hydrolysis products, or as the anion CrO_2^-. Chromium (VI) exists under strongly oxidizing conditions, occurs as dichromate $Cr_2O_7^{2-}$, or chromate $HCrO_4^-$ and CrO_4^{2-} anions. Plutonium exists in the III to VI oxidation states as Pu^{3+}, Pu^{4+}, PuO_2^+ and PuO_2^{2+} in strong acid conditions. Chromium (III) and Pu (III/IV) cations are sorbed to soil constituents, and therefore, immobile in most aqueous and soil environments. In contrast, Cr(VI) and Pu(VI) are quite mobile in soils and associated aqueous environments, because they are not sorbed by temperate soils to any extent. These elements in the hexavalent form are readily bioavailable and extremely toxic (Amacher and Baker, 1982) and, thus, of concern in food-chain contamination and ecosystem health. Manganese (III/IV) oxides can oxidize Cr(III) and Pu(III/IV) (Cleveland, 1970; Amacher and Baker, 1982). Therefore, Mn oxides can enhance the mobility and toxicity of Cr and Pu in soils and related ecosystems.

The kinetics of Cr(III) oxidation by Mn oxide is very rapid with most of the conversion of Cr(III) to Cr(VI) occurring during the first hour. The oxidation of Cr(III) increases with increasing temperature. The kinetics of Cr(III) oxidation on Mn(III/IV) oxides shows a trend similar to that observed for soils (Amacher and Baker, 1982).

Manganese oxides and oxyhydroxides can also catalyze the oxidation of other trace metals such as Co, Pb, Ni and Cu (Hem, 1978). The catalytic transformation of these metal ions may be directly influenced by redox processes coupled to disproportionation of Mn mixed-valence oxide, to catalyzed oxidation by aqueous O_2, or to other redox reactions involving changes from one Mn oxide species to another. When the oxidized form of the element has a lower solubility than the reduced form, this effect can be of major significance.

X-ray photoelectron spectroscopy measurements of Co adsorbed on MnO_2 reveal strong evidence that Co(II) is oxidized to Co(III) in the presence of the strong electric field at the MnO_2- solution interface (Murray and Dillard, 1979). Nickel (II), however, cannot be oxidized at the interface except at very high concentrations. Strong experimental evidence for the oxidation of other trace metals catalyzed by Mn oxides still remains to be attained. Further, more information on the kinetics and mechanisms of redox reactions of these trace metals on the surfaces of Mn oxides is needed.

Other than Mn oxides, heterogeneous oxidation/reduction reactions involving electron transfer between transition metals and Fe-containing minerals have been investigated in numerous studies. Vanadium (II) and V(IV) can be oxidized by Fe(III) oxyhydroxides (Wehrli and Stumm, 1989). Reduction of Cr(VI) by biotite has been shown by Eary and Rai (1989) and Ilton and Veblen (1994). Direct evidence of Cr(VI) reduction on magnetite surfaces has been recently documented by Peterson et al. (1996) using X-ray adsorption fine structure spectroscopy. Magnetite and ilmenite are the most common Fe(II)-containing oxide minerals in the Earth's crust and potentially important in controlling heterogeneous redox reactions involving aqueous transition metals in soil environments. Recently, experimental evidence demonstrates that structural Fe(II) in magnetite and ilmenite heterogeneously reduces aqueous Cu(II), V(IV), and Cr(VI) ions at the oxide surfaces over a pH range of 1 to 7 at 25 °C (White and Peterson, 1996). Compared with the catalytic oxidation of transition metals, far fewer

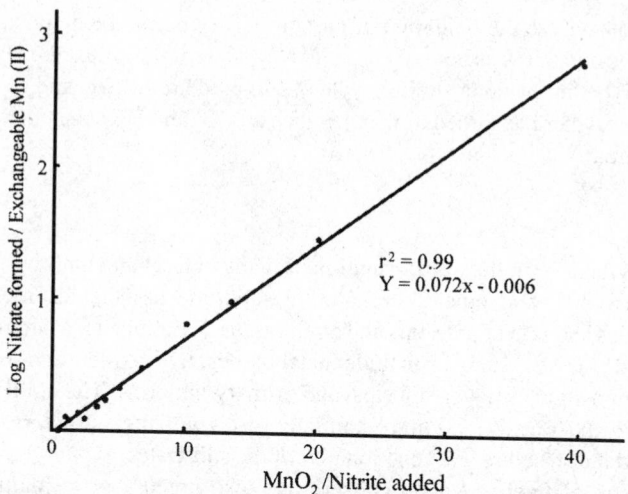

Fig. 9.6 The logs of the ratios of nitrate formed from nitrite over Mn reduced to the exchangeable form plotted against the MnO_2/nitrite ratios in the original suspensions [From Bartlett, 1981. Soil Sci. Soc. Am. J. 45:1054-1058 with permission of the Soil Science Society of America].

studies have addressed the reduction of aqueous transition metals on the surfaces of natural Fe(II)-containing minerals (White, 1990).

9.4.2 Transformations of Other Inorganics

Nitrate formation in soils from NO_2^- is related to soil level of reactive Mn oxides which catalyze the oxidation (Bartlett, 1981). Nitrite oxidation, NO_3^- formation, and MnO_2 reduction are stoichiometrically related reactions both in the presence and absence of atmospheric O_2. The relationship shown in Fig. 9.6 suggests that, at high ratios of MnO_2/NO_2^-, most of the Mn(IV) is reduced only to non-exchangeable solid Mn(III) oxide, rather than to exchangeable Mn(II) ions. When the MnO_2/NO_2^- ratio is high, little Mn(II) becomes exchangeable relative to the NO_3^- formed, as indicated by high values of $NO_3^-/Mn(II)$. At low MnO_2/NO_2^- ratios, more exchangeable Mn(II) relative to NO_3^- is formed. Therefore, the more MnO_2 relative to NO_2^-, the greater is the efficiency of the NO_2^- to NO_3^- transformation and the less exchangeable Mn(II) formed relative to NO_3^-. High MnO_2/NO_2^- ratios likely prevail in many soils. Equation [9.10], which does not involve a pH change, may describe the NO_2^- to NO_3^- transformation in high Mn oxide soils better than Equation [9.11]:

$$NO_2^- + 2MnO_2 = Mn_2O_3 + NO_3^- \tag{9.10}$$

$$NO_2^- + MnO_2 + 2H^+ = Mn^{2+} + NO_3^- + H_2O \tag{9.11}$$

Such a process of substantial reduction of Mn in the solid phase relative to the release of exchangeable Mn(II) might be described better as the insertion of electrons into the overlapping electronic orbitals of the solid, such that the electron excess is delocalized over the solid, rather than the reduction of one Mn^{4+} atom to Mn^{2+} at a particular surface site. The overlapping atomic orbitals form bands of many energy levels rather than a single energy level.

Manganese oxides also catalyze the oxidation of gaseous CO (Klier and Kuchynka, 1966). This reaction occurs by consumption of the structural O_2^{2-} of MnO_2. The catalyst must then be regenerated by chemisorption of O_2. Whether or not a similar cycle of Mn oxide reduction and O_2 consumption occurs in soil environments, where adsorbed molecules such as CO_2 and H_2O can poison the catalyst surface, remains unanswered.

9.5   Conclusions

Abiotic catalysis plays a vital role in the transformation of many natural and anthropogenic organic compounds, metals, metalloids, and other inorganics in soils and associated environments. Its significance in the dynamics and fate of nutrients and environmental pollutants is far greater than had previously been perceived. Abiotic catalysts include metal oxides, hydroxides, and oxyhydroxides, short-range ordered aluminosilicates, layer silicates, and primary minerals. The ability of inorganic soil constituents to catalyze the transformation substantially varies with their structural configuration and surface chemistry and the structure and functionality of the substrates.

Phenolic compounds are of environmental concern and also precursors of humic substances. Abiotic catalysis promotes the formation of humic macromolecules from phenolic compounds especially in the presence of amino acids through oxidative polymerization, polycondensation, ring cleavage, decarboxylation, dealkylation, and/or deamination. A series of mineral colloids promote such catalytic reactions. Among abiotic catalysts, Mn oxides are most reactive in the transformation of phenolic compounds. Many other organics can also be oxidatively decomposed by Mn oxides.

Hydrolysis of many organic pollutants can be catalyzed by mineral colloids. The catalytic properties of mineral colloids can be ascribed to the ability of the metal-containing surfaces to behave as Brønsted acids and donate protons or to act as Lewis acids and to accept electron pairs. The catalytic effectiveness of a metal ion depends on its ability to complex reactant molecules and shift electron density and conformation in ways favorable to the reactions. Such catalytic reactions may lead to detoxification of pesticides.

Among metalloids, the toxic As(III) is oxidized to the less toxic As(V) by Mn oxides. The ability of Mn oxides to catalyze the conversion of As(III) to As(V) varies with their structure and surface properties and also with the kinds and levels of surface coatings. Trace metals such as Cr(III), Pu(III), and Co(II) can be oxidized by disproportionation of Mn oxides. Besides Mn oxides, heterogeneous oxidation/reduction reactions involving electron transfer between transition metals such as V, Cr, and Cu and Fe-containing minerals have been demonstrated. Compared with the catalytic oxidation of trace metals, far fewer studies have addressed the reduction of trace metals on the surfaces of mineral colloids. Catalytic transformations of trace metals can greatly influence their solubility, mobility, toxicity, and food-chain contamination.

Transformations of Fe and Mn are also influenced by catalysis of Mn oxides. Manganese oxides, which have different structural and surface properties, vary substantially in their ability to catalyze the precipitation and crystallization of Fe oxides and oxyhydroxides. The oxidation of Mn(II) by soils has been shown to be proportional to the level of existing reactive Mn oxides. The oxidation of Mn(II) by Mn oxides is apparently autocatalytic in nature.

Furthermore, Mn oxides have been shown to catalyze the conversion of NO_2^- to NO_3^-. The reaction is theorized as mainly the insertion of electrons from NO_2^- into the overlapping electron orbitals of the Mn oxides. Manganese oxides also catalyze the oxidation of gaseous CO through consumption of structural O_2^{2-}.

Abiotic catalysis, thus, merits close attention in sustaining ecosystem health. Further, abiotic and biotic catalysts co-exist in soils and associated environments. Abiotic catalysts can influence microbial formation of enzymes (biotic catalysts) and enzymatic activity. The interactions of abiotic and biotic catalysts and the impact on terrestrial ecosystem health should, thus, be an issue of intense interest for years to come.

Acknowledgment

This study is supported by Research Grant GP 2383-Huang of the Natural Sciences and Engineering Research Council of Canada.

9.6   References

Amacher, M.L., and D.E. Baker, 1982. Redox reactions involving chromium, plutonium, and manganese in soils. DOE/DP/04515-1. Pennsylvania State University, University Park, PA.

Armstrong, D.E., G. Chesters, and R.F. Harris. 1967. Atrazine hydrolysis in soils. Soil Sci. Soc. Am. Proc. 31:61–66.

Barnhisel, R.I., and P.M. Bertsch. 1989. Chlorites and hydroxyl-interlayered vermiculite and smectite. p. 729–788. *In* J.B. Dixon and S.B. Weed (ed.) Minerals in soil environments. Soil Science Society of America, Madison, WI.

Bartlett, R.J. 1981. Nonmicrobial nitrite-to-nitrate transformation in soils. Soil Sci. Soc. Am. J. 45:1054–1058.

Bartlett, R.J. 1986. Soil redox behavior. p. 179–207. *In* D.L. Sparks (ed.) Soil physical chemistry. CRC Press, Boca Raton, FL.

Bartlett, R.J., and B. James. 1979. Behavior of chromium in soils. III. Oxidation. J. Environ. Qual. 8:31–35.

Bases, C.F. Jr., and R.E. Mesmer. 1986. The hydrolysis of cations. Krieger Publishing Company, Malabar, FL.

Basolo, F., and R.G. Pearson. 1968. Mechanisms of inorganic reactions. 2nd Ed. John Wiley and Sons, New York, NY.

Bell, R.P. 1941. Acid-base catalysis. Oxford University Press, Oxford, UK.

Blanchet, P.-F., and A. St.-George. 1982. Kinetics of chemical degradation of organophosphorus pesticides: Hydrolysis of chlorpyrifos and chlorpyrifos-methyl in the presence of copper (II). Pestic. Sci. 13:85–91.

Bollag, J.-M., S.Y. Liu, and R.D. Minard. 1982. Enzymatic oligomerization of vanillic acid. Soil Biol. Biochem. 14:157–163.

Bricker, O. 1965. Some stability relations in the system $Mn-O_2-H_2O$ at 25 C and one atmosphere total pressure. Am. Mineral. 50:1296–1354.

Brønsted, J.N. 1928. Acid and base catalysis. Chem. Rev. 5:231–338.

Brown, C.B., and J.L. White. 1969. Reactions of 12 *s*-triazines with soil clays. Soil Sci. Soc. Am. Proc. 33:863–867.

Butcher, W.W., and F.H. Westheimer. 1955. The lanthanum hydroxide gel promoted hydrolysis of phosphate esters. J. Am. Chem. Soc. 77:2420–2424.

Chang, H.M., and G.G. Allan. 1971. Oxidation. p. 433–485. *In* K.V. Sarkanen and C.H. Ludwig (ed.) Lignins. Wiley Interscience, New York, NY.

Cheney, M.A., G. Sposito, A.E. McGrath, and R.S. Criddle. 1996. Abiotic degradation of 2,4-D (dichlorophenoxyacetic acid) on synthetic birnessite: a calorespirometric method. Coll. Surf. A. Phy. Eng. Asp. 107:131–140.

Cheng, H.H. 1991. Pesticides in the soil environment: Processes, impacts, and modeling. Soil Science Society of America, Madison, WI.

Cleveland, J.M. 1970. The chemistry of plutonium. Gordon and Breach, New York, NY.

Daintith, J. 1990. A concise dictionary of chemistry. New edition. Oxford University Press, Oxford, UK.

Davis, J.A., and K.F. Hayes. 1986. Geochemical processes at mineral surfaces. ACS Symp. Ser. 323. American Chemical Society, Washington, DC.

Dixon, J.B. and S.B. Weed (ed.). 1989. Minerals in soil environments. Soil Science Society of America, Madison, WI.

Eary, L.E., and D. Rai. 1989. Kinetics of chromate reduction by ferrous ions derived from hematite and biotite at 25 °C. Amer. J. Sci. 289:180–213.

Faust, G.T. 1940. Staining of clay minerals as a rapid means of identification in natural and beneficiation products. Report of Investigation 3522. Bureau Mines, Washington, DC.

Fife, T.H., and T.J. Przystas. 1985. Divalent metal ion catalysis in the hydrolysis of ester of picolinic acid. Metal ion promoted hydroxide ion and water catalyzed reactions. J. Am. Chem. Soc. 107:1041–1047.

Filip, Z., W. Flaig, and E. Rietz. 1977. Oxidation of some phenolic substances as influenced by clay minerals. p. 91–96. *In* Isot. Radiat. Soil Org. Matter Stud., II. IAEA Bull., Vienna, Austria.

Flaig, W., H. Beutelspacher, and E. Rietz. 1975. Chemical composition and physical properties of humic substances. p. 1–211. *In* J.E. Gieseking (ed.) Soil components. Vol. 1, Organic components. Springer-Verlag, New York, NY.

Fowker, F.M., H.A. Benesi, R.B. Ryland, W.M. Sawyer, K.D. Detling, E.S. Loeffler, F.B. Folckemer, M.R. Johnson, and Y.P. Sun. 1960. Clay catalyzed decomposition of insecticides. J. Agric. Food Chem. 8:203–210.

Furukawa, T., and G.W. Brindley. 1973. Adsorption and oxidation of benzidine and aniline by montmorillonite and hectorite. Clays Clay Miner. 21:279–288.

Haider, K., J.P. Martin, and Z. Filip. 1975. Humus biochemistry. p. 175–234. *In* E.A. Paul and A.D. McLaren (ed.) Soil biochemistry, Vol. 4. Marcel Dekker, New York, NY.

Hannien, K.I., R. Klocking, and B. Helbig. 1987. Synthesis and characterization of humic acid-like polymers. Sci. Total Environ. 62:201–210.

Hatcher, P.G., I.A. Breger, and M.A. Mattingly. 1980. Structural characteristics of fulvic acids from continental shelf sediments. Nature 285:560–562.

Hatcher, P.G., M. Schnitzer, L.W. Dennis, and G.E. Maciel. 1981. Aromaticity of humic substances in soils. Soil Sci. Soc. Am. J. 45:1089–1094.

Hauser, E.A., and M.B. Leggett. 1940. Color reactions between clays and amines. J. Am. Chem. Soc. 62:1811–1814.

Hayes, M.H.B., and R.S. Swift. 1978. The chemistry of soil organic colloids. *In* D.J. Greenland and M.H.B. Hayes (ed.) The chemistry of soil constituents. John Wiley and Sons, New York, NY.

Hem, J.D. 1978. Redox processes at surfaces of manganese oxide and their effects on aqueous metal ions. Chem. Geol. 21:199–218.

Hightower, J.W., W.N. Delgass, E. Iglesia, and A.T. Bell. 1996. Studies in surface science and catalysis. Elsevier, Amsterdam, Netherlands.

Hoffmann, M.R. 1980. Trace metal catalysis in aquatic environments. Environ. Sci. Technol. 14:1061–1066.

Hoffmann, M.R. 1990. Catalysis in aquatic environments. p. 71–111. *In* W. Stumm (ed.) Aquatic chemical kinetics. John Wiley and Sons, New York, NY.

Houghton, R.P. 1979. Metal complexes in organic chemistry. Cambridge University Press, Cambridge, UK.

Huang, P.M. 1990. Role of soil minerals in influencing transformations of natural organics and xenobiotics in the environment. p. 29–115. *In* J.M. Bollag and G. Stotzky (ed.) Soil biochemistry, Vol. 6. Marcel Dekker, New York, NY.

Huang, P.M. 1991a. Kinetics of redox reactions on surfaces of Mn oxides and its impact on environmental quality. p. 191–230. *In* D.L. Sparks and D.L. Suares (ed.). Rates of chemical processes in soils. Soil Sci. Soc. Am. Spec. Pub. 27.

Huang, P.M. 1991b. Ionic factors affecting the formation of short-range ordered aluminosilicates. Soil Sci. Soc. Am. J. 55:1172–1180.

Huang, P.M. 1995. The role of short-range ordered mineral colloids in abiotic transformations of organic compounds in the environment. p. 135–167. *In* P.M. Huang, J. Berthelin, J.-M. Bollag, W.B. McGill, and A.L. Page (ed.), Environmental impact of soil component interactions: Vol. 1, Natural and anthropogenic organics. CRC Press/ Lewis Publishers, Boca Raton, FL.

Huang, P.M., and R. Fujii. 1996. Selenium and arsenic. p. 793–831. *In* D.L. Sparks (ed.) Methods of soil analysis: Part 3. Chemical methods. Soil Science Society of America, Madison, WI.

Iler, R.K. 1979. The chemistry of silica. John Wiley and Sons, New York, NY.

Ilton, E.S., and D.R. Veblen. 1994. Chromium sorption by phlogopite and biotite in acidic solutions at 25 °C: Insights from X-ray photoelectron spectroscopy and electron microscopy. Geochim. Cosmochim. Acta 58:2777–2788.

Jauregui, M.A., and H.M. Reisenauer. 1982. Dissolution of oxides of manganese and iron by root exudate components. Soil Sci. Soc. Am. J. 46:314–317.

Jenne, E.A. 1977. Trace element sorption by sediments and soils-sites and processes. p. 425–553. *In* W.R. Chappell and K.K. Peterson (ed.) Molybdenum in the environment, Vol. 2. Marcel Dekker, New York, NY.

Ketelaar, J.A.A., H.R. Gersmann, and M.M. Beck. 1956. Metal-catalysed hydrolysis of thiophoshporic esters. Nature 177:392–396.

Klier, K., and K. Kuchynka. 1966. Carbon monoxide oxidation and adsorbate-gas exchange reactions on MnO_2-based catalysts. J. Catal. 6:62–71.

Kolthoff, I.M. 1921. Iodometric studies. VII. Reactions between arsenic trioxide and iodine. Anal. Chem. 60:393–406.

Kresge, A.J. 1975. Water makes proton transfer fast. Acc. Chem. Res. 8:354–360.

Krishnamurti, G.S.R., and P.M. Huang. 1987. The catalytic role of birnessite in the transformation of iron. Can. J. Soil Sci. 67:533–543.

Krishnamurti, G.S.R., and P.M. Huang. 1988. Influence of manganese oxide minerals on the formation of iron oxides. Clays Clay Miner. 36:467–475.

Kroll, H. 1952. The participation of heavy metal ions in the hydrolysis of amino acid esters. J. Am. Chem. Soc. 74:2036–2039.

Kumada, K., and H. Kato. 1970. Browning of pyrogallol as affected by clay minerals. Soil Sci. Plant Nutr. 16:195–200.

Kummert, R., and W. Stumm. 1980. The surface complexation of organic acids on hydrous γ-Al_2O_3. J. Coll. Interf. Sci. 75:373.

Kung, K.-H., and M.B. McBride. 1988. Electron transfer processes between hydroquinione and hausmannite (Mn_3O_4). Clays Clay Miner. 36:297–302.

Kyuma, K., and K. Kawaguchi. 1964. Oxidative changes of polyphenols as influenced by allophane. Soil Sci. Soc. Am. Proc. 28:371–374.

Lee, J.S.K., and P.M. Huang. 1995. Photochemical effect on the abiotic transformation of polyphenols as catalyzed by Mn(IV) oxide. p. 177–189. *In* P.M. Huang, J. Berthelin, J.-M. Bollag, W.B. McGill, and A.L. Page (ed.) Environmental impact of soil component interactions: Vol. 1, Natural and anthropogenic organics. CRC Press/ Lewis Publishers, Boca Raton, FL.

Lehmann, R.G., and H.H. Cheng. 1988. Reactivity of phenolic acids in soil and formation of oxidation products. Soil Sci. Soc. Am. J. 52:1304–1309.

Lehmann, R.G., H.H. Cheng, and J.B. Harsh. 1987. Oxidation of phenolic acids by soil iron and manganese oxides. Soil Sci. Soc. Am. J. 51:352–356.

Liu, S.-Y., R.D. Minard, and J.-M. Bollag. 1981. Oligomerization of syringic acid, a lignin derivative, by a phenoloxidase. Soil Sci. Soc. Am. J. 45:1100–1105.

Loganathan, P., R.G. Burau, and D.W. Fuerstenau. 1977. Influence of pH on the sorption of Co^{2+}, Zn^{2+}, and Ca^{2+} by a hydrous manganese oxide. Soil Sci. Soc. Am. J. 41:57–62.

McBride, M.B. 1979. Reactivity of adsorbed and structural iron in hectorite as indicated by oxidation of benzidine. Clays Clay Miner. 27:224–230.

McBride, M.B. 1989a. Oxidation of dihydroxybenzenes in aerated aqueous suspensions of birnessite. Clays Clay Miner. 37:341–347.

McBride, M.B. 1989b. Oxidation of 1,2- and 1,4-dihydroxybenzene by birnessite in acidic aqueous suspension. Clays Clay Miner. 37:479–486.

McBride, M.B. 1989c. Surface chemistry of soil minerals. p. 35–88. *In* J.B. Dixon and S.B. Weed (ed.) Minerals in soil environments. Soil Science Society of America, Madison, WI.

McBride, M.B. 1994. Environmental chemistry of soils. Oxford University Press, Oxford, UK.

McKenzie, R.M. 1981. The surface charge on manganese dioxides. Aust. J. Soil Res. 19:41–50.

McKenzie, R.M. 1989. Manganese oxides and hydroxides. p. 439–465. *In* J.B..Dixon and S.B. Weed (ed.) Minerals in soil environments. 2nd Ed. Soil Science Society of America, Madison, WI.

Mill, T., and W. Mabey. 1988. Hydrolysis of organic chemicals. p. 71–111. *In* O. Hutzinger (ed.) The handbook of environmental chemistry, Vol. 2D: Reactions and processes. Springer-Verlag, Berlin, Germany.

Mingelgrin, U., and S. Saltzman. 1979. Surface reactions of parathion on clays. Clays Clay Miner. 27:72–78.

Mingelgrin, U., S. Saltzman, and B. Yaron. 1977. A possible model for the surface-induced hydrolysis of organophosphorus pesticides on kaolinite clays. Soil Sci. Soc. Am. J. 41:519–523.

Moore, J.W., and R.G. Pearson. 1981. Kinetics and mechanism. 3rd Ed. John Wiley and Sons, New York, NY.

Mortland, M.M. 1970. Clay-organic complexes and interactions. Adv. Agron. 22:75–117.

Mortland, M.M. 1986. Mechanisms of adsorption of nonhumic organic species by clays. p. 59–76. *In* P.M. Huang and M. Schnitzer (ed.) Interactions of soil minerals with natural organics and microbes. Soil Science Society of America, Madison, WI.

Mortland, M.M., and L.J. Halloran. 1976. Polymerization of aromatic molecules on smectites. Soil Sci. Soc. Am. J. 40:367–370.

Murray, J.W., and J.G. Dillard. 1979. The oxidation of cobalt(II) adsorbed on manganese dioxide. Geochim. Cosmochim. Acta 43:781–787.

Naidja, A., P.M. Huang, and J.-M. Bollag. 1998. Comparison of reaction products from the transformation of catechol catalyzed by birnessite or tyrosinase. Soil Sci. Soc. Am. J. 62:188–195.

Nannipieri, P., and L. Gianfreda. 1998. Kinetics of enzyme reactions in soil environments. p. 449–479. *In* P.M. Huang, N. Senesi, and J. Buffle (ed.) Environmental particles. Vol. 4: Structure and surface reactions of soil particles. IUPAC Ser. on Analytical and Physical Chemistry of Environmental Systems. John Wiley and Sons, New York, NY.

Niemeyer, J., Y. Chen, and J.-M. Bollag. 1992. Characterization of humic acids, composts and peat by diffuse reflectance Fourier-transform infrared spectroscopy. Soil Sci. Soc. Am. J. 56:135–140.

Oscarson, D.W., P.M. Huang, C. Defosse, and A. Herbillon. 1981a. Oxidative power of Mn (IV) and Fe (III) oxides with respect to As (III) in terrestrial and aquatic environments. Nature 291:50–51.

Oscarson, D.W., P.M. Huang, and W.K. Liaw. 1980. The oxidation of arsenite by aquatic sediments. J. Environ. Qual. 9:700–703.

Oscarson, D.W., P.M. Huang, and W.K. Liaw. 1981b. The role of manganese in the oxidation of arsenite by freshwater lake sediments. Clays Clay Miner. 28:219–225.

Oscarson, D.W., P.M. Huang, and W.K. Liaw. 1981c. The kinetics and components involved in the oxidation of arsenite by freshwater sediments. Verh. Int. Ver. Theor. Angew. Limnol. 21:181–186.

Oscarson, D.W., P.M. Huang, W.K. Liaw, and U.T. Hammer. 1983a. Kinetics of oxidation of arsenite by various manganese dioxides. Soil Sci. Soc. Am. J. 47:644–648.

Oscarson, D.W., P.M. Huang, U.T. Hammer, and W.K. Liaw. 1983b. Oxidation and sorption of arsenite by manganese dioxide as influenced by surface coatings of iron and aluminum oxides and calcium carbonate. Water Air Soil Pollut. 20:233–244.

Pal, S., J.-M. Bollag, and P.M. Huang. 1994. Role of abiotic and biotic catalysts in the transformation of phenolic compounds through oxidative coupling reactions. Soil Biol. Biochem. 26:813–820.

Pearson, R.G. 1966. Acids and bases. Science 151:172–177.

Pearson, R.G. 1976. Symmetry rules for chemical reactions. Wiley-Interscience, New York, NY.

Penrose, W.R. 1974. Arsenic in the marine and aquatic sediments: analysis, occurrence and significance. CRC Crit. Rev. Environ. Control 4:465–482.

Peterson, M.L., G.E. Brown, and G.A. Parks. 1996. Direct XAFS evidence for heterogeneous redox reaction at the aqueous chromium/magnetite interface. Colloid Surf. A. 107:77–88.

Pinnavaia, T.J., P.L. Hall, S.S. Cady, and M.M. Mortland. 1974. Aromatic radical cation formation on the intracrystal surfaces of transition metal layer lattice silicates. J. Phys. Chem. 78:994–999.

Plastourgou, M., and M.R. Hoffmann. 1984. Transformation and fate of organic esters in layered-flow systems: The role of trace metal catalysis. Environ. Sci. Technol. 18:756–764.

Pohlman, A.A., and J.G. McColl. 1989. Organic oxidation and manganese and aluminum mobilization in forest soils. Soil Sci. Soc. Am. J. 53:686–690.

Rai, D., and R.J. Serne. 1977. Plutonium activities in soil solutions and the stability and formation of selected plutonium mineals. J. Environ. Qual. 6:89–95.

Ribault, L.L. 1971. Présence d'ne pellicule de silice amorphe à la surface de cristaux de quartz des formations sableuses. C.R. Acad. Sci. Paris Ser. D 272:1933–1936.

Ross, D.S., and R.J. Bartlett. 1981. Evidence for non-microbial oxidation of manganese in soil. Soil Sci. 132:153–160.

Russell, J.D., M. Cruz, and J.L. White. 1968a. The adsorption of 3-aminotriazole by montmorillonite. J. Agric. Food Chem. 16:21–24.

Russell, J.D., M. Cruz, and J.L. White. 1968b. Model of chemical degradation of s-triazines by montmorillonite. Science 160:1340–1342.

Saltzman, S., B. Yaron, and E. Mingelgrin. 1974. The surface-catalyzed hydrolysis of parathion on kaolinite. Soil Sci. Soc. Am. Proc. 38:231–234.

Sanchez-Camazano, M., and M. J. Sanchez-Martin. 1983. Montmorillonite-catalyzed hydrolysis of phosmet. Soil Sci. 136:89–93.

Scheffer, F., B. Meyer, and E.A. Niederbudde. 1959. Huminstoffbildung unter katalyischer Einwirkung natürlich vorkommender Eisenverbindungen im Modellversuch. Z. Pflanzenernäehr. Bodenkd. 87:26–44.

Schnitzer, M. 1977. Recent findings on the characterization of humic substances extracted from soils from widely differing climatic zones. p. 117–130. In Soil Organic Matter Studies, Vol. II. IAEA-SM-211/7. IAEA, Vienna, Austria.

Schnitzer, M. 1978. Humic substances: chemistry and reactions. p. 1–64. In M. Schnitzer and S.U. Khan (ed.) Soil organic matter. Elsevier Science Publishing, New York, NY.

Schnitzer, M. 1982. Quo vadis soil organic matter research. Trans. 12th Int. Congr. Soil Sci. 5:67–78.

Schnitzer, M., M. Barr, and R. Hartenstein. 1984. Kinetics and characteristics of humic acids produced from simple phenols. Soil Biol. Biochem. 16:371–376.

Schnitzer, M., and K. Ghosh. 1982. Characteristics of water-soluble fulvic acid-copper and fulvic acid-iron complexes. Soil Sci. 134:354–363.

Schnitzer, M., and M. Lévesque. 1979. Electron spin resonance as a guide to the degree of humification of peats. Soil Sci. 127:140–145.

Senesi, N., and M. Schnitzer. 1977. Effect of pH, reaction time, chemical reduction and irradiation on ESR spectra of fulvic acids. Soil Sci. 123:224–234.

Shindo, H., and P.M. Huang. 1982. Role of Mn(IV) oxide in abiotic formation of humic substances in the environment. Nature 298:363–365.

Shindo, H., and P.M. Huang. 1984a. Catalytic effects of manganese (IV), iron (III), aluminum, and silicon oxides on the formation of phenolic polymers. Soil Sci. Soc. Am. J. 48:927–934.

Shindo, H., and P.M. Huang. 1984b. Significance of Mn(IV) oxide in abiotic formation of organic nitrogen complexes in natural environments. Nature 308:57–58.

Shindo, H., and P.M. Huang. 1985a. The catalytic power of inorganic components in the abiotic synthesis of hydroquinone-derived humic polymers. Appl. Clay Sci. 1:71–81.

Shindo, H., and P.M. Huang. 1985b. Catalytic polymerization of hydroquinone by primary minerals. Soil Sci. 139:505–511.

Shindo, H., and P.M. Huang. 1992. Comparison of the influence of Mn(IV) oxide and tyrosinase on the formation of humic substances in the environment. Sci. Total Environ. 117/118:103–110.

Siegel, H. 1976. Metal ions in biological systems. Marcel Dekker, New York, NY.

Smolen, J.M., and A.T. Stone. 1998. Organophosphorus ester hydrolysis catalyzed by dissolved metals and metal-containing surfaces. p. 157–171. *In* P.M. Huang et al. (ed.) Soil chemistry and ecosystem health. Soil Sci. Soc. Am. Spec. Pub. 52.

Solomon, D.H. 1968. Clay minerals as electron acceptors and/or electron donors in organic reactions. Clays Clay Miner. 16:31–39.

Solomon, D.H., and D.G. Hawthorne. 1983. Chemistry of pigments and fillers. John Wiley and Sons, New York, NY.

Solomon, D.H., B.C. Loft, and J.D. Swift. 1968. Reactions catalyzed by minerals. IV. The mechanism of the benzidine blue reaction on silicate minerals. Clay Miner. 7:389–397.

Sposito, G. 1984. The surface chemistry of soils. Oxford University Press, Oxford, UK.

Stanton, D.T. 1987. Glass-catalyzed decomposition of oxycarboxin in aqueous solution. J. Agric. Food Chem. 35:856–859.

Steinberger, R., and F.H. Westheimer. 1949. The metal ion catalyzed decarboxylation of dimethyloxaloacetic acid. J. Am. Chem. Soc. 71:4158–4159.

Steinberger, R., and F.H. Westheimer. 1951. Metal ion-catalyzed decarboxylation: A model for an enzyme system. J. Am. Chem. Soc. 73:429–435.

Stevenson, I.L., and M. Schnitzer. 1982. Transmission electron microscopy of extracted fulvic and humic acids. Soil Sci. 133:179–185.

Stone, A.T. 1987. Reductive dissolution of manganese (III)/(IV) oxides by substituted phenols. Environ. Sci. Technol. 21:979–988.

Stone, A.T., and J.J. Morgan. 1984a. Reduction and dissolution of manganese(III) and manganese (IV) oxides by organics. 1. Reaction with hydroquinone. Environ. Sci. Technol. 18:450–456.

Stone, A.T., and J.J. Morgan. 1984b. Reduction and dissolution of manganese(III) and manganese (IV) oxides by organics. 2. Survey of the reactivity of organics. Environ. Sci. Technol. 18:617–624.

Stone, A.T., and A. Torrents. 1995. The role of dissolved metals and metal-containing surfaces in catalyzing the hydrolysis of organic pollutants. p. 275–298. *In* P.M. Huang, J. Berthelin, J.-M. Bollag, W.B. McGill, and A.L. Page (ed.) Environmental impact of soil component interactions. Vol. 1. Natural and anthropogenic organics. CRC Press/Lewis Publishers, Boca Raton, FL.

Stumm, W. 1992. Chemistry of the solid-water interface. John Wiley and Sons, New York, NY.

Stumm, W., R. Kummert, and L. Sigg. 1980. A ligand exchange model for the adsorption of inorganic and organic ligands at hydrous interfaces. Croat. Chem. Acta 53:291.

Suflita, J.M. and J.-M. Bollag. 1981. Polymerization of phenolic compounds by a soil-enzyme complex. Soil Sci. Soc. Am. J. 45:297–302.

Tennakoon, D.T.B., J.M. Thomas, and M.J. Tricker. 1974. Surface and intercalate chemistry of layer silicates. Part II, An iron-57 Mössbauer study of the role of lattice-substituted iron in the benzidine blue reaction of montmorillonite. J. Chem. Soc. Dalton Trans. 1974:2211–2215.

Theng, B.K.G. 1974. The chemistry of clay-organic reactions. John Wiley and Sons, New York, NY.

Theng, B.K.G. 1979. Formation and properties of clay-polymer complexes. Elsevier Science Publishing, New York, NY.

Theng, B.K.G. 1982. Clay-activated organic reactions. p. 197–238. *In* H. van Olphen and F. Veniale (ed.) Proc. Int. Clay Conf. 1981. Elsevier, Amsterdam, Netherlands.

Thompson, T.D., and W.F. Moll, Jr. 1973. Oxidative power of smectites measured by hydroquinone. Clays Clay Miner. 21:337–350.

Torrents, A. 1992. Hydrolysis of organic esters at the mineral/water interface. Ph.D. Thesis, Johns Hopkins University, Balimore, MD.

Torrents, A., and A.T. Stone. 1991. Hydrolysis of phenyl picolinate at the mineral/water interface. Environ. Sci. Technol. 25:143–149.

Twigg, M.V. 1989. Catalyst handbook. 2nd Ed. Wolfe Publishing Ltd., London, UK.

Voudrias, E.A., and M. Reinhard. 1986. Abiotic organic reactions at mineral surfaces. p. 462–486. In J.A. Davis and K.F. Hayes (ed.) Geochemical processes at mineral surfaces. ACS Symp. Ser. 323. American Chemical Society, Washington, DC.

Wang, M.C., and P.M. Huang. 1986. Humic macromolecular interlayering in nontronite through interaction with phenol monomers. Nature 323:529–531.

Wang, M.C., and P.M. Huang. 1987. Polycondensation of pyrogallol and glycine and the associated reactions as catalyzed by birnessite. Sci. Total Environ. 62:435–442.

Wang, M.C., and P.M. Huang. 1991. Nontronite catalysis in polycondensation of pyrogallol and glycine and the associated reactions. Soil Sci. Soc. Am. J. 55:1156–1161.

Wang, M.C., and P.M. Huang. 1992. Significance of Mn(IV) oxide in the abiotic ring cleavage of polyphenol in natural environments. Sci. Total Environ. 113:147–157.

Wang, M.C., and P.M. Huang. 1994. Structural role of polyphenols in influencing the ring cleavage and related chemical reactions as catalyzed by nontronite. p. 173–180. In N. Senesi and T.M. Miano (ed.) Humic substances in the global environment and implications on human health. Elsevier, Amsterdam, Netherlands.

Wang, M.C., and P.M. Huang. 1997. Catalytic power of birnessite in abiotic formation of humic polycondensates from glycine and pyrogallol. p. 59–65. In J. Drozd, S.S. Gonet, N. Senesi and J. Weber (ed.) Proc. 8th Int. Conf. Int. Humic Substances Society, Wroclaw, Poland.

Wang, T.S.C., J.-H. Chen, and W.-M. Hsiang. 1985. Catalytic synthesis of humic acids containing various amino acids and dipeptides. Soil Sci. 140:3–10.

Wang, T.S.C., P.M. Huang, C.-H. Chou, and J.-H. Chen. 1986. The role of soil minerals in the abiotic polymerization of phenolic compounds and formation of humic substances. p. 251–281. In P.M. Huang and M. Schnitzer (ed.) Interactions of soil minerals with natural organics and microbes. Soil Science Society of America, Madison, WI.

Wang, T.S.C., M.M. Kao, and P.M. Huang. 1980. The effect of pH on the catalytic synthesis of humic substances by illite. Soil Sci. 129:333–338.

Wang, T.S.C., and S.W. Li. 1977. Clay minerals as heterogeneous catalysts in preparation of model humic substances. Z. Pflanzenernaehr. Bodenkd. 140:669–676.

Wang, T.S.C., M.C. Wang, and P.M. Huang. 1983. Catalytic synthesis of humic substances by using aluminas as catalysts. Soil Sci. 136: 226–230.

Wang, T.S.C., K.L. Yeh, S.Y. Cheng, and T.K. Yang. 1971. Behavior of soil phenolic acids. p. 113–120. In Biochemical interaction among plants. National Academy of Science, Washington, DC.

Weast, R.C. (ed.) 1978. CRC handbook of chemistry and physics. 58th Ed. CRC Press, Inc., Boca Raton, FL.

Wehrli, B. 1990. Redox reactions of metal ions at mineral surfaces. In W. Stumm (ed.) Aquatic chemical kinetics. Wiley-Interscience, New York, NY.

Wehrli, B., and W. Stumm. 1989. Vanadyl in natural waters: adsorption and hydrolysis promote oxygenation. Geochim. Cosmochim. Acta 53: 69–77.

Wehrli, B., E. Wieland, and G. Furrer. 1990. Chemical mechanism in the dissolution kinetics of minerals: The aspect of active sites. Aquat. Sci. 52:3–31.

White, A.F. 1990. Heterogeneous electrochemical reactions associated with oxidation of ferrous oxide and silicate surfaces. Rev. Mineral. 23:467–505.

White, A.F., and M.L. Peterson. 1996. Reduction of aqueous transition metal species on the surfaces of Fe(II)-containing oxides. Geochim. Cosmochim. Acta 60:3799–3814.

Wilkins, R.G. 1991. Kinetics and mechanisms of reactions of transition metal complexes. 2nd Ed. VCH Publishers, Weinheim, Germany.

Williams, L.P. 1965. Michael Faraday. Chapman and Hall, London, UK.

Wilson, M.A., and K.M. Goh. 1977. Proton-decoupled pulse Fourier-transform ^{13}C magnetic resonance of soil organic matter. J. Soil Sci. 28:645–652.

Ziechmann, W. 1959. Die Darstellung von Huminsäuren im heterogenen System mit neutraler Reaktion. Z. Pflanzenernäehr. Dueng. 84:155–159.

10

Soil pH and pH Buffering

Paul R. Bloom
University of Minnesota

10.1 Introduction

Soil pH is probably the single most important chemical characteristic of a soil. In soil, pH is a "master variable" (McBride, 1994) and knowledge of pH is needed to understand important chemical processes such as ion mobility, precipitation and dissolution equilibria, precipitation and dissolution kinetics, and oxidation-reduction equilibria. A knowledge of pH is also needed to understand nutrient availability to plants and the negative response of many plant species to soil acidity.

Soils are a complex mixture of solid-phase components that react simultaneously to yield a measured pH value. These components also buffer soils against pH changes caused by natural and anthropic inputs of acids and bases. Knowledge of the reactions that buffer soil pH is necessary for an understanding of natural soil weathering and the response of a soil to inputs of lime, acid-forming N fertilizers, acid mine wastes, and acid rain. The discussion in this chapter focuses on the reactions of components that buffer soil pH in upland soils. A discussion of the factors affecting pH and buffering in flooded soils is given in Section B, Chapter 5 and Section G, Chapter 4.

10.2 Definition and Determination of Soil pH

Because pH is a term that is only defined for solutions, in a strict sense, it cannot be applied to a solid-phase material like soil. However, the chemical properties of the solid phase components in soil define the pH of the soil solution, the solution in the soil pores (Section B, Chapter 3).

10.2.1 Definition of pH

The term pH was defined by Sorenson (1909) to provide a convenient way of representing the H^+ or OH^- concentrations in aqueous solutions. As currently defined, pH is the negative logarithm, base 10, of H^+ activity $[-\log(H^+)]$ where activity, ( ), is the concentration adjusted for nonideality caused by charge-charge interactions with other ions in solution. In all except saline soils, the ionic interaction correction is small and the difference between H^+ concentration and activity is small. The determination of H^+ also provides a measurement of OH^- in solution. At 25 °C in pure water, the relation between (OH^-) and (H^+) is given by:

$$K_w = (H^+)(OH^-) = 1.0 \times 10^{-14} \qquad\qquad [10.1]$$

where K_w is the ion ionization constant for water. Expressing Equation [10.1] as negative logarithms yields:

$$pH + pOH = 14 \qquad\qquad [10.2]$$

where pOH is the negative logarithm of the OH^- activity.

10.2.2   Determination of pH with a Glass Electrode

Soil pH is determined by measuring the pH after equilibrating soil with pure water or a salt solution (Glossary of Soil Science Terms, 1997). The pH of a solution in equilibrium with a soil varies with the composition and concentration of the salts in the solution. This is because cations in solution displace H^+ and Al^{3+} ions from soil surfaces. Three common standard methods for determining pH in soils involve the suspension in distilled water, 0.01 M $CaCl_2$, or 1 M KCl solutions (Thomas, 1996). In the United States, the most commonly reported value is for a 1:1 (w:v) suspension of air-dried soil in distilled water. A suspension of soil is prepared (10 g soil and 10 mL distilled water) and stirred vigorously and allowed to stand for 10 min. The suspension is stirred again and the pH is measured with a glass electrode pH meter (Thomas, 1996). This measure of soil pH_{H_2O} is expected to give a higher value than the pH of the soil solution at natural soil water contents. The quantity of water added results in a dilution of the salts in the soil solution and a lower equilibrium salt content. This effect is illustrated by the 0.4 unit increase in pH obtained for an increase in water content in the soil suspension from 1:1 to 1:10 (Thomas, 1996). The 1:1 water determination of pH is influenced by the natural seasonal variation in soil solution salt concentrations and the effects of soil management such as recent fertilization.

At low salt concentrations, an error due to the junction potential effect can occur in determination of pH_{H_2O}. A pH measurement requires contact with the solution by a glass electrode and a porous liquid junction of a reference electrode. This is necessary to complete the circuit between the pH sensitive glass electrode and the reference electrode. The reference electrode is often built in the same body as the glass electrode to form a combination electrode. The filling solution for the reference electrode is a concentrated KCl solution which slowly leaks into the test solution through a porous liquid junction. Because soils are cation exchangers, they can have a disproportionate effect on the rate of K^+ diffusion compared to Cl^- diffusion into solution which sets up a potential across the junction (Coleman et al., 1951). This can cause a reduction in the measured pH (Thomas, 1996).

The method for measurement of pH in 0.01 M $CaCl_2$ is the same as for measurement of pH_{H_2O} (Thomas, 1996). Because of the displacement of H^+ and Al^{3+} from soil materials by the Ca^{2+}, pH_{CaCl_2} is on average about 0.5 units lower than pH_{H_2O} (Lathwell and Peech, 1964). Because the pH_{CaCl_2} solution is sufficiently concentrated to overwhelm variations in soil solution concentrations, one of the major problems in measuring soil pH is minimized in all except very saline soils. The 0.01 M $CaCl_2$ also eliminates the junction potential effect. Calcium is used in the determination of pH because it is the predominant soil solution cation in soils of temperate regions. The 0.01 M concentration, however, is somewhat higher than found in most soil solutions, and thus the pH values obtained by this method are usually lower than in soil solutions at natural water contents.

The method for measurement of pH in M KCl is the same as for pH_{H_2O} (Thomas, 1996). The pH_{KCl} is generally lower than pH_{CaCl_2}. For a range of Nigerian soils, pH_{KCl} is 1 unit lower value than pH_{H_2O} at $pH_{H_2O} = 4$ and the difference is even larger at higher pH (Okusami et al., 1987). The KCl solution also eliminates problems due to solution concentration and junction potential.

For some subsurface horizons of highly weathered soils of the tropics, pH_{KCl} can actually be greater than pH_{H_2O} (van Raij and Peech, 1972), especially in subsoils low in permanent charge silicate clays and low in organic matter. Ion exchange in these soils is dominated by oxides and hydroxides of Fe and Al and at acid pH values, they are anion exchangers (Section 10.7). Thus, on KCl addition more OH⁻ is displaced by Cl⁻ than H⁺ by K⁺ resulting in an increase in pH (McBride, 1994). The problems associated with pH measurements using glass electrode systems are discussed by Sumner (1994).

10.2.3 Determination of Soil pH Using pH Sensitive Dyes

Rapid determination of soil pH can be made colorimetrically with pH indicator dyes. Colorimetric methods are based on the change in color that takes place upon dissociation of a weak acid , or weak base organic dye; thus for a weak acid dye:

$$HI = H^+ + I^-$$

[10.3]

where raising the pH will cause a change in color as the undissociated dye (HI) changes to the dissociated form (I⁻). Typically, a mixture of dyes is used to yield color changes that take place over a wide range of pH. A dye solution is mixed with soil to make a slurry and the dye is decanted from the soil and the color is visually compared with a color chart. The pH measured in this manner corresponds with pH_{H_2O} measured with a glass electrode but is not as precise and can vary from the glass electrode results by as much as 0.4 units (Mason and Obenshain, 1939). Dye impregnated pH papers also can be used to get quick pH measurement within 0.5 units of the true pH_{H_2O} (Thomas, 1996).

10.3 Acids and Bases in Soil Solutions

Soil solutions contain a mixture of soluble weak acids and weak bases that can buffer the pH of these solutions when soil solutions are extracted from soils. This buffering is small compared to the quantity of acidity or basicity associated with the solid phases in soil and has little impact on soil pH buffering. However, a discussion of the weak acids and bases in soil solutions can lead to a better understanding of the reactions of acids and bases in soils. The acid and base components in soil solutions directly impact plant roots and soil microbes and thus directly influence the vitality of the biotic community in a soil. Also, the acids and bases in the soil solution directly impact weathering and mineral resynthesis reactions involved in soil genesis.

In this chapter, Brønsted's definition of acids and bases, where an acid is an H⁺ donor and a base is an H⁺ acceptor, will be used (Stumm and Morgan, 1996). Because an H⁺ ion is just a proton, acids can be described as proton donors and bases as proton acceptors. According to this definition, reaction of an acid to lose a proton produces a conjugate base that can accept a proton, if the reaction is reversed. Conversely, a base that accepts a proton yields a conjugate acid. For an example, when HNO_3 donates protons to water, the base is H_2O. The reaction is

$$HNO_3 + H_2O = NO_3^- + H_3O^+$$

[10.4]

The NO_3^- anion is the conjugate base of HNO_3 and the H_3O^+ is the conjugate acid of water. In water, H⁺ ions hydrate to form H_3O^+ ions. Nitric acid is a strong acid because at all except very high H⁺ concentrations, Reaction [10.4] goes all the way to the right and all of the HNO_3 is ionized. The NO_3^- anion is a very weak base because it has very little tendency to accept protons. The above reaction in aqueous solution is often abbreviated to:

$$HNO_3 = NO_3^- + H^+ \tag{10.5}$$

10.3.1 Weak Acids

Soil solutions contain a mixture of inorganic and organic weak acid components. The weak acids include organic carboxylic acids, H_2CO_3, and Al^{3+}, a metal ion that undergoes hydrolysis. Carboxylic acids are organic compounds with one or more R-COOH groups that have the potential to ionize to form the carboxylate anion (RCOO⁻), where R represents any organic compound that might have a carboxyl group. Organic acids in soil solutions include the simple acids such as malic [(COOH)CH$_2$ CH(OH)CH$_2$(COOH)] and citric acids [(COOH)CH$_2$C(OH)(COOH)CH$_2$(COOH)] that are exuded by plant roots or soil microbes. These acids generally have a short life time in soils because they can be consumed by soil microbes. In addition, soil solutions contain macromolecular humified components produced by degradation of plant debris (Stevenson, 1994). The soluble humic fraction in soil solutions is dominated by fulvic acid. Both fulvic and the simple acids contain carboxyl groups that can donate protons to water according to the reaction:

$$RCOOH = RCOO^- + H^+ \tag{10.6}$$

The strength of the acid (ability to donate protons) is defined by the equilibrium constant:

$$K_a = (RCOO^-)(H^+)/(RCOOH) \tag{10.7}$$

where K_a is an acidity constant which is greater for stronger acids. Often the strength of a weak acid is given in terms of $-\log K_a = pK_a$. The lower the pK_a the greater the strength of the acid. When pH = pK_a, an acid is half ionized, being more and less than half ionized above and below this value. Thus, when solution pH $< pK_a$, a weak acid has a greater ability to donate protons. Carboxylic organic acids have pK_a values generally in the range of 3 to 5. Soluble fulvic acid also contains phenolic groups (ArOH) where Ar is an aromatic 6-C-ring structure. Phenolic components are generally much less concentrated than the carboxyl components and are much less acidic having pK_a values in the range 7–11.

The Al^{3+} ion, which is an important weak acid component in acid soil solutions, contributes to soil solution acidity and the acid toxicity response experienced by many plants in acid soils. In water, Al^{3+} undergoes a series of hydrolysis reactions (Nordstrom and May, 1996):

$$Al^{3+} + H_2O = AlOH^{2+} + H^+ \qquad pK_{a1} = 5.0 \tag{10.8}$$

$$AlOH^{2+} + H_2O = Al(OH)_2^+ + H^+ \qquad pK_{a2} = 5.1 \tag{10.9}$$

$$Al(OH)_2^+ + H_2O = Al(OH)_3^o + H^+ \qquad pK_{a3} = 6.7 \tag{10.10}$$

$$Al(OH)_3^o + H_2O = Al(OH)_4^- + H^+ \qquad pK_{a4} = 6.2 \tag{10.11}$$

Although the second, third, and fourth hydrolysis reactions are important in determining the chemistry of Al, only the first is a major source of protons in the soil solution. The low solubility of $Al(OH)_3$ limits the quantity of Al ions. A solution at pH 5.0 in equilibrium with $Al(OH)_3$ has a total concentration of Al^{3+} plus hydrolyzed Al of less than 10^{-6} M (Nordstrom and May, 1996).

When CO_2 dissolves in water, it produces carbonic acid (H_2CO_3) according to the reaction:

$$CO_2 + H_2O = H_2CO_3 \quad pK = 1.51 \quad \text{(Stumm and Morgan, 1996)} \tag{10.12}$$

and ionizes to produce bicarbonate and H^+:

$$H_2CO_3 = HCO_3^- + H^+ \quad pK_a = 6.35 \quad \text{(Stumm and Morgan, 1996)} \tag{10.13}$$

Thus, H_2CO_3 is a weaker acid than R–COOH or Al^{3+}. The concentration of H_2CO_3 is fixed by the pressure of CO_2 and at a constant content of CO_2, the concentration of HCO_3^- increases with increasing pH. In the Earth's atmosphere, P_{CO_2} is 3.5×10^{-4} atm (McBride, 1994) but in soils where root and microbial respiration are active, P_{CO_2} can easily be 100-fold greater (Fernandez and Kosian, 1987; Magnusson, 1992) which has a great effect on the content of HCO_3^- in soils.

10.3.2 Weak Bases

All of the anions of the weak acids above, as well as the hydrolyzed Al cations, are weak bases that can accept protons in the pH range found in soils. Because of the low pK_a values for organic acids and $AlOH^{2+}$, these ions are only important proton acceptors in very acid soils. For most neutral and alkaline soils, the concentration of HCO_3^- determines the alkalinity of soil solutions, where alkalinity is defined as the quantity of strong acid needed to lower the pH to the endpoint for completely titrating HCO_3^- (pH ~ 4.8) (Stumm and Morgan, 1996). The CO_3^- ion is important only in very high pH soils because the pK_a for the ionization of HCO_3^{2-} to CO_3^{2-} is 10.33 (Stumm and Morgan, 1996).

In addition, when fertilizer is applied as NH_3 (gas) or urea, NH_3 (aq) can be a significant base in the soil solution. When NH_3 (gas) is injected into a soil, it reacts to lower (H^+) thus raising the pH. Ammonia is a conjugate base of NH_4^+ and if the reaction in water is written like the weak acid reactions above:

$$NH_4^+ = H^+ + NH_3 \text{ (aq)} \quad pK_a = 9.2 \quad \text{(Stumm and Morgan, 1996)} \tag{10.14}$$

In all except the highest pH soils, the reaction goes to the left consuming protons. Because of this reaction, localized pH values near anhydrous NH_3 injection bands can be > 9.5 (Tisdale et al., 1985) and NH_3 (aq) contributes significantly to soil solution basicity.

The localized pH near urea fertilizer granules can also be high because in water in the presence of urease enzymes, which are ubiquitous in soils, urea reacts to form NH_3:

$$(NH_2)_2CO + 2H_2O = HCO_3^- + NH_4^+ + NH_3 \text{ (aq)} \tag{10.15}$$

The basicity effect of these fertilizers is short lived because biological oxidation of NH_4^+ to nitrite produces $2H^+$ ions with the net effect being acidification (Section 10.9.3).

10.4 Overview of Reactions with Soil Solids Controlling pH and pH Buffering

Reactions of the solid-phase components control soil pH, and consequently, buffer soil pH against rapid change when natural or anthropogenic inputs of acids and bases are made. Solid-phase soil materials have a much greater capacity to accept or donate protons than the soluble acids and bases in soil solution. Magdoff and Bartlett (1985) investigated the buffering in 51 surface soils of Vermont by reacting these soils with H_2SO_4 or $CaCO_3$ for 30 days and determining the final pH in 0.01 M $CaCl_2$.

They found maximal buffering (minimal slope of pH versus added acid or $CaCO_3$) to occur at both pH 3.5 and pH 7.5 (Fig. 10.1). The upper bound of the pH in the buffering shown in Fig. 10.1 is fixed by the solubility of $CaCO_3$ under the conditions in these soils (Magdoff and Bartlett, 1985). They also showed that the buffer plot was generally linear in the pH range of 4.5 to 6.5, a range that encompasses most acid agricultural soils.

The pH buffering reactions in soils include proton desorption and adsorption reactions by mineral and organic materials as well as ion exchange, dissolution, and precipitation reactions. Some of the soil components are effective in buffering over a wide range of pH values, while others are effective over a limited pH range (Table 10.1). The buffer ranges in Table 10.1 are somewhat similar to those shown by Ulrich (1991) with major modifications and additions. For example, Ulrich (1991) did not include SOM and surface reactions of oxides and hydrous oxides. The buffer reactions in Table 10.1 will be discussed in more detail in Sections 10.5 to 10.9.

10.5   Buffering by Soil Organic Matter

Soil organic matter (SOM) is a very important component of pH buffering in surface soils, even in typical upland soils that contain only little SOM. Like the soluble organic components discussed in Section 10.3.1, solid-phase SOM contains carboxyl and phenolic groups that can donate protons. Most of the acidity in SOM is contributed by the humified components, humic and fulvic acids. These acids can bind Al^{3+} ions, altering the proton donation behavior of SOM. Based on chemical analysis of humic and fulvic acids extracted from SOM (Section B, Chapter 2), various organic acids have been proposed to represent the acidity of carboxyl and phenolic sites in SOM. Most of the acidity is thought to arise from carboxyl and OH groups bound to aromatic ring structures. The most simple model compounds are benzoic acid (Ar–COOH, pK_a = 4.2) and phenol (Ar-OH, pK_a = 10.0) (Martell and Smith, 1974).

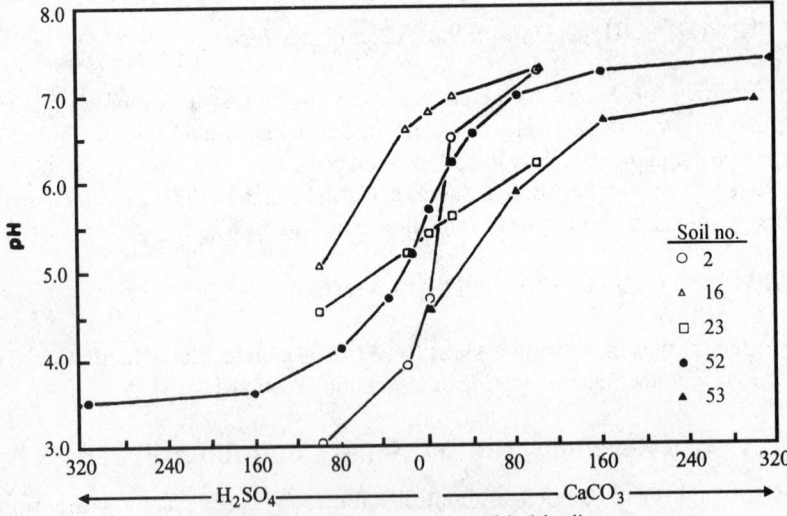

Fig. 10.1   Titration of surface soils from Vermont with $CaCO_3$ and H_2SO_4. The acid and base additions are reported per gram of soil. The pH($CaCl_2$) was measured after incubation for 20 days at field capacity water content [Reprinted from Magdoff and Bartlett, 1985. Soil Sci. Soc. Am. J. 49:145-148, with permission of the Soil Science Society of America].

Table 10.1   Reactions of solid-phase soil components that buffer pH in the range of 3.0 to 10.0

Buffer substance	pH range	Proton accepting and donating reactions
$MgCO_3$	> 9.5	Precipitation and dissolution
Soil organic matter	whole pH range	Ionization of carboxyl and phenolic groups
Oxides and hydroxides of Fe and Al, allophane, imogolite and silicate clay edges	whole pH range	Ionization and protonation of surface hydroxyl groups
Limestone, $CaCO_3$	7 to 9.5	Dissolution and precipitation
Permanent charge sites on silicate clay minerals	3.5 to 5	Ion exchange of H^+ and Al^{3+} with base cations
$Al(OH)_3$ - soil organic matter	5 to 8	Precipitation of organic bound Al^{3+} as $Al(OH)_3$ or dissolution of $Al(OH)_3$ by organic acids
Interlayer Al in 2:1 clays	5 to 7	Precipitation or dissolution of interlayer $Al(OH)_3$
Al - soil organic matter	< 5	Carboxyl H^+ exchange with Al^{3+}
Irreversible dissolution of high activity silicate clays and allophane	< 3.5	Consumption of H^+ upon dissolution of Al^{3+}
Very slow irreversible weathering of primary silicate minerals	whole pH range	Consumption of H^+ upon dissolution of Ca^{2+}, Mg^{2+}, K^+ and Na^+

Soil organic matter buffers pH over a wider range of pH values than predicted by a simple mixture of benzoic acid and phenol. The titration of humic acid with an alkali metal hydroxide shown in Fig. 10.2 illustrates the buffering over a very wide pH range. This broadening of the buffering occurs because the aromatic ring structures of the humic acid are polysubstituted with -COOH and phenolic -OH groups. This alters the pK_a values, resulting in aromatic ring structures with wide ranges of individual pK_a values. Also, with the ionization of the acid sites, negative charges accumulate on the macromolecular structures reducing the proton donation ability of humic and fulvic acids. The ionization of -COOH and phenolic -OH groups is generally complete at pH 8 and 11, respectively. Thus, the common methods for determination of -COOH and Ar-OH acidity involve titrating to these pH values (Stevenson, 1994). The negative charge sites produced by the ionization of the weak acid sites on humic and fulvic acids bind exchangeable cations. Organic matter contributes much of the variable charge (CEC) to soils (Section B, Chapter 7).

In acid soils, the binding of Al to acid sites in SOM is important both for Al chemistry and for acidity buffering. In acid mineral soils, many of the -COOH sites in SOM are satisfied by Al^{3+}. Only a small fraction of the SOM bound Al can be extracted with $1M$ KCl (Bloom et al., 1979), with most considered to be nonexchangeable (Bertsch and Bloom, 1996). However, this strongly bound Al has a large effect on buffering. Hargrove and Thomas (1982) found that the pH buffering of Al^{3+} treated SOM occurred at higher pH values than for H^+ saturated SOM (Fig. 10.3) indicating that Al-SOM is a weaker acid than the H^+ analogue.

As the pH of Al-SOM is raised above 5, solution Al^{3+} exceeds the solubility of $Al(OH)_3$ which then precipitates in an amorphous form (Walker et al., 1990) illustrated by:

$$Al(RCOO)_3 + 3H_2O = 3RCOO^- + 3H^+ + Al(OH)_3 \qquad\qquad [10.16]$$

This reaction is reversible and when the pH is lowered $Al(OH)_3$ dissolves and Al becomes bound to SOM. As with H-SOM, increasing pH by adding a base increases surface charge on SOM which becomes more saturated with the cation of the added base. Except under very acid conditions, soil pH is correlated with the saturation of the CEC by Ca^{2+}, Mg^{2+}, Na^+ and K^+; all cations of strong bases are commonly referred to as "base cations". Base saturation is {(sum of exchangeable bases)/CEC} where CEC is determined at a reference pH of 7 or 8 (Bloom and Grigal, 1985). In most soils, the most abundant base cation is Ca^{2+}.

At pH < 5, proton exchange with Al^{3+} is important in buffering pH and as pH is decreased to < 4.5, Al^{3+} hydrolysis becomes insignificant, and exchange can add very significant quantities of Al^{3+} to the soil solution. At these low pH values, which are typical of the A and O horizons in many forest soils, plots of base saturation versus pH show a better correlation when Al^{3+} is included as a base cation rather than as an acid (Skyllberg, 1994).

Soil organic matter provides much of the pH buffering in most surface soils. Helling et al. (1964) found the mean CEC of SOM, at pH 8, for 60 mineral soils of Wisconsin is 2,000 $mmol_c$ kg^{-1}. This value provides a reasonable estimate for the capacity of SOM in mineral soils to buffer pH in the pH range of 3 to 8. With the exception of $CaCO_3$, the pH buffering capacity of SOM is equal or greater than other soil components (Table 10.2). After normalizing data to equivalent SOM contents, Magdoff and Bartlett (1985) found a single relationship between pH and acid or base added to 51 surface soils, further supporting the importance of SOM in buffering pH (Fig. 10.4). Under very acid conditions, buffering increased partially due to the displacement of organic bound Al^{3+} by H^+.

10.6   Proton and Al Exchange in Silicate Clays

Ion exchange reactions on permanent charge sites in silicate clays are important in pH buffering. (Section B, Chapter 7). The pH buffering by permanent charge sites involves the exchange of H_3O^+ but

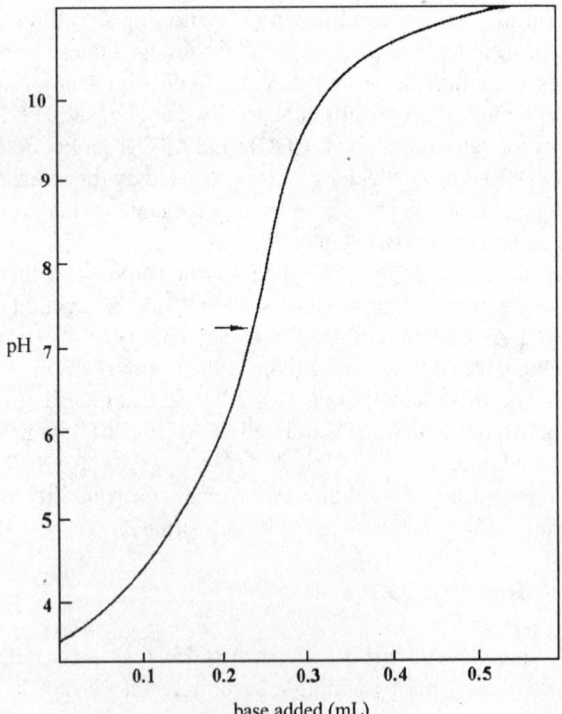

Fig. 10.2   Titration of a humic acid with a strong base

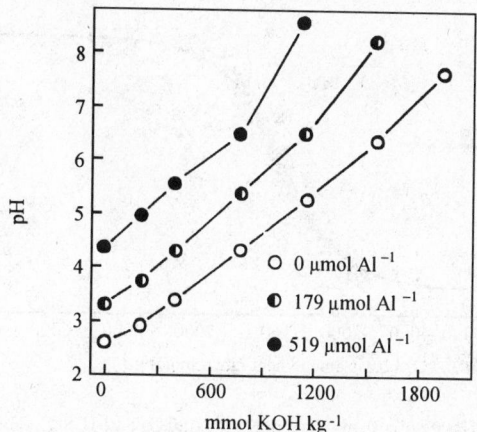

Fig. 10.3 Titration of Al-SOM containing different contents of Al^{3+} [Reprinted from Hargrove and Thomas, 1982. Titration properties of Al-organic matter. Soil Sci. 134:216-225, with permission of Williams & Wilkins, Baltimore, MD]

more importantly, the exchange of Al^{3+}. Aluminum clays are proton donors because of the ability of Al^{3+} to hydrolyze (Section 10.3) and to precipitate as $Al(OH)_3$. Extraction of acid soils with neutral salts solutions such as $1M$ KCl, yields mostly Al^{3+} not H^+ (Thomas and Hargrove, 1984).

High activity clays such as smectites, vermiculites and illites when saturated with H^+ have very low pH values and are not stable. When montmorillonite is saturated with H^+, the pH is < 3.5, but after 24 h of aging, acid dissolution of the clay produces Al^{3+} ions resulting in an increase in pH and replacement of the exchangeable H^+ ions by Al^{3+} (Coleman and Craig, 1961). Titration of Al^{3+} montmorillonite (Fig.10.5) shows that at pH_{H_2O} values in the range of 5 to 6, Al^{3+} on permanent charge sites strongly buffers the pH (Turner and Nichol, 1962). As in soil, pH_{CaCl_2} is lower than pH_{H_2O}.

When base is added to an Al^{3+} saturated smectite or vermiculite, the initial reaction is the hydrolysis of Al cations on exchange sites (Bloom et al., 1977), but with sufficient addition of strong base, the Al^{3+} ions hydrolyze and eventually precipitate as $Al(OH)_3$ in the interlayer space in the clay, accounting for much of the buffering at pH > 5. Because of OH^- deficiencies in interlayer $Al(OH)_3$, it can neutralize the negative charge on the clay producing hydroxy interlayered smectite (HIS) and hydroxy interlayered vermiculite (HIV).

Table 10.2 Approximate maximum proton donation or adsorption capacity of soil materials in the pH range of 3.5 to 8

Soil Material	Capacity, mmol kg^{-1}	References
Silicate Clays		
smectites	800-1500	McBride (1994)
vermiculite	1,500-2,000	McBride (1994)
illite	200-400	McBride (1994)
Kaolinite	10-50	Thomas and Hargrove et al. (1964)
SOM	2000	Helling et al. (1964)
Allophane and imogolite	200-500	Wada (1989)
Hydroxides and oxides of Fe and Al*	50-400	Borggaard (1983)
CaCO$_3$**	20,000	Stumm and Morgan (1996)

* Based on linear extrapolation between 8 and 3.5 of the data for hematite and goethite.
** CaCO$_3$ equilibrium occurs at pH 7 or above.

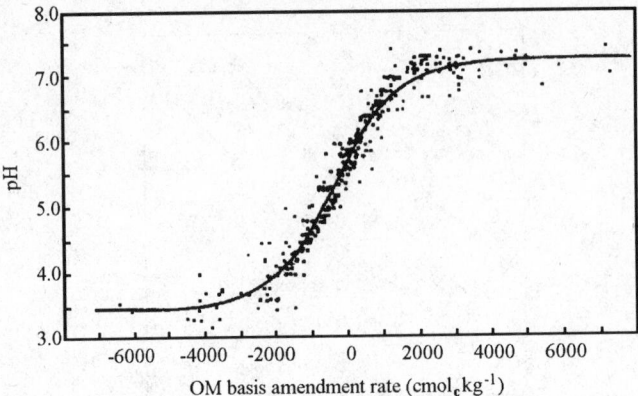

Fig. 10.4 Titration data for 51 surface soils from Vermont with $CaCO_3$ and H_2SO_4. The acid and base addition are reported per gram of organic matter. The zero amendment data point was transposed to fit soil 51 in Fig. 10.1. The $pH(CaCl_2)$ was measured after incubation for 20 days at field capacity water content [Reprinted from Magdoff and Bartlett, 1985. Soil Sci. Soc. Am. J. 49:145-148, with permission of the Soil Science Society of America].

As with SOM, buffering in Al^{3+}-saturated clays can be described as a function of base saturation, assuming that both H^+ and Al^{3+} are acid cations and that Al^{3+} represents 3 moles of acidity per mole of Al^{3+}. This is a usable generalization at pH values high enough for Al^{3+} to hydrolyze. Recent research with very acid surface soils from northern temperate forests has shown that at pH values < 4.5, Al^{3+} should be treated as a base cation (Skyllberg, 1994).

The capacity of silicate clays to buffer pH varies widely because of the large differences in clay CEC (Table 10.2). Kaolinite, a low activity clay, has little or no permanent charge and has little capacity to buffer pH while illite and smectite have a much greater CECs and are much more effective buffers. Among the silicate clays, only the highest charge vermiculites approach the buffer capacity of SOM.

10.7 Variable Charge Buffering by Mineral Components

Oxides and hydroxides of Fe(III) and Al and poorly crystalline aluminosilicates (allophane) contribute to variable charge in soil and to pH buffering while the edges of crystalline silicate clays make a minor contribution (Section B, Chapter 7). Unlike SOM, these materials have surfaces that are both proton donors and acceptors.

10.7.1 Oxides and Hydroxides of Iron and Aluminum

These minerals, that accumulate in soils upon weathering, are important in the pH buffering of soils, particularly in highly weathered soils (Uehara and Gillman, 1982). Common Fe minerals in soil include hematite (α–Fe_2O_3), goethite (α–$FeOOH$), and ferrihydrite ($Fe_5HO_8 \cdot 4H_2O$). By far the most common Al hydroxide mineral is gibbsite, [$Al(OH)_3$]. In water, the hydrated surfaces of these oxides and (oxy)hydroxides contain strongly bound OH groups that can act both as proton donors and acceptors. For hematite, the surface hydration reaction is

$$\text{surface } Fe_2O_3 + 3H_2O \rightarrow 2Fe(OH)_3 \qquad [10.17]$$

The strong bonding of the O atom to Fe(III) and Al weakens the bond of the proton to the O atom, making the surface OH groups weakly acidic, and at high pH, the metal-bound hydroxide groups can donate protons creating negative surface charges:

$$\text{surface FeOH} \rightarrow \text{FeO}^- + \text{H}^+ \tag{10.18}$$

This reaction can also be written as a adsorption of OH^-:

$$\text{surface FeOH} + \text{OH}^- \rightarrow \text{FeO}^- + \text{H}_2\text{O} \tag{10.19}$$

These surface OH groups can accept protons at low pH to become positively charged:

$$\text{surface FeOH} + \text{H}^+ \rightarrow \text{FeOH}_2^+ \tag{10.20}$$

Surface adsorption of H^+ and OH^- is quantified by titrating oxide suspensions in Na^+ or K^+ salts of anions that do not strongly adsorb to the oxide surfaces (e.g., NO_3^-). The data of Parks and de Bruyn (1962), which illustrate typical titrations for oxide suspensions (Fig. 10.6), show the effects of increasing KNO_3 concentration on the quantity of H^+ or OH^- adsorbed. McBride (1994) explains this effect as due to K^+ displacement of H^+ at high pH and NO_3^- displacement of OH^- at low pH. At pH values < 7, the oxides and hydroxides of Al and Fe are positively charged and these surfaces are anion

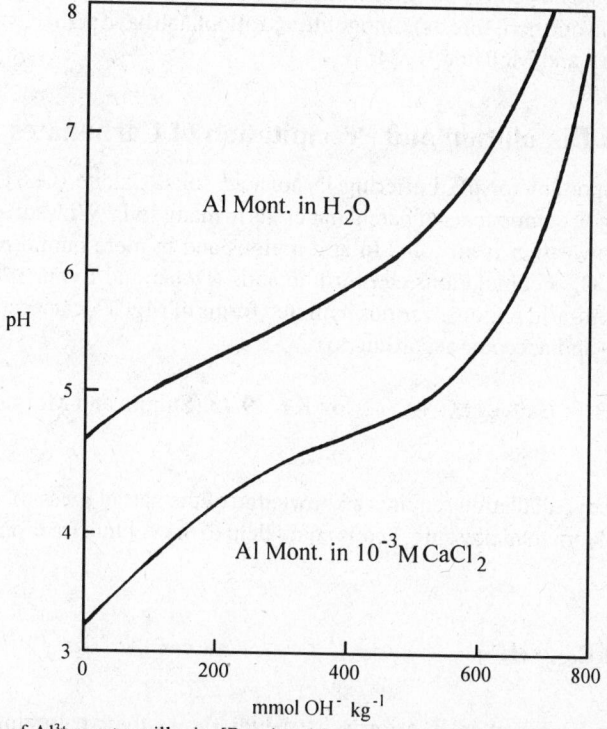

Fig. 10.5 Titration of Al^{3+} montmorillonite [Reprinted from Turner and Nichol, 1962. A study of the lime potential: 2. Relation between lime potential and percent base saturation of negatively charged clays in aqueous salt suspensions. Soil Sci. 94:58-63, with permission of Williams & Wilkins, Baltimore, MD].

exchangers while at pH > 9, they are cation exchangers. At pH values in the range of 7 to 9, a point is found where salt has no effect on the pH. This is the point at which the net charge on the surface is zero, the point of zero net charge (p.z.n.c.). For oxides and hydroxides of Fe and Al, almost all measured p.z.n.c. values are in the range 7–9 (Uehara and Gillman, 1982), meaning that unlike silicate clays and organic matter, the net surface charge of these materials is positive in all except alkaline soils.

10.7.2   Variable Charges on Silicate Clay Edges

The layer structure of silicate clays includes an AlOH layer with exposed hydroxyl groups on the edges (McBride, 1994). These AlOH groups can adsorb or desorb protons, but because of the influence of the Si^{4+} ions in the structure, the p.z.n.c. is lower than for gibbsite. For example, Sposito (1989) estimates the p.z.n.c. for the edge of kaolinite at 4–5 which means the edge charge can contribute to the negative charge in soils with slightly acid and higher pH values. For highly charged clays like smectites and vermiculite, this edge charge is small compared to the permanent structural charge. For example, for a smectite at pH 8, the variable charge is about 50 $mmol_c$ kg^{-1} compared to a permanent charge of about 750 $mmol_c$ kg^{-1} (McBride, 1994).

10.7.3   Imogolite and Allophane

Allophane and imogolite are aluminosilicates that, like silicate clay edges, have variable charge surfaces. Allophane is amorphous, not having a regular repeating structure, and imogolite, which has a threadlike morphology, is ordered only in one dimension (Wada, 1989). Both of these materials are found in abundance in young volcanic ash soils and also in some young soils in temperate regions (e.g., Spodosols in northern forests). Imogolite and allophane have p.z.n.c. values in the range 5.5–7 (Wada, 1980; Clark and McBride, 1984).

10.8   Buffering by Dissolution and Precipitation of Carbonates

Carbonates are very important for pH buffering in nonacid soils. Calcite ($CaCO_3$) and dolomite [$CaMg(CO_3)_2$] are common components of parent materials in many soils. With sufficient rainfall and drainage, these minerals weather from soils. In arid regions and in more humid regions with poor drainage, however, $CaCO_3$ accumulations can form in soils (Doner and Lynn, 1989). Under some conditions in arid and semi-arid regions, various hydrated forms of $MgCO_3$ can accumulate. Calcium carbonate acts as proton and acceptor according to:

$$CaCO_3 + 2H^+ \rightleftharpoons Ca^{2+} + H_2O + CO_2 \quad \log K = -9.73 \text{ (Stumm and Morgan, 1996)}$$

$$[10.21]$$

The solubility equilibrium calculation requires a knowledge of the partial pressure of CO_2, (P_{CO_2}), in the soil atmosphere. At 1 atm total pressure, P_{CO_2} is equivalent to the volume fraction. The equilibrium equation is

$$K = \left(Ca^{2+}\right)\left(P_{CO_2}\right) / \left(H^+\right)^2 \qquad [10.22]$$

Calcite equilibrium can be rewritten in a form that better shows the weathering of $CaCO_3$ from calcareous soils:

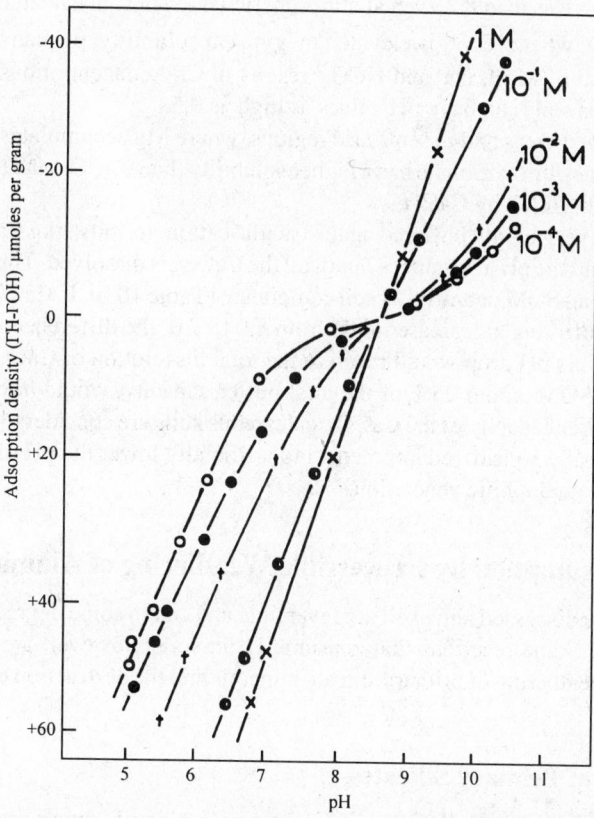

Fig. 10.6 Titration of Fe_2O_3 in varying concentrations of KNO_3 [Reprinted with permission from Parks and De Bruyn, 1962. J. Phys. Chem. 66:967–973. Copyright American Chemical Society].

$$CaCO_3 + CO_2 + H_2O \rightleftharpoons Ca^{2+} + 2HCO_3^- \quad \log K = -5.91$$
(Stumm and Morgan, 1996)

[10.23]

In soils, the CO_2 produced by soil microbial and root respiration dissolves $CaCO_3$ to produce soluble $Ca(HCO_3)_2$.

The pH buffer range for soils in equilibrium with $CaCO_3$ varies widely. A $CaCO_3$ suspension in distilled water in equilibrium with the ambient atmosphere ($P_{CO_2} = 3.5 \times 10^{-4}$ atm) will contain two HCO_3^- ions in solution for every Ca^{2+} ion, and the pH will be 8.3 (Stumm and Morgan, 1996). In soils, however, the P_{CO_2} is greatly increased due to microbial and root respiration. The increase in CO_2 pressure is greatly affected by the rate of gas exchange with the atmosphere, a process influenced by the quantity of air-filled pores in a soil. Inskeep and Bloom (1986) showed that the filling of soil pores with water greatly decreases CO_2 exchange with the air and increases P_{CO_2}. In pots with growing soybean plants, the P_{CO_2} ranged from 0.0006–0.031 atm in dry to moist soils approaching saturation. By the principle of LeChatalier (the mass action principle), Reaction [10.21] will be shifted to the left when P_{CO_2} is raised, resulting in an increase in (H^+) and a decrease in pH.

In soils, HCO_3^- concentrations are commonly not equal to twice the concentration of Ca^{2+}. When the chemistry of soil is dominated by Ca salts other than bicarbonate, $[Ca^{2+}]$ can be much greater than

[HCO$_3^-$] and pH is much less than 8.3 even at atmospheric P$_{CO_2}$. Thus, in calcareous soils containing gypsum (CaSO$_4$•2H$_2$O), where [Ca^{2+}] is elevated by gypsum solubility, pH can be as low as 7.0. In some arid regions, input of Mg^{2+}, Na$^+$ and HCO$_3^-$ results in Ca^{2+} concentrations much less than $^1/_2$ [HCO$_3^-$], and calcareous soils can have pH values as high as 9.5.

In some poorly drained areas and in semi-arid regions, where Mg accumulates, hydrated forms of MgCO$_3$ can form. Magnesium carbonate has a higher solubility than CaCO$_3$, and the pH of these soils is higher than for soils buffered by CaCO$_3$.

Calcareous soils are very highly buffered against acidification. In soils, 1 kg CaCO$_3$ can neutralize 20,000 mmol H$^+$ and hold the pH at a value > 7 until all the CaCO$_3$ is dissolved. Thus, CaCO$_3$ is a much more effective buffer than SOM or any other soil component (Table 10.2). If CaCO$_3$ is compared with SOM on the basis of buffering a decrease in pH from 8.0 to 7.0, the difference is even greater than shown in Table 10.2. This pH drop would result in the total dissolution of CaCO$_3$ but at pH 7, only about 500 mmol$_c$ kg^{-1} SOM, about 25% of the total buffer capacity, would have been neutralized. Because of the high buffer capacity of the CaCO$_3$, calcareous soils are considered to be unaffected by acid rain and it is generally considered impractical to artificially lower the pH of calcareous soils to allow for the growth of acidophilic vegetation.

10.9   Proton Consumption by Irreversible Weathering of Aluminous Minerals

The buffering reactions discussed above all are reversible and can to some degree be described using equilibrium equations. Some reactions that consume H$^+$ in soils, however, are irreversible. These reactions include the weathering of primary silicate minerals and the destruction of high activity clays in very low pH soils.

10.9.1 Weathering of Primary Silicates

Over the very long term, proton consumption by the weathering of primary minerals from parent material can partially counteract the effect of the natural acidification reactions that are discussed in Section 10.10. These reactions are irreversible because primary minerals only form at the high temperatures involved in rock forming processes. The weathering of the feldspar anorthite to form kaolinite serves as an example and consumes 2 protons for every mole of Ca mobilized:

$$CaAl_2Si_2O_8 + H_2O + 2H^+ \rightarrow Al_2Si_2O_5(OH)_4 + Ca^{2+} \qquad [10.24]$$
$$\text{anorthite} \qquad\qquad\qquad\qquad \text{kaolinite}$$

The rates of weathering reactions depend on the type of primary mineral and the surface area exposed to the soil solution (Bloom and Nater, 1991) which means that, per unit mass, silt sized particles weather much more rapidly than sand sized particles. The rates of these reactions are pH dependent with greater rates at lower pH (Bloom and Nater, 1991). In time periods less than decades, weathering of primary minerals is generally not significant in buffering pH.

10.9.2   Proton Consumption by Destruction of High Activity Clay and Allophane

When the pH$_{H2O}$ is decreased to values < 3.5, high activity clays and allophane can dissolve quite rapidly, consuming H$^+$ and releasing Al^{3+} into solution. As discussed in Section 10.6, H$^+$ clays have pH values < 3.5 and dissolve to release Al^{3+}. At pH values < 3.5, Al(OH)$_3$ does not form in soils and kaolinite will not dissolve nearly as rapidly as the high activity clays and allophane dissolve. For a Ca^{2+} montmorillonite:

$$Ca_{0.165}(Al_{1.67}Mg_{0.33})Si_4O_{10}(OH)_2 + 6H^+ + 4H_2O \rightarrow$$
$$0.165Ca^{2+} + 0.33Mg^{2+} + 1.67Al^{3+} + 4Si(OH)_4 \qquad [10.25]$$

where 5 of the 6 protons consumed in the dissolution are due to the release of Al^{3+}. The dissolution of allophane also involves the consumption of 3 moles of H^+ for every mole of Al^{3+} released. The dissolution of silicate clays likely accounts for much of the very strong buffering at pH 3.5 that was observed by Magdoff and Bartlett (1985) (Figs. 10.1, 10.4).

10.10   Determination of Buffer Capacities

The determination of soil pH buffer capacity has long been an interest to soil chemist. Many crops respond positively to the addition of lime to acid soils, but because of the differences in buffer capacity, soils of similar pH may require vastly different quantities of lime to yield the same increase in pH. Horticulturalists have also been interested in decreasing the pH of nonacid soils using alum $[KAl(SO_4)_2]$ or elemental S to allow for the vigorous growth of acidophilic plants such as azaleas, rhododendrons, and blueberries. As with liming, the quantity of acidifying agent necessary to lower soil pH by a given amount depends on the buffer capacity. More recently, the impact of acid rain on poorly buffered forest soils has raised the question of the capacity of soils to resist the long-term input of very dilute HNO_3 and H_2SO_4 (Bloom and Grigal, 1985; Ulrich, 1991).

10.10.1   Total Titratable Acidity

The total acidity in acid soils is operationally defined as consisting of two components, exchangeable (salt extractable) and residual (nonextractable) (Glossary of Soil Science Terms, 1997). Salt extractable acidity is the H^+ and Al^{3+} extractable with $1M$ KCl, while the residual acidity is the acidity titratable, but not easily exchangeable acidity (Bertsch and Bloom, 1996). In practice, the residual acidity is determined by the difference between the total acidity neutralized by raising the pH to a reference level (7.0 or 8.0) and the salt extractable acidity.

Total acidity can be determined by the titration of a soil suspension in a salt solution to a reference pH using a strong base or addition of increments of lime as shown in Fig. 10.1. This produces a plot of the pH versus the quantity of base consumed. However, this is a very slow procedure because of the slow reaction of some of the residual acidity.

A standard method for the determination of total acidity is to react a soil for several hours or overnight with a solution containing $0.5M$ $BaCl_2$ plus a triethanolamine (TEA) buffer adjusted to pH 8.0 or 8.2 (Thomas, 1982). Triethanolamine is well buffered at pH 8. The Ba^{2+} is included to displace acidity from soil components. A reference pH of 8.0 or 8.2 was chosen to represent the pH attained when a soil is limed with excess lime. From a more fundamental point of view pH 8 is a good choice because at this pH organic carboxyl acidity, as well as the Al^{3+} acidity bound to clays is neutralized. With the $BaCl_2$-TEA method the acidity is determined by the difference in the quantity of H^+ needed to titrate the $BaCl_2$-TEA solution to a pH of about 5 before and after reaction with the soil. This difference represents the quantity of acidity that reacts with the TEA buffer.

An alternative method for the calculation of total acidity is the difference between the CEC determined at pH 7.0 with NH_4OAc and the sum of exchangeable bases. Determination of CEC at pH 7.0 is a standard method used in the classification of soils (Soil Survey Staff, 1992). In this method, soil is reacted with an NH_4OAc (pH 7) solution, the excess NH_4OAc is leached from the soil, and the NH_4^+ ions bound to cation exchange sites are extracted with a salt solution and quantified (Sumner and

Miller, 1996). Subtraction of the exchangeable bases from the pH 7.0 CEC represents the total acidity referenced to pH 7.0.

10.10.2   Lime Requirement

Development of the recommendations to growers for the liming of a soil to produce maximal crop production requires a quick and easy method for determining the quantity of base needed to raise the soil pH to a value required for a given crop. Because crops differ in sensitivity to soil acidity, recommendations for liming may differ with crop. The most common methods used in the United States involves the reaction of a buffer solution with a soil and, after a period of reaction, measurement of the equilibrium pH. The Shoemaker-McLean-Pratt (SMP) method uses a mixed acetate, triethanolamine, p-nitrophenol and chromate buffer that provides for a linear buffer response to acidity in the pH range from 5 to 7.5 (McLean et al., 1977; McLean and Brown, 1984). The pH of the SMP solution is adjusted to 7.5 and the lime requirement acidity is determined by the decrease in pH after reaction with a soil. Possible equilibrium pH values versus the quantity of lime needed to raise the soil pH to a desired value have been tabulated (Sims, 1996). Other buffer methods for determination of lime requirement include the methods of Adams and Evans (1962), Woodruff (1948), and Nômmik (1983).

In areas of the world where highly weathered soils are common and the crop mix on farms is such that it is not necessary to raise the pH above a value of 6, the quantity of $1M$ KCl extractable Al^{3+} can be used to determine lime requirement (Sims, 1996). The Al^{3+} acidity is calculated as three times the molar quantity of extracted Al and the lime requirement is calculated by multiplying this acidity by a factor of 1.5 or 2.0. These procedures have been critically reviewed by Sumner (1997).

10.10.3   Buffering by Calcium Carbonate

The quantity of $CaCO_3$ in soils is generally determined by methods that measure the CO_2 evolved after adding acid or the weight loss due to CO_2 loss. For soils that have less than 2 or 3% carbonate the weight loss method cannot be used (US Salinity Lab Staff, 1954; Loeppert and Suarez, 1996). Carefully weighed soils samples are added to a carefully weighed flask containing $3M$ HCl and loosely covered with a rubber stopper (Loeppert and Suarez, 1996). The flask is swirled and uncovered periodically until all of the CO_2 is evolved and the flask is carefully reweighed. The content of carbonate is calculated assuming all of the carbonate is $CaCO_3$ with a formula weight of 100 g mol^{-1}.

Determination of the evolved CO_2 provides a more accurate and precise measure of $CaCO_3$. Several methods have been developed for the determination of evolved CO_2 (Loeppert and Suarez, 1996). Commonly used methods involve the determination of the pressure increase in a closed cell using an Hg manometer or pressure transducer CO_2 (Williams, 1949; Martin and Reeve, 1955; Skinner et al., 1959; Evangelou et al., 1984) or the gravimetric determination of CO_2 collected in a magnesium perchlorate trap (Loeppert and Suarez, 1996).

10.10.4   Determination of the Buffering of H$^+$ Inputs in Noncalcareous Soils

Routine laboratory methods have not been developed for the determination of the capacity of soils to buffer against acidification. The demand for this type of information is much less than the demand for lime requirement. However, addition of increments of acid and long-term incubation, as used by Magdoff and Bartlett (1985), can be used to determine the response of soils to acid additions (Figs. 10.1 and 10.3). An alternative method is to equilibrate soil suspensions repeatedly with dilute acid. Bloom and Grigal (1985) suspended soils in $2.5 \times 10^{-4} M$ H_2SO_4 for 24 h, decanted, and repeated 13 times. By measuring the pH of the equilibrated solutions they were able to determine the quantity of protons adsorbed and the Al^{3+} and base cation losses in each treatment. This method much more closely simulates the response to acidic deposition with the resultant loss of bases and Al^{3+}. Bloom and Grigal (1985) used their buffer plots to test a numerical model for the response of soils to acidic deposition.

10.11   Soil Acidification

Soil acidification occurs naturally during the very slow process of soil formation. With sufficient inputs of acid rain to poorly buffered soils natural acidification can be very significantly accelerated (Bloom and Grigal, 1985; Ulrich, 1991). A more rapid acidification is produced by the addition of NH_4^+-forming N fertilizers. An even more rapid acidification is produced by the oxidation of S and reduced S compounds added to soils, or deposited by natural processes in coastal wetlands where drainage results in oxidation.

10.11.1   Natural Acidification

Under conditions of good drainage and rainfall sufficient to produce leaching, H_2CO_3 and soluble organic acids can cause the very slow acidification of soils. At pH values > 6, the donation of protons by the reaction of H_2CO_3 to produce HCO_3^- (Reaction [10.13]) is the predominant acidifying reaction. As the pH is decreased to pH values much lower than the pK_a (6.35), H_2CO_3 is no longer an effective proton donor and organic acids, having lower pK_a values, are the predominant proton donors. This is a slow process that is well buffered by $CaCO_3$ dissolution in calcareous soils until the $CaCO_3$ is depleted (Section 10.8) and is counteracted to a lesser extent in noncalcareous soils by the slow weathering of primary silicate minerals (Section 10.9.1).

10.11.2   Acid Rain

The dry aerosol and wet acid deposition that makes up what is popularly called acid rain contains H_2SO_4 and HNO_3 (Grennfelt et al., 1980). Under natural conditions, these two strong acids are of minor importance in determining the acid composition of soil solutions but long-term deposition of acid rain can have negative impacts on soils (Bloom and Grigal, 1985; Ulrich, 1991). Because the concentration of acid rain is not sufficient to cause observable pH decreases over time periods of several decades, even in poorly buffered soils, soil pH is not significantly impacted in the short term. In the long term, however, in very poorly buffered soils, base cation depletion can be significant, decreasing the base saturation and the pH (Bloom and Grigal, 1985; Ulrich, 1991).

10.11.3   Acid-Producing Fertilizers

Nitrogen fertilizers that produce NH_4^+ in soils: ammonium sulfate $[(NH_4)_2SO_4]$, ammonium nitrate (NH_4NO_3), anhydrous ammonia (NH_3), and urea $[CO(NH_2)_2]$, acidify soils. In soil, bacteria oxidize NH_4^+ to NO_3^- producing acidity (Section C, Chapter 5) :

$$2NH_4^+ + 4O_2 \rightarrow 2NO_3^- + 4H^+ + 2H_2O \tag{10.26}$$

This reaction can produce sufficient acidification to require periodic addition of lime to crop land (Adams, 1984).

10.11.4   Acidification by Sulfur Compounds

Sulfuric acid produced in acid mine wastes or when the soils of coastal swamps are drained can have a large impact on soils. When pyrite (FeS_2) associated with a seam of coal or a coastal organic soil is exposed to air, bacterial oxidation results in the production of H_2SO_4 (Sanchez, 1976):

$$FeS_2 + 15/4O_2 + 7/2H_2O \rightarrow Fe(OH)_3 + 2H_2SO_4 \qquad [10.27]$$
pyrite

Under these conditions, soil pH values < 3.5 are possible. This reaction has produced large areas of acid sulfate soils in Vietnam, Malaysia, and Indonesia (Sanchez, 1976).

Acidification of nonacid soils for the growth of acid-loving (acidophilic) plants can be accomplished using elemental S. Bacterial oxidation produces an effect similar to the oxidation of pyrite:

$$S + 3/2O_2 + H_2O \rightarrow H_2SO_4 \qquad [10.28]$$

Growers using elemental S must be very careful because pH values of 3.5 are easily attainable when excess S produces excess H_2SO_4 (Fig. 10.1).

An alternative method is to add alum [$KAl(SO_4)_2$], which produces acidity according to the reaction:

$$2KAl(SO_4)_2 + 6H_2O \rightarrow 2Al(OH)_3 + 3H_2SO_4 + K_2SO_4 \qquad [10.29]$$

Acidification with alum is limited by the solubility of $Al(OH)_3$, and pH will not be decreased below about 4. This more expensive option is preferable for acidification in pots and small plots.

10.12   References

Adams F., and C.E. Evans, 1962. A rapid method for measuring of lime requirement of Red-Yellow Podzolic soils. Soil. Sci. Soc. Am. Proc. 26:355–357.

Adams, F. 1984. Crop response to lime in the southern United States. p. 211–265. In F. Adams (ed.) Soil acidity and liming. American Society of Agronomy, Madison, WI.

Bertsch, P.M., and P.R. Bloom. 1996. Aluminum. p. 517–549. In D.L. Sparks (ed.) Methods of soil analysis 3. Chemical methods. Soil Science Society of America, Madison, WI.

Bloom, P.R., and D.F. Grigal. 1985. Modeling soil response to acidic deposition in non-sulfate adsorbing soils. J. Environ. Qual. 14:481–495.

Bloom, P.R., and E.A. Nater. 1991. Kinetics of dissolution of oxide and silicate minerals. p. 151–189. In D.L. Sparks and D.L. Suarez (ed.) The kinetics of physicochemical processes in soils. Soil Science Society of America, Madison, WI.

Bloom, P.R., M.B. McBride, and B. Chadbourne. 1977. Adsorption of aluminum by smectite: I. Surface hydrolysis during Ca^{2+}– Al^{3+} exchange. Soil Sci. Soc. Am. J. 41:1068–1073.

Bloom, P.R., M.B. McBride, and R.M. Weaver. 1979. Aluminum organic matter in acid soils: Salt extractable aluminum. Soil Sci. Soc. Am. J. 43:813–815.

Borggaard, O.K. 1983. Effect of surface area and mineralogy of iron oxides on their surface charge and anion-adsorption properties. Clays Clay Miner. 31:230–232.

Clark, C.J., and M.B. McBride. 1984. Cation and anion retention by natural and synthetic allophane and imogolite. Clays Clay Miner. 32:291–299.

Coleman, N.T., and D. Craig. 1961. The spontaneous alteration of hydrogen clay. Soil Sci. 91:14–18.

Coleman, N.T., D.E. Williams, T.R. Nielson, and H. Jenny. 1951. On the validity of interpretations of potentiometrically measured soil pH. Soil. Sci. Soc. Am. Proc. 15: 106–110.

Doner, H.E., and W.E. Lynn. 1989 Carbonate, halide, sulfate, and sulfide minerals. p. 279–330. *In* Dixon, J.B. and S.B. Weed (ed.) Minerals in soil environments. Soil Science Society of America, Madison, WI.

Evangelou, V.P., L.D. Whittig, and K.K. Tanji. 1984. An automated manometric method for the quantitative determination of calcite and dolomite. Soil. Sci. Soc. Am. J. 48:1236–1239.

Fernandez, I.J., and P.A. Kosian. 1987. Soil air carbon dioxide concentrations in a New England spruce-fir forest. Soil Sci. Soc. Am. J. 51:261–263.

Glossary of Soil Science Terms. 1997. Soil Science Society of America, Madison, WI.

Grennfelt, P., C. Bengtson, and L. Skaby. 1980. An estimation of the atmospheric input of acidifying substances to a forest ecosystem. p. 29–40. *In* T. C. Hutchinson and M. Havas (ed.) Effects of acid precipitation on terrestrial ecosystems. Plenum Press, New York, NY.

Hargrove, W.L., and G. W. Thomas. 1982. Titration properties of Al-organic matter. Soil Sci. 134:216–225.

Helling, C.S., G. Chesters, and R.B. Corey. 1964. Contribution of organic matter and clay to soil cation-exchange capacity as affected by the pH of the saturating solution. Soil Sci. Soc. Am. Proc. 28:517–520.

Inskeep, W.P., and P.R. Bloom. 1986. Kinetics of calcite precipitation in the presence of water soluble organic ligands. Soil Sci. Soc. Am. J. 50:1167–1172.

Lathwell, D L., and M. Peech. 1964. Interpretation of chemical soil tests. Cornell Agric. Exp. Stn. Bull. 995.

Loeppert, R.H., and D.L. Suarez. 1996. Carbonate and gypsum. p. 437–474. *In* D.L Sparks (ed.) Methods of soil analysis. Part 3. Chemical methods. Soil Science Society of America, Madison, WI.

Magdoff, F.R., and R.J. Bartlett. 1985. Soil pH buffering revisited. Soil. Sci. Soc. Am. J. 49:145–148.

Magnusson, T. 1992. Studies of the soil atmosphere and related physical site characteristics in mineral forest soils. J. Soil Sci. 43:767–790.

Martell, A.E., and R.M. Smith. 1974. Critical stability constants. Vol. 3: Other organic ligands. Plenum, New York, NY.

Martin, S.E., and R. Reeve. 1955. A rapid manometric method for the determining soil carbonate. Soil Sci. 79:187–197.

Mason, D.D., and S.S. Obershain. 1939. A comparison of methods for the determination of soil reaction. Soil. Sci. Soc. Am. Proc. 3:129–137.

McBride, M.B. 1994. Environmental chemistry of soils. Oxford University Press, New York, NY.

McLean, E.O., J.F. Terwiler, and D.J. Eckert. 1977. Improved SMP buffer method for determination of lime requirement of acid soils. Commun. Soil Sci. Plant Anal. 8:667–675.

McLean, E.O., and J.R. Brown. 1984. Crop response to lime in midwestern United States. p. 267–304. *In* F. Adams. 1984. Soil acidity and liming. American Society of Agronomy, Madison, WI.

Nômmik, H., 1983. A modified procedure for the rapid determination of titratable acidity and lime requirements in soils. Act. Agric. Scand. 33:337–348.

Nordstrom, D.K., and H.M. May. 1996. Aqueous equilibrium data for mononuclear aluminum species. p. 30–80. *In* G. Spositio (ed.) The environmental chemistry of aluminum. Lewis Publishers, Boca Raton, FL.

Okusami, T.A., R.H. Rust, and A.S.R. Juo. 1987. Reactive characteristics of certain soils from south Nigeria. Soil. Sci. Soc. Am. J. 51:1256–1262.

Parks, G.A., and P.D. de Bruyn. 1962. The zero point of charge of oxides. J. Phys. Chem. 66:967–973.

Sanchez, P.A. 1976. Properties and management of soils in the tropics. John Wiley and Sons, New York, NY.

Sims, J.T., 1996. Lime requirement. p. 491–515. *In* D.L Sparks (ed.) Methods of soil analysis. Part 3. Chemical methods. Soil Science Society of America, Madison, WI.

Skinner, S.I.M., R.L. Halstead, and J.E. Brydon. 1959. Quantitative manometric determination of calcite and dolomite in soils and limestones. Can. J. Soil Sci. 39:197–204.

Skyllberg, U. 1994. Aluminum associated with a pH-increase in the humus layer of a boreal Haplic Podzol. Interciencia 19:356–365.

Soil Survey Staff. 1992. Soil survey laboratory methods manual. Soil Surv. Invest. Reps. 42. USDA-SCS, Washington, DC.

Sorensen, S.P.L. 1909. Enzyme studies: II. The measurement and importance of the hydrogen ion concentration in enzyme reaction. Compt. Rend. Trav. Lab. (Carlsberg). 8:1–168.

Sposito, G. 1989. The chemistry of soils. Oxford University Press, New York, NY.

Stevenson, F.J. 1994. Humus chemistry: Genesis, composition, reactions. 2nd Ed. John Wiley and Sons, New York, NY.

Stumm, W., and J.J. Morgan. 1996. Aquatic chemistry: Chemical equilibria and rates in natural waters. 3rd Ed. John Wiley and Sons, New York, NY.

Sumner, M.E. 1994. Measurement of soil pH: Problems and solutions. Commun. Soil Sci. Plant Anal. 25:859–879.

Sumner, M.E. 1997. Procedures used for diagnosis and correction of soil acidity: A critical review. In A.C. Moniz et al. (ed.) Soil-plant interactions at low pH: Sustainable agriculture and forestry production. Brazilian Soil Science Society, Campinas, SP, Brazil.

Sumner, M.E., and W.P. Miller. 1996. Cation exchange capacity and exchange coefficients. p. 1201–1229. In D.L Sparks (ed.) Methods of soil analysis. Part 3. Chemical methods. Soil Science Society of America, Madison, WI.

Thomas, G.W. 1982. Exchangeable acidity. p. 159–165. In A.L. Page (ed.) Methods of soil analysis. Part 2. Chemical and microbiological properties. American Society of Agronomy, Madison, WI.

Thomas, G.W. 1996. Soil pH and soil acidity. p. 475–490. In D.L. Sparks (ed.) Methods of soil analysis. Part 3. Chemical methods. Soil Science Society of America, Madison, WI.

Thomas, G.W., and W.L. Hargrove. 1984. the chemistry of soil acidity. p 3–56. In F. Adams (ed.) Soil acidity and liming. 2nd Ed. American Society of Agronomy, Madison, WI.

Tisdale, S.L., W.L. Nelson, and J.D. Beaton. 1985. Soil fertility and fertilizers. 4th Ed. Macmillan, New York, NY.

Turner, R.C., and W.E. Nichol. 1962. A study of the lime potential: 2. Relation between the lime potential and percent base saturation of negatively charged clays in aqueous salt suspensions . Soil Sci. 94:58–63.

US Salinity Laboratory Staff. 1954. Diagnosis and improvements of saline and alkali soils. USDA Handbook 60. US Government Printing Office, Washington, DC.

Uehara G., and G.P. Gillman. 1982. The mineralogy, chemistry and physics of tropical soils with variable charge clays. Westview Press, Boulder, CO.

Ulrich, B. 1991. An ecosystem approach to soil acidification. p. 28–79. In B. Ulrich and M.E. Sumner (ed.) Soil acidity. Springer-Verlag, Berlin, Germany.

van Raij, B., and M. Peech. 1972. Electrochemical properties of some oxisols and alfisols of the tropics. Soil Sci. Soc. Am. Proc. 36:446–451.

Wada, K. 1980. Mineralogical characteristics of Andosols. p. 87–107. In B.K.G. Theng (ed.) Soils with variable charge. New Zealand Soil Science Society, Lower Hutt, New Zealand.

Wada, K. 1989. Allophane and imogolite. p. 1051–1087. In J.B. Dixon and S.B. Weed (ed.) Minerals in soil environments. Soil Science Society of America, Madison, WI.

Walker, W.T., C.S. Cronan, and P.R. Bloom. 1990. Aluminum solubility in organic soil horizons from northern and southern forested watersheds. Soil Sci. Soc. Am. J. 54:369–374.

Williams, D.E., 1949. A rapid manometric method of the determination of carbonate in soil. Soil Sci. Soc. Am. Proc. 13:127–129.

Woodruff, C.M. 1948. Testing of lime requirement by means of a buffer solution and the glass electrode. Soil Sci. 66:53–63.

The soil was deep, it absorbed and kept the water in the loamy soil, and the water that soaked into the hills fed springs and running streams everywhere

-Plato

Soil Biology and Biochemistry

E.A. Paul
Michigan State University

C.1 The Soil Biota and its Processes

Soil organisms play a central role in soil formation, plant growth and the C cycle. Decomposition in returning CO_2 to the atmosphere effectively reverses the effects of photosynthesis, but is not 100% efficient, resulting in the production of soil organic matter (SOM) so important in nutrient release, soil physical attributes and ecosystem stability. The C cycle is normally tightly coupled. Global soil C storage is equivalent to 10 yr of photosynthesis (140×10^{15} g C yr^{-1}). Photosynthesis has in the past exceeded decomposition resulting in the present stocks of oil and gas that contain stored C equivalent to about 3 yr and coal that represents 35 yr of present day photosynthesis. Man is now acting as a decomposition agent in the use of fossil fuels and burning of forests, thus interfering with the C cycle and contributing to global climate change. The conversion of stored hydrocarbons to various toxic or recalcitrant chemicals that are new to the environment has resulted in the need to aid natural processes (bioremediation) in instances where the soil organisms are exposed to chemical structures with which they are unfamiliar.

Advances in molecular biology that make it possible to study the many organisms that do not grow on laboratory media should make it possible to better understand soil populations, and eventually, manage them for our benefit (Paul and Clark, 1996). Soil microbial populations are generally resistant to change and capable of persisting for many years. Those scientists that call soil a nutrient-starved, stressful situation do not understand the many safety factors that nature has built into soils during the billions of years of evolution. One should be most pleased that introduced organisms usually cannot persist in the highly competitive, diverse, multi-organism associations that exist within the many habitats and niches within soil. These are responsible for the soil self-cleansing that provides protection against the many plant and animal pathogens introduced to this milieu by both natural and anthropogenic means.

Ecosystem stability depends on the fact that soil C accumulates and is not easily degraded. The decomposition process results in a rapid degradation of most of the plant residues in a period of months. The remaining plant residues and associated biota, not easily separated from soil, constitute the small, active fraction of SOM that decomposes within one growing season (Paul et al., 1998). Microbial intermediates and plant residues, protected within soil aggregates, take years to decompose. These represent the slow fraction that accounts for up to half the soil C and constitutes the seat of soil fertility and ecosystem stability. The old, recalcitrant fraction can be identified with acid hydrolysis in conjunction with ^{14}C dating. It represents the other half of the soil C and is on the average 1,200 yr older than the total soil; subsurface ages reach many thousands of years (Paul et al., 1997). Understanding the complex interaction of the soil biota and organic C is most important to our understanding of ecosystem stability and sustainable agriculture. It is equally important in our attempts to manage nature (Brussaard and Ferrera-Cerrato, 1997) by introducing organisms into soil.

C.1.1   The Commercialization of Soil Organisms

Commercialization is defined as the use of soil organisms for specific processes not normally associated with their activity in that location (Alexander, 1994). This could include the use of yeast in fermentation, streptomyces in antibiotic production or soil bacilli as a source of Bt genes for insect control, but these are not carried out in soil. The most successful soil commercialization that actually occurs in soil has involved the inoculation of introduced legumes with N-fixing symbionts. Various forms of bioremediation and biocontrol constitute the major other forms of commercialization.

C.1.1.1   Bioremediation

Bioremediation, in its broadest sense, includes the age-old composting techniques as well as the more modern applications that involve as much engineering as biology. Interactions and biotic controls in bioremediation are shown in Fig. C.1. Biodegradation is more narrowly defined as the biologically catalyzed reduction in the complexity of chemicals (Alexander, 1994). This can lead to mineralization, the conversion of much of the C, N, S, P and other elements to inorganic products. Microorganisms degrade substrates in environments as diverse as soils, groundwater, surface water, the vadose zone between soils and groundwaters, sediments and geological structures as well as sewage facilities. This requires (1) the presence of an organism or organisms with the appropriate growth characteristic and enzymes, (2) accessibility of the substrate, and (3) appropriate abiotic and nutrient factors for microbial growth and activity. The degradation process usually involves an acclimation period that, if lengthy, can affect both distribution and toxicity of the chemical. Once a soil is acclimated, a second exposure can result in rapid degradation. This is desirable for detoxification, but can result in the too rapid breakdown of herbicides and insecticides before they have had time to act. Detoxification of a chemical involves changes in its structure to render it less harmful to one or more susceptible organisms such as humans, plants, other animals or the soil biota. Conversely, microbial activity can produce more toxic intermediates by the process known as activation of either organics or metals. The microbial methylation of Hg, Sn and As increases their toxicity and distibution.

Cometabolism is the transformation of an organic compound by an organism or organisms incapable of using that compound as a source of energy or cell constituent. While now being studied relative to detoxification of anthropogenics (Focht, 1993), perhaps, the most important cometabolism

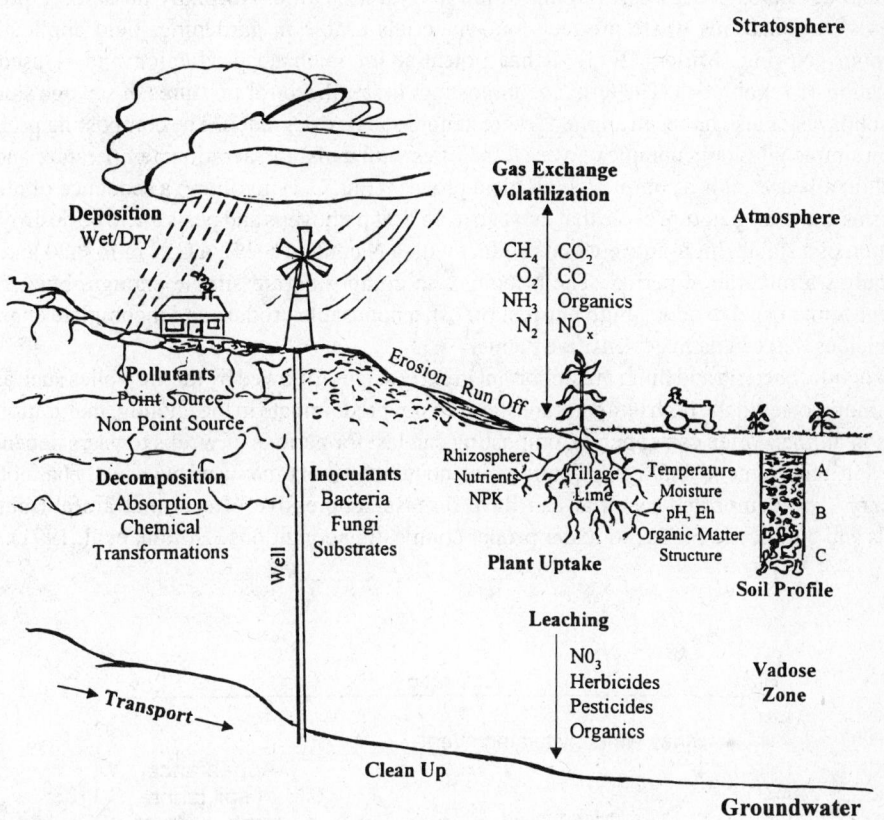

Fig. C.1 Soil factors affecting bioremediation

in nature is the degradation of the lignin molecule and soil humic constituents. Very little of the C of ^{14}C lignin added to soil enters the soil biomass; external energy supplies are required for its breakdown (Stevenson, 1994). Soil aromatics most likely behave similarly. The aromatics are decomposed to gain access to the associated carbohydrates and nitrogenous compounds.

Bioremediation in its simplest form involves the application of nutrients such as N and P to *in situ* spills and cultivation to provide aeration and mixing. Slightly more complex is land farming where the contaminant is removed and transported to another site for nutrient addition and mixing. The process is said to involve a bioreactor if it includes more complex engineering such as moisture and temperature control or the use of clay or plastic liners to control runoff and leaching. Contaminants at depth can be degraded by aeration via access wells and a vacuum system that can also collect volatiles in a bioventing system. Still more complex are systems that degrade groundwater contaminants via addition of substrate, electron acceptors and possibly added microorganisms (Bewley, 1996).

C.1.1.2 Composting

The controlled biological conversion of solid organics into a stable, humic-like substance involves two separate, but related processes, decomposition and humification. Normally an aerobic process, it decreases carbonaceous waste products into materials usable in gardening, field application and mushroom growing (Miller, 1993). It has potential for methane production and is used in the degradation of xenobiotics. The term cocomposting, first utilized for mixtures of sewage sludge and other solid wastes also has been applied where xenobiotics are degraded. The composting process can occur in simple piles or in complex covered facilities with moisture, aeration, temperature and runoff control. In all cases, it is a complex, biological process (Fig. C.2) involving a sequence of numerous organisms and a temperature cycle that must go to 65 °C if pathogens and pests are to be destroyed. The formation of a stable, high-nutrient end product with a N content > 2%, a C:N ratio < 20 and neutral pH requires a maturation period. The principles in composing are simple enough, but errors and mismanagement lead to odors, high nutrient runoff, a nonusable product and incomplete degradation of herbicides that can damage sensitive plants.

Mesophilic bacteria and fungi are important in early stages followed by thermophiles such as bacilli and actinomycetes in the high temperature stage. Fungi predominate in the cooling, maturation period when soil animals often can appear. Composting can last for periods of weeks to years depending on the mixing management employed. Commercial inoculation is at times used but usually has not proven necessary. The composted material can have disease suppressive effects that are attributable to phenols and antibiotics produced under proper composting conditions (Hoitink et al.,1991).

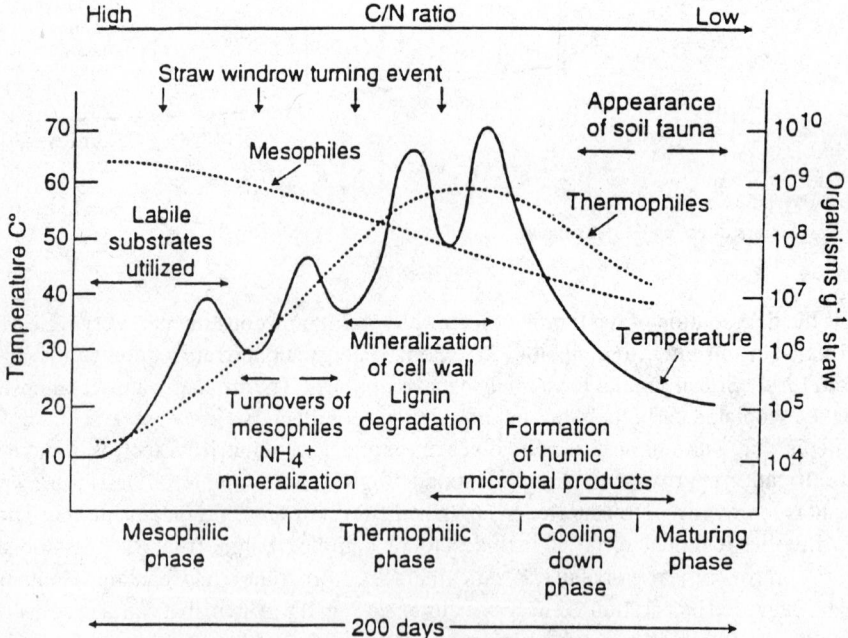

Fig. C.2 Organisms and processes in composting [Reprinted from Paul and Clark, 1996. Soil microbiology and biochemistry. Copyright Academic Press, San Diego, CA with permission]

C.1.2 Ecology of Soil Microorganisms

Microorganisms, next to living plants, constitute the largest biomass on our planet. They also carry out the greatest range of physiological processes ranging from decomposition to the many reactions in the N, S, P and other nutrient element cycles in a wide array of environments. Their small size and the difficulties in their isolation have inhibited both the study and the use of information on soil biota in the development of ecological theory (Coleman and Crossley, 1996). In turn, soil microbiologists have not been part of the mainstream in ecological thinking; concepts developed with larger plants and animals are difficult to test in the much more constricted, multihabitat, competitive soil environment (Wardle and Giller, 1996).

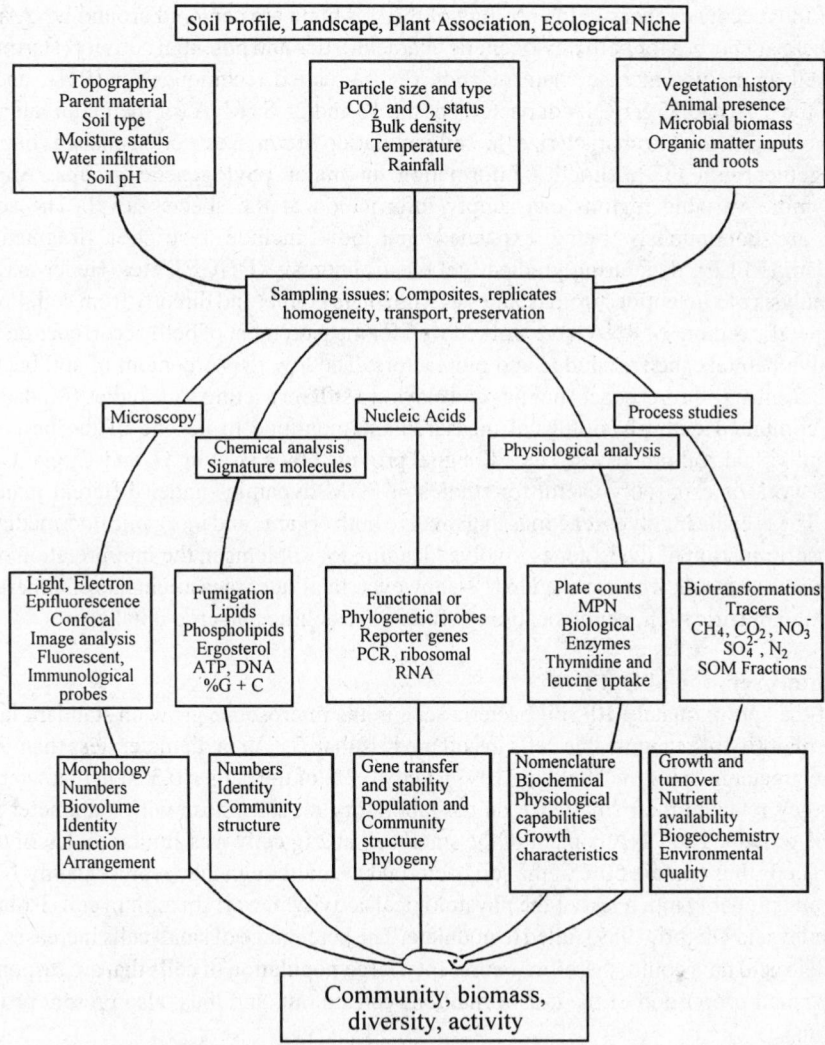

Fig. C.3 Methods for studying the soil biota

C.1.2.1   Methodology

Techniques that include improved, automated microscopy, chemical analysis of signature molecules, the study of nucleic acids (molecular biology) and tracers (Fig. C.3) are providing rapid advances in the study of soil organisms and their processes. Microscopy while slow with a requirement for expensive equipment, provides vital information on morphological features such as size and shape, some characteristics useful in identification and biovolume (Bloem et al., 1995). The small size and greater density of microorganisms in soil than in pure culture (Paul et al., 1998) are related to their adaptation to the soil habitat. Chloroform fumigation with either incubation (CFI) or extraction (CFE) has made possible extensive comparisons of biomass to activity in many habitats and management conditions (Powlson, 1994). Phospholipid, ergosterol, and lipid measurements of isolates and of the soil directly give population abundance and community structure estimates (Zelles and Alef, 1995).

Many of the recent advances in the ecology of the soil biota are centered around DNA and RNA analysis. There are probes for both phylogenetic characteristics and potential activity (Hartman et al., 1997) as well as the polymerase chain reaction (PCR)-related techniques for rRNA and rDNA. Analysis of the 5,16 and 23 S rRNA of bacteria or the 18 and 25 S rRNA (or their equivalent rDNA) of fungi have the potential to characterize the soil population *in situ*. Restriction analysis of preserved regions, together with PCR, supplies information on major phylogenetic groups. Analysis of associated, more variable regions can supply information at the species level. The separation techniques are continuously being expanded and now include restriction fragment length polymorphism (RFLP), denaturing gradient gel electrophoresis (DGGE), etc. (Heuer and Smalla, 1997) for analysis of nucleotides from isolates as from plate counts and directly from soil. Probes that react with specific regions of RNA have worked well for measurement of both occurrence and activity in more active habitats such as sludge and bioreactors. The low rRNA content of soil bacteria and interference from autofluorecsence, humics and clays are still restricting soil studies (Madsen, 1996).

Tracers combined with physiological measurements continue to be one of the best ways for following individual soil processes. The ^{13}C signal provided by a shift in C_3 and C_4 plants in many parts of the world is especially useful for studies of SOM dynamics under different management conditions. The N cycle involves reactions that involve both organic and inorganic intermediates. This results in discrimination of the isotopes involved leading to problems in the interpretation of natural abundance measurements. The use of added ^{15}N, however, facilitates measurements of mineralization, immobilization, nitrification, denitrification, N fixation, leaching and crop uptake.

C.1.2.2   Interpretation of Data

Only 1% of the approximately 10^9 soil bacteria seen in the microscope grow on standard laboratory media. The majority of recognizable cells in soil are less than 0.3 m in diameter; less than 50% have a biovolume greater than 0.1 m^3(Bakken, 1997). Only 0.2% of the cells < 0.3 m in diameter grew on plates, but they retained their small size on the laboratory media. Those with a diameter > 0.8 m showed 38% growth. The DNA content of the small cell at 2 fg cell^{-1} was similar to that of the larger cells. Dwarf cells thus contain 50% of the soil bacterial DNA although they represent only 10% of the biovolume and probably much less of the physiological activity. Dwarf rhizobium cells isolated from soil (Bottomley and Dughri, 1989) failed to nodulate. The percentage of small cells increases with soil depth. Nucleic acid data could, therefore, represent a large population of cells that are responsible for only a very small proportion of the transformations carried out, and thus, also present problems in interpretation.

The soil fungi present special problems. Plate counts represent primarily spores if special washing techniques and hyphal separations are not conducted. Their DNA is difficult to isolate; most DNA extractions of soil probably represent primarily bacterial DNA. This is related to the observation that most fungal cells have their cytoplasm concentrated in 2-5% of the hyphae at the growing tips. A comparison of fungal and bacterial DNA in one soil showed that, although the two had similar biomass, bacterial DNA was 10 times as prevalent as that of the fungi (Harris, 1994). Fungal nucleotides may thus be more closely related to function in soil than are the more easily extractable, more common bacterial nucleotides.

Diversity is an index of the number of different sequences or species in a habitat. A more useful index is that of community structure that includes quantitative information on numbers of different individuals or physiological groups. Extraction of bacterial DNA followed by denaturation/reassociation studies show 4,000 different genomes that could represent even more species in one soil (Torsvik et al., 1994). The available PCR analyses together with appropriate sequencing, RFLP and DGGE measurements (Liesack et al., 1997) show that the DNA isolated from nature only very seldom is similar to that obtained from isolated cultures. DNA representing whole new groups of organism is often found, leading to conjectures on the number of new organisms yet to be found. This has led to questions concerning the relation of soil diversity to ecological fitness and possible redundancy relative to physiological function. Microorganisms appear to follow the concepts on diversity established with larger organisms. In macroecological systems, an increase in diversity appears to enhance productivity in low nutrient environments. High nutrient environments, such as those often found in agriculture, appear to increase dominance and reduce diversity (Odum, 1998).

Ammonia oxidizers in soil represent a select group of soil bacteria of importance in N cycling and responsive to environmental stress (Prosser, 1989). Agricultural soils with high potential nitrification rates had only one genus (*Nitrosospira*). An adjacent grassland had very little nitrification capacity, but contained *Nitrosopira,* two *Nitrosomonos,* and a yet to be identified sequence (Bruns et al., 1998). One must ask, were the grassland species outcompeted in the disturbed soils by the stronger *Nitrosospira* that had a greater capacity to grow at variable NO_3 concentrations, or did the uncultivated grassland provide more microhabitats for otherwise less competitive organisms? The data indicate that, in this case, diversity was highest in the grassland site with the most competition for the available mineral N. The agricultural soil displayed dominance. The activity of keystone and cornerstone species (Wardle and Giller, 1996; OhTonen et al., 1997) must be further tested in microbial interactions within soils such as these.

The isolation of soil DNA has made possible the application of ^3H thymidine-growth incorporation to measure soil bacterial growth rates. Harris and Paul (1994) found bacterial turnover rates of 100 to 160 days. These were longer than those reported by Bakken (1997). The measured turnover rates in soil of the Harris and Paul (1994) study closely corresponded to calculation of growth potential based on rates of substrate incorporation and CO_2 evolution. They estimated 35% growth efficiency with a proportion of the available energy attributed to maintenance energy of the bacterial population.

No one approach answers all of today's questions. The best understanding comes from the application of a multiple, polyphasic approach. In addition to the techniques discussed above, this could include the use of marker genes (Liesack et al., 1997) such as the fluorescing *lux* gene, the physiological *gus* genes, antibiotic resistance, green fluorescent proteins, fatty acids or immunological techniques as well as MPN-PCR (van Elsas et al., 1997).

C.1.3 Management and Inoculation of Soil Organisms

Specific crops, rotations, cultivation, fertilizers and pH control have in the past been the primary ways to influence soil microbial activity. Dropping the pH of soils growing potatoes controls *Streptomyces* induced potato scab. Crop rotation deters the build up of both fungal pathogens and nematodes and has been said to control undesirable mycorrhyzal dominance (Johnson and Pfloger, 1992). Burning and residue incorporation are regularly used for pathogen control. Leaving residues on the surface of other sites or environments allows a build up of SOM necessary for aggregation and erosion control. The seeding of legumes in rotation or as cover crops makes it possible to selectively enhance symbiotic N fixation. Management of the soil system for agronomic or bioremediation sometimes involves inoculation of new organisms. The commercial use of genetically engineered soil biota has been slow because society is more willing to accept new genes incorporated in plants than in microorganisms where they can more readily escape to the environment. Placing a microbial gene into a plant also overcomes soil inoculation problems. Roundup-resistant crops and commercial crops containing the Bt gene from soil bacilli are now part of regular management.

The suppression of one organism by another can be of use in biocontrol of pests and pathogens (Killham, 1994). It continuously operates naturally to maintain our soils in a relatively pathogen-free situation considering all the human and plant pathogens that are routinely dumped on them. Certain soils show suppressiveness to one disease, but are conducive to another. Principles involved include those of antagonism and stimulation where one organism detracts or enhances the growth of another. The competitive inhibition of newly inoculated rhizobia in many legume systems may be a negative outcome from the viewpoint of a soil microbiologist, but not from that of the long time survival of the indigenous rhizobia.

Antibiotics are strong antagonistic agents *in vitro* and are active *in vivo*, but have not proven amenable to management in the multi-organism, multi-habitat, soil system in which many organisms have an active demand for a limiting resource. This includes not only organic substrates, but also competition for space as in the competition for root entry sites in mycorrhyzal-root infection interactions that help prevent plant diseases. Parasitism and predation are part of the normal food web of soil; soil structure (Ladd et al.,1996) plays a predominant role in protecting organisms from predation. Attempts to alter population in such an environment require a large inoculum and conditions that give a nutrient advantage and protective habitat to the introduced organism.

Inoculation has achieved its widest and most successful application with rhizobia for symbiotic N fixation in the nodulation and effective fixation by legumes where there is not extensive competition from indigenous rhizobia. The soybeans in their native habitat in China are often nodulated with ineffective strains and require fertilizer N for growth. Those brought to North America were initially inoculated with strains good enough for growth without N fertilization, but not for maximum fixation. Attempts to add more efficient fixers have largely failed because of the competitive ability of the original introductions. These persist in soil for up to 12 yr in the absence of soybeans by developing in the rhizosphere of other plants and may even have weak fixation capacities in the free-living state. The introduction of new symbionts to such situations must await the selection and engineering of plants that give the new introductions a competitive edge in the soil. Similarly, the use of inoculants in composting and bioremediation often produces results that are no different than amelioration of the soil by fertilizers, aeration, etc. and development of the indigenous flora.

The stability of the SOM is as equally important as the resistance to change and stability of the soil biota in sustainable ecosystem functioning and long-term agriculture. The active and slow pool can be increased with additional inputs and restricted cultivation. The large resistant pool that supplies physical stability must generally remain intact. The consequences must be carefully be considered before one drastically attempts to alter either the soil biota or SOM.

C.2   References

Alexander, M. 1994. Biodegradation and remediation. Academic Press, San Diego, CA.

Bakken, L.R. 1997. Culturable and non culturable bacteria in soil. p. 47-62. *In* J. van Elsas, J. Trevors, and E. Wellington (ed.) Modern soil microbiology. Marcel Dekker, New York, NY.

Bewley, R.F. 1996. Field implementation of *in situ* bioremediation: Key physiochemical and biological factors. p. 473-543 *In* J.M.Bollag and G.Stotsky (ed.) Soil biochemistry. Vol. 9. Marcel Dekker, New York, NY.

Bloem, J., P.R. Bolhuis, M.R. Veninga, and J. Weiringa. 1995. Microscopic methods for counting bacteria and fungi in soil. p. 162-192. *In* K. Alef and P. Nannipieri (ed.). Methods in applied soil microbiology and biochemistry. Academic Press, San Diego, CA.

Bottomley, P.J., and M. H. Dughri. 1989. Population size and distribution of *Rhizobium leguminosarum* bv *Trifolii* in relation to total soil bacteria and depth. Appl. Environ. Microbiol. 55:959-964.

Bruns, M.A., M.R. Fries, J. Tiedje, and E.A. Paul. 1998. Functional gene hybridization patterns of terrestrial ammonia oxidizing bacteria. Microbial Ecol. 36:293–302.

Brussaard L., and R. Ferrera-Cerrato. 1997. Soil ecology in sustainable agriculture. CRC/Lewis Publishers, Boca Raton, FL.

Coleman D.C., and D.A. Crossley. 1996. Fundamentals of soil ecology. Academic Press, San Diego, CA .

Focht, D. 1993. Microbial degradation of chlorinated byphenyls. p. 341-408. *In* J.M. Bollag and G. Stotsky (ed.) Soil biochemistry. Vol. 8. Marcel Dekker, New York. NY.

Harris, D. 1994. Analysis of DNA extracted from microbial communities. p. 111-119 *In* K. Ritz, J. Dighton, and K.E. Giller (ed.) Beyond the biomass. John Wiley and Sons, New York, NY.

Harris D., and E.A. Paul. 1994. Measurement of bacterial growth rates in soil. Appl. Soil Ecol. 1:277-290.

Hartman, A., B. Assmus, G. Kirchof, and M. Schoeter. 1997. Direct approaches for studying soil microbes. p. 279-310 *In* J. van Elsas, J. Trevors, and E. Wellington (ed.) Modern soil microbiology. Marcel Dekker, New York, NY.

Heuer, H., and K. Smalla. 1997. Application of denaturing gradient gel electrophoresis and temperature gradient gel electrophoresis for studying soil microbial communities. p. 353 374. *In* J. van Elsas, J. Trevors, and E. Wellington (ed.) Modern soil microbiology. Marcel Dekker, New York, NY.

Hoitink, H.A., Y. Imbar, and M. Boehm. 1991. Status of compost amended potting mixes naturally suppressive to soil borne diseases of floricultural crops. Plant Dis. 75: 869-873.

Johnson, N.C., and P.L. Pfloger. 1992. Mycorrhizae in sustainable agriculture. p. 45-70. *In* G.J. Bethlanfalvay and R. G. Linderman (ed.). Mycorrhizae in sustainable agriculture: proceedings of a symposium sponsored by Divisions S-3 and S-4 of the Soil. ASA Pub. 54. American Society of Agronomy, Madison, WI.

Killham, K. 1994. Soil ecology. Cambridge University Press, Cambridge, UK.

Ladd, J.M., R.C. Foster, P. Nannipieri, and J.M. Oades. 1996. Soil structure and biological activity p. 23-78. *In* J.M.Bollag and G.Stotsky (ed.) Soil biochemistry. Vol. 9. Marcel Dekker, New York, NY.

Liesack, W., P.H. Janssen, F.A. Rainey, N.L. Ward-Rainey, and E. Stackebrandt. 1997. Microbial diversity in soils: The need for a combined approach using molecular and cultural techniques. p. 375-426. *In* J.D. van Elsa, J.K. Trevors, and E.M. Wellington (ed.). Modern soil microbiology. Marcel Dekker Inc., New York, NY.

Madsen, E.L. 1996. A critical analysis of methods for determining the composition and biogeochemical activities of soil microbial communities *in situ*. p. 287-370. *In* J.M.Bollag and G.Stotsky (ed.) Soil biochemistry. Vol. 9. Marcel Dekker, New York, NY.

Miller, F.C. 1993. Composting as a process based on the control of ecologically selective factors. p. 515-544. *In* F. Metting Jr. (ed.) Soil microbial ecology. Applications in agriculture and environment management. Marcel Dekker, Inc., New York, NY.

Odum, E.P. 1998. Productivity and diversity: A two way relationship. Bull. Ecol. Soc. Amer. 79: 125.

OhTonen, R., S. Aiko, and V. Häre. 1997. Ecological theories in soil ecology. Soil Biol. Biochem. 29:1613-1619.

Paul, E.A., and F.E. Clark. 1996. Soil microbiology and biochemistry. 2nd Ed. Academic Press, Inc., San Diego, CA.

Paul, E.A., R.F. Follett, S.W. Leavitt, A. Halvorson, G.A. Peterson, and D.J. Lyon. 1997. Radiocarbon dating for determination of soil organic matter pool sizes and fluxes. Soil Sci. Soc. Amer. J. 61:1058-1067.

Paul, E.A., D. Harris, M. Klug, and R. Ruess. 1998. The determination of microbial biomass. *In* G.P. Robertson, C.S. Bledsoe, D.C. Coleman, and P. Sollins (ed.) Methods of soil analysis. Oxford University Press, New York.

Paul, E.A., H.P. Collins, and S. Haile-Mariam. 1998. Analytical determination of soil C dynamics. Trans. 16th Int. Congr. Soil Sci., Montpellier, France.

Powlson, D.S. 1994. The soil microbial miomass; before beyond and back. p. 3-20. *In* K. Ritz, J. Dighton, and K.E. Giller (ed.) Beyond the biomass. John Wiley and Sons, New York, NY.

Prosser, J.L. 1989. Autotrophic nitrification in bacteria. Adv. Microb. Physiol. 30:125-181.

Stevenson, F.J. 1994. Humus chemistry. John Wiley and Sons, New York, NY.

Torsvik, V., J. Goksoyr, F.L. Daae, R. Sorheim, J. Michalsen, and K. Salte. 1994. Use of DNA analysis to determine the diversity of microbial communities. p. 39-48. *In* K. Ritz, J. Dighton, and K.E. Diller (ed.) Beyond the biomass. Wiley Sayce, London, UK.

van Elsas, J.P., J.T. Trevors, and E.M. Wellington. 1997. Modern soil microbiology. Marcel Dekker, Inc., New York, NY.

Wardle, D.A., and K. E. Giller. 1996. The quest for a contemporary ecological dimension to soil biology. Soil Biol. Biochem. 28:1549-1554.

Zelles, L. and K. Alef. 1995. Biomakers. p. 422-438. *In* K. Alef and P. Nannipieri (ed.) Methods in applied soil microbiology and biochemistry. Academic Press, San Diego, CA.

1

Microbiota

1.1 Viruses

J.S. Angle
University of Maryland at College Park

Viruses are the most numerous of all organisms in soil. Populations may reach as high as 10^{10} g^{-1} soil. To say that viruses are living organisms, however, and thus comparable to higher organisms, is somewhat of a misnomer. Most viruses contain a core of DNA or RNA surrounded by a protein coat (Fig. 1.1). Because they contain little if any metabolic machinery, they are unable to autonomously replicate and must first invade a living host cell. The nucleic acid from the virus then directs the host to replicate new viral particles. Furthermore, once produced by the host, virions, or complete viral particles, do not grow in size. Since the only characteristic suggesting that viruses are living is their ability to maintain genetic continuity, few scientists consider viruses to be living organisms.

The overall importance of viruses is not easy to determine since so little is known about the vast majority of viruses present in the soil. Some viruses of agriculturally important bacteria (e.g., *Rhizobium* and *Bradyrhizobium*) have been shown to cause a significant decline in bacterial populations in soil to the point where crop growth is affected. However, viruses never appear to completely eliminate specific bacteria from soil (Basit et al., 1992). When the population of the target bacterial species declines to a low level, the virus no longer can maintain its population and subsequently also declines in number. The low viral population then allows for increased numbers of the bacterial host.

Viruses are also routinely used for biocontrol of insects. Since viruses are specific to individual genera and species, they can be applied to a field and infect only the desired target species of insect. The nuclear polyhidrosis viruses are most often used for biocontrol. Although still under development, viruses may one day also be used to control fungal, bacterial, nematode and even other viral pathogens with the use of specific viral antagonists.

1.1.1 Taxonomy

Viruses are classified according to the hosts they infect but are not grouped into a distinct class of organisms nor included in any of the three domains of living organisms. To a large extent, viruses are grouped into animal, plant, fungal or bacterial parasites. However, these relationships become blurred when individual species are able to infect hosts in different domains.

Bacterial viruses are also called bacteriophages or simply phages. Bacteriophages, which have been isolated for nearly every known species of soil bacteria, are classified according to their

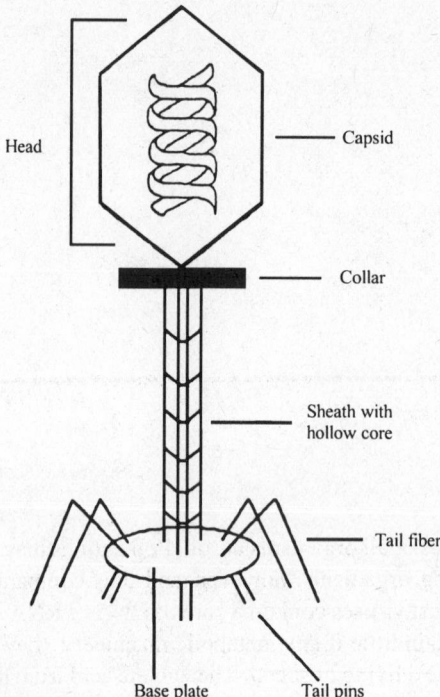

Head

Capsid

Collar

Sheath with
hollow core

Tail fiber

Base plate   Tail pins

Fig. 1.1   Structural anatomy of a typical soil virus

morphology and are found in many shapes and sizes. Fungal viruses which are called mycoviruses or mycophages, have been inadequately studied and many seem to have no apparent adverse effect on fungi. Currently, no cohesive taxonomic system exists for the plant viruses which, *per se*, are not indigenous inhabitants of the soil since they only infect living plant cells. Most plant viruses have no known economic importance, although a few are very important plant pathogens. Transmission of viruses between plants is most often via an insect vector.

Animal viruses are also not indigenous inhabitants of the soil, but are added to soil through a number of human activities (Bitton et al., 1987). The application of waste products, especially biosolids, can incorporate a very high number of human viruses, including pathogens, into soil. Polio, hepatitis A and adeno viruses are important human pathogens that have been detected in soil following application of wastes. Because of their small size, viruses have the potential to infiltrate through the soil profile and contaminate ground and other drinking water supplies. This has been an important mode of transmission in past viral epidemics.

1.1.2   Isolation and Enumeration

Viruses in soil are typically assayed using the soft agar overlay technique (Angle, 1994) which is based on the ability of a virus to attack and lyse a specific bacterial species or other host. Hence, the soft agar overlay technique enumerates viruses that attack specific bacterial species. It is not currently possible to enumerate all viruses in soil, although future refinements in immunofluorescent techniques and polymerase chain reaction (PCR) amplification of specific gene sequences may make this possible. In the soft agar overlay technique, a soil suspension is passed through a 0.45 μm filter. Bacteria and all higher organisms are retained on the filter, while only viruses pass through the membrane. An appropriate dilution of the filtrate is then mixed with a log phase culture of target bacteria and 0.5% molten agar. The mixture is then overlaid onto solidified agar and incubated for several days. During this time, each viable viral particle capable of attacking the bacteria will create a plaque, a clear zone of dead bacteria within a lawn of visible bacterial growth. The number of plaques represent the number of viruses present in the soil. Similar procedures for animal and plant viruses are available using specific cell cultures as the host organism.

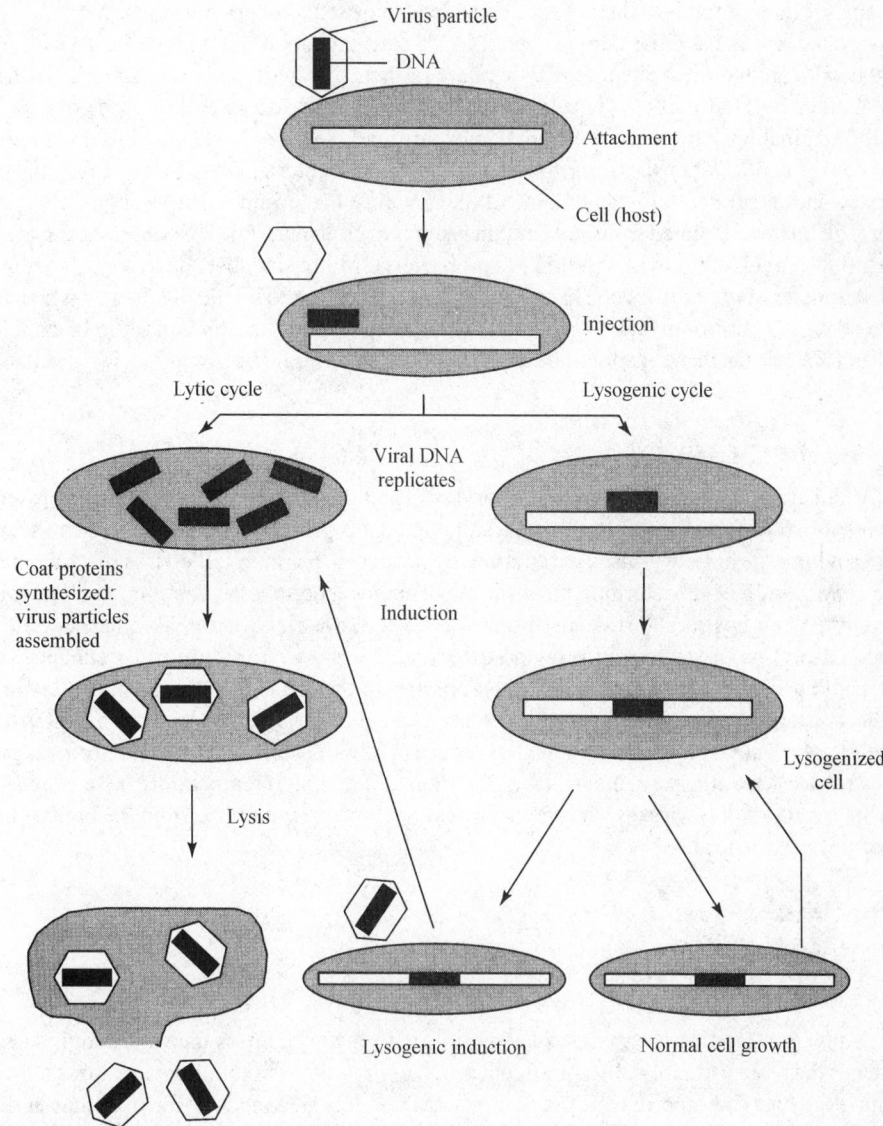

Fig. 1.2   Lytic and lysogenic viral replication

1.1.3   Reproduction

Viruses have two main types of reproduction, lysogenic and lytic (Fig 1.2). During lysogenic reproduction, the virus injects its DNA or RNA into a host cell. The viral genetic material then codes for the production of new viral particles, all of which are produced by the metabolic machinery of the host. Several hundred viruses per single host cell can be produced by this method of reproduction. In lysogenic reproduction, the viral genome does not immediately lyse the host but rather is stably integrated into the genome of the host. The viral genome may be maintained within the host for many generations. Occasionally, external stimuli (such as starvation of the host, exposure to UV radiation) cause the viral genome to be excised from the host genome and enter into a lytic phase. Viruses that are capable of integrating into the host genome are also called temperate viruses.

Temperate viruses have the unique capability of carrying small amounts of host DNA from the original host to another host during the lytic phase of their life cycle. The transfer of DNA from one host to another by viral particles is called transduction. During the assembly phase of new viruses within the original host, the new viruses may pick up small pieces of host DNA that are incorporated into the viral genome. When the virus infects a new host, the bacterial DNA from the original host can be released and integrated into the genome of the new host bacterium during lysogeny.

Nearly all bacteria isolated from soil contain lysogenic viral material. This suggests that the viruses are potentially involved in the evolution of soil bacteria. Many scientists have speculated that viral mediated transfer of genes from one bacterium to another is a powerful mechanism by which bacteria adapt to changing environments. The overall importance of transduction has yet to be conclusively proved, but it is known that transduction can occur between different bacterial species in soil (Stotzky, 1989).

1.1.4   Survival in Soil

Many factors affect the ability of viruses to survive in soil, foremost among them is the presence of a susceptible host for further propagation. If a suitable host is present in soil, the virus can survive for a long period of time as new viruses are continually produced. Further, if the virus is stably integrated into the genome of a host bacterium, the virus may become a permanent component of the microbial community. When host organisms are not present in soil, a variety of soil abiotic properties affect survival. Adsorption of viruses onto clay particles has been shown to significantly enhance survival. Tightness of adsorption is determined by the properties of the electric double layer on clay minerals, CEC, electrostatic and hydrophobic interactions, and cation bridging (Lipson and Stotzky, 1987). Adsorption of viruses onto clays and other solid surfaces protects the viruses from degradative enzymes released by other organisms. Low soil moisture and high temperature have been shown to reduce the length of time viruses survive in soil. Soil pH is also important, with the highest numbers found in soil near neutrality.

1.2   Bacteria

J.S. Angle
University of Maryland at College Park

Although fewer in number than viruses and constituting lower biomass than soil fungi, bacteria are nonetheless the most metabolically significant group of organisms in soil. Bacteria are assigned to the kingdom Protista. To some extent, the Protista are a catch-all kingdom for nonplant and animal species. Bacteria and cyanobacteria (blue green algae) all belong to what are referred to as the lower group of Protista called prokaryotes. Prokaryotes lack a true nucleus and the chromosome is found as

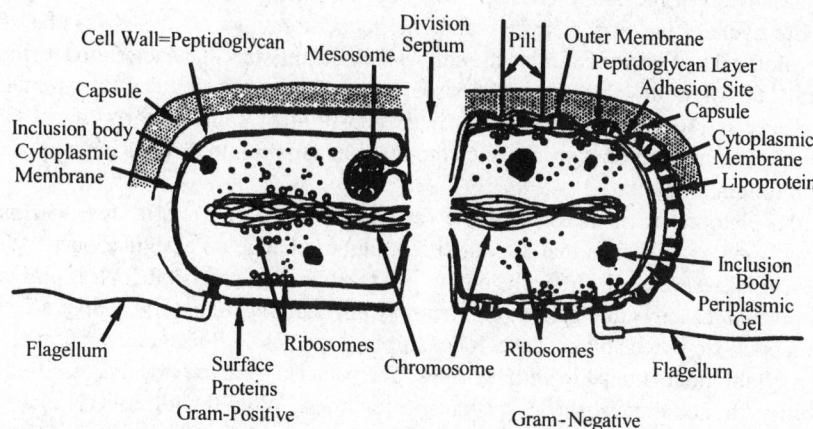

Fig. 1.3 Structural anatomy of a gram-positive and gram-negative bacterium

a closed circular loop within the cytoplasm (Fig. 1.3). Chromosomal DNA is essential to survival and replication of bacteria. However, additional smaller circular DNA molecules called plasmids that replicate independently of the chromosome may also be found. Many genes on a plasmid are cryptic; that is, they do not code for any known function. While it is recognized that cryptic genes are not essential for survival, the cryptic nature probable results from the fact that their function is yet unknown, rather than that they have no function. In addition, bacteria are separated from higher organisms because many of their organelles are not bound to a membrane. Many metabolic functions do not occur within discrete organelles. Ribosomes in bacteria are relatively small (70S [Svedberg units]) compared to higher organisms. Nearly all bacteria are surrounded by a cell wall (with the exception of the Mycoplasma) which contains peptidoglycan, a polymer not found in higher organisms.

1.2.1 Taxonomy

Given the wide diversity of bacteria found in soil and the frequency with which they exchange genetic material, it is not surprising that the taxonomy of these organisms is less than a perfect science. Early attempts at classification used morphology and function as primary discriminating criteria but more recent attempts using molecular approaches have been effective and have shown many previously unknown relationships between bacteria.

The standard for taxonomic classification of bacteria is *Bergey's Manual of Systematic Bacteriology* (1995). This multi-volume text recognizes four methods for classification of bacteria. Numerical taxonomy considers morphological, physiological and antigenic properties of an isolate and groups organisms into homogeneous clusters based upon like characteristics. A complete review of numerical taxonomy has been presented by Colwell (1973). Studying and comparing DNA from isolates has also proved useful for classification of bacteria. Genetic relatedness tends to be a more stable trait compared to morphological and physiological properties. Homology of DNA between known and unknown strains can be used to determine the level of similarity between these strains (Johnson, 1973). Isolates that are related to one another at rates greater than 60 to 70% are typically considered to be of the same species. More recently, 16S rRNA (ribosomal) has been studied and appears to be highly conserved within a species and thus has proved to be a reliable indicator for identification (Giovannoni et al., 1996). Use of extrachromosomal elements (plasmids, transposons and phages) also appears to offer some potential for classification. Although not presently used for systematic classification, the ubiquitous nature of genetic exchange suggests that this may be a normal part of the life cycle of bacteria. Serology and chemotaxonomy are also used for classification of bacteria. Serological techniques examine the cell wall characteristics of a bacterium and the ability of the cell wall to elicit an antigenic response (production of antibodies in animal species). Chemotaxonomy is the study of the chemical composition of cells or parts of cells. Cell wall composition, lipid composition, cytochrome composition and protein profiles have all proved to be highly useful for classification.

Despite the plethora of methods available for classification of bacteria, few soil bacteria are presently identified and of those that are classified, names change on a regular basis. Why has the classification of bacteria been so difficult? Numerous reasons are responsible for this observation, but the most important reason is that genetic elements within bacteria are mobile. Genes are continually flowing from one species to another. Take, for example, a species of *Rhizobium*, an organism capable of forming a symbiotic relationship with legumes. The bacteria were previously classified according to their ability to nodulate specific species of legumes. With the discovery that nodulation determinants were capable of moving from one bacterium to another, it became subjective as to what constitutes a specific species. Transduction (the viral mediated transfer of genes), transformation (uptake of naked DNA) and conjugation (plasmid transfer) have all been shown to be responsible for the movement of genes within the ecosystem. Analysis and comparison of 16S rRNA are presently considered to be the most powerful tools for classification of bacteria, although not the most commonly used method. Use of 16S rRNA has generated a phylogenetic tree of the all living organisms (Fig. 1.4).

As noted, the prokaryotes are divided into two large groups, the Archaebacteria and Eubacteria. The Eubacteria, which are further divided into 11 large groups, contain most known soil bacteria

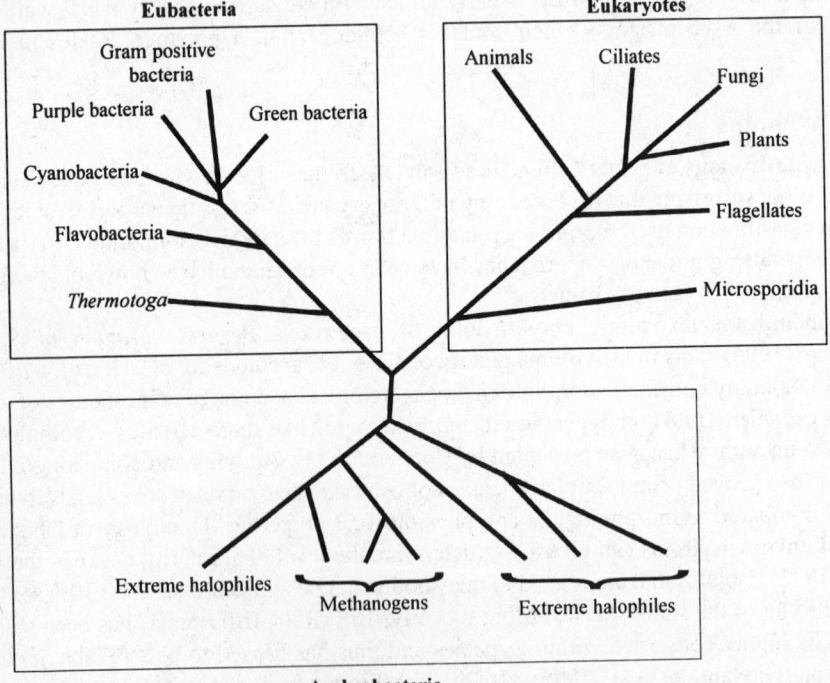

Fig. 1.4 Phylogenic classification of living organism

Table 1.1 Bacterial phyla and their subdivisions [Adapted from Paul and Clark, 1996]

Purple bacteria
- α subdivision
  Purple nonsulfur bacteria, rhizobacteria, agrobacteria, rickettsia, *Nitrobacter*
- β subdivision
  Rhodocyclus, (some) *Thiobacillus, Alcaligenes, Spirillum, Nitrosovibrio*
- γ subdivision
  Enterics, fluorescent pseudomonads, purple sulfur bacteria, *Legionella*
- δ subdivision
  Sulfur and sulfate reducers (*Desulfovibrio*), myxobacteria, bedellovibrios

Gram-positive eubacteria
- A. High-G+C species
  Actinomyces, Streptomyces, Arthrobacter, Micrococcus, Bifidobacterium
- B. Low-G+C species
  Clostridium, Peptococcus, Bacillus, mycoplasma
- C. Photosynthetic species
  Heliobacterium
- D. Species with Gram-negative walls
  Megashaera, Sporomusa

Cyanobacteria and chloroplasts
  Aphanocapsa, Oscillatoria, Noctoc, Synechococcus, Gleobacter, Prochloron

Spirochetes and relatives
- A. Spirochetes
  Spirochaeta, Treponema, Borrelia
- B. Leptospiras
  Leptospira, Leptonema

Green sulfur bacteria
  Chlorobium, Chloroherpeton

Bacteriods, Flavobacteria, and relatives
- A. Bacteroides
  Bacteroides, Fusobacterium
- B. Flavobacterium group
  Flavobacterium, Cytophaga, Saprospira, Flexibacter

Planctomyces and relatives
- A. Planctomyces group
  Planctomyces, Pasteuria
- B. Thermophiles
  Isocystis pallida

Chalmydiae
  Chlamydia psittaci, C. trachomatis

Radioresistant micrococci and relatives
- A. Deinococcus group
  Deinococcus radiodurans
- B. Thermophiles
  Thermus aquaticas

Green nonsulfur bacteria relatives
- A. Chloroflexus group
  Chloroflexus, Herpetosiphon
- B. Thermomicrobium group
  Thermomicrobium roseum

(Table 1.1). The Archaebacteria have been studied as ecological curiosities for many years, but have only been classified within this group for several years. The Archaebacteria have also been referred to as extremeophiles, a term that describes their proliferation in extreme environments. The Archaebacteria are thought to be remnants from organisms that survived from the earliest times when conditions were much more extreme than today. Archaebacteria are divided into three broad groups: methanogenic, halophilic and thermo acidophilic bacteria.

To some extent, the continuum of phenotypic traits and ability for modification of the genome make both traditional and molecular approaches to classification somewhat artificial. Many scientists prefer to describe bacteria not by their binomial taxonomy but rather by describing their physiological characteristics within the environment. Carbon and energy sources are two of the most important criteria for describing bacteria. Carbon is used for energy and new cellular material. Organisms that obtain energy from photosynthesis are referred to as phototrophic while those that obtain energy from the oxidation of existing chemicals are called chemotrophic. When C is obtained from preformed carbonaceous materials, the organism is called heterotrophic and those that obtain C from the atmosphere as CO_2 are autotrophic (lithotrophic). Various combinations of energy and C utilization are used to describe nearly all living organisms. Photoautotrophic organisms include all higher plants, cyanobacteria and purple and green sulfur bacteria. Photoheterotrophic bacteria are relatively rare and include members of the nonpurple sulfur bacteria; this group is relatively insignificant in the soil ecosystem. Chemoautotrophic bacteria use diverse energy sources, including NH_4, NO_3 and SO_4. While their numbers in soil are few, the metabolic processes they mediate are extremely important in the soil ecosystem. Many of these organisms are found in environments such as groundwater, where preformed C sources are lacking (Stevens and McKinley, 1995). Finally, the great majority of soil bacteria studied to date are chemoheterotrophic. Most higher organisms, excluding plants, but including fungi and animals, are also found within this group.

Since most soil bacteria require preformed organic C sources for growth, the C requirement can also be used to describe an organism. Oligotrophy, which is used to describe organisms that can live in environments with very low nutrient concentrations, is important within the soil ecosystem which is often nutrient limited. Environments such as groundwater are even more C limited. Copiotrophic bacteria, on the other hand, require higher concentrations of carbon in the environment for growth. Typically, these organisms cannot compete with the oligotrophs at low nutrient concentrations, but grow very rapidly when ample supplies of C are available.

Soil bacteria can also be characterized and described by their tolerance of temperature extremes. Mesophilic bacteria have an optimum temperature requirement of 30 °C but can survive within the range of 15 to 45 °C. Psychrophilic bacteria prefer lower temperatures and have been found to be metabolically active even below the freezing temperature of water. Thermophilic bacteria which live at temperatures ranging from 45 to 75 °C, are more common in soils exposed to intense solar radiation and in warm environments such as compost piles where they are responsible for the production of the intense heat generated. A few extreme theromphiles have been isolated from volcanic environments, living at temperatures > 750 °C.

The requirement for atmospheric O_2 has also been used to describe bacteria. Aerobic organisms require molecular O_2 for growth and reproduction where anaerobic organisms cannot tolerate O_2, because even low concentrations are thought to oxidize and inactivate O_2-sensitive enzymes with high sulphydryl contents. Most organisms in soil are facultative anaerobes. They can live either with or without molecular O_2, which is an important attribute for survival since the concentration of O_2 in soil is constantly changing.

Huq and Colwell (1996) reported that, in the 1970s, *Vibrio cholerae* isolated from water were viable and capable of causing disease but were nonculturable on agar media. This phenomena was termed viable but nonculturable. Since that time, it has been reported than many soil bacteria also exist in the viable but nonculturable state (Manahan and Steck, 1997). The exact reason for this state is unknown but it is believed to be conducive to survival when conditions are unfavorable for growth and reproduction. Further, it is generally recognized that most bacteria in soil cannot be cultured on artificial media since media are lacking in essential elements required for growth. No universal medium presently exists that can support the growth of all soil bacteria. Best estimates are that only 1–10% of the total bacterial population in soil is culturable. This estimate is based upon a comparison of numbers estimated by direct microscopy of soil and numbers calculated by direct plating on artificial media. With < 10% of the total bacterial population in soil having been studied, soil biology is considered to still be in its infancy. As molecular techniques allow for a more complete assessment of bacterial populations in soil, our understanding of the significance of bacteria in soil will rapidly expand. Current best estimates of the number of bacteria in soil suggest that there may be as many as 10^9 bacteria g^{-1} soil comprising > 20,000 species.

1.2.2 Isolation and Enumeration

Numerous methods for isolation of bacteria from soil have been reported (Parkinson and Coleman, 1991). Unfortunately, because each method has many limitations, there is no standard for isolation. Traditional methods of isolation and enumeration involve the culture of bacteria on media capable of supporting bacterial growth. However, since no single medium can support the growth of all, or even a fraction of all soil bacteria, results from this method must be interpreted with caution. Zuberer (1994) has discussed in detail most of the media currently used for enumeration of soil bacteria. Typically, a small sample of soil is throughly mixed with a diluent, further diluted to approximately 2,000 CFU (colony forming units) mL^{-1}, then a small aliquot of the sample is placed onto agar media. The agar media can be selective for specific bacteria or nonselective to support a wide range of bacteria. The advantage of this procedure is that individual bacteria can be isolated and studied further. The most probable number technique (MPN), which is a direct culture procedure, is used when it is not possible or appropriate to culture bacteria on agar media. In this technique, the highest dilution that supports the growth of bacteria in replicate inocula is used to estimate the population in the soil being enumerated (Woomer, 1994). Statistical tables or computer programs allow for the calculation of the best estimate of the population. The theoretical aspects of viable counts have been summarized by Hattori (1988).

The problem of noncultureability can be overcome with the use of molecular methods for enumeration of soil microbes. Because these techniques cannot be used for isolation of soil bacteria, their potential use is restricted. It is possible to extract bacteria from soil then extract DNA from this group of bacteria. Alternatively, bacteria can be lysed directly in soil and DNA subsequently extracted from the soil. The DNA can then be increased in concentration using the PCR technique. The amplified DNA can then be examined to estimate the total number of species in soil. Alternatively, by probing for gene sequences specific to an individual species or group of organisms, and examining dilutions of the DNA extract, it is possible to estimate the actual number of these bacteria in soil. PCR-MPN is also a powerful tool that shows that the genes for specific reactions are more common than can be determined by normal MPN techniques.

1.2.3 Occurrence in Soil

Numbers of bacteria in soil are not evenly distributed within the ecosystem with the highest numbers being found just below the soil surface. Solar radiation, which extends several mm into the soil, reduces the number of bacteria at the surface. Bacterial numbers increase rapidly below this zone, and then slowly decline deeper in the soil. The subsequent decline with depth is related to the lack of organic C associated with increasing depth. Higher numbers of bacteria in soil are also associated with the rhizosphere (portion of the soil that is influenced by the root). The root affects the nutrient and moisture contents and toxicant balance of the soil. This zone may extend several mm out from the root (Angle et al, 1996). Typical populations within the rhizosphere are often 10- to 100-fold higher than in nonrhizosphere soil. The rhizosphere population is composed of a higher proportion of gram negative bacteria compared to the nonrhizosphere population because they are able to grow quickly and rapidly in response to the presence of organic materials released by the root. The rhizosphere bacteria provide benefits to the plant such as buffering against invasion by pathogenic bacteria, solubilization of inorganic nutrients and degradation of toxic plant wastes.

Although the soil contains a wide diversity of bacteria, a few species often predominate in typical agricultural soils, each of which possesses a specific ecological character that allows this dominance. The genus *Arthrobacter* is gram positive and may be the most common of all (culturable) bacteria in soil. *Arthrobacter* is catabolically diverse because it can use a wide variety of organic C sources and is able to store glycogen. Glycogen storage is important because it is an energy storage material that is critical to survival during times of starvation. *Streptomyces* and *Bacillus* spp. also predominate in many soils due to their abilities to produce large numbers of spores and to survive under a wide range of environmental extremes. *Bacillus* sp. have been reported to survive temperatures from < 0–$80\ ^\circ C$, soil pHs from 2–9 and a very high soluble salt content. The nutritional flexibility of the genus *Pseudomonas* allows this organism to use many different and complex sources of organic C which probably accounts for its predominance in soil. *Pseudomonas* spp., which degrade many toxic organic materials, is the most useful in bioremediation.

1.2.4   Survival in Soil

The single most important factor affecting survival of bacteria in soil is the concentration of nutrients, especially C. Because C concentrations in soil are generally low, the soil is an environment where the organisms are most often in a state of starvation. During this period when C is lacking, bacteria are in a resting state or state of low metabolic activity which is why the ability to produce spores or enter a viable but nonculturable state is critical for the long-term success of an organism. When nutrients become available, for example, following the incorporation of organic residues, bacteria respond with rapid growth and reproduction. Once the supply of available C is exhausted, the bacteria again return to a state of reduced activity. Nitrogen, P and several micronutrients have also been shown to limit bacterial growth in soil, but the effects are much less dramatic compared to limitation in the C supply.

Soil pH also affects the number and distribution of bacteria in soil. The normal range for maximum numbers of bacteria in soil is from pH 4 to 10 with the optimum being ~7.2. Above pH 10, micronutrients become limiting as a result of immobilization and precipitation. At pH < 4, excess Al^{+3} and Mn^{+2} become toxic. Nevertheless, at pHs above or below the normal range, bacteria can still be found. For example, *Thiobacillus ferrooxidans*, which can grow in an environment where the pH ~1, is partially responsible for acid mine drainage. On the other hand, a number of species of *Streptomyces* can live at soil pH > 10, while a number of Archaebacteria (halophiles) can tolerate pH ≤ 13.

Soil moisture is critical for survival of bacteria in soil. The highest number of bacteria in soils are found at a matric potential of ~0.01 MPa. Both bacterial number and bacterial activity decrease as the soil moisture content increases or decreases from the optimum. At low soil moisture contents, bacterial numbers and activity are limited as a result of desiccation and accumulation of toxic products within the bacterial niche. At higher moisture contents, a reduction in gas exchange with the atmosphere causes an increase in CO_2 and concurrent reduction in O_2 concentrations. Most organisms in soil are capable of tolerating nonoptimal matric potentials, an important attribute since the moisture content of the soil is rarely optimum. Most soil bacteria have developed mechanisms for surviving short periods when conditions are far from optimal.

Soil moisture is directly related to the O_2 content of soil. Although most soil bacteria are facultative anaerobic organisms, a reduction in soil O_2 results in reduced numbers and activity. Oxygen in soil can be reduced either by restricted exchange with the atmosphere (due to a high soil moisture content) or rapid depletion due to degradation and respiration of organic C in soil. The O_2 content of the soil atmosphere averages 20.5%, slightly lower than that found in the atmosphere. Bacterial numbers are not affected until the O_2 concentration is < 10% and all aerobic activity ceases at concentrations < 1%.

Because soil temperature changes rapidly, both diurnally and seasonally, bacteria must be able to tolerate a wide range of temperatures in order to survive. As previously noted, bacteria have been described that can live between −10 and 110 °C. Organisms that can grow near freezing contain lipids that prevent the formation of ice crystals that would otherwise structurally damage the cell. The ability to survive at low temperature has been explored as a mechanism for the prevention of freezing of plants. Aside from the Archaebacteria that can grow at exceptionally high temperatures, most soil bacteria tolerant of high temperature exhibit the ability to produce heat resistant spores during times of temperature stress. Growth and reproduction are much more sensitive to the adverse effects of high temperature.

A number of recent studies have examined the influence of toxicants on soil bacterial populations. Heavy metals and toxic organics have both been shown to reduce the number of bacteria in highly contaminated soil. For example, Beyer (1988) showed that soil contaminated with Zn and Cd has a 75% lower number of bacteria than similar soil without contaminants. Contaminants in soil can also affect the genetic characteristics of bacteria in soil. Selection due to heavy metal or organic contaminants can cause the population to change over time. Hirsch et al. (1993) demonstrated that *Rhizobium leguminosarum* bv. *trifolii* lost its ability to fix atmospheric N_2 following long-term exposure to metals in soil. Whether or not this change was caused by selection for an ineffective subpopulation or actual mutation in normal cells remains unknown. The end result, however, was that growth of white clover on affected soils was much poorer than on noncontaminated soils. Chaudri et al. (1992) found that *Rhizobium leguminosarum* isolated from sewage-treated soil lost its ability to to fix N_2, while that isolated from farmyard manure-treated plots lost its ability to survive in metal contaminated soils because of the lack of tolerance to heavy metals.

In conclusion, understanding the role of bacteria and viruses in soil is in its infancy. The diversity of the population combined with unequal distribution makes this group of soil organisms very difficult to study. Future studies of soil bacteria and viruses will take two directions. First, specific organisms that may be adapted for useful agronomic and environmental purposes will be studied. Bioremediation, biocontrol of diseases, insects and weeds, and N_2 fixation may all one day be enhanced through the application of bacteria and viruses to soil. The second avenue of study is to examine community structure in order to understand how organisms are related to one another. Basic ecological studies, while long term, may elucidate mechanisms that allow us to better understand the complex interrelationships between all living organisms.

1.2.5 References

Angle, J.S. 1994. Viruses. p. 109–115. *In* R. Weaver, J.S. Angle, and P. Bottomley (ed.) Methods of soil analysis: Microbiological and biochemical properties. Soil Science Society of America, Madison, WI.

Angle, J.S. , J.V. Gagliardi, M.S. McIntosh, and M.A. Levin. 1996. Enumeration and expression of bacterial counts in the rhizosphere. Soil Biochem. 9:233–251.

Basit, H.A., J.S. Angle, S. Salem, and E.M. Gewaily. 1992. Phage coating of soybean seed reduces nodulation by indigenous soil bradyrhizobia. Can. J. Microbiol. 38:1264–1269.

Bergey's Manual of Systematic Bacteriology. 1995. Williams and Wilkins Co, Baltimore, MD.

Beyer, W.N. 1988. Damage to the forest ecosystem on Blue Mountain from zinc smelting. Trace Substances. Environ. Health. 12:249–262

Bitton, G., J.E. Maruniak, and F.W. Zettler. 1987. Virus survival in natural ecosystems. p. 301–332. *In* Y. Henis (ed.) Survival and dormancy of microorganisms. John Wiley and Sons, New York, NY.

Chaudri, A.M., S.P. McGrath, and K.E. Giller. 1992. Metal tolerance of isolates of *Rhizobium leguminosarum* biovar *trifolii* from soil contaminated by past applications of sewage sludge. Soil Biol. Biochem. 24:83–88.

Colwell, R.R. 1973. Genetic and phonetic classification of bacteria. Adv. Appl. Bacteriol. 104:410–433.

Giovannoni, S.J., M.S. Rappe, D. Gordon, E. Urbach, M. Suzuki, and K.G. Field. 1996. Ribosomal RNA and the evolution of bacterial diversity. p. 63–85. *In* D. McL. Roberts, P. Sharp, G. Alderson, and M. Collins (ed.) Evolution of microbial life. Cambridge University Press, Cambridge, UK.

Hattori, T. 1988. The viable count: Quantitative and environmental aspects. Springer-Verlag, Berlin, Germany.

Hirsch, P.R., M.J. Jones, S.P. McGrath, and K.E. Giller. 1993. Heavy metal from past applications of sewage sludge decrease the genetic diversity of *Rhizobium leguminosarum* biovar *trifolii* populations. Soil Biol. Biochem. 25:1485–1490.

Huq, A., and R.R. Colwell. 1996. A microbiological paradox: Viable but nonculturable bacteria with special reference to *Vibrio cholerae*. J. Food Protect. 59:96–101.

Johnson, J.L. 1973. Use of nucleic acid homologies in the taxonomy of anaerobic bacteria. Int. J. Syst. Bact. 23:308:315.

Lipson, S.M., and G. Stotzky. 1987. Interactions between viruses and clay minerals. p. 197–230. *In* V.C. Rao and J.L. Melnick (ed.). Human viruses in sediments, sludges and soils. CRC Press, Boca Raton, FL.

Manahan, S.H., and T.R. Steck. 1997. The viable but nonculturable state in *Agrobacterium tumefaciens* and *Rhizobium meliloti*. FEMS Microb. Ecol. 22:29–37.

Parkinson, D., and D.C. Coleman, 1991. Methods for assessing soil microbial populations, activity and biomass. Agric. Ecosyst. Environ. 34:3–33.

Paul, E.A., and F.A. Clark. 1996. Soil microbiology and biochemistry. Academic Press, San Diego, CA.

Stevens, T.O., and J.P. McKinley. 1995. Lithautotrophic microbial ecosystems in deep basalt aquifers. Science 270:450–454.

Stotzky, G. 1989. Gene transfer among bacteria in soil. p. 165–222. *In* S.B. Levy and R.V. Miller (ed.) Gene transfer in the environment. McGraw Hill, New York, NY.

Woomer, P.L. 1994. Most probable number counts. p. 59–79. *In* R.W. Weaver, J.S. Angle, and P.S. Bottomley. Methods of soil analysis. Part 2. Microbiological and biochemical properties. Soil Science Society of America, Madison, WI.

Zuberer, D.A. 1994. Recovery and enumeration of viable bacteria. p. 119–142. *In* R.W.Weaver, J.S. Angle, and P.S. Bottomley (ed.). Methods of soil analysis. Part 2. Microbiological and biochemical properties. Soil Science Society of America, Madison, WI.

1.3　Soil Fungi

R. Greg Thorn
University of Wyoming

1.3.1　Introduction

Soil by definition includes the L, F, H, and O organic horizons and the mineral horizons from A to C (Bridges, 1978; Fanning and Balluff-Fanning, 1989), and could be expanded to include the aerial soil that forms on the surfaces of trees that are covered with mosses, lichens, and vascular plant epiphytes in humid regions (Nadkarni, 1984; Ingram and Nadkarni, 1993). The activities of plants and soil organisms make a soil what it is. The dominant soil organisms, both in terms of processes and biomass, are the fungi. It is difficult to overemphasize the importance of fungi in soil. In many soils, the

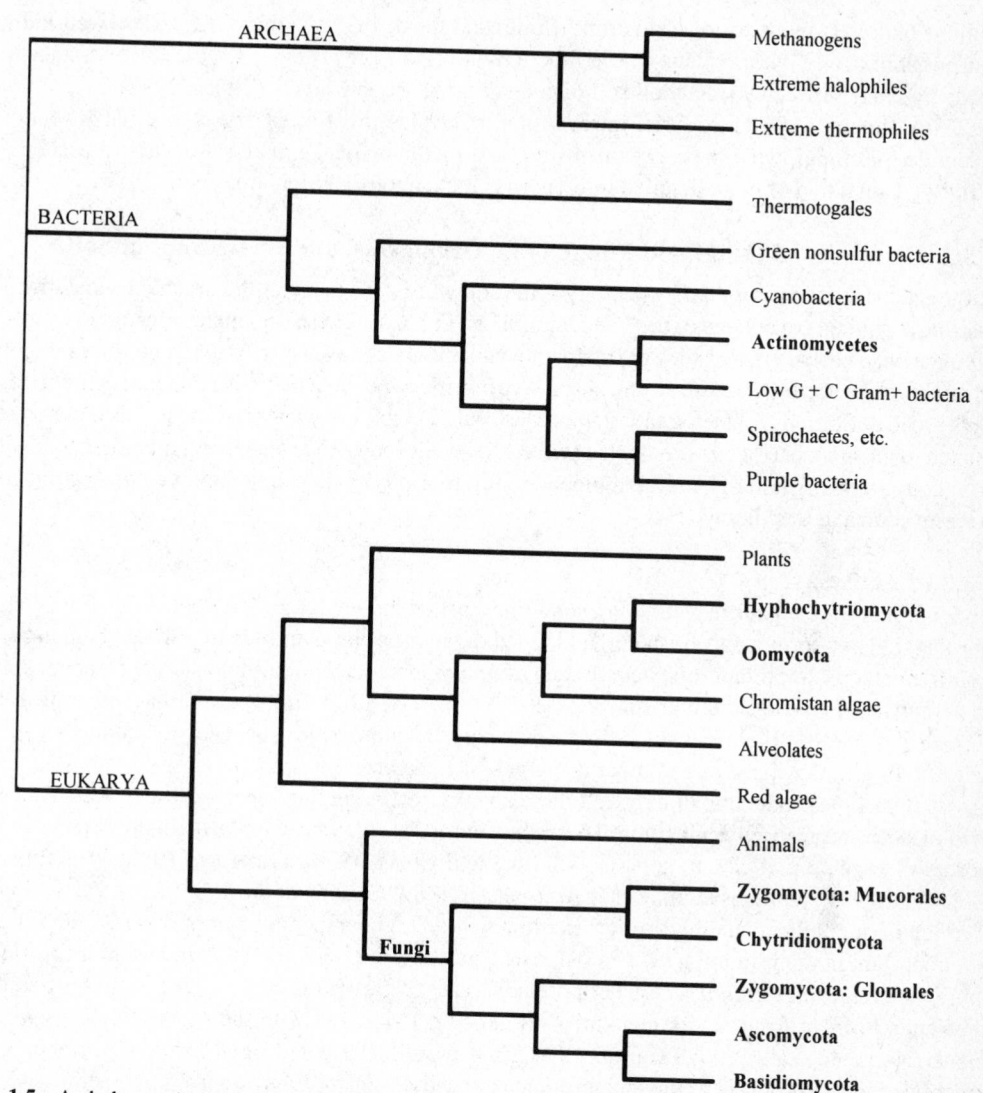

Fig. 1.5 A phylogenetic tree indicating the place of various groups of fungi and fungus-like organisms (indicated in bold font), based on Olsen et al. (1994) and Van de Peer and De Wachter (1997). Many groups of organisms are left out, branch lengths are not proportional to sequence divergence, and the tree is not rooted. The relationships of the three Domains: Archaea, Bacteria and Eukarya (Wheelis et al., 1992) are not resolved.

biomass of fungi exceeds that of all other soil organisms combined (excluding plant roots) by a factor of 10 to 1. In soil, as elsewhere, the dominant and best known role of fungi is the decomposition and mineralization of complex, recalcitrant compounds of plant and animal origins, such as cellulose, hemicellulose, lignin, and chitin (Cooke and Rayner, 1984; Rayner and Boddy, 1988). Other major roles of soil fungi include their involvements in beneficial and detrimental symbioses with plant roots. These range from the mutually beneficial mycorrhizae that enable land plants to exist in the face of nutrient and water limitations and other stresses to the plant pathogens that annually cause billions of dollars in losses to world crops (Smith and Read, 1997; Agrios, 1997). Less well-known, but possibly of great importance in the functioning of soil ecosystems, are the many connections made by fungi between virtually all soil organisms, at all trophic levels (Thorn, 1997).

This chapter will provide a brief introduction to the classification of organisms that are fungi or resemble true fungi, what these organisms do, and methods for the analysis of their presence and activity. A selection of more detailed references is included for particular aspects.

1.3.2   Diversity of Fungi and Fungus like Organisms, and their Roles in Soil

Although the strict definition of mycology is the study of fungi, many of the organisms that mycologists have studied are outside of the Kingdom Fungi (Fig. 1.5). These organisms are united by a suite of fungus-like characteristics: they are nonphotosynthetic chemoheterotrophs that reproduce by spores. Many have extracellular digestion and absorptive nutrition and most have a filamentous growth form. This broad definition includes some filamentous bacteria (actinomycetes), slime molds more closely related to the amoebae, water molds that prove to be achlorophyllous algae, and true members of the Kingdom Fungi. True fungi can be divided among four phyla: the Chytridiomycota, Zygomycota, Ascomycota, and Basidiomycota.

1.3.2.1   Actinomycetes

These prokaryotic organisms, although sometimes called the ray fungi, are Bacteria (= Eubacteria), not Fungi. They form a part of the high G-C subdivision of the Gram-positive lineage, along with non-filamentous, important soil bacteria such as *Arthrobacter* and human and animal pathogens such as *Mycobacterium* (Stackebrandt and Woese, 1981; Woese, 1987; Embley and Stackebrandt, 1994; Eppard et al., 1996; Rheims et al., 1996). As the name actinomycetes suggests (the ending -mycetes refers to fungi), they possess a number of fungus-like characteristics, including the hyphal (tubular) growth form, the production of extracellular enzymes (including lignin peroxidases) (Ramachandra et al., 1988), and absorptive nutrition. These characteristics combine to make them important decomposers of organic particles in soil (Goodfellow and Cross, 1974; Crawford, 1978; Williams and Robinson, 1981; Williams et al., 1984; Ames et al., 1989; Godden et al., 1992; Li and Gao, 1996). Certain thermophilic actinomycetes are predominant in high temperature composts (40–60 ºC), and are important in the formation of the composted substrate for commercial cultivation of the button mushroom, *Agaricus brunnescens* (Fermor and Grant, 1985; Iiyama et al., 1994). Actinomycetes in the genus *Frankia* form N -fixing symbioses called actinorhizae with the roots of various woody plants (Normand et al., 1996; Swensen, 1996; Huss-Danell, 1997). Nitrogen fixation by true fungi or other eukaryotes has not been convincingly demonstrated (Postgate, 1988). Saprotrophic actinomycetes are abundant in soil (Williams et al., 1984) and produce the compound geosmin, which is responsible for the characteristic odor of freshly tilled soil (Gerber and Lechevalier, 1965; Gerber, 1968). Actinomycetes, particularly *Streptomyces* species, are commonly recovered during the isolation of soil fungi, but are readily recognized by their tufted growth form and extremely fine (≤ 0.5 μm) hyphae. Media for the selective isolation of actinomycetes frequently make use of the antifungal antibiotic cycloheximide (actidione) and antibacterial antibiotics, and may contain colloidal chitin as the sole C and N source (Goodfellow and Cross, 1974). There is no realistic estimate available of the number of species of actinomycetes inhabiting soil (Rheims et al., 1996). However, over 400 described species of the important genera *Actinomyces*, *Frankia*, *Geodermatophilus*, *Streptomyces*, *Streptosporangium*, *Streptoverticillium*, and *Thermoactinomyces* are accepted in the Approved Lists of Bacterial Names (Skerman et al., 1989).

1.3.2.2 Protistan Fungi or Mycetozoa

The fungus-like protists include the slime molds and their relatives (myxomycetes, acrasids, dictyostelids, and plasmodiophoromycetes) (Alexopoulos et al., 1996). The informal name mycetozoa (fungus animals) refers to their mix of fungus-like and animal-like characters: reproduction by spores and ingestion of food by phagocytosis. The phylogeny of these organisms is not well-known enough to assign them formal taxonomic ranks (Baldauf and Doolittle, 1997). The acrasids, dictyostelids and myxomycetes occur in soil as amoebae, and may be abundant in the root rhizosphere (Amewowor and Madelin, 1991). Dictyostelids and myxomycetes also form motile, multicellular pseudoplasmodia or multinucleate plasmodia (slime stages) that continue to feed on bacteria, other microbes, and small organic particles; those of certain myxomycetes can be conspicuous on lawns and garden beds of woodchip mulch (Martin and Alexopoulos, 1969; Martin et al., 1983; Raper, 1984). The plasmodiophoromycetes include important root pathogens of cultivated crucifers (cabbage and cole crops) and potatoes, and may also serve as vectors of plant-pathogenic viruses (Karling, 1968; Buczacki, 1983). Owing to their abundance in soils and their lack of cell walls during their trophic phase, the protistan fungi may contribute a substantial proportion of the eukaryotic DNA recovered during lysis procedures for the isolation of bacterial DNA from soil. This group includes just over 800 described species, many of which live in habitats other than soil, such as rotting wood, bark, and dung (Hawksworth et al., 1996).

1.3.2.3 Stramenopile Fungi

Fungus-like members of the Kingdom Stramenopila include the Oomycota and Hyphochytriomycota, which have often gone by the name of water molds (Fuller and Jaworski, 1987; Alexopoulos et al., 1996). The molecular phylogeny of these organisms has been the subject of much recent research, unfortunately based on a limited sample of taxa (Van der Auwera et al., 1995; Kumar and Rzhetsky, 1996; Sogin et al., 1996; Van de Peer et al., 1996; Van de Peer and De Wachter, 1997). Like the actinomycetes, the Oomycota have a number of fungus-like characteristics, including the hyphal growth form, the production of extracellular enzymes, and absorptive nutrition (Alexopoulos et al., 1996). Unlike most true fungi, members of both the Oomycota and Hyphochytriomycota have motile stages known as zoospores. These are powered by two different types of flagella, the whiplash and the tinsel flagellum, in common with the motile stages of chromistan algae (the diatoms and the golden and brown algae). The Hyphochytriomycota are a small group known primarily as parasites of Oomycota; the Oomycota are a fairly large group that include the important soil-borne plant pathogens *Pythium* and *Phytophthora* (Buczacki, 1983; Erwin et al., 1983; Martin, 1992). There are approximately 760 described species of stramenopile fungi of which more than half occur in freshwater and marine habitats (Hawksworth et al., 1996).

1.3.2.4 Kingdom Fungi, Phylum Chytridiomycota

Soil chytrids are a relatively small group, including plant pathogens such as *Synchytrium* and *Olpidium*, nematode parasites such as *Catenaria*, parasites of algae such as *Chytridium*, and saprobes such as *Allomyces* and *Chytriomyces* (Barron, 1977; Karling, 1977; Fuller and Jaworski, 1987; Barr, 1990; Powell, 1993). The soil chytrids possess motile zoospores powered by a single, posterior flagellum, and are thought to be close to the common ancestor of fungi and animals. It has also been suggested that the most recent common ancestor of the uniflagellate chytrids and the biflagellate Stramenopila may have been a biflagellate zoosporic fungus (Van der Auwera and De Wachter, 1996). Recent phylogenetic analyses that have included more than one representative of each of the chytrids and Zygomycota have indicated that the chytrids may form a monophyletic group, splitting the Zygomycota

(Van de Peer and De Wachter, 1997) or that both groups may be polyphyletic and interspersed one among the other (Nagahama et al., 1995). Together, the chytrids and Zygomycota differ considerably from the Ascomycota and Basidiomycota in that they are typically coenocytic, with multiple genetically different nuclei per cell, whereas the Ascomycota and Basidiomycota typically have uninucleate primary mycelia and binucleate (dikaryotic) secondary mycelia. Many of the approximately 1,000 described species of chytrids (Barr, 1990) occur in freshwater or marine environments, and there is no known estimate of the number of species occurring in soil.

1.3.2.5   Kingdom Fungi, Phylum Zygomycota

Soil inhabiting members of the Zygomycota belong in three groups of very different morphology and ecology: the predominantly saprobic Zygomycetes (Mucorales and relatives), the endomycorrhizal Glomales, and the Zoopagales and Entomophthorales, which are parasitic on insects and other invertebrates, fungi, and humans (Hesseltine and Ellis, 1973; O'Donnell, 1979; Morton, 1988; Humber, 1989; Morton and Benny, 1990; Alexopoulos et al., 1996). It appears that the lineage of the Glomales goes back at least to the early Cambrian, when they may have been symbiotic with cyanobacteria and, later, the ancestors of modern land plants (Pirozynski and Malloch, 1975; Pirozynski, 1981; Pirozynski and Dalpé, 1989; Simon et al., 1993; Gehrig et al., 1996; Simon, 1996; Selosse and Le Tacon, 1998). Phylogenetic analyses that have included multiple examples of both chytrids and Zygomycota suggest that the latter are not a monophyletic group. The saprobic and mycoparasitic Zygomycetes (Mucorales and relatives) are more basal and cluster together with some or all of the chytrids, while the Glomales appear to be the sister group to the higher fungi, the Ascomycota and Basidiomycota (Nagahama et al., 1995; Van de Peer and De Wachter, 1997; Fig. 1.5). However, both groups share general morphological features such as broad, tubular, thin-walled hyphae, the coenocytic condition, and positive (red) staining of their hyphal walls with diazonium blue B (Summerbell, 1985).

In the earlier literature, the saprobic Mucorales were often referred to as sugar fungi. The accompanying view held that these organisms are rapid colonists making use of the soluble sugars before secondary colonists become established during the succession of different fungi on rich and freshly deposited substrates such as dung and fallen fruits (Garrett, 1963). In reality, the rapid appearance of Mucorales on such substrates reflects their rapid rates of growth and sporulation. In many cases, the secondary colonists are already there and digesting the more recalcitrant substrates, but are slower to sporulate and come to our notice (Pugh, 1974; Cooke and Rayner, 1984). Still, the term sugar fungi accurately portrays the typical growth substrate of these fungi. They are not noted for their extracellular degradative enzymes and may frequently depend on the breakdown products provided by lignocellulose degrading ascomycetes or basidiomycetes (Cooke and Rayner, 1984). From here, it may have been a short step to direct, biotrophic parasitism of the decomposer fungi, a nutritional mode found among some Mucorales and related groups (O'Donnell, 1979).

The Glomales are arguably among the most important of all of the fungi, approximately 80% of the world's land plants depend upon mycorrhizal symbioses with these fungi for survival in soils that are nutrient limiting or prone to drought or other stresses (Malloch et al., 1980; Smith and Read, 1997). It was recently discovered on the basis of ribosomal DNA sequence analysis that *Geosiphon*, an organism that forms an endosymbiosis with a photosynthetic cyanobacterium (the cyanobacterium is inside the *Geosiphon*), is a member of the Glomales (Gehrig et al., 1996). This provides new support for the speculated origin of the Glomales some 350–550 million years ago based on fossil evidence (Pirozynski, 1981; Pirozynski and Dalpé, 1989) or DNA sequence divergence (Berbee and Taylor, 1993; Simon et al., 1993), given that the cyanobacteria date back at least 3.4 billion years (Schopf, 1993).

Of the 867 described species of Zygomycota (including Entomophthorales, Glomales, Mucorales, and Zoopagales) accepted in the Dictionary of the Fungi (Hawksworth et al., 1996), the majority might be found in soil or close to soil in various organic substrates.

1.3.2.6   Kingdom Fungi, Phylum Ascomycota

This Phylum includes most yeasts, and most asexual forms previously included in the Deuteromycetes. The Ascomycota include species of diverse nutritional modes, from saprotrophs through plant and animal pathogens, mycoparasites, and mycorrhizal associates. The Ascomycota include fungi that reproduce sexually through the production of meiotic spores in a cell called an ascus (plural asci), as well as their phylogenetic relatives that are only known to reproduce asexually, or rarely sexually. The inclusion of asexual relatives may seem obvious to the non-mycological reader, but the tradition in mycology has been to recognize separate taxa for the sexual and asexual forms (Webster, 1981; Hawksworth et al., 1983). Filamentous asexual forms have been called imperfect fungi, fungi imperfecti, deuteromycetes, hyphomycetes, or molds. There have been separate names and classifications all the way from genus and species (e.g., asexual *Aspergillus nidulans* versus sexual *Emericella nidulans*) up to class (Deuteromycetes versus Ascomycetes). The vast majority of deuteromycete species have not been connected with any sexual state (indeed, it is thought that many may not have a sexual state) (Kendrick, 1979; Domsch et al., 1980). However, there are a number of morphological, physiological, and molecular criteria by which these fungi can be recognized as belonging to the Ascomycota (see below), so there seems little justification for retaining separate, artificial classifications above the level of the genus. *Aspergillus*, *Cladosporium*, *Fusarium*, and *Penicillium* will remain familiar genera of soil fungi, even if some species are connected with sexual states that have different names. The Ascomycota also include the majority of species of the yeasts, predominantly cellular (non-filamentous) organisms that usually reproduce asexually by budding or fission (Kreger-van Rij, 1984; Barnett et al., 1990). However, the cellular growth form of yeasts is an adaptation to life in moist or liquid environments, and some members of the Basidiomycota have independently developed the yeast life form (see below). Thus, yeasts and molds are artificial categories of fungi, useful to describe their appearance and ecology but not their classification.

A particular disadvantage of this system has been that a single genus name used for asexual forms (e.g., *Sclerotium* or *Rhizoctonia*) may include species of both Ascomycota and Basidiomycota; two phyla in one genus! Likewise, maintenance of a separate class Deuteromycetes has hidden the fact that one does not know the correct phylum for many of the species included. With new knowledge from ultrastructural, physiological and molecular studies, this situation is gradually improving. The genus *Rhizoctonia* has recently been subdivided along phylogenetic lines (Moore, 1987; Andersen, 1996; Stalpers and Andersen, 1996). A number of recent molecular phylogenetic analyses have included the asexual forms among their sexually reproducing relatives (Bruns et al., 1992; Untereiner et al., 1995; LoBuglio et al., 1996; Okada et al., 1997). Likewise, our level of knowledge of the broad relationships within the Ascomycota has been improved by recent molecular phylogenetic studies (Berbee and Taylor, 1992; Spatafora and Blackwell, 1993; Landvik et al., 1996; Sjamsuridzal et al., 1997; Wedin and Tibell, 1997).

The Ascomycota and their asexual relatives are certainly the largest group of fungi, with over 46,000 described species (Hawksworth et al., 1996). Over 20,000 described species form lichens, stable symbiotic associations between fungi and green algae or cyanobacteria, in which the fungal partner (mycobiont) forms a characteristic structure called a thallus that encloses and protects the alga or cyanobacterium (photobiont) (Hawksworth, 1988; Ahmadjian, 1993). As with mycorrhizae, the lichen habit or symbiosis has multiple separate evolutionary origins amongst Zygomycota,

Ascomycota and Basidiomycota. These separate origins, long recognized on the basis of morphology, are also suggested by molecular phylogenetic analysis (Gargas et al., 1995). Lichens are important in soil formation by the colonization and physicochemical degradation of bare rocks (Hale, 1983), and form part of the cryptogamic crusts stabilizing and providing nutrients to desert soils (West, 1990; Johansen, 1993), but they are not normally considered soil fungi. Many other species of Ascomycota are parasites or endophytes of aerial plant parts, are human or animal pathogens, or inhabit dung or other specialized substrates. Some of these have a soil borne stage or a significant reservoir in soil. Perhaps 15,000 or more described species of Ascomycota might be found in soil or in litter and other organic substrates on soil.

Persons isolating fungi from substrates such as soil, rotting wood or litter, or diseased plants or animals frequently recover cultures of morphologically undistinguished fungi that lack the sexual reproductive structures necessary to identify them as members of the Ascomycota or Basidiomycota. These fungi may sporulate asexually, but belong to a group that has never been connected with any sexual state, or may be non-sporulating in pure culture. The latter, sterile forms are often referred to in the literature as sterile white mycelium, sterile dark mycelium, or *mycelium radicis atrovirens* (L. greenish black mycelium from roots). These terms are unhelpful, in terms of physiological, taxonomic, and ecological precision, and should no longer be necessary or accepted in publication, given the many tools with which one can now quite simply make a more accurate identification.

Hyphae of Ascomycota have simple septal pores, that is, their septa thin to a circular pore that provides a cytoplasmic connection between adjacent cells. Almost all members of the Basidiomycota that grow in culture as filamentous colonies (not yeasts) have what are called dolipore septa, a donut-like thickening around the septal pore that may be surrounded by a perforate or imperforate, membranous septal pore cap (Moore and Patton, 1975; Wells, 1978; Moore, 1985; Wells, 1994). Some Basidiomycota also form clamp connections (looping connections across septa) in their dikaryotic mycelia, and these are diagnostic if present. For those fungi without clamp connections, the dolipore septum or lack of it can usually be seen with a good light microscope in light-colored hyphae more than 1 μm in diameter if they are stained with trypan blue (Sneh et al., 1991) or with Congo Red in an aqueous solution of ammonia or sodium dodecyl sulfate (Nawawi et al., 1977; Clemençon, 1998). Laser confocal microscopy should also provide a good method for detection of dolipore or simple septal pores in living or fixed materials. Dolipore septa and simple septal pores are readily identified with transmission electron microscopy (TEM) in appropriate sections. If material is fixed, sectioned, and examined with TEM, Ascomycota can usually be distinguished from Basidiomycota even if septal pores are not seen. Most Ascomycota have bilayered hyphal walls, with a narrow, electron dense outer layer and a broad, pale inner layer. In contrast, most Basidiomycota have multilamellate hyphal walls, with multiple narrow light and dark layers as seen in TEM (Alexopoulos et al., 1996). The hyphal walls of Ascomycota and Basidiomycota (especially yeasts, and less dependably filamentous forms) also differ in their reaction to staining with diazonium blue B. The hyphal walls of Basidiomycota stain red or purplish (a pretreatment with KOH is necessary for filamentous forms), whereas the walls of most Ascomycota are nonstaining (Hagler and Ahearn, 1981; Summerbell, 1985). As mentioned above, hyphae of Zygomycetes also stain with diazonium blue B, but these can easily be distinguished from members of the Basidiomycota by their large diameter and coenocytic nature. Physiologically, cultures can often be determined as Ascomycota or Basidiomycota by their pattern of tolerance or sensitivity to sodium chloride, benomyl, and cycloheximide (Hutchison, 1990). Finally, great progress has been made in the extraction and sequencing of fungal DNA, enabling the rapid and certain identification of an unknown strain by comparison with available sequences of known strains (Bruns et al., 1991; Maidak et al., 1997; Bruns et al., 1998).

1.3.2.7   Kingdom Fungi, Phylum Basidiomycota

This Phylum includes some yeasts, and some asexual forms previously included in the Deuteromycetes. The Basidiomycota include two of the most important groups of soil fungi: species forming ectomycorrhizae of forest trees, and species that are the dominant decay organisms of plant polymers such as lignin, hemicellulose and cellulose. A few basidiomycetes, such as *Moniliopsis solani* (= *Rhizoctonia solani*), are very important plant pathogens of field and forage crops ranging from wheat to rice to alfalfa. Others are root pathogens of tropical tree crops such as rubber and oil palms, or are root pathogens or cause rot and butt rots of tropical and temperate forest trees and turf (Cook, 1981; Farr et al. 1989; Smiley et al., 1992; Ploetz, 1994; Tainter, 1996; Hansen and Lewis, 1997). The Ascomycota and their asexual relatives are usually thought to be the major group of soil fungi, both in number of species and in biomass. This may be true in agricultural soils and in native tropical soils, but in many temperate and boreal forest soils, the basidiomycetes probably outnumber and outweigh the ascomycetes in at least the organic rich soil horizons (Frankland, 1982; Entry et al., 1992). Many form extensive mycelia with considerable biomass. The most famous of these is *Armillaria*, the humongous fungus of which one individual mycelium was calculated to occupy 15 ha and have a weight of 10 Mg (Smith et al., 1992). Another individual *Armillaria* thought to occupy 40 ha has been reported (Shaw and Roth, 1976). Clearly, the Basidiomycota include the fungi with the best ligninolytic abilities (Orth et al., 1993; Worrall et al., 1997) and are, therefore, likely to be the most important in decomposition of plant litter in most soils. Approximately 10,000 or more described species of Basidiomycota may be found in soil or litter (Hawksworth et al., 1996).

As with the situation in the Ascomycota, a number of asexual or predominantly asexual fungi belong in the Basidiomycota. Their natural relationships have been hidden (perhaps especially to non-mycologists) by their inclusion in the artificial class Deuteromycetes. Formerly orphaned fungi include molds such as the *Sporotrichum* (= *Chrysosporium*) state of *Phanerochaete chrysosporium* and many yeasts frequently encountered in soil, such as *Rhodotorula* (Stalpers, 1987; Sugiyama and Suh, 1993; Swann and Taylor, 1995; Sugita and Nakase, 1998; Okada et al., 1998). Molecular phylogenetic analyses of Basidiomycota (Swann and Taylor, 1993; Hibbett et al., 1997) are improving our knowledge of the relationships of these fungi, and providing a framework into which the asexual forms and unidentified isolates from soil or mycorrhizae may be placed and identified (Bruns et al., 1998).

Despite their numbers and biomass in soil, the Basidiomycota are rarely recovered in surveys of soil fungi based on isolation into culture, because few produce significant numbers of propagules in soil in the way that the asexual Ascomycota do. On isolation plates, the few hyphal propagules of Basidiomycota are tremendously outnumbered by mold spores and outcompeted by the more rapidly growing mold fungi. Because of this, the Basidiomycota remain the missing link in soil mycology (Chesters, 1949), 50 years after he first made that claim. To overcome our inability to routinely isolate decomposer basidiomycetes from soil, Thorn et al. (1996) developed a method of plating washed organic particles from soil on a medium made selective for basidiomycetes with benomyl (Edgington et al., 1971; Worrall, 1991; Johnson, 1994) and with guaiacol as an indicator of laccase(s) and peroxidase(s). The addition of dicloran (2,6-dichloro-4-nitroaniline) to the medium should provide control of Mucorales (which are tolerant to benomyl), for soils or litters in which these fungi are a problem (Worrall, 1991); hymexazol (3-hydroxy-5-methylisoxazole) would be an alternative (Tsao and Guy, 1977; W. Gams, 1997, personal communication).

1.3.2.8   Methods for Characterization of Communities and Activities of Soil Fungi

Among the methods available for the characterization of communities and activities of soil fungi are direct counts (Bloem et al., 1995) and immunological detection (Frankland, 1984; Clausen, 1997), differential inhibition with antibacterial and antifungal biocides and measurement of CO_2 evolution or other indicators of physiological activity (Anderson and Domsch, 1975; Kjøller and Struwe, 1982; Stamatiadis et al., 1990; Landi et al. 1993; but see Velvis, 1997), chloroform fumigation and incubation or extraction methods for total microbial activity (Jenkinson and Powlson, 1976; Voroney and Paul, 1984; Vance et al., 1987; Harris et al., 1997), linkage of microbial populations to biogeochemical processes by ^{13}C labeling of biomarkers (Boschker et al., 1998), activity stains applied to fungi in soil smears (Tsuji et al., 1995), assays on enzymes extracted from soil (Wood, 1979; Matcham et al., 1985; Frankenberger and Johanson, 1986; Falih and Wainwright, 1996; Kiem and Kandeler, 1997), isolation and quantification of fungal biomarkers such as ergosterol or chitin derived glucosamine (Swift, 1973; Tunlid and White, 1992; Sancholle and Dalpé, 1993; Stahl and Klug, 1996; Eckblad et al., 1998), isolation of fungi into culture, and isolation and characterization of fungal DNA or RNA from soil. A brief overview of methods for isolating fungi from soil and isolation and characterization of soil fungal DNA and RNA are all that can be included here.

1.3.2.9 Methods for Isolating Fungi from Soil

There exist a number of reviews of methods for the isolation of various fungi from soil into culture (Barron, 1971; Stevens, 1974; Gams, 1992; Singleton et al., 1992; Parkinson, 1994). Keys for the identification of common genera and species of soil fungi of temperate areas are available (Barron, 1968; O'Donnell, 1979; Domsch et al., 1980; Gams, 1993; Watanabe, 1994), but identification of many isolates from tropical soils or from temperate or tropical litter must be sought in the large mycological literature, to which Hawksworth et al. (1996) provide an entry. In choosing a method for isolating fungi from soil, one should first define the objective: is it to obtain a complete list for one sample, or a suite of predominant species by which to compare two or more samples, or to recover a particular species or ecological group of interest? If it is to obtain a complete list, be prepared for a challenge. Christensen (1989), using just one method and medium, found that 8 to 15 previously unseen species of microfungi were added with every 100 incremental isolates past 1,100. No one method, medium, or incubation temperature will recover all fungi in any sample (Carreiro and Koske, 1992). A number of sources discuss methods and media for the selective recovery of particular groups of fungi (Barron, 1977; Singleton et al., 1992; Thorn et al., 1996). The greatest difficulty with any of these methods is the inability to relate the species and numbers of isolates obtained to the identities and importance of fungi in the native soil.

1.3.2.10 Isolation and Characterization of Nucleic Acids from Soil

Methods for the isolation of DNA and RNA from soil have seen great improvements in recent years (Holben et al, 1988; Porteous and Armstrong, 1991; Tsai and Olson, 1991; Claassen et al., 1996; Purdy et al., 1996; Yeates et al., 1997). Perhaps most notable is the hydroxyapatite spin column method developed by Purdy et al. (1996), which allowed the rapid and separate isolation of high quality DNA and rRNA. Most of these methodological improvements have been directed toward characterization of the soil bacterial community. However, Yeates et al. (1997) reported successful amplification of fungal beta-tubulin genes following isolation of DNA from soil by bead beating and SDS lysis followed by cleaning and precipitation with polyethylene glycol, potassium acetate, phenol and isopropanol. Claassen et al. (1996) were successful in PCR amplification of endomycorrhizal (Glomalean) rDNA from DNA extracted from bulk soil, using standard and Glomales specific (Simon et al., 1992) PCR primers. The development of the extraction techniques for soil community nucleic acids allows for many methods for their characterization, from probing with taxon specific probes (Liesack and Stackebrandt, 1992; Frischer et al., 1996) to PCR-amplification of ribosomal RNA or other genes with conserved fungal primers and subsequent cloning and sequencing or RFLP analysis of amplified products (Erland, 1995; Zhou et al., 1997; Lloyd-Jones and Lau, 1998; Dykhuizen, 1998), or with taxon specific primers (PCR-based detection assays) (Gardes and Bruns, 1993; O'Gorman et al., 1994). In a simple substrate with presumably low fungal diversity, these approaches may finally allow us to answer the question of what proportion of the fungal community can be isolated into culture by comparing results with both methods (Rossman et al., 1998). The future of soil mycology just got more interesting.

1.3.3 References

Agrios, G.N. 1997. Plant pathology. 4th Ed. Academic Press, San Diego, CA.

Ahmadjian, V. 1993. The lichen symbiosis. John Wiley, New York, NY.

Alexopoulos, C.J., C.W. Mims, and M. Blackwell. 1996. Introductory mycology. 4th Ed. John Wiley, New York, NY.

Ames, R.N., K.L. Mihara, and H.G. Bayne. 1989. Chitin-decomposing actinomycetes associated with a vesicular-arbuscular mycorrhizal fungus from a calcareous soil. New Phytol. 111:67–71.

Amewowor, D.H.A.K., and M.F. Madelin. 1991. Numbers of myxomycetes and associated microorganisms in the root zones of cabbage (*Brassica oleracea*) and broad bean (*Vicia faba*) in field plots. FEMS Microbiol. Lett. 86:69-82.

Andersen, T.F. 1996. A comparative taxonomic study of *Rhizoctonia sensu lato* employing morphological, ultrastructural and molecular methods. Mycol. Res. 100:1117–1128.

Anderson, J.P.E., and K.H. Domsch. 1975. Measurement of bacterial and fungal contributions to respiration of selected agricultural and forest soils. Can. J. Microbiol. 21:314-322.

Baldauf, S.L., and W.F. Doolittle. 1997. Origin and evolution of the slime molds (Mycetozoa). Proc. Natl. Acad. Sci. USA 94:12007–12012.

Barnett, J.A., R.W. Payne, and D. Yarrow. 1990. Yeasts: characteristics and identification. 2nd Ed. Cambridge University Press, Cambridge, UK.

Barr, D.J.S. 1990. Phylum Chytridiomycota. p. 454–466. *In* L. Margulis, J.O. Corliss, M. Melkonian and D.J. Chapman (ed.) Handbook of Protoctista. Jones and Bartlett, Boston, MA.

Barron, G.L. 1968. The genera of hyphomycetes from soil. Williams and Wilkins, Baltimore, MD.

Barron, G.L. 1971. Soil fungi. p. 405-427. *In* C. Booth (ed.) Methods in microbiology. Vol. 4. Academic Press, London, UK.

Barron, G.L. 1977. The nematode-destroying fungi. Canadian Biological Publications, Guelph, Ontario, Canada.

Berbee, M.L., and J.W. Taylor. 1992. Two Ascomycete classes based on fruiting body characters and ribosomal DNA sequences. Mol. Biol. Evol. 9:278–284.

Berbee, M.L., and J.W. Taylor. 1993. Dating the evolutionary radiations of the true fungi. Can. J. Bot. 71:1114–1127.

Bloem, J., M. Veninga, and J. Shepherd. 1995. Fully automatic determination of soil bacterium numbers, cell volumes, and frequencies of dividing cells by confocal laser scanning microscopy and image analysis. Appl. Environ. Microbiol. 61:926-936.

Boschker, H.T.S., S.C. Nold, P. Wellsbury, D. Bos, W. de Graaf, R. Pel, R.J. Parkes, and T.E. Cappenberg. 1998. Direct linking of microbial populations to specific biogeochemical processes by ^{13}C-labelling of biomarkers. Nature 392:801–805.

Bridges, E.M. 1978. World soils. 2nd Ed. Cambridge University Press, Cambridge, UK.

Bruns, T.D., T.M. Szaro, M. Gardes, K.W. Cullings, J.J. Pan, D.L. Taylor, T.R. Horton, A. Kretzer, M. Garbelotto, and Y. Li. 1998. A sequence database for the identification of ectomycorrhizal basidiomycetes by phylogenetic analysis. Mol. Ecol. 7:257–272.

Bruns, T.D., R. Vilgalys, S.M. Barns, D. Gonzalez, D.S. Hibbett, D.J. Lane, L. Simon, S. Stickel, T.M. Szaro, W.G. Weisburg, and M.L. Sogin. 1992. Evolutionary relationships within the fungi: analyses of nuclear small subunit rRNA sequences. Mol. Phylog. Evol. 1:231–241.

Bruns, T.D., T.J. White, and J.W. Taylor. 1991. Fungal molecular systematics. Annu. Rev. Ecol. Syst. 22:525–564.

Buczacki, S.T. (ed.) 1983. Zoosporic plant pathogens. Academic Press, London, UK.

Carreiro, M.M., and R.E. Koske. 1992. Room temperature isolations can bias against selection of low temperature microfungi in temperate forest soils. Mycologia 84:886-900.

Chesters, C.G.C. 1949. Concerning fungi inhabiting soil. Trans. Br. Mycol. Soc. 32:197-216.

Christensen, M. 1989. A view of fungal ecology. Mycologia 81:1-19.

Claassen, V.P., R.J. Zasoski, and B.M. Tyler. 1996. A method for direct soil extraction and PCR amplification of endomycorrhizal fungal DNA. Mycorrhiza 6:447–450.

Clausen, C.A. 1997. Immunological detection of wood decay fungi-an overview of techniques developed from 1986 to the present. Intl. Biodet. Biodegrad. 39:133–143.

Clemençon, H. 1998. Observing the dolipore with the light microscope. Inoculum 49:(2)3.

Cook, A.A. 1981. Diseases of tropical and subtropical field, fiber and oil plants. Hafner Press, New York, NY.

Cooke, R.C., and A.D.M. Rayner. 1984. Ecology of Saprotrophic Fungi. Longman, London, UK.

Crawford, D.L. 1978. Lignocellulose decomposition by selected *Streptomyces* strains. Appl. Environ. Microbiol. 35:1041–1045.

Domsch, K.H., W. Gams, and T.-H. Anderson. 1980. Compendium of soil fungi. 2 vol. Academic Press, London, UK.

Dykhuizen, D.E. 1998. Santa Rosalia revisited: Why are there so many species of bacteria? Ant. van Leeuw. 73:25–33.

Eckblad, A., H. Wallander, and T. Näsholm. 1998. Chitin and ergosterol combined to measure total and living fungal biomass in ectomycorrhizas. New Phytol. 138:143–149.

Edgington, L.V., K.L. Khew, and G.L. Barron. 1971. Fungitoxic spectrum of benzimidazole compounds. Phytopath. 61:42-44.

Embley, T.M., and E. Stackebrandt. 1994. The molecular phylogeny and systematics of the Actinomycetes. Annu. Rev. Microbiol. 48:257–289.

Entry, J.A., C.L. Rose, and K. Cromack. 1992. Microbial biomass and nutrient concentrations in hyphal mats of the ectomycorrhizal fungus Hysterangium setchellii in a conifer forest soil. Soil Biol. Biochem. 24:447–453.

Eppard, M., W.E. Krumbein, C. Koch, E. Rhiel, J.T. Staley, and E. Stackebrandt. 1996. Morphological, physiological, and molecular characterization of actinomycetes isolated from dry soil, rocks, and monument surfaces. Arch. Microbiol. 166:12–22.

Erland, S. 1995. Abundance of *Tylospora fibrillosa* ectomycorrhizas in a south Swedish spruce forest measured by RFLP analysis of the PCR-amplified rDNA ITS region. Mycol. Res. 99:1425–1428.

Erwin, D.C., S. Bartnicki-Garcia, and P.H. Tsao (ed.) 1983. *Phytophthora*: Its biology, taxonomy, ecology, and pathology. APS Press, St. Paul, MN.

Falih, A.M.K., and M. Wainwright. 1996. Microbial enzyme activity in soils amended with a natural source of easily available carbon. Biol. Fertil. Soils 21:177–183.

Fanning, D.S., and M.C. Balluff-Fanning. 1989. Soil: morphology, genesis, and classification. John Wiley, New York, NY.

Farr, D.F., G.F. Bills, G.P. Chamuris, and A.Y. Rossman. 1989. Fungi on plants and plant products in the United States. APS Press, St. Paul, MN.

Fermor, T.R., and W.D. Grant. 1985. Degradation of fungal and actinomycete mycelia by *Agaricus bisporus*. J. Gen. Microbiol. 131:1729–1734.

Frankenberger, W.T., Jr., and J.B. Johanson. 1986. Use of plasmolytic agents and antiseptics in soil enzyme analysis. Soil Biol. Biochem. 18:209–213.

Frankland, J.C. 1982. Biomass and nutrient cycling by decomposer basidiomycetes. p. 241-261. *In* J.C. Frankland, J.N. Hedger and M.J. Swift (ed.) Decomposer basidiomycetes. Cambridge University Press, Cambridge, UK.

Frankland, J.C. 1984. Autecology and the mycelium of a woodland litter decomposer. p. 241-260. *In* D.H. Jennings and A.D.M. Rayner (ed.) The ecology and physiology of the fungal mycelium. Cambridge University Press, Cambridge, UK.

Frischer, M.E., P.J. Floriani, and S.A. Nierzwicki-Bauer. 1996. Differential sensitivity of 16S rRNA targeted oligonucleotide probes used for fluorescence in situ hybridization is a result of ribosomal higher order structure. Can. J. Microbiol. 42:1061–1071.

Fuller, M.S., and A. Jaworski. 1987. Zoosporic fungi in teaching and research. Southeastern Publishing, Athens, GA.

Gams, W. 1992. The analysis of communities of saprophytic microfungi with special reference to soil fungi. p. 183-223. *In* W. Winterhoff (ed.) Fungi in vegetation science. Kluwer Academic, Dordrecht, Netherlands.

Gams, W. 1993. Supplement and corrigendum to the compendium of soil fungi. IHW-Verlang, Eching, Germany.

Gardes, M., and T.D. Bruns. 1993. ITS primers with enhanced specificity for basidiomycetes-application to the identification of mycorrhizae and rusts. Mol. Ecol. 2:113–118.

Gargas, A., P.T. DePriest, M. Grube, and A. Tehler. 1995. Multiple origins of lichen symbioses in fungi suggested by SSU rDNA phylogeny. Science 268:1492–1495.

Garrett, S.D. 1963. Soil fungi and soil fertility. Pergamon Press, Oxford, UK.

Gehrig, H., A. Schüßler, and M. Kluge. 1996. *Geosiphon pyriforme*, a fungus forming endocytobiosis with *Nostoc* (Cyanobacteria), is an ancestral member of the Glomales: evidence by SSU rRNA analysis. J. Mol. Evol. 43:71–81.

Gerber, N.N. 1968. Geosmin from microorganisms is *trans* 1,10-dimethyl-*trans*-9-decalol. Tetrahedron Lett. 25:2971–2974.

Gerber, N.N., and H.A. Lechevalier. 1965. Geosmin, an earth-smelling substance isolated from actinomycetes. Appl. Microbiol. 13:935–938.

Godden, B., A.S. Ball, P. Helvenstein, A.J. McCarthy, and M.J. Penninckx. 1992. Towards elucidation of the lignin degradation pathway in actinomycetes. J. Gen. Microbiol. 138:2441–2448.

Goodfellow, M., and T. Cross. 1974. Actinomycetes. p. 269–302. *In* C.H. Dickinson and G.J.F. Pugh (ed.) Biology of plant litter decomposition. Vol. 2. Academic Press, London, UK.

Hagler, A.N., and D.G. Ahearn. 1981. Rapid diazonium blue B test to detect basidiomycetous yeasts. Intl. J. Syst. Bact. 31:204-208.

Hale, M. 1983. Biology of lichens. 3rd Ed. E. Arnold, London, UK.

Hansen, E.M., and K.J. Lewis. 1997. Compendium of conifer diseases. APS Press, St. Paul, MN.

Harris, D., R.P. Voroney, and E.A. Paul. 1997. Measurement of microbial biomass N:C by chloroform fumigation-incubation. Can. J. Soil Sci. 77:507–514.

Hawksworth, D.L. 1988. The variety of fungal-algal symbioses, their evolutionary significance, and the nature of lichens. Bot. J. Linn. Soc. (Lond.) 96:3-20.

Hawksworth, D.L., B.C. Sutton, and G.C. Ainsworth. 1983. Ainsworth & Bisby's dictionary of the fungi. 7th Ed. Commonwealth Mycological Institute, Kew, UK.

Hawksworth, D.L., P.M. Kirk, B.C. Sutton, and D.N. Pegler. 1996. Ainsworth & Bisby's dictionary of the fungi. 8th Ed. CAB International, Wallingford, UK.

Hesseltine, C.W., and J.J. Ellis. 1973. Mucorales. p. 187–217. *In* G.C. Ainsworth, F.K. Sparrow and A.F. Sussman (ed.) The fungi. Vol. IVB. Academic Press, New York, NY.

Hibbett, D.S., E.M. Pine, E. Langer, G. Langer, and M.J. Donoghue. 1997. Evolution of gilled mushrooms and puffballs inferred from ribosomal DNA sequences. Proc. Natl. Acad. Sci. USA 94: 2002–2006.

Holben, W.E., J.J. Jansson, B.K. Chelm, and J.M. Tiedje. 1988. DNA probe method for the detection of specific microorganisms in the soil bacterial community. Appl. Environ. Microbiol. 54:703–711.

Humber, R.A. 1989. Synopsis of a revised classification for the Entomophthorales (Zygomycotina). Mycotaxon 34:441–460.

Huss-Danell, K. 1997. Actinorhizal symbioses and their N$_2$ fixation. New Phytol. 136:375–405.

Hutchison, L.J. 1990. Studies on the systematics of ectomycorrhizal fungi in axenic culture. IV. The effect of some selected fungitoxic compounds upon linear growth. Can. J. Bot. 68:2172–2178.

Iiyama, K, B.A. Stone, and B.J. Macauley. 1994. Compositional changes in compost during composting and growth of *Agaricus bisporus*. Appl. Environ. Microbiol. 60:1538–1546.

Ingram, S.W., and N.M. Nadkarni. 1993. Composition and distribution of epiphytic organic matter in a neotropical cloud forest, Costa Rica. Biotropica 25:370–383.

Jenkinson, D.S., and D.S. Powlson. 1976. The effects of biocidal treatments on metabolism in soil. V. A method of measuring soil biomass. Soil Biol. Biochem. 8:209–213.

Johansen, J.R. 1993. Cryptogamic crusts of semiarid and arid lands of North America. J. Phycol. 28:139-147.

Johnson, G.C. 1994. A comparison of the efficacy of six media for isolating wood-decaying Hymenomycetes. Can. J. Microbiol. 41:104–107.

Karling, J.S. 1968. The Plasmodiophorales. 2nd Ed. Hafner Press, New York., NY

Karling, J.S. 1977. Chytridiomycetarum iconographia. J. Cramer, Vaduz, Liechtenstein.

Kendrick, W.B. (ed.) 1979. The whole fungus. 2 vol. National Museum of Natural Sciences, Ottawa, Canada.

Kiem, R., and E. Kandeler. 1997. A simple method for the determination of trehalase activity in soils. Microbiol. Res. 152:19–25.

Kjøller, A., and S. Struwe. 1982. Microfungi in ecosystems: fungal occurrence and activity in litter and soil. Oikos 39:389-422.

Kreger-van Rij, N.J.W. (ed.) 1984. The yeasts: a taxonomic study. 3rd Ed. Elsevier, Amsterdam, Netherlands.

Kumar, S., and A. Rzhetsky. 1996. Evolutionary relationships of eukaryotic kingdoms. J. Mol. Evol. 42:183–193.

Landi, L., L. Badalucco, F. Pomare, and P. Nannipieri. 1993. Effectiveness of antibiotics to distinguish the contribution of fungi and bacteria to net nitrogen mineralization, nitrification and respiration. Soil Biol. Biochem. 25:1771-1778.

Landvik, S., N.F.J. Shailer, and O.E. Eriksson. 1996. SSU rDNA sequence support for a close relationship between the Elaphomycetales and the Eurotiales and Onygenales. Mycosci. 37:237–241.

Li, X., and P. Gao. 1996. Isolation and partial characterization of cellulose-degrading strain of *Streptomyces* sp. LX from soil. Lett. Appl. Microbiol. 22:209–213.

Liesack, W., and E. Stackebrandt. 1992. Unculturable microbes detected by molecular sequences and probes. Biodiv. Conserv. 1:250–262.

Lloyd-Jones, G., and P.C.K. Lau. 1998. A molecular view of microbial diversity in a dynamic landfill in Québec. FEMS Microbiol. Lett. 162:219–226.

LoBuglio, K.F., M.L. Berbee, and J.W. Taylor. 1996. Phylogenetic origins of the asexual mycorrhizal symbiont *Cenococcum geophilum* Fr. and other mycorrhizal fungi among the Ascomycetes. Mol. Phylog. Evol. 6:287–294.

Maidak, B.L., G.J. Olsen, N. Larsen, R. Overbeek, M.J. McCaughey, and C.R. Woese. 1997. The RDP (ribosomal database project). Nuc. Acids Res. 25:109–110.

Malloch, D., K.A. Pirozynski, and P.H. Raven. 1980. Ecological and evolutionary significance of mycorrhizal symbioses in vascular plants (a review). Proc. Natl. Acad. Sci. USA 77:2113-2118.

Martin, F.N. 1992. Pythium. p. 39–49. *In* L.L. Singleton, J.D. Mihail and C.M. Rush (ed.) Methods for research on soilborne phytopathogenic fungi. APS Press, St. Paul, MN.

Martin, G.W., and C.J. Alexopoulos. 1969. The Myxomycetes. University of Iowa Press, Iowa City, IA.

Martin, G.W., C.J. Alexopoulos, and M.L. Farr. 1983. The genera of Myxomycetes. University of Iowa Press, Iowa City, IA.

Matcham, S.E., B.R. Jordan, and D.A. Wood. 1985. Estimation of fungal biomass in a solid substrate by three independent methods. Appl. Microbiol. Biotech. 21:108-112.

Moore, R.T. 1985. The challenge of the dolipore / parenthosome septum. p. 175–212. *In* D. Moore, L. Casselton, D.A. Wood and J.C. Frankland (ed.) Developmental biology of higher fungi. Cambridge University Press, Cambridge, UK.

Moore, R.T. 1987. The genera of *Rhizoctonia*-like fungi: *Ascorhizoctonia, Ceratorhiza gen. nov., Epulorhiza gen. nov., Moniliopsis,* and *Rhizoctonia.* Mycotaxon 29: 1–99.

Moore, R.T., and A.M. Patton. 1975. Parenthosome fine structure in *Pleurotus cystidiosus* and *Schizophyllum commune.* Mycologia 67:1200–1205.

Morton, J.B. 1988. Taxonomy of VA mycorrhizal fungi: classification, nomenclature, and identification. Mycotaxon 32:267-324.

Morton, J.B., and G.L. Benny. 1990. Revised classification of arbuscular mycorrhizal fungi (Zygomycetes): a new order, Glomales, two new suborders, Glomineae and Gigasporineae, and two new families, Acaulosporaceae and Gigasporaceae, with an emendation of Glomaceae. Mycotaxon 37:471-491.

Nadkarni, N. 1984. Epiphyte biomass and nutrient capital of neotropical elfin forest. Biotropica 16:249–26.

Nagahama, T., H. Sato, M. Shimazu, and J. Sugiyama. 1995. Phylogenetic divergence of the entomophthoralean fungi: evidence from nuclear 18S ribosomal RNA gene sequences. Mycologia 87:203–209.

Nawawi, A., J. Webster, and R.A. Davey. 1977. *Dendrosporomyces prolifer, gen. et sp. nov.,* a basidiomycete with branched conidia. Trans. Br. Mycol. Soc. 68:59-63.

Normand, P., S. Orso, B. Cournoyer, P. Jeannin, C. Chapelon, J. Dawson, L. Evtushenko, and A.K. Misra. 1996. Molecular phylogeny of the genus *Frankia* and related genera and emendation of the family Frankiaceae. Intl. J. Syst. Bacteriol. 46:1–9.

O'Donnell, K. L. 1979. Zygomycetes in culture. Department of Botany, University of Georgia, Athens, GA.

O'Gorman, D., B. Xue, T. Hsiang, and P.H. Goodwin. 1994. Detection of *Leptosphaeria korrae* with the polymerase chain reaction and primers from the ribosomal internal transcribed spacers. Can. J. Bot. 72:342–346.

Okada, G., A. Takematsu, I. Gandjar, and T. Nakase. 1998. Morphology and molecular phylogeny of *Tretopileus sphaerosprous,* a synnematous hyphomycete with basidiomycetous affinities. Mycosci. 39:21–30.

Okada, G., A. Takematsu, and Y. Takamura. 1997. Phylogenetic relationships of the hyphomycete genera *Chaetopsina* and *Kionochaeta* based on 18S rDNA sequences. Mycosci. 38:409–420.

Olsen, G.J., C.R. Woese, and R. Overbeek. 1994. The winds of (evolutionary) change: Breathing new life into microbiology. J. Bacteriol. 176:1–6.

Orth, A.B., D.J. Royse, and M. Tien. 1993. Ubiquity of lignin-degrading peroxidases among various wood-degrading fungi. Appl. Environ. Microbiol. 59:4017-4023.

Parkinson, D. 1994. Filamentous fungi. p. 329-350. *In* R.W. Weaver et al. (ed.) Methods of soil analysis. Part 2. Microbiological and biochemical properties. Soil Soil Science Society of America, Madison, WI.

Pirozynski, K.A. 1981. Interactions between fungi and plants through the ages. Can. J. Bot. 59:1824-1827.

Pirozynski, K.A., and Y. Dalpé. 1989. Geological history of the Glomaceae with particular reference to mycorrhizal symbiosis. Symbiosis 7:1-36

Pirozynski, K.A., and D.W. Malloch. 1975. The origin of land plants: a matter of mycotrophism. BioSyst. 6:153-164.

Ploetz, R.C. 1994. Compendium of tropical fruit diseases. APS Press, St. Paul, MN.

Porteous, L.A. and J.L. Armstrong. 1991. Recovery of bulk DNA from soil by a rapid, small-scale extraction method. Curr. Microbiol. 22:345–348.

Postgate, J. 1988. The ghost in the laboratory. New Sci. 4:49–52.

Powell, M.J. 1993. Looking at fungi with a Janus face: a glimpse at Chytridiomycetes active in the environment. Mycologia 85:1-20.

Pugh, G.J.F. 1974. Terrestrial fungi. p. 303–336. *In* C.H. Dickinson and G.J.F. Pugh (ed.) Biology of plant litter decomposition. Vol. 2. Academic Press, London, UK.

Purdy, K.J., T.M. Embley, S. Takii, and D.B. Nedwell. 1996. Rapid extraction of DNA and rRNA from sediments by a novel hydroxyapatite spin-column method. Appl. Environ. Microbiol. 62:3905–3907.

Ramachandra, M., D.L. Crawford, and G. Hertel. 1988. Characterization of an extracellular lignin peroxidase of the ligninolytic actinomycete *Streptomyces viridosporus*. Appl. Environ. Microbiol. 54:3057–3063.

Raper, K.B. 1984. The Dictyostelids. Princeton University Press, Princeton, NJ.

Rayner, A.D.M., and L. Boddy. 1988. Fungal decomposition of wood: its biology and ecology. Wiley, Chichester, UK.

Rheims, H., C. Spröer, F.A. Rainey, and E. Stackebrandt. 1996. Molecular biological evidence for the occurrence of uncultured members of the actinomycete line of descent in different environments and geographical locations. Microbiol. 142:2863–2870.

Rossman, A.Y., R.E. Tuloss, T.E. O'Dell, and R.G. Thorn. 1998. Protocols for an all taxa biodiversity inventory of fungi in a Costa Rican conservation area. Parkway Publishers, Boone, NC.

Sancholle, M., and Y. Dalpé. 1993. Taxonomic significance of fatty acids of arbuscular mycorrhizal fungi and related species. Mycotaxon 49:187-193.

Schopf, J.W. 1993. Microfossils of the Early Archean Apex chert: new evidence of the antiquity of life. Science 260:640–646.

Selosse, M.-A., and F. Le Tacon. 1998. The land flora: a phototroph-fungus partnership? Trends Ecol. Evol. 13:15–20.

Shaw, C.G. III, and L.F. Roth. 1976. Persistence and distribution of a clone of *Armillaria mellea* in ponderosa pine forest. Phytopath. 66:1210–1213.

Simon, L. 1996. Phylogeny of the Glomales: deciphering the past to understand the present. New Phytol. 133:95–101.

Simon, L., J. Bousquet, R.C. Lévesque, and M. Lalonde. 1993. Origin and diversification of endomycorrhizal fungi and coincidence with vascular land plants. Nature 363:67–69.

Simon, L., M. Lalonde, and T.D. Bruns. 1992. Specific amplification of 18S fungal ribosomal genes from vesicular-arbuscular endomycorrhizal fungi colonizing roots. Appl. Environ. Microbiol. 58:291–295.

Singleton, L.L., J.D. Mihail, and C.M. Rush. 1992. Methods for research on soilborne phytopathogenic fungi. APS Press, St. Paul, MN.

Sjamsuridzal W., Y. Tajiri, H. Nishida, T.B. Thuan, H. Kawasaki, A. Hirata, A. Yokota, and J. Sugiyama. 1997. Evolutionary relationships of members of the genera *Taphrina*, *Protomyces*, *Schizosaccharomyces*, and related taxa within the Archiascomycetes: integrated analysis of genotypic and phenotypic characters. Mycosci. 38:267-280.

Skerman, V.B.D., V. McGowan, and P.A. Sneath. 1989. Approved lists of bacterial names. American Society of Microbiology, Washington, DC.

Smiley, R.W., P.H. Dernoeden, and B.B. Clarke. 1992. Compendium of turfgrass diseases. APS Press, St. Paul, MN.

Smith, M.L., J.N. Bruhn, and J.B. Anderson. 1992. The fungus *Armillaria bulbosa* is among the largest and oldest living organisms. Nature 356:428-431.

Smith, S.E., and D.J. Read. 1997. Mycorrhizal symbiosis. Rev. Ed. Academic Press, San Diego, CA.

Sneh, B., L. Burpee, and A. Ogoshi. 1991. Identification of *Rhizoctonia* species. APS Press, St. Paul, MN.

Sogin, M.L., H.G. Morrison, G. Hinkle, and J.D. Silberman. 1996. Ancestral relationships of the major eukaryotic lineages. Microbiología 12:17–28.

Spatafora, J.W., and M. Blackwell. 1993. Molecular systematics of unitunicate perithecial ascomycetes: the Clavicipitales-Hypocreales connection. Mycologia 85:912–922.

Stackebrandt, E., and C.R. Woese. 1981. Towards a phylogeny of the actinomycetes and related organisms. Curr. Microbiol. 5:197–202.

Stahl, P.D., and M.J. Klug. 1996. Characterization and differentiation of filamentous fungi based on fatty acid composition. Appl. Environ. Microbiol. 62:4136–4146.

Stalpers, J.A. 1987. Pleomorphy in Holobasidiomycetes. p. 201–220. *In* J. Sugiyama (ed.) Pleomorphic fungi: The diversity and its taxonomic implications. Kodansha, Tokyo, Japan.

Stalpers, J.A. and T.F. Andersen. 1996. A synopsis of the taxonomy of teleomorphs connected with *Rhizoctonia* s.l. p. 49–63. *In* B. Sneh, S. Jabaji-Hare, S. Neate and G. Dijst (ed.) *Rhizoctonia* species: Taxonomy, molecular biology, ecology, pathology and disease control. Kluwer Academic, Dordrecht, Netherlands.

Stamatiadis, S., J.W. Doran, and E.R. Ingham. 1990. Use of staining and inhibitors to separate fungal and bacterial activity in soil. Soil Biol. Biochem. 22:81-88.

Stevens, R.B. 1974. Mycology guidebook. University of Washington Press, Seattle, WA.

Sugita, T., and T. Nakase. 1998. Molecular phylogenetic study of the basidiomycetous anamorphic yeast genus *Trichosporon* and related taxa based on small subunit ribosomal DNA sequences. Mycosci. 39:7–13.

Sugiyama, J., and S.-O. Suh. 1993. Phylogenetic analysis of basidiomycetous yeasts by means of 18S ribosomal RNA sequences: relationship of *Erythrobasidium hasegawianum* and other basidiomycetous yeast taxa. Ant. van Leeuw. 63:201–209.

Summerbell, R.C. 1985. The staining of filamentous fungi with diazonium blue B. Mycologia 77:587-593.

Swann, E.C., and J.W. Taylor. 1993. Higher taxa of basidiomycetes: an 18S rRNA gene perspective. Mycologia 85:923–936.

Swann, E.C., and J.W. Taylor. 1995. Toward a phylogenetic systematics of the Basidiomycota: integrating yeasts and filamentous basidiomycetes using 18S rRNA gene sequences. Stud. Mycol. 38:147–161.

Swensen, S.M. 1996. The evolution of actinorhizal symbioses: evidence for multiple origins of the symbiotic association. Amer. J. Bot. 83:1503–1512.

Swift, M.J. 1973. The estimation of mycelial biomass by determination of hexosamine content of wood decayed by fungi. Soil Biol. Biochem. 5:321-332.

Tainter, F.H. 1996. Principles of forest pathology. John Wiley, New York, NY.

Thorn, R.G. 1997. The fungi in soil. p. 63–127. *In* J.D van Elsas, J.T. Trevors and E.M.H. Wellington (ed.) Modern soil microbiology. Marcel Dekker, New York, NY.

Thorn, R.G., C.A. Reddy, D. Harris, and E.A. Paul. 1996. Isolation of saprophytic basidiomycetes from soil. Appl. Environ. Microbiol. 62:4288–4292.

Tsai, Y.-L., and B.H. Olson. 1991. Rapid method for direct extraction of DNA from soil and sediments. Appl. Environ. Microbiol. 57:1070–1074.

Tsao, P.H., and S.O. Guy. 1977. Inhibition of *Mortierella* and *Pythium* in a *Phytophthora*-isolation medium containing hymexazol. Phytopath. 67:796–801.

Tsuji, T., Y. Kawasaki, S. Takeshima, T. Sekiya, and S. Tanaka. 1995. A new fluorescence staining assay for visualizing living microorganisms in soil. Appl. Environ. Microbiol. 61:3415-3421.

Tunlid, A., and D.C. White. 1992. Biochemical analysis of biomass, community structure, nutritional status, and metabolic activity of microbial communities in soil. p. 229-262. *In* G. Stotsky and G.M. Bollag (ed.) Soil biochemistry. Vol 7. Marcel Dekker, New York, NY.

Untereiner, W.A., N.A. Straus, and D. Malloch. 1995. A molecular-morphotaxonomic approach to the systematics of the Herpotrichiellaceae and allied black yeasts. Mycol. Res. 99:897–913.

Van de Peer, and R. De Wachter. 1997. Evolutionary relationships among the eukaryotic crown taxa taking into account site-to-site rate variation in 18S rRNA. J. Mol. Evol. 45:619–630.

Van de Peer, Y., G. Van der Auwera, and R. De Wachter. 1996. The evolution of Stramenopiles and Alveolates as derived by "substitution rate calibration" of small ribosomal subunit RNA. J. Mol. Evol. 42:201–210.

Van der Auwera, G., and R. De Wachter. 1996. Large-subunit rRNA sequence of the Chytridiomycete *Blastocladiella emersonii*, and implications for the evolution of zoosporic fungi. J. Mol. Evol. 43:476–483.

Van der Auwera, G., R. De Baere, Y. Van de Peer, P. De Rijk, I. Van den Broeck, and R. De Wachter. 1995. The phylogeny of the Hyphochytriomycota as deduced from ribosomal RNA sequences of *Hyphochytrium catenoides*. Mol. Biol. Evol. 12:671–678.

Vance, E.D., P.C. Brookes, and D.S. Jenkinson. 1987. An extraction method for measuring soil microbial biomass C. Soil Biol. Biochem. 19:703-707.

Velvis, H. 1997. Evaluation of the selective respiratory inhibition method for measuring the ratio of fungal:bacterial activity in acid agricultural soils. Biol. Fertil. Soils. 25:354–360.

Voroney, R.P., and E.A. Paul. 1984. Determination of k_C and k_N *in situ* for calibration of the chloroform fumigation incubation method. Soil Biol. Biochem 16:9–16.

Watanabe, T. 1994. Pictorial atlas of soil and seed fungi. Lewis Publishers, Boca Raton, FL.

Webster, J. 1981. Introduction to fungi. 2nd Ed. Cambridge University Press, Cambridge, UK.

Wedin, M., and L. Tibell. 1997. Phylogeny and evolution of Caliciaceae, Mycocaliciaceae, and Sphinctrinaceae (Ascomycota), with notes on the evolution of the prototunicate ascus. Can. J. Bot. 75:1236–1242.

Wells, K. 1978. The fine structure of septal pore apparatus in the lamellae of *Pholiota terrestris*. Can. J. Bot. 56:2915-2924

Wells, K. 1994. Jelly fungi, then and now! Mycologia 86:18-48

West, N.E. 1990. Structure and function of microphytic soil crusts in wildland ecosystems of arid to semi-arid regions. Adv. Ecol. Res. 20:179-223.

Wheelis, M.L., O. Kandler, and C.R. Woese. 1992. On the nature of global classification. Proc. Natl. Acad. Sci. USA 89:2930–2934.

Williams, S.T., S. Lanning, and E.M.H. Wellington. 1984. Ecology of Actinomycetes. p. 481–528. *In* M. Goodfellow, M. Mordarski and S.T. Williams (ed.) Biology of Actinomycetes. Academic Press, London, UK.

Williams, S.T., and C.S. Robinson. 1981. The role of Streptomycetes in decomposition of chitin in acid soils. J. Gen. Microbiol. 127:55–63.

Woese, C.R. 1987. Bacterial evolution. Microbiol. Rev. 51:221–271.

Wood, D.A. 1979. A method for estimating biomass of *Agaricus bisporus* in a solid substrate, compost wheat straw. Biotechnol. Lett. 1:255-260.

Worrall, J.J. 1991. Media for the selective isolation of Hymenomycetes. Mycologia 83:296–302.

Worrall J.J., S.E. Anagnost, and R.A. Zabel. 1997. Comparison of wood decay among diverse lignicolous fungi. Mycologia 89:199-219

Yeates, C., M.R. Gillings, and D.A. Veal. 1997. PCR amplification of crude microbial DNA extracted from soil. Lett. Appl. Microbiol. 25:303–307.

Zhou, J., M.E. Davey, J.B. Figuera, E. Rivkina, D. Gilichinsky, and J.M. Tiedje. 1997. Phylogenetic diversity of a bacterial community determined from Siberian tundra soil DNA. Microbiol. 143:3913–3919.

1.4 Mycorrhizae

John Klironomos

University of Guelph

1.4.1 What is a Mycorrhiza?

Mycorrhizae (Gr. fungus root) are mutualistic symbiotic associations between plant roots and soil fungi. The plant provides the fungus with C in the form of photosynthate, and in return, the fungus provides the plant with inorganic nutrients. Recently, other benefits to the plant by the fungal symbiont have also been identified, such as protection from soil borne pathogens (Newsham et al., 1995). However, not all fungus-root interactions are labeled as mycorrhizal, even if both organisms benefit. In the soil, there is a gradient of increasing association between plants and soil microorganisms. At one extreme, one finds the general soil microorganisms (Barron, 1972). There are many free-living fungi in the rhizosphere, such as *Trichoderma* spp and *Penicillium* spp that can stimulate plant growth. They are responsible for the mineralization of soil organic matter. Nutrients that are produced from mineralization are not always available to plants. It depends on how nutrient-limited the microorganisms are, as well as the proximity of the nutrients to active roots. Mycorrhizal fungi, on the other hand, are intimately associated with roots of plants and have coevolved over millions of years. They develop intraradical structures, but can also extend into the soil, farther than the rhizosphere, to capture nutrients that were otherwise not available to the plant. The vast majority of plants on Earth are believed to be mycorrhizal (Trappe, 1987), so such associations are not oddities, but rather the norm.

1.4.2 Types of Mycorrhizae

Several types of mycorrhizae have been described based on morphological characteristics (Table 1.2). They can also be divided into two groups based on the types of fungi that are involved. Arbuscular mycorrhizae (AM) involve fungi from the order Glomales (Division Zygomycota). Their hyphae are typically aseptate, and they reproduce asexually. All other types involve fungi from the Division

Table 1.2 Morphological characteristics of mycorrhizal types

Type	Arbuscules	Intracellular colonization	Fungal mantle	Hartig net
Arbuscular	+	+	–	–
Ecto-	–	–	+	+
Ectendo-	–	+	+/–	+
Arbutoid	–	+	+/–	+
Monotropoid	–	+	+	+
Ericoid	–	+	–	–
Orchid	–	+	–	–

Dikaryomycota. This includes both Ascomycetes and Basidiomycetes. Their hyphae are regularly septate, and often produce macroscopic, sexual fruit bodies.

The main characteristic of arbuscular mycorrhizal fungi is that hyphae can penetrate through the walls of root cortical cells to form arbuscules. Arbuscules are modified hyphae that have branched many times to increase their surface area, which makes them effective structures for nutrient transfer. This type of mycorrhiza was formerly called vesicular-arbuscular mycorrhiza (VAM). One no longer uses this term because many Glomalean species do not form vesicles. This is the most common type of mycorrhiza. Approximately 300,000 plant species form arbuscular mycorrhizae from all plant divisions, including Bryophytes, Pteridophytes, Gymnosperms and Angiosperms. There is hardly an ecosystem that does not contain arbuscular mycorrhizal symbioses.

The six other types of mycorrhizae collectively involve fewer plant species. Ecto-, ectendo-, arbutoid, and ericoid mycorrhizae are found mainly on some shrubs and trees. Monotropoid mycorrhizae are associated with the Monotropaceae, a group of achlorophyllous plants. Orchids also contain a unique type of mycorrhiza (Orchid mycorrhiza). In these groups, the type of mycorrhiza is defined by the presence of specific structures in plant roots, and is thus influenced by the identity of specific fungi and plants. Certain fungal species can form different types of mycorrhizae with different host plants. In ectomycorrhizae, the fungus forms a thick mantle around most active feeder roots. Hyphae from the mantle penetrate the roots and form an extensive branched mycelium that envelopes each cortical cell. This is referred to as the hartig net. Penetration of plant cells is typically absent. Other hyphae extend away from the mantle into the soil matrix.

In ectendomycorrhizae, the mantle is reduced or absent. There is a hartig net, but some hyphae also penetrate into cortical cells. In arbutoid mycorrhizae, the mantle and hartig net are well developed, and cortical cells are heavily colonized by coiled hyphae. Monotropoid mycorrhizae are also similar in structure; the mantle and hartig net are well developed, but in this case, the plants are achlorophyllous, and C flows from the fungus to plant. The fungi extend out into the soil and form an ectomycorrhizal association with a nearby tree. Carbon is then shunted from the tree to the monotropoid plant. These plants are essentially fungal mediated plant parasites. Orchids are also achlorophyllous early in their development and are thus heterotrophic. They also depend on carbohydrate translocated to them via fungal mycelia. The source of C is usually dead organic matter.

Arbuscular and ectomycorrhizae are the most commonly found types, and also the most economically important so they are discussed in more detail below. For more information on all types of mycorrhiza, refer to the excellent books by Allen (1991) and Smith and Read (1997). There are a number of good methods-based books (Brundrett et al., 1994; Norris et al., 1994; Brundrett et al., 1996; Smith and Dickson, 1997).

1.4.3 Arbuscular Mycorrhizae

1.4.3.1 Plant Symbionts

Arbuscular mycorrhizae are found in all divisions of land plants, and on every biome on earth. The range of host plants is extremely wide. Only 3% of plant species have been examined as to their mycorrhizal status, but most plant families have been included in this survey (Trappe, 1987). Of the entire terrestrial flora, over 85% of plant species belong to families that are typically arbuscular mycorrhizal. Even in those few families that are considered to be non-mycorrhizal (Brassicaceae, Caryophyllaceae, Chenopodiaceae, Juncaceae, Polygonaceae), arbuscular mycorrhizal structures have been reported, albeit not very frequently. With over 95% of plants remaining to be surveyed, much work remains to be done if one is to generalize with more confidence.

The majority of trees, shrubs and herbs that have been surveyed have tested positive for arbuscular mycorrhizae (Baylis, 1962; Janos, 1987). Except for members of the Pinaceae, conifer families are dominantly arbuscular mycorrhizal, as are most angiosperms. Most experimental work has been performed on herbs (Klironomos and Kendrick, 1993). They are easier to grow under controlled settings, and their relatively short life cycles make them more manageable. Crop plants, in particular, have been well studied.

Some plant species are not always colonized by mycorrhizal fungi, yet they can still proliferate in nature. These are referred to as facultative mycorrhizal plants. In contrast, obligate mycorrhizal (or mycorrhizal dependent) plants are always colonized and cannot typically survive in the absence of a suitable mycobiont. Very few habitats on Earth are predominantly non-mycorrhizal, and these normally include heavily disturbed areas (Allen, 1991). Mycorrhizal dependent plants are not found in such habitats, at least until successful migration of mycorrhizal fungal inoculum (Allen, 1989).

1.4.3.2 Fungal Symbionts

The fungi that form arbuscular mycorrhizae all belong to one fungal order (Glomales) in the Division Zygomycota (Morton and Benny, 1990). There are six genera in total: *Acaulospora*, *Entrophospora*, *Gigaspora*, *Glomus*, *Sclerocystis*, and *Scutellospora*, containing approximately 150 species. More than 80 of those species belong to *Glomus*. Their taxonomy is based on subcellular characters of asexually produced spores. They are the largest of all fungal spores (< 1mm diam), but are difficult to identify with confidence, since many characters disappear with age. Thus, it is recommended that fungi are not identified from field collected spores, but rather from young spores produced in dual cultures. The majority of these fungal taxa are ubiquitous, and are non-host specific. They can colonize the roots of most plants. They are all, however, obligate biotrophs. They have no saprobic abilities and, thus, cannot be cultured under axenic conditions. They can only be maintained in dual culture with a suitable living host plant.

There is fossil evidence that these fungi arose during the Devonian period (~ 400 million yr BP). This is the same period in which plants were making the transition from water to land, and it has been hypothesized that this newly evolved mycorrhizal symbiosis was largely responsible for their success (Pirozynski and Malloch, 1975). Early plants did not possess true roots, but they did possess rhizoids, and these were infected by fungal hyphae, arbuscules and vesicles. Early soils were harsh environments for plants and largely deficient in nutrients. Fungi are absorbotrophs, efficient nutrient capturing machines. They, in turn, were limited by C availability. Fossil structures have been observed from the Phynie chert flora and from the Triassic flora of Antarctica (Stubblefield et al., 1987; Remy et al., 1994).

1.4.3.3   Morphology and Function

Arbuscular mycorrhizal fungi occupy two matrices in the soil, the intra- and extraradical matrices. Colonization of plant root can be initiated by spores, root fragments, and extraradical hyphae. Development of mycorrhizae is highly structured and under the control of plant and fungal genes, which act in a coordinated manner (Peterson and Bonfante, 1994). To enter the root, the fungus penetrates the plant epidermis, sometimes through root hairs, and then penetrates into the cortical layers. It then grows between and within cortical cells, but cannot penetrate the endodermis, and thus never enters into the vascular tissue. Intraradical hyphae will often penetrate cortical cells to form haustorial structures called arbuscules. These highly branched structures represent the plant/fungus interface and the sites of inorganic nutrient transfer from fungus to plant. Arbuscules are usually short lived (days to weeks). Carbon transfer from plant to fungus occurs between intercellular hyphae and plant cortical cells. Up to 20% of net photosynthate is transferred to the fungus. Glucose, which can be absorbed by the fungus, is then incorporated into trehalose and becomes unavailable to the plant.

Periodically, hyphal tips will swell to form vesicles. These can be intra- or intercellular. They are typically thick walled, rich in lipid, and are reconsidered to be resting structures that can reinfect uncolonized plant roots. Vesicles are not formed by species of *Gigaspora* and *Scutellospora*, and hence, the term vesicular arbuscular mycorrhiza (VAM) was recently changed to arbuscular mycorrhiza. Colonization inside the root is not systemic, but rather is restricted to localized infection units.

The extraradical matrix is composed of a complex mycelial network and spores (Friese and Allen, 1991). There are typically two types of hyphae, runner and absorptive hyphae. Runner hyphae grow along root systems initiating new points of infection. They can also bridge different root segments from the same or different plants. Absorptive hyphae develop from runner hyphae, are highly branched and extend into the soil. They can penetrate soil pores normally inaccessible to roots. Their hyphal tips actively absorb inorganic nutrients, particularly P, which are then translocated back to the plant. Transfer to the plant occurs at the arbuscule. The extraradical hyphal network is active throughout the seasons and can survive extremes in temperature and moisture. Spores are also formed in the extraradical network. They are the largest spores in the fungal kingdom. They range from 20 to almost 1,000 μm in diameter. They are not products of sexual reproduction, and thus not zygospores. Spores contain thousands of nuclei, are effective long-term resting structures and units of dispersal.

Fungi dominate in soils, and a large proportion of that belongs to mycorrhizal fungi. For example, a deciduous forest dominated by arbuscular mycorrhizal plants contained an average of 4 km hyphae g^{-1} dry soil (Klironomos and Kendrick, 1995). Fifty percent of that was estimated to belong to arbuscular mycorrhizal fungi with the remainder being saprobes and pathogens. Mycorrhizal fungi, thus, represent a significant C pool in soil ecosystems and are involved in other functions such as soil aggregation.

1.4.4   Ectomycorrhizae

1.4.4.1   Plant Species Distribution

The species diversity of plants that forms ectomycorrhizae is much more restricted than that of arbuscular mycorrhizae (Smith and Read, 1997). There are ~ 2,000 plant species, mainly woody, belonging to families such as the Pinaceae, Salicaceae, Betulaceae, Fagaceae, Caesalpinioideae, Dipterocarpaceae, Myrtaceae, Papilionoideae, Rosaceae and a few others. Although only a small number of plant species are ectomycorrhizal, they are very commonly found, covering large areas of the globe and are of high economic value. For example, the ectomycorrhizal Pinaceae are dominant in boreal ecosystems of the northern hemisphere, the Fagaceae dominate in northern temperate forests, and the Myrtaceae dominate in temperate and tropical forests of the southern hemisphere.

1.4.4.2 Fungal Species

In contrast to arbuscular mycorrhizae, a large number of fungal taxa are involved in ectomycorrhizal associations. These include more than 90 genera and 5,000 species (Molina et al., 1992). They are members of the division Dikaryomycota. Most of these are basidiomycetes and some are ascomycetes. Most form large fruit bodies, such as mushrooms and truffles, some of which are poisonous, others prized edibles. Most are epigeous (above ground), many are hypogeous (below ground). Most of the ascomycetes are hypogeous, including the black truffle (*Tuber melanosporum*). Many of these mycobionts belong to the cosmopolitan agaric genera like *Russula*, *Lactarius*, *Cortinarius*, *Amanita*, *Tricholoma*, and *Laccaria*. *Cortinarius* alone is estimated to have 2,000 species. Some families are entirely mycorrhizal, for example: Boletaceae, Gomphidiaceae, Russulaceae, and Cantherellaceae. Most known hypogeous basidiomycetes such as *Rhizopogon*, *Hymenogaster* are also mycorrhizal.

Compared to Glomalean fungi, ectomycorrhizal fungi are host specific. Many fungi are associated with only a select group of plants, often within the same genus. For example, *Suillus grevillei* associates only with *Larix*, and *Suillus lackei* only with Douglas fir. Also, there is usually a succession of fungi that are associated with a plant throughout its lifespan. Several hundred fungal species can be found associated with a single plant species. There are other fungi, however, that have a broad host range with a worldwide distribution such as *Pisolithus tinctorius*, *Thelephora terrestris*, and *Cenococcum geophilum* (Molina et al., 1992).

1.4.4.3 Morphology and Function

The morphology of an ectomycorrhiza differs dramatically from that of an arbuscular mycorrhiza (Peterson and Bonfante, 1994). Macroscopically, ectomycorrhizal roots are thicker and have altered branching patterns. Lateral roots are typically surrounded by a dense mass of fungal mycelium called a mantle. Depending on the plant and fungal species involved, the thickness of the mantle can vary from 1 to 30 cells thick. The mantle is often a different color, so it is easily distinguishable from non-colonized root segments. In contrast, roots infected by arbuscular mycorrhizal fungi do not look any different at a macroscopic level compared to non-mycorrhizal roots. It is necessary to stain the roots and observe them under a microscope for any intraradical evidence of a symbiosis.

From the developing mantle, hyphae penetrate the root but not the cells. They develop a mycelial network, forming a fungal envelope around each individual cortical cell. This unique highly branched network is called a hartig net. Cortical cells are virtually separated from each other and can only communicate with each other via plasmodesmata. The hartig net is the plant/fungus interface, where solutes are transferred between the partners, C from plant to fungus, and inorganic nutrients from fungus to plant.

Ectomycorrhizal fungi also produce an extensive extraradical hyphal network that extends into the soil from the mantle. These fungi are typically found in highly organic soils (Read, 1991). They can easily utilize NH_4-N, NO_3-N as well as complex organic compounds. Ammonium N tends to be the preferred inorganic source. In addition, the fungal mantle is an important structure in absorption, storage, and transfer of P that is captured from organic and inorganic sources. Ectomycorrhizal fungi are facultative biotrophs. They have limited saprobic abilities, but the majority of their C is obtained directly from the phytobiont. Up to 20% of plant photosynthate is allocated to ectomycorrhizal fungi. This is used to maintain the extensive hyphal network and the production of fruiting bodies.

1.4.5   Fungal Plant Pathogens

Fungi are responsible for approximately 60% of all plant diseases, and many of these are soil borne (Kendrick, 1992). Plant pathogens can be classified into two groups: obligate biotrophs and facultative saprobes. Obligate biotrophs have lost their independence, and can only grow on or in a suitable host. They typically have a very narrow host range, and are usually pathogenic on ephemeral host parts. They do not survive well in the off season, when suitable host plants die back. In contrast, facultative saprobes can live saprobically or parasitically. They typically have a broad host range, and many attack annual herbaceous plants and survive between growing seasons as members of the normal soil mycota. They are usually difficult to control. A good example is *Fusarium oxysporum*, which can survive in the soil for decades as a saprobe. The majority of soil-borne plant pathogens are facultative saprobes (Bruehl, 1987).

Both Protoctistan and Eumycotan fungi contain important plant pathogens. Members of the Chytridiomycota are typically obligate parasites, but most have a relatively broad host range. They are primitive fungi that do not produce hyphae. Zoospores infect plant tissues, and their vegetative thallus typically consists of one cell which is then converted into a reproductive sporangium producing many more zoospores. Many zoosporic fungi are not pathogens themselves but are dispersal vectors for viruses. A good example is *Olpidium brassicae*, which has a very wide host range.

In the Oomycota, the order Peronosporales contains obligate and facultative plant parasites. Most are soil borne and cause epidemics on important crops. For example, species of *Pythium* produce pectinases that dissolve plant tissues, or cause damping off. Species of *Phytophthora* have a worldwide distribution. They can either have a broad host range, infecting plants from different families, or a specific host range focusing on one or two plant species. A famous example of the latter is *Phytophthora infestans* that wiped out the Irish potato crop in 1845-1847, causing more than a million human deaths.

The kingdom Eumycota (the higher fungi), however, contains the highest diversity of plant pathogens. Many teleomorphic fungi are important pests such as *Ceratocystis* spp (Ascomycotina) causing vascular wilts in deciduous trees, *Sclerotinia* spp (Ascomycotina) producing their characteristic sclerotia and causing diseases on a wide variety of crops worldwide, and *Armillaria* spp (Basidiomycotina) which occur on a wide range of angiosperms and gymnosperms. The most widespread and diverse group of plant pathogens, however, are the dikaryomycotan anamorphs, such as *Cylindrocarpon*, *Fusarium, Rhizoctonia* and *Verticillium*, among many others. For a few recommended books, Farr et al. (1989) provide a reference list of fungi found on different plants and plant products, Domsch et al. (1980) provide a comprehensive literature review of commonly found soil fungi, including pathogens, and Singleton et al. (1993) provide detailed methods for studying various soil-borne pathogenic fungi.

1.4.6   References

Allen, M.F. 1991. The ecology of mycorrhizae. Cambridge University Press, Cambridge, UK.

Allen, M.F. 1989. Mycorrhizae and rehabilitation of disturbed arid soils: processes and practices. Arid Soil Res. Rehabil. 3:229-241.

Barron, G.L. 1972. The genera of hyphomycetes from soil. Krieger Publishing Company, Huntington, NY.

Baylis, G.T.S. 1962. Rhizophagus: The catholic symbiont. Austr. J. Sci. 25:195-200.

Bruehl, G.W. 1987. Soilborne plant pathogens. McMillan Publishing Co, New York, NY.

Brundrett, M., L. Melville, and L. Peterson. 1994. Practical methods in mycorrhiza research. Mycologue Publications, Waterloo, Canada.

Brundrett, M., N. Bougher, B. Dell, T. Grove, and N. Malajczuk. 1996. Working with mycorrhizas in forestry and agriculture. ACIAR Monogr. 32, Canberra, Australia.

Domsch, K. H., W. Gams, and T.-H. Anderson. 1980. Compendium of soil fungi. Vol 1. Academic Press, New York, NY.

Farr, D.F., G.F. Bills, G.P. Chamuris, and A.Y. Rossman. 1989. Fungi on plants and plant products in the United States. APS Press, St. Paul, MN.

Friese, C.F., and M.F. Allen. 1991. The spread of VA mycorrhizal fungal hyphae in the soil: Inoculum types and external hyphal architecture. Mycologia 83:409-418.

Janos, D.P. 1987. VA mycorrhizae in humid tropical ecosystems. p. 107-134. *In* G.R. Safir (ed.) Ecophysiology of VA mycorrhizal plants. CRC Press, Boca Raton, FL.

Kendrick, B. 1992. The fifth Kingdom. 2nd Ed. Mycologue Publications, Waterloo, Canada.

Klironomos, J.N., and W.B. Kendrick. 1993. Research on mycorrhizas: Trends in the past 40 years as expressed in the 'Mycolit' database. New Phytol. 125:595-600.

Klironomos, J.N., and B. Kendrick. 1995. Relationships among microarthropods, fungi, and their environment. Plant Soil 170:183-197.

Molina, R., H. Massicotte, and J.M. Trappe. 1992. Specificity phenomena in mycorrhizal symbiosis: community-ecological consequences and practical implications. p. 357-423. *In* M.F. Allen (ed.) Mycorrhizal functioning. Chapman and Hall, London, UK.

Morton, J.B., and G.L. Benny. 1990. Revised classification of arbuscular mycorrhizal fungi (Zygomycetes): A new order, Glomales, two new suborders, Glominae and Gigasporineae, and two new families, Acaulosporaceae and Gigasporaceae, with an emendation of Glomaceae. Mycotaxon 37:471-491.

Newsham, K.K., A.H. Fitter, and J.W. Merryweather. 1995. Multifunctionality and biodiversity in arbuscular mycorrhizas. Trends Ecol. Evol. 10:407-411.

Norris, J.R., D. Read, and A.K. Varma. 1994. Techniques for mycorrhizal research. Academic Press, San Diego, CA.

Peterson, R.L., and P. Bonfante. 1994. Comparative structure of vesicular-arbuscular mycorrhizas and ectomycorrhizas. Plant Soil 159:79-88.

Pirozynski, K.A., and D.W. Malloch. 1975. The origin of land plants: A matter of mycotropism. Biosyst. 6:153-164.

Read, D.J. 1991. Mycorrhizas in ecosystems. Experimentia 47:376-391.

Remy, W., T.N. Taylor, H. Haas, and H. Kerp. 1994. Four hundred-million-year-old vesicular-arbuscular mycorrhizae. Proc. Nat. Acad. Sci. 91:11841-11843.

Singleton, L.L., J.D. Mihail, and C.M. Rush. 1993. Methods for research on soilborne phytopathogenic fungi. APS Press, St. Paul, MN.

Smith, S., and S. Dickson. 1997. VA mycorrhizas: Basic research techniques. Cooperative Research Centre for Soil and Land Management, Glen Osmond, Australia.

Smith, S.E., and D.J. Read. 1997. Mycorrhizal symbiosis. 2nd Ed. Academic Press, San Diego, CA.

Stubblefield, S.P., T.N. Taylor, and J.M. Trappe. 1987. Antarctic VAM fossils. Am. J. Bot. 74:1904-1911.

Trappe, J.M. 1987. Phylogenetic and ecologic aspects of mycotropy in the angiosperms from an evolutionary standpoint. p. 5-25. *In* G.R. Safir (ed.) Ecophysiology of VA mycorrhizal plants. CRC Press, Boca Raton, FL.

2

Soil Fauna

Paul F. Hendrix, Editor
University of Georgia

The soil fauna consist of a broad grouping of animals that spend all or part of their lives in soil, and in many situations have significant effects on structural and functional properties of soils. In total numbers and biomass, the soil fauna are dominated by invertebrates, representing many animal phyla and constituting a rich diversity of organisms and adaptations to life in the soil. This chapter covers the groups of soil fauna most often considered important in soil processes and most commonly encountered in the literature or in the field. Although sometimes important in local situations, several groups of animals are not considered in this chapter, including the rotifers and tardigrades, flatworms, molluscs (snails and slugs), and vertebrates (burrowing reptiles and mammals). Information on these animals can be found in the general references cited in the following section.

2.1 Protozoa

Stuart S. Bamforth
Tulane University

2.1.1 Introduction

Protozoa and nematodes, the microbial feeding microfauna, are pivotal organisms in nutrient recycling. They live in the water films covering soil aggregates and filling their pores, where bacteria and fungi decompose organic matter and immobilize the extracted nutrients into their bodies. Microfaunal grazing regulates the size and composition of the microbial community, and enhances microbial and plant growth through microfaunal excretions. Protozoa, especially small flagellates and amoebae, can graze bacteria in tiny pore spaces unavailable to nematodes. Estimates of numbers of protozoa in soils are presented in Table 2.1. Organic matter, organisms, and water are scattered discontinuously in soils, producing intermittent activity and inactivity; consequently, many protozoa (like bacteria) exist in a dormant (encysted) state most of the time. Activity is increased by comminution of organic matter and redistribution of microsites by microarthropods and larger fauna, especially earthworms.

2.1.2 Basic Taxonomy

Soil protozoa comprise four groups: the "naked" flagellates, amoebae and ciliates, and the more slowly growing shelled amoebae or testacea (Fig. 2.1). Ciliates constitute a well-defined taxon, but the

Table 2.1 Numbers of protozoa in soils and litters, from selected sources. The number in parentheses refers to the table describing the method employed.

Soil type and region	Nos. x 10^3 g^{-1}	Source
Total Protozoa		
Hardwood forest, New Zealand	24.00	Yeates et al. (1991)
Pasture, New Zealand	60.00	Yeates et al. (1991)
Permanent grassland, Scotland	1.13	Griffiths and Ritz (1988)
Arable field soil, Scotland	14.73	Griffiths and Ritz (1988)
Flagellates		
Trop. rain forest, Costa Rica		
ground litter	156.83	Bamforth and Lousier (1995)
suspended soils	66.15	Bamforth and Lousier (1995)
Spruce litter, Austria	2.00–10.00	Foissner (1991)
Savanna, Kenya	6.18	Bamforth and Lousier (1995)
Cropland, Scotland	0.57–13.95	Darbyshire and Greaves (1967)
Desert soils, Arizona	0.14–2.32	Bamforth (1984)
Amoebae		
Trop. rain forest, Costa Rica		
ground litter	972.93	Bamforth and Lousier (1995)
suspended soils	316.35	Bamforth and Lousier (1995)
Pine forest, Sweden	100–2,000	Clarholm (1981)
Savanna, Kenya	57.61	Bamforth and Lousier (1995)
Cropland, Scotland	10.58–50.05	Darbyshire and Greaves (1967)
Desert soils, Arizona	0.76–9.40	Bamforth (1984)
Ciliates		
Trop. rain forest, Costa Rica		
ground litter	68.89	Bamforth and Lousier (1995)
suspended soils	8.54	Bamforth and Lousier (1995)
Spruce forest, Austria	0.60	Aescht and Foissner (1993)
Savanna, Kenya	0.21	Bamforth and Lousier (1995)
Wheat field, tilled, Austria	1.28	Foissner (1987)
Wheat field, organic, Austria	1.02	Foissner (1987)
Cropland, Scotland	0.03–0.28	Darbyshire and Greaves (1967)
Desert soils	0.11–0.46	Bamforth (1984)
Testacea		
Aspen woodland, Canada	93.48	Lousier (1982)
Oak forest, Belgium	4.30–32.50	Couteaux (1975)
Alpine forest, Austria	1.29	Berger et al. (1986)
Wheat field, tilled, Austria	0.75	Foissner (1987)
Wheat field, organic, Austria	0.87	Foissner (1987)
Fellfields, Signy Island	0.25–0.45	Smith and Tearle (1985)

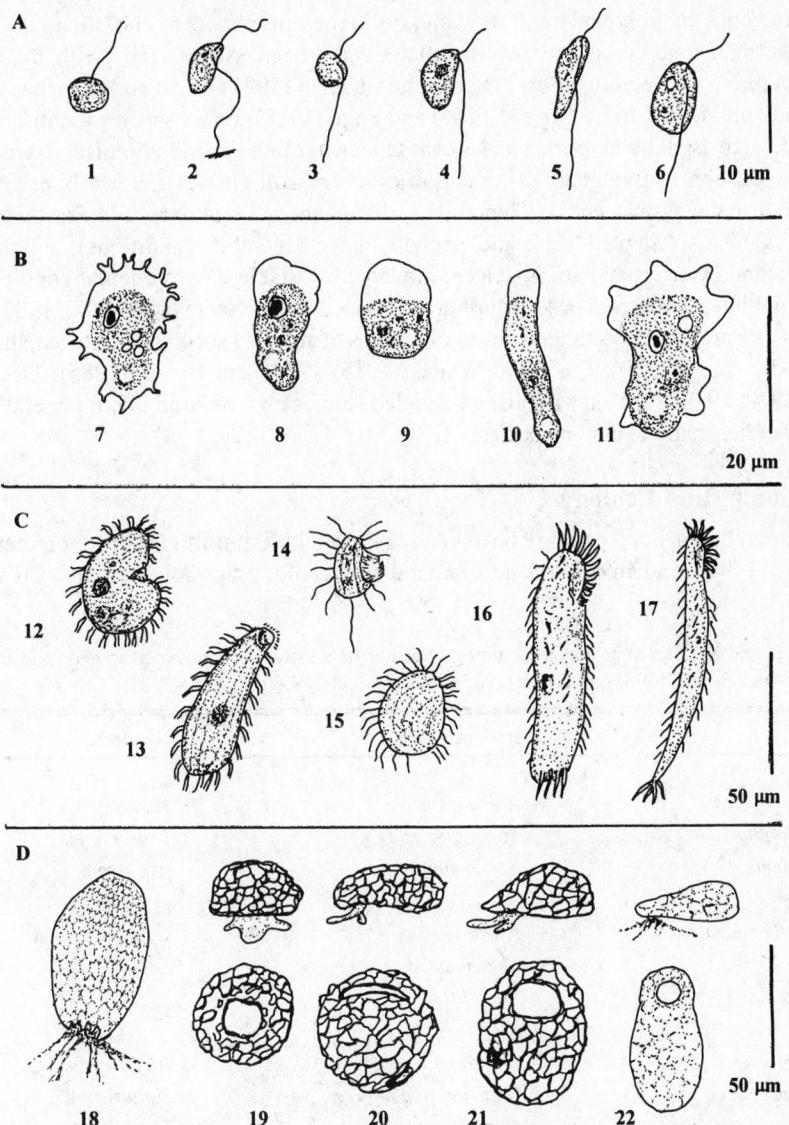

Fig. 2.1 Some common soil protozoa. For the 3 naked groups, r-selected species are arranged on the left, intermediate and K-selected species on the right (see Table 2.2). (A) flagellates: (1) *Oikomonas termo*, (2) *Pleuromonas jaculans* (= *Bodo saltans*), (3) *Heteromita globosa*, (4) *Bodo edaz*, (5) *Cercobodo* sp., (6) *Anisonema ovale*. (B). Gymnamoebae: (7) *Acanthamoeba* sp., (8) *Vahlkampfia* sp., (9) *Platyamoeba placida*, (10) *Hartmanella* sp., (11) *Mayorella* sp. (C) Ciliates: (12) *Colpoda inflata*, (13) *Platyophyra vorax*, (14) *Cyclidium muscicola*, (15) *Leptopharynx costatus*, (16) *Holosticha* sp., (17) *Hemisincirra gracilis*. (D). Testacea, arranged by shell type (acrostome, globose, wedge): (18) *Euglypha* sp., (19) *Phryganella* sp., (20) *Plagiopyxis callida*, (21) *Centropyxis aerophila*, (22) *Trinema enchelys*.

other three groups are polyphyletic. Lee et al. (1985) presents a modern classification. The scattered (and obsolete) taxonomic studies of flagellates have been synthesized (with a review of soil flagellates) into a single volume by Patterson and Larsen (1991). Most soil amoebae belong to the Lobosea and are described by Bovee (1985) and Page (1988). Exact species identifications among major taxa, such as Vahlkampfidae and *Acanthamoeba*, often depend upon isoenzyme techniques. With the exception of two genera, the classification of soil ciliates follows both taxonomic and ecological groupings (taxocenes). Colpodida and the genera *Keronopsis* and *Paraholosticha* are r-selected organisms (adapted for rapid growth under favorable conditions); Polyhymenophora (hypotrichs and heterotrichs) are K-selected (adapted to stable environmental conditions); and all other taxa are intermediates. These groups are described by Foissner (1980, 1982, 1987) (Table 2.2). Although there are several detailed studies on groups of selected species of testacea, the best overall works are by Cash (1909), Cash and Wailes (1915, 1919) and Bovee (1985). Decloitre (1976, 1977a,b, 1978, 1979, 1981) has published detailed studies on five large major genera, and Foissner (1987) illustrates many common species.

2.1.3 Biology and Ecology

Soil protozoan biology and ecology have been reviewed by Bamforth (1985), Foissner (1987), and Darbyshire (1994), and integrated into general soil ecology by Coleman and Crossley (1996).

Table 2.2 r- and K-selected protozoa including indicator taxa of (a) acid moder-mor; and mild mull (m) humus conditions [After Bonnet (1964); Foissner (1987); Wodarz et al. (1992)]

r-selected	intermediate	K-selected
Flagellates		
Bodo	*Anisonema*	*Heteronema*
Heteromita	*Mastigamoeba*	*Notosolenus*
Pleuromonas	*Scytomonas*	*peranema*
Amoebae		
Acanthamoeba	*Hartmanella*	*Gephyramoeba*
Platyamoeba	*Thecamoeba*	*Leptomyxa*
Vahlkampfia	*Vanella*	
Ciliates		
(a) *Bryometopus*	*cyclidium*	*Blepharisma*
colpoda	(a) *Frontonia depress.*	*Halteria*
Cyrtolophosis	*Homalogastra*	*Hemisincirra*
Grossglockneria	*Leptopharynx*	*Holosticha*
Platyophrya	*Sathrophilus*	*Oxytricha*
Pseudocyrtolophosis	*Spathidium*	*Urosomoida*
Testacea		
Cryptodifflugia	*Arcella*	(m) *Bullinularia*
difflugiella vulgaris	(a) *Assulina*	some *Centropyxis*
many *Euglypha*	(a) *Corythion*	many *Heleopera*
Schoenbornia	(a) many *Nebela*	many *Plagiophyxis*
Trinema enchelys	*Phryganella acro.*	(m) *Pseudawerintzewia*
T. Lineare	(a) *Trignonopyxis*	(m) *Schwabia terricola*

Protozoa spend less energy on locomotion and maintenance and more on growth than metazoa. Their rapid encystment-excystment abilities enable them to adapt to the fluctuating conditions of soil life, and their cysts also facilitate transport to new sites via plant root and animal movements, and passage through animal guts.

The distribution of protozoan groups is determined by pore sizes of the aggregates in which they live. There is a protozoan domain in the smaller pore spaces, inhabited by small flagellates and amoebae; and a protozoan-nematode domain in the larger spaces, shared by all protozoan groups, rotifers, and nematodes (Bamforth, 1985).

Small size and flexibility allows the raptorially feeding small flagellates and amoebae to ingest bacteria in pores down 8 μm in diameter, where the tiny protozoa are protected from predation and water fluctuations. Most of these small protozoa are highly competitive r-selected organisms (Table 2.2), furnishing the first colonists to soils (Yeates et al., 1991) and the majority of protozoan numbers in soils (Table 2.1). The less abundant, but morphologically more diverse ciliates and testacea have adjusted with varying degrees of success to the fluctuations of moisture and CO_2 in the larger pore spaces, and show a wide spectrum of r/K selected species (Table 2.2; Fig. 2.1); hence they can serve as bioindicators of soil conditions. Many soil ciliates have small laterally compressed or elongate worm-like bodies to produce a largely autochthonous fauna (Foissner, 1987). Although many ciliates feed on bacteria, about 50% of species prey on other protozoa also. The colpodid family Grossglockneridae feed on fungal spores (Foissner, 1987).

The ventral surface of the shell of testacea is flattened for better adhesion to the substrate (during periods of desiccation) and the round aperture (pseudostome) is reduced to a slit in some edaphic forms (Fig. 2.1). Testacean shells show an evolution of four types: flat ("Arcella"), vaulted, globose, and wedge-shaped (Fig. 2.1), the latter two adapted to more xeric habitats. Different assemblages of testacea describe soil types (Table 2.3). Some specimens with a few ciliates can help distinguish acid "mor" from more alkaline "mull" soils, and many species reflect r/K selection (Table 2.2). Thus, testacea, like ciliates, can serve as indicators of soil conditions. Most testacea appear to feed on humus particles but the genus *Nebela* is carnivorous.

Table 2.3 Ecological classification of testacean communities with soil types [After Bonnet, 1964]

I.   Azonal soils low in organic matter: *Centropyxis*
     A.   Acid soils: *Corythion dubium*
         1.   Soils with saxicole vegetation and roots: *Centrophyxis deflandriana*
         2.   Skeletal soils: *Centropyxis vandeli*
     B.   Calcareous soils: *Bullinaria gracilis*
         1.   Soils very low in organic matter: *Paraquadrula* and *Hyalosphenia insecta*
         2.   Soils with accumulated organic matter: *Cento. Plagiostoma*
         3.   Decalcified clays: *Pseudawerintzeia calcicola*
         4.   White rendzinas: *Geopyxella sylvicola*
         5.   Skeletal soils: *arcella arenaria*
II.   Evolved soils rich in organic matter: *Plagiopyxis callida*
     A.   Brown forest soils: *Plagiopyxis Callida*
         1.   Mor: *Plagiopyxis callida*
         2.   Mull: *Plagio. Penardi, Euglypha polylepis*
     B.   Grassland soils: *Tracheleuglypha acolla*
         1.   Humic alpine soils: *Tracheleuglypha acolla*
         2.   Calcareous grassland: *Trach. Acolla* and *Centro. Elongata*
III   Saline soils: *Centropyxis halophila*

2.1.4   Importance to Soil Processes

Protozoa are found in the water films of microsite "hot spots" of activity, such as litter, organic matter, soil aggregates, the rhizosphere, and the "drillosphere" of worm tunnels and casts (Coleman and Crossley, 1996). Here mineralization occurs through the predation of protozoa and nematodes on the microflora. Protozoa have a greater assimilation efficiency and turnover time than other soil fauna, and consume several times their mass of living microbial tissues. Amoebae are especially important in N mineralization (Darbyshire, 1996; Coleman and Crossley, 1996). Protozoa are also ingested by higher fauna, thus serving as transformers of bacterial protoplasm into higher trophic levels.

Protozoa can serve as bioindicators of soil conditions because their exposed cell membranes ensure close association with the environment; they do not migrate much in soil; and their short life span enables them to react quickly to environmental changes. They are abundant (Table 2.1), and in stressed environments are the only fauna. Consequently, they can help evaluate soil conditions and effects of perturbations such as impoverishment leading to an increase of opportunistic (r-selected) species (Table 2.2) (Darbyshire, 1996).

2.1.5   Methods

The tiny size and large numbers of protozoa in litters and soils dictate that only a small sample can be analyzed, and since these materials are mosaics of microhabitats, composite samples must be collected from an area and thoroughly mixed. Most protozoa adhere closely to soil particles, and must be shaken off in order to be found. Another problem is distinguishing active from encysted forms. There is no universal method for soil protozoa identification and enumeration, so investigators must select the methods best addressing their objectives, recognizing the limitation of the techniques. Using a spatula or a 15-mm-diameter cork borer, 6–12 samples are collected from 0–3 cm soil depth from several 4–16 m^2 areas. Samples are combined, thoroughly mixed, and subsampled for study.

All methods described in this section require a compound microscope with oil immersion objectives, and desirably phase contrast or interference optics, since observations of living specimens enable definition of important taxonomic characters, e.g., locomotor patterns, organelles (nucleus, contractile vacuole), and cysts (amoebae). Slides, cover slips, and pipettes are needed in all methods.

Table 2.4   Species richness: Supplemented nonflooded Petri dish for protozoa in general [Bamforth, 1992, 1995; Foissner, 1987, 1992]

Materials:	
1.	Sterile Petri dishes, 10–15 cm diameter
2.	Cover slips and lens paper
3.	Distilled water
Procedure:	
1.	Place 10–50 cm of material at least 1 cm deep in a Petri dish. Saturate, but do not flood with distilled water, until 5–20 mL will drain off when material is gently pressed with a finger
2.	Place two cover slips, with a piece of lens paper under each, on the surface of the material.
3.	On day 1, transfer each cover slip (but not the lens paper) onto a drop of 25% Ringers saline on a slide and examine for flagellates. Replace cover slips onto the material and repeat examination on day 2.
4.	Examine the run-off fluid for protozoa on the following days:

	2–3	Flagellates, colpodids, some other ciliates
	5–6	Small ciliates, hypotrichs, some amoebae
	8–10	Hypotrichs, testacea
	15–20	Mainly testacea
	30	Mainly testacea

Although Page (1988) describes several isotonic media to study amoebae, 25% frog Ringers solution is more convenient to study all soil protozoa.

The naked protozoa must be "enticed" from their habitats. The nonflooded Petri dish method (Table 2.4) creates a closed system where larger, and a few smaller, protozoa appear in a succession. Supplementing the dish with cover slips (Table 2.4) and using bacterized agar culture methods (Table 2.5) entice more species of flagellates and amoebae, respectively, from their small pore spaces.

The shell of testacea allows them to be examined like larger fauna, and the direct count and stained slide methods discern living from empty tests, for both species richness and enumeration studies (Couteaux, 1967; Korgonova and Geltser, 1977; Aescht and Foissner, 1993). Because the examined quantity is so small, these methods should be supplemented by the nonflooded Petri dish method (Table 2.4) to find rare large species which may not appear on the slides.

Traditionally, the Singh most probable number (MPN) culture method and its modifications (Table 2.6) have been used to estimate numbers. A number of inaccuracies occur in this procedure (Foissner, 1987) and the reader should refer to Berthold and Palzenberger (1995) before selecting a method. To meet these criticisms, three alternate methods have been developed: (1) Density centrifugation (Griffiths and Ritz, 1988) fixes cells quickly after soil sampling. The percoll phosphate gradient centrifugation separates more protozoa from soil particles than other methods, and the "DAPI" stains only active cells. However, the method is more applicable to agricultural than forest soils (Griffiths and Ritz, 1988). (2) Direct counts of fresh soil suspensions record active ciliates by their motility, and stained testacea suspensions distinguish active from encysted forms (Aescht and Foissner, 1992; Foissner, 1992). (3) The third approach dilutes a soil suspension to isolate single cells, either spreading 1 mL replicates onto agar plates (plaque count) or inoculating aliquots of suspension into wells of tissue culture or microwell plates ensuring the dilution is high enough that some wells will be negative (Butler and Rogerson, 1995; Bamforth, 1992, 1995). This culture method measures total populations, and has been used for amoebae and ciliates, and can be used for flagellates if diluted to higher levels.

Taxonomic keys are listed under Section 2.1.2.

Table 2.5 Species richness: agar plate methods for amoebae [Adapted from Bamforth, 1995]

Materials and apparatus:
1.   Sterile Petri dishes, 10–15 cm diameter
2.   Scalpel, 10 mm cork borer, L-shaped glass rod (hockey stick)
3.   Bunsen burner

Reagents and bacteria:
1.   1.5% water (non-nutrient) agar
2.   Distilled water, 95% ethyl alcohol
3.   Culture of *Escherichia coli*, *Klebsiella aerogenes*, or a suitable soil bacterium

Procedure A. Streak plate method:
1.   Streak two "sine waves" of a bacterial suspension over the agar in a Petri dish. Add two straight streaks of a sample suspension bisecting the bacterial streaks.
2.   Incubate at room temperature (20 ºC), and examine for amoebae migrating into the bacterial streaks from 4th day and thereafter.

Procedure B. Well plate method:
1.   Using alcohol flamed instruments, cut a 5 x 1 cm rectangular trough and 3 1-cm diameter wells in the agar in a Petri dish. Spread a bacterial suspension over the agar with the hockey stick.
2.   Fill 4 wells with material and add distilled water to moisten but not overflow the wells. Incubate at room temperature and examine at 4 d and after for amoebae migrating out of the wells over the agar.

Table 2.6   Enumeration: Most probable number (MPN) methods

Materials and apparatus for all MPN methods:
1.   Sterile Petri dishes
2.   Glass or polypropylene rings (for A, 0.5 mL; for B, 1.0 cap.)

Reagents and bacteria for all MPN methods:
1.   1.5% water (non-nutrient) agar
2.   Soil extract: boil 300 g soil in 1 L distilled water for 10 min, decant, filter, and autoclave. Dilute 1:5
        extract: distilled water. Adjust to pH 7 with HCl or NaOH.
3.   Culture of *Escherichia coli, Klebsiella aerogenes*, or a suitable soil bacterium.

Method A. Singh (1955) method:
1.   Embed 8 rings in agar per Petri dish for 15 dishes (= levels). Prepare an initial 1:5 soil suspension,
        using soil extract as the diluent; then make successive two-fold dilutions (1:10, 1:20, 1:40, etc.).
2.   Pipette 0.5 mL of each dilution into each of 8 rings to obtain 15 levels.
3.   Incubate at room temperature and examine at 4-day intervals. Add drop of bacterial suspension to
        each well after day 8 to promote protozoan excystment.

Method B. Stout (1962) modification:
1.   Embed 10 1-mL capacity rings in agar per Petri dish for 3 dishes (=3 levels)
2.   Prepare 3 10-fold dilutions: 10^1-10^3 for nutrient poor, 10^4-10^6 for very rich soils. Pipette 1 mL of
        each dilution into each of 10 rings to obtain 3 levels.
3.   Incubate, examine, and calculate numbers as in Section A.

Method C. Darbyshire et al. (1974) modification
Additional materials:
        96-well microtiter plates (substitute for rings)

Procedure:
1.   Prepare two-fold dilution series as in Section A.
2.   Pipette 0.05 mL of each dilution into 8 wells of a 96-well microtiter plate.
3.   Incubate, examine, and calculate numbers as in Section A.

2.2   Nematodes
Robert McSorley
University of Florida

2.2.1   Introduction

Nematodes, or roundworms, which are common in most habitats, are a diverse group of invertebrates, including forms which inhabit marine, freshwater, or terrestrial habitats, as well as parasites of man and other vertebrates and of insects and other invertebrates. Many different kinds of nematodes inhabit soil, but while often present in large numbers, they often go unnoticed because of their cryptic habits and microscopic size, usually 0.5 to 2.0 mm in length. Over 15,000 species have been described (Poinar, 1983), but many more species remain unknown and it is common to encounter undescribed species in most soil habitats.

2.2.2   Basic Taxonomy and Identification

Nematodes belong to the Phylum Nematoda (Poinar, 1983), also referred to by some authors as the Phylum Nemata (Thorne, 1961; Maggenti, 1991). Formerly, the Nematoda were considered a class within the phylum Aschelminthes, along with rotifers and related organisms (Barnes, 1965). The phylum Nematoda is divided into two classes: Adenophorea and Secernentea (Thorne, 1961; Poinar,

1983; Freckman and Baldwin 1990). Although other authors may recognize a slightly different arrangement (Andrassy, 1976; Maggenti, 1991; Lorenzen, 1994), Poinar (1983) divides the two classes into 16 orders (Table 2.7).

Some orders of nematodes are not common in soil, but certain stages of animal parasites may sometimes be found (Levine, 1968; Anderson, 1992) and marine nematodes may occur in soil from estuaries or other coastal habitats. Many nematodes are associates of insects (referred to as entomogenous nematodes), and are occasionally found in soil (Poinar, 1979; Nickle, 1984; Gaugler and Kaya, 1990). These include insect-parasitic, or entomopathogenic, nematodes. The genera *Steinernema* and *Heterorhabditis* in the order Rhabditida have been used frequently in the biological control of soil insects (Gaugler and Kaya, 1990). Plant-parasitic nematodes are common constituents of the soil fauna. Nonparasites of all types are referred to collectively as "free-living" nematodes.

Specific references are useful for the identification of some of the orders commonly occurring in soil (Table 2.7). More general texts (Thorne, 1961; Goodey and Goodey, 1963; Ferris, 1971;

Table 2.7   Habitats and taxonomic references for nematode orders, based on classification scheme of Poinar (1983)

Order	Main Habitats	Other Habitats	Taxonomic References to Soil-Inhabiting Forms
Class Adenophorea			
Araeolaimida	Marine, freshwater	Soil	Freckman and Baldwin (1990), Andrassy (1976), Ferris, et al. (1973), Goodey and Goodey (1963)
Chromadorida	Marine, freshwater	Soil	Freckman and Baldwin (1990), Ferris et al. (1973), Goodey and Goodey (1963)
Desmodorida	Marine, freshwater	Soil	Freckman and Baldwin (1990), Ferris et al. (1973)
Desmoscolecida	Marine		
Dorylaimida	Soil	Freshwater	Ferris (1971), Thorne (1974)
Enoplida	Marine, freshwater	Soil	Freckman and Baldwin (1990), Andrassy (1976), Ferris et al. (1973), Goodey and Goodey (1963)
Mermithida	Invertebrates		
Monhysterida	Marine, freshwater	Soil	Andrassy (1983)
Mononchida	Soil	Freshwater	Jairajpuri and Khan (1982), Jensen and Mulvey (1968)
Class Secernentea			
Aphelenchida	Soil	Plants, invertebrates	Hunt (1993)
Ascaridida	Vertebrates	Invertebrates	---
Oxyurida	Invertebrates, vertebrates		
Rhabditida	Soil, freshwater	Invertebrates, vertebrates	Andrassy (1983)
Spirurida	Vertebrates, invertebrates	---	
Strongylida	Vertebrates	---	
Tylenchida	Soil	Plants, invertebrates	Mai and Mullins (1996), Siddiqi (1985)

Andrassy, 1976; Freckman and Baldwin 1990) include these as well as genera from orders which occur less frequently in soil. In most texts, identification is to genus level; species identification is difficult and many species are undescribed.

2.2.3 Biology and Ecology

2.2.3.1 Morphology

Nematodes are unsegmented roundworms and most are wormlike in appearance, although some modifications exist. They have a complete digestive system, nervous, excretory and reproductive systems, and longitudinal muscles, but lack respiratory and circulatory systems. Sexes are separate, although intersexes occur in some species (Bird and Bird, 1979; Wharton, 1986). Usually reproduction is sexual and amphimictic, although some species reproduce parthenogenetically or hermaphroditically (Bird and Bird, 1979; Poinar, 1983; Wharton, 1986). With substage lighting under a microscope, nematodes appear transparent, so that internal anatomical and morphological features important in taxonomy can be examined. The characteristics of the stoma and esophagus are particularly important in separating some of the nematode orders present in soil (Fig. 2.2). In some groups, the stoma is modified into a needle-like stylet to aid in certain types of feeding.

2.2.3.2 Life Cycle

A typical nematode life cycle consists of the egg, four juvenile stages, and the adult (Bird and Bird, 1979; Wharton, 1986). The juvenile stages differ in size but are similar morphologically. Modifications and exceptions occur in some groups, such as the plant-parasitic genus *Meloidogyne*, in which latter juvenile stages become swollen, developing into a saccate female (Poinar, 1983).

The lengths of nematode life cycles vary from a few days to more than a year, with those of most common plant parasites probably averaging 3–6 weeks and those of some free-living forms only 1–2 weeks (Norton, 1978; Bird and Bird, 1979). In all cases, the length of the life cycle is strongly dependent on temperature (Norton, 1978).

Various types of delayed development, such as the formation of dauer larvae, may occur in some species in response to adverse environmental conditions (Bird and Bird, 1979; Wharton, 1986). In extreme cases, nematodes of some species may enter anabiosis, an extended state of metabolic dormancy which may last for years (Wharton, 1986; Norton, 1978). Anhydrobiosis is the anabiotic response to low soil moisture and desiccation (Wharton, 1986).

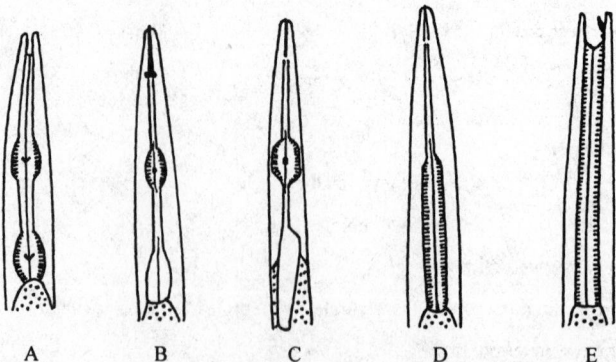

Fig. 2.2 A. Rhabditida. B. Tylenchida. C. Aphelenchida. D. Dorylaimida. E. Mononchida. The most anterior region of the digestive system (stoma) is modified into a stylet in B, C, and D.

2.2.3.3 Nematodes in Belowground Food Webs

Nematodes are an integral part of food webs in soil ecosystems, appearing at several positions and levels due to their diverse feeding habits which form the basis for subdivisions into a number of trophic groups (Elliott et al., 1988; Moore and de Ruiter, 1991; Yeates et al., 1993). Five of these groups are frequent in soil Table 2.8): bacterivores, or microbivores, feed on soil bacteria; fungivores feed on fungal mycelia; herbivores, or phytophages, are the plant parasites; predators feed on nematodes and other small invertebrates; and omnivores feed on a variety of foods such as fungi, algae, small invertebrates, bacteria, etc. (Freckman, 1982; Freckman and Baldwin, 1990). Because the biology of many nematodes is relatively unstudied, the food habits of some groups are unknown (Yeates et al., 1993). Some of these groups, such as the family Tylenchidae in the order Tylenchida, which are common in soil, are sometimes grouped as "plant associates" because it is unclear whether their food source is discarded plant material or the fungi associated with its decomposition. Feeding habits vary within the nematode orders occurring in soil.

2.2.3.4 Abundance, Distribution, Diversity

A diverse nematode fauna of typically two or three dozen different genera are present in most soils, with numbers often in the range of 10^6 to 10^7 m^{-2} (Table 2.8). When dry biomass is reported, values in the range of 100–200 mg m^{-2} are common, except in communities containing a high percentage of dorylaimid omnivores (Table 2.9). The percent composition of the nematode fauna by trophic group varies, depending on the plant species present and the geographic location. Relatively wide ranges in trophic composition or abundance within a site may result from seasonal variation.

2.2.4 Importance to Soil Processes

2.2.4.1 Decomposition and Nutrient Cycling

Sohlenius (1980) reviewed the allocation of energy for production or respiration in nematode communities at various sites. Although nematodes accounted for < 1% of total soil respiration, they released about 20–60 kcal m^{-2} yr^{-1} in productive sites and 7–10 kcal m^{-2} yr^{-1} in cooler or unproductive sites. Nevertheless, the role of nematodes in energy cycling and flow is significant because their consumption rates are high. Estimates of nematode consumption of root biomass from several studies in Colorado and South Dakota prairies ranged from 34.8 to 57 g m^{-2} (more than vertebrate herbivores),

Table 2.8 Feeding habits of nematode orders found in soil ecosystems

Order	Bacterivores	Fungivores	Herbivores	Omnivores	Predators	Other
Class Adenophorea						
Araeolaimida	All soil forms					
Chromadorida	Few					Mostly algivores
Desmodorida	Some?					
Dorylaimida		Some	Some	Some	Some	
Enoplida	Some			Few	Some	
Monhysterida	Most soil forms					
Mononchida	Few				Most	
Class Secernentea						
Aphelenchida		Many	Some		Few	Many insect associates
Rhabditida	Most				Some	Some insect parasites
Tylenchida		Some	Most			Some plant associates, some insect parasites

Table 2.9   Examples of total nematode numbers and biomass from selected studies

Ecosystem	Location	Genera per Site	Nematodes m^{-2} ($\times 10^6$) (mean or range)	Biomass (mg dry wt m^{-2}) (mean or range)	Reference
Tundra	Summary, 15 sites		3.5		Sohlenius (1980)
Grassland	Summary, 20 sites		9.2		Sohlenius (1980)
Grassland	California	26	2.7-3.2	108-130	Freckman et al. (1979)
Grassland	New Zealand	29-32	0.8-4.5		Yeates (1981)
Grassland	New Zealand	21-35	2.3		Yeates (1984)
Grassland	South Australia	17-19	0.8-3.4		Yeates and Bird (1994)
Grassland	Poland		1.9-12.0		Wasilewska (1994)
Grassland	Sweden	38-41	7.6		Bostrom and Sohlenius (1986)
Cultivated field, barley	Sweden	38-41	6.2		Bostrom and Sohlenius (1986)
Cultivated field, soybean	Florida	26-30	1.7-5.9		McSorley and Frederick (1996)
Cultivated fields, various	South Australia	10-22	0.6-10.6		Yeates and Bird (1994)
Shrub	South Australia	21-26	0.6-2.8		Yeates and Bird (1994)
Deciduous forest	Summary, 15 sites		6.3		Sohlenius (1980)
Deciduous forest	Indiana		1.2-2.5	260-530	Ferris and Ferris (1974)
Deciduous forest	Tennessee		1.1-6.9	490-2100	Ferris and Ferris (1974)
Forests, various	Poland		0.5-7.0	192-691	Wasilewska (1971)
Pine forest	Summary, 14 sites		3.3		Sohlenius (1980)
Pine forest	Sweden	23	4.1	116	Sohlenius (1979)
Pine forest	Florida	34-37	0.4-0.5		McSorley (1993)
Tropical forest	Summary, 2 sites		1.7		Sohlenius (1980)
Tropical forest	New Caledonia		0.1-0.3		Yeates (1991)
Moss	Signy I. (Antarctic)	9	0.4-0.7	105-355	Caldwell (1981)
Bog	Summary, 9 sites		1.7		Sohlenius (1980)
Desert	Summary, 2 sites		0.8		Sohlenius (1980)

and in one study amounted to 5.8–12.6% of annual net primary production (Ingham and Detling, 1984).

Since herbivores and decomposers (bacterivores plus fungivores) make up the most abundant nematode groups in most systems, energy flow through the soil nematode community takes two major pathways. Nematode herbivores function as primary consumers, removing energy directly from the plant, whereas energy flow from plant to decomposers is indirect. Nematode decomposers do not feed directly on organic matter of plant origin, but on the bacteria and fungi which break down this material. In one study, 34–44% of the energy in organic matter from crop residues was passed through bacteria and fungi and consumed by nematodes within a year (Wasilewska and Bienkowski, 1985). Changes in the relative proportions of nematode bacterivores and fungivores over time reflect changes in the pathway of decomposition (Wasilewska and Bienkowski, 1985). Nematodes accelerate the decomposition process by dispersing relatively immobile microflora to new sites and by their feeding, which regulates bacterial growth and decomposition rates (Ingham et al., 1985; Wasilewska and Bienkowski, 1985; Freckman, 1988). Overgrazing of bacteria is avoided as nematodes in turn are regulated by mites and other invertebrate predators (Whitford et al., 1982; Elliott et al., 1988).

The increased decomposition rates resulting from nematode feeding increase the recycling and mineralization of C and other elements, and CO_2 evolution is a consequence of nematode activities (Ingham et al., 1985; Freckman, 1988). Nematodes are particularly important in recycling N, which can become immobilized in bacterial populations during decomposition (Freckman, 1988). Since nematodes have a higher C:N ratio (8:1 to 12:1) than their bacterial food source (3:1 to 4:1) (Wasilewska and Bienkowski, 1985), their feeding results in the excretion of N, mostly as NH_3 (Freckman, 1988). Numerous studies confirm the increase of NH_4^+ and other inorganic N sources in soil with nematodes present, and increased N levels in plant tissue have resulted in some instances (Ingham et al., 1985; Freckman, 1988). Nematodes are also important in the enhanced mineralization of P(Ingham et al., 1985; Freckman, 1988), and even S in some systems (Freckman, 1988).

2.2.4.2 Economic Importance to Agriculture

Nematode herbivores are of particular importance in agriculture because of their potential to damage the roots of crop plants (Norton, 1978; Poinar, 1983). Ectoparasites feed at the root surface but do not enter roots, while endoparasites may feed and live within root tissue. Feeding by large numbers of plant-parasitic nematodes can result in yield losses, wilting, nutrient deficiency symptoms, and other problems associated with a damaged root system. Nematode problems of various crops are reviewed by several texts (Webster, 1972; Nickle, 1984; Luc et al., 1990; Evans et al., 1993); some of the more important examples are presented (Table 2.10).

Table 2.10 Major nematode pests of selected agricultural crops

Crop	Nematodes
Alfalfa (*Medicago sativa*)	*Ditylenchus dipsaci, Meloidogyne* spp.
Banana (*Musa* spp.)	*Radopholus similis, Helicotylenchus multicintus*
Citrus (*Citrus* spp.)	*Tylenchulus semipenetrans, Radopholus citrophilis*
Coconut (*Cocos nucifera*)	*Bursaphelenchus (Rhadinaphelenchus) cocophilus*
Coffee (*Coffea arabica*)	*Meloidogyne* spp., *Pratylenchus* spp.
Cotton (*Gossypium hirsutum*)	*Meloidogyne incognita, Rotylenchulus reniformis*
Grape (*Vitis vinifera*)	*Meloidogyne* spp., *Xiphinema* spp.
Maize (*Zea mays*)	*Pratylenchus* spp., *Belonolaimus longicaudatus*
Peanut (*Arachis hypogaea*)	*Meloidogyne* spp., *Pratylenchus brachyurus*
Pine (*Pinus* spp.)	*Bursaphelenchus xylophilus*
Potato (*Solanum tuberosum*)	*Globodera* spp., *Meloidogyne* spp., *Pratylenchus* spp.
Rice (*Oryza sativa*)	*Hirschmanniella* spp., *Aphelenchoides besseyi*
Soybean (*Glycine max*)	*Heterodera gylcines, Meloidogyne* spp.
Sugarbeet (*Beta vulgaris*)	*Heterodera schachtii, Meloidogyne* spp.
Sugarcane (*Saccharum* spp.)	*Pratylenchus* spp.
Sweetpotato (*Ipomoea batatas*)	*Meloidogyne* spp., *R. reniformis*
Vegatables	
fam. Cruciferae (cabbages, etc.)	*H. schachtii, Meloidogyne* spp.
fam. Leguminosae (beans, peas)	*Meloidogyne* spp., *R. reniformis, Heterodera* spp.
fam. Solanaceae (tomatoes, etc.)	*Meloidogyne* spp.
Wheat (*Triticum aestivum*)	*Heterodera avenae*

Table 2.11 Outline of methods for sampling and extraction of nematodes from soil

Outline of procedures			Remarks	References
Collection of samples			Field collection	Barker and Nusbaum (1971), Barker and Campbell (1981), Goodell (1982), McSorley (1987), Freckman and Baldwin (1990)
	Sampling tools		Usually removes a core of soil	Barker and Campbell (1981), Goodell (1982)
	Depth of sampling		Most nematodes in top 15-30 cm of soil	Barker and Nusbaum (1971), Norton (1978), Barker and Campbell (1981), McSorley (1987)
	Timing of sampling		Seasonal peaks in populations	Barker and Nusbaum (1971), Barker and Campbell (1981), Goodell (1982)
	Spatial distribution		Nematodes highly aggregated	Barker and Nusbaum (1971), Barker and Campbell (1981), Goodell and Ferris (1981), Goodell (1982), McSorley (1987)
	Sampling plans		No of separate cores needed to form a composite sample (e.g., 20 cores/1.6 ha in N.C.)	Barker and Nusbaum (1971), Barker and Campbell (1981), Goodell and Ferris (1981), McSorley (1987), Freckman and Baldwin (1990)
	Sampling pattern		Spatial arrangement of cores in field	Barker and Campbell (1981), Goodell and Ferris (1981), McSorley (1987)
Handling of samples			Transport and storage of samples	Barker and Nusbaum (1971), Barker and Campbell (1981), Goodell and Ferris (1981), Goodell (1982), McSorley (1987)
Extraction of nematodes from soil			Laboratory methods	Ayoub (1980), Goodell (1982), Southey (1986), McSorley (1987), Hooper (1990), Freckman and Baldwin (1990)
	Subsample preparation		Usually 50-500 cm³ soil removed for extraction	McSorley (1987), Freckman and Baldwin (1990)
	Extraction of vermiform stages			Barker and Nusbaum (1971), Ayoub (1980), Goodell (1982), Southey (1986), McSorley (1987), Hooper (1990), Freckman and Baldwin (1990), McSorley and Walter (1991)
		Active techniques	Dependent on nematode motility	McSorley (1987), Freckman and Baldwin (1990), McSorley and Walter (1991)
			Diagram of Baermann method	Ayoub (1980), Southey (1986), Hooper (1990)
		Passive techniques	Independent of nematode motility	McSorley (1987), McSorley and Walter (1991)
			Diagram of sieving	Ayoub (1980), Southey (1986)
			Diagram of centrifugation method	Ayoub (1980)
	Extraction efficiency		All methods <100% efficient	McSorley (1987)
	Extraction of resistant stages		Cysts, anabiotic forms	Freckman et al. (1975), Ayoub (1980), Southey (1986)

Table 2.11 (cont.)

Outline of procedures	Remarks	References
Extraction from roots	Illustrations and diagrams	Ayoub (1980), Southey (1986), Hooper (1990)
Assessment of biological control	Special methods for predators and parasites	Stirling (1991)
Enumeration	Counting nematodes in laboratory	Southey (1986), Hooper (1990)
Analysis: energetics	Calculation of biomass, respiration, production	Klekowski et al (1972), Yeates (1979), Freckman (1982), Freckman and Baldwin (1990)
Analysis: community structure	Calculation of indices	Yeates (1979), Bostrom and Sohlenius (1986), Bongers (1990), Freckman and Ettema (1993), De Goede and Dekker (1993), McSorley (1993), Yeates and Bird (1994), Wasilewska (1994), Neher and campbell (1994), Yeates (1994), McSorley and Frederick (1996)

2.2.5 Methods

2.2.5.1 Collection, Sampling, and Extraction of Nematodes

Methods for collection of samples and extraction are reviewed in Table 2.11. The irregular horizontal spatial distribution of nematodes presents particular problems in obtaining precise samples from the field (Barker and Nusbaum, 1971; Barker and Campbell, 1981; Goodell and Ferris, 1981; Goodell, 1982; McSorley, 1987) Since many different methods are used for extracting nematodes from samples, comparison of counts between laboratories can be difficult, and should be standardized (McSorley, 1987).

2.2.5.2 Community Structure Analysis and Bioindicators

Data on the population density of a key genus or species are often collected in nematology. However, interest is increasing in the development and use of measures to assess community structure rather than individual taxa. Environmental changes impact both nematode numbers and community structure (Freckman and Ettema, 1993). The ability of nematodes to respond to changes, as well as their diversity, widespread distribution, size, abundance, and other features make nematodes useful as bioindicators of ecological disturbances (Freckman 1988; Bongers, 1990; Freckman and Ettema, 1993; Neher and Campbell, 1975). Much current work focuses on the utility of various indices of community structure for this purpose (Freckman and Ettema, 1993; Neher and Campbell, 1975; Wasilewska, 1994; Yeates and Bird, 1994; McSorley and Frederick, 1996).

2.3 Microarthropods

D.A. Crossley, Jr. and David C. Coleman
University of Georgia

2.3.1 Introduction

Soil microarthropods are a major fraction of the mesofauna, namely, those arthropods with body widths ranging between approximately 0.1 and 2 mm, and body lengths between 0.2 mm and 10 mm

(Fig. 2.3). This scheme of classification, although imprecise, is practical, defined by the method of sampling. Microarthropods are sampled by collecting a fragment of habitat (e.g., a soil core) and extracting them from it, while macroarthropods (Section C, Chapter 2.4) are collected by hand sorting, pitfall trapping, or other methods dealing with individuals. Microarthropods are dominated by two groups: the mites (Acari) and the springtails (Insect order Collembola). Together, mites and springtails account for about 90% of the microarthropods in most soil systems. Also included in this group, among others, are the Protura, Pauropoda, dipteran larvae, small spiders, pseudoscorpions, some Homoptera and Coleoptera, and thrips. Immature stages of many insect Orders are collected from soil samples, and some may be considered microarthropods for purposes of a particular study. These minor groups typically constitute less than 10% of the total number of microarthropods.

Numbers of microarthropods in soil systems range upwards to 200,000 m^{-2} or more (Table 2.12). Forested systems generally support higher microarthropod population densities than do grasslands, deserts, or agricultural systems, with densities being higher in soils from temperate than tropical forests, and coniferous than deciduous forests. Soils in agroecosystems may have sparse populations, although numbers increase under conservation tillage management. Together with protozoans, nematodes and other small soil fauna, the microarthropods make up a food web of several trophic levels, driven by energy sources from decomposing residues and mobilizing nutrient elements.

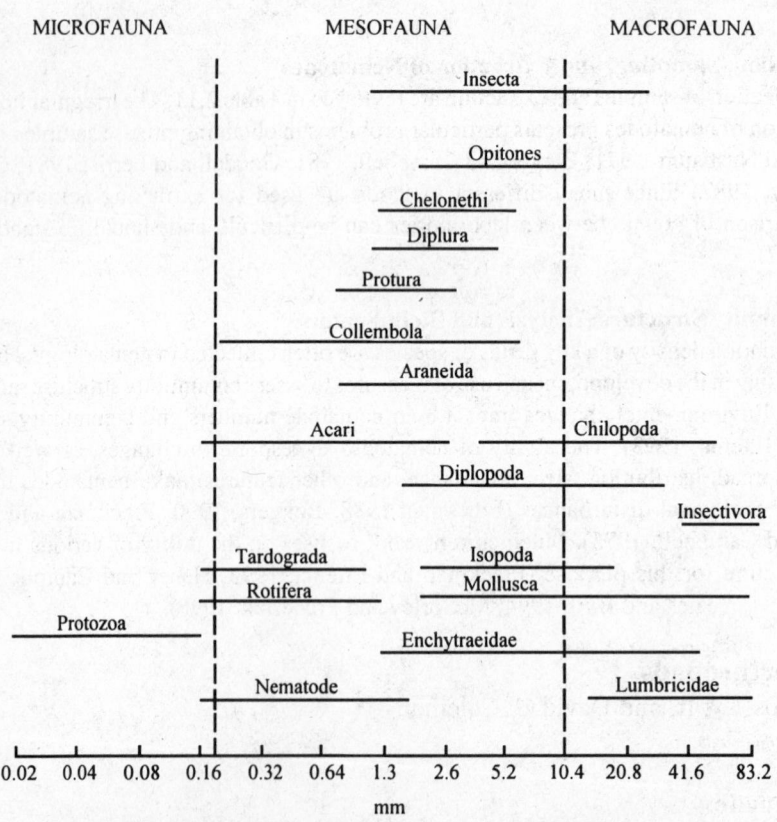

Fig. 2.3 A generalized classification of soil fauna by body length [Reprinted from Wallwork, 1983. Earthworm biology. Camelot Press, Southampton, UK with permission of McGraw-Hill]

2.3.2   Biology and Ecology

2.3.2.1 Collembolans

These minute insects (Fig. 2.4) are commonly called "springtails" in reference to an abdominal appendage, the furcula, which enables them to leap through the air. They possess a characteristic ventral tube with eversible sacs, important in moisture balance. Collembolans are most obvious when they swarm on the soil surface, sometimes on snowbanks, leading to the name "snowfleas." But springtime swarms are common in North American forest, grassland and agricultural soils as well. Eighteen families of Collembola are currently recognized (Hopkin, 1997) but the taxonomy is in a state of flux. General reference works on Collembola are presented in Maynard (1951) and Christiansen and Bellinger (1980).

Collembolan numbers in terrestrial ecosystems range between 10^4 and 10^5 m^{-2} and typically constitute 20–50% of the soil microarthropods. They are opportunistic species and dominate microarthropod communities in springtime during population blooms. Their habitat extends from the litter layers of soil down into deeper substrata. Their morphology reflects their habitat in that surface dwellers are larger, often colored, with well-developed furcula and long antennae, while those inhabiting the soil are white, with reduced furcula and shorter antennae (Fig. 2.4).

Although most collembolans are fungivores or detritivores, a few species are economic pests, and some have proved to be predaceous. Like other microarthropods, collembolans may feed opportunistically upon nematodes or other periodically abundant resources in the soil. Collembolans appear to be attracted to plant roots and may be important in rhizosphere dynamics. In experiments, collembolans grazed selectively on fungal root pathogens, thereby protecting cotton plants (Curl and Truelove, 1986).

2.3.2.2   Soil Mites (Acari or Acarina)

Being members of the Class Arachnida, mites are eight-legged, chelicerate relatives of spiders and phalangids. Conservatively, mites are placed in the Order Acari (or Acarina) and are divided into several suborders. While many species of mites are plant feeders or parasites, only the soil forms will be considered here. For a general treatment of the Acari, see Evans (1961) or Krantz (1978). Most soil

Table 2.12   Representative densities of soil microarthropods from various ecosystems

Region	Ecosystem type	Density (m^{-2})	Reference
North Carolina	Deciduous forest	133,000	Seastedt and Crossley (1981)
North Carolina	Deciduous forest	88,000	Lamoncha and Crossley (1997)
Tennessee	Pine forest	102,000	Bohnsack and Crossley (1960)
Belgium	Oak forest	70,000 - 180,000 (mites only)	Lebrun (1965)
Canada	Aspen forest	123,000 (mites only)	Mitchell (1977)
Australia	Tropical rain forest	33,000 - 49,000	Holt (1985)
Africa	Tropical rain forest	22,000 (dry season) 65,000 (wet season)	Madge (1969)
Georgia	Agroecosystem (Sorghum)	119,000 (no-tillage) 49,000 (conventional)	House and Parmelee (1985)
South Carolina	Pine plantation	166,000	Johnston (1996)
Colorado	Short-grass prairie	102,000	Walter et al. (1987)

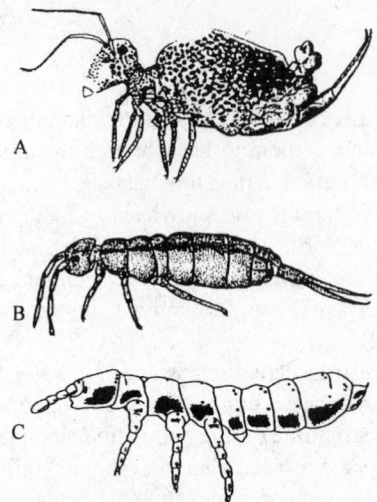

Fig. 2.4   Collembolan life forms: (A). Family Sminthuridae, typically found at the soil surface. (B). Family Isotomidae, from forest leaf litter. (C). Family Onychuridae, from mineral soil layers [Reprinted from Coleman and Crossley, 1996. Fundamentals of soil ecology with permission of Academic Press, Inc, Orlando, FL]

mites fit into one of four suborders, separation of which is easily performed using a dissecting microscope at 20x. Each suborder is considered below.

2.3.2.3   Acari: Prostigmata

Representatives of this very diverse suborder are often among the most numerous of the soil mites. In general, they are delicate, white to colorless, and subject to desiccation. While most species are predaceous, some are fungivorous and these species may become abundant in decomposing organic litter. Many Prostigmata are opportunistic species, able to reproduce rapidly when food resources become abundant. Large populations may build in disturbed situations, such as forest clearings, drained marshes, and so forth. Larger Prostigmata are predators, including the bright red "velvet mites" seen walking on the soil surface in the spring, or following rains in desert systems. Smaller species with piercing stylet chelicerae are generally fungal feeders in soil and litter layers, but some are effective predators on other mites or nematodes. Some species may switch between fungal hyphae and nematodes, or perhaps supplement a fungal diet with occasional protein.

Recent comprehensive treatments of the Prostigmata are those of Kethley (1990). He gives extensive tables of abundance and biomass for the Prostigmata in a range of habitats, as well as a key to families from edaphic habitats.

2.3.2.4   Acari: Mesostigmata

This suborder consists of generally flattened, tick-like mites. Although not as numerous as other mite groups, mesostigmatic mites are important predators in soil systems, particularly on nematodes and other arthropods. The smaller species are mainly nematophagous. Larger ones are active predators of collembolans, other small arthropods and arthropod eggs. One species in the family Macrochelidae has been used successfully to control housefly populations in manure (Krantz, 1978). In forest floors, litter species tend to be larger, and mineral soil inhabitants are generally smaller and colorless. In agricultural soils, Mesostigmata are attracted to decomposing roots. Members of the family Uropodidae feed upon fungal hyphae and associated organic debris.

A large and comprehensive literature discusses the soil mesostigmatic mites of North America, but the group is in need of a comprehensive revision (Evans and Till, 1979; Krantz and Ainscough, 1990).

2.3.2.5 Acari: Oribatei

Next to collembolans, oribatid mites (Fig. 2.5) are the most numerous of microarthropods (Balogh and Balogh, 1992; Marshall et al., 1987; Moldenke and Fichter, 1988). They are usually brown and beetle-like in form and, like other mites, are octopod with the body essentially unsegmented except for some secondary sutures. Their rate of reproduction is slower than that of collembolans, but they maintain high populations by longer survivorship. Adult oribatids have a heavily sclerotized exoskeleton containing deposits of $CaCO_3$ evidently accumulated from calcium oxalate crystals in their fungal food sources. Together with snails, millipedes and isopods, oribatids may play a significant role in Ca metabolism in soil systems. The immature stages of oribatid mites often do not resemble the adults, to the extent that recognition of species based on immatures is difficult.

Densities of oribatid mites in soils, as with Collembola, range from 10^4 to 10^5 m^{-2}. In deciduous forests, peaks of abundance occur in autumn and again in spring, with numbers remaining high during summer months. Year-to-year variation may be large, but fluctuations are not as wide as with Collembola. Coniferous forest floors typically support the largest populations of oribatids, followed by deciduous hardwood forest, grassland, and tundra (Coleman and Crossley, 1996). Cultivation of agricultural soils reduces oribatid population sizes.

Oribatids are primarily fungivorous, but will eat a variety of foods. Those species with stout chelate chelicerae can fragment and ingest decomposing leaf litter and wood, but many species can be reared on algae or lichens. Although oribatids have been observed to ingest nematodes, it seems clear that fungi are the major resource base. The major importance of oribatids in soils is their effect on the decomposition process; they are able to break down organic litter by fragmenting in or tunneling within materials such as woody residues. Their activities may stimulate microbial immobilization of nutrient elements, but at the same time, ingestion of fungal hyphae may destroy hyphal bridges (Lussenhop, 1992).

There are over 1,000 genera of oribatid mites, with newly discovered ones being published annually. Perhaps as few as 20% of the world fauna of oribatid species have been named (Behan-Pelletier and Bissett, 1993). The identification of oribatids to genus has been greatly facilitated by the

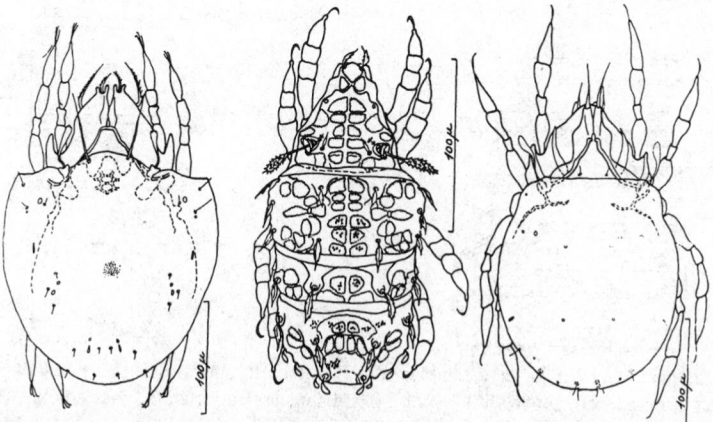

Fig. 2.5 Illustrations of oribatid mites. Left, Family Achipteriidae. Center, Family Brachychthoniidae. Right, Family Astegistidae [After D. L. Dindal, State University of New York, Syracuse with permission]

herculean efforts of Janos Balogh, who has authored a series of well-illustrated works during the past two decades (Balogh and Balogh, 1992). Recent publications, especially those which include scanning electron micrographs of oribatids, will aid in identification, but validation by an expert taxonomist is essential (Marshall et al., 1987; Moldenke and Fichter, 1988).

2.3.2.6 Acari: Astigmata

This suborder is occasionally found in soil samples. These mites, commonly called cheese mites, seem to be associated with highly organic, decomposing materials such as manure. Buried agricultural residues may support Astigmata, and they may become pests of root crops on occasion.

2.3.3 Sampling and analysis

Microarthropods are not sampled directly. Rather, samples of habitat are collected and microarthropods extracted from them. Soil cores (5 cm dia x 5 cm deep) are extracted on micro-Tullgren apparati (Fig. 2.6) or by flotation methods (Walter et al., 1986). An alternate sampling method employs litterbags containing leaf litter, which are placed in the field at the beginning of the season and then retrieved through time (Crossley and Hoglund, 1962). Microarthropods are extracted from them using Tullgren funnels ("Berlese funnels") (Evans, 1961). Using this technique, it is possible to identify microarthropod groups associated with various stages of litter decay. With either Tullgren or micro-Tullgren funnels, animals are collected in containers of 70% ethyl alcohol.

Samples of microarthropods may be sorted into major taxa with the use of a dissecting microscope at 20 times magnification. Suborders of mites and many families of Collembolans are readily identified, once their basic morphology and characteristics are learned. Further identification requires slide mounted material. Slide mounts can be made directly from alcohol (Table 2.13), although heavily sclerotized specimens may require preliminary clearing. Identification of North American microarthropod species is problematic because of the large number of poorly described or unnamed species (Behan-Pelletier and Bissett, 1993). While it is essential that preliminary identifications be

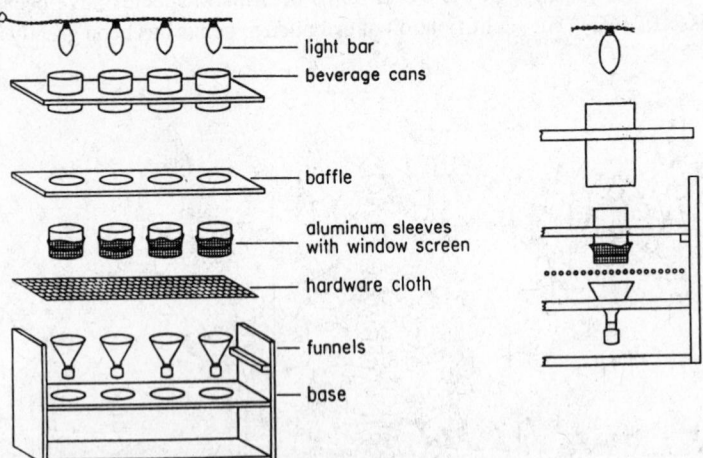

Fig. 2.6 A high-efficiency extractor for core samples of soil microarthropods. Soil cores contained in aluminum sleeves are inverted and heated from above with 5-watt lights. Fauna are collected in funnels below [Reprinted from Crossley and Blair, 1991. Agric. Ecosys. Environ. 34:187. Copyright with kind permission of Elsevier Science, Amsterdam, Netherlands].

verified by an expert taxonomist specializing in the taxon, enumeration of the major taxa often is adequate for comparative studies of soil microarthropods.

2.4 Macroarthropods

D.A. Crossley, Jr. and D.C. Coleman
University of Georgia

2.4.1 Introduction

The macroarthropods are those large enough to be sampled as individuals, in contrast to the microarthropods which are sampled by extraction from a fragment of habitat (Section 2.3) (Borrer et al, 1992; Dindal, 1990; Arnett, 1993). Although smaller macroarthropods overlap in size with the larger microarthropods (Fig. 2.3), the distinction between them is a practical one, based on method of sampling. A functional difference lies in their impact on soils. Macroarthropods are capable of restructuring soil profiles or relocating large amounts of soil, whereas microarthropods must take advantage of existing pore spaces (Coleman and Crossley, 1996). Two insect groups, the ants and termites, are responsible for major disruptions of soil profiles, while other macroarthropods may cause some disturbance. Examples include emergence tunnels of periodical cicadas (Insecta: Homoptera) or chimneys made by crayfish (Crustacea: Decapoda) in marshy areas.

The macroarthropods in soil systems are a highly diverse group. Most terrestrial insect orders contain species that live in the soil at some phase of their life cycle. Transient species (some Lepidoptera), are those that overwinter or pupate in surficial soil layers. Other temporary residents such as midges and other flies, spend their immature stages in the soil but emerge as adults to feed elsewhere. Permanent soil residents, such as predaceous beetles, remain in the soil or on soil surfaces. Spiders (Araneae) and centipedes (Chilopoda) are numerous and important predators in soil systems (Kastan, 1978; Camatini, 1979; Kevan and Scudder, 1989; Foelix, 1996). Detritivores include

Table 2.13 Procedures for preparation of slide mounts of microarthropods (Krantz, 1978)

Storage media:	70% ethyl alcohol
	95% ethyl alcohol
	Oudeman's fluid:
	Glycerin.........................5 parts
	70% alcohol.................87 parts
	Glacial acetic acid.........8 parts
Clearing agents:	Lactophenol:
	Lactic acid...............50 parts
	Phenol crystals.........25 parts
	Distilled water.........25 parts
Mounting media:	Hoyer's medium:
	Distilled water.........50 mL
	Gum Arabic.............30 g
	Chloral hydrate........200 g
	Glycerin..................20 mL
	CMC-10 medium*

*available from Masters Chemical Company, 200 Wilson Court, Bensenville, Il 60106

millipedes (Diplopoda) and sowbugs (Isopoda) (Shachak et al., 1976; Camatini, 1979; Snider and Shaddy, 1980). Scorpions (Scorpionida) and windscorpions (Solifugae) are important predators in desert systems (Crawford, 1981; Williams, 1987).

2.4.2   Biology and Ecology

The major groups of macroarthropods likely to be found in soil and litter samples are listed in Table 2.14. The list is not inclusive because, aside from representatives of the minor orders of arachnids and insects not listed, samples may also include representatives of major orders not usually considered to be soil fauna. Grasshoppers and crickets are frequently found on the soil surface and in pitfall traps. Even caterpillars which have descended from plant canopies to pupate in soil will be sampled. In fact, nearly every free-living group of terrestrial arthropods may occasionally enter soil food webs as prey items.

The majority of the scientific literature deals with individual taxa in detail, rather than providing overviews of entire macroarthropod faunas. Where general overviews do exist, they attempt broad syntheses, often without detailed information about macroarthropods (Borrer et al., 1992; Dindal, 1990). The biology and importance of some ecological groups, such as root feeders, remain poorly known (Coleman and Crossley, 1996).

Macroarthropod faunas vary considerably between and within types of ecosystems. Forested ecosystems in general contain macroarthropod faunas dominated by millipedes, spiders and beetles (Table 2.15). Numbers and biomass tend to be maximal in hardwood as compared to evergreen forests, in which microarthropod abundance is high (Section 2.3). Spiders, carabid beetles and crickets are abundant in pitfall traps placed in agricultural areas (House and Stinner, 1983; Blumberg and Crossley 1983). Most species of macroarthropods may be more sensitive to cultivation than other soil fauna, and as a result, investigations have neglected the sensitive species in favor of the abundant ones (Wolters and Ekschmitt, 1997). Spiders are the most abundant and probably the most important of the predaceous macroarthropods in terms of their impact on food webs (Ekschmitt et al., 1997). The ranges of macroarthropod population sizes vary widely. Considerable overlap in abundance of macroarthropod taxa among different ecosystems has been found for millipedes and centipedes, although arable land tends to contain the lower part of the range. The ranges listed in Tables 2.15 and 2.16 illustrate the differences which occur between habitat types and seasons. Adjacent forest, grassland and agricultural lands often have markedly different species of taxa such as spiders (Reichert and Lockly, 1984; Draney 1992).

Table 2.14   Major groups of macroarthropods

Class	Order	Common name
Arachnida	Araneae	Spiders
	Scorpiones	Scorpions
	Opiliones	Phalangids, harvestmen
	Pseudoscorpiones	Pseudoscorpions
	Solifugae	Windscorpions
Malacostraca	Isopod	Sowbugs, pillbugs
Diplopoda	(Ten Orders)	Millipedes
Chilopoda	(Four Orders)	Centipedes
Hexapoda	(Thirty-one Orders)	Insects

Table 2.15 Abundance and biomass of macroarthropod taxa in forest ecosystems

Forest type	Diplopoda		Orthoptera		Araneae		Coleoptera		Source
	N m^{-2}	g m^{-2}	N m^{-2}	g m^{-2}	N m^{-2}	g m^{-2}	N m^{-2}	g m^{-2}	
Mixed Hardwood	14.0	6.10	0.9	0.21	202.5	1.59	11.0	0.30	Gist and Crossley (1975)
Tulip-poplar	39.0	1.50	4.0	1.01	258.0	0.28	102.0	0.17	Moulder et al. (1970)
Oak	433.0	5.59	0.5	0.10	148.0	0.06	194.0	0.33	Bornebusch (1930)
Beech	33.0	1.07	2.0	0.04	66.0	0.09	16.0	0.92	Bornebusch (1930)
Spruce	51.0	0.84	---	---	34.0	0.03	180.0	0.47	Bornebusch (1930)
Beech	11.0	0.11	---	---	186.0	0.74	17.0	0.60	Kitizawa (1967)
Beech	80.0	1.30	---	---	163.4	---	510.7	1.50	van der Drift (1950)

2.4.2.1 Social Insects

Collectively, ants and termites are responsible for major modifications of soil. Termites are typical insects of tropical and subtropical regions and may be dominant in arid or semiarid ecosystems (Lee and Wood, 1971; Bryan, 1978; Holldobler and Wilson, 1990; Stork and Eggleton, 1992; Bolton, 1994). The following soil modifications are brought about by the activities of termites: (1) physical changes of soil profiles, (2) changes in soil structure, (3) changes in the nature and distribution of organic matter, (4) changes in the distribution of plant nutrients, and (5) construction of subterranean galleries (macropores). Ants are more widely distributed than termites, occurring in most terrestrial habitats, and are usually predaceous. Their colonies are smaller than those of termites, but they are responsible for the same kinds of soil modifications. As predators, ants may have a considerable impact on herbivorous insects. The importance of ants as predators of insect pests has been well demonstrated in a variety of stable forest ecosystems. Close, evidently coevolved, relationships exist between some plant and ant species, but in relatively unstable, annual agroecosystems, less is known of the importance of ants.

2.4.2.2 Myriapods

The many-legged arthropods (myriapods) are abundant in undisturbed soils of many types, but less abundant in agricultural systems (Table 2.15). Millipedes are major saprovores, feeding upon decomposing organic matter in a variety of ecosystems. Although moisture dependent, they are among the macroarthropods of desert ecosystems (Crawford, 1979). Millipedes, which are important in Ca cycling in forests, have a calcareous exoskeleton and may process 15–25% of Ca inputs into forest floors (Coleman and Crossley, 1996). Some millipedes are obligate coprophages, feeding on their own microbially enriched fecal matter, while others secrete noxious chemicals as a defense mechanism against predation.

Centipedes are ubiquitous and active predators in soil and litter habitats, able to run rapidly and capture small prey such as microarthropods (especially collembolans). Most are 3–5 cm in length, but tropical centipedes may exceed 30 cm. Typically, centipedes constitute about 20% of the predaceous macroarthropods in temperate forests, but the percentage is lower in subarctic, boreal and dry forests

Table 2.16 Abundance and biomass of macroarthropod taxa in three ecosystems. (Numbers in parentheses indicate ranges of abundance estimates)[After Wolters and Ekschmitt, 1997]

Taxon	Arable land		Temperate grassland		Temperate deciduous forest	
	N m^{-2}	mg m^{-2}	N m^{-2}	mg m^{-2}	N m^{-2}	mg m^{-2}
Isopoda	5 (0 - 25)	15	1200 (500-7900)	1600	286 (96 - 1850)	93
Diplopoda	200 (70-400)	---	--- (500 - 7900)	1250	55 (210 - 700)	618
Chilopoda	100 (40 - 220)	---	60 (63 - 387)	140	187 (50 - 790)	265

(Albert, 1979). Species diversity is lower for centipedes than for other predators such as spiders and staphylinid beetles.

2.4.2.3 Spiders

Spiders are the most numerous of the predaceous macroarthropods in ecosystems ranging from forest to grassland to agroecosystems. The taxonomy of soil and litter dwelling spiders remains unsettled, especially for the numerous species in the family Linyphiidae, which contains many small soil species. Spiders, which are strictly carnivorous, are generalist feeders that attack insects but also feed upon other invertebrates, including other spiders (Wise, 1993). Despite their numbers, there is no consensus on the abilities of spiders to control insect populations (Riechert and Lockly, 1984). Spiders do not reproduce rapidly enough to keep pace with exploding prey populations. Also, many species are territorial. In forest habitats, spiders may act as a stabilizing influence on populations of forest floor invertebrates by maintaining a continual predation pressure.

2.4.2.4 Beetles

In terms of numbers of species, beetles are the largest order of insects, being found in every habitat except the oceans (Richter, 1958; Thiele, 1977). In soil systems, beetles include species that are phytophagous, saprophagous and predaceous. Carabid, tenebrionid and staphylinid beetles are numerous predators in disturbed and undisturbed systems alike. Together with spiders, carabids are the typical ground-surface macroarthropod predators taken in pitfall traps in agroecosystems. Other active predators include the tiger beetles (Cicindelidae) whose larvae construct belowground retreats from which they capture prey. Phytophagous beetles include the Scarabaeidae (June bugs) whose larvae feed extensively on roots. Larvae of elaterid beetles (wireworms) are important root feeders in cropping systems and in forests. Although predaceous beetles are conspicuous, especially on the soil surface in agricultural fields, the phytophagous species are probably more important. The predaceous carabids may exert some control on caterpillars such as armyworms and similar species. Gypsy moth caterpillars, descending to the soil of the forest to pupate, may fall prey to carabid beetles in large numbers.

2.4.3 Sampling and Analysis

Macroarthropods are sampled in several ways. The most basic method is simply to delineate an area—a square meter or less—and, by hand sorting, collect all the macroarthropods. If large Tullgren funnels are available, bulk samples of soil and litter may be extracted (Section 2.3 and Fig. 2.7). Subterranean

Fig. 2.7 Large Tullgren funnels for extraction of macroarthropods from soil cores and litter

macroarthropods are sampled by taking soil cores (10–15 cm dia) and sieving them, either dry or with the use of a wet sieving apparatus. A third method in widespread use involves pitfall traps–cans set flush with the soil surface and containing a preservative (Fig. 2.8). Pitfall traps are an inexpensive sampling technique, but are not entirely quantitative, since captures depend upon the mobility of

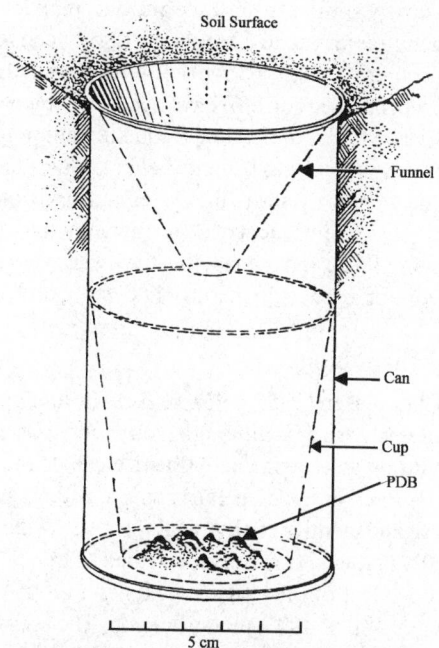

Fig. 2.8 A pitfall trap for surface-active macroarthropods. PBD, *p*-dichlorobenzene preservative [Reprinted from Reichle and Crossley, 1965. Radiocesium dispersion in a cryptozoan food web. Health Phys. 11:1375-1384, with permission of Williams & Wilkins, Baltimore, MD]

animals as well as their density. In sampling programs designed for comparison of areas or seasons, pitfall traps are a preferred method (Bater, 1996).

Because ants and termites are social insects, population estimation for these insects requires special techniques. Species diversity of ants is readily sampled with baited pitfall traps (Romero and Jaffe, 1989), while subterranean termites are sampled using soil cores (Lee and Wood, 1971). Termite mounds may require complete destruction. A comparative sampling technique for desert termites uses rolls of toilet paper, which are placed on the soil surface and shielded from the sun with aluminum foil (Whitford et al., 1982).

Samples may be preserved in 70% alcohol prior to sorting. While alcoholic storage is satisfactory for storage of specimens, entomologists prefer that insects be pinned if possible. Insect orders are pinned in different ways. Instructions for pinning and labeling of specimens are given in most entomology textbooks (Borror et al., 1992).

Sorting of macroarthropods into major taxa is straightforward. Reference to a general entomology textbook will allow the novice to make identifications of major taxa to family levels. More detailed sorting of samples may be done with the aid of literature guides (Dindal, 1990). Identifications to generic and species levels will require the services of a specialist in the taxonomy of the group.

2.5 Enchytraeids

P.C.J. van Vliet
University of Western Australia

2.5.1 Introduction

The enchytraeids constitute a family (Enchytraeidae) of Oligochaetes that occurs in terrestrial, littoral and aquatic habitats with a total of some 600 species worldwide (Dash, 1990). Enchytraeids are mostly pale colored and are anatomically similar to earthworms but smaller. Their length ranges from 5 cm to less than 1 mm. Larger enchytraeids (up to 60 mm long) have been found in subarctic soils from the unglaciated portion of the northern Yukon (Canada) (Smith et al., 1990).

Of the Oligochaeta, earthworms (Section 2.6) have been the subject of most studies. Enchytraeidae (also known as 'potworms') have been studied less frequently although they are distributed globally. The biology and ecology of enchytraeids has become better known during the last 30 years, especially in Europe. This intensification was partly due to the systematic revision for European species (Nielsen and Christensen, 1961, 1963), the development of efficient methods of extraction of enchytraeids from soil (Nielsen, 1953; O'Connor, 1955), and the fact that they were very abundant in certain ecosystems and were expected to have an important role in nutrient cycling and soil structure formation.

2.5.2 Taxonomy and Morphology

Currently the Nielsen and Christensen (1959, 1961, and 1963) monographs for European species are used for the identification of enchytraeids, although many more genera and species have been found since. A key to the most common genera of Enchytraeidae can be found in Dash (1990). Taxonomic literature on North American Enchytraeidae is fairly sparse and no good key is available.

The easiest way to observe and identify enchytraeids is by allowing them to clear their gut in a Petri dish in a cool place. A sexually mature worm (alive) can then be transferred to a drop of water on a slide and covered with a coverslip. Most of the structures shown in Fig. 2.9 are fairly transparent in live worms and can be examined at 100 to 400 x magnification. The known genera of enchytraeids, two-thirds of which are generally found in soil, are presented in Table 2.17. Information about the habitats of the other genera is mentioned as well.

2.5.3 Biology and Ecology

2.5.3.1 Life History and Reproduction

Several types of reproductive strategies are known in enchytraeids (Table 2.18). Parthenogenesis, self-fertilization and fragmentation are often used by cosmopolitan and widespread species. These strategies allow animals to survive under particularly unfavourable conditions and enhance their chances to successfully colonize new habitats (Dósza-Farkas, 1996). Reproduction occurs mostly through the formation of cocoons which contain 1 or more eggs. Depending on species and temperature the duration from cocoon hatching to maturity is about 65 to 120 days. The maximum lifespan of enchytraeids is assumed to be one year (Dash, 1990).

2.5.3.2 Food Preferences

Enchytraeids ingest mineral particles, leaf litter, fungi, bacteria, oats and yeast, seaweeds and sewage. Dash et al. (1981) and Urbášek and Chalupský (1991) found that enchytraeids were capable of digesting di- and some polysaccharides in food, but that most complex plant substances became available through microbial decomposition. In food preference experiments, Dósza-Farkas (1982) showed that enchytraeids preferred old and microbially decomposed over fresh leaves. However, consumption of the partially decomposed leaves differed by species. According to Toutain et al. (1982), enchytraeids only consumed the epidermis and parenchyma cells of leaves, while avoiding the veins. After gut passage, the material had changed little in structure, only bacteria and fungi being digested.

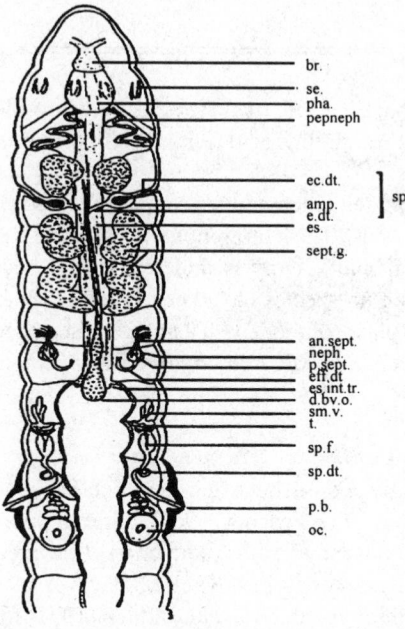

Fig. 2.9 Morphological characters of an enchytraeid worm. amp., ampulla; an.sept, ante-septal; br., brain; d.bv.o., dorsal blood vessel origin; ec.dt., ectal duct; e.dt. ental duct; eff.dt., efferent duct; es., esophagus; es.int.tr., esophageal intestinal transition; neph., nephridia; oc., oocyte; pha., pharynx; p.b., penial bulb; pepneph., peptonephridia; p.sept., postseptal; se., setae; sept.g., septal gland; sm.v., seminal vesicle; sp. spermatheca; sp.dt., sperm duct; sp.f., sperm funnel; t., testes (From the "Soil Biology Guide", Dindal, D.L. Copyright ©1990 John Wiley & Sons, Inc. Reprinted with permission of John Wiley & Sons, Inc.)

Table 2.17 Enchytraeid genera and their occurrence in soil [Nielsen and Christensen, 1959; Römbke, 1992; Römbke and Dósza-Farkas, 1996; Dósza-Farkas, pers. com.]

Genera occurring in soil	Other genera	Environment
Achaeta	Aspidodrilus	epizoic on earthworms
Bryodrilus	Barbidrilus	freshwater
Buchholzia	Enchylea	only found in Enchytraeid culture
Cernosvitoviella	Enchytraeina	marine
Cognettia	Grania	marine
Enchytraeus	Pelmatodrilus	epizoic on earthworms
Enchytronia	Propappus	freshwater
Fridericia	Randidrilus	marine
Guaranidrilus	Stephensoniella	marine
Hemienchytraeus		
Hemifridericia		
Henlea		
Isosetosa		
Lumbricillus		
Marionina		
Mesenchytraeus		
Oconnorella		
Stercutus		
Tupidrilus		

Brockmeyer et al. (1990) found that two enchytraeid species were not able to break down the cell walls of the fresh plant material, while the muramic acid cell walls of the microorganisms were no barrier to the worms' digestive capabilities.

Persson et al. (1980) considered enchytraeid communities to be 50% saprovorous, 25% fungivorous and 25% bacterivorous. This differentiation is often used in studies of production ecology. It seems, however, that microbivory is underestimated, because some species are well adapted to digest microbes, while no species have been found that produce sufficient amounts of enzymes to decompose complex plant compounds. Therefore, it seems to be more realistic to assume that an enchytraeid community contains 80% microbivorous and 20% saprovorous feeders (Didden, 1993).

2.5.3.3 Abundance and Distribution

Enchytraeidae are common in cultivated soil but reach their highest abundances in acid soils with high organic matter contents (O'Connor, 1957; Peachey, 1963). Large populations have been found in cold and temperate moist habitats, but this may be more a function of the greater amount of data available from Europe than of environmental conditions. Table 2.19 gives average annual abundances of enchytraeids in various habitats throughout the world. Didden (1991) found no correlations between enchytraeid densities and annual precipitation, annual temperature and acidity of the soil.

Enchytraeids occur in aggregated distribution patterns (Nielsen, 1954: Peachey, 1963; O'Connor, 1967; Abrahamsen, 1969; Abrahamsen and Strand, 1970; Nakamura, 1979, 1982), sometimes in multispecies aggregation centers (Chalupský and Lepš, 1985) in relation to environmental conditions, food availability, etc. The vertical distribution of enchytraeids in soil is related to the organic matter

Table 2.18 Reproductive strategies in enchytraeids [After Dósza-Farkas, 1996]

Reproduction	Description	Genera/species
Only sexually, all during the year possible	irregular reproduction periods, specimens with eggs can be found any time of the year	many *Fridericia* sp., most *Enchytraeus* sp., *Henlea* species, *Enchytronia* sp., *Marionina* sp., *Achaeta* sp.
Only sexually, become mature once a year	thereafter they regress and stay in rejuvenile state till the next reproduction period	*Mesenchytraeus* species *Stercutus niveaus*
Parthenogenesis	egg development without fertilization	*Lumbricillus* sp. and *Fridericia* sp.
Fragmentation		*Buchholzia* sp. and *Cognettia* sp.
Self-fertilization		*Enchytraeus buchholzi*, E. *bulbosis*

distribution in the soil profile. In most natural habitats (e.g., coniferous forests), up to 90% of the population occurs in the upper layers where most of the organic matter is located (Persson et al. 1980; O'Connor, 1957). The importance of organic matter for the vertical distribution of enchytraeids is apparent in agricultural ecosystems. In pastures and no-tillage fields, where organic matter is not plowed in, enchytraeids occur mostly in the top layer, whereas in tilled soils they are more evenly distributed (Lagerlöf et al. 1989; Didden, 1991; van Vliet et al., 1997). The vertical distribution of enchytraeids is also influenced by soil moisture content and temperature. Enchytraeids are known to react rapidly to changes in soil moisture (Nielsen, 1955a; O'Connor, 1967). Springett et al. (1970) showed that enchytraeids were able to move vertically as the soil water content changed, covering up to 6 cm in a few hours. Apart from the distribution of organic material and environmental variables, the vertical distribution of enchytraeids may also be species specific.

Nielsen (1955a) determined that seasonal trends in population densities of enchytraeids were caused directly or indirectly by external factors and not by inherent rhythmic fluctuations. According to O'Connor (1957), seasonal trends in enchytraeid numbers can be predicted from climatological data. In a temperate oceanic climate, soil moisture would be sufficient for reproduction during the entire year, resulting in a generally high density of enchytraeids with a winter minimum due to low temperatures. A temperate continental climate would have low abundances in summer and fall when temperature is too high and moisture too low for reproduction. Maximum densities would occur in winter and spring. In a New Zealand pasture, Yeates (1986) found significant correlations between enchytraeid abundances and soil moisture and temperature at sampling or three weeks prior. Abrahamsen (1972) found that enchytraeid abundances were correlated with soil moisture but also with vegetation type. Environmental conditions can cause changes in the active state of the animals (cocoon production, diapause) and/or accelerate or retard the reproductive cycle (Didden, 1993).

2.5.4 Importance to Soil Processes

2.5.4.1 Organic Matter Decomposition and Nutrient Cycling
Abrahamsen (1990) found that temperature, moisture and the presence of enchytraeids significantly influenced N mineralization in humus. Didden (1995) also found that the decomposition of litter and N mobilization were enhanced in the presence of enchytraeids. Enchytraeids have also been shown to stimulate nitrogenase activity in soil (Šimek et al., 1991).

Table 2.19 Enchytraeid abundances (annual average number m^{-2}) in different biotypes and countries

Biotope	Country	Abundance (number m^{-2})	Reference
Forest			
Douglas Fir	Wales	134,300	O'Connor (1967)
Pinus radiata, 50 stems ha^{-1}	New Zealand	64,002	Yeates (1988)
Pacific Silver Fir, mature stand	WA, USA	49,400	Piper (1982)
Pinus radiata, 200 stems ha^{-1}	New Zealand	39,270	Yeates (1988)
Spruce	Norway	34,700	Abrahamsen (1972)
Rhododendron-Oak, 1160 m altitude	NC, USA	32,630	van Vliet et al. (1995)
Rhododendron-Oak, 750 m altitude	NC, USA	26,811	van Vliet et al. (1995)
Pine	Norway	22,900	Abrahamsen (1972)
Pinus radiata, 100 stems ha^{-1}	New Zealand	21,391	Yeates (1988)
Scotch pine forest	Sweden	16,200	Persson et al. (1980)
Deciduous forest	UK	14,590	Phillipson et al. (1979)
Spruce	South Finland	13,400	Huhta and Koskenniemi (1975)
Pacific Silver Fir, young stand	WA, USA	11,400	Piper (1982)
Pinus radiata, 0 stems ha^{-1}	New Zealand	10,647	Yeates (1988)
Spruce	South Finland	8,200	Huhta and Koskenniemi (1975)
Spruce	North Finland	4,000	Huhta and Koskenniemi (1975)
Arable land			
Sugarbeet	The Netherlands	30,000	Didden (1991)
Winterwheat	The Netherlands	19,437	Didden (1991)
NT corn-clover	GA, USA	16,830	van Vliet et al. (1995)
CT corn-clover	GA, USA	15,270	van Vliet et al. (1995)
Potato field	Poland	13,200	Ryl (1977)
Barley, no N	Sweden	10,000	Lagerlöf et al. (1989)
Rye field	Poland	9,800	Ryl (1977)
Barley, 120 kg N	Sweden	8,100	Lagerlöf et al. (1989)
Rice/wheat/barley (organic)	Japan	4,940	Nakamura (1989)
Rice/wheat/barley (conven.)	Japan	525	Nakamura (1989)
Moor			
Juncus peat	UK	145,000	Peachey (1963)
Nardus	UK	71,000	Peachey (1963)
Blanket bog	UK	40,000	Standen (1973)
Fen	Canada	5,600	Dash and Cragg (1972)
Grassland			
Grassland soil	Sweden	24,000	Persson & Lohm (1977)
Lucerne ley	Sweden	9,900	Lagerlöf et al. (1989)
Grassland 10 sheep ha^{-1}	Australia (NSW)	6,000	King and Hutchinson (1976)
Grassland 30 sheep ha^{-1}	Australia (NSW)	2,300	King and Hutchinson (1976)

Enchytraeids can influence decomposition processes by affecting soil microorganisms in several ways. Grazing can (1) lead to a younger microbial population with a higher metabolic activity (Sparling et al., 1981); (2) release nutrients from the microbial biomass (Williams and Griffiths, 1989); (3) change the species composition of the microbial community and inhibit the growth of fungal hyphae. In a laboratory experiment Forster et al. (1995) showed that enchytraeids increased CO_2 production, and decreased microbial biomass.

Wolters (1988) demonstrated that enchytraeids promoted decomposition through an increase in the inoculation capacity of the microflora and through alteration of physiochemical conditions of the substrate, but decreased mineralization rates by reducing the number of microbial inhabitants in the environment and probably by occluding in their faeces the products synthesized by soil fungi. Therefore, the net effect of enchytraeids on litter decomposition must be regarded as the result of both promotion and inhibition of soil microflora.

Golebiowska and Ryszkowski (1977) found that tillage shifted the proportional importance of the Annelida from earthworms to the metabolically more active enchytraeids. They concluded that in conventionally tilled fields Enchytraeidae were important in the process of organic matter mineralization due to their high respiration. Hendrix et al. (1986) came to a similar conclusion that enchytraeids were more important in the detritus food web in conventionally tilled than in no-tillage fields. Kasprzak (1982) concluded that in cultivated fields in Poland the contribution of enchytraeids to the process of organic matter mineralization was much greater than that of earthworms. The percentage of total soil respiration that can be attributed to enchytraeid activity varied from 0.4% in a potato field in Poland (Golebiowska et al., 1974) to 3.2% in a sugar beet field in The Netherlands (Didden, 1991).

In summary, enchytraeids can influence the decomposition of organic matter by (1) comminution of the material which enhances microbial decomposition, (2) selective grazing on microbial populations, (3) incorporation of SOM into soil and soil aggregates (see below), (4) dispersal of microbial spores and propagules, and (5) immobilization of nutrients in their biomass.

2.5.4.2 Soil Structure

Enchytraeids excrete fecal pellets consisting of fine particles with little cellulosic plant residues. In coniferous forests, these pellets contain a considerable portion of humus; while in mineral soils, the fecal pellets are more sponge-like structures with humus and loamy material combined (Kasprzak, 1982). Didden (1990) concluded that enchytraeids have a positive effect on aggregate stability in the 600–1,000 μm fraction. These aggregates might have been partly composed of enchytraeid excrements with a size of about 200 μm. However, he observed no significant effects of enchytraeids on the distribution of water stable aggregates.

In the O horizon of forest soils, enchytraeids are secondary decomposers of the larger excrements of soil macrofauna (Rusek, 1985). They make narrow channels in large, spongy earthworm casts. When high densities of enchytraeids are present, their fecal matter may be important for SOM dynamics and changes in soil structure. Encapsulation of organic materials into fecal pellets protects it from extensive decomposition. Also, these microaggregates may in turn serve as building blocks for macroaggregates (Tisdall and Oades, 1982).

van Vliet et al. (1995) calculated the amount of soil ingested by enchytraeids in an agricultural field and in rhododendron-oak forest ecosystems to be 1% and 0.5% of the soil volume (0–15 cm), respectively. The enchytraeid community in the agricultural field was composed of more than 50% large *Fridericia* which can contain large quantities of soil in their guts. Thus, due to species-specific differences between the forest and agricultural field, the effect of enchytraeids on soil structure was greater in agricultural fields. However, Didden (1990) found that the amount of soil transported by *Enchytraeus buchholzi* through ingestion was only about 0.001–0.01% of the total soil volume. The large difference between these two measurements is due to different methods of determining the amount of ingested soil, and to species-specific differences.

The burrowing activity of enchytraeids in soil is poorly known. Rusek (1985) concluded that enchytraeids can make active microtunnels in the soil matrix. Didden (1990) suggested that they increased pore continuity and volume corresponding to their body size (50–200 μm). From his analysis of thin sections he concluded that enchytraeids removed pore necks. Observations of enchytraeids in small microcosms led to the conclusion that they do make channels and increase the hydraulic conductivity of soil. The impact of enchytraeids on hydraulic conductivity, however, depended not only on the incorporation of organic matter in the soil, but also on the number of enchytraeids present and the duration of the experiment (van Vliet et al., 1993, 1997).

Overall, enchytraeids affect soil structure through their fecal pellets and their burrowing capacity. However, the total effect of enchytraeids on soil structure depends on the species composition of the enchytraeid community, on their abundance and on environmental conditions.

2.5.5   Collection and Extraction Methods

For quantitative purposes, enchytraeids can best be sampled with a cylindrical soil corer which keeps the soil intact; optimum sampler size is 5 to 7.5 cm in diameter (Lal et al., 1981; Didden et al., 1995). Because enchytraeids are clustered in soil (Section 2.5.3.3), sufficient replicates need to be taken, but the size and number of the sampling units are mostly chosen as a compromise between accuracy of the abundance estimates and the amount of work involved (Didden, 1993).

Handsorting of enchytraeids from soil is possible, but the efficiency of this labor intensive method is very low. Enchytraeids are often covered with small sand grains and organic particles (Ponge, 1984), making them hard to distinguish from soil material. Small animals are also hard to recover with the handsorting method.

The most commonly used method to extract enchytraeids from soil is the wet funnel method, using an extractor such as that shown in Fig. 2.10 (O'Connor, 1955). In this method, a soil slice (max. 3 cm thick), resting on a sieve in a funnel filled with water, is exposed to light and heat. In 3 hours, the light intensity is increased gradually until the soil surface has reached a temperature of 45 °C. Enchytraeids respond to the light and heat by moving away from the source and pass through the sieve into the water below. A modified extraction method has been proposed (Graefe, 1973; Schauermann, 1983) in which enchytraeids are extracted from soil without heat. The total extraction time is extended from 3 hours to several days for soils rich in organic matter and up to 2 weeks for mineral soils. The length of the extraction period is determined by the risk of an oxygen deficit and the death of larger individuals.

A comparative study of the two methods (Didden et al., 1995) has shown that a better extraction is obtained without heating. The length of the cold extraction period and the total extraction time had a significant positive influence on the extraction efficiency. The Graefe method is cheap and easy to set up, but due to its long extraction time, is not very appropriate for studies where large numbers of samples are taken or where samples are taken at short time intervals. In those cases, the faster O'Connor (1955) method, modified with a longer settling time before the heated extraction starts, might be more appropriate (Didden et al., 1995).

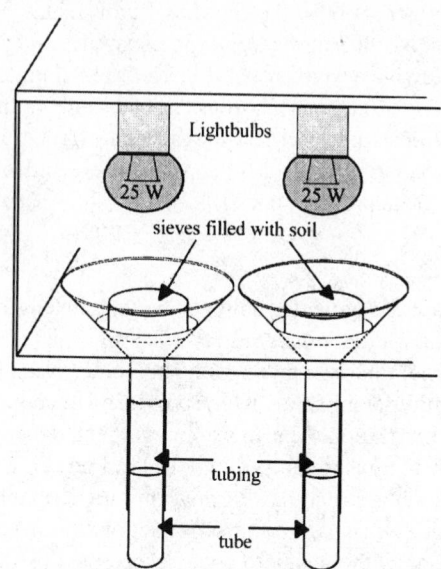

Fig. 2.10 Enchytraeid extractor: The sieves with soil are completely submerged in water; after 4 hr in the dark, the light intensity increased in 3 h from very dim to full bright [After O'Connor, 1955].

2.6   Earthworms
Paul F. Hendrix
University of Georgia

2.6.1   Introduction

Among the soil fauna, earthworms are perhaps the most widely recognized and, in addition to ants and termites, have the most significant biological effects on soil structure and function. For these reasons earthworms have been intensively studied for their potential benefits in agriculture, waste management and land reclamation. The literature on earthworms dates back over 200 years to the taxonomic description of *Lumbricus terrestris* by Linnaeus (1758). The modern era of earthworm research, in the context of soil science, began with Darwin (1881) and a vast literature has accumulated since then. Reviews of much of the recent literature can be found in Lee (1985), Hendrix (1995), and Edwards and Bohlen (1996). This chapter draws from these works to give a brief overview of earthworm biology, ecology and methods of collection.

2.6.2   Basic Taxonomy

Earthworms are classified within the phylum Annelida and class Oligochaeta which consists of some 36 families worldwide (Reynolds and Cook, 1993). Many of these families consist of aquatic or semiaquatic worms, whereas the others are mostly or exclusively terrestrial forms. Twenty terrestrial families are recognized by Jamieson (1988) and 23 by Reynolds and Cook (1993). It is reported that there are 7254 species, both terrestrial and aquatic, in 739 genera (Edwards and Bohlen, 1996). Earthworm families along with their geographic distributions are presented in Table 2.20. This table also includes the family Enchytraeidae, discussed in Section 2.5. Further information on earthworm taxonomy and biogeography can be found in Reynolds (1977), Gates (1982), Sims and Gerard (1985), Edwards and Bohlen (1996), Satchel (1983), Dindal (1990) and Hendrix (1995).

According to Edwards and Bohlen (1996), the families Lumbricidae and Megascolecidae are ecologically the most important in North America, Europe, Australia and Asia. Species from both familes have been introduced worldwide by human activities, and now dominate the earthworm fauna in many areas. Such "peregrine" or "anthropochorous" species are highly successful in many agricultural or otherwise disturbed areas, and often show significant effects on soil processes. Whether introduced earthworms displace native species or occupy areas devoid of native species due to disturbance is a subject of debate (Kalisz and Wood ,1995).

2.6.3   Biology and Ecology

Earthworms are elongated, cylindrical, segmented animals, ranging in length from a few millimeters to 1.4 m, such as the giant Australian *Megascolides australis*. They consist of a relatively simple, tube-within-a-tube body plan, the internal tube comprising the alimentary canal. The body segments are separated by septa and are filled with coelomic fluid which provides a dynamic, hydrostatic "skeleton" for locomotion. Respiration occurs through the moist integument, where blood in subcuticular capillaries absorbs oxygen which is transported throughout the body in a closed vascular system driven by a series of muscular heart-like structures. Earthworms are hermaphroditic, each individual carrying male and female reproductive organs. During reproduction, sperm is exchanged between two individuals and later released, along with eggs, into cocoons secreted by the glandular clitellum, a characteristic thickening along several anterior segments of sexually mature individuals. Further details of earthworm biology can be found in Wallwork (1983), Lee (1985) and Edwards and Bohlen (1996).

Earthworms occur worldwide in most areas where water and temperatures are favorable for at least part of the year (all but desert and polar conditions). Across this range of habitats, earthworms display

Table 2.20   Systematics and global distribution/origin of major families of the terrestrial oligochaeta [Summarized from Wallwork (1983) and Reynolds and Cook (1993)]

Phylum: Annelida	
Class: Oligochaeta	
Order: Haplotaxida	
Suborders: Enchytraeina	
Family: Enchytraeidae	NH, SH
Suborder: Lumbricina	
Family: Lumbricidae	NH--NA, EU
Komarekionidae	NH--NA
Sparganophilidae	NH--NA
Lutodrilidae	NH--NA
Megascolecidae	NH, SH--SA, OC, AS
Glossoscolecidae	SH–SA
Eudrilidae	SH–AF
Acanthodrilidae	SH–AS
Octochaetidae	SH–OC
Ocnerodrilidae	SH–SA, AF, AS, MA
Ailoscolecidae	NH--EU
Hormogastridae	NH--ME
Kynotidae	SH–MA
Microchaetidae	SH–AF
Almidae	SH– SA, AF, AS
Biwadrilidae	NH--JA

NH = northern hemisphere, SH = southern hemisphere; AF = Africa, AS = Asia, EU = Europe, JA = Japan, MA = Madagascar, ME = Mediterranean, NA = North America, OC = Oceania, SA = South America

a wide array of morphological, physiological and behavioral adaptations to environmental conditions. During unfavorable periods, many species are able to enter a temporary dormant state (aestivation or diapause) or produce resistant cocoons which hatch when conditions improve.

The abundance of earthworms across habitats is highly variable, depending on climatic and edaphic conditions, ecosystem type, and the degree to which the habitat has been altered, for example, by agriculture. Under otherwise suitable conditions, soil C concentration has been shown to be highly correlated with earthworm population density and biomass (Edwards, 1983; Hendrix et al., 1992). Earthworm density and biomass in a variety of habitats worldwide are presented in Table 2.21. Densities range from < 10 to > 2,000 m^{-2}, with the highest values occurring in grasslands (especially fertilized) and the lowest in acid or arid soils (coniferous or sclerophyllous forest). Typical densities from temperate deciduous or tropical forests and certain arable systems range from ~ 100–400 m^{-2}, the latter depending on intensity and nature of management. Earthworm biomass tends to track density, but biomass estimates are problematic due to different methods used by various investigators (Section 2.6.5.1). Because earthworm populations often show seasonal variation in abundance (especially in temperate regions), time of sampling also affects density and biomass estimates.

Within habitats, earthworms often show heterogeneous spatial distributions. "Single-tree-influence" (Boettcher and Kalisz, 1991) or spatial distributions of other controlling environmental factors such as soil texture or organic matter (SOM) often result in clumping of earthworm populations, while habitat and feeding preferences dictate vertical distributions of species within the soil profile. A series of earthworm life forms or functional categories based on habitat and feeding ecology is presented in Table 2.21. These categories describe niche separation of earthworm species within a soil volume. Polyhumic epigeic and epiendogeic species utilize litter and organically enriched surface layers; poly-, meso- and oligohumic endogeic species inhabit mineral soil within the rhizosphere and beyond; and anecic species exploit both the surface litter as a source of food and the mineral soil as a refuge. Lee (1985) summarizes data showing that within a particular soil, commonly less than a half-dozen earthworm species are found. The species in a given earthworm association often effectively partition the soil volume according to the functional categories mentioned above. Furthermore, the activities of earthworms within these categories influence biogeochemical processes in various ways. For example, epigeic species facilitate the breakdown and mineralization of surface litter, while anecic species incorporate SOM deeper into the soil profile and enhance aeration and water infiltration through burrow formation (Lee, 1985; Edwards et al., 1990).

For management of earthworms in agroecosystems, Lee (1991, 1995) recommends that "target earthworm communites" consist of one or more anecic/epigeic species that make deep vertical burrows and that cast on the surface and bury residues, and one or more endogeic species that feed below ground on dead roots and SOM and that make horizontal burrows. Diverse assemblages of earthworms may more effectively exploit soil resources and influence a wider array of processes, such as SOM turnover, in addition to soil structural properties, than a single species.

2.6.4   Importance to Soil Processes

Where earthworms are abundant, they can exert significant influence on soil processes through effects on: (1) soil structure, (2) SOM and nutrient cycling, and (3) other soil organisms, especially microbes. These topics are reviewed in Lee (1985), Lavelle (1988), Blair et al. (1995), Tomlin et al. (1995) and Edwards and Bohlen (1996).

2.6.4.1   Organic Matter Dynamics and Nutrient Cycling
As with soil structure, effects of earthworms on SOM and nutrient cycling are closely linked with the life form and feeding ecology of earthworms (Table 2.21). Epigeic species typically live in the O and

upper A soil horizons where, through feeding and casting activities, they mix mineral soil and plant litter, fragment organic particles, inoculate them with microbes, and thereby accelerate organic matter decomposition rates. Anecic forms pull surface litter into their burrows, thus transporting organic material deeper into the soil profile. They cast on the soil surface, mixing organic and mineral particles in the litter layer. The activities of both epigeic and anecic earthworms produce "mull" soil horizons, defined as those in which organic matter is intimately incorporated into the upper mineral soil of a well-developed A horizon overlain with litter or humus layers < 2 cm thick. The extreme case is termed "vermimull," in which the Ah horizon is granular and characterized by strong organomineral complexes consisting of earthworm casts (Green et al. 1993). Endogeic earthworms feed within the soil on SOM and microbes associated with the rhizosphere or mineral soil. They are termed oligo-, meso- or polyhumic, depending on the level of organic enrichment of their substrate. Casts and burrows of endogeic earthworms are also sites of increased microbial activity and organic matter decomposition. Mineralization of SOM in earthworm casts and burrow linings produces zones of nutrient enrichment compared to bulk soil. These "hot spots" (the "drilosphere") are often sites of enhanced activity of plant roots and other soil biota (Lavelle, 1988; Beare et al., 1995).

2.6.4.2 Soil Structure

Soil structure is affected by earthworms principally through: (1) production of casts which can form stable aggregates upon and within the soil, and (2) formation of burrows which produce macropores that may increase water infiltration and aeration within the soil. Casts are produced by ingestion of mineral and organic particles, mixing, organic enrichment, and microbial stimulation in the gut, and egestion of the material as a slurry or as discrete pellets (depending on earthworm species), which harden into stable aggregates. Mechanisms of cast stabilization include organic bonding of particles by polymers secreted by earthworms and microbes, mechanical stabilization by plant fibers and fungal hyphae, and stabilization due to wetting and drying cycles and age hardening/thixotrophic effects (Tomlin et al., 1995). Earthworm casts are usually enriched with plant-available nutrients and thus may enhance soil fertility.

Earthworms create burrows of various sizes, depths and orientations, depending on species and soil type. Burrows tend to be similar in diameter to that of the body, ranging from 1 to > 10 mm diameter and constituting among the largest of soil pores (Lee, 1985; Tomlin et al., 1995). Geophagous species (Table 2.21) may form networks of variously oriented macropores, as the earthworms consume the soil and cast behind them as they burrow. Although such networks may form continuous pores for some depth, casting within the burrows may impede free water movement. Anecic earthworms may create vertical burrows which form continuous macropores to depths of up to one meter or more. Such burrows are often highly stable since their walls are lined with organic metter drawn in or secreted by earthworms, and they tend to have higher bulk density than surrounding soil. Continuous macropores resulting from earthworm burrowing may greatly enhance water infiltration by functioning as bypass flow pathways through saturated soils (Lee, 1985; Tomlin et al., 1995). These pores may or may not be important in solute transport depending on antecedent soil water, nature of the solute, and exchange properties of the burrow linings (Edwards et al., 1990, 1993).

2.6.5   Methods

2.6.5.1   Earthworm Collection

Techniques for field sampling of earthworms are reviewed in Lee (1985) and Edwards and Bohlen (1996). Unless otherwise given, methodological details and specific reference citations can be found in these works. Collection techniques are passive, behavioral, and indirect (Table 2.23).

Table 2.21 Abundance and biomass of earthworms in selected habitats from various parts of the world [Reproduced with permission from Lee, 1985. Earthworms: Their ecology and relationships with soils and land use. Copyright Academic Press, Sydney, Australia]

Habitat	Location	Collection method	Earthworm Taxa	Abundance no. m^{-2}	Biomass g m^{-2}
Sown pastures	New Zealand	Hand sorting	Lumbricidae	208–775	60–241
				740–1235	146–303
				690–2020	305 (mean)
Sown pastures	South Australia	Hand sorting	Lumbricidae	460–625	62–78
Sown pastures	South Africa	Hand sorting	Lumbricidae	72–112	—
Fertilized pasture	Argentina	Hand sorting	Lumbricidae, Megascolecidae and Glossoscolecidae	27	—
Pastures heavily fertilized	Ireland	Hand sorting	Lumbricidae	400–500	100–200
Old pasture	Sweden	Hand sorting	Lumbricidae	109	59
Old pasture	England	Hand sorting	Lumbricidae	390–470	52–110
Old pasture	Wales	Hand sorting	Lumbricidae	646	149
Old pasture	France	Washing/sieving	Lumbricidae	288	125
Fallow	South Australia	Hand sorting	Lumbricidae	210–460	16–76
Fallow	Wales	Hand sorting	Lumbricidae	226	79
Cropland	South Australia	Hand sorting	Lumbricidae	20–25	2–2.5
Cropland	Rumania	Hand sorting	Lumbricidae	5–100	0.5–20
Natural grassland	Rumania	Hand sorting	Lumbricidae	200 (mean)	10–60
Natural grassland	Wales	Hand sorting	Lumbricidae	22	8
Natural grassland	Tennessee, USA	Hand sorting	Lumbricidae	13–41	3.2–7.5
Natural grassland	South Africa	Hand sorting	Glossoscolecidae	74	96
Natural grassland	India	Hand sorting	Megascolecidae and Ocnerodrilidae	64–800	6–60
Tropical savannas	Ivory Coast	Hand sorting and washing/sieving	Megascolecidae and Eudrilidae	230	49
Natural grassland	New Zealand	Hand sorting	Megascolecidae	250–750	—
Orchard	Netherlands	Hand sorting	Lumbricidae	300–500	75–122
Orchard	Australia	Hand sorting	Lumbricidae	150	—
Mulched and irrigated orchards	Australia	Hand sorting	Lumbricidae	2000	—
Garden	Egypt	Hand sorting	Megascolecidae	420	153

Table 2.21   (cont.)

Habitat	Location	Collection method	Earthworm Taxa	Abundance no. m^{-2}	Biomass g m^{-2}
Gardens	Argentina	Hand sorting	Lumbricidae, Megascolecidae and Glossoscolecidae	73	—
Taiga	Finland	Hand sorting	Lumbricidae	17.4	2.8
	Siberia			23.0	8.4
	USSR			3–7	—
Northern European and Asian coniferous forests	Finland	Hand sorting	Lumbricidae	14–68	30–35
	Sweden			103–167	—
	USSR			12	—
	Japan			27–72	—
Spruce forest with lime topdressing	USSR	Hand sorting	Lumbricidae	1000	—
European deciduous forests	England	Hand sorting	Lumbricidae	118–138	—
	USSR			136	68.3
	Czechoslovakia			106	98.1
North American deciduous forsts	Canada	Hand sorting	Lumbricidae	240–780	38–109
	Tennessee, USA			2–96	1.3–14
	Indiana, USA			14–124	26.3–280.3
				7–38	1.3–25.5
Dry schlerophyll forest		Hand sorting	Megascolecidae	34–76	12.3–47.9
Wet schlerophyll forest	Australia	Hand sorting	Megascolecidae	15–106	5.7–35.7
Subalpine woodland				70–130	3.4–6.8
Gallery forests	Ivory Coast	Hand sorting and washing/sieving	Megascolecidae and Eudriliidae	34	10.2
Tropical forests	Nigeria		Eudrilidae	61.7	2.5
Tropical forest	Nigeria	Hand sorting	Eudrilidae	37–92	0.7–1.3
Lowland dipterocarp forest			Moniligastridae and Megascolecidae	55	3.1
Lower Montane forest			Moniligastridae and Megascolecidae	47–108	1.8–2.7
Upper Montane forest	Sarawak	Hand sorting	Moniligastridae and Megascolecidae	2–24	0.2–2.1
Upper Montane low forest			Megasolecidae		

Table 2.22 Ecological strategies of earthworms [Summarized from Lee (1959); Bouché (1977); and Lavelle (1983)]

Epigeic (litter dweller) - mesophage; detritivore.
    Lives in and consumes litter; small size; uniformly pigmented.
    (*Lumbricus rubellus, Bimastos* spp., *Dendrobaena octaedra, D. rubida, Eisenia foetida*)
Endogeic (subsoil dweller) – microphage; geophage; (epiendogeic or hypoendogeic; oligohumic, mesohumic,
    or polyhumic).
    Lives in horiontal, branching burrows in organo-mineral layer; consumes soil; small to large in size;
    weakly pigmented.
        (*Aporrectodea caliginosa, Octolasion cyaneum, Diplocardia* spp.)
Anecic (topsoil dweller) – macrophage; detritivore.
    Lives in deep vertical burrows, casting on surface; emerges at night to draw down organic matter (plant
    residue, etc.); large as adults (200-1,100 mm); brown pigment anteriorly and dorsally.
    (*Lumbricus terrestris, Allolobophora longa*)

Hand digging and sorting, which is the most commonly used method for quantitative sampling of earthworms, involves digging pits of known dimensions (such as 25 x 25 x 25 cm), breaking the soil by hand, and collecting all earthworms found. Often the collected specimens are immediately preserved in 70% ethanol or 5% formalin for later counting and identification.

Washing and sieving is an elaboration of hand sorting, in that the soil is dispersed in water (or a dispersing agent), poured through a sieve or nest of sieves, and the earthworms and cocoons hand picked from the sieve contents. Mechanized approaches to washing and sieving are described by Bouché and Beugnot (1972). Flotation of sieve contents in a high density solution, such as 1.16–1.20 SG $MgSO_4$, is an additional means of separating earthworms and other soil fauna.

Several approaches have been taken to extracting earthworms from soil based on their behavioral response to certain stimuli. A number of chemical irritants have been used, including $HgCl_2$, $KMnO_4$, mustard, and formalin. Aqueous solutions of 0.165–0.550% formalin are most commonly used and have been shown to be effective on *L. terrestris* when applied in three sequential doses of 18 L m^{-2} (Raw, 1959; Lee, 1985); but formalin may be less effective on other species (Satchell, 1969; Callaham and Hendrix, 1997). Effectiveness varies with earthworm species and activity, soil water content, porosity and temperature. Comparisons with hand sorting should be done before adopting extraction techniques for quantitative sampling.

The heat extraction method is a modification of that used for enchytraeids (Section 2.5.5). Soil cores or blocks are placed in pans of water, exposed to heat from overhead lightbulbs, and earthworms are collected from the water after several hours. This technique was more effective than hand sorting or formalin extraction on small earthworms in dense root mats (Satchell, 1969). As with hand sorting, it is not effective on deep burrowing, anecic species such as *L. terrestris*.

Mechanical vibration employs a rod or stake driven into the soil, vibration for a few minutes with a bow or flat piece of metal, such as an automobile leaf spring, and collection of earthworms that emerge onto the soil surface. Some megascolecid species have been sampled with this technique (Reynolds, 1973; Hendrix et al., 1994), but it is not effective on lumbricids and probably only useful for selective or comparative sampling of certain populations.

Electrical extraction of earthworms involves inserting metal rods into the soil, connecting them to a source of alternating current and collecting earthworms that come to the soil surface. Different voltages and amperages have been used with varying degrees of success; effectiveness of the technique is highly dependent on soil water content, electrolyte concentration and temperature. As with mechanical vibration, the soil volume sampled is not known and therefore this method is best suited for qualitative or comparative sampling. The method is potentially very dangerous and should only be used with extreme caution.

Two earthworm trapping techniques have been described. Pitfall traps (Section 2.4.3; open top containers buried level with the soil surface and containing a fixative solution, such as picric acid) may be useful for sampling surface active species in diurnal or seasonal studies. Arrays of traps are installed and sampled at 12 h, 24 h or longer intervals. Baited traps, such as perforated clay pots containing manure or other attractants and inserted into the soil may also be useful for collecting certain species.

Table 2.23 Descriptions of methods for collecting earthworms [From Lee (1985) and Edwards and Bohlen (1996)]

Method	Description	Advantages	Disadvantages
Passive			
Hand sorting	Known volume of soil cut with spade or corer, broken and worms removed by hand	Simple, reliable in the field; low cost	Laborious; may not collect deep burrowing species, small earthworms and cocoons
Washing and sieving	Known volume of soil cut with spade or corer, soaked in dispersant/preservative, and washed through sieve(s) by hand or mechanical device	Higher recovery of cocoons and small individuals	Laborious; may not collect deep burrowing species
Flotation	Material from hand sorting or washing/sieving floated in high-density solution (e.g., $MgSO_4$)	Separates earthworms from soil and plant debris; cocoons and small individuals collected	Laborious; may not collect deep burrowing species
Behavioral			
Chemical extraction	Soil saturated with chemical irritant (e.g., 0.2% formalin) causing earthworms to emerge onto soil surface	Simple; effective on deep burrowing anecic species	Not effective on all species, in all soils or under all conditions.
Heat extraction	Soil blocks or cores suspended under heat lamps in water into which earthworms migrate	Effective on dense root mats	Not effective on all species; inconvenient for field use
Electrical extraction	Metal rods inserted into soil and connected to AC electrical source	Useful for selective or comparative sampling	Highly variable; not convenient in the field; dangerous
Mechanical vibration	Stake or rod inserted into soil and vibrated with bow or flat iron	Simple; useful for selective or comparative sampling	Not effective on all species
Trapping	Pitfall or baited traps placed in soil and sampled at desired intervals	Simple; useful for selective or comparative sampling	Not effective on all species
Mark-recapture	Individuals tagged, released, and population sampled at intervals	Useful for estimating population density, dispersal and mortality	Laborious
Indirect			
Cast counting	Surface castings enumerated and identified	Simple	Not a quantitative estimate of population density

As with other behavioral methods, trapping is probably highly selective and best suited for qualitative or comparative sampling.

Mark, release and recapture techniques have been widely used to study population dynamics of animals, including earthworms. Large numbers of individuals of desired species are collected, marked (e.g., with brands or nontoxic dyes) and released into the population of interest. Sampling over time and distance from the target site, and enumeration of tagged relative to untagged individuals, yield information on dispersal, mortality and population density. Radioisotope and, more recently, immunofluorescent antibody techniques have been employed in earthworm mark-recapture studies.

For earthworm species that cast on the soil surface, such as *Aporrectodea longa*, numbers and identity of castings may be a useful index of population activity. Because casting is dependent on soil temperature and moisture, this technique is highly variable and not a quantitative estimate of population density.

In summary, digging and hand sorting or washing are probably the most reliable means of sampling earthworms. However, no single method will be adequate to sample earthworm populations in all situations. Combinations of methods will probably achieve reasonable results. For example, formalin solution can be applied to the bottom of pits previously excavated for hand sorting, to extract deep burrowing anecic forms not sampled by digging (Edwards and Bohlen 1996). Combinations of various methods may be useful in other situations.

2.6.5.2 Identification

Many earthworm species in the family Lumbricidae can be identified from external body characteristics if the specimens are sexually mature. Taxonomic keys by Reynolds (1977), Sims and Gerard (1985), and Schwert (1990) are very useful for the common lumbricids found worldwide. Reynolds (1977) also includes a key to North American Sparganophilidae, a limicolous or semiaquatic group.

Most earthworms other than the Lumbricidae require dissection for accurate taxonomic identification. Locations and characteristics of sexual organs, the gut and associated glands, and other structures are required. The procedures must be done carefully and require a degree of skill and practice. At the family level, several keys are available: Jamieson (1988) includes a key and diagrammatic comparison of characteristic internal structures of most families; Edwards and Bohlen (1996) review major characteristics of the families; Sims and Gerard (1985) provide keys and species descriptions for seven families found in Great Britain; Fender and McKey-Fender (1990) give keys to the families of the United States and the genera of Megascolecidae from the western United States; James (1990) provides a key to the genus *Diplocardia*, a group of megascolecids found in the eastern United States, Mexico and the Caribbean. Moreno (1996) includes keys and generic descriptions for the Glossoscolecidae in Central and South America. Many of the works cited in Gates (1982) include keys and species descriptions for families and genera found in North and Central America.

2.7 References

Abrahamsen, G. 1969. Sampling design in studies of population densities in Enchytraeidae (Oligochaeta). Oikos 20:45–66.

Abrahamsen, G. 1972. Ecological study of Enchytraeidae (Oligochaeta) in Norwegian coniferous forest soils. Pedobiol. 12:26–82.

Abrahamsen, G. 1990. Influence of *Cognettia sphagnetorum* (Oligochaeta) on nitrogen mineralization in homogenized mor humus. Biol. Fertil. Soils 9:159–162.

Abrahamsen, G., and L. Strand. 1970. Statistical analysis of population density data of soil animals, with particular reference to Enchytraeidae (Oligochaeta). Oikos 21:276–284.

Aescht, E., and W. Foissner. 1992. Enumerating soil testate amoebae by direct counting, p. B-6.1–B-6.4. *In* J.J. Lee and A.T. Soldo (ed.) Protocols in protozoology. Allen Press, Lawrence, KS.

Aescht, E., and W. Foissner. 1993. Effects of organically enriched magnesite fertilizers on the soil ciliates of a spruce forest. Pedobiol. 37:321–335.

Albert, A.M. 1979. Chilopoda as part of the predatory macroarthropod fauna in forests: abundance, life-cycle, biomass and metabolism. p. 215–231 *In* M. Camatini (ed.) Myriapod biology. Academic Press, London, UK.

Anderson, R. C. 1992. Nematode parasites of vertebrates. CAB International, Wallingford, UK.

Andrassy, I. 1976. Evolution as a basis for the systematization of nematodes. Pitman Publishing, San Francisco, CA.

Andrassy, I. 1983. A taxonomic review of the suborder Rhabditina (Nematoda: Secernentia). ORSTOM, Paris, France.

Arnett, R. H. Jr. 1993. American insects. A handbook of the insects of America north of Mexico. Sandhill Crane Press, Gainesville, FL.

Ayoub, S.M. 1980. Plant nematology, an agricultural training aid. NemaAid Publications, Sacramento, CA.

Balogh, J., and P. Balogh. 1992. The Oribatid mites genera of the world. Vol. 1, 2. Hungarian National Museum Press, Budapest, Hungary.

Bamforth, S.S., and J.D. Lousier. 1995. Protozoa in tropical litter decomposition, p. 59–73. *In* M.K. Reddy (ed.) Soil organisms and litter decomposition in the tropics. IBH, Oxford, UK.

Bamforth, S.S. 1992. Sampling and enumerating soil protozoa. p. B-5.1–B-5.3. *In* J.J. Lee and A.T. Soldo (ed.) Protocols in protozoology. Allen Press, Lawrence, KS.

Bamforth, S.S. 1985. The role of protozoa in litters and soils. J. Protozool. 32:404–409.

Bamforth, S.S. 1984. Microbial distributions in Arizona deserts and woodlands. Soil Biol. Biochem. 16:133–137.

Bamforth, S.S. 1995. Isolating and counting of protozoa, p. 174–180. *In* K. Alef and P. Nannipieri (ed.) Methods in applied soil microbiology and biochemistry. Harcourt Brace, London, UK.

Barker, K.R., and C.L. Campbell. 1981. Sampling nematode populations. *In* B.M. Zuckerman, W.F. Mai and R.A. Rohde (ed.) Plant parasitic nematodes. Vol. III, Academic Press, New York, NY.

Barker, K.R., and C.J. Nusbaum. 1971. Diagnostic and advisory programs. *In* B.M. Zuckerman, W.F. Mai, and R.A. Rohde (ed.) Plant parasitic nematodes. Vol. I. Academic Press, New York, NY.

Barnes, R. D. 1965. Invertebrate zoology. W. B. Saunders Co., Philadelphia, PA.

Bater, J. 1996. Micro- and macro-arthropods. p. 163–174. *In* G.S. Hall (ed.) Methods for the examination of organismal diversity in soils and sediments. CAB International, New York, NY.

Beare, M.H., D.C. Coleman, D.A. Crossley, Jr., P.F. Hendrix, and E.P. Odum. 1995. A hierarchical approach to evaluating the significance of soil biodiversity to biogeochemical cycling. Plant Soil 170:5–22.

Behan-Pelletier, V.M., and B. Bisset. 1993. Biodiversity of nearctic soil arthropods. Canad. Biodiv. 2:5–14.

Berger, H., W. Foissner, and H. Adam. 1986. Field experiments on the effects of fertilizers and lime on the soil microfauna of an alpine pasture. Pedobiol. 29:261–272.

Berthold, A., and M. Palzenberger. 1995. Comparison between direct counts of active soil ciliates (Protozoa) and most probable number estimates obtained by Singh's dilution culture method. Biol. Fertil. Soils 19:348–356.

Bird, A.F., and J. Bird. 1979. The structure of nematodes. 2nd Ed. Academic Press, San Diego, CA.

Blair, J.M., R.W. Parmelee, and P. Lavelle. 1995. Influences of earthworms on biogeochemistry. *In* P.F. Hendrix (ed.) Ecology and biogeography of earthworms in North America. Lewis Publishers, Boca Raton, FL.

Blumberg. A.Y., and D.A. Crossley, Jr. 1983. Comparison of soil surface arthropod populations in conventional tillage, no-tillage and old field systems. Agroecosys. 8: 247–253.

Boettcher, S.E., and P.J Kalisz. 1991. Single-tree influence on earthworms in forest soils in eastern Kentucky. Soil Sci. Soc. Am. J. 55:862–865.

Bohnsack, K.K. and D.A. Crossley, Jr. 1960. Long-term ecological study in the Oak Ridge area. III. The oribatid mite fauna in pine litter. Ecol. 41:785–790.

Bolton, B. 1994. Identification guide to the ant genera of the world. Harvard University Press, Cambridge, MA.

Bongers, T. 1990. The maturity index: An ecological measure of environmental disturbance based on nematode species composition. Oecolog. 83:14-19.

Bonnet, L. 1964. Le peuplement thecamoebien des sols. Rev. Ecol. Sol 1:123–408.

Bornebusch, C.H. 1930. The fauna of the soil. Forstl. Forsogavias Dan. 11:1–224.

Borror, D.J., C.A. Triplehorn, and N.F. Johnson. 1992. An introduction to the study of insects. Saunders College Publishers, Fort Worth, TX.

Bostrom, S., and B. Sohlenius. 1986. Short-term dynamics of nematode communities in arable soil. Influence of a perennial and an annual cropping system. Pedobiol. 29:345-357.

Bouché, M.B., and M. Beugnot. 1972. Contribution à l'approache méthodologique de l'étude des biocénoses. II. L'extraction des macroéléments du sol par lavage-tamisage. Ann. Zool. Ecol. Anim. 4:537–544.

Bouché, M.B. 1977. Stratégies lombriciennes. *In* Soil organisms as components of ecosystems. Biol. Bull. 25:122–133.

Bovee, E.C. 1985. Class Lobosea Carpenter 1861, p. 159–215. Class Filosea Leidy 1879, p. 228–245. *In* J.J. Lee, S.H. Hunter, and E.C. Bovee (ed.) An illustrated guide to the protozoa. Allen Press, Lawrence, KS.

Brockmeyer, V., R. Schmid, and W. Westheide. 1990. Quantitative investigations of the food of two terrestrial enchytraeid species (Oligochaeta). Pedobiol. 34:151–156.

Bryan, M. V. 1978. Production ecology of ants and termites. Cambridge University Press, Cambridge, UK.

Butler, H., and A. Rogerson. 1995. Temporal and spatial abundance of naked amoebae (Gymnamoebae) in marine benthic sediments of the Clyde Sea area, Scotland. J. Euk. Microbiol. 42:724–730.

Caldwell, J.R. 1981. Biomass and respiration of nematode populations in two mass communities at Signy Island, maritime Antarctic. Oikos 37:160-166.

Callaham, M.A., Jr., and P.F. Hendrix. 1997. Relative abundance and seasonal activity of earthworms (Lumbricidae and Megascolecidae) as determined by hand-sorting and formalin extraction in forest soils on the southern Appalachian Piedmont. Soil Biol. Biochem. 29:317–322.

Camatini, M. (ed.) 1979. Myriapod biology. Academic Press, London, UK.

Cash, J. 1909. The British freshwater Rhizopoda and Heliozoa, Vol. 2. The Ray Society, London, UK.

Cash, J., and G. H. Wailes. 1915, 1919. The British freshwater Rhizopoda and Heliozoa. Vol. 3, 4. The Ray Society, London, UK.

Chalupský Jr., J., and J. Lepš. 1985. The spatial pattern of Enchytraeidae (Oligochaeta) Oecolog. 68:153–157.

Christensen, B. 1956. Studies on Enchytraeidae 6. Technique for culturing Enchytraeidae, with notes on cocoon types. Oikos 7:303–307.

Christiansen, K., and P. Bellinger. 1980. The Collembola of North America, north of the Rio Grande. Grinnell College, Grinnell, IA.

Clarholm, M. 1981. Protozoan grazing of bacteria in soil-impact and importance. Microb. Ecol. 7:343–350.

Coleman, D.C. and D. A. Crossley, Jr. 1996. Fundamentals of soil ecology. Academic Press, San Diego, CA.

Couteaux, M.M. 1967. Une technique d'observation des thecamoebiens du sol pour l'estimation de leur densite absolue. Rev. Ecol. Biol. Sol 4:593–596.

Couteaux, M.M. 1975. Ecologie des Thecamoebiens de quelques humus bruts forestiers: l'especie dans la dynamique de l'equilibre. Revue Ecol. Biol. Sol, 12: 421–444.

Crawford, C.S. 1979. Desert millipedes: A rationale for their distribution. p. 171–181. *In* M. Camatini. (ed.) Myriapod biology. Academic Press, London, UK.

Crawford, C. S. 1981. Biology of desert invertebrates. Springer-Verlag, Berlin, Germany.

Crossley, D.A. Jr., and M.P. Hoglund. 1962. A litter-bag method for the study of microarthropods inhabiting leaf litter. Ecol. 43:571–573.

Curl, E.A., and B. Truelove. 1986. The rhizosphere. Springer-Verlag, New York, NY.

Darbyshire, J.F. (ed.) 1996. Soil protozoa. CAB International, London, UK.

Darbyshire, J.F., and M.P. Greaves. 1967. Protozoa and bacteria in the rhizosphere of *Sinapis alba* L., *Trofolium repens* L., and *Lolium perenne* L. Can. J. Microbiol. 13:1057–1068.

Darbyshire, J.F., R.E. Wheatley, M.P. Greaves, and R.H.E. Inkson. 1974. A rapid micromethod for estimating bacterial and protozoan populations in soil. Rev. Ecol. Biol. Sol, 11: 465–475.

Darwin, C.R. 1881. The formation of vegetable mould through the action of worms, with observations on their habits. Murray, London, UK.

Dash, M.C. 1990. Enchytraeidae. p. 311-340. *In* D.L. Dindal (ed.). Soil biology guide. John Wiley and Sons, New York, NY.

Dash, M.C., and J.B. Cragg. 1972. Ecology of Enchytraeidae (Oligochaeta) in Canadian Rocky Mountain soils. Pedobiol. 12:323–335.

Dash, M.C., B. Nanda, and P.C. Mishra. 1981. Digestive enzymes in three species of Enchytraeidae (Oligochaeta). Oikos 36:316–318.

Decloitre, L. 1976. Le genre Euglypha. Arch. Protistenk. 118:18–33.

Decloitre, L. 1977a. Le genre Cyclopyxis. Arch. Protistenk. 119:31–53.

Decloitre, L. 1977b. Le genre Nebela. Arch. Protistenk. 119:325–352.

Decloitre, L. 1978. Le genre Centropyxis I. Arch. Protistenk. 120:63–85.

Decloitre, L. 1979. Le genre Centropyxis II. Arch. Protistenk. 121:162–192.

Decloitre, L. 1981. Le genre Trinema Dujardin. Arch. Protistenk. 124:193–218.

Didden, W. 1990. Involvement of Enchytraeidae (Oligochaeta) in soil structure evolution in agricultural fields. Biol. Fertil. Soils 9:152–158.

Didden, W. 1991. Population ecology and functioning of Enchytraeidae in some arable farming systems. Ph.D. dissertation. Agricultural University Wageningen, Netherlands.

Didden, W. 1993. Ecology of Enchytraeidae. Pedobiol. 37:2–29.

Didden, W. 1995. The effect of nitrogen deposition on enchytraeid-mediated decomposition and mobilization-a laboratory experiment. Acta Zool. Fennica 196:60–64.

Didden, W., H. Born, H. Domm, U. Graefe, M. Heck, J. Kühle, A. Mellin, and J. Römbke. 1995. The relative efficiency of wet funnel techniques for the extraction of Enchytraeidae. Pedobiol. 39:52–57.

Dindal, D. L. (ed.). 1990. Soil biology guide. Wiley Interscience, New York, NY.

Dósza-Farkas, K. 1996. Reproduction strategies in some enchytraeid species. p. 25–33. In K. Dósza-Farkas (ed.) Newsletter on Enchytraeidae No. 5. Eötvös Loránd University. Budapest, Hungary.

Dósza-Farkas, K. 1982. Konsum verschiedener Laubarten durch Enchytraeiden (Oligochaeta). Pedobiol. 23:145–156.

Draney, M.L. 1992. Biodiversity of ground-layer macroarachnid (Araneae and Opiliones) assemblages from four Piedmont floodplain habitats. MS Thesis, University of Georgia, Athens, GA.

Edwards, C.A., and P.J. Bohlen. 1996. Biology and Ecology of earthworms. 3rd Ed. Chapman and Hall, New York, NY.

Edwards, C.A. 1983. Earthworm ecology in cultivated soil. p. 123–138. In J.E. Satchell (ed.) Earthworm ecology from Darwin to vermiculture. Chapman and Hall, London, UK.

Edwards, W.M., M.J. Shipitalo, L.B. Owens, and L.D. Norton. 1990. Effect of Lumbricus terrestris L. burrows on hydrology of continuous no-till corn fields. Geoderma 46:73–84.

Edwards, W.M., M.J. Shipitalo, L.B.Owens, and W.A. Dick. 1993. Factors affecting preferential flow of water and atrazine through earthworm burrows under continuous no-till corn. J. Environ. Qual. 22:453–457.

Ekschmitt, K., V. Wolters, and M. Weber. 1997. Spiders, carabids and staphylinids: the ecological potential of predatory macroarthropods. p. 307–362. In G. Benckiser (ed.) Fauna in soil ecosystems. Marcel Dekker, New York, NY.

Elliott, E.T., H.W. Hunt, and D.E. Walter. 1988. Detrital foodweb interactions in North American grassland ecosystems. Agric. Ecosys. Environ. 24:41-56.

Evans, G.O. 1961. Terrestrial Acari of the British Isle. Vol 1. Introduction and biology. British Museum of Natural History, London. UK.

Evans, G.O. 1992. Principles of acarology. CAB International, Wallingford, UK.

Evans, K., D.L. Trudgill, and J.M. Webster. 1993. Plant parasitic nematodes in temperate agriculture. CAB International, Wallingford, UK.

Fender, W.M., and D. McKey-Fender. 1990. Oligochaeta: Megascolecidae and other earthworms from western North America. p. 357–378. In D.L. Dindal (ed.) Soil biology guide. John Wiley and Sons, New York, NY.

Fender, W.M. 1995. Native earthworms of the Pacific Northwest: An ecological overview. p. 53–66. In P.F. Hendrix (ed.) Earthworm ecology and biogeography in North America. Lewis Publishers, Boca Raton, FL.

Ferris, V. R. 1971. Taxonomy of the Dorylaimida. In B.M. Zuckerman, W.F. Mai, and R.A. Rohde (ed). Plant parasitic nematodes. Academic Press, New York, NY.

Ferris, V. R., J.M. Ferris, and J.P. Tjepkema. 1973. Genera of freshwater nematodes (Nematoda) of eastern North America. US Environmental Protection Agency, Washington, DC.

Ferris, V.R., and J.M. Ferris. 1974. Inter-relationships between nematode and plant communities in agricultural ecosystems. Agroecosyst. 1:275-299.

Fisher, R.A., and F. Yates. 1963. Statistical tables in biological, agricultural, and medical research. Oliver and Boyd, Edinburgh, UK.

Foelix, R.F. 1996. Biology of spiders. Oxford University Press, New York, NY.

Foissner, W. 1980. Colpodidae Ciliaten (Protozoa: Ciliophora) aus alpinen Boden. Zool. Jb. Syst. 107:391–432.

Foissner, W. 1982. Okologie und Taxonomie der Hypotrichida (Protozoa: Ciliophora) einiger osterreichischer Boden. Arch. Protistenk. 126:19–143.

Foissner, W. 1987. Soil protozoa: fundamental problems, ecological significance, adaptations in ciliates and testaceans, bioindicators, and guide to the literature. Progr. Protistol. 2:69–212.

Foissner, W. 1991. Diversity and ecology of soil flagellates. p. 93–112. In D.J. Patterson and J. Larsen (ed.) The biology of free-living heterotrophic flagellates, Clarendon Press, Oxford, UK.

Foissner, W. 1992. Enumerating active soil ciliates by direct counting, p.B-7.1–B-7.4 In J.J. Lee and A.T. Soldo (ed.) Protocols in protozoology. Allen Press, Lawrence, KS.

Forster, B., J. Römbke, T. Knacker, and E. Morgan. 1995. Microcosm study of the interactions between microorganisms and enchytraeid worms in grassland soil and litter. Europ. J. Soil Biol. 31:21–27.

Freckman, D.W. 1982. Parameters of the nematode contribution to ecosystems. *In* D.W. Freckman (ed.) Nematodes in soil ecosystems. University of Texas Press, Austin, TX.

Freckman, D.W. 1988. Bacterivorous nematodes and organic-matter decomposition. Agric. Ecosys. Environ. 24:195-217.

Freckman, D. W., and J.G. Baldwin. 1990. Nematoda. p. 155. *In* D.L. Dindal (ed.) Soil biology guide. John Wiley and Sons, New York, NY.

Freckman, D.W., and C.H. Ettema. 1993. Assessing nematode communities in agroecosystems of varying human intervention. Agric. Ecosys. Environ. 45:239-261.

Freckman, D.W., R. Mankau, and H. Ferris. 1975. Nematode community structure in desert soils: Nematode recovery. J. Nematol. 7:343-346.

Freckman, D.W., D.A. Duncan, and J.R. Larson. 1979. Nematode density and biomass in an annual grassland system. J. Range Manag. 32:418-421.

Gates, G.E. 1982. Farewell to North America Megadriles. Megadrilog. 4:12–77.

Gaugler, R., and H.K. Kaya. 1990. Entomopathogenic nematodes in biological control. CRC Press, Boca Raton, FL.

Gist, C.S., and D.A. Crossley, Jr. 1975. The litter arthropod community in a southern Appalachian hardwood forest: Numbers, biomass and mineral element content. Amer. Midl. Nat. 93:107–122.

Golebiowska, J., Z. Margowski, and L. Ryszkowski. 1974. An attempt to estimate the energy and matter economy in the agroeconoses. p. 19–40. In: L. Ryszkowski (ed.) Ecological effects of intensive agriculture. Polish Scientific Publishers, Warszawa, Poland.

Golebiowska, J., and L. Ryszkowski. 1977. Energy and carbon fluxes in soil compartments of agroecosystems. *In* U. Lohm and T. Persson (ed.) Soil organisms as components of ecosystems. Ecol. Bull. 25:274–283.

Goodell, P.B. 1982. Soil sampling and processing for detection and quantification of nematode populations for ecological studies. *In* D.W. Freckman (ed.) Nematodes in soil ecosystems. University of Texas Press, Austin, TX.

Goodell, P.B., and H. Ferris. 1981. Sample optimization for five plant-parasitic nemtodes in an alfalfa field. J. Nematol. 13:304-313.

Goodey, T., and J.B. Goodey, J. B. 1963. Soil and freshwater nematodes. John Wiley and Sons, New York, NY.

Graefe, U. 1973. Systematische Untersuchungen an der Gattung *Achaeta* (Enchytraeidae, Oligochaeta). Diplomarbeit, Universität Hamburg, Hamburg, Germany..

Green, R.N., R.L. Trowbridge, and K. Klinka. 1993. Towards a taxonomic classification of humus forms. Society of American Foresters, Bethesda, MD.

Griffiths, B.S., and K. Ritz. 1988. A technique to extract, enumerate and measure protozoa from mineral soils. Soil Biol. Biochem. 20:163–173.

Hendrix, P.F. 1995. Earthworm ecology and biogeography in North America. Lewis Publishers, Boca Raton, FL.

Hendrix, P.F., B.R. Mueller, R.R. Bruce, G.W. Langdale, and R.W. Parmelee. 1992. Abundance and distribution of earthworms in relation to landscape factors on the Georgia Piedmont, USA. Soil Biol. Biochem. 24:1357-1361.

Hendrix, P.F., M.A. Callaham, Jr., and L. Kirn. 1994. Ecological studies of nearctic earthworms in the southern USA: II. Effects of bait harvesting on *Diplocardia* populations in Apalachicola National Forest in north Florida. Megadrilog. 5:73–76.

Hendrix, P.F., R.W. Parmelee, D.A. Crossley, Jr., D.C. Coleman, E.P. Odum, and P.M. Groffman. 1986. Detritus food webs in conventional and no-tillage agroecosystems. Biosci. 36:372–380.

Holldobler, B., and E. O. Wilson. 1990. The ants. Harvard University Press, Cambridge, MA.

Holt, J.A. 1985. Acarina and collembola in the litter and soil of three North Queensland rain forests. Austr. Jour. Ecol. 10:57–65.

Hooper, D.J. 1990. Extraction and processing of plant and soil nematodes. *In* M. Luc, R.A. Sikora, and J. Bridge (ed.) Plant parasitic nematodes in subtropical and tropical agriculture. CAB International, Wallingford, UK.

Hopkin, S.P. 1997. Biology of the springtails. Insecta: Collembola. Oxford University Press, New York, NY.

House, G.J., and R.W. Parmelee. 1985. Comparison of soil arthropods and earthworms from conventional and no-tillage agroecosystems. Soil Till. Res. 5:351–360.

House, G.J., and B.R. Stinner. 1983. Arthropods in no-tillage soybean agroecosystems: Community composition and ecosystem interactions. Environ. Manag. 7:23–28.

Huhta, V., and A. Koskenniemi. 1975. Numbers, biomass and community respiration of soil invertebrates in spruce forests at two latitudes in Finland. Ann. Zool. Fennici 12:164-182.

Hunt, D.J. 1993. Aphelenchida, Longidoridae, and Trichodoridae. CAB International, Wallingford, UK.

Ingham, R.E., and J.K. Detling. 1984. Plant-herbivore interactions in a North American mixed-grass prairie. III. Soil nematode populations and root biomass on *Cynomys ludovicianus* colonies and adjacent uncolonized areas. Oecolog. 63:307-313.

Ingham, R.E., J.A. Trofymow, E.R. Ingham, and D.C. Coleman. 1985. Interactions of bacteria, fungi, and their nematode grazers: Effects on nutrient cycling and plant growth. Ecol. Monogr. 55:119-140.

Jairajpuri, M.S., and W.U. Khan. 1982. Predatory nematodes (Mononchida). Associated Publishing Co., New Delhi, India.

James, S.W. 1995. Systematics, biogeography and ecology of nearctic earthworms from eastern, central, southern and southwestern United States. p. 29–52. *In* P.F. Hendrix (ed.) Earthworm ecology and biogeography in North America. Lewis Publishers, Boca Raton, FL.

James, S.W. 1990. Oligochaeta: Megascolecidae and other earthworms from southern and midwestern North America. p. 379–386. *In* D.L. Dindal (ed.) Soil biology guide. John Wiley and Sons, New York, NY.

Jamieson, B.G.M. 1988. On the phylogeny and higher classification of the Oligochaeta. Cladistics 4:367–401.

Jensen, H.J., and R.H. Mulvey. 1968. Predaceous nematodes (Mononchidae) of Oregon. Oregon State University Press, Corvallis, OR.

Johnston, J.M. 1996. Microarthropod ecology in managed loblolly pine (*Pinus taeda* L.) forests: Relations of oribatid diversity and microarthropod community structure to forest management practices. Ph.D. dissertation, University of Georgia, Athens, GA.

Kalisz, P.J. and Wood, H.B. 1995. Native and exotic earthworms in wildland ecosystems. p. 117–126. *In* P.F. Hendrix (ed.) Earthworm ecology and biogeography in North America. Lewis Publishers, Boca Raton, FL.

Kasprzak, K. 1982. Review of enchytraeid (Oligochaeta, Enchytraeidae) community structure and function in agricultural ecosystems. Pedobiol. 23:217–232.

Kastan, B. J. 1978. How to know the spiders. William C. Brown, Dubuque, IA.

Kethly, J.B. 1990. Acarina: Prostigmata (Actinedida). P667-756. *In* D.L. Dindal (ed.) Soil biology guide. Wiley Interscience, New York, NY.

Kevan, D. K. McE., and G. G. E. Scudder. 1989. Illustated keys to the families of terrestrial arthropods of Canada. 1. Myriapods (Millipedes, Centipedes, etc.). Biol. Surv. Canada Tax. Series 1. Biological Survey of Canada, Ottawa, Canada.

King, K.L., and K.J. Hutchinson. 1976. The effects of sheep stocking intensity on the abundance and distribution of mesofauna in pastures. J. Appl. Ecol. 13:41-55.

Kitizawa, Y. 1967. Community metabolism of soil invertebrates in forest ecosystems in Japan. p. 649–661. *In* K. Petrusewicz (ed.). Secondary production of terrestrial ecosystems. Polish Academy of Sciences, Warsaw, Poland.

Klekowski, R.Z., L. Wasilewska, and E. Paplinska. 1972. Oxygen consumption by soil inhabiting nematodes. Nematolog. 18:391-403.

Korgonova, G.A., and J.C. Geltser. 1977. Stained smears for the study of Testacida (Protozoa, Rhizopoda). Pedobiol. 12:222–225.

Krantz, G.W. 1978. A manual of acarology. 2nd Ed. Oregon State University Bookstore, Corvallis, OR.

Krantz, G.W., and B.D. Ainscough. 1990. Acarina: Mesostigmata (Gamasida). p. 583-665. *In* D.L. Dindal (ed.) Soil biology guide. Wiley Interscience, New York, NY.

Lagerlöf, J., O. Andrén, and K. Paustian. 1989. Dynamics and contribution to carbon flows of Enchytraeidae (Oligochaeta) under four cropping systems. J. Appl. Ecol. 26:183–199.

Lal, V.B., J. Singh, and B. Prasad. 1981. A comparison of different size samplers for collection of Enchytraeidae (Oligochaeta) in tropical soil. J. Soil Biol. Ecol. 1:21–26.

Lamoncha, K.L., and D.A. Crossley, Jr. 1998. Oribatid mite diversity along an elevation gradient in a southeastern Appalachian forest. Pedobiol. 42:43–55.

Lavelle, P. 1983. The structure of earthworm communities. p. 449–466. *In* J.E. Satchell (ed.) Earthworm ecology: From Darwin to vermiculture. Chapman and Hall, London, UK.

Lavelle, P. 1988. Earthworms and the soil system. Biol Fertil. Soils 6:237–251.

Lawson, D.L., and R.W. Merritt. 1979. A modified Ladell apparatus for the extraction of wetland invertebrates. Can. Entomol. 111:1389–1393.

Lebrun, P. 1965. Contribution a l'etude ecologique des Oribatides de la litiere dans une foret de Moyenne-Belgique. Mem. Roy. Soc. Nat. Belg. 133:1–96.

Lee, J.J., S.H. Hunter, and E.C. Bovee (ed.) 1985. An illustrated guide to the Protozoa. Allen Press, Lawrence, KS.

Lee, K.E. 1959. The earthworm fauna of New Zealand, Wellington: NZ Dept. Sci. Ind. Res. Bull. 130.

Lee, K.E. 1961. Interactions between native and introduced earthworms. Proc. NZ Ecol. Soc 8:60–62.

Lee, K.E., and T.G. Wood. 1971. Termites and soils. Academic Press, London, UK.

Lee, K.E. 1985. Earthworms: Their ecology and relationships with soils and land use. Academic Press, Sydney, Australia.

Lee, K.E. 1991. The diversity of soil organisms. p. 73–87. *In* D.L. Hawksworth (ed.) The biodiversity of microorganisms and invertebrates: Its role in sustainable agriculture. CAB International, Wallingford, UK.

Lee, K.E. 1995. Earthworms and sustainable land use. p. 215–234. *In* P.F. Hendrix (ed.) Earthworm ecology and biogeography in North America. Lewis Publishers, Boca Raton, FL.

Levine, N.D. 1968. Nematode parasites of domestic animals and of man. Burgess Publishing Co., Minneapolis, MN.

Linnaeus, C. 1758. Systema Naturae. Regnum Animale. 10th ed. Stechert-Hafner Service Agency, New York, NY.

Lorenzen, S. 1994. The phylogenetic systematics of free-living nematodes. Intercept Ltd., Andover, Hants., UK.

Lousier, J.D. 1982. Colonization of decomposing leaf litter by Testacea (Protozoa: Rhizopoda): Species succession, abundance and biomass. Oecolog. 52:381–388.

Luc, M., R.A. Sikora, and J. Bridge. 1990. Plant parasitic nematodes in subtropical and tropical agriculture. CAB International, Wallingford, UK.

Lussenhop, J. 1992. Mechanisms of microarthropod-microbial interactions in soil. Adv. Ecol. Res. 23:1–33.

Madge, D.S. 1969. Litter disappearance in forest and savannah. Pedobiol. 9:288–299.

Maggenti, A.R. 1991. Nemata: higher classification. p. 147. *In* W.R. Nickle (ed.) Manual of agricultural nematology. Marcel Dekker, New York, NY.

Mai, W.R., and P.G. Mullins. 1996. Plant parasitic nematodes-A pictorial key to Genera. 5th Ed. Cornell University Press, Ithaca, NY.

Marshall, V.G., R.M. Reeves, and R.A. Norton. 1987. Catalog of the Oribatida (Acari) of continental United States and Canada. Mem. Entom. Soc. Can. 139:1–148.

Maynard, E.A. 1951. A monograph of the Collembola or springtail insects of New York State. Comstock Publishers, Ithaca, NY.

McSorley, R. 1993 Short-term effects of fire on the nematode community in a pine forest. Pedobiol. 37:39-48.

McSorley, R. 1987. Extraction of nematodes and sampling methods. *In* R.H. Brown and B.R. Kerry (ed.) Principles and practices of nematode control in crops. Academic Press, Sydney, Australia.

McSorley, R., and D.E. Walter. 1991. Comparison of soil extraction methods for nematodes and microarthropods. Agric. Ecosys. Environ. 34:201–208.

McSorley, R., and J.J. Frederick. 1996. Nematode community structure in rows and between rows of a soybean field. Fund. Appl. Nematol. 19:251–260.

Mitchell, M.J. 1977. Population dynamics of oribatid mites in an aspen woodland. Pedobiol. 17:305–319.

Moldenke, A.R., and B.L. Fichter. 1988. Invertebrates of the H.J. Andrews Experimental Forest, Western cascade Mountains, Oregon. IV. The oribatid mites (Cryptostigmata). US For. Serv. Gen. Tech. Rep. PNW-GTR-217.

Moore, J.C., and P.C. de Ruiter. 1991. Temporal and spatial heterogeneity of trophic interactions within below-ground food webs. Agric. Ecosys. Environ. 34:371-397.

Moreno, A.G. 1996. Catalogo de la Familia Glossoscolecidae. Unpublished.

Moulder, B.C., D.E. Reichle, and S.I. Auerbach. 1970. Significance of spider predation in the dynamics of forest floor.

Nakamura, Y. 1979. Microdistribution of Enchytraeidae in *Zoysia* grassland (Studies on Japanese enchytraeids I). Bull. Nat. Grassl. Res. Inst. 14:21–27.

Nakamura, Y. 1982. Distribution pattern of *Achaeta camerani* (Cogenetti) in grazing and cutting grasslands (Studies on Japanese enchytraeids III). Edaphologia 25/26: 41–47.

Nakamura, Y. 1989. Oribatids and enchytraeids in ecofarmed and conventionally farmed dryland grainfields of central Japan. Pedobiol. 33:389–398.

Neher, D.A., and C.L. Campbell. 1975. Nematode communities and microbial biomass in soils with annual and perennial crops. Appl. Soil Ecol. 1:17-28.

Nickle, W.R. 1984. Plant and insect nematodes. Marcel Dekker, New York, NY.

Nielsen, C.O. 1953. Studies on Enchytraeidae. 1. A technique for extracting Enchytraeidae from soil samples. Oikos 4:187–196.

Nielsen, C.O. 1954. Studies on Enchytraeidae. 3. The microdistribution of Enchytraeidae. Oikos 5:167–178.

Nielsen, C.O. 1955a. Studies on Enchytraeidae 5: Factors causing seasonal fluctuations in numbers. Oikos 6:153–169.

Nielsen, C.O. 1955b. Studies on Enchytraeidae 2: Field studies. Natura Jutl. 4–5: 1–58.

Nielsen, C.O., and B. Christensen. 1959. The Enchytraeidae: Critical revision and taxonomy of European species. Natura Jutl. 8–9:1–160.

Nielsen, C.O., and B. Christensen. 1961. The Enchytraeidae: Critical revision and taxonomy of European species. Natura Jutl. (Suppl. 1) 10:1–23.

Nielsen, C.O., and B. Christensen. 1963. The Enchytraeidae: Critical revision and taxonomy of European species. Natura Jutl. (Suppl. 2) 10:1–19.

Norton, D.C. 1978. Ecology of plant-parasitic nematodes. John Wiley and Sons, New York, NY.

O'Connor, F.B. 1955. Extraction of enchytraeid worms from a coniferous forest soil. Nature 175:815–816.

O'Connor, F.B. 1957. An ecological study of the enchytraeid worms from a coniferous forest soil. Oikos 8:161–169.

O'Connor, F.B. 1967. The Enchytraeidae. p. 213–257. In A. Burgess and F. Raw (ed.) Soil biology. Academic Press, New York, NY.

Page, F.C. 1988. A new key to freshwater and soil Gymnamoebae. Freshwater Biology Association, Ambleside, UK.

Patterson, D.J., and J. Larsen. (ed.) 1991. The biology of free-living heterotrophic flagellates. Clarenden Press, Oxford, UK.

Peachey, J.E. 1963. Studies on Enchytraeidae (Oligochaeta) of moorland soils. Pedobiol. 2:81–95.

Persson, T., and U. Lohm. 1977. Energetical significance of the annelids and arthropods in a Swedish grassland soil. Ecol. Bull. 23:1-211.

Persson, T., E. Bååth, M. Clarholm, H. Lundkvist, B.E. Söderström, and B. Sohlenius. 1980. Trophic structure, biomass dynamics and carbon metabolism of soil organisms in Scots pine forest. Ecol. Bull. 32:419–459.

Phillipson, J., R. Abel, J. Steel, and S.R.J. Woodell. 1979. Enchytraeid numbers, biomass and respiration metabolism in a Beech woodland-Wytham Woods, Oxford. Oecologia 43:173-193.

Piper, S.R. 1982. Enchytraeid (Oligochaeta) worms: Species composition, distribution, abundance, respiration and production in two Pacific Silver fir stands. MSc. Thesis. University of Washington, Seattle, WA.

Poinar, G.O., Jr. 1983. The natural history of nematodes. Prentice-Hall. Englewood Cliffs, CA.

Poinar, G.O., Jr. 1979. Nematodes for biological control of insects. CRC Press, Boca Raton, FL.

Ponge, J.F. 1984. Étude écologique d'un humus forestier par l'observation d'un petit volume, premiers résultats. I. La couche L1 d'un moder sous pin sylvestre. Rev. Écol. Biol. Sol 21:161–187.

Raw, F. 1959. Estimating earthworm populations by using formalin. Nature 184:1661–1662.

Reichle, D.E., and D.A. Crossley, Jr. 1965. Radiocesium dispersion in a cryptozoan food web. Health Phys. 11:1375–1384.

Reynolds, J.W. 1973. Earthworm (Annelida: Oligochaeta) ecology and systematics. p. 95–120. In D.L. Dindal (ed.) Proc. 1st Soil Microcomm. Conf. US Atomic Energy Commission, Washington, D.C.

Reynolds, J.W. 1977. The earthworms (Lumbricidae and Sparganophilidae) of Ontario. Life Sci. Misc. Publ., Roy. Ont. Mus., Toronto, Canada.

Reynolds, J.W. 1995. Status of exotic earthworm systematics and biogeography in North America. p. 1–28. In P.F. Hendrix (ed.) Earthworm ecology and biogeography in North America. Lewis Publishers, Boca Raton, FL.

Reynolds, J.W., and D.G. Cook. 1993. Supplementum Tertium: A catalogue of names, descriptions and type specimens of the Oligochaeta, New Brunswick: New Brunswick Museum Monographic Series (Natural Science) No. 9.

Richter, P. O. 1958. Biology of Scarabaeidae. Ann. Rev. Entomol. 3:25–58.

Riechert, S.E., and T. Lockly. 1984. Spiders as biological control agents. Ann. Rev. Entomol. 29:299–320.

Römbke, J. 1992. New taxa since 1985. In J. Römbke (ed.) Newsletter on Enchytraeidae No. 3. Institut für Angewandte Bodenbiologie. Hamburg, Germany.

Römbke, J. and K. Dósza-Farkas. 1996. New taxa since 1992. p. 5-9. In K. Dósza-Farkas (ed.) Newsletter on Enchytraeidae No. 5. Eötvös Loránd University, Budapest, Hungary.

Romero, H., and K. Jaffe. 1989. A comparison of methods for sampling ants (Hymenoptera, Formicidae) in savannas. Biotrop. 21:348–352.

Rusek, J. 1985. Soil microstructures-contributions of specific soil organisms. Quaest. Entomol. 21:497–514.

Ryl, B. 1977. Enchytraeids (Oligochaeta, Enchytraeidae) on rye and potato fields in Turew. Ekol. Pol. 25:519-529.

Satchell, J.E. 1969. Methods of sampling earthworm populations. Pedobiol. 9:20–25.

Satchell, J.E. 1983. Earthworm ecology from Darwin to vermiculture. Chapman hall, London, UK.

Schauermann, J. 1983. Eine Verbesserung der Extraktionsmethode für terrestrische Enchytraeiden. p. 669–670. In P. Lebrun, H.M. André, A. De Medts, C. Grégoire-Wibo and G. Wauthy (ed.) New trends in soil biology. Dieu-Brichart, Louvain-la-Neuve, Belgium.

Schwert, D.P. 1990. Oligochaeta: Lumbricidae. p. 341–356. In D.L. Dindal (ed.) Soil biology guide. John Wiley and Sons, New York, NY.

Seastedt, T.R., and D.A. Crossley, Jr. 1981. Microarthropod response to cable logging and clear-cutting in the southern Appalachians. Ecol. 62:126–135.

Shachak. M. E., A. Chapman and Y. Steinberger. 1976. Feeding, energy flow and soil turnover in the desert isopod, Hemilepistus reamuri. Oecologia 24:57-69.

Siddiqi, M.R. 1985. Tylenchida parasites of plants and insects. CAB, Farnham Royal, UK.

Šimek, M., V. Pizl, and J. Chalupský. 1991. The effect of some terrestrial oligochaeta on nitrogenase activity in the soil. Plant Soil 137:161–165.

Sims, R.W. 1983. The scientific names of earthworms. *In* J.E. Satchell (ed.) Earthworm ecology from Darwin to vermiculture. Chapman Hall, London, UK.

Sims, R.W., and B.M. Gerard. 1985. Earthworms: Synopses of British fauna. Nat. Hist. Mus. 31:1–171.

Singh, B.N. 1955. Culturing soil protozoa and estimating their numbers in soil, p. 403–411. *In* Kevan, D.K. McE. (ed.) Soil zoology. Butterworths, London, UK.

Smith, C.A.S., A.D. Tomlin, J.J. Miller, L.V. Moore, M.J. Tynen, and K.A. Coates. 1990. Large enchytraeid (Annelida: Oligochaeta) worms and associated fauna from unglaciated soils of the northern Yukon, Canada. Geoderma 47:17–32.

Smith, H.G., and P.V. Tearle. 1985. Aspects of microbial and protozoan abundances in Sigany Island fellfields. Bull. Brit. Antarct. Surv. 68:83–90.

Snider, R. M. and J. H. Shaddy. 1980. The ecobiology of Trachelipus rathkei (Isopoda). Pedobiol. 20:394-410.

Sohlenius, B. 1979. A carbon budget for nematodes, rotifers and tardigrades in a Swedish coniferous forest soil. Holarctic Ecol. 2:30-40.

Sohlenius, B. 1980. Abundance, biomass and contribution to energy flow by soil nematode in terrestrial ecosystems. Oikos 34:186-194.

Southey, J.F. 1986. Laboratory methods for work with plant and soil nematodes. Her Majesty's Stationery Office, London, UK.

Sparling, G., B.G. Ord, and D. Vaughan. 1981. Microbial biomass and activity in soils amended with glucose. Soil Biol. Biochem. 13:99–104.

Springett, J.A., J.E. Brittain and B.P. Springett. 1970. Vertical movement of Enchytraeidae (Oligochaeta) in moorland soils. Oikos 21:16–21.

Standen, V. 1973. The production and respiration of an enchytraeid population in blanket bog. J. Anim. Ecol. 42:219-245.

Stirling, G.R. 1991. Biological control of plant parasitic nematodes. CAB International, Wallingford, UK.

Stork, N.E., and P. Eggleton. 1992. Invertebrates as determinants of soil quality. American J. Altern. Agric. 7:38–47.

Stout, J.D. 1962. An estimation of microfaunal populations in soils and forest litter. J. Soil Sci. 13:314–320.

Thiele, H. U. 1977. Carabid beetles in their environments. Springer-Verlag, Berlin, Germany.

Thorne, G. 1961. Principles of nematology, McGraw-Hill Book Co., New York, NY.

Thorne, G. 1974. Nematodes of the northern Great Plains. Part II. Dorylaimoidea in part (Nemata: Adenophorea). SD Agric. Exp. Sta., Brookings, SD.

Thorne, G., and R.B. Malek. 1968. Nematodes of the northern Great Plains. Part I. Tylenchida (Nemata: Secernentea). SD Agric. Exp. Sta., Brookings, SD.

Tisdall, J.M., and J.M. Oades. 1982. Organic matter and water stable aggregates in soils. J. Soil Sci. 33:141–163.

Tomlin, A.D., M.J. Shipitalo, W.M. Edwards, and R. Protz. 1995. Earthworms and their influence on soil structure and infiltration. *In* P.F. Hendrix (ed.) Ecology and biogeography of earthworms in North America. Lewsi Publishers, Boca raton, FL.

Toutain, F., G. Villemin, A. Albrecht and O. Reisinger. 1982. Etude ultrastructurale des processes de biodégradation. II Modèle enchytraeides-litière de feuillus. Pedobiol. 23:145–156.

Urbášek, F., and J. Chalupský. 1991. Activity of digestive enzymes in 4 species of Enchytraeidae (Oligochaeta). Rev. Ecol. Biol. Sol 28:145–154.

van der Drift, J. 1950. Analysis of the animal community in a beech forest floor. Tijdscr. Entomol. 94:1–168

van Vliet, P.C.J., M.H. Beare and D.C. Coleman. 1995. A comparison of the population dynamics and functional roles of Enchytraeidae (Oligochaeta) in hardwood forest and agricultural ecosystems. p. 237–245. *In* H.P Collins, G.P. Robertson and M.J. Klug (ed.) The significance and regulation of soil biodiversity. Kluwer Academic, Dordrecht, Netherlands.

van Vliet, P.C.J., D.C. Coleman, and P.F. Hendrix 1997. Population dynamics of Enchytraeidae (Oligochaeta) in different agricultural systems. Biol. Fertil. Soils 25:123-129.

van Vliet, P.C.J., L.T. West, P.F. Hendrix, and D.C. Coleman. 1993. The influence of Enchytraeidae (Oligochaeta) on the soil porosity of small microcosms. Geoderma 56:287–299.

van Vliet, P.C.J., D.E. Radcliffe, P.F. Hendrix, and D.C. Coleman. 1998. Hydraulic conductivity and pore size distribution in small microcosms with and without enchytraeids (Oligochaeta). Appl. Soil Ecol. 9:277–282..

Wallwork, J.A. 1983. Earthworm biology. Camelot Press, Southampton, UK.

Walter, D.E., and J. Kethley. 1987. A heptane flotation method for recovering microarthropods from semiarid soils. Pedobiol. 30:221–232.

Walter, D.E., H.W. Hunt, and E.T. elliott. 1987. The influence of prey type on the development and reproduction of some predatory soil mites. Pedobiol. 30:419–424.

Wasilewska, L. 1994. The effect of age of meadows on succession and diversity in soil nematode communities. Pedobiol. 38:1-11.

Wasilewska, L., and P. Bienkowski. 1985. Experimental study on the occurrence and activity of soil nematodes in decomposition of plant material. Pedobiol. 28:41-57.

Wasilewska, L. 1971. Nematodes of the dunes in the Kampinos Forest. II. Community structure based on numbers of individuals, state of biomass and respiratory metabolism. Ekol. Polska 19:651-688.

Webster, J.M. 1972. Economic nematology. Academic Press, London, UK.

Wharton, D.A. 1986. A functional biology of nematodes. The Johns Hopkins University Press, Baltimore, MD.

Whitford, W.G., Y. Steinberger, and G. Ettershank. 1982. Contributions of subterranean termites to the "economy" of Chihuahuan desert ecosystems. Oecolog. 55:298–302.

Whitford, W.G., D.W. Freckman, P.F. Santos, N.Z. Elkins, and L.W. Parker. 1982. The role of nematodes in decomposition in desert ecosystems. In D.W. Freckman (ed.) Nematodes in soil ecosystems. University of Texas Press, Austin, TX.

Williams, B.L., and B.S. Griffiths. 1989. Enhanced nutrient mineralization and leaching from decomposing Sitka spruce litter by enchytraeid worms. Soil Biol. Biochem. 21:183–188.

Williams, S. C. 1987. Scorpion bionomics. Ann. Rev. Entomol. 32: 275-295.

Wise, D.H. 1993. Spiders in ecological webs. Cambridge University Press, Cambridge, UK.

Wodarz, D., E. Aescht, and W. Foissner. 1992. A weighted coenotic index (WCI): description and application to soil animal assemblages. Biol. Fertil. Soils 14:5-13.

Wolters, V. 1988. Effects of *Mesenchytraeus glandulosus* (Oligochaeta, Enchytraeidae) on decomposition processes. Pedobiologia 32:387–398.

Wolters, V., and K. Ekschmitt. 1997. Gastropods, isopods, diplopods and chilopods: Neglected groups of the decomposer food web. p. 265–306. In G. Benckiser (ed.) Fauna in soil ecosystems. Marcel Dekker, New York, NY.

Yeates, G.W. 1981. Populations of nematode genera in soils under pasture. IV. Seasonal dynamics at five North Island sites. NZ J. Agric. Res. 24:107-121.

Yeates, G.W. 1984. Variation in soil nematode diversity under pasture with soil and year. Soil Biol. Biochem. 16:95-102.

Yeates, G.W. 1986. Enchytraeidae-some population estimates for grasslands and a New Zealand bibliography. NZ Soil Bur. Sci. Rep. 77. Lower Hutt, New Zealand.

Yeates, G.W. 1988. Earthworm and enchytraeid populations in a 13-year-old agroforestry system. NZ J. For. Sci. 18:304-310.

Yeates, G.W. 1991. Nematode populations at three forest sites in New Caledonia. J. Trop. Ecol. 7:411-413.

Yeates, G.W. 1994. Modification and qualification of the nematode maturity index. Pedobiol. 38:97-101.

Yeates, G.W., and A.F. Bird. 1994. Some observations on the influence of agricultural practices on the nematode faunae of some South Australian soils. Fund. Appl. Nematol. 17:133-145.

Yeates, G.W., S.S. Bamforth, D.J. Ross, K.R. Tate, and P. Sparling. 1991. Recolonization of methyl bromide sterilized soils under four different field conditions. Biol. Fertil. Soils 11:181–189.

Yeates, G.W., T. Bongers, R.G.M. de Goede, D.W. Freckman, and S.S. Georgieva. 1993. Feeding habits in soil nematode families and genera-an outline for soil ecologists. J. Nematol. 25:315-331.

3

Microbially Mediated Processes

3.1 Phosphorus, Sulfur and Metal Transformations

J.J. Germida and S.D. Siciliano

University of Saskatchewan

3.1.1 Introduction

Microorganisms play an important role in the cycling of biosphere elements (Schlesigner, 1991), most of which are essential for growth of living organisms (Table 3.1). Major bio-elements such as phosphorus (P) and sulfur (S) are required in high concentrations ($>10^{-4}$ M) and form important constituents of cell components or, through chemical bonds, store energy. Some major bio-elements, such as S or iron (Fe), may be used by bacteria to generate energy, e.g., oxidation of Fe^{2+} to Fe^{3+}. Minor bio-elements such as manganese (Mn), selenium (Se) and zinc (Zn) are required at low concentrations ($<10^{-4}$ M) to act as important cofactors in cellular metabolism, but are toxic at higher concentrations. All bio-elements are subject to various microbial transformations that can alter the element's bioavailability, ecological flux or toxicity.

3.1.2 General Overview of Microbial Transformations

Oxidation and reduction reactions are microbial processes that provide energy and reducing equivalents necessary for cell growth and replication. The element acts either as an electron donor or acceptor, undergoing a redox reaction which alters the element's speciation and valency, and hence, its physicochemical characteristics. As cells grow and increase their biomass, they assimilate (i.e., biologically immobilize) elements by utilizing these elements for the construction of cellular components such as phospholipids or proteins. During growth, metabolic byproducts (e.g., organic acids) exuded by microorganisms alter the surrounding environment and often solubilize elements. In some cases, microorganisms produce metabolites (e.g., siderophores, enzymes, etc.) that function to specifically acquire or increase the bioavailability of a certain element. During degradation of dead biomass, microorganisms with a surplus of nutrients will convert biologically immobilized elements back into an inorganic form, i.e., mineralization.

Microbial transformations of P, S, and metals regulate the bioavailability, toxicity, and environmental impact of these elements in the biosphere. All of these transformation are closely linked to microbial metabolism and provide the organism with either an essential nutrient, energy or detoxification process. For more detailed information the reader is referred to excellent books with extensive references on P (Tiessen, 1995), S (Howarth et al., 1992a,b), and metals (Ehrlich, 1996;

Table 3.1 Bio-elements, their sources, and some of their functions in microorganisms [Reprinted from G. Gottschalk, 1979. Bacterial Metabolism. Copyright Springer-Verlag GmbH & Co with permission]

Major Bio-Elements (required at $>10^{-4}$ M)		
Element	Source	Role in Organism
C	organic compounds, CO_2	main constituents of cellular material
O	O_2, H_2O, organic compounds, CO_2	
H	H_2, H_2O, organic compounds	
N	NH_4^+, NO_3^-, N_2, organic compounds	
S	SO_4^{2-}, HS^-, S_0, $S_2O_3^{2-}$	constituent of cysteine, methionine, thiamin pyrophosphate, coenzyme A, biotin, and α-lipoic acid
P	HPO_4^{2-}	constituent of nucleic acids, phospholipids, and nucleotide
K	K_+	cofactor of many enzymes (e.g., kinases): present in cell walls, membranes, and phosphate esters
Ca	Ca_{2+}	cofactor of enzymes; present in exoenzymes (amylases, proteases): Ca-dipicolinate is an important component of endospores
Fe	Fe_{2+}, Fe_{3+}	present in cytochromes, ferredoxins; and other iron-sulfur proteins; cofactor of enzymes (some dehydratases)
Zn	Zn_{2+}	present in alcohol dehydrogenase, alkaline phosphatatase, adolase, RNA and DNA polymerase
Mn	Mn_{2+}	present in bacterial superoxide dismutase; cofactor of some enzymes (PEP carboxykinase, re-citrate synthase)
Mo	MoO_4^{2-}	present in nitrate reductase, nitrogenase, and formate dehydrogenase
Se	SeO_3^{2-}	present in glycine reductase and formate dehydrogenase
Co	Co_{2+}	present in coenzyme B_{12}-containing enzymes (glutamate mutase, methylmalonyl-CoA mutase)
Cu	Cu_{2+}	present in cytochrome oxidase and oxygenases
W	WO_4^{2-}	present in some formate dehydrogenases
Ni	Ni_{2+}	present in urease, required for autotrophic growth of hydrogen-oxidizing bacteria

Hughes and Poole, 1989). Addtional information on transformations of P, S, and metals in the rhizosphere may be found in Lynch (1990a) and Smith and Read (1997).

3.1.3 Phosphorus Transformations

The microbially mediated transformations of P in soil are depicted in Fig. 3.1. Phosphorus is solubilized from inorganic sources and enters the microbial or plant biomass. This organic P can itself be solubilized by microbial action and be made available to plants or microbes. At any one time, very little P is available in the soil solution. Microbial reactions that solubilize inorganic P by acidification and chelation are important due to the limited solubility of P in soil systems. There are approximately 10^2 to 10^8 bacteria capable of solubilizing inorganic P per gram of soil (Kucey et al., 1989), although the mechanism by which microorganisms increase P bioavailability is not clear. Phosphorus bioavailability increases with decreasing pH until solubilization of Al limits root growth, or until Al

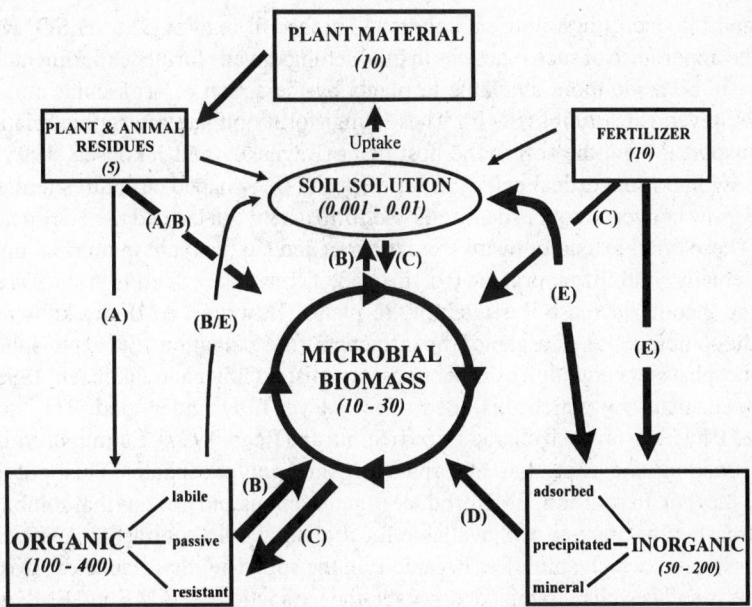

Fig. 3.1 Schematic representation of P fractions and flows in soil. The numbers in parentheses indicate amounts of P (kg ha^{-1}, 0-10 cm) in these fractions. A, decomposition; B, mineralization; C, immobilization; D, solubilization; E, inorganic adsorption-desorption, precipitation-solubilization [Courtesy of A.E. Richardson, 1994]

or Fe oxides/hydroxides complex P (Frossard et al., 1995). Thus several investigators have suggested that microbial production of organic acids such as citric, oxalic or fumaric acid which acidify soil solution increases P bioavailability. However, Kucey et al. (1989) argue that if an increase in acidity is the only mechanism by which microorganisms solubilize P, then the amount of acid required to solubilize the P taken up by the plant could not be produced by rhizosphere bacteria. Thus, Lahuerte and Berthelin (1988) suggest that the role of microorganisms is not to solubilize P directly, but instead to alter the plant's carbon allocation such that the rhizosphere is enriched with organic acids that solubilize P.

There is evidence that besides reducing solution pH, microbially produced organic acids also chelate counter ions to P and thereby solubilize P. Kucey (1988) found that the addition of EDTA to solutions containing insoluble copper and zinc had the same solubilizing effect as inoculation with *Penicillium bilaii*, but that reduction of the solution pH to 4.0 by 0.1 N HCl did not result in metal ion solubilization. This organism was later found to produce citric and oxalic acids which solubilize P by binding Ca^{2+} resulting in dissolution of $CaPO_4$ (Cunningham and Kuiack, 1992; Kucey et al., 1989). In addition to organic acid production, microorganisms may solubilize P by acidifying the soil solution. Illmer et al. (1995) found that *Penicillium aurantiogriseum, P. simplicissimum* and *Pseudomonas* sp. PI18/89 release protons into the soil during respiration or cation uptake (e.g., NH_4^+), and thereby solubilize P from hardly soluble $AlPO_4$ without producing organic acids.

Microbial processes such as dissimilatory Fe reduction or H_2S production can lead to P solubilization in anaerobic systems (Ehrlich, 1996), but the importance of these reactions in controlling the bioavailability of P in agricultural soils is not fully understood (Frossard et al., 1995). Ghani et al. (1994) found that extractable P from reactive phosphate rock was increased nine-fold by including S^o and thiobacilli in a nutrient solution. Thiobacilli oxidize S^o to H_2SO_4 which, in turn, dissolves P. The importance of such reactions in field settings awaits further experimental verification.

Phosphate can be made more available to plants by the action of arbuscular mycorrhiza fungi (AMF) and ectomycorrhiza fungi (EMF). These fungi form obligate biotrophic relationships with plants and transport P from the soil to the host plant (Ravnskov and Jakobsen, 1995). Arbuscular mycorrhiza grow into root cortical cells and form special tree-shaped structures termed arbuscules whereas EMF grow between root cortical cells and form a network termed the Hartig net (Smith and Read, 1997). These fungi extend outward from the root and aid the plant in nutrient uptake. It is not clear if AMF actually solubilize inorganic P or, instead, increase the effective surface area of a plant's root system and thereby increase P availability to plants. However, AMF are known to indirectly contribute to the solubilization of organic P by influencing the exudation of acid phosphatase by roots and alkaline phosphatase production by other microorganisms (Joner and Jakobsen, 1995; Joner et al., 1995). The mechanisms by which this occurs are not yet fully understood. The acquisition and translocation of P by EMF are well characterized (Smith and Read, 1997). Ectomycorrhiza fungi form a mantle surrounding the root that absorbs inorganic polyphosphate. This polyphosphate is translocated to the root. In addition, EMF produce organic anions and protons that solubilize inorganic P, thereby increasing the amount of P available for root uptake. In contrast to AMF, EMF produce phosphomonoesterases which solubilize organic P in the soil. The importance of AMF and EMF P transformations in soil is well recognized; however, the interactions of AMF and EMF with other soil microorganisms and the resulting effect on P transformations are still unclear.

Microorganisms immobilize a significant portion (1–5%) of P in soil (Paul and Clark, 1996). This biologically immobilized P may take up to a year to be mineralized depending upon nutrient states (Frossard et al., 1995). Phosphorus in microorganisms has two distinct forms: (1) P associated with cell constituents and (2) P in storage granules (i.e., volutin). As microorganisms multiply and grow, their cellular biomass increases. This requires additional P to construct cellular components such as phospholipids and proteins. Microbial growth can result in P immobilization or mineralization depending upon the elemental composition of the substrate. Microbial respiration of organic substrates with a C:P ratio less than 200:1 will result in P being mineralized, i.e., released into the environment. In contrast, substrate C:P ratios above 300 result in immobilization, i.e., the microorganism utilizes all the P from the substrate and additional P from the environment to form its cellular components (Mullen, 1998). Microorganisms can also store inorganic phosphate as volutin granules which are formed by the sequential addition of P residues to pyrophosphate with ATP serving as the donor (Stanier et al., 1986). These volutin granules are formed during the removal of P from waste water (Heymann et al., 1989). However, the role of volutin in a soil setting has received little attention. Mycorrhiza fungi also immobilize P by forming polyphosphate granules in their vacuoles. These are later translocated to the plant (Smith and Read, 1997).

Phosphate stored in dead biomass can be a significant pool of P in the soil. Scarcity of nutrients for microbial use will result in microorganisms excreting phosphatase that can hydrolyse organic P to inorganic P, and thereby make it bioavailable. Phosphatase is produced by over 75% of soil microorganisms assessed, including bacteria from the genera: *Bacillus, Serratia, Proteus, Arthrobacter, Streptomyces* and fungi from the genera: *Aspergillus, Penicillium, Rhizopus,* and *Cunninghamella* (Ehrlich, 1996). Regulation of phosphatase expression by bacteria is coupled with the P status of the microbial cell. For example, Schweizer (1994) found that expression of an alkaline phosphatase by *Erwinia carotovora* subsp. *carotovora* was increased by P starvation. However, alkaline phosphatase activity is also linked to metal ion concentration with Mg^{2+} increasing and Na^+ decreasing activity. Only bacteria produce alkaline phosphatase. Roots, fungi as well as bacteria, produce acid phosphatase which has an optimal pH < 7 (Dinkelaker and Marschner, 1992). Both

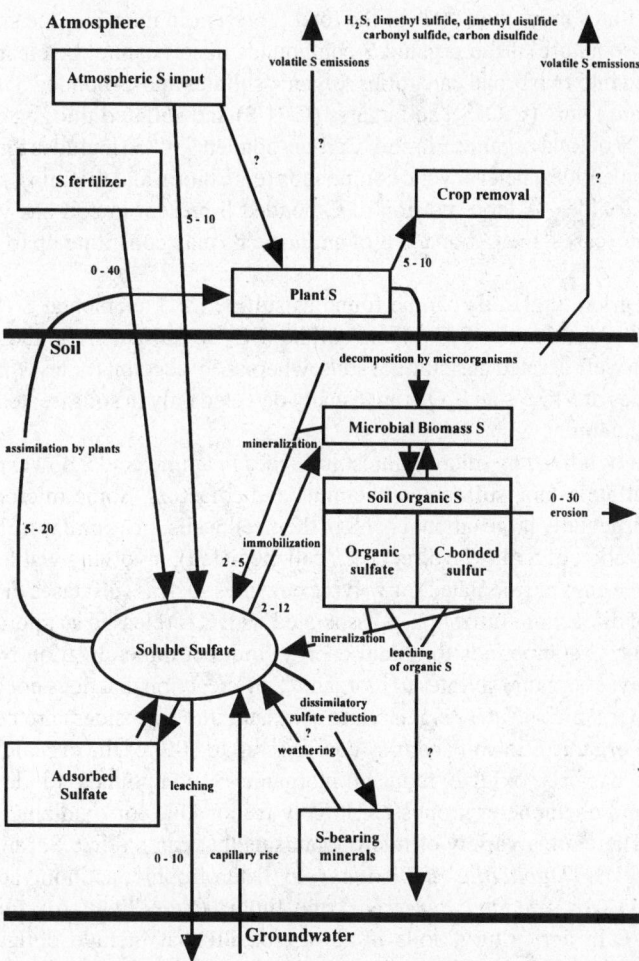

Fig. 3.2 Conceptual sulfur cycle in agroecosystems. Numbers represent flux estimated (kg ha^{-1} yr^{-1}) for sulfur transformations in western Canadian soils [Reprinted from Schoenau and Germida, 1992. Sulphur cycling in upland agriculture systems. R.W. Howarth et al. (ed.) Sulphur cycling on the continents. SCOPE 48. Copyright John Wiley and Sons]

alkaline and acid phosphatases have the potential to solubilize organic P but the reaction rate is controlled by organic P sorption to surfaces which increases organic P resistance to enzymatic attack (McLachlan et al., 1988).

3.1.4 Sulfur Transformations

Total S in soil ranges from 20 to 100,000 mg kg^{-1}, with the highest levels in tidal flats, and in saline, acid sulphate, and organic soils. Figure 3.2 depicts the S cycle in agroecosystems. Microbial transformations of S include the decomposition of organic S found in plant matter. This is converted in part to soil organic sulfate or C-bonded S, both of which can be eroded, leached or mineralized to soluble sulfate. In turn, soluble sulfate can be immobilized back into organic S, incorporated into plant biomass, or sorbed into S-bearing minerals by dissimilatory sulfate reduction.

Organic S constitutes more than 90% of the total S present in most surface soils (Freney, 1986). However, the precise nature of the organic S compounds in soil cannot be clearly identified. Thus organic S is grouped into two broad categories: organic sulfates and C-bonded S. Organic sulfate (R-O-S) includes sulfate esters (C-O-S), sulfamates (C-N-S) and sulfated thioglycosides (N-O-S) and constitutes 30 to 75% of total organic S in soil. Carbon-bonded S (C-S) includes the S present in amino acids, proteins, polypeptides, heterocyclic compounds (e.g., biotin and thiamin), sulfinates, sulfones, sulfonatees and sulfoxides. A large portion of C-bonded S present in soil has yet to be identified; nevertheless, in some cases, the C-bonded S of amino acids may constitute up to 30% of the organic S in soil.

Inorganic S in agricultural soils can be found as sulfide (S^{2-}), elemental S (S^o), sulfite (SO_3^{2-}), thiosulfate ($S_2O_3^{2-}$), tetrathionate ($S_4O_6^{2-}$) and sulfate (SO_4^{2-}). Sulfate is the most common form of inorganic S found in well-aerated agricultural soils, whereas S^{2-} account for less than 1% of total S and measurable quantities of $S_2O_3^{2-}$ and $S_4O_6^{2-}$ are usually detected only in soils treated with S fertilizer or those receiving pollutants.

Most of the S immobilized by microorganisms resides in amino acids; however, microorganisms also accumulate sulfate esters, sulfonates, vitamins and cofactors. Some microorganisms, such as fungi, accumulate especially large amounts of S in their cell walls. In a similar fashion to that seen for organic P, mineralization of S may be direct (i.e., cell mediated), involving viable microbial cells, or indirect (i.e., cell-free enzyme mediated), involving enzymes such as sulfatases that hydrolyse sulfate esters. In the case of direct mineralization, S associated with C is released as microorganisms oxidize S-containing organic C compounds to obtain energy. Indirect mineralization by enzymes such as arylsulfatase hydrolyses organic sulfates to inorganic S in a reaction that does not generate energy for the microbial cell. Organic sulfates (e.g., sulfate ester and thioglucosides) are considered to be the most labile form of organic S in soil and may comprise up to 70% of the organic S in surface soils.

Many different bacteria oxidize reduced inorganic S compounds to derive energy. Both chemoautotrophs and chemoheterotrophs are largely responsible for oxidizing S in most aerobic, agricultural soils. There are a variety of microorganisms that can oxidize S, ranging from obligate chemolithotrophs like *Thiobacillus thiooxidans* to heterotrophic actinomycetes (*Streptomyces sioyaensis*), bacteria (*Arthrobacter aurescens*) and fungi (*Penicillium* sp.). One major group of bacteria oxidizing S in agricultural soils is the thiobacilli that include obligate and facultative chemolithotrophs as well as chemolithotrophic heterotrophs. These organisms can oxidize a variety of S compounds and excrete SO_4^{2-}, but the metabolic pathway can vary greatly depending upon the initial S substrate and the thiobacilli species. The role thiobacilli play in S oxidation and the S cycle in soil is still not fully understood. Marked increases in thiobacilli are found in soils amended with reduced forms of S, but no consistent correlation between S-oxidation rates and the incidence of thiobacilli exists, except that rates of S-oxidation are generally low in soils that lack these organisms and are accelerated by inoculation. A consortia of heterotrophs and autotrophs probably work to bring about the oxidation of S^o in agricultural soils (Lawrence and Germida, 1991). In this scenario, organic matter coating S^o is colonized by heterotrophic bacteria which produce $S_2O_3^{2-}$, $S_4O_6^{2-}$, SO_4^{2-} and/or H_2S. These reduced inorganic S compounds induce the development of autotrophic thiosulfate-oxidizing communities in association with S^o.

The reduction of SO_4^{2-} to H_2S is mediated mainly by anaerobic bacteria such as *Desulfovibrio* spp. (Table 3.2), but is usually not important in well-aerated agricultural soils. The formation of S^{2-} in rice paddy soils is dependent upon the levels of Fe^{+3} and MnO_2 present in the soil as the metabolic pathways which use Fe^{+3} and Mn^{+4} as electron acceptors generate more energy than metabolic pathways that reduce S^{+6}. However, the low abundance of Fe^{+3} and MnO_2 and the precipitation of FeS_2 suggest that SO_4^{2-} reduction may occur in rice paddy soils. The reduction of SO_4^{2-} is also involved in

Table 3.2 Sulfur-using bacteria occurring in soil and aquatic habitats [Reprinted from Cook and Kelly, 1992. Sulfur cycling and fluxes in temperate dimictic lakes. R.W. Howarth et al. (ed.) Sulfur cycling on the continents. Copyright John Wiley and Sons with permission]

Group	S-Conversion	Habitat Requirements	Habitat Example	Examples of Genera
Heterotrophs that use oxidized S species as electron acceptors	$SO_4^{2-} \rightarrow HS^-$ $S_2O_3^- \rightarrow HS^-$ or S^O $S^O \rightarrow HS^-$ $SO_3 \rightarrow HS^-$	anaerobic; organic substrates available; light not required	anoxic sediments and soils	*Desulfomonas, Desulfovibrio, Desulfotomaculum, Desulfuromonas, Campylobacter*
Obligate and facultative autotrophs that use reduced S as an energy source	$HS^- \rightarrow S^O$ $S^O \rightarrow SO_4^{2-}$ $S_2O_3^- \rightarrow SO_4^{2-}$	H_2S-O_2 interface light not required	mud; hotsprings; mine drainage; soils	*Thiobacillus, Thiomicrospira, Achromatium, Beggiatoa*
Phototrophs that use reduced S as an electron donor	$HS^- \rightarrow S^O$ $S^O \rightarrow SO_4^{2-}$	anoxia; H_2S, light	shallow water; anoxic sediments; metalimion or hypolimnion; anoxic water	*Chlorobium, Chromatium, Ectothiorhodospira, Thiopedia, Rhodopseudomonas*
Heterotrophs that use organic S compounds as energy sources or that hydrolyze esters	org S $\rightarrow HS^-$ org S $\rightarrow$ volatile org S ester-$SO_4 \rightarrow SO_4^{2-}$	source of organic S compounds	sediments; soils; water column	Many
Microorganisms that use SO_4^{2-} or H_2S in biosynthesis	$SO_4^{2-} \rightarrow$ protein $HS^- \rightarrow$ protein $SO_4^{2-} \rightarrow$ DMSP	nonspecific	sediments; soils; water column	Many

methane production and perhaps organic compound degradation. Methane is oxidized during SO_4^{2-} reduction with the net result that methane production in anoxic soils (e.g., rice paddies) is inversely correlated with SO_4^{2-} concentrations. Few measurements of the flux or controls of H_2S production have been made in rice paddy soils and more field measurements are required before the cycling of inorganic S in rice paddy soils is fully understood (Lefroy et al., 1992).

The capacity to reduce SO_4^{2-} is widespread among anaerobic bacteria including the Gram-negative eubacteria (*Desulfovibrio, Desulfobacter, Desulfobacterium, Desulfobulbus, Desulfococcus, Desulfosarcina, Desulfomonas, Thermosulfobacterium, Desulfonema*) as well as in Gram-positive eubacteria (*Desulfotomaculum*) and a group of archaebacteria (*Archaeoglobus*) (Kelly, 1989). These organisms typically use H_2, lactate and acetate as electron donors. The initial stage of SO_4^{2-} reduction requires that SO_4^{2-} be transported into the cell. This is accomplished in an energy-dependent process which transforms SO_4^{2-} to SO_3^{2-} and requires 1-2 mol of ATP per mole of SO_4^{2-} reduced. The SO_3^{2-} is reduced to H_2S in an electron-transport-linked phosphorylation which yields 2 to 3 mol ATP. This process generates between 1-2 mol of ATP for every mole of SO_4^{2-} reduced.

3.1.5 Metal Transformations

Some trace elements have biogeochemical cycles similar to those of P and S. For example, the behavior of Arsenic (As) closely mimics that of P whereas Se mimics that of S. Furthermore, SO_4^{2-}-

reducing bacteria are capable of reducing SeO_4^{2-} to Se^{2-} which, in turn, can be substituted for S^{2-} in proteins. Such similarities can also be detrimental for trace elements such as molybdenum (Mo). The biological uptake of MoO_4^{2-}, which has an atomic structure similar to SO_4^{2-}, can be inhibited by SO_4^{2-} which is typically found at concentrations a thousand- to million-fold greater than MoO_4^-. Consequently, Mo bioavailability in natural waters can be quite low (Howarth and Stewart, 1992).

As shown in Table 3.1 a number of metals are essential for life, especially as micronutrients and co-factors for enzymes. Metals are also important as a source or acceptor of electrons. In these cases, the metal plays an integral function in the synthesis of ATP (source of energy for metabolic processes) or the formation of NADH (source of a reductant for anabolic metabolism) (Table 3.3). As shown in Fig. 3.3, various metals can act as electron acceptors during the degradation of organic matter. However, the sequence in which each element is used as an acceptor is dependent upon the O_2 status of the environment with O_2, NO_3^-, Mn^{4+}, Fe^{3+}, SO_4^{2-} and finally H_2 used in order of declining O_2 status. The metabolic pathway used to reduce each element further along in the progression imparts less energy to the microorganism.

The Fe and S cycles are closely interwoven. Some S-oxidizing bacteria also oxidize Fe to Fe^{+3} that acts as an oxidizing agent to chemically react with FeS_2 to form Fe^{2+} and SO_4^{2-} (Kuenen and Bos, 1989). The oxidation of S results in a drop in pH that favors certain Fe^o-oxidizing bacteria such as *T. ferroxidans* and *Sulflobus* spp., both of which can also oxidize S. Not all bacteria that oxidize Fe are acidophilic, as neutrophilic bacteria of the genera *Leptothrix, Sphaerotilus, Planctomyces, Hyphomicrobium* and *Pedomicrobium* are capable of oxidizing Fe^o in anoxic environments containing low levels of Fe^{3+} (Kuenen and Bos, 1989). The reduction of Fe^{3+} is an important reaction in the anoxic soils found in rice paddies because bacteria-degrading organic matter can use Fe^{3+} as a terminal electron acceptor. Furthermore, the precipitation of sulfides as pyrite (FeS_2) reduces the toxicity of H_2S to crops (Lefroy et al., 1992).

Above pH 4, Fe^{3+} is extremely insoluble (Lynch, 1990b). Consequently, microorganisms have evolved the ability to secrete Fe-chelating compounds called siderophores that are large organic molecules (1,000–1,500 Daltons) that strongly and specifically bind Fe^{3+}. Fungi and bacteria secrete

Table 3.3 Elements subject to microbial oxidation-reduction reactions in soils and sediments and examples of bacterial genera involved with each reaction [From Mullen, 1998. D. Sylvia et al. (ed.) Principles and applications of soil microbiology. Copyright. Reprinted with permission of Prentice-Hall, Inc, Upper Saddle River, NJ]

Elements and their common oxidation states	Reaction, significance, and redox couple		Some bacterial genera reported to be involved
Cr^{6+}, Cr^{3+}	Oxidation-NR		*Aeromonas, Bacillus,*
	Reduction-AR, D	$Cr^{6+} + 3e^- \rightarrow Cr^{3+}$	*Chlorella, Pseudomonas*
Fe^{3+}, Fe^{2+}	Oxidation-E	$2Fe^{2+} \rightarrow 2Fe^{3+} + 2e^-$	*Thiobacillus*
	Reduction-AR	$2Fe^{3+} + 2e^- \rightarrow 2Fe^{2+}$	*Geobacter, Desulfovibrio, Pseudomonas, Thiobacillus*
Hg^{2+}, Hg^o	Oxidation-NE	$Hg^o \rightarrow Hg^{2+} + 2e^-$	*Bacillus, Pseudomonas*
	Reduction-D	$Hg^{2+} + 2e^- \rightarrow Hg^o$	*Chlorellas, Pseudomonas, Streptomyces*
Mn^{4+}, Mn^{2+}	Oxidation-E,D	$Mn^{2+} \rightarrow Mn^{4+} + 2e^-$	*Arthrobacter, Pseudomonas*
	Reduction-Ar	$Mn^{4+} + 2e^- \rightarrow Mn^{2+}$	*Bacillus, Geobacter, Pseudomonas*
Se^{6+}, Se^{4+}, Se^o, Se^{2-}	Oxidation-E	$Se^{2-} \rightarrow Se^o + 2e^-$	*Bacillus, Thiobacillus*
	Reduction-AR	$SeO_4^{2-} + 8e^- \rightarrow Se^{2-}$	*Clostridium, Desulfovibrio, Micrococcus*

AR, element used as a terminal-electron acceptor in anaerobic respiration; D, detoxification mechanisms; E, energy source, NE, nonenzymatic reaction, microorganism alters the physiochemical environment.
*NR, not reported to be biologically mediated.

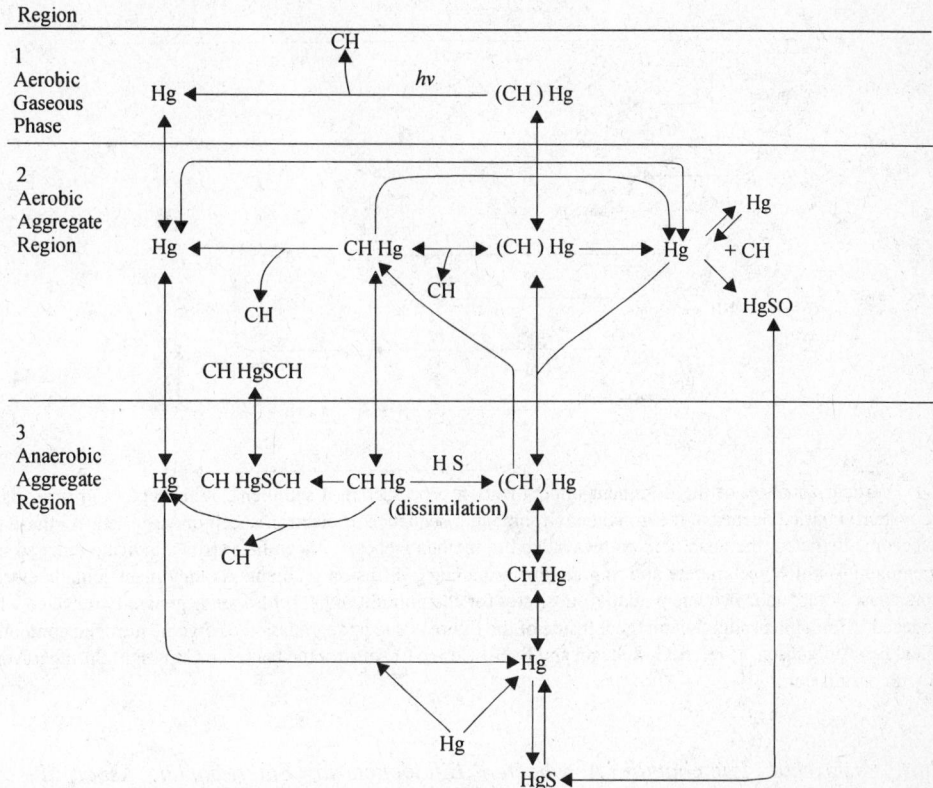

Fig. 3.3 Major mercury transformations in the soil environment, including interactions between the aerobic bulk gaseous phase (region 1), the aerobic aggregate region (region 2), and the anaerobic aggregate phase (region 3) [Reprinted from Klein and Thayer, 1990. H.G. Schegel and B. Bowien (ed.) Autrtrophic bacteria. Copyright Springer-Verlag GmbH with permission]

hydroxymate siderophores (containing amide functional groups) whereas catecholate siderophores (containing aromatic functional groups) are produced only by bacteria (Lynch, 1990b). These siderophores are recognized by specific receptors on the cell surface, transported back into the cell and the Fe^{3+} is released into the cytoplasm by highly elaborate transport systems. Furthermore, the uptake of siderophores by bacteria can be highly selective with one bacterial species unable to use another's siderophores and vice versa. Microbial competition for Fe^{3+} has been implicated in suppression of phytopathogens and root colonization by rhizobacteria.

Not all microbially mediated transformations of metals result in the production of energy for microbial use. The oxidation of Mn^{2+} to MnO_2 is thought to be catalyzed by bacteria. However, it is not clear if bacteria obtain energy from this process. The enthalpy change of Mn^{2+} to Mn^{4+} is positive, but due to the insolubility of MnO_2, the overall reaction has a negative free energy change (Kuenen and Bos, 1989). To date, there has been only one unconfirmed report of a bacterial species (*Pseudomonas*) that was able to fix CO_2 while oxidizing Mn^{2+}, indicating that it was obtaining energy from the process (Kepkay and Nealson, 1987). As in the case of Fe, MnO_2 can act as an electron acceptor during S^{2-} oxidation.

Many microbially mediated transformations of toxic elements can result in the entry of elements into the food chain. For example, a wide range of bacterial genera (i.e., *Pseudomonas, Microbacter,*

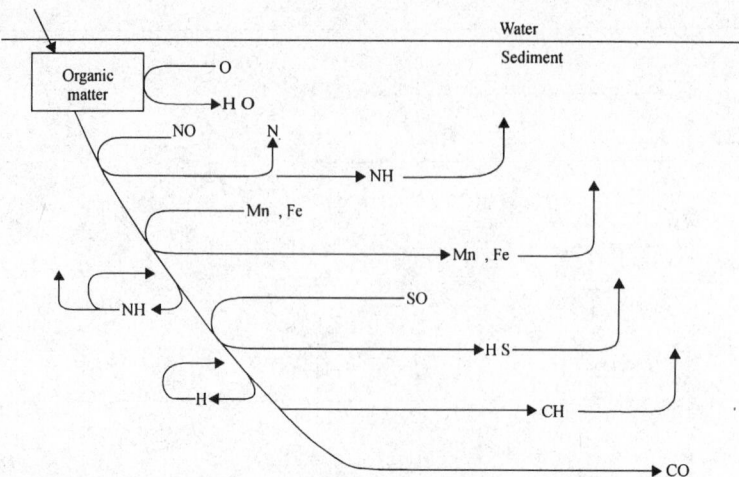

Fig. 3.4 Vertical zonation of the dominant mineralization processes in a sediment. With depth and time, as the organic matter is buried deeper in the anoxic environment, a sequence of electron acceptors are used. As these ultimately become depleted, the anaerobic pathways lead to methanogenesis. Ascending arrows indicate reduced inorganic compounds which accumulate and migrate upward along a diffusion gradient. As they meet suitable electron acceptors, these compounds become potential substrates for chemolithotrophs. Hydrogen is generally recycled where it is produced. Ammonia production on the left side of the figure is due to degradation of organic nitrogen compounds [Reprinted from Jorgensen, 1989. H.G. Schegel and B. Bowien (ed.) Autrtrophic bacteria. Copyright Springer-Verlag GmbH with permission].

Bacillus, Escerichia, Lactobacilli, Aerobacter, Bifidobacteria, Enterobacter, Aspergillus and *Scopulariopsis*) can methylate Hg^{2+} which increases the lipophilicity of the Hg and thereby its toxicity (Craig, 1986). In addition, other bacteria can further methylate CH_3Hg^+ by utilizing S^{2-} in a dismutation process to produce $(CH_3)_2Hg$ and HgS. A wide number of bacterial genera are capable of demethylating Hg by reduction. Further complicating the picture, these processes can also occur by other pathways such as extracellular enzymes. Abiotic methylation of Hg has been observed where the donor molecule has been added to Hg in distilled water. These enzymes (e.g., methionine synthetase, acetate synthetase and methane synthetase) will methylate Hg in the absence of organisms. There is some evidence that Hg can be methylated by methyl cobalamin (CH_3CoB_{12}) by carbanion transfer. Thus, the environmental fate of Hg is extremely complex. Mercury transformations in three phases are summarized in Fig. 3.4. Beginning from Hg^{2+}, bacteria can methylate or dimethylate mercury. This dimethylmercury can be reduced back to Hg^{2+} or demethylated to CH_3Hg^+. Methylmercury, in turn, can be reduced to produce methane and volatile Hg^o. Thus, it is important to take into account the physical chemical environment surrounding microbial transformations as this can influence the net effect of microbially mediated transformations of elements.

In addition to direct microbially mediated transformations, microorganisms can facilitate the entry of other elements into the food chain by the production of chelation agents. Many of these chelators are the same compounds implicated in the solubilization of inorganic P (i.e., oxalic and citric acid). In a manner similar to that seen with P solubilization, these low molecular weight organic acids are thought to work not only by acidifying the soil solution but also by chelating the metal directly (Dynes and Huang, 1995; Berthelin et al., 1995). The exact chemical mechanism by which this occurs is not completely understood. However, such low molecular weight acids can have a significant impact on the bioavailability of heavy metals such as Cd (Krishnamurti et al., 1997). Cadmium found as a trace contaminant in certain fertilizers is accumulated by wheat and can bioaccumulate in the food chain.

3.1.6 Summary and Conclusions

Microbial transformations of P, S, and metals regulate the bioavailability, toxicity, and environmental impact of these elements in the biosphere. All of these transformations are closely linked to microbial metabolism and provide the organism with an essential nutrient, energy, or detoxification process. For more detailed information, the reader is referred to excellent books with extensive references on P (Tiessen, 1992), S (Howarth et al., 1992), and metals (Hughes and Poole, 1989; Ehrlich, 1996). Additional information on transformations of P, S. and metals in the rhizosphere may be found in Lynch (1990a), Smith and Read (1997), and Sylvia et al. (1998).

3.1.7 References

Berthelin, J., C. Munier-Lamy, and C. Leyval. 1995. Effect of microorganisms on mobility of heavy metals in soils. p. 3–16. *In* P.M. Huang, J. Berthelin, J.-M. Bollag, W.B. McGill, and A.L. Page (ed.). Environmental impact of soil component interactions: Metals, other inorganics, and microbial activities. Lewis Publishers, Boca Raton, FL.

Cook, R.B., and C.A. Kelly. 1992. Sulphur cycling and fluxes in temperate dimictic lakes. p. 145–188. *In* R.W. Howarth, J.W.B. Stewart, and M.V. Ivanov (ed.) Sulphur cycling on the continents: wetlands, terrestrial ecosystems and associated water bodies. John Wiley and Sons, New York, NY.

Craig, P.J. 1986. Organomercury compounds in the environment. p. 65–101. *In* P.J. Craig (ed.) Organometallic compounds in the environment: Principles and reactions. John Wiley and Sons, New York, NY.

Cunningham, J.E., and C. Kuiack. 1992. Production of citric and oxalic acids and solubilization of calcium phosphate by *Penicillium bilaji*. Appl. Environ. Microbiol. 58:1451–1458.

Dinkelaker, B., and H. Marschner. 1992. *In vivo* demonstration of acid phosphatase activity in the rhizosphere of soil-grown plants. Plant Soil 144:199–205.

Dynes, J.J., and P.M. Huang. 1995. Influence of citrate on selenite sorption-desorption on short-range ordered aluminum hydroxides. p. 47–60. *In* P.M. Huang, J. Berthelin, J.-M. Bollag, W.B. McGill, and A.L. Page (ed.) Environmental impact of soil component interactions: Metals, other inorganics, and microbial activities. Lewis Publishers, Boca Raton, FL.

Ehrlich, H.L. 1996. Geomicrobiology. 3rd Ed. Marcel Dekker, Inc, New York, NY.

Freney, J.R. 1986. Forms and reactions of organic sulfur compounds in soils. p. 207–232. *In* M.A. Tabatabai (ed.). Sulfur in agriculture. American Society of Agronomy, Madison, WI.

Frossard, E., M. Brossard, M.J. Hedley, and A. Metherell. 1995. Reactions controlling the cycling of P in soils. p. 107–138. *In* H. Tiessen (ed.). Phosphorus in the global environment: Transfers, cycles and management. John Wiley and Sons, Chichester, UK.

Ghani, A., S.S.S. Raan, and A. Lee. 1994. Enhancement of phosphate rock solubility through biological processes. Soil Biol. Biochem. 26:127–136.

Gottschalk, G. 1979. Bacterial metabolism. Springer-Verlag, New York, NY.

Heymann, J.B., I.M. Eagle, H.A. Greben, and D.J.J. Potgieter. 1989. The isolation and characterization of volutin granules as subcellular components involved in biological phosphorus removal. Water Sci. Tech. 21:397–408.

Howarth, R.W., J.W.B. Stewart, and M.V. Ivanov. 1992. Sulphur cycling on the continents: Wetlands, terrestrial ecosystems, and associated water bodies. John Wiley and Sons, Chichester, UK.

Howarth, R.W., and J.W.B. Stewart. 1992. The interactions of sulphur with other element cycles in ecosystems. p. 67–79. *In* R.W. Howarth, J.W.B. Stewart, and M.V. Ivanov (ed.) Sulphur cycling on the continents: Wetlands, terrestrial ecosystems and associated water bodies. John Wiley and Sons, New York, NY.

Hughes, M.N., and R.K. Poole. 1989. Metals and microorganisms. Chapman and Hall, London, UK.

Illmer, P., A. Barbato, and F. Schinner. 1995. Solubilization of hardly soluble $AlPO_4$ with P-solubilizing microorganisms. Soil Biol. Biochem. 27:265–270.

Joner, E.J., and I. Jakobsen. 1995. Growth and extracellular phosphatase activity of arbuscular mycorrhizal hyphae as influenced by soil organic matter. Soil Biol. Biochem. 27:1153–1159.

Joner, E.J., J. Magid, T.S. Gahoonia, and I. Jakobsen. 1995. P depletion and activity of phosphatases in the rhizosphere of mycorrhizal and non-mycorrhizal cucumber (*Cucumis sativus* L.). Soil Biol. Biochem. 27:1145–1151.

Jorgensen, B.B. 1989. Biogeochemistry of chemoautotrophic bacteria. p. 117–146. *In* H.G. Schlegel and B. Bowien (ed.). Autotrophic Bacteria. Springer Verlag, Berlin, Germany.

Kelly, D.P. 1989. Physiology and biochemistry of unicellular sulfur bacteria. p.193–218. *In* H.G. Schlegel, and B. Bowien (ed.). Autotrophic bacteria. Springer Verlag, Berlin, Germany.

Kepkay, P.E., and K.H. Nealson. 1987. Growth of a manganese oxidizing *Pseudomonas* sp. in continuous culture. Arch. Microbiol. 148:63–67.

Klein, D.A., and J.S. Thayer. 1990. Interactions between soil microbial communities and organometallic compounds. p. 431–481. *In* H.G. Schlegel and B. Bowien (ed.). Autotrophic Bacteria. Springer Verlag, Berlin, Germany.

Krishnamurti, G.S.R., G. Cieslinski, P.M. Huang, and K.C.J. Van Rees. 1997. Kinetics of cadmium release from soils as influenced by organic acids: implication in cadmium availability. J. Environ. Qual. 26:271–277.

Kucey, R.M.N. 1988. Effect of *Penicillium bilaji* on the solubility and uptake of P and micronutrients from soil by wheat. Can. J. Soil Sci. 68:261–270.

Kucey, R.M.N., H.H. Janzen, and M.E. Leggett. 1989. Microbially mediated increases in plant-available phosphorus. Adv. Agron. 42:199–226.

Kuenen, J.G., and P. Bos. 1989. Habitats and ecological niches of chemolitho(auto)trophic bacteria. p. 53–80. *In* H.G. Schlegel, and B. Bowien (ed.) Autotrophic bacteria. Springer Verlag, Berlin, Germany.

Lahuerte, F., and J. Berthelin. 1988. Effect of a phosphate solubilizing bacteria on maize growth and root exudation over four levels of labile phosphorus. Plant Soil 105:11–17.

Lawrence, J.R., and J.J. Germida. 1991. Microbial and chemical characteristics of elemental sulfur beads in agricultural soils. Soil Biol. Biochem. 23:617–622.

Lefroy, R.D.B., C.P. Mamaril, G.J. Blair, and P.B. Gonzales. 1992. Sulphur cycling in rice wetlands. p. 279–296. *In* R.W. Howarth, J.W.B. Stewart and M.V. Ivanov (ed.) Sulphur cycling on the continents: Wetlands, terrestrial ecosystems and associated water bodies. John Wiley and Sons, New York, NY.

Lynch, J.M. 1990a. The rhizosphere. John Wiley and Sons, New York, NY.

Lynch, J.M. 1990b. Microbial metabolites. p. 177–206. *In* J.M. Lynch (ed.). The rhizosphere. John Wiley and Sons, New York, NY.

McLachlan, M.J., A.M. Alston, and J.K. Martin. 1988. Phosphorus cycling in wheat pasture rotations. III. Organic phosphorus turnover and phosphorus cycling. Aust. J. Soil Res. 26:343–353.

Mullen, D.D. 1998. Transformations of other elements. p. 369–386. *In* D.M. Sylvia, J.J. Fuhrman, P.G. Hartel, and D.A. Zuberer (ed.). Principles and applications of soil microbiology. Prentice Hall, Upper Saddle River, NJ.

Paul, E.A., and F.E. Clark. 1996. Soil microbiology and biochemistry. 2nd Ed. Academic Press, San Diego, CA.

Ravnskov, S., and I. Jakobsen. 1995. Functional compatibility in arbuscular mycorrhizas measured as hyphal P transport to the plant. New Phytol. 129:611–618.

Richardson, A.E. 1994. Soil microorganisms and phosphorus availability. p. 50–62. *In* C.E. Pankhurst, B.M. Doube, V.V.S.R. Gupta, and P.R. Grace (ed.) Soil biota. CSIRO Publications, East Melbourne, Australia.

Schlesigner, W.H. 1991. Biogeochemistry: An analysis of global change. Academic Press, San Diego, CA.

Schoenau, J.J., and J.J. Germida. 1992. Sulphur cycling in upland agriculture systems. p. 261–277. *In* R.W. Howarth, J.W.B. Stewart, and M.V. Ivanov (ed.). 1992. Sulphur cycling on the continents: Wetlands, terrestrial ecosystems, and associated water bodies. SCOPE 48. John Wiley and Sons, Chichester, UK.

Schweizer, H.P. 1994. Coordinate derepression of alkaline phosphatase, and binding protein dependent transport systems for phosphate and *sn*-glycerol-3-phosphate by phosphate starvation in *Erwinia carotovora* subsp. *carotovora*. Can. J. Microbiol. 40:310–313.

Smith, S.E., and D.J. Read. 1997. Mycorrhizal symbiosis. 2nd Ed. Academic Press, San Diego, CA.

Stanier, R.Y., E.A. Adelberg, and J.L. Ingraham. 1986. The microbial world. 4th Ed. Prentice Hall, Englewood Cliffs, NJ.

Sylvia, D.M., J.J. Fuhrmann, P.G. Hartel, and D.A. Zuberer. 1998. Principles and applications of soil microbiology. Prentice Hall, Upper Saddle River, NJ.

Tiessen, H. 1995. Phosphorus in the global environment: Transfers, cycles and management. John Wiley and Sons, Chichester, UK.

3.2 Decomposition

E.G. Gregorich and H.H. Janzen

Agriculture and Agri-Food Canada, Research Branch,
Ottawa, Ontario and Lethbridge, Alberta

3.2.1 Overview

Decomposition is the progressive dismantling of organic materials, ultimately into inorganic constituents. In nature, decomposition is mediated mainly by soil microorganisms, who derive energy and nutrients from the process. The net effect is the release of carbon and nutrients back into biological circulation, both on a local and global scale.

Decomposition usually occurs not as a single step, but as a cascade. Fresh material (R_1), usually plant litter, is converted to altered form (R_2), often with the release of CO_2, NH_4^+, and other inorganic compounds (Fig. 3.5). The revised organic material (R_2), in turn, is susceptible to further decomposition. Although the overall trend with each step is always toward simpler compounds, some of the products (R_2) may be more complex than their precursors. Indeed, a tiny fraction of elements entering the soil is eventually converted, via successive steps, into material of high complexity (humus).

The process of decomposition is governed by the interaction of three components (Fig. 3.5): the soil organisms (O), the quality or composition of the organic substrate (Q), and the physical-chemical environment (P). These components not only determine the rate of the process, but also, to some extent, the final products of decomposition.

Decomposition is central to biogeochemical cycles encompassing terrestrial, aquatic, and atmospheric systems (Fig. 3.6). It completes the C cycle, reverting the C fixed by photosynthesis back into CO_2. At the same time, decomposition releases N, P, S, and micronutrients for plant uptake. This continual recycling of organically bound nutrients drives the biogeochemical cycles; were decomposition to stop, the essential biological elements would soon all be tied up, and life would cease within months (Jenkinson, 1981).

The substrates for decomposition include a wide range of materials, forming a continuum from recently added plant litter to very stable, humified organic matter. For simplicity, decomposition can

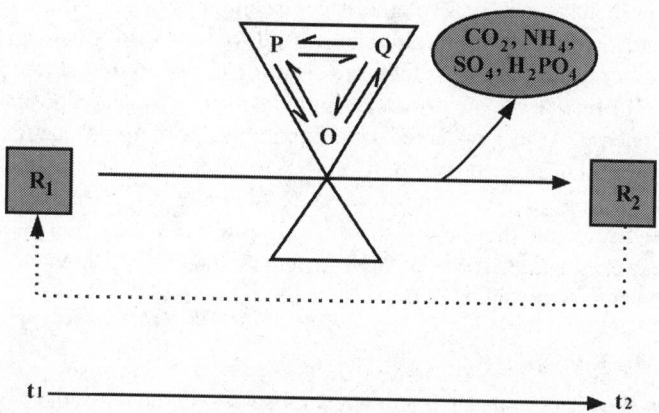

Fig. 3.5 The decomposition process is regulated by three groups of factors: the physical-chemical environment (P), and the quality of the resource (Q) acting through decomposer organisms (O). The organic resource is changed in this process from R_1 to R_2 over time, t_1 to t_2. The changed resource enters the decomposition cascade where it is further decomposed and redistributed through comminution, catabolism and leaching [Adapted from Swift et al., 1979].

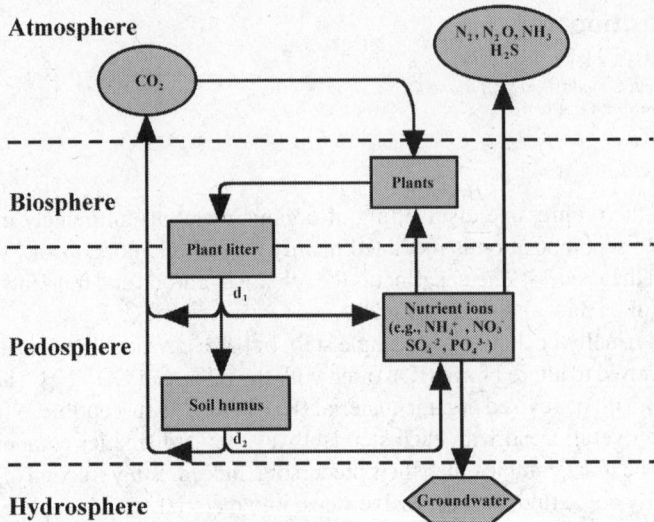

Fig. 3.6 The process of decomposition of organic residues is central to biogeochemical cycles encompassing the atmosphere, biosphere, pedosphere and the hydrosphere

be subdivided into two processes: primary decomposition (d_1, Fig. 3.6), involving the breakdown of fresh litter; and secondary decomposition (d_2), involving the progressive breakdown of decomposition products. Fresh plant litter decomposes very quickly; consequently, though it represents only a small fraction of C in soil, about half of the CO_2 output from soil, globally, comes from decomposition of the annual litter fall (Coûteaux et al., 1995). Stable organic matter, at the other extreme, decomposes very slowly over centuries or millennia (Campbell et al., 1967), but the size of this pool is very large.

The rate of decomposition, relative to that of other ecosystem processes, influences soil, water, and atmospheric composition. In 'native' climax ecosystems, decomposition is synchronized with plant growth, so that carbon and nutrients are used with maximum efficiency. Disturbance of these ecosystems, however, may retard or accelerate decomposition, imposing effects on other components of the system. For example, the initial cultivation of soils of the North American Great Plains stimulated decomposition relative to photosynthesis, resulting in the accumulation of CO_2 in the atmosphere (Fig. 3.7). The flow of other nutrients is similarly affected by decomposition rate. If decomposition is slowed, release of nutrients is reduced, limiting net primary production of that ecosystem. Alternatively, acceleration of decomposition, relative to plant assimilation, results in the accumulation of inorganic forms which can be transformed to volatile forms, affecting air quality (e.g., N_2O, NH_3, H_2S), or leached below the rooting zone, affecting water quality (e.g., NO_3^-) (Table 3.4, Fig. 3.6). Thus, decomposition rate affects land, water, and air quality, and an understanding of this process will help to ensure preservation of ecosystems.

3.2.2 Biotic Mechanisms

Decomposition occurs largely through biological processes. Of course, abiotic processes cannot be discounted entirely; in harsh environments, appreciable mass loss from plant litter may occur by abiotic mechanisms: fragmentation, physical abrasion, photochemical breakdown, and leaching (Coûteaux et al., 1995; Dormaar, 1991). Moorhead and Reynolds (1989) proposed that abiotic mechanisms might account for as much as 50 to 75% of the total annual loss from litter in a desert environment. But even where abiotic processes of mass loss predominate, the eventual oxidation to CO_2 and other inorganic constituents probably is mediated by soil organisms.

Of the soil organisms, fauna play an important role by physically fragmenting and mixing organic residues into the soil, thereby increasing the surface area of substrates and their exposure to microbial activity. Soil fauna also influence decomposition by inoculating substrates with microbes, inhibiting or stimulating microbial activity (e.g., by deposition of mucus or wastes), and by altering the composition of the decomposer community. Although soil fauna are important agents of litter comminution and redistribution, microorganisms are the primary agents of mineralization (Juma and McGill , 1986). Consequently, research into decomposition has focused largely on microbial activity.

The type of microorganisms present in soil affects the biochemical process and products of decomposition. Bacteria, the most numerous microorganisms, derive most of their carbon from organic substrates. They are sometimes classified based on their activity and response to substrates in soil: *Authochthonous* organisms grow slowly in soil, and predominate when there is little oxidizable substrate; *Zymogenous* organisms respond rapidly, often with high levels of activity when substrates are made available to microbial attack. Another way to classify microorganisms is based on the ecological theory that relates the density of a species to its soil food supply (Paul and Clark, 1989). According to this approach, a *K*-selected species is adapted to a more continuous supply of energy source, where competition is intense. An *r*-selected organism, in contrast, lives under conditions of normally sparse food supply with minimal competition, but exhibits rapid growth per unit of food to take advantage of sporadic flushes of substrate.

Most of the soil microbial community is dormant because of limited mobility or restricted access to food. The community is extremely diverse and, as a whole, is capable of surviving a wide range of environmental and food-related stresses. A notable difference between bacteria and fungi is their mode of growth (Coleman and Crossley, 1997). Bacteria depend on water films in soil for activity

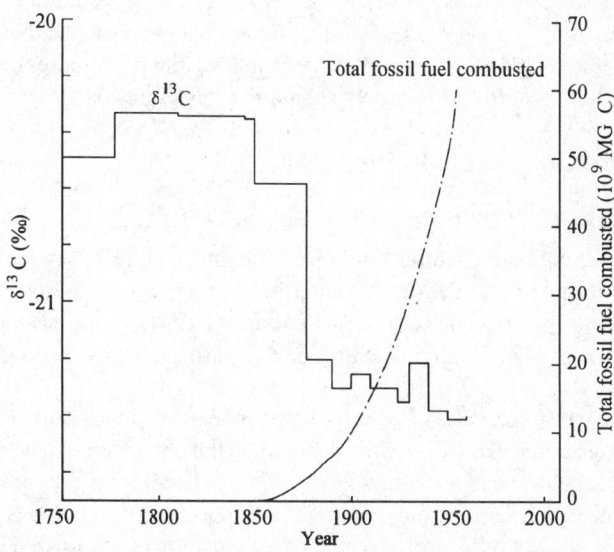

Fig. 3.7 Plot of $\delta^{13}C$ values of the cellulose across a section of Bristlecone Pine. Note the sharp change in values between 1860 and 1890 which occurred before the combustion of fossil fuel had contributed significant quantities of CO_2 to the atmosphere [Adapted from Wilson, 1978].

Table 3.4   Some of the major end products of decomposition and their transformation in soil

End product	Remarks
CO_2	formed during aerobic decomposition of organic compounds by heterotrophic organisms; used by plants in photosynthesis; a greenhouse gas.
CH_4	formed during anaerobic decomposition of organic matter by methanogens; stable in the absence of O_2; under aerobic conditions used as substrate by methanotrophic bacteria; a greenhouse gas.
NH_3	formed during ammonification; NH_4^+ formed in soil environment; used by plants, microorganisms; is readily oxidized to NO_3^- in moist, well-aerated soils.
NO_3^-	formed during nitrification; used by plants, aerobic and anaerobic microorganisms - assimilatory and dissimilatory reduction; affects water quality.
N_2O	formed during nitrification and anaerobic decomposition (dissimilatory NO_3 reduction); used by microorganisms; a greenhouse gas; affects ozone depletion.
SO_4^{2-}	formed during aerobic decomposition of organic compounds; used by plants, aerobic and anaerobic microorganisms - assimilatory and dissimilatory reduction.
H_2S	formed during anaerobic decomposition (dissimilatory SO_4 reduction); used as substrate by microorganisms; affects air quality.

and motility. They are usually clustered in colonies occupying a volume of soil measuring only a few cubic millimeters. Their movement is mostly episodic and related to such factors as rainfall, root growth, tillage, or possibly ingestion by soil fauna. In contrast, fungi have hyphae that can grow over relatively large distances and that penetrate into small spaces where they can decompose organic matter by secreting enzymes and translocating the nutrients back through the hyphae.

Enzymes catalyze many of the reactions necessary for the life processes of microorganisms, as well as for decomposition of litter and cycling of nutrients. In soil, enzymes occur both within living microorganisms or fauna and outside of living cells, as extracellular excretions, or in association with organic and inorganic colloids. The enzymes most commonly linked to decomposition are hydrolases, which catalyze hydrolysis, and transferases, which catalyze the transfer of chemical groups from one molecule to another (Dick, 1994). Other enzymes widely studied to date include those prominent in carbon cycling (e.g., amylase, cellulase, lipase, glucosidases, and invertase), those catalyzing nitrogen processes (e.g., proteases, amidases, ureases, and deaminases), and enzymes which promote the release of other nutrients (e.g., phosphatase, arylsulfatase).

3.2.3   Factors Affecting Decomposition

3.2.3.1   Residue Quality
Organic residues in soils consist of complex mixtures of lignin, cellulose, hemicellulose, and proteins from plant, animal, and microbial debris. Although the carbon concentration of these materials is relatively constant, usually between 40 to 50% (Jenkinson, 1981), the concentration of other constituents varies widely; for example, N concentration varies from < 0.5% in wood to >10% in protein and bacterial cells.

Various indices have been proposed to describe the effect of residue composition on decomposition. The most common of these, the C:N ratio, is based on the observation that N is the nutrient most limiting to decomposition. In residues with wide C:N ratios, decomposition may be slowed because N immobilization exceeds mineralization, resulting in a shortage of available N for the decomposer organisms. The threshold C:N ratio, above which decomposition is suppressed, is often about 20 to 30 (corresponding to an N concentration of about 15 to 25 g kg^{-1}).

The C:N ratio, however, is not an unerring measure of N availability for decomposition (Jenkinson 1981). Any factor that increases the rate of decomposition, and hence the N demand, will tend to increase the threshold N concentration (narrow the threshold C:N ratio). For example, more favourable moisture or temperature regimes, higher rates of residue application, and a higher proportion of readily available C in the substrate will all stimulate greater microbial activity, increase N demand, and increase the threshold N concentration. Laboratory studies, generally performed under very favourable conditions, may therefore provide misleading estimates of threshold C:N ratios, or even overstate the impact of residue N-content field decomposition rates (Dendooven et al., 1990).

Decomposition depends not only on N content, but also on the lignin, polyphenol, and carbohydrate content of residues (Herman et al., 1977; Melillo et al., 1989; Palm and Sanchez, 1991). Consequently, various other ratios have been proposed as predictors of the decomposition rate, based on observed interactions between N and organic constituents like lignin, carbohydrates, or polyphenols. Berg (1986) proposed that different factors predominate at various stages of decomposition: in early stages, decomposition rate is determined by concentration of nutrients and readily degradable C; while at later stages, the rate is determined by lignin content.

3.2.3.2 Physical/Chemical Environment

Water
Water potential affects decomposition because soil microorganisms depend on water for survival and mobility. As well, water potential can influence decomposition indirectly through effects on O_2 supply, amount of soluble substrate, and pH of the soil solution.

Optimum soil water potentials for decomposition of residue are between –0.01 and –0.05 MPa. With decreasing water potential, decomposition slows because of reduced microbial activity. The activity of bacteria becomes limited first, because of a reduction in cell mobility and limited availability of substrate (Sommers et al., 1981). Fungi tend to tolerate greater water stress and are capable of surviving at very low water potentials (–4 to –10 MPa). At water potentials higher than about – 0.01 MPa, decomposition is slowed by low oxygen tension, because the rate of oxygen diffusion in water is much lower than that in air.

Oxygen Supply
The complete decomposition of organic substrates to CO_2 and other inorganic constituents is dependent on an adequate supply of oxygen. Hence, rates of aerobic decomposition decline when oxygen concentration falls below a threshold level, about 10% according to Schachtschabel et al., 1984 (cited by Verburg et al., 1995). Whether or not oxygen depletion affects decomposition, however, depends on the rate of oxygen replenishment relative to that of oxygen consumption at the site of decomposition. Thus any factor that reduces rates of oxygen diffusion, like high moisture content, poor soil structure, or increasing depth within soil will tend to limit oxygen availability. Similarly, oxygen deficiency may be induced by high rates of decomposition (i.e., oxygen consumption) at localized sites. Even well-drained soils, therefore, may show appreciable anaerobic activity, presumably in local pockets of low oxygen concentration within soil aggregates (Tiedje et al., 1984).

Although fungi, algae, and many bacteria are considered aerobic organisms, the microorganisms present in a particular soil may range from obligate aerobes to facultative forms to obligate anaerobes. Some fungi may have the ability to survive anaerobic conditions, though anaerobic activity in soil is dominated by bacteria.

Temperature
Decomposition rate is highly responsive to temperature, increasing several-fold over the range found in unfrozen soils (0 to 35 $^\circ$C). Laboratory incubations, for example, suggest that decomposition rate

increases about 2-fold for a 10 °C increase (i.e., $Q_{10} = 2$) between 10 and 40 °C (Stanford et al., 1973), though Q_{10} values may range from 1.6 to 3.2 (Anderson, 1991). Near freezing, microbial activity is constrained more by the absence of liquid water than by low temperature *per se*. Some activity can occur below 0 °C, probably in unfrozen water films around soil colloids.

At a regional level, the effects of temperature on decomposition are reflected in the accumulation of organic matter in soil. Transects of soils in the semi-arid, semi-humid, and humid regions in the central United States and Canada show that soils from warm climates contain less organic matter than those in cool regions, in part because of faster decomposition (Jenny, 1941).

pH

Soil pH affects decomposition by restricting the activity of soil microorganisms. Highly acidic or alkaline soil conditions tend to depress the size of the bacterial community, which prefers near-neutral (pH 6 to 7.5) soil conditions (Alexander, 1977). Fungi are less sensitive to low pH and usually predominate in acidic soils.

Jenkinson (1977) found that organic material decomposed more slowly in strongly acid soils than in neutral soils. After one year, soils with pH 3.7 had more residue than soils with pH 6.9. However, in the long term (> 5 yr), this difference disappeared, suggesting that the effects of soil acidity on decomposition were predominant in the early stages of decomposition.

3.2.3.3   Accessibility

Degree of Physical Protection

Organic substrates can be physically protected from microbial and enzyme attack in two ways. First, clay minerals can adsorb large organic molecules directly, reducing their availability to decomposition. Enzymes that attack the organic molecules may be adsorbed to clay surfaces and become inactivated. Second, organic material may be located in pores too small for microorganisms to enter. The majority of habitable space in soil may consist of pores so small that they prevent entry of most bacteria (Juma, 1993).

Evidence for the physical protection of organic matter is the flush of CO_2 released when aggregates are crushed (Elliott, 1986). Chemical analysis of organic material located within aggregates indicates that some of it is relatively undecomposed (Golchin et al., 1994; Gregorich et al., 1997). The capacity of a soil to physically protect organic matter may be limited, and it has been postulated (Hassink, 1996) that this protective capacity is saturated in soils that have not been cultivated or that have been under grass for a long time. According to this hypothesis, the potential for preserving soil organic matter against microbial activity is directly related to the proportion of the soils' potential protective capacity not yet occupied.

The effect of physical protection may be mitigated by other factors. For example, Skene et al. (1997) suggested that physical protection may be the primary deterrent to decomposition of high quality substrates, but that chemical protection (i.e., chemical composition) is the predominant limiting factor in low quality substrates.

Particle Size

Finely divided organic matter usually decomposes more quickly than coarse organic matter because of increased surface area and greater exposure to microbial activity. Although soil fauna account for only a small proportion (often < 10%) of soil respiration (Juma and McGill 1986; Anderson 1991), they help to increase the rate of decomposition by comminuting and redistributing organic matter through the soil profile, making it more accessible to microbial attack. In one study, for example, decomposition of straw in an 8–10 month period was 26 to 47% faster in mesh bags allowing entry of earthworms than in bags excluding earthworms (Curry and Byrne, 1992).

Position

Plant litter at the soil surface decomposes more slowly than that within the matrix of the soil, though the effect of placement diminishes with time (Christensen, 1986; Cogle et al., 1987). For example, Cogle et al. (1989) observed that incorporated straw decomposed faster than surface straw during the first 15 days, but the difference abated thereafter. The initial lag in decomposition of residues on the soil surface is probably attributable to reduced contact with decomposer organisms, and less favorable temperature and moisture conditions.

3.2.4 Human Impact on Decomposition

Human intervention in agricultural and forested systems can affect decomposition by altering some of the factors described above. One of the most widespread and pronounced effects on decomposition was the original clearing or cultivation of land for intensive agriculture (Wilson, 1978). In most cases, this action resulted in accelerated decomposition because of changes to soil moisture, aeration, temperature, and accessibility of SOM.

In agricultural systems, where soil and plant residues are often intensively manipulated, human impact on decomposition is especially pronounced (Campbell, 1978). Management practices like tillage, selection of crops and cropping sequences, and fertilization can alter decomposition rates by their effects on soil moisture, soil temperature, aeration, composition and placement of residues. The rate of decomposition in agricultural systems is generally more rapid than in forested systems; temperature, moisture and aeration conditions tend to enhance decomposition in managed soils (Fig. 3.8).

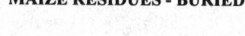

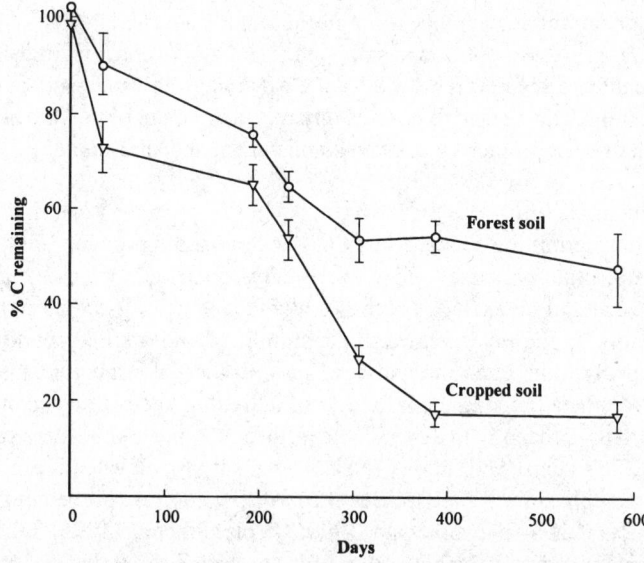

Fig. 3.8 Carbon loss rates for buried (15 cm) maize residues in agricultural and forested systems [From Gregorich and Ellert, 1994]

Human activity can also influence decomposition indirectly by altering global conditions. For example, increases in global temperature resulting from an enhanced 'greenhouse effect' could accelerate decomposition, resulting in depletion of soil organic matter reserves (Jenkinson et al. 1991). Furthermore, increasing concentrations of atmospheric CO_2 may affect decomposition by altering the composition of plant litter. For example, Van Ginkel et al. (1996) showed that decomposition of grass roots grown under 700L L^{-1} CO_2 was slower than that of roots grown under 350L L^{-1} CO_2. Another anthropogenic effect that may influence decomposition is the change in nitrogen cycling; the transfer of nitrogen from the atmosphere to terrestrial ecosystems has been approximately doubled by fertilizer manufacture, use of legume crops, and other human activities (Vitousek et al., 1997).

Human activities that alter decomposition rates can, however, conceivably also have environmental benefits. For example, adoption of agricultural practices that slow decomposition, relative to primary production, may result in increased storage of C in soil and net removal of CO_2 from the atmosphere (e.g., Dumanski et al., 1999). With growing concern about the depletion of SOM and potential global warming, there have been renewed efforts to adopt practices that retard decomposition, thereby favouring accumulation of SOM and reducing emissions of CO_2.

3.2.5 Methods for Measuring Decomposition

3.2.5.1 Field Methods

The most widely used method of measuring decomposition in the field is the litter bag technique. In this approach, litter is placed in nylon bags, incubated in the field, and mass loss is determined by periodic sampling. This technique has a number of limitations (Heal et al., 1997). It measures only net changes, and ignores the fate of material leaving the bag. This material may be more labile and contain more nutrients than the residue remaining and may also play an important role in other ecosystem processes affecting water and atmospheric quality. As well, the exclusion of fauna and vegetation and different climatic conditions in the bag from those of the surrounding soil may affect decomposition.

Another method to determine decomposition in the field is adding labelled (e.g., ^{14}C, ^{15}N) plant residues to soil. These residues are placed in small cylinders to prevent contamination by unlabelled material. At periodic intervals, soil is removed from the cylinders and analyzed to measure the quantity of the label remaining. This approach permits direct measurement of the fate of plant litter, but is usually limited to small-scale studies by cost of establishment and analysis.

3.2.5.2 Laboratory Methods

Laboratory incubations permit the detailed study of decomposition products and generate information about the decomposition processes. They are usually conducted under favourable temperature and moisture regimes, and often exclude processes like faunal activity and nutrient leaching which influence decomposition in the field. Although the simplified and optimal conditions under which they are conducted prevent direct extrapolation of rates to the field, laboratory measurements are nevertheless useful for characterizing the influence of individual factors on decomposition. For example, they have been used recently to describe the influences on decomposition of inorganic matrices (Skene et al., 1997), sodicity and salinity (Nelson et al., 1996), elevated levels of CO_2 (Gorissen et al., 1995), residue composition (Vanlauwe et al., 1996), N addition (Green et al., 1995), and the presence of living roots (Cheng and Coleman, 1990; Nicolardot et al., 1995). Furthermore, laboratory incubation studies remain a valuable method for measuring the flows of carbon and nutrients among various SOM fractions, including microbial biomass (e.g., Wang and Bakken, 1997; Hassink, 1995).

3.2.6 Modeling

3.2.6.1 Decomposition Rate and Turnover Time
The rate of disappearance of plant residues can be described using kinetic analysis. First-order rate kinetics are usually used to characterize decomposition of plant residues, assuming that the annual input of plant residues is independent of the rate of their decomposition. Using first-order kinetics to describe decomposition implies that the metabolic potential of the soil microbial biomass exceeds the substrate supply. Therefore, changes in microbial biomass are not associated with changes in the rates of decomposition.

3.2.6.2 Simulation Models
Early studies describing N loss in cultivated soils used a first-order decay model with a single component (Jenny, 1941):

$$\frac{dX}{dt} = A - kX$$

where X is the soil organic C or N content, A is the addition rate, and k is the first-order rate constant (i.e., the fraction of soil C or N decomposed each year). This model was fitted to data from several field experiments and with calculated turnover times (1/k) varying from 18 to 36 yr. The shortcoming of this model is that all plant residue components are assumed to decompose at the same rate (i.e., have the same k). Radiocarbon dating of various soil fractions showed that this was not true and also indicated that the age of soil organic matter ranged from hundreds to thousands of years (Campbell, 1978).

Other studies have used first-order models with multiple components to describe the loss of C (or N):

$$C = C_1 e^{-k_1 t} + C = C_2 e^{-k_2 t} + \dots$$

where C_x is the C content of a particular component and k_x is the rate constant for that component. In field studies using ^{14}C-labelled plant material, a double exponential model best described the data (Jenkinson, 1977; Voroney et al., 1989). In these models, the fitted data predicted that about 70% of the plant residue has a turnover time of 0.35 yr, while the remainder has a turnover time of about 12 yr. This model oversimplifies decomposition in soil, because it does not account for the formation and decomposition of either the microbial biomass or inert material (Jenkinson, 1977).

More recently, multicompartment models (Fig. 3.9) have been developed using data from long-term field experiments, the amount of microbial biomass in soil, the age of soil organic matter, and incubation experiments on the decomposition of labeled plant material in soil (e.g., Jenkinson and Raynor, 1977). Most of these models are process oriented, operating in the time scale of months to centuries (McGill, 1996). Material in each compartment decays by first-order kinetics, so that each compartment has a unique rate constant. Many of the models include environmental driving variables such as daily air temperature, incoming radiation, and precipitation. The rate constants are multiplied by factors for the different environmental variables to alter the speed of decomposition. Soil texture

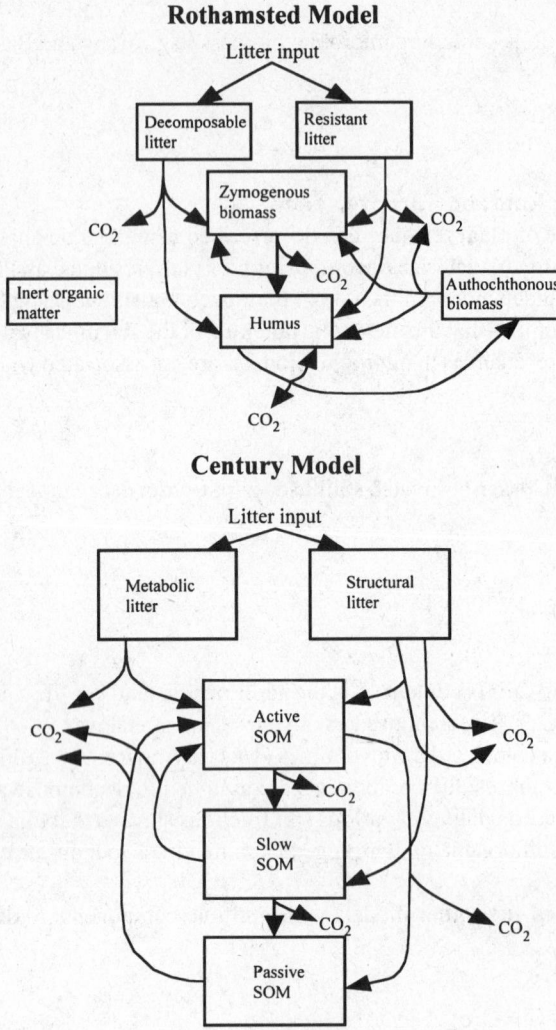

Fig. 3.9 Carbon pools and mass transfer pathways in the Rothamsted (Jenkinson and Raynor, 1977) and Century (Parton et al., 1987) models of soil organic matter dynamics [Adapted from Paustian et al., 1997]

is often used as a control to both protect or slow decomposition of soil organic components and to regulate the partitioning of C among pools.

Noncompartmental models have been developed to describe the decomposition process as a continuum, with organic matter changing "quality" as it decays (Bosatta and Agren, 1985). Quality is defined as a measure of substrate accessibility and ability to promote microbial growth (Bossata and Agren, 1991). Microbial activity is the principal mechanism behind decomposition in this model, and the quality of each fresh litter cohort is defined by a continuous variable (q) which determines its decay rate. For each litter cohort, q varies over time according to a dispersion function representing the reduction in substrate quality as the organic matter is transformed by microorganisms.

Research models have been developed recently which incorporate algorithms for decomposition and crop production for assessing changes in ecosystem behavior under defined conditions (e.g.,

Parton et al., 1987; Grant, 1997). In these models, processes such as C fixation and nutrient uptake are coupled with those that determine decomposition and C transformations. An example of this type of model is the mathematical model, *ecosys*, which simulates the transformations and transfers of water, heat, oxygen, C, N, P and salts within the plant-microbe community under specified soil, climate and management conditions (Grant, 1997).

The complexity of the interactions among the components regulating the decomposition process and the uncertainty in measuring them have been a driving force in the development of simulation models. Models help to study the process of decomposition under different soils/climates and management practices and attempt to develop a predictive capability for future land-use or climate-change scenarios. While they may not always be successful in predicting the future, models are important in a more basic way. They help to organize knowledge, identify gaps in knowledge and test hypotheses on the decomposition process. They also help to view decomposition in a holistic perspective to improve our understanding of how ecosystems operate.

3.2.7 Conclusions

Decomposition is central to the biogeochemical cycles in terrestrial, aquatic and atmospheric systems. It releases nutrients and energy tied up in organic materials and feeds them back into local and global cycles, thereby affecting land, air and water quality. Three interrelated factors regulate decomposition: the quality of the residue, the physical-chemical environment in which decomposition occurs and the type of organisms in the decomposer community. Humans can affect decomposition by altering some of these factors, especially in agricultural systems which occupy large portions of the terrestrial landscape. Tillage practices, selection and sequence of crops, and fertilization can alter the rate of decomposition by changing moisture content, temperature, and aeration of soil or the composition and placement of residues.

Decomposition rates, which integrate the effects of organisms, physical-chemical environment and litter quality, can be estimated by various field and laboratory techniques. Field methods include litter bag studies, which may underestimate the rate of decomposition because soil fauna are excluded, and isotopic techniques, which are limited in scale by their cost. Laboratory incubations are usually conducted under simplified and optimal conditions, thus preventing direct extrapolation of decomposition rates to the field. However, if their limitations are recognized and appropriate comparisons are made, these methods provide reliable data for predicting the rate of mass loss and nutrient release during decomposition.

The current understanding of the decomposition processes, learned from field and laboratory studies, is embodied in simulation models. Often models are the only practical means to extrapolate information in time and space; equally important, they also help to identify gaps in that knowledge.

Decomposition processes will be influenced by impending changes in climate, global nutrient cycles, and demands for food production. To ensure the preservation of ecosystems under these pressures, we need a broader understanding of decomposition and the factors that govern its rate.

3.2.8 References

Alexander, M. 1977. Introduction to soil microbiology. John Wiley and Sons, New York, NY.

Anderson, J.M. 1991. The effects of climate change on decomposition processes in grassland and coniferous forests. Ecol. Appl. 1:326–347.

Berg, B. 1986. Nutrient release from litter and humus in coniferous forest soils – a mini review. Scand. J. For. Res. 1:359–369.

Bosatta, E., and G.I. Ågren. 1985. Theoretical analysis of decomposition of heterogeneous substrates. Soil Biol. Biochem. 17:601–610.

Bosatta, E., and G.I. Ågren. 1991. Dynamics of carbon and nitrogen in the organic matter of the soil: A generic theory. Am. Nat. 138:227–245

Campbell, C.A. 1978. Soil organic carbon, nitrogen and fertility. p. 173–271. *In* M. Schnitzer and S.U. Khan (ed.) Soil organic matter. Elsevier Scientific Pub. Co., Amsterdam, Netherlands.

Campbell, C.A., E.A. Paul, D.A. Rennie, and K.J. McCallum. 1967. Applicability of the carbon-dating method of analysis to soil humus studies. Soil Sci. 104:217–224.

Cheng, W., and D.C. Coleman. 1990. Effect of living roots on soil organic matter decomposition. Soil Biol. Biochem. 22: 781–787.

Christensen, B.T. 1986. Barley straw decomposition under field conditions: Effect of placement and initial nitrogen content on weight loss and nitrogen dynamics. Soil Biol. Biochem. 18:523–529.

Cogle, A.L., W.M. Strong, P.G. Saffigna, J.N. Ladd, and M. Amato. 1987. Wheat straw decomposition in subtropical Australia. II. Effect of straw placement on decomposition and recovery of added ^{15}N urea. Aust. J. Soil Res. 25: 481–490.

Cogle, A.L., P.G. Saffigna, and W.M. Strong. 1989. Carbon transformations during wheat straw decomposition. Soil Biol. Biochem. 21: 367–372.

Coleman, D.C. and D.A. Crossley, Jr. 1997. Fundamentals of soil ecology. Academic Press Inc., San Diego, CA.

Coûteaux, M.-M., P. Bottner, and B. Berg. 1995. Litter decomposition, climate and litter quality. Tree 10:63–66.

Curry, J.P., and D. Byrne. 1992. The role of earthworms in straw decomposition and nitrogen turnover in arable land in Ireland. Soil Biol. Biochem. 24:1409–1412.

Dendooven, L., L. Verstraeten, and K. Vlassak. 1990. The N-mineralization potential: An undefinable parameter. p.170–181. *In* R. Merckx, H. Vereecken, and K. Vlassak (ed.) Fertilization and the environment. Leuven University Press, Leuven, Belgium.

Dick, R.P. 1994. Soil enzyme activities as indicators of soil quality. p. 107–124. *In* J.W. Doran, D.C. Coleman, D.F. Bezdicek, and B.A. Stewart (ed.) Defining soil quality for a sustainable environment. Soil Sci. Soc. Am. Spec. Pub. 35.

Dormaar, J.F. 1991. Decomposition as a process in natural grasslands. p. 121–136. *In* R.T. Coupland (ed.) Ecosystems of the world 8a: Natural grasslands - Introduction and western hemisphere. Elsevier Science Publishers, Amsterdam, Netherlands.

Dumanski, J., R.L. Desjardins, C. Tarnocai, C.M. Monreal, E.G. Gregorich, C.A. Campbell, and V. Kirkwood. 1999. Possibilities for future carbon sequestration in Canadian agriculture in relation to land use changes. Clim. Change (in press).

Elliott, E.T. 1986. Aggregate structure and carbon, nitrogen, and phosphorus in native and cultivated soils. Soil Sci. Soc. Am. J. 50:627–633.

Golchin, A., J.M. Oades, J.O. Skjemstad, P. Clarke. 1994. Study of free and occluded particulate organic matter in soils by solid state ^{13}C CP/MAS NMR spectroscopy and scanning electron microscopy. Aust. J. Soil Res. 32:285–309.

Gorissen, A., J.H. Van Ginkel, J.J.B. Keurentjes, and J.A. Van Veen. 1995. Grass root decomposition is retarded when grass has been grown under elevated CO_2. Soil Biol. Biochem. 27:117–120.

Grant, R.F. 1997. Changes in soil organic matter under different tillage and rotation: Mathematical modeling in *ecosys*. Soil Sci. Soc. Am. J. 61:1159–1175.

Green, C.J., A.M. Blackmer, and R. Horton. 1995. Nitrogen effects on conservation of carbon during corn residue decomposition in soil. Soil Sci. Soc. Am. J. 59:453–459.

Gregorich, E.G., C.F. Drury, B.H. Ellert, and B.C. Liang. 1997. Fertilization effects on physically protected light fraction organic matter. Soil Sci. Soc. Am. J. 61:482–484.

Gregorich, E.G., and B.H. Ellert. 1994. Decomposition of plant residues in soils under different management. p. 13–18. *In* Proc. 13th Int. Soil Till. Res. Org. Conf., Aalborg, Denmark.

Hassink, J. 1995. Decomposition rate constants of size and density fractions of soil organic matter. Soil Sci. Soc. Am. J. 59:1631–1635.

Hassink, J. 1996. Preservation of plant residues in soils differing in unsaturated protective capacity. Soil Sci. Soc. Amer. J. 60:487–491.

Heal, O.W., J.M. Anderson, and M.J. Swift. 1997. Plant litter quality and decomposition: An historical overview. p. 3–30. *In* G. Cadisch and K.E. Giller (ed.) Driven by nature: Plant litter quality and decomposition. CAB International, Wallingford, UK.

Herman, W.A., W.B. McGill, and J.F. Dormaar. 1977. Effects of initial chemical composition on decomposition of roots of three grass species. Can. J. Soil Sci. 57:205–215.

Jenkinson, D.S. 1977. Studies on the decomposition of plant material in soil V. The effects of plant cover and soil type on the loss of carbon from ^{14}C labelled ryegrass decomposing under field conditions. J. Soil Sci. 28:424–434.

Jenkinson, D.S. 1981. The fate of plant and animal residues in soil. p. 505–561. *In* D.J. Greenland and M.H.B. Hayes (ed.) The chemistry of soil processes. John Wiley and Sons Ltd., New York, NY.

Jenkinson, D.S., and J.H. Raynor. 1977. The turnover of soil organic matter in some of the Rothamsted classical experiments. Soil Sci. 123:298–305.

Jenkinson, D.S., D.E. Adams, and A. Wild. 1991. Model estimates of CO_2 emissions from soil in response to global warming. Nature 351:304–306.

Jenny, H. 1941. Factors of soil formation. McGraw-Hill, New York, NY.

Juma, N.G. 1993. Interrelationships between soil structure/texture, soil biota/soil organic matter and crop production. Geoderma 57:3–30.

Juma, N.J., and W.B. McGill. 1986. Decomposition and nutrient cycling in agro-ecosystems. p. 74–136. *In* M.J. Mitchel and J.P. Nakas (ed.) Microfloral and faunal interactions in natural and agro-ecosystems. Martinus Nijhoff/ Dr. W. Junk Publishers, Rotterdam, Netherlands.

McGill, W.B. 1996. Review and classification of ten soil organic matter (SOM) models. p. 111–132. *In* D.S. Powlson, P. Smith, and J.U. Smith. Evaluation of soil organic matter models. NATO Series, Vol. I 38, Springer-Verlag, Berlin, Germany.

Melillo, J.M., J.D. Aber, A.E. Likens, A. Ricca, B. Fry, and K.J. Nadelhoffer. 1989. Carbon and nitrogen dynamics along the decay continuum: Plant litter to soil organic matter. p. 53–62. *In* M. Clarholm and L. Bergstrom (ed.) Ecology of arable land. Kluwer Academic Pub., Dordrecht, Netherlands.

Moorhead, D.L. and J.F. Reynolds. 1989. Mechanisms of surface litter mass loss in the northern Chihuahan desert: a reinterpretation. J. Arid Environ. 16:157–163.

Nelson, P.N., J.N. Ladd and J.M. Oades. 1996. Decomposition of ^{14}C-labelled plant material in a salt-affected soil. Soil Biol. Biochem. 28: 433–441.

Nicolardot, B., D. Denys, B. Lagacherie, D. Cheneby, and M. Mariotti. 1995. Decomposition of ^{15}N-labelled catch-crop residues in soil: Evaluation of N mineralization and plant-N uptake potentials under controlled conditions. Europ. J. Soil Sci. 46: 115–123.

Palm, C.A., and P.A. Sanchez. 1991. Nitrogen release from the leaves of some tropical legumes as affected by their lignin and polyphenolic contents. Soil Biol. Biochem. 23:83–88.

Parton, W.J., D.S. Schimel, C.V. Cole, and D.S. Ojima. 1987. Analysis of factors controlling soil organic matter levels in Great Plains grasslands. Soil Sci. Soc. Am. J. 51:1173–1179.

Paul, E.A., and F.E. Clark. 1989. Soil microbiology and biochemistry. Academic Press, San Diego, CA.

Paustian, K., G.I. Ågren, and E. Bosatta. 1997. Modelling litter quality effect on decomposition and soil organic matter dynamics. *In* G. Cadisch and K.E. Giller (ed.) Driven by nature: Plant litter quality and decomposition. CAB International, Wallingford, UK.

Skene, T.M., J.O. Skjemstad, J.M. Oades, and P.J. Clarke. 1997. The influence of inorganic matrices on the decomposition of *Eucalyptus* litter. Aust. J. Soil Res. 35:73–87.

Sommers, L.E., C.M. Gilmour, R.E. Wildung, and S.M. Beck. 1981. The effect of water potential on decomposition processes in soils. p. 97–117. *In* J.F. Parr, W.R. Gardner, and L.F. Elliott (ed.) Water potential relations in soil microbiology. Soil Sci. Soc. Am. Spec. Pub. 9.

Stanford, G., M.H. Frere, and D.H. Schwaninger. 1973. Temperature coefficient of soil nitrogen mineralization. Soil Sci. 115:321–323.

Swift, M.J., O.W. Heal, and J.M. Anderson. 1979. Decomposition in terrestrial ecosystems. University of California Press, Berkeley, CA.

Tiedje, J.M., A.J. Sexstone, T.B. Parkin, N.P. Revsbech, and D.R. Shelton. 1984. Anaerobic processes in soil. Plant Soil 76:197–212.

Van Ginkel, J.H., A. Gorissen, and J.A. van Veen. 1996. Long-term decomposition of grass roots as affected by atmospheric carbon dioxide. J. Environ. Qual. 25:1122–1128.

Vanlauwe, B., O.C. Nwoke, N. Sanginga, and R. Merckx. 1996. Impact of residue quality on the C and N mineralization of leaf and root residues of three agroforestry species. Plant Soil 183: 221–231.

Verburg, P.S.J., D. van Dam, J.C.Y. Marinissen, R. Westerhof, and N. van Breemen. 1995. The role of decomposition in C sequestration in ecosystems. p. 85–112. *In* M.A. Beran (ed.) Carbon sequestration in the biosphere, NATO ASI Series. Series I, Global Environmental Change, Vol. 33. Springer-Verlag, Berlin.

Vitousek, P.M., J. Abner, R.W. Howarth, G.E. Likens, P.A. Matson, D.W. Schindler. W.H. Schlesinger, and G.D. Tilman. 1997. Human alteration of the global nitrogen cycle: Causes and consequences. Issues in Ecology, Number 1, Spring 1997. Published by the Ecological Society of America.

Voroney, R.P., E.A. Paul, and D.W. Anderson. 1989. Decomposition of wheat straw and stabilization of microbial products. Can. J. Soil Sci. 69:63–77.

Wilson, W.T. 1978. Pioneer agriculture explosion and CO_2 levels in the atmosphere. Nature 273:40-41.

Wang, J. and L.R. Bakken. 1997. Competition for nitrogen during mineralization of plant residues in soil: Microbial response to C and N availability. Soil Biol. Biochem. 29:163–170.

3.3 Anaerobic Microbially Mediated Processes

Zhengping Wang and Wm. H. Patrick, Jr.

Louisiana State University

3.3.1   Methanogenesis

3.3.1.1   Introduction

Methanogenesis is a microbiological process that occurs in O_2 free environments. Methanogenic bacteria or methanogens are strict anaerobic archaeobacteria existing in different habitats including anaerobic soils and water environments, intestines of ruminant animal, insects and humans. In the natural C cycle, methanogens function at the end of the anaerobic food chain and play important roles in other nutrient cycles as well.

The product of methanogenesis is methane (CH_4), which is considered as one of the most important greenhouse gases because of its relatively high potential for thermal absorption (Bouwman, 1990). The steady increase in atmospheric CH_4 in the past 200 years has attracted much attention (Rasmussen and Khalil, 1983; Blake and Rowland, 1988). The main sources of atmospheric CH_4 are microbial mineralization of SOM under strictly anaerobic conditions, such as enteric fermentation in the digestive tract of ruminants, landfills, paddy fields and natural wetland ecosystems, as well as biomass burning and coal mining (Schütz et al., 1990).

Rice paddies and natural wetlands are considered to be the most important biotic sources of CH_4, both sources contributing 20-25% of the total (Neue, 1993). From the agricultural perspective, rice paddies contribute significantly to atmospheric CH_4 and increased yield and harvest area may give rise to increased CH_4 emissions.

Flooded soil is a unique physio-biochemical soil environment with the existence of both aerobic and anaerobic soil conditions simultaneously. A thin aerobic soil surface layer and an underlying reduced soil layer develop after soil submergence. Because of the transmission of atmospheric O_2 through rice, its rhizosphere can be highly aerobic. These coexisting aerobic and anaerobic soil environments make paddy soil function as both a source and sink for CH_4.

Methane production in flooded rice soils is a microbiological process controlled by many biological, chemical, and physical factors.

3.3.1.2   Taxonomy and Characteristics of Methanogens

Oremland (1988) reviewed the microbiological aspects of methanogens while the most recent and popular taxonomy of methanogens was proposed by Balch et al. (1979) differing from the earlier work by Bryant (1974) and Wosese and Fox (1977). This proposal was a major revision of the taxonomy of methanogens based on the characteristics of the 13 species (17 strains) recognized at the time but was later modified because of several new isolates. The modified taxonomy classifies methanogens into 3 orders (Methanobacteriales, Methanomicrobiales, and Methanococcales) comprised of 6 families (Methanobacteriaceae, Methanothermacea, Methanococcaceae, Methanomicrobiaceae, Methanoplanaceae and Methanosarcinaceae), 9 genera and 29 species.

Different shapes and sizes of methanogens were found in different natural habitats including living organisms, aquatic and extreme environments such as anoxic, hypersaline lakes, and hot springs as well as high temperature seafloor sites. Morphologically, the methanogens exhibit a wide variety of shapes and sizes including various rods, cocci, long-chained spirilla, sarcina and unusual flattened plates.

Methanogens can be both Gram-positive and Gram-negative. Physiologically, they are very different from other bacteria because of the differences in the composition and structure of cell envelopes, the composition of cell membranes, C supply and coenzymes (Oremland, 1988).

3.3.1.3 Organic Carbon Transformation and Redox Processes in Anaerobic Soils

Carbon Flow

Studies using C isotope techniques demonstrate the two distinct metabolic pathways of biological CH_4 formation: (1) CO_2 reduction that utilizes H_2, fatty acids or alcohols as a H donor, and (2) transmethylation of acetic acid or methyl alcohol, which does not involve CO_2 as an intermediate. These are illustrated as follows (Vogels et al., 1988):

$$CO_2 + 4 H_2 \rightarrow CH_4 + 2H_2O \qquad [3.3.1]$$
$$CH_3COO^- + H^+ \rightarrow CH_4 + CO_2 \qquad [3.3.2]$$

Simple compounds such as various organic acids formed during organic matter fermentation provide energy and C for the growth of methanogenic bacteria. Currently, recognized substrates include H_2 plus CO_2, formate, acetate, methanol and methylated amines. In flooded rice soil, three main carbon sources are involved in supplying C for methanogenic bacteria: (1) the original SOM, (2) the exogenous supply of organic material to the soil, and (3) root litter and exudates from growing rice plants.

Exogenous supply of organic material to the soil, whether it be for the disposal of crop residues or as a source of fertilizer, appears to be the most important contributor to CH_4 production. Yagi and Minami (1990) found that application of rice straw at rates of 6-9 Mg ha^{-1} increased CH_4 emission by 1.8- to 3.5-fold. However, application of compost prepared from rice straw did not appreciably enhance CH_4 fluxes, indicating the importance of labile C in CH_4 formation. Methane production rates have been shown to be linearly correlated with soil water soluble or readily mineralizable C in both laboratory and field experiments (Yagi and Minami, 1990; Van Cleemput et al., 1991; Wang et al., 1992). In addition, CH_4 production also depends on the soil oxidant contents (Wang et al., 1993a). The method of application (i.e., depth of placement) also affects methanogenesis (Sass et al., 1991). More CH_4 is released and the emission is more sporadic when organic matter is placed at a greater depth.

Root exudates and plant litter consisting of carbohydrates, organic acids, amino acids and phenolic compounds (Martin, 1977) provide substrates for both methanogenic and methanotrophic bacteria. Root exudation has been reported to enhance CH_4 emission in laboratory experiments with 3-week-old rice plants (Raimbault et al., 1977). Kludze et al. (1993) reported a small decrease in CH_4 production in unvegetated compared to vegetated pots. This disparity was attributed to the additional contribution made by root exudates and root autolysis products to the organic substances in the soil medium.

In addition to root exudation, deteriorating plants contribute to the organic carbon pool by aboveground plant and root litter (Schütz et al., 1989a, 1991). Such additional sources of substrates are reported to be the cause of temporal and spatial variation in CH_4 production and emission.

Electron Flow and Critical Initial Eh of Methanogenesis

Flooded soils are characterized by a lack of sufficient O_2 in the soil atmosphere to act as the sole electron acceptor for microbial, plant and animal respiration (Reddy and Patrick, 1984). As long as O_2 is present, other soil oxidants such as NO_3^-, Mn^{4+}, Fe^{3+}, SO_4^{2-}, and CO_2 are not used as electron acceptors in biological reduction. However, after submergence, O_2 dissolved in the flooded water is consumed quickly. The need for electron acceptors by facultative anaerobic and true anaerobic organisms results in the reduction of several oxidized components. Reductions of NO_3^- to NO_2^-, N_2O and N_2, Mn^{4+} to Mn^{2+}, Fe^{3+} to Fe^{2+}, SO_4^{2-} to S^{2-} and CO_2 to CH_4 will occur sequentially in the soil due to thermodynamic principles as long as available C sources exist (Patrick and Delaune, 1977). A corresponding decrease in soil Eh indicates the depletion of the subsequent oxidants. For instance, when NO_3^- is reduced to N_2O and N_2 the corresponding Eh range is +250 to +350 mV. Manganic forms are reduced in a slightly lower range. Ferric iron reduction occurs in the range of +120 to +180 mV. Sulfate reduction was reported to occur when the soil Eh was as low as −150 mV (Connell and Patrick, 1969; Jakobson et al., 1981).

Methane production does not reach appreciable rates until most SO_4^{2-} is removed from soil and water systems by SO_4^{2-}-reducing bacteria (Martens and Berner, 1974; Jakobsen et al., 1981). Although the initiation of methanogenesis was reported at an Eh as low as −300 mV (Cicerone and Oremland, 1988), methanogenesis can occur following the reduction of SO_4^{2-} which has been shown by Connell and Patrick (1969) to occur at a higher Eh (~ −150 mV). The CH_4 production rate is very low until all SO_4^{2-} in the soil is reduced and soil Eh reaches a certain low level. The critical Eh for CH_4 production was recently observed in the range of −140 to −160 mV (Wang et al., 1993b). An exponential relationship between CH_4 production and soil Eh was observed when soil Eh was lower than −150 mV. This indicates that any small drop in soil Eh is accompanied by a huge increase in CH_4 production.

3.3.1.4   Other Factors Affecting Methanogenesis

Soil pH

Methanogens are pH-sensitive organisms (Alexander, 1977; Oremland, 1988). Most of them grow over a relatively narrow pH range (~ 6-8) and the optimal pH is about 7. Nevertheless, CH_4 production does occur in acidic environments such as peat bogs (Crawford, 1984), although at a slow rate. There are also a few strains of alkaliphilic (pH optima 8.1 to 9.7) CH_4-producing bacteria (Oremland et al., 1982).

The application of rice straw and chemical fertilizers might have some influence on soil pH, resulting in a temporary change in CH_4 production rate. Fertilizer application, especially urea, changes soil pH significantly due to hydrolysis. Conflicting observations on the effect of urea fertilization on CH_4 emission have been reported. Schütz et al. (1989b) reported a reduction in CH_4 emission after the incorporation of $(NH_4)_2SO_4$ or urea into the soil while a stimulatory effect of urea on CH_4 emission was observed by Lindau et al. (1991). The cause of these conflicts may be related to the effect of urea on soil pH. Wang et al. (1993b) studied 16 flooded rice soils (pH 4.8 - 8.1). In most acidic soils, the addition of urea-stimulated CH_4 emission, while in all neutral and alkaline soils, soil pH also increased, but CH_4 emission was inhibited. Apparently, a critical effect of fertilizer applications is to shift the soil pH toward or away from the range for CH_4 production.

Soil Temperature

Contradictory effects of soil temperature on the activity of methanogens have been reported. Cicerone et al. (1983) did not observe a significant correlation between soil temperature and CH_4 emission rate. In their experiment, CH_4 fluxes rose 10- to 100-fold during July and August with soil temperature at 10 cm depth always close to 23 °C. However, Seiler et al. (1984) found a positive correlation between CH_4 emissions and soil temperature at 0.5 cm depth. A positive correlation between CH_4 flux and soil temperature was also observed in a field study conducted in China with a Q_{10} value of 3 (Khalil et al., 1991).

A study conducted in peat slurry concluded that the microflora involved in CH_4 metabolism are not well adapted to low temperature (Dunfield et al., 1993). By incubating soil samples at different temperatures (0–35 °C), they found the optimum temperature for both CH_4 production and consumption to be 25 °C. However, CH_4 production was much more temperature sensitive (activation energy 123-271 kJ mol^{-1}, $Q_{10} = 5.3$–16) than was CH_4 consumption (20-80 kJ mol^{-1}, $Q_{10} = 1.4$–2.1). In the 0–10 °C range, CH_4 production was negligible but CH_4 consumption was 13–38% of maximum.

Westerman (1993) reported that temperature sensitivity of methanogenesis decreased with decreasing substrate concentrations paralleling results with axenic methanogenic cultures. The most stimulatory effect on methanogenesis was found at 2 and 10 °C, but not at 37 °C in two temperate, permanently waterlogged swamps.

Fertilization

Fertilization, especially application of N, is essential for intensive rice cultivation. Since fertilizer application may alter soil pH, microbial populations, plant litter and root exudate inputs, changes in CH_4 production are also likely. Contradictory results have been obtained by different researchers following chemical fertilizer application. Yagi and Minami (1991) did not observe a significant difference in CH_4 emission between fertilized and control treatments in a field experiment. In a different study, a significant reduction in CH_4 emission was found when $(NH_4)_2SO_4$ or urea was incorporated into the soil (Schütz et al., 1989b).

Because the most common N fertilizer used in flooded rice production is urea, constituting 75% of the total (Maene et al., 1987), much attention has been paid to its effect on CH_4 production. In a field experiment, Kimura et al. (1992) compared the effects of broadcasting $(NH_4)_2SO_4$, NH_4Cl and urea, on CH_4 emission rate. Urea-treated rice produced the most CH_4, followed by NH_4Cl and $(NH_4)_2SO_4$ treatments. The effect of urea fertilization on CH_4 production might to be related to its effects on changing microbial activity, increasing plant litter, roots exudates and altering soil chemical characteristics, e.g., pH in the short term.

3.3.1.5 Inhibitors of Methanogenesis

Oxygen, alternate electron acceptors and some chemicals inhibit methanogenesis by different mechanisms. A central electron carrier in CH_4 production biochemistry is coenzyme F420. Oxygen causes an irreversible disassociation of this coenzyme (Vogels et al., 1988). Exposure to low levels of O_2 (e.g., a few mg kg^{-1}) lowers the adenylate charge of methanogens and causes death (Roberton and Wolfe, 1970).

Alternate electron acceptors other than O_2 (e.g., NO_3^-, Fe^{3+}, Mn^{+4}, SO_4^{2-}) inhibit methanogenesis in mixed microbial ecosystems, by channeling electron flow to microorganisms that are thermodynamically more efficient than methanogens (e.g., denitrifiers or SO_4 reducers). The sequential reductions of N, Mn, Fe and SO_4 have been described above.

The use of inhibitors such as 2-bromoethanesulfonic acid, which is a structural analog of coenzyme M in methanogens has recently gained popularity. Chlorinated CH_4 compounds (e.g., chloroform, methylene chloride, etc.) are competitive inhibitors of CH_4 formation (Cicerone and Oremland, 1988).

Some metal ions, such as Cu and Cd, may inhibit methanogenesis (Drauschke and Neumann, 1992). Inubushi et al. (1990a) showed that addition of Cd at 1.9 (Cd^+) or 6.9 mg kg^{-1} (Cd^{2+}) on a dry soil basis to paddy soils suppressed CH_4 formation in the early stage of anaerobic incubation.

3.3.1.6   Methane Oxidation

Emission of CH_4 from a particular source is the result of both production and consumption of CH_4 in an ecosystem. In the presence of O_2, some ecosystems that produce CH_4 can also function as sinks for atmospheric CH_4 (Seiler, 1984; Schütz et al., 1990; Frenzel et al., 1992). Bacteria that consume CH_4 for growth are known as methanotrophs and are part of a larger group of organisms termed methylotrophs. All methanotrophs isolated and studied to date are obligate aerobes since the enzyme responsible for the initial step in CH_4 oxidation is a monooxygenase that requires molecular O_2. Three main soil characteristics affecting CH_4 oxidation are discussed below.

Surfacial Layer

Water saturation for an extended period of time usually results in changes in chemical properties of soil as well as microbial populations. In flooded rice soils, the dissolved O_2 is maintained within the top thin layer (Frenzel et al., 1992). Methane generated in anaerobic zones of the soil profile can be utilized by methanotrophic bacteria as it passes through the upper oxidized layer (Wagatsuma et al., 1992). A recent study has shown that CH_4 oxidation in surface soil of a flooded rice field consumed about 80% of the potential diffusive CH_4 flux (Conrad and Rothfuss, 1991).

Rhizosphere

In rice and other hydrophytes, atmospheric O_2 is transported to the submerged organs from leaves and stems above water through plant aerenchyma and intercellular gas space systems by diffusion or mass flow. An oxidized environment surrounding the root system exists in the rhizosphere where CH_4 oxidation is considered to be the most important internal sink for CH_4 produced in the soil profile (Holzapfel-Pschorn and Seiler, 1986). Methane fluxes increased after changing the gas phase of an incubation chamber from air to N_2 (Frenzel et al., 1992) indicating that CH_4 oxidation was decreased due to reduced O_2 supply transported through plants from the atmosphere into the rhizosphere. Consequently, O_2 transported to the root rhizosphere is a decisive factor in the magnitude of CH_4 emissions.

Algae Effect

Algae on the surface of a flooded soil may also affect CH_4 emission because of the release of O_2 during photosynthesis. Harrison (1914) observed that the gases which occur in rice soils consist mainly of CH_4 and N_2 together with small amounts of CO_2 and H_2. Aiyer (1920) showed that, with the exception of N_2, the other gases leaving a rice soil were almost undetectable as long as the surface was not disturbed. An "organized film" which covered the soil surface was considered to possess the power of arresting and assimilating these gases. The possible composition of this organized film could be a thin layer of algae which is commonly observed in most rice fields. Undoubtedly, the O_2 released during algal photosynthesis affects the biochemical characteristics of soils. Consequently, CH_4 emission is affected by increased CH_4 oxidation.

Methane emission decreased substantially in the presence of a thin layer of algae in microcosms without rice plants (Wang et al., 1995). In the presence of rice, an algal layer did not reduce CH_4 emission substantially. This effect occurred because CH_4 emission through rice plant aernchyma was greater than that by diffusion and ebullition.

3.3.1.7 Factors Affecting Methane Emission

Processes regulating CH_4 emission from paddy fields into the atmosphere include ebullition, molecular diffusion and vascular transport by plants.

Ebullition

Ebullition of CH_4 gas occurs when the partial pressure within the soil exceeds the hydrostatic pressure; this results in an upward surge into the atmosphere. Bartlett et al. (1988) attributed some 49-64% of the total CH_4 flux to ebullition, while Crill et al. (1988) reported values of up to 70%. The presence of vegetation moderates ebullition. For example, Takai and Wada (1990) observed that CH_4 ebullition is important during the early stage of flooding when rice plants were small, whereas vascular transport became more important as the rice plants grew. Possible factors that may influence ebullition include wind speed, water temperature, atmospheric pressure, solar radiation, water level and local watertable (Mattson and Likens, 1990).

Molecular Diffusion

Dissolved CH_4 gas may diffuse from paddy soils through the soil-water and air-water interface as observed by Bartlett et al. (1985). The diffusion of gases in water is about 10^4 times slower than in air, so that the exchange of gases almost stops when soils are waterlogged. However, CH_4 emission by diffusion does occur. Conrad and Rothfuss (1991) observed that subsurface microbial CH_4 oxidation is important in controlling CH_4 emission by diffusion.

Vascular Transport

Rice plants develop aerenchyma, an intercellular gas space system which provides the roots with O_2. This gas space system enables the transport of other gases, including CH_4 and CO_2, from the soil/sediment to the atmosphere. Much higher CH_4 emission was found in vegetated than in non-vegetated paddy fields. A seasonal pattern of the highest CH_4 fluxes in the last 2-3 weeks before harvest suggested either greater gas permeability at ripening or a greater C input due to root death or leaf litter input (Cicerone et al., 1983).

Rice aerenchyma tissue facilitates the emission of CH_4 produced in the anaerobic zones of flooded soil. About 95% of the CH_4 emitted by rice soils may be transported through the aerenchyma system (Cicerone et al., 1983, Seiler et al., 1984). Inubushi et al. (1990a) reported that soil pore CH_4 and plant stem CH_4 concentrations are correlated with CH_4 distribution as well as root biomass. In this context, Kludze and DeLaune (1995) found that stressful conditions which curtailed root growth resulted in significant reductions in CH_4 emission, even though production rates and individual root air space formation were high. This is believed to be due to a reduction in the total pathway for gas escape. The growth stage of the plant (Inubushi et al., 1990b) and diurnal fluctuations in photosynthesis and respiration rates (Bouwman, 1990) have also been reported to affect plant-mediated CH_4 emissions. However, Seiler et al. (1984) noted that CH_4 emission from plants does not seem to be under stomatal control.

Methane Emission and Rice Cultivars

Variations exist among rice cultivars in root aerenchyma formation (Kludze et al., 1994) and root density (Kludze and DeLaune, 1994), both factors which collectively determine the total pathway for the transport and oxidation capabilities of CH_4 gas. In addition, cultivar differences in the formation of root exudates and litter would influence CH_4 formation and its net emission. It is, therefore, imperative that morphological differences in gas conduction paths and variations in root exudation and litter formation among rice cultivars be considered in any breeding program aimed at developing rice cultivars with enhanced CH_4 oxidation and reduced CH_4 production.

3.3.1.8   Mitigation of Methane Emissions

Comprehensive global measurements of atmospheric CH_4 concentrations show that CH_4 levels are increasing substantially. Although knowledge of CH_4 sources and sinks is still limited, a protocol on worldwide CH_4 emission control, especially from the industrial sector, focusing on technology transfer and increasing research opportunities in developing countries is considered to be important (Rotmans et al., 1992).

Manipulation of rice floodwater can provide a means of mitigating CH_4 emission without reducing yields. A field experiment by Sass et al. (1992) showed that 88% less CH_4 was emitted from a multiple aeration treatment than from the normal irrigation treatment without reducing rice yields. However, more water was necessary to reach the same yields.

Minimizing incorporation of crop residue prior to planting can decrease CH_4 emission from flooded rice and reduce the potential for yield loss, particularly with some cultivars and in soils with low rates of seepage and percolation (Sass et al., 1991).

Fertilization methods affect CH_4 emission in flooded rice fields. A remarkable suppression in methanogenesis by foliar application of $(NH_4)_2SO_4$, NH_4Cl and urea instead of broadcasting has been observed but with reduced yields (Kimura et al., 1992). A recent field study showed that encapsulated CaC_2, dicyandiamide, $(NH_4)_2SO_4$ and Na_2SO_4 had a mitigating effect on CH_4 emission (Lindau et al., 1993).

3.3.1.9   Measurement

Methane production potential reflects the CH_4 production capacity of a soil under predetermined incubation conditions. Laboratory CH_4 production potential measurement is a common incubation technique to study the production potential of any individual soil. The static chamber technique is the most popular way to assess seasonal fluxes in the field. Pore water and ebullition measurements in the field are generally used to collect data to support and explain field flux measurements. Establishment of a standardized methodology is urgently needed to compare data from different sources.

3.3.2   References

Aiyer, P.A.S. 1920. The gases of swamp rice soils V: A methane-oxidizing bacterium for rice soils. Mem. Dept. Agri. India 5:173–180.

Alexander, M. 1977. Introduction to soil microbiology. 2nd Ed. John Wiley and Sons, New York, NY.

Balch, W.E., G.E. Fox, L.J. Magrum, C.R. Woese, and R.S. Wolfe. 1979. Methanogens: Reevaluation of a unique biological group. Microbiol. Rev. 43:260–296.

Bartlett, K.B., R.C. Harris, and D.I. Sebacher. 1985. Methane flux from coastal salt marshes. J. Geophys. Res. 90:5710–5720.

Bartlett, K.B., P.M. Crill, D.I. Sebacher, R.C. Harris, J.O. Wilson, and J.M. Melack. 1988. Methane fluxes from the central Amazonian floodplain. J. Geophys. Res. 93:1571–1582.

Blake D.R., and F.S. Rowland. 1988. Continuing worldwide increase in tropospheric methane 1978-1987. Science 239:1129–1131.

Bouwman A.F. 1990. Soils and the greenhouse effect. John Wiley and Sons, New York, NY.

Bryant, M.P. 1974. Methane-producing bacteria, p. 472–477. In R.E. Buchanan and N.E. Gibbons (ed.) Bergey's manual of determinative bacteriology. 8th Ed. Williams and Wilkins, Baltimore, MD.

Cicerone, R.J., J.D. Shetter, and C.C. Delwiche. 1983. Seasonal variation of methane flux from a California rice paddy. J. Geophys. Res. 88:11022–11024.

Cicerone, R.J., and R.S. Oremland. 1988. Biogeochemical aspects of atmospheric methane. Global Biogeochem. Cycl. 2:299–327.

Connell, W.E., and W.H. Patrick, Jr. 1969. Reduction of sulfate to sulfide in waterlogged soil. Soil Sci. Soc. Am. Proc. 33:711–715.

Conrad, R., and F. Rothfuss. 1991. Methane oxidation in the soil surface layer of a flooded rice field and the effect of ammonium. Biol. Fert. Soil 12:28–32.

Crawford, W. 1984. Methane production in Minnesota peat lands. Appl. Environ. Microbiol. 47:1266–1271.

Crill, P.M, K.B. Bartlett, R.C. Harris, E.S. Verry, D.I. Sebacher, L. Madzar, and W. Sanner. 1988. Methane flux from Minnesota peatlands. Global Biogeochem. Cycl. 2:371–384.

Crozier, C.R., R.D. DeLaune, and W.H. Patrick, Jr. 1995. Methane production in Mississippi Deltaic Plain wetland soils as a function of soil redox species. p. 247–255. *In* R. Lal, J. Kimble, E. Levine, and B.A. Stewart (ed.) Soils and global change. CRC Press, Boca Raton, FL.

Drauschke, G., and W. Neumann. 1992. Investigation on influences of toxic substances on microbial methane production from cattle wastes. Zentrabl. Mikrobiol. 147:308–318.

Dunfield, P., R. Knowles, R. Dumont, and T.R. Moore. 1993. Methane production and consumption in temperate and subartic peat soils: Response to temperature and pH. Soil Biol. Biochem. 25:321–326.

Frenzel, P., F. Rothfuss, and R. Conrad. 1992. Oxygen profiles and methane turnover in a flooded rice microcosm. Biol. Fertil. Soils 14:84–89.

Harrison, W.H. 1914. The gases of swamp rice soils II. Their utilization for the aeration of roots of the crop. Mem. Dept. Agric. India 4:1–17.

Holzapfel-Pschorn, A., and W. Seiler. 1986. Methane emission during a cultivation period from an Italian rice paddy. J. Geophy. Res. 90:11803–11814.

Inubushi, K., D. Taja, H. Obata, and M. Umebayashi. 1990a. Suppression of methane emission from paddy field by heavy metal. Researches related to the UNESCO's Man and the Biosphere Program in Japan 1989-1990.

Inubushi, K., M. Umebayashi, and H. Wada. 1990b. Methane emission from paddy fields. Trans. 14th Int. Congr. Soil Sci. II:249–254.

Jakobsen, P., W.H. Patrick, Jr., and B.G. Williams. 1981. Sulfide and methane formation in soils and sediments. Soil Sci. 132:279–287.

Khalil, M.A.K., R.A. Rasmussen, M.X. Wang, and L. Ren. 1991. Methane emission from rice fields in China. Environ. Sci. Technol. 25:979–981.

Kimura, M., K. Asai, A. Watanabe, J. Murase, and S. Kuwatsuka. 1992. Suppression of methane fluxes from flooded paddy soil with rice plants by foliar spray of nitrogen fertilizers. Soil Sci. Plant Nutr. 38:735–740.

Kludze, H.K., R.D. DeLaune, and W.H. Patrick, Jr. 1993. Aerenchyma formation and methane and oxygen exchange in rice. Soil Sci. Soc. Am. J. 57:386–391.

Kludze, H.K., and R.D. DeLaune. 1994. Methane emissions and growth of *Spartina patens* in response to soil redox intensity. Soil Sci. Soc. Am. J. 58:1838–1845.

Kludze, H.K., R.D. DeLaune, and W.H. Patrick, Jr. 1994. A colorimetric method of assaying dissolved oxygen loss from container-grown rice roots. Agron J. 86:483–487.

Kludze, H.K., and R.D. DeLaune. 1995. Gaseous exchange and wetland plant response to soil redox intensity and capacity. Soil Sci. Soc. Am. J. 59:939–945.

Lindau, C.W., P.K. Bollick, R.D. Delaune, A.R. Mosier, and K.F. Bronson. 1993. Methane mitigation in flooded Louisiana rice fields. Biol. Fertil. Soils. 15:174–178.

Lindau, C.W., P.K. Bollich, R.D. Delaune, W.H. Patrick, Jr., and V.J. Law. 1991. Effect of urea fertilizer and environmental factors on CH_4 emission from a Louisiana USA, rice field. Plant Soil 136:195–203.

Maene, L.M., A. De Vuyst, and S.B. Pradhan. 1987. The importance of urea as a source of nitrogen for crop production in the Asian and Pacific region. Urea-Tech '87 16-19 March, Kuala Lumpur, Malaysia.

Martens, C.S., and R.A. Berner. 1974. Methane production in the interstitial water of sulfate-depleted marine sediments. Science 185:1167–1169.

Martin, J.K. 1977. Factors influencing the loss of organic carbon from wheat roots. Soil Biochem. 9:1–7.

Mattson, M.D., and G.E. Likens. 1990. Air pressure and methane fluxes. Nature 347:718–719.

Neue, H.U. 1993. Methane emission from rice fields. Bio Sci. 43:466-474.

Oremland, R., L.M. Marsh, and S. Polein. 1982. Methane production and simultaneous sulfate reduction in anoxic marsh sediments. Nature 296:143–145.

Oremland, R.S. 1988. Biogeochemistry of methanogenic bacteria. p. 641–706. *In* A.J.B. Zehnder (ed.) Biology of anaerobic microorganisms. John Wiley and Sons, New York, NY.

Patrick, W.H. Jr., and R.D. DeLaune. 1977. Chemical and biological redox systems affecting nutrient availability in the coastal wetlands. Geosci. Man 18:131–137.

Raimbault, M., G. Rinaudo, J.L. Garcia, and M. Boureau. 1977. A device to study metabolic gases in the rice rhizosphere. Soil Biol. Biochem. 2:193–196.

Rasmussen, R.A., and M.A.K. Khalil. 1983. Global production of methane by termites. Nature 301:700–702.

Reddy, K.R., and W.H. Patrick, Jr. 1984. Nitrogen transformations and loss in flooded soils and sediments. CRC Crit. Rev. Environ. Contr. 13:273–309.

Roberton, A.M., and R.S. Wolfe. 1970. ATP pools in Methanobacterium. J. Bacteria. 102:43–51.

Rotmans, J., M.G.J. den Elzen, M.S. Krol, R.J. Swart, and H.J. van der Woerd. 1992. Stabilizing atmospheric concentrations: Towards international methane control. Ambio 21:404–413.

Sass, R.L., F.M. Fisher, P.A. Harcombe, and F.T. Turner. 1991. Mitigation of methane emission from rice fields: Possible adverse effects of incorporated rice straw. Global Biogeochem. Cycl. 5:275–287.

Sass, R.L., F.M. Fisher, Y.B. Wang, F.T. Turner, and M.F. Jund. 1992. Methane emission from rice fields: The effect of flood water management. Global Biogeochem. Cycl. 6:249–262.

Schütz, H., W. Seiler, and H. Rennenberg. 1990. Soil and land use related sources and sinks of methane (CH_4) in the context of the global methane budget. p. 269–285. In A.F. Bouwman. (ed.) Soils and the greenhouse effect. John Wiley and Sons, New York, NY.

Schütz H., W. Seiler, R. Conrad. (1989a). Processes involved in formation and emission of methane in rice paddies. Biogeochem. 7:33–53.

Schütz, H., A. Holzapfel-Pschorn, R. Conrad, H. Rennenberg, and W. Seiler. 1989b. A 3-year continuous record on the influence of daytime, season, and fertilizer treatment on methane emission rates from an Italian rice paddy. J. Geophys. Res. 94:16405–16416.

Schütz, H., P. Schroder, and H. Rennenberg. 1991. Role of plants in regulating the methane flux to the atmosphere. p. 29–61. In Trace gas emission by plants. Academic Press Inc., New York, NY.

Seiler, W. 1984. Contribution of biological processes to the global budget of CH_4 in the atmosphere. p. 468–477. In M.J. Klug and C.A. Reddy. (ed.) Current perspectives in microbial ecology. American Society for Microbiology, Washington, DC.

Seiler, W., A. Holzapfel-Pschorn, R. Conrad, and D. Scharffe. 1984. Emission of methane from rice paddies. J. Atom. Chem. 1:241–268.

Takai, Y., and E. Wada. 1990. Methane formation in waterlogged paddy soils and its controlling factors. p. 101–107. In H.W. Scharpenseel, M. Schomaker and A. Ayoub. (ed.) Soils in a warmer earth. Elsevier, Amsterdam, Netherlands.

van Cleemput, O., H. Ramon, and A. Vermoessen. 1991. Emission of C1-C3 hydrocarbons from soils. p. 189–191. In Proc. EUROTRAC Symp. '90. The Hague, Netherlands.

Vogels, G.D., J.T. Keltjens, and C. van der Drift. 1988. Biochemistry of methane production. p. 707–770. In A.J.B. Zehnder (ed.) Biology of anaerobic microorganisms. John Wiley and Sons, New York, NY.

Wagatsuma, T., K. Jujo, K. Tawaraya, T. Sato, and A. Ueki. 1992. Decrease of methane concentration and increase of nitrogen gas concentration in the rhizosphere by hygrophytes. Soil Sci. Plant Nutr. 38:467–476.

Wang, Z., R.D. DeLaune, C.W. Lindau, and W.H. Patrick, Jr. 1992. Methane production from anaerobic soil amended with rice straw and nitrogen fertilizers. Fert. Res. 33:115–121.

Wang, Z., C.W. Lindau, R.D. DeLaune, and W.H. Patrick, Jr. 1993a. Methane emission and entrapment in flooded rice soils as affected by soil properties. Bio Fertil. Soils 16:163–168.

Wang, Z., R.D. DeLaune, P.H. Masscheleyn, and W.H. Patrick, Jr. 1993b. Soil redox and pH effects on methane production in a flooded rice soil. Soil Sci. Soc. Am. J. 57:382–385.

Wang, Z., C.R. Crozier, and W.H. Patrick, Jr. 1995. Methane emission as affected by the presence of rice and algae in a flooded rice field. p. 245–250. In R. Lal, J. Kimble, E. Levine and B.A. Stewart (ed.) Soil management and the greenhouse effect. CRC Press, Boca Raton, FL.

Westerman, R.L. 1993. Temperature regulation of methanogenesis in wetlands. Chemosphere 26:321–328.

Wosese, C.R., and G.E. Fox. 1977. Phylogenetic structure of the procaryotic domain: Primary kingdoms. Proc. Natl. Acad. Sci.74:5088–5090.

Yagi, K., and K. Minami. 1990. Effect of organic matter application on methane emission from some Japanese paddy fields. Soil Sci. Plant Nutr. 36:599–610.

Yagi, K., and K. Minami. 1991. Emission and production of methane in the paddy fields of Japan. JARQ 25:165–171.

3.4 Soil Enzymes

Paolo Nannipieri and Loretta Landi
Dip. della Scienza del Suolo e Nutrizione della Pianta, Firenze, Italy

3.4.1 Introduction

Chemical, physiochemical and biochemical reactions are involved in nutrient cycling in soil. Biochemical reactions are catalyzed by enzymes (protein molecules) which markedly increase the reaction rate and act in a precise (only products and no byproducts are formed in the enzyme-catalyzed reactions) and specific (distinguish between substrate stereoisomers) way (Lehninger et al., 1993). For example, proteolytic enzymes hydrolyze peptide bonds in proteins containing L- but not D-amino acids. As with any catalyst, enzymes are unaltered at the end of the reaction. An enzyme-catalyzed reaction occurs within the confines of a pocket of the enzyme molecule called the active site, consisting of a binding and catalytic site. Enzymes have molecular weights ranging from 12,000 to over 1 million and some of them are activated only when they interact with an additional chemical component called a cofactor. One or more inorganic ions (Fe^{3+}, Fe^{2+}, Mn^{2+}, Zn^{2+}, etc.) or a complex organic or metallorganic molecule, also called a coenzyme, can function as cofactors. When the coenzyme or the metal ion is covalently bound to the enzyme protein, it is called a prosthetic group. Enzymatic activity is markedly affected by pH, ionic strength, temperature, and the presence or absence of inhibitors or activators. Enzymes are denatured by elevated temperature and extremes of pH.

Interest in soil enzymatic activities derives from the notion that their potential activities, analyzed under artificial but relatively controlled conditions, relate to their activities in the natural soil and environment, with varying pH, moisture and substrate supply. During the past thirty years many reviews (Skujins, 1976; Burns, 1982; Ladd, 1985; Mayaudon, 1986; Boyd and Mortland, 1990; Nannipieri et al., 1990; Tabatabai and Fu, 1992; Dick and Tabatabai, 1992; Dick, 1994; Nannipieri, 1992; Schaffer, 1993; Gianfreda and Bollag, 1996; Nannipieri et al., 1996) have detailed the activities and properties of soil enzymes and a book devoted entirely to the subject has been published (Burns, 1978).

Oxidoreductases (catalyzing electron transfer reactions), transferases (catalyzing reactions with transfer of molecular groups, such as $-NH_2$, R-CO-, etc.) and hydrolases (catalyzing bond hydrolysis) have been the most studied enzyme systems in soil because of their role in the oxidation and release of inorganic nutrients from soil organic matter (SOM) (Dick, 1994; Nannipieri, 1994).

3.4.2 Location and Origin

The activity of any particular enzyme in soil is the result of activities associated at various biotic and abiotic components (Fig. 3.10). Enzymes may be intracellularly located in proliferating and non-proliferating cells or with dead cells and cell debris (Burns, 1982). They may have leaked from cells or lysed cells and they may exist temporarily in enzyme-substrate complexes or adsorbed to clay minerals or be associated with humic colloids. The same enzyme may change location in soil; it may be intracellular in a viable cell, become associated with cell debris after cell death, be released in the aqueous phase as cell membranes are broken and eventually be adsorbed by soil colloids. Enzymes present in some of the locations may not contribute significantly to the overall enzymatic activity measured in soil. Activities due to free enzymes in the soil solution are probably negligible because these enzymes are short-lived (Burns, 1982). In addition, enzymes in spores and cysts are unlikely to exhibit any activity in enzyme assays.

Specific terms have been introduced to indicate a group of locations. Kiss et al. (1975) and Sequi (1974) coined the term accumulated enzymes to indicate active enzymes soluble or bound to inorganic

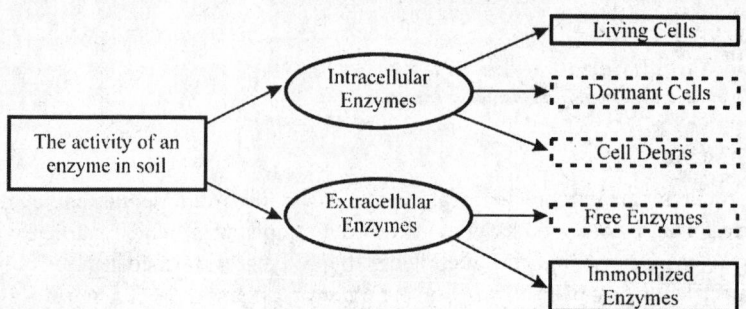

Fig. 3.10   The activity of an enzyme in soil depends on intra- and extracellular enzymes. The contribution of enzymes surrounded by dashed lines to the overall activity is generally insignificant.

and organic soil colloids, associated with particulate cell remains and dead cells, and on viable but nonproliferating cells. Skujins (1976) introduced the term abiotic to describe all these enzymes plus those secreted by proliferating cells during normal growth (truly extracellular enzymes) but excluding those either within or attached to the outer surface of living cells. The term immobilized, which has appropriate associations with industrial enzyme technology, has been used to define active enzymes bound to clays and humic molecules.

Enzymes in soil may be synthetized by microbial, vegetal and animal cells but probably microorganisms are the most important sources (Ladd, 1985). The contribution of enzymes to soil from soil fauna has been poorly investigated (Dick, 1994). Plants can positively affect enzymatic activity in soil directly or indirectly. Roots can release extracellular enzymes into soil, or stimulate microbial growth and thus the microbial enzymatic activities through the release of root exudates, mucilages and root remains (Ladd, 1985; Dick, 1994; Nannipieri, 1994). As a consequence of the plant effect, enzymatic activities of the rhizosphere are often higher than those of the bulk soil. It is difficult to conclusively discriminate between sources of enzymes in soil. Ultracytochemical and ultrahistochemical methods applied to soil with examination of preparations by electron microscope have allowed the detection of enzymes, such as acid phosphatases in root fragments, mycorhizae, soil microorganisms and fragments of microbial membranes as small as 7-20 nm (Ladd et al., 1996). Unfortunately, these ultracytochemical tests used to locate enzymes have limits in electron-transparent materials, such as microbial and root cells or cell fragments but not in naturally electron-dense components, such as soil minerals.

3.4.3   Methodology and Interpretations of Measurements

The most serious problem in assaying enzymatic activities in soil is to establish the contribution of each according to the state in which it exists in soil. Differentiating between the contributions of extra- and intracellular components is important because it would enable the activity of stabilized extracellular enzymes made resistant to thermal denaturation and proteolysis by association with soil colloids to be measured. In addition, this extracellular enzymatic activity being unrelated to microbial activity is not subject to repression or induction and probably not sensitive to environmental conditions affecting the physiological state of the microorganisms. The use of bacteriostatic agents to inhibit intracellular enzymatic activity is unsatisfactory due to various factors such as wall permeability, the possible induction of plasmolysis with release of intracellular enzymes, the promotion of microbial growth, and their variable effect depending on soil and enzyme assay conditions (Ladd, 1985; Frankenberger and Johanson, 1986; Nannipieri, 1994).

An innovative approach to study the contribution of extracellular components to the overall enzymatic activity of soil was developed by McLaren and Pukite (1973) based on the work of Paulson and Kurtz (1969) who monitored urease activity and the number of ureolytic microorganisms in a soil treated with glucose to promote microbial growth. McLaren and Pukite (1973) observed that both variables were significantly correlated. By plotting urease activity against the number of ureolytic microorganisms and extrapolating to zero population, the positive intercept gave an estimate of extracellular urease activity. By following the same approach with measurement of microbial biomass by ATP (more accurate method than the plate count), extracellular phosphatase at pH 6.5 was measured in fresh but not in air-dried soils (Nannipieri et al., 1996). Further research is needed to verify the validity of this approach for enzymes other than phosphatase in soils with different properties.

Enzymatic activity in soil was first demonstrated some 90 years ago, and today routine assays have been devised for a wide range of soil enzymes representing all classes. Research in soil enzymology has concentrated on the effects of pollutants on enzymatic activities and their role in nutrient cycling (Nannipieri, 1994). The fact that enzyme methods have been collected in two recent books (Weaver et al., 1994; Alef and Nannipieri, 1995) devoted to analytical procedures in soil microbiology and biochemistry indicates the importance of enzyme assays to complement chemical and microbiological analyses. Usually, methods for determining enzymatic activities in soil are based on the efficient extraction of the substrate or the reaction products from soil before the determination of the relative concentration (Tabatabai, 1994). Buffers unable to extract SOM are preferred because most procedures for determining either product formation or substrate disappearance involve colorimetric reactions. In addition, the assay is valid when the substrate concentration does not limit the reaction rate and the period of time selected is such that the product formation or substrate disappearance is linear. The shortest incubation time necessary to produce a measurable value of activity should be chosen so as to avoid complications associated with microbial growth, intracellular catalysis, and new enzyme production. The effect of temperature and pH on enzymatic activity must be investigated to establish the optimal conditions for the activity; if buffers are used (and not all assays include them), they must maintain the pH required for optimum activity throughout the incubation period. Non-biological conversion of substrate and product can occur during the assay and must be taken into consideration by designing proper controls.

In interpreting the results of soil enzyme measurements, it should be remembered that the maximum potential rather than the actual activity is measured because the incubation conditions of enzyme assays are chosen to ensure optimum rates of catalysis. The activity of an enzyme in the field is subject to conditions different than those characterizing an enzyme assay *in vitro*. *In situ* substrates are not generally in excess, temperature and pH are rarely at an optimum, and soil moisture fluctuates widely. In addition, the activities of some soil enzymes (e.g., phosphatase and sulphatase) are conveniently determined by the use of substrates (*p*-nitrophenyl phosphate and sulphate) not found in nature.

The fact that the activity of any particular enzyme in soil is a composite of activities associated with various components, and enzyme assays give the maximum potential activity rather than the actual values, have often led to the conclusion that soil enzyme assays are meaningless at least in ecological and agricultural terms. However, it is appropriate to note that other long-established methods in soil microbiology are founded on the use of unrealistic conditions. The many problems associated with the measurement of actual nitrification rates *in situ* are an example (Berg and Rosswall, 1985). Enzyme measurements answer qualitative questions regarding specific metabolic processes; therefore, even if enzyme measurements reflect potential rather than actual activities, they (along with CO_2 evolution, ATP content, biomass measurements, etc.) are of great value in screening procedures for determining the susceptibility of soil processes to agrochemical amendments, cultivation practices, and environmental and climatic factors (Nannipieri et al., 1990; Nannipieri, 1994).

3.4.4   Kinetics of Enzyme Reactions in Soil

In a homogeneous and isotropic system, enzymes exert their catalytic action in a single phase where all components (enzymes, hydrogen ions, substrates, products, inhibitors, activators, cofactors, etc.) are simultaneously present. By assuming that the concentration of the active complex [ES] (where E is the enzyme and S the substrate) is constant with time, enzyme kinetics follow the Michaelis-Menten equation:

$$v = V_{max}[S] / (K_m + [S])$$ [3.4.1]

where v is the observed reaction rate at a given substrate concentration [S], K_m is the Michaelis-Menten constant and V_{max} is the maximum reaction rate at saturating concentrations of the substrate. Both K_m and V_{max} values vary with soil type, assay conditions, soil texture, changes in microbial populations following organic amendments, etc. (Tabatabai, 1973; Nannipieri and Gianfreda, 1997).

The Michaelis-Menten theory is founded on basic assumptions, including the presence of a homogeneous system, which is not fulfilled in soil (Nannipieri and Gianfreda, 1997). Therefore, it is surprising that K_m and V_{max} values have been calculated for some enzymatic activities in soil slurries and soil columns. Soil being inhabited by different microbial animal and plant organisms may include many enzymes catalyzing the same reaction (for example, different phosphatases) but characterized by different properties including different kinetic constants. The measured rate values, therefore, only represent average reaction ratio whose weighting factors are unknown. However, under controlled conditions (pH, temperature, buffer, etc.) and with care, these values can be used to compare soil characteristics (Nannipieri and Gianfreda, 1997).

Most soil enzymes act in a heterogeneous system which affects the relative kinetics (McLaren and Packer, 1970; Nannipieri and Gianfreda, 1997; Tabatabai, 1994). Particularly, the immobilization of enzymes on soil colloids may affect the conformation of the protein molecule, and consequently, the kinetic values. Both K_m and V_{max} values can be affected by enzyme immobilization due to diffusional limitations of substrates and reaction products, steric and microenvironmental effects. Soil enzyme kinetics can also be influenced by nonuniform enzyme distribution in the soil matrix as well as by the presence of multiple enzymes catalyzing the same reaction (Nannipieri et al., 1988).

3.4.5   Enzymes Immobilized in Soil: Properties and Role

The presence of humic or clay enzyme complexes has attracted much attention because, as mentioned above, they are indicative of the soil enzymatic capacity which is independent of the extent of microbial activity. Adsorption and binding of proteins to clays and the stability of enzyme-clay complexes to microbial attack have been extensively studied (Stotzky, 1986; Boyd and Mortland, 1990; Nannipieri et al., 1996). According to Stotzky (1986), a protein is bound to the clay when no protein is released from the protein-clay complex by water washing. No direct evidence exists for the occurrence in soil of clay-protein or clay-enzyme complexes, and the failure to detect their presence is attributable to practical difficulties in isolating such complexes (Ladd, 1985). The indirect proof of the occurrence of clay-protein complexes in soil is derived from the fact that the highest overall enzymatic activity is observed in clay sized particles (< 50 μm) (Nannipieri et al., 1996).

Clay enzyme interaction is a complex process depending on several factors such as moisture, nature of the exchangeable cation, clay type and surface area, pH of bulk phase and clay water interface. It also depends on the characteristics of the protein molecule: solubility, molecular weight, shape, isoelectric point (pZ), polar and ionizable functional groups. Ionic binding, ion-dipole interaction, hydrogen bonds and physical forces are the mechanisms involved in clay enzyme interactions. Studies with clay minerals covered with mixtures of polymeric oxides, and hydroxides of Fe, Al and Mn (the so-called "dirty clays") or with synthetic clay organic complexes have been conducted because they reflect *in situ* conditions better than pure homoionic mineral clays (Stotzky, 1986; Fusi et al., 1989; Gianfreda et al., 1992).

The indirect proof of the presence of extracellular humic protein complexes in soil comes from the fact that amino acids are the main identifiable organic N compounds after acidic hydrolysis (Stevenson, 1982). Active humic fractions have been extracted from soil using different solutions and shaking for periods ranging from a few min to 1 day (Tabatabai and Fu, 1992; Nannipieri et al., 1996). Generally conditions are chosen to avoid damage to living cells, to favor extraction of enzymes bound to soil colloids and to prevent the formation of artifacts such as the adsorption or entrapment of the extracted enzymes within the dispersed soil colloids. In addition, large quantities of natural enzyme humus complexes must be extracted before investigation of their properties (Nannipieri et al., 1980). Most of the buffers currently used in conventional biochemistry (phosphate, *tris*-borate, bicarbonate, etc.) are not effective in extracting large amounts of humic enzyme complexes (Ceccanti et al., 1978; Nannipieri et al., 1978). However, sodium pyrophosphate extracts extracellularly stabilized urease in large quantities (Nannipieri et al., 1974), and is effective in extracting laccase (Ruggiero and Radogna, 1984), amylase (Scherbakova et al., 1981), catalase (Perez Mateos et al., 1988), phosphatase, casein and benzoylarginamide-hydrolysing proteases (Nannipieri et al., 1980).

Purification studies of extracted enzymes that have attempted to elucidate the mechanisms by which enzymes are attached to humic molecules (Tabatabai and Fu, 1992; Nannipieri et al., 1996) and show that the enzymatic activity is dispersed throughout a range of molecular weight fractions and that most persistent (against thermal and proteolytic denaturation) enzymes are likely to be associated with high molecular weight (> 50,000) polycondensates (Nannipieri et al., 1996). The association of the enzyme with the humic moiety probably stabilizes the tertiary structure of the protein thus improving the resistance of the enzyme to thermal denaturation. Many mechanisms are proposed to explain the association of enzymes with humic molecules; these include ion exchange, H-bonding, lyophilic reactions and covalent bonding to the humic moiety during synthesis (Nannipieri et al.,1996). It has been postulated that humic-enzyme complexes of higher molecular weight are more likely to possess molecular arrangements which permit the movement of substrates and products towards the enzyme but not that of a large molecule such as protease (Burns, 1982). According to this hypothesis, enzymes (for example, casein-hydrolysing activity) acting on high molecular weight substrates are short-lived in soil. Bonding and entrapment of these enzymes in humic complexes may protect the enzyme protein but may render it inaccessible and inactive toward high molecular weight substrates (Ladd, 1985). Higher molecular weight humic-enzyme complexes extracted from soil have been shown to contain two enzymes catalyzing the same reaction (Nannipieri et al., 1996). According to Mayaudon (1986), humic-enzyme complexes are composed of glycoenzymes of fungal origin associated to bacterial lipolysaccharides, which are linked through Ca^{2+} bridges to the humic moiety.

Another way of understanding the relationship between enzymes and humic matter is to attempt to complex enzymes with either extracted humates or synthetic humic analogues. Usually stable complexes are not formed by simple adsorption of the enzyme by humates (Maignan, 1982; Sarkar, 1986); entrapment within a rigid humic matrix and/or chemical binding seems essential to form stable humic-enzyme complexes. Model humic-enzyme complexes have been prepared by oxidative coupling or condensation reactions of phenolic compounds with enzymes. Terminal amino groups as well as sulphydryl, phenolic and imidazole groups of the latter molecules bind to carboxyl and carbonyl groups associated with reactive quinones formed from phenols through both enzymatic (monophenol monoxygenases, peroxidases and laccasases) and abiotic catalysis (Section B, Chapter 9). The synthetic humic-enzyme complexes are more resistant to thermal and proteolytic degradation than the respective free enzymes (Rowell et al., 1973; Sarkar and Burns, 1984; Grego et al., 1990). Enzymes are incorporated after reactive quinone formation and during the subsequent condensation stage. This, and the failure to immobilize enzymes on or within preformed humic analogues suggest that chemical bonding, perhaps due to co-crosslinking, is of greater significance than mechanisms dependent on ionic forces or entrapment (Sarkar and Burns, 1984). In addition, this confirms that the mere physical mixing of enzymes and humates does not produce a stable complex.

Burns (1982) suggested that enzymes immobilized on organic colloids or occluded by humic molecules play an important role in soil microbial ecology. A particular nutritional problem is faced by microorganisms that rely on the extracellular enzymes for the supply of nutrients and energy sources. Most of the organic substrates (lignin, cellulose, hemicellulose, proteins, etc.) entering the soil system are not suitable for uptake until they are depolymerized by extracellular enzymes. These enzymes may not survive long enough to detect and degrade their exogenous substrates because they can be adsorbed and inactivated by soil colloids or used as substrates for proteolytic microorganisms. Even if the enzyme survives long enough to detect the exogeneous polymer and the microenvironment surrounding the substrate is suitable for catalysis, the product may not be taken up by the enzyme-producing microorganisms due to competitive factors such as adsorption by soil particles, non-biological degradation and uptake by opportunistic microorganisms (Burns, 1982). Immobilized enzymes in soil may act as stable catalysts for the detection of potential substrates making unnecessary the continuous synthesis and secretion of the extracellular enzymes by microorganisms (Burns, 1982). The products of the reaction catalyzed by immobilized enzymes may act as "effector molecules". Thus, the microorganisms can synthesize and release the extracellular enzyme only if the substrate is present and the environmental conditions are suitable for catalysis. The hypothesis proposed by Burns (1982) has not been verified yet because models such as this are at the moment difficult to investigate.

3.4.6 Biotechnological Applications

The manipulation of the activity of an enzyme in soil can be useful for practical purposes. The enzymatic activity can be decreased by using enzyme inhibitors, such as phenylphosphorodiossidate (PPD) and N-(n-butyl) thiophosphoryltriamide (NBPT), used for the urease activity (Martens and Bremner, 1984; Bremner and Chai, 1986). Under certain conditions (e.g., high urease activity in alkaline soils) when urea is applied, soil urease activity must be decreased to limit NH_3 volatization losses.

The increase in the activity of an immobilized enzyme may be needed for transforming an organic pollutant to a less toxic intermediate in polluted soil (Nannipieri and Bollag, 1991) or to release a soluble nutrient from the insoluble form (Burns, 1987).

Nannipieri and Bollag (1991) have discussed the use of either enzymes or microorganisms for degrading pesticides in pesticide-polluted soils. Microorganisms containing the desired enzymes are preferred over enzyme preparations when the detoxification treatment requires multistep transformations with different enzymes acting sequentially and/or where the regeneration of cofactors or coenzymes is required. Enzymes are preferred if the single-step reaction catalyzed by the enzyme produces a byproduct that is less toxic than the pollutant molecule. However, soil inoculation with microorganisms containing the desired enzyme presents several problems. Usually, inoculated microorganisms do not survive long enough to successfully perform the reaction for which they have been used in soil (Nannipieri and Bollag, 1991). The added microorganisms are weak competitors with the indigenous population and are easily attacked by predators. The situation of the added microorganisms can be further made problematic by unfavorable conditions such as chemical shocks, unsuitable pH and temperature, presence of toxins and high concentration of pollutants. The activity of an enzyme can be increased in soil by adding either the enzyme immobilized on a solid support or by promoting its microbial synthesis. Immobilization is required because free enzymes in soil are short-lived. Increases in enzymatic activities due to the addition of organic compounds promoting microbial growth in soil are also short-lived (Nannipieri et al., 1983). Microbial synthesis of catabolic enzymes is normally regulated at the cellular level by induction or repression due to the exposure of the cell to a substrate or a product of the pathway, respectively (Gottschalk, 1979). The intimate contact of the inducer or repressor with the cell wall is required for the success of this strategy. Both the inducer and the repressor may not survive long enough to interact with the microbial cell due to adsorption by soil particles or non-biological and biological degradation (Burton and McGill, 1991). Soil is mainly an oligotrophic environment and with the exception of a few microsites, it is likely that inducers will not reach concentrations high enough to induce enzyme synthesis; thus constitutive enzyme synthesis is probably the dominant microbial source of enzymatic activity in soil (Burton and McGill, 1991). This hypothesis seems to be supported by the observations on protease and cellulase activity which are limited in soil by substrate availability and not enzyme content (Tateno, 1988). Therefore, the present evidence indicates that induction or the catabolic repression of the enzymatic activity associated with microbial cells is not a practical approach for the manipulation of the soil enzymatic activity.

In spite of the promising potential applications, only a few investigations have been carried out to manipulate soil enzymatic activity by applying immobilized enzymes. Further research is needed to compare the efficacy of enzymes immobilized on different matrices in accelerating reaction rates *in situ*. The best result is probably a compromise between moderate enzyme protection against proteolytic degradation and the limitation of diffusion problems between the enzyme and the substrates. In addition to laboratory studies, investigations in the field are needed so as to ascertain the feasibility of these preparations on a large scale.

3.4.7 References

Alef, K., and P. Nannipieri. 1995. Methods in applied soil microbiology and biochemistry. Academic Press, London, UK.

Berg, P., and T. Rosswall. 1985. Ammonium oxidizer numbers, potential and actual exudation rates in two Swedish arable soils. Biol. Fert. Soil 1:131–140.

Boyd, S.A., and M.M. Mortland. 1990. Enzymes interactions with clays and clay-organic matter complexes. p. 1–28. *In* J-M- Bollag and G. Stotzky, (ed.) Soil biochemistry, Vol. 6. Marcel Dekker, New York, NY.

Bremner, J.M., and H.S. Chai. 1986. Evaluation of N-butyl phosphorothioic triamide for retardation of urea hydrolysis Commun. Soil Sci. Plant Anal. 17:337–351.

Burns, R.G. 1978. Soil enzymes. Academic Press, London, UK.

Burns, R.G. 1982. Enzyme activity in soil: location and a possible role in microbial ecology. Soil Biol. Biochem. 14:423–427.

Burns, R.G. 1987. Interactions of humic substances with microbes and enzymes in soil and possible implications for soil fertility. Anal. Edaf. Agrobiol. 46:1247–1259.

Burton, D.L., and W.B. McGill. 1991. Inductive and repressive effects of carbon and nitrogen on L-histidine ammonia-lyase activity in a black chernozemic soil. Soil Biol. Biochem. 23:939–946.

Ceccanti, B., P. Nannipieri, S. Cervelli, and P. Sequi. 1978. Fractionation of humus urease complexes. Soil Biol. Biochem. 10:39–45.

Dick, R.P. 1994. Soil enzyme activities as indicators of soil quality. p. 107–124. *In* J.W. Doran, D.C. Coleman, D.F. Berdicek and B.A. Stewart (ed.) Defining soil quality for a sustainable environment. Soil Sci. Soc. Am. Spec. Pub. 35, Soil Science Society of America, Madison, WI.

Dick, W.A., and M.A. Tabatabai. 1992. Significance and potential uses of soil enzymes. p. 95–127. *In* F. Blaine (ed.) Soil microbial ecology. Marcel Dekker, New York, NY.

Frankenberger, W.T., and J.B. Johanson. 1986. Use of plasmolytic agents and antiseptics in soil enzyme assays. Soil Biol. Biochem. 18:209–213.

Fusi, P., G.G. Ristori, L. Calamai, and G. Stotzky. 1989. Adsorption and binding of protein on "clean" (homoionic) and "dirty" (coated with Fe oxyhydroxides) montmorillonite, illite and kaolinite. Soil Biol. Biochem. 21:911–920.

Gianfreda, L., M.A. Rao, and A. Violante. 1992. Adsorption, activity and kinetic properties of urease on montmorillonite, aluminium hydroxide and Al(OH)$_x$ montmorillonite complexes. Soil Biol. Biochem. 24:51–58.

Gianfreda, L., and J.M. Bollag. 1996. Influence of natural and anthropogenic factors on enzyme activity in soil. p. 123–193. *In* J.M. Bollag and G. Stotzky, (ed.) Soil biochemistry. Vol. 9. Marcel Dekker, New York, NY.

Gottschalk, G. 1979. Regulation of bacterial metabolism. p. 178–207. *In* G. Gottschalk (ed.) Bacterial metabolism. 2nd Ed. Springer Verlag, New York, NY.

Grego, S., A. D'Annibale, M. Luna, L. Badalucco, and P. Nannipieri. 1990. Multiple forms of synthetic pronase-phenolic copolymers. Soil Biol. Biochem. 22:721–724.

Kiss, S., M. Dragan-Bularda, and D. Radulescu. 1975. Biological significance of enzymes accumulated in soil. Adv. Agron. 27:25–87.

Ladd, J.N. 1985. Soil enzymes. p. 175–221. *In* D. Vaughan and R.E. Malcom (ed.) Soil organic matter and biological activity. Martinus Nijhoff/Dr. W. Junk Publishers, Dordrecht, Netherlands.

Ladd, J.N., R.C. Foster, P. Nannipieri, and J.M. Oades. 1996. Soil structure and biological activity. p. 23–78. *In* G. Stotzky and J.M. Bollag (ed.) Soil biochemistry. Vol 9. Marcel Dekker, New York, NY.

Lehninger, A.L., D.L. Nelson, and M.M. Cox. 1993. Principles of biochemistry. 2nd Ed. Worth Publishers, New York, NY.

Maignan, C. 1982. Activite' des complexes acides humques-invertases: Influence du mode de preparation. Soil Biol. Biochem. 14:439–445.

Martens, D.A., and S.M. Bremner. 1984. Effectiveness of phosphoroamides for retardation of urea hydrolysis in soils. Soil Sci. Soc. Am. J. 48:302–305.

Mayaudon, C. 1986. The role of carbohydrates in the free enzymes in soil. p. 263–309. *In* C. H. Fuchsman (ed.) Peat and water. Elsevier Applied Science Publishers, London, UK.

McLaren, A.D., and L. Packer. 1970. Some aspects of enzyme reactions in heterogeneous systems. Adv. Enzymol. 33:245–308.

McLaren, A.D., and A. Pukite. 1973. Ubiquity of some soil enzymes and isolation of soil organic matter with urease activity. p. 187–193. *In* D. Povoledo and M.L. Goltermaan (ed.) Humic substances and function in the biosphere. Pudoc, Wageningen, Netherlands.

Nannipieri, P., B. Ceccanti, S. Cervelli, and P. Sequi. 1974. Use of 0.1 M sodium pyrophosphate to extract urease from a soil. Soil Biol. Biochem. 6:355–362.

Nannipieri, P., B. Ceccanti, S. Cervelli, and P. Sequi. 1978. Stability and kinetic properties of humus-urease complexes. Soil Biol. Biochem. 10:143–147.

Nannipieri, P., B. Ceccanti, S. Cervelli, and E. Matarese. 1980. Extraction of phosphatase, urease, protease, organic carbon and nitrogen from soil. Soil Sci. Soc. Am. J. 44:1011–1016.

Nannipieri, P., L. Muccini, and C. Ciardi. 1983. Microbial biomass and enzyme activities: production and persistence. Soil Biol. Biochem. 15:679–685.

Nannipieri, P., B. Ceccanti, and D. Bianchi. 1988. Characterization of humus-phosphatase complexes extracted from soil. Soil Biol. Biochem. 20:683–691.

Nannipieri, P., B. Ceccanti, and S. Grego. 1990. Ecological significance of the biological activity in soil. p. 293–366. *In* J.M. Bollag and G. Stotzky (ed.) Soil biochemistry. Vol. 6. Marcel Dekker, New York, NY.

Nannipieri, P., and J.M. Bollag. 1991. Use of enzymes to detoxify pesticide-contaminated soils and waters. J. Environ. Qual. 20:510–517.

Nannipieri, P. 1994. The potential use of soil enzymes as indicators of productivity, sustainability and pollution. p. 238–244. *In* C.E. Pankhurst, B.M. Doube, V.V.S.R. Gupta and P.R. Grace (ed.) Soil biota. Management in sustainable farming systems. CSIRO, Adelaide, Australia.

Nannipieri, P., P. Fusi, and P. Sequi. 1996. Humus and enzyme activity. p. 293–328. *In* A. Piccolo (ed.) Humus substances in terrestrial ecosystems. Elsevier, Amsterdam, Netherlands.

Nannipieri, P., I. Sastre, L. Landi, M.C. Lobo, and G. Pietramellara. 1996. Determination of extracellular neutral phosphomonoesterase activity in soil. Soil Biol. Biochem. 28:107–112.

Nannipieri, P., and L. Gianfreda.1997. Kinetics of enzyme reactions in soil environments. *In* P.M. Huang, N. Senesi and J. Buffle (ed.) Environmental particles. Vol. 5. Analytical and physical chemistry of soils. John Wiley and Sons, New York, NY.

Paulson, K.N., and L.T. Kurtz. 1969. Locus of urease activity. Proc. Am. Soil Sci. Soc. 33:897–901.

Perez Mateos, M., S. Gonzales Carcedo, and M.D. Busto Nunez. 1988. Extraction of catalase from soil. Soil Sci. Soc. Am. J. 52:408–411.

Rowell, M.J., J.N. Ladd, and E.A. Paul. 1973. Enzymatically active complexes of proteases and humic acid analyses. Soil Biol. Biochem. 5:699–700.

Ruggiero, P., and V.M. Radogna. 1984. Properties of laccase in humic-enzyme complexses. Soil Sci. 138:74–87.

Sarkar, J.M., and R.G. Burns. 1984. Synthesis and properties of D-glucosidase phenolic copolymers as analogues of soil humic-enzyme complexes. Soil Biol. Biochem. 16:619–625.

Sarkar, J.M. 1986. Formation of (^{14}C) cellulase-humic complexes and their stability in soil. Soil Biol. Biochem. 18:251–254.

Schaffer, A. 1993. Pesticide effects on enzyme activities in the soil ecosystems. p. 273–340. *In* J.M. Bollag and G. Stotzky. (ed.) Soil biochemistry. Vol. 8. Marcel Dekker, New York, NY.

Sequi, P. 1974. Gli enzimi del terreno. L'Italia Agricola. 112: 80–101.

Shcherbakova T.A., A.A. Mas'ko, and V.F. Galushko. 1981. Extraction of organo-mineral complexes with enzymatic activity from soil. Soviet Soil Sci. (English translation) 13:45–52.

Skujins, J.S. 1976. Extracellular enzymes in soil. CRC Crit. Rev. Microbiol. 4:383–427.

Stevenson, F.J. 1982. Humus chemistry. Genesis, composition, reactions. John Wiley and Sons, New York, NY.

Stotzky, G. 1986. Influence of soil mineral colloids on metabolic processes, growth, adhesion, and ecology of microbes and viruses p. 305–428. *In* P.M. Huang and M. Schnitzer (ed.) Interactions of soil minerals and natural organics and microbes. Soil Science Society of America, Madison, WI.

Tabatabai, M.A. 1973. Michaelis constant of urease in soils and soil fractions. Soil Sci. Soc. Am. Proc. 37:707–710.

Tabatabai, M.A., and M. Fu. 1992. Extraction of enzymes from soils. p. 197–227. *In* G. Stotzky and J.M. Bollag (ed.) Soil biochemistry. Vol. 7. Marcel Dekker, New York, NY.

Tabatabai, M.A.1994. Soil enzyme. p. 775–833. *In* R.W. Weaver, S. Angle, P. Bottomley, D. Bezdicek., S. Smith, A. Tabatabai and A. Wollem (ed.) Methods of soil analysis. Part 2. Soil Science Society of America, Madison, WI.

Tateno, M. 1988. Limitations of available substrates for the expression of cellulase and protease activities in soil. Soil Biol. Biochem. 20:117–118.

Weaver, R.W., S. Angle, P. Bottomley, D. Bezdicek., S. Smith, A. Tabatabai, and A. Wollem. 1994. Methods of soil analysis. Part 2. Microbiological and biochemical properties. Soil Science Society of America, Madison, WI.

4

Nitrogen Transformations

4.1 Dinitrogen Fixation
Peter H. Graham
University of Minnesota

4.1.1 Importance of Dinitrogen Fixation within the Global N Cycle

Biological dinitrogen (N_2) fixation is, after photosynthesis, the second most important biological process on earth. Estimates of total N_2 fixation vary, but the figure of 175 Tg yr^{-1} (Burns and Hardy, 1975) is still widely accepted. N_2 fixation accounts for 65% of the N currently required for agriculture, and in Brazil alone is equivalent to applying some 2.5 million Mg fertilizer N yr^{-1}, a saving of $1.8 billion yr^{-1} (Dobereiner et al., 1995). Initially emphasized predominantly for grain and pasture legumes, N_2 fixation is now of increasing interest for forest (Dommergues, 1995), sugar cane (Boddey, 1995; Boddey and Dobereiner, 1995) and rice (Roger, 1995) production. Its contribution to oceanic and tidal basin N flux is also under study (Lipschultz and Owens, 1996; Bergman et al., 1997). Progress in our understanding of this field has been rapid, but a number of available technologies have still to be applied at the field level. Adoption of these technologies will be key to sustaining or improving agricultural production at a time when population pressures again threaten food self-sufficiency in third world countries, and pollution, fossil energy and cost concerns limit fertilizer N production and use.

4.1.2 Organisms Capable of N_2 Fixation

The ability to fix N_2 is restricted to prokaryotic organisms, but within this group, occurs in organisms that are taxonomically diverse including heterocystous (*Anabaena*, *Nostoc*) and nonheterocystous (*Trichodesmium*, *Gloeocapsa* [Bergman et al., 1997]) cyanobacteria, actinomycetes (*Frankia*), and heterotrophic (*Azotobacter*, *Bacillus*), autotrophic (*Thiobacillus*), aerobic (*Pseudomonas*, *Methylosinus*), anaerobic (*Clostridium*, *Desulfovibrio*) and phototrophic (*Chlorobium*, *Rhodospirillum*) bacteria.

Young (1992) provides a detailed listing of N_2-fixing species. The diversity of organisms known to fix N_2, the plasmid location of *nif* genes in some organisms, and evidence of *nif* gene conservation in organisms as different as *Anabaena* and *Frankia*, led Normand and Bousquet (1989) to propose horizontal transfer of this trait among species of bacteria. More recent studies provide evidence for both horizontal and vertical transfer of *nif* genes (Hirsch et al., 1995). N_2-fixing prokaryotes can (1) live free in nature (*Azotobacter*, *Trichodesmium*), (2) occur in loose or associative symbiosis with other plants or animals (*Acetobacter* or *Herbaspirillum* and sugar cane (Boddey and Dobereiner, 1995), or (3) enter into symbiosis with their host and be housed within specialized structures such as *Rhizobium* and the legume nodule (Graham, 1997) and *Anabaena* and the water fern, *Azolla* (Wagner, 1997).

4.1.3 Measurement of Dinitrogen Fixation

Four methods are in common use for the measurement of N_2 fixation; the N difference method, acetylene reduction assay, xylem solute analysis, and isotopic procedures based on ^{15}N. These differ in sensitivity, appropriateness for species other than plants, integration of N_2 fixation over time, and cost. Weaver and Danso (1994) and Danso (1995) review these methods, but tend to emphasize their application in agricultural situations. This account will emphasize those methods used for other organisms and environments.

The only method to directly measure N_2 fixation employs $^{15}N_2$, and evaluates incorporation of ^{15}N into test organisms using emission or mass spectrometry. This defines whether an organism can or cannot fix N_2, but is expensive and inappropriate for field experimentation.

Other methods based on ^{15}N depend on the remarkably constant ratio of $^{15}N:^{14}N$ in the atmosphere (0.3663% ^{15}N:99.6337% ^{14}N), and apply fertilizer, organic material or nutrient solutions with a ^{15}N:^{14}N ratio very different from that of the atmosphere. Specimens that do not fix N_2 will have $^{15}N:^{14}N$ ratios close to that of their substrate, while the ability to fix N_2 will result in sample $^{15}N:^{14}N$ ratios closer to that of the atmosphere. A variant of this approach, the natural abundance method, uses the fact that soil has a higher $^{15}N:^{14}N$ ratio than the atmosphere. Thus, vascular rainforest epiphytes from Brazil, Australia and the Solomon Islands, were generally lower in ^{15}N than the tree species with which they were associated (Stewart et al., 1995). This difference was consistent with a dependence on N_2 fixation as the principal source of N. Carpenter et al. (1997) have also used this approach to demonstrate significant N_2 fixation by the cyanobacterium *Trichodesmium* in the Sargasso and Caribbean seas.

The acetylene reduction assay is also widely used to measure N_2 fixation in biological samples other than plants. The assay depends on the ability of the enzyme nitrogenase to reduce acetylene to ethylene (Hardy et al., 1968), with the ethylene produced then measured by gas chromatography. Equivalent rates of N_2 fixation can then be calculated. Problems with this technique limit its use in symbiotic N_2 fixation (Vessey, 1994; Minchin et al., 1994; Danso, 1995), but the assay is widely employed in identifying N_2-fixing organisms, and in the measurement of N_2 fixation in cyanobacterial blooms (Lipschultz and Owens, 1996) or extreme environments (Solheim et al., 1996).

4.1.4 Nitrogen Fixation *in vitro* and *in vivo*

Levels of nonsymbiotic N_2 fixation achieved by organisms in pure culture vary. Thus, Azotobacter can fix up to 350 μg N mL^{-1} d^{-1}, and grows well with N_2 as the only source of N, whereas *Clostridium butyricum* fixes only 13.6 μg N mL^{-1} d^{-1}, and needs combined N for normal growth (Alexander, 1977). Rhizobia, the group of bacteria responsible for symbiotic N_2 fixation in legumes such as clovers and peas, fix no N_2 in pure culture, though N_2 fixation has been demonstrated in *Bradyrhizobium* and *Azorhizobium* (Kurz and LaRue, 1975; Pagan et al., 1975; Gebhardt et al., 1984). As evident from the equation (Allen et al., 1994):

$$N_2 + 10H^+ + nMgATP + 8e^- \rightarrow 2NH_4 + H_2 + nMgADP + nP_I (n \geq 16) \qquad [4.1.1]$$

N_2 fixation is an energy requiring process, with fixation by nonsymbiotic organisms under field conditions often energy limited. Thus, Vitousek (1994) found nonsymbiotic N_2 fixation in primary succession in the Hawaii Volcanoes National Park to vary from only 0.3 to 2.8 kg N ha^{-1} yr^{-1}, while Yegorov (1996) found only 3 kg N$_2$ fixed ha^{-1} yr^{-1} in cultivated podsolic soils. The situation in rice culture, tidal flats or intermittently inundated areas, where photosynthetic bacteria and cyanobacteria can play a significant role, is not as limiting (Joye and Paerl, 1994; Roger, 1995; Grimm and Petrone, 1997). Joye and Paerl (1994) obtained N$_2$ fixation rates in exposed mud flats and marsh surface sediments of 6-79 mg N m^{-2} d^{-1}, while traditional wetland rice culture allows moderate but sustainable yields because of N$_2$ fixation rates that average 30 kg N ha^{-1} crop cycle^{-1} (Roger, 1995). Approximately two-thirds of the fixation in rice paddies is due to photosynthetic bacteria and cyanobacteria. In a review of 634 studies in which rice was inoculated with cyanobacteria, Roger (1995) noted an average yield increase of 11.3%, but found that wide variation in response limited acceptance of this practice on the part of farmers.

A number of N$_2$-fixing bacteria can establish in the rhizosphere or within the host tissue of monocotyledonous plants (Dong et al., 1994; James et al., 1994; Olivares et al., 1996). Rates of N$_2$ fixation in such associative symbioses are extremely variable, and assessing N benefits to the host can sometimes be complicated by plant hormones released from the microsymbiont (Kucey, 1988; Bashan et al., 1989). Bremer et al. (1995) found N$_2$ fixation in wheat to be of little significance, but Okon and Labandera-Gonzalez (1994) in reporting on 20 years of experimentation with *Azospirillum* inoculation of cereals noted inoculation responses of from 5-30% in almost two-thirds of the trials. Boddey and Dobereiner (1995) cite similar results for corn and *Brachiaria*. The most striking results, however, have been with sugar cane, where some cultivars have been shown to fix as much as 170 kg N ha^{-1} yr^{-1} (Urquiaga et al., 1992). *Acetobacter diazotrophicus* and *Herbaspirillum seropedicae*, the microsymbionts in sugar cane (Baldani et al., 1986: Li and MacRae, 1992), have now been detected in other hosts (Olivares et al., 1996).T riplett (1996) has now suggested the exploitation of these endophytes in corn as the easiest approach to the establishment of N$_2$ fixation in this crop.

Rates of symbiotic N$_2$ fixation in legumes vary with plant species and cultivar, growing season, soil fertility and crop fertilization. Rates reported for forage legumes have reached 600 kg ha^{-1} yr^{-1} (Sears et al., 1965: Stevenson, 1986), but more commonly range between 100 and 300 kg ha^{-1} yr^{-1} (Vance, 1997). Rates achieved by grain legumes are generally lower. Average rates of fixation for lupine, field pea and chickpea in Southern Australia were 165, 83 and 70 kg N ha^{-1} growing season^{-1} (Unkovich et al., 1997), while in Fairbanks, Alaska, fixation rates for lupine, *Lens*, *Vicia faba* and *Pisum sativum* were 140-202, 21-91, 179-215, and 65-144 kg ha^{-1} growth season^{-1} respectively (Sparrow et al., 1995). Generally similar rates are reported in the review by Peoples et al. (1995). N$_2$ fixation rates in soybean are strongly affected by maturity group (Hardy et al., 1973), but range from 80 to more than 200 kg N ha^{-1} growth cycle^{-1} (Neves et al., 1985; Herridge et al., 1990; Keyser and Li, 1992).

The extensive rooting of tree species can complicate measurement of N$_2$ fixation. In *Prosopis*, nodule formation can occur to a depth of 4-6 m below the soil surface, with the rhizobia recovered from surface nodules markedly different from those occurring in nodules at depth (Jenkins et al., 1987). Dommergues (1995) reports nodule mass in tree species of from 34-500 g tree^{-1}, with rates of N$_2$ fixation of from 4.5 to 42.2 g tree^{-1} yr^{-1}.

4.1.5 Recent Progress in N$_2$ Fixation Research

4.1.5.1 Improving Levels of N$_2$ Fixation in Crop Plants
Cultivar variation affects levels of N$_2$ fixation in many crop species. This includes rice (App et al., 1986; Ladha et al., 1987), corn (Boddey and Dobereiner, 1995), sugarcane (Urquiarga et al., 1992)

and a number of legume species (Buttery et al., 1992; Keyser and Li, 1992; Herridge and Danso, 1995), including some trees (Toky et al., 1995). In the study of Urquiarga et al. (1992) the cane varieties CB45-3 and SP70-1143 maintained yields equivalent to 200 Mg cane ha^{-1} yr^{-1} for three years without N supplementation, and with 60% plant N derived from fixation, while most other varieties showed a decrease in yield with time. Dobereiner et al. (1995) discuss the impact of indirect selection for N$_2$ fixation on rates achieved by sugarcane and soybean in Brazil, but provide no genetic basis for this program. No parallel focus has been attempted in the United States, though some gains have been made in improving soybean N$_2$ fixation simply because yield selection was carried out on soils low in N (Coale et al., 1985; Cregan and Yaklich, 1986). N$_2$ fixation in crop and pasture legumes appears moderately heritable (Burias and Planchon, 1990; Pazdernik et al., 1997; Vance, 1997) and a number of programs are beginning to emphasize this trait. The challenge will be to identify and combine different factors contributing to overall N$_2$ fixation, in what is usually considered a quantitatively inherited trait. Molecular markers are likely to play an important role in such research.

4.1.5.2   Infection of Legumes and Other Hosts

Nodulation in legumes results from molecular signaling between host and rhizobia. This signaling has been the subject of intensive research in recent years, and has already been extensively reviewed (Denarie et al.,1996; Long, 1996; Pueppke, 1996; Spaink, 1996). This research focus culminated recently in the sequencing of the sym plasmid from the promiscuous *Rhizobium* strain NGR234 (Freiberg et al., 1997). Both common and host specific nodulation genes have been identified, with the majority of them only expressed in the presence of an appropriate host. Expression requires the presence of flavonoids exuded from the root of the homologous host, and results in the formation of lipo-oligosaccharides or nod-factors by the homologous *Rhizobium* or *Bradyrhizobium*. The nod-factors not only induce root hair deformation and curling at concentrations less than 10^{-9} *M*, but as well effect a number of morphological changes associated with nodule formation. Specificity in nodulation arises from differences in the flavonoid(s) able to induce nod-gene expression in particular rhizobia, and from structural differences in the lipo-oligosaccharide nod factors produced. Infection with *Rhizobium* also results in the formation of numerous nodulins, proteins not present in significant quantity in either host or microbe, but expressed during symbiosis. Early nodulins function in infection, and in nodule development and morphogenesis; later nodulins (e.g., hemoglobin and the enzyme uricase) play a role in nodule function and N$_2$ fixation (Pawlowski, 1997).

While much of the initial research on infection emphasized rhizobia, parallels and differences with other symbiotic or associative symbioses are emerging (Pawlowski and Bisseling, 1996; Akkermans and Hirsch, 1997). Many actinorhizal plants also undergo root hair infection. Germinating seeds of red alder produce flavonoid-like compounds which influence nodulation by *Frankia* (Benoit and Berry, 1997) and culture filtrates from *Frankia* elicit root hair deformation (Ghelue et al., 1997). Nodulation mutants of *Phaseolus vulgaris*, *Pisum sativum* and *Vicia faba* may also be resistant to infection with mycorrhizal fungi (Duc et al., 1989; Shirtliffe and Vessey, 1996). Nodulins expressed during legume/*Rhizobium* symbiosis have also been detected in mycorrhizal plants (Wyss et al., 1990). *Acetobacter diazotrophicus* is normally present within the setts used to propagate sugarcane, but can also colonize the rhizosphere of its host, then gain entry between cells in the root cap, or through wounds left by lateral root emergence. Peanut rhizobia also use this mode of entry. Within this host, *A. diazotrophicus* appears surrounded by a mucilaginous material (Dong et al., 1994; James et al., 1994). This perhaps serves to prevent recognition by the host and the initiation of defense responses, as occurs in legume nodules.

4.1.5.3   The Ecology of N₂-Fixing Bacteria

Recent developments in molecular biology have resulted in improved techniques for studying the diversity of N_2 fixing organisms in soil and water. Techniques include restriction fragment length polymorphism (RFLP), the polymerase chain reaction (PCR), and multilocus enzyme electrophoresis (MLEE) (Sadowsky and Graham, 1998). Again, emphasis in such studies has been on the rhizobia, but increasingly is being directed toward other N_2-fixing organisms. In one study with rhizobial isolates from *Phaseolus vulgaris* in the Mesoamerican center of origin of this crop, Pinero et al. (1988) found a mean genetic distance per enzyme locus of 0.691 among the 51 isolates tested, indicating considerable biodiversity. Many of the rhizobia in these soils were noninfective (Segovia et al., 1991) but survived saprophytically. The diversity revealed in molecular studies has led in a number of cases to modifications in the taxonomy of the rhizobia, as detailed by Young and Haukka (1996). Host and environment can also influence soil populations. Thus, Hirsch (1996) found *R. leguminosarum* to survive better than *Sinorhizobium meliloti* in Rothamsted soil in the absence of the homologous host, while Anyango et al. (1995) found *R. etli* to occupy 95% of the nodules formed on beans in a Kenyan soil of pH 6.8, but to be essentially replaced by *R. tropici* in a soil of pH 4.5.

Less diversity has been found in strains of *Acetobacter diazotrophicus* (Caballero-Mellado and Martinez-Romero, 1994; Caballero-Mellado et al., 1995), *Frankia* (Nazaret et al, 1991; Rouvier et al., 1996; Nalin et al., 1997) and *Anabaena azollae* (Caudales et al., 1995; van Coppenolle et al., 1995). Caballero-Mellado et al. (1995) identified only 7 electrophoretic types among 55 strains of *Acetobacter diazotrophicus*, but noted that Brazilian strains were more variable than those from Mexico, perhaps because of the lower dependence there on fertilizer N. In both *Frankia* and *Anabaena*, phylogenetic groupings parallel host plant relationships (Caudales et al., 1995; Rouvrier et al., 1996), and are suggestive of co-evolution between host and microbe (Caudales et al., 1995). Additional studies with other plant species and in other environments will undoubtedly expand the level of diversity in these organisms.

4.1.5.4   Chemistry of Dinitrogen Fixation

The enzyme, nitrogenase, which catalyses the reduction of atmospheric N_2 to NH_4 comprises two O labile proteins (Allen et al., 1994; Vance, 1997): (1) dinitrogenase (MoFe protein or component 1), an $\alpha_2\beta_2$ tetramer of MW ~220,000, which contains two [MoFe^{6}S^8] clusters per molecule; and (2) dinitrogenase reductase (Fe protein or component 2), an α_2 dimer of MW ~60,000, and containing a single [4Fe-4S] cluster plus two MgATP binding sites. The Fe protein donates electrons to the MoFe protein.

Nine *nif* genes are needed for the synthesis of the MoFe and Fe proteins; additional *fix* genes provide both positive and negative regulation of N_2-fixing activity (Fischer, 1994).

While dinitrogenase in most organisms contains Mo, alternate dinitrogenases in which V replaces Mo, or Fe alone is present have been identified. Homocitrate is also an integral component of dinitrogenase (Hoover et al., 1989). Dinitrogenase reduction occurs one electron at a time (Lowe and Thorneley, 1984; Thorneley and Lowe, 1984), with concomitant hydrolysis of MgATP (Equation [4.1.1]).

Elucidation of the crystal structure of dinitrogenase and dinitrogenase reductase (Kim and Rees, 1992; Bolin et al., 1993) has not shed light on the exact functioning of this system, though Leigh (1995) does propose a model for nitrogenase function, and more recently has commented on the possibility of N_2 caging within nitrogenase (Leigh, 1997)

Both dinitrogenase and dinitrogenase reductase are O labile, the latter having a half-life in air of 0.5 to 0.75 sec. (Zuberer, 1997). N_2-fixing organisms have developed a number of mechanisms to avoid exposure of these enzymes to O_2. These include barriers to O transport and use of leghemoglobin as an O carrier in legumes (Witty and Minchin, 1994; Iannetta et al., 1995; James et al., 1996), heterocyst formation, dark phase N_2 fixation and other structural changes in cyanobacteria (Wolk, 1996; Fredrickson and Bergman,1997), and respiratory protection in *Azotobacter* and other organisms (Prosperi, 1994). The need for protection against oxygen can have interesting ecological consequences. Thus Joye and Paerl (1994) found that N_2 fixation in mud flats of the Tomales Bay region of California only exceeded denitrification during the daylight hours, suggesting an important role for heterocystous species in the N economy of that environment.

4.1.6   Constraints to the Implementation of N_2 Fixation Technologies

Many of the techniques needed to make N_2 fixation an integral part of sustainable agricultural production are already available. In Brazil, as stated earlier, inoculation with bradyrhizobia has been a major factor in the successful expansion of soybean production, while N fertilizer usage in sugar cane production is only one-third that of countries such as Mexico, Venezuela and the United States (Dobereiner et al., 1995). Few countries have been as successful. Hall and Clark (1995) have reviewed the impact of rhizobial inoculants for soybean in three areas of Thailand. They note high adoption rates in new areas of production, but also report significant use of chemical N in older production areas where indigenous rhizobia have established, and inoculation is of little benefit. Surprisingly in this study, farmers sometimes chose to use inoculants for reasons very different from those extolled by scientists and extension workers. In the midwestern United States, soybean plants may satisfy less than 50% of their N needs from symbiosis.

Similar problems affect all of the biological N_2 fixation technologies discussed in this section and include (1) poor quality inoculant strains limited in ability to fix N_2, or with marked varietal or species specificities. This is a particular problem with *Frankia* where cross inoculation is still under study, and inoculants are often prepared from crushed nodules. Graham (1997), amongst others, has listed desirable traits in inoculant rhizobia; (2) inoculants with cell counts affected by contaminant organisms or additives, inappropriate carrier material or packaging, or exposed to adverse conditions during shipment; (3) the use of fertilizer N application rates likely to interfere with nodulation and/or nitrogen fixation; (4) competition between inoculant and indigenous organisms for establishment in the soil. Roger (1995) comments on the nonestablishment of inoculant cyanobacteria, and Streeter (1994) on the parallel problem with rhizobia; (5) limited strain persistence in soil. *Acetobacter diazotrophicus* survives poorly in soil; the saprophytic competence of some clover and bean rhizobia has been questioned (Brockwell et al., 1982; Vlassak et al., 1996); and (6) soil or environmental conditions not appropriate for active N_2 fixation. Phosphorus deficiency is a major factor in both symbiotic and nonsymbiotic N_2 fixation; soil acidity and drought are common concerns.

Even population pressures and social change can impact N_2 fixation technologies. Thus, Roger (1995) notes a decline in the area sown to *Azolla* in China from 6.5 million ha before 1979 to only 0.7 million ha in 1982. This was attributed to cheap sources of urea, and to economic policies leading to the disbanding of communes and the redistribution of labor. In the United States, the development of sustainable agricultural systems that depend on N_2 fixation rather than N fertilization will require a significant change in research orientation, and better integration of research, extension and legislative programs.

4.1.7   References

Akkermans, A.D.L., and A.M.Hirsch. 1997. A reconsideration of terminology in *Frankia* research: A need for congruence. Physiol. Plant 99:574–578.

Alexander, M.A. 1977. Introduction to soil microbiology. Wiley, New York, NY.

Allen, R.M., R. Chatterjee, M.S. Madden, P.W. Ludden, and V.K. Shar. 1994. Biosynthesis of the iron-molybdenum cofactor of nitrogenase. Crit. Rev. Biotech. 14:225–249.

Anyango, B., K.J. Wilson, J.L. Beynon, and K.E. Giller. 1995. Diversity of rhizobia nodulating *Phaseolus vulgaris* L. in two Kenyan soils with contrasting pHs. Appl. Environ. Microbiol. 61:4016–4021.

App, A.A., I. Watanabe, T. Santiago-Ventura, M. Bravo, and C. Daez-Jurey. 1986. The effect of cultivated and wild rice varieties on the nitrogen balance of flooded soil. Soil Sci. 141:448–452.

Baldani, J.I., V.L.D. Baldani, L. Seldin, and J. Dobereiner. 1986. Characterization of *Herbaspirillum seropedicae* gen. nov., sp. nov., a root-associated nitrogen-fixing bacterium. Intern. J. Syst. Bacteriol. 36:86–93.

Bashan, Y., M. Singh, and H. Levanony. 1989. Contribution of *Azospirillum brasilense* Cd to growth of tomato seedlings is not through nitrogen fixation. Can. J. Bot. 67:2429–2434.

Benoit, L.F., and A.M.Berry. 1997. Flavonoid-like compounds from seeds of red alder (*Alnus rubra*) influence host nodulation by *Frankia* (Actinomycetales). Physiol. Plant. 99:588–593.

Bergman, B., J.R. Gallon, A.N. Rai, and L.J. Stal. 1997. N_2 fixation by non-heterocystous cyanobacteria. FEMS Microbiol. Revs. 19:139–185.

Boddey, R.M. 1995. Biological nitrogen fixation in sugar cane: A key to energetically viable biofuel production. Crit. Rev. Plant Sci. 14:263–279.

Boddey, R.M., and J. Dobereiner. 1995. Nitrogen fixation associated with grasses and cereals: Recent progress and perspectives for the future. Fert. Res. 42:241–250.

Bolin, J.T., N. Campobasso, S.W. Muchmore, T.V. Morgan, and L.E. Mortenson. 1993. The structure and environment of the metal clusters in the nitrogenase MoFe protein from *Clostridium pasteurianum*. p. 186–195. *In* E.I. Stiefel et al. (ed.) Molybdenum enzymes, cofactors and model systems. American Chemical Society, Washington, DC.

Bremer, E., H.H. Janzen, and C. Gilbertson. 1995. Evidence against associative N_2 fixation as a significant N source in long-term wheat plots. Plant Soil 175:13–19.

Brockwell, J., R.R. Gault, M. Zorin, and M.J. Roberts. 1982. Effects of environmental variables on the competition between inoculum strains and naturalized populations of *Rhizobium trifolii* for nodulation of *Trifolium subterraneum* L. and on rhizobia persistence in the soil. Aust. J. Agric. Res. 33:803–815.

Burias, N., and C. Planchon. 1990. Increasing soybean productivity through selection for nitrogen fixation. Agron. J. 82:1031–1034.

Burns, R.C., and R.W.F. Hardy. 1975. Nitrogen fixation in bacteria and higher plants. Springer Verlag, Berlin, Germany.

Buttery, B.R., S.J. Park, and D.J. Hume. 1992. Potential for increasing nitrogen fixation in grain legumes. Can. J. Plant Sci. 72:323–349.

Caballero-Mellado, J., L.E. Fuentes-Ramirez, V.M. Reis, and E. Martinez-Romero. 1995. Genetic structure of *Acetobacter diazotrophicus* populations and identification of a new genetically distant group. Appl. Environ. Microbiol. 61:3008–3013.

Caballero-Mellado, J., and E. Martinez-Romero. 1994. Limited genetic diversity in the endophytic sugarcane bacterium *Acetobacter diazotrophicus*. Appl. Environ. Microbiol. 60:1532–1537.

Carpenter, E.J., H.R. Harvey, B. Fry, and D.G. Capone. 1997. Biogeochemical tracers of the marine cyanobacterium *Trichodesmium*. Deep sea research. Oceanogr. Res. Papers 44:27–38.

Caudales, R., J.M. Wells, A.D. Antoine, and J.E. Butterfield. 1995. Fatty acid composition of symbiotic cyanobacteria from different host plant (*Azolla*) species: Evidence for coevolution of host and microsymbiont. Intern. J. Syst. Bacteriol. 45:364–370.

Coale, F.J., J.J. Meisinger, and W.J. Wiebold. 1985. Effects of plant breeding and selection on yields and nitrogen fixation in soybean under two soil nitrogen regimes. Plant Soil 86:357–367.

Cregan, P.B., and R.W. Yaklich. 1986. Dry matter and nitrogen accumulation and partitioning in selected soybean genotypes of different derivation. Theor. Appl. Gen. 72:782–786.

Danso, S.K.A. 1995. Assessment of biological nitrogen fixation. Fert. Res. 42:33–41.

Denarie, J., F. Debelle, and J.C. Prome. 1996. *Rhizobium* lipo-chito oligosaccharide nodulation factors: Signaling molecules mediating recognition and morphogenesis. Ann. Rev. Biochem. 65:503–535.

Dobereiner, J, S. Urquiaga, and R.M. Boddey. 1995. Alternatives for nitrogen nutrition of crops in tropical agriculture. Fert. Res. 42:339–346.

Dommergues, Y.R. 1995. Nitrogen fixation by trees in relation to soil nitrogen economy. Fert. Res. 42:215–230.

Dong, Z., M.J. Canny, M.E. McCully, M.R. Roboredo, C.F. Cabadilla, E. Ortega, and R. Rodes. 1994. A nitrogen-fixing endophyte of sugarcane stems. Plant Physiol. 105:1139–1147.

Duc, G., A. Trouvelot, V. Gianinazzi-Pearson, and S. Gianinazzi. 1989. First report of non-mycorrhizal plant mutants (myc⁻) obtained in pea (*Pisum sativum* L.) and fababean (*Vicia faba* L.). Plant Sci. 60:215–222.

Fischer, H.M. 1994. Genetic regulation of nitrogen fixation in rhizobia. Microbiol. Revs. 58:352–386

Fredrickson, C., and B. Bergman. 1997. Ultrastructural characterization of cells specialized for nitrogen fixation in a non heterocystous cyanobacterium, *Trichodesmium* spp. Protoplasma 197:76–85.

Freiberg, C., R. Fellay, A. Belroch, W.J. Broughton, A. Rosenthal, and X. Perret. 1997. Molecular basis of symbiosis between *Rhizobium* and legumes. Nature 387:394–401.

Gebhardt, C., G.L. Turner, A.H. Gibson, B.L. Dreyfus, and F.J. Bergersen. 1984. Nitrogen-fixing growth in continuous culture of a strain of *Rhizobium* sp isolated from stem nodules of *Sesbania rostrata*. J. Gen. Microbiol. 130:843–848.

Ghelue, M.V., E. Lovaas, E. Ringo, and B. Solheim. 1997. Early interactions between *Alnus glutinosa* and *Frankia* strain ArI3. Production and specificity of root hair deformation factor(s). Physiol. Plant. 99:579–587.

Graham, P.H. 1997. Biological dinitrogen fixation: Symbiotic. p, 322–345. *In* D. Sylvia et al. (ed.) Principles and applications of soil microbiology. Prentice Hall, New York, NY.

Grimm, N.B., and K.C. Petrone. 1997. Nitrogen fixation in a desert stream ecosystem. Biogeochem. 37:33–61.

Hall, A., and N. Clark. 1995. Coping with change, complexity and diversity in agriculture: The case of *Rhizobium* inoculants in Thailand. World Devel. 23:1601–1614.

Hardy, R.W.F., R.C. Burns, and R.D. Holsten. 1973. Applications of the acetylene- ethylene assay for measurement of nitrogen fixation. Soil Biol. Biochem. 5:47–81.

Hardy, R.W.F., R.D. Holsten, E.K. Jackson, and R.C. Burns. 1968. The acetylene-ethylene assay for N_2 fixation: Laboratory and field evaluation. Plant Physiol. 43:1185–1207.

Herridge, D.F., F.J. Bergersen, and M.B. Peoples. 1990. Measurement of nitrogen fixation by soybean in the field using the ureide and natural ^{15}N abundance methods. Plant Physiol. 93:708–716.

Herridge, D.F., and S.K.A. Danso. 1995. Enhancing crop legume N_2 fixation through selection and breeding. Plant Soil 174:51–82.

Hirsch, A.M., H.I. McKhann, A. Reddy, J. Liao, Y. Fang, and C.R. Marshall. 1995. Assessing horizontal transfer of *nif*HDK genes in Eubacteria: Nucleotide sequence of *nif*K from *Frankia* strain HFPCc13. Mol. Biol. Evol. 12:16–27.

Hirsch, P.R. 1996. Population dynamics of indigenous and genetically modified rhizobia in the field. New Phytol. 133:159–171.

Hoover, T.R., J. Imperial, P.W. Ludden, and V.K. Shah. 1989. Homocitrate is a component of the iron-molybdenum cofactor of nitrogenase. Biochem. 28:2768–2771.

Iannetta, P.P.M., E.K. James, J.I. Sprent, and F.R. Minchin. 1995. Time course of changes involved in the operation of the oxygen diffusion barrier in white lupin nodules. J. Exp. Bot. 46:565–575.

James, E.K., P.P.M. Iannetta, P.J. Nixon, A.J. Whiston, L. Peat, R.M.M. Crawford, J.I. Sprent, and N.J. Brewin. 1996. Photosystem II and oxygen regulation in *Sesbania rostrata* stem nodules. Plant Cell Environ. 19:895–910.

James, E.K., V.M. Reis, F.L. Olivares, J.I. Baldani, and J. Dobereiner. 1994. Infection of sugar cane by the nitrogen-fixing bacterium *Acetobacterium diazotrophicus*. J. Exp. Bot. 45:757–766.

Jenkins, M.B., R.A. Virginia, and W.M. Jarrell. 1987. Rhizobial ecology of the woody legume mesquite (*Prosopis glandulosa*) in the Sonoran desert. Appl. Environ. Microbiol. 53:36–40

Joye, S.B., and H.W. Paerl. 1994. Nitrogen cycling in microbial mats- rates and patterns of denitrification and nitrogen fixation. Marine Biol. 119:285–295.

Keyser, H.H., and F. Li. 1992. Potential for increasing biological nitrogen fixation in soybean. Plant Soil 141:119–135.

Kim, J., and D.C. Rees. 1992. Structural models for the metal centers in the nitrogenase molybdenum-iron protein. Science 257:1677–1682.

Kucey, R.M.N. 1988. Plant growth-altering effects of *Azospirillum brasilense* and *Bacillus* C-11-25 on two wheat cultivars. J. Appl. Bacteriol. 64:187–196.

Kurz, W.G.W., and T.A. LaRue. 1975. Nitrogenase activity in rhizobia in absence of host plant. Nature 256:407–409.

Ladha, J.K., A. Tirol, G.C. Punzalan, and I. Watanabe. 1987. N_2- fixing (C_2H_2-reducing) activity and plant growth characters of 16 wetland rice varieties. Soil Sci. Plant. Nut. 33:187–200.

Leigh, G.J. 1995. The mechanism of dinitrogen reduction by molybdenum nitrogenases. Europ. J. Biochem. 229:14-20.

Leigh, G.J. 1997. Biological nitrogen fixation and model chemistry. Science 275:1442.

Li, R., and I.C. MacRae. 1992. Specific identification and enumeration of *Acetobacter diazotrophicus* in sugar cane. Soil Biol. Biochem. 24:413–419.

Lipschultz, F., and N.J.P. Owens. 1996. An assessment of nitrogen fixation as a source of nitrogen to the North Atlantic Ocean. Biogeochem. 35:261–274.

Long, S.R. 1996. *Rhizobium* symbiosis: Nod factors in perspective. Plant Cell 8:1885–1898.

Lowe, D.J., and R.N.F.Thorneley. 1984. The mechanism of *Klebsiella pneumoniae* nitrogenase action. Pre-steady-state kinetics of H_2 formation. Biochem. J. 224:877–886.

Minchin, F.R., J.F. Witty, and L.R. Mytton. 1994. Reply "Measurement of nitrogenase activity in legume root nodules: In defense of the acetylene reduction" by J.K.Vessey. Plant Soil 158:163–167.

Nalin, R., P. Normand, and A.M. Domenach. 1997. Distribution and N_2-fixing activity of *Frankia* strains in relation to soil depth. Physiol. Plant.99:732–738.

Nazaret, S., B. Cournoyer, P. Normand, and P. Simonet. 1991. Phylogenetic relationships among *Frankia* genomic species determined by use of amplified 16S rDNA sequences. J. Bacteriol. 173:4072–4078.

Neves, M.C.P., A.D. Didonet, F.F. Duque, and J. Dobereiner. 1985. *Rhizobium* strain effects on nitrogen transport and distribution in soybeans. J. Exp. Bot. 36:1179–1192.

Normand, P., and J. Bousquet. 1989. Phylogeny of nitrogenase sequences in *Frankia* and other nitrogen-fixing microorganisms. J. Mol. Evol. 29:436–444.

Okon, Y., and C.A. Labandera-Gonzalez. 1994. Agronomic applications of *Azospirillum*: An evaluation of 20 years worldwide field inoculation. Soil Biol. Biochem. 26:1591–1601.

Olivares, F.L., V.L.D. Baldani, V.M. Reis, J.I. Baldani, and J. Dobereiner. 1996. Occurrence of the endophytic diazotrophs *Herbaspirillum* spp. in roots, stems and leaves, predominantly of Gramineae. Biol. Fertil. Soils. 21:197–200.

Pagan, J.D., J.J. Child, W.R. Scowcroft, and A.H. Gibson. 1975. Nitrogen fixation by *Rhizobium* cultured on a defined medium. Nature 256:406–407.

Pawlowski, K. 1997. Nodule-specific gene expression. Physiol. Plant. 99:617–631.

Pawlowski, K., and T. Bisseling. 1996. Rhizobial and actinorhizal symbioses: What are the shared features? Plant Cell 8:1899–1913.

Pazdernik, D.L., P.H. Graham, and J.H. Orf. 1997. Heritability in the early nodulation of F_3 and F_4 soybean lines. Can. J. Plant Sci. 77:201–205.

Peoples, M.B., D.F. Herridge, and J.K. Ladha. 1995. Biological nitrogen fixation: An efficient source of nitrogen for sustainable agricultural production. Plant Soil 174:3–28.

Pinero, D., E. Martinez, and R.K. Selander. 1988. Genetic diversity and relationships among isolates of *Rhizobium leguminosarum* biovar phaseoli. Appl. Environ. Microbiol. 54:2825–2832.

Prosperi, C.H. 1994. The role of respiration as a protection for nitrogenase activity in the cyanobacterium *Anabaena flos-aquae*. Phyton. 55:159–162.

Pueppke, S.G. 1996. The genetic and biochemical basis for nodulation of legumes by rhizobia. Crit. Rev. Biotech. 16:1–51.

Roger, P.A.. 1995. Biological N_2 fixation and its management in wetland rice cultivation. Fert. Res. 42:261–276.

Rouvier, C., Y. Prin, P. Reddell, P. Normand, and P. Simonet. 1996. Genetic diversity among *Frankia* strains nodulating members of the family Casuarinaceae in Australia revealed by PCR and restriction fragment length polymorphism analysis with crushed nodules. Appl. Environ. Microbiol. 62:979–985

Sadowsky, M.J., and P.H. Graham. 1998. Soil biology of the rhizoshpere. p. 155–172. *In* H.P. Spank, A. Kondorosi and P.J.J. Hooykaas (ed.) The Rhizobiaceae. Kluwer Academic Publishers, Dordrecht, Netherlands.

Sears, P.D., V.C. Goodall, R.H. Jackman, and G.S. Robinson. 1965. Pasture growth and soil fertility. 8. The influences of grasses, white clover, fertilizers, and the return of herbage clippings on pasture production of an impoverished soil. NZ J. Agric. Res. 8:270–283.

Segovia, L., D. Pinero, R. Palacios, and E. Martinez-Romero. 1991. Genetic structure of a soil population of nonsymbiotic *Rhizobium leguminosarum*. Appl. Environ. Microbiol. 57:426–433.

Shirtliffe, S.J., and J.K.Vessey. 1996. A nodulation (Nod+/Fix-) mutant of *Phaseolus vulgaris* L. has nodule-like structures lacking peripheral vascular bundles (Pvb-) and is resistant to mycorrhizal infection (Myc-). Plant Sci. 118:209–220.

Solheim, B., A. Endahl, and H. Vigstadt. 1996. Nitrogen fixation in arctic vegetation and soils from Svalbard, Norway. Polar Biol. 16:35–40.

Spaink, H.P. 1996. Regulation of plant morphogenesis by lipo-chitin oligosaccharides. Crit. Rev. Plant Sci. 15:559-582.

Sparrow, S.D., V.L. Cochran, and E.B. Sparrow. 1995. Dinitrogen fixation by seven legume crops in Alaska. Agron. J. 87:34–41.

Stevenson, F.J. 1986. Cycles of soil. Wiley Interscience, New York, NY.

Stewart, G.R., S. Schmidt, L.L. Handley, M.H. Turnbull, P.D. Erskine, and C.A. Joly. 1995. ^{15}N natural abundance of vascular rainforest epiphytes: Implications for nitrogen source and acquisition. Plant Cell Environ. 18:85–90.

Streeter, J.G. 1994. Failure of inoculant rhizobia to overcome the dominance of indigenous strains for nodule formation. Can. J. Microbiol. 40:513–522.

Thorneley, R.N.F., and D.J. Lowe. 1984. The mechanism of *Klebsiella pneumoniae* nitrogenase action. Pre-steady-state kinetics of an enzyme-bound intermediate in N_2 reduction and of NH_3 formation. Biochem. J. 224:887–894.

Toky, O.P., N. Kaushik, and P.K. Sharma. 1995. Genetic variability in progenies of *Acacia nilotica* (L) ex del ssp indica (Benth.) Brenan for nitrogen fixing ability. Silvae Genet. 44:161–165.

Triplett, E.W. 1996. Diazotrophic endophytes: progress and prospects for nitrogen fixation in monocots. Plant Soil 186:29–38.

Unkovich, M.J., J.S. Pate, and P. Sanford. 1997. Nitrogen fixation by annual legumes in Australian Mediterranean agriculture. Aust. J. Agric. Res. 48:267–293.

Urquiaga, S., K.H.S. Cruz, and R.M. Boddey. 1992. Contribution of nitrogen fixation to sugar cane: Nitrogen-15 and nitrogen balance estimates. Soil Sci. Soc. Amer. J. 56:105–114

van Coppenolle, B., S.R. McCouch, I. Watanabe, N. Huang, and C. van Hove. 1995. Genetic diversity and phylogeny analysis of *Anabaena azollae* based on RFLPs detected in *Azolla-Anabaena azollae* DNA complexes using *nif* gene probes. Theor. Appl. Genet. 91:589–597.

Vance, C.P. 1997. Legume symbiotic nitrogen fixation: Agronomic aspects. p. 509–530. *In* H.P.Spaink, A. Kondorosi and P.J.J. Hooykaas (ed.) The Rhizobiaceae. Kluwer Academic Publishers, Dordrecht, Netherlands.

Vessey, J.K. 1994. Measurement of nitrogenase activity in legume root nodules: In defense of the acetylene reduction assay. Plant Soil 158:151–162.

Vitousek, P.M. 1994. Potential nitrogen fixation during primary succession in Hawaii Volcanoes National Park. Biotropica 26:234–240

Vlassak, K., J. Vanderleyden, and A. Franco. 1996. Competition and persistence of *Rhizobium tropici* and *Rhizobium etli* in tropical soil during successive bean (*Phaseolus vulgaris* L.) cultures. Bio. Fertil. Soil 21:61–68.

Wagner, G.M. 1997. *Azolla*: A review of its biology and utilization. Bot. Rev. 63:1–26.

Weaver, R.W., and S.K.A. Danso. 1994. Dinitrogen fixation. p. 1019–1045. *In* R.W.Weaver et al. (ed.) Methods of soil analysis, Part 2. Microbiological and biochemical properties. Soil Science Society of America, Madison, WI.

Witty, J.F., and F.R. Minchin. 1994. A new method to detect the presence of continuous gas-filled pathways for oxygen diffusion in legume nodules. J. Exp. Bot. 45:967–978.

Wolk, C.P. 1996. Heterocyst formation. Ann. Rev. Genet. 30:59–78.

Wyss, P., M.B. Mellor, and A. Wiemken. 1990. Vesicular arbuscular mycorrhizas of wild-type soybeans and non-nodulating mutants with *Glomus mosseae* contain symbiosis specific polypeptides (mycorrhizins) immunologically cross reactive with nodulins. Planta 182:22–26.

Yegorov, V.I. 1996. Nitrogen regime in cultivated podzolic soils of the Kola north. Europ. J. Soil Sci. 26:98–107.

Young, J.P.W. 1992. Phylogenetic classification of nitrogen-fixing organisms. p. 43–86. *In* G. Stacey et al. (ed.) Biological nitrogen fixation. Chapman and Hall, New York, NY.

Young, J.P.W., and K.E. Haukka. 1996. Diversity and phylogeny of rhizobia. New Phytol. 133:87–94.

Zuberer, D.A. 1997. Biological dinitrogen fixation: Introduction and nonsymbiotic. p. 295–321. *In* D. Sylvia et al. (ed.) Principles and applications of soil microbiology. Prentice Hall, New York, NY.

4.2   Nitrogen Mineralization Immobilization Turnover

Jeanette M. Norton

Utah State University

4.2.1   Introduction and Definitions

The transformations between organic and inorganic N form a central part of the internal soil N cycle. Mineralization is the general term for the conversion of organic to inorganic N as either NH_4^+ or NO_3^-, ammonification is the conversion of organic N to NH_4^+, while nitrification (Section 4.3) is the conversion of NH_4^+ to NO_3^-. Immobilization is the assimilation of inorganic N to organic N mediated by microorganisms. An overview of the soil N-cycle transformations is given in Fig. 4.1. Mineralization-immobilization turnover (MIT) refers to the combined transformations between

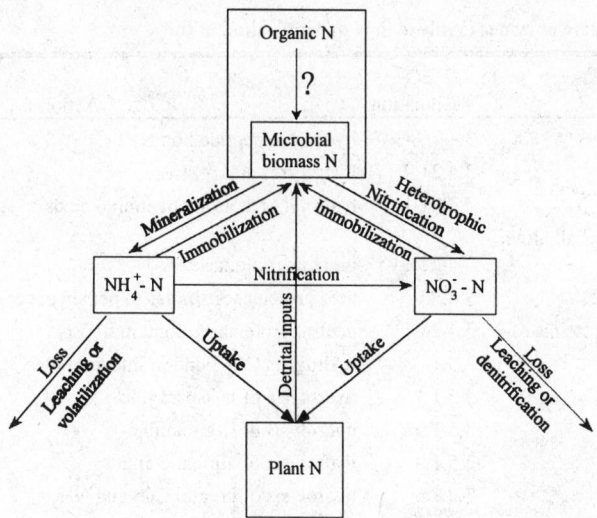

Fig. 4.1 Nitrogen transformations in the soil environment

organic and inorganic N that accompany the growth and death of the soil microbial biomass. To what extent MIT in soils necessitates passage through the NH_4^+ pool is the subject of current debate (Barraclough, 1997). However, microbial physiology would indicate that small N containing compounds (amino acids, amines, etc.) can be transported into microbial cells when available without prior release of the N to the NH_4^+ pool. The amount of N produced by mineralization is a major control on N availability to plants especially in unfertilized systems (Binkley and Hart, 1989; Powlson and Barraclough, 1993; Mengel, 1996).

4.2.2 Biochemistry and Physiology of Mineralization

As organic N, including decaying microorganisms, is assimilated by the microbial biomass, N in excess of microbial requirements is released as NH_4^+ or mineralized. N can be released from organic N by the action of extracellular enzymes (primarily microbial in origin) including the proteinases, chitinases, kinases, amidases and amidohydrolases. The substrates for mineralization include a broad range of proteins, microbial cell wall constituents (amino sugars and their polymers, chitin and peptidoglycan) and nucleic acids. Nitrogen is also mineralized from SOM which contains various N compounds including a large amount of heterocyclic and phenolic N which is only slowly decomposed.

Mineralization of organic N is basically a sequence of enzymatic reactions. Some of the specific enzymes that are involved in mineralization and their reactions are summarized in Table 4.1. Since proteins and peptides are a major source of mineralizable N (Mengel, 1996), the enzymes that can degrade these compounds are produced abundantly and in a wide diversity by soil microorganisms. Microorganisms produce both endo- and exopeptidases such as trypsin and chymotrypsin which cleave preferentially at specific amino acid residues. The bacterial subtilisins are alkaline proteinases that were originally characterized from *Bacillus subtilis*. Subtilisins and related enzymes belong to the class of proteinases in which a serine residue is involved at the active site and are also known as alkaline serine proteinases. Fungi produce many carboxyl or acid proteinases (formerly proteases) and many of these are useful commercially. The purified acid proteinases have pH optima between pH 2

Table 4.1   Representative enzymes involved in N mineralization in soil

Enzyme	EC designation	Action
Proteinases/peptidases	3.4	hydrolyze peptide bonds
trypsin	3.4.21.4	hydrolyzes at Arg, Lys
chymotrypsin	3.4.21.1	hydrolyzes at aromatic amino acids
subtilisin (microbial alkaline proteinase)	3.4.21.14	alkaline proteinase
carboxyl proteinases	3.4.23	acid proteinases similar to pepsin or rennin
microbial metaloproteinases	3.4.24	neutral proteinases containing Zn
Amidohydrolases	3.5.1	hydrolysis C-N bond in linear amides
L-asparaginase	3.5.1.1	hydrolysis of L-asparagine
L-glutaminase	3.5.1.2	hydrolysis of L-glutamine
amidase	3.5.1.4	hydrolysis of aliphatic amides
urease	3.5.1.5	hydrolysis of urea to CO_2 and NH_3
peptidoglycan hydrolase	3.5.1.28	hydrolysis of peptidoglycan linkages
Amidinohydrolases	3.5.3	hydrolysis C-N bond of linear amidines
arginase	3.5.3.1	hydrolysis of arginine to ornithine and urea
Transaminases	2.6.1	transfers amine to a-ketoglutarate
Dehydrogenases which deaminate	1.4	
Glutamate dehydrogenase	1.4.1.3	deaminates glutamate
Glycosidases	3.2	important in hydrolysis of amino sugars and their polymers
chitinase	3.2.1.14	hydrolyze chitin linkages
muramidase	3.2.1.17	hydrolyze muropeptide (peptidoglycan) or mucopolysaccharide
Nucleases	3.1.11-16	degrade DNA and RNA
DNase	3.1.11	exodeoxyribonuclease
RNase	3.1.13	exoribonuclease

and 5. The release of N from nonpeptide C-N bonds in amino acids and urea is mediated by amidohydrolases, dehydrogenases such as glutamate dehydrogenase (Fig. 4.2) and transaminases. The polymers of amino sugars that are important components of bacterial and fungal cell walls are hydrolyzed by glycosidases. The D-isomers of alanine, glutamine and glutamic acid which serve as crosslinks in peptidoglycan are degraded more slowly than the common L-isomers and may accumulate in soil (Hopkins, 1996). The observation that free nucleic acids are degraded rapidly substantiates the ubiquitous presence of the nucleases in the soil environment.

Specific enzyme activities may be measured in soils by the addition of the substrate in excess, incubation under optimal conditions with inhibition of microbial growth and determination of product accumulation or substrate depletion (Tabatabai, 1994). These assays measure the enzymatic potential or V_{max} under the conditions chosen, not necessarily the process rate *in situ* (Section 4.2.5). Details for enzyme assays of L-asparaginase, L-glutaminase, amidase and urease in soil are given in Tabatabai (1994).

The activity of urease in soil has been widely studied because urea is the most important N fertilizer in world agriculture (Bremner, 1995) and a major component of animal wastes. Urease catalyzes the following reaction:

$$CO(NH_2)_2 + H^+ + 2H_2O \rightarrow 2NH_4^+ + HCO_3^- \qquad [4.2.1]$$

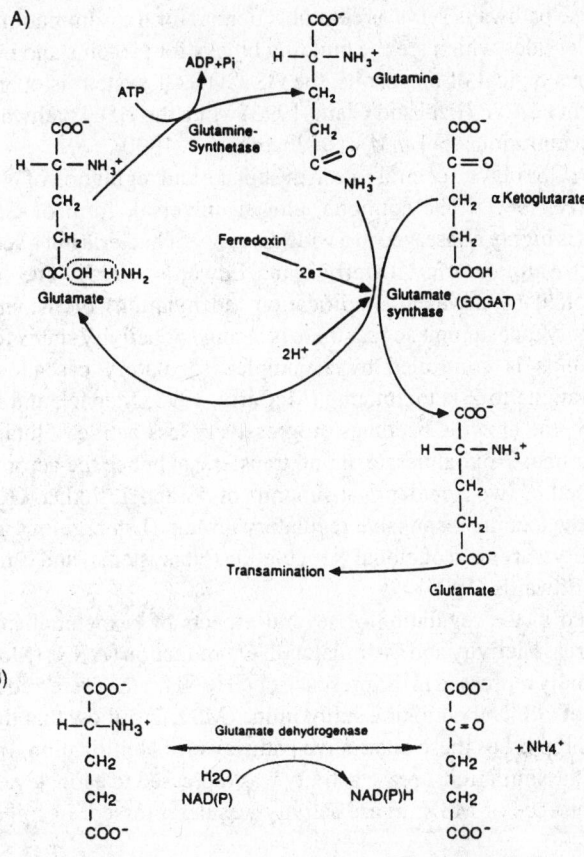

Fig. 4.2 Nitrogen assimilation pathways (A) Glutamine sythetase-glutamine synthase (GS-GOGAT) high affinity pathway (B) Glutamate dehydrogenase reversible low affinity pathway for NH_4^+ incorporation or release [Reprinted from Paul and Clark, 1996. Soil microbiology and biochemistry, with permission from Academic Press, Inc.]

In warm moist soils, urea will be hydrolyzed to NH_4^+ within several days. The NH_4^+ released is then readily available for immobilization, nitrification or for loss by NH_3 volatilization. The high potential for N loss from urea has emphasized the importance of urease inhibitors in N management (Bremner, 1995).

4.2.3 Biochemistry and Physiology of Immobilization

The incorporation of N into the microbial biomass and organic N occurs through numerous enzymatic and abiotic pathways. The preferred inorganic N source for assimilation by bacteria and fungi is NH_3/NH_4^+ although NO_3^- is also used under appropriate conditions. Typically, NH_3 enters microbial cells by rapid diffusion across cytoplasmic membranes although there is now evidence of NH_4^+ active transport in several bacteria (Merrick and Edwards, 1995). Whether active transport of NH_4^+ is important in the soil environment remains to be investigated. Biological NH_4^+ immobilization by soil microorganisms is mainly accomplished by two enzymatic pathways: glutamine synthetase/glutamate synthase (GS-GOGAT system) and glutamate dehydrogenase (GDH) (Fig. 4.2). The glutamate and glutamine formed by these pathways serve as central N donors for transamination reactions yielding the amino acids and nucleotides which are the building blocks for proteins and nucleic acids. At the lower NH_4^+ concentrations typical of most soils, the GS-GOGAT system is operative requiring the input of energy in the form of ATP (Paul and Clark, 1996) while the GDH pathway immobilizes N at relatively high NH_4^+ concentrations (> 1 mM) (Neidhardt et al., 1990).

Glutamine synthetase (GS) plays a central role in the uptake and regulation of N assimilation in soil microorganisms (McCarty, 1995). The common, almost universal, form of GS composed of 12 identical 55 kDa subunits is highly conserved in a wide diversity of bacteria but a second form has been found in *Rhizobium* and *Agrobacterium* (Merrick and Edwards, 1995). The NH_4^+-incorporating activity of GS is controlled by covalent modification (adenylation) of the individual subunits; adenylation inactivates only that subunit so that there is a range of activity states for the enzyme. The adenylation of the subunits is controlled by a complex regulatory cascade which senses the intracellular ratio of glutamine to 2-ketoglutarate (McCarty, 1995; Merrick and Edwards, 1995) so that, under N sufficiency, the enzyme becomes progressively less active. Glutamate synthase was formerly known as glutamine: 2-oxoglutarate amino transferase; hence the acronym GOGAT. In *E. coli*, GOGAT is comprised of two nonidentical subunits of 53 and 135 kDa. GOGAT synthesis is subject to repression by the leucine responsive regulatory protein (Lrp regulon) under conditions of high amino acid availability. A review of global N regulation (Ntr systems) and N transport in bacteria is given by Merrick and Edwards (1995).

GS activity is involved in the regulation of several aspects of N metabolism in soil microbial communities including urease activity and assimilatory NO_3^- reduction (ANR) (McCarty, 1995). Both urease and ANR are normally repressed in the presence of NH_3/NH_4^+, the preferred N source. McCarty (1995) used an inhibitor of GS, L-methionine sulfoximine (MSX), to show that the production of L-glutamine by GS activity represses these alternative pathways for N utilization. In soils treated with glucose to stimulate NH_4^+ assimilation, urease activity was repressed to a low level in the presence of added NH_4^+ but in the presence of MSX urease activity was derepressed. In soil slurries, ANR was

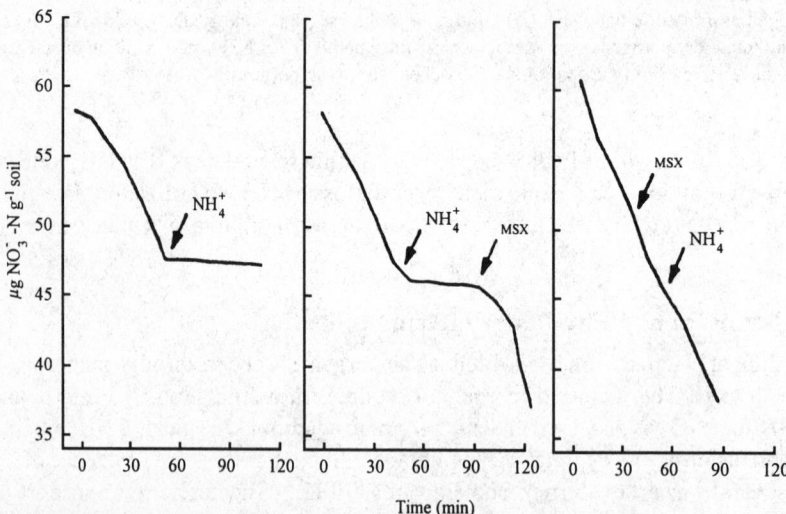

Fig. 4.3 Effect of L-methonine sulfoximine (MSX) on assimilatory reduction of NO_3^- in aerated slurries of Harps soil treated with glucose. Slurries treated with NH_4^+ (0.12 mg N g^{-1} soil) and MSX (1 mol g^{-1} soil) at the times indicated [Reprinted from McCarty, 1995. Plant Soil 170:141-147. Copyright with kind permission from Kluwer Academic Publishers]

repressed in response to NH_4^+ addition, but when MSX was added, ANR resumed (Fig. 4.3). Both of these studies emphasize the key role of L-glutamine production by GS as an important regulatory point for N assimilation in soil microorganisms.

Glutamate dehydrogenase can function bidirectionally to immobilize or mineralize NH_4^+ (Fig. 4.2). The direction of catalysis is controlled by N availability; GDH typically functions to immobilize N only under high NH_3 availability due to the relatively high Km (1mM) for this enzyme. Inside the cell, GDH can catalyze the oxidative deamination of L-glutamate leading to the mineralization of N. In *E. coli*, GDH is a hexamer of identical 50 kDa subunits which are encoded by the *gdhA* gene.

Nitrate may be immobilized directly by both bacteria and fungi by ANR (Paul and Clark, 1996). Energy is required for the transport of NO_3^- across the cell membrane prior to reduction to NH_4^+ by the ANR system. The enzymes responsible for reduction are assimilatory nitrate reductase and assimilatory nitrite reductase. The pathway of ANR is

$$NO_3^- \rightarrow NO_2^- \rightarrow NH_2OH \rightarrow NH_4^+ \rightarrow R-NH_2 \text{ (Paul and Clark, 1996)} \qquad [4.2.2]$$

The synthesis of the ANR enzymes (nitrate and nitrite reductases) is regulated by the global Ntr system and inducible by NO_3^- (Merrick and Edwards, 1995) but repressed by NH_4^+ (Rice and Tiedje, 1989; Recous et al., 1990), or glutamine (McCarty, 1995).

Recent soil investigations have shown that both direct assimilation of the amino acids, leucine and glycine, and extracellular deamination and subsequent immobilization can occur in parallel (Barraclough, 1997). The mechanisms of organic N assimilation into the soil microbial biomass are a subject for current investigation.

4.2.4   Substrate Quality Effects on Mineralization-Immobilization Turnover

Substrates for decomposition are of varying chemical constitution and complexity. These differences affect the rate at which microorganisms process the material and the amount of N that is required or released during microbial growth on the substrate. The substrates which are decomposed include materials that are inputs to the soil such as plant debris, animal manure and industrial wastes and those which are internal to the soil system such as SOM and microbial decay products. The availability of the substrate for decomposition, the efficiency of the microbial C assimilation and the C/N ratio of the microbial biomass determine the requirement for N during substrate decomposition (Janssen, 1996). If excess N is present in the substrate relative to the microbial demand, then net N mineralization will result; conversely, when the N required by microorganisms is greater than that present in the substrate, then microorganisms will immobilize inorganic N from the soil solution. While the actual situation during decomposition is extremely complex with numerous fractions of varying composition and availability, there is a long history of the use of simple characteristics as indicators for substrate quality and in estimation of the N demand during decomposition. The C/N ratio and the N content (N%) of the substrate have often served as indicators of the potential for immobilization. Table 4.2 gives some representative C/N ratios for various organic materials and Fig. 4.4 illustrates a simplified calculation of the first round of decomposition of two contrasting substrates, alfalfa residues and wheat straw. Generally, substrates with C/N ratios < 20 or N contents > 1.5% N result in net N mineralization while those higher than 30/1 or with N contents < 1.5% tend to immobilize soil inorganic N (Tisdale et al., 1993). This oversimplification does not represent substrates with large proportions of resistant materials such as lignin which may mineralize N at C/N ratios up to 50 due to their slow decay rates (Paul and Clark, 1996). As decomposition of a substrate proceeds, the C/N ratio of the material remaining in the soil (a combination of the remaining substrate and the microbial products) will

Table 4.2 C/N ratios in various organic materials [Tisdale et al., 1993; Hyvönen et al., 1995]

Organic material	C/N ratio
Soil microorganisms	8/1
Soil organic matter	10/1
Sewage sludge	9/1
Alfalfa residues	16/1
Farmyard manure	20/1
Corn stover	60/1
Grain straw	80/1
Oak	200/1
Pine	300/1
Crude oil	400/1
Conifer sawdust	625/1

decrease and N will eventually be mineralized. Models of C and N turnover in soil have used the lignin to N ratio of the substrate as an indicator of substrate quality. Substances with high lignin to N ratios are considered to contain more structural materials which degrade more slowly with lower microbial efficiencies (Paustian et al., 1992).

4.2.5 Process Rates and Kinetics

4.2.5.1 Net, Gross, and Potential Mineralization and Immobilization Rates

The common uses of the term mineralization rate include at least three different definitions. The net mineralization rate is the rate of NH_4^+ and NO_3^- accumulation being the difference between the rates of conversion and consumption of organic N to NH_4^+/NH_3 and NO_3^-. Net mineralization refers to the case where the inorganic N pool is increasing with time due to higher gross rates of mineralization than immobilization. The gross mineralization rate is the actual rate of conversion of organic N to NH_4^+/NH_3 and is the same as the sum of the rates of NH_4^+ accumulation and consumption. Determination of gross rates requires the use of N isotopes to separate simultaneous production and consumption. The potential mineralization rate is the net rate of mineralization in the absence of plant uptake and leaching (the potential rate is defined by the actual method chosen for its determination). Previously, many investigators assumed that measured net mineralization rates in the absence of plant uptake and leaching were adequate indicators of the importance and rate of mineralization in the environment. This assumption neglected the potential role of inorganic N assimilation by microorganisms and denitrification. Significant rates of inorganic N assimilation are often measured simultaneously with net mineralization (Powlson and Barraclough, 1993).

Immobilization is defined as the assimilation of inorganic N (NH_4^+/NO_3^-) into the microbial biomass. Net immobilization refers to the case where the inorganic N pool is decreasing with time due to higher gross rates of immobilization than mineralization. The gross immobilization rate is the actual rate of conversion of inorganic N to organic N and requires the use of N isotopes to separate simultaneous production and consumption. Gross immobilization is only a portion of the total NH_4^+ consumption which also includes nitrification and plant root uptake. Inorganic N consumption includes NH_4^+ and NO_3^- immobilization by heterotrophic microorganisms, plant root uptake, NO_3^- leaching and denitrification. The distinction between total consumption of inorganic N and NH_4^+ consumption and immobilization becomes particularly important in the determination of gross immobilization rates. A schematic diagram of the processes involved in the internal N cycle is given in Fig. 4.1.

4.2.5.2 Measurements of Mineralization and Immobilization Rates

Although numerous measurements of soil mineralization rates have been made in a variety of ecosystems, the majority are net rates; relatively few estimates are available for gross mineralization/immobilization rates on undisturbed soil cores. The basic principles and application of N isotopes for determination of gross mineralization and immobilization rates were developed several decades ago (Kirkham and Bartholomew, 1954, 1955; Jansson, 1958) but recent innovations in direct combustion isotope ratio mass spectrometry (Hauck et al., 1994) have facilitated the use of ^{15}N for estimating gross rates.

The determinations of NH_4^+, NO_2^- and NO_3^- concentrations in soils (Hart et al., 1994a; Bundy and Meisinger, 1994) are the starting points for almost all the rate measurements. The inorganic N is extracted using a strong salt solution, filtered and NH_4^+, NO_2^-/NO_3^- determined using colorimetric methods (Bundy and Meisinger, 1994). Glassware should be acid washed and all solutions made with high quality deionized water to prevent N contamination. The soil should be extracted as soon as possible after sampling as inorganic N levels change rapidly in disturbed samples. For isotopic work, samples should be extracted in the field rather than after transport to the laboratory.

4.2.5.3 Laboratory Methods for Net and Potential Mineralization Rates

Laboratory incubation experiments have a long history of being used to examine net mineralization in the absence of plant uptake and under conditions that minimize gaseous and leaching losses (Stanford and Smith, 1972). These types of experiments are most useful for comparisons of the N supplying potential of closely related soils or a soil which has received various amendments or treatments. Typically, the soils are sieved and homogenized before incubation; however, intact cores may be a more realistic estimate of field mineralization rates (Cabrera and Kissel, 1988). The artificial conditions of constant temperature and controlled moisture and the reliance on net rates minimize the

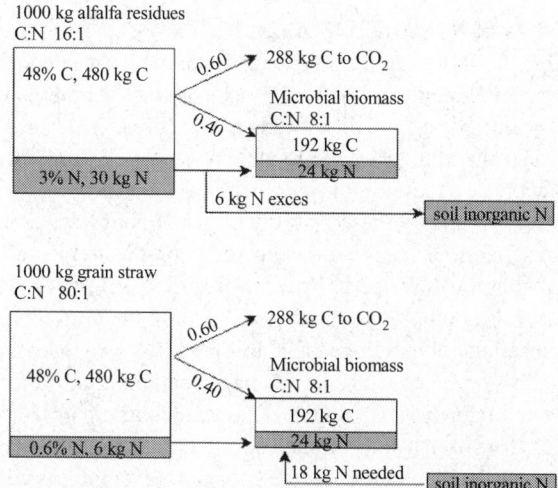

Fig. 4.4 Idealized C and N relationship in the first round of decomposition of a high N alfalfa residue versus a low N wheat straw. Assume a microbial efficiency of 40% and a C/N ratio of the microbial biomass is 8/1.

relation of mineralization potential measurements to gross mineralization rates in the field. The many variations of the laboratory incubation method (Stanford and Smith, 1972; Hart et al., 1994a; Bundy and Meisinger, 1994) may give contrasting results on different soils. The basic premise is that initial soil inorganic N determinations are made, the soil is incubated under a defined temperature and moisture regime, and the amount of inorganic N produced during a specified time period is determined. Details of incubation methods which determine NH_4^+ production during waterlogged incubation and inorganic N production during long-term aerobic incubation with leaching are given by Bundy and Meisinger (1994), while an aerobic incubation method under nonleaching conditions is described by Hart et al. (1994a). The inorganic N produced may be measured once or several times over the course of a long-term incubation. Interpretations may involve first order modeling of the time course of released inorganic N to determine the pool size of the mineralizable N according to the equation (Stanford and Smith, 1972):

$$N_m = N_o\left(1 - e^{-kt}\right)$$ [4.2.3]

where N_m (mg kg^{-1}) is the mineralized N at time t (d), N_o is the mineralization potential (mg kg^{-1}) and k (d^{-1}) is the mineralization rate constant. Parameter estimation is made by a nonlinear least squares method (Smith et al., 1986). Representative values for net mineralization rates and first order mineralization rate constants from a variety of ecosystems are summarized by Smith and Paul (1990). Recent analysis using simulation modeling more realistically represents mineralizable N as several pools (up to 7) with different mineralization rates (Richter and Benbi, 1996). However, the description and modeling of numerous pools of mineralizable N may result in models that are overparameterized with regard to the available data (Richter and Benbi, 1996).

The arginine ammonification potential (AAP) has also been used as a laboratory indicator of a soil's ability to process organic N (Fuller and Scow, 1996). The release of NH_4^+ from arginine is measured in a short-term shaken slurry incubation. Although the AAP assay is in effect an enzyme potential assay for argininase, it has also been used as a general indicator of the size of the soil microbial biomass (Alef and Kleiner, 1986).

4.2.5.4 Field Measurements of Net Mineralization

Several methods are available for the determination of net mineralization and nitrification in the field; all include a barrier to prevent root uptake of NH_4^+ and NO_3^- and to either prevent or capture the NO_3^- in the mass flow of water. The buried bag and the closed top soil core methods are suitable procedures for field estimates of net mineralization (Hart et al., 1994a). Typically intact soil cores are used to minimize disturbance artifacts (i.e., increased mineralization) caused by sieving and mixing. The number of cores per site is determined by the variability of the site but Hart et al. (1994a) suggested minimum numbers are of 8 separate determinations on soils composited from 3 samples. Both the buried bag and the solid core method described below may result in differences in the moisture status of the soil inside the container as compared to the external soil. In the buried bag method, soil cores are removed intact and enclosed in polyethylene bags. Initial samples are taken in adjacent positions and extracted immediately for inorganic N. The buried bags containing the cores are incubated for 1 month or sometimes longer and then extracted in the same manner as the initial determinations. This technique is most appropriate for surface soils. The closed top solid core method is a variation of this approach in which the cores are incubated intact inside PVC or steel cylinders with polyethylene film or a styrofoam cup to prevent water entry (Hart et al., 1994a). Another possible variation involves the use of open topped intact cores with the bottom of the core containing a ring filled with mixed bed ion exchange resin. This technique allows water entry to the top of the core and captures N leaching (Hart et al., 1994a). Net mineralization rates are generally in the range of 0.1 to 1.0 mg N kg^{-1} d^{-1} for surface mineral soils. Net mineralization and nitrification rates have been related to N loss from forest ecosystems (Vitousek et al., 1982) and used to predict plant available N in agroecosystems (Robertson et al., 1997). Net N mineralization determinations confound the simultaneous production and consumption of inorganic N and may not be well correlated with gross rates of N processes (Hart et al., 1994b; Davidson et al., 1992).

4.2.5.5 Gross Rates of Mineralization and Immobilization: Use of Isotopes

The measurement of actual or gross rates of mineralization and immobilization requires the use of N isotopes with ^{15}N being used almost exclusively. In general, ^{15}N can be used in two contrasting ways to examine MIT: (1) tracer methods in which ^{15}N is added as the substrate for the process of interest, and (2) pool dilution methods in which ^{15}N is added to the product pool of the transformation of interest. Mineralization rates can be assessed by tracer techniques in which ^{15}N-labeled plant material or other ^{15}N organic substrates are added to the source pool and the fate of the added ^{15}N is monitored. In the pool dilution approach, ^{15}NH$_4^+$ is added and the dilution of its enrichment by the production of ^{14}NH$_4^+$ by mineralization is monitored over time. Immobilization rates may be estimated by the tracer method or by a combination of the isotope dilution and tracer techniques (Fig. 4.5).

The tracer approach for mineralization involves the preparation and use of ^{15}N-labeled organic substrates and allows the investigator to assess the mineralization potential of the material and follow the fate of the added ^{15}N. The data from these studies are particularly useful for parameterizing simulation models of MIT. Methods for labeling plants and the microbial biomass with ^{15}N are described by Wolf et al. (1994) and the preparation of organic materials for ^{15}N analysis is described by Hauck et al. (1994) and Harris and Paul (1989). The tracer approach for estimating immobilization rates involves adding ^{15}NH$_4^+$ or ^{15}NO$_3^-$ and then measuring the appearance of ^{15}N in the sink pool, usually the microbial biomass or SOM (Powlson and Barraclough, 1993). In N-limited systems, the addition of substrate may cause stimulation, and therefore an overestimate of immobilization rates (Hart et al., 1994a; Norton and Firestone, 1996). The tracer approach is particularly useful for estimating immobilization rates in the presence of several consumptive processes (i.e., plant uptake,

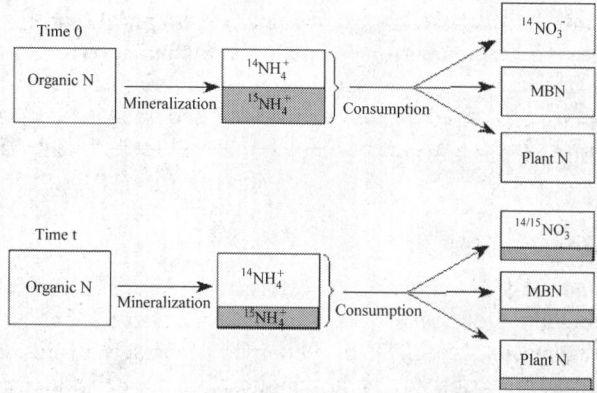

Fig. 4.5 Combined pool dilution and tracer approach for estimation of mineralization and immobilization rate. At the beginning of incubation (Time 0), ^{15}NH$_4^+$ is added and the initial pool sizes and enrichments are determined. At the end of the incubation (Time t) the final pool sizes and enrichments are determined. The mineralization and NH$_4^+$ consumption rates are determined by pool dilution calculation, NH$_4^+$ immobilization rate is determined by tracer analysis.

nitrification, NH_3 volatilization) and is often used in combination with the pool dilution approach described below to answer mechanistic questions on MIT (Schimel et al., 1989; Norton and Firestone, 1996).

The pool dilution approach for estimating mineralization and NH_4^+ consumption rates has been used for laboratory and field studies (Davidson et al., 1991, 1992; Wessel and Tietema, 1992; Hart et al., 1994a, 1994b; Norton and Firestone, 1996; Watkins and Barraclough, 1996). The pool dilution approach is ideally suited for measurements of gross mineralization rates while the estimates of NH_4^+ immobilization are less reliable due to stimulation of rates by the addition of the $^{15}NH_4^+$ substrate and the presence of several additional consumptive processes. In the typical pool dilution experiment $^{15}NH_4^+$ is added to the soil at the beginning of the incubation (time 0) and the soil is harvested and extracted after incubation (time t). Typically, one sample is harvested at the beginning of the incubation (near time 0) to correct for incomplete recoveries and rapid abiotic consumption such as that due to NH_4^+ fixation by clay minerals (Davidson et al., 1991). The rate at which the enrichment of the NH_4^+ pool declines and the changes in pool size are used to calculate the rate of production of NH_4^+ and gross NH_4^+ consumption. The rate determination can be done either by analytical solution (Kirkham and Bartholomew, 1954; Hart et al., 1994a) or by dynamic simulation modeling (Myrold and Tiedje, 1986; Wessel and Tietema, 1992; Powlson and Barraclough, 1993). The pool dilution approach does not distinguish the pathways of production and consumption. While the gross production of NH_4^+ is nearly equivalent to gross mineralization, the gross consumption of NH_4^+ is a combination of several significant processes including immobilization, nitrification, plant uptake and NH_3 volatilization. A combined pool dilution and tracer approach (Fig. 4.5) can be used to partition the total consumption into its component processes. Immobilization rates can be estimated by determining the ^{15}N incorporated into the microbial biomass or the soil organic N and appropriate modeling (Davidson et al., 1991, 1992; Norton and Firestone, 1996). The rate of nitrification may be determined from parallel NO_3^- pool dilution experiments (Section 4.3.5.5) or nitrification may be blocked using acetylene (Section 4.3.4). Plant uptake can be calculated from the ^{15}N content of the plant tissue or blocked by preventing root uptake with a barrier method. Ammonia volatilization, which is important only in alkaline soils, can be measured by including an acid trap for the volatilized NH_3 in the experiment or calculated by difference. It is recommended that total recoveries of ^{15}N in the plant-soil system be determined for both time 0 and t samples to detect any unexplained loss of ^{15}N. The specific recommendations for sampling and techniques for field pool dilution experiments in intact cores without plants are discussed by Hart et al. (1994a) and Davidson et al. (1991). The background and calculations for estimating mineralization and assimilation in the plant-soil system are given in Powlson and Barraclough (1993). Some representative rates of NH_4^+ production and consumption determined by pool dilution from agricultural and wildland ecosystems are given in Table 4.3. Few estimates for agricultural systems that are from field experiments using intact cores are available.

4.2.6 Simulation Models of MIT

Kinetic and simulation models of N turnover in the plant-soil system can be useful for understanding the interactions of the multiple processes involved. Models can also be of assistance in identifying gaps in our knowledge, typically the failure of models to adequately express the realities of the observed system is the first step towards new understanding of a neglected process. An example of this is the identification of the importance of microbial immobilization occurring simultaneously with net mineralization for the simulation of short-term inorganic N dynamics (Richter and Benbi, 1996). Nitrogen process models have become more crucial for the prediction of NO_3^- leaching after

Table 4.3   Gross mineralization and NH_4^+ immobilization rates in suface mineral soils

Location	Ecosystem type	Treatment	Gross Min.	Gross NH_4^+ Immob.	Source
			mg N kg^{-1} soil day^{-1}		
CA	young mixed conifer forest	greenhouse study	1.2	0.8	1
CA	annual grassland-open	intact cores spring	4.9	7.3*	2
CA	annual grassland canopy	intact cores spring	9.1	11.7*	2
CA	annual grassland-open	intact cores early spring	1.4	1.9	3
CA	young mixed conifer forest	intact cores spring	1.4	1.0	4
CA	>100 yr. old conifer forest	intact cores spring	5.0	4.8	4
OR	mixed conifer old growth	mixed soil laboratory	5.5	5.4	5
La Selva, Costa Rica	old growth tropical forest	intact cores 7 day	5.6	1.3	6
Hokkaido, Japan	agricultural soil	mixed soil laboratory	4.0	2.9	7
MI	northern hardwood	mixed soil laboratory	1.3	0.5-1.0	8

* probably overestimates due to abiotic consumption by clay fixation
1 = Norton and Firestone (1996), 2 = Davidson et al. (1990), 3 = Davidson et al. (1991), 4 = Davidson et al. (1992), 5 = Hart et al. (1994b), 6 = Zou et al. (1992), 7 = Nishio et al. (1985), 8 = Myrold and Tiedje (1986)

application of organic materials and mineral fertilizers; because it is no longer sufficient to predict plant needs and economic returns, environmental impacts of excess N must be considered. Models vary in their treatment of organic matter pools, number of distinct pools involved in MIT, treatment explicitly of the microbial biomass and its turnover and whether nitrification and denitrification are considered. Current N turnover models have been critically reviewed and compared for their treatment of the above factors (Willigen, 1991; Richter and Benbi, 1996). A general consensus is that improved submodels dealing with the maintenance and growth of the microbial biomass (Fig. 4.6) and particularly seasonal variation in the microbial biomass N are needed to adequately predict inorganic N dynamics and, therefore, the potential for N leaching. In general, most models link the cycling of N to the C cycle processes; C availability controls microbial biomass growth and turnover (Paustian et al., 1992). While the abiotic effects of temperature and moisture on mineralization and decomposition rates can be simulated adequately, abiotic factor effects on N immobilization by the soil microbial biomass deserve further model development. The detailed aspects of the different simulation models are beyond the scope of this chapter and the reader is referred to Willigen (1991) and Paustian et al (1992). While our conceptual models and the computer simulations of MIT in soils may never converge, the process of quantitatively formulating the complex processes in the soil environment improves our predictive ability and identifies our misconceptions.

4.2.7   Concluding Remarks

The mineralization and immobilization of N in soils process many times the net inputs and outputs of inorganic N from the soil ecosystem. Unfortunately, surprisingly few reliable estimates of MIT rates in the field and even less information on their seasonal and spatial variability are available. Identification of the controlling factors on the individual processes, not their combined outcome, will be necessary to develop an improved mechanistic understanding of N flow through the soil microbial biomass.

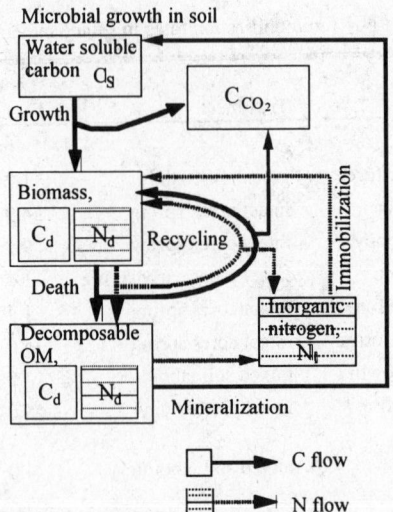

Microbial growth in soil

Fig. 4.6 An example of a submodel of microbial growth in soil and its relationship to N mineralization and immobilization [Reprinted from Richter and Benbi, 1996. Plant Soil 181:109-121, with permission from Kluwer Academic Publishers using the data and equations of Yevdokimov et al., 1993]

4.3 Nitrification

Jeanette M. Norton
Utah State University

4.3.1 Introduction and Definitions

Nitrification is defined as the biological conversion of reduced N in the form of NH_3 or NH_4^+ to oxidized N in the form of NO_2^- or NO_3^-. The conversion of the cation (NH_4^+) to an anion (NO_2^- or NO_3^-) determines the movement of N through the generally negatively charged soil matrix, and therefore, strongly influences its fate in the ecosystem. Nitrate is more likely than NH_4^+ to move rapidly by mass flow to plant roots, leach out of the root zone or be lost from the soil by denitrification (Section 4.4). For these reasons, agricultural soils are often managed to reduce nitrification in order to increase N use efficiency. In waste processing, nitrification is often a desirable process as NO_3^- is less toxic to aquatic systems than NH_4^+/NH_3 and is easily removed from the waste stream by denitrification.

In most soil environments, nitrification is mediated primarily by autotrophic bacteria which gain energy from the oxidation of N, although heterotrophic nitrification may be of importance in specialized situations (Section 4.3.2.3) (van Neil et al., 1993). In autotrophic nitrification, the conversion of N takes place in two steps: the NH_3-oxidizing bacteria such as *Nitrosomonas* convert NH_3 to NO_2^- according to:

$$NH_3 + O_2 + 2H^+ + 2e^- \rightarrow NH_2OH + H_2O \rightarrow NO_2^- + 5H^+ + 4e^- \qquad [4.3.1]$$

while the NO_2^--oxidizing bacteria such as *Nitrobacter* convert NO_2^- to NO_3^-

$$NO_2^- + H_2O \rightarrow NO_3^- + 2H^+ + 2e^- \qquad [4.3.2]$$

No bacteria have been found which can convert NH_3 to NO_3^- directly (Hooper et al., 1997). The recently described anaerobic oxidation of NH_4^+ to N_2 gas in wastewater treatment systems termed the anammox process (Jetten et al., 1997) is not considered further in this discussion.

The difficulties associated with the isolation and maintenance of pure cultures of nitrifying bacteria and the relative ease of quantifying the substrates (NH_3 /NH_4^+) and products (NO_2^-/NO_3^-) have resulted in nitrification generally being studied at the process level. Comparatively few investigations have effectively examined the community ecology or diversity of the bacteria responsible (Belser and Schmidt, 1978a; Stanley and Schmidt, 1981; Both et al., 1992; Stephen et al., 1996). The selective nature of enrichment and isolation techniques (Section 4.3.3.3) may have resulted in a biased representation of the soil nitrifier community in pure cultures. Current development of molecular techniques for directly identifying and quantifying nitrifiers in the soil environment (Section 4.3.3.4) is beginning to rectify this situation (Degrange and Bardin, 1995; Stephen et al., 1996).

The process of nitrification is a major control point in the N cycle and rates have been intensively studied in a variety of ecosystems (Section 4.3.5.1). However, the techniques used have often relied on measurements of substrate and product pool sizes, and therefore, measure the net rate of nitrification. More recently, isotopic techniques (Section 4.3.5.2) have allowed investigators to examine the turnover of NO_3^- and separate the confounding processes of NO_3^- production (nitrification) and consumption (Davidson et al., 1992; Stark and Hart, 1997). Mechanistic models of nitrification will require kinetic parameters suitable for the soil nitrifiers from a range of ecosystems. Kinetic constants from available pure cultures are unlikely to represent the values for the soil community (Stark and Firestone, 1996). Understanding various aspects of the biochemistry, physiology, and microbial ecology of the nitrifying bacteria will lead to an integrative approach to the management and interpretation of nitrification in the soil environment.

4.3.2 Biochemistry, Physiology and Molecular Biology of Nitrification

4.3.2.1 Ammonia Oxidation by Chemolithoautotrophic Bacteria

Obligate chemolithoautotrophic bacteria such as *Nitrosomonas europaea* convert NH_3 to NO_2^- via hydroxylamine and in so doing obtain all the energy needed for growth. The energy yield from this reaction is relatively small, most of which is expended in CO_2 fixation by the Calvin cycle (Wood, 1986). Thus NH_3 oxidizers live on the metabolic edge, and typically, have long generation times and slow growth rates (Prosser, 1989).

The substrate for the first enzyme in the nitrification pathway, NH_3 monooxygenase, is thought to be NH_3 rather than NH_4^+ (Suzuki et al., 1974). Urea may also serve as primary substrate for NH_3

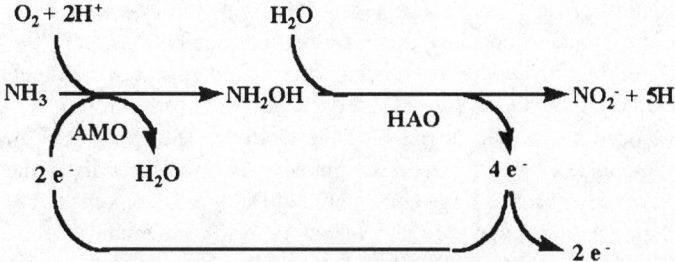

Fig. 4.7 Pathway of NH_3 oxidation to NO_2^- via the enzymes ammonia monooxygenase (AMO) and hydroxylamine oxidoreductase (HAO) in *Nitrosomonas eurpoaea* [Reprinted from Arp et al., 1996. M.E. Lidstrom and F.R. Tabita. Microbial growth on Cl compounds, with kind permission from Kluwer Academic Publishers]

oxidation in those NH_3 oxidizers which produce urease (Holt et al., 1994). A schematic of the process of NH_3 oxidation and the enzymes involved is given in Fig. 4.7 (Arp et al., 1996).

Ammonia oxidation to hydroxylamine is catalyzed by the membrane bound enzyme, ammonia monooxygenase (AMO) (Arp et al., 1996) which has not been purified in an active form so that its subunit composition and tertiary structure are not fully elucidated (Julliette et al., 1995). The ability of ^{14}C-labeled acetylene to bind to the active site of AMO was used to identify and purify the peptide containing the active site of the complex (AmoA) and the copurifying 45 kDa peptide (AmoB) (Arp et al., 1996). The genes which encode amoA and amoB have been sequenced from *N. europaea* (McTavish et al., 1993; Bergmann and Hooper, 1994) and from several other NH_3 oxidizers (Klotz and Norton, 1994, 1995; Norton et al., 1996; Alzerreca et al., 1997). Recently, a gene designated as *amoC* which encodes a membrane bound polypeptide of approximately 36 kDa has been identified in *Nitrosospira* (Klotz et al., 1997). AMO and the related particulate methane monooxygenase (pMMO) comprise a new class of Cu-containing monooxygenases (Holmes et al., 1995). AMO which is the target for several nitrification inhibitors (Section 4.3.4), catalyzes the oxidation of a wide variety of nonpolar compounds (Hooper et al., 1997); however, these reactions are cooxidations incapable of supporting growth of the organism.

The oxidation of hydroxylamine is the reaction which yields the reducing equivalents necessary for activity of ammonia monooxygenase and for the energy yielding reaction at the terminal aa_3 oxidase. This reaction is catalyzed by the highly complex periplasmic enzyme, hydroxylamine oxidoreductase (HAO) (Hooper et al., 1997). The crystal structure of purified HAO reveals that it is a trimeric molecule with a total of 24 haems arranged in a complex structure that may serve to direct electrons into the two pathways (Prince and George, 1997; Igarashi et al., 1997). The gene encoding the 63 kDa peptide of HAO from *N. europaea* has been sequenced (Sayavedra-Soto et al., 1994). Additional electron carriers involved in NH_3 oxidation include several c-cytochromes, cytochrome P460 (P460), and the Cu containing terminal aa_3 oxidase (the site of O_2 reduction) (Yamanaka, 1996; Hooper et al., 1997). The electrons which pass down this electron transport chain generate a proton motive force (Prosser, 1989). *N. europaea* can also denitrify using NO_2^- as the terminal electron acceptor under low O_2 availability (Jetten et al., 1997).

4.3.2.2   Nitrite Oxidation

The NO_2^--oxidizing bacteria are chemolithotrophs with members capable of autotrophic, mixotrophic or even heterotrophic growth (Bock and Koops, 1992; Teske et al., 1994). *Nitrobacter* strains can grow under a wide variety of conditions including aerobic/lithotrophic, aerobic/mixotrophic or anoxic/heterotrophic (Bock et al., 1988) and can grow heterotrophically on soluble organic substrates such as acetate, formate, pyruvate and glycerine while using NO_3^- as a terminal electron acceptor (Bock et al., 1988). Therefore, NO_2^- oxidation should be considered a reversible reaction (Wood, 1986) and accumulation of NO_3^- may not represent its production rates especially under anaerobic conditions. While important to consider these alternative metabolic modes, the primary energy-generating mechanism for *Nitrobacter* and related bacteria in the soil environment is still considered to be the aerobic oxidation of NO_2^- to NO_3^-. Under aerobic conditions, the oxidation of NO_2^- to NO_3^- (Reaction [4.3.2]) occurs without detectable intermediates catalyzed by the enzyme nitrite oxidoreductase (Yamanaka, 1996) with oxygen supplied by H_2O and O_2 acting as the terminal electron acceptor. Nitrite oxidoreductase is a large enzyme (254 kDa) which contains two Mo atoms, an Fe-S cluster (Fe_4S_4), a heme a, and a heme c (Yamanaka, 1996). A schematic of the electron transport system in *Nitrobacter winogradsky* is shown in Fig. 4.8.

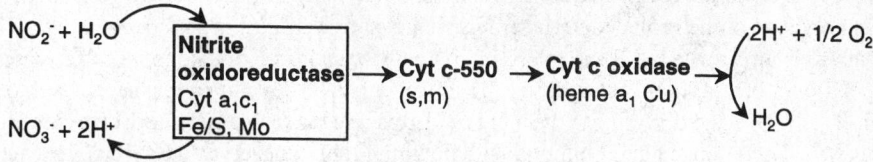

Fig. 4.8 Pathway of NO_2^- oxidation to NO_3^- catalyzed by nitrite oxidoreductase and coupling to electron transfer system in *Nitrobacter winogradsky* [Modified from Yamanaka, 1996. Plant Cell Physiol. 37:569-574 with permission]

4.3.2.3 Heterotrophic Nitrification

Heterotrophic nitrification may be broadly defined as the oxidation of reduced N compounds (including organic N) producing NO_2^- or NO_3^-. Heterotrophic nitrification is catalyzed by a variety of microorganisms including fungi, actinomycetes and bacteria presumably using a wide variety of metabolic pathways (Prosser, 1989). However, in no case has heterotrophic nitrification been shown to be associated with energy production or growth of the organism (Prosser, 1989; Jetten et al., 1997). The rates of production of NO_2^- or NO_3^- by heterotrophic nitrification have generally been much lower than autotrophic rates (Prosser, 1989; Jetten et al., 1997) but this may need to be reevaluated as heterotrophic nitrifiers such as *Thiosphaera pantotropha* (also known as *Paracoccus denitrificans*) are capable of denitrification (Jetten et al., 1997). The substrate for heterotrophic nitrification is generally amino N although some organisms have been identified that can oxidize NH_4^+ through an inorganic pathway (Prosser, 1989). The postulated inorganic and organic pathways for heterotrophic nitrification are shown in Fig. 4.9. In many forest soils, fungi have been identified as the major mediators of heterotrophic nitrification (Killham, 1986) so the ecological importance of the *T. pantropha* pathway has not been substantiated.

4.3.3 Nitrifying Organisms

4.3.3.1 Taxonomy

The two groups of autotrophic nitrifying bacteria were formerly included in the family Nitrobacteriaceae (Watson et al., 1981), defined by their ability to grow as chemolithotrophs by the oxidation of NH_3 to NO_2^- and NO_2^- to NO_3^-. Recent investigations which have examined the phylogeny of these bacteria based on their 16S ribosomal sequences have shown that the family Nitrobacteriaceae is polyphyletic with members belonging to four different subdivisions of the proteobacteria. Therefore, while convenient to group the nitrifying bacteria together, there is no phylogenetic basis for doing so except that they are all members of the large group of the division Proteobacteria (Head et al., 1993; Teske et al., 1994). A functional distinction separates the obligate autotrophic NH_3 oxidizing bacteria from the bacteria which oxidize NO_2^-, some of which can grow mixotrophically or heterotrophically. The genus names of the NH_3 oxidizers begin with *Nitroso*, while

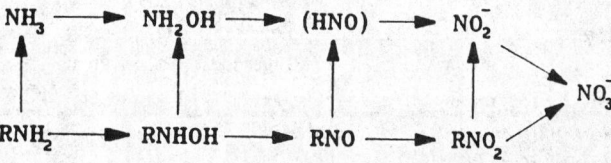

Fig. 4.9 Proposed parallel inorganic and organic pathways for heterotrophic nitrification [Reprinted from Prosser, 1989. Adv. Microbial Physiol 30:125-181, with permission from Academic Press Ltd]

those of the NO_2^- oxidizers begin with *Nitro*. The soil NH_3 oxidizing bacteria are generally restricted to the ß subdivision of the Proteobacteria, although the marine strain *Nitrosococcus oceanus* belongs in the γ subdivision. Nitrite oxidizers are found in the α, γ, and δ subdivisions. For this reason while group specific probes based on 16S ribosomal sequences can be developed for the soil NH_3 oxidizers (Stephen et al., 1996), the diversity of the NO_2^- oxidizers will make this task more difficult. However, the genus *Nitrobacter* appears to be monophyletic with highly conserved 16S ribosomal sequences (Orso et al., 1994). Formerly, classification to the genus level was based on cell shape and the arrangement of intracytoplasmic membranes and has resulted in the reported genera and species shown in Table 4.4. The taxonomy of the nitrifiers is changing rapidly and is likely to continue to be redefined as additional environmental isolates are characterized. Any attempt at identification of isolates or dominant members of enrichments should use 16S rDNA as the basis for discrimination, and possibly the intergenic spacer region between the 16S and 23S genes in the ribosomal operon for distinguishing between very closely related strains (Navarro et al., 1992).

4.3.3.2   Methods of Enumeration, Isolation and Maintenance of Nitrifying Bacteria

Nitrifying bacteria have slow growth rates and small colony size on plates; therefore, the traditional plate count technique alone is seldom successful for their isolation or enumeration. Most studies have used liquid cultures either for enrichment culture or for most probable number (MPN) enumeration and counting. Both of these cultural methods suffer from serious limitations due to the selective nature of media composition, and incubation time and conditions. Although direct molecular methods (Section 4.3.3.4) may avoid some of these problems, progress on the ecophysiology of nitrifiers will

Table 4.4   Type strains of chemolithotrophic nitrifiers

Genus	Species	Source	Cell Shape	Intracytoplasmic Membranes	Source
Ammonia oxidizers					
Nitrosomonas	*europaea*	soil, water sewage	straight rods	peripheral	3,7,8
	eutropha	sewage	straight rods	peripheral	4,7
	oligotropha	freshwater	straight rods	peripheral	4,7
	cryotolerans	marine	straight rods	peripheral	3,7,8
	marina	marine	straight rods	peripheral	4,7,8
Nitrosospira	*briensis*	soil	spirals	minimal	3,8
Nitrosolobus[1]	*multiformis*	soil	lobate	compartmental	3,5,8
Nitrosovibrio[2]	*tenuis*	soil	curved rods	minimal	3,5,8
Nitrosococcus	*oceanus*	marine	spheres	central stack	3,6,8
	halophilus	salt lakes	spheres	central stack	3,8
	mobilis	brackish	spheres	peripheral	3,6,7,8
Nitrite oxidizers					
Nitrobacter	*winogradsky*	soil, water	pleomorphic rods	polar cap	3,6,9
Nitrobacter	*hamburgensis*	soil	pleomorphic rods	polar cap	3,6,9
Nitrospira	*marina*	marine	spirals	minimal	3,6,9
Nitrospina	*gracilis*	marine	slender rods	minimal	3,6,9
Nitrococcus	*mobilis*	marine	sphere	tubular	3,6,9

[1]May be reclassified as *Nitrosospira multiformis* (Head et al., 1993).
[2]May be reclassified as *Nitrosospira tenuis* (Head et al., 1993).
3 = Holt et al. (1994), 4 = Koops and Möller (1992), 5 = Head et al. (1993), 6 = Teske et al. (1994), 7 = Pommerening-Roser et al. (1996), 8 = Koops and Möller (1992), 9 = Bock and Koops (1992).

continue to rely on the isolation and characterization of representative pure cultures. Immunofluorescent techniques also rely on pure cultures as the starting point for the production of specific antibodies to either the NH_3- or NO_2^--oxidizing bacteria. For successful growth of nitrifiers in the laboratory it is important to remember that nitrifying bacteria are photosensitive (strongly inhibited even by room fluorescent lights), sensitive to carbon disulphide (from rubber seals, tubing) and inhibited by low levels of some organic compounds. All glassware should be acid washed, silcone or cotton closures used and strict attention given to sterile procedures for the maintenance and isolation of pure cultures. Relatively large inocula (5–10% v/v) are used for the transfer of nitrifier cultures; this practice is an attempt to avoid the prolonged lag times observed with more dilute suspensions of cells. Extended lag times in dilute suspensions of cells may be explained by the phenomenon of cell density regulated recovery of nitrifiers when grown in suspension cultures in contrast to shorter recoveries under the biofilm or surface growth conditions which are prevalent in native habitats (Batchelor et al., 1997).

Soil suspensions (typically 10 g soil in 95 ml of 1 mM phosphate buffer, pH 7.2) are the starting point for the enrichment or MPN enumeration of soil nitrifiying bacteria. This suspension is dispersed by blending for 1 min; Tween 80 (6 drops) and antifoam (2 drops) may be added to aid in dispersion (Belser and Schmidt, 1978a). Traditional enrichment cultures are prepared by adding a large inoculum (1%) to a medium favoring the growth of either NH_3 or NO_2^- oxidizers and incubating for several weeks. These cultures serve as inoculum for the next serial passage in the same medium until a rapidly nitrifying culture is established. This enrichment procedure will tend to select for the fastest growing organisms under the given conditions although these strains may not represent dominant members of the original source. Enrichment cultures may still be useful tools for selecting for and isolating a specific physiological type of nitrifier. Alternatively, isolations made from the highest positive MPN tubes (dilution to extinction) will tend to isolate the predominant nitrifiers from the environment which are able to grow on the medium chosen (Underhill, 1990; Schmidt and Belser, 1994).

The composition of the media used for isolation and growth will determine the numbers and diversity of the nitrifiers examined. Representative media for use in enrichments or MPN cultures for NH_3 and NO_2^- oxidizers are shown in Table 4.5. The concentration of the N substrate and the pH are the key aspects that may determine media selectivity.

For the NH_3 oxidizers reported substrate concentrations range from approximately 0.7 mM (Donaldson and Henderson, 1989; Suwa et al., 1994) to 36 mM NH_4^+(Suwa et al., 1994). Substrate (NH_4^+/NH_3) levels > 1 mM have been shown to be inhibitory to some NH_3-sensitive strains. Soil NH_3 oxidizer enrichment cultures from oak grassland and mixed conifer sites showed maximum NH_4^+ oxidation rates at approximately 0.4 mM NH_4^+, higher concentrations resulted in substrate inhibition (Stark and Firestone, 1996). The NH_3 tolerance of soil NH_3 oxidizers from wildland, unfertilized systems are likely to be somewhat intermediate between those in heavily fertilized soils or sewage and the oligotrophic NH_3 oxidizers of the open ocean (Stark and Firestone, 1996). For the NO_2^- oxidizers, reported substrate concentrations range from 0.05 mM (Both et al., 1990) up to 30 mM NO_2^- (Bock and Koops, 1992) . While most pure culture representatives can be grown at the higher range of NO_2^-, 1 mM or less would appear to be more appropriate for soil enumeration or isolation. The recommendation is to adjust NH_4^+ and NO_2^- concentrations in the media to reflect environmentally relevant concentrations or to use several concentrations to select for different subgroups of the community. Interpretation of MPN counts must always be made in the context of the selective nature of the medium.

The dilution procedures and recommended NO_2^- spot test for determining positive MPN tubes for both NH_3 and NO_2^- oxidizers are described in detail by Schmidt and Belser (1994). Nitrite can be detected simply by using Greiss-Ilosvay reagents or may be measured quantitatively using these

Fig. 4.5   Media composition for isolation, enumeration, and growth of ammonia- and nitrite-oxidizing bacteria

Chemical constituent	Ammonia Oxidizer					Nitrite Oxidizer	
	Soriano and Walker (1968)	Watson (1971)	Watson et al. (1981)	Gerna et al. (1992) #929	Schmidt and Belser (1994)	Bock et al. (1983)	Schmidt and Belser (1994)
				L^{-1}			
$(NH_4)_2SO_4$	500 mg	130 mg	2,000 mg	1,320 mg	500 mg		
$NaNO_2$						2,000 mg	
KNO_2							8.5 mg
$MgSO_4 \cdot 7H_2O$	40 mg	200 mg	200 mg	380 mg	40 mg	50 mg	4 mg
$CaCl_2 \cdot 2H_2O$	40 mg	20 mg	20 mg	20 mg	134 mg		134 mg
KH_2PO_4	200 mg				204 mg	150 mg	27 mg
K_2HPO_4		87 mg	16 mg	87 mg			139 mg
NaCl						500 mg	
$CaCO_3$						3 mg	
Fe (Chelated)[*]	500 µg	1,000 µg	1,000 µg	1,000 µg	500 µg		500 µg
$FeSO_4 \cdot 7H_2O$						150 µg	
$Na_2MoO_4 \cdot 2H_2O$		100 µg	100 µg	100 µg	100 µg		100 µg
$(NH_4)_6Mo_7)_{24} \cdot 4H_2O$						50µg	
$MnCl_2 \cdot 4H_2O$		200 µg	200 µg	200 µg	200 µg		200 µg
$CoCl_2 \cdot 6H_2O$		2 µg	2 µg	2 µg	2 µg		2 µg
$CuSO_4 \cdot 5H_2O$		20 µg	20 µg		20 µg		20 µg
$ZnSO_4 \cdot 7H_2O$		100 µg	100 µg	100 µg	100 µg		100 µg
Phenol red	500 µg			130 µg			
Bromothymol blue					2 mg		
Final pH 7.5-8.0							

* Fe added either as the 13% Geigy chemical, Fe-EDTA or as $FeSO_4 \cdot 7H_2O$ with Na-EDTA

reagents and spectroscopy. Incubation periods are from a minimum of 3 weeks to 3 months (Both and Laanbroek, 1991). The tubes showing product (NO_2^-/NO_3^-) formation for NH_3 oxidizers or substrate depletion for the NO_2^- oxidizers are considered positive and scored using MPN tables (Woomer, 1994) to give cell numbers per g of material. MPN counts may underestimate cell numbers of nitrifiers due to the tendency of both NH_3 and NO_2^- oxidizers to form exopolysaccharides and flocs and problems with media selectivity (Belser, 1979). Reported MPN numbers from various soil habitats are shown in Table 4.6. Typically, MPN counts are around 10^4 g^{-1} soil but have notoriously high variability spatially, seasonally and with media composition (Belser, 1979; Both et al., 1992). For this reason, large numbers of replicate determinations are usually required to make any statistical conclusions as to treatment effects. Replicates at the field level are more useful than additional replicate MPN determinations on the same soil sample.

The MPN tubes at the highest dilutions showing positives are the recommended starting point for the isolation of pure cultures representing dominant members of the community capable of growth under the chosen conditions. These tubes can also be examined by immunofluorescence if strain-specific antibodies are available. Pure cultures may be attained by further dilution of the MPN tubes (Schmidt and Belser, 1994) or alternatively by plating onto solid media made with purified agar or silica gel (Underhill, 1990). On solid media, nitrifier colonies are initially visible only microscopically, and typically remain small (~ 0.2 mm). All isolates should be rigorously checked for heterotrophic contamination. Pure cultures of nitrifiers are generally maintained by serial transfers

Table 4.6 MPN population estimates of ammonia-oxidizing (AO) and nitrite-oxidizing (NO) bacteria in surface soils

Habitat / Soil	AO numbers Log_{10}	NO numbers $Log_{10} x$	Comments	Source
Agricultural soil, Minnesota	5.0-5.7		varied w/ media	1
Loamy sand, Texas unmined surface soil	3.2	3.1	pre-fertilization	2
	4.2	3.9	post N fertilization	2
Pasture and agricultural soils, central Sweden	4.4-4.7	5.5-6.0	varied with crop and seasonally	3
Acid forest litter, central Netherlands	4.9		media pH 7.5	4
Seasonal dry tropical forest (Ultisol), India	6.2	6.1	rainy season	5
	3.9	3.9	dry season	5
Seasonal dry tropical cropland (Ultisol),	5.8	5.6	rainy season	5
India	3.7	3.8	dry season	5

1 = Belser and Schimdt (1978a), 2 = Waggoner and Zuberer (1996), 3 = Berg and Rosswall (1987), 4 = de Boer et al. (1993), 5= Jha et al. (1996).

(Schmidt and Belser, 1994). Limited success has been achieved by freezing cultures with glycerol (10% final concentration) although this procedure may result in reduced viability. The time-consuming and selective nature of the MPN and isolation process have promoted the development of the immunofluorescent and molecular techniques described in the next section.

4.3.3.3 Immunofluorescent and Molecular Methods for Community Characterization

Pure cultures of nitrifiers can be used for the production of strain specific antibodies for both NH_3-(Belser and Schmidt, 1978b) and NO_2^--oxidizing bacteria (Bohlool and Schmidt, 1980; Hankinson and Schmidt, 1988). Polyclonal antibodies are prepared from the blood of rabbits immunized with a pure culture of the nitrifier. The antibodies produced are reactive only to a limited number of serologically closely related strains (Belser and Schmidt, 1978b). Antibodies may be purified and then labeled with a fluorescent dye such as fluorescein isothiocyanate. The fluorescent antibody (FA) can then be used to specifically stain serologically related nitrifiers from mixed cultures or environmental samples. The stained bacterial cells can be viewed using epifluorescent microscopy or alternatively the FA can be used in an enzyme-linked immunosorbent assay (ELISA) (Saraswat et al., 1994). The strong specificity means that several FA may be needed to assess the community composition of even a single soil (Belser and Schmidt, 1978b). The need for pure cultures and the great specificity of the FA reagent are the major limitations of this approach. If the research interest is to follow a specific nitrifier strain in the environment, this technique would still be a useful tool. Future developments of this technique may involve the production of monoclonal antibodies with a wider specificity for either NH_3- or NO_2^--oxidizing bacteria (Saraswat et al., 1994).

Molecular methods for the detection of nitrifiying bacteria can target either the genes for the 16S ribosomal subunit (16S rDNA) or a specific functional gene such as those encoding ammonia monooxygenase or nitrite oxidoreductase (Hastings et al., 1997). The database and techniques for the use of 16S rDNA are more advanced although functional gene probes have the potential to examine gene expression in response to environmental regulators. The diversity of soil NH_3-oxidizing bacteria in the subdivision has been examined in agricultural soils using 16S rDNA sequencing of polymerase chain reaction (PCR) products directly from soil DNA or from enrichment cultures with subsequent DNA extraction and amplification (Stephen et al., 1996). This study revealed a greater diversity of soil NH_3 oxidizers than is represented by the pure cultures available and also showed the selection for *Nitrosomonas* that occurs during laboratory enrichments. The general approach for examining the species richness of the community is to isolate DNA from soil (Picard et al., 1992; Holben, 1994; Zhou et al., 1996), use PCR to amplify 16S rDNA from bacteria closely related to either known NH_3 or NO_2^- oxidizers, then clone and sequence these PCR products (Stephen et al., 1996). The dominant sequence types identified are thought to be representative of the community of nitrifiers although there are some questions on the quantification of gene abundance due to possible bias in the PCR amplification (Suzuki and Giovannoni, 1996). The design of primers is a key step in the process, broad specificity in the primers may result in amplification of sequences from nontarget bacteria which are not nitrifiers but are phylogenetically related. In contrast, a primer very specific to known sequences may eliminate some of the diversity of the uncultured nitrifying bacteria. Therefore, new primers should be thoroughly evaluated with pure cultures of nitrifiers and their close relatives. The sequences obtained are analyzed with standard phylogenetic techniques (Felsenstein, 1993; Swofford et al., 1996) to yield trees showing clusters of related sequences. The novel sequences are presumed to be from nitrifiers based on their close association with known nitrifier sequences. Further work is necessary after the construction of a clone library (i.e., obtain isolates or use PCR on the MPN tubes from the highest dilution) to definitively link novel 16S rDNA sequences to NH_3- or NO_2^--oxidizing activity.

The diversity of 16S rDNA PCR products from soil NH_3-oxidizing bacteria has been examined using profiling by denaturing gradient gel electrophoresis (DGGE) (Kowalchuk et al., 1997). The DGGE technique involves polyacrylamide gels which contain a gradient of DNA denaturants, thus allowing differentiation of PCR products with identical sequence length but divergence in sequence composition. The patterns of bands in a DGGE gel can be compared for soil treatments and with known representatives of the clusters. Individual bands can be excised, cloned and sequenced to augment the existing sequence database. PCR has also been used in conjunction with MPN serial dilution to give a more rapid estimate of the population of NO_2^- oxidizers containing the specific target sequence (Degrange and Bardin, 1995). Dilutions of extracted soil DNA can be screened for positive amplification of the 16S rDNA target and related back to an estimate of population size. This technique requires calibration of the number of cells with pure culture representatives. PCR can also be used to amplify the intergenic spacer region of the ribosomal operon, with one primer in the 16S rDNA and the other in the 23S rDNA. The intergenic region is longer and more variable than that within the 16S rDNA gene and can be profiled by restriction fragment length polymorphisms (RFLP). This approach was used to examine the diversity of isolates belonging to the *Nitrobacter* from a variety of ecosystems (Navarro et al., 1992).

An alternative approach to PCR-based techniques is to design oligonucleotide probes specific to the 16S rRNA of the NH_3- or NO_2^--oxidizing bacteria and to use these probes to perform an *in situ* hybridization (Wagner et al., 1995; Mobarry et al., 1996; Schramm et al., 1996). The probe is typically an oligonucleotide of 16–17 nucleotides in length which is labeled with a fluorochrome. The choice of target and design of the probe sequence determines its range and selectivity. *In situ* hybridization conditions must be strictly controlled to permeabilize the cells, maintain stringency of the probe binding and eliminate artifacts (Amann et al., 1995). After hybridization, the cells in which rRNA has hybridized to the probe sequence can be visualized using epifluorescent microscopy or confocal laser scanning microscopy (Schramm et al., 1996). The same preparation of cells can be examined with multiple probes which are labeled with different fluorochromes. Using these techniques, NH_3- and NO_2^--oxidizing bacteria were visualized in a biofilm preparation simultaneously monitored for NO_3^- production using a microelectrode (Schramm et al., 1996). Spatial relationships of the cells in the biofilm were maintained revealing a close association between NH_3- and NO_2^--oxidizing bacteria. Further method development may be necessary before these elegant techniques can be fully exploited for the soil environment. In contrast to the use of labeled oligonucleotides as probes, Guschin et al. (1997) immobilized an array of oligonucleotide probes (including those targeting NH_3- and NO_2^--

oxidizing bacteria) to slides and then hybridized these to labeled environmental DNA or RNA. This approach allows for a quantitative analysis of the signal from 16S rDNA which hybridizes to nitrifier-specific targets versus the signal from a eubacteria (general) target.

4.3.4 Nitrification Inhibitors

Inhibition of nitrification is an important consideration in three main areas: (1) agricultural management to increase N use efficiency and minimize N loss, (2) ecological and physiological studies of the nitrifying organisms, and (3) nontarget effects of inhibitory substances in waste treatment systems and in soil ecology. There are three principal groups of nitrification inhibitors: reduced nitrogenous compounds such as nitrapyrin, acetylenic compounds and phosphoramides (Okey et al., 1996). A summary of the common nitrification inhibitors is given in Table 4.6. While some compounds are formulated and used specifically to inhibit nitrification, many agricultural chemicals may have a nontarget effect on the nitrifying bacteria. As current wastewater treatment systems are dependent upon the nitrifiers, knowledge of these nontarget effects may be important for preventing unintentional failures of nitrification in waste treatment and the resulting release of toxic NH_4^+/NH_3 to the surface waters (Okey et al., 1996). The specificity is the primary concern for the choice of inhibitors in scientific investigations while ease of application, persistence and expense determine the usefulness of inhibitors for agriculture.

Nitrification inhibitors have been used to estimate the contribution of nitrifiers to NH_4^+/NH_3-oxidizing activity in the environment or in environmental samples. Typically, a soil sample is equally divided into two portions; one receives the inhibitor while the other receives only the carrier. The disappearance of endogenous NH_4^+ and/or the production of NO_3^- are compared for the two sets of samples. Alternatively, labeled substrates may be added and their transformation rates determined using isotope techniques. The assumption made is that the inhibitor is specific for the nitrification process and that inhibition is 100% effective. Since neither of these assumptions is completely valid (Bedards and Knowles, 1989), results of studies using inhibitors must be critically examined for potential artifacts. Most nitrification inhibitors are targeted to prevent NH_3 oxidation so as to avoid the build up of potentially toxic NO_2^- in the soil. Acetylene (C_2H_2) and nitrapyrin (2-chloro-6-(trichloromethyl)pyridine) are the two most commonly used nitrification inhibitors for ecological studies. Both of these substances act on NH_3 oxidation at the level of the NH_3 monooxygenase enzyme.

Acetylene is inhibitory to soil autotrophic NH_3 oxidation at 10 Pa partial pressure (compared to denitrification inhibition at 10 kPa) (Berg et al., 1982) and has the advantage of being distributed easily in the soil gas phase and soluble in the soil water. The C_2H_2 used for inhibition of NH_3 oxidation should be acetone free, as there are potential side effects on N cycling from adding the rather large quantities of acetone found in commercial grade C_2H_2. Acetone-free purified C_2H_2 may be purchased (Matheson Gas Products) or conveniently made in small quantities from calcium carbide (CaC_2) by the addition of water (Weaver and Danso, 1994). Acetylene is typically added into the headspace of a jar containing soil and, if the only objective is to block autotrophic nitrification, excess C_2H_2 (1% vol/vol) is typically used to ensure complete inhibition. Typically the C_2H_2 is left in the soil or in the headspace for the duration of the experiment, although in studies differentiating N_2O production from nitrifiers and denitrifiers, 24-h exposure to 100 Pa C_2H_2 with subsequent removal of C_2H_2 resulted in 90% inhibition of nitrification over 6 days with recovery of N_2O reduction by denitrifiers after 24 h (Kester et al., 1996). Acetylene is a suicide substrate for ammonia monooxygenase and the inactivation of the enzyme is irreversible (Hyman and Arp, 1992). This results in the requirement for the NH_3 oxidizers to synthesize new enzyme before they can recover their oxidizing activity usually taking more than one week (Kester et al., 1996). While autotrophic nitrification is strongly inhibited by C_2H_2, heterotrophic nitrification is relatively unaffected, making C_2H_2 a useful tool to differentiate autotrophic from heterotrophic nitrification in soils (Killham, 1986). Consumption of C_2H_2 in soil has been observed (Terry and Duxbury, 1985; de Boer et al., 1993; Klemedtsson and Mosier, 1994) and may be responsible for decreased inhibition in longer experiments. Coated CaC_2 has been used to slowly release C_2H_2 and maintain inhibitory concentrations in the soil (Klemedtsson and Mosier, 1994), although consumption of C_2H_2 eventually increased to the point that nitrification inhibition was no longer effective. Acetylene also acts as an inhibitor of CH_4 oxidation by pMMO (Bedards and Knowles, 1989).

Nitrapyrin (2-chloro-6-(trichloromethyl)pyridine) is commercially available under the trade name of N-serve (Dow Chemical) and is the most widely used commercial nitrification inhibitor with approximately 10^6 kg of nitrapyrin manufactured and sold each year in the United States (Okey et al., 1996). Nitrapyrin is only sparingly soluble in water, and therefore, is typically dissolved in an organic solvent forming an emulsifiable concentrate before addition to soil. The carrier itself may have an inhibitory effect on nitrification or may affect the availability of NH_4^+ through increasing immobilization. These effects may be important for research applications although not for agricultural use. An alternative is to prepare an aqueous solution by shaking nitrapyrin in a stoppered bottle for 1 h. Care must be taken to use glass bottles and stoppers, to minimize shaking time and to store solutions at 4 °C (Bremner et al., 1978). Inhibitory concentrations used vary from 0.5 to 20 g nitrapyrin g^{-1} soil (Keeney, 1986; McCarty and Bremner, 1990a; Bauhaus et al., 1996) or approximately 1–2 kg ha^{-1} (Chalk et al., 1990). Nitrapyrin is degraded in the soil with a half-life generally around two weeks at 20 °C, although this varies with soil type, organic matter content and especially temperature (Keeney, 1986; McCarty and Bremner, 1990a). Gross rates of N mineralization and immobilization have been observed to be only slightly inhibited by nitrapyrin (Chalk et al., 1990) while nitrapyrin also acts to inhibit CH_4 oxidation by pMMO (Bedards and Knowles, 1989).

Several other potentially important nitrification inhibitors are given in Table 4.7. Their effectiveness, as determined by laboratory incubation at 25 °C for 28 days, decreased in the order of 2-ethynylpyridine, etridiazole, nitrapyrin or 4-amino-1,2,4 triazole to 2,4,diamino-6-trichloromethyl triazine (McCarty and Bremner, 1990b). Only etridiazole (Dwell) and nitrapyrin (N-

Table 4.7 Chemical inhibitors of nitrification

Chemical	Common or commercial name	Company	Comments	Source
Nitrapyrin) (2-chloro-6-(trichloro-methyl) pyridine $C_6H_3NCl_4$	N-Serve	Dow Chemical	most widely used in US	1,3,4
Dicyanamide $C_2H_4N_4$	DCD or Dycan	Showa Denko	also a slow release N source	1,3,4
Etridiazole (5-ethoxy-3-(trichloro-methyl)-1,2,4 thiadiazole	Dwell	Olin Corp.	also a fungicide	1,4
2-ethnylpyridine		Farchan Laboratories	acetylenic pyridine recently developed	1
2,4-diamino-6-trichloromethyl triazine	CL-1580	American Cyanamid		1,4
Ammonium thiosulfate $(NH_4)_2S_2O_3$	ATS		also a N & S fertilizer, urease inhibitor	1,4
		Ishihada		
4-amino-1,2,4-triazole HCl	ATC	Industry	less effective than nitrapyrin	1,4
2-amino-4-chloro-6-methylpyridine	AM	Toyo-Koatsu		1
Coated calcium carbide			releases C_2H_2	2,3,4

1 = McCarty and Bremner (1990b), 2 = Klemedtsson and Mosier (1994), 3 = Sarawat et al. (1994), 4 = Keeney (1986)

Serve) are currently registered as nitrification inhibitors for commercial use in the United States. Both 2-ethynylpyridine and etridiazole are more persistent in a variety of soils and conditions than nitrapyrin (McCarty and Bremner, 1990a). Allylsulfide may be a promising research tool for differentially inhibiting NH_3 oxidation at concentrations which do not substantially affect CH_4 oxidation (Roy and Knowles, 1995).

The quantitative structure activity relationship model derived by Okey et al. (1996) is a systematic approach to relating chemical structure to inhibitory effectiveness which should prove useful for identifying new nitrification inhibitors and potential side effects of nitrification inhibition by other agricultural chemicals. This may be particularly important for screening of new herbicides, many of which contain heterocyclic, amine or amide N resembling the reduced nitrogenous nitrification inhibitors. Because herbicides are generally applied to soils at relatively high levels, nontarget effects on soil nutrient cycling and on downstream waste treatment systems should be evaluated.

Chlorate is commonly used to inhibit NO_2^- oxidation although there are potential nontarget effects on NH_3 oxidation if NO_2^- is allowed to accumulate to high levels especially in unmixed systems (Berg and Rosswall, 1985). Chlorate is used at 10 mM concentration to inhibit NO_2^- oxidation in a mixed slurry assay (see nitrification potential below) so that NO_2^- rather than NO_3^- accumulation can be measured (Belser and Mays, 1980). Chlorate has also been observed to inhibit heterotrophic nitrification (Bauhaus et al., 1996). The true inhibitor of NO_2^- oxidation may actually be chlorite (ClO_2^-) produced by the reduction of chlorate (ClO_3^-) by the NO_3^- reductase enzyme in *Nitrobacter* (Prosser, 1989). The inhibition of NO_2^- oxidation by chlorate is decreased at high NO_2^- concentrations (Belser and Mays, 1980).

4.3.5 Nitrification Rates and Kinetics in Soil Environments

4.3.5.1 Net, Gross, and Potential Nitrification Rates: Definitions and Introduction

The term nitrification rate can be defined in at least three ways. The net nitrification rate is the rate of NO_3^- accumulation and is equal to the rate of conversion of NH_4^+/NH_3 to NO_3^- minus any consumption of NO_3^-. The gross nitrification rate is the actual rate of conversion of NH_4^+/NH_3 to NO_3^- and is equal to the rate of NO_3^- accumulation (net rate) plus the rate of NO_3^- consumption. Determination of gross rates requires the use of stable or radioactive isotopes of N to separate simultaneous production and consumption. Nitrate consumption is primarily accounted for by plant root uptake, NO_3^- immobilization, leaching and denitrification. The potential nitrification rate is the maximum rate of nitrification in a mixed system with nonlimiting substrate (NH_4^+) supply (although the potential rate is defined by the actual method chosen). Previously, many investigators assumed that measured net nitrification rates in the absence of plant uptake and leaching were adequate indicators of the importance and rate of nitrification in the environment. This assumption neglected the potential role of NO_3^- assimilation by microorganisms and denitrification. While in well-aerated soils the assumption of low denitrification rates is generally justified, significant rates of NO_3^- assimilation have been measured even in the presence of measurable NH_4^+ pools (Davidson et al., 1990; Stark and Hart, 1997) indicating it may be of greater significance than formerly recognized.

4.3.5.2 Measurements of Nitrification Rates

Numerous measurements of soil nitrification rates have been made in ecosystems throughout the world; however, relatively few estimates are available for gross nitrification rates on undisturbed soil cores. Most information is available for net rates although the basic principles and application of N isotopes for determination of gross rates in soils were developed several decades ago (Kirkham and

Bartholomew, 1954, 1955; Jansson, 1958). The determinations of NH_4^+, NO_2^-, and NO_3^- concentrations in soils (Hart et al., 1994a; Bundy and Meisinger, 1994) which are the starting points for almost all the rate measurements described below are described in Section 4.2.5.2.

4.3.5.3   Laboratory Methods for Potential and Net Nitrification Rates

The objective of laboratory measurements of nitrification potentials is to characterize nitrifying activity under nonlimiting conditions of substrate and O_2 for a short period thus avoiding population growth of the nitrifying bacteria. Nitrification potentials generally reflect NH_3/NH_4^+ oxidation, although NO_2^- oxidation potentials may also be determined. The measurement of NH_3/NH_4^+ oxidation potentials generally follow the shaken slurry method of Belser and Mays (1980) although recently some modifications have been suggested (Hart et al., 1994a). Soil samples are collected, sieved and stored at 4 °C until the assay can be performed. While NH_3 oxidizer populations are thought to be relatively stable, storage should be minimized to prevent changes in potential rates. The assay for NH_3 oxidation is performed in a rapidly shaken slurry (200 RPM) (Stark, 1996a) to facilitate O_2 and substrate diffusion and prevent denitrification. The substrate is added in excess to achieve the maximum rate of NH_3/NH_4^+ oxidation under the specified conditions of temperature and pH. A method outline for determining NH_3/NH_4^+ oxidation potential is given in Table 4.8. The original method included chlorate (10 mM) to inhibit NO_2^- oxidation so that its production could be measured with increased sensitivity especially in the presence of high background NO_3^- (Belser and Mays, 1980; Schmidt and Belser, 1994). The suggested method measures both NO_2^- and NO_3^- production in the absence of inhibitors to avoid problems with incomplete inhibition of NO_2^- oxidation (Hart et al., 1994a) and possible inhibition of NH_3/NH_4^+ oxidation by chlorite (Schmidt and Belser, 1994). In a

Table 4.8   Methods for nitrification potential assay [Adapted from Hart et al., 1994a]

Equipment and supplies:
   orbital shaker with platform and clamps for 250 mL flasks
   250 mL flasks
   15 mL centrifuge tubes or filters (Whatman no. 40) and funnels
   5 or 10 ml pipette with wide mouth tips (cut to 0.5 cm orifice)
   freezer tubes (10 mL) for samples
Reagents:
   0.2 M K_2HPO_4
   0.2 M KH_2PO_4
   0.05 M NH_4SO_4
   Sieved (2 or 4 mm) field moist soil (determine gravimetric soil water content on separate subsample)
Procedure:
   1.   Combine 1.5 mL 0.2 M KH_2PO_4, 3.5 mL 0.2 M K_2HPO_4 and 15 mL 0.05 M NH_4SO_4 in a 1 L volumetric, adjust to pH 7.2, bring to volume with dH_2O. This solution is 1.5 mM NH_4^+ and 1 mM phosphate.
   2.   Place 15 g field moist soil into a 250 mL flask, add 100 ml of solution above.
   3.   Place on shaker, cover with a vented cap (rubber stopper with hole or other closure). Shake at 200 RPM for a total of 24 h.
   4.   Sample 4 times (Suggested at 2, 4, 22, 24 h). At least 1 h is necessary at the beginning of the incubation for equilibration. Shake the sample immediately before sampling, use pipetter with the wide mouth tip to sample an aliquot with the same soil/solution ratio as original. Centrifuge sample (8000 x g for 8 min) or filter through preleached Whatman #40. Cap and freeze tubes.
   5.   Analyze for NO_2^-/NO_3^- using colorimetric methods (Bundy and Meisinger, 1994). Calculate rate of NO_2^-/NO_3^- production by linear regression of solution concentration over time, convert to a soil dry weight basis by correcting for the moisture content of the soil.

similar fashion, NO_2^- oxidation potential can be determined by determining the consumption of NO_3^- in a shaken slurry to which NO_2^- has been added (8 μg $NaNO_2$ mL^{-1}) and NH_3/NH_4^+ oxidation is inhibited by nitrapyrin (40 μg nitrapyrin mL^{-1}) (Fuller and Scow, 1996; Schmidt and Belser, 1994). Results from nitrification potentials can be used to estimate theoretical cell numbers using maximum activities from pure cultures (Schmidt and Belser, 1994); these may be related to MPN counts if desired. However, the maximum activity values from pure cultures may be overestimates for the soil NH_3 oxidizer community (Stark and Firestone, 1996). Typically, nitrification potentials are in the range of 100–1000 ng N g^{-1} h^{-1} (Berg and Rosswall, 1985; Stienstra et al., 1994; Stark and Firestone, 1996). The lower values are generally from unfertilized wildland ecosystems while higher values are found in heavily fertilized or manured agricultural soils.

Net nitrification in soil laboratory incubations is measured by a procedure similar to that for soil mineralization (Section 4.2.5.3). The accumulation of NO_3^- is measured over the course of the incubation and is interpreted as a net nitrification potential under the soil conditions specified (Hart et al., 1994a; Schmidt and Belser, 1994). The laboratory incubation may be useful in situations where a comparison of treatment effects under tightly controlled conditions is desirable although its relevance to the field situation is limited because of the absence of plants and temperature and moisture fluctuations.

4.3.5.4 Field Measurements of Net Nitrification

Several methods are available for the determination of net nitrification in the field; all include a barrier to prevent root uptake of NO_3^- and to either prevent or monitor NO_3^- in the mass flow of water. The buried bag and the closed top soil core methods are described under N mineralization and immobilization (Section 4.2.5.4). Net nitrification is calculated as the change in soil NO_3-N per day either on a mass or an area basis and has been related to NO_3^- leaching potential following disturbance (Vitousek et al., 1982). Field measurements of net nitrification can be combined with the use of C_2H_2 to block autotrophic nitrification (Section 4.3.4). The difference between net NO_3^- production in paired samples without and with C_2H_2 is an estimate of gross nitrification (Stark and Hart, 1997):

$$\text{net } NO_3^- \text{production}_{(\text{no } C_2H_2)} - \text{net } NO_3^- \text{production}_{(\text{w/}C_2H_2)} = \text{gross nitrification} \quad [4.3.3]$$

4.3.5.5 Gross Rates of Nitrification: Use of Isotopes

The measurement of actual or gross rates of nitrification requires the use of isotopes of N; the stable isotope (^{15}N) is used almost exclusively because ^{13}N has an extremely short half-life (~10 min) that is not conducive to ecological studies. Gross nitrification rates are generally measured by one of two contrasting isotope approaches: (1) tracer techniques in which $^{15}NH_4^+$ is added and the production of $^{15}NO_3^-$ is monitored, or (2) pool dilution approach where $^{15}NO_3^-$ is added and the dilution from $^{14}NO_3^-$ production by nitrification is monitored over time (Mosier and Schimel, 1993). These two approaches can both yield valid estimates of nitrification rates but are subject to different limitations and potential problems. Heterotrophic nitrification is not measured by tracer approaches but is included in the measurement of gross nitrification by pool dilution. The major limitation of the tracer technique is that ^{15}N is added as $^{15}NH_4^+$, the substrate for nitrification, and therefore, rates can be stimulated in substrate-limited systems (Hart et al., 1994a). The consumption of the product pool (NO_3^-) and decreasing enrichment of the substrate NH_4^+ pool due to mineralization will both cause underestimation of the nitrification rate if the rate estimates are not properly corrected. While simulation modeling may account for some of these problems, most can be avoided by use of the pool dilution approach.

In the pool dilution approach where $^{15}NO_3^-$ is added to the product pool, stimulation of the nitrification rate by substrate addition is avoided. The rate at which the enrichment of the NO_3^- pool declines due to production of $^{14}NO_3^-$ is determined over the specified time period and related to the gross nitrification rate by either analytical solution (Kirkham and Bartholomew, 1954; Hart et al., 1994a) or simulation modeling (Myrold and Tiedje, 1986; Nason and Myrold, 1991). Depending on the complexity of the rate equations used, several assumptions are required. The simplest rate equations based on those of Kirkham and Bartholomew (1954) as applied by Hart et al. (1994a) require the following assumptions: (1) ^{15}N and ^{14}N react similarly (isotopic discrimination is considered negligible), (2) production and consumption rates are constant during incubation, and (3) no recycling of ^{15}N occurs from the sink (NO_3^-) to the source pool (NH_4^+). These assumptions can be most easily accommodated in short-term experiments lasting 1 or 2 days. More complex cases where one or more of the above assumptions are expected to be violated, require multiple time points over a longer incubation period as input for simple simulation models (Myrold and Tiedje, 1986; Barraclough and Smith, 1987; Nason and Myrold, 1991). The pool dilution approach to estimate nitrification rates has been used in laboratory incubation experiments (Hart et al., 1994b), greenhouse studies (Norton and Firestone, 1996) and in intact cores in the field (Davidson et al., 1990, 1992; Stark and Hart, 1997). Some representative values for gross nitrification rates and comparison to net nitrification are given in Table 4.9. The rates measured in disturbed samples may not be directly comparable to those from intact cores although representative values are given so that the range of values can be evaluated. Unfortunately, few examples of gross nitrification rates are available for undisturbed cores from agricultural systems. Many greenhouse and laboratory incubation experiments can benefit from the application of ^{15}N pool dilution methods to estimate gross nitrification. These situations have the added advantage that paired matching samples can be set from the beginning of the experiment to be injected with $^{15}NO_3^-$ and one harvested within a short time (< 2 h) and the other after 24- or 48-h incubation. A detailed discussion of a field pool dilution method for estimating gross nitrification rates in intact cores is given in Hart et al. (1994a).

For both the pool dilution or tracer methods, the concentrations of NO_2^- and NO_3^- in the soil extracts are determined and the appropriate amount of solution is prepared for mass spectrometry generally by

Table 4.9   Nitrification rates in surface mineral soils

Location	Ecosystem type	Treatment	Gross rate	Net rate	Source
			mg N kg^{-1} soil day^{-1}		
California	young mixed conifer forest	greenhouse study	1.3	−0.7	1
Oregon	red alder/ Douglas fir	intact cores	4.3	0.2	4
Oregon	Douglas fir	intact cores	1.7	0.0	4
Oregon	w. hemlock/ sitka spruce	intact cores	1.9	−1.9	4
New Mexico	Ponderosa pine	intact cores	0.3	0.3	4
California	young mixed conifer forest	intact cores spring	0.5	0.1	3
California	>100-yr-old conifer forest	intact cores spring	0.5	0.06	3
California	annual grassland-open	intact cores spring	0.8	−0.8	2
California	annual grassland-canopy	intact cores spring	3.5	−0.2	2
Oregon	mixed conifer old growth	mixed soil laboratory	1.1	0.03	5
La Selva, Costa Rica	old growth tropical forest	intact cores 7 day	5.0	2.1	6
Hokkaido, Japan	agricultural soil	mixed soil laboratory	3.2	2.6	7
Michigan	northern hardwood	mixed soil laboratory	1.1	ND	8

1 = Norton and Firestone (1996), 2 = Davidson et al. (1990), 3 = Davidson et al. (1992), 4 = Stark and Hart (1997), 5 = Hart et al. (1994b), 6 = Zou et al. (1992), 7 = Nishio et al. (1985), 8 = Myrold and Tiedje (1986)

the diffusion method (Hart et al., 1994a, Stark and Hart, 1996). The increased availability of direct combustion isotope ratio mass spectrometry which can analyze relatively small N quantities (< 10 μg) (Hauck et al., 1994) has made the pool dilution method for estimating nitrification rates more attractive. The values of the NO_2^-/NO_3^- pool size and its enrichment (atom % excess ^{15}N) are used in analytical solutions (Hart et al., 1994a) or simulation models (Nason and Myrold, 1991). Gross nitrification rates can be used to calculate the mean residence time for N in the NO_2^-/NO_3^- pool and as an indicator of the availability of NO_3^- for plant and microbial consumption. Determination of gross nitrification rates in conjunction with pool dilution estimates of NH_4^+ production and consumption (Section 4.2.5.5) can be used to partition NH_4^+ consumption into immobilization and nitrification. together these processes comprise the internal N cycle in soils.

4.3.5.6 Kinetics and Models of Nitrification

The kinetics, growth and product formation of nitrifiers have been studied in pure and mixed enrichment cultures, and modeled in the soil environment (Prosser, 1989; Nishio and Fujimoto, 1990; Grant, 1994; Stark and Firestone, 1996). Most studies involve Michaelis-Menten kinetics and/or the Monod growth curve. The maximum specific growth rate (μ_{max}) for balanced growth of nitrifiers is quite low compared to that for most heterotrophic bacteria; reported values for μ_{max} typically range from 0.014 to 0.087 h^{-1}, equivalent to generation times of 50 to 8 h (Prosser, 1989). The highest growth rates are for *N. europaea* which are very slow compared to the typical μ_{max} of 1.8 h^{-1} (generation time of 23 min) for *Eschericia coli* growing on complex rich media (Neidhardt et al., 1990). However, in the soil environment the ability to use substrates at low concentrations may be a more important factor than a high specific growth rate (Prosser, 1989). The relationship of substrate utilization or product formation to substrate concentration is typically analyzed using a hyperbolic equation defined by the half saturation constant (K_s) or the Michaelis constant (K_m) and the maximum rate of substrate utilization (V_{max}). The Haldane equation can be used to account for substrate inhibition at high NH_4^+ levels which has been observed both in pure and enrichment cultures (Suwa et al., 1994; Stark and Firestone, 1996). The general observation is that K_m reflects the ambient NH_4^+ concentration in the environment (Prosser, 1989) indicating that K_m values for wildland, unfertilized systems are likely to be lower than those from agricultural soils. The K_m for NH_3 oxidation in a variety of soils generally ranges from 5 to 50 μM NH_4^+ based on the solution concentration (Hart et al., 1994a; Stark and Firestone, 1996; Low et al., 1997) while sewage K_m values are generally higher (up to 700 μM). The effects of temperature, moisture and osmotic potential on gross nitrification rates have been examined for several soils (Stark and Firestone, 1995, 1996; Low et al., 1997). The combination of all of these environmental parameters with submodels for NH_4^+ supply and diffusion in the soil environment will allow for more effective simulation modeling of nitrification in the future (Grant, 1994).

4.3.6 Concluding Remarks

The combination of traditional methods and innovative techniques will continue to provide new insights into the functioning and management of the unique group of soil microorganisms, the nitrifiers. With this information, insight into plant N availability, NO_3^- pollution of surface and groundwater, N saturation of native ecosystems and the production of atmospherically active N trace gases will be obtained. Clearly, the controlling factors of nitrification will continue as a key area of research for understanding human impact on the terrestrial N cycle.

4.3.7 References

Alef, K., and D. Kleiner. 1986. Arginine ammonification, a simple method to estimate microbial activity potentials in soils. Soil Biol. Biochem. 18:233–235.

Alzerreca, J.J., J.M. Norton, and M.G. Klotz. 1997. *Nitrosococcus oceanus* ammonia monooxygenase subunit A (amoA). GenBank Accession U96611.

Amann, R.I., L. Krumholz, and K.H. Schleifer. 1995. Phylogenetic identification and in situ detection of individual microbial cells without cultivation. Microbiol. Rev. 59:143–169.

Arp, D.J., N.G. Hommes, M.R. Hyman, L.Y. Juliette, W.K. Keener, S.A. Russel, and L.A. Sayavedra-Soto. 1996. Ammonia monooxygenase from *Nitrosomonas europaea*. p.159–166. *In* M.E. Lidstrom and F.R. Tabita (ed.) Microbial growth on C1 compounds. Kluwer Academic Publishers, Dordrecht, Netherlands.

Barraclough, D. 1997. The direct or MIT route for nitrogen immobilization: a ^{15}N mirror image study with leucine and glycine. Soil Biol. Biochem. 29:101–108.

Barraclough, D., and M.J. Smith. 1987. The estimation of mineralization, immobilization and nitrification in nitrogen-15 field experiments using computer simulation. J. Soil Sci. 38:519–530.

Batchelor, S.E., M. Cooper, S.R. Chhabra, L.A. Glover, G.S.A.B. Stewart, P. Williams, and J.I. Prosser. 1997. Cell density-regulated recovery of starved biofilm populations of ammonia-oxidizing bacteria. Appl. Env. Microbiol. 63:2281–86.

Bauhus J., A.C. Meyer, and R. Brumme. 1996. Effect of the inhibitors nitrapyrin and sodium chlorate on nitrification and N_2O formation in an acid forest soil. Biol. Fert. Soil. 22:318–325.

Bedards C., and R. Knowles. 1989. Physiology, biochemistry and specific inhibitors of CH_4^+, NH_4^+ and CO oxidation by methanotrophs and nitrifiers. Microbiol. Rev. 53:68–84.

Belser, L.W.1979. Population ecology of nitrifying bacteria. Ann. Rev. Microbiol. 33:309–333.

Belser, L.W., and E.L. Mays. 1980. Specific inhibition of nitrite oxidation by chlorate and its use in assessing nitrification in soils and sediments. Appl. Environ. Microbiol. 39:505–510.

Belser, L.W., and E.L. Schmidt. 1978a. Diversity in the ammonia-oxidizing nitrifier population of a soil. Appl. Environ. Microbiol. 36:584–588.

Belser, L.W., and E.L. Schmidt. 1978b. Serological diversity within a terrestrial ammonia-oxidizing population. Appl. Environ. Microbiol. 36:589–593.

Berg, P., L. Klemedsston, and T. Rosswall. 1982. Inhibitory effects of low partial pressures of acetylene on nitrification. Soil. Biol. Biochem. 14:301–303.

Berg, P., and T. Rosswall. 1985. Ammonium oxidizer numbers, potential and actual oxidation rates in two Swedish arable soils. Biol. Fert. Soils 1:131–140.

Berg, P., and T. Rosswall. 1987. Seasonal variations in abundance and activity of nitrifiers in four arable cropping systems. Microb. Ecol. 13:75–87.

Bergmann, D.J., and A.B. Hooper. 1994. Sequence of the gene, *amoB*, for the 43-kDa polypeptide of ammonia monooxygenase of *Nitrosomonas europaea*. Biochem. Biophys. Res. Commun. 204:759–762.

Binkley, D., and S.C. Hart. 1989. The components of nitrogen availability assessments in forest soils. Advances in Soil Science 10:55–112.

Bock, E., H. Sundermeyer-Klinger, and E. Stackebrandt. 1983. New facultative lithoautotrophic nitrite-oxidizing bacteria. Arch. Microbiol. 136:281–284.

Bock, E., and H.-P. Koops. 1992. The genus *Nitrobacter* and related genera. p. 2302–2309. *In* A. Balows, H.G. Truper, M. Dworkin, W. Harder, K. Schleifer (ed.) The Prokaryotes. 2nd Ed. Springer-Verlag, New York, NY.

Bock, E., P.A. Wilderer, and A. Frietag. 1988. Growth of *Nitrobacter* in the absence of dissolved oxygen. Wat. Res. 22:245–250.

Bohlool, B.B., and E.L. Schmidt. 1980. The immunofluorescent approach in microbial ecology. Adv. Microbiol. Ecol. 4:204–241.

Both, G.J., S. Gerards, and H.J. Laanbroek. 1990. Enumeration of nitrite-oxidizing bacteria in grassland soils using a most probable number technique: effect of nitrite concentration and sampling procedure. FEMS Microbiol. Ecol. 74:277–286.

Both, G.J., S. Gerards, and H.J. Laanbroek. 1992. Temporal and spatial variation in the nitrite-oxidizing bacterial community of a grassland soil. FEMS Microbiol. Ecol. 101:99–112.

Both, G.J., and H.J. Laanbroek. 1991. The effect of incubation period on the result of MPN enumerations of nitrite-oxidizing bacteria: theoretical considerations. FEMS Microbiol. Ecol. 85:335–344.

Bremner, J.M. A.M. Blackmer, and L.G. Bundy. 1978. Problems in the use of nitrapyrin (N-Serve) to inhibit nitrification in soils. Soil. Biol. Biochem. 10:441–442.

Bremner, J.M. 1995. Recent research on problems in the use of urea as a nitrogen fertilizer. Fert. Res. 42:321–329.

Bundy, L.G., and J.J. Meisinger. 1994. Nitrogen availability indices. p. 951–984. *In* R.W. Weaver (ed.) Methods of soil analysis: Microbiological and biochemical properties. 3rd Ed. Soil Science Society of America, Madison, WI.

Cabrera, M.L., and D.E. Kissel. 1988. Evaluation of a method to predict nitrogen mineralized from soil organic matter under field conditions. Soil Sci. Soc. Am. J. 52:1186–1187.

Chalk P.M., R.L. Victoria, T. Muraoka, and M.C. Piccolo. 1990. Effect of a nitrification inhibitor on immobilization and mineralization of soil and fertilizer nitrogen. Soil Biol. Biochem. 22:533–538.

Davidson, E.A., J.M. Stark, and M.K. Firestone. 1990. Microbial production and consumption of nitrate in an annual grassland. Ecol. 7:1968–1975.

Davidson, E.A., S.C. Hart, and M.K. Firestone. 1992. Internal cycling of nitrate in soils of a mature coniferous forest. Ecol. 73:1148–1156.

Davidson, E.A., S.C. Hart, C.A. Shanks, and M.K. Firestone. 1991. Measuring gross nitrogen mineralization, immobilization, and nitrification by ^{15}N isotope dilution in intact soil cores. J. Soil Sci. 42:335–349.

de Boer, W., P.J.A. Klein-Gunnewiek, and R.A. Kester. 1993. The effect of acetylene on N transformations in an acid oak-beech soil. Plant Soil 149:292–296.

Degrange, V., and R. Bardin. 1995. Detection and counting of *Nitrobacter* populations in soil by PCR. Appl. Environ. Microbiol. 61:2093–2098.

Donaldson, J.M., and G.S. Henderson. 1989. A dilute medium to determine population size of ammonium oxidizers in forest soils. Soil Sci. Soc. Am. J. 53:1608–1611.

Felsenstein, J. 1993. Phyllip (Phylogeny Inference Package). Ver. 3.5c. Department of Genetics, University of Washington, Seattle, WA.

Fuller M.E., and K.M. Scow 1996. Effects of toluene on microbially mediated processes involved in the nitrogen cycle. Microb. Ecol. 32:171–184.

Gerna, R., R. Cote, and P. Pienta. 1992. American type culture collection catalogue of bacteria and pPhages. American Type Culture Collection, Rockville, MD.

Grant, R.F. 1994. Simulation of ecological controls on nitrification. Soil Biol. Biochem. 26:305–315.

Guschin, D.Y., B.K. Mobarry, D. Proudnikov, D.A. Stahl, B.E. Rittmann, and A.D. Mirzabekov. 1997. Oligonucleotide microchips as genosensors for determinative and environmental studies in microbiology. Appl. Env. Microbiol. 63:2397–2402.

Hankinson, T.R., and E.L. Schmidt. 1988. An acidophilic and a neutrophilic *Nitrobacter* strain isolated from the numerically predominant nitrite-oxidizing population of an acid forest soil. Appl. Environ. Microbiol. 54:1536–1540.

Harris D., and E.A. Paul 1989. Automated analysis of ^{15}N and ^{14}C in biological samples. Commun. Soil Sci. Plant Anal. 20: 935–947.

Hart, S.C., J.M. Stark, E.A. Davidson, and M.K. Firestone. 1994a. Nitrogen mineralization, immobilization and nitrification. p. 985–1018. *In* R.W. Weaver (ed.) Methods of soil analysis: Microbiological and biochemical properties. 3rd Ed. Soil Science Society of America, Madison, WI.

Hart S.C., G.E. Nason, D.D. Myrold, and D.A. Perry. 1994b. Dynamics of gross nitrogen transformations in an old growth forest: the carbon connection. Ecol. 75:880–891.

Hastings, R.C., M.T. Ceccherini, N. Miclaus, J.R. Saunders, M. Bazzicalupo, and A.J. McCarthy. 1997. Direct molecular biological analysis of ammonia oxidising bacteria populations in cultivated soil plots treated with swine manure. FEMS Microbiol. Ecol. 23:45–54.

Hauck, R.D. J.J. Meisinger, and R.L. Mulvaney. 1994. p. 907–950. *In* R.W. Weaver (ed.) Methods of soil analysis: Microbiological and biochemical properties. 3rd Ed. Soil Science Society of America, Madison, WI.

Head, I.M., W.D. Hiorns, T.M. Embley, A.J. McCarthy, and J.R. Saunders. 1993. The phylogeny of autotrophic ammonia-oxidizing bacteria as determined by analysis of 16S ribosomal RNA gene sequences. J. Gen. Microbiol. 139: 1147–1153.

Holben, W.E. 1994. Isolation and purification of bacterial DNA from soil. p. 729–751. *In* R.W. Weaver (ed.) Methods of soil analysis: Microbiological and biochemical properties. 3rd Ed. Soil Science Society of America, Madison, WI.

Holmes, A.J., A. Costello, M.E. Lidstrom, and J.C. Murrell. 1995. Evidence that particulate methane monooxygenase and ammonia monooxygenase may be evolutionarily related. FEMS Microbiol. Lett. 132:203–208.

Holt, J.G., N.R. Krieg, P.H. Sneath, J.T. Staley, and S.T. Williams. 1994. Bergey's manual of determinative Bacteriology. 9th Ed. Williams and Wilkins, Baltimore, MD.

Hooper, A.B., T. Vannelli, D.J. Bergmann, and D.M. Arciero. 1997. Enzymology of the oxidation of ammonia to nitrite by bacteria. Antonie van Leeuwenhoek 71:59–67.

Hopkins, D.W. 1996. D- and L-amino acid metabolism in soil. p. 57–61. *In* P. van Cleemput et al. (ed.). Progress in nitrogen cycling studies. Kluwer Academic Publishers, Dordrecht, Netherlands.

Hyman, M.R., and D.J. Arp. 1992. $^{14}C_2H_2$ and $^{14}CO_2$-labeling studies of the *de Novo* synthesis of polypeptides by *Nitrosomonas europaea* during recovery from acetylene and light inactivation of ammonia monooxygenase. J. Biol. Chem. 267:1534–1545.

Hyvönen, R., G.I. Ågren and O. Andrén. 1996. Modelling long-term carbon and nitrogen dynamics in an arable soil receiving organic matter. Ecol. Appl. 6:1345–1354.

Igarashi, N., H. Moriyama, T. Fujiwara, Y. Fukumori, and N. Tanaka. 1997. The 2.8 Å structure of hydroxylamine oxidoreductase from a nitrifying chemoautotrophic bacterium, *Nitrosomonas europaea*. Nature Struct. Biol. 4:276–284.

Janssen B.H. 1996. Nitrogen mineralization in relation to C:N ratio and decomposibility of organic materials. Plant Soil 181:39–45.

Jansson, S.L. 1958. Tracer studies on nitrogen transformations in soil with special attention to mineralisation-immobilization relationships. Ann. Roy. Agric. Coll. Sweden 24:101–361.

Jetten, M.S.M., S. Logemann, G. Muyzer, L.A. Robertson, S. de Vries, M.C.M. van Loosdrecht, and J.G. Kuenen. 1997. Novel principles in the microbial conversion of nitrogen compounds. Antonie van Leeuwenhoek. 71:75–93.

Jha, P.B., J.S. Singh, and A.K. Kashyap. 1996. Dynamics of viable nitrifier community and nutrient availability in dry tropical forest habitat as affected by cultivation and soil texture. Plant Soil 180:277–285.

Juliette, L.Y., M.R. Hyman, and D.J. Arp. 1995. Roles of bovine serum albumin and copper in the assay and stability of ammonia monooxygenase activity in vitro. Appl. Environ. Microbiol. 177:4908–4913.

Keeney, D.R. 1986. Inhibition of nitrification in soils. p. 99–115. *In* J.I. Prosser (ed.) Nitrification. IRL Press, Oxford, UK.

Kester, R.A., W. de Boer, and H. Laanbroek. 1996. Short exposure to acetylene to distinguish between nitrifier and denitrifier nitrous oxide production in soil and sediment samples. FEMS Microbiol. Ecol. 20:111–120.

Killham, K. 1986. Heterotrophic nitrification. p. 117–126. *In* J.I. Prosser (ed.) Nitrification. IRL Press, Oxford, UK.

Kirkham, D., and W.V. Bartholomew. 1954. Equations for following nutrient transformation in soil, utilizing tracer data. Soil Sci. Soc. Am. Proc. 18:33–34.

Kirkham, D., and W.V. Bartholomew. 1955. Equations for following nutrient transformation in soil, utilizing tracer data: II. Soil Sci. Soc. Am. Proc. 19:189–192.

Klemedtsson, L.K., and A.R. Mosier. 1994. Effect of long-term field exposure of soil to acetylene on nitrification, denitrification and acetylene consumption. Biol. Fert. Soils 18:42–48.

Klotz, M.G., and J.M. Norton. 1994. Sequence of *amoA*, a gene encoding the catalytic subunit of ammonia monooxygenase in *Nitrosolobus multiformis* ATCC 25196. GenBank accession number U15733.

Klotz, M.G., and J.M. Norton. 1995. Sequence of an ammonia monooxygenase subunit A-encoding gene from *Nitrosospira sp.* NpAV. Gene 163:159–160.

Klotz, M.G., J. Alzerreca, and J.M. Norton. 1997. A gene encoding a membrane protein exists upstream of the *amoA/amoB* genes in ammonia-oxidizing bacteria; a third member of the *amo* operon? FEMS Microbiol. Lett. 150:65–73.

Koops, H.-P., and U.C. Möller. 1992. The lithotrophic ammonia-oxidizing bacteria. p. 2625–2637. *In* A. Balows, H.G. Truper, M. Dworkin, W. Harder, K. Schleifer (ed.) The prokaryotes. 2nd Ed. Springer-Verlag, New York, NY.

Kowalchuk, G.A., J.R. Stephen, W. de Boer, J.I. Prosser, T. M. Embley, and J.W. Woldendorp. 1997. Analysis of ammonia-oxidizing bacteria of the subdivision of the class Proteobacteria in coastal sand dunes by denaturing gradient gel electrophoresis and sequencing of PCR-amplified 16S ribosomal DNA fragments. Appl. Environ. Microbiol. 63:1489–1497.

Low, A.P., J.M. Stark, and L.M. Dudley. 1997. Effects of soil osmotic potential on nitrification, ammonification, N-assimilation, and nitrous oxide production. Soil Sci. 162:16–27.

McCarty, G.W. 1995. The role of glutamine synthetase in regulation of nitrogen metabolism within the soil microbial community. Plant Soil 170:141–147.

McCarty, G.W., and J.M. Bremner. 1990a. Persistence of effects of nitrification inhibitors added to soils. Commun. Soil Sci. Plant Anal. 21:639–648.

McCarty, G.W., and J.M. Bremner. 1990b. Evaluation of 2-ethynylpyridine as a soil nitrification inhibitor. Soil Sci. Soc. Am. J. 54:1017–1021.

McTavish, H., J.A. Fuchs, and A.B. Hooper 1993. Sequence of the gene coding for ammonia monooxygenase in *Nitrosomonas europaea*. J. Bacteriol. 175: 2436–2444.

Mengel, K. 1996. Turnover of organic nitrogen in soils and its availability to crops. Plant Soil 181:83–93.

Merrick, M.J., and R.A. Edwards. 1995. Nitrogen control in bacteria. Microbiol. Rev. 59:604–622.

Mobarry, B.K., M. Wagner, V. Urbain, B.E. Rittmann, and D.A. Stahl. 1996. Phylogenetic probes for analyzing abundance and spatial organization of nitrifying bacteria. Appl. Env. Microbiol. 62:2156–2162.

Mosier, A.R., and D.S. Schimel. 1993. Nitrification and denitrification. p. 181–208. *In* R. Knowles and T.H. Blackburn (ed.) Nitrogen isotope techniques. Academic Press, San Diego, CA.

Myrold, D.D., and J.M. Tiedje. 1986. Simultaneous estimation of several nitrogen cycle rates using ^{15}N: Theory and application. Soil Biol. Biochem. 18:559–568.

Nason, G.E., and D.D. Myrold. 1991. ^{15}N in soil research: appropriate application of rate estimation procedures. Agri. Ecosystems Environ. 34: 427–441.

Navarro, E., P. Simonet, P. Normand, and R. Bardin. 1992. Characterization of natural populations of *Nitrobacter* spp. using PCR/RFLP analysis of the ribosomal intergenic spacer. Arch-Microbiol. 157:107–115.

Neidhardt, F.C., J.L. Ingraham, and M. Schaechter. 1990, Physiology of the bacterial cell. Sinauer Associates, Inc. Sunderland, MA.

Nishio, T., and T. Fujimoto. 1990. Kinetics of nitrification of various amounts of ammonium added to soils. Soil Biol. Biochem. 22:51–55.

Nishio, T., T. Kanamori, and T. Fujimoto. 1985. Nitrogen transformations in an aerobic soil as determined by a ^{15}NH$_4$$^+$ dilution technique. Soil Biol. Biochem. 17:149–154.

Norton, J.M., and M.K. Firestone. 1996. Nitrogen dynamics in the rhizosphere of *Pinus ponderosa* seedlings. Soil Biol. Biochem. 28:351–362.

Norton, J.M., J. M. Low, and M.G. Klotz. 1996. The gene encoding ammonia monooxygenase subunit A exists in three nearly identical copies in *Nitrosospira sp.* NpAV. FEMS Microbiol. Lett. 139:181–188.

Okey, R.W., H.D. Stensel, and M.C. Martis. 1996. Modeling nitrification inhibition. Wat. Sci. Tech. 33:101–107.

Orso, S., M. Gouy, E. Navarro, and P. Normand. 1994. Molecular phylogenetic analysis of *Nitrobacter* spp. Int. J. Sys. Bacteriology 44:83–86.

Paul, E.A., and F.E. Clark. 1996. Soil microbiology and biochemistry. 2nd Ed. Academic Press, San Diego, CA.

Paustian K., W.J. Parton, and J. Persson. 1992. Modeling organic matter in organic-amended and nitrogen-fertilized long-term plots. Soil Sci. Soc. Am. J. 56:476-488.

Picard, C., C. Ponsonnet, and E. Paget. 1992. Detection and enumeration of bacteria in soil by direct DNA extraction and polymerase chain reaction. Appl. Environ. Microbiol. 58:2717–22

Pommerening-Roser, A., G. Rath, and H.-P. Koops. 1996. Phylogenetic diversity within the Genus *Nitrosomonas*. System. Appl. Microbiol. 19:344–351.

Powlson D.S., and D. Barraclough. 1993. Mineralization and assimilation in soil-plant systems. p. 209–242. *In* R. Knowles and T.H. Blackburn (ed.) Nitrogen isotope techniques. Academic Press, San Diego, CA.

Prince, R.C., and G.N. George. 1997. The remarkable complexity of hydroxylamine oxidoreductase. Nature Struct. Biol. 4:247–250.

Prosser, J.I. 1989. Autotrophic nitrification in bacteria. Adv. Microbial Physiol. 30:125–181.

Recous, S., B. Mary, and G. Faurie. 1990. Microbial immobilization of ammonium and nitrate in cultivated soils. Soil Biol. Biochem. 22:913–922.

Rice, C. W., and J. M. Tiedje. 1989. Regulation of nitrate assimilation by ammonium in soils and in isolated soil microorganisms. Soil Biol. Biochem. 21:597–602.

Richter J., and D.K. Benbi. 1996. Modeling of nitrogen transformations and translocations. Plant Soil 181:109–121.

Robertson, G.P., K. M. Klingensmith, M.J. Klug, E.A. Paul, J.R. Crum, and B.G. Ellis. 1997. Soil resources, microbial activity, and primary production across an agricultural ecosystem. Ecol. Appl. 7:158–170.

Roy, R., and Knowles, R. 1995. Differential inhibition by allylsulfide of nitrification and methane oxidation in fresh-water sediment. Appl. Env. Microbiol. 61: 4278–83

Saraswat, N., J.E. Alleman, and T.J. Smith. 1994. Enzyme immunoassay detection of *Nitrosomonas europaea*. Appl. Env. Microbiol. 60:1969–73.

Sayavedra-Soto, L. A., N.G. Hommes, and D.J. Arp. 1994. Characterization of the gene encoding hydroxylamine oxidoreductase in *Nitrosomonas europaea*. J. Bacteriol. 176:504–510.

Schimel, J. P., L.E. Jackson, and M.K. Firestone. 1989. Spatial and temporal effects on plant-microbial competition for inorganic nitrogen in a California annual grassland. Soil Biol. Biochem. 21:1059–1066.

Schmidt, E. L., and L..W. Belser. 1994. Nitrifying bacteria. p. 159–177. *In* R.W. Weaver (ed.) Methods of soil analysis: Microbiological and biochemical properties. 3rd Ed. Soil Science Society of America, Madison, WI.

Schramm, A., L.H. Larsen, N.P. Revsbech, N.B. Ramsing, R. Amann, and K.H. Schleifer. 1996. Structure and function of a nitrifying biofilm as determined by in situ hybridization and the use of microelectrodes. Appl. Env. Microbiol. 62:4641–47.

Smith, J. L., B. L. McNeal; H. H. Cheng, and G. S. Campbell. 1986. Calculation of microbial maintenance rates and net nitrogen mineralization in soil at steady state. Soil Sci. Soc. Am. J. 50:332–338.

Smith J. L., and E. A. Paul. 1990. The significance of soil microbial biomass estimations. p. 357–396. *In* J. M. Bollag and G. Stotzky (ed.) Soil biochemistry. Vol. 6. Marcel Dekker, Inc., New York, NY.

Soriano, S., and N. Walker. 1968. Isolation of ammonia-oxidizing autotrophic bacteria. J. Appl. Bacteriol. 31:493–497.

Stanford G., and S.J. Smith.1972. Nitrogen mineralization potential of soils. Soil Sci. Soc. Am. Proc. 38:103–107.

Stanley P.M., and E.L. Schmidt. 1981. Serological diversity of *Nitrobacter* from soil and aquatic habitats. Appl. Environ. Microbiol. 41:1069–1071.

Stark, J.M. 1996a. Shaker speeds for aerobic soil slurry incubations. Comm. Soil Sci. Plant Anal. 27:2625–2631.

Stark, J.M. 1996b. Modeling the temperature response of nitrification. Biogeochem. 35:433–445.

Stark, J.M., and M.K. Firestone. 1995. Mechanisms for soil moisture effects on activity of nitrifying bacteria. Appl. Environ. Microbiol. 61:218–221.

Stark, J.M., and M.K. Firestone. 1996. Kinetic characteristics of ammonium-oxidizer communities in a California oak woodland-annual grassland. Soil Biol. Biochem. 28:1307–1317.

Stark, J.M., and S.C. Hart. 1996. Diffusion technique for preparing salt solutions, Kjeldahl digests, and persulfate digests for nitrogen-15 analysis. Soil. Sci. Soc. Am J. 60:1846–1855.

Stark, J.M., and S.C. Hart. 1997. High rates of nitrification and nitrate turnover in undisturbed coniferous forests. Nature 385:61–64.

Stephen J. R., A. McCaig, Z. Smith, J.I. Prosser, and T. M. Embley. 1996. Molecular diversity of soil and marine 16S rRNA gene sequences related to subgroup ammonia-oxidizing bacteria. Appl. Environ. Microbiol. 62:4147–4154.

Stienstra, A.W., P.K. Gunnewiek, and H.J. Laanbroek. 1994. Repression of nitrification in soils under a climax grassland vegetation. FEMS Microbiol. Ecol. 14:45–52.

Suwa, Y., Y. Imamura, T. Suzuki, T. Tashiro, and Y. Urushigawa. 1994. Ammonia-oxidizing bacteria with different sensitivities to $(NH_4)_2SO_4$ in activated sludges. Water Res. 28:1523–1532.

Suzuki, I., U. Dular, and S.C. Kwok. 1974. Ammonia or ammonium ions as substrates for oxidation by *Nitrosomonas europaea* cells and extracts. J. Bacteriol. 120:556–558.

Suzuki, M.T., and S.J. Giovannoni. 1996. Bias caused by template annealing in the amplification of mixtures of 16S rRNA genes by PCR. Appl. Env. Microbiol. 62:625–630.

Swofford, D.L., G.J. Olsen, P.J. Waddell, and D.M. Hillis. 1996. Phylogenetic inference. p. 407–514. *In* D.M. Hillis, C. Moritz and B.K. Mable (ed.) Molecular systematics. 2nd Ed. Sinauer Associates, Inc., Sunderland, MA.

Tabatabai, M.A. 1994. Soil enzymes. p. 775–833. *In* R.W. Weaver (ed.) Methods of soil analysis: Microbiological and biochemical properties. 3rd Ed. Soil Science Society of America, Madison, WI.

Terry, R.E., and J.M. Duxbury. 1985. Acetylene decomposition in soils. Soil Sci. Soc. Am. J. 49:90–94.

Teske, A., E. Alm, J.M. Regan, S. Toze, B.E. Rittmann, and D.A. Stahl. 1994. Evolutionary relationships among ammonia and nitrite-oxidizing bacteria. J. Bacteriol. 176:6623–6630.

Tisdale, S.L., W.L. Nelson, J.D. Beaton, and J. L. Havlin. 1993. Soil fertility and fertilizers. 5th Ed. MacMillan, New York, NY.

Underhill, S.E. 1990. Techniques for studying the microbial ecology of nitrification. Methods Microbiol. 22:417–445.

van Neil, E.W.J., P.A.M. Arts, B.J. Wesselink, L.A. Robertson, and J.G. Kuenen. 1993. Competition between heterotrophic and autotrophic nitrifiers for ammonia in chemostat cultures. FEMS Microbiol. Ecol. 102:109–118.

Vitousek P. M., J.R. Gosz, C.C. Grier, J.M. Melillo, and W.A. Reiners. 1982. A comparative analysis of potential nitrification and nitrate mobility in forest ecosystems. Ecol. Mon. 52:155–177.

Waggoner, P.J., and D.A. Zuberer. 1996. Response of nitrification and nitrifying bacteria in mine spoil to urea or ammonium sulfate. Soil Sci. Soc. Am. J. 60:477–486.

Wagner, M., G. Rath, R. Amann, H.P. Koops, and K. H. Schleifer. 1995. In situ identification of ammonia oxidizing bacteria. Syst. Appl. Microbiol. 18:251–264.

Watkins, N., and D. Barraclough. 1996. Gross rates of N mineralization associated with the decomposition of plant residues. Soil Biol. Biochem. 28:169–175.

Watson, S.W. 1971. Reisolation of *Nitrosospira briensis*. S. Winogradsky and H. Winogradsky. 1933. Arch. Mikrobiol. 75:179–188.

Watson, S.W., F.W. Valois, and J.B. Waterbury. 1981. The family Nitrobacteriaceae. p. 1005–1022. *In* M.P. Starr, H. Stolt, H.G. Trhpen, A. Bolows, and H.G. Schlegel (ed.) The prokaryotes, a handbook of habitats, isolation and identification of bacteria. Vol. 1. Springer-Verlag, Berlin, Germany.

Weaver, R.W., and S.K. Danso. 1994. Dinitrogen fixation. p. 1019–1045. *In* R.W. Weaver (ed.) Methods of soil analysis microbiological and biochemical properties. 3rd Ed. Soil Science Society of America, Madison, WI.

Wessel, W.W., and A. Tietema. 1992. Calculating gross nitrogen transformations in ^{15}N pool dilution experiments with acid forest litter: analytical and numerical approaches. Soil Biol. Biochem. 24:931–942.

Willigen, P. 1991. Nitrogen turnover in the soil-crop system; comparison of fourteen simulation models. Fert. Res. 27:141–149.

Wolf, D.C., J.O. Legg, and T. W. Boutton. 1994. Isotopic methods for the study of soil organic matter dynamics. p. 865–906. *In* R.W. Weaver (ed.) Methods of soil analysis: Microbiological and biochemical properties. 3rd Ed. Soil Science Society of America, Madison, WI.

Wood, P.M. 1986. Nitrification as a bacterial energy source. p. 39–62. *In* J.I. Prosser (ed.) Nitrification. IRL Press, Oxford, UK.

Woomer, P.L. 1994. Most probable number counts. p. 59–80. *In* R.W. Weaver (ed.) Methods of soil analysis: Microbiological and biochemical properties. 3rd Ed. Soil Science Society of America, Madison, WI.

Yamanaka, T. 1996. Mechanisms of oxidation of inorganic electron donors in autotrophic bacteria. Plant Cell Physiol. 37:569–574.

Yevdokimov, I.V., S.A. Blagodatski, and V.N. Kudeyarov. 1993. Microbiological immobilization, remineralization and plant uptake of fertilizer nitrogen. Eurasian Soil Sci. 25(8):64–72.

Zhou, J.Z., M.A. Bruns, and J.M. Tiedje. 1996. DNA recovery from soils of diverse composition. Appl. Environ. Microbiol. 62: 316–322.

Zou X., D.W. Valentine, R.L. Sanford, and D. Binkley. 1992. Resin-core and buried-bag estimates of nitrogen transformations in Costa Rican lowland rainforests. Plant Soil 139:275–283.

4.4 Denitrification

G. Philip Robertson
Michigan State University

4.4.1 Introduction

Denitrification is the dissimilatory reduction of soil NO_3^- to the nitrogen gases NO, N_2O, and N_2. It is carried out by a wide variety of mainly heterotrophic bacteria that use NO_3^- as a terminal electron acceptor when O_2 is unavailable. Hence, denitrification occurs in soil environments where C and NO_3^- are available during periods of restricted O_2 availability. In wetland soils, these conditions may exist most of the time. In upland soils, these conditions occur mainly following rainfall and within soil aggregates and decomposing litter.

At a global scale, denitrification is a crucial part of the overall N cycle. It is the only point in the cycle in which fixed N reenters the atmosphere as N_2. Thus, denitrification closes the N cycle. Without denitrification, atmospheric N_2 would eventually be drawn down to nil by N fixers (Section 4.1), and the biosphere would be awash in NO_3^-. Globally, denitrification in soil may account for > 60% of total $N_2 + N_2O$ production (Bowden, 1986; Aulakh et al., 1992).

Denitrification is important for other reasons as well. It is a major source (arguably the major source) of atmospheric N_2O, an important, radiatively active greenhouse gas that also consumes ozone once it reaches the stratosphere. It is a significant source of NO, another trace gas important to atmospheric chemistry. That concentrations of N_2O in the atmosphere are increasing annually is an important global change issue (IPCC, 1996) and N_2O is one of the three biogenic trace gases targeted for reduction at the recent International Framework Convention on Climate Change in Kyoto (Bolin, 1998).

At an ecosystem scale, denitrification can rival leaching as a vector for N loss in upland soils, and can exceed hydrologic losses of N both in wetlands and in arid and semiarid environments following rainfall or irrigation events. From a management perspective, denitrification can be a positive attribute when it is desirable to remove excess NO_3^- from soil prior to its movement to surface or groundwater (Lowrance et al., 1984). More often, however, managers seek to minimize denitrification in order to further conserve N for plant uptake. Another ecosystem level consequence of denitrification is its tendency to counter soil acidification. By removing a NO_3^- anion from the soil solution, denitrification effectively consumes acidity, an effect that can be particularly important in highly weathered, variable charge soils, in which CEC is largely governed by soil pH (Sollins et al., 1988; Robertson et al., 1988). Many tropical soils are dominated by variable charge minerals (Uehara and Gillman 1981), and denitrification appears to be an especially active process in the humid tropical soils examined thus far (Robertson and Tiedje, 1988; Groffman, 1995).

Complicating our understanding of denitrification in ecosystems with well-drained soils is denitrification's extreme spatial variability. Not only does denitrification appear to cluster temporarily around rain events (e.g., Sexstone et al., 1985a), it also is highly variable spatially on the order of centimeters to meters (Folorunso and Rolston, 1985; Robertson et al., 1988).

4.4.2 Denitrifiers

Denitrification is carried out by a broad array of bacteria, including mostly organotrophs, but also chemo- and photolithotrophs, N_2 fixers, thermophiles, halophiles, and various pathogens. It is remarkable that denitrification occurs in so broad an array of microbial taxa; over 50 genera with over 125 denitrifying species have been identified (Zumft, 1992). In soil, most culturable denitrifiers are facultative anaerobes from only 3–6 genera, principally *Pseudomonas* and *Alcaligenes*, and to a lesser extent *Bacillus*, *Agrobacterium*, and *Flavobacterium* (Tiedje, 1994). Typically, denitrifiers constitute 0.1 to 5% of the total culturable soil population (Tiedje, 1988).

Organisms denitrify to obtain energy (ATP) by electron transport phosphorylation via the cytochrome system. The general pathway is

$$2NO_3^- \xrightarrow{N_{ar}} 2NO_2^- \xrightarrow{N_{ir}} 2NO \xrightarrow{N_{or}} N_2O \xrightarrow{N_{os}} N_2 \qquad [4.4.1]$$

Each step is catalyzed by individual enzymes, namely, nitrate reductase (N_{ar}), nitrite reductase (N_{ir}), nitric oxide reductase (N_{or}), and nitrous oxide reductase (N_{os}). Each is inhibited by O_2, and the organization of these enzymes in the cell membrane has been worked out for Gram-negative bacteria as described in Fig. 4.10. At any step in this process intermediate products can be exchanged with the soil environment, making denitrifiers a significant source of NO_2^- in soil solution, and important sources of NO and N_2O gas fluxes.

Each denitrification enzyme is inducible, primarily in response to O_2 partial pressure (P_{O2}), soil pH, and substrate C:N ratio. Because enzyme induction is sequential and substrate dependent, there is usually a lag between the production of an intermediate substrate and its consumption by the next

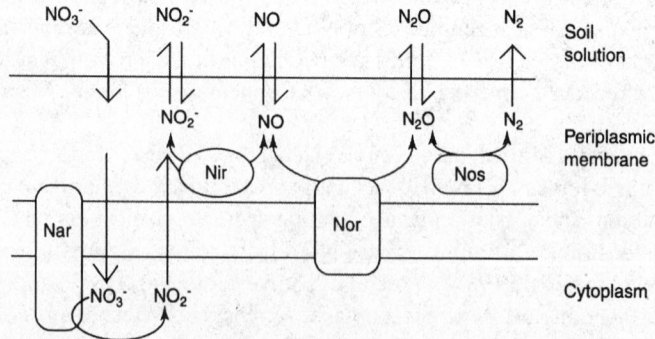

Fig. 4.10 The organization of denitrification enzymes in the cell membrane for Gram-negative bacteria [Reprinted from Ye et al., 1994. Appl. Environ. Microbiol 60:1053-1058, with permission from American Society for Microbiology]

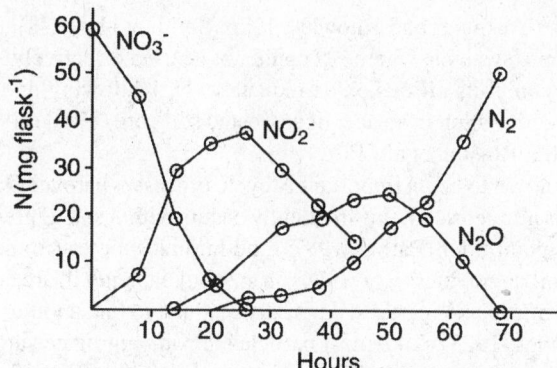

Fig. 4.11 The sequence of products formed during denitrification [Redrawn from Cooper and Smith, 1963. Soil Sci. Soc. Am Proc. 27:659-662, with permission of the Soil Science Society of America]

enzyme. In pure culture, these lags can be on the order of hours (Fig. 4.11); in bulk soil, lags can be substantially longer, and differences in lags among groups comprising the complex soil microbial community, together with substrate abundance, may significantly affect the contribution of denitrifiers to fluxes of the intermediates NO and N_2O to the atmosphere. The fact that induced enzymes degrade at different rates, and more slowly than they are induced, also leads to a hysteretic response around environmental conditions that induce denitrification; whether a soil has denitrified recently, i.e., denitrifying enzymes are present, may largely determine the soil's response to newly favorable conditions for denitrification. Rainfall onto soil that is moist, for example, will, all else being equal, lead to a faster, and perhaps stronger denitrification response than will rainfall onto the same soil when it is dry (Groffman and Tiedje, 1988).

By the same token, denitrifier N_2O production following rainfall on a previously dry soil may be initially stronger than N_2O production in a previously wet soil, because in the dry soil there will be a lag between the onset of rainfall and the induction and synthesis of nitrous oxide reductase (N_{os} in Fig. 4.10). This means that the moisture regime immediately prior to rainfall may partly determine the mole fraction of gas product that is N_2O (or the $N_2O:N_2$ ratio). Whether or not these lags are significant contributors to differences in N_2O production among different habitats is unknown and under active investigation.

Also contributing to differences in the molar ratio of $N_2O:N_2$ flux in soil may be population level differences among the bacterial species comprising the denitrifier community. Denitrifiers differ in their ability to reduce N_2O to N_2 (Abou Seada and Ottow, 1985; Munch, 1991), and recent evidence suggests that these differences, coupled with differences in denitrifier community composition among different habitats, may help to explain differences among habitats in rates of N_2O flux that do not appear to be directly related to environmental conditions (Cavigelli, 1998).

4.4.3 Controls on Denitrification

For decades after its discovery as an important microbial process, denitrification was assumed to be unimportant in nonsaturated soil environments. Only in sediments and wetland soils was denitrification recognized to be important to ecosystem function. In fact, it was not until the advent of whole ecosystem N budgets and the use of ^{15}N to trace the fate of fertilizer N that denitrification was considered potentially important in soils other than hydrosols. It was these studies (Allison, 1955) that suggested a role for denitrification in upland soils, though until the development of the acetylene block technique in the 1970s (Yoshinari and Knowles, 1976; Smith et al., 1978), the importance of denitrification in nonagronomic systems, where ^{15}N could not be used effectively, could not be confirmed. Acetylene selectively inhibits nitrous oxide reductase (N_{os}), allowing one to assess N_2 production by following N_2O accumulation in an acetylene treated soil core (Tiedje, 1994) or soil monolith (Ryden and Dawson, 1982; Rolston et al., 1982).

Today, denitrification is known to be an important N cycle process wherever O_2 is limiting and C and NO_3^- are available. In unsaturated soils, this frequently occurs within soil aggregates (Sexstone et al., 1985b), in decomposing plant litter (Parkin, 1987), and in rhizospheres (Prade and Trolldenier, 1990; Nieder et al., 1989). Soil aggregates vary widely in size but, in general, are comprised of small mineral particles and pieces of organic matter < 2 mm diam. glued to one another with biologically derived polysaccharides (Oades, 1993). Like most particles in soil, aggregates are surrounded by a thin water film that impedes gas exchange. Modeling efforts in the 1970s and 1980s (Arah and Smith, 1989) suggested that the centers of these aggregates ought to be anaerobic owing to a higher respiratory demand in the aggregate center than could be satisfied by O_2 diffusion from the bulk soil atmosphere. This was experimentally confirmed in 1985 (Fig. 4.12) (Sexstone et al., 1985b), providing a logical explanation for active denitrification in soils that appeared otherwise to be aerobic, and an explanation for the almost universal presence of denitrifiers and denitrification enzymes in soils worldwide (Gamble et al., 1977).

Three major environmental controls act in concert to regulate denitrification fluxes in soil, namely, C, NO_3^-, and O_2. Carbon is important because in most ecosystems most denitrifiers are heterotrophs (Poth and Focht, 1985) and require C as the electron donor. Nitrate serves as the electron acceptor, and must be provided via nitrification (Section 4.3) or from external sources such as atmospheric deposition or fertilization (Section 4.5). Oxygen is a preferred electron acceptor because of its high energy yield as compared to alternate acceptors such as NO_3^- or SO_4^{2-}, and thus O_2 must be depleted before denitrification occurs. In most soils, the majority of the denitrifiers present are facultative

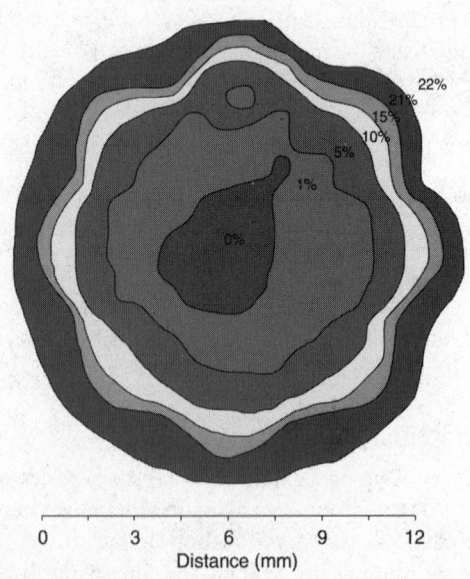

Fig. 4.12 Measured oxygen profile through a 1.2 cm soil aggregate [Redrawn from Sexstone et al., 1985b. Soil Sci. Soc. Am. J. 49:645-651, with permission of the Soil Science Society of America]

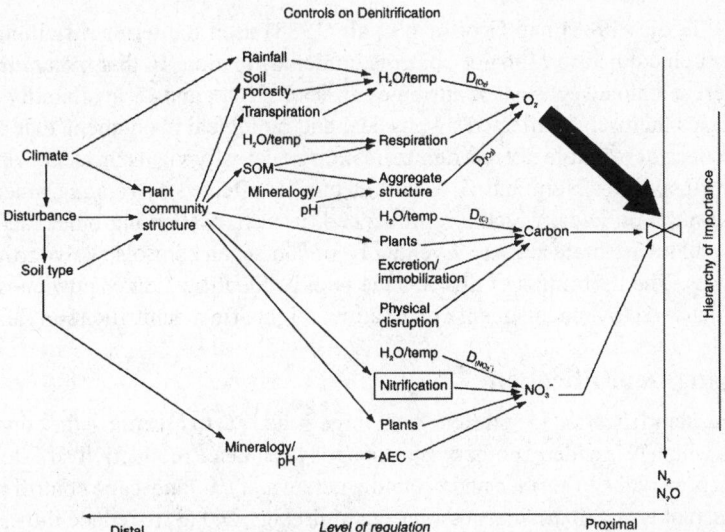

Fig. 4.13 Major controls on denitrification from the cellular (right) to landscape scales [Reprinted from Robertson, 1989. Nitrification and denitrification in humid tropical ecosystems. J. Proctor (ed.) Mineral nutrients in tropical forest and savanna ecosystems with permission of Blackwell Science, Oxford, UK]

anaerobes (Tiedje, 1988; Knowles, 1981) that will simply avoid synthesizing denitrifier enzymes until O_2 drops below some critical threshold.

In the field, O_2 is by far the dominant control on denitrification ($N_2O + N_2$) flux from soils. It is an easy matter to induce denitrification in an aerobic soil by removing O_2, and to reduce denitrification in saturated soil by drying or otherwise aerating it. The relative importance of C and NO_3^-, the other major controls, will vary by ecosystem. Under saturated conditions, such as those found in wetlands and lowland rice paddies, NO_3^- limits denitrification because nitrifiers, the principal source of NO_3^- in unfertilized ecosystems, are inhibited at low O_2 concentrations. Consequently, denitrification occurs only in the slightly oxygenated rhizosphere and at the sediment-water interface, places where nitrifiers have sufficient O_2 to oxidize NH_4^+ to NO_3^-, which can then diffuse to the denitrifiers in the increasingly anaerobic zones away from the root surface or sediment-water interface. It is often difficult to find NO_3^- in persistently saturated soils not only because of low nitrification, but also because of the tight coupling between nitrifiers and denitrifiers (Patrick, 1982; Reddy and Patrick, 1986; Mosier et al., 1990). In wetlands with fluctuating water tables or with significant inputs of NO_3^- from groundwater, NO_3^- may be less limiting.

In unsaturated soils, on the other hand, C availability more often limits denitrification. In these soils, carbon supports denitrification both directly, by providing donor electrons to denitrifiers, and indirectly, by stimulating O_2 heterotroph consumption. It can be difficult to distinguish between these two effects experimentally; from a management perspective, there probably is no need to. It is well recognized that exogenous C stimulates denitrification (Bremner and Shaw, 1958a,b; Knowles, 1981), although the C added must be an available form and must not lead to N immobilization sufficient to deplete NO_3^- availability (Firestone, 1982).

Where the effect of C on denitrification is indirect, through its effect on O_2 availability, C can be considered a secondary control, similar in effect to other controls on denitrification that serve to affect denitrification rates via their effects on the availability of the three primary controls of substrate

C, NO_3^- and O_2. Tiedje (1988) and Groffman et al. (1988) used the terms distal and proximal to describe the hierarchical nature of major controls on denitrification. In this scalar characterization (Fig. 4.13), different soil and ecosystem attributes affect C, NO_3^-, and O_2 availability differentially, with these attributes themselves affected by physical and biological phenomena that occur at larger spatial scales. Water, for example, affects denitrification principally via its influence on O_2 availability and on the diffusivity of NO_3^- and C; water, in turn, is affected by (among other factors) soil porosity and transpiration rates, which are influenced, in turn, by (among other factors) the plant community and soil invertebrate activity. Eventually, of course, all controls evolve from climate and land use influences. The usefulness of this scheme is in its identification of how ecosystem disturbance, whether delivered by management or by nature, might affect denitrification fluxes.

4.4.4   Managing Denitrification

Efforts to manage denitrification in soil stem from three goals: (1) to maximize the conservation of N in ecosystems to which N is added or otherwise managed to enhance productivity, (2) to remove NO_3^- from hydrologic flow paths to surface and groundwater, usually at landscape control points such as sewage treatment plants and riparian zones adjacent to streams, and (3) to reduce fluxes of N_2O to the atmosphere. This latter goal is relatively new and may well dominate efforts to control denitrification in the coming decades as legislation and policy are developed to meet the provisions of international agreements to reduce greenhouse gas emissions (Bolin, 1998). All three goals are interrelated, but their concomitant objectives are quite distinct from one another: the first goal can be met only by inhibiting or otherwise attenuating denitrification *in situ*, the second goal can be met principally by stimulating denitrification, and the third by either inhibiting denitrification or by skewing the molar end product ratio $N_2O:N_2$ toward N_2. Goals (1) and (2) are not mutually exclusive only because they are addressed in different parts of the landscape, which emphasizes the need to embrace a management perspective that is greater than the scale of individual fields.

Cropping systems receive most of the 80 Tg of fertilizer N now produced globally each year (Bumb, 1995). In general, about half of the fertilizer N applied to any given cropping system is lost as NO_3^- or N_2 gas in the year it is applied (Legg and Meisinger, 1982; Robertson, 1997). The portion lost as NO_3^- is that portion that is targeted for NO_3^- removal downstream (Goal 2 above); thus, conserving N loss in general, rather than simply inhibiting denitrification *in situ*, obviates the need to control denitrification both upstream and downstream.

How does one conserve N, and thereby, reduce denitrification at the field scale? Fertilizer composition and the rate and timing at which it is applied have a substantial impact on N loss from cropping systems. Myriad techniques are available for improving crop N use efficiency (Keeney, 1982; Myers, 1988; Peoples et al., 1995; Robertson, 1997). The most important of these is to synchronize N supply with plant N demand. Typically fertilizer is added to crops prior to the period of maximum crop growth, and prior to uptake this N is available to denitrifiers for transformation to N_2O and N_2 and is available to wetting fronts for hydrologic transport from the field. Moreover, much of the mineralizable N in crop residues from the previous year is mineralized (transformed from organic to inorganic form ([Section 4.2]) prior to crop growth, and this N also increases the soil solution pool available for loss. It is not unusual for the soil NO_3^- pool under annual field crops to increase by an order of magnitude prior to significant crop uptake.

The release of N from crop residues can be manipulated directly by tillage, by managing residue quality and quantity, and by the use of cover crops during periods when the field would otherwise be fallow. Cover crops remove inorganic N from the soil solution, and thereby, act as a temporary buffer against N loss during that portion of the cropping cycle when mineralized N accumulates in the soil solution and is available for uptake or loss. Properly managed, the N retained by the cover crop can be slowly released by mineralization during the growing season after it is killed when the principal crop is planted.

Unless fertilizer can be added to irrigation water, it is also difficult to manage the timing of fertilizer application. Usually, agronomic considerations dictate fertilizer application in a single pulse at a time when farm machinery can access the field without harming the growing crop or soil structure. Alternately, slow-release fertilizers such as sulfur-coated urea (Allen, 1984) can improve crop nutrient synchrony, though they generally are not economically practical under most cropping regimes.

In situ inhibitors have also been developed for microbial N transformations, and can effectively block denitrification in the field. Various nitrification inhibitors, for example, block NO_3^- formation, and thereby, reduce both denitrification and leaching losses of N. They are not widely used, however, owing to their expense relative to the lower expense of excess N use, and to their inconsistent behavior (Meisinger et al., 1983). They do, however, have the added advantage of blocking nitrifier-derived N_2O (Peoples et al., 1995; Klemedtsson and Mosier, 1994).

Denitrification downstream of the agricultural field, in particular at the soil-stream interface, can effectively remove NO_3^- from the hydrologic flow path (Lowrance et al., 1984; Peterjohn and Correll, 1986; Hedin et al., 1998). Together with NO_3^- uptake by vegetation, denitrification in riparian zones and wetlands provides an effective filter to protect surface and groundwater from NO_3^- contamination. Thus the design and management of riparian buffer strips have become a particularly important consideration for managers seeking to reduce watershed level NO_3^- inputs to coastal waters (Hubbard and Lowrance, 1997). Likewise, wetland restoration and water level management may also become an important management tool for reducing downstream NO_3^- inputs.

Moving into an era in which fluxes of greenhouse gases become increasingly regulated, one may see a call for reducing N_2O fluxes irrespective of reductions in denitrification *in toto* ($N_2O + N_2$). Management to favor the further reduction of N_2O to N_2 will, in cropping systems, likely focus on available soil NO_3^- concentrations and on water management. For energetic considerations, NO_3^- is the preferred electron acceptor when both NO_3^- and N_2O are present (Harris, 1982), which leaves N_2O as a more likely denitrifier end product when NO_3^- is plentiful (Weier et al., 1993; Linn and Doran, 1984). Likewise, N_2O will be a more likely end product when soil conditions that favor denitrification are separated by periodic droughts or other conditions that inhibit denitrification activity, periods during which denitrifiers lose N_2O reductase (N_{os} in Fig. 4.10), and therefore, their ability to reduce N_2O to N_2 immediately following the onset of NO_3^- reduction. Management that favors small NO_3^- pools (e.g., smaller, more frequent fertilizer applications), more consistently moist soils (e.g., conservation tillage or irrigation), or more consistently anaerobic soil aggregates (e.g., no-till management) may skew the $N_2O:N_2$ ratio toward N_2.

4.4.5 Conclusions

Denitrification is a crucial part of the global N cycle and can be a major pathway of N loss from terrestrial ecosystems. As a major source of the atmospheric trace gas N_2O, and a significant source of the trace gas NO, denitrifiers also play an important role in atmospheric chemistry. Recent advances in our understanding of denitrification have clarified its biochemical pathway, have provided knowledge of its importance in even well-aerated, upland soils, have demonstrated its ability for helping to mitigate watershed level NO_3^- fluxes, and have clarified ecosystem and watershed level controls on rates of denitrification in both natural and managed habitats. Efforts to model denitrification quantitatively have been stymied by an incomplete understanding of how the factors that affect denitrification rates interact *in situ*, including the influence of denitrifier community composition. Calls to better understand denitrification and especially the factors regulating end product composition (the $N_2O:N_2$ ratio) may grow more intense as policymakers respond to international pressure to reduce greenhouse gas emissions globally.

Acknowledgments

I thank T.T. Bergsma, J.C. Boles, and especially M.A. Cavigelli and an anonymous reviewer for valuable comments on earlier drafts of this chapter. Support for this effort was provided in part by the DOE-NIGEC Program, the NSF LTER Program, and the Michigan Agricultural Experiment Station.

4.4.6   References

Abou Seada, M.N.I., and J.C.G. Ottow. 1985. Effect of increasing oxygen concentration on total denitrification and nitrous oxide release from soil by different bacteria. Biol. Fert. Soils 1:31–38.

Allen, S.E. 1984. Slow-release nitrogen fertilizers. p. 195–206. *In* R.D. Hauck (ed.) Nitrogen in crop production. American Society of Agronomy, Madison, WI.

Allison, F.E. 1955. The enigma of soil nitrogen balance sheets. Adv. Agron 7:213–250.

Arah, J.R.M., and K.A. Smith. 1989. Steady-state denitrification in aggregated soil: A mathematical model. J. Soil Sci. 40:139–149.

Aulakh, M. S., J. W. Doran, and A. R. Mosier. 1992. Soil denitrification–significance, measurement, and effects of management. Adv. Soil Sci. 18: 1–52.

Banerjee, N.K., and A.R. Mosier. 1989. Coated calcium carbide as a nitrification inhibitor in upland and flooded soils. J. Ind. Soc. Soil Sci. 37:306–313.

Bolin, B. 1998. The keys to negotiations on climate change: A science perspective. Science 279:330–331.

Bowden, W. B. 1996. Gaseous nitrogen emissions from undisturbed terrestrial ecosystems: an assessment of their impacts on local and global nitrogen budgets. Biogeochem. 2: 249–279.

Bremner, J.M., and K. Shaw. 1958a. Denitrification in soil. I. Methods of investigation. J. Agric. Sci. 51:22–39.

Bremner, J.M., and K. Shaw. 1958b. Denitrification in soil. II. Factors affecting denitrification. J. Agric. Sci. 51:39–52.

Bumb, B.L. 1995. World nitrogen supply and demand: an overview. p. 1–41. *In* P.E. Bacon (ed.) Nitrogen fertilization in the environment. Marcel Dekker, New York, NY.

Cavigelli, M. 1998. Ecosystem consequences and spatial variability of soil microbial community structure. Ph.D. Dissertation, MI State University, East Lansing, MI.

Cooper, G.S., and R. Smith. 1963. Sequence of products formed during denitrification. Soil Sci. Soc. Am. Proc. 27:659–662.

Firestone, M.K. 1982. Biological denitrification. p. 289–326. *In* F.J. Stevenson (ed.) Nitrogen in agricultural soils. American Society of Agronomy, Madison, WI.

Folorunso, O. A., and D. E. Rolston. 1985. Spatial variability of field-measured denitrification gas fluxes. Soil Sci. Soc. Am. J. 48: 1214–1219.

Gamble, T.N., M.R. Betlach, and J.M. Tiedje. 1977. Numerically dominant denitrifying bacteria from world soils. Appl. Environ. Microbiol. 33:926–939.

Groffman, P.M., J.M. Tiedje, G.P. Robertson, and S. Christensen. 1988. Denitrification at different temporal and geographical scales: proximal and distal controls. p. 174–192. *In* J.R. Wilson (ed.) Advances in nitrogen cycling in agricultural ecosystems. CAB International, Wallingford, UK.

Groffman, P. M. 1995. Assessment and importance of denitrification as a source of soil nitrogen loss in tropical agroecosystems. Fert. Res. 42:139–148.

Groffman, P.M., and J.M. Tiedje. 1988. Denitrification hysteresis during wetting and drying cycles in soil. Soil Sci. Soc. Am. J. 52:1626–1629.

Groffman, P.M., and J.M. Tiedje. 1989. Denitrification in north temperate forest soils: spatial and temporal patterns at the landscape and seasonal scales. Soil Biol. Biochem. 21:613–620.

Harris, R.F. 1982. Energetics of nitrogen transformations. p. 833–890. *In* F.J. Stevenson (ed.) Nitrogen in agricultural soils. American Society of Agronomy, Madison, WI.

Hedin, L.O., J.C. von Fischer, N.E. Ostrom, M.G. Brown, and G.P. Robertson. 1998. Thermodynamic constraints on nitrogen transformation and other biogeochemical processes at soil-stream interfaces. Ecol. 79:684–703.

Hubbard, R.K., and R. Lowrance. 1997. Assessment of forest management effects on nitrate removal by riparian. Trans. ASAE 40:383.

IPCC. 1996. Climate change 1995-The science of climate change. Cambridge University Press, Cambridge, MA.

Keeney, D.R. 1982. Nitrogen management for maximum efficiency and minimum pollution. p. 605–650. *In* F.J. Stevenson (ed.) Nitrogen in agricultural soils. American Society of Agronomy, Madison, WI.

Keeney, D.R. 1983. Factors affecting the persistence and bioactivity of nitrification inhibitors. p. 33–46. *In* J.J. Meisinger, G.W. Randall, and M.L. Vistosh (ed.) Nitrification inhibitors: Potentials and limitations. American Society of Agronomy, Madison, WI.

Klemedtsson, L. K., and A. R. Mosier. 1994. Effect of long-term field exposure of soil to acetylene on nitrification, denitrification, and acetylene consumption. Biol. Fert. Soils 18:42–48.

Knowles, R. 1981. Denitrification. p. 315–329. *In* F.E. Clark and T. Rosswall (ed.) Terrestrial nitrogen cycles. Swedish Natural Science Research Council, Stockholm, Sweden.

Legg, J.O., and J J. Meisinger. 1982. Soil nitrogen budgets. p. 503–566. *In* F.J. Stevenson (ed.) Nitrogen in agricultural soils. American Society of Agronomy, Madison, WI.

Linn, D.M., and J.W. Doran. 1984. Effect of water-filled pore space on CO_2 and N_2O production in tilled and non-tilled soils. Soil Sci. Soc. Am. J. 48:1267–1272.

Lowrance, R.R., R.L. Todd, J. Fail, O. Hendrickson, R. Leonard, and L. Asmussen. 1984. Riparian forests as nutrient filters in agricultural watersheds. BioSci. 34:374–377.

Meisinger, J.J., G.W. Randall, and M.L. Vitosh, eds. 1983. Nitrification inhibitors-Potentials and limitations. American Society of Agronomy, Madison, WI.

Mosier, A. R., S. K. Mohanty, A. Bhadrachalam, and S. P. Chakravorti. 1990. Evolution of dinitrogen from the soil to the atmosphere through rice plants. Biology and Fertility of Soils 9:61–67.

Munch, J.C. 1991. Nitrous oxide emission from soil as determined by the composition of denitriifying microbial population. p. 309–316. *In* J. Berthelin (ed.) Diversity of environmental biogeochemistry. Elsevier, Amsterdam, Netherlands.

Myers, R.J.K. 1988. Nitrogen management of upland crops: from cereals to food legumes to sugarcane. p. 257–273. *In* J.R. Wilson (ed.) Advances in nitrogen cycling in agricultural ecosystems. CAB International, Wallingford, UK.

Nieder, R., G. Schollmayer, and J. Richter. 1989. Denitrification in the rooting zone of cropped soils with regard to methodology and climate: a review. Biol. Fert. Soils 8: 219–226.

Oades, J.M. 1993. The role of biology in the formation, stabilization and degradation of soil structure. Geoderma 56:377–400.

Parkin, T. B. 1987. Soil microsites as a source of denitrification variability. Soil Sci. Soc. Am. J. 51:1194–1199.

Patrick, W.H. 1982. Nitrogen transformation in submerged soils. p. 449–465. *In* F.J. Stevenson (ed.) Nitrogen in agricultural soils. American Society of Agronomy, Madison, WI.

Peoples, M.B., A.R. Mosier, and J.R. Freney. 1995. Minimizing gaseous losses of nitrogen. p. 565–602. *In* P.E. Bacon (ed.) Nitrogen fertilization in the environment. Marcel Dekker, New York, NY.

Peterjohn, W.T., and D.L. Correll. 1986. The effect of riparian forest on the volume and chemical composition of baseflow in an agricultural watershed. p. 244–262. *In* D.L. Correll (ed.) Watershed research perspectives. Smithsonian Institution Press, Washington, DC.

Poth, M., and D.D. Focht. 1985. ^{15}N kinetic analysis of N_2O production by *Nitrosomonas europaea* and examination of nitrifier denitrification. Appl. Environ. Microbiol 49:1134–1141.

Prade, K., and G. Trolldenier. 1990. Denitrification in the rhizosphere of plants with inherently different aerenchyma formation: wheat (*Triticum aestivum*) and rice (*Oryza sativa*). Biol. Fert. Soils 9:215–219.

Reddy, K.R., and W.H. Patrick Jr.. 1986. Denitrification losses in flooded rice fields. Fert. Res. 9:99–116.

Robertson, G.P. 1989. Nitrification and denitrification in humid tropical ecosystems. p. 55–70. *In* J. Proctor (ed.) Mineral nutrients in tropical forest and savanna ecosystems. Blackwell Scientific, Cambridge, MA.

Robertson, G.P. 1997. Nitrogen use efficiency in row-crop agriculture: crop nitrogen use and soil nitrogen loss. p. 347–365. *In* L. Jackson (ed.) Ecology in agriculture. Academic Press, New York, NY.

Robertson, G. P., and J. M. Tiedje. 1988. Denitrification in a humid tropical rainforest. Nature 336: 756–759.

Robertson, G., M. Huston, F. Evans, and J. Tiedje. 1988. Spatial variability in a successional plant community: patterns of nitrogen availability. Ecol. 69: 1517–1524.

Rolston, D.E., A.N. Sharpley, D.W. Toy, and F.E. Broadbent. 1982. Field measurement of denitrification: III. Rates during irrigation cycles. Soil Sci. Soc. Am. J. 46:289–296.

Ryden, J.C., and K.P. Dawson. 1982. Evaluation of the acetylene inhibition technique for the measurement of denitrification in grassland soils. J. Sci. Food Agric. 33:1197–1206.

Sexstone, A. J., T. P. Parkin, and J. M. Tiedje. 1985a. Temporal response of soil denitrification rates to rainfall irrigation. Soil Sci. Soc. Am. J. 49:99–103.

Sexstone, A.J., N.P. Revsbech, T.P. Parkin, and J.M. Tiedje. 1985b. Direct measurement of oxygen profiles and deni-
    trification rates in soil aggregates. Soil Sci. Soc. Am. J. 49:645–651.
Smith, M.S., M.K. Firestone, and J.M. Tiedje. 1978. The acetylene inhibition method for short-term measurement of
    soil denitrification and its evaluation using nitrogen-13. Soil Sci. Soc. Am. J. 42:611–615.
Sollins, P., G.P. Robertson, and G. Uehara. 1988. Nutrient mobility in variable- and permanent-charge soils. Biogeochem.
    6:181–199.
Tiedje, J.M. 1988. Ecology of denitrification and dissimilatory nitrate reduction to ammonium. p. 179–244. In A.J.B.
    Zehnder (ed.) Biology of anaerobic microorganisms. John Wiley and Sons, New York, NY.
Tiedje, J.M. 1994. Denitrifiers. p. 245–267. In R.W. Weaver, J.S. Angle, P.S. Bottomley, D. Bezdecik, S. Smith, A.
    Tabatabai, and A. Wollum (ed.) Methods of soil analysis, Part 2. Microbiological and biochemical properties. Soil
    Science Society of America, Madison, WI.
Uehara, G. and G.P. Gillman. 1981. The mineralogy, chemistry and physics of tropical soils with variable charge.
    Westview Press, Boulder, CO.
Weier, K.L., J.W. Doran, J.F. Power, and D.T. Walters. 1993. Denitrification and the dinitrogen/nitrous oxide ratio as
    affected by soil water, available carbon, and nitrate. Soil Sci. Soc. Am. J. 57:66–72.
Ye, R.W., B.A. Averill, and J.M. Tiedje. 1994. Denitrification of nitrite and nitric oxide. Appl. Environ. Microbiol.
    60:1053–1058.
Yoshinari, T., and R. Knowles. 1976. Acetylene inhibition of nitrous oxide reduction by denitrifying bacteria. Biochem.
    Biophys. Res. Commun. 69:705–710.
Zehnder, A.J.B., and W. Stumm. 1988. Geochemistry and biogeochemistry of anaerobic habitats. p. 1–38. In A.J.B.
    Zehnder (ed.) Biology of anaerobic microorganisms. Wiley, New York, NY.
Zumft, W.G. 1992. The denitrifying procaryotes. p. 555–582. In A. Balows et al. (ed.) The Prokaryotes. Springer-
    Verlag, New York, NY.

4.5   Nitrogen in the Environment

Peter M. Groffman
Institute of Ecosystem Studies

4.5.1   Introduction

While N is an extremely valuable and important nutrient within the soils of nearly all ecosystems on
the globe (Vitousek and Howarth, 1991), it becomes a highly problematic pollutant once it leaves one
ecosystem and moves into another. Given that human alteration of the global N cycle is extreme
(Tamm, 1991), that there are multiple forms of N that can be considered as pollutants, and that there

Table 4.10   Forms of N of concern in the environment

N form	Sources	Dominant transport vectors	Environmental effects
Nitrate (NO_3^-)	Fertilizer Disturbance Combustion (acid rain)	Groundwater	Toxic Eutrophication
Ammonia (NH_3) Ammonium (NH_4^+)	Fertilizer Animal waste	Surface runoff Atmosphere	Toxic Eutrophication
Nitrous oxide (N_2O)	Byproduct of multiple transformations	Atmosphere Groundwater	Greenhouse gas Ozone destruction
Nitric oxide (NO)	Byproduct of multiple transformations	Atmosphere	Ozone precurser
Dissolved organic N	Byproduct of natural decomposition	Surface runoff Groundwater	Eutrophication(?)

are multiple transport vectors for these forms of N, it is not surprising that the topic of N in the environment is one of the most well studied and controversial in soil and environmental science.

Most of the concern and research about N in the environment is about nitrate (NO_3^-), which is the most mobile form of N and is the most common groundwater pollutant in the United States (Keeney, 1986; USEPA, 1990). However, there is also environmental concern about other forms of N that can be transported from one ecosystem to another, including ammonia (NH_3), ammonium (NH_4^+), nitrous oxide (N_2O), nitric oxide (NO) and dissolved organic N (DON). This section provides a brief review of how each of these species of N is produced, transformed and transported in the environment and a framework for site or ecosystem scale evaluations: (1) is this site (ecosystem) a source of N to the environment (surrounding ecosystems), and (2) does this site (ecosystem) have the potential to absorb N, i.e., function as a sink? This framework is based on the idea that N pollution problems can be best addressed in a landscape scale context, where a landscape consists of a range of ecosystems (different cropping systems, different forest types on different soils, wetlands, pastures) that can be viewed as either N sources or sinks, and that are connected by hydrologic and atmospheric transport pathways (National Research Council, 1993). A landscape approach allows for reconciliation of the within system benefits and external pollution problems that are inherent in analysis of N pollution problems, and provides a strong basis for evaluation and management of these problems in the field. The goal is to provide the reader with an approach to evaluation of key N source, sink and transport processes at any point, in any landscape.

4.5.2 Forms of Nitrogen in the Environment

Detailed descriptions of the forms and transformations of N can be found elsewhere (Section B, Chapter 1; Section D, Chapters 1, 3, 4, 5; Section G, Chapters 7, 10, 11) and earlier in this chapter. A brief discussion of why certain forms of N are of concern in the environment, how they are produced and/or consumed within ecosystems, and how they are transported between ecosystems (Table 4.10) will be presented.

4.5.2.1 Nitrate

Nitrate is regulated as a drinking water pollutant (USEPA, 1990) and is considered to be a prime agent of eutrophication in estuarine and coastal ecosystems (Ryther and Dunstan, 1971; Howarth et al., 1996). This ion is added directly to ecosystems in some fertilizer types and is produced by mineralization (conversion of organic N to NH_4^+) and nitrification (conversion of NH_4^+ to NO_3^-) of other fertilizer and organic materials (e.g., plant detritus, manures, organic wastes). Nitrate is highly soluble and does not have a strong affinity for soil particles, except in highly weathered sesquioxidic subsoils (Bellini et al., 1996); therefore, it is readily transported in surface runoff and percolation. It is frequently the dominant form of N in percolating water and aquifers.

Ecosystems tend to be sources of NO_3^- to the environment if they are heavily fertilized and/or highly disturbed. As described earlier, there is intense competition for N between and among plants and microbes. In many ecosystems, NO_3^- losses are prevented by this competition, primarily because nitrifying organisms do not compete well with plants and other microbes for NH_4^+ (Vitousek et al., 1982). Therefore, NO_3^- loss occurs only where N inputs from fertilization, atmospheric or hydrologic sources are high enough to satisfy plant and microbial demands for N or where some type of disturbance disrupts plant and microbial competition, allowing nitrification to occur (Aber et al., 1989). There are many types of disturbance that can cause this disruption. Disturbances to the plant community that induce NO_3^- losses include crop harvest (Gold et al., 1990), forest clear cutting

(Bormann and Likens, 1979), insect outbreaks (Swank et al., 1981) and hurricanes (Bowden et al., 1993). The dominant soil disturbance that can stimulate NO_3^- losses is tillage, but more subtle disturbances (e.g., earthworm activity) can also be important (Edwards and Bohlen, 1996).

In recent years considerable attention has been paid to atmospheric transport of NO_3^- (Agren and Bosatta, 1988; Aber et al., 1989). This NO_3^- is formed in the atmosphere from N oxides emitted by combustion processes and returns to earth in rainfall (wet deposition) or by surface deposition as particles and/or vapor (dry deposition). Rates of atmospheric NO_3^- deposition range from 5 to more than 20 kg N ha^{-1} yr^{-1} worldwide (Johnson, 1992; Lovett, 1994). Recognition of the importance of NO_3^- deposition has led to concern about N saturation of natural (i.e., unfertilized) ecosystems, which has been linked to a variety of deleterious chemical and biological changes in temperate ecosystems (e.g., soil acidification) (van Breemen et al., 1982; Christ et al., 1995).

Many ecosystems function as sinks for NO_3^- in the landscape. Ecosystems where the growth of plants and/or microbes is strongly N limited will take up NO_3^- that is added in fertilizer or that enters by atmospheric or hydrologic transport (Vitousek and Howarth, 1991). Ecosystems with high actual or potential rates of denitrification (i.e., with significant volumes of anaerobic soil) function as very effective sinks for NO_3^-. The location of sink ecosystems relative to sources (e.g., wet riparian zones located between intensively farmed uplands and streams) is often a critical determinant of the nature and extent of NO_3^- pollution problems in a landscape (Lowrance et al., 1984). The capacity of groundwater to support denitrification is another critical (and highly variable) determinant of landscape scale NO_3^- dynamics (Madsen, 1995). There also appear to be abiotic sinks for N in soil (Davidson et al., 1991; Johnson, 1992; Hart et al., 1993), likely caused by chemical and physical condensation reactions between old organic matter, e.g., humus and NO_3^- (Haider et al., 1975; Johnson, 1992).

The challenge in evaluations of NO_3^- pollution problems is to determine the location, strengths and hydrologic and atmospheric connections between sources and sinks in the landscape. Approaches to making these determinations are described in Section 4.5.3.

4.5.2.2 Ammonia and Ammonium

Ammonia is likely the most common inorganic form of N in soils. Most N fertilizers are NH_3 (or urea, which rapidly degrades to NH_3) based and NH_3 is the first inorganic N form released in decomposition (mineralization). However, free NH_3 is rarely present in soils of low and moderate pH (< 8.0). Under these conditions, NH_3 exists in a protonated form, NH_4^+, which has a strong affinity for negatively charged clay and organic matter particles and is, therefore, much less mobile than NO_3^-. Free NH_3 is quite toxic however, and both NH_3 and NH_4^+ can readily cause eutrophication in N limited aquatic ecosystems. There is a vast flux of NH_4^+ in most soils from mineralization-immobilization turnover, but most of this NH_4^+ never leaves the surface horizons of the soil. In contrast to NO_3^-, NH_4^+ is seldom found in groundwater.

The dominant vector of NH_3 transport in the environment is atmospheric. In areas of intensive animal production, rates of NH_3 volatilization are high enough to cause significant loading (e.g., 40–50 kg N ha^{-1} yr^{-1}) of adjacent ecosystems by atmospheric deposition (Berendse et al., 1993). Some of the most dramatic concerns about N saturation of natural ecosystems have arisen in areas of intensive animal production (e.g., The Netherlands). Ammonium is frequently transported in surface runoff, most commonly in soils that have been treated with organic wastes (Vellidis et al., 1996).

Similar to NO_3^-, many ecosystems have a high capacity to function as sinks for NH_4^+. In most cases, NH_4^+ is more conservative than NO_3^- because it is less mobile and because in addition to biological uptake by plants and microbes, there can be strong physicochemical sinks for NH_4^+ on cation exchange sites and in the interlayer spaces of some clay minerals.

4.5.2.3 Nitrous Oxide

Nitrous oxide is produced and/or consumed by several N transformation processes described earlier, including nitrification, denitrification and dissimilatory NO_3^- reduction to NH_3. There is environmental concern about N_2O because it is a greenhouse gas which can influence the earth's radiative budget and it plays a role in stratospheric ozone destruction (Mooney et al., 1987; Prather et al., 1995).

Identifying and quantifying sources of N_2O in the landscape is hindered by high spatial and temporal variability in the dynamics of this gas (Folonoruso and Rolston, 1984). This variability is caused by the fact that there are multiple processes that affect this gas and there are multiple factors influencing each of these processes. However, it is clear that there is a generic relationship between N richness and N_2O flux. This relationship applies to both the inherent N richness of natural ecosystems (Matson and Vitousek, 1987; Firestone and Davidson, 1989) as well as to fertilizer inputs (Mosier et al., 1996). Nitrous oxide flux tends to be particularly high in N-rich ecosystems with low pH (Firestone and Davidson, 1989). Wet tropical soils, which tend to be N-rich and acidic, are considered to be the strongest N_2O sources (Matson and Vitousek, 1990).

Although N_2O is readily consumed by several processes in soils, most notably by denitrification in wet soils, it is quite stable once it gets into the atmosphere. This stability contributes greatly to the environmental concern about this gas. High concentrations of N_2O are also common in groundwater (Dowdell et al., 1979; Ronen et al., 1988; Ueda et al., 1993). Similar to NO_3^-, wet soils with a high potential for denitrification should be considered to be potential sinks for N_2O. In areas where N_2O-rich groundwater interacts with such soils, this potential could be important to landscape N_2O fluxes.

4.5.2.4 Nitric Oxide

Similar to N_2O, NO is a gas that is produced as a byproduct of several soil N cycle processes, especially nitrification and denitrification. In contrast to N_2O, however, NO is quite reactive and short lived in the atmosphere. There is environmental concern about NO because it is a precursor to ozone in the lower atmosphere and contributes to the formation of atmospheric NO_3^- deposition (Logan, 1983). The highly reactive nature of NO makes it very difficult to predict ecosystems that will be strong NO sources. However, similar to N_2O, ecosystems that are N rich are likely to produce relatively large amounts of NO (Hutchinson and Davidson, 1993). Wet ecosystems are likely to have a high potential to consume NO.

4.5.2.5 Dissolved Organic Nitrogen

Dissolved organic N has generally not been considered to be of environmental concern. However, recent studies have shown that DON is a significant pathway of N loss from many natural ecosystems (Hedin et al., 1995) that increases in response to N inputs (Currie et al., 1996). Moreover, much of the DON that leaves ecosystems is highly labile (Qualls and Haines, 1992) and can decompose to inorganic N and cause eutrophication problems in N-limited waters (Carlsson et al., 1993). It is likely that DON will receive increased attention as an environmental concern in the near future.

4.5.3 Evaluating N Sources and Sinks and Landscape Hydrologic Connections

A series of questions that allow for evaluation of key N source, sink and transport processes at any point, in any landscape will be posed which serves two purposes. First, they are a convenient vehicle for summarizing the diverse sources of information on N dynamics that are relevant to evaluating N in the environment and second, they provide a practical basis for managing N pollution problems. Consider the landscape depicted in Fig. 4.14. To evaluate N in this landscape, identification of which

Table 4.11 Criteria for determining if a site is a source of sink of N in the landscape

Criteria	Determinants
Is the site N rich?	Fertilized
	Fine texture
	Legumes
	Wet tropics
Is the site highly disturbed?	Disturbance of plant uptake (e.g., harvest)
	Stimulation of mineralization (e.g., tillage)
	Disturbance of links between plant and microbial processes (e.g., tillage)
Does the site have a high potential to consume NO_3^-?	Wet soil
	High organic matter
Does the site have a high potential for NH_3 volatilization?	High pH (> 8.0)

ecosystems are sources of N, how different forms of N move from one ecosystem to the other, and which ecosystems function as sinks for N, is needed. Note that Fig. 4.14 does not show atmospheric transport. The following questions are designed to provide a basis for thinking about N in a landscape context, and to point to key landscape scale factors that should be considered in evaluating N in the environment.

4.5.3.1 Is this Ecosystem Potentially a Sink or Source of N?

There are several relatively simple factors (Table 4.11) that can be used to determine if a particular site is a source of N in the landscape. The primary factor to determine is if this is an inherently N-rich site.

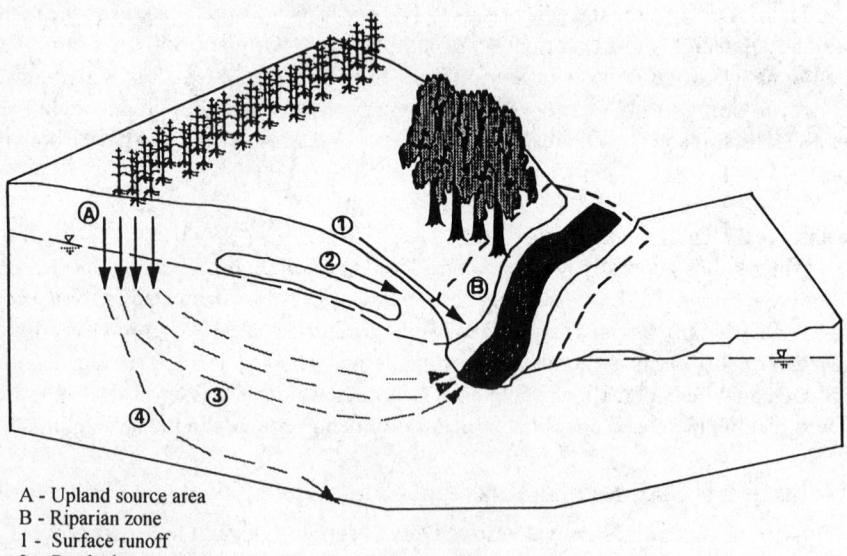

A - Upland source area
B - Riparian zone
1 - Surface runoff
2 - Perched water
3 - Groundwater flow
4 - Underflow or deep recharge

Fig. 4.14 Conceptual diagram of a landscape showing different hydrologic flowpaths from uplands to riparian zones to streams [Reprinted from Schnabel et al., 1994. Riperian ecosystems in the humid U.S., with permission of the National Association of Conservation Districts]

In natural ecosystems, inherent N richness is strongly linked to soil texture and to a lesser extent to parent material. Ecosystems developed on coarse textured soils tend to have vegetation that exhibits adaptations to frequent water stress. These adaptations (e.g., thick cuticle) result in the production of low quality plant litter that results in low rates of N availability (Pastor et al., 1984). Coarse textured soils also hold less organic matter and support lower microbial biomass than fine textured soils (Paul and Clark, 1996), which also reduces the inherent N richness of a site.

All fertilized ecosystems can be considered to be N rich, but there are marked differences in the N source strength of different fertilized ecosystems; e.g., fertilized forests will tend to be much weaker N sources in the landscape than fertilized agricultural fields. Much of this difference is related to the nature and extent of fertilization and to the disturbance regime (discussed below) of the site, e.g., agricultural sites lose more N than forests because they are more highly disturbed. In addition to fertilizer, sites subject to N inputs through symbiotic fixation by legumes, high rates of N deposition, or high rates of groundwater or surface runoff N loading should also be considered to be potential sources of N in the landscape. Tropical soils tend to be more N rich than temperate zone soils (Matson and Vitousek, 1990).

Next to N richness, the dominant factor controlling whether a site is an N source in the landscape is disturbance regime. There is a large body of literature describing the effects of disturbances ranging from clear cutting to tillage to grazing to acid rain on N losses from natural and agricultural ecosystems (Bormann and Likens, 1979; Mooney and Godron, 1983). It is important to note that not all disturbances foster N losses from all ecosystems. The effect of disturbance on N cycling is dependent both on the nature and extent of the disturbance and on properties of the ecosystem being disturbed; for example, clear cutting is more likely to produce N losses from inherently N-rich forests than from inherently N-poor forests (Vitousek et al., 1982). Disturbances that reduce plant uptake (e.g., clear cutting, harvest, herbiciding, grazing, fire), stimulate microbial mineralization activity (e.g., tillage), or disrupt linkages between microbial mineralization and plant uptake (e.g., tillage, harvest, earthworm activity, fire) are particularly likely to increase N losses from an ecosystem. It is important to stress that the effects of disturbance on N loss are often transient and/or difficult to predict. For example, many disturbances that reduce plant uptake in the short term (e.g., clear cutting reduces plant uptake for several months; fire decreases plant uptake for several weeks) increase uptake in the mid and long term (years for clear cutting; months for fire). Ecosystems that have not had any major disturbance for a long time (e.g., late successional or old growth forests) tend to have higher N losses than younger, more actively growing mid-successional ecosystems (Bormann and Likens, 1979).

In addition to criteria for predicting if a site is a source of N in the landscape (N richness, disturbance regime), there are several factors that can be evaluated to determine if a site is a potential sink for N. Obviously, if a site is inherently N poor (e.g., N poor soils, young successional), it will have a high potential to absorb inorganic N and transform it into organic N forms that are not of environmental concern. In addition to considering N richness, it is important to consider if a site is particularly supportive of key processes (e.g., denitrification, NH_3 volatilization) that consume N forms of environmental concern. Sites with high volumes of anaerobic soil (e.g., wetlands) have a high potential to support denitrification. Sites with high pH (> 8.0) have a high potential to support NH_3 volatilization.

Evaluation of site sink potential is complicated in several ways, however. First, not all inherently N-poor sites will be effective sinks for N; for example, sites on coarse textured soils transmit water so fast that N is not taken up by plants or microbes very effectively (Lajtha et al., 1995) (Section 4.5.3.2). Second, it is important to recognize that consumption of one N form of environmental concern can lead to production of another form of concern (e.g., denitrification can convert NO_3^- into N_2O). Finally,

Table 4.12 Critical variables affecting the movement of N into, through and out of a site

Variable	Status	Results
Texture	Coarse	Infiltration-movement to groundwater
	Fine	Surface runoff
Structure	Well	Infiltration-movement to groundwater
	Poor	Surface runoff
Drainage	Well	Infiltration-movement to groundwater
	Poor	Surface runoff
Soil Cover	High	Infiltration-movement to groundwater
	Low	Erosion by wind and water
pH	High	NH_3 volatilization
	Low	NH_3 stabilization as NH_4^+
		Potential for high yield of N_2O and NO

there can be a big difference between the actual and potential sink strength of an ecosystem within a landscape. If an ecosystem with a high potential to act as an N sink is not situated in a position to receive N from landscape N sources, it will not function as a sink (Section 4.5.3.3)

While it is clear that there are several readily observable criteria (N richness, disturbance regime, wet soil) that can be used to predict if a site is an N source or sink in the landscape, it is also clear that there are numerous complicating factors influencing the nature and extent of source and sink strengths that motivate more sophisticated evaluation techniques. Simulation models that integrate and synthesize the diverse factors affecting N dynamics can be used to produce an integrated depiction of N dynamics at a site. There are numerous site or ecosystem scale models of N dynamics for both agricultural (Rose et al., 1990; De Willigen, 1991; Bouma et al., 1993) and natural (Power, 1993) ecosystems. A review of these models, which vary widely in their approach, emphasis, complexity and ease of application is well beyond the scope of this chapter. The key challenge for landscape scale assessments of N in the environment is the ease with which a particular model can be used to identify sources and sinks within the landscape, and then to link them through hydrologic or atmospheric flows. Identifying multiple sources and sinks in a landscape is challenging to site or ecosystem scale models because soil, plant and management conditions can vary widely across the landscape (e.g., few models are capable of simulating N dynamics in agricultural and natural components of the landscape). Linking sources and sinks through hydrologic or atmospheric flows with site or ecosystem scale models is difficult because these models seldom consider transport processes at scales not relevant to internal N dynamics. Some efforts at true landscape scale modeling of N dynamics are described in Section 4.5.3.3.

4.5.3.2   How does N Enter and/or Leave the Ecosystem by Hydrologic or Atmospheric Vectors?

Once the inherent potential of a site to function as a sink or source of N in the landscape has been determined, the nature and extent of the connections between this site and adjacent landscape components need to be examined. The first part of this examination relates to the way that water moves into, through and out of a site and to the potential of a site to absorb or transmit N to the atmosphere.

Hydrologic transport into, through and out of a site is influenced primarily by soil texture, structure, drainage and slope, properties that are discussed extensively elsewhere in this volume. For the purposes of this chapter, the objective is only to highlight key properties that predict the hydrologic nature of a site as the basis for more detailed investigations (Table 4.12).

Sites with coarse soil texture have a high capacity for infiltration relative to sites with fine texture. As a result, coarse textured soils are likely to produce leaching and movement of N to groundwater, and fine textured soils are more likely to produce surface runoff. Texture effects on infiltration are strongly modified by soil structure and cover, both of which increase the capacity for infiltration. Soil drainage class, which is controlled by the height of the water table in the soil profile, is a good predictor of water movement into and out of a site to the extent that saturated (more poorly drained) soils are less able to absorb incoming water. Slope and nature of the soil surface (crusting) have obvious effects on the transmission of incoming water.

The soil variables that most strongly influence the potential of a site to transmit N by atmospheric vectors are soil cover and pH. Soil cover strongly influences the potential of a site to have wind erosion. Sites with high soil pH (> 8.0) have a high potential to produce NH_3 gas by volatilization. Sites with low soil pH (< 5.0) have a relatively high potential to produce N_2O and NO.

4.5.3.3 Where does this Ecosystem Sit in the Landscape?

The juxtaposition of N sources and sinks in the landscape is clearly a major determinant of overall landscape N flux. For example, Fig. 4.14 illustrates why riparian forests have been a major focus of research on controlling N outputs from agricultural watersheds (Lowrance et al., 1984; Peterjohn and Correll, 1984; Jacobs and Gilliam, 1985; Gilliam, 1994; Hill, 1996). The presence of an ecosystem with a high potential to act as an N sink (an unfertilized forest, with wet soil), in a position to absorb N moving in both surface runoff and groundwater, can have a big impact on watershed N output. Connections between landscape components can be complex and counter intuitive; for example, groundwater flow can bypass riparian zones.

Quantitative landscape scale evaluations of N flux are difficult. These evaluations require models capable of depicting internal and input/output N dynamics in diverse ecosystems, as well as landscape-scale hydrologic processes. These requirements are a challenge to modelers who need to balance the need for detail in their models with the need to be comprehensive. These requirements also demand large amounts of high resolution input data on soil, hydrologic and ecological conditions. Linking simulation models with geographic information systems to carry out landscape scale assessments of N flux is a rapidly developing field that will likely become a major tool for managing N in the environment in the next few decades (Robinson and Ragan, 1993; Tim and Jolly, 1994; Corwin and Wagenet, 1996).

4.5.3.4 What are Subsurface Conditions in this Site?

Landscape scale assessments of N flux must consider subsurface processes. The most common concern about N in the environment is NO_3^- pollution of groundwater. The importance of considering the physical, chemical and biological conditions of the subsurface material that NO_3^- laden groundwater passes through as it moves through the landscape is illustrated in Fig. 4.14. There is considerable controversy about N dynamics in groundwater. For example, some studies have found high potential and/or actual denitrification in subsurface material (Trudell et al., 1986; Slater and Capone, 1987, Smith and Duff, 1988; Francis et al., 1989; Obenhuber and Lowrance, 1991; Haycock and Pinay, 1993), while others have found little or no activity (Parkin and Meisinger, 1989; Groffman et al., 1996; Bradley et al., 1992; Yeomans et al., 1992; Lowrance, 1992; Starr and Gillham, 1993; Groffman et al., 1996).

Most of the uncertainty around subsurface processes concerns the presence of energy sources to support microbial processes. A major challenge for the next few years is to develop a basis for predicting N dynamics in the subsurface based on factors such as soil type, geologic setting or regional hydrologic conditions. There is also an important need to determine if there are functional links between groundwater processes and surface soil processes (e.g., is it possible to influence subsurface denitrification with surface management practices?).

4.5.4   Conclusions

It is clear that much of the information needed to assess the dynamics of N in the environment is available. Detailed knowledge of N-cycle processes (Sections 4.1–4.4) forms a basis for predicting and managing internal N dynamics in ecosystems. The challenge, however, is to apply this knowledge at the landscape and watershed scales, which is where N pollution problems must be assessed and managed. In a conceptual sense, it is necessary to develop the ability to think at larger scales, to consider landscape units such as those depicted in Fig. 4.14 as the objects of study rather than the single crop fields or forest stands that have been the traditional focus of soil research. In a practical sense, it is necessary to modify existing models and/or develop new models that are capable of dealing with the wide range of sites and ecosystems within a landscape, and that are capable of incorporating true landscape-scale hydrologic and atmospheric transport phenomena. These conceptual and practical challenges will require contributions from many areas of soil science over the next couple of decades.

4.5.5   References

Aber, J.D., K.J. Nadelhoffer, P. Steudler, and J.M. Melillo. 1989. Nitrogen saturation in northern forest ecosystems. BioSci. 39:378–386.

Agren, G.I., and E. Bosatta. 1988. Nitrogen saturation of terrestrial ecosystems. Environ. Pollut. 54:185–197.

Bellini, G., M.E. Sumner, D.E. Radcliffe, and N.P. Qafoku. 1996. Anion transport through columns of highly weathered acid soil: Adsorption and retardation. Soil Sci. Soc. Am. J. 60:132–137.

Berendse, F., R. Aerts, and R. Bobbink. 1993. Atmospheric nitrogen deposition and its impact on terrestrial ecosystems. p. 104–121. In C.C. Vos and P. Opdam (ed.) Landscape Ecology of a Stressed Environment. Chapman and Hall, London, UK.

Bormann, F.H., and G.E. Likens. 1979. Pattern and process in a forested ecosystem. Springer-Verlag, New York, NY.

Bouma, J., R.J. Wagenet, M.R. Hoosbeek, and J.L. Hutson. 1993. Using expert systsems and simulation modelling for land evaluation at farm level: A case study from New York state. Soil Use Manag. 9:131–139.

Bowden, R.D., M.S. Castro, J.M. Melillo, P.A. Steudier, and J.D. Aber. 1993. Fluxes of greenhouse gases between soils and the atmosphere in a temperate forest following a simulated hurricane blowdown. Biogeochem. 21:61–71.

Bradley, P.M., M. Fernandez, Jr., and F.H. Chapelle. 1992. Carbon limitation of denitrification in an anaerobic groundwater system. Environ. Sci. Technol. 12:2377–2381.

Carlsson, P., A. Segatto, and E. Granéli. 1993. Nitrogen bound to humic matter of terrestrial origin – a nitrogen pool for coastal phytoplankton? Mar. Ecol. Prog. Ser. 97:105–116.

Christ, M., Y. Zhang, G.E. Likens, and C.T. Driscoll. 1995. Nitrogen retention capacity of a northern hardwood forest soil under ammonium sulfate additions. Ecol. Appl. 5:802–812.

Corwin, D.L., and R.J. Wagenet. 1996. Applications of GIS to the modeling of nonpoint source pollutants in the vadose zone: A conference overview. J. Environ. Qual. 25:403–411.

Currie, W.S., J.D. Aber, W.H. McDowell, R.D. Boone, and A.H. Magill. 1996. Vertical transport of dissolved organic C and N under long-term amendments in pine and hardwood forests. Biogeochem. 35:471–505.

Davidson, E.A., S.C. Hart, C.A. Shanks, and M.K. Firestone. 1991. Measuring gross nitrogen mineralization, immobilization, and nitrification by ^{15}N isotopic pool dilution in intact soil cores. J. Soil Sci. 42:335–349.

DeWilligen, P. 1991. Nitrogen turnover in the soil-crop system: Comparison of fourteen models. Fert. Res. 27:141–149.

Dowdell, R.J., J.R. Burford, and R. Crees. 1979. Losses of nitrous oxide dissolved in drainage water from agricultural land. Nature 278:342–343.

Edwards, C.A., and P.J. Bohlen. 1996. Biology and ecology of earthworms. Chapman and Hall, London, UK.

Firestone, M.K., and E.A. Davidson. 1989. Microbiological basis of NO and N_2O production and consumption in soil. p. 7–21. In M.O. Andreae and D.S. Schimel (ed.) Exchange of trace gases between terrestrial ecosystems and the atmosphere. John Wiley and Sons, Chichester, UK.

Foloronuso, O.A., and D.E. Rolston. 1984. Spatial variability of field-measured denitrification gas fluxes and soil properties. Soil Sci. Soc. Am. J. 49:1087–1093.

Francis, A.J., J.M. Slater, and C.J. Dodge. 1989. Denitrification in deep subsurface sediments. Geomicro. J. 7:103–116.

Gilliam, J.W. 1994. Riparian wetlands and water quality. J. Environ. Qual. 23:896–900.

Gold, A.J., W.R. DeRagon, W.M. Sullivan, and J.L. Lemunyon. 1990. Nitrate-nitrogen losses to groundwater from rural and suburban land uses, J. Soil Water Conserv. 45:305–310.

Groffman, P.M., G. Howard, A.J. Gold, and W.M. Nelson. 1996. Microbial nitrate processing in shallow groundwater in a riparian forest. J. Environ. Qual. 25:13091316.

Haider, K., J.P. Martin, and Z. Filip. 1975. Humus biochemistry. p. 195–244. *In* E.A. Paul and A.D. McLaren (ed.). Soil biochemistry. Marcel Dekker, Inc., New York, NY.

Hart, S.C., M.K. Firestone, E.A. Paul, and J.L. Smith. 1993. Flow and fate of soil nitrogen in an annual grassland and a young mixed-conifer forest. Soil Biol. Biochem. 25:431–442.

Haycock, N.E., and G. Pinay. 1993. Groundwater nitrate dynamics in grass and poplar vegetated riparian buffer strips during the winter. J. Environ. Qual. 22:273–278.

Hedin, L.O., J.J. Armesto, and A.H. Johnson. 1995. Patterns of nutrient loss from unpolluted, old-growth temperate forests: Evaluation of biogeochemical theory. Ecol. 76:493–509.

Hill, A.R. 1996. Nitrate removal in stream riparian zones. J. Environ. Qual. 25:743–755.

Howarth, R.W., G. Billen, D. Swaney, A. Townsend, N. Jaworski, K. Lajtha, J.A. Downing, R. Elmgren, N. Caraco, T. Jordan, F. Berendse, J. Freney, V. Kudeyarov, P. Murdoch, and Z. Zhao-Liang. 1996. Regional nitrogen budgets and riverine N and P fluxes for the drainages to the North Atlantic Ocean: Natural and human influences. Biogeochem. 35:75–139.

Hutchinson, G.L., and E.A. Davidson. 1993. Processes for production and consumption of gaseous nitrogen oxides in soil. p. 79–83. *In* Agricultural ecosystem effects on trace gases and global climate change. Am. Soc. Agron Spec. Pub. 55. American Society of Agronomy, Madison, WI.

Jacobs, T.C., and J.W. Gilliam. 1985. Riparian losses of nitrate from agricultural drainage water. J. Environ. Qual. 14:472–278.

Johnson, D.W. 1992. Nitrogen retention in forest soils. J. Environ. Qual. 21:1–12.

Keeney, D.R. 1986. Sources of nitrate to ground water. CRC Crit. Rev. Environ. Contr. 16:257–304.

Lajtha, K., B. Seely, and I. Valiela. 1995. Retention and leaching losses of atmospherically derived nitrogen in the aggrading coastal watershed of Waquoit Bay, MA. Biogeochem. 28:33–54.

Logan, J.A. 1983. Nitrogen oxides in the troposphere: Global and regional budgets. J. Geophys. Res. 88:10785–10807.

Lovett, G.M. 1994. Atmospheric deposition of nutrients and pollutants in North America: An ecological perspective. Ecol. Appl. 4:629–650.

Lowrance, R. 1992. Groundwater nitrate and denitrification in a coastal plain riparian forest. J. Environ. Qual. 2:401–405.

Lowrance, R., R. Todd, J. Fail, Jr., O. Hendrickson, Jr., R. Leonard, and L. Asmussen. 1984. Riparian forests as nutrient filters in agricultural watersheds. BioSci. 34:374–377.

Madsen, E.L. 1995. Impacts of agricultural practices on subsurface microbial ecology. Adv. Agron. 54:1–67.

Matson, P.A., and P.M. Vitousek. 1987. Cross-system comparisons of soil nitrogen transformations and nitrous oxide flux in tropical forest ecosystems. Global Biogeochem. Cycl. 1:163–170.

Matson, P.A., and P.M. Vitousek. 1990. Ecosystem approach to a global nitrous oxide budget. BioSci. 40:667–672.

Mooney, H.A., and M. Godron. 1983. Disturbance and ecosystems. Springer-Verlag, Berlin, Germany.

Mooney, H.A., P.M. Vitousek, and P.A. Matson. 1987. Exchange of materials between terrestrial ecosystems and the atmosphere. Science 238:926–932.

Mosier, A.R., J.M. Duxbury, J.R. Freney, O. Heinemeyer, and K. Minami. 1996. Nitrous oxide emissions from agricultural fields: Assessment, measurement and mitigation. Plant Soil 181:95–108.

National Research Council. 1993. Soil and water quality: An agenda for agriculture. National Academy Press, Washington, DC.

Obenhuber, D.C., and R. Lowrance. 1991. Reduction of nitrate in aquifer microcosms by carbon additions. J. Environ. Qual. 20:255–258.

Parkin, T.B., and J.J. Meisinger. 1989. Denitrification below the crop rooting zone as influenced by surface tillage. J. Environ. Qual. 18:12–16.

Pastor, J., J.B. Aber, C.A. McClaugherty, and J.M. Melillo. 1984. Aboveground production and N and P cycling along a nitrogen mineralization gradient on Blackhawk Island, Wisconsin. Ecol. 65:256–268.

Paul, E.A., and F.E. Clark. 1996. Soil microbiology and biochemistry. Academic Press, New York, NY.

Peterjohn, W.T., and D.L. Correll. 1984. Nutrient dynamics in an agricultural watershed. Observations on the role of a riparian forest. Ecol. 65:1466–1475.

Power, M. 1993. The predictive validation of ecological and environmental models. Ecol. Model. 68:33–50.

Prather, M., R. Derwent, D. Ehhalt, P. Fraser, E. Sanhueza, and X. Zhou. 1995. Other trace gases and atmospheric chemistry. p. 77–126. In J. Houghton, L.G. Meira, E. Haites, N. Harris, and K. Maskell (ed.). Climate change 1994: Radiative forcing of climate changes and an evaluation of the IPCC IS92 emission scenarios. Cambridge University Press, New York, NY.

Qualls, R.G., and B.L. Haines. 1992. Biodegradability of dissolved organic matter in forest throughfall, soil solution, and stream water. Soil Sci. Soc. Am. J. 56:578–586.

Robinson, K.J., and R.M. Ragan. 1993, Geographic information system based nonpoint pollution modeling. Water Res. Bull. 29:1003–1008.

Ronen, D., M. Magaritz, and E. Almon. 1988. Contaminated aquifers are a forgotten component of the global N_2O budget. Nature 335:57–59

Rose, C.W., W.T. Dickinson, S.E. Jorgensen, and H. Ghadiri. 1990. Agricultural nonpoint source runoff and sediment yield water quality (NPSSWQ) models: Modeler's perspective. p. 145–170. In D.G. Decoursey (ed.) Proc. Int. Symp. Water Qual. Model. Agric. Nonpoint Sour. USDA-ARS, Washington, DC.

Ryther, J.H., and W.M. Dunstan. 1971. Nitrogen, phosphorus and eutrophication in the coastal marine environment. Sci. 171:1008–1013.

Schnabel. R.R., W.J. Gburek, and W.L. Stout. 1994. Evaluating riparian zone control on nitrogen entry into Northeast Streams. p. 432–445. In Riparian ecosystems in the humid. US National Association of Conservation Districts, Washington, D.C.

Slater, J.M., and D.G. Capone. 1987. Denitrification in aquifer soil and nearshore marine sediments influenced by groundwater nitrate. Appl. Environ. Microbiol. 53:1292–1297.

Smith, R.L., and J.H. Duff. 1988. Denitrification in a sand and gravel aquifer. Appl. Environ. Microbiol. 54:1071–1978.

Starr, R.C., and R.W. Gillham. 1993. Denitrification and organic carbon availability in two aquifers. Groundwater 31:934–947.

Swank, W.T., J.B. Waide, D.A. Crossley, Jr., and R.L. Todd. 1981. Insect defoliation enhances nitrate export from forest ecosystems. Oecologia 51:297–299.

Tamm, C.O. 1991. Nitrogen in terrestrial ecosystems. Springer-Verlag, Berlin, Germany.

Tim, U.S., and R. Jolly. 1994. Evaluating agricultural nonpoint-source pollution using integrated geographic information systems and hydrologic/water quality model. J. Environ. Qual. 23:25–35.

Trudell, M.R., R.W. Gillham, and J.A. Cherry. 1986. An in-situ study of the occurrence and rate of denitrification in a shallow unconfined sand aquifer. J. Hydrol. 83:251–268.

Ueda, S., N. Ogura, and T. Yoshinari. 1993. Accumulation of nitrous oxide in aerobic groundwaters. Water Res. 27:1787–1792.

USEPA. 1990. National pesticide survey: Nitrate. Office of Water, Office of Pesticides and Toxic Substances, US Environmental protection Agency, Washington, DC.

van Breemen, N., P.A. Burrough, E.J. Velthorst, H.F. van Dobben, T. de Witt, T.B. de Ridder, and H.F.R. Reijnders. 1982. Acidification from atmospheric ammonium sulphate in forest canopy throughfall. Nature 299:548–550.

Vellidis, G., R.K. Hubbard, J.G. Davis, R. Lowrance, R.G. Williams, J.C. Johnson, and G.L. Newton. 1996. Nutrient concentrations in the soil solution and shallow groundwater of a liquid dairy manure land application site. Trans. ASAE 39:1357–1365.

Vitousek, P.M., and R.W. Howarth. 1991. Nitrogen limitation on land and in the sea: How can it occur? Biogeochem. 13:87–115.

Vitousek, P.M., J.R. Gosz, C.C. Grier, J.M. Melillo, W.A. Reiners, and R.L. Todd. 1982. A comparative analysis of potential nitrification and nitrate mobility in forest ecosystems. Ecol. Monog. 52:155–177.

Yeomans, J.C., J.M. Bremner, and G.W. McCarty. 1992. Denitrification capacity and denitrification potential of subsurface soils. Commun. Soil Sci. Plant Anal. 23:919927.

And he gave it for his opinion, that whoever could make two ears of corn or two blades of grass to grow upon a spot of ground where only one grew before, would deserve better of mankind, and do more essential service to his country, than the whole race of politicians put together.

-Jonathan Swift

Soil Fertility and Plant Nutrition

Eugene J. Kamprath
North Carolina State University

The section on soil fertility and plant nutrition examines the factors which control the availability of the 13 essential mineral elements, soil and tissue tests for assessing nutrient sufficiency, interactions which affect availability, application methods, and soil management practices for efficient use of nutrients.

The 13 mineral elements are divided into macro- and micronutrients based on amounts required for growth. Nitrogen, P, K, Ca, Mg and S are classified as macronutrients and Fe, Mn, Cu, Zn, B, Mo and Cl are classified as micronutrients.

The availability of mineral nutrients is controlled by the chemical and physical properties of the soil. The cation exchange properties of the soil clay and organic matter regulate the availability of the cation nutrients, while the hydrated oxides of Fe and Al and soil Ca affect to various degrees the availability of the relatively immobile anions, P, S, and Mo. Availability of organic sources of N is dependent upon mineralization of the N to the inorganic forms, ammonium and nitrate. These reactions are controlled by the chemical, physical and biological properties of the soils. The solubility of most micronutrients is affected by soil pH and organic matter content.

To assess the sufficiency of the mineral nutrients for optimum plant growth, various soil and tissue tests have been developed. Interpretation of soil tests must take into account the soil chemical properties. Interpretation of tissue tests requires knowledge of plant requirements and nutrient interactions.

For sustained high crop yields, the application of nutrients is required. Efficient use of applied nutrients depends upon the timing and methods of nutrient application. The chemical and physical properties of the soil determine which methods of application and soil management practices are best suited for a given soil.

The principles of soil fertility and plant nutrition affecting the growth of plants and their response to application of mineral nutrients are discussed in the following six chapters.

1

Bioavailability of Major Essential Nutrients

1.1 Introduction

The major essential mineral elements, N, P, K, Ca, Mg, and S are discussed in this chapter. Nitrogen, P, and K are discussed in separate sections because of their importance in soil fertility and the large amount of research done with these three elements. Calcium and Mg, which behave similarly in soils and are applied as dolomitic lime or sulfate salts, are discussed with S in the final section of this chapter.

1.2 Bioavailability of Nitrogen

Alfred M. Blackmer
Iowa State University

1.2.1 Nitrogen as a Plant Nutrient

Plant tissues usually contain more N than any other nutrient normally applied as a fertilizer. This N is needed to form chlorophyll, proteins, and many other molecules essential for plant growth. Although some plants acquire N from air by forming symbiotic relationships with N-fixing bacteria, most plants rely on their roots to acquire N by taking up NO_3^- and NH_4^+ from the soil solution.

Nitrogen has been recognized as an important plant nutrient for more than a century. The past 50 years have brought marked advances in the capacity to manufacture and apply plant available N as commercial fertilizers. These advances, however, have not diminished the importance of problems related to N management but have created a greater appreciation for the importance of avoiding yield limiting N deficiencies. Also, marked increases in fertilizer N rates applied to agricultural soils have raised concern about the environmental impacts of N that escapes from the rooting zone (Aldrich, 1980; Pratt, 1985; Hallberg, 1989; NRC, 1993).

Research during the past few decades has provided a wealth of new information about the many factors and processes that influence N availability in soils. Many indices for measuring or predicting N availability in soils have been described. The objective of this section is to summarize current knowledge about N availability in soils. Included are discussions of definitions of availability, factors affecting N availability in soils, and commonly used N availability indices and their use.

1.2.2 Defining Terms Related to Availability

Availability of N, which refers to the quantity of available N, seems like an easy concept to understand in general discussions. In general, N is considered available if it is susceptible to uptake by plants. Such definitions, however, provide insufficient information to guide quantitative measurements of available N or N availability.

Quantitative discussions of available N and availability of N are hindered by several problems: (1) these terms have been defined and used in a variety of ways (Table 1.1); (2) the position of N relative to roots is important but difficult to quantify; (3) uptake involves a flow of N into the root, so distinctions between available and unavailable N cannot be made without specifying the time period involved and conditions that influence the N flow; and (4) soil microorganisms are continuously altering forms of N, so it often is difficult to distinguish between N that is available to plants and microor-

Table 1.1 Examples illustrating the variety of ways in which terms related to availability have been defined

Source	Definition
Bray (1954)	Available forms of nutrients are those forms whose variations in amount are responsible for significant variations in yield and response. The availability of these forms, however, involves not only their chemical and physical nature but also the ability of the plant to forage for them with its root system.
Scarsbrook (1965)	Available N is N in the root zone in a form readily absorbed by plant roots.
Brady (1974)	Available means the portion of any element or compound in soil that can be absorbed and assimilated by growing plants.
Barber (1984)	An available or bioavailable nutrient is one that is present in a pool of ions in the soil and can move to the plant root during plant growth if the root is close enough.
Dahnke and Johnson (1990)	Nitrogen availability indices are a measurement of the potential of a soil to supply N to plants when conditions are ideal for mineralization. They do not include existing inorganic N present when the soil is sampled.
Peck and Soltenpour (1990)	By plant-available nutrient, one usually means the chemical form or forms of an essential plant nutrient in the soil whose variation in amount is reflected in variations in plant growth and yield.
Troeh and Thompson (1993)	Available nutrient means the small portion of each essential plant nutrient present in the soil that is available to plants.
Black (1993)	Available means susceptible to absorption to plants, availability means effective quantity. Biological indices of nutrient availability estimate ratios of availabilities or values that are proportional to availabilities; actual availabilities cannot be measured.
Bundy and Meisinger (1994)	Available N in soils is that N present in forms, concentrations, and spatial position that allow utilization by plants growing in the soil. Nitrogen availability indices are N analyses or chemical or biological tests to measure or predict the amounts of available N released from soil under a specified set of conditions.
SSSA (1997)	Available nutrients are (1) the amount of a soil nutrient in chemical forms accessible to plant roots or compounds likely to be convertible to such forms during the growing season, and (2) the contents of legally designated available nutrients in fertilizers determined by specified laboratory procedures, which in most states constitute the legal basis for guarantees.

ganisms and N that is available only to microorganisms. The terms available N and N availability, therefore, refer to nebulous concepts unless used to refer to a specific N availability index.

The term N availability index is used here to denote measured values intended to estimate amounts of N that were, are, or will be susceptible to plant uptake under specified conditions. The conditions specified can be a unique situation or a range of conditions normally expected during growth of a specific crop or group of similar crops in a particular region. A variety of commonly used N availability indices are described in Section 1.2.6.

Problems relating to terminology can be simplified by assuming that, even when not stated, available N denotes N that is measured by a particular N availability index and N availability denotes the quantity of N measured when a particular N availability index is used. Such substitution of terms greatly simplifies syntax, and is often necessary for clear discussions of data and concepts.

1.2.3   Pools of Available N

The concept of pools is essential when discussing N availability in soil-plant systems. A pool of N can be considered to be N in forms and locations that are similar enough that they need not be treated differently. An N availability index, for example, measures the amount of N in a specified pool. The concept of pools is useful because the chemical form and location of N in soils can change rapidly with time. By dividing soil N into appropriate pools, availability can be described as a function of time and position as root systems grow and as N is transformed and moved in the soil.

There is no established convention for deciding how N in soils should be grouped into pools, and the most appropriate method varies greatly with type of study. A pool of N, for example, can be all N present as a specific chemical form (e.g., NO_3^-) within the entire rooting zone, a layer of soil in the rooting zone, or a small cell in a three-dimensional matrix of cells near a single root. A pool of N also can be a group of N atoms in unknown chemical forms that will be converted (e.g., mineralized) to a single chemical form under specified conditions. Fig. 1.1 illustrates how soil N can be divided into pools by N form without regard to its position. Such a figure is useful when discussing N forms and transformations, but not when discussing N movement or the effects of N position on availability to roots or microorganisms in soils.

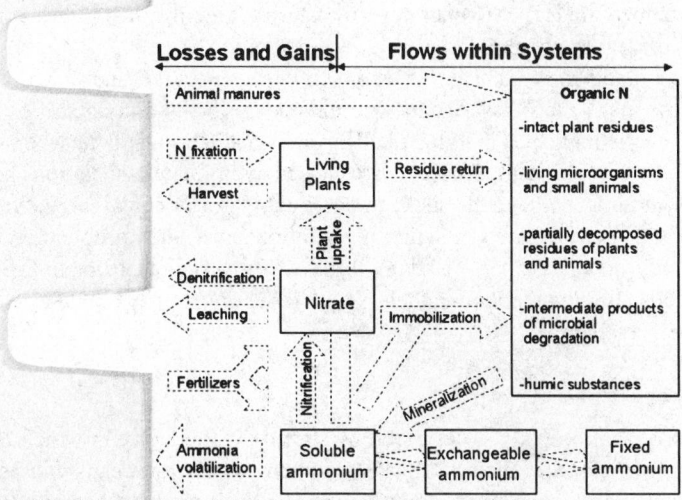

Fig. 1.1 Schematic diagram showing major pools and flows of N in soil-plant systems. The rectangles indicate pools of N. Arrows indicate the transformations and movements that cause N to flow among pools.

1.2.4   Forms of Nitrogen in Soil-Plant Systems

Important factors affecting the susceptibility of N to uptake by roots are how the N is bonded with other atoms to form molecules and how these molecules are held in or on soil minerals. These two factors provide a basis for dividing soil N into the six pools shown in Fig. 1.1.

Nitrate is the preferred form of N for uptake by most plants, and it usually is the most abundant form that can be taken up in well-aerated soils. The quantities of NO_3^- found in soil at any time, however, usually represent only enough N to support uptake for a short period. Nitrate usually remains in the soil solution and, therefore, is relatively free to move with water flows. Water flows induced by transpiration (i.e., movement of water from the soil through the plant to the atmosphere) facilitate uptake (Wild, 1981; Nye and Tinker, 1977; Barber, 1984). Drainage of excess water often moves NO_3^- downward in the soil profile and out of the rooting zone. For these reasons, quantities of NO_3^- in soil usually provide poor estimates of the quantities of N susceptible to uptake during extended periods.

Ammonium in the soil solution is susceptible to uptake by roots and is in equilibrium with a much larger quantity of exchangeable NH_4^+ (Nommik and Vahtras, 1982). The exchangeable NH_4^+ often equilibrates with nonexchangeable NH_4^+, which is not rapidly displaced by neutral salt solutions. Concentrations of NH_4^+ in the soil solution usually are low enough that relatively little movement occurs with flows of water. Quantities of NH_4–N can exceed those of NO_3–N in anaerobic soils or in soils recently treated with NH_4^+-containing or producing fertilizers. This can also occur immediately after precipitation has leached ambient NO_3^- but nitrification has had insufficient time to convert NH_4^+ to NO_3^-.

Most (usually > 95%) of the N in soils is bonded to C in humus or in living and decaying cells of plants, microorganisms, or small animals (Allison, 1973; Stevenson, 1982a,b). This organic N is not susceptible to plant uptake until it is mineralized to NO_3^- or exchangeable NH_4^+ by soil microorganisms. However, mineralization usually occurs as an ongoing process that supplies a flow of new N that is susceptible to uptake by plants. Mineralization makes organic N the major source of plant-available N under many conditions of practical interest. Because the N-containing molecules within this pool differ greatly in resistance to decomposition, it is reasonable to consider that this N shows continuous gradation from easily mineralizable to unavailable N. The rate of mineralization from this pool is determined by environmental factors (temperature, moisture availability, aeration status, etc.) and by types or amounts of organic N present.

Much of the N taken up by plants remains in above- and belowground plant tissues for some time. Most of this N is organic, but many plant tissues can accumulate significant amounts of NO_3^-. Although it would be desirable to distinguish N in living plants from N not in living plants when analyzing soils, clear distinctions often are difficult. Problems are caused by intimate associations between fine roots and soil particles and microorganisms, leakage or sloughing of N-containing compounds from healthy root systems and tissues shared in symbiotic relationships. Such problems often are addressed by considering only plant parts large enough to be easily separated from soil. Because small root members are usually short lived, pooling their N with other organically bound N usually is not a problem in estimating N availability.

1.2.5   Flows of N Between Pools

The behavior of N on a global scale usually is described by starting with a cycle in which N flows between the soil and the atmosphere (Stevenson, 1982a,b). Numerous subcycles can be imposed on the major cycle. The cycles describe flows (transformations and/or movement) of N among various pools of N.

The behavior of N in field scale soil-plant systems tends to be dominated by transient rather than by cyclic or steady state flows. Flows are transient because they often are driven by events initiated outside the system and because rates of flow vary greatly with time. The flows can move N among pools within the system and they can move N into or out of the system. Processes (Fig. 1.1) that affect N availability in soils are briefly discussed but more information is presented by Allison (1973), Stevenson (1982a,b), and Hayes (1986) .

Biological fixation takes N_2 from air and incorporates it into organic molecules produced during growth of plants and microorganisms. Nitrogen can be fixed by free-living microorganisms and by microorganisms loosely associated with plants, but fixation through symbiotic relationships between plants and microorganisms has the greatest effect on N availability to plants. Biological N fixation produces the primary flow of N into some soil-plant systems, but is unimportant in others.

Denitrification is the biological reduction of NO_3–N (or NO_2–N) to N_2O–N or N_2. On a global scale, denitrification balances N fixation. In small soil-plant systems, however, denitrification is note-worthy because it causes losses of plant-available N. Conditions needed for denitrification include absence of O_2, and presence of NO_3^- and an electron-donating substrate (e.g., organic residues) to support microbial respiration. Although absence of one or more of these conditions usually prevents rapid denitrification, a short period of favorable conditions can result in substantial losses of plant-available N from most systems.

Leaching refers to the movement of N in water moving downward through the soil profile and out of the rooting zone. Nitrate is the form of N most likely to be lost by this process because it usually is the most abundant form of N in the water that moves. The importance of N losses by leaching varies greatly with factors that determine how much and when water flows downward through soils. Losses can be substantial in systems where mineralization or fertilization results in high concentrations of NO_3^- during periods when leaching is likely. In most soil-plant systems, losses of N by leaching are associated with short-term weather events that can be neither controlled nor predicted. Flows of water often increase N availability by moving soluble N from the soil surface to depths where roots are active, but this usually is not considered leaching. In many highly weathered soils, NO_3^- leaching can be retarded by positive charge present in acid subsoils (Bellini et al., 1996) leading to substantial storage capacity for NO_3–N in the soil profile.

Ammonium in soil (and plant) solutions tends to equilibrate with NH_3 in air near the solutions, and this equilibration often results in losses of N by NH_3 volatilization. Such losses are favored by high pH and short distances between the solution and moving air. Significant losses from soils are most likely to occur when NH_4^+ from fertilizers or animal manure remains on the surface. Plants can lose N as NH_3 through stomata. Soils and plants at other sites may absorb some of the NH_3 volatilized from one site, and such absorption may provide a source of plant-available N under some conditions.

Nitrification, which is the biological oxidation of NH_4–N via NO_2–N to NO_3–N, is carried out primarily by a small group of chemoautotrophic bacteria. These bacteria keep concentrations of ex-changeable NH_4^+ relatively low in aerobic soils, but their activity is inhibited by lack of O_2 in anaero-bic soils. This process does not result in losses of N from soils, but it alters the susceptibility of N to loss by various mechanisms (e.g., ammonia volatilization, leaching, denitrification). The use of nitri-fication inhibitors is discussed in detail in Section D, Chapter 5.

Mineralization is the transformation of organic to mineral N (i.e., NO_3^- or exchangeable NH_4^+) and immobilization is the opposite process. Mineralization occurs when soil microorganisms release surplus N as they break down organic materials. Immobilization occurs as plants and microorganisms take up mineral N and utilize it to build tissues during growth. Both processes can occur simulta-neously, but one usually dominates the other. Which process dominates often changes with time, so these processes result in a net production of plant-available N during some and a net consumption during other parts of the year.

Ammonium fixation is induced by additions of relatively high concentrations of NH_4^+ to soils with 2:1 layer clays (Nommik and Vahtras, 1982). This recently fixed NH_4^+ is released after concentrations of ammonium are decreased by nitrification or plant uptake (Green et al., 1994). Field studies have shown that concentrations of fixed NH_4^+ are temporarily decreased during periods of plant uptake of N (Li et al., 1990). Such observations suggest that fixation and release of this NH_4^+ may serve as a buffer for soil NH_4^+ concentrations, and thereby, increase availability of N during periods of intense demand by plants.

Applications of animal manures or commercial fertilizers often add more N than is taken up by plants during a year. The amount of N added for production of many field crops, for example, is the amount of N expected to be taken up by the crop plus enough extra N to compensate for losses of N expected to occur (Stanford, 1973; Bock and Hergert, 1991).

Uptake of N is usually considered to have occurred when it crosses the plasma membrane of root cells (Nye and Tinker, 1977). The overall process of N uptake in soils, however, usually is considered to involve movement to and across the plasma membrane as separate processes (Nye and Tinker, 1977; Russell, 1977; Barber, 1984). Ammonium moves to the membrane primarily by diffusion, and NO_3^- often moves primarily by mass flow in the transpiration stream. Movement of these ions across the membrane can occur by active transport and more than one mechanism may be involved. Rates of N uptake by plants tend to vary greatly with stage of growth. Many annual plants take up most of the N they need within a few weeks during the growing season, and growth of these plants is influenced primarily by N availability only during this period.

1.2.6 Indices of N Availability

Many different methods have been used to estimate N availability in soils. Following is a brief description of several indices that have been used and that illustrate the diversity in approach to estimating N availability.

1.2.6.1 Mitscherlich b Values

Mitscherlich et al. (1925) [see Black (1993) for recent discussions in English] conducted extensive studies on the effect of quantities of nutrients in soils on dry matter yields and found that a simple exponential function (Fig. 1.2) could relate one to the other. The function can be fitted to observed yields resulting from incremental applications of fertilizer, and it provides a yield response curve that describes expected yields as a continuous and smooth function of nutrient availability. Extrapolation of the curve to the x axis gives the Mitscherlich b value, which estimates quantities of plant-available N in the soil. Such estimates can be made from data collected in field or greenhouse studies.

Quantities of nutrient in soil are expressed in terms of fertilizer equivalents rather than absolute units (e.g., mass of N/quantity of soil). In Fig. 1.2, for example, the soil in each pot contained a quantity of plant-available N that was equivalent to 42 mg of fertilizer N. The absolute quantity of soil N represented by a unit of fertilizer equivalent, however, varies with the percentage of the added fertilizer that was taken up by the plant. The percentage taken up varies with the amounts of fertilizer N lost or immobilized soon after application and the proportion of the fertilized soil actually foraged by plant roots. Mitcherlich b values, therefore, can be expressed in absolute units only by making assumptions about the fate of N applied and the volume of soil foraged by the roots, which depend on how the study was conducted. Use of this availability index, however, gives an assessment of the ability of the soil to supply N under the conditions studied, and the results can be used to predict soil nutrient availability (in fertilizer equivalents) under similar conditions.

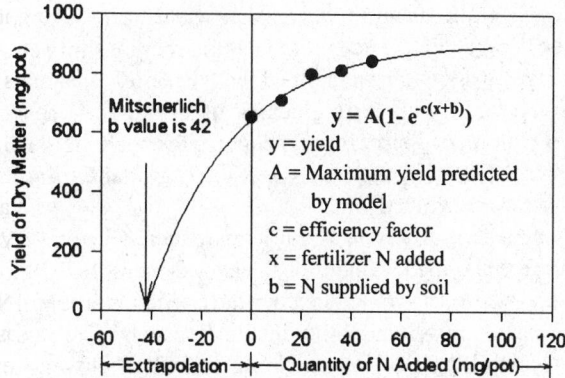

Fig. 1.2 Example of the Mitscherlich model fit to data collected in a greenhouse study involving the growth of corn seedlings in pots to which various amounts of N had been applied

1.2.6.2 The Percentage Sufficiency Concept

Mitscherlich c values (Fig. 1.2) describe the efficiency with which added nutrient is converted to plant dry matter. This efficiency depends on the percentage of the applied nutrient that is actually taken up by plants and on the amount of dry matter produced per unit of nutrient taken up. If both of these factors remain reasonably constant across a range of conditions (i.e., if the c value is reasonably constant), then the problem of relating soil nutrient availability indices to yield level can be greatly simplified by transforming absolute to relative yields. Relative yields are expressed as a percentage of the highest yield that can be attained by adding that nutrient under otherwise similar conditions. Any given level of nutrient availability, therefore, corresponds to a given percentage of the highest yield that can be attained by adding that nutrient. Plants that essentially reach the 100% relative yield level are assumed to have sufficient N for growth. Use of this relationship avoids the need for knowing absolute yield levels and, therefore, enables data collected from a variety of conditions to be described as a single relationship. The ideas involved are discussed by Black (1993).

Bray (1954, 1963) noted that Mitscherlich c values should be reasonably constant across a reasonable range of conditions if a plant's ability to forage for the nutrient was proportional to yield level (i.e., if higher yielding plants had greater ability to extract nutrients from a soil). He reasoned that this relationship, often called proportionate feeding, should apply to relatively immobile nutrients (e.g., P, K, and NH_4^+) but not to relatively mobile nutrients (e.g., NO_3^-). The basic idea is that relatively few roots were required to take up all the mobile nutrients in a volume of soil, but the amounts of immobile nutrients taken up varied with density of rooting within the volume of soil. His reasoning suggests that, unlike quantities of available P and K, the quantities of available N needed by plants should be expected to vary with absolute yield level of the plant.

An important notion that emerges from the ideas of Mitscherlich and Bray is that measured quantities of available N in soils can be related to levels of N sufficiency for plant growth. The minimum measurable quantity of N needed to avoid yield limiting deficiencies is an important reference point in any discussion of N availability to plants and is important whether or not proportionate feeding occurs.

1.2.6.3 Tissue Tests to Evaluate N Sufficiency

Plant tissues often are analyzed to evaluate the N sufficiency (i.e., supply of N relative to the needs of plants). The basis for such evaluations was provided by Macy (1936), who found useful relationships between tissue nutrient concentrations and relative plant yields. Use of this method requires that fertilizer response trials be conducted to discover the natural relationship between tissue concentrations and relative yields. Fig. 1.3 shows an example of the type of relationship expected. Once such a relationship has been established, tissue testing can be used to determine if N availability under specific conditions was less, more, or approximately equal to the needs of the plant (Ulrich and Hills, 1973; Cerrato and Blackmer, 1990). Because plants measure N availability, problems associated with sampling and analysis of soils are avoided.

New variations on tissue testing to assess N sufficiency include use of chlorophyll meters and remote sensing. The meters are clamped on individual leaves to measure light transmittance at wavelengths of 650 and 940 nm, where transmittance correlates with chlorophyll content. Such meters enable rapid in field assessments of N availability in situations where the amount of chlorophyll is largely determined by N availability (Piekielek and Fox, 1992). Remote sensing involves measuring the light reflectance from many leaves simultaneously and offers advantages when trying to characterize spatial patterns of N sufficiency within fields (Blackmer and Schepers, 1996; Blackmer and White, 1996). Both of these techniques are most effective when measurements are referenced to plants having adequate N under otherwise similar conditions.

1.2.6.4 Fried and Dean A Values

Fried and Dean (1952) used isotope dilution techniques to estimate N availability in soils. Their method involves adding known quantities of ^{15}N labeled N to the soil, growing plants, and analyzing the plants to determine relative amounts of soil and labeled N taken up. A key assumption is that plants take up soil and labeled N in proportion to their effective quantities in soils. Isotope dilution techniques are used to estimate availability of soil N in absolute units (e.g., mass of N/quantity of soil).

A limitation of this method is that it usually is difficult to assess the validity of important assumptions made during calculation of soil N availability in absolute terms (Edwards, 1978). The assumptions are that (1) the labeled fertilizer is mixed uniformly within the exact volume of soil foraged by plant roots; (2) added N is not selectively lost from the soil soon after application; and (3) transforma-

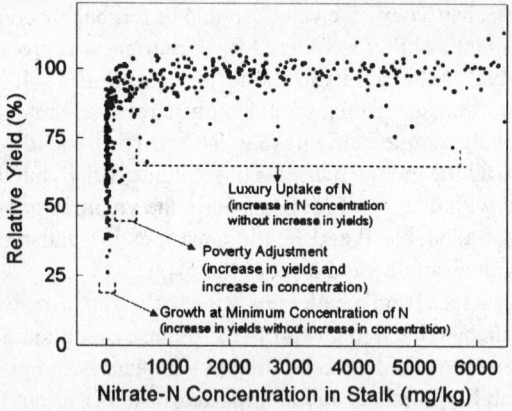

Fig. 1.3 Example of the type of relationship Macy observed between concentrations of N in plant tissues and relative yields (yields expressed as a percentage of the highest yields that could be attained by adding N under otherwise similar conditions). Data are from 45 different trials in which ten rates of N were applied shortly before the crop was planted. Nitrate concentrations were measured in the lower portion of the cornstalk at the end of the growing season. Nitrogen sufficiency occurs at the upper end of the zone of poverty adjustment [Data from Binford et al., 1992. Agron. J. 84:886 with permission of the American Society of Agronomy].

tions such as immobilization and mineralization do not occur significantly during the study (Jenkinson et al., 1985; Blackmer and Green, 1995).

1.2.6.5 Dean a Values

Dean (1954) studied relationships between nutrient quantities in soils and accumulated by plants growing on the soils. He observed that, over a limited range of conditions, the quantities of nutrient accumulated tend to be linearly related to amounts of nutrient added to soils. Plotting this linear relationship and extrapolating it to the x axis provides an estimate of N availability in the soil (Fig. 1.4). This index is analogous to Mitscherlich b values, but the relationships observed are more likely to be linear.

Units for Dean a values are often expressed as quantity of nutrient per quantity of soil. As with Mitscherlich b values, however, Dean a values are in units of fertilizer equivalents rather than absolute quantities of N. Relationships between quantities of N in soils and quantities of N accumulated in plants often are not linear, so the assumption that observed relationships should be linear can introduce significant errors when estimating availability in soils.

1.2.6.6 Tests for Mineralizable Nitrogen

Many tests have been developed to estimate the quantities of plant-available N that a soil can supply by mineralization (Keeney, 1982; Campbell et al., 1993) and these can be divided into biological and chemical tests. Biological tests involve incubating soil samples under specified laboratory conditions favorable for mineralization and measuring amounts of exchangeable NH_4–N plus NO_3–N produced. The rates of mineralization observed often are used to predict quantities of mineralizable N in the soil (Stanford and Smith, 1972). Chemical tests, which involve chemically treating soil samples and measuring the amounts of N released by the treatments, are of interest because they enable rapid testing and reporting of results. It must be shown, however, that the chemical tests are reasonable predictors of the amounts of N mineralized by soil microorganisms.

Indices of N availability based on mineralization tests are most useful in situations where an important portion of the N taken up by plants is supplied by N mineralization. However, use of such indices relies on the assumption that the measured rates of mineralization can be used to predict the rates that occur under field conditions. Moreover, it must be assumed that plants take up a relatively constant fraction of the N mineralized. These assumptions cause problems in situations where vari-

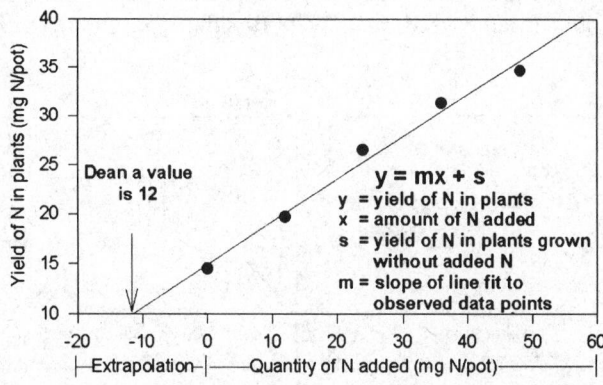

Fig. 1.4 Example showing Dean a values calculated from data collected in a greenhouse study involving the growth of corn seedlings in pots to which various amounts of N had been applied

ability in weather causes great variability in rates of mineralization, losses of N from soils, and depth of rooting. In many soil-plant systems of practical interest, the amounts of N mineralized are small compared to the amounts of N supplied by fertilizers.

1.2.6.7 Carryover of Nitrogen from Legumes

Nitrogen fixed by legumes provides a major input of plant-available N into many soil-plant systems. The fixed N supports growth of the legumes and part of this N remains in the soil for use by other plants. The quantity of plant-available N that remains in the soil varies greatly with legume and the amounts and types of legume residues left to decompose in the soil (Voss and Shrader, 1984; Hesterman, 1988).

Quantities of plant-available N left by legumes often are estimated from trials in which yield responses to incremental increases in rate of N application are compared on plots with and without a preceding legume crop. Examples of such studies are illustrated in Fig. 1.5, where the apparent amount of N carried over is measured by differences in amounts of fertilizer N required to avoid yield limiting N deficiencies. The quantity of N left is estimated in fertilizer equivalents rather than in units of mass of N/quantity of soil.

A key assumption is that legume-induced differences in yields at a single rate of N application are caused solely by the quantity of N left by the legume. This assumption often is not valid because previous crops alter many different factors, some of which (extent of damage by insects or diseases, moisture availability, etc.) influence the ability of roots to forage for N. These factors can cause substantial differences in yields even when large amounts of fertilizer N are applied (Fig. 1.5). The finding that factors other than N availability are important under some conditions gives reason to question the assumption that differences in yield can be attributed solely to differences in N availability under other conditions.

1.2.6.8 Available Nitrogen from Animal Manure

Much of the manure generated during livestock production is applied to land used for crop production (Smith and Peterson, 1982; NRC, 1993). The primary reason for land application often is inexpensive waste disposal, but manure also is a source of nutrients needed for crop production. Applications of animal manures can result in relatively large inputs of N into some soil-plant systems.

Estimates of the quantities of plant-available N supplied by animal manure usually are based on estimates of quantities of N in plant-available forms after adjustments are made for N losses likely to occur soon after application (Mathers and Goss, 1979; Bouldin et al., 1984). The amounts of N in the

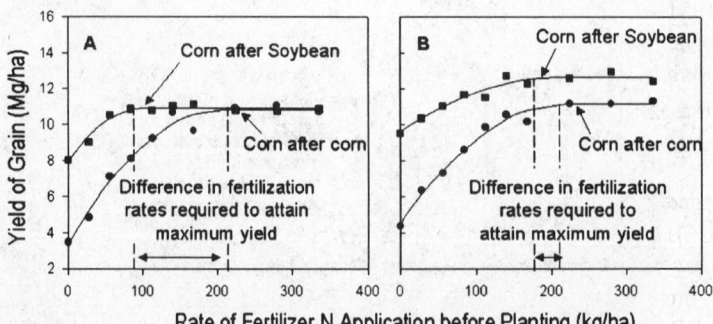

Fig. 1.5 Examples showing a commonly used method for estimating quantities of plant-available N left by legumes for the following crop. The data shown in A and B were collected at different sites [Data from Meese, 1993].

manure are most often estimated by considering the type of manure and methods by which the manure was handled. Adjustments for losses soon after application are made because substantial N losses can occur by commonly used methods. If manure has been stored under anaerobic conditions, for example, much of the N is converted to the NH_4^+ form which can be lost by NH_3 volatilization if the manure is not immediately incorporated into the soil. Because much of the N is contained in organic forms that must be mineralized before it is available for uptake by roots, estimated rates of mineralization often are used to predict the amounts of N that will be available to the first and each succeeding crop.

1.2.6.9 Guaranteed Analysis of Fertilizers

Fair marketing practices require that commercial fertilizers be sold on the basis of their nutrient supplying power as estimated by a standardized procedure. These estimations usually are based on availability indices indicating the quantities of N present as various chemical forms. Discussions of such availability indices usually assume that N present in forms such as NO_3^-, NH_4^+, and urea are 100% available to plants. This definition of availability should not be confused with definitions that consider proximity to roots to be an important factor affecting availability.

1.2.6.10 Tests for Mineral Nitrogen

Availability of N in soils often is estimated by measuring concentrations of NO_3^- and/or exchangeable NH_4^+ in samples collected to various depths. Such tests are commonly described as residual N (Hergert, 1987; Dahnke and Johnson, 1990), N-min (Neeteson, 1985; Greenland, 1986), preplant N (Bundy and Meisinger, 1994), or pre-sidedress N tests (Bundy and Meisinger, 1994). The depth and time of soil sampling tends to vary with climate, soil properties, and crop grown. Testing for mineral N tends to be most effective in regions where rainfall patterns and soil characteristics limit the potential for losses of NO_3^- by leaching or denitrification. The tests tend to be more useful for annual than perennial crops because NO_3^- is more likely to accumulate between growing seasons for annual crops. For additional information on N soil tests, see Section D, Chapter 4.

Results of the soil test can be interpreted by considering mass balance or sufficiency levels. Methods based on mass balance assume that measured concentrations can be converted to absolute quantities of plant-available N (e.g., kg N ha^{-1} in the surface 60 cm layer of soil). This conversion is accomplished by using an efficiency factor that describes the ability of the plant to take up N from the pool measured (Section 1.2.6.11).

Interpretations based on sufficiency levels relate measured concentrations of soil N to those needed to avoid yield limiting N deficiencies. Fig. 1.6 shows an example of a soil nitrate test used in this way. Observations from many different field trials are pooled to form a single relationship between soil NO_3^- concentrations found early in the growing season and relative yield levels observed at harvest. A concentration of 25 mg NO_3–N kg^{-1} is the lowest needed to avoid yield limiting N deficiencies under the conditions shown. This critical concentration has been found to be remarkably constant for corn grown across a wide range of conditions (Bundy and Meisinger, 1994). This index is used without estimating absolute quantities of plant available N in the soil profile, uptake efficiency by plants, or quantities of N taken up by plants. Quantities of N are assessed only relative to the soil test values usually needed by plants to avoid N deficiencies within the range of conditions studied.

1.2.6.11 Efficiency of Nitrogen Uptake

A mass balance approach commonly is used to describe the transformations and movement of N in soil-plant systems (Bock, 1984; Meisinger, 1984; Meisinger and Randall, 1991). By this method,

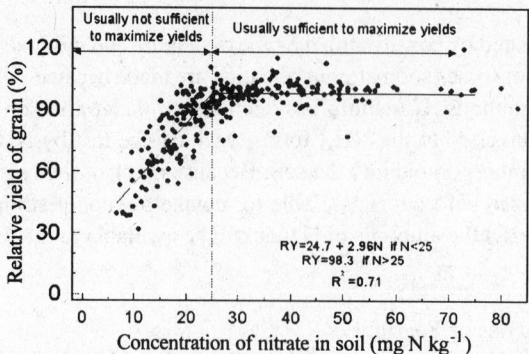

Fig. 1.6 Example of an observed relationship between concentrations of nitrate found in the surface 30 cm layer of soil when corn plants were 15-30 cm tall and relative yield levels at harvest. Data shown are from more 34 sites at which N was applied at 10 different rates shortly before the crop was planted [Data from Binford et al., 1992a. Agron. J. 84:58 with permission of the American Society of Agronomy].

selected pools of N (e.g., N applied as a commercial fertilizer, NO_3^- in the surface 60 cm layer of soil, mineralizable N in the surface 60 cm layer of soil, etc.) are considered to be 100% available for plant uptake. However, the efficiency with which this N is taken up is considered to vary with soil characteristics, plant type and condition, weather conditions, and many other factors. Efficiency is not measured directly; it is calculated from the difference between measured quantities of N in soils and measured quantities of N taken up by plants. Use of uptake efficiency requires an assumption that plants do not take up N from pools of N other than those measured. If roots forage deeper than expected, promote release of fixed NH_4^+, or stimulate mineralization, for example, then uptake efficiency from the measured pool is overestimated.

When the mass balance approach is used, availability is defined so that the unavoidable uncertainty often associated with availability is shifted to the efficiency factor. By this definition, the proximity of NO_3^- to a root influences its uptake efficiency rather than its availability. Nitrate present in the soil before planting is considered to be 100% available for plant uptake even (1) if it will be lost from the system before the seed germinates, (2) when the quantity present greatly exceeds the amounts that can be taken up, or (3) when uptake of much of this N is prevented by lack of water, insect damage to roots, or other factors. Problems associated with definition of terms can be avoided by recognizing that N availability has no specific meaning unless it refers to a specific index of N availability and that N uptake efficiency is a commonly used index of N availability.

1.2.7 Unavoidable Uncertainty

Common use of availability indices involves the assumption that susceptibility of N to plant uptake during a period can be described by using a single number. This assumption is flawed because uptake of N by plants occurs as a flow of N during a period and because many soil, plant, and environmental factors interact to influence the rate of this flow during the period of uptake. Estimates of N availability, therefore, are necessarily confounded with the factors affecting its availability, if only a single estimate is used to characterize N availability during periods lasting days, weeks, or months. It is impossible, therefore, to develop a meaningful definition for availability that describes all that is connoted by this term and applies across a range of soil-plant systems.

Defining availability by using dictionary synonyms (e.g., susceptibility) does not provide adequate direction for measurement of quantities of plant-available N in soils. It is impossible, for example, to describe the susceptibility of NO_3^- to uptake in a small volume of soil surrounding a single root unless assumptions are made concerning length of the period of uptake and many other factors. Rate of N uptake can be expressed as a function of time in simulation models [e.g., CERES-Maize) (Jones and Kiniry, 1986) or LEACHM (Wagenet and Hutson, 1989)] that describe transformations and movement of N, plant growth and demand for N, and many other important factors as they normally influence N uptake in soil-plant systems. Even when such models are used, however, susceptibility to uptake is a difficult concept to quantify.

1.2.8 Use of Nitrogen Availability Indices

Various N availability indices are used in many different ways for many different purposes. Directly or indirectly, most of these uses relate to the practical need to identify appropriate rates of N fertilization for plant growth. For economic and environmental reasons, N application rates should be selected to complement the quantities of available N already in soils. Various indices are used to estimate the amounts of plant-available N that will be supplied by soil, organic materials, or inorganic fertilizers.

Problems associated with defining N availability are avoided by establishing relationships between appropriate indices of N availability. The problem of confounding availability with factors that affect availability is reduced by recognizing that relationships between various indices of N availability should be expected to apply only across a limited range of conditions (e.g., a single crop and selected soil types within a region). A process that can be used to determine how soils and crops should be grouped into meaningful soil categories has been described (Waugh et al., 1973).

The reliability of any N availability index can be evaluated by degree of agreement with other such indices when measurements are made over an appropriate range of conditions. Indices that involve direct measurement of quantities of N should be correlated with indices that rely on plants to assess N availability. If N quantities are measured with the objective of assessing N fertilizer needs, then these quantities should be related to an index that indicates N sufficiency in plants. Degree of agreement between indices can be quantified by amount of variability that can be explained by models (e.g., the R^2 value in Fig. 1.6).

Comparisons of degree of agreement between appropriate N availability indices can be used to evaluate alternative hypotheses about basic processes involved. Finding that a concentration of ~25 mg NO_3–N kg^{-1} is the minimum amount needed to maximize yields across a wide range of conditions (Fig, 1.6), for example, suggests that N tends to follow Bray's expectation for immobile nutrients. Contrary to Bray's expectation for mobile nutrients, the concentration of soil NO_3^- needed to avoid N deficiencies often does not seem to vary in proportion to corn yield (Blackmer et al., 1989). Such observations indicate that, for one reason or another, the ability of plants to take up N from the soil often tends to be proportional to the yield levels attainable with adequate N.

The high mobility and biological reactivity of NO_3^- in soil-plant systems complicates the task of relating measured quantities of soil N to the quantities of N taken up by plants. Complications occur because substantial amounts of NO_3^- can be lost from the system due to transient and unpredictable events such as rainfall. Complications occur because many factors interact to influence flows of N between forms and locations within the rooting zone. Studies of relationships between various N availability indices provide the practical information needed to manage N more efficiently. These studies also provide basic knowledge that helps to understand and describe the complex interaction of factors that control N availability in soils.

1.2.9 References

Aldrich, S.R. 1980. Nitrogen in relation to food, environment, and energy. Univ. IL Agri. Exp. Sta. Spec. Pub. 61.

Allison, F.E. 1973. Soil organic matter. Elsevier North Holland, New York, NY.

Barber, S.A. 1984. Soil nutrient bioavailability. A mechanistic approach. John Wiley and Sons, New York, NY.

Bellini, G., M.E. Sumner, D.E. Radcliffe, and N.P. Qafoku. 1996. Anion transport through columns of highly weathered acid soils: Adsorption and retardation. Soil Sci. Soc. Am. J. 60:132–137.

Binford, G.D., A.M. Blackmer, and M.E. Cerrato. 1992a. Relationships between corn yields and soil nitrate in late spring. Agron. J. 84:53–59.

Binford, G.D., A.M. Blackmer, and B.G. Meese. 1992b. Optimal concentrations of nitrate in cornstalks at maturity. Agron. J. 84:881–887.

Black, C.A. 1993. Soil fertility evaluation and control. Lewis Publishers, Boca Raton, FL.

Blackmer, A. M., and C.J. Green. 1995. Nitrogen turnover by sequential immobilization and mineralization during residue decomposition in soils. Soil Sci. Soc. Am. J. 59:1052–1058.

Blackmer, A.M., D. Pottker, M.E. Cerrato, and J. Webb. 1989. Correlations between soil nitrate concentrations in late spring and corn yields in Iowa. J. Prod. Agric. 2:103–109.

Blackmer, A.M., and S.E. White. 1996. Remote sensing to identify spatial patterns in optimal rates of N fertilization. p. 33–41. In P.C. Robert, R.H. Rust and W.E. Larsen (ed.) Proc. 3rd Int. Conf. Site Specific Manage. Agric. Sys. American Society of Agronomy, Madison, WI.

Blackmer, A.M., and J.S. Schepers. 1996. Aerial photography to detect nitrogen stress in corn. J. Plant. Physiol. 148:440–444.

Bock, B.R. 1984. Efficient use of nitrogen in cropping systems. p. 273–294. In R.D. Hauck (ed.) Nitrogen in crop production. American Society of Agronomy, Madison, WI.

Bock, B.R., and G.W. Hergert. 1991. Fertilizer nitrogen management. p. 139–164. In R.F. Follett, D.R. Keeney and R.M. Cruse (ed.) Managing nitrogen for groundwater quality and farm profitability. Soil Science Society of America, Madison, WI.

Bouldin, D.R., S.D. Klausner, and W.S. Reid. 1984. Use of nitrogen from manure. p. 221–245. In R.D. Hauck (ed.) Nitrogen in crop production. American Society of Agronomy, Madison, WI.

Brady, N.C. 1974. The nature and properties of soils. MacMillan, New York, NY.

Bray, R.H. 1963. Confirmation of the nutrient mobility concept of soil-plant relationships. Soil Sci. 95:124–130.

Bray, R.H. 1954. A nutrient mobility concept of soil-plant relationships. Soil Sci. 78:9–22.

Bundy, L.G., and J.J. Meisinger. 1994. Nitrogen availability indices. p. 951–984. In R.W. Weaver (ed.) Methods of soil analysis. Part 2. Microbiological and biochemical properties. Soil Science Society of America, Madison, WI.

Campbell, C.A., B.H. Ellert, and Y.W. Jame. 1993. Nitrogen mineralization potential in soils. p. 341–349. In M.R. Carter (ed.) Soil sampling and methods of soil analysis. Lewis Publishers, Boca Raton, FL.

Cerrato, M.E., and A.M. Blackmer. 1990. Relationships between grain nitrogen concentrations and the nitrogen status of corn. Agron. J. 82:744–749.

Dahnke, W.C., and G.V. Johnson. 1990. Testing soils for available nitrogen. p. 128–139. In R.L. Westerman (ed.) Soil testing and plant analysis. 3rd Ed. Soil Science Society of America, Madison, WI.

Dean, L.A. 1982. Yield-of-phosphorus curves. Soil Sci. Soc. Am. Proc. 18:462–466.

Edwards, A.P. 1978. Real vs. imaginary sources of error in agronomic research involving ^{15}N. Soil Sci. Soc. Am. J. 42:174–175.

Fried, M., and L.A. Dean. 1952. A concept concerning the measurement of available soil nutrients. Soil Sci. 73:263–271.

Green, C.J., A.M. Blackmer, and N.C. Yang. 1994. Release of fixed ammonium during nitrification in soils. Soil Sci. Soc. Am. J. 58:1411–1415.

Greenland, D.J. 1986. Prediction of nitrogen fertilizer needs of arable crops. Adv. Plant Nutr. 2:1–61.

Hallberg, G.R. 1989. Nitrate in ground water in the United States. p. 35–138. In R.F. Follett (ed.) Nitrogen management and groundwater protection. Elsevier, Amsterdam, Netherlands.

Haynes, R.J. 1986. Mineral nitrogen in the soil-plant system. Academic Press, Orlando, FL.

Hergert, G.W. 1987. Status of residual nitrate-nitrogen soil tests in the United States of America. p. 73–88. In J.R. Brown (ed.) Soil testing: Sampling, correlation, calibration, and interpretation. Soil Sci. Soc. Am. Spec. Pub. 21, Soil Science Society of America, Madison, WI.

Hesterman, O.B. 1988. Exploiting forage legumes for nitrogen contribution in cropping systems. p. 155–166. *In* W.L. Hargrove (ed.) Cropping strategies for efficient use of water and nitrogen. ASA Spec. Pub. 51. American Society of Agronomy, Madison, WI.

Jenkinson, D.S., R.H. Fox, and J.H. Rayner. 1985. Interactions between fertilizer nitrogen and soil nitrogen: The so-called "priming" effect. J. Soil Sci. 36:425–444.

Jones, C.A., and J.R. Kiniry. 1986. CERES-Maize: A simulation model for maize growth and development. TX A&M University Press, College Station, TX.

Keeney, D.R. 1982. Nitrogen availability indices. p. 711–733. *In* A.L. Page (ed.) Methods of soil analysis. Part 2. Chemical methods. 2nd Ed. American Society of Agronomy, Madison, WI.

Li, C., X. Fan, and K. Mengel. 1990. Turnover of interlayer ammonium in loess-derived soils grown with winter wheat in the Shanxi Province of China. Biol. Fert. Soil 9:211–214.

Macy, P. 1936. The quantitative mineral nutrient requirements of plants. Plant Physiol. 11:749–764.

Marshner, H. 1995. Mineral nutrition of higher plants. Academic Press, London, UK.

Mathers, A.C., and D.W. Goss. 1979. Estimating animal waste applications to supply crop nitrogen requirements. Soil Sci. Soc. Am. J. 43:364–366.

Meese, B.G. 1993. Nitrogen management for corn after corn and corn after soybean. Ph.D. Dissertation, IA State University, Ames, IA.

Meisinger, J.J. 1984. Evaluating plant-available nitrogen in soil-crop systems. p. 391–416. *In* R.D. Hauck (ed.) Nitrogen in crop production. American Society of Agronomy, Madison, WI.

Meisinger, J.J., and G.W. Randall. 1991. Estimating nitrogen budgets for soil-crop systems. p. 85–124. *In* R.F. Follett, D.R. Keeney and R.M. Cruse (ed.) Managing nitrogen for groundwater quality and farm profitability. Soil Science Society of America, Madison, WI.

Mitscherlich, E.A., F. Durhring, S.V. Saucken, and C. Bohm. 1925. Die Pflanzenphysiologische Losung der Chemischen Bodenanalyse. Landwirtsch. Jahrb. 58:601–626.

Neeteson, J.J. 1985. Effectiveness of the assessment of nitrogen fertilizer requirement for potatoes on the basis of soil mineral nitrogen. p. 15–24. *In* J.J. Neeteson and K. Dilz (ed.) Assessment of nitrogen fertilizer requirement. Proc. 2nd Meet. NW-Europ. Study Group Assess. N Fert. Requ. Netherlands Fertilizer Institute, Haren, Netherlands.

Nommik, H., and K. Vahtras. 1982. Retention and fixation of ammonium and ammonia in soils. p. 123–171. *In* F.J. Stevenson (ed.) Nitrogen in agricultural soils. American Society of Agronomy, Madison, WI.

NRC. 1993. Soil and water quality: An agenda for agriculture. National Research Council, National Academy Press, Washington, DC.

Nye, P.H., and P.B. Tinker. 1977. Solute movement in the soil-root system. University of California Press, Berkeley, CA.

Peck, T.R., and P.N. Soltanpour. 1990. The principles of soil testing. p. 1-9. *In* R.L. Westerman (ed.) Soil testing and plant analysis. 3rd Ed. Soil Science Society of America, Madison, WI.

Piekielek, W.P., and R.H. Fox. 1992. Use of a chlorophyll meter to predict sidedress nitrogen requirements for maize. Agron. J. 84:59-65.

Pratt, P.F. 1985. Agriculture and groundwater quality. CAST Rep. 103.

Russell, R.S. 1977. Plant root systems. Their function and interaction with the soil. McGraw-Hill, London, UK.

Scarsbrook, C.E. 1965. Nitrogen availability. p. 486-502. *In* W.V. Bartholomew and F.E. Clark (ed.) Soil nitrogen. American Society of Agronomy, Madison, WI.

Smith, J.H., and J.R. Peterson. 1982. Recycling of nitrogen through land application of agricultural, food, and municipal wastes. p. 791-831. *In* F.J. Stevenson (ed.) Nitrogen in agricultural soils. American Society of Agronomy, Madison, WI.

SSSA. 1997. Glossary of soil science terms. Soil Science Society of America, Madison, WI.

Stanford, G. 1973. The rationale for optimum nitrogen fertilization in corn production. J. Environ. Qual. 2:159-166.

Stanford, G., and S.J. Smith. 1972. Nitrogen mineralization potentials of soils. Soil Sci. Soc. Am. Proc. 36:465-472.

Stevenson, F.J. 1982a. Nitrogen in agricultural soils. American Society of Agronomy, Madison, WI.

Stevenson, F.J. 1982b. Origin and distribution of nitrogen in soil. p. 1-42. *In* F.J. Stevenson (ed.) Nitrogen in agricultural soils. American Society of Agronomy, Madison, WI.

Troeh, F.R., and L.M. Thompson. 1993. Soils and soil fertility. 5th Ed. Oxford University Press, New York, NY.

Ulrich, A., and F.J. Hills. 1973. Plant analysis as an aid in fertilizing sugar crops: Part 1. Sugar beets. p. 271-288. *In* L.M. Walsh and J.B. Beaton (ed.) Soil testing and plant analysis. American Society of Agronomy, Madison, WI.

Voss, R.D., and W.D. Shrader. 1984. Rotation effects and legume sources of N for corn. p. 61-68. *In* D.F. Bezdicheck (ed.) Organic farming: Current technology and its role in a sustainable agriculture. ASA Spec. Pub. 46. American Society of Agronomy, Madison, WI.

Wagenet, R.J., and J.L. Hutson. 1989. LEACHM:Leaching estimation and chemistry model: A process-based model of water and solute movement, transformations, plant uptake and chemical reactions in the unsaturated zone, continuum. Vol. 2, Water Resources Institute, Cornell University, Ithaca, NY.

Waugh, D.L., R.B. Cate. and L.A. Nelson. 1973. Discontinuous models for rapid correlation, interpretation, and utilization of soil analysis and fertilizer response data. Int. Soil Fert. Eval. Improv. Prog. Tech. Bull. 7. NC State University, Raleigh, NC.

Wild, A. 1981. Mass flow and diffusion. p. 37-80. *In* D.J. Greenland and M.H. Hayes (ed.) The chemistry of soil processes. John Wiley and Sons, New York, NY.

1.3   Phosphorus Availability

Andrew Sharpley
USDA-ARS, University Park, PA

1.3.1   Introduction

Phosphorus (P) is an essential nutrient for plant growth. The low concentration (100–$3,000$ mg P kg^{-1}) and solubility (< 0.01 mg P L^{-1}) in soils, however, make it a critical nutrient limiting plant growth. In natural ecosystems, P availability is controlled by sorption, desorption, and precipitation of P released during weathering and dissolution of rocks and minerals of low solubility. Thus, P availability is generally inadequate for crop needs in production agriculture. To meet these needs, P is added as fertilizers or animal manures to build up or maintain P availability at predetermined optimum levels (Section D, Chapter 4). In this section, P availability is defined as that P in soil or water that is available by desorption and dissolution processes for uptake by plants in terrestrial and aquatic ecosystems.

The components, forms, availability, and cycling of P in soil are conceptualized in Fig. 1.7. Complex and interrelated processes determine the amounts and availability of several inorganic and organic forms of soil P. This section will describe these processes, how they are affected by agricultural management, and how P availability for crop production can be optimized.

1.3.2   Forms and Amounts in Soil

Soil P exists in inorganic (P_i) and organic (P_o) forms (Fig. 1.7). Inorganic P forms are associated with amorphous and crystalline sesquioxides, and calcareous compounds (Fig. 1.7 and Table 1.2). Organic P forms include relatively labile phospholipids and fulvic acids and more resistant humic acids. Forms generalized in Fig. 1.7 are not discrete entities, as intergrades and dynamic transformations between forms occur continuously to maintain equilibrium conditions. These approximized forms of P are assigned based on the extent to which sequential extractants of increasing acidity or alkalinity can dissolve soil P (Hedley et al., 1982).

Topsoil P content is usually greater than that in subsoil due to the sorption of added P and greater biological activity and accumulation of organic material. However, soil P content varies with parent material, extent of pedogenesis, soil texture, and management factors, such as rate and type of P applied and soil cultivation. These factors also influence the relative amounts of P_i and P_o. In most soils 50 to 75% is P_i, although this fraction can vary from 10 to 90% (Table 1.2). Generally, Ca-P_i decreases with weathering, whereas amorphous Al and Fe-P_i and P_o forms tend to increase, due in part to changes in soil clay fraction from basic primary minerals to Al and Fe dominated oxides.

Phosphorus additions are usually needed to maintain adequate available P for plant uptake. The level of these additions varies with both soil and plant type (Pierzynski and Logan, 1993). Once applied, P is either taken up by the crop, becomes weakly (physical) or strongly (chemical) adsorbed

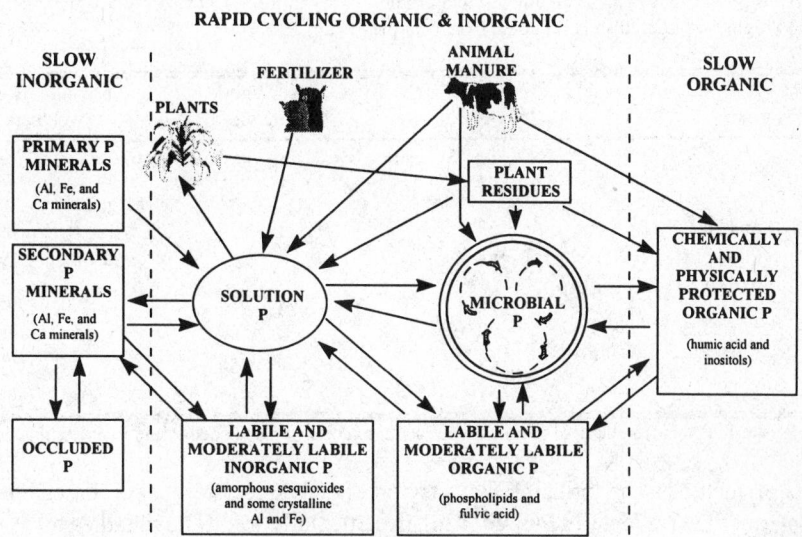

RAPID CYCLING ORGANIC & INORGANIC

Fig. 1.7 The soil P cycle: its components, forms, and flows [Adapted from Stewart and Sharpley, 1987]

onto Al, Fe and Ca surfaces, or incorporated into organic P (McLaughlin et al., 1988; Syers and Curtin, 1989) (Fig. 1.8). As P_i generally supplies most of the P taken up by crops in productive agricultural soils, more attention has been focused on the availability of P_i rather than P_o following P additions.

Overall, soil pH is the main property controlling P_i forms, although Al, Fe, Mn, and Ca contents determine the amounts of these forms (Fig. 1.8). In acid soils, amorphous and hydrous oxides of Al, Fe, and Mn dominate P sorption processes, while Ca compounds dominate P sorption and precipitation reactions in alkaline soils. As a result, P availability is greatest at soil pH values between 6 and 7 (Fig. 1.8). Immobilization of P_i by these processes renders a portion of the added P unavailable for plant uptake (Fig. 1.9). Mehlich III P decreased with time after application of P to a Kingsbury clay (Aeric Ochraqualfs; pH 5.7) and Hagerstown silt loam (Typic Hapludalfs; pH of 6.8) incubated at room temperature (25 °C) and field moisture (~ 30% water). At the same time, P_i becomes more tightly bound with Al and Fe complexes (Fig. 1.9). This illustrates why the removal of P_i from soil by crops is generally low. For the United States, an average 29% of P added in fertilizer and manures is removed by harvested crops, ranging from < 1% in Hawaii to 71% in Wyoming (National Research Council, 1993). The low recovery reflects the predominance of high P-fixing soils in Hawaii.

Even though P_i has generally been considered the major source of plant-available P in soils, the mineralization of labile P_o has also been shown to be important in both low and high fertility soils (Stewart and Tiessen, 1987; Tate et al., 1991). Amounts of P_o mineralized in temperate dryland soils range from 5 to 20 kg P ha^{-1} yr^{-1} (Stewart and Sharpley, 1987). Mineralization of soil P_o tends to be higher in the tropics (67 to 157 kg P ha^{-1} yr^{-1}), where distinct wet and dry seasons and higher soil temperatures enhance microbial activity. In contrast, P_o compounds may also become resistant to hydrolysis by phosphatase through complexation with Al and Fe (Tate, 1984).

1.3.3 Principles of Analysis

The various forms and amounts of P_i and P_o in soil can be estimated by extraction with acids and alkalis that dissolve specific complexes binding P (Olsen and Sommers, 1982; Tiessen and Moir, 1993). A

Table 1.2 Average amounts of inorganic P (P_i) and organic P (P_o) in the A1 horizon of virgin calcareous, slightly weathered, and highly weathered soils [adapted from Sharpley et al., 1987]

P form	Calcareous (n = 41)	Slightly weathered (n = 40)	Highly weathered (n = 39)
	------------------------------------mg kg^{-1}------------------------------------		
Bioavailable P_i [†]	11	21	22
Amorphous Al and Fe-P_i	37	74	109
Ca-P_i	285	85	16
Labile P_o	8	18	34
Protected P_o	28	60	78
Residual P [†]	152	254	179
Total P	521	512	438

[†]Bioavailable P_i is resin P_i and residual P is chemically resistant, mineral, and occluded P_i and P_o as designated on Fig. 1.8

wide range of methods and principles exists for effective recovery of P based on soil type, environment, and level of detail required. This is particularly true for available soil P_i estimation, which has traditionally been based on acid dissolution (acetic, citric, hydrochloric, lactic, nitric, and sulfuric), anion exchange (acetate, bicarbonate, citrate, lactate, and sulfate), cation complexation (citrate, fluoride, and lactate), or cation hydrolysis (buffered bicarbonate). Several excellent reviews of these methods are available for further reading (Kamprath and Watson, 1980; Fixen and Grove, 1990). In the United States, the most common P tests have been Mehlich I, Bray 1, and Olsen P; however, a number of laboratories have converted to the multielement Mehlich III or AB-DTPA extractants in the past decade (Kuo, 1996).

It is unlikely that an extractant would exclusively measure a single pool of soil P_i, although some components of extractants are aimed at specific pools (Fardeau et al., 1988). For example, F in the Bray I extractant exchanges with Al-bound P_i, with the assumption that this Al-bound P_i contributes to available P in acid soils. The success of any extractant to estimate available P depends on the appropriateness of the chemical used relative to soil properties. Alternative methods utilize P sinks, such as anion exchange resins, ion exchange membranes, and Fe oxide impregnated paper, to determine the quantity of soil P available to plants with negligible chemical extraction (Sharpley et al., 1994a). These methods more closely mimic rhizosphere conditions and often provide comparable or better correlations with crop response than chemical extractants (Table 1.3).

1.3.3.1 Anion Exchange Resins

Anion exchange resins (AER) are the most common P sink methods for assessing available P_i. The procedure typically involves the use of chloride saturated resin at a 1:1 resin-to-soil ratio in 10 to 100 mL of water or weak electrolyte for 16 to 24 h (Kuo, 1996). To prevent the diffusion of P from soil to resin from being the rate limiting step, resins should be intimately mixed with the soil, which creates difficulties in separating resin from soil for P analysis. Soil can be ground to a smaller size than the resin, but this probably changes P release characteristics. Resin and soil may also be separated by enclosing the resin in a mesh bag, which may limit resin-soil contact or float resin from soil in a sucrose solution (Thien and Myers, 1991). Skogley et al. (1990) encapsulated a mixture of anion and cation exchange resins in a mesh sphere. Greenhouse studies indicated that the correlation between P uptake by sorghum-sudangrass and resin sphere results were as good or better than those with the Olsen P soil test (Table 1.3).

1.3.3.2 Ion Exchange Membranes

A similar approach using ion exchange resin impregnated membranes facilitates separation of the resin beads from the soil, may eliminate the soil grinding step, and an extraction time as short as 15 min can be used without reducing the accuracy of predicted P availability for a wide range of soils (Qian et al., 1992; Saggar et al., 1992). In pot studies, the resin membranes have provided a better index of P availability than conventional chemical extraction methods for canola and ryegrass (Table 1.3).

1.3.3.3 Iron Oxide Impregnated Paper

Another P sink that has received attention is that of Fe oxide-impregnated filter paper (Fe O strip), which has successfully estimated available P_i in a wide range of soils and management systems (Sharpley, 1993; Menon et al., 1997) (Table 1.3).

Widespread adoption of P sink methods for routine soil testing has not yet occurred in the United States, although parts of Brazil have used the method for the last decade (Raij et al., 1986). As the P sink methods operate with limited chemical extraction, they are more suited to a wide range of soils, irrespective of management history (Sharpley, 1993; Yang et al., 1991; Qian et al., 1992; Somasiri and Edwards, 1992). Where fertilizer history is unknown and frequent changes in fertilizer type, including rock phosphate, may have been made, it is difficult to choose the appropriate soil test. For example, Olsen P can underestimate and Bray 1 P overestimate P availability in soils amended with rock P, while P sink methods have provided accurate estimates when KCl rather than $CaCl_2$ is used as the support medium (Raij et al., 1986; Menon et al., 1989a; Saggar et al., 1992). Even so, detailed field calibration and improvement in standardized methodology will be essential before any of the P sink approaches can be used routinely to estimate available P and make reliable fertilizer recommendations.

Phosphorus isotopes ^{32}P and ^{33}P have also been widely used to characterize P availability. Using laboratory incubation, greenhouse pot, and limited field plot studies, valuable insights have been gained into P availability in terms of exchange kinetics, plant-available forms, and the rate, extent, and direction of P cycling in soils. Readers are directed to excellent reviews by Di et al. (1997), Fardeau (1996), and Frossard and Sinaj (1997) for more information.

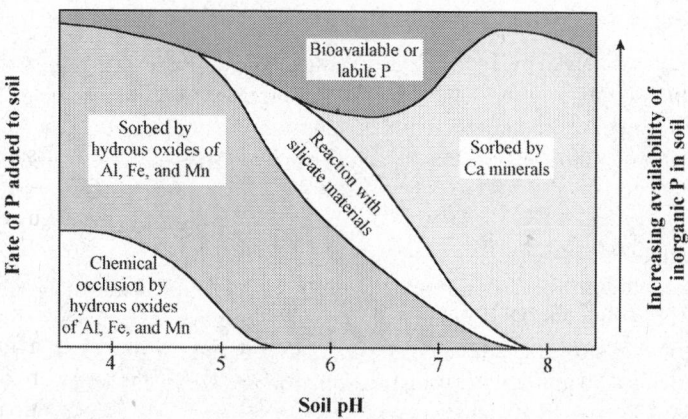

Fig. 1.8 Approximate representation of the fate of P added to soil by sorption and occlusion in inorganic forms, as a function of soil pH [Adapted from Buckman and Brady, 1970]

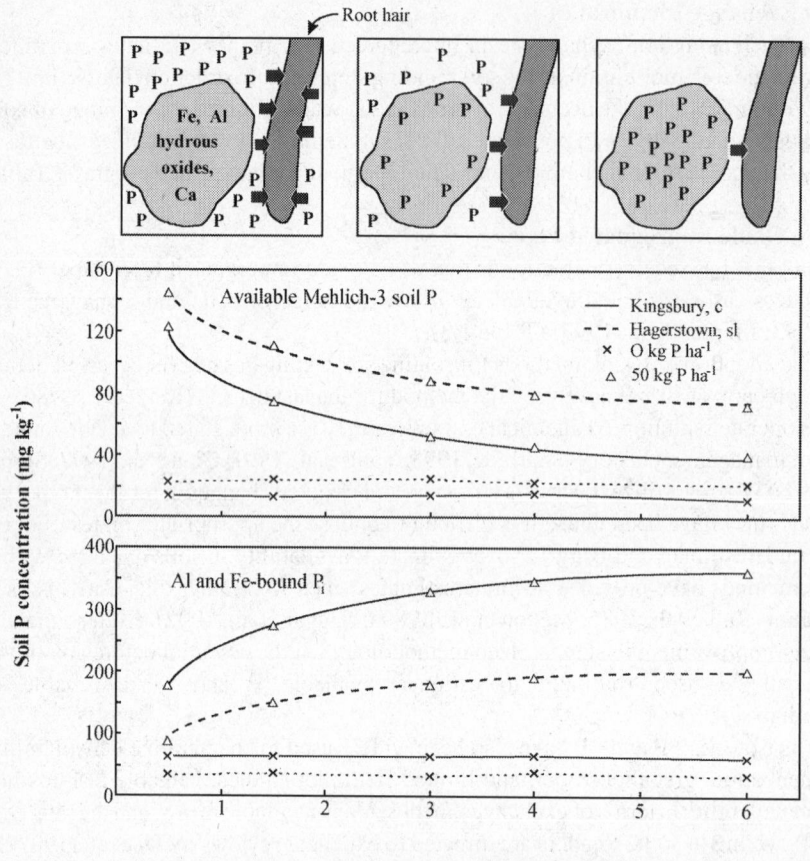

Fig. 1.9   Change in available (Mehlich III-P) and adsorbed soil P (Al and Fe-bound) with time after application of P

1.3.4   Cycling in Soil

Dynamic processes involved in P cycling are chemical and biological (Fig. 1.7). Chemical processes include precipitation/dissolution and sorption/desorption while biological processes involve immobilization/mineralization initiated by P uptake and decay of plants. Also, soil faunal and floral activities often modify the direction, extent, and rate at which these chemical and biological processes occur.

1.3.4.1   Chemical Processes

Precipitation/dissolution processes differ from sorption/desorption in that the solubility product of the least soluble P compound in the solid phase controls dissolution and, thus solution P concentration, whereas solution P controls the amount of P sorbed (Syers and Curtin, 1989). In reality, retention of P by soil material is a continuum between precipitation and surface reactions.

Precipitation/Dissolution

In general, Ca controls these reactions in neutral or calcareous environments, while Al and Fe are the dominant controlling cations in acidic environments. Apatite is the most common primary P mineral

$[Ca_{10}(PO_4)_6 X_2$, where X is F or OH]. Apatite dissolution requires a source of H^+ from soil or biological activity and a sink for Ca and P (Mackay et al., 1986):

$$Ca_{10}(PO_4)_2 X_2 + 12H^+ = 10\ Ca^{2+} + 6H_2PO_4^- + 2X^- \qquad\qquad [1.3.1]$$

Table 1.3 Relationship between crop uptake of P and bioavailable soil P determined by P sink and chemical extraction

Crop	Location	Number of soils	Soil P test[†]	Correlation coefficient	Source
Canola	Saskatchewan	135	IEM	0.92	Qian et al. (1992)
			Olsen	0.78	
Cotton	Brazil	28	AER	0.85	Raij et al. (1986)
			$0.02\ M\ H_2SO_4$	0.68	
Maize	Alabama	32	Fe-O strip	0.87	Menon et al. (1989b)
			AER	0.62	
			Olsen	0.81	
			Bray 1	0.74	
Maize[‡]	Australia	2	Fe-O strip	0.91	Kumar et al. (1992)
			IEM	0.91	
			Bray 1	0.87	
Maize	Egypt	10	Fe-O strip	0.90	Monem and Gadalla (1992)
			Olsen	0.91	
Maize	Uganda	2	Fe-O strip	0.85	Butegwa et al. (1996)
			Bray 1	0.77	
			Mehlich I	0.64	
Rice	Brazil	8	AER	0.98	Raij et al. (1986)
			AB-DTPA	0.41	
Ryegrass[‡]	New Zealand	56	IEM	0.92	Saggar et al. (1992)
			Olsen	0.87	
Ryegrass	Finland	32	Fe-O strip	0.93	Yli-Halla (1990)
			Acetic acid	0.88	
Sorghum/sudangrass	Montana	75	PST	0.91	Schaff et al. (1992)
			Olsen	0.87	
Sudangrass/sorghum/ barley	Colorado	23	AER	0.92	Bowman et al. (1978)
			Olsen	0.88	
Wheat[‡]	Australia	2	Fe-O strip	0.96	Kumar et al. (1992)
			IEM	0.97	
			Bray 1	0.98	
Wheat	China	39	Fe-O strip	0.84	Lin et al. (1991)
			AER	0.83	
			Olsen	0.83	
			Bray 1	0.56	

[†] AEM represents anion exchange membrane; IEM, ion exchange resin; AB-DTPA, ammonium bicarbonate diethylene triamine pentaacetic acid; and PST resin phytoavailability soil test.
[‡] Relationship between relative crop yield and soil P test.

and occurs during soil development due to weathering (Walker and Syers, 1976). Even though the rate of apatite weathering or dissolution will vary with rainfall and temperature, it is still difficult to predict this first step in the P cycle (Pierzynski, 1991).

Precipitates in Ca systems occur in the following sequence: monocalcium phosphate $[Ca(H_2PO_4)_2]$; dicalcium phosphate dihydrate $(CaHPO_4 \cdot 2H_2O)$; octacalcium phosphate $[Ca_8 \, H_2(PO_4)_6 \cdot 5H_2O]$; and finally, hydroxyapatite $[Ca_{10}(PO_4)_6 (OH)_2]$ or fluorapatite $[Ca_{10}(PO_4)_6 F_2]$, which have low solubilities and should, thus, control soil solution P concentration (Lindsay et al., 1989; Syers and Curtin, 1989).

In Al and Fe dominated soils, few well-crystallized precipitates have been observed. Generally, P reacts with Al oxides to form amorphous Al-P or organized phases such as sterretite $[Al(OH_2)_3HPO_4 \, H_2PO_4]$ and with Fe oxides to such precipitates as tinticite $[Fe_6 (PO_4)_4(OH)_6 \cdot 7H_2O)]$ or griphite $[Fe_3 Mn_2 (PO_4)_2 \cdot 5(OH)_2]$ (Hsu, 1982; Lindsay et al., 1989). Many other amorphous mixed Al-Fe-Si-P compounds have been observed in soils with high P concentrations from phosphatic parent material or large fertilizer or manure applications (Pierzynski et al., 1990).

Sorption/Desorption

In most soils, sorption/desorption processes describe P uptake and release by soils that control P availability better than precipitation/dissolution reactions. In this section, the term sorption covers surface adsorption and subsequent penetration of P into the retaining component or absorption. Sorption curves or isotherms have been used extensively to describe the relationship between the amount of P sorbed and that remaining in solution (Fig. 1.10). For a given solution P concentration, there is a large difference in the amount of P sorbed by the three soil types. In general, clay content approximates the reactive surface area of a soil responsible for P sorption (Syers et al., 1971; Sharpley et al., 1984a; Hedley et al., 1995). This surface reactivity is a function of the amount and type of hydrous oxides of Al and Fe and reactive Ca components present, other ions (Ca or Na dominated), pH of the system, and reaction kinetics.

Even in calcareous soils, hydrous Fe oxides can influence P sorption reactions (Holford and Mattingly, 1975a,b). The types of bonding associated with hydrous oxides and pH dependency have been described in detail in Section B, Chapters 7 and 8. Ligand exchange of P at hydrous OH and Fe

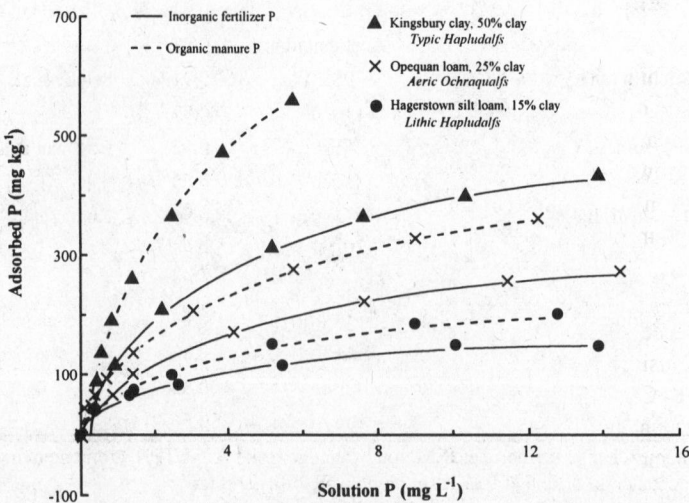

Fig. 1.10 Phosphorus sorption isotherms as a function of the type of soil and P source

Table 1.4 Soil P sorption properties calculated from the Langmuir isotherm as a function of soil type and source of added P (triple superphosphate and dairy manure)

Soil type	P sorption maximum		Binding energy		Equilibrium P concentration	
	Fertilizer	Manure	Fertilizer	Manure	Fertilizer	Manure
	mg kg^{-1}		L mg^{-1}		mg L^{-1}	
Hagerstown, sl Typic Hapludalf	172	245	2.17	2.78	0.019	0.044
Kingsbury, c Aeric Ochraqualf	476	909	2.24	3.82	0.069	0.113
Opequan, l Lithic Hapludalf	303	455	1.97	3.09	0.034	0.114

oxide surfaces results in the formation of monodentate, bidentate, or binuclear complexes. As soil solution pH increases, P sorption is decreased by the greater negative charge at the oxide surface and reduced polarization of the Al- or Fe-O bond.

Soil P sorption has been characterized by parameters calculated from the Langmuir equation:

$$S = \left(\frac{kS_{max}C}{1+kC} \right)$$

where S is sorbed P, C is solution P concentration, S_{max} is the maximum amount of P that can be sorbed, and k is an affinity constant describing binding energy (Section B, Chapter 8). Phosphorus sorption maximum is calculated as the reciprocal of the slope of the plot C/S and C and binding energy as the slope/intercept of the same plot (Table 1.4). Equilibrium P concentration is the solution P concentration at which no net sorption or desorption occurs (Fig. 1.10 and Table 1.4). These parameters have been widely used to quantify the extent of specific soil reactions and effects of soil types, counter ion and background electrolyte, P source, and soil management on P sorption (Fixen and Grove, 1990) (Fig.1.11). However, the theoretical assumptions of the Langmuir equation have been questioned (Barrow, 1989) as sorbed P carries charges which decrease the surface charge and potential of the sorbing surface, leading to large errors in the estimates of S_{max} (Kuo, 1988). The Freundlich equation often fits data better than the Langmuir equation but does not estimate S_{max}.

Organic anions can compete with P for similar sorption sites on soil surfaces (Hue, 1991; Ohno and Crannell, 1996). However, P sorption can be increased when the addition of organic compounds to amorphous oxides in soil impedes their crystallization and increases specific surfaces (Table 1.4). Also, humic compounds can complex with Fe, Al, and to a lesser extent Ca and sorb P (Frossard et al., 1995).

The release of sorbed P into solution or desorption is not completely reversible. Thus, desorption curves are displaced to the left of the sorption curves. This hysteresis effect is the result of precipitation, occlusion, solid-state diffusion, and bidentate or binuclear bonding with the colloid surface (Section B, Chapter 8).

Several studies have reported that P desorption during a short period of time is a low activation energy process (2 and 3 kcal mol^{-1}) (Sharpley et al., 1981). The low activation energies suggest that P desorption during short reaction times may be limited more by the diffusion of the desorbed P

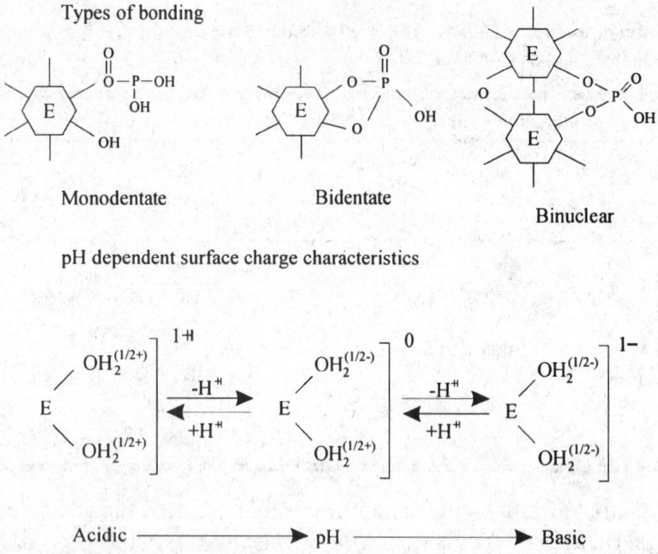

Types of bonding

Monodentate   Bidentate

Binuclear

pH dependent surface charge characteristics

Acidic $\longrightarrow$ pH $\longrightarrow$ Basic

Fig. 1.11 Types of P bonding associated with hydrous oxides where E is Al or Fe [Adapted from Fixen and Grove, 1990]

through the stagnant water films present around the soil particles and within the soil aggregates than by the chemical reaction. Higher activation energies (20 kcal mol^{-1}) have been reported (Barrow and Shaw, 1975) for a longer reaction period (up to 100 days) and represent the transfer of P between desorbable and fixed forms. A continuous range of activation energies for P desorption has been suggested (Posner and Bowden, 1980). During the initial stages, P held at low activation energies is desorbed followed by P held at higher activation energies.

More detailed analysis and review of these chemical processes controlling P availability are given by Barrow (1985), Syers and Curtin (1989), Fixen and Grove (1990), Frossard et al. (1995) (Section B, Chapter 8).

1.3.4.2 Biological Processes

Uptake of soluble P by bacteria and fungi, stimulated by the addition of microbial substrates such as litter and crop residues, and release of P as the result of cell lysis or predation are represented in Fig. 1.7 as a revolving wheel. This is done deliberately to emphasize the central role of the microbial population in P cycling and availability. For example, if the wheel is stopped or slowed down by lack of C inputs, the supply of P to plants will be limited to the quantity of labile P_i. If the wheel is operating, then solution P is constantly being replenished from labile P_i and P_o forms. In a study of P cycling through soil microbial biomass in England, Brookes et al. (1984) measured annual P fluxes of 5 and 23 kg P ha^{-1} yr^{-1} in soils under continuous wheat and permanent grass, respectively. Although biomass P flux under continuous wheat was less than P uptake by the crop (20 kg P ha^{-1} yr^{-1}), annual P flux in the grassland soils was much greater than P uptake by the grass (12 kg P ha^{-1} yr^{-1}).

Within the microbial cell, P exists in a wide variety of compounds, principally RNA (30–50%), acid soluble P_i and P_o (15–20%), phospholipids (< 10%), and DNA (5–10%) (Stewart and McKercher, 1982). If the microbial cell is ruptured or lysed, all these compounds will be released to the soil solution to react with both inorganic and organic soil components to form a host of P_i and P_o compounds of differing solubility or susceptibility to mineralization. The rate of mineralization of P_o

forms depends largely on phosphatase activity which, in turn, can be controlled by solution P concentration (McGill and Cole, 1981). Stable P_o accumulates in both chemically resistant and aggregate protected forms (Tisdale and Oades, 1982) (Section C, Chapter 4).

Chemically or physically protected P_o may be slowly mineralized as a byproduct of overall soil organic matter mineralization or by specific enzyme action in response to the need for P. Therefore, organic matter turnover, as well as solution P_i concentration and the demand for P by microbial and plant components, will be factors controlling the lability of P_o (McGill and Cole, 1981). A continuous drain on soil P pools by cultivation and crop removal will rapidly deplete labile P_i and P_o forms and thereby reduce available P (Sharpley, 1985).

1.3.4.3 Fauna and Flora Processes

The effects of soil fauna (e.g., earthworms and termites) and flora (actinomycetes, bacteria, and fungi) on soil physical, chemical, and biological processes have been extensively studied and reviewed (Lee, 1985). This section outlines the role of the more important of these, earthworms and mycorrhizae, in soil P cycling and their impact on P availability.

Earthworms

Under favorable soil temperature and moisture conditions, *Pheretima* has been found annually to consume 100% (8 Mg ha^{-1} yr^{-1}) of the litter of an evergreen oak forest in Japan (Sugi and Tanaka, 1978) and *Lumbricus terrestris* of mixed deciduous forest in England (3 Mg ha^{-1} yr^{-1}) (Satchell, 1967), about 30% of the litter decomposed each year in grass savanna (Lavelle, 1978) and 100% of added cattle manure (17 to 30 Mg ha^{-1} yr^{-1}) (Guild, 1955). During this consumption, earthworms commonly ingest 100–500 Mg soil ha^{-1} yr^{-1} (equivalent to 0.5–3.0 cm of topsoil). Consequently, earthworms can have a major influence on soil physical, chemical, and biological properties through incorporation and assimilation of plant litter and associated ingestion of soil material (Lee, 1985; Lavelle, 1988). Egestion as surface and subsurface cast material can rapidly redistribute P in a soil profile, increasing availability and potential for uptake. Also, earthworm burrowing can allow a greater soil volume to be exploited by plant roots, which decreases susceptibility of plants to water stress. Opening of these burrows at the soil surface increases soil infiltration rates and aeration (Edwards et al., 1989).

The main source of P affected by earthworms is soil organic matter (SOM), which includes plant litter in various stages of decomposition, roots, and organic matter in the soil with its complement of microflora and fauna. Ingestion, maceration, and intimate mixing of organic matter with soil increases the rate of humification and mineralization. Although most organic matter undergoes little chemical change during passage through the earthworm gut, it is finely ground with the increased surface area exposed to microbial activity, facilitating further decomposition. The five- to tenfold increase in P content and availability in earthworm casts results from enhanced mineralization of organic P, enrichment of clay-sized particles in casts, and a reduction in P sorption capacity of soil by organic matter blockage (Sharpley et al., 1992). Thus, most of the additional P present in casts is held in more physically sorbed than chemisorbed forms, which are readily available to plants.

Both microbial population and activity are increased during passage through the earthworm gut as a function of the organic matter content of the initial food source (Scheu, 1987). Parle (1963) observed that numbers of bacteria and actinomycetes increased 1,000-fold during passage through the gut, and O_2 consumption remained higher in earthworm casts than in soil for 50 days, indicating increased microbial activity. This enhanced activity is probably responsible for the increased phosphatase activity found in earthworm casts compared to underlying soil.

Mycorrhizal Associations

Mycorrhizae are widespread under natural conditions, with vesicular-arbuscular mycorrhizae (VAM) most common in agricultural soils. Three primary mechanisms by which VAM enhances P availability are increased physical exploration of the soil, chemical modification of the rhizosphere, and physiological differences between VAM and plant roots. For example, extensive hyphal growth of VAM reduces the distance for diffusion of P in soil, thereby increasing uptake. This effect is greater when diffusion limits uptake (Gerdemann, 1968). Consequently, a greater response to VAM infection has been exhibited in coarse than fine rooted plant species (Crush, 1973), in high than low P sorbing soils (Yost and Fox, 1979), and in soils than solution culture (Howeler et al., 1982). In addition, the generally smaller diameter of VAM hyphae (2–4 μm) compared to root hairs (7–10 μm) affords a greater absorptive surface area for hyphae and enables their entry into soil pores and organic matter that cannot be entered by root hairs.

Vesicular-arbuscular mycorrhizae may chemically modify the rhizosphere through exudation of chelating compounds (Jayachandran et al., 1989) or phosphatases (Harley, 1989), which could solubilize poorly soluble soil P. It is clear, however, that VAM utilize the same sources of P as nonmycorrhizal plants (Kucey et al., 1989), but do so more efficiently.

On a unit weight basis, mycorrhizal plants can rapidly absorb larger amounts of P than nonmycorrhizal plants (Bolan et al., 1987). Because this difference cannot be accounted for by increased hyphal surface area alone, a greater affinity for P in mycorrhizal plants may increase absorption rates and the critical or minimum P concentration, below which there is limited net absorption of P, may be lower for mycorrhizal than nonmycorrhizal plants because of an increase in physical contact between hyphae and P, thereby reducing P diffusion distance.

1.3.5 Optimizing Phosphorus Availability

Critical available soil P concentrations are required to maximize crop yield and can be optimized by P additions, liming, and cultivation. This section discusses how these factors influence P availability; the soil testing and recommendation process that quantifies P additions is discussed in Section D, Chapter 4.

1.3.5.1 Critical Concentrations for Plant Production

Estimates of soil P concentrations, above which little or no crop response to P additions is obtained, vary with the extractant used (Kamprath and Watson, 1980) and are 10 and 30 mg P kg^{-1} for Olsen P and Bray 1 P, respectively, and 10 and 25 mg P kg^{-1} for clayey and sandy soils, respectively.

Several additional factors influence P availability. These include temperature, compaction, moisture, aeration, pH, type and amount of clay, and nutrient (including P) status of soil. When soil temperatures are low during early plant growth, P uptake is reduced. Compaction reduces pore space decreasing the amount of water and O_2 which, in turn, reduces P uptake. The use of liming materials to increase topsoil pH and thereby P availability has been adequately reviewed, but reducing subsoil acidity has not been highly successful due to physical limitations of incorporating lime. However, surface applications of gypsum with sufficient time for transport into the subsoil have been shown to increase crop yields and subsoil Ca, and decrease subsoil Al (Sumner et al., 1986). Soils with high clay content tend to fix more P than sandy soils with a low clay content. Thus, more P needs to be added to raise the soil test level of clay than loamy and sandy soils. In addition, the presence of NH_4^+ enhances P uptake by creating an acid environment due to nitrification and NH_4^+ uptake. High concentrations of NH_4–N and P fertilizer may interfere with and delay normal P fixation reactions, prolonging the availability of fertilizer P (Murphy, 1988).

1.3.5.2 Phosphorus Additions

To optimize P availability through P additions, many factors must be considered. Because of the immobility of P in most soils, the timing of fertilizer P application is less critical than placement. Even so, small amounts of placed starter fertilizer for vegetable crops have successfully reduced the need for much larger broadcast applications of P (Costigan, 1988), and a similar strategy (e.g., foliar applications) may be appropriate for other crops. In efforts to minimize P inputs in sustainable or low input management systems, there has been renewed interest in the estimation and utilization of residual P availability from fertilizer or manure amendments (McCollum, 1991).

Rate

The application of P either as mineral fertilizer or animal manure increases available P (Table 1.5). In many areas of intensive crop and livestock operations, the application of P at rates greater than crop removal has increased available P above critical concentrations for crop production (Sharpley et al., 1994b). The increase in available P ranged from 5 to 31 mg kg^{-1} for every 100 kg ha^{-1} of P added (an average 18% increase; Table 1.5). McCollum (1991) observed similar increases in available P following mineral fertilizer P application. These values are similar to proportional increases following beef (7–23 mg kg^{-1}; an average 14%), poultry (14–28 mg kg^{-1}; an average 20%), and swine manure application (5–20 mg kg^{-1}; an average 11%) (Table1.5).

Type

Ways to broaden the use of slow release P fertilizers, such as partially acidulated rock P (RP), beyond soils which have low pH, Ca, and P contents have been evaluated (Hedley et al., 1989; Muchovej et al., 1989). For example, in soils of neutral pH, it may be possible to apply a heavy initial dressing of finely ground RP and include a rotation of fine rooted legumes to generate a low pH rhizosphere with low Ca concentrations and thus increase RP dissolution. Other methods designed to increase acidity in the immediate RP soil environment, and thereby its dissolution, include addition of elemental S (Muchovej et al., 1989), NH_4^+ fertilizers, or organic matter such as animal manure and crop residues (Hedley et al., 1989).

Animal manure itself is a valuable source of P for crop production. The availability of manurial P in some soils may differ from fertilizer P (Table 1.4; Fig. 1.10) (Frossard et al., 1995; Hue, 1991). As a major proportion of manurial P can be organic (25 to 50%), biological processes in soil will play a greater role in determining P availability than for fertilizer P. The slower release of P from manure may make it a longer term source of P to crops than more readily soluble fertilizer P. For more information on the fertilizer value of manures, readers are directed to the reviews of Overcash et al. (1983), Sharpley et al. (1998), and Sims (1995).

Placement

Due to the general immobility of P in the soil profile, fertilizer placement is generally more critical for P than N. Depending on soil and environmental factors, band applications may or may not be better than broadcast incorporated applications of P. The effect of P application also varies with soil type. For six soils having a hundredfold variation in P sorptivity, Holford (1989) found that fertilizer P effectiveness, as measured by yield response in the first crop (wheat), residual effect in the second crop (clover), or cumulative recovery of applied P, was consistently greater for shallow banding at 5 cm depth compared to banding at 15 cm or broadcast applications. The almost equal effect obtained by mixing P throughout the soil, regardless of P sorptivity, suggested that the positional availability of P in the root zone is important in maximizing fertilizer effectiveness in addition to reducing P sorption.

Table 1.5 Available soil P of soil treated with fertilizer or manure for several years and untreated soil in the United States and United Kingdom studies

Soil[†]	Crop	Added P kg ha⁻¹ yr⁻¹	Time yr	Available soil P Method	Untreated mg kg⁻¹	Treated mg kg⁻¹	Source and Location
Fertilizer							
Portsmouth, fsl Typic Umbraquult	Mixed veg.	20	9	Mehlich I	18	73	Cox et al. (1981) North Carolina and Rothamsted, UK
Batcombe, cl Typic Haploboroll		27	19	Olsen	16	44	
Richfield, scl Aridic Argiudoll	Mixed veg.	20	14	Bray 1	12	54	Hooker et al. (1983) Kansas
		40	14		12	56	
Pullman, cl Torrertic Paleustoll	Sorghum	56	8	Bray 1	15	76	Sharpley et al. (1984b) Texas
Keith, sil Aridic Argiudoll	Wheat	11	6	Bray 1	22	31	McCallister et al. (1987) Nebraska
		33	6		24	47	
Rosebud, sil Aridic Argiustoll	Wheat	11	6	Bray 1	10	28	
		33	6		10	48	
Beef Manure							
Lethbridge, cl Typic Haploboroll	Barley	160	11	Bray 1	22	424	Chang et al. (1991) Alberta
		320	11		22	736	
		480	11		22	893	
Pullman, cl Torrertic Paleustoll	Sorghum	90	8	Bray 1	15	63	Sharpley et al. (1984b) Texas
		273	8		15	230	
Poultry Litter							
Cahaba, vfsl Typic Hapludult	Grass	130	12	Bray 1	5	216	Sharpley et al. (1993) Oklahoma
Ruston, fsl Typic Paleudult		100	12		12	342	
Stigler, sl Aquic Paleudalf		35	35		14	239	
Swine Manure							
Norfolk, l Typic Kandiudult	Grass	109	11	Mehlich I	80	235	King et al. (1990) North Carolina
		218	11		80	310	
		437	11		80	450	
Captina, sl Typic Fragiudult	Grass	101	9	Bray 1	5	121	Sharpley et al. (1991) Oklahoma
Sallisaw, l Typic Paleudalf	Grass	81	15		6	147	
Stigler, sl Aquic Paleudalf	Wheat	37	9		15	82	
Cecil, sl Typic Kanhapludult	Grass	160	3	Mehlich I	19	45	Reddy et al. (1980) North Carolina
		320	3		19	100	

[†]vfsl - very fine sandy loam, fsl - fine sandy loam, sl - sandy loam, sil - silt loam, l - loam, scl - silty clay loam, cl - clay loam, c - clay.

Positional availability is also influenced by crop type. In order for banding or restricted fertilizer placement to increase potential root extraction of P, the rate of P absorption and growth of roots in fertilized soil must increase to compensate for roots in unfertilized soil. Increased root growth and P uptake in the P fertilized volume of soil compared to unfertilized soil has been observed for corn (Anghioni and Barber, 1980), soybeans (Borkert and Barber, 1985), and wheat (Yao and Barber, 1986). In contrast, several studies have shown that flax does not respond to banded fertilizer due to an inability of its root system to expand and proliferate into and efficiently absorb P from high concentrations in the fertilized zone (Strong and Soper, 1974). In the case of flax, increased P uptake and yield response were obtained when fertilizer P was placed 2–5 cm directly below the seed, ensuring adequate P levels during early growth (Bailey and Grant, 1989).

Residual Availability

Halvorson and Black (1985) showed that soil test P levels were increased above the initial available-P level for more than 16 yr, by a one-time P application on a Williams loam (Typic Argiborolls) in Montana (Fig. 1.12). After the initial increase, available-P levels declined for about 12 yr and then stabilized at a higher available level than was initially present, thus establishing what appears to be a new equilibrium level of available P. Fixen (1986) reported similar changes in available-P levels with time. Crop yields reported by Halvorson and Black (1985) were also improved by the residual P fertilizer for a period of 16 yr (Fig. 1.12).

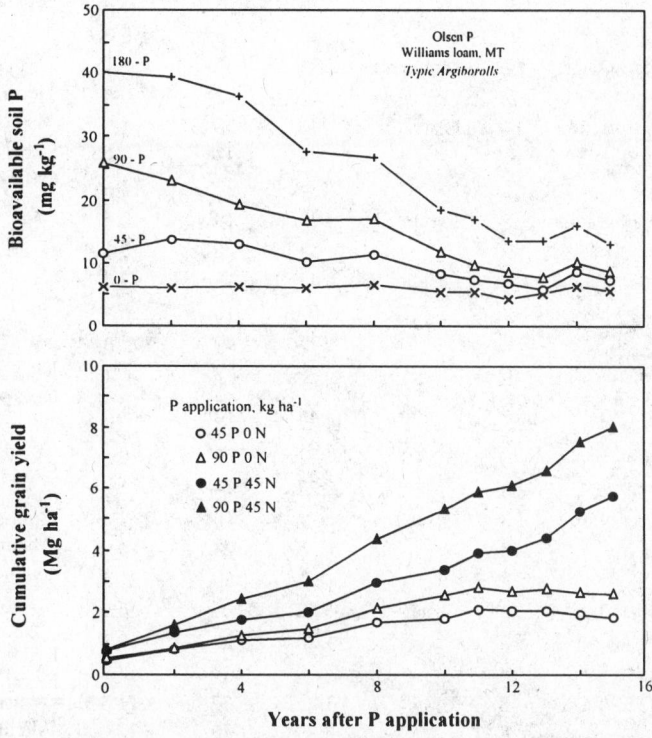

Fig. 1.12 Changes in bioavailable soil P (Olsen P) and cumulative wheat yields following a single application of fertilizer P [Data adapted from Halvorson and Black, 1985]

The rate of decline in available P in high P soils when no further P is added varies with soil type and management (Table 1.6). The rate of decline in available P ranged from 0.1 to 30 mg kg^{-1} yr^{-1}. McCollum (1991) estimated that without further P, 16 to 18 yr of corn (*Zea mays* L.) or soybean (*Glycine max* (L.) Merr.) production would be needed to deplete soil P (Mehlich III P) in a Portsmouth fine sandy loam (Typic Umbraquults) from 100 mg P kg^{-1} to the agronomic threshold level of 20 mg P kg^{-1}. Several authors have found the rate of decrease in available P with depletion by cropping when no P is added, is inversely related to the soil's P buffering capacity (Aquino and Hanson, 1984) or P sorption saturation (available soil P/ P sorption maximum) (Sharpley, 1996).

1.3.5.3   Cultivation

Cultivation and associated fertilizer applications can influence the amount of soil P_i and P_o and should be considered in optimizing P availability. Thompson et al. (1954) measured a decrease in P_i and P_o content of 25 surface (0–15 cm) soils from Iowa, Texas, and Colorado, with cultivation. Similarly, Adepetu and Corey (1977) reported that 25% of the P_o content of the surface of a Nigerian soil was

Table 1.6   Decrease in available soil P after P application is stopped in North American studies

Soil[†]	Crop	Time yr	Method	Available soil P Initial mg kg^{-1}	Final	Decline mg kg^{-1} yr^{-1}	Source and Location
Thurlow, l	Small grains	9	Olsen	13	4	1.0	Campbell (1965)
Ustollic Haplargid		9		20	4	1.8	Montana
		9		60	6	6.0	
Georgeville, scl	Small grains	7	Mehlich I	3	1	0.1	Cox et al. (1981)
Typic Hapludult		7		7	2	0.6	North Carolina
Haverhill, cl	Wheat/fallow	14	Olsen	40	25	1.1	Saskatchewan
Typic Epiaquoll		14		74	33	2.9	
		14		134	69	4.6	
Portsmouth, fsl	Small grains	8	Mehlich I	23	18	0.6	
Typic Umbraquult		9		54	26	3.1	
Sceptre, c	Wheat/fallow	8	Olsen	45	18	3.4	
Typic Haploboroll		8		67	18	6.1	
		8		147	40	13.4	
Williams, l	Wheat/barley	16	Olsen	26	8	1.1	Havlorson and Black (1985)
Typic Argiboroll		16		45	14	1.9	Montana
Richfield, scl	Corn	8	Bray 1	12	8	0.5	Hooker et al. (1983)
Aridic Argiustoll		8		22	14	1.0	Kansas
Carroll, cl	Wheat/flax	8	Olsen	71	10	7.6	Spratt et al. (1980)
Typic Haploboroll		8		135	23	14.0	Manitoba
		8		222	50	21.5	
Waskada, l	Wheat/flax	8	Olsen	48	9	4.9	
Typic Haploboroll		8		88	23	8.1	
		8		200	49	18.9	
Waskada, cl	Wheat/flax	8	Bray	140	50	11.3	Wagar et al. (1986)
Typic Haploboroll		8		320	80	30.0	Manitoba

[†]fsl - fine sandy loam, l - loam, scl - silty clay loam, cl - clay loam, c - clay.

mineralized during the first two cropping periods following cultivation. Where no fertilizer P is added, there is a general P_i consistency, although available levels will gradually decline. The net loss from the system through removal in the harvested crop is primarily accounted for by a decrease in P_o. For example, 60 years (1913–1973) of cotton growth on a Mississippi Delta soil, Dundee silt loam (Typic Endoaqualfs), with no reported P added, resulted in no appreciable effect on P_i (Sharpley and Smith, 1983). However, the P_o content of cultivated (93 mg kg^{-1}) compared to virgin analogue (223 mg kg^{-1}) surface soil (0–15 cm) decreased. Mineralization of P_o replenished the P_i pool and thus, provided adequate amounts of available P. Where cultivation involved P application, P_i and P_o increased 130 and 227%, respectively, while available P decreased 33% on average for 7 agricultural soils (Sharpley and Smith, 1983).

1.3.6 Conclusions

The amount and availability of soil P are determined by physical, chemical, and biological processes, which are often managed in attempts to increase or optimize crop uptake of P and yields. In many areas, P cycles have been fragmented by the specialization of agricultural production systems in specific regions so that mineral fertilizer P is imported to areas of crop production from continental United States and overseas deposits, which have been treated to varying degrees to increase P solubility. The harvested grain is used to meet human or animal demands. In both cases, major population areas and confined animal operations are geographically removed from areas of crop production. For example, most of the corn produced in the Midwest is used as feed in the eastern United States. As a result, P is moving from areas of ore deposits, through crop producing regions and accumulating in areas of confined animal operations.

The fragmentation of P cycling has emphasized that many areas of the United States require P inputs to maintain adequate available-P levels for crop production, while localized areas are accumulating P and have soil P levels in excess of crop requirements. Thus, P availability issues must consider the P imbalances in large-scale cycling that are occurring. Future management of P availability must address the impacts that this specialization of agricultural systems is having on regional P requirements and production on a national scale. At the small scale, however, efforts should continue to find ways of enhancing P availability, through soil testing, appropriate rates of P application, and utilization of organic or manure sources of P in an increasing number of agricultural production systems. If these goals are not met, P availability may become an environmental rather than soil fertility issue in more areas.

1.3.7 References

Adepetu, J.A., and R.B. Corey. 1976. Organic phosphorus as a predictor of plant available phosphorus in soils of Southern Nigeria. Soil Sci. 122:159–164.

Anghioni, I., and S.A. Barber. 1980. Predicting the most efficient phosphorus placement for corn. Soil Sci. Soc. Am. J. 44:1016–1020.

Aquino, B.F., and R.G. Hanson. 1984. Soil phosphorus supplying capacity evaluated by plant removal and available phosphorus extraction. Soil Sci. Soc. Am. J. 48:1091–1096.

Bailey, L.D., and C.A. Grant. 1989. Fertilizer phosphorus placement studies on calcareous and noncalcareous chernozemic soils: Growth, P-uptake and yield of flax. Commun. Soil Sci. Plant Anal. 20:635–654.

Barrow, N.J. 1985. Reactions of anions and cations with variable charge soils. Adv. Agron. 38:183–230.

Barrow, N.J. 1989. Surface reactions of phosphate in soil. Agric. Sci. 2:33–37.

Barrow, N.J., and T.C. Shaw. 1975. The low reactions between soil and anions: 2. Effect of time and temperature on the decrease in phosphate concentration in the soil solution. Soil Sci. 119:167–177.

Bolan, N.S., A.D. Robson, and N.J. Barrow. 1987. Effects of phosphorus application and mycorrhizal inoculation on root characteristics of subclover and ryegrass in relation to phosphorus uptake. Plant Soil 104:294–298.

Borkert, C.M., and S.A. Barber. 1985. Soybean shoot and root growth and phosphorus concentration as affected by phosphorus placement. Soil Sci. Soc. Am. J. 49:152–155.

Bowman, R.A., S.R. Olsen, and F.S. Watanabe. 1978. Greenhouse evaluation of residual phosphate by four phosphorus methods in neutral and calcareous soils. Soil Sci. Soc. Am. J. 42:451–454.

Brookes, P.C., D.S. Powlson, and D.S. Jenkinson. 1984. Phosphorus in the soil microbial biomass. Soil Biol. Biochem. 16:169–175.

Buckman, H.O., and N.C. Brady. 1970. The nature and properties of soils. Collier-Macmillan Limited, London, UK.

Butewaga, C.N., G.L. Mullins, and S.H. Chien. 1996. Induced phosphorus fixation and the effectiveness of phosphate fertilizers derived from Sukulu Hills phosphate rock. Fert. Res. 44:213–240.

Campbell, R.E. 1965. Phosphorus fertilizer residual effects on irrigated crops in rotation. Soil Sci. Soc. Am. Proc. 29:67–70.

Chang, C., T.G. Sommerfeldt, and T. Entz. 1991. Soil chemistry after eleven annual applications of cattle feedlot manure. J. Environ. Qual. 20:475–480.

Costigan, P. 1988. The placement of starter fertilizers to improve early growth of drilled and transplanted vegetables. Proc. Fert. Soc. No. 274. The Fertilizer Society, Peterborough, UK.

Cox, F.R., E.J. Kamprath, and R.E. McCollum. 1981. A descriptive model of soil test nutrient levels following fertilization. Soil Sci. Soc. Am. J. 45:529–532.

Crush, J.R. 1973. The effect of *Rhizophagus tenius* mycorrhizas on rye grass, cocksfoot and sweet vernal. New Phytol. 72:965–973.

Di, H.J., L.M. Condron, and E. Frossard. 1997. Isotope techniques to study phosphorus cycling in agricultural and forest soils: A review. Bio. Fert. Soils 27:1–12.

Edwards, W.M., M.J. Shipitalo, L.B. Owens, and L.D. Norton. 1989. Water and nitrate movement in earthworm burrows within long-term no-till cornfields. J. Soil Water Conserv. 44:240–243.

Fardeau, J.C. 1996. Dynamics of phosphate in soils: An isotopic outlook. Fert. Res. 45:91–100.

Fardeau, J.C., C. Morel, and R. Boniface. 1988. Pourquoi choisirla methode Olsen pour estimer le phosphore "assimilable" des sols? Agronomie 8:577–584.

Fixen, P.E. 1986. Residual effects of P fertilization: Lessons for the '80's. p. 1–8. *In* Proc. 16th North Central Extension-Industry Soil Fertility Workshop. Potash and Phosphate Institute, Atlanta, GA.

Fixen, P.E., and J.N. Grove. 1990. Testing soils for phosphorus. p. 141–180. *In* R.L. Westerman (ed.) Soil Testing and Plant Analysis, 3rd Ed. Soil Science Society of America, Madison, WI.

Frossard, E., and S. Sinaj. 1997. The isotopic exchange kinetic technique: A method to describe the availability of inorganic nutrients. Applications to K, PO_4, SO_4, and Zn. Isotop. Environ. Health Stud. 33:61–67.

Frossard, E., M. Brossard, M.J. Hedley, and A. Metherell. 1995. Reactions controlling the cycling of phosphorus in soils. p. 107–138. *In* H. Tiessen (ed.) Phosphorus in the global environment: Transfers, cycles and management. J. Wiley and Sons, New York, NY.

Gerdemann, J.W. 1968. Vesicular-arbuscular mycorrhiza and plant growth. Ann. Rev. Phytopath. 6:397–418.

Guild, W.J. 1955. Earthworms and soil structure. p. 83–98. *In* D. K. McE. Kevan (ed.) Soil zoology. Butterworth, London, UK.

Guttay, A.J.R., and L.M.C. Dandurand. 1989. Interaction of the vesicular-arbuscular mycorrhizae of maize with extractable soil phosphorus levels and nitrogen-potassium fertilizers. Biol. Fertil. Soils 8:307–310.

Halvorson, A.D., and A.L. Black. 1985. Long-term dryland crop responses to residual phosphorus fertilizer. Soil Sci. Soc. Am. J. 49:928–933.

Harley, J.L. 1989. The significance of mycorrhizae. Mycol. Res. 92:129–139.

Hedley, M.J., J.M. Mortvedt, N.S. Bolan, and J.K. Syers. 1995. Phosphorus fertility management in agroecosystems. p. 59–92. *In* H. Tiessen (ed.) Phosphorus in the global environment: Transfers, cycles and management. J. Wiley and Sons, New York, NY.

Hedley, M.J., J.W.B. Stewart, and B.S. Chanhan. 1982. Changes in inorganic and organic soil phosphorus fraction induced by cultivation practices and by laboratory incubations. Soil Sci. Soc. Am. J. 46:970–976.

Hedley, M.J., R.W. Tillman, and G. Wallace. 1989. The use of nitrogen fertilizers for increasing the suitability of reactive phosphate rocks for use in intensive agriculture. p. 311–320. *In* R.E. White and L.D. Currie (ed.) Nitrogen fertilizer use in New Zealand agriculture and horticulture. Occasional Report 3, Fertilizer and Lime Research Centre, Massey University, New Zealand.

Holford, I.C.R. 1989. Efficacy of different phosphate application methods in relation to phosphate sorptivity in soils. Aust. J. Soil Res. 27:123–133.

Holford, I.C.R., and G.E.G. Mattingly. 1975a. The high- and low-energy phosphate adsorbing surfaces in calcareous soils. J. Soil Sci. 26:407–417.

Holford, I.C.R., and G.E.G. Mattingly. 1975b. Surface areas of calcium carbonate in soils. Geoderma 13:247–255.

Hooker, M.L., R.E. Gwin, G.M. Herron, and P. Gallagher. 1983. Effects of long-term annual applications of N and P on corn grain yields and soil chemical properties. Agron. J. 75:94–99.

Howeler, R.H., C.J. Asher, and D.G. Edwards. 1982. Establishment of an effective mycorrhizal association on cassava in flowing solution culture and its effects on phosphorus nutrition. New Phytol. 90:229–238.

Hsu, P.H. 1982. Crystallization of iron (III) phosphate at room temperature. Soil Sci. Soc. Am. J. 46:928–932.

Hue, N.V. 1991. Effects of organic acids/anions on P sorption and phytoavailability in soils with different mineralogies. Soil Sci. 152:463–471.

Jayachandran, J., A.P. Schwab, and B.A.D. Hetrick. 1989. Mycorrhizal mediation of phosphorus availability: Synthetic iron chelate effects on phosphorus solubilization. Soil Sci. Soc. Am. J. 53:1701–1706.

Kamprath, E.J., and M.E. Watson. 1980. Conventional soil and tissue tests for assessing the phosphorus status of soils. p. 433–469. *In* F.E. Khawsaneh et al. (ed.) The role of phosphorus in agriculture. Soil Science Society of America, Madison, WI.

King, L.D., J.C. Burns, and P.W. Westerman. 1990. Long-term swine lagoon effluent applications on 'Coastal' Bermudagrass: II. Effects on nutrient accumulations in soil. J. Environ. Qual. 19:756–760.

Kucey, R.M.N., H.H. Janzen, and M.E. Leggett. 1989. Microbially mediated increases in plant-available phosphorus. Adv. Agron. 42:199–228.

Kumar, V., R.J. Gilkes, and M.D.A. Bolland. 1992. Comparison of seven soil-P tests for plant species with different external-P requirements grown on soils containing rock phosphate and superphosphate residues. Fert. Res. 33:35–45.

Kuo, S. 1988. Application of a modified Langmuir isotherm to phosphate sorption by some acid soils. Soil Sci. Soc. Am. J. 52:97–102.

Lavelle, P. 1978. Les vers de terre de la savane de Lamto (Cote d'Ivoire). Peuplements, populations et fonctions de Pecosysteme. Publ. Lab. Zool. E.N.S. 12:1–301.

Lavelle, P. 1988. Earthworm activities and the soil system. Biol. Fert. Soils 6:237–251.

Lee, K.E. 1985. Earthworms: Their ecology and relationships with soils and land use. Academic Press, London, UK.

Lin, T.H., S.B. Ho, and K.H. Houng. 1991. The use of iron oxide-impregnated filter paper for the extraction of available phosphorus from Taiwan soils. Plant Soil 133:219–226.

Lindsay, W.L., P.L.G. Vlek, and S.H. Chien. 1989. Phosphate minerals. p. 1089–1130. *In* J.B. Dixon and S.B. Weed (ed.) Minerals in soil environments. 2nd Ed. Soil Science Society of America, Madison, WI.

Mackay, A.D., J.K. Syers, R.W. Tillman, and P.E.H. Gregg. 1986. A simple model to describe the dissolution of phosphate rock in soils. Soil Sci. Soc. Am. J. 50:291–296.

McCallister, D.L., C.A. Shapiro, W.R. Raun, F.N. Anderson, G.W. Rhem, O.P. Engelstadt, M.O. Russelle, and R.A. Olson. 1987. Rate of phosphorus and potassium buildup/decline with fertilization for corn and wheat on Nebraska Mollisols. Soil Sci. Soc. Am. J. 51:1646–1652.

McCollum, R.E. 1991. Buildup and decline in soil phosphorus: 30-year trends on a Typic Umbraquult. Agron. J. 83:77–85.

McGill, W.B., and C.V. Cole. 1981. Comparative aspects of cycling of organic C, N, S, and P through soil organic matter. Geoderma 26:267–286.

McLaughlin, M.J., A.M. Alston, and J.K. Martin. 1988. Phosphorus cycling in wheat-pasture rotations. III. Organic phosphorus turnover and phosphorus cycling. Aust. J. Soil Res. 26:343–353.

Menon, R.G., S.M. Chien, and W.J. Chardon. 1007. Iron oxide-impregnated filter paper (Pi): II. A review of its applications. Nut. Cycl. Agroecosys. 47:7–18.

Menon, R.G., S.H. Chien, and L.L. Hammond. 1989a. Comparison of Bray 1 and P$_i$ tests for evaluating plant-available phosphorus from soils treated with different partially acidulated phosphate rocks. Plant Soil 114:211–216.

Menon, R.G., L.L. Hammond, and H.A. Sissingh. 1989b. Determination of plant-available phosphorus by the iron hydroxide-impregnated filter paper (P$_i$) soil test. Soil Sci. Soc. Am. J. 52:110–115.

Menon, R.G., S.H. Chien, L.L. Hammond, and B.R. Arora. 1990. Sorption of phosphorus by the iron oxide-impregnated filter paper (P$_i$ soil test) embedded in soils. Plant Soil 126:287–294.

Monem, M.A., and A.M. Gadalla. 1992. Evaluation of the Pi soil test in Egyptian soils and its development as a rapid field test. p. 139–144. *In* J. Ryan and A. Matar (ed.) Fertilizer use efficiency under rain-fed agriculture in West Asia and North Africa. Proc. 4th Regional Workshop, May 1991, Morocco.

Muchovej, R.M.C., J.J. Muchovej, and V.H. Alvarez. 1989. Temporal relations of phosphorus fractions in an Oxisol amended with rock phosphate and *Thiobacillus thiooxidans*. Soil Sci. Soc. Am. J. 53:1096–1100.

Murphy, L.S. 1988. Phosphorus management strategies for MEY of spring wheat. *In* Proc. Profitable Spring Wheat Production Workshop, Jan. 6–7, 1988, Fargo, ND.

National Research Council. 1993. Soil and water quality: An agenda for agriculture. National Academy Press, Washington, DC.

Ohno, T., and B.S. Crannell. 1996. Green and animal manure-derived dissolved organic matter effects on phosphorus sorption. J. Environ. Qual. 25:1137–1143.

Olsen, S.R., and L.E. Sommers. 1982. Phosphorus. p. 403–429. *In* A.L. Page et al. (ed.) methods of soil analysis, Part 2. 2nd Ed. American Society of Agronomy, Madison, WI.

Olsen, S.R., and F.S. Watanabe. 1963. Diffusion of phosphorus as related to soil texture and plant uptake. Soil Sci. Soc. Am. Proc. 27:648–653.

Overcash, M.R., F.J. Humenik, and J.R. Miner. 1983. Introduction to livestock waste management. p.114–182. *In* M.R. Overcash et al. (eds.), CRC Livestock Waste Management. Volume 2. CRC Press, Boca Raton, FL.

Parle, J.N. 1963. Microorganisms in the intestines of earthworms. J. Gen. Microbiol. 31:1–11.

Pierzynski, G.M. 1991. The chemistry and mineralogy of phosphorus in excessively fertilized soils. Crit. Rev. Environ. Contr. 21:265–295.

Pierzynski, G.M., and T.J. Logan. 1993. Crop, soil, and management effects on phosphorus soil test levels. J. Prod. Agric. 6:513–520.

Pierzynski, G.M., T.J. Logan, S.J. Traina, and J. M. Bigham. 1990. Phosphorus chemistry and mineralogy in excessively fertilized soils: descriptions of phosphorus-rich particles. Soil Sci. Soc. Am. J. 54:1583–1589.

Posner, A.M., and J.W. Bowden. 1980. Adsorption isotherms: Should they split? J. Soil Sci. 31:1–10.

Qian, P., J.J. Schoenau, and W.Z. Huang. 1992. Use of ion exchange membranes in routine soil testing. Commun. Soil Sci. Plant Anal. 23:1791–1804.

Raij, B. van, J.A. Quaggio, and N.M. de Silva. 1986. Extraction of phosphorus, potassium, calcium, and magnesium from soils by an ion-exchange resin procedure. Commun. Soil Sci. Plant Anal. 17:547–566.

Reddy, K.R., M.R. Overcash, R. Kahled, and P.W. Westerman. 1980. Phosphorus absorption-desorption characteristics of two soils utilized for disposal of animal manures. J. Environ. Qual. 9:86–92.

Saggar, S., M.J. Hedley, R.E. White, P.E.H. Gregg. K.W. Perrot, and I.S. Cornforth. 1992. Development and evaluation of an improved soil test for phosphorus. 2. Comparison of the Olsen and mixed cation-anion exchange resin tests for predicting the yield of ryegrass growth in pots. Fert. Res. 33:135–144.

Satchell, J.E. 1967. Lumbricidae. p. 259–322. *In* A. Burges and F. Raw (ed.) Soil biology. Academic Press, London, UK.

Schaff, B.E., E.O. Skogley, J.W. Bauder, and D.J. Sieler. 1992. Resin capsule adsorption of P for predicting plant response and P uptake. Agron. Abs. 84:290.

Scheu, S. 1987. Microbial activity and nutrient dynamics in earthworm casts (Lumbricidae). Biol. Fertil. Soils 5:230–234.

Sharpley, A.N. 1985. Phosphorus cycling in unfertilized and fertilized agricultural soils. Soil Sci. Soc. Am. J. 49:905–911.

Sharpley, A.N. 1991. Soil phosphorus extracted by iron-aluminum-oxide-impregnated filter paper. Soil Sci. Soc. Am. J. 55:1038–1041.

Sharpley, A.N. 1996. Availability of residual phosphorus in manured soils. Soil Sci. Soc. Am. J. 60:1459–1466.

Sharpley, A.N., and S.J. Smith. 1983. Distribution of phosphorus forms in virgin and cultivated soil and potential erosion losses. Soil Sci. Soc. Am. J. 47:581–586.

Sharpley, A.N., and S.J. Smith. 1985. Fractionation of inorganic and organic phosphorus in virgin and cultivated soils. Soil Sci. Soc. Am. J. 49:1276–130.

Sharpley, A.N., J.T. Sims, and G.M. Pierzynski. 1994a. Innovative soil phosphorus indices: Assessing inorganic phosphorus. p. 115–142. *In* J. Havlin, J. Jacobsen, P. Fixen, and G. Hergert (ed.) New directions in soil testing for nitrogen, phosphorus and potassium. Soil Science Society of America, Madison, WI.

Sharpley, A.N., H. Tiessen, and C.V. Cole. 1987. Soil phosphorus forms extracted by soil tests as a function of pedagenesis. Soil Sci. Soc. Am. J. 51:362–365.

Sharpley, A.N., L.R. Ahuja, M. Yamamoto, and R.G. Menzel. 1981. The kinetics of phosphorus desorption from soil. Soil Sci. Soc. Am. J. 45:493–496.

Sharpley, A.N., S.J. Smith, and W.R. Bain. 1993. Nitrogen and phosphorus fate from long-term poultry litter applications to Oklahoma soils. Soil Sci. Soc. Am. J. 57:1131–1137.

Sharpley, A.N., C.A. Jones, C. Gray, and C.V. Cole. 1984. A simplified soil and plant phosphorus model. II. Prediction of labile, organic, and sorbed phosphorus. Soil Sci. Soc. Am. J. 48:805–809.

Sharpley, A.N., J.J. Meisinger, J. Power, and D. Suarez. 1992. Root extraction of nutrients associated with long-term soil management. Adv. Soil Sci. 19:151–217.

Sharpley, A.N., S.J. Smith, B.A. Stewart, and A.C. Mathers. 1984. Forms of phosphorus in soil receiving cattle feedlot waste. J. Environ. Qual. 13:211–215.

Sharpley, A.N., B.J. Carter, B.J. Wagner, S.J. Smith, E.L. Cole, and G.A. Sample. 1991. Impact of long-term swine and poultry manure applications on soil and water resources in eastern Oklahoma. OK State Univ. Tech. Bull. T169.

Sharpley, A.N., S.C. Chapra, R. Wedepohl, J.T. Sims, T.C. Daniel, and K.R. Reddy. 1994. Managing agricultural phosphorus for protection of surface waters: Issues and options. J. Environ. Qual. 23:437–451.

Sharpley, A.N., J.J. Meisinger, A. Breeuwsma, T. Sims, T.C. Daniel, and J.S.Schepers. 1997. Impacts of animal manure management on ground and surface water quality. p.173–242. *In* J. Hatfield (ed.). Effective management of animal waste as a soil resource. Lewis Publishers, Boca Raton, FL.

Sims, J.T. 1993. Environmental soil testing for phosphorus. J. Prod. Agric. 6:501–507.

Sims, J.T. 1995. Animal waste management. p. 185–201. *In* Encyclopedia of agricultural science. Academic Press, New York, NY.

Skogley, E.O., S.J. Georgitis, J.E. Yang, and B.F. Schaff. 1990. The phytoavailability soil test - PST. Commun. Soil Sci. Plant Anal. 21:1229–1243.

Somasiri, L.L.W., and A.C. Edwards. 1992. An ion exchange resin method for nutrient extraction of agricultural advisory soil samples. Commun. Soil Sci. Plant Anal. 23:645–657.

Spratt, E.D., F.G. Warder, L.D. Bailey, and D.W.L. Read. 1980. Measurement of fertilizer phosphorus residues and its utilization. Soil Sci. Soc. Am. J. 44:1200–1204.

Stewart, J.W.B., and R.B. McKercher. 1982. Phosphorus cycle. p. 221–238. *In* R.G. Burns and J.H. Slater (ed.) Experimental microbial ecology. Blackwell Scientific Publications, Oxford, UK.

Stewart, J.W.B., and A.N. Sharpley. 1987. Controls on dynamics of soil and fertilizer phosphorus and sulfur. p. 101–121. *In* R.F. Follett, J.W.B. Stewart, and C.V. Cole (eds.) Soil fertility and organic matter as critical components of production systems. Soil Sci. Soc. Am. Spec. Pub. 19, Soil Science Society of America, Madison, WI.

Stewart, J.W.B., and H. Tiessen. 1987. Dynamics of soil organic phosphorus. Biogeochem. 4:41–60.

Strong, W.M., and R.J. Soper. 1974. Utilization of pelleted phosphorus by flax, wheat, rape and buckwheat from a calcareous soil. Agron. J. 65:18–21.

Sugi, Y., and M. Tanaka. 1978. Number and biomass of earthworm populations. p. 171–178. *In* T. Kira, Y. Ono, and T. Hosokawa (ed.) Biological production in a warm-temperature evergreen oak forest of Japan. University of Tokyo Press, Tokyo, Japan.

Sumner, M.E., H. Shahandah, J. Bouton, and J. Hammel. 1986. Amelioration of an acid soil profile through deep liming and surface application of gypsum. Soil Sci. Soc. Am. J. 50:1254–1258.

Syers, J.K., and D. Curtin. 1989. Inorganic reactions controlling phosphorus cycling. p. 17–29. *In* H. Tiessen (ed.) Phosphorus cycles in terrestrial and aquatic ecosystems. Saskatchewan Institute of Pedology, Saskatoon, Canada.

Syers, J.K., T.D. Evans, J.D.H. Williams, and J.T. Murdock. 1971. Phosphate sorption parameters of representative soils from Rio Grande Do Sul, Brazil. Soil Sci. 112:267–275.

Tate, K.R. 1984. The biological transformation of P in soil. Plant Soil 76:245–256.

Tate, K.R., T.W. Spier, D.J. Ross, R.L. Parfitt, K.N. Whale, and J.C. Cowling. 1991. Temporal variations in some plant and soil P pools in two pasture soils of different P fertility status. Plant Soil 132:219–232.

Thien, S.J., and R. Myers. 1991. Separating ion-exchange resin from soil. Soil Sci. Soc. Am. J. 55:890–892.

Thompson, L.M., C.A. Black, and J.A. Zoellner. 1954. Occurrence and mineralization of organic phosphorus in soils with particular reference to associations with nitrogen, carbon, and pH. Soil Sci. 77:185–196.

Tiessen, H., and J.O. Moir. 1993. Characterization of available P in sequential extraction. p. 75–86. *In* M.R. Carter (ed.) Soil sampling and methods of analysis. Lewis Publishers, Boca Raton, FL.

Tisdale, J.M., and J.M. Oades. 1982. Organic matter and water stable aggregates in soils. J. of Soil Sci. 33:141–163.

Wagar, B.I., J.W.B. Stewart, and J.L. Henry. 1986. Comparison of single large broadcast and small annual seed-placed phosphorus treatments on yield and phosphorus and zinc content of wheat on chernozemic soils. Can. J. Soil Sci. 66:237–248.

Walker, T.W., and J.K. Syers. 1976. The fate of phosphorus during pedogenesis. Geoderma 15:1–19.

Yang, J.E., E.O. Skogley, S.J. Georgitis, B.F. Schaff, and A.H. Ferguson. 1991. Phytoavailability soil test: Development and verification of theory. Soil Sci. Soc. Am. J. 55:1358–1365.

Yao, J., and S.A. Barber. 1986. Effect of one phosphorus rate placed in different soil volumes on P uptake and growth of wheat. Commun. Soil Sci. Plant Anal. 17:819–827.

Yli-Halla, M. 1990. Comparison of a bioassay and three chemical methods for determination of plant-available P in cultivated soils of Finland. J. Agric. Sci. Finl. 62:213–319.

Yost, R.S., and R.L. Fox. 1979. Contribution of mycorrhizae to P nutrition of crops growing on an oxisol. Agron. J. 71:903–908.

1.4 Bioavailability of Soil Potassium

Donald L. Sparks

University of Delaware

1.4.1 Introduction

The role of K in soils is prodigious. Of the many plant nutrient-soil mineral relationships, those involving K are of major if not prime significance (Sparks and Huang, 1985).

Since the middle of the 17th century, when J. R. Glauker in The Netherlands first proposed that saltpeter (KNO_3) was the principle of vegetation, K has been recognized as being beneficial to plant growth (Russell, 1961). Glauker obtained large increases in plant growth from addition of saltpeter to the soil that was derived from the leaching of coral soils. The essentiality of K to plant growth has been known since the work of von Liebig published in 1840.

Of the major nutrient elements, K is usually the most abundant in soils (Reitemeier, 1951). Igneous rocks of the Earth's crust have higher K contents than sedimentary rocks. Of the igneous rocks, granites and syenites contain 46 to 54, basalts 7, and peridotites 2.0 g K kg^{-1}. Among the sedimentary rocks, clayey shales contain 30, whereas limestones have an average of only 6 g K kg^{-1} (Malavolta, 1985).

Mineral soils generally range between 0.04 and 3% K. Total K contents in soils range between 3000 and 100,000 kg ha^{-1} in the upper 0.2 m of the soil profile. Of this total K content, 98% is bound in the mineral form, whereas 2% is in soil solution and exchangeable phases (Schroeder, 1979; Bertsch and Thomas, 1985).

Potassium, among mineral cations required by plants, is the largest in nonhydrated size (r = 0.133 nm) and the number of oxygen atoms surrounding it in mineral structures is high (8 or 12), which suggests that the strength of each K-O bond is relatively weak (Sparks and Huang, 1985). Potassium has a polarizability equal to 0.088 nm^3, which is higher than for Ca^{2+}, Li^+, Mg^{2+}, and Na^+ but lower than for Ba^{2+}, Cs^+, NH_4^+, and Rb^+ ions (Rich, 1968, 1972; Sparks and Huang, 1985). Ions with higher polarizability are preferred in ion exchange reactions. Potassium has a hydration energy of 142.5 kJ g^{-1} ion^{-1}, which indicates little ability to cause soil swelling (Helfferich, 1962).

1.4.2 Forms of Soil K

Soil K exists in four forms in soils: solution, exchangeable, fixed or nonexchangeable, and structural or mineral (Fig. 1.13). Quantities of exchangeable, nonexchangeable, and total K in the surface layer (0–20 cm) of a variety of soils are shown in Table 1.7. Exchangeable K and nonexchangeable K levels comprise a small portion of the total K. The bulk of total soil K is in the mineral fraction (Sparks and Huang, 1985). There are equilibrium and kinetic reactions between the four forms of soil K that affect the level of soil solution K at any particular time, and thus, the amount of readily available K for plants. The forms of soil K in the order of their availability to plants and microbes are solution > exchangeable > fixed (nonexchangeable) > mineral (Sparks and Huang, 1985; Sparks, 1987).

1.4.2.1 Solution K

Soil solution K is the form of K that is directly taken up by plants and microbes and also is the form most subject to leaching in soils. Levels of soil solution K are generally low, unless recent amendments

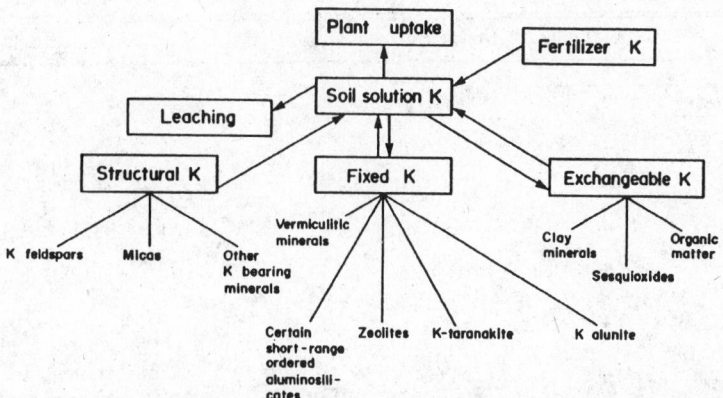

Fig. 1.13 Interrelationships of various forms of soil K [Reprinted from Sparks and Huang, 1985. Munson. (ed.) Potassium in agriculture with permission of the Soil Science Society of America]

of K have been made to the soil. The quantity of K in the soil solution varies from 2 to 5 mg K L^{-1} for normal agricultural soils of humid regions and is an order of magnitude higher in arid region soils (Haby et al., 1990). Levels of solution K are affected by the equilibrium and kinetic reactions that occur between the forms of soil K, the soil moisture content, and the concentrations of bivalent cations in solution and on the exchanger phase (Sparks and Huang, 1985).

1.4.2.2 Exchangeable K
Exchangeable K is the portion of the soil K that is electrostatically bound as an outer-sphere complex to the surfaces of clay minerals and humic substances. It is readily exchanged with other cations and also is readily available to plants.

1.4.2.3 Nonexchangeable K
Nonexchangeable or fixed K differs from mineral K in that it is not bonded within the crystal structures of soil mineral particles. It is held between adjacent tetrahedral layers of dioctahedral and trioctahedral micas, vermiculites, and intergrade clay minerals such as chloritized vermiculite (Rich, 1972; Sparks and Huang, 1985; Sparks, 1987). Potassium becomes fixed because the binding forces between K and the clay surfaces are greater than the hydration forces between individual K$^+$ ions. This results in a partial collapse of the crystal structures and the K$^+$ ions are physically trapped to varying degrees, making K release a slow, diffusion controlled process (Sparks, 1987). Nonexchangeable K also can be found in wedge zones of weathered micas and vermiculites (Rich, 1964). Only ions with a size similar to K$^+$, such as NH$_4^+$ and H$_3$O$^+$, can exchange K from wedge zones (Section F, Chapter 2). Large hydrated cations, such as Ca^{2+} and Mg^{2+}, cannot fit into the wedge zones. Release of nonexchangeable K to the exchangeable form occurs when levels of exchangeable and soil solution K are decreased by crop removal and/or leaching and perhaps by large increases in microbial activity (Sparks, 1980).

Nonexchangeable K is moderately to sparingly available to plants (Mengel, 1985; Sparks and Huang, 1985; Sparks, 1987). Mortland et al. (1956) showed that biotite could be altered to vermiculite by plant removal of K. Schroeder and Dummler (1966) showed that the nonexchangeable K associated with some German soil illites was an important source of K to crops. The ability of plants to take up nonexchangeable K appears to be related to the plant species. Steffens and Mengel (1979) found that ryegrass (*Lolium perenne*) could take up nonexchangeable K longer without yield

Table 1.7 Potassium status of selected soils[†]

Origin of Soil	Exchangeable K	Nonexchangeable K	Total K	Source
	------------------------ (cmol$_c$ kg^{-1}) ------------------------			
Alfisols				
Nebraska, USA	0.40	--	--	*Soil Taxonomy* (1975)
West Africa	0.46	--	3.07	Juo (1981)
Inceptisols				
California, USA	0.40	--	--	*Soil Taxonomy* (1975)
Maryland, USA	0.20	--	--	*Soil Taxonomy* (1975)
Mollisols				
Iowa, USA	0.27	--	--	*Soil Taxonomy* (1975)
Nebraska, USA	0.40	--	--	*Soil Taxonomy* (1975)
Ultisols				
Delaware, USA	0.33	0.49	22.5	Parker et al. (1989)
Florida, USA	0.14	0.25	2.71	Yuan et al. (1976)
Virginia, USA	0.11	0.17	6.5	Sparks et al. (1980)
West Africa	0.24	--	8.06	Juo (1981)

† Data are for surface soils (0-20 cm depth).

reductions, while red clover (*Trifolium pratense*) could not. This was attributed to the ryegrass having longer root length which would allow it to grow at a relatively low K concentration. A similar concentration would result in a K deficiency in red clover. It may be that the difference in root mass, root length, and root morphology between monocots and dicots explains why monocots feed better from nonexchangeable K than dicots (Mengel, 1985).

1.4.2.4   Mineral K

As noted earlier, most of the total K in soils is in the mineral form, mainly as K-bearing primary minerals such as muscovite, biotite, and feldspars. For example, in some Delaware soils, Sadusky et al. (1987) found that mineral K comprised about 98% of the total K (Table 1.8). Most of the mineral K was present as K feldspars in the sand fractions.

Common soil K-bearing minerals, in the order of availability of their K to plants, are biotite, muscovite, orthoclase, and microcline (Huang et al., 1968; Sparks, 1987). Mineral K is generally assumed to be only slowly available to plants; however, the availability is dependent on the level of K in the other forms, and the degree of weathering of the feldspars and micas constituting the mineral K fraction (Sparks and Huang, 1985; Sparks, 1987). Sadusky et al. (1987) and Parker et al. (1989a,b) have found that a substantial amount of K is released from the sand fractions of Delaware soils that are high in K feldspars. This finding, along with the large quantities of mineral K in these and other Atlantic Coastal Plain soils, could help in explaining the often observed lack of crop response to K amendments on these soils (Liebhardt et al., 1976; Yuan et al., 1976; Sparks et al., 1980; Woodruff and Parks, 1980; Parker et al., 1989b).

1.4.3   Factors Affecting Potassium Availability

1.4.3.1   Solution-Exchangeable K Dynamics

The rate and direction of reactions between solution and exchangeable forms of K determine whether applied K will be leached into lower horizons, taken up by plants, converted into unavailable forms, or released into available forms.

The reaction rate between soil solution and exchangeable phases of K is strongly dependent on the type of clay minerals present (Sivasubramaniam and Talibudeen, 1972; Sparks et al., 1980; Sparks and Jardine, 1981, 1984; Jardine and Sparks, 1984) and the method employed to measure kinetics of K exchange (Sparks, 1989,1995; Amacher, 1991; Sparks et al., 1996). Vermiculite, montmorillonite, kaolinite, and hydrous mica vary drastically in their ionic preferences, ion binding affinities, and types of ion exchange reactions. Such fundamental differences in these clay minerals account for the varying kinetics of K exchange.

Kinetics of K exchange on kaolinite and montmorillonite are usually quite rapid (Malcolm and Kennedy, 1969; Sparks and Jardine, 1984). An illustration of this is shown in Fig. 1.14. In the case of kaolin clays, the tetrahedral layers of adjacent clay layers are held tightly by H bonds; thus, only planar

Table 1.8   Potassium status of Delaware soils and sand fractions [From Sadusky et al. 1987. Soil Sci. Soc. Am. J. 51:1460-1465 with permission of the Soil Science Society of America]

Horizon	Depth	Soils				Sand fractions	
		$CaCl_2$ extractable	HNO_3 extractable	Mineral $K^‡$	Total K	Total $K^ƒ$	K feldspars[¶]
	cm	---------------------------------- cmo$_c$l kg^{-1} ----------------------------					frequency %
Kenansville loamy sand							
Ap	0-23	0.25	0.42	35.02	35.69	30.88	9.5
Bt2	85-118	0.25	0.49	45.30	46.04	33.86	12.0
Rumford loamy sand							
Ap	0-25	0.33	0.49	21.67	22.51	18.62	6.7
BC	89-109	0.21	0.54	23.39	23.96	16.76	8.2
Sassafras fine loamy sand							
Ap	0-20	0.35	0.56	43.54	44.45	28.95	16.0
C1	84-99	0.13	0.36	45.99	46.68	36.69	24.0

‡   Mineral K = [(total K) - (CaCl$_2$ ext. K + HNO$_3$ ext. K)].
ƒ   These data represent the amount of total K in the sand based on a whole soil basis.
¶   Determined through petrographic analyses of the whole sand fractions and represents the percentage of total point counts in a given sample that were K feldspars. The remaining minerals in the sand fractions were quartz, plagioclase, and opaques.

external surface and edge sites are available for ionic exchange. With montmorillonite, the inner peripheral space is not held together by H bonds, but instead is able to swell with adequate hydration, and thus allow for rapid passage of ions into the interlayer space. Malcolm and Kennedy (1969) found that the rate of Ba exchange on kaolinite and montmorillonite was rapid with 75% of the total exchange occurring in three s.

Kinetics of K exchange on vermiculitic and micaceous minerals tend to be extremely slow. Both are 2:1 phyllosilicates with peripheral spaces that impede many ion exchange reactions. Micaceous minerals typically have a more restrictive interlayer space than vermiculite since the area between layer silicates of the former is selective for certain types of cations (e.g., K^+, Cs^+). Bolt et al. (1963) theorized the existence of three types of binding sites for K exchange on a hydrous mica. The authors hypothesized that slow kinetics were due to internal exchange sites, rapid kinetics to external planar sites, and intermediate kinetics to edge sites.

1.4.3.2   Potassium Fixation

The phenomenon of K fixation or retention significantly affects K availability. The fact that fixation processes are limited to interlayer ions such as K^+ has been explained in terms of the good fit of K^+ ions (the crystalline radius and coordination number are ideal) in an area created by holes and adjacent oxygen layers (Barshad, 1951). The important forces involved in interlayer reactions in clays are electrostatic attractions between the negatively charged layers and the positive interlayer ions, and expansive forces due to ion hydration (Kittrick, 1966).

The degree of K fixation in clays and soils depends on the type of clay mineral and its charge density, the degree of interlayering, the moisture content, the concentration of K^+ ions as well as the concentration of competing cations, and the pH of the ambient solution bathing the clay or soil (Rich, 1968; Sparks and Huang, 1985).

The major clay minerals responsible for K fixation are montmorillonite, vermiculite, and weathered micas. In acid soils, the principal clay mineral responsible for K fixation is dioctahedral

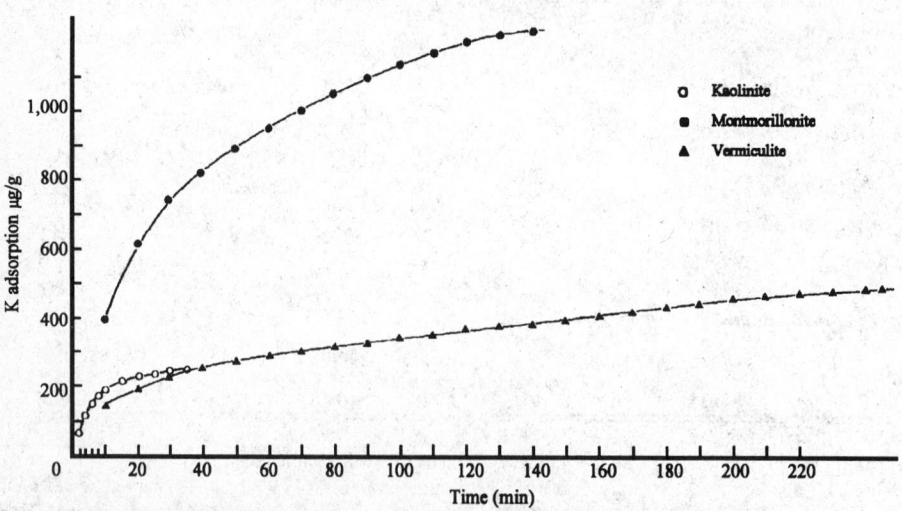

Fig. 1.14   Potassium adsorption versus time in pure systems [Reprinted from Sparks and Jardine, 1984. Comparison of kinetic equations to describe K-Ca exchange in pure and mixed systems. Soil Sci. 138:115-122 with permission of Williams and Wilkins, Baltimore, MD]

vermiculite. Weathered micas fix K under moist as well as dry conditions, whereas some montmorillonites fix K only under dry conditions (Rich, 1968).

The degree of K fixation is strongly influenced by the charge density on the layer silicate. Those with high charge density fix more K than those with low charge density (Walker, 1957). Weir (1965) noted that K fixation by montmorillonites is limited unless the charge density of the clays is high. Low charge montmorillonite (Wyoming) stays at 1.5 nm when K saturated unless it is heated (Laffer et al., 1966). Schwertmann (1962a,b) noted that soil montmorillonites have a greater capacity to fix K than do many specimen montmorillonites. Soil montmorillonites have higher charge density and a greater probability of having wedge positions near mica-like zones where the selectivity for K is high (Rich, 1968).

The importance of interlayer hydroxy Al and hydroxy Fe^{3+} material on K fixation was first noted in the classic work of Rich and Obenshain (1955). They theorized that hydroxy Al and hydroxy Fe^{3+} interlayer groups acted as props to decrease K fixation. This theory was later corroborated in the work of Rich and Black (1964) who found that the introduction of hydroxy Al groups into Libby vermiculite increased the Gapon selectivity coefficient (k_G) from 5.7 to 11.1 x 10^{-2} L mmol$^{-\frac{1}{2}}$.

Wetting and drying and freezing and thawing can significantly affect K fixation (Hanway and Scott, 1957; McLean and Simon, 1958; Cook and Hutcheson, 1960). The degree of K fixation or release on wetting or drying is dependent on the type of colloid present and the level of K^+ ions in the soil solution.

Potassium fixation by 2:1 clay minerals may be strongly influenced by the kind of adsorbed cations or the anions within the system. In studies with the silicate ion, Mortland and Gieseking (1951) found that montmorillonite clays dried with K_2SiO_3 were altered in their swelling properties and fixed K in large amounts. Hydrous mica clays also fixed large amounts of K that could not be removed with boiling HNO_3.

Volk (1934) observed a marked increase in K fixation in soils where pH was raised to about 9 or 10 with Na_2CO_3. Martin et al. (1946) showed at pH values up to 2.5 there was no fixation; between pH 2.5 and 5.5, the amount of K fixation increased very rapidly. Above pH 5.5, fixation increased more slowly. These differences in K fixation with pH were discussed by Thomas and Hipp (1968). At pH values > 5.5, Al^{3+} cations precipitate as hydroxy polycations, which increase in the number of OH groups as pH increases until they have a form like gibbsite (Thomas, 1960). At this pH (~ 8), Al^{3+} does not neutralize the charge on the clay and cannot prevent K fixation. Below pH 5.5, Al^{3+} and $Al(OH)_x$ species dominate. Below pH 3.5, H_3O^+ predominates (Coleman and Harward, 1953; Thomas and Hipp, 1968).

The increase in K fixation between pH 5.5 and 7.0 can be ascribed to the decreased numbers of $Al(OH)_x$ species which decrease K fixation (Rich and Obenshain, 1955; Rich, 1960, 1964; Rich and Black, 1964). At low pH, the lack of K fixation is probably due to large numbers of H_3O^+ and their ability to replace K as well (Rich, 1964; Rich and Black, 1964).

1.4.3.3 Potassium Release

The release of K from micas proceeds by two processes: (1) the transformation of K-bearing micas to expansible 2:1 layer silicates by exchanging the K with hydrated cations, and (2) the dissolution of the micas followed by the formation of weathering products. The relative importance of these two mechanisms depends on the stability of micas and the nature of soil environments (Sparks and Huang, 1985).

Release of K from feldspars appears to involve a rapid exchange with H, which creates a thin layer of hydrolyzed aluminosilicate. This residual layer ranges in thickness from several to a few tens of nm and seems to cause the initial nonstoichiometric release of alkali and alkaline earths relative to Si and Al. Following this step, there is continued dissolution, which removes hyperfine particles. After these are removed, further dissolution breaks down the outer surface of the residual layer at the same rate that alkalis are replaced by H at the interface between fresh mineral surfaces and the residual layer. This releases all constituents to the solution. Release is now stoichiometric. Thus, the weathering of feldspars appears to be a surface-controlled reaction (Sparks, 1989).

A number of physiochemical and mineralogical factors govern the release of K from micas by both cation exchange reactions and dissolution processes. These include tetrahedral rotation and cell dimensions, degree of tetrahedral tilting, hydroxyl orientation, chemical composition, particle size, structural imperfections, degree of K depletion, layer charge alterations and associated reactions, hydronium ions, biological activity, inorganic cations, wetting and drying, and other factors (Sparks and Huang, 1985). This review will focus on the latter four factors.

Biological activity promotes K release from micas (Mortland et al., 1956; Boyle et al., 1967; Weed et al., 1969; Sawhney and Voight, 1969). The organisms deplete the K in the soil solution, and their action may be analogous to that of tetraphenylboron (TPB) in artificial weathering of micas. Furthermore, the overall action of organisms is more complex when organic acids are produced (Boyle et al., 1967; Spyridakis et al., 1967; Sawhney and Voight, 1969).

The importance of organic acids in weathering of rock-forming minerals has been recognized for a long time (Sprengel, 1826; Bolton, 1882; Huang and Keller, 1970). All soils contain small but measurable amounts of biochemical compounds such as organic acids. Furthermore, since the time required for soil formation can extend over a period of centuries, the cumulative effect in a soil of even very small quantities of chelating agents will be considerable. The influence of oxalic and citric acids on the dynamics of K release from micas and feldspars was studied by Song and Huang (1988). They found that the sequence of K release from K-bearing minerals by oxalic and citric acids is biotite > microcline orthoclase > muscovite.

The activity of K^+ ions in soil solution around mica particles greatly influences the release of K from micas by cation exchange. When the K level is less than the critical value, K is replaced from the interlayer by other cations from the solution. On the contrary, when the K level is greater than the critical value, the mica expansible 2:1 mineral takes K from the solution. The critical K level is highly mineral dependent, being much higher for the trioctahedral minerals (Scott and Smith, 1967; Newman, 1969; von Reichenbach, 1973; Henderson et al., 1976). The critical levels for muscovite are so low that even the K impurities in laboratory chemicals or dissolved from glassware are often sufficient to prevent any K release (Scott and Smith, 1967).

The nature and concentration of the replacing cations also influences the critical K level of the cations tested in Cl^- solutions. Rausell-Colom et al. (1965) found the critical K levels decreased in the order $Ba^{2+} > Mg^{2+} > Ca^{2+} \approx Sr^{2+}$ for the same concentration of these ions and with a constant mica particle size. The activity of all of these replacing ions in the solution phase must be much greater than that of the K for significant K release to occur. The activity of various cations in the soil solution is governed by other minerals in the soil systems, by pedogenic processes, and by anthropogenic activities.

The release of K upon drying a soil is related to the clay fraction (Scott and Hanway, 1960). When a soil is dried, the degree of rotation of weathered soil minerals, such as micas, may be changed. Thus, the K-O bond may be modified. Dehydration of interlayer cations may permit a redistribution of interlayer cations, because Ca could now compete with K for wedge sites. This seems to account for the release of K from soils upon drying. Rich (1972) found that Virginia soils, which contain hydroxy Al interlayers and appreciable amounts of K, did not release K upon drying. The presence of the hydroxy Al interlayers may block or retard the interlayer diffusion of K ions and may change the b dimension of micas, the degree of tetrahedral rotation, and the length and strength of the K-O bond.

Other factors that can affect K release from soils are leaching, redox potential (Eh), and temperature. Leaching promotes the K release from K-bearing minerals by carrying away the reaction products. Therefore, leaching accelerates the transformation of minerals, e.g., micas, to expansible 2:1 layer silicates and other weathering products if the chemistry of leaching water favors the reaction.

Redox potential of soils could influence K release from micas since it has been pointed out that the tenacity with which K is held by biotite is greater after oxidation of its structural Fe. It appears that, other factors being equal, the extent of the K release from biotite should be less in soil environments that oxidize Fe than in soil environments that reduce it. Major elements in K-bearing feldspars do not exist in more than one valence state, thus the prevailing Eh of a soil may not be of direct concern to chemical weathering of the feldspars. However, the weatherability of feldspars can be affected by complexing organic acids that are vulnerable to oxidation. Therefore, the stability of feldspars may be indirectly related to the prevailing Eh of a soil.

Increasing temperature has been shown to increase the rate of K release from biotite (Rausell-Colom et al., 1965) and K feldspars (Rasmussen, 1972). Under conditions of leaching of biotite with 0.1 mol NaCl L^{-1}, the rate of K release appears directly proportional to temperature in the range of 293 to 323 K (Mortland, 1958). Under similar leaching conditions, Mortland and Ellis (1959) observed that the log of the rate constant for K release from fixed K in vermiculite was directly proportional to the inverse of the absolute temperature.

Preheating of micas to high temperatures (1273 K) prior to TPB extraction (Scott et al., 1973) was found to enhance the rate of K extraction from muscovite, to decrease the rate for biotite, and to have little effect on phlogopite, except at very high temperatures. The decrease in K release from biotite by preheating is presumably because of oxidation of Fe at high temperatures. The more rapid rate with muscovite following heating remains unexplained.

1.4.3.4 Leaching of Potassium in Soils

Soil solution K is either leached or sorbed by plants or soils. A number of factors influence the movement of K in soils, including the CEC, soil pH and liming, method and rate of K application, and K absorption by plants (Terry and McCants, 1968; Sparks, 1980).

The ability of a soil to retain applied K is very dependent on the CEC of the soil. Thus, the amount of clay and SOM in the soil strongly influences the degree of K leaching. Soils with higher CEC have a greater ability to retain added K, whereas leaching of K is often a problem on sandy soils (Sparks and Huang, 1985).

Retention of K can often be enhanced in sandy, Atlantic Coastal Plain soils after application of lime, since in such variable charge soils, the CEC is increased as soil pH is increased. Nolan and Pritchett (1960) found that liming a Lakeland fine sand soil (thermic, coated Typic Quartzipsamment) to pH 6 to 6.5 caused maximum retentivity of applied K. Potassium was replaced by Ca on the exchange complex at higher levels of limestone application. Less leaching of K occurred at pH 6.0 to 6.5 due to enhanced substitution of K for Ca than for Al, which was more abundant at low pH. Lutrick (1963) found that K leaching occurred on unlimed but not on limed areas when 112 to 224 kg K ha^{-1} was applied on a Eustis loamy fine sand (sandy, siliceous, thermic Psammentic Paleudult).

Movement of applied K has been related to the method of application. Nolan and Pritchett (1960) compared banded and broadcast placement of KCl applied at several rates to an Arredondo fine sand (loamy, siliceous, hyperthermic Grossarenic Paleudult) in lysimeters under winter and summer crops. For the low rate of application, cumulative K removal for both placements was only about 5.0 kg ha^{-1}.

A number of investigations have been conducted to determine the relationship of crop uptake and rate of K application to leaching of K. Jackson and Thomas (1960) applied up to 524 kg K ha^{-1} prior to planting sweet potatoes (*Ipomoea batatas* L.) on a Norfolk sandy loam (fine loamy, siliceous, thermic Typic Paleudult). At harvest time, soil and plant K exceeded applied K at the 131 and 262 kg K ha^{-1} rates. However, at the 524 kg K ha^{-1} rate, 38 kg K was unaccounted for by soil and plant K. This deficiency of K was attributed to leaching below sampling depths. During a two-year study with corn (*Zea mays* L.) on two Dothan (Typic Paleudult) soils of Virginia, Sparks et al. (1980) found that 83 and 249 kg of applied K ha^{-1} increased the exchangeable K in the E and B21t horizons of the two soils. These increases were ascribed to leaching of applied K. The magnitude of leaching varied directly with rate of K application. No accumulation of K was found in the top 0.76 m of a Leon sand (sandy, siliceous, thermic Aeric Haplaquod) after 40 yr of heavy K fertilization (Blue et al., 1955).

1.4.4 Assessing Potassium Extractability and Availability

The extractability and availability of soil and plant K can be assessed by using chemical extractants to quantify the various forms of soil K, soil test extractants, quantity/ intensity (Q/I) analyses, and plant analysis.

1.4.4.1 Chemical Extractants for Soil K Forms

Methods to determine total K and the other alkali elements in soils use acids or a high temperature fusion to decompose the soil. The most widely employed digestion techniques for total elements in soils and minerals have used combinations of HF and either H_2SO_4 or $HClO_4$ (Helmke and Sparks, 1996).

Exchangeable K is that K which is typically extracted with a neutral normal salt, usually 1.0 M NH_4OAc minus the water soluble K (Knudsen et al., 1982). In soils that are not saline, levels of water soluble K are minimal and can be ignored. However, in saline soils, the levels of water soluble K should be determined from a saturated extract or some similar extract and subtracted from the amount of K determined using NH_4OAc.

It should be noted that in soils that contain weathered vermiculitic and micaceous minerals wedge zones can be present that contain K. This K is not accessible to large index cations such as Ca and Mg, but can be extracted by NH_4, which is of similar size to K. For example, in soils that contain wedge zones, NH_4OAc will extract more K than an extractant like 1 M $CaCl_2$. It is debatable whether this K is truly exchangeable. Thus, in soils containing wedge zones exchangeable K could be overestimated with NH_4OAc (Sparks and Huang, 1985; Helmke and Sparks, 1996).

There are a number of chemical methods that can be employed to extract nonexchangeable K. These include boiling HNO_3, H_2SO_4, hot HCl, electroultrafiltration, Na tetraphenylboron with EDTA, and ion exchange resins such as H and Ca saturated resins (Hunter and Pratt, 1957; Martin and Sparks, 1985; Helmke and Sparks, 1996).

The most commonly used method for extraction of nonexchangeable K is the boiling HNO_3 technique. Most researchers that use this method boil the soil in M HNO_3 for 10 min over a flame, transfer the slurry to a filter, leach the soil with dilute HNO_3, and then determine the concentration in the filtrate. One of the problems with boiling for only 10 minutes over a flame is that it is difficult to be precise about the correct boiling time, the time it takes for boiling to occur, and the vigor of boiling (Martin and Sparks, 1985). Some workers have attempted to diminish these problems by using a 386 K oil bath for 25 minutes, including heating time (Pratt, 1965). This modification is particularly useful when large numbers of samples are being analyzed. Of course, one of the major concerns with using a boiling HNO_3 procedure is the potential to cause partial dissolution of mineral forms of K.

Other researchers have used continuous leaching of the soil with dilute acids such as 0.01 M HCl or with dilute salts such as 0.1 M NaCl, repeated extractions with 3, 0.3 and 0.03 M NaCl, Sr salts, hot $MgCl_2$, and sodium cobaltinitrite (Martin and Sparks, 1985).

Cation exchange resins saturated with H or Ca also have been widely used to measure nonexchangeable K. These resins have high CECs and when saturated with an appropriate cation and mixed with soil and with a dilute solution, they will adsorb released K. One of the major advantages of using cation exchange resins to extract K is that they act as a sink for the released K and thus prevent an inhibition of further K release. This is a problem with many batch methods that employ dilute acids and electrolytes. One major disadvantage of cation exchange resins for extracting K is that the resins are expensive and the procedure is time consuming and tedious.

In order for electrolyte and acid solutions and cation exchange resins to be effective in extracting K, the K concentration in the solution phase must be kept very low, or K release is inhibited (Rausell-Colom et al., 1965; Wells and Norrish, 1968; Feigenbaum et al., 1981; Martin and Sparks, 1983, 1985). The critical concentration above which release is inhibited is 4 mg L^{-1} for soils in general, 2.3 to 16.8 mg L^{-1} for trioctahedral micas in dilute solution, and as low as < 0.1 mg L^{-1} for muscovite and illite (Smith and Scott, 1966; Martin and Sparks, 1985). A low enough concentration of solution K can be maintained by employing a continuous flowing extracting or exchanging solution, cation exchange resins, or Na tetraphenylboron (Scott et al., 1960).

One can quantitatively analyze for mineral K (K feldspars and micas) by using a selective dissolution method employing $Na_2S_2O_7$ fusion. The technique and method for calculating the quantities of K feldspars and micas can be found in Helmke and Sparks (1996). A semiquantitative approach for measuring mineral K is to subtract the quantity of nonexchangeable K, using the boiling HNO_3 procedure, from the quantity of total K, using the HF digestion method. One also can quantify K feldspars in the sand fraction of soil using petrographic analyses (Parker et al., 1989a).

1.4.4.2 Soil Tests for Potassium
Soil test extractants for K were developed to easily and rapidly measure K in soils and to estimate K availability. Based on the amounts of extractable K, recommendations that are based on field test calibrations can then be made on the amount of K that is needed to maximize plant yields. Soil tests for K usually estimate the quantity of solution and exchangeable K, and since acids are usually employed as extractants, some nonexchangeable and mineral K also is extracted (Wolf and Beegle, 1991). The soil tests used to measure extractable K in the United States include Mehlich I and Mehlich III procedures in the northeastern and southeastern United States, the Morgan and modified Morgan procedures in parts of the northeastern United States, the 1 M NH_4OAc at pH 7 procedure in the north central United States, and the ammonium bicarbonate-DTPA extraction in the western United States. Procedures for these soil tests are fully described in Helmke and Sparks (1996) and Section D, Chapter 4.

1.4.4.3   Q/I Analysis
Schofield (1947) proposed that the ratio of the activity of cations such as K and Ca was defined by the relation $a_K/(a_{Ca})^{1/2}$ where a is the ion activity. He appears to have been the first person to apply the concepts of quantity (Q) and intensity (I) to the mineral nutrient status of soils (Schofield, 1955).

Beckett (1964a), following a consideration of the Ratio law (Schofield, 1947), suggested that the I of K in a soil at equilibrium with its soil solution could best be defined by the ratio $a_K/(a_{Ca} + a_{Mg})^{1/2}$ of the soil solution. This equilibrium activity ratio for K or AR^K (Beckett, 1964a) has often been used as a measure of K availability to plants (Sumner and Marques, 1966; le Roux and Sumner, 1968).

Beckett (1964c) suggested that exchangeable K is held by two distinct mechanisms. The majority is held by general force fields comparable with those that hold exchangeable Na or Ca. A small proportion is held at sites offering a specific binding force for K but not for Ca and Mg. The electrochemical potential of exchangeable K in the diffuse double layer dictates the chemical potential of K in the soil solution. The K activity is also affected by the difference in electrical potential across the diffuse double layer that surrounds the exchange complex. Thus, no simple relationship exists between the activity of K in soil solution and quantity of K on the exchange phase (San Valentin et al., 1973). Moss (1967) and Lee (1973) note that a soil with a given complement of exchangeable K, Ca, and Mg gives rise to an activity ratio for K (AR^K) in the equilibrium soil solution that will be characteristic of that soil and independent of the soil-to-solution ratio and total electrolyte concentration. Moss (1967) noted the ratio depends only on K saturation and the strength of adsorption of cations.

However, the relation of the amount of exchangeable K to AR^K must be specified to accurately describe the K status of a soil. Beckett (1964b) noted that different soils showing the same value of AR^K may not possess the same capacity for maintaining AR^K while K is removed by plant roots. Therefore, one must include not only the current potential of K in the labile pool but also the form of Q/I or the way in which potential depends on quantity of labile K present. These findings brought about the classic Q/I curves where the ratio of $a_K/(a_{Ca} + a_{Mg})^{1/2}$ is related to the change in exchangeable K to obtain the effect of quantity (exchangeable K) on intensity. The Q/I concept has been widely promulgated in the scientific literature to investigate the K status of soils (Evangelou et al., 1994).

The traditional method for Q/I analyses involves equilibrating a soil with solutions containing a constant amount of $CaCl_2$ and increasing the amounts of KCl (Beckett, 1964a). The soil gains or loses K to achieve the characteristic AR^K of the soil or remains unchanged if its AR^K is the same as the equilibrating solution. The AR^K values are then plotted versus the gain or loss of K to form the characteristic Q/I curve. From the Q/I plot, one can obtain several parameters to characterize the K status of a soil. The AR^K when the Q factor or ΔK equals zero is a measure of the degree of K availability at equilibrium or AR^K_e. The value of ΔK when $AR^K = 0$ is a measure of labile or exchangeable K in soils (ΔK°). The slope of the linear portion of the curve gives the potential buffering capacity of K (PBC^K) and is proportional to the CEC of the soil. The number of specific sites for K (K_x) is the difference between the intercept of the curved and linear portion of the Q/I plot at $AR^K = 0$ (Beckett, 1964b; Sparks and Liebhardt, 1981; Evangelou et al., 1994).

The traditional method described above is too time consuming for routine analyses. Advances in ion selective electrode technology (ISE) have allowed for more rapid Q/I analysis (Evangelou et al., 1994). Parra and Torrent (1983) developed a ISE simplified Q/I method whereby a single K-ISE in an electrochemical cell with liquid junction was employed to quantitate the K concentration (C_K) in equilibrated soil suspensions based on a successive addition procedure. The values of AR^K were estimated based on the expression, $AR^K = (11.5 - 0.3b) C_K + 22 \times 10^{-6}$, where b is the CEC (cmol$_c$ kg^{-1}) based on the weight of the soil samples used. The method of Parra and Torrent (1983) is quicker than the traditional Q/I method because equilibration time is reduced to 10 min compared to 24 h for the traditional method. Parra and Torrent (1983) achieved results with their modified procedure that were comparable to the traditional method. Wang et al. (1988) modified the procedure of Parra and Torrent (1983) by making direct measurements of CR_K [concentration ratio: $C_K/(C_{Ca+Mg})^{1/2}$] values with Ca and K-ISEs in an electrochemical cell with or without liquid junction.

1.4.4.4 Plant K Analysis

Plant K analysis will not be discussed in any detail in this review. For extensive discussions on plant analysis, the reviewer is referred to chapters in Westerman (1990). Table 1.9 lists the critical level or concentration of K for various agronomic crops. These levels are usually determined by relating yield (e.g., % of maximum yield or growth rate) to nutrient concentration (g kg^{-1}) for a specific plant part sampled at a given stage of development (Munson and Nelson, 1990). This method is based on the principle that if an element such as K is deficient in a plant, growth rates and yield will be decreased.

Table 1.9 Critical K concentrations in agronomic crops [From Westerman, 1990. Soil testing and plant analysis. Soil Science Society of America with permission]

Crop	Time of Sampling	Plant Part	Critical Concentration[‡] g kg^{-1}
Sugarbeet		Blade	10
Cotton		Leaves	< 9-15
Wheat	Jointing (GS6)	Total tops	20-25
	Early Boot (GS9)	Total tops	15-20
Rice	Flag Leaf to Mid-Tillering		10-14
Corn	At tassel	Ear leaf	19
	At tassel	Leaves	17-27
	At silk	Sixth leaf from base	13
	At silk	Leaf opposite and just below ear shoot	20
Grain sorghum	Full heading	Second blade below apex	18
Alfalfa		Whole top	8-22
Red Clover		Tops	15-22.5
Bermudagrass		Tops	13-15
Orchardgrass		Tops	23-25
Tall Fescue		Tops	24-38
Kentucky Bluegrass		Tops	16-20

[‡] Critical concentration is that nutrient concentration at which plant growth begins to decrease in comparison with plants above the critical concentration.

If one adds increasing amounts of K, the concentration of the element in the plant or plant part increases until an optimum level is attained. Using this approach, growth or yield is expressed as a percentage of the maximum. The zone between the deficient and optimum concentration can be referred to as the transition zone (Ulrich and Hills, 1967). Ulrich and Hills (1967) referred to the transition zone as the zone between nutrient concentrations that produced a 20% reduction in growth or yield and continues to those that cause optimum or 100% in the maximum yield. Dow and Roberts (1982) refer to this latter zone as the critical range where researchers select the yield reduction and nutrient concentrations that are acceptable.

1.4.5 References

Amacher, M.C. 1991. Methods of obtaining and analyzing kinetic data. p. 19–59. *In* D.L. Sparks, and D.L. Suarez (ed.) Rates of soil chemical processes. Soil Sci. Soc. Am. Spec. Publ. 27. Soil Science Society of America, Madison, WI.

Barshad, I. 1951. Cation exchange in soils: I. Ammonium fixation and its relation to potassium fixation and to determination of ammonium exchange capacity. Soil Sci. 77:463–472.

Beckett, P.H.T. 1964a. Studies on soil potassium: I. Confirmation of the ratio law: Measurement of potassium potential. J. Soil Sci. 15:1–8.

Beckett, P.H.T. 1964b. Studies on soil potassium: II. The 'immediate' Q/I relations of labile potassium in the soil. J. Soil Sci. 15:9–23.

Beckett, P.H.T. 1964c. K-Ca exchange equilibria in soils: Specific sorption sites for K. Soil Sci. 97:376–383.

Bertsch, P.M., and G.W. Thomas. 1985. Potassium status of temperate region soils. p. 131–162. *In* R.E. Munson (ed.) Potassium in agriculture. American Society of Agronomy, Madison, WI.

Blue, W.G., C.F. Eno, and P.J. Westgate. 1955. Influence of soil profile characteristics and nutrient concentrations on fungi and bacteria in Leon fine sands. Soil Sci. 80:303–308.

Bolt, G.H., M.E. Sumner, and A. Kamphorst. 1963. A study between three categories of potassium in an illitic soil. Soil Sci. Soc. Am. Proc. 27:294–299.

Bolton, H.C. 1882. Application of organic acids to the examination of minerals. Proc. Am. Assoc. Adv. Sci. 31:3–7.

Boyle, J.R., G.K. Voight, and B.L. Sawhney. 1967. Biotite flakes: Alteration by chemical and biological treatment. Science 155:193–195.

Coleman, N.T., and M.E. Harward. 1953. The heats of neutralization of acid clays and cation-exchange resins. J. Am. Chem. Soc. 75:6045–6046.

Cook, M.G., and T.B. Hutcheson, Jr. 1960. Soil potassium reactions as related to clay mineralogy of selected Kentucky soils. Soil Sci. Soc. Am. Proc. 24:252–256.

Dow, A.I., and S. Roberts. 1982. Proposal: Critical nutrient ranges for crop diagnosis. Agron. J. 74:401–403.

Evangelou, V.P., J. Wang, and R.E. Phillips. 1994. New developments and perspectives on soil potassium quantity/ intensity relationships. Adv. Agron. 52:173–227.

Feigenbaum, S., R. Edelstein, and I. Shainberg. 1981. Release rate of potassium and structural cations from micas to ion exchangers in dilute solutions. Soil Sci. Soc. Am. J. 45:501–506.

Haby, V.A., M.P. Russelle, and E.O. Skogley. 1990. Testing soils for potassium, calcium, and magnesium. p. 181–228. *In* R.L. Westerman (ed.) Soil testing and plant analysis. Soil Science Society of America, Madison, WI.

Hanway, J.J., and A.D. Scott. 1957. Soil potassium-moisture relations: II. Profile distribution of exchangeable K in Iowa soils as influenced by drying and rewetting. Soil Sci. Soc. Am. Proc. 20:501–504.

Helfferich, F. 1962. Ion exchange. McGraw-Hill Book Co., New York, NY.

Helmke, P.A., and D.L. Sparks. 1996. Lithium, sodium, potassium, rubidium, and cesium. p. 551–574. *In* D.L. Sparks (ed.) Methods of soil analysis: Chemical methods. Soil Science Society of America, Madison, WI.

Henderson, J.H., H.E. Doner, R.M. Weaver, J.K. Syers, and M.L. Jackson. 1976. Cation and silica relationships of mica weathering to vermiculite in calcareous Harps soils. Clays Clay Miner. 24:93–100.

Huang, P.M., L.S. Crossan, and D.A. Rennie. 1968. Chemical dynamics of K-release from potassium minerals common in soils. Trans. 9th Int. Congr. Soil Sci. 2:705–712.

Huang, W.H., and W.D. Keller. 1970. Dissolution of rock forming minerals in organic acids. Am. Mineral. 55:2076–2094.

Hunter, A. H., and P. F. Pratt. 1957. Extraction of potassium from soils by sulfuric acid. Soil Sci. Soc. Am. Proc. 21:595–598.

Jackson, W. A., and G. W. Thomas. 1960. Effects of KCl and dolomitic limestones on growth and ion uptake of the sweet potato. Soil Sci. 89:347–352.

Jardine, P. M., and D. L. Sparks. 1984. Potassium-calcium exchange in a multireactive soil system: I. Kinetics. Soil Sci. Soc. Am. J. 47:39–45.

Juo, A. S. R. 1981. Chemical characteristics. p. 51–79. *In* D. J. Greenland (ed.). Characterization of soils in relation to their classification and management for crop production: Examples from some areas of the humid tropics. Clarendon Press, Oxford, UK.

Kittrick, J. A. 1966. Forces involved in ion fixation by vermiculite. Soil Sci. Soc. Am. Proc. 30:801–803.

Knudsen, D., G.A. Peterson, and P.F. Pratt. 1982. Lithium, sodium, and potassium on a Coastal Plain soil. Commun. Soil Sci. Plant Anal. 7:265–277.

Laffer, B.G., A.M. Posner, and J.P. Quirk. 1966. Hysteresis in the crystal swelling of montmorillonite. Clay Miner. 6:311–321.

Lee, R. 1973. The K/Ca Q/I relation and preferential adsorption sites for potassium. NZ Soil Bur. Sci. Rep. II.

le Roux, J., and M. E. Sumner. 1968. Labile potassium in soils: I. Factors affecting the quantity-intensity (Q/I) parameters. Soil Sci. 106:35–41.

Liebhardt, W.C., L.V. Svec, and M.R. Teel. 1976. Yield of corn as affected by potassium on a Coastal Plain soil. Commun. Soil Sci. Plant Anal. 7:265–277.

Lutrick, M. C. 1963. The effect of lime and phosphate on downward movement of potassium in Red Bay fine sandy loam. Proc. Soil Crop Sci. Soc. Fla. 23:90–94.

Malavolta, E. 1985. Potassium status of tropical and subtropical region soils. p. 163–200. *In* R.E. Munson (ed.) Potassium in agriculture. American Society of Agronomy, Madison, WI.

Malcolm, R.L., and V.C. Kennedy. 1969. Rate of cation exchange on clay minerals as determined by specific ion electrode techniques. Soil Sci. Soc. Am. Proc. 33:245–253.

Martin, H.W., and D.L. Sparks. 1983. Kinetics of nonexchangeable potassium release from two Coastal Plain soils. Soil Sci. Soc. Am. J. 47:883–887.

Martin, H.W., and D.L. Sparks. 1985. On the behavior of nonexchangeable potassium in soils. Commun. Soil Sci. Plant Anal. 16:133–162.

Martin, J.C., R. Overstreet, and D.R. Hoagland. 1946. Potassium fixation in soils in replaceable and nonreplaceable forms in relation to chemical reactions in the soil. Soil Sci. Soc. Am. Proc. 10:94–101.

McLean, E. O., and R. H. Simon. 1958. Potassium status of some Ohio soils as revealed by greenhouse and laboratory studies. Soil Sci. 85:324–332.

Mengel, K. 1985. Dynamics and availability of major nutrients in soils. Adv. Soil Sci. 2:65–131.

Mortland, M.M. 1958. Kinetics of potassium release from biotite. Soil Sci. Soc. Am. Proc. 22:503–508.

Mortland, M.M., and B.G. Ellis. 1959. Release of fixed potassium as a diffusion-controlled process. Soil Sci. Soc. Am. Proc. 23:363–364.

Mortland, M.M., and J.E. Gieseking. 1951. Influence of the silicate ion on potassium fixation. Soil Sci. 71:381–385.

Mortland, M.M., K. Lawton, and G. Uehara. 1956. Alteration of biotite to vermiculite by plant growth. Soil Sci. 82:477–481.

Moss, P. 1967. Independence of soil quantity/intensity relationships to changes in exchangeable potassium: Similar potassium exchange constants for soils within a soil type. Soil Sci. 103:196–201.

Munson, R.D., and W.L. Nelson. 1990. Principles and practices in plant analysis. p. 359–387. *In* R.L. Westerman (ed.) Soil testing and plant analysis. Soil Science Society of America, Madison, WI.

Newman, A.C.D. 1969. Cation exchange properties of micas: I. The relation between mica composition and potassium exchange in solutions of different pH. J. Soil Sci. 20:357–373.

Nolan, C.N., and W.L. Pritchett. 1960. Certain factors affecting the leaching of potassium from sandy soils. Proc. Soil Crop Sci. Soc. Fla. 20:139–145.

Parker, D.R., D.L. Sparks, G.J. Hendricks, and M.C. Sadusky. 1989a. Potassium in Atlantic Coastal Plain soils: I. Soil characterization and distribution of potassium. Soil Sci. Soc. Am. J. 53:392–396.

Parker, D.R., G.J. Hendricks, and D.L. Sparks. 1989b. Potassium in Atlantic Coastal Plain soils: II. Crop responses and changes in soil potassium under intensive management. Soil Sci. Soc. Am. J. 53:397–401.

Parra, M.A., and J. Torrent. 1983. Rapid determination of the potassium quantity-intensity relationships using a potassium-selective ion electrode. Soil Sci. Soc. Am. J. 47:335–337.

Pratt, P.F. 1965. Potassium. p. 1023–1031. *In* C.A. Black et al. (ed.) Methods of soil analysis. Part 2. American Society of Agronomy, Madison, WI.

Rasmussen, K. 1972. Potash in feldspars. Proc. Colloq. Int. Potash Inst. 9:57–60.

Rausell-Colom. J.A., T.R. Sweetman, L.B. Wells, and K. Norrish. 1965. Studies in the artificial weathering of micas. p. 40–70. *In* E.G. Hallsworth, and D.V. Crawford (ed.) Experimental pedology. Butterworths, London, UK.

Reitemeier, R.F. 1951. The chemistry of soil potassium. Adv. Agron. 3:113–164.

Rich, C.I. 1960. Aluminum in interlayers of vermiculite. Soil Sci. Soc. Am. Proc. 24:26–32.

Rich, C.I. 1964. Effect of cation size and pH on potassium exchange in Nason soil. Soil Sci. 98:100–106.

Rich, C.I. 1968. Mineralogy of soil potassium. p. 79–91 *In:* V.J. Kilmer et al. (ed.) The role of potassium in agriculture. American Society of Agronomy, Madison, WI.

Rich, C.I. 1972. Potassium in minerals. Proc. Colloq. Int. Potash Inst. 9:15–31.

Rich, C.I., and W.R. Black. 1964. Potassium exchange as affected by cation size, pH, and mineral structure. Soil Sci. 97:384–390.

Rich, C.I., and S.S. Obenshain. 1955. Chemical and clay mineral properties of a red-yellow podzolic soil derived from mica schist. Soil Sci. Soc. Am. Proc. 19:334–339.

Russell, E.W. 1961. Soil conditions and plant growth. Longmans, London, UK.

Sadusky, M.C., D.L. Sparks, M.R. Noll, and G.J. Hendricks. 1987. Kinetics and mechanisms of potassium release from sandy soils. Soil Sci. Soc. Am. J. 51:1460–1465.

San Valentin, G.O., L.W. Zelazny, and W. K. Robertson. 1973. Potassium exchange characteristics of a Rhodic Paleudult. Proc. Soil Crop Sci. Soc. Fla. 32:128–132.

Sawhney, B.L., and G.K. Voight. 1969. Chemical and biological weathering in vermiculite from Transvaal. Soil Sci. Soc. Am. Proc. 33:625–629.

Schofield, R.K. 1947. A ratio law governing the equilibrium of cations in the soil solution. Proc. Int. Congr. Pure Appl. Chem. 11:257–261.

Schofield, R.K. 1955. Can a precise meaning be given to "available" soil phosphorus. Soils Fert. 18:373–375.

Schroeder, D. 1979. Structure and weathering of potassium containing minerals. Proc. Congr. Int. Potash Inst. 11:43–63.

Schroeder, D., and H. Dummler. 1966. Kalium-Nachlieferung, Kalium-Festlegung und Tonmineralbestand schleswigholsteinischer Boden. Z. Pflanzenernäehr Dueng. Bodenkd. 113:213–215.

Schwertmann, U. 1962a. Die selective Kationensorption der Tonfraktion einiger Boden aus Sedimenten. Z. Pflanzenernäehr. Dueng. Bodenkd. 97:9–25.

Schwertmann, U. 1962b. Eigenschaften und beldung aufwertbarer (quellbarer)Dreischichttonminerale in Böden aus Sedimenten. Beitr. Mineral. Petrogr. 8:199–209.

Scott, A.D., A.P. Edwards, and J.M. Bremner. 1960. Removal of fixed ammonium from clay minerals by cation exchange resins. Nature 185:792.

Scott, A.D. and J.J. Hanway. 1960. Factors influencing the change in exchangeable soil K observed on drying. Trans. 7th Int. Congr. Soil Sci. 4:72–79.

Scott, A.D., F.T. Ismail, and R.R. Locatis. 1973. Changes in interlayer potassium exchangeability induced by heating micas. p. 467–479. In J. M. Serratosa (ed.) Proc. 4th Int. Clay Conf., Div. de Ciencias, Madrid, Spain.

Scott, A.D., and S.J. Smith. 1967. Visible changes in macro mica particles that occur with potassium depletion. Clays Clay Miner. 15:367–373.

Sivasubramaniam, S., and O. Talibudeen. 1972. Potassium-aluminum exchange in acid soils: I. Kinetics. J. Soil Sci. 23:163–176.

Smith, S.J., and A.D. Scott. 1966. Extractable potassium in Grundite illite: 1. Method of extraction. Soil Sci. 102:115–122.

Soil Survey Staff. 1975. Soil Taxonomy. USDA Agric. Handb. 436, US Government Printing Office, Washington, DC.

Song, S.K., and P.M. Huang. 1988. Dynamics of potassium release from potassium-bearing minerals as influenced by oxalic and citric acids. Soil Sci. Soc. Am. J. 52:383–390.

Sparks, D.L. 1980. Chemistry of soil potassium in Atlantic Coastal Plain soils: A review. Commun. Soil Sci. Plant Anal. 11:435–449.

Sparks, D.L. 1987. Potassium dynamics in soils. Adv. Soil Sci. 6:1–63.

Sparks, D.L. 1989. Kinetics of soil chemical processes. Academic Press, San Diego, CA.

Sparks, D.L. 1995. Environmental soil chemistry. Academic Press, San Diego, CA.

Sparks, D.L., and P.M. Huang. 1985. Physical chemistry of soil potassium. p. 201–276. In R.D. Munson (ed.) Potassium in agriculture. American Society of Agronomy, Madison, WI.

Sparks, D.L., and P. M. Jardine. 1981. Thermodynamics of potassium exchange in soil using a kinetics approach. Soil Sci. Soc. Am. J. 45:1094–1099.

Sparks, D.L., and P.M. Jardine. 1984. Comparison of kinetic equations to describe K-Ca exchange in pure and in mixed systems. Soil Sci. 138:115–122.

Sparks, D.L., T.H. Carski, S.E. Fendorf, and C.V. Toner, IV. 1996. Kinetic methods and measurements. p. 1275–1307. In D.L. Sparks (ed.) Methods of soil analysis: Chemical methods, Soil Science Society of America, Madison, WI.

Sparks, D.L., and W.C. Liebhardt. 1981. Effect of long-term lime and potassium applications on quantity-intensity (Q/I) relationships in sandy soil. Soil Sci. Soc. Am. J. 45:786–790.

Sparks, D.L., D.C. Martens, and L.W. Zelazny. 1980. Plant uptake and leaching of applied and indigenous potassium in Dothan soils. Agron. J. 72:551–555.

Sprengel, C. 1826. Uber Pflanzenhumus, Humussaure und Humussaure Salze. Kastners Arch. Naturlehre 8:145–220.

Spyridakis, D.E., G. Chesters, and S.A. Wilde. 1967. Kaolinization of biotite as a result of coniferous and deciduous seedling growth. Soil Sci. Soc. Am. Proc. 31:203–210.

Steffens, D., and K. Mengel. 1979. Das Aneignungsvermögen von Lolium perenne im Vergleich zu Trifolium pratense für Zwischenschicht-Kalium der Tonminerale. Landw. Forsch. SH 36:120–127.

Sumner, M.E., and J.M. Marques. 1966. Ionic equilibria in a ferrallitic clay: Specific adsorption sites for K. Soil Sci. 102:187–192.

Terry, D.L., and C.B. McCants. 1968. The leaching of ions in soils. NC Agric. Expt. Stn. Tech. Bull. 184.

Thomas, G.W. 1960. Forms of aluminum in cation exchangers. Trans. 7th Int. Congr. Soil Sci. 2:364–369.

Thomas, G.W., and B.W. Hipp. 1968. Soil factors affecting potassium availability. p. 269–291. In V. J. Kilmer et al. (ed.) The role of potassium in agriculture. American Society of Agronomy, Madison, WI.

Ulrich, A., and F.J. Hills. 1967. Principles and practices of plant analysis. p. 11–24. In G. W. Hardy (ed.) Soil testing and plant analysis. Part II. Soil Sci. Soc. Am. Spec. Publ. 2. Soil Science Society of America, Madison, WI.

Volk, N.J. 1934. The fixation of potash in difficultly available forms in soils. Soil Sci. 37:267–287.

von Reichenbach, H.G. 1973. Exchange equilibria of interlayer cations in different particle size fractions of biotite and phlogopite. p. 480–487. *In* J.M. Serratosa (ed.) Proc. 4th Int. Clay Conf., Div. de Ciencias, Madrid, Spain.

Walker, G.F. 1957. On the differentiation of vermiculites and smectites in clays. Clay Miner. Bull. 3:154–163.

Wang, J., R.E. Farrell, and A.D. Scott. 1988. Potentiometric determination of potassium Q/I relationships. Soil Sci. Soc. Am. J. 52:657–662.

Weed, S.B., C.B. Davey, and M.G. Cook. 1969. Weathering of mica by fungi. Soil Sci. Soc. Am. Proc. 33:702–706.

Weir, A.H. 1965. Potassium retention in montmorillonite. Clay Miner. 6:17–22.

Wells, C.B., and K. Norrish. 1968. Accelerated rates of release of interlayer potassium from micas. Trans. 9th Int. Congr. Soil Sci. 2:683–694.

Westerman, R.L. (ed.) 1990. Soil testing and plant analysis. Soil Science Society of America, Madison, WI.

Wolf, A., and D. Beegle. 1991. Recommended soil tests for macronutrients: Phosphorus, potassium, calcium and magnesium. p. 25–34. *In* J.T. Sims and A. Wolf (ed.) Recommended soil testing procedures for the Northeastern Region. DE Agric. Exp. Stn. Bull. 493.

Woodruff, J.R., and C.L. Parks. 1980. Topsoil and subsoil potassium calibration with leaf K for fertility rating. Agron. J. 72:392–396.

Yuan, L.L., L.W. Zelazny, and A. Ratanaprasotporn. 1976. Potassium status of selected Paleudults in the lower Coastal Plain. Soil Sci. Soc. Am. Proc. 40:229–233.

1.5 Bioavailability of Calcium, Magnesium, and Sulfur

J.J. Camberato
Clemson University

W.L. Pan
Washington State University

Calcium, Mg, and S are essential mineral elements that are classified as macronutrients. While these three elements generally accumulate in plant tissues in higher concentrations (1.5 to 35 g kg^{-1}) (Table 1.10) than the micronutrients, they are not as commonly limiting in crop production as N, P, and K. Yet, when the soil bioavailability of Ca, Mg, and S is low, crop yield and quality can be suboptimal, depending on crop species and environmental conditions.

The bioavailability of Ca, Mg, and S is governed by the following factors: (1) parent material, (2) ion exchange reactions, (3) biological transformations, (4) losses from the crop root zone by leaching and crop removal from the field, and (5) replenishment via atmospheric deposition, fertilizer, and soil amendments. Factors influencing Ca and Mg bioavailability are collectively discussed due to the similarity in soil chemical and plant nutritional behavior of these two elements. In contrast, S chemistry in soils and plant nutrition is distinctly different, and is discussed in the latter half of the section.

1.5.1 Calcium and Magnesium

1.5.1.1 Overview of Calcium and Magnesium Nutrition

General reviews of the function of Ca and Mg in plants have been written by Clark (1984), Hanson (1984), and Pooviah and Reddy (1996). A large proportion of Ca is found as Ca pectates in cell walls for structural rigidity, or as Ca oxalate and other organic acids in the apoplasm or vacuoles (Borchert, 1990; Kinzel, 1989). Calcium functions in plants include maintenance of membrane integrity (Caldwell and Haug, 1981) and cell wall stability (Konno et al., 1984), root (Pooviah and Reddy, 1996) and shoot (Slocum and Roux, 1983) gravitropism, callose deposition (Lerchl et al., 1989), and regulation of several enzymes including α-amylase (Mitsui et al., 1984), protein kinases (Raghothama et al., 1987) and ATPases (Clarkson and Hanson, 1980).

Table 1.10   Calcium and magnesium concentrations in plant tissue considered adequate for plant growth [Ca and Mg data adapted from Clark, 1984; S data compiled from Asher et al., 1983; Fox and Blair, 1986; Kamprath and Jones, 1986; Jones, 1986; Tiwari, 1990]

Plant	Tissue	Range of Concentration		
		Ca	Mg	S
		$g\ kg^{-1}$ tissue		
Maize	Ear Leaf	2.1-10.0	2.0-4.0	1.2-2.5
Rice	Leaves (Tillering)	1.6-3.9	1.6-3.9	0.9-3.8
Wheat	Leaves (Heading)	2.0-5.0	1.5-5.0	1.5-4.0
Soybean	Leaves	3.6-20.0	2.6-10.0	2.0-4.0
Peanut	Leaves & Stem	7.5-20.0	3.0-7.5	2.0-3.0
Cotton	Leaves	19.0-35.0	3.0-7.5	2.9-15.0
Fruit trees	Leaves	14.0-21.0	4.1-6.8	1.3-3.0

Magnesium functions as an enzyme activator, in addition to serving as the coordinating central cation in the chlorophyll molecule. Magnesium activates numerous plant enzymes by bridging enzymes with ligand groups of substrates to optimize geometric conformation of enzyme systems involving the transfer of phosphate or carboxyl groups (Marschner, 1995). Systems activated by Mg include chlorophyll biosynthesis (Walker and Weinstein, 1991), chlorophyll degradation (Langmeier et al., 1993), photosynthesis (Pierce, 1986), protein synthesis (Cammarano et al., 1972), ATP synthesis (Lin and Nobel, 1971), and ATPases (Balke and Hodges, 1975) some of which are involved in phloem loading of sucrose (Williams and Hall, 1987).

1.5.1.2   Occurrence of Soil Calcium and Magnesium
Calcium and Mg occur in soils primarily in mineral forms and as ions on the exchange complex or in the soil solution. Little Ca or Mg (Mokwunye and Melsted, 1972) occurs in organic complexes in soils. The relative amounts of Ca derived from parent materials rank as follows: calcareous sedimentary rock > basic igneous rock > acid igneous rock (Jenny, 1941). Parent materials rank with respect to Mg contents: basic igneous rock > acid igneous rock > sedimentary rock (Metson, 1974). Generally, soils with the highest 2:1 lattice clay content have the highest Mg contents (Mokwunye and Melsted, 1972). Contents of Ca and Mg in soils are presented in Table 1.11.

1.5.1.3   Additions and Losses of Soil Calcium and Magnesium
Calcium and Mg accession by soils from atmospheric deposition may be significant in comparison to that removed in crop and forest plants (Johnson and Todd, 1987). Calcium and Mg depositions from precipitation in a Tennessee forest were 6.3 and 0.7 kg ha^{-1} yr^{-1}, respectively (Johnson and Todd, 1987). Hedin et al. (1994) estimated atmospheric Ca depositions for Sweden, the northeastern United States, and The Netherlands in the 1990s to be 0.5, 1.0, and 4.2 kg Ca ha^{-1} yr^{-1}, respectively.

Considerable amounts of Ca and Mg are added to soils in the form of limestone (21–32% Ca, 3–12% Mg), fertilizers, and animal manures. Fertilizers used to specifically supply Ca to soils include gypsum (22% Ca) and calcium nitrate (19% Ca). Calcium may also be added to the soil incidently

Table 1.11 Contents of calcium and magnesium in various soils

	Content		
Soil type	Ca	Mg	Source
	g kg^{-1}		
Temperate soils	0.7-36	1.2-15	Brady (1974)
Humid region soil	4	3	Brady (1974)
Arid region soil	10	6	Brady (1974)
Peat soils	1.1-48.3		Brady (1974)
Muck soils		0.7-5.7	Millar (1955)

while supplying P as superphosphates (14 to 20% Ca). Common Mg-containing fertilizers include epsom salts (10% Mg) and sulfate of potash magnesia (11% Mg).

Loss of Ca and Mg from the soil occurs through leaching and crop removal. Leaching losses of Ca and Mg from an aspen forest (*Populus tremuloides* Michx.) on Typic Fragiochrept soils averaged 39 and 9 kg ha^{-1} yr^{-1}, respectively (Silkworth and Grigal, 1982). Cropping with perennial species reduces leaching losses in comparison to annuals. Calcium and Mg leaching losses were 193 and 104 kg ha^{-1}, respectively, from a continuous corn rotation but only 64 and 48 kg ha^{-1} from a bluegrass (*Poa pratensis*) sod (Bolton et al., 1970). The application of Ca generally increases the leaching of Mg (Pratt and Harding, 1957), in part because the adsorption affinity for Ca is greater than that for Mg in nonvermiculitic soils (Hunsaker and Pratt, 1971). Leaching of Ca and Mg is accelerated by acidification resulting from the nitrification of ammoniacal N sources (Schwab et al., 1989; Darusman et al., 1991). Under acid conditions, Al^{3+} and H$^+$ ions displace Ca and Mg from the exchange complex enabling leaching to occur.

Plant removals of Ca and Mg from the soil are dependent on the plant species and genotype and the harvested fraction that is removed from the field. The Ca concentration of plants varies considerably, but one generalization can be made: the Ca concentration of dicots, legumes, and crucifers (12–18 g kg^{-1}) is far greater than the 4 g kg^{-1} Ca typically found in grasses (Parker and Truog, 1920). In contrast, the Mg concentration of these groups of plants differs only slightly (~ 2–4 g kg^{-1}). Several authors have provided compilations of plant mineral composition that may be useful in estimating crop removal of Ca and Mg (Truog, 1918; Beeson, 1941). Magnesium removals were 5, 7, 8, and 16 kg Mg ha^{-1} for snap beans (*Phaseolus vulgare* L.), okra (*Hibiscus esculentus* L.), white potatoes (*Solanum tuberosum* L.), and sweet potatoes (*Ipomoea batatas* (L.) Lam.), respectively (Prince, 1951). More recently representative uptakes and removals of Ca and Mg for several crops were reported by Buol (1995). Crop removal was in the range of 1 to 30 kg Ca or Mg ha^{-1} yr^{-1}. Spring wheat (*Triticum aestivum* L.) accumulated 35 kg Ca ha^{-1} and 13 kg Mg ha^{-1}, whereas sugar beets (*Beta vulgaris* L.) accumulated 104 kg Ca ha^{-1} and 44 kg Mg ha^{-1} (Strebel and Duynisveld, 1989). Barber (1984) reported the total uptake of Mg by maize (*Zea mays* L.) to be 45 kg ha^{-1}.

Hardwood tree species accumulate substantial amounts of Ca. For example, a mixed oak (*Quercus* sp.) stand grown on Typic Paleudult soils contained 1090 kg Ca ha^{-1} (16 kg Ca ha^{-1} yr^{-1}) in the above ground plant parts when harvested (Johnson and Todd, 1987). Loblolly pine (*Pinus taeda* L.) grown on the same soils removed only 196 kg Ca ha^{-1} (6 kg Ca ha^{-1} yr^{-1}). Both species removed about 45 kg Mg ha^{-1} from the soil (0.6 and 1.5 kg Mg ha^{-1} yr^{-1} for mixed oak and pine, respectively). In a more temperate climate, aspen removed 1,034 kg Ca ha^{-1} and 95 kg Mg ha^{-1} from Typic Fragiochrept soils in northeastern Minnesota (Silkworth and Grigal, 1982).

1.5.1.4 Factors Affecting Availability of Soil Calcium and Magnesium

The activities of Ca and Mg in the soil solution, which directly influence the rates of uptake into plant roots, are dependent on the level of exchangeable cations (Albrecht, 1941) and the type of colloid. Organic matter and 1:1 clays retain Ca less tightly than 2:1 clay minerals, resulting in greater Ca availability at any given level of Ca saturation (Mehlich and Colwell, 1945; Allaway, 1945; Mehlich, 1946).

The affinity of cations for the exchange complex is dependent on the mineralogy of the colloid. For instance, the affinities of Ca and Mg for montmorillonite are similar, but the affinity of Mg for vermiculite is much greater than the affinity of Ca (Wild and Keay, 1964). Soils with exchange complexes arising from organic matter, peat, allophane, kaolinite, and oxides of Fe and Al have a higher affinity for Ca than for Mg (Hunsaker and Pratt, 1971).

The relative abundance of basic cations on the exchange complex affects the plant availability of these ions. Bear et al. (1945) asserted that the ideal base saturation of the exchange complex was 65% of Ca, 10% of Mg, and 5% of K. This concept was termed the basic cation saturation ratio. Although this ideology of soil test interpretation was widely adopted in the midwestern United States for making fertilizer and lime recommendations (McLean and Brown, 1984), subsequent research demonstrated that a fairly broad range of basic cation saturations would produce equivalent crop yields (Hunter, 1949; Giddens and Toth, 1951; Key et al., 1962; Simson et al., 1979).

The degree of Ca saturation needed to provide sufficient plant-available Ca to crops is dependent on the colloid. Kamprath (1984) concluded from the results of a number of studies that a Ca saturation of about 25–30% and an exchangeable Ca level of $1.0 \, cmol_c \, kg^{-1}$ in highly weathered soils, dominated by kaolinitic clays and Fe/Al oxides, was adequate for supplying the Ca requirement of most plants. However, other factors such as Al may still limit plant growth at this level of Ca saturation. Soil solution Ca levels in acid soils ranging from 0.38 to 9.3 mM were reported by Kamprath (1978). The soil solution Ca level required for optimum growth of tropical legumes was 0.125 mM, much lower than the 2.0 mM required by temperate legumes (Kamprath, 1978). The soil solution Ca concentration necessary for maximizing cotton (*Gossypium hirsutum* L.) root growth was between 0.04 and 0.34 mM (Adams and Moore, 1983; Howard and Adams, 1965).

1.5.1.5 Calcium and Magnesium Soil-Plant Relationships

The transport of Ca and Mg ions to the root surface occurs by mass flow and diffusion. The soil solution composition and transpiration have been used to estimate the contribution of mass flow to the movement of these ions to plant roots (Barber, 1962). The difference between total nutrient accumulation in the plant and the estimate of that supplied by mass flow gives the relative contribution of diffusion and root interception. From these estimates, Barber (1962) recognized that mass flow could supply more than the required amount of Ca to plant roots of some soil-plant systems, resulting in a buildup of Ca in the rhizosphere that was observed in subsequent experiments (Barber and Ozanne, 1970). He also surmised that in situations where plant Ca demand is high and soil solution Ca is low, the supply of Ca to plant roots can be diffusion limited (Barber, 1962), resulting in a depletion of soluble Ca around the roots (Barber and Ozanne, 1970). Similarly, the mode of Mg transport is cropping system dependent. While mass flow could account for the Mg accumulated by wheat, a majority of Mg moving to sugar beet roots occurred by diffusion (Strebel and Duynisveld, 1989).

The presence of Ca is required throughout the root profile to maintain the integrity of root cell membranes and nutrient uptake (Haynes and Robbins, 1948). Insufficient Ca limits rooting into the subsoil of many Ultisols (Adams and Moore, 1983) and Oxisols (Ritchey et al., 1982) limiting the volume of soil explored and crop access to soil moisture supplies (Sumner et al., 1986; Sumner, 1995). Subsoil horizons in these soils may also be Al and H toxic (Adams and Moore, 1983). The Ca concentration of the soil solution needed to obtain maximum soybean root growth increases as pH decreases and Al activity increases (Lund, 1970).

The effects of liming on soil Ca levels are initially confined to the zone of incorporation (Ritchey et al., 1980; Pavan et al., 1984, 1987). However, gypsum, phosphogypsum, and soluble Ca applications are effective at providing Ca deep into the soil profile. Forty-four weeks after application, soil Ca was increased in the upper 20 cm of a tropical Inceptisol, but phosphogypsum and $CaCl_2$ increased exchangeable Ca to a depth of 80 cm (Pavan et al., 1987). Growth of alfalfa (*Medicago sativa* L.) was increased in an Appling coarse sandy loam (Typic Hapludult) by a surface application of gypsum which increased subsoil Ca and decreased soluble Al (Sumner et al., 1986; Sumner and Carter, 1988; Sumner, 1995).

Low supplies of Ca inhibit the nodulation, growth, and N fixation of bacteria associated with the roots of legumes (Albrecht, 1931; McCalla, 1937). No N fixation occurs unless the Ca saturation is > 40% (Albrecht, 1937). The number of nodules and the amount of N fixed increased to 97% Ca saturation. The number of nodules per plant was positively correlated with the total Ca content of the primary root (Sartain and Kamprath, 1978). The Ca requirement for nodulation is greater at low pH and greater than that for the host plant (Alva et al., 1990). Calcium concentration of 0.5 mM maximized nodule number and weight at pH > 5, but 2.5 mM Ca was required at pH 4.5.

The ratio of Ca and Mg to other cations in the soil solution is important in Ca and Mg sufficiency to the plant. Maximum root length of cotton occurred when the activity of Ca exceeded approximately 15% of the total cation activity in the soil solution (Adams, 1966) and when exchangeable Ca was 13% or more of the total exchangeable cations (Howard and Adams, 1965; Adams and Moore, 1983). These relationships were nearly identical for the two soils examined even though the clay mineralogy of one soil was kaolinitic and that of the other was vermiculitic. Other crops respond similarly to the relative activity of Ca in the soil solution. Root growth of soybean (*Glycine max* (L.) Merr.) (Lund, 1970) and loblolly pine (Lyle and Adams, 1971) were maximized when the ratio of Ca activity to total cation activity in the soil solution exceeded 10 to 20%. Magnesium activity expressed as a function of total cation activity in the soil solution was closely correlated with ryegrass Mg concentration (Salmon, 1964).

Calcium has low phloem mobility, which results in low Ca redistribution from older plant tissues to growing meristems (Jeschke and Pate, 1991), thereby imposing heavy reliance on concurrent Ca uptake and xylem transport to support new growth and development of vegetative and reproductive tissues (Morard et al., 1996). Plant organs with low rates of transpiration (e.g., new leaves, fruits, tubers) and low rates of xylem flow exhibit Ca deficiency disorders such as pod rot in peanut (*Arachis hypogaea* L.) (Cox et al., 1982; Sumner et al., 1988), internal brown spot in potato (Tzeng et al., 1986), blossom end rot in tomato (*Lycopersicon esculentum* Mill.) (Geraldson, 1957), bitter pit in apple (*Malus* sp.) (Perring, 1986) and cork spot of pear (*Pyrus communis* L.) (Raese and Drake, 1993), and blackheart in celery (*Apium graveolens* L.) (Geraldson, 1954). Although low soil Ca levels may increase the frequency and severity of Ca deficiency in these crops, the deficiency is more a function of poor Ca translocation within the plant than of low soil Ca levels.

Calcium uptake and translocation occur at greater rates in actively growing meristematic regions than older root sections (Clarkson, 1984; Marschner, 1995). As a result, soil Ca availability must be optimized in the root zone at the time of active root development (Kratzke and Palta, 1986). Since environmental stresses that curtail root meristematic development can inhibit Ca uptake and translocation to shoot meristems, temperature and moisture stresses can exacerbate low soil Ca availability by imposing transient Ca deficiencies during critical periods of crop development. These environmental stress by soil Ca interactions explain why these Ca-related disorders are only displayed in some growing seasons. Foliar and fruit sprays, as well as maintenance of adequate soil Ca, are required to completely alleviate Ca deficiency disorders in sensitive crops (Geraldson, 1954,1957; Raese and Drake, 1993).

1.5.1.6   Crop Response to Calcium and Magnesium Availability

Optimum soil Ca levels for maximizing yield and quality of peanuts are quite high, 538 kg ha^{-1} for small seeded cultivars, and 1,600 kg ha^{-1} for large seeded cultivars (Walker et al., 1979; Gaines et al., 1989). Adsorption of Ca for kernel development occurs through the shell. Small seeded cultivars have relatively more surface area per unit mass for Ca adsorption than large seeded cultivars, hence the lower soil Ca requirement (Sumner et al., 1988). Reduced peanut yield and quality with low soil Ca are, in part, due to destruction of pods by *Pythium* and *Rhizoctonia* fungal pathogens (Hallock and Garren, 1968).

Low Mg uptake by forage grasses can induce grass tetany in ruminants (Grunes et al., 1970), while low Ca in wheat forage leads to wheat pasture poisoning (Bohman et al., 1983). Low rates of Mg and Ca uptake due to suboptimal soil Mg and Ca availability can be exacerbated by high K (Thill and George, 1975; Ohno and Grunes, 1985), low soil temperatures (Miyasaka and Grunes, 1990), low P availability (Reinbott and Blevins, 1994) and wet soil conditions (Karlen et al., 1980). A survey of the incidence of grass tetany in the United States showed greater appearance of the problem in areas where soil parent material is naturally low in Mg and in cooler climates (Kubota et al., 1980). This study concurred with earlier findings of a greater incidence of grass tetany when the Mg concentration in forages is less than 2.0 g Mg kg^{-1} and the K/(Ca + Mg) equivalent ratio is above 2.2 (Kemp and Hart, 1957).

Exchangeable Mg levels are often poorly related to crop yield response to applied Mg. Incongruities between exchangeable Mg and crop response primarily arise from two sources: a pool of available Mg in the A horizon soil that is not exchangeable, and an accumulation of available Mg below the A horizon. Nonexchangeable Mg may be an important source of plant-available Mg in some soils. Prince and Toth (1937) noted that only about 3 to 4% of the total Mg content of a Sassafras loam soil was exchangeable, suggesting fixation of Mg in an insoluble form in the colloidal fraction of the soil. Substantial plant uptake of nonexchangeable Mg occurred in five Coastal Plain soils where only 4 to 9% of the total Mg content was in the exchangeable form (Rice and Kamprath, 1968). An additional 4 to 10% of the total soil Mg content was extractable with dilute acid and was termed nonexchangeable, but plant available.

Significant fixation of Mg occurred in Oxisols and Ultisols when limed to pH above 7 (Sumner et al., 1978; Grove et al., 1981). Magnesium fixation was nearly 80% in a Bradson soil (Typic Hapludult) limed to a pH of 7.2 (Grove et al., 1981). Subsurface accumulations of Mg may also satisfy crop requirements for Mg. Adams and Hartzog (1980) noted that Mg accumulated in the subsoil of Ultisols could explain the lack of crop growth response to applied Mg on soils with extremely low exchangeable Mg in the topsoil. Exchangeable Mg levels were tenfold greater in the subsoil (60 to 120 cm) of a Wagram loamy sand (Arenic Kandiudult) than in the surface 40 cm (Schmidt and Cox, 1992).

Magnesium deficiencies are intensified by high levels of K or Ca (Welte and Werner, 1963). Severe Mg deficiency of potato and tobacco (*Nicotiana tabacum* L.) and mild Mg deficiencies of sugar beet and barley (*Hordeum vulgare* L.) occurred with heavy applications of K, even though exchangeable Mg levels were considered adequate (Walsh and O'Donohoe, 1945). Magnesium deficiency of citrus occurred when the Mg:K ratio was less than 2.5:1 (Pratt et al., 1957). Magnesium deficiency of alfalfa occurred when Mg comprised less than 6% of the exchangeable cations (Prince et al., 1947). Similarly, Mg deficiency of sudangrass (*Sorghum sudanense* (Piper) Stapf) and clover (*Trifolium repens* L.) occurred when soils had < 4% of the CEC occupied by Mg (Adams and Henderson, 1962), and in citrus when Mg was < 4 to 8% of the CEC (Martin and Page, 1969). Exchangeable Mg in excess of 10% of the CEC generally ensures Mg availability to crops will not be limiting.

When Mg fertilizer applications are made to the soil surface, as in orchards or pastures, or to crops with high Mg requirements, high solubility fertilizers are more effective than low solubility materials (Boynton, 1947; Camp, 1947). For example, $MgSO_4$ and MgO increased grapefruit (*Citrus paradisi*) leaf Mg and alleviated visual symptoms of Mg deficiency within 9 months, whereas effects of Mg CO_3 were not detected for more than 24 months (Koo, 1971). McMurtrey (1931) reported that considerably more dolomitic limestone was needed than soluble Mg fertilizer to prevent Mg deficiency in tobacco. Foliar applications of $MgSO_4$ can be more effective than soil applications of Mg in correcting Mg deficiency (Scott and Scott, 1951).

1.5.2 Sulfur

1.5.2.1 Overview of Sulfur Nutrition
Sulfur functions in plants have been reviewed by Duke and Reisenauer (1986), DeKok et al. (1993), and Marschner (1995). Sulfur is a structural constituent of organic compounds, some of which are uniquely synthesized by plants, providing animals with essential amino acids (methionine and cysteine) required to synthesize S-containing proteins. Disulfide bonding plays an important role in regulating the three-dimensional conformation of proteins, affecting enzyme function. Sulfur is also contained in vitamins and coenzymes (thiamin, biotin, coenzyme A, and lipoic acid) (Mengel and Kirkby, 1982; Marschner, 1995). Other S-containing compounds include glutathione, an antioxidant (Bergmann and Rennenberg, 1993) and phytochelatin (Grill et al., 1987), as well as glucosinolates that can serve as natural plant protectants (Schnug, 1993).

1.5.2.2 Occurrence of Soil Sulfur
Soil S exists in organic compounds, and adsorbed and soil solution SO_4^{2-}. Because plants absorb SO_4^{2-} from the soil solution, replenishment from organic and adsorbed sources is important in determining S supply to the plant. Temperate region soils ordinarily contain between 0.1 and 2.0 g S kg^{-1} (Brady, 1974). Representative soils from arid regions contain more S than those from humid regions (0.8 and 0.4 mg kg^{-1}, respectively) (Brady, 1974). The predominant mineral form of S in arid soils is gypsum. Organic S comprises most of S in most soils (> 90% of the total) (Table 1.12). The C:S and N:S ratios of mineral soils vary. The organic pool is composed of three major sources of S (ester sulfates, amino acid S, and C bonded S). Ester sulfate is considerably more labile than C bonded S and is considered an important source of plant-available S. Ester sulfate and C bonded S accounted for 53 and 14% of total organic S, respectively, in 48 surface soils of Ghana (Acquaye and Kang, 1987). Ester sulfate ranged from 35 to 52% of total S in 54 Canadian soils (Bettany et al., 1973), from 20 to 65% in 6 Brazilian soils, and 43 to 60% in 6 Iowa soils (Neptune et al., 1975). Commonly, more than 30% of the total organic S in soils is not identified as either ester sulfate or C bonded S (Neptune et al., 1975; Acquaye and Kang, 1987). Amino acid S has been reported to account for 21 to 31% of the total organic S content of two Australian soils (Freney et al., 1972) and 19 to 31% of C bonded S in four mineral soils of Scotland (Scott et al., 1981).

Adsorbed SO_4^{2-} occurs on the positively charged exchange sites at edges of clay minerals, organic matter, and Fe and Al oxides (Chao et al., 1962). The SO_4^{2-} adsorption capacity of a soil is decreased by increased soil pH and P content (Ensminger, 1954; Kamprath et al., 1956). Ultisols, Alfisols, and Oxisols have a high adsorption capacity due to an abundance of Fe and Al oxides and typically low pH. Binding is due both to electrostatic attraction and ligand exchange mechanisms (Marsh et al., 1987;

Table 1.12   Total-S, organic-S, and C:N:S ratio of selected mineral soils

Origin	Total-S	Organic-S	C:N:S[†]	Reference
	------ mg kg^{-1}------			
Saskatchewan, Canada	296	291	79:7.3:1.0	Bettany et al. (1973)
Oregon, USA	247	235	144:10:1.0	Harward et al. (1962)
Brazil, South America	166	146	139:7.1:1.0	Neptune et al. (1975)
Ghana, Africa	129	121	66:6.8:1.0	Acquaye and Kang (1987)
New Zealand	717	640	91:8.0:1.0	Perrot and Sarathchandra (1987)

[†] Organic C and S for all entries. Total N for all entries, except organic N for Bettany et al. (1973).

Marcano-Martinez and McBride, 1989) and highly dependent on the occurrence of positive charge on the soil colloids. Sulfate adsorption occurs rapidly, with nearly complete adsorption in minutes (Rajan, 1978) to days (Marcano-Martinez and McBride, 1989). However, a slower second stage of SO_4^{2-} adsorption also exists, rendering initially adsorbed SO_4^{2-} less available over time (Barrow and Shaw, 1977). Sulfate retention in acid soils is enhanced by Ca (Barrow, 1972; Ryden and Syers, 1976). Marcano-Martinez and McBride (1989) proposed that the stimulation in SO_4 adsorption arose from the bonding of a SO_4-Ca complex to the oxide surface by an O ligand and attraction of the Ca to a negatively charged hydroxyl.

1.5.2.3   Additions and Losses of Sulfur

Major sources of atmospheric S are oceans (24% of the total), soils (35%), volcanic activity (7%), and industry (35%) (Noggle et al., 1986). Sulfur accretion in soils occurs through direct adsorption of S gases and as rainfall. Distance from the source of S determines the amount of deposition. Sulfur in rainfall in Hawaii was exponentially related to distance from the ocean, averaging from 24 to 1 kg S ha^{-1} yr^{-1} from the coast to 24 km inland (Hue et al., 1990). As much as 168 kg S ha^{-1} was deposited on soils near industry, whereas urban and rural locations generally received < 15 kg S ha^{-1} (Olson and Rehm, 1986). Areas distant from the sea and devoid of industrial sources may have rainfall concentration of < 0.1 mg L^{-1} (< 0.1 kg S ha^{-1} per 100 mm rainfall) (Fox and Blair, 1986). Industrial S emissions continue to be reduced in industrialized countries thus reducing S deposition (Schnug, 1991).

The replacement of ordinary superphosphate (12% S) with concentrated superphosphate (1% S) beginning in the late 1940s as the most common P source resulted in an increase in the occurrence of crop S deficiencies in the southern United States (Mehring and Lundstrom, 1938; Jordan, 1964). The use of gypsum as a Ca source for peanuts is another example of incidental S applications to soils. Nutritive applications of S are most frequently applied with N sources. The primary source of S in both liquid and solid N fertilizers is ammonium sulfate (24% S). Ammonium bisulfite (32% S) and thiosulfate (26% S) are also used to provide S in liquid fertilizers. Sulfur sources used in solid fertilizers are potassium sulfate (18% S), potassium magnesium sulfate (23% S), magnesium sulfate (14% S), and ordinary superphosphate. Elemental S is also used sometimes as a slow release source of nutritive S. The S becomes available to plants when the elemental S is oxidized by bacteria to SO_4^{2-} (Starkey, 1966). Oxidation rate is dependent on the size of the S particles, temperature, and moisture (Burns, 1967).

The retention of fertilizer S in surface soils is dependent on rainfall. Nearly 50% of the SO_4^{2-} added to maize was lost from the soil profile by leaching in the subhumid savannah of West Africa (1500 mm rainfall), whereas 100% of that added to a soil in the semiarid savannah (640 mm rainfall) remained in the soil profile to a depth of 105 cm (Friesen, 1991). Nearly all of 56 kg SO_4-S added to a Wagram loamy sand (Arenic Paleudult) was leached from the upper 0.45 m of soil with 445 mm of rainfall in 180 days (Rhue and Kamprath, 1973). Leaching losses (24.1 kg S ha^{-1} yr^{-1}) exceeded crop removal (15 kg S ha^{-1} yr^{-1}) even with S inputs from the atmosphere, crop seed, and fertilizer of 37.1 kg S ha^{-1} yr^{-1} and rainfall of 660 mm yr^{-1} in a clay loam Typic Eutrochrept in central Sweden (Kirchmann et al., 1996). Liming and P fertilization decrease S adsorption and increase SO_4-S leaching (Bolan et al., 1988).

Crop removal of S varies by species. Cruciferous forages, alfalfa, and rapeseed (*Brassica napus* L.) accumulate large amounts of S, about 70 kg S ha^{-1} (Spencer, 1975). Sugarcane *(Saccharum officinarum* L.), coffee (*Coffea arabica* L.), and coconut (*Cocos nucifera*) accumulate moderate amounts of S (< 50 kg S ha^{-1}) (Spencer, 1975), while field crops and forages accumulate between 20 to 30 kg S ha^{-1} annually (Kamprath and Jones, 1986; Hoeft and Fox, 1986).

1.5.2.4 Factors Affecting the Availability of Soil Sulfur

Sulfate uptake by plants from the rooting solution can be saturated at relatively low concentrations of SO_4^{2-}. Soybean growth was optimized with 0.23 mM SO_4-S in nutrient solution (Elkins and Ensminger, 1971). Wheat achieved maximum S accumulation at 0.01 mM SO_4-S in solution (Reisenauer, 1969), while Fox and Blair (1986) concluded that approximately 0.14 mM SO_4-S was required in the soil solution to optimize growth of some tropical and subtropical crops. Soil reactions that buffer SO_4^{2-} concentrations above these critical soil solution levels are required to optimize S availability.

Extractable S fluctuates during the year. Tan et al. (1994b) reported that SO_4^{2-} and C-bonded S were inversely related throughout the year in three New Zealand pasture soils (Inceptisols), with the highest proportions of C-bonded S occurring in the winter and SO_4-S predominating in the spring. Although Watkinson and Kear (1996b) found SO_4-S to be constant throughout the year, in contrast to Tan et al. (1994b), they found extractable organic S concentrations to increase in the fall and winter in agreement with the prior study.

Sulfate adsorption is negligible in surface soils because little adsorption occurs at soil pH and phosphate levels conducive to crop plant growth (Kamprath et al., 1956). Therefore, most of the effective SO_4^{2-} adsorption in cultivated soils occurs in acid argillic horizons, which often retain enough S to support plant growth. The range in water soluble SO_4-S concentrations of surface horizons of Ultisols was 0.09 to 0.16 mM and in the argillic horizon, 0.20 to 0.36 mM (Camberato and Kamprath, 1986). The depth of the argillic horizon, in part, determines the plant availability of S accumulated in that soil layer. When the argillic horizon was > 0.45 m deep, S fertilization increased the yield of tobacco in two of ten site years on four Ultisols, but S deficiency did not occur when the argillic horizon was within 0.30 m of the soil surface (Smith et al., 1987).

The depth of crop rooting also influences the amount of S accessible to the crop in soils with subsoil accumulations of SO_4^{2-}. The amount of SO_4-S available to tobacco increased during the growing season and was greater than that of cotton on a Durham coarse sandy loam soil (Paleudult) with < 18 kg SO_4-S ha^{-1} in the upper 0.45 m of soil, but > 72 kg S ha^{-1} below 0.45 m (Kamprath et al., 1957). Differences in rooting between these crops explained the difference in S supply. Tobacco was previously shown to have 19.8% of its root activity below 0.45 m at 7 weeks after transplanting, whereas cotton had a much shallower root system with only 1.5% of its root activity at this soil depth 11 weeks after planting.

If root growth is prevented by impeding soil layers or high Al^{3+} levels, response to S may occur on soils with high levels of subsoil SO_4^{2-}. Failure to disrupt the tillage pan in a Typic Kandiudult prevented rooting into a SO_4^{2-} rich subsoil and wheat grain yield was increased by S fertilization (Oates and Kamprath, 1985). On an adjacent soil that was subsoiled to allow rooting below the tillage pan, wheat obtained sufficient S for maximum crop productivity. Aluminum saturations in excess of 50% were implicated in reducing rooting in the SO_4^{2-} rich subsoil of an Aquic Hapludult, resulting in an increase in maize yield with S fertilization (Kline et al., 1989). Limited crop rooting due to shallow soils causes S deficiency of maize and alfalfa in the midwestern United States (Hoeft and Fox, 1986).

Sulfur commonly limits crop production in the subhumid and humid regions of the Pacific Northwest. Low atmospheric deposition of S (< 6 kg ha^{-1} yr^{-1}), low S-containing surface irrigation water and basalt, granite, and volcanic ash parent materials have resulted in soils that have low total S content and low S bioavailability (Rasmussen and Kresge, 1986). Sulfur responses in cereals are dependent on having adequate N availability (Koehler, 1965) and high yield potential (Ramig et al., 1975). High rainfall and winter leaching potential in the humid regions of the Western coastal areas, coupled with low S deposition create S deficient conditions in cereal, forage, fruit, and vegetable production (Rasmussen and Kresge, 1986). In contrast, arid soils in the western United States with high accumulations of soluble SO_4^{2-}, low leaching potential under nonirrigated conditions, or that are irrigated with high SO_4^{2-}-containing groundwater respond less frequently to S fertilization.

1.5.2.5 Assessing Levels of Soil Sulfur

Many studies have demonstrated that total S or organic S is poorly correlated to crop response to applied S. Measures of extractable S, which may include soil solution, exchangeable, and organically bound SO_4^{2-} and S, have been reasonably successful at predicting response to S fertilization. Differences in the chemical extractant, concentration, temperature, and other procedural factors affect the quantity and form of S extracted from the soil (Anderson et al., 1992). The analytical procedure used to determine S in the extract also influences the quantity of S measured. Turbidimetric and anion exchange chromotography methods measure only solution SO_4^{2-} levels, whereas inductively coupled argon plasma spectrophotometry measures total S in solution, including solution SO_4^{2-}, ester SO_4^{2-}, and soluble organic S (Anderson et al., 1992). The contribution of S in rainfall to the crop (Hoeft et al., 1973) and subsoil SO_4, which are not often quantified, may be reasons why soil S analysis for predicting crop response has not been totally successful.

Extractable SO_4-S was not a good indicator of S sufficiency on 7 silt loam Mollisols in Minnesota (O'Leary and Rehm, 1991). However, the amount of S mineralized during a 4- or 12-week aerobic incubation was related to maize yield response to S fertilization, indicating the importance of organic S to plant-available S in some soils. Approximately 30% of the S extracted from 8 field-moist New Zealand soils was in the soluble organic form, while the remainder was SO_4^{2-} (Tan et al., 1994a). Extractable organic S is a good predictor of the amount of mineralizable S. Organic S extracted by potassium phosphate was a better predictor of maximum yield in pastoral soils of New Zealand than initial SO_4^{2-}, SO_4^{2-} mineralized during short-term incubation, organic S extracted by calcium phosphate, or organic S extracted by sodium bicarbonate (Watkinson and Kear, 1996a). Potassium phosphate extracted more organic S from high organic matter soils than the commonly used calcium phosphate extractant. Blair et al. (1991) found that extraction with KCl was superior to either calcium phosphate or sodium bicarbonate as an indicator of pasture response to fertilizer S in 18 soils from northern New South Wales, Australia. Critical levels for the three extractants ranged from 6.5 to 8.4 mg kg^{-1}. The bicarbonate extractant in this case was probably inferior to the others because it overestimated the contribution of available S arising from the ester SO_4^{2-} fraction.

1.5.2.6 Crop Response to Sulfur Availability

Sulfur deficiencies occur worldwide, but are most prevalent in areas where S accretions from atmospheric deposition, the S contents of fertilizers and pesticides, and irrigation water are low, and soils are sandy with low SOM content, and rainfall is plentiful (Tisdale et al., 1986). Numerous examples of crop S deficiencies and responses to S fertilization are presented by Tabatabai (1986).

Sulfur deficiencies decrease crop yields and in certain instances also reduce crop nutritional value and quality. Sulfur fertilization of subirrigated meadow vegetation (Nichols et al., 1990) and forage maize (O'Leary and Rehm, 1990) increased dry matter accumulation, but did not affect crop quality. In other instances, S fertilization increased both the protein- and S-containing amino acid content of forages, resulting in increased animal performance (Rendig, 1986). In mixed pastures, S application up to 90 kg S ha^{-1} increased total forage production and the proportion of clover to grass (Jones, 1964). Numerous yield responses of coastal bermudagrass (*Cynodon dactylon* (L.) Pers.) to S fertilization have been documented (Kamprath and Jones, 1986). Alfalfa is the most S responsive crop grown in the Midwest and Northeast regions of the United States (Hoeft and Fox, 1986).

Cyst(e)ine and methionine levels of seeds are increased by S fertilization resulting in more nutritious foods (Rendig, 1986). Baking quality of wheat flour is highly dependent on the S concentration of the grain and is increased by S fertilization (Haneklaus et al., 1992). Bread dough made from S-deficient grain resists extension and has lower extensibility (Moss et al., 1981), which is related to low albumin proteins (Wrigley et al., 1984). Sulfur-containing metabolites influence the flavor of several crops including asparagus (*Asparagus officinalis* L.) and various *Allium* sp. (Schnug, 1990).

1.5.3 References

Acquaye, D.K., and B.T. Kang. 1987. Sulfur status and forms in some surface soils of Ghana. Soil Sci. 144:43–52.

Adams, F. 1966. Calcium deficiency as a causal agent of ammonium phophate injury to cotton seedlings. Soil Sci. Soc. Am. Proc. 30:485–488.

Adams, F., and D.L. Hartzog. 1980. The nature of yield responses of Florunner peanuts to lime. Peanut Sci. 7:120–123.

Adams, F., and J.B. Henderson. 1962. Magnesium availability as affected by deficient and adequate levels of potassium and lime. Soil Sci. Soc. Am. Proc. 26:65–68.

Adams, F., and B.L. Moore. 1983. Chemical factors affecting root growth in subsoil horizons of Coastal Plain soils. Soil Sci. Soc. Am. J. 47:99–102.

Albrecht, W.A. 1931. The function of calcium in the growth of certain legumes. J. Am. Soc. Agron. 23:1052–1053.

Albrecht, W.A. 1937. Physiology of root nodule bacteria in relation to fertility levels of the soil. Soil Sci. Soc. Am. Proc. 2:315–327.

Albrecht, W.A. 1941. Plants and the exchangeable calcium of the soil. Am. J. Bot. 28:394–402.

Allaway, W.H. 1945. Availability of replaceable calcium from different types of colloids as affected by degree of calcium saturation. Soil Sci. 59:207–217.

Alva, A.K., C.J. Asher, and D.G. Edwards. 1990. Effect of solution pH, external calcium concentration, and aluminum activity on nodulation and early growth of cowpea. Aust. J. Agric. Res. 41:359–365.

Anderson, G., R. Lefroy, N. Chinoim, and G. Blair. 1992. Soil sulphur testing. Sulphur Agric. 16:6–14.

Asher, C.J., F.P.C. Blamey, and C.P. Mamaril. 1983. Sulfur nutrition of tropical annual crops. p. 54–64. *In* G.J. Blair and A.R. Till (ed.) Sulfur in south east Asian and south Pacific agriculture. The University of New England, Armidale, New South Wales, Australia.

Balke, N.E., and T.K. Hodges. 1975. Plasma membrane adenosine triphosphatase of oat roots. Plant Physiol. 55:83–86.

Barber, S.A. 1962. A diffusion and mass-flow concept of soil nutrient availability. Soil Sci. 93:39–49.

Barber, S.A. 1984. Soil nutrient bioavailability. A mechanistic approach. John Wiley and Sons, New York, NY.

Barber, S.A., and P.G. Ozanne. 1970. Autoradiographic evidence for the differential effect of four plant species in altering the calcium content of the rhizosphere soil. Soil Sci. Soc. Am. Proc. 34:635–637.

Barrow, N.J. 1972. Influence of solution concentration of calcium on the adsorption of phosphate, sulfate, and molybdate by soils. Soil Sci. 113:175–180.

Barrow, N.J., and T.C. Shaw. 1977. The slow reactions between soil and anions: 7. Effect of time and temperature of contact between an adsorbing soil and sulfate. Soil Sci. 124:347–354.

Bear, F.E., A.L. Prince, and J.C. Malcom. 1945. Potassium needs of New Jersey soils. NJ Agric. Exp. Stn. Bull. 721.

Beeson, K.C. 1941. The mineral composition of crops with particular reference to the soils in which they were grown: A review and compilation. USDA Misc. Pub. 369.

Bergmann, L., and H. Rennenberg. 1993. Glutathinone metabolism in plants. p. 109–123. In L.J. DeKok et al. (ed.) Sulfur nutrition and assimilation in higher plants. SPB Academic Publishing, The Hague, Netherlands.

Bettany, J.R., J.W.B. Stewart, and E.H. Halstead. 1973. Sulfur fractions and carbon, nitrogen, and sulfur relationships in grassland, forest, and associated transitional soils. Soil Sci. Soc. Am. Proc. 37:915–918.

Blair, G.J., N. Chinoim, R.D.B. Lefroy, G.C. Anderson, and G.J. Crocker. 1991. A soil sulfur test for pastures and crops. Aust. J. Soil Res. 29:619–626.

Bohman, V.R., F.P. Horn, E.T. Littledike, J.G. Hurst, and D. Griffin.1983. Wheat pasture poisoning. II. Tissue composition of cattle grazing cereal forages and related to tetany. J. Anim. Sci. 57:1364–1373.

Bolan, N.S., J.K. Syers, R.W. Tillman, and D.R. Scotter. 1988. Effect of liming and phosphate additions on sulphate leaching in soils. J. Soil Sci. 39:493–504.

Bolton, E.F., J.W. Aylesworth, and F.R. Hore. 1970. Nutrient losses through tile drains under three cropping systems and two fertility levels on a Brookston clay soil. Can. J. Soil Sci. 50:275–279.

Borchert, R. 1990. Ca^{2+} as developmental signal in the formation of Ca-oxalate crystal spacing patterns during leaf development in *Carya ovata*. Planta 182:339–347.

Boynton, D. 1947. Magnesium nutrition of apple trees. Soil Sci. 63:53–58.

Brady, N.C. 1974. The nature and properties of soils., 8th Ed. Macmillan, New York, NY.

Buol, S.W. 1995. Sustainability of soil use. Ann. Rev. Ecol. Syst. 26:25–44.

Burns, G.R. 1967. Oxidation of sulphur in soils. Sulphur Inst. Tech. Bull. 13.

Caldwell, C.R., and A. Haug. 1981. Temperature dependence of the barley root plasma membrane-bound Ca^{2+} and Mg^{2+}-dependent ATPase. Physiol. Plant. 53:117–124.

Camberato, J.J., and E.J. Kamprath. 1986. Solubility of adsorbed sulfate in Coastal Plain soils. Soil Sci. 142:211–213.

Cammarano, P., A. Felsani, M. Gentile, C. Gualerzi, C. Romeo, and G. Wolf. 1972. Formation of active hybrid 80-S particles from subunits of pea seedlings and mammalian liver ribosomes. Biochim. Biophys. Acta 281:625–642.

Camp, A.F. 1947. Magnesium in citrus fertilization in Florida. Soil Sci. 63:43–52.

Chao, T.T., M.E. Harward, and S.C. Fang. 1962. Soil constituents and properties in the adsorption of sulfate ions. Soil Sci. 94:276–283.

Clark, R. 1984. Physiological aspects of calcium, magnesium, and molybdenum deficiencies in plants. p. 99–170. In F. Adams (ed.) Soil acidity and liming. Soil Science Society of America, Madison, WI.

Clarkson, D.T. 1984. Calcium transport between tissues and its distribution in the plant. Plant Cell Environ. 7:449–456.

Clarkson, D.T, and J.B. Hanson. 1980. The mineral nutrition of higher plants. Ann. Rev. Plant Physiol. 31:239–298.

Cox, F.R., F. Adams, and B.B. Tucker. 1982. Liming, fertilization, and mineral nutrition. In H.E. Pattee and C.T. Young (ed.) Peanut science and technology. American Peanut Research and Education Society Inc., Yoakum, TX.

Darusman, L.R.S., D.A. Whitney, K.A. Janssen, and J.H. Long. 1991. Soil properties after twenty years of fertilization with different nitrogen sources. Soil Sci. Soc. Am. J. 55:1097–1100.

DeKok, L.J., I. Stulen, H. Rennenberg, C. Brunold, and W.E. Rauser. 1993. Sulfur nutrition and sulfur assimilation in higher plants. SPB Academic Publishing, The Hague, Netherlands.

Duke, S.H., and H.M. Reisenauer. 1986. Roles and requirements of sulfur in plant nutrition. p. 123–168. In M.A. Tabatabai (ed.) Sulfur in agriculture. American Society of Agronomy, Madison, WI.

Elkins, D.M., and L.E. Ensminger. 1971. Effect of soil pH on the availability of adsorbed sulfate. Soil Sci. Soc. Am. Proc. 35:931–934.

Ensminger, L.E. 1954. Some factors affecting the adsorption of sulfate by Alabama soils. Soil Sci. Soc. Am. Proc. 18:259–264.

Fox, R.L., and G.J. Blair. 1986. Plant response to sulfur in tropical soils. p. 405–434. In M.A. Tabatabai (ed.) Sulfur in agriculture. American Society of Agronomy, Madison, WI.

Freney, J.R., F.J. Stevenson, and A.H. Beavers. 1972. Sulphur-containing amino acids in soil hydrolysates. Soil Sci. 114:468–476.

Friesen, D.K. 1991. Fate and efficiency of sulfur fertilizer applied to food crops in West Africa. Fert. Res. 29:35–44.

Gaines, T.P., M.B. Parker, and M.E. Walker. 1989. Runner and Virginia type peanut response to gypsum in relation to soil calcium level. Peanut Sci. 16:116–118.

Geraldson, C.M. 1954. The control of blackheart of celery. Proc. Am. Soc. Hort. Sci. 63:353–358.

Geraldson, C.M. 1957. Control of blossom-end rot of tomatoes. Proc. Am. Soc. Hort. Sci. 69:309–317.

Giddens, J., and S.J. Toth. 1951. Growth and nutrient uptake of ladino clover grown on red and yellow and grey-brown podzolic soils containing varying ratios of cations. Agron. J. 43:209–214.

Grill, E., E.L. Winnacker, and M.H. Zenk. 1987. Phytochelatins, a class of heavy metal binding peptides from plants are functionally analogous to metallothioneins. Proc. Nat. Acad. Sci. 84:439–443.

Grove, J.H., M.E. Sumner, and J.K. Syers. 1981. Effect of lime on exchangeable magnesium in variable surface charge soils. Soil Sci. Soc. Am. J. 45:497–500.

Grunes, D. L., P.R. Stout, and J.R. Brownell. 1970. Grass tetany of ruminants. Adv. Agron. 22:331–374.

Hallock, D.L., and K.H. Garren. 1968. Pod breakdown, yield, and grade of Virginia type peanuts as affected by Ca, Mg, and K sulfates. Agron. J. 60:253–257.

Haneklaus, S., E. Evans, E. Schnug. 1992. Baking quality and sulphur content of wheat. I. Influence of grain sulphur and protein concentrations on loaf volume. Sulphur Agric. 16:31–34.

Hanson, J.B. 1984. The function of calcium in plant nutrition. Adv. Plant Nutr. 1:149–208.

Harward, M.E., T.T. Chao, and S.C. Fang. 1962. The sulfur status and sulfur supplying power of Oregon soils. Agron. J. 54:101–106.

Haynes, J.L., and W.R. Robbins. 1948. Calcium and boron as essential factors in the root environment. J. Am. Soc. Agron. 40:707–715.

Hedin, L.O., L. Granat, G.E. Likens, T.A. Buishand, J.N. Galloway, T.J. Butler, and H. Rodhe. 1994. Steep declines in atmospheric base cations in regions of Europe and North America. Nature 367:351–354.

Hoeft, R.G., and R.H. Fox. 1986. Plant response to sulfur in the Midwest and Northeastern United States. p. 345–356. *In* M.A. Tabatabai (ed.) Sulfur in agriculture. American Society of Agronomy, Madison, WI.

Hoeft, R.G., L.M. Walsh, and D.R. Keeney. 1973. Evaluation of various extractants for available soil sulfur. Soil Sci. Soc. Am. Proc. 37:401–404.

Howard, D.D., and F. Adams. 1965. Calcium requirement for penetration of subsoils by primary cotton roots. Soil Sci. Soc. Proc. 29:558–562.

Hue, N.V., R.L. Fox, and J.D. Wolt. 1990. Sulfur status of volcanic ash-derived soils in Hawaii. Commun. Soil Sci. Plant Anal. 21:299–310.

Hunsaker, V.E., and P.F. Pratt. 1971. Calcium magnesium exchange equilibria in soils. Soil Sci. Soc. Am. Proc. 35:151–152.

Hunter, A.S. 1949. Yield and composition of alfalfa as affected by variations in the calcium-magnesium ratio in the soil. Soil Sci. 67:53–62.

Jenny, H. 1941. Calcium in the soil: III. Pedologic relations. Soil Sci. Soc. Am. Proc. 6:27–35.

Jeschke, W.D., and J.S. Pate. 1991. Cation and chloride partitioning through xylem and phloem within the whole plant of *Ricinus communis* L. under conditions of salt stress. J. Exp. Bot. 42: 1105–1116.

Johnson, D.W., and D.E. Todd. 1987. Nutrient export by leaching and whole-tree harvesting in a loblolly pine and mixed oak forest. Plant Soil 102:99–109.

Jones, M.B. 1964. Effect of applied sulfur on yield and sulfur uptake of various California dryland pasture species. Agron. J. 56: 235–237.

Jones, M.B. 1986. Sulfur availability indexes. p. 549–566. *In* M.A. Tabatabai (ed.) Sulfur in agriculture. American Society of Agronomy, Madison, WI.

Jordan, H.V. 1964. Sulfur as a plant nutrient in the southern United States. USDA Tech. Bull. 1297.

Kamprath, E.J. 1984. Crop response to lime on soils in the tropics. p. 349–368. *In* F. Adams (ed.) Soil acidity and liming. American Society of Agronomy, Madison, WI.

Kamprath, E.J. 1978. Lime in relation to Al toxicity in tropical soils. p. 233–245. *In* C.S. Andrew and E.J. Kamprath (ed.) Mineral nutrition of legumes in tropical and subtropical soils. The Dominion Press, North Blackburn, Victoria, Australia.

Kamprath, E.J., and U.S. Jones. 1986. Plant response to sulfur in the Southeastern United States. p. 323–343. *In* M.A. Tabatabai (ed.) Sulfur in agriculture. American Society of Agronomy, Madison, WI.

Kamprath, E.J., W.L. Nelson, and J.W. Fitts. 1957. Sulfur removed from soils by field crops. Agron. J. 49:289–293.

Kamprath, E.J., W.L. Nelson, and J.W. Fitts. 1956. The effect of pH, sulfate and phosphate concentrations on the adsorption of sulfate by soils. Soil Sci. Soc. Am. Proc. 20:463–466.

Karlen, D.L., R. Ellis, Jr., D.A. Whitney, and D.L. Grunes. 1980. Influence of soil moisture on soil solution cation concentrations and the tetany potential of wheat forage. Agron. J. 72:73–78.

Kemp, A., and M.L. t'Hart. 1957. Grass tetany in grazing milk cows. Neth. J. Agric. Sci. 5:4–17.

Key, J.L., L. T. Kurtz, and B.B. Tucker. 1962. Influence of ratio of exchangeable calcium-magnesium on yield and composition of soybeans and corn. Soil Sci. 93:265–270.

Kinzel, H. 1989. Calcium in the vacuoles and cell walls of plant tissue. Forms of deposition and their physiological and ecological significance. Flora 182:99–125.

Kirchmann, H., F. Pichlmayer, and M.H. Gerzabek. 1996. Sulfur balances and sulfur-34 abundance in a long-term fertilizer experiment. Soil Sci. Soc. Am. J. 59:174–178.

Kline, J.S., J.T. Sims, and K.L. Schilke-Gartley. 1989. Response of irrigated corn to sulfur fertilization in the Atlantic Coastal Plain. Soil Sci. Soc. Am. J. 53:1101–1108.

Koehler, F.E. 1965. Fertilizer interactions in wheat producing areas of eastern Washington. p.19–21. In Proc. 10th Northwest Ann. Fert. Conf. Tacoma, WA.

Konno, H., T. Yamaya, Y. Yamasaki, and H. Matsumoto. 1984. Pectic polysaccharide break-down of cell walls in cucumber roots grown with calcium starvation. Plant Physiol. 76:633–637.

Koo, R.C.J. 1971. A comparison of magnesium sources for citrus. Soil Crop Sci. Soc. Fla. Proc. 31:137–140.

Kratzke, M.G., and J.P. Palta. 1986. Calcium accumulation in potato tubers: role of the basal roots. Hort. Sci. 21:1022–1024.

Kubota, J., G.H. Oberly, and E.A. Naphan. 1980. Magnesium in grasses of three selected regions in the United States and its relation to grass tetany. Agron. J. 72:907–914.

Langmeier, M., S. Ginsburg, and P. Matile. 1993. Chlorophyll breakdown in senescent leaves: Demonstration of Mg-chelatase activity. Physiol. Plant. 89:347–353.

Lerchl, D., S. Hillmer, R. Grotha, and D.G. Robinson. 1989. Ultrastructural observations on CTC-induced callose formation in Riella helicophylla. Bot. Acta 102:62–72.

Lin, D.C., and P.S. Nobel. 1971. Control of photosynthesis by Mg^{2+}. Arch. Biochem. Biophys. 145:622–632.

Lund, Z.F. 1970. The effect of calcium and its relation to several cations in soybean root growth. Soil Sci. Soc. Amer. Proc. 34:456–459.

Lyle, E.S., Jr., and F. Adams. 1971. Effect of available soil calcium on taproot elongation of loblolly pine (Pinus taeda L.) seedlings. Soil Sci. Soc. Am. Proc. 35:800–805.

Marcano-Martinez, E., and M.B. McBride. 1989. Calcium and sulfate retention by two Oxisols of Brazilian Cerrado. Soil Sci. Soc. Am. J. 53:63–69.

Marschner, H. 1995. Mineral nutrition of higher plants, 2nd Ed. Academic Press, Inc., San Diego, CA.

Marsh, K.B., R.W. Tillman, and J.K. Syers. 1987. Charge relationships of sulfate sorption by soils. Soil Sci. Soc. Am. J. 51:318–323.

Martin, J.P., and A.L. Page. 1969. Influence of exchangeable Ca and Mg and of percentage base saturation on growth of citrus plants. Soil Sci. 107:39–46.

McCalla, T.M. 1937. Behavior of legume bacteria (Rhizobium) in relation to exchangeable calcium and hydrogen ion concentration of the colloidal fraction of the soil. MO Agr. Exp. Sta. Res. Bul. 256.

McLean, E.O., and J.R. Brown. 1984. Crop response to lime in the midwestern United States. p. 267–303. In F. Adams (ed.). Soil acidity and liming. American Society of Agronomy, Madison, WI.

McMurtrey, J.E. Jr. 1931. Relation of calcium and magnesium to the growth and quality of tobacco. J. Am. Soc. Agron. 23:1051–1052.

Mehlich, A. and W.E. Colwell. 1945. Absorption of calcium by peanuts from kaolin and bentonite at varying levels of calcium. Soil Sci. 60: 369–374.

Mehlich, A. 1946. Soil properties affecting the proportionate amounts of calcium, magnesium, and potassium in plants and in HCl extracts. Soil Sci. 62:393–409.

Mehring, A.L., and F.O. Lundstrom. 1938. The calcium, magnesium, sulphur and chlorine contents of fertilizers. Am. Fert. 88:5–10.

Mengel, K., and E. Kirkby. 1982. Principles of plant nutrition, 3rd Ed. International Potash Institute, Bern, Switzerland.

Metson, A.J. 1974. Magnesium in New Zealand soils. I. Some factors governing the availability of soil magnesium: A review. NZ J. Exp. Agric. 2:277–319.

Millar, C.E. 1955. Soil fertility. John Wiley and Sons, New York, NY.

Mitsui, T., J.T. Christeller, I. Hara-Nishimura, and T. Akazawa. 1984. Possible roles of calcium and calmodulin in the biosynthesis and secretion of α-amylase in rice seed scutellar epithelium. Plant Physiol. 75:21–25.

Miyasaka, S.C., and D.L. Grunes. 1990. Root temperature and calcium level effects on winter wheat forage. II. Nutrient composition and tetany potential. Agron. J. 82:242–249.

Mokwunye, A.U., and S.W. Melsted. 1972. Magnesium forms in selected temperate and tropical soils. Soil Sci. Soc. Am. Proc. 36:762–764.

Morard, P., A. Pujos, A. Bernadac, and G. Bertoni. 1996. Effect of temporary calcium deficiency on tomato growth and mineral nutrition. J. Plant Nutr. 19:115–127.

Moss, H.J., C.W. Wrigley, R. MacRitchie, and P.J. Randall. 1981. Sulfur and nitrogen fertilizer effects on wheat. II. Influence on grain quality. Aust. J. Agric. Res. 32: 213–226.

Neptune, A.M.L., M.A. Tabatabai, and J.J. Hanway. 1975. Sulfur fractions and carbon-nitrogen-phosphorus-sulfur relationships in some Brazilian and Iowa soils. Soil Sci. Soc. Am. Proc. 39:51–55.

Nichols, J. T., P.E. Reece, G.W. Hergert, and L.E. Moser. 1990. Yield and quality response of subirrigated meadow vegetation to nitrogen phosphorus and sulfur fertilizer. Agron. J. 82:47–52.

Noggle, J.C., J.F. Meagher, and U.S. Jones. 1986. Sulfur in the atmosphere and its effects on plant growth. p. 251–278. *In* M.A. Tabatabai (ed.) Sulfur in agriculture. American Society of Agronomy, Madison WI.

Oates, K.M., and E.J. Kamprath. 1985. Sulfur fertilization of winter wheat grown on deep sandy soils. Soil Sci. Soc. Am. J. 49:925–927.

Ohno, T., and D.L. Grunes. 1985. Potassium-magnesium interactions affecting nutrient uptake by wheat forage. Soil Sci. Soc. Am. J. 49:685–690.

O'Leary, M.J., and G.W. Rehm. 1991. Evaluation of some soil and plant analysis procedures as predictors of the need for sulfur for corn production. Commun. Soil Sci. Plant Anal. 22:87–98.

O'Leary, M.J., and G.W. Rehm. 1990. Nitrogen and sulfur effects on the yield and quality of corn grown for grain and silage. J. Prod. Agric. 3:135–140.

Olson, R.A., and G.W. Rehm. 1986. Sulfur in precipitation and irrigation waters and its effects on soils and plants. p. 279–294. *In* M.A. Tabatabai (ed.) Sulfur in agriculture. American Society of Agronomy, Madison, WI.

Parker, F.W., and E. Truog. 1920. The relation between the calcium and the nitrogen content of plants and the function of calcium. Soil Sci. 10:49–56.

Pavan, M.A., F.T. Bingham, and F.J. Peryea. 1987. Influence of calcium and magnesium salts on acid soil chemistry and calcium nutrition of apple. Soil Sci. Soc. Am. J. 51:1526–1530.

Pavan, M.A., F.T. Bingham, and P.F. Pratt. 1984. Redistribution of exchangeable calcium, magnesium, and aluminum following lime or gypsum applications to a Brazilian Oxisol. Soil Sci. Soc. Am. J. 48:33–38.

Perring, M.A. 1986. Incidence of bitter pit in relation to the calcium content of apples: Problems, and paradoxes, a review. J. Sci. Food Agric. 37:591–606.

Perrott, K.W., and S.U. Sarathchandra. 1987. Nutrient and organic matter levels in a range of New Zealand soils under established pasture. NZ J. Agric. Res. 30:249–259.

Pierce, J. 1986. Determinants of substrate specificity and the role of metal in the reaction of ribolosebisphosphate carboxylase/oxygenase. Plant Physiol. 81:943–945.

Poovaih, B.W., and A.S.N. Reddy. 1996. Calcium and geotropism. p. 307–321. *In* Y. Waisel et al. (ed.) Plant roots: The hidden half. Academic Press, New York, NY.

Pratt, P.F., and R.B. Harding. 1957. Decreases in exchangeable magnesium in an irrigated soil during 28 years of differential fertilization. Agron. J. 49:419–421.

Pratt, P.F., W.W. Jones, and F.T. Bingham. 1957. Magnesium and potassium content of orange leaves in relation to exchangeable magnesium and potassium in the soil at various depths. Proc. Am. Soc. Hort. Sci. 70:245–251.

Prince, A.L. 1951. Magnesium economy in the Coastal Plain soils of New Jersey. Soil Sci. 71:91–98.

Prince, A.L., and S.J. Toth. 1937. Effects of long-continued use of dolomitic limestone on certain chemical and colloidal properties of a Sassafras loam soil. Soil Sci. Soc. Am. Proc. 2:207–214.

Prince, A.L., M. Zimmerman, and F.E. Bear. 1947. The magnesium-supplying powers of 20 New Jersey soils. Soil Sci. 63:69–78.

Raese, J.T., and S.R. Drake. 1993. Effects of preharvest calcium sprays on apple and pear quality. J. Plant Nutr. 16:1807–1819.

Raghothama, K.G., A.S.N. Reddy, M. Friedmann, and B.W. Poovaiah. 1987. Calcium-regulated *in vivo* protein phosphorylation in *Zea mays* L. root tips. Plant Physiol. 83:1008–1013.

Rajan, S.S.S. 1978. Sulfate adsorbed on hydrous alumina, ligands displaced, and changes in surface charge. Soil Sci. Soc. Am. J. 42:39–44.

Ramig, R.E., P.E. Rasmussen, R.R. Allmaras, and C.M. Smith. 1975. Nitrogen-sulfur relations in soft white winter wheat. I. Yield response to fertilizer and residual sulfur. Agron. J. 67:219–224.

Rasmussen, P.E., and P.O. Kresge. 1986. Plant response to sulfur in the Western United States. p. 357–374. *In* M.A. Tabatabai (ed.) Sulfur in agriculture. American Society of Agronomy, Madison, WI.

Reinbott, T.M, and D.G. Blevins. 1994. Phosphorus and temperature effects on magnesium, calcium, and potassium in wheat and tall fescue leaves. Agron. J. 86:523–529.

Reisenauer, H.M. 1969. A technique for growing plants at controlled levels of all nutrients. Soil Sci. Soc. Am. Proc. 27: 553–555.

Rendig, V.V. 1986. Sulfur and crop quality. p. 635–652. In M.A. Tabatabai (ed.) Sulfur in agriculture. American Society of Agronomy, Madison, WI.

Rhue, R.D., and E.J. Kamprath. 1973. Leaching losses of sulfur during winter months when applied as gypsum, elemental S or prilled S. Agron. J. 65:603–605.

Rice, H.B., and E.J. Kamprath. 1968. Availability of exchangeable and nonexchangeable Mg in sandy Coastal Plain soils. Soil Sci. Soc. Am. Proc. 32:386–388.

Ritchey, K. D., D. M. G. Souza, E. Lobato, and O. Correa. 1980. Calcium leaching to increase rooting depth in a Brazilian Savannah Oxisol. Agron. J. 78: 40–44.

Ritchey, K.D., J.E. Silva, and U.F. Costa. 1982. Calcium deficiency in clayey B horizons of Savanna Oxisols. Soil Sci. 133:378–382.

Ryden, J.C., and J.K. Syers. 1976. Calcium retention in response to phosphate sorption by soils. Soil Sci. Soc. Am. J. 40:845–846.

Salmon, R.C. 1964. Cation-activity ratios in equilibrium soil solutions and the availability of magnesium. Soil Sci. 98:213–221.

Sartain, J.B., and E.J. Kamprath. 1978. Aluminum tolerance of soybean cultivars based on root elongation in solution culture compared with growth in acid soil. Agron. J. 70:17–20.

Schmidt, J.P., and F.R. Cox. 1992. Evaluation of the magnesium soil test interpretation for peanuts. Peanut Sci. 19:126–131.

Schnug, E. 1990. Sulphur nutrition and quality of vegetables. Sulphur Agric. 14:3–7.

Schnug, E. 1991. Sulphur nutritional status of European crops and consequences for agriculture. Sulphur Agric. 15:7–12.

Schnug, E. 1993. Physiological functions and environmental relevance of sulfur-containing secondary metabolites. p. 179–190. In L.J. deKok et al. (ed.) Sulfur nutrition and assimilation in higher plants. SPB Academic Publishing, The Hague, Netherlands.

Schwab, A.P., M.D. Ransom, and C.E. Owensby. 1989. Exchange properties of an Argiustoll: Effects of long-term ammonium nitrate fertilization. Soil Sci. Soc. Am. J. 53:1412–1417.

Scott, L.E., and D.H. Scott. 1951. Response of grapevines to soil and spray applications of magnesium sulfate. Proc. Amer. Soc. Hort. Sci. 57:53–58.

Scott, N.M., W. Bick, and H.A. Anderson. 1981. The measurement of sulphur-containing amino acids in some Scottish soils. J. Sci. Food Agric. 32:21–24.

Silkworth, D.R., and D.F. Grigal. 1982. Determining and evaluating nutrient losses following whole-tree harvesting of aspen. Soil Sci. Soc. Am. J. 46:626–631.

Simson, C.R., R.B. Corey, and M.E. Sumner. 1979. Effect of varying Ca:Mg ratios on yield and composition of corn (Zea mays) and alfalfa (Medicago sativa). Commun. Soil Sci. Plant Anal. 10:153–162.

Slocum, R. D., and S. J. Roux. 1983. Cellular and subcellular localization of calcium in gravistimulated oat coleoptiles and its possible significance in the establishment of tropic curvature. Planta 157:481–492.

Smith, W.D., G.F. Peedin, W.K. Collins, M.R. Tucker, G.S. Miner, and E.J. Kamprath. 1987. Tobacco response to sulfur on soils differing in depth to the argillic horizon. Tob. Sci. 31:36–39.

Spencer, K. 1975. Sulphur requirements of plants. p. 98–116. In K.D. McLachlan (ed.) Sulphur in Australasian agriculture. Sydney University Press, Sydney, Australia.

Starkey, R.L. 1966. Oxidation and reduction of sulfur compounds in soils. Soil Sci. 101:297–306.

Strebel, O., and W.H.M. Duynisveld. 1989. Nitrogen supply to cereals and sugar beet by mass flow and diffusion on a silty loam soil. Z. Pflanzenernahr. Bodenk. 152:135–141.

Sumner, M.E. 1995. Amelioration of subsoil acidity with minimum disturbance. p. 147–186. In N.S. Jayawardane and B.A. Stewart (ed.) Subsoil management techniques. Lewis Publishers, Boca Raton, FL.

Sumner, M.E., and E. Carter. 1988. Amelioration of subsoil acidity. Commun. Soil Sci. Plant Anal. 19:1309–1318.

Sumner, M.E., P.M.W. Farina, and V.J. Hurst. 1978. Magnesium fixation: A possible cause of negative yield responses to lime applications. Commun. Soil Sci. Plant Anal. 9:995–1007.

Sumner, M.E., H. Shahandeh, J. Bouton, and J. Hammel. 1986. Amelioration of an acid soil profile through deep liming and surface application of gypsum. Soil Sci. Soc. Am. J. 50:1254–1258.

Sumner, M.E., C.S. Kvien, H. Smal, and A.S. Csinos. 1988. On the Ca nutrition of peanuts (*Arachis hypogaea* L.) I. Conceptual model. J. Fert. Issues 5:97–102.

Tabatabai, M.A. 1986. Sulfure in agriculture. American Society of Agronomy, Madison, WI.

Tan, Z., R.G. McLaren, and K.C. Cameron. 1994a. Forms of sulfur extracted from soils after different methods of sample preparation. Aust. J. Soil Res. 32:823–834.

Tan, Z., R.G. McLaren, and K.C. Cameron. 1994b. Seasonal variations in forms of extractable sulfur in some New Zealand soils. Aust. J. Soil Res. 32:985–993.

Thill, J.L., and J.R. George. 1975. Cation concentration and K to Ca + Mg ratio of nine cool-season grasses and implications with hypomagnesaemia. Agron. J. 67:89–91.

Tisdale, S.L., R.B. Reneau, Jr., and J.S. Platou. 1986. Atlas of sulfur deficiencies. p. 295–322. *In* M.A. Tabatabai (ed.) Sulfur in agriculture. American Society of Agronomy, Madison, WI.

Tiwari, K.N. 1990. Sulphur research and agricultural production in Uttar Pradesh, India. Sulphur Agric. 14:29–34.

Truog, E. 1918. Soil acidity: I. Its relation to the growth of plants. Soil Sci. 5:169–195.

Tzeng, K.C., A. Kelman, K.E. Simmons, and K.A. Kelling. 1986. Relationship of calcium nutrition to internal brown spot of potato tubers and sub-apical necrosis of sprouts. Am. Pot. J. 63:87–97.

Walker, C.J., and J.D. Weinstein. 1991. Further characterization of the magnesium chelatase in isolated developing cucumber chloroplasts. Plant Physiol. 95:1189–1196.

Walker, M.E., R.A. Flowers, R.J. Henning, T.C. Keisling, and B.G. Mullinix. 1979. Response of early bunch peanuts to calcium and potassium fertilization. Peanut Sci. 6:119–123.

Walsh, T., and T.F. O'Donohoe. 1945. Magnesium deficiency in some crop plants in relation to the level of potassium nutrition. J. Agric. Sci. Camb. 35:254–263.

Watkinson, J.H., and M.J. Kear. 1996a. Sulfate and mineralisable organic sulfur in pastoral soils of New Zealand. II. A soil test for mineralisable organic sulfur. Aust. J. Soil Res. 34:405–412.

Watkinson, J.H., and M.J. Kear. 1996b. Sulfate and mineralisable organic sulfur in pastoral soils of New Zealand. I. A quasi equilibrium between sulfate and mineralisable organic sulfur. Aust. J. Soil Res. 34:385–403.

Welte, E., and W. Werner. 1963. Potassium-magnesium antagonism in soils an crops. J. Sci. Food Agric. 44: 180–186.

Wild, A., and J. Keay. 1964. Cation exchange equilibria with vermiculite. J. Soil Sci. 15:135–144.

Williams, L., and J.L. Hall. 1987. ATPase and proton pumping activities in cotyledons and other phloem-containing tissues of *Ricinus communis*. J. Exp. Bot. 38:185–202.

Wrigley, C.W., D.L. duCros, J.G. Fullington, and D.D. Kasarda. 1984. Changes in polypeptide composition and grain quality due to sulfur deficiency in wheat. J. Cereal Sci. 2:15–24.

2

Bioavailability of Micronutrients

John J. Mortvedt
Colorado State University

2.1 Introduction

Micronutrients (B, Cu, Fe, Mn, Mo, Zn, Cl) are elements which are essential for plant growth, but are required in much smaller amounts than the major nutrients (N, P, K, Ca, Mg, S). While Cl is a micronutrient, deficiencies rarely occur in nature so most discussions on supplying micronutrient fertilizers are confined to the other six micronutrients. Micronutrients have also been called minor elements, indicating that their content in plant tissues is minor in relation to the macronutrients. Another term used for micronutrients has been trace elements, since only traces of these elements are found in plant tissues.

With the exception of Fe and Mn, which are among the 12 most abundant elements, the other micronutrients occur at concentrations of less than 0.1% in the lithosphere, another reason for their being named minor or trace elements (Harmsen and Vlek, 1985). Several other elements which are related to plant nutrition (Al, Co, Na, Ni and Si) are known as beneficial elements. They are defined as those elements which have positive effects on plant growth, but do not fit the current criterion of essentiality (Asher, 1991).

Deficiencies of micronutrients have been increasing in some crops. Some reasons are higher crop yields which increase plant nutrient demands, use of high analysis NPK fertilizers containing lower quantities of micronutrient contaminants, and decreased use of farmyard manure on many agricultural soils. Micronutrient deficiencies have been verified in many soils through increased use of soil testing and plant analysis. Methods of evaluating the micronutrient status of soils through soil testing and plant analysis are discussed in Section D, Chapter 4.

2.2 Iron, Manganese, Copper, and Zinc

2.2.1 Forms in Soil

The micronutrient elements are a diverse group of elements geochemically (Chesworth, 1991). While the four metallic micronutrient cations occur mainly in the divalent form in soils, differences in the ionic character of their chemical bonding are great enough so that only Fe^{2+} and Mn^{2+} can substitute extensively for each other. Chemical forms of micronutrients have been defined by pools determined by various chemical extractions. Pickering (1981) termed these pools as ion exchangeable, absorbed, organically bound, hydrous oxide segment and lattice component micronutrients. These pools and extraction techniques for each micronutrient were discussed in detail by Shuman (1991). The most common minerals containing the micronutrient elements are shown in Table 2.1.

Iron is the most abundant element in soils, ranging from 7,000 to 500,000 mg kg^{-1}, with a mean concentration of 38,000 mg kg^{-1} (3.8%) in soils (Lindsay, 1979). Iron occurs mainly in the Fe^{2+} and Fe^{3+} forms as iron oxides, silicates, sulfates and carbonates in the Earth's crust. The most abundant form is Fe_2O_3, which is very stable and insoluble in water. Ferrous sulfides occur under permanent reducing conditions, such as in bogs and other wetlands.

Manganese is similar to Fe in both its chemistry and geology. Concentrations of Mn in the Earth's crust, igneous and sedimentary rocks, and soil are second only to those of Fe. Concentrations of total Mn in soils range from 20 to 6,000 mg kg^{-1}, with a mean concentration of 600 mg kg^{-1} in soils. Neither total nor exchangeable Mn in soil is correlated with the composition of the bedrock, which indicates the mobility of Mn in the Earth's crust. Manganese has three common valences (Mn^{2+}, Mn^{3+}, and Mn^{4+}) of which the divalent form is the most common, especially with reference to plant nutrition. The trivalent ion is unstable in solution, while the divalent ion forms stable compounds in reducing environments. The most stable compound is MnO_2 in strongly oxidizing environments.

Table 2.1   Minerals containing micronutrient elements [After Krauskopf, 1972]

Boron

Borates (hydrous):	borax $Na_2B_4O_7 \cdot 10H_2O$, kernite $Na_2B_4O_7 \cdot 4H_2O$, colemanite $Ca_2B_6O_{11} \cdot 5H_2O$, ulexite $NaCaB_5O_9 \cdot 4H_2O$
Borates (anhydrous):	ludwigite Mg_2FeBO_5, kotoite $Mg_3(BO_3)_2$
Complex borosilicates:	tourmaline, axinite

Chloride

Sylvite:	KCl
Kainite:	$KCl. MgSO_4 \cdot 3H_2O$
Langbeinite:	$K_2SO_4 \cdot 2MgSO_4$

Copper

Carbonates:	malachite $Cu_2(OH)_2CO_3$ azurite $Cu_3(OH)_2(CO_3)_2$
Oxides:	cuprite Cu_2O, tenorite CuO
Simple sulfides:	chalcocite Cu_2S, covellite CuS
Complex sulfides:	chalcopyrite $CuFeS_2$, bornite Cu_5FeS_4

Iron

Carbonates:	siderite $FeCO_3$
Oxides:	hematite Fe_2O_3, goethite $FeOOH$, magnetite Fe_3O_4
Sulfides:	pyrite FeS_2, pyrrhotite $Fe_{1-x}S$
Sulfates:	jarosite $KFe_3(OH)_6(SO_4)_4$

Manganese

Carbonates:	rhodochrosite $MnCO_3$
Simple oxides:	pyrolusite MnO_2, hausmannite Mn_3O_4, manganite $MnOOH$
Complex oxides:	braunite $(Mn, Si)_2O_3$, psilomelane $BaMg_9O_{18} \cdot 2H_2O$
Silicates:	rhodanate $MnSiO_3$

Molybdenum

Oxides:	ilsemanite $Mo_3O_8 \cdot 8H_2O$,
Molybdates:	wulfenite $PbMoO_4$, powellite $CaMoO_4$, ferrimolybdite $Fe_2(MoO_4)_3 \cdot 8H_2O$
Sulfides:	molybdenite MoS_2

Zinc

Carbonates:	smithsonite $ZnCO_3$
Silicates:	hemimorphite $Zn_4(OH)_2 Si_2O_7 \cdot H_2O$
Sulfides:	sphalerite ZnS

Copper occurs in the Earth's crust mainly as sulfide minerals. Less stable forms include oxides, silicates, sulfates and carbonates. Copper occurs in the cuprous (Cu^+) and cupric (Cu^{2+}) forms, but can also occur in metallic form in some ores. The concentrations of total Cu in soils range from 2 to 100 mg kg^{-1}, with a mean concentration of 30 mg kg^{-1}. Copper is present mainly in the adsorbed form and as organic complexes in soils in the Cu^{2+} form, which is also of primary interest in plant nutrition.

The principal Zn ore is sphalerite (ZnS) but Zn also occurs as silicates and carbonates in the Earth's crust. Total Zn concentrations in soils range from 10 to 300 mg kg^{-1}, with a mean concentration of 50 mg kg^{-1}. Zinc occurs as the divalent cation (Zn^{2+}) in soils and does not undergo reduction in nature, due to its electropositive nature (Krauskopf, 1972). In comparison with Cu, Zn migrates further in soil and is known as the most mobile of the heavy metals. Divalent Zn is sorbed less strongly than Cu, probably because the covalent bonds holding Cu^{2+} are stronger than those of Zn^{2+}.

2.2.2 Factors Affecting Bioavailability

2.2.2.1 Soil pH

Bioavailability of all four of the metallic micronutrients is significantly affected by soil pH, decreasing with increasing pH. Solubility of Fe decreases a thousandfold for each unit increase in soil pH in the range 4 to 9 (Lindsay, 1979), and consequently, most Fe deficiencies occur on calcareous soils. This decrease in solubility is much greater for Fe than for Mn, Cu or Zn. The activity (and consequent bioavailability) of Mn, Cu and Zn decreases 100-fold for each unit increase in soil pH.

Complex ions of Mn, such as $MnHCO_3^-$ and $MnOH^-$, in addition to Mn^{2+}, exist in soil solution and their relative concentrations are affected by pH (Lindsay, 1979). Ion species of Cu in soil solution are Cu^{2+} at pH values below 7.3, while $CuOH^-$ is most common above that pH. The most common form of Zn in solution is Zn^{2+} below pH 7.7, while $Zn(OH)_2$ is most common above that level. Both Ca and Mg can replace Zn from solution complexes and from adsorption sites on soil solids (Barak and Helmke, 1993). Amounts of exchangeable metals in soils are related to their concentrations in soil solution, so soil pH affects exchangeable Fe, Mn, Cu and Zn similarly.

2.2.2.2 Soil Organic Matter

Reactions with soil organic matter (SOM) significantly affect bioavailability of these metallic micronutrients (Stevenson, 1991). Copper reacts with SOM to form very stable complexes, especially with carboxyl and phenolic groups. Some of these complexes are so stable that most Cu deficiencies have been associated with organic soils. Reactions of Zn with SOM are also important in providing bioavailable Zn, but the strength of these bonds is not so great as with Cu.

Manganese also forms stable complexes with organic ligands, and Mn deficiencies also occur in organic soils. The stability of these complexes is such that the incidence of Mn deficiency above pH 6.5 is much lower in soils with appreciable levels of SOM than in low organic matter soils. While Fe also complexes with SOM, its bioavailability is more affected by soil pH than by SOM content.

2.2.2.3 Redox Reactions

Redox reactions are very common in soils, and these relationships affect micronutrient availability. A comprehensive discussion of redox chemistry is presented in Section B, Chapter 5. Redox potentials have been expressed in terms of pe (– log of electron activity) (Lindsay, 1979). The redox potential of soil is dependent upon soil pH, soil aeration and soil microbial activity. Iron is significantly affected by redox potential. The reduced form (Fe^{2+}) is found under waterlogged conditions in soil.

Manganese can occur in three oxidation states, so it is significantly affected by soil redox status. Trivalent Mn is unstable in aqueous solutions because it reduces to Mn^{2+} or it disproportionates to

Mn^{2+} and MnO_2 (Harmsen and Vlek, 1985). The reduction of Mn^{+4} to Mn^{2+} is greater at lower soil pH levels, and very acid soil conditions can lead to Mn toxicities in some sensitive plant species. Although Cu^{2+} can be reduced to Cu^+ ions, neither Cu nor Zn are affected by oxidation-reduction reactions which occur under most soil conditions.

2.2.3   Sources

Micronutrient sources vary considerably in their physical state, chemical reactivity, cost per unit micronutrient, and availability to plants. The four main classes of micronutrient sources are inorganic products, synthetic chelates, organic complexes, and fritted glass products (Mortvedt, 1991a).

The most common sources of Cu, Fe, Mn and Zn are shown in Table 2.2. Most of the inorganic sources are sulfates and oxides, with lesser amounts of chlorides and carbonates being sold. The main

Table 2.2   Sources commonly used as micronutrient fertilizers [After Mortvedt, 1991a]

Source	Solubility in H_2O	Element content (%)
Boron		
H_3BO_3	Soluble	17
$Na_2B_4O_7 \cdot 5H_2O$	Soluble	20
$Na_2B_4O_7 \cdot 10H_2O$	Soluble	11
$Ca_2B_6O_{11} \cdot 5H_2O$	Slightly soluble	10
Chloride		
$CaCl_2$	Soluble	50
KCl	Soluble	48
Copper		
$CuSO_4 \cdot 5H_2O$	Soluble	25
CuO	Insoluble	50-75
Iron		
$FeSO_4 \cdot 7H_2O$	Soluble	20
FeHEDTA	Soluble	5-9
FeEDDHA	Soluble	6
Manganese		
$MnSO_4 \cdot 4H_2O$	Soluble	24
MnO	Insoluble	41-68
Mn oxysulfate	Variable	30-50
Molybdenum		
$Na_2MoO_4 \cdot 2H_2O$	Soluble	39
$(NH4)_2MoO_4$	Soluble	49
MoO_3	Soluble	66
Zinc		
$ZnSO_4 \cdot H_2O$	Soluble	36
$ZnSO_4$-NH_3 complex	Soluble	10-15
ZnO	Insoluble	60-78
Zn oxysulfate	Variable	18-50
ZnEDTA	Soluble	6-14

synthetic chelate is EDTA (ethylenediamine tetraacetate), although other chelates are also used as micronutrient fertilizers.

2.2.3.1 Inorganic Sources

Inorganic sources include oxides and carbonates, and metallic salts such as sulfates, chlorides, and nitrates. The sulfates are the most common of the metallic salts and are sold in crystalline or granular form. An ammoniated $ZnSO_4$ solution also is used in polyphosphate starter fertilizers. Oxides of Mn and Zn are commonly sold in granular form and as fine powders, but because they are insoluble, their immediate effectiveness for crops is rather low in granular form. Also, the available divalent form of Mn in MnO will oxidize to the unavailable tetravalent form, so there is very little residual availability for succeeding crops. Thus, agronomic effectiveness of granular MnO may be rather low. Since Mn in MnO_2 is already in an unavailable form, it should not be used as a Mn fertilizer.

Oxysulfates (usually industrial by-products) are oxides which have been partially acidulated with sulfuric acid, and generally are sold in granular form. The percentage of water soluble Mn or Zn in oxysulfates is directly related to the degree of acidulation by sulfuric acid. Research results have shown that about 35 to 50% of the total Zn in granular Zn oxysulfate or other Zn fertilizers should be in water soluble form to be immediately effective for crops (Mortvedt, 1992). Similar results would be expected for Mn oxysulfate fertilizers. Inorganic sources usually are the least costly per unit of micronutrient, but they may not always be the most effective for crops.

2.2.3.2 Synthetic Chelates

These sources are formed by combining a chelating agent with a metal through coordinate bonding. Stability of the metal chelate bond affects the bioavailability of the micronutrient metals (Cu, Fe, Mn, Zn). An effective chelate is one in which the rate of substitution of the chelated micronutrient for other cations in the soil is quite low, thus maintaining the applied micronutrient in chelated form (Mortvedt, 1991b). Relative effectiveness for crops per unit of micronutrient as soil-applied chelates may be from two to five times greater than that of inorganic sources, but chelate costs per unit of micronutrient may be five to 100 times higher. Several types of chelates are sold, so relative effectiveness depends on the types of chelates and inorganic products being compared.

2.2.3.3 Organic Complexes

These complexes are made by reacting metallic salts with some organic byproducts (lignosulfonates, polyflavonoids, phenols) of the wood pulp or related industries. The chemical bonding of the metals to the organic components is not well understood. Some of the bonds may be coordinate as in the chelates, but other types may be present. While organic complexes are less costly per unit of micronutrient, they usually are also less effective and more readily decomposed by microorganisms in soil than synthetic chelates. These sources are more suitable for foliar sprays and mixing with fluid fertilizers.

2.2.3.4 Frits

Fritted glassy products (frits) are not as commonly used as the other types of micronutrient fertilizers. Micronutrient solubility in frits is controlled by particle size and changes in matrix composition (Mortvedt, 1991a). Micronutrient concentrations vary from 2 to 25%, and more than one micronutrient may be included in a fritted product. Fritted micronutrients generally are used only on sandy soils in regions of high rainfall where leaching occurs. This material class is more appropriate for maintenance programs than for correcting micronutrient deficiencies.

2.2.4   Rates of Application

The most common method for Cu, Mn and Zn application for crops is soil application. Recommended application rates usually are less than 10 kg ha^{-1} (on an elemental basis), so uniform application of micronutrient sources separately in the field is difficult. Therefore, both granular and fluid NPK fertilizers are commonly used as carriers of micronutrients. Including micronutrients with mixed fertilizers is a convenient and cost effective method of application and allows more uniform distribution with conventional equipment.

Foliar sprays are widely used to apply micronutrients, especially Fe and Mn, for many crops. Because soluble inorganic salts generally are as effective as synthetic chelates in foliar sprays, they are usually chosen because of lower costs. Suspected micronutrient deficiencies may be diagnosed with foliar spray trials with one or more micronutrients. Correction of deficiency symptoms usually occurs within the first several days after spraying, allowing the entire field to be sprayed with the appropriate micronutrient source thereafter. Inclusion of sticker spreader agents in the spray is suggested to improve micronutrient adherence to the foliage. Caution should be used because of leaf burn due to high salt concentrations or inclusion of certain compounds in foliar sprays.

Advantages of foliar sprays are (1) rates are much lower than for soil application; (2) a uniform application is easily obtained; and (3) response to the applied nutrient is almost immediate so deficiencies can be corrected during the growing season. Low residue foliar sprays of Mn and Zn have been used to correct deficiencies of citrus and other fruit crops, but sprays which will discolor the fruit should be avoided. Disadvantages of foliar sprays are (1) leaf burn may result if salt concentrations in the spray are too high; (2) nutrient demand often is high when plants are small and their leaf surface is insufficient for foliar absorption; (3) maximum yields may not be possible if spraying is delayed until deficiency symptoms appear; and (4) there is little residual effect from foliar sprays thus requiring multiple sprayings for season-long correction. Application costs will be higher if more than one spray is needed, unless they can be combined with pesticide spray applications. Principles of placement and timing of fertilizer application are discussed in Section D, Chapter 5.

2.2.5   Crop Responses

Responses to micronutrient fertilizers have been reported on many crops around the world. As a result, the relative needs for each micronutrient on each crop are well-known. Therefore, diagnosis of possible micronutrient deficiencies is simplified with this knowledge (Martens and Westermann, 1991). For example, young maize plants with yellow leaf midribs growing on a high pH soil may be Zn deficient, but yellow wheat plants growing on soil at the same pH may be more likely to be N or S deficient. Crop responses to Cu, Fe, Mn and Zn application generally occur on high pH soils on those crops having higher requirements of the particular micronutrient.

2.2.5.1   Copper

Most Cu deficiencies are found on organic and sandy soils. Recommended Cu application rates range from 1 to 10 kg ha^{-1}. Residual effects of Cu are very common, with responses being noted up to 12 years after application, so annual applications usually are not necessary (Gartrell, 1981). Soil tests should be used to monitor the Cu buildup in soils where Cu is applied. Some Cu-sensitive crops are cereals, corn, clover, and some fruits and vegetables.

Soil applications of Cu sources are more commonly used than foliar sprays, because the applied Cu remains bioavailable for some time after application. Inorganic sources are effective, and soluble $CuSO_4$ is suggested instead of CuO where Cu deficiencies have been reported, because immediate bioavailability is important (Karamanos et al., 1986). Cereals respond significantly to $CuSO_4$

Table 2.3 Effects of copper fertilizer application on yield and kernel plumpness of wheat and barley in Canada [Adapted from Solberg et al., 1993]

Crop	Copper fertilization	Yield kg ha^{-1}	Kernel plumpness %
Barley	No	3333	48
	Yes	4946	82
Wheat	No	1010	59
	Yes	2420	72

applications on Cu-deficient soils, especially those with relatively high SOM contents (Table 2.3). Broadcast applications of $CuSO_4$ at rates up to 10 kg ha^{-1} were ineffective for dryland wheat in Australia (Grundon, 1980). Low Cu uptake was attributed to low soil moisture conditions in the zone of Cu application. Since $CuSO_4$ is hygroscopic, it may not blend well with granular NPK fertilizers because it could react with P to form insoluble compounds with low Cu bioavailability (Gilkes, 1977). Foliar applications of Cu are effective for cereals at the late tillering stage, but research results show that crop response to soil-applied Cu is more consistent than to foliar applications (Solberg et al., 1993).

Knowledge of the residual bioavailability of Cu is needed to determine when reapplication to a particular field is necessary. This is usually done through soil analysis. Residual effectiveness of Cu also is affected by Cu source and rate. Results in Table 2.4 show that spring CuO (water insoluble) applications at 5 and 10 kg Cu ha^{-1} were not effective in the year of application, but corrected Cu deficiencies in crops grown in the following year. In contrast, the organic complex, Cu lignosulfonate, applied at 0.5 and 1.0 kg ha^{-1} was effective for the immediate crop but not for that grown a year later.

2.2.5.2 Iron

Although the most abundant element in soils, levels of available Fe often are limiting in calcareous soils, and to a much lesser extent in sandy soils in high rainfall areas. Plant species vary considerably in their susceptibility to Fe chlorosis. Growing Fe-tolerant species, or varieties within a species, is one of the main methods of controlling Fe chlorosis on low Fe soils (Chen and Barak, 1982).

Soil applications of soluble inorganic Fe sources usually are very ineffective in controlling Fe chlorosis. Banded applications are more effective than broadcast applications of $FeSO_4$, because soil

Table 2.4 Differential crop response to Cu sourcse with time after application [Adapted from Karamanos et al., 1986]

Fertilizer source	Cu applied kg ha^{-1}	Yield (kg ha^{-1})			
		Spring wheat		Canola	
		1983	1984	1983	1984
Control	0	2185	2142	1794	779
CuO	5	1910	2568	1766	1159
	10	2037	2656	1978	1340
Cu lignosulfonate	0.5	2775	2012	1979	831
	1.0	3065	1991	2224	926

fertilizer contact is more limited with band application. Two methods which have been tested to decrease the rate of conversion of applied Fe^{2+} to less soluble forms are to use spot or mixed placement of $FeSO_4$ with H_2SO_4 (Wallace, 1988) and inclusion of $FeSO_4$ with organic residues such as biosolids (Mostaghimi et al., 1988). Inclusion of K_2SO_4 with $FeSO_4$ resulted in correction of Fe chlorosis in peanuts on a highly calcareous soil in Israel (Table 2.5).

Foliar sprays of soluble Fe sources, mainly $FeSO_4$, are recommended for correction of Fe chlorosis (Martens and Westermann, 1991). Both inorganic and organic Fe sources are effective. Inclusion of urea and NH_4NO_3 in foliar sprays enhanced Fe uptake by corn (Hsu and Ashmead, 1984). Raese and Staiff (1988) compared several Fe compounds as foliar sprays on pear trees. The least amount of phytotoxicity to the fruit was from Fe lignosulfonate, but all Fe sprays increased Fe concentrations in the leaves and reduced Fe chlorosis symptoms.

Four foliar $FeSO_4$ sprays were more effective in producing grain sorghum on a calcareous soil (pH 8.3) than soil application of FeEDDHA (Table 2.6). Grain sorghum dry matter production was greater with soil application of farmyard manure (containing 2.1 g kg^{-1} Fe) than with FeEDDHA. The same amount of Fe was applied in each treatment. Residues of *Amaranthus* spp. (an Fe-efficient weed) enriched with Fe by spraying $FeSO_4$ on the plants just prior to harvest, and an acidified Fe-rich mining residue of the Cu industry also were more effective than soil application of FeEDDHA. Application of $FeSO_4$ mixed with manure, cottonseed cake or other organic substances in 8–10 holes around apple trees in China corrected Fe chlorosis (Zhen-ging and Chang-zhen, 1982). These results suggest that application of Fe sources complexed by organic matter may be more effective than Fe sources applied alone.

2.2.5.3 Manganese

Deficiencies of Mn are found mainly on well-drained neutral or calcareous or organic soils. Recommended Mn application rates range from 2 to 20 kg ha^{-1}. There are few residual effects from applied Mn because oxidation of the available Mn^{2+} to MnO_2 occurs rapidly in aerated soils (Mortvedt and Cox, 1985). Foliar sprays of $MnSO_4$ are very effective in correcting Mn deficiencies in small

Table 2.5 Dry matter production and chlorophyll content of peanuts as affected by Fe and K sources [Adapted from Shaviv and Hagin, 1987]

Fertilizer applied	Fe applied kg ha^{-1}	K applied mg pot^{-1}	Yield g pot^{-1}	Chlorophyll mg cm^{-2}
Control	0	0	3.3	0.64
K_2SO_4	0	300	4.3	0.85
	0	600	5.4	0.97
	0	900	4.4	0.97
$FeSO_4$-K_2SO_4	105	340	10.3	3.15
	210	680	10.2	3.33
	315	1020	14.3	5.93
$FeSO_4$-KCl	105	179	4.2	1.38
	210	358	8.3	2.59
	315	537	8.8	3.89
FeEDDHA	1.5	0	11.8	5.73
$LSD_{0.05}$			2.0	0.23

Table 2.6 Effects of soil or foliarly applied Fe treatments on dry matter and grain production and Fe uptake by grain sorghum [Adapted from Matocha, 1984]

| | April 6 | | June 3 | |
| | Dry matter kg ha^{-1} | Fe uptake mg ha^{-1} | Dry matter kg ha^{-1} | Grain yield kg ha^{-1} |
Treatment				
Control	19.5 a[a]	3454	0 a	0 a
FeEDDHA (2.2kg Fe ha^{-1})	25.3 a	4774	567 abc	43 ab
Acidified Fe-rich mine residue (4480 kg ha^{-1})	39.6 a	5171	2421 c	821 c
Untreated *Amaranthus*	26.9 a	4764	0 a	0 a
Fe-treated *Amaranthus*	34.6 a	5816	4092 d	548 b
Manure (33.6 Mg ha^{-1})	29.5 a	4413	540 ab	102 ab
Four foliar sprays (2.5% FeSO$_4$)	25.0 a	4260	4094 d	1531 c

[a] Treatment means within columns followed by the same letter are not significantly different at the 0.05 level

grain crops (Reuter et al.,1973). Some Mn sensitive crops are cereals, cotton, peanuts, soybean, sugarcane, fruits and vegetables (Martens and Westermann, 1991).

Rates of soil-applied Mn are higher than those for band application because of soil reactions which reduce Mn bioavailability. Mascagni and Cox (1984) found that MnSO$_4$ banded with acid-forming starter fertilizers was more effective than broadcasting an equal rate of the fertilizer mixture. The acidity in the fertilizer band retards the rate of oxidation of Mn^{2+}, decreasing reversion to less available forms (Randall et al., 1975). Foliar sprays and band applications of MnSO$_4$ were both effective in supplying available Mn for soybeans (Table 2.7). Alley et al. (1978) reported that banding 16.8 kg Mn ha^{-1} as MnSO$_4$ with the seed corrected Mn deficiency in soybean with adequate soil moisture but not under drought conditions.

Foliar Mn applications generally are recommended for correcting Mn deficiencies of most Mn-sensitive crops. Multiple applications of MnSO$_4$ spray are more effective than a single spray for barley (Reuter et al., 1973), peanut (Hallock, 1979), soybean (Cox, 1968), and wheat (Nayyar et al., 1985). Data in Table 2.8 show that one foliar spray at the V6 or R1 growth stage was less effective for soybean production than equivalent Mn foliar applications at both growth stages. Mascagni and Cox (1984)

Table 2.7 Effects of band and broadcast application of MnSO$_4$ on soybean yields, [Adapted from Mascagani and Cox, 1985]

| Mn applied kg ha^{-1} | Method of application | Soybean grain yield (kg ha^{-1}) Location | |
		a	b
0	Broadcast	485	2050
3		510	2080
10		1220	2400
30		1280	2480
0	Banded	1730	2010
3		2010	2130
7		1970	2160
13		2030	2290

Table 2.8 Effect of single or multiple $MnSO_4$ spray applications on soybean yields [adapted from Gettier et al., 1984]

| | Soybean grain yield (kg ha^{-1}) | | |
Foliar Mn application	Location a	Location b	Location c
Control	2863 a[a]	1308 a	950 a
V6 stage	3402 a	2969 b	2703 c
R 1 stage	2985 a	2694 b	1670 b
Both stages	3488 a	3826 c	2940 c

[a] Treatment means with the same letters in each column are not significantly different at the 0.05 level.

reported higher corn grain yields with $MnSO_4$ spray applications at the 4 and 8 leaf stages than one application at either growth stage.

2.2.5.4 Zinc

Deficiencies of Zn are more widespread than those of other micronutrients. Deficiencies occur in many soil types with pH levels > 6.0, especially in low organic matter soils (Mortvedt and Gilkes, 1993). As with Cu, residual effects of applied Zn are substantial, with responses reported at least 5 years after Zn application. Recommended Zn application rates range from 1 to 10 kg ha^{-1} (Mortvedt and Cox, 1985). Some Zn sensitive crops are corn, citrus, field beans, rice, and some fruits and vegetables.

Crop response to Zn has been documented in many countries and on numerous crops. Broadcast applications of Zn sources should be incorporated into the soil because Zn movement in soil is restricted. Schulte and Walsh (1982) showed that $ZnSO_4$, ZnO and Zn frits were equally effective for corn, edible bean and vegetables with band and broadcast applications. However, about twice as much Zn was required for broadcast as banded applications. Zinc oxide must be finely ground to be effective for correcting Zn deficiencies (Mengel, 1980). Boawn (1973) reported that the chelate, ZnEDTA, was 2.0–2.5 times more effective per unit of Zn than $ZnSO_4$ for edible beans and sweet corn.

Hergert et al. (1984) banded five Zn sources with liquid ammonium polyphosphate at four Zn rates as a starter fertilizer for corn (Table 2.9). Zinc EDTA was the most effective source at the three lowest rates, but crop yields decreased at the highest Zn rate with this source. The suggested reason was that Fe uptake was limited because of increased Zn uptake, resulting in decreased crop yields. All sources were equally effective at the highest Zn rate.

Table 2.9 Response of corn to Zn sources band-applied with a 10-15-0 starter fertilizer [Adapted from Hergert et al., 1984]

| | Corn grain yield (Mg ha^{-1}) | | | | |
| | Zn applied (kg ha^{-1}) | | | | |
Zn source	0	0.11	0.33	1.12	3.36
	3.9				
ZnEDTA		8.6	8.7	9.5	8.8
ZnO		8.2	8.2	8.6	9.1
$ZnSO_4$		8.3	8.9	8.7	9.1
$ZnSO_4$ - NH_3 complex		7.8	8.6	8.5	8.8
$Zn(NO_3)_2$ - UAN		8.2	8.2	8.9	8.8

Table 2.10 Effect of zinc sources and methods of application on rice yields, West Pakistan [Adapted from Yoshida, 1970]

Zn source	Zn rate kg ha^{-1}	Method of application	Rice grain yield Mg ha^{-1}
Control	0		4.32 a[a]
ZnSO$_4$	10	Preplant	5.98 bcd
ZnSO$_4$	100	Preplant	6.52 cd
ZnSO$_4$	10	Broadcast after puddling	6.01 cd
ZnSO$_4$	100	Broadcast after puddling	6.92 d
ZnSO$_4$	10	Broadcast after first symptoms	5.69 bcd
ZnSO$_4$	100	Broadcast after first symptoms	6.17 bcd
ZnO	0.1	Dip seedlings in 1% suspension	5.86 bcd

[a] Treatment means with the same letters in each column are not significantly different at the 0.05 level.

Zinc deficiency in rice was not verified until 1966 (Yoshida and Tanaka, 1969). Prior to that time this physiological disorder was thought to be a disease. Chlorosis of young rice is often associated with heavily limed or calcareous soils, land leveling operations and use of flood water high in bicarbonates (Westfall et al., 1971). Yoshida and Tanaka (1969) studied several methods for correcting Zn deficiency in rice. They concluded that soil or foliar applications of ZnSO$_4$ were as effective as dipping rice seedlings in a 1% ZnSO$_4$ solution, but the cost of the latter method was less. Yoshida (1970) reported that preplant or broadcast applications of ZnSO$_4$ (10 kg Zn ha^{-1}) after puddling were as effective as dipping roots of rice plants in a 1% ZnO suspension at transplanting from nurseries (Table 2.10).

2.3 Boron, Molybdenum, and Chloride

2.3.1 Forms in Soil

Boron always occurs as an ion in combination with O in nature. Although B is found in some insoluble silicate minerals such as tourmaline, the Na and Ca borates are its primary minerals in terms of abundance. The common borates such as borax (Na$_2$B$_4$O$_7$·10 H$_2$O) and kernite (Na$_2$B$_4$O$_7$·4 H$_2$O) are quite soluble and they occur mainly as evaporite deposits in the Earth's crust. Concentrations of B range from 2 to 100 mg kg^{-1}, with a mean concentration of 10 mg kg^{-1} in soils where it generally occurs as boric acid (H$_3$BO$_3$) (Goldberg, 1993).

Molybdenum is the least abundant of all the micronutrients in the Earth's crust and soils. Concentrations of total Mo range from 0.2 to 5 mg kg^{-1}, with a mean concentration of 2 mg kg^{-1} in soils. The most common form of Mo in igneous rocks is molybdenite (MoS$_2$), but Mo occurs as an anion (MoO$_4^{2-}$) in minerals of the Earth's crust and in alkaline solution (Reddy et al., 1997). The mobility of the molybdate anion in soils is high, compared to that of the metallic micronutrient cations

Chlorine, or more correctly the chloride ion (Cl$^-$), is common in nature and is the most recent element to be designated as a micronutrient. Its concentration range in soils is 20 to 900 mg kg^{-1}, with a mean concentration of 100 mg kg^{-1}. The majority of Cl$^-$ in soils is considered to have originated from salts trapped in parent materials, and from volcanic emissions and marine aerosols. Most of the soil Cl$^-$

commonly occurs in soluble salts such as NaCl, $CaCl_2$ and $MgCl_2$ and is the principal anion in extracts of saline or sodic soils. The concentration of Cl^- in soil solutions ranges from less than 0.5 to more than 6,000 mg kg^{-1}.

2.3.2 Factors Affecting Bioavailability

2.3.2.1 Soil pH

Boron occurs in soil solution mainly as undissociated H_3BO_3, which dissociates only slightly below pH 9.2. Boron is more strongly adsorbed by soil components than is Cl^- or NO_3^-. Sorption of B to Fe and Al oxides is pH dependent and is highest at pH 6 to 9 (Goldberg, 1993). Bioavailability of B is highest between pH 5.5 and 7.5, decreasing below and above this range mainly due to these pH-dependent reactions.

Molybdenum is usually present as adsorbed molybdate ions (MoO_4^{2-}) in soils, but $HMoO_4^-$ may also exist under very acid soil conditions (Reddy et al., 1997). Solubility of $CaMoO_4$ and molybdic acid (H_2MoO_4) increases with increasing pH while the Mo sorption to hydrous Fe oxides increases with decreasing soil pH in the range 7.8 to 4.5 (Hodgson, 1963; Harmsen and Vlek, 1985). Consequently, Mo bioavailability increases with increasing soil pH.

The Cl^- anion is bound only very slightly by most soils under acid conditions and becomes negligible at pH levels above 7.0. Appreciable amounts of Cl^- can be adsorbed, particularly on kaolinitic and oxidic soils which have significant positive charge.

2.3.2.2 Soil Organic Matter

Most of the bioavailable B in soils is found in SOM. The nature of reactions of B with organic matter are not well understood, but may involve hydroxyl groups on organic complexes (Offiah and Axley, 1993). Soil conditions favoring decomposition of SOM, such as warm, moist soil, good aeration and optimum microbial activity, will result in improved B bioavailability.

While Mo concentrations have been correlated with high levels of SOM, Hodgson (1963) stated that such accumulations did not necessarily indicate the presence of Mo organic complexes in these soils. Chloride bioavailability is not related to SOM content with which it does not complex.

2.3.3 Sources and Rates of Application

The most common sources of B, Cl^- and Mo are shown in Table 2.2. Recommended B rates are lower than those for Cu, Fe, Mn and Zn, generally < 2 kg ha^{-1}, because the range between deficiency and

Table 2.11   Effects of Mo seed treatment on nodulation and yield parameters for peanuts and soybean [Adapted from Shivashankar and Hagstrom, 1991]

Seed Mo g kg^{-1}	Nodule weight plant^{-1}	Pods plant^{-1}	Pod yield kg ha^{-1}	Stover kg ha^{-1}
Peanuts				
0	0.14	17.3	1690	2060
4	0.16	18.0	1750	2140
8	0.19	20.9	2190	2280
Soybean				
0	0.12	62	2330	1940
4	0.15	80	2830	2050
8	0.16	82	2940	2170

toxicity is rather narrow. Because many plant species are sensitive to B applications, care must be used in applying boronated fertilizers (Mortvedt and Woodruff, 1993). Molybdenum application rates are even lower, generally < 0.2 kg ha^{-1} for seed treatment or foliar sprays, but slightly higher rates are used as soil applications (Adams, 1997). Deficiencies of Mo are found on acidic soils and may be corrected by liming soils to pH 6.0. Deficiencies are mainly reported in legume crops, because Mo is essential for N_2 fixation. However, some vegetable crops are also sensitive to Mo deficiencies.

Seed treatment is the most common method of Mo application. This method ensures a more uniform application of the low rates required for correction of Mo deficiencies (Mortvedt and Cox, 1985). Including Mo sources with bacterial inoculants on legume seed is a common practice but care must be taken not to apply too much Mo. Data in Table 2.11 show that nodule weight, number of pods plant^{-1} and yield of peanuts and soybean were increased with application of 4 and 8 g Mo kg^{-1} seed as Na_2MoO_4. Increasing the rate of Mo above 8 g kg^{-1} reduced seed germination rate and nodulation (data not shown). Soaking red clover seed in a 1% solution of Na_2MoO_4 significantly increased the yield of harvested seed (Hagstrom and Berger, 1965). Most soils contain sufficient levels of Cl$^-$ for adequate plant nutrition. However, Cl$^-$ deficiencies have been reported mainly on cereal crops grown on sandy soils in high rainfall areas, those derived from low Cl$^-$ parent materials or soils in low rainfall areas located away from oceans. The recommended application rate of Cl$^-$ is 10-30 kg ha^{-1} when Cl$^-$ deficiencies are suspected.

2.3.4 Crop Responses

2.3.4.1 Boron

Crop species differ considerably in their B requirement as well as to their tolerance to over applications. Residual effects of applied B also vary with soil conditions, with the least effects found on acidic sandy soils in areas of high rainfall. Some crops with high B requirements are alfalfa, cotton, peanuts, irrigated corn, root crops, soybean, and some fruits and vegetables. Hollow heart of peanut kernels was attributed to B deficiency in 33% of the fields inspected in northern Thailand (Rerkasem et al., 1988).

Both soil and foliar applications are effective for crops. Soil applications generally are used for field crops, but foliar sprays are more common on perennial crops such as fruit and nut trees. Foliar applications of B at the prebloom or early bloom stages are effective in increasing fruit and nut yields (Nyomora et al., 1997). Murphy and Lancaster (1971) reported maximum yields of cotton when 0.5

Table 2.12 Response of cotton to boron in Arkansas [Murphy and Lancaster, 1971]

B applied kg ha^{-1}	Method of application	Seed cotton (kg ha^{-1})		
		Location a	Location b	Location c
Control		1350 a[a]	1945 a	2316 a
0.29	Drilled	2539 b	2342 b	2377 a
0.57	Drilled	2642 b	2377 b	2354 a
1.14	Drilled	2684 b	2445 b	2429 a
1.14	Broadcast	2536b	2357 b	2416 a
	Foliar[b]	2779 b	2302 b	2462 a
0.57	Drilled and foliar[b]	2700 b	2262 b	2375 a

[a] Means within each column with the same letter are not significantly different at the 0.05 probability level.
[b] 0.1 kg ha^{-1} of B applied at weekly intervals for five weeks.

Table 2.13   Effect of B applied to a previous crop of rutabaga on the grain yield of barley and wheat [Adapted from Gupta and Cutliff, 1982]

Crop	Method of B application	Grain yield (kg ha^{-1})				
		B applied in previous year (kg ha^{-1})				
		0	2.24	4.48	8.96	Significance
Barley	Broadcast	2931	2999	2979	2679	NS
	Banded	3201	3221	3136	2902	NS
Wheat	Broadcast	3446	3463	3110	3029	NS
	Banded	3855	3720	3780	3628	NS

kg ha^{-1} was applied as a foliar spray (five times at 0.1 kg ha^{-1}) or at a soil application rate > 0.3 kg B ha^{-1} (Table 2.12).

Boron leaches quite readily in most soils, but especially in sandy soils under high rainfall. Therefore, B should be applied for the immediate crop. Gupta (1984) reported that a broadcast application of $Na_2B_4O_7$ at a rate of 2 kg B ha^{-1} on a loam soil provided sufficient B for alfalfa and red clover for 2 years in eastern Canada, while Shorrocks (1987) stated that soil applications of 50 g H_3BO_3 per citrus tree were effective up to 5 years in Taiwan. Gupta and Cutliffe (1982) applied B at rates up to 8.96 kg ha^{-1} for rutabaga on several sandy loam soils in eastern Canada. The following year they found no significant reductions in grain yields of barley and wheat grown on these fields (Table 2.13). Since the recommended B rates for rutabaga are 2-4.5 kg ha^{-1}, they concluded any residual effects of B applied for high B-requiring crops should not be detrimental to succeeding cereal crops.

2.3.4.2   Molybdenum
Production of forage legumes is significantly affected by soil pH. Liming of many soils may provide sufficient bioavailable Mo so that Mo fertilization may not be required. Some soils are so acid that legume production is not possible without at least a low rate of lime application. Liming resulted in increased alfalfa forage yields in the absence of Mo fertilization in the southeastern United States (Table 2.14). Annual applications of Mo resulted in increased forage production at soil pH 5.3 and 5.5, but not at pH 6.5.

Table 2.14   Alfalfa responses to Mo applications, as affected by soil pH level [Adapted from Mortvedt and Anderson, 1982]

Lime Mg ha^{-1}	Soil pH	Alfalfa yield (kg ha^{-1})			
		Mo rate[a] (g ha^{-1})			
		0	50	100	400
0	5.3	770	1520	1290	1580
2.2	5.5	2240	2620	2320	3240
9.0	6.5	3210	3370	3140	3300

[a] Mo applied annually

Table 2.15 Effects of KCl, KNO_3 and $CaCl_2$ on barley and spring wheat yields at several locations [Adapted from Fixen et al., 1986]

Product	Yield ($Mg\ ha^{-1}$)					
	Wheat				Barley	
	1[a]	2	3	4	1	2
Control			5.0	3.0		
KCl	2.4	2.7	5.4	3.4	3.0	3.0
KNO_3	1.9	2.5	5.1	3.1	2.7	2.9
$CaCl_2$	2.4	2.4	5.3	3.4	3.0	3.0
$LSD_{0.05}$	0.1	0.1	0.2	0.2	0.2	0.1

[a] The Cl^- rate was 85 kg ha^{-1} for locations 1 and 2, and 61 kg ha^{-1} for locations 3 and 4.

Results of a long-term corn-soybean experiment in northern Alabama showed no response to Mo applied biennially to corn (Adams et al.,1990). Observation of Mo deficient symptoms on soybean leaves led to a study which showed Mo response by soybean. The lack of Mo response to corn had led to the erroneous conclusion that these soils were not Mo deficient for soybean.

2.3.4.3 Chloride

Barley and spring wheat grain yields were increased by KCl applications on high K testing soils in 7 of 12 locations in eastern South Dakota (Fixen et al., 1986). Oat yields were not affected by KCl applications at all five locations in this study. Data in Table 2.15 show that barley and wheat responses were due to Cl^- and not to K in the products applied. Results also indicated that the Cl^- concentrations in wheat plants were highly correlated with soil Cl^- in the top 60 cm of the soil profile. There are few areas of Cl^- deficient soils in the United States, so this micronutrient generally is not considered in most fertilizer programs. In addition, Cl^- is applied to soils with KCl, the dominant K fertilizer.

The role of Cl^- in decreasing the incidence of various diseases in small grains is perhaps more important than its nutritional role from a practical viewpoint. Data in Table 2.16 show that wheat yields were 1.19 Mg ha^{-1} higher with a spring application of NH_4Cl instead of $(NH_4)_2SO_4$ on an acid soil (pH 5.5), but were similar on a limed soil (pH 6.6). The severity index of disease caused by the take-all fungus [*Gaeumannomyces graminis* (Sacc.) Arx & Oliv.] also was reduced by application of the Cl-containing N fertilizer on the acid soil but not on the limed soil.

Table 2.16 Wheat yields and take-all index, as affected by soil pH and N source [Adapted from Christensen et al., 1986]

Soil pH	Nitrogen source	Wheat yield Mg ha^{-1}	Take all index
5.5	$(NH_4)_2SO_4$	3.50	3.48
	NH_4Cl	4.69	3.16
6.6	$(NH_4)_2SO_4$	3.86	3.72
	NH_4Cl	3.78	3.68
S_x		0.30	

2.4 References

Adams, J.F. 1997. Yield responses to molybdenum by field and horticultural crops. p. 182–201. *In* U. C. Gupta (ed.) Molybdenum in agriculture. Cambridge University Press, Cambridge, UK.

Adams, J.F., C.H. Burmester, and C.C. Mitchell. 1990. Long term fertility treatments and molybdenum availability. Fert. Res. 21:167–170.

Alley, M.M., C.I. Rich, G.W. Hawkins, and D.C. Martens. 1978. Correction of Mn deficiency of soybeans. Agron. J. 70:35–38.

Asher, C.J. 1991. Beneficial elements, functional nutrients and possible new essential elements. p. 703–724. *In* J.J. Mortvedt et al. (ed.) Micronutrients in agriculture. 2nd Ed. Soil Science Society of America, Madison, WI.

Barak, P., and P.A. Helmke. 1993. The chemistry of zinc. p. 1–14. *In* A. D. Robson (ed.) Zinc in soils and plants. Kluwer Academic Publishers, Dordrecht, Netherlands.

Boawn, L.C. 1973. Comparison of zinc sulfate and zinc EDTA as zinc fertilizer sources. Soil Sci. Soc. Am. Proc. 37:111–115.

Chen, Y., and P. Barak. 1982. Iron nutrition of plants in calcareous soils. Adv. Agron. 35:217–240.

Chesworth, W. 1991. Geochemistry of micronutrients. p. 1-30. *In* J.J. Mortvedt et al. (ed.) Micronutrients in agriculture. 2nd Ed. Soil Science Society of America, Madison, WI.

Christensen, N.W., and M. Brett. 1985. Chloride and liming effects on soil nitrogen form and take all disease of wheat. Agron. J. 77:157–163.

Cox, F.R. 1968. Development of a yield response prediction and manganese soil test interpretation for soybeans. Agron. J. 60:521–524.

Fixen. P.E., R.H. Gelderman, J.K. Gerwing, and F. A. Cholick. 1986. Response of spring wheat, barley, and oats to chloride in potassium chloride fertilizers. Agron. J. 78:664–668.

Gartrell, J.W. 1981. Distribution and correction of copper deficiency in crops and pastures. p. 313–349. *In* J. F. Loneragan et al. (ed.) Copper in soils and plants. Academic Press, New York, NY.

Gettier, S.W., D.C. Martens, D.L. Hallock, and M.J. Stewart. 1984. Residual Mn and associated soybean yield response from MnSO₄ application on a sandy loam soil. Plant Soil 81:101–110.

Gilkes, R.J. 1977. Factors influencing the release of copper and zinc additives from granulated superphosphate. J. Soil Sci. 28:103–111.

Goldberg, S.A. 1993. Chemistry and mineralogy of boron in soils. p. 3–44. *In* U.C. Gupta (ed.) Boron and its role in crop production. CRC Press, Inc., Boca Raton, FL.

Grundon, N.J. 1980. Effectiveness of soil dressings and foliar sprays of copper sulphate in correcting copper deficiency in wheat (*Triticum aestivum*) in Queensland. Aust. J. Exp. Agric. Anim. Husb. 20:717–723.

Gupta, U.C. 1984. Boron nutrition of alfalfa, red clover, and timothy grown on podzol soils of eastern Canada. Soil Sci. 133:16–22.

Gupta, U.C., and J.A. Cutliffe. 1982. Residual effect of boron applied to rutabaga on subsequent cereal cops. Soil Sci. 133:155–159.

Hagstrom, G. R., and K.C. Berger. 1965. Molybdenum deficiencies of Wisconsin soils. Soil Sci. 100:52–56.

Hallock, D.L. 1979. Relative effectiveness of several Mn sources on Virginia-type peanuts. Agron. J. 71:685–688.

Harmsen, K., and P.L.G. Vlek. 1985. The chemistry of micronutrients in soil. p. 1–42. *In* P.L.G. Vlek (ed.) Micronutrients in tropical food crop production. Martinus Nijhoff / Dr. W. Junk Publishers, Dordrecht, Netherlands.

Hergert, G.W., G.W. Rehm, and R. A. Wiese. 1984. Field evaluations of zinc sources band applied in ammonium polyphosphate suspension. Soil Sci. Soc. Am. J. 46:1190–1192.

Hodgson, J.F. 1963. Chemistry of micronutrients in soils. Adv. Agron. 15:119–159.

Hsu, H.H., and H. D. Ashmead. 1984. Effect of urea and ammonium nitrate on the uptake of iron through leaves. J. Plant Nutr. 7:291–300.

Karamanos, R.E., G.A. Kruger, and J.W.B. Stewart. 1986. Copper deficiency in cereal and oilseed crops in northern Canadian prairie soils. Agron. J. 78:317–323.

Krauskopf, K.B. 1972. Geochemistry of micronutrients. p. 7–40 *In* J.J. Mortvedt, P.M. Giordano and W. L. Lindsay (ed.) Micronutrients in agriculture, 1st Ed. Soil Science Society of America, Madison, WI.

Lancaster, J.D. 1971. Response of cotton to boron. Agron. J. 63:539–540.

Lindsay, W.L. 1979. Chemical equilibria in soils. Wiley Interscience, New York, NY.

Martens, D.L., and D.T. Westermann. 1991. Fertilizer applications for correcting micronutrient deficiencies. p. 549–592. *In* J.J. Mortvedt et al. (ed.) Micronutrients in agriculture. 2nd Ed. Soil Science Society of America, Madison,

WI.

Mascagni, H, J., Jr., and F.R. Cox. 1984. Diagnosis and correction of manganese deficiency in corn. Commun. Soil Sci. Plant. Anal. 15:1323–1333.

Mascagni, H.J., Jr., and F.R. Cox. 1985. Effective rates of fertilization for correcting manganese deficiency in soybeans. Agron. J. 77:363–366.

Matocha, J.E. 1984. Grain sorghum response to plant residue-recycled iron and other iron sources. J. Plant Nutr. 7:259–270.

Mengel, D.B. 1980. Role of micronutrients in efficient crop production. IN Coop. Ext. Serv. AY–239.

Mortvedt, J.J. 1991a. Micronutrient fertilizer technology. p. 523–548. *In* J.J. Mortvedt et al. (ed.) Micronutrients in agriculture. 2nd Ed. Soil Science Society of America, Madison, WI.

Mortvedt, J.J. 1991b. Sequestration and chelation. p. 177–188. *In* D.A. Palgrave (ed.) Fluid fertilizer science and technology. Marcel Dekker, Inc., New York, NY.

Mortvedt, J.J. 1992. Crop response to level of water-soluble zinc in granular zinc fertilizers. Fert. Res. 33:249–255.

Mortvedt, J.J. 1997. Sources and methods for molybdenum fertilization of crops. p. 171–181. *In* U. C. Gupta (ed.) Molybdenum in agriculture. Cambridge University Press, Cambridge, UK.

Mortvedt, J.J., and O.E. Anderson (ed.). 1982. Forage legumes: Diagnosis and correction of molybdenum and manganese problems. Univ. GA Coop. Series Bull. 278.

Mortvedt, J.J., and F.R. Cox. 1985. Production, marketing and use of calcium, magnesium and micronutrient fertilizers. p. 455–481. *In* O.P. Engelstad (ed.) Fertilizer technology and use. 3rd Ed. Soil Science Society of America, Madison, WI.

Mortvedt, J.J., and R.J. Gilkes. 1993. Zinc fertilizers. *In* A. D. Robson (ed.) Zinc in soils and plants. Kluwer Academic Publishers, Dordrecht, Netherlands.

Mortvedt, J. J., and J. R. Woodruff. 1993. Technology and application of boron for crops. p 157–176. *In* U.C. Gupta (ed.) Boron and its role in crop production. CRC Press, Inc., Boca Raton, FL.

Mostaghimi, S., J.E. Matocha, and C.C. Crenshaw. 1988. Effects of sewage sludge on iron chlorosis and yields of grain sorghum grown on calcareous soils. J. Plant Nutr. 11.1397–1415.

Murphy, B.C., and J.D. Lancaster. 1971. Response of cotton to boron. Agron. J. 63:539–540.

Nayyar, V. K., U.S. Sadana, and T.N. Takkar. 1985. Methods and rates of application of Mn and its critical levels for wheat following rice on coarse-textured soils. Fert. Res. 8:173–178.

Nyomora, A.M.S., P.H. Brown, and M. Freeman. 1997. Fall foliar-applied boron increases tissue boron concentration and nut set of almond. J. Am. Soc. Hort. Sci. 122:405–410.

Offiah, O.O., and J.H. Axley. 1993. Soil testing for available boron in soils. p. 106–123. *In* U.C. Gupta (ed.) Boron and its role in crop production. CRC Press, Inc., Boca Raton, FL.

Pickering, W.F. 1981. Selective chemical extraction of soil components and bound metal species. CRC Crit. Rev. Anal. Chem. 12:233–266.

Raese, J.T., and D.C. Staiff. 1988. Chlorosis of "Anjou" pear trees reduced with foliar sprays of iron compounds. J. Plant Nutr. 11:1379–1385.

Randall, G.W, E.E. Schulte, and R.B. Corey. 1975. Effects of soil and foliar-applied manganese on the micronutrient content and yield of soybeans. Agron. J. 67:502–507.

Reddy, K.R., L.C. Munn, and L. Wang. 1997. Chemistry and mineralogy of molybdenum in soils. p. 4–22. *In* U.C. Gupta (ed.) Molybdenum in agriculture. Cambridge University Press, Cambridge, UK.

Rerkasem, B., R. Netsangtip, R.W. Bell, J.F. Loneragan, and N. Hiranburana. 1988. Comparative species responses to boron on a Typic Tropaqualf in northern Thailand. Plant Soil 106:15–22.

Reuter, D.G., T.G. Heard, and A.M. Alston. 1973. Correction of manganese deficiency in barley crops on calcareous soils. 1. Manganese sulphate applied at sowing and as foliar sprays. Aust. J. Exp. Agric. Anim. Husb. 13:434–439.

Schulte, E.E., and L.M. Walsh. 1982. Soil and applied zinc. Univ. WI Coop. Ext. Serv. A2528.

Shaviv, A., and J. Hagin. 1987. Correction of lime-induced chlorosis by application of iron and potassium sulfates. Fert. Res. 33:43.

Shivashankar, K., and G. Hagstrom. 1991. Molybdenum fertilizer sources and their use in crop production. p. 297–305. *In* S. Portch (ed.) Proc. Int. Symp. Role of sulphur, magnesium and micronutrients in balanced plant nutrition. Sulphur Institute, Washington, DC.

Shorrocks, V.M. 1987. Citrus-residual boron effects. Micronutrient News 8:7.

Shuman, L.M. 1991. Chemical forms of micronutrients. p. 113–144. *In* J.J. Mortvedt et al. (ed.) Micronutrients in agriculture, 2nd Ed. Soil Science Society of America, Madison, WI.

Solberg, E., D. Penny, and L. Evans. 1993. Copper deficiency in cereal crops. Agdex 532–2. Agrifax, Edmonton, Canada.

Stevenson, F.J. 1991. Organic matter-micronutrient reactions in soil. p. 145–186. *In* J. J. Mortvedt et al. (ed.) Micronutrients in agriculture, 2nd Ed. Soil Science Society of America, Madison, WI.

Wallace, A. 1988. Acid and acid-iron fertilizers for iron deficiency control in plants. J. Plant Nutr. 11:1311–1319.

Westfall, D. G., W. B. Anderson, and R. J. Hodges. 1971. Iron and Zn response of chlorotic rice grown on calcareous soils. Agron J. 63:702–705.

Yoshida, S., G.W. McLean, M. Shafi, and K.E. Mueller. 1970. Effects of different methods of zinc application on growth and yields of rice in a calcareous soils, West Pakistan. Soil Sci. Plant Nutr. 16:147–149.

Yoshida, S., and A. Tanaka. 1969. Zinc deficiency of the rice plant in calcareous soils. Soil Sci. Plant Nutr. 15:57–80.

Zheng-ging, Z., and L. Chang-zhen. 1982. Studies on the application of ferrous sulfate for controlling chlorosis of apple trees on calcareous soils. J. Plant. Nutr. 5:883–896.

3

Nutrient Interactions in Soil and Plant Nutrition

S.R. Wilkinson and D.L. Grunes
USDA

Malcolm E. Sumner
University of Georgia

3.1 Introduction

Although interactions between nutrients have been studied in great detail in hydroponic or sand culture (McDonald, 1994; Marschner, 1995) such cases will not be discussed here. As this chapter falls in the section on Soil Fertility, only those interactions that pertain to field situations will be investigated in the discussion to follow.

3.2 What are Interactions?

Russell (1950) stated it this way "If two nutrients are limiting, or nearly limiting growth or composition (concentration), where adding only one of the nutrients has little effect while adding both gives a considerable effect, the effect is said to be a positive interaction. Similarly, if adding the two together has less effect than when each is added separately, the effect is said to be negative interaction." When factors acting in concert result in a positive growth response that is greater than the sum of their individual effects, the interaction is positive (synergistic); when less than the sum of the individual effects, the interaction is negative (antagonistic); and when the same, no interaction. Important to this concept is the Law of the Minimum, or Principle of Limiting Factors which states that the amount of growth for a given environment depends on the quantity and balance of growth-determining factors with the least optimum factor-limiting growth the most. However, two additional aspects must also be considered. First, Liebig's Law of the Minimum applies to conditions where inflows of energy, minerals, and other factors are balanced (steady-state condition). Second, factors may interact in positive or negative ways to modify effects of individual factors. Thus, when solar radiation, temperature and soil water availability are nonlimiting, plant nutrient requirements will be higher, for which Wallace (1990) proposed the Law of the Maximum in contrast to the Law of the Minimum. The Law of Maximum states that when the need is fully satisfied for every factor involved in the process, the rate of the process can be at its maximum potential, which is greater than the sum of its parts because of sequentially additive interaction (Wallace, 1990). Mineral nutrient responses may be of the

D-89

Law of the Minimum or Mitcherlich type where responses are proportional to the level of the input and its degree of deficiency. He used a multiple fractional yield plot to illustrate how this relationship might work (Fig. 3.1). The large increase in yield represented by regions A and B indicates strong positive interactions typical of Liebig responses; region C represents additive yield effects as multiple yield fractions increase; region D represents an area of no response; while region E illustrates antagonism, or negative interaction. The importance of distinguishing Liebig and Mitscherlich type responses is that, with the latter, one has options to control yield levels; while with the former, one has fewer options.

Interactions occur when the supply of one nutrient affects the absorption, distribution, or function of another nutrient (Robson and Pitman, 1983). The result may be induced deficiencies, toxicities, modified growth responses, and/or modified nutrient composition. Interactions may be specific or nonspecific. Nonspecific interactions become important when the contents of both nutrients are near the deficiency range or excessive in total or proportion. When the supply of one nutrient is increased, dilution may induce a deficiency of another nutrient especially when the supply of the other nutrient is limiting (or excessive in total or proportion). Such nonspecific interactions are theoretically possible for any mineral nutrient combination. Specific nutrient interactions may occur when: (1) competition occurs between ions which have similar physicochemical properties (valency and diameter) or which form chemical bonds; (2) ions with sufficiently similar chemical properties compete for adsorption (on clay minerals, oxides, cell walls and SOM), or absorption (influx and/or efflux) sites, or transport within xylem and phloem and in metabolic functions (at active sites) (Robson and Pitman, 1983). Interactions may also be associated with absorption, adsorption, translocation, and/or precipitation at any of the sites at the soil-root interface, which can affect uptake by roots and translocation. Plant germplasm interactions at the root-soil interface can cause impacts of one nutrient upon the availability of another, and can also affect root morphology which may, in turn, affect shoot growth and composition.

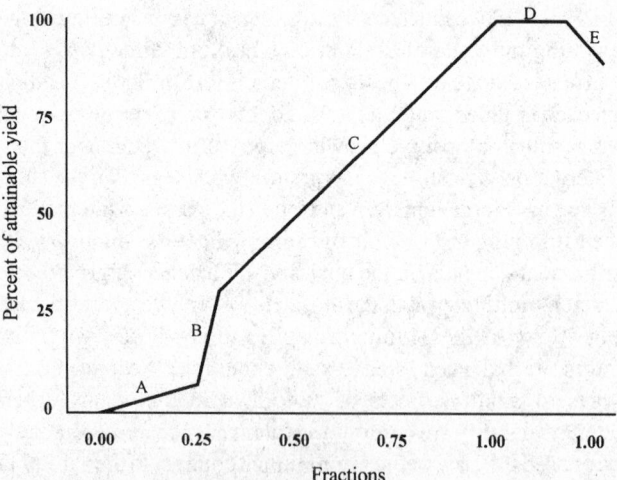

Fig. 3.1 Fractions of multiple fraction yield plots or the total of all stresses as fractions multiplied together. Regions A and B represent responses typical of Liebig type factors, region C represents additive yield effects as multiple yield fractions increase, region D represents an area of no effect, while region E represents antagonisms or negative multiplying effects [Reprinted from Wallace, 1990. J. Plant Nutr. 13:313-325 by courtesy of Marcel Dekker, Inc.]

A conceptual model (Sumner and Farina, 1986) illustrating how a crop responds to the alleviation of multiple nutrient stresses is presented in Fig. 3.2, which is an extension of the pattern in Fig. 3.1 to a multinutrient situation. As the number of limiting factors (n) is decreased, the permissible variation in the growth factor(s) being plotted on the X axis becomes narrower with a concomitant increase in yield, highlighting the need for improved nutritional balance in order to obtain a high yield. When the growth factor is very limiting, yield is also limited but increases (A→B) as the level of the growth factor is increased until one of the other n growth factors becomes limiting. When that growth factor is identified and remedied, a further increase in yield (B→C) becomes possible until one of the n-1 growth factors begins to limit growth. In this way as limiting growth factors are identified and remedied, yield continues to increase (C→D→E→F→G→H). Similarly, when toxic factors or excesses come into play, yields would decrease (I→O). In other words, the composition of a plant which is performing perfectly can only vary within certain limits. When composition falls outside those limits, yield decreases (H→A or I→O). This relationship is useful in interpreting interactions.

3.3 Nutrient Interactions Involving Nitrogen

Nitrogen, which is required in the greatest quantity of all mineral nutrients and absorbed by plant roots either as a cation or anion, is an essential component of proteins. Because N function in plant growth and nutrition is closely connected to C, the C/N ratio controls N availability, and potentially affects

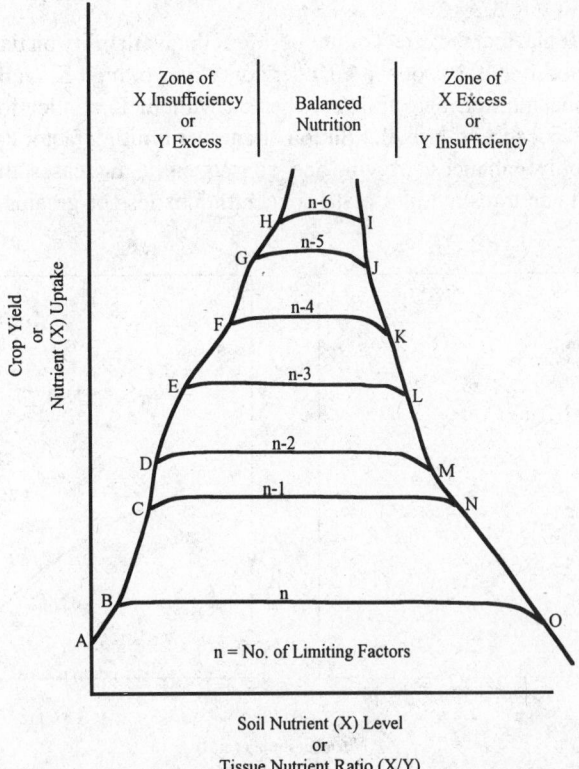

Fig. 3.2 Interptetive model illustrating the response of a crop to a number of limiting factors [Reprinted from Sumner and Farina, 1986. Adv. Soil Sci. 5:201-236 with permission of Springer-Verlag New York, Inc.]

interactions through the processes of SOM formation, biomass production through photosynthesis and respiration, and energy flow in and through all levels of the ecosystem. Values for C/N ratios range from 8/1 for soil microorganisms to 400/1 for sawdust. Ratios > 30/1 generally result in immobilization of N (Fixen, 1996) and potential deficiencies. When growth is limited by carbohydrate availability, increases in C assimilation lead to proportional increases in N uptake when N resources are available in the root zone (Gastal and Saugier, 1989).

Botanical cycling between grass and legume components, caused by variations in soil N, constitutes an important interaction in livestock production (Fales et al., 1996). The ability of grasses with their extensive fibrous root systems to exploit soil supplies of nutrients such as P and K can also bring about deficiencies in the associated legume which may reduce its competitiveness in multiple species production situations.

3.3.1   Nitrogen*Potassium Interactions

Nitrogen and K are macronutrients which are required in the greatest quantity for most plants. Consequently, availabilities vary greatly in soils which have been cropped over extended periods. Interactions having economic significance occur when one of these two nutrients is present at near deficiency levels, and the other at high or toxic levels.

The importance of K in promoting barley grain and straw yield responses to N in hydroponic culture together with the depressive effect at deficient levels is shown in Fig. 3.3 (MacLeod, 1969). The relationships vary depending on whether grain or straw yield is being considered. These results are in accord with the model in Fig. 3.2.

Kemp (1971, 1983) found that the effect of increasing N bioavailability on tissue K concentrations depends on K bioavailability in the root zone. Under conditions of high K availability, increasing N supply increases K concentration and uptake; whereas without K application, K concentrations decrease at high N rates because of growth dilution or another limiting factor coming into play (Fig. 3.4). Increasing N supply enhances growth, and consequently, increases the demand for other nutrients. This demand can translate into plant concentrations less or greater than that needed for

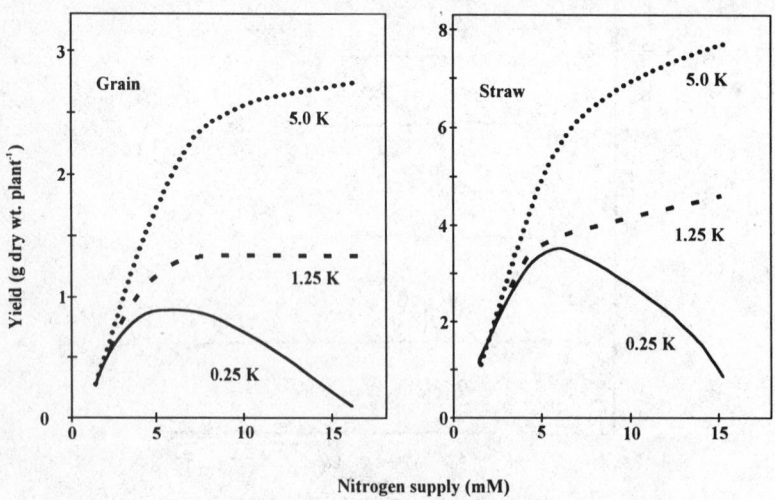

Fig. 3.3   Interactive effects of increasing N and K supply on barley grain and straw yield [From MacLeod, 1969. Agron. J. 61:26-29 with permission of the American Society of Agronomy]

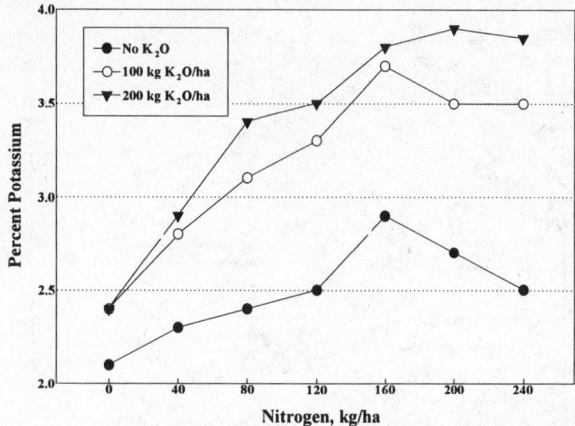

Fig. 3.4 Effect of N and K fertilization of the K concentrations of grass [From Wilkinson and Stuedemann, 1979. Am. Soc. Agron. Spec. Pub. 35:93-121 with permission of the American Society of Agronomy]

sufficiency, depending on the nutrient supply in the root zone (dilution-accumulation effect). Low temperature survival of field grown bermudagrass (*Cynodon dactylon*) depends on the balance between N and K. As the ratio of applied N to K widens, bermudagrass survival decreases at an increasing rate (Fig. 3.5). Increasing N fertilization depletes soil available K, which lowers plant K and results in mortality during the dormant winter season (Carreker et al., 1977).

A practical application of the competitive interaction between ions with similar properties is illustrated in Fig. 3.6 in which N applications increased and K applications decreased ^{137}Cs uptake by oats on a site contaminated by Cs during the Chernobyl disaster. Combined N and K applications resulted in increased yields (data not shown) with lower levels of ^{137}Cs uptake. The net effect of combined N and K applications was antagonistic to ^{137}Cs concentration. Westermann et al. (1994) reported a three-way interaction involving N and K rate and source in potato yield at low available soil N, and specific gravity at adequate soil N levels. Without applied N, KCl reduced specific gravity of potatoes while K_2SO_4 did not.

3.3.2 Nitrogen*Phosphorus*Potassium Interactions

An N*P*K interaction in Coastal bermudagrass illustrates the yield response surfaces for N and K applications to Cecil soil with (Fig. 3.7B) and without (Fig. 3.7A) P fertilization. Highest yields were obtained at the highest rates of N, P and K. Working with *Festuca ovina* in an N*P*K factorial, Wilson (1993) confirmed the generalization that the response to one nutrient depends on the sufficiency level of other nutrients. Yield depressions were found when high levels of one nutrient were combined with low levels of the other nutrients. An N*P*K interaction resulting in toxicity was reported. Alleviating the yield depressing effect of excessive macronutrient supply involved removing the limitation of a low supply of other nutrients. Teng and Timmer (1994) reported that white spruce seedlings receiving NH_4NO_3 alone increased in biomass by 34% while monocalcium phosphate addition stimulated growth by 107%. Applying N and P together increased dry matter production by 362%, illustrating the benefit that can be realized by exploiting interactions.

It is important to make the distinction that there are no toxic mineral nutrients, but that there are nutrient element concentrations which are toxic. Comments by Sumner and Farina (1986) reiterated by Marschner (1995) are worth repeating in this regard. Interactions between two nutrients are

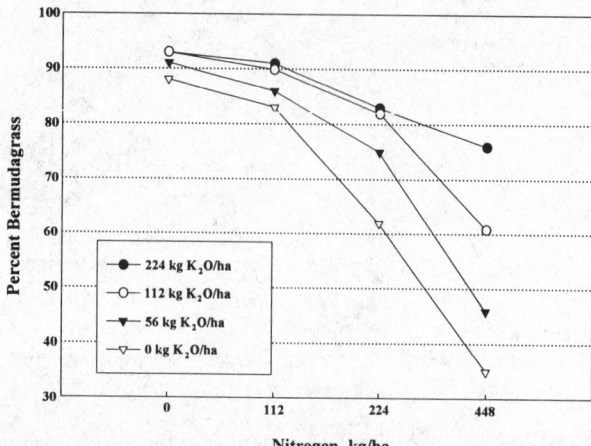

Fig. 3.5   Effect on N and K fertilization on the ability of coastal bermudagrass to survive winter cold temperatures [Carreker et al., 1977]

important when the contents of both are near the deficiency range. Increased supply of only one nutrient will stimulate growth which may, in turn, induce a deficiency of the other nutrient. In principle, these nonspecific interactions hold true for any mineral nutrient with contents at or near critical deficiency levels. Using optimal ratios alone as references has limitations because such ratios can be obtained when both nutrients are at deficient or at toxic levels. With this limitation in mind, optimal ratios indicate only that the two nutrients involved are not limiting.

3.3.3   Form of Nitrogen (NH_4^+ versus NO_3^-) Interactions

Ammonium and NO_3^- forms make up as much as 80% of the total cations and anions taken up by plants. Consequently, the concentration of these ions in the root uptake zone has a great effect on the uptake of other cations and anions (van Beusichem et al., 1988). The uptake of these N ions greatly impacts rhizosphere soil pH, and cellular pH regulation. Uptake of NH_4^+ and NO_3^- creates increased and decreased root medium acidity, respectively, as follows:

$$3NH_4^+ \rightarrow 3R - NH_2 + 4H^+ \tag{3.1}$$

$$3NO_3^- \rightarrow 3NH_2 + 2OH^- \tag{3.2}$$

Similarly, biologically fixed N_2 also produces acidity as follows:

$$1.5N_2 \text{ fixed} \rightarrow R - NH_2 + H^+ \tag{3.3}$$

From the standpoint of interactions, changes in soil pH occur naturally whatever the form of N taken up by the root system. Marschner (1995) generalized that calcifuges (plants adapted to acid soils and to low redox potentials) may prefer NH_4^+ sources while calcicoles (plants adapted to calcareous soils) may prefer NO_3^- sources. Detrimental effects of NH_4^+ on uptake of K^+ and Ca^{2+} are more likely when the root medium is acid.

Nitrification of NH_4^+ decreases bulk soil pH while preferential NH_4^+ uptake and release of H^+ reduce rhizosphere pH causing the dissolution of other plant nutrients, particularly in soils of high pH. The NH_4^+ effect is particularly important in the case of P, Mn, Zn, and Cu where bulk soil and rhizosphere acidification increases their solubility (Thomson et al., 1993). Plant nutrient interactions with soil acidity are an important aspect of the effect of N form on plant growth and chemical composition. Engels and Marschner (1992) found that nutrient demand in maize (defined as shoot growth rate per unit weight of roots) was markedly lower at 12 than 24 °C shoot base temperature. As a consequence of the lower nutrient demand imposed on the root system, uptake rates of NO_3^- and NH_4^+ declined by more than 50%. Ammonium depressed uptake of K^+ and Ca^{2+}, particularly in the shoot, both in plants with high and low nutrient demand. Control of cation concentrations in shoots may be the result of internal demand rather than the capacity of roots to absorb nutrients.

Magalhaes and Wilcox (1983) provided NH_4^+ or NO_3^- to tomato plants either as foliar sprays, or to the roots. Ammonium sulfate reduced growth when supplied through the roots relative to foliar sprays, indicating that NH_4^+ translocation to shoots can be harmful. Barber et al. (1994) observed greater efficiency in converting radiation into dry matter when corn was supplied with high proportions of NH_4^+ than when supplied primarily with NO_3^-. Variable responses to N form by sorghum genotypes, mediated through effects on Al toxicity, were reported by Tan et al. (1992).

Most crop plants perform best when supplied with a combination of both N forms. Plants supplied with NO_3^- tend to accumulate more cations (K^+, Ca^{2+}, Mg^{2+}), while those supplied with NH_4^+ accumulate more anions (SO_4^{2-}, $H_2PO_4^-$, Cl^-).

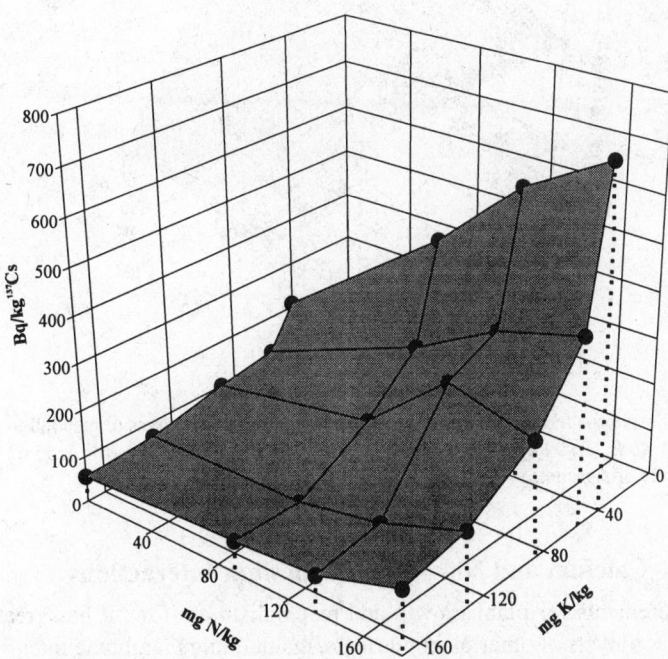

Fig. 3.6 Effect of N and K additions on the concentration of cesium in oat shoots [Reprinted from Tulin et al., 1995. Potash Rev. 2:1-2 with permission of the International Potash Institute]

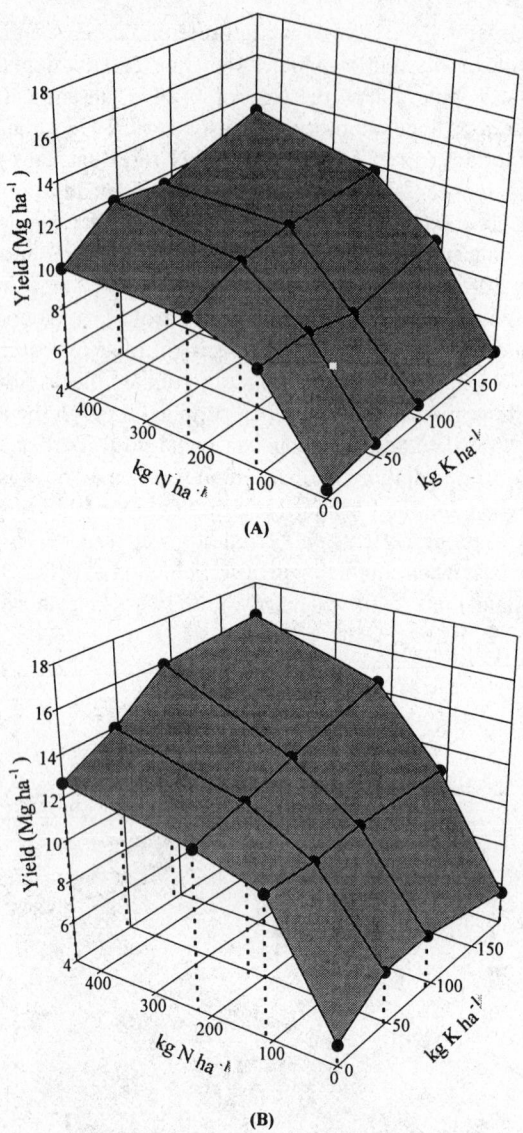

Fig. 3.7 Effect of phosphorus fertilization on the ability of coastal bermudagrass to respond to nitrogen and potassium fertilization. (A) No P; (B) 99 kg P ha^{-1} [Reprinted from Welch et al., 1963. Agron J. 55:63-67 with permission of the American Society of Agronomy]

3.3.4 Nitrogen*Calcium and Nitrogen*Magnesium Interactions

Although Ca requirements for plant growth and metabolism are low, it has great significance in providing balance for levels of other plant nutrients, maintaining membrane integrity, and reducing potential for toxicity of other elements (Wallace et al., 1966). The Ca concentration in solution is important in meeting plant Ca demand, because Ca is supplied primarily by mass flow, and once in the plant, is not readily transported to other tissues during senescence. Potassium, NH_4^+, Mg^{2+}, Al^{3+}, Mn^{2+}, H^+ and heavy metals can reduce Ca uptake by binding at the exterior surface of the plasma membrane which increases the Ca requirement. Marschner (1995) concluded that, while increased external Ca concentration may lead to increased leaf Ca, this may not occur in nontranspiring tissue such as fruit because there is no driving force. Consequently, as fruit develop, the Ca concentration becomes diluted, raising the risk of Ca deficiency. Some known Ca-related disorders are tip burn in lettuce, blackheart in celery, blossom end rot in tomatoes and watermelon, and bitter pit in apple. Induced Mg deficiency is very common, particularly on soils of low CEC heavily fertilized with N and K. Fertilization with acid-forming fertilizers may cause preferential leaching of K, Ca, and Mg (Adams et al., 1967; Wilkinson et al., 1987). The strength with which Ca and Mg are held on exchange sites is an important factor in the interaction between K, Ca, and Mg. On 2:1 layer silicates, Mg is held less strongly than Ca, and both are held less strongly than K (Barber, 1995), but this is not true for kaolinite and organic matter. Fertilization with NO_3^- generally enhances Ca and Mg concentrations in plants driven by the need for cation-anion balance. When N is supplied as NO_3^-, electrical neutrality is maintained internally by its reduction in synthesizing organic acids by release from roots of anions such as OH^- or HCO_3^- or by uptake of cations. When N is supplied as NH_4^+, internal electrical neutrality is maintained by release of H^+ or by uptake of anions.

3.3.5 Nitrogen*Micronutrient Interactions

Micronutrient interactions with N occur frequently due to the effect of N form on pH either in the rhizosphere, and/or bulk soil. Application of N as NO_3^- increases soil pH, whereas NH_4^+ decreases it (Section B, Chapter 10).

The effect of acidification on plant nutrient uptake and utilization is site and nutrient dependent. That is, if the soil is alkaline and Fe deficient, the application of NO_3^- may reduce Fe availability, whereas NH_4^+, or $R-NH_2$ sources which release H^+, increase it. Other interactions of N with micronutrients occur when N-stimulated growth causes increased demand for a nutrient and possible deficiency. Such effects may occur with B, Cu, Mn, and Mo where cation competition between NH_4^+ and Cu^{2+}, Mn^{2+}, and Zn^{2+} for absorption can occur. Manganese uptake can be stimulated by either or both forms of N. The NO_3^- form may lead to greater uptake of cations, i.e., Mn^{2+}, while the NH_4^+ form may increase the bioavailability of Mn by acidification. In neutral to alkaline soils with low available Fe, increased acidity from NH_4^+ forms may enhance availability of Fe^{2+} by promoting the reduction of the unavailable Fe(III). The N*Fe interaction is of greatest economic significance. Because most soil Fe is present as Fe(III), it must be reduced to the Fe^{2+} form before uptake can occur. Plants that are Fe efficient may release H^+ ions or reducing agents into the rhizosphere to increase Fe availability. Graminaceous plant species are capable of releasing phytosiderophores which can chelate Fe, but dicotyledonous plants do not have this ability (Barber, 1995; Marschner, 1995). Dijkshoorn et al. (1983) reported that differences in accumulation of Mn, Ni, Zn and Cd in ryegrass grown on sludge amended soil were associated with differences in soil pH, brought about by the N form applied.

3.3.6 Nitrogen*Sulfur Interactions

Nitrogen and S uptake and pathways of assimilation must be balanced for efficient protein synthesis and optimal crop quality. In winter oilseed rape, yield responses to N were limited by S deficiencies as N fertilization rates increased from 180 to 230 kg N ha^{-1} (McGrath and Zhao, 1996), but not at lower rates (50-100 kg N ha^{-1}). The N*S interaction was a result of growth stimulation from N causing dilution of S in plant tissue stemming from insufficient bioavailable S in the soil. In corn in North

Carolina, Cassel et al. (1996) found that subsoiling marginal droughty soils to encourage root growth into the B horizon increased S supply. On the other hand, Morris et al. (1994) did not obtain yield responses to S fertilization of annual ryegrass at N rates of 280 kg ha⁻¹ and relatively low levels of atmospheric S deposition (5.8 kg ha⁻¹ yr⁻¹), because S was being supplied from sources other than rainfall, most likely from the subsoil where large quantities were adsorbed by Fe and Al oxides. This subsoil S storage is the most likely reason for the lack of yield response to added S observed in industrialized countries in the past; but with increased air quality regulations, potential interactions with N in shallow rooted annual crops are likely in the future. Soliman et al. (1992) found that the acidifying effect of S applications to a highly calcareous soil was important in the mobilization of Fe, Mn, Zn and P and availability of micronutrients to subsequent crops.

3.4   Interactions Primarily Involving Potassium

Potassium requirements for higher plants are second to N. The importance of K*Ca*Mg interactions as they impact plant, ruminant animal, and human nutrition has been reviewed extensively (Wilkinson et al., 1990). Hypomagnesemic grass tetany is a conditioned Mg deficiency in grazing ruminants associated with inadequate daily intake of bioavailable Mg in herbage or hay. High rates of applied K to cool season grass pastures, whether from manures or inorganic fertilizers, increase the incidence of grass tetany (Wilkinson et al., 1987). Consequently, any interactions between Ca and K that depress forage Mg contents exacerbate this problem. The economic problem of grass tetany is complicated because, for most cool season grasses, the sufficiency level for maximum crop yields is about half the safe level for prevention of grass tetany in grazing, lactating ruminants such as cattle and sheep. Although supplying Mg as mineral supplements is generally helpful, complete protection is only obtained when adequate bioavailable Mg in forage is consumed daily. Consequently, it is important to evaluate the interactions which contribute to this syndrome.

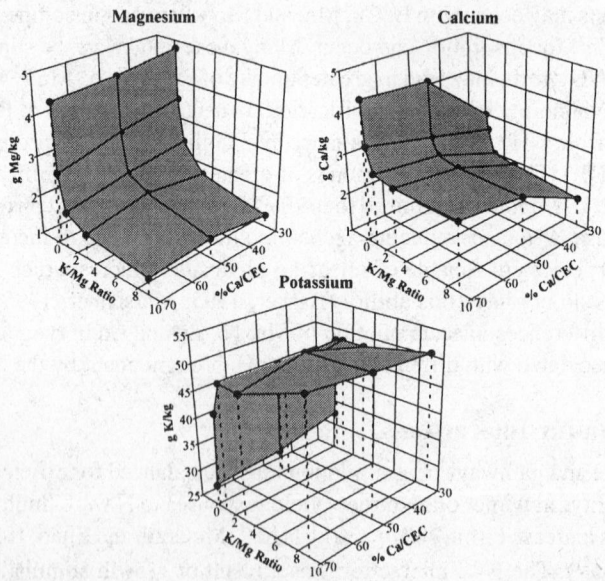

Fig. 3.8   Effect of ratios of K/Mg saturation and Ca/CEC ratios on uptake of K, Ca and Mg by oats [Data from Johannsonn and Hahlin, 1977]

3.4.1 Potassium*Calcium*Magnesium Interactions

Johannsonn and Hahlin (1977) illustrated the effect of various K/Mg/Ca percentage saturations of the CEC on Mg concentrations in oats (Fig. 3.8). Over wide ranges in K/Mg ratios, oat yields were unaffected (data not shown) but tissue Mg and Ca concentrations increased while those of K decreased markedly at low K/Mg ratios. Magnesium saturation did not affect yields for 30 and 50% Ca saturation, but at 70% Ca saturation, yield decreased at 18% K and 2% Mg saturation. Consequently, when the balance between K, Ca and Mg is upset, tissue composition is impacted, which can carry through to the grazing animal's health. Hannaway et al. (1980, 1982) showed that increasing the level of K decreased the Mg concentration and accumulation in tall fescue shoots, but did not affect its influx into the roots. Reduced translocation from the roots to the shoots appeared to be the source of the K/Mg antagonism. Ohno and Grunes (1985) and Ohno et al. (1985) confirmed this pattern of behavior.

Huang et al. (1990) reported that net Mg translocation from roots to shoots was depressed by increased K concentrations in roots. When the Mg concentration in the soil solution is low and K is high, fertilization with NO_3^- may decrease shoot Mg concentration because of the competitive effect of high K in the roots. This interaction resulted from NO_3^- stimulating K uptake, which, in turn, inhibited Mg translocation to the shoots. Cold, wet soils may also exhibit this same effect. The presence of high Ca in the soil may also reduce Mg concentration in the shoots.

Grunes et al. (1992) found that K fertilization significantly increased K tissue concentrations at the expense of Mg and Ca concentrations in 3 cool season grasses accompanied by increased concentrations of organic acids which may impact the bioavailability of Mg to livestock. Because delayed nitrification depressed Mg and Ca uptake more than that of K, and resulted in higher K/(Ca + Mg) ratios, Mathers et al. (1982) recommended that nitrification inhibitors not be used on grazed wheat pastures. Seasonal variations in the K/(Ca + Mg) ratio are quite common with values reaching very high levels in early spring growth (Wilkinson and Stuedemann, 1979). Syedomar and Sumner (1991) found that low rates of gypsum increased alfalfa yield, but at higher rates reduced topsoil exchangeable Mg and K causing yield reductions. In addition, Mg was more susceptible to loss by leaching than K. Because highly hydrated Mg^{2+} is bound weakly in the cell walls and possibly at the binding sites on the plasma membrane, other cations (K^+ and Ca^{2+}) compete quite effectively with Mg^{2+} and strongly depress its uptake. Competition from K generally has a greater effect on Mg translocation to the shoot than on Mg absorption by the root (Hannaway et al., 1980, 1982; Ohno and Grunes, 1985). Potassium and Mg interactions are extremely important from an economic standpoint in maintaining animal health and performance, and production of high quality leafy vegetables. Fertilization with Mg to increase tissue Mg concentrations is often ineffective because, in order to double tissue Mg, a fourfold increase in Mg activity in the soil solution is required (Wild, 1988).

Table 3.1 Interactive effects of N and P on yield, and leaf N and P contents of corn grown on a South African Oxisol [From Sumner and Farina, 1986]

Soil test P (mg L^{-1})	Corn yield (Mg ha^{-1}) N applied (kg ha^{-1})				Tissue P content (%) N applied (kg ha^{-1})				Tissue N content (%) N applied (kg ha^{-1})			
	0	60	120	180	0	60	120	180	0	60	120	180
3	2.31	2.22	3.11	3.41	0.13	0.12	0.15	0.15	2.17	2.35	2.70	2.89
16	4.74	6.68	7.91	7.84	0.19	0.24	0.25	0.26	2.19	2.73	2.90	3.21
27	4.06	7.14	9.12	9.74	0.20	0.26	0.32	0.32	2.09	2.64	3.17	3.24
46	4.54	8.17	9.42	9.96	0.24	0.32	0.34	0.35	2.06	2.88	3.08	3.27
LSD$_{0.05}$	0.82	0.82	0.82	0.82	0.03	0.03	0.03	0.03	0.16	0.16	0.16	0.16

Table 3.2 Least cost N and P combinations for specified corn yields [From Sumner and Farina, 1986. Adv. Soil Sci. 5:201-236 with permission of Springer-Verlag New York, Inc.]

Target corn yield (Mg ha^{-1})	Fertilizer N rate (kg ha^{-1})	Soil test P level (mg L^{-1})
3.0	10	5
4.0	20	7
5.0	30	10
6.0	60	14
6.5	80	16
7.0	110	19
7.5	150	22

3.5 Interactions Involving Phosphorus

A detailed discussion of interactions between P and other nutrients has been presented by Sumner and Farina (1986). Phosphorus intercations of greatest economic interest involved N or Zn.

3.5.1 Phosphorus*Nitrogen Interactions

Nitrogen can increase P concentrations in plants by increasing root growth, by increasing the ability of roots to absorb and translocate P, and by decreasing soil pH as a result of absorption of NH_4^+ and thus increasing solubility of fertilizer P (Grunes, 1959; Cole et al., 1963; Miller, 1974; Adams, 1980). The P*N interaction for corn illustrated in Table 3.1 is typical. Responses to both N and P are small at low levels of the other nutrient, but increase markedly for combinations of N and P at higher rates. The highest yield was obtained at the highest rates of N and P. The contents of N and P in leaves followed the same pattern, which is in accordance with that in Fig. 3.1. Nitrogen stimulated the uptake of P and *vice versa*. Because increased growth requires more nutrients to maintain tissue composition within acceptable limits, mutually synergistic effects as shown here for N and P promote growth even more. The economic importance of interactions such as these becomes clear in less developed areas where fertilizers are in short supply, costs are high and soils are frequently deficient in N and P. Because similar yields are possible at different combinations of N and P (Table 3.1), calculation of least cost inputs for a chosen yield level allows practical exploitation of the consequences of interaction (Table 3.2). An interesting question arises as to whether or not this interaction could have been further exploited by varying other limiting factors which also interact with N or P. The model (Fig. 3.1) would certainly so suggest.

3.5.2 Phosphorus*Zn Interactions

There appear to be many contradictions concerning the P*Zn interaction. Sumner and Farina (1986) demonstrated that when the model in Fig. 3.2 is used to interpret the results, much of this apparent contradiction disappears. For example, Boawn et al. (1954) and Ellis et al. (1964) failed to induce Zn deficiency by increasing levels of P. However, reinterpretation of their data indicates that they did not add sufficient P to induce this deficiency (Sumner and Farina, 1986). On the other hand, in an experiment in which the P*Zn interaction was significant (Takkar et al., 1976), the data followed the model very well (Fig. 3.9). At low P/Zn ratios in the tissue (insufficient P or excessive Zn), yield increased steeply as the ratio increased until the optimum was reached (P/Zn = 100), after which yield declined again at high P/Zn ratios. At low and high P/Zn ratios, P and Zn deficiency symptoms, respectively, were observed. These data clearly indicate the importance of nutrient balance in obtaining high yields.

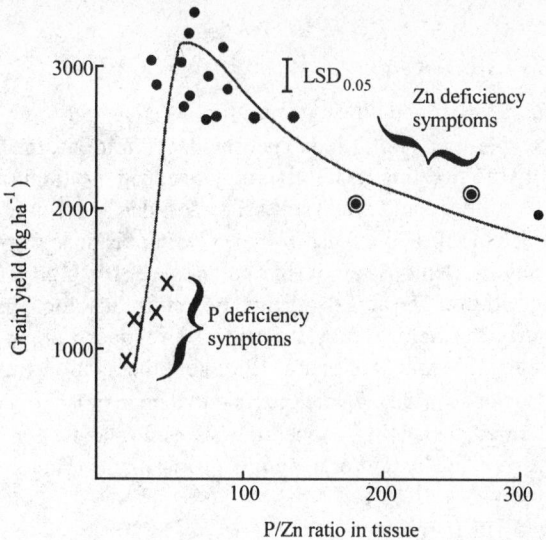

Fig. 3.9 Effect of P∗Zn interaction on the yield of corn and appearance of deficiency symptoms [Data from Takkar et al., 1976]

Peck et al. (1980) found that increasing levels of concentrated superphosphate without Zn decreased Zn concentrations in vegetable crops, while with Zn fertilizer, concentrations increased. On the other hand, Norvell et al. (1987) reported that N and P fertilization had little effect on soluble Zn concentrations in the soil, and probably was only important in situations where Zn levels were marginally deficient in alkaline soils. Similarly, Mallarino and Webb (1995) showed that large P applications failed to produce economically important P∗Zn interactions in corn. Loneragan et al. (1982) observed that P accumulated to very high levels in Zn-deficient okra leaves accompanied by accumulation of Mg at all levels of Zn supply. Deiorio et al. (1996) found a negative P∗Zn and positive P∗Cu interaction in lettuce. Increased retention of Zn and Cu in roots occurred when P was supplied at high levels. Welch and Norvell (1993) found that Zn-deficient roots leaked greater quantities of K, Mn, Ca and Al than roots supplied with adequate supplies of Zn.

3.5.3 Phosphorus∗Magnesium Interactions

Positive interrelationships between P and Mg are expected since Mg is an activator of kinase enzymes and activates most reactions involving phosphate transfer. Although significant P∗Mg interactions involving growth were found in the literature, positive correlations and interactions between P and Mg concentrations were found in tall fescue leaves, wheat seedlings, bermudagrass and grape vines suggesting the existence of a synergism (Skinner and Matthews, 1990; Reinbott and Blevins, 1991, 1994, 1997; Hillard et al., 1992). When fertilized with NH_4^+, P or P and K, Follett et al. (1977) showed that bromegrass had significantly greater Mg content than that receiving only K; while with tall fescue, fertilization with broiler litter resulted in increased P content without concomitant increases in Mg concentration or in ability to prevent grass tetany in animals. This conflicts with the recommendation that additional P fertilization may reduce the risk of grass tetany. Perhaps there is a threshold P level below which additions of P increase Mg, or the effect of broiler litter in increasing tissue K and decreasing tissue Mg concentrations may override any P effects (Wilkinson and Stuedemann, 1979).

3.5.4   Phosphorus*Iron Interactions

Although Sumner and Farina (1986) found little evidence for P*Fe interactions which affected growth or P or Fe concentrations except under controlled experimental conditions, the ability of P and Fe to form insoluble products suggests that this interaction may occur in the field. Haleem et al. (1992) demonstrated that banded P fertilizer could affect soil pH by as much as 0.5 unit as far as 2.0 cm from the band, which caused extractable Fe to decrease with increasing distance from the band. Chandler and Scarseth (1941) were able to demonstrate in the greenhouse that high P inputs decreased Fe concentrations in peanuts and alfalfa. The ability of P and Fe to form insoluble products suggests that such interactions may occur in the field. Heavy P fertilization induced apparent Fe deficiency in macadamia nut trees in Hawaii (Hue and Nakamura, 1988), but the relationship was also thought to involve low Fe/Mn ratios (Hue et al, 1988). While the field evidence of P*Fe interactions that are of economic significance is limited, balancing concentrations of P and Fe for best performance is necessary, especially for hydroponically and/or container grown plants (Handreck, 1991).

3.5.5   Phosphorus*Lime Interactions

Liming has been found to increase (Ryan and Smillie, 1975; Parfitt, 1977; Smythe and Sanchez, 1980), decrease (Munns and Fox, 1976; Friesen et al., 1980), or not affect (Reeve and Sumner, 1970; Jones

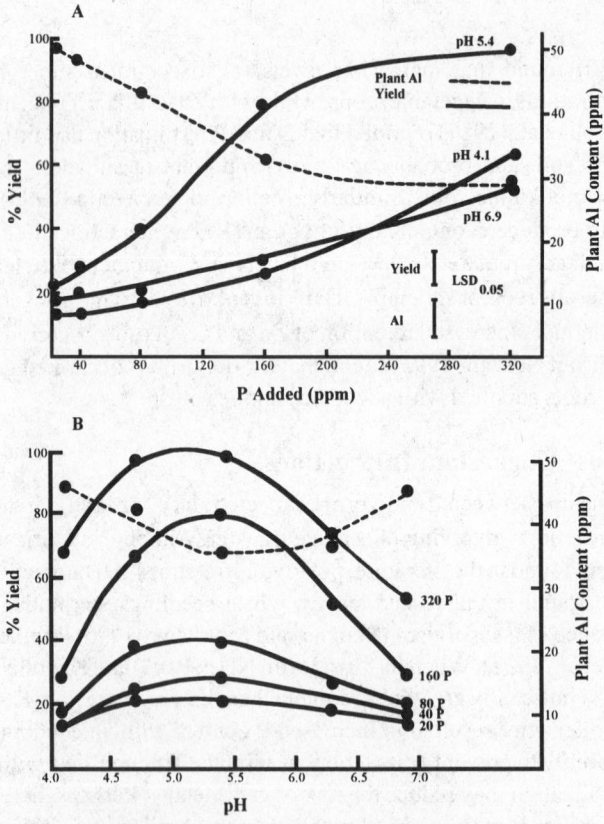

Fig. 3.10   Effect of added P (A) and lime (B) on the relative yield and Al content of maize grown in the greenhouse on a Natal Oxisol [Reprinted from Sumner and Farina, 1986. Adv. Soil Sci. 5:201-236 with permission of Springer-Verlag New York, Inc.]

and Fox, 1978) soil P availability. Similarly, P uptake has been increased (Ryan and Smillie, 1975; Friesen et al., 1980), decreased (Sumner 1979; Farina et al., 1980) or not affected (Brams, 1971). Most P*lime interactions are intimately associated with toxic soil Al which limits root growth and proliferation, and nutrient uptake. Once in the root, Al can precipitate P (Foy et al., 1978) preventing its translocation to the tops.

An example of a P*lime interaction is illustrated in Fig. 3.10 (Sumner and Farina, 1986) where lime produced ranges of pH values in factorial combination with levels of P. With increasing pH_{KCl} up to 5.5, yield response was much greater at high than low levels of applied P. Above pH_{KCl} 5.5, high P rates had relatively greater depressive effects on yield than low rates. The mechanism involved was difficult to establish, but the highest P rate in the absence of lime reduced exchangeable Al by 18%. Both lime and P reduced tissue Al.

3.6   Soil Acidity and Related Interactions

The major plant nutrition problems on acid soils are toxicities of Al and Mn, and deficiencies of Ca. Forest decline has been associated with increased stress from declining Ca/Al ratios (Shortle et al., 1997) which are a result of the depletion of basic cations, and mobilization of Al from soil minerals.

3.6.1   Calcium*Aluminum Interactions

The results of solution culture experiments show that the negative effects of Al^{3+} on root growth can be partially or totally offset by increasing the level of Ca^{2+} in solution (Noble et al., 1988; Nichol et al., 1993; Zysset et al., 1996), but sometimes, this depressed Mg uptake leading to deficiency symptoms (Tan et al., 1992). Aluminum disrupts the transport of Ca^{2+} at the root apex leading to Al toxicity in sensitive wheat cultivars. Because Al in its various forms can inhibit Ca^{2+} uptake within the root apex and possibly block Ca^{+2} channels, the dramatic differences in sensitivity to Al within wheat varieties may lie in their ability to prevent the inhibition of Ca^{2+} uptake or homeostasis processes in the root apex areas. Other cations may also reduce the impact of Al^{3+} on Ca^{2+} movement into the xylem and translocation to shoots (Huang et al., 1992). Consequently, because a favorable Ca/Al balance must be maintained in the root apex region to allow for root elongation, it is essential that a continuous supply of Ca be available (Huang et al., 1993, 1995).

Cronan and Grigal (1995) concluded that the risk of adverse impact on root growth and function increased from 50 to 75 to 100% as the soil solution Ca/Al ratio decreased from 1.0 to 0.5 to 0.2. Threshold conditions for impacts on forest growth from Al stress were indicated by four successive measurement endpoints: (1) base saturations < 15% of the effective CEC; (2) Ca/Al ratios < 1; (3) fine root tissue Ca/Al ratios < 0.2; and (4) foliar tissue Ca/Al molar ratios ≤ 125. On Brazilian Oxisols where both Al toxicity and Ca deficiency occur, Smyth and Cravo (1992) obtained maximum corn and soybean yields at 27% Al saturation whereas that for peanuts was 54%. Baligar et al. (1992) found that although surface horizons had lower pHs and similar Al levels to subsoils, they exhibited less toxicity to soft red winter wheat, probably because of higher levels of exchangeable cations which ameliorated Al toxicity. Other results also suggest that other cations (K, Mg) may ameliorate the effects of Al in the soil solution (Huang and Grunes, 1992), while Pellet et al. (1995) indicated that differential Al tolerance in corn cultivars was partly related to Al-induced release of citrate, phosphate and malate from cells which complexed Al and helped these plants to survive in soils containing high concentrations of Al. Delafuente et al. (1997) reported that improved Al tolerance in transgenic plants was associated with enhanced citrate synthesis. Pellet et al. (1997) found that in wheat Al tolerance appeared to be mediated by multiple exclusion mechanisms controlled by different genes. These mechanisms included phosphate and malate exudation by roots. Aluminum toxicity in yeast was associated with Mg deficiency (Macdiarmid and Garner, 1996). Calcium/Al balance in the soil-tree root interface was found to generally improve with closeness to the root surface, and suggests that trees growing on nutrient poor, acid soils invest energy around their roots to create a more favorable microenvironment for both roots and microorganisms (Gobran and Clegg, 1996).

3.6.2   Interactions Involving Manganese

Manganese availability is controlled by the total quantity of Mn, pH, SOM, and redox potential, but in most soils, redox potential is the most important. Manganese interactions with Mg are involved in the tolerance of wheat plants to Mn toxicity. The ratio of Mg/Mn in the soil solution predicted the corresponding ratio in plants (Goss et al., 1992). In four diverse soils, Shuman et al. (1990) found that reductions in yield were more strongly related with soil Al than Mn levels. Schomberg and Weaver (1991) demonstrated that Mn decreased N fixation in arrowleaf clover more than mineral N uptake. Consequently, the potential impact of excess soluble Mn in the root zone was partially offset by availability of mineral N for uptake and metabolism. Rhizobial strain by pH interactions in arrowleaf has been reported (Coll et al., 1989) in which low soil pH (5.6) affected yield less when plants were dependent on mineral N than when inoculated and dependent on N_2 fixation. There was also a significant effect of Mn on N_2 fixation. Interactions between Mn and Al have not been described. Whether the effects are additive, antagonistic, or synergistic is unknown. Changes in soil aeration and pH at the root-soil interface may create conditions where Mn may interact with Fe. Noble and Sumner (1988) reported that increasing Al levels in solution significantly depressed Ca, Mg, P, and Mn concentrations in soybean shoots over multiple Ca levels.

3.7   Other Interactions

3.7.1   Interactions involving Selenium, Molybdenum, and Arsenic

Selenium uptake in alfalfa (*Medicago sativa* L.) from coal fly ash could be reduced by gypsum applications probably as a result of competition with S for uptake (Arthur et al., 1993), while S fertilization of grazed herbage reduced herbage and cattle blood Se contents (Murphy and Quirke, 1997). Bell et al. (1992) suggested that primary Se accumulators such as *Astragalus bisulcatus* have the unique ability to accumulate Se in the presence of SO_4^{2-}, but SeO_4^{2-} increased alfalfa Se content only on the verge of incipient toxicity indicating that alfalfa may have the ability to discriminate against Se while *A. bisulcatus* preferentially absorbs it. Khattak et al. (1991) found interactions between As, Se, Mo, and P at relatively low concentrations in alfalfa in sand culture even though yield was not affected. Arsenic stimulated Se uptake, Se depressed As uptake, while P depressed both Se and As uptake.

Although Nass et al. (1993) found that plants grown on pure fly ash took up enough Mo to possibly induce Cu deficiency in animals fed the material, sheep have grazed pastures fertilized with fly ash without detrimental effects (Vona et al., 1992). Fertilization or dietary supplementation of animals may be a possible solution when problems arise. Gupta (1997) summarized the interactive effects of P, S, N, Mn and Cl with Mo as follows: (1) soluble P enhances plant uptake of Mo; (2) soluble S decreases Mo uptake; (3) N applications over time may decrease Mo uptake; (4) there is an inverse relationship between leaf Mn and Mo; (5) on acid soils with high Mn, poor growth may be associated with deficient Mo or toxic Mn; and (6) Mo is the only element whose availability increases with pH. Part of the N*Mo interaction may arise from the opposite effects of NH_4^+ and NO_3^- on soil pH which would change Mo availability. Molybdenum is also essential for NO_3^- reduction.

Other ions may also play a role in micronutrient interactions. For example, P can replace Mo from anion exchange sites, while SO_4^{2-} may compete directly at absorption sites on the roots (Khattak et al., 1991).

3.7.2   Silicon Interactions

In most soils, Si as $Si(OH)_4$ is readily available for uptake and plays a role in growth, mineral nutrition, mechanical strength, resistance to disease, herbivory and tolerance to adverse soil or rooting conditions (Epstein, 1994). Galvez et al. (1989) found that Si could alleviate Mn toxicity in sorghum grown in solutions containing up to 1800 μM Mn. While Li et al. (1996) found little effect of Si on Al toxicity to sorghum, Hammond et al. (1995) reported that Si reduced Al toxicity and promoted Ca uptake in barley in solutions which could not have been due to growth stimulation alone. However, Epstein (1994) pointed out that plants grown in solutions may produce misleading results as far as Si is concerned. Liang et al. (1996) suggested that Si enhanced K uptake and inhibited Na intake in salt stressed barley mitigating salt toxicity and increasing salt tolerance. Hodson (1995) found that plant species vary considerably in the amount of Al and Si that they can transport in their tissues, and that high accumulations of Si and Al appear to be mutually exclusive. He suggested that codeposition of Al and Si within the plant, internal biochemical effects, and indirect effects were responsible for the ameliorative effect. Wallace (1993) has suggested that soluble Si might tip cation/anion balance in favor of excess anions, thereby releasing equivalent OH^- from roots which could raise rhizosphere pH, reduce soluble Al, and ameliorate its toxicity. A similar effect might account for some of the effect of Si on Mn toxicity. Silicate amendments may increase P availability by displacing it from sesquioxide surfaces.

3.7.3   Salinity Interactions with Plant Nutrition

High levels of NaCl compete for uptake with cations such as K, Ca, Mg, and anions such as NO_3^-, and $H_2PO_4^-$ (Curtin et al., 1993). Zhong and Lauchli (1994) concluded that supplemental Ca could

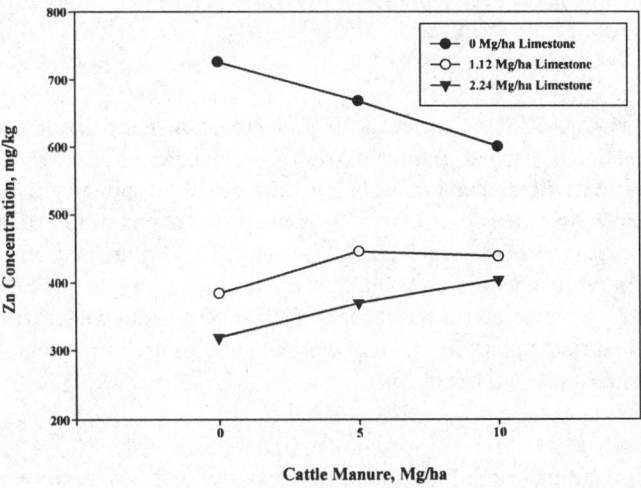

Fig. 3.11 Interactive effects of cattle manure and lime on the zinc concentration of soybeans grown on metal contaminated soil [From Pierzynski and Schwab, 1993. J. Environ. Qual. 22:247-254 with permission of the American Society of Agronomy]

alleviate the inhibitory effect of salinity on root growth by maintaining membrane selectivity for K over Na. Nitrogen fertilization particularly as NH_4^+ has been found to help overcome yield reducing effects of salinity (Langdale et al., 1971). Addition of Ca to saline irrigation water affects nutrient uptake, at least partly by modifying root function resulting in reduced uptake of heavy metals even though their solubility may not be appreciable in saline waters (Helal et al., 1996). There is evidence that when Cl^- rather than SO_4^{2-} is the anion in saline soils, Ca deficiency could be alleviated (Curtin et al., 1993).

3.7.4   Nutrient Interactions with Heavy Metals

Bioavailability of heavy metals increases with decreasing soil pH and with the decomposition of organic matter (Hani, 1996). Application of lime and manures has been suggested as a measure to reduce heavy metal bioavailability as illustrated in Fig. 3.11 (Pierzynski and Schwab, 1993) where increasing the lime rate decreased the Zn concentration in soybean tissue. On the other hand, manure only reduced tissue Zn in the absence of lime.

Zinc/Cd interactions have been studied perhaps more extensively than those of other heavy metals because of the potential threat of Cd to human health. Low rates of Zn (5.0 kg ha^{-1}) can reduce Cd concentrations in wheat (Oliver et al., 1994) and potato tubers (McLaughlin et al., 1997). Choudhary et al. (1993) found that N and P fertilization reduced tissue Cd in wheat by promoting growth (dilution). McLaughlin et al. (1995) showed, that in saline soils, Cd interacts with Cl^- to form a $CdCl^+$ complex which is readily taken up and transported in plants. In solution culture, Cd has been shown to reduce the uptake of other heavy metals (Yang et al., 1996).

3.8   Contributing Factors in Plant Nutrient Interactions

3.8.1   Nonplant Nutrient Factors

Factors which affect plant growth such as temperature, soil water availability, and light intensity, as well as biological factors such as species, cultivars, strains, and ecotypes, all influence nutrient demand, and consequently, nutrient requirements. Response to nutrient supply is impacted whether that supply is indigenous or from fertilizer.

3.8.1.1   Temperature
Chemical reactions generally double for each 10 °C increase in temperature until the optimum temperature for the reaction is reached. Temperature affects plant growth similarly and consequently, affects demand for nutrients. Plant membrane composition, structure and integrity are also impacted by temperatures particularly as they deviate from the optimum (Marschner, 1995). Loss of membrane integrity may mean loss of control of normal ion uptake while temperature may affect nutrient uptake differently for different plant species. For example, P uptake is generally depressed more than other nutrients by low root zone temperatures (Marschner, 1991). Temperature also affects the growth of roots which may limit nutrient uptake to shoots and growth of tops at low nutrient supply, but not be limiting at high nutrient supply in the root zone.

3.8.1.2   Light Intensity
Marschner (1995) reported that high light intensities may increase sensitivity to B deficiency by raising the amount of B required to detoxify the phenolics produced at the higher light intensities. Bates (1971) indicated that critical deficiency concentrations of B and Zn were higher under low light

intensity. Adequate light energy stimulates growth and nutrient demand. Low light intensity can result in accumulation of nitrates in plant tissue and inefficient use of N.

3.8.1.3   Soil Physical Conditions

Soil aeration is a very important factor controlling soil microbial activity, nutrient availability, and plant root growth (Walters et al., 1992). When soils remain wet and become reduced, the availability of Fe, Mn, Cu , and P usually increases and this is reversed on drying. The balance between water and air-filled porosity is crucial in plant nutrition. Oxygen consumption by microbial and root respiration and slow O_2 diffusion in water leads to poor aeration, which reduces root growth. Consequently, because soil compaction reduces pore space and O_2 supply, root growth and nutrient interception are impaired.

Water availability influences nutrient availability through its effect on mass flow and diffusion, although this effect is likely to be less than its effect on top growth. Increasing water availability facilitates nutrient diffusion, while soil texture also interacts with water content. For example, at constant soil water potential, the soil solution P concentration required to supply the same amount of P to the root by diffusion decreases with increasing volumetric moisture content (Cox and Barber, 1992).

3.8.1.4   Biological Factors Involved in Interactions

Plant nutrient interactions vary between and within species, cultivars, strains, and ecotypes. The large number of plant species, and their genotypic variability including annuals, perennials, C3 or C4 type metabolism, etc. suggests that the potential for plant germplasm interaction with nutrients is very great. Agricultural technology has normally stressed changing the soil to meet the needs of the crop. More recently, plant breeding and biotechnology have been used to manipulate the crop to adapt it to various soil conditions (Barrow, 1993; Marschner, 1995). Organisms can have profound effects on plant nutrient uptake through their effect on nutrient accumulation, depletion, immobilization, mineralization, and pH of the rhizosphere, where they reside in much greater numbers than in the bulk soil. These organisms may be either detrimental (root pathogens, subclinical pathogens, detrimental rhizobacteria, cyanide producers, and nematodes) or beneficial (rhizobia, mycorrhiza, antagonists of detrimental microorganisms, hormone producer, and plant growth stimulating bacteria) to plant growth and health (Marschner, 1995). Mycorrhizal associations which may be either mutualistic (beneficial), neutral, or parasitic with higher plants are most often beneficial in terms of plant nutrition, especially with P and sometimes other nutrients. Occasionally, they may not be beneficial under very wet, or high nutrient conditions (Clark and Zeno, 1996).

3.8.1.5   Cultural Management

According to Sharpley et al. (1992) management variables controlling soil water content, nutrient availability, root growth and development, and thereby root extraction of nutrients are interactive, complex and dynamic. These include crop production system (field, container with soil or artificial media, hydroponics), level and placement of fertilizers and water, and residue management.

Cultural practices can interact with specific nutrient management plans for a crop or pastoral production system. Holistic approaches to soil-crop-animal nutrition management offer opportunities to take advantage of the interconnection of these components in the ecosystem. Improved models are needed which predict nutrient interactions, and quantify their impacts on yield and composition as affected by other contributing factors in the environment.

3.9   References

Adams, F. 1980. Interactions of phosphorus with other elements in soils and in plants. p. 655–680. *In* R.C. Dinauer (ed.) The role of phosphorus in agriculture. American Society of Agronomy, Madison, WI.

Adams, W.E., A.W. White, Jr., and R.N. Dawson. 1967. Influence of lime sources and rates on 'Coastal' bermudagrass production-soil profile reaction, exchangeable Ca and Mg. Agron. J. 59:147–149

Arthur, M.A., G. Rubin, P.B. Woodbury, H.L. Weinstein. 1993. Gypsum amendment to soil can reduce selenium uptake by alfalfa grown in presence of coal fly-ash. Plant Soil 148:83–90.

Baligar, V.C., R.J. Wright, and K.D. Ritchey. 1992. Soil acidity effects on wheat seedling growth. J. Plant Nutr. 15:845–856.

Barber, S.A. 1995. Soil nutrient availability. A mechanisitic approach. John Wiley and Sons, New York, NY.

Barber, K.L., G.M. Pierzynski, and R.L. Vanderlip. 1994. Ammonium/nitrate ratio effects on dry-matter partitioning and radiation use efficiency of corn. J. Plant Nutr. 17:869–882.

Barrow, N.J.1993. Plant nutrition proceedings: From genetic engineering to field practice. Kluwer Academic Publishers, Dordrecht, Netherlands.

Bates, T.E. 1971. Factors affecting critical nutrient concentrations in plants and their evaluation: A review. Soil Sci. 112:116–126.

Bell, P. F., D.R. Parker, and A. L. Page. 1992. Contrasting selenate sulfate interactions in selenium-accumulating and nonaccumulating plant-species. Soil Sci. Soc. Amer. J. 56:1818–1824.

Boawn, L.C., F.G. Viets, and C.L. Crawford. 1954. Effect of phosphate fertilizer on zinc nutrition of field beans. Soil Sci. 78:1–7.

Brams, E.A. 1971. Continuous cultivation of West African soils: Organic matter diminution and effects of applied lime and phosphorus. Plant Soil 35:401–414.

Carreker, J.R., S.R. Wilkinson, A.P. Barnett, and J.E. Box, Jr. 1977. Soil and water management systems for sloping lands. ARS-USDA Spec. Pub. ARS-S-160.

Cassel, D.K., E.J. Kamprath, and F.W. Simmons. 1996. Nitrogen-sulfur relationships in corn as affected by landscape attributes and tillage. Agron. J. 88:133–140.

Chandler, W.V., and G.D. Scarseth. 1941. Iron starvation as affected by overphosphating and sulfur treatment on Houston and Sumter Clay soils. J. Am. Soc. Agron. 33:93–104.

Choudhary, M., D. Bailey, and C.A. Grant. 1993. Effect of zinc on cadmium concentration in the tissue of durum wheat. Can. J. Plant Sci. 74:549–552.

Clark, R.B., and S.R. Zeno. 1996. Mineral acquisition by mycorrhizal maize grown on acid and alkaline soil. Soil Biol. Biochem. 28:1495–1503

Cole, C.V., D.L. Grunes, L.K. Porter, and S.R. Olsen. 1963. The effects of nitrogen on short term phosphorus absorption and translocation in corn (*Zea mays*). Soil Sci. Soc. Amer. Proc. 27:671-674.

Coll, J.J., H.H. Schomberg, and R.W. Weaver. 1989. Effectiveness of rhizobial strains on arrowleaf clover grown in acidic soil containing manganese. Soil Biol. Biochem. 21:755–758.

Cox, M.S., and S.A. Barber. 1992. Soil phosphorus levels for equal P uptake from four soils with different water contents at the same potential. Plant Soil 143:93–98.

Cronan, C.S., and D. F. Grigal. 1995. Use of calcium/aluminum ratios as indicators of stress in forest ecosystems. J Environ. Qual. 24:209–226.

Curtin, D., H. Stepphuhn, and F. Selles. 1993. Plant growth responses to sulfate and chloride salinity-growth and ionic relations. Soil Sci. Soc. Am. J. 57:1304–1310.

Deiorio, A.F., L. Gorgoschide, A. Rendina, and M.J. Barros. 1996. Effect of phosphorus, copper and zinc addition on the phosphorus/copper and phosphorus/zinc interaction in lettuce. J. Plant Nutr.19:481–491.

Delafuente, J.M., V. Ramirezrodiquez, J.L. Caberaponce, and L. Herreralestrella. 1997. Aluminum tolerance in transgenic plants by alteration of citrate synthesis. Science 276: 1566–1568.

Dijkshoorn, W., E.J. Lampe, and L.W. Broekhaven. 1983. Effect of soil pH and ammonium nitrate treatments on heavy metals in ryegrass from sludge amended soil. Neth. J. Agric. Sci. 31:181–188.

Ellis, R., J.F. Davis, and D.L. Thurlow. 1964. Zinc availability in calcareous Michigan soils as influenced by phosphorus level and temperature. Soil Sci. Soc. Am. Proc. 28:83–86.

Engels, C., and H. Marschner. 1992. Root to shoot translocation of macronutrients in relation to shoot demand in maize (*Zea mays* L.) grown at different root zone temperatures. Z. Pflanzenäehr. Bodenk. 155:121–128.

Epstein, E. 1994. The anomaly of silicon in plant biology. Proc. Nat. Acad. Sci. 91:11–17.

Fales, S.L., A.S. Laidlaw, and M.G. Lambert. 1996. Cool-season grass ecosystems. p. 267–296. *In* L.E. Moser, D.R. Buxton, and M.D. Casler (ed.) Cool-season forage grasses. American Society of Agronomy, Madison, WI.

Farina, M.P.W., M.E. Sumner, C.O. Plank, and W.S. Letsch. 1980. Aluminum toxicity in corn at near neutral pH levels. J. Plant Nutr. 2:683–697.

Fixen, P. 1996. Nutrient management following conservation reserve program. Better Crops 80:16–19.

Follett, R.F., J.F. Power, D.L. Grunes, and C.A. Klein. 1977. Effect of N, K, and P fertilization, N source and clipping on potential tetany hazard of bromegrass. Plant Soil 48:485–508.

Foy, C.D., R.L. Chaney, and M.C. White. 1978. The physiology of metal toxicity in plants. Ann. Rev. Plant Physiol. 29:511–566.

Friesen, D.K., A.S.R. Juo, and M.H. Miller. 1980. Liming and lime phosphorus-zinc interactions in two Nigerian Ultisols. Soil Sci. Soc. Am. J. 44:1221–1226.

Galvez, L., R.B. Clark, L.M. Gourley, and J.W. Maranville. 1989. Effects of silicon on mineral composition of sorghum grown with excess manganese. J. Plant Nutr. 12:547–561.

Gastal, F., and B. Saugier. 1989. Relationship between nitrogen uptake and carbon assimilation in whole plants of tall fescue. Plant Cell Environ. 12:407–418.

Gobran, G.R. and S. Clegg. 1996. A conceptual model for nutrient availability in the soil-root system. Can. J. Soil Sci. 76:125–131.

Goss, M.J., G.P.R. Carvalho, V. Cosimini, and M.L. Fearnhead. 1992. An approach to the identification of potentially toxic concentrations of manganese in soils. Soil Use Manag. 8:40–44.

Grunes, D.L. 1959. Effect of nitrogen on the availability of soil and fertilizer phosphorus to plants. Adv. Agron. 11:369-396.

Grunes, D.L., J.W. Huang, F.W. Smith, P.K. Joo, and D.A. Hewes. 1992. Potassium effects on minerals and organic acids in three cool season grasses. J. Plant Nutr.15:1007–1025.

Gupta, U.C. 1997. Soil and plant factors affecting molybdenum uptake by plants. p. 71–91. *In* U.C. Gupta (ed.) Molybdenum in agriculture. Cambridge University Press, Cambridge, UK.

Haleem, A.A., W.B. Anderson, M.K. Sadik, and A.A. Salam. 1992. Iron movement from band placed ferrous-iron fertilizer in an iron-deficient soil as influenced by phosphorus fertilizers. J. Plant Nutr. 15:1983–1994.

Hammond, K.E., D.E. Evans, and M.J. Hodson. 1995. Aluminum silicon interactions in barley (*Hordeum vulgare* L.) seedlings. Plant Soil 173:89–95.

Handreck, K.A. 1991. Interactions between iron and phosphorus in the nutrition of *Banksia ericifolia* LF var Ericifolia (Proteaceae) in soil-less potting media. Austr. J. Bot. 39:373–384.

Hani, H. 1996. Soil analysis as a tool to predict effects on the environment. Commun. Soil Sci. Plant Anal. 27:289–306.

Hannaway, D.B., L.P. Bush, and J.E. Leggett. 1980. Plant nutrition: magnesium and hypomagnesemia in animals. KY Agric. Exp. Stn. Bull. 716.

Hannaway, D.B., L.P. Bush, and J.E. Leggett. 1982. Mineral composition of Kenhy Tall Fescue as affected by nutrient solution concentrations of Mg and K. J. Plant Nutr. 5:137–151.

Helal, H.M., S.A. Haque, A.B. Ramadan, and E. Schnug, 1996. Salinity-heavy metal interactions as evaluated by soil extraction and plant analysis. Commun. Soil Plant Anal. 27:1355–1361,

Hillard, J.B., V.A. Haby, and F.M. Hons. 1992. Phosphorus effects on magnesium uptake by forage grasses. Better Crops Plant Food 76:22–23.

Hodson, M.J. 1995. Aluminum/silicon interactions in higher plants. J. Expt. Bot. 46:161–171.

Huang, J.W., J.E. Shaff, D.L. Grunes, and L.V. Kochian. 1992. Aluminum effects on calcium fluxes at the root apex of aluminum-tolerant and aluminum-sensitive wheat cultivars. Plant Physiol. 98:230–237.

Huang, J.W., D.L. Grunes, and L.V. Kochian. 1993. Aluminum effects on calcium uptake and translocation in wheat forages. Agron. J. 85:867–873.

Huang, J.W., D.L. Grunes, and R. M. Welch. 1990. Magnesium, nitrogen form, and root temperature effects on grass tetany potential of wheat forage. Agron. J. 82:581–587.

Huang, J.W., and D.L. Grunes. 1992. Potassium/magnesium ratio effects on aluminum tolerance and mineral composition of wheat forage. Agron. J. 84:643–650.

Huang, J.W., D.L. Grunes, and L.V. Kochian. 1995. Aluminum and calcium transport interactions in intact roots and root plasmalemma vesicles from aluminum sensitive and tolerant Wheat Cultivars. Plant Soil 171:131–135.

Huang, J.W., D.L. Grunes, and L.V. Kochian. 1993. Aluminum effects on calcium ($^{45}Ca^{2+}$) translocation in aluminum tolerant and aluminum-sensitive wheat (*Triticum aestivum* L.) cultivars. Plant Physiol. 102:85–93.

Huang, J.W, D.L. Grunes, and L.V. Kochian. 1992. Aluminum effects on the kinetics of calcium uptake into the cells of the wheat root apex. Planta 88:414–421.

Hue, N.V., and E.T. Nakamura. 1988. Iron chlorosis in Macadamia as affected by P-Fe interactions. J. Plant Nutr. 11:1635–1648.

Hue, N.V., R.L. Fox, and W.W. McCall. 1988. Chorosis in Macadamia as affected by phosphate fertilization and soil properties. J. Plant Nutr. 11:161–173.

Johannsonn, O.A.H., and J.M. Hahlin. 1977. Potassium/magnesium balance for maximum yield. p. 487–495. *In* Proc. Int. Sem. Soil Environ. Fert. Manag. Intens. Agric., Tokyo, Japan.

Jones, J.P., and R.L. Fox. 1978. Phosphorus nutrition of plants influenced by manganese and aluminum uptake from an Oxisol. Soil Sci. 126:230–236.

Kemp, A. 1971. The effects of K and N dressings on the mineral supply of grazing animals. p. 1–14. *In* Potassium and systems of grassland farming. Potassium Institute, Henley-on-Thames, UK.

Kemp, A. 1983. The effect of fertilizer treatment of grassland on the biological availability of magnesium to ruminants. p.143–157. *In* J.P. Fontenot, G.E. Bunce, K.E. Webb, Jr., V.G. Allen (ed.). Role of magnesium in animal nutrition. VPISU, Blacksburg, VA.

Khattak, R.A., A. L. Page, D.R. Parker, and D. Bakhtar. 1991. Accumulation and interactions of arsenic, selenium, molybdenum, and phosphorus in alfalfa. J. Environ. Qual. 20:165–168.

Langdale, G.W., and J.R. Thomas. 1971. Soil salinity effects on absorption of nitrogen, phosphorus, and protein synthesis by Coastal bermudagrass. Agron. J. 63:708–711.

Li, Y.C., M.E. Sumner, W.P. Miller, and A.K. Alva. 1996. Mechanism of silicon induced alleviation of aluminum phytotoxicity. J. Plant Nutr. 19:1075–1087.

Liang, Y., Q. Shen, Z. Shen, and T. Ma. 1996. Effects of silicon on salinity tolerance of two barley cultivars. J. Plant Nutr. 19:173–183.

Loneragan, J.F., D.L. Grunes, R.M. Welch, E.A. Aduayi, A. Tengah, V.A. Lazar, and E.E. Cary. 1982. Phosphorus accumulation and toxicity in leaves in relation to zinc supply. Soil Sci. Soc. Am. J. 46:345–352.

Macdiarmid, C.W. and R.C. Gardner. 1996. Al toxicity in yeast-a role for Mg. Plant Physiol. 112:1101–1109.

MacLeod, L.B. 1969. Effects of N, P, and K and their interactions on the yield and kernel weight of barley in hydroponic culture. Agron. J. 61:26–29.

Magalhaes, J.R., and G.E. Wilcox. 1983. Tomato growth, nitrogen fraction and mineral composition in response to nitrate and ammonium foliar sprays. J. Plant Nutr. 6:911–939.

Mallarino, A.P., and J.P. Webb. 1995. Long-term evaluation of phosphorus and zinc interactions in corn. J. Prod. Agric. 8:52–55.

Marschner, H. 1995. Mineral nutrition of higher plants. 2nd Ed. Academic Press, London, UK.

Marschner, H. 1991. Plant-soil relationships: Acquisition of mineral nutrients by roots from soils. p.125–155. *In* J.R. Porter and D.W. Lawlor (ed.) Plant growth: Interactions with nutrition and environment. Cambridge University Press, Cambridge, UK.

Mathers, A.C., B.A.Stewart, and D.L. Grunes. 1982. Effect of a nitrogen inhibitor on the K, Ca, and Mg composition of winter wheat forage. Agron. J. 74:569–573.

McDonald, J.S. 1994. Nutrient supply and plant growth. p. 47–57. *In* P.J. Lunsden, J.R. Nicholas, and W.J. Davies (ed.) Physiology, growth, and development of plants in culture. Kluwer Academic Publishers, Dordrecht, Netherlands.

McGrath, S.P. and J. F. Zhao. 1996. Sulfur uptake, yield responses and the interactions between nitrogen and sulfur in winter oilseed rape (*Brassica napus*). J. Agric. Sci. 126:53–62.

McLaughlin, M.J., K G. Tiller, and M K. Smart. 1997. Speciation of cadmium in soil solutions of saline/sodic soils and relationship with cadmium concentrations in potato tubers (Solanum-tubersum L). Aust. J. Soil Res. 35:83–198.

McLaughlin, M.J., N.A. Maier, K. Freeman, K.G. Tiller, C.M.J. Williams, and M.K. Smart. 1995. Effect of potassic and phosphatic fertilizer type, fertilizer Cd concentration, and zinc rate on cadmium uptake by potatoes. Fert. Res. 40:63–70.

Miller, M.H. 1974. Effects of nitrogen on phosphorus absorption by plants. p. 643–683. *In* E.W. Carson (ed.) The plant root and its environment. Proc. Institute, Southern Regional Education Board, Virginia Polytechnic Institute and State University, 5-16 July, 1971. University Press of Virginia. Charlottesville, VA.

Morris, D.R., T.F. Brown, V.C. Baligar, D.L. Corkern, L.K. Zeringue, and L.F. Mason. 1994. Ryegrass forage yield and quality response to sulfur and nitrogen-fertilizer on Coastal plain soil. Commun. Soil Sci. Plant Anal. 25:3035–3046.

Munns, D.N., and R.L. Fox. 1976. Depression of legume growth by liming. Plant Soil 45:701–705.

Murphy, M. D., and W.A. Quirke. 1997. The effect of sulphur/nitrogen/selenium interactions on herbage yield and quality. Irish J. Agric. Food Res. 36:31–38.

Nass, M.M., T.M. Lexmond, M.L. van Beusichem, and M. Janssenenjurkkovicova.1993. Long-supply and uptake by plants of elements from coal fly-ash. Commun. Soil Sci. Plant Anal. 24:899–913.

Nichol, B.E., L.A. Oliveira, A.D. Glass, and M.Y. Siddiqi. 1993. The effects of aluminum on the influx of calcium, potassium, ammonium, nitrate, and hosphate in an aluminum sensitive cultivar of barley (*Hordeum vulgare* L.). Plant Physiol. 101:1263–1266.

Noble, A.D., M.V. Fey, and M.E. Sumner. 1988. Calcium-aluminum balance and the growth of soybean roots in nutrient solutions. Soil Sci. Soc. Am. J. 52:1651–1656.

Noble, A.D., and M.E. Sumner. 1988. Calcium and Al interactions and soybean growth in nutrient solutions. Commun. Soil Sci. Plant Anal. 19:1119–1131.

Norvell, W.A., H. Dabkowska-Naskret, and E.E. Cary. 1987. Effect of phosphorus and zinc fertilization on the solubility of Zn^{+2} in two alkaline soils. Soil Sci. Soc. Amer. J. 46:345–352.

Ohno, T., and D.L. Grunes. 1985. Potassium-magnesium interactions affecting nutrient uptake by wheat forage. Soil Sci. Soc. J. 49:685–690.

Ohno, T., D.L. Grunes, and C.A. Sanchirico. 1985. Nitrogen and potassium fertilization and environmental factors affecting the grass tetany hazard of wheat forage. Plant Soil 86:173–184.

Oliver, D.P, R. Hannum, K.G. Tiller, N.S. Wilhelm, R.H. Merry, and G.D. Cozens. 1994. The effects of zinc fertilization on cadmium concentration in wheat grain. J. Environ. Qual. 23:705–711.

Parfitt, R.L. 1977. Phosphate adsorption by an Oxisol. Soil Sci. Soc. Am. J. 41:1064–1067.

Peck, N.H., D.L. Grunes, R.M. Welch, and G.E. MacDonald. 1980. Nutritional quality of vegetable crops as affected by phosphorus and zinc fertilizers. Agron. J. 72:528–534.

Pellet, D.M., D.L. Grunes, and L.V. Kochian. 1995. Organic acid exudation as an aluminum-intolerance mechanism in maize (*Zea mays* L.). Planta 196:798–795.

Pellet, D.M., L.A. Papernick, D.L. Jones, P.R. Darrah, D.L. Grunes, and L.V. Kochian. 1997. Involvement of multiple aluminum exclusion mechanisms in aluminum tolerance in wheat. Plant Soil 192:63–68.

Pierzynski, G.M., and A.P. Schwab. 1993. Bioavailability of zinc, cadmium, and lead in a metal contaminated alluvial soil. J. Environ. Qual. 22:247–254.

Reeve, N.G., and M.E. Sumner. 1970. Effects of aluminum toxicity and phosphorus fixation on crop growth on Oxisols in Natal. Soil Sci. Soc. Am. Proc. 34:263–267.

Reinbott, T.M., and D.G. Blevins. 1991. Phosphate interaction with uptake and leaf concentration of magnesium, calcium and potassium in winter wheat seedlings. Agron. J. 83:1043–46.

Reinbott, T.M., and D.G. Blevins. 1997. Phosphorus and magnesium fertilization interaction with soil phosphorus level: Tall Fescue yield and mineral element content. J. Prod. Agric. 10:260–265.

Reinbott, T.M., and D.G. Blevins. 1994. Phosphorus and temperature effects on magnesium, calcium, and potassium in wheat and tall fescue leaves. Agron. J. 86:523–529.

Robson, A.D., and M.G. Pitman. 1983. Interactions between nutrients in higher plants. p. 147–180. *In* A. Lauchli and R.L. Bielski (ed.) Encyclopedia of plant physiology. Springer-Verlag, Berlin, Germany.

Russell, E.J. 1950. Soil conditions and plant growth. Longmans, Green and Co., London, UK.

Ryan, J., and G.W. Smillie. 1975. Liming in relation to soil acidity and P fertilizer efficiency. Commun. Soil Sci. Plant Anal. 16:409–420.

Schomberg, H.H., and R.W. Weaver. 1991. Growth and N_2 fixation response of Arrowleaf clover to manganese and pH in solution culture. Develop. Plant Soil Sci. 45:641–647.

Sharpley, A.N., J.J. Meisinger, J.F. Power, and D.L. Suarez. 1992. Root extraction of nutrients associated with long-term soil management. Adv. Soil Sci. 19:151–217.

Shortle, W.C., K.T. Smith, R. Minodia, G.B. Lawrence, and M.B. David. 1997. Acidic deposition, cation mobilization, and biochemical indicators of stress in healthy Red Spruce. J. Environ. Qual. 26:871–876.

Shuman, L.M., E.L. Ranseur, and R.R. Duncan. 1990. Soil aluminum effects on the growth and aluminum concentration of Sorghum. Agron. J. 82:313–318.

Skinner, P.N., and M.A. Matthews. 1990. A novel interaction of magnesium translocation with the supply of phosphorus to roots of grape vines (*Vitis vinifera* L.). Plant Cell Environ. 13:821–826

Smyth, T.J., and M.S. Cravo. 1992. Aluminum and calcium constraints to continous crop production in a Brazilian Amazon Oxisol. Agron. J. 84:843–850.

Smyth, T.J., and P.A. Sanchez. 1980. Effects of lime, silicate and phosphorus applications to an Oxisol on phosphorus sorption and iron retention. Soil Sci. Soc. Am. J. 44:500–505.

Soliman, M.F., S.F. Kostandi, and M.L. van Beusichem. 1992. Influence of sulfur and nitrogen fertilizer on the uptake of iron, manganese, and zinc by corn plants grown in calcareous soil. Comm. Soil Sci. Plant Anal. 23:1289–1300.

Sumner, M.E., and M.P.W. Farina. 1986. Phosphorus interactions with other nutrients and lime in field cropping systems. Adv. Soil Sci. 5:201–236.

Sumner, M.E. 1979. Response of alfalfa and sorghum to lime and P on highly weathered soils. Agron. J. 71:763–766.

Syedomar, S.R., and M.E. Sumner. 1991. Effect of gypsum on soil potassium and magnesium status. Commun. Soil Plant Anal. 22:2017–2028.

Takkar, P.N., M.S. Mann, R.L. Bansal, N.S. Randhawa, and H. Singh. 1976. Yield and uptake response of corn to zinc as influenced by phosphorus fertilization. Agron. J. 68:942–946.

Tan, K., W.G. Keltjens, and G.R. Findenegg. 1992. Effect of nitrogen form on aluminum toxicity in sorghum genotypes. J. Plant Nutr. 15:1383–1394.

Teng, Y., and V.R. Timmer. 1994. Nitrogen and phosphorus interactions in an intensively managed nursery soil-plant system. Soil Sci. Soc. Am. J. 58:232–238.

Thomson, C.J., H. Marschner, and V. Romheld. 1993. Effect of nitrogen fertilizer form on pH of the bulk soil and rhizosphere and on the growth, phosphorus and micronutrient uptake of bean. J. Plant Nutr. 16:493–506.

Tulin, S.A., N.G. Stavrova, S.O. Koroviakovskaya, and A.S. Tulina. 1995. The effects of potassium fertilizers on ^{137}Cs and yield of crops on Bryansk Soddy Podzolic sandy soil contaminated by the Chernobyl disaster. Potash Review, International Potash Institute, Basel, Switzerland 2:1–2.

van Beusichem, M.L., E.A.Kirkby, and R. Baas. 1988. Influence of nitrate and ammonium on the uptake, assimilation and distribution of nutrients in *Ricinus communis*. Plant Physiol. 86:914–921.

Vona, L.C., C. Merideth, R.L. Reid, J.C. Hern, H.D. Perry, and O.L. Bennett. 1992. Effect of fluidized bed combustion residue on the health and performance of sheep grazing hill pastures. J. Environ. Qual. 21:335–40.

Wallace, A. 1993. Participation of silicon in cation-anion balance as a possible mechanism for aluminum and iron tolerance in some Gramineae. J. Plant Nutr.16:547–553.

Wallace, A., E. Frolich, and O.R.Lunt. 1966. Calcium requirements of higher plants. Nature 209:634.

Wallace, A. 1990. Crop improvement through multi-disciplinary approaches to different types of stresses: Law of the Maximum. J. Plant Nutr. 13:313–325.

Walters, D.T., M.S. Aulakh, and J. W. Doran. 1992. Effects of soil aeration, legume residue, and soil texture on transformations of macronutrients and micronutrients in soils. Soil Sci. 153:100–107.

Welch, R.M., and W.A. Norvell. 1993. Growth and nutrient uptake by barley (*Hordeum vulgaris*, cv Herta): Studies using an N(2-hydroxyethyl)ethylene dinitrilotriacetic acid buffered solution technique. II. Role of zinc in the root uptake and leakage of mineral nutrients. Plant Physiol. 101:6277–631.

Welch, L.F., W.E. Adams, and J.L. Carmon. 1963. Yield response surface isoquants and economic fertilizer optima for Coastal bermudagrass. Agron. J. 55:63–67.

Westerman, D.T., T.A. Tindall, D.W. James, and R. L. Hurst. 1994. Nitrogen and potassium fertilization of potatoes: Yield and specific gravity. Amer. Potato J. 71:417–431.

Wild, A. 1988. Potassium, sodium, calcium, magnesium, sulphur, silicon. p. 743–780. *In* A. Wild (ed.) Russell's soil conditions and plant growth. 11th Ed. Longman, Scientific and Technical, London, UK.

Wilkinson, S.R., R.N. Dawson, O. Devine, and J.B. Jones, Jr. 1987. Influence of dolomitic limestone and NPK fertilization on the mineral composition of Coastal bermudagrass grown under very acid conditions. Commun. Soil Sci. Plant Anal. 18:1191–1215.

Wilkinson, S.R., J.A. Stuedemann, D.L. Grunes, and O.J. Devine. 1987. Relation of soil and plant magnesium to nutrition of animals and man. Magnesium 6:74–90.

Wilkinson, S.R., R.M. Welch, H.F. Mayland, and D.L. Grunes. 1990. Magnesium in plants: Uptake, distribution, function, and utilization by animals and man. p. 33–56. *In* H. Siegel and A. Siegel (ed.) Compendium on magnesium and its role in biology, nutrition, and physiology. Marcel Dekker, Inc, New York, NY.

Wilkinson, S.R., and J.A. Stuedemann. 1979. Tetany hazard of plants as affected by fertilization with N, K, or poultry litter and methods of grass tetany prevention. p. 93–121. *In* V.V. Rendig and D.L. Grunes (ed.) Grass tetany. Am. Soc. Agron. Spec. Pub. 35. American Society of Agronomy, Madison, WI.

Wilson, J.B. 1993. Macronutrient (NPK) toxicity and interactions in the grass *Festuca ovina*. J. Plant Nutr. 16:1151–1159.

Yang, X., V.C. Baligar, D.C. Martens, and R.B. Clark. 1996. Cadmium effects on influx and transport of mineral nutrients in 4 plant-species. J. Plant Nutr.19:643–656.

Zhong, H.L., and A. Lauchli. 1994. Spatial distribution of solutes, K, Na, Ca and their deposition rates in the growth zone of primary cotton roots: Effects of NaCl and CaCl$_2$. Planta 194:34–41.

Zysset, M., I. Brunner, B. Frey, and P. Blaser. 1996. Response of european chestnut to varying calcium/aluminum ratios. J. Environ. Qual. 25:702–708.

4

Soil Fertility Evaluation

J. Thomas Sims
University of Delaware

4.1 Introduction

4.1.1 Soil Fertility: A Modern Definition

Soil fertility is a scientific discipline that integrates the basic principles of soil biology, chemistry, and physics to develop the practices needed to manage nutrients in a profitable, environmentally sound manner. Historically, the study of soil fertility has focused on managing soil nutrient status to create optimal conditions for plant growth. Fertile, productive soils are vital components of stable societies to ensure that the plants needed for food, fiber, animal feed and forage, medicines, industrial products, and for an aesthetically pleasing environment can be grown. Two other fundamental principles underlay the study of soil fertility. First is the recognition that optimum nutrient status alone will not ensure soil productivity. Other factors, such as soil moisture and temperature, soil physical condition, soil acidity and salinity, and biotic stresses (disease, insects, weeds) can reduce the productivity of even the most fertile soils. Second is the realization that modern soil fertility practices must stress environmental protection as well as agricultural productivity.

4.1.2 Soil Fertility Evaluation: Purpose, General Principles and Practices

The fundamental purpose of soil fertility evaluation has always been to quantify the ability of soils to supply the nutrients required for optimum plant growth. Knowing this, the nutrient management practices needed to achieve economically optimum plant performance can be optimized. Related, equally important goals, are (1) to identify other factors that reduce soil productivity (e.g., acidity, salinity, elemental phytotoxicity), and (2) to determine if the intended use of the soil may negatively impact environmental quality.

Basically, soil fertility is evaluated by observations and tests which are used to predict the response of plants, and the larger environment, to nutrient management (Black, 1993; Tisdale et al., 1993; Foth and Ellis, 1997). Soil fertility evaluation involves an impressive array of field and laboratory diagnostic techniques and a series of increasingly sophisticated empirical and/or theoretical models that quantitatively relate these indicators of soil fertility to plant response. Diagnostic techniques include chemical and biological soil tests, visual observations of plant growth for nutrient deficiency or toxicity symptoms, and chemical analysis of plant tissues. New approaches include remote-sensing technologies and geographic information systems (GIS) that facilitate landscape-scale, site-specific assessments of soil fertility. Computerized expert systems enable these indicators of soil fertility to be related to quantitative or qualitative assessments of plant performance (yield, composi-tion, quality, color, health), and thus, to rapidly adjust soil management practices for the most efficient use of nutrients.

4.1.3   Soil Fertility Evaluation for Agricultural and Nonagricultural Systems

The study of soil fertility evolved within ecosystems devoted primarily to the production of agricultural crops. The importance of soil fertility to world agriculture continues today as a spiraling world population and a diminishing arable land base create unprecedented pressures on scientists and practicing agriculturalists to produce more food per unit area of land than ever before. Advances in plant genetics and breeding and other agricultural technologies (e.g., irrigation) are increasing agricultural productivity. However, higher crop yields mean greater depletion of soil nutrient supplies which eventually must be balanced by increased nutrient inputs to maintain fertile soils. Soil fertility evaluation will play an increasingly important role in the future of global agriculture in identifying new lands that can be brought into production and to maximize production from existing soils. Other land uses also require a thorough, in-depth evaluation of soil fertility for maximum economic and environmental efficiency. Examples are horticultural systems, disturbed lands needing reclamation, and soil conservation and remediation practices.

Horticulture includes an extremely diverse range of situations where nutrients must be managed, often quite intensively. Land reclamation can be equally diverse in terms of the nature of the growth media and the types of plants. Soils at land reclamation sites (e.g., surface mining, highway construction, landfills) are often highly disturbed by human activity and may possess extremely unfavorable chemical and physical characteristics, including very low soil fertility.

Related, but slightly different problems are faced when evaluating the fertility of soils used for soil conservation purposes, such as grassed waterways, terraces, buffer strips, constructed wetlands and wildlife habitats. In cases such as these, where the goal is not maximum yield but a stable vegetative cover, the objective may often be low to moderate soil fertility, not agronomically optimum nutrient values. Finally, the need to ensure optimum soil fertility in soil remediation programs is a new, but increasingly important aspect of soil fertility evaluation, whether the goal is enhancing microbial degradation of an organic contaminant, such as an oil spill, or phytoremediation (plant-based remediation of a contaminated soil) of an inorganic contaminant, such as Cd, Pb, or Zn from the soil near an industrial site (Berti and Cunningham, 1994).

4.1.4   Soil Fertility Evaluation: Environmental Issues

Environmental quality is inextricably linked with soil fertility. Soils must be managed to optimize plant productivity, and to avoid or minimize pollution of water, atmosphere, and the food chain. Some essential plant nutrients contribute to environmental problems. Nitrogen may cause human and animal health problems if NO_3-N leaches to groundwaters used for drinking water supplies. Ammonia-N volatilized from fertilizers and animal manures causes soil acidification, particularly in forest ecosystems on redeposition. Nitrogen oxides (NO_x) produced by denitrification have been implicated in ozone depletion and global warming. Eutrophication of surface waters is caused by entry of P and N in runoff, erosion and aerial deposition. Soil salinity problems arise where salts accumulate, particularly in arid regions.

Recycling of wastes or byproducts as nutrient sources can also, directly or indirectly, affect environmental quality. Increasing pressure exists to land apply wastes to avoid the costs and undesirable environmental impacts of landfilling and incineration. Less developed countries often use wastes and wastewaters as fertilizers and for irrigation because of a lack of resources, equipment, and infrastructure. Use of municipal biosolids (sewage sludges, composts) as soil amendments, which is beneficial in recycling nutrients and building soil organic matter (SOM), is carefully regulated in most countries to limit nonessential and potentially toxic elements from impacting human, animal, or ecosystem health. Land application for biosolids (and animal manures) is usually based on the amount of N needed for optimum yield. However, the unfavorable N:P ratio in most organic wastes, relative to that in crops, means that P accumulates above required levels in waste amended soils. This creates an environmental dilemma because organic wastes used at beneficial N rates cause the buildup of P which can impact surface waters by losses in runoff and erosion. Other byproducts such as papermill sludges, municipal composts, wood ashes, flue gas desulfurization gypsum, and coal fly ash have beneficial effects, but often create similar dilemmas in terms of acidity/alkalinity, soluble salts, and immobilization of plant nutrients. Soil fertility evaluation is more complex today because of the need to balance productivity and environmental protection for a wider and more diverse range of land uses (Fig. 4.1).

4.2   Soil Testing

4.2.1   Introduction

Soil testing is defined as "...rapid chemical analyses to assess the plant-available nutrient status, salinity, and elemental toxicity of a soil .. a program that includes interpretation, evaluation, fertilizer

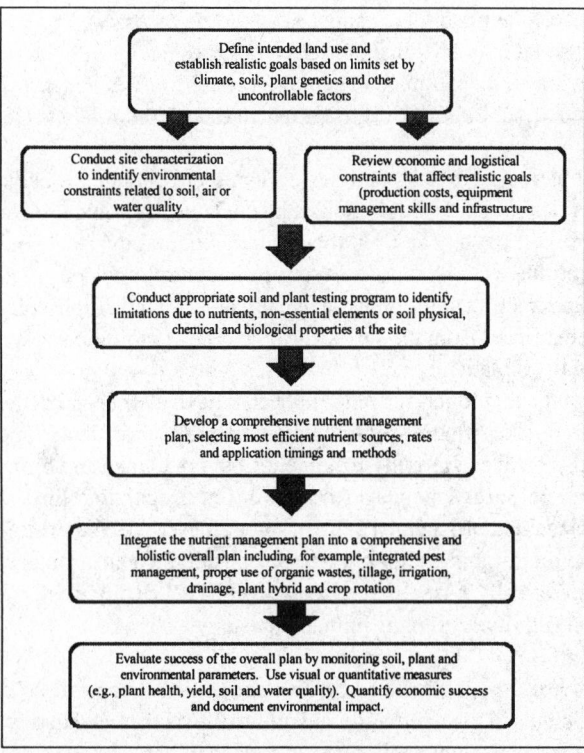

Fig. 4.1 Conceptual summary of the process of soil fertility evaluation

and amendment recommendations based on results of chemical analyses and other considerations" (Peck and Soltanpour, 1990). The use of soil testing represents perhaps the most significant practical application of our knowledge of soil science to land use management and should be viewed fundamentally as an interpretive process, not simply as a series of laboratory methods. The purpose of soil testing is to provide a quantitative basis for soil management decisions, usually, but not always, for agricultural systems where yield and quality are the ultimate measures of success. Soil testing also has applications to other systems in which the goal is protection of human health or the environment where the measure of success is not productivity.

4.2.2   Soil Testing: Assessing Elemental Availability in Soils

One of the fundamental tenets of soil testing is that only a proportion of the total quantity of an element in a soil will be available for assimilation by a biological organism. This means that measuring total elemental concentration in soils is usually of little value. Instead, testing methods that can extract (complex, dissolve, desorb, exchange, hydrolyze) a portion of the total that is proportional to the quantity that will become available to the organism of interest during the time period of concern are needed. The term labile is often used to describe the chemical and biological forms of an element that are in rapid equilibrium with the soil solution and are thus most likely to be available for biological assimilation. Much of soil testing research has focused on the development of chemical extracting solutions that can selectively remove labile elements from the soil in a rapid and reproducible manner.

4.2.2.1   Assessing Essential and Nonessential Element Availability

For soil fertility purposes, availability of the 13 essential mineral elements (N, P, K, Ca, Mg, S, B, Cl, Cu, Fe, Mn, Mo, Zn) for plant growth in soils needs to be assessed. For environmental purposes, essential plant nutrients that may be transported to other ecosystems and the fate of essential and non essential elements that may impact human or ecosystem health, are of concern. Nonessential elements of greatest concern are Al, As, Ba, Cd, Cr, Pb, Hg, Ni, and Se. Potentially toxic essential elements are Cu, Mn, Mo, and Zn.

The availability of elements in soils is assessed using chemical extractants, which are typically dilute solutions or mixtures of acids, bases, salts, and chelates. Biological techniques (e.g., bioassays based on microbial growth) have been used to a limited extent, but are generally too expensive and time consuming for routine use. The most effective concentrations and relative proportions of the reagents in a chemical extracting solution were usually determined empirically by comparison of the amount of an element extracted from the soil with some type of biological response (usually yield) by a target organism (usually a plant).

Soil testing extractants developed for plant nutrients have also been used, with some success, to assess the risk of plant uptake of nonessential elements (O'Connor, 1988; Risser and Baker, 1990). Other methods have also been used for these elements, but not to measure biological availability such as that used to measure total sorbed metals up to some defined regulatory limit (USEPA, 1986) and the toxicity characteristic leaching procedure (TCLP), sometimes used to determine if a soil is sufficiently polluted with an element or organic compound to be considered a hazardous waste. For some metals (Cd, Pb, Zn), a physiologically based extraction test (PBET) (Ruby et al., 1996) has been used to simulate the biological activity within the human digestive system.

4.2.2.2   Influence of Soil Properties and Environment on Elemental Availability

One of the major challenges in the evaluation of soil fertility is the need to develop tests (chemical or otherwise) that can estimate nutrient availability in soils of widely differing biological, chemical, and physical properties. The availability of elements in soils depends both on immutable soil properties (e.g., texture, sesquioxides, carbonates, and organic matter) and those that are more sensitive to natural and anthropogenic inputs (pH, CEC, Eh). Elemental availability can also be markedly affected by broader soil properties (drainage class, nature of soil horizons, slope) and by the soil environment (aeration, moisture, temperature). Therefore, to estimate the present and future element availability, a good understanding of the basic properties of that soil and of the environmental conditions likely to be present during the time interval of interest is required.

Two approaches are used to integrate soil properties and environmental conditions with elemental availability provided by soil testing. First, soils can be tested for some of these properties such as pH and SOM content and this information can be used to modify our assessment of biological availability; other parameters (e.g., CEC, texture, oxides, carbonates) are usually too time consuming to measure routinely. Second, information on difficult to measure soil properties and on environmental conditions can be obtained by simply asking a few key questions of the individual submitting the soil sample for analysis. Knowledge of geographic location, soil series, drainage class, slope, and historical information on previous soil management practices (e.g., fertilization, liming, crop rotation) can be invaluable when evaluating soil productivity, and thus crop nutrient requirements. The ability to integrate information such as this into management recommendations has improved in the past few years with the increased use of computers and the advent of geographic information systems (GIS) that are capable of layering different databases to provide a more holistic view of the relationship between soil fertility and land use.

4.2.3 Soil Testing Program

All modern soil testing programs have four basic components: (1) sample collection, handling, and preparation; (2) analysis; (3) interpretation of analytical results; and (4) recommendations for action. For soil testing to be successful, each component must be conducted properly, keeping in mind the overall objective (production or environmental protection), and an awareness of the potential sources of error at each step.

4.2.3.1 Soil Sample Collection and Handling
Collection of a sample that is representative of the entire area of interest, whether it is a farm field, a lawn or garden, or a severely disturbed soil at a construction or mining site, is the most important step in any soil testing program. Proper handling of the sample, once collected, is also important to avoid contamination or changes in elemental concentrations due to improper storage and/or the use of incorrect techniques to prepare the soil sample for analysis (drying, grinding, sieving). An effective soil sampling and handling program must be based on an understanding of the natural and anthropogenic sources of soil variation, the proper method of sample collection (depth, time of year, sampling tools), and the sources of error in sample handling and preparation.

Understanding and Compensating for Variability
A high degree of natural variability in soil chemical and physical properties can exist even within a very small area. Sample collection should reflect this natural variability to the extent that it is likely to significantly influence the intended use of the soil. In some cases, the differences between adjacent soil series in a large field may be pronounced and require not only different soil samples, but entirely different soil and crop management practices. In other situations, differences are minor and collecting additional samples and/or altering management practices is not economically justifiable. Information on the spatial distribution of soil series in an area is available in soil surveys, which should be consulted as a first step in identifying natural sources of variability. A simple followup is the visual inspection of the areas of interest, particularly during periods of plant growth or major seasonal changes in

weather, to determine how natural soil variability may affect land use. New technologies (GIS, Global Positioning Systems [GPS]) are now available to integrate soil survey information and qualitative data on soil spatial variability and to link soil sample location to soil series. For agronomic crops, yield monitoring devices can be installed on harvesting equipment, providing a spatially based data set that relates soil fertility, soil series, and plant performance (Section 4.4).

Natural variation can often be overshadowed by variability in soil fertility caused by human activities. Many soil management practices, such as methods of fertilizer or organic waste application, tillage, land leveling, terracing, and even the year to year selection of what plants will be grown and where, can produce marked differences in soil nutrient status and other soil properties. These differences can be even more pronounced in nonagricultural settings, such as land reclamation projects, where soil disturbance can be severe and byproducts may be used as soil amendments at unusually high rates. It is also important to remember that, as with natural soil variability, anthropogenic activities can produce three-dimensional variability. For example, changing from conventional, plow-tillage to no-tillage means that fertilizers and other soil amendments will no longer be physically incorporated in the soil. Many studies have shown that this often results in an accumulation of P and a decrease in soil pH in the upper few cm of the topsoil, factors that must be accounted for when designing a soil sampling program. Similarly, nutrient management practices that involve subsoil tillage or injection of soil amendments (e.g., lime, fertilizer, manures) can extend spatial variability below the topsoil, a factor which should not be ignored in the soil sampling process.

Many other examples could be cited to make the point that spatial variability in soil properties, natural or anthropogenic, is inevitable. What is more important, however, are the practices that can be used to compensate for any known source of variability. James and Wells (1990) suggested that soil sample collection basically occurs under either uniform or nonuniform conditions and that sampling techniques should reflect this. Uniform fields are those that are reasonably similar in physical properties (slope, aspect, drainage, soil series, etc.) and management practices (crop rotation, fertilizer/manure management history, tillage, irrigation, etc.). In many agricultural settings this is a rather common scenario. The proper approach to use for uniform fields is to collect a random, composite sample. This is normally done by following a zig-zag path across the area to be sampled, collecting enough soil samples to minimize the influence of any localized nonuniformity (Fig. 4.2). Typically, this involves collecting about 25–30 separate soil cores from a uniform area < 8 ha, ensuring that field corners and edges are included in the sample. Individual soil cores are combined into one composite sample that represents the entire area by crushing and mixing and the composite sample is submitted for analysis.

Nonuniform fields are those with a high degree of either macro- (significant variation between sample points separated by > 2 m) or mesovariation (significant variation between sample points separated by 0.05 to 2 m) (James and Wells, 1990). For areas with high macrovariation, a nonrandom sampling process must be followed to characterize the average value for the soil properties of interest and to understand the spatial location of extreme values. This prevents skewing of average soil test results for an area by samples that are extremely high or low in some soil property. It also allows for more site-specific application of fertilizers, organic wastes, lime, etc. at appropriate rates, thus avoiding under or over application of these soil amendments. Nonrandom sampling requires a large number of soil samples, and is usually done by establishing a field sample grid with spacings of 15 to 30 m between grid intersection points. A composite soil sample is collected at each grid point by combining 8 to 10 soil cores collected from a 1 m diameter circle placed around the sample point. Grid spacing is a function of the intended land use and the anticipated degree of macrovariation. Note that 15 and 30 m grid spacings will result in about 45 and 12 composite soil samples per ha, respectively, considerably more than the one sample per 8 ha associated with random sampling. Grid sampling has

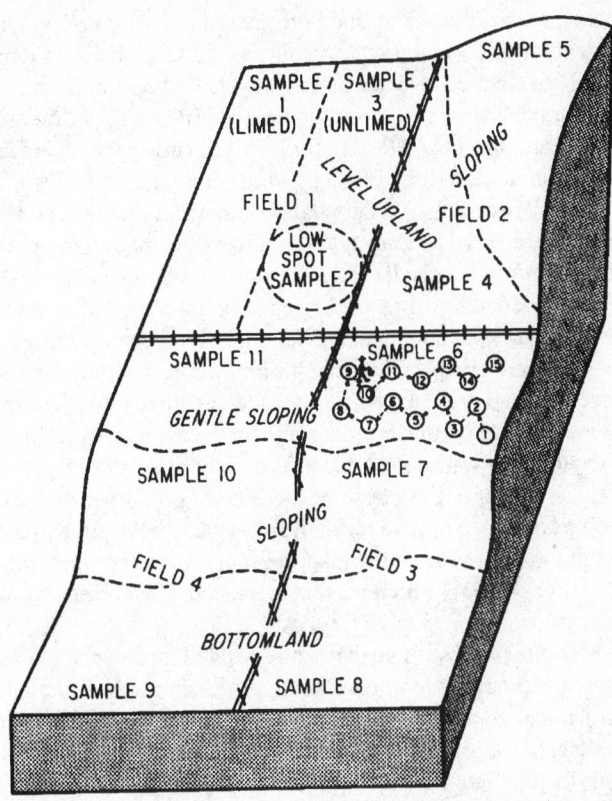

Fig. 4.2 Illustration of soil sampling practices for uniform fields [Courtesy of Nebraska Agricultural Extension Service]

become more common in recent years with the advent of GIS and GPS technology and its use in precision agriculture (Section 4.4). For areas with significant and identifiable mesovariation, as might occur when fertilizers are consistently placed in bands across a field, a more intensive random sampling pattern can be used to ensure that the average value is not skewed by either the very high value present in the fertilizer band or the lower value for the bulk soil located between the bands. The number of soil cores to be collected in this situation will usually be four or five times as great as those needed for a random, composite sample (i.e., as many as 1–2 soil samples per ha). Finally, it should be noted that random and nonrandom sampling techniques are not appropriate for all aspects of soil fertility evaluation, particularly for elements that are more mobile in the soil profile (e.g., NO_3-N and SO_4-S).

Soil Sampling Methods

The most important factors to consider are sampling depth and frequency, and time of year. Depth depends on crop to be grown and type of soil test to be performed. Soil samples for routine soil tests are usually obtained from topsoils (0–20 cm depth), but important exceptions are when soil tests are conducted for mobile soil nutrients (NO_3-N and SO_4-S) to estimate the effect of pH on herbicide activity in no-till cropping systems, or for soil fertility evaluation for shallow or deep rooted crops, or to monitor potential leaching of pollutants. Shallow soil sampling is most often recommended for conservation tillage systems where nutrients and lime are not incorporated with the soil by plowing, for permanent pastures and/or turf where root systems rarely extend below a depth of 10 cm, and to estimate the potential for P loss in erosion or runoff in watersheds where eutrophication is an important environmental concern (sample to 0–5 cm). The recommended depth for subsoil sampling varies with the intent of the test. For example, subsoil testing for NO_3-N in arid regions may require sampling to 0.6–1 m. Testing sandy soils in humid regions for SO_4-S may require a subsoil sample from the B horizon (usually 20–60 cm) where SO_4-S accumulates by sorption to clays and Fe/Al oxides. When collecting subsoil samples care should be taken to minimize contamination of the subsoil sample by topsoil which can seriously influence results and subsequent recommendations.

Soil samples can be collected at any time of the year when the ground is not frozen, although the ideal time is shortly before making a land management decision because this gives the most current indication of soil properties (fertility, pH). In general, for most agricultural systems routine soil tests for lime and fertilizer recommendations are normally collected three to six months prior to planting a crop. This usually provides sufficient time for management decisions to be made and implemented in a timely manner. For example, if limestone is needed to correct soil acidity, this should be known several months in advance of planting because of the time required for the limestone to react and correct the problem. Soil testing well in advance of planting also allows time to change the crop if soil test results indicate that growing conditions are inappropriate. An exception is the pre-sidedress NO_3^- test for corn (PSNT) (Section 4.2.4.4) which must be collected when corn plants are about 30 cm in height.

The frequency of soil testing varies somewhat with intended land use, i.e., the plants to be grown and the nutrient management practices required. Ideally, soils should be sampled at the same time of year (e.g., spring or fall) and at no more than two- to three-year intervals. Sampling at the same time of year minimizes the effect of seasonal variations on soil pH, which decreases during the summer as soluble salts increase in the soil from fertilization and mineralization of SOM. Some studies have shown similar decreases in P and K during the year, with lower values reported in the fall than spring.

Soil Sample Handling

Proper handling is necessary to prevent contamination and to minimize extreme changes in elemental concentrations or pH between the soil in the field and the sample to be analyzed. Handling includes collecting soil cores, mixing to prepare a composite sample, transporting the sample from the field to the laboratory, and drying, grinding, sieving, and storing in the laboratory. A variety of tools are available to collect soil cores, including handheld soil probes and augurs, hydraulic soil coring devices, shovels, and hand trowels. The major consideration is to use sampling and mixing tools made of materials that will not contaminate the soil (e.g., stainless steel or plastic). Once collected, individual soil cores from the area of interest are composited by combining them in a clean container, preferably made of plastic to avoid contamination, mixing well, and removing a subsample that represents the entire mixture. During the sampling process, avoid contamination from dirty or rusty sampling and mixing devices, fertilizer materials, galvanized metals (Zn), and paper bags that may contain B. Soil samples should be delivered to the testing laboratory as soon as possible after collection to minimize any changes in elemental concentration that may take place prior to drying. The greatest concern is with NO_3-N which can increase when soils are kept for extended periods in a warm, moist state (from mineralization of organic N and nitrification of exchangeable NH_4-N), or decrease when soils are maintained in a warm, wet state (from denitrification of NO_3-N). Once received by the laboratory, soil samples are normally dried at low temperatures (ambient to 50 °C; avoid higher temperatures), ground and sieved, typically to pass a 2 mm screen prior to analysis.

4.2.3.2 Soil Analyses: Chemical, Physical, and Biological Methods

Chemical analysis of soils is based on the principle that chemical solutions can rapidly, reproducibly, and inexpensively assess soil nutrient-supplying capacity and other soil properties that affect plant growth (pH, soluble salts, SOM). The most common chemical methods used are extraction and equilibration; others include titration (for acidity) and chemical or thermal oxidation (for SOM).

Chemical extraction is almost always conducted with dried, ground, and sieved soil samples. For most soil tests, the process involves scooping or weighing a small representative portion of the soil sample (from 1 to 10 cm^3 or g) into an extracting vessel (flask, beaker, extraction bottle), adding a known volume of chemical extracting solution (from 10 to 100 mL), shaking rapidly for a short time period (from 5 to 30 minutes), filtering the sample, and analyzing the filtrate for the elements of interest (Sparks, 1996).

Soil chemical properties are not always assessed by extraction methods. Another common chemical testing method is equilibration, in which a solution is added to the soil, the resulting suspension is shaken (or sits) for a short time period, and some property of the soil suspension is measured. This approach is used to measure soil pH, lime requirement and soluble salts (Section 4.2.5). Some soil testing laboratories use titrimetric techniques to measure soil acidity by first extracting soil with a neutral salt solution (e.g., M KCl), followed by titration of the acidic extractant with a dilute base (e.g., 0.1 M NaOH). Soil organic matter tests originally used wet chemical oxidation to estimate SOM from the amount of C that could be oxidized by $K_2Cr_2O_7$. Because of environmental concerns on the use and disposal of Cr, high temperature oxidation (360 °C) is now used to estimate SOM from weight loss upon ignition.

Most soil testing methods have been standardized by regional and national soil testing organizations (NCR-13, 1988; SRIEG-18, 1992; SPAC, 1992; NEC-67, 1995; Sparks, 1996). Closely following recommended methods is essential in accurately assessing soil fertility.

4.2.3.3 Interpretation of Soil Testing Results

Interpretation of soil testing results may be defined as quantitatively relating the results of a soil analysis to the probability that a soil management activity will have the desired result. For soil fertility evaluation, this means using soil test results to accurately predict crop yield without nutrient addition and the probability of a profitable plant response when fertilizers or other soil amendments are added. Interpreting soil test results for land use where the desired result is not an economically optimum yield, requires that measures of success must be quantitatively related to a soil test value and to soil management practices.

For agricultural systems, a statistical correlation must exist between a soil test value and some aspect of plant response (e.g., nutrient concentration in the plant, crop yield) which can be used as the basis for soil test calibration. The main goal of soil test calibration is to rate soils in terms of the probability that nutrient additions will be profitable (e.g., to divide the population of soil test values into responsive and nonresponsive categories) (Table 4.1 and Section 4.2.6).

4.2.3.4 Recommendations

Soil test recommendations must integrate test data with many other factors such as climate, economics, soil and crop management practices, management ability of the soil test user, and any considerations imposed by environmental protection. Because soil test values provide no information on these other factors, other sources of information, basic scientific principles, practical experience, and professional judgment must be accessed to integrate them with soil test results into a reasonable recommendation.

Table 4.1 Generalized soil test categories and recommendations based on crop response and environmental impact [Adapted from Beegle, 1995]

Category name	Category Definition	Recommendations
Crop Response		
Below Optimum (Very Low, Low, Medium)	The nutrient is considered deficient and will probably limit crop yield. There is a high to moderate probability of an economic yield response to adding the nutrient.	Nutrient recommendations are based on crop response and will build soil fertility into the optimum range over time. Starter fertilizer may be recommended for some crops
Optimum (Sufficient, Adequate)	The nutrient is considered adequate and will probably not limit crop growth. There is a low probability of an economic yield response to adding the nutrient.	If soils are tested annually, no nutrient additions are needed for the current crop. For other than annual soil testing, nutrient applications are often recommended to maintain the soil in the optimum range. Starter fertilizer may be recommended for some crops.
Above Optimum (High, Very High, Excessive)	The nutrient is considered more than adequate and will not limit crop yield. There is a very low probability of an economic yield response to adding the nutrient. At very high levels, there is the possibility of a negative impact on the crop if nutrients are added.	No nutrient additions are recommended. At very high or excessive levels, remedial action may be needed to prevent phytotoxicity or environmental problems.
Environmental Response		
Potential Negative Environmental Impact (Very High, Excessive)	Soils testing at this level or above have higher potential to cause environmental degradation and should be monitored closely. The likelihood of environmental problems depends on other site-specific characteristics (e.g., slope, hydrology, rainfall). This soil test level is independent of the crop response categories above and may be above or below the optimum level based on crop response.	If other site factors minimize environmental impact, some nutrient additions may be recommended according to crop response guidelines. If other site factors indicate a potential environmental impact is likely, nutrient additions including starter fertilizer are not recommended. Remedial actions may be required to protect the environment.

The recommendation process starts with knowledge of the intended land use and factors that affect recommendations. For production agriculture, recommendations usually include the rate, timing, and method of application of fertilizers, liming materials, and other soil amendments such as biosolids and manures. In this situation, where the goals are economically optimum crop yields and minimal environmental impact, recommendations are based on: (1) current soil test values and any other soil characteristics that affect a recommendation, such as soil drainage class; (2) the crop to be grown and the realistic yield potential for that crop; (3) soil test calibration data that indicate the degree of response expected to any soil amendment; (4) the source of nutrients and/or lime to be used and any restrictions on application method and timing that exist; (5) soil management history, such as the use of animal manures, biosolids, or growth of a leguminous cover crop; and (6) environmental parameters that may require modification of standard recommendations, such as soil leaching potential and depth to groundwater and/or soil erosion and runoff potential and proximity to surface waters sensitive to eutrophication.

Not surprisingly, given the rather subjective nature of many components of the soil test recommendation process, several different recommendation philosophies have evolved for agricultural systems. These philosophies (Section 4.2.6) vary mainly in the approach used to maintain soil fertility in the range needed for optimum crop production. Some recommend more liberal and more frequent applications of nutrients to ensure that nutrient deficiency does not limit crop production. Others are more conservative, relying heavily on soil tests as the basis for any nutrient addition, often recommending that no nutrients be added when soils are in the optimum or excessive range. The decision on which philosophy to follow is primarily based on the economics of production at this time. However, for nutrients known to have environmental impacts (N, P), the need for environmentally based recommendations to protect ground and surface water quality is now being considered in some areas even if the result is suboptimum crop performance.

4.2.4 Soil Testing Methods for Plant Nutrients

4.2.4.1 Phosphorus

Soil P

Phosphorus occurs in inorganic and organic forms with ~30–50% of the total P in most soils in the organic fraction. Chemical weathering of soil minerals and mineralization of SOM releases P into the soil solution where it exists in very low concentrations (0.003 to 0.3 mg P L^{-1}, mean ~0.05 mg P L^{-1}) almost exclusively as orthophosphate. In acid soils, $H_2PO_4^-$ predominates while HPO_4^{2-} is the main species above pH 7.2. Once in solution, P can be assimilated by biological organisms, sorbed to sesquioxides and $CaCO_3$, precipitate as an insoluble compound, or be lost in surface or subsurface runoff. Each solid phase can also release P back into the soil solution by mineralization, desorption, or dissolution.

Plant-available forms of soil P are mainly those found in the soil solution, sorbed by soil colloids, or precipitated as relatively soluble minerals. Mineralization of organic P and dissolution of very stable P minerals proceeds too slowly in most soils to provide sufficient available P for plant growth. Desorption (and dissolution) of P occurs when plant uptake decreases the P concentration in the soil solution, thermodynamically favoring the release of sorbed or solid phase P. Soil tests for P are designed to simulate this process, extracting P from sorbed forms and metastable precipitates by four processes: (1) acid dissolution, (2) anion exchange, (3) cation complexation, and (4) cation hydrolysis (Kamprath and Watson, 1980). For example, the Bray P$_1$ soil test, developed for slightly acid and neutral soils where Al- and Ca-P are major sources of plant-available P, is a mixture of 0.025 M HCl + 0.03 M NH$_4$F. The F$^-$ ion forms a strong complex with Al^{+3} in solution, causing the dissolution of Al-P compounds, and the dilute HCl dissolves a proportion of the Ca-P and lesser amounts of Al-P and Fe-P. Together the NH$_4$F and the HCl cause the labile pools of soil P to release P into solution, similar to what would occur in soils in response to the depletion of soil solution P by plant uptake. In near neutral soils, F$^-$ forms CaF$_2$, enhancing the dissolution of Ca-P. The Olsen soil test (0.5M NaHCO$_3$, pH 8.5) functions similarly in calcareous soils where HCO$_3^-$ precipitates soluble Ca as CaCO$_3$ causing the release of P from CaHPO$_4$. Soil P tests may also extract some organic P either by direct acid hydrolysis of organic P esters or by enhancing P release from organometallic complexes (Al-OM, Fe-OM).

Soil tests based on soluble P alone are not an accurate index of P-supplying capacity because they do not reflect the capacity of the soil to replenish solution P as uptake occurs. Similarly, soil tests using extremely strong chemical reagents that cause extensive dissolution of mineral phases or oxidation of SOM are also inappropriate.

Current Soil Testing Methods for P

The soil testing methods commonly used for P today are shown in Table 4.2 and were reviewed extensively by Fixen and Grove (1990). The Bray P_1, Ca lactate, Morgan, and Olsen soil tests are only used to extract P while Mehlich I, Mehlich III, and AB-DTPA (NH_4HCO_3+DTPA) are multielement soil tests used to extract P, K, Ca, Mg, and some micronutrients (Cu, Fe, Mn, Zn). As a general rule, the acidic extractants (Bray, Mehlich I, Mehlich III, Morgan) are used on acid soils and the alkaline extractants (AB-DTPA, Olsen) on calcareous soils. A number of studies, however, have shown that the Olsen P test may be an accurate test for a broader range of soils.

Two soil testing approaches for P, the P_i soil test and ion exchange resins, do not remove P from soils by chemical extraction. In the P_i test which shows considerable promise (Chardon et al., 1996; Menon et al., 1997), soil is equilibrated with 0.01 M $CaCl_2$ in the presence of an Fe oxide-coated filter paper strip, which acts as an infinite sink for soil P. Phosphorus in solution is sorbed by the strip, causing desorption of labile P from the soil. After a specified equilibration period (usually 2 h) the strip

Table 4.2   Summary of soil testing methods currently used for phosphorus (P)

Soil test	Extractant composition	Comments, critical values,[†] and sources
AB-DTPA	M NH_4HCO_3 + 0.005M DTPA - pH 7.5	Multinutrient extractant primarily used with alkaline soils. Critical value: $\geq$ 8mg kg^{-1} (Soltanpour and Schwab, 1977).
Bray P_1	0.03M NH_4F + 0.025M HCl	Used only to extract P on acid soils with moderate CEC. Critical value: $\geq$ 30 mg kg^{-1} (Bray and Kurtz, 1945).
Mehlich I	0.05M HCl + 0.0125M H_2SO_4	Multinutrient extractant used on acidic, low CEC soils. Critical value: $\geq$ 25 mg kg^{-1} (Mehlich, 1953).
Mehlich III	0.2M CH_3COOH + 0.25M NH_4NO_3 + 0.015M NH_4F + 0.013M HNO_3 + 0.001M EDTA - pH 2.5	Multinutrient extractant suitable for wide range of soils. Well correlated with Bray P_1, Mehlich I, and Olsen P. Critical value: $\geq$ 50 mg kg^{-1} (Mehlich, 1984).
Morgan and Modified Morgan	Morgan: 0.7M $NaC_2H_3O_2$ + 0.54M CH_3COOH + - pH 4.8 Modified Morgan: 0.62M NH_4OH + 1.25M CH_3COOH - pH 4.8	Multinutrient extractant primarily used in the northeast United States for acid, low CEC soils. Not suitable for calcareous soils. Critical value: $\geq$ 4-6 mg kg^{-1} (Morgan, 1941).
Olsen	0.5M $NaHCO_3$ - pH 8.5	Originally developed as P extractant for alkaline soils in the western United States; now also used for acid and neutral soils. Critical value: $\geq$ 10 mg kg^{-1} (Olsen et al., 1954).
Egner	P-CAL: 0.01M Ca lactate + 0.02M HCl P-AL: 0.10M NH_4 lactate + HOAc - pH 3.75	Multinutrient extractant used in Europe and Scandinavia but not in the United States. (Egner et al., 1960).

[†]Critical value is defined as the soil test concentration above which the soil test level is considered optimum for plant growth and responses to additions of the nutrient are unlikely to occur. Critical values cited in this table are approximate, can be affected by soil type and crop, and were obtained from several sources.

is removed, rinsed lightly to remove soil particles, and sorbed P is extracted by shaking for 2 h with M H_2SO_4. Anion exchange resins saturated with Cl^- or HCO_3^- and suspended with soils in aqueous suspensions accurately estimate labile P in soils of widely differing properties (Amer et al., 1955; Sibbesen, 1978; Wolf et al., 1985; van Raij, et al., 1986; Abrams and Jarrell, 1992). Resins simulate root uptake by removing P from solution by surface sorption processes, with the rate of P sorption controlled by diffusion (Kuo, 1996). Ion exchange resins have not been widely adopted primarily because of practical difficulties in separating resin from soil after equilibration. An encapsulated ion exchange resin technique for use as a multinutrient soil test (phytoavailability soil test [PST] (Skogley et al., 1990; Skogley, 1994) and a resin strip method (Leal et al., 1994) have been developed to overcome these problems. van Raij (1997) successfully adapted ion exchange resin beads for routine use as multinutrient soil tests. Ion exchange resins and chemical extractions can both be used as environmental soil P tests to identify soils with higher potential for P loss in erosion or runoff (Section 4.2.7).

Considerations in Soil Testing for P

Soil samples for P can be collected at any time during the year because P is relatively immobile in most soils. The standard sample depth is 0–20 cm, except for permanent pastures and turf (0–10 cm). In fields where the use of banded fertilizers containing P is a common practice, the number of soil samples collected per unit area should be increased to overcome the high mesovariation (Section 4.2.3.1). Where P transport to surface waters is a concern, samples should be collected from 0–5 cm, since runoff interacts with the upper few cm of soil (Sharpley and Smith, 1989; Sharpley et al., 1996). There are no special handling or storage requirements for samples collected for P analysis. Usually, P is analyzed as orthophosphate colorimetrically (Murphy and Riley, 1962) or by ICP-AES, which measures both orthophosphate and some organic P causing ICP-AES values to be higher.

Interpretation of P tests must consider that P moves to roots primarily by diffusion, a process highly dependent on moisture and temperature. Early season P deficiency in crops grown in soils that are rated optimum or excessive in P can occur sometimes. Low soil temperatures and dry soil conditions can inhibit diffusion, root growth, and plant uptake of P, creating a temporary P deficiency that often disappears as soils warm and receive rainfall or irrigation. For the most part, however, soil P tests are interpreted using the sufficiency level approach described in Table 4.1.

4.2.4.2 Potassium, Calcium and Magnesium

Soil K, Ca and Mg

The cycling and plant availability of K, Ca, and Mg are sufficiently similar that these three nutrients can be considered together for the purposes of soil fertility evaluation. The primary sources of plant available K, Ca, and Mg are soil minerals from which they are released during weathering.

Major K-bearing minerals include the feldspars (orthoclase, microcline, sanidine) and micas (biotite, muscovite, phlogopite). Total K concentrations in soils average 1.9%, but can range from 0.03% in organic soils (peats, mucks) to 0.3% in very sandy soils to as high as 3.0% in mineral soils derived from feldspars and micas. Once mineral dissolution has occurred soil K is primarily found in soluble, exchangeable, and nonexchangeable forms. Soil solution K ranges from 1 to 80 mg K L^{-1} (mean 2–5 mg K L^{-1}) and is in rapid equilibrium with exchangeable K which accounts for < 5% of the CEC. Plant-available forms are soluble and exchangeable K (< 1–2% of total K). Nonexchangeable K (fixed) occurs in the interlayer positions of micaceous minerals and is a slowly available reserve. Soluble K is < 1% of exchangeable K and can be quickly depleted by plant uptake or leaching. The ability of soils to maintain an adequate concentration of K in solution by releasing K from exchangeable and nonexchangeable forms is referred to as the K buffer capacity and is an important measure of soil fertility. The K buffer capacity of soils depends largely upon the amount and types of clay minerals present and to a lesser extent on SOM content. Fine textured soils have higher K buffer capacities than sandy or organic soils and more ability to maintain soluble K in an optimum range for plant growth.

Major mineral sources of soil Ca are carbonates, feldspars, and phosphates. The most important Ca-bearing mineral is anorthite ($CaAl_2Si_2O_8$; a plagioclase feldspar), except in calcareous soils where calcite ($CaCO_3$) and dolomite [$CaMg(CO_3)_2$] predominate. Total soil Ca varies widely, from < 0.1% in highly weathered, tropical soils to as high as 25% in calcareous soils. Typical total Ca values in noncalcareous, humid, temperate soils are from 0.7 to 1.5% with values > 3%, indicating the presence of $CaCO_3$. In most soils, Ca is the dominant exchangeable cation, occupying, from 20–80% of the CEC. Consequently, soil solution concentrations of Ca are quite high relative to most other plant nutrients, ranging from 30 to 300 mg Ca L^{-1} in noncalcareous soils.

Plant available Mg originates from the weathering of minerals such as biotite, dolomite, hornblende, olivine, and serpentine. Total soil Mg varies from < 0.1% in coarse, sandy soils of humid regions to 4% in fine textured soils formed from Mg bearing minerals. As with K and Ca, soluble and exchangeable Mg are most important to plant growth. Exchangeable Mg occupies from 4 to 20% of the CEC and soluble Mg ranges from 50 to 120 mg Mg L^{-1} in temperate region soils.

Current Soil Testing Methods

Soil chemical extraction is by far the most common approach used today to test soils for plant-available K, Ca, and Mg and most extracting methods simultaneously extract and assay soluble and exchangeable K, Ca, and Mg (Table 4.3). Haby et al. (1990) reported that the most common soil test for K, Ca, and Mg in the United States and Canada was ammonium acetate (M NH_4OAc, pH 7.0) with some newer, multi-element soil testing extractants, such as AB-DTPA and Mehlich III gaining in popularity to reduce costs and time for analysis. Most soil test extractants for K, Ca, and Mg displace these cations from exchange sites on soil colloids with a replacing cation, usually NH_4^+ (NH_4OAc, AB-DTPA), Na^+ (Morgan), H^+ (Mehlich I) or a combination of cations (Mehlich II, Modified Morgan). All extractants also remove any solution K, Ca, and Mg present in the soil. Acidic soil tests (e.g., Mehlich I, Mehlich III) may also extract some nonexchangeable K from micaceous clays because the H^+ ion can penetrate the interlayers and displace K. Acidic extractants may also overestimate exchangeable Ca and Mg in calcareous soils due to dissolution of Ca- and Mg-bearing minerals.

Ion exchange resins (Raij et al., 1986; Skogley, 1994) and electroultrafiltration (EUF)(Nemeth, 1979; Haby et al., 1990) have also been used to evaluate plant-available K, Ca and Mg. The basic principles of ion exchange resin techniques were discussed earlier (Section 4.2.4.1). The EUF method combines electrodialysis and ultrafiltration and has been used as a multi-element extractant for NH_4^+, NO_3^-, P, K, Ca, Mg, Na, S, B, Mn, and Zn. In this approach, a soil suspension is stirred in a central compartment separated by microfiber filters from cells on each side containing Pt electrodes; voltage and vacuum are applied to withdraw water and dissolved ions after specified time intervals that characterize the different forms of plant nutrients (e.g., 0–5, 5–10 and 10–30 min extractions giving soluble and exchangeable K, and K buffering capacity, respectively). Solutions collected at the anode and cathode sides of the EUF device are combined and analyzed by standard instrumental techniques. While EUF has been shown to be an effective method to simultaneously extract plant-available nutrients, it is rather slow and expensive and not as widely used as chemical soil tests (Haby et al., 1990; van Lierop and Tran, 1985).

Table 4.3   Summary of soil testing methods currently used for potassium (K), calcium (Ca) and magnesium (Mg)

Soil test extractant	Extractant composition	Comments, critical values[†], and sources
Ammonium acetate	M NH$_4$OAc, pH 7.0	Used for > 50 yr as soil test for K, Ca, and Mg (Chapman and Kelley,1930; Schollenberger and Simon, 1945). Suited for wide range of soils, but primarily used in midwestern and western United States states. Critical values for K, Ca, and Mg vary widely based on soil type (pH, CEC, clay mineralogy) and crop and reportedly range from K:110 to 200 mg kg^{-1}; Ca: 250 to 500 mg kg^{-1}; and, Mg: 30 to 60 mg kg^{-1} (Haby et al., 1990).
AB-DTPA	M NH$_4$HCO$_3$ + 0.005M DTPA - pH 7.5	Multinutrient extractant primarily used with alkaline soils. (Soltanpour and Schwab. 1977).
Mehlich I	0.05M HCl + 0.0125M H$_2$SO$_4$	Multinutrient extractant used on acidic, low CEC soils. (Mehlich, 1953).
Mehlich III	0.2M CH$_3$COOH + 0.25M NH$_4$NO$_3$ + 0.015M NH$_4$F + 0.013M HNO$_3$ + 0.001M EDTA - pH 2.5	Multinutrient extractant suitable for wide range of soils. (Mehlich, 1984).
Morgan and Modified Morgan	Morgan: 0.7M NaC$_2$H$_3$O$_2$ + 0.54M CH$_3$COOH + - pH 4.8 Modified Morgan: 0.62M NH$_4$OH + 1.25M CH$_3$COOH - pH 4.8	Multinutrient extractant primarily used in the northeast United States for acid, low CEC soils. Not suitable for calcareous soils. (Morgan, 1941).

[†]Critical value is defined as the soil test concentration above which the soil test level is considered optimum for plant growth and responses to additions of the nutrient are unlikely to occur. Critical values cited in this table are approximate, can be affected by soil type and crop, and were obtained from several sources.

Considerations in Soil Testing for K, Ca, and Mg

Soil samples for K, Ca, and Mg analysis are collected following the standard approaches described in Section 4.2.3.1. The most important exception is with K where subsoil samples are sometimes recommended for soils with a very sandy surface horizon and a shallow B horizon (zone of clay accumulation). In these situations subsoil K has been shown to be an important source of plant available K and testing the surface horizons alone may underestimate the true K-supplying capacity. This is particularly true if the B horizon is a reservoir of slowly available, nonexchangeable K. Subsoil samples are rarely tested for Ca and Mg. Samples should be collected at the same time each year to minimize the effect of natural, seasonal changes in K concentrations caused by processes such as leaching, freezing and thawing, biological transformations (uptake, mineralization, biocycling of K from subsoils to topsoils), and seasonal differences in soil moisture content.

Soil sample handling, particularly drying, can markedly and unpredictably alter extractable K. Air drying will usually cause an increase in exchangeable K except in soils that have very high K values, where K fixation can decrease soil test K. Changes caused by drying are greatest in fine textured soils dominated by 2:1 clays. While it can be argued that soil tests for K would be conducted best on field moist soils, virtually all laboratories air dry soils to ease handling, grinding, sieving, mixing, and weighing of a representative subsample. These advantages outweigh the changes in extractable K that occur during drying; however, soils to be analyzed for K should only be air dried at moderate temperatures (< 50 °C) because oven drying, particularly at > 60 °C greatly enhances the release of K. Analyses for K, Ca, and Mg can be conducted by either AAS or ICP-AES.

Interpretation of soil test results for K, Ca, and Mg follows the sufficiency level approach (Table 4.1) with only a few minor modifications related to soil type, plant to be grown, and soil/crop management. Soil test K interpretations are often modified based on CEC. Soils with higher CEC values will often have a higher critical value (point above which crop response is not expected and thus no fertilizer is recommended). As an example, the critical value for soil test K in Alabama (Mehlich I soil test) increases from 40 to 80 mg K kg^{-1} as CEC increases from 4.5 to 9.0 $cmol_c$ kg^{-1}. Subsoil K is occasionally considered when interpreting the results of a soil test for K, usually by the use of indirect information on subsoil properties, such as soil survey data on horizonation (e.g., depth to B horizon), texture, and clay mineralogy. Other factors that may alter a soil test based K recommendation include crop and yield goal which affect K removal, tillage practices, and climate, which affects release of K from unweathered K bearing minerals. Modification of the results of a soil test for Ca or Mg by inclusion of other information is unusual. In general, a soil liming program will supply adequate levels of Ca and Mg in most soils.

4.2.4.3 Sulfur

Soil S

The total S content of temperate zone soils ranges from 0.005 to 0.04%, more than 90% of which is found in organic forms. Total S values can be much higher in arid and semiarid regions where soils can accumulate soluble and mineral forms of SO_4-S, such as gypsum ($CaSO_4 \cdot 2H_2O$), epsomite ($MgSO_4 \cdot 7H_2O$), and mirabilite ($Na_2SO_4 \cdot 10H_2O$). Important S-bearing minerals in humid regions are pyrite (Fe_2S), sphalerite (ZnS), and chalcopyrite ($CuFeS_2$). Sulfur originating from burning fossil fuels or from volcanic activity accrues from wet or dry atmospheric deposition.

The plant-available form of S is SO_4^{2-} which originates from the dissolution of soluble salts and minerals containing S, oxidation of elemental S, and mineralization of organic S. Solution SO_4-S can be taken up by plants, immobilized in microbial biomass, sorbed by soil colloids, precipitated in an insoluble mineral form by reaction with Ca, Mg, or Na, or leached to subsoils. If plant roots penetrate subsoils, sorbed SO_4-S can be released and absorbed. Under reducing conditions, SO_4-S can be converted to H_2S gas and lost from the soil by volatilization or precipitated as metal sulfide minerals such as FeS_2. Concentrations of SO_4-S in the soil solution (A horizon) of temperate zone soils range from 5–20 mg SO_4-S L^{-1}, higher than the 3–5 mg SO_4-S L^{-1} required for the optimum growth of most plants. Hence, plant S deficiency is uncommon except with high yielding crops grown on deep sandy soils with low SOM contents or on soils developed from parent materials low in S.

Current Soil Testing Methods for S

Because crop response to S fertilization is uncommon, less effort has been directed toward the development and calibration of soil tests for S than for P and K. Soils tested for S are normally sampled from the A horizon, except in sandy, low SOM soils with shallow B horizons that can be a significant reserve of plant available SO_4-S. In these situations, a subsoil sample may be required because the subsoil may have enough available S for plant growth.

Soil testing for S relies on chemical extraction. More than 20 extracting solutions have been developed and evaluated as S soil tests, including water and various concentrations of dilute acids, dilute salts, acetates, and phosphates (Johnson and Fixen, 1990). Most extractants remove soluble and sorbed forms of SO_4-S, along with a small percentage of organic S, as these are the soil fractions regarded as plant-available. In arid regions where the concentration of SO_4-S is often quite high due to the accumulation of sulfate salts, extraction with deionized or distilled water is used to identify S deficiency. In humid regions, the use of an extractant that contains a replacing anion such as phosphate is often more successful. One of the most widely used extracting solutions for S is a 500 mg P L^{-1} solution of Ca(H$_2$PO$_4$)$_2$, sometimes in combination with 2M HOAc. The phosphate ions displace sorbed SO$_4$-S, the HOAc extracts some organic S and the Ca^{2+} causes flocculation of the soil allowing for ease of analysis of S by either colorimetric or turbidimetric means. Recently, there has been interest in using Mehlich III as an S soil test to eliminate the need for a separate extraction to determine available S. Once extracted, SO$_4$-S can be analyzed by colorimetry, titrimetry, ion chromatography, and ICP-AES. The most common analytical techniques are turbidimetry (if S is determined alone), and ICP-AES (for S alone and in multi-element analyses). A review of the advantages and disadvantages of analytical methods for S is given by Tabatabai (1996).

Interpretation of S Soil Tests
Sulfur soil testing is a moderately reliable approach to determine S fertility status but is best used in conjunction with other information, such as plant analysis, knowledge of soil type and plant yield potential, SOM content, and inputs of S from sources other than fertilizers (e.g., manures, crop residues, rainfall and irrigation waters). Testing for S does not account for the S inputs from the atmosphere or irrigation waters, which can exceed S recommendations. In situations where S deficiency is likely, the cost of applying a small amount of S fertilizer (10 to 15 kg S ha^{-1}) may be cheaper than soil testing.

4.2.4.4 Nitrogen
Nitrogen deficiency is the most common soil fertility problem for nonlegumes. Nitrogen losses from soils are also known to negatively impact water and air quality. However, despite the importance of N in agricultural production and environmental quality, a widely accepted method to test soils for plant-available N, particularly in humid regions, has not been developed. The reasons for this center around the complex N cycle transformations (Tisdale et al., 1993; Pierzynski et al., 1994; Foth and Ellis, 1997) (Section C, Chapter 5).

Most of the N available for biological assimilation by plants and animals originated from the atmosphere. Only a small fraction of global N is in soils where total N values typically range from 0.05 to 0.15% and most of the N (> 98%) is organic in nature. Atmospheric N is converted by symbiotic or nonsymbiotic biological N fixation to forms of N that can be directly or indirectly used by plants. *Rhizobium* and *Bradyrhizobium* spp. form symbiotic relationships with plants to assimilate N$_2$ from the atmosphere while nonsymbiotic N fixation occurs in free-living algae, bacteria, and actinomycetes. Electrical discharges and industrial processes synthesize NH$_3$ while the burning of fossil fuels and volcanic eruptions also contribute atmospheric sources of N to soils.

The key components of the soil N cycle are (1) mineralization, in which organic N is converted to NH$_4$-N by microbial decomposition; (2) immobilization, in which microorganisms assimilate NH$_4$-N and NO$_3$-N from the soil solution for growth and biomass production which is essentially the reverse of mineralization; (3) nitrification, in which certain soil bacteria convert NH$_4$-N to NO$_3$-N, a rapid process in most well-aerated soils; (4) ion exchange in which NH$_4$-N is retained by cation exchange sites on soil clays or SOM (including fixation within micaceous clays as nonexchangeable NH$_4$-N) and NO$_3$-N is retained by any positively charged sites present on soil colloids (rarely of consequence in most soils); (5) denitrification, in which anaerobic bacteria convert NO$_3$-N to forms of N (N$_2$O, NO, N$_2$) which are lost from soil as gases; (6) volatilization, in which NH$_4$-N is converted to gaseous NH$_3$-N under certain conditions (high pH) and lost to the atmosphere; and (7) leaching, in which NO$_3$-N moves downward with percolating waters.

Plants absorb NH$_4$-N and NO$_3$-N from the soil solution. An accurate soil test for N, therefore, must integrate all components of the soil N cycle that affect the availability of NH$_4$-N and NO$_3$-N, which is a complex task given the dependence of N cycling on biological activity and environmental conditions.

Current Soil Testing Methods for N

Soil testing for N differs markedly between arid and humid regions. In arid or semiarid regions, evapotranspiration usually exceeds precipitation and inorganic N is leached or denitrified to a lesser extent. For this reason, a sample collected from the rooting zone shortly before the start of the growing season and analyzed for residual inorganic N (NH_4-N, NO_3-N) can accurately measure plant-available N. Nitrogen inputs are then reduced accordingly. In most cases, samples for residual inorganic N must be collected to deeper depths (60–180 cm) than for standard soil testing (20 cm). Soil samples tested for residual inorganic N are often only analyzed for NO_3-N since this is usually the dominant form of inorganic N in most soils. Samples for residual inorganic N are usually taken just before planting or early in the growing season, although in very cold and dry areas (minimal mineralization and leaching), samples can be collected the preceding fall or winter.

After sample collection, proper handling is critical to avoid changes during storage. Moist and wet soils stored under warm conditions can mineralize, immobilize or lose a significant amount of inorganic N. To avoid problems, samples should be rapidly air dried at ambient temperatures by spreading the soil in a thin layer. Extraction of inorganic N is usually accomplished by shaking a dried, ground soil sample for 30 min to 1 h with a salt solution [e.g., 2 M KCl, 0.01 M $CaSO_4$, 0.04 M $(NH_4)_2(SO_4)$], followed by filtration. Automated colorimetry and ion chromatography are usually used to determine NH_4-N and NO_3-N in soil extracts, but steam distillation, specific ion electrodes, and microdiffusion techniques are also used (Bundy and Meisinger, 1996).

Because soil testing for N in humid regions is a more complex process and a less accurate predictor of soil N fertility, routine soil tests are seldom offered. The greater rainfall and warmer temperatures in humid areas cause rapid, seasonal changes in the amount of inorganic N in the profile, making direct measures of residual inorganic N less reliable estimates of plant-available N.

The general consensus on soil testing for residual inorganic N in humid regions is that this practice has value if conducted at or near planting and if samples are collected to a reasonable depth in the soil profile (not just the topsoil). If residual inorganic N values are high, reductions in N inputs should be made. Bundy et al. (1992) reported on the use of the preplant soil profile nitrate test (PPNT) in the upper Midwestern United Sates, where moderate rainfall and cooler winter temperatures make this approach more likely to be successful than in warmer, higher rainfall humid regions. The economically optimum N fertilizer rate for corn (*Zea mays* L.) was shown to decrease in a near linear manner with increasing PPNT values in soil samples collected to a depth of 1 m. Another situation where residual inorganic N testing has been successful in humid regions has been with short season crops where there is less likelihood that significant losses of residual inorganic N will occur before plant N uptake.

In general, the most promising recent advance in soil N testing in the humid regions has been the pre-sidedress soil nitrate test (PSNT) originally developed for corn but now being investigated for a wider range of agronomic and vegetable crops (Magdoff et al., 1984; Bock and Kelley, 1992). The PSNT was conceived and evaluated to address the problem of overfertilization of corn with N in the Northeastern United States, particularly in fields with histories of manure and legume use where residual organic N would likely provide an appreciable percentage of the total N requirement for many nonleguminous crops. The PSNT has four basic tenets: (1) all fertilizer N for corn except for a small amount banded at planting, should be applied by sidedressing when the crop is beginning its period of maximum N uptake, usually early June; (2) soil and climatic conditions prior to sampling integrate the factors influencing the availability of N from the soil, crop residues and from previous applications of organic wastes; (3) a rapid sample turn around (< 14 d) by a testing laboratory is possible, thus allowing time for farmers to collect the soil sample, submit it to the laboratory, have it analyzed, receive the results and recommendations, and then apply (or not apply) sidedress N before the corn crop becomes too large for equipment to move through the field; and (4) farmers will normally only sample to a depth of 30 cm, the recommended depth for the PSNT sample. In practice, a PSNT sample is collected early during corn growth (~30 cm tall). The sample is rapidly air or oven dried (< 60 °C) after spreading the soil in a thin (< 1 cm) layer. Extraction and analysis for NO_3-N proceed as described above for residual inorganic N. However, while the PSNT actually measures inorganic N, it is not a measure of residual inorganic N, but an indirect, field-based expression of the capacity to provide an adequate supply of inorganic N during the growing season (i.e., of the soil N mineralization potential). The PSNT has been successfully evaluated in over 300 field studies in the Northeastern and Midwestern Unites States in identifying N sufficient soils (Magdoff et al., 1990; Bock and Kelley, 1992; Meisinger et al., 1992; Sims et al., 1995). Some of the logistical difficulties associated with the need for rapid sample analysis have been overcome by the development of quick test kits and specific ion electrodes that can be used in the field (Jemison and Fox, 1988).

Considerations in Interpretation of N Soil Tests
Most N recommendations are not based on soil testing, but on field calibration studies quantifying plant performance in response to N inputs (fertilizers, manures, biosolids, etc.). Widespread, commercial scale use of N testing today is confined to certain areas and crops, such as the PSNT for corn in humid regions or residual inorganic N testing for grain crops in arid regions. Recommended N rates for the major grain crops, such as corn, wheat (*Triticum aestivum* L.), and sorghum (*Sorghum bicolor* (L.) Moench) are initially determined from an equation using the expected, realistic yield goal and a conversion factor appropriate to that crop. For example, for corn, the fertilizer N rate is arrived at by directly multiplying the realistic yield goal by an empirically determined factor that ranges from 17–18 kg N Mg^{-1} of expected yield. Modifications (reductions) to this recommendation are then made based on residual soil inorganic N, the previous or intended use of other N sources (animal manures), documented N inputs from other sources (high NO_3-N irrigation waters), and credits for N supplied by a previous legume crop in the rotation (e.g., alfalfa, soybeans). Tisdale et al. (1993) summarized the general approach used to make N recommendations as follows:

$$N_{fertilizer} = N_{crop} - N_{soil} - (N_{organic\,matter} + N_{previous\,crop} + N_{organic\,waste})$$

$N_{fertilizer}$ = amount of N needed from fertilizers, manures, biosolids, etc.
N_{crop} = crop N requirement at realistic yield goal
N_{soil} = residual soil inorganic N (NH_4-N + NO_3-N)   [4.1]
$N_{organic\,matter}$ = N mineralized from soil organic matter
$N_{previous\,crop}$ = residual N available from previous legume crops
$N_{organic\,waste}$ = residual N available from previous organic waste use such as animal manures, biosolids, wastewater irrigation, etc.

In some cases, an N availability index, based on SOM content, texture (indication of leachability and moisture-holding capacity), and climate (indicated by crop being grown) is used to estimate soil N-supplying capacity. Fertilizer N recommendations are adjusted accordingly with fine textured soils with higher SOM contents presumed to provide more plant-available N from mineralization, and thus, to need less fertilizer N. Mathematical models have also been developed to predict crop N requirements, but with only limited success because of the amount of site-specific information required to function with any degree of accuracy (Tanji, 1982).

Interpretations of N soil tests are done routinely and with reasonable success. However, soil N testing may be improved in conjunction with plant analysis, remote sensing or leaf chlorophyll meters (Sections 4.2.7.3 and 4.3).

4.2.4.5 Soil Testing for Micronutrients

Micronutrients are essential elements normally present in plants at very low concentrations (< 100 mg kg^{-1}) and include B, Cl, Cu, Fe, Mn, Mo, and Zn. Four micronutrients exist in soils as cations (Cu^{2+}, Zn^{2+}, $Fe^{2+, 3+}$ and Mn^{2+}) while three are found as an uncharged molecule ($H_3BO_3^0$) or anions [($B(OH)_4^-$, Cl^-, MoO_4^{2-}].

Plant-available Cu originates from the weathering of igneous and sedimentary rocks containing chalcopyrite ($CuFeS_2$), chalcocite (Cu_2S), and bornite ($CuFeS_4$). Total Cu concentrations typically range from 1–40 mg kg^{-1} and concentrations in the soil solution are quite low, from 10^{-8} to 10^{-9} M. More than 99% of the Cu^{2+} in the soil solution is complexed with dissolved organic matter (DOM); above pH 6.9, the dominant inorganic form of Cu is $Cu(OH)_2^0$. Soluble Cu is in equilibrium with Cu complexed by SOM, exchangeable Cu, and Cu sorbed, occluded, or coprecipitated on soil oxides. Sorption of Cu by SOM primarily controls plant availability of Cu, although Cu solubility is also highly pH dependent, decreasing 100-fold for each unit increase in pH.

Plant-available Zn also originates from the weathering of igneous and sedimentary rocks and total Zn concentrations usually range from 10 to 300 mg kg^{-1}. The major Zn bearing minerals include franklenite ($ZnFe_2O_4$), smithsonite ($ZnCO_3$), and willemite ($ZnSiO_4$). Zinc concentrations in the soil solution range from 2 to 70 g L^{-1} and Zn^{2+} is the major species below pH 7.7. Approximately 50 to 60% of soluble Zn is complexed with SOM. Plant-available Zn also includes exchangeable Zn and Zn sorbed by clays, oxides, and carbonates. As with Cu, the solubility of Zn is highly dependent upon soil pH, decreasing markedly above pH 6.0–6.5.

Iron is one of the major constituents of the Earth's crust (~5%) and total Fe contents in most soils are quite high, ranging from 1,000 to 50,000 mg kg^{-1}. Major mineral forms include olivene [$(Mg,Fe)_2SiO_4$], pyrite (FeS), siderite ($FeCO_3$), hematite (Fe_2O_3), and goethite (FeOOH). Plant-available forms of Fe include that sorbed by clays and SOM. Soil solution concentrations of Fe in equilibrium with these minerals are very low and depend greatly upon soil pH, ranging from 10^{-6} M in very acid soils to $< 10^{-20}$ M in soils above pH 7.0. The form and solubility of Fe in the soil solution also depend upon soil redox potential; in well-aerated, oxidized soils, Fe^{3+} or $Fe(OH)_2^+$ predominate, while in reduced, waterlogged soils the major inorganic species is Fe^{2+}. Each unit increase in pH decreases the solubility of Fe^{3+} by 1,000-fold, but only decreases soluble Fe^{2+} by 100-fold. Total concentrations of soluble inorganic Fe in most soils are too low to meet the nutritional needs of most plants, even under very acid soil conditions, yet plants are able to obtain adequate Fe for growth. Natural organic compounds (chelates) exuded from plant roots play an important role in preventing the precipitation of Fe as insoluble compounds by forming Fe-chelate complexes that can move to plant roots by mass flow or diffusion. At the root surface Fe dissociates from the chelate and is taken up by the plant.

Plant-available Mn originates from the weathering of minerals such as pyrolusite (MnO_2), hausmannite (Mn_3O_4), manganite (MnOOH), rhodochrosite ($MnCO_3$), and rhodonite ($MnSiO_3$). Total soil Mn ranges from 20 to 3000 mg kg^{-1} while soluble Mn is usually between 0.01 and 1.0 mg L^{-1}, existing primarily as Mn^{2+} in equilibrium with MnO_2 in oxidized soils and $MnCO_3$ in reduced soils. Exchangeable, sorbed, and organically complexed Mn are the plant-available forms. As much as 80% of soluble Mn in some soils is complexed with SOM. Soil properties and processes affecting Mn solubility include soil pH, complexation/chelation, and redox potential. Manganese solubility decreases about 100-fold as pH increases by one unit, and also increases markedly when soils become reduced and Mn oxides (e.g., MnO_2) dissolve.

Plant-available B originates in most soils from the weathering of sedimentary rocks and tourmaline. Total soil B concentrations usually range from 2 to 200 mg kg^{-1} with < 5% available to plants. Boron is found as undissociated $H_3BO_3^0$ (pH 5 to 9) or as the $B(OH)_4^-$ anion (at pH > 9.2). Major sources of plant-available B are those sorbed by soil clays and hydrous oxides and SOM. The $H_3BO_3^0$ molecule is highly mobile, particularly in soils low in clay, oxides, and SOM. However, B availability decreases at pH > 6.5–7.0 because of an increased affinity of clays and oxides for the $B(OH)_4^-$ anion.

Molybdenum is present in soils at very low levels, with total Mo values ranging from 0.2 to 5.0 mg kg^{-1} and soil solution concentrations < 0.5 g L^{-1}. Molybdenum is available as $HMoO_4^-$ or MoO_4^{2-}, which are strongly sorbed by Fe/Al oxides under acidic conditions and/or complexed by SOM. Unlike all other micronutrients, Mo availability increases with pH, due to the greater solubility of several Mo-bearing minerals at pH >7.0 and to a decreased affinity of most soils for Mo.

Chloride occurs mostly in igneous and metamorphic rocks and, once weathered, is found as soluble salts such as NaCl, $MgCl_2$, and $CaCl_2$ and in the soil solution as the Cl$^-$ anion. Dissolution of these salts is the primary process controlling Cl$^-$ availability to plants. Soil solution concentrations of Cl vary widely as a function of soil type and geographic location, ranging from < 0.5 - > 6,000 mg L^{-1}. Chloride is very mobile in most soils and may be retained in highly acid soils with variable charge.

Current Soil Testing Methods for Micronutrients

Soil tests for micronutrients have historically been special tests, restricted to situations where soil properties or crop characteristics indicated an economic response to micronutrient fertilization was likely. Consequently, separate soil tests were used for individual, or groups of micronutrients with similar properties. However, the advent of multi-element extractants (Mehlich III, DTPA, and AB-DTPA) and analytical techniques (ICP-AES) has increased micronutrient soil testing.

Soil sampling, handling, and storage are conducted following the standard techniques outlined in Section 4.2.3. Soil samples are almost always collected from topsoils (0–15 or 0–20 cm) but precautions should be taken to avoid contamination. For example, galvanized sampling tools and mixing buckets and some rubber stoppers contain Zn; metal grinding and sieving equipment may contain Cu, Fe, and Zn; borosilicate glassware, paper bags and boxes contain B; and many common laboratory reagents contain Cl. Sample drying, duration and intensity of grinding, speed of shaking during extraction, type of extraction vessel and soil:solution ratio affect the quantity of extractable Cu, Fe, Mn, and Zn in some soils (Soltanpour et al., 1976; Soltanpour et al., 1979). Given these somewhat unpredictable potential sources of error, following standardized methods for soil sampling, handling, preparation, and extraction is crucial to micronutrient soil testing. Only a small amount of contamination or a slight alteration in procedure can markedly skew a soil analysis, resulting in an erroneous recommendation.

Chemical extraction using dilute acids or chelating agents is the standard approach to assess Cu, Fe, Mn, and Zn availability. Less commonly used extractants include neutral salts and reducing agents (hydroquinone) for Mn (Table 4.4). In general, soil tests have been designed to remove soluble micronutrient forms (including organically complexed) by solvent action, to displace exchangeable and sorbed forms by ion exchange and desorption reactions with other cations or with H_3O^+, to partially dissolve soil minerals or oxides that contain precipitated and occluded forms, and to dissociate or chelate micronutrient cations that are complexed by solid phases of SOM. In most cases, all four cations are extracted and analyzed simultaneously by AAS and/or ICP-AES.

Dilute acids (0.025 to 0.1 M) are used as extractants for micronutrient cations, particularly on acid soils because they lack buffer capacity for use on calcareous soils. These extractants work primarily by dissociation, displacement, and partial acidic dissolution of cations from soil clays, oxides, and SOM. The most common dilute acid soil tests are the Mehlich I and 0.1 M HCl.

Table 4.4 Summary of major soil testing methods, interacting factors, and references for micronutrient soil tests [From Martens and Lindsay, 1990 and Sims and Johnson, 1991][†]

Micronutrient	Soil test and critical range	Comments and interacting factors used in soil test interpretation
Boron	Hot water: 0.1-2.0 mg kg^{-1} Mehlich III: 0.7-3.0 mg kg^{-1}	Hot water is the most widely used method. Interacting factors include crop yield goal, pH, soil moisture, texture, organic matter and soil type
Copper	AB-DTPA: 0.5-2.5 mg kg^{-1} DTPA: 0.1-2.5 mg kg^{-1} Mehlich I: 0.1-10 mg kg^{-1} Mehlich III: 0.3-15 mg kg^{-1} 0.1M HCl: 0.1-2.0 mg kg^{-1}	AB-DTPA and DTPA are used for alkaline soils, Mehlich III for alkaline and acid soils, and Mehlich I and 0.1M HCl for acid, low CEC soils. Interacting factors include crop type, organic matter, pH, and % $CaCO_3$.
Iron	AB-DTPA: 4.0- 5.0 mg kg^{-1} DPTA: 2.5-5.0 mg kg^{-1}	AB-DTPA and DTPA are used for alkaline soils (Fe deficiency is very rare with acid soils). Interacting factors include pH, % $CaCO_3$, CEC, organic matter and soil moisture.
Manganese	AB-DTPA: 0.5-5.0 mg kg^{-1} DTPA: 1.0-5.0 mg kg^{-1} Mehlich I: 5.0 mg kg^{-1} @ pH 6 10 mg kg^{-1} @ pH 7 Mehlich III: 4.0 mg kg^{-1} @ pH 6 8.0 mg kg^{-1} @ pH 7 0.1M HCl: 1.0-4.0 mg kg^{-1}	AB-DTPA and DTPA are used for alkaline soils, Mehlich III for alkaline and acid soils, and Mehlich I for acid, low CEC soils. Interacting factors include pH, texture, organic matter, and % $CaCO_3$.
Molybdenum	Ammonium oxalate - pH 3.3: 0.1-0.3 mg kg^{-1}	Soil testing for Mo is rarely done. Interacting factors are pH and crop.
Zinc	AB-DTPA: 0.5-1.0 mg kg^{-1} DTPA: 0.2-2.0 mg kg^{-1} Mehlich I: 0.5-3.0 mg kg^{-1} Mehlich III: 1.0-2.0 mg kg^{-1} 0.1M HCl: 1.0-5.0 mg kg^{-1}	AB-DTPA and DTPA are used for alkaline soils, Mehlich III for alkaline and acid soils, and Mehlich I and 0.1M HCl for acid, low CEC soils. Interacting factors are pH, % $CaCO_3$, P, organic matter, % clay, and CEC.

[†]References: Hot water B, Berger and Truog (1940); Mehlich 3 B, Shuman et al. (1992). DTPA (0.005M DTPA + 0.01M CaCl$_2$ + 0.1M TEA - pH 7.3) Lindsay and Norvell 1978); Ammonium oxalate [$(NH_4)_2C_2O_4$] for Mo, Griggs (1953); 0.1M HCl for Zn, Wear and Evans (1968). References for AB-DTPA, Mehlich I and Mehlich III are in Tables 4.2 and 4.3.

Chelating agents (DTPA, EDTA) are commonly used to extract micronutrient cations. Chelating agents reduce the activity of free metal ions in the soil solution by forming metal-chelate complexes, which promote replenishment by release from solid phases (clays, oxides, SOM). Consequently, the amount of micronutrient extracted by a chelate based soil test reflects both the initial quantity present in the soil solution and the ability of the soil to maintain that concentration. Chelate-based extractants thus simulate nutrient removal from the soil by plant uptake and replenishment of the soil solution from labile solid phases. Because most chelate-based soil tests were developed for specific physiographic regions and soil types and are buffered at specific pH and ionic strength values to avoid the release of micronutrients from nonlabile solid phases, they should only be used for the soil type and conditions for which they were originally calibrated. For example, the DTPA soil test, commonly used for calcareous soils, is buffered at pH 7.3 and contains 0.01 M CaCl$_2$ to prevent the dissolution of carbonate minerals that might contain occluded or precipitated Cu, Fe, Mn, and Zn.

Clearly, since the DTPA was developed for calcareous soils, it would be inappropriate for highly acid soils without careful calibration and perhaps modification of the extractant composition (Norvell, 1984; O'Connor, 1988). EDTA has been successfully used on a wide range of soils either alone or in multi-element soil tests (i.e., 0.001 M EDTA is in the Mehlich III soil test and the modified Olsen soil test is $0.5M$ NaHCO$_3$ + 0.01 M EDTA, pH 8.6). Similarly, DTPA is included in the AB-DTPA extractant (M NH$_4$HCO$_3$ + 0.005M DTPA, pH 7.6) now widely used in the western United States.

Soil tests for the anionic or uncharged micronutrients (B, Mo, and Cl) have received less attention than those for micronutrient cations because of the relatively rare responses to fertilization. Most soil tests have focused on methods that remove soluble, sorbed, or organically complexed forms of these micronutrients.

The most common soil test used for B has been the hot water extraction method of Berger and Truog (1940) in which soil is boiled with water or 0.01 M CaCl$_2$, using a reflux condenser, removing soluble and organically complexed B. Although shown to be a reasonably good predictor of plant response to B, its cumbersome and time-consuming nature has made routine use difficult. Mahler et al. (1984) using boiling plastic pouches and Shuman et al. (1992) with the Mehlich III extractant have identified some practical alternatives to the original hot water method. Molybdenum is usually extracted with acid ammonium oxalate, primarily by a desorption reaction with the added oxalate, while deionized water or any dilute salt solution [e.g., 0.01M Ca(NO$_3$)$_2$, 0.5M K$_2$SO$_4$] can be used as an extractant for Cl because of its high solubility.

Considerations in Interpretation of Micronutrient Soil Tests

Micronutrient deficiencies are, for the most part, associated with specific combinations of soil and plant factors that are reasonably well understood (Tisdale et al., 1993). Copper deficiencies are most common on soils that are extremely high in SOM (peats, mucks) or are alkaline, while deficiencies of Fe, Mn, Zn are almost always confined to calcareous or overlimed soils and sensitive crops. Boron deficiency is most frequently observed on sands with low SOM or following extremely dry periods that inhibit mineralization of SOM, and thus, the release of organically bound B. Molybdenum deficiency rarely occurs except on very acid soils and then only with crops that are highly sensitive to low concentrations (e.g., legumes, crucifers, citrus). Chloride deficiencies are very unusual and confined to certain physiographic regions, such as the Northern Great Plains of the United States.

The critical value approach is the most widely used method to interpret the results of a micronutrient soil test (Tables 4.1, 4.4). However, the predictive value of a micronutrient soil test can be improved by evaluation of more than one soil property or by knowledge of the crop to be grown. For instance, soil tests for Mn and Zn are much more accurate when soil pH is known; other examples of interacting factors that can improve soil test interpretation for micronutrients are given in Table 4.4. In some cases, quantitative availability indices based on multiple regression analysis of the soil test result and another property are calculated and used in place of the critical value approach.

Finally, several micronutrients can be toxic to plants if present in soils at high concentrations, the most common being Mn in highly acid soils (pH < 5.2) and B where only a slight overapplication of fertilizer B can cause phytotoxicity. Although unusual, B, Cu and Zn can occasionally become phytotoxic in soils amended with agricultural, municipal, and industrial waste products, such as animal manures (pig and poultry), municipal biosolids, and some byproducts of mining industries, if recommended or mandated management practices are not strictly followed. Critical phytotoxic levels based on micronutrient soil tests are much more difficult to establish than deficiency values and are usually highly specific to the plant that is grown and the soil type.

4.2.5 Methods for Soil Chemical, Physical, and Biological Properties

The availability of essential and nonessential elements to plants, their potential to become phytotoxic, or to cause environmental problems through leaching, erosion, runoff, and/or volatilization depend upon soil physical, chemical, and biological properties. Some of these properties are routinely measured by soil testing laboratories; others are only measured on selected samples. Some are rarely measured at all but can be inferred from other soil properties or from information in soil surveys. Only the soil properties that are most relevant to plant growth will be discussed.

4.2.5.1 Soil pH

Soil pH is an index of acidity/alkalinity in the soil solution in equilibrium with soil colloids (van Lierop, 1990), commonly measured using a pH meter (Thomas, 1996). Soil pH is most useful in soil fertility evaluation and management because it provides information on the solubility, and thus potential availability or phytotoxicity of some plant nutrients and nonessential elements, and the relative biological activity of plants and microorganisms. The solubility of most micronutrients, and several nonessential trace elements (e.g., Cd, Ni, Pb) is highly pH dependent. For most elements, solubility increases as the soil becomes more acidic; exceptions include P, which is most available at pH 5.5–7.5 and Ca and Mo, which are more available at pH > 7.0. Other processes (e.g., cation exchange, sorption/desorption) important to nutrient and nonessential element retention in soils also vary with pH. Acid soil infertility is most severe at pH values < 5.5, and is caused by increased solubility and toxicity of Al and Mn and by the decreased plant availability of Ca, Mg, Mo, and P. Soil N availability is lower under acid conditions because the bacteria responsible for nitrification are more active under neutral or slightly acid conditions. Alkaline soil infertility is most common in calcareous or overlimed soils, resulting from reduced availability due to the decreased elemental solubility (P, Cu, Fe, Mn, Zn) or greater sorption (B).

4.2.5.2 Lime Requirement

Soil pH is an index of the suitability of the soil chemical environment for plant growth, but provides no information on the amount of amendment needed to correct acid or alkalkine soil infertility. To determine the rate of lime (or acidulent) needed, soil buffer capacity must be measured. For acid soils, lime requirement is defined as the amount of agricultural limestone or other basic material needed to increase soil pH to a value that is optimum for the desired use (Sims, 1996). The acidification requirement is similar to the lime requirement and refers to the amount of acidulent (usually elemental S or $[Al_2(SO_4)_3]$) needed to decrease soil pH to an optimum value. Soil lime requirement can be measured by a number of methods, the most common being the measurement of buffer pH (Sims, 1996) by adding a buffered solution (pH 7.5–8.0), allowing time for equilibration, and then measuring the pH of the soil buffer suspension. The measured decrease in buffer pH is an index of the amount of soil acidity that must be neutralized to adjust the soil to the desired pH. Field calibrations between buffer pH measurements and changes in soil pH on liming are essential to the development of a buffer pH test such as those of Shoemaker-McLean-Pratt (SMP, pH 7.5), Adams-Evans (pH 8.0), and Mehlich (pH 6.6). Another lime requirement technique involves extraction of exchangeable acidity or exchangeable Al^{3+} from a soil with a salt solution (e.g., M KCl), followed by titration of the extract with a standardized base. This is a rapid, inexpensive method that is well adapted to highly acidic, aluminous soils. Application of sufficient limestone to neutralize 1.5 to 2.0 times the amount of exchangeable acidity is often adequate to eliminate some of the more serious limitations associated with acid soil fertility (e.g., Al toxicity). The procedures for diagnosis and correction of soil acidity have been critically reviewed by Sumner (1997).

4.2.5.3 Organic Matter

Because SOM content is extremely important in soil fertility, it is frequently used as a standard component in the routine soil test. Although present at low levels in topsoils (1 to 5%) and subsoils (< 0.5%), SOM (1) provides plant nutrients, especially N, B, P, and S during decomposition; (2) acts as a chelate, maintaining micronutrient cations in a plant-available form and complexing Al in a less phytotoxic form; (3) has a high water holding capacity by weight minimizing the effects of moisture stress; (4) is the predominant source of variable charge in most soils contributing to CEC which retains nutrient and nonessential cations; (5) is a significant source of pH buffering, preventing marked and often undesirable changes in soil pH due to anthropogenic inputs; and (6) acts as an aggregating agent in improving soil structure, resulting in better aeration and more prolific root growth. Despite its many important roles, it is only recently that soil testing laboratories have begun to measure and routinely report this soil property. The long-standing, traditional approach used to estimate SOM content was to measure organic C by wet chemical oxidation using dichromate ($Cr_2O_7^{2-}$) as the oxidant. Soil organic matter was then calculated from an empirically derived relationship with organic C. Because this method was time consuming and generated a high Cr waste, it was primarily used as a special test. Most soil testing laboratories now estimate SOM content by the loss on ignition (LOI) method in which a soil sample is combusted at ~360 to 400 °C for several hours or overnight. The weight loss upon ignition is assumed to be proportional to SOM content. Usually an empirical relationship between LOI and some direct measure of SOM (e.g., dichromate oxidation) is used to calculate estimated SOM content. The LOI method is well suited to modern soil testing laboratories that wish to include an estimate of SOM content in the routine soil test. Advances in electronic weighing and data acquisition and processing have resulted in LOI becoming an efficient, reasonably accurate approach to estimate SOM.

4.2.5.4 Soluble Salts

Soil salinity is a global problem and directly affects soil fertility. Soils high in soluble salts negatively affect plant growth in several ways. In addition to specific ion toxicities (Na, Cl, B) causing direct injury to plants, salinity makes it more difficult for plants to extract water from soils, even to the point of causing plant injury and death. Measuring soluble salts is a fairly easy task and is usually done as a special test involving the measurement of the electrical conductivity of a soil:water extract (1:2 or 1:5 soil:solution). A more time-consuming test, but one that better represents the soluble salts concentration in the soil solution is the saturated paste extract, obtained by mixing soil and deionized water to the point of saturation, followed by filtration and analysis of the extract for EC.

4.2.6 Interpretation and Recommendations

Interpretation systematically uses statistical relationships between soil test parameters and plant performance, or other indices of the success of a land use program. Soil test recommendations although quantitative, also include professional judgment since all the factors that control plant performance cannot be estimated from analysis of a single soil sample. Individuals responsible for nutrient management recommendations must be thoroughly familiar with the process of soil testing and with all aspects of the intended land use, including soil types, plants to be grown, climate, crop management practices, and any economic or environmental factors that may restrict a recommendation.

4.2.6.1   Correlation and Calibration

Interpretation of soil analyses begins with soil test correlation, defined as "... the process of determining whether there is a relationship between plant uptake of a nutrient or yield and the amount of nutrient extracted by a particular soil test" (Corey, 1987). To be of value a soil test must be statistically correlated with some measure of plant performance, preferably in the field. Greenhouse studies are usually the first step in soil test correlation and can rapidly and inexpensively assess the potential value of a soil test for widely differing soils and plants. The standard approach is to obtain soil samples representative of variation where the soil test will be used, measure the amount of extractable nutrient (or nonessential element) in each soil, grow plants under controlled greenhouse conditions, where moisture, light, and spatial variability are minimized, and then measure plant yield and elemental composition. Statistical correlation and regression methods describe the relationship between soil test level and plant response. Correlation analysis determines if the change in plant yield or nutrient composition is proportional to the amount of nutrient extracted by a soil test. If a high degree of correlation exists, regression analysis will provide a predictive equation that reliably estimates plant yield or elemental composition at each soil test value. In some cases, multiple correlation and regression analyses are used to develop a predictive equation that quantifies the relationship between plant performance and more than one soil property (e.g., soil test value and pH, SOM, texture, etc.).

If a soil test is significantly correlated with plant performance in the well-controlled greenhouse environment, field experiments are then conducted to determine how accurate the test will be under normal growing conditions. For greatest reliability, multiyear field trials should be conducted with multiple rates of the nutrient(s) being investigated at many locations with a range of soil types and soil test values. Replicated experiments in which other variables affecting plant performance are optimized should be used. Correlation analysis is again used to obtain a statistically significant relationship between the soil test value and plant response. Lower correlation coefficients are obtained in the field than greenhouse because of uncontrolled variability. Because of the costs and time required to conduct field and greenhouse studies greater reliance has been placed on laboratory correlations as a means of evaluating new soil test extractants. For example, several laboratory studies have compared the Mehlich III multi-element soil test with current soil tests (Hanlon and Johnson, 1984; Wolf and Baker, 1985; Sims, 1989). Many of these studies reported highly significant correlations between nutrients extracted by Mehlich III and those extracted by the Mehlich I, Bray P_1, M NH_4OAc, and EDTA, indicating that the Mehlich III could be as reliable a means of evaluating soil fertility as the existing soil test. However, while laboratory-based correlations may be acceptable for the preliminary evaluation of new soil tests, they should not be used as the sole means of determining soil test reliability. Soil test correlations must relate the amount of an element extracted by the soil test to plant yield or elemental composition (Fixen and Grove, 1990).

The next step in soil test development is calibration, defined as "... ascertaining the degree of limitation to crop growth or the probability of a growth response to applications of a nutrient at a given soil test level" (Dahnke and Olson, 1990). The purpose of calibration is to categorize soil test levels in terms of the probability of economic response to nutrient applications or their potential effect on the environment. Traditionally, soil test results have been categorized as very low, medium, high, and very high. More recently, optimum and excessive categories and crop response and environmental impact classes have been added (Beegle, 1995). Tisdale et al. (1993) suggested that probabilities of economic responses to P and K fertilization for soil rated as low, medium, high, and very high would be 70–95%, 40–70%, 10–40%, and < 10%, respectively. Several of the more important approaches used in soil test calibration are briefly described below; Dahnke and Olson (1990) and Black (1993) present more thorough reviews.

The calibration process is essentially an effort to mathematically model the relationship between soil test level and plant response to nutrient additions, which is almost always nonlinear. Consequently, curvilinear regression models are often used to identify the point where plant performance is optimal (e.g., the plateau yield, usually associated with 93–95% of maximum attainable yield) and then to determine the soil test value associated with optimum yield, which is referred to as the critical level or critical value (Fig. 4.3a). Curvilinear models are often based on relative yield or percent yield, defined as the yield obtained without addition of the nutrient being studied divided by the yield attained at that location when no other factors are limiting. Relative yield data from field trials in the region of interest are combined and plotted against the soil test values and the critical level is determined mathematically or graphically. Use of relative yields minimizes the influence of uncontrolled variables and allows for more effective interpretation of data collected over many different years, locations, soils, climates, and management settings. Once the critical value has been identified, soil test values below this level are subdivided into categories that are associated with the probability of crop response (e.g., low, medium, optimum) and the nutrient rate required for an optimum yield. Critical soil test levels vary between soils, crops, climatic regions and soil test extractants. For instance, critical soil test P values by the Bray P_1, Olsen, Mehlich I, and Mehlich III soil tests are about 30, 12, 25, and 50 mg kg^{-1}, respectively (SPAC, 1992).

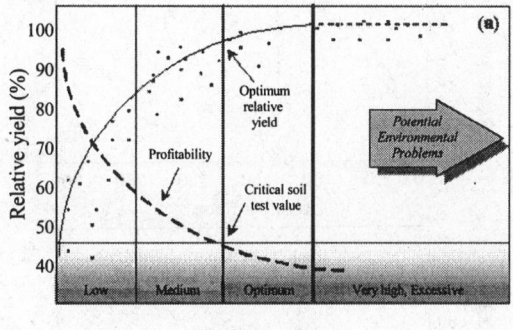

Soil test value

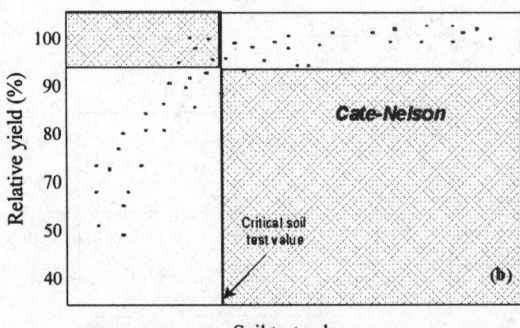

Soil test value

Fig. 4.3   Generalized illustration of the principles of soil test correlation and calibration using (a) curvilinear models and (b) the Cate-Nelson approach

Another calibration approach used to identify critical soil test levels is the Cate-Nelson method (Figure 4.3b) (Cate and Nelson, 1965). In this approach, relative yield is plotted against soil test value and the data are subdivided graphically into four quadrants either visually, using professional judgement, or mathematically (Nelson and Anderson, 1977) by placing a horizontal line at the optimum relative yield (93–95% of maximum) and a vertical line at the soil test value that minimizes the number of points in the upper left and lower right quadrants indicating that the soil test is reliable. Points in the upper left quadrant would mean that a low soil test value was associated with a high relative yield; points in the lower right quadrant are those where a high soil test value was associated with a low yield. Both of these are inconsistent with the basic premise of soil testing, that a soil test can accurately and reliably separate responsive from nonresponsive sites. If most points are in the lower left or upper right quadrants, then the soil test accurately predicts plant performance; low soil tests have low relative yields and high soil tests have high relative yields. As shown in Fig. 4.4, the mathematical approach used to model the relationship between plant performance (relative yield) and rate of nutrient added can affect the determination of the economically optimum nutrient rate. Thorough discussions of the mathematical models used to interpret soil test results are provided by Black (1993).

4.2.6.2   Recommendation Philosophies for Soil Test Interpretation

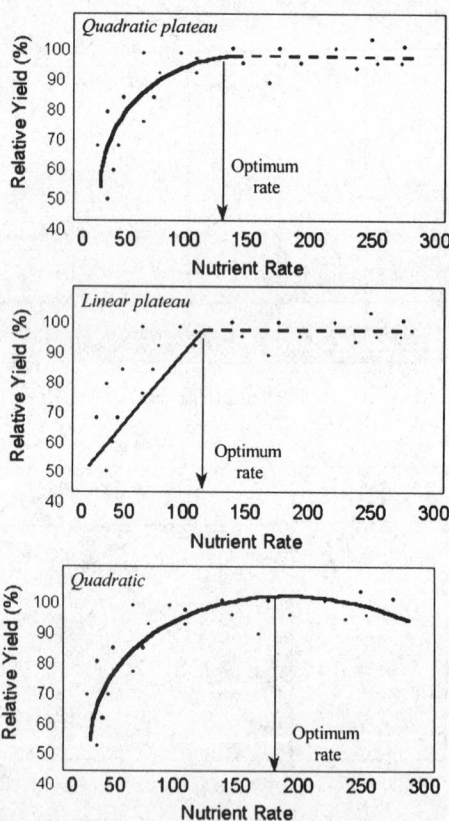

Fig. 4.4   Illustration of the influence of type of mathematical model selected on the identification of the nutrient rate required for optimum yield

The final phase in soil test development is the recommendation process in which the actual amount of nutrient to be applied, application method, and timing are specified. Individuals making a nutrient recommendation must integrate the quantitative information from soil test calibrations (e.g., the probability and magnitude of response likely to occur at a given soil test value of the nutrient) with other, more subjective aspects of nutrient management. Individuals skilled at making efficient recommendations are able to combine soil testing information with their professional experience, judgment, and scientific understanding of the system of interest in a manner that optimizes the profitability of nutrient use while minimizing any potential impacts on environmental quality. Given the subjective nature of this process, several different recommendation philosophies have evolved and now receive widespread use. The two most common philosophies for agricultural crops are rapid buildup and maintenance and sufficiency level. Both philosophies are more commonly used with immobile (P, K, Ca, Mg, Cu, Fe, Mn, Mo, Zn) than mobile nutrients (B, Cl, NO_3-N, SO_4-S) where more comprehensive soil test approaches, such as subsoil testing, must be relied upon.

The rapid buildup and maintenance approach recommends that soil fertility be built to an optimum level as rapidly as possible, usually within two years. Following this initial buildup of soil fertility, maintenance nutrient applications are made annually at rates equal to the amounts removed in the harvested portion of the crop. This philosophy is sometimes referred to as fertilization of the soil and proceeds somewhat independent of soil testing since nutrient applications are made each year, regardless of soil test results. As noted by Dahnke and Olson (1990), "... the rapid buildup and maintenance concept discounts the inherent nutrient delivery capacity of a soil's native mineral reserves which, with most soils other than sands is large for most nutrients" and "Complete adherence to this system ... would eliminate the need for further soil testing". While still used, economic and environmental questions about the appropriateness of this recommendation philosophy have grown in recent years (Olson et al., 1987).

The most widely used recommendation philosophy today is the sufficiency level approach which "... promotes the idea that a measurable soil test level exists below which responses to added fertilizer are probable and above which they are not" (Eckert, 1987). This approach is also sometimes referred to as fertilization of the crop and is a more conservative approach than rapid buildup and maintenance because nutrients are only recommended when soil test values are below the critical soil test level and are applied in proportion to the soil test category (i.e., more nutrients are added to soils that are rated low than those that are rated medium). The sufficiency level approach also inherently includes a buildup phase, but once soil fertility is in the optimum range, nutrient additions cease, or are minimal, until subsequent soil testing indicates that soil fertility has declined to the point where an economic response to further nutrient inputs is likely. There is no evidence to support the contention that the sufficiency level approach causes unnecessary depletion of soil nutrient reserves; in fact, most studies have shown that adhering to this philosophy results in a slight buildup of soil fertility. Consequently, most soil test calibration research supports the use of the sufficiency level concept for soil test recommendations.

4.2.7 Environmental Soil Testing

4.2.7.1 Principles and Purposes of Environmental Soil Tests
In recent years, interest has grown in environmental soil testing, defined as quantitative analysis of soils to determine if environmentally unacceptable levels of nutrients, nonessential elements or organic compounds are present. Environmental soil testing is a much more ambiguous process than agricultural soil testing because the meaning of the term environmentally unacceptable is difficult to quantify. In the absence of a clear, quantitative measure of success, the entire process of soil testing

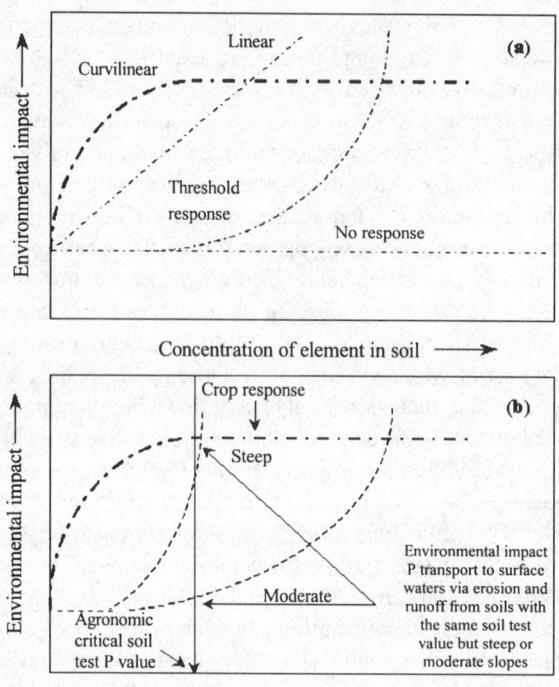

Fig. 4.5   Illustration of points to consider in interpretation of environmental soil tests. As seen in (a) increasing the concentrations of soil nutrients of nonessential elements can induce no environmental response, cause a response once a threshold is reached, or have a linear or curvilinear effect on the environmental parameter of interest [adapted from Pierzynski et al., 1994]; in (b) crop response to increasing soil concentrations of a nutrient is contrasted with the environmental impact of that element, which can be modified by other soil properties. For example, crop response to inputs of phosphorus (P) is normally curvilinear, while the impacts of P on surface waters due to erosion and runoff can follow a threshold response. Soils with the same soil test P value, but varying slopes (steep or moderate), will have differing environmental impacts.

becomes more diffuse and complex. Nevertheless, the rising interest in environmental protection has prompted an increased effort to use soil testing to assess risks posed by soils to other sectors of the environment, particularly ground and surface waters. Consequently, soil testing practices must be restructured to meet environmental goals (Sims et al., 1997a).

In the broadest sense, the goals of environmental soil testing are the same as those of routine, agricultural soil testing, namely, rapid, accurate, appropriate and reproducible evaluation, and interpretation of results related to environmental risk. In both cases, the conceptual differences between environmental (Fig. 4.5) and agricultural interpretation (Fig. 4.3) must be considered.

4.2.7.2   Soil Testing for Potentially Toxic Trace Elements

Soil testing for potentially toxic trace elements is an environmental issue because some plant nutrients (Cu, Mo, and Zn) and nonessential elements (As, Cd, Cr, Hg, Ni, Pb, Se) are toxic to either plants, animals, or humans. Naturally, high concentrations of one or more of these elements (very unusual) may occur, or may increase as a result of anthropogenic activity such as land application of wastes as beneficial soil amendments (e.g., animal manures, municipal sewage sludges, industrial byproducts). Some soils may be highly polluted with toxic elements due to mismanagement of potentially beneficial wastes, by an accidental spill or discharge, or as a result of an industrial activity such as mining or smelting. Note that contamination and pollution are not the same. Contamination occurs when concentrations exceed natural levels, but no adverse effect on an organism is apparent while pollution implies elevated concentrations, and clearly documented adverse effects on some organism (Pierzynski et al., 1994).

Soil testing for potentially toxic trace elements begins with an understanding of the nature of the risk involved; what organisms may be affected, by what pathway, and by which elements. Primary areas of concern are direct soil ingestion, phytotoxicity, plant uptake and food chain contamination, and water pollution from erosion, runoff, or leaching. Direct ingestion is normally only a concern for Pb and with young children who are most likely to ingest contaminated soil or inhale dust. Phytotoxicity is rarely an issue with Pb except in highly polluted soils at industrial sites but is a greater concern with Cu, Cr, Ni, and Zn. Food chain contamination and human health effects are most often associated with Cd and Hg and water quality concerns with As, Cu, Hg, and Se. Knowledge of the nature of the risk and the element of concern helps to determine the most effective soil sampling protocol. In the case of Pb in urban soils where human health is the concern, very shallow soil samples (< 2 cm) must be used since this is the soil depth most likely to be ingested. A similar depth would be useful if runoff or erosion are the issue since rainfall primarily interacts with only the uppermost few cm of the soils. However, in terms of groundwater pollution, sampling into subsoil horizons (1–2 m) would be recommended to determine if elemental leaching is occurring. For elements where the main concerns are phytotoxicity and food chain contamination, the normal sampling depths are usually acceptable (e.g., 0–20 cm). If site remediation is the goal, either by soil removal, washing, or amendment, then systematic deep sampling (e.g., 0–5, 5–10, 10–20, 20–40, 40–60, 60–100 cm) is recommended to determine the depth of soil contamination and thus the extent of remediation required. Soil sample handling and preparation must avoid contamination from equipment and protect personal health and safety. The method of analysis to be used varies with the intent of the test. If the goal is to assess biological availability (e.g., plant uptake), then many dilute acid- or chelate-based soil test extractants may be suitable, providing due consideration is given to the most appropriate test for the intended land use (O'Connor, 1988). However, if the goal is to quantify the extent of accumulation of an element, relative to normal soils or natural background levels, or to monitor this accumulation over time, methods that determine or approximate total elemental content are recommended. One example is USEPA Method 3050 which successively digests a soil sample with concentrated HNO_3, H_2O_2, and HCl to measure total sorbed metals by acid dissolution of clays, oxides, and carbonates and oxidation of organic matter; elements associated with silicates are not dissolved. For true total elemental content, complete soil digestion of the soil with strong acids (e.g., HNO_3-$HClO_4$ for Cd, Hg, and Pb), by carbonate fusion (Cr, Ni), or by alkaline oxidation techniques (As, Se) is required. Given the costs and difficulty of measuring total elemental content of trace elements in soils, use of rapid soil testing methods as surrogate monitoring techniques may be advisable (Sims and Johnson, 1991).

Interpretations and recommendations for soil tests for potentially toxic trace elements are considerably more difficult than for agricultural systems and are often very site and element specific. As mentioned above, the main difficulty lies with soil test calibration, establishing a quantitative relationship between an agreed upon measure of environmental risk and the amount of an element measured by the soil test. In most cases, this has been done by the use of complex risk assessment models that first identify the target organism of concern (e.g., human versus plant) and then evaluate all possible pathways by which the target organism may be exposed to the risk factor (toxic trace element). If possible, a most sensitive pathway is identified, defined as the lowest soil concentration level at which an adverse effect on the organism would be likely to occur (e.g., soil ingestion versus consumption of contaminated drinking water). Regulatory upper limits may then be established for that pathway, which can be monitored by the soil testing methods described above. This approach was used by USEPA in the formulation of the national rule for the disposal and utilization of municipal sewage sludges which established regulatory limits for the total amount of several trace elements that could be land applied to agricultural soils (Ryan and Chaney, 1993). Additional information on the methods used to interpret soil testing results is found in USEPA (1989), Risser and Baker (1990), and Pierzynski et al. (1994).

4.2.7.3   Soil Testing for Plant Nutrients with Water Quality Impacts

Nitrogen and P are the primary nutrients that degrade ground (N) and surface (N, P) water quality. The methods described in Section 4.2.4.4 (e.g., the PSNT), which focus on identifying sites with an adequate N supply, and thus, avoiding unnecessary applications of fertilizers or organic sources of N (manures, sludges) are appropriate environmental soil testing. The role of P in the eutrophication of surface waters has stimulated much effort in developing environmental soil testing for P. The focus of this effort has been the use of soil testing alone, or as a component of indices and models, to identify soils that are most likely to be significant nonpoint sources of P pollution of surface waters. The changes needed when surface water protection and not agricultural production is the primary goal are discussed below. More detailed information is available from Sims (1993, 1997), Sharpley et al. (1996), Sibbesen and Sharpley (1997), and Tunney et al. (1997).

Establishing Upper Critical Limits for Soil Test P

One approach proposed for environmental soil testing for P is to simply establish an upper critical limit for soil P using currently available agronomic soil testing methods (e.g., Bray P_1, Mehlich I, Olsen P). Soils that exceeded the upper critical limit would no longer receive P inputs from any source and would be targeted as priority areas for soil and water conservation practices to prevent P loss in erosion, runoff, and leaching. Two reasons are usually given to justify the need for this upper critical limit. First, soils which are overfertilized with P relative to crop requirements create increased risk of nonpoint source pollution of surface waters (Sharpley et al., 1994; Sims et al., 1997; Tunney et al., 1997). Second, continued P application to soils beyond values that are needed for crop production contradicts the principles of sustainable development and sustainable agriculture. Because P is obtained from a finite natural resource, at a cost to society, agricultural practices which waste this resource are inconsistent with sustainability. Despite these concerns there has been a reluctance to establish upper critical limits for soil test P presumably because (1) agronomic soil tests were not designed or calibrated for environmental purposes; and (2) there may be an unjustified reliance upon soil test P alone by environmental regulatory agencies attempting to control nonpoint source pollution of surface waters, ignoring the complex interaction between soil P and the transport processes that move P from soil to water.

Other approaches to environmental soil testing for P are direct measurement of soluble or biologically available P, estimating potentially desorbable P by Fe oxide strips, and characterizing the degree of P saturation of soils (Gartley and Sims, 1994; Sibbesen and Sharpley, 1997; Sims, 1997). While research has been promising, none of these has received widespread acceptance in quantifying the environmental risk associated with high P soils.

4.3   Plant Testing

Soil fertility evaluation does not rely upon soil analysis alone. Many techniques assess soil fertility by characterizing the growth and elemental composition of plants, including visual diagnosis, in-field evaluation, tissue analysis, and remote sensing. Plant analysis includes both rapid in field tissue testing and laboratory total elemental analysis of plant samples. In both cases the underlying premise is that the plant concentration of an element is proportional to the availability of the nutrient in the soil and thus is an index of soil fertility. Visual diagnosis and remote sensing do not actually determine plant nutrient concentration, but instead rely upon changes in plant color or growth as indices of soil fertility.

4.3.1 Visual Diagnosis of Deficiency Symptoms

With experience, visual symptoms that result from nutrient deficiencies can be identified, including severely stunted growth and purpling of older leaves (P), chlorosis and necrosis of leaf margins (K), interveinal yellowing of young (Fe, Mn) and older leaves (Mg), and distorted meristems and blackened internal tissues (B). In general, visual diagnosis should be verified by soil and plant analysis, prior to taking corrective actions because (1) symptoms may be caused by a deficiency of more than one nutrient; (2) deficiency of one nutrient may be caused by an excess of another; (3) other factors, such as insect and disease injury, can create symptoms similar to nutrient deficiencies; and (4) symptoms may be caused by more than one growth factor (e.g., P deficiency can occur in cold, wet soils that have optimum P fertility levels but disappear when soils warm) (Tisdale et al., 1993). In addition, the same symptom may be caused by different nutrients in differing plants. For example, purpling of older leaves, typically due to P, can be caused in crucifers by S deficiency. The location of a deficiency on the plant can assist in proper identification of the cause. Nitrogen and S can both cause yellowing of plant leaves, but because N is more mobile than S, its deficiency symptoms usually occur on older leaves, while S deficiencies appear as a more overall yellowing. Because Ca, B, Cu, Fe, Mn, and Zn are rather immobile in plants, their deficiencies usually occur on new growth, while those of more mobile N, P, K, and Mg appear on older plant parts. Finally, by the time a nutrient deficiency symptom can be accurately diagnosed, it is either too late for corrective action (unless nutrients can be added by fertigation or sidedressing) or unwise because the deficiency has already damaged the plant beyond the point where nutrient additions can profitably correct the problem.

4.3.2 In-Field Evaluation of Plant Nutritional Status

In addition to visual diagnosis, several in-field diagnostic techniques can be used to provide semi-quantitative information on plant nutritional status. The most common are rapid tissue testing kits for the colorimetric analysis of plant sap (usually for N, P, and K). The entire process takes only a few minutes. Because of their semi-quantitative nature, rapid tissue tests should only be used as guides, and not as the sole basis for nutrient management recommendations. Tisdale et al. (1993) also cautioned that rapid tissue testing must: (1) only be done with the correct plant part, sampled at the proper time of year and, for some nutrients (e.g., NO_3-N), at the proper time of day because nutrient concentrations change during the day; (2) be done in a comprehensive manner, continuing the testing throughout the growing season, and not as a one time activity; (3) focus on periods of maximum vegetative growth and reproductive stages to best assess a fertilizer program; and (4) be the average value from the analysis of at least 10–15 plants and be collected from areas of normal and deficient plant growth.

A recent advance in field plant N testing is the leaf chlorophyll meter (LCM), a small, hand-held spectrometer that essentially measures leaf greenness which is correlated with leaf chlorophyll content, plant N, and the likelihood of an economic response to N fertilization. The LCM allows for rapid, ongoing monitoring of plant N nutrition, and is particularly well suited to irrigated crops where N can be added by the irrigation system. Since other nutrients affect leaf greenness, especially S, the LCM must be used with caution and with soil N testing where possible to avoid recommending N fertilization when some other factor is responsible for the observed chlorosis (Schepers et al., 1992).

Nevertheless, the LCM is an important advance in the field scale evaluation of plant N nutrition, particularly for agronomic crops.

4.3.3 Plant Analysis

4.3.3.1 Basic Principles

Plant analysis is defined as "... the determination of the elemental composition of plants, or a portion of the plant, for elements essential for growth. It can also include determining elements that are detrimental to growth or animals or humans through our food chain" (Munson and Nelson, 1990). The components of a plant analysis program are similar to those used for soil testing: (1) collection of a representative sample from a plant, or the whole plant, at the proper stage of plant development; (2) proper handling of the sample to avoid contamination or damage that could affect interpretation of results, followed by accurate sample analysis using standardized laboratory methods; (3) correlation and calibration studies that establish quantitative relationships between plant analysis and plant performance; and (4) economically and environmentally sound nutrient management recommendations to correct any nutrient deficiencies. Key factors to consider in each step are described below. Detailed descriptions of the protocols used for plant sampling, handling, and laboratory analysis have been presented by Plank (1989), Jones and Case (1990), Westerman (1990), and Mills and Jones (1996).

4.3.3.2 Interpretation of Analytical Results

The two most widely used methods for interpretation of plant analysis are (1) the critical nutrient concentration or range (CNC or CNR), and (2) the diagnosis and recommendation integrated system (DRIS) which relies primarily on nutrient ratios, emphasizing the importance of nutrient balance in the plant for optimum plant performance.

Interpretation by the critical nutrient range (CNR) approach relies, as with soil testing, mainly on correlation and calibration studies that show a statistically significant relationship between plant nutrient concentration and plant yield to nutrient additions (Fig. 4.6). In most cases, plant analyses are

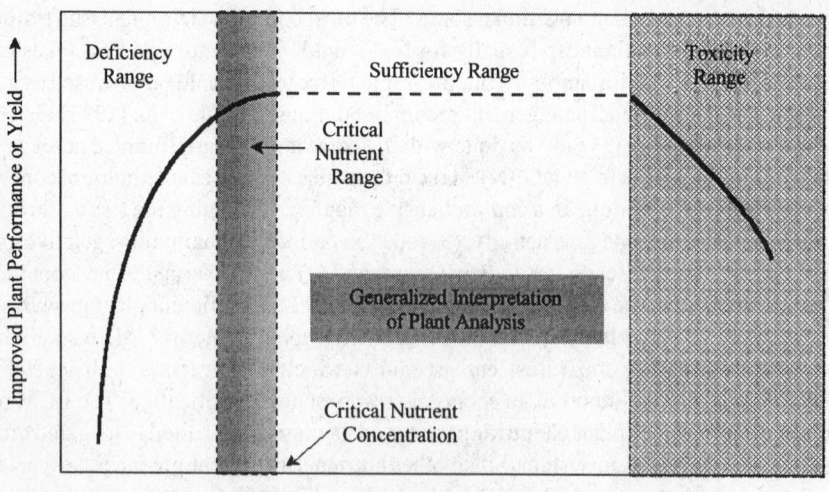

Fig. 4.6 Illustration of the critical nutrient range approach for plant analysis interpretation

compared to optimum CNRs. Nutrient concentrations below the CNR are considered deficient, those above it adequate or optimum; the CNR represents a transition zone between these two categories. The critical nutrient concentration (CNC) is defined as the nutrient concentration where plant performance changes from suboptimum and unsatisfactory to optimum. Because the CNC is a single value that can be difficult to determine experimentally, given the uncertainties and variations associated with field studies, and because it can vary somewhat within a plant species due to genetic differences among hybrids, the CNR approach is more commonly used to interpret plant analysis results. Extensive research conducted to determine CNRs for most plants with economic value clearly shows that the CNR for a plant depends upon growth stage and plant part that is sampled (Mills and Jones, 1996). Therefore, comparisons of plant analysis results with CNR values are only valid if the plant sample is taken at the same growth stage and from the same plant part used in the calibration studies conducted to determine CNRs. Interpretation of plant analysis results by this method, therefore, requires careful adherence to recommended protocols for sampling, handling, and analysis. Extensive listings of CNC and CNR values for different plants are to be found in Reuter and Robinson (1997).

The balance of nutrients within a plant is often more important than the concentration of any individual nutrient. The DRIS system focuses on nutrient balance as an alternative to the CNR approach to plant analysis interpretation. One advantage of the DRIS system, compared to the CNR, is that nutrient ratios in plants tend to be more constant throughout the growing season than individual nutrient concentrations. Use of DRIS, therefore, allows for greater flexibility in the time of plant sample collection as samples do not necessarily have to be taken at a specific growth stage. The DRIS system first establishes norms for all nutrient ratios (N:P, N:K, P:K, N:S, P:S, etc.) associated with maximum crop yield. These norms can be developed from reviews of the scientific literature or by widespread sampling of a crop in a given physiographic region. As much as possible, all factors that affect crop yield are measured at the time of plant sample collection (soil test values, pest pressure, climate, cultural practices, etc.). In this way an integrated relationship can be established between the DRIS norms, crop yield, and the other growth-limiting factors. Interpretation of plant analysis by the DRIS system can be done graphically or mathematically by the calculation of DRIS indices. The DRIS system has received intense interest and considerable research in the past decade. Computer programs that can rapidly calculate DRIS indices have been developed, resulting in wider use of DRIS by plant analysis laboratories. Readers are referred to Beaufils (1973), Escano et al. (1981), Walworth and Sumner (1988), Sumner (1989), and Sumner and Barbour (1992) for examples of the use of DRIS.

4.4 Soil Fertility Evaluation: The Future

The principles and practices used to assess soil fertility continue to evolve, and it is clear that the process of soil fertility evaluation will take on new dimensions in the next decade. Given this, two important questions should be considered now. First, how should newly emerging technologies for the evaluation of soil fertility be assessed and integrated with current practices? Second, how can soil fertility evaluation expand beyond production agriculture to address new challenges, especially those related to environmental protection?

4.4.1 Precision Agriculture and Remote Sensing

The past decade has seen the development of several new technologies that have the potential to significantly alter soil fertility evaluation and management. The two most important and closely related examples are precision agriculture and remote sensing (Hergert, 1997). Precision agriculture, also referred to as site-specific management, is a relatively new approach to farming that permits: (1) precise mapping of soil fertility in a field or on a farm in great detail (1 ha grids) through the use of hand-held or equipment mounted global positioning systems (GPS) that use signals from a network of satellites to instantaneously locate a position on the Earth's surface; (2) use of variable rate application equipment, equipped with GPS and on-the-go sensors of soil nutrient status, to supply nutrients in accordance with these maps; and (3) generation of GPS-based maps of crop yields. Factors related to yield that cannot easily be measured by soil and plant testing can be obtained from other computerized databases (e.g., soil series from soil surveys) or be assessed using remote sensing (e.g., plant health, soil moisture, irrigation and drainage, pest pressure). The most commonly used remote sensing devices include cameras and other imaging systems mounted on aircraft or satellites. Other, more recent examples are sensors mounted on ground equipment (e.g., tractors, pesticide applicators) or on large permanent structures (irrigation systems, electrical towers). Remote sensing devices acquire electromagnetic energy that is emitted or reflected from plants (or bare soils) and convert this energy into data that can be used in soil fertility evaluation. Each combination of soils, plants, environmental conditions, and management practices has a characteristic spectral signature (a specific spectrum of radiation) that can be used to diagnose soil fertility problems. As an example, infrared aerial photography has been used to assess plant performance because healthy green plants reflect large amounts of infrared radiation while plants damaged by insects, drought, or nutrient deficiencies do not. As with traditional approaches to soil fertility evaluation, the ultimate value of these new technologies lies in our ability to interpret the results. Indeed, one of the problems with precision agriculture is that vast quantities of spatially located data can be generated very quickly, creating an information overload that can confuse or obscure interpretation. Research is now underway to develop expert systems that use computerized GIS to integrate the various layers of data in a meaningful way, thus guiding a more holistic approach to crop management, of which soil fertility evaluation is only a part. Some areas in precision agriculture where research is needed include (1) an economic analysis of the value of the additional information obtained by extensive, GPS-based grid soil sampling, relative to traditional, less intensive techniques; (2) a reevaluation of current soil test calibration and correlation models given the enormous, large-scale databases that are rapidly becoming available; (3) the most effective and rapid means to obtain ground truth (verification of the accuracy of remote sensing devices) and on-the-go sensors as indicators of soil fertility; and (4) the value of site-specific management techniques for applications other than crop production, such as minimizing N and P losses by leaching and runoff, irrigation scheduling, erosion control practices, and more efficient use of wastes and byproducts. Readers are referred to Pierce and Sadler (1997) and Robert et al. (1995) for reviews of precision agriculture and site-specific management.

4.4.2 New Directions and Uses for Soil Fertility Evaluation Techniques

The formal, institutionalized practice of soil fertility evaluation originated to serve the needs of production agriculture, a function that continues to be of unquestioned importance today. However, some of the needs of the agricultural sector are changing as, for example, in developed countries where the potential environmental impacts of soils that have become overenriched with nutrients, has begun to supersede the need to identify and correct nutrient deficiency problems. At the same time, those responsible for nonagricultural land uses are beginning to see the potential value of practices originally developed to optimize soil fertility for crop production. There are several challenges that must be overcome if the process of soil fertility evaluation is to evolve and respond to needs such as these. First is the establishment of better and more direct interfaces between nontraditional users of soil and plant testing and researchers with expertise in these areas. This will help to clarify when, where, and how it is appropriate to use current soil fertility evaluation techniques for purposes other than those for which they were originally designed. It will also provide insight into the advances in soil science needed to more effectively address the problems faced by those charged with protecting air and water quality, preventing further damage or restoring soil quality to lands disturbed by erosion, salinization, construction, surface mining, and similar activities, and remediation of soils polluted by anthropogenic activities. Second is the need for public and private organizations responsible for soil fertility evaluation to recognize the contribution they can make to solving land management problems that are not solely directed at profitable crop production. This will require better interactions with: (1) researchers who have developed many new soil testing methods that have value in these areas but that have not been adopted by routine soil testing laboratories, and (2) with new clientele who often have unusual problems and limited understanding of the principles and practices of soil and plant testing. Strengthening these interfaces will provide the research base essential to support nontraditional uses of soil fertility evaluation techniques and the educational component needed to ensure that only the appropriate analytical methods are used and that proper interpretations of the results are made.

In conclusion, soil fertility evaluation is a vital, integral part of global agriculture and plays important, but somewhat limited roles in most nonagricultural land uses. Successful application of the principles and practices described here will increase the profitability and minimize the environmental impacts of nutrient use, the fundamental goal of soil fertility management. Appropriate integration of newly emerging technologies with current practices will further enhance our ability to evaluate soil fertility and make sound nutrient management decisions. Expanding the process of soil fertility evaluation to more fully include nonagricultural settings, and especially to situations where environmental protection or land restoration is the goal, is perhaps the greatest challenge we face today.

4.5 References

Abrams, M.M., and W.M. Jarrell.1992. Bioavailability index for phosphorus using ion exchange resin impregnated membranes.Soil Sci. Soc. Am. J. 56:1532–1537.

Amer, F., D.R. Bouldin, C.A. Black, and F.R. Duke. 1955. Characterization of soil phosphorus by anion exchange resin adsorption and ^{32}P-equilibration. Plant Soil 6:391–408.

Beaufils, E.R. 1973. Diagnostic and recommendation integrated system (DRIS). A general scheme for experimentation and calibration based on principles developed from research in plant nutrition. Soil Sci. Bull. 1. University of Natal, Pietermaritzburg, South Africa.

Beegle, D. 1995. Interpretation of soil test results. p. 84–91. *In* J. T. Sims and A. M. Wolf (ed.) Recommended soil testing procedures for the Northeastern United States. University of DE Bull. 493.

Berger, K.C., and E. Truog. 1940. Boron deficiency as revealed by plant and soil tests. .J. Am. Soc. Agron. 32:297–301.

Berti, W.R., and S.D. Cunningham. 1994. Remediating soils with green plants. p. 43–51. *In* C. R. Cothern (ed.) Trace substances, environment, and health. Sci. Rev., Northwood, Middlesex, UK.

Black, C.A. 1993. Soil fertility evaluation and control. Lewis Publishers, Boca Raton, FL.

Bock, B.R., and K.R. Kelley. 1992. Predicting nitrogen fertilizer needs for corn in humid regions. National Fertilizer and Environmental Research Center, Muscle Shoals, AL.

Bray, R.H., and L.T. Kurtz. 1945. Determination of total organic and available forms of phosphorus in soil. Soil Sci. 59:39–46.

Bundy, L.G., and J.J. Meisinger. 1996. Nitrogen availability indices. p. 951–984. *In* R. W. Weaver et al. (ed.) Methods of soil analysis. Part 2: Microbiological and biochemical properties. Soil Science Society of America, Madison, WI.

Bundy, L. G., M. A. Schmitt, and G. W. Randall. 1992. Advances in the upper midwest. p. 73–90. *In* Bock, B.R. and K.R. Kelley (ed.). Predicting nitrogen fertilizer needs for corn in humid regions. National Fertilizer and Environmental Research Center, Muscle Shoals, AL.

Cate, R.B., Jr., and L.A. Nelson. 1965. A rapid method for correlation of soil test analyses with plant response data. NC Agric. Exp. Sta. Int. Soil Test. Ser. Tech. Bull. 1.

Chapman, H.D., and W.P. Kelley. 1930. The determination of the replaceable bases and base-exchange capacity of soils. Soil Sci. 30:391–406.

Chardon, W.J., R.G. Menon, and S.H. Chien. 1996. Iron oxide impregnated filter paper (Pi test): A review of its development and methodological research. Nutr. Cycl. Agroecosys. 46:41–51.

Corey, R.B. 1987. Soil test procedures: Correlation. p. 15–22. In J.R. Brown (ed.). Soil testing: Sampling, correlation, calibration, and interpretation. Soil Sci. Sov. Am. Spec. Pub. 21. Soil Science Society of America, Madison, WI.

Dahnke, W.C., and R.A. Olson. 1990. Soil test correlation, calibration, and recommendation. p. 45–71. In R.L. Westerman (ed.). Soil testing and plant analysis. 3rd Ed. Soil Science Society of America, Madison, WI.

Eckert, D. J. 1987. Soil test interpretations: Basic cation saturation ratios and sufficiency levels. p. 53–64. In Soil testing: Sampling, correlation, calibration, and interpretation. Soil Sci. Soc. Am. Spec. Pub. 21, Soil Science Society of America, Madison, WI.

Egner, H., H. Riehm, and W.R. Domingo. 1960. Untersuchungen uber die chemishe bodenanalyse als grundlage fur die beurteilung des nahrstoffzustandes der boden. II. Chemische extraktions-methoden zur phosphor-und kalimbestimmung kungl.Lantbrukshoegsk. Ann. 26:204–209.

Escano, C.R., C.A. Jones, and G. Uehara. 1981. Nutrient diagnosis in corn grown on Hydric Dystrandepts. II. Comparison of two systems of tissue diagnosis. Soil Sci. Soc. Am. J. 45:1140–1144.

Fixen, P.E., and J.H. Grove. 1990. Testing soils for phosphorus. p. 141–180. In R.L. Westerman (ed.) Soil testing and plant analysis. 3rd Ed. Soil Science Society of America, Madison, WI.

Foth, H. D., and B. G. Ellis. 1997. Soil fertility. 2nd Ed. CRC Press, Boca Raton, FL.

Gartley, K. L., and J. T. Sims. 1994. Phosphorus soil testing: Environmental uses and implications. Commun. Soil Sci. Plant Anal. 25:1565–1582.

Grigg, J.L. 1953. Determination of the available molybdenum of soils. NZ J. Sci. Technol. 34:405–414.

Haby, V.A., M.P. Russelle, and E.O. Skogley. 1990. Testing soils for potassium, calcium, and magnesium. p. 181–227. In R.L. Westerman (ed.) Soil testing and plant analysis. 3rd Ed. Soil Science Society of America, Madison, WI.

Hanlon, E.A., and G.V. Johnson. 1984. Bray/Kurtz, Mehlich III, AB/D and ammonium acetate extractions of P, K, and Mg in four Oklahoma soils. Commun. Soil Sci. Plant Anal. 15:277–294.

Hergert, G.W. 1997. A futuristic view of soil and plant analysis and nutrient recommendations. p. 24–35. Proc. 5th Intl. Symp. Soil Plant Anal., Minneapolis, MN.

James, D.W., and K.L. Wells. 1990. Soil sample collection and handling: Technique based on source and degree of field variability. p. 25–44. In R.L. Westerman (ed.) Soil testing and plant analysis. 3rd Ed. Soil Science Society of America, Madison, WI.

Jemison, J.M., Jr., and R.H. Fox. 1988. A quick-test procedure for soil and plant tissue nitrates using test strips and a hand-held reflectometer. Commun. Soil Sci. Plant Anal. 19:1569–1582.

Johnson, G.V., and P.E. Fixen. 1990. Testing soils for sulfur, boron, molybdenum, and chlorine. p. 265–273. In R.L. Westerman (ed.) Soil testing and plant analysis. 3rd Ed. Soil Science Society of America, Madison, WI.

Jones, J.B., Jr., and V.W. Case. 1990. Sampling, handling, and analyzing plant tissue samples. p. 389–427. In R.L. Westerman (ed.) Soil testing and plant analysis. 3rd Ed. Soil Science Society of America, Madison, WI.

Kamprath, E.J., and M.E. Watson. 1980. Conventional soil and tissue tests for assessing the phosphorus status of soils. p. 433–470. In F.E. Khawsaneh (ed.) The role of phosphorus in agriculture. Soil Science Society of America, Madison, WI.

Keeney, D.R. 1982. Nitrogen-availability indices. p. 711–733. In A.L. Page et al. (ed.) Methods of soil analysis. Part 2. 2nd Ed. Soil Science Society of America, Madison, WI.

Keeney, D.R., and D.W. Nelson. 1982. Nitrogen: Inorganic forms. p. 643–698. In A.L. Page et al. (ed.) Methods of soil analysis. Part 2. 2nd Ed. Soil Science Society of America, Madison, WI.

Kuo, S. 1996. Phosphorus. p. 869–919. In D.L. Sparks (ed.) Methods of soil analysis. Part 3. Chemical methods. Soil Science Society of America, Madison, WI.

Leal, J.E., M.E. Sumner, and L. T. West. 1994. Evaluation of available P with different extractants on Guatemalan soils. Commun. Soil Sci. Plant Anal. 25:1161–1196.

Lindsay, W.L., and W.A. Norvell. 1978. Development of a DTPA soil test for zinc, iron, manganese and copper. Soil Sci. Soc. Am. J. 42:421–428.

Magdoff, F.R., D. Ross, and J. Amadon. 1984. A soil test for nitrogen availability to corn. Soil Sci. Soc. Am. J. 48:1301–1304.

Magdoff, F.R., W.E. Jokela, R.H. Fox, and G.F. Griffin. 1990. A soil test for nitrogen availability in the Northeastern United States. Commun. Soil Sci. Plant Anal. 21:1103–1115.

Mahler, R.L., D.V. Naylor, and M.K. Fredrickson. 1984. Hot water extraction of boron from soils using sealed plastic pouches. Commun. Soil Sci. Plant Anal. 15:479–492.

Martens, D.C., and W.L. Lindsay. 1990. Testing soils for copper, iron, manganese, and zinc. p. 229–264. *In* R.L. Westerman (ed.) Soil testing and plant analysis. 3rd Ed. Soil Science Society of America, Madison, WI.

Mehlich, A. 1953. Determination of P, Ca, Mg, K, Na and NH_4. North Carolina Soil Test Division mimeo.

Mehlich, A. 1984. Mehlich III soil extractant: A modification of Mehlich II extractant. Commun. Soil Sci. Plant Anal. 15:1409–1416.

Meisinger, J.J., V.A. Bandel, J.S. Angle, B.E. O'Keefe, and C.M. Reynolds. 1992. Pre-sidedress soil nitrate test evaluation in Maryland. Soil Sci. Soc. Am. J. 56:1527–1532.

Menon, R.G., S.H. Chien, and W.J. Chardon. 1997. Iron oxide impregnated filter paper (Pi test): II. A review of its application. Nutr. Cycl. Agroecosys. 47:7–18.

Mills, H.A., and J.B. Jones, Jr. 1996. Plant analysis handbook. II. MicroMacro Publishing, Inc. Athens, GA.

Morgan, M.F. 1941. Chemical soil diagnosis by the Universal Soil Testing System. CT Agric. Exp. Sta. Bull. 450.

Murphy, J., and J. P. Riley. 1962. A modified single solution method for the determination of phosphorus in natural waters. Anal. Chim. Acta. 27:31–36.

NCR-13. 1988. Recommended chemical soil test procedures for the North Central region. ND Agric. Expt. Sta. Bull. 499.

NEC-67. 1995. Recommended soil testing procedures for the Northeastern United States. DE Agric. Expt. Sta. Bull 493.

Nelson, L.A., and R.L. Anderson. 1977. Partitioning of soil test-crop response probability. p. 19–38. *In* T.R. Peck et al. (ed.) Soil testing: Correlating and interpreting the analytical results. Soil Sci. Soc. Am. Spec. Publ. 29. Soil Science Society of America, Madison, WI.

Nemeth, K. 1979. The availability of nutrients in the soil as determined by electro-ultrafiltration (EUF). Adv. Agron. 31:155–188.

Norvell, W.A. 1984. Comparison of chelating agents as extractants for metals in diverse soil materials. Soil Sci. Soc. Am. J. 48:1285–1292.

O'Connor, G.A. 1988. Use and misuse of the DTPA soil test. J. Environ. Qual. 17:715–718.

Olsen, S.R., C.V. Cole, F.S. Watanabe, and L.A. Dean. 1954. Estimation of available P in soils by extraction with sodium bicarbonate. USDA Circ. 939.

Olson, R.A., F.N. Anderson, K.D. Frank, P.H. Grabouski, G.W. Rehm, and C.A. Shapiro. 1987. Soil test interpretations: Sufficiency vs. build-up and maintenance. p. 41–52. *In* J.R. Brown (ed.) Soil testing: Sampling, correlation, calibration, and interpretation. Soil Sci. Soc. Am. Spec. Publ. 21, Soil Science Society of America, Madison, WI.

Peck, T.R., and P.N. Soltanpour. 1990. The principles of soil testing. p. 1–10. *In* R. L. Westerman (ed.) Soil testing and plant analysis. Soil Science Society of America, Madison, WI.

Pierce, F., and E.J. Sadler. 1997. The state of site specific management for agriculture. Soil Science Society of America, Madison, WI.

Pierzynski, G.M., J.T. Sims, and G.F. Vance. 1994. Soils and environmental quality. Lewis Publishers, Boca Raton, FL.

Plank, C.O. 1989. Plant analysis handbook for Georgia. University of Georgia, Athens, GA.

Power, J.F., and J.S. Schepers. 1989. Nitrate concentration of ground water in North America. Agric. Ecosys. Environ. 26:165–188.

Reuter, D., and J.B. Robinson. 1997. Plant Analysis: An interpretation manual. 2nd Ed. CSIRO Publishing, Umbrae, South Australia.

Risser, J.A., and D.E. Baker. 1990. Testing soils for toxic metals. p. 275–298. *In* R.L. Westerman (ed.) Soil testing and plant analysis. 3rd Ed. Soil Science Society of America, Madison, WI.

Robert, P.C., R.H. Rust, and W.E. Larson. 1995. Site-specific management for agricultural systems. Soil Science Society of America, Madison, WI.

Ruby, M.V., A. Davis, R. Schoof, S. Eberle, and C.M. Sellstone. 1996. Estimation of lead and arsenic bioavailability using a physiologically based extraction test. Environ. Sci. Technol. 30:422–430.

Ryan, J.A. and R.L. Chaney. 1993. Regulation of municipal sewage sludge under the Clean Water Act Section 503: A model for exposure and risk assessment for MSW-compost. *In* H.A. Hoitnik and H.M. Keener (ed.) Science and engineering of compost: Design, environmental, microbiological, and utilization aspects. Renaissance Publishers, Worthington, OH.

Schepers, J.S., T.M. Blackmer, and D.D. Francis. 1992. Using chlorophyll meters. p. 105–114. *In*. Bock, B.R. and K.R. Kelley (ed.) Predicting nitrogen fertilizer needs for corn in humid regions. National Fertilizer and Environmental Research Center, Muscle Shoals, AL.

Schollenberger, C.J., and R.H. Simon. 1945. Determination of exchange capacity and exchangeable bases in soil-ammonium acetate method. Soil Sci. 59:13–24.

Sharpley, A.N., and S.J. Smith. 1989. Prediction of soluble phosphorus transport in agricultural runoff. J. Environ. Qual. 18:313–316.

Sharpley, A.N., S.C. Chapra, R. Wedepohl, J.T. Sims, T.C. Daniel, and K.R. Reddy. 1994. Managing agricultural phosphorus for protection of surface waters: Issues and options. J. Environ. Qual. 23:437–451.

Sharpley, A.N., T.C. Daniel, J.T. Sims, and D.H. Pote. 1996. Determining environmentally sound soil phosphorus levels. J. Soil Water 51:160–166.

Sharpley, A.N., U Singh, G. Uehara, and J. Kimble. 1989. Modeling soil and plant phosphorus dynamics in calcareous and highly weathered soils. Soil Sci. Soc. Am. J. 53:153–158.

Shuman, L.M., V.A. Bandel, S.J. Donahue, R.A. Isaac, R.M. Lippert, J.T. Sims, and M.R. Tucker. 1992. Comparison of Mehlich I and Mehlich III extractable soil boron with hot-water extractable boron. Comm. Soil Sci. Plant Anal. 23:1–14.

Sibbesen, E. 1978. An investigation of the anion-exchange resin method for soil phosphate extraction. Plant Soil 50:305–321.

Sibbesen, E., and A.N. Sharpley. 1997. Setting and justifying upper critical limits for phosphorus in soils. p. 151–176. In H. Tunney et al. (ed.) Phosphorus loss from soil to water. CAB International, London, UK.

Sillanpaa, M. 1982. Micronutrients and the nutrient status of soils: A global study. FAO Soils Bull. 48.

Sims, J.T. 1989. Comparison of Mehlich 1 and Mehlich III extractants for P, K, Ca, Mg, Mn, Cu, and Zn in Atlantic coastal plain soils. Commun. Soil Sci. Plant Anal. 20:1707–1726.

Sims, J.T. 1993. Environmental soil testing for phosphorus. J. Prod. Agric. 6:501–507.

Sims, J.T. 1996. Lime requirement. p. 491–515. In D.L. Sparks (ed.) Methods of soil analysis. Part 3. Chemical methods. Soil Science Society of America, Madison, WI.

Sims, J.T. 1998. Phosphorus soil testing: Innovations for water quality protection. Commun. Soil Sci. Plant Anal. 29:1471–1489.

Sims, J.T., and G.V. Johnson. 1991. Micronutrient soil tests. In (ed.) Micronutrients in agriculture. 2nd Ed. Soil Science Society of America, Madison, WI.

Sims, J.T., S.D. Cunningham, and M.E. Sumner. 1997. Assessing soil quality for environmental purposes: Roles and challenges for soil scientists. J. Environ. Qual. 26:20–25.

Sims, J.T., R.R. Simard, and B.C. Joern. 1998. Phosphorus losses in agricultural drainage: Historical perspective and current research. J. Environ. Qual. 27:277–293.

Sims, J.T., B.L. Vasilas, K.L. Gartley, B. Milliken, and V. Green. 1995. Evaluation of soil and plant nitrogen tests for maize on manured soils of the Atlantic coastal plain. Agron. J. 87:213–222.

Skogley, E.O. 1994. Reinventing soil testing for the future. p. 187–201. In J.L. Havlin et al. (ed.) Soil testing: Prospects for improving nutrient recommendations. Soil Science Society of America, Madison, WI.

Skogley, E.O., S.J. Georgitis, J.E. Yang, and B.E. Schaff. 1990. The phytoavailability soil test. Commun. Soil Sci. Plant Anal. 21:1229–1243.

Soil and Plant Analysis Council, Inc. 1992. Handbook on reference methods for soil analysis. Council on Soil Testing and Plant Analysis, Athens, GA.

Soltanpour, P.N., and A.P. Schwab. 1977. A new soil test for simultaneous extraction of macro- and micronutrients in alkaline soils. Commun. Soil Sci. Plant Anal. 8:195–207.

Soltanpour, P.N., A. Khan, and W.L. Lindsay. 1976. Factors affecting DTPA-extractable Zn, Fe, Mn and Cu. Commun. Soil Sci. Plant Anal. 7:797–821.

Soltanpour, P.N., A. Khan, and A.P. Schwab. 1979. Effect of grinding variables on the NH_4HCO_3-DTPA soil test values for Fe, Zn, Mn, Cu, P, and K. Commun. Soil Sci. Plant Anal. 10:903–909.

Sparks, D.E. 1996. Methods of soil analysis, Part 2: Chemical properties. Soil Science Society of America, Madison, WI.

SRIEG-18. 1992. Reference soil and media diagnostic procedures for the Southern region of the United States. VA Agric. Expt. Sta. Bull. 374.

Sumner, M.E. 1989. Advances in the use and application of plant analysis. Commun. Soil Sci. Plant Anal. 21:13–16.

Sumner, M.E. 1997. Procedures used for diagnosis and correction of soil acidity: A critical review. p.195–204. In A.C. Moniz et al. (ed.) Plant-soil interactions at low pH: Sustainable agriculture and forestry production. Brazilian Soil Science Society, Campinas, Brazil.

Sumner, M.E., and N.W. Barbour. 1992. The Diagnosis and Recommendation Integrated System (DRIS) as an ecological indicator. p. 663–674. In D.H. MacKenzie (ed.) Ecological indicators, Elsevier Publishers, New York, NY.

Sumner, M.E., 1979. Interpretation of foliar analysis for diagnostic purposes. Agron. J. 71:343–348.

Sumner, M.E. 1981. Diagnosing the sulfur requirement of corn and wheat using foliar analysis. Soil Sci. Soc. Am. J. 45:87–90.

Tabatabai, M.A. 1996. Sulfur. p. 921–960. In D.L. Sparks (ed.) Methods of soil analysis. Part 3. Chemical methods.

Soil Science Society of America, Madison, WI.

Tanji, K.K. 1982. Modeling of the soil nitrogen cycle. p. 721–772. *In* F.J. Stevenson (ed.) Nitrogen in agricultural soils. American Society of Agronomy, Madison, WI.

Thomas, G.W. 1996. Soil pH and soil acidity. p. 475–490. *In* D.L. Sparks (ed.) Methods of soil analysis. Part 3. Chemical methods. Soil Science Society of America, Madison, WI.

Tisdale, S.L., W.L. Nelson, J.D. Beaton, and J.L. Havlin. 1993. Soil fertility and fertilizers. 5th Ed. MacMillan Publishers, New York, NY.

Tunney, H., O.T. Carton, P.C. Brookes, and A.E. Johnston. 1997. Phosphorus loss from soil to water. CAB International, Wallingford, UK.

USEPA. 1986. Acid digestion of sediment, sludge and soils. *In* Test methods for evaluating solid waste SW-846. US Environmental Protection Agency, Cincinnati, OH.

USEPA. 1989. Development of risk assessment methodology for land application and distribution and marketing of municipal sludges. EPA/600/6-89/001. US Government Printing Office, Washington, DC.

van Lierop, W. 1990. Soil pH and lime requirement. p. 73–126. *In* R.L. Westerman (ed.). Soil testing and plant analysis. 3rd Ed. Soil Science Society of America, Madison, WI.

van Lierop, W. and T.S. Tran. 1985. Comparative potassium levels removed from soils by electro-ultrafiltration and some chemical extractants. Can. J. Soil Sci. 65:25–34.

van Raij, B. 1994. New diagnostic techniques, universal soil extractants. Commun. Soil Sci. Plant Anal. 25:799–816.

van Raij, B., J.A. Quaggio, and N.M. da Silva. 1986. Extraction of phosphorus, potassium, calcium and magnesium from soils by an ion-exchange resin procedure. Commun. Soil Sci. Plant Anal. 17:544–566.

Viro, P.J. 1955. Use of ethylenediaminetetraacetic acid in soil analysis: I. Experimental. Soil Sci. 79:459–465.

Walworth, J.L., and M.E. Sumner. 1987. The Diagnosis and Recommendation Integrated System. Adv. Soil Sci. 6:149–185.

Walworth, J.L., and M.E. Sumner. 1988. Foliar diagnosis: A review. Adv. Plant Nutr. 3:193–241.

Wear, J.I., and C.E. Evans. 1968. Relationships of zinc uptake of corn and sorghum to soil zinc measured by three extractants. Soil Sci. Soc. Am. Proc. 32:543–546.

Westerman, R.L. 1990. Soil testing and plant analysis. 3rd Ed. Soil Science Society of America, Madison, WI.

Wolf, A.M., and D.E. Baker. 1985. Comparison of soil test phosphorus by Olsen, Bray P_1, Mehlich I and Mehlich III methods. Commun. Soil Sci. Plant Anal. 16:467–484.

Wolf, A.M., D.E. Baker, H.B. Pionke, and H.M. Kunishi. 1985. Soil tests for estimating labile, soluble, and algae-available phosphorus in soils. J. Environ. Qual. 14:341–348.

5

Fundamentals of Fertilizer Application

David Mengel
Kansas State University

George Rehm
University of Minnesota

5.1 Introduction

Efficiency of nutrient application is one of the major keys which determines the overall effectiveness of a fertilizer program. There are a number of ways that the efficiency of fertilizer use can be measured, but regardless of the measure used, a number of interacting factors determine the efficiency of a given fertilizer application. The relative mobility of the nutrient of concern in the soil is foremost in importance. Nitrogen is an excellent example of a nutrient which is mobile in the soil and easily lost from the root zone. Devising an application system which can provide N to the plant when it is needed is the challenge in some situations. Sulfur, B, and Cl are also considered to be mobile in soils. Compared to N usage, uptake of these nutrients is rather small and fewer management decisions are needed.

Phosphorus, on the other hand, is generally considered to be relatively immobile in soils. While P may remain in the soil, it is subject to a number of reactions and transformations which can reduce P availability to the plant. Placement techniques that can minimize the effects of these transformations and enhance P availability are valuable to crop managers.

In choosing appropriate application technology a number of characteristics of the system beyond nutrient mobility must also be considered. These include the nature of the crop being fertilized, weather and climate, soil properties, the form in which the nutrient is applied, the method of application utilized and the timing of the fertilizer application. By understanding how these individual factors interact, one can devise an efficient system of nutrient delivery requiring the least possible amounts of fertilizer, yet achieving optimum growth and subsequent yield.

To arrive at useful and effective decisions with respect to fertilizer application, it is necessary to have some understanding of how plant nutrients get to the root system. There is general agreement that nutrients reach a root by: (1) mass flow, and/or (2) diffusion, and/or (3) root interception.

Mass flow is somewhat self descriptive. The nutrients flow to the surface of the root in the water present around soil particles, i.e., mobile nutrients travel or move to the root with soil water in mass flow. Diffusion is a process whereby nutrients move from an area of high concentration to an area of lower concentration. The immobile nutrients move to the surface of the root by diffusion which takes place over relatively short distances (< 1 mm). In the root interception process, the root intercepts plant nutrients as it grows through the soil. Both mobile and immobile nutrients are taken into plants by this process.

5.2 Application of Mobile Nutrients

Nitrogen, S, B, and Cl are the essential nutrients which are considered to be mobile in soils. Because of the differences in magnitude of uptake and sources available for use in a fertilizer program, the application of each will be discussed separately.

5.2.1 Nitrogen

Application of N is a major concern in the production of most nonlegume crops. When nitrogen application is considered, the modern grower is faced with decisions about: (1) rate, (2) source, (3) time of application, (4) method of application, and (5) use of a nitrification inhibitor. Decisions revolving around rate of application are covered in a Section D, Chapter 4.

5.2.1.1 Sources

There are numerous sources of N that have been used for crop production over the years. The N sources most commonly used for crop production are anhydrous ammonia (82-0-0), urea-ammonium nitrate (28-0-0), urea (46-0-0), ammonium nitrate (33-0-0), and ammonium sulfate (21-0-0-24). Some information about the most commonly used sources is summarized in Table 5.1. The choice of the source of fertilizer N is dependent on several factors such as soil texture, rooting patterns of the intended crop, moisture relationships in the field, cost of the product, and method of application. Slow-release N products are also marketed but the relatively high cost precludes their use for production of most agronomic crops and use is limited to special cropping situations.

The selection of a source of fertilizer N has always been a major decision for crop producers. Consequently, claims of effectiveness have been both debated and researched. In most studies, when care was taken to prevent N loss, all sources had an equal effect on yield when used at rates to supply the same amount of N. There are probably numerous situations where one N source is preferred or recommended over the others. Examples will be used to illustrate some of them.

Rooting depth can dictate a choice of N source. For example, the edible bean crop has a very shallow root system when grown on irrigated sandy soils. The application of 82-0-0 may place a

Table 5.1 Common sources of N fertilizers, their analysis and form applied to soil

Fertilizer	Analysis	Form applied to the soil
Anhydrous ammonia	822-0-0	high pressure liquid
Aqua ammonia	20-0-0 to 24-0-0	low pressure liquid
Nitrogen solutions	28-0-0 to 32-0-0	liquid without pressure
Ammonium sulfate	21-0-0-24	granules
Urea	46-0-0	granules
Ammonium nitrate	33.5-0-0	granules
Diammonium phosphate	18-46-0	granules
Potassium nitrate	13-0-44	granules
Ammonium phosphate	10-34-0 or 11-37-0	liquid without pressure

substantial amount of N below the root zone which could reduce the effectiveness of this N source for this crop. Some herbicides can be impregnated on dry materials or mixed with 28-0-0 or 32-0-0. This combination eliminates the possibility of using 82-0-0 if a grower wishes to apply the herbicide together with the N fertilizer. There are, obviously, many other situations that dictate the N fertilizer source that can be used. Space does not allow for a listing of all situations.

5.2.1.2 Time of Application

The optimum time for N applications to agronomic crops is dependent on several factors including (1) intended crop, (2) its root architecture, (3) its N uptake characteristics, (4) climate, (5) soil texture, and (6) the rate of N fertilizer needed for optimum yield. Many suggestions for optimum time of application are strongly influenced by potential for loss. Management of N fertilizers to avoid loss will be discussed in detail in the following section.

Growth and development of the root system are important factors to consider in timing decisions. Because of the downward mobility of NO_3-N in soils, more frequent applications of fertilizer N have been most effective for shallow rooted crops such as edible beans and potatoes. If soils are not sandy, timing of the application of fertilizer N is rather flexible for deep rooted crops such as corn. If timing itself has little effect on yield, the suggestions for most appropriate time of application are governed by N loss and environmental considerations.

Considering soil texture, split applications have been shown to be most effective for crops such as corn grown on irrigated sandy soils. When grown on nonirrigated sandy soils, N fertilizers are usually applied during the growing season for optimum use of the fertilizer N.

Nitrogen uptake by crops is not constant throughout the growing season. For crops such as wheat, barley, oats, and corn, the rate of N uptake is slow during the early part of the growing season, accelerates before anthesis, and decreases after pollination is completed. Therefore, the application of fertilizer N should be completed before grain fill is started.

Timing of needed fertilizer N is also an important consideration in the production of forage grasses. Because cool season species produce most of their dry matter in late spring and early summer, fertilizer N should be applied before this season of rapid growth. In contrast, the majority of the growth of warm season species occurs when temperatures are high and therefore, application of fertilizer N should be delayed beyond the early season.

5.2.1.3 Managing Fertilizer N to Prevent Losses

Nitrogen is lost from agricultural soils primarily by: (1) leaching, (2) denitrification, and (3) NH_3 volatilization. In addition, N can be transformed to unavailable forms through immobilization and ammonium fixation. Detailed reviews of each of these processes can be found in Sections C, D, Hauck (1984) and Scharf and Alley (1988). However, some basic fundamental concepts will be discussed to facilitate an understanding of how application techniques may minimize the impact of these processes.

Leaching is simply the downward transport of NO_3^- (and NH_4^+ in low CEC soils) with water. Leaching occurs when water reaching the soil surface from precipitation and/or irrigation exceeds evapotranspiration (Nelson and Uhland, 1955). Leaching is a special concern in the more humid climate of the Eastern US, where precipitation exceeds evapotraspiration by 150 to 500 mm annually. Soil water-holding capacity also plays an important role in determining the importance of leaching losses. Soils with a high water-holding capacity, such as silt loams, can accumulate large quantities of water before NO_3^- is transported below the root zone. Thus, soils with high water-holding capacities are much less prone to leaching losses than coarse textured, low water-holding capacity soils. Tile drainage, such as is commonly used in the Midwest, can enhance leaching by providing a shallow outlet for drainage water to surface water sources.

Denitrification is a microbial process by which soil bacteria (facultative anaerobes) utilize the O in NO_3 to support metabolism. In the process, NO_2 and N_2 gases are released into the soil atmosphere and are subsequently lost. A number of factors or conditions interact to impact the rate of denitrification. These include energy sources such as soil organic matter (SOM), crop residue, and animal manure, moisture/O availability, temperature and pH (Wijler and Delwhiche, 1954; Bailey and Beauchamp, 1973; Burford and Bremner, 1975; Rolston et al., 1978). Denitrification is a primary N loss process and is a major concern for finer textured, poorly drained soils, or soils with high seasonal watertables.

Ammonia volatilizaton is the gaseous loss of free NH_3 to the atmosphere from soils or fertilizers. There are three general situations where NH_3 volatilization is a significant loss problem. When NH_4-based fertilizers such as ammonium sulfate (21-0-0-24) are broadcast on the surface of high pH or alkaline soils and not incorporated, some NH_4 is converted to NH_3 which can volatilize to the atmosphere. This only occurs in significant amounts when pH > 7. When urea-based fertilizers are applied to soils, the naturally present urease enzyme hydrolyzes the urea to NH_3. The hydrolysis reaction generates enough OH ions to temporarily raise the pH of the soil around the urea fertilizer which, in turn, causes NH_3 volatilization. Volatilization of N from urea is enhanced if urea is applied to a soil surface covered with crop residue or vegetation, as in a no-till corn field or a pasture or hay field. The third situation deals with the addition of urea- or NH_4-based fertilizers to flooded rice paddies. A diurnal fluctuation in pH due to algal utilization of CO_2 results in a rapid rise in water pH reaching a maximum of 7 to 10 during midday. The high pH favors the conversion of NH_4 to NH_3, and its volatilization. In all cases, incorporation of urea- or NH_4-based fertilizers into the soil effectively stops volatilization (Ernst and Massey, 1960). Unfortunately, incorporation is not always possible so that alternative management strategies are sometimes utilized.

There are a number of alternative application strategies which can be used to reduce leaching loss of N from agricultural soils. Probably the most commonly used is timing of fertilizer applications to match periods of rapid utilization by crops. The use of sidedressing, delaying application of N fertilizers until the crop is established and taking up N, or the use of split applications to avoid N leaching losses is used for crops such as corn, wheat, and cotton, particularly on sandy soils. An example of how time of N application has been used to enhance N use efficiency on coarse textured soils by reducing leaching losses can be found in Evanylo (1991).

The use of nitrification inhibitors and slow-release fertilizers is an alternative strategy to application timing as a means of reducing N loss from leaching. While the use of nitrification inhibitors has proven to be a successful means of reducing N loss on fine textured soils in the Eastern Corn Belt (Huber et al., 1982; Stehouwer and Johnson, 1990; Mann, 1995), they have shown mixed results in irrigated areas and the western Corn Belt (Maddux et al., 1985; Cerrato and Blackmer, 1990). A number of slow-release fertilizer products have been studied and have potential for reducing N loss from leaching in coarse textured soils. However, the cost of these products has limited their use at this time. A review of these products is presented by Hauck (1985).

Many of the same strategies that are used to avoid leaching losses can be used to avoid losses from denitrification, namely, timing of application, the use of slow-release fertilizers, and nitrification inhibitors. However, one important difference among soils where these problems occur must be emphasized. Soils prone to high rates of leaching loss tend to be well drained, with low water-holding capacity. These soils dry rapidly after a leaching event and a farmer has a high probability of being able to drive across these fields in a short period of time. Many of these fields may be irrigated, which allows for the application of fertilizer N with irrigation water (fertigation). Soils prone to high rates of loss by denitrification are not usually irrigated and tend to be poorly drained and slow to dry and trafficability is reduced. Thus, while sidedressing or split N applications are excellent tools to reduce

N loss from denitrification, they entail risk to the grower who chooses to use them. This is a risk that additional N may not be applied in a timely manner.

Nitrification inhibitors are useful tools for reducing N loss in soils prone to denitrification. A number of studies in fine textured soils of the eastern Corn Belt have shown responses to the use of inhibitors with fall or spring preplant applied N (Fry et al., 1978; Huber et al., 1982; Stehouwer and Johnson, 1990).

Incorporation of the fertilizer N into the soil is the most effective strategy for reducing N loss from NH_3 volatilization (Ernst and Massey, 1960). While this is not always possible with no tillage systems, a number of tools have been developed to allow injection or banding of N fertilizers below the surface in these production systems. A number of studies have demonstrated that injection of N fertilizers into the soil and below the surface residue in corn and grain sorghum greatly enhances N use efficiency (Mengel et al., 1982; Lamond et al., 1991). Irrigation water can also be used to incorporate urea into the soil to reduce ammonia volatilization (Mengel and Wilson, 1988).

Choosing a non-volatile N fertilizer source is another alternative (Bandel et al., 1980; Fox and Hoffman, 1981). When compared to urea, ammonium nitrate and ammonium sulfate are preferred N sources for use in high residue situations (Bandel et al., 1980). Ammonia volatilization losses from these products are negligible compared to those from urea or urea-based products (Keller and Mengel, 1986).

Urease inhibitors are a recent addition to the management tools available to reduce NH_3 volatilization from surface applied urea-based fertilizers. The rationale for using a urease inhibitor to reduce NH_3 volatilization is to limit hydrolysis and subsequent NH_3 production until after the urea has been incorporated into the soil by rain or irrigation water. At present, one urease inhibitor, N-(n-butyl) thiophosphoric triamide (NBPT) is commercially available in the United States. Research has shown that urea impregnated with NBPT is a viable alternative to a non-volatile N source such as ammonium nitrate, in no-till corn and for summer applications to pastures and hay fields.

5.2.2 Sulfur

Sulfur-like N, is mobile in soils, yet its transformations and chemistry differ considerably from many of the reactions associated with N. Because S is mobile, many of the best management practices suggested for use of fertilizer N are also appropriate for the management of fertilizer S. Compared to N, S is taken up in relatively small amounts by actively growing crops. Various research projects have evaluated the effect of rate and management of S fertilizers on production and quality of a variety of crops. Specific management practices associated with optimum production vary with the crops that are grown.

The legumes, (alfalfa, clovers, soybeans) and canola remove relatively large amounts of S from the soil system while other crops (corn, small grains, forage grasses) remove smaller amounts. Options for placement of fertilizer S also vary with the intended crop.

Approximately 90% of the total S in soils is found in SOM, which can supply substantial amounts of S for crop production. Research has shown that crops grown on fine textured soils with a high SOM content do not respond to the application of fertilizer S. On the other hand, use of fertilizer S has produced dramatic increases in production when applied to crops grown on sandy soils with a low SOM content (Section D, Chapter 1).

5.2.2.1 Sulfur Sources and Methods of Application

Application of S in a fertilizer program is most frequently associated with the production of alfalfa and other perennial legumes, corn, and small grains. As reported by Hoeft and Walsh (1975), alfalfa responds favorably to annual topdress applications of fertilizer S. Products containing elemental S (0-

0-0-90) or SO_4 form can be used for this method of application. The most common dry sources of SO_4-S are ammonium sulfate (21-0-0-24) and the double salt of potassium and magnesium sulfate (0-0-22-22-11). If K is needed in a fertilizer program for alfalfa, the double salt is a logical choice for topdress applications. Perennial grasses have also responded to annual topdress applications in situations where S is needed in a fertilizer program (Lamond et al., 1995). In general, annual applications of fertilizer S are suggested for all perennial crops whenever there is a need for S in a fertilizer program. Although it is a legume which removes substantial amounts of S from soils, soybeans have not responded consistently to S fertilization. Yield increases have been noted when fertilizer S is applied on very sandy soils (Matheny and Hunt, 1981) but not on fine textured soils (Brown et al., 1981; Sweeney and Grande, 1993). Sulfur needed for soybeans can be either broadcast before planting or applied in a band at planting. There is no research information to suggest that one method of application is superior.

When corn is grown on sandy soils where responses to S fertilization might be expected, the needed S can either be broadcast and incorporated before planting or applied as a starter fertilizer at planting (Hoeft et al., 1985; Kline et al., 1989). In general, rates of applied S are doubled if S fertilizers are broadcast and incorporated before planting rather than applied in a band near the seed at planting.

There is general agreement that the use of split applications of N fertilizers is a best management practice for corn production on irrigated sandy soils. In evaluating the timing of S applications for sandy soils, however, a single application at planting has been as effective as split applications during the first half of the growing season (Rehm, 1993). When wheat and other small grains are considered, responses to S fertilization are not frequently reported. The majority of responses reported have been for the silt loam soils of the Pacific Northwest where broadcast applications have been popular (Ramig et al., 1975; Mahler and Maples, 1987).

For annual crops, S can be applied in the form of elemental S or SO_4-S. These dry sources can be either broadcast or applied in a band at planting. Because crops do not absorb elemental S from soils, microbes must convert it to SO_4-S before it can be used by crops (Mahler and Maples, 1987).

When fluid fertilizers are used for production of annual crops, S can be supplied as ammonium thiosulfate (12-0-0-26) or potassium thiosulfate (0-0-25-17). Although these fluid materials are best suited for a band application at planting, caution should be used because germination can be impaired if 12-0-0-26 is placed in contact with the seed. The safety of potassium thiosulfate, when placed in contact with the seed, has not been fully documented. Fluid fertilizer containing S can also be injected into irrigation water. This is not a preferred practice and should be limited to correcting identified S deficiencies.

5.2.2.2 Managing Sulfur to Prevent Losses

When present as SO_4-S, downward movement through soils can occur during either heavy rainfall or over-irrigation. The rate of downward movement of SO_4-S is, however, not as rapid as that of NO_3-N. Unless excessive amounts of irrigation water are applied to sandy soils, leaching of SO_4-S should be of minor concern. Because SO_4-S is not associated with problems of water quality, its loss by leaching is of economic concern which can be minimized by applying fertilizer S, when needed, in a band close to the seed at planting.

5.2.3 Boron

This essential nutrient is also mobile in soils. Boron, which is found in soils as uncharged H_3BO_3, is not strongly sorbed by soil particles and is susceptible to leaching. Boron is classified as a micronutrient because small amounts are required for optimum crop production. The majority of the research conducted with B fertilization has focused on the production of alfalfa (Brown, 1972), corn (Touchton and Boswell, 1975), and small grains (Gupta et al., 1976). Positive responses have not been consistent and are limited to unique soils and/or situations. As with N and S, annual applications of B are suggested often as topdress applications as for alfalfa. Since there is some indication that B applied with the seed may cause germination damage, broadcast applications incorporated before planting are suggested for annual crops and placement of B in contact with the seed should be avoided.

5.2.4　Chloride

Chloride, like B, is mobile in soils behaving like NO_3-N in terms of leaching (Endelman et al., 1974). Although not studied intensively, responses to this nutrient have been documented for corn (Heckman, 1995) and small grain production in limited situations (Fixen et al., 1986a,b; Engel et al., 1994). Because Cl is mobile, placement should have little effect on crop response to this nutrient. In contrast to the management of N, there is no indication that split applications would be superior to a single application at or before planting.

5.3　Application of Immobile Nutrients

Nutrients such as P, K, Ca, Mg, Zn, Cu, Fe and Mn that are strongly sorbed by soil components are immobile and do not move easily through soils (Section B, Chapter 8). Unless there are substantial losses of soil from the landscape, there are usually no significant losses of immobile nutrients from the soil system other than through crop removal. There are situations, however, where nutrients normally considered to be immobile can move through a soil profile. Two examples are leaching of K and Mg in soils with a low cation exchange capacity (CEC) and leaching of Mn^{2+} and Fe^{2+} through very sandy soils that are poorly drained.

5.3.1　Placement Considerations

Fertilizer placement becomes a major management consideration in the overall management of immobile nutrients with the following options available.

5.3.1.1　Broadcasting

Broadcasting with or without soil incorporation is the commonly used method of application for the immobile nutrients. Throughout the United States, a very high percentage of the P and K fertilizers are applied in this manner. With this placement option, essentially 100% of the top few cm of topsoil comes in contact with the applied fertilizer. Some of the advantages of broadcast applications are (1) it is fast and easy; (2) there are several opportunities for application which can reduce the workload at critical times during the year; (3) there is a high probability that crop roots will come in contact with fertilized soil especially when incorporated with some form of tillage; and (4) when soils have a low CEC, there are more potential sites for adsorption and subsequent retention of K and Mg. Some of the disadvantages of this placement option are (1) thorough mixing with tillage increases the probability of fixation and reduced availability; and (2) it is difficult to achieve a uniform application of low rates of fertilizers which supply the immobile nutrients.

5.3.1.2　Banding

Banding of fertilizer is a mechanical technique in which some type of tine is used to open a furrow or trench and the fertilizer is applied in a band below the soil surface. With most band applications, < 1% of the topsoil volume (15 cm) comes in contact with the fertilizers. Three common band placement options are (1) place a low rate of fertilizer in close proximity to the seed to enhance availability of nutrients to young plants (starter); (2) place a low rate of fertilizer in direct contact with the seed at planting (pop-up); and (3) place various amounts of fertilizer in positions throughout the root zone to overcome fixation reactions and create zones of high fertility.

In North Carolina, Nelson et al. (1959) compared various band placement options to broadcast applications of P for corn and cotton production. Using low rates of P on P-responsive sites, they found that the band placement enhanced early growth and resulted in better utilization of the applied P. There were, however, no differences in yield. Barber (1958) established a long-term study in Indiana and evaluated both the direct and residual responses to both broadcast and banded placement of P fertilizers. He found that band applications provided the greatest response when the P soil test was low and decreased as the soil test for P increased. This study also showed that production of optimum corn yields required both broadcast and banded application of P fertilizer.

Welch et al. (1966a,b) found similar results with the application of both P and K in Illinois. With low fertilizer rates on responsive soils, the band applications resulted in more efficient utilization of immobile nutrients. However, highest yields were obtained by combining both low rates in a band and broadcast application at higher rates. For soils with higher soil test values, response to fertilizer was smaller and placement had only limited effects on use efficiency.

The primary advantages of banded applications of immobile nutrients are (1) it creates a zone of enhanced nutrient availability by minimizing contact between soil and fertilizer; and (2) it is a simple and efficient method of applying small amounts of fertilizer. Disadvantages include (1) only a limited portion of the root system has a high probability of coming in contact with the fertilizer; (2) damage to germination can occur when high rates of some fertilizer products are placed too close to the seed; and (3) the cost of equipment used in banding can be high.

5.3.1.3 Strip Application

In strip application, a compromise alternative to band and broadcast applications, the fertilizer supplying the immobile nutrients is placed in a band on the soil surface which is then incorporated with some tillage. By varying the width of the surface band and the type of tillage used for incorporation, 5% to 15% of the volume of the surface soil is mixed with the fertilizer. While the level of nutrient availability in the treated zone may be lower than with other banding options, a much larger portion of the root system can potentially come in contact with the fertilizer. This has enhanced uptake of immobile nutrients. Nelson et al. (1959) were the first to test this concept with corn and cotton and found that a combination of in-row or seed-placed phosphate with stripping resulted in the best utilization of P by young seedlings. In a 5-yr study with P and K, Barber (1974) found that strip application produced significantly higher corn yields when compared to either banding or broadcasting equivalent rates. Barber and Cushman (1981) later developed a mathematical model capable of predicting optimum fertilizer placement under a wide range of soil conditions.

5.3.1.4 Point Fertilization

In its simplest form, this consists of opening a hole in the soil with a stick or hoe, and placing a quantity of fertilizer into the soil near the crop to be fertilized. An early study conducted by Coe (1926) in Iowa compared broadcast applications to hill placement, or short bands of fertilizer 7.5 to 15 cm long and 3 to 4 cm wide at the same depth as the seed. He concluded that banding the fertilizer in the hills was equal or superior to broadcasting, if contact between fertilizer and seed was avoided, and suggested that the fertilizer should be placed to the side of and below the seed. When used at moderate rates, fertilizer could be placed in the hill with the seed, but at high rates, the fertilizer should be split with part applied in the hill and part broadcast.

The point system is commonly used today in the production of perennial shrub or tree crops. Holes are dug near the outer edge of the canopy (drip line) of trees and fertilizer placed in those holes. Tree stakes, solid blocks of fertilizer shaped to facilitate pushing them into the ground, can also be placed at points along the outer edge of the canopy. In many developing countries, a hole is dug near or in a hill of corn or sorghum and a small quantity of fertilizer is added to maximize nutrient utilization and to minimize competition from weeds. Large individual granules of fertilizer (super granules) have been developed specifically for this purpose. With a point application system, only a very small portion of the soil, generally < 1%, will come in contact with the fertilizer.

5.3.1.5 Application with Irrigation Water
Fluid fertilizer can be injected into irrigation water but in the case of immobile nutrients, particularly P, this method of application has limited use. Research evaluating this placement has been reviewed by Mikkelsen (1989).

5.3.1.6 Foliar Application
Although the majority of nutrients needed for growth and development of plants enter through the root system, beneficial effects of foliar fertilization have been reported. For example, Shafer and Reed (1986) studied the foliar absorption of 31 organic and inorganic compounds. Much of the research effort on foliar fertilization has been summarized by Alexander and Schroeder (1987).

5.4 Placement of P and K Fertilizers

The efficiency of use of the immobile nutrients has been the focus of field and greenhouse research projects for many years. In most cases, these studies have focused on the relative advantages or disadvantages of banding versus broadcasting of P and K fertilizers. The impact of the placement of immobile nutrients on uptake by crops and subsequent yield has been studied in both the laboratory and the field. A mechanistic model to explain some of the complex relationships which influence crop response to fertilizer placement was developed by Claassen and Barber (1976), Anghinoni and Barber (1980) and Barber and Cushman (1981).

Using 33 diverse soils to test the model, Kovar and Barber (1987, 1988, 1989) found that the concentration of P in the soil solution was the soil parameter which had the greatest effect on P uptake. When P fertilizer was placed in contact with small volumes of soil, banding produced the largest increase in soil solution P for each unit of applied fertilizer. However, the number of roots which can come in contact with the increased solution P is limited by banding. Since roots have a finite capacity to take up nutrients, it is possible that although banding increases the concentration of P in the fertilized zone, restricted root volume in contact with the high P zone could limit uptake. Using the Barber-Cushman model to predict the optimum volume of soil which should be fertilized, they found that the optimum volume varied substantially from a low of about 3% to a maximum of 15% to 20%. Borkert and Barber (1985) found a very close relationship between predicted and measured P uptake by soybeans as a function of soil volume fertilized in a pot experiment. Barber (1995) concludes "Placement of P is most important in soils low in available P that adsorb or fix large quantities of added P." Phosphorus placement also becomes more important as the rate of applied P decreases.

Although the quantity of P taken up by crops is only a fraction of that of K, fertilizer placement effects are similar for both. The use efficiency of banded K for corn is greater at low soil test values for K with the difference between banded and broadcast K diminishing as soil test K increases (Welch et al. 1966a,b).

5.4.1 Field Comparison of Placement Options

Field comparisons between band and broadcast applications of P and K fertilizers show that banding, especially for P fertilizers, produced greater yield increases when equal rates were used for both. The results summarized by Welch et al. (1966a,b) are typical of many of the results reported.

Working with several forage crops, Sheard (1980) concluded that the banded application of 30 kg P ha^{-1} increased seeding growth as much as fivefold regardless of the species and soil test for P. Working with oats grown on contrasting soils in a greenhouse, Sleight et al. (1984) concluded that the beneficial effects of banding are obtained mainly from placing all the immobile nutrients where contact by active roots is likely, rather than from any increase in availability that may be obtained from the decreased soil/fertilizer contact associated with banding.

Some research utilizing a wide range of application rates has shown that rate can influence the effect of placement. Barber (1958, 1959) has shown that at high fertilization rates, broadcast applications may give higher yields than band applications while Welch et al. (1966a,b) have demonstrated that the optimum combination of banding and broadcasting can be a complex interaction of application rate and soil test level.

With P and K fertilizers, the most frequent placement is to the side of and below the seed at planting (starter fertilizer). In the Midwest, starter fertilizer has consistently affected early season growth more than yield (Randall and Hoeft, 1988). However, starter fertilizer has consistently been more beneficial in no-till than conventional or clean till corn (Reeves and Touchton, 1986; Mengel et al., 1992). This benefit, in part, may be attributed to the fact that in no-till the soil may be cooler and/or have a higher bulk density and thus early root growth is inhibited. Starter fertilizer effects are generally found to be greatest when soil temperatures are lowest (Kitcheson, 1968).

Earlier planting dates typically coincide with cooler soil temperatures resulting in slower root growth and metabolism. Although crops typically respond to starter fertilizer under these planting conditions, it can increase growth and grain yield on late planted corn, when soil temperatures were very warm (Farber and Fixen, 1986). This was also true on the Coastal Plain soils where inherent soil strength inhibits root growth and exploration (Karlen et al., 1984).

Much debate has focused on which nutrients in a starter fertilizer are responsible for the increased early season growth and subsequent increased grain yield, particularly under no-till. In Wisconsin, Motavalli et al. (1993) obtained a response to starters containing N and P in only one of three years where manure had been applied. In Illinois under conventional tillage on a soil with high residual P, N and P starter fertilizer increased early season growth without increasing final plant dry weight or yield (Bullock et al., 1993). In Indiana, N has been the single most important nutrient in the starter solution on soils containing adequate levels of P and K (Mengel, 1992). In Alabama, Karim and Touchton (1983) found N primarily responsible for increased growth 14 days after emergence, but P appeared to have the greatest effect at 28–42 days. Wright (1985) in Florida found improved yields and earlier maturity when using a starter fertilizer containing N and P on a high P soil.

While N and P are the two nutrients most commonly used in starter fertilizers, a number of others are routinely applied including K, S, Mg, and Zn. Starter fertilizer bands provide a convenient delivery system for micronutrients in many crops. For example, Miner et al. (1986) obtained a response to Mn applied in a starter on an Atlantic Coastal Plain soil that was Mn deficient due to excessive liming.

Occasionally, fertilizer will be broadcast at planting specifically to enhance early growth. With no-till corn at several locations, Mann (1995) showed that broadcast applications of 44 kg N ha^{-1} as urea-ammonium nitrate solution negated the response to a N starter in no-till corn. This broadcast application of fluid N has become a common practice in some areas of the Midwest where many corn planters do not have starter fertilizer attachments.

In general, the band application of immobile nutrients has proven to be a cost effective method of fertilizer placement, particularly when soil test values are in the low or very low range. Either band or broadcast applications have been effective when the P or K soil test values are in the medium or high range. A combination of broadcast and band applications seems to be most valuable at lower soil test values and with higher rates of fertilizers.

The effect of placement of immobile nutrients has also been evaluated for small grain production. As with corn, band applications have been superior to broadcasting, particularly at lower rates of fertilizer (Sander et al., 1990). The distance between bands is not a concern when immobile nutrients are applied with a drill but could pose a problem when phosphate is applied in a subsurface band before planting small grains. Comparing spacings of 25, 30, and 50 cm, Maxwell et al. (1984) reported that spacing affected P uptake by young plants but had no effect on grain yield.

The residual effects of both broadcast and banded applications of phosphate are also important for small grain production. Following the harvest of spring wheat, Selles (1993) measured greater availability of P in soil samples collected directly over the band. When P was banded in soil having a range of soil test levels established by previous broadcast applications, Alessi and Power (1980) reported that the effects of the combination of band and residual effects of previous broadcast applications were additive.

Starter fertilizer usually involves the application of a small amount of fertilizer, particularly those containing N and P close to the seed at planting for the purpose of enhancing the early season growth of crops. In some cases, a small quantity of fertilizer is placed directly in the seed furrow. This pop-up or in-furrow placement should only be used with crops such as corn and small grains which are relatively salt tolerant. There are a number of reports of injury to corn seedlings from fertilizer placed in direct contact with the seed (Coe, 1926; Allred and Ohlrogge, 1963; Creamer and Fox, 1980), sometimes resulting in stand reductions (Mengel, 1992). The amount of fertilizer that can be placed safely on or in contact with seed varies with crop, row spacing, soil moisture, and climate. For corn in Indiana, a maximum fertilizer rate of 5.6 kg N plus K ha^{-1} on silt loam or heavier soils is recommended for seed placement (Mengel 1992). The use of seed-placed fertilizer is a common practice for small grain production and no damage has been reported unless excessive, uneconomical fertilizer rates are used. Soybeans are much more sensitive to pop-up fertilizer and this practice has severely reduced germination and subsequent yield (Clapp and Small, 1970; Hoeft et al., 1975).

5.4.2 Fertilizer Placement in Conservation Tillage Systems

The principles for placement of P and K fertilizers (Section 5.4.1) were developed primarily in production systems using a moldboard plow for primary tillage. Because of the emphasis on soil conservation, tillage systems are changing to minimize soil disturbance resulting in less incorporation of broadcast P and K fertilizers with a subsequent reduction in the redistribution of nutrients in crop residues (Larson, 1964).

The stratification of immobile nutrients in the conservation tillage production systems has been documented for a variety of soils (Cruz, 1982; Weil et al., 1988; Robbins and Voss, 1991; Rehm et al., 1995). In no-till or reduced till systems, extremely high levels of fertilizer can accumulate in the top 4 or 5 cm of soil because of a lack of incorporation of fertilizers and re-incorporation of nutrients cycling through the vegetation. Interestingly, this stratification is not restricted to no-till and ridge-till but is also found in chisel plow systems (Cruz, 1982; Hollanda et al., 1998).

In addition to vertical stratification, horizontal stratification can also occur. Work in Indiana clearly shows that nutrients can accumulate near the old row area when the crop row is consistently placed in the same area, a controlled traffic system. Thus nutrient stratification can occur in both vertical and horizontal planes. Unless a substantial portion of the root system develops near the soil surface and in the zone of nutrient enrichment, and the root environment is conducive to nutrient uptake, stratification can reduce the amount of nutrients available for uptake by the actively growing crop. A number of studies have shown that roots do concentrate in fertilized zones of soil, particularly in areas of high P (Barber, 1995) and this can be near the soil surface, especially in more humid climates (Cruz, 1982; Kaspar et al., 1991; Hollanda et al., 1998). Thus, the effect of nutrient stratification near the soil surface could be either a negative or positive factor influencing nutrient uptake and is highly dependent on the environment.

The advantages of band applications of starter fertilizer for corn in no-till planting systems have been discussed earlier. However, a number of studies have documented the importance of subsurface or deep banding for ridge-till and no-till planting in the northern Corn Belt (Rehm et al., 1995).

A similar response to placement of immobile nutrients has been found for other crops grown in conservation tillage production systems. Yields of grain sorghum were improved when P and K fertilizers were knifed in below the soil surface rather than broadcast on the surface (Sweeney, 1989). Yield increases from this knife placement were larger where soil test values for P and K were in the low rather than the medium range.

Nutrient stratification has not always been associated with reduced yields and response to alternative fertilizer placement techniques, however. In a long-term tillage in Indiana, moldboard plow, chisel plow, ridge-till, and no-till tillage systems were included, in both a continuous corn monoculture and a corn/soybean rotation (Griffith et al., 1988). Extensive stratification of P, K, and acidity, both vertical and horizontal, was found within the surface 30 cm of soil in both rotations (Cruz, 1982; Hollanda et al., 1998). However, differences in corn yields among tillage systems were relatively small in the corn/soybean rotation (< 4%), but rather large differences (> 17%) were found in continuous corn yields from 1980 to 1995, indicating that stratification of nutrients alone was not a significant yield limiting factor.

Yibirin et al. (1993) evaluated the effect of mulch on the response of corn to banded K and concluded that its benefits decreased as the amount of mulch on the soil surface increased. The mulch apparently increased soil moisture and reduced soil temperature near the soil surface, thereby increasing root development and subsequent K uptake from that zone.

In contrast to other crops, yield of winter wheat grown in no till planting systems in the Great Plains was not affected by P placement when soil test values for P were in the medium range (Halvorson and Havlin, 1992). The absence of a placement effect may be a consequence of the distribution of roots closer to the soil surface when compared to the corn and grain sorghum crops, and a root environment more conducive to nutrient uptake during key periods of growth.

5.5   Calcium and Magnesium

Like P and K, these nutrients are relatively immobile in most agricultural soils. In contrast to K, however, uptake of Ca and Mg by crops is considerably less. As a result, the necessity for these nutrients in a fertilizer program is diminished to special or localized situations. Therefore, research which has focused on the application of these two nutrients has not been extensive.

In most agricultural soils, Ca dominates the CEC which satisfies the Ca requirements of most crops (Mortvedt and Cox, 1985). Under acid conditions, the Ca content of the soil decreases, while the H and Al content increases. The addition of calcitic or dolomitic limestone to correct acidity problems is the primary means by which Ca is supplied to crops. With the exception of the peanut crop, Ca deficiency of field crops is generally rare under slightly acid to neutral soil conditions. Calcium deficiency is much more common in fruit and vegetable crops such as tomato, apple, and lettuce (Shear, 1975).

The most common Ca deficiency found in field crops is in peanuts. While lime may be a somewhat effective source of Ca, gypsum is preferred (Walker, 1975). Gypsum at a rate of 600 to 800 kg ha^{-1} is normally applied in a broad band directly over the row at early flowering. While full yield response has been obtained at rates as low as 200 kg ha^{-1}, higher rates are normally used as a means of improving seed quality and germination and reducing pod diseases (Cox, 1972). Broadcast applications of gypsum require that high rates of material be applied.

Like Ca, availability of Mg to plants decreases under acid soil conditions due to both a pH effect and loss of Mg through leaching from the root zone (Mortvedt and Cox, 1985). Increasing the pH of the soil, even with calcitic lime, can increase Mg availability (Christenson et al., 1973). Dolomitic limestone is the preferred Mg source in areas where it is available.

Applications of $MgSO_4$ or $KMg(SO_4)_2$ fertilizers are made on low Mg soils when dolomitic lime is not available. Soil tests are commonly used to estimate the need for Mg. In the Midwest, Mg applications for most field crops are recommended when the exchangeable Mg levels are < 50 mg kg^{-1} (Vitosh et al., 1995). General recommendations are to apply dolomitic lime if the soil is acid or to apply either 55 kg ha^{-1} soluble Mg, or 0.5 ton dolomitic lime ha^{-1} if the soil pH is adequate.

Since both Ca and Mg are not routinely added to fertilizer programs, the effectiveness of placement of these nutrients for agronomic crops has not been investigated. If lime is routinely used in a crop rotation, these nutrients should not generally be needed in a fertilizer program.

5.6 Micronutrients

The management of the remaining five micronutrients (Zn, Fe, Mn, Cu and Mo), which are taken up in small amounts by crops, will be discussed individually. Application of B, a mobile nutrient, was discussed in Section 5.2.3.

5.6.1 Zinc

A deficiency of Zn is most likely to occur on sandy soils formed from parent material low in Zn (Krauskopf, 1972), highly weathered tropical soils, highly calcareous soils, and organic soils (Schulte and Walsh, 1982). Overliming (Rehm and Penas, 1982), application of high rates of P (Murphy et al., 1981), and removal of surface soil by erosion, land leveling, and terracing (Frye et al., 1978; Gerwing et al., 1982) are management practices that can enhance Zn deficiency.

Crops differ in their sensitivity to Zn deficiency. Corn, edible beans, and sorghum are highly sensitive; barley, sugar beet, soybean, and sudangrass are moderately sensitive; and wheat, alfalfa, and most forage grasses are not sensitive to Zn deficiency (Robertson et al., 1981). While Zn deficiency can be corrected through either soil or foliar application, soil applications are more common for agronomic crops (Martens and Westerman, 1992) and foliar applications in the citrus industry.

Relatively high rates of Zn, broadcast as $ZnSO_4 \cdot 7H_2O$, are commonly used to correct a Zn deficiency in agronomic crops. Other inorganic sources include ZnO and Zn frits which are not readily available for crop uptake unless finely ground. Application rates vary depending on the demands of the crop being grown, soil properties which could affect Zn availability, and the native supply of Zn in the soil. Higher application rates are commonly needed on calcareous than noncalcareous soils (Wiese and Penas, 1979). Higher rates are also recommended for soils with low levels of extractable Zn (Killorn, 1984). Lower rates of Zn are normally applied in band than broadcast applications because of the reduced Zn - soil contact which slows the reversion of Zn to less available forms and the lowered pH in the band when Zn is applied in combination with N or N-P fertilizers. In addition to inorganic sources, a number of organic Zn sources are used for band applications. These include Zn EDTA, Zn lignosulfate, Zn acetate, and Zn citrate. The chelated materials are commonly used for foliar applications.

5.6.2 Manganese

Manganese deficiencies in crops are found in organic, calcareous, and acid poorly drained sandy soils. Soybeans, wheat, barley, and oats are highly sensitive to Mn deficiency while corn, sugar beets, alfalfa, and forage grasses are moderately sensitive (Robertson and Lucas, 1981). The availability of Mn is highly influenced by the total Mn content of the soil, drainage, oxidation state, SOM content, and pH. pH is probably the most important soil property controlling the availability of Mn to plants which decreases as pH increases. Overliming soil frequently produces Mn deficiencies (Gilbert et al., 1926; Blair and Prince, 1936; Sherman et al., 1942; Snider, 1943; Sanchez and Kamprath, 1959). The fact that Mn availability is influenced by so many factors has led to confusion concerning the effectiveness of various methods of Mn application. Positive yield responses from broadcast application of $MnSO_4$ or MnO, banding of Mn products at planting, banding of Mn in conjunction with an acid forming fertilizer, and foliar application of Mn have all been reported (Mederski et al., 1960). For soils low in Mn, broadcast applications of $MnSO_4 \cdot 7H_2O$ have been used successfully as a means of correcting Mn deficiencies in a number of crops (Gilbert and McLean, 1928; Evans and Purvis, 1948; Anderson and Carstens, 1973; Alley et al., 1978; Gettier et al., 1984; Mascagni and Cox, 1985a,b). Broadcast applications were not found to be effective on all soils. Harner (1942), Steckel (1946), Gupta (1986), Randall et al. (1975) and Eck (1995) all found broadcast applications of Mn to be ineffective or less effective than alternative methods of Mn fertilization. Banding Mn has generally been more effective than broadcasting as a means of correcting Mn deficiency across a broad range of soils. Mascagni and Cox (1985) found that optimum yields in soybeans were obtained on Atlantic Coastal Plains soils by banding 3 kg Mn ha^{-1} compared to broadcasting 14 kg Mn ha^{-1}. Randall et al. (1975) found similar results in Wisconsin, although an additional foliar application was required for optimum yield when deficiencies were severe. Excellent results have also been obtained when Mn is band applied in conjunction with acid forming fertilizers such as 21-0-0-24, 11-52-0, and 18-46-0 (Steckel, 1946; Mederski et al., 1960; Petrie and Jackson, 1984). Banding an acid forming fertilizer alone can also increase the availability of native Mn (Steckel et al., 1948).

Foliar applications of Mn have also been used to correct Mn deficiency. Although applications of < 1.0 to 5.0 kg Mn ha^{-1} as $MnSO_4 \cdot 7H_2O$ are commonly recommended (Mengel, 1980; Vitosh et al., 1981), the number of applications required has not been unequivocally established. Eck (1995) found a single application of 1.1 kg Mn ha^{-1} adequate for correcting Mn deficiency in soybeans in Indiana while Mascagni and Cox (1985a,b) found that up to three applications were required.

5.6.3 Iron

Like Mn, Fe availability to plants is controlled by a number of soil factors including pH, free $CaCO_3$ content, SOM content, and redox potential. The reader is referred to reviews by Moraghan and Mascagni (1992) and Lindsay (1992) for a detailed discussion of these factors. Iron deficiency (chlorosis) is common in soybeans on calcareous soils (Martens and Westerman, 1992). Because of large differences in tolerance to Fe chlorosis among and within plant species, development of tolerant cultivars has been possible. Because broadcast applications of inorganic sources of Fe to the soil are not effective in controlling or correcting Fe chlorosis, foliar sprays using chelates such as FeEDDHA are generally recommended in field crops but multiple applications may be required.

5.6.4 Copper

Copper deficiency occurs most commonly on organic soils, but can also occur where sandy soils are highly weathered, on mineral soils with a high SOM content, and on calcareous mineral soils (Martens and Westerman, 1992). Crops differ greatly in their susceptibility to Cu deficiency with wheat, oats, sudan grass, and alfalfa being highly sensitive, barley, corn, and sugar beet moderately sensitive and soybeans and most forage grasses tolerant to Cu deficiency (Robertson et al., 1981).

While a number of fertilizers to supply Cu are available, $CuSO_4 \cdot 5H_2O$ is the most common fertilizer material used because of low cost and high water solubility. Karamanos et al. (1986) evaluated a number of Cu fertilizers and their effect on crop production. Copper oxide was generally ineffective in the year of application but residual effects alleviated Cu deficiency, while $CuSO_4 \cdot 5H_2O$ and chelated products were effective in the year of application. However, there was no residual effect from the application of chelated materials. Applications of animal manure or biosolids containing Cu will also correct deficiencies. Swine manure is an excellent source of Cu in many areas because of the high levels of Cu fed to growing pigs. Soil application is the preferred method of correcting Cu deficiency because of the good residual effects from Cu fertilization. Copper fertilizers are most commonly broadcast at or before planting but can also be banded. Common rates for soil application for field crops are 2.2 to 3.3 kg Cu ha^{-1} banded or 33 to 66 kg Cu ha^{-1} broadcast (Mengel, 1980). Foliar applications can be used during the growing season to correct Cu deficiencies with chelated materials being favored.

5.6.5 Molybdenum

Molybdenum is required for NO_3 reduction reactions in plants and the symbiolic fixation of N by legumes. Deficiencies of this nutrient have been reported for a number of crops grown on acid soils in the Great Lakes region, coastal areas of the United States, as well as in New Zealand and Australia. Deficiencies are usually observed with the legume crops.

Molybdenum deficiencies can be corrected by liming, by soil or foliar application of Mo or with seed treatment. While Mo toxicity in plants is rare, high concentrations of Mo in forages may induce Cu deficiency in animals (Miltmore and Mason, 1971). Therefore, care must be exercised in making Mo applications. Application methods which utilize the lowest effective Mo rate, generally foliar or seed treatment, are preferred.

5.7 References

Alessi, J., and J.F. Power. 1980. Effects of banded and residual fertilizer phosphorus on dryland spring wheat yield in the Northern Plains. Soil Sci. Soc. Am. J. 44:792–796.

Alexander, A., and M. Schroeder. 1987. Modern trends in foliar fertilization. J. Plant Nutr. 10:1391–1399.

Alley, M.M., C.I. Rich, G.W. Hawkins, and D.C. Martens. 1978. Correction of Mn deficiency of soybeans. Agron. J. 70:35–38.

Allred, S.E., and A.J. Ohlrogge. 1963. Principles of nutrient uptake from fertilizer bands. VI. Germination and emergence of corn as affected by ammonia and ammonium phosphate. Agron. J. 56:309–313.

Anderson, W.C., and J.B. Carstens. 1973. Effect of manganese soil and seed treatments on growth and yield of peas. Soc. Hort. Sci. J. 98:581–582.

Anghinoni, I., and S.A. Barber. 1980. Predicting the most efficient phosphorus placement for corn. Soil Sci. Soc. Am. J. 44:1016–1020.

Bailey, L.D., and E.G. Beauchamp. 1973. Effects of temperature on NO_3 and NO_2 reduction, nitrogenous gas production and redox potential in a saturated soil. Can. J. Soil Sci. 53:213–218.

Bandel, V.A., S. Dzienia, and G. Stanford. 1980. Comparison of N fertilizers for no-till corn. Agron. J. 72:337–341.

Barber, S.A. 1959. Relation of fertilizer placement to nutrient uptake and crop yield. II. Effects of row potassium, potassium soil-level and precipitation. Agron. J. 51:97–99.

Barber, S.A. 1958. Relation of fertilizer placement to nutrient uptake and crop yield. I. Interaction of row phosphorus and soil level of phosphorus. Agron. J. 50:535–539.

Barber, S.A. 1974. A program for increasing the efficiency of fertilizers. Fert. Soln. 18:24–25.

Barber, S.A. 1995. Soil nutrient bioavailability: a mechanistic approach. 2nd Ed. John Wiley and Sons, New York, NY.

Barber, S.A., and J.H. Cushman. 1981. Nitrogen uptake model for agronomic plants. p. 382–409. In I.K. Iskander (ed.) Modeling waste water renovation-land treatment. Wiley-Interscience, New York, NY.

Blair, A.W., and A.L. Prince. 1936. Manganese in New Jersey soils. Soil Sci. 42:327–333.

Borkert, C.M., and S.A. Barber. 1985. Predicting the most efficient phosphorus placement for soybeans. Soil Sci. Soc. Am. J. 49:901–904.

Brown, J.R. 1972. Micronutrient topdressing of alfalfa (Medicago sativa L.) on Udollic Albaqualf. Commun. Soil Sci. Plant Anal. 3:211–221.

Brown, J.R., W.O. Thom, and L.L. Wall, Sr. 1981. Effects of sulfur application on the yield and composition of soybeans and soil sulfur. Commun. Soil Sci. Plant Anal. 12:247–261.

Bullock, D.G., F.W. Simmons, I.M. Chung, and G.I. Johnson. 1993. Growth analysis of corn with and without starter fertilizer. Crop Sci. 33:112–117.

Burford, J.R., and J.M. Bremner. 1975. Relationships between denitrification capacities of soils and total, water-soluble, and readily decomposable soil organic matter. Soil Biol. Biochem. 7:389–394.

Cerrato, M.E., and A.M. Blackmer. 1990. Effects of nitrapyrin on corn yields and recovery of ammonium-N at 18 site-years in Iowa. J. Prod. Agric. 3:513–521.

Christenson, D.R., R.P. White, and E.C. Doll. 1973. Yields and magnesium uptake by plants as affected by soil pH and calcium levels. Agron. J. 65:205–206.

Claassen, N., and S.A. Barber. 1976. Simulation model for nutrient uptake from soil by a growing plant root system. Agron. J. 68:961–964.

Clapp, J.G., Jr., and H.G. Small. 1970. Influence of "pop-up" fertilizers on soybean stands and yield. Agron. J. 62:802–803.

Coe, D.G. 1926. The effects of various methods of applying fertilizers on crop yields. Soil Sci. 21:127–141.

Cox, F.R. 1972. Effect of calcium sources and fungicide on peanut production. J. Am. Peanut Res. Educ. Assoc. 4:122–129.

Creamer, F.L., and R.H. Fox. 1980. The toxicity of banded urea or di-ammonium phosphate to corn as influenced by soil temperature, moisture and pH. Soil Sci. Soc. Am. J. 44:298–300.

Cruz, J.C. 1982. The effect of crop rotation and tillage system on some soil properties root distribution and crop production. Ph.D. Thesis, Purdue University, West Lafayette, IN.

Eck, K.J. 1995. Diagnosing and correcting manganese deficiencies of soybeans in Indiana. MS Thesis, Purdue University, West Lafayette, IN.

Endelman, F.J., D.R. Keeney, J.T. Gilmour, and P.G. Saffigna. 1974. Nitrate and chloride movement in the Plainfield loamy sand. J. Environ. Qual. 3:295–298.

Engel, R.E., J. Eckhoff, and R.K. Berg. 1994. Grain yield, kernel weight, and disease responses of winter wheat cultivars to chloride fertilization. Agron. J. 86:391–396.

Ernst, J.W., and H.F. Massey. 1960. The effects of several factors on volatilization of ammonia formed from urea in the soil. Soil Sci. Soc. Am. Proc. 24:87–90.

Evans, H.J., and E.R. Purvis. 1948. An instance of manganese deficiency of alfalfa and red clover in New Jersey. Agron. J. 40:1046–1047.

Evanylo, G.W. 1991. No-till corn response to nitrogen rate and timing in the middle Atlantic Coastal Plain. J. Prod. Agric. 4:180–185.

Farber, B.G., and P.E. Fixen. 1986. Phosphorus response of late planted corn in three tillage systems. J. Fert. Issues. 3:46–51.

Fixen, P.E., G.W. Buchenan, R.H. Gelderman, T.E. Shumacher, J.R. Gerwing, F.A. Cholick, and B.G. Farber. 1986b. Influence of soil and applied chloride on several wheat parameters. Agron. J. 78:736–740.

Fixen, P.E., R.H. Gelderman, J. Gerwing, and F.A. Cholick. 1986a. Response of spring wheat, barley, and oats to chloride in potassium chloride fertilizers. Agron. J. 78:664–668.

Fox, R.H., and L.D. Hoffman. 1981. The effect of N fertilizer source on grain yield, N uptake, soil pH and lime requirement in no-till corn. Agron. J. 73:891–895.

Frye, W.W., H.F. Miller, L.W. Murdock, and D.E. Peaslee. 1978. Zinc fertilization of corn in Kentucky. KY Coop. Ext. Serv. Agron. Notes 11:1–4.

Gettier, S.W., D.C. Martens, D.L. Hallock, and M.J. Stewart. 1984. Residual Mn and associated soybean yield response from MnSO₄ application on a sandy loam soil. Plant Soil 81:101–110.

Gerwing, J., P. Fixen, and R. Gelderman. 1982. Zinc rate and source studies. SD Exp. Stn. Prog. Rep. 13.

Gilbert, B.E., and F.T. McLean. 1928. A "deficiency disease": the lack of available manganese in a lime-induced chlorosis. Soil Sci. 26:27–31.

Gilbert, B.E., F.T. McLean, and L.J. Hardin. 1926. The relation of manganese and iron to be lime-induced chlorosis. Soil Sci. 22:437–446.

Griffith, D.R., E.J. Kladivko, J.V. Mannering, T.D. West, and S.D. Parsons. 1988. Long-term tillage and rotation effects on corn growth and yield on high and low organic matter poorly drained soils. Agron. J. 80:599–605.

Gupta, U.C., J.A. MacLeod, and J.D.E. Sterling. 1976. Effects of boron and nitrogen on grain yield and boron and nitrogen concentration of wheat and barley. Soil. Sci. Soc. Am. J. 40:723–726.

Gupta, U.C. 1986. Manganese nutrition of cereals and forages grown in Prince Edward Island. Can. J. Soil Sci. 66:59–65.

Halvorson, A.D., and J.L. Havlin. 1992. No-till winter wheat response to phosphorus placement and rate. Soil Sci. Soc. Am. J. 56:1635–1639.

Harner, P.M. 1942. The occurrence and correction of unproductive alkaline organic soil. Soil Sci. Soc. Am. Proc. 7:378–386.

Hauck, R.D. 1984. Nitrogen in crop production. American Society of Agronomy, Madison, WI.

Hauck, R.D. 1985. Slow-release and bioinhibitor-amended nitrogen fertilizers. *In* O.P. Engelstad (ed) Fertilizer technology and use. 3rd Ed. Soil Science Society of America, Madison, WI.

Heckman, J.R. 1995. Corn responses to chloride in maximum yield research. Agron. J. 84:415–419.

Hoeft, R.G., and L.M. Walsh. 1975. Effect of carrier, rate, and time of application of S on the yield, and S and N content of alfalfa. Agron. J. 67:427–430.

Hoeft, R.G., L.M. Walsh, and E.A. Liegel. 1975. Effect of seed-placed fertilizer on the emergence (germination) of soybeans (*Glycine max* L.) and snapbeans (*Phaseolus vulgaris* L.). Commun. Soil Sci. Plant Anal. 6:655–664.

Hoeft, R.G., J.E. Sawyer, R.M. Van Den Heuvel, M.A. Schmitt, and G.S. Brinkman. 1985. Corn response to sulfur on Illinois soils. J. Fert. Issues 2:95–104.

Hollanda, F.S.R., D.B. Mengel, M.B. Paula, J.G. Caruaha, and J.C. Bertoni. 1998. Influence of crop rotations and tillage systems on phosphorus and potassium stratification and root distribution in the soil profile. Commun. Soil Sci. Plant Anal. 29:2383–2394.

Huber, D.M., H.L. Warren, D.W. Nelson, C.Y. Tsai, M.A. Ross, and D.B. Mengel. 1982. Evaluation of nitrification inhibitors for no-till corn. Soil Sci. 134:388–394.

Karamanos, R.E., G.A. Kruger, and J.W.B. Stewart. 1986. Copper deficiency in cereal and oilseed crops in northern Canadian Prairie soils. Agron. J. 78:317–323.

Karim, F., and J.T. Touchton. 1983. Response of corn seedlings to high concentrations of ammonium phosphates. Commun. Soil Sci. Plant Anal. 14:847–858.

Karlen, D.L., P.G. Hunt, and R.B. Campbell. 1984. Crop residue removal effects on corn yield and fertility of a Norfolk sandy loam. Soil Sci. Soc. Am. J. 48:868–872.

Kaspar, T.C., H.J. Brown, and E.M. Kassmeyer. 1991. Corn root distribution as affected by tillage wheel traffic and fertilizer placement. Soil Sci. Soc. Am. J. 55:1390–1394.

Keller, G.D., and D.B. Mengel. 1986. Ammonia volatilization from nitrogen fertilizers surface applied to no-till corn. Soil Sci. Soc. Am. J. 50:1060–1063.

Killorn, R. 1984. Zinc-an essential nutrient. IA Coop. Ext. Serv. Bull. PM-1129.

Kitcheson, J.W. 1968. Effect of controlled air and soil temperature and starter fertilizer on the growth and nutrient composition of corn (*Zea mays* L.). Soil Sci. Soc. Am. J. 32:531–534.

Kline, J.S., J.T. Sims, and K.L. Schike-Gartley. 1989. Response of irrigated corn to sulfur fertilization in the Atlantic Coastal Plain. Soil Sci. Soc. Am. J. 53:1101–1108.

Kovar, J.L., and S.A. Barber. 1987. Placing phosphorus and potassium for greatest recovery. J. Fert. Issues 4:1–6.

Kovar, J.L., and S.A. Barber. 1988. Phosphorus supply characteristics of 33 soils as influenced by seven rates of P addition. Soil Sci. Soc. Am. J. 52:160–165.

Kovar, J.L., and S.A. Barber. 1989. Reasons for differences among soils in placement of phosphorus for maximum predicted uptake. Soil Sci. Soc. Am. J. 53:1733–1736.

Krauskopf, K.B. 1972. Geochemistry of micronutrients. p. 7–40. *In* J.J. Mortvedt et al. (ed.) Micronutrients in agriculture. Soil Science Society of America, Madison, WI.

Lamond, R.E., D.A. Whitney, and B.H. Marsh. 1995. Sulfur fertilization of smooth bromegrass in Kansas. Agron. J. 87:13–16.

Lamond, R.E., D.A. Whitney, J.S. Hickman, and L.C. Bonczkowski. 1991. Nitrogen rate and placement for grain sorghum production in no-tillage systems. J. Prod. Agric. 4:531–535.

Lindsay, W.L. 1992. Inorganic equilbria affecting micronutrients in soils. *In* J.J. Mortvedt et al. (ed.) Micronutrients in agriculture. 2nd Ed. Soil Science Society of America, Madison, WI.Larson, W.E. 1964. Soil parameters for evaluating tillage needs and operations. Soil Sci. Soc. Am. Proc. 18:118–122.

Maddux, L.D., D.E. Kissel, J.D. Ball, and R.J. Raney. 1985. Nitrification inhibition by nitrapyrin and volatile sulfur compounds. Soil Sci. Soc. Am. J. 49:239–242.

Mahler, R.J., and R.L. Maples. 1987. Effect of sulfur additions on soil and the nutrition of wheat. Commun. Soil Sci. Plant Anal. 18:653–673.

Mann, C.L. 1995. Efficiency of nitrogen management systems in no-till corn production. MS Thesis. Purdue University, West Lafayette, IN.

Martens, D.C., and D.T. Westerman. 1992. Fertilizer applications for correcting micronutrient deficiencies. *In* Mortvedt et al. (ed.) Micronutrients in agriculture. 2nd Ed. Soil Science Society of America, Madison, WI.

Mascagni, H.J., Jr., and F.R. Cox. 1985a. Effective rates of fertilization for correcting manganese deficiency in soybeans. Agron. J. 77:363–366.

Mascagni, H.J., Jr., and F.R. Cox. 1985b. Critical levels of manganese in soybean leaves at various growth stages. Agron. J. 7:373–375.

Matheny, T.A., and P.G. Hunt. 1981. Effects of irrigation and sulphur application on soybeans grown on a Norfolk loamy sand. Commun. Soil Sci. Plant Anal. 12:147–159.

Maxwell, T.M., D.E. Kissel, M.G. Wagger, D.A. Whitney, M.L. Cabrera, and H.C. Moser. 1984. Optimum spacing of preplant bands of N and P fertilizer for winter wheat. Agron. J. 76:243–247.

Mederski, H.J., D.J. Hoff, and J.H. Wilson. 1960. Manganese oxide and sulfate as fertilizer sources for correcting Mn deficiency in soybeans. Agron. J. 52:667.

Mengel, D.B. 1980. Role of micronutrients in efficient crop production. Purdue Univ. Coop. Ext. Serv. Agron. Guide AY-239.

Mengel, D.B. 1992. Fertilizing corn grown using conservation tillage. Purdue Univ. Coop. Ext. Serv. Agron. Guide AY-268.

Mengel, D.B., and F.E. Wilson. 1988. Timing of nitrogen for rice in relation to paddy flooding. J. Prod. Agric. 1:90–92.

Mengel, D.B., J.F. Moncrief, and E.E. Schulte. 1992. Fertilizer management. p. 83–87. *In* Conservation tillage systems and management. Midwest Planning Service, Agricultural and Biosystems Engineering, Iowa State University, Ames IA.

Mengel, D.B., D.W. Nelson, and D.M. Huber. 1982. Placement of nitrogen fertilizers for no-till and conventional till corn. Agron. J. 74:515–518.

Mikkelsen, R.L. 1989. Phosphorus fertilization through drip irrigation. J. Prod. Agric. 2:279–286.

Miltmore, J.E., and J.L. Mason. 1971. Copper to molybdenum rations and molybdenum and copper concentrations in ruminent feeds. Can. J. Anim. Sci. 51:193–200.

Miner, G.S., S.Traoe, and M.R. Tucker. 1986. Corn response to starter fertilizer acidity and manganese materials varying in water solubility. Agron. J. 78:291–295.

Moraghan, J.T., and H.J. Mascagni, Jr. 1992. Environmental and soil factors affecting micronutrient deficiencies and toxicities. *In* J.J. Mortvedt et al. (ed.) Micronutrients in agriculture. 2nd Ed. Soil Science Society of America, Madison, WI.

Motavalli, P.P., K.A. Kelling, T.D. Syverud, and R.P. Wolkowski. 1993. Interaction of manure and nitrogen or starter fertilizer in northern corn production. J. Prod. Agric. 6:191–194.

Mortvedt, J.J., and F.R. Cox. 1985. Production, marketing and use of calcium, magnesium and micronutrient fertilizers. *In* O.P. Engelstad (ed.) Fertilizer technology and use. 3rd Ed. Soil Science Society America, Madison, WI.

Murphy, L.S., R. Ellis, Jr., and D.C. Adriano. 1981. Phosphorus-micronutrient interaction effects on crop production. J. Plant Nutr. 3:593–613.

Nelson, L.B., and R.E. Uhland. 1955. Factors that influence loss of fall applied fertilizers and their probable importance in different section of the US. Soil Sci. Soc. Am. Proc. 19:492–496.

Nelson, W.L., B.A. Krantz, C.D. Welch, and N.S. Hall. 1959. Utilization of phosphorus as affected by placement: II. Cotton and corn in North Carolina. Soil Sci. 68:139–144.

Petrie, S.E., and T.L. Jackson. 1984. Effect of fertilization on soil solution pH and manganese concentration. Soil Sci. Soc. Am. J. 48:315–318.

Ramig, R.E., P.E. Rasmussen, R.R. Allmaras, and C.M. Smith. 1975. Nitrogen-sulfur relations in soft white winter wheat. I. Yield response to fertilizer and residual sulfur. Agron. J. 67:219–224.

Randall, G.W., and R.G. Hoeft. 1988. Placement methods for improved efficiency of P and K fertilizers: A review. J. Prod. Agric. 1:70–79.

Randall, G.W., E.E. Schulte, and R.B. Corey. 1975. Effect of soil and foliar applied manganese on the micronutrient content and yield of soybeans. Agron. J. 67:502–507.

Reeves, D.W., and J.T. Touchton. 1986. Relative phytotoxicity of dicyandiamide and availability of its nitrogen to cotton, corn and grain sorghum. Soil Sci. Soc. Am. J. 50:1352–1357.

Rehm, G.W., and E.J. Penas, 1982. Use and management of micronutrient fertilizers in Nebraska. NB Coop. Ext. Serv. Bull. NebGuide G82-596.

Rehm, G.W. 1993. Timing sulfur applications for corn (*Zea mays* L.) production on irrigated sandy soil. Commun. Soil Sci. Plant Anal. 24:285–294.

Rehm, G.W., G.W. Randall, A.J. Scobbie, and J.A. Vetsch. 1995. Impact of fertilizer placement and tillage system on phosphorus distribution in soil. Soil Sci. Soc. Am. J. 59:1661–1665.

Robbins, S.G., and R.D. Voss. 1991. Phosphorus and potassium stratification in conservation tillage systems. J. Soil Water Cons. 46:300.

Robertson, L.S., and R.E. Lucas. 1981. Manganese: an essential plant micronutrient. MI Coop. Ext. Serv. Bull. E-1031.

Robertson, L.S., D.D. Warncke, and B.D. Knezek. 1981. Copper: an essential plant micronutrient. MI Coop. Ext. Serv. Bull. E-1159.

Rolston, D.E., D.L. Hoffman, and D.W. Toy. 1978. Field measurements of denitrification: I. Flux of N_2 and N_2O. Soil Sci. Soc. Am. J. 42:863–689.

Sanchez, C., and E.J. Kamprath. 1959. Effect of liming and organic matter content on the availability of native and applied manganese. Soil Sci. Soc. Am. Proc. 23:302–304.

Sander, D.H., E.J. Penas, and B. Eghball. 1990. Residual effects of various phosphorus application methods on winter wheat and grain sorghum. Soil Sci. Soc. Am. J. 54:1473–1478.

Scharf, P.C., and M.M. Alley. 1988. Nitrogen loss pathways and nitrogen loss inhibitors: A review. J. Fert. Issues. 5:109–125.

Schulte, E.E., and L.M. Walsh. 1982. Soil and applied zinc. WI Coop. Ext. Serv. Bull. A2528.

Selles, P. 1993. Residual effect of phosphorus fertilizer when applied with the seed or banded. Commun. Soil Sci. Plant Anal. 24:951–960.

Shafer, W.E., and D.W. Reed. 1986. The foliar application of potassium from organic and inorganic potassium carriers. J. Plant Nutr. 9:143–157.

Shear, C.B. 1975. Calcium related disorders of fruits and vegetables. Hort. Sci. 10:361–365.

Sheard, W.W. 1980. Nitrogen in the P band for forage establishment. Agron. J. 72:89–97.

Sherman, G.D., J.S. McHargue, and W.S. Hodgkiss. 1942. Determination of active manganese in soil. Soil Sci. 54:253–257.

Sleight, D.W., D.H. Sander, and G.A. Peterson. 1984. Effect of fertilizer phosphorus placement on the availability of phosphorus. Soil Sci. Soc. Am. J. 48:336–340.

Snider, H.J. 1943. Manganese in some Illinois soils and crops. Soil Sci. 56:186–195.

Steckel, J.E. 1946. Manganese fertilization of soybeans in Indiana. Soil Sci. Soc. Am. Proc. 11:346–348.

Steckel, J.E., B.R. Bertramson, and A.J. Ohlrogge. 1948. Manganese nutrition of plants as related to applied superphosphate. Soil Sci. Soc. Am. Proc. 13:108–111.

Stehouwer, R.C., and J.W. Johnson. 1990. Urea and anhydrous ammonia management for conventional tillage corn production. J. Prod. Agric. 3:507–513.

Sweeney, D.W. 1989. Suspension N-P-K placement methods for grain sorghum in conservation tillage systems. J. Fert. Issues 6:83–88.

Sweeney, D.W., and G.V. Grande. 1993. Yield, nutrient, and soil sulfur response to ammonium sulfate fertilization of soybean cultivars. J. Plant Nutr. 16:1083–1098.

Touchton, J.T., and F.C. Boswell. 1975. Boron application for corn growth on selected southeastern soils. Agron. J. 67:197–200.

Vitosh, M.L., J.W. Johnson, and D.B. Mengel. 1995. Tri-state fertilizer recommendations for corn, soybeans, wheat, and alfalfa. MI State Univ. Ext. Bull. E-2567.

Vitosh, M.L., D.D. Warncke, D.B. Knezek, and R.E. Luca. 1981. Secondary and micronutrients for vegetables and field crops. MI Coop. Ext. Serv. Bull. E-486.

Walker, M.E. 1975. Calcium requirements of peanuts. Commun. Soil Sci. Plant Anal. 6:299–313.

Weil, R.R., P.W. Bennedetto, L.J. Sikora, and V.A. Bandel. 1988. Influence of tillage practices on phosphorus distribution and forms in three ultisols. Agron. J. 80:503–509.

Welch, L.F., P.E. Johnson, G.E. McKibben, L.V. Boone, and J.W. Pendleton. 1966a. Relative efficiency of broadcast versus banded potassium for corn. Agron. J. 58:618–621.

Welch, L.F., D.L. Mulvaney, L.V. Boone, G.E. McKibben, and J.W. Pendleton. 1966b. Relative efficiency of broadcast versus banded phosphorus for corn. Agron. J. 58:283–287.

Wiese, R.A., and E.J. Penas. 1979. Fertilizer suggestions for corn. NB Coop. Ext. Serv. NebGuide G74-174.

Wijler, J., and C.C. Delwhiche. 1954. Investigations on the denitrifying process in soil. Plant Soil 2:155–169.

Wright, D.L. 1985. No-till corn response to starter fertilizer and starter placement. p. 137–140. *In* Proc. 1985 South. Region No-Till Conf., Griffin, GA.

Yibirin, H., J.W. Johnson, and D.J. Eckert. 1993. No-till corn production as affected by mulch, potassium placement and soil exchangeable potassium. Agron. J. 85:636–644.

6

Nutrient and Water Use Efficiency

Robert L. Westerman, William R. Raun, and Gordon V. Johnson
Oklahoma State University

6.1 Introduction

Nutrient use efficiencies seldom exceed 50% in most grain crop production systems, yet few research/extension programs in the United States or abroad have demonstrated significant improvement in this area. Use efficiencies are generally understood to mean the fraction of a substance found in the used form divided by the amount of that substance available or presented for use. Unless otherwise noted, nutrient and water use efficiencies will be used here in that manner. An often encountered difficulty in calculating use efficiencies relates to the nutrient transformation dynamics that occur with time in both the soil and plant. Consequently, the total amount of a nutrient (such as N) available for plant uptake just after fertilizer addition may decrease with time by an amount greater than that taken up by the plant. Conversely, under some conditions (priming), the total amount of N available to the plant may be greater than the sum of soil-available N plus fertilizer N at the time of fertilizer addition.

Poor use efficiencies are especially true for N, which in many cases represents the most expensive input for crop production around the globe. In the past, applying excess fertilizer N was affordable in the developed world since the economic risk was low and the environmental risk was thought to be insignificant. However, this approach is no longer practical.

To a large extent, universities, industry and international centers have focused on increased production. Similar importance should be placed on combining high yields with increased nutrient use efficiency (NUE) and water use efficiency (WUE). This is especially true for N, which represents in many cases, the most expensive input used by farmers for nonlegume production. This chapter highlights ideas and concepts that have focused on improving NUE and WUE in production agriculture. It also challenges some of the present knowledge and provides an independent view for improvements.

6.2   Nutrient Use Efficiency

6.2.1   Bray's Mobility Concept

Bray (1954) whose work has been pivotal in the management of mobile and immobile nutrients in soils, was the first to distinguish between root system sorption zones (mobile nutrients) and root surface sorption zones (immobile nutrients) (Fig. 6.1). Yield response to mobile nutrients was found to depend on the total amount available in the soil (root system sorption zone). If the total amount of available mobile nutrient is sufficient to produce only a yield of 2 Mg ha^{-1} and midway through the season better than average weather conditions raise the yield potential to 3 Mg ha^{-1}, Bray's mobility concept implies that the yield will be limited to 2 Mg ha^{-1} because the total supply of nutrient will be consumed to produce 2 Mg ha^{-1}. Thus, for a mobile nutrient like N, deficiencies are more prevalent in advanced growth stages. Additional yield can only be obtained if more of the mobile nutrient is added.

Alternatively, for immobile nutrients such as P, Bray (1954) showed that the addition of soluble P fertilizers results in yield responses that are independent of the size of the yield obtained (root surface sorption zone), but depend on the kind of crop, planting pattern, seeding rate, and the form and distribution pattern of the nutrient relative to the planting pattern. The supply of immobile nutrients to plants takes place by diffusion to the roots within a given soil volume. As a plant grows and roots extend out into the soil, roots explore a greater volume of soil from which they can extract P with the amount extracted being limited by the concentration at (or very near) the root-soil interface and the P buffer capacity. Many of these concepts are discussed in Section D, Chapter 5. If the concentration of P available to the plant at the root-soil interface is inadequate to meet the needs of the plant, then the plant will be deficient in P throughout its development. The deficiency will always be present, and plant growth and crop yield will be limited by the degree to which the immobile nutrient is deficient. Another, perhaps more common way of expressing this nutrient limitation is to state that yield will be obtained according to the sufficiency of the nutrient supply. When this is expressed as a percentage of the possible yield, then the term percent sufficiency may be used. Whenever percent sufficiency < 100, plant performance will be lower than the yield potential provided by the growing environment. Consequently, if the percent sufficiency is 80, it does not matter whether the yield potential is 2 or 3 Mg ha^{-1} because the actual yield obtained (theoretically) will only be 80% of the potential. Consequently, sufficiency (immobile nutrient fertilization indices) is considered to be independent of

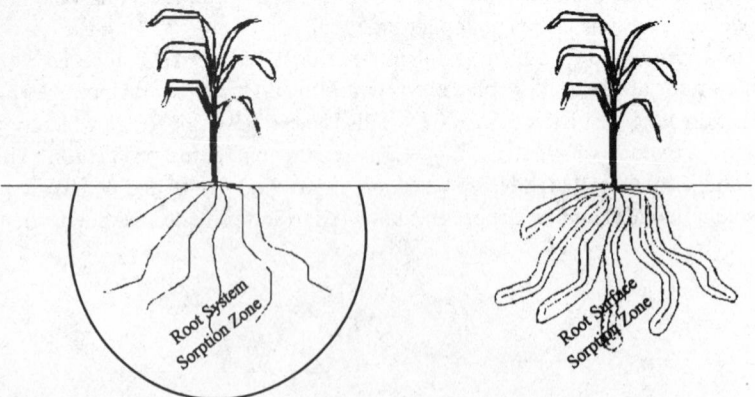

Fig. 6.1   The root system (mobile nutrients) and root surface sorption zones (immobile nutrients) as proposed by Bray (1954) for nurtient management depending on mobility.

the environment since, as roots grow, they extend into soil with concentrations of the element in question and the total amount present in the soil is not greatly affected (Fig. 6.1).

6.2.2 Mobile Nutrients

Moll et al. (1982) were the first to evaluate the contributions of N uptake and utilization processes to the variation in N use efficiency in corn hybrids. The utility of this work has been reflected in numerous articles dealing with N use efficiency in many crops. There are two primary components of N use efficiency: (1) the efficiency of absorption or uptake (Nt/Ns); and (2) the efficiency with which the N absorbed is utilized to produce grain (Gw/Nt) where Nt is the total N in the plant at maturity (grain + stover), Ns is the N supply or rate of fertilizer N, and Gw is the grain weight, all expressed in the same units. Other parameters defined in their work together with minor modifications subsequently introduced are presented in Table 6.1.

Additional N use efficiency parameters have been incorporated which reflect the recent improvements in the understanding of plant N losses. Harper et al. (1987) documented that N was lost as volatile NH_3 from wheat plants after fertilizer application and during flowering. Maximum N accumulation has been found to occur at or near flowering in wheat and corn and not at harvest. In order to estimate plant N loss without the use of labeled N forms, the stage of growth where maximum N accumulation is known to occur needs to be identified. The amount of N remaining in the grain plus straw or stover is subtracted from the amount at maximum N accumulation to estimate potential plant N loss (difference method). However, even the use of difference methods for estimating plant N losses is flawed since continued uptake is known to take place beyond flowering or the point of maximum N accumulation.

Table 6.1 Components of nitrogen use efficiency as reported by Moll et al. (1982) and modifications (in italics) for grain crops

Component	Abbreviation	Unit
Grain weight	Gw	kg ha^{-1}
Nitrogen supply (rate of fertilizer N)	Ns	kg ha^{-1}
Total N in the plant at maturity (grain + stover)	Nt	kg ha^{-1}
N accumulation after silking	Na	kg ha^{-1}
N accumulated in grain at harvest	Ng	kg ha^{-1}
Stage of growth where N accumulated in the plant is at a maximum, at or near flowering	*Nf*	*kg ha^{-1}*
Total N accumulated in the straw at harvest	*Nst*	*kg ha^{-1}*
Estimate of gaseous loss of N from the plant	*Nl =Nf-(Ng+Nst)*	*kg ha^{-1}*
Flowering uptake efficiency	*Eup=Nf/Ns*	
Harvest uptake efficiency (Uptake efficiency)	*Eha=Nt/Ns*	
Utilization efficiency	Gw/Nt	
Efficiency of use	Gw/Ns	
Grain produced per unit of grain N	Gw/Ng	
Fraction of total N translocated to grain	Ng/Nt	
Fraction of total N accumulated after silking	Na/Nt	
Ratio of N translocated to grain to N accumulated after silking	Ng/Na	

Francis et al. (1993) recently demonstrated that as much as 73% of the unaccounted N in ^{15}N balance calculations was due to plant N losses which were greater when N supply was increased. Kanampiu et al. (1997) and Francis et al. (1993) found that maximum N accumulation in wheat and corn occurred soon after flowering as illustrated in Fig. 6.2. In addition, Francis et al. (1993) highlighted the importance of plant N losses on the development and interpretations of strategies to improve N fertilizer use efficiencies. Harper et al. (1987) reported that 21% of the N fertilizer applied to wheat was lost as NH_3, of which 11.4% was from both soil and plant soon after fertilization and 9.8% from leaves between anthesis and maturity. Francis et al. (1993) concluded that failure to include direct plant N losses when calculating N budgets leads to overestimation of N losses by denitrification, leaching and NH_3 volatilization.

The quotient Nt/Ns is really the harvest or physiological maturity uptake efficiency (Eha), whereas uptake efficiency should be estimated at the stage of maximum N accumulation as Eup = Nf/Ns and not at maturity after some N has been lost (Nt/Ns) as proposed by Moll et al. (1982). Uptake efficiency as proposed by Moll et al. (1982) could be partitioned into two separate components since plant N losses (from flowering to maturity) can be significant (Daigger et al., 1976; Harper et al., 1987; Francis et al., 1993). The fraction of N translocated to the grain should be estimated as Ng/Nf instead of Ng/Nt (Table 6.1) since more N is accumulated in the plant at an earlier stage of growth. Plants losing significant quantities of N as NH_3 would have very high fractions of N translocated to the grain when calculated using Nt instead of Nf. In terms of plant breeding efforts, this could be a highly misleading statistic.

6.2.2.1   Increasing Nitrogen Use Efficiency

Split applications of N (Sowers et al., 1994) and low levels of applied N (Fowler et al., 1990) could increase N use efficiency in wheat and rye. Forage production systems have higher N use efficiencies than grain and/or forage/grain combinations, which do not decrease with increasing N applied as in grain production systems (Altom et al., 1996). This may be due to lower plant N losses (improved N use efficiency) because forage is harvested prior to anthesis, thus avoiding the period after anthesis when NH_3 losses are greater (Parton et al., 1988; Hooker et al., 1980). Ammonia has the potential to be lost because it must be formed as a precursor in the assimilation of NO_3^- into an amino form (Bidwell, 1979).

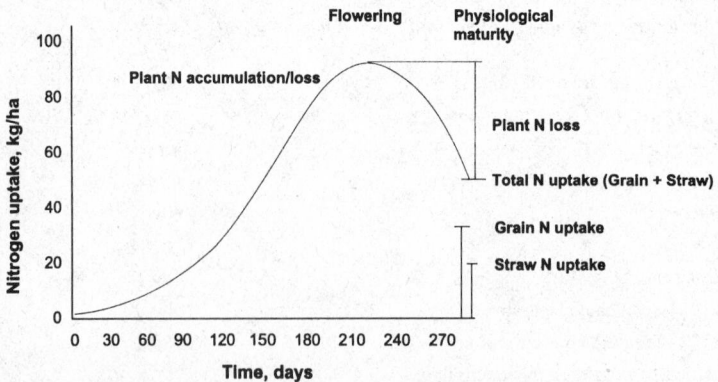

Fig. 6.2   Illustration of total N uptake in winter wheat and estimated loss following flowering in winter wheat [From Raun et al., 1996]

Russelle et al. (1981) found that maximum fertilizer use efficiency in corn was obtained with a low sidedressed N rate with light frequent irrigation. Wuest and Cassman (1992a) reported that recovery of N applied at anthesis (0–60 kg N ha^{-1}) ranged from 55 to 80% compared to 30–50% when all N was applied at planting. They showed that late N applications could be efficiently taken up by irrigated wheat and would not decrease N uptake from soil. Rao and Dao (1992) reported that N placement in narrow bands increased use efficiency by increasing N assimilation and reduction, but had little effect on grain and straw yields in conventional and no-till wheat. Water quality can be improved by modifying N management (combination of split, banded and spring topdress applications), which also increased winter wheat grain yield and N use efficiency (Mahler et al., 1994).

Improved fertilizer use efficiency has also been achieved by injecting anhydrous NH_3 below the surface instead of broadcasting (without incorporation) urea, NH_4NO_3 and UAN in no-till corn (Mengel et al., 1982). Sidedress N applications in no-till corn improved use efficiency compared to preplant N, particularly when urea was used (Fox et al., 1986). Method of placement has often been confounded with source in these and others studies (e.g., NH_3 cannot be broadcast), thus restricting direct source comparisons.

Because soil and plant tissue testing are reliable tools for detecting nutrient deficiencies, they are indispensible in improving NUE in virtually all production systems. A sound soil testing program is one of the best ways of determining what constitutes adequate, but not excessive fertilizer use for efficient crop production (Peck and Soltanpour, 1990). Despite the advent of nondestructive analytical methods that evaluate spectral radiance from crop canopies and soils, correlation and calibration with known soil testing procedures are still required. Because of the many factors which affect crop growth, and which are confounded using present day spectral measurements, soil testing will be required for nutrient recommendations well into the future.

Nitrogen use efficiency is improved when either the source or the method of application is selected to reduce, or eliminate, a process in the soil-plant system known to lose N or otherwise decrease its accumulation in harvested plant material. Thus, in comparison to preplant N, side- and topdress applications limit N losses (NH_3 volatilization, denitrification or leaching) and increase plant availability by reducing microbial immobilization early in the growing season. Sidedress and topdress applications can also be timed to coincide with peak crop N demand. Similarly, N use efficiency may be improved by physical incorporation or irrigation following broadcast application of urea or urea-based materials, thus minimizing N losses by NH_3 volatilization from urea hydrolysis when the soil or residue surface dries. Injection or band application of N eliminates the potential for surface NH_3 losses and reduces soil-fertilizer contact and subsequent risk of fertilizer N immobilization by soil microorganisms. Foliar applications at or near peak crop demand totally eliminate the soil-plant interactions that reduce N use efficiency.

Research has identified many of the conditions and components of the fertilizer-soil-plant system that lead to poor NUE. Implementation of practices that improve NUE are largely dependent on associated cost/benefit ratios. For this reason, farmers willingly accept potentially lower N use efficiency when they inject anhydrous NH_3 preplant instead of split applications of urea because anhydrous NH_3-N costs much less. Also, the lower N use efficiency of preplant than split, top- and sidedress N applications which must be made within narrow windows of opportunity in row crops is acceptable because of the reduced hassle factor. Sadly, many farmers/producers are unaware that N use efficiency can be compromised by one practice relative to another.

6.2.2.2 Decreasing Nitrogen Use Efficiency

Decreased N use efficiency in winter wheat with increasing N is common (Wuest and Cassman, 1992b; Sowers et al., 1994). Management systems designed for high protein cereal production in which N rates exceed those for maximum yield have very low use efficiencies (Fowler et al., 1990) due to increased grain N uptake without a corresponding yield increase (Fowler and Brydon, 1989; Raun and Johnson, 1995). However, increased grain protein is often associated with delayed N availability (Fowler and Brydon, 1989). The highest rates of plant NH_3 volatilization were found to occur during grain fill (Morgan and Parton, 1989), which would help explain why late N applications would tend to be more inefficient. Alternatively, Wuest and Cassman (1992a) showed that late N applications (anthesis) can be efficiently taken up by spring wheat due to a more extensive root system, better photosynthate supply, and a larger sink capacity of the fully developed crop at anthesis.

In lettuce production, Thompson and Doerge (1996) found that when water rates were exceeded, up to 50% of the fertilizer N was unutilized and unaccounted for at the high N rate thus decreasing use efficiency, which was probably due to the increased N losses by leaching and/or denitrification associated with the wetter conditions.

Because adding excess N at planting reduces overall partitioning efficiency, early season N must be managed to optimize grain yield, whereas applications late in the season can be adjusted to increase grain protein levels without reducing partitioning efficiency (Wuest and Cassman, 1992b). Under dryland corn, leaf water and osmotic potentials which have been associated with improved WUE, decreased significantly as N rates increased (Eghball and Maranville, 1991). For three C3 species, Polley et al. (1995) found that WUE increased at elevated CO_2 levels while N use efficiency was less affected .

Recently, soil-plant inorganic N buffering in dryland winter wheat production systems where inorganic N accumulation in the profile did not increase until N rates exceeded those for maximum yield has been demonstrated (Johnson and Raun, 1995; Raun and Johnson, 1995). When applied N exceeded maximum yield requirements, the system was able to buffer against soil N accumulation by increases in plant protein, plant N volatilization and denitrification. Therefore, when N rates exceeded those required for maximum yield, N removal from the soil continued to increase (plant protein, plant N loss and denitrification). Johnson and Raun (1995) also reported that the same mechanisms which prevent 100% fertilizer N use efficiency by plants being reached, also delay and prevent inorganic N accumulation in soils and the associated risk of subsequent leaching to groundwater. In other words, improving N use efficiencies in crop production systems may have adverse environmental consequences, especially in relation to NO_3^- contamination of groundwater. This is also likely to occur for other environmentally important mobile nutrients.

6.2.2.3 Variable Rate Technology and Spatial Variability

Spectral radiance readings for red and near infrared (NIR) wavelengths collected between Feekes stages 4 and 6 (Large, 1954) have been used successfully by Stone et al. (1996) to predict total N uptake and correct N deficiencies in early wheat growth and increase N use efficiency, demonstrating that variable rate technologies are likely to decrease the risk that overfertilization poses to the environment. Similar work in corn (Blackmer et al., 1994) showed that at later stages of vegetative growth, reflected radiation was correlated with relative grain yield. The improved N use efficiency which resulted was largely due to being able to detect and treat microvariability (1 m^2) in plant growth within larger areas (> 1 ha) that normally would receive one fixed N rate.

Significant spatial variability in soil test and plant biomass parameters has also been documented at the 1m^2 resolution. Solie et al. (1996) found that field element sizes (area which provides the most precise measure of the available nutrient where the level of that nutrient changes with distance) > 1.96 m^2 are unlikely to optimize fertilizer N inputs and have the potential for fertilizer misapplication because of too coarse a grid. Chancellor and Goronea (1994), who found similar effects of spatial variability on site specific applications in wheat, reported a field element size of 1 m^2 while Fiez et al. (1994) indicated that spatially variable N management programs are limited by their ability to predict site-specific yield potentials and the resultant N requirements. These findings raise the question of the potential environmental hazards associated with fixed fertilizer rates applied over large areas, which is the norm today. Because microvariability in fields makes intensive soil testing uneconomical, inexpensive, indirect plant tissue assays (spectral radiance) that reliably detect nutrient deficiencies are needed to replace soil testing on small scales.

Intensive sampling of a large number of small plots in Oklahoma demonstrated great spatial variability in bermudagrass yields (< 1,300 to > 10,000 kg ha^{-1}) and soil pH (4.37 to 6.29) within a 45 m^2 area. Recommended fertilizer rates varied from 0 to 31 kg P ha^{-1} and 0 to 108 kg K ha^{-1} with the soil test results and associated P recommendations for individual samples taken 0.30 m apart being illustrated in Fig. 6.3. This microvariability demonstrates one of the potential problems associated with grid-based soil samples since more than 10,000 soil samples ha^{-1} would theoretically be required. At a cost of $10 per sample, funds in excess of $100,000 ha^{-1} would be required to accurately map nutrient variability. Although such soil variability is real, soil sampling is not an option due to the excessive time and expense required to obtain this information. Because of this, cheaper indirect methods of sensing nutrient variability in plants and soils are currently being developed. This work indicates that sound environmental stewardship and high nutrient use efficiencies can only be attained where applicator resolution and rate are matched to soil test recommendations so that small areas requiring different fertilizer rates can be accurately treated in the field.

6.2.3 Immobile Nutrients

Phosphorus fertilizer use efficiency varies with placement. Sander et al. (1991) found that P use efficiencies in winter wheat production systems varied from 2% for broadcast to 18% for knifed P. In general, P use efficiencies, which are ~25% that of N, are heavily dependent on method of placement and soil type (Sander et al., 1991). Peterson et al. (1981) demonstrated that increased fertilizer was required when P was broadcast rather than band placed in conventional tillage but this depended on the soil test P level; less additional fertilizer would be required in a broadcast application as soil test levels increased. Increased use efficiency of band compared to broadcast P is due to less fertilizer contact with the soil in the former preventing decreased availability. Soluble P in fertilizer is adsorbed or precipitated on the surfaces of Fe and Al oxides in which forms it is less available (Bell and Black,

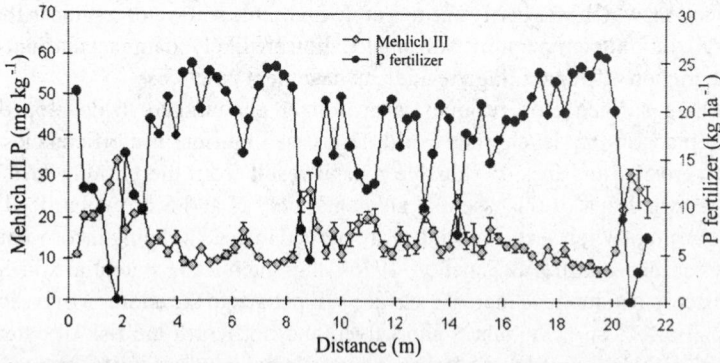

Fig. 6.3 Mean surface soil P from duplicate samples collected from consecutive 0.30 by 0.30 m plots, and fertilizer P recommendations based on soil test analyses, Burneyville, OK [Reprinted from Raun et al., 1998. Soil Sci. Soc. Am. J. 62:683-690 with permission of the Soil Science Society of America]

1970). Randall and Hoeft (1988) suggested that small grain crops generally respond better to band than broadcast applications because of increased fertilizer P efficiency with seed placement. On the other hand, Halvorson and Havlin (1992) reported that broadcast applications of P without incorporation for no-till winter wheat were equally as effective as band applications on moderately P-deficient soils. Similar to band applications in conventional tillage, broadcasting P in no-till decreases fertilizer-soil contact and subsequent P fixation. Broadcast P in no-till can be viewed as a horizontal band on the surface where increased moisture and root activity are often found.

Sloan et al. (1995) reported that banded P fertilizer is a viable alternative to liming for reducing Al likely to be toxic to seeds in strongly acid soils. Rehm et al. (1995) found that when band applications are used, soil test values become unreliable largely due to the spatial variability induced by banding (large differences in soil test values between band and untreated soil). Because soil testing is essential in assessing P fertilizer requirements, use efficiency will be determined by a method of sample collection and whether or not it adequately reflects P availability. Welch et al. (1966) obtained improved fertilizer K use efficiency when K was banded compared to broadcast, while Randall and Hoeft (1988) improved fertilizer use efficiency by banding both P and K when soil tests were low and soil moisture limiting.

In the Sahel, Payne et al. (1995) found that pearl millet (*Pennisetum glaucum* L.) production was limited by the inefficient use of N, P and water. Nitrogen and P use efficiencies (calculated as dry matter mg^{-1} of N and P uptake) increased and decreased with added P, respectively, from which they developed the phosphate root uptake efficiency (PRE) index (mg shoot P g^{-1} of root dry matter), which was found to be positively correlated with WUE and yield.

6.2.4   Tillage/Residue Management

When converting from conventional to no-till systems, the increased potential for immobilization of surface-applied N and lower levels of available N must be taken into account (Doran, 1980). If the soil potential to supply N during periods of peak demand is equal for conventional and no till systems, N demands would be no greater under no-till if an immobilization-mineralization equilibrium is established. However, increased N demand under no-till would suggest that an N equilibrium was not achieved, or that N losses (volatilization, denitrification or leaching) were greater compared to conventional tillage. Meek et al. (1995) found that soil NO_3-N in the 1.35–3.3 m depth was 21 kg ha^{-1} higher after 2 years of a corn-wheat rotation under conventional tillage than no-till, while Eck and Jones (1992) found that NO_3-N moved deeper into the profile under no-till compared to stubble mulch. Legg et al. (1979) reported greater uptake and fertilizer N recovery from no-till than conventional tillage corn production, which was partly due to a more favorable soil water regime in the untilled than tilled plots. Overall, rainfall, temperature, soil and location are likely to impact nutrient use efficiency differently depending on different tillage/residue management practices.

Although burning surface crop residues is no longer encouraged in developed countries, it continues to take place in the developing world. Decreased nutrient use efficiencies as a result of increased surface erosion and loss of valuable topsoil result from the burning of marginal lands. Boerner (1982) reported that of the essential elements, only N and S are volatilized in significant amounts during burning. When surface residues are burned instead of being incorporated, up to 75% of the S present can be volatilized (Sanchez, 1976), thus increasing potential soil S deficiencies. Although burning crop residues can increase surface soil pH due to the addition of basic cations in the ash, this beneficial effect on acid soils is short lived and not worth the risk of extensive erosion. Recently, Saa et al. (1994) showed that burning increases total P losses due to removal of particulate matter in runoff, which increases the risk of eutrophication in receiving waters.

6.2.5   Crop Rotation/Green Manures

Karlen and Doran (1991) noted that cover crops have not been widely accepted by farmers because of potential agronomic problems leading to lower yields and profits compared to primary grain crop production. Owens et al. (1995) found that a corn-soybean rotation in which no N is applied to the soybeans so that they can forage for the residual N from the corn crop provided a potential strategy for reducing NO_3-N loading to groundwater. However, much of the NO_3-N leaching occurred during the winter months (February-April) when no crop was actively growing. Oberle and Keeney (1990) found that N contributions from soybeans or alfalfa to first year corn were largely related to soil type and frequency of legumes in the rotation. Their work indicated that soil series information provided essential knowledge for N management evaluation.

Staley et al. (1991), who evaluated N use efficiency in cool (fescue) and warm (switchgrass) season grasses, found that although N recoveries for both species were low (23–33%), increased use of switchgrass could be recommended based on relatively high production levels at low N rates (90 kg N ha^{-1}) when compared to fescue. This work conducted on fairly shallow soils and on marginal landscapes with low water-holding capacities applies to many forage production systems.

6.2.6   Crop/Variety/Hybrid Selection

Eghball and Maranville (1991) reported that N use efficiency closely paralleled WUE in hybrid corn, suggesting that selecting for one component of efficiency may well improve the other. May et al. (1991) reported that N use efficiency traits are strongly influenced by the environment, and would be difficult to modify in a breeding program. Moll et al. (1982) showed that breeding for improved N use efficiency requires selection for both uptake and utilization efficiencies which resulted in the release a yellow dent corn inbred with improved NUE (Moll et al., 1991). Dhugga and Waines (1989) suggested specific crosses in two bread wheat genotypes based on large genotypic differences in N accumulation/loss between anthesis and maturity in an attempt to recombine superior N accumulation capacity and superior grain yield.

Consequently, it would appear that it is possible to develop varieties and hybrids that have improved nutrient use efficiencies; however, it should probably be carried out simultaneously with selection for WUE which largely controls yield potential.

6.2.7   Micronutrients

Although micronutrient use efficiency has received relatively little attention largely because the rates required to correct known deficiencies are low and widespread deficiencies are not common, improved Fe efficient soybeans have been bred by recurrent selection for this trait (Beeghly and Fehr, 1989). This approach has also been used for crop tolerance to Al and Mn toxicities. Nevertheless, much of the micronutrient deficiency work has focused on plant symptoms and plant diagnosis.

6.3   Water Use Efficiency

As the world population increases, the efficient use of fresh water will become increasingly important. This is not to say that it is not important today, but rather that it is recognized as a critical resource that demands judicious and prudent management. In rainfed regions, water runoff, surface evaporation, low soil water-holding capacity and random precipitation are major problems constraining efficiency (Kemper, 1993). Many of the management practices discussed in this chapter address these and other issues associated with improved WUE. However, it should be noted that the most efficient method of managing water supplies will not apply to all regions primarily because technological capacities differ from one region/country to the next. When water is the only, or most limiting growth factor, WUE will be maximized when water supply and potential evapotranspiration (humid, cool environment) are low. Conversely, WUE will be low when potential evapotranspiration exceeds water supply (plants dying in an inadequately irrigated desert environment).

Following Erie et al.'s (1965) observation that water is a limiting factor in the expansion of irrigated areas and in the production of food, WUE is a worldwide problem which needs urgent attention.

6.3.1   Soil Water/Physical Properties

Water use efficiency in irrigated agriculture has been increased significantly by the use of irrigation scheduling based on soil moisture determinations. A common approach is to irrigate when available water in the rooting depth reaches 50% depletion. Initially, soil water content was determined by tedious gravimetric methods which have been replaced by more convenient techniques such as resistance blocks, neutron probes, tensiometers, granular matrix sensors (GMS) and time domain reflectrometry (TDR) (Tollner et al., 1991; Eldredge et al., 1993; Robinson et al., 1994).

Many interacting factors affect sustainable use of soil resources, which is largely determined by the balance between soil destructive and restorative processes (Hornick and Parr, 1987). Crop residues returned to the soil can maintain or enhance soil quality and productivity through favorable effects on soil properties (Lal, 1995). Crop residues contribute to the maintenance of soil organic C, add plant nutrients and reduce fertilizer requirements. Fiez et al. (1994) found that improved N management could be achieved by site-specific yield estimates calculated from soil water availability and wheat spike density. They also noted that the large differences in grain yield among landscape positions may justify spatially variable N applications. Thus, improving soil physical properties will enhance WUE as well as NUE.

6.3.2   Source of Water and Water Quality

The most desirable irrigation system should not apply water and associated materials uniformly throughout the field, but rather should apply precisely the amounts required depending on the production capacity of each parcel (Hoffman and Martin, 1993). This is consistent with goals of many variable rate technology teams who find that estimated yield potentials are needed on a fine scale to refine recommended fertilizer rates. If water and nutrients based on yield potential could be applied simultaneously, synergistic improvements in the efficiencies of both could be expected.

Because large quantities of saline water frequently exist in irrigated areas, various strategies have been proposed for their use, including blending with good quality water (Bradford and Letey, 1992). Cyclic strategies use water of various salinities separately either during one season or in a crop rotation as a function of crop salt tolerance. Models predicts higher yields of salt-sensitive corn from cyclic than blending strategies. Andreu et al. (1996) showed that with reduced irrigation water more frequent applications give marginally better yields for the same quantity, but at the expense of slightly increased salt concentrations in the root zone.

Many current irrigation systems are open and have relatively low efficiencies of water application (Hagin and Lowengart, 1995). In addition, water quality relative to management practices has been largely ignored. Erie et al. (1965) noted that for optimum production at a specific site, particular crops require definite amounts of water during the growing season.

6.3.3 Crop Rotations/Cover Crops/Fallow

Biederbeck and Bouman (1994) reported that while a fallow phase is the most common practice to store soil water to make small grain farming feasible in south central Saskatchewan, partial fallow replacement with legumes could reduce the risk of erosion and nutrient leaching and minimize the hazard of salinization and eutrophication of downstream ecosystems. As dry matter production increased, so did WUE, especially on marginal lands.

Norwood (1994) indicated that a wheat-sorghum-fallow system (two crops in 3 years) was superior to wheat-fallow where the fallow period was no-till. In general, no-till fallow systems have resulted in increased water storage and decreased soil erosion, while clean fallow practices (conventional tillage with no crop) have led to increased soil erosion and decreased soil fertility, leading Dumanski et al. (1986) to conclude that clean fallow is an inefficient means of water conservation.

Tanaka (1989) showed that grain crop WUE was greater for spring wheat under stubble mulch than chemical fallow but under severe drought stress, chemical fallow was superior. Nevertheless, over 4 years, stubble mulch and reduced tillage resulted in greater WUE in grain production than chemical fallow.

Various methods have been evaluated to reduce evaporation by covering the soil surface with layers of different materials, but most have been surface mulches (Willis, 1962; Unger and Parker, 1968) whose effectiveness increases with the amount of mulch and percentage surface covered. However, Sembiring et al. (1995) have demonstrated that placing a wheat straw layer (3 Mg ha^{-1}) 3 cm beneath the surface reduced water loss compared to no-till on silty clay and sandy loam soils but not on sandy clay loam soil.

6.3.4   Crop/Variety/Hybrid Selection

In selecting for N efficient and/or drought tolerant corn genotypes, the root system should be considered because applied N slows root growth, and moderate water stress significantly increases root length which is positively correlated with WUE (Eghball and Maranville, 1993). On the other hand, although rice is highly inefficient in water use, wet seeded rice is superior to the traditional transplanted rice in terms of WUE, has greater drought tolerance, and requires less labor for establishment and weed control (Bhuiyan et al., 1995).

6.3.5   Soil Water Balance/Models

Recent technological advances have allowed researchers to simulate water and nutrient transport, dry matter production, and other dependent variables as a function of variable water, radiation energy, nutrient inputs, etc. Modeling allows the experimenter to change input parameters and evaluate the potential outcome without having to conduct experiments that are costly and time consuming. Although this is an extremely valuable tool, models developed and validated under one set of controlled conditions are not necessarily applicable under different conditions. Nevertheless, models are useful in improving WUE.

6.3.6   Nutrient Interactions

Selles et al. (1992) suggested that soil testing laboratories should modify their fertilizer recommendations according to different levels of available water, rather than the traditional dry, normal and wet classes used. Heitholt (1989) and Heitholt et al. (1991) found that optimal leaf N concentrations promoted higher WUE, and low leaf N led to poor WUE in N-limited and drought-stressed wheat in the southern Great Plains. Under low water supply, Sandhu and Sidhu (1995) recommended that fertilizer N be broadcast before the preseeding irrigation to obtain higher crop yields. Both WUE and NUE were improved when nutrients were placed beneath the surface of the soil, as deep as 15 cm in turfgrass (Murphy and Zaurov, 1994), while WUE in irrigated winter wheat increased with increasing N up to 140 kg N ha^{-1} under nonstressed conditions (Eck, 1988). In West Africa, Bationo et al. (1993) reported that the addition of crop residues and fertilizer increased soil water use over the control for pearl millet production. Ryan and Tabbara (1989) reported that urea phosphate may enhance WUE in some irrigated soils as well as serving as an effective source of N and P. Their work showed that infiltration rates increased due to H_3PO_4 solubilizing $CaCO_3$ for exchange reactions of Ca with soil Na in initially sodic soils. In general, much of the work conducted, where both WUE and NUE have been evaluated, suggests that increasing one will increase the other.

6.4   Concluding Remarks

Agricultural development and production have focused on NUE and WUE from the time crops were first cultivated. Significant improvements have come from soil and plant tissue testing, residue management, crop rotation, method, source and time of fertilizer application, and improved hybrids and varieties. In crop production systems, soil conservation and appropriate residue management practices are critical to future improvements in NUE and WUE. This is especially true in the developing world where soil erosion losses lead to devastating land degradation.

Future improvements in NUE and WUE will likely come from the management of small scale variability in soil. At present, indirect spectral radiance measurements to detect differences in the nutrient status over short distances allow fertilizer applications based on need. Similar approaches with irrigation systems will likely take place in the near future.

6.5   References

Altom, W., J.L. Rogers, W.R. Raun, G.V. Johnson, and S.L. Taylor. 1996. Long-term rye-wheat-ryegrass forage yields as affected by rate and date of N application. J. Prod. Agric. 9:510–516.

Andreu, L., N. J. Jarvis, F. Moreno, and G. Vachaud. 1996. Simulating the impact of irrigation management on the water and salt balance in drained march soils (Marismas, Spain). Soil Use Manag. 12:109–111.

Bationo, A., C.B. Christianson, and M.C. Klaij. 1993. The effect of crop residue and fertilizer use on pearl millet yields on Niger. Fert. Res. 34:251–258.

Beeghly, H. H., and W. R. Fehr. 1989. Indirect effects of recurrent selection for Fe efficiency in soybean. Crop Sci. 29:640–643.

Bell, L. C., and C. A. Black. 1970. Crystalline phosphates produced by interaction of orthophosphate fertilizers with slightly acid and alkaline soils. Soil Sci. Soc. Am. J. 34:735–740.

Bhuiyan, S. I., M. A. Sattar, and M. A. K. Khan. 1995. Improving water use efficiency in rice irrigation through wet-seeding. Irrig. Sci. 16:1–8.

Bidwell, R.G.S. 1979. Plant physiology. MacMillan Publishing Co. New York, NY.

Biederbeck, V. O., and O. T. Bouman. 1994. Water use by annual green manure legumes in dryland cropping systems. Agron. J. 86:543–549.

Blackmer, T.M., J.S. Schepers, and G.E. Varvel. 1994. Light reflectance compared with other nitrogen stress measurements in corn leaves. Agron. J. 86:934–938.

Boerner, R.E. J. 1982. Fire and nutrient cycling in temperate ecosystems. Biosci. 32:187–192.

Bradford, S., and J. Letey. 1992. Cyclic and blending strategies for using nonsaline and saline waters for irrigation. Irrig. Sci. 13:123–128.

Bray, R.H. 1954. A nutrient mobility concept of soil-plant relationships. Soil Sci. 78:9–22.

Chancellor, W.J., and M.A. Goronea. 1994. Effects of spatial variability of nitrogen, moisture, and weeds on the advantages of site-specific applications for wheat. Trans. ASAE 37:717–724.

Daigger, L.A., D.H. Sander, and G.A. Peterson. 1976. Nitrogen content of winter wheat during growth and maturation. Agron. J. 68:815–818.

Dhugga, K.S., and J.G. Waines. 1989. Analysis of nitrogen accumulation and use in bread and durum wheat. Crop Sci. 29:1232–1239.

Doran, J.W. 1980. Soil microbial and biochemical changes associated with reduced tillage. Soil Sci. Soc. Am. J. 44:765–771.

Dumanski, J., D.R. Coote, G. Luciuk, and C. Lok. 1986. Soil conservation in Canada. J. Soil Water Conserv. 41:204–210.

Eck, H.V., and O.R. Jones. 1992. Soil nitrogen status as affected by tillage, crops, and crop sequences. Agron. J. 84:660–668.

Eck, H.V. 1988. Winter wheat response to nitrogen and irrigation. Agron. J. 80:902–908.

Eghball, B., and J.W. Maranville. 1991. Interactive effects of water and nitrogen stresses on nitrogen utilization efficiency, leaf water status and yield of corn genotypes. Commun. Soil Sci. Plant Anal. 22:1367–1382.

Eghball, B., and J.W. Maranville. 1993. Root development and nitrogen influx of corn genotypes grown under combined drought and nitrogen stresses. Agron. J. 85:147–152.

Eldredge, E.P., C.C. Shock, and T.D. Stieber. 1993. Calibration of granular matrix sensors for irrigation management. Agron. J. 85:1228–1232.

Erie, L.J., O.F. French, and K. Harris. 1965. Consumptive use of water by crops in Arizona. AZ Agric. Exp. Sta. Tech. Bul. 169.

Fiez, T.E., B.C. Miller, and W.L. Pan. 1994. Winter wheat yield and grain protein across varied landscape positions. Agron. J. 86:1026–1032.

Fowler, D.B., and J. Brydon. 1989. No-till winter wheat production on the Canadian prairies: timing of nitrogen fertilization. Agron. J. 81:817–825.

Fowler, D.B., J. Brydon, B.A. Darroch, M.H. Entz, and A.M. Johnston. 1990. Environment and genotype influence on grain protein concentration of wheat and rye. Agron. J. 82:655–664.

Fox, R.H., J.M. Kern, and W.P. Piekielek. 1986. Nitrogen fertilizer source, and method and time of application effects on no-till corn yields and nitrogen uptakes. Agron. J. 78:741–746.

Francis, D.D., J.S. Schepers, and M.F. Vigil. 1993. Post-anthesis nitrogen loss from corn. Agron. J. 85:659–663.

Gaston, L.A., R.S. Mansell, and H.M. Selim. 1992. Predicting removal of major soil cations and anions during acid infiltration: model evaluation. Soil Sci. Soc. Am. J. 56:944–950.

Hagin, J., and A. Lowengart. 1995. Fertigation for minimizing environmental pollution by fertilizers. Fert. Res. 43:5–7.

Halvorson, A.D., and J.L. Havlin. 1992. No-till winter wheat response to phosphorus placement and rate. Soil Sci. Soc. Am. J. 56:1635–1639.

Harper, L.A., R.R. Sharpe, G.W. Langdale, and J.E. Giddens. 1987. Nitrogen cycling in a wheat crop: Soil, plant and aerial nitrogen transport. Agron. J. 79:965–973.

Heitholt, J.J. 1989. Water use efficiency and dry matter distribution in nitrogen- and water stressed winter wheat. Agron. J. 81:464–469.

Heitholt, J.J., R.C. Johnson, and D.M. Ferris. 1991. Stomatal limitation to carbon dioxide assimilation in nitrogen- and drought-stressed wheat. Crop Sci. 31:135–139.

Hoffman, G.J., and D.L. Martin. 1993. Engineering systems to enhance irrigation performance. Irrig. Sci. 14:53–63.

Hooker, M.L., D.H. Sander, G.A. Peterson, and L.A. Daigger. 1980. Gaseous N losses from winter wheat. Agron. J. 72:789–792.

Hornick, S.B., and J.F. Parr. 1987. Restoring the productivity of marginal soils with organic amendments. Am. J. Altern. Agric. 2:64–68.

Hutson, J.L., and R.J. Wagenet. 1993. A pragmatic field-scale approach for modeling pesticides. J. Environ. Qual. 22:494–499.

Johnson, G.V., and W.R. Raun. 1995. Nitrate leaching in continuous winter wheat: use of a soil-plant buffering concept to account for fertilizer nitrogen. J. Prod. Agric. 8:486–491.

Kanampiu, F.K., W.R. Raun, and G.V. Johnson. 1997. Effect of nitrogen rate on plant nitrogen loss in winter wheat varieties. J. Plant Nutr. 20:389–404.

Karlen, D.L., and J.W. Doran. 1991. Cover crop management effects on soybean and corn growth and nitrogen dynamics in an on-farm study. Amer. J. Altern. Agric. 6:71–82.

Kemper, W.D. 1993. Effects of soil properties on precipitation use efficiency. Irrig. Sci. 14:65–73.

Lal, R. 1995. The role of residues management in sustainable agricultural systems. J. Sust. Agric. 5:51–78.

Large, E.C. 1954. Growth stages in cereals, illustration of the Feekes scale. Plant Pathol. 3:128–129.

Legg, J.O., G. Stanford, and O.L. Bennett. 1979. Utilization of labeled-N fertilizer by silage corn under conventional and no-till culture. Agron. J. 71:1009–1015.

Mahler, R. L., F.E. Koehler, and L.K. Lutcher. 1994. Nitrogen source, timing of application, and placement: effects on winter wheat production. Agron. J. 86:637–642.

Mallawatantri, A.P., and D.J. Mulla. 1996. Uncertainties in leaching risk assessments due to field averaged transfer function parameters. Soil Sci. Soc. Am. J. 60:722–726.

May, L., D.A. van Sanford, C.T. MacKown, and P.L. Cornelius. 1991. Genetic variation for nitrogen use in soft red x hard red winter wheat populations. Crop Sci. 31:626–630.

Meek, B.D., D.L. Carter, D.T. Westerman, J.L. Wright, and R.E. Peckenpaugh. 1995. Nitrate leaching under furrow irrigation as affected by crop sequence and tillage. Soil Sci. Soc. Am. J. 59:204–210.

Mengel, D.B., D.W. Nelson, and D.M. Huber. 1982. Placement of nitrogen fertilizers for no-till and conventional till corn. Agron. J. 74:515–518.

Moll, R.H., E.J. Kamprath, and W.A. Jackson. 1982. Analysis and interpretation of factors which contribute to efficiency of nitrogen utilization. Agron. J. 74:562–564.

Moll, R.H., E.J. Kamprath, and W.A. Jackson. 1991. Registration of NC201 parental line of maize. Crop Sci. 31:857.

Morgan, J.A, and W.J. Parton. 1989. Characteristics of ammonia volatilization from spring wheat. Crop Sci. 29:726–731.

Murphy, J.A., and D.E. Zaurov. 1994. Shoot and root growth response of perennial ryegrass to fertilizer placement depth. Agron. J. 86:828–832.

Nofziger, D.L., and A.G. Hornsby. 1987. Chemical movement in layered soils: User's manual. Univ. FL Circ. 780.

Norwood, C. 1994. Profile water distribution and grain yield as affected by cropping system and tillage. Agron. J. 86:558–563.

Oberle, S.L., and D.R. Keeney. 1990. Factors influencing corn fertilizer N requirements in the northern U.S. corn belt. J. Prod. Agric. 3:527–534.

Owens, L.B., W.M. Edwards, and M.J. Shipitalo. 1995. Nitrate leaching through lysimeters in a corn-soybean rotation. Soil Sci. Soc. Am. J. 59:902–907.

Parton, W.J., J.A. Morgan, J.M. Altenhofen, and L.A. Harper. 1988. Ammonia volatilization from spring wheat plants. Agron. J. 80:419–425.

Payne, W.A., L.R. Hossner, A.B. Onken, and C.W. Wendt. 1995. Nitrogen and phosphorus uptake in pearl millet and its relation to nutrient and transpiration efficiency. Agron. J. 87:425–431.

Peck, T.R., and P.N. Soltanpour. 1990. The principles of soil testing. p. 1–9. In R.L. Westerman (ed.) Soil testing and plant analysis. 3rd Ed. Soil Science Society of America, Madison, WI.

Peterson, G.A., D.H. Sander, P.H. Grabouski, and M.L. Hooker. 1981. A new look at row and broadcast phosphate recommendations for winter wheat. Agron. J. 73:13–17.

Polley, H.W., H.B. Johnson, and H.S. Mayeux. 1995. Nitrogen and water requirements of C3 plants grown at glacial to present carbon dioxide concentrations. Funct. Ecol. 9:86–96.

Ragab, R., F. Beese, and W. Ehlers. 1990. A soil water balance and dry matter production model. II. Dry matter production of oat. Agron. J. 82:157–161.

Randall, G.W., and R.G. Hoeft. 1988. Placement methods for improved efficiency of P and K fertilizers: A review. J. Prod. Agric. 1:70–79.

Rao, S.C., and T.H. Dao. 1992. Fertilizer placement and tillage effects of nitrogen assimilation by wheat. Agron. J. 84:1028–1032.

Raun, W.R., and G.V. Johnson. 1995. Soil-plant buffering of inorganic nitrogen in continuous winter wheat. Agron. J. 87:827–834.

Raun, W.R., G.V. Johnson, S.L. Taylor, and R.L. Westerman. 1996. Soil-plant relationships. Dept. Plant and Soil Science, OK State Univ., Stillwater, OK.

Raun, W.R., J.B. Solie, G.V Johnson, M.L. Stone, R.W. Whitney, H.L. Lees, H. Sembiring, and S.B. Phillips. 1998. Micro-variability in soil test, plant nutrient and yield parameters in bermudagrass. Soil Sci. Soc. Am. J. 62:683–690.

Rehm, G.W., G.W. Randall, A.J. Scobbie, and J.A. Vetsch. 1995. Impact of fertilizer placement and tillage system on phosphorus distribution in soil. Soil Sci. Soc. Am. J. 59:1661–1665.

Robinson, D.A., J.P. Bell, and C.H. Batchelor. 1994. Influence of iron minerals on the determination of soil water content using dielectric techniques. J. Hydrol. 161: 169–181.

Russelle, M.P., E.J. Deibert, R.D. Hauck, M. Stevanovic, and R.A. Olson. 1981. Effects of water and nitrogen management on yield and 15N-depleted fertilizer use efficiency of irrigated corn. Soil Sci. Soc. Am. J. 45:553–558.

Ryan, J., and H. Tabbara. 1989. Urea phosphate effects on infiltration and sodium parameters of a calcareous sodic soil. Soil Sci. Soc. Am. J. 53:1531–1536.

Saa, A., M.C. Trasar-Cepeda, F. Gil-Sotres, and F. Diaz-Fierros. 1994. Forms of phosphorus in sediments eroded from burnt soils. J. Environ. Qual. 23:739–746.

Sanchez, P. A. 1976. Properties and management of soils in the tropics. John Wiley and Sons, New York, NY.

Sander, D.H., E.J. Penas, and D.T. Walters. 1991. Winter wheat phosphorus fertilization as influenced by glacial till and loess soils. Soil Sci. Soc. Am. J. 55:1474–1479.

Sandhu, K.S., and A.S. Sidhu. 1995. Response of dryland wheat to supplemental irrigation and rate and method of N application. Fert. Res. 45:135–142.

Selles, F., R.P. Zentner, D.W.L. Read, and C.A. Campbell. 1992. Prediction of fertilizer requirements for spring wheat grown on stubble in southwestern Saskatchewan. Can. J. Soil Sci. 72:229–241.

Sembiring, H., W.R. Raun, G.V. Johnson, and R.K. Boman. 1995. Effect of wheat straw inversion on soil water conservation. Soil Sci. 159:81–89.

Sloan, J.J., N.T. Basta, and R.L. Westerman. 1995. Aluminum transformations and solution equilibria induced by banded phosphorus fertilizer in acid soil. Soil Sci. Soc. Am. J. 59:357–364.

Solie, J.B., W.R. Raun, R.W. Whitney, M.L. Stone, and J.D. Ringer. 1996. Optical sensor based field element size and sensing strategy for nitrogen application. Trans. ASAE 39:1983–1992.

Sowers, K.E., W.L. Pan, B.C. Miller, and J.L. Smith. 1994. Nitrogen use efficiency of split nitrogen applications in soft white winter wheat. Agron. J. 86:942–948.

Staley, T.E., W.L. Stout, and G.A. Jung. 1991. Nitrogen use by tall fescue and switchgrass on acidic soils of varying water holding capacity. Agron. J. 83:732–738.

Stone, M.L., J.B. Solie, W.R. Raun, R.W. Whitney, S.L. Taylor, and J.D. Ringer. 1996. Use of spectral radiance for correcting in-season fertilizer nitrogen deficiencies in winter wheat. Trans. ASAE 39:1623–1631.

Tanaka, D.L. 1989. Spring wheat plant parameters as affected by fallow methods in the northern great plains. Soil Sci. Soc. Am. J. 53:1506–1511.

Thompson, A.L., and T.A. Doerge. 1996. Nitrogen and water interactions in subsurface trickle-irrigated leaf lettuce. II. Agronomic, economic and environmental outcomes. Soil Sci. Soc. Am. J. 60:168–173.

Tollner, E.W., A.W. Tyson, and R.B. Beverly. 1991. Estimating the number of soil-water measurement stations required for irrigation decisions. Appl. Eng. Agric. 7:198–204.

Unger, P.W., and J.J. Parker, Jr. 1968. Residue placement effects on decomposition, evaporation and soil moisture distribution. Agron. J. 60:469–472.

van Sanford, D.A., and C.T. MacKown. 1987. Cultivar differences in nitrogen remobilization during grain fill in soft red winter wheat. Crop Sci. 27:295–300.

Welch, L.F., P.E. Johnson, G.E. McKibben, L.V. Boone, and J.W. Pendleton. 1966. Relative efficiency of broadcast versus banded potassium for corn. Agron. J. 58:618–621.

Willis, W.O. 1962. Effect of partial surface covers on evaporation from soil. Soil Sci. Soc. Am. Proc. 27:586–589.

Wuest, S.B., and K.G. Cassman. 1992a. Fertilizer-nitrogen use efficiency of irrigated wheat: I. Uptake efficiency of preplant versus late-season application. Agron. J. 84:682–688.

Wuest, S.B., and K.G. Cassman. 1992b. Fertilizer-nitrogen use efficiency of irrigated wheat: II. Partitioning efficiency of preplant versus late-season application. Agron. J. 84:689–694.

A nation that destroys its soils destroys itself
-Franklin D. Roosevelt

E

Pedology

Larry P. Wilding
Texas A & M University

Pedology is the earth science that quantifies the factors and processes of soil formation including the quality, extent, distribution and spatial variability of soils from microscopic to megascopic scales (Sposito and Reginato, 1992; Wilding et al., 1994). Spatial variability of soils in landforms is governed by the processes of soil formation which are, in turn, interactively conditioned by lithology, climate, biology, and relief through geologic time. Soils are welded together into a continuum like chains; processes and impacts on higher topographic surfaces directly affect adjacent lower lying surfaces. This is because transfer of energy flow and mass flux, the driving forces of pedogenesis, occur within and over three-dimensional soil landform bodies. Renewal vectors of biomass production, rainfall and dusts counter constituent losses via drainage waters, lateral interflow and downslope migration of erosion products.

The development of open versus closed drainage patterns during landform evolution strongly governs energy and mass flux in the system. In the closed drainage network, dispersal of chemical and erosion products is distributed to adjacent local sinks and depressions in the area. In contrast, for open drainage systems, dispersal is mostly external to source areas. In this case, distribution occurs via upland drainage ways to tributary streams, rivers, lakes and oceans. Differences in drainage network are paramount when one considers the effect of landscape on soil moisture and nutrient regimes, pollution of the environment, recharge of groundwater aquifers, crop production potentials, etc. Hence, to adequately comprehend, interpret, and transfer knowledge of soil resources from one area to the next, a soil/geomorphic systematic landscape model must be applied.

Through extensive knowledge of soil/landform relationships, pedologists have verified the occurrence, configuration, depth and pedogenic formation of root and water restrictive layers; documented the origin and distribution of cracking and fissuring patterns in soils and geologic materials that govern bypass flow of nutrients, chemicals solutes and fluids; identified the scale, mode, and occurrence of systematic spatial variability fundamental to the design efficiency and sampling of soil units; and utilized soil color patterns on a macro- and microscale to infer major periods of soil aeration/reduction and relative periods of excess, sufficiency and deficiency of soil moisture contents for specific land uses. In most soil systems, reaction kinetics and diffusion rates rather than thermodynamic equilibria control chemical reactions, solute movement and precipitation of chemicals and minerals. Reactive aggregate and fissure surfaces rather than bulk soil or geologic media are the significant phases. These and other impacts of geomorphology and pedogenic processes reflected in soil landscape patterns are covered in Chapters 1 and 2, respectively.

Qualitative landscape models are heavily premised on the state-factor analysis approach of Jenny (1941), but more recently, computer simulation models after Hoosbeek and Bryant (1992) have gained considerable interest. A thorough evaluation of the pros and cons of both modeling approaches and appropriate applications are considered in Chapter 3.

Pedology is an integrative and extrapolative science. It provides an organizational framework to catalogue modes, mechanisms and magnitudes of spatial variability, and to generalize this knowledge base for synthesis of models. Pedology provides a vehicle for extrapolation and scaling of spatial variability from components of soils (hand specimens and horizons) to the population of soils within the continuum as a whole (pedons, toposequences, physiographic entities and the pedosphere) (Fig. 6.1). Fig. 6.2 illustrates hierarchical levels in this continuum of soil organization relative to tools used to generalize information content at multiple scales of resolution.

Various taxonomic systems of soil classification have been developed to accommodate this cataloging of soil attributes. *Soil Taxonomy* (Soil Survey Staff, 1975, 1998) is such a system developed using morphogenetic indicators (diagnostic horizons and properties) as class criteria.

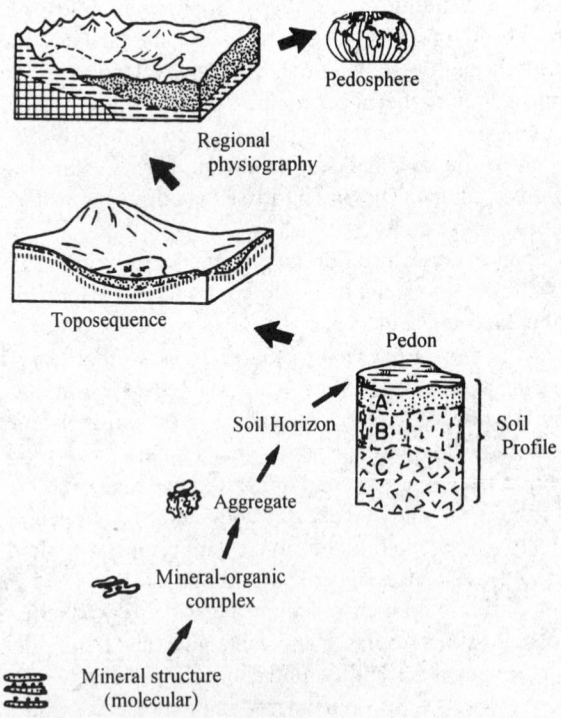

Fig. 6.1 Scaling of spatial variability from soil components through pedons and toposequences to the pedosphere

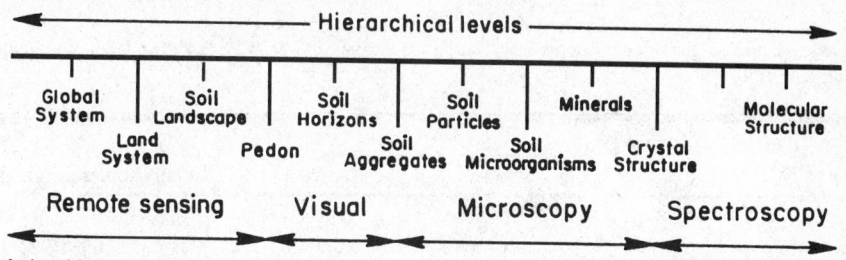

Fig. 6.2 Relationship between hierarchical levels of the soil continuum and the tools used for their study

Chapter 4 discusses the history leading to the development of *Soil Taxonomy*, an overview of class criteria, nomenclature and interpretational inferences. The functions and objectives of these systems have nationalistic overtones, but strong attempts are underway to develop correlative relationships that will further accommodate exchange of information pertinent to pedogenesis, land use and quality of land resources. Chapter 5 addresses these issues and uses the following systems to illustrate these concepts: the FAO-UNESCO Soil Map of the World, World Reference Base for Soil Resources, French system of soil classification, the system of classification used in the Former Soviet Union (USSR), and the recently developed Chinese soil taxonomy.

Spatial variability is the change in a soil property as a function of time and space. It may be temporal or of a more permanent nature, but it is a real landscape attribute, the norm (Wilding and Drees, 1983; Wilding et al., 1994). The soil variability dilemma is that soils are a continuum; many properties are not single valued; many properties are temporal; and properties are systematically and time spatially dependent. In short, the medium is vertically and laterally anisotropic. Examples of magnitudes of spatial variability in soil attributes within landscape units of a given mapping unit portraying a series concept are illustrated in Table E.1.

Spatial variability in soil systems belongs to two broad categories: systematic (structured) and random (unstructured and unknown causes). Systematic variability is a gradual or marked change in soil properties as a function of physiography, geomorphology and interactions of soil-forming factors (Wilding and Drees, 1983). Systematic variability permits pedologists to partition spatial variability in soils by subsets of properties that constitute soil survey map units corresponding to geomorphic landscape elements (summit, shoulder, backslope, footslope, etc.). The purpose of soil surveys is to partition the spatial variability of landforms into stratified subsets that are less variable than the medium as a whole and to remove systematic components of error. Information gained from soil surveys on the quality and distribution of soils when correlated with their classification provides a powerful vehicle for technology transfer. Chapter 6 illustrates characteristics of the 12 soil orders in *Soil Taxonomy*, their properties, processes and distribution patterns on a global scale. Each of the 12 soil orders are covered as subsets of this chapter. The importance of spatial variability of soils on classification, use and management is nicely illustrated in this chapter.

From a land use and interpretational base, there continues to be strong inertia to develop more comprehensive models of land evaluation driven largely by the era of information and technology explosion, availability of computer hardware and software, client demand and a systems analysis approach. Pedologists have long recognized spatial variability as the mainstay of their profession, but are being strongly challenged to better integrate this information for assessments of land quality, environmental quality and risk avoidance. Further, simulation models of plant growth, water and chemical solute transport, land quality, climatic change and environmental quality are widely used and heavily based on soil attributes and pedogenic processes. Chapter 7 considers developments in land evaluation both using classical methodology and nontraditional approaches. This chapter illustrates

Table E.1 Means and ranges of coefficients of variation for various soil properties within landscape units of a given mapping unit

Soil Property	CV (%)		Relative order of soil variability
	Mean	Range	
Bulk density	7	5-13	
Soil color-hue	9	2-20	
Soil color-value	10	4-12	Least variable
Soil pH	10	5-15	
Plasticity limit	15	5-28	
Liquid limit	17	8-31	
A horizon thickness	18	8-31	
Water retention (1/3 bar)		10-31	
Base saturation	25	17-33	
total sand content	25	8-46	
Total clay content	25	10-61	Moderately variable
Calcium carbonate equiv.	28	20-30	
Soil color-chroma	28	15-50	
Depth to carbonates	30	20-49	
Cation exchange capacity	32	20-40	
Depth to mottling	35	20-50	
Organic matter content	39	20-61	
Plasticity index	41	20-63	
Soil thickness	43	25-58	
Exchangeable Ca	48	30-73	Most variable
Exchangeable K	57	7-160	
Exchangeable Mg	58	31-121	
Water soluble salt ext.	48	---	
Hydraulic conductivity	75	31-150	

how land evaluation assessments may be used in data-rich and data-poor environments, the role of land, case studies on field and farm scale land evaluation, and the world food crisis.

References

Hoosbeek, M.R., and R.B. Bryant. 1992. Towards the quantitative modeling of pedogenesis-A review. Geoderma 55:183–210.

Jenny, H. 1941. Factors of soil formation. McGraw-Hill Book Co., Inc., New York, NY.

Soil Survey Staff. 1975. Soil taxonomy; a basic system of soil classification for making and interpreting soil surveys. USDA-SCS Agric. Handb. 436. U.S. Gov. Print Office, Washington, DC.

Soil Survey Staff. 1998. Keys to soil taxonomy, 8th Ed. Govt. Printing Office, Washington, DC.

Sposito, G., and R. Reginato. 1992. Opportunities in basic soil research. Soil Science Society of America, Madison, WI.

Wilding, L.P., J. Bouma, and D. Goss. 1994. Impact of spatial variability on modeling. pp. 61–75. *In* R. Bryant and M.R. Hoosbeek (eds.). Quantitative modeling of soil forming processes. Soil Sci. Soc. Am. J., Special Publ. No. 39. Madison, WI.

Wilding, L.P., and L.R. Drees, 1983. Spatial variability and pedology. P. 83–116. *In* L. P. Wilding et al. (ed.) Pedogenesis and soil taxonomy: I. Concepts and interactions. Elsevier Publ. Co., Amsterdam, Netherlands.

1

Geomorphology of Soil Landscapes

Douglas A. Wysocki and Philip J. Schoeneberger
National Soil Survey Center, Natural Resources Conservation Service
Lincoln, NE

Hannan E. LaGarry
Conservation and Survey Division, University of Nebraska-Lincoln

1.1 Introduction

Soils form a continuum across the Earth's land surface that is the interface of atmospheric, biological, and geological processes. Soils are more than a veneer of surficial alteration on landscapes or sediments. Soils, landscapes, and surficial materials or rocks together comprise three-dimensional systems that co-evolve through the interaction of physical and chemical weathering, erosion and deposition. To fully understand soils, successfully predict soil patterns, and anticipate soil behavior, one must comprehend the relationships among soils, landscapes, and surficial sediments. Soil geomorphology is the scientific study of the origin, distribution, and evolution of soils, landscapes, and surficial deposits and the processes that create and alter them. As a science, soil geomorphology is directly linked to Pedology, Geology, Hydrology, Archeology, Geomorphology, Physical Geography, and Geotechnical Engineering (Fig. 1.1). Soil geomorpholgy relies primarily, but not solely, upon geological principles and techniques (Daniels and Hammer, 1992). Principles and techniques drawn from geology or from other sciences, however, have applications or expressions that are unique to soils and soil landscapes. This chapter summarizes major geomorphic principles and techniques used to understand the relationships among soils, landscapes, and surficial sediments.

1.1.1 Goals of Soil Geomorphology

Soil geomorphology serves two important functions: (1) it provides basic principles for understanding the geomorphic history of landscapes (e.g., spatial and time relationships among soils, landscapes, and surficial sediments); (2) geomorphic-based landscape models segregate the soil continuum into meaningful soil bodies with a minimum of effort and can explain soil distribution from local hillslope to continental scales.

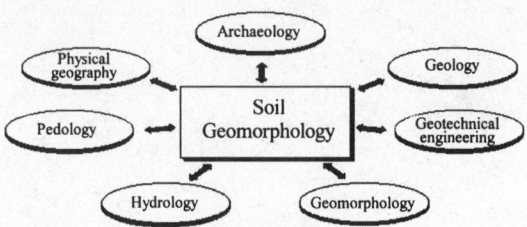

Fig. 1.1   Conceptual diagram showing the relationships of Soil Geomorphology to other sciences

1.2   Key Terminology

Before discussing the principles of soil geomorphology in detail, a few key definitions are needed. Although some terms may have alternate definitions, this discussion uses those following:

Depositional Surface: Part of a geomorphic surface formed by the deposition of sediments (alluvial, colluvial, eolian, etc.) derived from an erosional surface or erosional processes (Ruhe, 1975). Water, wind, ice, and gravity are active agents that construct and shape depositional surfaces.

Erosional Surface: Part of a geomorphic surface formed by the removal of rock, soil, or sediment under the wearing and transport action of water, wind, ice, and gravity (Ruhe, 1975). Running water is the dominant agent that forms and shapes most erosional surfaces.

Geomorphic Surface: A definable part of the land surface that forms during a given time and under a common set of erosional and depositional processes. A geomorphic surface may include many landforms, but it is a mappable feature with borders that can be identified (Ruhe, 1975).

Landform: Any physical, recognizable feature on the Earth's surface, having a characteristic shape (and range of composition and internal structure), that is produced by natural processes (modified from Soil Science Society of America, 1997). Landforms are the features of the Earth that together comprise the land surface (Ruhe, 1975).

Landscape: The portion of the land surface that the eye can comprehend in a single view (Ruhe, 1969). A collection or population of landforms that can be observed in a single view (Ruhe, 1975). Individual landforms in a single landscape can differ in both age and origin.

Pedon: A three-dimensional body of soil with lateral dimensions large enough to permit the complete study of horizon shapes and relations and commonly ranges from 1 to 10 m² in area (Soil Science Society of America, 1997).

Soil: A natural, three-dimensional body with definable boundaries that commonly, but not always, consists of horizons made up of mineral and organic materials, contains living matter, and can support vegetation (Soil Survey Staff, 1996).

Soil Delineation: An individual polygon shown on a soil map with a map unit symbol and name that defines a three-dimensional soil body of specified area, shape, and location on the landscape (Soil Science Society of America, 1997).

Soil Map Unit: An aggregate of all soil delineations in a soil survey area that have a similar, established set of characteristics (van Wambeke and Forbes, 1986). Each delineation (polygon) of a map unit is identified by the same symbol and name on the soil map.

1.3   Soil as a Landscape Unit or Body

Soil Geomorpholgy is a field science that studies soils on landscapes. The soil continuum on a landscape, for ease of comprehension, is divided into discrete units. The pedon is the soil unit most commonly described, sampled, and classified by pedologists. The relative size or scale of a pedon is an intellectual construct useful for description and classification. A pedon, however, lacks distinct lateral boundaries and is not a natural landscape unit. It is more helpful to consider natural groupings or clusters of pedons, rather than individual pedons, in order to understand landscapes. Geological processes, which are driven by the atmospheric agents of water, wind, ice, and gravity impact landscapes as a continuum and thereby affect entire groups of pedons. Landscapes possess natural boundaries that restrict or control mass and energy transfer. Examples of landscape boundaries are topographic divides, contacts between different rocks or sediment bodies, inflections in slope gradient, and contacts between landforms of different age, origin, and internal structure.

1.3.1 Landscape Scale and Function

Soil delineations depicted on a soil map at or near a scale of 1:24,000 are more closely linked to geomorphic processes than are individual pedons. Accordingly, soil delineations are better suited to geomorphic studies than pedons. Soil delineations depicted in a soil survey are landscape units (Fig. 1.2). Boundaries between delineations are established by the soil surveyor, but not in an arbitrary fashion. Boundaries on a soil map mark observable differences in soil morphology such as horizon type and thickness, and soil color, texture, and structure across the landscape. Changes in soil morphology across a landscape generally result from differences in the transfer of mass and energy driven by ecological, geomorphic, or atmospheric processes. Soil surveyors use observation, experience, and geomorphic landscape models to help create soil maps.

1.4 Models of Soil Formation

Soils are complex systems that defy easy comprehension. Therefore, scientific models are used to help understand or explain soils, but no single model provides complete understanding. Models of various form, function, and design are used to understand different aspects of soils (Dijkerman, 1974). Two important and well-known soil models are the factorial approach popularized by Jenny (1941) and Simonson (1959). An extensive review of pedological modeling is presented in Section E, Chapter 3.

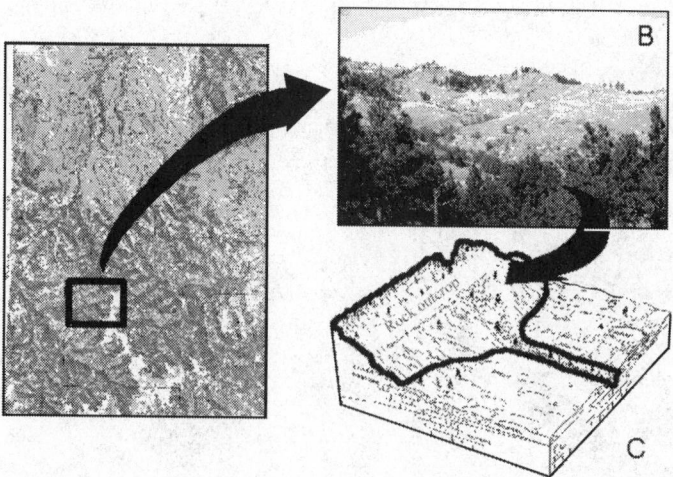

Fig. 1.2   Image of landscape and soil map showing soil as a landscape unit

1.4.1   Factors of Soil Formation

The eloquent, conceptual model of soil forming factors (Dokuchaev, 1883; Jenny, 1941) describes soil as a function of climate, biological influences, topography, parent material, and time (Fig. 1.3). Implicit in this model are the distinct relationships among ecosystems (biological factor), landscapes (topography), surficial sediments (parent material), and landscape evolution (time). Stratigraphy of sediments or bedrock and surface contours strongly influence the movement of water within and over the landscape. Topography and parent materials have a strong control on both local (e.g., soil watertables, water holding, nutrient capacity, salt content, soil temperature) and regional soil environments (e.g., rain shadows, elevational induced climate zones, adiabatic winds), and therefore, impact ecosystem form and function. All five soil-forming factors are linked either, directly or indirectly, to landscapes, surficial sediments, and landscape evolution.

1.4.2   Simonson's Process Model

Simonson (1959) explained soil formation by the interaction of four processes: additions, removals, translocations, and transformations (Fig. 1.4). This model is more helpful than Jenny's (1941) for understanding the spatial relationships and dynamics within soil landscapes. Geologic or geomorphic processes cause additions, removals, translocations, and transformations on a landscape scale that create and modify landforms, sediments, and soils. For example, sediment eroded from the flank of a hillslope is deposited as slope alluvium or colluvium at the base of the slope, or as alluvium in nearby drainageways or flood plains. The sediment is incorporated into the uppermost horizons of an existing soil or becomes fresh parent material in which a new soil begins to form.

1.5   Soil Landscape Models

1.5.1   The Catena

The models of Simonson (1959) and Jenny (1941) provide an important conceptual framework for understanding soil formation. Neither model, however, establishes functional boundaries for segregating the soil continuum into natural landscape units. The *catena* is a fundamental concept that

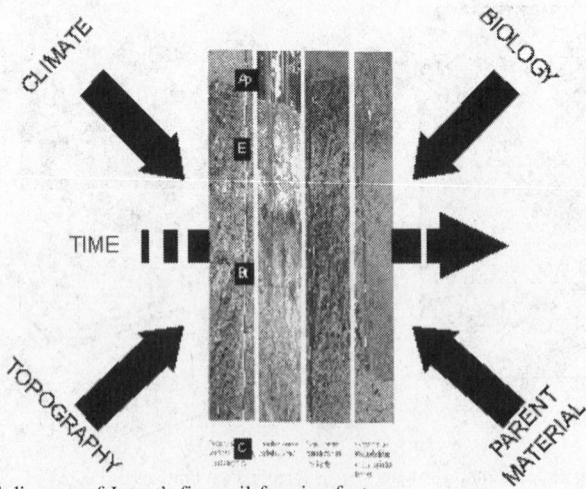

Fig. 1.3   Conceptual diagram of Jenny's five soil-forming factors

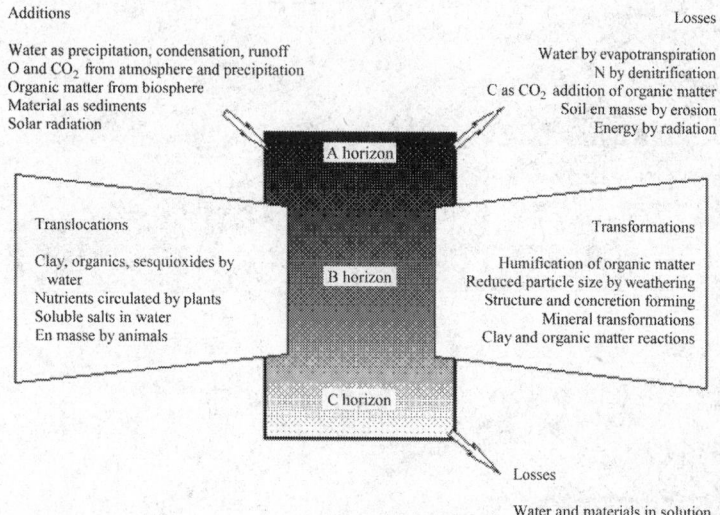

Additions

Water as precipitation, condensation, runoff
O and CO_2 from atmosphere and precipitation
Organic matter from biosphere
Material as sediments
Solar radiation

Losses

Water by evapotranspiration
N by denitrification
C as CO_2 addition of organic matter
Soil en masse by erosion
Energy by radiation

Translocations

Clay, organics, sesquioxides by
    water
Nutrients circulated by plants
Soluble salts in water
En masse by animals

Transformations

Humification of organic matter
Reduced particle size by weathering
Structure and concretion forming
Mineral transformations
Clay and organic matter reactions

A horizon

B horizon

C horizon

Losses

Water and materials in solution
or suspension

Fig. 1.4 Simonson's (1959) process model of soil genesis showing the interactions of additions, removals, transfers, and transformations [Modified from Foth, 1984]

explains the pattern of soils on hillslopes. Milne (1936a,b) coined the term to describe a repeating sequence of soils that occurs from the top of a hillslope to the adjacent valley bottom and distinguished two types of catenas. The first type occurred on hillslopes developed over a single kind of parent rock. Despite the uniformity in parent rock, he observed a sequential change in soils along the slope gradient and attributed the sequence of soils to variations in subsurface drainage, the lateral transport of sediments, and the translocation of materials at or beneath the soil surface (Fig. 1.5a).

In Milne's (1936a,b) second example, the hillslope contained more than one type of parent rock (Fig. 1.5b). An observable sequence of soils also occurred on this hillslope. Variations in drainage and lateral transport also contributed to this catena, but stratigraphic differences in the parent rock increased the complexity of the soil pattern. In this example, the surficial sediments form a drape on the landscape that is not coincident with the underlying rock strata. The catena concept includes both surficial stratigraphy and internal hillslope structure or lithology.

Furthermore, the catena concept is both a soil landscape model and a geomorphic model. Milne (1936a) recognized that lateral sorting contributed to the sequential variation of soils down the hillslope. Erosion and sedimentation processes driven by relief and water movement, redistribute sediments across hillslopes creating subtle lateral differences in soil parent materials (Kleiss, 1970). Recall that both parent material and topography are factors in Jenny's (1941) model of soil formation. The same erosional and depositional processes that drive landscape evolution influence the soil pattern on landscapes. The sequential change in soil morphology across a landscape is linked by process to landscape evolution on hillslopes both in time and space. In Jenny's model, landscape evolution means that parent material and topography are not independent variables, but are dependent on and vary in time.

A well-studied catena consists of the Clarion-Nicollet-Webster soils that occupy about 31,000 km² of the Des Moines Lobe in south central Minnesota and north central Iowa (Fig. 1.6). The Des Moines Lobe represents the last Late Wisconsinan glacial advance into Iowa about 14,000 yr BP. Major topographic areas include hummocky, high relief end moraines separated by undulating areas of low to moderate relief ground moraine (Ruhe and Scholtes, 1959; Ruhe, 1969; Kemmis et al., 1981).

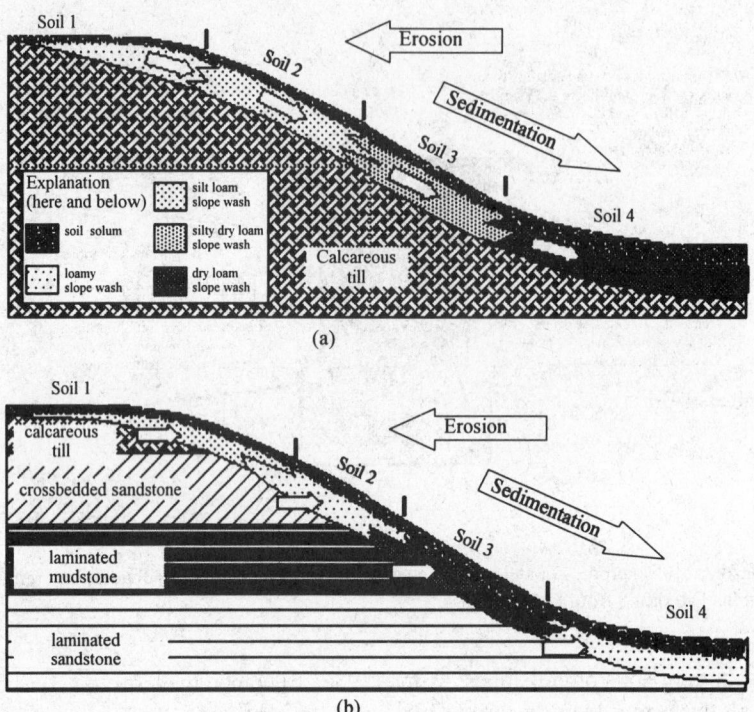

Fig. 1.5 Two-dimensional diagram based on Milne's (1936) catena showing idealized landscape relations for two different situations (a) and (b)

Closed, semi-closed, and linked (open) depressions occur throughout the Des Moines lobe with closed depressions most abundant in the end moraines (Kemmis, 1991). The Clarion-Nicollet-Webster soils are Mollisols formed in a stratigraphic sequence composed of hillslope sediments, supraglacial sediments, and loam textured till. The catenary relationships on the Des Moines Lobe landscape result from surficial sorting during deglaciation (Kemmis, 1991; Steinwand and Fenton, 1995), post-glacial hillslope sorting (Walker, 1966; Burras and Scholtes, 1982; Steinwand and Fenton, 1995) and subsurface flow relationships (Steinwand and Fenton, 1995).

Numerous studies (Dan and Yaalon, 1964; Blume and Schlichting, 1965; Blume, 1968; Walker and Ruhe, 1968; Huggett, 1975; Pennock and Vreeken, 1986; Pennock and Acton, 1989) have confirmed that catena relationships occur in various climates and landscapes. Conacher and Dalrymple (1977) and Dalrymple et al. (1968) provided a quantitative description of the catena. They defined the soil hillslope relationship as a three-dimensional unit having arbitrary lateral dimensions extending from the hilltop to the valley bottom and from the soil surface to the base of the solum. They segmented the hillslope into nine land-surface units (Fig. 1.7) based on soil morphology, mobilization and transport of soil constituents, redeposition of soil constituents by overland flow and throughflow, or by gravity as mass movements.

1.5.2   The Toposequence

Bushnell (1942) studied morphological differences in soils differing mainly in color across a hillslope gradient. He attributed the changes in morphology to elevational position and local hydrology. This concept is commonly referred to as a toposequence. Unlike Milne (1936a,b), he did not recognize the influence of hillslope erosion and sedimentation. The terms catena and toposequence are often used as synonyms, but the original meanings are not identical.

1.5.3 The Valley Basin

Huggett (1975) expanded on the catena concept, proposing that the basic three-dimensional unit of the soil landscape is the first-order valley basin. The functional boundaries of this soil landscape are defined as the atmosphere-soil interface, the weathering front at the base of the soil, and the drainage divides of the basin. The topographic boundaries of a drainage basin define the physical limits and direction of overland flow, and thus, control geomorphic processes such as erosion, transport, and deposition. In humid climates, groundwater divides are generally coincident with topographic divides that are boundaries for overland flow. Thus, the first-order basin forms the natural boundary

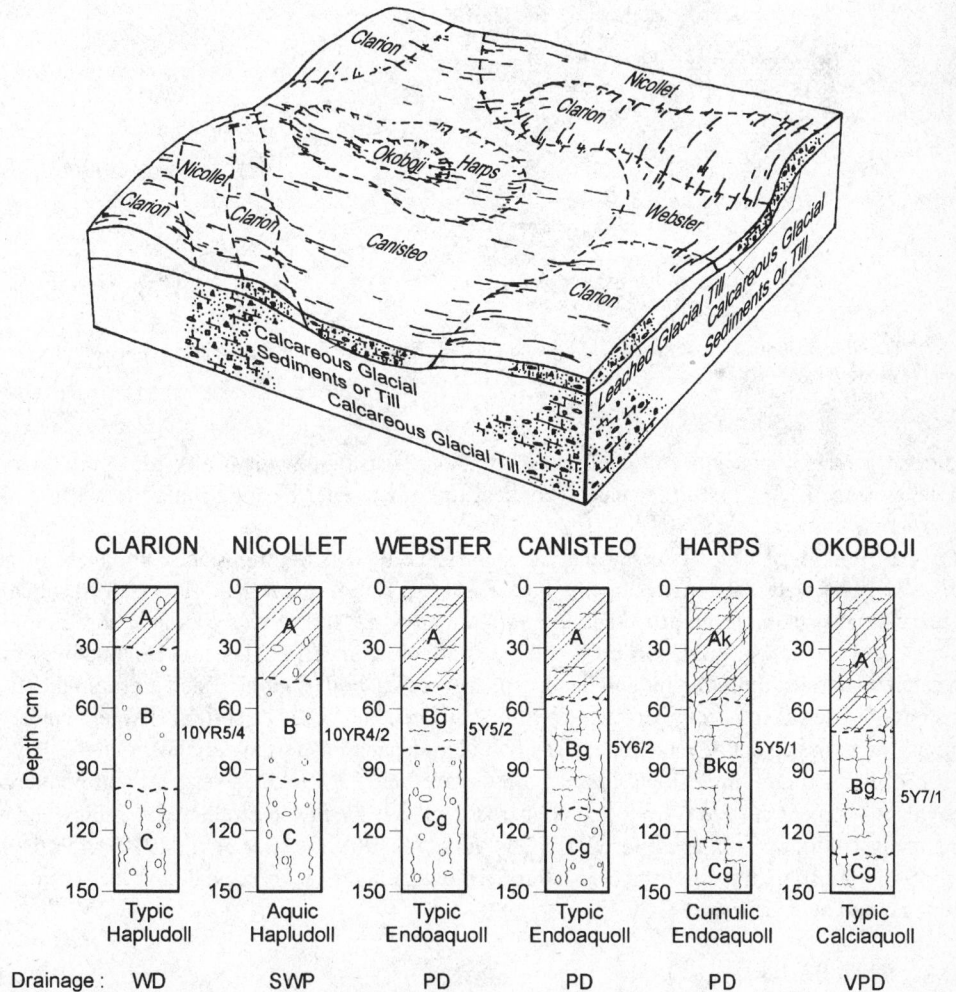

Fig. 1.6 Three-dimensional block diagram showing the soil and landscape relations of the Clarion-Nicollet-Webster catena and stylized soil profiles [Andrews and Dideriksen, 1981]

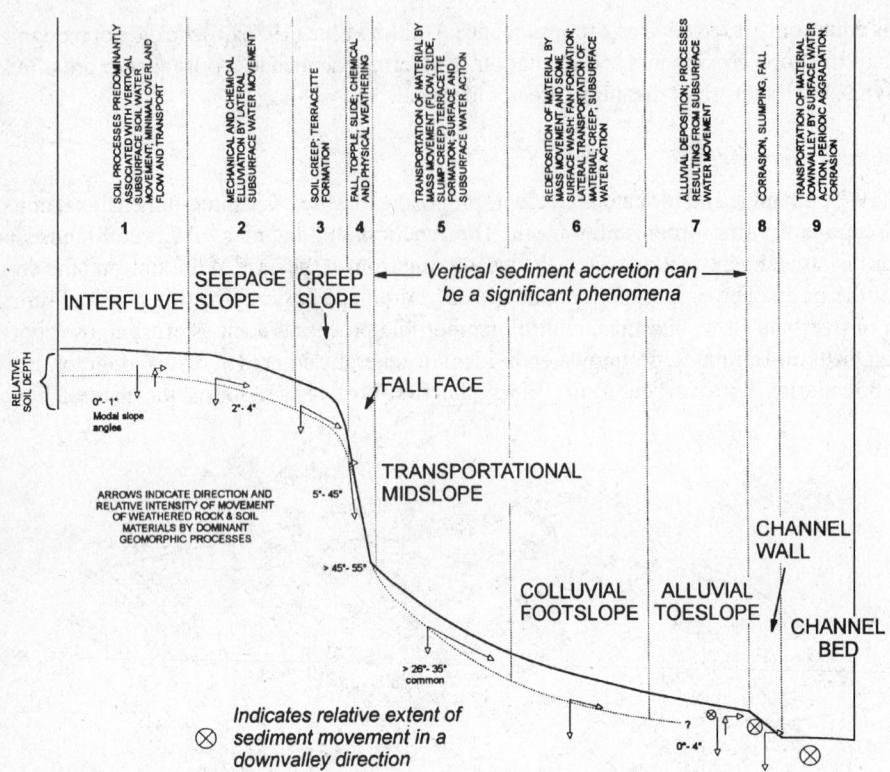

Fig. 1.7 Two-dimensional, nine-unit soil and hillslope model showing soil hillslope relationships [Modified from Dalrymple et al., 1968]

conditions for chemical and colloidal transport and redistribution via water transport on most landscapes (Fig. 1.8a). First order basins connect to higher order basins forming a larger, linked system across landscapes.

The partitioning of water into surface and subsurface flow is an important component of both Huggett's (1976) watershed model and the catena. Surface runoff or overland flow is the mechanism that drives many geomorphic processes. Water that enters the soil, drives chemical and biological reactions, and via throughflow, transports soluble or colloidal materials. The movement of water and associated materials through the soil landscape occurs as both saturated and unsaturated flow. Transport of material can occur between or within soil horizons, between soils at different landscape positions, or across an entire watershed. Deep percolation can transport material beyond the depth of a soil profile. A three-dimensional approach must account for lateral, divergent, and convergent throughflow. Recent studies (Arndt and Richardson, 1989; Knuteson et al., 1989; Steinwand and Richardson, 1989), as well as earlier work (Glazovskaya, 1968; Cleaves et al., 1970; Crabtree and Burt, 1983) attest to the importance of throughflow and the geochemical link between soils, hydrology, and landscapes.

1.5.3.1 Open Basins
Drainage basins can be categorized as either open or closed (Ruhe, 1969). An open drainage basin is confined by the head of the watershed and the perimeter divide. The mouth of the basin is open (Fig.

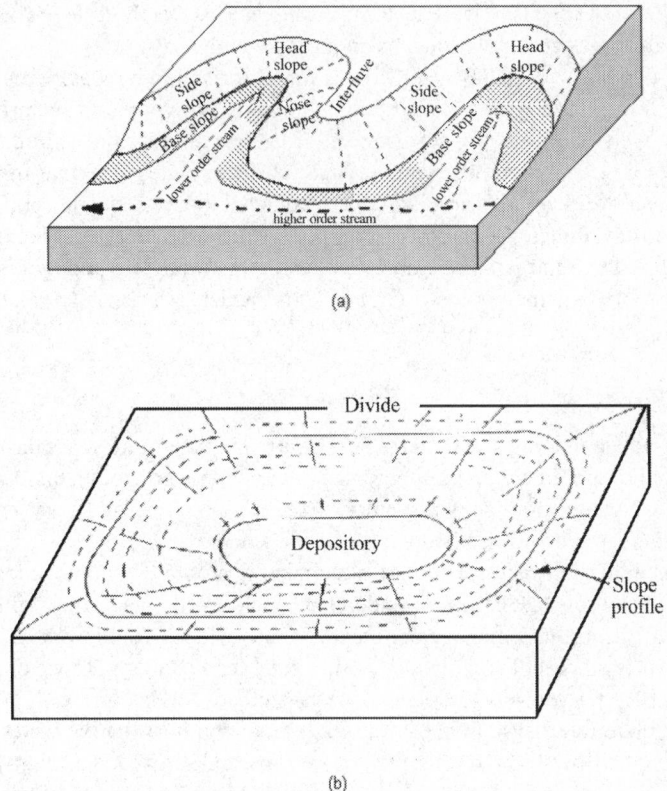

Fig. 1.8 Three-dimensional diagram of (a) open [Schoeneberger and Wysocki, 1997] and (b) closed drainage basins as soil landscape models [Modified from Walker and Ruhe, 1968]

1.8a). Surface water collected within the basin during precipitation events discharges through the outlet at the mouth. Eroded sediment can be transported and redeposited either within the basin or removed entirely from it. Both mass and energy can enter and leave an open basin by surface flow or runoff. The sediment retained within an open basin represents only a portion of the total sediment produced by erosion. The sediment package in an open basin is an incomplete record of the geologic history. A study of the sediment stratigraphy may yield only a partial understanding of the geomorphic evolution of the basin.

1.5.3.2 Closed Basins
A closed drainage basin lacks a surface outlet. The drainage pattern within the basin flows to a central area or point (Fig. 1.8b); no water leaves the basin as surface runoff. Water from precipitation is lost by evaporation, transpiration, or infiltrates and becomes subsurface water. Mass and energy transfers by surface flow are contained entirely within the basin. Surface flow does not remove mass from the basin in the form of clastic sediment. Mass is primarily lost as soluble or fine colloidal material in subsurface water. Mass can be added or removed from the basin by eolian processes. Alluvial sediment contained in a closed basin forms a complete record of its erosional history.

The retainment of surface runoff that distinguishes closed basins from open basins is independent of scale. Open basins span the range from tiny, erosional rills on a hillslope to continental scale drainage basins like those of the Amazon or Mississippi rivers. Likewise, closed basins vary in size from small depressions like tree tip pits to large tectonic basins such as Death Valley. The magnitude and complexity of the transfer processes within a basin increase with size.

The lack of a surface outlet in closed basins precludes initial formation by water erosion. Closed basins originate through several mechanisms including deposition or removal of material by ice or wind, dissolution of relatively soluble rock, subsidence due to structural failure of underlying rock often related to dissolution or groundwater fluctuations, and subsidence caused by faulting or tectonic downwarping. Closed basins are common in landscapes produced by these geomorphic processes. Examples include the prairie potholes and chain lakes of the glaciated midcontinent, depression lakes in dune topography such as the Sand Hills of Nebraska, karst topography created by dissolution of limestone, and the playa lakes of the southern Great Plains, which are partially related to wind deflation and deposition.

1.6 Soil Hydrology

A major advancement in understanding soil systems has occurred in the last 20 years through insight into how water moves through landscapes: where the water goes, so goes soil development. Historically, the study of water flow in soils emphasized either overland flow, erosion, and sedimentation, or water movement within a pedon.

Erosion and sedimentation control were foci of national soil programs and an integral part of agronomic applications of soil science. Efforts to understand and quantify water flow within a pedon or within fields (artificially delimited management areas versus naturally soil bodies) such as for irrigation, have been an integral part of soil physics studies for over a century. However, subsurface water movement has not been pervasively integrated into the study of soil landscapes. To understand natural soil systems and thereby ecosystem behavior, conceptual and quantitative water movement models must encompass water flow through soils, the vadose zone, and across and through landscapes.

Milne's (1936a,b) concept of the catena recognized the existence of soils with different drainage classes across landscapes. His early recognition of lateral transport as a contributing factor to landscape evolution has been deemphasized over time. The main focus shifted to relatively static soil water conditions and vertical water flow (deep percolation) at a given position on a landscape. Lateral transport was primarily recognized as overland flow with concomitant sediment sorting across the surface (Walker and Ruhe, 1968; Ruhe, 1975). The equally important processes of subsurface water flow and transport through soil landscapes were substantially ignored or forgotten. Soil hydrology studies in the last 20 years have expanded the catena concept to include water flow through landscapes (Arndt and Richardson, 1989; Steinwand and Fenton, 1995) with a simultaneous recovery from the historical emphasis on soil erosion caused by overland flow. The study of soil systems now includes the dynamic flow of water through as well as across soil and landscapes (Fig. 1.9). This approach of dynamic water flow considers the entire vadose zone, not just the ground surface or the soil solum.

A variety of soil hydrologic terms has been developed or adopted which partition water flow in soil systems. The terms can be placed in an idealized schematic to demonstrate relationships (Fig. 1.10). These terms portray the potential fate of precipitation onto, into, and through soils. The emphasis is on water at or above the permanent groundwater table (i.e., the vadose zone). Water dynamics and aquifer conditions below the water table generally are the purview of groundwater hydrology. Water flow can be predominantly downward, lateral, or upward toward the ground surface, depending upon prevailing energy dynamics. This can be demonstrated by looking at near-surface water movement patterns in different climatic settings.

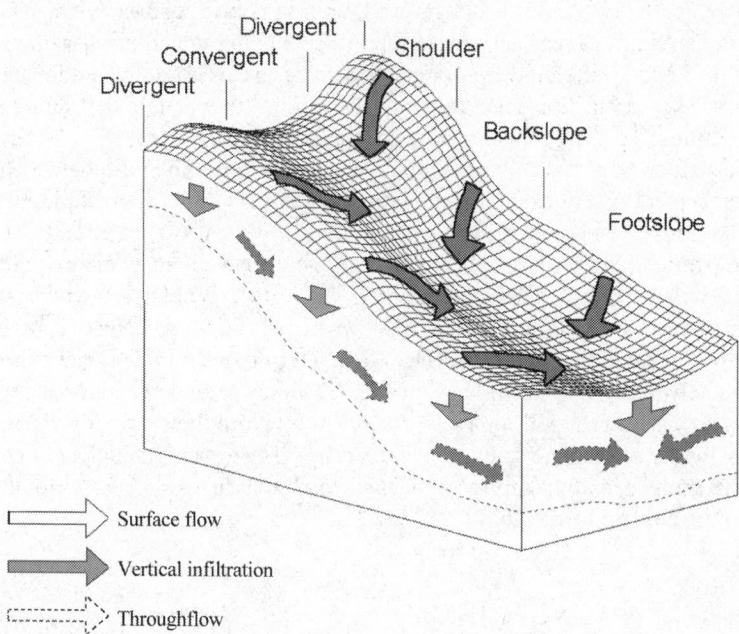

Fig. 1.9   Three-dimensional hillslope with flow directions [Modified from Pennock et al., 1987]

1.6.1   Climate Influence

Regional climate conditions control precipitation levels and patterns and evapotranspiration dynamics (either directly via solar radiation or indirectly through vegetation populations). All else being equal, regional climate determines the dominant water flow direction through the vadose zone. Within a given temperature range, precipitation levels can vary with predictable, generalized results. In humid environments, watertables and flow paths tend to loosely mimic the local topography with water generally infiltrating (i.e., recharge) in topographically high areas, flowing to and discharging from

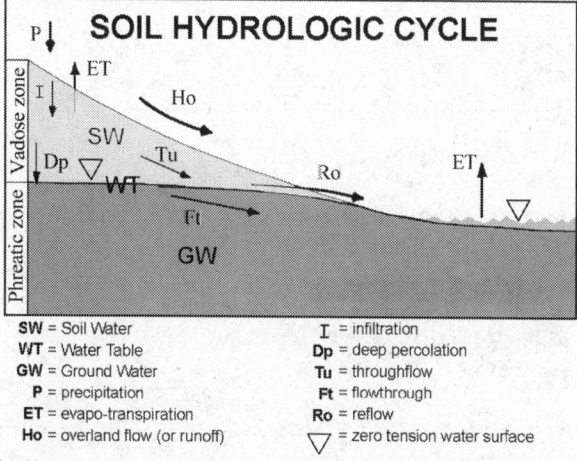

Fig. 1.10   Soil hydrologic cycle with terminology

lower lying areas (Fig. 1.11a). In arid to semi-arid environments, conditions are reversed, with recharge sites dominant in low areas and subsurface water moving to and discharging from higher elevation areas (Fig. 1.11c). Subhumid environments behave as tension or transition zones between these two extremes. These transition zones temporarily follow either pattern depending upon short-term climatic conditions (e.g., seasonal climatic variations or annual cycles) (Fig. 1.11b).

Temperature also exerts a controlling climatic influence on soil hydrologic behavior both permanently and ephemerally. For example, permafrost effectively limits or precludes vertical water flow. If local conditions warm (human induced or natural), permafrost melts and ceases to be a restriction to water flow. A seasonally ephemeral example occurs in areas that experience annual ground frost. The internal water flow behavior in soil is vastly different between winter, spring thaw, and summer (Emerson et al., 1990). Ground frost when present is an effective barrier to both subsurface water flow and the infiltration of surface water. During several weeks in the spring, thawed soil layers immediately above a remnant frost layer can become saturated and highly erosive (e.g., Willamette Valley, OR). After the soil completely thaws, subsurface flow is greatly enhanced and the surface layer is no longer as prone to saturation and erosion. The transport and fate of contaminants can be radically different depending upon seasonal soil conditions. This is the basis for discouraging land application of manure on frozen soil.

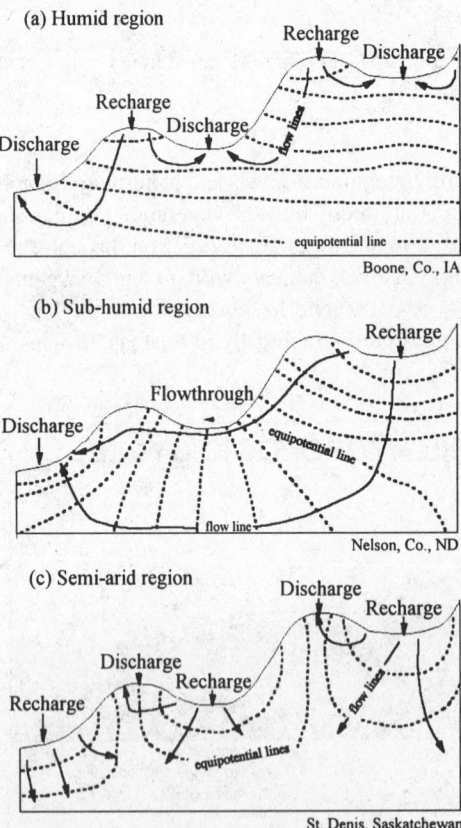

Fig. 1.11 Two-dimensional landscape diagrams with recharge-discharge, flowlines, and equipotential lines in different climatic settings [Modified from Richardson et al., 1991]

1.6.2   Geostratigraphic Influence

Water restrictive (aquitard) or conductive (aquifer) sediments or rock strata can fundamentally alter subsurface water flow patterns. Tilted rock strata can preferentially redirect water such that infiltration in one location is shunted (laterally displaced) to an unlikely recharge or discharge site. Where bedrock strata approach or breach the ground surface, the vertical movement of subsurface water can be restricted, forcing water to flow back to the surface (i.e., reflow) (Fig. 1.10) to form a spring, seep, or moist area. The composition of subsurface stratigraphic sequences and the extent to which they are connected to conductive materials also influences subsurface water flow. For example, a porous stratum that is normally conductive will be nonconductive, if no outlets to other conductive material are available (e.g., a sand layer in a flood plain deposit confined by clay sediments) (Fig. 1.12). Similarly, a topographic position that precludes or minimizes water inputs may cause a stratum that is potentially conductive to be functionally nonconductive.

1.6.3   Pedostratigraphic Influence

Some features unique to soil or derived from soil processes emulate geostratigraphic influences on water flow. Some pedogenically derived layers (e.g., argillic layer, duripan, etc.) can restrict vertical water flow and enhance lateral flow (Fig. 1.13). Water repellent materials or layers (e.g., fire induced hydrophobic layer in chaparral) can reduce infiltration of water into the soil or function as an aquitard and restrict vertical movement. Conversely, pedogenic features or related phenomena, such as pedogenic structure and biotic activity, can greatly enhance water flow in upper soil horizons (Schoeneberger and Wysocki, 1996). Substantial differences in infiltration or internal water movement in a soil can result directly from different management practices (Franks et al., 1993, 1995).

1.6.4   Vegetative Pumping

Biotic activities, particularly plant respiration, can directly impact water flow patterns in soils. Plant communities can consume substantial volumes of water during transpiration. Water consumption may be sufficient to change local flow dynamics. Consider the case of phreatophytes fringing a marsh or

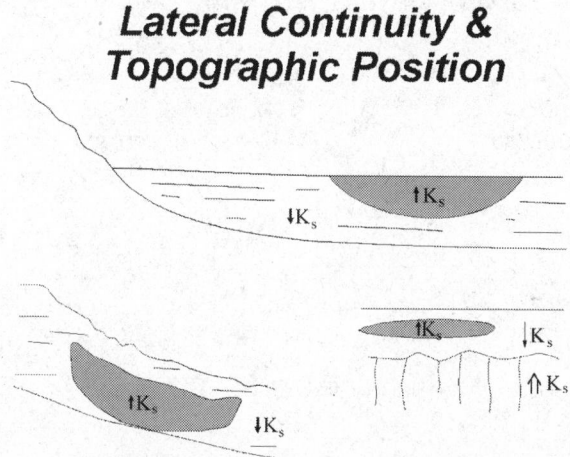

Lateral Continuity & Topographic Position

Fig. 1.12   Stratigraphic isolation of otherwise conductive strata; K_s indicates relative magnitude of saturated hydraulic conductivity.

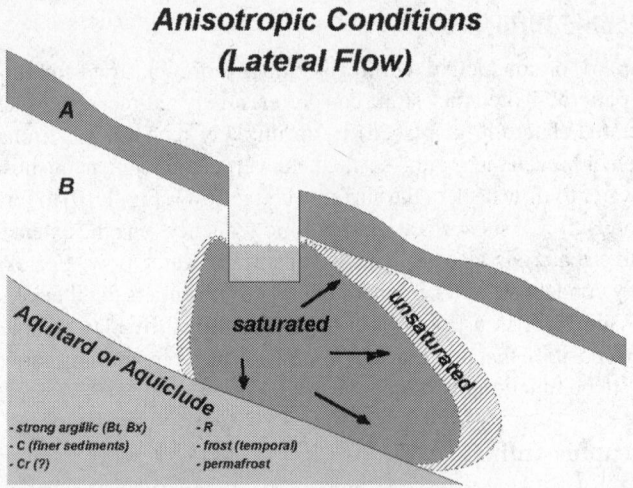

Fig. 1.13   Pedogenic or geogenic aquitard restricting water flow

riparian system (Fig. 1.14). Minor differences in elevation (e.g., 10 cm) can promote plant establishment resulting in a disproportionately large reduction in the watertable, and subsequent differences in soils (e.g., the soil drainage class).

1.6.5   The Soil Sponge

An important but often overlooked function of soil is its role as a hydrologic sponge. The porous character of soil typically allows some to much of the local precipitation to infiltrate. Water that does not infiltrate, evaporate, or pond, runs off into local drainage networks. Infiltrated water is gradually transmitted vertically to recharge groundwater or laterally to be released as discharge to surface waters via throughflow. Obviously, the rate of water flow through the porous medium of soil is less than that

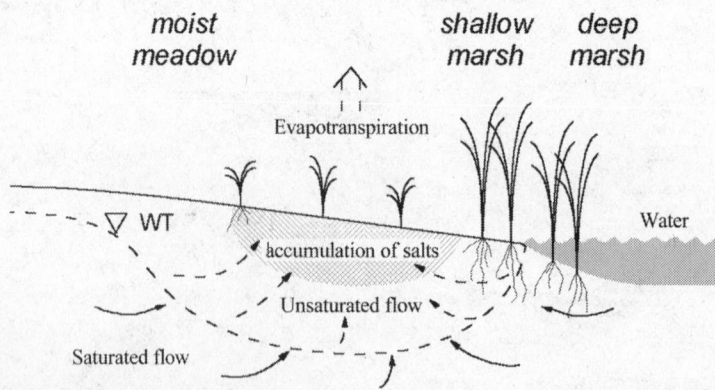

Fig. 1.14   Diagram of vegetative pumping

above ground. Infiltrated water is thereby slower to discharge to surface waters. The hydrologic result is a delayed and dampened streamflow hydrograph. The importance of streamflow response can be seen from a practical, human perspective. The streamflow response for areas from which substantial amounts of soil are removed or sealed over (e.g., construction sites in urban areas) typically results in much higher and quicker peak flow compared to the original conditions. This pattern can have a profound financial impact if structures are underdesigned (culverts, bridges, etc.) or if the new hydrologic dynamics initiate a new cycle of stream incisement or substantial change in sedimentation patterns (Hammer, 1995; Hammer et al., 1998; Lickerman et al., 1998).

1.6.6   Soil Morphology, Landscapes, and Water

Water is the dominant catalyst, mediator, and transport agent in most natural systems on earth. This is true for soils. Although flora, fauna, and geologic substrate all affect soil and landscape development, water is the controlling force providing both chemical and potential energy. The presence and flow of water in a landscape can be observed directly by excavation or piezometers. This is a tedious and time-consuming task given the seasonal or transient presence of water. The presence of saturated conditions and the movement of water in landscapes can be predicted from soil morphology (e.g., redoximorphic features). Soil morphology forms in response to long-term prevailing water state conditions. Soil patterns on a landscape result from differing water conditions and the prevailing water flow dynamics. Soil patterns on landscapes can explain where the water is and how it flows. Soil morphology and soil landscape models can explain and/or predict soil water dynamics and subsequent soil geography.

1.7   Geomorphic Description of Landscapes

1.7.1   Hillslopes

Soil patterns or sequences on hillslopes will generally follow a catenary relationship. The soil and landscape relationship inherent in the catena can be used to predict and describe soil occurrence. A fairly simple set of geometric or morphometric descriptors can be used to define hillslopes, and therefore, associated soil patterns. These descriptors include slope gradient, aspect, shape, complexity and position, and geomorphic position.

1.7.1.1   Slope Gradient

Slope gradient, which is measured along the vertical profile of a slope, is the angle of inclination of the ground surface from the horizontal plane. Commonly, slope gradient is expressed in degrees from the horizontal plane (e.g., 45°) or as a percent, which for a 45° slope is one unit of drop or rise per one unit of distance (slope gradient equals 100%). Slope gradient is a proxy expression of potential energy that drives mass movements and the erosive force of surface runoff on a slope.

Soil map units in a soil survey include a typical range for the slope gradient (e.g., Chemawa Loam, 8 to 15% slopes) (van Wambeke and Forbes, 1986). Inflection points or changes in slope gradient that repeat on a landscape are readily discernible, and generally, correspond to differences in the internal structure (underlying lithology), past erosion events (stream incision), or contacts between landforms or sediment bodies (Fig. 1.15). Inflection points on slopes, therefore, often mark natural boundaries between soils. Soil mappers use slope inflections as visible clues to changes in soils on landscapes.

1.7.1.2   Slope Aspect

Aspect is the direction that a slope faces. Slope aspect is usually expressed as a compass azimuth (e.g., 215°) or as a cardinal direction (e.g., SW). Soil microclimates are strongly influenced by the amount

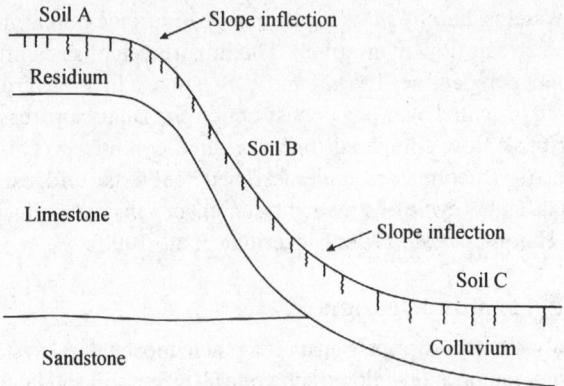

Fig. 1.15 Diagram of natural landscape boundaries

of direct solar radiation, which is a function of slope aspect. In the northern hemisphere, north and east facing slopes are cooler and moister than south and west facing slopes. Distinct differences in soils occur as a result of aspect (Lotspeich and Smith, 1953; Finney et al., 1962; Franzmeier et al., 1969). The effects of slope aspect are more pronounced in mountainous or high relief terrain than in low relief areas. The influence of slope aspect is also more pronounced in temperate than in equatorial latitudes (Buol et al., 1989).

1.7.1.3 Slope Shape

Slope shape is the three-dimensional geometry of a slope. The geometric form is obtained by combining the shapes of both the vertical profile (up and down slope) and the elevation contours (across slope). A two-dimensional shape is either linear or curved. If curved, the shape can be convex or concave. This yields nine possible geometric forms to describe all slopes (Fig. 1.16).

Slope shape is a property of hillslopes that strongly influences the lateral movement of water across the surface as overland flow and internally as throughflow. For example, a slope that is linear in both vertical profile and contour shape creates parallel, lateral flow (Fig. 1.16a) whereas a slope that is convex in both profile and contour causes divergent flow (Fig. 1.16e), and a slope concave in both

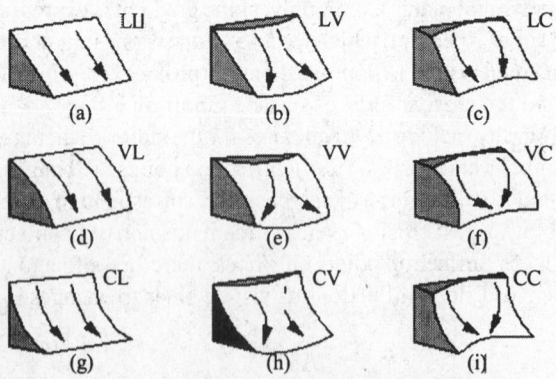

Fig. 1.16 Three-dimensional diagram of the nine hillslopes shapes with flowlines shown; L = linear, V = convex, C = concave [Schoeneberger and Wysocki, 1997]

profile and contour causes convergent flow (Fig. 1.16i). Slope shape redistributes moisture received by precipitation, creating distinct microenvironments on the landscape. Areas of convergent flow are moister than the general soil microclimate whereas areas of divergent flow are drier than the general soil microclimate. The influences of slope shape on water movement and soil moisture on landscapes, in turn, control soil formation and vegetation.

In addition to its role in the redistribution of runoff, slope shape is a visual clue to the internal structure of a hillslope. Hillslopes that have a steeply convex, vertical profile are inherently resistant to erosion. Convex slopes occur predominately in landscapes where erosion is controlled by resistant bedrock. Soils on these slopes tend to be shallow and usually display limited horizon development. In contrast, linear or concave slopes generally occur on unconsolidated material or weakly resistant rocks. Soils on these slopes tend to be thicker and display greater horizon development.

Concave slopes or concave portions of slopes denote a decrease in slope gradient. Both potential energy and the velocity of overland flow decrease at the inflection of a concave slope. An accumulation of colluvium or sediment derived from slope wash often occurs on concave slopes. Soils on these slopes often form by cumulic processes and commonly have thicker and less poorly developed horizons compared to soils on adjacent uplands.

1.7.1.4 Slope Complexity

Slope complexity is a simplistic description of the ground surface in respect to the downslope path of overland flow (Fig. 1.17). Simple slopes are relatively smooth, with few obstructions to surface flow or decreases in slope gradient, and thus, nominal opportunities for sediment deposition except at the base. Complex slopes contain substantial irregularities in slope conditions. Surface water flow is apt to be interrupted and largely nonparallel with considerable changes in flow velocity and subsequent erosion, sediment transport capacity and deposition.

1.7.1.5 Slope Position

Slopes can be divided into segments or elements along a two-dimensional, cross-sectional profile based on slope shape, the degree of erosion or deposition, the presence or absence of sediment, and the nature of the sediment. Wood (1942) identified four segments across a fully developed slope for a bedrock-controlled landscape. These elements are the waxing slope, free face, debris slope, and pediment. King (1957) elaborated on the geomorphic processes active on these four slope elements. The waxing slope is the convex crest of a hillslope dominated by chemical weathering rather than erosional removal. The free face is an outcrop of bare bedrock on the upper reach of a hill. Erosion is most active on this element. The debris slope occurs below the free face and is composed of material eroded and transported from the free face. Below the debris slope is the pediment, which is an inclined ramp extending from the hillslope base to an alluvial basin.

Ruhe (1960) modified Wood's (1942) two-dimensional hillslope elements and applied them to the study of soil landscapes. His hillslope profile elements, which include the summit, shoulder, backslope, footslope, and toeslope, can be distinguished by inflections in slope gradient and line segment shape (Fig. 1.17). The summit is the relatively level, uppermost portion of a hillslope profile. It is the most geomorphically stable and least erosive part of a hillslope. The main vector of water flow is downward through the soil and erosional transport is minimal. Soils in this position display the greatest degree of profile development. The shoulder is the convex portion of the hillslope below the summit. The break between summit and shoulder is identified by an increase in slope gradient. The shoulder is subject to a greater degree of erosion, and greater lateral flow of water compared to the summit. Soils tend to be similar to, but thinner than those on the summit, and may appear to be vertically compressed or truncated. The shoulder descends to the steepest and more linear portion of

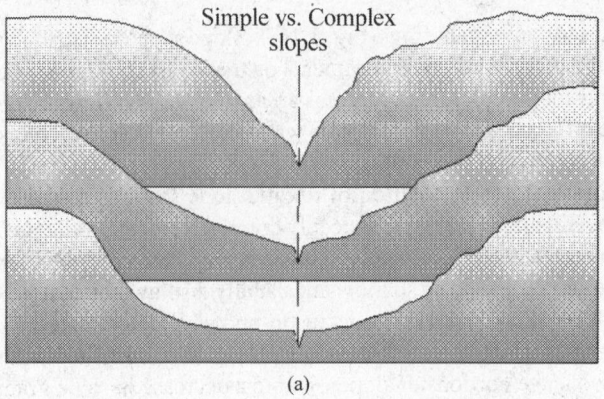

(a)

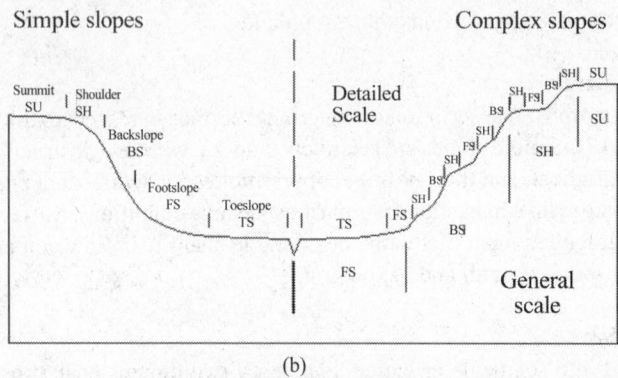

(b)

Fig. 1.17 Two-dimensional diagram of simple vs. complex slopes

the slope, the backslope, where surface runoff and erosional transport are greatest. The vector of water flow may be more lateral than vertical depending on the slope gradient. Soils may reflect inputs of less weathered parent material than soils on the summit and shoulder. On long slopes, some degree of lateral sorting may be evident. The backslope descends to the concave portion of a hillslope, the footslope. The decrease in slope gradient reduces the carrying capacity of flowing water and increases sediment accumulation. Water flow vectors may be primarily lateral, but the position is characterized by the concentration of water from upslope. The footslope merges downslope with the toeslope, which is predominantly linear or slightly concave. The comparatively low slope gradient and low lying position at the toeslope allows alluvial processes to dominate, given sufficient surface water from upslope or adjacent streams. Soils tend to be deep, comparatively moist, and composed of or strongly influenced by alluvial sediments. Toeslope sediments from lower order streams reflect short distance transport and a lower degree of fluvial modification (sorting, rounding, stratification, etc.) than toeslope sediments derived from higher order streams.

1.8   Geomorphic Components

A major focus of soil geomorphology is the movement of water through soil bodies and the way that a landscape sheds or concentrates water. This focus is the result of the impact of surficial fluvial processes on major agricultural areas and population centers. Consequently, terms have evolved to describe portions of the Earth's surface that share a common location, form, and geomorphic process and the practical emphasis on surficial water movement.

In addition to the two-dimensional hillslope elements (Figs. 1.17 and 1.18), Ruhe (1960, 1968) refined and popularized geomorphic descriptors for three-dimensional pieces of landforms (Ruhe, 1969, 1975). These area descriptors, called geomorphic components, are based in part, upon the convergent, linear (or parallel), or divergent nature of overland water flow (Fig. 1.16) and associated sediment transport. The most widely used suite of geomorphic components was developed for, and is most appropriately applied to hills. Historically, this set of terms has been applied to most landscapes, including mountains, due more to the lack of alternatives than to their utility. Other geomorphic settings (mountains, terraces, flat plains) have unique dynamics or complexity and warrant different area descriptors.

1.8.1 Geomorphic Components

1.8.1.1 Hills
The interfluve is the uppermost area of a hill, and represents the oldest, most stable part of the landscape, typically, with the most developed soils. A noted exception is where opposing hillslopes have narrowed an interfluve or merged (e.g., a saddle or crest) to the extent that erosion begins to lower the crest. Hillsides are areas characterized by active backwearing (erosion), in general, and at the heads of streams, in particular.

Hillsides can be divided into discrete parts based on the dominant behavior of overland water flow (Fig. 1.8a): converging (head slope), linear or parallel (side slope), or diverging (nose slope). Soils on hillsides are dominated by colluvial sediments in gradual transport down the slope and modified to varying degrees by slope wash (nonchannel, overland flow). Unless masked by changes in parent materials, some degree of lateral sorting downslope is usually present. Soil profiles on hillsides can range from thin to thick depending on the rate of erosion (high or low, respectively), the magnitude of the slope gradient, and the extent to which the bedrock is resistant to weathering and erosion. In many landscapes, the traditional assumption of a prevalence of shallow soils over hard rock on hillsides and mountainsides is erroneous (Graham, 1986). The base slope is commonly an apron of colluvium and/ or slope alluvium at the bottom of hillslopes (Schoeneberger and Wysocki, 1996). This drape of transported material can range from coarser, unsorted gravity deposits to finer sediments that have been winnowed or sorted by slope-wash processes. The base slope does not typically include sorted

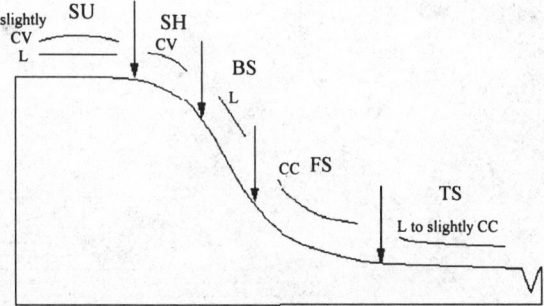

Fig. 1.18 Hillslope profile elements and associated slope shapes; SU = summit, BS = backslope, FS = footslope, TS = toeslope, L = linear, CV = convex, CC = concave

and stratified alluvium associated with stream (fluvial) processes. Distal base slope sediments commonly grade into or interfinger with alluvial fills.

1.8.1.2 Mountains

Mountains represent a unique geomorphic setting due to their scale and slope complexity. Mountainsides commonly have complex backslopes sometimes kilometers long, steep slope gradients, highly diverse sediment mantles and complex, near-surface hydrology. Mass movement processes and features are more prevalent than in hills. Consequently, new area descriptors have been recently proposed (Schoeneberger and Wysocki, 1996) to identify geomorphic components for mountains.

The mountaintop (Fig. 1.19) is the summit or crest of a mountain and is commonly characterized by comparatively short, simple slopes composed of bare rock, residuum, or short transport (angular) colluvial sediments. In humid climates, soils on mountaintops can be quite thick (Oliver et al., 1997). The side of a mountain, the mountainflank, is characterized by comparatively long, complex slopes dominated by mantles of long transport (subangular) colluvium. Residuum, if present, is usually buried by 1 to 2 m of colluvium. Incised drainageways, structural benches, and mass movement features can be common. Rock outcrops, while visually prominent, are not a major portion of the land surface. The mountainbase is an apron of colluvium at the bottom of a mountain slope and analogous to the baseslope in hills. It is marked by a substantial decrease in slope gradient compared to the mountain flank. The mountain base is characterized by a thick mantle or wedge of colluvium, and commonly, contains a comparatively high percentage of coarse rock fragments. The colluvium can extend out onto more level land surfaces, and ultimately interfingers with or is buried by alluvium, or thins and joins reemergent residuum (as on a pediment).

1.8.1.3 Terraces

Terraces are landforms that form a relatively level or gently inclined surface, a constructional strip, or plain that borders a stream, lake, or sea. Terraces are a unique geomorphic setting and have dynamics quite different from those of hills or mountains. Stream terraces and flood-plain steps (Fig. 1.20) originally developed by alluvial processes and sediments, rather than the dynamic slope processes and colluvial sediments that characterize hills and mountains or the localized, low energy processes typical of flat plains.

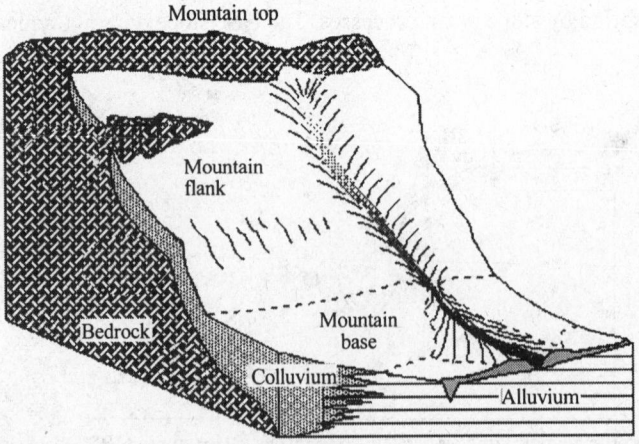

Fig. 1.19 Geomorphic components for mountain landscapes

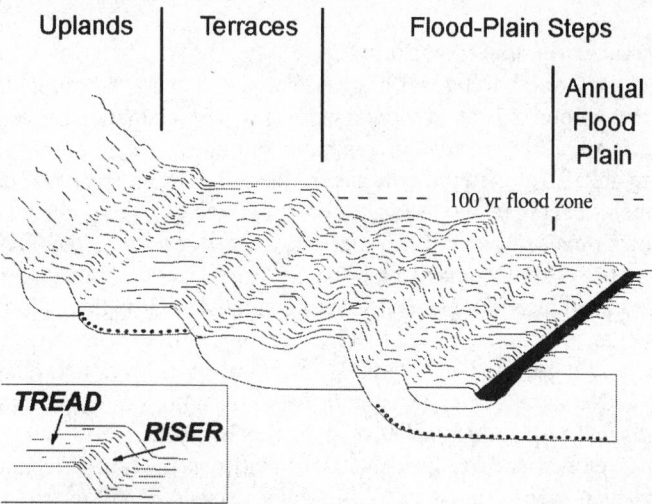

Fig. 1.20   Geomorphic components for terraces and flood plains

The tread is the comparatively broad, generally level part of a terrace or flood-plain step. Treads can extend laterally for many km (Saucier, 1994). Treads are level or gently inclined (low slope gradients) and underlain by alluvial, lacustrine, or marine sediments. Strath terraces are similar in form to other stream terraces, but are erosional landforms characterized by thin alluvial sediments over an eroded bedrock bench or platform.

The riser is an escarpment that separates terrace or flood-plain levels. The riser commonly consists of a short, steep, planar slope cut into the sediments that underlie an adjacent higher tread. The areal extent of a riser rarely exceeds tens of m, and therefore, is typically depicted on a soil map with a spot symbol rather than a delineation. Geomorphically, risers represent an abrupt change to a lower hydrologic base level, which suppresses the watertable along the edge of the adjacent, higher surface. This directly affects soil processes and subsequent soil geography. Soils above and adjacent to a riser tend to be better drained than those farther away from the escarpment. Daniels et al. (1984) called a similar relationship along drainage ways the dry edge effect (Fig. 1.21). Soils at the base of a riser tend to be as wet or wetter than those farther from the scarp. This is especially true in flood plains.

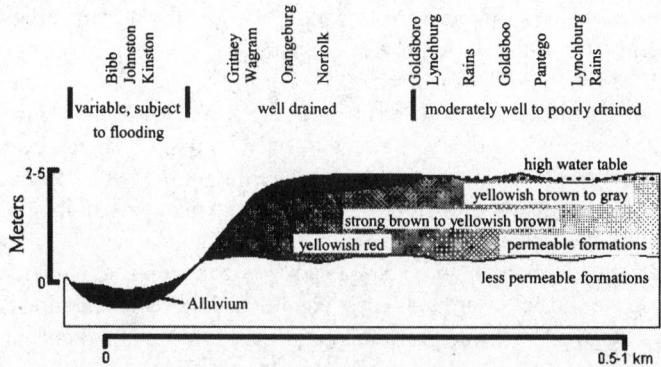

Fig. 1.21   Two-dimensional landscape diagram show drainage relationships on low relief divides [Modified from Daniels et al., 1984]

1.8.1.4   Stream Terraces versus Flood-Plain Steps

Fluvial landforms include the active channel, adjacent surfaces that are intermittently inundated, and higher areas that no longer flood or receive alluvial sediments. The distinction between stream terraces and flood plain steps is an arbitrary break in a natural continuum. There are practical pedological reasons for separating areas that are within the active flood plain from areas that no longer flood or receive alluvial sediments except in rare instances. For most land uses (e.g., home site), it is valuable to know the relative occurrence of flooding. Terraces are defined as a part of the fluvial system that no longer actively experiences fluvial modification (Ruhe, 1975), i.e., no longer aggrading (receiving additional alluvial sediments), no longer experiencing significant flooding. Flood-plain steps are morphometrically similar to terraces, except that they occur within the fluvially active portion of a flood plain and are subject to relatively frequent modification (experience relatively regular or significant flooding and alluvial sediment inputs). Pedologically, terraces commonly have more extensively developed soils (e.g., Alfisols, Ultisols; Section E, chapters 6.11 and 6.12, respectively) compared to lesser developed soils (e.g., Inceptisols, Entisols; Section E, chapters 6.5 and 6.4, respectively) of flood-plain steps. The practical pedological separation of terraces from flood-plain steps varies. Stream terraces which, by definition, no longer flood, could be inundated during rare, catastrophic events (e.g., 500-yr flood). Soil development on such terraces is usually controlled more by precipitation or groundwater than by fluvial processes of the adjacent river. Terrace soils in cold or dry climates, despite a greater age and lack of new sediment, may closely resemble soils on adjacent flood-plain steps that periodically receive fresh sediment. Conversely, soils on flood-plain steps in hot, humid settings may exhibit extensive development and resemble soils on the older, adjacent terraces. In temperate North America a practical separation between terraces and flood-plain steps (Fig. 1.20) is the 100-yr flood stage (Soil Survey Staff, 1998).

1.8.1.5   Flat Plains

Broad, flat plains such as proglacial lake plains, low lying coastal plains, and low gradient glacial till plains are a geomorphic setting distinct from hills, mountains, or terraces. The primary differences are geomorphic origin (nonfluvial) and the low potential energy to drive water flow. Historically, generic terms (e.g., broad interstream divide, rise, flat, etc.) or terms developed for other settings have been applied to flat plains with unsatisfactory results. Recent interest and recognized value of wetlands and hydric soils (Mausbach and Richardson, 1994) highlight the need for unique descriptors for flat areas. Small differences in elevation (e.g., 15 cm or less) on a landscape can mark the change from moderately well drained to very poorly drained soils (Knuteson et al., 1989). Presently, the search is on for better, unique descriptors to express the geomorphic components (pieces) of flat plains. The following terms are proposed as provisional contenders. Even though relief materials are only on the magnitude of a few tens of cm, this difference in elevation has a strong impact on hydrology, as discussed in section 1.8.1.6.

Flat plains are subdued landscapes dominated by broad areas with low slope gradients (e.g., 0–1%) called Talfs (Fig. 1.22), and closed depressions, with only nominal changes in local relief (i.e., microhighs, microlows). A rise is a slightly elevated area (i.e., microhigh) which tends to be broad with low slope gradients (e.g., 1–3%). The low slope gradients of flat plains result in minimal erosion by running water, especially if vegetated. Consequently, fluvial drainage networks tend to be poorly developed, i.e., nonintegrated, incipient, or deranged. Precipitation tends to pond locally and lateral transport is slow both above and below ground. These conditions favor accumulation of organic matter (e.g., pocosins, upland bogs) and retention of water and fine earth (< 2 mm diameter) sediments.

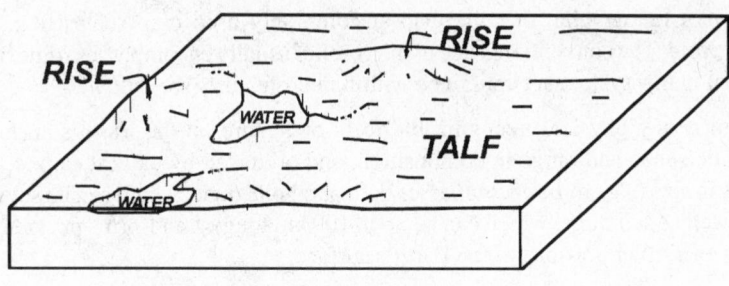

- very low gradients (e.g., slope 0-1%)

- deranged, non-integrated, or incipient drainage network

- "high areas" are broad and low (e.g., slope 1-3%)

- sediments commonly lacustrine, alluvial, or till

Fig. 1.22   Geomorphic components for flat low relief plains

1.8.1.6   Microrelief

As used here, microrelief refers to slight variations in the height of a land surface that are too small to delineate on a topographic or soils map at commonly used scales (e.g., 1:24,000 and 1:10,000). Examples of microrelief include microhigh, microlow relief. These minor elevational differences can have a surprisingly large impact on vegetation distribution, soil hydrologic dynamics (Hopkins, 1996), and sedimentation patterns. This approach to microrelief is strictly morphometric. A distinction is made between small elevational differences (microrelief) and small features (microfeature) with which these elevational differences are associated. An example is gilgai (a kind of microfeature), with the associated microrelief (mound areas = microhighs, bowl areas = microlows). It is an unfortunate yet common practice to confuse the two entities and describe both as microrelief.

1.9   Landscapes, Landforms, Microfeatures, and Anthropogenic Features

Armed with the ability to describe the most detailed morphometric nuances of the land surface, one is inevitably faced with some variation of the question: "Taken all together, what is this feature that we see and what should it be called?" A potpourri of names have been applied over time that connote internal composition, form, arrangement, collective relationships and origin. These terms range in scale from small, human scale features to continental scale assemblages. Confronted with such complexity, some organization of terminology is reasonable and, in fact, necessary. Various schema have been developed to achieve this, typically from a particular geographical perspective: physiography (Fenneman, 1931, 1938, 1946; Thornbury, 1965, 1969; Hunt, 1967), land use (USDA-SCS, 1981) and the currently popular ecological (Omernick, 1987, 1995; Bailey et al., 1994; USDA-USEPA, 1996). A different approach is to deemphasize the geographical context and focus on geomorphic processes (Peterson, 1981). One pseudohierarchy of terms has been assembled specifically from and for soil survey and geographic applications (Schoeneberger and Wysocki, 1997). In addition to morphometric terminology and physiographic location, this system loosely arrays land surface features in a progression of scale in the following fashion.

1. A microfeature is a small, local, natural form (feature) on the land surface that is too small to delineate on a topographic or soils map at commonly used map scales (e.g., 1:24,000 to 1:10,000). *Note*: The conventional use of microrelief usually encompasses some of the terms or features that in this system are contained within microfeature (see below).

2. A landform is any physical, recognizable form or feature on the Earth's surface, having a characteristic shape and range in composition, and produced by natural causes; it can span a wide range in size (e.g., dune encompasses both parabolic dune, which can be several tens of m across, as well as seif dune, which can be up to 100 km long). Landforms provide an empirical description of similar portions of the Earth's surface.

3. A landscape is a collection of spatially related, natural landforms, usually the collective land surface that the eye can comprehend in a single view (Soil Survey Staff, 1998).

4. An anthropogenic feature is an artificial feature on the land surface, having a characteristic shape and range in composition, composed of unconsolidated earthy or organic materials, artificial materials, or bedrock, that is the direct result of human manipulation or activities. Historically, "landforms" have been defined as, and restricted to, natural features. A population of natural features (e.g., sand dunes) has relatively consistent formational processes with common compositional, structural, stratigraphic results. In contrast, human-made features may have a common theme in intent (e.g., water impoundment), but can have an almost limitless variety of formational procedures, material composition and internal structure.

1.10   Age Assessment of Soil Landscapes

Age assessment of landscapes is the single most important contribution that geomorphology makes to the understanding of soils. The law of superposition and the concept of geomorphic surfaces are the key principles for defining relative age relationships in a landscape (Daniels et al., 1971; Hall, 1983). Radiometric, isotopic, or palentological dating combined with field studies using geomorphic and stratigraphic principles can establish the absolute age of a deposit, soil, or a landscape.

The longer a landscape is exposed to subaerial weathering the greater the potential for soil development. The terms weathered, well developed, and old are relative indicators of time in soil formation, but none are accurate descriptors of age. Weathered refers to the relative stability stage (Goldich, 1938; Jackson, 1968) of the minerals contained in a soil. The weathering stage is a function of age, weathering intensity, and mineral stage of the parent material. A chronologically young soil may be composed of highly weathered minerals if the parent material is preweathered.

Soil development is a function of both age and the rate of horizon formation. Consequently, soil development has been used as a relative measure of soil age based on the type and degree of horizonation present. Rate of horizon formation depends on the intensity of soil-forming processes (Section E, Chapter 2), and composition and resistance of the parent material to change. Soils of similar morphology can differ substantially in age. For example, Burgess and Drover (1953) reported that Spodic horizons in Australia formed in sandy beach deposits within 1,000 to 3,000 yr. Franzmeier and Whiteside (1963) found that Spodic horizons took 8,000 yr to form on dunes in northern Michigan. Holzhey et al. (1975) suggested 21,000 to 28,000 yr for development of thick Spodic horizons in sandy soils in North Carolina. Well-developed, weathered soils are commonly referred to as "old". In this context, old is a proxy for age based on weathering and degree of soil formation. Geomorphic principles are a means to assess soil and landscape age independent of horizon development or weathering stage.

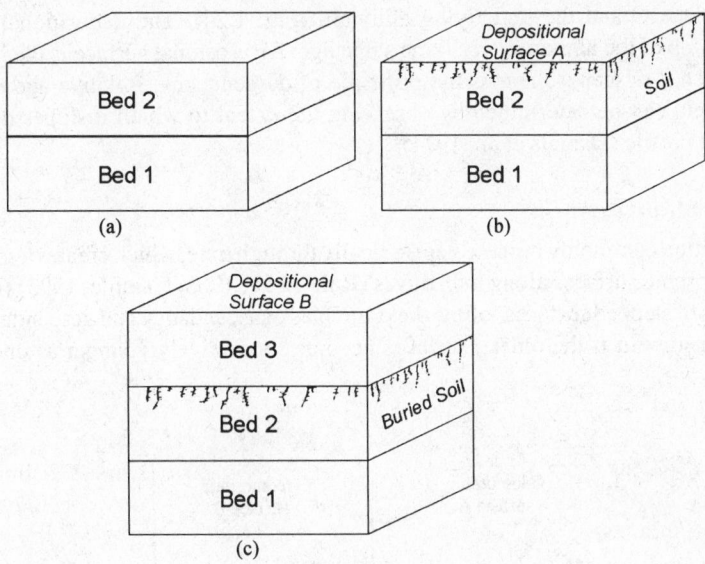

Fig. 1.23 Three-dimensional block diagram depicting the Law of Superposition

1.10.1 The Law of Superposition

This specifies the sequence of deposition of sedimentary rocks and unconsolidated sedimentary deposits. Sediments can only be deposited atop preexisting materials. Younger beds or strata invariably overlie older beds, if not overturned or tectonically displaced. Bed 2 (Fig. 1.23a) is younger than bed 1. Once deposition ceases, the top of bed 2 represents a depositional surface (surface A) equal in age to the youngest strata in the bed. Stabilization of the surface allows a soil to form in the sediment (Fig. 1.23b). This is time zero for subaerial weathering and soil formation. The surface and soil age are the same. The passage of time is recorded by an increasing degree of soil development (Fig. 1.23c). Soil formation requires a period of nondeposition, and therefore, indicates an unconformity in the sedimentary sequence. A buried soil represents an unconformity in the sedimentary sequence. For example, deposition of bed 3 buries the older units and the soil formed on them (Fig. 1.23d). Bed 3 is the youngest deposit. The top of bed 3 is a new, younger depositional surface B. The stratigraphic record contains three depositional beds and a buried soil.

1.10.2 Geomorphic Surfaces

Geomorphic surfaces can be composed of two parts, an erosional component upslope that cuts into existing materials, and a lower depositional component to which the eroded material is transported and deposited. In a landscape, an erosion cycle or event (e.g., stream incision, hillslope retreat, pedimentation) that cuts into existing beds forms a geomorphic surface B of erosional origin (Fig. 1.24). Surface B is younger than any stratum, soil, or surface that it cuts. An erosional surface can only cut preexisting materials, stratas, or surfaces. As one ascends a landscape, the geomorphic surfaces that one crosses increase in age. This is the principle of ascendancy. Surface B is younger than surface A and younger than beds 1, 2, and 3. The cutting of an erosional surface produces sediment that is deposited downslope or down valley. The erosional surface descends or grades to a corresponding depositional surface C and the underlying alluvium (Fig. 1.24). The depositional and erosional surfaces and the top of the alluvium are all the same age. An erosional surface is the same age as the alluvium to which it descends. This is the principle of descendency. Relative age of an erosional surface in the field can be determined by observing the extent to which it slopes downward in a smooth, concave profile (Daniels et al., 1971).

1.10.3   Stepped Surfaces

Landscape evolution commonly proceeds episodically through time, which creates a series of stepped or faceted geomorphic surfaces along interfluves (Ruhe et al., 1968; Gamble, 1993) (Fig. 1.25). The age relationships of stepped surfaces follow the principles of ascendancy and descendency. Surface A on the interfluve summit is the oldest; surfaces become successively younger as one descends the

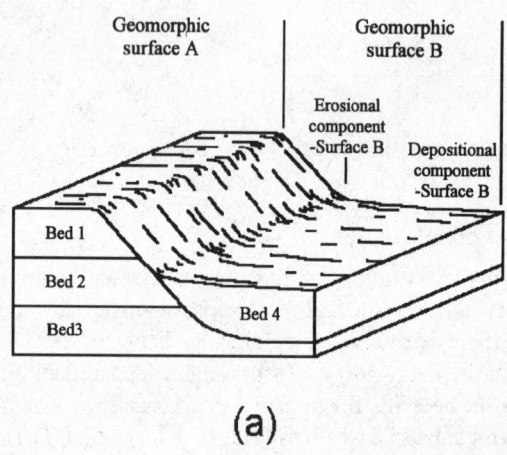

(a)

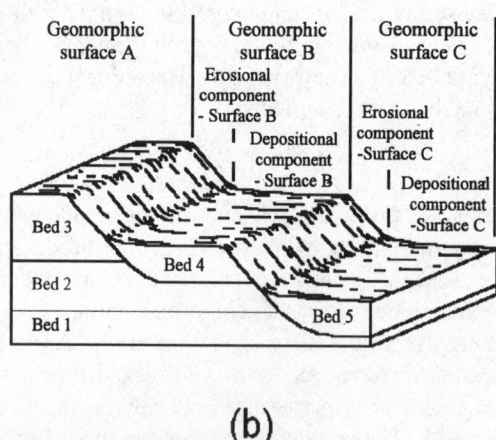

(b)

Fig. 1.24 Three dimensional block diagram showing geomorphic surfaces and associated time relationships

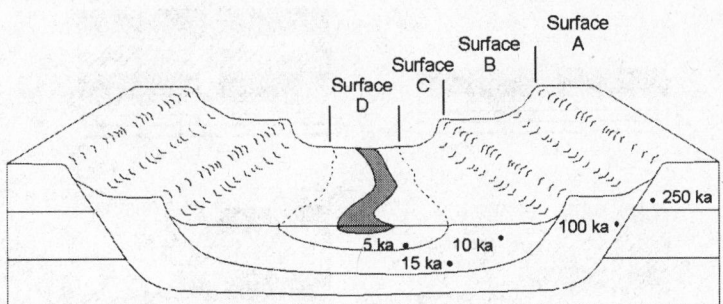

Fig. 1.25   Three-dimensional block diagram showing geomorphic surfaces and dating of sediments

landscape. The age differences between surfaces depend on the timing of the erosional processes that produced them.

1.10.4   Absolute Age Assessment

Dating techniques along with the geomorphic relationships provide a framework for establishing a landscape chronology. For example, ages shown in Fig. 1.25, which could be obtained by ^{14}C, luminescence techniques, or fossil analysis, establish the periods for deposition. Surface D is underlain by sediment with an age of 5 ka at the base. This unit is inset into an alluvium that has an age of 15 ka at the base and 10 ka near the top. Stream incision and back filling, therefore, occurred between 5 and 10 ka that deposited the alluvium under surface D. Surface C includes both an erosional and depositional element. Cutting of the erosional surface occurred between 10 to 15 ka based on the alluvium to which it descends. The alluvium is a time transgressive deposit as is the cutting of the erosional surface. Alluvium under surface B has an age of 100 ka and the erosional element ascends to a surface of 250 ka. The erosional processes that created surface B began later than 250 ka and ended by 100 ka. Given this scenario, the absolute age differences between the surfaces and the soils on them can be established.

1.10.5   Geomorphic History

The concepts as displayed in Figs. 1.23 to 1.25 are straightforward. Application to soil landscapes requires detailed stratigraphic control obtainable only by field observations from natural (e.g., stream banks) or man-made (e.g., road embankments) exposures or drill cores and the ability to think in three dimensions. The geomorphic history of a region must sometimes be established through detailed stratigraphic investigation before soil patterns are completely understood (Ruhe et al., 1967; Daniels et al., 1970). Consider the geomorphic history of a hypothetical landscape in Fig. 1.26. A landscape is composed of soil s1 on geomorphic surface A, which is formed in a single deposit (Fig. 1.26a). The deposit is buried by a younger material (e.g., alluvium, glacial till, loess) with subsequent development of soil s2 on surface B (Fig. 1.26b). The landscape now consists of two stacked stratigraphic units separated by a buried soil s1. Stream incision cuts an erosion surface across both units (Fig. 1.26c). The landscape becomes a two-stepped sequence comprised of surfaces B and C. Soil s3 forms in surface C (Fig. 1.26d). Two soils, s2 and s3 of different age, make up the landscape. Valley alluviation buries surface C and reduces relief in the landscape (Fig. 1.26e). Soil s4 forms in surface D on the alluvium (Fig. 1.26f). The soil landscape at this point consists of two soils, s2 and s4, which again differ considerably in age. Stream incision and backfilling insert a deposit into the

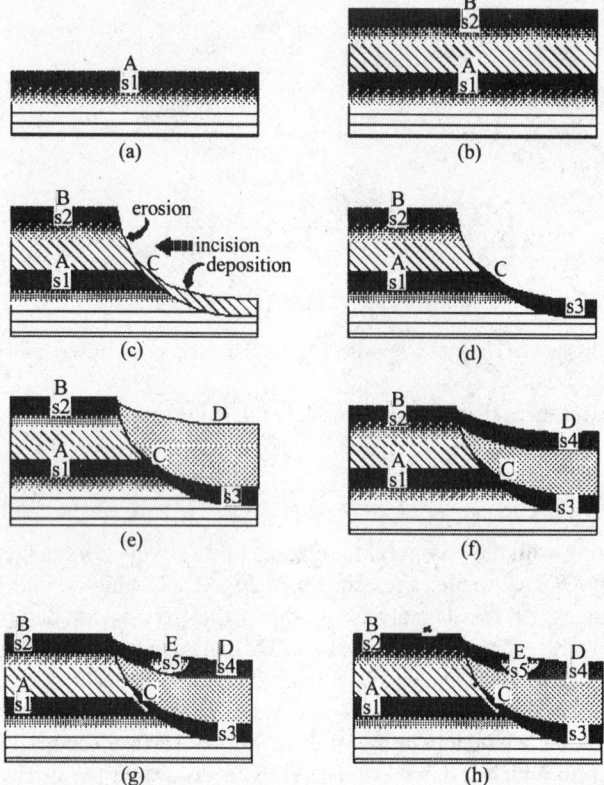

Fig. 1.26   Two-dimensional diagram explaining hypothetical landscape evolution

preexisting alluvium (Fig. 1.26g). The insert is denoted by surface E and soil s5. The landscape now includes three surface soils s2, s4, and s5 all of different ages. Soil distribution on the landscape, soil age, and the geomorphic history of this landscape can only be established from multiple field observations and drill cores at critical locations that identify all the stratigraphic units and buried soils.

1.10.6   Soils and Geomorphic Surfaces

Despite the strong association between soils and geomorphic surfaces, soil patterns and morphology cannot be used to initially define or recognize geomorphic surfaces. Geomorphic surfaces must first be defined using geomorphic principles, and then, the linkages made to soil patterns. A soil or suites of soils may occur on more than one geomorphic surface and soil boundaries do not always correspond to geomorphic boundaries on a landscape. Consider Fig. 1.27 which shows a stratigraphic sequence and soil horizons cut by an erosional surface. Soils 2, 3, and 4 on erosional surface B form over beds of varying composition and age and from preexisting soil horizons. Soils 2, 3, and 4, however, are all the same age (the age of surface B). In this case, three different soils occur on the erosional surface. The sequence of soils is determined by the stratigraphic units present and erosional sorting along the surface. The same erosional surface in its geographic occurrence can cut different strata or sediments. The soils in that case would differ.

The boundary between soil 1 and soil 2 (Fig. 1.27d) lies below the slope break that denotes the erosional surface. Erosion has not removed enough of the original profile to cause a change in the soil.

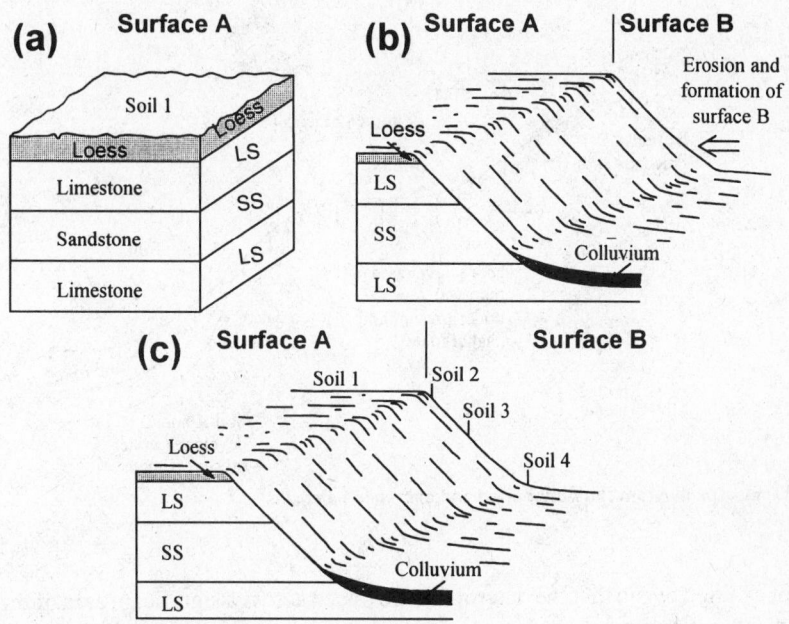

Fig. 1.27 Two-dimensional diagram relating soils and geomorphic surfaces

The boundary between the surfaces and the boundary between soils is not coincident. A soil map and a geomorphic surface of the same landscape would not have coincident boundaries.

1.11 Paleosols, Geosols, and Climate Interpretation

1.11.1 Paleosols

Geologic processes that create and destroy soils and landscapes vary in time and space. Continental scale processes like glaciation can create or destroy entire landscapes. More commonly, landscapes evolve by erosion and deposition on only parts of them. Erosion removes soils and sediments from the active parts of landscapes, while soils continue to form on the stable parts. Sediment from erosion buries preexisting surfaces, soils, and sediments. Buried soils and soils that endure on stable landscapes for long periods become part of the geologic record and are valuable for interpreting Earth history. Paleosols are soils that formed on landscapes of the past (Yaalon, 1971; Valentine and Dalrymple, 1976). Age alone, however, does not denote a paleosol. The process or processes responsible for the soil morphology must no longer operate due to a change in climate, local environment, or because of burial.

Ruhe (1965) recognized three types of paleosols: buried, exhumed, and relict. Buried paleosols form at the land surface, but later are covered by sediment, which removes the paleosol from the dominant soil-forming zone. Burial must be deep enough to suspend the soil-forming processes, and rapid enough to prevent formation of a cumulic soil profile. Post-burial and diagenetic changes in soil properties and horizons are common (Olson and Nettleton, 1999; Yaalon, 1971) and must be considered when studying paleosol composition.

Exhumed paleosols form at the land surface, are buried, and later reexposed on younger landscapes by erosion of the covering sediment (Ruhe, 1969). Erosion exposes both the paleosol and the former

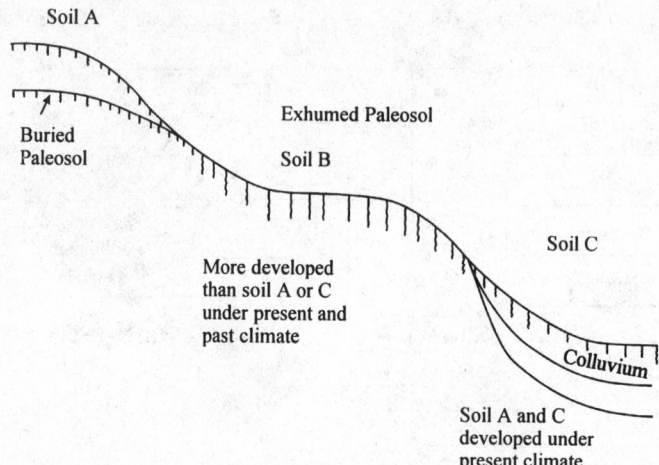

Fig. 1.28   Landscape diagram showing buried and exhumed paleosols

differs from a buried one in that the outcrop can be traced across a significant area of the present land surface. An exhumed paleosol can be out of balance with both the existing climate and the adjacent soils on the landscape. Landscapes that include exhumed paleosols may not conform to the expected catena relationship in a geographic area (Fig. 1.28) (Ruhe, 1969).

Relict paleosols are soils formed on preexisting landscapes and were never buried (Ruhe, 1965). Therefore, identification of relict paleosols is somewhat problematic. A relict paleosol must have endured one or more shifts in regional climate or local environment (e.g., lowering of a watertable by stream incision or base level change) such that it is no longer in balance with the existing conditions. This concept requires that the age of the landscape be known or inferred and that one has a precise understanding of the processes responsible for soil morphology (Section E, Chapter 2). Soils on old, easily distinguishable geomorphic surfaces such as terraces (Fig. 1.29) are the most commonly recognized relict paleosols.

The present knowledge of soil processes is inadequate to determine relict versus nonrelict soil conditions in all situations. Soil-forming processes can be reversible self terminating, steady state or metastable, or irreversible self-terminating (Yaalon, 1971). This means that all morphological features do not have equal utility as indicators of climate change. For example, the formation of silica-cemented soil horizons (duripans) occurs in arid or semi-arid climates (Soil Survey Staff, 1996). Once formed, a duripan is resistant to destruction even in humid conditions. Thus, the presence of a duripan in a humid climate is strong evidence of a relict condition. In contrast, some morphological features, such as A horizon thickness and color, which have a strong correlation to present climatic regimes, can adjust rapidly to a shift in climate. Despite the difficulties in defining and recognizing relict paleosols, this concept is valuable because it focuses on the relationships of landscape age and soil processes.

1.11.2   Geosols

Buried paleosols are integral components of the stratigraphic record. Formation of a soil requires a period of landscape stability with exposure to subaerial weathering. Therefore, a buried paleosol (1) represents an unconformity in a sedimentary sequence, and (2) marks a consistent stratigraphic position at the top of a sedimentary body or bodies. Buried paleosols that are laterally traceable with

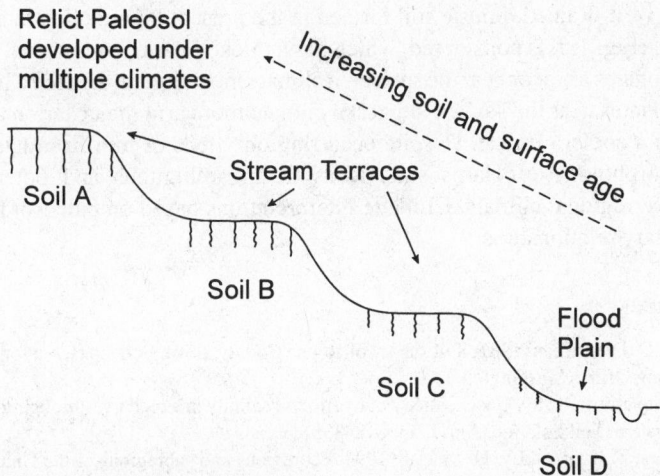

Fig. 1.29 Landscape diagram depicting relict paleosol

three-dimensional form and have mappable distribution at a scale of 1:24,000 can be formally defined as pedostratigraphic units (Morrison, 1967) called geosols (NACSN, 1983). Geosols include only pedogenic horizons formed in preexisting rock or sediment bodies. Buried O horizons and some C horizons of paleosols are excluded from geosols. A geosol may form in parent materials of diverse ages and composition. Formal naming and establishing of geosols follow the same criteria of other stratigraphic units (NACSN, 1983). A geosol is named after a geographic feature in the type area.

1.11.3 Climate Change and Paleosols

Jenny (1941) originally proposed a mathematical model that can be solved for an individual soil-forming factor. For example, the equation

$$\text{Soil} = f'(c)_{p,r,b,t} \tag{1.1}$$

ascertains the climate during soil formation, where the subscripts denote a constancy in the factors of parent material, relief, biota, and time. However, there are problems with this approach; Equation [1.1] defines the soil-forming factors as independent variables, but, in reality, the factors covary. It is also difficult to assign quantitative values to some factors, particularly parent material and the biota. Nonetheless, this approach allows useful, conceptual and semiquantitative comparisons between soil morphology and individual soil-forming factors.

Processes of soil formation occur over time scales of tens to thousands of years. Soil morphology forms a physical record of the climate, vegetation, and/or environment during the time of formation. Recent emphasis on global climate change has renewed interest in paleosols as indicators of past climates. Numerous studies have used soil morphology to interpret past climates or environments. The interpretation of climate based on paleosols requires one or more of the following: (1) that the paleosol existed on a stable landscape and had reached a steady state within a particular climate; (2) identification of the ancient soil environment (e.g., flood plain); and/or (3) identification of the catenary relationship or position of the paleosol on the ancient landscape.

For example, a climatic interpretation based solely on the properties of the Clarion profile (Fig. 1.6), such as A horizon thickness, organic C content, depth to inorganic carbonate, and B horizon color are indicative of a well-drained, prairie soil formed in the present humid climate of central Iowa. If only the Harps soil (Fig. 1.6) is considered, which has a thicker, darker A horizon, a gray B horizon, and inorganic carbonates are at or near the surface, a climate more humid than the Clarion morphology is indicated. Carbonates near the surface suggest a climate more arid than Clarion and the A horizon thickness suggests a cooler climate. Despite occurring only tens of mm from Clarion soils on the landscape, the morphology of Harps soils leads to an ambiguous and potentially erroneous interpretation of the regional climate. Climatic interpretations based on paleosol morphology must include paleocatenary relationships.

1.12 References

Andrews, W.F., and R.O. Dideriksen. 1981. Soil survey of Boone County, Iowa. Soil Conservation Service, USDA. US Government Printing Office, Washington, DC.

Arndt, J.L., and J.L. Richardson. 1989. Geochemistry of hydric soil salinity in a recharge-throughflow-discharge prairie-pothole wetland system. Soil Sci. Soc. Am. J. 53: 848–855.

Bailey, R.G., P.E. Avers, T. King, and W.H. McNab. 1994. Ecoregions and subregions of the United States (map scale 1:7,500,000) (supplementary table of map unit descriptions compiled and edited by McNab, W.H. and Bailey, R.G.) USDA- Forest Service, Washington, DC.

Blume, H. 1968. Die pedogenetische Deutung einer Catena durch die Untersuchung der Bodendynamik. Trans. 9th. Int. Cong. Soil Sci. 4:441–449.

Blume, H.P., and E. Schlichting. 1965. The relationship between historical and experimental pedology. In E.G. Hallsworth and D.V. Crawford (ed.) Experimental pedology. Butterworths, London, UK.

Buol, S.W., F.D. Hole, and R.J. McCracken. 1989. Soil genesis and classification. Iowa State University Press, Ames, IA.

Burgess, A., and D.P. Drover. 1953. The rate of Podzol development in sands of the Woy Woy District N.S.W. Aust. J. Bot. 1:83–94.

Burras, C.L., and W.H. Scholtes. 1987. Basin properties and postglacial erosion rates of minor moraines in Iowa. Soil Sci. Soc. Amer. J. 51:1541–1547.

Bushnell, T.M. 1942. Some aspects of the soil catena concept. Soil Sci. Soc. Am Proc. 7:466–476.

Cleaves, E.T., A.E. Godfrey, and O.P. Bricker. 1970. Geochemical balance of a small watershed and its geomorphic implications. Geol. Soc. Am Bull. 81:3015–3032.

Conacher, A.J., and J.B. Dalrymple. 1977. The nine-unit land surface model: an approach to pedogeomorphic research. Geoderma 18:1–154.

Crabtree, R.W., and T.P. Burt. 1983. Spatial variation in solutional denudation and soil moisture over a hillslope hollow. Earth Surf. Process. Landf. 8:151–160.

Dalrymple, J. B., R.J. Blong, and A.J. Conacher. 1968. A hypothetical nine-unit landsurface model. Z. Geomorphol. 12:60–76.

Dan, J. and D.H. Yaalon. 1964. The application of the catena concept of pedogenesis in Mediterranean and desert fringe regions. Trans 8th. Int. Cong. Soil Sci. p. 751–758.

Daniels, R.B., E.E. Gamble, and J.G. Cady. 1971. The relation between geomorphology and soil morphology and genesis. Adv. Agron. 23:51–88.

Daniels, R.B. and R.D. Hammer. 1992. Soil Geomorphology. John Wiley and Sons, New York, NY.

Daniels, R.B., E.E. Gamble, and J.G. Cady. 1970. Some relations among coastal Plain soils and geomorphic surfaces North Carolina. Soils Sci. Soc. Am. Proc. 34:648–653.

Daniels, R.B., H.J. Kleiss, S.W. Buol, H.J. Byrd, and J.A. Phillips. 1984. Soil systems in North Carolina. NC State Univ. Agric. Res. Serv. Bull. 467.

Dijkerman, J.C., 1974. Pedology as science: The role of data models and theories in the study of natural systems. Geoderma 11:73–93.

Dokuchaev, V.V. 1883. Russian Chernozems (Russkii Chernozem) Israel Prog. Sci. Trans., Jerusalem, Israel, 1967, translated by N. Kaner.

Emerson, D.G., M.D. Sweeney, V.M. Dressler, and S.W. Norbeck. 1990. Instrumentation and data for a study of seasonally frozen soil in southeastern North Dakota. US Geol. Surv. Open-File Rep. 90–107.

Fenneman, N.M. 1931. Physiography of the western United States. McGraw-Hill Co., New York, NY.

Fenneman, N.M. 1938. Physiography of the eastern United States. McGraw-Hill Co., New York, NY.

Fenneman, N.M. 1946. Physical divisions of the United States. US Geological Survey, US Government Printing Office, Washington, DC.

Finney, J.R., N. Holowaychuk, and M.R. Heddleson. 1962. The influence of micro-climate on the morphology of certain soils of the Allegheny Plateau of Ohio. Soil Sci. Soc Amer. Proc. 26:287–292.

Foth, H.D. 1984. Fundamentals of soil science. 7th Ed. John Wiley and Sons, New York, NY.

Franks, C.D., L.C. Brockman, P.M. Whited, and S.W. Waltman. 1993. Using the benchmark site selection process to enhance modeling. USDA-NRCS Internal Report, National Soil Survey Center, Lincoln, NE.

Franks, C.D., F.B. Pierson, A.G. Mendenhall, K.E. Spaeth, and M.A. Weltz. 1995. Interagency rangeland water erosion project report and state data summaries. NWRC 98-1. USDA-Agricultural Research Service Northwest Watershed Research Center, Boise ID.

Franzmeier, D.P. and E.P. Whiteside. 1963. A chronosequence of podsols in northern Michigan. I. Ecology and description of pedons, II. Physical and chemical properties. MI State Univ. Agric. Exp. Stn. Quart. Bull. 46:1–36.

Franzmeier, D.P., E.J. Pedersen, T.J. Longwell, J.G. Byrne, and C.K., Loshe. 1969. Properties of some soils in the Cumberland Plateau related to slope aspect and position. Soil Sci. Soc. Am. Proc. 33:755–761.

Gamble, E.E. 1993. Geomorphic study in the upper Gasconade River Basin, Laclede and Texas Counties, Missouri. Soil Survey Investigations Report 43.

Glazovskaya, M.A. 1968. Geochemical landscapes and types of geochemical soil sequences. Trans. 9th Int. Congress Soil Sci. 4:303–312.

Goldich, S.S. 1938. A study in rock weathering. J. Geol. 46:17–58.

Graham, R.C. 1986. Geomorphology, mineral weathering, and pedology in an area of the Blue Ridge Front, North Carolina. Ph.D. dissertation, North Carolina State Univesity, Raleigh, NC.

Hall, G.F. 1983. Pedology and geomorphology. p. 117–140. *In* L. P. Wilding et al., (ed) Pedogenesis and soil taxonomy. I. Concepts and interactions. Elsevier, New York, NY.

Hammer, R.D. 1995. An environmental report to the City of Wildwood. p. 65–137. *In* Master plan for the City of Wildwood, MD. S.H. Lickerman, Chair, Planning and Zoning Committee.

Hammer, R.D., F.E. Heisner, and S.H. Lickerman. 1998. Soil and geomorphology in an Ozark Border forested watershed. Agron. Abst. p. 300.

Holzhey, C.S., R.B. Daniels, and E.E. Gamble. 1975. Thick Bh horizons in the North Carolina coastal plain: II. Physical and chemical properties and rates of organic additions from surface sources. Soil Sci. Soc. Am. Proc. 39:1182–1187.

Hopkins, D.H. 1996. Hydrologic and abiotic constraints of soil genesis and natural vegetation patterns in the sandhills of North Dakota. Ph.D. dissertation, North Dakota State University, Fargo, ND.

Huggett, R.J. 1975. Soil landscape systems: a model of soil genesis. Geoderma 13:1–22.

Huggett, R.J. 1976. Lateral translocation of soil plasma through a small valley in the northern Great Wood Hertforshire. Earth Surf. Process. 1:99–109.

Hunt, C.B. 1967. Physiography of the United States. W.H. Freeman and Co., London, UK.

Hunt, C.B. 1986. Surficial deposits of the United States. Reinhold Co., Inc., New York, NY.

Jackson, M.L. 1968. Weathering of primary and secondary minerals in soils. Trans. 9th Intern. Cong. Soil Sci. 4:281–292.

Jenny, H.J., 1941. Factors of soil formation, McGraw-Hill Co., New York, NY.

Kemmis. T.J. 1991. Glacial landforms, sedimentology, and depositional environments of the Des Moines Lobe northern Iowa. Ph.D. Diss., University of Iowa, Iowa City, IA.

Kemmis, T.J., G.R. Hallberg, and A.J. Lutenegger. 1981. Depositional environments of glacial sediments and landforms on the Des Moines Lobe. IA Geol. Survey Guidebook 6.

King, L.C. 1957. The uniformitarian nature of hillslopes. Trans. Edinb. Geol. Soc. 17:81–102.

Kleiss, H.J. 1970. Hillslope sedimentation and soil formation in northeastern Iowa. Soil Sci. Soc Amer. Proc. 34:287–290.

Knuteson, J.A., J.L. Richardson, D.D. Patterson, and L. Prunty. 1989. Pedogenic carbonates in a Calciaquoll associated with a recharge wetland. Soil Sci. Soc. Am J. 53: 495–499.

Lickerman, S.H., R.D. Hammer, and F.E. Heisner. 1998. Bringing science to urban planning in Ozark Border watersheds: Problems and prespectives. Agron. Abst. p. 300.

Lotspeich, F.B., and H.W. Smith. 1953. Soils of the Palouse loess: I. The Palouse catena. Soil Sci. 76:467–480.

Mausbach, M.J., and J.L. Richardson. 1994. Biogeochemical processes in hydric soil formation. *In*: Current topics in wetland biogeochemistry. Wetland Biogeochemistry Institute, Louisiana State University 1:68–127.

Milne, G. 1936a. A provisional soil map of East Africa. East African Agriculture Research Station, Amani Memoirs, Tanganyika Territory.

Milne, G. 1936b. Normal erosion as a factor in soil profile development. Nature 138:548–541.

Morrison, R.B. 1967. Principles of Quaternary soil stratigraphy. p 1–69. *In* R.B. Morrison and R. E. Wright (ed.) Quaternary soils. Center for Water Resources Research, Desert Research Institute, University of Nevada, Reno, NV.

North American Commission on Stratigraphic Nomenclature. 1983. North American Stratigraphic Code. Amer. Assoc. Petrol. Geol. Bull. 67:841–875.

Oliver, M.C. 1997. An investigation of two landscapes on the Holston Limestone Formation in McMinn Count, Tennessee. MS Thesis, University of tennessee, Knoxville, TN.

Olson, C.G., and W.D. Nettlton. 1999. Paleosols and the effects of alteration. Quat. Int. 51/52:185–194.

Omernik, J.M. 1987. Ecoregions of the coterminous US (map supplement). Ann. the Assoc. Am. Geogr. 77:118–125.

Omernik, J.M. 1995. Ecoregions - a framework for environmental management. p. 49–62. *In*: W.S. Davis and T.P. Simon (ed.) Biological assessment and criteria - tools for water resource planning and decision making. Lewis Publishers, Boca Raton, FL.

Pennock D.J., and D.F. Acton. 1989. Hydrological and sedimentological influences on Boroll catenas, Central Saskatchewan. Soil Sci. Soc. Am. J. 53:904–910.

Pennock, D.J. and W.J. Vreeken. 1986. Soil-geomorphic evolution of a Boroll catena in southwestern Alberta Soil Sci. Soc. Amer. J. 50:1520–1526.

Pennock, D.J., B.J. Zebarth, and E. de Long. 1987. Landform classification and soil distribution in hummocky terrain, Sasketchewan, Canada. Geoderma 40:297–315.

Peterson, F.F. 1981. Landforms of the Basin and Range Province for soil survey. NV Agric. Exp. Sta. Tech. Bull. 28.

Richardson, J.L., J.L. Arndt, and R.G. Eilers. 1991. Soils in three prairie pothole wetland systems. Annual manitoba Soil Science Society Meeting, Winnepeg, Canada.

Ruhe, R.V. 1960. Elements of the soil landscape. Trans. 7th Int. Congr. Soil Sci. 23:165–169.

Ruhe, R.V. 1965. Quaternary paleopedology. p. 755–764. *In* H.E. Wright and D.E. Frey. The Quaternary of the United States. Princeton University Press, Princeton, NJ.

Ruhe, R.V. 1969. Quaternary landscapes in Iowa. IA State Univ. Press, Ames, IA.

Ruhe, R.V. 1975. Geomorphology. Houghton Mifflin, Boston, MA.

Ruhe, R.V., and W.H. Scholtes. 1959. Important elements in the classification of the Wisconsin glacial stage. J. Geol. 67:585–593.

Ruhe, R.V., and P.H. Walker. 1968. Hillslope models and soil formation. I. Open systems. Trans. 9th Int. Congr. Soil Sci. 4:551–560.

Ruhe, R.V., R.B. Daniels, and J.G. Cady. 1967. Landscape evolution and soil formation in southwestern Iowa. USDA Tech. Bull. 1349.

Ruhe, R.V., W.P. Dietz, T.E. Fenton, and G.F. Hall. 1968. The Iowan problem northeastern Iowa. Report of Invest. 7, IA Geol. Surv., Iowa City, IA.

Saucier, R.T. 1994. Geomorphology and Quaternary geologic history of the lower Mississippi Valley. US Army Engineer Waterways Experiment Station, US Army Corps of Engineers, Vicksburg, MS.

Schoeneberger, P.J., A. Amoozegar, and S.W. Buol. 1995. Physical properties of a soil and saprolite continuum at three geomorphic positions. Soil Sci. Soc. Am. J. 59:1389–1397.

Schoeneberger, P.J., and D.A. Wysocki. 1996. Geomorphic descriptors for landforms and geomorphic components: effective models, weaknesses and gaps. American Society of Agronomy, Annual Meetings, Indianapolis, IN.

Schoeneberger, P.J., and D.A. Wysocki. 1997. Geomorphic description system. *In* Schoeneberger, P. J., D. A. Wysocki, E. C. Benham, and W. D. Broderson, 1998. Field book for describing and sampling soils. USDA - Natural Resource Conservation Service, Lincoln, NE.

Simonson, R.W. 1959. Outline of a generalized theory of soil genesis. Soil Sci Soc. Am. Proc. 23:152–156.

Soil Science Society of America., 1997. Glossary of Soil Science Terms. Soil Science Society of America. Madison, WI

Soil Survey Staff. 1975. Soil taxonomy. US Government Printing Office, Washington, DC.

Soil Survey Staff. 1996. Keys to soil taxonomy. US Government Printing Office, Washington, DC.

Soil Survey Staff. 1998. National soil survey handbook. USDA - Natural Resources Conservation Service, Lincoln, NE.

Steinwand, A.L., and T.E. Fenton. 1995. Landscape evolution and shallow groundwater hydrology of a till landscape in central Iowa. Soil Sci. Soc. Amer. J. 59:1370–1377.

Steinwand, A.L., and J.L. Richardson. 1989. Gypsum occurrence in soils on the margin of semipermanent prairie pothole wetlands. Soil Sci Soc. Am. J. 54:836–842.

Thornbury, W.D. 1965. Regional geomorphology of the United States. John Wiley and Sons, Inc., New York, NY.

Thornbury, W.D. 1969. Principles of geomorphology. 2nd Ed. John Wiley and Sons, Inc., New York, NY.

USDA-Soil Conservation Service. 1981. Land resource regions and major land resource areas of the United States. Agricultural Handbook 296.

USDA-USEPA. 1996. Developing a spatial framework of ecological units for the United States. *p. 46–54. In* GIS/LIS 96 Ann. Conf. Expo. Proc., Denver, CO.

Valentine, K.W.G., and J.B. Dalrymple. 1976. Quaternary buried paleosols: A critical review. Quat. Res. 6: 209–222.

van Wambeke, A., and T. Forbes. 1986. Guidelines for using soil taxonomy in the names of soil map units. Soil Management Support Services, Technical Mono. 10.

Walker, P.H., 1966. Postglacial environments in relation to landscape and soils on the Cary drift, Iowa. IA Agric. Exp. Sta. Res. Bull. 549:838–875.

Walker, P.H., and R.V. Ruhe. 1968. Hillslope models and soil formation. I. Closed systems. Trans. 9th Int. Congr. Soil Sci. 4:561–568.

Wood, A. 1942. The development of hillside slopes. Geol. Assoc Proc. 53:128–138.

Yaalon, D.H. 1971. Soil-forming processes in time and space. p. 29–40. *In* D.H. Yaalon (ed) Paleopedology. Israel University Press, Jerusalem, Israel.

2

Pedogenic Processes

Oliver A. Chadwick
University of California, Santa Barbara

Robert C. Graham
University of California, Riverside

2.1 Introduction

Soil is a mixture of geological parent material, living organisms and the residue of their interaction. The residue imparts soil with its unique characteristics. Soil is a heterogeneous, polyphasic, particulate, disperse, and porous system. The colloidal nature of soil exposes a large surface area which gives rise to adsorption of water and chemicals, ion exchange, adhesion and capillarity, swelling and shrinking, and dispersion and flocculation. Soil colloids are composed of interbonded inorganic (clay minerals) and organic (humus) molecules. The nature and amount of these colloids vary in response to environmental stimuli and organic input. Their variation is responsible for lateral differentiation in soil properties. Soils are separated vertically into horizons that indicate differences in the internal soil environment which, in turn, determines the amount and character of soil colloid accumulation. This chapter provides an overview of the way in which soil properties vary as soils develop under varying environmental stimuli.

Soil formation is powered by gradients in chemical and physical potential as Earth's atmosphere and biosphere interact with rocks and minerals. Rock minerals are formed at high temperature and pressure which makes them susceptible to attack by water and the biospherically derived acids existing in soils. The fundamental mechanisms of physical, chemical, and biological processes in soils are discussed in Sections A, B, and C. Here the focus will be on the matrix of processes that produce three-dimensional soil bodies called pedons which have differentiated horizons and lateral variability.

2.2 Environmental Factors that Drive Pedogenesis

The concept of soil-forming factors is one of the earliest and most important of soil science. It defines soil as a component of ecosystems that must be characterized in terms of both geological substrate and biological input (Jenny, 1941; Amundson and Jenny, 1997). It provides a far reaching description of the controls on soil processes that allows prediction of general soil properties based on the interaction of a relatively small number of driving variables. In the words of Hans Jenny (1980), "pedogenic order in a landscape is unraveled by stratified random sampling along vectors of the state factors." Knowledge of the soil properties produced by the interaction of environmental variables allows prediction of soil characteristics in detail, locally, or in general, regionally and globally. The fact that characteristics of soil properties can be predicted based on environmental variables is critical to soil resource mapping because it is impossible to observe soil profiles everywhere.

The interaction of environmental variables establishes the soil-forming processes whose actions on parent material manifest themselves in characteristic soil morphologies which may, in turn, alter the nature of the ongoing processes. Environmental variables are broadly grouped as geological, climatological, biological and topographical. The geological characteristics constitute a site factor that sets the initial condition for soil formation whereas the climatological and biological factors represent energy input that drives soil development. The energy factors share a strong spatial covariance because of the dependence of organisms on climate. Within any of these groupings, local differences in topography modify the activity of the more broadly defined variables. Human manipulation of edaphic environments is significant enough to warrant close consideration of their effect on soil properties (Amundson and Jenny, 1991).

2.2.1   The Geologic Foundation: Lithology, Mineralogy and Topography

Geological processes ranging from plate tectonics and volcanism to deep ocean carbonate sedimentation to glaciations create conditions that eventually provide substrate for soil development. Plate tectonics drive global processes that differentiate lighter, silica-rich rocks from the mantle into the crust and create relief that produces erosion, accumulation of alluvium, and formation of sedimentary rocks. Continental areas that are actively uplifted, such as the Tibetan plateau and the cordillera of South and North America, support thin, young soils. In contrast, those areas on stable cratons, such as central West Africa, Brazil and Australia, have deep soils that are superimposed on thick, highly weathered saprolite zones (Paton et al., 1995). In these ancient cratons, depth to unweathered rock can exceed 100 m. Volcanism brings locally melted silica-rich rock in subduction zones to the surface in areas such as Japan, Indonesia, and Washington State. Hot spots, such as Hawaii, the Canary Islands, or Yellowstone, provide a direct conduit from mantle to surface for Fe- and Mg-rich rocks. A dramatic example of mantle-fed volcanism is exhibited by flood basalts on the Columbia and Daccan plateaus. On ocean floors, far from continental inflow, deposition of carbonate produces limestone that, when subaerially exposed, has distinct parent material properties when compared with igneous rock. In polar and temperate zones, glacial advances deposit moraines and outwash composed of the mix of lithologies crossed by the glacier. The more finely ground glacial drift is lofted into the atmosphere and deposited as loess. Numerous Pleistocene glaciations have left distinctly younger soils than the deeply weathered soils of stable craton areas in the tropics. Even within temperate zones, there are strong differences in soil properties between recently glaciated areas and those that have not been overrun. Globally, much of the variation in soil properties can be attributed directly to differences in parent material mineralogy and the dynamics of Earth history. Locally they are important as well. For instance, stream channel and associated hillslope erosion may cut through an extensive surficial geological unit into an underlying lithology that has different mineralogical and chemical properties which affect soil properties (Section E, Chapter 1).

Parent material lithology determines the physical and mineralogical nature of soil. For example, soil formed in marine clays or shale inherits a large quantity of colloids that lead to poor water infiltration, relatively low plant available water, and substantial shrink/swell behavior. In contrast, soils formed on quartz sandstone or weathered granite are likely to be sandy with little inherited clay. Commonly, they are excessively drained and often have low fertility and water-holding capacity. Fracture patterns in the underlying bedrock provide preferential lines of water flow which enhance weathering and soil formation. The stratigraphic juxtaposition of soft and hard lithologies provides gentler and steeper relief and less or more rock fragments in the overlying soil.

Parent material mineralogy determines ecological soil properties, such as nutrient supply and retention, and water movement. For example, mafic mantle-derived rocks typically weather to a smectite and Fe oxide-rich colloidal fraction with the simultaneous release of Ca and Mg for use by plants and soil organisms. The smectites provide high CEC which can easily retain nutrient cations. Felsic crust-derived rocks tend to weather to a kaolin- and gibbsite-rich colloidal fraction with release of Ca and K for use by the biota. Kaolin and gibbsite provide only a small amount of cation exchange capability leading to relatively rapid leaching of nutrient cations. The presence of quartz in the felsic rocks provides a skeletal framework of resistant sand size grains that tends to promote water percolation. In contrast, the smectites derived from mafic rocks tend to form high clay content horizons and the clays tend to swell when wet, which can reduce water flow by closing off macropores. Often these igneous rocks contain a broad suite of minerals that weather to provide a broad mix of nutrients and a diverse array of colloid mineralogy. In contrast, limestone is quite pure calcite which should only release Ca and H_2CO_3 when it reacts with water. Soils forming on limestone often have inherited their colloidal fraction from weathering of impurities in the rock or from eolian addition. Similarly, soils forming on quartz sandstone or quartzite must inherit their colloidal fraction from eolian input.

2.2.2   Energy: Water, Temperature, and Biology

The developing soil profile is strongly dependent on energy that is ultimately provided by the sun and Earth's gravitational field. Solar heating is by direct energy absorption at the surface and indirect transfer from the atmosphere. These processes set up superimposed diurnal and annual waves of energy moving through soil profiles. Energy can be lost through latent heat transfers or through long wavelength emissions. Soil provides a waystation in the hydrological cycle which moves water from oceans to the atmosphere and back. Rain and snow melt provide water to soils through downward infiltration under the influence of gravity and suction. Within soil profiles, water is held by adhesive and cohesive forces against gravitational leaching. Once the water-holding capacity is exceeded, water can move downward to the groundwater table or laterally to river systems. Removal of water requires solar energy to evaporate water from the surface of soil and leaves. Plant roots collect water from much greater depths than are accessed by evaporation, thus extending greatly the role of latent heat flux in driving land-atmosphere hydrologic transfers.

During Earth history, evolution of photosynthesis and establishment of rooted plants on land masses initiated a powerful source of energy to drive pedogenesis. Photosynthesis produces complex organic molecules which are shed to the soil either directly or after they have cycled through animals. In the soil, organic matter is metabolized by microorganisms releasing the CO_2 and water that was utilized in photosynthesis (often called soil respiration). The paired photosynthesis and soil respiration reactions set up a giant planetary reduction-oxidation cycle involving carbon and oxygen. Carbon in CO_2 is reduced to organic compounds; O in water is oxidized to molecular form. In turn, microbial breakdown of organic compounds oxidizes C back to CO_2 with O_2, acting as an electron acceptor, being reduced to water. Photosynthesis captures massive amounts of solar energy and much of the ensuing organic matter is transferred to or below Earth's surface. Its breakdown, effected by vast microbial populations, provides most of the energy that drives pedogenesis. Without water, nutrients and an appropriate temperature range, biological reactions cannot function, but once these environmental conditions are met, it is the C-O redox cycle that releases organic acid into soil which, in turn, drives specific pedological transformations.

Organic acids lower the pH of soil solution thereby enhancing mineral weathering, but the release of cations during weathering acts to neutralize the acidity. The amount of water percolation and cation removal determines the extent to which weathering buffers soil pH against acidification and determines the nature of secondary mineral synthesis. In well-drained humid environments, dilute soil solution leads to strong chemical gradients, intense weathering, efficient leaching of soluble components (basic cations and silicic acid), and secondary mineral synthesis involving only the least soluble components (Fe and Al oxyhydroxides and kaolin minerals). In poorly-drained sites, soil solution accumulates highly soluble ions as well as sparingly soluble ions leading to synthesis of smectite amd other minerals. In these soils, Fe and Mn may be reduced and slowly leached away. In contrast, arid environments often have soil solutions that contain large concentrations of ions, leading to minor chemical gradients, short leaching distances for only the most soluble ions, and secondary mineral synthesis that is dominated by ionic salts (calcite, gypsum, halite, etc.). Thus, long-term average soil solution concentrations determine most pedochemical properties. Further, they help to condition the terrestrial portion of many biogeochemical cycles because elements released by weathering and not utilized biologically or assimilated into pedogenic minerals are rapidly leached from soil into groundwater, rivers and lakes or oceans (Berner and Berner, 1996).

The annual amount and seasonal distribution of precipitation and solar energy input define broad life zones, such as boreal forest, temperate grasslands, tropical seasonal forest, and tropical rainforest, based on ecosystem composition and productivity (Aber and Melillo, 1991). Up to some maximum, increasing rainfall leads to greater net primary production of C which, in turn, leads to greater production of organic matter and more rapid rates of nutrient cycling (Schlesinger, 1997). Similarly, highly productive ecosystems often have high levels of CO_2 and enhanced organic acid production which lead to low soil solution pH values. These same environments sustain high weathering rates that release more nutrients until their store of primary minerals is exhausted. Temperature controls the rate of both biological and mineral weathering/synthesis processes. At 0 °C, and below, little water is available to mediate chemical reactions. In extremely cold environments, physical disintegration of rock material often predominates over chemical breakdown of minerals. For each 10 °C increase, biologically mediated reaction rates double until the upper limit of enzymatic functioning is reached. This implies that, all other things being equal, soils in tropical environments will receive far more yearly heat energy and will be more chemically weathered than those in arctic environments (van Wambeke, 1992; Ugolini and Spaltenstein, 1992). In hilly and mountainous terrain, precipitation and solar energy inputs are modified by local topographic position. Soil moisture varies along hillslopes because of redistribution from convex to concave positions with consequent impact on vegetation type and productivity. Locally, sites are drier or moister than are predicted by average climatic parameters. Similarly, angle and duration of solar illumination, and hence temperature, are modified by local relief.

In addition to capturing solar energy by photosynthesis, plants have more direct influences on soil formation. The locus of deposition of dead plant material is a factor determining the nature of near-surface horizons. In forests, annual or periodic litter fall deposits organic matter above the mineral soil. The organic matter in the resultant O horizons is partly decomposed in situ and the residues are carried downward either in solution or by soil fauna. Downward movement of organic acids provides chelating power which can produce horizons of depletion in the mineral surface soil (E horizon) where one might expect to find organic accumulation (A horizon). In contrast, grasslands tend to be dominated by fire which releases nutrients as ash at the soil surface, and they shed more organic matter as dead roots and root exudates in the mineral surface soil horizons (producing A horizons). When trees fall, soil and underlying geologic material are often pulled up with the root mass, acting to mix and destroy existing horizons. Root growth breaks up geological substrate through physical expansion and enhances chemical weathering by exuding acidic molecules. In this way, they produce preferential flow paths where pedogenic processes are intensified. Roots fuel nutrient cycling by providing a direct conduit for nutrient movement from soil to aboveground plant parts which are then returned to the upper soil horizons. Mycorrhizal fungi associated with the roots of many plant species are particularly good at decomposing mineral substrate to release P (Section C, Chapter 3).

Roots are themselves an important source of food for animals that burrow either explicitly to find food or primarily for shelter. Fossorial mammals are very effective at mixing parent material fabric and soil horizons. They can sort particle sizes and create macropores that direct preferential water flow. Smaller animals are equally effective at driving processes that either enhance or retard specific vectors of soil formation. Ants, termites, springtails, and earthworms move organic matter from the surface downward, mix soil horizons, and create macropores in different environments. On hillslopes, animals that deposit soil material on the surface can significantly enhance downslope redistribution of mass through their diggings.

2.3 Pedogenic Processes

In combination, environmental variables drive processes that act on geological or preexisting soil substrate to effect pedogenic alteration at scales ranging from the microscopic parts of soil fabric to watershed soil mantles. Commonly, there will be many processes acting partly in conjunction and partly in opposition. In their totality, these processes produce a soil body with recognizable horizons. In some cases, continued enhancement of horizon properties can, itself, modify ongoing processes, changing the vector of soil formation, and initiating horizon deterioration. Understanding soil formation requires measurement of present processes, prediction of future trends based on these processes, and interpretation of past processes based on present, relict morphology. A large body of pedological research informs this chapter. The authors have not referenced most of it, choosing instead to focus on essential concepts and specific examples.

2.3.1 Conceptual Process Model

A given body of *in situ* soil material, regardless of size, can be altered by additions, losses, transformations, or translocations (Simonson, 1959). These processes interact differently depending on the depth from the soil surface and the mix of environmental variables at a specific location. Some of these processes lend themselves to mass balance calculations, others to use of specific tracers or extractions, but, in general, there is no single, universal approach that will provide knowledge of these four processes.

2.3.1.1 Additions
Soil is formed when C (and N) are added from the atmosphere by biological fixation processes. The role of biological enzymes in plants and associated microorganisms is to reduce atmospheric components into forms that are available for building living structures. Dead aboveground structures are shed to the soil surface and roots are sloughed into the soil. Much of this material supports a host of microorganisms which utilize the C and N for their protoplasmic needs and oxidize the organic material for energy. Residues from these processes accumulate in soils in complex forms collectively known as soil organic matter (SOM). Since addition of organic residues is greatest near the surface, soil functioning can often be conceptualized into two compartments: an upper part where organic acids and microbial dynamics hold sway and a lower part where inorganic processes dominate but are still influenced by organic processes. It is the mixture of SOM with the porous matrix of parent material minerals and newly synthesized minerals that fundamentally defines Earth's soils.

Solutes and particulate matter are added to soils either directly from the atmosphere (Simonson, 1995) or by movement from topographically higher points in the soil landscape (Paton et al., 1995). Although these added components are often incorporated into the soil with little outward evidence of their influence, their role can be significant, such as when ecosystems growing on highly weathered soils are sustained by nutrients added from atmospheric sources. When additions are rapid and large, existing soils may be buried and new soils developed on the added material. An intermediate result is formation of cumulic soils where soil formation continues as new matter is added at the surface, thickening the soil. Cumulic soils are often found along the lower portions of hillslopes where matter eroded from topographically higher positions accumulates and in areas receiving moderate amounts airborne dust (Section E, Chapter 1).

2.3.1.2   Losses

Mass is lost from soils through erosion by wind and water at the surface and by leaching of solubilized components and colloids downward or laterally from the soil column in percolating water. Eroded soils often lie on the upper slope portions of hillslopes. Microbes break down SOM and minerals weather, releasing ions to the soil solution. In the organically dominated soil compartment, organic chelates facilitate movement of ions deeper into the profile. When excess water moves through soil pores, some of these elements are carried further down profile and out of the soil, either moving into local stream water or into groundwater. Typical cations lost through leaching are Na^+, Ca^{2+}, K^+, Mg^{2+}, and as soil acidity increases, smaller amounts of Al are leached in variously charged oxyhyroxide forms. Except at pH values > 9, $Si(OH)_4$ is leached as a nonionized compound. Anions leached from the soil are dominated by HCO_3^- which leads to significant amounts of atmospheric carbon sequestration on geological time scales (Chadwick et al., 1994). Lesser amounts of NO_3^- and SO_4^{2-} are leached from soils but those receiving acid rain input may show large losses of these anions (Sposito, 1989). Under aerobic conditions, little Fe is leached from soils and both Fe and Al tend to accumulate by residual enrichment because of leaching of other elements. Iron and Mn can be leached from soils under anaerobic conditions because their reduced forms are soluble.

2.3.1.3   Transformations

Organic matter added to soil is modified by microbial respiration leading to accumulation of more complex SOM. Weathering of parent material minerals provides not only nutrients for biological activity, but also building blocks for synthesis of secondary minerals. Compared with mantle- and crust-derived minerals that form soil parent material, secondary minerals are enriched in stable elements such as Al and Fe and depleted in easily solubilized elements such as base cations and Si. Under humid, well-drained conditions, where leaching is maximized, secondary mineral formation is characterized by synthesis of aluminosilicate clays and Al and Fe oxyhydroxides. In arid environments, leaching is negligible and even basic cations accumulate in the lower parts of soil profiles as soluble and semisoluble salts. These chemical transformations can lead to formation of secondary minerals that are particularly responsive to physical gradients in soil. For example, wetting and drying results in swelling and shrinking that transform the soil fabric, producing a structure reflective of its mineralogical, textural and organic matter status. Smectite, vermiculite and noncrystalline clays hold much more hydroscopic water than do the primary minerals from which they form. Chemical and mineralogical transformations cannot only release nutrients for biological use but also sequester nutrients. Nitrogen can be immobilized by NH_4^+ fixation in clay minerals and P can be fixed by sorption onto Fe and Al oxides at low to neutral pH values and precipitation with $CaCO_3$ at high pH values.

2.3.1.4 Translocations

Colloids composed of low molecular weight organic matter, small clay particles, and dissolved constituents are moved by water in soil profiles. Commonly, the location of mobilization and deposition are controlled by complex chemical interactions between soil solution and the colloidal fraction. Translocation in humid environments usually is effected by downward flow of water. In arid environments, it is often downward as well but salts can be carried upward by capillary flow of water in response to evaporation. These processes lead to formation of zones of colloid depletion such as E horizons and enrichment such as Bhs ($\approx$spodic), Bt ($\approx$argillic), Bk ($\approx$calcic), and Bqm ($\approx$duripan) horizons. The activity of larger fauna and flora often acts as a counterbalance to colloid redistribution because they stir soil profiles and slow or prevent horizon formation.

2.3.2 Dominant Processes

Locally, soil processes reflect a balance among gains, losses, internal redistribution, and chemical and physical changes. The resulting soil properties represent the long-term effect of these processes acting on a three-dimensional reaction column that is open to exchange of matter and energy with the environment.

Below, important soil-forming processes are described beginning with two universal aspects of soil formation: organic matter accumulation and formation of soil structure. Subsequent sections are organized to cover processes that are driven by water flux and to some degree, reflect the influence of progressively greater leaching: accumulation of soluble salts and gypsum, accumulation of $CaCO_3$, accumulation of opaline silica, redistribution of clay, complexation and redistribution of Fe and Al, leaching of Si and concentration of resistant oxides. Finally, the processes in soil environments where the presence of water is so prevalent that it greatly limits the supply of O are discussed. In these soils, microbially induced reduction controls important aspects of the chemistry, mineral transformation and morphology.

2.3.2.1 Organic Matter Accumulation and Alteration

The primary source of organic matter in soils is vegetation, with leaves, stems, and floral parts added to the surface as they drop from the plant, and roots added directly into the soil itself as they grow. Animals and microorganisms (Section C, Chapters 2 and 3) feed on the vegetable matter, and each other, decomposing the organic matter parent material to yield gases (e.g., CO_2, CH_4) and humus. The influence of SOM is out of proportion to its weight, a reality that is recognized in the designation of soils as organic when they contain a minimum of only 12 to 20% organic C. The organic fraction of soils can be extremely important in determining many aspects of soil processes. For example, even in soils with as little as 0.5 to 1% organic C, SOM controls pesticide adsorption.

Organic Matter Decomposition

Soil organisms use organic tissues as a source of energy and C. Each type of organism has its own role in the decomposition process, from primary decomposers such as beetles, ants, termites, earthworms, and fungi, to decomposers that feed on the feces or tissue of the primary decomposers such as bacteria, fungi and various meso- and macrofauna. A succession of populations operates, each using the altered material from the previously active population and further altering the organic substrate itself. The result is a successive depletion of chemical energy sources, increased resistance to microbial attack, and a lowering of the C:N ratio of the SOM. The rate of decomposition is controlled by a number of factors related to the quality of the organic matter and environmental conditions that affect biological activity. The amount and kind of the various compounds in the organic matter are critical (Melillo et al., 1989). Simple organic acids, sugars, and starches are very quickly (on the order of days to months) utilized by microorganisms. Protein, chitin, cellulose and hemicellulose are used less rapidly and are listed in order of increasing resistance to decomposition. Lignin, fats, and waxes are most resistant to decomposition. Biological activity in soils is greatest when the soil pH is between 6 and 8, soil temperature is 20 to 40 °C, soil water potential is between field capacity and about -1.5 MPa, and there are no deficiencies in essential nutrients. On the other hand, there are microbes adapted for virtually every condition that may come about naturally in soils (Atlas and Bartha, 1981).

Characteristics of Soil Organic Matter

Soils may contain organic matter ranging across the entire spectrum of decomposition, products, particularly if a mixing mechanism, such as burrowing animals or surface cracks, operates to incorporate fresh organic material into the soil (Quideau et al., 1998). In the absence of aggressive mixing, much decomposition occurs within a litter layer. In general, SOM is increasingly decomposed with depth. The final product is humus: the dark brown, complex, microbially resistant, colloidal compounds that are highly modified original materials or materials synthesized by microbes. Humus has high surface area and chemical reactivity (CEC ranges from 100 to 300 $cmol_c$ kg^{-1}), is very hydroscopic, and readily forms complexes with the inorganic fraction (Section B, Chapter 2). Humus is so complex that its composition is usually defined using the following operational criteria (Oades, 1989): fulvic acid fraction which is soluble in both alkali and acid, humic acid fraction which is soluble in alkali but insoluble in acid, and the humin fraction which is insoluble in both alkali and acid. Fulvic acid is composed of the lowest molecular weight compounds and has the greatest number of acid groups and the highest cation exchange capacity. In contrast, humin is composed of the highest molecular weight compounds and has the least amount of acid groups and the lowest CEC.

Relationship of Organic Matter to Morphology and Function

The accumulation of humified organic matter is key to formation of an A horizon which, if it becomes thick enough, may evolve to qualify as a mollic or umbric epipedon. A horizons are an important link between soils and plants, since most plant roots exist within the A horizon and most of the nutrients are extracted from there. Decomposition of organic matter in the A horizon, including the roots themselves, biocycles essential nutrients from the vegetation back into the plant-available pool. High levels of organic acids lead to maximum rates of primary mineral decomposition in A horizons as well. Addition of organic matter also tends to promote development of soil structure and porosity, thereby enhancing water infiltration and water-holding capacity.

2.3.2.2 Development of Soil Structure

An essential component of soil development is the substitution of soil structure for the geologic structure of parent material. While geologic structure is inherited from jointing cracks and bedding planes, soil structure forms as a function of the downward movement of water, organic matter and mineral components along preferential flow lines. Soil structural units, known as peds, become more strongly expressed as their faces are differentiated from their interiors. The surface/interior differentiation comes about through several mechanisms. Wetting-induced expansion forms pressure faces or stress argillans by orienting platy silt and clay particles along contact planes between peds. Subsequent drying causes contraction which opens voids between peds. Upon rewetting, water flows preferentially through the contraction voids depositing clays and organic matter on the ped faces, further differentiating the outer from the inner parts of the peds. Other compounds, including calcite, silica, and Fe and Mn oxides can also be deposited along ped faces. Conversely, under some conditions, outer margins of peds become depleted by preferential leaching of clay and soluble compounds. As roots grow, they follow paths of least resistance and concentrate in voids between structural units. These roots enrich ped surfaces with organic matter and enhance preferential drying of ped exteriors relative to interiors.

Size, Degree of Expression, and Reinforcement of Peds

A number of factors influence the size and degree of expression of soil structural units. The amount of volume change associated with wetting and drying cycles (swelling and shrinking) is very important. Fine-textured soils and those with smectitic mineralogy typically have relatively large volume changes, and therefore, develop the most strongly expressed structural units. Fine-textured soils also have less resistance to shear and so tend to have more cracks and finer structural units. Frequent shrink/swell cycles tend to reinforce the expression of structural units (Southard and Buol, 1988). Slow drying allows a relatively uniform shrinkage and contraction of the soil mass, producing larger peds. Crystallization of salts in saline soils yields a finer structure than is found in similar nonsaline soils (Reid et al., 1993).

Granular Structure in Surface Horizons

The structure of surface horizons is often most strongly influenced by biological factors, including organic matter, microbes, plant roots, and soil fauna (Fig. 2.1). Fecal casts of soil macrofauna, such as earthworms (Lee and Foster, 1991) and snails (Anderson et al., 1975), often form compact, stable aggregates which dominate in some A horizons. Dense networks of fine grass roots and fungal hyphae cause uniform fine-scale shrinkage cracks and enmesh soil materials into granular or crumb-like units

Type of Structure	Soil Environment	Formation processes
Granular 	A horizon	Aggregation by biological agents: organic compounds, fungal hyphae, fine roots, fecal casts
Platy 	A and other surface horizons Petrocalcic horizons Duripans	Vertical compression from compaction
Prismatic 	B horizons with uniform texture, fine roots, few coarse fragments, relatively slow drying	Horizontal compression from shrink-swell
Blocky 	B horizons with coarse fragments, heterogeneous root sizes	Shrink-swell interrupted by nonhomogeneous matrix

Fig. 2.1 Types of soil structure, their occurrence within soils, and the processes responsible for their formation

(Oades, 1993). At the most fundamental level, soil aggregates in surface horizons are usually held together by soil organic matter-mineral complexes (Section A, Chapter 7; Section B, Chapter 2).

Platy Structure

Platy structure typically forms in response to unidirectional compressional forces (Fig. 2.1). It is most often produced in surface soils through compaction by animals or machines, or by raindrop impact. Vertical compaction yields horizontally oriented, platy structural units. Gravity can contribute to the necessary vertical force for platy structure development in the silty vesicular A horizons that are common in arid and semi-arid regions. In these surface horizons, vesicles are formed by escaping gases as the soil material is wetted. Repeated wetting and drying cycles enlarge the vesicles until they can no longer support the weight of the overlying material and collapse, forming planes that bound platy structural units (Miller, 1971). Platy structure is also common in highly developed petrocalcic horizons and duripans. The origin of platy structure in these subsoil horizons is less clear, but may be caused by vertical compression generated as precipitation of secondary calcite and opal increases subsoil volume. In pergelic soils (Gelisols), platy structure is associated with the formation of lenticular ice (vein ice) (Section E, Chapter 6.6)

Prismatic and Blocky Structure

Shrinkage of moist soil material as it dries results in multidirectional compression centered around numerous loci within the mass. When shrinkage forces are largely resolved in lateral directions, the soil material may contract uniformly toward more or less equally spaced centers forming prismatic or columnar peds bounded by vertical cracks (Fig. 2.1). In formation of prismatic structure, uniform shrinkage is important because it develops the consistently spaced and arranged cracks that form ped boundaries. Uniform shrinkage is favored by slow drying and homogeneous soil materials. It is often associated with uniformly fine-textured sediments in poorly drained, low-lying landscape positions. Prismatic or columnar structure is common in natric horizons because of enhanced swelling and poor drainage associated with Na-saturated clays. In contrast to the elongated character of prismatic structure, blocky structure forms as a result of combined lateral and vertical shrinkage forces (Fig. 2.1), with neither predominating. Conditions favoring blocky structure development are rapid drying and factors of nonhomogeneity, such as woody roots, and coarse fragments. Also, bedded moderately fine and fine-textured fluvial or marine deposits tend to form blocky structure in B horizons because horizontal planes of weakness, inherited from parent materials, are bisected by vertical shrinkage cracks.

Vertic Features

Vertisols and other clayey, expansive soils develop characteristic structural features as a result of their dynamic shrink/swell behavior (Section E, Chapter 6.6). Upon wetting, the fabric of these soils swells, commonly increasing its volume by about 10 to 15%. In surface horizons, this increase in volume is accommodated by upward movement, but in the subsoil, overburden pressures confine vertical movement so that lateral swelling pressures are at least four times as great as the vertical pressures and exceed the soil shear strength (Wilding and Tessier, 1988). Under these conditions, shear failure takes place at angles between 10 and 60 degrees from horizontal. The shear planes present slick, grooved surfaces, known as slickensides, and bound characteristic wedge-shaped aggregates.

Aggregation by Fe and Al Oxides

Highly weathered soils, such as Oxisols, have a fine (< 2 mm diam.) granular structure that is the result of aggregation by Fe and Al oxyhydroxides (van Wambeke et al., 1983) (Section E, Chapter 6.13).

Aggregation is effected by very small positively charged Fe and Al oxyhydroxide particles and surface coatings that form bridges between negatively charged clay particles to build up small aggregates (Hsu, 1989; Schwertmann and Taylor, 1989).

2.3.2.3 Accumulation and Redistribution of Salts

Salts of Na, Mg, K, and sulfates, carbonates, and chlorides occur in soils of arid environments (Table 2.1). They are more than 100 times more soluble in water than gypsum and move readily with saturated and unsaturated water flow in soil. Salts precipitate when their solubility is exceeded, usually as the soil solution is concentrated by evapotranspiration. Salt content and the ionic concentration in soil solution impact plant growth and soil properties such as structure and infiltration. The ultimate source of salts is weathering of primary minerals (Fig. 2.2) which produces dilute soil solution. Atmospheric deposition contributes appreciable quantities of salts to soils, particularly as dryfall (dust) in arid regions and as sea spray along coastal margins. Most saline soils receive salts from sources of salt that have been concentrated by geological processes. For instance, dissolution of marine or closed basin lake sediments provides preconcentrated salts to near-surface waters.

The distribution of salts in soil landscapes is controlled primarily by subsurface hydrology and the balance between evapotranspiration and leaching. The exfiltration of saline water through seeps, often following outcrops of impermeable strata that act as aquitards, causes localized occurrences of highly saline soils (Fig. 2.2). In well-drained soils where leaching is greater than evapotranspiration, salts do not accumulate because the constituent ions are leached to groundwater. On the other hand, salts accumulate when leaching is minimal (Section E, Chapter 1). Low leaching results from high evapotranspiration rates and/or low rainfall, convex topography that disperses water flow, and soil conditions such as crusting that yield low infiltration rates. In each of these cases, the areal and vertical distribution of salts reflects a balance between a downward or lateral flux of leaching water that removes salts, and the upward flux of evapotranspiration that helps retain salts within the soil. Although the dominant influence of plants on salt distribution is through their role in evapotranspiration, some plants have a localized effect on soil salinity and sodicity through biocycling. For example, surface soils under the canopy of greasewood have electrical conductivity (EC) values six times those of soils in interspaces between the shrubs, and exchangeable Na percentage (ESP) values are more than 10 times greater (Fireman and Harward, 1952).

Table 2.1 Some common salts in soils and their solubilities in water at 25 °C

Name	Formula	Solubility
		mol L^{-1}
Halite	NaCl	6.15
Thenardite	Na_2SO_4	1.97
Mirabilite	$Na_2SO_4 \cdot 10H_2O$	2.74
Nahocolite	$NaHCO_3$	1.22
Soda	$Na_2CO_3 \cdot 10H_2O$	2.77
Trona	$Na_3H(CO_3)_2 \cdot 2H_2O$	2.56
Bloedite	$Na_2Mg(SO_4)_2 \cdot 4H_2O$	2.31
Hexihydrite	$MgSO_4 \cdot 6H_2O$	3.17
Epsomite	$MgSO_4 \cdot 7H_2O$	3.03
Gypsum	$CaSO_4 \cdot 2H_2O$	0.005
Calcite	$CaCO_3$	0.0006 (pH 8 & $CO_2 = 10^{-4}$ MPa)

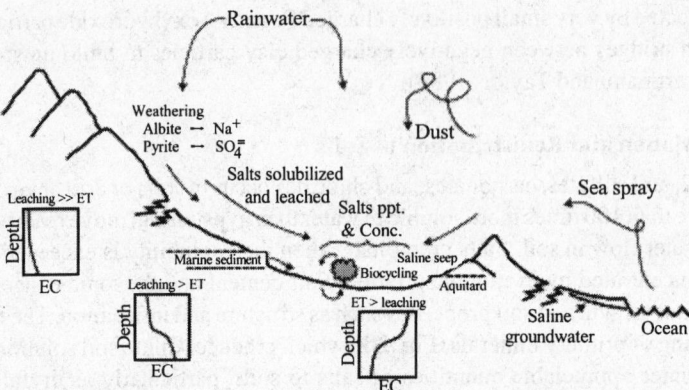

Fig. 2.2 Sources and redistribution of soluble salts in soils in relation to hydrology, lithology, and landscape position. When the downward leaching flux of water greatly exceeds upward flux due to evapotranspiration (ET), soluble salts, as indicated by electrical conductivity (EC) of saturation extracts, are minimal throughout the soil profile. When leaching is slightly greater than ET, EC increases with depth, since salts are leached from the surface soil. When ET is greater than leaching, EC is greatest near the surface due to the predominating upward flux of water, carrying salts, to the evaporative surface.

Salt-affected soils have major impacts on the flocculation or dispersion of clay minerals and organic matter (Section G, Chapters 1 and 2). In general, high electrolyte concentrations keep soils flocculated, whereas colloidal dispersion occurs when the electrolyte concentration is low and Na predominates on exchange sites. Thus, sodic soils (ESP > 15%, EC < 4 dS m^{-1}) are dispersed, resulting in clogged pores and very low infiltration rates, whereas saline (ESP < 15%, EC > 4 dS m^{-1}) and saline-sodic (ESP > 15%, EC > 4 dS m^{-1}) soils are flocculated, promoting porosity and higher infiltration rates. Sodic soils also have high pH values (9–12) resulting from the production of hydroxyls by dissolution of Na_2CO_3 and, to a lesser extent, by hydrolysis of exchangeable Na (Section G, Chapter 2). Sodium carbonate can form in soils through several mechanisms (Whittig and Janitzky, 1963). Weathering of Na aluminosilicates releases Na, which reacts with HCO_3^- in water to form $NaHCO_3$, which concentrates to Na_2CO_3 under the influence of evaporation. Neutral Na salts can react with $CaCO_3$ to yield Na_2CO_3:

$$Na_2SO_4 + CaCO_3 \rightleftharpoons Na_2CO_3 + CaSO_4 \qquad [2.1]$$

and Na can be replaced from the exchange by H or Ca from carbonates:

$$[CLAY]-2Na + CaCO_3 \rightleftharpoons [CLAY]-Ca + Na_2CO_3 \qquad [2.2]$$

Examples

The prairie pothole wetlands of North Dakota illustrate the strong links between hydrology and soil chemistry and mineralogy (Fig. 2.3) (Arndt and Richardson, 1989). The highest topographic positions are known as recharge areas because they receive water only from precipitation, resulting in seasonally ponded basins where leaching predominates. This yields a dilute soil solution with Ca as the dominant cation, HCO_3^- and SO_4^{2-} as the dominant anions, and no solid-phase calcite or salts. The intermediate elevations have basins that are semipermanently ponded with input from both precipitation- and from

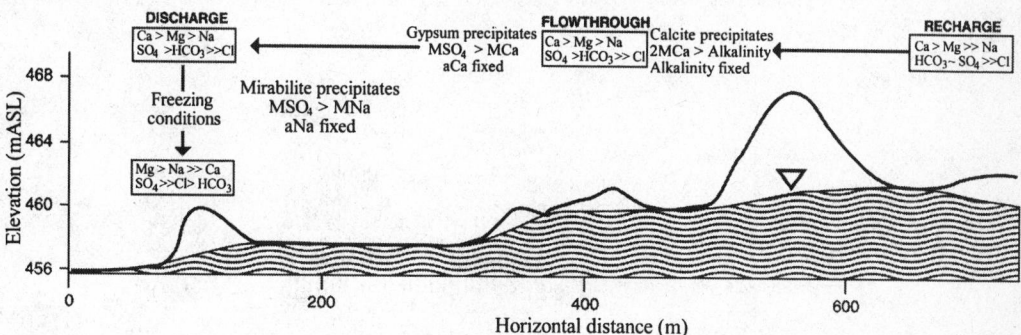

Fig. 2.3 Linkages between hydrology, soil solution chemistry, and mineralogy in the prairie-pothole wetlands of North Dakota. Alkalinity as HCO_3^- molarity (M) and activity (a) as designated [Reprinted from Arndt and Richardson, 1989. Soil Sci. Soc. Am. J. 53:848-855 with permission of the Soil Science Society of America]

recharge-derived groundwater. In these parts of the system, known as throughflow areas, salinity levels are higher. Coming into the throughflow areas, Ca^{2+} is still the predominant cation in solution, but the precipitation of calcite has taken HCO_3^- from solution, thereby fixing alkalinity levels. The dominant anion is then SO_4^{2-}, some of which, together with most of the remaining Ca is precipitated as gypsum. Soils in the throughflow areas may contain both calcite and gypsum. The lowest landscape positions are permanently ponded with most input coming from the discharge of groundwater from the upslope areas. In these discharge areas, SO_4^{2-} remains the dominant anion and Mg^{2+} and Na^+ are the dominant cations. The soils still contain some calcite and gypsum, but also may contain more soluble salts, such as mirabilite. Thus, the chemical and mineralogical characteristics of soils in this landscape are controlled by the hydrologic system and the relative solubilities of minerals that can precipitate from the solution.

A second example illustrates the feedback relationships among salts, soil properties and processes, and vegetation. The Carrizo Plain is a valley with no external drainage in the semi-arid central Coast Ranges of California (Reid et al., 1993). Within 1 km to the east of the soda lake that fills the lowest part of the valley, the grass covered plain is broken by spots of bare soil, known as slickspots, ranging from 0.1 to 15 m in diameter and generally depressed 10 cm below the surrounding grassed surfaces (Fig. 2.4). The grass-cover traps dust, including salts, blown up from the soda lake playa, and protects the surface from wind and water erosion. The grass-covered soils have clayey Bt horizons with strong prismatic structure which allows rapid infiltration of rainwater and leaching of salts. In these soils, solubility of the evaporite minerals increases with depth in the profile (calcite-gypsum-soluble salts), indicating a regime in which leaching predominates over evapotranspiration. In contrast, < 1 m away, the bare surface of the slickspot is crusted from raindrop impact and the upper few centimeters are sufficiently leached of salts so that the soil disperses. Crusting and dispersion result in very low infiltration rates so that leaching of the profile is minimal. The predominance of evaporation over leaching is indicated by the decreasing solubility of pedogenic minerals with depth in the profile (soluble salts-gypsum-calcite), a trend opposite that in the grassed areas. Under the slickspots, the minimal infiltration at the surface causes the subsoils to remain dry even during the wet season. This, in turn, establishes a matric potential gradient from the moist subsoils under grass so that salts are carried laterally in solution as water is drawn by capillarity into the slickspot soils. The buildup of salts in the slickspot soils eventually has a toxic effect on the grasses at the slickspot margin, which exposes bare soil to raindrop impact and wind erosion, thereby promoting expansion of the slickspot. Under the influence of these processes, the landscape is becoming increasingly saline and barren.

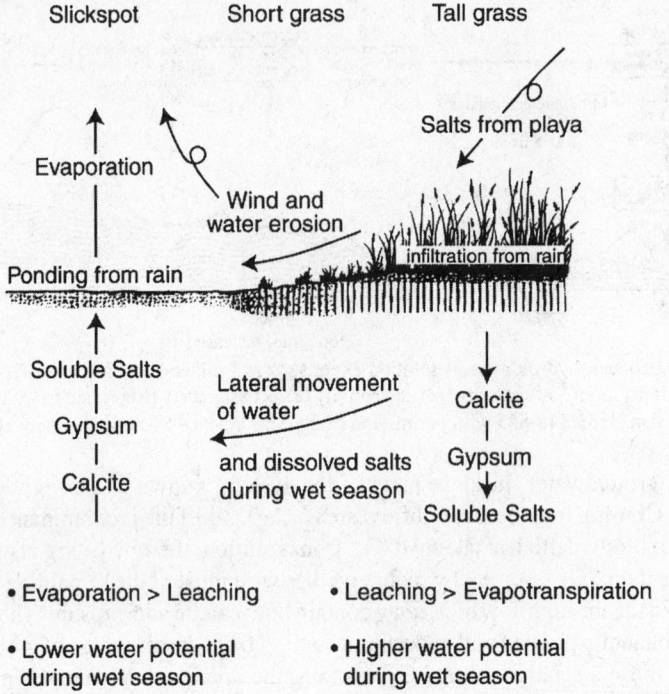

Fig. 2.4 Diagrammatic representation of processes contributing to the salinization and expansion of slickspots on the Carrizo Plain, California [Modified from Reid et al., 1993. Soil Sci. Soc. Am. J. 57:162-168 with permission of the Soil Science Society of America]

2.3.2.4 Accumulation and Redistribution of Calcite

Where soil pH values are above about 7.2, $CaCO_3$ can precipitate in soils to form the mineral species calcite. Calcite contains < 5 mol % Mg. At 25 °C, pH 8, and Pco_2 of 10^{-4} MPa, calcite has a solubility of 0.0006 mol L^{-1} (0.06 g L^{-1}), which is much more soluble than silicate minerals, but is considerably less soluble than the salts commonly found in soils. In well-drained soils, calcite is found in soil formed in arid, semi-arid, and subhumid environments. With progressively greater leaching, it is found at greater depths below the soil surface. It can also accumulate in poorly drained soils where calcium and bicarbonate are concentrated in ponded water.

Natural Sources of Calcite in Soils

While calcite may be inherited in soils from parent materials, such as limestone, pedogenic calcite precipitates from soil solution and requires sources of Ca^{2+} and CO_3^{2-} ions. Calcium ions may be derived from the weathering of parent materials. Calcareous rocks provide the most abundant Ca, with limestone containing 30 to 40% Ca and dolomite containing about 20% Ca. Calcareous tills, loess, and other sediments also provide a ready supply of Ca for pedogenic calcite formation. Calcium is present in primary minerals, particularly Ca-rich plagioclase, but also in some amphiboles, pyroxenes, garnets, and epidote. Among igneous rocks, Ca is most abundant in the more mafic (but not ultramafic) rocks, such as gabbro and basalt, which contain about 7% Ca. Atmospheric deposition can be a substantial, even dominant, source of Ca, both in ionic form and as particulate $CaCO_3$. For example, in southern Nevada and California, dust contains 10 to 30% $CaCO_3$, equating to a deposition rate of

1 to 6.6 g $CaCO_3$ m^{-2} yr^{-1} (Reheis and Kihl, 1995). In southern New Mexico, dust contributes about 0.4 g $CaCO_3$ m^{-2} yr^{-1}, whereas rain delivers sufficient Ca to precipitate 1.2 g m^{-2} yr^{-1} (Birkeland, 1984). On the Edwards Plateau in Texas, no $CaCO_3$ is delivered as dust, but rain supplies Ca equivalent to 2.3 g $CaCO_3$ m^{-2} yr^{-1} (Rabenhorst et al., 1984). Calcium is also derived from biocycled plant material and groundwater which may be highly calcareous, depending on the lithologies exposed in the aquifer. When Ca laden groundwater lies within or just below the soil zone, calcite can form in the soil as it is precipitated within the capillary fringe.

The source of the carbonate anion in calcite is the dissolution of CO_2 in water, which yields the following species:

$$CO_2 + H_2O \rightleftharpoons H_2CO_3 \qquad\qquad [2.3]$$

$$H_2CO_3 \rightleftharpoons HCO_3^- + H^+ \quad pK = 6.4 \qquad\qquad [2.4]$$

$$HCO_3^- \rightleftharpoons CO_3^{2-} + H^+ \quad pK = 10.3 \qquad\qquad [2.5]$$

The amount of CO_2 in the soil solution depends on the relationships expressed in Henry's Law:

$$K_H = (H_2CO_3)/P_{CO_2} = 10^{-1.5} \qquad\qquad [2.6]$$

An increase in the proportion of CO_2 in the gas phase increases the CO_2 concentration in the solution in contact with the gas. The P_{CO_2} in soils is on the order of 0.003 to 0.03, compared to 0.00033 in the atmosphere. The distribution of CO_2 in soils depends on a balance between production of CO_2 within the soil by microbes and plant roots and diffusion losses of CO_2 out of the soil to the atmosphere. As an example, the CO_2 concentration in a coarse-textured soil in the southern Nevada desert was greatest in spring when biologic activity was at its peak (Fig. 2.5) (Terhune and Harden, 1991). Even at this time, the CO_2 concentration in the surface soil was less compared to the subsoil due to diffusional losses to the atmosphere. Biologic activity and CO_2 production were low during the winter due to cold temperatures and the summer due to dryness.

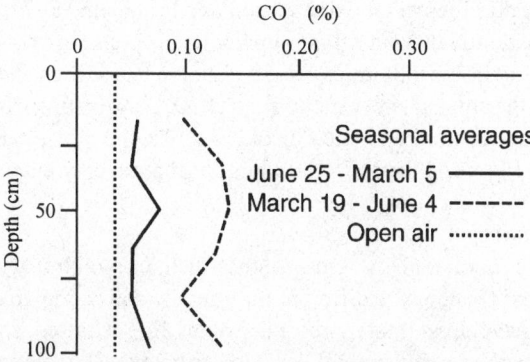

Fig. 2.5 Seasonal change of CO_2 with depth in an Entisol in southwestern Nevada [Reprinted from Terhune and Harden, 1991. Soil Sci. 151:417-429. Copyright Williams & Wilkins, Baltimore, MD with permission]

Calcite Precipitation in Soils

The reaction for calcite precipitation can be expressed as:

$$Ca^{2+} + 2HCO_3^- \rightleftharpoons CaCO_3 + H_2O + CO_2 \qquad\qquad [2.7]$$

Precipitation of calcite is promoted by a number of factors that influence this relationship. An increase in pH drives the reaction to the right by supplying more HCO_3^- as a result of $CO_2 + OH^-$. Decreasing the P_{CO_2} through the loss of CO_2 to the atmosphere also drives the reaction to the right. Loss of water through evapotranspiration increases the ionic concentration of the soil solution to the point where it exceeds the solubility product of calcite and results in precipitation. Precipitation of calcite in soils is not simply an inorganic chemical phenomenon. In fact, the role of microorganisms seems to be ubiquitous and perhaps essential in pedogenic calcite formation. Microbial influence is revealed by calcified biological structures, such as fungal hyphae, and experimental evidence that shows that both bacteria and, especially fungi, produce calcite as a byproduct of their metabolism. In Ca-rich soils, microbes excrete excess Ca, which concentrates on their external surfaces and reacts with HCO_3^- in the soil solution. In the laboratory, calcite precipitates rapidly in inoculated soil columns, but not at all in sterile columns that were otherwise identical (Monger et al., 1991).

Precipitation of calcite in soils occurs preferentially in certain types of morphologic sites. These favored sites include the vicinity of roots, where microbes and nucleation sites are abundant and in large pores where drying is relatively rapid and the P_{CO_2} is lower than in the matrix due to effective gas exchange with the atmosphere. In gravelly soils, calcite tends to precipitate preferentially on the undersides of gravels, perhaps because water tends to collect and is last to dry in those sites. In any case, calcite is preferentially precipitated on preexisting calcite crystals.

Examples of Calcite Distribution and Precipitation in Soils

The processes governing precipitation of calcite in a well-drained soil where water flux is downward, are illustrated in Fig. 2.6a. Calcium in the soil originates from weathering, rain, and/or dissolution of dust. Organic matter decomposition in the A horizon produces a high P_{CO_2}, although it decreases substantially near the soil surface due to gas exchange with the atmosphere. The high P_{CO_2} yields a relatively large amount of CO_2 dissolved in the soil solution, producing HCO_3^-. The Ca^+ and HCO_3^- ions are leached with soil water to a depth at which calcite precipitates. This depth is determined by decreased P_{CO_2} below the zone of major biological activity and, perhaps more importantly, by the soil solution being concentrated as it enters the dry subsoil and is depleted by evapotranspiration.

In poorly drained soils, dominant water flux is upward in response to evaporation (Fig. 2.6b). The relatively high water content in subsoil limits gas diffusion, so P_{CO_2} is high and decreases near the soil surface as CO_2 diffuses to the atmosphere. Calcium and HCO_3^-, originating from groundwater sources or *in situ* from the soil, move upward with the flux of water. Calcite precipitates in an upper soil zone where the P_{CO_2} is relatively low and the soil solution is concentrated by evaporation.

Relation to Soil Morphology

Soils with pedogenic calcite commonly progress through distinct evolutionary stages (Table 2.2). The stages and morphologic expressions are different for gravelly soils compared to fine-textured soils. Because gravelly soils have less total pore space than fine-textured soils, segregated calcite accumulates more rapidly, precipitating readily on pebbles bounded by relatively large pores. In fine-textured soils, calcite precipitates first as filaments in root pores, then as soft masses. Eventually a calcite-cemented (petrocalcic) horizon forms where carbonate plugs nearly all the pores and laminar plates of carbonate build up at the interface between the overlying horizons and the plugged ones.

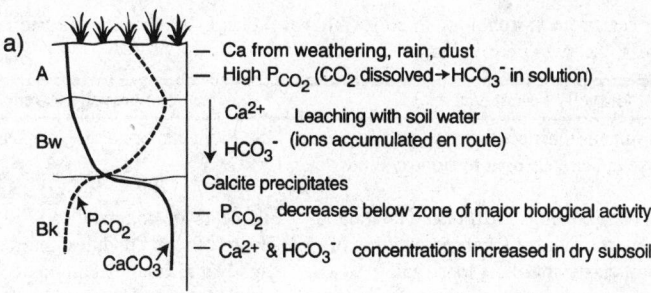

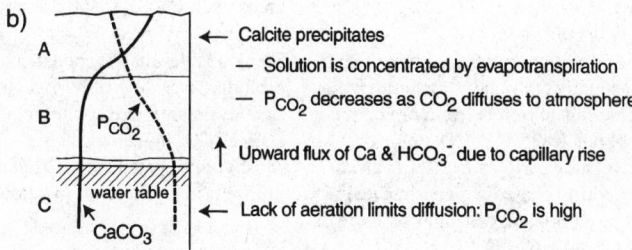

Fig. 2.6 Idealized diagram of processes involved in calcite precipitation in (a) a well-drained soil (water flux is downward), and (b) a poorly drained soil (water flux is upward).

Formation of a petrocalcic horizon (stage IV or greater) has major implications for the direction of further development. Plugged horizons form a barrier to most root penetration. They also resist erosion and the overlying horizons may be stripped off, leaving the petrocalcic exposed at the surface. Burrowing animals and erosion commonly bring fragments of petrocalcic horizons to the surface, where the calcite is dissolved and recycled back into the soil (Egbhal and Southard, 1993).

2.3.2.5 Accumulation and Redistribution of Silica

After O, Si is the most abundant element in Earth's crust. It exists in many primary mineral forms and is incorporated into soil-formed minerals as well. When Si is released by weathering, it is leached and can be lost or can be incorporated into aluminosilicate clay minerals, such as kaolinite or smectite. In humid environments, most Si is either leached or consumed during clay mineral synthesis. In subhumid and semi-arid environments where there is enough water to support mineral weathering but not enough to completely leach Si, it can also precipitate as amorphorus silica that cements soil horizons.

Characteristics of Silica in Soils

Silica refers to compounds consisting of SiO_2 in crystalline, poorly crystalline, or amorphous forms, and is sometimes hydrated to some degree. Quartz is a common form of silica in many soils. It is highly crystalline SiO_2 that is inherited from parent materials and usually does not form pedogenically. Opal-A, the common pedogenic and biogenic silica is a hydrated, X-ray amorphous, form of SiO_2. A somewhat more crystalline form of opal, opal-CT, may form in some very old soils, but it is not nearly as common as opal-A.

Table 2.2 Stages of carbonate morphology in soils (After Birkeland, 1984. Soils and geomorphology. Copyright Oxford University Press, Inc. with permission]

Stage	Gravelly Parent Material	Nongravelly Parent Material
I	Thin discontinuous clast coatings; some filaments; matrix can be calcareous next to stones; about 4% $CaCO_3$	Few filaments or coatings on sand grains; < 10% $CaCO_3$
I+	Many or all clast coatings are thin and continuous	Filaments are common
II	Continuous clast coatings; local cementation of few to several clasts; matrix is loose and calcareous enough to give somewhat whitened appearance	Few to common nodules; matrix between nodule is slightly whitened by carbonate (15-50% by area), and the latter occurs in veinlets and as filaments; some matrix can be noncalcareous about 10-15% $CaCO_3$ in whole sample, 15-75% in nodules
II+	Same as stage II, except carbonate in matrix is more pervasive	Common nodules; 50-90% of matrix is whitened; about 15% $CaCO_3$ in whole sample
Continuity of fabric high in carbonate		
III	Horizon has 50-90% fine-grained with carbonate forming an essentially continuous medium; color mostly white; carbonate rich layers more common in upper part; about 20-25% $CaCO_3$	Many nodules, and carbonate coats many grains such that over 90% of horizon is white; carbonate-rich layers are more common in upper part; about 20% $CaCO_3$ in whole sample
III+	Most clasts have thick carbonate coats; matrix particles continuously coated with carbonate or pores plugged by carbonate; cementation more or less continuous; > 40% $CaCO_3$	Most grains coated with carbonate; most pores plugged; > 40% $CaCO_3$ in whole sample
Partly or entirely cemented		
IV	Upper part of horizon is nearly pure cemented carbonate (75-90% $CaCO_3$) and has a weak platy structure due to the weakly expressed laminar depositional layers of carbonate; the rest of the horizon is plugged with carbonate (50-75% $CaCO_3$)	
V	Laminar layer and platy structure are strongly expressed; incipient brecciation and pisolith (thin, multiple layers of carbonate surrounding particles) formation	
VI	Brecciation and recementation, as well as pisoliths, are common	

Sources of Soluble Silicon

The weathering of silicate minerals releases Si to solution that can be precipitated, under appropriate conditions, as opal-A. Soils rich in easily weathered silicates, such as olivine ($[Mg, Fe]_2SiO_4$) and anorthite ($CaAl_2Si_2O_8$), release abundant Si into solution. More resistant silicates, such as quartz, release Si very slowly. Amorphous silica is much more soluble than crystalline forms (Fig. 2.7). Volcanic glass is a primary form of amorphous silica which is abundant in many soils or parent materials, sometimes even in those far from current volcanic activity, since volcanic ash can travel great distances downwind from eruptions. Another source of amorphous silica is biogenic opal, which is actually opal-A produced as part of plant structures, known as phytoliths, or as part of aquatic organisms, such as diatom tests or sponge spicules. Phytoliths are a biocycled form of silica in soils and can be very abundant in A horizons, particularly in grasslands. In soils that are depleted of primary mineral silica, weathering of phytoliths provides a significant portion of the $Si(OH)_4$ in solution (Alexander et al., 1997). Diatoms and sponge spicules are most often found in soils derived from lacustrine, deltaic, or fl-podplain sediments.

Dissolution of Silica

Release of Si into solution from a solid is controlled by a number of factors. The inherent solubility of the mineral or amorphous compound varies widely, with opal-A being the most soluble source of silica

(Fig. 2.7). Solubility also increases dramatically for very small particles (< 0.01 μm). External conditions play a large role, with Si solublization increasing when the soil solution has a low ionic strength, high pH (> 9), and relatively high temperature (Drees et al., 1989). Organic acids promote dissolution by complexing $Si(OH)_4$ released into solution, resulting in highest silica dissolution potentials in the root zone. While coatings of organic materials or Fe/Al oxides can retard dissolution by isolating the reactive surface from the soil solution, sorption of Si onto Fe/Al oxides can make them a sink for soluble Si, thereby keeping solution concentrations of Si low and increasing dissolution. Solubility of silica increases with increased pressure, a situation that may arise in petrocalcic horizons as calcite crystallization increases the subsoil volume and generates grain-to-grain pressure contacts (Monger and Daugherty, 1991).

Translocation and Precipitation of Silicon

In solution, Si exists mostly as silicic acid [$Si(OH)_4$]. It moves with percolating water and precipitates when conditions are favorable. The soil pH influences the way in which silica precipitates. At pH < 7, $Si(OH)_4$ precipitates as SiO_2 on adsorption sites as individual molecules or low molecular weight polymers, whereas at pH 7 to 10, Si in solution takes the form of high molecular weight polymers and these are flocculated by cations in solution. Other factors may alter, or override, the effect of pH on Si precipitation in soils. Silica precipitation is promoted in soil solutions with high ionic strength and adsorption is promoted by a solid phase with a high surface area.

In soils, $Si(OH)_4$ is adsorbed on exposed hydroxyl groups of clays, oxides, and primary silicates (Fig. 2.8). A high surface area (e.g., argillic horizon) or ionic strength (e.g., calcic horizon) promotes rapid adsorption of silica as individual or low molecular weight polymers. Adsorption is greatest at pH 7 to 9. Drying dehydrates the adsorbed $Si(OH)_4$ causing amorphous SiO_2 to precipitate on the surface (Fig. 2.8A). The strong Si-O bond inhibits rehydration and desorption. The precipitated silica acts as a template for further precipitation (Fig. 2.8B) and, eventually, opaline silica forms bridges between the grains on which it precipitates (Fig. 2.8C) (Chadwick et al., 1987).

In fine-textured soil horizons, silica diffuses into smaller voids in the soil matrix, where there is more surface area for adsorption, and precipitates as polymerized layers. The zone of accumulation tends to correspond to the zone of clay accumulation because of the abundant sites for precipitation there. The silica polymers are found as flocs (~1 μm diam.) on grains in the soil matrix and interlaminated with clay skins. The same mechanisms apply for coarse-textured soils, but silica not in

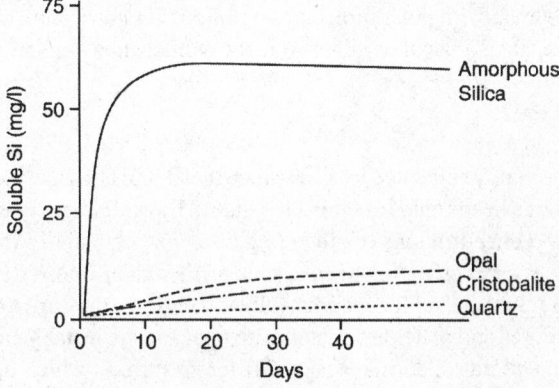

Fig. 2.7   Dissolution of silicon in water from various silica phases as a function of time [Reprinted from Drees et al., 1989. J.B. Dixon and S.B. Weed. Minerals in soil environments, with permission of the Soil Science Society of America]

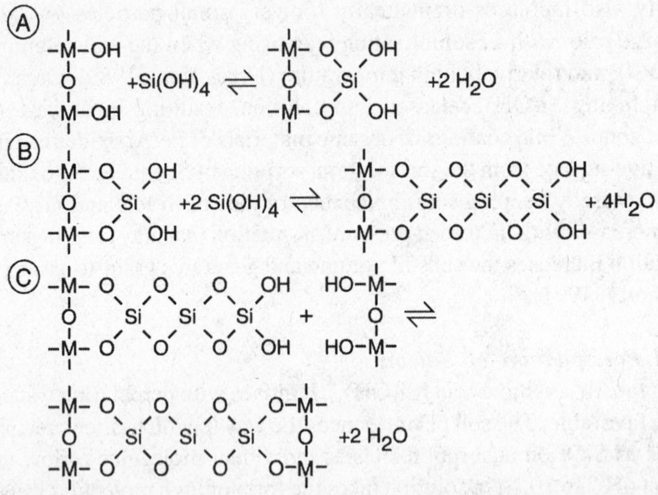

Fig. 2.8 Model of silica precipitation in soils (A) reversible Si(OH)$_4$ adsorption on Fe and Al oxides and oxyhydroxides, silicates, and aluminosilicates (M may be Al, Fe, Mg, or Si); (B) further Si(OH)$_4$ adsorption followed by dehydration to SiO$_2$ during soil drying decreases the reversibility of the original adsorption process; (C) two soil components with Si(OH)$_4$ adsorption surfaces may be bonded together by opaline SiO$_2$ bridges [Reprinted from Chadwick et al., 1987. Soil Sci. Soc. Am. J. 51:975-982 with permission of the Soil Science Society of America]

direct contact with adsorption surfaces may precipitate by flocculation if the ionic strength is sufficiently high (Chadwick et al., 1987). Opaline silica may also precipitate preferentially on the undersides of gravels, even forming pendants (Munk and Southard, 1993).

In many soils of arid and semi-arid regions, the depth distributions of silica and calcite overlap, but these pedogenic compounds are concentrated in different microsites by nature of their mode of precipitation. Calcite precipitates by self-nucleation, particularly in larger voids because these dry early, concentrating the soil solution. They often have the most effective contact with the atmosphere so the P$_{CO_2}$ is relatively low. Silica specifically bonds with Al and Fe hydroxy compounds whereas under most soil conditions, calcium does not. Thus, silica precipitates primarily by adsorption on surfaces and so is found in the matrix, coating grains and clay skins, rather than in voids. Furthermore, there is no direct chemical bonding between Ca and Si(OH)$_4$ at pH < 9, since Si(OH)$_4$ is neutral and nonpolar and cannot compete with polar molecules or anions for adsorption on calcite (Chadwick et al., 1987). Coatings of opal also occur as silans in macrovoids when the soil matrix is plugged with calcite and silica cements.

Relation to Soil Morphology

A soil horizon that is thoroughly cemented by silica, such that it will not slake in water or HCl, is known as a duripan. Since the silica needs only to cement the matrix by bridging between grains, cementation can occur with very low concentrations of silica (e.g., ~ 4%), particularly in coarse-textured soils. Duripans often have accessory cementing agents, such as Fe oxides or calcite. Duripans exhibit two forms corresponding largely to the regional climate. In arid regions, the duripan tends to be platy, with plates 1 to 15 cm thick. Pores and plates are coated with opal and are usually engulfed with pedogenic calcite. In Mediterranean climates, duripans take a different form in which opal coats the faces and pores of very coarse to extremely coarse (0.3–3 m diam) prisms that make up the subsoil. The upper boundary is abrupt and the pan may have an opal coating on top in the strongly developed cases. Water often perches on top of the pan during the rainy season, so that Fe/Mn oxide nodules and illuvial clay accumulate there. The matrix of this kind of duripan is extremely hard when dry, but is brittle when moist and can be penetrated, with some difficulty, with a hand auger. Pedogenic calcite may, or may not, be present, but it is not a dominant component. Reversible, weakly expressed silica cementation seems to play a role in the formation of fragipans. These horizons are dense and appear cemented when dry but will slake in water (Franzmeier et al., 1989; Marsan and Torrent, 1989).

Silica cementation is often enhanced in soils near scarps or incisions, such as at terrace edges. Evaporation at the scarp surface concentrates the soil solution and promotes silica precipitation. The resulting induration impacts water movement and slope retreat, and renders the soil profile exposed along the scarp atypical of the soil under the broader geomorphic surface (Moody and Graham, 1997).

2.3.2.6 Accumulation and Redistribution of Clay

Clay (inorganic particles < 2 μm diam) is produced by weathering of primary mineral grains such as feldspars, micas, pyroxenes, and amphiboles. The ions released by weathering are carried in solution to other parts of the soil profile or the landscape and precipitated as colloids such as smectite, gibbsite, kaolin, or Fe oxides. Commonly, as primary grains are weathered, secondary minerals replace them pseudomorphically. Thus, the secondary minerals are not initially dispersed as colloids in the soil (Nahon, 1991); they enter the clay size fraction when pseudomorph grains are crushed by turbation processes (Graham and Buol, 1990). Clay contained in some parent materials, especially sediments (e.g., alluvium, colluvium, lacustrine deposits, loess) and sedimentary rocks (e.g., shale, limestone, mudstone), is simply inherited in the soils derived from them. Another significant source of clay that is particularly important in arid regions is aerosolic input (Reheis and Kihl, 1995; Simonson, 1995).

Mobilization and Transport of Clay

As the smallest particle size fraction in soils, clay is most susceptible to suspension in and transport by water percolating through the soil. Mobilization of a clay particle by water is effected by slaking or by dispersion. Slaking is the physical detachment of soil particles in water which is enhanced by wetting-drying cycles. As a soil rewets, moving water can dislodge loose particles and carry them in suspension (Hudson, 1977). Physical detachment of clay is also enhanced by increased pore water flow rates because it increases shear stress needed to detach particles from the matrix (Kaplan et al., 1993). Slaking is inhibited by processes that produce stable aggregates. Thus interparticle bonding that is enhanced by organic matter and noncrystalline inorganic compounds minimizes slaking and clay redistribution.

Dispersion, the chemical repulsion of colloids in suspension, greatly enhances clay movement. Conditions that favor dispersion in soils are those that result in an expanded diffuse double layer (Section B, chapter 6), most notably Na-saturated clay and absence of soluble salts or partially soluble minerals (e.g., calcite, gypsum) that would increase the electrolyte concentration in solution or put Ca on exchange sites at the expense of Na. Dispersion is also enhanced by high pH because it imparts a negative charge to variable charge materials (e.g., oxyhydroxides, kaolin), which are then repulsed by each other and by the permanently negatively charged minerals. An example of this in natural systems is the effect of woodash from forest fires, which may raise the surface soil pH above 10 (Ulery et al., 1993), causing kaolinite to disperse (Durgin and Vogelsang, 1984). Low molecular weight organic acids can be effective dispersive agents (Jenny and Smith, 1935; Kaplan et al., 1993; Kretzschmar et al., 1993, 1995). They complex cations in solution, thereby keeping them from flocculating clays, and they can specifically adsorb to positively charged mineral edges, preventing edge-to-face bonding with negatively charged minerals (Durgin and Chaney, 1984; Heil and Sposito, 1993a,b).

In general, the finest clay particles are more readily transported, moving even at low pore water flow rates (Kaplan et al., 1993), but even silt particles move in suspension if conditions are favorable. Clay-sized minerals with relatively strong negative charge, such as smectite, are most mobile, but preferential movement of certain mineral species may result if they occur as very small particles. This is true of smectites precipitated from solution, and of Fe oxides and gibbsite (Kaplan et al., 1993). Kaolin particles are less mobile because they are larger and have a low negative charge. Micas, vermiculites, quartz, and feldspars can be transported if they are small (usually <1 µm) and if flow rates are sufficient to carry them. Most clay movement is driven by macroscopic flow, although local redistribution can occur in micropores. Water flow in macropores, while less frequent than in micropores, is at much greater velocities. As a result, it can move more clay and larger particles to greater depths.

Deposition of Clay

Clay in suspension is deposited by cessation of water flow, flocculation by high electrolyte concentrations as in saline or calcareous subsoils, adsorption of phyllosilicate clays with permanent negative charge to Fe and Al oxyhydroxides in acid subsoils, and by sieving and abundant adsorption sites in fine-textured horizons (Jenny and Smith, 1935). Deposition of clay frequently occurs as the suspension travels into a drier part of the soil where the moving water is imbibed and retained by capillary forces. For example, when a suspension of clay flows through a macropore (Fig. 2.9), water is pulled by matric forces into the soil fabric and the clay in suspension is deposited in or on the pore wall. Deposition of this type occurs in tubular, interpedal, and intergranular pores, and results in the characteristic parallel orientation of platy phyllosilicate clays on pore walls. These morphologic features, known as clay skins, clay films, clay linings, or argillans, are typically most abundant in subsoils (Bt horizons), but are often best preserved and expressed within the fractures of underlying saprolite or bedrock (Graham et al., 1994). They form rapidly in the laboratory (Dalrymple and Theocharopoulos, 1984) and develop within several decades in the field if pedoturbation is minimal (Graham and Wood, 1991). Clay skins are often, but not always, enriched in finer clay sizes than the soil matrix as a whole. Clay deposition is a strong contributor to soil structure formation by creating distinct differences between the surface and interior of peds. Deposited clays on ped surfaces are strongly oriented and can be discerned easily using a petrographic microscope and cross-polarized light. Clay skins are destroyed in soils having extensive shrink/swell activity; they are replaced by shiny slickensides or pressure faces on ped surfaces which can be distinguished microscopically from clay skins by a different pattern of clay orientation (Nettleton et al., 1969) (Section E, Chapters 6.8, 6.10, 6.11, and 6.12).

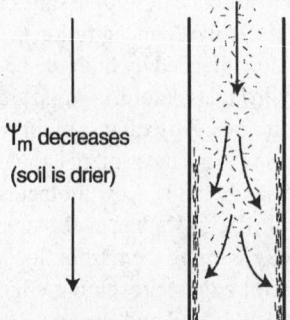

Ψ_m decreases
(soil is drier)

1. Water flows down pore carrying clay in suspension.
2. Drier subsoil is encountered.
3. Matric forces pull infiltrating water out of pore, into matrix.
4. Clay from suspension is plastered on pore walls.
5. Phyllosilicates have c-axis perpendicular to pore wall.

Fig. 2.9 Illustration of clay deposition from suspension in a tubular pore to form an oriented clay lining (i.e., clay film, channel illuviation argillan)

2.3.2.7 Complexation and Redistribution of Fe and Al

In many soils, particularly Spodosols, dissolved organic matter plays a critical role in pedogenic processes by complexing metals, predominantly Fe and Al, in surface horizons, translocating them, and depositing them in subsoils. This process enhances a distinctive style of mineral weathering, in which chelation removes weathering products, and produces a characteristic soil morphology, epitomized by an albic horizon overlying a spodic horizon (Section E, chapter 6.9).

Reactive Agents and Sources

Dissolved organic acids act as the carriers of metal cations. Some are simple acids derived directly from leachates of relatively fresh plant material, either from the vegetative canopy or from the leaf litter at the soil surface. Typically, the more important dissolved organic acids in soils are the byproducts of microbial decomposition of organic matter produced in the O or A horizons. They are complex, heterogeneous, relatively low molecular weight organic acids referred to as fulvic acids (Section B, Chapter 2). Organic acids chelate and remove cations from the surface of mineral grains. This type of weathering is very effective, since it leaves a fresh grain surface with no coatings of secondary minerals to impede solution access to the surface, a condition that inhibits further weathering. Since the organic acids keep the soil solution pH below the pK of H_2CO_3 (6.4), bicarbonate weathering is not involved in these systems (Ugolini and Spaltenstein, 1992). Iron and Al are the cations preferentially removed by organic complexation since, being relatively small cations with high valence, they form the most stable chelates (Schnitzer, 1969). At a given pH, Fe and Al will remain in solution at a much higher concentration if organically complexed than if simply inorganic.

Translocation and Accumulation

Chelates, forming primarily in the O and A horizons, move in solution with percolating water. During catastrophic leaching episodes, organometal colloids, as well as dissolved metal chelates, may be flushed downward in the profile (Stoner and Ugolini, 1988). The dissolved chelates precipitate when the metal:organic carbon ratio exceeds a critical value at which all polar-bonding sites are full. For Fe, precipitation occurs when the Fe^{3+}:fulvic acid ratio equals 6.1 (Schnitzer, 1969). When the metal:organic C ratio is low, chelates may still be arrested by adsorbing on positively charged material, such as Fe, Al oxyhydroxides or high metal:organic C material already precipitated. Such reactions cause polymerization of soluble low molecular weight compounds into insoluble forms (Fig. 2.10a). Chelates may also be deposited by desiccation (Fig. 2.10b) or by aggregation in a zone of relatively high ionic strength, and low H^+ activity, and exposed negative charges (Fig. 2.10c). Spodic horizons are often identified by microscopic examination of thin sections that reveal the presence of organic and Fe oxide-rich silt-size aggregates, many with cracked coatings indicative of post-depositional dehydration (DeConnink, 1980).

Relation to Soil Morphology

The O horizon is the major source of dissolved organic acids, which strip Fe and Al from the mineral soil as they move downward (Table 2.3). This produces the bleached E (~ albic) horizon, which exhibits the colors of the fresh primary mineral grains. The zone in which the chelates are deposited is the Bhs (or Bh or Bs; ~ spodic) horizon. As time passes, weatherable minerals in the E horizon may become deeply pitted by dissolution while the B horizon is enriched with humus and metals. Microbial activity within the B horizon oxidizes organic C, increasing the metal:organic C ratios and releasing metals from the organic complexes to precipitate as poorly crystalline oxyhydroxides. Thus, the B horizon takes on a dark reddish brown color reflecting the humus and Fe oxyhydroxide components. Metal-humus complexes can accumulate in such a way as to produce cemented horizons. One such

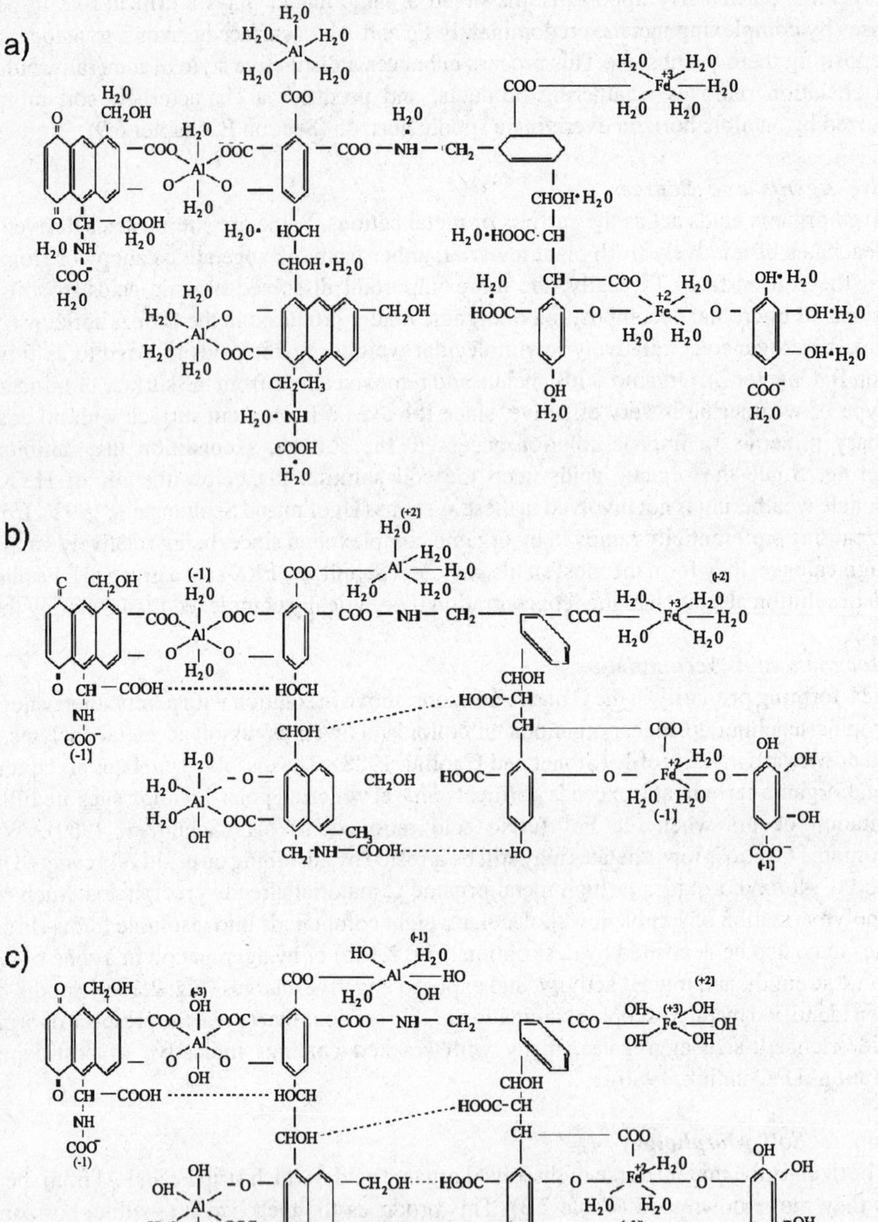

Fig. 2.10 Three configurations of a molecular model describing metal ion chelation and subsequent precipitation: a) hypothetical polymerized hydrated compound, having several Al, Fe^{2+}, and Fe^{3+} cations in 6-fold coordination; b) dehydrated version with several protonic bridges indicated by dotted lines, the overall molecular charge is zero; c) version showing response to increased system pH, more -COOH are dissociated and part of the metal coordinated water is dissociated [Modified from DeConnink, 1980. Major mechanisms in formation of spodic horizons. Geoderma 24:101-128 with permission from Elsevier Science].

Table 2.3 Chemical variables and their roles in the complexation and redistribution of Fe and Al in soils [After Marrett, 1988]

Variable	Major role and interaction with other variables	Trend, interaction with compartment			
		O	E and/or A	B	C
DOC	Major driving variable mobile anion, acidity source, metal complexing agent	Major source	Minor source	Major sink	No trend
pH	Low pH controlled by DOC; major variable	Lowered greatly	Lowered slightly	Rises greatly	Rises slightly
HCO_3	Controlled by pH, P_{CO_2}	Lowered	Insignificant	Insignificant	Rises significantly
Fe	Complexed and mobilized by DOC, causes DOC immobilization in B	Source	Major source	Major sink	Insignificant
Al	Complexed and mobilized by DOC, causes DOC immobilization in B; may be mobilized inorganically at low pH	Minor source	Major source	Major sink	Insignificant or minor
Basic cations	Leached in association with DOC (upper horizons) and with HCO_3 (C horizons)	Major source	Source	Sink	Deep leaching (sink)

feature is the placic horizon, which is a thin (2–10 mm), hard, brittle, and wavy zone cemented by Fe/Mn humus. Quite commonly these features accumulate at a hydrological boundary, such as a change in particle size. A more massive cemented horizon is ortstein, which is essentially a spodic horizon cemented by Fe/Al humus. Both of these features require soil that is relatively free of physical disruption in order to form and persist.

Relation to Environmental Conditions
The process of Fe and Al translocation and accumulation is associated with specific conditions of climate and vegetation. In general, the effective climate is wet, to provide a strong leaching environment, and cool, to produce a low decomposition rate of organics. In these cool, humid environments, coniferous forests and ericaceous shrubs often prevail and are particularly effective in promoting chelation since they have acidic foliage. As a consequence of this close relationship between climate, vegetation, and the process of Fe/Al translocation and accumulation in soils, Spodosols and similar soils are found in vast areas north of 45° latitude and at lower latitudes where high precipitation and low temperature prevail, such as coastal regions or high elevations. On the other hand, Spodosols are common in warm humid regions where soils are poorly drained. Organic acids in the groundwater chelate Fe and Al, which are concentrated by the fluctuating watertable to precipitate in a Bh horizon. The Bh horizons typically contain chelates but little or no free Fe as oxides or oxyhydroxides since it is reduced and leached away. Soils formed in this way have very pronounced albic and spodic horizons. They are common in the coastal plains of the southeastern United States and tropical forests such as the Orinoco basin in Brazil.

Parent material composition is also very influential in the process of metal-humus translocation and accumulation. The process is promoted by relatively low levels of Ca, Fe, and Al. High levels of these cations prevent mobilization because a high metal:organic C complex forms quickly which is not soluble and readily translocated. Furthermore, high Ca contents promote microbial activity which decomposes soluble organics so they are not available for chelation and leaching. As a rule, translocation of Fe- and Al-humus complexes is favored by silicic or felsic parent materials, but not carbonate or mafic materials. It is further favored by coarse-textured materials (sand to coarse loamy) where surface area is low, water infiltration is rapid and cation release is slower.

2.3.2.8 Desilication and Concentration of Resistant Oxides

Weathering of primary minerals occurs to different degrees in all soil environments. It releases highly mobile basic cations, moderately mobile $Si(OH)_4$, and relatively immobile Al^{3+} and Fe^{3+} into soil solution. The fate of these ions depends on soil leaching intensity, organic matter composition, the amount of reactive surface area, and pH. In this section, the cumulative effect of leaching on soil volume change, mineral composition, and soil fabric is discussed.

Leaching and Mineral Composition

In humid environments, well-drained soils lose many of the mobile constituents in a process that lowers pH and changes solution ionic composition from bases and $Si(OH)_4$ to Al. At neutral pH, Si is more soluble than Al or Fe (Fig. 2.11). In the pH range from 5 to 7, much silica can be lost by leaching. Below pH 5, Al is leached also, but commonly at slower rates than Si because Al can be strongly sorbed by organic matter. Usually there is enough Al in soil minerals that the pH is buffered near 5 and seldom drops to levels (< 4) where Fe is dissolved due to acidity alone (van Breemen et al., 1983). Reducing conditions are required to solubilize Fe as described in the next section. Thus, leaching changes soil mineral stability fields in favor of minerals composed of Fe and Al and relatively small amounts of Si. A typical soil mineral assemblage accumulated after intense weathering in a humid environment is shown in Fig. 2.12. The secondary mineral assemblage includes kaolin containing Si and Al, gibbsite containing Al, and hematite and goethite containing Fe. There is also quartz which is inherited from the parent material or added by dust, and concentrated by the dissolu-

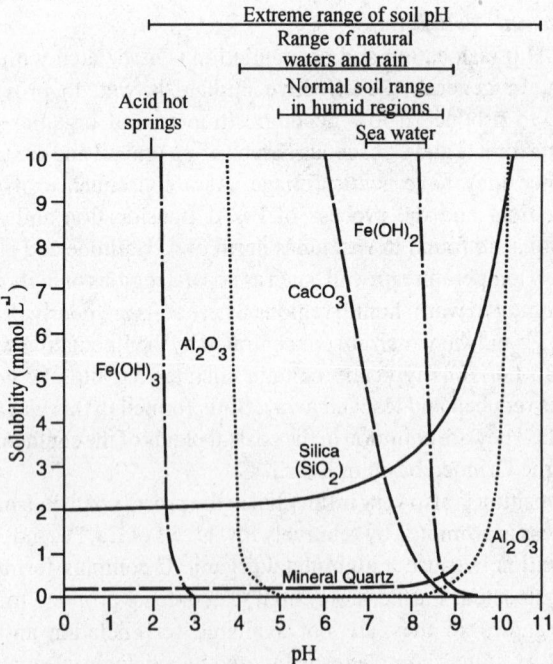

Fig. 2.11 Mineral solubility as a function of pH

tion of other primary minerals. These minerals have low nutrient retention and supply capabilities which means that nearly all plant nutrients must be derived from breakdown of organic matter or from atmospheric deposition. Because Si has been dramatically depleted and that remaining resides in quartz and kaolinite, these minerals are subject to weathering to a greater extent than in soils that still contain weatherable primary minerals. Even though kaolinite is considered to be a stable end product of weathering, it can decompose (and form) quite rapidly in highly leached soils (Giral-Kacmarcik et al., 1998). Silicon is conserved by biocycling between rainforest vegetation, where it forms opal phytoliths, and soils, where it is released by weathering of the phytoliths (Lucas et al., 1993; Alexander et al., 1997).

Leaching and Collapse

As primary rock minerals are dissolved, they commonly lose volume (Chadwick et al., 1990; Brimhall et al., 1992). This process is countered by any mass addition to soil which will serve to expand or dilate it. Quantitative studies of leaching losses during soil formation account for these changes through the use of index minerals (or elements) which are extremely resistant to weathering (Haseman and Marshall, 1945; Brewer, 1964; Smeck and Wilding, 1980; Brimhall and Dietrich, 1987). Dilation is indicated when the quantity of an index element in a soil horizon is less than in the parent material and soil collapse is indicated when the opposite it true. As depicted in Fig. 2.13, young soils are often dilated because accumulation of organic matter is more rapid than mineral weathering. Older soils show progressively greater collapse because mineral weathering becomes the dominant control as organic C accumulation is balanced by microbial respiration (Chadwick et al., 1994). Fig. 2.13 was

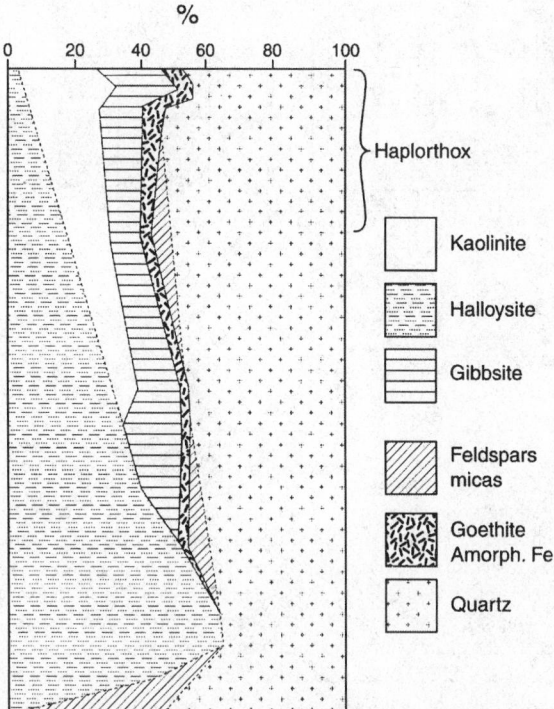

Fig. 2.12 Mineral distribution in an Oxisol and associated saprolite on granite [From Eswaran and Bin, 1978. Soil Sci. Soc. Am. J. 42:144-149 with permission of the Soil Science Society of America]

constructed by analyzing soils developing in beach sand on a suite of progressively older uplifted marine terraces (Merritts et al., 1992; Brimhall et al., 1992); similar patterns of early C driven dilation followed by weathering driven collapse occur in soils on lava flows in Hawaii (Vitousek et al., 1997).

Accumulation of Atmospherically Derived Constituents

Soils that reside on old stable geomorphic positions accumulate atmospherically transported minerals. For example, dust from Africa augments soils in the Amazon rainforest (Swap et al., 1992) and dust from Asia can be found in Hawaiian soils (Jackson et al., 1971). In parts of Africa and Australia, deeply weathered bauxite deposits are produced partly by the accumulation of chemically mature Fe and Al compounds that are blown in from other regions (Brimhall et al., 1988). These eolian additions are translocated into the top few meters of a deposit through macropores created by biological activity where they effectively dilate previously collapsed horizons (Brimhall et al., 1992). Below the lower limit of root growth, translocation is no longer accommodated by smaller pores in the saprolite so leaching has led to collapse. Because these soils have accumulated eolian additions of highly weathered Fe and Al compounds, Si and basic cation losses could be overestimated when comparing soil horizons to underlying material. In contrast, when unweathered primary minerals are added to soils, it is possible that leaching losses will be underestimated. Old soils commonly have complex histories.

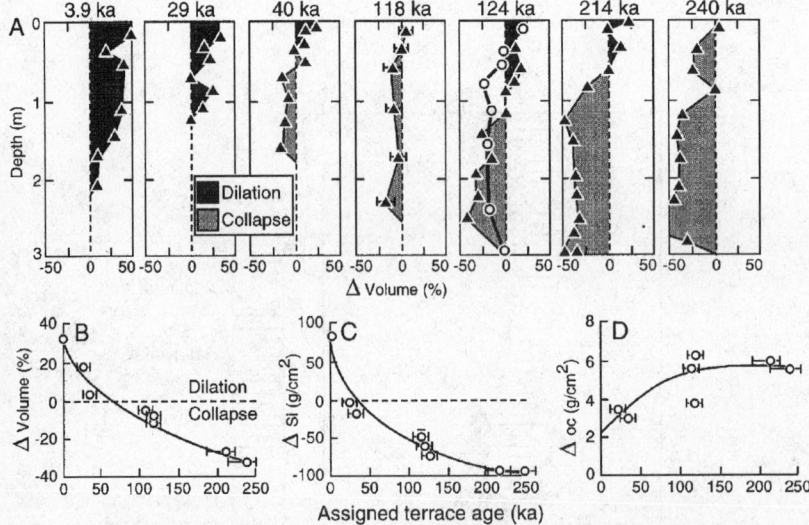

Fig. 2.13 Parent material and soil values for bulk density and Zr are used to compute volume change that occurs as soils develop from beach sand to 240 ky on uplifted marine terraces [Modified from Brimhall et al., 1992. Science 255:695-702. Copyright American Association for the Advancement of Science]: (A) volume change plotted as a function of depth for each profile sampled; the 124 ky terrace was sampled in two different locations; (B) volume change integrated to the sampling depth (depth-weighted mean) to provide an average value for each profile plotted as a function of terrace age; (C) the quantity of Si leached from each profile as a summation over the soil sampling depth plotted as a function of terrace age. For these soils, base cations are leached more rapidly and Al is not leached to a significant extent. Given the mass dominance of Si in the arkosic beach sand, Si leaching is the main control on soil collapse; (D) the quantity of organic carbon summed over the sampling depth plotted as a function of terrace age. Organic matter accumulation is the dominant factor driving early dilation and partly offsetting desilication in the older profiles.

Relation to Soil Morphology

Horizons in the top 1-2 m of old, highly weathered soils, display fine micropeds (< 2 mm) which can agglomerate into larger structural units, but they rarely exceed 50 mm in diamter (van Wambeke et al., 1983). The peds are composed of strongly interbonded kaolinite and Al and Fe oxyhydroxides that have low CEC, which defines, in part, the existence of an oxic or kandic diagnostic horizon. In oxic horizons, clay films are rare both because of high aggregate stability, and intense bioturbation caused by ants and termites. The soil has high microporosity within peds and relatively large interped pores which produce excellent permeability.

In Fe-rich soils that are imperfectly drained, localized reduction and oxidation allows Fe to become mobile for short distances. Its redistribution produces localized accumulations of hematite or goethite which appear as reddish or yellowish mottles. The location of initial deposition can be in small pores where it is thought that precipitation is favored because the chemical potential of water in close association with the solid phase is at a minimum (Tardy and Nahon, 1985), although precipitation can also occur in large pores where high P_{O_2} can lead to oxidation (Bouma, 1983). Initially, the accumulated Fe forms a weakly defined glaebule, or segregated body, that includes primary mineral grains and secondary clays as well as Fe oxides. Slowly, the glaebule becomes more clearly defined because the engulfed minerals decompose and are replaced by Fe oxide (Nahon, 1991). Mineral decomposition is driven by alternating oxidation and reduction of Fe, which creates acidity for enhanced hydrolysis (Brinkman, 1978). Continued growth of the glaebule produces an abrupt boundary between it and the surrounding soil matrix. In time, glaebules grow into each other producing a reticulate pattern of reddish Fe oxide called plinthite, interspersed with less red, kaolin and gibbsite rich matrix. Plinthite in perennially moist soils remains soft, but under wetting and drying conditions, it can solidify into iron stone nodules or continuously cemented ferricrete. Wetting and drying can be the result of natural climate, climatic drying, erosional dissection of a plateau, or excavation of a road (Daniels et al., 1971; Nahon, 1991). There is no certain evidence on how long it takes for plinthite to harden when exposed to wetting and drying conditions.

2.3.2.9 Reduction and Oxidation Leading to Depletions and Concentrations

Environmental factors, particularly climate, topography, and the chemical and physical nature of the substratum create the drainage properties of soils, which, in turn, influence the intensity, duration, and spatial occurrence of anoxic conditions. Soils are susceptible to anoxia because O_2 is consumed by belowground microbial respiration but it can only be supplied to soil pores by diffusion from the overlying planetary atmosphere. Soils with few macropores and many water-filled pores often consume O_2 more rapidly than can be resupplied by diffusion. Many soils will be anoxic for short periods right after intense wetting events and the interiors of peds may be anoxic even in otherwise well-aerated soils. When O_2 is depleted, microbially induced reduction reactions result in the dissolution of redox sensitive compounds with the common result being an increase in their solubility, leading to selective elemental loss from the anoxic area with subsequent precipitation during exposure to higher O_2 levels. These processes leave long-lasting visible imprints on soil morphology.

Reactive Agents in Redox Processes

The reactive agents in redox processes include organic matter, Fe and Mn oxides, nitrates, sulfides and sulfates, and microbes (Section B, Chapter 5). Microbial activity is the key to reduction in soils. Microbes require a C source, supplied by solid or dissolved organic matter, and electron acceptors. In well-aerated soils, O_2 is the electron acceptor, but as it is used up, nitrates, Mn and Fe oxides, and sulfates are used by different microbial populations. Such anaerobic conditions are usually associated with saturated or very wet soils in which there is little free pore space for the influx of O_2 from the atmosphere. Each electron acceptor compound is associated with a characteristic range of redox potentials (Section B, Chapter 5). Oxygen levels and redox potentials within a soil may show extreme variation at any given time, even on a scale of millimeters, such as from the exterior to the interior of an aggregate (Zausig et al., 1993).

Oxidation of a Reduced Soil

As a wet soil drains, macropores are the first to lose water. Often shrinkage occurs during the drying of soils so that cracks form between peds. Oxygen penetrates through the interpedal pores, root channels, macrofaunal burrows, and other macropores. In response to the higher Eh that develops within these voids, Fe and Mn oxides precipitate from the soil solution remaining within the matrix near the pore wall surface. Precipitation of these oxides removes the metal from solution, thereby establishing a diffusion gradient causing Fe^{2+} and Mn^{2+} to migrate from the still reduced soil matrix to the oxidizing zone adjacent to the macropores (Fanning and Fanning, 1989). Redox sensitive elements such as Fe and Mn often accumulate along macropores or the exterior of peds (Fig. 2.14). If Fe oxides precipitate, the color is red, orange, or yellow, depending on the species of Fe oxide formed (Section F, Chapter 3). Manganese oxides are a strong pigment and if they are present, even at levels of tenths of a percent, a pore wall will be blackened.

Reduction of an Oxidized Soil

Soils that are generally well-aerated and oxidized periodically become so wet that macropores are filled with water and O_2 is excluded. Organic matter from roots typically persists in old root channels and along ped faces of strongly structured soils. This organic matter provides an abundant source of C for microbes, and Fe oxides in the soil adjacent to the macropore serve as electron acceptors and are reduced to yield Fe^{2+} in solution. At the same time, particularly in highly structured, fine-textured soils, ped interiors and soil matrix at some distance from the macropores may contain sites that are not water saturated and retain relatively high Eh conditions. In this case, Fe^{2+} mobilized from soil adjacent to the pores diffuses into the higher Eh environments within the matrix where it precipitates as Fe oxides (Fanning and Fanning, 1989; Vepraskas, 1996). Soil near the pore is the site of redox depletion and is generally a gray or white color, whereas redox concentrations of Fe (red, orange, or yellow) occur

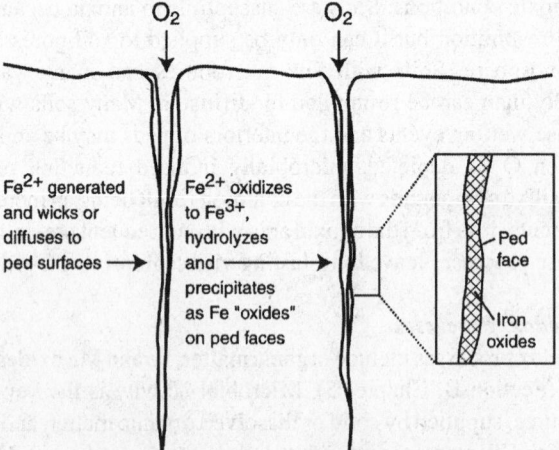

Fig. 2.14   Macropores (e.g., cracks) facilitate O_2 entry into a reduced soil, resulting in the precipitation of Fe oxides on macropore wall surfaces (e.g., ped faces) [Modified from Fanning and Fanning, 1989. Soil morphology, genesis, classification. Copyright. Reprinted with permission of John Wiley & Sons, Inc.]

within the soil matrix (Fig. 2.15). Because Mn reduction and oxidation occur at higher Eh values than Fe, Mn oxides should also be dissolved near the water-saturated pores and precipitated within the soil matrix, but not as far from the pore as the Fe.

Redox Soil Systems

A classic expression of the effects of varying redox conditions is found in the catena concept, where upslope soils are usually well drained and oxidized while soils at the base of the slope are, at least seasonally, poorly drained and reduced (Fanning and Fanning, 1989). As a result of these topographically induced pedochemical conditions, the upslope soils contain relatively abundant Fe oxides and are reddish, whereas soils at the base of the slope contain few Fe oxides, are enriched in Mn oxides, and generally have low chroma colors (Weitkamp et al., 1996). Redox potentials in the upper slope soils are sufficiently high so that Fe oxides are not reduced, although they are low enough occasionally to reduce Mn oxides. Mn^{2+} is mobilized and transported to the base of the slope where Eh is often low enough to limit the persistence of Fe oxides in quantity, but is usually high enough for Mn oxides to precipitate and become stable.

Soils which contain a slowly permeable layer, such as a dense argillic horizon or fragipan, may develop a seasonally perched watertable. The resulting epiaquic conditions may not substantially deplete Fe or Mn from the soil as a whole, since there is little downward leaching. The redox sensitive elements are redistributed above and within the restrictive layer, largely by processes described above for the reduction of oxidized soils (Fanning and Fanning, 1989). An E horizon, depleted of Fe and Mn oxides by reduction and lateral transport during the period of saturation, may develop above the restrictive layer (McDaniel and Falen, 1994). Macropore surfaces have low chroma colors, the matrix has high chroma reddish or yellowish colors. Fe/Mn nodules are formed by precipitation of Fe in

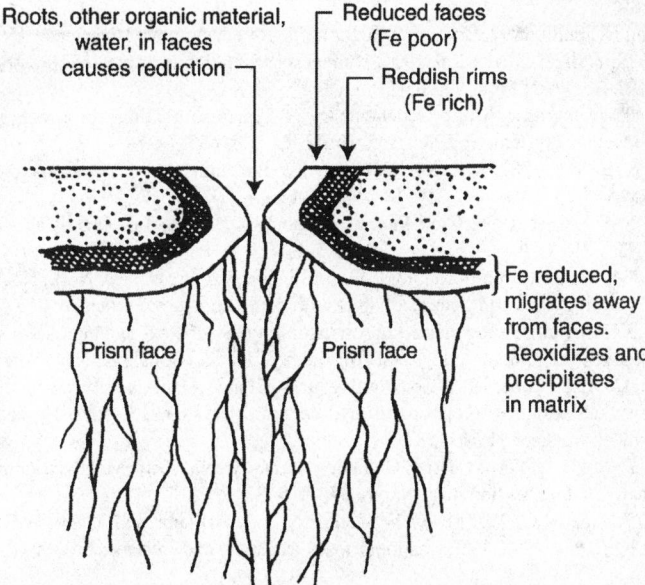

Fig. 2.15 Occasional, localized reduction occurs in well-drained soils when macropores (e.g., cracks between peds) are temporarily filled with water. Fe oxides in rims of peds are dissolved and removed or reprecipitated in more oxidized interior zones of the peds [After Fanning and Fanning, 1989. Soil morphology, genesis, classification. Copyright. Reprinted with permission of John Wiley & Sons, Inc.]

oxidized microsites and Mn oxides precipitate preferentially on the Fe oxides. The initial precipitate acts as a template for further precipitation and the microsite continues to be conducive to oxidation.

2.4   From Property to Process

For convenience, specific soil-forming processes have been identified and discussed separately. This leaves the impression that one can easily interpret soil-forming processes from soil properties. To a degree this inversion can be done quite well, but care is required. Many processes occur in the same pedon simultaneously which imprints a less than clear suite of properties. Local differences in processes can be dramatic depending on soil–landscape position. For example, reducing conditions can prevail in a valley bottom soil while silica and carbonate accumulation and clay redistribution are modifying pedons on surrounding hillslopes, or lithologies can change from one location to another. Soils are often polygenetic, their properties developed under different climatic conditions. Past climate change can superimpose different processes on a pedon at different times during soil formation. The dominant processes operating on a pedon today could be considerably different than in the past. It is important to be alert to the possibility that the present set of soil properties is a composite derived from a series of temporally varying processes. Interpretation of soil processes requires as full an Earth history context as possible. Smart sampling, keen observation and an open mind are prerequisites to successful understanding.

2.5   References

Aber, J.D., and J.M. Melillo. 1991. Terrestrial ecosystems. Saunders College Publishing, Tokyo, Japan.

Alexander, A., J.D. Meunier, F. Colin, and J.M. Mathias Koud. 1997. Plant impact on the biogeochemical cycle of silicon and related weathering processes. Geochim. Cosmochim. Acta. 61:677–682.

Amundson, R., and H. Jenny. 1991. The place of humans in the state factor theory of ecosystems and their soils. Soil Sci. 151:99–109.

Amundson, R., and H. Jenny. 1997. On a state factor model of ecosystems. Bioscience 47:536–543.

Anderson, J.U., D. Silberman, and D. Rai. 1975. Humus accumulation in a forested Haploboroll in south-central New Mexico. Soil Sci. Soc. Am. Proc. 39:905–908.

Arndt, J.L., and J.L. Richardson. 1989. Geochemistry of hydric soil salinity in a recharge-throughflow-discharge prairie-pothole wetland system. Soil Sci. Soc. Am. J. 53:848–855.

Atlas, R.M., and R. Bartha. 1981. Microbial ecology: Fundamentals and applications. Addison-Wesley Publishing Co., Reading, MA.

Berner, E.K., and R.A. Berner. 1996. Global environment: Water, air, and geochemical cycles. Prentice-Hall, Upper Saddle River, NJ.

Birkeland, P.W. 1984. Soils and geomorphology. Oxford University Press, New York, NY.

Bouma, J. 1983. Hydrology and soil genesis of soils with aquic moisture regimes. p. 253–281. In L.P. Wilding, N.E. Smeck, and G.F. Hall, (ed.). Pedogenesis and soil taxonomy. I. Concepts and interactions. Elsevier, New York, NY.

Brewer, R. 1964. Fabric and mineral analysis of soil. John Wiley and Sons, Inc. New York, NY.

Brimhall, G.H., and W.E. Dietrich. 1987. Constitutive mass balance relations between chemical composition, volume, density, porosity, and strain in metasomatic hydrochemical systems: Results on weathering and pedogenesis. Geochim. Cosmochim. Acta 51:567–587.

Brimhall, G.H., C.J. Lewis, C. Ford, J. Bratt, G. Taylor, and O. Warin. 1988. Metal enrichment in bauxites by deposition of chemically mature aeolian dust. Nature 333:819–824.

Brimhall, G.H., O.A. Chadwick, C.J. Lewis, W. Compston, I.S. Williams, K.J. Danti, W.E. Dietrich, M.E. Power, D. Hendricks, and J. Bratt. 1992. Deformational mass transport and invasive processes in soil evolution. Science 255:695–702.

Brinkman, R.H. 1978. Ferrolysis. Elsevier, New York, NY.

Chadwick, O.A., G.H. Brimhall, and D.M. Hendricks. 1990. From a black to a gray box-a mass balance interpretation of pedogenesis. Geomorphology 3:369–390.

Chadwick, O.A., D.M. Hendricks, and W.D. Nettleton. 1987. Silica in duric soils: I. A depositional model. Soil Sci. Soc. Am. J. 51:975–982.

Chadwick, O.A., E.F. Kelly, D.M. Merritts, and R.G. Amundson. 1994. Carbon dioxide consumption during soil development. Biogeochem. 24:115–127.

Dalrymple, J.B., and S.P. Theocharopoulos. 1984. Intrapedal cutans-an experimental production of depositional (illuviation) channel argillans. Geoderma 33:237–1683.

Daniels, R.B., E.E. Gamble, and J.C. Cady. 1971. The relation between geomorphology and soil morphology and genesis. Adv. Agron. 23:51–88.

DeConnink, A. 1980. Major mechanisms in formation of spodic horizons. Georderma 24:101–128.

Drees, L.R., L.P. Wilding, N.E. Smeck, and A.L. Senkayi. 1989. Silica in soils: Quartz and disordered silica polymorphs. p. 913–974. *In* J.B. Dixon and S.B. Weed (ed.) Minerals in soil environments. Soil Science Society of America, Madison, WI.

Durgin, P.B., and J.G. Chaney. 1984. Dispersion of kaolinite by dissolved organic matter from Douglas fir roots. Can. J. Soil Sci. 64:445–455.

Durgin, P.B., and P.J. Vogelsang. 1984. Dispersion of kaolinite by water extracts of Douglas fir ash. Can. J. Soil Sci. 64:439–443.

Egbhal, M.K., and R.J. Southard. 1993. Micromorphological evidence of polygenesis of three Aridisols, western Mojave Desert, California. Soil Sci. Soc. Am. J. 57:1041–1050.

Eswaran, H., and W.C. Bin. 1978. A study of a deep weathering profile on granite in peninsular Malaysia: I. Physicochemical and micromorphological properties. Soil Sci. Soc. Am. J. 42:144–149.

Fanning, D.S., and M.C.B. Fanning. 1989. Soil morphology, genesis, classification. John Wiley and Sons, New York, NY.

Fireman, H., and H.E. Harward. 1952. Indicator significance of some shrubs in the Escalante Desert, Utah. Bot. Gaz. 114:143–154.

Frazmeier, D.P., L.D. Norton, and G.C. Steinhardt. 1989. Fragipan formation in loess of the midwestern United States. p. 69–95. *In* N.E. Smeck and E.J. Ciolkosz (ed.) Fragipans: Their occurrence, classification, and genesis. SSSA Spec. Publ. 24., Soil Science Society of America, Madison, WI.

Giral-Kacmarcik, S., S.M. Savin, D.B. Nahon, J-P. Girard, Y. Lucas, and L.J. Abel. 1998. Oxygen isotope geochemistry of kaolinite in laterite-forming processes, Manaus, Amazonas, Brazil. Geochim. Cosmochim. Acta 62:1865–1879.

Graham, R.C. and S.W. Buol. 1990. Soil-geomorphic relations on the Blue Ridge Front: II. Soil characteristics and pedogenesis. Soil Sci. Soc. Am. J. 54:1367–1377.

Graham, R.C., and H.B. Wood. 1991. Morphologic development and clay redistribution in lysimeter soils under chaparral and pine. Soil Sci. Soc. Am. J. 55:1638–1646.

Graham, R.C., W.R. Guertal, and K.R. Tice. 1994. The pedologic nature of weathered rock. p. 21–40. *In* D.L. Cremeens et al. (ed.) Whole regolith pedology. Spec. Publ. 34., Soil Science Society of America, Madison, WI.

Haseman, J.F., and C.E. Marshall. 1945. The use of heavy minerals in studies of the origin and development of soils. MO Agric. Exp. Stn. Res. Bull. 387.

Heil, D., and G. Sposito. 1993a. Organic matter role in illitic soil colloids flocculation: I. Counter ions and pH. Soil Sci. Soc. Am. J. 57:1241–1246.

Heil, D., and G. Sposito. 1993b. Organic matter role in illitic soil colloids flocculation: II. Surface charge. Soil Sci. Soc. Am. J. 57:1241–1246.

Hsu, P.H. 1989. Aluminum oxides and oxyhydroxides. p. 331–378. *In* J.B. Dixon and S.B. Weed (ed.) Minerals in soil environments. Soil Science Society of America, Madison, WI.

Hudson, B.D. 1977. Cohesive water films as a factor in clay translocation. Soil Surv. Horiz. 18:9–15.

Jackson, M.L., T.W.M. Levelt, J.K. Syers, R.W. Rex, R.N. Clayton, G.D. Sherman, and G. Uehara. 1971. Geomorphological relationships of tropospherically derived quartz in the soils of the Hawaiian Islands. Soil Sci. Soc. Am. Proc. 35:515–525.

Jenny, H. 1941. Factors of soil formation. McGraw-Hill Book Co., New York, NY.

Jenny, H. 1980. The soil resource, origin and behavior. Springer-Verlag, New York, NY.

Jenny, H., and G.D. Smith. 1935. Colloid chemical aspects of clay pan formation in soil profiles. Soil Sci. 39:377–389.

Kaplan, D.I., P.M. Bertsch, D.C. Adriano, and W.P. Miller. 1993. Soil-borne mobile colloids as influenced by water flow and organic carbon. Environ. Sci. Technol. 27:1193–1200.

Kretzschmar, R., W.P. Robarge, and S.B. Weed. 1993. Flocculation of kaolinitic soil clays: Effects of humic substances and iron oxides. Soil Sci. Soc. Am. J. 57:1277–1283.

Kretzschmar, R., W.P. Robarge, and A. Amoozegar. 1995. Influence of natural organic matter on colloid transport through saprolite. Water Resour. Res. 31:435–445.

Lee, K.E., and R.C. Foster. 1991. Soil fauna and soil structure. Aust. J. Soil Res. 29:745–775.

Lucas, Y., F.J. Luizao, A. Chauvel, J. Rouiller, and D. Nahon. 1993. The relation between biological activity of the rain forest and mineral composition of soils. Science 260:521–523.

Marrett, D.J. 1988. Acid soil processes in the Okpilak Valley, artic Alaska. PhD Dissertation, University of washington, Seattle, WA.

Marsan, F., and J. Torrent. 1989. Fragipan bonding by silica and iron oxides in a soil from northwestern Italy. Soil Sci. Soc. Am. J. 53:1140–1145.

McDaniel, P.A., and A.L. Falen. 1994. Temporal and spatial patterns of episaturation in a Fragixeralf landscape. Soil Sci. Soc. Am. J. 58:1451–1457.

Melillo, J.M., J.D. Aber, A.E. Linkins, A. Ricca, B. Fry, and K.J. Nadelhoffer. 1989. Carbon and nitrogen dynamics along the decay continuum: plant litter to soil organic matter. Plant Soil 115:189–198.

Merritts, D.J., O.A. Chadwick, D.M. Hendricks, G.H Brimhall, and C.J. Lewis. 1992. The mass balance of soil evolution on late Quaternary marine terraces, northern California. Geol. Soc. of Am. Bull. 104:1456–1470.

Miller, D.E. 1971. Formation of vesicular structure in soil. Soil Sci. Soc. Am. Proc. 35:635–637.

Monger, H.C., and L.A. Daugherty. 1991. Pressure solution: Possible mechanism for silicate grain dissolution in a petrocalcic horizon. Soil Sci. Soc. Am. J. 55:1625–1629.

Monger, H.C., L.A. Daugherty, W.C. Lindemann, and C.M. Liddell. 1991. Microbial precipitation of pedogenic calcite. Geol. 19:997–1000.

Moody, L.E., and R.C. Graham. 1997. Silica cemented edges, central California coast. Soil Sci. Soc. Am. J. 61:1723–1729.

Munk, L.P., and R.J. Southard. 1993. Pedogenic implications of opaline pendants in some California late-Pleistocene Palexeralfs. Soil Sci. Soc. Am. J. 57:149–154.

Nahon, D.B. 1991. Introduction to the petrology of soils and chemical weathering. John Wiley and Sons, New York, NY.

Nettleton, W.D., K.W. Flach, and B.R. Brasher. 1969. Argillic horizons without clay skins. Soil Sci. Soc. Am. Proc. 33:121–125.

Oades, J.M. 1989. An introduction to organic matter in mineral soils. p. 89–159. *In* J.B. Dixon and S.B. Weed, (ed.) Minerals in soil environments. Soil Science Society of America, Madison, WI.

Oades, J.M. 1993. The role of biology in the formation, stabilization, and degradation of soil structure. Geoderma 56:377–400.

Paton, T.R., G.S. Humphreys, and P.B. Mitchell. 1995. Soils: A new global view. Yale University Press, New Haven, CT.

Quideau, S.A., R.C. Graham, O.A. Chadwick, and H.B. Wood. 1998. Carbon sequestration under chaparral and pine after four decades of soil development. Geoderma 83:227–242.

Rabenhorst, M.C., L.P. Wilding, and C.L. Girdner. 1984. Airborne dusts in the Edwards Plateau Region of Texas. Soil Sci. Soc. Am. J. 48:621–627.

Reheis, M.C., and R. Kihl. 1995. Dust deposition in southern Nevada and California, 1984–1989: Relations to climate, source area, and source lithology. J. Geophys. Res. 100:D5 8893–8918.

Reid, D.A., R.C. Graham, R.J. Southard, and C. Amrhein. 1993. Slickspot soil genesis in the Carrizo Plain, California. Soil Sci. Soc. Am. J. 57:162–168.

Schlesinger, W.H. 1997. Biogeochemisty: An analysis of global change. Academic Press, San Diego, CA.

Schnitzer, M. 1969. Reactions between fulvic acids, a soil humic compound and inorganic soil consituents. Soil Sci. Soc. Am. Proc. 33:75–81.

Schwertmann, U., and R.M. Taylor. 1989. Iron oxides. p. 379–438. *In* J.B. Dixon and S.B. Weed (ed.) Minerals in soil environments. Soil Science Society of America, Madison, WI.

Simonson, R.W. 1959. Outline of a generalized theory of soil genesis. Soil Sci. Soc. Am. Proc. 23:152–156.

Simonson, R.W. 1995. Airborne dust and its significance to soils. Geoderma 65:1–43.

Smeck, N.E., and L.P. Wilding. 1980. Quantitative evaluation of pedon formation in calcareous glacial deposits in Ohio. Geoderma 24:1–16.

Southard, R.J., and S.W. Buol. 1988. Subsoil blocky structure formation in some North Carolina Paleudults and Paleaqults. Soil Sci. Soc. Am. J. 52:1069–1076.

Sposito, G. 1989. The chemistry of soils. Oxford University Press, New York, NY.

Stoner, M.G., and F.C. Ugolini. 1988. Arctic Pedogenesis. II. Threshold-controlled subsurface leaching episodes. Soil Sci. 145:46–51.

Swap, R., M. Garstang, S. Greco, R. Talbot, and P. Kallberg. 1992. Saharan dust in the Amazon basin. Tellus 44B:133–149.

Tardy, Y., and D. Nahon. 1985. Geochemistry of laterites, stability of Al-goethite, Al-hematite, and Fe3+-kaolinite in bauxite and ferricretes. An approach to the mechanism of concretion formation. Am. J. Sci. 285:865–903.

Terhune, C.L., and J.W. Harden. 1991. Seasonal variations of carbon dioxide concentrations in stony, coarse-textured soils of southern Nevada, USA. Soil Sci. 151:417–429.

Ugolini, F.C., and H. Spaltenstein. 1992. Pedosphere. p. 123–153. *In* F.C. Butcher et al. (ed.) Global biogeochemical cycles. Academic Press, San Diego, CA.

Ulery, A.L., R.C. Graham, and C. Amrhein. 1993. Wood-ash composition and soil pH following intense burning. Soil Sci. 156:358–364.

van Breemen, N., J. Mulder, and C.T. Driscoll. 1983. Acidification and alkalization of soils. Plant and Soil 75:283–308.

van Wambeke, A., H. Eswaran, A.J. Herbillon, and J. Comerma. 1983. Oxisols. p. 325–354. *In* L.P. Wilding, N.E. Smeck, and G.F. Hall (ed.) Pedogenesis and soil taxonomy. II. The soil orders. Elsevier Science Publishers, Amsterdam, Netherlands.

van Wambeke, A. 1992. Soils of the tropics. McGraw-Hill, New York, NY.

Vepraskas, M.J. 1996. Redoximorphic features for identifying aquic conditions. NC Agric. Res. Serv. Tech. Bull. 301.

Vitousek, P.M., O.A. Chadwick, T.E. Crews, J.H. Fownes, D.M. Hendricks, and D. Herbert. 1997. Soil and ecosystem development across the Hawaiian Islands. GSA Today 7:1–8.

Weitkamp, W.A., R.C. Graham, M.A. Anderson, and C. Amrhein. 1996. Pedogenesis of a vernal pool Entisol-Alfisol-Vertisol catena in southern California. Soil Sci. Soc. Am. J. 60:316–323.

Whittig, L.D., and P. Janitzky. 1963. Mechanisms of formation of sodium carbonate in soils. I. Manifestations of biological conversions. J. Soil Sci. 14:322–333.

Wilding, L.P., and D. Tessier. 1988. Genesis of Vertisols: Shrink/swell phenomena. p. 55–81. *In* L.P. Wilding, and R. Puentes (ed.) Vertisols: Their distribution, properties, classification, and management. Texas A&M University Printing Center, College Station, TX.

Zauzig, J., W. Stepniewski, and R. Horn. 1993. Oxygen concentration and redox potential gradients in unsaturated model soil aggregates. Soil Sci. Soc. Am. J. 57:908–916.

3

Pedological Modeling

Marcel R. Hoosbeek
Wageningen Agricultural University

Ronald G. Amundson
University of California

Ray B. Bryant
Cornell University

3.1 Introduction

Since the early history of soil science pedologists have been trying to organize and capture their observations in models. These models include maps representing the distribution of soils, descriptions of soils, models that relate environmental factors with the genesis of soils, and models that aim to simulate soil-forming processes. The general objective for modeling is to better organize information related to the understanding of soils and ultimately improve predictions of the consequences of human interactions with soils.

Dokuchaev in Russia and Hilgard in the USA are generally given credit for developing the first multi-factor pedogenetic models (Jenny, 1961; Fanning and Fanning, 1989; Tandarich and Sprecher, 1994). In 1882 Dokuchaev presented the following equation (Dokuchaev, 1948; Volobuyev, 1984):

$$S = f(cl, o, p)t^0 \tag{3.1}$$

where S is soil, cl is the climate of a given region, o is plants and organisms, p is the geologic substratum (parent material), and t^0 is the relative age. Relief was mentioned as important for some azonal soils. Jenny (1941) published a soil-forming factor equation that became known to most soil scientists:

$$S = f(cl, o, r, p, t, ...) \tag{3.2}$$

where s denotes any soil property, cl is environmental climate, o is organisms and their frequencies referring to species germules, r is topography including certain hydrologic features, p is parent material defined as state of soil at soil formation time zero, t is the absolute age of the soil, and ... are additional unspecified factors. Jenny sees the factors not as formers, creators or forces, but as state factors that define the state of the soil system.

In a subsequent paper, Jenny (1961) broadened the scientific base of the soil-forming equations by using an open systems analysis approach. This approach is still the basis to most modeling projects in the environmental sciences and is described in any textbook on systems analysis and modeling (e.g., Leffelaar, 1993). The following definitions by De Wit (1993) will be used throughout this chapter. A system is a limited part of reality that contains interrelated elements. The totality of relations within the system is known as the system structure; both systems and models have a structure. A model is a simplified representation of a system.

As an example and introduction to this chapter, we will quote parts of Jenny's (1961) original text in which several basic questions are covered that are the basis of any modeling project, namely: (1) what part of the environment will be included in the model? (2) what are the major state variables in the selected compartment? (3) what are the major driving variables acting on the selected compartment? (4) which fluxes are entering or leaving the compartment? and (5) how many subcompartments are needed to model the internal processes?

Let us examine, as in [Fig. 3.1], a portion of a natural landscape, an ecosystem, the totality of soil, vegetation and animal life. The boundaries as dashed lines, are placed arbitrarily. Their enclosure delineates the scope of interest of an ecosystem analyst. A pedologist might draw the boundaries around the soil per se, a student of microbial ecology might place them around a minute volume of soil or vegetation.

The ecosystem is given the label L (from *l*arger system). Its individual properties and those of its parts are designated with small letters. Specifically, l is any property of the ecosystem in its totality, such as the nitrogen content of the entire ecosystem. Soil properties are denoted by s, vegetation properties by v, and animal properties by a.

The ecosystem, whatever its size, is open. Throughout its existence it gains and loses energy and matter. A partial list of gains or influxes from outside the assigned boundaries includes the following variables: (1) energy (solar radiation, heat transfer, entropy transfer from the outside heat reservoir), and

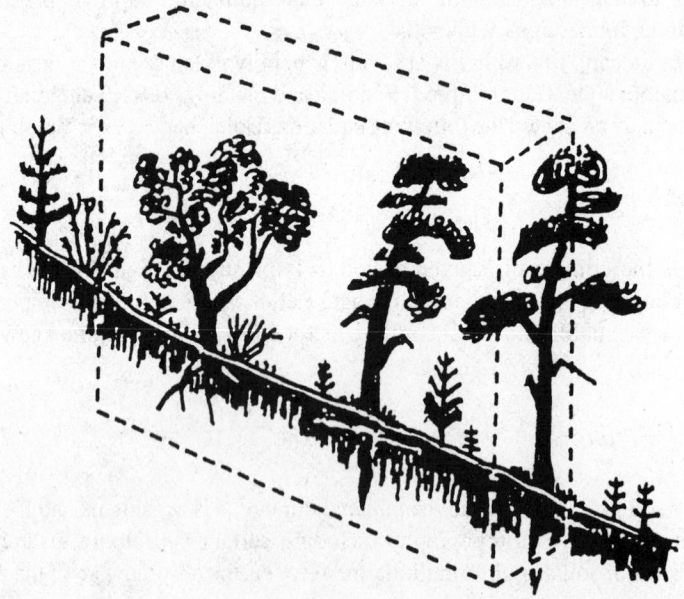

Fig. 3.1 Ecosystem with arbitrary boundaries [Reprinted from Jenny, 1961. Soil Sci. Soc. Am. Proc. 25:385-388 with permission of the Soil Science Society of America]

(2) matter: gases (H_2O, N_2, O_2, CO_2, etc. entering by diffusion or as mass flow), water (precipitation, runoff and floods, and capillary), solids dissolved or dispersed in water (e.g. NO_3^- and Sr^{90} in precipitation, nutrients added as fertilizers, sediments deposited by runoff and floods), solids carried by wind, and immigration of organisms.

A partial list of losses or outfluxes includes the following variables: (1) energy (heat radiation, light reflection), and (2) matter: gases (especially H_2O from evapotranspiration, CO_2, O_2, NH_3, etc. from photosynthesis, respiration, decomposition of organic matter), water (runoff, vertical percolation, side-seepage), solids (dissolved or dispersed in water, such as cations and anions and colloidal substances (e.g. humus, clay) in percolates and runoff), mass movement (solifluction), solids carried off by wind (wind erosion), and emigration of organisms.

Within the ecosystem its properties are subject to positional changes. Carbon assimilated in the leaves is translocated to the roots. Calcium from soil minerals is transported to the leaves and returned to the soil in litter fall (cycling of elements). Quite generally, growth of organisms implies movement of matter. Translocations within the soil lead to horizon formation and development of the soil profile.

So far, Jenny described the structure of the system. The next step involves the quantification of the fluxes, which is primarily a matter of bookkeeping through time. In the real world, time is a continuum, but for most modeling approaches it is necessary to chop up time into discrete time intervals. During each time interval the system is kept constant in the model, i.e., the state variables are only affected by the driving variables going from one to the next time interval. This implies that the length of the time interval may have quite an influence on the simulation results. A more theoretical treatise on this issue and on modeling approaches in general will be presented in Section 3.2.

In the following quote, Jenny (1961) assumed a time interval of one year, which for the processes he had in mind was an appropriate time scale. The student of microbial ecology, mentioned earlier by Jenny, would have been more interested in time intervals of weeks, days or even hours to describe the dynamics of microbial populations.

"During any interval of time Δt the changes in the properties of the total ecosystem are given by the gains minus the losses, or the influxes minus the outfluxes. The change in any l-property (Δl) is given, on a yearly basis, by

$$\frac{\Delta l}{\Delta t} = \frac{\Delta l}{1} = l_{in} - l_{out} \qquad [3.3]$$

where l_{in} and l_{out} denote quantities, large or small, of influx and outflux during one year. The difference l may be positive or negative.

Starting with an ecosystem at time zero, denoted by L_0, having any one of its l-properties labeled as l_0, the value of l after one year will be:

$$l_1 = l_0 + (\Delta l)_1 = l_0 + (l_{in} - l_{out})_1 \qquad [3.4]$$

After t years it will be:

$$l_t = l_0 + (l_{in} - l_{out})_1 + (l_{in} - l_{out})_2 + \dots (l_{in} - l_{out})_t \qquad [3.5]$$

In abbreviated form:

$$l_t = l_0 + \sum_1^t \left(l_{in} - l_{out}\right) \qquad [3.6]$$

The yearly net change ($l_{in} - l_{out}$) may have a different magnitude each year.

Equation [3.6] may be evaluated with lysimeters and phytotrons or biotrons which permit determination of influxes and outfluxes. If such studies were carried out during centuries and millennia – a utopia, of course – an accurate picture of ecosystem evolution could be obtained."

This touches on the subject of data availability, or, in more practical terms, the scarcity of data. At this point the modeler will have to make choices with respect to how to continue his or her modeling approach. For instance, is there enough process knowledge available and are there enough data available to pursue a mechanistic approach, or is an empirical approach more feasible? Are the differential equations that summarize the knowledge of a system simple enough to be solved analytically, or are numerical methods needed? These and other questions will be discussed in the next section in which we will provide some basic elements of dynamic simulation.

In the quotes presented above, Jenny defined the spatial scale to be at the soil-landscape level and the temporal scale to be at the century to millennium level. These are the extent scale levels, i.e., the scale levels to be explained by the model. Based on these scales and his knowledge of the processes involved he chose a time step of one year to run the model. In the remainder of his 1961 paper, Jenny worked out a modeling approach that allowed him to deal with the data scarcity at the spatial and temporal scales he had defined. This modeling approach became the well-known State Factor Analyses approach. Based on different spatial and temporal scales, different computation methods, and different degrees of model complexity, modelers have developed a variety of pedogenetic modeling approaches. Jenny's approach and many other approaches will be discussed in Section 3.3.

3.2 Basics of Modeling

Scientific knowledge is based on the interaction between observations and the hypotheses that correspond with those observations (Popper, 1959) and the hypotheses or theories adopted even before the observations are made (Kuhn, 1962). The scientific method used by most pedologists was described by Dijkerman (1974) in seven stages: (1) selection of a system (e.g., soil-landscape, pedon, or horizon); (2) measurement of properties; (3) ordering and condensing of data; (4) explanation of the data by hypotheses; (5) testing of these hypotheses against new data; (6) structuring of confirmed hypotheses into scientific laws which together form a body of well-established formal theory; and (7) use of scientific laws in predicting new unknown phenomena. This last stage of using scientific laws to predict new unknown phenomena may also apply to the beginning of a research activity. To be able to use the several stages, it is necessary to reduce the complex natural soil system to a level of abstraction that can be dealt with by the pedologist. This abstract reduction or simplification results in a conceptual model. In this chapter, conceptual models, as opposed to concrete models (real physical objects, e.g., a soil column in the laboratory), will be referred to as models.

With this wide definition of models in mind, we will present in the following sections a diagram for the classification of all kinds of models. With this classification diagram we will discuss several modeling approaches for the major groups of models. The organization of the sections dealing with existing pedogenetic models is also based on this diagram.

3.2.1 Modeling at Which Scale?

The first question posed in Section 3.1 was "What part of the environment will be included in the model?" The ideal situation would be that the system of interest is isolated from the rest of the environment. Since this is hardly possible, choosing a border should be attempted so that the environment influences the system, but the system affects the environment as little as possible (De Wit, 1993).

The scale of the defined system is then the extent of the model, i.e., the model aims to explain at this scale. A number of extent scales are indicated along the vertical axis in Fig. 3.2 (Hoosbeek and Bryant, 1992; Bouma and Hoosbeek, 1996). Each scale level can be regarded as a system by itself, with its own terminology, and can be seen as a combination of subsystems at lower levels or as a subsystem of higher level systems (France and Thornley, 1984). Each level integrates the knowledge of subsystems at lower levels, which means that investigations at a subsystem level, e.g., at the horizon level, provide a mechanistic understanding of a model that aims to explain at the pedon level (the extent).

We placed the pedon at the central i level of the presented scale hierarchy. Soil Taxonomy (Soil Survey Staff, 1975) defined a pedon as a three-dimensional natural body large enough to represent the nature and arrangements of its horizons and variability. The other scale levels were defined with the same idea in mind, i.e., three-dimensional bodies large enough to represent the nature and variability at a certain level.

The spatial variability of a soil system needs to be investigated at a level of resolution appropriate for the extent of the model. Spatial variability in soils forms a continuum from megascopic to microscopic levels of resolution (Wilding, 1985). The size of a sample that is needed for a certain resolution is called the support size in geostatistics. Therefore, we call the scale level at which the sampling takes place the support scale. For instance, many pedogenetic models aim to describe soil-forming processes that affect a pedon. To be able to mechanistically model horizonation (e.g., eluviation and illuviation), samples are take within each horizon. The extent is then at the pedon level and the support scale is at the soil structure level. The support scale depends on the extent scale, the soil properties used in the model, and on the sampling methods.

There is no scientific reasoning for the number of scale levels presented in Figure 3.2. The original figure by Hoosbeek and Bryant (1992) distinguished only 7 scale levels. The present list is based on what seems to be useful in practical terms. In Section 3.4 we will discuss the usefulness of Figure 3.2 to projects in which several models are used. The scale levels can be changed to fit any project. For instance, for an agronomic project it may be useful to distinguish the following scales: leaf (leaf area index) - plant - plot - field - farm - county.

3.2.2 Qualitative and Quantitative Models

Next to the vertical axis indicating the scale levels, two other axes are distinguished (Fig. 3.2). The axis labeled degree of computation distinguishes between qualitative and quantitative models.

Mental models are concepts and ideas that exist in the human mind to enable a soil scientist to work with the complex natural soil system (Dijkerman, 1974). For instance, in Section 3.3.2 we will discuss how a soil surveyor uses a mental version of the soil-landscape model to extrapolate soil information across a landscape in order to delineate map units. A soil profile description like 'Ap: Dark grayish brown light silty clay loam ...' is a descriptive model. Mental, verbal and descriptive models are placed at the qualitative side of the axis.

Mathematical models express abstractions in the form of algorithms using either a deterministic or a stochastic approach. Deterministic models presume that a simulated system will return one uniquely defined outcome as the result of an input data set. The outcome of a stochastic model is not one determined outcome, but rather a distribution around an average. The stochastic model presupposes an uncertain outcome and therefore uses a simulation structure to take this uncertainty into account. Quantitative models are placed at the opposite side of the axis. Actually, many models fall somewhere between the qualitative and quantitative extremes. For instance, the Munsell color scheme has both descriptive and numerical characteristics.

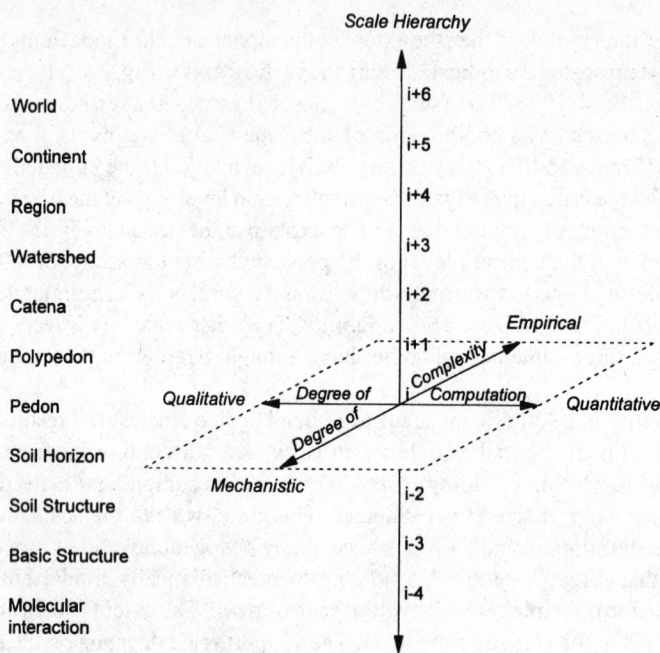

Fig. 3.2 Scale hierarchy and knowledge type diagram (Hoosbeek and Bryant, 1992; Bouma and Hoosbeek, 1996). Model classification based on: (1) scale hierarchy, (2) degree of computation, and 3) degree of complexity

3.2.3 Empirical and Mechanistic Models

The third axis of Figure 3.2 classifies models based on the degree of complexity. Empirical models use a simplified relation between observed and simulated data without making a claim to fundamentalness with regard to the processes involved. In these models usually having a low degree of complexity, cause and effect are related through a black box. Since the empirical model only describes the observations, the extent and support are at the same scale level (France and Thornley, 1984).

Mechanistic models are based on available process knowledge. The behavior of a system at the extent scale (i level) is described in terms of attributes at a lower i-1 scale level. The two scale levels are connected by a process of analyses and resynthesis, accompanied by hypotheses or assumptions (France and Thornley, 1984). The description of level i-1 behavior may be purely empirical, not containing any elements that refer to level i-2, or it may be partly empirical and partly mechanistic with some reference to i-2 or lower levels. Eventually, every mechanistic model is based on empiricism.

3.2.4 The Scale Hierarchy and Knowledge Type Diagram

The combination of the above three characteristics: scale hierarchy, degree of computation, and degree of complexity results in the Scale Hierarchy and Knowledge Type Diagram (Fig. 3.2). At each scale level a knowledge type plane can be constructed with the horizontal axis (degree of computation, and degree of complexity) illustrated at pedon level. For example, at any scale in the hierarchy a regression equation would have the knowledge type Quantitative-Empirical.

3.2.5 Time

Time is not considered as a fourth dimension in the Scale Hierarchy and Knowledge Type diagram, but is indirectly present in the 'scale hierarchy' and the 'degree of complexity'. Spatial and temporal scales are frequently related. For instance, in the introduction, Jenny combined a spatial extent scale at the soil-landscape level (or catena) with a temporal extent scale at the century to millennium level and a temporal support of one year. The student of microbial ecology mentioned by Jenny would have combined spatial with temporal extent scales at the soil horizon and year levels and spatial with temporal support scales at the soil structure and day levels. Generally speaking, processes at higher spatial scales have higher temporal scales and *vice versa*.

Qualitative and quantitative models may or may not take the time dimension into account. The time dimension does, however, have an influence on the complexity of a model. A static model represents a system at one time, for instance, a soil map. If the time dimension is taken into account during data collection and incorporated into the model, the model becomes dynamic.

Capacity models simulate changes of quantities without time as a direct variable, as opposed to rate models in which the change of a variable is defined as a function of time. The latter are more complex because they require the use of differential equations, and frequently, the use of iterative procedures to solve for the individual variables.

3.2.6 Dynamic Simulation Modeling

In pedology we are frequently interested in how soil characteristics and soil processes change over time. Dynamic simulation models are based on the assumption that the state of each system at any moment can be quantified, and that changes in the state can be described by rate or differential equations. This leads to models in which state, rate, and driving variables can be distinguished (Leffelaar, 1993). State variables are variables like OM content, volumetric water content, or amount of biomass. Generally, state variables have dimensions of length, number, volume, weight, energy or heat content. Driving variables quantify the effect of external factors on the system, and are not influenced by the processes within the system. Their value must be monitored continuously, e.g., meteorological variables like rain, wind and radiation. Rate variables indicate the rate at which state variables change. Their values are determined by the state and driving variables according to equations that are based on process knowledge.

In the following paragraphs we will introduce the method of constructing models according to the state variable approach by drawing in part on a chapter by Leffelaar (1993). The following examples of elementary systems will be used: (1) the amount of water draining through the bottom of a peat bog at a constant rate; (2) a number of anaerobic bacteria that increase each year by a certain fraction; (3) a dry polder being filled with water by seepage from a nearby river (Fig. 3.3) which may occur when the active drainage of the polder stops, and eventually, the water level in the polder will reach the water level of the river. The state variables in these examples are the volume of drained water, number of bacteria, and the amount of water in the polder. The rate variables appear on the left-hand side of the following equations:

$$dD/dt = c \qquad\qquad [3.7]$$
$$dA/dt = c \cdot A \qquad\qquad [3.8]$$
$$dW/dt = c \cdot (W_m - W) \qquad\qquad [3.9]$$

where D is the drained volume, t is time, and c is a constant, A is the number of bacteria, W is the amount of water in the polder, and W_m is the maximum water level that can be reached, i.e., the water level of the river (Fig. 3.3).

The rate variables have dimensions of a state variables per unit of time, namely, $m^3 \cdot t^{-1}$, $number \cdot t^{-1}$, and $m^3 \cdot t^{-1}$, respectively. The rate variables in Equations [3.8] and [3.9] are functions of the state variables, whereas in Equation [3.7] the state has no effect on the rate (the amount of water in the bog is assumed to have no effect on the rate of deep drainage in this example). The proportionality coefficient c in Equations [3.8] and [3.9] is important for the behavior of the state variables and often has a special name. In biological systems, such as in Equation [3.8], c is called the relative growth rate, while in technical or hydrological systems, such as in Equation [3.9], the inverse of c is called the time coefficient.

When a rate variable, dX/dt, depends on the state variable X, a feedback loop exists, namely, the state of the variable determines its rate of change, and hence its subsequent state. This feedback continuously adjusts the system through either positive or negative feedback loops. In a negative feedback loop the rate of change may be either positive or negative, but it is a negative function of the value of the state variable. For instance, in Equation [3.9] the rate is positive, but it decreases linearly with increasing water level in the polder. Negative feedback causes disturbed systems to return to some stable equilibrium. A positive feedback loop exists when there is a positive relation between the values of rate and state variables, so that both continuously increase. The exponential growth of the bacteria in Equation [3.8] is the result of positive feedback. In reality, however, factors like food shortage will put a limit to the growth.

When the knowledge of a system is summarized in differential equations, and the state of the model is known at a certain moment, the state of the model at other times can be calculated. Therefore, the differential equations need to be solved with respect to time, either by analytical or numerical integration. The following derivations provide the analytical solutions to:

Equation [3.7]
$$\int dD = c \cdot \int dt$$
$$D = c \cdot t + k \qquad \text{with } D = D_0 \text{ at } t = 0, \text{ then } k = D_0$$
$$D = c \cdot t + D_0 \qquad \text{(Fig. 3.4a)}$$
[3.10]

Equation [3.8]
$$\int (dA / A) = c \cdot \int dt$$
$$\ln A = c \cdot t + k \qquad \text{with } A = A_0 \text{ at } t = 0, \text{ then } k = \ln A_0$$
$$\ln(A / A_0) = c \cdot t$$
$$A = A_0 \cdot e^{c \cdot t} \qquad \text{(Fig. 3.4b)}$$
[3.11]

Equation [3.9]
$$\int [d(W - W_m) / (W - W_m)] = c \cdot \int dt$$
$$\ln (W - W_m) = -c \cdot t + k \qquad \text{with } W = W_0 \text{ at } t = 0, \text{ then } k = \ln(W_0 - W_m)$$
$$\ln[(W - W_m)/(W_0 - W_m)] = -c \cdot t$$
$$W = W_m - (W_m - W_0) \cdot e^{-c \cdot t} \text{(Fig. 3.4c)}$$
[3.12]

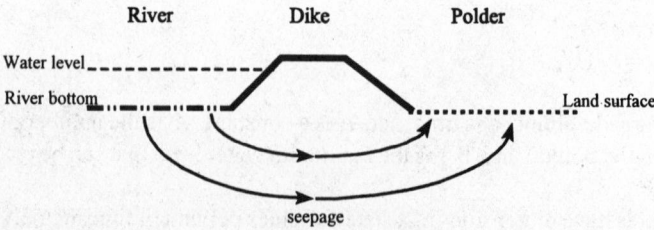

Fig. 3.3 A dry polder being filled with water by seepage from a nearby river. This situation may occur when the drainage of the polder stops. Eventually the water level in the polder will reach the water level of the river.

In case of the inundation of the polder, the rate of inflow of water decreases linearly with the difference between a maximum water level (W_m) and the actual water level (W). The solved differential equation shows that the water level in the polder approaches W_m exponentially (Fig. 3.4c). The rate equations are now obtained by differentiating the solved (integrated) equations:

Equation [3.10]
$$dD/dt = c \qquad\qquad\qquad [3.13]$$
Equation [3.11]
$$dA/dt = A_0 \cdot c \cdot e^{c \cdot t} \qquad\qquad\qquad [3.14]$$
Equation [3.12]
$$dW/dt = (W_m - W_0) \cdot c \cdot e^{-c \cdot t} \qquad\qquad\qquad [3.15]$$

These examples were chosen because they could be described by relatively simple differential equations. However, minor additions or changes to the equations, for instance, adding a temperature dependence to the biological system, make an analytical solution impossible. In practical applications, differential equations are frequently too complex to solve analytically.

Numerical integration is based on the assumption that the rate of change of a state variable is constant over a short period of time (Δt). The rate at t ($rate_t$) is calculated from $state_t$ with the rate equation. The new state, $state_{t+\Delta t}$, of the system is calculated according to:

$$state_{t+t} = state_t + \Delta t \cdot rate_t \qquad\qquad\qquad [3.16]$$

From this new state ($state_{t+\Delta t}$) a new rate ($rate_{t+\Delta t}$) is calculated, multiplied by Δt, and so on. This calculation scheme is called the rectangular or Euler's integration method.

Numerical integration always yields some approximation of the true value. This deviation depends on the calculation scheme and the length of Δt. With the above rectangular integration scheme, the rate is held constant during Δt. In reality this rate will increase or decrease during Δt, and as a result, error is introduced. Therefore, a shorter t will be more accurate, but the computation time will increase. For relatively simple systems like the three examples, the c coefficient may be used to choose an appropriate Δt. The inverse of c is called the time coefficient (τ). In models containing more than one rate variable, a first approximation of Δt is obtained by taking Δt smaller than one-tenth of the smallest τ in the model. Next to this rule of thumb, the modeler's knowledge of the system is also important in choosing an appropriate Δt. More elaborate integration schemes may improve the accuracy of approximation by, for instance, estimating an average state and rate during the time interval. These methods are beyond the scope of this introduction to dynamic modeling.

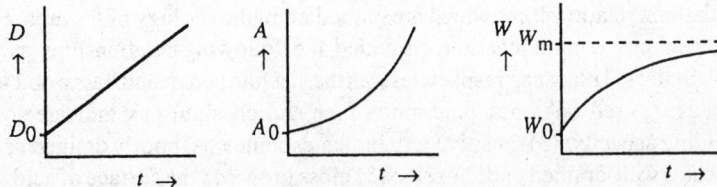

Fig. 3.4 Analytical solutions to the differential equations [Reprinted with kind permission of Kluwer Academic Publishers from Leffelaar, 1993. On systems analysis and simulation of ecological processes. (p. 16, Fig. 2.4) Copyright Kluwer Academic Publishers]

3.3   Pedological Models

Pedological models include all models that are used to describe or predict the distribution of soils in
a landscape and models that describe or simulate processes that occur in soils. In the remainder of this
chapter we will discuss existing pedological models. For organizational purposes we will classify the
models based on the Scale Hierarchy and Knowledge Type Diagram by grouping the extent scales
indicated in Figure 3.2 into two groups: (1) the landscape scales including polypedon up to world
scales, and (2) the pedon and subpedon scales including the pedon down to molecular interaction
scales.

The distinction between qualitative versus quantitative and empirical versus mechanistic does not
include a judgement with respect to the quality or sophistication of a model. A descriptive, qualitative,
model of a complex soil-landscape system may be more accurate and beneficial than a quantitative
model that can necessarily only take a limited number of processes into account. On the other hand,
quantitative models provide information that may be critical for planning purposes or that serve to
test the degree of understanding that is embodied in our qualitative models. Also, the purpose for
which a model is developed may dictate the approach. For management or policy making an empirical
approach may be the most suitable. However, when system knowledge needs to be tested or further
developed, a mechanistic approach is needed.

3.3.1   Qualitative-Empirical Models at Landscape Scales: Soil Maps

The number of qualitative pedological soil-landscape models (both empirical and mechanistic) that
have been developed probably exceeds the combined number of soils and soil scientists. Many of
these models have been documented in the literature, but the vast majority have never been recorded.
All of them contributed to our current understanding of soils to a greater or lesser degree, but failure
to record these models has hindered the advancement of pedology. Perhaps a better understanding of
the nature of these models and their contribution to the advancement of pedology will encourage
better documentation.

Soil maps are qualitative empirical models of how soils are distributed in a landscape. The map itself
does not provide information on the processes that lead to the presented distribution. This process
knowledge may be added in the legend as a descriptive, qualitative, mechanistic model. Additional
information, like the surface area of each map unit, may make the model less qualitative and more
quantitative. The information needed to make the soil map was collected by a surveyor using either a
mental version of the soil-landscape model (a qualitative model which may be empirical or
mechanistic) or a geostatistical approach (a quantitative empirical model).

Pedologists frequently use ecological indicators in the field to observe the distribution of soil
conditions in the landscape. Vegetative relationships with soils and soil conditions are qualitative
and empirical because the mechanisms are hidden in an extremely complex mix of factors such as
competition and adaption relationships, which are embodied in the ecology of the area. For example,
S. Carlisle (1993, personal communication) provided the following relationships as observed in
northern New York State: (1) blackcap raspberries, northern white cedar, and basswood indicate soils
with high base status; (2) red oak, jack pine, pitch pine and chestnut oak indicate soils with low
available water holding capacity; (3) sensitive fern indicates somewhat poorly drained or wetter soils;
(4) sweet fern indicates well-drained or drier soils; (5) moss grows on the surface of acid soils; and (6)
tag alder and swamp grass sharply coincide with the boundaries of flood plains.

3.3.2 Qualitative-Mechanistic Models at Landscape Scales: Soil Survey Paradigm

Soil scientists frequently develop and use qualitative models of this kind to facilitate soil survey by reducing the required number of soil observations and improving the placement of soil boundaries. Unfortunately, these models are seldom recorded and are lost once the soil survey is completed. Experience shows that soil surveys require periodic update and maintenance. Models of this kind would be a valuable tool for this purpose.

It was suggested earlier that Equation [3.2] could be solved in different ways. In this section this equation will be applied in its qualitative (mental descriptive) form. In subsequent sections the same soil-forming factors equation will be solved quantitatively both empirically and mechanistically.

Qualitative mechanistic models derived from soil-landscape models can be used to extrapolate point data based on conceptual relationships between observations of the soil property being mapped and easily observable landscape features. A conceptual soil-landscape model may be present in the head of a soil scientist as thoughts relative to the association of soils with various observable landforms that allow the extrapolation of a point observation to a landscape segment. This association of soils with landforms is largely based on process knowledge of the soil-landscape system. Soil-landscape models of this kind have been widely accepted as the foundation for soil survey. The paradigm of soil survey is based on the soil forming factors (Hudson, 1992), and can be described by Equation [3.2].

A qualitative predictive model for mapping, e.g., soil degradation, can be derived by adding a management variable as the sixth factor to Equation [3.2] (Hoosbeek et al., 1997):

$$S = f(cl, o, r, p, t, m) \qquad [3.17]$$

where m represents recent and historical land use and management practices. Just as climate and organisms are factors of soil formation, management practices associated with the history of man's use of the land define the factor of soil degradation. The mechanism of the interactions between specific land management practices and soil and climate characteristics that result in erosion and soil degradation is well understood. The model embodies that knowledge.

The qualitative application of Equation [3.2] provides a predictive conceptual model that can be used to extrapolate point data to spatial areas. It is important to note that landscape features that define land segments having relatively uniform characteristics for the five soil-forming factors must be visible in order to allow delineation of the segment. The conceptual model allows one to extrapolate observations to the limits of the delineation in which those observations occur. The exact location of where the line between map units will be drawn is subjective, but is based on perceived changes in landscape features.

Application of the qualitative soil-forming factors model is based on tacit knowledge of the user. Hudson (1992) explained that students can only learn to use the model by working in the field guided by experienced pedologists. Early in the learning process, a *gestalt* shift takes place, i.e., the student starts to see the soil-landscape as a collection of geomorphic attributes, followed by a gradual extension of the tacit knowledge.

3.3.3 Quantitative-Empirical Models at Landscape Scales

The following four modeling approaches will be discussed: (1) state factor model as regression equation; (2) statistical soil-landscape models; (3) spatial and temporal variance models; and (4) learning algorithms. At first it may seem counterintuitive to place these approaches in one section.

However, all four approaches are characterized by the use of statistical modeling techniques without the use of mechanistic process knowledge.

3.3.3.1   State Factor Model as Regression Equation: Chronofunctions

To illustrate the use of quantitative empirical models we continue to quote from Jenny's (1961) paper: "To render state factor equations useful for acquiring specific pedologic knowledge they must be "solved". Soil properties must be related to individual state factors. To cite an example, in the great valleys of northern India the average nitrogen content in surface soils under native vegetation varies from 0.05 to 0.20%. This range is caused by the many state factors which operate in that area. In general, the contribution of any state factor $\underline{F}$ to the variation of a soil property, such as nitrogen N, is given by:

$$\int_{c}^{d}\frac{\delta N}{\delta \underline{F}}d\underline{F}$$ [3.18]

This mathematical expression consists of two parts. First, the range (c d) of the state factor $\underline{F}$ in the area, second, the slope ($\delta N/\delta \underline{F}$) which expresses the effectiveness of the state factor in the area. In northern India the precipitation factor ranges from 10 to 120 inches (254 to 3048 mm·y^{-1}), as seen on the horizontal axis in Fig. 3.5. The effectiveness of this state factor, that is, the slope of the N-curve is variable. A rise in rainfall of 1 inch (25.4 mm) greatly affects soil nitrogen in the precipitation range of 5 to 40 inches (127 to 1016 mm·y^{-1}), but it has very little influence in the range 60 to 120 inches (1524 to 3048 mm·y^{-1}), where the curve is almost horizontal, and the slope ($\delta N/\delta \underline{F}$) is nearly zero. Here, variation in rainfall is inefficient."

With this model Jenny described how one factor is related to the other. The model does not explain why the factors behave as described. In his book *The Soil Resource,* Jenny (1980) provided many more examples in which soil properties were related to individual state factors. Richardson and Edmonds (1987) used linear regression to describe segments of a curve for soil properties plotted

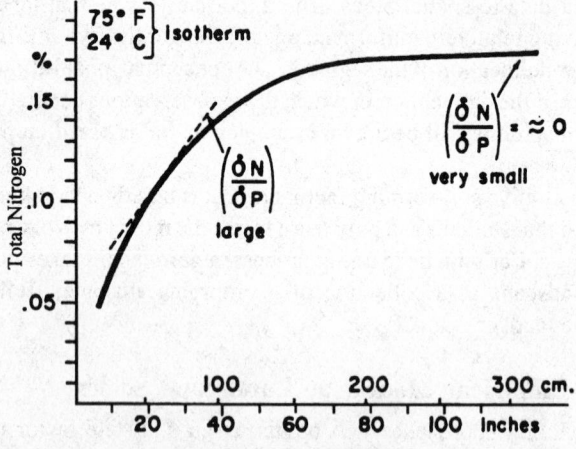

M. A. Precipitation

Fig. 3.5   Relationship between soil nitrogen and mean annual precipitation in northern India. The slopes ($\delta N/\delta P$) are large in the dry regions and small in the wet regions [Reprinted from Jenny, 1961. Soil Sci. Soc. Am. Proc. 25:385-388 with permission of the Soil Science Society of America].

against a state factor. The relative effectiveness as indicated by the slope was examined for specific examples.

It is often possible to select in a region a series of pedons for which the contribution of one single state factor outranks the combined contribution of all the other factors (Jenny, 1961). Such cases of a dominant factor provide pedogenic sequences. The chronofunction is defined as:

$$S = f(t)_{\,cl,\,o,\,r,\,p,\,t,\,...} \qquad\qquad\qquad\qquad\qquad [3.19]$$

where S, any soil property, is a function of time, while the other state factors are assumed constant. Climo-, bio-, topo-, and litho-function were defined likewise with S as a function of the dominant factor. A chronosequence is constituted by a group of experimental sites where the primary distinction is the age of the soils. To the degree possible, differences in other soil-forming factors are kept to a minimum. Locating sites that meet these conditions is often difficult, in part because accurate determination of the ages of soils is troublesome (Rabenhorst, 1997). Nevertheless, common types of chronosequences that have been studied are series of alluvial (Chittleborough et al., 1984) and marine (Merritts et al., 1991) terraces of differing ages.

Recently, Rabenhorst (1997) provided an overview of how chronofunctions were solved: "Before 1980, there were few instances of mathematical solutions to chronofunctions (Hay, 1960; Ruxton, 1968). More recently, Bockheim (1980) calculated best fit computer regression equations of previously published chronosequence data. Others have subsequently described mathematical chronofunctions for various soil properties or soil formation indices (Birkenland, 1984; Harden, 1990; Harden et al., 1991; Mellor, 1986; Muhs, 1982). Several models have emerged to describe different soil properties and include linear (Muhs, 1982), logarithmic (Bockheim, 1980), power (Muhs, 1982), and exponential functions (Birkeland, 1984). Schaetzl et al. (1994) have argued persuasively for the necessity of applying sound pedological theory when selecting mathematical models for soil chronofunctions and have discouraged the blind application of statistics in choosing "best fit" models." The plea by Schaetzl et al. (1994) for a less empirical approach would involve the examination of the system at one or more lower scale levels to acquire the needed process knowledge to allow for a more mechanistically guided mathematical solution.

In addition, regression equations based on a clay accumulation index calculated by Levine and Ciolkosz (1983) and the use of the soil development index by Harden and Taylor (1983) should be mentioned. The clay accumulation index values of eleven well-drained soils formed in glacial till parent materials of varying ages in northeastern Pennsylvania were calculated as follows (Levine and Ciolkosz, 1983):

$$\text{clay accumulation index (per pedon)} = (B_c - C_c) \times T \qquad\qquad [3.20]$$

where B_c is the clay content (%) of the B2 horizon, C_c is the clay content (%) of the C horizon and T is the thickness of the B2 horizon (cm). A regression equation was derived based on soils of known Woodfordian (15,000 yr B.P.) and Holocene ages to predict the age of soils formed in tills of different ages:

$$\log Y = 1.80 + 0.992\,(\log X) \qquad\qquad r^2 = 0.913 \qquad\qquad [3.21]$$

where Y is the age in years and X is the clay accumulation index.

Harden and Taylor (1983) quantified soil development over long periods of time by investigating four chronosequences at the catena scale level in different climatic zones. Points were assigned to properties of a horizon as compared to the reference state of the parent material (e.g., parent material has hue 7.5YR, then a horizon with hue 5YR scored 10 points because of rubification). Quantified properties were presented in time plots to determine whether they developed at similar rates. The soil development index was found to be useful and was presented as a semi-quantitative method.

Generally, two major problems have affected chronofunction studies (Rabenhorst, 1997). First, most studies are limited by the availability of suitable sites (2–6 pedons). Second, the spatial variability, induced by the other state variables, usually increases rapidly with increasing distances between the sampling points. In addition, soil chronosequences that extend beyond the Holocene period are affected by significant climatic variations related to glacial/interglacial effects. Jenny (1980) was very hesitant about fitting smooth curves to these sequences given that we will never know the pathway as to how those old soils acquired their properties; consequently, he only plotted the points and did not connect them, or at least, did not assign real meaning to the line.

Rabenhorst (1997) described a theoretical approach to develop more accurate chronofunctions in transgressive estuarine pedogenic systems by minimizing these obstacles and called it a chrono-continuum. The encroachment of the sea on upland soils during the most recent coastal transgression caused a dramatic shift in pedogenic processes, which induced significant changes in soil properties. The slowly rising sea level has permitted the establishment of tidal marsh vegetation and its subsequent vertical accretionary growth. The rate of sea-level rise during an interval of marsh soil formation can be determined by various dating techniques, which can be related to the elevation of the submerged land surface through a topographic survey. Thus, the continuum in space, along the topographic surface, can be translated into a continuum in time. This permits one to sample a large number of points along the temporal continuum within a relatively small distance, minimizing the two problems mentioned earlier with respect to chronofunction studies. Changes in soil properties along the chrono-continuum can be quantified between adjacent sampling points (P_1 and P_2) and related to the corresponding temporal points (t_1 and t_2):

$$\delta P = P_1 - P_2 \qquad\qquad\qquad\qquad [3.22]$$

and

$$\delta t = t_1 - t_2 \qquad\qquad\qquad\qquad [3.23]$$

The rate of that particular process is then equal to $\delta P/\delta t$. Rabenhorst acknowledged that chronofunctions obtained as described above are formulated under a narrow range of conditions (intended to suppress the other state factors) and are therefore, like any other empirical state factor function, limited to the soil-landscape for which they were developed. Nevertheless, once reliable chronofunctions are developed for the area under investigation, such as the coastal area under ongoing transgression, they may be applied to predict changes in soil properties and to quantify pedogenic processes.

3.3.3.2 Statistical Soil-Landscape Models

Shovic and Montagne (1985) developed a statistical model of soil-landscape relations to serve as an aid in making maximum use of limited data. Thematic mapper data and topographic information from a digital elevation model were combined by Lee et al. (1988) to determine soil characteristics of hilly terrain in southwestern Wisconsin. Havens (1988) developed a GIS-based statistical model which predicts the percentage of the soil surface covered with rock fragments. Although the model failed to predict for fragments smaller than 25 cm in diameter, the probability of encountering a particular stoniness class was simulated correctly in 77% of the cases during validation in the field. A similar approach of soil-landscape modeling was used by Bell (1990) to produce soil drainage class maps.

The model had a 74% overall agreement rate with field observations in the area of development (calibration), but failed when applied to another site within the same physiographic province. The approach is innovative in the use of new technology and appears to be useful when properly calibrated for a specific area; however, empirical models are inherently restricted to areas for which they are calibrated.

Slater et al. (1994) reviewed different types of Geographic Information Systems that may be of use for pedological modeling. They also described the idea of using a 3-dimensional version of raster pixels called *voxels*, that can be thought of as stacked boxes each occupying a fixed volume and containing certain attributes. The voxel structure can be flexibly applied at any data scale and may be useful for modeling soil systems from thin sections to the entire pedosphere. In order to represent short-distance variability, a large number of small voxels are needed with consequent high data storage requirements. The proposed GIS based on voxels would be a 3-dimensional bookkeeping system that is suitable for both empirical or mechanistic models at all scales.

3.3.3.3 Spatial and Temporal Variance Models

Regionalized variable theory (Journel and Huijbregts, 1978; Burgess and Webster, 1980a,b; Webster, 1985; Davis, 1986; Burrough, 1986) assumes that the spatial variability of any variable can be expressed as the sum of three major components:

$$Z(x) = m(x) + \epsilon'(x) + \epsilon'' \qquad [3.24]$$

where $Z(x)$ is a spatial variable at location x; $m(x)$ is a deterministic function describing the systematic variability of $Z(x)$ at x; $\epsilon'(x)$ denotes the random locally varying spatially dependent residuals from $m(x)$; and ϵ'' is a spatially independent residual having a zero mean and variance σ^2 (Burrough, 1991). Observations obtained close to each other are more likely to be similar than observations at a larger distance from each other. This spatial correlation of $\epsilon'(x)$ is described by the semivariance γ. If γ is plotted as a function of the lag distance h, the semivariogram $\gamma(h)$ is obtained. $\gamma(h)$ is estimated as (Journel and Huijbregts, 1978; Davis, 1986; Burrough, 1986, 1991):

$$\gamma(h) = \frac{1}{2N(h)} \sum_{i=1}^{N(h)} \left[Z(x_i) - Z(x_i + h) \right]^2 \qquad [3.25]$$

where N is the total number of pairs of data points.

A typical semivariogram may have a nugget variance indicating the variance at distance zero. This nugget variance represents the measurement error. With increasing lag distance the variance will increase up to the sill variance. The distance at which the sill is reached is called the range. Beyond this range the observations are not spatially correlated. Several standard models are available to fit the experimental semivariogram, e.g., linear, spherical, exponential, Gaussian and power models (Deutsch and Journel, 1992).

Regionalized variable theory may also be applied to temporal variables. Temporal variability can be expressed as the sum of three major components:

$$Z(t) = m(t) + \epsilon'(t) + \epsilon'' \qquad [3.26]$$

where $Z(t)$ is a temporal variable at time t; $m(t)$ is a deterministic function describing the systematic variability of $Z(t)$ at t; $\epsilon'(t)$ denotes the random locally varying temporally dependent residuals from

$m(t)$; and $\in''$ is a temporally independent residual having a zero mean and variance σ^2 (adapted from Burrough, 1991).

Spectral analyses is the partitioning of the variation in a time series into components according to the duration or length of the intervals within which the variation occurs (Davis, 1986). Time series are considered the sum of many regular sinusoids with different amplitudes, wavelengths and starting points. Once the deterministic part, $m(t)$, is removed by time series analysis, $\in'(t)$ and $\in''$ can be described by a temporal semivariogram.

Some observations are more variable in space and relatively less variable in time or *vice versa*. Many observations in soil science tend to belong to one of these two groups. However, some variables cannot be placed convincingly in either group. Organic matter fractions with a short turn over time may exhibit both characteristic spatial and temporal variabilities. In addition to variables that can be measured directly in the field or laboratory, many environmental studies involve the use of georeferenced data as input to computer models that simulate processes over periods of time (Bouma and Hoosbeek, 1996; Hoosbeek and Bouma, 1998). For the proper interpretation of the output variables resulting from these simulations at each georeferenced point, it is important to describe both their spatial and temporal variability.

Several methods have been proposed to integrate both spatial and temporal variability into one model (Hoosbeek, 1998). Journel (1986) and Rouhani and Myers (1990) discussed possible problems associated with integrating spatial and temporal dimensions into one semivariogram model and pointed out that, although some directional dependence (i.e., anisotropy) may exist, spatial phenomena in general show no ordering, whereas a notion of past, present, and future exists for temporal phenomena. Measured data may represent a unique realization for the past and present, but in the case of the future a truer stochastic dependence exists. Because of this inequality, the additive space-time model cannot be used for interpolation; however, it is useful for describing and depicting the combined spatial and temporal variability of soil data.

Hoosbeek (1998) treated time as a third dimension added to a two-dimensional space. Each datum obtained as output from a model, e.g., a θ-value, has two spatial coordinates, x and y, and one temporal coordinate, t, i.e., $\theta(x_1, y_1, t_1)$. For each combination of two data points in space and time, the squared difference between the two data values, the lag distance in space and the lag distance in time were calculated. The lag distance in space, h, was calculated as the euclidian distance between (x_1, y_1) and (x_2, y_2). The lag distance in time, τ, was calculated as the absolute difference between t_1 and t_2. The following spherical model was used to model the space-time semivariograms:

$$\gamma(0,0) = 0$$

$$\gamma(h,\tau) = c_0 + c_1 * \left[1.5\left(\frac{h}{a}\right) - 0.5\left(\frac{h}{a}\right)^3 \right] +$$

$$c_2 * \left[1.5\left(\frac{\tau}{b}\right) - 0.5\left(\frac{\tau}{b}\right)^3 \right] \qquad for\ h \leq a \wedge \tau \leq b$$

$$\gamma(h,\tau) = c_0 + c_1 * \left[1.5\left(\frac{h}{a}\right) - 0.5\left(\frac{h}{a}\right)^3 \right] + c_2 \qquad for\ h \leq a \wedge \tau > b$$

$$\gamma(h,\tau) = c_0 + c_1 + c_2 * \left[1.5\left(\frac{\tau}{b}\right) - 0.5\left(\frac{\tau}{b}\right)^3 \right] \qquad for\ h > a \wedge \tau \leq b$$

$$\gamma(h,\tau) = c_0 + c_1 + c_2 \qquad for\ h > a \wedge \tau > b$$

[3.27]

where c_0 is the nugget variance, c_1 is the spatial variance contribution, c_2 is the temporal variance contribution, a is the spatial range, and b is the temporal range. For instance, the spatio-temporal variogram of leaching simulated with DSSAT (Tsuji et al., 1994) is presented in Fig. 3.6 (Hoosbeek and Bouma, 1998; Hoosbeek, 1998).

Spatial and/or temporal variance models may be used for the optimization of sampling schemes or for interpolation techniques like kriging. These and other geostatistical techniques are further discussed in Section A, Chapter 9.

3.3.3.4 Learning Algorithms

Artificial Neural Networks (ANNs) are computer algorithms that are composed of a network of connected nodes (Fu, 1994). The mechanics of these networks are inspired by biological neural systems. ANNs have the ability to learn patterns in large data sets given a fixed set of inputs. Input and output variables are related without any knowledge about the underlying processes. Single layer ANNs with linear transfer functions are analogous to linear regression models. Multilayered networks with nonlinear transfer functions can model complex nonlinear relationships as well as feedback mechanisms which provide the net with a memory of previous input relationships.

Levine et al. (1996) tested the feasibility of using ANNs as a method to predict a qualitative soil property like soil structure (granular, blocky, and massive) from standard soil survey data (all textural classes, C%, CEC/clay ratio, etc.). The analyses were limited to Ustolls from the National Cooperative Soil Survey (USDA) database in order to control the effects of climate and natural vegetation. The pedon descriptions included soil structure classes as observed in the field. One half of the 390 available samples were used to train various net structures with 3 to 12 input nodes (characterization variables), 0 to 3 hidden nodes, and 3 output nodes (the 3 soil structure classes). The other samples were used to test the networks. The best network appeared to have 3 input nodes (C%, clay% and silt%), 2 hidden nodes and the 3 (fixed) output nodes. This network predicted soil structure with an accuracy of 79%, while linear regression yielded an accuracy of 46%.

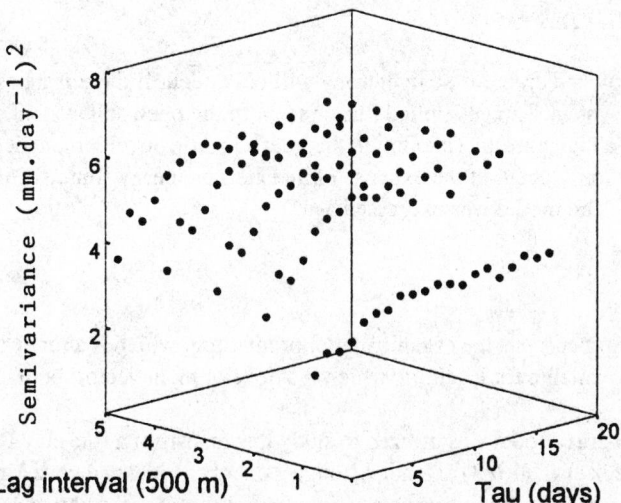

Fig. 3.6 Spatio-temporal semivariogram of leaching simulated with DSSAT [Hoosbeek and Bouma, 1998; Hoosbeek, 1998]

Lam and Pupp (1993) used ANNs to estimate missing data in a database. They trained the network with available data and then used it to predict the missing values. Pachepsky et al. (1996) found that ANNs gave better results than linear regression for predicting soil water retention in Ustolls from easily measurable data (texture and bulk density).

Artificial Neural Networks need, depending on their complexity, large data sets consisting of hundreds to thousands of data points to be trained and tested. In pedology, data sets of this size are usually limited to standard soil characterization data from national soil surveys or are remotely sensed data. Another limitation is that ANNs can only be used as an interpolation tool (only knowledge that was input during training is available).

3.3.4 Quantitative-Mechanistic Models at Landscape Scales

After defining the boundaries of his system, Jenny (1961) provided a list of in and out fluxes which are the result of many processes. The mechanistic approach involves the careful definition, based on process knowledge, of differential equations that describe transport processes (in and out fluxes per compartment) and transformation processes (e.g., oxidation-reduction reactions).

In this section, we will discuss quantitative mechanistic models at the landscape scales. For most regions, landscape scales include hillslope processes. Models based on one-dimensional vertical transport processes cannot, or only roughly, describe processes in hilly terrain. Therefore, in this section we will discus models that include multi-dimensional processes, while the one-dimensional models will be discussed in Section 3.3.8.

3.3.4.1 Runge's Energy Model

Runge (1973) developed an energy model in which gravity is the source of energy. Soil development at a landscape segment is dependent on the relative amount of water running off (no net soil development) versus the amount of water infiltrating into the soil (producing soil development). Runge based his ideas on the first two laws of thermodynamics. A closed system will approach a state of minimum energy and maximum entropy at equilibrium. The Gibb's free energy, G, is defined as the difference between the change in enthalpy, H, and entropy, S, at constant temperature and pressure:

$$\Delta G = \Delta H - T\Delta S \tag{3.28}$$

A decrease in enthalpy and an increase in entropy will only occur if there is a positive energy input, ΔG, into the closed system. Runge applied this concept to the open soil system. A loess soil parent material has initially a high entropy (maximum disorder), profile development decreases the entropy (ordering) which will only occur at the expense of an external energy source. The leaching of water provides this energy. The model was expressed as:

$$s = f(o, w, t) \tag{3.29}$$

where s is soil development; o is the organic matter production, which is a renewing vector; and w is the amount of water available for leaching, which is a development vector; both vectors are intensity factors.

Runge's (1973) energy model was utilized to study the genesis of a fine clay B horizon in a stable loess landscape. Smeck et al. (1983) extended Runge's energy model and gave a graphical representation of entropy changes on a relative scale as a function of the contribution of selected soil forming processes to the development of the ten taxonomic soil orders. Both Runge and Smeck used laws of thermodynamics applicable to closed systems. Although the energy model is useful in a qualitative way, soils are open systems, therefore, actual quantitative thermodynamical calculations may not be possible. An approach based on nonequilibrium thermodynamics would be more desirable (Hoosbeek and Bryant, 1992). Another point is that gravity is not the only source of energy. Solar energy dries soils and also provides energy for the plants and organisms that modify soils.

3.3.4.2   Soil Landscape Systems Approaches

Huggett (1975) used the soil landscape system as an approach to model the soil system. The valley basin is defined as the basic organizational unit of the soil system and consists of a three-dimensional open system whose boundaries are defined as drainage divides, the surface of the land, and the weathering front at the base of the soil profile. Two subsystems are defined within the organizational unit, the soil skeleton and the soil solution. The soil skeleton is comprised of relatively large and stable clastic sediments (including loess) and is part of a geomorphic sediment-transporting system. The very mobile and unstable soil solution is part of the hydrological system. The model is based on the examination of the material fluxes of the subsystems through the valley basin.

Kirkby (1985) developed a mathematical model for soil profile development with a strong link to geomorphology. A weathering profile, organic profile, and inorganic profile are simulated based on the following processes: percolation, equilibrium solution, leaching, ionic diffusion, organic mixing, leaf fall, organic decomposition, and mechanical denudation. The weathering profile describes the proportion of substance remaining (p), at any depth (Fig. 3.7). The accumulated deficit of weathered material is

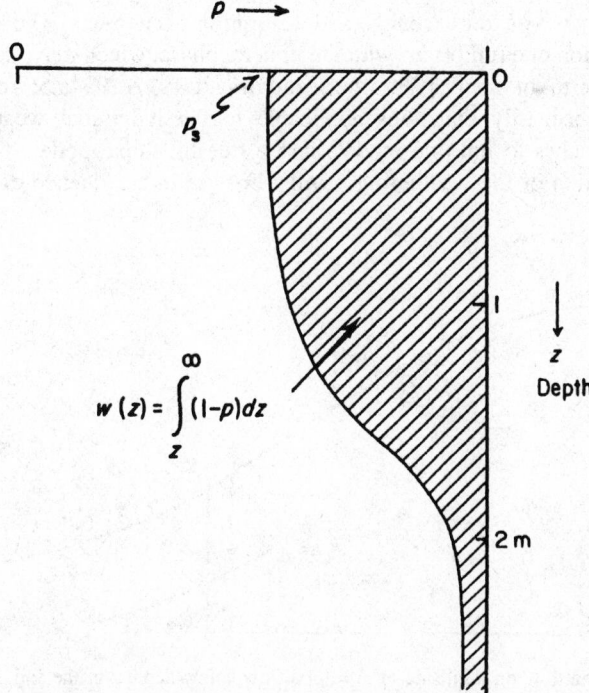

Fig. 3.7   Definition sketch for soil profile terminology [Reprinted from Kirkby, 1985. J. Soil Sci. 36:97-121 with permission of Blackwell Science Ltd]

$$w(z) = \int_{z=0}^{\infty} (1-p)dz \qquad\qquad\qquad [3.30]$$

where z is depth below the soil surface. A simple linear flow law for a convex slope (dg/dx constant) with uniform permeability, K, is written as:

$$q(z) = K.g.w(z) \qquad\qquad\qquad [3.31]$$

where $q(z)$ is the total flow below depth z in the soil, g is the surface gradient, and K is the flow velocity on unit hydraulic gradient (Fig. 3.8). Maximum percolation at depth z is

$$F(z) = F_0 + d\big((K.g)w(z)\big)/dx = F_0 + \lambda w(z) \qquad\qquad\qquad [3.32]$$

where F_0 is the limiting percolation rate at large depths, λ is a constant for any given site. The process of weathering is expressed as:

$$-\delta p / \delta t = c(p)q' \qquad\qquad\qquad [3.33]$$

where $c(p)$ is the total solute concentration at p, and q' is the rate of flow through the soil in $l\,mg^{-1}\,yr^{-1}$. These equations and equations for the other processes are combined to second-order linear partial differential equations for each of the three profiles. Kirkby (1985) mentioned the neglect of processes such as physical translocation of clay, complexing, and cheluviation as simplifications. Other limitations are the neglect of ion exchange and adsorption phenomena. The solution chemistry is based on the assumption of equilibrium with the mineral phase, which may, however, be described more successfully in terms of kinetics (Hoosbeek and Bryant, 1992, 1994a,b; Addiscott, 1994).

Kirkby provided primarily equations applicable to the individual weathering profiles and suggested two approaches to link the models into a true hillslope model. The first involves the extension and correction of the weathering profiles for use in a sequence of profiles; the second

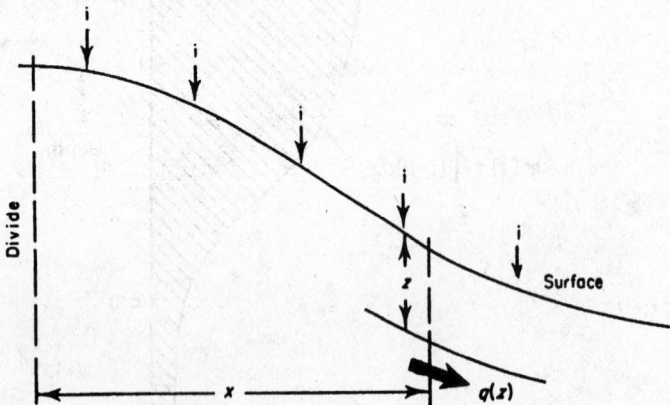

Fig. 3.8 Percolation and flow on a hillside. $q(z)$ = total flow below depth z in the soil. i = mean infiltration rate (rainfall - evapotranspiration - overland flow) [Reprinted from Kirkby, 1985. J. Soil Sci. 36:97-121 with permission of Blackwell Science Ltd]

method is to incorporate the profile model more directly in a hillslope model with two independent space dimensions, i.e. distance downslope and depth within the soil. The second method for combining slope and soil models requires an explicit hydrological model which routes subsurface flow downslope. The two models may then be formally linked, calculating flows and solute transport within the two-dimensional soil layer on the hillside. Sediment transport may also then be modeled in terms of calculated overland flows. This type of combined model has theoretical advantages, but Kirkby acknowledged the considerable cost in added computation and complexity. Moreover, we expect that the data availability would be the most serious limitation to the application of Kirkby's model to a field situation.

3.3.4.3 Watershed Study

Chen et al. (1982) produced an early acid rain model (ILWAS) to simulate lake-watershed acidification processes. The model is comprised of several modules including watershed hydrology, stream and lake hydraulics, canopy, snow, soil, stream and lake chemistry. Incoming precipitation is routed through a layered soil profile to the streams and ultimately to the lake. Field studies indicated the importance of the several possible flow-paths (surface, shallow subsurface, and groundwater flow) in determining lake water acidity. ILWAS can be useful to pedogenetic modeling in providing a shell consisting of nonsoil compartments, leaving the soil compartment for further improvements.

3.3.5 Qualitative-Empirical Models at Pedon and Subpedon Scales

A large part of the information in soil databases around the world is present in the form of qualitative empirical horizon and pedon descriptions, such as the first horizon of the first pedon described in Appendix IV of *Soil Taxonomy* (Soil Survey Staff, 1975): "Ap 0–15 cm. Dark grayish brown (10YR 4.5/2) light silty clay loam, very dark grayish brown (10YR2.5/2) moist; moderate fine granular structure; hard (dry), friable (moist); clear boundary." In this description the horizon depth is quantitative, the textural class and Munsell color code are semi-quantitative, and the rest is a qualitative model of a soil horizon. This model allows a pedologist to get a mental picture of this horizon without having seen the real thing. A pedon model is comprised of one or more horizon models.

Similarly, a thin-section description portrays the character of the soil fabric within a horizon. Although digital imaging technology allows partial quantification of a thin-section description, the vast majority of these models are qualitative. The micromorphologist may develop a more mechanistic model of the soil forming processes that led to the observed characteristics of one or more thin sections, but the qualitative empirical model must precede the development of a mechanistic model.

In both examples above, the tools of model development are the terms which are defined and adapted for standard use by soil scientists. Without widely recognized terminology, these models could not fulfill their purpose of communicating a mental image of the soil. *The Soil Survey Manual* (Soil Survey Division Staff, 1993) and the *Handbook for Soil Thin Section Description* (Bullock et al., 1985) are examples of defined terminologies widely recognized among soil scientists.

3.3.6 Qualitative-Mechanistic Models at Pedon and Subpedon Scales

The qualitative empirical models presented in the previous section may lead to a description of soil-forming processes which may result in the observed soil characteristics. By describing the processes within the horizons (at one or two lower scale levels), the formation of the horizons can be explained. By explaining the formation of the horizons, the formation of the pedon is explained. This explanatory model has then become mechanistic. Soil horizon nomenclature is based on these models. Horizon names represent the best guess of soil changes relative to the parent material.

There are many good examples in the literature in which macro- and micro-morphological techniques are used to build qualitative empirical models, which in turn are used to deduce qualitative mechanistic models of soil formation. For instance, Jongmans et al. (1991) used macro- and micromorphological, mineralogical, chemical and physical characteristics to build a qualitative mechanistic model of soil formation in a terrace chronosequence of nine gravelly Quaternary terraces of the Allier River, Limagne, France.

3.3.7 Quantitative-Empirical Models at Pedon and Subpedon Scales

Simonson's (1959) Generalized Theory of Soil Genesis can be regarded as the archetype of quantitative models at pedon and subpedon scales. Most models, both empirical and mechanistic, are in some way derived from Simonson's model. The models in the following subsections are derived by applying statistical techniques to observations.

3.3.7.1   Generalized Theory of Soil Genesis Pedon Model

Simonson's (1959) Generalized Theory of Soil Genesis is primarily concerned with the differentiation of horizons in a profile at the pedon level. Four kinds of changes are described: additions, removals, transfers, and transformations. Differences between soils and their horizons are explained by differences in relative intensity of the four changes which are not defined in terms of actual processes (Huggett, 1975), making the model empirical. However, it can easily be converted into a mechanistic model by describing the additions, removals, transfers, and transformations in terms of differential equations representing actual processes. For instance, the pedodynamic approach (Section 3.3.8.5) is such an application.

3.3.7.2   Dynamic Soil-Forming Processes Model

Yaalon (1971) noted that Simonson's (1959) model does not attempt to indicate how the system will develop with time and suggested that soil-forming processes can be viewed as belonging to one of two large groups, each with two subgroups: (1) those approaching a state of dynamic equilibrium at (a) a fast or (b) a slower initial rate; and (2) the irreversible or self-terminating processes, where the balance of input and output is not maintained, with (c) gains greater than losses (I>E) or (d) gains less than losses (I<E). The two major groups are not mutually exclusive, since the size of the system will often determine whether a certain attribute resulted from one or the other group of processes.

Paleosols can play an important role in interpreting past and ongoing climatic changes (Valentine and Yaalon, 1990; Yaalon, 1991). This arrangement of dynamic soil-forming processes was used by Yaalon (1971) to determine the persistence of various soil features and horizons in Paleosols. Features resulting from processes belonging to the equilibrium group (1) that relatively slowly (> 10^3 years) attain steady state are relatively persistent in paleosols and are thus good diagnostic features of paleopedogenesis. The irreversible, self-terminating features resulting from group (2) are the best indicators of paleopedogenic conditions.

3.3.7.3   Evolutional Soil Genesis Model

Johnson and Watson-Stegner (1987) claimed that their evolutional model of pedogenesis meets the following conditions: "Such a model should be consistent with the observation that soils are complex open systems, continuously adjusting by variable degrees, scales, and rates to variable energy and mass fluxes, thermodynamic gradients, and other changing exogenous environmental conditions." The model is written as:

$$S = f(P, R)$$ [3.34]

where S is the soil; P stands for progressive pedogenesis, or soil progression, which includes those processes, factors, and conditions that promote differentiated profiles leading to physicochemical stability, namely, horizonation/leaching processes, developmental (assimilative) upbuilding, and/or deepening; and R stands for regressive pedogenesis, or soil regression, which includes those processes, factors, and conditions that promote simplified profiles leading to physicochemical instability, namely, haploidization/rejuvenation processes, surface removals, and/or retardant (non-assimilative) upbuilding.

In fact, this model is a rephrasing of earlier models developed by Jenny (1941), Simonson (1959), Runge (1973), and others, with added emphasis on certain aspects. Three years later, Johnson et al. (1990) introduced a dynamic-rate model of soil evolution, which depends heavily on philosophical redefinitions of existing concepts. This model uses dynamic and passive pedogenetic vectors and their derivatives:

$$S = f(D, P, (dD/dt), (dP/dt)) \qquad [3.35]$$

where S is the state of the soil or profile evolution, D a set of dynamic vectors, P a set of passive vectors (both sets of vectors vary spatially and fluctuate through time), and dD/dt and dP/dt are vector changes through time. For most applications the 'vectors' will be similar to the factors of Jenny's state factor model.

3.3.7.4 Mass Balance Interpretation of Pedogenesis

Haseman and Marshall (1945) and Brewer (1964) developed techniques of quantitative mineral analysis that can be used to estimate gains and losses of weight, thickness (or volume), and particular constituents in the various soil horizons due to soil formation. The calculations of soil formation are based on quantitative estimations of the constituents of the soil materials of the horizon in relation to the amount of some substance considered to be essentially stable to the soil-forming processes which have been operative in the particular profile. The accuracy of these calculations depends on the validity of 3 basic assumptions: (1) uniformity of the parent material; (2) selection of the true parent material; and (3) the properties of the stable constituent. On the basis of estimation of a stable constituent (e.g., zircon or tourmaline), the percentage of each mineral species that has been weathered in the formation of each horizon can be plotted against depth in the profile. In general, smooth curves without inflections indicate a true parent material and uniformity of this parent material. Zircon and quartz have been used most frequently as a stable constituent because of their apparent relative stability under a wide range of conditions, although under certain conditions (e.g., laterite formation) they may weather to some extent. Smeck and Wilding (1980) evaluated the assumptions of mass balance reconstructions and errors in these reconstructions using elemental Zr as the stable constituent for soil developed from glacial tills in Ohio.

Van Wambeke (1972, 1976), Smeck and Wilding (1980) and later Brimhall and Dietrich (1987) and Brimhall et al. (1991a,b) extended Brewer's (1964) techniques and presented a methodology, referred to as "a mass balance interpretation of pedogenesis", to estimate the rates of weathering and pedogenesis. Chadwick et al. (1990) combined a set of analytical mass-balance functions consisting of basic conservation equations, strain equations (to account for volumetric changes), and flux equations with traditional selective extraction and particle-size separation procedures to investigate the overall long-term pedogenetic processes that have turned beach sand into an Alfisol.

Mass-balance analysis was also used by Langley-Turnbaugh and Bockheim (1998) to quantify elemental losses, gains and transformations for a soil chronosequence developed on elevated marine terraces in south coastal Oregon. The major soil-forming processes and the net gains and losses of the major soil forming elements (Si, Al, Fe, Na) of four soils ranging in age from 80 to 250 10^3 yr were determined.

3.3.7.5   Regression Model Using Standard Soil Data

Levine and Ciolkosz (1986) developed a Computer Simulation Model for Soil Genesis Applications which uses Kline's (1973) compartment design to represent the zone above the soil surface, and an upper and lower horizon. Each compartment has an input and output flow and a parameter that describes its content. Input consists of annual precipitation volume and chemistry (the major ions) and standard soil characterization information. The simulation of the major soil processes is based on empirical relations, such as pH values calculated from base saturation using linear regression equations derived from The Pennsylvania State University soil database:

$$A\ horizons: pH = (0.04 \times BS) + 3.48 \qquad (r^2 = 0.92) \qquad\qquad [3.36]$$
$$B\ horizons: pH = (0.04 \times BS) + 3.34 \qquad (r^2 = 0.82) \qquad\qquad [3.37]$$

The model was tested with soil analysis data from 21-week and 5-year samplings. T-test analysis indicated an effective simulation for 1- and 5-year periods. However, 1- and 5-year test periods are very short compared to the 1-year time increments of the model. A major advantage of the model is its ability to simulate based on standard soil data. Levine and Ciolkosz (1988) used the same model to determine the sensitivity of Pennsylvania soils to acid deposition.

3.3.7.6   Soil Moisture Regimes

The Newhall simulation model is a simple empirical model useful in determining the soil moisture regimes used in Soil Taxonomy (van Wambeke et al., 1986). It approximates daily moisture conditions in a soil based on limited monthly climatic data. Eight layers are divided by eight compartments (8 x 8 matrix) in which water is held at a tension between field capacity and permanent wilting point. The water-holding capacity is fixed, accretion takes place after precipitation, and depletion depends on the potential evapotranspiration. The model was tested in the U.S.A. and proved to be useful in parts of the world where data are sparse.

3.3.7.7   Pedotransfer Functions

Bouma (1989a) discussed "the challenge for soil science to translate data we have (soil survey) to data we need." Pedotransfer functions (PTFs) define relationships between different soil characteristics and properties such as the moisture retention characteristic and the pressure head-hydraulic conductivity (Bouma and van Lanen, 1987; Wösten et al., 1990). These hydraulic properties are essential for the description of water and solute movement in unsaturated soils. There are two main approaches that can be distinguished: point estimation and parameterization (Vereecken, 1988). In the point estimation method, specific points of the moisture retention characteristic [$h(\theta)$], or the hydraulic conductivity-pressure head relation [$K(\theta)$] are estimated from basic soil properties using linear statistical models. A continuous curve through the individual points is derived either through interpolation, smoothing, or data-fitting techniques. The parameterization method estimates the parameters of models, describing $h(\theta)$, $K(\theta)$, or $K(h)$ data. The model parameters are then related to basic soil properties using linear statistical models referred to as PTFs by Bouma (1989b).

The goodness of fit of PTFs is usually evaluated by the coefficient of determination or by visual inspection of the estimated and measured curves. Other evaluation methods have been discussed by Wösten et al. (1986), Wösten and van Genuchten (1988), Mishra and Parker (1989), and Vereecken et al. (1992).

3.3.7.8 Transfer Function Models for Pedogenic Transport Processes

Jury and Sposito (1985) mentioned the limitations of one-dimensional mechanistic models based on the convection-dispersion equation when applied to solute transport in a three-dimensional experiment: "... if the area scale is large and pronounced local variations exist in the pore-water velocity, the convection-dispersion equation may not describe well the situation wherein area-averaged soil solute concentrations develop from a distributed surface input-source." The stochastic (empirical) transfer-function model (TFM), developed by Jury (1982), takes the spatial variability of water flux into account and can be applied to the upper part of the unsaturated zone. Solute movement is described by a travel-time probability density function $f_L(t)$, so that $f_L(t)dt$ yields the probability of a solute reaching a certain depth ($z=L$) in time interval (t to $t+dt$) after it was initially at the surface ($z=0$) at time zero. The model parameters can be determined by applying a narrow pulse of a nonreactive solute at the soil surface ($z=0$) with concentration C_{in}. The flux concentration at $z = L$ is

$$C(L,t) = \int_0^\infty C_{in}(t-t')f_L(t')dt' \qquad [3.38]$$

With the travel-time density function represented in parametric form, the result of integration can be fitted to the experimental concentration data. Area-averaged field-scale solute concentration data were used to determine the shape of the two-parameter log normal density function:

$$f_L(t) = \left[(2\pi)^{1/2}\sigma t\right]^{-1}\exp\left[(\ln t - \mu)^2/2\sigma^2\right] \qquad [3.39]$$

where μ and σ are mean and variance parameters.

Jury et al. (1986) generalized the transfer function model to describe solute movement that may undergo physical, chemical, and biological transformations in a natural soil. A soil unit with a transport volume (V_{st}) and its boundary surface (S_{st}) was defined in which the flow of water may vary in space and time. The general transfer function equation is written as:

$$Q_{out}(t) = {}_0\int^\infty g(t-t'/t')Q_{in}(t')dt' \qquad [3.40]$$

where $Q_{out}(t)$ is the rate of solute loss from the soil unit, $Q_{in}(t')$ is the marginal probability density function for the solute input time and can be interpreted as the rate at which solute mass enters V_{st} for the first time divided by the total mass of solute input, $g(t-t'/t')$ is the lifetime density function which represents the net effect of soil processes (convection, dispersion, sorption, transformations, etc.).

In order to solve the lifetime density function for a particular solute transport, the transfer function equation needs to be solved with the use of experimental field data that relate solute mass input and loss rates as functions of time. White et al. (1986) applied the transfer function model to solute transport through undisturbed soils with Br and Cl as tracers. Travel times varied with soil type, initial water content, and rate of water input. The log normal distribution of travel times, as initially assumed by Jury (1982), described reasonably well the probability density function of some soils, but was found inadequate when applied to well-structured soils. The existence of a component of fast solute transport through large, interped voids and a component of slow transport through intraped voids, resulted in probability density functions with different shapes. This phenomenon should be incorporated in the transfer function model of a well-structured soil.

Both mechanistic and empirical (stochastic) approaches to modeling solute movement are of interest for pedogenetic modeling. The mechanistic approach, based on the Richard's equation, for example, might be more accessible when standard soil data are available. The empirical approach has the theoretical advantage of incorporating spatial variability, but the intensity of specific experimental field data which are needed to solve the transfer function equations for each soil might be a disadvantage. However, the lifetime density function which represents the net effect of soil processes (convection, dispersion, sorption, transformations, etc.) may be useful to model a suite of soil-forming processes that are too complex to model individually.

3.3.8 Quantitative-Mechanistic Models at Pedon and Subpedon Scales

Kline (1973) discussed a linear, constant-coefficient, compartment structure for quantitative mechanistic models which can be used for simulation of a soil system represented by interconnected compartments. The behavior of a compartment is described by: (1) the flow rate (F) through the compartment, and 2) its content (C). From these two measurable parameters, the rate constant or transfer coefficient (λ) is defined as: $\lambda = F/C$. Many other, more complicated methods describing compartmental transfers are available. De Wit and van Keulen (1972) presented an early treatise on the transport of heat, salts, ions and water in the unsaturated phase. Campbell (1985) described the transport of water, gases, heat, and nutrients in soil and plant systems with the use of example programs written in Basic. Richter (1987) also described how to model the major soil processes. Anlauf et al. (1990) presented many applications and programs based on the book by Richter (1987).

Bryant and Olson (1987) called for orientation of soil genesis modeling efforts toward finding solutions to anthropogenically created environmental problems. They proposed using a systems approach to combine subsystems and compartments of various levels of resolution into a pedogenetic model. The compartments dealing with soil processes would have a greater concentration of interior detail than, for example, a compartment dealing with the plant canopy. But, from a mathematical point of view, one has to keep in mind that "the chain is as strong as the weakest link". The accuracy of input to a soil compartment (e.g., the O horizon) heavily depends on the accuracy of output from a neighboring compartment (e.g., the canopy). Pedological modelers should not try to reinvent the wheel, but rather incorporate established models in soil physics, chemistry and biology (Sections A, B, and C) as subsystems into larger systems.

3.3.8.1 Simplified Water and Solute Movement Models

CALSOIL (Mayer, 1985) simulates the development of calcic horizons. The simulation of water movement is similar to the Newhall model. Precipitation enters the first compartment and fills it to field capacity, excess water "flows over" to the next lower compartment. Carbonate can cross compartment boundaries only in solution and is modeled as a function of calcite equilibria, which depend on temperature and soil P_{CO_2}.

Another model of interest for pedogenesis is the Trace Element Transport Model (TETrans)(Corwin and Waggoner, 1990). TETrans is a one-dimensional capacity model which defines changes in amounts of solute driven by the amounts of rainfall, irrigation, or evapotranspiration with time as an indirect variable. The model utilizes a mass-balance approach to determine solution and adsorbed (or exchangeable) solute concentration distributions. Model options include: plant water uptake, hydraulic bypass, exchange, and adsorption. Hydraulic bypass occurs when water moves through pores where stagnant areas of immobile water exist (immobile water films, dead-end pores) and when water moves through large pores thereby bypassing water in smaller pores. Several simplifying assumptions are made in TETrans; however, this approach has the advantage that the model is applicable to situations where available data are limited.

3.3.8.2 The CALDEP and CALGYP Models

Marion et al. (1985) developed a regional soil genesis model for $CaCO_3$ deposition in desert soils of the southwestern United States. The CALDEP model consists of five major components: (1) a stochastic precipitation model based on monthly data resulting in three precipitation seasons; (2) an evapotranspiration model (Thornthwaite's equation $\rightarrow$ pan evaporation $\rightarrow$ actual evapotranspiration); (3) chemical thermodynamic relationships of the carbonate system; (4) soil parameterization (Pco_2 per horizon, water-holding capacity); and (5) soil water and $CaCO_3$ fluxes (only saturated flow was considered, influx of Ca is through weathering and from dust and precipitation). The model was run using the present climate and three Pleistocene climate scenarios and was highly sensitive to the frequency of storm events, water holding capacity, and biotic control of Pco_2. Predicted $CaCO_3$ deposition rates agreed with the rates for most field studies (1 to 5 g m^{-2} yr^{-1}).

Marion and Schlesinger (1994) provided an overview of published $CaCO_3$ models and their major characteristics. The CALDEP and CALGYP models share the same basic structure except that CALGYP includes additional routines to describe sulfate chemistry. The CALGYP model and its application are discussed in detail by Marion and Schlesinger (1994).

3.3.8.3 Soil Organic Matter Model: CENTURY

The CENTURY model which simulates the formation of soil organic matter (SOM) considers three SOM fractions (Parton et al., 1987): (1) active SOM with a short turnover time (1–5 yr); (2) slow SOM with an intermediate turnover time (20–40 yr); and (3) passive SOM with the longest turnover time (200–1500 yr). The system is defined by a set of four driving variables: (1) annual precipitation affects the decomposition and production of SOM; (2) temperature controls decomposition; (3) soil texture influences turnover rates; and (4) plant lignin content controls decomposition rates of above and below ground material as a function of climate. The parameters for these relations were estimated from published data using a nonlinear data fitting procedure. Many of these relations are empirical, others are of a more mechanistic type, such as those based on laboratory incubation experiments and data. The model was validated by comparing soil C and N to mapped values at 24 sites on the Great Plains where regional trends in SOM were predicted with an overall error of 15%. The model provides insight into the factors controlling SOM levels which is valuable to taxonomic and pedogenetic studies. Later versions of CENTURY include a revised submodel for surface litter decomposition and long-term soil incubation data were used to improve the effect of soil texture on soil organic matter dynamics (Parton et al., 1994).

3.3.8.4 Isotope Models

The C isotope composition ($^{13}C/^{12}C$) of pedogenic $CaCO_3$ was long believed to be related to the C isotope composition of the flora since CO_2 released by plant respiration or decomposition reacts with dissolved Ca to form $CaCO_3$ in arid or semi-arid environments. However, measurement of C isotope ratios with depth was not found to be constant, but instead commonly showed decreasing ratios of $^{13}C/^{12}C$ with depth. Over the years, researchers (Thorstenson et al., 1983; Cerling, 1984) recognized that the pattern could be explained by considering the processes which produce and transport CO_2 to the overlying atmosphere, namely, molecular diffusion of each isotopic species of CO_2 along its own concentration gradient. Furthermore, for both simplicity and ease of computation, as well as a means to best represent real soils, it was assumed that the CO_2 production and transport was at steady state for the soil as a whole or for any soil layer:

$$D_s \frac{\delta^2 C}{\delta z^2} = \phi \qquad\qquad\qquad [3.41]$$

where D_s is the diffusion coefficient of CO_2 in soil, C is the concentration of CO_2, and ϕ is the biological production of CO_2. By choosing atmospheric CO_2 as an upper boundary, and setting a no-flux lower boundary, a steady-state equation for both $^{13}CO_2$ and $^{12}CO_2$ could be derived, and the isotopic composition of CO_2 at any soil depth could be predicted by the following expression:

$$R_s^{13} = \frac{\dfrac{\phi R_p^{13}}{D_s^{13}}\left(Lz - \dfrac{z^2}{2}\right) + C_{atm}R_{atm}^{13}}{\dfrac{\phi R_p^{13}}{D_s^{13}}\left(Lz - \dfrac{z^2}{2}\right) + C_{atm}} \qquad\qquad\qquad [3.42]$$

where R^{13} is the C isotope ratio of soils (s), plants (p), or the atmosphere (atm), L is the total soil thickness and z is the soil depth.

This model has multiple practical uses: (1) it explains the observed variation in the $^{13}C/^{12}C$ of $CaCO_3$ versus depth (which ranges from near atmospheric values near the soil surface to values close to those of plants with increasing depth); (2) it has served as the basis for subsequent models of the ^{14}C content of $CaCO_3$ and usefulness of $CaCO_3$ as a substrate for ^{14}C dating (Wang et al., 1994; Amundson et al., 1994); and (3) when rearranged and solved for atmospheric CO_2, it has proven to be a very useful means of estimating past atmospheric CO_2 concentrations from the C isotope composition of paleosol carbonates (Cerling, 1991).

3.3.8.5 Pedodynamic Approach
In the introduction of this chapter we presented Jenny's (1961) open systems analysis approach. Jenny was interested in a time span of centuries to millennia and was limited by data availability, which resulted in the derivation of the empirical state factor approach. The pedodynamic approach can be seen as the mechanistic solution to the open systems analysis approach. Pedodynamics is defined as the quantitative integrated simulation of physical, chemical and biological soil processes acting over short time increments in response to environmental factors (Hoosbeek and Bryant, 1994a).

Hoosbeek and Bryant (1992, 1994a, 1994b, 1995) applied the open systems analyses approach to several pedons in a boreal forest and were interested in a temporal extent of months to several years to capture seasonal fluctuations in temporal properties. The spatial and temporal extent in combination with an intensive monitoring program allowed for a mechanistic solution. Process knowledge and data needed to be collected at spatial scales of one or two levels below the extent (pedon) at the horizon and soil structure levels. The pedodynamic approach will be illustrated in the remainder of this section by the study of Hoosbeek and Bryant (1994a, 1994b, 1995).

To obtain data for the development and validation of various submodels of the pedodynamic model, three detailed monitoring sites were installed in a boreal forest in the Adirondack Mountains near Tupper Lake, New York. The soil is a well-expressed Typic Haplorthod developed in sandy glacial outwash deposits and has never been disturbed by logging or agriculture. The site is nearly level, and there is no measurable runoff or runon. Tensiometers, soil solution samplers, and thermistors were installed in the Oi, Oa, E, Bh/Bhs, Bs, BC, and C horizons. Redox potentials were measured in the Oi, Oa, and E horizons. Soil solutions, precipitation, and canopy throughfall samples were collected every four weeks over 2 years and were chemically analyzed for organic and inorganic components.

Each submodel within the pedodynamic model will be described. Taken together, the submodels represent the complex processes that govern the chemistry and movement of aluminum, silica, organic components, and other chemical species in Spodosols. The ORTHOD model was applied to the simulation of C dynamics and to test several hypotheses on Spodosol formation. However, this section focuses only on the development of the model.

The movement of water in the soil is fundamental to any quantitative mechanistic pedogenetic model since water plays a major role in transfer and transformation processes. Realistic representation of saturated and unsaturated flow in a typically non-isotropic medium like soil is a complex problem. Several mechanistic models have been developed that simulate the flow of water in the unsaturated zone based on Darcy's law and the continuity principle:

$$q = -K \frac{\delta H}{\delta z} \quad \left(cm \cdot d^{-1} \right)$$

[3.43]

$$\frac{\delta \theta}{\delta t} = -\frac{\delta q}{\delta z} \quad \left(d^{-1} \right)$$

[3.44]

where q is the volumetric fluid flux density, K is the hydraulic conductivity, H is the pressure head of soil water, z is the vertical space coordinate, θ is the volume fraction of liquid, and t is time. Combination yields a partial differential equation in terms of hydraulic head, called the Richard's equation:

$$\frac{\delta \theta}{\delta t} = \frac{\delta}{\delta z} \left(K \frac{\delta H}{\delta z} \right)$$

[3.45]

Using the pressure head form of the flow equation, where ψ is the potential of soil water, and definition of the differential moisture capacity C, $C = d\Theta/d\psi$, yields the flow equation for predicting water movement in layered soils:

$$\frac{\delta \psi}{\delta t} = \frac{1}{C(\psi)} \frac{\delta}{\delta z} \left[K(\psi) \left(\frac{\delta \psi}{\delta t} - 1 \right) \right]$$

[3.46]

The sandy soil at the study site permitted the use of a model based on the Richard's equation. The LEACHM model (Hutson and Wagenet, 1992) uses a numerical solution to the Richard's equation and was selected for use in the ORTHOD model. Measured and simulated tensiometer values gave a good fit. Given precipitation input, this submodel gave a relatively good representation of soil moisture conditions in all horizons at any point in time and over the full range of soil moisture conditions.

The LEACHM model (Hutson and Wagenet, 1992) uses a numerical solution to the convection-dispersion equation to estimate chemical fluxes. Solute movement consists of three components: convective, diffusive, and dispersive transport. Combining the equations of these three processes leads to the following expression for solute flux (J_s) (van Genuchten and Wierenga, 1986):

$$J_s = -\theta D(\delta C / \delta x) + qC \tag{3.47}$$

where θ is the volumetric water content, C is the volume-averaged solute concentration, x is distance, q is the volumetric fluid flux density, and D is the summation of the molecular diffusion and mechanical dispersion coefficients. Both the diffusion and dispersion coefficients can be assumed to be negligible for a sandy soil with a relative large downward convective flow throughout the year. The solute movement between two layers is then dominantly convective flow, which is the product of the volumetric fluid flux density and the volume-averaged solute concentration (Wagenet, 1986):

$$J_s = qC \tag{3.48}$$

This simplification significantly reduced computation time and did not cause significant error when using relatively thin compartmental layers (25 mm).

Heat transfer and soil profile temperatures were simulated as described by Tillotson et al. (1980). The heat flow equation is

$$\rho C_p \frac{\delta T}{\delta t} = \frac{\delta}{\delta z}\left(K_t(\theta) \frac{\delta t}{\delta z} \right) \tag{3.49}$$

where ρ is bulk density (kg.m^{-3}), C_p is gravimetric heat capacity of the soil (J m^{-3} C^{-1}), T is temperature (C), t is time (s), z is depth (m), and $K_t(\theta)$ is the thermal conductivity of the soil (J.m^{-1} s^{-1} $^{\circ}$C^{-1}) at water content θ (m^3 m^{-3}). The volumetric heat capacity is calculated from:

$$\rho C_p = \rho_s C_s + \theta \rho_w C_w \tag{3.50}$$

where ρ_s and C_s are the bulk density and the gravimetric heat capacity of the solid phase and ρ_w and C_w are the density and the gravimetric heat capacity of the liquid phase.

The rate of dissolution of clay minerals and metal hydrous oxides is surface controlled and is observed to follow zero-order kinetics and can be expressed by the equation (Stumm et al., 1985; Sposito, 1989):

$$\frac{d[A]}{dt} = k \tag{3.51}$$

where $[A]$ is the aqueous-phase concentration of an ion, and t is time. The parameter k is a function of temperature, pressure, mineral surface area, proton concentration $[H^+]$, and, in the presence of a strong chelator, the ligand concentration $[L^-]$. In a controlled laboratory experiment at constant temperature, pressure, and for a given soil material of some specific mineralogy and mineral surface area, k' is then a function of $[H^+]$ and $[L^-]$:

$$k = k'\left[H^+\right]^m\left[L^-\right]^n \tag{3.52}$$

where m and n are fractional order constants. Combination of the two equations yields:

$$\frac{d[A]}{dt} = k'[H^+]^m[L^-]^n \tag{3.53}$$

Microbial decomposition is considered to take place in several pools (Bohn et al., 1985; Parton et al., 1987), such as structural and metabolic plant remains, active soil organic carbon (SOC) (decomposing plant residues, live microbes), slow SOC (microbial metabolites), passive SOC (humified material), and dissolved organic carbon (DOC). The general equation for the rate of decomposition per pool within a soil environment is

$$\frac{dC_x}{dt} = K_x \cdot \theta \cdot T \cdot C_x \tag{3.54}$$

where C_x is the carbon state variable of pool x, K_x is the decomposition rate for pool x, θ is the volumetric water content, and T is the soil temperature. The decomposition products are CO_2, H_2O, SOC flowing to other pools, and DOC. The ratio in which these products are produced depends on the volumetric water content.

Ion exchange was recognized as an important process controlling the chemistry of the soil solution. Many different exchange equations have been proposed to describe the exchange between cations of unequal charge. The most general form of an exchange equation is based on the mass action equation (Bohn, 1985), for instance:

$$3CaX + 2Al^{3+} = 2AlX + 3Ca^{2+} \tag{3.55}$$

with the selectivity coefficient:

$$K = \frac{(AlX)^2(Ca^{2+})^3}{(CaX)^3(Al^{3+})^2} \tag{3.56}$$

where X denotes the ion in the adsorbed phase. It is assumed that the ions in the adsorbed phase behave as if being in solution. Vanselow (1932) substituted the mole fractions of the exchangeable ions in the above exchange equation. The mole fraction, e.g., $N(Al)$, is defined as:

$$N(Al) = n_{Al} / (n_{Al} + n_{Ca}) \tag{3.57}$$

where n is mole of exchangeable ions per g soil. The Gapon equation is widely used for Na^+–Ca^{2+} exchange (Bohn et al., 1985). The mass action equation for Ca^{2+}–Al^{3+} exchange, which is an important exchange reaction in a Spodosol, is written as:

$$CaX + 2/3Al^{3+} = (Al)_{2/3}X + Ca^{2+} \tag{3.58}$$

with the exchange coefficient:

$$K_{Gapon} = \frac{[Al_{2/3}X][Ca^{2+}]}{[CaX][Al^{3+}]^{2/3}}$$

[3.59]

Exchange equations based on mass action equations assume that the activities of the adsorbed ions are proportional to their equivalent or mole fraction in the adsorbed phase. However, data from a Ca^{2+}–Al^{3+} exchange experiment on montmorillonite at low pH showed a large preference of the clay for the trivalent ion (McBride and Bloom, 1977). An exchange model based on a statistical thermodynamic approach was derived to better describe the activity of Al^{3+} at the mineral surface. The equation relates the Al^{3+} in solution and the equivalent fraction of adsorbed Al^{3+} as:

$$(Al^{3+}) = K_{McBride}N(Al)/(1-N(Al))$$

[3.60]

This equation emphasizes the lack of dependence of Al^{3+} adsorption on the $[Ca^{2+}]$ in solution, a result of the more solution-like nature of adsorbed Ca^{2+} in comparison to the strongly adsorbed Al^{3+}.

Due to the concentration charge effect (a preference of adsorption of tri- and divalent ions as compared to monovalent ions at low electrolyte concentrations), which applies to the study site (excessive leaching, low electrolyte concentrations), Al^{3+}, and to a lesser degree Ca^{2+} and Mg^{2+}, will dominate the exchange sites (McBride, 1994). Based on the experimental results, the system was simplified to an exchange between Al^{3+} and the summation of the base cations for the B horizons.

The chemical equilibrium program MINEQL+ (Westall et al., 1976; Schecher and McAvoy, 1991) was selected to calculate chemical speciation and the precipitation of solid phases.

The submodels presented in this section were used as building blocks in the pedodynamic ORTHOD model. The model was applied successfully to the simulation of carbon dynamics in three Typic Haplorthod soil profiles (Hoosbeek and Bryant, 1995). As an example, the measured and simulated DOC concentrations of the E and Bhs horizons are presented in Fig. 3.9. CO_2 and DOC fluxes were simulated per day. The net DOC balance for each layer over a 365-day period shows a large DOC flux into the Bh horizon and relatively small fluxes in the BC and C horizons (Fig. 3.10).

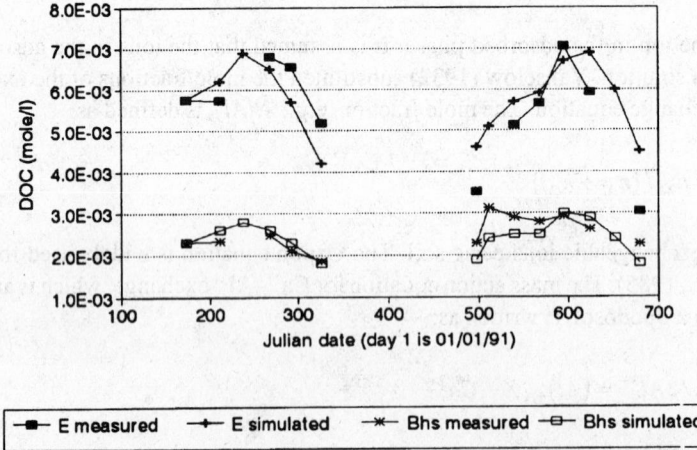

Fig. 3.9 Measured and simulated DOC concentrations of the E and Bhs horizons [Reprinted from Hoosbeek and Bryant, 1994. Soil Sci. Soc. Am. Spec. Pub. 39:111-128 with permission of the Soil Science Society of America]

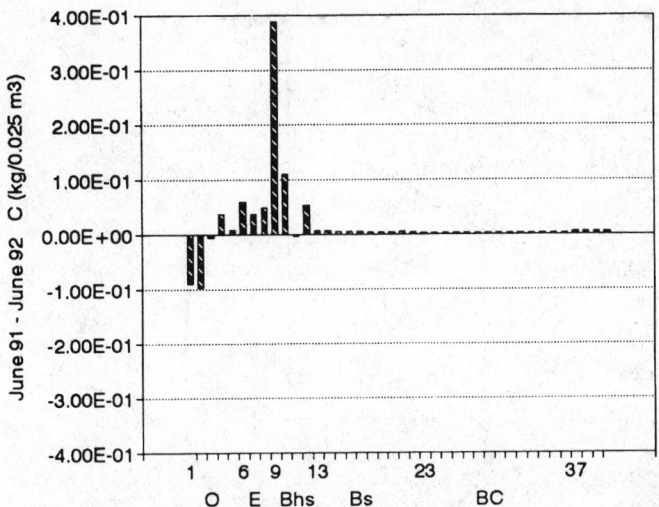

Fig. 3.10 Simulated net DOC fluxes for each layer over a 365-day period [Reprinted from Hoosbeek and Bryant, 1994. Soil Sci. Soc. Am. Spec. Pub. 39:111-128 with permission of the Soil Science Society of America]

Several podzolization theories applicable to the investigated Typic Haplorthod were tested with the ORTHOD model (Hoosbeek and Bryant, 1994a, 1994b). The model simulated the chemistry and movement of Al, Si, DOC, and other chemical species involved in the podzolization process. The net Al balance over a 365-day period per layer showed significant losses of Al from the E horizon, losses from the Bh and Bhs, and accumulations of Al in the Bs and BC horizons (Fig. 3.11).

Some pedological or environmentally related questions cannot be answered by applying a state factor model. For instance, how much DOC, produced in the O horizon, will be sequestered in the B horizons, and how much DOC will leave the profile per year? Or, which mechanisms regulate Al

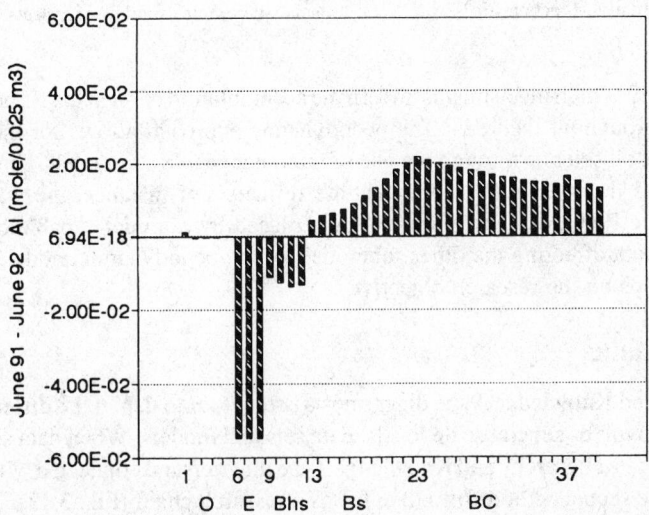

Fig. 3.11 Simulated net Al balances for each layer over a 365-day period [Reprinted from Hoosbeek and Bryant, 1994. Soil Sci. Soc. Am. Spec. Pub. 39:111-128 with permission of the Soil Science Society of America]

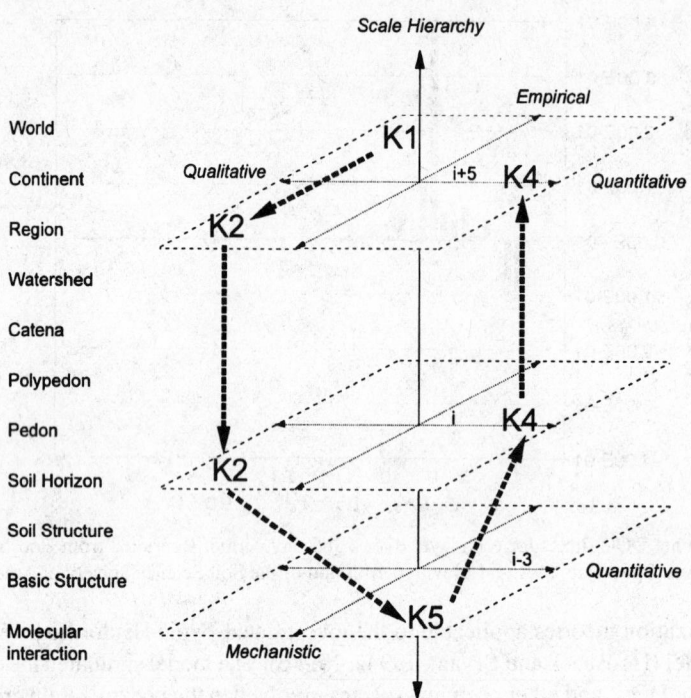

Fig. 3.12 Scale hierarchy and knowledge type diagram [Hoosbeek and Bryant, 1992; Bouma and Hoosbeek, 1996]. Research chain indicating the sequence of knowledge types used at different scale hierarchies for a study to estimate critical loads on forest soils in Europe.

Knowledge types:

K1 = Qualitative-Empirical	User
K2 = Qualitative-Mechanistic	Expert
K3 = Quantitative-Empirical	Specialist: simple comprehensive models
K4 = Quantitative-Mechanistic/Empirical	Specialist: complex comprehensive models
K5 = Quantitative-Mechanistic	Specialist: complex models of aspects

activity in the rootzone? Which mechanisms govern the accumulation of Al in the Bs horizon and how do these rates vary throughout the year? The pedodynamic approach allows for the utilization of current knowledge in soil science through the use of interchangeable submodels. The model can be gradually improved as the individual submodels are refined. For instance, the water movement submodel based on the Richards equation might be replaced by a model capable of dealing with preferential flow without affecting the other submodels of the pedodynamic model. The choice of submodels also depends on the research objectives.

3.4 Research Chains

The Scale Hierarchy and Knowledge Type diagram is a useful tool to depict the different stages of a research project that involves several scale levels, data sets and models. When data sets and models are positioned in the Scale Hierarchy and Knowledge Type diagram and connected with arrows in the order of execution, the sequence of arrows then forms a research chain (Fig. 3.12). These research chains are a valuable aid in determining which scales and knowledge types are involved in one project. Vertical arrows pointing downward indicate downscaling while arrows pointing upward indicate upscaling or aggregation of data. Arrows within a knowledge type plane indicate a change of knowledge type (e.g., grouping of numerical values into descriptive classes). In practice, many arrows indicate a combination of scale and knowledge type transitions.

Figure 3.12 depicts the research chain of a study estimating critical loads of N and S on forest soils in Europe by De Vries et al. (1992) (Bouma and Hoosbeek, 1996). Soil data were derived from the FAO soil map (1 : 1,000,000) at continent scale level. Each map unit was defined in terms of soil type, texture class, and slope class at the pedon level. The qualitative-mechanistic understanding of soils represented in the soil survey (based on the pedogenetic soil-landscape model) made this jump from continent to pedon level possible. Laboratory studies at the basic structure level yielded quantitative-mechanistic weathering rate coefficients of base ions. These data were, in turn, used as input to the simulation model PROFILE (Sverdrup and Warfvinge, 1988; Warfvinge and Sverdrup, 1992) to estimate quantitative-semimechanistic field weathering rates based on soil mineralogy at the pedon level. A class-pedotransfer function was used to upscale these field weathering rates from pedon level to the soil types of the soil map at continent level.

3.5   Model Selection and Application

In this chapter we have tried to provide a wide overview of the many different approaches available to pedological modeling. However, we do not claim to have provided a comprehensive listing of published pedological models. Our aim was to provide some guidance and examples to the pedological modeler.

The scale issue is important, but frequently intuitively dealt with. The research question or application frequently dictates the extent scale. Then the data availability and the nature of the research question, be it applied or fundamental, call for an empirical or mechanistic approach. In case of a mechanistic approach, data and process knowledge will need to be collected at a support scale of one or more levels below the extent.

For studies that involve different data sets and models, it may be useful to position these items in the Scale Hierarchy and Knowledge Type diagram. When the sequence of execution is indicated in the diagram by arrows, an overview is obtained of the different scales and knowledge types that are involved in the study. Awareness of the different scales and knowledge types and subsequently dealing with them in an appropriate way by up- or downscaling and transformations may improve the quality of the overall study.

3.6   References

Addiscott, T.M. 1994. Simulation, prediction, foretelling or prophecy? Some thoughts on pedogenetic modeling. p. 1–15. *In* R.B. Bryant and R.W. Arnold (ed.) Quantitative modeling of soil forming processes. Soil Sci. Soc. Am. Spec. Pub. 39.

Amundson, R., Y. Wang, O. Chadwick, S. Trumbore, L. McFadden, E. McDonald, S. Wells, and M. DeNiro. 1994. Factors and processes governing the ^{14}C content of carbonate in desert soils. Earth Planet. Sci. Lett. 125:385–405.

Anlauf, R., K. Ch. Kersebaum, Liu Ya Ping, A. Nuske-Schüler, J. Richter, G. Springob, K.M. Syring, and J. Utermann. 1990. Models for processes in the soil: Programs and exercises. Catena Verlag, Cremlingen, Germany.

Bell, J.C. 1990. A GIS-based soil-landscape modeling approach to predict soil drainage Class. Ph.D. Thesis. PA State University, University Park, PA.

Birkeland, P.W. 1984. Holocene soil chronofunctions, Southern Alps, New Zealand. Geoderma 34:115–134.

Bockheim, J.G. 1980. Solution and use of chronofunctions in studying soil development. Geoderma 24:71–85.

Bohn, H.L., B.L. McNeal, and G.A. O'Connor. 1985. Soil Chemistry. Wiley-Interscience, New York, NY.

Bouma, J. 1989a. Using soil survey data for quantitative land evaluation. p.177–213. *In* B.A. Stewart (ed.) Advances in Soil Science, Volume 9. Springer-Verlag, New York, NY.

Bouma, J. 1989b. Land qualities in space and time. p.3–13. In J. Bouma and A.K. Bregt (ed.) Land qualities in space and time. PUDOC, Wageningen, Netherlands.

Bouma, J., and H.A.J. van Lanen. 1987. Transfer functions and threshold values: From soil characteristics to land qualities. p. 106–111. In K.J. Beek, P.A. Burrough, and D.E. McCormack (ed.) Proc. ISSS/SSSA Worksh. Quant. Land Eval. Proced. Int. Inst. Aerospace Surv. and Earth Sci. Publ. 6. Enschede, Netherlands.

Bouma, J., and M.R. Hoosbeek. 1996. The contribution and importance of soil scientists in interdisciplinary studies dealing with land. p. 1–15. In R.J. Wagenet and J. Bouma (ed.) The role of soil science in interdisciplinary research. Soil Sci. Soc. Am. Spec. Pub. 45.

Brewer, R. 1964. Fabric and mineral analysis of soils. Wiley, New York, NY.

Brimhall, G.H., O.A. Chadwick, C.J. Lewis, W. Compston, I.S. Williams, K.J. Danti, W.E. Dietrich, M.E. Power, D. Hendricks, and J. Bratt. 1991a. Deformational mass transport and invasive processes in soil evolution. Science 255:695–702.

Brimhall, G.H., and W.E. Dietrich. 1987. Constitutive mass balance relations between chemical composition, volume, density, porosity, and strain in metasomatic hydrochemical systems: results on weathering and pedogenesis. Geochim. Cosmochim. Acta 51:567–587.

Brimhall, G.H., C.J. Lewis, C. Ford, J. Bratt, G. Taylor, and O. Warin. 1991b. Quantitative geochemical approach to pedogenesis: importance of parent material reduction, volumetric expansion, and eolian influx in laterization. Geoderma 51:51–91.

Bryant, R.B., and C.G. Olson. 1987. Soil genesis: Opportunities and new directions for research. In L.L. Boersma (ed.) Future developments in soil science research. Soil Science Society of America, Madison, WI.

Bullock, P., N. Fedoroff, A. Jongerius, G. Stoops, and T. Tursina. 1985. Handbook for soil thin section description: Prepared under the auspices of the International Society of Soil Science. Waine Research Publications, Wolverhampton, U.K.

Burgess, T.M., and R. Webster. 1980a. Optimal interpolation and isarithmic mapping of soil properties: 1. The semivariogram and punctual kriging. J. Soil Sci. 31:315–331.

Burgess, T.M., and R. Webster. 1980b. Optimal interpolation and isarithmic mapping of soil properties: 2. Block kriging. J. Soil Sci. 31:333–341.

Burrough, P.A. 1986. Principles of geographical information systems for land resources assessment. Clarendon Press, Oxford, UK.

Burrough, P.A. 1991. Sampling designs for quantifying map unit composition. In M.J. Mausbach and L.P. Wilding (ed.) Spatial variabilities of soils and landforms. Soil Sci. Soc. Am. Spec. Pub. 28.

Campbell, G.S. 1985. Soil physics with BASIC - Transport models for soil-plant systems. Elsevier, Amsterdam, Netherlands.

Cerling, T.E., 1984. The stable isotopic composition of modern soil carbonate and its relationship to climate. Earth Planet. Sci. Lett. 71:229–240.

Cerling, T.E., 1991. Carbon dioxide in the atmosphere: Evidence from Cenozoic and Mesozoic paleosols. Amer. J. Sci. 291:377–400.

Chadwick, O.A., G.H. Brimhall, and D.M. Hendricks. 1990. From a black to a gray box - a mass balance interpretation of pedogenesis. Geomorph. 3:369–390.

Chen, C.W., J.D. Dean, S.A. Gherini, and R.A. Goldstein. 1982. Acid rain model: Hydrologic module. J. Environ. Eng. Div., Proc. Am. Soc. Civil Eng. 108:EE3.

Chittleborough, D.J., P.H. Walker, and J.M. Oades. 1984. Textural differentiation in chronosequences from eastern Australia: I. Descriptions, chemical properties and micromorphologies of soils. Geoderma 32:181–202.

Corwin, D.L., and B. Waggoner. 1990. Trace element transport model: Solute transport modeling software, IBM-compatible version 1.5 USDA-ARS, US Salinity Laboratory, Riverside, CA.

Davis, J.C. 1986. Statistics and data analysis in geology. 2nd Ed. John Wiley, New York, NY.

Deutsch, C.V., and A.G. Journel. 1992. GSLIB: geostatistical software library. Oxford University Press, New York, NY.

De Vries, W., M. Posch, G.J. Reinds, and J. Kamari. 1992. Critical loads and their exceedance on forest soils in Europe. Report 58 Winand Staring Center, Wageningen, Netherlands, and Water and Environment Res. Inst., Helsinki, Finland.

De Wit, C.T. 1993. Philosophy and terminology. p.3–9. In P.A. Leffelaar (ed.) On systems analysis and simulation of ecological processes. Kluwer Academic Publishers, Dordrecht, Netherlands.

De Wit, C.T. and H. van Keulen. 1972. Simulation of transport processes in soils. Pudoc, Wageningen, Netherlands.

Dijkerman, J.C. 1974. Pedology as a science: The role of data, models and theories in the study of natural soil systems. Geoderma 11:73–93.

Dokuchaev, V.V. 1948. Russian Chernozem - Selected works of V.V. Dokuchaev, Volume I. Moskva, 1948. Translated from Russian by Israel Program for Scientific Translations, Jerusalem, 1967.

Fanning, D.S., and M.C.B. Fanning. 1989. Soil, genesis, and classification. John Wiley and Sons, New York, NY.

France, J. and J.H.M. Thornley. 1984. Mathematical models in agriculture. Butterworths, London, UK.

Fu, L. 1994. Neural networks in computer intelligence. McGraw-Hill, New York, NY.

Harden, J.W. 1990. Soil development on stable landforms and implications for landscape studies. Geomorph. 3: 391–398.

Harden, J.W., and E.M. Taylor. 1983. A qualitative comparison of soil development in four climatic regimes. Quart. Res. 20:342–359.

Harden, J.W., E.M. Taylor, C. Hill, R.K. Mark, L.D. MacFadden, M.C. Reheis, J.M. Sowers, and S.G. Wells. 1991. Rates of soil development from four soil chronosequences in the Southern Great Basin. Quat. Res. 35:383–399.

Haseman, J.F., and C.E. Marshall. 1945. The use of heavy minerals in studies of the origin and development of soils. MO Agric. Exp. Sta. Res. Bull. 387.

Havens, M.W. 1988. A GIS-based soil-landscape modeling approach to predict surface rock fragment distributions. M.S. Thesis. PA State University, University Park, PA.

Hay, R.L. 1960. Rate of clay formation and mineral alteration in a 4000 year old volcanic ash soil on St. Vincent, BWI. Am. J. Sci. 258:354–368.

Hoosbeek, M.R. 1998. Incorporating scale into spatio-temporal variability: applications to soil quality and yield data. Geoderma 85:113–131.

Hoosbeek, M.R., and Bouma, J., 1998. Obtaining soil and land quality indicators using research chains and geostatistical methods. Nutr. Cycl. Agroecosys. 50:35–50.

Hoosbeek, M.R., and R.B. Bryant. 1992. Towards the quantitative modeling of pedogenesis - a review. Geoderma 55:183–210.

Hoosbeek, M.R., and R.B. Bryant. 1994a. Developing and adapting soil process submodels for use in a pedodynamic model. p.111–128. *In* R.B. Bryant and R.W. Arnold (ed.) Quantitative modeling of soil forming processes. Soil Sci. Soc. Am. Spec. Pub. 39.

Hoosbeek, M.R., and R.B. Bryant. 1994b. Pedodynamics: An approach to study and quantify soil forming processes. Proc. 15th World Congr. Soil Sci., Acapulco, Mexico, July 1994. International Society of Soil Science.

Hoosbeek, M.R., and R.B. Bryant. 1995. Modeling the dynamics of organic carbon in a Typic Haplorthod. p.415–431. *In* R. Lal, J.M. Kimble, E.R. Levine, and B.A. Stuart (ed.) Soils and global change. CRC Lewis Publishers, Chelsea, MI.

Hoosbeek, M.R., A. Stein, and R.B. Bryant. 1997. Mapping soil degradation. p.407–422. *In* R. Lal, W.H. Blum and C. Valentin (ed.) Methodology for assessment of soil degradation. CRC Press LLC, Boca Raton, FL.

Hudson, B.D. 1992. The soil survey as paradigm-based science. Soil Sci. Soc. Am. J. 56:836–841.

Huggett, R.J. 1975. Soil landscape systems: A model of soil genesis. Geoderma 13:1–22.

Hutson, J.L., and R.J. Wagenet. 1992. LEACHM: leaching estimation and chemistry model. Res. Series No. 92–3. Dep. Soil, Crop and Atmos. Sci., Cornell University, Ithaca, NY.

Jenny, H. 1941. Factors of soil formation - A system of quantitative pedology. McGraw-Hill, New York, NY.

Jenny, H. 1961. Derivation of state factor equation of soils and ecosystems. Soil Sci. Soc. Am. Proc. 25:385–388.

Jenny, H. 1980. The soil resource, origin and behavior. Springer Verlag, New York, NY.

Johnson, D.L., E.A. Keller, and T.K. Rockwell. 1990. Dynamic pedogenesis: New views on some key soil concepts, and a model for interpreting quaternary soils. Quat. Res. 33:306–319.

Johnson, D.L., and D. Watson-Stegner. 1987. Evolution model of pedogenesis. Soil Sci. 143:349–366.

Jongmans, A.G., T.C.J. Feijtel, R. Miedema, N. van Breemen, and A. Veldkamp. 1991. Soil formation in quaternary terrace sequence of the Allier, Limagne, France: Macro- and micromorphology, particle size distribution and chemistry. Geoderma 49:215–239.

Journel, A.G. 1986. Geostatistics, models and tools for the earth sciences. Math. Geol. 18:119–140.

Journel, A.G., and C.J. Huijbregts. 1978. Mining geostatistics. Academic Press, London, UK.

Jury, W.A. 1982. Simulation of solute transport using a transfer function model. Water Resour. Res. 18:363–368.

Jury, W.A. and G. Sposito. 1985. Field calibration and validation of solute transport models for the unsaturated zone. Soil Sci. Soc. Am. J. 49:1331–1341.

Jury, W.A., G. Sposito, and R.E. White. 1986. A transfer function model of solute transport through soil 1. Fundamental concepts. Water Resour. Res. 22:243–247.

Kirkby, M.J. 1985. A basis for soil profile modeling in a geomorphic context. J. Soil Sci. 36:97–121.

Kline, J.R. 1973. Mathematical simulation of soil-plant relationships and soil genesis. Soil Sci. 115:240–249.

Kuhn, T.S. 1962. The structure of scientific revolutions. The University of Chicago Press, Chicago, IL.

Lam, D. and C. Pupp. 1993. Integration of GIS, expert system and modeling for state of environment reporting. In: Proc. 2nd Int. Conf. Integr. GIS Environ. Model. NCGIA, Breckenridge, CO.

Langley-Turnbaugh, S.J., and J.G. Bockheim. 1998. Mass balance of soil evolution on late Quaternary marine terraces in coastal Oregon. Geoderma 84:265–288.

Lee, K-S., G.B. Lee, and E.J. Tyler. 1988. Thematic mapper and digital elevation modeling of soil characteristics in hilly terrain. Soil Sci. Soc. Am. J. 52:1104–1107.

Leffelaar, P.A. 1993. Basic elements of dynamic simulation. p.11–27. In P.A. Leffelaar (ed.) On systems analysis and simulation of ecological processes. Kluwer Academic Publishers, Dordrecht, Netherlands.

Levine, E.R., and E.J. Ciolkosz. 1983. Soil development in till of various ages in northeastern Pennsylvania. Quat. Res. 19:85–99.

Levine, E.R., and E.J. Ciolkosz. 1986. A computer simulation model for soil genesis applications. Soil Sci. Soc. Am. J. 50:661–667.

Levine, E.R., and E.J. Ciolkosz. 1988. Computer simulation of soil sensitivity to acid rain. Soil Sci. Soc. Am. J. 52:209–215.

Levine, E.R., D.S. Kimes, and V.G. Sigillito. 1996. Classifying soil structure using neural networks. Ecol. Model. 92:101–108.

Marion, G.M., and W.H. Schlesinger. 1994. Quantitative modeling of soil forming processes in deserts: The CALDEP and CALGYP models. p.129–145. In R.B. Bryant and R.W. Arnold (ed.) Quantitative modeling of soil forming processes. Soil Sci. Soc. Am. Spec. Pub. 39.

Marion, G.M., W.H. Schlesinger, and P.J. Fonteyn. 1985. CALDEP: A regional model for soil $CaCO_3$ (Caliche) deposition in southwestern deserts. Soil Sci. 139:468–481.

Mayer, L. 1985. The distribution of calcium carbonate in soils: A computer simulation using program CALSOIL. US Geol. Surv. Bull. 975.

McBride, M.B. 1994. Environmental chemistry of soils. Oxford University Press, New York, NY.

McBride, M.B., and P.R. Bloom. 1977. Adsorption of aluminum by a smectite: II. An Al^{3+}-Ca^{2+} exchange model. Soil Sci. Soc. Am. J. 41:1073–1077.

Mellor, A. 1986. A micromorphological examination of two alpine soil chronosequences, southern Norway. Geoderma 39:41–57.

Merritts, D.J., O.A. Chadwick, and D.M. Hendricks. 1991. Rates and processes of soil evolution on uplifted marine terraces, northern California. Geoderma 51:241–275.

Mishra, S., and J.C. Parker. 1989. Effects of parameter uncertainty on predictions of unsaturated flow. J. Hydrol. 108:19–33.

Muhs, D.R. 1982. A soil chronosequence on Quaternary marine terraces, San Clemente Island, California. Geoderma 28:257–283.

Pachepsky, Y.A., D. Timlin, and G. Varallyay. 1996. Artificial neural networks to estimate soil water retention from easily measurable data. Soil Sci. Soc. Am. J. 60:727–733.

Parton, W.J., D.S. Schimel, C.V. Cole, and D.S. Ojima. 1987. Analysis of factors controlling soil organic matter levels in Great Plain grasslands. Soil Sci. Soc. of Am. J. 51:1173–1179.

Parton, W.J., D.S. Schimel, D.S. Ojima, and C.V. Cole. 1994. A general model for soil organic matter dynamics: Sensitivity to litter chemistry, texture and management. p.147–167. In R.B. Bryant and R.W. Arnold (ed.) Quantitative modeling of soil forming processes. Soil Sci. Soc. Am. Spec. Pub. 39.

Popper, K.R. 1959. The logic of scientific discovery. Hutchinson, London, UK.

Rabenhorst, M.C. 1997. The chrono continuum: An approach to modeling pedogenesis in march soils along transgressive coastlines. Soil Sci. 162:2–9.

Richardson, J.L., and W.J. Edmonds. 1987. Linear regression estimations of Jenny's relative effectiveness of state factors equation. Soil Sci. 144:203–208.

Richter, J. 1987. The soil as a reactor. Catena Verlag, Cremlingen, Germany.

Rouhani, S., and D.E. Myers. 1990. Problems in space-time kriging of geohydrological data. Math. Geol. 22:611–623.

Runge, E.C.A. 1973. Soil development sequences and energy models. Soil Sci. 115:183–193.

Ruxton, B.P. 1968. Rates of weathering of Quaternary volcanic ash in northeast Papua. Trans. 9th Int. Congr. Soil Sci., Adelaide. 4:551–560.

Schaetzl, R.J., L.R. Barret, and J.A. Winkler. 1994. Choosing models for soil chronofunctions and fitting them to data. Europ. J. Soil Sci. 45:219–232.

Schecher, W.D., and D.C. McAvoy. 1991. MINEQL: A chemical equilibrium program for personal computers. User's manual, version 2.1. Environmental Research Software, Edgewater, MD.

Shovic, H.F., and C. Montagne. 1985. Application of a statistical soil-landscape model to an order III wildland soil survey. Soil Sci. Soc. Am. J. 49:961–968.

Simonson, R.W. 1959. Outline of a generalized theory of soil genesis. Soil Sci. Soc. Am. Proc. 23:152–156.

Slater, B.K., K. McSweeney, A. McBratney, S.J. Ventura, and B.J. Irvin. 1994. A spatial framework for integrating soil-landscape and pedogenetic models. p.169–185. *In* R.B. Bryant and R.W. Arnold (ed.) Quantitative modeling of soil forming processes. Soil Sci. Soc. Am. Spec. Pub. 39.

Smeck, N.E., and L.P. Wilding. 1980. Quantitative evaluation of pedon formation in calcareous glacial deposits in Ohio. Geoderma 24:1–16.

Smeck, N.E., E.C.A. Runge, and E.E. Mackintosh. 1983. Dynamics and genetic modelling of soil systems. p. 51–81. *In* L.P. Wilding, N.E. Smeck, and G.F. Hall (ed.) Pedogenesis and soil taxonomy I. Concepts and interactions. Elsevier, Amsterdam, Netherlands.

Soil Survey Division Staff. 1993. Soil survey manual. USDA Handbk. 18.

Soil Survey Staff. 1975. Soil taxonomy. USDA-SCS, Washington DC.

Sposito, G. 1989. The chemistry of soils. Oxford University Press, New York, NY.

Stumm, W., G. Furrer, E. Wieland, and B. Zinder. 1985. The effects of complex-forming ligands on the dissolution of oxides and aluminosilicates. p. 55–74. *In* J.I. Drever (ed.) The chemistry of weathering. D. Reidel Publishing Company Dordrecht, Netherlands.

Sverdrup, H.U., and P.G. Warfinge. 1988. Chemical weathering of minerals in the Gardsjon catchment in relation to a model based on laboratory rate coefficients. p. 81–129. In: J. Nillson and P. Grennfelt (ed.) Critical loads for sulphur and nitrogen. Copenhagen Nordic Council of Ministers. Miljo rapport 15.

Tandarich, J.P., and S.W. Sprecher. 1994. The intellectual background for the factors of soil formation. p.1–13. *In* R. Amundson, J. Harden and M. Singer. Factors of soil formation: A fiftieth anniversary retrospective. Soil Sci. Soc. Am. Spec. Pub. 33.

Thorstenson, D.C., E.P. Weeks, H. Haas, and D.W. Fisher. 1983. Distribution of gaseous $^{12}CO_2$, $^{13}CO_2$, and $^{14}CO_2$ in sub-soil unsaturated zone of the western US Great Plains. Radiocarb. 25:315–346.

Tillotson, W.R., C.W. Robbins, R.J. Wagenet, and R.J. Hanks. 1980. Soil water, solute, and plant growth simulation. UT State Agr. Exp. Stn. Bull. 502.

Tsuji, G.Y., G. Uehara, and S. Balas. (ed.) 1994. DSSAT (Decision Support System for Agrotechnology Transfer) v.3. University of Hawaii, Honolulu, HI.

Valentine, K.W.G., and D.H. Yaalon. (ed.) 1990. Climatic and lithostratigraphic significance of Paleosols. Geoderma 45:103–195.

Van Genuchten, Th., and P. Wierenga. 1986. Solute dispersion coefficients and retardation factors. p. 1025–1054. *In* A. Klute (ed.) Methods of soil analysis, Part 1, Physical and mineralogical methods. Soil Science Society of America, Madison, WI.

Vanselow, A.P. 1932. Equilibria of the base-exchange reactions of bentonites, permutites, soil colloids, and zeolites. Soil Sci. 33:95–113.

van Wambeke, A. 1972. Mathematical expression of eluviation-illuviation processes and the computation of the effects of clay migration in homogeneous soil parent materials. J. Soil Sci. 23:325–332.

van Wambeke, A. 1976. A mathematical model for the differential movement of two soil constituents into illuvial horizons: application to clay ratios in argillic horizons. J. Soil Sci. 27:111–120.

van Wambeke, A., P. Hastings, and M. Tolomeo. 1986. Newhall simulation model: A BASIC program for the IBM PC. Cornell University, Ithaca, NY.

Vereecken, H. 1988. Pedotransfer functions for the generation of hydraulic properties of Belgian soils. Ph.D. dissertation, Katholieke Universiteit Leuven, Belgium.

Vereecken, H., J. Diels, J. van Orshoven, J. Feyen, and J. Bouma. 1992. Functional evaluation of pedotransfer functions for the estimation of hydraulic properties. Soil Sci. Soc. Am. J. 56:1371–1378.

Volobuyev, V.R. 1984. Two key solutions of the energetics of soil formation. *In* Genesis and geography of soils. 1985. Scripta Publishing Co. Translated from: Pochvovedeniye, 1984, 7:5-11.

Wagenet, R.J. 1986. Water and solute flux. p.1055–1088. *In* Methods of soil analysis, Part I. Soil Science Society of America, Madison,WI.

Wang, Y., R. Amundson, and S. Trumbore. 1994. A model for soil $^{14}CO_2$ and its implications for using ^{14}C to date pedogenic carbonate. Geochim. Cosmochim. Acta 58:393–399.

Warfvinge, P. and H. Sverdrup. 1992. Calculating critical loads of acid deposition with PROFILE: A steady state soil chemistry model. Water Air Soil Pollut. 63:119–143.

Webster, R., 1985. Quantitative spatial analysis of soil in the field. Adv. Soil Sci. 3:2–70.

Westall, J.C., J.L. Zachary, and F.M.M. Morel. 1976. MINEQL, A computer program for the calculation of chemical equilibrium composition of aqueous systems. MA Inst. Technol. Dept. Civil Eng., Tech. Note 18.

White, R.E., J.S. Dyson, R.A. Haigh, W.A. Jury, and G. Sposito. 1986. A transfer function model of solute transport through soil 2. Illustrative applications. Water Resour. Res. 22:248–254.

Wilding, L.P. 1985. Spatial variability: its documentation, accommodation and implication to soil surveys. *In* D.R. Nielsen and J. Bouma (ed.) Soil spatial variability. Pudoc Wageningen, Netherlands.

Wösten, J.H.M., M.H. Bannink, J.J. De Gruijter, and J. Bouma. 1986. A procedure to indentify different groups of hydraulic conductivity and moisture retention curves for soil horizons. J. Hydrol. 86:133–145.

Wösten, J.H.M., C.H.J.E. Schuren, J. Bouma, and A. Stein. 1990. Functional sensitivity analysis of four methods to generate soil hydraulic functions. Soil Sci. Soc. Am. J. 54:832–836.

Wösten, J.H.M., and M. Th. van Genuchten. 1988. Using texture and other soil properties to predict the unsaturated soil hydraulic conductivity. Soil Sci. Soc. Am. J. 52:1762–1770.

Yaalon, D.H. 1971. Soil-forming processes in time and space. p. 29–39. *In* D.H. Yaalon (ed.) Paleopedology - Origin, nature and dating of Paleosols. Int. Soc. Soil Sci. and Israel Universities Press, Jerusalem, Israel.

Yaalon, D.H. 1991. Soil processes and global change. p. 196–199. *In* M. Graber, A. Cohen and M. Magaritz (ed.) Regional implications of future climate change. Proc. Int. Worksh., Weizmann Institute of Science, Rehovot, Israel.

4

Soil Taxonomy

Robert J. Ahrens and Richard W. Arnold
USDA-NRCS

4.1 Conditions Favoring the Development of *Soil Taxonomy*

By 1960, the United States had 60 years of experience with a soil survey program which mapped and interpreted soils in various parts of the country. Ever since the earliest mapping of specialty croplands and those of problem saline soils, the primary purpose of the soil survey program was to predict the consequences of alternative uses of soils.

During that time, significant events had taken place that affected the production and delivery of consistent products and services of the soil survey program. Among these were World War I, new and improved industries, enhanced energy distribution, gas motors and automobiles, mechanization of agriculture and a shift from mainly family farms to more commercial enterprises. The Dust Bowl devastated the lives of many farmers as land degradation and unfavorable climatic conditions collided. New federal and state parks and forest reserves were established, World War II spanned the globe, followed by the Korean War. Economic reconstruction was promoted, and global markets expanded. Agriculture was also changing from a mode of unbridled production to one of increased emphasis on conservation of soil and water resources in more responsible ways. Many of these changes had impacts on the prediction of the consequences of alternative soil uses.

The model of soil changed during those times. When the soil survey started, it was perceived that soils were derived from the rocks, or from the transported materials, on which they rested, a classic geological conclusion. This concept gradually changed to that proposed by the Russian, Dokuchaev, where soils were considered to be the result of processes that were influenced by the interactions of soil-forming factors, namely, climate, biota, parent material, topography, and time. Soils were recognized as independent natural bodies worthy of study by, and for themselves, and thus the course of history of soil science in the United States began to change in the 1920s. Marbut promoted the independence of soils, presented his ideas of soil classification, and helped America become recognized in international affairs of soil science.

Similar climates existed over fairly large areas, as did vegetation groups such as forests and grasses, although both had microvariations. On more local scales, there were differences in parent materials and landform topography that comprised landscapes. The overlapping of the soil-forming factors in space was crucial to applying the model of soil as the result of processes that were influenced by the interactions of the soil forming factors. Soil mappers discovered empirical relations linking sets of soil properties (generally called soil profiles) to specific landscape features which represented soil factors. The correlation of soils with landscape segments was found to be consistent enough to be delineated on base maps and to be satisfactory for the purpose of the soil survey.

Soils, therefore, had a certain predictability and so did their expected behavior. Where conditions had been the same, the soils would be the same; and where the soils were the same, those responses that depend on soil properties would be the same. Where similar but not identical conditions or soils occurred, the soil responses would be similar, but not identical to those where the soils were identical.

Soil, which was considered to be a continuum covering the Earth's terrestrial surface, could be subdivided into classes in a variety of ways, thereby creating a population or collection of individual soil bodies. Emphasis changed from thinking about the whole with loosely defined and indistinct parts to the concept in which the parts were sharply in focus and the whole was an organized collection of parts.

Locally, the individual soils were called soil types. The members of the same series, that is soils having a similar sequence of the same kinds of horizons, were separated according to the general texture of the profile. This was a way to recognize different groups of parent material such as glacial till and loess. Initially, the soil types were grouped together into a soil series which had the same horizonation and commonality of properties. Later, soil type referred only to the texture of the surface soil and was considered as a phase of a soil series. The differences in parent materials became a basis for establishing separate series.

Mapping of soil types proceeded rapidly as there were hundreds of survey parties mapping in all parts of the country. Many new soil series were set up; however, an adequate system of correlation and classification lagged behind. It became more and more difficult to keep track of all the information being collected by the soil scientists, to compare soils from one region to another, and to communicate about soil properties and characteristics.

A number of differences in the responses to management and in land uses were found to vary geographically. Some regional variations were related to climate and age, and locally many variations were related to drainage condition and parent material. These observations supported the concept of important geographic differences and similarities and led to the need for some way of recognizing these similarities and differences. The significance of factors and factor interactions varied from region to region and if reasonable ways to group or separate them could be found, they might serve as a basis for classification.

Over time the question about soil had changed from "How much yield can be expected from this soil with this amount of input?" to "How much input must be used on this soil to produce a given volume of produce?" This recognized that soil was dynamic and capable of modification and manipulation rather than being a static responder to management. Humans and their activities were accepted as major factors of soil formation; in fact, it was recognized that humans were the dominant influence on temporal soil quality. The changing concepts and attitudes about soil and the evolving world economic development were important contributors to the decision to develop a new soil classification system.

4.2 Recognition of Guiding Principles for a Soil Classification System

The leaders of the National Cooperative Soil Survey agreed that a classification system must serve the soil survey program of the United States. The program was undertaken for very practical reasons, namely, to identify and locate soils, and predict the consequences of alternative uses of soils.

They indicated that a system should provide a basis for developing principles of soil genesis and behavior that would enable them to continue to provide predictions of soil behavior and responses to management and manipulation. Because soil genesis attempts to find cause and effect relationships rather than merely identifying empirical relationships, it was felt that genetic principles should serve well as a means of spatial extrapolation and to assist in interpreting soils for use (Smith, 1963).

It was clear to them that classes of a system must have real counterparts in mappable soil bodies. Some small geographic body was the entity to be classified; thus, the criteria of classification should not be applied independently of the values and variability within natural landscapes. This need to link with identifiable geographic bodies separated the practical from the purely hypothetical organization of soil property information (Cline, 1949).

It was implied that the component soil bodies in nature that were to be classified must be relevant to applied objectives. Geographic size and shape and soil attributes included within those areas were considered important for making relevant interpretations. Experience had demonstrated that this would be a problem in a classification with mutually exclusive classes; however, some of the concerns could be handled with mapping conventions.

Although the system would serve the practical uses of the soil survey, it was not implied that the classes themselves had to serve directly as the technical interpretive units. The leaders realized that valid interpretations of behavior and response likely would involve an additional step, level, or reasoning which would rely on the known or assumed relationships between soil properties and behavior. This was also true for valid interpretations about soil genesis. It was recognized that no set of classes, or basis for groupings, or basis of subdivision, could provide units homogeneous enough for direct application to multiple objectives. Therefore, the classes had to be capable of being regrouped or subdivided as needed to satisfy different applied purposes. This was most important for the classes of the lowest category, which was the soil series (Smith, 1963).

There was agreement that a classification system must also serve to understand nature by revealing our understanding of soils as natural bodies. This departed from classification schemes of plants and animals where lineage relationships were assumed. The accurate mapping of soils depended on this understanding. This meant that science was to be basic to the process. In as much as such a system would be able to satisfy these ideas, it would be a reflection of the knowledge at a given point in time.

It was desired to have a system that could be applied uniformly by competent soil scientists. It would have to be objective in the sense that it should be based on properties of soils and not on beliefs of the classifier. The taxa should be defined, or recognized by observable or measurable properties selected to group soils of similar genesis, but also capable of providing groups with similar properties, even if the genesis was uncertain or unknown. Soil genesis would be used to help select those properties that would provide useful groups of soils; however, the properties themselves would be the criteria for actual placement.

4.3 Science and Classification

Several lessons had been learned in trying to modify the 1938 classification. The concepts of some Great Soil Groups were not clear and seemed to overlap with other groups. The methods of determining properties varied from place to place, or from time to time, and the correlations were not always certain. The mapping of soil types and soil series at the lower end of the spectrum, and the division of the pedosphere into regions and provinces and groups at the other end left an undecided middle ground with no good connections. There was great interest in not overly disturbing the thousands of soil series that had been established and used throughout the United States; yet there was a need to be able to link the series to classes in the higher categories of a classification system.

Mapping of soils and detailed laboratory studies of their properties gave rise to many consistent and repeatable correlations of soil features with landscape features. To maintain this consistency the nature of the operations and methods used to obtain the data need to be specified and used. It is known that explanation is primarily a recognition of familiar correlations among phenomena in nature and that explanations are critical to understanding. It had been suggested that to go beyond empirical correlations to hypothesize the reasons for them will prejudice the future; therefore, there would be an effort to try to minimize using *a priori* principles which determine or limit possibilities of new experiences. This meant that concepts of soil genesis would be useful to help select soil properties of interest, but that the operational character of soil facts would be used so as to not limit new experiences in the future.

There already were several thousand series considered to be classes of a lower category. They wanted some broader, more inclusive classes for making generalized maps, and an orderly organization of current understanding. It was also thought that they required flexibility to better serve other needs relying on soil information. Due to the magnitude and complexity of information, a hierarchical system of organization seemed most appropriate. For the most part, the logic of J. S. Mill as explained by Cline (1963) was followed. Some of the requirements follow.

Each category in a multicategoric system contains all members of the population of interest and is defined by abstracting the concepts that group the member classes together. Higher level categories are more abstract than those of the lower level.

The definition of a category suggests criteria that may be appropriate to separate the classes of that category. The criteria are then recognized by properties or features which are called differentiating characteristics. These characteristics are observable or measurable properties of individuals that are grouped into a class.

Classes separated at a higher categorical level remain separated throughout the lower categories of the system. For ease of operation (placement of objects into the system), the classes are mutually exclusive, that is, without overlap. Operationally, the listing of the categories and the listing of the classes within each category need to be by priority because only a few of the characteristics of individuals are used to obtain placement within the scheme (Smith, 1983). This means that the scheme can be presented as a key or set of keys for consistent application of the definitions and the criteria (differentiating criteria).

The logic of the hierarchy represented by *Soil Taxonomy* is one of its major strengths, yet at the same time it retains the weaknesses of such organizational frameworks. The efficiency with which accessory characteristics are carried along with the differentiating ones is very powerful indeed. On the other hand, the uncertainty of the degree of accessory property relationships and the rigid class structure of the system limit the precision of statements that can be made for soils as they are observed in the field.

Recognition of a soil individual is crucial to the proper functioning of the system. Because the system was to support the soil survey, the individual had to be a recognizable geographic entity and one that could be characterized with limited sampling. The sample unit became the pedon and the soil individual became the polypedon. Their application has not been without some difficulty.

The nomenclature of a hierarchical system is meaningful to the extent that there is consistent composition of the classes. A mnemonic scheme of phrases and combinations seemed a worthy goal. It was thought desirable to have a system of naming that would minimize previous biases and confusion and which might be translatable or usable in other languages. Consequently, roots of words from Latin and Greek and their combinations were proposed and tested.

And finally, it was agreed that the definitions must be continually tested by the nature of functioning of the soils grouped in a taxon. A taxonomy for the use of the soil survey needed to be tested by the nature of the interpretations that could be made (Smith, 1983). These interpretations included those for genesis and the mapping of soils as well as for the behavior of soils under multiple uses.

4.4 Definitions of Categories of *Soil Taxonomy*

A category is the aggregate of classes that is formed by differentiation within a population on a single basis. A category includes the entire population; it includes all classes differentiated on one basis; it is distinguished from class which is only one part of a category; and it is definable only in terms of the basis of differentiation. In a hierarchy the higher categories have fewer classes and they are more inclusive than the classes of the lower categories which have accumulated attributes from all of the higher categories. Thus high categorical levels are associated with high level generalizations or abstractions. These abstractions are used as the bases of differentiation; however, these ideas are expressed in terms of attributes that are assumed to be their consequences. Thus soil attributes that are thought to be the results of such processes are the criteria used to segregate soils.

The definitions of categories, orders through series, although not stated very clearly, are intended to guide the selection and testing of properties and features used to characterize and classify soils. Soil properties that are thought to reflect processes or control processes are of great importance in this system. Although the processes of formation are influential in the selection of properties, it is the properties themselves that are used to determine the placement of soils into respective classes. The applications of quantified definitions and observations are the tests of adequacy of the theories behind this model of soil. Hence, *Soil Taxonomy* is considered a morphogenetic classification system.

4.4.1 Order

Classes at the order level are separated on the basis of properties resulting from the major processes and pathways of soil formation. Neither the genetic processes nor the courses of development are precisely known but the accepted concepts have influenced the selection of soil properties that are used to recognize and define the twelve classes currently considered. Many of the features are thought to have taken a reasonably long time to develop, are stable in a pedological sense, and are mainly static historical.

4.4.2 Suborder

Classes at the suborder level are separated within each order on the basis of soil properties that are major controls, or reflect such controls on the current set of soil-forming processes. Most of the properties selected are dynamic such as soil moisture regime or cold soil temperatures. Other properties relate to materials or processes that retard horizon development, such as sand or alluvial sedimentation.

4.4.3 Great Group

Classes at the great group level are differentiated within each suborder on the basis of properties that constitute subordinate or additional controls, or reflect such controls on the current set of soil-forming processes. The properties selected are generally static, such as layers that retard percolation of water or root extension, but some are dynamic, such as the moisture regime where it was not a criterion at the suborder level.

4.4.4 Subgroup

Classes at the subgroup level are differentiated within each great group on the basis of properties resulting from either (1) a blending or overlapping of sets of processes in space or time that cause one kind of soil to develop from, or toward another kind of soil that has been recognized at the great group, suborder or order level; or (2) sets of processes or conditions that have not been recognized as criteria for any class at a higher level. A third kind of subgroup fits neither (1) nor (2) but is considered to typify the central concept of the great group.

4.4.5   Family

Classes at the family level are separated with the subgroup on the basis of properties that reflect important conditions affecting behavior or the potential for further change. Particle size, mineralogy, and soil depth are mainly capacity factors, whereas soil temperature and exchange activity are mainly intensity factors.

4.4.6   Series

Classes at the series level are separated within the family on the basis of properties that reflect relatively narrow ranges of soil-forming factors and of processes that transform parent materials into soils. Some properties are indicative of parent materials such as coarse fragments, sand or silt content, color, and horizon thickness or expression. Others reflect influences on processes such as differences in intensity or amount of precipitation, and depth to the presence or concentration of soluble compounds.

4.5   Differentiating Characteristics

The differentiating characteristics mentioned above are referred to in *Soil Taxonomy* as diagnostic horizons and characteristics. They are the building blocks of *Soil Taxonomy*. The diagnostic horizons and characteristics help define the criteria for the various taxa of *Soil Taxonomy*. They are also terms used by scientists to communicate the language of soil.

4.5.1   Diagnostic Horizons

Diagnostic horizons are defined in the *Keys to Soil Taxonomy* (Soil Survey Staff, 1998). The intent in this chapter is not to reiterate the complete definition of the diagnostic horizons, but to provide a brief summary.

4.5.2   Diagnostic Epipedons

Diagnostic horizons that form at or near the surface are referred to as epipedons. There are 8 epipedons recognized in *Soil Taxonomy* which are briefly described below.

4.5.2.1   Anthropic Epipedon (Gr. *anthropos*, human being)
This is formed during long continuous use by humans, either as a place of residence (kitchen middens) or as a site for irrigated crops. Anthropic epipedons often have the dark color of an umbric or a mollic epipedon, but were formed largely due to the actions of humans. Anthropic epipedons occur in a variety of soil orders.

4.5.2.2   Folistic Epipedon (L. *folia*, leaf)
This is 20 cm or more thick and contains high amounts of organic C, but is not saturated with water for more than a few days after heavy rains. These epipedons occur mostly in cool, humid regions under forest vegetation. Folistic epipedons occur most commonly in Spodosols and Inceptisols, but can occur in other orders.

4.5.2.3 Histic Epipedon (Gr. *histos*, tissue)
This is 20 to 40 cm thick and contains high amounts of organic C. It differs from the folistic epipedon in that it is commonly saturated with water for long periods. Histic epipedons can occur in many soil orders.

4.5.2.4 Melanic Epipedon (Gr. *melas*, *melan-*, black)
This is a thick, black horizon that contains high concentrations of organic C and short-range order minerals or Al-humus complexes. It commonly occurs in areas associated with a volcanic influence under forest vegetation. Melanic epipedons are associated with Andisols.

4.5.2.5 Mollic Epipedon (L. *mollis*, soft)
This is a relatively thick, dark colored, humus-rich surface horizon in which divalent cations are dominant on the exchange complex. It forms from the underground decomposition of organic residues, chiefly grasses. Mollic epipedons have good structure and porosity and are associated with some of the richest agricultural soils in the world. Mollic epipedons are most commonly associated with Mollisols; they very rarely occur in other soil orders.

4.5.2.6 Ochric Epipedon (Gr. *ochros*, pale)
This fails to meet the definition of any of the other diagnostic epipedons. It can be light colored, thin and dark colored, or even thin with high amounts of organic C. It has few or no accessory characteristics, but was defined to serve as a means of placing a name on the surface horizons that failed the criteria of any of the other seven epipedons. Ochric epipedons can occur in a variety of soil orders.

4.5.2.7 Plaggen Epipedon (Ger. *Plaggen*, sod)
Plaggen epipedons are rare and occur mostly in Europe where in medieval times sod or other materials were used for bedding livestock, and the manure was spread on cultivated fields. The gradual additions of bedding materials raised the level of the fields. The plaggen epipedon is associated with Inceptisols.

4.5.2.8 Umbric Epipedon (L. *umbra*, shade, hence dark)
Umbric epipedons resemble mollic epipedons, but formed largely under a forest vegetation and have a lower natural fertility (base saturation) than the mollic epipedon. The umbric epipedon is most commonly associated with Inceptisols, Andisols, Alfisols, Ultisols, and Gelisols.

4.5.3 Diagnostic Subsurface Horizons

The subsurface horizons form below the surface, but can be at or near the surface in eroded soils. Brief descriptions follow.

4.5.3.1 Agric Horizon (L. *ager*, field)
This is an illuvial horizon which has formed under cultivation and contains significant amounts of illuvial silt, clay, and humus. This horizon is relatively rare and is often more associated with old world agriculture. Agric horizons are associated with Alfisols.

4.5.3.2 Albic Horizon (L. *albus*, white)
This is an eluvial horizon 1.0 cm or more thick with a color that is largely determined by the color of the primary sand and silt grains, rather than the color of their coatings. The clay and Fe oxides have

been removed by pedogenesis. Albic horizons are most commonly associated with Spodosols, Alfisols, Ultisols, and Mollisols.

4.5.3.3 Argillic Horizon (L. *argilla*, white clay)
An argillic horizon is commonly a subsurface horizon with a significantly higher content of phyllosilicate clay than the overlying horizon. There is also evidence of clay illuviation. Argillic horizons occur on stable landscapes and are most commonly associated with Alfisols, Ultisols, Aridisols, and Mollisols.

4.5.3.4 Calcic Horizon (L. *calcis*, lime)
This is an illuvial horizon in which secondary $CaCO_3$ has accumulated to a significant extent. In arid environments, precipitation is insufficient to move carbonates to great depths. In soils with water near the surface, capillary rise, evaporation, and transpiration concentrate $CaCO_3$ toward the surface. Calcic horizons occur in Aridisols, Alfisols, Mollisols, Inceptisols, and Gelisols.

4.5.3.5 Cambic Horizon (L. *cambiare*, to exchange)
The cambic horizon forms as a result of physical alterations, chemical transformations, accumulations, removals, or a combination of these processes. They are commonly identified by structure and a higher clay content, redder hue, or higher chroma than an overlying horizon. Cambic horizons are most commonly associated with Inceptisols, but they can occur in Aridisols, Mollisols, Vertisols, Andisols, and Gelisols.

4.5.3.6 Duripan (L. *durus*, hard)
A duripan is cemented by illuvial silica. Duripans limit the downward growth of roots and movement of water. They are most commonly associated with the Aridisols, but are known to occur in Alfisols, Mollisols, Andisols, Inceptisols, and Spodosols.

4.5.3.7 Fragipan (L. *fragilis*, brittle, and pan)
This is a subsurface horizon that is noncemented, but restricts the entry of water and roots into the soil matrix. Fragipans most commonly are associated with Ultisols, Alfisols, Spodosols, and Inceptisols.

4.5.3.8 Glossic Horizon (Gr. *glossa*, tongue)
The glossic horizon develops as a result of the degradation of an argillic, kandic, or natric horizon. Clay and Fe oxides are removed starting from the exterior of peds. These horizons are transitional from argillic, natric, and kandic horizons to albic horizons. Glossic horizons occur most commonly in the Alfisols and Ultisols.

4.5.3.9 Gypsic Horizon (L. *gypsum*)
This is an illuvial horizon in which secondary gypsum has accumulated to a significant extent. Most gypsic horizons occur in arid environments where the parent materials are rich in gypsum. In soils with groundwater close to the surface, gypsum can accumulate by capillary rise, evaporation, and transpiration. Gypsic horizons mostly commonly occur in the Aridisols, but a few Inceptisols and Gelisols also have gypsic horizons.

4.5.3.10 Kandic Horizon (modified from kandite)
This is a subsurface horizon with a significantly higher content of clay than an overlying horizon. Kandic horizons are dominated by low activity clays (1:1) and, therefore, have low CEC. Kandic

horizons occur in the Ultisols and Oxisols.

4.5.3.11 Natric Horizon (L. *natrium*, sodium)

The natric horizon has all the characteristics of the argillic horizon (significant clay increase) and, in addition, has an accumulation of Na. Sodium adversely affects the physical properties of a soil. Of the soil orders natric horizons are most commonly associated with the Aridisols, but they also occur in the Alfisols and Mollisols and to a very limited extent in the dry Gelisols.

4.5.3.12 Ortstein

Ortstein is a cemented horizon that consists of complexes of Al and organic matter with or without Fe (spodic materials). Ortstein limits root growth and the downward movement of water. Ortstein occurs in Spodosols.

4.5.3.13 Oxic Horizon (F. *oxide*)

This is a subsurface horizon of sandy loam or a finer texture with a low CEC and low amount of weatherable minerals. Oxic horizons occur in old, weathered soils or soils derived from highly weathered parent materials. Oxic horizons occur in Oxisols.

4.5.3.14 Petrocalcic Horizon (Gr. *petra*, rock and calcic)

This is cemented or indurated by $CaCO_3$ or Ca and Mg carbonates. It is a barrier to roots and the downward movement of water. Petrocalcic horizons are most commonly associated with the Aridisols, but also occur in the Alfisols, Mollisols, and Inceptisols.

4.5.3.15 Petrogypsic Horizon (Gr. *petra*, rock and gypsic)

This is cemented or indurated by gypsum. It is a barrier to roots and downward movement of water. Petrogypsic horizons are most commonly associated with Aridisols.

4.5.3.16 Placic Horizon

This is a thin (< 25 mm), dark colored horizon that is cemented by either Mn or Fe and Mn and organic matter. Commonly, they occur in moist, cool climates and are associated with the Spodosols and Inceptisols.

4.5.3.17 Salic Horizon (L. *sal*, salt)

This is an accumulation of salts that are more soluble than gypsum in water. Many plants are intolerant of high concentrations of salt. Most salic horizons are associated with the Aridisols; a few occur in the drier Gelisols.

4.5.3.18 Sombric Horizon (F. *sombre*, dark)

This is an illuvial horizon that contains humus that is not associated with Al nor dispersed by Na. It is largely confined to the cool, moist soils of high plateaus or mountains of the tropics and subtropics. Sombric horizons are associated with Inceptisols, Oxisols, and Ultisols.

4.5.3.19 Spodic Horizon (Gr. *spodos*, wood ash)

This is an illuvial horizon that contains active amorphous materials composed of organic matter and Al with or without Fe. Spodic horizons are associated with Spodosols.

4.5.3.20 Sulfuric Horizon (L. *sulfur*, sulfur)

The sulfuric horizon forms when sulfide rich and organic materials are oxidized as a result of drainage, most commonly artificial. Sulfuric horizons are toxic to most plants. They occur in Entisols, Histosols, and Inceptisols.

4.5.4 Other Diagnostic Soil Characteristics

The diagnostic soil characteristics are features of the soil that are used repeatedly in *Soil Taxonomy*.

Abrupt textural change is a considerable increase in clay within a short distance, a very important feature used to help predict water movement in soils.

Andic soil properties result from the presence of significant amounts of allophane, imogolite, ferrihydrite, or Al-humus complexes.

Aquic conditions arise in soils that currently experience continuous or periodic saturation and reduction. The presence of these features is indicated by redoximorphic features. There are several types of saturation: episaturation describes a perched watertable, endosaturation describes a ground watertable, and anthraquic saturation describes controlled flooding such as that used to grow rice and cranberries.

Coefficient of linear extensibility (COLE) is the ratio of the difference between the moist and dry lengths of a clod to its dry length. COLE is used to determine the shrink/swell potential of a horizon.

Densic contact occurs between soil and densic materials and is a barrier to roots.

Densic materials are noncemented, relatively unweathered, mostly earthy materials such as till, volcanic mudflows, and noncemented rocks.

Durinodes are cemented to indurated nodules with SiO_2, presumably opal and microcrystalline forms of silica.

Fragic soil properties are similar to a fragipan, but do not have the required thickness or volume for a fragipan.

Identifiable secondary carbonates refer to translocated authigenic $CaCO_3$ that has been precipitated in place from the soil solution rather than precipitated in place from the parent material.

Lamellae are thin (< 7.5 cm) illuvial horizons that contain evidence of illuvial clay and occur only in coarse-textured soils.

Lithic contact occurs between soil and a coherent underlying material that is in a strongly or more cemented class. Lithic contacts occur at the boundary between soil and hard bedrock.

Paralithic contact occurs between soil and paralithic materials (defined below) such as between soil and unconsolidated or weathered bedrock.

Paralithic materials are relatively unaltered (by pedogenesis) materials that are in a moderately or less cemented class.

Petroferric contact is a boundary between soil and a continuous layer of indurated material in which Fe is an important cement and organic matter is either absent or present in trace amounts.

Plinthite is an Fe-rich, humus-poor material that irreversibly hardens when exposed to wetting and drying, especially if heated by the sun.

Slickensides are polished and grooved surfaces produced by one mass sliding past another. They are a feature of soils with a high capacity to shrink and swell.

4.5.5 Soil Moisture and Temperature Regimes

Soil Taxonomy is unique among many systems used to classify soils because it recognizes the temperature of a soil and the moisture status of a soil over time. Both temperature and moisture are important properties of the soil and convey essential information about soils. Because complete definitions for the soil moisture and temperature regimes can be found in the *Keys to Soil Taxonomy* (Soil Survey Staff, 1998), only brief descriptions are presented here.

4.5.5.1 Soil Moisture Regimes

For most soil orders, moisture regime is used to determine placement of a soil at the suborder level.

Aquic moisture regime signifies a reducing regime virtually free of dissolved O_2 from saturation with water. Most soils that have an aquic moisture regime are not saturated with water to the soil surface year round.

Aridic (torric) moisture regime applies to soils which commonly occur in arid climates and are dry in all parts more than half the time during the growing season and moist in some or all parts for less than 90 consecutive days during the growing season.

Udic moisture regime applies to soils which occur in climates with well-distributed rainfall or sufficient summer rain so that rainfall plus stored moisture equals or exceeds the amount of evaporation.

Ustic moisture regime is moister than aridic and drier than udic. The concept of ustic is one of limited moisture, but at least some moisture at a time when conditions are suitable for plant growth.

Xeric moisture regime applies to soils which occur in climates with cool, moist winters and warm, dry summers.

Perudic moisture regime applies to soils which occur in areas where precipitation exceeds evaporation in all months when the soil is not frozen.

4.5.5.2 Soil Temperature Regimes

The following soil temperature regimes are based on the mean annual temperature of a soil measured at 50 cm or at a lithic, paralithic, or densic contact, whichever is shallowest.

Cryic soil temperature regime applies to soils which have a mean annual soil temperature of < 8 °C, but no permafrost and a summer temperature cooler than soils in a frigid regime.

Frigid soil temperature regime applies to soils which have a mean annual soil temperature of < 8 °C.

Mesic temperature regime applies to soils which have a mean annual soil temperature ≥ 8 °C but < 15 °C.

Thermic temperature regime applies to soils which have a mean annual soil temperature ≥ 15 °C but < 22 °C.

Hyperthermic temperature regime applies to soils which have a mean annual soil temperature ≥ 22 °C.

When the difference between mean summer and winter soil temperatures is less than 6 °C, *iso* is added to the name. *Isofrigid, isomesic, isothermic*, and *isohyperthermic* are the only classes used. Soil temperature is most commonly used at the family level in *Soil Taxonomy*.

4.6 Categories of *Soil Taxonomy*

The information that follows is largely updated from the first edition of *Soil Taxonomy* (Soil Survey Staff, 1975). The diagnostic horizons and characteristics help define the categories of *Soil Taxonomy*.

A category of *Soil Taxonomy* is a set of classes that is defined approximately at the same level of generalization or abstraction and includes all soils. There are six categories or levels in *Soil Taxonomy*. In order of decreasing rank and increasing number of differentiae and classes, the categories are order, suborder, great group, subgroup, family, and series.

The nomenclature of *Soil Taxonomy* is based on the following premises: each taxon requires a name if it is to be used in speech; a good name is short, easy to pronounce, and distinctive in meaning; a name is connotative, that is, capable of mnemonic attachment to the concept of the thing itself (Heller, 1963); it is useful if the name of a taxon indicates its position in the classification, if similarities in important properties are reflected by similarities in names, if the mnemonic attachments hold in many languages, and if the name fits into many languages without translation.

The name of each taxon above the category of series indicates its class in all categories of which it is a member. The name of a soil series indicates only the category of series. Thus, a series name may be recognized as a series, but one cannot tell from the name to what order, suborder, and so on, it belongs.

Table 4.1 gives a few examples of names of taxa in each category from order to series. Because of the system for assigning names and because most formative elements carry the same meaning in any combination, a name can convey a great deal of information about a soil.

4.7 Recognition of the Categories

4.7.1 Orders

The name of each order ends in sol (L. *solum*, soil) with the connecting vowel *o* for Greek roots and *i* for other roots. Each name of an order contains a formative element that begins with the vowel immediately preceding the connecting vowel and ends with the last consonant preceding the connecting vowel. In the order name, Entisol, the formative element is *ent*. In Aridisol, it is *id*. These formative elements are used as endings for the names of suborders, great groups, and subgroups. Thus, the names of all taxa higher than the series that are members of the Entisol order end in *ent* and can be recognized as belonging to the order of Entisol. Names ending in *id* are names of taxa belonging to the order of Aridisols. Table 4.2 lists all the soil orders and their formative elements.

4.7.2 Suborders

Names of suborders have exactly two syllables. The first syllable connotes something of the diagnostic properties of the soils or the soil moisture regime. The second is the formative element from the name of the order. Thirty-one formative elements (Table 4.3) are used with the 12 formative elements from names of the orders to make names of over 60 suborders. The suborder of Entisols that has aquic conditions throughout is called Aquents, (L. *aqua*, water, plus *ent* from Entisol). The formative element *aqu* is used with this meaning in 9 of the 12 orders. The suborder of Entisols that consists of very young sediments is called Fluvents (L. *fluvius*, river, plus *ent* from Entisol).

4.7.3 Great Groups

The name of a great group consists of the name of a suborder and a prefix that consists of one or two formative elements suggesting something of the diagnostic properties. Names of great groups, therefore, have three or four syllables and end with the name of a suborder. Fluvents that have a cryic temperature regime are called Cryofluvents (Gr. *kyros*, icy cold, plus fluvent). Fluvents that have a torric moisture regime are called Torrifluvents (L. *toridus*, hot and dry). The formative elements for the great groups are listed in Table 4.4.

Table 4.1 Example of naming systems for soils used in *Soil Taxonomy*

Order	Suborder	Great Group	Subgroup	Family	Series
Aridisols	Argids	Calciargids	Ustic Calciargids	fine-loamy, mixed, superactive, mesic	Barx Highland

Table 4.2 Formative elements in names of soil orders in *Soil Taxonomy*

Formative element in name	Name of order	Derivation of formative element	Pronunciation of formative element
Alf	Alfisol	Meaningless syllable	Pedalfer
And	Andisol	Modified from ando	Ando
Id	Aridisol	L. *aridus, dry*	Arid
Ent	Entisol	Meaningless syllable	Recent
El	Gelisol	L. *gelare*, to freeze	Jell
Ist	Histosol	Gr. *histos,* tissue	Histology
Ept	Inceptisol	L. *inceptum*, beginning	Inception
Oll	Mollisol	L. *mollis*, soft	Mollify
Ox	Oxisol	F. *oxide*, oxide	Oxide
Od	Spodosol	Gr. *spodos*, wood ash	Odd
Ult	Ultisol	L. *ultumus*, last	Ultimate
Ert	Vertisol	L. *verto*, turn	Invert

4.7.4 Subgroups

The name of a subgroup consists of the name of a great group modified by one or more adjectives. The adjective, typic, represents in some instances what is thought to typify the great group and in other instances, typic subgroups simply do not have any of the characteristics used to define the other subgroups in a great group. Each typic subgroup has, in clearly expressed form, all the diagnostic properties of the order, suborder, and great group to which it belongs. Typic subgroups also have no additional properties indicating a transition to another great group. Typic subgroups are not necessarily the most extensive subgroup of a great group.

Intergrade subgroups are those that belong to one great group but that have some properties of another order, suborder, or great group. They are named by using the adjectival form of the name of the appropriate taxon as a modifier of the great group name. Thus, the Torrifluvents that have some of the properties or properties closely associated with Vertisols are called Vertic Torrifluvents. Vertic Torrifluvents have some of the properties of Vertisols superimposed on the complete set of diagnostic properties of Torrifluvents.

Extragrade subgroups are those that have important properties that are not representative of the great group but that do not indicate transitions to any other known kind of soil. They are named by modifying the great group name with an adjective that connotes something of the nature of the aberrant properties. Thus, a Cryorthent that has bedrock that is at least strongly cemented within 50 cm of the mineral soil surface is called a Lithic Cryorthent (lithic, Gr. *lithos*, stone). This subgroup is listed as an example in Table 4.5.

4.7.5 Families

Names of families are polynomial. Each consists of the name of a subgroup and descriptive terms, generally three or more, to indicate the particle-size class (or combinations thereof if strongly contrasting), the mineralogy (26 classes), the cation exchange activity (4 classes), the calcareous and reaction (4 classes), the temperature (8 classes), and, in a few families, depth of soil (3 classes), rupture resistance (2 classes), and classes of coatings and classes of cracks (3 classes). Names of most families have three to five descriptive terms that modify the subgroup name, but a few have only one or two and

Table 4.3 Formative elements in names of suborders in *Soil Taxonomy*

Formative element	Derivation	Connotation
Alb	L. *albus*, white	Presence of albic horizon
Anth	Gr. *anthropos*, human	Modified by humans
Aqu	L. *aqua*, water	Aquic conditions
Ar	L. *arare*, to plow	Mixed horizon
Arg	L. *argilla*, white clay	Presence of argillic horizon
Cal	L. *calcis*, lime	Presence of calcic horizon
Camb	L. *cambiare*, to exchange	Presence of cambic horizon
Cry	Gr. *kryos*, icy cold	Cold
Dur	L. *durus*, hard	Presence of duripan
Fibr	L. *fibra*, fiber	Least decomposed stage
Fluv	L. *fluvius*, river	Flood plain
Fol	L. *folia*, leaf	Mass of leaves
Gyps	L. *gypsum*, gypsum	Presence of gypsic horizon
Hapl	Gr. *haplos*, simple	Minimum horizon development
Hem	Gr. *hemi*, half	Intermediate stage of decomposition
Hist	Gr. *histos*, tissue	Presence of organic materials
Hum	L. *humus*, earth	Presence of organic matter
Ochr	Gr. *ochros*, color	Presence of an ochric horizon
Per	L. *per*, throughout time	Perudic moisture regime
Psamm	Gr. *psammos*, sand	Sandy texture
Rend	Modified from Rendzina	High carbonate content
Sal	L. *sal*, salt	Presence of salic horizon
Sapr	Gr. *saprose*, rotten	Most decomposed stage
Torr	L. *torridus*, hot and dry	Torric moisture regime
Turb	L. *turbidis*, disturbed	Presence of cryturbation
Ud	L. *udus*, humid	Udic moisture regime
Umbr	L. *umbra*, shade	Presence of an umbric horizon
Ust	L. *ustus*, burnt	Ustic moisture regime
Vitr	L. *vitrum*, glass	Presence of glass
Xer	Gr. *xeros*, dry	Xeric moisture regime

a few as many as six. An example given in Table 4.1 is a family of fine-loamy (particle-size), mixed (mineralogy), superactive (cation exchange activity), and mesic (soil temperature) Ustic Calciargids.

4.7.6 Series

Names of series as a rule are abstract place names. The name usually is taken from a place near where the series was first recognized. It may be the name of a town, a county, or some local feature. Some series have coined names while many have been carried over from earlier classifications with some having been in use since 1900. The name of a series carries no meaning to people who have no other source of information about the soils in it. There are over 19,000 soil series currently recognized in the United States.

Table 4.4　Formative elements in names of great groups in *Soil Taxonomy*

Formative element	Derivation	Connotation
Acr	Gr. *akros*, at the end	Extreme weathering
Al	Modified from aluminum	High aluminum, low iron
Alb	L. *albus*, white	Presence of an albic horizon
Anhy	Gr. *anhydros*, waterless	Very dry
Anthr	Gr. *anthropos*, human	An anthropic epipedon
Aqu	L. *aqua*, water	Aquic conditions
Arg	L. *argilla*, white clay	Presence of an argillic horizon
Calc	L. *calcis*, lime	A calcic horizon
Cry	Gr. *kryos*, icy cold	Cold
Dur	L. *durus*, hard	A duripan
Dystr (Dys)	Gr. *dys*, ill; dystrophic, infertile	Low base saturation
Endo	Gr. *endo*, within	Implying a groundwater table
Epi	Gr. *epi*, on, above	Implying a perched water table
Eutr (Eu)	Gr. *eu*, good; eutrophic, fertile	High base saturation
Ferr	L. *ferrum*, iron	Presence of iron
Fibr	L. *fibra*, fiber	Least decomposed stage
Fluv	L. *fluvius*, river	Flood plain
Fol	L. *folia*, leaf	Mass of leaves
Frag	L. *fragilis*, brittle	Presence of fragipan
Fragloss	Compound of fra(g) and gloss	See Frag and Gloss
Fulv	L. *fulvus*, dull	Dark brown color, presence of organic C
Glac	L. *glacialis*, icy	Ice lenses or wedges
Gyps	L. *gypsum*, gypsum	Presence of gypsic horizon
Gloss	Gr. *glossa*, tongue	Presence of glossic horizon
Hal	Gr. *hals*, salt	Salty
Hapl	Gr. *haplos*, simple	Minimum horizon
Hem	Gr. *hemi*, half	Intermediate stage of decomposition
Hist	Gr. *histos*, tissue	Presence of organic materials
Hum	L. *humus*, earth	Presence of organic matter
Hydr	Gr. *hydor*, water	Presence of water
Kand	Modified from kandite	1:1 layer silicates
Luv	Gr. *louo*, to bath	Illuvial
Melan	Gr. *melasanos*, black	Black, presence of organic C
Molli	L. *mollis*, soft	Presence of a mollic epipedon
Natr	L. *natrium*, sodium	Presence of a natric horizon
Orth	Gr. *orthos*, true	The common ones
Pale	Gr. *paleos*, old	Excessive development
Petro	Gr. *petra*, rock	A cemented horizon
Plac	Gr. *plax*, flat stone	Presence of a thin pan
Plagg	Ger. *plaggen*, sod	Presence of a plaggen epipedon
Plinth	Gr. *plinthos*, brick	Presence of plinthite
Psamm	Gr. *psammos*, sand	Sand texture
Quartz	Ger. *quarz*, quartz	High quartz content
Rhod	Gr. *rhodon*, rose	Dark red color
Sal	L. *sal*, salt	Presence of a salic horizon
Sapr	Gr. *saprose*, rotten	Most decomposed stage
Somb	F. *sombre*, dark	Presence of sombric horizon
Sphagn	Gr. *sphagnos*, bog	Prescence of sphagnum
Sulf	L. *sulfur*, silfur	Presence of sulfides or their oxidation products
Torr	L. *torridus*, hot	Torric moisture regime and dry
Ud	L. *udus*, humid	Udic moisture regime
Umbr	L. *umbra*, shade	Presence of an umbric horizon
Ust	L. *ustus*, burnt	Ustic moisture regime
Verm	L. *vermes*, worm	Wormy or mixed by animals
Vitr	L. *vitrum*, glass	Presence of volcanic glass
Xer	Gr. xeros, dry	Xeric moisture regime

Table 4.5 Adjectives in names of extragrades in *Soil Taxonomy* and their meaning

Adjective	Derivation	Connotation
Abruptic	L. *abruptum*, torn off	Abrupt textural change
Aeric[a]	Gr. *aerios*, air	Aeration
Albic	L. *albus*, white	Presence of an albic horizon
Alic	Modified from aluminum	High Al^{3+} status
Anionic	Gr. *anion*	Positively charged colloid
Anthraquic	Gr. *anthropos*, human and L. *aqua*, water	Controlled flooding
Anthropic	Gr. *anthropos*, human	An anthropic epipedon
Arenic	L. *arena*, sand	Sandy between 50 and 100 cm thick
Calcic	L. *calcis*, lime	Presence of a calcic horizon
Chromic	Gr. *chroma*, color	High chroma
Cumulic	L. *cumulus*, heap	Thickened epipedon
Durinodic	L. *durus*, hard	Presence of durinodes
Dystic	Gr. *dys*, ill	Low base status
Eutric	Gr. *eu*, good; eutrophic, fertile	High base status
Fragic	L. *fragilis*, brittle	Presence of fragic properties
Glacic	Gr. *glacialis*, icy	Presence of ice lenses or wedges
Glossic	Gr. *glossa*, tongue	Tongued horizon boundaries
Grossarenic	L. *grossus*, thick; *arena*, sand	Thick, sandy layer
Gypsic	L. *gypsum*, gypsum	Presence of a gypsic horizon
Halic	Gr. *hals*, salt	Salty
Humic	L. *humus*, earthy	Presence of organic matter
Hydric	Gr. *hydor*, water	Presence of water
Kandic	Modified from kandite	Presence of 1:1 layer silicates
Lamellic	L. *lamella*, dim	Presence of lamellae
Leptic	Gr. *leptos*, thin	A thin soil
Limnic	Gr. *limn*, lake	Presence of a limnic layer
Lithic	Gr. *lithos*, stone	Presence of a shallow lithic contact
Natric	L. *natrium*, sodium	Presence of sodium
Nitric	Modified from *nitron*	Presence of nitrate salts
Ombroaquic	Gr. *ombros*, rain; L. *aqua*, water	Surface wetness
Oxyaquic	Oxy representing oxygen and aquic	Aerated
Pachic	Gr. *pachys*, thick	A thick epipedon
Petrocalcic	Gr. *petra*, rock; L. *calcis*, lime	Presence of a petrocalcic horizon
Petroferric	Gr. *petra*, rock; L. *ferrum*, iron	Presence of a petroferric contact (ironstone)
Petrogypsic	Gr. *petra*, rock; L. *gypsum*, gypsum	Presence of a petrogypsic horizon
Petronodic	Gr. *petra*, rock; L. *nodulus*, a little knot	Presence of concretions and/or nodules
Placic	Gr. *plax*, flat stone	Presence of a thin pan (placic horizon)
Plinthic	Gr. *plinthos*, brick	Presence of plinthite
Rhodic	Gr. *rhodon*, rose	Dark red color
Ruptic[a]	L. *ruptum*, broken	Intermittent or broken horizons
Sodic	Modified from sodium	Presence of sodium
Sombric	F. *sombric*, dark	Presence of sombric horizon
Sulfic	L. *sulfur*, sulfur	Presence of sulfides or their oxidation products
Terric	L. *terra*, earth	A mineral substratum
Thapto(ic)[a]	Gr. *thapto*, buried	A buried soil
Ultic	L. *ultimus*, last	Low base status
Umbric	L. *umbra*, shade	Presence of umbric epipedon
Xanthic	Gr. *xanthos*, yellow	Yellow color

[a] Not strictly an extragrade. Name used to indicate a special departure from the Typic subgroup

The Barx and Highland series (Table 4.1) are two members of the fine-loamy, mixed, superactive, mesic family of Ustic Calciargids. The meaning of each of these terms is defined later, but in a general way the name tells us the following.

Fine-loamy means that from a depth of 25 to 100 cm there is no marked contrast in particle-size class, the content of clay is between 18 and 35%, 15% or more of the material is coarser than 0.1 mm in diameter (fine sand to very coarse sand plus gravel), but less than 35% by volume of the material is rock fragments 2.0 mm or more in diameter (less than about 50% by weight). The average texture, then, is more likely to be loam, clay loam, or sandy clay loam. *Mixed* means a mixed mineralogy that is less than 40% of any one mineral other than quartz in the fraction between 0.02 mm and 2.0 mm in diameter; less than 20% (by weight) glauconitic pellets in the fine earth fraction; a total iron plus gibbsite (by weight) in the fine earth fraction of 5% or less; a fine earth fraction that has at least one of the following: free carbonates, the pH of a suspension of 1 g soil in 50 mL 1 M NaF of 8.4 or less after two minutes, or a ratio of −1,500 kPa water to measured clay of 0.6 or less. *Superactive* means the CEC divided by the clay content (%) is 0.60 or more. *Mesic* indicates a mesic temperature regime, that is, the mean annual soil temperature is between 8 and 15 °C (47 and 59 °F) and the soil temperature fluctuates more than 6 °C between summer and winter. In other words, the soil is somewhere in the midlatitudes, summer is warm or hot, and winter is cool or cold. *Depth of soil* when no class is used, in Ustic Calciargids, means the soil is 50 cm or more deep.

It is not necessary to know the exact criteria in the *Keys to Soil Taxonomy* in order to communicate useful information about a soil. It is necessary to know the language of *Soil Taxonomy*. For example, from Ustic Calciargids one knows the following.

Ustic at the subgroup level of an Aridisol means that the soil is moister than what is typical for an Aridisol or aridic soil moisture regime and close to having an ustic moisture regime. *Calci* means the soil has a calcic horizon within 150 cm of the soil surface. *Argi* denotes presence of an argillic horizon. The *id* means it is an Aridisol. Therefore, one can say that the soil occurs in an arid, temperate climate. Presence of an argillic horizon implies that the soil occurs on stable, older landscape. It has a calcic horizon and under irrigation, Fe chlorosis may be a problem in sensitive plants. If the soil is not irrigated, it can be used only for limited grazing.

4.8 Forming Names

4.8.1 Names of Orders, Suborders, and Great Groups

The names of the orders, suborders, and great groups that are currently recognized are presented in Section E, Chapter 6 under each order.

4.8.1.1 Names of Subgroups

The name of a subgroup consists of the name of a great group modified by one or more adjectives. As explained earlier, the adjective typic is used for the subgroup that is thought to typify the central concept of the great group, or for soils that fail to meet the criteria of the other subgroups defined for a great group.

Intergrade subgroups that have, in addition to the properties of their great group, some properties of another taxon carry the name of the other taxon in the form of an adjective. The names of orders, suborders, or great groups or any of the prior (first) formative elements of those names may be used in the form of an adjective in subgroup names. A few soils may have aberrant properties of two great groups that belong in different orders or suborders. For these, it is necessary to use two names of taxa as adjectives in the subgroup name.

Extragrade subgroups carry the name of one or more special descriptive adjectives (Table 4.5) to modify the name of the great group and connote the nature of the aberrant properties.

4.8.1.2 Names of Intergrades Toward Other Great Groups in the Same Suborder

If the aberrant property of a soil is one that is characteristic of another great group in the same suborder, only the distinctive formative element of the great group name is used to indicate the aberrant properties. Thus, Typic Argidurid is defined, in part, as having an indurated or very strongly cemented duripan. If the only aberrant feature of an Argidic Argidurid is that the duripan is strongly cemented or less cemented throughout, it is considered to intergrade toward the Argids. The name, however, is Argidic Argidurids, not Haplargidic Argidurids. Only the prior (first) formative element is used in adjectival form if the two great groups are in the same suborder.

4.8.1.3 Names of Intergrades Toward a Great Group in the Same Order but Different Suborder

Two kinds of names have been chosen here. If the only aberrant features are color and moisture regime and the hue is too yellow or the chroma is too high or too low for the typic subgroup, the adjectives aeric and aquic are used.

If an Epiaquult has chroma too high for the typic subgroup, but has no other aberrant feature, it is placed in an aeric subgroup. Using an adjective taken from the suborder, udic, would not suggest that the difference is one of aeration alone.

If the only aberrant feature of a Hapludult is redoximorphic features that are too shallow for a Typic Hapludult, the adjective aquic is used in the subgroup name. If redox depletions (accompanied by aquic conditions, unless artificially drained) appear within the upper 60 cm of the argillic horizon, the soil is called an Aquic Hapludult, not an Aquultic Hapludult.

In other instances, the adjective in the subgroup name is made from the first two formative elements of the appropriate great group name in that suborder. For example, if a Paleudult has both shallow redoximorphic features and some plinthite, it is called a Plinthaquic Paleudult, not a Plinthaquultic Paleudult. Note that the formative element for the order is not repeated in the adjective if the two great groups are in the same order.

4.8.1.4 Names of Subgroups not Intergrading Toward Any Known Kind of Soil (Extragrade)

Some soils have aberrant properties that are not characteristic of a class in a higher category of any order, suborder, or great group. One example might be taken from the concave pedons that are at the base of slopes, in depressions, or in other places where new soil material accumulates slowly on the surface. In these soils, material is added to the A horizon. The presence of an overthickened A horizon is not used to define any great group, but the soils lie outside the range of the typic subgroup and there is no class toward which they intergrade. Hence, a descriptive adjective is required. For this particular situation, the adjective cumulic (L. *cumulus*, heap, plus ic, Gr. *ikos*) is used to form the subgroup names. Pachic is used to indicate an overthickened epipedon if there is no evidence of new material at the surface.

Other soils lie outside the range of typic subgroups in an opposite direction. Such soils are, in effect, truncated by hard rock and are shallow or are intermittent between rock outcrops. They are, in effect, intergrades to nonsoil and are called lithic subgroups. The names of formative elements in groups of this sort, which are called extragrades, are listed together with their derivation in Table 4.5.

4.8.2 Names of Families

Each family requires one or more names. The technical family name consists of a series of descriptive terms modifying the subgroup name. For these terms the class names that are given later for particle-size class, mineralogy, and so on, in family differentiae are used. To have consistent nomenclature, the order of descriptive terms in names of families is particle-size class, mineralogy class, cation exchange activity class, calcareous and reaction class, soil temperature class, soil depth class, rupture resistance class, classes of coatings, and classes of cracks.

Redundancy in names of families is avoided. Particle-size class and temperature classes should not be used in the family name if they are specified above the family. Psamments, by definition, all have a sand or loamy sand texture and are in a sandy particle-size class, unless they are ashy. It is, therefore, redundant to use a particle-size class for Psamments, unless they are ashy.

4.9 References

Cline, M.G. 1949. Basic principles of soil classification. Soil Sci. 67:81–91.

Cline, M.G. 1963. Logic of the new system of soil classification. Soil Sci. 96:17–22.

Heller, J.L. 1963. The nomenclature of soils or what's in a name? Soil Sci. Soc. Amer. Proc. 27:216–220.

Smith, G.D. 1963. Objectives and basic assumptions of the new classification system. Soil Sci. 96:6–16.

Smith, G.D. 1983. Historical development of Soil Taxonomy: Background. p. 23–49. *In* L.P. Wilding, N.E. Smeck, and G.F. Hall (ed.) Pedogenesis and *Soil Taxonomy*. I. Concepts and interactions. Elsevier, New York, NY.

Soil Survey Staff. 1975. *Soil Taxonomy*: A basic system of soil classification for making and interpreting soil surveys. USDA Handb. 436. US Government Printing Office, Washington, DC.

Soil Survey Staff. 1998. Keys to Soil Taxonomy. 8th Ed. US Government Printing Office, Washington, DC.

5

Other Systems of Soil Classification

Otto C. Spaargaren
International Soil Reference and Information Centre
Wageningen, The Netherlands

5.1   Introduction

Soil classification is probably as old as farming. The fact that, around 8,000 BP, the first farming communities in Europe settled on the better loess soils indicates that, during these times, farmers were already capable of distinguishing between the more and less productive soils. The oldest historical record of soil classification is most likely the Chinese book, *Yugong*, in which the soils of China were classified into three categories and nine classes, based on soil color, texture and hydrological features (Gong, 1994). Even today, such criteria are still in use by farmers to differentiate soils. Studies on indigenous soil knowledge in northern Ghana, for example, have shown that farmers use texture, color, stoniness and soil depth to stratify the soils (Asiamah et al., 1997).

This chapter will describe the Legend of the Soil Map of the World (FAO-UNESCO, 1974), its revised version (FAO, 1988), and the World Reference Base for Soil Resources (ISSS-ISRIC-FAO, 1994; 1998) as international systems of soil classification; the French system of soil classification (CPCS, 1967), which is still the system in use in large parts of West and Central Africa; the new *Référentiel Pédologique Français* (AFES, 1990; 1992), the system of soil classification used in the former USSR (VASKhNIL, 1986); the recently developed soil classification system of China (Gong, 1994); and the newly developed system of soil classification for Australia (Isbell, 1996) as classification systems which are covering a range of ecological regions (from the arctic to the (sub)tropics). In addition, some attention will be paid to the soil classification systems in use in Brazil, Canada, England and Wales, New Zealand and South Africa, which are more oriented towards one ecological region.

5.2   The FAO-UNESCO Legend of the Soil Map of the World 1:5,000,000

5.2.1   Introduction

The Soil Map of the World is a response to recommendations made by the international soils community in the 1950s to give special attention to developing the classification and correlation of the soils of great regions of the world (FAO-UNESCO, 1974). It was the first attempt to prepare, on the basis of international cooperation, a soil map covering all continents of the world using a uniform legend, thus enabling the correlation of soil units and the comparison of soils on a global scale. It should be borne in mind that the Soil Map of the World project produced a legend accompanying the map and not a global soil classification system.

5.2.2 History

The Food and Agriculture Organization (FAO) and United Nations Educational, Scientific, and Cultural Organization (UNESCO), in association with the International Society of Soil Science (ISSS), jointly took up the recommendations made during the Sixth and Seventh ISSS Congresses in 1956 and 1960 to prepare a soil map of the world at scale 1:5,000,000. The project started in 1961 and was based on the compilation of available soil survey material and field correlation. A scientific advisory panel was convened to study the scientific and methodological problems relative to the preparation of such a soil map of the world.

During a number of meetings, the advisory panel worked out the organization of the field correlation, selected the scale of the map and its topographic base, and prepared the first draft definitions of soil units. These were presented to the Eighth ISSS Congress in 1964. In 1966, a general agreement was reached on the principles for constructing the international legend, on the preparation of the definitions of soil units, and on the adoption of a unified nomenclature. The first draft was presented in 1968 to the Ninth ISSS Congress, which approved the outline of the legend, the definitions and the nomenclature. Moreover, it recommended that the Soil Map of the World be published as soon as possible.

5.2.3 Objectives

The objectives of the Soil Map of the World project (FAO-UNESCO, 1974) were to: (1) make a first appraisal of the world's soil resources, (2) supply a scientific basis for the transfer of experience between areas with similar environments, (3) promote the establishment of a generally accepted soil classification and nomenclature, (4) establish a common framework for more detailed investigations in developing areas, (5) serve as a basic document for educational, research, and development activities, and (6) strengthen international contacts in the field of soil science.

5.2.4 The Soil Units

The soil units which form the basis of the FAO-UNESCO Legend have been defined in terms of measurable and observable properties of the soil itself. They form a monocategorical and not a taxonomic system with different levels of generalization. However, for the sake of logical presentation, they can be grouped on generally accepted principles of soil formation (FAO-UNESCO, 1974). Based on soil development status, material, and major geographical zone, 24 major soils and 106 soil units are distinguished (Table 5.1).

The soil units are characterized by the presence or absence of diagnostic horizons and properties. Key properties have been selected on the basis of generally accepted principles of soil formation so as to correlate with as many other characteristics as possible. Clusters of properties have been combined into so-called diagnostic horizons which have been adopted to formulate the definitions of the soil units. The definitions and nomenclature of the diagnostic horizons and properties used are drawn from those adopted in Soil Taxonomy (Soil Survey Staff, 1975), but the definitions have been summarized, and sometimes, simplified to serve the purpose of the legend. For full background and details, the user is referred to Soil Taxonomy (Soil Survey Staff, 1975, 1996; Section E, Chapter 4, this volume). A brief overview of the diagnostic horizons and properties used in the FAO-UNESCO Legend and their meaning are given in Table 5.2.

Table 5.1 List of 106 soil units of the FAO-UNESCO Legend of the Soil Map of the World 1:5,000,000 [FAO-UNESCO, 1974]

Non or weakly developed soils

J	**Fluvisols**	**G**	**Gleysols**	**R**	**Regosols**	**I**	**Lithosols**
Je	Eutric Fluvisols	Ge	Eutric Gleysols	Re	Eutric Regosols		
Jc	Calcaric Fluvisols	Gc	Calcaric Gleysols	RC	Calcaric Regosols		
Jd	Dystric Fluvisols	Gd	Dystric Gleysols	Rd	Dystric Regosols		
Jt	Thionic Fluvisols	Gm	Mollic Gleysols	Rx	Gelic Regosols		
		Gh	Humic Gleysols				
		Gp	Plinthic Gleysols				
		Gx	Gelic Gleysols				

Soils conditioned by parent material

Q	**Arenosols**	**E**	**Rendzinas**	**U**	**Rankers**	**T**	**Andosols**	**V**	**Vertisols**
Qc	Cambic Arenosols					To	Ochric Andosols	Vp	Pellic Vertisols
Ql	Luvic Arenosols					Tm	Mollic Andosols	Vc	Chromic Vertisols
Qf	Ferralic Arenosols					Th	Humic Andosols		
Qa	Albic Arenosols					Tv	Vitric Andosols		

Soils from (semi-)arid regions

Z	**Solonchaks**	**S**	**Solonetz**	**Y**	**Yermosols**	**X**	**Xerosols**
Zo	Orthic Solonchaks	So	Orthic Solonetz	Yh	Haplic Yermosols	Xh	Haplic Xerosols
Zm	Mollic Solonchaks	Sm	Mollic Solonetz	Yk	Calcic Yermosols	Xk	Calcic Xerosols
Zt	Takyric Solonchaks	Sg	Gleyic Solonetz	Yy	Gypsic Yermosols	Xy	Gypsic Xerosols
Zg	Gleyic Solonchaks			Yl	Luvic Yermosols	Xl	Luvic Xerosols
				Yt	Takyric Yermosols		

Soils from the steppe regions

K	**Kastanozems**	**C**	**Chernozems**	**H**	**Phaeozems**	**M**	**Greyzems**
Kh	Haplic Kastanozems	Ch	Haplic Chernozems	Hh	Haplic Phaeozems	Mo	Orthic Greyzems
Kk	Calcic Kastanozems	Ck	Calcic Chernozems	Hc	Calcaric Phaeozems	Mg	Gleyic Greyzems
Kl	Luvic Kastanozems	Cl	Luvic Chernozems	Hl	Luvic Phaeozems		
		Cg	Glossic Chernozems	Hg	Gleyic Phaeozems		

Moderately developed soils mainly from temperate regions

B	**Cambisols**	**L**	**Luvisols**	**D**	**Podzoluvisols**	**P**	**Podzols**	**W**	**Planosols**
Be	Eutric Cambisols	Lo	Orthic Luvisols	De	Eutric Podzoluvisols	Po	Orthic Podzols	We	Eutric Planosols
Bd	Dystric Cambisols	Lc	Chromic Luvisols	Dd	Dystric Podzoluvisols	Pl	Leptic Podzols	Wd	Dystric Planosols
Bh	Humic Cambisols	Lk	Calcic Luvisols	Dg	Gleyic Podzoluvisols	Pf	Ferric Podzols		
Bg	Gleyic Cambisols	Lv	Vertic Luvisols			Ph	Humic Podzols	Wm	Mollic Planosols
Bx	Gelic Cambisols	Lf	Ferric Luvisols			Pp	Placic Podzols	Wh	Humic Planosols
Bk	Calcic Cambisols	La	Albic Luvisols			Pg	Gleyic Podzols	Ws	Solodic Planosols
Bc	Chromic Cambisols	Lp	Plinthic Luvisols						
Bv	Vertic Cambisols	Lg	Gleyic Luvisols					Wx	Gelic Planosols
Bf	Ferralic Cambisols								

Strongly weathered soils mainly from the tropical regions

A	**Acrisols**	**N**	**Nitosols**	**F**	**Ferralsols**
Ao	Orthic Acrisols	Ne	Eutric Nitosols	Fo	Orthic Ferralsols
Af	Ferric Acrisols	Nd	Dystric Nitosols	Fx	Xanthic Ferralsols
Ah	Humic Acrisols	Nh	Humic Nitosols	Fr	Rhodic Ferralsols
Ap	Plinthic Acrisols			Fh	Humic Ferralsols
Ag	Gleyic Acrisols			Fa	Acric Ferralsols
				Fp	Plinthic Ferralsols

Organic soils

O	**Histosols**
Oe	Eutric Histosols
Od	Dystric Histosols
Ox	Gelic Histosols

Table 5.2 Diagnostic horizons and properties in the FAO-UNESCO Legend of the Soil Map of the World 1:5,000,000 [FAO-UNESCO, 1974)]

Horizons

Albic E	light colored eluvial horizon generally associated with argic and spodic horizons
Argillic B	subsurface horizon with distinct clay accumulation
Calcic	horizon with accumulation of calcium carbonate
Cambic B	subsurface horizon showing evidence of alteration relative to the underlying horizon(s)
Gypsic	horizon with accumulation of gypsum
Histic H	poorly aerated, waterlogged, highly organic surface horizon
Mollic A	thick, dark colored surface horizon with high base saturation and moderate to high organic matter content
Natric B	subsurface horizon with distinct clay accumulation and a high exchangeable sodium percentage
Ochric A	weakly developed surface horizon, either light colored, or thin, or having a low organic matter content
Oxic B	strongly weathered subsurface horizon with low cation exchange capacity
Spodic B	dark colored subsurface horizon with illuvial alumino-organic complexes, with or without iron
Sulfuric	extremely acid subsurface horizon with sulfuric acid resulting from oxidation of sulfides
Umbric A	thick, dark colored surface horizon with low base saturation and moderate to high organic matter content

Properties and materials

Abrupt textural change	sharp increase in clay content within a limited depth range
Albic material	light colored mineral soil material
Aridic moisture regime	no available water in any part of the moisture control section for as long as 90 consecutive days, or more than half the time when soil temperature (at 50 cm depth) is above 5 °C
Exchange complex	CEC (pH 8.2) more than 150 $cmol_c$ kg^{-1} clay; if 15-bar water content is 20% or more, pH NaF is more than 9.4; ratio 15-bar water content to clay more than 1.0; more than 0.6% organic carbon; DTA shows low temperature endotherm; bulk density is 0.85 g cm^{-3} at 1/3 bar tension
Ferralic	low cation exchange capacity (<24 $cmol_c$ kg^{-1} clay by NH_4Cl)
Ferric	iron concentrated in large mottles or concretions, or low cation exchange capacity (<24 $cmol_c$ kg^{-1} clay by NH_4Cl)
Gilgai microrelief	succession of enclosed microbasins and microknolls in level areas, or microvalleys and microridges on slopes
High organic matter content	organic matter content of 1.35% or more averaged to a depth of 100 cm, or 1.5% organic matter in the upper part of the B horizon (Acrisols only)
High salinity	EC of saturation extract more than 15 dS m^{-1} at 25 °C at specified depths, or more than 4 dS m^{-1} within 25 cm of the surface if pH (H_2O, 1:1) exceeds 8.5
Hydromorphic properties	saturation with groundwater; occurrence of a histic H horizon; dominant neutral (N) hues or hues bluer than 10Y; saturation with water at some period of the year (unless artificially drained) with evidence of reduction or of reduction and segregation
Interfingering	penetrations of an albic E horizon into an underlying argillic or natric B horizon, not wide enough to constitute tonguing
Permafrost	perennial temperature at or below 0 °C
Plinthite	iron-rich, humus-poor soil material which hardens irreversibly upon repeated wetting and drying
Slickensides	polished and grooved ped surfaces that are produced by sliding past another
Smeary consistence	presence of thixotropic soil material
Soft powdery lime	accumulation of translocated calcium carbonate in soft powdery form
Sulfidic materials	waterlogged deposit containing 0.75% or more sulfur, and less than three times as much carbonates ($CaCO_3$ equivalent) as sulfur
Takyric features	combination of heavy texture, polygonal cracks and platy or massive surface crust
Thin iron pan	black to dark reddish layer cemented by iron, by iron and manganese, or by iron-organic matter complexes
Tonguing	penetrations of an albic E horizon into an argillic B horizon with specified dimensions
Vertic properties	cracks 1 cm or more wide within 50 cm of the upper boundary of the B horizon, extending to the surface or at least to the upper part of the B horizon
Weatherable minerals	presence of minerals considered to be unstable relative to other minerals such as quartz and 1:1 lattice clays, and which produce plant nutrients upon weathering

Volume I of the Legend of the Soil Map of the World provides a key to the soil units which can be used to identify the soil (FAO-UNESCO, 1974). An abbreviated version listing the major soil units is reproduced in Table 5.3.

5.3 The Revised Legend of the FAO-UNESCO Soil Map of the World

5.3.1 Introduction

In 1988, a revised version of the Legend of the Soil Map of the World was issued (FAO, 1988) which assessed to what extent the objectives of the original Legend of the Soil Map of the World (FAO-UNESCO, 1974; Section 5.2.3) were met, and analyzed its present day function. It was realized that, in order to keep the maps and accompanying legend up to date, revisions were necessary. New knowledge on soils, particularly from the developing world, had emerged and more recent soil surveys had yielded better insight into the distribution of soils in the world. The Revised Legend did not replace the 1974 Legend, which continues to serve as reference for the Soil Map of the World. The Revised Legend is used for updating the Soil Map of the World, contained in the UNEP-ISSS-ISRIC-FAO sponsored programme of SOTER (Section H, Chapter 1), and served as framework for the establishment by the ISSS of the World Reference Base for Soil Resources (Section 5.4).

5.3.2 Amendments to the 1974 Legend of the Soil Map of the World

The monocategorial character of the 1974 Legend, using only soil units, was transformed into a multicategorial system with Major Soil Groupings (MSG) (e.g., Fluvisols), soil units (e.g., Dystric Fluvisols), and soil subunits (e.g., Gleyi-dystric Fluvisols) (FAO, 1988). The soil subunit level is described in more detail by Nachtergaele et al. (1994), providing guidelines for distinguishing soil subunits.

The introduction of a multicategorical system was necessitated by the increasing use of the 1974 Legend in more detailed surveys, particularly in Africa, where during the 1980s 1:1,000,000 soil maps were produced of, for example, Kenya, Botswana and Zambia.

The Revised Legend of the Soil Map of the World distinguishes 28 MSGs, four more than the 1974 Legend, and 153 soil units, 47 more than in 1974. Major changes are (1) amalgamation of the Lithosols, Rendzinas and Rankers into one major soil grouping (Leptosols), since the three had been difficult to show on the map; (2) deletion of the Xerosols and Yermosols, which were characterized by an aridic moisture regime (as a general principle, climatic criteria were not to be used in separating the soil units); (3) division of the Acrisols and Luvisols, in 1974 distinguished by base saturation, into four MSGs, introducing clay activity as an additional separating criterion; (4) introduction of new MSGs, (Alisols, Calcisols, Gypsisols, Lixisols, Plinthosols and Anthrosols); and (5) renaming of Nitosols (Nitisols). A listing of the MSGs and the soil units is given in Table 5.4.

The diagnostic horizons and properties have also been largely adapted. In 1974, they were fairly similar to those in Soil Taxonomy (Soil Survey Staff, 1975), but in 1988, many were redefined and renamed, and new additions made. The argillic and oxic B horizons were renamed argic and ferralic B horizons, respectively. The argic B horizon now includes both the argillic and the kandic horizon as defined in Soil Taxonomy (Soil Survey Staff, 1996), while the ferralic B horizon includes additional criteria such as silt/clay ratio and water dispersible clay content. An addition is the thick, manmade, fimic A horizon which includes both the anthropic and plaggen epipedons of Soil Taxonomy.

Newly defined diagnostic properties include continuous hard rock, fluvic, geric, nitic, and sodic properties. The 1974 hydromorphic properties were split into gleyic (wetness conditioned by groundwater) and stagnic (wetness conditioned by surface water) properties. Definitions of albic material,

Table 5.3   Key to major soil units of the 1974 FAO-UNESCO Legend [FAO-UNESCO, 1974)]

Soils having an H horizon of 40 cm or more (60 cm or more if the organic material consists mainly of sphagnum or moss or has a bulk density of less than 0.1 g cm^{-3}) either extending down from the surface or taken cumulatively within the upper 80 cm of the soil; the thickness of the H horizon may be less when it rests on rock or on fragmental material of which the interstices are filled with organic matter.

HISTOSOLS (O)

Other soils which are limited in depth by continuous coherent and hard rock within 10 cm of the surface.

LITHOSOLS (I)

Other soils which, after the upper 20 cm have been mixed, have 30 percent or more clay in all horizons to at least 50 cm from the surface; at some period in most years have cracks at least 1 cm wide at a depth of 50 cm, unless irrigated, and have one or more of the following characteristics: gilgai microrelief, intersecting slickensides or wedge-shaped or parallelepiped structural aggregates at some depth between 25 and 100 cm from the surface.

VERTISOLS (V)

Other soils developed from recent alluvial deposits, having no diagnostic horizons other than (unless buried by 50 cm or more new material) an ochric or an umbric A horizon, a histic H horizon, or a sulfuric horizon.

FLUVISOLS (J)

Other soils having high salinity and having no diagnostic horizons (unless buried by 50 cm or more new material), an A horizon, an H horizon, a cambic B horizon, a calcic or a gypsic horizon.

SOLONCHAKS (Z)

Other soils showing hydromorphic properties within 50 cm of the surface; having no diagnostic horizons other than (unless buried by 50 cm or more new material) an A horizon, an H horizon, a cambic B horizon, a calcic or gypsic horizon.

GLEYSOLS (G)

Other soils having either a mollic or an umbric A horizon possibly overlying a cambic B horizon, or an ochric A horizon and a cambic B horizon; having no other diagnostic horizons (unless buried by 50 cm or more new material); having to a depth of 35 cm or more one or both of: (a) a bulk density (at 1/3 bar water retention) of the fine earth (less than 2 mm) fraction of the soil of less than 0.85 g cm^{-3} and an exchange complex dominated by amorphous material; (b) 60 percent or more vitric volcanic ash, cinders, or other vitric pyroclastic material in the silt, sand and gravel fractions.

ANDOSOLS (T)

Other soils of coarse texture consisting of albic material occurring over a depth of at least 50 cm from the surface, or showing characteristics of argillic, cambic or oxic B horizons which, however, do not qualify as diagnostic horizons because of textural requirements; having no diagnostic horizons other than (unless buried by 50 cm or more new material) an ochric A horizon.

ARENOSOLS (Q)

Other soils having an umbric A horizon, which is not more than 25 cm thick; having no other diagnostic horizons (unless buried by 50 cm or more new material).

RANKERS (U)

Other soils having a mollic A horizon which contains or immediately overlies calcareous material with a calcium carbonate equivalent of more than 40 percent (when the A horizon contains a high amount of finely divided calcium carbonate, the color requirements of the mollic A horizon may be waived).

RENDZINAS (E)

Other soils having a spodic B horizon.

PODZOLS (P)

Other soils having an oxic B horizon.

FERRALSOLS (F)

Other soils having an albic E horizon overlying a slowly permeable horizon (for example, an argillic or natric B horizon showing an abrupt textural change, a heavy clay, a fragipan) within 125 cm of the surface; showing hydromorphic properties at least in a part of the E horizon.

PLANOSOLS (W)

Other soils having a natric B horizon.

SOLONETZ (S)

Other soils having a mollic A horizon with a moist chroma of 2 or less to a depth of at least 15 cm, showing bleached coatings on structural ped surfaces.

GREYZEMS (M)

Other soils having a mollic A horizon with a moist chroma of 2 or less to a depth of at least 15 cm; having one or more of the following: a calcic or a gypsic horizon, or concentrations of soft powdery lime within 125 cm of the surface when the weighted average textural class is coarse, within 90 cm for medium textures, within 75 cm for fine textures.

CHERNOZEMS (C)

Other soils having a mollic A horizon with a moist chroma of more than 2 to a depth of at least 15 cm; having one or more of the following: a calcic or a gypsic horizon, or concentrations of soft powdery lime within 125 cm of the surface when the weighted average textural class is coarse, within 90 cm for medium textures, within 75 cm for fine textures.

KASTANOZEMS (K)

Other soils having a mollic A horizon.

PHAEOZEMS (H)

Other soils having an argillic B horizon showing an irregular or broken upper boundary resulting from deep tonguing of the E into the B horizon or from the formation of discrete nodules (ranging from 2 to 5 cm up to 30 cm in diameter) the exteriors of which are enriched and weakly cemented or indurated with iron and having redder hues and stronger chroma than the interiors.

PODZOLUVISOLS (D)

Table 5.3 Cont.

Other soils having no diagnostic horizons or none other than (unless buried by 50 cm or more new material) an ochric A horizon. **REGOSOLS (R)**	Other soils having a weak ochric A horizon and an aridic moisture regime; lacking permafrost within 200 cm of the surface. **XEROSOLS (X)**
Other soils having a very weak ochric A horizon and an aridic moisture regime; lacking permafrost within 200 cm of the surface **YERMOSOLS (Y)**	Other soils having an argillic B horizon; having a base saturation which is less than 50 percent (by NH_4OAc) in at least some part of the B horizon within 125 cm of the surface. **ACRISOLS (A)**
Other soils having an argillic B horizon with a clay distribution where the percentage of clay does not decrease from its maximum amount by as much as 20 percent within 150 cm of the surface; lacking plinthite within 125 cm of the surface; lacking vertic and ferric properties **NITOSOLS (N)**	Other soils having an argillic B horizon. **LUVISOLS (L)** Other soils having a cambic B horizon or an umbric A horizon which is more than 25 cm thick. **CAMBISOLS (CM)**

aridic moisture regime, high organic matter content and thin iron pan were deleted as they are no longer used in defining soil units. Andic properties have replaced the 1974 exchange complex dominated by amorphous material.

A brief overview of the diagnostic horizons and properties used in the Revised Legend and their meaning is given in Table 5.5. A key to the Major Soil Groupings in the Revised Legend is given in Table 5.6.

5.4 The World Reference Base for Soil Resources

5.4.1 Introduction

The World Reference Base for Soil Resources (WRB) was initiated by the ISSS, FAO, and ISRIC to "... provide scientific depth and background to the 1988 FAO-UNESCO-ISRIC Revised Legend of the Soil Map of the World, so that it incorporates the latest knowledge relating to global soil resources and interrelationships." (ISSS-ISRIC-FAO, 1994).

5.4.2 History

The history of the initiative is closely related to the Soil Map of the World 1:5,000,000 project of FAO and UNESCO. After its completion in the early 1980s, it was realized that 20 years had elapsed since the project had started. During this period, numerous new soil surveys, often using local or national soil classification systems, had been carried out both in developing and developed countries, generating new knowledge and insight on the distribution and potential of our soil resources. If the Soil Map of the World were to retain its value as a global soil resource inventory, it needed regular updating with the most recent information.

Preliminary discussions were started on the establishment of an International Reference Base for Soil Classification (IRB), an initiative undertaken by FAO and UNESCO, supported by the United Nations Environmental Program (UNEP) and the ISSS. The intention of the IRB project was to work toward the establishment of a framework through which existing soil classification systems could be correlated and ongoing soil classification work could be harmonized (Dudal, 1990). Meetings were organized in Sofia, Bulgaria, to commence such an international program. The outcomes were draft definitions of 16 major soil groups occurring globally.

Table 5.4 Major soil groupings (MSG) and soil units, 1988 Revised Legend of the Soil Map of the World [FAO, 1988]

FL	**FLUVISOLS**	**AR**	**ARENOSOLS**	**CM**	**CAMBISOLS**	**CL**	**CALCISOLS**
FLe	Eutric Fluvisols	ARh	Haplic Arenosols	CMe	Eutric Cambisols	CLh	Haplic Calcisols
FLc	Calcaric Fluvisols	ARb	Cambic Arenosols	CMd	Dystric Cambisols	CLc	Luvic Calcisols
FLd	Dystric Fluvisols	ARl	Luvic Arenosols	CMu	Humic Cambisols	CLp	Petric Calcisols
FLm	Mollic Fluvisols	ARo	Ferralic Arenosols	CMc	Calcaric Cambisols		
FLu	Umbric Fluvisols	ARa	Albic Arenosols	CMv	Vertic Cambisols		
FLt	Thionic Fluvisols	ARc	Calcaric Arenosols	CMo	Ferralic Cambisols		
FLs	Salic Fluvisols	ARg	Gleyic Arenosols	CMg	Gleyic Cambisols		
				CMi	Gelic Cambisols		
GL	**GLEYSOLS**	**AN**	**ANDOSOLS**			**GY**	**GYPSISOLS**
GLe	Eutric Gleysols	ANh	Haplic Andosols			GYh	Haplic Gypsisols
GLk	Calcic Gleysols	ANm	Mollic Andosols			GYk	Calcic Gypsisols
GLd	Dystric Gleysols	ANu	Umbric Andosols			GYl	Luvic Gypsisols
GLa	Andic Gleysols	ANz	Vitric Andosols			GYp	Petric Gyopsisols
GLm	Mollic Gleysols	ANg	Gleyic Andosols				
GLu	Umbric Gleysols	ANi	Gelic Andosols				
GLt	Thionic Gleysols						
GLi	Gelic Gleysols						
RG	**REGOSOLS**	**VR**	**VERTISOLS**			**SN**	**SOLONETZ**
RGe	Eutric Regosols	VRe	Eutric Vertisols			SNh	Haplic Solonetz
RGc	Calcaric Regosols	VRd	Dystric Vertisols			SNm	Mollic Solonetz
RGy	Gypsiric Regosols	Vrk	Calcic Vertisols			SNk	Calcic Solonetz
RGd	Dystric Regosols	VRy	Gypsic Vertisols			SNy	Gypsic Solonetz
RGu	Umbric Regosols					SNj	Stagnic Solonetz
RGi	Gelic Regosols					SNg	Gleyic Solonetz
LP	**LEPTOSOLS**					**SC**	**SOLONCHAKS**
LPe	Eutric Leptosols					SCh	Haplic Solonchaks
LPd	Dystric Leptosols					SCm	Mollic Solonchak
LPk	Rendzic Leptosols					SCk	Calcic Solonchaks
LPm	Mollic Leptosols					SCy	Gypsic Solonchaks
LPu	Umbric Leptosols					SCn	Sodic Solonchaks
LPq	Lithic Leptosols					SCg	Gleyic Solonchaks
LPi	Gelic Leptosols					SCi	Gelic Solonchaks
KS	**KASTANOZEMS**	**LV**	**LUVISOLS**	**LX**	**LIXISOLS**	**HS**	**HISTOSOLS**
KSh	Haplic Kastanozems	LVh	Haplic Luvisols	LXh	Haplic Lixisols	HSl	Folic Histosols
KSl	Luvic Kastanozems	LVf	Ferric Luvisols	LXf	Ferric Lixisols	HSs	Terric Histosols
KSk	Calcic Kastanozems	LVx	Chromic Luvisols	LXp	Plinthic Lixisols	HSf	Fibric Histosols
KSy	Gypsic Kastanozems	LVk	Calcic Luvisols	LXa	Albic Lixisols	HSt	Thionic Histosols
		LVv	Vertic Luvisols	LXj	Stagnic Lixisols	HSi	Gelic
		LVa	Albic Luvisols	LXg	Gleyic Lixisols		
		LVj	Stagnic Luvisols				
		LVg	Gleyic Luvisols				
CH	**CHERNOZEMS**	**PL**	**PLANOSOLS**	**AC**	**ACRISOLS**	**AT**	**ANTHROSOLS**
CHh	Haplic Chernozems	PLe	Eutric Planosols	ACh	Haplic Acrisols	ATa	Aric Anthrosols
CHk	Calcic Chernozems	PLd	Dystric Planosols	ACf	Ferric Acrisols	ATc	Cumulic Anthrosols
CHl	Luvic Chernozems	PLm	Mollic Planosols	ACu	Humic Acrisols	ATf	Fimic Anthrosols
CHw	Glossic Chernozems	PLu	Umbric Planosols	ACp	Plinthic Acrisols	ATu	Urbic Anthrosols
CHg	Gleyic Chernozems	PLi	Gelic Planosols	ACg	Gleyic Acrisols		
PH	**PHAEOZEMS**	**PD**	**PODZOLUVISOLS**	**AL**	**ALISOLS**		
PHh	Haplic Phaeozems	PDe	Eutric Podzoluvisols	ALh	Haplic Alisols		
PHc	Calcaric Phaeozems	PLd	Dystric Podzoluvisols	ALf	Ferric Alisols		
PHl	Luvic Phaeozems	PLj	Stagnic Podzoluvisols	ALu	Humic Alisols		
PHj	Stagnic Phaeozems	PDg	Gleyic Podzoluvisols	ALp	Plinthic Alisols		
PHg	Gleyic Phaeozems	PDi	Gelic Podzoluvisols	ALj	Stagnic Alisols		
				ALg	Gleyic Alisols		
GR	**GREYZEMS**	**PZ**	**PODZOLS**	**NT**	**NITISOLS**		
GRh	Haplic Greyzems	PZh	Haplic Podzols	NTh	Haplic Nitisols		
GRg	Gleyic Greyzems	PZb	Cambic Podzols	ATr	Rhodic Nitisols		
		PZf	Ferric Podzols	NTu	Humic Nitisols		
		PZc	Carbic Podzols				
		PZg	Gleyic Podzols				
		PZi	Gelic Podzols				
				FR	**FERRALSOLS**		
				FRh	Haplic Ferralsols		
				FRx	Xanthic Ferralsols		
				FRr	Rhodic Ferralsols		
				FRu	Humic Ferralsols		
				FRg	Geric Ferralsols		
				FRp	Plinthic Ferralsols		
				PT	**PLINTHOSOLS**		
				PTe	Eutric Plinthosols		
				PTd	Dystric Plinthosols		
				PTu	Humic Plinthosols		
				PTa	Albic Plinthosols		

Table 5.5 Diagnostic horizons and properties, 1988 Revised Legend of the Soil Map of the World [FAO, 1988]

Horizons

Albic E	light colored eluvial horizon generally associated with argic and spodic horizons
Argic B	subsurface horizon with distinct clay accumulation
Calcic	horizon with accumulation of calcium carbonate
Cambic B	subsurface horizon showing evidence of alteration relative to the underlying horizon(s)
Ferralic B	strongly weathered subsurface horizon with low cation exchange capacity
Fimic A	surface and subsurface horizons resulting from long continued cultivation
Gypsic	horizon with accumulation of gypsum
Histic H	poorly aerated, waterlogged, highly organic surface horizon
Mollic A	thick, dark colored surface horizon with high base saturation and moderate to high organic matter content
Natric B	subsurface horizon with distinct clay accumulation and a high exchangeable sodium percentage
Ochric A	weakly developed surface horizon, either light colored, or thin, or having a low organic matter content
Petrocalcic	continuous cemented or indurated calcic horizon
Petrogypsic	continuous cemented or indurated gypsic horizon
Spodic B	dark colored subsurface horizon with illuvial alumino-organic complexes, with or without iron
Sulfuric	extremely acid subsurface horizon with sulfuric acid resulting from oxidation of sulfides
Umbric A	thick, dark colored surface horizon with low base saturation and moderate to high organic matter content

Properties

Abrupt textural change	sharp increase in clay content within a limited depth range
Andic	high acid oxalate extractable Al and Fe content, low bulk density, high phosphate retention, high amount of coarse volcanoclastic material
Calcareous	strong effervescence with 10% HCl (more than 2% calcium carbonate)
Calcaric	presence of calcareous soil material between 20 and 50 cm depth
Continuous hard rock	presence of coherent rock, practically impermeable for roots
Ferralic	low cation exchange capacity (< 24 $cmol_c$ kg^{-1} clay or 4 $cmol_c$ kg^{-1} fine earth)
Ferric	presence of many coarse mottles or large iron concretions
Fluvic	presence of fresh fluviatile, lacustrine or marine sediments at the surface
Geric	extremely low to negative effective cation exchange capacity
Gleyic	wetness producing reduced conditions caused by groundwater
Gypsiferous	presence $\geq$ 5% gypsum
Interfingering	narrow penetrations of an albic E horizon into an argic or natric B horizon
Nitic	presence of strongly developed, nut-shaped structure and shiny pedfaces
Organic material	material containing a very high amount of organic debris
Permafrost	perennial temperature at or below 0 °C
Plinthite	presence of iron-rich, humus-poor material which hardens irreversibly upon repeated wetting and drying
Salic	high soluble salt content (electrical conductivity > 15 dS m^{-1}, or > 4 if pH exceeds 8.5)
Slickensides	presence of polished and grooved surfaces produced by one mass sliding past another
Smeary consistence	presence of thixotropic soil material
Sodic	exchangeable sodium percentage $\geq$ 15%, or exchangeable Na + Mg $\geq$ 50%
Soft powdery lime	accumulation of translocated calcium carbonate in soft powdery form
Stagnic	wetness producing reduced conditions caused by stagnating surface water
Strongly humic	high organic matter content
Sulfidic material	waterlogged deposit containing sulfides, and only moderate amounts of calcium carbonate
Tonguing	wide penetrations of an albic E horizon into an argic B horizon, or penetrations of a mollic A horizon into an underlying cambic B horizon or into a C horizon (Chernozems only)
Vertic	presence of cracks, slickensides, wedgeshaped or parallel piped structural aggregates
Weatherable minerals	presence of minerals unstable in a humid climate relative to other minerals

Table 5.6   Key to major soil groupings of the 1988 Revised Legend of the Soil Map of the World [FAO, 1988)]

Soils having an H horizon, or an O horizon, of 40 cm or more (60 cm or more if the organic material consists mainly of sphagnum or moss or has a bulk density of less than 0.1 Mg m^{-3}) either extending down from the surface or taken cumulatively within the upper 80 cm of the soil; the thickness of the H or O horizon may be less when it rests on rocks or on fragmental material of which the interstices are filled with organic matter.

HISTOSOLS (HS)

Other soils in which human activities have resulted in a profound modification or burial of the original soil horizons, through removal or disturbance of surface horizons, cuts and fills, secular additions of organic materials, long-continued irrigation, etc.

ANTHROSOLS (AT)

Other soils, which are limited in depth by continuous hard rock or highly calcareous materials (calcium carbonate equivalent > 40%) or a continuous cemented layer within 30 cm of the surface or having < 20% of fine earth over a depth of 75 cm from the surface. Diagnostic horizons may be present.

LEPTOSOLS (LP)

Other soils having, after the upper 18 cm have been mixed, 30 percent or more clay in all horizons to a depth of 50 cm; developing cracks from the soil surface downward which at some period in most years (unless the soil is irrigated) are at least 1 cm wide to a depth of 50 cm; having one or more of the following: intersecting slickensides or wedge-shaped or parallelepiped structural aggregates at some depth between 25 and 100 cm from the surface.

VERTISOLS (VR)

Other soils showing fluvic properties and having no diagnostic horizons other than an ochric, mollic, an umbric A horizon, or a histic H horizon, or a sulfuric horizon, or sulfidic material within 125 cm of the surface.

FLUVISOLS (FL)

Other soils showing salic properties and having no diagnostic horizons other than an ochric, umbric or mollic A horizon, a histic H horizon, a cambic B horizon, a calcic or a gypsic horizon.

SOLONCHAKS (SC)

Other soils, exclusive of coarse textured materials (except when a histic H horizon is present), showing gleyic properties within 50 cm of the surface; having no diagnostic horizons other than an A horizon, a histic H horizon, a cambic B horizon, a sulfuric horizon, a calcic or a gypsic horizon; lacking plinthite within 125 cm of the surface.

GLEYSOLS (GL)

Other soils showing andic properties to a depth of 35 cm or more from the surface and having a mollic or an umbric A horizon possibly overlying a cambic B horizon, or an ochric A horizon and a cambic B horizon; having no other diagnostic horizons.

ANDOSOLS (AN)

Other soils which are coarser than sandy loam to a depth of at least 100 cm from the surface, having < 35% of rock fragments or other coarse fragments in all subhorizons within 100 cm of the surface, having no diagnostic horizons other than an ochric A horizon or an albic E horizon.

ARENOSOLS (AR)

Other soils having no diagnostic horizons other than an ochric or umbric A horizon; lacking soft powdery lime.

REGOSOLS (RG)

Other soils having a spodic B horizon.

PODZOLS (PZ)

Other soils having ≥ 25% plinthite by volume in a horizon which is at least 15 cm thick within 50 cm of the surface or within a depth of 125 cm when underlying an albic E horizon or a horizon which shows stagnic properties within 50 cm of the surface or gleyic properties within 100 cm of the surface.

PLINTHOSOLS (PT)

Other soils having a ferralic B horizon.

FERRALSOLS (FR)

Other soils having an E horizon showing stagnic properties at least in part of the horizon and abruptly overlying a slowly permeable horizon within 125 cm of the surface, and lacking a natric or a spodic B horizon.

PLANOSOLS (PL)

Other soils having a natric B horizon.

SOLONETZ (SN)

Other soils having a mollic A horizon with a moist chroma of 2 or less to a depth of at least 15 cm, showing uncoated silt and sand grains on structural pedfaces; having an argic B horizon.

GREYZEMS (GR)

Other soils having a mollic A horizon with a moist chroma of 2 or less to a depth of at least 15 cm; having a calcic or petrocalcic horizon, or concentrations of soft powdery lime within 125 cm of the surface, or both.

CHERNOZEMS (CH)

Other soils having a mollic A horizon with a moist chroma of more than 2 to a depth of at least 15 cm; having one or more of the following: a calcic, petrocalcic or gypsic horizon, or concentrations of soft powdery lime within 125 cm of the surface.

KASTANOZEMS (KS)

Other soils having a mollic A horizon; having a base saturation (by NH$_4$OAc) of ≥ 50% throughout the upper 125 cm of the soil.

PHAEOZEMS (PH)

Table 5.6 Cont.

Other soils having an argic B horizon showing an irregular or broken upper boundary resulting from deep tonguing of the A into the B horizon or from the formation of discrete nodules larger than 2 cm, the exteriors of which are enriched and weakly cemented or indurated and have redder hues and stronger chromas than the interiors **PODZOLUVISOLS (PD)**	Other soils having an argic B horizon which has a cation exchange capacity equal to or more than 24 $cmol_c$ kg^{-1} clay and a base saturation (by NH_4OAc) of less than 50 percent in at least some part of the B horizon within 125 cm of the surface. **ALISOLS (AL)**
Other soils having a gypsic or petrogypsic horizon within 125 sm of the surface; having no diagnostic horizons other than an ochric A horizon, a cambic B horizon or an argic B horizon permeated with gypsum or calcium carbonate, a calcic or petrocalcic horizon **GYPISOLS (GY)**	Other soils having an argic B horizon which has a cation exchange capacity of less than 24 $cmol_c$ kg^{-1} clay and a base saturation (by NH_4OAc) of less than 50 percent in at least some part of the B horizon within 125 cm of the surface. **ACRISOLS (AC)**
Other soils having a calcic or a petrocalcic horizon, or a concentration of soft powdery lime, within 125 cm of the surface; having no diagnostic horizons other than an ochric A horizon, a cambic B horizon, or an argic B horizon which is calcareous. **CALCISOLS (CL)**	Other soils having an argic B horizon which has a cation exchange capacity equal to or more than 24 $cmol_c$ kg^{-1} clay and a base saturation (by NH_4OAc) of 50 percent or more throughout the B horizon within 125 cm of the surface. **LUVISOLS (LV)**
Other soils having an argic B horizon with a clay distribution which does not show a relative decrease from its maximum of more than 20 percent within 150 cm of the surface; showing gradual to diffuse horizon boundaries between the A and B horizons; having nitic properties in some subhorizon within 125 cm of the surface. **NITISOLS (NT)**	Other soils having an argic B horizon which has a cation exchange capacity of less than 24 $cmolc$ kg^{-1} clay and a base saturation (by NH_4OAc) of 50 percent or more throughout the B horizon within 125 cm of the surface. **LIXISOLS (LX)** Other soils having a cambic B horizon. **CAMBISOLS (CM)**

During the ISSS congresses in 1982 and 1986 and expert consultations in 1987 and 1988, the IRB took form, and as a result, 20 major soil groupings were identified and agreed upon as being representative of the principal components of the world's soil cover.

Subsequently, it became clear that some of the proposed 20 major soil groupings were too broad to be defined consistently and, consequently, had to be subdivided. By doing so, the list of major soil groupings became very close to those of the Revised Legend of the Soil Map of the World (FAO, 1988). As a result, it was decided in 1992 to adopt the Revised Legend as the frame for further IRB work. This was also prompted by the fact that both the Revised Legend and the International Reference Base for Soil Classification were supported by the ISSS and that it would be inappropriate to pursue two programs which essentially had the same goal, namely, to arrive at a rational inventory of global soil resources (ISSS-ISRIC-FAO, 1994). The two programs were, therefore, merged under the name World Reference Base for Soil Resources (WRB), an ISSS/FAO/ISRIC undertaking.

5.4.3 Objectives

The specific objectives of the WRB are to: (1) develop an internationally acceptable framework for delineating soil resources to which national classifications can be attached and related, using the FAO Revised Legend as a guideline; (2) provide this framework with a sound scientific base so that it can also serve different applications in related fields, such as agriculture, geology, hydrology and ecology; (3) acknowledge in the framework important lateral aspects of soils and soil horizon distribution as characterized by topo- and chronosequences; and (4) emphasize the morphological characterization of soils rather than to follow a purely analytical approach.

5.4.4   Concepts and Principles

For describing and defining the reference soil groups of the WRB, use is made of characteristics, properties and horizons, which when combined, define soils and their relationships. Soil characteristics are single parameters which are observable or measurable in the field or laboratory, or can be analyzed using microscopic techniques. They include such characteristics as color, texture and structure of the soil, features of biological activity, arrangement of voids and pedogenic concentrations (mottles, cutans, nodules, etc.) as well as analytical determinations (soil reaction, particle size distribution, CEC, exchangeable cations, amount and nature of soluble salts, etc.).

Soil properties are combinations (assemblages) of soil characteristics, which are known to occur in soils and indicative of present or past soil-forming processes (e.g., vertic properties, which are a combination of fine texture, smectitic mineralogy, slickensides, hard consistence when dry, sticky when wet, shrinking when dry, and swelling when wet).

Soil horizons are three-dimensional pedological bodies which are more or less parallel to the Earth's surface. Each horizon exhibits one or more properties, occurring over a certain depth, which characterizes it and permits its recognition. The thickness varies from a few cm to several m. The upper and lower limits (boundaries) are more or less clear, and progressive, or abrupt. Laterally, the extension of a soil horizon varies greatly, from a m to several km. However, a soil horizon is never infinite. Laterally, it disappears or grades into another horizon.

Soils are defined by the vertical combination of horizons, occurring within a defined depth, and by the lateral organization (sequence) or lack thereof, of the soil horizons at a scale reflecting the relief of a land unit. Soil horizons and properties are intended to reflect the expression of genetic processes, which are widely recognized as occurring in soils. They are considered to be diagnostic when they reach a minimum degree of expression, which is determined by visibility, prominence, measurability, importance and relevance for soil formation and soil use. To be diagnostic, soil horizons also require a minimum thickness, which must be appraised in relation to bioclimatic factors (e.g., a spodic horizon in boreal regions can be expected to be thinner than in the tropics).

The relationships between soil characteristics, properties, horizons and soils in terms of reference soil groups are shown in Table 5.7. A brief description of the diagnostic horizons, properties and materials recognized in the WRB is presented in Table 5.8.

Table 5.7   Relationships between soil characteristics, soil properties, soil horizons and soils in terms of reference soil groups

Soil characteristics	Limiting value	Diagnostic property	Thickness	Diagnostic horizon	Reference Soil Group
CEC	≤ 24 cmol$_c$ kg^{-1} clay ≤ 16 cmol$_c$ kg^{-1} clay	Ferralic (protic) $\longrightarrow$	(≥ 15 cm) $\longrightarrow$	(Cambic) $\longrightarrow$	Ferralic Cambisols
ECEC*	< 12 cmol$_c$ kg^{-1} clay < 1.5 cmol$_c$ kg^{-1} clay	Geric $\longrightarrow$	≥ 30 cm $\longrightarrow$	Ferralic $\longrightarrow$	Geric Ferralsols
Soil reaction	pH$_{KCl}$ - pH$_{water}$ ≥ 0				
Water-dispersible clay*	$< 10\%$				
Gravel content**	$< 90\%$				
Weatherable minerals*	$< 10\%$	Ferralic (orthic) $\longrightarrow$	≥ 30 cm $\longrightarrow$	Ferralic $\longrightarrow$	Other Ferralsols
Texture***	Sandy loam or finer				
Structure	Weak to moderate				

* Failing one or more of these, the soil classifies as Ferralic Cambisols

** Failing this the soil classifies as Leptosol

*** Failing this the soil classifies as Ferralic Cambisol

Table 5.8 Diagnostic horizons, properties and materials in the World Reference Base (WRB) [ISSS-ISRIC-FAO, 1998]

Horizons

Albic	light colored eluvial horizon generally associated with argic and spodic horizons
Andic	moderately weathered horizon in pyroclastic material dominated by short-range order minerals
Anthropedogenic	surface and subsurface horizons resulting from long continued cultivation
Argic	subsurface horizon with distinct clay accumulation
Calcic	horizon with accumulation of calcium carbonate
Cambic	subsurface horizon showing evidence of alteration relative to the underlying horizon(s)
Chernic	thick, well-structured, black, base-saturated surface horizon, rich in organic matter content and biological activity
Cryic	perennially frozen mineral or organic soil horizon
Duric	subsurface horizon with weakly cemented to indurated nodules cemented by silica ("durinodules")
Ferralic	strongly weathered subsurface horizon with low cation exchange capacity
Ferric	horizon in which iron is concentrated in large mottles or concretions
Folic	well-aerated, highly organic surface horizon
Fragic	natural, non-cemented subsurface horizon in which access for roots and water is restricted to interped faces
Fulvic	thick, black horizon rich in organic matter associated with pyroclastic deposits and vegetation other than grassland
Gypsic	horizon with accumulation of gypsum
Histic	poorly aerated, waterlogged, highly organic surface horizon
Melanic	thick, black horizon rich in organic matter associated with pyroclastic deposits and grassland vegetation
Mollic	thick, dark colored surface horizon with high base saturation and moderate to high organic matter content
Natric	subsurface horizon with distinct clay accumulation and a high exchangeable sodium percentage
Nitic	clayey subsurface horizon with strongly developed, nut-shaped structure and shiny pedfaces
Ochric	weakly developed surface horizon, either light colored, or thin, or having a low organic matter content
Petrocalcic	continuous cemented or indurated calcic horizon
Petroduric	continuous cemented or indurated duric horizon
Petrogypsic	continuous cemented or indurated gypsic horizon
Petroplinthic	continuous cemented or indurated plinthic horizon
Plinthic	iron-rich, humus-poor subsurface horizon which hardens irreversibly upon repeated wetting and drying
Salic	surface or shallow subsurface horizon with a high soluble salt content
Spodic	dark colored subsurface horizon with illuvial alumino-organic complexes, with or without iron
Sulfuric	extremely acid subsurface horizon with sulfuric acid resulting from oxidation of sulfides
Takyric	heavy textured crusted surface horizon occurring under arid conditions
Umbric	thick, dark colored surface horizon with low base saturation and moderate to high organic matter content
Vertic	clayey subsurface horizon dominated by shrink-swell clays
Vitric	horizon dominated by volcanic glass and other primary minerals derived from volcanic ejecta
Yermic	surface horizon with desert pavement or a loamy vesicular crust covered by windblown deposits

Properties

Abrupt textural change	sharp increase in clay content within a limited depth range
Albeluvic tonguing	penetrations of clay and iron-depleted material into an argic horizon
Alic	very acid mineral soil material with a high amount of exchangeable aluminum
Aridic	presence of properties (low organic matter, aeolian activity, light colors, high base saturation) in surface horizons, characteristic of arid environments
Continuous hard rock	presence of coherent rock, practically impermeable for roots
Ferralic	low cation exchange capacity
Geric	extremely low to negative effective cation exchange capacity
Gleyic	wetness producing reduced conditions caused by groundwater
Permafrost	perennial temperature at or below 0 °C
Soft powdery lime	accumulation of translocated calcium carbonate in soft powdery form
Stagnic	wetness producing reduced conditions caused by stagnating surface water
Strongly humic	high organic matter content

Soil material

Anthropogeomorphic	unconsolidated mineral or organic material produced by human activity
Calcaric	material containing calcium carbonate
Fluvic	fresh fluviatile, lacustrine or marine sediments
Gypsiric	material containing gypsum
Organic	material containing a very high amount of organic debris
Sulfidic	waterlogged deposit containing sulfides, and only moderate amounts of calcium carbonate
Tephric	unconsolidated, non- or only slightly weathered pyroclastic products

The general principles governing the construction of the WRB can be summarized as follows: (1) classification of soils is based on soil properties defined in terms of diagnostic horizons, properties and materials, which, to the greatest extent possible, should be measurable and observable in the field; (2) selection of diagnostic horizons, properties and materials takes into account their relationship with soil-forming processes; however, at a high level of generalization, it also attempts to select, to the extent possible, diagnostic features which are of significance for management purposes; (3) climatic parameters are not applied; (4) the Reference Base is limited to the highest level only and comprises 30 reference soil groups; listing of adjectives to the reference soil group names, each with a unique meaning, is provided to enable precise characterization and classification of individual soil bodies; (5) the reference soil groups identified are representative of major soil regions so as to provide a comprehensive overview of the world's soil cover; (6) the Reference Base is not meant to substitute for national classification systems, but rather, to serve as a common denominator for communication at the international level and as a link between existing soil classification systems; in addition, WRB may also serve as a consistent communication tool for compiling global soil databases for inventory and monitoring of the world's soil resources; (7) the FAO-UNESCO-ISRIC Revised Legend of the Soil Map of the World is used as base to develop the WRB to take advantage of international soil correlation work, which has already been conducted in the course of the project; (8) where possible, definitions and descriptions of soils reflect both vertical and lateral variations in soil characteristics to account for spatial linkages in the landscape; and (9) the nomenclature used to distinguish reference soil groups and soil units will have terms which are traditionally used, or which can easily be introduced in current language.

5.4.5 The WRB Reference Soil Groups

Thirty reference soil groups are now identified in Table 5.9 (ISSS-ISRIC-FAO, 1994). Compared to the FAO Revised Legend, one major soil grouping has been omitted (Greyzems), and three new ones are introduced (Cryosols, Durisols and Umbrisols). Greyzems were deleted as they constitute the smallest major soil grouping and were amalgamated with the Phaeozems. Cryosols were newly introduced to identify a group of soils which occur under the unique environmental conditions of thawing and freezing. Durisols have been added to group soils together, which have accumulation of secondary silica, analogous to the Calcisols and Gypsisols. Umbrisols constitute the group of soils which have a thick accumulation of desaturated organic matter at the surface, and are the natural counterpart of Chernozems, Kastanozems and Phaeozems. The key to the reference soil groups of the WRB is presented in Table 5.9.

5.5 The French Systems of Soil Classification

5.5.1 Introduction

The *Commission de Pédologie et de Carthographie des Sols* (CPCS, 1967) issued the French soil classification, building on previous work published by Aubert and Duchaufour (1956) which has been the basis for many soil surveys during the 1970s and 1980s, not only in France, but also in many of the former French colonies, notably in Africa. It was replaced by the Pedological Reference Base (*Référentiel Pédologique Français*)(AFES-INRA, 1990; 1992).

5.5.2 The 1967 CPCS System

The CPCS soil classification system comprises four main levels: the class (*classe*), the subclass (*sous-classe*), the group (*groupe*) and the subgroup (*sous-groupe*); followed by four minor levels: the fam-

Table 5.9 Key to the reference soil groups of the World Reference Base (WRB) [ISSS-ISRIC-FAO, 1998]

Soils having a *histic* or *folic* horizon,
1. *either* a. 10 cm or more thick from the soil surface to a lithic or paralithic contact;
   or b. 40 cm or more thick and starting within 30 cm from the soil surface; *and*
2. lacking an *andic* horizon starting within 30 cm from the soil surface.

HISTOSOLS

Other soils having one or more *cryic* horizons within 100 cm from the soil surface.

CRYOSOLS

Other soils having *either*
1. a *hortic, irragric, plaggic* or *terric* horizon 50 cm or more thick; *or*
2. an *anthraquic* horizon and an underlying *hydragric* horizon with a combined thickness ≥ 50 cm.

ANTHROSOLS

Other soils, which are *either*
1. limited in depth by *continuous hard rock* within 25 cm from the soil surface; *or*
2. overlying material with a calcium carbonate equivalent ≥ 40%, both within 25 cm from the soil surface; *or*
3. containing less than 10% (by weight) fine earth to a depth of 75 cm or more from the soil surface; *and*
4. having no diagnostic horizons other than a *mollic, ochric, umbric, yermic* or *vertic* horizon.

LEPTOSOLS

Other soils having
1. a *vertic* horizon within 100 cm from the soil surface; *and*
2. after the upper 20 cm have been mixed, ≥ 30% clay in all horizons to a depth of 100 cm or more, or to a contrasting layer (lithic or paralithic contact, *petrocalcic, petroduric* or *petrogypsic* horizons, sedimentary discontinuity, etc.) between 50 and 100 cm; *and*
3. cracks which open and close periodically.

VERTISOLS

Other soils having
1. *fluvic* soil material within 25 cm from the soil surface; *and*
2. no diagnostic horizons other than a *histic, mollic, ochric, takyric, umbric, yermic, salic* or *sulfuric* horizon.

FLUVISOLS

Other soils having
1. a *salic* horizon starting within 50 cm from the soil surface; *and*
2. no diagnostic horizons other than a *histic, mollic, ochric, takyric, yermic, calcic, cambic, duric, gypsic* or *vertic* horizon.

SOLONCHAKS

Other soils having
1. *gleyic* properties within 50 cm from the soil surface; *and*
2. no diagnostic horizons other than a *histic, mollic, ochric, takyric, umbric, andic, calcic, cambic, gypsic, plinthic, salic* or *sulfuric* horizon within 100 cm from the soil surface.

GLEYSOLS

Other soils having
1. *either* a *vitric or andic* horizon, both starting within 25 cm from the soil surface; *and*
2. having no diagnostic horizons (unless buried deeper than 50 cm) other than a *histic, fulvic, melanic, mollic, umbric, ochric, duric* or *cambic* horizon.

ANDOSOLS

Other soils having a *spodic* horizon starting within 200 cm from the soil surface, underlying an *albic, umbric* or *ochric* horizon, or an *anthropedogenic* horizon less than 50 cm thick.

PODZOLS

Other soils having *either*
1. a *petroplinthic* horizon starting within 50 cm from the soil surface; *or*
2. a *plinthic* horizon starting within 50 cm from the soil surface; *or*
3. a *plinthic* horizon starting within 100 cm from the soil surface when underlying either an *albic* horizon or a horizon with *stagnic* properties.

PLINTHOSOLS

Other soils
1. having a *ferralic* horizon at some depth between 25 and 200 cm from the soil surface; *and*
2. lacking a *nitic* horizon; *and*
3. lacking a layer which fulfills the requirements of an *argic* horizon and which has in the upper 30 cm ≥ 10% water-dispersible clay (unless the soil material has *geric* properties or ≥ 1.4% organic carbon).

FERRALSOLS

Other soils having
1. an eluvial horizon, the lower boundary of which is marked, within 100 cm from the soil surface, by an *abrupt textural change* associated with *stagnic* properties above that boundary; *and*
2. no *albeluvic tonguing.*

PLANOSOLS

Other soils having a *natric* horizon within 100 cm from the soil surface.

SOLONETZ

Other soils having
1. a *chernic* horizon or a *mollic* horizon with a moist chroma of 2 or less to a depth of at least 20 cm or directly below any plow layer; *and*
2. concentrations of *soft powdery lime* starting within 50 cm of the lower limit of the Ah horizon but within 200 cm from the soil surface; *and*
3. no *petrocalcic* horizon between 25 and 100 cm from the soil surface; *and*
4. no secondary gypsum; *and*
5. no uncoated silt and sand grains on structural ped surfaces.

CHERNOZEMS

Table 5.9   cont.

Other soils having
1. a *mollic* horizon with a moist chroma of more than 2 to a depth of at least 20 cm or directly below any plough layer; *and*
2. concentrations of *soft powdery lime* within 100 cm from the soil surface; *and*
3. no diagnostic horizons other than an *argic*, *calcic*, *cambic*, *gypsic* or *vertic* horizon.

KASTANOZEMS

Other soils having
1. a *mollic* horizon; *and*
2. a base saturation (by 1 *M* NH$_4$OAc, pH 7) of 50% or more at least to a depth of 100 cm from the soil surface, or to a contrasting layer (lithic or paralithic contact, *petrocalcic* horizon) between 25 and 100 cm; *and*
3. no concentrations of *soft powdery lime* within 200 cm from the soil surface; *and*
4. no diagnostic horizons other than an *albic*, *argic*, *cambic* or *vertic* horizon or a *petrocalcic* horizon in the substratum.

PHAEOZEMS

Other soils having
1. *either* a *gypsic* or *petrogypsic* horizon within 100 cm from the soil surface, *or* 15% (by volume) or more gypsum, which is accumulated under hydromorphic conditions, averaged over a depth of 100 cm within 1.5 m from the soil surface; *and*
2. no diagnostic horizons other than an *ochric* or *cambic* horizon, an *argic* horizon permeated with gypsum or calcium carbonate, or a *calcic* or *petrocalcic* horizon underlying the gypsic horizon.

GYPSISOLS

Other soils having a *duric* or *petroduric* horizon within 100 cm from the soil surface.

DURISOLS

Other soils having
1. a *calcic* or *petrocalcic* horizon within 100 cm of the surface; *and*
2. no diagnostic horizons other than an *ochric* or *cambic* horizon, an *argic* horizon which is calcareous, or a *gypsic* horizon underlying a petrocalcic horizon.

CALCISOLS

Other soils having an *argic* horizon within 100 cm from the soil surface with an irregular upper boundary resulting from *albeluvic tonguing* into the argic horizon.

ALBELUVISOLS

Other soils having
1. *alic* properties in the major part between 25 and 100 cm from the soil surface; *and*
2. an *argic* horizon, which has a cation exchange capacity (by 1 *M* NH$_4$OAc, pH 7) of 24 cmol$_c$ kg^{-1} clay or more in some part, either starting within 100 cm from the soil surface, or within 200 cm from the soil surface if the argic horizon is overlain by loamy sand or coarser textures throughout; *and*
3. no diagnostic horizons other than an *ochric*, *umbric*, *albic*, *andic*, *ferric*, *nitic*, *plinthic* or *vertic* horizon.

ALISOLS

Other soils having
1. a *nitic* horizon starting within 100 cm from the soil surface; *and*
2. gradual to diffuse horizon boundaries between the surface and the underlying horizons; *and*
3. no *ferric*, *plinthic* or *vertic* horizon within 100 cm from the soil surface.

NITISOLS

Other soils having
1. a base saturation (by 1 *M* NH$_4$OAc, pH 7) of less than 50 % in the major part between 25 and 100 cm; *and*
2. an *argic* horizon, which has a cation exchange capacity (by 1 *M* NH$_4$OAc, pH 7) of less than 24 cmol$_c$ kg^{-1} clay in some part, either starting within 100 cm from the soil surface, or within 200 cm from the soil surface if the argic horizon is overlain by loamy sand or coarser textures throughout.

ACRISOLS

Other soils having an *argic* horizon with a cation exchange capacity (by 1 *M* NH$_4$OAc, pH 7) equal to or more than 24 cmol$_c$ kg^{-1} clay throughout, either starting within 100 cm from the soil surface, or within 200 cm from the soil surface if the argic horizon is overlain by loamy sand or coarser textures throughout.

LUVISOLS

Other soils having an *argic* horizon, either starting within 100 cm from the soil surface, or within 200 cm from the soil surface if the argic horizon is overlain by loamy sand or coarser textures throughout.

LIXISOLS

Other soils having
1. an *umbric* horizon; *and*
2. no diagnostic horizons other than an *anthropedogenic* horizon less than 50 cm thick, or an *albic* or *cambic* horizon.

UMBRISOLS

Other soils having *either*
1. a *cambic* horizon; *or*
2. a *mollic* horizon overlying a subsoil which has a base saturation (by 1 *M* NH$_4$OAc, pH 7) of less than 50% in some part within 100 cm from the soil surface; *or*
3. one of the following diagnostic horizons within the specified depth from the soil surface:
   a.   an *andic* or *vitric* horizon between 25 and 100 cm;
   b.   a *plinthic*, *petroplinthic* or *salic* horizon between 50 and 100 cm, in absence of loamy sand or coarser textures to a depth of at least 100 cm.

CAMBISOLS

Other soils having
1. a texture which is loamy sand or coarser to a depth of at least 100 cm from the surface; *and*
2. less than 35% (by volume) of rock fragments or other coarse fragments within 100 cm from the soil surface; *and*
3. no diagnostic horizons other than an *ochric*, *yermic* or *albic* horizon, or a *plinthic*, *petroplinthic* or *salic* horizon below 50 cm from the soil surface.

ARENOSOLS

Other soils.

REGOSOLS

ily (*famille*), the series (*série*), the type (*type*) and the phase (*phase*). However, due to the limited knowledge at the time of design of the system, the four minor levels have not been developed for all classes.

The class comprises soils which have main characteristics in common such as a certain degree of profile development, weathering mode, composition and distribution of organic matter, and predominant soil-forming factors (e.g., wetness). The subclass differentiation is related to criteria resulting from climatic factors which influence, among others, the pedoclimate. The groups are defined according to morphological characteristics corresponding to soil development, while the subgroup is differentiated either on degree of intensity of the fundamental evolutionary characteristics, or on the presence of important secondary soil-forming processes.

In total, 12 classes are distinguished, namely, nondeveloped mineral soils (*sols minéraux bruts*), slightly developed soils (*sols peu évolués*), vertisols, andosols, Ca/Mg-saturated soils (*sols calcimagnésiques*), humus-rich soils (*sols isohumiques*), brunified soils (*sols brunifiés*), podzolized soils (*sols podzolisés*), soils rich in Fe (*sols à sesquioxydes de fer*), ferralitic soils, hydromorphic soils, and sodic soils.

Nondeveloped mineral soils show very little trace of soil development apart from some accumulation of organic matter at the surface. They are characterized by an (A)C, (A)R or R horizon sequence. Included in this class are eroded soils (Lithosols, Régosols), alluvial, colluvial and eolian accumulations, volcanic deposits, manmade soils, nondeveloped soils in the arctic regions (Cryosols) and nondeveloped desert soils.

The class of slightly developed soils has higher organic matter content than the nondeveloped soils and is characterized by an AC or AR horizon sequence. No B horizon is permitted in this class. Included are soils from the arctic regions, with or without permafrost, soils with high organic matter content directly overlying hard rock (rankers, soils over limestone and slightly weathered soils on volcanic ashes), slightly developed desert soils and soils resulting from erosion and deposition.

Vertisols are described as clayey soils which are homogenized or irregularly differentiated as a result of internal movement and which are dominated by swell/shrink clays. The normal horizon sequence is A(B)C, A(B)$_g$C or A(B)C$_g$. Subdivision into subclasses and groups is based on external drainage factors and type of structure in the surface horizon (rounded or angular).

Andosols are defined as soils in which the mineral fraction is dominated by poorly crystalline minerals and/or metal humus complexes, associated with variable, but usually high amounts of organic matter. Their diagnostic characteristics are described as follows: (1) morphologically, AC or A(B)C horizon sequence; (2) physically, low bulk density (generally < 0.8 g cm^{-3}), high water retention; (3) chemically, organic matter content between 3 and 30%, pH between 5 and 6 when high in allophane, but between 4 and 5 when metal humus content becomes high; CEC > 25 cmol$_c$ kg^{-1} fine earth; and (4) mineralogically, presence of (a) primary minerals (orthosilicates, pyroxenes, hornblendes, feldspars) and glass particles, (b) amorphous weathering products (allophanes, ferrihydrite, imogolite), and (c) crystalline secondary minerals (halloysite, smectite, gibbsite). Andosols are divided according to climate (tropical or temperate), which is related to their organic matter content, and acidity.

The Ca/Mg-saturated soils have an exchange complex 90% of which is saturated with Ca and/or Mg, and have a pH above 6.8. They are generally associated with calcareous or basic rocks, and have an AR, AC, A(B)R or A(B)C horizon sequence. They are subdivided into: (1) carbonate rich soils (rendzinas, cryptorendzinas and brown calcareous soils), (2) saturated soils (soils containing only traces of primary CaCO$_3$ in the fine earth fraction), and (3) gypsiferous soils.

The class of humus-rich soils (*sols isohumiques*) comprises soils characterized by a moderate to high accumulation of well-humified, polymerized organic matter. Normally, base saturation is high with Ca as dominant cation, followed by Mg and, sometimes, Na. If the base saturation is only moderate (50 – 80%) in the upper part of the soil, it increases with depth. Profile evolution is slight to moderate, with an A(B)C or ABC, rarely AC horizon sequence. Subdivision of these soils is based at subclass level on pedoclimatic characteristics. Where the pedoclimate is humid, Brunizems (soils with a moderate base saturation) are found, while in cold regions, Chernozems (non- or slightly calcareous soils with an organic matter content of > 5%), Chestnut soils (decalcified soils with an organic matter content between 3 and 6%), and Brown soils (soils with an organic matter content between 1 and 3%) are formed. If the pedoclimate is cool during the rainy season, Chestnut soils are formed, but with a lower organic matter content and a more strongly weathered mineral fraction, as well as Sierozems. Finally, subarid Brown soils with a relatively low amount of organic matter and a relatively high content of free Fe occur where the pedoclimate is warm during the rainy season. The requirements for organic matter content do not take into account the loss in organic matter content upon cultivation. Thus it is possible that, for example, soils classified as Chernozems may become Chestnut soils if the amount of organic matter becomes less than 5%.

Brunified soils are well-developed soils with an A(B)C or ABC horizon sequence and are characterized by the presence of dominantly mull type humus. They may have a structural or textural B horizon. Subdivision is based on climate (humid temperate, continental temperate, boreal or tropical) and the morphology of the profile, giving rise to the groups of brown soils (with a structural B horizon), eluvial soils (with a textural B horizon), gray wooded soils, dernopodzolic soils, eluvial boreal soils, and eutrophic brown tropical soils.

The class of podzolic soils is characterized by the processes of alteration and destruction of the silicate minerals by fulvic acids and complexation of liberated Fe and Al. These processes result morphologically in a strongly depleted and light colored eluvial horizon and an illuvial horizon, which has a higher organic matter content than the eluvial horizon and a sesquioxide content, which is higher than the original material. Division at subclass level is based on climatic or pedoclimatic characteristics. Podzolized soils of the temperate climates, podzols of the cold climate, and hydromorphic podzolized soils are distinguished. Division into groups is based on horizon differentiation, notably the presence or absence of an eluvial A_2 horizon and the expression of the eluvial horizon.

Soils rich in Fe have an ABC or A(B)C profile characterized by the presence of Fe and/or Mn (hydr)oxides giving the soils characteristic red, yellow, rusty brown or even black (in the case of Mn) colors, an SiO_2/Al_2O_3 ratio of > 2, a base saturation of > 50%, and a low amount of organic matter. Two subclasses are recognized, one in which the role of sesquioxides is dominant (ferruginous tropical soils), and another in which the behavior of the clay fraction predominates (fersiallitic soils).

The ferruginous tropical soils have an A(B)C, ABC or AB_gC horizon sequence, yellow colors (hue of 7.5–10YR with value > 5 and chroma > 4), a massive structure, moderate base saturation and a clay fraction dominated by kaolinite. Groups distinguished are weakly leached ferruginous tropical soils (actual clay increase < 5%), leached ferruginous tropical soils (actual clay increase > 5%), and impoverished ferruginous tropical soils (clay increase present, but the clay content does not decrease between the B and C horizons and no visible clay films).

The second subclass, the fersiallitic soils have redder colors (hues of 10R, 2.5 or 5YR), are decalcified, have a clay mineralogy dominated by inherited minerals (mainly illite), and base saturation is moderate to high. Two groups are distinguished, one with reserves of Ca, and one without.

The ferralitic soils are characterized by complete weathering of primary minerals, residual enrichment of resistant minerals (quartz, rutile, etc.), loss of nutrients, and the presence of neoformations such as kaolinite, gibbsite, goethite, hematite, etc. Subclasses are distinguished on the degree of leaching: (1) strongly leached ferralitic soils with a low amount of exchangeable bases (< 1 $cmol_c$ kg^{-1} fine earth), low base saturation and low pH (< 5.5); (2) moderately leached ferralitic soils with a moderate amount of exchangeable bases (1–3 cmol$_c$ kg^{-1} fine earth) and moderate base saturation; and (3) weakly leached ferralitic soils with a fair amount of exchangeable bases (2–8 cmol$_c$ kg^{-1} fine earth), moderate to high base saturation and a slightly acid soil reaction (pH 5.0–6.5).

A large number of groups are distinguished in each subclass. Distinction is made on a number of criteria, namely, the typical, humiferous, illuvial, leached, reworked and rejuvenated or young soils. At subgroup level, important secondary characteristics are introduced, such as induration, hydromorphism, eolian influence and color.

The class of hydromorphic soils comprises both organic and mineral soils. The three subclasses are separated on organic matter content. The first subclass (hydromorphic organic soils) is defined as having > 30% organic matter over a depth of at least 40 cm if the mineral component is clayey, or > 20% if the mineral component is sandy. The second subclass (moderately organic hydromorphic soils) has between 8 and 30% organic matter over at least 20 cm depth; while the third subclass, the mineral soils (weakly organic hydromorphic soils) have < 8% organic matter over a depth of at least 20 cm. The organic hydromorphic soils are at group level separated on decomposition rate (weak or fibric, moderate or hemic, or strong or sapric), while the two other subclasses are divided on the character of hydromorphism (gley or stagnogley) and accumulation of Fe ($CaCO_3$ or gypsum).

The sodic soils comprise both soils which have a high amount of soluble salts as well as soils with a high exchangeable sodium percentage (ESP). This difference is used to separate the subclasses into sodic soils with a nondegraded structure, comprising the saline soils (solonchak) with an AC horizon sequence, and sodic soils with a degraded structure, having an A(B)C or ABC horizon sequence and comprising alkaline saline soils (high ESP), sodic soils with a textural B horizon (solonetz), and so-called solodized solonetz, which are acid at the surface.

5.5.3 The Pedological Reference Base (*Référentiel Pédologique Français*)(PRB)

This differs basically from the older CPCS system which it replaces, in that it is being presented as a reference system, not a hierarchical classification (AFES-INRA, 1992). It considers the soil mantles as objects of study for which three sets of data are required: (1) the composition of the soil mantle (mineral, organic, etc.); (2) the internal arrangement of the individual constituents (e.g., structure); and (3) the soil dynamics (e.g., evolution over time).

At the highest level, the PRB recognizes the pedological system which comprises several associated horizons grouped in a three-dimensional space pattern. A horizon is defined as a part of the soil mantle which can be considered homogenous. Because dimensions of horizons and pedological systems are not infinite, vertically and laterally, they merge into other systems (e.g., bedrock or other pedological systems).

The authors of the PRB have tried to design a system which, at the same time, is both scientific and practical as well as precise but flexible. An example of this is the depth indications used in the descriptions of the horizons; the RPB starts off with the tolerance limits concerning depth or thickness requirements: for example, 10 cm must be considered as 5 to 15 cm, 40 cm means 30 to 50 cm, etc. Therefore, only two categories are distinguished (the References and the Types), the latter being indicated by one or more qualifiers.

The reference horizons form the basis of the system. So far, 50 have been proposed (AFES-INRA, 1992), defined and described by several of the following: (1) morphological characteristics (constituents, pedological features, etc.), (2) analytical data, (3) pedogenetic significance, (4) major possible variations of the characteristics, and (5) most common positions within soil mantles. A succession or combination of reference horizons identifies a diagnostic solum and permits the association of such a solum with a Reference.

The PRB has proposed some 90 References (Table 5.10). Several of these are closely associated with each other because, for example, they may have the same reference horizons. Such References are described together as Major Groupings of References to avoid duplication and to associate the References with traditional pedological concepts. For example, the Podzosols Major Grouping of References comprise 7 References characterized by a process of podzolization. However, Major Groupings of References do not form part of the PRB.

5.6 The Soil Classification System of the Former USSR

5.6.1 Introduction

This system was published in Russian in 1977 and in English in 1986 (VASKhNIL, 1986). This classification serves as a manual for soil examination and survey in the former USSR, for governmental assessment of land records and evaluation, as well as a reference book on agricultural and industrial planning for agronomists, land use planners, reclamation specialists, and others. It was amended (Shishov and Sokolov, 1990) to correct obvious errors and to integrate new knowledge and data. The taxonomic levels of the classification system are expanded and new names have been introduced for a number of soils.

5.6.2 Structure of the System

It does not have a hierarchical structure as in other systems in which a soil is classified using a key. Profound knowledge of the soil-forming processes is required in order to recognize and classify the soils. The higher levels are known as types, subtypes, genera and species. At subtype level, the facies modifier is added to indicate the thermal regime. Twenty-seven of these facies are recognized, ranging from arctic permafrost to subtropical hot nonfreezing.

5.6.3 Brief Description of the Types

Some 71 soil types are distinguished at the highest level and characterized as follows:
   Podzolic soils are characterized by downward movement of organic acids facilitating decomposition of primary and secondary minerals, and removal of weathering products and clay particles.
   Bog-podzolic soils have stagnating water conditions in the upper part of the soil, resulting in a bleached eluvial horizon and hydromorphic properties in both the eluvial and illuvial horizons. Most of these soils have a peaty layer at the surface.
   Sod-calcareous soils have a dark colored, base-saturated and humus-rich surface horizon overlying calcareous parent material.
   Sod-gley soils are poorly drained with a dark, humus-rich surface horizon and hydromorphic features at shallow depth.
   Gray forest soils have a dark colored, humus-rich surface horizon of variable thickness with conspicuous white powdery spots overlying a clay illuviated subsurface horizon.
   Gley gray forest soils are similar to those above but with distinct hydromorphic features below the surface horizon.
   Brown forest soils (or Burozems) are weakly developed apart from an enrichment with organic matter in the surface horizon(s).
   Gley brown forest soils (or Gley Burozems) are similar to those above except for clear hydromorphic features below the surface horizon.

Table 5.10 Major Groupings of References (MGR) and References of the AFES-INRA Pedological Reference Base [AFES-INRA, 1992]

MGR	Reference	Brief description
ALOCRISOLS	Typic Alocrisols	Very acid, brown or yellow soils with a high amount of exchangeable aluminum (2-8 cmol$_c$ kg^{-1} fine earth and Al saturation of 20-50%)
	Humic Alocrisols	Very acid, brown or yellow soils with a thick, dark colored surface horizon high in organic matter and a high amount of exchangeable aluminum
ALU-ANDISOLS	Humic Alu-Andisols	Non-allophanic soils in weathered volcanic deposits having a thick surface horizon rich in organic matter
	Typic Alu-Andisols	Soils in volcanic deposits or strongly weathered ferralitic material having an aluminum-rich surface horizon and an allophane-dominated subsurface horizon
ANDOSOLS	Humic Andosols	Allophane-rich soils having a thick surface horizon rich in organic matter
	Eutric Andosols	Allophane-rich soils having a surface horizon with base saturation > 50%
	Dystric Andosols	Allophane-rich soils having a surface horizon with base saturation < 50%
	Perhydrated Andosols	Allophane-rich soils having a high irreversible water content
ANTHROPOSOLS	Transformed Anthroposols	Soils modified by intensive or long-continued human activities
	Artificial Anthroposols	Man-made soils consisting of non-soil material (mine refuse, urban debris, etc.)
	Reshaped Anthrosols	Man-made soils consisting of transported soil material
ARENOSOLS	Arenosols	Deep (> 120 cm) sandy soils
BRUNISOLS	Saturated Brunisols	Non-calcareous soils with a structural B horizon and 80-100 % base saturation
	Meso-saturated Brunisols	Non-calcareous soils with a structural B horizon and 50-80 % base saturation
	Oligo-saturated Brunisols	Non-calcareous soils with a structural B horizon and 20-50 % base saturation
	Resaturated Brunisols	Non-calcareous soils with a structural B horizon and > 80 % base saturation due to cultivation
CALCARISOLS	Calcarisols	Soils with a calcic horizon at least 10 cm thick, starting within 20 cm depth
CALCISOLS	Calcisols	Soils with non-calcareous, base-saturated (mainly Ca^{2+}) A and B horizons
CALCOSOLS	Calcosols	Soils with calcareous A and B horizons (CaCO$_3$ >5 %)
CASTANOSOLS	Castanosols	Soils with a moderately thick to thick, dark colored, base-saturated surface horizon rich in organic matter
CHERNOSOLS	Chernosols	Soils with a thick, very dark colored, base-saturated surface horizon rich in organic matter
COLLUVIOSOLS	Colluviosols	Soils in colluvial deposits
CRYOSOLS	Histic Cryosols	Soils with permafrost within 1 m depth and a histic surface horizon
	Mineral Cryosols	Soils with permafrost within 2 m depth lacking a histic surface horizon
DOLOMITOSOLS	Dolomitosols	Soils with dolomitic A and B horizons (molar ratio of CaCO$_3$/MgCO$_3$ < 1.5)
FERRALLISOLS	Soft Ferrallisols	Strongly weathered soils with a ferralitic or oxidic mineralogy
	Nodular Ferrallisols	Strongly weathered soils with a high amount of sesquioxide nodules
	Petroxydic Ferrallisols	Strongly weathered soils with indurated sesquioxide layers (e.g. cuirasses)
FERSIALSOLS	Carbonated Fersialsols	Calcareous soils with significant amounts of 2:1 clays and "free iron"
	Saturated Fersialsols	Base-saturated soils with significant amounts of 2:1 clays and "free Fe"
	Desaturated Fersialsols	Desaturated soils with significant amounts of 2:1 clays and "free Fe"
	Xanthic Fersialsols	Yellowish soils with significant amounts of 2:1 clays and "free Fe"
FLUVIOSOLS & THALASSOSOLS	Raw Fluviosols	Soils in fluviatile deposits lacking any horizon development
	Typical Fluviosols	Soils in fluviatile deposits with one ore more not fully developed reference horizons
	Brunified Fluviosols	Soils in fluviatile deposits with a well-developed structural B horizon
	Thalassosols	Non- or weakly developed soils in marine or fluvio-marine deposits
GYPSOSOLS	Gypsosols	Soils with accumulation of gypsum
HISTOSOLS	Leptic Histosols	Shallow organic soils with consolidated or unconsolidated rock within 40 cm
	Fibric Histosols	Organic soils with weakly decomposed organic material more than 60 cm thick
	Mesic Histosols	Organic soils with moderately decomposed organic material more than 40 cm thick
	Sapric Histosols	Organic soils with strongly decomposed organic material more than 40 cm thick
	Composite Histosols	Organic soils without dominance of either fibric, mesic or sapric materials between 40 and 120 cm
	Covered Histosols	Organic soils with a cover of mineral soil material 10-40 cm thick
	Floating Histosols	Organic soils on water occurring between 40 and 160 cm depth
LITHOSOLS	Lithosols	Shallow soils (< 10 cm) over continuous hard rock or indurated layer
LUVISOLS	Neoluvisols	Soils with a moderately developed eluvial and well developed textural B horizon
	Typic Luvisols	Soils with a well-developed eluvial and textural B horizon
	Degraded Luvisols	Soils with a well-developed, partially light colored and hydromorphic eluvial horizon penetrating a gleyed textural B horizon
	Dernic Luvisols	Soils with a well-developed, partially light colored eluvial horizon penetrating a textural B horizon
	Truncated Luvisols	Soils with a textural B horizon but lacking an eluvial horizon

Table 5.10 cont.

MGR	References	Brief description
MAGNESISOLS	Magnesisols	Soils with non-calcareous, base-saturated (Ca^{2+}/Mg^{2+} <2) A and B horizons
ORGANOSOLS	Calcaric Organosols	Well-drained, organic matter rich (>8% organic C), calcareous soils directly overlying an unconsolidated or consolidated substratum
	Saturated Organosols	Well-drained, organic matter rich (>8% organic C), base-saturated (Ca^{2+}/Mg^{2+} >5) soils directly overlying an unconsolidated or consolidated substratum
	Undersaturated Organosols	Well-drained, organic matter rich (>8% organic C), undersaturated (BS <80%) soils directly overlying an unconsolidated or consolidated substratum
	Tangelic Organosols	Well-drained, organic matter rich (>8% organic C), base-saturated soils with a thick, greasy horizon consisting of soil animal casts ("tangel horizon")
PELOSOLS	Typic Pelosols	Clay-rich, slightly weathered soils lacking coloration in the B horizon
	Brunified Pelosols	Clay-rich, slightly weathered soils with a brown colored B horizon
	Differentiated Pelosols	Clay-rich, slightly weathered soils with a clear eluvial horizon
PEYROSOLS	Stony Peyrosols	Soils which have throughout the upper 50 cm 40% or more stones plus 20% or more other coarse fragments
	Gravelly Peyrosols	Soils which have throughout the upper 50 cm 60% or more gravel, stones and boulders, but less than 40% stones
PLANOSOLS	Typic Planosols	Soils with abrupt textural change and temporary perched watertable within 50 cm
	Distal Planosols	Soils with abrupt textural change and temporary perched watertable below 50 cm
	Structural Planosols	Soils with a temporary perched watertable within 50 cm of the surface caused an impermeable layer which is not texturally induced (e.g., fragipan, duripan)
PODZOSOLS	Duric Podzosols	Soils with an eluvial horizon and a cemented podzol B horizon
	Humo-Duric Podzosols	Soils with an indurated podzol B horizon, lacking an eluvial horizon
	Soft Podzosols	Soils with an eluvial horizon and a soft podzol B horizon
	Placic Podzosols	Soils with podzol B horizon and a placic horizon
	Ochric Podzosols	Soils with a weakly developed humic podzol B horizon, lacking an eluvial horizon
	Humic Podzosols	Soils with a soft humic podzol B horizon, lacking an eluvial horizon
	Post_Podzosols	Man-modified soils in which (remnants of) the podzol B horizon can be recognized
	Eluvial Podzosols	Soils lacking a podzol B horizon, but having lateral linkage to a podzol B horizon
RANKOSOLS	Rankosols	Soils with a moderately thick A horizon with non-calcareous coarse fragments overlying consolidated or unconsolidated rock
REDUCTISOLS & REDOXISOLS	Typic Reductisols	Hydromorphic soils conditioned by saturation of fluctuating groundwater table
	Stagnic Reductisols	Hydromorphic soils conditioned by a perched water table
	Duplex Reductisols	Hydromorphic soils conditioned by groundwater and stagnating surface water
	Redoxisols	Soils with a textural discontinuity and a perched water table
REGOSOLS	Regosols	Shallow soils (<10 cm) over unconsolidated material or only slightly coherent rock
RENDISOLS	Rendisols	Soils with non-calcareous, base-saturated (mainly Ca^{2+}) A horizon over consolidated or unconsolidated calcareous rock
RENDOSOLS	Rendosols	Soils with deep (>30 cm) calcareous A horizon ($CaCO_3$ >5 %) over consolidated or unconsolidated calcareous rock
SALISOLS	Chloridi-Sulfatic Salisols	Neutral soils with a high amount of sodium, magnesium or calcium salts
	Carbonatic Salisols	Alkaline soils with a high amount of carbonate/bicarbonates
SODISOLS	Undifferentiated Sodisols	Alkaline soils with a high amount of exchangeable sodium
	Solonetzic Sodisols	Soils with clay illuviation and moderate leaching of sodium in the upper part
	Solodic Sodisols	Soils with clay illuviation and strong leaching of sodium in the upper part
SULFATOSOLS	Sulfatosols	Very acid soils with jarosite within 50 cm depth
THIOSOLS	Thiosols	Waterlogged soils containing sulfide minerals, rapidly acidifying upon aeration
VERACRISOLS	Veracrisols	Soils with a thick (50-150 cm), acid, dark colored surface horizon with a high biological activity overlying a slowly permeable horizon (e.g., textural B horizon)
VERTISOLS	Topovertisols	Deep, clayey soils in level, low-lying positions which crack and show gilgai microrelief, slick
VITRANDISOLS	Vitrandisols	Soils in slightly weathered pyroclastic material
YERMOSOLS	Yermosols	Hot desert soils

Podzolic brown forest soils (or Podzolic Burozems) have a clearly developed eluvial horizon and weak hydromorphism due to seasonal surface waterlogging.

Gley podzolic brown forest soils (or Gley podzolic Burozems) are similar to those above, but with seasonal wetness more pronounced. The process of acidic hydrolysis may take place in the upper part of these soils.

Bleached meadow soils (Podbels) are seasonally waterlogged with a bleached horizon near the surface, in which removal of Fe in concretions is the main ongoing process.

Meadow chernozem-like soils have a thick, dark colored and humus-rich surface horizon and distinct features of hydromorphism (gray and rusty colors, Fe/Mn concretions, white powdery coatings) in the lower part of the soils.

Chernozem-like dark meadow soils are waterlogged with a peaty or mucky surface horizon overlying a dark colored, humus-rich mineral horizon with rust colored mottles.

Chernozems are well-drained, base-saturated or only slightly undersaturated soils with a thick, dark colored surface horizon rich in organic matter.

Meadow-chernozem soils are similar to those above, but with distinct features of wetness in the lower part of the solum.

Chestnut soils are well drained and base saturated with a dark colored surface horizon which is less thick and less rich in organic matter than in Chernozems. The lower part of the solum often contains accumulations of calcium carbonate or gypsum.

Meadow chestnut soils are similar as the soils above, but with distinct features of wetness in the lower part of the solum.

Meadow soils are conditioned by a brief period of surface waterlogging and a longer period of saturation by groundwater, resulting in humus-rich surface horizons overlying a gleyed subsoil.

Semi-desert brown soils have an accumulation of calcium carbonate, possibly overlying accumulations of gypsum and a crusty surface horizon.

Semi-desert meadow brown soils are similar to those above, but with a higher organic matter content, weak signs of hydromorphism and deeper $CaCO_3$ accumulations.

Desert gray brown soils are calcareous with a low organic matter content and a variable degree of salinization.

Desert takyr-like soils are weakly developed with a friable porous surface crust.

Takyrs have a hard, porous but crusted surface horizon cracking into polygonal patterns.

Desert sandy soils are coarse textured with little horizon differentiation apart from some accumulation of organic matter and segregation of calcium carbonate at depth.

Meadow desert soils are poorly differentiated, characterized by an enrichment with organic matter at the surface and signs of hydromorphism in the subsoil.

Serozems have a surface enrichment in organic matter and a calcareous illuvial layer in the subsoil.

Meadow-serozem soils are similar to those above, but signs of wetness occur in the deeper subsoil.

Semi-desert and desert meadow soils have periodic or permanent wetness through capillary rise reaching the surface, giving rise to a well-developed, humus-rich surface horizon and a gleyed subsoil, in which carbonate concentrations are linked to the groundwater level.

Irrigated soils are all related to the original soil (irrigated serozems, irrigated brown soils, irrigated meadow brown soils, irrigated gray brown soils, irrigated takyr like soils, irrigated meadow desert soils, irrigated meadow soils and irrigated bog soils), in which irrigation has caused considerable modification including enhancement of biological activity, leaching, accumulation of sediments from irrigation water, enrichment in carbonates and soluble salts, etc.

Gray-cinnamon brown soils have a well-developed surface horizon with a low amount of organic matter and a clay-enriched subsurface horizon. The soils are calcareous throughout and differ from chestnut soils in that they do not freeze during wintertime.

Meadow gray-cinnamon brown soils are similar to those above, but with clear indications of increased wetness in the subsoil.

Cinnamon brown soils are deep with a moderate amount of organic matter and a clay-enriched subsurface horizon with a characteristic cinnamon brown color.

Meadow cinnamon brown soils are similar to those above, but with clear indications of increased wetness in the subsoil.

Gray meadow forest soils have a thick, humus-enriched surface horizon and hydromorphic features starting at or near to the surface.

Zheltozems are leached subtropical soils with only weak textural differentiation and rich in sesquioxides.

Gley zheltozems are similar to those above, but with pronounced gleying throughout the profile.

Podzolic-zheltozem soils are leached subtropical soils with clear textural differentiation and a high content in sesquioxides. Gley features are common in the transition zone between the eluvial and illuvial horizons.

Podzolic-zheltozem gley soils are similar to those above, but with pronounced gleying throughout the profile, resulting in the accumulation of Fe in the illuvial horizon.

Krasnozems are strongly weathered subtropical soils with a high amount of sesquioxides. The clay fraction mainly contains kaolinite, halloysite, goethite and gibbsite.

Peat-bog soils are waterlogged organic soils, both in upland and lowland positions.

Reclaimed peat soils are drained peat soils with a plow layer.

Meadow-bog soils are waterlogged mineral soils with or without a shallow organic layer at the surface.

Bog soils of the semi-deserts and deserts have shallow groundwater (usually < 50 cm) and an organic matter-rich surface horizon in desert or semi-desert conditions.

Solods are degraded solonetzes and solonetzic soils of which the upper horizons are acidified, resulting in a well-differentiated soil with a humus-rich surface horizon, a white eluvial horizon and a brownish colored, compact illuvial B horizon.

Solonetzes have a high amount of exchangeable Na, an (near) absence of readily soluble salts, resulting in a well-expressed eluvial horizon and a compact illuvial horizon. Automorphic (related to the parent material), semihydromorphic and hydromorphic (related to groundwater influence) solonetzes are distinguished.

Solonchaks have a high amount of soluble salts. Automorphic and hydromorphic solonchaks are distinguished.

Alluvial soils are divided into three main types: alluvial soils with deep groundwater and only a brief period of flooding (sod alluvial soils), alluvial soils influenced by both flooding and groundwater at moderate (1–2 m) depth (meadow alluvial soils), and alluvial soils which are conditioned by a long period of flooding, or shallow groundwater in combination with surface flooding (bog alluvial soils). In addition, the soil reaction is used to further subdivide the types.

Mountain meadow soils occur in cold and moist high mountains which receive a large excess of moisture resulting in a strongly leaching regime, an acid soil reaction and a considerable accumulation of organic matter in the surface horizon.

Chernozem-like mountain meadow soils occur in high mountains which, although receiving an excess in moisture, have only a moderately leaching regime, resulting in a weakly acid to weakly alkaline soil reaction as well as a considerable accumulation of organic matter in the surface horizon.

Mountain meadow-steppe soils develop under similar conditions as above, but have a much lower exchange capacity than the chernozem-like mountain meadow soils.

An overview of the types and subtypes distinguished in the 1986 USSR soil classification is given in Table 5.11.

5.7   The Chinese Soil Taxonomic Classification

5.7.1   Introduction

Until 1949, the Chinese soil classification was based on that of the United States, but then was replaced by that of the USSR geographical classification (ISS-AS, 1990), and integrating locally important soils, such as long continued cultivated soils, paddy soils among others. In 1994, a drastically renewed first proposal for a new Chinese Soil Taxonomic Classification (CSTC) based largely on Soil Taxonomy (Section E, chapter 4) was issued (Gong, 1994), which is still under review and subject to modification. In developing the CSTC system, many elements of the Legend of the Soil Map of the World (FAO-UNESCO, 1974) and other soil classifications were incorporated.

5.7.2   Structure

The CSTC is a hierarchical system with seven categories: order, suborder, group, subgroup, genus, species and variety. The first four levels are used to construct mapping units for small scale maps, the lower three for more detailed maps.

The order is based on soil properties which result from or reflect major soil-forming processes. The suborder is defined according to soil properties, which either control recent soil-forming processes or which reflect dominant limiting factors. At group level, intensities of major or secondary soil-forming processes are used to differentiate among the soils, while the subgroup reflects the deviation from the central concept of the group.

The system presently has defined 13 orders, 33 suborders, 78 groups and 301 subgroups. The nomenclature used is a mixture of older and more recent names, as well as local names for typical Chinese soils. The names of the order and suborder are linked, the group and subgroup nomenclature is different, to avoid names becoming too long. This is illustrated in Table 5.12 with Aquisols (soils saturated with water within 1 m depth, having a (per)aquic soil moisture regime and a layer with redoxic features or a gleyic horizon within 50 cm depth) as an example. An overview of the orders, suborders and groups of the Chinese Soil Taxonomic System is given in Table 5.13.

5.7.3   Diagnostic Horizons and Characteristics

Like many other soil classification systems, the Chinese Soil Taxonomic System uses diagnostic horizons and characteristics to identify the soil. A number of the diagnostic horizons are directly taken from other systems, such as the argillic horizon from Soil Taxonomy, while others are newly defined to suit Chinese conditions. Thirty diagnostic horizons are defined, 8 surface horizons or epipedons, 10 subsurface horizons, and 12 horizons which may occupy any position in the soil profile.

Three categories of diagnostic surface horizons are recognized: histic, humus (isohumic, umbrihumic and ochrihumic epipedons) and anthropic epipedons. Particularly the latter category has a number of diagnostic epipedons which do not occur in other systems, namely, the warpic (finely stratified epipedon resulting from long continued irrigation), cumulic (thick epipedon resulting from additions of manure and organic rich soil material), mellowic (epipedon resulting from long and intensive cultivation and applying night soil, organic trash and manure), and hydragric (epipedon resulting from cultivating the soil under wet conditions) epipedons.

The ten diagnostic subsurface horizons are the albic (white), weathering B, humilluvic (illuvial organic matter), argillic (clay illuviated), clayific (*in situ* clay accumulation through weathering of

Table 5.11 Types and subtypes of the 1986 USSR soil classification [VASKhNIL, 1986]

Types	Subtypes
Podzolic soils	Gley-podzolic soils
	True podzolic soils
	Sod-podzolic soils
Bog-Podzolic soils	Surface gleyed peaty podzolic soils
	Surface gleyed soddy-podzolic soils
	Surface gleyed mucky-podzolic soils
	Subsoil gleyed peaty-podzolic soils
	Subsoil gleyed soddy-podzolic soils
	Subsoil gleyed mucky-podzolic soils
Sod-Calcareous soils	Typical sod-calcareous soils
	Leached sod-calcareous soils
	Podzolized sod-calcareous soils
Sod-Gley soils	Sod surface gleyey soils
	Mucky surface gley soils
	Sod subsurface gleyey soils
	Mucky subsurface gley soils
Gray Forest soils	Light gray forest soils
	Gray forest soils
	Dark gray forest soils
Gley Gray Forest soils	Surface gleyey (and surface meadow gray forest) soils
	Subsurface gleyey gray forest soils
	Subsurface gley gray forest soils
Brown Forest soils	Acid mor brown forest soils
	Acid mor podzolized brown forest soils
	Acid brown forest soils
	Acid podzolized brown forest soils
	Slightly unsaturated brown forest soils
	Slightly unsaturated podzolized brown forest soils
Gley Brown Forest soils	Podzolized surface gleyey brown forest soils
	Podzolized surface gley brown forest soils
	Gleyey brown forest soils
	Gley brown forest soils
Podzolic Brown Forest soils	Unsaturated podzolic brown forest soils
	Slightly unsaturated podzolic brown forest soils
Gley Podzolic Brown Forest soils	Surface gleyey podzolic brown forest soils
	Surface gley podzolic brown forest soils
	Gleyey podzolic brown forest soils
	Gley podzolic brown forest soils
Bleached Meadow soils	Podzolized bleached meadow soils
	Podzolized gley bleached meadow soils
Meadow Chernozem-like soils	Meadow chernozem-like soils (surface-wet)
Chernozem-like Dark Meadow Soils	Dark meadow prairie soils
	Dark moist-meadow prairie soils
Chernozems	Podzolized chernozems
	Leached chernozems
	Typical chernozems
	Ordinary chernozems
	Southern chernozems
	Mountain chernozems
Meadow-Chernozem soils	Meadowy chernozemic soils
	Meadow-chernozem soils
Chestnut soils	Dark chestnut soils
	Chestnut soils
	Light chestnut soils
	Mountain chestnut soils
Meadow Chestnut soils	Meadowy chestnut soils
	Meadow chestnut soils
Meadow soils	Meadow soils
	Moist-meadow soils

Table 5.11 cont.

Types	Subtypes
Semi-Desert Brown soils	Semi-desert brown soils
Semi-Desert Meadow Brown soils	Semi-desert meadowy brown soils
	Semi-desert meadow brown soils
Desert Gray Brown soils	Desert gray brown soils
Desert Takyr-like soils	Desert takyr-like soils
Takyrs	Takyr
Desert Sandy soils	Desert sandy soils
Meadow Desert soils	Meadow desert (meadowy takyr-like) soils
	Meadow desert (meadow takyr-like) soils
	Meadow desert soils with complementary surface moistening
	Gray brown meadow desert soils
	Sandy meadow desert soils
Serozems Light colored	Typical serozems
	Dark serozems
Meadow-Serozem soils	Meadowy serozem
	Meadow serozem
Semi-Desert and Desert Meadow soils	Semi-desert and desert meadow
	Semi-desert and desert moist-meadow soils
Irrigated Serozems	Irrigated light coloured serozem soils
	Irrigated typical serozem soils
	Irrigated dark serozem soils
	Old irrigated serozem soils
Irrigated Meadow Serozems	Irrigated meadow serozem soils
	Irrigated serozem meadow soils
Irrigated Brown soils	
Irrigated Meadow Brown soils	
Irrigated Gray Brown soils	
Irrigated Takyr-like soils	Irrigated takyr-like soils
	Ancient irrigated takyr-like soils
Irrigated Meadow Desert soils	Irrigated meadow desert soils
	Ancient irrigated meadow desert soils
Irrigated Meadow soils	Irrigated meadow soils
	Irrigated moist meadow soils
	Ancient irrigated meadow soils
Irrigated Bog soils	
Gray-Cinnamon Brown soils	Dark gray-cinnamon brown soils
	Common gray-cinnamon brown soils
	Light gray-cinnamon brown soils
Meadow Gray-Cinnamon Brown	Surface-meadowy gray-cinnamon brown soils
	Meadowy gray-cinnamon brown soils
	Meadow gray-cinnamon brown soils
Cinnamon Brown soils	Leached cinnamon brown soils
	Typical cinnamon brown soils
	Calcareous cinnamon brown soils
Meadow Cinnamon Brown soils	Surface-meadowy cinnamon brown soils
	Meadowy cinnamon brown soils
	Meadow cinnamon brown soils
Gray Meadow-Forest soils	Gray meadow-forest soils
	Gray wet meadow-forest soils
Zheltozems	Unsaturated zheltozems
	Weakly unsaturated zheltozems
	Unsaturated podzolized zheltozems
	Weakly unsaturated podzolized zheltozems

Table 5.11 cont.

Types	Subtypes
Gley Zheltozems	Surface gleyey zheltozems
	Gleyey zheltozems
	Gley zheltozems
Podzolic-Zheltozem soils	Unsaturated podzolic zheltozem soils
	Slightly unsaturated podzolic zheltozem soils
Podzolic-Zheltozem Gley soils	Podzolic-zheltozem surface-gleyey soils
	Podzolic-zheltozem gleyey soils
	Podzolic-zheltozem gley soils
Krasnozems	Typical krasnozems
	Podzolized krasnozems
Upland Peat-Bog soils	Upland peat-gley bog soils
	Upland peat-bog soils
Lowland Peat-Bog soils	Depleted lowland peat-gley bog soils
	Lowland (typical) peat-gley bog soils
	Depleted lowland peat-bog soils
	Lowland (typical) peat-bog soils
Reclaimed Upland Peat soils	
Reclaimed Lowland Peat soils	Reclaimed depleted lowland peat-gley soils
	Reclaimed depleted lowland peat soils
	Reclaimed lowland muck-gley soils
	Reclaimed lowland mucky-peat soils
Meadow Bog soils	Mucky meadow-bog soils
	Clayey meadow-bog soils
Bog soils (semi-desert/deserts)	Peat-bog soils
	Clayey-bog soils
Solods	Meadow-steppe solods
	Solods meadow
	Meadow-bog solods
Automorphic Solonetzes	Chernozemic solonetzes
	Chernozemic solonchak-solonetzes
	Solonchakic chernozemic solonetzes
	Deep-solonchakic chernozemic solonetzes
	Deep-salinized chernozemic solonetzes
	Chestnut solonetzes
	Semi-desert solonetzes
Semihydromorphic Solonetzes	Meadow-chernozemic solonetzes
	Meadow-chernozemic solonchak-solonetzes
	Solonchakic meadow-chernozemic solonetzes
	Deep-solonchakic meadow-chernozemic solonetzes
	Meadow-chestnut solonetzes
	Meadow semi-desert solonetzes
	Semihydromorphic cryogenic solonetzes
Hydromorphic Solonetzes	Chernozemic-meadow solonetzes
	Chestnut-meadow solonetzes
	Meadow-bog solonetzes
	Cryogenic-meadow solonetzes
Automorphic Solonchaks	Typical automorphic solonchaks
	Takyrized automorphic solonchaks
Hydromorphic Solonchaks	Typical solonchaks
	Meadow-solonchaks
	Bog-solonchaks
	Sor-solonchaks
	Mud-volcanic solonchaks
	Hummocky solonchaks
Sod Acidic Alluvial soils	Primitive stratified sod acidic alluvial soils
	Stratified sod acidic alluvial soils
	Typical sod acidic alluvial soils
	Podzolized sod acidic alluvial soils

Table 5.11 cont.

Types	Subtypes
Saturated Sod Alluvial soils	Primitive stratified saturated sod alluvial soils
	Stratified saturated sod alluvial soils
	Typical stratified sod alluvial soils
	Saturated steppe sod alluvial soils
Calcareous Sod Desertified Alluvial soils	Primitive stratified calcareous sod desertified alluvial soils
	Stratified calcareous sod desertified alluvial soils
	Typical calcareous sod desertified alluvial soils
Meadow Acid Alluvial soils	Primitive stratified acid meadow alluvial soils
	Stratified acid meadow alluvial soils
	Typical acid meadow alluvial soils
Meadow Saturated Alluvial soils	Primitive stratified saturated meadow alluvial soils
	Stratified saturated meadow alluvial soils
	Typical saturated meadow alluvial soils
	Dark colored saturated meadow alluvial soils
Calcareous Meadow Alluvial soils	Stratified calcareous meadow alluvial soils
	Tugai calcareous meadow alluvial soils
	Typical calcareous meadow alluvial soils
Meadow-Bog Alluvial soils	Typical meadow-bog alluvial soils
	Peaty meadow-bog alluvial soils
Clayey-Muck-Gley Bog Alluvial soils	Clayey-peat-gley bog alluvial soils
	Muck-gley-bog alluvial soils
Clayey-Peat Bog Alluvial soils	Clayey-peat-gley bog alluvial soils
	Clayey-peat bog alluvial soils
Mountain Meadow soils	Alpine mountain meadow soils
	Subalpine mountain meadow soils
Chernozem-like Mountain Meadow soils	Typical chernozem-like mountain meadow soils
	Leached chernozem-like mountain meadow soils
	Calcareous chernozem-like mountain meadow soils
Mountain Meadow-Steppe soils	Alpine mountain meadow-steppe soils
	Subalpine mountain meadow-steppe soils

primary minerals), claypan (slowly permeable, clay rich horizon), alkalic (clay-illuviated horizon with high Na percentage), spodic (illuviation of amorphous organic compounds in combination with Al and/or Fe), agric (illuviated humus clay or humus silt clay under conditions of cultivation) and hydragric-redoxic (oxidation-reduction horizon related to wet cultivation) horizons.

The other diagnostic horizons comprise the calcic, hypercalcic, gypsic, hypergypsic, salic, hypersalic, sulfuric, phosphic and gleyic horizons and the calci-, gypsi- and salipans.

Twenty-three diagnostic characteristics (characters, properties, features, contacts, saturation, materials, regimes or layers) are defined. These are lithologic character (features inherited from the parent material), vertic features (shrink/swell phenomena), desertic features (surface and topsoil features related to arid conditions), takyric features (cracking of surface soil in arid conditions induced by wetting and drying), redoxic features (hydromorphism), frost/thaw features, regosolic features (< 35% coarse fragments), skeletic features (between 35 and 70% coarse fragments), lithic features (> 70% coarse fragments), lithic contact, paralithic contact, siallic properties (weathering B horizon of which the clay fraction is dominated 2:1 or 2:1:1 clays), fersiallic properties (weathering B horizon of which the clay fraction is dominated 2:1 or 2:1:1 clays and which has $\geq$ 2% free Fe oxides), ferralic properties (weathering B horizon of which the clay fraction is dominated 1:1 clays and which has $\geq$ 2% free Fe oxides), andic properties (related to weathering of pyroclastic deposits), base saturation, Al saturation, calcaric properties ($\geq$ 1% $CaCO_3$), humic properties (high organic matter content), organic soil materials, soil moisture regimes, soil temperature regimes and permafrost layer (perennial temperature at or below 0 °C).

Table 5.12 Subdivision of the Aquisol order in the Chinese Soil Taxonomic System [Gong, 1994]

Order	Suborder	Group	Subgroup
Aquisols	Gelic Aquisols	Geli-gley soils	Histic geli-gley soils
			Haplic geli-gley soils
	Peric Aquisols	Gley soils	Sulfuric gley soils
			Salic gley soils
			Histic gley soils
			Aquic gley soils
			Haplic gley soils
	Orthic Aquisols	Chao soils	Warpic Chao soils
			Alkalic Chao soils
			Medihumic Chao soils
			Salific Chao soils
			Salinic Chao soils
			Aquic Chao soils
			Haplic Chao soils
		Umbrihumic Chao soils	Gleyic umbrihumic Chao soils
			Albic umbrihumic Chao soils
			Haplic umbrihumic Chao soils
		Shajiang black soils	Alkalic Shajiang black soils
			Salic Shajiang black soils
			Vertic Shajiang black soils
			Albic Shajiang black soils
			Haplic Shajiang black soils
		Foliaged-Chao soils	Alkalic-salic foliaged-Chao soils
			Salic foliaged-Chao soils
			Medihumic foliaged-Chao soils
			Haplic foliaged-Chao soils

When important and obvious properties of soil horizon do not fulfill the combination required for a diagnostic horizon, the CSTC uses the term evidence. They are used to identify soil taxa, particularly at subgroup level. Twelve diagnostic evidences are defined so far, namely, histic, warpic, cumulic, mellowic, clayific (all meeting the requirements of the related diagnostic horizons apart from thickness), alkalic (lower ESP), spodic (lower ratio of pyrophosphate extractable Al+Fe to clay), calcic (either shallower thickness or lower $CaCO_3$ content), gypsic (either shallower thickness or lower gypsum content), salic (lower salt content), gleyic (discontinuous gleyic parts), and vertic (less wider cracks and a clayey subhorizon within 50 cm depth) evidences.

5.8   The Australian Soil Classification

5.8.1   Introduction

In 1996, a new soil classification system for Australia was issued (Isbell, 1996). Until then, two different soil classification systems were widely used (Stace et al., 1968; Northcote, 1979). A soil classification committee has worked over 15 years to develop the new system, taking advantage of the previously used systems, the numerous soil surveys carried out all over the country, and the large database containing data of some 14,000 soil profiles, which was constructed during the time of the committee's work.

Table 5.13   Orders, suborders and groups of the Chinese Soil Taxonomic System [Gong, 1994]

Order	Suborder	Group
Histosols	Gelic Histosols	Geli-peat soils
	Orthic Histosols	Peat soils
Anthrosols	Hydragric Anthrosols	Paddy soils
	Dryagric Anthrosols	Cumulated soils
		Lou soils
		Irrigation-warping soils
		Mellow soils
Andisols	Andisols	Andisols
Spodosols	Orthic Spodosols	Podzols
Vertisols	Aquic Vertisols	Black clay soils
	Udic Vertisols	Pell clay soils
		Chrom clay soils
Halosols	Alkalic Halosols	Solonetz
	Salic Halosols	Solonchaks
		Arid solonchaks
Aridisols	Altocryic Aridisols	Frost desert soils
		Cold desert soils
		Frost-calc soils
		Cryo-calc soils
	Gypsi-salic Aridisols	Brown desert soils
		Grey desert soils
	Calcic Aridisols	Brown calc soils
		Sierozems
	Orthic Aridisols	Takyr soils
		Haplo-calc soils
		Haplo-desert soils
Aquisols	Gelic Aquisols	Geli-gley soils
	Peric Aquisols	Gley soils
	Orthic Aquisols	Chao soils
		Umbrihumic Chao soils
		Shajiang black soils
		Foliaged Chao soils
Isohumisols	Altocryic Isohumisols	Cryo-black soils
		Frost-sod soils
		Cryo-sod soils
	Lithomorphic Isohumisols	Phospho-calc soils
		Rendzinas
	Udic Isohumisols	Black soils
		Thermo-black soils
	Ustic Isohumisols	Greyzems
		Grey-cinnamon soils
		Chernozems
		Castanozems
		Heilu soils
Ferrallisols	Perudic Ferrallisols	Lato-yellow soils
		Latored-yellow soils
		Yellow soils
	Ustic Ferrallisols	Dry red soils
	Udic Ferrallisols	Latosols
		Latored soils
		Red soils
Fersiallisols	Perudic Fersiallisols	Yellow limestone soils
		Grey-yellow brown soils
		Para-yellow soils

Table 5.13 Orders, suborders and groups of the Chinese Soil Taxonomic System [Gong, 1994] (cont.)

Order	Suborder	Group
	Ustic Fersiallisols	Yellow-cinnamon soils
		Red-cinnamon soils
	Udic Fersiallisols	Brown limestine soils
		Red limestone soils
		Yellow brown soils
		Brown-red soils
		Para-red soils
Siallisols	Perchic Siallisols	Albisols
	Perudic Siallisols	Grey-brown soils
		Humus-brownified soils
	Ustic Siallisols	Cinnamon soils
	Udic Siallisols	Brown soils
		Acid brown soils
		Dark brown soils
		Cryo-brown soils
Primarosols	Anthropic Primarosols	Disturbance soils
	Lithic Primarosols	Leptisols
		Skeletisols
	Regosic Primarosols	Red-bed soils
		Purple soils
		Loessal soils
		Blown sand soils
		Alluvial soils

5.8.2 Structure and Nomenclature

The new system is multicategorical, comprising orders, suborders, great groups, subgroups and families. In total, 14 orders are recognized (Anthroposols, Calcarosols, Chromosols, Dermosols, Ferrosols, Hydrosols, Kandosols, Kurosols, Organosols, Podosols, Rudosols, Sodosols, Tenosols and Vertosols), distinguished by fairly straightforward criteria which can easily be recognized in the field. Nomenclature is often based on Latin or Greek roots, as with many other modern soil classification systems, but the names are clearly different from Soil Taxonomy, the FAO-UNESCO Legend or the World Reference Base for Soil Resources, in order to avoid confusion.

Unlike many other soil classification systems, there is no fixed soil depth for classification purposes in the new Australian system. The concept of pedologic organization (McDonald et al., 1990) is used to define the soil to be classified. Isbell (1996) describes this as "... a broad concept used to include all changes in soil material resulting from the effect of the physical, chemical and biological processes that are involved in soil formation." Consequently, soil studies for classification purposes may need to go to considerable depth. At family level, this is recognized since one of the family criteria, soil depth, includes a class Giant for those soils which are deeper than 5 m.

Division of seven orders (Chromosols, Dermosols, Ferrosols, Kandosols, Kurosols, Sodosols, and Vertosols) into suborders is based on color criteria (e.g., red, brown, yellow, grey and black). Suborders of Anthroposols are distinguished on the kind and nature of the human activity, in Calcarosols the degree and kind (shells, gypsum, $CaCO_3$) of the accumulation is used for the subdivision, suborders of the Hydrosols reflect the type of tidal inundation, the degree of salinity, and the reduction/oxidation regime, Organosols are classically divided according to the degree of decomposition of the organic material, suborders of the Podosols reflect the degree of wetness, Rudosols are divided according the nature of the material, while, in Tenosols, the type of surface horizon, the occurrence of a bleached horizon, and the presence or absence of a weakly developed B horizon are used to characterize the suborders.

Subdivision of the great groups and subgroups is still incomplete for a number of suborders. For those suborders that are developed into great groups and subgroups, division is generally made on the presence or absence of important diagnostic horizons and materials, and on the degree of importance of certain characteristics.

Criteria considered at family level are A horizon thickness (not in Organosols, Rudosols and Vertosols), gravel content of the surface and A1 horizon (not in Organosols), A1 horizon texture (not in Organosols and Vertosols), B horizon maximum texture (not in Organosols and Rudosols), soil depth (not in Organosols), thickness of the soils above the upper boundary of the Bk horizon (in Calcarosols only), nature of the uppermost organic materials and cumulative thickness of the organic materials (in Organosols only), and clay content of the upper 0.1 m (in Vertosols only).

The full soil name is constructed as follows: subgroup, great group, suborder, order, family. A unique coding system for all levels is provided for recording the classification of soil profiles. Included in this coding system is the opportunity to indicate a confidence level to the soil classification.

5.8.3 Description of the Orders

The 14 orders now recognized in the Australian soil classification, and their characteristerics are, in alphabetical order:

Anthroposols (Gr. *anthropos*, man): soils resulting from human activities;

Calcarosols (L. *calcis*, lime): soils having pedogenetic carbonate accumulations throughout the solum, or at least directly below the A1 or Ap horizon;

Chromosols (Gr. *chroma*, color): soils having a clear or abrupt textural B horizon in which the major part of the upper 0.2 m of the B2 horizon is not strongly acid [pH_{H_2O} (1:5) $\geq$ 5.5 and pH_{CaCl_2} (1:5) $\geq$ 4.6];

Dermosols (L. *dermis*, skin): soils having a B2 horizon which has a structure more developed than weak throughout the major part of the horizon;

Ferrosols (L. *ferrum*, iron): soils having a B2 horizon in which the major part contains > 5% free Fe oxide in the fine earth fraction;

Hydrosols (Gr. *hydor*, water): soils which are saturated with water in the major part of the solum for at least 2–3 months in most years. Reducing conditions are not an essential requirement;

Kandosols (kandite, 1:1 clay minerals): soils having a well-developed B2 horizon of which the major part is massive or has only a weak grade of structure, and having a maximum clay content in some part of the B2 horizon >15%;

Kurosols (no root): soils having a clear or abrupt textural B horizon in which the major part of the upper 0.2 m of the B2 horizon is strongly acid [pH_{H_2O} (1:5) < 5.5 and pH_{CaCl_2} (1:5) < 4.6];

Organosols (no root): soils having more than 0.4 m of organic materials within the upper 0.8 m, or having organic materials from the surface to a minimum depth of 0.1 m overlying either hard rock or other hard layers, partially weathered or decomposed rock or fragmental material with interstices (partially) filled;

Podosols (R. *pod*, under, and *zola*, ash): soils having a Bs, Bhs or Bh horizon (horizons with illuvial accumulations of amorphous organic matter Al and Al silica complexes, with or without Fe);

Rudosols (L. *rudimentum*, a beginning): soils having neglegible (or rudimentary) pedological organization apart from the minimal development of an A1 horizon, or the presence of less than 10% of B horizon material in fissures in the parent rock or saprolite;

Sodosols (E. *sodium*): soils having a clear or abrupt textural B horizon, in which the major part of the upper 0.2 m of the B2 horizon is sodic (ESP of the fine earth soil material $\geq$ 6);

Tenosols (L. *tenius*, weak, slight): soils having weak expression of pedological organization;

Vertosols (L. *vertere*, to turn): soils having ≥ 35% clay, and developing open cracks which are at least 5 mm wide and extend upward to the surface or to the base plough layer at some time in most years self-mulching horizon, or a thin, surface crust, and having slickensides or lenticular peds.

5.9 Classification Systems of Brazil, Canada, England and Wales, New Zealand and South Africa

5.9.1 Introduction

The soil classification systems of Brazil, Canada, England and Wales, New Zealand and South Africa have in common that they are developed in one, or at the most, two major world ecological regions, and therefore, focus on the characteristics important in that particular region. The Brazilian soil classification specializes in tropical soils, the Canadian system of soil classification focuses mainly on soils of the (sub)arctic and drier boreal regions, that of England and Wales aims at classifying soil conditions characteristic of the humid temperate regions, as does that of New Zealand, but with focus on volcanic regions, while the South African system is designed to classify soils in dry subtropical areas with emphasis on land use and management aspects.

5.9.2 The Brazilian System

The Brazilian system finds its origin in the older United States soil classifications of Baldwin et al. (1938) and Thorp and Smith (1949). Over the years, adaptations of these systems necessary to suit the Brazilian environmental conditions were introduced, and thus, the present day Brazilian soil classification took shape. Adaptations included modifications of the criteria, new subdivisions and intergrades.

The development of the system is still ongoing. The latest version dates from 1996 (Do Prado, 1996) and adaptation and enlargements are still being discussed. It is intended to arrive at a multi-categorical, open system (to accommodate new classes), which has a morphogenetic basis. It centers around soil horizon characteristics which result from pedogenetic processes, and consequently, morphological, physical, chemical and mineralogical characteristics are used to differentiate the soils. Many of the new elements of the classification system are drawn from soil classification systems currently in use in other countries, such as from Soil Taxonomy (Soil Survey Staff, 1975) and the Legend of the Soil Map of the World (FAO-UNESCO, 1974).

An important element in the Brazilian soil classification is the activity of the clay. It is calculated from the CEC (pH 7), and corrected for the contribution of organic matter. A graphic method as developed by Bennema (1966) is used, especially for the Latosols. Two classes are distinguished, with subclasses for the latosolic B horizon, as follows: (1) high activity soils with a CEC ≥ 24 cmol$_c$ kg^{-1} clay, and (2) low activity soils with a CEC < 24 cmol$_c$ kg^{-1} clay. For the latosolic B horizon (thick subsurface horizon with low CEC, low in weatherable minerals, and a silt/clay ratio < 0.7), the following CEC subdivisions are used: (1) upper limit for latosolic B horizons of 13 cmol$_c$ kg^{-1} clay, and (2) upper limit for highly weathered latosolic B horizons of 6.5 cmol$_c$ kg^{-1} clay.

The following 15 classes are presently recognized at the highest level of the Brazilian soil classification system:

Latosols (*Latossolos*): soils with a latosolic B horizon;

Structured earths (*Terra estruturada*): non-hydromorphic mineral soils with a well-structured, clayey textural B horizon;

Podzolic soils (*Solos podzólicos*): non-hydromorphic mineral soils with a textural B horizon,
Podzols (*Podzolos*): mineral soils with a podzol B horizon;
Rubrozems (*Rubrozéms*): non-hydromorphic, clayey soils with a humic A horizon (equivalent to umbric epipedon), a textural B horizon and an extremely high aluminium saturation;
Non-calcic brown soils (*Solos brunos não cálcico*): non-hydromorphic mineral soils with a weak A horizon and a textural B horizon;
Planosols (*Planossolos*): mineral soils with an abrupt textural change and a textural B horizon,
Solodized Solonetz (*Solonetz-solodizado*): mineral soils with a natric B horizon;
Solonchaks (*Solonchaks*): mostly hydromorphic mineral soils with a salic horizon, or a saline C horizon;
Cambisols (*Cambissolos*): mostly non-hydromorphic mineral soils with an incipient B horizon,
Plinthosols (*Plintossolos*): soils with plinthite (soft or hardened);
Hydromorphic soils (*Solos hidromórficos*): hydromorphic mineral soils;
Vertisols (*Vertissolos*): soils with ≥ 30% clay, developing cracks upon drying and having intersecting slickensides, wedge-shaped aggregates or gilgai microrelief;
Rendzinas, Lithosolic soils, Regosols, Quartzose sands, Alluvial soils (*Rendzinas, Solos litólicos, Regossolos, Areias quartzosas, Solos aluviais*): weakly developed soils without diagnostic subsurface horizon;
Organic soils (*Solos orgânicos*): hydromorphic organic soils.

5.9.3 The Canadian System

The Canadian system (Canada Soil Survey Committee, Subcommittee on Soil Classification, 1978) comprises a number of taxa which are defined on the basis of soil properties; however, the system is genetically biased with respect to the definition of the higher categories. The system is differentiated as follows: orders, suborders, great groups, subgroups, families and series. The following 9 orders are recognized:
Brunosolic order: weak to moderately developed soils;
Chernozemic order: soils with a dark colored, high base-saturated (dominantly Ca) A horizon which is at least 10 cm thick and which have an organic carbon content between 1 and 17% (depending on clay content);
Cryosolic order: soils which have permafrost within 1 m of the surface, or 2 m if they are strongly cryoturbated;
Gleysolic order: soils permanently or temporarily saturated with water and which experience reducing conditions;
Luvisolic order: soils with a subsurface horizon showing evidence of clay accumulation;
Organic order: soils dominated by organic horizons (horizons > 17% organic C);
Podzolic order: soils having a podzolic B horizon (horizon with accumulation of amorphous materials);
Regosolic order: soils showing little or no soil development;
Solonetzic order: soils having a solonetzic B horizon (subsurface horizon with prismatic or columnar structures and a Ca/Na ratio ≥ 10).

5.9.4 The Soil Classification System of England and Wales

The differentiating criteria which build the soil classification for England and Wales (Avery, 1980) comprise ... the composition of the soil material, the presence or absence of particular horizons or sequences of horizons, or other specified differentiating criteria... Ten major soil groups are recognized which are further subdivided in soil groups and soil subgroups. The 10 major soil groups are

Brown soils: soils having a weathered (cambic) or argillic B horizon;

Groundwater gley soils: soils influenced by the presence of a shallow fluctuating ground watertable,

Lithomorphic soils: shallow soils (mineral substratum starting within 40 cm depth) with a distinct, humose or peaty topsoil;

Manmade soils: soils having a thick manmade A horizon or which are profoundly disturbed to depths exceeding 40 cm;

Peat soils: soils with > 40 cm organic material, or > 30 cm if directly overlying bedrock, and lacking overlying nonhumose mineral horizons extending below 30 cm;

Pelosols: slowly permeable clayey soils;

Podzolic soils: soils having a podzolic B horizon (subsurface horizon with accumulations of organic matter and Al, Fe, or both);

Raw gley soils: soils with a gleyed subsurface horizon lacking a distinct topsoil;

Surface-water gley soils: soils having a gleyed subsurface horizon which can be attributed to saturation by surface water;

Terrestrial raw soils: soils consisting of little altered mineral material having no diagnostic surface or subsurface horizons.

5.9.5 The New Zealand System

The newly published New Zealand soil classification (Hewitt, 1998) comprises 15 orders, which are subdivided into groups and subgroups. They are distinguished from each other by the presence or absence of diagnostic horizons and other differentia, some of which are unique to New Zealand. An important element in the classification is a listing of accessory properties of the order, some of which are directly related to soil management and conservation. The 15 orders are

Allophanic soils: soils strongly influenced by minerals with short-range order (esp. allophane, imogolite and ferrihydrite);

Anthropic soils: soils formed by the direct action of people;

Brown soils: soils having a weathered B, argillic or cutanic horizon (horizon containing translocated material which, however, fails to meet the requirements for argillic or Bh horizon), generally having a low base status;

Gley soils: poorly and very poorly drained soils;

Granular soils: clayey soils dominated by kaolinite group minerals;

Melanic soils: soils with highly base-saturated, well-structured, very dark A horizons;

Organic soils: soils occurring in partly decomposed remains of wetland plants or forest litter;

Oxidic soils: soils containing low activity clays and secondary oxides, which give rise to variable charge properties;

Pallic soils: soils with moderate to high base status and low contents of secondary Fe oxides;

Podzols: acid soils with low base saturation and subsurface accumulation of organo Al complexes, with or without Fe;

Pumice soils: soils having properties dominated by a pumiceous and glassy skeleton with a low content of clay;

Raw soils: soils lacking distinct topsoil development;

Recent soils: soils showing only incipient marks of soil-forming processes;

Semi-arid soils: soils having a high base status and soil water deficit during the growing season;

Ultic soils: acid soils with clayey and/or organic illuvial features.

5.9.6 The South African System

The South African taxonomic system of soil classification (Soil Classification Working Group, 1991) comprises only two categories, soil forms and soil families. Soil forms are defined by a unique vertical sequence of diagnostic horizons and materials, while the soil families are separated on the basis of other properties. Nomenclature is South African in the sense that both soil forms and soil families bear local, geographical names (e.g., Escourt Form, Haarlem Family). So far 73 soil forms are defined, divided into 406 soil families. The definitions of the diagnostic horizons and materials fit the unique South African conditions, and are very difficult to correlate with other existing classification systems. Important elements for classification of the soil forms are the various topsoil horizons (organic O, humic A, vertic A, melanic A and orthic A horizons), the pedality (apedal, structural grade weaker than moderate, versus structured B horizons) various clay-enriched horizons (prismacutanic, pedocutanic, lithocutanic and neocutanic B horizons), signs of wetness (e.g., soft and hard plinthic B horizons), carbonate accumulation (neocarbonate B horizon, soft carbonate horizon and hardpan carbonate horizon), and nature of the underlying material (regic sand, stratified alluvium, saprolite, hard rock). Distinguishing criteria at soil family level comprise the thickness, colors, base status structure, consistency and wetness of horizons.

5.10 References

AFES-INRA (Association Française pour l'Etude du Sol) (Institut National de la Recherche Agronomique). 1990. Référentiel Pédologique Français, 3^éme proposition. AFES, Plaisir, France.

AFES-INRA (Association Française pour l'Etude du Sol) (Institut National de la Recherche Agronomiqie). 1992. Référentiel Pédologique, principaux sols d'Europe. INRA, Paris, France.

Asiamah, R.D., J.K. Senayah, T. Adjei-Gyapong, and O.C. Spaargaren. 1997. Ethno-pedology surveys in the semi-arid savanna zone of northern Ghana. Soil Research Institute, Kwadaso-Kumasi, Ghana and International Soil Reference and Information Centre, Wageningen, Netherlands.

Aubert, G., and P. Duchaufour. 1956. Projet de classification des sols. *In* Rapp. VIe Congr. Int. Sc. Sol, V. E:597–604.

Avery, B.W. 1980. Soil classification for England and Wales. Technical Monograph 14. Soil Survey for England and Wales, Harpenden, UK.

Baldwin, M., C.E. Kellogg, and J. Thorp. 1938. Soil classification. *In* Soils and men. USDA Yearbk. p. 979–1002. US Government Printing Office, Washington, DC.

Canada Soil Survey Committee, Subcommittee on Soil Classification. 1978. The Canadian system of soil classification. Can. Dep. Agric. Pub. 1646. Supply and Services Canada, Ottawa, Canada..

CPCS (Commission de Pédologie et de Cartographie des Sols). 1967. Classification de sols. Mimeogr. 96 p., Thivernal-Grignon, France.

Do Prado H. 1996. Manual de classificacão de solos do Brasil. 3a ed. Jaboticabal: FUNEP, Brasilia, Brazil.

Dudal R. 1990. Progress in IRB preparation. p. 69–70. *In* B.G. Rozanov (ed.) Soil classification. Reports of the International Conference on Soil Classification. Centre for International Projects, USSR State Committee for Environmental Protection, Moscow, Russia.

FAO. 1988. FAO-UNESCO Soil map of_the world, Revised Legend. World Soil Resources Report 60. Food and Agriculture Organization, Rome, Italy.

FAO-ISRIC-ISSS. 1998. World reference base for soil resources. World Soil Resources Report 84. Food and Agriculture Organization, Rome, Italy.

FAO-UNESCO. 1974. Soil map of the world 1:5,000,000. Volume I, Legend. United Nations Educational, Scientific and Cultural Organization, Paris, France.

Gong, Z. 1994. Chinese soil taxonomic classification (First Proposal). Institute of Soil Science, Academia Sinica, Nanjing, China.

Hewitt A.E. 1998. New Zealand soil classification. Landcare Research Science Series 1. Manaaki Whenua Press, Lincoln, New Zealand.

Isbell R.F. 1996. The Australian soil classification. Australian Soil and Land Survey Handbook. CSIRO, Collingwood, Australia.

ISS-AS (Institute of Soil Science-Academia Sinica). 1990. Soils of China. Science Press, Beijing, China.

ISSS-ISRIC-FAO. 1994. World reference base for soil resources. Draft. Compiled and edited by O.C. Spaargaren. International Society of Soil Science, International Soil Reference and Information Centre, and the Food and Agriculture Organization of the United Nations. Wageningen/Rome, Netherlands/Italy.

ISSS-ISRIC-FAO. 1998. World reference base for soil resources. Acco Press, Leuven, Belguim.

McDonald R.C., R.F. Isbell, J.G. Speight, J. Walker, and M.S. Hopkins. 1990. Australian soil and land survey field handbook. 2nd Ed. Inkata Press, Melbourne, Australia.

Nachtergaele, F., A. Remmelzwaal, J. Hof, J. van Wanbeke, A. Souirji and R. Brinkman. 1994. Guidelines for distinguishing soil subunits. Trans. 15th World Congr. Soil Sci. 6a:818–833.

Northcote K.H. 1979. A factual key for the recognition of Australian soils. 4th Ed. Rellim Technical Publications, Glenside, South Australia.

Shishov L.L. and I.A. Sokolov. 1990. Genetic classification of soils in the USSR. p. 77–93. *In* B.G. Rozanov (ed.) 1990. Soil classification. Reports of the International Conference on Soil Classification. USSR State Committee for Environmental Protection, Moscow, Russia.

Soil Classification Working Group. 1991. Soil Classification. A taxonomic system for South Africa. Memoirs on the Agricultural Natural Resources of South Africa 15. Department of Agricultural Development, Pretoria, South Africa.

Soil Survey Staff. 1975. Soil taxonomy. A basic system of soil classification for making and interpreting soil surveys. Agriculture Handbk. 436. SCS-USDA, Washington, DC.

Soil Survey Staff. 1996. Keys to soil taxonomy. 7th Ed. Natural Resources Conservation Service, USDA, Washington, DC.

Stace H.C.T., G.D. Hubble, R. Brewer, K.H. Northcote, J.R. Sleeman, M.J. Mulcahy, and E.G. Hallsworth. 1968. A Handbook of Australian Soils. Rellim Technical Publications Glenside, South Australia.

Thorp, J. and G.D. Smith. 1949. Higher categories of soil classification: Order, suborder, and great groups. Soil Sci. 67:117–126.

VASKhNIL (V.V. Dokuchaev Institute of Soil Science). 1986. Classification and diagnostics of soils of the USSR. Translated from Russian by S. Viswanathan. Amerind Publishing Co., New Delhi, India.

6

Classification of Soils

6.1 Introduction: General Characteristics of Soil Orders & Global Distributions

Larry P. Wilding
Texas A & M University

The purpose of Chapter 6 is to present a discussion of the distribution, attributes, pedogenesis, classification, use and management of the 12 Soil Orders recognized by *Soil Taxonomy* (Soil Survey Staff, 1998). In this chapter, soil orders have been arrayed in a sequence to consider the organic soils first, followed next by soils which are weakly expressed or lack subsoil diagnostic horizons, and closing with soils exhibiting the greatest evidence of pedogenic weathering and/or accumulation of sesquioxides and translocated phyllosilicates. Only general statements can be made at this categorical level, but the information content becomes more precise as one descends to lower taxa (Sections 6.2–6.13). Because *Soil Taxonomy* is a dynamic classification system, several of the orders are under revision or have been revised recently; hence, subject matter included herein is subject to change. This is particularly true of the Histosols, Mollisols, Alfisols, Inceptisols, and Gelisols.

Because of space limitations, data sets verifying the physical, chemical, mineralogical and biological attributes of given orders are abridged. However, access to complete soil characterization databases, engineering properties, pedon descriptions, soil interpretations, and other information for different soil taxa are available online at the following address: http://www.statlab.iastate.edu/soils/soildiv/. For example, through the Soil Survey Laboratory, National Soil Survey Center, Lincoln, Nebraska, analytical data for more than 20,000 pedons in the United States and 1,100 pedons from other countries have been archived. Standard morphological pedon descriptions are available for about 15,000 of these pedons. Further, these databases include listings of published soil surveys in the United States, Puerto Rico, Virgin Islands, Trust Territories and some foreign countries, areal extent and classification placements of soil series, ongoing soil survey investigations, including soil-related studies on climatic change, and field and laboratory methodologies.

Fig. 6.1 illustrates the global distribution of soil orders at 1:1 million scale and Table 6.1 provides the areal extent of suborders associated with these orders. Tables 6.2 and 6.3 present proportional percentages of soil orders occurring in different soil temperature (Table 6.2) and soil moisture (Table 6.3) regimes as defined in *Soil Taxonomy* (Soil Survey Staff, 1998). Broad geographical distribution patterns (Fig. 6.1) reflect regional conditions of climate, vegetation, geology, and topography that function interactively over time. Soils found at any point on the land surface are considered products of multiple sets of pedogenic processes with state factors serving as controls (Section E, Chapter 4). Generally, soils are developed along polygenetic pathways, on dynamically evolving landforms under the influence of paleoclimates, in nonuniform parent materials. Hence, rarely can bio-, litho-, climo-,

Global Soil Regions

U.S. Department of Agriculture
Natural Resources Conservation Service
Soil Survey Division
World Soil Resources

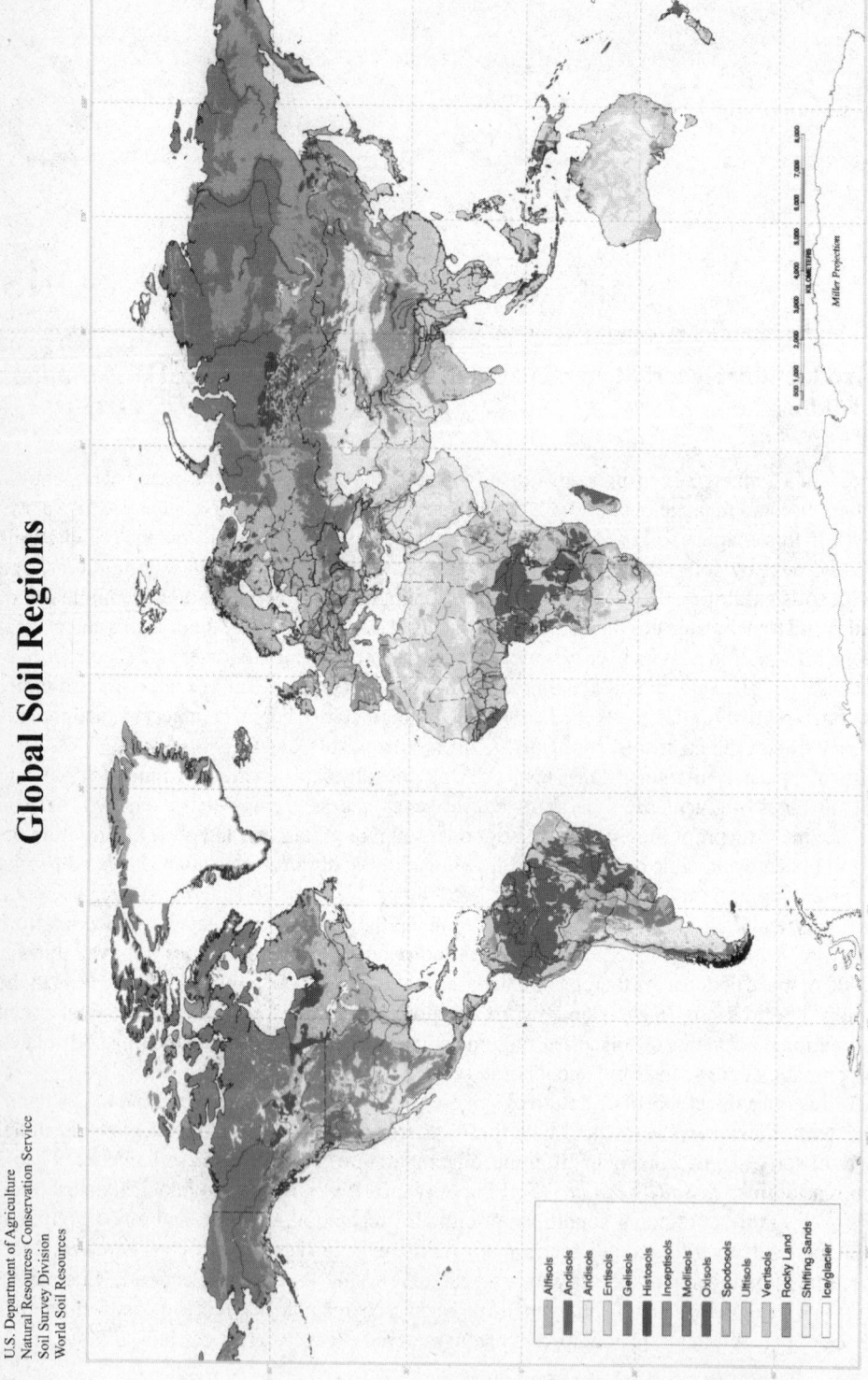

Country boundaries are not authoritative

Fig. 6.1 Illustration of global distribution of soil orders at 1:1 million scale. See color insert following page E-204. [Courtesy of the USDA-NRCS, Soil Survey Division, World Soil Resources, 1998]

Table 6.1　Global areas and percentages of suborders and miscellaneous land units based on ice-free land area [Courtesy of the USDA-NRCS, Soil Survey Division, World Soil Resources, 1998]

Order	Suborder		Area km^2 (x 10^3)	Proportion %
Alfisols	Aqualfs		836	0.7
	Cryalfs		2,518	1.9
	Ustalfs		5,664	4.3
	Xeralfs		897	0.7
	Udalfs		2,706	0.2
		Subtotal	**12,621**	**9.6**
Andisols	Cryands		255	0.2
	Torrands		2	« 0.1
	Xerands		32	« 0.1
	Vitrands		281	0.2
	Ustands		63	0.1
	Udands		279	0.2
		Subtotal	**912**	**0.7**
Aridisols	Cryids		943	0.7
	Salids		890	0.7
	Gypsids		683	0.5
	Argids		5,408	4.1
	Calcids		4,873	3.7
	Cambids		2,931	2.3
		Subtotal	**15,728**	**12.0**
Entisols	Aquents		116	0.1
	Psamments		4,428	3.4
	Fluvents		2,860	2.2
	Orthents		13,733	10.5
		Subtotal	**21,137**	**16.2**
Gelisols	Histels		1,013	0.8
	Turbels		6,333	4.8
	Haplels		3,914	3.0
		Subtotal	**11,260**	**8.6**
Histosols	Folists		« 1	« 0.1
	Fibrists		197	0.1
	Hemists		988	0.8
	Saprists		341	0.3
		Subtotal	**1,527**	**1.2**
Inceptisols	Aquepts		3,199	2.5
	Anthrepts		< 1	« 0.1
	Cryepts		457	0.3
	Ustepts		4,241	3.2
	Xerepts		685	0.5
	Undepts		4,247	3.3
		Subtotal	**12,830**	**9.8**
Mollisols	Albolls		28	« 0.1
	Aquolls		118	0.1
	Rendolls		266	0.2
	Xerolls		924	0.7
	Cryolls		1,164	0.9
	Ustolls		5,245	4.0
	Udolls		1,261	1.0
		Subtotal	**9,006**	**6.9**

Table 6.1   (cont.)

Order	Suborder		Area km^2 (x 10^3)	Proportion %
Oxisols	Aquox		320	0.2
	Torrox		31	« 0.1
	Ustox		3,096	2.4
	Perox		1,162	0.9
	Udox		5,201	4.0
		Subtotal	**9,810**	**7.5**
Spodosols	Aquods		169	0.1
	Cryods		2,460	1.9
	Humods		58	< 0.1
	Orthods		667	0.5
		Subtotal	**3,354**	**2.5**
Ultisols	Aquults		1,281	1.0
	Humults		344	0.3
	Udults		5,540	4.2
	Ustults		3,870	3.0
	Xerults		19	« 0.1
		Subtotal	**11,054**	**8.5**
Vertisols	Aquerts		5	« 0.1
	Cryerts		15	« 0.1
	Xererts		99	0.1
	Torrerts		889	0.7
	Usterts		1,768	1.3
	Uderts		384	0.3
		Subtotal	**3,160**	**2.4**
Miscellaneous	Shifting		13,076	10.0
	Other		5,322	4.0
		Subtotal	**18,398**	**14.0**
Total ice-free land area			**130,797**	**100.0**

chrono- and toposequence functions be developed with scientific rigor. Clearly, the distribution of soil orders in which local conditions strongly govern pedogenic dynamics over short-range distances is not well illustrated in (Fig. 6.1). In particular, such comments are germane to distributions of Histosols, Spodosols, Vertisols and Gelisols. In these cases, the distributions presented in subsections of this chapter may differ somewhat from those presented herein (Fig. 6.1). This uncertainty will also impact the accuracy of global distributions and areal data given for Histosols, Andisols, and Inceptisols found interspersed with Gelisols.

The human impact on soils is recognized in some taxa of Entisols and Inceptisols, but, in general, *Soil Taxonomy* does not adequately handle the anthropogenic effects on soils. For example, Mollisols can be transformed to Alfisols upon severe erosion; paddy soils can undergo secondary salinity and waterlogging under long-term rice culture; soils in urban/industrial environments can be markedly modified by landfills, landfarming, earth movement and heavy metal contamination; and drastically disturbed soils are common in regions where precious metals, rock aggregate and fossil fuels have been mined. The International Committee on Anthropic Soils (ICOMANTH) where ANTH stands for anthropogenic, is currently addressing this matter. For information on this activity refer to the ICOMANTH web-site address: http://wwwscas.cit.cornell.edu/icomanth.

Soil Taxonomy currently does not accommodate systematic cataloging of paleosols formed under remarkably different paleoenvironments. With increasing geomorphic age, properties of soils commonly reflect a welding of contemporaneous and paleoenvironments. Especially when paleosols

Table 6.2 Extent of soil orders found in different temperature regions [Modified from data supplied through the courtesy of the USDA-NRCS, Soil Survey Division, World Soil Resources, 1998]

| Soil Order | Soil temperature regimes* | | | |
	Tropical	Temperate	Boreal	Tundra
		%		
Alfisols	38	39	23	0
Andisols	49	23	28	0
Aridisols	12	74	14	« 1
Entisols	28	68	4	0
Gelisols	0	0	0	100
Histosols	21	8	71	0
Inceptisols	47	42	11	0
Mollisols	4	50	46	0
Oxisols	98	2	0	0
Spodosols	« 1	18	82	0
Ultisols	69	31	« 1	0
Vertisols	47	52	1	0

*Soil temperature regimes defined by *Soil Taxonomy* (Soil Survey Staff, 1998, http://www.statlab.iastate.edu/soils/soildiv)

Tropical: isomesic, isothermic and isohyperthermic (MAST > 8 °C in which the difference in mean summer and mean winter temperatures is < 6 °C at 50 cm soil depth)

Temperate: mesic, thermic, and hyperthermic (MAST is > 8 °C at 50 cm soil depth).

Boreal: frigid, isofrigid, and cryic (MAST is 0-8 °C at 50 cm soil depth).

Tundra: pergelic (MAST < 0 °C at 50 cm soil depth)

are well preserved, they are valuable proxies of biological and physicochemical evolution of the Earth. This is an area for *Soil Taxonomy* development that will likely gain greater attention in the future. With multiple uses of soil surveys for connectivity between the land surface and geological substrata (e.g., for environmental, hydrological, archaeological and biogeoscience synergisms), there are increasing driving forces to observe soils beyond the 2 m depth limit currently in vogue. This will likely enhance the interests in revising Soil Taxonomy to better accommodate recognition and cataloging of paleosols as important morphogentic markers of paleoenvironments (Section E, chapter 1).

In perusing the global distributions of soil orders (Fig. 6.1), and aerial extents given in Tables 6.1–6.3, several noteworthy relationships become apparent. These are summarized in a cursory overview in the following paragraphs. The organic-rich kingdom of soils [Histosols (Chapter 6.2)] comprises about 1.2 % of the Earth's ice-free land surface (Table 6.1). They are generally in cool, humid, high latitude, boreal regions (or in microhabitats) that favor the balance of organic matter accumulation over decomposition (Tables 6.2 and 6.3). This is because of saturated and reduced conditions, biological inhibitors, and/or cool, short summer periods with low evapotranspiration rates. Many of the Histosols that could be illustrated on this general map are now considered Gelisols or Andisols with the introduction of these two new soil orders. Histosols are important wildlife and wetland habitats. They also serve extensively for cranberry, citrus, vegetable, rice, and sugarcane production. Major constraints are subsidence, nutrient deficiencies and wind erosion. Effective watertable management is critical to minimize subsidence. These land resources are also critically limiting to construction activities because of their low soil strength and load bearing capacities.

Entisols (Chapter 6.3) comprise about 16.2% of the Earth's ice-free land surface (Fig. 6.1). Over two thirds of the Entisols are in the temperate region and the remainder mostly in the tropics (Table 6.2). Sixty percent of the Entisols have aridic soil moisture regimes with ustic and udic, the next most extensive (Table 6.3). These are mineral soils which do not have expression of diagnostic subsurface horizons within a specified depth of the soil surface, generally 1–2 m. Subsoil materials of these soils

Table 6.3 Extent of soil orders found in different soil moisture regimes [Modified from data supplied through the courtesy of the USDA-NRCS, Soil Survey Division, World Soil Resources,1998]

Soil Order	Soil moisture regimes*			
	Aridic	Xeric	Ustic	Udic**
		%		
Alfisols	0	8	56	36
Andisols	4	3	30	63
Aridisols	95	1	3	1
Entisols	60	4	22	14
Gelisols	N/A[†]	N/A	N/A	N/A
Histosols	8	1	23	68
Inceptisols	0	6	42	52
Mollisols	0	13	66[‡]	22
Oxisols	« 1	0	35	65
Spodosols	« 1	1	9	90
Ultisols	0	« 1	42	58
Vertisols	28	3	56	13

* Soil moisture regimes defined by *Soil Taxonomy* (Soil Survey Staff, 1998, http://www.statlab.iastate.edu/soils/soildiv)
** Udic soil moisture regime includes perudic (continuously moist) and aquic conditions (saturated, reduced and with redoximorphic features)
[†] N/A - not appropriate for the Gelisol order
[‡] A few percentage of Mollisols placed in Ustic soil moisture regimes may in fact occur under Aridic soil moisture regimes, especially in steppe zones of Eurasia that border on Ustic-Aridic soil moisture regimes.

reflect the nature of the geological materials from which they were derived. These weakly expressed soils are commonly found on geomorphic surfaces which are unstable because of frequent flooding, erosion/truncation, or human impact (drastically disturbed lands). They are formed also in coarse-textured, resistant mineral parent materials (e.g., quartzose sands) that are subject to little pedogenic development over time. Entisols are common along flood pains of river and stream valley systems, sand dunes in desert regions, on high gradient mountainous terrain, and associated with recently mined or disturbed lands.

Andisols (Chapter 6.4) comprise < 0.7% of the Earth's ice-free land surface (Table 6.1), but these soils are so unique physically, chemically and mineralogically that a separate soil order was established for them. They are found along the tectonically active Pacific Ring of Fire, Central Atlantic Ridge, North Atlantic rift, the Caribbean, and the Mediterranean regions where volcanic or pyroclasic deposits are common (Fig. 6.1). About half of them are located in tropical regions and the remainder split between temperate and boreal climates (Table 6.2). About two thirds are in udic soil moisture regimes and one third in ustic regions (Table 6.3). These weakly developed, fertile soils are texturally undifferentiated and characterized by short-range order (amorphous) aluminosilicates that have not been translocated from upper to lower horizons. Andisol land resources are used extensively for crop production in developing countries employing slash/burn traditional agricultural practices.

Inceptisols (Chapter 6.5) comprise about 9.8% of the Earth's ice-free land surface (Fig. 6.1). These soils occur indiscriminately globally because they lack a sharply defined central concept. Over 90% occur in tropical and temperate climates (Table 6.2) under ustic and udic soil moisture regimes (Table 6.3). Inceptisols serve to make Soil Taxonomy fully bifurcated ("junk basket" category). Soils classified as Inceptisols generally contain light colored surface horizons and weakly expressed diagnostic subsoil horizons (cambic), but under certain circumstances, may contain more strongly developed subsoil diagnostic horizons (e.g., petrocalcic, petrogypsic, duripan, and fragipan). Where these soils reflect youthfulness, they occur in high gradient mountainous regions, along major river systems as terraces and deltaic/fluvial plains, and as soils developed from carbonate rich bedrocks or sediments in Mediterranean environments. They also occur interspersed among more strongly developed Alfisols, Ultisols and Oxisols in tropical and subtropical regions.

Gelisols, the newest of the soil orders (Chapter 6.6), comprise about 8.6% of the Earth's ice free land surface (Fig. 6.1). These are soils of the high latitude polar regions underlain by permafrost (materials wet/dry that remain below 0 °C for 2 consecutive years) at depths of 1–2 m depending on cryoturbation (frost churning) activity (Table 6.2). Cryoturbation in Gelisols is driven by the physical volume change from water to ice and subsequent moisture migration along thermal, hydrostatic, chemical, and electrical gradients. Cryopedogenic processes include freezing/thawing, cryoturbation, frost heaving, cryogenic sorting, thermal cracking, and ice segregation. Remarkable spatial variability occurs in soil properties over distances of a few meters or less giving rise to pattern ground in the form of stone stripes, stone circles, high center polygons, low center polygons and barren frost boils. Construction activities on these land resources require management of thermal balances so the insulation properties of the surface are not disturbed. If the organic rich surface layers are compacted, compressed or removed, substrata ice wedges will melt causing the surface to become a highly irregular thermokarst topography. Because Gelisols are the most recent soil order established, and because the depth and presence (or absence) of permafrost is locally controlled, the precise global distribution of Gelisols as presented in Fig. 6.1 is preliminary and approximate. Likewise, the areal extent of Gelisol suborders presented in Table 6.1 is still under development and refinement.

Vertisols (Chapter 6.7) comprise about 2.4% of the Earth's ice-free land surface (Fig.6.1). About half the Vertisols are found in tropical environments and about half in temperate (Table 6.2). Over 50% of the Vertisols occur in regions of ustic soil mosture regimes where seasonal desiccation is common. The remainder are found primarily under aridic, udic and xeric environments; here the oscillation from wet to dry soil conditions is less frequent (Table 6.3) and gilgai expression less extreme. These clayey, shrink/swell soils are highly diverse in physical, chemical, biological and mineralogical properties. They are concentrated in subtropical and tropical environments, but occur across broad climatic, vegetative, topographic, and geologic regions of Africa, India, Australia, China and North and South America (Fig. 6.1). Vertisols are active and subject to soil failure when swelling pressures exceed shear strength upon wetting a dry soil. Soil failure is usually expressed as slickensides occurring as diagonal, polished and grooved slip surfaces, wedge-shaped structural units, and microtopography in the form of gilgai. Commonly these soils have self-swallowing and self-mulching surface features where strongly aggregated surface materials infill vertical structural cracks. While inversion by turbation is commonly invoked as a major pedogenic processes, most movement in these soils is believed to be diagonally along inclined major slickenside fault planes. Major limitations for using these highly productive soils for agriculture, especially in developing countries, are their high energy requirements for tillage and narrow favorable soil moisture range for workability. Construction activities are constrained by their propensity for soil failure and high shrink/swell activity.

Mollisols (Chapter 6.8) comprise about 7% of Earth's ice-free land surface (Fig. 6.1). Essentially all the Mollisols are in temperate and boreal regions, approximately equally split (Table 6.2). Approximately two thirds of the Mollisols are in ustic soil moisture regimes (Ustolls); most of the remainder occur under udic regions (Udolls) and the rest are in climates with winter precipitation and summer periods of soil moisture deficit (Xerolls) (Table 6.3). A few Mollisols may occur under aridic soil moisture regimes in the southern steppe zones of Eurasia; however, clear definition of the extent of these soils is still wanting. Mollisols are the dark colored, high base status soils commonly found under prairie grass vegetation in mid latitudes of North and South America and Eurasia. The distribution pattern of these soils in Eurasia is in east-west trending belts between Alfisols to the north and Aridisols and Inceptisols to the south. This reflects decreasing precipitation and increasing temperature gradients in traversing from north to south. In contrast, in North America the precipitation isohyets and isotherms are set normal to each other with precipitation decreasing from east to west and temperature increasing from north to south. These climatic controls result in Mollisols being bordered on the east by Alfisols and Ultisols, and to the west by Aridisols, Entisols and drier Alfisols. Parent materials also control Mollisol distribution; Mollisols are commonly associated with calcareous or high base status glacial drift deposits, limestones and loess. Mollisols (especially Udolls) are the "bread basket" soils of the Americas and Eurasia because of their high native fertility, favorable climate, and excellent physical and chemical properties for high crop production potentials.

Spodosols (Chapter 6.9) comprise about 4% of Earth's ice-free land area (Fig. 6.1). Over 80% of the Spodosols occur under boreal environments (Table 6.2) with udic soil moisture regimes (Table 6.3). Spodosols occur between Mollisols of mid to high latitudes and Alfisols in lower latitudes. These soils are very photogenic, and commonly, have prominent expression of eluvial E horizons over black to red spodic and placic horizons. The latter are formed from translocated illuvial organic-metal (Al and Fe) complexes. These soils have many properties in common with Andisols. The major difference is that Spodosols have formed for a sufficient period of time for the weathered organic-metal complexes to be translocated to subsoils. In Andisols little or no translocation of these complexes has taken place. Formation of Spodosols is favored by environments with abundant acid litter, coarse-textured, resistant parent materials, and strong leaching potentials. These soils are very strongly acid and extremely phosphorus deficient. Native vegetation reflects acid tolerant plants. They are little used for agricultural production because of the short, cool growing seasons, high management imputs, and low nutrient and water retention. Major land use is forests for lumbering and recreation. Pasture and forages are used for livestock production and dairy animals.

Aridisols (Chapter 6.10) comprise about 12.0% of the Earth's ice-free land area (Fig. 6.1). Aridisols are mostly in hot temperate deserts with aridic soil moisture regimes, dry coastal regions, or on rain-shadow plains leeward of high mountains (Tables 6.2 and 6.3). Locally, they also occur in more humid regions where salts have concentrated at or near the soil surface through evaporative pumping from shallow saline ground waters. Aridisols range from base-rich to base-poor soils and exhibit a wide diversity in physical, chemical and mineralogical properties. Commonly, they represent contemporaneous xerophytic environments, but some Aridisols have well-expressed subsoil diagnostic properties such as argillic, natric, calcic, petrocalcic, etc. horizons. These likely reflect more pluvial paleoclimates; under such conditions, these diagnostic horizons would be, in part, relict. Aridisols are used extensively for rangelands, military bases, and nomadic agrarian activities. These soils are not generally considered arable land resources without irrigation. However, they are valuable land resources where an adequate quantity of high quality subsurface water is available for intensive agriculture and urban development. Irrigation and drainage must be managed for a favorable salt balance in Aridisols.

Alfisols (Chapter 6.11) and Ultisols (Chapter 6.12) are closely related soil orders that comprise, respectively, about 9.6% and 8.5% of the Earth's ice-free land area (Fig. 6.1). These soils, developed mostly under deciduous forested or savannah environments, are extensive in mid to lower latitudes. Alfisols are distributed approximately equally across tropical, temperate and boreal environments whereas the distribution of Ultisols is skewed towards tropical conditions (Table 6.2). For both Alfisols and Ultisols most of their extent is associated with ustic and udic soil moisture regimes (Table 6.3). Alfisols and Ultisols are differentiated from other orders on the basis of textural differentiation resulting from translocation and/or *in situ* neoformation of clays that form argillic and kandic horizons. Alfisols differ from Ultisols in having higher base saturation (> 35%) at a specified depth in the subsoil. Rationale for this differentiation was that Alfisols could sustain traditional slash/burn subsistence agriculture without significant external fertilizer inputs, while Ultisols would require applied nutrients for crop production. Both orders are used extensively for forest and crop production. Major limitations are nutrient deficiencies, wind and water erosivity, and seasonal soil moisture deficits, especially for those in xeric and ustic regions. Alfisols are used extensively in developing countries of Africa, Central America, South America and India for subsistence traditional slash/burn agriculture.

Sesquioxide-rich Oxisols (Chapter 6.13) comprise about 8% of the Earth's ice-free land surface. They occur in equatorial regions intermixed with Ultisols and Inceptisols (Fig. 6.1). Nearly all the Oxisols occur in tropical environments (Table 6.2) with two-thirds under udic and one-third under ustic soil moisture regimes (Table 6.3). Oxisols have low activity clays and are nutrient poor, weakly buffered systems that have little or no horizon differentiation. This reflects (1) their high weathering intensities in low latitude, humid environments, (2) their geomorphic stability and age associated with mafic-rich, Precambrian platforms, or (3) development from polcycled, preweathered parent materials derived from highly weathered terrestrial source areas. These soils are often considered residual carcasses, but many are very productive for agriculture and forestry if properly managed for water conservation, soil erosion and nutrient inputs. With high nutrient imputs and management to conserve soil and water, Oxisols can be very productive agricultural soil systems.

6.2 Histosols

Martin C. Rabenhorst
University of Maryland

David Swanson
USDA-NRCS, Alaska

6.2.1 Introduction

While most soils of the world are comprised primarily of mineral materials, a small but important group of soils are formed from organic materials derived from plants, or less frequently, from animals. Organic soil materials contain a minimum of 12–18% organic C, depending on the particle size of the mineral component (Soil Survey Staff, 1996). Generally speaking, soils within which at least 40 cm of the upper 80 cm are organic materials, and which do not have permafrost within 1 m of the soil surface, are Histosols. Prior to 1997, organic soils with permafrost were included in the Histosols order, but are now placed in the Histel suborder of Gelisols (Soil Survey Staff, 1998). Therefore, earlier discussions of the worldwide distribution of Histosols include organic soils of permafrost regions and might overestimate the distribution by today's concepts (Everett, 1983). Organic soil materials are commonly referred to as peat, and land covered by Histosols or Histels is known as peatland. The term mire is a synonym of peatland that is more commonly used in Europe. Histosols also include a narrowly distributed group of soils, the Folists, that consist of well-drained organic soil materials that directly overlie bedrock or coarse fragments with little or no intervening fine soil. The peat layer in Folists may be (and usually is) thinner than the 40 cm required for other Histosols.

Because of their high organic C content, many Histosols have been utilized as a combustible energy resource. Mankind has mined and burned peat since prehistoric times, and peat is still an important fuel in a number of northern countries, although it has a lower energy rating than oil or coal. In Russia, Germany, and Ireland, peat is not only utilized for domestic heating, but also on a large scale, in electricity generation. In 1991, peat burning was the source of 14% of Ireland's electricity (McNally, 1997), and as recently as the late 1980s, when fuel costs were higher, a peat-burning electric power plant was built in Deblois, ME, although the subsequent fall in fuel prices precluded the plant from ever becoming operational. In addition to being mined as an energy source, peat is mined for use as a soil amendment in agriculture and horticulture. In addition, the agricultural value of mineral-rich Histosols has long been recognized. Provided that the watertables can be effectively managed, high yields of vegetables and other specialty crops can be produced on Histosols in such different climatic regions as Michigan and Florida (Lucas, 1982).

Because most Histosols occur in wetlands, their utilization as agricultural and energy resources has come under intense scrutiny, particularly in the United States. Histosols perform many of the beneficial functions of wetlands, and they are negatively impacted by mining, drainage or other practices associated with agriculture; thus, there are benefits derived from the preservation of Histosols in their natural state. Histosols are perhaps uniquely fragile and are highly vulnerable to degradation. When drained or dry, organic soil material is highly susceptible to wind erosion (Lucas, 1982). Organic soils also have very low strength, are highly compressible (MacFarlane, 1969; MacFarlane and Williams, 1974), and gradually disappear by decomposition, if drained. Furthermore, within the framework of current discussions of global climate change and C budgeting, Histosols contribute significantly to the terrestrial C pool. On an areal basis, C storage in Histosols is often greater than mineral soils by an order of magnitude (Rabenhorst, 1995.) A listing of great groups and subgroups in the Histosols order is presented in Table 6.4.

6.2.2 Distribution

Histosols occur at all latitudes, but are most prevalent in the boreal forest regions of northern North America, Europe, and Asia (Fig. 6.2). The world's largest expanses of Histosols occur in the West Siberian lowland (Walter, 1977) and the Hudson Bay lowland of central Canada (Sjors, 1963; Canada Committee on Ecological (Biophysical) Land Classification, 1988). At lower latitudes, Histosols occur locally on humid coastal plains, notably southeast Asia and Indonesia (Anderson, 1983).

Histosols in the United States are most widespread in lowlands of the Great Lakes region, the northeast, the Atlantic Coastal plain and Florida, the Pacific Northwest, and Alaska (Fig. 6.2). The largest expanses of Histosols in the continental United States are on the Lake Agassiz plain in north central Minnesota (Wright et al., 1992). Coastal and estuarine areas inundated by tidal water are also sites for Histosol formation, most notably along the Atlantic and Gulf coastlines. Drained Histosols are

Table 6.4 Listing of suborders, great groups and subgroups in the Histosols Order [Soil Survey Staff, 1998]

Suborder	Great Group	Subgroup
Folists	Cryofolists	Lithic, Typic
	Torrifolists	Lithic, Typic
	Ustifolists	Lithic, Typic
	Udifolists	Lithic, Typic
Fibrists	Cryofibrists	Hydric, Lithic, Terric, Fluvaquentic, Sphagnic, Typic
	Sphagnofibrists	Hydric, Lithic, Limnic, Terric, Fluvaquentic, Hemic, Typic
	Haplofibrists	Hydric, Lithic, Limnic, Terric, Fluvaquentic, Hemic, Typic
Saprists	Sulfosaprists	Typic
	Sulfisaprists	Terric, Typic
	Cryosaprists	Lithic, Terric, Fluvaquentic, Typic
	Haplosaprists	Lithic, Limnic, Halic Terric, Halic, Terric, Fluvaquentic, Hemic, Typic
Hemists	Sulfohemists	Typic
	Sulfihemists	Terric, Typic
	Luvihemists	Typic
	Cryohemists	Hydric, Lithic, Terric, Fluvaquentic, Typic
	Haplohemists	Hydric, Lithic, Limnic, Terric, Fluvaquentic, Fibric, Sapric, Typic

widely used in agriculture in the Great Lakes region, southern Florida, and the Sacramento-San Joaquin delta region of California. In the semi-arid Great Plains and mountainous west, Histosols are very rare and occur only in areas of steady groundwater discharge (Mausbach and Richardson, 1994) or humid areas at high elevations (Cooper and Andrus, 1994). Organic soils are widespread in lowlands throughout Alaska, although most of the organic soils in central and northern Alaska have permafrost, and hence, are classified as Gelisols rather than Histosols.

6.2.3   Formation

6.2.3.1   Parent Material

In contrast to the wide variety of mineral materials which may serve as parent materials for other soils, the parent materials from which Histosols are formed are organic in nature. The conditions which cause the accumulation of organic parent materials are very closely tied to the processes which form various organic soil horizons. The unique properties of Histosols result from the nature of the organic parent material.

Some of the factors which affect the nature of organic parent materials include hydroperiod, water chemistry, and vegetation type and will be discussed in more detail later. The net accumulation of organic materials occurs when rates of additions (usually as primary plant production) exceed those of decomposition. In natural soils of most ecosystems, a steady state exists between these two processes, which maintains the quantity of organic C in the surface horizons somewhere between 0.5% and 10%, although some forest soils have thin layers of organic soil materials (O horizons). In Histosols, the rates of decomposition are slowed and organic matter accumulates to significant depth. In most cases, this is caused by saturation leading to anaerobic conditions which causes soil organic

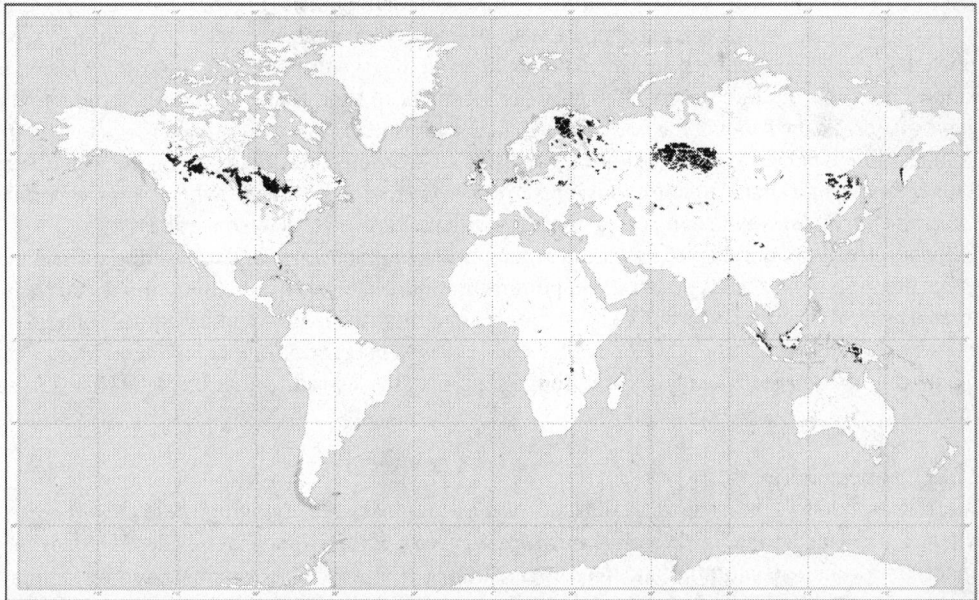

Fig. 6.2 Global distribution of Histosols. The extent of Histosols is probably significantly greater in Alaska than shown on this map due to an overestimation in the extent of permafrost (Gelisols) in Alaska used in developing this map [Courtesy of USDA-NRCS, Soil Survey Division, World Soil Resources, 1998].

matter (SOM) decomposition to be less efficient than under aerobic conditions. Occasionally, usually under cool and moist conditions, organic parent materials may accumulate without prolonged saturation, producing soils in the unique Folist suborder of Histosols. Under certain conditions and landscapes, mineral materials can be added to accumulating organic parent materials, typically by alluviation or eolian deposition. In such cases, the balance between mineral inputs and organic accumulation will determine whether a Histosol or mineral soil will form.

The accumulation of organic parent materials often occurs over long periods of time and under changing conditions. Thus, the stratigraphy of a bog may reflect many thousands of years of organic accumulation. Rarely does vegetation remain constant over such long periods. Microscopic examination of the partially decomposed peat, or evaluation of pollen or plant microfossils can provide information concerning the types of plants which have contributed to the organic parent material during various stages of accumulation.

The organic parent material of Histosols is a major source of acidity. Acids produced by the partial decomposition of organic matter cause the organic horizons of Histosols to be highly acidic unless they have been neutralized by bases that were dissolved from mineral soils or rocks and transported into the peat by groundwater. Highly acidic peatlands are called bogs and commonly described as ombrotrophic (rainfed, because all nutrients are derived from atmospheric sources). The less acid Histosols that receive base-rich groundwater or runoff are called fens (if treeless) or swamps (if forested) and usually described as minerotrophic (fed with mineral-derived nutrients). The term swamp is also applied to forested wetlands on mineral soils. Peatlands are also sometimes divided by their acidity into oligotrophic (very acid and mineral poor), mesotrophic (intermediate), and eutrophic (weakly acidic to neutral and mineral rich) classes (Moore and Bellamy, 1974; Gore, 1983).

6.2.3.2 Climate

Histosol formation is favored by wet or cold climates by hindering decomposition of organic matter. The northern Histosol-dominated regions of North America and Eurasia (Fig. 6.2) have temperate and boreal climates in which average precipitation exceeds annual potential evapotranspiration (Trewartha, 1968). Further to the north, in Greenland and the islands bordering the Arctic Ocean, for example, Histosols are rare because the growing season is so short that there is little production of organic matter. To the south of the boreal zone, Histosol formation is apparently constrained by rapid decomposition of SOM. At lower latitudes, Histosols are restricted mostly to coastal plains with very flat topography, high annual precipitation, and no dry season, or to those areas where a high watertable is maintained by tidal waters and coastal marshes with mangroves (Anderson, 1983).

Climate affects not only where Histosols occur, but also the chemistry of the resulting soils. Where average precipitation exceeds potential evapotranspiration, Histosols in suitable settings can remain saturated by rainfall alone (Ivanov, 1981). In such regions, both the highly acidic Histosols of bogs and the less acid Histosols of groundwater-fed fens and swamps may form. In less humid regions, where groundwater is required to maintain saturation of soils, only the less acid, minerotrophic Histosols can form.

6.2.3.3 Topography

While climate controls the occurrence of Histosols on a regional scale, topography dictates where they occur on a given landscape. The classical soil-forming factor of topography is considered broadly here to incorporate relief, geomorphic, and hydrologic settings. Histosols occur where the setting facilitates concentration of runoff, discharge of groundwater, or retention of precipitation. (Fig. 6.3A,B,C). These conditions are most often satisfied in topographic depressions or very flat areas. On plains with low permeability substrates in the boreal zone, Histosols may cover entire interfluves (Sjors, 1963;

Walter, 1977), and in extremely humid climates, such as the British Isles and southeastern Alaska, Histosols may also occur on gentle slopes (Fig. 6.3B) (Moore and Bellamy, 1974; Sjors, 1985). Histosols of flood plains receive suspended mineral matter during floods (Fig. 6.3C), while other Histosols obtain only dissolved material from source waters or inputs of eolian materials.

In lower latitude coastal areas, continually rising sea level has caused brackish or saline waters to engulf drowned river valleys or to extend over formerly upland soils. This has led to the formation of coastal marsh Histosols in several geomorphic settings (Darmody and Foss, 1979). A schematized cross-section of a submerged upland type marsh where organic soil materials are accumulating over what were previously upland soils, forming Histosols is presented in Fig. 6.4. As sea level continues to rise, the margin of the marsh is pushed landward, and the organic materials continue to thicken so that older and deeper Histosols generally exist nearer to the open water (Rabenhorst, 1997).

The formation of Histosols, like other soils, is a function of topography, but Histosols are unique among soils in that their formation also modifies the topography. Accumulation of organic soil material can fill depressions and create gentle topographic highs that were once level or depressed (Fig. 6.5) which can prevent base-rich water from moving onto the peatland and thereby facilitate formation of a highly acidic bog. Peat accumulation also produces intriguing microtopography on some peatlands, notably a pattern of ridges (or strings) and pools (Foster et al., 1983; Seppälä and

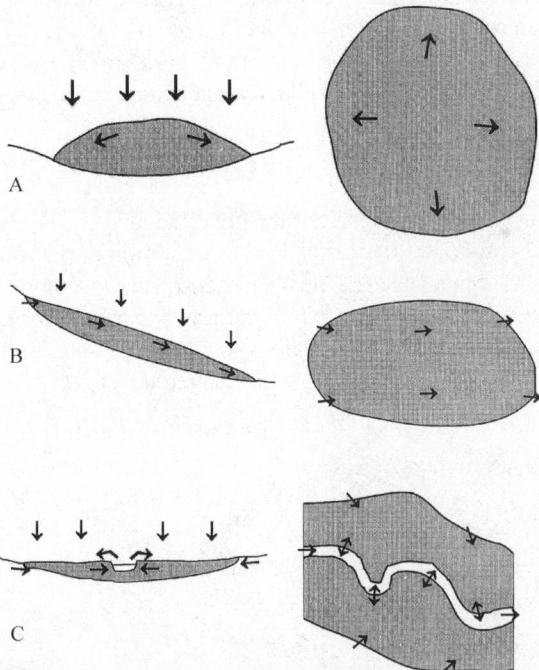

Fig. 6.3 Settings of peatlands. Peat is shown as shaded areas, and arrows indicate the direction of water movement. (A) Bog. The domed surface of the bog precludes input of runoff or groundwater containing bases dissolved from minerals. All water is derived from the atmosphere and evaporates or runs radially off the bog. (B) Fen or swamp. Runoff and groundwater from mineral soils surrounding the peatland supplement precipitation on the peatland. (C) Floodplain fen or swamp. The peatland receives precipitation and runoff or groundwater from mineral soils adjacent to the peatland. Water seeps from the peatland into the stream during periods of low water, while during floods suspended mineral matter is deposited on the peatland, increasing the content of mineral matter in the peat.

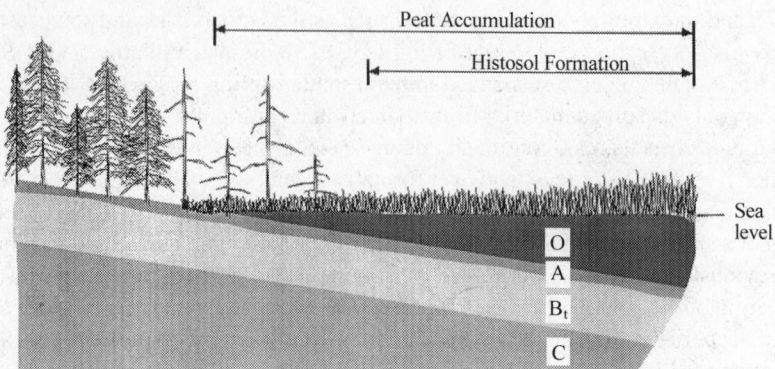

Fig. 6.4   A schematized cross-section of a submerged upland type marsh showing development of O horizons and a Histosol (probably a sulfihemist) over what was formerly an upland soil with an argillic (Bt) horizon (The vertical scale and slope are exaggerated in this diagram.)

Koutaniemi, 1985; Swanson and Grigal, 1988) (or flarks), the cause of which is still debated (Washburn, 1980).

Histosols frequently form by accumulation of organic matter in basins of lakes or ponds, a process known as lakefill or terrestrialization and by paludification, the expansion of wetland onto what was originally drier soils. Both processes operated in the formation of the peatland depicted in Fig. 6.5. Folists, unlike other Histosols, generally occur in mountainous regions (Reiger, 1983; Wakeley et al., 1996) which provide the high precipitation and underlying bedrock or fragmental material required for Folist formation.

6.2.3.4   Vegetation

The flora of Histosols vary widely as a result of the great range in climates over which they occur. Moreover, while most Histosols owe their existence to saturation of the soil by water, the depth at which saturation occurs is variable, which has a major effect on the vegetation. The vegetation of bogs in the worldwide circumboreal zone is remarkably uniform, apparently due to the limited number of plants that can tolerate the poor nutrient conditions, cold climate, and high watertable (Fig. 6.6). *Sphagnum* mosses cover the ground, along with scattered sedges (*Carex*) and cottonsedges

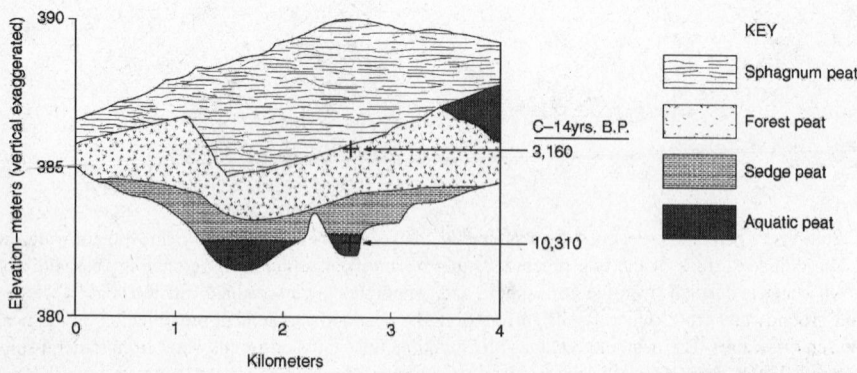

Fig. 6.5   Stratigraphic cross-section of a bog in the Myrtle Lake peatland, northern Minnesota [Modified from Heinselman, 1970]

(*Eriophorum*). Low shrubs from the family Ericaceae are common, and trees are usually present but stunted; black spruce (*Picea mariana*) is most widespread in North America and Scot's pine (*Pinus sylvestris*) in Eurasia. The vegetation of bogs in more southerly climates includes different species, but shares with northern bogs, the *Sphagnum* moss and prevalence of nutrient-conserving evergreen plants (Anderson, 1983; Hofstetter, 1983).

The less acid conditions of groundwater-fed Histosols permit a greater diversity of plants, many of which also occur on wet mineral soils. On minerotrophic peatlands with the watertable continuously near the surface (fens), most vascular plants are aerenchymous grass-like plants (mainly sedges, family Cyperaceae) that can transport O_2 downward to their roots through hollow stems, allowing root metabolism in anaerobic soils (Crawford, 1983). Swamps occur on soils where an aerobic surface horizon is present at least during part of the growing season, allowing growth of nonaerenchymous plants, which includes most woody plants (Gill, 1970; Kozlowski, 1984). Though nutrient conditions for tree growth on these Histosols are superior to those of bogs, trees may be stunted due to short duration of aerobic conditions or thinness of the aerobic rooting zone.

Many of the plants that occur on Histosols are highly adapted to specific conditions of pH with nutrients and high watertable becoming useful indicators of these conditions (Sjors, 1963; Heinselman, 1963, 1970; Jeglum, 1971; Vitt and Slack, 1975; Vitt and Bayley, 1984; Andrus, 1986; Janssens and Glaser, 1986; Swanson and Grigal, 1989,1991; Glaser, 1992; Janssens, 1992). Because the high water table restricts rooting of most plants to near surface soil horizons, vegetation is a useful indicator of pH and nutrient conditions near the surface but not at depth.

Because the vegetation actually creates most of the soil material in Histosols, composition of the vegetation that formed the soil exerts a strong control over its physical and chemical properties. Peats

Fig. 6.6 Typical bog vegetation on a highly acidic Histosol in boreal North America. Trees are stunted black spruce (*Picea mariana*); the largest trees visible are about 4 m tall. Understory plants include ericaceous dwarf shrubs, sedges (*Carex* spp.), and a continuous cover of *Sphagnum* spp. moss. The soil is a Typic Borohemist. Toivola peatland, northeastern Minnesota.

are commonly divided into three broad groups on the basis of botanical composition: moss, herbaceous (sedge), and woody peat (Kivinen, 1977), which are generally deposited in bogs, fens, and swamps, respectively. The botanical origin of organic soil material may be determined by examination of plant remains (Birks and Birks, 1980; Janssens, 1983; Lévesque et al., 1988).

The changes in soil drainage and trophic conditions on peatlands that accompany peat accumulation affect the vegetation, and hence, botanical composition of the peat. As peat accumulates, the rooting zone often becomes more and more isolated from mineral nutrient sources, and portions of the peatland may become drier as the ground surfaces rises. A cross-section of a bog in Minnesota shows how a lake was filled in with aquatic peat (i.e., limnic material) (Fig. 6.4). Then, the peatland expanded onto what was originally dryland of the lakeshore as minerotrophic sedge peat accumulated, followed by minerotrophic forest peat as the surface rose and became drier. By about 3,000 years BP, the center of the peatland became isolated from minerotrophic water, allowing accumulation of highly acidic *Sphagnum* moss peat, which subsequently expanded over the entire peatland and continues to accumulate today. Such changes in vegetation over time in peatlands make it difficult to predict subsurface peat properties from surface vegetation (Swanson and Grigal, 1989).

6.2.3.5   Time
Essentially, all extant Histosols have formed since the end of the Pleistocene epoch. Most northern latitude Histosols occupy regions which were covered by glaciers during the last ice age and have formed following the glacial retreat. Reported average rates of peat accumulation in northern bogs and fens have been as high as > 1 mm yr^{-1}, but more typically fall in the range of 0.2–0.7 mm yr^{-1} (Table 6.5). These average rates usually are based on basal ^{14}C dates and actual rates may have been higher or lower during particular periods.

Although distant from the glacial activity, coastal Histosols at lower latitudes were also impacted by the glaciation. During the glacial maximum (~ 20,000 yr BP) when large quantities of water were tied up in the glacial ice, sea level worldwide was approximately 150 m below the present level. Melting of the ice and concurrent ocean warming caused sea level to rise at such a rapid rate (10–20 mm yr^{-1}) that initially vegetation could not colonize the tidal regions. Approximately 3,000–5,000 years ago, sea-level rise slowed to a more modest pace so that marsh vegetation could become established and organic parent materials began to accumulate (Bloom and Stuvier, 1963; Redfield, 1972). As sea level has continued to rise, organic materials have accumulated in Histosols, and coastal marshes and mangroves generally have been thought to have accreted at approximately the rate of sea-level rise. In addition to the eustatic sea-level rise, sediments in transgressing coastal regions are subsiding (for example, along the Atlantic and Gulf coasts). The combination of rising sea level (presently estimated at 1 mm yr^{-1} worldwide) and coastal subsidence can be joined to yield an apparent sea-level rise, which is substantially greater. Current estimates of peat accretion in coastal areas generally range from 3–8 mm yr^{-1}, which are much higher than in noncoastal regions, with even higher rates reported in rapidly subsiding areas (Table 6.5). Current evidence suggests that the highest rates of sea-level rise may be too great for marsh systems to maintain, and that some of these areas are suffering marsh loss.

6.2.4   Morphological Properties
Most organic soil material is derived from terrestrial plants, and soil particles initially resemble the plants from which they were derived. As decomposition progresses, the organic matter is converted into a homogenous, dark colored mass. Some organic soil materials are derived from aquatic plants and animals that accumulate on the bottom of water bodies, producing limnic materials (Finney et al., 1974; Soil Survey Staff, 1996).

Table 6.5 Rates of peat accretion/accumulation in various organic soils

Location	Site characteristics	Peat accretion - accumulation rate mm yr⁻¹	Source
			Dereived from Almquist-Jacobson and Foster
Bergslagen, Sweden	Raised Bog	0.3 - 1.0	(1995)
Slave Lake, Canada	Sphagnum peat	0.3 - 0.6	Kuhry and Vitt (1996)
West Greenland	Nearshore peat	0.43	Bennike (1992)
Subarctic and Boreal Canada	Based on 138 basal ^{14}C dates	0.31 - 0.54	Gorham (1991)
S. Sweden and N. Germany		0.70	Tolonen (1979) (after Gorham, 1991)
S. and Central Finland		0.75	Tolonen (1979) (after Gorham, 1991)
N. Europe		0.60	Aaby (198 6) (after Gorham, 1991)
Boreal USSR	Raised bogs	0.6 - 0.8	Botch and Masing (1983)
Siberian USSR	Palsa Province	0.2 - 0.4	Botch and Masing (1983)
Eurasia		0.52	Zurek (1976) (after Gorham, 1991)
Maine		0.35 - 0.75	Tolonen et al. (1988) (After Gorham, 1987)
Minnesota	Red Lake, Minnesota	0.85 - 1.15	Gorham (1987)
Louisiana	Coastal Marsh		Derived from Nyman and DeLaune (1991)
	Fresh	6.5 - 8.5	
	Brackish	5.9 - 9.5	
	Saline	7.5 - 7.6	
Chesapeake Bay, MD	Coastal Marsh, Brackish	3.3 - 7.8	Derived from Kearney and Stevenson (1991)
Chesapeake Bay, MD	Coastal Marsh, Brackish	3.5 - 7.5	Griffin and Rabenhorst (1989)
Chesapeake Bay, MD	Coastal Marsh, Brackish		Hussein (1996)
	Current (^{210}Pb)	1.4-3.2	
	Long term (^{14}C)	0.5-1.1	
Louisiana	Barataria Basin, Coastal Marsh	7-13	Hatton et al. (1983)
Massachusetts	Barnstable Coastal Marsh	1.1 - 2.6	Redfield and Rubin (1962)

The most obvious morphological properties of organic soil horizons are related to their degree of decomposition. Master horizons Oi, Oe, and Oa are used to designate fibric, hemic and sapric horizons, respectively (Table 6.6), and are defined by the portion of the soil material which retains plant fibers after rubbing and by the color of a Na pyrophosphate-extracting solution (Soil Survey Staff, 1996). Other methodologies and rating scales have been developed and utilized for evaluating degree of decomposition; the one most broadly used outside of the soil science community and in Europe is the Von Post scale, which ranks organic materials on a scale of decomposition from 1 to 10 based on soil color, quantity of recognizable fibers, and the proportion of material remaining in one's hand when the sample is squeezed. Von Post scale numbers 1–4 correspond approximately to fibric material, 5–7 to hemic, and 8–10 sapric (Von Post and Granlund, 1926; Clymo, 1983).

There are a number of soil properties which are related to the degree of decomposition of the organic materials, including color and a variety of physical and chemical properties. Field moist colors

Table 6.6   Defining morphological criteria for organic horizons [Soil Survey Staff, 1996]

Horizon designation	Type of material	Common descriptor	Volumetric rubbed fiber (RF) content		Color of pyrophosphate extract
Oi	Fibric	Peat	RF > 3/4 or R F > 2/5	and	7/1, 7/2, 8/1, 8/2, 8/3
Oe	Hemic	Mucky Peat	2/5 > RF > 1/6	or	does not otherwise qualify for either Fibric or Sapric materials
Oa	Sapric	Muck	RF < 1/6	and	below or right of line drawn to exclude blocks 5/1, 6/2, 7/3

of sapric organic soil material are usually nearly black. Fibric peat is lighter colored and often reddish in hue, while hemic peat has an intermediate color (Table 6.7). The relation between the degree of decomposition and various physical properties is discussed later under the section on physical properties.

Undrained Histosols typically lack pedogenic structure as it is usually defined, although flattened plant remains and stratification commonly produce a plate-like structure (Lee and Manoch, 1974). This sedimentary structure often results in Histosols having different shear strength and hydraulic conductivity in horizontal and vertical directions (MacFarlane and Williams, 1974; Rycroft et al., 1975). Drained Histosols may develop pedogenic structure in the anthropogenic aerobic zone, such as granular structure due to earthworm casts and blocks or prisms due to cycles of wetting and drying (Lee and Manoch, 1974).

The mineral soil material that occurs under the peat in Histosols is typically chemically reduced as a result of saturation by water and the abundance of organic matter above it. Where paludification has occurred, soil horizons from prior mineral soils may be buried beneath the peat. In some cases, the preceding mineral soil pedogenesis may facilitate paludification by producing low permeability horizons such as placic horizons (Ugolini and Mann, 1979; Klinger, 1996). The depth to mineral soil underlying drained Histosols has been measured with some success by ground-penetrating radar (Shih and Doolittle, 1984; Collins et al., 1986). The morphology of Folists differs from that of other Histosols in that the peat is thinner and underlain by fragmental material or bedrock (Witty and Arnold, 1970; Everett, 1971; Lewis and Lavkulich, 1972).

6.2.5   Micromorphology

Micromorphological observations can provide a direct examination of the structural integrity of plant fragments and components in organic soil materials. Fox (1985) and Lévesque and Dinel (1982) have

Table 6.7   Color of organic soil material in relation to degree of decomposition[1]

Degree of decomposition	Median Munsell color (hue value/chroma)	N
Sapric	10YR 2/1	69
Hemic	10YR 3/2	49
Fibric	7.5YR 3/2	18

[1]Data include all organic horizons of Histosols in the USDA Natural Resources Conservation Service, National Soil Survey Laboratory characterization database.

summarized the characteristics of organic soil materials at various stages of decomposition (fibric, hemic and sapric materials). Fibric materials mainly show unaltered or slightly altered plant tissues without appreciable darkening, and with little organic fine material. The plant fragments, which are only slightly decomposed, appear to be loosely arranged with a porous and open structure. Partially decomposed (hemic) materials also possess a fibrous appearance, and most fragments show incomplete degradation. The development of brown or black colors in the plant tissues is typical. Fine organic material is also present intermixed with, or adhering to, the coarser fragments of plant tissue (Fig. 6.7). Fecal pellets, which are evidence of faunal activity, can also be common. In the most highly decomposed (sapric) materials, organic fragments are sufficiently darkened and decomposed that identification of botanical origin is not possible. Fine organic material is usually the dominant component although faunal excrement is also common.

The effects of draining organic soils can sometimes be seen during microfabric examination. When an undrained sphagnum peat profile in Ireland was compared with those which had been drained for between 10 and 100 yr, the drained profiles had undergone substantial alteration and decomposition leaving little of the original tissue structures (Hammond and Collins, 1983). The fine organic material was dominant, showing some biological granulation (Pons, 1960; Lee, 1983). The change in microfabric directly corresponded to an increase in density of the material.

In another study, Lee and Manoch (1974) concluded that 50 yr of drainage and cultivation of organic soils led to significant decomposition and the development of pedogenic structure in the subsoil, whereas in the lower portion of the profile where the soil remained saturated, sedimentary structure persisted and the peat was more highly fibrous and less decomposed. The activity of soil fauna in the drained portions of the soil contributed to biological granulation and the formation of two

Fig. 6.7 Thin section showing organic material from hemic Oe horizon (80-88 cm) of a Typic Sulfihemist in a coastal marsh of Chesapeake Bay; the organic material reflects an intermediate degree of decomposition with some discernible plant structures and cell components intermixed with decomposed organic material; frame length = 5mm.

distinct types of surface horizons. The *moder* mostly consists of faunal excrement and usually forms in oligotrophic peats, while the *mull* is formed by an intense mixing and binding of organic with mineral particles by larger organisms such as earthworms, and usually forms in mesotrophic or eutrophic peats.

While not widely reported, following drainage and cultivation of organic soils, illuvial humus termed humilluvic material (Soil Survey Staff, 1996) may accumulate in the lower parts of acid organic soils (van Heuveln et al., 1960). Both the lower pH of the oligotrophic peat and the disturbance by cultivation apparently contribute to the dispersion of the organic fraction which can then be translocated within the soil, and accumulates within the lower horizons of the peat at the peat-mineral soil contact, or within the underlying mineral soil material.

6.2.6 Classification

The definition of organic materials for saturated soils requires a minimum of 12% organic C, if there is no clay, and a minimum of 18% organic C if the soil contains $\geq 60\%$, with a sliding scale for intermediate textures. Those soils, which are not saturated, must contain at least 20% organic C to be considered organic soil materials (Soil Survey Staff, 1998, 1996). In order for a soil to be classified as a Histosol, at least 40 cm of the upper 80 cm must be comprised of organic materials, and it must not have permafrost within 1 m of the surface. However, if the organic materials are especially low in density ($< 0.1\,\mathrm{g\,cm^{-1}}$), then at least 60 cm of the upper 100 cm must be organic materials. Histosols may be buried by as much as 40 cm of overlying mineral soil materials and still be considered Histosols.

The types of differentiating characteristics used to discriminate between classes of soils at the various categorical levels are presented in Table 6.8. Basically, organic soils which are not saturated for extended periods are classified as Folists, while the saturated organic soils are differentiated according to the degree of decomposition of the organic materials in the subsurface tier (the zone approximating 40–100 cm) into Fibrists, Hemists, or Saprists. Within the United States, some 234 soil series have been established for Histosols distributed among the classes as shown in Table 6.9. The number of series that exist within a particular class may result from many factors and should not be taken to represent the areal extent of those soils. Histosols classified at the family level are differentiated into classes based upon particle size and mineralogy (used only for terric subgroups or for those containing limnic materials), reaction (pH_{CaCl_2}), temperature regime, and soil depth (only used if < 50 cm deep.)

Table 6.8 Criteria utilized in the classification of Histosols

Suborder	Great group	Subgroup	Family
Degree of saturation with water	Soil temperature regime	Thickness of organic materials (Terric vs. Typic)	Particle size and mineralogy (used only for Terric subgroups or for those containing limnic materials)
Degree of decomposition of the subsurface tier	Special components (sphagnum fibers, sulfidic materials or sulfuric horizon, humiluvic materials)	Underlying materials	
		Special materials contained (Limnic)	Reaction (pH_{CaCl_2})
			Temperature regime
		Intergrades to other great groups (Cyric, S phagnic)	soil depth (only used if < 50 cm deep

Table 6.9 Number of soil series in the United States which are classified into particular taxonomic groups

Suborders		Great Groups		Subgroups	
	Number of Series	Formative Element	Number of Series	Formative Element	Number of Series
Fibrists	17	Boro[†]	60	Fluvaquentic	11
Folists	25	Cryo	22	Limnic	14
Hemists	61	Medi[‡]	98	Lithic	29
Saprists	131	Sphagno	5	Terric	88
		Sulfi	18	Typic	76
		Tropo	31	Others	16
Total	234	Total	234	Total	234

[†] This great group is no longer used in *Soil Taxonomy* and these series have been recorrelated to alternative taxa.
[‡] This great group was essentially replaced by the Haplo great group in the 1998 version of Soil Taxonomy (Soil Survey Staff, 1998)

6.2.7 Biological and Chemical Properties

Organic C contents are generally higher and N contents lower in less decomposed than highly decomposed peats giving high C:N ratios and ash contents for less and more highly decomposed peats, respectively (see also Table 6.12) (Lévesque et al., 1980; Lee et al., 1988).

Thus, more highly decomposed peats are generally more fertile than less decomposed peats. Some drained, sapric peats may, in fact, supply N in excess of crop requirements without fertilization. However, nutrients derived from minerals, such as P, K, and most micronutrients, are frequently deficient in Histosols (Lucas, 1982; Yefimov, 1986).

The chemical and physical properties of peats are also related to their botanical composition. Peats derived mostly from *Sphagnum* mosses tend to be more acid, less decomposed, contain less ash, have lower CEC and bulk density than woody peats; sedge peats typically have intermediate properties (Farnham and Finney, 1965; Lévesque et al., 1980).

6.2.7.1 Soil Carbon

Recent interest in the global C cycle has focused attention on the high proportion (75%) of terrestrial C stored in soils (Lal et al., 1995). As a group, wetland soils maintain a disproportionately high level of soil C, and Histosols, which are composed largely of organic matter, clearly store the highest quantities. While typical agricultural soils may contain between 2–10 kg C m^{-2}, reported values for Histosols typically are an order of magnitude greater, with some values as high as nearly 200 kg C m^{-2} (Table 6.10). The quantity of C stored in some very deep Histosols is undoubtedly even higher.

Histosols are very dynamic and may be particularly significant in the overall terrestrial C budget. Many Histosols continue to sequester C at significant rates. This is particularly true for soils of coastal marshes, where rising sea level continues to power the engines of marsh accretion and C storage. Therefore, Histosols are generally viewed as a significant C sink. However, because Histosols contain such a high concentration of soil C, if they are drained or in some other way exposed to an aerobic environment, they may begin to oxidize and in the process yield large quantities of C to the atmosphere.

Most of the discussion of possible global warming and greenhouse gas emission has focused on rising levels of CO_2 in the atmosphere. Methane, however, is 32 times more efficient than CO_2 in trapping infrared radiation. Because many Histosols are strongly reducing (low Eh), they represent an

Table 6.10 Carbon storage values for organic soils

Site Characteristics	Location	Carbon Accumulation Rate $kg\ m^{-2}\ yr^{-1}$	Quantity of stored C $kg\ m^{-2}$	Reference

Coastal Marsh
  Fresh
  Brackish
  Saline
	Louisiana	
		0.17 - 0.22
		0.17 - 0.27
		0.21 - 0.22
				Derived from Nyman and DeLaune (1991)

Coastal Marsh Brackish	Chesapeake Bay, Maryland	0.12 - 0.42		Derived from Kearney and Stevenson (1991)
Coastal Marsh Brackish	Chesapeake Bay, Maryland		59 (range 18-166)	Derived from Griffin and Rabenhorst (1989)
Barataria Basin, Coastal Marsh	Louisiana	0.18 - 0.30		Smith et al. (1983)
Coastal Marshes	Atlantic and Gulf Coasts		64 (range 9 - 191)	Rabenhorst (1995)
Sphagnum peat	Slave Lake, Canada	0.014 - 0.035		Kuhry and Vitt (1996)
Based on 138 basal ^{14}C dates	Subartic and Boreal Canada	0.023 - 0.029		Gorham (1991)

ideal environment for the formation of CH_4. In systems where SO_4^{2-} is more abundant in the soil solution, such as in coastal or estuarine environments, SO_4^{2-} reduction is favored over methanogenesis and CH_4 production may be more limited (Bartlett et al., 1987; Widdell, 1988) (Section C, Chapter 4.3) However, in many freshwater or inland areas, Histosols may be the locus of significant CH_4 emission to the atmosphere (Table 6.11). Minerotrophic fens have higher CH_4 emissions than ombrotrophic bogs. Methane emission from Histosols seems to be directly related to the location of the watertable, with greater generation when soils are saturated to or above the surface. This is due to the fact that soil microbes in the aerobic zone utilize CH_4 as it passes through on its way to the surface. Carbon dioxide emissions, which are produced by oxidation of SOM, are generally greater when watertables drop.

6.2.7.2 Sulfides

The biogeochemical environment in which Histosols form can also be conducive to the formation of sulfides. The occurrence of Fe sulfides can, in some circumstances, lead to the generation of extreme acidity and acid sulfate soils (van Breemen, 1982) (Section B, Chapter 5). Sulfate reduction generally requires the presence of organic matter (energy source), low Eh, SO_4^{2-} (electron acceptor), and SO_4^{2-} reducing bacteria (Rickard, 1973). If sulfide is formed in the presence of a reactive Fe source, then such minerals as pyrite (FeS_2) can form. The C source and anaerobic conditions are almost always present in Histosols, but SO_4^{2-} levels may vary dramatically among environments. Many inland Histosols receive SO_4^{2-} only in small amounts as atmospheric deposition, and under these circumstances, sulfidization (Rabenhorst and James, 1992) is insignificant, and most of the S in those soils is bound in organic forms (Novak and Wieder, 1992). Some inland peats have developed acid sulfate conditions, although usually they are associated with deposits of cuprogeneous earth (Lucas, 1982). On the other hand, SO_4^{2-} reduction is common in coastal Histosols which contain an abundance of SO_4^{-2} from sea water. Extensive areas of these Sulfihemist soils have been identified along the Atlantic coast of the United States. The distribution of FeS_2 within coastal Histosols can be highly

Table 6.11 Reported fluxes of CO_2 and CH_4 emissions from Histosols

Location	Site Details	Notes	CO₂ emission rate Mean	CO₂ emission rate Range	CH₄ emission rate Mean	CH₄ emission rate Range
			\-\-\-\-\-\-\-\-\-\-\-\-\-\-\-\-\-mmol m^{-2} d^{-1}\-\-\-\-\-\-\-\-\-\-\-			
Quebec Canada[1]	Gatineau Park			1.97–7.24		1.15–2.18
Wales[2]	Peat Monoliths	Laboratory study	14.7	9.6–21.0	14.4	4.7–34.4
Finland[3]	Natural Fen	Annual	11.3		3.49	
	Drained Fen		30.8		0.03	
Finland[4]	Ombrotrophic	12 °C	88.2	42.5–141.3		
Alaska[5,6]		Summer measurements			9.2	
Boreal Canada[7]	Swamp (n = 20)	Annual Averages			0.21	
	Fen (n = 6)				0.57	
	Bog (n = 13)				0.39	
Canada[8]	Bog flooded	19–23 °C	0.005		0.012	
(lab study)	Bog drained		0.19		0.006	
	Fen flooded		0.009		0.58	
	Fen drained		0.14		0.025	
Alaska[9]	Moist tundra	August			0.3	
	Waterlogged tundra				7.4	
	Wet meadows				2.5	
	Alpine fen				18	
Northern Sweden[10]	Ombrotrophic bog	Summer				
	Hummocks		10.12		0.05	
	Between humocks		14.92		0.14	
	Shallow depressions		11.61		0.77	
	Deeper depressions		12.45		1.21	
	Ombro-Minerotrophic		12.49		2.73	
	Minerotrophic Fen		11.48		16.89	
Minnesota[11]	Bog	Sampled during August			8.2	1.2–29.2
	Fen				0.25	0.19–0.31
Georgia[5,6]		During midsummer			6.6	
West Virginia[5,6]	Mountain Bog	During midsummer			11.7	
Minnesota[6,12]	Forest bog	Summer measurement			3.6	0.5–5.33
	Forest fen				6.7	3.2–12
	Open bog				14	0.9–41
	Neutral fen				15	7.1–33
	Acid fen				4.8	
Minnesota[13]	Open poor feen	Winter measurement			3.0	
	Open bog				0.7	
	Forest bog hollow				0.8	
	Hummock				0.3	
Virginia[14]	Coastal Marsh	Summer				
	York River	Creek bank			0.45	0.13–0.82
	Chesapeake Bay	Short spartina			0.29	0.05–0.87
	Estuary	High marsh			0.09	0–0.36
Virginia[15]	Coastal Marsh	Summer				
	Tidal creek in	Low salinity			11	5–16
	Chesapeake Bay	Moderate salinity			7.5	4–11
	Estuary	High salinity			2	0.5–2.5
West Virginia[16]	Appalachian Bog		1?7	75–250	17.0	0–53
Maryland[16]	Appalachian Bog		152	100–250	4.4	0–12
Louisiana[17]	Barataria Basin, Coastal Marsh	Annual averages	41–141			
Florida[5,6]		During midsummer			6.0	
Malaysia[18]			170			
Malaysia[19]	Drained & cultivated peatland			139–727		

[1]Buttler et al. (1994), [2]Freeman et al. (1993), [3]Nykanen et al. (1995), [4]Silvola et al. (1996), [5]After Gorham (1991), [6]after Crill et al. (1988), [7]Derived from Moore and Roulet (1995), [8]Derived from Moore and Knowles (1989), [9]Derived from Sebacher et al. (1986), [10]Svensson and Rosswall (1984), [11]After Harriss et al. (1985), [12]After Mitsch and Wu (1995), [13]Dise (1992), [14]Derived from Bartlett et al. (1985), [15]Derived from Bartlett et al. (1987), [16]Wieder et al. (1990), [17]Smith et al. (1983), [18]Wosten et al. (1997), [19]Murayama and Bakar, 1996

variable and is often related to microsite differences in the availability of either reactive Fe or S^{2-} (Rabenhorst and Haering, 1989).

6.2.7.3 Acidity and Base Saturation

In Histosols of bogs, where inputs of bases are minimal because soil water is derived from rainfall that has never contacted mineral soil, base saturation is low (at least in the rooting zone) and the organic acids produced by partial decomposition of organic matter typically buffer soil water pH near 4; the soil pH_{CaCl_2} is typically 3–4 (Figs. 6.8 and 6.9) (Gorham et al., 1985; Urban, 1987). Minerotrophic (fen and swamp) peats have higher base saturation and water pH is generally above 5 ($pH_{CaCl_2} > 4.5$). The pH of minerotrophic peats is buffered by cation exchange with the soil (Bloom et al., 1983) or by carbonates if they are present. The Al^{3+} ion comprises a small proportion of the total acidity in most Histosols, because organic soil material is rich in H^+ but contains little Al which is derived from silicate minerals, and has a low solubility at pH > 5.5 (Table 6.12). The pH is generally lower in less than more highly decomposed peats (Table 6.12) (Lee et al., 1988).

The distinction between euic and dysic reaction classes at the family level in *Soil Taxonomy* separates the highly acid, bog peats from less acid fen and swamp peats (Farnham and Finney, 1965). Because a $pH_{CaCl_2} > 4.5$ anywhere in the control section (i.e., anywhere within 130 or 160 cm of the surface) places the soil into the euic class, a bog with highly acidic near-surface rooting zone may classify as euic rather than dysic due to the presence of higher pH horizons at depth in the soil.

Histosols which contain sulfide minerals such as FeS_2 have the potential to develop extreme acidity. Under saturated and anaerobic conditions, such sulfide-bearing soils have circumneutral pH. If they are drained, dredged, or in some other way exposed to oxidizing conditions, acid sulfate weathering takes place (van Breemen, 1982). The oxidation of pyrite can lead both to the extreme acidification of the soil (pH < 3.5) and also to the generation of acidity, which can be moved offsite through mobilization of acid-generating soluble salts such as $FeSO_4$.

6.2.7.4 Cation Exchange Capacity

The CEC of Histosols is quite high as a result of the high CEC of SOM (Table 6.12). The CEC is higher for more highly decomposed peats than fibric peats (Table 6.12) (Lévesque et al., 1980). The difference between the CEC of sapric and fibric peats is even more marked if the CEC is expressed on a volume rather than mass basis, as a result of the very low bulk densities of fibric materials (Table

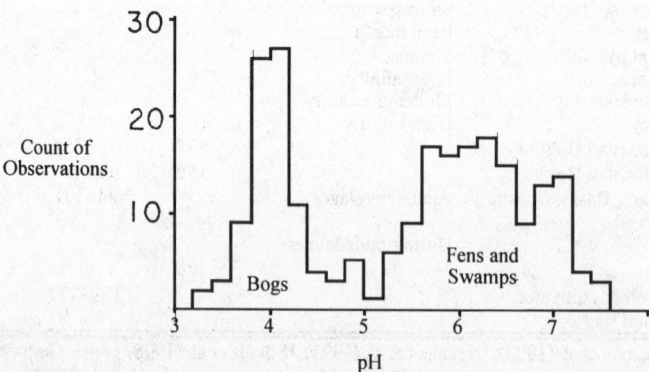

Fig. 6.8 Frequency distribution of pH in peatland surface water for 232 sites in Minnesota. Samples with surface water pH near 4 are from bogs and those with pH above 5 are from fens (unforested) and swamps (forested) [Swanson and Grigal, 1989].

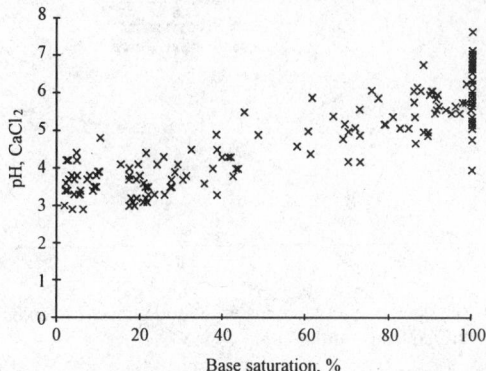

Fig. 6.9 Relationship between base saturation and pH of organic soil horizons of Histosols. Peat samples with pH in $CaCl_2$ of less than 4.5 (presumably deposited in bogs) generally have less than 50% base saturation. Peat samples with pH greater than 4.5 (deposited in fens and swamps) typically have high base saturation. Data are for all organic horizons of Histosols in the USDA Natural Resources Conservation Service, National Soil Survey Laboratory characterization database.

6.12). The CEC per unit soil volume is a more useful measure than CEC per unit soil mass in Histosols, because their bulk densities are very low and because roots occupy a volume rather than mass of soil. The CEC per unit volume of fibric peats (~ 20 cmol$_c$ L^{-1}) is comparable to that of most mineral soils (5–20 cmol$_c$ L^{-1}) assuming mineral soil bulk density of 1.0–1.5 kg L^{-1} (Holmgren et al., 1993). Even when expressed per unit volume, the average CEC of sapric and hemic peats (Table 6.12) is much higher than the 5–20 cmol$_c$ L^{-1} of most mineral soils.

6.2.8   Physical Properties

6.2.8.1   Bulk Density
The physical properties of Histosols differ greatly from those of mineral soils. Bulk densities for organic soil materials generally are quite low, ranging from as little as 0.02 up to 0.30 g cm^{-3}. Bulk

Table 6.12   Chemical properties of organic soil material as related to degree of decomposition: mean (std. deviation, N)[1]

Property	Sapric	Hemic	Fibric	AOV Probability[2]
Organic Carbon, g kg^{-1}	313 (128, 129)	347 (135, 61)	372 (130, 26)	0.055
Total Nitrogen, g kg^{-1}	18 (9, 131)	16 (6, 54)	14 (5, 23)	0.058
C:N ratio	21 (10, 113)	25 (11, 48)	27.5 (10, 20)	0.007
CEC, cmol$_c$ g kg^{-1}	101 (44, 129)	88 (41, 61)	83 (33, 28)	0.046
CEC, cmol$_c$ L^{-1}	76 (42, 28)	44 (25, 25)	21 (2, 5)	0.000
Ash, g kg^{-1}	250 (nd)	178 (110, nd)	100 (50, nd)	nd
pH	5.1 (1.2, 143)	4.9 (1.2, 59)	4.5 (1.2, 27)	0.024
Al, mol Al mol^{-1} TEA[3]	0.074 (0.076, 49)	0.022 (0.024, 23)	0.038 (0.050, 16)	0.004

[1]Data (except for ash) are for all organic horizons of Histosols in the USDA Natural Resources Conservation Service, national Soil Survey laboratory characterization database. Ash content is taken from Lee et al., 1988; data for 1300 samples of Wisconsin Histosols, nd - no data.
[2]F-test probability from one-way analysis of variance between fibric, hemic, and sapric peats
[3]KCl extractable Al divided by total NH$_4$-acetate extractable acidity

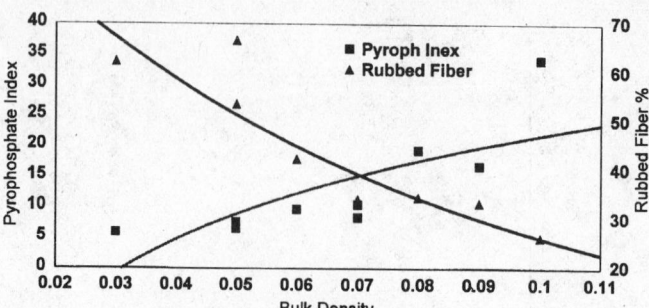

Fig. 6.10   Relationship between bulk density of organic materials and indices of decomposition such as rubbed fiber content or pyrophosphate index [Derived from Buttler et al., 1994]

density is clearly related to the degree of decomposition of the organic materials generally increasing as the materials become more highly decomposed (Fig. 6.10). Because organic materials contain varying amounts of mineral matter, this also can affect the bulk density of organic soil horizons. For coastal marsh peat, the organic matter content is generally about twice the content of organic C, with the remainder representing the mineral fraction. The relationship between the mineral content (roughly the difference remaining from twice the organic C content) and the bulk density for Oe and Oa horizons from Sulfihemists along the Atlantic Coast is illustrated in Fig. 6.11 (Griffin and Rabenhorst, 1989).

6.2.8.2   Porosity, Hydraulic Conductivity, and Water Retention

Histosols have very high porosity levels which can reach over 80% as shown by the water content at saturation (Fig. 6.12) (Boelter, 1969). The high porosity and low bulk density of organic soils would lead one to expect high rates of water transmission through Histosols. Weakly or undecomposed peat often has a fairly high hydraulic conductivity. However, as the material becomes more decomposed the hydraulic conductivity decreases (Table 6.13). The hydraulic conductivity of sapric and some hemic peats is quite low, comparable to that of clay (i.e., 10^{-5} m s^{-1}). Because the peat in the lower part of Histosol profiles is often sapric or hemic, little deep percolation occurs and water tends to evaporate or move laterally through the less decomposed surface horizons (Ivanov, 1981). Thus peat

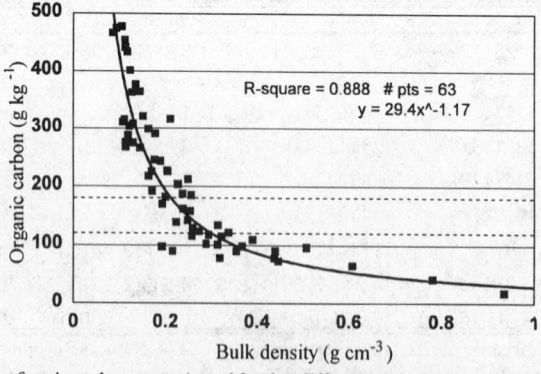

Fig. 6.11   The effect of mineral content (roughly the difference remaining from 2X the OC content) on the bulk density of Oe and Oa horizons from Sulfihemists along the Atlantic Coast. Dashed lines represent 12 and 18% OC which is necessary for soil materials to be considered organic [Derived from Griffin and Rabenhorst, 1989].

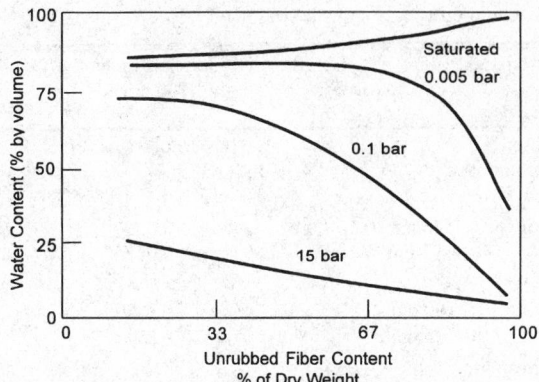

Fig. 6.12 Volumetric soil water content as a function of unrubbed fiber content at various moisture tensions [Reprinted from Boelter, 1969. Soil Sci. Soc. Am. Proc. 33:606-609 with permission of the Soil Science Society of America]

accumulation helps to maintain the watertable near the surface by limiting water loss to percolation. While the increased density corresponds to a decrease in porosity, this change alone is insufficient to explain the dramatic reduction in conductivity. Evidently, the pore continuity is reduced in more strongly decomposed materials.

Water retention in Histosols is also closely related to the degree of decomposition (Fig. 6.12). The large pores of slightly decomposed peats drain readily at low suction. In contrast, the fine pores of more well-decomposed peats retain considerable water at low suctions (Boelter, 1974). The water retention difference (water content at --0.01 MPa (–0.1 bar) minus that at –1.5 MPa (–15 bar); approximates plant-available soil water capacity) of most Histosols is very high; it approaches 50% in some peats and exceeds 25% in all except the least decomposed peats (Fig. 6.12).

6.2.9   Utilization and Management

6.2.9.1   Interest in Preservation of Coastal and Nontidal Wetlands
Apart from any benefits which may be achieved from managed systems, such as commercial forestry, grazing or intensive agriculture, Histosols perform a number of environmental and ecological functions. Because most Histosols occur in wetland environments, they typically provide wetland functions including wildlife habitat, flood-water control, groundwater recharge, nutrient and biogeochemical cycling, ion sorption, purification of surface and shallow groundwater, as well as functioning as an important sink for terrestrial C. The importance of peatlands as paleoenvironmental and archeological archives has also been documented (Godwin, 1981.) Throughout the years, the value which society has placed on these functions has been minimal, and the Histosols of peatlands and coastal marshes have been exploited and extensive areas destroyed. More recently people have recognized that the functions which Histosols perform in a natural setting have significant benefit for society, and legislation has been passed in the United States and elsewhere to preserve Histosols and other wetlands.

6.2.9.2   Histosols as Agricultural Resources
As was mentioned earlier, Histosols are utilized as important agricultural soils in many areas, so long as watertables can be effectively managed. For example, in Japan, over 70,000 ha of rice are grown on

Table 6.13 Physical and hydraulic properties of organic soils

Location	Site Characteristics		Bulk Density g cm^{-1}	Hydraulic Conductivity 10^{-5} cm s^{-1}		Reference
Northern Minnesota	Undecomposed Partially decomposed Decomposed			3,810 - 15,000 13.9 - 132 0.9 - 15	Ā[†] = 8,650 Ā = 63 Ā = 5.1	Boelter (1965)
Wyoming	mountain bog	46 cm 91 cm	0.16 - 0.22	0.0277 0.0185		Sturges (1968)
Ottawa and St. Lawrence River Valleys, Canada	Swamp and bog	0 -60 cm 0-100 cm 0-125 cm		624 366 269		Mathur and Lévesque (1985)
Northern Minnesota	Lost River Peatland Bog Fen Margin Spring Fen		0.06 - 0.14	25 - 560 150 - 2,600 67 - 1,600		Chason and Siegel (1986)
Quebec, Canada	Gatineau park		0.03 - 0.10			Buttler et al. (1994)
Eastern New Brunswick, Canada	Sphagnum peat from raised bogs - Von Post scale	1 - 2 3 - 4 5 - 6 7 - 8	0.02 - 0.08 0.03 - 0.10 0.06 - 0.11 0.09 - 0.13	90 - 175 6.9 - 56 1.4 - 17 0.14 - 2.8		Korpijaako and Radforth (1972)
Wisconsin	Fibric Hemic Sapric		0.13 0.17 0.20, 0.24			Lee et al. (1988)
Minnesota	Bog 0 - 10 cm Von Post 1 10 - 20 cm Von Post 2 - 3 20 - 30 cm Von Post 4 - 5 30 - 40 cm Von Post 5 - 6 40 - 50 cm Von Post 5 - 7 Fen 0 - 10 cm 10 - 20 cm 20 - 30 cm			23,495 7,697 5,498 799 995 31,597 5,000 1,400		Gafni and Brooks (1990)

[†] Ā = average

peatlands. Peatlands in a number of tropical countries such as Cuba, Guyana, Malayasia and Indonesia have been reclaimed from mangroves and are used for the production of sugarcane (Moore and Bellamy, 1974). Extensive vegetable production on peatlands is practiced in the northern and southern United States. The agricultural use of Histosols presents some special challenges regarding nutrient management and fertility, but the most significant problem is the high watertable requiring drainage.

6.2.9.3 Impacts of Drainage

Where Histosols have been converted for higher intensity land uses, such as in agriculture, horticulture, or silvaculture, they are almost always drained to lower the watertable. Such drainage results in a number of short-term effects, such as shrinkage and consolidation due to desiccation, the loss of the buoyant force of groundwater, and compaction. Ongoing consolidation and fabric alteration are caused by the enhanced decomposition of the organic materials following the shift from an anaerobic to an oxidizing regime (Stephens and Speir, 1969). Reported rates of long-term peat subsidence range up to 10 cm yr^{-1} but most reports are in the range of 2–5 cm yr^{-1} (Table 6.14). This

Table 6.14 Reported rates of subsidence after drainage of organic soils

Location	Site Characteristics	Subsidence rate cm yr^{-1}	Length of record yr	Reference
Florida	Everglades, Muck and Peat	3.2	41	Thomas (1965)
Hunts, England	Holme marsh	3.4	103	Nickolson (1951)
California	Sacramento-San Joaquin Delta, muck and peat 1 - 3 cm deep	6.4 - 9.8	26	Weir (1950)
Indiana		3.0	30+	Ellis and Morris (1945)
Northern Indiana	Muck	1.2 - 3.0 dependent on WT level	6	Jongedyk et al. (1950)
Michigan		0 - 3.6	5	Davis and Engberg (1955) [after Thomas (1965)]
Southern Ontario, Canada	Holland march, deep loose muck	3.3	19	Mirza and Irwin (1964)
Minnesota		5.1	3	Row (1940)
Florida	Everglades	3	55	Stephens and Speir (1969)
Florida	Everglades	2.3 - 1.8	20	Shih et al. (1981)
California	Sacramento-San Joaquin Delta	2.3	78	Rojstaczer and Deverel (1995)
Quebec, Canada		2.1	38	Parent et al. (1982)
Malaysia		2		Wosten et al. (1997)

Table 6.15 Peat mining by country, 1994 (in millions of tons)[1]

Country	Agriculture and Horticulture Use	Fuel Use
Belarus	10	6
Canada	1	
Estonia	4.5	1
Finland	0.6	8
Germany	2.8	0.2
Ireland	0.3	6.4
Latvia	4.5	0.5
Lithuania	4.5	0.2
Russia	60	4
Sweden	0.3	1.4
Ukraine	20	1
United States	0.5	Nd[2]

[1]Data from Cantrell (1994); includes countries with 500,000 tons or more production
[2]No data

consolidation is accompanied by changes in the physical properties of the peat, including higher bulk densities and lower moisture contents.

6.2.9.4   Histosols as Energy Resources

Peat is mined in northern Europe and used as fuel (Table 6.15), but a large amount is used as a horticultural amendment and as an ingredient in potting soils, mixed fertilizers, components of mushroom beds, as a seed inoculant, as a material for packing of flowers, plants etc., as well as a general soil amendment to increase organic matter content in gardens, golf courses, etc (400,000 Mg in the United States) (Cantrell, 1994).

6.2.9.5   Engineering Properties

Histosols are notorious for their low strength and great compressibility, which make them poor foundation materials for roads, buildings, and other structures. Compression and settlement of peats may continue for years after loading (MacFarlane, 1969; MacFarlane and Williams, 1974; Dhowian and Edil, 1981). Special engineering techniques, such as removal of the peat or precompression before construction, are thus required. The high water content and acidity of most Histosols also make corrosion of concrete and metal structures a potential problem (MacFarlane, 1969; MacFarlane and Williams, 1974). While Histosols are poor foundation materials, their high porosity and great adsorption capacity make them very useful for treatment of wastewater. Peats have potential for treatment of municipal effluent and removal of heavy metals and hydrocarbon pollutants from wastewater (Malterer et al., 1996).

6.2.10   References

Aaby, B. 1986. Paleoecological studies of mires. p. 145–164. *In* B. E. Berglund, (ed.) Handbook of Holocene palaeoecology and palaeohydrology. John Wiley and Sons, New York, NY.

Almquist-Jacobson, H., and D.R. Foster. 1995. Toward an integrated model for raised-bog development: Theory and field evidence. Ecol. 76:2503–2516.

Anderson, J.A.R. 1983. The tropical peat swamps of western Malesia. p. 181–199. *In* A.J.P. Gore (ed.) Mires: Swamp, bog fen, and moor. Elsevier, Amsterdam, Netherlands.

Andrus, R.E. 1986. Some aspects of Sphagnum ecology. Can. J. Bot. 64:416–426.

Bartlett, K.B., R.C. Harriss, and D. Sebacher. 1985. Methane flux from coastal salt marshes. J. Geophys. Res. 90:5710–5720.

Bartlett, K.B., D.S. Bartlett, R.C. Harriss, and D.I. Sebacher. 1987. Methane emissions along a slat marsh salinity gradient. Biogeochem. 4:183–202.

Bennike, O. 1992. Paleoecology and paleoclimatology of a late Holocene peat deposit from Broendevinsskaer, Central West Greenland. Arctic Alp. Res. 24: 249–252.

Birks, H.J.B., and H.H. Birks. 1980. Quaternary paleoecology. Edward Arnold, London, UK.

Bloom, P.R., W.E. Elder, and J. Grava. 1983. Chemistry and mineralogy of mineral elements in Minnesota Histosols. p. 29–43. *In* Proc. 26th Ann. Meet. Manitoba Soc. Soil Sci., Winnipeg, Manitoba, Canada.

Bloom, A.L., and M. Stuvier. 1963. Submergence of the Connecticut coast. Science 139:333–34.

Boelter, D.H. 1965. Hydraulic conductivity of peats. Soil Sci. 100:227–231.

Boelter, D.H. 1969. Physical properties of peats as related to degree of decomposition. Soil Sci. Soc. Amer. Proc. 33:606–609.

Boelter, D.H. 1974. The hydrologic characteristics of undrained organic soils in the Lake States. p. 33–46. *In* A.R. Aandahl (ed.) Histsols: Their characteristics, classification, and use. Soil Sci. Soc. Amer. Spec. Pub. 6, Soil Science Society of America, Madison, WI.

Botch, M.S. and V.V. Masing. 1983. Mire ecosystems in the USSR. p. 95–152. *In* A.J.P. Gore (ed.) Mires: Swamp, bog, fen and moor. Elsevier, Amsterdam, Netherlands.

Buttler, A., H. Dinel, and P.E.M. Lévesque. 1994. Effects of physical, chemical and botanical characteristics of peat on carbon gas fluxes. Soil Sci. 158:365–374.

Canada Committee on Ecological (Biophysical) Land Classification, National Wetlands Working Group. 1988. Wetlandes of Canada. Polyscience, Montreal, Canada.

Cantrell, R. L. 1994. Peat. p. 583–585. *In* Minerals yearbook: Metals and minerals. Vol. I. US Government Printing Office, Washington, D.C.

Chason, D.B., and D.I. Siegel, 1986. Hydraulic conductivity and related physical proprties of peat, Lost River Peatland, northern Minnesota. Soil Sci. 142:91–99.

Clymo, R.S. 1983. Peat. p. 159–224. *In* A.J.P. Gore (ed.) Mires: Swamp, bog, fen, and moor. Elsevier, Amsterdam, Netherlands.

Collins, M.E., G.W. Schellentrager, J.A. Doolittle, and S.F. Shih. 1986. Using ground-penetrating radar to study changes in soil map unit composition in selected Histosols. Soil Sci. Soc. Amer. J. 50:408–412.

Cooper, D.J., and R.E. Andrus. 1994. Patterns of vegetation and water chemistry in peatlands of the west-central Wind River Range, Wyoming, U.S.A. Can. J. Bot. 72:1586–1597.

Crawford, R.M.M. 1983. Root survival in flooded soils. p. 257–283. *In* A.J.P. Gore (ed.) Mires: Swamp, bog fen, and moor. Elsevier, Amsterdam, Netherlands.

Crill, P.M., K.B. Bartlet, R.C. Harriss, E. Gorham, E.S. Verry, D.I. Sebacher, L. Mazdar, and W. Sanner. 1988. Methane flux from Minnesota peatlands. Global Biogeochem. Cycl. 2:317–384.

Darmody R.G., and J.E. Foss, 1979. Soil-landscape relationships of the tidal marshes of Maryland. Soil. Sci. Soc. Amer. J. 43:534–541.

Davis, J.F., and C.A. Engberg. 1955. A preliminary report of investigations of subsidence of organic soil in Michigan. MI Agr. Exp. Sta. Quart. Bull 37:498–505.

Dhowian, A.W., and T.B. Edil. 1981. Consolidation behaviour of peats. Geochem. Test. J. 2:105–114.

Dise, N.B. 1992. Winter fluxes of methane from Minnesota peatlands. Biogeochem. 17:71–83.

Ellis, N.K., and R. Morris. 1945. Preliminary observations on the relation of yield of crops grown on organic soil with controlled water table and the area of aeration in the soil and subsidence of the soil. Soil Sci. Soc. Amer. Proc. 10:282–283.

Everett, K.R. 1971. Composition and genesis of the organic soils of Amchitka Island, Aleutian Islands, Alaska. Arctic Alp. Res. 3:1–6.

Everett, K.R. 1983. Histosols. p. 1–53. *In* L.P. Wilding, N.E. Smeck, and G.F. Hall. (ed.) Pedogenesis and Soil Taxonomy. II. The Orders. Elsevier, Amsterdam, Netherlands.

Farnham, R.S., and H.R. Finney. 1965. Classification and properties of organic soils. Adv. Agron. 17:115–162.

Finney, H.R., E.R. Gross, and R.S. Farnham. 1974. Limnic materials in peatlands. p. 21–31 *In* A.R. Aandahl (ed.) Histosols: Their characteristics, classification, and use. Soil Sci. Soc. Amer. Spec. Pub. 6, Soil Science Society of America, Madison, WI.

Foster, D.R., C.A. King, P.H. Glaser, and H.E. Wright Jr. 1983. Origin of string patterns in boreal mires. Nature 306:256–258.

Fox, C.A. 1985. Micromorphological characterization of Histosols. p. 85–104. *In* Soil micromorphology and soil classification. Soil Sci. Soc. Am. Spec. Pub. 15, Soil Science Society of America, Madison, WI.

Freeman, C.M., A. Lock, and B. Reynolds. 1993. Fluxes of CO_2, CH_4 and N_2O from a Welsh peatland following simulation of water table draw-down: Potential feedback to climatic change. Biogeochem. 19:51–60.

Gafni, A. and K.N. Brooks. 1990. Hydraulic characteristics of four peatlands in Minnesota. Can. J. Soil Sci. 70:239–253.

Gill, C.J. 1970. The flooding tolerance of woody species. For. Abst. 31:671–678.

Glaser, P.H. 1992. Vegetation and water chemistry. p. 15–26. *In* H.E. Wright, B.A. Coffin, and N.E. Aaseng. (ed.) The patterned peatlands of Minnesota. University of Minnesota, Minneapolis, MN.

Godwin, H. 1981. The archives of the peat bogs. Cambridge University Press. New York, NY.

Gore, A.J.P. 1983. Mires: Swamp, bog, fen, and moor. Elsevier, Amsterdam, Netherlands.

Gorham, E. 1987. The ecology and biogeochemistry of *Sphagnum* bogs in central and eastern North America. p. 3–15. *In* A.D. Laderman (ed.) Atlantic white cedar wetlands. Westview Press, Boulder, CO.

Gorham, E. 1991. Northern peatlands: Role in the carbon cycle and probable responses to climatic warming. Ecolog. Applic. 1:182–195.

Gorham, E.S., J. Eisenreich, J. Ford, and M.V. Santelmann. 1985. The chemistry of bog waters. p. 339–363. *In* W. Stumm (ed.) Chemical processes in lakes. John Wiley and Sons, New York, NY.

Griffin, T.M., and M.C. Rabenhorst. 1989. Processes and rates of pedogenesis in some Maryland tidal marsh soils. Soil Sci. Soc. Am. J. 53:862–870.

Hammond, R.F., and J.F. Collins. 1983. Microfabrics of a sphagnofibrist and related changes resulting from soil amelioration. p. 689–697. *In* P. Bullock and C.P. Murphy (ed.) Soil micromorphology 2. Soil genesis. A B Academic Publishers, Berkhamsted, UK.

Harriss, R.C., E. Gorham, D.I. Sebacher, K.B. Bartlett, and P.A. Glebbe. 1985. Methane flux from northen peatlands. Nature 315:652–654.

Hatton, R.S., R.D. DeLaune, and W.H. Patrick, Jr. 1983 Sedimentation, accretion, and subsidence in marshes of Barataria Basin Louisiana. Limnol. Oceanogr. 28:494–502.

Heinselman, M.L. 1963. Forest sites, bog processes, and peatland types in the glacial Lake Agassiz region, Minnesota. Ecol. Monogr. 33:327–374.

Heinselman, M.L. 1970. Landscape evolution, peatland types and the environment in the Lake Agassiz Peatlands Natural Area, Minnesota. Ecol. Monogr. 40:235–261.

Hofstetter, R.H. 1983. Wetlands in the United States. p. 201–244. *In* A.J.P. Gore (ed.) Mires: Swamp, bog, fen, and moor. Elsevier, Amsterdam, Netherlands.

Holmgren, G.G.S., M.W. Meyer, R.L. Chaney, and R.B. Daniels. 1993. Cadmium, lead, zinc, copper, and nickel in agricultural soils in the United States of America. J. Environ. Qual. 22:335–348.

Hussein, A.H. 1996. Soil chronofunctions in submerging coastal areas of Chesapeake Bay. Ph.D. Thesis. University of Maryland, College Park, MD.

Ivanov, K.E. 1981. Water movement in mirelands. Academic Press, London, UK.

Janssens, J.A. 1983. A quantitative method for stratigraphic analysis of bryophytes in Holocene peat. J. Ecol. 71:189–196.

Janssens, J.A. 1992. Bryophytes. p. 43–57. *In* H.E. Wright, B.A. Coffin, and N.E. Aaseng. (ed.) The patterned peatlands of Minnesota. University of Minnesota, Minneapolis, MN.

Janssens, J.A., and P.H. Glaser. 1986. The bryophyte flora and major peat-forming mosses at Red Lake peatland, Minnesota. Can. J. Bot. 64:427–442.

Jeglum, J.K. 1971. Plant indicators of pH and water level in peatlands at Candle Lake, Saskatchewan. Can. J. Bot. 49:1661–1672.

Jongedyk, H.A., R.B. Hickok, I.D. Mayer, and N.K. Ellis. 1950. Subsidence of muck soil in the Northern Indiana. IN Agri. Exp. Sta. Cir. 366.

Kearney, M.S., and J.C. Stevenson. 1991. Island land loss and marsh vertical accretion rate evidence for historical sea-level changes in Chesapeake Bay. J. Coastal Res. 7:403–415.

Kivinen, E. 1977. Survey, classification, ecology, and conservation of peatlands. Bull. Int. Peat Soc. 8:24–25.

Klinger, L.F. 1996. Coupling of soils and vegetation in peatland succession. Arctic Alp. Res. 28:380–387.

Korpijaako, M., and N.W. Radforth. 1972. Studies of the hydraulic conductivity of peat. *In* Proc. 4th Int. Peat Congr. 3:323–334.

Kozlowski, T.T. 1984. Responses of woody plants to flooding. p. 129–163. *In* T.T. Kozlowski. (ed.) Flooding and plant growth. Academic Press, London, UK.

Kuhry, P., and D.H. Vitt. 1996. Fossil carbon/nitrogen ratios as a measure of peat decomposition. Ecol. 77:271–275.

Lal, R. J. Kimble, E. Levine, and C. Whitman. 1995. World soils and greeenhouse effect: An overview. p. 1–8. *In* R. Lal, J. Kimble, E. Levine, and B.A. Stewart (ed.) Soils and global change. Lewis Publishers, Boca Raton, FL.

Lee, G.B. 1983. The micromorphology of peat. p. 485–501. *In* P. Bullock and C. P. Murphy (ed.) Soil micromorphology 2. Soil genesis. A B Academic Publishers, Berkhamsted, UK.

Lee, G.B., and B. Manoch. 1974. Macromorphology and micromorphology of a Wisconsin saprist. p. 47–62. *In* A. R. Aandahl (ed.) Histosols: Their characteristics, classification, and use. Soil Sci. Soc. Am. Spec. Pub. 6, Soil Science Society of America, Madison, WI.

Lee, G.B., S.W. Bullington, and F.W. Madison. 1988. Characteristics of Histic materials in Wisconsin as arrayed in four classes. Soil Sci. Soc. Amer. J. 52:1753–1758.

Lévesque, M., H. Dinel, and R. Marcoux. 1980. Evaluation des critères de différentiation pour la classification de 92 matériaux tourbeaux du Québec et de l'Ontario. Can. J. Soil Sci. 60:479–486.

Lévesque, M.P. and H. Dinel. 1982. Some morphological and chemical aspects of peats applied to the characterization of Histosols. Soil Sci. 133:324–333.

Lévesque, P.E.M., H. Dinel, and A. Larouche. 1988. Guide to the identification of plant macrofossils in Canada peatlands. Land Resource Centre, Research Branch, Agriculture Canada, Ottawa.

Lewis, T. and L.M. Lavkulich. 1972. Some Folisols in the Vancouver area, British Columbia. Can. J. Soil Sci. 52:91–98.

Lucas, R.E. 1982. Organic soils (Histosols). Formation, distribution, physical and chemical properties, and management for crop production. MI Sta. Univ. Agric. Expt. Sta. Res. Rep. 435.

Global Soil Regions

U.S. Department of Agriculture
Natural Resources Conservation Service
Soil Survey Division
World Soil Resources

Alfisols
Andisols
Aridisols
Entisols
Gelisols
Histosols
Inceptisols
Mollisols
Oxisols
Spodosols
Ultisols
Vertisols
Rocky Land
Shifting Sands
Ice/glacier

0 500 1,000   2,000   3,000   4,000   5,000   6,000   7,000   8,000
KILOMETERS

Miller Projection

Country boundaries are not authoritative

MacFarlane, I.C. 1969. Muskeg engineering handbook. National Research Council of Canada, University of Toronto Press, Toronto, Canada.

MacFarlane, I.C., and G.P. Williams. 1974. Some engineering aspects of peat. p. 79–93. *In* A.R. Aandahl (ed.) Histosols: Their characteristics, classification, and use. Soil Sci. Soc. Am. Spec. Publ. 6, Soil Science Society of America, Madison, WI.

Malterer, T., B. McCarthy, and R. Adams. 1996. Use of peat in waste treatment. Min. Eng. 48:53–56.

Mathur, S.P., and M. Lévesque. 1985. Negative effect of depth on saturated hydraulic conductivity of Histosols. Soil Sci. 140:462–466.

Mausbach, M.J., and J.L. Richardson. 1994. Biogeochemical processes in hydric soil formation. Cur. Top. Wetl. Biogeochem. 1:68–127.

McNally, G. 1997. Peatlands, power and post-industrial use. p. 245–251. *In* L. Parkyn, R.E. Stonemen, and H.A.P. Ingram. (ed.) Conserving peatlands. CAB International, New York, NY.

Mirza, C. and R.W. Irwin. 1964. Determination of subsidence of an organic soil in Southern Ontario. Can. J. Soil Sci. 44:248–253.

Mitsch W.J., and X. Wu. 1995. Wetlands and global change. p. 205–230. *In* R. Lal, J. Kimble, E. Levine, and B.A. Stewart. (ed.) Soil management and greenhouse effect. Lewis Publishers, Boca Raton, FL.

Moore, P.D., and D.J. Bellamy. 1974. Peatlands. Springer Verlag, New York, NY.

Moore, T.R., and R. Knowles. 1989. The influence of water table levels on methane and carbon dioxide emissions from peatland soils. Can. J. Soil Sci. 69:33–38.

Moore, T.R., and N.T. Roulet. 1995. Methane emissions from Canadian peatlands. p.153–164. *In* R. Lal, J. Kimble, E. Levine, and B.A. Stewart (ed.) Soils and global change. Lewis Publishers, Boca Raton, FL.

Murayama, S., and Z. A. Bakar. 1996. Decomposition of tropical peat soils. 2. Estimation of *in situ* decomposition by measurement of CO_2 flux. JARQ 30:153–158.

Nickolson, H.H. 1951. Groundwater control in reclaimed marshlands. World Crops 3: 251–254.

Novak, M., and R.K. Wieder. 1992. Inorganic and organic sulfur profiles in nine Sphagnum peat bogs in the United States and Czechoslovakia. Water Air Soil Poll. 65:353–369.

Nykanen, H., J. Alm, K. Lang, J. Silvola, and P.J. Martikainen. 1995. Emissions of CH_4, N_2O and CO_2 from a virgin fen and a fen drained for grassland in Finland. J. Biogeogr. 22:351–357.

Nyman, J.A. and R.D. DeLaune. 1991. CO_2 emission and soil Eh responses to different hydrological conditions in fresh, brackish, and saline marsh soils. Limnol. Oceanogr. 36:1406–1414.

Parent, L.E., J.A. Millette, and G.R. Mehuys. 1982. Subsidence and erosion of a Histosol. Soil Sci. Soc. Am. J. 46:404–408.

Pons, L.J. 1960. Soil Genesis and classification of reclaimed peat soils in connection with initial soil formation. Proc. 7th Int. Congr. Soil Sci. IV:205–211.

Rabenhorst, M.C. 1995. Carbon storage in tidal marsh soils. p. 93–103. *In* R. Lal, J. Kimble, E. Levine, and B.A. Stewart (ed.) Soils and global change. Lewis Publishers, Boca Raton, FL.

Rabenhorst, M.C. 1997. The chrono-continuum: An approach to modeling pedogenesis in marsh soils along transgressive coastlines. Soil Sci. 167:2–9.

Rabenhorst, M.C., and K.C. Haering. 1989. Soil micromorphology of a Chesapeake Bay tidal marsh: Implications for sulfur accumulation. Soil Sci. 147:339–347.

Rabenhorst, M.C., and B.R. James. 1992. Iron sulfidization in tidal marsh soils. Catena Suppl. 21:203–217.

Redfield, A.C. 1972. Development of a New England salt marsh. Ecol. Mongr. 42:201–237.

Redfield, A.C., and M. Rubin. 1962. The age of salt marsh peat and its relation to recent changes in sea level at Barnstable, MA. Proc. Natl. Acad. Sci. 48:1728–1735.

Reiger, S. 1983. The genesis and classification of cold soils. Academic Press, New York, NY.

Rickard, D.T. 1973. Sedimentary iron sulphide formation. *In* H. Dost (ed.) Acid sulfate soils. I. ILRI Publ. 18, Wageningen, Netherlands.

Rojstaczer, S., and S.J. Deverel. 1995. Land subsidence in drained Histosols and highly organic mineral soils of California. Soil Sci. Soc. Am. J. 59:1162–1167.

Row, H.B. 1940. Some soil changes resulting from drainage. Soil Sci. Soc. Amer. Proc. 4:402–409.

Rycroft, D.W., D.J. Williams, and H. A. Ingram. 1975. The transmission of water through peat: 1. Review. J. Ecol. 63: 535–556.

Sebacher, D.I., R.C. Harriss, K.B. Bartlet, S.M. Sebacher, and S.S. Grice. 1986. Amospheric methane sources: Alaskan tundra bogs, and alpine fens, and a subarctic boreal marsh. Tellus 38B:1–10.

Seppälä, M., and L. Koutaniemi. 1985. Formation of a string and pool topography as expressed by morphology, stratigraphy, and current processes on a mire in Kuusamo, Finland. Boreas 14:287–309.

Shih, S.F., and J.A. Doolittle. 1984. Using radar to investigate organic soil thickness in the Florida Everglades. Soil Sci. Soc. Amer. J. 48:651–656.

Shih, S.F., D.E. Vandergrift, D.L. Myhre, G.S. Rahi, and D.S. Harrison. 1981. The effect of land forming on subsidence in the Florida Eveglades' organic soil. Soil Sci. Soc. Am. J. 45:1206–1209.

Silvola, J., J. Alm, U. Ahlholm, H. Nykanen, and P.J. Martikainen. 1996. CO_2 fluxes from peat in boreal mires under varying temperature and moisture conditions. J. Ecol. 84:219–228.

Sjors, H. 1963. Bogs and fens on the Attawapiskat River, northern Ontario. Mus. Can. Bull. Contr. Bot. 186:43–133.

Sjors, H. 1985. A comparison between mires of southern Alaska and Fennoscandia. Aquilo Ser. Bot. 21:89–94.

Smith, C.J., R.D. DeLaune, and W.H. Patrick, Jr. 1983. Carbon dioxide emission and carbon accumulation in coastal wetlands. Est. Coastal Shelf Sci. 17:21–29.

Soil Survey Staff. 1996. Keys to Soil Taxonomy. 7th Ed. USDA-Natural Resources Conservation Service. US Government Printing Office, Washington, DC.

Soil Survey Staff. 1998. Soil Taxonomy: A basic system of soil classification for making and interpreting soil surveys. USDA Agric. Handb. 436, US Government Printing Office, Washington, DC.

Stephens, J.C. and W.H. Speir. 1969. Subsidence of organic soils in the USA. Assoc. Intern. Hydrol. Sci. Spec. Pub. 89:523–534.

Sturges, D.L. 1968. Hydrologic properties of peat from a Wyoming mountain bog. Soil Sci. 106:262–264.

Svensson, B.H., and T. Rosswall. 1984. *In situ* methane production from acid peat in plant communities with different moisture regimes in a subarctic mire. Oikos 43: 341–350.

Swanson, D.K., and D.F. Grigal. 1988. A simulation model of mire patterning. Oikos 53:309–314.

Swanson, D.K., and D.F. Grigal. 1989. Vegetation indicators of organic soil properties in Minnesota. Soil Sci. Soc. Am. J. 53:491–495.

Swanson, D.K. and D.F. Grigal. 1991. Biomass, structure, and trophic environment of peatland vegetation in Minnesota. Wetlands 11:279–302.

Thomas, F.H. 1965. Subsidence of peat and muck soils in Florida and other parts of the United States: A review. Soil Crop Sci. Soc. Fl. Proc. 25:153–160.

Tolonen, K. 1979. Peat as a renewable resource: long-term accumulation rates in North European mires. p. 282–296. *In* Proc. Int. Symp. Classif. Peat Peatl., International Peat Society, Helsinki, Finland.

Tolonen, K., R.B. Davis, and L.S. Widoff. 1988. Peat accumulation rates in selected Maine peat deposits. ME Geol. Surv. Bul. 33.

Trewartha, G.T. 1968. An introduction to climate. McGraw-Hill, New York, NY.

Ugolini, F.C., and D.H. Mann. 1979. Biopedological orgin of petlands in south east Alaska. Nature 281:366–368.

Urban, N.R. 1987. The nature and origins of acidity in bogs. Ph.D. Thesis, University of Minnesota, Minneapolis, MN.

USDA-NRCS. 1996. Field indicators of hydric soils in the United States. USDA, NRCS, Fort Worth, TX.

van Breemen, 1982. Genesis, morphology and classification of acid sulfate soils in Coastal Plains. p. 95–108. *In* J.A. Kittrick, D.S. Fanning, and L.R. Hossner (ed.) Acid sulfate weathering. Soil Sci. Soc. Amer. Spec. Pub. 10., Soil Science Society of America, Madison, WI.

van Heuveln, B., A. Jongerius, and L.J. Pons. 1960. Soil formation in organic soils. Proc. 7th Int. Congr. Soil Sci. IV:195–204.

Vitt, D.H., and N.G. Slack. 1975. An analysis of the vegetation of Sphagnum dominated kettle-hole bogs in relation to environmental gradients. Can. J. Bot. 53:332–359.

Vitt, D.H., and S. Bayley. 1984. The vegetation and water chemistry of four oligotrophic basin mires in northwestern Canada. Can. J. Bot. 62:1485–1500.

Von Post, L., and E. Granlund. 1926. Sodra Sveriges torvtillgångar I. Sven. Geol. Unders. C. 335.

Wakeley, J.S., S.W. Sprecher, and R.W. Lichvar. 1996. Relationships among wetland indicators in Hawaiian rain forest. Wetlands 16:173–184.

Walter, H. 1977. The oligotrophic peatlands of Western Siberia: The largest peinhelobiome in the world. Vegetatio 34:167–178.

Washburn, A.L. 1980. Geocryology. John Wiley and Sons, New York, NY.

Weir, W.W. 1950. Subsidence of peat lands in the Sacramento-San Juaquin Delta of California. Hilgardia 20:37–55.

Widdell, F. 1988. Microbiology and ecology of sulfate- and sulfur-reducing bacteria. p. 469–585. *In* A. J. B. Zehnder (ed.) Biology of anaerobic microorganisms. John Wiley and Sons, New York, NY.

Wieder, R.K., J.B. Yavitt, and G.E. Lang. 1990. Methane production and sulfate reduction in two Appalachian peatlands. Biogeochem. 10:81–104.

Witty, J.E., and R.W. Arnold. 1970. Some Folists on Whiteface Mountain, New York. Soil Sci. Soc. Am. Proc. 34:653–657.

Wosten, J.H.M, A.B. Ismail, and A.L.M. van Wijk. 1997. Peat subsidence and its practical implications: A case study in Malaysia. Geoderma 78:25–36.

Wright, H.E., B.A. Coffin, and N.E. Aaseng. 1992. The patterned peatlands of Minnesota. University of Minnesota, Minneapolis, MN.

Yefimov, V.N. 1986. Peat soils and their fertility. Agropromizdat, Leningrad, Russia.

Zurek, S. 1976. The problem of growth of the Eurasia peatland in the Holocene. Proc. 5th Int. Peat Congr. 2:99–122.

6.3 Andisols

J.M. Kimble
USDA-NRCS

C.L. Ping
University of Alaska Palmer

Malcolm E. Sumner
University of Georgia

Larry P. Wilding
Texas A & M University

6.3.1 Introduction

The Andisol order was added to *Soil Taxonomy* as a result of a proposal developed by Smith (1978) who provided the rationale for separating these soils into an order category. The name Ando soil (J. *An,* dark, and *do*, soil) connoting acid soils derived from volcanic ash which have thick, dark horizons was introduced in 1947 during reconniassance soil surveys by United States soil scientists in Japan (Simonson and Rieger, 1967). Part of the name was retained as a formative element in the systems of nomenclature used in different global soil classification systems. The term Ando soil is still used in Japan (Wada, 1986) but evolved to Andosols in the FAO soil classification system (FAO-UNESCO, 1989) and Andepts in *Soil Taxonomy* (Soil Survey Staff, 1975). The term Andisol was introduced in *Soil Taxonomy* in 1990 (Soil Survey Staff, 1990).

The central concept of Andisols is one of deep soils with depositional stratification developing from volcanic ash, pumice, cinders or other volcanic ejecta and volcaniclastic materials, with an exchange complex dominated by amorphous materials composed of Al, Si and organic matter, or a matrix-dominated by glass, having one or more diagnostic horizons other than an ochric epipedon. Bulk density is always comparatively low in most horizons, although the actual values may vary with degree of weathering, moisture regime, and in a very few cases, with the degree of cementation by silica or other cements. The most common diagnostic horizon sequences are an umbric or mollic (rare) epipedon and a cambic horizon, or an ochric epipedon and a cambic horizon. In the driest regimes, a duripan may be found while in the wettest, a placic horizon is not uncommon. A melanic (black) epipedon was added during the testing of the proposal. While soils which are not formed on volcanic parent materials sometimes are included in the Andisol order, Andisols are classified on selected chemical, physical and mineralogical properties acquired through weathering and not on parent material alone, which is sometimes a common misconception.

Andisols possess a suite of common properties including amorphous (short-range order) mineralogy, low permanent and high variable charge, low bulk density, high C content and high P and water retention relative to other mineral soils of similar field texture. Field texture has been used because laboratory particle size analyses of Andisols are often erroneous due to difficulties in obtaining complete soil dispersion (Warkentin and Maeda, 1980). Andisols pose unique engineering problems because of the fragility of pumice and their thixotropic properties (Section 6.3.6.4). In addition to Andisols, many other soils (Spodosols, Mollisols, Ultisols, Alfisols, Entisols, Inceptisols and Oxisols) can form in parent materials of volcanic origin.

6.3.2 Geographic Distribution

Andisols cover more than 120 million ha or nearly 1% of the Earth's surface. By far, the largest concentration is to be found along the Pacific Ring of Fire, a zone of tectonic activity and volcanoes stretching from South through Central and North America via the Aleutian Islands to the Kamchatka Peninsula of Russia through Japan, the Philippines and Indonesia to Papua New Guinea and New Zealand. Other areas include the Carribean, central Atlantic ridge, northern Atlantic rift (Iceland) and the Mediterranean (Arnold, 1988). The largest area of Andisols in the United States occurs in Alaska (10 million ha) and in the Pacific Northwest (WA, OR, ID, CA) (Péwé, 1975; Rieger et al., 1979; Southard and Southard, 1991; Ugolini and Dahlgren, 1991; Goldin et al., 1992; Takahashi et al., 1993; McDaniel et al., 1993). Worldwide distribution of Andisols is depicted in Fig. 6.13. Volcanoes constructed during the period of intense tectonic activity in the Holocene, Pleistocene and later Tertiary periods produced about 1,000 m^3 of lava and associated ejecta from which Andisols have developed (Leamy, 1984).

Andisols can occur in all soil moisture and temperature regimes (Dudal and Suparpthardjo, 1961). In the Pacific Northwest, very few Andisols occur in the thermic temperature regime because the summers are too hot and dry to allow sufficient weathering to produce the required allophane and oxalate extractable Fe and Al for placement in the order. In other parts of the world, Andisols are found in thermic and even hyperthermic temperature regimes but under wetter moisture regimes (udic and

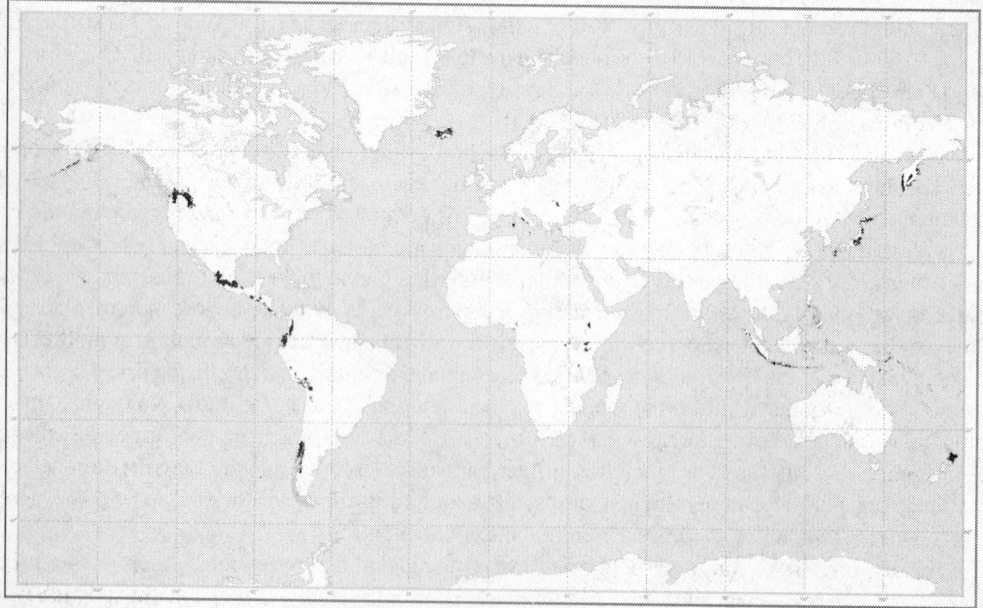

Fig. 6.13 Global distribution of Andisols [Courtesy of USDA-NRCS, Soil Survey Division, World Soil Resources, 1998]

perudic) (Derking, 1987). Because of their unique physical and chemical properties (Section 6.3.6), Andisols support a much greater population than their limited extent would suggest (Leamy, 1984). This is because they frequently occur in humid regions, which together with their excellent physical condition and high fertility made the establishment of nonshifting, early traditional agriculture possible.

Soils which classify as Andisols are also found in areas not associated with volcanic activity such as in the southern Appalachian mountains, Kyushu (Japan) and the Alps. These soils have high levels of Al associated with organic matter and similar management constraints as soils formed from volcanic ejecta, and consequently, key out as Andisols. This further highlights the importance of realizing that Andisols are not classified on parent material, but on the properties acquired during weathering.

6.3.3 Formation of Andisols

6.3.3.1 Parent Material
In most cases, the parent materials from which Andisols have formed are of volcanic origin which accounts for their close association with areas of volcanic activity. Parent materials vary from ultrafine dusts to lava flows, any of which may contain large boulders and volcanic bombs ejected by explosive eruptions. Because sequential eruptions laid down a variety of materials (ash, pumice, boulders), many Andisols are layered which allows chronological dating of the soil mass based on the known or datable age of the various layers. Andisol properties are affected by the chemical, mineralogical, textural and depositional properties of the volcanic ash which can be divided into dark and light colored varieties depending on contaminants from other rock types (Shoji, 1986). Light colored ash is acidic (andesitic and rhyollitic) containing quartz, and smaller amounts of hornblende and biotite, while dark colored ash contains mafic rich minerals, mainly olivine. Intermediate color groups are richer in plagioclases and pyroxenes. Opaque minerals are present in all three groups. Soil formation takes place faster in dark colored glassy (basaltic) than light colored glassy tephras (Shoji et al., 1993). In southcentral Alaska and the Aleutian Islands, and southeast Alaska, most Andisols are derived from light and darker colored tephras, respectively (Riehle, 1985; Ping et al., 1989), with the former being more acidic than the latter. The Vitrands suborder is the only one in which the vitric parent material plays a dominat role in the classification (Soil Survey Staff, 1998). In Iceland, Andisols form in mixtures of volcanic ash and loessial deposits with distinct layers of tephra that permit dating by radiocarbon and from historical records (Arnalds et al., 1995). These soils are highly subject to wind and water erosion when surface cover is destabilized. Accretions of volcanic ash from large cataclysmic events lead to buried soils.

6.3.3.2 Climate
Volcanic ejecta occur in all moisture and temperature regimes; consequently, the climates under which Andisols are found are very varied. Climate, which strongly controls Andisol formation, is recognized at all suborder levels except Vitrands. Combinations of temperature and moisture result in the different great groups of Andisols. Because Torrands form in arid climates, they are the least weathered and exhibit calcite accumulation in the soil profile. Ustands form in semi-arid climates and experience incomplete leaching, particularly of silica, which is translocated from the upper to lower profile and forms a duripan in some cases. Xerands, which form in Mediterranean climates with dry hot summers and cool, wet winters, have greater accumulation of organic matter than in the above less humid environments. Udands which form under humid or very humid climates are highly leached, very acidic and have high organic matter contents. Cryands which form in climates with mean annual temperatures

< 8 °C with cool summers, frequently occur in Alaska, New Zealand, Iceland, and mountainous regions of the Pacific Ring of Fire countries. Aquands which occur in localized wet environments commonly have organic surface horizons and evidence of soil reduction independent of the broad climatic bias of other suborders.

To be classified as an Andisol, sufficient weathering must have taken place to produce specific quantities of acid oxalate extractable Fe and Al (Soil Survey Staff, 1998). Thus, in very dry environments, one would not expect to find Andisols. Parfitt (1990) suggested that a minimum precipitation of 1,200 mm yr^{-1} would be required to bring about this weathering. However, in Iceland, Cryands form under as little as 700 mm yr^{-1} (Arnalds et al., 1995). Because the occurrence of Torrands is limited, the question arises as to whether they really exist. Do they occur in seep or other areas where moisture is concentrated by other means? Because of their limited extent, this may well be the case.

The classic Andisols of Japan, New Zealand and other areas along the Pacific Ring of Fire are formed in humid environments which also tend to be fairly cool (Wada, 1986; Parfitt, 1990). They are deep organic rich soils which suggests that the preferred climate for Andisol formation is one with cool temperatures and high rainfall. Even in the tropics, the better expressed Andisols are found in isomesic temperature regimes, but that does not mean that they do not occur in warmer areas. However, more rapid weathering in the warmer areas leads to the rapid transformation to Oxisols, Ultisols, and Inceptisols.

6.3.3.3 Topography

Although Andisols are found on all types of topography at a wide range of elevations (0 to > 3,000 m), they tend to occur most frequently on mountainous to hilly terrain on the slopes of volcanoes. On the other hand, fine ash can be carried long distances and deposited on any type of landscape, while water can rework ash, both of which result in extensive flat areas of Andisols in intermountain valleys. In addition, large lava flows may result in fairly flat areas on which Andisol formation can take place. Because of the different modes of depostion of volcanic materials, no particular type of topography favors Andisol formation. However, Aquands which form in areas with restricted drainage mainly occur at low points in the landscape. These soils tend to be dark colored because of the accumulation of organic matter which forms in these anaerobic conditions.

6.3.3.4 Vegetation

Because Andisols occur in all moisture and temperature regimes on different landscapes at different elevations, vegetation is accordingly highly variable. In general, Andisols are moderately acid so vegetation associated with acid soils will predominate. Because of their relatively high water-holding capacities, Andisols, even in xerophytic areas, are likely to support more luxuriant vegetation than other nonandic soils in the same environment.

Although vegetation on Andisols is quite variable, it has a major influence on the type of Andisol formed according to Shoji and Otowa (1988). In the Pacific Northwest and Alaska, Udands generally form under forest vegetation. However, they are known to form under other vegetation types such as Japanese pampus grass (*Miscanthus sinensis*) (Melanoudands), Sitka spruce *(Picea sitchensis)* (Hydrudands) and blue joint grass *(Calmagrostis canadiansis)* (Fulvudands). The organic matter is dominated by humic acids in the first cases and by fulvic acids in the last (Ping et al., 1989, 1995; Shoji, et al., 1993). Most Xerands, such as those in Washington and Oregon, form under mixed vegetation of grass and forest while Torrands form under shrubby or grass vegetation and Vitrands under grass or forest. When the vegetation is grass as in Mollisols, a more humus-rich material develops. In Andisols, these soils are placed in great groups containing melanic (black) epipedons. In Iceland and other high latitude polar regions, Cryands form under moss, heath and grass covers (Arnalds et al.,

1995). In the tropics, the vegetation is often rain forest, particularly on the slopes of volcanoes where grassland and savanna types also occur.

6.3.3.5 Time

As with all soils, time is a major factor in the formation of Andisols. The time at which volcanic ash was deposited is regarded as time zero. Andisols generally are found on young parent materials in humid climates because, on older and more stable landscapes, weathering would have taken place for longer periods of time resulting in the transformation to other soil orders. The drier and cooler the environment, the longer it will take for an Andisol to form. Volcanic materials, either very young, in extremely dry areas, or that are recently deposited which have not weathered sufficiently to form at least the minimum requirements for andic materials, are placed in the Entisol order. According to Shoji et al. (1993), a minimum of several hundred years is required for an A horizon to form in newly deposited ash whereas for an A–B sequence, several hundred to thousands of years may be necessary. In Japan, most Andisols are 4,000–7,000 yr old whereas in SE Alaska, a Cryand was dated at 12,000 yr using ^{14}C dating. In Iceland, most of the Andisols range in age from a few hundred yr to over 10,000 yr (Arnalds et al., 1995). Episodic accumulation of volcanic and admixed eolian materials follows glacial retreat. With progressive age, A horizons formed in eolianandic materials are transformed to a series of Bw cambic horizons, interspersed by coarse-textured tephra in marker layers (Fig. 6.14).

6.3.4 Pedogenic Processes in Andisols

Interactions of all the soil-forming factors (Jenny, 1941) are required for an Andisol to form. In general, the most common environment where Andisols form in a humid climate on young parent materials under forest vegetation. Similar processes are at work in the formation of Andisols and other soil orders, particularly Spodosols. The end products of weathering for both podzolization and andolization are very similar, namely, Fe and Al associated with organic matter with or without the presence of Si. Andisols are found on all types of landscapes and slopes.

Acid oxalate extracts Al and Fe from organic complexes and poorly crystalline oxides. Andisols contain high amounts of acid oxalate extractable Al (Al_o) and/or Fe (Fe_o). Spodosols are often high in Al_o and Fe_o as well. Pyrophosphate extractable Al (Al_p), or the amount of Al contained in organic complexes, is often proportionally higher in spodic than andic materials. Mizota and van Reeuwijk

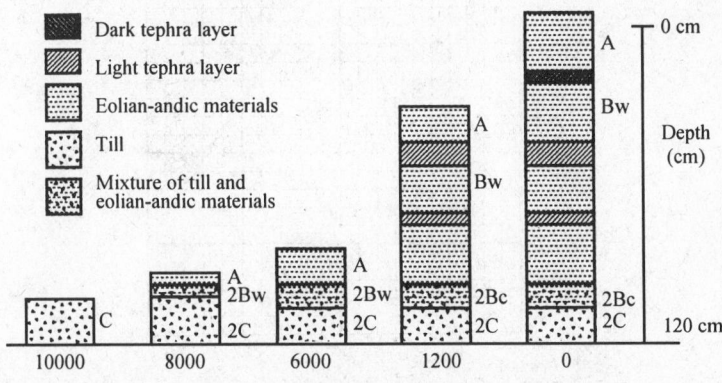

Fig. 6.14 Chronological development of Andisols in Iceland formed from tephra and eolian-andic materials [Reprinted from Arnalds et al., 1995. Soil Sci. Soc. Am. J. 59:161-169 with permission of the Soil Science Society of America]

(1989) have proposed a binary ratio (Al_p/Al_o) as a means of separating allophanic from nonallophanic soils. Soils with higher binary ratios are nonallophanic, are dominated by Al organic complexes, and would more likely be associated with spodic types of material. Morphologically, Spodosols often contain albic (bleached) horizons formed by the complexation and vertical translocation of metal organic complexes. Because this process typically does not occur in Andisols, they do not possess albic horizons.

The genetic concept of an Andisol is a soil formed in a humid environment on volcanic ash. A minimum amount of weathering to produce andic soil properties (Soil Survey Staff, 1996) is required. As most Andisols are formed in humid environments on easily weathered ash, weathering is quite rapid (Flach et al., 1980). Allophane which is a coprecipitate of Al, Fe and Si oxides is the first mineral to form from the weathering of volcanic glass together with organic complexes of Fe and Al (acid oxalate extractable). Under more intense weathering, halloysite and other crystalline minerals may form (Wada, 1989).

Because Andisols form under a wide range of environmental conditions, their properties vary quite widely. Common properties are low bulk densities, high levels of organic C to depth and high P fixation, all of which are used to define andic soil material. The overall processes lead to the formation of stable complexes of Al, Fe and Si with organic matter and the formation of allophane, ferrihydrite, and imogolite minerals (Southard and Southard, 1991; Ugolini and Dahlgren, 1991; Wada et al., 1992; Takahashi et al., 1993). If weathering continues, minerals become more crystalline leading to the formation of soils which would fall in other soil orders.

6.3.5   Classification of Andisols

Andisols are classified on the basis of andic soil properties which are defined by the Soil Survey Staff (1998) (Fig. 6.15):

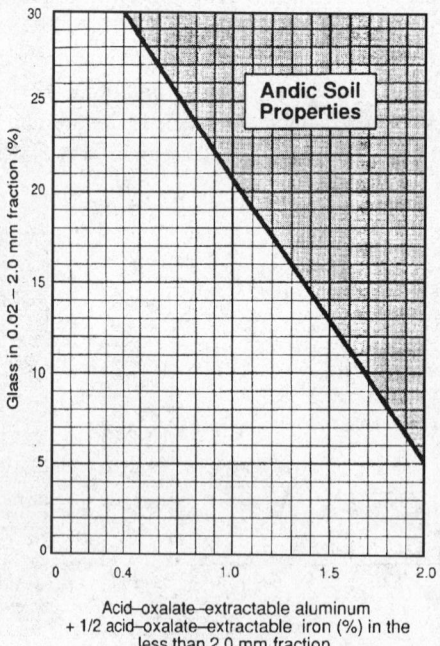

Fig. 6.15   Classification of andic versus nonandic materials as a function of glass content and ammonium oxalate extractable Al and Fe contents assuming 25% phosphate fixation [Soil Survey Staff, 1998]

Soil material must contain < 25% (by weight) of organic C and meet one or both of the following requirements in the fine earth fraction: (1) Al plus ½Fe (%) by ammonium oxalate ≥ 2.0%, a bulk density at 33 kPa ≥ 0.90 g cm^{-3}, and a P retention ≥ 85%, (2) a P retention ≥ 25%, ≥ 30% of particles between 0.02 and 2 mm, and one of the following: (a) Al plus ½ Fe (%) by ammonium oxalate ≥ 0.40% and ≥ 30% volcanic ash in the 0.02 to 2 mm fraction, or (b) Al plus ½ Fe (%) by ammonium oxalate ≥ 2.0% and ≥ 5% volcanic glass in the 0.02 to 2 mm fraction, or (c) Al plus ½ Fe (%) by ammonium oxalate between 0.40 and 2.0% and enough volcanic glass in the 0.02 to 2 mm fraction so that the glass (%) when plotted against the value obtained by adding Al plus ½ Fe (%) falls within prescribed limits (Soil Survey Staff, 1998).

Table 6.16   Listing of suborders, great groups and subgroups in the Andisols order [Soil Survey Staff, 1998]

Suborder	Great Group	Subgroup
Aquands	Cryaquands	Lithic, Histic, Thapic, Typic
	Placaquants	Lithic, Duric Histic, Duric, Histic, Thaptic, Typic
	Duraquands	Histic, Acraquoxic, Thaptic, Typic
	Vitraquants	Lithic, Duric, Histic, Thaptic, Typic,
	Melanaquands	Lithic, Acraquoxic, Hydric Pachic, Hydric, Pachic,Thaptic, Typic
	Epiaquands	Duric, Histic, Alic, Hydric, Thaptic, Typic
	Endoquands	Lithic, Duric, Histic, Alic, Hydric, Thaptic, Typic
Cryands	Duricryands	Aquic, Typic
	Hydrocryands	Lithic, Placic, Aquic, Thaptic, Typic
	Melanocryands	Lithic, Vitric, Typic
	Fulvicryands	Lithic, Pachic, Vitric, Typic
	Vitricryands	Lithic, Aquic, Oxyaquic, Thaptic, Humic Xeric, Xeric, Ultic, Alfic, Humic, Typic
	Haplocryands	Lithic, Alic, Aquic, Acrudoxic, Vitric, Thaptic, Xeric, Typic
Torrands	Duritorrands	Petrocalcic, Vitric, Typic
	Vitritorrands	Lithic, Duric, Aquic, Calcic, Typic
	Haplotorrands	Lithic, Duric, Calcic, Typic
Xerands	Vitrixerands	Lithic, Aquic, Thaptic, Alfic Humic, Ultic, Alfic, Humic, Typic
	Melanoxerands	Pachic, Typic
	Haploxerands	Lithic, Aquic, Thaptic, Calcic, Ultic, Alfic Humic, Alfic, Humic, Typic
Vitrands	Ustivitrands	Lithic, Aquic, Thaptic, Calcic, Humic, Typic
	Udivitrands	Lithic, Aquic, Thaptic, Ultic, Alfic, Humic, Typic
Ustands	Durustands	Aquic, Thaptic, Humic, Typic
	Haplustands	Lithic, Aquic, Dystric Vitric, Vitric, Pachic, Thaptic, Calcic, Dystric, Oxic, Ultic, Alfic, Humic, Typic
Udands	Placudands	Lithic, Aquic, Acrudoxic, Hydric, Typic
	Durudands	Aquic, Acrudoxic, Hydric, Pachic, Typic
	Melanudands	Lithic, Anthraquic, Aquic, Acrudoxic Vitric, Acrudoxic Hydric, Acrudoxic, Pachic Vitric, Vitric, Hydric Pachic, Pachic, Hydric, Thaptic, Ultic, Eutric, Typic
	Hydrudands	Lithic, Aquic, Acrudoxic Thaptic, Acrudoxic, Thaptic, Eutric, Ultic, Typic
	Fulvudands	Eutric Lithic, Lithic, Aquic, Hydric, Acrudoxic, Ultic, Eutric Pachic, Eutric, Pachic, Thaptic, Typic
	Hapludands	Lithic, Anthraquic, Aquic Duric, Duric, Alic, Aquic, Acrudoxic Hydric, Acrudoxic Thaptic, Acrudoxic Ultic, Acrudoxic, Vitric, Hydric Thaptic, Hydric, Eutric Thaptic, Thaptic, Eutric, Oxic, Ultic, Alfic, Typic

Suborders are based on moisture or temperature regimes and the great groups and subgroups on other soil properties, which is evident from the names in Table 6.16. The melanic (dark humus rich soils generally formed under grass) and fulvic (browner, generally formed under forest) great groups are the most common. Cryands, which are Andisols formed under cryic or pergelic soil temperature regimes, were recently added to the Andisol order. The classification is based on low bulk density, high P fixation and the presence of amorphous minerals (allophane and imogolite).

Under specified conditions where andic materials have influenced surface horizons of other soil orders, but where they are insufficiently thick or do not meet the andic requirements, these soils are placed in andic subgroups of other orders.

6.3.6 Andisol Properties

6.3.6.1 Morphological Features

Most Andisols have striking morphological features. They usually have multiple sequences of horizons (Fig. 6.16) as a result of intermittent volcanic fallout (Ping et al., 1988). In the field, one can easily identify dark A horizons overlying reddish brown or dark yellowish brown Bw cambic horizons with one or more buried A–Bw sequences below (Tables 6.17 and 6.19). Quite often, a thin gray or whitish layer of recent ashfall may be found at or near the surface as in the El Sueño series (Bw horizon). Because of the nature of the ashfall, the horizons are generally parallel to the surface and the transition from one horizon to the next is distinct. Such tephra or ash layers provide good markers of historical or datable volcanic ejection chronology.

Fig. 6.16 Typical profile of a Cryand from Iceland showing a multiple stratified horizon sequence [Photo courtesy of L.P. Wilding]

Table 6.17 Selected chemical properties of some Andisols from Guatemala [Sumner et al., 1992]

Soil horizon	Depth cm	pH H₂O	pH *M* KCl	pH *M* NaF	Organic matter %	ECEC	CEC pH 7	AEC pH 4	Ca	Mg	Na	K	Al	P retention %
		H_2O	*M* KCl	*M* NaF	%				Ca	Mg	Na	K	Al	%
						------------------------cmol$_c$ kg^{-1}-----------------------								
medial, isothermic Acrudoxic Hapludand: Las Cruces series (Guatemala)														
A1	0-30	4.7	4.6	11.5	7.2	2.2	26.8	1.2	0.55	0.09	0.34	0.06	1.20	90
Bw1	30-60	4.6	4.8	10.9	2.6	1.3	29.5	1.6	0.50	0.08	0.37	0.07	0.30	95
Bw2	60-92	4.8	5.1	11.5	1.3	1.4	28.4	2.0	0.50	0.12	0.35	0.03	0.40	96
Bw3	92-185	4.3	5.0	11.4	0.9	1.3	26.9	2.2	0.50	0.03	0.26	0.03	0.50	96
ashy over medial, isothermic Thaptic Udivitrand: El Sueño series (Guatemala)														
A	0-16	5.0	4.4	10.8	4.0	3.3	13.1	0.4	1.52	0.09	0.29	0.13	1.30	88
Bw	16-75	5.7	4.9	10.7	0.2	0.8	2.2	nd	0.50	0.03	0.22	0.06	0.00	68
2Ab	75-102	5.7	5.1	11.7	6.5	4.7	39.7	nd	3.70	0.16	0.38	0.43	0.00	98
2Bwb	102-163	5.8	5.3	11.3	2.8	4.7	39.8	nd	3.67	0.21	0.37	0.41	0.00	94
2C	163-170	5.6	5.4	11.4	1.7	2.1	30.3	nd	1.33	0.09	0.52	0.11	0.00	60
medial-skeletal, isothermic Typic Haplustand: Tecuamburro series (Guatemala)														
A	0-15	6.0	5.1	10.2	5.8	17.5	39.8	0.0	15.17	1.57	0.45	0.35	0.00	92
Bw1	15-47	5.6	4.8	10.1	0.8	15.9	35.4	nd	13.60	1.80	0.38	0.07	0.00	90
Bw2	47-112	5.6	0.6	10.2	0.6	17.8	36.0	nd	15.17	1.86	0.65	0.07	0.00	95
Bw3	112-200	5.8	4.8	10.0	0.4	14.8	32.6	nd	12.15	2.17	0.47	0.03	0.00	95

In the field when excavated, Andisols give a feeling of puffiness due to the low bulk density which allows roots to penetrate to great depths. Andisols generally have granular structure in the A horizon due to pertubations caused by roots and soil fauna. Structure in the Bw horizon is generally weak subangular blocky. Some Udands formed in areas of high rainfall have higher clay contents while soils that are subjected to wet and dry cycles form columnar structure. Soils high in allophane are sticky under field moist conditions.

6.3.6.2 Mineralogical Properties

The mineralogy of Andisols is responsible for their unique physical and chemical properties. The major minerals are allophane, allophane-like minerals, imogolite and crystalline aluminosilicate clay minerals. All types of clay minerals can be found in Andisols, but those of greatest interest are allophane, imogolite and poorly crystalline Fe hydroxides (ferrlhydrite) because they confer andic properties (Wada, 1986; Ping et al., 1988; Yoshinga, 1988; Shoji et al., 1993).

Volcanic ash consists of glass-containing aluminosilicates with minor amounts of ferromagnesian minerals. Upon weathering, ash releases Al and Fe into solution which then form stable complexes with organic matter derived from plant decomposition in the surface soil. These Al and Fe complexes, which are immobile, accumulate at the surface to form a dark or darkish brown A horizon. These complexes dominate the fine fraction. Normally in humid environments, dissolved Al is complexed with organic matter, but the soil solution may contain large amounts of dissolved Si which leads to the formation of secondary Si minerals (laminar opaline silica) from the saturated solution (Shoji and Masui, 1971; Ping et al., 1988). This material is circular or elliptical in shape (0.2–0.5 μm diam) and extremely thin (Fig. 6.17). Opaline silica is also found in the surface horizons of some Andisols (Aleutian Islands and Kenai Peninsula).

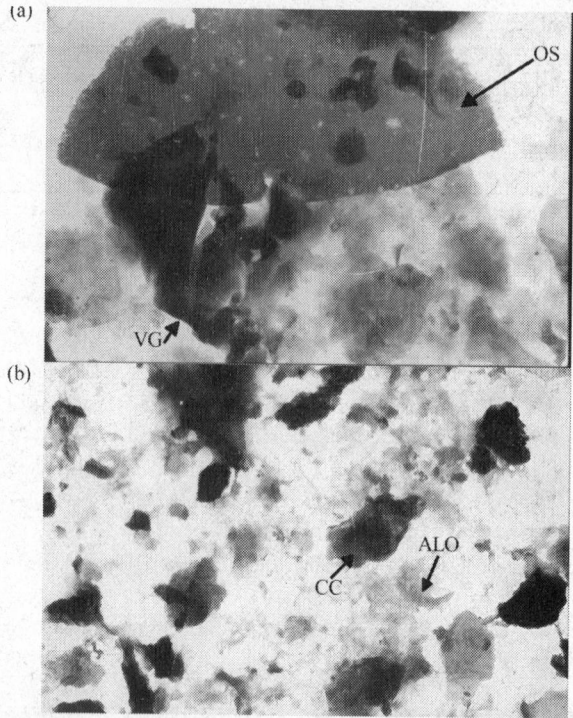

Fig. 6.17 Electron micrographs of clay minerals in Andisols from Alaska: (a) Laminar opaline silica (OS) in the A horizon of a Melanocryand, (b) allophane (ALO) and crystalline clay minerals (CC) in the Bs horizon of a Fulvucryand

As depth to Bw horizon increases, the supply of organic matter becomes more limited, and as a result, the dissolved Al and Si from volcanic glass weathering precipitate to form non- and para-crystalline minerals, mainly allophane and imogolite. Allophane has been used as a synonym for amorphous clay because it is nearly X-ray amorphous. However, under the electron microscope, it appears as hollow spheres (3.55 nm diam) which have the chemical composition $(1-2)SiO_2 \cdot Al_2O_3 \cdot (2.5-3.0)H_2O$ (Wada, 1989). Imogolite is para-crystalline and gives characteristic X-ray diffraction, IR and differential thermal analysis patterns. Under the electron microscope, it appears as smooth and curved threads or tubes with inner and outer diameters of 1.1 and 2 nm, respectively. Chemically it has the composition $(OH)SiO_3 \cdot Al_2(OH)_3$ (Wada, 1989). Allophane content increases with depth in the profile, usually being concentrated in the Bw and buried horizons; across the landscape, it increases with increasing age of the volcanic deposit and annual precipitation. Allophane and imogolite are soluble in acid oxalate solution and the Si dissolved is used to estimate their contents in soils (Parfitt and Henmi, 1982). Allophane in B horizons ranges from 2–12, 4–14 and 10–20% in southcentral Alaska, the Aleutian Islands and Baranof Island, respectively (Ping et al., 1989).

In contrast, the soils from Guatemala (Table 6.18) can contain halloysite, gibbsite, kaolinite, vermiculite and crystobalite in addition to amorphous allophanic materials. When the pH_{NHF} is above 9.5, Fieldes and Perrott (1966) showed that it was diagnostic for the presence of allophane. In all pedons from Guatemala, this was the case which confirms the presence of this mineral. All pedons contained substantial amounts of free Fe oxides.

Ferrihydrite $(5Fe_2O_3 \cdot 9H_2O)$ (Schwertmann and Taylor, 1989), also an amorphous mineral (short-range order) found in Andisols under high rainfall or in soils with restricted drainage, is a widespread

Table 6.18 Selected physical and mineralogical properties of some Andisols from Guatemala [Sumner et al., 1992]

Soil horizon	Depth cm	Clay content	Gravel content	Water content 30 kPa	1500 kPa	Bulk density g cm^{-3}	Surface area m^2 g^{-1}	Free Fe$_2$O$_3$ %	Mineralogy[†]
			% (w/w)						
medial, isothermic Acrudoxic Hapludand: Las Cruces series (Guatemala)									
A1	0-30	7.6	0	53.0	29.0	0.56	51	3.64	AM,HA,GI,CT
Bw1	30-60	6.0	0	56.9	40.5	0.53	217	5.19	AM,HA,GI,CT
Bw2	60-92	6.4	0	60.7	41.3		244	4.44	
Bw3	92-185	6.9	0	64.6	43.6	0.62	250	3.51	AM,HA,GI,CT
ashy over medial, isothermic Thaptic Udivitrand: El Sueño series (Guatemala)									
A	0-16	5.8	2	38.7	10.8	0.86	7	0.88	AM,HA,VR,CT
Bw	16-75	2.8	4	10.0	2.9	0.92	1	0.23	AM,HA,VR,CT
2Ab	75-102	22.4	0	79.1	33.6		69	3.76	
2Bwb	102-163	23.0	0	71.5	34.6	0.53	181	3.89	
2C	163-170	26.6	0	95.2	41.4		179	2.38	AM,CT
medial-skeletal, isothermic Typic Haplustand: Tecuamburro series (Guatemala)									
A	0-15	23.8	14	44.4	27.8	0.58	44	0.82	HA,KK,AM,CT
Bw1	15-47	18.5	34	47.8	29.1		71	3.59	HA,KK,AM,CT
Bw2	47-112	18.0	32	41.1	28.8		71	4.14	
Bw2	112-200	24.5	31	49.3	32.1		80	5.34	HA,KK,AM,CT

[†] AM = amorphous, HA = halloysite, GI = gibbsite, CT = crystobalite, VR = vermiculite, KK = kaolinite

and characteristic component of young Fe-oxide accumulations precipitated from Fe-rich solutions. Ferrihydrite content can be estimated from oxalate or dithionite extractable Fe (Parfitt, 1983). In B horizons, allophane with or without ferrihydrite dominates the colloidal fraction with Al and Fe humus complexes as minor components. The lower B and C horizons can be rich in allophane and ferrihydrite but usually contain only small amounts or no Al and Fe humus complexes (Ping et al., 1989).

In the silt and sand fractions of Andisols, the dominant components are volcanic glass and other primary minerals. The glass particles (shards) which, like shattered glass, have sharp angles and edges, are very abrasive. However, in soils, the glass particles are usually coated with amorphous materials including allophane, Al and Fe oxides and their humus complexes which all contribute to aggregate formation. The oxide and organic matter coatings on glass or other mineral surfaces can be easily removed using HCI and NaOH solutions, respectively, prior to the measurement of glass content.

Mineralogical characteristics of Andisols have been reviewed by Wada (1980) and a detailed discussion of short-range order minerals in presented in Section F, Chapter 4.

6.3.6.3 Chemical Properties

Most Andisols are acid to very acid (Tables 6.17 and 6.19) and many contain elevated levels of exchangeable Al and low amounts of exchangeable basic cations. All pedons presented have high P-retention capacities due to the presence of allophane and Fe oxyhydroxides (Parfitt, 1989). Organic matter contents are high relative to nonandic soils and are concentrated in materials which have been weathered. Note the very low organic matter content in a recent ash layer (Bw) in the El Sueño series (Table 6.17).

Variable charge arising from the amorphous minerals and Fe and Al organic complexes is high in most Andisols as illustrated in Table 6.17 by the large difference between ECEC and CEC values. In

Table 6.19   Selected chemical and physical properties of some Andisols from Alaska

Soil horizon	Depth cm	pH H$_2$O	pH KCl	pH NaF	Org. C %	CEC pH 7 cmol$_c$ kg^{-1}	Exch. Al[1]	Base sat. %	P reten. %	Bulk density g cm^{-3}	Allophane content	1,500 kPa water[2] %	Volcanic glass[3]	USDA texture
colspan Island series (Melanocryand)
A1	0-23	5.2			9.1	37	3.1	7	90	0.65	0	30	36	sil
A2	23-40	5.6			6.1	33	1.3	10	94	0.73	2.1	27	34	vfsl
A3	40-60	6.0			3.5	13	0.1	16	82	0.64	3.6	19	34	sil
Bw	60-101	6.0			2.2	13	0.3	15	81	0.71	4.0	18	20	sil
C	101-152	5.9			0.2	5	0.6	15	20	1.00	0.5	4	2	lfs
Kachemak series (Fluvucryand)
Bs	0-8	4.5			9.5	39	7.1	15	94	0.63	0	23	90	sil
Ab	8-20	4.8			12.4	42	7.6	10	96	0.69	0	37	86	sil
Bhs	20-30	4.9			7.3	45	7.5	5	97	0.67	1.5	36	50	sil
Bw	30-56	5.2			3.0	23	3.8	5	95	0.77	3.0	24	25	sil
2C	56-114	5.3			0.9	17	4.5	30	70	1.21	1.0	14	5	sil
Kashwitna series (Haplocryand)
E	0-3	4.3			6.9	35	6.0	17	54	0.63	0	16	51	sil
Bs	3-20	5.5			4.7	27	1.9	6	94	0.74	4.2	20	57	sil
Eb	20-25	5.6			3.2	11	0.7	5	55	---	2.8	---	---	sil
Bsb	20-30	5.8			1.4	14	0.8	8	66	---	4.3	11	14	sil
BC	30-35	5.6			0.8	8	0.2	9	50	1.03	2.8	7	13	sil
2BC	35-46	5.9			0.2	4	---	6	34	---	0.7	3	2	gvlq
2C	46-102	5.7			---	2	---	24	16	---	0	2	2	gxs

[1]Exchangeable Al extracted with M KCl
[2] On field moist samples, wt%
[3] In very fine sand fraction

fact, one pedon (Las Cruces series) contains substantial AEC and is poised near its point of zero charge (p.z.c.) which can be inferred from the higher pH$_{KCl}$ than pH$_{water}$ values. For Papua New Guinea Andisols (Hydrudands), Radcliffe and Gillman (1984) demonstrated that the p.z.c. decreased with increasing organic C and increased with increasing Fe$_2$O$_3$ content. Parfitt (1988) found that ECECs could range from 1–8 cmol$_c$ kg^{-1} and CECs (pH 7) from 20–40 cmol$_c$ kg^{-1} for allophanic A horizons. Shoji et al. (1993) demonstrated that AEC was detected in allophanic but not in nonallophanic soils.

A review of the chemical properties of variable charge soils is presented by Parfitt (1980) and the chemistry of variable charge surfaces is presented in Section B, Chapter 7.

6.3.6.4   Physical Properties

Unique physical attributes of Andisols are related to compound structural assemblages of hollow spheres and tubular threads as mineral entities into progressively larger aggregated domains. Hence, close packing of spherules and threads from 1–2 nm basic units to 0.1–1 μm microaggregated domains to aggregates of several mm in size results in low density, high porosity, high surface area, high soil water retention even at low water potentials, and the disparity in water movement from within basic units (intraaggregate pores) to interparticle porosity associated with aggregated assemblages of different sizes. This phenomenon also accounts for the low thermal conductivity of andic materials which is 3 to 4 times less than that of the layer lattice phyllosilicates in other mineral soils. It accounts for the thixotropic character of these soils (discussed later), and the irreversible changes in bulk density (increased), water retention (decreased) and cohesive strength (increased) upon drying andic materials.

Andisols have low bulk density, usually < 0.9 g cm^{-3} due to the high organic matter and amorphous mineral contents. This makes them excellent rooting media but on the other hand, they have low bearing capacity, are highly susceptible to wind and water erosion when surface cover is removed or degraded, and present problems from an engineering standpoint. Because of the nature of ejecta, many contain appreciable amounts of gravel and stones.

Most Andisols have a smeary feeling when rubbed between the fingers when field moist. Even moist soil after prolonged rubbing can become greasy and wet. However, once the pressure is released, the excess water is reabsorbed by the soil. This property is called thixotrophy and is due to their unusually high water retention capacities (Tables 6.18 and 6.19). This high water retention is due to the allophane present which has a very high surface area (~ 1,000 m^2 g^{-1}) due to its hollow sphere shape (Wada, 1989). Exchange of water between micropores associated with these spheres and macropores is incomplete and very slow. Field capacity moisture contents of many Andisols can exceed 100%, particularly where weathering has been intense. The Guatemalan Andisols all exhibited high surface areas in weathered horizons (Table 6.18).

Many Andisols exhibit irreversible changes on drying. On dehydration, allophane aggregates to silt or larger size particles which do not breakdown on resetting. This phenomenon can also lead to crust formation at the soil surface during hot dry periods. This also results in particle size analyses that are commonly unreliable because clay content is underestimated and sand and silt contents overestimated (Ping et al., 1988). This is well illustrated in Table 6.18 for the Las Cruces pedon where measured clay content was ~ 6% but the surface area was 50–250 m^2 g^{-1}, clearly an incongruity.

Plasticity is a physical property which distinguished allophanic from all other soils containing crystalline clay minerals. Generally, field moist Andisols have high liquid (60–350%) and plastic limits (70–180%) (Warkentin and Maeda, 1980). Plasticity measurements can be used as an index of physical behavior in Andisols as a substitute for grain size distribution which is unreliable. Air-dry samples, on the other hand, show little plasticity because of the irreversible changes on drying and behave like sandy soils. The physical and mechanical characteristics of Andisols have been discussed in detail by Warkentin and Maeda (1980).

6.3.7 Management of Andisols

In many parts of the world, crop productivity on Andisols is very high, leading to population concentrations such as on Java where the soils are particularly fertile (Sutanto et al., 1988) and in Japan, the Andean portion of South America, and also Central America. Fresh deposits of ash resupply nutrients to many Andisols as they weather, thus maintaining fertility. Consequently, few inputs with the exception of P have been required for successful agricultural production on many Andisols.

However, Andisols are all acid and where acidity is particularly high, addition of lime may be required to overcome Al toxicity or alternatively acid-tolerant crops must be selected (Michaelson and Ping, 1987). Where Andisols have been acidified by N inputs, serious consequences for crop production can arise because as pH decreases, CEC declines towards the p.z.c. As a result, such soils are able to hold few basic cations for plant uptake (Sumner and Hylton, 1993). Because of their high P-fixing capacities, most Andisols require inputs of P for successful production (Michaelson and Ping, 1990). Such inputs in the form of rock phosphate are effective on very acid Andisols but in any event, soluble P sources should be band placed or otherwise protected from intimate mixing with the soil (Sumner and Hylton, 1993). Requirements for N vary according to the ease with which organic matter is mineralized in these soils. Liming often promotes the release of substantial amounts of N from non-available reserves. Other nutrients may sometimes be deficient in Andisols depending of the chemical composition of the parent material.

Andisols, in general, are noted for their good physical properties (high water-holding capacity, good tilth, resistance to water erosion and stable aggregation). Consequently, they are excellent media for root growth and proliferation, but under heavy grazing can become compacted (Schlichting, 1988). In high latitudes, Cryands may be overgrazed, denuded, and subject to wind erosion (Arnalds, et al., 1995). Likewise, Vitrands have high potential to be eroded by wind and have low nutrient and water holding capacities due to their sandy nature. In areas where volcanoes are or have been active, stratified ash deposits can cause textural discontinuities resulting in problems with water transmission within the profile, root growth and poor tilth. Deep plowing can often remedy these situations.

6.3.8   References

Arnalds, O., C.T. Hallmark, and L.P. Wilding. 1995. Andisols from four different regions of Iceland. Soil Sci. Soc. Am. J. 59:161–169.

Arnold, R.W. 1988. The world wide distribution of Andisols and the need for an Andisol order in Soil Taxonomy. *In* D.I. Kinloch, S. Shoji, F.H. Beinroth, and H. Eswaran (ed.) Proc. 9th Int. Soil Classif. Workso. USDA Soil Management Support Services, Washington, DC.

Derking, R. 1987. A review of the classification of Andepts, their intergrades and related families in the northwestern United States. *In* J. Kimble and W.D. Nettleton (ed.) Characterization, classification and utilization of Andisols. Proc. 1st Int. Soil Correl. Meet. USDA-SCS and USAID Soil Management Support Service, Washington, DC.

Dudal, R., and M. Suparpthardjo. 1961. Some considerations of the genetic relationships between Latosols and Andosols in Java (Indonesia). Trans. 7th Int. Cong. Soil Sci. 4:229–233.

FAO-UNESCO. 1989. Soils of the world. UNESCO, Paris, France.

Fassbender, H.W. 1969. Panel on volcanic ash soils in Latin America. Training and Research Center of the IAAIS, Turrialba, Costa Rica.

Fieldes, M., and K.W. Perrott. 1966. The nature of allophane in soils. Part 3. A rapid field and laboratory test for allophane. NZ J. Sci. 9:623–629.

Flach, K.W., C.S. Holzhey, F. DeConinck, and R.J. Bartlett. 1980. Genesis and classification of Andepts and Spodosols. p. 411–426. *In* B.K.G. Theng (ed.) Soils with variable charge. New Zealand Society of Soil Science, Lower Hutt, New Zealand.

Goldin, A., W.D. Nettleton, and R.J. Engel. 1992. Pedogenesis on outwash and glacial marine drift, northwestern Washington. Soil Sci. Soc. Am. J. 56:1545–1552.

Jenny, H. 1941. Factors of soil formation. McGraw-Hill Book Co., New York, NY.

Leamy, M.L. 1979. ICOMAND Circ. Let. 1. NZ Soil Bureau, Lower Hutt, New Zealand.

Leamy, M.L. 1984. Andisols of the world. p. 369–387. *In* Comm. Congr. Int. Suelos Volcan. Secretariado de Publicaciones Serie Informes 13, Universidad de la Laguna, La Laguna, Spain.

Leamy, M.L. 1988. ICOMAND Circ. Let 1 0. NZ Soil Bureau, Lower Hutt, New Zealand.

Maeda, T., H. Takenaka, and B.P. Warkentin. 1977. Physical properties of allophane soils. Adv. Agron. 29:229–264.

McDaniel, P.A., M.A. Fosberg, and A. L. Falen. 1993. Expression of andic and spodic properties in tephra-influenced soils of northern Idaho. Geoderma 58:79–94.

Michaelson, G.J., and C.L. Ping. 1987. Effect of P, K and liming on soil pH, Al, Mn, K, matter yield and forage barley dry matter yield and quality for a newly cleared Cryorthod. Plant Soil 104:155–161.

Michaelson, G.J., and C. L. Ping. 1990. Mehlich 3 extractable P of Alaska soils as affected by P fertilizer application. Appl. Agric. Res. 5:255–260.

Mizota, C., and L.P. van Reeuwijk. 1989. Clay mineralogy and chemistry of soils formed in volcanic material in diverse climatic regions. International Soil Reference and Information Centre, Wageningen, Netherlands.

Parfitt, R.L. 1983. Identification of allophane in Inceptisols and Spodosols. Soil Taxonomy News 5:11 and 18. New Zealand Soil Bureau, Lower Hutt, New Zealand.

Parfitt, R.L. 1980. Chemical properties of variable charge soils. p. 167–194. *In* B.K.G. Theng (ed.) Soils with variable charge. NZ Society of Soil Science, Lower Hutt, New Zealand.

Parfitt, R.L. 1988. Variable charge in Andisols. p. 60–73. *In* D.I. Kinloch, S. Shoji, F.H. Beinroth, and H. Eswaran (ed.) Proc. 9th Int. Soil Classif. Worksh. USDA Soil Management Support Services, Washington, DC.

Parfitt, R.L. 1989. Phosphate reactions with natural allophane, ferrihydrite and goethite. J. Soil Sci. 40:359–369.

Parfitt, R.L. 1990. Soils formed in tephra in different climatic regimes. Trans. 14th. Int. Congr. Soil Sci. VII:134–139.

Parfitt, R.L., and T. Henmi. 1982. Comparison of an oxalate extraction method and an infrared spectroscopic method for determining allophane in soil clays. Soil Sci. Plant Nutr. 28:183–190.

Péwé, T.L. 1975. Quaternary geology of Alaska. US Geol. Surv. Prof. Pap. 835.

Ping, C.L., S. Shoji, and T. Ito. 1988. The properties and classification of three volcanic ash derived pedons from Aleutian Islands and Alaska Peninsula. Soil Sci. Soc. Am. J. 52:455–462.

Ping, C.L., S. Shoji, T. Ito, T. Takahashi, and J.P. Moore. 1989. Characteristics and classification of volcanic ash-derived soils in Alaska. Soil Sci. 148:8–28.

Ping, C.L., G.J. Michaelson, and R.L. Malcolm. 1995. Fractionation and carbon balance of soil organic matter in selected cryic soils in Alaska. p. 307–314. *In* R. Lal, J. Kimble, E. Levine and B.A. Stewart (ed.) Soils and global change. Lewis Publishers, Boca Raton, FL.

Radcliffe, D.J., and G.P. Gillman. 1984. Surface charge characteristics of volcanic ash soils from the southern highlands of Papua New Guinea. p. 164–192. *In* Comm. Congr. Int. Suelos Volcan. Secretariado de Publicaciones Serie Informes 13, Universidad de la Laguna, La Laguna, Spain.

Rieger, S., D.B. Schoepherster, and C.E. Furbush. 1979. Exploratory soil survey of Alaska. USDA-SRC. US Government Printing Office, Washington, DC.

Riehle, J.R. 1985. A reconnaissance of the major Holocene tephra deposits in the Upper Cook Inlet Region, Alaska. J. Volcanol. Geotherm. Res. 26:37–74.

Schlichting, E. 1988. Physical properties of Andisols conducive to treading damage. *In* D.I. Kinloch, S. Shoji, F.H. Beinroth, and H. Eswaran (ed.) Proc. 9th lnt. Soil Classif. Worksh. USDA Soil Management Support Services, Washington, DC.

Schwertmann, U., and R.M. Taylor. 1989. Iron oxides. p. 379–438. *In* J.B. Dixon and S.B. Weed (ed.) Minerals in soil environments. Soil Science Society of America, Madison, WI.

Shoji, S. 1983. Mineralogy of volcanic ash soils. p. 31–47. *In* N. Yoshinaga (ed.) Volcanic ash soils: Genesis, properties, classification. Hakuyusha, Tokyo, Japan (In Japanese).

Shoji, S. 1986. Mineralogical characteristics. p. 21–40. *In* K. Wada (ed.) Ando soils in Japan. Kyushu University Press, Fukuoka, Japan.

Shoji, S., and J. Masui. 1971. Opaline silica of recent ash soils in Japan. J. Soil Sci. 22:101–108.

Shoji, S., and Y. Fujlwara. 1984. Active aluminum and iron in the humus horizons of Andisols from northeastern Japan: Their forms, properties, and significance in clay weathering. Soil Sci. 137:216–226.

Shoji, S., M. Nanzyo, and R.A. Dahigren. 1993. Volcanic ash soils: Genesis, properties and utilization. Elsevier, Amsterdam, Netherlands.

Shoji, S., and M. Otowa. 1988. Distribution and significance of Andisols in Japan. p. 13–24. *In* D.I. Kinloch, S. Shoji, F.H. Beinroth and H. Eswaran (ed.) Proc. 9th Int. Soil Classif. Worksp. Soil Management Support Services, Washington, DC.

Simonson, R.W., and S. Rieger. 1967. Soils of the Andept suborder in Alaska. Soil Sci. Soc. Am. Proc. 31:692–699.

Smith, G.D. 1978. The Andisol proposal–A preliminary proposal for reclassification of the Andepts and some Andic subgroups. p. 1–165. *In* M.L. Leamy, D.I. Kinloch, and R.L. Parfitt (ed.) International Committee on Andisols-Final report. USDA Soil Manag. Supp. Serv. Tech. Monogr. 20.

Soil Survey Staff. 1975. Soil Taxonomy-A basic system of soil classification for making and interpreting soil surveys. USDA-SCS Agric. Handb. 436. US Government Printing Office, Washington, DC.

Soil Survey Staff. 1990. Keys to Soil Taxonomy. 4th Ed. Soil Manag. Supp. Serv. Tech. Monogr. 6. Virginia State University, Blacksburg, VA.

Soil Survey Staff. 1998. Keys to Soil Taxonomy. 8th Ed. USDA-NRCS. Washington, DC.

Southard, S.B., and R.J. Southard. 1991. Mineralogy and classification of Andic soils in northeastern California. Soil Sci. Soc. Am. J. 53:1784-1791.

Sumner, M.E., and K. Hylton. 1993. A diagnostic approach to solving soil fertility problems in the tropics. p. 215–234. *In* J.K. Syers and D.L. Rimmer (ed.) Soil science and sustainable land management in the tropics. CAB International, Wallingford, UK.

Sumner, M.E., L.T. West, and J.E. Leal. 1992. Suelos de la agroindustria cafetalera de Guatemala: Región Sur. University of Georgia, Athens, GA.

Sutanto, R., F. DeConinck, and M. Doube. 1988. Mineralogy, charge properties and classification of soils on volcanic materials and limestone in central Java (Indonesia). State University Ghent, Ghent, Belgium.

Takahashi, T., R. Dahlgren, and P. van Suteren. 1993. Clay mineralogy and chemistry of soils formed in volcanic materials in the xeric moisture regime of northern California. Geoderma 59:131–150.

Ugolini, F.C., and R.A. Dahlgren. 1991. Weathering environments and occurrence of imogolite/allophane in selected Andisols and Spodosols. Soil Sci. Soc. Am. J. 55:1166–1171.

Wada, K. 1980 Mineralogical characteristics of Andisols. p. 87–108. In B.K.G. Theng (ed.) Soils with variable charge. NZ Society of Soil Science, Lower Hutt, New Zealand.

Wada, K. 1986. Ando soils in Japan. Kyushu University Press, Fukuoka, Japan.

Wada, K. 1989. Allophane and imogolite. p. 1051–1087. In J.B. Dixon and S.B. Weed. Minerals in soil environments. 2nd Ed. Soil Science Society of America, Madison, WI.

Wada, K., O. Arnalds, Y. Yakuto, L.P. Wilding, and C.T. Hallmark. 1992. Clay minerals of four soils formed in eolian and tephra materials in Iceland. Geoderma 52:351–365.

Warkentin, B.P., and T. Maeda. 1980. Physical and mechanical characteristics of Andisols. In B.K.G. Theng (ed.) Soils with variable charge. New Zealand Society of Soil Science, Lower Hutt, New Zealand.

Yoshinaga, N. 1988. Mineralogy of Andisols. p. 45–59. In D.I. Kinloch, S. Shoji, F.H. Beinroth and H. Eswaran. Proc. 9th Int. Soil Classif. Worksp. Soil Management Support Services, Washington, DC.

6.4   Entisols

L.C. Nordt
Baylor University

M.E. Collins
University of Florida

D.S. Fanning
University of Maryland

H.C. Monger
New Mexico State University

6.4.1   Introduction

Entisols are mineral soils that exhibit minimal pedogenic alteration (Soil Survey Staff, 1975, 1998). The ability to support plants distinguishes Entisols from areas that are not considered soil, such as rock outcrops or some highly disturbed lands. Entisols typically have AC, ACr, or AR profiles and inherit many physical and chemical properties from their parent material. The A horizon can be classified as any taxonomic epipedon, except mollic. Subsoil horizon symbols may be used when a pedogenic pathway is inferred, but the properties of the horizon must not meet the minimum requirements of any subsurface diagnostic horizon.

Entisols form in areas where pedogenesis has been restricted or retarded. Examples include landscapes with recent deposition or erosion, resistant bedrock, inert parent material, saturation, or human activity (Buol et al., 1980). The world distribution of soil orders reveals where many of these conditions exist (Fig. 6.18). The most common areas where Entisols form are mountains, sand dunes, flood plains, coastal plains, and urban areas. Entisols comprise approximately 16% of the world distribution of soils, and are the most widespread of all orders (Soil Survey Staff, 1997).

In contrast to most other soil orders, suborders of Entisols are not defined by precipitation or temperature regime. Rather, Entisol suborders are designed to differentiate the conditions under which subsurface horizon development is impeded (Smith, 1986). The five Entisol suborders are Orthents, Psamments, Fluvents, Aquents, and Arents (Figs. 6.19 and 6.20).

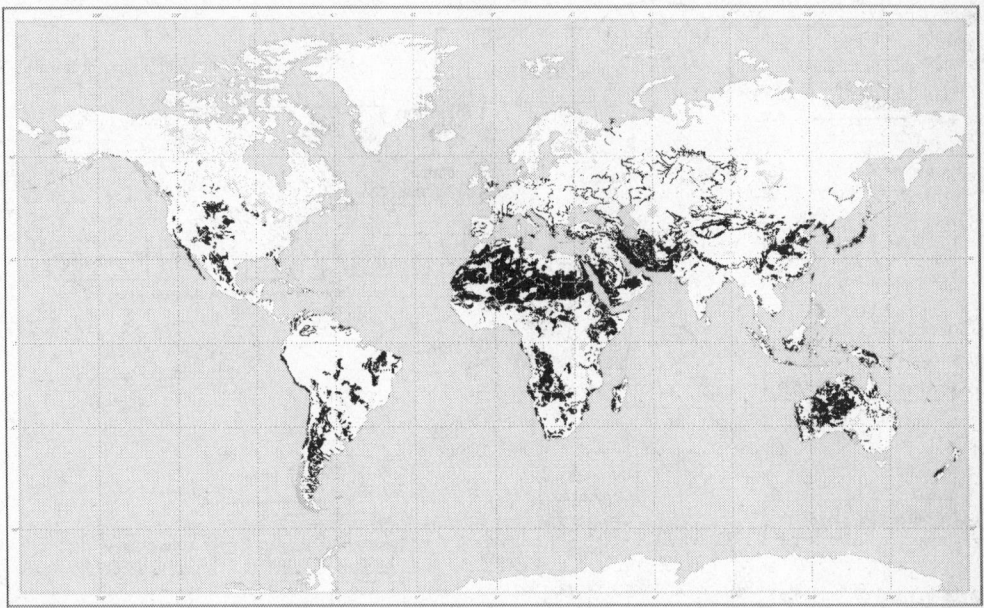

Fig. 6.18   Global distribution of Entisols [Courtesy of USDA-NRCS, Soil Survey, Division, World Soil Resources, 1998]

Parent material lithology is one factor that may inhibit pedogenesis. Resistant bedrock increases runoff and erosion, and decreases weathering. These processes promote the formation of the suborder Orthents, particularly in mountain regions (Figs. 6.19 and 6.20). Orthents may also form in sandy parent materials that have an abundance of rock fragments. When parent materials consist of quartzitic sands without coarse fragments, such as in many sand dunes, profile development is also retarded. Here, the suborder Psamments may form (Figs. 6.19 and 6.20). Both Orthents and Psamments typically have ochric epipedons.

Fluvents may develop in flood plains, fans, deltas, or mudflows when sedimentation rates exceed pedogenic rates (Section E, Chapter 1). Rapid deposition limits the effects of pedogenesis and subsurface horizon development. These soils typically have ochric epipedons (Figs. 6.19 and 6.20).

Another condition that slows pedogenesis is anaerobiosis. Persistent high watertables may prevent leaching of soil constituents, inhibit redox cycles, and preclude subsurface horizon development. These soils make up the Aquent suborder (Figs. 6.19 and 6.20). The geographical distribution of these soils may include poorly drained flood plains, hillslope seeps, and coastal marshes. Ochric or histic epipedons may form in these areas.

Lastly, Arents and other disturbed soils are becoming more common with intensified human activity. These soils typically form in mine spoil, dredge spoil, and in other recently deposited urban or industrially related materials (Fig. 6.19). These soils may have ochric or anthropic epipedons.

6.4.1.1   Engineering and Land Use
Many engineering and land use problems arise in areas where Entisols form (Buol et al., 1980; Grossman, 1983). Erosion may occur with Orthents, rapid sedimentation with Fluvents, wind erosion or deposition with Psamments, unpredictable behavior of physical and chemical properties with Arents, and poor drainage with Aquents. In contrast, some of the most productive agricultural soils in the world occur in flood plains. Flooding provides much needed nutrients and a water supply for crop

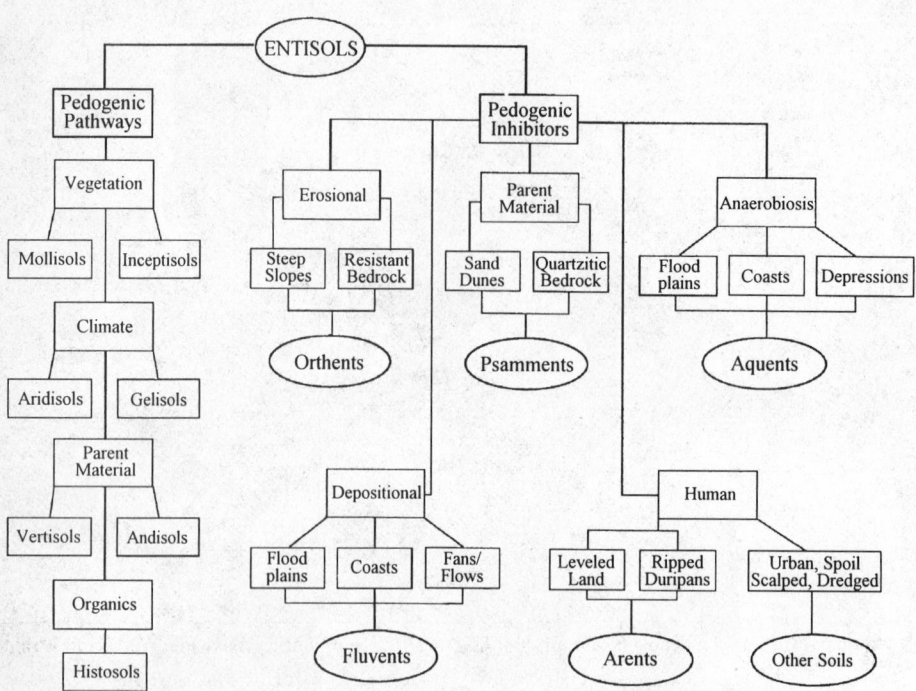

Fig. 6.19 Flow diagram illustrating (a) pedogenic pathways of soil orders from Entisols; and (b) landscape factors used to differentiate the suborders of Entisols (not included for Orthents are sandy parent materials with more than 35% rock fragments.)

production in areas where fertilization and irrigation practices are not used. Further, some of the most populated regions of the world occur in flood plains.

6.4.1.2 Genesis

The influence of the soil-forming factors (Jenny, 1941) on the development of Entisols is limited. Properties of Entisols may not reflect climate conditions because of recent deposition, erosion, or human disturbance. The primary role of biota is that of initiating the physical and chemical weathering process of freshly exposed substrate. Topography influences drainage and can contribute to the rapid formation of wet Entisols, or induce erosion in steeply sloping areas. Many of these factors also reduce the passive effects of time on soil development. Regardless of the influence of the other soil-forming factors, the properties of Entisols are largely inherited from the parent material.

Melanization (Buol et al., 1980) is the principal soil-forming process that initiates the development of Entisols. This process contributes organic matter to the parent material, promotes structural development, and may be associated with minor leaching of soluble constituents. Grossman (1983) presents data showing that bulk densities in alluvial settings initially increase in Entisols because of frequent wetting and drying cycles, which lead to compaction. Eventually, however, melanization will decrease bulk densities in surface horizons as organic matter inputs increase. In resistant bedrock, melanization rapidly decreases bulk densities as rock material is weathered and replaced by organic material or pore space. The process of melanization in the formation of Entisols can proceed as long as mollic epipedons or subsurface diagnostic horizons do not form.

6.4.1.3 Pedogenic Pathways

Entisols serve as the precursor to all other soil orders during their evolutionary history (Fig. 6.19) (Soil Survey Staff, 1998, 1975). In grassland ecosystems, Mollisols may form when steady state has been attained in the accumulation of organic matter. In forest settings, ochric epipedons and Inceptisols may form. Entisols develop into Aridisols outside of flood plains in areas with arid moisture regimes. Vertisols and Andisols are influenced more by parent material characteristics than any other soil order except Entisols. Vertisols form from Entisols in high shrink/swell parent materials where wet/dry cycles are prevalent. Andisols commonly form from Entisols in parent materials with large quantities of volcanic ash in cool and humid climates. Entisols will develop into Histosols with the accumulation of significant amounts of organic matter in association with high water tables and anaerobic conditions. Lastly, the new Gelisol order now encompasses areas formerly mapped as Entisols in cold regions. A listing of great groups and subgroups in the Entisol order is presented in Table 6.20.

6.4.2 Orthents

Orthents are loamy and clayey Entisols (better drained than Aquents), with a regular decrease in organic C with depth. They occupy approximately 10.5% of the Earth's land surface. They are more extensive than the combined area of the other Entisol suborders, which together occupy an area of about 5.7% (Soil Survey Staff, 1997). Orthents also have more taxonomic subdivisions, reflecting properties related to climate, depth to bedrock, mineralogy, wetness, fertility, shrink/swell, and bioturbation (Soil Survey Staff, 1998).

Taxonomically, Orthents commonly grade into Fluvents and Psamments. The rationale for separating Orthents from Fluvents was to maintain the old concept of flood plain soils (Fluvents), which have immense agricultural importance (Smith, 1986). The rationale for separating Orthents from Psamments was to identify soils (the Psamments) that are readily susceptible to wind erosion (Gile et al., 1981).

Although Orthents are most commonly associated with Psamments and Fluvents, they may also be related to Arents. For example, where soils have been badly gullied, such as in some loess deposits of the southern United States, Alfisols occur on narrow ridges and Orthents in eroded areas between the ridges. When the areas are reclaimed by leveling with bulldozers, Arents are formed from Alfisols

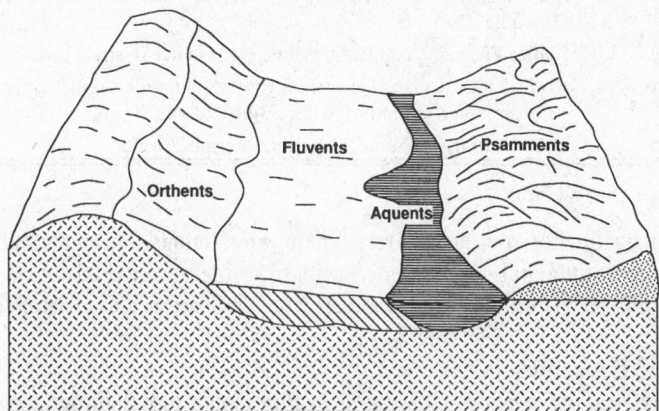

Fig. 6.20 Block diagram illustrating typical landscape positions where Entisols suborders form. Orthents on steep slopes associated with bedrock, Fluvents in floodplains, Aquents in poorly drained riparian conditions, and Psamments in sand dunes

Table 6.20   Listing of suborder, great groups and subgroups in the Entisol order [Soil Survey Staff, 1998]

Suborder	Great Group	Subgroup
Aquents	Sulfaquents	Haplic, Histic, Thapto-Histic, Typic
	Hydraquents	Sulfic, Sodic, Thaptic-Histic, Typic
	Cryaquents	Aquandic, Typic
	Psammaquents	Lithic, Sodic, Spodic, Humaqueptic, Mollic, Typic
	Fluvaquents	Sulfic, Vertic, Thapto-Histic, Aquandic, Aeric, Humaqueptic, Rhodic, Mollic, Typic
	Epiaquents	Aeric, Humaqueptic, Mollic, Typic
	Endoaquents	Sulfic, Lithic, Sodic, Aeric, Humaqueptic, Mollic, Typic
Arents	Ustarents	Haplic
	Xerarents	Sodic, Duric, Alfic, Haplic
	Torriarents	Sodic, Duric, Haplic
	Udarents	Alfic, Ultic, Mollic, Haplic
Psamments	Cryopsamments	Lithic, Aquic, Oxyaquic, Vitrandic, Spodic, Lamellic, Typic
	Torripsamments	Lithic, Vitrandic, Haploduridic, Ustic, Xeric, Rhodic, Typic
	Quartzipsamments	Lithic, Aquodic, Aquic, Oxyaquic, Ustoxic, Udoxic, Plinthic, Lamellic Ustic, Lamellic, Ustic, Xeric, Spodic, Typic
	Ustipsamments	Lithic, Aquic, Oxyaquic, Aridic, Lamellic, Rhodic, Typic
	Xeropsamments	Lithic, Aquic Durinodic, Aquic, Oxyaquic, Vitrandic, Durinodic, Lamellic, Dystric, Typic
	Udipsamments	Lithic, Aquic, Oxyaquic, Spodic, Lamellic, Plaggeptic, Typic
Fluvents	Cryofluvents	Andic, Vitrandic, Aquic, Oxyaquic, Mollic, Typic
	Xerofluvents	Vertic, Aquandic, Andic, Vitrandic, Aquic, Oxyaquic, Durinodic, Mollic, Typic
	Ustifluvents	Aquertic, Torrertic, Vertic, Anthraquic, Aquic, Oxyaquic, Aridic, Udic, Mollic, Typic
	Torrifluvents	Ustertic, Vertic, Vitrixerandic, Vitrandic, Aquic, Oxyaquic, Duric Xeric, Duric, Ustic, Xeric, Anthropic, Typic
	Udifluvents	Aquertic, Vertic, Andic, Vitrandic, Aquic, Oxyaquic, Mollic, Typic
Orthents	Cryorthents	Lithic, Vitrandic, Aquic, Oxyaquic, Lamellic, Typic
	Torriorthents	Lithic Ustic, Lithic Xeric, Lithic, Xerertic, Ustertic, Vertic, Vitrandic, Aquic, Oxyaquic, Duric, Ustic, Xeric, Typic
	Xerorthents	Lithic, Vitrandic, Aquic, Oxyaquic, Durinodic, Dystric, Typic
	Ustorthents	Aridic Lithic, Lithic, Torrertic, Vertic, Anthraquic, Aquic, Oxyaquic, Durinodic, Vitritorrandic, Vitrandic, Aridic, Udic, Vermic, Typic
	Udorthents	Lithic, Vitrandic, Aquic, Oxyaquic, Vermic, Typic

because fragments of argillic horizons are scattered in the smooth land left by the bulldozers (Smith, 1986). A variety of environmental conditions contribute to the absence of subsurface diagnostic horizons in Orthents. The most common factors are erosion, deposition, and resistant parent material.

6.4.2.1   Erosion and Deposition
Because steep slopes promote erosion, Orthents are common in mountainous terrains, especially where vegetation is sparse. An example of this landscape occurs in the Basin and Range area of southern New Mexico (Fig. 6.21). Mountain slopes in that area are characteristically composed of

intermixed Orthents and rock outcrops, where the Orthents are composed of A overlying R horizons (Gile and Grossman, 1979). As with steep mountain slopes, erosional scarps that separate pleistocene fan terraces (an alluvial fan deposited on a flood plain that was subsequently abandoned with channel downcutting) are also areas where Orthents form.

In some dissected arid areas, Orthents develop by episodic erosion. Here well-developed Aridisols with argillic and calcic subsurface diagnostic horizons are eroded to the extent that the parent material becomes surficially exposed. As a result, soil classification changes from Aridisols to Entisols (Argid to Calcid to Orthent).

On the scarps of Pleistocene fan terraces, another type of Orthent occurs not as the result of erosion, but as the result of deposition. Orthents in this area form in colluvial footslopes because the rate of sedimentation is greater than the rate of pedogenesis (Fig. 6.21). Consequently, no subsurface diagnostic horizons occur in this landscape position.

Orthents also commonly occupy fan terraces of Holocene age (Fig. 6.21). Although the rates of sedimentation and erosion are low, soils are too young to have yet formed subsurface diagnostic horizons. In contrast, topographically higher fan terraces of Pleistocene age have developed calcic or petrocalcic horizons, and thus classify as Aridisols (Calcids).

Many arroyo channels in southern New Mexico are also areas of Orthents (Fig. 6.21). Although these channels ephemerally carry runoff water and are in a nearly level flood plain position, they typically do not contain organic matter irregularly distributed with depth. This is because arid climates are not conducive to organic matter production or preservation. Therefore, instead of Fluvents that are common on the larger flood plains with frequent flooding, such as along the Rio Grande, the smaller arroyo flood plains have Orthents (Gile et al., 1981).

6.4.2.2 Resistant Parent Material

Exceptional resistance to weathering of some parent materials, such as quartzite-rich bedrock, prolongs the period of undistinguished horizonation (Buol et al., 1980). A comparison of soils formed

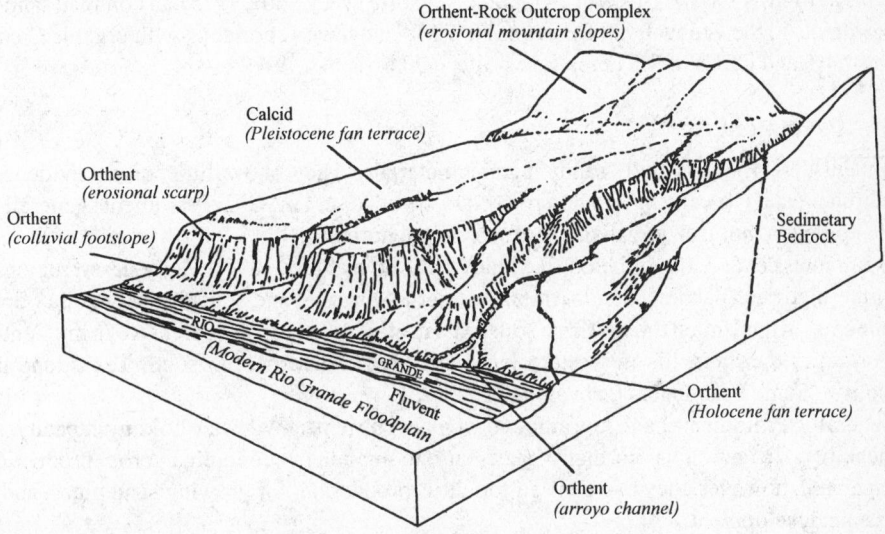

Fig. 6.21 Block diagram of a Basin and Range terrain in southern New Mexico illustrating landforms where Orthents occur [Modified from Gile and Grossman, 1979]

in monzonite versus rhyolite parent material, for instance, reveals a large difference in the nature of the soils (Gile et al., 1981). Monzonite disintegrates along grain mineral contacts and may produce a fine-textured soil, while rhyolite resists weathering because it lacks planes of weakness along which water can enter. As a result, monzonite weathers into soil (Orthents) much sooner than rhyolite.

The amount of detrital carbonate in alluvial parent materials can, like resistant bedrock, curtail the formation of diagnostic horizons and prolong the time a soil remains an Entisol. For example, soils in southern New Mexico that formed in calcareous alluvium deposited in the late Holocene often do not have diagnostic subsurface horizons. Because these soils do not have an irregular depth distribution of organic matter, they classify as Orthents (Gile et al., 1981). In contrast, neighboring soils of similar age that formed in alluvium derived from igneous rocks have faint subsurface coatings of oriented clay, which is enough development to qualify as cambic horizons and Aridisols. It appears that the formation of cambic and argillic horizons is delayed in carbonate-rich parent materials because the large amount of Ca keeps the clay flocculated and immobile.

6.4.2.3 Lower Taxonomic Levels

Orthents are subdivided at the great group level based on temperature and moisture regimes (Soil Survey Staff, 1998). For example, Torriorthents occur in desert environments and Udorthents in humid environments. At the subgroup level, Orthents are subdivided according to many features of the soils. These subgroup categories include: lithic, pergelic (frozen), vitrandic, aquic, oxyaquic, alfic (lamellae), ustic, ustertic, xerertic, vertic, haploduridic, oxyaquic, xeric, vermic, anthraquic, and dystric.

6.4.2.4 Physical and Chemical Properties

In contrast to soils that are classified according to their pedogenically produced diagnostic horizons, the physical and chemical properties of Orthents are inherited largely from their parent materials. As a result, the range of Orthent properties is as wide as the properties of the unconsolidated material in which they occur. For example, coarse-silty Udorthents in Iowa can have A horizons with organic C contents as high as 1% and CECs of 14 $cmol_c$ kg^{-1} (Soil Survey Staff, 1975). In contrast, sandy skeletal Torriorthents in the Chihuahuan Desert of New Mexico have A horizons with organic C contents as low as 0.4% and CECs of 7.5 $cmol_c$ kg^{-1} (Gile and Grossman, 1979).

6.4.3 Psamments

Psamments are Entisols with sandy parent materials. They show little or no evidence of soil development and typically have a horizon sequence of AC or A/AC-C. Psamments generally have an ochric epipedon, but lack any diagnostic subsurface horizons.

Psamments cover 3.4% of the Earth's land (Soil Survey Staff, 1997). Major desert regions, such as in northern Africa (Sahara), Saudi Arabia, western Australia, and southern Mongolia (Gobi), have Psamments. In the United States, these soils exist in all states. They are extensive in the Central Ridge of Florida, in the Sandhills in western Nebraska, in the sandsheet in south Texas, and in central Wisconsin. Some Psamments form along beaches.

Generally, Psamments have low amounts of plant nutrients, low water-holding capacity, and rapid permeability. Irrigation is normally necessary to maintain economical crop production. With management, however, they can be used for citrus production, for growing sand pines, and even for real estate development.

Natural vegetation associated with Psamments in Florida consists of bluejack (*Quercus incana*), blackjack oak (*Quercus marilandica*), turkey oak (*Quercus laevis*), longleaf threeawn (*Aristida stricta*), and bluestem (*Andropogon* sp.). In the major cattle-growing states of Texas and Nebraska, Psamments are used for spring and summer grazing of native grasses. In central Wisconsin, truck crops are an important use.

6.4.3.1 Lower Taxonomic Levels

Psamments are restricted to soils that have a texture of loamy fine sand or coarser in the control section (family particle size class) and < 35% by volume of particles > 2 mm diameter (Soil Survey Staff, 1998). The latter distinguishes Psamments from sandy Orthents.

Psamments occur in a wide range of climates. Six of the seven great groups reflect a climatic/geographic bias. Only Quartzipsamments do not indicate a geographic location, but rather the dominant mineralogy, quartz. Psamments exist from cryic or pergelic (Cryopsamments) soil temperature regimes to areas in the world where temperature at a depth of 50 cm from the soil surface does not change more than 6 °C between mean summer and winter soil temperatures (Tropopsamments). Also, they are located in torric (Torripsamments), ustic (Ustipsamments), xeric (Xeropsamments), and udic (Udipsamments) soil moisture regimes.

Subsurface features present in some Psamments include a lithic contact (Lithic), aquic or oxyaquic conditions, lamellae (Argic), plinthite, or durinodes. Thus, subsoil features vary greatly and the variation is noted in the subgroup taxonomic name. Examples include Lithic Cryopsamments, Plinthic Quartzipsamments, and Aquic Durinodic Xeropsamments.

At the subgroup taxonomic level, climatic/geographic conditions are also described. As an example, Aridic Ustipsamments are located in several soil temperature regimes, but must have a period of time in which the soil is dry in all parts. The length of time that the soil must be dry is defined by the soil temperature regime. Xeric Torripsamments are defined by both soil temperature (Torri) and moisture (Xeric) regimes.

6.4.3.2 Genesis

Genetically, Psamments develop in areas high in sand content (> 85%). The parent material can be geologically young, as in eolian or alluvial materials or old, as in sandy residual bedrock. Sand dunes of either marine (Florida) or eolian (Saudia Arabia or Nebraska) origin are common parent materials. Some Psamments form in materials weathered from sandstone bedrock (e.g., South Dakota or Montana). Depending on the mineralogy of the sand, Psamments (Quartzipsamments) can be very resistant to weathering.

6.4.3.3 Representative Psamments

Psamments from Florida (Astatula series: hyperthermic, uncoated, Typic Quartzipsamments), Nebraska (Valentine series: mixed, mesic Typic Ustipsamments), and Saudi Arabia (Torripsamments) were selected to represent different geographic areas and, therefore, different genetic pathways. But morphologically, these soils are very similar. The soils are also excessively drained.

Astatula, Valentine, and the Torripsamments from Saudi Arabia have thick C horizons with thin A horizons. If an A horizon is present, it is light colored and < 15 cm thick. The A horizon is subject to continual mixing by wind action. Astatula soils form in eolian and marine sands and are extensive in peninsular Florida on upland slopes that range from 0 to 30%. These soils occur in high rainfall areas (about 1,270 mm yr^{-1}), are highly leached, and are typically strongly acid (pH < 5.5). Valentine soils form in sand dunes in the sandhills of Nebraska on slopes that range from 0 to 60%. The sand dunes generally are oriented in a northwest to southeast direction (prevailing wind direction). In addition to Nebraska, Valentine soils occur extensively in Kansas, Montana, New Mexico, South Dakota, Texas, and Wyoming (> 2 million ha). The Torripsamments in Saudi Arabia form in sand dunes and also in alluvial sandy deposits on plains and stream terraces. The dunes range in height from < 2 to > 10 m. The Torripsamments in these desert areas (< 100 mm yr^{-1} precipitation) typically have an accumulation of bases, and are alkaline in reaction (pH > 8.5).

6.4.3.4 Physical and Chemical Properties of Representative Psamments
Physical properties and chemical properties are presented for two selected Psamments (Table 6.21). Organic C contents typically are very low with pH ranging from strongly acid in the Psamments of Florida that occur in high rainfall areas (Astatula series), to strongly alkaline soils in Saudi Arabia (data not shown) that occur in the very dry desert regions. The sand size fraction of the Astatula and Valentine series typically is dominated by fine sand. Clay content of these Psammments is typically < 5%.

6.4.4 Fluvents

Fluvents are Entisols that typically form in alluvium of rapidly aggrading flood plains, fans, deltas, and in some cases, mudflows. In the United States, the greatest abundance of Fluvents occur within the numerous and large floodplains of the Midwest (Lindbo, 1997). World coverage of Fluvents is 2.2% (Soil Survey Staff, 1997).

When rates of deposition exceed rates of pedogenesis in alluvial settings, A-C soil profiles develop (Table 6.22). Fluvents typically have ochric epipedons, but no subsurface diagnostic horizons (Soil Survey Staff, 1975, 1998). Organic C content in Fluvents must decrease irregularly with depth, or be above 0.2% at a depth of 125 cm (Table 6.22). The rationale of the Fluvent suborder is that deposition inhibits diagnostic horizon formation and is accompanied by fluctuations in organic C levels associated with stratification.

Fluvents are differentiated from other Entisol suborders by: (1) having textures that are finer than loamy very fine sand and with less than 35% rock fragments (Psamments), (2) occurring on slopes less than 25% and not having bedrock within 25 cm (Orthents), (3) not having aquic conditions (Aquents), (4) not having fragments of diagnostic horizons (Arents), and (5) having mean annual soil temperatures greater than 0 °C (Soil Survey Staff, 1998, 1975).

6.4.4.1 Lower Taxonomic Levels
Great groups of the Fluvents are differentiated by either soil temperature or soil moisture regime (Soil Survey Staff, 1996). Fluvents occur in udic (Udifluvents), ustic (Ustifluvents), xeric (Xerofluvents), and torric (Torrifluvents) soil moisture regimes. They also occur in cryic (Cryofluvents), isomesic, isothermic, and isohyperthermic (Tropofluvents) soil temperature regimes. At the subgroup taxonomic level, Entisols intergrade to the Andisol (andic), Mollisol (mollic), and Vertisol (vertic) soil orders, and to numerous other soil moisture and temperature regimes not used at the suborder categorical level (Soil Survey Staff, 1998).

6.4.4.2 Climate, Vegetation, and Topography
Fluvents are soils that are not in equilibrium with the effects of climate, vegetation, and in some cases, topography. Thus, Fluvents can occur in most climates associated with most vegetation communities, and on a variety of slopes. Climate can, however, indirectly influence the chemical and physical properties of Fluvents. Alluvium associated with flood plains, fans, deltas, or mudflows may be inherited from upland soils where properties are more likely to reflect equilibrium with regional climate, vegetation, and topography.

Table 6.21 Soil characterization data for the Astula fine sand [Sodek et al., 1990] and Valentine fine sand [Soil Survey Staff, 1966]

Horizon	Depth	Very Coarse Sand	Coarse Sand	Medium Sand	Fine Sand	Very Fine Sand	Silt	Clay	pH[a]	Organic Carbon	Extractable Ca	Extractable Mg	Extractable Na	Extractable K	Cation Exchange Capacity	Base Saturation
	cm	------------------------------------ % ------------------------------------								%	----------------cmol$_c$ kg^{-1}----------------					%
Astatula fine sand (Typic Quartzipsamments)																
A	0-13	--	0.1	18.8	76.7	2.1	0.9	1.4	5.5	0.5	0.5	0.1	0.0	0.0	2.7	23
C1	13-64	--	0.1	18.4	76.5	2.4	1.6	1.5	5.2	0.2	0.0	0.0	0.0	--	1.5	5
C2	64-175	--	0.1	19.7	75.8	2.2	0.6	1.6	5.0	0.1	0.0	0.0	0.0	--	0.7	5
C3	175-203	--	0.1	19.4	76.2	2.0	0.6	1.7	5.0	0.0	0.0	0.0	0.0	--	0.9	6
Valentine fine sand (Typic Ustipsamments)																
A	0-10	0.2	4.2	12.7	50.2	24.7	4.1	3.9	6.0	0.1	3.6	0.6	0.1	0.3	5.9	78
AC	10-25	0.1	4.6	13.3	55.1	21.6	2.2	3.1	6.2	0.5	2.1	0.4	0.1	0.3	3.7	78
C1	25-45	--	3.6	11.2	56.2	24.0	1.8	3.2	6.5	0.2	2.0	0.6	0.1	0.1	0.1	88
C2	45-81	--	4.4	18.6	58.9	14.7	1.0	2.4	6.7	0.1	1.5	0.2	0.1	0.2	2.4	79
C3	81-137	0.1	3.8	10.9	59.1	22.1	1.3	2.9	6.8	0.0	1.8	0.8	0.1	0.1	2.8	100

[a]Determined in a 1:1 H$_2$O solution for the Astatula fine sand and in a 1:5 CaCl$_2$ solution for the Valentine fine sand

Table 6.22   Soil characterization data for Fluvents from Iowa (Udifluvents) and New Mexico (Torrifluvents)

Horizon	Depth	Sand	Silt	Clay	Organic Carbon
	cm	--%----------------------------------			
		Udifluvents[a]			
Ap	0-14	5.1	78.4	16.5	1.28
C1	14-31	6.9	78.6	14.5	0.57
C2	31-42	4.1	81.0	14.9	0.60
C3	42-56	16.5	73.0	10.5	0.45
C4	56-66	8.2	75.3	16.5	0.57
C5	66-98	34.1	55.3	10.6	0.42
C6	98-125	13.0	71.6	15.4	0.52
		Torrifluvents[b]			
A1	0-5	74.8	16.2	9.0	0.66
A2	5-13	66.9	21.1	12.0	0.71
2A	13-18	73.1	16.1	10.8	0.49
3A	18-38	60.4	25.2	14.4	0.54
4C	38-43	69.5	19.2	11.3	0.33
5C	43-64	62.6	25.1	12.3	0.41
6C	64-81	77.8	13.5	8.7	0.19
7C	81-119	81.7	11.5	6.8	0.24

[a] Missouri River flood plain, Monona County, Iowa. Modified from Soil Survey Staff, 1975, p. 614
[b] Gardner Springs Arroy, Dona Ana County, New Mexico. Modified from Gile and Grossman, 1979, p. 874

6.4.4.3   Hydrology

Flood frequency, flood magnitude, and sediment load control the amount of deposition that occurs on flood plains. Prior to 1,000 yr BP, many North American rivers were forming cumulative Mollisols and Inceptisols in response to moderate frequency and low magnitude floods (Hall, 1990; Waters, 1991; Blum and Valastro, 1994; Arbogast and Johnson, 1994; Waters and Nordt, 1995). Between 1,500 and 500 yr BP there was widespread regional river channel trenching. This event transformed many broad flood plains into infrequently flooded surfaces, herein called flood terraces. Flood terraces flood on occasion, but the frequency is low enough that pedogenesis is beginning to transform the associated soils into orders beyond Entisols. In Australia, Walker and Coventry (1976) reported that flood plains that flooded once each year maintained soil development in the Entisol stage, whereas when flood frequencies were on the order of 1 to 10 yr, Mollisols began forming. Reservoirs built along major rivers during the last 100 yr have also reduced the flood frequency of many flood plains.

Countering late Holocene flood reduction by channel trenching or reservoirs, historic flood magnitudes are increasing in some areas because of intensified land use, increased runoff and channel peak discharge. As a consequence, it appears that some flood plains that were flooded infrequently during the last 500 to 1,000 yr, may be transforming back to more frequently flooded flood plains because of historic land use practices. This has resulted in deposition of post-settlement alluvium across numerous flood plains (Knox, 1987; Grossman, 1983; Scully and Arnold, 1981; Daniels and Jordan, 1966).

Deposition of post-settlement alluvium has buried many late Holocene soils in flood terrace positions. In most cases, if a recent flood deposit is < 50 cm thick, it is treated as a mapping phase and the buried soil is classified (Soil Survey Staff, 1975, 1998). If a recent flood plain deposit is over 50 cm thick, the recent material is classified, which may be a Fluvent (Fig. 6.22).

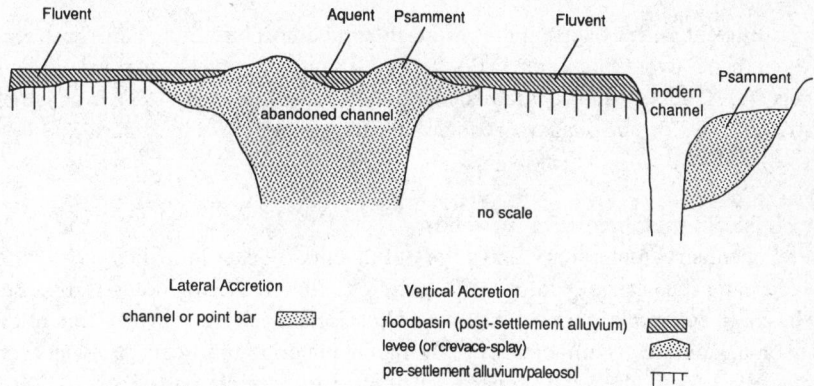

Fig. 6.22 Soil stratigraphic cross-section of an idealized alluvial floodplain

6.4.4.4 Time

Under conditions of landscape stability, the evolution of Entisols into other soil orders typically occurs in association with one or more properties approaching steady state. Surface horizons, and consequently Fluvents, can develop almost immediately after deposition because of the inheritance of detrital organic matter (Nordt and Hallmark, 1990). In general, organic matter inputs attain steady state in < 1,000 yr as Entisols evolve into other soil orders (Dickson and Crocker 1953a, 1953b; Parsons et al., 1970; Walker and Coventry, 1976; Hallberg et al., 1978; Schafer et al., 1980; Nordt and Hallmark, 1990). In some humid regions, mollic epipedons and Mollisols can develop in as little as 100 yr (Ruhe et al., 1975). In other humid regions, Entisols can develop into Inceptisols within 500 to 600 yr (Bilzi and Ciolkosz, 1977; Scully and Arnold, 1981). Fluvents in arid regions can persist for a thousand years in low carbonate parent material and up to several thousand years in carbonatic parent material (Gile, 1975). After this time, subsurface diagnostic horizons and Aridisols form.

6.4.4.5 Parent Material

Flood plains, fans, and deltas are constructed by the alluvial processes of vertical and lateral accretion (Reeding, 1982). Lateral accretion results from lateral channel migration and deposition of bar and channel facies (Fig. 6.22). Vertical accretion results from overbank deposition, which entails levee, floodbasin, and sometimes crevasse splay facies. The distribution of facies in a flood plain system is dependent on stream type (Chorely et al., 1985). Braided channels in either floodplains or alluvial fans are dominated by laterally accreted channel and bar deposits. Straight to meandering streams tend to deposit thick vertical accretion facies. Anastomosing streams have multiple bifurcating and reconnecting channels with a high proportion of fine grained lateral accretion deposition. Deltas are dominated by a distributary channel network with a complex array of lateral and vertical accretion deposits. Fluvents are most likely to form in lateral accretion deposits and in rapidly aggrading vertical accretion facies.

Sedimentological packages of meandering streams in humid regions typically display a fining upward sequence consisting of lower channel, intermediate point bar, and upper flood basin or levee facies. In some cases, Fluvents may form entirely within only one of these facies and not exhibit a noticeable textural change within the solum. If a Fluvent develops in both a point bar and flood basin facies, for example, a fining upward sequence may exist in the profile. In other situations, reverse grading may be encountered. For example, a levee may be deposited over a floodbasin facies, thus producing a Fluvent with a coarsening upward textural distribution. In addition, other suborders of Entisols other than Fluvents may form in flood plains. Psamments may form in sandy channel and point bar deposits and Aquents in poorly drained topographic positions (Fig. 6.22). In sum, recognizing the alluvial stratigraphy of flood plains can enhance the understanding of the genesis of Fluvents (Section E, chapter 1).

6.4.4.6 Physical and Chemical Properties

The texture, chemistry, mineralogy, and color of Fluvents are determined largely by the character of the soils and parent materials (bedrock, till, loess, etc.) that are being eroded from the surrounding drainage basin. For example, permeability, water-holding capacity, and nutrient retention vary greatly depending on the texture and mineralogy of the parent material. In essence, Fluvents can have a wide variety of physical and chemical properties that are for the most part inherited from their parent materials. Most alluvial soils have high fertility levels and support some of the most important agricultural areas of the world (Edelman and van der Voorde, 1963; Radwanski, 1968; Sidhu et al., 1977). In the tropics, however, upland soils tend to be depleted in nutrients such that their alluvial counterparts produce low fertility flood plain soils (Edelman and van der Voorde, 1963).

6.4.5 Aquents

Aquents are wet Entisols. They exist widely in the United States, except in Arizona and New Mexico with the largest concentrations in south Florida (Soil Survey Staff, 1990). Worldwide, Aquents are not extensive, covering only approximately 0.1% of the Earth's land (Soil Survey Staff, 1997). Areas where Aquents occur include salt marshes of coastal tidal areas, freshwater marshes and swamps, alluvial flats and backswamps of flood plains, and outwash plains that receive new deposits of sediment at frequent intervals. Depending on where Aquents are located, some can be used for agricultural production when drained. Many undrained areas of Aquents are used for pasture or wildlife habitat.

Aquents are defined as having aquic conditions and sulfidic materials within 50 cm of the mineral soil surface, or permanent saturation and a reduced matrix below a depth of 25 cm from the mineral soil surface, or texture and color requirements if a densic, lithic, or paralithic contact exists or aquic conditions are present for some time in some years in a layer between 40 and 50 cm from the mineral soil surface (Soil Survey Staff, 1998). The soil properties of Aquents are highly variable. One common feature of all Aquents is their saturation with water at or near the surface for much of the year. Aquents can range in texture from sandy to clayey.

Wetness in Aquents is normally inferred by the color distribution generated by the redox state of Fe. In most Aquents, this process produces redoximorphic features (Fe concentrations or depletions) within the upper 50 cm of the surface. Redoximorphic features are most common in nonsandy Aquents. The sandy Aquents (Psammaquents) often lack redoximorphic features because of the very low Fe content of the quartz parent materials. Many Aquents have hydric soil indicators (thin layer of muck on the soil surface, strong sulfidic odor, or a gray and reduced soil matrix in the upper 15 cm of the soil). In Florida according to F.S. 17–340 (Florida Statutes, 1995), all Sulfaquents and Hydraquents are considered to be hydric soils. Many Psammaquents are also hydric according to the county listings of hydric soils.

6.4.5.1 Lower Taxonomic Levels

Aquents are associated with landscapes that accumulate soil water. They exist in cryic to isohyperthermic soil temperature regimes. Sulfaquents and Hydraquents are located along coastal areas in tidal marshes. Sulfaquents, as the name implies, have sulfidic materials. Hydraquents are thixotropic soils, meaning that they cannot support appreciable overburden pressures (n value > 0.7). Psammaquents have a texture of loamy fine sand or coarser in all layers within the control section for the family particle size class. Fluvaquents occur on flood plains and are defined by the amount of organic C and its distribution in the soil. Cryaquents form in the colder climatic areas of Alaska and other higher latitudes that receive significant snowfalls, and soil development is slow.

6.4.5.2 Genesis

Aquents form where pedogenetic development does not occur or is very slow under wet conditions. The soil must be saturated for long periods of time, and in some areas, organic matter can accumulate to create a histic epipedon. The saturation in Aquents can be episaturated (perched watertable) or endosaturated (apparent watertable). The time needed for Aquents to develop can be short. On flood plains and other areas in which water flows across the area, Aquents commonly have stratified layers of sand, silt, clay and organic matter. Aquents of flood plains may also have buried histic epipedons, and Aquents with a high clay content may have cracks and slickensides.

Aquents from Florida (Pellicer series: fine, smectitic, nonacid, hyperthermic Typic Sulfaquents), Egypt (Edko series: fine-loamy, mixed, nonacid, hyperthermic Mollic Fluvaquents), and Alaska (Beluga series: coarse-loamy, mixed, nonacid, Typic Cryaquents) were selected to represent Aquents from different geographic areas and to reflect different genetic pathways. The Pellicer soils form in clayey marine sediments in tidal marshes along the Atlantic Ocean coastal areas of Florida (Carlisle et al., 1981). These soils are very poorly drained and are subject to daily tidal flooding. The representative Aquent from Egypt (Edko area) has anthric saturation and oximorphic features in the form of brownish Fe accumulations. The soil was sampled in a depressional area south of Lake Burullus near Resheed City in northern Egypt (Rasheed et al., 1992). It developed in recent fluvio-lacustrine deposits in an aridic climate. The Beluga soils are extensive in flood plains and alluvial fans in southern Alaska and form in stratified alluvium and colluvium derived from soft clayey shale and sandstone. These soils are poorly drained and are saturated with water at the surface for portions of the year.

6.4.5.3 Physical and Chemical Properties

Physical properties such as particle size distribution, and chemical properties such as organic C content, pH, extractable bases, and CEC are presented for the Pellicer and Edko soils (Table 6.23). The sand, silt, and clay content of the Pellicer soil varies considerably within the profile, reflecting the different conditions under which the materials were deposited. Organic C content of the Pellicer soils is high in the surface horizon, the result of the accumulation of organic materials under wet conditions. The Pellicer soils are high in S content (1 to 4%), and this is reflected in the extremely low pH values (< 3.5). The sand, silt, and clay content of the Edko soil is fairly evenly distributed in each horizon. The pH of this soil is very high (> 8.5) due to the high Na content.

6.4.6 Arents

Humans affect soils in many ways. Some Entisols have been created where humans have deposited or exposed soil parent material by earth-moving operations. Some of these operations include highway construction, surface mining, landfill construction, or dredging. In other cases, as in Arents described below, Entisols have been created by humans because of severe physical disturbance of previously existing soils.

Table 6.23 Soil characterization data for the Pellicer clay [Carlisle et al., 1981] and Edko soil [Rasheed et al., 1992]

Horizon	Depth	Very Coarse Sand	Coarse Sand	Medium Sand	Fine Sand	Very Fine Sand	Silt	Clay	pH[a]	Organic Carbon	Extractable Ca	Extractable Mg	Extractable Na	Extractable K	Cation Exchange Capacity	Base Saturation
	cm	%								%	$cmol_c\ kg^{-1}$					%
Pellicer clay (Typic Sulfaquents)																
A	0-51	0.1	0.5	2.1	3.5	9.3	23.6	60.9	2.7	14.3	11.7	24.3	27.7	1.0	129.6	50
C1	51-89	0.1	1.1	15.9	21.6	11.3	16.6	33.4	3.2	7.1	6.8	15.7	20.7	1.1	78.6	56
C2	89-140	0.1	2.6	38.7	34.1	10.9	7.5	6.1	3.1	0.8	1.3	3.3	5.0	0.2	15.4	63
Edko soil (Mollic Fluvaquents)																
A1	0-6		4.6		37.4		26.4	31.6	8.6	3.3	12.0	16.0	64.4	1.1		
A2	6-23		4.0		29.1		36.0	30.9	8.4	1.6	14.0	24.0	67.2	0.9		
C	23-56		0.4		29.9		34.8	29.9	8.6	0.9	10.0	14.0	65.2	4.5		
2C	56-76		0.4		29.9		34.8	29.9	8.6	0.9	10.0	14.0	65.2	4.5		

[a] Determined in a 1:1 solution of distilled water

Arents are a suborder of Entisols that are physically disturbed by the actions of humans (Soil Survey Staff, 1998). Arents are restricted to those soils which have, in one or more layers between 25 and 100 cm from the mineral soil surface, 3% or more by volume of fragments of diagnostic horizons that are not arranged in any discernible order (Soil Survey Staff, 1998). This suborder was placed in *Soil Taxonomy* for situations where previously existing soils had been significantly physically modified. One situation where this is known to have happened is in California where duripans have been ripped by heavy equipment to improve the soils for the growth of almond trees, based on observations in the 1960s (Simonson, 1969). The ripping was possible only with moderately developed duripans because thick, strongly developed duripans were impossible to rip. The value of the land was sufficiently improved to permit successful commercial almond production. As a result of the duripan ripping, the great group classification of the soils changed from Durixeralfs to Xerarents.

In formerly gullied lands in loess bluffs in western Tennessee, land leveling was used to improve the soils for agricultural purposes. Fragments of argillic horizons from the original Alfisols permitted the new soils to be identified as Arents after the land disturbance. Thus, Hapludalfs (e.g., soils of the Memphis soil series) were reclassified to Udarents following the leveling. In Texas, reclamation of lignite surface mine lands with mixed spoil materials gives rise to Arents, especially near the margins of the mine lands. Where the overburden is thin relative to thick Paleudalfs in premine condition.

6.4.6.1 Soils in Anthromorphic Parent Materials

In many soils with human-deposited or human-exposed parent materials, fragments of diagnostic horizons are not identifiable; thus a classification of Arents would not be appropriate. Such soils commonly have been classified in the suborder of Orthents and subdivided into great groups according to the soil moisture regime (e.g., into Udorthents in the eastern United States) (Soil Survey Staff, 1998). This classification has seemed insufficient to many soil scientists because it does not reflect the role of humans in the genesis of the soils (Short et al., 1986a; Strain and Evans, 1994). Although these soils are commonly thought to be highly variable, studies have shown a degree of variability similar to that of some soils in naturally deposited parent materials (Short et al., 1986b).

Many soils in materials deposited by bulldozers or other earth-moving equipment have an irregular distribution of organic C with depth. In theory, these soils would classify as Fluvents at the suborder taxonomic level (Soil Survey Staff, 1998). In contrast to Fluvents, however, human-disturbed soils are not stratified by natural processes, are not flooded in most years, and are more dense. Because these soils behave more like Orthents than Fluvents, they have been grouped, during soil surveys such as that of Washington, DC (Smith, 1976), into Orthents by ignoring their irregular organic-C depth distribution.

Because of the inadequacies of the present classification system, new suborders and/or subgroups of Entisols have been proposed for soils in human-deposited or human-exposed parent materials (Fanning and Fanning, 1989). At present, there is considerable interest in an international committee for anthropogenic soils (ICOMANTH) for the addition of new taxonomic classes (Section E, Chapter 6.1). However, the taxonomic level at which the recognition should take place has yet to be firmly decided. The suborder of Arents, discussed above, established a precedence for recognition at the suborder level. Fanning and Fanning (1989) have supported this viewpoint. However, some others (Strain and Evans, 1994; Kosse, 1988) have suggested that the recognition be at the order level called Anthrosols.

For soils in earth spoil, such as those deposited during surface mining operations, a suborder of Spolents (for spoil) was proposed by West Virginia workers (Sencindiver, 1977; Smith and Sobek, 1978). During the soil survey of Washington, DC (Smith, 1976), a suborder of Urbents was considered for soils in miscellaneous urban fill that had object artifacts such as brick or concrete in the particle size control section (generally between a depth of 25 and 100 cm). For soils with garbage of an organic nature within 2 m of the soil surface, which would readily subside and generate methane under anaerobic conditions, a suborder of Garbents was proposed (Fanning and Fanning, 1989; Fanning, 1991). Proposals for special diagnostic characteristics (urbic, garbic, spolic, and dredged material, and the scalped land surface) for *Soil Taxonomy* were proposed by Fanning and Fanning (1989). These characteristics were employed to propose special subgroup classes, such as Urbic Garbic Udorthents for sanitary landfills where the cover material qualified as urbic material (Fanning and Fanning, 1989). It was proposed that soils on scalped land surfaces (an engineering perspective) could be placed in Scalpic subgroups.

Soils created by dredging operations commonly qualify as Sulfaquents immediately after sulfidic-dredged materials have been deposited (Fanning and Fanning, 1989). Under aerobic conditions at dredged material sites, sulfuric horizons commonly form from sulfidic materials in a few weeks or months. This causes the soils to become Sulfaquepts, a great group of Inceptisols. Thus, the period that such soils remain as Entisols, before becoming a soil of a different order, may be very brief.

With the present strong interest in improving knowledge about the classification of human-influenced soils, much progress may be expected in the near future. This is needed as soil surveys of lands with human-influenced soils are becoming more widespread. A soil survey of the City of New York is presently underway, and many of the problems associated with the soils discussed in this section will be encountered. For example, sulfide-bearing clays have been used as landfill cover materials, which can lead to the formation of acid sulfate soils (Kargbo et al., 1993; Fanning and Burch, 1997). A more resolving classification system can enhance future interpretations and mapping techniques of soil surveys in these areas. Further, it has been suggested (Fanning, 1990) that a discipline of pedotechnology is needed to improve the design of soils brought into existence through the action of humans.

6.4.7 References

Arbogast, A.F., and W.C. Johnson. 1994. Climatic implications of the Late Quaternary alluvial record of a small drainage basin in the central Great Plains. Quat. Res. 41:298–305.

Bilzi, A.F., and E.J. Ciolkosz. 1977. Time as a factor in the genesis of four soils developed in recent alluvium in Pennsylvania. Soil Sci. Soc. Am. J. 41:122–127.

Blum, M.D., and S. Valastro, 1994. Late Quaternary sedimentation, lower Colorado River, Gulf Coastal Plain of Texas. Geol. Soc. Am. Bull. 106:1002–1016.

Buol, S.W., F.D. Hole, and R.J. McCracken. 1980. Soil genesis and classification. 2nd Ed. Iowa State University Press, Ames, IA.

Carlisle, V.W., C.T. Hallmark, F. Sodek III, R.E. Caldwell, L.C. Hammond, and V.E. Berkheiser. 1981. Characterization data for selected Florida soils. Univ. Fl. Soil Sci. Res. Rep. 81–1.

Chorely, R.J., S.A. Schumm, and D.E. Sugden. 1985. Geomorphology. Methuen and Co., London, UK.

Daniels, R.B., and R.H. Jordan. 1966. Physiographic history and the soils, entrenched stream systems, and gullies, Harrison County, Iowa. USDA Tech. Bull. 1348.

Dickson, B.A., and R.L. Crocker. 1953a. A chronosequence of soils and vegetation near Mt. Shasta, California, I. Definition of the ecosystem investigated and features of the plant succession. J. Soil Sci. 4:123–141.

Dickson, B.A., and R.L. Crocker. 1953b. A chronosequence of soils and vegetation near Mt. Shasta, California, II. The development of the forest floors and the carbon and nitrogen profiles of the soils. J. Soil Sci. 4:142–154.

Edelman, C.H., and P.K.J. van der Voorde. 1963. Important characteristics of alluvial soils in the tropics. Soil Sci. 95:258–263.

Fanning, D.S. 1991. Human-influenced and disturbed soils: Overview with emphasis on classification. p. 3–14. *In* C.V. Evans (ed.) Human-influenced and disturbed soils. Proc. Conf. Univ. NH Dep. Natural Res., University of New Hampshire, Durham, NH.

Fanning, D.S. 1990. Pedotechnology-soil genetic engineering: How and why soils scientists should be involved. Soil Surv. Hor. 31:29–32.

Fanning, D.S., and S.N. Burch. 1997. Acid sulphate soils and some associated environmental problems. p. 145–158. *In* K. Auerswald, H. Stanjek, and J.M. Bigham (ed.) Soils and environments. Catena Verlag, Reiskirchen, Germany.

Fanning, D.S., and M.C.B. Fanning. 1989. Soil: Morphology, genesis, and classification. John Wiley and Sons, New York, NY.

Florida Statutes. 1995. Wetlands delineation, Chapter 373–4211. Ratification of Chapter 17–340, Florida Administrative Code on Delineation of the Landward Extent of Wetlands and Surface Waters. State of Florida, Tallahassee, FL.

Gile, L.H. 1975. Holocene soils and soil-geomorphic relations in an arid region of southern New Mexico. Quat. Res. 5:321–360.

Gile, L.H., and R.B. Grossman. 1979. The desert project soil monogrograph. Doc. PB80-135304, National Technical Information Service, Springfield, VA.

Gile, L.H., J.W. Hawley, and R.B. Grossman. 1981. Soil and geomorphology in the Basin and Range area of southern New Mexico-Guidebook to the Desert Project. NM Bur. Mines Min. Res. Mem. 39.

Grossman, R.B. 1983. Entisols. *In* L.P. Wilding, N.E. Smeck, and G.F. Hall (ed.) p. 55–90. Pedogenesis and Soil Taxonomy, II. The soil orders. Elsevier, Amsterdam, Netherlands.

Hall, S.A. 1990. Channel trenching and climatic change in the southern U.S. Great Plains. Geol. 18:342–245.

Hallberg, G.R., N.C. Wollenhaupt, and G.A. Miller. 1978. A century of soil development in spoil derived from loess in Iowa. Soil Sci. Soc. Am. J. 42:399–343.

Jenny, H. 1941. Factors of soil formation. McGraw-Hill, New York, NY.

Kargbo, D.M., D.S. Fanning, H.I. Inyang, and R.W. Duell. 1993. Environmental significance of acid sulfate "clays" as waste covers. Env. Geol. 22:218–226.

Knox, J.C. 1987. Historical valley floor sedimentation in the Upper Mississippi Valley. Ann. Assoc. Am. Geog. 77:224–244.

Kosse, A. 1988. Anthrosols: Proposal for a new soil order. Agron. Abst. 80:260.

Lindbo, D.L. 1997. Entisols-Fluvents and Fluvaquents: Problems recognizing aquic and hydric conditions in young, flood plain soils. p. 133–152. *In* M.J. Vepraskas and S.W. Sprecher (ed.) Aquic conditions and hydric soils: The problem soils. Soil Sci. Soc. Am. Spec. Publ. 50, Soil Science Society of America, Madison, WI.

Nordt, L.C., and C.T. Hallmark. 1990. Soils-Geomorphology Tour Guidebook. Dep. Soil Crop Sci. Tech. Rep. 90-7, Texas A&M University, College Station, TX.

Parsons, R.B., C.A. Balster, and A.O. Ness. 1970. Soil development and geomorphic surfaces, Willamette Valley, Oregon. Soil Sci. Soc. Am. Proc. 34:485–491.

Radwanski, S.A. 1968. Field observations of some physical properties in alluvial soils of arid and semi-arid regions. Soil Sci. 106:314–316.

Rasheed, M.A., F.B. Labib, and Th.K. Ghabour. 1992. The wet soils of Egypt. p. 206–211. *In* J.M. Kimble (ed.) Proc. 8th Int. Soil Correl. Meet. (VIII ISCOM): Characterization, classification, and utilization of wet soils. USDA-SCS, National Soil Survey Center, Lincoln, NE.

Reeding, H.G. 1982. Sedimentary environments and facies. 2nd Ed. Blackwell Scientific Publications, Oxford, UK.

Ruhe, R.V., T.E. Fenton, and L.L. Ledesma. 1975. Missouri River history, flood plain construction, and soil formation in southwestern Iowa. IA Agric. Home Econ. Exp. Sta. Res. Bull. 580:38–791.

Schafer, W.M., G.A. Nielsen, and W.D. Nettleton. 1980. Mine soil genesis and morphology in a spoil chronosequence in Montana. Soil Sci. Soc. Am. J. 44:802–807.

Scully, R.W., and R.W. Arnold. 1981. Holocene alluvial stratigraphy in the upper Susquehanna River basin, New York. Quat. Res. 15:327–344.

Sencindiver, J.A. 1977. Classification and Genesis of Minesoils. Ph.D. dissertation. University of West Virginia, Morgantown, WV.

Short, J.R., D.S. Fanning, M.S. McIntosh, J.E. Foss, and J.C. Patterson. 1986a. Soils of the mall in Washington, DC. I. Statistical summary of properties. Soil Sci. Soc. Am. J. 50:699–705.

Short, J.R., D.S. Fanning, M.S. McIntosh, J.E. Foss, and J.C. Patterson. 1986b. Soils of the mall in Washington, DC. II. Genesis, classification, and mapping. Soil Sci. Soc. Am. J. 50:705–710.

Sidhu, P.S., J.L. Sehgal, and N.S. Randhawa. 1977. Elemental distribution and associations is some alluvium-derived soils of the Indo-Gangetic Plain of Punjab (India). Pedologie 27:225–235.

Simonson, R.W. 1969. Personal communication.

Smith, G.D. 1986. The Guy Smith interviews: Rationale for concepts in Soil Taxonomy. Soil Manage. Supp. Serv. Tech. Monogr. 11, US Government Printing Office, Washington, DC.

Smith, R.M., and A.A. Sobek. 1978. Physical and chemical properties of overburden, spoils, wastes, and new soils. p. 149–172. *In* F.W. Schaller and P. Sutton (ed.) Reclamation of drastically disturbed lands. American Society of Agronomy, Madison, WI.

Smith, H. 1976. Soil survey of the District of Columbia. USDA-SCS. US Government Printing Office, Washington, DC.

Sodek III, F., V.W. Carlisle, M.E. Collins, L.C. Hammond, and W.G. Harris. 1990. Characterization data for selected Florida soils. Univ. Fl. Soil Sci. Res. Rep. 90-1.

Soil Survey Staff. 1997. Global soil regions. USDA-NRCS, Lincoln, NE.

Soil Survey Staff. 1998. Keys to soil taxonomy. 8th Ed. USDA-NRCS, Lincoln, NE.

Soil Survey Staff. 1990. Soil series of the United States, including Puerto Rico and the U.S. Virgin Islands. USDA-NCRS Pub. 1483. US Government Printing Office, Washington, DC.

Soil Survey Staff. 1975. Soil taxonomy: A basic system of soil classification for making and interpreting soil surveys. USDA-SCS Handb. 436. US Government Printing Office, Washington, DC.

Soil Survey Staff. 1966. Soil descriptions for some soils of Nebraska. USDA-SCS. Soil Surv. Invest. Rep. 5. Lincoln, NE.

Strain, M.R., and C.V. Evans. 1994. Map unit development for sand- and gravel-pit soils in New Hampshire. Soil Sci. Soc. Am. J. 58:147–155.

Walker, P.H., and R.J. Coventry. 1976. Soil profile development in some alluvial deposits of eastern New South Wales. Aust. J. Soil Res. 14:305–317.

Waters, M.R. 1991. The geoarchaeology of gullies and arroyos in southern Arizona. J. Field Arch. 18:141–159.

Waters, M.R., and L.C. Nordt. 1995. Late Quaternary floodplain history of the Brazos River in east-central Texas. Quat. Res. 43:311–319.

6.5   Inceptisols

Wayne H. Hudnall and Lois M. West
Louisiana State University Agricultural Center

Ellis C. Benham
USDA-NRCS

Larry P. Wilding
Texas A & M University

6.5.1   Introduction

Inceptisols serve two important functions in *Soil Taxonomy*. First, they group soils with incipient soil development (L. *Inceptum*, beginning); and second, they ensure that the classification system is fully bifurcated—that is, all soils fall within some taxa of the classification system. Because of the latter, Smith (1986) referred to Inceptisols as the wastebasket order. They are the repository for all soils that do not meet class differentiae of other orders. While the central concept of Inceptisols is that of soils in cool to very warm humid and subhumid regions that have a weakly developed subsoil horizon (cambic) and a light colored surface horizon (orchric), the order contains a wide variety of soils (Soil Survey Staff, 1998a,b). In some areas they contain soils with minimal development, while in other areas they represent soils with well-expressed diagnostic horizons that fail the criteria of other orders.

Inceptisols, because of their broad and inclusive nature, served as the precursor for new soil orders (Andisols and Gelisols) and new taxa in other orders. For example, shrink/swell soils with aquic conditions and those with cryic soil temperature regimes previously placed within Inceptisols are now Vertisols (Aquerts and Cryerts). Inceptisols with isomesic or warmer *iso* temperature regimes, previously called Tropepts, are now distributed among other great groups and families of Inceptisols with *iso* temperature regimes; Inceptisols as markers of past cultural habitats and anthropogenic activities (Plaggepts) are now termed Anthrepts; and Ochrepts, a previous suborder of Inceptisols, has now been incorporated into Inceptisol suborders reflecting soil moisture bias. Table 6.24 illustrates the dynamics of these changes and how suborders of Inceptisols have evolved with increased knowledge over the past 40 years.

The central concept of Inceptisols includes soils that have undergone modifications of the parent material by structural development and alteration sufficient to differentiate them from Entisols. Pedogenesis may include eluviation (losses of constituents); the translocation of clay, iron, silica, alumi-

Table 6.24 Inceptisol suborders reflecting the evolution of *Soil Taxonomy*

7th Approximation (1960)[a]	*Soil Taxonomy* (1975)[b]	*Keys to Soil Taxonomy* (1990)[c]	*Soil Taxonomy* (1998)[d]
Andepts	Andepts	Aquepts	Anthrepts
Aquepts	Aquepts	Ochrepts	Aquepts
Ochrepts	Ochrepts	Plaggepts	Cryepts
Umbrepts	Plaggepts	Tropets	Udepts
	Tropets	Umbrepts	Ustepts
	Umbrepts		Xerepts

[a] Soil Survey Staff (1960)
[b] Soil Survey Staff (1975)
[c] Soil Survery Staff (1990)
[d] Soil Survey Staff (1998a,b)

num, carbonates, bases, and organic matter; and the formation of redoximorphic features by aquic (hydromorphic) conditions. Inceptisols occur in all known climates, except under aridic and pergelic conditions, and have many kinds of diagnostic horizons and epipedons. Hence, while the central concept expresses soils with incipient development, Inceptisols are an order with inordinate diversity. No attempt will be made herein to consider their full diversity. Rather, examples of the great range in physical, chemical, biological and mineralogical attributes of this order pertinent to land use and behavior will be illustrated. For a more extensive coverage of these attributes the reader is referred to online resources (http://www.statlab.iastate.edu/soils/soildiv/).

6.5.2 Distribution

The worldwide distribution of Inceptisols is illustrated in Fig. 6.23. Because these soils lack a sharply focused central concept, they occur indiscriminately globally. Where Inceptisols reflect youthfulness, they occur in high gradient mountainous regions, along major river systems as deltaic/fluvial plains, and as soils developed from carbonate-rich bedrocks or sediments in positions of geomorphic instability. Where Inceptisols serve to bifurcate *Soil Taxonomy* there is no particular pattern to their occurrence, but they are commonly dispersed among more strongly developed Alfisols, Ultisols and Oxisols in tropical and subtropical regions (Fig. 6.23).

6.5.3 Formation of Inceptisols

Because soils placed in Inceptisols are varied and represent many different pedogenic processes and combinations of soil-forming factors, examples will be given that typify central concepts or exhibit unique characteristics.

6.5.3.1 Parent Material

Foss et al. (1983) state that Inceptisols develop on geologically young sediments or landscapes and/ or under environmental conditions that inhibit soil development (e.g., coarse-textured siliceous deposits that are resistant to weathering or parent materials within cool climates that inhibit pedogenesis). Given the above constraints, parent materials include almost all types of igneous, metamorphic, and sedimentary materials (residuum, alluvium, colluvium, loess/eolian, glacial drift, etc.). Inceptisols are excluded from soils having a sandy texture throughout because it is believed that pedological features indicative of eluviation, weathering, translocation, and transformation (e.g., development of color Bw horizons) take a very short time to form in subsoils with such low specific surface area

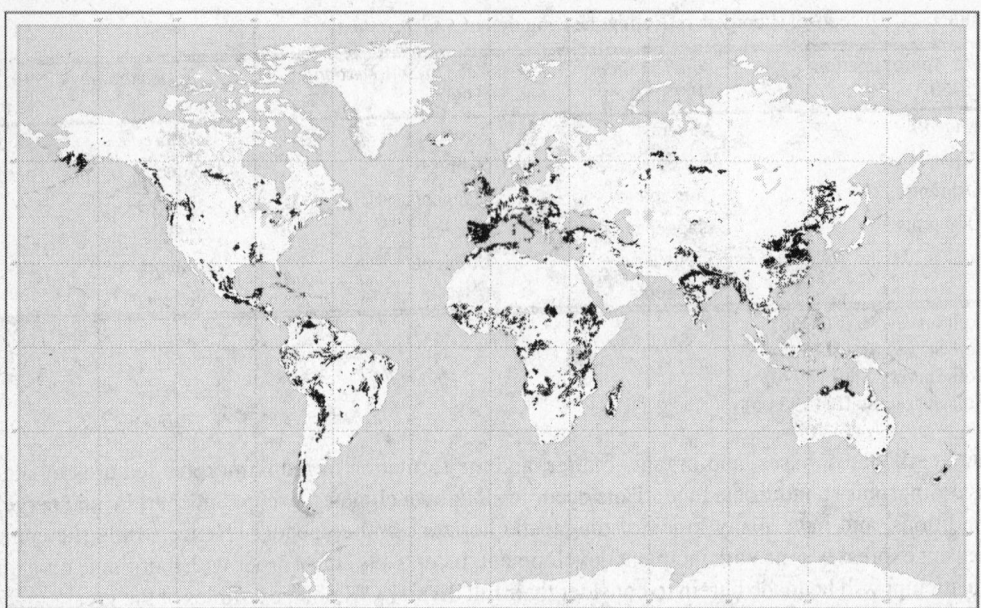

Fig. 6.23   Global distribution of Inceptisols [Courtesy of USDA-NRCS, Soil Survey Division, World Soil Resources, 1998]

coarse-textured materials. The kinds and arrangement of horizons, and their chemical, physical and mineralogical properties, are controlled to a major degree by the kinds of parent material from which Inceptisols have developed. For example, fine-textured parent materials give rise to clayey Inceptisols, acidic parent materials result in base-poor (*Dystr*) Inceptisols, and calcareous parent materials form neutral or alkaline, base-rich (*Eutr*) Inceptisols. Depth, color, organic carbon content and drainage frequently reflect the combination of geomorphic position, landscape instability, and nature of the parent materials.

Table 6.25 illustrates the diversity in selected physical, chemical and mineralogical properties of soils developed from six different parent materials [e.g., Mississippi River alluvium (*Fluvaquentic Epiaquepts*); noncalcareous glacial outwash (*Aeric Fragiaquepts*); Permian red bed shales (*Vertic Calciustepts*); basic igneous rock (*Oxic Haplustepts*); colluvium weathered from sandstone, siltstone and shale (*Typic Dystrudepts*); and Coastal Plains alluvium (*Typic Sulfaquepts*)]. Abridged attributes of these soils follow with detailed morphological, physical, chemical and mineralogical data found online at http://www.statlab.iastate.edu.soils.soildiv/:

Fluvaquentic Epiaquepts (Conciene series): These soils developed from loamy alluvial sediments that have weathered into soils with moderately expressed cambic horizons. Their nearly level to depressional landforms favor restricted drainage, especially under humid climates. Under slightly better drained conditions, soils developed in this alluvium may form argillic horizons in a relatively short time (hundreds to several thousand years). This is because the eluviation-illuviation process proceeds relatively rapidly in such permeable, base-rich, medium textured parent sediments.

Vertic Calciustepts (Vernon series): The calcareous clays and shales of the Permian red beds restrict water percolation and profile development. Well-drained soils that develop in these sedimentary materials have a calcic or salic horizon whereby lithogenic carbonates and soluble salts have been translocated to subsoils and precipitated as pedogenic products at the terminus of the wetting

Table 6.25 Physical and chemical properties of selected pedons in the Inceptisol order. See http://www.statlab. iastat.e.edu/soils/soildiv/ for complete morphological and characterization data sets

Horizon	Depth	Clay	Sand	OC[H]	Extr. Fe	CEC/ clay	WRD[I]	CEC	BS*	CaCO₃	pH_H₂O	COLE[I]
	cm	----%----			cm cm⁻¹	cm cm⁻¹		cmol_c kg⁻¹	----%----			cm cm⁻¹
Conciene series (S88-LA-047-001)(fine-silty, mixed, superactive, hyperthermic, Fluvaquentic Epiaquepts) Louisiana, USA												
Ap1	0-14	13.8	20.4	1.94	0.6	1.25	0.25	17.3	100	---	6.8	0.024
Bg2	51-78	15.9	17.6	0.30	0.4	0.91	0.22	14.5	100	1	7.7	0.009
Bgssb	180-215	50.4	4.3	0.71	1.0	0.69	0.19	34.9	100	1	7.5	0.116
Vernon soil series (S80-TX-253-001) (fine, mixed, semiactive, thermic, Vertic Calciustepts) Texas, USA												
A1	0-6	31.5	23.7	1.46	0.9	0.65	---	20.6	100	7	8.3	---
Bca2	45-65	40.5	13.8	0.21	1.7	0.34	0.08	14.8	100	23	8.5	0.042
Cr2	160-190	52.9	5.2	0.04	1.9	0.48	---	25.4	100	11	8.7	---
Kahana series (S89-HI-009-004) (very fine, kaolinitic, semiactive, isohypothermic Oxic Haplustepts) Hawaii, USA												
Ap	0-38	71.9	3.3	1.20	5.7	0.21	---	14.8	26	---	4.7	---
Bo2	56-87	59.8	7.1	0.45	5.9	0.23	---	14.0	71	---	6.2	---
Bo4	120-155	48.3	12.1	0.33	5.9	0.28	---	13.7	80	---	6.3	---
Volusia series (S86-NY-025-001) (fine-loamy, mixed, mesic, Aeric Fragaquepts) New York, USA												
Ap	0-20	---	---	4.50	2.0	---	---	22.5	44	---	5.4	---
Bw	20-29	---	---	0.95	1.8	---	---	11.3	33	---	5.5	---
Eg	29-42	---	---	0.27	1.2	---	---	6.3	30	---	5.2	---
2Bx3	90-145	---	---	0.16	1.5	---	---	10.9	68	---	6.2	---
Fedscreek series (S83-KY-195-018) (coarse-loamy, mixed, semiactive, mesic, Typic Dystrudepts) Kentucky, USA												
A	0-9	13.5	38.0	2.17	1.2	0.75	0.28	10.1	37	---	5.0	1.3
Bw2	42-76	14.7	43.6	0.20	1.2	0.34	0.17	5.0	20	---	4.8	0.4
BC	122-152	17.4	36.8	0.43	1.5	0.41	0.13	7.2	31	---	4.9	0.5
Gimhae series (S85-FN-515-012) (fine-silty, mixed, active, mesic, Typic Sulfaquepts) Korea												
Ap1	0-9	33.2	4.6	1.97	1.5	0.34	0.19	11.3	80	---	5.1	0.2
Bg1	15-32	34.4	5.2	1.59	1.9	0.38	---	13.2	42	---	4.0	---
Bg3	66-125	19.4	29.2	1.64	0.9	0.58	0.28	11.3	100	---	3.6	1.4
Vidhas soil (S94-FN-120-009) (fine-silty, mixed, superactive, thermic, Fluventic Haploxerepts) Albania												
Ap	0-25	21.4	12.4	0.98	0.9	0.75	0.17	16.1	100	20	8.1	0.6
Bw1	49-83	31.1	0.9	0.59	0.9	0.66	0.18	20.5	100	20	8.3	1.6
Bw2	83-116	59.8	1.9	0.69	1.3	0.52	0.09	31.2	100	15	8.2	1.4
Moran series (S85-CO-051-005) (loamy-skeletal, mixed, superactive, Humic Dystrocryepts) Colorado, USA												
Oi	8-0	---	---	16.6	---	---	---	---	---	---	---	---
A1	0-19	17.5	46.1	5.07	2.3	1.33	---	23.3	33	---	4.8	---
A2	19-64	13.1	54.0	3.36	2.3	1.40	---	18.4	24	---	5.2	---
Bw	64-98	14.5	50.7	3.05	2.3	1.32	---	19.2	22	---	5.2	---

[H] OC = organic C, [I] WRD = water retention difference, * BS = base saturation, [I] COLE = coefficient of linear expandability

front. The depth of carbonate and salt translocation, degree of calcic or salic horizon expression, and thickness of the cambic horizons are conditioned by the landform position, texture and permeability of the parent materials.

Oxic Haplustepts (*Kahana series*): These well-drained, clayey soils have developed in basic igneous rocks in Hawaii. Under a warm, tropical environment, the weathering of nonresistant mafic-rich minerals and fine-grained to amorphous materials has been rapid. An oxic-like (sesquioxide rich, subactive clay) cambic subsoil has developed. On a stable landform position, under an ustic moisture regime, subsequent weathering over time will result in an Oxisol with low activity clay.

Aeric Fragiaquepts (*Volusia series*): These soils have developed from weakly or noncalcareous glacial outwash of stratified sand, silt, and gravel. Upon weathering, these poorly sorted materials form loamy pans that have a brittle or fragic rupture upon deformation (fragipans). Weathering discontinuities in these materials enhanced by textural stratification are believed to result in weak cementation between closely packed skeleton grains (Smeck et al., 1989). The bonding agents between the grains are believed to be hydrous oxides of Al and Si and/or translocated clay coatings or bridges. Restricted drainage would complement transport discontinuities in these soils to form weathering discontinuities.

Typic Dystrudepts (*Fedscreek series*): These well-drained soils form under udic moisture regimes in loamy colluvium weathered from sandstone, siltstone and shale. The parent materials consist of acid channery-sized skeletal materials. Because the chemical and physical buffering capacity of the colluvium is low, and the parent materials are siliceous clasts, very acid conditions develop upon weathering. Profile development within the solum, however, is minimal because the preweathered clasts are resistant to further decomposition and disintegration. Hence, release of clay and other products of weathering for subsequent translocation and transformation is limited. Again, the nature of the parent material has been pivotal in governing the intensity, modes, and mechanisms of pedogenesis in these coarse-textured, highly permeable, acid upland soils.

Typic Sulfaquepts (*Gimhae series*): These soils are common along coastlines with tidal influence. Here the parent materials are medium textured alluvial deposits, usually associated with deltas or depositional regions of rivers that empty into the ocean. These wet environments are favorable for the accumulation of metal sulfides (e.g., pyrite) which upon drainage (natural or artificial) oxidize to form sulfuric acid and other acid sulfate weathering products (van Breeman, 1982). Sulfaquepts are important soils because of their land use, distribution, and behavior are conditioned by their acidic character.

6.5.3.2   Climate

Inceptisols occur in subhumid to humid climates from equatorial to tundra regions where, if they have permafrost within 2 m, they are replaced by Gelisols (Soil Survey Staff, 1998a,b). Inceptisols cannot have an aridic (torric) moisture regime unless they have an Anthropic epipedon upon which they become Anthrepts.

Soil temperature and moisture are used at the suborder level in *Soil Taxonomy* to reflect major controls on current soil-forming processes. Such controls not only influence rates, modes and mechanisms of physical and chemical weathering, but biological and geomorphic processes that influence relief, landscape stability, topography and soil drainage. Because Inceptisols form over a wide range of climatic conditions, excluding aridic and pergelic, the climatic impact on Inceptisols is similar to that of most other orders.

Soil moisture regime is used as a differentiaton for four Inceptisol suborders (Aquepts, Udepts, Ustepts and Xerepts), soil temperature regime for one suborder (Cryepts), and past human cultural impacts for one suborder (Anthrepts). For Cryepts it was judged that temperature constraints on

Table 6.26   Listing of suborders, great groups and subgroups for Inceptisols Order (Soil Survey Staff, 1998a)

Suborder	Great Group	Subgroups
Aquepts	Sulfaquepts	Salidic, Hydraquentic, Typic
	Petraquepts	Histic Placic, Placic, Plinthic,Typic
	Halaquepts	Vertic, Aquandic, Duric, Aeric, Typic
	Fragiaquepts	Aeric, Humic, Typic
	Cryaquepts	Sulfic, Histic Lithic, Lithic, Vertic, Histic, Aquabdic, Fluvaquentic, Aeric Humic, Aeric, Humic,Typic
	Vermaquepts	Sodic, Typic
	Humaquepts	Hydraquentic, Histic, Aquandic, Cumulic, Fluvaquentic, Aeric, Typic
	Epiaquepts	Vertic, Aquandic, Fluvaquentic, Fragic, Aeric, Humic, Mollic, Typic
	Endoaquepts	Sulfic, Lithic, Vertic, Aquandic, Fluvaquentic, Fragic, Aeric, Humic, Mollic, Typic
Anthrepts	Plagganthrepts	Typic
	Haplanthrepts	Typic
Cryepts	Eutrocryepts	Humic Lithic, Lithic, Andic, Vitrandic, Aquic, Oxyaquic, Lamellic, Xeric, Ustic, Humic, Typic
	Dystrocryepts	Humic Lithic, Lithic, Andic, Vitrandic, Aquic, Oxyaquic, Lamellic, Spodic, Xeric, Ustic, Humic, Typic
Ustepts	Durustepts	Typic
	Calciustepts	Lithic Petrocalcic, Lithic, Torrertic, Vertic, Petrocalcic, Gypsic, Aquic, Aridic, Udic, Typic
	Dystrustepts	Lithic, Andic, Vitrandic, Aquic, Fluventic, Oxic, Humic, Typic
	Haplustepts	Aridic Lithic, Lithic, Udertic, Torrertic, Vertic, Andic, Vitrandic, Anthraquic, Aquic, Oxyaquic, Oxic, Lamellic, Torrifluventic, Udifluventic, Fluventic, Gypsic, Haplocalcidic, Calcic Udic, Calcic, Aridic, Dystric, Udic, Typic
Xerepts	Durixerepts	Aquandic, Andic, Vitrandic, Aquic, Entic, Typic
	Calcixerepts	Lithic, Vertic, Petrocalcic, Sodic, Vitrandic, Aquic, Typic
	Fragixerepts	Andic, Vitrandic, Aquic, Humic, Typic
	Dystroxerepts	Humic Lithic, Lithic, Aquandic, Andic, Vitrandic, Fragiaquic, Fluvaquentic, Aquic, Oxyaquic, Fragic, Fluventic Humic, Fluventic, Humic, Typic
	Haploxerepts	Humic Lithic, Lithic, Vertic, Aquandic, Andic, Vitrandic, Gypsic, Aquic, Lamellic, Fragic, Fluventic, Calcic, Humic, Typic
Udepts	Sulfudepts	Typic
	Durudepts	Aquandic, Andic, Vitrandic, Aquic, Typic
	Fragiudepts	Andic, Vitrandic, Aquic, Humic, Typic
	Eutrudepts	Humic Lithic, Lithic, Aquertic, Vertic, Andic, Vitrandic, Anthraquic, Fragiaquic, Fluvaquentic, Aquic Dystric, Aquic, Oxyaquic, Fragic, Lamellic, Dystric Fluventic, Fluventic, Arenic, Dystric, Rendollic, Humic, Ruptic-Alfic, Typic
	Dystrudepts	Humic Lithic, Lithic, Vertic, Aquandic, Andic, Vitrandic, Fragiaquic, Fluvaquentic, Aquic Humic, Aquic, Oxyaquic, Fragic, Lamellic, Humic Psammentic, Fluventic Humic, Fluventic, Spodic, Oxic, Humic Pachic, Humic, Ruptic-Alfic, Ruptic-Ultic, Typic

pedogenesis, use and management were more important than moisture constraints, even though both are important; the soil moisture bias was introduced at the great group level (Table 6.26). Soil temperature regime was used as a differentiation at the family level for all orders except Gelisols and those soils that have a cryic soil temperature regime. Anthrepts have no particular climatic bias and are of limited geographical extent.

Aquepts form under aquic or excess wetness conditions (saturation with water, periodic reduction, and formation of contemporaneous redoximorphic features, redox concentrations and/or depletions). Udepts form under better drained, strongly leaching environments, where precipitation exceeds evapotranspiration. These soils generally lack salts and carbonates in the solumn and have undergone moderate to strong weathering. Generally, summer periods of soil moisture deficit are insufficient to negatively impact mesophytic agricultural crops. Ustepts are formed under drier climates with significant periods of summer soil moisture deficit. Precipitation is less than evapotranspiration, at least during significant periods of the summer. They commonly exhibit incomplete leaching, weak to moderate weathering intensities and pedogenic salts and/or carbonates in the lower column. Moisture deficit is sufficiently severe to limit crop growth during summer months. Xerepts are similar to Ustepts but are formed in climates where winter precipitation is dominant. These soils are common in Mediterranean regions of Europe and along the Pacific coastal zones of western United States. The Cryepts serve as an interface between the cold soils of high latitudes with pergelic conditions (permafrost within 2m) and cold soils of temperate regions or orographic mountainous zones without permafrost within 2m.

In Table 6.25 the climatic effects on soil classification and corresponding properties are illustrated for a few pedons. Vertic Calciustepts (*Vernon series*) and Oxic Haplustepts (*Kahana series*) are examples of soils with an ustic moisture regime and contrasting cambic horizon composition, development and chemical characteristics. The Calciustepts are incompletely leached, have carbonates throughout the pedon and have a clay-mineral composition that favors shrink/swell activity with changing moisture content. Comparisons between Fluventic Haploxerepts (*Vidhas series*) and Humic Dystrocryepts (*Moran series*) reflect contrasting temperature and precipitation regimes due to elevational differences. Although parent materials are different, climate is the major determinant governing increased organic matter contents in the Humic Dystrocryepts, which formed under udic moisture and cryic temperature regimes about 1050m elevation. This is compared to the Fluventic Haploxerepts formed under xeric moisture and thermic temperature regimes at an elevation of about 90 m. The Dystrocryepts have accumulated much greater organic carbon contents in the histic layer (Oi) overlying an umbric epipedon and a low base subsoil. In contrast, the Haploxerepts have accumulated much less organic matter in the ochric epipedon and the presence of carbonates distributed throughout the profile implies weak leaching potentials. Climate, in this example, is considered an intensity factor whereby the rate of soil development increases as temperature and rainfall increase, but decreases as either decreases. Limited soil moisture during periods of optimal soil weathering limits profile development in the Haploxerepts. In contrast, the Dystrocryepts have optimal moisture for pedogenesis, but the soil temperature is limiting.

6.5.3.3 Topography

The topography associated with many Inceptisols is a sloping to strongly sloping landscape subject to erosion rates that are equal to or greater than soil formation rates. Many Inceptisols occur on hillsides of major mountains throughout the world. Profile development here is constrained by unstable landforms and cool climates. Youthful Inceptisols are common also on nearly level fluvial, lacustrine and deltaic plains (or associated depressions) where profile development is limited by age of the sediments and/or shallow ground water tables. Coastal estuaries are also probable locations for Inceptisols, especially those where metallic sulfides and sulfates are likely to accumulate.

6.5.3.4 Vegetation

It is difficult to isolate vegetation as an independent soil-forming factor governing the synthesis of Inceptisols because vegetation is dependent on climate and topography. Interdependency of these factors is easily demonstrated by comparing the Humic Dystrocryepts (*Moran series*) and Vertic Calciustepts (*Vernon series*) of Table 6.25. Dystrocryepts support mountain meadow and alpine vegetative communities such as Kobres (*Koberesia sp.*), Asters (*Aster sp.*), and Yarro (*Achillea sp.*) under cold, moist climates. Under hotter, drier conditions of the Calciustepts, native vegetation is mixed, midgrass prairie consisting of buffalograss (*Buchloe dactyloides*), grama grasses [blue (*Bouteloua gracilis*), hairy (*B. hirsuta*) and sideoats (*B. curtipendula*)] and tobosa grass (*Hilaria mutica*). Primary productivity, accumulation of soil organic matter, and leaching potentials are governed by effective soil moisture and temperature regimes. Effects on soil properties are remarkably evident in Table 6.25 for these two soils. Primary productivity for the Dystrocryepts is limited by soil temperature regime while a seasonal soil moisture deficit limits primary productivity for Calciustepts. Higher organic matter accumulation, low turnover rates, and moderate to strong leaching potentials (acid subsoils) are consequences in Dystrocryepts. In contrast, lower organic matter contents, high turnover rates and less effective leaching potentials (calcareous solumn) occur in Calciustepts.

6.5.3.5 Time

Many examples are available in the literature to demonstrate that time is a soil-forming factor (Jenny, 1941, 1980). Some might argue that factors other than time have more influence on soil formation. While this may be true in some cases, time in combination with the intensity factor (climate) has been responsible for many attributes of Inceptisols. Inceptisols are usually considered immature soils, but some have advanced stages of development and are just short of completing their evolutionary cycle from Entisols to Oxisols. For example, Oxic Haplustepts are similar to Oxisols but either have oxic conditions too deep in the soil (greater than 150 cm) or have CEC charge characteristics just a little too high for oxic horizon requirements—a prerequisite for all Oxisols. Upon erosion and truncation, some Inceptisols have soil properties at depths associated with old stable landforms prior to dissection. Further, several taxonomic classes of Inceptisols have morphological features that could be interpreted either as youthful or an advanced stage of pedogenesis. Arenic Eutrudepts is one such example in which the soils may have developed from recent sandy sediments or may be residuum from weakly consolidated sandstone on old stable landforms.

6.5.4 Morphology

Inceptisols have many kinds of diagnostic horizons and epipedons but the most common sequence of horizons is an ochric or umbric epipedon overlying a cambic or fragipan horizon. However, a cambic horizon is not required if an umbric, histic, or plaggen epipedon is present or if there is a fragipan, duripan, placic, calcic, petrocalcic, gypsic, petrogypsic, salic or sulfuric horizon (Soil Survey Staff, 1998a,b). They can also have anthropic and mollic epipedons but not argillic, kandic, or natric horizons unless buried below recent sediments. Oxic horizons are permitted only if the upper boundary is deeper than 150 cm. Spodic horizons are permitted only if they are less than 10 cm thick or if the upper boundary is deeper than 50 cm.

It is clear from the above discussion that the number and kinds of processes active in Inceptisols and the definition of Inceptisols are unavoidably complex because the order serves dual functions in grouping soils of minimal pedogenic development and bifurcating *Soil Taxonomy*. In summary, they are youthful soils in rejuvenated landscapes but pedogentically advanced on more stable landforms. Their attributes are highly variable and seldom uniquely definitive.

6.5.5 Pedogenic Processes

Pedogenic processes common to Inceptisols have been briefly considered under Section 6.5.3. Such processes are covered in greater detail in Section E, Chapters 2 and 3. The processes germane to Inceptisols are combinations and complexes of subprocesses and reactions found in most other soil orders. The nature of these processes, including their terminology, are reviewed by Marbut (1921), Kellogg (1936), Byers et al. (1938), Simonson (1959), Arnold (1983), Buol et al. (1980), Fitzpatrick (1986), and Fanning and Fanning (1989).

Biological processes including establishment of macro and micro fauna and flora, microbial and fungal mineralization, decomposition (humification) and accumulation of organic matter, biotic transformations (e.g., ammonification, nitrification, denitrification and nitrogen fixation), and pedoturbation are important in epipedon and cambic horizon genesis. Chemical processes including hydration, hydrolysis, solution, mineral synthesis, and oxidation/reduction reactions are responsible for both epipedon and cambic horizon synthesis. Physical processes contribute to the formation of pedogenic structure (obliteration of at least 50% of the "rock" structure) and transport of materials within the solum. They include aggregation, expansion/contraction (shrink/swell), freeze-thaw, and mass transport by solution or suspension (translocation).

Formation of a cambic horizon requires conversion of rock structure to soil structure (peds) and evidence of at least one of the following: oxidation/reduction (aquic conditions); neoformation of clay; translocation of salts, carbonates, clay and/or organic metal complexes; or liberation of free iron oxides. The cambic horizon is a prerequisite for most Inceptisols, but exceptions occur for certain epipedons or subsurface diagnostic features (Section 6.5.3).

Illuviation and/or neosynthesis of clay and organic metal complexes cannot be of such magnitude that argillic, natric, kandic, or spodic horizons are formed, unless buried. Processes responsible for formation of oxic horizons include desilication, neoformation of low activity clays and/or residual concentration of secondary oxides and oxyhydroxides of iron and aluminum (Kellogg, 1936; Sivarajasingham et al., 1962; Cline, 1975 and Jenny, 1980). Acid sulfate soils are formed by processes termed sulfidization and sulfuricazation. These processes are responsible for the accumulation of metal sulfides in soils and sediments, oxidation of these sulfides, formation of sulfuric acid, ferrolysis, soil acidification and mineral dissolution (Kittrick et al., 1982; van Breeman, 1982; Fanning and Fanning, 1989).

6.5.6 Classification

In the 1938 classification scheme (Baldwin et al., 1938) and in most subsequent revisions in the U.S. prior to *Soil Taxonomy*, Inceptisols were included in a number of great soil groups including Ando, Hydrol Humic Latosols, Humic Gley, Tundra, Half Bog, Sols Bruns Acides, Brown Forest, Regosols, Lithosols, Aluvial, and numerous other minor groups (Foss et al., 1983). Even at that time this was a very extensive and diverse taxa of soils. In the Canadian soil classification system, the wet Inceptisols (Aquents) were classified as Gleysols but most of the better drained members were Brunisols. In the legend of the FAO/UNESCO soil map of the world (Section E, Chapter 5), most well-drained Inceptisols would be classed as Cambisols or Phaeozems where the base saturation was low. The wet Inceptisols would be classed mostly as Gleysols except those with acid sulfate conditions, which would be grouped with Fluvisols.

Inceptisol suborders, great groups and subgroups are presented in Table 6.26. Rationale for these taxa have been considered previously in this chapter and in Section E, Chapter 4. The following discussion of suborders has been taken directly from *Soil Taxonomy* (1998b) to illustrate the general nature and characteristics of Inseptisols suborders.

6.5.6.1 Anthrepts

These are the Inceptisols that are the more or less freely drained that have either an anthropic or a plaggen epipedon. Most of these soils have been used as cropland or places of human occupation for many years. They can have almost any temperature regime, and almost any vegetation. Most of them have a cambic horizon. Only two subgroups are recognized at this time, those with plaggen epipedons indicative of ancient cultural sites (*Plaggenanthrepts*), and those with anthropic epipedons reflecting other long-term human impacts on soils such as irrigated farmlands or housing areas. Most of these soils are in Eurasia or Northern Africa.

6.5.6.2 Aquepts

These are the wet Inceptisols. Their natural drainage is poor or very poor and, if they have not been artificially drained, groundwater is at or near the soil surface at some time during normal years but typically not during at all seasons. They mostly have a gray to black surface horizon and a gray subsurface horizon with redox concentrations that begin at a depth of less than 50 cm. A few have a brownish surface horizon that is less than 50 cm thick.

Most Aquepts have developed in late Pleistocene or younger deposits in depressions, nearly level plains, or flood plains. They occur from the Equator to latitudes with discontinuous permafrost. The common features of most of these soils are the gray and rusty colors of redoximorphic features at a depth of 50 cm or less and the shallow groundwater or artificial drainage. They may have almost any particle size class except fragmental, and any reaction class, any temperature regime except pergelic, and almost any vegetation. Most of them have a cambic horizon, and some have a fragipan. It is possible that some have a plaggen epipedon.

Table 6.26 lists the 9 Aquepts. These suborders are based upon limiting features or horizons that impact use, management and behavior. Specifically, these include soils with acid sulfuric horizons within 50 cm of the surface associated with oxidation of metal sulfides (*Sulfaquepts*); soils with a restrictive cemented subsurface horizon that forms a continuous phase within 100 cm of the soil surface (*Petraquepts*); salty (salic) or alkali (natric) subsoil horizons (*Halaquepts*); soils with a fragipan (*Fragiaquepts*); soils with a cryic temperature regime (*Cryaquepts*); soils with strongly bioturbated layers by macrofauna such as crayfish, worms, and mammals (*Vermaquepts*); soils with darkened, organic-enriched surface horizons (*Humaquepts*); soils with a perched water table (*Epiaquepts*); soils that are saturated from a groundwater source (*Endoaquepts*).

6.5.6.3 Cryepts

Cryepts are the cold Inceptisols of high mountains or high latitudes. They cannot have permafrost. The vegetation is mostly conifers or mixed conifers and hardwood trees. Few of them are cultivated. These soils may be formed in loess, drift, alluvium, or solifluction deposits, mostly late Pleistocene or Holocene in age. They commonly have a thin, dark brownish ochric epipedon and a brownish cambic horizon. Some have bedrock within 100 cm of the surface. In the United States these soils are moderately extensive in the high mountains of the West as well as other mountainous areas of southern Alaska and the world.

Two suborders of Cryepts are recognized (Table 6.26). Those that are calcareous or have high base saturation are *Eutocryepts*, and Cryepts with low base saturation are *Dystrocryepts*.

6.5.6.4 Udepts

Udepts are mainly the more or less freely drained Inceptisols that have a udic or perudic moisture regime. They formed on nearly level to steep surfaces mostly of late Pleistocene or Holocene age. Some, where the soil moisture regime is perudic, formed in older deposits. Most of them had or now have a forest vegetation, but some have shrub or grass. A few have been formed from Mollisols by truncation of the mollic epipedon (*Eutrudepts*), mostly under cultivation. Most of them have an ochric or umbric epipedon and a cambic horizon with low base saturation (*Dystrudepts*). These were the Sols Bruns Acides of earlier classification schemes (Foss et al., 1983). Some also have a sulfuric horizon (*Sulfudepts*), a fragipan (*Fragiudepts*), or a duripan (*Durudepts*). In the United States, Udepts are most extensive on the Appalachian Mountains, the Allegheny Plateau, and the West Coast; they also occur extensively in Eurasia,

6.5.6.5 Ustepts

Ustepts are the more or less freely drained Inceptisols that have a ustic moisture regime. They have dominantly summer precipitation or an isomesic, hyperthermic or warmer temperature regime. They formed mostly in Pleistocene or Holocene deposits or on steep slopes. Many of these soils are calcareous at a shallow depth and have a Bk or a calcic horizon (*Calciustepts*). A few have formed from Mollisols by truncation of the mollic epipedon, mostly under cultivation. Most of them have an ochric or umbric epipedon and a cambic horizon (*Haplustepts*). Some have a duripan (*Durustepts*), especially in areas where a labile silica source is associated with pyroclastic deposits or volcanic ash. Ustepts with low base saturation (*Dystrustepts*) occur in areas with polycycled preweathered acidic sediments or outcrops of acid bedrocks common in West Africa and isolated areas of the United States. The native vegetation commonly was grass, but some supported trees and savannas. Ustepts are of moderate extent in the United States. They are most common on the Great Plains mostly in Montana, Texas, and Oklahoma. They occur extensively in ustic sectors of the Americas, Eurasia, West Africa, Australia, and several island countries.

6.5.6.6 Xerepts

Xerepts are the more or less freely drained Inceptisols that have a xeric moisture regime, dominantly winter precipitation. They have a frigid, mesic, or thermic temperature regime. They formed mostly in Pleistocene or Holocene deposits or on steep slopes. Many of these soils are calcareous at a shallow depth and have a Bk, a calcic or a petrocalcic horizon (*Calcixerepts*). Others have low base saturation (*Dystrixerepts*). Most of them have an ochric or umbric epipedon and a cambic horizon (*Haploxerepts*). Some have a duripan (*Durxerepts*) and a few have a fragipan (*Fragixerepts*). The native vegetation was commonly coniferous forest on those with a thermic temperature regime. These soils are of moderate extent in the United States. They are most common near the West Coast in the states of California, Oregon, Washington, Idaho, and Utah. They are major soils of the Mediterranean regions of Eurasia and Northern Africa.

6.5.7 Physical, Chemical and Mineralogical Properties

Because of the diverse nature of Inceptisols, few generalized statements can be made about their physical, chemical, biological, and mineralogical properties. Their properties are nearly as inclusive as all the other soil orders collectively. Inceptisols span the global regions from intensively weathered to minimally developed soils. For example, they are acidic to alkaline in reaction, weak to strongly physico-chemically buffered, have low to high organic matter contents, low to high hydraulic transport functions, low to high water retention values, fertile to infertile in nutrient status, etc. The Soil

Survey manual (Soil Survey Staff, 1993) identify class ranges for the above classes and for many other soil attributes described and measured. Table 6.25 illustrates the kind and magnitude of diversity vested in many of these Inceptisol attributes. The database on-line for Inceptisols (http://www.statlab.iastate.edu/soils/soildiv/) further documents this aspect.

The reader is referred to discussions of other orders in Sections 6.2-6.13 for physical, chemical, biological, and mineralogical attributes likely to be associated with Inceptisols. For example, those Inceptisols with andic surface materials are similar to Andisols; those with vertic shrink/swell features to Vertisols; those with calcic, gypsic, salic, petrocalcic and petrogypsic horizons to Aridisols; those with translocation and/or neosynthesis of clay and organic metal colloids to Alfisols, Ultisols and Spodosols; those with high sesquioxide contents and/or low activity clay mineral suites to Oxisols; and those with cryic temperature regimes to Gelisols.

The sand and silt mineralogy commonly consists of quartz, mica and feldspars with minor components of weatherable heavy minerals (opaques). The opaque minerals in soils derived from basic bedrocks are feldspars, hematite, magnetite, ilmenite and rutile. Inceptisols rich in sequioxides (Oxic subgroups) are comprised of oxyhydroxides of Fe, Al and Ti including gibbsite, hematite, goethite, boehmite, rutile and anatase. Those soils associated with drier climates (Ustepts and Xerepts) contain soluble salts (e.g., $NaCO_3$, Na_2SO_4), gypsum and carbonates (calcite and dolomite). Inceptisols with andic and vitrandic materials (pyroclastics) contain amorphous or short-range order minerals such as allophane, ferrihydrite and glasses. Phyllosilicate clay minerals range from smectite, mica, kaolinite, chlorite, vermiculate and mixed layer assemblages to rather unique suites dominated by serpentine and glauconite. In acid sulfate soils (e.g., Sulfaquepts and Sulfudepts), jarosite is a common constituent that marks very acid conditions associated with metal sulfide oxidation. Pedogenic gypsum is also common in these soils.

6.5.8 Management

For management purposes, Inceptisols can be subdivided into three land uses, namely forestry, pasture production and agronomic cropping.

6.5.8.1 Forestry
Most Inceptisols under forested land use occur in mountainous regions on slopes ranging from 3 to 90%. On steep terrains, management systems other than natural regrowth are environmentally unacceptable and practically impossible. Most forested Inceptisols have carbon contents sufficiently high as long as the surface is not eroded, that indigenous nutrient recycling is sufficient to sustain growth without external fertilizer amendments. On less sloping terrain, many large, commercial timber companies find fertilizer amendments, especially phosphorous applications, to be economically beneficial. Management techniques to control competition from unwanted species and disease suppression commonly involve controlled burns or physical removal of dead or unwanted species. Harvesting methods depend on the slope gradient; they range from surface skidding to aerial removal by helicopter or cable lifts. Harvest schemes may involve clear cutting or selective harvesting. Damages from erosion and compaction of harvest operations on less sloping Inceptisols are commonly ameliorated with surface tillage operations to break up surface crusts, compacted zones and to establish soil and water conservation buffers.

6.5.8.2 Pasture Production
Most of the Inceptisols under pasture management are planted to improved pastures that respond to N, P and K fertilization, especially if the pasture is intensely managed for animal or forage produc-

tion. Areas, which occur under native range, or under shifting traditional agriculture as in developing countries, should be managed using best management practices for stocking rate, pasture rotation and fallow period. These vary from one region to another, not only in terms of soil conditions but the ability of the operator to provide investment inputs. Commonly under traditional agriculture, the inputs are minimum and the soil dictates the best management and utilization practices to be followed. Inceptisols with an ustic or xeric moisture regime are limited by insufficient rainfall. Irrigation may be used to supplement natural precipitation on some Inceptisols, but the cost of this investment for improved pastures may make it economically unattractive. However, many examples of irrigated improved pastures occur in Europe and the United States where the economics are favorable and adequate high quality aquifers are available. Surface or subsurface drainage may be required to remove surface water or lower the watertable for optimal forage production, especially for Aquepts.

6.5.8.3 Agronomic Cropping

Inceptisols are the major soils on which agronomic crops are produced in some parts of the world. Aquepts, Cryepts, Udepts, Ustepts, and Xerepts are very productive and valuble agricultural land resources, if properly managed. Crops produced depend on climate (length of growing season, dependability of precipitation, seasonal periods of soil moisture excess or deficit, photoperiodism, degree days, etc.), but because of Inceptisol diversity, most major food, feed and fiber crops are included. Irrigation is used on vegetable, citrus, and other important cash or speciality crops when summer periods of moisture deficit become extreme. The highly weathered Inceptisols within the tropics are used for sugarcane, pineapple, cotton and some upland rice production with many of these crops being produced under irrigation. Most of these soils respond to N, P, and K fertilizers and some soils require lime in order to neutralize acidity and sustain production.

Cryepts are used mostly for cereal crops because of the short growing season, photoperiodism and few number of degree days. Sometimes these soils are used for cool season vegetables and forage production. Where these soils interface with Gelisols, they are constrained by cold soil temperatures, freeze-thaw heaving, and erosion induced by wind, water and mass movement (solifluction). They may be under alpine meadow.

Aquepts used for agronomic production usually require surface or internal drainage but depending on soil temperature, a wide range of crops are grown including cotton, sugarcane, rice and corn in the subtropical and tropical climates and corn, soybeans, sorghum and some cereal crops in temperate climates. Sulfaquepts (and Sulfudepts) require specialized management because when they are drained mainly for lowland rice production, the sulfuric horizon produces large amounts of sulfuric acid. Because the reactions involved were not well understood when most of these soils were drained, vast wastelands were created. Management schemes, which reduce acid production, have been developed by Ponnamperuma et al. (1973), van Breeman (1982) and Coly (1996). Reclamation and management practices for acid sulfate soils have been proposed by van Breeman (1982), Rimwanich and Suebsuri (1983), Cisse et al. (1993) and Coly (1996). They are discussed in detail in Section G, Chapter 6. Briefly the formation of a sulfuric horizon should be prevented or minimized by controlling the depth of the water table upon drainage (keeping the soils saturated to prevent pyrite oxidation). Sulfaquepts can be reclaimed by inducing reducing conditions within the sulfuric horizon and/ or flushing the soil to remove the sulfuric acid. The flushing can be accomplished using a combination of saline and fresh water to remove the soluble salts. The addition of $CaCO_3$ is required to neutralize any acidity produced and $CaSO_4$ is required when saline water has been used for reclamation to remove excess sodium (Cisse et al., 1993; Coly, 1996).

6.5.9 References

Arnold, R.W. 1983. Concepts of soils and pedology. *In* L.P. Wilding, N.E. Smeck, and G.F. Hall (ed.) Pedogenesis and soil taxonomy I. Concepts and interactions. Elsevier Science Publishers, Amsterdam, Netherlands.

Baldwin, M., C.E. Kellogg, and J.Thorp. 1938. Soil classification. Soils and men. p. 979–1001. 1938 Yearbook of Agriculture. USDA. Government Printing Office, Washington, DC.

Buol, S.W., F.D. Hole, and R.J. McCracken. 1980. Pedogenic process: Internal, soil-building processes. Soil genesis and classification, 2nd Ed. IA State Press, Ames, IA.

Byers, H.G., C.E. Kellogg, and J.Thorp. 1938. Soil classification. Soils and men. p. 948–978. 1938 Yearbook of Agriculture. USDA. Government Printing Office, Washington, DC.

Cisse, S., W.H. Hudnall, and A.A. Szogi. 1993. Influence of land reclamation on coastal acid sulfate soils in the Republic of Guinea. p. 101–112. *In* J.M. Kimble (ed.) Proc. 8th Int. Soil Manag. Worksh. Util. Soil Surv. Info. Sustain. Land Use. USDA-NSSC, Lincoln, NB.

Cline, M.C. 1975. Origin of the term latosol. Soil Sci. Soc. Am. Proc. 39:162.

Coly, L. 1996. Management of acid sulfate saline soils in Casamance, Seneal. MS Thesis. Louisiana State University, Baton Rouge, LA.

Fanning, D.S., and M.C.B. Fanning. 1989. Soil morphology, genesis, and classification. John Wiley and Sons, New York, NY.

Fitzpatrick, E.A.. 1986. An introduction to soil science. 2nd Ed. Longman Scientific & Technical, Longman Group UK Limited, Essex, UK.

Foss, J.E., F.R. Morman, and S. Reiger. 1983. Inceptisols. p. 355–381. *In* L.P. Wilding, N.E. Smeck and G. F. Hall (ed.) Pedogenesis and soil taxonomy. II. The soil orders. Elsevier Science Publishers, Amsterdam, Netherlands.

Jenny, H. 1941. Factors of soil formation. McGraw-Hill, New York, NY.

Jenny, H. 1980. The soil resource: Origin and behavior. Springer Verlag, New York, NY.

Kellogg, C.E. 1936. Development and significance of the great group in the United States. USDA Misc. Publ. 229.

Kittrick, J.A., D.S. Fanning, and L.R. Hossner (ed.). 1982. Acid sulfate weathering. Soil Sci. Soc. Am. Spec. Pub. 10, Soil Science Society of America, Madison, WI.

Marbut, C.F. 1921. The contribution of soil survey to soil science. Soc. Prom. Agri. Sci. Proc. 41:116–142.

Ponnamperuma, F.N., T. Attanandana, and G. Beye. 1973. Amelioration of three acid sulfate soils for lowland rice. p. 391-406. *In* H. Dost (ed.) Acid sulfate soils. Proc. Int. Symp. Vol 2. ILRI, Wageningen, Netherlands.

Rimwanich, S., and B. Suebsuri. 1983. Nature and management of problem soils in Thailand. p. 1–23. *In* Int. Sem. Ecol. Manag. Prob. Soils Asia. Kasetsart University, Bangkok, Thailand.

Simonson, R.W. 1959. Outline of a generalized theory of soil genesis. Soil Sci. Soc. Am. Proc. 23:152–156.

Sivarajasingham, S., L.T. Alexander, J.G. Cady, and M.G. Cline. 1962. Laterite. Adv. Agron. 14:1–60.

Smeck, N.E., J.L. Thompson, L.D. Norton, and M.J. Shipitalo. 1989. Weathering discontinuities: A key to fragipan formation. p. 99-112. *In* N.E. Smeck and E. Ciolkosz (ed). Fragipans: Their occurrence, classification, and genesis. Soil Sci. Soc. Amer. Spec. Pub. 24. Soil Science Society of America, Madison, WI.

Smith, G. D. 1986. The Guy Smith interviews: Rationale for concepts in soil taxonomy. SMSS Tech. Monogr. 11. USDA Soil Management Support Services, Soil Conservation Service, Washington, DC.

Soil Survey Staff. 1960. Soil classification, a comprehensive system, 7th approximation. Soil Conservation Service. USDA. US Government Printing Office, Washington DC.

Soil Survey Staff. 1975. Soil taxonomy. Basic system of soil classification for making and interpreting soil surveys. USDA-SCS Handbook 436. US Government Printing Office, Washington, DC.

Soil Survey Staff. 1990. Keys to soil taxonomy. US Government Printing Office, Washington, DC.

Soil Survey Staff. 1993. Soil survey manual. USDA Agric. Handbook 18. USDA, Washington, DC.

Soil Survey Staff. 1998a. Keys to soil taxonomy, 8th Ed. US Dept. Agric. Natural Resources Conservation, Washington, DC.

Soil Survey Staff. 1998b. Soil Taxonomy. U.S. Dept. of Agric. Natural Resources Conservation Service, Washington, DC. Online (http://www.statlab.iastate.edu/soils/soildiv/)

van Breeman, N. 1982. Genesis, morphology and classification of acid sulfate soils in coastal plains. p. 98–105. *In* J.A. Kittrick, S. Fanning, and L.R. Hossner (ed.) Acid sulfate weathering. Soil Sci. Soc. Am. Spec. Pub. 10. Soil Science Society of America, Madison, WI.

6.6   Gelisols

J. G. Bockheim
University of Wisconsin

C. Tarnocai
Agriculture and Agri-Food Canada, Ottawa

6.6.1   Introduction

Gelisols, which are the permafrost-affected soils, constitute the twelfth and newest soil order. They comprise 18 million km^2 or about 13% of the Earth's land surface and occur in the Arctic, Antarctic, Subarctic, Boreal, and some alpine regions under cold continental, subhumid or semi-arid, and arid conditions (Bockheim et al., 1994). They support unvegetated to continuously vegetated tundra, subarctic and boreal forest, and some alpine tundra. Gelisols are of global concern because they contain many protected areas, support numerous indigenous populations who depend on the land and surrounding oceans for sustenance, and may be subject to considerable impacts from human development (oil, coal and gas exploration and mining) and global warming.

Gelisols are defined as soils having permafrost within 100 cm of the soil surface, or gelic materials within 100 cm of the soil surface and permafrost within 200 cm of the soil surface (Soil Survey Staff, 1998). Gelic materials, in turn, are seasonally or perennially frozen mineral or organic soil materials that have evidence of cryoturbation (frost churning), ice segregation, or cracking from cryodesiccation. Gelic materials contrast with other kinds of materials, such as andic or spodic materials, in being defined entirely on the basis of physical and thermal characteristics, rather than chemical properties.

A representative Gelisol is shown in Fig. 6.24. Gelic materials occur in both the active layer and the upper part of the permafrost as evidenced by cryoturbation, denoted in soil descriptions by the subscript *jj* (*y* in the Canadian system).

6.6.2   Permafrost and the Occurrence of Gelisols

Gelisols only occur in areas containing permafrost within 100 cm of the soil surface if the soil is not cryoturbated or 200 cm of the surface if the soil is cryoturbated. Permafrost is defined here as a thermal

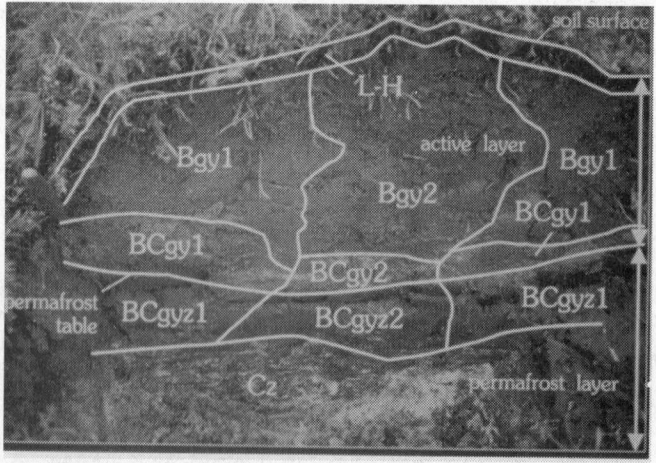

Fig. 6.24   An Aquaturbel developed on an earth hummock in northern Canada. The Canadian soil horizon nomenclature is used, where L-H = O, y = jj, and z = f in *Soil Taxonomy*.

condition in which a material (including soil materials) remains below 0 °C for two or more years in succession. Permafrost may be ice-cemented (designated in soil descriptions with the subscript f) or in the case of insufficient interstitial water, dry (designated as ff). In the frozen layer, a variety of ice lenses, vein ice, segregated ice crystals, and ice wedges are evident. An important consideration is that the permafrost is in dynamic equilibrium with the environment.

Permafrost, which comprises about 26% of the Earth's surface, including the Antarctic and Greenland ice sheets (Black, 1954), is differentiated into two broad zones on land in the circumpolar regions: the continuous and the discontinuous (Péwé, 1991). In the continuous zone, permafrost is present nearly everywhere except under large lakes and rivers that do not freeze to the bottom (Fig. 6.25). The discontinuous zone is divided into the widespread discontinuous permafrost subzone and the sporadic discontinuous subzone in the northern and southern parts, respectively.

The distribution and thickness of permafrost are influenced by natural surface features such as snow and vegetation cover, topography, and bodies of water, but as Fig. 6.25 shows, are most affected by regional climate.

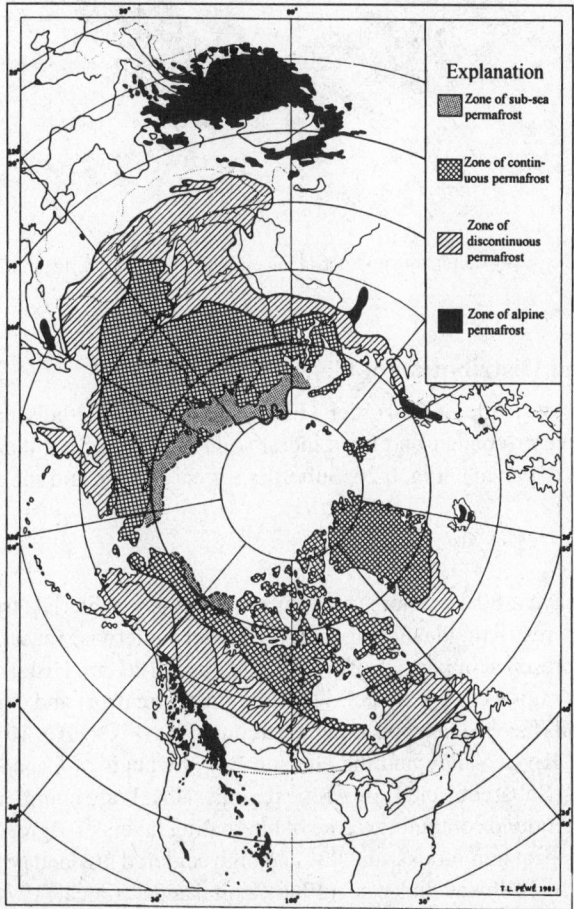

Fig. 6.25   Distribution of permafrost in the Northern Hemisphere [From Péwé, 1983. Artic Alpine Res. 15:146 with permission from the Regents of the University of Colorado]

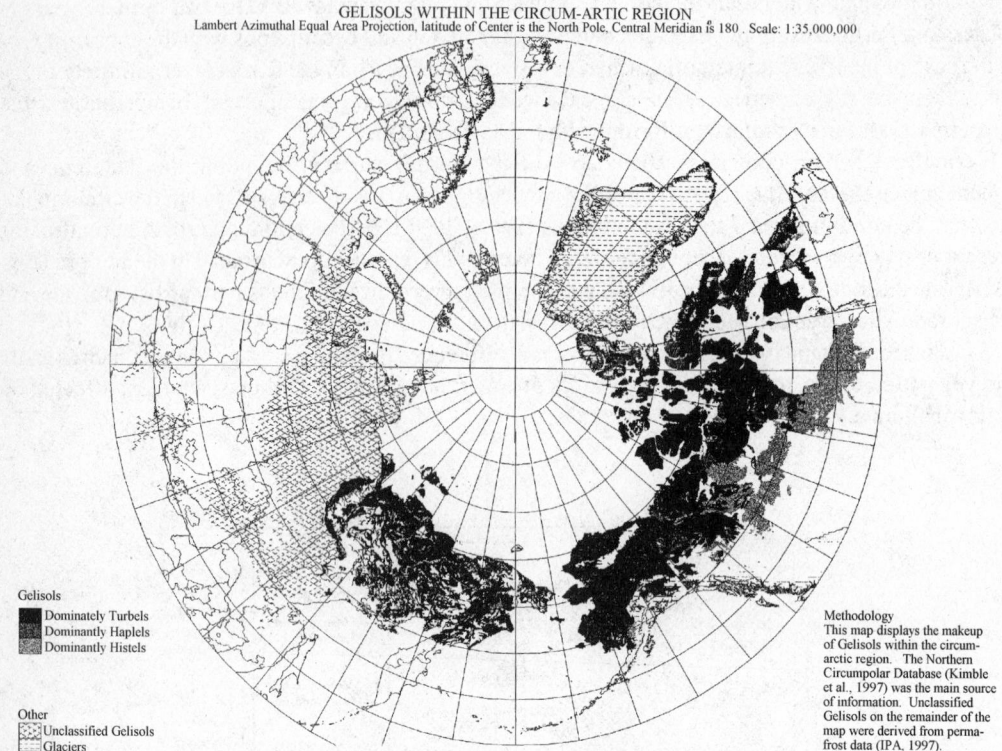

GELISOLS WITHIN THE CIRCUM-ARTIC REGION
Lambert Azimuthal Equal Area Projection, latitude of Center is the North Pole, Central Meridian is 180 . Scale: 1:35,000,000

Gelisols
Dominately Turbels
Dominantly Haplels
Dominantly Histels

Other
Unclassified Gelisols
Glaciers

Methodology
This map displays the makeup
of Gelisols within the circumarctic region. The Northern
Circumpolar Database (Kimble
et al., 1997) was the main source
of information. Unclassified
Gelisols on the remainder of the
map were derived from permafrost data (IPA, 1997).

Fig. 6.26   Distribution of Gelisols by dominant suborder in the circum-arctic region [From Kimble et al., 1997 with permission]

6.6.3   Description and Distribution of Gelisols

There are three suborders within the Gelisol order: Histels, Turbels, and Orthels that are differentiated on the basis of organic matter content and for mineral soils whether or not there is cryoturbation. Distribution of Gelisols is depicted in Fig. 6.26. Suborders, great groups, and subgroups are shown in Table 6.27.

6.6.3.1   Histels

Histels are Gelisols that have $\geq$ 80% or more (by volume) of organic materials from the soil surface to a depth of 50 cm, or to a restricting layer (Fig. 6.27A). Histels otherwise meet the requirements of a Histosol except for the presence of permafrost within the upper 100 cm. Histels occur primarily in Subarctic and Low Arctic regions in the area of discontinuous permafrost and comprise 5.3% of the Gelisols in areas where soils have been mapped in the circumarctic (Fig. 6.26). Most of these soils are located in the Mackenzie River Valley and the Hudson Bay lowlands of Canada. In general, they mostly occur in the Boreal, Subarctic, and Low Arctic regions. Histels are commonly associated with palsas (a peaty permafrost mound containing a core of alternating layers of segregated ice and peat or mineral soil), peat plateau, peat hummocks, and low and high centered lowland polygons. These soils have been discussed fully by Holowaychuk et al. (1966), Zoltai and Tarnocai (1971), Tarnocai (1972, 1973), Zoltai and Tarnocai (1974, 1975), Everett (1979), and the National Wetlands Working Group (1988).

Table 6.27 Listing of suborders, great groups and subgroups of Gelisols order [Soil Survey Staff, 1998]

Suborder	Great Group	Subgroups
Histels	Folistels	Lithic, Glacic, Typic
	Glacistels	Hemic, Sapric, Typic
	Fibristels	Lithic, Terric, Fluvaquentic, Sphagnic, Typic
	Hemistels	Lithic, Terric, Fluvaquentic, Typic
	Sapristels	Lithic, Terric, Fluvaquentic, Typic
Turbels	Histoturbels	Lithic, Glacic, Ruptic, Typic
	Aquiturbels	Lithic, Glacic, Sulfuric, Ruptic-Histic, Psammentic, Typic
	Anhyturbels	Lithic, Glacic, Petrogypsic, Gypsic, Nitric, Salic, Calcic, Typic
	Molliturbels	Lithic, Glacic, Vertic, Andic, Vitrandic, Cumulic, Aquic, Typic
	Umbriturbels	Lithic, Glacic, Vertic, Andic, Vitrandic, Cumulic, Aquic, Typic
	Psammoturbels	Lithic, Glacic, Spodic, Typic
	Haploturbels	Lithic, Glacic, Aquic, Typic
Orthels	Historthels	Lithic, Glacic, Ruptic, Typic
	Aquorthels	Lithic, Glacic, Sulfuric, Ruptic-Histic, Andic, Virandic, Salic, Psammentic, Typic
	Anhyorthels	Lithic, Glacic, Petrogypsic, Gypsic, Nitric, Salic, Calcic, Typic
	Mollorthels	Lithic, Glacic, Vertic, Andic, Vitrandic, Cumulic, Aquic, Typic
	Umbrorthels	Lithic, Glacic, Vertic, Andic, Vitrandic, Cumulic, Aquic, Typic
	Argorthels	Lithic, Glacic, Natric, Typic
	Psammorthels	Lithic, Glacic, Spodic, Typic
	Haplorthels	Lithic, Glacic, Aquic, Typic

Histels are divided into five great groups, including the Folistels, Glacistels, Fibristels, Hemistels, and Sapristels. With the exception of the Glacistels, these great groups follow the suborders in the Histosol order (Section E, Chapter 6.2). The Glacistels contain a glacic layer within 50 cm of the surface that is 30 cm or more thick and contains 75% or more ice (by volume).

6.6.3.2 Turbels
Turbels are mineral soils that occur in areas with patterned ground. Patterned ground is a general term for any ground surface with a discernibly ordered, more or less symmetrical, morphological pattern of ground and, where present, vegetation (Washburn, 1980). They show marked influence of cryoturbation (Fig. 6.27B). These soils occur throughout the circumpolar regions and are the dominant suborder, accounting for more than 57% of the Gelisols mapped in the circumarctic (Fig. 6.26). Field and laboratory data for these soils are contained in publications by Douglas and Tedrow (1960), Day and Rice (1964), Holowaychuk et al. (1966), Everett (1968), Tedrow et al. (1968), Tedrow (1970), Zoltai and Tarnocai (1974), Pettapiece (1975), and Tarnocai and Smith (1992).

Most horizons and layers within the active layer of Turbels are strongly affected by cryoturbation. Turbels commonly contain irregular or broken horizons, involutions, organic matter that usually accumulates on the surface of the permafrost, oriented rock fragments, and silt caps and silt-enriched subsoil horizons. Permafrost occurs within 200 cm of the soil surface. These soils are differentiated into seven great groups that link them with other orders not containing permafrost, including the Histoturbels (40 to 80% organic materials by volume in the upper 50 cm), Aquiturbels (aquic conditions), Anhyturbels (anhydrous conditions) [Anhydrous conditions refer to soils of cold deserts

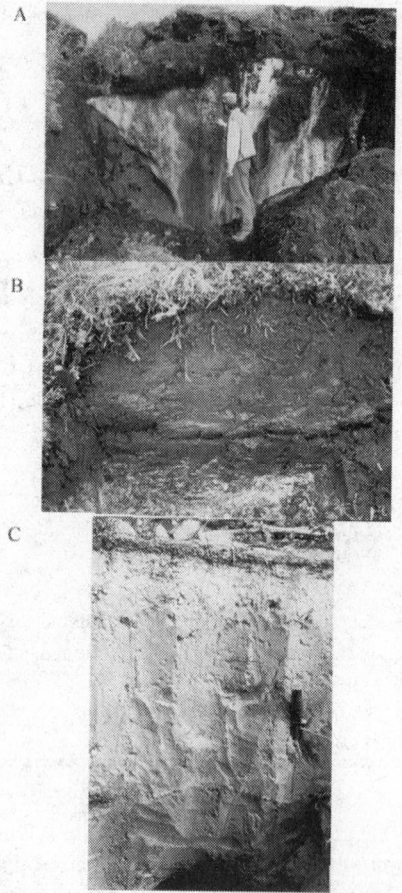

Fig. 6.27 Examples of the three suborders of Gelisols: (A) Histel, (B) Turbel, and (C) Orthel

and other areas with permafrost (often dry permafrost) and low precipitation (usually < 50 mm yr^{-1} water equivalent). Anhydrous soil conditions are similar to the aridic and torric soil moisture regimes except that the soil temperature is < 0 °C (Soil Survey Staff, 1998).], Molliturbels (mollic epipedon), Umbriturbels (umbric epipedon), Psammoturbels (sandy texture), and Haploturbels (other Turbels).

6.6.4 Orthels

Orthels are mineral soils containing permafrost within the upper 100 cm, but that lack cryoturbation (Fig. 6.27C). These soils comprise < 3.6% of the Gelisols mapped in the circumarctic and occur primarily in areas with dry permafrost such as floodplains and the dry valleys of Antarctica. They are divided into great groups that parallel the Turbels, i.e., Historthels, Aquorthels, etc. Subgroups within each great group of Gelisols are listed in Table 6.27.

6.6.5 Cryopedogenic Processes

Cryopedogenic processes that lead to gelic materials are driven by the physical volume change of water to ice, moisture migration along (1) thermal, (2) hydrostatic pressure, (3) solute concentration, and (4) electrical potential gradients in the frozen (or unfrozen) system (Marion, 1995), or thermal contraction of the frozen material by continued rapid cooling. These processes include freezing and thawing, cryoturbation, frost heaving, cryogenic sorting, thermal cracking, and ice segregation (Tedrow, 1977; Washburn, 1980) (Fig. 6.28).

It should be emphasized that cryopedogenic processes are soil-forming processes characteristic of soils with permafrost and should not be viewed as operating against the other soil-forming processes in lower latitude soils; rather, they are distinctive processes producing horizons and properties that are uncommon to other soil orders. Processes common to the other soil orders operate in Gelisols but at a lesser magnitude because of the dominance of cryopedogenic processes.

6.6.6   The Pedon as a Basic Soil Unit

The pedon is the basic soil unit for sampling in *Soil Taxonomy* (1975); it is an especially important concept for describing, classifying, and sampling Gelisols. The pedon is defined so as to encompass the full cycle of patterned ground with a 1 or 2 m linear interval or a half cycle with a 2 to 7 m cycle (Fig. 6.29A). This interval is suitable for most patterned ground features such as earth hummocks, circles, nets, and nonsorted polygons. In the case of large scale (> 7 m) ice-rich polygons, such as those that occur along the Alaskan Coastal Plain, two pedons are delineated: one within the center of the polygon and the other within the ice wedge (Fig. 6.29B). If no patterned ground exists, the pedon is arbitrarily selected but approximates about 1 m² in area.

Scaled sketches of a pedon showing soil horizons, including patches of cryoturbated material, should be drawn on graph paper in the field, or Polaroid photographs should be taken and the horizons annotated directly on the print (Fig. 6.24). Samples should be collected from each horizon across the pedon and composited for subsequent laboratory characterization. In the case of highly cryoturbated soils, the areal percentage of each horizon is reported in soil descriptions rather than depth intervals (Kimble et al., 1993).

6.6.7   Historical Approaches to Classification of Permafrost-Affected Soils

Tedrow (1977) provided a comprehensive review of early approaches to classification of soils in the cold regions, emphasizing the zonal systems. Of the soil taxonomic systems used today, only the Canadian (Agriculture Canada Expert Committee on Soil Survey, 1987) and United States (Soil Survey Staff, 1998) approaches recognize soils with permafrost in a separate order. The World Reference Base (WRB) for Soil Resources (Section E, chapter 5.4) is considering permafrost-affected soils in a separate soil unit (Tarnocai, 1997, personal communication). The proposed WRB system has received the preliminary approval of the international soil community, and is remarkably parallel to the Canadian and United States taxonomic systems with regard to permafrost-affected soils (Table 6.28).

Whereas permafrost-affected soils are identified as Gelisols (from the Greek word, *gelid*, meaning very cold) in *Soil Taxonomy*, they are called Cryosols (from the Greek word, *kraios*, meaning cold or ice) in the other two systems. Each system divides permafrost-affected soils into three categories: organic soils, cryoturbated mineral soils, and other mineral soils. The soils are further delineated on the basis of key properties that link them with soils of lower latitudes.

6.6.8   Properties of Gelisols

Gelisols/Cryosols encompass a vast array of soils in terms of chemical and mineralogical properties. However, with the exception of the Histels, Gelisols are uniform in terms of physical and thermal properties.

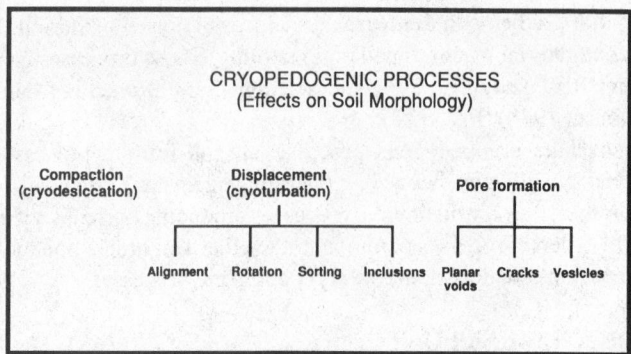

Fig. 6.28 Conceptual framework showing the interrelationships of the effects of cryopedogenic processes on development of specific fabric types in Gelisols [From Fox, 1994 with permission]

6.6.8.1 Macromorphology

The most common macroscopic soil features are due to cryoturbation and include irregular or broken horizons and incorporation of organic matter in lower horizons, especially along the top of the permafrost. Oriented stones and displacement of soil materials are common in Gelisols. Freezing and thawing produce granular and platy structures in surface horizons and blocky, prismatic or massive structures in subsurface horizons. The massive structure is due to cryostatic pressure and desiccation that develop when the two freezing fronts, one from the surface and the other from the permafrost, merge during freeze back in the autumn. The perennially frozen layer commonly contains ground ice in the form of segregated ice crystals, vein ice, ice lenses and wedges, and thick ground ice.

The granular, platy, or blocky structures of the surface mineral horizons are also the result of cryopedogenic processes such as the freeze-thaw process and vein-ice formation (ice segregation process). The subsurface horizons often have massive structures and are associated with higher bulk densities, especially in fine-textured soils.

Almost all Gelisols contain ice in the form of crystals, lenses, layers (vein ice), wedges, or massive ground ice, often to a thickness of several m. Soil texture is one of the factors controlling ice content in mineral soils with fine-textured Gelisols generally having higher ice contents than coarse-textured soils. Although coarse-textured soils have a relatively low ice content, they are often associated with ice wedges in the form of polygons. Histels have an ice content of 60 to 90% on a volume basis (Zoltai and Tarnocai, 1975).

The active layer that is subject to annual thawing and refreezing lies above the permafrost and not only supports biological life, but also protects the underlying permafrost. The thickness of the active layer is controlled by soil texture and moisture, thickness of the surface organic layer, vegetation cover, aspect and latitude.

Dilitancy, often confused with thixotropy, is common in soils with high silt content, greatly affects the trafficability of the soil, and is frequently present during the thaw period. Dilitancy is a property of granular masses expanding due to the increase of space between rigid particles upon displacement of the particles. When dilitant soils dry out, a characteristic vesicular porosity develops.

Salt crusts, patches, and pans are common in polar desert soils of the high arctic and cold desert soils of Antarctica (Tedrow, 1977; Bockheim, 1997).

6.6.8.2 Micromorphology

When viewed in thin sections, Gelisols contain a variety of fabrics resulting from desiccation and displacement due to alignment, rotation, sorting, and inclusions (Fig. 6.28). These features are accompanied by planar voids, cracks and vesicles during pore formation. More specifically, the fabric of Gelisols varies from granular (granic and granoidic) in the surface horizons to mainly porphyroskelic with fragmic and fragmoidic components in subsurface mineral horizons (Pawluk and Brewer, 1975; Fox, 1985; Smith et al., 1991). The micromorphology of Gelisols can show evidence of matrix displacement and movement, with resultant reorganization of skeleton grains into circular or elliptical patterns, producing an orbicular fabric (Fox and Protz, 1981). Ice lensing and vein-ice development lead to the formation of lenticular fabrics, while cryoturbation and cryodesiccation can lead to granic or granoidic fabrics (Smith et al., 1991). In addition, suscitic and conglomeric fabrics also are common in Gelisols (Fox and Protz, 1981).

6.6.8.3 Thermal Characteristics

Soil temperatures for a Gelisol located north of the arctic treeline in Canada are shown in Fig. 6.30. The unique thermal signature that separates Gelisols from all other soils is the presence of a perennially frozen layer, usually below 50 cm. Because of this frozen layer, Gelisols have a steep

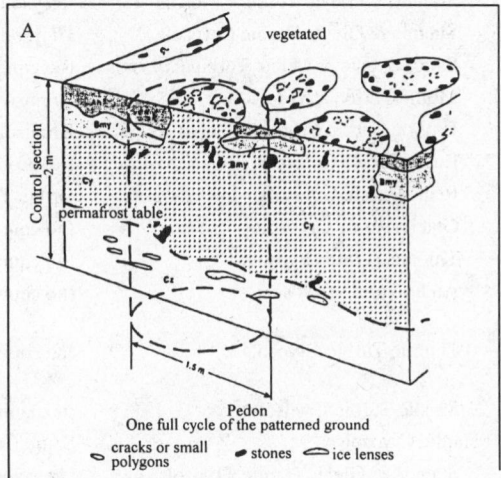

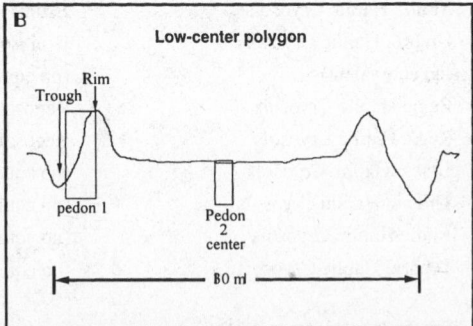

Fig. 6.29 The pedon concept as it applies to Gelisols and form of patterned ground, including (A) a nonsorted circle [Tarnocai and Smith, 1992] and (B) a large-scale, low-centered nonsorted polygon. The Canadian soil horizon nomenclature is used, where Ah = A, Bmy = Bwjj, Cy = Cjj, and Cz = Cf.

Table 6.28 Comparison of soil taxa among *Soil Taxonomy*, the Canadian system, and the proposed World Reference Base (FAO)

Soil Taxonomy (Soil Survey Staff, 1998)	World Reference Base	Canadian (Agriculture Canada Expert Committee on Soil Survey, 1987)
Gelisol (order)	Cryosol (soil unit)	Cryosol (order)
Histels (suborder)	Histic Cryosols (soil subunit)	Organic Cryosols (great group)
Folistels	(no equivalent)	(no equivalent)
Glacistels	Glacic Histic Cryosols	Glacic Organic Cryosols
Fibristels	Fibric Histic Cryosols	Fibric Organic Cryosols
Hemistels	Mesic Histic Cryosols	Mesic Organic Cryosols
Sapristels	Humic Histic Cryosols	Humic Organic Cryosols
[Terric Fibristels]	(no equivalent)	Terric Fibric Organic Cryosols
[Terric Hemistels]	(no equivalent)	Terric Mesic Organic Cryosols
[Terric Sapristels]	(no equivalent)	Terric Humic Organic Cryosols
Turbels	Turbic Cryosols	Turbic Cryosols
Histoturbels	Stagnic & Gleyic Turbic Cryosols	Gleysolic Turbic Cryosols
Aquiturbels	Stagnic & Gleyic Turbic Cryosols	Gleysolic Turbic Cryosols
Anhyturbels	Salic, Gypsic & Calcic Turbic Cryosols	(no equivalent)
Molliturbels	Mollic Turbic Cryosols	Brunisolic Turbic Cryosols
Umbriturbels	Umbric Turbic Cryosols	Brunisolic Turbic Cryosols
Psammoturbels	Regic Turbic Cryosols	Regosolic Turbic Cryosols
Haploturbels	Regic Turbic Cryosols	Orthic Turbic Cryosols
[Glacic subgroups]	Glacic Turbic Cryosols	(no equivalent)
[Sulfuric Aquaturbels]	Thionic Turbic Cryosols	(no equivalent)
[Andic Molli- & Umbriturbels]	Andic Turbic Cryosols	(no equivalent)
[Vitrandic Molli- & Umbriturbels]	Tephric Turbic Cryosols	(no equivalent)
[Spodic Psammoturbels]	Spodic Turbic Cryosols	(no equivalent)
Orthels	Haplic Cryosols	Static Cryosols
Historthels	Stagnic & Gleyic Haplic Cryosols	Gleysolic Static Crysols
Aquorthels	Stagnic & Gleyic Haplic Cryosols	Gleysolic Turbic Cryosols
Anhyorthels	Salic, Gypsic & Calcic Haplic Cryosols	(no equivalent)
Mollorthels	Mollic Haplic Cryosols	Brunisolic Static Cryosols
Umbrorthels	Umbric Haplic Cryosols	Brunisolic Static Cryosols
Argiorthels	(no equivalent)	(no equivalent)
Psammorthels	Regic Haplic Cryosols	Regosolic Static Cryosols
Haplorthels	Regic Haplic Cryosols	Regosolic Static Cryosols
[Glacic subgroups]	Glacic Haplic Cryosols	(no equivalent)
[Sulfuric Aquaorthels]	Thionic Haplic Cryosols	(no equivalent)
[Andic Molli- & Umbrorthels]	Andic Haplic Cryosols	(no equivalent)
[Vitrandic Molli- & Umbrorthels]	Tephric Haplic Cryosols	(no equivalent)
[Spodic Psammorthels]	Spodic Haplic Cryosols	(no equivalent)

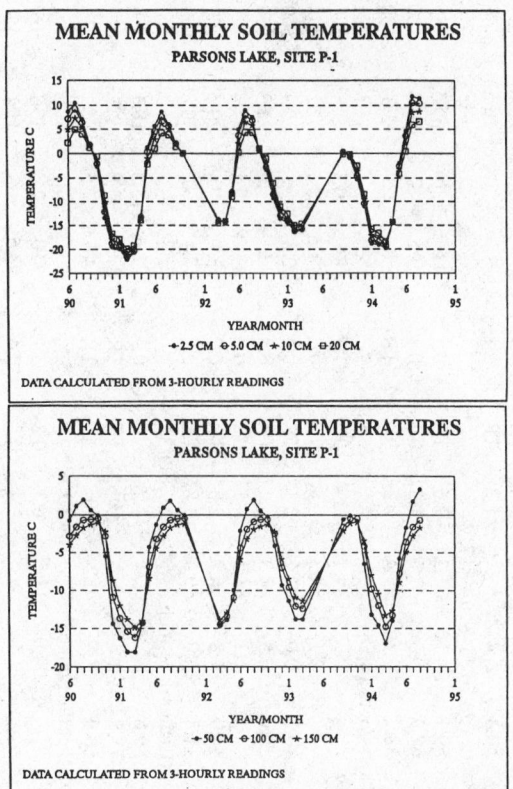

Fig. 6.30   Soil temperature variation in a Turbel from Parson Lake, Canada (68°58′09′N, 133°3254′W)

vertical temperature gradient. If these soils are associated with certain types of patterned ground, the horizonal temperature gradient can also be large. For example, in the case of Gelisols associated with earth hummocks, the soil temperature at the center of the hummock can decrease from 12 °C at the surface to 0 °C at 50 cm during the summer months. Soil temperatures at comparable depths under the interhummock depression, < 1 m away and at equivalent depths, can be 5 to 7 °C lower (Tarnocai and Zoltai, 1978).

6.6.8.4   Physical and Chemical Characteristics
Histels contain weakly to well-decomposed moss, sedge, woody and amorphous peats. The pH of these soils ranges from 2.5 to 7.0 (Table 6.29). Histels derived from moss peats (Fibristels) tend to have lower pH values than other organic soils with permafrost. Turbels and Orthels developed from calcareous parent materials have a pH_{CaCl_2} of 7 and a high base saturation. On the other hand, where these soils are developed from acidic parent materials, $pH_{CaCl_2} < 5.5$.

Anhyturbels and Anhyorthels in polar and cold deserts often have high salt contents and electrical conductivities. Although these soils have many features common to Aridisols (Section E, Chapter 6.10), they have patterned ground, dry permafrost in the upper meter over ice-cemented permafrost, and show evidence of thermal cracking (Bockheim, 1997).

Gelisols, especially the Histels and Turbels, often contain large amounts of organic C in the upper m, with values commonly ranging from 3 kg C m^{-3} in Haplorthels to over 100 kg C m^{-3} in Sapristels

Table 6.29 Analytical characteristics of selected Gelisols

Horizon	Depth	pH	BD	C	N	Clay	Silt	Sand	Extr. P	Ca	Mg	Na	K	Ex. Acid.	Ex. Al	CEC NH$_4$OAc	Base Sat.
	cm		Mg m^{-3}				%		mg L^{-1}				cmol$_c$ kg^{-1}				%
Typic Haploturbel; Dry Acidic Tundra (Empetrum-Betula-Dryas-Arctostaphylos); till/congelifractate; 68°37'N, 149°18'W																	
Ajj	2-46	4.91	0.56	8.68	0.53	9.8	53.2	37.0	0.7	2.4	1.2	TR†	0.1	39.4	11.5	36.1	10
Bwjj	4-40	4.83	0.87	2.3	0.13	18.4	46.2	35.4	1.2	0.7	0.8	TR	TR	25.2	9.1	20.5	8
BCjj	0-34	5.14	0.68	9.31	0.26	11.0	55.3	33.7	3.5	0.7	0.4	0.1	0.1	34.5	9.7	26.6	5
C	8-90	6.94	nd	0.58	0.05	5.2	44.4	50.4	5.4	7	1.2	TR	0.1	3.1	0	10.8	77
Typic Molliturbel; Dry Nonacidic Tundra (Carex-Dryas-Tomentypnum); loess/outwash; 70°11'N, 149°17'W																	
A	1-21	6.93	nd	5.66	0.37	14.0	26.8	59.1	0.1	61.3	2.6	TR	0.1	11.2	0	69.3	92
Oa/Ckjj	21-42	7.55	nd	15	0.96	16.0	16.1	67.9	0.5	47.8	1.2	0.1	0.1	3.6	0	30.3	100
2Ck	42-61	7.88	nd	2.37	0.04	5.7	6.1	88.2	0.1	35.4	0.8	0	TR	0	0	3.3	100
Ruptic-Histic Aquiturbel; Moist Nonacidic Tundra (Dryas-Salix-Eriophorum); loess/outwash; 69°27'N, 148°40'W																	
Oi	0-8	8.37	0.12	35.7	1.25				15.2	84.1	15.4	0.2	1.5	16	0.1	113	89
Oeij1	0-40	7.81	0.20	24.2	1.25				2.1	84.3	6.8	0.3	0.2	11.6	0.1	102	89
Bw	0-12	8.01	1.07	3.55	0.24	19.8	70.0	10.2	0.0	47.1	3.1	TR	0.1	0.8	0	22.1	100
Bg	0-88	7.92	1.19	3.8	0.29	23.5	66.9	9.6	0.5	36.3	2	TR	0.2	3	0	24.3	100
Oeij2	0-50	7.55	0.43	20.12	1.29				1.1	81.1	5.1	0.1	0.3	13.7	0.1	97.8	89
2Ajj	0-85	7.73	1.09	7.27	0.39	16.7	69.3	14.0	1.8	55.1	2.3	TR	0.2	3.7	0	29.5	100
2Cg/Oeijfm	65-85	7.40	nd	10.93	0.74	18.1	70.1	11.8	3.0	69.5	3.8	TR	0.2	7.4	0	50.8	100
Typic Umbrorthel; Moist Acidic Prostrate Shrub-Grass Tundra (Salix-Carex-Polytrichum); residuum; 68°46'N, 149°35'W																	
A	0-8	4.21	0.57	7.13	0.48	3.8	15.6	80.6	10.3	1.9	1.2	0	0.3	21.4	3.3	18.1	19
Bw	8-22	4.05	1.30	2.05	0.12	5.2	15.9	78.9	1.8	0.5	0	TR	TR	13.9	3.9	9.2	7
BC	22-31	4.64	1.38	0.44	0.05	1.2	11.9	86.9	7.5	0.2	0	0	TR	4.6	1.5	3	7
Typic Histoturbel; Moist Acidic Tussock Tundra (Eriophorum-Sphagnum-Betula); colluvium; 68°37'N, 149°19'W																	
Oi	0-9	4.46	0.04	44.34	0.85				34.0	11.5	6.3	0.8	3.1	110	2.5	136	16
Oe1	9-17	5.06	0.10	39.8	1.40				2.4	9.3	2.4	0.3	0.4	93.5		104	12
Oe2	17-23	5.37	0.37	24.3	1.34				1.2	3.9	1.6	0.1	0.3	63.2		59.9	10
Bg	23-40	5.17	1.46	3.04	0.11	20.0	41.9	38.1	0.4	0.5	0.4	TR	TR	14.7	4.2	15.8	6
BCg	40-48	5.70	1.09	3.46	0.14	20.5	39.8	39.7	0.6	0.5	0.4		TR	15	3.9	14.7	6
Cg/Oefm	48-80	6.39	nd	4.24	0.19	19.0	38.9	42.1	1.5	0.7	0.4	TR	TR	16	3.3	14	9
Typic Fibristel; Wet Acidic Tundra (Eriophorum-Sphagnum); organic basin deposits; 68°37'N, 149°19'W																	
Oi1	0-13	4.48	0.03	43.56	0.65				10.4	1.9	0.8	0.1	0.3	57.2	4.1	109	3
Oi2	13-33	4.52	0.19	44.82	2.58				4.0	8	2.6	0.1	0.2	107	4.8	77.1	14
Oe	33-44	4.66	0.35	25.2	1.17				0.6	3.4	0.7	TR	TR	75.2	6.9	51.3	8
Oa	44-62	4.99	0.52	18.5	1.02				0.5	3.6	0.7	TR	TR	58.2	9	46.6	9

†Tr = trace

and averaging about 50 kg C m^{-3} (Michaelson et al., 1996; Tarnocai, 1997). This is similar to Histosols of other regions (Section E, Chapter 6.2). Less than half of the organic C in Turbels is in the active layer, with the remainder in the upper 30 to 50 cm of permafrost (Michaelson et al., 1996).

The particle size distribution of Gelisols varies from clayey to coarse gravelly sand. The composition of the fine earth fraction is commonly dependent on the composition and age of the parent materials.

6.6.9   Special Problems in Managing Gelisols

Gelisols present special problems in terms of management, not only because of frost churning, heaving, sorting and cracking, but also because melting of segregated ice following a disturbance to the thermal regime leads to subsidence, or thermokarst. To preserve the integrity of structures (buildings, roads and pipelines) in permafrost soil, it is important to maintain the negative thermal balance of the soil. This is achieved by using special construction methods. For agricultural development it is important to determine the ice content of the soil; otherwise, after clearing of the land or within a few years after cultivation begins, severe subsidence and thermokarst can develop (Péwé, 1982).

6.6.9.1   Global Warming and Trace Gas Emissions

General circulation models (GCM) predict that with a projected twofold increase in atmospheric CO$_2$ by the year 2050, the mean air temperature of the Earth's surface could increase by 1.5 to 4.5 °C (Maxwell and Barrie, 1989); however, warming at the high northern latitudes could be on the order of 4 to 5 °C. Indeed, sea ice variations over the past several decades are compatible with a distinct warming of air temperatures in the arctic, especially during the winter and spring (Chapman and Walsh, 1993). In addition, permafrost temperatures in northernmost Alaska have increased by 2 to 4 C during the last few decades (Lachenbruch and Marshall, 1986).

As mentioned previously, Gelisols are large C sinks. The concern is that warming in the circumpolar regions could increase the thickness of the active layer and enhance heterotrophic respiration, releasing additional CO$_2$ to the atmosphere (Oechel and Billings, 1992; Shaver et al., 1992; Waelbroeck et al., 1997) and Gelisols would become a C source.

6.6.9.2   Human-Caused Disturbances

Arctic regions contain vast energy reserves, including fossil fuels (coal, oil and gas), biomass, and hydropower. The extraction of fossil fuels and minerals and deforestation may have dramatic long term effects on arctic ecosystems as they have a low resistance and resilience to disturbance (Reynolds and Tenhunen, 1996). As the world's population continues to expand, there will be increased pressure for development at the high latitudes. For example, Siberia already contains several cities that were developed in permafrost and have in excess of 500,000 inbabitants.

The circumarctic contains numerous indigenous peoples who are dependent on the terrestrial and marine ecosystems for food, fuel, and shelter. Special efforts must be made to ensure that these ecosystems remain sustainable.

The polar regions are less diverse biologically than other life zones. However, ancient (~ 2.5 million years old) permafrost contains viable microorganisms that may give clues to the evolution of microbes (Gilichinsky, 1993).

6.6.10   References

Agriculture Canada Expert Committee on Soil Survey. 1987. The Canadian System of Soil Classification. 2nd Ed. Research Branch, Agriculture Canada. Publ. 1646.

Black, R.F. 1954. Permafrost: a review. Geol. Soc. Am. Bull. 65:839–855.

Bockheim, J.G. 1997. Properties and classification of cold desert soils from Antarctica. Soil Sci. Soc. Am. J. 61:224–231.

Bockheim, J.G., C.L. Ping, J.P. Moore, and J.M. Kimble. 1994. Gelisols: a new proposed order for permafrost affected soils, p. 25–44. *In* J.M. Kimble and R.J. Ahrens (ed.) Proc. Mtg. on the Classification, Correlation, and Management of Permafrost-affected Soils. USDA-SCS, National Soil Survey Center, Lincoln, NE.

Chapman, W.L., and J.E. Walsh. 1993. Recent variations of sea-ice and air temperatures in high latitudes. Bull. Am. Meteor. Soc. 74:33–47.

Day, J.H., and H.M. Rice. 1964. The characteristics of some permafrost soils in the Mackenzie Valley, N.W.T. Arctic 17:223–236.

Douglas, L.A., and J.C.F. Tedrow. 1960. Tundra soils of arctic Alaska. Trans. 7th Intern. Congr. Soil Sci. IV:291–304.

Everett, K.R. 1968. Soil development in the Mould Bay and Isachsen areas, Queen Elizabeth Islands, Northwest Territories, Canada. OH State Univ. Inst. of Polar Studies, Rep. No. 24.

Everett, K.R. 1979. Evolution of the soil landscape in the sand region of the Arctic Coastal Plain as exemplified at Atkasook, Alaska. Arctic 32:207–223.

Fox, C.A. 1985. Micromorphology of an Orthic Turbic Cryosol-A permafrost soil, p. 699–705. *In* Bullock, P. and C.P. Murphy (ed.) Soil micromorphology. Vol. 2. Soil Genesis. AB Academic Publishers, Berkhamsted, UK.

Fox, C.A. 1994. Micromorphology of permafrost-affected soils, p. 51–62. *In* J.M. Kimble and R.J. Ahrens (ed.) Proc. Mtg. on the Classification, Correlation, and Management of Permafrost-affected Soils. July 1994. USDA-SCS, National Soil Survey Center, Lincoln, NE.

Fox, C.A., and R. Protz. 1981. Definition of fabric distributions to characterize the rearrangement of soil particles in the Turbic Cryosols. Can. J. Soil Sci. 61:29–34.

Gilichinsky, D.A. 1993. Viable microorganisms in permafrost: the spectrum of possible applications to new investigations, p. 268–270. *In* D.A. Gilichinsky (ed.), Proc. 1st Int. Conf. on Cryopedology. Russian Academy of Science, Pushchino, Russia.

Holowaychuk, N., J.H. Petro, H.R. Finney, R.S. Farnham, and P.L. Gersper. 1966. Soils of Ogotoruk Creek watershed, p. 221–273. *In* N.J. Wilmovsky and J.N. Wolfe (ed.) Environment of Cape Thompson Region, Alaska. U.S. Atomic Energy Commission, Oak Ridge, TN.

Kimble, J.M., B. Lacelle, Y.M. Naumov, D. Swanson, C. Tarnocai, and S. Waltman. 1997. Northern circumpolar database. Agriculture and Agri-Food Canada, USDA and V.V. Dokuchaev Soils Institute.

Kimble, J.M., C. Tarnocai, C.L. Ping, R. Ahrens, C.A.S. Smith, J. Moore, and W. Lynn. 1993. Determination of the amount of carbon in highly cryoturbated soils, p. 277–291. *In* Gilichinsky, D.A. (ed.) Post Seminar Proceedings of the Joint Russian-American Seminar on Cryopedology and Global Change. Nov. 1992. Russian Academy of Science, Pushchino, Russia.

Lachenbruch, A.H., and B.V. Marshall. 1986. Changing climate: geothermal evidence from permafrost in the Alaskan Arctic. Science 234:689–696.

Marion, G.R. 1995. Freeze-thaw processes and soil chemistry. U.S. Army Corps of Engineers, Cold Regions Research & Engineering Laboratory, Spec. Rep. 95–12, Hanover, NH.

Maxwell, J.B., and L.A. Barrie. 1989. Atmospheric and climate change in the Arctic and Antarctic. Ambio 18:42–49.

Michaelson, G.J., C.L. Ping, and J.M. Kimble. 1996. Carbon storage and distribution in tundra soils of arctic Alaska, U.S.A. Arctic Alpine Res. 28:414–424.

National Wetlands Working Group. 1988. Wetlands of Canada. Ecol. Land Classif. Ser. No. 24. Polyscience Publications, Montreal, Quebec.

Oechel, W.C., and W.D. Billings. 1992. Effects of global change on the carbon balance of arctic plants and ecosystems, p. 139–168. *In* Chapin, F.S. III, R.L. Jeffries, J.F. Reynolds, G.R. Shaver, and J. Svoboda (ed.) Arctic Ecosystems in a Changing Climate. Academic Press, San Diego, CA.

Pawluk, S., and R. Brewer. 1975. Micromorphological and analytical characteristics of some soils from Devon and King Christian Islands, N.W.T. Can. J. Soil Sci. 55:349–361.

Pettapiece, W.W. 1975. Soils of the subarctic in the lower Mackenzie basin. Arctic 28:35–53.

Péwé, T.L. 1982. Geological hazards of the Fairbanks area, Alaska. Alaska Geol. Surv. Spec. Rep. 15, Alaska Div. Geol. Geophys. Surv., College, AK.

Péwé, T.L. 1983. The periglacial environment in North America during Wisconsin time, p. 157–189. *In* Porter, S.C. (ed.) The Late Pleistocene-Late Quaternary Environment of the United States. University of Minnesota Press, Minneapolis, MN.

Reynolds, J.F, and J.D. Tenhunen. 1996. Ecosystem response, resistance, resilience, and recovery in arctic landscapes: progress and prospects, p. 419–428. *In* Reynolds, J.F. and J.D. Tenhunen (ed.) Landscape Function and Disturbance in Arctic Tundra. Ecol. Studies, Vol. 120. Springer-Verlag, Berlin, Germany.

Shaver, G.R., W.D. Billings, F.S. Chapin, A.E. Giblin, K.J. Nadelhoffer, W.C. Oechel, and E.B. Rastetter. 1992. Global change and the carbon balance of arctic ecosystems. BioSci. 42:433–441.

Smith, C.A.S., C.A. Fox, and A.E. Hargrave. 1991. Development of soil structure in some Turbic Cryosols in the Canadian Low Arctic. Can. J. Soil Sci. 71:11–29.

Soil Survey Staff. 1975. Soil taxonomy: A basic system of soil classification for making and interpreting soil surveys. USDA-SCS, Agric. Handb. 436, U.S. Government Printing Office, Washington, DC.

Soil Survey Staff. 1998. Keys to soil taxonomy. 8th Ed. USDA-NRCS, Washington, DC.

Tarnocai, C. 1972. Some characteristics of cryic organic soils in northern Manitoba. Can. J. Soil Sci. 52:485–496.

Tarnocai, C. 1973. Soils of the Mackenzie River area. Rep. No. 73–26, Task Force on Northern Oil Development QS-1528-000-EE-A1, Information Canada, Winnepeg, Canada.

Tarnocai, C. 1997. The amount of organic carbon in various soil orders and ecological provinces of Canada. *In* R. Lal, J.M. Kimble, and R.F. Follet (ed.) Soils and Global Changes. CRC Press, Boca Raton, FL.

Tarnocai, C., and C.A.S. Smith. 1992. The formation and properties of soils in the permafrost regions of Canada, p. 21–42. *In* Gilichinsky, D.A. (ed.), Proc. 1st Int. Conf. on Cryopedology. Nov. 1992. Russian Academy of Science, Pushchino, Russia.

Tarnocai, C., and S.C. Zoltai. 1978. Earth hummocks of the Canadian Arctic and Subarctic. Arctic Alpine Res. 10:581–594.

Tedrow, J.C.F. 1970. Soil investigations of Inglefield Land, Greenland. Meddelelser om Grønland 188:1–93.

Tedrow, J.C.F. 1977. Soils of the Polar Landscapes. Rutgers University Press, New Brunswick, NJ.

Tedrow, J.C.F., P.F. Bruggeman, and G.F. Walton. 1968. Soils of Prince Patrick Island. Res. Pap. 44, Arctic Institute of North America, Washington, D.C.

Waelbroeck, C., P. Monfray, W.C. Oechel, S. Hastings, and G. Vourlitis. 1997. The impact of permafrost thawing on the carbon dynamics of tundra. Geophys. Res. Let. 24:229–232.

Washburn, A.L. 1980. Geocryology. Wiley and Sons, Halstead Press, New York, NY.

Zoltai, S.C., and C. Tarnocai. 1971. Properties of a wooded palsa in northern Manitoba. Arctic Alpine Res. 3:115–129.

Zoltai, S.C., and C. Tarnocai. 1974. Soils and vegetation of hummocky terrain. Rep. No. 74–5, Task Force on Northern Oil Development, QS-1552-000-EEE-A1. Information Canada, Winnepeg.

Zoltai, S.C., and C. Tarnocai. 1975. Perennially frozen peatlands in the western arctic and subarctic of Canada. Can. J. Earth Sci. 12:28–43.

6.7 Vertisols

Clement E. Coulombe
Resource Management Group, Inc.

Larry P. Wilding
Texas A & M University

Joe B. Dixon
Texas A & M University

6.7.1 Introduction

Vertisols are clay soils which exhibit impressive volume change due to shrink/swell processes when soil moisture conditions change. The intrinsic shrink/swell behavior is considered the dominant process involved in Vertisols and results in significantly greater spatial and temporal variability of soil properties than in any other soil order. Thus, it is no surprise that Vertisols remain the most difficult and challenging type of land to successfully manage worldwide for natural resource and engineering applications.

Despite a voluminous literature, some aspects of the Vertisol order are still under debate or not fully understood. Nonetheless, considerable changes in the knowledge base of Vertisols have occurred, particularly in the last two decades. This subsection provides a condensed overview of Vertisols characteristics. For a more detailed discussion about any specific aspect of Vertisols, the reader is referred to Ahmad (1982), Probert et al. (1987), Wilding and Puentes (1988), Kimble (1991), Ahmad and Mermut (1996), and Coulombe et al. (1996b).

6.7.2 Distribution and Formation

Geographically, Vertisols have been reported in more than 100 countries. They are estimated to occupy 308.5 Mha (10^6 ha) which corresponds to 2.36% of the ice-free land area (Table 6.1) (USDA-SCS, 1994). World extent by suborder is discussed in Section 6.7.5 with Australia and India having the greatest extent with a respective estimate of 80 Mha each (Dudal and Eswaran, 1988; Isbell, 1991). Sudan follows with approximately 50 Mha and then the United States, China, and Ethiopia with 12–15 Mha each. The global distribution of Vertisols is illustrated in Fig. 6.31.

In the United States, Vertisols are reported in 25 states and territories (Coulombe et al., 1996b). More than half (~6.5 Mha) occurs in Texas with South Dakota, California, and Montana having 1.5 Mha, 1 Mha, and 0.6 Mha, respectively. The remaining states have less than 0.25 Mha of Vertisols and vertic intergrades (Fig. 6.32).

Vertisols can be derived from a wide variety of parent materials of varying age and under differing climatic conditions (Coulombe et al., 1996a,b). Most Vertisols have developed in recent times (Quaternary). Vertisols begin to develop at the time when the geomorphic surface is exposed and undergo surficial weathering processes. While most Vertisols typically occur at low elevation and on footslope to depressional positions in the landscape, some can occur at higher elevations and on higher slope gradients and positions on the geomorphic landscape. Natural vegetation of Vertisols is typically grassland and savanna, but mixed/deciduous forested and scrub/shrub can occur, depending on the ecological succession of the vegetation and human impact.

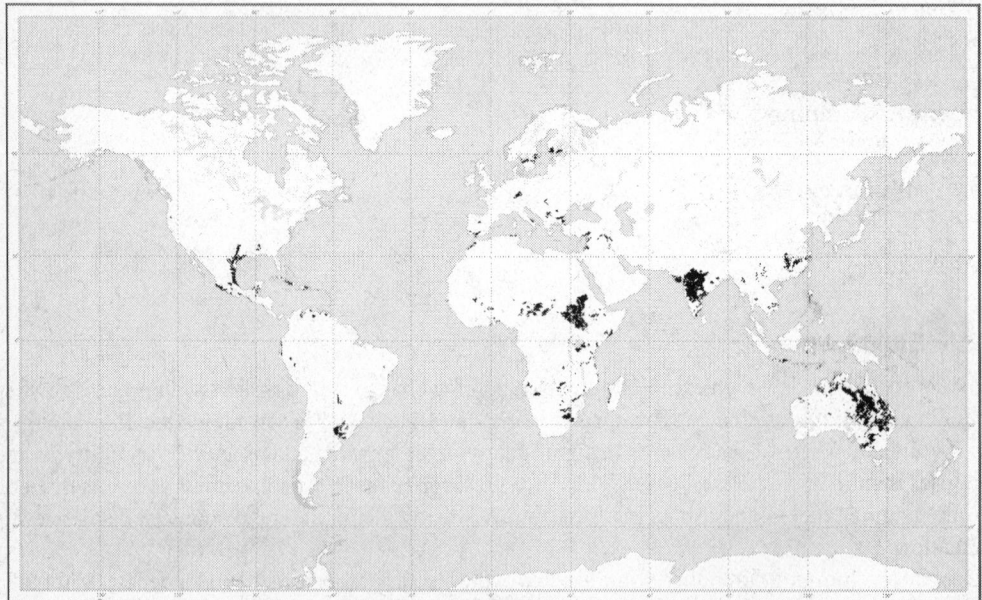

Fig. 6.31 Global distribution of Vertisols [Courtesy of USDA-NRCS, Soil Survey World Soil Resources, 1998]

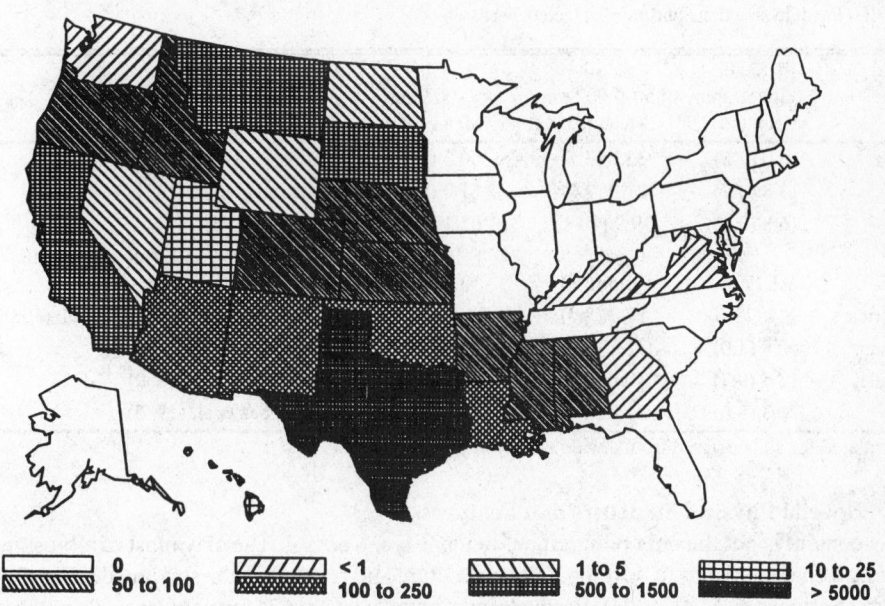

Vertisol Hectares x 1000

Fig. 6.32   Distribution of Vertisols in the United States [Reprinted from Coulombe et al., 1996. Overview of Vertisols. Adv. Agron. 57:289-375. Copyright Academic Press with permission]

6.7.3   Morphology

6.7.3.1   Color

In general, Vertisols are dark in color and commonly lack distinct horizonation. Vertisols with gray, brown, or red pigmentation have also been reported in various regions of the world. In the original version of *Soil Taxonomy* (Soil Survey Staff, 1975), color was considered at the great group category as *Pellic* and *Chromic* for low (< 1.5) and high (≥ 1.5) chroma color, respectively. In the most recent *Keys to Soil Taxonomy* (Soil Survey Staff, 1998), the color criterion is at the subgroup category by considering the chromic intragrade only, dark Vertisols being the norm. Chromic Vertisols require a color value of 4 or more (moist) or 6 or more (dry) or a chroma of 3 or more. The chroma requirement does not apply to the subgroup of Aquerts.

6.7.3.2   Texture

Vertisols are fine-textured soils. According to *Soil Taxonomy* (Soil Survey Staff, 1975; 1998), Vertisols must have: "a weighted average of 30% or more clay in the fine earth fraction either between the mineral soil surface and a depth of 18 cm or in an Ap horizon, whichever is thicker, and 30% or more clay in the fine earth fraction of all horizons between a depth of 18 cm and either a depth of 50 cm, or a densic, lithic, or paralithic contact, duripan, or petrocalcic horizon if shallower." Clay content of Vertisols can be as high as 90%, particularly for those derived from pyroclastic deposits. If the clay content requirement is not met, but the soil type exhibits considerable shrink/swell activities as measured by the coefficient of linear extensibility (COLE), these soils would be considered as vertic intergrades to other soil orders. Particle size distribution for selected Vertisols is listed in Table 6.30.

Table 6.30   Particle size distribution of selected Vertisols

Regions	% Sand 2.00-0.50 mm Mean (sd)	% Silt 0.50-0.002 mm Mean (sd)	% Clay < 0.002 mm Mean (sd)	References
Australia	16.0 (2.5)	33.6 (3.7)	50.3 (5.5)	Boettinger (1992)
India	13.8 (3.7)	21.9 (2.6)	61.0 (5.2)	Hirekurubar et al. (1991)
Sudan	16.8 (8.4)	19.9 (10.1)	63.1 (9.7)	Blokhuis (1993)
Texas	9.6 (7.3)	36.3 (6.9)	53.5 (3.7)	Kunze et al. (1963); Yule and Ritchie (1980a); Hallmark et al. (1986)
West Africa	18.4 (11.6)	27.7 (8.0)	54.0 (15.4)	Beavington (1978); Yerima et al. (1988)
El Salvador	10.8 (4.0)	31.2 (7.3)	58.1 (8.5)	Yerima et al. (1985)
West Indies	17.5 (9.7)	10.6 (3.0)	74.9 (11.4)	Ahmad and Jones (1969)
Uruguay	26.3 (5.6)	27.0 (8.1)	46.7 (13.3)	Lugo-Lopez et al. (1985)

Data calculated for soils derived from a wide variety of parent materials.

6.7.3.3   Special Physical Structure and Features

The clay content is not the only requirement to identify a Vertisol. The clay must exhibit significant shrink/swell characteristics to develop unique structure and features in the soil profile (Fig. 6.33). By definition (Soil Survey Staff, 1998), Vertisols must have "a layer of 25 cm or more thick, with an upper boundary within 100 cm of the mineral soil surface that has either slickensides close enough to intersect" or "wedge shaped aggregates which have their long axes tilted 10 to 60 degrees from the horizontal." Slickensides are features which exhibit shiny and smooth surfaces at the interface of the peds (Fig. 6.34). The presence of slickensides in a soil horizon is designated in the field as Bss and is the diagnostic characteristic common to all Vertisols.

Wedge-shaped aggregates are structural units that generally result from the intersection of slickensides. The presence of cracks (Fig. 6.35) that open and close in response to drying and wetting is also a requirement for Vertisols (Soil Survey Staff, 1975; 1998). Since cracking occurs in soils other than Vertisols, criteria for depth and duration when cracks remain open are the basis to discriminate and infer soil moisture regimes of Vertisols.

Gilgai and diapir are other features which may be present in Vertisols. Gilgai is an Australian aboriginal term to describe a microtopography that consists of mounds, shelves, and/or depressions, either in circular, linear, or complex patterns (Fig. 6.36). Since gilgai does not occur in all Vertisols, it was abandoned as a facultative criterion in *Soil Taxonomy* (Soil Survey Staff, 1998).

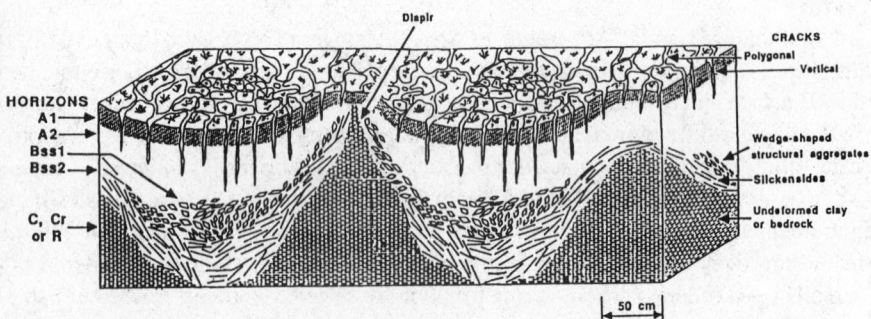

Fig. 6.33   Schematic of morphological structure and features of vertisols [Reprinted from Coulombe et al., 1996. Overview of Vertisols. Adv. Agron. 57:289-375. Copyright Academic Press with permission]

Fig. 6.34 Slickenside in a Vertisol of Texas [Reprinted from Coulombe et al., 1996. Overview of Vertisols. Adv. Agron. 57:289-375. Copyright Academic Press with permission]

Diapir, also called mukkara by the Australians, is another feature which may occur in Vertisols. This feature is a protrusion of soil material which penetrates from the subsoil through the upper layers and can be identified by a contrast in color and/or texture (Fig. 6.37). Diapir may occur independent of gilgai; however, when a gilgai is present, the tip of the diapir appears under the mound.

6.7.4 Genesis

Genesis of Vertisols which leads to the formation of characteristic features and structure is attributed to shrink/swell processes induced by changes in soil moisture conditions and can be seen at macroscopic to submicroscopic scales of resolution (Fig. 6.38). For decades, shrink/swell phenomena were attributed to the paradigm of the interlayer expansion/collapse of crystallographic structure of

Fig. 6.35 Typical polygonal cracking pattern of a Vertisol of Mississipi [Courtesy of Dr. Dennis Mengel, CH2M HILL]

Fig. 6.36 Circular gilgai under ponded conditions in a Vertisol of Texas [Reprinted from Coulombe et al. 1996. Overview of Vertisols. Adv. Agron. 57:289-375. Copyright Academic Press with permission]

clay minerals such as smectites. However, changes in the interlayer spacing of smectites contribute only 10–30% of the change in volume when saturated with Ca (Greene-Kelly, 1974; Tessier, 1984). Tessier (1994) reported that the type of microstructure, e.g., crystallite, domains, or quasicrystals (Fig. 6.38) and the surface area of the clay mineral phases govern the shrink/swell potential. The microstructure and the porosity will accommodate the changes in water content and potential of the clay water systems. Other determinant factors involved in the shrink/swell phenomenon include, but are not limited to, predominance of cations present on the exchange sites, the electrolyte concentration of the soil solution, the amount and location of the charge in the crystallographic structure of clay minerals which may impact the structural organization of the layers and the interlayer distance which may differ from smectite to smectite, particle rigidity, and stress history, etc. Models and factors involved in shrink/swell processes have been extensively reviewed by Tessier (1984), Quirk (1994), and Coulombe et al. (1996a).

Despite advances in the knowledge of clay-water behavior, controversy remains regarding the theoretical models that can account for the genesis of various structures and features (Coulombe et al., 1996b). Coulombe et al. (1996b) also reviewed some of the models proposed for the genesis of

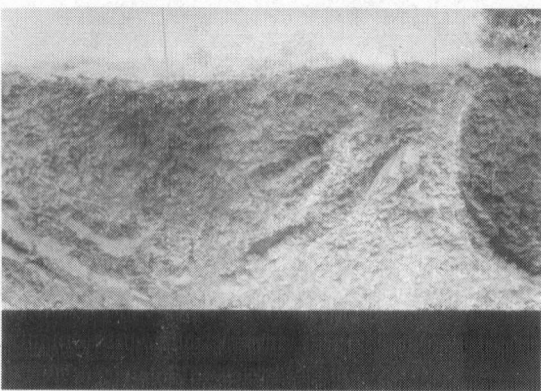

Fig. 6.37 Cross-sectional view of a Vertisol of Texas showing the diapiric structure [Reprinted from Coulombe et al. 1996. Overview of Vertisols. Adv. Agron. 57:289-375. Copyright Academic Press with permission]

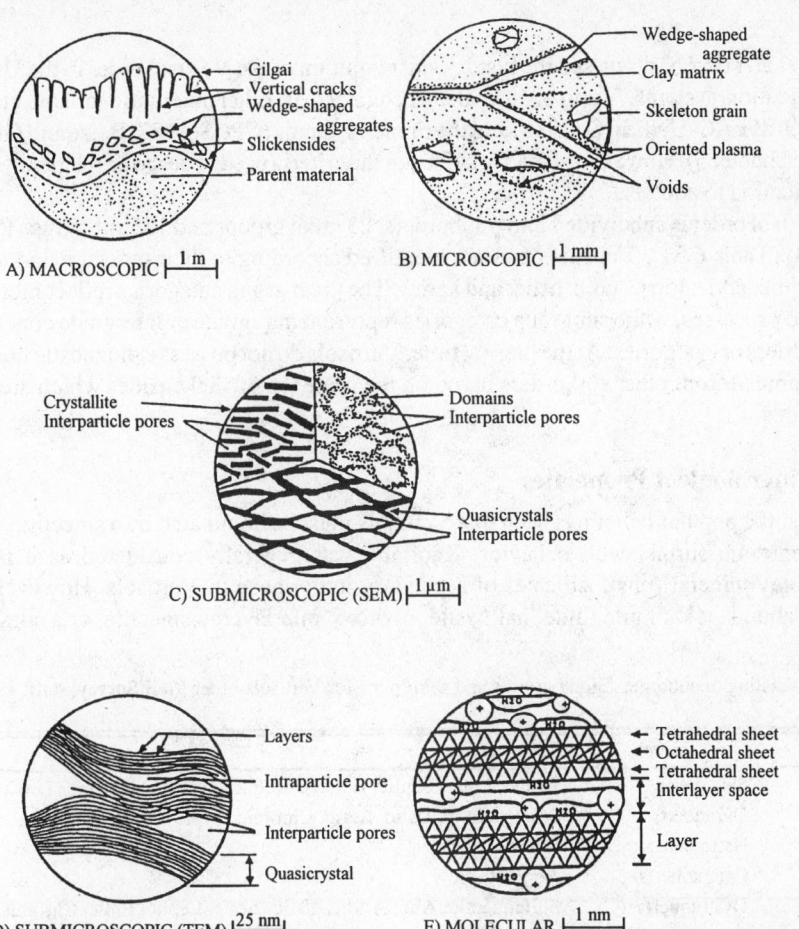

Fig. 6.38 Shrink/swell features which occurred at different scales of resolution [Reprinted from Coulombe et al., 1996. Overview of Vertisols. Adv. Agron. 56:289-375. Copyright Academic Press with permission]

features and structure in Vertisols. At the present time, the application of the Coulomb-Mohr theory of shear failure constitutes the best model to account for the formation and occurrence of features and structure in Vertisols. Shear failure is more likely to occur under moist or wet than dry conditions, since a lower normal stress is required to promote shear failure. Even though shrink/swell phenomena are significant in Vertisols, other pedogenic processes occur such as accumulation of salts (e.g., salic, natric, calcic, and gypsic great groups), silica (duric great group) and depth of saturation by water (epi and endo great groups).

6.7.5 Classification

Vertisol is a globally accepted term to designate fine-textured soils with extensive shrink/swell behavior. The term originates from *vertere* (L) meaning to churn or turn over and *sol* for soils (Dudal and Eswaran, 1988). It was first introduced as a soil order in the 7th Approximation (Soil Survey Staff, 1960) which was superceded by *Soil Taxonomy* as an international soil classification system.

Coulombe et al. (1996b) discussed the history and evolution of the Vertisol order in the United States soil classification systems. Vertisols are also recognized in other international soil classification systems such as FAO-UNESCO (1988) and the French systems (CPCS, 1987; Baize and Girard, 1996) (Section E, Chapter 5). However, some variations in the criteria used to describe them may occur from one classification to another.

The Vertisol order is subdivided into 6 suborders, 23 great groups and 155 subgroups (Soil Survey Staff, 1998) (Table 6.31). The suborders are identified according to their moisture and temperature regimes (aquic, cryic, torric, udic, ustic, and xeric). The great group categories reflect morphogenetic markers and processes, while subgroup categories represent intragrade or intergrade concepts toward other soil orders or categories. At the present time, Vertisols do not possess a diagnostic horizon. They are discriminated from other soil orders based on the presence of slickensides which are diagnostic characteristics.

6.7.6   Mineralogical Properties

For decades, the popular belief has been that Vertisols must be dominated by a smectitic mineralogy based on extreme shrink/swell behavior. Kaolinite was generally considered as a nonswelling accessory clay mineral (phyllosilicate) of secondary importance in Vertisols. However, Vertisols containing abundant kaolinite, illite, halloysite, hydroxy-interlayered smectite, or a combination of

Table 6.31   Listing of suborders, great groups and subgroups for Vertisols order [Soil Survey Staff, 1998]

Suborders	Great groups	Subgroups
Aquerts	Salaquerts	Aridic, Ustic, Leptic, Entic, Chromic, Typic
	Duraquerts	Aridic, Xeric, Ustic, Aeric, Chromic, Typic
	Natraquerts	Typic
	Calciaquerts	Aeric, Typic
	Dystraquerts	Sulfaqueptic, Alic, Aridic, Ustic, Aeric, Leptic, Entic, Chromic, Typic
	Epiaquerts	Halic, Sodic, Aridic, Xeric, Ustic, Aeric, Leptic, Entic, Chromic, Typic
	Endoaquerts	Halic, Sodic, Aridic, Xeric, Ustic, Aeric, Leptic, Entic, Chromic, Typic
Cryerts	Humicryerts	Sodic, Typic
	Haplocryerts	Sodic, Chromic, Typic
Xererts	Durixererts	Halic, Sodic, Aquic, Aridic, Udic, Haplic, Chromic, Typic
	Calcixererts	Lithic, Aridic, Petrocalcic, Leptic, Entic, Chromic, Typic
	Haploxererts	Lithic, Halic, Sodic, Aridic, Aquic, Udic, Leptic, Entic, Chromic, Typic
Torrerts	Salitorrerts	Aquic, Leptic, Entic, Chromic, Typic
	Gypsitorrerts	Chromic, Typic
	Calcitorrerts	Petrocalcic, Leptic, Entic, Chromic, Typic
	Haplotorrerts	Halic, Sodic, Leptic, Entic, Chromic, Typic
Uderts	Dystruderts	Lithic, Salic, Aridic, Udic, Aquic, Leptic, Entic, Chromic, Typic
	Hapluderts	Lithic, Aquic, Oxyaquic, Leptic, Entic, Chromic, Typic
Usterts	Dystrusterts	Lithic, Aquic, Aridic, Udic, Leptic, Entic, Chromic, Typic
	Salusterts	Lithic, Sodic, Aquic, Aridic, Leptic, Entic, Chromic, Typic
	Gypsiusterts	Lithic, Halic, Sodic, Aridic, Udic, Leptic, Entic, Chromic, Typic
	Calciusterts	Lithic, Halic, Sodic, Petrocalcic, Aridic, Udic, Leptic, Entic, Chromic, Typic
	Haplusterts	Lithic, Halic, Sodic, Petrocalcic, Aridic, Leptic-Udic, Entic-Udic, Chromic-Udic, Udic, Leptic, Entic, Chromic, Typic

interstratified clay minerals have been also found in various regions of the world (Coulombe et al., 1996a). Recognition of Vertisols with these characteristics has caused geoscientists to reconsider the Vertisol paradigm and mechanisms involved in the shrink/swell processes of soils not dominated by smectite. As previously mentioned, comprehension of shrink/swell behavior is beyond the crystallographic level, involving changes in microstructure and porosity associated with changes in water content and potential. Only under very specific conditions will the expansion/collapse of the interlayer space of clay minerals play a role in shrink/swell phenomena (Tessier, 1984; Wilding and Tessier, 1988; Coulombe et al., 1996a).

Nonetheless, smectites remain a universal and significant component of most Vertisols. Characteristics of smectites in Vertisols, which may also apply to other clay minerals and other soil orders, include (1) a greater thermodynamic stability than reference minerals; (2) a higher layer charge (> 0.45 per half unit cell) which may be as high as vermiculite or illite and in return impact the availability of some ions such as K and NH_4; and (3) a higher Fe content in their crystallographic structure, particularly when derived from basaltic parent materials containing abundant ferromagnesian minerals (Coulombe et al., 1996a). Despite the importance of clay minerals, a broad variety of minerals have been reported in Vertisol profiles and parent materials (Coulombe et al., 1996a).

6.7.7 Chemical Properties

6.7.7.1 pH

Vertisols are commonly known as being neutral to alkaline in reaction, since the majority are derived from calcareous or base-rich parent materials. However, acid Vertisols (Ahmad, 1985) exist and have been recognized as the Dystric great group (Soil Survey Staff, 1998). Dystric Vertisols must exhibit pH values of 4.5 or less in $0.01M$ $CaCl_2$ (5 or less in 1:1 water) and an electrical conductivity of the saturation extract of less than 4.0 dS m^{-1} at 25 °C with a total thickness of 25 cm or more within 50 cm of the mineral soil surface. Processes leading to acidification include removal of carbonates and bases by leaching or due to chemical processes such as (1) hydrolysis, (2) acid sulfate weathering, which is the result of the oxidation of sulfidic minerals such as pyrite to sulfuric acid and sulfate minerals, and (3) ferrolysis, which involves destabilization of crystallographic structure of clay minerals by proton attack. Ranges in pH of selected Vertisols are listed in Table 6.32.

6.7.7.2 Cation Exchange Capacity and Exchangeable Cations

Cation exchange capacity (CEC) of Vertisols generally ranges from 20 to 45 cmol$_c$ kg^{-1} (soil) or higher depending on the amount and type of mineral phases and soil organic matter (SOM) (Table 6.32). Calcium and Mg, and to a lesser extent, K and Na are the main exchangeable cations. When exchangeable Na percentage (ESP) > 15, structural deterioration occurs. Soil dispersion may occur at low ESP values when electrolyte concentration is low while at higher electrolyte concentrations, and even ESP, soil particles may remain flocculated (Section G, Chapter 2). However, in the case of high electrolyte concentration, the osmotic potential gradient between soil and plant roots increases and induces wilting and eventually death of the plants that are not tolerant to these saline conditions. Presence of some minerals such as gypsum, even in trace concentrations, can be sufficient to overcome the dispersing power of Na and flocculate soil particles.

Other cations such as Al, Mg, and H may increase and replace exchangeable Ca, particularly under acid conditions due to dissolution of minerals (Ahmad, 1985). An increase in the concentration of these cations and a decrease in CEC indicates that chemical degradation has occurred.

Table 6.32 Chemical characteristics of selected Vertisols

Regions	pH (1:1water) Mean (sd)	CEC (cmol$_c$ kg^{-1} soil) Mean (sd)	Extractable Bases (cmol$_c$ kg^{-1} soil)				References
			Ca Mean (sd)	Mg Mean (sd)	K Mean (sd)	Na Mean (sd)	
Australia	7.0 (0.6)	33.7 (2.5)	22.4 (3.4)	15.7 (3.4)	0.1 (0.1)	0.9 (0.4)	Boettinger (1992)
India	8.2 (0.2)	55.8 (4.7)	53.4 (5.6)	*	0.9 (0.1)	1.4 (0.5)	Hirekurubar et al. (1991)
Sudan	7.3 (0.4)	40.6 (11.0)	n/a	n/a	n/a	2.95 (3.9)	Blokhuis (1993)
Texas	7.4 (0.9)	41.0 (5.4)	52.9 (23.0)	6.8 (3.5)	0.7 (0.3)	2.7 (3.4)	Kunze et al. (1963); Hallmark et al. (1986)
West Africa	7.4 (0.8)	30.3 (8.0)	17.9 (7.9)	7.5 (3.1)	0.7 (0.4)	3.2 (3.3)	Beavington (1978); Yerima (1986)
El Salvador	6.7 (0.6)	45.5 (3.4)	33.9 (10.3)	11.6 (2.5)	0.5 (0.2)	1.9 (1.5)	Yerima et al. (1985)
West Indies	7.3 (0.6)	48.0 (11.7)	51.8 (23.6)	8.1 (5.8)	0.4 (0.3)	2.4 (2.1)	Ahmad and Jones (1969)
Uruguay	6.5 (0.5)	33.9 (7.9)	23.1 (7.5)	8.0 (2.3)	0.3 (0.1)	0.3 (0.1)	Lugo-Lopez et al. (1985)

n/a: not available; * Extractable Mg was compiled with data for extractable calcium.
Data were calculated for soils derived from a wide variety of parent materials.

6.7.8 Biological Properties

Biological properties involve the quantity and type of organic matter as they relate to pedofauna and land use (Table 6.33). Generally, organic C content ranges from 3 to 60 g kg^{-1} but most commonly from 3 to 35 g kg^{-1} (Coulombe et al., 1996a). The SOM varies with (1) climatic conditions (higher under humid and cooler regimes), (2) microtopography (higher under depressional positions than under mound positions of a gilgai) (Kunze and Templin, 1956; Wilding and Tessier, 1988), (3) depth (decreases with depth in the soil profile, which strongly refutes the argument that complete self-mixing occurs in Vertisols) (Wilding and Tessier, 1988; Southard and Graham, 1992; Coulombe, 1997), and (4) tillage (higher in virgin than cultivated areas) (Kunze and Templin, 1956; Skjemstad and Dalal, 1987; Coulombe, 1997).

Inferences of SOM content based on soil color can be misleading. Black Vertisols may contain as low as 3 g kg^{-1} of organic C. The dark color is the result of strong clay organic complexes in which smectites are coated with humic materials in the presence of Ca (Singh, 1954, 1956). The resistance to degradation is due to this complex rather than the recalcitrance of the humic materials to degradation (Skjemstad et al., 1986; Skjemstad and Dalal, 1987).

Pedofauna in Vertisols are generally diversified (Stephan et al., 1983). Humus type is a mull with a C/N ratio ranging from 10–25 (Table 6.33), unless the soil conditions are poorly drained which may lead to the formation of a hydromull or hydromoder. Microbial biomass may represent up to 5% of total organic C (Jocteur Monrozier et al., 1991). The loci of the organic matter such as polysaccharides and fungal hyphae are at interparticle positions in the domains and quasicrystals of clay minerals (Chenu, 1989; Chenu and Jaunet, 1990).

Loss of organic C can occur shortly after a change in land use systems from native/virgin conditions to cultivation. After a few decades or more of continuous cultivation, loss of organic C can be significant (Skjemstad et al., 1986; Skjemstad and Dalal, 1987; Puentes, 1990; Puentes and Wilding, 1990; Coulombe, 1997).

Table 6.33 Biochemical and biological characteristics of surface horizons of selected Vertisols

Regions	Soil Series (and/or Parent Material)	Land Use	OC (g kg^{-1})	Total N (g kg^{-1})	C/N ratio	Total P (mg kg^{-1})	References
Australia	Langlands-Logie clays	Virgin	23.0	2.41	9.5	n/a	Skjemstad et al. (1986)
	(Alluvial clayey sediments)	20 yr cultivation	11.4	1.12	10.2	n/a	Skjemstad et al. (1986)
		35 yr cultivation	8.9	0.89	10.0	n/a	Skjemstad et al. (1986)
		45 yr cultivation	7.8	0.77	10.1	n/a	Skjemstad et al. (1986)
	Black Earth	10 yr pasture	14.2	0.9-2.6	5.8-12.2	n/a	Jocteur Monrozier et al. (1991)
Texas	Houston Black	Virgin	20.9-48.2	1.37-3.53	13.69-21.10	0.35-0.55	Coulombe (1997)
		20 yr restoration	13.7-24.0	0.87-1.45	18.37-19.81	0.23-0.29	Coulombe (1997)
		Pasture	15.5-33.3	0.97-2.02	18.18-25.83	0.29-0.44	Coulombe (1997)
		Cereal rotation	12.2-15.6	0.83-1.03	14.74-20.28	0.34-0.43	Coulombe (1997)
		Row crop rotation	15.6-19.3	0.96-1.45	13.36-19.39	0.29-0.48	Coulombe (1997)
Nigeria	Lacustrine Plain	Cereal	0.26-0.58	0.06-0.08	5.2-7.2	5-10	Beavington (1978)
Trinidad	Princes Town Clay	> 100 yr Sugar cane	31.0-38.0	2.7-3.5	11	29-58	Ahmad and Jones (1969)
Jamaica	Carron Hall Clay	Unimproved pasture	42.0-60.0	4.5-5.7	9-11	13-15	Ahmad and Jones (1969)
Barbados	Black Soil	> 300 yr Sugar cane	12.0-15.0	0.8-1.0	15	250-500	Ahmad and Jones (1969)
Antigua	Fitches Clay	Mixed scrubs/grasses	32.0-38.0	2.2-2.3	15-17	13-17	Ahmad and Jones (1969)
Dominica	Black Soil	Scrub	11.0-28.0	1.0-2.1	11-13	2-4	Ahmad and Jones (1969)

n/a: not available

6.7.9 Physical Properties

6.7.9.1 Moisture Retention and Shrink/Swell Behavior

Soil moisture conditions are extremely critical in Vertisols because they govern all physical properties and behavior. Due to their clay content, Vertisols exhibit high moisture retention characteristics at all water potentials. Plant-available water in Vertisols is determined as the difference between water held at −0.033 and −1.5 MPa. Ranges in moisture retention values for selected Vertisols are presented in Table 6.34. Moisture content can vary considerably depending on the amount and kind of clay, climatic conditions, and land use.

Shrink/swell behavior severely limits the utilization of Vertisols. Under normal field conditions, most of the volume change occurs at water potentials > −1.5 MPa (Yule and Ritchie, 1980a; Wilding and Tessier, 1988), but it is not clear whether volume change is isotropic (equidimensional) or anisotropic. Current knowledge suggests that individual undisturbed peds shrink and swell isotropically (Yule and Ritchie, 1980a; Cabidoche and Voltz, 1995). Yule and Ritchie (1980b) reported that shrinkage in large (73 cm diam, 140 cm long) and small (10 cm diam) cores of a Texas Vertisol was similar. However, under field conditions, Cabidoche and Voltz (1995) reported anisotropic volume change in a Vertisol from Guadeloupe (French West Indies) with vertical slightly larger than the horizontal movements (anisotropy ratio ≈ 0.8). They attributed the anisotropy to movement of the peds along oblique slickensides, which highlights the importance of spatial arrangement or geometry of the peds in a soil profile on shrink/swell behavior.

6.7.9.2 Bulk Density and Coefficient of Linear Extensibility

Bulk density values for selected Vertisols are presented in Table 6.34. Factors that influence bulk density include (1) moisture content, (2) clay content, but more specifically the particle density of the mineral fraction, and (3) methodology used to determine bulk density (core vs. Saran® coated clods). Bulk density is routinely used to estimate total porosity and assess structural conditions of soils, but in several cases, bulk density has not been a sensitive indicator of structural changes in Vertisols. Bulk density should be coupled with morphological characteristics and other properties for a thorough assessment of soil structural conditions.

COLE, which is an index of soil shrinkage, is correlated with total and fine clay contents, surface area, water retention at field capacity, and ESP (Anderson et al., 1973; Yule and Ritchie, 1980a; Yerima et al., 1989). COLE values for selected Vertisols are presented in Table 6.34.

6.7.9.3 Consistence and Atterberg Constants

Vertisols exhibit very narrow ranges of soil moisture and consistence favorable for workability and trafficability. Consistence is estimated from the pressure required to deform a soil clod placed between the index finger and thumb. Consistence of Vertisols varies from friable to firm when moist, and from firm to extremely firm when dry. When wet, Vertisols become very plastic and sticky. Thus, determination of optimal moisture conditions is extremely critical for proper timing of cultivation and the energy/power required.

The Atterberg constants (liquid limit, plastic limit, and plasticity index) (Section A, Chapter 2) are empirical tests used by geoscientists and engineers to determine the mechanical characteristics of a

Table 6.34 Physical and mechanical characteristics of selected Vertisols

Regions	% Water 0.033 MPa Mean (sd)	% Water 1.5 Mpa Mean (sd)	Bulk Density 0.033 MPa ($Mg\ m^{-3}$) Mean (sd)	Bulk Density Oven Dry ($Mg\ m^{-3}$) Mean (sd)	COLE ($cm\ cm^{-1}$) Mean (sd)
India	38.3 (2.9)	17.8 (0.9)	n/a	1.3 (0.1)	n/a
Texas	42.2 (3.2)	23.4 (2.3)	n/a	1.8 (0.2)	n/a
West Africa	31.4 (6.5)	n/a	1.4 (0.1)	1.4 (0.1)	0.10 (0.03)
El Salvador	n/a	n/a	1.1 (0.1)	1.1 (0.1)	0.16 (0.03)

	Liquid Limit (%) Mean (sd)	Plastic Limit (%) Mean (sd)	Plasticity Index (%) Mean (sd)	References
India	61.0 (2.4)	34.0 (1.2)	26.9 (1.3)	Hirekurubar et al. (1991)
Texas	67.1 (3.8)	22.7 (4.3)	44.3 (3.6)	Kunze et al. (1963); Yule and Ritchie (1980a)
West Africa	n/a	n/a	n/a	Yerima (1986)
El Salvador	n/a	n/a	n/a	Yerima et al. (1985)

n/a: not available
Data calculated for soils derived from a wide variety of parent materials.

soil. Values for Vertisols are presented in Table 6.34. In general, values of Atterberg constants (1) increase as the clay content increases, and (2) decrease in the order: smectite > illite > kaolinite.

6.7.9.4 Structure/Porosity and Related Properties

Structure and porosity are extremely critical components of Vertisols. Soil processes such as gas exchange, heat transfer, water flow, and movement of solutes and organics are dependent on structure and porosity. One of the typical characteristics of Vertisols is the propensity to form large pores (macro- and mesopores) through wetting/drying cycles and/or biotic activities. The presence of macro/mesopores can be fairly significant for various processes that occur in Vertisols. For instance, they allow greater nitrification rates than in the rest of the soil mass (Kissel et al., 1974) and promote movement of water and solutes through preferential or bypass flow (Kissel et al., 1973, 1974; Bouma, 1983; Bouma and Loveday, 1988). Lin et al. (1996) reported, for a Texas Vertisol, that macropores (> 0.5 mm effective diam.) and mesopores (0.06 mm – < 0.5 mm effective diam.) contributed 89% and 10% of the water flow, respectively. Numerous factors influence infiltration. Infiltration rates reported for selected Vertisols are presented in Table 6.35. Because in shrink/swell clay soils, hydraulic conductivity (K) does not meet the assumption of uniform flow in a homogenous soil (Darcy's law), infiltration rates are preferred over K (Lin, 1995; Lin and McInnes, 1995). Nonetheless, infiltration rates can be highly variable in space and time and remain dependent on a variety of factors including land utilization and infiltration methods (Table 6.35).

6.7.10 Land Use and Management Considerations

6.7.10.1 Natural Resources

Agriculture
For many regions of the world, Vertisols are highly productive and sustainable resources for various crop production systems. However, despite their resilience, Vertisols in developed as well as developing countries are undergoing degradation after only decades of cultivation due to intensive and/or inappropriate management practices. Soil properties may partially recover from the effects of cultivation but it is unlikely that native conditions can be rejuvenated even after a restoration period of 20 years or more (Puentes, 1990; Puentes and Wilding, 1990; Coulombe, 1997). One of the difficulties in succesfully managing Vertisols involves technology transfer. Vertisols require site-specific management which is discussed in detail by Ahmad and Mermut (1996).

Wetlands
Vertisols subjected to aquic and udic soil moisture conditions can potentially be used for the creation or restoration of wetland habitats (Section G, Chapter 4). Besides the presence of hydric conditions, a requirement for successful establishment of desired wetland functions includes development of wetland hydrology and the presence of hydrophytic vegetation. Hydrological conditions can be supported by a permanent or seasonally perched watertable, or surface water. This causes the clay to swell and, consequently, drastically reduce permeability. In the process, an organic mat will develop on the floor of the wetland over time, further contributing to the decrease in permeability. Alternatively, the soil surface could be mechanically puddled such as for rice production, which alters the soil structure and breaks pore continuity and connectivity. Wetland hydrology is critical for development of hydric soil conditions and establishment of hydrophytic vegetation.

Table 6.35 Infiltration values under saturated conditions for surface horizons of selected Vertisols with reference to land use and methodology

Regions	Soil Series (and/or Parent Material)	Land Use	Methodology (Reference)	Infiltration ($\mu m\ s^{-1}$)	References
India	(Chlorite schist)	n/a	Double ring infiltrometer (Marshall and Stirk, 1950)	3.3	Hirekurubar et al. (1991)
	(Shale)	n/a	Double ring infiltrometer (Marshall and Stirk, 1950)	5.6	Hirekurubar et al. (1991)
	(Granite gneiss)	n/a	Double ring infiltrometer (Marshall and Stirk, 1950)	0.8	Hirekurubar et al. (1991)
	(Deccan trap)	n/a	Double ring infiltrometer (Marshall and Stirk, 1950)	1.2	Hirekurubar et al. (1991)
	(Limestone)	n/a	Double ring infiltrometer (Marshall and Stirk, 1950)	11.1	Hirekurubar et al. (1991)
Texas	Heiden	15-25 yr Pasture	Double-ring infiltrometer (Bouwer, 1986)	0.70-0.94	Puentes (1990)
	Heiden	15-25-yr cultivation	Double-ring infiltrometer (Bouwer, 1986)	0.14	Puentes (1990)
	Houston Black	Virgin	Column method (Bouma, 1982)	1.86	Coulombe (1997)
			Ponded infiltrometer (Prieksat et al., 1992)	1.77	Potter (pers. comm.)
			Disc permeameter (Perroux and White, 1988)	2.13	Lin (pers. comm.)
		20-yr restoration	Column method (Bouma, 1982)	1.04	Coulombe (1997)
			Ponded infiltrometer (Prieksat et al., 1992)	1.77	Potter (pers. comm.)
			Disc permeameter (Perroux and White, 1988)	2.29	Lin (pers. comm.)
		Pasture	Column method (Bouma, 1982)	0.61	Coulombe (1997)
			Ponded infiltrometer (Prieksat et al., 1992)	1.26	Potter (pers. comm.)
			Disc permeameter (Perroux and White, 1988)	1.86	Lin (pers. comm.)
		Cereal rotation	Column method (Bouma, 1982)	0.39	Coulombe (1997)
			Ponded infiltrometer (Prieksat et al., 1992)	1.75	Potter (pers. comm.)
			Disc permeameter (Perroux and White, 1988)	1.40	Lin (pers. comm.)
		Row crop rotation	Column method (Bouma, 1982)	0.88	Coulombe (1997)
			Ponded infiltrometer (Prieksat et al., 1992)	1.65	Potter (pers. comm.)
			Disc permeameter (Perroux and White, 1988)	1.37	Lin (pers. comm.)

n/a: not available

6.7.10.2   Engineering

Environmental Engineering

Due to their extensive shrink/swell behavior, Vertisols are generally not suitable for waste and wastewater disposal. Water would generally percolate too fast when dry and too slow when wet. In either case, this would contribute to potential contamination of surface and ground water and potential environmental health hazards. For similar reasons, Vertisols that exhibit a permanent (endoaquic) or seasonally perched (epiaquic) watertable would not be a suitable filter for wastewater disposal.

On the other hand, Vertisols could be used as constructed treatment wetlands for final disposal of tertiary effluent provided that wetland hydrology is maintained at saturation to enhance low soil permeability and to reduce groundwater recharge. Treatment wetlands are effective technologies for water quality improvement of various wastewaters while being beneficial for wildlife habitats (Kadlec and Knight, 1995).

Civil Engineering

The shrink/swell nature of Vertisols severely limits civil engineering applications. Vertisols must be stabilized for construction of buildings, roads, pipelines and utility corridors, to name a few. For example, several major cities in Texas (Dallas, Houston, San Antonio, and Corpus Christi) have been built on Vertisols. The swell/shrink properties of Vertisols make effective stabilization very difficult. The most economic choice may be to select stable soil for construction sites. If construction on a Vertisol is necessary, on-site investigation is required. Morphological properties of these soils often indicate movement at depths between 2–4 m due to natural moisture variability (M. E. Bloodworth, personal communication). These depths of movement indicate that prevention is the best way to minimize damage to structures on Vertisols and other expansive soils. Homeowners should maintain moist soils around the foundations of their dwellings that are built on steel reinforced concrete slabs to minimize movement. Application of gypsum to reduce ESP minimizes excessive swelling of smectite clay and maintains a favorable pH for plant growth in a climate with a seasonal water deficit.

Prior to construction of roads, streets and many structures that are to be built on Vertisols, lime $[Ca(OH)_2]$ stabilization of the soil is required to a depth of about 0.3 m or less (Eades and Grim, 1960; National Lime Association, 1991). Although this is beneficial, it obviously is only a partial treatment and complete stabilization would be extremely expensive. Larger multistory buildings sited on vertic soils and deep smectitic clays typically are supported by concrete piers with large bell-shaped feet at about 4 m depth to avoid contact with expansive soil.

Many other commercial methods to stabilize Vertisols are being promoted, some of which may be as successful as lime stabilization. Persons unfamiliar with these soils should proceed with caution because of the magnitude of the task and the unproven methods that have been marketed. One such treatment with a sulfonated napthalene was tested under laboratory conditions on three Vertisols with some success (Marquart, 1995), but the results were inconsistent, depending on soil conditions and level of treatment. The method is not recommended without proof of its effectiveness on a given soil.

It is imperative for construction workers to seriously consider the use of shoring materials and other safety precautions when a Vertisol is excavated. In Texas, from 1980 to 1985, 50 construction workers were killed due to destabilization of Vertisols that occurred during excavation (USDA-SCS, 1986).

6.7.11 References

Ahmad, N. 1982. Vertisols. p. 91–124. *In* L.P. Wilding, N.E. Smeck and G.F. Hall (ed.) Pedogenesis and Soil Taxonomy. Elsevier Scientific Publishing Co., New York, NY.

Ahmad, N. 1985. Acid Vertisols of Trinidad. p. 141–151. *In* Proc. 5th Int. Soil Classification Workshop, Sudan 1982. Soil Survey Administration, Khartoum, Sudan.

Ahmad, N., and R.L. Jones. 1969. Genesis, chemical properties and mineralogy of Carribean Grumusols. Soil Sci. 107:166–174.

Ahmad, N., and A.R. Mermut. 1996. Vertisols and technologies for their management. Elsevier Scientific Publishers, Amsterdam, Netherlands.

Anderson, J.U., K.E. Fadul, and G.A. O'Connor. 1973. Factors affecting the coefficient of linear extensibility in Vertisols. Soil Sci. Soc. Am. Proc. 37:296–299.

Baize, D., and M.C. Girard. 1996. Referentiel Pedologique 1995, Principaux sols d'Europe. INRA Editions, Versailles, France.

Beavington, F. 1978. Studies of some cracking clay soils in the Lake Chad basin of north east Nigeria. J. Soil Sci. 29:575–583.

Blokhuis, W.A. 1993. Vertisols of the central clay plain of the Sudan. Doctoral Thesis, Wageningen Agricultural University, Wageningen, Netherlands.

Boettinger, J.L. 1992. Genesis, mineralogy and geochemistry of a red-black (Alfisol-Vertisol) complex, northeastern Queensland, Australia. Ph.D. dissertation, University of California, Davis, CA.

Bouma, J. 1982. Measuring the hydraulic conductivity of soil horizons with continuous macropores. Soil Sci. Soc. Am. J. 46:438–441.

Bouma, J. 1983. Hydrology and soil genesis of soils with aquic moisture regimes. p. 253–281. In L.P. Wilding, N.E. Smeck, and G.F. Hall (ed.) Pedogenesis and Soil Taxonomy: I. Concepts and interactions. Elsevier Scientific Publishers, Amsterdam, Netherlands.

Bouma, J., and J. Loveday. 1988. Characterizing soil water regimes in swelling clay soils. p. 83–96. In L.P. Wilding and R. Puentes (ed.) Vertisols: Their distribution, properties, classification and management. Tech. Mono. 18, Texas A&M Printing Center, College Station, TX.

Bouwer, H. 1986. Intake rate: Cylinder infiltrometer. p. 825–844. In A. Klute (ed.) Methods of soil analysis, Part. I. Physical and mineralogical methods, 2nd Ed. Soil Science Society of America, Madison, WI.

Cabidoche, Y.M., and M. Voltz. 1995. Non-uniform volume and water content changes in swelling clay soil: II. A field study on a Vertisol. Europ. J. Soil Sci. 46:345–355.

Chenu, C. 1989. Influence of a fungal polysaccharide, scleroglucan, on clay microstructures. Soil Biol. Biochem. 21:299–305.

Chenu, C., and A.M. Jaunet. 1990. Modifications de l'organisation texturale d'une montmorillonite calcique liees a l'adsorption d'un polysaccharide. C.R. Acad. Sci. Paris 310:975–980.

Coulombe, C.E. 1997. Surface properties of Vertisols of Texas and Mexico under different land use systems. Ph.D. dissertation, Texas A&M University, College Station, TX.

Coulombe, C.E., J.B. Dixon, and L.P. Wilding. 1996a. Mineralogy and chemistry of Vertisols. p.115–200. In N. Ahmad and A.R. Mermut (ed.) Vertisols and technologies for their management. Elsevier Scientific Publishers, Amsterdam, Netherlands.

Coulombe, C.E., L.P. Wilding, J.B. Dixon. 1996b. Overview of Vertisols: Characteristics and impacts on society. Adv. Agron. 57:289–375.

CPCS. 1987. Classification des sols. E.N.S.A., Grignon, France.

Dudal, R., and H. Eswaran. 1988. Distribution, properties and classification of Vertisols. p. 1–22. In L.P. Wilding and R. Puentes (ed.) Vertisols: Their distribution, properties, classification and management. Tech. Mono. 18, Texas A&M Printing Center, College Station, TX.

Eades, J.L., and R.E. Grim. 1960. The reaction of hydrated lime with pure clay minerals in soil stabilization. 39th Ann. Meet. Hwy. Res. Bd., National Academy of Sciences, Washington, DC.

FAO-UNESCO. 1988. Soil map of the World: Revised legend. World Soil Resources Report 60, Food and Agriculture Organization, Rome, Italy.

Greene-Kelly, R. 1974. Shrinkage of clay soils: A statistical correlation with other soil properties. Geoderma 11:243–257.

Hallmark, C.T., L.T. West, L.P. Wilding, and L.R. Drees. 1986. Characterization data for selected Texas soils. TAES, USDA-SCS. College Station, TX.

Hirekurubar, B.M., V.S. Doddamani, and T. Satyanarayana. 1991. Some physical properties of Vertisols derived from different parent materials. J. Ind. Soc. Soil Sci. 39:242–245.

Isbell, R.F. 1991. Australian Vertisols. p. 73–80. In J.M. Kimble (ed.) Characterization, classification, and utilization of cold Aridisols and Vertisols, Proc. VI ISCOM, USDA-SCS, National Soil Survey Center, Lincoln, NB.

Jocteur Monrozier, L., J.N. Ladd, R.W. Fitzpatrick, R.C. Foster, and M. Raupach. 1991. Components and microbial biomass content of size fractions in soils of contrasting aggregation. Geoderma 50:37–62.

Kadlec, R.H., and R.L. Knight. 1995. Treatment wetlands. CRC Press, Boca Raton, FL.

Kimble, J. M. 1991. Characterization, classification, and utilization of cold Aridisols and Vertisols. In Proc. VI ISCOM, USDA-SCS, National Soil Survey Center, Lincoln, NB.

Kissel, D.E., J.T. Ritchie, and E. Burnett. 1973. Chloride movement in undisturbed swelling clay soil. Soil Sci. Soc. Am. Proc. 37:21–24.

Kissel D.E., J.T. Ritchie, and E. Burnett. 1974. Nitrate and chloride leaching in a swelling clay soil. J. Environ. Qual. 3:401–404.

Kunze, G.W., and E.H. Templin. 1956. Houston Black Clay, the type Grumusol: II. Mineralogical and chemical characterization. Soil Sci. Soc. Am. Proc. 27:412–421.

Kunze, G.W., H. Oakes, and M.E. Bloodworth. 1963. Grumusols of the Coast Prairie of Texas. Soil Sci. Soc. Amer. Proc. 27: 412–421.

Lin, H.S. 1995. Hydraulic properties and macropore flow of water in relation to soil morphology. Ph.D. dissertation, Texas A&M University, College Station, TX.

Lin, H.S., and K.J. McInnes. 1995. Water flow in clay soil beneath a tension infiltrometer. Soil Sci. 159:375–382.

Lin, H.S., K.J. McInnes, L.P. Wilding, and C.T. Hallmark. 1996. Effective porosity and flow rate with infiltration at low tensions into a well-structured subsoil. Trans. ASAE 39:131–135.

Lugo-Lopez, M.A., J.P. Carnelli, and G. Acevedo. 1985. Morphological, physical and chemical properties of major soils from Calagua in northwestern Uruguay. Soil Sci. Soc. Am. J. 49:108–113.

Marquart, D.K. 1995. Chemical stablization of three Texas Vertisols with sulfonated naphthalene. M.S. Thesis, Texas A&M University, College Station, TX.

Marshall, J.T., and G.B. Stirk. 1950. The effect of lateral movement of water in soil on infiltration measurement. Aust. J. Agric. Res. 1:253–257.

National Lime Association. 1991. Lime stabilization construction manual. Bull. 326. National Lime Association, Arlington, VA.

Perroux, K.M., and I. White. 1988. Design for disc permeameters. Soil Sci. Soc. Am. J. 52:1205–1215.

Prieksat, M.A., M.D. Ankeny, and T.C. Kaspar. 1992. Design for an automated, self-regulating, single ring infiltrometer. Soil Sci. Soc. Am. Proc. 36:874–879.

Probert, M.E., I.F. Fergus, B.J. Bridge, D. McGarry, C.H. Thompson, and J.S. Russel. 1987. The properties and management of Vertisols. CAB International, Wallingford, UK.

Puentes, R. 1990. Soil structure restoration in Vertisols under pastures in Texas. Ph.D. dissertation, Texas A&M University, College Station, TX.

Puentes, R., and L.P. Wilding. 1990. Structural restoration in Vertisols under pastures in Texas. Proc. 14th Int. Soil Sci. Soc. Cong. VII: 244–249.

Quirk, J.P. 1994. Interparticle forces: a basis for the interpretation of soil physical behavior. Adv. Agron. 53:121–183.

Singh, S. 1954. A study of the black cotton soils with special reference to their coloration. J. Soil Sci. 5:289–299.

Singh, S. 1956. The formation of dark coloured clay-organic complexes in Black Soils. J. Soil Sci. 7:43–58.

Skjemstad, J.O., and R.C. Dalal. 1987. Spectroscopic and chemical differences in organic matter of two Vertisols subjected to long periods of cultivation. Aust. J. Soil Res. 25:323–335.

Skjemstad, J.O., R.C. Dalal, and P.F. Barron. 1986. Spectroscopic investigations of cultivation effects on organic matter of Vertisols. Soil Sci. Soc. Am. J. 50:354–359.

Soil Survey Staff. 1960. Soil classification, a comprehensive system. 7th Approximation, USDA. US Government Printing Office, Washington, DC.

Soil Survey Staff. 1975. Soil Taxonomy: A basic system of soil classification for making and interpreting soil surveys. USDA-SCS, Agric. Handb. 436, US Government Printing Office, Washington, DC.

Soil Survey Staff. 1998. Keys to Soil Taxonomy. 8th Ed. USDA-NRCS. US Government Printing Office, Washington, DC.

Southard, R.J., and R.C. Graham. 1992. Cesium-137 distribution in a California Pelloxerert: Evidence of pedoturbation. Soil Sci. Soc. Am. J. 56:202–207.

Stephan, J., J. Berrier, A.A. De Petre, C. Jeanson, M.J. Kooistra, H.W. Scharpenseel, and H. Schiffmann. 1983. Characterization of *in situ* organic matter constituents in Vertisols from Argentina, using submicroscopic and cryochemical methods. Geoderma 30:21–34.

Tessier, D. 1984. Etude experimentale de l'organisation des materiaux argileux; Hydratation, gonflement et structuration au cours de la dessication et de la rehumectation. These D.Sc., University of Paris, Paris, France.

USDA-SCS. 1986. Soil cave-in: A fatal slip. Fact sheet. USDA, Washington, DC.

USDA-SCS. 1994. Global soil regions. World Soil Resources, Soil Survey Division, USDA, Washington, DC.

Wilding, L.P., and R. Puentes. 1988. Vertisols: Their distribution, properties, classification, and management. Tech. Mono. 18, Texas A&M Printing Center, College Station, TX.

Wilding, L.P. and D. Tessier. 1988. Genesis of Vertisols: Shrink/swell phenomena. p. 55–81. *In* L.P. Wilding and R. Puentes (ed.) Vertisols: Their distribution, properties, classification and management. Tech. Mono. 18, Texas A&M Printing Center, College Station, TX.

Yerima, B.P.K. 1986. Soil genesis, phosphorus and micronutrients of selected Vertisols and associated Alfisols of northern Cameroon. Ph.D. dissertation, Texas A&M University, College Station, TX.

Yerima, B.P.K., F.G. Calhoun, A.L. Senkayi, and J.B. Dixon. 1985. Occurrence of interstratified kaolinite-smectite in El Salvador Vertisols. Soil Sci. Soc. Am. J. 49:462–466.

Yerima, B.P.K., L.P. Wilding, C.T. Hallmark, and F.G. Calhoun. 1989. Statistical relationships among selected properties of northern Cameroon Vertisols and associated Alfisols. Soil Sci. Soc. Am. J. 53:1758–1763.

Yerima, B.P.K., L.R. Hossner, L.P. Wilding, and F.G. Calhoun. 1988. Forms of phosphorus and phosphorus sorption in northern Cameroon Vertisols and associated Alfisols. p. 147–164. *In* L.P. Wilding and R. Puentes (ed.) Vertisols: Their distribution, properties, classification and management. Tech. Mono.18, Texas A&M Printing Center, College Station, TX.

Yule, D.F., and J.T. Ritchie. 1980a. Soil shrinkage relationships of Texas Vertisols: I. Small cores, Soil Sci. Soc. Am. J. 44:1285–1291.

Yule, D.F., and J.T. Ritchie. 1980b. Soil shrinkage relationships of Texas Vertisols: II. Large cores. Soil Sci. Soc. Am. J. 44:1291–1295.

6.8 Mollisols

J.C. Bell
University of Minnesota

P.A. McDaniel
University of Idaho

6.8.1 Introduction

Mollisols are generally characterized as soils with thick dark surface horizons (mollic epipedons) resulting from organic C incorporation. The terms mollic and Mollisol are derived from the Latin *mollis*, soft. While the soil taxon, Mollisols, is defined by an exact set of soil property criteria, these criteria arose from conceptual ideas and empirical observations of soil landscapes with thick, dark, and often fertile surface horizons. Mollisols can form under multiple environmental conditions that facilitate accumulation of organic C in the upper soil profile. Although Mollisols correspond to the Chernozem soils of the Russian and older U.S. classification systems, the Mollisol soil order includes soils that are beyond the central concept of Chernozems (Fanning and Fanning, 1989).

Mollisols are among the most important soils for food and fiber production due to relatively high levels of native fertility coupled with climatic conditions conducive to plant growth. For this reason, the characteristics of Mollisols have been studied extensively, with early efforts focusing on the Chernozems of the Russian Steppes. During the late 19th century, droughts and crop failures afflicted the Russian Chernozem region resulting in widespread famine. In an effort to determine if the droughts could be prevented or moderated, Dokuchaev (1883, 1886) initiated an extensive study of the climate, soils, vegetation, and topography of the Chernozem Steppes. Based on the physical evidence that he and his colleagues collected, Dokuchaev theorized that the Chernozems of the Russian Steppes formed under specific conditions of climate, vegetation, and topography. These ideas contradicted the in vogue theories that Chernozems originated under either aquatic or marsh conditions. Dokuchaev's (1883, 1886) classic studies not only provided insight into the potential processes responsible for the formation of Chernozems, and subsequently Mollisols, but also articulated a conceptual framework for soil genesis that is still one of the foundations of modern pedology.

6.8.2 Geography and General Characteristics

6.8.2.1 Geographic Distribution

Extensive areas of Mollisols are distributed throughout the midlatitudes of the world predominantly on subhumid steppes and prairies. (Fig. 6.39). Mollisols occur under xeric, ustic, udic, and aquic soil

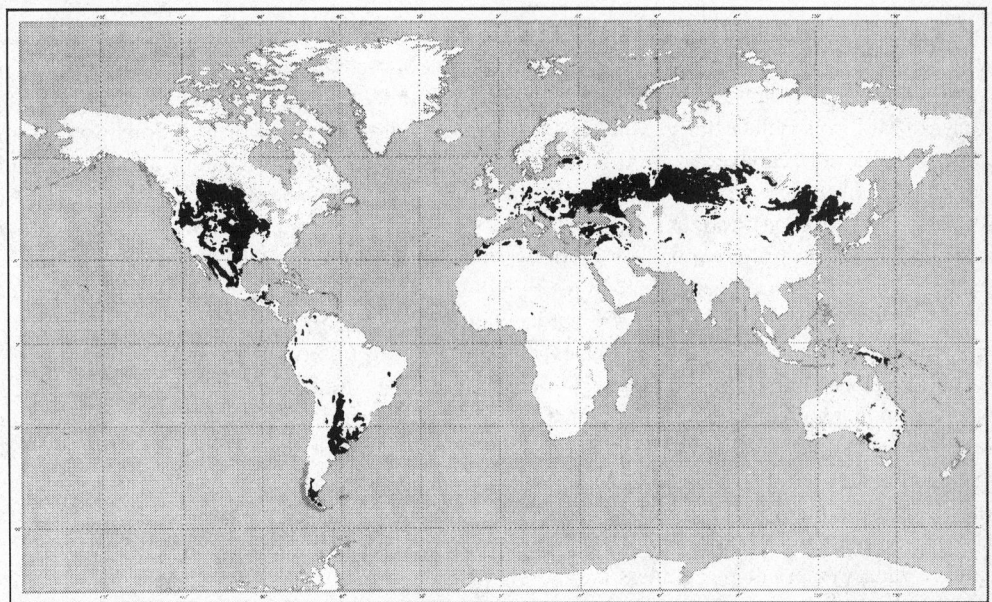

Fig. 6.39 Global distribution of Mollisols [Courtesy of USDA-NRCS, Soil Survey Division, World Soil References, 1998]

moisture regimes and cryic, frigid, mesic, thermic, and hypothermic soil temperature regimes. Spatially, Mollisols are the 8th (of 12) most common soil order and are estimated to cover 9,005,433 km² or approximately 7% of the ice-free land area on earth (Table 6.1). The distribution of Mollisol suborders can be distinguished by differences in climate with the exception of wet Mollisols (Aquolls and Albolls), which typically occur in localized areas subject to saturated and reduced soil conditions (Table 6.36).

6.8.2.2 General Characteristics of Mollisols

The general concept of Mollisols is that of dark colored soils of semi-arid to subhumid grassland ecosystems. The dark color reflects soil organic matter (SOM) enrichment in the upper portion of the

Table 6.36 Estimated occurrence of Mollisol suborder by general climatic condition (based on 1997 estimates by the USDA-NRCS)

Climatic conditions	Major geographic locations	Mollisol suborder	Approximate land area (km²)
Semi-Arid	Great Plains of United States and Canada, Southern Russia, Pampas Region of Argentina, Yucatan Peninsula of Mexico	Ustolls	5,244,636
Humid	Black Sea Region of Russia, Coastal Regions of United States, North-Central United States, Western Europe	Udolls, Rendolls	1,526,878
Cold	Mountainous regions of the Western US, Canadian Plains, higher latitudes across Kasikstan, Ukraine, and Russia	Borolls	1,163,797
Mediterranean	Turkey, Argentina, Palouse Region of United States	Xerolls	924,394
Seasonally Saturated	Red River Valley of North, riparian zones, Gulf coastal plain, glaciated regions of the US	Aquolls, Albolls	145,728

profile (Fig. 6.40). The formation of dark surface horizons is termed melanization, which is actually a combination of several processes involving the addition of organic matter to the soil in the form of plant residues and its subsequent transformation into humus (Buol et al., 1989). Distinguishing features of Mollisols include the presence of a mollic epipedon and high base status (Soil Survey Staff, 1998). A mollic surface layer by itself is not diagnostic for Mollisols, as it can be a feature of soils belonging to other orders. Mollisols are differentiated in these cases by the presence of high base status (> 50% base saturation) horizons that underlie the mollic epipedon, or by the lack of features associated with high shrink/swell clays (some Vertisols). A wide variety of subsurface horizons can occur beneath the mollic epipedon for Mollisols, including albic, cambic, argillic, calcic, petrocalcic, natric, duripans, and gleyed horizons. The nature of any diagnostic horizon is usually expressed at the great group level in classification with the exception of Albolls (albic) and Aquolls (gleyed).

A mollic epipedon is defined in *Soil Taxonomy* (Soil Survey Staff, 1975) by several quantitative criteria including horizon thickness, organic C content, color, consistence, structure, and base status (Soil Survey Staff, 1975; 1996). Although a mollic epipedon must contain at least 1% SOM (0.6% organic C) on a weight basis, many Mollisols have higher SOM contents. A mollic epipedon must generally be a minimum of 18 cm thick for shallow and 25 cm thick for deep soils. Soils which are very shallow require only a 10 cm thick mollic epipedon (Soil Survey Staff, 1998). Color requirements include a value (a measure of lightness/darkness) that is 3 or less for a moist sample and 5 or less for a dry sample. In both cases, the chroma (intensity or purity of color) is 3 or less. Thus, the colors of a mollic epipedon range from black or very dark brown when moist and are somewhat lighter (higher value) when dry. The color criteria were established to separate high organic matter soils formed under grasslands from their lower organic matter counterparts under forest. Mollic epipedons often have well-developed granular structure that is quite friable. By definition, a mollic epipedon cannot be both hard and massive when dry. A minimum of 50% base saturation is required for a mollic epipedon indicating a dominance of basic cations, primarily Ca and Mg, occupying the CEC.

Fig. 6.40   Soil profile of a Mollisol showing the thick, dark mollic epipedon. The mollic epipedon extends to depth of almost 60 cm (2 feet); scale is in feet.

The high base status of Mollisols generally translates into a high level of native fertility. Calcium and Mg are the dominant exchangeable cations, and both of these are required in fairly large quantities for plant growth. The pH values for Mollisols may be quite variable ranging from strongly acid (5.1–5.5) to strongly alkaline (8.5–9.0). However, many Mollisols have pH values somewhere in the middle of this range and are generally considered to be favorable for plant growth without widespread use of liming agents. Their high native fertility is one of the main reasons for their extensive use in agriculture throughout the world.

Mollisols generally have not undergone intensive weathering and their mineralogy is, therefore, often dominated by minerals inherited from parent material. Many Mollisols have formed in recently deposited parent materials such as glacial till and loess, which has not allowed sufficient time for significant mineral weathering to occur. Mineral weathering is also limited by the cooler temperatures and/or lack of moisture that are associated with many Mollisols. Clay mineralogy of Mollisols is typically dominated by 2:1 layer silicates, including clay mica (also referred to as illite), smectite, and vermiculite (Allen and Fanning, 1983). Generally, kaolinite, a 1:1 layer silicate, is commonly found in Mollisols although in small quantities (Allen and Hajek, 1989) but exceptions such as the Waialua soil from Hawaii (dominated by kaolinite) are sometimes found (USDA-NRCS, 1997).

Many of the physical characteristics of Mollisols are influenced by SOM and the associated biological processes. The relatively high SOM content of Mollisols is conducive to the formation of water stable aggregates which typically have crumb or granular structure, and are formed and maintained by the interaction of plant roots, microbial structures, and biological molecules (Tate, 1992). This type of stable structure is important for good water infiltration and reduced susceptibility of soils to erosion. Organic matter plays a considerable role in enhancing the water-holding capacity of Mollisols, as it can hold up to 20 times its weight of water (Stevenson, 1994). The dark colors associated with Mollisols are attributable to the humic fractions of soil SOM which are black and mask the lighter colors associated with fulvic fractions (Schultze et al., 1993).

6.8.3 Classification

Soils classified as Mollisols in *Soil Taxonomy* can be divided into seven suborders. While all of these soils have a mollic epipedon, they are separated into the various suborders largely on the basis of their climatic regimes. Mollisols include soils that were formerly classified as Chernozems, Brunizems, Chestnut and Rendzina soils and some soils formerly classified as Brown, Brown Forest, Humic Gley, and Planosols (Fenton, 1983). Table 6.37 lists the suborders, great groups , and subgroups in the Mollisol order.

6.8.3.1 Albolls

Albolls have a well-developed, light colored horizon (albic) and a fluctuating watertable. The albic horizon forms as pigmenting agents such as SOM, clays, and Fe oxides are removed from mineral grains (primarily quartz) exposing their light color. In Albolls, the albic horizon typically occurs just below the mollic surface layer and directly above a less permeable horizon of clay accumulation. Albolls often have a watertable at or near the surface for several months during the winter and spring. In the United States, Albolls are most extensive in loess deposits of the midwest (Soil Survey Staff, 1975).

6.8.3.2 Aquolls

Aquolls are the wettest Mollisols and are characterized by the presence of redoximorphic features below a very dark brown or black epipedon. Depending on the duration of wet conditions and Fe

Table 6.37 Listing of suborders, great groups, and subgroups in the Mollisols order [Soil Survey Staff, 1998]

Suborder	Great Group	Subgroups
Albolls	Natralbolls	Leptic, Typic
	Argialbolls	Xerertic, Vertic, Argiapuic Xeric, Argiaquic, Xeric, Aquandic, Typic
Aquolls	Cryaquolls	Vertic, Histic, Thapto-Histic, Aquandic, Argic, Calcic, Cumulic, Typic
	Duraquolls	Natric, Vertic, Argic, Typic
	Natraquolls	Vertic, Typic
	Calciaquolls	Petrocalcic, Aeric, Typic
	Argiaquolls	Arenic, Grossarenic, Vertic, Abruptic, Typic
	Epiaquolls	Cumulic Vertic, Fluvaquentic Vertic, Vertic, Histic, Thapto-Histic, Aquandic, Duric, Cumulic, Fluvaquentic, Typic
	Endoaquolls	Lithic, Cumulic Vertic, Fluvaquentic Vertic, Vertic, Histic, Thapto-Histic, Aquandic, Duric, Cumulic, Fluvaquentic, Typic
Rendolls	Cryendolls	Lithic, Typic
	Haprendolls	Lithic, Vertic, Inceptic, Entic, Typic
Cryolls	Duricryolls	Argic, Typic
	Natricryolls	Typic
	Palecryolls	Aquic, Oxyaquic, Abruptic, Pachic, Ustic, Xeric, Typic
	Argicryolls	Lithic, Vertic, Andic, Vitrandic, Abruptic, Aquic, Oxyaquic, Pachic, Alfic, Ustic, Xeric, Typic
	Calcicryolls	Lithic, Petrocalcic, Pachic, Ustic, Xeric, Typic
	Haplocryolls	Lithic, Vertic, Andic, Vitrandic, Aquic Cumulic, Cumulic, Fluvaquentic, Aquic, Oxyaquic, Calcic Pachic, Pachic, Fluventic, Calcic, Ustic, Xeric, Typic
Xerolls	Durixerolls	Vertic, Vitritorrandic, Vitrandic, Aquic, Paleargidic, Abruptic Argiduridic, Cambidic, Haploduridic, Argidic, Argiduridic, Haplic Palexerollic, Palexerollic, Haplic Haploxerollic, Haploxerollic, Haplic, Typic
	Natrixerolls	Vertic, Aquic Duric, Aquic, Aridic, Duric, Typic
	Palexerolls	Vertic, Vitrandic, Aquic, Pachic, Petrocalcidic, Duric, Aridic, Petrocalcic, Ultic, Haplic, Typic
	Calcixerolls	Lithic, Vertic, Aquic, Oxyaquic, Pachic, Vitrandic, Aridic, Vermic, Typic
	Argixerolls	Lithic Ultic, Lithic, Torrertic, Vertic, Andic, Vitritorrandic, Vitrandic, Aquultic, Aquic, Oxyaquic, Alfic, Calcic Pachic, Pachic Ultic, Pachic, Argiduridic, Duric, Calciargidic, Aridic, Calcic, Ultic, Typic
	Haploxerolls	Lithic Ultic, Lithic, Torrertic, Vertic, Vitritorrandic, Vitrandic, Aquic Cumulic, Cumulic Ultic, Cumulic, Fluvaquentic, Aquic Duric, Aquultic, Aquic, Oxyaquic, Calcic Pachic, Pachic Ultic, Pachic, Torrifluventic, Duridic, Calcidic, Torripsammentic, Torriorthentic, Aridic, Duric, Psammentic, Fluventic, Vermic, Calcic, Entic Ultic, Ultic, Entic, Typic
Ustolls	Durustolls	Natric, Haploduridic, Argiduridic, Entic, Haplic, Typic
	Natrustolls	Leptic Torrertic, Torrertic, Leptic vertic, Glossic Vertic, Vertic, Aridic leptic, Leptic, Aquic, Aridic, Duric, Glossic, Typic
	Calciustolls	Salidic, Lithic Petrocalcic, Lithic, Torrertic, Udertic, Vertic, Petrocalcic, Gypsic, Pachic, Aquic, Oxyaquic, Aridic, Udic, Typic
	Paleustolls	Torrertic, Udertic, Vertic, Pachic, Petrocalcic, Calcidic, Aridic, Udic, Calcic, Entic, Typic
	Argiustolls	Aridic Lithic, Alfic Lithic, Lithic, Torrertic, Udertic, Vertic, Andic, Vitritorrandic, Vitrandic, Pachic, Aquic, Oxyaquic, Alfic, Calcidic, Aridic, Udic, Duric, Typic
	Vermustolls	Lithic, Aquic, Pacic, Entic, Typic
	Haplustolls	Salidic, Rutic-Lithic, Lithic, Torrertic, Pachic Udertic, Urdertic, Vertic, Torroxic, Oxic, Andic, Virtitorrandic, Vitrandic, Aquic Cumulic, Cumulic, Anthraquic, Fluvaquentic, Pachic, Aquic, Oxyaquic, Torrifluventic, Torriorthentic, Aridic, Fluventic, Duric, Udorthentic, Udic, Entic, Typic
Udolls	Natrudolls	Petrocalcic, Leptic Vertic, Glossic vertic, Vertic, Leptic, Glossic, Calcic, Typic
	Calciudolls	Lithic, Vertic, Aquic, Fluventic, Typic
	Paleudolls	Vertic, petrocalcic, Aquic, Pachic, Oxyaquic, Calcic, Typic
	Argiudolls	Lithic, Aquertic, Oxyaquic Vertic, Pachic Vertic, Alfic Vertic, Vertic, Andic, Vitrandic, Aquic, Pachic, Oxyaquic, Lamellic, Psammentic, Arenic, Abruptic, Alfic, Oxic, Calcic, Typic
	Vermudolls	Lithic, Haplic, Typic
	Hapludolls	Lithic, Aqertic, Vertic, Andic, Vitrandic, Aquic Cumulic, Cumulic, Fluvaquentic, Aquic, Pachic, Oxyaquic, Fluventic, Vermic, Calcic, Entic, Typic

reduction, expression of Fe-depleted zones can range from entire horizons with predominantly dull gray colors to horizons that have a strongly contrasting mottled gray color pattern. They typically occupy the lower lying areas of the landscape where water accumulates and a high watertable exists. The most extensive areas of Aquolls in the United States occur in the Red River Valley of the North, along flood plains and terraces of major rivers in the central part, and along the Gulf coastal plain (Soil Survey Staff, 1975). Most uses of Aquolls, whether urban or agricultural, are limited by wetness. Many Aquolls are suitable for agriculture if artificially drained. Aquolls typically have black surface horizons underlain by gray colors indicating reduction and/or depletion of Fe (Table 6.38). Organic C accumulates in the surface horizon due to reduced rates of organic matter decomposition resulting from seasonal anaerobic conditions. The presence or absence of carbonates in Aquoll profiles is quite variable and can be used as an indicator of wetland hydrology. Carbonates tend to be removed from recharge wetlands by leaching, whereas discharge wetlands are enriched often to the soil surface. Depressional basins in subhumid climates often have recharge hydrology resulting in leaching of carbonates and formation of argillic horizons (Mausbach and Richardson, 1994).

6.8.3.3 Rendolls

Rendolls form in highly calcareous parent materials of humid or cold regions, often under forest vegetation. These soils typically consist of a mollic surface layer overlying mineral material that contains > 40% $CaCO_3$ by weight (Soil Survey Staff, 1998). Rendolls are associated with parent materials such as chalk, limestone, highly calcareous glacial till, and shell deposits. These soils are not extensive in the United States where only a few soils with this classification have been recognized. Rendolls are extensive in western Europe (Smith, 1986).

6.8.3.4 Xerolls

Xerolls form in areas that receive the majority of precipitation during the winter and spring months when temperatures are cool (Mediterranean climates). Little precipitation is received during the growing season and, as a result, Xerolls frequently become dry at some point during the summer. This type of moisture regime is very effective for leaching $CaCO_3$ and other soil constituents in higher precipitation zones. Conversely, Xerolls of the drier areas are subject to long periods of soil moisture deficits during the growing season. In the United States, Xerolls comprise < 5 % of the land area occupied by Mollisols and are restricted to the western states. An example of a Xeroll profile from the Palouse region of northern Idaho is shown in Fig. 6.41, with selected data in Table 6.39. Important

Table 6.38 Selected morphological, physical, and chemical properties of an Aquoll from southern Minnesota; Location: Waseca County, Minnesota; Mean annual air temperature: 6.2 °C; Mean annual precipitation: 836 mm Vegetation: Reed Canary Grass (*Phalaris* Parent material: *arundinacea* L.), Cattails (*Typha* sp.), Wisconsinan-age glacial till; Landscape position: depression, 0-1% slope; Classification: fine-loamy, mixed, mesic, Typic Endoaquoll

Horizon	Depth	Texture	Moist Color	Clay	pH	Organic C	Cation Exchange Capacity	$CaCO_3$
	(cm)			(%)		(%)	(cmol$_c$ kg^{-1})	(%)
Ap	0-28	clay loam	Black	37.5	7.5	4.5	45.5	1
A	28-41	clay loam	Black	35.9	7.2	1.3	34.9	trace
AB	41-55	clay loam	Very dark gray	32.8	7.4	0.5	28.9	trace
Bg	55-71	clay loam	Olive gray	28.6	7.5	0.4	23.0	7
Cg1	71-105	loam	Olive gray	26.3	7.5	0.2	20.5	11
Cg2	105-125	loam	Gray	25.3	7.6	0. 2	18.8	12

Fig. 6.41 Soil profile of a Xeroll from northern Idaho. Accompanying data are presented in Table 6.39; scale is in decimeters.

features of this soil include the lack of CaCO$_3$ in the profile, acidic conditions, and dense subsoil horizons (94-152+ cm) in the lower part of the profile. Effective leaching by winter precipitation is responsible for the lack of CaCO$_3$ and the low pH. The dense subsoil horizons restrict downward water movement and result in the formation of perched watertables during the winter and spring months. Xerolls are used primarily for production of wheat, hay, and some timber.

6.8.3.5 Cryoborolls

Cryoborolls occur in cold climates having a mean annual soil temperature < 8 °C with cool summers. Cryoborolls are generally associated with a short growing season that limits their use for agricultural

Table 6.39 Selected morphological, physical, and chemical properties of a Xeroll (Southwick series, 93-ID-29151) from northern Idaho; Location: 460°44′N Lat., 116° 50′ W long., Latah County, Idaho; 13 km east of Moscow; Elevation: 825 m; Mean annual air temperature: 7 °C; Mean annual precipitation: 635 mm; Vegetation: ponderosa pine, Idaho fescue, snowberry; Parent material: Late Pleistocene and Holocene loess; Landscape position: near summit of gently sloping (4%) spur ridge; Classification: fine-silty, mixed, mesic Boralfic Argixerolls

Horizon	Depth	Moist Color	Sand	Silt	Clay	Bulk Density	pH	Organic C
	(cm)		----	(%)	----	(g cm^{-3})		(%)
A	0-18	very dark brown	6.9	72.2	21.0	--†	6.5	3.09
AB	18-38	very dark grayish brown	7.9	72.7	19.3	1.25	5.9	1.69
Bw	36-71	dark brown	7.7	70.0	22.3	1.30	5.8	1.20
BE	71-58	brown	7.3	74.0	18.7	1.37	5.6	0.46
E	58-94	grayish brown	8.7	78.3	13.0	1.48	5.5	0.46
Btb1	94-114	brown	5.5	63.2	31.3	1.76	5.2	0.42
Btb2	114-152+	brown	6.6	63.8	29.6	1.75	5.6	0.17

† not sampled

production. They can, however, be productive for small grains and hay production. Data from a representative Cryoboroll profile from Montana are presented in Table 6.40. Large quantities of organic C are associated with such soils giving a strong black color (~ 10% C in the 0–6 cm layer) (Table 6.40). Limited leaching has resulted in the removal of $CaCO_3$ from the upper 61 cm of this profile and the development of acid pH values. Base saturation increases with depth and is generally ~100% in lower horizons where $CaCO_3$ has accumulated. These soils are mainly used for winter wheat and hay production.

6.8.3.6 Ustolls

Ustolls occur in subhumid or semi-arid climates receiving significant amounts of spring and summer rain (350–900 mm annual precipitation), which is characteristic of the ustic soil moisture regime (Soil Survey Staff, 1975). As a result, the potential productivity of these soils for agriculture varies considerably. In the higher rainfall areas, drought sensitive crops, such as corn and soybeans, can be produced, while in the drier regions, summer fallow which allows recharge of soil moisture in a fallowed field (typically every other year), is often necessary for crop production (Soil Survey Staff, 1975). Ustolls are extensive throughout the central Great Plains and are the dominant suborder in the United States.

6.8.3.7 Udolls

Udolls occur in humid regions having typically formed in late Pleistocene or Holocene deposits under tallgrass prairie. Udolls are characterized by having moderate precipitation, well-distributed throughout the year (Soil Survey Staff, 1975). Because extended soil moisture deficits are not experienced, most Udolls have been placed into agricultural production. Udolls are extensive in the Midwest and comprise one of the most productive grain-producing regions of the world (Soil Survey Staff, 1975). Corn and soybeans, which are among the principal crops produced on Udolls, are generally not produced or produced with higher risks on other Mollisols because of climatic limitations. Data for a Udoll from Iowa are presented in Table 6.41. Although the parent material is calcareous loess, the relatively high annual precipitation has completely removed $CaCO_3$ from the upper 175 cm. Calcium is still the dominant soil cation and, with Mg, occupies most of the CEC.

Table 6.40 Selected morphological, physical, and chemical properties of a Boroll (S71 MT 31-1) from Montana (data courtesy of Dr. G.A. Nielsen, Montana State University); Location: Gallatin County, Montana; Elevation: 1585 m; Mean annual air temperature: 5 °C; Mean annual precipitation: 510 mm; Vegetation: bluebunch wheatgrass, Idaho fescue, prairie junegrass, sagebrush; Parent material: mixed alluvium/colluvium; Landscape position: alluvial/colluvial fan, 4% slope

Horizon	Depth	Moist Color	Clay	pH	Organic C	Cation Exchange Capacity	$CaCO_3$	Base Saturation
	(cm)		(%)		(%)	(cmol$_c$ kg^{-1})	(%)	(%)
A1	0-6	black	28.1	5.5	9.88	43.7	nd†	70
A2	6-18	black	28.6	5.8	5.27	36.8	nd	80
Bt1	18-29	dark brown	32.8	6.1	1.88	32.1	nd	84
Bt2	29-42	brown	28.9	6.4	1.11	32.4	nd	100
Bt3	42-61	brown	27.1	7.0	0.87	34.0	nd	100
Bk1	61-86	brown	25.0	7.7	0.55	27.2	11	100
Bk2	86-130	brown	26.2	8.0	0.28	26.8	8	100
2C	130-175	brown	26.3	8.2	0.21	25.6	2	100

† not detected

Table 6.41 Selected morphological, physical, and chemical properties of a Udoll (S63 Iowa-83-2) from Iowa (data from Appendix IV, Soil Survey Staff, 1975); Location: Shelby County, Iowa; 135 km west of Des Moines; Mean annual air temperature: 9 °C; Mean annual precipitation: 720 mm; Vegetation: cropland; Parent material: Late Pleistocene loess; Landscape position: axis of short interfluve, 3% slope; Classification: fine-silty, mixed, mesic Typic Hapludoll

Horizon	Depth	Texture	Moist Color	Clay	pH	Organic C	Cation Exchange Capacity	Exch.Ca
	(cm)			(%)		(%)	(cmol$_c$ kg^{-1})	-----
Ap	0-18	silty clay loam	very dark brown	30.4	5.6	2.20	22.0	13.9
A	18-33	silty clay loam	very dark brown	33.5	5.7	1.87	22.9	14.7
AB	33-46	silty clay loam	very dark grayish brown	32.8	5.8	1.11	21.6	14.8
Bw1	46-69	silty clay loam	dark brown/ brown	30.4	5.8	0.58	20.0	14.8
Bw2	69-86	silty clay loam	dark brown/ brown	28.2	5.9	0.33	20.7	14.7
BC1	86-110	silt loam	yellowish brown	26.9	5.9	0.21	20.4	14.6
BC2	110-125	silt loam/ silty clayloam	yellowish brown/ olive gray	28.0	6.0	0.17	20.7	15.2
C1	125-145	silt loam	brown/ olive gray	26.9	6.0	0.11	20.8	14.6
C2	145-175	silt loam	yellowish brown/ olive gray	25.7	6.2	0.10	19.7	13.8

6.8.4 Pedogenic Processes

Soils classified as Mollisols develop under a variety of environmental conditions through several distinct genetic pathways. Soil genesis typically occurs over time frames where environmental conditions, and hence, the theoretical equilibrium state are variable. The morphological imprints left by grassland vegetation are ephemeral in pedogenic time frames, and hence, are probably indicative of the soil climate over the past few hundreds to thousands of years. The genesis of Mollisols will be discussed from two different perspectives (factors and processes of soil formation) in an attempt to elucidate the development of these unique soils.

6.8.4.1 Factors of Soil Formation

The factor approach applied by Jenny (1941, 1980) to the soil-forming factor model is predicated on the hypothesis that soil properties result from the interaction of at least five site and flux factors. This model proposes that soil bodies result from the action of climate, organisms, and relief acting upon parent material over time and provides a useful conceptual framework by which soil differences can be attributed to state or site factors.

Organisms

To a large degree, Mollisols are distinguished from other soils on the basis of organisms, or more specifically, the overriding effects of grassland vegetation. The reason why grassland vegetation imparts distinctive soil morphologies is discussed later. While some Mollisols have developed under consistent vegetative conditions, abundant evidence suggests that many were forested during glacial and post glacial periods (Walker, 1966; Ruhe, 1970; Wright, 1970). Similar evidence also suggests that the climate has changed sufficiently during the Holocene to allow fluctuations between forest and prairie vegetation, especially along humid to subhumid ecotones (Walker, 1966; Geis and Boggess, 1968). Most soils classified as Mollisols (with the possible exception of Rendolls, Aquolls, and Albolls) have probably had grassland vegetation at some time during their development. Because soil characteristics reflect the integration of the soil forming factors over long time periods (centuries to millennia), the contemporary environment may or may not accurately reflect the conditions under which a soil has developed. Mollisols are common in transitional areas between humid and subhumid environments such as the north central United States. Changes in climate during the Holocene have probably resulted in the formation and subsequent degradation of Mollisols as forest encroached on areas that were previously prairies (Fenton, 1983).

The mixing and incorporation of organic matter within the mollic epipedon have also been attributed to soil faunal activity. Of particular importance are burrowing organisms, such as gophers, prairie dogs, worms, and ants (Thorp, 1948; Munn, 1993). The net effect of burrowing activities is to move organic matter lower in the soil profile and to mix soil materials, resulting in a rather homogeneous epipedon.

Climate

As previously suggested, climate is also a key factor in the formation of Mollisols, primarily through its effect on vegetation. Grassland ecosystems are associated with broad ranges in mean annual precipitation and temperature. Consequently, Mollisols occur under a wide range of soil temperature and moisture regimes. In the United Sates, these soils are found from as far south as Texas to the Canadian border. Mollisols are most common in the zone between the arid deserts of the western and the forests of the midwestern states, a range of approximately 300 mm to more than 900 mm in annual precipitation.

The effects of climate on Mollisol formation can be demonstrated by comparing soil morphology along climatic gradients. Munn et al. (1978) examined a warm/dry to cool/moist sequence of grassland soils in Montana. They observed an increase in annual aboveground biomass production with increasing precipitation and decreasing temperature (Fig. 6.42). Similarly, SOM content, which increased as well, is expressed by increasing thickness of the mollic epipedon. This suggests that the dark, organically enriched mollic epipedon provides a record of the balance between annual site production and microbial decomposition. As production increases with increasing precipitation, decreasing temperature favors the accumulation of SOM by decreasing microbial decomposition rates. Leaching of $CaCO_3$ in the soils was greater with increasing annual precipitation, indicating that depth to carbonate serves as an indicator of rainfall within a geographic region.

Relief

The local topography or hillslope position affects hydrologic processes, especially erosion and sedimentation and the timing, duration, and depth of seasonal soil saturation. Many glacial till landscapes of the eastern portion of the central United States contain both Mollisols and Alfisols. Aquolls and Albolls occur in the wetter, lower lying portions of the landscape (darker soil surface) (Fig. 6.43). Conversely, Alfisols occupy the adjacent higher positions having better drainage (Allen and Fanning, 1983).

Richardson et al. (1992) proposed a framework to view wetland hydrology based on differences in how water flows through depressional basins (recharge, flowthrough, or discharge). Differences in basin hydrology result in distinctive differences in soil morphology along hillslopes (Thompson and Bell, 1996; Bell and Richardson, 1997). Examination of a toposequence of Mollisols from south central Minnesota illustrates typical catena relationships (Fig. 6.44). Along this hillslope continuum, the following changes occur from upper to lower landscape positions: (1) increased darkness (organic

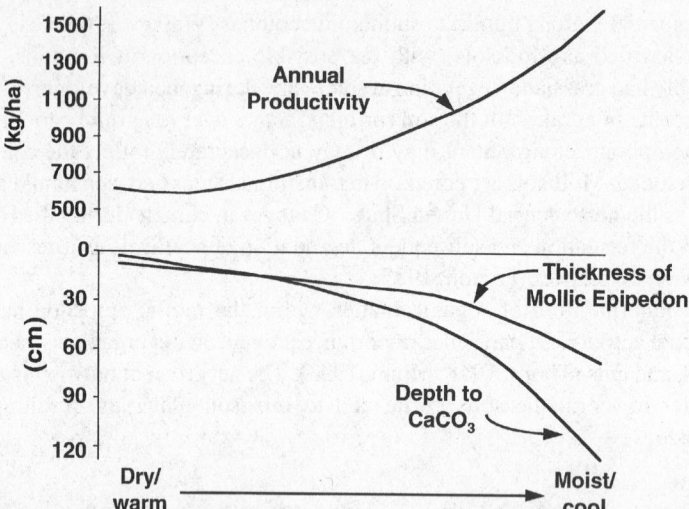

Fig. 6.42 Generalized relationship between annual aboveground productivity of rangelands in western Montana, depth to CaCO₃, and thickness of mollic epipedon. [Adapted from Munn et al., 1978. Soil Sci. Soc. Am. J. with permission of the Soil Science Society of America]

C) and thickness of the surface horizon, (2) decreased depth to redoximorphic features, and (3) decreased chroma of the subsurface horizons. These morphological differences can often be traced back to differences in soil processes, such as erosion, deposition, and reduction, that are affected by hydrology and the shape and position on the hillslope. Numerous studies have documented increased duration of saturation in lower landscape positions for similar landscapes (Khan and Fenton, 1994; Bell et al., 1995; Thompson and Bell, 1996). For the example from Waseca, Minnesota (Fig. 6.44); higher carbonate concentrations between 45 to 90 cm at the toeslope (Fig. 6.45) are probably indicative of discharge hydrology and subsequent enrichment of the soil profile from carbonate laden groundwaters. The distribution of organic C with depth along the hillslope clearly indicates a trend of

Fig. 6.43 Glacial till landscape in central Indiana showing patterns of Alfisols and Mollisols. Wet Mollisols (Aquolls) occupy wetter, lower lying portions of the landscape and appear as the dark areas. Alfisols are found in the higher, better drained positions and appear as the lighter areas in the photograph [Photo used with permission of the Soil Science Society of America].

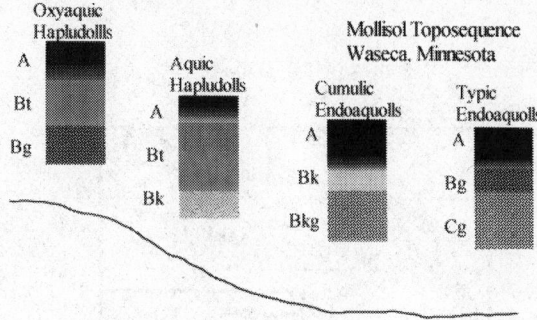

Fig. 6.44   Toposequence of soil profiles along a hillslope in glacial tills near Waseca, Minnesota

increasing organic C in the surface horizons from upper to lower hillslope positions (Fig. 6.46). Organic C concentrations tend to converge at a depth of approximately 75 cm for all landscape positions where concentrations fall below 0.5%. These same spatial relationships cannot be assumed for all hillslopes as differences in soil stratigraphy, regional hydrology, and climate can cause different spatial patterns to develop.

Parent Material

Mollisols occur on a variety of substrate materials, but are most commonly associated with unconsolidated sediments from coastal, riverine, or glacial depositional environments, including loess (Soil Survey Staff, 1975). Loess is a common parent material of Mollisols in southern and central Russia, and Duchaufour (1982) has suggested that the permeability and $CaCO_3$ content of loess is especially conducive to the formation of Mollisols. Many Mollisols have formed in calcareous parent materials resulting in appreciable quantities of $CaCO_3$ within the soil profile, particularly those Mollisols in lower precipitation areas. Rendolls form exclusively in highly calcareous parent materials such as chalk, limestone, and shell deposits.

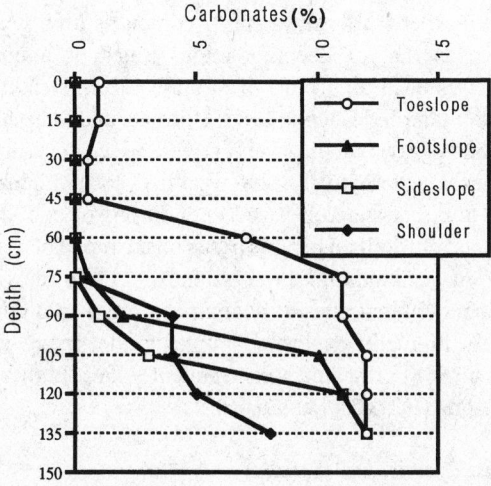

Fig. 6.45   Depth distribution of soil carbonate concentrations at four hillslope positions in glacial till near Waseca, Minnesota

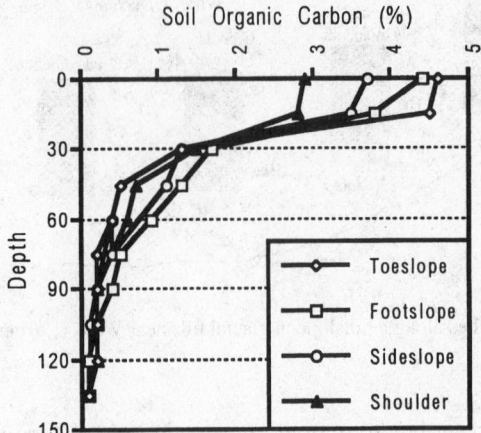

Fig. 6.46 Depth distribution of soil organic carbon at four hillslope positions in glacial till near Waseca, Minnesota

Time

Most Mollisols in the United States and Russia appear to be relatively young and have formed on late Pleistocene and Holocene deposits and surfaces (Soil Survey Staff, 1975). However, the recognition of Mollisols in intertropical areas may warrant re-examination of this concept. It has been shown that accumulation of SOM occurs relatively fast. Schafer et al. (1980) who compared organic C levels in newly created mine soils with those of adjacent undisturbed grassland soils of eastern Montana, found that only 30 years was required for C to build up in the top 10 cm to the level in undisturbed soils. In Iowa, the rate of formation of a Udoll surface horizon was estimated at 0.08 cm yr^{-1} (Buol et al., 1989). Foss et al. (1985) suggested that mollic epipedons developed in < 900 yr in the Red River Valley of the North in Minnesota and North Dakota. From these and other studies, it is clear that the formation of a mollic surface horizon can occur in relatively short time spans.

6.8.4.2 Genetic Pathways for Mollisol Formation (Soil Processes)

The accumulation of organic C in the upper soil profile is the salient morphological feature distinguishing Mollisols from other soil orders. The C content is high when the long-term rate of addition and retention exceeds that of decomposition. Rates of annual root production and decomposition in grasslands result in high rates of organic C turnover in the soil to depths often approaching 50 to 100 cm (Dahlman and Kucera, 1965; Dormaar and Sauerbeck, 1983). Thorp (1948) estimated that 113 to 409 kg ha^{-1} of raw OM (dry weight) were added annually for short grass and 136–500 kg ha^{-1} for tall grass prairies. Alternatively, soil organic C accumulates when rates of microbial decomposition are low due to either anaerobic (wet) conditions or cool soil temperatures (Jenny, 1930). Three primary pathways for Mollisol genesis based on the mode of organic C accumulation can be identified: (1) high rates of accumulation in grasslands, (2) low rates of decomposition under anaerobic (wet) conditions, and (3) low rates of decomposition in cold climates (Fig. 6.47). These pathways are not necessarily mutually exclusive. For example, many wet soils may have also developed under grassland, but wet grassland soils frequently have higher organic C contents than drier soils in the same landscape (Figs. 6.44 and 6.46).

Grassland Mollisols (Udolls, Ustolls, and Xerolls)

A comparison of the chemical composition and growth forms of forests and grasslands explains why soils under these covers have distinctive morphologies. In forests, organic C tends to accumulate on

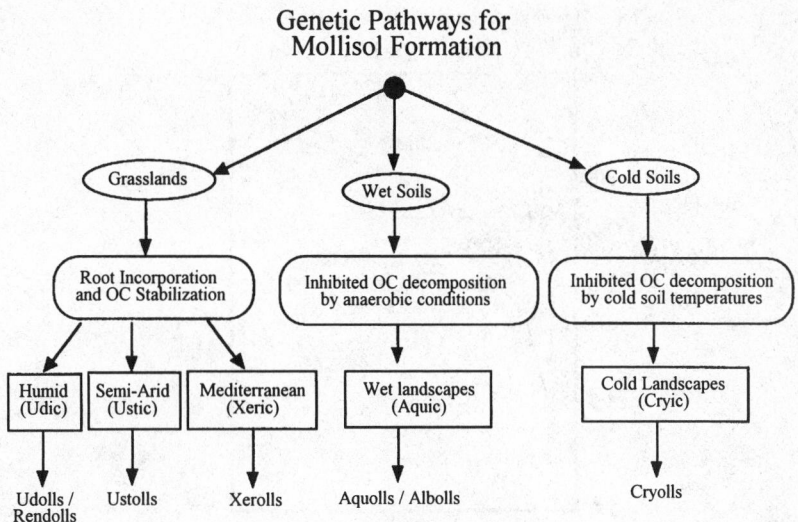

Genetic Pathways for
Mollisol Formation

Grasslands · Wet Soils · Cold Soils

Root Incorporation and OC Stabilization · Inhibited OC decomposition by anaerobic conditions · Inhibited OC decomposition by cold soil temperatures

Humid (Udic) · Semi-Arid (Ustic) · Mediterranean (Xeric) · Wet landscapes (Aquic) · Cold Landscapes (Cryic)

Udolls / Rendolls · Ustolls · Xerolls · Aquolls / Albolls · Cryolls

Fig. 6.47 Theoretical pathways for the genesis of Mollisols

the soil surface from annual decomposition cycles of forest floor litter. Despite the fact that the forested soils (Alfisols) receive more annual precipitation, they contain less organic C throughout the mineral soil and lack the dark soil colors associated with Mollisols. Much of the organic matter associated with the forest soils is contained in litter layers lying on top of the mineral soil. Leachate from the decomposition of forest floor litter is often composed of carbonic, fluvic, and/or tannic acids facilitating translocation (eluviation) of soil materials deeper within the soil profile resulting in morphological features usually associated with Alfisols and Spodosols. These conditions and processes inhibit the accumulation and stabilization of organic C in the upper soil layers.

By contrast, photosynthesis and other metabolic processes in grassland vegetation quickly transport organic C to dense and fibrous root systems. Estimates indicate that grasslands have the highest annual additions of C to soil of any of the terrestrial ecosystems, including tropical forests (Bolin et al., 1979). Related research has demonstrated that the greatest biomass production in grassland ecosystems is below ground (Caldwell, 1975; Lauenroth and Whitman, 1977; Jenny, 1980). As such, Mollisols have distinctive organic matter profiles with depth that differ markedly from Alfisols, where most of the biomass is produced aboveground (Fig. 6.48). The annual proliferation of roots and their subsequent decomposition is responsible for the deep accumulation of SOM leading to the formation of mollic epipedons whose thickness is largely determined by the depth and amount of grass roots (Hole and Nielsen, 1968; Cannon and Nielsen, 1984). Incorporation of organic C is also facilitated by the mixing of near surface soil horizons by ants (Formicidae), earthworms (Lumbricus), and other soil fauna (Curtis, 1959; Baxter and Hole, 1967). Schlesinger (1977, 1991) has estimated that temperate grassland soils have a mean organic matter content of 19.2 kg C m^{-2}, ranking them behind only soils of wetlands and tundra/alpine ecosystems on a global scale.

Once added to the soil, organic materials undergo further decomposition by complex, microbially mediated processes that results in the formation of a relatively stable organic fraction. The large annual additions of C to Mollisols and subsequent cycling results in the formation of an active (bioavailable) as well as a very stable (labile) organic fraction. The humus contained in Mollisols appears to be more stable than that found in other soils (Stevenson, 1994), which is possibly related to the chemical

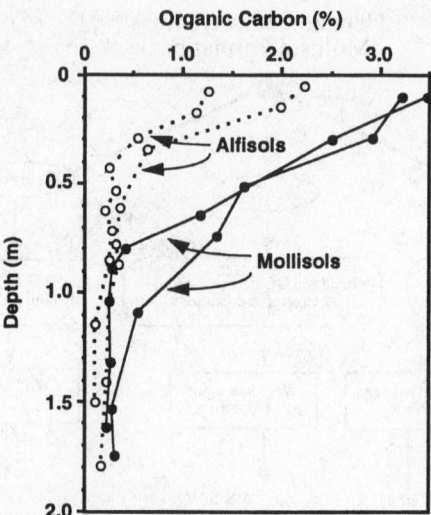

Fig. 6.48 Depth distribution of organic C in Mollisols and Alfisols of the Palouse region of northern Idaho. The Mollisols have formed under Idaho fescue grassland with approximately 580 mm of annual precipitation; the Alfisols have formed under grand fir forest with approximately 830 mm of annual precipitation [Data courtesy of Soil Characterization Laboratory, University of Idaho].

composition of grasses and soil parent material. Grasses have a higher ratio of humic to fluvic acids (Glazovskaya, 1985). Novak and Smeck (1991) found higher concentrations of humic substances in the surface horizons of Mollisols compared to Alfisols in southwestern Ohio where these soils are adjacent in the same landscape. Minimal differences were found in fluvic acid content. The combination of humic substance in the presence of calcareous soil parent material results in the formation of Ca humates that are thought to bind silicate grains to organic C which becomes stabilized (Evans and Russell, 1959; Stevenson, 1994). Formation of these complexes increases the resistance of SOM to physical disintegration, chemical extraction, and further biological change through microbial decomposition (Oades, 1989; Stevenson, 1994). Estimates as to the extent of organic matter-mineral associations in grassland soils vary. As much as 80% of the organic C in a mollic epipedon may be so closely associated with the mineral fraction that it cannot be separated by physical means (Bremmer, 1965; Greenland, 1965; McKeague et al., 1986). Researchers have been able to determine mean residence time (MRT) of organic fractions of Mollisols in several different environments. The MRT represents the average length of time that an organic fraction has been present in the soil. Oldest MRT values measured in soils are associated with Mollisols and Histosols (Oades, 1989), for example, between 1255 and 2973 years for Mollisols in Canada and the United States (Anderson and Paul, 1984; Hsieh, 1992).

The stabilization of organic C in Mollic epipedons is also facilitated by soil clays. Organic C associated with fine clays is protected from further rapid decomposition lengthening the turnover time of otherwise labile humic substances from days to months, years, or even decades (Anderson and Paul, 1994). Anderson (1979) found that as much as 50% of the total humus in some grasslands is associated with clay. From these and other studies, it appears that texture is an important factor in determining the stable level of organic C in Mollisols as well as other soils. Texture influences the water-holding capacity of a soil and the quantity of clays available to form complexes with organic matter. Nichols (1984) observed a good correlation (r = 0.86) between soil organic C and clay content in 65 Mollisols and associated soils of Texas, Oklahoma, and New Mexico where mean annual temperatures exceed ~15 °C (Fig. 6.49). These data indicate that water-holding capacity and the protection afforded to humus through formation of mineral organic complexes may control the equilibrium organic C contents in Mollisols of warmer regions. In contrast, clay content does not appear to exert such strong control of organic C contents in Mollisols of cooler regions. McDaniel and Munn (1985) found little correlation between SOM and clay in 137 Mollisols and associated soils in Montana and Wyoming having mean annual soil temperatures < 8 °C (Fig. 6.50). Comparison of these data with those of Nichols (1984) suggests that it is temperature rather than moisture that becomes the determining factor in establishing an equilibrium SOM content. Furthermore, organic matter may be able to persist in cold soils without the stabilizing influence afforded by complexation with clays (McDaniel and Munn, 1985).

Wet Mollisols (Albolls, Aquolls)

An alternate genetic pathway for the development of Mollisols involves organic C accumulation under anaerobic soil conditions due to reduced rates of organic matter decomposition (Ponnamperuma, 1972; Gambrell and Patrick, 1978; Tate 1980). These Mollisols have an aquic soil moisture regime and are classified as Aquolls or Albolls. The decomposition of organic matter is less efficient under anaerobic conditions and a net accumulation of SOM will occur if rates of biomass production are sufficiently high. If this rate of organic matter accumulation is high enough, organic soils (Histosols) will develop. In many landscapes, Mollisols are found at the transition between organic and mineral soils.

In general, Aquolls have black (N 2/0) surface horizons with predominant gleyed or depleted horizons directly beneath the surface horizon (Table 6.38). Increases in the duration of soil saturation and anaerobic conditions are usually associated with higher soil organic C contents (Khan and Fenton, 1994; Bell et al., 1995). Topographically, organic C increases downslope evidenced by the darker soil surfaces in lower landscape positions (Fig. 6.44) associated with higher surface concentrations of organic C in depressional areas (Fig. 6.46). Thompson and Bell (1996) found good agreement between a profile darkness index based on the thickness and darkness of the surface horizons and the duration of saturation for soil hydrosequences at several locations in the United States. They also

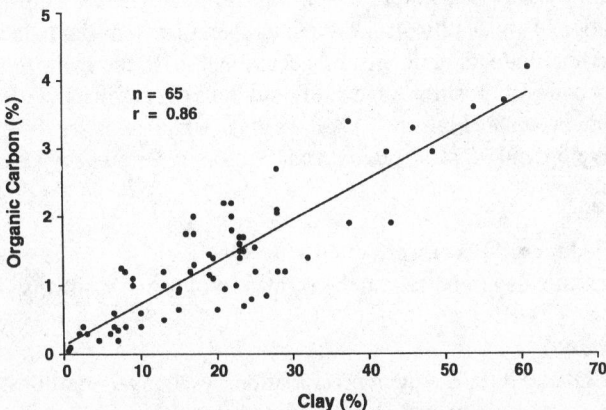

Fig. 6.49 Relationship between clay and organic matter content in warm grassland soils of the southern United States [Adapted from Nichols, 1984. Soil Sci. Soc. Am. J. 48:1382-1384 with permission of the Soil Science Society of America]

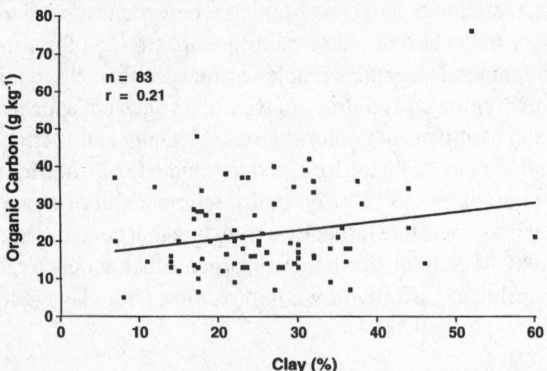

Fig. 6.50 Relationship between clay and organic matter content in cool grassland soils of the northern United States [Adapted from McDaniel and Munn, 1985. Soil Sci. Soc. Am. J. 49:1486-1489 with permission of the Soil Science Society of America]

found that the profile darkness index was highly correlated with soil organic C for the soils that were derived from glacial deposits.

Soil saturation, the development of anaerobic conditions and subsequent reduction of Fe(III) to Fe(II), causes the formation of distinctive color patterns (redoximorphic features) in the horizons beneath the mollic epipedon. Biochemical reduction results in the translocation of Fe compounds, which are the primary determinants of soil color in low organic matter horizons. The processes involved in the development of redoximorphic features are discussed in Section B, Chapter 5. If the soil water is stagnant, Fe(II) which has a distinctive bluish or greenish color creating a reduced (or gleyed) soil matrix is often present. If the soil water is moving, Fe(III) coatings on mineral grains will be removed as Fe^{2+} ions, leaving the mineral grains the dull, gray color of the minerals, known as depletions. Under conditions of fluctuating water tables, Fe(II) may be reoxidized creating discrete concentrations of orange or reddish Fe(III) (Vepraskas and Sprecher, 1997). Bell and Richardson (1997) discuss specific soil process and morphologies associated with Aquolls and Albolls.

Cold Mollisols (Cryolls)

The formation of Mollisols under cold conditions (mean annual soil temperature < 8 °C) is somewhat analogous to the formation of wet Mollisols. Low soil temperatures (similar to lack of soil O_2) reduce microbial activity and facilitate organic matter accumulation if the long term rate of biomass production exceeds that of decomposition. Again, organic soils (Borofolists) will develop where rates of organic matter accumulation are high and Cryolls in areas where rates are lower. Cryolls, Aquolls, and Albolls usually develop under grassland, but can develop under other types of vegetation where conditions are favorable.

6.8.4.3 Associated Pedogenic Processes

The following processes are also important in the formation of some Mollisols.

Carbonate Translocation

The dissolution of $CaCO_3$ and its subsequent precipitation as secondary carbonates lower in the soil profile is common in many calcareous parent materials where sufficient leaching takes place. Carbonates in layers of maximum accumulation are more finely divided than those associated with the parent material supporting secondary carbonate formation (Redmond and McClelland, 1959). Total precipitation is important in determining the depth at which $CaCO_3$ is deposited (Jenny, 1941). In many Xerolls where precipitation is low but exceeds evapotranspiration especially in winter, $CaCO_3$ is readily mobilized and reprecipitated at depth. In the Palouse region of eastern Washington and northern Idaho, $CaCO_3$ has been removed from the upper 1.5 m of the profile in Xerolls receiving more than 530 mm of rainfall (Barker, 1981). In contrast, Ustolls of the Great Plains receiving comparable amounts of total rainfall typically have well-developed zones of $CaCO_3$ accumulation within 1 m of the soil surface. The Ustolls receive much of their precipitation during the growing season when evapotranspirational demand is high, resulting in less moisture being available for mobilization of $CaCO_3$. As previously discussed, spatial patterns of carbonate removal and accumulation along hillslopes can often be used to interpret soil hydrology.

Clay Translocation

The movement and accumulation of clay are common soil-forming processes that occur in soils occupying relatively stable landscape positions and depression-focused recharge depressional basins. Clay translocation causes differences in soil texture that can affect soil water movement and the subsequent genesis of certain soil morphological features. This process is thought to be important in the formation of Albolls. Appreciable clay movement does not occur until carbonates have been removed from the upper portions of the soil profile (Duchaufour, 1982; Fanning and Fanning, 1989). Carbonate removal releases clay previously cemented which may be subject to further chemical alteration prior to movement (Fenton, 1983). In calcareous parent materials, subsoil clay accumulation commonly occurs immediately above the zone of $CaCO_3$ accumulation. In Mollisols, subsurface clay accumulation is highly variable and is not always observed.

In a study of Mollisols and associated soils in the forest steppe-dry steppe transition in Eastern Europe and the forest-prairie transition in the United States Great Plains, Bronger (1991) was unable to find micromorphological evidence of clay illuviation in Mollisols commonly described as having argillic horizons. While Nettleton et al. (1969) proposed that the lack of illuvial clay skins in some fine-textured soils was due to disruption by shrink/swell clays, Bronger (1991) suggested that increased clay contents in Mollisol subsoils are either lithologic discontinuities or are polygenetic in origin and, as such, are relict features that formed under older and moister climates. Sobecki and Wilding (1983) investigated Texas Coast Prairie Mollisols and found that argillic horizons were confined to microtopographic lows, suggesting depression-focused recharge. Weakly developed Bt horizons on microtopographic highs were determined to be relict features based on micromorphological analysis. They suggested that processes leading to both carbonate accumulation and clay illuviation in the same horizon were incompatible under the current humid climate.

Erosion and Sedimentation

Erosional processes result in the redistribution of surface soil material from upper to lower hillslope positions. This process often results in thickening of the mollic epipedons in concave slope positions. Soils with these overthickened surface horizons are often classified as cumulic at the subgroup level. Anthropogenically accelerated erosion causes significant reduction in the thickness of A upland horizons which no longer meet the surface horizon thickness criteria for Mollisols. These eroded soils will typically be classified as Inceptisols or Alfisols (if an argillic horizon is present). Erosional losses as a genetic pathway for Alfisols or Inceptisols which is contrary to the conceptual models for the genesis of these soils, has been a long-standing taxonomic dilemma in the classification of Mollisols (Fenton, 1983).

6.8.5   Land Use

Human impact on the use of Mollisols has increased dramatically over the past century. Early cultures had little influence on grassland soils other than the use of fire as a management tool. In the plains of North America, burning was used to stimulate the subsequent year's production and attract buffalo (Warkentin, 1969; Anderson, 1987). With the advent of the moldboard plow, Mollisols became widely used as arable soils. During the early 1900s, artificial drainage (tile drains and surface ditching) and flood protection measures lowered watertables sufficiently for the conversion of extensive areas of Aquolls and other marginally wet Mollisols to productive cropland. In the United States, where Mollisols make up ~25% of the land area, much of the agriculturally based, westward settlement was a direct result of the ease with which productive Mollisols could be cleared by fire. The high level of native fertility and favorable physical properties meant that most areas of Mollisols were converted to row cropping. Virgin areas of Mollisols and associated prairies are now quite rare in the United States.

Mollisols are among the most agriculturally productive soils in the world. Although Mollisols make up only ~7% of the Earth's ice free land area, a large part of the world's wheat and other small grains are grown on Ustolls and Xerolls (Troeh and Thompson, 1993). Because Mollisols are commonly found in subhumid climates, lack of sufficient soil moisture can limit the production of traditional agricultural crops. Soil moisture conservation strategies, such as fallowing and residue management, must be implemented for sustainable dryland farming. Unfortunately, the use of unsustainable farming practices not designed to protect the soil from erosion or to conserve soil organic C, has led to a decline in soil quality and subsequent productivity in some regions of the world. Long-term cultivation has reduced the organic C content of Mollisols; reductions of ~35 % in C contents in 60 to 70 yr of cultivation have been documented (Tiessen et al., 1982). These reductions are often associated with degradation of soil structure and increased susceptibility to erosion. Additionally, the conversion of prairie to cropland further increases the erosion potential and has resulted in devastating wind erosion during cyclic droughts in portions of North America and Asia. These processes of soil degradation result in loss of organic matter and organically bound nutrients and subsequent declines in soil productivity. The use of appropriate soil conservation practices can greatly reduce soil degradation and maintain soil productivity.

As population pressure has increased, conversion of Mollisols from arable to urban uses increased. Many of the same characteristics that make Mollisols preferable for cropland also favor many non-agricultural uses. The conversion of these prime agricultural soils and the associated decline in the resource base for food and fiber production has generated much debate in the United States and stimulated farmland preservation efforts.

6.8.6   References

Allen, B.L., and D.S. Fanning. 1983. Composition and soil genesis. p. 141–192. *In* L.P. Wilding et al. (ed.) Pedogenesis and soil taxonomy. I. Concepts and interactions. Elsevier, Amsterdam, Netherlands.

Allen, B.L., and B.F. Hajek. 1989. Mineral occurrence in soil environments. p. 199–278. *In* J.B. Dixon and S.B. Weed (ed.) Minerals in soil environments. 2nd Ed. Soil Science Society of America, Madison, WI.

Anderson, D.W. 1987. Pedogenesis of grassland and adjacent forests of the Great Plains. Adv. Soil Sci. 7:53–93.

Anderson, D.W., and E.A. Paul. 1984. Organo-mineral complexes and their study by radiocarbon dating. Soil Sci. Soc. Am. J. 48:298–301.

Anderson, D. W. 1979. Processes of humus formation and transformations in soils of the Canadian Great Plains. J. Soil Sci. 30:79–84.

Barker, R.J. 1981. Soil survey of Latah County area, Idaho. USDA-SCS. US Government Printing Office, Washington, DC.

Baxter, F.P., and F.D. Hole. 1967. Ant (*Formica cinerea*) pedoturbation in a prairie soil. Soil Sci. Soc. Am. Proc. 31:425–28.

Bell, J.C., and J.L. Richardson. 1997. Aquic conditions and hydric soil indicators for Aquolls and Albolls. p. 23–40. *In* M.J. Vepraskas and S. Sprecher (ed.) Aquic conditions and hydric soils: The problem soils. Soil Science Society of America, Madison, WI.

Bell, J.C., J.A. Thompson, and C.A. Butler. 1995. Morphological indicators of seasonally saturated soils for a hydrosequence in southeastern Minnesota. J. Minn. Acad. Sci. 59:25–34.

Bolin, B.E., T. Degens, P. Duvigneaud, and S. Kempe. 1979. The global biogeochemical carbon cycle. p. 1–56. *In* B. Bolin et al. (ed.) The global carbon cycle. John Wiley and Sons, Ltd, Chichester, UK.

Bremner, J. M. 1965. Organic nitrogen in soils. p. 93–149 *In* W.V. Bartholomew and F.E. Clark (ed.) Soil nitrogen. Soil Science Society of America, Madison, WI.

Bronger, A. 1991. Agrillic horizons in modern loess soils in an ustic soil moisture regime: Comparative studies in forest-steppe and steppe areas from eastern Europe and the United States. Adv. Soil Sci. 15:41–90.

Buol, S.W., F.D. Hole, and R.J. McCracken. 1989. Soil genesis and classification. 3rd Ed. Iowa State University Press, Ames, IA.

Caldwell, M. 1975. p. 41–73. *In* J.P. Cooper (ed.) Photosynthesis and productivity in different environments. Cambridge University Press, New York, NY.

Cannon, M.E., and G.A. Nielsen. 1984. Estimating production of range vegetation from easily measured soil characteristics. Soil Sci. Soc. Am. J. 48:1393–1397.

Curtis, J.T. 1959. The vegetation of Wisconsin: An ordination of plant communities. University of Wisconsin Press, Madison, WI.

Dahlman, R.C. and C.L. Kucera. 1965. Root productivity and turnover in native prairie. Ecol. 46:84–89.

Dokuchaev, V.V. 1883. Russian Chernozems (Russkii Chernozem). Israel Prog. Sci. Trans., Jerusalem, 1967. Transl. from Russian by N. Kaner. US Department of Commerce, Springfield, VA.

Dokuchaev, V.V. 1886. Report to the provincial zenstvo (local authority) of Nizhnii-Novgorod (Gorki), No. 1. Main phases in the history of land assessment in European Russia, with classification of Russian soils. In Acad. Sci., USSR. Collected writings (Sochineniya), Vol. 4, Moscow, 1950.

Dormaar, J.F. and D.R. Sauerbeck. 1983. Seasonal effects on photoassimilated carbon-14 in the root system of blue gramma and associated soil organic matter. Soil Biol. Biochem. 15:475–479.

Duchaufour, P. 1982. Pedology: Pedogenesis and classification. George Allen and Unwin, London, UK.

Evans, L.T., and E.W. Russell. 1959. The adsorption of huic and fluvic acids by clays. J. Soil Sci. 10:119–132.

Fanning, D.S., and Fanning, M.C.B. 1989. Soil morphology, genesis, and classification. John Wiley and Sons, New York, NY.

Fenton, T.E. 1983. Mollisols. p. 125–163. *In* L.P. Wilding, N.E. Smeck and G.G. Hall (ed.) Pedogenesis and soil taxonomy. II. The soil orders. Elsevier, Amsterdam, Netherlands.

Foss, J.E., M.G. Michlovic, J.L. Richardson, J.L. Arndt, and M.E. Timpson. 1985. Pedologic study of archaelogical sites along the Red River. ND Acad. Sci. Proc. 39:51–69.

Gambrell, R.P., and W.H. Patrick, Jr. 1978. Chemical and microbilogical properties of anaerobic soils and sediments. p. 375–423. *In* D.D. Hook and R.M.M. Crawford (ed.) Plant life in anaerobic environments. Ann Arbor Science Publishers Inc., Ann Arbor, MI.

Geis, J.W., and W.R. Burgess. 1968. The Prairie Peninsular: Its origin and significance in the vegetational history of Central Illinois. Univ. IL Coll. Agr. Spec. Pub. 14.

Glazovskaya, M.A. 1985. Soils of the world. Vol. I. A.A. Balkema Publishers, Rotterdam, Netherlands.

Graham, R.C., and A.R. Southerd. 1983. Genesis of a Vertisol and associated Mollisol in Northern Utah. Soil Sci. Soc. Am. J. 47:552–559.

Greenland, D.J. 1965. Interaction between clays and organic compounds in soils. Part 2. Soils Fert. 28:521–532.

Hole, F.D. and G.A. Nielsen. 1968. Some processes of soil genesis under prairie. p. 28–34. *In* P. Schramm (ed.) Proc. Symp. of Prairie and Prairie Restoration. Spec. Pub. 3. Know College Field Station, Galesburg, IL.

Hsieh, Y.P. 1992. Pool size and mean age of stable soil organic carbon in cropland. Soil Sci. Soc. Am J. 56:460–464.

Jenny, H. 1930. A study on the influence of climate upon the nitrogen and organic matter content of soils. Univ. Missouri Agric. Exp. Stn. Res. Bull. 52.

Jenny, H. 1941. Factors of soil formation: A system of quantitative pedology. McGraw-Hill, New York, NY.

Jenny, H. 1980. The soil resource: Origin and behavior. Ecol. Stud. 37:1–377.

Khan, F.A., and T.E. Fenton. 1994. Saturated zones and soil morphology in a Mollisol catena of central Iowa. Soil Sci. Soc. Am. J. 58:1457–1464.

Kodama, H. 1979. Clay minerals in Canadian soils: Their origin, distribution, and alteration. Can. J. Soil Sci. 59:37–58.

Lauenroth, W.K., and W.C. Whitman. 1977. Dynamics of dry matter production in a mixed-grass prairie in western North Dakota. Oecologia 27:339–351.

Mausbach, M.J., and J.L. Richardson. 1994. Biogeochemical processes in hydric soil formation. Cur. Topics Wetld. Biogeoch. 1:68–127.

McDaniel, P.A., and L.C. Munn. 1985. Effect of temperature on organic carbon-texture relationships in Mollisols and Aridisols. Soil Sci. Soc. Am. J. 49:1486–1489.

McKeague, J.A., M.V. Cheshire, F. Andreux, and J. Berthelin. 1986. Organo-mineral complexes in relation to pedogenesis. p. 549–592. *In* P.M. Huang and M. Schnitzer (ed.) Interactions of soil minerals with natural organics and microbes. Soil Science Society of America, Madison, WI.

Munn, L.C. 1993. Effect of prairie dogs on physical and chemical properties of soils. *In* J.L. Oldemeyer, D.E. Biggins, B.J. Miller and R. Crete (ed.) Management of prairie dog complexes for reintroduction of the black-footed ferret. USDA Fish and Wildlife Service, Washington, DC.

Munn, L.C., G.A. Nielsen, and W.F. Mueggler. 1978. Relationships of soils to mountain and foothill range habitat types and production in western Montana. Soil Sci. Soc. Am. J. 42:135–139.

Nettleton, W.D., K.W. Flatch, and B.R. Brasher. 1969. Argillic horizons without clay skins. Soil Sci. Soc. Am. Proc. 33:121–125.

Nichols, J.D. 1984. Relation of organic carbon to soil properties and climate in the southern Great Plains. Soil Sci. Soc. Am. J. 48:1382–1384.

Novak, J.M., and N.E. Smeck. 1991. Comparison of humic substances extracted from contiguous Alfisols and Mollisols in southwestern Ohio. Soil Sci. Soc. Am. J. 55:96–102.

Oades, J.M. 1989. An introduction to organic matter in mineral soils. p. 89–159. *In* J.B. Dixon and S.B. Weed (ed.) Minerals in soil environments. 2nd Ed. Soil Science Society of America, Madison, WI.

Ponnamperuma, F.N. 1972. The chemistry of submerged soils. Adv. Agron. 24:29–96.

Redmond, C.E., and J.E. McClelland. 1959. Occurrence and distribution of lime in calcium carbonate Solonchak and associated soils of eastern North Dakota. Soil Sci. Soc. Am. Proc. 23:61–65.

Richardson, J.L., L.P. Wilding, and R.B. Daniels. 1992. Recharge and discharge of groundwater in aquic conditions illustrated with flownet analysis. Geoderma 53:65–78.

Ruhe, R.V. 1970. Soils paleosols, and environment. p. 37–52. *In* W. Dort, Jr. and J.K. Jones, Jr. (ed.) Pleistocene and recent environments of the central Great Plains. University Press of Kansas, Lawrence, KS.

Schafer, W.M., G.A. Nielsen, and W.D. Nettleton. 1980. Minesoil genesis and morphology in a spoil chronosequence in Montana. Soil Sci. Soc. Am. J. 44:802–807.

Schlesinger, W.H. 1977. Carbon balance in terrestrial detritus. Ann. Rev. Ecol. Syst. 8:51–81.

Schlesinger, W.H. 1991. Biogeochemistry: An analysis of global change. Academic Press, San Diego, CA.

Schultze, D.G., J.L. Nagel, G.E. van Scoyoc, T.L. Henderson, M.F. Baumgarder, and D.E. Stott. 1993. Significance of organic matter in determining soil color. p. 71–90. *In* J.M. Bigahm and E.J. Ciolkosz (ed.) Soil color. Soil Science Society of America, Madison, WI.

Smith, G. 1986. The Guy Smith interviews: Rationale for concepts in Soil Taxonomy. T.R. Forbes (ed.) SMSS Tech. Monogr. 11.

Sobecki, T.M., and L.P. Wilding. 1983. Formation of calcic and argillic horizons in selected soils of the Texas coast prairie. Soil Sci. Soc. Am. J. 47:707–715.

Soil Survey Staff. 1975. Soil taxonomy: A basic system of soil classification for making and interpreting soil surveys. USDA Agric. Handb. 436. US Government Printing Office, Washington, DC.

Soil Survey Staff. 1998. Keys to soil taxonomy. 8th ed. USDA-NRCS. US Government Printing Office, Washington, DC.

Stevenson, F.J. 1994. Humus chemistry: Genesis, composition, reactions. 2nd Ed. John Wiley and Sons, New York, NY.

Tate, R.L., III. 1980. Microbial oxidation of organic matter of Histosols. Adv. Microbial Ecol. 4:169–201.

Tate, R.L., III. 1992. Soil organic matter: Biological and ecological effects. Krieger Publishing Co., Malabar, FL.

Thompson, J.A. and J.C. Bell. 1996. A color index for identifying hydric conditons for seasonally saturated conditions prairie soils in Minnesota. Soil Sci. Soc. Am. J. 60:1979–1988.

Thompson, J.A., J.C. Bell, and C.A. Butler. 1997. Quantitative soil-landscape modeling for estimating the areal extent of hydromorphic soils. Soil Sci. Soc. Am. J. 61:971–980.

Thorp, J. 1948. How soils develop under grass. p. 55–66. *In* Yearbook of Agriculture, US Government Printing Office, Washington, DC.

Tiessen, H., J.W.B. Stewart, and J.R. Bettany. 1982. Cultivation effects on the amounts and concentration of carbon, nitrogen, and phophorus in grassland soils. Agron. J. 74:831–835.

Troeh, F.R., and L.M. Thompson. 1993. Soils and soil fertility. Oxford University Press, New York, NY.

USDA-NRCS. 1997. Official soil series descriptions fron the Internet. http://www.statlab.iastate.edu:80/soils/osd.

Vepraskas, M.J., and S.W. Sprecher. 1997. Overview of aquic conditions and hydric soils. p. 1–22. *In* M.J. Vepraskas and S. Sprecher (ed.) Aquic conditions and hydric soils: The problem soils. Soil Sci. Soc. Am. Spec. Pub. 50. Soil Science Society of America, Madison, WI.

Walker, P.H.. 1966. Postglacial environments in relation to landscape and soils on the Cary Drift, Iowa. IA Ag. Home Ec. Exp. Sta. Res. Bul. 549:838–875.

Warkentin, J. 1969. The western interior of Canada. McClelland and Stewart Ltd., Toronto, Canada.

Wright, H.E. 1970. Vegetational history of the central plains. p. 157–172. *In* W. Dort, Jr. and J.K. Jones, Jr. (ed.) Pleistocene and recent environments of the central Great Plains. University Press of Kansas, Lawrence, KS.

6.9 Spodosols

Delbert L. Mokma
Michigan State University

Christine V. Evans
University of New Hampshire

6.9.1 Introduction

The formative element of the word, Spodosols, was derived from the Greek, *spodos*, meaning wood ashes. The name is connotative of the bleached eluvial (E) horizon that often overlies the distinctive reddish brown spodic (Bs or Bhs) horizon, which is characterized by an accumulation of active amorphous materials composed of organic matter and Al with or without Fe (Soil Survey Staff, 1998). Most Spodosols are sandy. The strongest grade of structure is in B horizons as a result of the accumulation of C, Al, and Fe. The humus and sesquioxides provide the cohesion in the sandy materials. Organic matter also produces structure in A horizons. Sandy E and C horizons which usually are single grained, have a high proportion of macropores, and are rapidly permeable. The sandy nature of Spodosols gives them a relatively low water-holding capacity. The available water-holding capacity increases as the amount of amorphous materials in B horizons increases (Shetron, 1974).

Typical Spodosol profiles (Fig. 6.51) have four major horizons. A or O horizons have a black or dark brown color (Table 6.42), as a result of accumulation of organic materials. E horizons have an ashy gray color, frequently albic (Soil Survey Staff, 1998), as a result of the weathering of nonresistant minerals and subsequent eluviation of amorphous materials. The Bs or Bhs horizons have dark reddish brown colors as a result of the illuvial accumulation of organometallic complexes; usually the reddest hue, lowest value and lowest chroma occur in the upper spodic subhorizon. The distinctive reddish brown color of Spodosol subsoils is related to amorphous material (Mokma, 1993). With depth, hues become more yellow, and value and chroma increase. C horizons usually have a yellower hue and higher value than the spodic horizon (Table 6.42; Fig. 6.51). The POD index is a numerical index of Spodosol development based on differences between eluvial and illuvial horizon colors (Schaetzl and Mokma, 1988). Redder Munsell hues and thicker illuvial horizons result in higher POD indices. The POD index of the Typic Haplorthod in Table 6.42 is 8.

The E horizons generally have uncoated sand grains and edge weathering on ferromagnesian minerals (Fig. 6.52). As organometallic complexes are immobilized in B horizons, they form coatings on most sand grains (Fig. 6.53). With time these coatings become thicker. As vegetation removes water from the horizon, desiccation occurs allowing van der Waals bonds and protonic bridges (De Coninck, 1980) to form, causing the horizon to become at least partially hydrophobic. Desiccation is also responsible for the development of cracked coatings (Fig. 6.54), the presence of which is a criterion for spodic materials (Soil Survey Staff, 1998). Many Spodosols also have thin Al and/or Fe

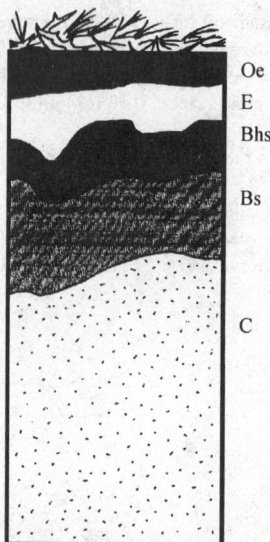

Fig. 6.51 Typical Spodosol profile

oxide coatings on sand grains in C horizons. For a more comprehensive treatment of Spodosols, the reader is referred to McKeague et al. (1983).

6.9.2   Distribution

Spodosols are found on all continents, but are most extensive in cool, humid climates (Fig. 6.56). They occupy about 2.5% of the ice-free land surface (Table 6.1). Spodosols occupy about $3, 354, 000$ km^2. Some Spodosols have also formed in hot, humid areas in quartz-rich sands that are nearly devoid of weatherable minerals. Spodosols are absent in arid regions. Most Spodosols in the United States are in New England, New York, Great Lakes states, Alaska, along the Atlantic coast in the South, and in high mountains in the West.

6.9.3   Formation of Spodosols

6.9.3.1   Parent Materials

Sand fractions are composed primarily of quartz with lesser amounts of orthoclase, plagioclase, mica, pyroxene, and amphibole (Haile-mariam and Mokma, 1995). Accessory minerals are often highly

Table 6.42   Selected morphological properties of a Typic Haplorthod

Horizon	Depth (cm)	Texture	Moist Color
O	0-4	s	N 2/0
A	4-8	s	10YR 2/1
E	8-26	s	10YR 5/2
Bhs	26-38	s	7.5YR 3/2
Bs	38-57	s	7.5YR 3/4
BC	57-96	s	10YR 4/4
C	96-157	s	10YR 5/4

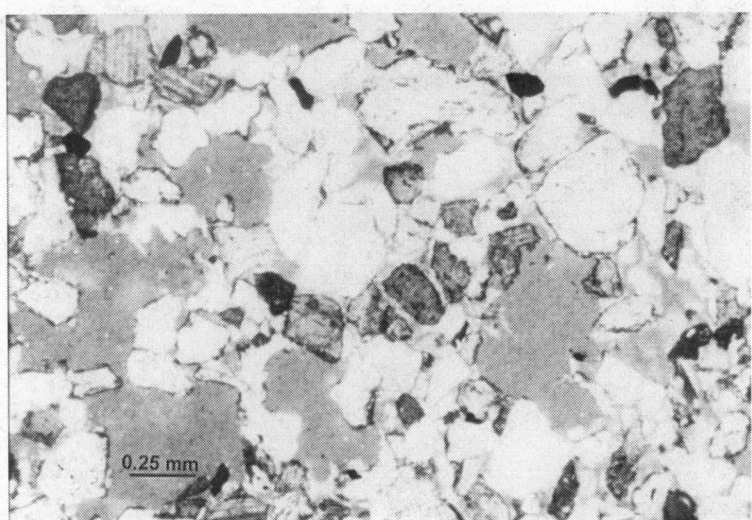

Fig. 6.52 Microphoto of E horizon in polarized light. Most grains are quartz. A few feldspar grains are conspicuous due to twinned diffraction.

weathered, opaque grains that are not easily identifiable. Less resistant minerals have weathered from E and B horizons. Parent materials contain little or no carbonate minerals and are either naturally acidic or intensively leached. Silt fractions are composed dominantly of quartz with lesser amounts of orthoclase and plagioclase. Most Spodosols have little silicate clay. Phyllosilicate minerals present in clay fractions of parent materials include mica, chlorite, and small amounts of kaolinite and vermiculite. Clay mineral distributions in Spodosols suggest that mica transforms to vermiculite and hydroxy interlayered vermiculite in B horizons, and to smectite through an intermediate vermiculite

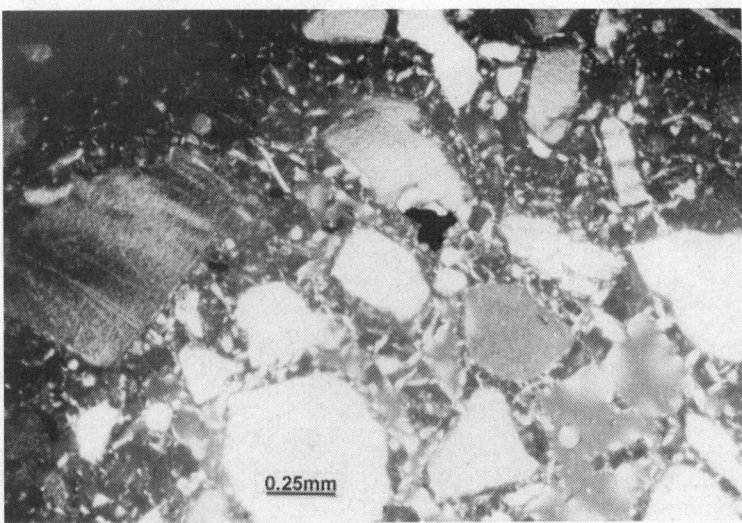

Fig. 6.53 Microphoto of Bs horizon showing distribution of sesquioxide coatings on sand grains in plain transmitted and reflected light

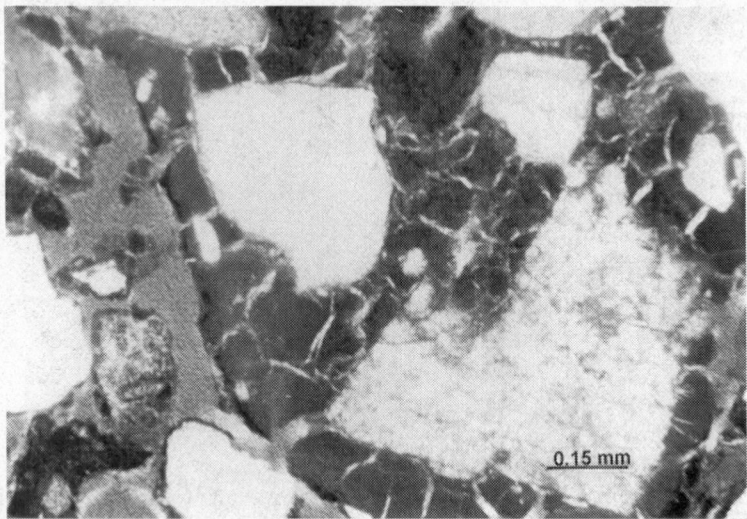

Fig. 6.54 Microphoto of Bsm horizon (ortstein) showing development of sesquioxide matrix with desiccation cracking in plain transmitted and reflected light

state, in E horizons. Nonphyllosilicate minerals present in the < 2 mm fraction are quartz and small
amounts of feldspars.

As the silicate clay content increases, the translocation of humus and sesquioxides decreases; the
thickness of E and B horizons decreases, and B horizon color is less red and less dark (Gardner and
Whiteside, 1952). Silicate clay tends to inhibit translocation of humus and sesquioxides by adsorbing

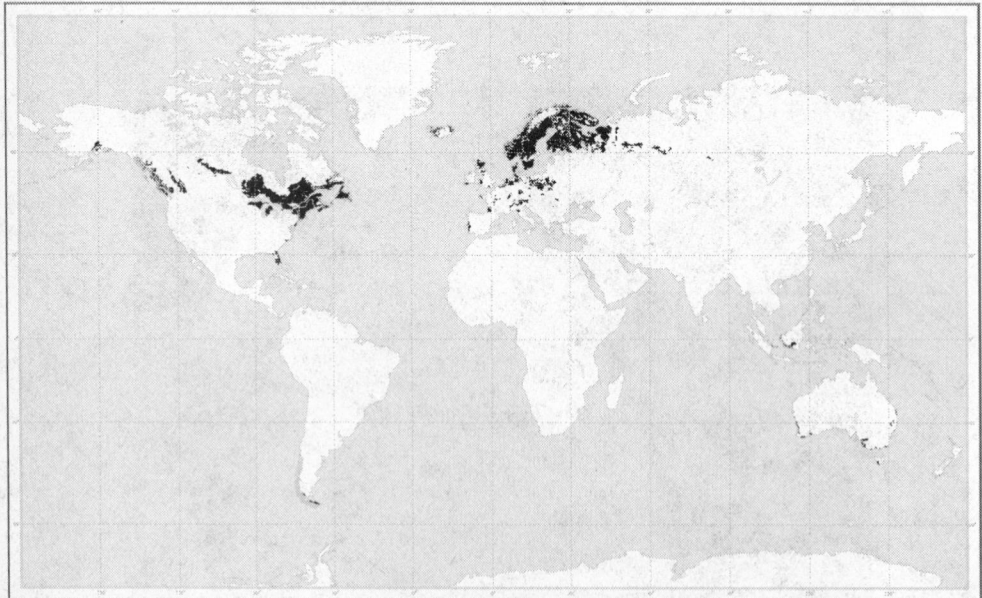

Fig. 6.55 Global distribution of Spodosols [Courtesy USDA-NRCS, Soil Survey Division, World Soil Resources,
1998]

these components. Also, weathering of silicate clays in acid, upper horizons may release many cations, immobilizing chelate complexes and inhibiting translocation.

6.9.3.2 Topography

Slope Position and Aspect
In Michigan Spodosols on north to northeast facing slopes are more strongly developed than those on south to southwest facing slopes (Table 6.43) (Hunckler, 1996).

Soil Drainage
Spodosols form on surfaces that have very deep to shallow watertables. Development of spodic horizons is influenced by depth and duration of water saturation (Mokma, 1991; Mokma and Sprecher, 1994). Spodic horizons do not appear to form in soils that are permanently saturated near the soil surface. Ratios of resistant minerals in the sand fraction of subsoils to those in parent materials tend to decrease from well (1.08) to somewhat poorly (1.06) to poorly drained (1.04) pedons. Ratios of quartz contents in the silt fractions of the subsoils to those in parent materials tend to be greater for horizons that were not water saturated (1.04) than for those that were saturated (0.93). This suggests that water saturation hinders weathering of sand and silt sized minerals, as well as translocation processes. Poorly drained soils in Spodosol landscapes tend to have less C, Al, and Fe in their B horizons than do better drained Spodosols. Some of these have particularly low Fe levels because it was chemically reduced and leached from the soil. In some low lying landscapes, hydrologic discharge may result in reversal of translocation, such that the spodic horizon may accumulate materials eluviated from the C horizon, as the water table moves upward (Evans and Mokma, 1996). These soils were formerly referred to as Groundwater Podzols (Baldwin et al., 1938).

Maximum accumulations of C, Al, and Fe occur usually in somewhat poorly drained Spodosols (Table 6.44) in which the saturated zone acts as a barrier to water movement, and organometallic complexes are carried from A and E horizons to B horizons, but rarely move through the B horizons in significant amounts. In these soils, intensity of oxalate extractable Fe illuviation into spodic horizons is strongly related ($R^2 = 0.7$) to the frequency of watertable fluxes (Evans and Mokma, 1996). Spodic horizon chemistry is unresponsive to duration of saturation and watertable fluctuation, but C

Table 6.43 Mean optical density of oxalate extract and extractable Al and Fe in paired pedons on opposing aspects [Adapted from Hunckler, 1996]

Aspect	Horizon	ODOE[#]	Al_o[†]	Al_p[†]	Fe_o[‡]	Fe_p[‡]
			-------------------------- g kg⁻¹ --------------------------			
N-NE	E	0.0*	0.2*	0.1*	0.2*	0.1*
	Bs1	0.3*	4.7	2.5*	5.1*	2.5*
	Bs2	0.1	3.5*	1.7*	2.3	1.1
S-SW	E	0.1	0.3	0.2	0.4	0.2
	Bs1	0.1	3.5	1.4	3.5	1.1
	Bs2	0.1	2.1	0.9	1.8	0.6

[#] ODOE = optical density of oxalate extract
[†] Al_o and Fe_o = oxalate extractable Al and Fe
[‡] Al_p and Fe_p = pyrophosphate extractable Al and Fe
* Indicates or promotes significantly stronger podzolization when comparing equivalent horizons of paired pedons on opposing aspects.

Table 6.44 Weighted index amounts of pyrophosphate extractable Al and Fe in soils of a Michigan hydrosequence

Drainage Class	Horizon	Al	Fe
		-------------g kg^{-1} x cm-------------	
Well	E	1	2
	Bs1	42	47
	Bs2	29	16
Moderately Well	E	1	1
	Bs1	48	19
	Bs2	24	18
Somewhat Poorly	E	2	2
	Bs1	75	112
	Bs2	76	36
Poorly	Bg	4	0

horizon chemistry is more responsive and that of E horizons is most responsive to watertable fluctuation patterns (Evans and Mokma, 1996).

In well-drained Spodosols, organometallic complexes can be carried by percolating waters into, and even through, B horizons. The brownish colored lakes and streams in Spodosol areas support this; however, spodic horizons are a significant sink for Fe, Al, and dissolved organic C (DOC) that effectively decreases stream content of these elements (McDowell and Wood, 1984).

Vegetation

Spodosols form under coniferous or deciduous forest but not under prairie vegetation. More Al and Fe occurs in B horizons formed under conifers than under deciduous trees (Messenger et al., 1972; De Kimpe and Martel, 1976). In mixed forests, Spodosols are more strongly expressed under some species, such as hemlock (*Tsuga canadensis*) and Kauri pine (*Agathis australis*), than under other species. Since organic C is involved in the translocation of C, Al, and Fe, the more leaf/needle litter produced, the more organic compounds to form organometallic complexes.

In some areas of sandy Spodosols, fire frequency affects the formation of B horizons by influencing forest successional events (Mokma and Vance, 1989). When the interval between severe fires is 5 to 50 yr, jack pine (*Pinus banksiana* L.) survives and regenerates. Formation of the Bs horizon is slow because much of the organic matter in the forest floor is destroyed by the frequent fires. Relatively small amounts of organic compounds are generated for complexing Al and Fe (Table 6.45). The E horizons are thin or nonexistent and B horizons have small amounts of C, Al, and Fe. The soils in this environment are weakly developed. When the fire interval is 50 to 350 yr, red (*Pinus resinosa* Ait.)

Table 6.45 Weighted index amounts of pyrophosphate extractable Al, Fe and C in B horizons of a developmental sequence of soils [Adapted from Vance et al., 1986]

Subgroup	Fire frequency	Al	Fe	C	Al + Fe	Al + Fe + C
	yr	---------------------------------- g kg^{-1} x cm ----------------------------------				
Typic Udipsamment	5 - 50	62	33	196	95	291
Entic Haplorthod	50-350	65	44	275	109	384
Typic Haplorthod	>350	68	59	388	127	515

and/or white (*Pinus strobus* L.) pine succeed the jack pine. More needle litter accumulates as a result of fewer fires; thus, more organic acids and chelating substances are available to translocate Al and Fe from E horizons to B horizons. The soils in this environment are primarily Entic Haplorthods. When fire intervals exceed 350 yr, succession will likely produce a northern hardwood forest [sugar maple (*Acer saccharum* Marsh) and beech (*Fagus grandifolia*)] which is more shade tolerant and more fire resistant. The accumulation of C, Al, and Fe in B horizons continues, but is likely to be slower under hardwoods because less Al and Fe and more Ca are being recycled (Messenger, 1975). The soils in this environment are primarily Typic Haplorthods.

Treethrow is another form of vegetative influence in Spodosol formation. Tree uprooting disrupts soil horizons and arrests and regresses soil development. Portions of the Spodosol profile are removed, producing a pit. The soil material is deposited on top of adjacent soil producing a mound. The resulting microtopography in an area of Haplorthods had Entic Haplorthods or Spodic Udipsamments in pits and Typic Udipsamments on mounds (Schaetzl, 1990).

Climate
Accumulations of C, Al and Fe in spodic horizons tend to be greater in cooler areas of the Appalachian Mountains than in warmer areas (Stanley and Ciolkosz, 1981). In New England, Spodosols are the dominant soil order in frigid and cryic temperature regimes, but occur less frequently in mesic areas, except when aquic conditions are present (Evans, 1996). Similarly, many Aquods also occur in sandy parent materials in Florida. In Michigan, strong spodic development is associated with belts of deep snowpack that were in place before the soil froze. Meltwater from the snow could then flush freely through the soil in the spring (Schaetzel and Isard, 1991).

Time
Spodic horizons can form in a few hundred years if conditions are optimum, but they usually form more slowly. E horizons have been reported in soils that were 300 yr old or less (Gjems, 1960; Ugolini, 1968). Weak Spodosol profiles have been reported after about 2000 yr of soil development (Burges and Drover, 1953; Franzmeier and Whiteside, 1963). Even though weak Bs horizons form in a few thousand yr, it takes 8000 to 10,000 yr of soil development for the majority of soils on a surface to qualify as Spodosols (Table 6.46) (Franzmeier and Whiteside, 1963; Barrett and Schaetzl, 1992).

6.9.4   Pedogenic Processes in Spodosols

6.9.4.1 Eluviation of Humus and Sesquioxides
Formation of the Spodosol profile is thought to involve the formation, translocation and immobilization of organometallic complexes. Organic acids are produced in forest canopies and fine roots and by incomplete decomposition of the forest floor. These acids form soluble organometallic complexes with the Al and Fe released during mineral weathering in A and E horizons. Translocation of clay, resulting in argillic horizon formation, may also occur in Spodosols. The clay moves downward to form E and Bt horizons, and when the E horizon becomes sufficiently coarse, a shallow E-Bhs or Bs horizon sequence forms in it. Before the clay eluviates, it adsorbs the Al and Fe released during mineral weathering and they are not available to form organmetallic complexes.

Acidification and Mineral Weathering
Organic acids move downward through O and A horizons with percolating waters. They acidify the upper mineral horizons initiating intense weathering of minerals resulting in the release of large amounts of sesquioxides, particularly Al, which often shows a strong depth correlation with pH

Table 6.46 Color, optical density of oxalate extract and extractable Al and Fe of pedons from different aged surfaces [Adapted from Barrett and Schaetzl, 1992]

Surface Age	Horizon	Color	ODOE[#]	Al_o[†]	Al_p[†]	Fe_o[‡]	Fe_p[‡]
yr BP					---g kg^{-1}---		
3,000	E	10YR 4/2	0.02	0.1	0.1	0.3	0.1
	Bs	10YR 4/4	0.01	0.2	0.1	0.1	0.1
	C	10YR 6/4	0.00	0.1	0.1	0.1	0.0
4,000	E1	10YR 4/1	0.01	0.2	0.1	0.1	0.0
	E2	10YR 5/3	0.03	0.2	0.1	0.3	0.1
	Bs1	10YR 5/6	0.02	0.7	0.3	0.7	0.3
	Bs2	7.5YR 5/8	0.02	2.3	0.5	0.9	0.2
	Bs3	10YR 5/4	0.02	0.9	0.3	0.4	0.2
	BC	10YR 5/6	0.02	1.3	0.3	0.5	0.1
	C	10YR 6/4	0.01	0.4	0.1	0.2	0.1
10,000	E1	10YR 4/2	0.01	0.1	0.0	0.1	0.0
	E2	7.5YR 4/2	0.05	0.5	0.2	0.7	0.4
	Bs1	7.5YR 4/6	0.09	3.6	1.2	2.3	1.2
	Bs2	7.5YR 3/4	0.22	5.6	1.7	1.4	0.6
	Bs3	10YR 5.6	0.04	3.1	0.8	0.6	0.2
	BC	10YR 6/6	0.02	2.0	0.6	0.4	0.1
	C	10YR 6/4	0.01	1.3	0.4	0.3	0.1
11,000	E	10YR 5/2	0.01	0.1	0.1	0.3	0.2
	Bhs	7.5YR 3/2	0.28	2.8	0.9	3.2	1.3
	Bs1	7.5YR 4/6	0.10	3.3	1.0	1.7	0.6
	Bs2	10YR 4/6	0.04	1.9	0.6	0.7	0.2
	BC	10YR 5/4	0.02	0.9	0.3	0.2	0.1
	C	10YR 6/4	0.02	0.8	0.3	0.3	0.1

[#] ODOE = optical density of oxalate extract
[†] Al_o and Fe_o = oxalate extractable Al and Fe
[‡] Al_p and Fe_p = pyrophosphate extractable Al and Fe
[*] Indicates or promotes significantly stronger podzolization when comparing equivalent horizons of paired pedons on opposing aspects.

(McDowell and Wood, 1984; Mokma and Vance, 1989; Freeland and Evans,1993). The Si, Al, Ca, and Mg in the leachate of granitic parent materials increased with increasing DOC concentration, which coincided with decreasing pH (McCracken et al., 1996). Iron release was also greatest in materials leached with the highest DOC concentration, but at low and medium concentrations, Fe was actually retained by the granitic materials, presumably because the organometallic complex precipitated out of solution (Burman, 1985).

Chelation and Eluviation
Organic compounds form chelate complexes with the Al and Fe released during weathering of minerals. Organic compounds, capable of complexing metal ions such as Al and Fe, are produced in forest canopies, canopy leachate and SOM. The kinds of compounds differ depending on tree species. The complexes are translocated downward by percolating waters.

6.9.4.2 Illuviation of Humus and Sesquioxides

C, Al and Fe
Immobilization of low molecular weight phenolic compounds, such as protocatechuic acids, may occur as a result of increases in pH. Immobilization of larger phenolic compounds is thought to occur

when the amount of Al and Fe taken up by the organic molecules exceeds a critical limit (McKeague et al., 1983). Morphological expression of podzolization has been related to the content of phenolic compounds. Seven phenolic compounds have been extracted from spodic horizons (Vance et al., 1986). Vanillic and *p*-hydroxybenzoic acids dominate in surface horizons, whereas protocatechuic acid is found in greater concentrations in B horizons (Table 6.47). The distribution of protocatechuic acid parallels that of Al and Fe in the Spodosol profiles suggesting that it plays a role in the podzolization process (Table 6.48). Of the seven phenolic compounds, only protocatechuic acid has *ortho*-dihydroxy functional groups which have the potential to form chelate complexes with Al and Fe. Content of low molecular weight phenolic compounds increases in A horizons as the proportion of coniferous to deciduous tree species in the forest overstory increases (Evans, 1980).

Ortstein

Ortstein is a spodic horizon in which sands are cemented, primarily by organic matter and Al with lesser amounts of Fe and little or no Si (McKeague et al., 1967; McKeague and Wang, 1980; Lee et al., 1988). Most roots do not penetrate ortstein. Ortstein may occur in a horizontal orientation which tends to be root restrictive, in vertical orientation which is less root restrictive, or as nodules. Ortstein in somewhat poorly drained Spodosols is more continuous and more strongly cemented with horizontal orientation, whereas that in better drained Spodosols is less continuous and less strongly cemented with vertical orientation (Mokma et al., 1994). Somewhat poorly drained soils tend to form ortstein more readily than well-drained soils, presumably due to increased accumulation of amorphous Fe and Al (Langley-Turnbaugh, 1995). Saturated zones provide a longer residence time in B horizons for organometallic complexes, thus the complexes have more time to become immobilized by attaching themselves to sand grains. Continued accumulation of the complexes may fill pores and the sand grains become cemented together. Roots are unable to penetrate ortstein, but water is able to slowly percolate through these horizons. Root and faunal activity occurs above the cemented B horizon. Ortstein reduces the permeability to moderate or moderately rapid rates (10 cm hr^{-1}), and Spodosols with ortstein tend to have lower potential forest productivity than those without ortstein. The site index for red pine on a Typic Haplorthod without ortstein is 72, whereas that with 60 to 100% ortstein is 59.

Table 6.47   Amounts of low molecular weight phenolic compounds in a Typic Haplorthod [Adapted from Vance et al., 1986]

Horizon	PA*	PHBA	VA	PHB	V	CA	FA
				mg g^{-1} x cm			
A	2.1	4.2	0.6	0.1	0.4	0.3	0.4
E	0.6	0.1	0.8	0.1	0.1	0	0
Bhs	5.6	3.8	7.0	0.5	0.5	0.1	0.1
Bs	13.4	4.6	8.0	1.3	1.7	0	0
BC	0.4	0.2	0.1	0.1	0.1	0	0
C	0.6	1.3	0.6	0.3	0.6	0	0

* PA = protocatechuic acid   V = vanillin
  PHBA = *p*-hydroxybenzoic acid   CA = *trans - p* - coumaric acid
  VA = vanillic acid   FA = ferulic acid.
  PHB = *p*-hydroxybenzaldehyde

Table 6.48 Weighted index contents of protocatechuic acid (PA) and pyrophosphate extractable Al_p and Fe_p in pedons of a sandy developmental sequence of Spodosols [Adapted from Vance et al., 1986]

Subgroup	Horizon	PA	Al_p	Fe_p
		mg kg^{-1} x cm	---------------g kg^{-1} x cm------------	
Typic Udipsamment	Oa			
	A/E	0.5	3	5
	Bw1	2.0	36	24
	Bw2	0.8	21	10
Entic Haplorthod	Oe			
	A	0.3	1	1
	E	1.2	3	6
	Bs	9.9	22	48
	BC	2.4	35	14
	C	1.3	26	13
Typic Haplorthod	Oe			
	A	2.1	3	4
	E	0.6	3	6
	Bhs	5.6	32	31
	Bs	13.4	46	17
	BC	0.4	4	2
	C	0.6	25	12

Silicon

Silicon has recently been proposed as a component of podzolization (Farmer et al., 1980; Childs et al., 1983; Parfitt and Saigusa, 1985; Wang et al., 1986; Jakobsen, 1989; Ugolini et al., 1991; Freeland and Evans, 1993) and Si data have been interpreted as an accumulation of imogolite or allophane. Imogolite and/or allophane are identified by various methods, namely, petrography, infrared spectroscopy, energy dispersive X-ray analysis, electron microscopy, or inference from chemical data as oxalate extractable Si is nearly exclusively derived from allophane or imogolite. In some Spodosols accumulations of apparent imogolite coincide with cemented horizons with lighter colors than those typical of the C-rich ortstein (McKeague and Sprout, 1975; McKeague and Kodama, 1981; Childs et al., 1983; Freeland and Evans, 1993). Imogolite-cemented horizons usually occur as a lower Bs horizon, whereas ortstein usually occurs as upper Bhs or Bs horizons. There are no specific provisions in *Soil Taxonomy* for imogolite- or allophane-cemented horizons; thus, these horizons are considered duripans to acknowledge the role of Si. Such duripans are very different, however, from those most commonly identified in arid to semi-arid climates in base-rich soils (Section E, Chapter 6.10). Furthermore, duripans are defined as cemented horizons due to an illuvial accumulation of silica (Soil Survey Staff, 1996), and there has been controversy as to whether Si in spodic horizons has been translocated or accumulates from *in situ* weathering (Ugolini et al., 1991). Other studies (Ugolini and Dahlgren, 1987; Gustafsson et al., 1995) indicated imogolite/allophane formed in B horizons after Al had been translocated as organometallic complexes. Strong positive correlations between oxalate extractable Al and Si, but not between Al and C or between Fe and Al, occur in Haplorthods (Freeland and Evans, 1993) and Aquods (Evans and Mokma, 1996). It is possible that, following translocation, organic complexes are broken down by soil microbes, thus oxidizing and decreasing soil C levels. In weathering studies of granitic materials, however, close correlations have been found between Al and Si release rates but not between Fe and Al release rates in response to leaching with DOC (McCracken et al., 1996).

6.9.5 Classification: From Marbut to *Soil Taxonomy*

Soils with translocation of organic matter Al and Fe were classified as Podzols by Marbut (1936). Podzols were Pedalfers that had developed under forest cover, dominated by conifers. Soils with Podzol-like profiles and shallow water tables were recognized as Groundwater Podzols but not normal Podzols. No great soil group was established for these wet soils. Baldwin et al. (1938) added the groundwater Podzol great soil group for soils with Podzol-like profiles and somewhat poor drainage. Also, the great soil group, Brown Podzolic, was added for soils that have morphological properties of an immature Podzol. No modifications were made in the classification of Podzols by Thorp and Smith (1949) before the new approach to soil classification was begun in 1951. The Spodosol order was created for soils that had a spodic horizon (Soil Survey Staff, 1960, 1975). Emphasis was placed on the presence of the spodic horizon rather than on the albic or E horizon. Spodosols include primarily the soils that had been classified as Podzols, Brown Podzols and Groundwater Podzols. Not all Podzols classified as Spodosols. Four suborders of Spodosols were recognized: Aquods, Ferrods, Humods and Orthods (Soil Survey Staff, 1960). The Suborder Ferrods was dropped and the suborder Cryods was added in 1992 (Soil Survey Staff, 1992). Aquods have aquic conditions, Cryods have a cryic or pergelic soil temperature regime, Humods have a layer with at least 6% organic C within the spodic horizon, and Orthods have a spodic horizon that contains C, Al and Fe with no one element dominating. The criteria for the spodic horizon also changed in 1992. Oxalate extractable Al and Fe and the optical density of the oxalate extract (Soil Survey Staff, 1992) are used rather than

Table 6.49 Listing of suborders, great groups, and subgroups in the Spodosols order [Soil Survey Staff, 1998]

Suborders	Great groups	Subgroups
Aquods	Cryaquods	Lithic, Placic, Duric, Andic, Entic, Typic
	Alaquods	Lithic, Duric, Histic, Alfic Arenic, Arenic Ultic, Arenic Umbric, Arenic, Grossarenic, Alfic, Ultic, Aeric, Typic
	Fragiaquods	Histic, Plaggeptic, Argic, Typic
	Placaquods	Andic, Typic
	Duraquods	Histic, Andic, Typic
	Epiaquods	Lithic, Histic, Andic, Alfic, Ultic, Umbric, Typic
	Endoaquods	Lithic, Histic, Andic, Argic, Umbric, Typic
Cryods	Placocryods	Andic, Humic, Typic
	Duricryods	Aquandic, Andic, Aquic, Oxyaquic, Humic, Typic
	Humicryods	Lithic, Aquandic, Andic, Aquic, Oxyaquic, Typic
	Haplocryods	Lithic, Aquandic, Andic, Oxyaquic, Aquic, Entic, Typic
Humods	Placohumods	Andic, Typic
	Durihumods	Andic, Typic
	Fragihumods	Typic
	Haplohumods	Lithic, Andic, Plaggeptic, Typic
Orthods	Placorthods	Typic
	Durorthods	Andic, Typic
	Fragiorthods	Aquic, Alfic Oxyaquic, Oxyaquic, Plaggeptic, Alfic, Ultic, Entic, Typic
	Alorthods	Oxyaquic, Arenic Ultic, Arenic, Entic Grossarenic, Entic, Grossarenic, Plaggeptic, Alfic, Ultic, Typic
	Haplorthods	Entic Lithic, Lithic, Fragiaquic, Aqualfic, Aquentic, Aquic, Alfic Oxyaquic, Oxyaquic Ultic, Fragic, Lamellic, Oxyaquic, Andic, Alfic, Ultic, Entic, Typic

pyrophosphate and dithionite-citrate extractable Al and Fe, extractable C, clay content, and CEC (Soil Survey Staff, 1975). The Spodosol order in the 8th Edition of the *Keys to Soil Taxonomy* (Soil Survey Staff, 1998) has 4 suborders, 20 great groups, and 108 subgroups (Table 6.49).

6.9.5.1 Relationship to Andisols

Both Spodosols and Andisols are characterized by significant amounts of amorphous, hydrated oxyhydroxide compounds of Fe, Al, and Si that are chemically active. In Spodosols, these materials are present as illuvial accumulations in Bs horizons, but in Andisols such compounds are typically distributed throughout the soil profile due to weathering of the volcanic parent material (Table 6.50) (Parfitt and Saigusa, 1985) (Section E, Chapter 6.3). Carbon contents are often greater in Spodosols than in Andisols, especially in subsurface horizons in which illuvial humus accumulates. In some cases, Spodosols also form in volcanic ash parent materials. These Spodosols may contain greater than usual amounts of clay but remain distinguished from Andisols on the basis of profile distributions of active amorphous materials (Takahashi et al., 1989).

6.9.6 Chemical Properties

Organic C is highest in O and A horizons followed by the upper B horizon (Table 6.49). E horizons have an intermediate amount of organic C and usually less Al and Fe than C horizons as a result of their eluviation. Organic C contents decrease with depth from the upper B horizon to the C horizon. The distributions of Al and Fe often parallel each other, presumably because Fe and Al are comobilized by organic acids (Mokma and Vance, 1989; Mokma, 1991, 1993). In these soils, the amounts of Al and Fe are greatest in upper B horizons and decrease with depth. In other studies, however, Al is more closely related to pH (McDowell and Wood, 1984; Freeland and Evans, 1993), while Fe and C distributions are parallel. In those instances, maximum accumulation of Al tends to occur below horizons in which C and Fe reach a maximum.

The levels of Al in B horizons of Spodosols may be toxic to plants. Root growth, stem growth and leaf development may be reduced by the Al toxicity (Hoyle, 1971). Reductions in root growth associated with high Al levels are often complicated by low levels of other nutrients which are common in Spodosols. Root tolerance differs widely among plants.

Table 6.50 Extractable Fe, Al, Si and organic C in horizons of a Spodosol and an Andisol [Adapted from Parfitt and Saigusa, 1985]

Sample	Horizon	Dithionite # Fe$_d$	Al$_d$	Acid Oxalate Fe$_o$	Al$_o$	Si$_o$	Pyrophosphate Fe$_p$	Al$_p$	C$_p$
						%			
Tihoi	HO	0.14	0.16	0.07	0.15	0	0.01	0.13	4.02
Spodosol	E	0.15	0.05	0.03	0.04	0	0.03	0.04	0.77
	Bhs	0.70	0.60	0.50	0.56	0	0.44	0.49	4.21
	Bhs	0.37	1.01	0.26	0.96	0.01	0.23	0.88	5.79
	C	0.82	1.56	0.63	1.60	0.36	0.58	1.30	3.92
Taupo	A1	0.33	0.56	0.21	0.47	0	0.19	0.48	2.90
Andisol	A3	0.37	0.47	0.28	0.83	0.30	0.15	0.27	1.03
	B	0.32	0.38	0.19	0.80	0.34	0.07	0.17	0.72
	C	0.22	0.24	0.13	0.74	0.37	0.02	0.11	0.31

Dithionite-citrate-bicarbonate extract.

Table 6.51 Selected chemical properties of a sandy Typic Haplorthod

Horizon	Org C	ODOE	pH	Al_d	Al_o	Al_p	Fe_d	Fe_o	Fe_p	Si_o	P Retention
	%						$g\ kg^{-1}$				%
A	3.4	0.06	3.6	0.4	0.3	0.2	0.3	0.4	0.3	tr	4.5
E	0.2	0.03	4.0	0.2	0.1	0.1	0.2	0.2	0.2	tr	5.6
Bhs	1.1	0.23	4.2	2.0	2.8	1.5	2.1	2.8	2.1	0.4	24.1
Bs	0.6	0.13	4.5	1.9	2.7	1.7	1.1	1.1	1.1	1.1	21.5
BC	0.2	0.04	4.8	0.8	1.2	0.7	0.3	0.4	0.3	0.5	11.2
C	-	0.03	4.9	0.4	0.5	0.4	0.2	0.2	0.2	0.3	7.2

Spodosols are acidic soils having low native fertility. A and E horizons are very strongly to extremely acid with gradual increases in pH with depth (Table 6.51).

Spodic horizons have large amounts of variable charge as a result of the accumulation of amorphous C, Al, and Fe which have the ability to retain large amounts of P (Table 6.51) not readily available to plants.

6.9.7 Management of Spoddsols

Many Spodosols are used for forestry and the major products are lumber, pulpwood, and Christmas trees. Forestry has been the dominant economic activity on these soils, but agriculture is also important. Major constraints to productivity are low available P levels, low CEC, acidity and drought susceptibility.

6.9.7.1 Low Input Cultural Systems

After loggers removed most of the timber, extensive areas of Spodosols were settled by farmers and converted to crop production. These soils produced a good first crop of fall-planted small grains, such as wheat. After two or three years, yields decreased and land was abandoned. The sandy nature of these soils makes them droughty which together with acidity and low CEC make them infertile. Some Spodosols with argillic (Bt) horizons which are not as infertile as the more sandy Spodosols are used to support dairy enterprises. The application of animal manure returns nutrients and the organic matter contributes to increased water- and nutrient-holding capacities.

6.9.7.2 High Input Cultural Systems

Because of the droughty and infertile nature of Spodosols, their utilization requires high inputs, especially P. Therefore, high value crops such as potatoes and blueberries are grown. Potatoes are grown on these acid soils to avoid potato scab disease. Blueberries, both high and low bush, are acid-loving plants that do well on Spodosols. Cranberries, another acid-loving crop, are produced on wet Spodosols.

6.9.8 References

Baldwin, M., C.E. Kellogg, and J. Thorp. 1938. Soil classification. p. 997–1001. *In* Soils and men. USDA Yearbook. US Government Printing Office, Washington, DC.

Barrett, L.R., and R.J. Schaetzl. 1992. An examination of podzolization near Lake Michigan using chronofunctions. Can. J. Soil Sci. 72:527–541.

Burges, A., and D.P. Drover. 1953. The rate of Podzol development in sands of the Woy Woy District N.S.W. Aust. J. Bot. 1:83–94.

Buurman, P. 1985. Carbon/sesquioxide ratios in organic complexes and the transition albic-spodic horizon. J. Soil Sci. 36:255–260.

Childs, C.W., R.L. Parfitt, and R. Lee. 1983. Movement of aluminum as an inorganic complex in some podzolized soils, New Zealand. Geoderma 29:139–155.

De Coninck, F. 1980. Major mechanisms in formation of spodic horizons. Geoderma 24:101–128.

De Kimpe, C.R., and Y.A. Martel. 1976. Effects of vegetation on the distribution of carbon, iron, and aluminum in the B horizons of northern Appalachian Spodosols. Soil Sci. Soc. Am. J. 40:77–80.

Evans, C. V. 1996. Preliminary investigations of hydric soil hydrology and morphology in New Hampshire. p. 114–126. In J.S. Wakely, S.W. Sprecher, and W.C. Lynn. (ed.). Preliminary investigations of hydric soil hydrology and morphology in the United States. Tech. Rp., WRP-DE-13, U.S. Army Engineer Waterways Exp. Stn. Vicksburg, MS.

Evans, C.V., and D.L. Mokma. 1996. Sandy wet Spodosols: Water tables, chemistry, and pedon partitioning. Soil Sci. Soc. Am. J. 60:1495–1501.

Evans, L. J. 1980. Podzol development north of Lake Huron in relation to geology and vegetation. Can. J. Soil Sci. 60:527–539.

Farmer, V. C., J.D. Russell, and M.L. Berrow. 1980. Imogolite and proto-imogolite allophane is spodic horizons: Evidence for a mobile aluminum silicate complex in Podzol formation. J. Soil Sci. 31:673–684.

Franzmeier, D. P., and E.P. Whiteside. 1963. A chronosequence of podzols in northern Michigan. I. Ecology and description of pedons. II. Physical and chemical properties. MI State. Univ. Agric. Exp. Sta. Quart. Bul. 46:1–36.

Freeland, J. A., and C.V. Evans. 1993. Genesis and profile development of Success soils, northern New Hampshire. Soil Sci. Soc. Am. J. 57:183–191.

Gardner, D. R., and E.P. Whiteside. 1952. Zonal soils in the transition region between the Podzol and Gray-Brown Podzolic regions in Michigan. Soil Sci. Soc. Am. Proc. 16:137–141.

Gjems, O. 1960. Some notes on clay minerals in podzol profiles in Fennoscandia. Clay Miner. Bul. 4:208–211.

Gustafsson, J. P., P. Bhattacharya, D.C. Bain, A.R. Fraser, and W.J. McHardy. 1995. Podzolisation mechanisms and the synthesis of imogolite in northern Scandinavia. Geoderma 66:167–184.

Haile-mariam, S., and D.L. Mokma. 1995. Mineralogy of two sandy Spodosol hydrosequences in Michigan. Soil Surv. Horiz. 36:121–132.

Hoyle, M. C. 1971. Effects of the chemical environment on yellow-birch root development and top growth. Plant Soil 35:623–633.

Hunckler, R. V. 1996. Spodosol development as affected by geomorphic aspect, Baraga County, Michigan. M.A. thesis. Michigan State University, East Lansing, MI.

Jakobsen, B. H. 1989. Evidence for translocations into the B horizon of a subarctic Podzol in Greenland. Geoderma 45:3–17.

Langley-Turnbaugh, S. J. 1995. Soil evolution and pedogenic processes on elevated marine terraces in Coastal Oregon. Ph.D. dissertation, University of Wisconsin, Madison, WI.

Lee, F. Y., T.L. Yuan, and V.W. Carlisle. 1988. Nature of cementing materials in ortstein horizons of selected Florida Spodosols: I. Constituents of cementing materials. Soil Sci. Soc. Am. J. 52:1411–1418.

Marbut, C. F. 1936. Soils of the United States. p. 1–93. In USDA atlas of American agriculture. Part III. US Government Printing Office, Washington, DC.

McCracken, K.L., C.V. Evans, and W.H. McDowell. 1996. Effects of dissolved organic carbon concentration on metal and silica release: Leaching and initial weathering. Agron. Abs. p. 262.

McDowell, W.H., and T. Wood. 1984. Podzolization: Soil processes control dissolved organic carbon concentrations in stream water. Soil Sci. 137:23–32.

McKeague, J.A., F. De Coninck, and D.P. Franzmeier. 1983. Spodosols. p. 217–252. In L.P. Wilding, N.E. Smeck, and G.F. Hall (ed.) Pedogenesis and soil taxonomy. II. The soil orders. Elsevier, New York, NY.

McKeague, J.A., and H. Kodama. 1981. Imogolite in cemented horizons of some British Columbia soils. Geoderma 25:189–197.

McKeague, J.A., and P.N. Sprout. 1975. Cemented subsoils (duric horizons) in some soils of British Columbia. Can. J. Soil Sci. 55:189–203.

McKeague, J.A., and C. Wang. 1980. Micromorphology and energy dispersive analysis of ortstein horizons of Podzolic soils from New Brunswick and Nova Scotia, Canada. Can. J. Soil Sci. 60:9–21.

McKeague, J. A., M. Schnitzer, and P.K. Heringa. 1967. Properties of an ironpan humic Podzol from Newfoundland. Can. J. Soil Sci. 47:23–32.

Messenger, A.S. 1975. Climate, time, and organisms in relation to Podzol development in Michigan sands: II. Relationships between chemical element concentrations in mature tree foliage and upper humic horizons. Soil Sci. Soc. Am. Proc. 39:698–702.

Messenger, A.S., E.P. Whiteside, and A.R. Wolcott. 1972. Climate, time, and organisms in relation to Podzol development in Michigan sands: I. Site descriptions and microbiological observations. Soil Sci. Soc. Am. Proc. 36:633–638.

Mokma, D.L. 1991. Genesis of Spodosols in Michigan, USA. Trends Soil Sci. 1:25–32.

Mokma, D.L. 1993. Color and amorphous materials in Spodosols from Michigan. Soil Sci. Soc. Am. J. 57:125–138.

Mokma, D.L., J.A. Doolittle, and L.A. Tornes. 1994. Continuity of ortstein in sandy Spodosols of Michigan. Soil Surv. Horiz. 35:6–10.

Mokma, D.L., and S.W. Sprecher. 1994. Water table depths and color patterns in Spodosols of two hydrosequences in northern Michigan, USA. Catena 22:275–286.

Mokma, D.L., and G.F. Vance. 1989. Forest vegetation and origin of some spodic horizons, Michigan. Geoderma 43:311–324.

Parfitt, R.L., and M. Saigusa. 1985. Allophane and humus-aluminum in Spodosols and Andepts formed from the same volcanic ash beds in New Zealand. Soil Sci. 139:149–155.

Schaetzl, R.J. 1990. Effects of treethrow microtopography on the characteristics and genesis of Spodosols, Michigan, USA. Catena 17:111–126.

Schaetzl, R.J., and S.A. Isard. 1991. The distribution of Spodosol soils in southern Michigan: A climatic interpretation. Ann. Assoc. Am. Geo. 81:425–442.

Schaetzl, R.J., and D.L. Mokma. 1988. A numerical index of Podzol and podzolic soil development. Phys. Geog. 9:232–246.

Shetron, S.G. 1974. Distribution of free iron and organic carbon as related to available water in some forested sandy soils. Soil Sci. Soc. Am. Proc. 38:359–362.

Soil Survey Staff. 1960. Soil classification, a comprehensive system, 7th Approximation. US Government Printing Office, Washington, DC.

Soil Survey Staff. 1975. Soil taxonomy. A basic system of soil classification for making and interpreting soil surveys. Agric. Handbk. 436, US Government Printing Office, Washington, DC.

Soil Survey Staff. 1992. Keys to soil taxonomy, 5th Ed. SMSS Technical Monograph 19, Pocahontas Press, Inc., Blacksburg, VA.

Soil Survey Staff. 1996. Keys to soil taxonomy, 7th Ed. USDA, NRCS, Lincoln, NE.

Stanley, S.R., and E. J. Ciolkosz. 1981. Classification and genesis of Spodosols in the central Appalachians. Soil Sci. Soc. Am. J. 45:912–917.

Takahashi, T., S. Shoji, and A. Sato. 1989. Clayey Spodosols and Andisols showing a biosequential relation from Shimokita Peninsula, northeastern Japan. Soil Sci. 148:204–218.

Thorp, J., and G.D. Smith. 1949. Higher categories of soil classification: Order, suborder, and great soil groups. Soil Sci. 67:117–126.

Ugolini, F.C. 1968. Soil development and alder invasion in a recently deglaciated area of Glacier Bay, Alaska. p. 115–140. *In* Biology of Alder. US Dept. Agric., Forest Serv., Portland, OR.

Ugolini, F.C., and R. Dahlgren. 1987. The mechanism of podzolization as revealed by soil solution studies. p. 195–203. *In* D. Righi and A. Chauvel (ed.) Podzols et Podzolisation. AFES et INRA, Paris, France.

Ugolini, F.C., R. Dahlgren, J. LaManna, W. Nuhn, and J. Zachara. 1991. Mineralogy and weathering processes in recent and Holocene tephra deposits of the Pacific Northwest, USA. Geoderma 51:277–299.

Vance, G. F., D.L. Mokma, and S.A. Boyd. 1986. Phenolic compounds in soils of hydrosequences and developmental sequences of Spodosols. Soil Sci. Soc. Am. J. 50:992–996.

Wang, C., J.A. McKeague, and H. Kodama. 1986. Pedogenic imogolite and soil environments: Case study of Spodosols in Quebec, Canada. Soil Sci. Soc. Am. J. 50:711–718.

6.10 Aridisols

Randal J. Southard
University of California, Davis

6.10.1 Concept of Aridisols

Aridisols are soils that do not have water available to mesophytic plants for long periods. Limited supplies of water may be caused by a large excess of evapotranspiration over precipitation, low

water-holding capacity due to a shallow soil depth or restricted infiltration, or to low osmotic potential due to salinity. Generally speaking, water for plant growth is available for fewer than 90 consecutive days, and there are long periods of water deficit. The lack of water also inhibits many other soil
processes, including mineral weathering and removal of weathering products by leaching. As a result,
Aridisols are characterized by the accumulation of relatively soluble weathering products, which are
more extensively leached, or removed entirely, from soils of more humid regions.

As the order name implies, Aridisols are associated with arid climates, but not all soils in regions
with arid climates are Aridisols. Approximately one-third of the Earth has an arid climate (Shantz,
1956), but Aridisols occupy less than half of the arid land area. Soils of arid regions are also classified
in the Gelisol, Oxisol, Vertisol, Andisol, Mollisol and Entisol orders.

6.10.2   Geographical Distribution

6.10.2.1   Global

Aridisols occupy approximately 12% (about 16.0 million km^2) of the land surface of the Earth (Fig.
6.56, Table 6.52), and rank second behind Entisols in areal extent (Table 6.1). Large areas of these
soils occur in southwestern and central Asia, throughout most of Australia, western North and South
America, and on the continental margins of Africa. The distribution of Aridisols coincides quite
closely with arid climates associated with the descending branches of the Hadley cell of global atmospheric circulation (Ahrens, 1991) at about 30° N and S latitudes (subtropical high pressure). Their
distribution is further influenced by orographic influences (rain shadows), particularly on the western
margins of North and South America and in central Asia, and by long distances from oceanic sources
of moisture, as in the case of the continental interiors of North America and Asia.

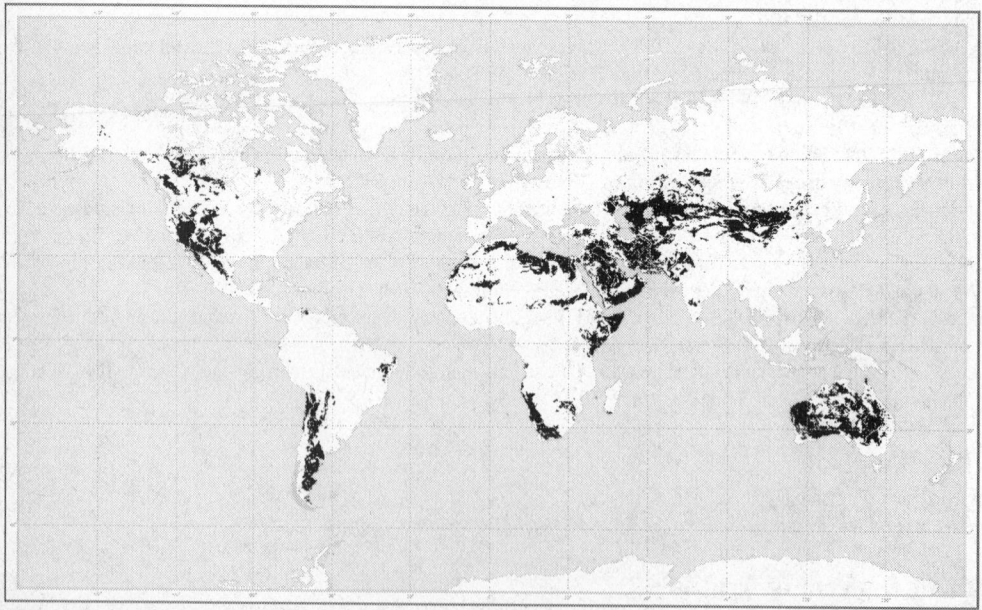

Fig. 6.56   Global distribution of Aridisols [Courtesy of USDA-NRCS, Soil Survey Division, World Soil Resources,
1998]

Table 6.52 Estimated extent (km²) of Aridisol suborders globally (USDA-NRCS, 1997a) and in the United States [Compiled from Fosberg et al., 1991; Hipple et al., 1990; Johnson, 1990; and Witty, 1990]

State	Cryids	Salids	Durids[a]	Gypsids	Argids	Calcids	Cambids	Total
WORLD	948,365	890,118	349,580	682,963	5,407,965	4,872,554	2,897,853	16,049,398
AZ	0	0	2,366	112	22,613	10,040	6,376	41,507
CA	0	45	914	0	4,616	346	350	6,271
CO	0	0	3,531	53	24,231	3,261	2,206	33,382
HA	0	5	0	0	0	0	82	87
ID	1,428	0	8,001	0	4,595	12,993	4,367	31,384
KS	0	0	1,319	0	0	0	29	1,348
MT	0	94	3,700	39	3,684	3,090	6,395	17,002
NE	0	0	74	0	26	0	137	237
NV	0	246	13,976	196	23,651	4,837	7,910	50,816
NM	0	115	1,557	3,425	34,132	36,189	7,986	83,404
ND	0	0	241	0	0	0	35	276
OK	0	0	481	0	0	808	0	1,289
OR	0	0	1,282	0	61	0	2,209	3,552
SD	0	12,052	463	17	4,396	0	1,839	18,767
TX	0	669	8,134	1,189	1,136	25,457	4,735	41,320
UT	0	696	2,311	264	1,987	9,611	1,876	16,745
WA	0	0	894	0	14	0	6,474	7,382
WY	0	0	336	0	2,403	317	350	3,406
USA TOTAL	1,428	13,922	49,580	5,295	127,545	106,949	53,356	358,075

[a] Includes Durids in the USA and red and brown hardpan soils in Australia (Hubble et al., 1983; Isbell, 1990), although some of these soils may be in other orders.

6.10.2.2 United States

Aridisols occupy approximately 358,000 km² in 18 of the western states, including Hawaii, generally west of the 100th meridian (Table 6.52). They are most extensive in New Mexico, Nevada, Arizona, Texas, Colorado, and Idaho, occupying more than 30,000 km² in each of those states.

The most southerly Aridisols in the United States occur in the subtropical high pressure belt and receive most of their precipitation in the summer from subtropical convective thunderstorms. The Aridisols of the intermountain west have more continental climates, but precipitation is still dominated by summer thunderstorms driven by orographic lifting in the Rocky Mountains. Farther west, the Aridisols of the Great Basin region receive most of their precipitation in the winter from Pacific frontal storms.

6.10.3 Soil-Forming Environment

The most important aspect of the Aridisol soil-forming environment is the limited water available for plant growth, mineral weathering, and leaching of soluble weathering products. A typical water budget (Fig. 6.57) is dominated by a long period of deficit during the summer when potential evapotranspiration greatly exceeds precipitation. The deficit period is followed by a partial recharge of water-holding capacity, then rapid utilization of that stored water during the brief growing season. The lack of a surplus of precipitation over evapotranspiration and water-holding capacity prevent wholesale leaching of weathering products from Aridisols. Nonetheless, many Aridisols have thin, but morphologically well-developed, soil profiles due to the concentration of the limited water in a relatively small volume of soil. Further, the typically erratic distribution of rainfall in arid regions (Archibold, 1995) causes occasional periods of relatively high precipitation in the winter months, when deeper

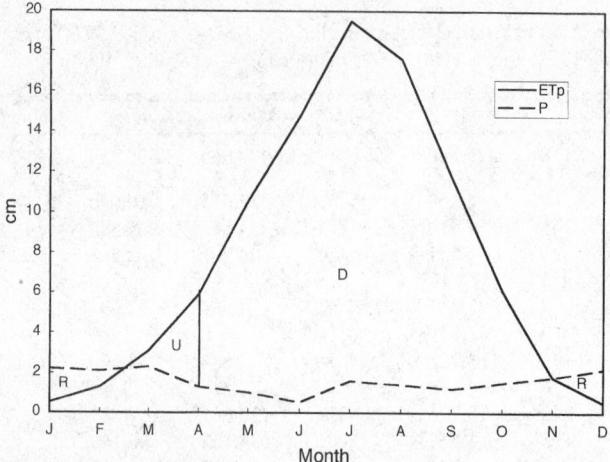

Fig. 6.57 Water budget for St. George, Utah, USA, calculated by the Thornthwaite (1948) method, assuming a soil water-holding capacity of 200 mm. R = recharge of soil water-holding capacity, U = utilization of stored water, D = deficit of precipitation (P) with respect to potential evapotranspiration (ETp)

leaching could occur. In this case, the depth of leaching and horizon development may reflect the rainfall of the extreme years, rather than that of average years.

Soil temperature regimes of Aridisols range from cryic to isohyperthermic (Soil Survey Staff, 1998).

Vegetation on Aridisols is dominated by drought-escaping annual grasses and forbs and drought-enduring evergreen shrubs (Archibold, 1995). Aridisols transitional to xeric or ustic soil moisture regimes, or in frigid or cryic soil temperature regimes, often have a significant component of perennial grasses (Passey et al., 1982). The spatial distribution of plants on the landscape is generally very

Fig. 6.58 An arid region landscape with desert shrub vegetation (foreground) and active, high-gradient alluvial fans (background), in Owens Valley, California. Coppice dunes around the base of creosote bush (*Larrea tridentata*) often support microbiotic crusts.

heterogeneous (Eckert et al., 1978; Schlesinger and Jones, 1984), giving rise to islands of fertility (Schlesinger et al., 1990) occupied by the shrubs and understory grasses, forbs, and microbiotic crusts (Johnston, 1997). These islands tend to be relatively stable, resistant to erosion (Otterman and Gornitz, 1983), and often trap eolian very fine and fine sand to form a coppice dune (Gile, 1966; Reid et al., 1993) at the base of the shrub (Fig. 6.58). These coppice dunes range in height from a few cm, around the base of shrubs, to 1 m or more, in which case they dominate the landscape microtopography.

Most Aridisols form from sediments transported by wind or water on relatively low gradient landscapes. Lithology of the sediments varies widely. Truncation of steep landforms and slow weathering rates retard the formation of subsurface soil horizons and result in Entisols on most erosional landforms. Some Aridisols have formed in residual rock material on erosional landforms, for example, on old and highly weathered granitic pediments in the Mojave Desert (Boettinger and Southard, 1991; 1995) and on stable, steep hills where episodic erosion has been minimal (Nettleton and Peterson, 1983).

Landscapes with Aridisols range in age from late Holocene to early Pleistocene or older (Nettleton and Peterson, 1983; Machette, 1985). The youngest Holocene landforms generally have not experienced sufficient weathering to produce Aridisols and are dominated by Entisols. Most Aridisols on Pleistocene landscapes have probably experienced fluctuating climatic conditions (cooler, moister Pleistocene pluvials) that significantly affected the rate and intensity of weathering.

6.10.4   Soil Morphology and Physical Properties

6.10.4.1   Surface Horizons
Pedogenic processes in Aridisols have produced numerous surface soil features. Of special interest are (1) surface crusts, (2) vesicular horizons, and (3) desert pavement. A soil crust, widespread, but not present everywhere, is a surficial layer of uncemented, coherent fine earth that can be broken free from underlying soil. The crust is generally less than 10 to 20 mm thick and is often platy in the upper part and massive below. Water infiltration of the crust is slow, in contrast to the rapid infiltration in uncrusted coppice sand dunes. Crusts form by raindrop impact, localized sheet erosion, and by repeated wetting and drying of loamy soil material (Nettleton and Peterson, 1983).

Beneath the crust, especially in silty eolian material, a horizon with many fine, vesicular pores (Fig. 6.59) often forms (sometimes referred to as an Av horizon). The vesicles result from repeated saturation and desiccation and progressive enlargement of air bubbles trapped beneath the thin, saturated surface layer (Springer, 1958; Miller, 1971).

Desert pavement, a surficial layer of rock fragments, usually one or two fragments thick, occurs on many Aridisols (Fig. 6.60). The origin of the pavement has been attributed to loss of fine particles by wind deflation or water erosion, surficial weathering and splitting of some stones (Cooke, 1970), vertical sorting of coarse fragments by repeated wetting and drying (Springer, 1958), and by rafting of surface fragments by infiltrating eolian material (McFadden et al., 1987). The pavement itself serves as a dust trap (Yaalon and Ganor, 1973; Gile, 1975a; Peterson, 1977), whose effectiveness decreases as the surface fragments become more interlocked, until the land surface is nearly completely covered by the pavement. At this point, the surface probably neither retains much new eolian material, nor loses much material to erosion, unless disturbed.

Desert varnish, typically yellowish to reddish brown on the undersides of desert pavement rock fragments, and nearly black on upper surfaces, includes 2:1 silicate clay minerals and oxyhydroxides of Fe and Mn. These constituents are probably of eolian origin and appear to be microbially precipitated (Dorn and Oberlander, 1981). The varnish occurs most frequently on stable, mafic rock frag-

Fig. 6.59 A vesicular surface horizon (an Av horizon) rafts desert-varnished basalt pebbles of a desert pavement in the Cima volcanic field, in the Mojave Desert, California. The lens cap is about 60 mm in diameter.

ments and generally is thickest (as much as 100 μm thick) and darkest on Pleistocene pavements (Nettleton and Peterson, 1983).

The low organic C content and relatively light colors of the surface soil horizon are characteristic of the ochric epipedons of virtually all Aridisols (dark anthropic epipedons occur, but are very rare and localized).

Fig. 6.60 Desert varnished surface clasts interlock to form a nearly continuous desert pavement on Pleistocene basalt flows in the Mojave Desert. The clasts act as a dust trap to catch eolian materials derived from nearby playas and alluvial fans. The length of the tape measure scale is about 30 cm.

Table 6.53 Summary of properties of subsurface horizons in Aridisols [Soil Survey Staff, 1998]

Horizon	Properties
Argillic	Clay accumulation relative to overlying horizon, evidence of clay translocation, minimum thickness of 7.5 cm if loamy or clayey, 15 cm if sandy
Calcic	$CaCO_3$ accumulation, generally $\geq$ 15 cm thick, $\geq$ 15% $CaCO_3$ equivalent, $\geq$ 5% more $CaCO_3$ than underlying horizon
Cambic	Evidence of alteration, reddening, development of soil structure, loss of carbonates
Duripan	Silica-cemented, does not slake in water or acid, evidence of opal accumulation
Gypsic	Gypsum accumulation, $\geq$ 15 cm thick, $\geq$ 5% gypsum, $\geq$ 1% visible secondary gypsum, thickness in cm multiplied by % gypsum has product $\geq$ 150
Petrocalcic	$CaCO_3$-cemented, $\geq$ 10 cm thick, or $\geq$ 1 cm thick if a laminar cap is underlain directly by bedrock
Petrogypsic	Gypsum-cemented, $\geq$ 10 cm thick, $\geq$ 5% gypsum and thickness in cm multiplied by % gypsum has product $\geq$ 150
Natric	Properties of argillic horizon, plus columnar or prismatic structure, or degraded blocky structure, sodium adsorption ratio (SAR) $\geq$ 13 or exchangeable sodium percentage (ESP) $\geq$ 15
Salic	Accumulation of salts more soluble than gypsum, $\geq$ 15 cm thick, electrical conductivity (EC) $\geq$ 30 dS m^{-1}, thickness in cm multiplied by EC has product $\geq$ 900

6.10.4.2 Subsurface Horizons

Subsurface horizons include cambic, argillic, natric, calcic, petrocalcic, gypsic, petrogypsic and salic horizons, and duripans. None of these horizons is unique to Aridisols, although the occurrence of salic, gypsic, and petrogypsic horizons in other soil orders is relatively rare. A summary of properties of these horizons is given in Table 6.53.

6.10.5 Chemical Properties

6.10.5.1 Organic Carbon and Nitrogen

Organic C contents are among the lowest of all soils, often < 1% on a mass basis, due to low biomass production and rapid decomposition. Accumulation of organic C is confined to relatively shallow soil depths in the thin ochric epipedon. Organic C content generally is inversely related to mean annual soil temperature. In cool and cold arid regions (frigid and cryic soil temperature regimes), organic C contents may rival those of semi-arid to subhumid climates (Table 6.54).

Nitrogen occurs mostly in organic matter, and is usually quickly nitrified after the organic matter is decomposed. The NO_3^- produced is susceptible to some leaching, but more importantly, to runoff during overland flow and accumulation in a playa environment, where denitrification can occur (Schlesinger et al., 1990).

6.10.5.2 Soluble Salts

Limited leaching and concentration of soluble ions by evaporation near the soil surface causes accumulation of soluble salts (gypsum and more soluble evaporites) (Section 6.10.6). The distribution of salts in Aridisol profiles reflects the relative solubilities of the compounds and the relative importance of downward leaching from precipitation versus the upward movement of groundwater due to

Table 6.54 Selected properties[a] of representative Aridisol pedons [Data from Soil Survey Staff, 1975 and USDA-NRCS, 1997b]

Depth cm	Horizon	Sand %	Silt %	Clay %	OC[a] %	N[a] %	CCE[a] %	BD[a] g cm^{-3}	pH[b]	EC[a] dS m^{-1}	SAR[a]
Xeric Calcicryid											
0 - 10	A	32	56	12	2.98	0.275	15	0.95	7.7	1.7	tr
10 - 18	Bk1	30	55	15	1.83	0.189	80		8.2	1.0	1
18 - 33	Bk2	34	44	22	0.78	0.078	41	1.17	7.8	12.9	18
33 - 58	Bk3	35	41	24	0.37	0.037	73	1.27	8.0	19.0	21
58 - 84	Bkq1	31	34	35	0.38		54	1.11	7.9	19.6	19
84 - 107	Bkq2	31	30	39	0.46		88		7.7	30.2	19
107 - 152	Bkq3	38	32	30	0.23		62		7.8	24.3	18
Typic Haplosalid											
0 - 11	A	1	50	49	0.69	0.064	4		8.4[b]	22.5	67
11 - 15	Cyz	1	44	55	0.38	0.039	2		8.7	116.7	281
15 - 20	Czy	2	55	43	0.60	0.063	3		8.8	122.5	553
20 - 36	Cz	10	53	37	0.44	0.046	4		8.4	83.8	201
36 - 56	C1	17	54	29	0.27	0.030	6		8.4	66.7	110
56 - 76	C2	2	58	40	0.26	0.033	13		8.6	63.0	110
76 - 105	C3	4	78	18	0.18	0.019	11		8.6	37.7	74
105 - 150	C4	4	53	43	0.21	0.028	13		8.6	38.3	34
Typic Haplodurid											
0 - 3	A1	86	12	2	0.11	0.013	3		8.2[b]	0.4	1
3 - 10	A2	81	14	5	0.08	0.015	5		8.3	0.4	1
10 - 41	A3	75	14	11	0.15	0.021	7		8.3	0.3	1
41 - 58	Bqkm1	87	7	6	0.20	0.019	14		8.1	0.4	4
58 - 76	Bqkm2	81	13	6	0.03		9		8.4	0.4	10
76 - 94	C	87	9	4	0.01		3		8.5	0.5	15
94 - 140	Cqkm	84	12	4	0.01		14		8.3	1.0	7
140 - 175	C'	90	8	2			4		8.4	0.9	6
Typic Calcigypsid											
0 - 0.5	A1	31	43	26	1.27	0.112	12		7.9	2.5	
0.5 - 5	A2	28	42	30	0.96	0.094	12		7.9	3.8	3
5 -13	A3	29	42	29	0.93	0.088	12	1.37	7.9	4.0	3
13 - 33	Bk1	25	40	35	0.63	0.066	19		7.8	9.2	9
33 - 56	Bk2	22	40	38	0.63	0.068	26	1.22	7.4	29.1	8
56 - 84	Cy1	20	30	50	0.26	0.025	10	1.30	7.6	21.3	
84 - 115	Cy2	10	50	39	0.14		8		7.7	17.4	2
115 - 140	Cy3	15	41	44	0.16		14		7.8	16.6	6
140 - 180	Cy4	14	43	43	0.18		11		7.9	21.3	7
Typic Haplargid											
0 - 3	A1	58	28	14	0.98	0.095	-	1.49	6.9[b]	0.5	
3 - 12	A2	60	23	17	0.26	0.038	-	1.50	6.6	0.5	
12 - 23	BAt	57	22	21			-	1.63	6.9	0.3	
23 - 46	Bt	46	24	30			-	1.47	7.5	0.2	
46 - 66	Btk1	40	31	29			3	1.39	7.9	0.2	
66 - 84	Btk2	33	31	36			8	1.41	8.0	0.2	
84 - 110	Btk3	31	39	29			15	1.34	8.1	0.3	
110 - 130	BCtk1	34	40	26			16	1.45	8.1	0.2	
130 - 160	BCtk2	33	44	23			18	1.60	8.3	1.0	7
160 - 180	BCtk3	27	49	24			12	1.42	8.1	2.7	11
180 - 230	C	32	49	19			9	1.37	7.9	6.8	10

[a] OC = organic carbon, N = nitrogen, CCE = calcium carbonate equivalent, BD = bulk density at 33 kPa water potential, EC = electrical conductivity, and SAR = sodium adsorption ratio.
[b] pH of 1:1 soil:water mixture, unless noted by [b], then pH of saturated paste.

Table 6.54 cont.

Depth cm	Horizon	Sand %	Silt %	Clay %	OC^a %	N^a %	CCE^a %	BD^a g cm^{-3}	pH^b	EC^a dS m^{-1}	SAR^a
Argic Petrocalcid											
0 - 5	A	85	5	10	0.25		-				
5 - 18	BAt1	87	4	9	0.13		-	1.86^c			
18 - 25	BAt2	82	5	13	0.17		-	1.78			
25 - 36	Bt	80	5	15	0.14		1	1.70			
36 - 48	Btk	77	6	17	0.24		16	1.53			
48 - 74	Bkm1	68	9	23	0.15		75				
74 - 100	Bkm2	71	11	18	0.05		65	1.68			
100 - 150	Bkm3	75	6	19	0.05		51	1.66			
Typic Haplocambid											
0 - 6	A	47	43	10	0.52		tr	1.41	8.0		
6 - 15	Bw1	44	45	12	0.31		tr	1.26	8.3		
15 - 33	Bw2	46	43	11	0.27		tr	1.22	8.5		
33 - 48	Bk1	44	44	12	0.35		4	1.22	8.7	0.7	5
48 - 60	Bk2	65	29	6	0.28		tr	1.24	8.7	1.7	10
60 - 87	Bk3	49	44	7	0.22		tr	1.17	8.2	7.8	15
87 - 135	C1	39	52	9	0.22		tr	1.24	7.9	13.3	14
135 - 150	C2	66	25	9	0.12		tr	1.30	7.8	11.8	12
150 - 175	C3	62	29	9	0.12		tr	1.30	7.7	11.1	11
175 - 205	C4	64	22	14	0.10		tr	1.30	7.9	10.6	9

[a] OC = organic carbon, N = nitrogen, CCE = calcium carbonate equivalent, BD = bulk density at 33 kPa water potential, EC = electrical conductivity, and SAR = sodium adsorption ratio.
[b] pH of 1:1 soil:water mixture, unless noted by [b], then pH of saturated paste.
[c] Bulk density at 33 kPa water potential, unless noted by [c], then air-dry.

evaporation. Typically, well-drained Aridisols have accumulations of the most soluble salts (chlorides and sulfates) at greater depths than those that are less soluble (gypsum and calcium carbonate), whereas more poorly drained Aridisol have accumulations of soluble salts closest to the surface. The distribution of these salts within the soil profile is reflected in the electrical conductivity (EC) of the soil solution (Table 6.54)

6.10.5.3 Calcium Carbonate

Calcium carbonate is present in almost all Aridisols, except for some Aridisols on stable platforms in West Africa and Australia. Inputs of Ca from mineral weathering or from the atmosphere, either dissolved in rain water or as particles of Ca-bearing minerals, coupled with relatively low solubility, cause $CaCO_3$ to accumulate. Once in soil solution, Ca can react with HCO_3^- from rainwater or that produced by the reaction of soil CO_2 (from respiration) with water. Loss of water by evaporation, or reduction of the partial pressure of CO_2 (P_{CO_2}), drives the precipitation of $CaCO_3$ from solution.

Aridisols on Holocene or latest Pleistocene landscapes are often calcareous in all horizons, because the relatively dry conditions of the Holocene prevent leaching of carbonate from surface horizons (e.g., the Typic Haplodurid and Typic Calcigypsid in Table 6.54). In contrast, Aridisols on older Pleistocene landscapes are often carbonate-free in the upper horizons, presumably due to more effective leaching during the cooler, moister Pleistocene (e.g., the Typic Haplargid and Argic Petrocalcid in Table 6.54). Aridisols of West Africa are acidic and lack carbonates because they form on old pre-weathered platform surfaces or from more recent polycycled sediments derived therefrom (L.P. Wilding, 1999, personal communication).

6.10.5.4   Exchangeable Sodium and Alkalinity

The accumulation of Na on the cation exchange complex is a common phenomenon in Aridisols and is related to conditions in soil solution [Na adsorption ratio (SAR)] (Table 6.54). Hydrolysis of exchangeable Na, particularly when soil solution electrolyte concentrations are low (i.e., little or no soluble salts) can produce alkaline soil reactions near pH 10. Swelling and dispersion of soil clays can result, in some cases causing plugging of soil pores and reducing permeability (Section G, Chapter 2). Clay dispersion also enhances the mobility of clay particles in the soil solution and aids the accumulation of clay in subsoil horizons, particularly natric horizons.

6.10.6 Mineralogy

Most minerals in Aridisols are inherited from the parent material (Nettleton and Peterson, 1983; Allen, 1990). The arid climate provides little surplus water for mineral hydrolysis or for leaching of weathering products. As a result, primary silicates experience little chemical weathering and Si-rich smectites are expected to dominate the clay fraction. Some Aridisols are kaolinitic and reflect paleoenvironmental conditions in regions of West Africa and Australia (L.P. Wilding, 1999, personal communication).

In certain soil-forming environments, there is clear evidence that mineral weathering is an important process and that some minerals are pedogenic. For example, on the deeply weathered granitic pediments of the Mojave Desert, alteration of biotite to vermiculite is extensive. This alteration and neosynthesis of Al-rich smectites from feldspar weathering are the main sources of phyllosilicates in the clay fraction (Boettinger and Southard, 1995). It is likely that much of this weathering is a result of more humid conditions during the Pleistocene.

The pedogenic formation of the fibrous clays, palygorskite [ideal structure: $Si_8Mg_5O_{20}(OH)_2(OH_2)_4 \cdot 4H_2O$] and sepiolite [ideal structure: $Si_{12}Mg_8O_{30}(OH)_4(OH_2)_4 \cdot 8H_2O$] (Singer, 1989) has been documented in the alkaline, Si-preserving, environment of petrocalcic horizons.

In most other Aridisols, the sand and silt fractions reflect relatively little mineral weathering, and soil mineral composition is dominated by the mineralogy of the parent material. Generally, the silicate mineral suite is dominated by quartz, feldspar, amphiboles, and micas. A few Aridisols contain abundant inherited zeolites (Allen, 1990).

The least intensively leached Aridisols may have significant contents of relatively soluble, nonsilicate minerals, including calcite ($CaCO_3$), gypsum ($CaSO_4 \cdot 2H_2O$), halite (NaCl), thenardite (Na_2SO_4),

Table 6.55   Mineralogical classes of Aridisol soil series in the United States [USDA-NRCS, 1997a]

Mineralogy class	Sandy and loamy, including skeletal families	Clayey and clayey-skeletal families	Substitute class families	Contrasting particle-size class families
Mixed	1651	126		82
Smectitic		316		9
Carbonatic	136	4		
Gypsic	31			
Siliceous	11			
Glassy			7	2
Illitic		3		
Kaolinitic		2		

mirabilite ($Na_2SO_4 \cdot 10H_2O$), epsomite ($MgSO_4 \cdot 7H_2O$), nahcolite ($NaHCO_3$), and trona ($Na_2CO_3 \cdot 10H_2O$) (Doner and Lynn, 1989).

The mixture of minerals in the sand and silt fractions and the dominance of smectites (usually montmorillonite) in the clay fraction are reflected by the mineralogical classes of *Soil Taxonomy* (Table 6.55).

6.10.7 Soil-Forming Processes

6.10.7.1 Eolian Accretion

The significant role of eolian additions to many Aridisols is well documented (Yaalon and Ganor, 1973; Gile and Grossman, 1979; Reheis, 1987). The role of eolian dust is particularly important in many Holocene Aridisols, where nearly all of the carbonate, gypsum, and opaline silica are derived from an external source (Harden et al., 1991). In some cases, eolian material constitutes nearly the entire fine-earth fraction and accounts for the nearly stone-free vesicular horizons below many desert pavements (McFadden et al., 1987). It is clear that an external source of carbonates is required for the development of thick petrocalcic horizons in soils formed from noncalcareous parent materials (Gile et al., 1966). The large volume of primary rocks that would need to be weathered to produce the carbonates present in many Aridisols is unrealistic, given the limited availability of water for silicate mineral weathering (Birkeland, 1984). Landscapes derived from rock types that tend to weather to soils with few surface clasts are less likely to have significant eolian accretion because an effective dust trap is absent (Boettinger and Southard, 1991). Those Aridisols derived from limestone bedrocks or sediments derived therefrom may have some accumulation of eolian dusts, but dust impact in terms of carbonate source may be of lesser importance in these soils (Rabenhorst and Wilding, 1986).

6.10.7.2 Mineral Weathering

Mineral weathering in Holocene Aridisols is minimal, due to the aridity of the Holocene climate. Aridisols on Pleistocene landscapes show clear signs of weathering of primary silicate minerals to produce Ca- and Mg-rich clays (Harden et al., 1991), interstratified mica-vermiculite (Eghbal and Southard, 1993a), vermiculite and Al-rich smectite (Boettinger and Southard, 1995), and Ca and silica in calcareous duripans (Boettinger and Southard, 1991; Eghbal and Southard, 1993c).

6.10.7.3 Clay Accumulation

The processes by which clay accumulates in Aridisols are similar to those that occur in other soil orders (Section E, Chapters 6.11, 6.12). Argillic and natric horizons, ranging in texture from sand to clay, and in thickness from 7.5 to 75 cm, generally begin at shallow depths (4 to 25 cm) below the soil surface, and, thus, are shallower than argillic and natric horizons in other soils.

Major processes of clay accumulation in subsurface horizons include disaggregation and dispersion of clays in eolian dust accompanied by burial by additional increments of dust (eolian accretion), *in situ* weathering of primary minerals, illuviation of dispersed clay from surface horizons, selective loss of fine particles from surface horizons by deflation, sheet flow, or dissolution, and neosynthesis of clay minerals from soil solution. Mineral weathering to produce clays susceptible to dispersion and translocation is primarily a relict of wetter Pleistocene climatic conditions (Nettleton and Peterson, 1983). Further, wetter conditions of the Pleistocene pluvials were probably required to dissolve and leach carbonates from surface horizons, the removal of Ca being a prerequisite for dispersion and translocation of the clays.

Most Aridisols with argillic horizons occur on the earliest Holocene and older landscapes, and the minimum time required to form an argillic horizon is largely a function of the carbonate content of the parent material and the rate at which Ca is added by eolian dust (Gile, 1975b). Exchangeable Na increases the rate of subsurface clay accumulation. Enhanced dispersion of clays, even in calcareous soils, causes natric horizons to form in less than 6,600 years (Alexander and Nettleton, 1977; Peterson, 1980).

6.10.7.4   Calcium Carbonate and Soluble Salt Redistribution

The depth and magnitude of $CaCO_3$ accumulation in many Aridisols illustrate the importance of varying eolian input and Pco_2 (McFadden et al., 1991). Formation of pedogenic Bk horizons is by dissolution of carbonates from upper horizons and precipitation from soil solution in lower horizons. Biogenic CO_2 raises Pco_2 to levels as high as 1.2%, and the relative rates of CO_2 production and loss to the atmosphere by diffusion generally cause Pco_2 to increase with soil depth (McFadden et al., 1991). Plant roots take up water preferentially over dissolved HCO_3^-, causing $CaCO_3$ to precipitate due to desiccation.

The progressive accumulation of carbonate in many Aridisols (Table 6.56, Fig. 6.61) reflects variable, but more or less continuous, input of Ca in eolian material, or from indigenous calcareous parent materials, and a large expansion of the volume of the calcic and petrocalcic horizons as carbonate precipitates and forces matrix grains apart. Cementation of horizons occurs when carbonate content reaches 25 to 60% (Machette, 1985).

Gile et al. (1965, 1966) proposed that a master K horizon be used to designate the more advanced stages of carbonate accumulation. The proposal was not formally adopted (the formal designation is Bkm), but K horizon is still used (Machette, 1985), as are the terms caliche and calcrete, although these latter terms are not well defined. The preferred terms for horizons of $CaCO_3$ accumulation are calcic horizon (often designated Bk) if not cemented, and petrocalcic if cemented (Soil Survey Staff, 1996). Many petrocalcic horizons may be relicts of older land forms and not related to present soil profiles, although upper, laminar layers may be the result of present-day processes of localized dissolution and reprecipitation (Harden et al., 1991; Eghbal and Southard, 1993b).

Gypsic and salic horizons are typical of basins with playas (Driessen and Schoorl, 1973) where leaching is limited by a high watertable and where eolian salts are recycled from the playa to surrounding terrain. The distribution of these salts in soil profiles is often seasonally dynamic due to their high solubility and the ease with which they are moved downward by leaching during the wet season and upward by evapotranspiration during the dry season (Eghbal et al., 1989).

Table 6.56   Stages of calcium carbonate accumulation in Aridisols [Gile et al., 1966; Machette, 1985]

Stage	General character	---------------Carbonate morphology---------------	
		Gravelly soils	**Nongravelly soils**
I	Weakest expression of macroscopic carbonate	Thin, discontinuous pebble coating, usually on undersides.	Few filaments or faint coatings
II	Carbonates segregated	Continuous pebble coatings, some interpebble fillings.	Few to common nodules
III	Carbonate continuous, plugged horizon forms in last part	Many interpebble fillings.	Many nodules and internodular fillings
		Any soil	
IV	Thin laminar layer	Laminar layer < 1 cm thick over plugged horizon.	
V	Thick laminar layer	Laminar layer > 1 cm thick over plugged horizon. Pisolites. Vertical faces laminated.	
VI	Multiple laminae, recemented, relaminated	Multiple generations of laminae, breccia, and pisolites. Many case-hardened vertical faces.	

Figure 6.61 Profile of an Argic Petrocalcid in New Mexico, with a platy, Stage V petrocalcic horizon about two feet (60 cm) thick. The scale on the tape is in feet.

6.10.7.5 Silica Accumulation

Silica accumulates in the form of opal ($SiO_2 \bullet nH_2O$) (Flach et al., 1973; Chadwick et al., 1987a), and possibly as chalcedony and quartz (Flach et al., 1973; Smale, 1973). Incipient cementation produces horizons that are very hard when dry and brittle when moist. More complete cementation of the soil by opal produces a duripan. Unlike the cementation process in petrocalcic horizons, wherein silicate grains in the matrix are forced apart, cementation by opal tends to be less disruptive of the silicate grains in the soil fabric, and relatively small amounts of silica (about 10% by weight) cause signifi-

Table 6.57 Ratios of 1,500 kPa water content and CEC to clay in Aridisols with indurated (typic) duripans and less strongly cemented (haplic) duripans from Nevada and Arizona [After Southard et al., 1990]

--------------1500 kPa water/clay--------------		-------------------CEC/clay-------------------	
Bw or Bt horizon	Bqm or Bqkm horizon	Bw or Bt horizon	Bqm or Bqkm horizon
Typic (n = 9)			
0.56 ± 0.15[a]	1.15 ± 0.22	1.02 ± 0.39	2.03 ± 1.23
Haplic (n = 15)			
0.52 ± 0.14	1.47 ± 0.71	1.09 ± 0.46	3.24 ± 1.92

[a] Mean ± standard deviation.

cant cementation of clay particles (Nettleton and Peterson, 1983), significantly changing water retention characteristics and CEC (Table 6.57).

Duripans have been called silcretes and duricrust (Jackson, 1957), although some are largely of
geologic, not pedologic, origin. Most duripans in Aridisols have a considerable content of $CaCO_3$, as
much as 70% by weight (Southard et al., 1990), but at least half the volume of a duripan does not
slake in acid, as petrocalcic horizons do. Many of these duripans have indurated, laminar upper layers
as much as 2 cm thick that overlie massive and less strongly cemented, highly calcareous material.
These duripans can be distinguished from a petrocalcic horizon only by the acid treatment. In many
cases only the upper laminar layer survives the acid treatment intact. The silica cements are derived
by weathering of siliceous rocks (Boettinger and Southard, 1991), volcanic ash (Chadwick et al.,
1987b), and eolian dust (Harden et al., 1991).

6.10.7.6 Polygenesis

Evidence of leaching below the calculated average depth of water storage (Arkley, 1963) is often
observed in Aridisols and ascribed to more humid Pleistocene paleoclimates (Nettleton and Peterson,
1983), and to variation in the accretion of eolian material and rates of weathering attributable to
Quaternary climate change (Wells et al., 1987). Evidence for climatically driven shifts in the dominant soil-forming processes includes engulfment of carbonate-free argillic horizons, presumably leached
of carbonates during the Pleistocene, by eolian-derived carbonates during the more arid Holocene
(Nettleton and Peterson, 1983) and micromorphologic evidence of episodic deposition of carbonates, opal, and clay in argillic horizons and duripans (Eghbal and Southard, 1993c).

6.10.8   Classification

At the order level, Aridisols are differentiated from other soils by an aridic soil moisture regime, an
ochric or anthropic epipedon, and one or more of the following subsurface horizons within 100 cm of
the soil surface: argillic, cambic, natric, salic, gypsic, petrogypsic, calcic, petrocalcic, or duripan
(Table 6.58). These subsurface horizons, and the cryic soil temperature regime, are used to differentiate seven suborders. Prior to 1994, only two suborders existed: Argids, with an argillic or a natric
horizon, and Orthids. Great groups of the Orthids were differentiated on the basis of the other diagnostic horizons now used to identify the suborders (Witty, 1990), thus placing more emphasis on the
presence of a horizon of clay accumulation.

The Cryids have a cryic soil temperature regime and occur at high latitudes and high elevations.
These soils, in which soil processes are inhibited not only by lack of water, but also by low temperatures, are common in the intermountain western region of the United States (Hipple et al., 1990). The
Argids, Durids, and Petrocalcids (petrocalcic horizon) occur on the oldest geomorphic surfaces, usually late Pleistocene and older, as on the least eroded parts of dissected alluvial fans or on hillslopes
protected from significant erosion by vegetation. Cambids and other Calcids are found on geologically younger side slopes and surfaces of intermediate age, usually latest Pleistocene and younger.
Gypsids and Salids often occur near playa margins, where salts are concentrated at or near the soil
surface by upward flux from a watertable driven by evaporation. Less commonly, they occur in association with Cambids and Calcids where parent materials are saline or where gypsum is added with
eolian material (Reheis, 1987; Eswaran and Gong, 1991).

The great groups are differentiated on the basis of a number of other properties, including aquic
conditions, cementation of horizons by gypsum or carbonates, and presence of diagnostic subsurface
horizons not identified at the suborder level, generally at depths between 100 and 150 cm. Subgroups

Table 6.58 Listing of suborders, great groups and subgroups of Aridisols order [Soil Survey Staff, 1998]

Suborders	Great groups	Subgroups
Cryids	Salicryids	Aquic, Typic
	Petrocryids	Xereptic, Duric Xeric, Duric, Petrogypsic, Xeric, Ustic, Typic
	Gypsicryids	Calcic, Vitrixerandic, Vitrandic, Typic
	Argicryids	Lithic, Vertic, Natric, Vitrixerandic, Vitrandic, Xeric, Ustic, Typic
	Calcicryids	Lithic, Vitrixerandic, Vitrandic, Xeric, Ustic, Typic
	Haplocryids	Lithic, Vertic, Vitrixerandic, Vitrandic, Xeric, Ustic, Typic
Salids	Aquisalids	Gypsic, Calcic, Typic
	Haplosalids	Duric, Petrogypsic, Gypsic, Calcic, Typic
Durids	Natridurids	Vertic, Aquic Natrargidic, Aquic, Natrixeralfic, Natrargidic, Vitrixerandic, Vitrandic, Xeric, Typic
	Argidurids	Vertic, Aquic, Abruptic Xeric, Abruptic, Haploxeralfic, Argidic, Vitrixerandic, Vitrandic, Xeric, Ustic, Typic
	Haplodurids	Aquicambidic, Aquic, Xereptic, Cambidic, Vitrixerandic, Vitrandic, Xeric, Ustic, Typic
Gypsids	Petrogypsids	Petrocalcic, Calcic, Vitrixerandic, Vitrandic, Xeric, Ustic, Typic
	Natrigypsids	Lithic, Vertic, Petronodic, Vitrixerandic, Vitrandic, Xeric, Ustic, Typic
	Argigypsids	Lithic, Vertic, Calcic, Petronodic, Vitrixerandic, Vitrandic, Xeric, Ustic, Typic
	Calcigypsids	Lithic, Petronodic, Vitrixerandic, Vitrandic, Xeric, Ustic, Typic
	Haplogypsids	Lithic, Leptic, Sodic, Petronodic, Vitrixerandic, Vitrandic, Xeric, Ustic, Typic
Argids	Petroargids	Petrogypsic Ustic, Petrogypsic, Duric Xeric, Duric, Natric, Xeric, Ustic, Typic
	Natrargids	Lithic Xeric, Lithic Ustic, Lithic, Vertic, Aquic, Durinodic Xeric, Durinodic, Petronodic, Glossic Ustic, Haplic Ustic, Haploxeralfic, Haplic, Vitrixerandic, Vitrandic, Xeric, Ustic, Glossic, Typic
	Paleargids	Vertic, Aquic, Arenic Ustic, Arenic, Calcic, Durinodic Xeric, Durinodic, Petronodic Ustic, Petronodic, Vitrixerandic, Vitrandic, Xeric, Ustic, Typic
	Gypsiargids	Aquic, Durinodic, Vitrixerandic, Vitrandic, Xeric, Ustic, Typic
	Calciargids	Lithic, Xerertic, Ustertic, Vertic, Aquic, Arenic Ustic, Arenic, Durinodic Xeric, Durinodic, Petronodic Xeric, Petronodic Ustic, Petronodic, Vitrixerandic, Vitrandic, Xeric, Ustic, Typic
	Haplargids	Lithic Ruptic-Entic, Lithic Xeric, Lithic Ustic, Lithic, Xerertic, Ustertic, Vertic, Aquic, Arenic Ustic, Arenic, Durinodic Xeric, Durinodic, Petronodic Ustic, Petronodic, Vitrixerandic, Vitrandic, Xeric, Ustic, Typic
Calcids	Petrocalcids	Aquic, Natric, Xeralfic, Ustalfic, Argic, Calcic lithic, Calcic, Xeric, Ustic, Typic
	Haplocalcids	Lithic Xeric, Lithic Ustic, Lithic, Vertic, Aquic Durinodic, Aquic, Duric Xeric, Duric, Durinodic Xeric, Durinodic, Petronodic Xeric, Petronodic Ustic, Petronodic, Sodic Xeric, Sodic Ustic, Sodic, Vitrixerandic, Vitrandic, Xeric, Ustic, Typic
Cambids	Aquicambids	Sodic, Durinodic Xeric, Durinodic, Petronodic, Vitrixerandic, Vitrandic, Fluventic, Xeric, Ustic, Typic
	Petrocambids	Sodic, Vitrixerandic, Vitrandic, Xeric, Ustic, Typic
	Anthracambids	Typic
	Haplocambids	Lithic Xeric, Lithic Ustic, Lithic, Xerertic, Ustertic, Vertic, Durinodic Xeric, Durinodic, Petronodic Xeric, Petronodic Ustic, Petronodic, Sodic Xeric, Sodic Ustic, Sodic, Vitrixerandic, Vitrandic, Xerifluventic, Ustifluventic, Fluventic, Xeric, Ustic, Typic

are differentiated by a large number of properties. Some common criteria are shallow lithic contacts, soil properties that grade toward Andisols, and soil moisture regimes that border on ustic and xeric.

6.10.9 Management

Agricultural use of Aridisols is limited chiefly by the lack of water. Where water is not available, most Aridisols are used for grazing by livestock and wildlife. Where irrigation systems have been developed, most Aridisols present some problem with irrigation and generally are less well suited for irrigation than are the Entisols (e.g., Torriorthents and Torrifluvents) that are commonly associated with the Aridisols. Land leveling to allow flood, furrow, or sprinkler irrigation often exposes calcic, petrocalcic, natric, or argillic horizons, or duripans.

Under irrigation, control of salinity in the rooting zone is critical to the long-term success of any crop management system. Flushing of salts by an amount of water in excess of the crop evapotranspiration demand (a leaching fraction) helps reduce salinity in the root zone. The excess water and dissolved salts must be removed by subsurface drainage. The presence of shallow subsurface horizons with slow permeability (duripans and natric and petrocalcic horizons) compounds the irrigation and drainage management problems for crop production, as well as for septic tank leach fields and irrigation of lawns in urbanized areas. Soil subsidence, due to solution of gypsum, and corrosion of concrete are common problems where soils containing gypsum are irrigated (Nettleton et al., 1982). Concrete corrosion occurs when sulfate reacts with Na to form mirabilite or thenardite (section 6.10.6), or with Ca and Al to form ettringite [$Ca_6Al_2(SO_4)_3(OH)_{12}.26H_2O$]. Crystallization of these sulfate minerals increases the volume of solids, leading to the disintegration of the concrete.

Other than N, most major plant nutrients are abundant in Aridisols, although micronutrient availability is often limited by low solubility at alkaline pH. Careful management of P is also often needed due to sorption of P by $CaCO_3$.

6.10.10   References

Ahrens, C.D. 1991. Meteorology today: An introduction to weather, climate, and the environment. 4th Ed. West Publishing Co., St. Paul, MN.

Alexander, E.B., and W.D. Nettleton. 1977. Post-Mazama Natrargids in Dixie Valley, Nevada. Soil Sci. Soc. Am. J. 41:1210–1212.

Allen, B.L. 1990. Mineralogy of Aridisols. p. 191–196. *In* J.M. Kimble and W.D. Nettleton (ed.) Proc. 4th Int. Soil Correlation Meet. (ISCOM) Characterization, classification, and utilization of Aridisols. Part A: Papers. USDA, SCS, Lincoln, NE.

Archibold, O.W. 1995. Ecology of world vegetation. Chapman and Hall, London, UK.

Arkley, R. J. 1963. Calculation of carbonate and water movement in soil from climatic data. Soil Sci. 96:239–248.

Birkeland, P.W. 1984. Soils and geomorphology. Oxford University Press, New York, NY.

Boettinger, J.L., and R.J. Southard. 1991. Silica and carbonate sources for Aridisols on a granitic pediment, western Mojave Desert. Soil Sci. Soc. Am. J. 55:1057–1067.

Boettinger, J.L., and R.J. Southard. 1995. Phyllosilicate distribution and origin in Aridisols on a granitic pediment, western Mojave Desert. Soil Sci. Soc. Am. J. 59:1189–1198.

Chadwick, O.A., D.M. Hendricks, and W.D. Nettleton. 1987a. Silica in duric soils: II. Mineralogy. Soil Sci. Soc. Am. J. 982–985.

Chadwick, O.A., D.M. Hendricks, and W.D. Nettleton. 1987b. Silica in duric soils: I. A depositional model. Soil Sci. Soc. Am. J. 975–982.

Cooke, R.U. 1970. Stone pavements in deserts. Ann. Assoc. Am. Geogr. 60:560–577.

Doner, H.E., and W.C. Lynn. 1989. Carbonate, halide, sulfate, and sulfide minerals. p. 279–330. *In* J.B. Dixon and S.B. Weed (ed.) Minerals in soil environments. 2nd Ed. Soil Science Society of America, Madison, WI.

Dorn, R.I., and T.M. Oberlander. 1981. Microbial origin of desert varnish. Science 213:1245–1247.

Driessen, P.M., and R. Schoorl. 1973. Mineralogy and morphology of salt efflorescences on saline soils in the Great Konya Basin, Turkey. J. Soil Sci. 24:436–442.

Eckert, R.E., Jr., M.K. Wood, W.H. Blackburn, F.F. Peterson, J.L. Stephens, and M.S. Meurisse. 1978. Effects of surface-soil morphology on improvement and management of some arid and semi-arid rangelands. p. 299–302. *In* Proc. 1st. Int. Rangeland Congr. American Society for Range Management, Denver, CO.

Eghbal, M.K., and R.J. Southard. 1993a. Mineralogy of Aridisols on dissected alluvial fans, western Mojave Desert, California. Soil Sci. Soc. Am. J. 57:538–544.

Eghbal, M.K., and R.J. Southard. 1993b. Stratigraphy and genesis of Durorthids and Haplargids on dissected alluvial fans, western Mojave Desert, California. Geoderma 59:151–174.

Eghbal, M.K., and R.J. Southard. 1993c. Micromorphological evidence of polygenesis of three Aridisols, western Mojave Desert, California. Soil Sci. Soc. Am. J. 57:1041–1050.

Eghbal, M.K., R.J. Southard, and L.D. Whittig. 1989. Dynamics of evaporite distribution in soils on a fan-playa transect in the Carrizo Plain, California. Soil Sci. Soc. Am. J. 53:898–903.

Eswaran, H., and Gong, Z.-T. 1991. Properties, genesis, classification, and distribution of soils with gypsum. p. 89–119. *In* W.D. Nettleton (ed.) Occurrence, characteristics and genesis of carbonate, gypsum, and silica accumulations in soils. Soil Science Society of America, Madison, WI.

Flach, K.W., W.D. Nettleton, and R.E. Nelson. 1973. The micromorphology of silica-cemented soil horizons in western North America. p. 714–729. *In* G. K. Rutherford (ed.) Soil microscopy. Proc. 4th Int. Work. Meet. Soil Micromorph., Aug. 27–31, 1973, Queen's University, Kingston, Ontario, Canada.

Fosberg, M.A., A.L. Falen, R.R. Blank, and K.W. Hipple. 1991. Soil forming processes in soils with cryic and frigid soil temperature regimes in Idaho. p. 43–53. *In* J. M. Kimble (ed.) Proc. 6th Int. Soil Correlation Meet. (VI ISCOM) Characterization, classification, and utilization of cold Aridisols and Vertisols. USDA, SCS, National Soil Survey Center, Lincoln, NE.

Gile, L. H. 1966. Coppice dunes and the Rotura soil. Soil Sci. Soc. Am. Proc. 30:657–660.

Gile, L.H. 1975a. Holocene soils and soil-geomorphic relations in an arid region of southern New Mexico. Quatern. Res. 5:321–360.

Gile, L.H. 1975b. Causes of soil boundaries in an arid region: I. Age and parent material. Soil Sci. Soc. Am. Proc. 39:316–323.

Gile, L.H., and R.B. Grossman. 1979. The desert project soil monograph. USDA, SCS, Washington, DC.

Gile, L.H., F.F. Peterson, and R.B. Grossman. 1965. The K horizon: A master soil horizon of carbonate accumulation. Soil Sci. 99:74–82.

Gile, L.H., F.F. Peterson, and R.B. Grossman. 1966. Morphological and genetic sequences of carbonate accumulation in desert soils. Soil Sci. 101:347–360.

Harden, J.W., E.M. Taylor, M.C. Reheis, and L.D. McFadden. 1991. Calcic, gypsic, and siliceous soil chronosequences in arid and semi-arid environments. p. 1–16. *In* W.D. Nettleton (ed.) Occurrence, characteristics and genesis of carbonate, gypsum, and silica accumulations in soils. Soil Science Society of America, Madison, WI.

Hipple, K.W., G.H. Logan, and M.A. Fosberg. 1990. Aridisols with cryic soil temperature regimes. p. 99–109. *In* J.M. Kimble and W.D. Nettleton (ed.) Proc. 4th Int. Soil Correlation Meet. (ISCOM) Characterization, classification, and utilization of Aridisols. Part A: Papers. USDA, SCS, Lincoln, NE.

Hubble, G.D., R.F. Isbell, and K.H. Northcote. 1983. Features of Australian soils. p. 17–47. *In* Soils, an Australian viewpoint. Division of Soils, Academic Press, London, UK.

Isbell, R.F. 1990. Soil of the Australian arid zone. p. 67–72. *In* J.M. Kimble and W.D. Nettleton (ed.) Proc. 4th Int. Soil Correlation Meet. (ISCOM) Characterization, classification, and utilization of Aridisols. Part A: Papers. USDA, SCS, Lincoln, NE.

Jackson, E.A. 1957. Soil features in arid regions with particular reference to Australia. J. Aust. Inst. Agric. Sci. 23:196–208.

Johnson, W.M. 1990. The Argids. p. 15–20. *In* J.M. Kimble and W.D. Nettleton (ed.) Proc. 4th Int. Soil Correlation Meet. (ISCOM) Characterization, classification, and utilization of Aridisols. Part A: Papers. USDA, SCS, Lincoln, NE.

Johnston, R. 1997. Introduction to microbiotic crusts. USDA-NRCS, Soil Quality Institute and Grazing Lands Technology Institute. Fort Worth, TX.

Machette, M.N. 1985. Calcic soils of the southwestern United States. p. 1–21. *In* D. L. Weide (ed.) Soils and Quaternary geology of the southwestern United States. Geol. Soc. Am. Spec. Pap. 203.

McFadden, L.D., R.G. Amundson, and O.A. Chadwick. 1991. Numerical modeling, chemical and isotopic studies of carbonate accumulation in soils of arid regions. p. 17–35. *In* W.D. Nettleton (ed.) Occurrence, characteristics and genesis of carbonate, gypsum, and silica accumulations in soils. Soil Science Society of America, Madison, WI.

McFadden, L.D., S.G. Wells, and M.J. Jercinovic. 1987. Influences of eolian and pedogenic processes on the origin and evolution of desert pavements. Geology 15:504–508.

Miller, D.E. 1971. Formation of vesicular structure in soil. Soil Sci. Soc. Am. Proc. 35:635–637.

Nettleton, W.D., and F.F. Peterson. 1983. Aridisols. p. 165–215. *In* L.P. Wilding, N.E. Smeck, and G.F. Hall (ed.) Pedogenesis and soil taxonomy II. The soil orders. Elsevier, Amsterdam, Netherlands.

Nettleton, W.D., R.E. Nelson, B.R. Brasher, and P.S. Derr. 1982. Gypsiferous soils in the western United States. p. 147–168. *In* J.A. Kittrick, D.S. Fanninf and L.R. Hossner (ed.) Acid sulfate weathering. Soil Sci. Soc. Am. Spec. Pub. 10. Soil Science Society of America, Madison, WI.

Otterman, J., and V. Gornitz. 1983. Saltation vs. soil stabilization: Two processes determining the character of surfaces in arid regions. Catena 10:339–362.

Passey, H.B., V.K. Hugie, E.W. Williams, and D.E. Ball. 1982. Relationships between soil, plant community, and climate on rangelands of the Intermountain West. USDA Tech. Bul. 1669.

Peterson, F.F. 1977. Dust infiltration as a soil forming process in deserts. Agron. Abst. p. 172.

Peterson, F.F. 1980. Holocene desert soil formation under sodium salt influence in a playa-margin environment. Quatern. Res. 13:172–186.

Rabenhorst, M.C., and L.P. Wilding. 1986. Pedogenesis on the Edwards Plateau, Texas: II. Formation and occurrence of diagnostic horizons in a climosequence. Soil Sci. Soc. Am. J. 50:687–692.

Reheis, M.C. 1987. Gypsic soils on the Kane alluvial fans, Big Horn County, Wyoming. US Geol. Surv. Bull. 1590.

Reid, D.A., R.C. Graham, R.J. Southard, and C. Amrhein. 1993. Slickspot soil genesis in the Carrizo Plain, California. Soil Sci. Soc. Am. J. 57:162–168.

Schlesinger, W.H., and C.S. Jones. 1984. The comparative importance of overland runoff and mean annual rainfall to shrub communities of the Mojave desert. Bot. Gaz. 145:116–124.

Schlesinger, W.H, J.F. Reynolds, G.L. Cunningham, L. F. Huenneke, W. M. Jarrell, R.A. Virginia, and W.G. Whitford. 1990. Biological feedbacks in global desertification. Science 247:1043–1048.

Shantz, H.L. 1956. History and problems of arid lands development. p. 3–25. In G.F. White (ed.) The future of arid lands. Greenwood Press, Westport, CT.

Singer, A. 1989. Palygorskite and sepiolite group minerals. p. 829–872. In J.B. Dixon and S.B. Weed (ed.) Minerals in soil environments. 2nd Ed. Soil Science Society of America, Madison, WI.

Smale, D. 1973. Silcretes and associated silica diagenesis in southern Africa and Australia. J. Sediment. Petrol. 43:1077–1089.

Soil Survey Staff. 1975. Soil taxonomy: A basic system of soil classification for making and interpreting soil surveys. USDA-SCS, Agric. Handbk. 436. US Government Print Office, Washington, DC.

Soil Survey Staff. 1998. Keys to soil taxonomy. 8th ed. USDA-NRCS. US Government Printing Office, Washington, DC.

Southard, R.J., J.L. Boettinger, and O.A. Chadwick. 1990. Identification, genesis, and classification of duripans. p. 45–60. In J.M. Kimble and W.D. Nettleton (ed.) Proc. 4th Int. Soil Correlation Meet. (ISCOM) Characterization, classification, and utilization of Aridisols. Part A: Papers. USDA, SCS, Lincoln, NE.

Springer, M.E. 1958. Desert pavement and vesicular layer of some soils of the desert of the Lahontan Basin, Nevada. Soil Sci. Soc. Am. Proc. 22:63–66.

Thornthwaite, C.W. 1948. An approach toward a rational classification of climate. Geograph. Rev. 38:55–94.

USDA-NRCS. 1997a. Soils data bases. Survey Section, Statistical Laboratory, Iowa State University, Ames, IA.

USDA-NRCS. 1997b. Soil Survey Laboratory database. National Soil Survey Center, Lincoln, NE.

Wells, S.G., L.D. McFadden and J.C. Dohrenwend. 1987. Influence of late Quaternary climatic changes on geomorphic and pedogenic processes on a desert piedmont, eastern Mojave Desert, California. Quatern. Res. 27:130–146.

Witty, J.E. 1990. The Orthids past, present and future classification. p. 93–98. In J.M. Kimble and W.D. Nettleton (ed.) Proc. 4th Int. Soil Correlation Meet. (ISCOM) Characterization, classification, and utilization of Aridisols. Part A: Papers. USDA, SCS, Lincoln, NE.

Yaalon, D.H., and E. Ganor. 1973. The influence of dust on soils during the Quaternary. Soil Sci. 116:146–155.

6.11 Alfisols

C.T. Hallmark
Texas A & M University

D.P. Franzmeier
Purdue University

6.11.1 Introduction

The central concept of Alfisols embraces soils of the semi-arid to humid climates with light colored surfaces and subsoils moderately rich in basic cations, formed, at least in part, by movement of clay from overlying horizons into the subsoil. Most Alfisols developed under deciduous forest vegetation in humid climates and grass or savannah vegetation in drier regimes. In loamy and silty Alfisols of the humid regions, fragipans are not uncommon; in drier climates, incomplete leaching of salts and carbonates can result in concentrations of secondary carbonates, gypsum, amorphous silica or enrichment of exchangeable Na. As the processes of formation of Alfisols require both time and energy, most are found on geomorphic surfaces that are of Pleistocene age or older. They form on a wide variety of parent materials.

In previous classification systems in the United States, most Alfisols were in the Gray-Brown Podzolic Great Group, but others were in the Low-Humic Gley, Noncalcic Brown, Reddish-Brown Lateritic, Reddish Brown, Reddish Chestnut, Gray Wooded and Planosol Great Groups. Further, natric great groups or subgroups of Alfisols also include salt-affected soils that were previously classified as Solonetz and Soloth soils (Soil Survey Staff, 1975).

6.11.2 Distribution

Alfisols occur on all continents except Antarctica (Fig. 6.62). They occupy about 9.6% of the ice-free Earth's surface (Table 6.1). They are found between the latitudes of 65° N and 45° S and concentrated in the Northern Hemisphere, specifically North America, Europe and east Central Asia between latitudes of 30° N and 65° N; also, significant occurrences are found in subsaharan Africa, India, South America, and Australia. Cryalfs are found within the colder reaches such as Northern Europe and into Western Russia. Ustalfs are common in subsaharan Africa, Eastern Brazil, the eastern portion of India, and Southeastern Australia. Udalfs are prominent in Central Europe and Eastern Australia, while Xeralfs are found in countries bordering the Mediterranean Sea.

Within the United States, Cryalfs are found in the north central states and at higher elevations in the western states. Udalfs are prominent in the eastern portion of the Midwest, along the Mississippi and Ohio River valleys while Ustalfs occur to the south and west of the Great Plains as well as the southern portion of the High Plains. Most Xeralfs are restricted to the west coast states.

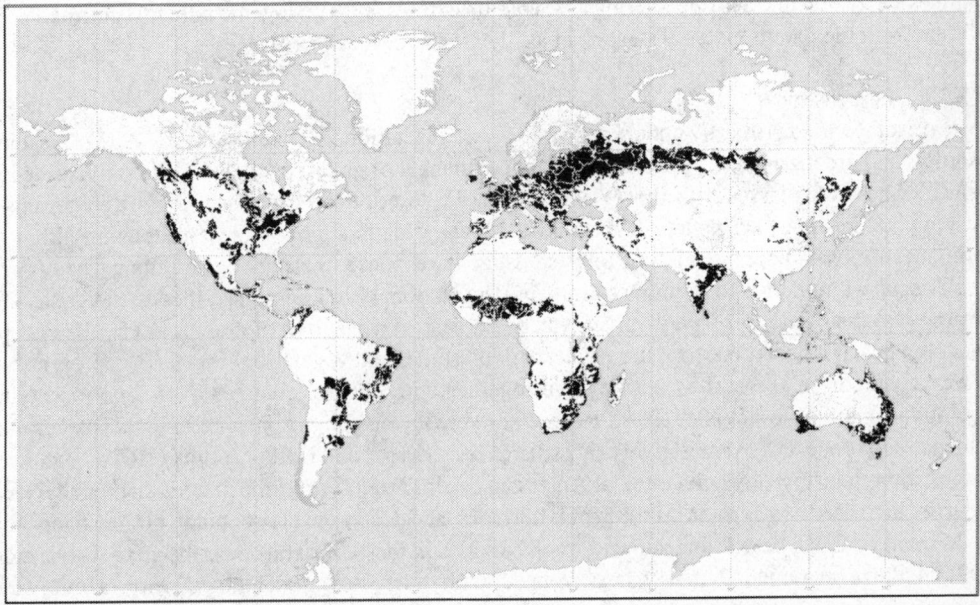

Fig. 6.62 Global distribution of Alfisols [Courtesy of USDA-NRCS, Soil Survey Division, World Soil Resources, 1998]

6.11.3 Factors of Soil Formation

6.11.3.1 Parent Material

Alfisols form on a wide variety of parent materials which includes glacial till (Borchardt et al., 1968; Smeck et al., 1968; Smith and Wilding, 1972; Ranney et al., 1975; Franzmeier et al., 1985), loess (Bouma et al., 1968; Grossman and Fehrenbacher, 1971), eolian sands (Davis, 1970; Allen et al., 1972; Miles and Franzmeier, 1981), residuum from limestone (Reynders, 1972), sandstone (Anderson et al., 1975; Stahnke et al., 1983; Chittleborough, 1989; Delgado et al., 1994), siltstones and shales (Gallez et al., 1975; Delgado et al., 1994), acid igneous and metamorphic rock (Goss and Allen, 1968; Ranney and Beatty, 1969; Yerima et al., 1989; Kooistra et al., 1990); mafic igneous and metamorphic rock (Gallez et al., 1975; Bhattacharyya et al., 1993; Juo and Wilding, 1996), lacustrine sediments (Borchardt et al., 1968; Ranney and Beatty, 1969), colluvium derived from acid shales and siltstones (Bailey and Avers, 1971), basalt (Weitkamp et al., 1996), coastal plain sediments (Vepraskas and Wilding, 1983; Juo and Wilding, 1996) and alluvium (Parsons et al., 1968; Ranney et al., 1975; Chittleborough et al., 1984; Klich et al., 1990; Delgado et al., 1994). The literature is replete with examples of Alfisols developing from two or more parent materials, for example, loess over till (Foss and Rust, 1968; Allan and Hole, 1968; Ransom et al., 1987), loess over basalt (Blank and Fosberg, 1991), loess over shale (Ranney et al., 1975), and colluvium over residuum (Weitkamp et al., 1996). Uniformity of parent material also continues to be a question in evaluating the influence of lithogenic inherited versus pedogenic and translocated clay in the argillic horizon (Ruhe, 1984a). In eolian sands, the argillic (zone of clay accumulation) horizon usually consists of a sequence of thin bands called lamellae that contain somewhat more clay than the soil material between the bands.

Lotspeich and Coover (1962) stated that "texture of the parent material controls the texture of the soil because texture is a nearly permanent characteristic of soil," particularly in arid and semi-arid regions. While this may be true in Alfisols for which the parent materials have undergone previous cycles of weathering and are rich in quartz and resistant minerals (Anderson et al., 1975), many examples are available where weathering gives rise to textures and a pedogenic mineral suite markedly different from the parent material (Gallez et al., 1975; Paeth et al., 1971).

6.11.3.2 Organisms

Most Alfisols formed primarily under hardwood forest vegetation (Soil Survey Staff, 1975), but some suborders and great groups support a significant component of grass vegetation. This is illustrated in Fig. 6.63 where soil classes with abbreviated vegetative descriptions from the literature are arranged in a temperature-moisture regime matrix. Aqualfs and Udalfs support forest vegetation although dominant species and composition of forests change in response to climatic conditions. Ustalfs and Xeralfs support moderate to dominant grassland vegetation, commonly described as savanna. The forest vegetation of moister Cryalfs gives way to a greater composition of grasses as Cryalfs occupy dryer climates. It is hazardous to assume present or presettlement vegetation (Fig. 6.63) represents the native vegetation of the soil as both evolve together with climate changes over time modifying vegetative communities (Wells, 1970). Even within broad vegetative zones, individual species may influence soil properties over relatively short distances. Gersper and Holowaychuk (1970a,b) showed precipitation stemflow down the trunk of American beech (*Fagus grandifolia*) trees thickened E (A2) horizons, increased low chroma mottling of B horizons, and decreased clay content, pH, exchangeable Ca, Mg and K, CEC, base saturation and free Fe oxides in the B horizons near the stem. Davis et al. (1995) believe available P that was correlated with microelevational changes on a cultivated

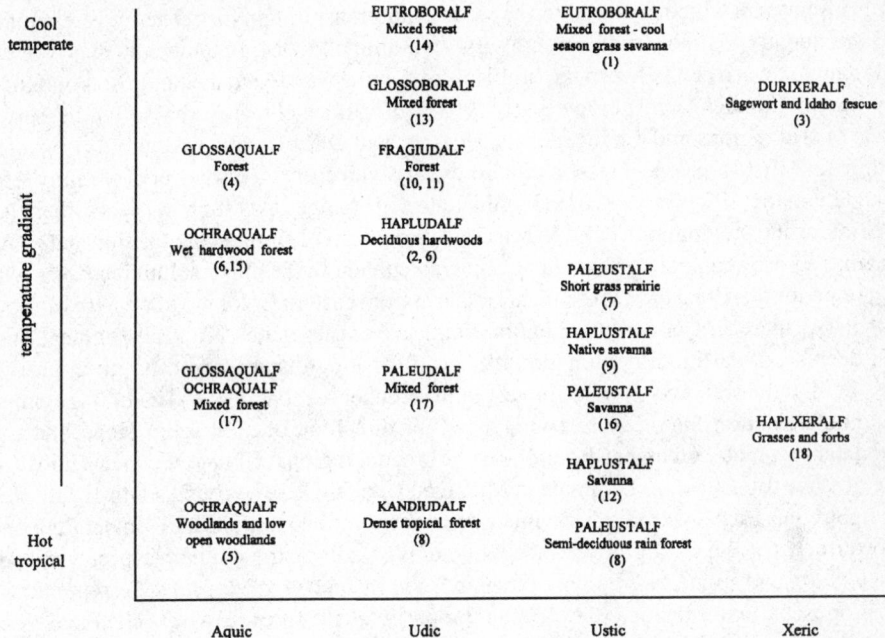

Fig. 6.63 Generalized native vegetation for selected Alfisols as a function of temperature and moisture regimes. [(1) Anderson et al., 1975; (2) Bailey and Avers, 1971; (3) Blank and Fosberg, 1991; (4) Bouma et al., 1969; (5) Coventry and Williams, 1984; (6) Cremeens and Mokma, 1986; (7) Davis, 1970; (8) Gallez et al., 1975; (9) Goss and Allen, 1968; (10) Lozet and Herbillon, 1971; (11) Miller et al., 1971; (12) Moberg and Esu, 1991; (13) Ranney and Beatty, 1969; (14) Schaetzl, 1996; (15) Smith and Wilding, 1972; (16) Stahnke et al., 1983; (17) Vepraskas and Wilding, 1983; (18) Weitkamp et al., 1996] Those soils listed as Ochraqualfs would be placed as Epiqualfs or Endoqualfs, and those as Boralfs as Cryalfs in the current 1998 *Keys to Soil Taxonomy* [Soil Survey Staff, 1998]

Psammmentic Paleustalf in Niger is related to location of bushes and trees in the savanna or to cultivation.

Fauna dwelling in Alfisols have also been credited with affecting soil formation. Nielson and Hole (1964) found earthworm population masses of 2218 kg ha^{-1} under a deciduous forest and earthworm middens numbering 274,000 ha^{-1}. Essentially all annual leaf fall was moved to the middens by earthworms. Wiecek and Messenger (1972) described calcite spheroids in acid A horizons of Hapludalfs under forest vegetation that are produced in earthworm calciferous glands and were responsible for close-range variability in soil pH of A horizons. Hugie and Passey (1963) describe burrowing activity of western cicadas, crediting them with development of cylindrical structure as they burrowed through loamy soils. Their numbers were greatly reduced on coarse- and fine-textured materials. Ant, termite, and crustacean activity have also been described (Thorp, 1949). Crayfish activity in Ochraqualfs of the Texas coast has been so intense that up to 75% of the subsoil volume are comprised of their krotovinas prompting the call for a new great group (Vermaqualfs) to accommodate these soils highly modified by fauna (Vepraskas and Wilding, 1983). Arkley and Brown (1954) observed sufficient rodent activity to suggest that mima mounds in western states are the result of pocket gophers.

6.11.3.3 Climate
In the process of clay eluviation, percolating water has mobilized silicate clays in upper horizons, carried them downward in the pedon with final deposition in subsoil horizons. This occurs most

readily in climates in which precipitation exceeds evapotranspiration for at least several months each year. Consequently, Alfisols occur in semi-arid to humid climates. In temperate and cool climates, Alfisols tend to form a belt between the Mollisols of the grasslands and the Spodosols and Inceptisols of very humid climates (Soil Survey Staff, 1975). In warmer climates, they often lie between the Aridisols of arid regions and the Inceptisols, Ultisols, and Oxisols of warm, humid climates.

Within the Alfisol areas, several soil properties vary with climate. Early works of Jenny (1935) and Jenny and Leonard (1934) serve to broadly illustrate soil changes over long transects where gradients of temperature and precipitation exist. When crystalline rocks of similar composition and geomorphic age weather across a temperature gradient, soil clay content in the upper solum increases with mean annual temperature. Along the 11 °C annual isotherm from eastern Colorado to western Missouri, they found that with increased annual precipitation, depth to free carbonates, N and clay contents increased, and pH decreased. Partitioning climatic influence over long distances is difficult at best. Climate implies both temporal and spatial fluxes in a number of parameters including temperature, precipitation, radiation, humidity, and wind speed and direction. Locally, relief, slope, and aspect can provide a microclimate somewhat dissimilar to the broader regional climate. Cooper (1960) illustrated the role of slope aspect on microclimate in Michigan Udalfs where he found south-facing slopes had higher light intensities, maximum air temperature, evaporation rate, and soil temperature and lower soil moisture than north-facing slopes. The B horizons of soils on the southern aspect were redder and finer in texture, and solums were thinner than those of the northern exposure. Torrent and Nettleton (1979) proposed a textural index (based upon fine silt:total silt ratios) to assess chemical weathering. Using soils, primarily Udalfs and Aqualfs developed from loess along the Mississippi Valley from Minnesota to Mississippi, they related the textural index describing weathering of silts to clays with increasing mean annual temperature. Subsequently, Ruhe (1984a), utilizing a smaller data set of Hapludalfs along a similar Minnesota to Mississippi transect, interpreted the trends as sedimentation dependent. In a companion study on the same Hapludalfs, Ruhe (1984b) related the more intensive and deeper leaching of bases in the southern sector to both historic and paleo climate differences in annual and effective precipitation.

6.11.3.4 Relief

Alfisols occur mostly on hillslopes that are of moderate gradient and are flat or convex in cross-section. On steeper slopes, clay translocation is limited because the soil is very shallow or stony or because material moves downslope, resulting in Inceptisols or Entisols. In depressional areas, downward water movement might be limited, or soil organic matter (SOM) content of upper horizons is high enough for the soils to be Mollisols.

Relief is responsible for the distribution of Aqualfs in many settings where water moves to lower topographic positions either by overland or throughflow. In these lower positions, epiaquic (perched water table) or endoaquic conditions may occur. Cremeens and Mokma (1986) described soil hydrosequences in Michigan where well- and moderately well-drained Hapludalfs yielded to somewhat poorly drained Ochraqualfs and poorly drained Haplaquepts and Argiaquolls. Argillans (clay films) and parameters related to movement of clay from eluvial to illuvial zones were greatest in the better drained soils. Working with seasonally saturated Alfisols in a toposequence on the Texas Coastal Plain, Vepraskas and Wilding (1983) found soil pH generally increased as did base saturation and exchangeable Na levels from upper to lower topographic positions. Also, smectite was the predominant clay mineral in the upper sola of the soils in the lower positions while kaolinite was dominant in higher positions. Cremeens and Mokma (1986) also found smectite in the wettest sites of their study with its absence in the better drained soils. Goss and Allen (1968) found appreciable accumulation of exchangeable Na downslope in a Natrustalf and evidence of accumulation of smectite due to relief.

6.11.3.5 Time

Alfisols in areas with moderate and cool temperatures are mostly on late Pleistocene surfaces as most of these surfaces were affected by glaciation. In warmer areas, they may be on older surfaces. Usually, there has not been enough clay translocation for the soils to be classified as Alfisols on surfaces younger than late Pleistocene. On old surfaces, leaching of bases commonly has progressed to the point where the soils are classified as Ultisols instead of Alfisols.

The length of time for development of soil and specific soil properties has long been of scientific interest (Jenny, 1941). In recent years, interest in soil age has been increasing as evidenced by the growing volume of literature (Bilzi and Ciolkosz, 1977; Bockheim, 1980; Little and Ward, 1981; Meixner and Singer, 1981; Harden, 1982; Muhs, 1982; Chittleborough et al., 1984; Catt 1991; Markewich and Pavich, 1991; Delgado et al., 1994). Part of this renewed interest is due to application of soil interpretation to subjects of current interest such as climate change, C sequestration rates, landscape stability, and archaeology. Improved and new methodologies to directly determine soil age such as radiocarbon assay, thermoluminesence, amino acid analysis, palaeomagnetic measurement and U-Th disequilibrium series have augmented dendrochronology and recorded historic events (mudflows, volcanic eruptions, etc.) and indirect methods of correlation of stratigraphic units. As soil (and landscape) age increases, so does the concern for significant changes in other factors of state, particularly climate and its effect on vegetation. The time required for the development of Alfisols is incumbent upon the formation of argillic and kandic horizons. Bilzi and Ciolkosz (1977) believe that 2,000 to 12,000 years are required to develop fragipan-like features and argillic horizons in alluvium in Pennsylvania under udic soil moisture and mesic soil temperature regimes. In northeast Iowa on earthen mounds of varying age built by Indians, Parsons et al. (1962) found that soil genesis had taken place with A horizon properties remaining relatively constant with increasing age beyond 1000 yr, E horizon platy structure developing within 1,000 yr (maximum expression required > 2500 yr) and the blocky structure plus clay translocation sufficient to meet argillic horizon definitions in the B horizon occurring between 1,000 and 2,500 yr BP. In comparison, the nearby Fayette soil (Typic Hapludalf) had developed over a 14,000 yr period. In preconditioned material used by Indians to construct mounds and ridges in Louisiana, argillic horizons in Ultisols and Alfisols had developed in 5,000 to 5,400 yr (Saunders et al., 1997).

In the Midwest, most Alfisols are found on surfaces affected by late Wisconsinian glaciation (12,000–25,000 yr BP). In the Upper Midwest, many Alfisols formed in late Wisconsinian till deposited by glacier ice, and in outwash and lacustrine deposits from the glacier when it melted. Also, much of the Midwest and the lower Mississippi Valley was blanketed with late Wisconsinian age loess that was calcareous when it was deposited. Soils formed in it are relatively high in basic cations. This loess also covered surfaces on which soils had formed previously and had been eroded. These buried soils (paleosols) were enriched with basic cations (especially Ca and Mg) that were leached from the loess and translocated downward.

Alfisols of ustic and xeric soil moisture regimes and developed from alluvium have been included in a number of chronosequence studies (Table 6.59). The annual precipitation ranged from 170 to 420 mm. The youngest Alfisol (Ultic Haplustalf) was 3,740 yr BP and the oldest (Paleustalf) was 763,000 yr BP, while the youngest Paleustalf was 9,220 yr BP. Much older Haploxeralfs were noted in the California studies, but in the xeric moisture regime sufficient development of the argillic horizon to permit classification as a Palexeralf had not occurred.

Table 6.59 Ages of Ustalfs and Xeralfs included in chronosequence studies. All soils developed from alluvium. Dating methods in the studies include amino acid, U-series, uplift rates, and stratigraphic correlations

Soil Class	Age, yrs BP	MAP (mm)	Location	Source
Typic Haploxeralf	250,000	310	California	Meixner and Singer (1981)
Typic Haploxeralf	40,000	410	California	Harden (1982)
Typic Haploxeralf	130,000			
Typic Haploxeralf	600,000			
Typic Natrixeralf	60-107,000	170	California	Muhs (1982)
Typic Natrixeralf	120-134,000			
Typic Natrixeralf	375-460,000			
Ultic Haplustalf	3,740	420	New South Wales, Australia	Chittleborough et al. (1984)
Rhodic Paleustalf	9,220			
Ultic Paleustalf	26,700			
Ultic Paleustalf	27,040			
Ultic Paleustalf	29,000			
Paleustalf	42,500	--	Victoria, Australia	Little and Ward (1981)
Paleustalf	210,000			
Paleustalf	227,000			
Paleustalf	763,000			

6.11.4 Morphology

The morphology of Alfisols ranges greatly across the broad spectrum of conditions under which they occur. In all Alfisols, however, there is an increase in clay content from the A and E horizons near the surface to the B horizons below. In most Alfisols, this clay increase meets the concept of an argillic horizon where the illuvial Bt (t from *ton* Gr, clay) horizon shows evidence of oriented clays. In older

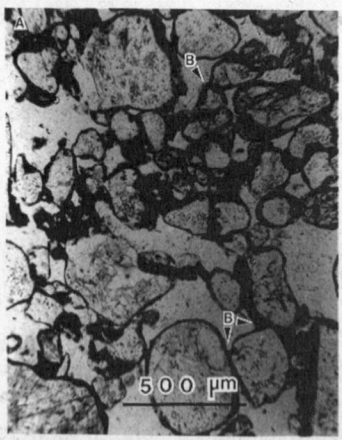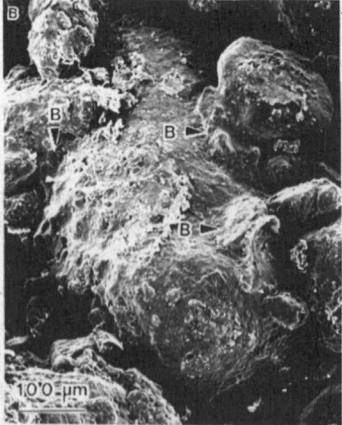

Fig. 6.64 The Bt2 horizon (51-72 cm depth) of a sandy clay, siliceous, subactive, isohyperthermic Psammentic Paleustalf from Niger. A. Plane polarized light micrograph of ferriargillans around quartz skeleton grains and clay bridging (B) of sand grains; B. Scanning electron micrograph of clay bridging (B) between sand grains [West et al. (1984)]

soils dominated by low activity clays (low CEC of the clay fraction), evidence of oriented clay may be absent but clay increase is sufficient for a kandic horizon.

Where some of the clay in the Bt horizon is oriented, that is, stacked like a pile of papers or playing cards, the oriented clay may be on ped surfaces where it forms clay films (thin layers of oriented clay) that often cover and may be a different color from sand grains in the ped (see Section E, Chapter 6.12 for electron micrograph of a clay film). In sandy soils, the oriented clay forms bridges between adjacent sand grains (Fig. 6.64) and may line root channels. Often, this clay mentioned above actually contains a fair amount of silt. In many Alfisols, the transitional EB or BE horizons have silt coatings instead of clay on the outside of peds. In these, and some other Alfisols, especially those formed in calcareous parent materials, the best developed clay films are in the lower Bt horizon. Apparently, clay is stripped from the ped coatings in the upper B or BE horizon and is deposited on ped surfaces deeper in the profile, illustrating that clay translocation is a dynamic process, and that the argillic horizon is moving down in the profile.

Before a soil can be classified in *Soil Taxonomy*, germaine diagnostic horizons and soil properties (defined in Section E, Chapter 4.5) must be identified. Several are essential or common in Alfisols and their occurrence in the various suborders of Alfisols is shown in Table 6.61.

6.11.5   Suborders: Their Morphology, Distribution, and Use

The Alfisol Order is subdivided into five suborders, one based on temperature and four based on soil moisture regime. Cryalfs are the Alfisols in cold climates. Of the four suborders based on soil moisture regime, three (Udalfs, Ustalfs, Xeralfs) follow broad geographical climatic belts. The distribution of Aqualfs depends on local landscape and parent material factors. A complete listing of the Great Groups and Subgroups in each suborder appears in Table 6.60.

6.11.5.1   Aqualfs
These are the wet Alfisols in which the watertable is at or near the surface (50 cm) for at least several weeks each year, but may be very deep during the dry season. The watertable may be continuous to the aquifer, or it may be held up by a layer with low permeability in or just beneath the soil surface. Their location in the landscape is controlled mainly by local conditions of parent material and relief. They occur as small areas in larger areas of Cryalfs, Udalfs, Ustalfs, Xeralfs or other soil suborders, but they are more common in zones of Udalfs than Ustalfs. They may have several kinds of diagnostic horizons (Table 6.61). Nearly all Aqualfs had forest vegetation at some time in the past (Soil Survey Staff, 1975). Most are cultivated after they were drained by underground tiles or ditches. Rice is a common crop in warmer regions.

Aqualfs display quantifiable redoximorphic features (gray colors) immediately below a depth of 25 cm from the mineral soil surface. In the absence of quantifiable redoximorphic features, a positive reaction to active Fe^{2+} (α, α-dipyridyl) can be used. In this process (gleying), soil horizons are saturated with water, microbes deplete available O_2, causing Fe(III) in brownish and reddish Fe oxide minerals to be reduced. The resulting Fe^{2+} is released into the soil solution, much of which is leached from the soil. The result is that the horizon, or much of it, takes on the gray color of the silicate minerals. Gleyed horizons are designated with a g as Btg.

Gleying in Aqualfs may be noted in the upper 25 cm of the soil. Gleyed soils are generally found on lower topographic positions receiving additional water, or on broad, level interfluves where surface drainage may be very slow. The A horizon, commonly overlain by O horizon material in varying stages of decomposition, is usually either too thin, or too low in organic C to meet requirements of a mollic epipedon. A few may qualify as an umbric or mollic epipeon, but have insufficient bases in the subsoil

Table 6.60 Listing of suborders, great groups and subgroups in the Alfisols order [Soil Survey Staff, 1998]

Suborder	Great Group	Subgroup
Aqualfs	Cryaqualfs	Typic
	Plinthaqualfs	Typic
	Duraqualfs	Typic
	Natraqualfs	Vertic, Vermic, Albic Glossic, Albic, Glossic, Mollic, Typic
	Fragiaqualfs	Vermic, Aeric, Plinthic, Humic, Typic
	Kandiaqualfs	Arenic, Grossarenic, Plinthic, Aeric Umbric, Aeric, Umbric, Typic
	Vermaqualfs	Natric, Typic
	Albaqualfs	Arenic, Aeric Vertic, Chromic Vertic, Vertic, Udollic, Aeric, Aquandic, Mollic, Umbric, Typic
	Glossaqualfs	Histic, Arenic, Aeric Fragic, Fragic, Aeric, Mollic, Typic
	Epiaqualfs	Aeric Chromic Vertic, Aeric Vertic, Chromic Vertic, Vertic, Aquandic, Aeric Fragic, Fragic, Arenic, Grossarenic, Aeric Umbric, Udollic, Aeric, Mollic, Umbric, Typic
	Endoaqualfs	Aquandic, Chromic Vertic, Vertic, Aeric Fragic, Fragic, Arenic, Grossarenic, Udollic, Aeric Umbric, Aeric, Mollic, Umbric, Typic
Cryalfs	Palecryalfs	Andic, Vitrandic, Aquic, Oxyaquic, Xeric, Ustic, Mollic, Umbric, Typic
	Glossocryalfs	Lithic, Vertic, Andic, Vitrandic, Aquic, Oxyaquic, Fragic, Xerollic, Umbric Xeric, Ustollic, Xeric, Ustic, Mollic, Umbric, Eutric, Typic
	Haplocryalfs	Lithic, Vertic, Andic, Vitrandic, Aquic, Oxyaquic, Lamellic, Psammentic, Inceptic, Xerollic, Umbric Xeric, Ustollic, Xeric, Ustic, Mollic, Umbric, Eutric, Typic
Ustalfs	Durustalfs	Typic
	Plinthustalfs	Typic
	Natrustalfs	Salidic, Leptic Torrertic, Torrertic, Aquertic, Aridic Leptic, Vertic, Aquic Arenic, Aquic, Arenic, Petrocalcic, Leptic, Haplargidic, Aridic, Mollic, Typic
	Kandiustalfs	Grossarenic, Aquic Arenic, Plinthic, Aquic, Arenic Aridic, Arenic, Aridic, Udic, Rhodic, Typic
	Kanhaplustalfs	Lithic, Aquic, Aridic, Udic, Rhodic, Typic
	Paleustalfs	Aquertic, Oxyaquic Vertic, Udertic, Vertic, Aquic Arenic, Aquic, Oxyaquic, Lamellic, Psammentic, Arenic Aridic, Grossarenic, Arenic, Plinthic, Petrocalcic, Calcidic, Aridic, Kandic, Rhodic, Ultic, Udic, Typic
	Rhodustalfs	Lithic, Kanhaplic, Udic, Typic
	Haplustalfs	Lithic, Aquertic, Oxyaquic Vertic, Torrertic, Udertic, Vertic, Aquic Arenic, Aquultic, Aquic, Oxyaquic, Vitrandic, Lamellic, Psammentic, Arenic Aridic, Arenic, Calcidic, Aridic, Kanhaplic, Inceptic, Calcic Udic, Ultic, Calcic, Udic, Typic
Xeralfs	Durixeralfs	Natric, Vertic, Aquic, Abruptic Haplic, Abruptic, Haplic, Typic
	Natrixeralfs	Vertic, Aquic, Typic
	Fragixeralfs	Andic, Vitrandic, Mollic, Aquic, Inceptic, Typic
	Plinthoxeralfs	Typic
	Rhodoxeralfs	Lithic, Vertic, Petrocalcic, Calcic, Inceptic, Typic
	Palexeralfs	Vertic, Aquandic, Andic, Vitrandic, Fragiaquic, Aquic, Petrocalcic, Lamellic, Psammentic, Arenic, Natric, Fragic, Calcic, Plinthic, Ultic, Haplic, Mollic, Typic
	Haploxeralfs	Lithic Mollic, Lithic Ruptic-Inceptic, Lithic, Vertic, Aquandic, Andic, Vitrandic, Fragiaquic, Aquultic, Aquic, Natric, Fragic, Lamellic, Psammentic, Plinthic, Calcic, Inceptic, Ultic, Mollic, Typic
Udalfs	Natrudalfs	Vertic, Glossaquic, Aquic, Typic
	Ferrudalfs	Aquic, Typic
	Fraglossudalfs	Andic, Vitrandic, Aquic, Oxyaquic, Typic
	Fragiudalfs	Andic, Vitrandic, Aquic, Oxyaquic, Typic
	Kandiudalfs	Plinthaquic, Aquic, Oxyaquic, Arenic Plinthic, Grossarenic Plinthic, Arenic, Grossarenic, Plinthic, Rhodic, Mollic, Typic
	Kanhapludalfs	Lithic, Aquic, Oxyaquic, Rhodic, Typic
	Paleudalfs	Vertic, Andic, Vitrandic, Fragiaquic, Plinthaquic, Glossaquic, Albaquic, Aquic, Anthraquic, Oxyaquic, Fragic, Arenic Plinthic, Grossarenic Plinthic, Lamellic, Psammentic, Arenic, Grossarenic, Plinthic, Glossic, Rhodic, Mollic, Typic
	Rhodudalfs	Typic
	Glossudalfs	Aquertic, Oxyaquic Vertic, Vertic, Aquandic, Andic, Vitrandic, Fragiaquic, Aquic, Oxyaquic, Fragic, Arenic, Haplic, Typic
	Hapludalfs	Lithic, Aquertic Chromic, Aquertic, Oxyaquic Vertic, Chromic Vertic, Vertic, Andic, Vitrandic, Fragiaquic, Fragic Oxyaquic, Aquic Arenic, Albaquultic, Albaquic, Glossaquic, Aquultic, Aquollic, Aquic, Anthraquic, Oxyaquic, Fragic, Lamellic, Psammentic, Arenic, Glossic, Inceptic, Ultic, Mollic, Typic

Table 6.61 Occurrence of soil horizons or properties in soils of the suborders of Alfisols

Suborder	Soil horizon or property									
	Argillic horizon	Natric horizon	Kandic horizon	Glossic horizon	Fragipan	Duripan	Plinthite	Very red argillic	Very deep argillic	Other
Aqualfs	X	X	X	X	X	X	X			Worm activity, very cold *
Cryalfs	X			X					X	
Udalfs	X	X	X	X	X			X	X	Fe nodules
Ustalfs	X	X	X			X	X	X	X	
Xeralfs	X	X			X	X	X	X	X	

* Also: abrupt clay increase at upper boundary of the argillic horizon, and perched water table

to be classified as Mollisols. Most Aqualfs have well-developed E horizons although the underlying Bt and Btg horizons often show only minimally expressed clay films. Boundaries between E and Bt and Btg horizons often show evidence of mixing by fauna (i.e., crayfish) or tongues and stringers of E horizon material between structural units of Bt and Btg material (glossic material).

6.11.5.2 Cryalfs
Cryalfs are restricted to frigid and cryic temperature regimes, and generally occur in more mountainous and sloping landscapes that are generally well drained. They formed in North America, eastern Europe, and Asia above 49° N latitude and in some high mountains of lower latitude where they occur at lower altitudes than Spodosols or Inceptisols. Most have been under coniferous forest because of the cool, short growing season. Commonly, parent materials are residuum or colluvium from local bedrock. Most are not cultivated, and thus, have thin to moderately thick (2–10 cm) O horizons in varying stages of organic matter decomposition. Many lack A horizons, but when present, they are thin (5 cm) with granular or subangular blocky structure. Most have E horizons of platy or subangular blocky structure that are relatively thick (15 to 30 cm). The Bt horizons range from thin to thick (15 to 135 cm), display a wide range of clay film features, and commonly have angular or subangular blocky structure. Many have lithic or paralithic contacts, and most contain significant quantities of coarse fragments throughout the solum.

6.11.5.3 Udalfs
These are the freely drained Alfisols of humid regions, extensive in the United States and Europe. They formed under forest vegetation at some time during their development. The forests were mainly deciduous, but in colder regions, may have been mixed coniferous and deciduous. Many Udalfs have been cleared and intensively farmed; while in some that have been severely eroded, the Bt horizon is immediately below or may be incorporated into the Ap horizon (Soil Survey Staff, 1975). Many have fragipans, and some have natric and other diagnostic horizons (Table 6.61). Corn and soybeans are common crops on Udalfs.

In well-drained, temperate humid environs under forest, uncultivated Udalfs have thin (2–5 cm) O horizons with organic matter in various stages of decomposition. Most also have a thin (10–15 cm) A horizon commonly of granular structure darkened by organic matter and underlain by a thin (10–15 cm), light colored E horizon with platy structure, a zone of maximum eluviation of silicate clays.

Because many Alfisols are cultivated, the A and E horizons have been mixed by plowing producing a light color. The intense eluviation in E horizons results in a concentration of sand and silt, and in extreme cases, all or most coloring compounds have been removed leaving skeletal grains whose color dominates the horizon. Often there is a transitional EB or BE horizon between the E and Bt horizons. As compared to the thinner A and E horizons, the Bt horizon is thick, possesses subangular or angular blocky structure, and contains significantly more silicate clay. At least a portion of the silicate clay occurs as clay films on structural or pores surfaces. In the Midwest where most of the parent material is calcareous and associated with the last glaciation, complete leaching of carbonates from the solum has occurred while C horizons remain calcareous.

6.11.5.4   Ustalfs
These are the Alfisols of subhumid to semi-arid regions. Moisture moves through these soils to deeper layers only in occasional years (Soil Survey Staff, 1975). If there are carbonates in the parent material or in dust, the soil may have a carbonate enriched horizon (calcic) below the argillic. Original vegetation may have been xerophytic trees, savanna, or grassland. Dryland crops for Ustalfs include drought tolerant selections such as sorghum, wheat, cotton, and millet. Ustalfs which are common in the southern Great Plains, do not have fragipans, but may have several other kinds of diagnostic horizons (Table 6.61).

In the United States, Ustalfs tend to be on older landforms than Udalfs, and have thicker sola; many are in the Paleustalf great group because of either an abrupt textural change between E and Bt horizons or a thick Bt horizon enriched in clay throughout. There is less tendency for E horizons to display platy structure; while with less rainfall, E horizons are often absent. The Bt horizons often have prismatic parting to angular or subangular blocky structure; at the drier end of Ustalfs, sufficient secondary carbonates commonly accumulate to form calcic horizons in the lower portion of the argillic horizon or just below it.

6.11.5.5   Xeralfs
These occur in regions that have a Mediterranean climate, where precipitation occurs during the cool season and the summers are hot and dry. In some winters water moves through the entire soil profile to deeper layers. Most Xeralfs border the Mediterranean Sea or lie to the east of an ocean, such as those in the western United States and Australia. Where there is no irrigation, grains are common crops but with irrigation, a variety of crops can be grown. Grapes and olives are grown in the warmer areas (Soil Survey Staff, 1975). Occurrence of E horizons is uncommon, and A horizons with weak subangular blocky structure to massive condition tend to range from 15 to 25 cm thick. Most argillic horizons are relatively thin (15 to 75 cm) with significant quantities of coarse fragments throughout the pedon.

6.11.6   Soil Formation Processes

6.11.6.1   Clay Translocation
Alfisols and Ultisols form mainly by eluviation of silicate clay from A and E to Bt horizons. In Alfisols, however, the base saturation of deep subsoil layers is greater than it is in Ultisols. Alfisols are similar to the Gray Brown Podzolic soils of previous soil classification systems. The name Alfisol was derived from the term Pedalfer, a coined word used in previous classifications derived from *ped*, as in pedology (the science of natural soils), *Al*, aluminum, and *Fe*, iron. The implication is that in Pedalfers, Al and Fe accumulate in the subsoil, in contrast to Pedocals, in which Ca accumulates.

The concept of Gray Brown Podzolic soils originated at a time when soil colloids (clays) were thought to be mainly amorphous. Then, soil scientists determined total chemical composition of the colloidal fraction, and interpreted the results by examining SiO_2/Al_2O_3 and $SiO_2/(Fe_2O_3 + Al_2O_3)$ ratios of the various horizons. These ratios showed that both Podzols (Spodosols) and Gray Brown Podzolic soils had B horizons enriched in Fe and Al, but that Podzol B horizons were also enriched in organic C. This led Marbut (1935) to conclude that the differences in the processes leading to the two kinds of soil were of degree, not kind, and that Gray Brown Podzolic soils were less developed than Podzols. The word Podzolic is a descriptor previously used for many Alfisols.

Hendricks and Fry (1930) reported two key discoveries that laid the groundwork for the current understanding of the processes by which Alfisols form. First, they showed that soil colloids are mainly crystalline silicate minerals, and that when clay suspensions are allowed to dry, the plate-shaped minerals lay flat to form masses that have a regular crystallographic, or at least optical, arrangement and are large enough to study with a polarizing microscope. This discovery led to an understanding that in Alfisols the material translocated is mainly silicate clay minerals (with Fe oxides on their surfaces), but in Spodosols, it is mainly amorphous complexes of organic matter with Al and usually Fe. Second, they provided the basis for interpreting what are now called clay films.

These ideas were gradually applied to studies of soil genesis. Jenny and Smith (1935) produced early stages of argillic horizon formation in columns of pure quartz sand by first coating the sand grains with electropositive iron hydroxide and then passing dispersed clay through the column. Brown and Thorp (1942) noted that the B horizons in the Miami soil (Hapludalf) from Indiana and similar soils contained much more clay than did the A and E horizons. They concluded, however, that the clay increase in the B horizons was only partly due to illuviation. Although the types of clay were identified by X-ray diffraction and thermal methods, soil formation processes were still discussed mainly in terms of $SiO_2/(Fe_2O_3 + Al_2O_3)$ ratios. Frei and Cline (1949) concluded that the apparent loss of clay from A and E horizons and the apparent gain in clay in B horizons of a New York soil could be accounted for in at least three possible ways: (1) clay migrates downward as a sol in percolating water; (2) clay is synthesized in the B horizon from soluble weathering products that were released there or moved down in the profile; or (3) the apparent increase in clay was really due to loss of other materials, such as $CaCO_3$. They showed, too, that there was a strong concentration of clays with a high degree of optical continuity on peds in the B horizon of a Gray Brown Podzolic soil, but in the upper B horizon, these coatings were highly degraded. According to their interpretation, clay that was mobilized in A and E horizons and stripped from ped surfaces in the upper B horizon, moved down the profile in suspension, and was deposited on ped surfaces as films of oriented clay particles. Thorp et al. (1959), in another study of the Miami soil of Indiana, showed that the small particle size of clays, their association with organic compounds, and shrink/swell cycles in upper horizons all enhanced their mobility.

In summary, the movement of clay from A and E horizons to Bt horizons can be viewed as the net result of three subprocesses: dispersion, transport, and deposition (Jenny, 1980). Dispersion, the release of individual clay particles from aggregates, is favored by the replacement of exchangeable Ca^{2+} with Na^+, by adsorption of humus molecules on clay particles, and by mechanical disruption of aggregates which may be caused by wetting and drying, freezing and thawing, or by mechanical disruption by faunal activity. Small particles move more readily than large particles, so that illuvial clay, as in clay films, contains more fine than immobile clay, such as that in ped interiors. Little clay is dispersed in calcareous soil materials because Ca^{2+} ions on exchange sites favor clay flocculation. Clay that is mobilized in A and E horizons moves as a percolating sol or as slurry creep along ped surfaces. Deposition is favored when Na^+ is replaced by Ca^{2+}, resulting in clay flocculation by an increase in pH, or by reaction of the silicate clay with oxides such as Fe oxide. When the water of a slurry enters into a ped, the clay it contains is deposited on the ped surface, like passing the slurry through filter paper. In some cases, the slurry moves down to the wetting front in the subsoil, and clay is deposited as water evaporates or is taken up by plants. Texture discontinuities may also arrest the percolation of suspensions.

Because many Alfisols formed in calcareous materials, $CaCO_3$ must be dissolved and the weathering products leached from the soil before much clay can move. Usually H^+ replaces a portion of the Ca^{2+} in these Alfisols, causing leached horizons to become acidic. The shallowest depth at which carbonate minerals are found, as detected by dilute HCl, serves as an index of the intensity and length of soil formation.

It should be stressed that, although illuviation of clays is diagnostic for the argillic horizon, much or even most clay in the Alfisol Bt horizons originates from other sources to include inheritance from parent material, clay neoformation, *in situ* weathering of lithogenic minerals, and dissolution of silt and sand sized carbonates (Smeck and Wilding, 1980).

6.11.6.2 Natric Horizons
Natric horizons are argillic horizons high in exchangeable Na^+. A natric horizon in an Alfisol seems to be an oxymoron; if rainfall has been high enough to move sufficient clay for the horizon to qualify as an argillic horizon, Na^+ should have long since been leached from the soil. However, natric horizons do exist in Alfisols indicating that there are circumstances in which they form. Some might be caused by a change in climate where the argillic horizon was formed during a previous wetter period, but rainfall is now less, and the leaching of Na^+ is reduced. Other soils with natric horizons occur in small areas where Na^+ has accumulated. Although the original source of much of the Na is feldspar minerals, Na^+ moves about in the landscape once it is released into solution. Often, soils with natric horizons are low in the landscape because solutions containing Na^+ move downslope either on the soil surface or through subsurface horizons, thus concentrating Na^+ in the low areas. In some soils, Na^+ is retained in the soil because the hydraulic conductivity of B horizons is very low, and leaching of Na^+ is retarded. This seems to be a self-perpetuating process; Na^+ accumulates because permeability is low, while Na^+ promotes dispersion, further slowing permeability. Furthermore, some soils with natric horizons were formerly saline soils. When soluble salts are leached from a soil, Na^+ may remain behind on the exchange sites.

6.11.6.3 Fragipan Formation
Fragipans occur in many Alfisols of the humid temperate regions, especially those of silty and loamy textures. These subsurface horizons restrict the entry of water and roots into the soil matrix (prisms) and often underlie argillic horizons. Commonly, the fragipan has a relatively low content of organic matter and high bulk density relative to horizons above. Most fragipans consist of prisms more than 10 cm across with light colored coatings on the prism faces. These light colored coatings form a polygonal pattern as shown in Fig 6.65. Material within prisms has a brittle fracture, a tendency for a clod to rupture suddenly rather than undergo slow deformation when pressure is applied. Fragments of the fragipan slake (fall apart) when submersed in water.

There are many theories on how fragipans form, summarized in two reviews (Smalley and Davin, 1982; Smeck and Ciolksz, 1989). One reason for the diversity in ideas is that there may be several kinds of horizons included under the name fragipan, each with its own set of processes of formation. Most soil scientists believe that they have formed, or are forming, under present day conditions. The main evidence that fragipans are genetic horizons related to the current land surface is that they are roughly parallel to the soil surface, and the upper boundary has a relatively narrow range of about 50 to 100 cm depth. Some scientists, however, believe that they formed much earlier. The following are some of the various theories of fragipan formation.

Paleosols

According to this theory, fragipans formed in earlier times and were buried by younger deposits in which fragipans are not forming. In western Tennessee, Buntley et al. (1977) observed that fragipans are mainly in older loess deposits that were buried by younger Peorian loess in which no fragipans form.

Rapid Initial Development

Some workers believe that fragipans formed quickly in their life history. Fragipan formation has been related to periglacial conditions (Payton, 1992) and to permafrost (FitzPatrick, 1956; van Vliet and Langohr, 1981). The polygonal pattern formed by coatings on the coarse prismatic structure is similar to that found in periglacial regions. Others present evidence that the structure of loess collapses when it first dries after deposition (Bryant, 1989), and that this denser material becomes the fragipan.

Much glacial till is very dense because it was compacted by the mass of glacier ice. Subsoil horizons that form in dense till may inherit some of their strength from the parent material (Lindbo and Veneman, 1989), but they might also develop fragipan properties through soil-forming processes. In

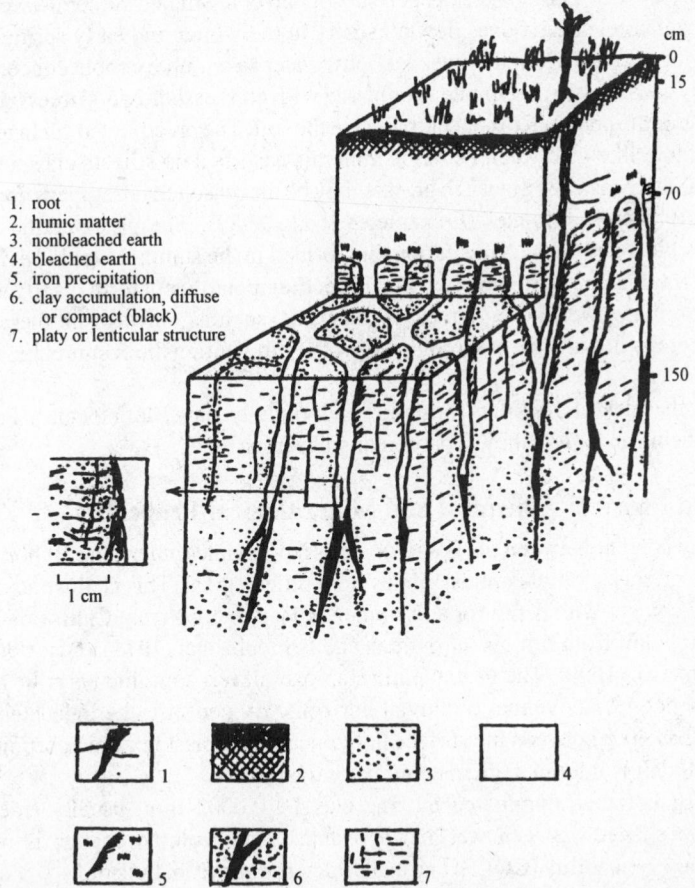

1. root
2. humic matter
3. nonbleached earth
4. bleached earth
5. iron precipitation
6. clay accumulation, diffuse or compact (black)
7. platy or lenticular structure

Fig. 6.65 Schematic of a fragipan in an imperfectly drained silty soil [Reprinted from van Vliet and Langohr, 1981. Correlation between fragipans and permafrost with special reference to silty deposits in Belgium and northern France. Catena 8:137-154. Copyright with permission of Elsevier Science]

these soils, which are common from eastern Ohio to New England, it is difficult to decide whether the dense material should be considered a fragipan or a slightly modified parent material. The great strength of these horizons is due in part to their high density, but it may also be due to cementation by Si or carbonates (McBurnett and Franzmeier, 1997).

Chemical Cementation

Fragipans could also be cemented by Fe-, Al-, or Si-rich materials. Iron-rich materials cause the cementation of ortstein layers, and silica bonds particles together to form duripans. Although duripans form mainly in aridic and xeric climates, in more humid climates, they grade into fragipans (Soil Survey Staff, 1998). The idea that fragipans are cemented by silica was first proposed in the 1930s and 1940s but was discredited and ignored for about 40 years before revitalized in the late 1970s. The proposal requires certain interactions of parent materials and climate (Franzmeier et al., 1989). In addition to undergoing periodic leaching, soils must also become seasonally dry, which commonly occurs in areas with a udic moisture regime under hardwood forest. Also, downward percolation of soil solutions must be arrested, rather than passing quickly through subsoil layers. This could be caused by the depth of penetration of the wetting front, by discontinuities in parent materials (Smeck et al., 1989), or by slowly permeable, deeper subsoil layers. Silicate minerals weather in upper horizons, and the silica released moves down, usually in the winter and early spring, to the subsoil where it remains for a time. When the trees leaf out, water taken up by roots concentrates silica in solution eventually causing it to precipitate. Compared with grasses, hardwood trees take up relatively little Si, and thus have the potential to concentrate Si in the soil. The precipitated silica bridges between clay particles, or more likely between Fe oxide minerals adsorbed on silicate clay surfaces, causing weak cementation. The greater Si uptake in grasses may be the reason that fragipans do not form under prairie vegetation in humid climates (Franzmeier et al., 1989). The precipitation of Si must be somewhat irreversible, or else any cementation that formed in the summer would be destroyed in the winter, and no net fragipan formation would take place. Fragipans do not form readily on steep slopes, in deep loess, or in loess over more permeable materials, such as outwash. In these soils, the soil solution moves laterally downslope or through the profile, preventing the accumulation of Si-rich soil solutions.

Most of these theories of fragipan formation individually have deficiencies. Perhaps various combinations of them can explain how a particular soil forms.

6.11.7   Selected Physical, Chemical and Mineralogical Properties

An increase in clay content between the E and/or A horizon(s) and underlying Bt horizons occurs in all Alfisols; however, this textural contrast ranges from little to great. This is illustrated (Fig. 6.66) by the change in clay content with depth for a Psammentic Haploxeralf from California (Torrent et al., 1980), a Typic Hapludalf from Illinois (Grossman and Fehrenbacher, 1971) and a Udertic Paleustalf from Texas (Klich et al., 1990). The Psammentic Haploxeralf has a minimal argillic horizon, with a clay increase of about 4% above that of eluvial horizons. By contrast, the Paleustalf shows a clay increase of more than 40% between the surface horizon and the finest textured portion of the argillic horizon. The Hapludalf is intermediate in textural contrast.

Alfisols with Bt horizons dominated by fine clay (< 0.0002 mm, usually smectitic) display significant volume change between wet and dry states. This volume change is reported as the coefficient of linear extensibility (COLE) and can be determined from bulk density values associated with moist and dry, natural clods (Grossman et al., 1968). The Paleustalf in Fig. 6.66 has COLE values > 0.07 (7% volume change between moist and dry states) for horizons containing 30% or more clay.

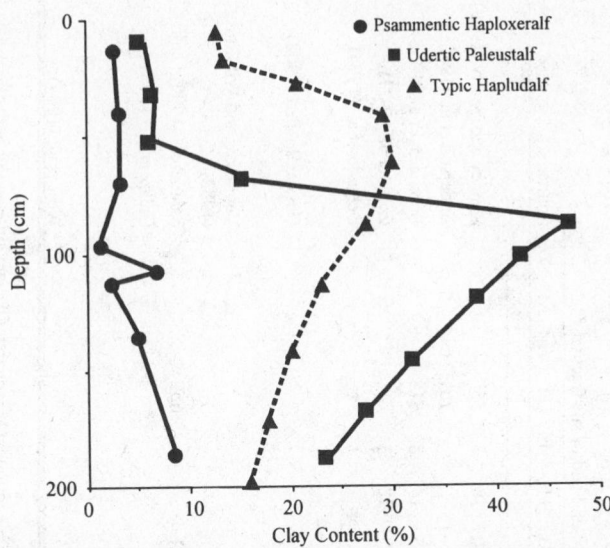

Fig. 6.66 Clay distribution in three Alfisols of contrasting degrees of argillic horizon development [Sources are Torrent et al., 1980; Grossman and Fehrenbacher, 1971; and Klich et al., 1990 for the Psammentic Haploxeralf, Typic Hapludalf and Udertic Paleustalf, respectively]

Because such soils are intergrades to Vertisols and possess similar features such as high subsoil clay content and shrink/swell potential, and slickensides, they present similar problems for use as construction materials (i.e., unstable subgrades for roads and houses).

Texture and clay content affect water-holding and water transmission properties of soils since water-related properties are dependent on void size and continuity. When the size of voids change greatly over a short distance, water movement is restricted. Alfisols with lamella type argillic horizons may exhibit slower drainage and better soil-plant relations due to greater available water retention than similar textured soils without lamella. Under more extreme conditions, strongly contrasting textures of eluvial and argillic horizons may result in water perched above the argillic horizons for several weeks as in Natraqualfs and Albaqualfs. In Alfisols with compact, clayey argillic horizons, root growth may be limited primarily to structural planes due to the high bulk density of ped interiors. Jones (1983) showed that root exclusion due to soil bulk density is a function of clay content.

Although organic C is greatest in the surface horizons, Alfisols tend to have greater CEC in the argillic horizon where clay and smectite contents are greatest. By definition, Alfisols must have a base saturation of 35% or more at the shallower of the following depths: (1) 1.25 cm below the top of the argillic horizon, (2) 1.8 m below the soil surface, (3) 75 cm below the top of a fragipan, or (4) immediately above a densic, petroferric, lithic or paralithic contact (just above continuous hard or soft bedrock). Most Alfisols have base saturation percentages well above 35%; those that approach low base status are recognized as intergrades to Ultisols by classification in Ultic subgroups. Generally, base saturation (and pH) decrease initially with depth in Alfisols, often reaching a minimum in the E horizons or upper portions of the argillic horizon, then increase again with depth. Alfisols under deciduous forest vegetation have relatively high base saturations in the surface horizon due to cycling of basic cations from depth to the surface in leaf fall. The relationship between soil pH and base saturation (Beery and Wilding, 1971; Blosser and Jenny, 1971; Ranney et al., 1974) is often used in

Table 6.62 Mineralogy of the argillic horizon of selected Alfisols

| | | | Minerals Reported | | | | | | | | | | | | | | | | |
| | | | Clay fraction *(< 0.002 mm) | | | | | | | | Sand and/or silt-fraction ** (%) | | | | | | | | |
Classification	Location	Parent material	Ch	Ve	In	Sm	Mi	Ka	Qz	GH	Ps	Kf	Pl	Mi	Qz	Vg	AP	Ze	Source
Glossoboralfs	Wisconsin	Lacustrine sediments	1	2		2	2	1				12	18	8	55				Ranney and Beatty (1969)
Eutroboralf	New Mexico	Sandstone				2	2	2					-18-		76	6			Anderson et al. (1975)
Hapludalf	Wisconsin	Thin loess over till	1	2		2	2	2	1			13	14	8	61				Borchardt et al. (1968)
Hapludalf	Oregon	Greenish tuff & breccia	1			3					22		52	3	1	t	6	11	Paeth et al. (1971)
Haploxeralf	Spain	Calcareous alluvium			1	2	3	1				- a -		t	t				Delgado et al. (1994)
Rhodoxeralf		Red sandstone			1	1	3	1				- a -		c					
Haplustalf	Texas	Granite				2	2	1											Goss and Allen (1968)
Natraustalf						2	2	1											
Paleudalf	Texas	Coastal plains sediments				1	1	3+	1	1									Vepraskas and Wilding (1983)
Haplustalf?	India	Shist/quartzite/phyllite				t	2	3	1	1		31			65				Sahu et al. (1990)
Kandiudalf (reclassified)	Nigeria	Olivine basalt						3	t	1									Gallez et al. (1975)
Plinthustalf		Basement complex			t		1	3		1									
Paleustalf	Niger	Eolian sands						3	1										West et al. (1984)

*Relative amounts: 3 = >50%; 2 = 10 to 50%; 1 = <10%; t = trace

Ch = chlorite; Vr = vermiculite; In = interstratified 2:1 minerals; Sm = smectite; Mi = hydrous mica or illite; Ka = kaolinite; Qz = quartz; GH - goethite and/or hematite.

** Values are for a major fraction of sand or silt.

Ps = Pseudomorphs of clay minerals; Kf = K feldspar; Pl = plagioclase feldspar; Mi = mica; Qz = quartz; Vg = volcanic glass; Ze = zeolite; a = abundant; c = common; t = trace. +; Chloritized vermiculite also noted

field operations to separate Alfisols from Ultisols, but care must be exercised as this relationship changes with mineralogy.

The mineralogy from a range of Alfisols illustrating variations in parent materials, climates, and degree of development is presented in Table 6.60. Most Alfisols contain substantial quantities of 2:1 clay minerals such as smectite and hydrous mica, and perhaps some vermiculite, chlorite and kaolinite. However, for more intensely weathered Alfisols (Kandiudalf, Paleudalf, Paleustalf, and Plinthustalf), kaolinite is the dominant clay mineral in the argillic horizon. Similarly, the most common mineral dominating the sand and silt fraction is quartz (Table 6.62) which comprised > 90% of the sand fraction in the Paleustalf and Paleudalf. The Udalf from Oregon which formed from tuff and breccia, was dominated by plagioclase feldspars. Most Alfisols contain significant quantities of weatherable minerals including K feldspars, plagioclase and mica in the sand and silt fractions.

6.11.8 References

Allan, R.J., and F.D. Hole. 1968. Clay accumulation in some Hapludalfs as related to calcareous till and incorporated loess on drumlins in Wisconsin. Soil Sci. Soc. Am. Proc. 32:403–408.

Allen, B.L., B.L. Harris, K.R. Davis, and G.B Miller. 1972. The mineralogy and chemistry of High Plains playa lake soils and sediments. TX Tech. Univ. Water Resour. Cent. Pub. 72–4.

Anderson, J.U., O.F. Bailey, and D. Rai. 1975. Effects of parent material on genesis of Borolls and Boralfs in south-central New Mexico mountains. Soil Sci. Soc. Am. Proc. 39:901–904.

Arkley, R.J., and H.C. Brown. 1954. The origin of mima mound (hogwallow) microrelief in the far western states. Soil. Sci. Soc. Am. Proc. 18:195–199.

Bailey, H.H., and P.E. Avers. 1971. Classification and composition of soils from mountain colluvium in eastern Kentucky and their importance in forestry. Soil Sci. 111:244–251.

Beery, M., and L.P. Wilding. 1971. The relationship between soil pH and base-saturation percentage for surface and subsoil horizons of selected Mollisols, Alfisols, and Ultisols in Ohio. Ohio J. Sci. 71:43–55.

Bhattacharyya, T., D.K. Pal, and S.B. Deshpande. 1993. Genesis and transformation of minerals in the formation of red (Alfisols) and black (Inceptisols and Vertisols) soils on Deccan basalt in the Western Ghats, India. J. Soil Sci. 44:159–171.

Bilzi, A.F., and E.J. Ciolkosz. 1977. Time as a factor in the genesis of four soils developed in recent alluvium in Pennsylvania. Soil Sci. Soc. Am. J. 41:122–127.

Blank, R.R. and M.A. Fosberg. 1991. Duripans of the Owyhee Plateau region of Idaho: Genesis of opal and sepiolite. Soil Sci. 152:116–133.

Blosser, D.L., and H. Jenny. 1971. Correlations of soil pH and percent base saturation as influenced by soil-forming factors. Soil Sci. Soc. Am. Proc. 35:1017–18.

Bockheim, J.G. 1980. Solution and use of chronofunctions in studying soil development. Geoderma 24:71–85.

Borchardt, G.A., F.D. Hole, and M.L. Jackson. 1968. Genesis of layer silicates in representative soils in a glacial landscape of southeastern Wisconsin. Soil Sci. Soc. Am. Proc. 32:399–402.

Bouma, J., L.J. Pons, and J. van Schuylenborgh. 1968. On soil genesis in temperature humid climate. VI. The formation of a Glossudalf in loess (silt loam). Neth. J. Agric. Sci. 16:58–70.

Bouma, J., and J. van Schuylenborgh. 1969. On soil genesis in temperate humid climate. VII. The formation of a Glossaqualf in a silt loam terrace deposit. Neth. J. Agric. Sci. 17:261–271.

Brown, I.C., and J. Thorp. 1942. Morphology and composition of some soils of the Miami family and the Miami catena. USDA Tech. Bull. 834.

Bryant, R.B. 1989. Physical processes of fragipan formation. p. 141–150. *In* N.E. Smeck and E.J. Ciolkosz (ed.) Fragipans: Their occurrence, classification, and genesis. Soil Sci. Soc. Am. Spec. Publ. 24. Soil Science Society of America, Madison, WI.

Buntley, G.J., R.B. Daniels, E.E. Gamble, and W.T. Brown. 1977. Fragipan horizons in soils of the Memphis-Loring-Grenada sequence in west Tennessee. Soil Sci. Soc. Am. J. 41:400–407.

Catt, J.A. 1991. Soils as indicators of Quaternary climatic change in mid-latitude regions. Geoderma 51:167–187.

Chittleborough, D.L. 1989. Genesis of a Xeralf on Feldspathic sandstone, South Australia. J. Soil Sci. 40:235–250.

Chittleborough, D.J., P.H. Walker, and J.M. Oades. 1984. Textural differentiation in chronosequences from eastern Australia. I. Description, chemical properties and micromorphologies of soils. Geoderma 32:181–202.

Cooper, A.W. 1960. An example of the role of microclimate in soil genesis. Soil Sci. 90:109–120.

Coventry. R.J., and J. Williams. 1984. Quantitative relationships between morphology and current soil hydrology in some Alfisols in semi-arid tropical Australia. Geoderma 33:191–218.

Cremeens, D.L., and D.L. Mokma. 1986. Argillic horizon expression and classification in the soils of two Michigan hydrosequences. Soil Sci. Soc. Am. J. 50:1002–1007.

Davis, J.G., L.R. Hossner, L.P. Wilding, and A. Manu. 1995. Variability of soil chemical properties in two sandy, dunal soils of Niger. Soil Sci. 159:321–330.

Davis, K.R. 1970. Mineralogy of a playa soil and underlying sediments from the High Plains medium-textured soil zone. M.S. Thesis, Texas Tech University, Lubbock, TX.

Delgado, R., J. Aguilar, and G. Delgado. 1994. Use of numerical estimators and multivariate analysis to characterize the genesis and pedogenic evolution of Xeralfs from southern Spain. Catena 23:309–325.

FitzPatrick, E.A. 1956. An indurated horizon formed in permafrost. J. Soil Sci. 7:248–254.

Foss, J.E., and R.H. Rust. 1968. Soil genesis study of a lithologic discontinuity in glacial drift in western Wisconsin. Soil Sci. Soc. Am. Proc. 32:393–398.

Franzmeier, D.P., R.B. Bryant, and G.C. Steinhardt. 1985. Characteristics of Wisconsin glacial tills in Indiana and their influence on argillic horizon development. Soil Sci. Soc. Am. J. 49:1481–86.

Franzmeier, D.P., L.D. Norton, and G.C. Steinhardt. 1989. Fragipan formation in loess of the midwestern United States. p. 69–98. In N.E. Smeck and E.J. Ciolkosz (ed.) Fragipans: Their occurrence, classification, and genesis. Soil Sci. Soc. Am. Spec. Publ. 24. Soil Science Society of America, Madison, WI.

Frei, E., and M.G. Cline. 1949. Profile studies of normal soils of New York: II. Micro-morphological studies of the Gray-Brown Podzolic soil sequence. Soil Sci. 68:333–344.

Gallez, A., A.S.R. Juo, A.J. Herbillon, and F.R. Moormann. 1975. Clay mineralogy of selected soils in southern Nigeria. Soil Sci. Soc. Am. Proc. 39:577–585.

Gersper, P.L., and N. Holowaychuck. 1970a. Effect of stemflow water on a Miami soil under a beech tree. I. Morphological and physical properties. Soil Sci. Soc. Am. Proc. 34:779–786.

Gersper, P.L., and N. Holowaychuk. 1970b. Effects of stemflow water on a Miami soil under a beech tree: II. Chemical properties. Soil Sci. Soc. Am. Proc. 34:786–794.

Goss, D.W. and B.L. Allen. 1968. A genetic study of two soils developed on granite in Llano County, Texas. Soil Sci. Soc. Am. Proc. 32:409–413.

Grossman, R.B., B.R. Brasher, D.P. Franzmeier, and J.L. Walker. 1968. Linear extensibility as calculated from natural-clod bulk density measurements. Soil Sci. Soc. Am. Proc. 32:570–573.

Grossman, R.B., and J.B. Fehrenbacher. 1971. Distribution of moved clay in four loess-derived soils that occur in southern Illinois. Soil Sci. Soc. Am. Proc. 35:948–951.

Harden, J.W. 1982. A quantitative index of soil development from field descriptions: Examples from a chronosequence in central California. Geoderma 28:1–28.

Hendricks, S.B., and W.H. Fry. 1930. The results of X-ray and microscopical examinations of soil colloids. Soil Sci. 29:457–479.

Hugie, V.K., and H.B. Passey. 1963. Cicadas and their effect upon soil genesis in certain soils in southern Idaho, northern Utah, and northeastern Nevada. Soil Sci. Soc. Am. Proc. 27:78–82.

Jenny, H. 1935. The clay content of the soil as related to climatic factors, particularly temperature. Soil Sci. 40:111–128.

Jenny, H. 1941. Factors of soil formation. McGraw-Hill, New York, NY.

Jenny, H. 1980. The soil resource. Springer-Verlag, New York, NY.

Jenny, H., and C.D. Leonard. 1934. Functional relationships between soil properties and rainfall. Soil Sci. 38:363–381.

Jenny, H., and G.D. Smith. 1935. Colloid chemical aspects of clay pan formation in soil profiles. Soil Sci. 39:377.

Jones, C.A. 1983. Effect of soil texture on critical bulk densities for root growth. Soil Sci. Soc. Am. J. 47:1208–1211.

Juo, A.S.R., and L P. Wilding. 1996. Soils of the lowland forests of West and Central Africa. Proc. Royal Soc. Edin. 104B:15–29.

Klich, I., L.P. Wilding, and A.A. Pfordresher. 1990. Close-interval spatial variability of Udertic Paleustalfs in East Central Texas. Soil Sci. Soc. Am. J. 54:489–494.

Kooistra, M.J., A.S.R. Juo, and D. Schoonderbeek. 1990. Soil degradation in cultivated Alfisols under different management systems in southwestern Nigeria. p. 61–69. In L.A. Douglas (ed). Soil micromorphology: A basic and applied science. Elsevier, New York, NY.

Lindbo, D.L., and P.L.M. Veneman. 1989. Fragipans in the northeastern United States. p. 11–31. In N.E. Smeck and E.J. Ciolkosz (ed.) Fragipans: Their occurrence, classification, and genesis. Soil Sci. Soc. Am. Spec. Publ. 24. Soil Science Society of America, Madison, WI.

Little, I.P., and W.T. Ward. 1981. Chemical and mineralogical trends in a chronosequence developed on alluvium in eastern Victoria, Australia. Geoderma 25:173–188.

Lotspeich, F.B., and J.R. Coover. 1962. Soil forming factors on the Llano Estacado: Parent material, time and topography. Texas J. Sci. 14:7–17.

Lozet, J.M., and A.J. Herbillon. 1971. Fragipan soils of Controz (Belgium): Mineralogical, chemical and physical aspects in relation to their genesis. Geoderma 5:325–343.

Marbut, C.F. 1935. Atlas of American agriculture. Part III: Soils of the United States. US Government Printing Office, Washington, DC.

Markewich, H.W., and M.J. Pavich. 1991. Soil chronosequence studies in temperature to subtropical, low-latitude, low-relief terrain with data from the eastern United States. Geoderma 51:213–239.

McBurnett, S.L., and D.P. Franzmeier. 1997. Pedogenesis and cementation in calcareous till in Indiana. Soil Sci. Soc. Am. J. 61:1098–1104.

Meixner, R.E., and M.J. Singer. 1981. Use of a field morphology rating system to evaluate soil formation and discontinuities. Soil Sci. 131:114–123.

Miles, R.J., and D.P. Franzmeier. 1981. A lithochronosequence of soils formed in dune sand. Soil Sci. Soc. Am. J. 45:362–67.

Miller, F.P., N. Holowaychuk, and L.P. Wilding. 1971. Canfield silt loam, a Fragiudalf: I. Macromorphological, physical and chemical properties. Soil Sci. Soc. Am. Proc. 35:319–324.

Moberg, J.P., and I.E. Esu. 1991. Characteristics and composition of some savanna soils in Nigeria. Geoderma 48:113–129.

Muhs, D.R. 1982. A soil chronosequence on Quaternary marine terraces, San Clemente Island, California. Geoderma 28:257–283.

Nielsen, G.A., and F.D. Hole. 1964. Earthworms and the development of coprogenous A1 horizons in forest soils of Wisconsin. Soil Sci. Soc. Am. Proc. 28:426–430.

Paeth, M.E., M.E. Harward, E.G. Knox, and C.T. Dryness. 1971. Factors affecting mass movement of four soils in the Western Cascades of Oregon. Soil Sci. Soc. Am. Proc. 35:943–947.

Parsons, R.B., W.H. Scholtes, and F. F. Riecken. 1962. Soils of Indian mounds in northeast Iowa as benchmarks of soil genesis. Soil Sci. Soc. Am. Proc. 26:491–496.

Parsons, R.B., G.H. Simonson, and C.A. Balster. 1968. Pedogenic and geomorphic relationships of associated Aqualfs, Albolls, and Xerolls in Western Oregon. Soil Sci. Soc. Am. Proc. 32:556–563.

Payton, R.W. 1992. Fragipan formation in argillic brown earths (Fragiudalfs) of the Milfield Plain, northeast England. I. Evidence for a periglacial stage of development. J. Soil Sci. 43:621–644.

Ranney, R.W., and M.T. Beatty. 1969. Clay translocation and albic tongue formation in two Glossoboralfs of west-central Wisconsin. Soil Sci. Soc. Am. Proc. 33:768–775.

Ranney, R.W., E.J. Ciolkosz, R.L. Cunningham, G.W. Petersen, and R.P. Matelski. 1975. Fragipans in Pennsylvania soils: Properties of bleached prism face materials. Soil Sci. Soc. Am. Proc. 39:695–698.

Ranney, R.W., E.J. Ciolkosz, G.W. Peterson, R.P. Matelski, L.J. Johnson, and R.L. Cunningham. 1974. The pH and base-saturation relationships in B and C horizons of Pennsylvania Soils. Soil Sci. 118:247–253.

Ransom, M.D., N.E. Smeck, and J.M. Bigham. 1987. Stratigraphy and genesis of polygenetic soils on the Illinoisan till plain of Southwestern Ohio. Soil Sci. Soc. Am. J. 51:135–141.

Reynders, J.J. 1972. A study of argillic horizons in some soils in Morocco. Geoderma 8:267–271.

Ruhe, R.V. 1984a. Loess-derived soils, Mississippi Valley Region: I. Soil-sedimentation system. Soil Sci. Soc. Am. J. 48:859–863.

Ruhe, R.V. 1984b. Loess-derived soils, Mississippi Valley Region: II. Soil-climate system. Soil Sci. Soc. Am. J. 48:864–867.

Sahu, G.C., S.N. Patnaik, and P.K. Das. 1990. Morphology, genesis, mineralogy and classification of soils of Northern Plateau Zone of Orissa. J. Ind. Soc. Soil Sci. 38:116–121.

Saunders, J.W., R.D. Mandel, R.T. Saucier, E.T. Allen, C.T. Hallmark, J.K. Johnson, E.H. Jackson, C.M. Allen, G.L. Stringer, D.S. Frink, J.K. Feathers, S. Williams, K.J. Gremillion, M.F. Vidrine, and R. Jones. 1997. A mound complex in Louisiana at 5400–5000 years before the present. Science 277:1796–1799.

Schaetzl, R.J. 1996. Spodosol-Alfisol intergrades: Bisequal soils in N. E. Michigan, USA. Geoderma 74:23–47.

Smalley, I.J., and J.E. Davin. 1982. Fragipan horizons in soils: A bibliographic study and review of some of the hard layers in loess and other materials. NZ Soil Bur. Bibl. Rep. 30. Department of Scientific and Industrial Research, Wellington, New Zealand.

Smeck, N.E., and E.J. Ciolkosz (ed.) 1989. Fragipans: Their occurrence, classification, and genesis. Soil Sci. Soc. Am. Spec. Publ. 24. Soil Science Society of America, Madison, WI.

Smeck, N. E., M. L. Thompson, L. D. Norton, and M. J. Shipitalo. 1989. Weathering disontinuities: A key to fragipan formation. p. 99–112. In N. E. Smeck and E. J. Ciolkosz (ed.) Fragipans: Their occurrence, classification, and genesis.Soil Sci. Soc. Am. Spec. Publ. 24. Soil Science Society of America, Madison, WI.

Smeck, N.E., and L.P. Wilding. 1980. Quantitative evaluation of pedon formation in calcareous glacial deposits in Ohio. Geoderma 24:1–16.

Smeck, N.E., L.P. Wilding, and N. Holowaychuk. 1968. Genesis of argillic horizons in Celina and Morley soils of western Ohio. Soil Sci. Soc. Am. Proc. 32:550–556.

Smith, H., and L.P. Wilding. 1972. Genesis of argillic horizons on Ochraqualfs derived from fine-textured till deposits of northwestern Ohio and southeastern Michigan. Soil. Sci. Soc. Am. Proc. 36:808–815.

Soil Survey Staff. 1998. Keys to soil taxonomy. 8th Ed. USDA-NRCS. US Government Printing Office, Washington, DC.

Soil Survey Staff. 1975. Soil taxonomy: A basic system of soil classification for making and interpreting soil surveys. USDA-SCS Agric. Handb. 436. US Government Printing Office, Washington, DC.

Stahnke, C.R., L.P. Wilding, J.D. Moore, and L.R. Drees. 1983. Genesis and properties of Paleustalfs of North Central Texas: Morphological, physical, and chemical properties. Soil Sci. Soc. Am. J. 47:728–733.

Thorp, J. 1949. Effects of certain animals that live in soils. Sci. Monthly 68:180–191.

Thorp, J., J.G. Cady, and E.E. Gamble. 1959. Genesis of Miami silt loam. Soil Sci. Soc. Am. Proc. 23: 156–161.

Torrent, J., and W.D. Nettleton. 1979. A simple textural index for assessing chemical weathering in soils. Soil Sci. Soc. Am. J. 43:373–377.

Torrent, J., W.D. Nettleton, and G. Borst. 1980. Clay illuviation and lamella formation in a Psammentic Haploxeralf in southern California. Soil Sci. Soc. Am. J. 44:363–369.

van Vliet, B., and R. Langohr. 1981. Correlation between fragipans and permafrost with special reference to silty deposits in Belgium and northern France. Catena 8: 137–154.

Vepraskas, M.J., and L.P. Wilding. 1983. Deeply weathered soils in the Texas Coastal Plain. Soil Sci. Soc. Am. J. 47:293–300.

Weitkamp, W.A., R.C. Graham, and M.A. Anderson. 1996. Pedogenesis of a vernal pool Entisol-Alfisol-Vertisol catena in southern California. Soil. Sci. Soc. Am J. 60:316–323.

Wells, P.V. 1970. Postglacial vegetational history of the Great Plains. Science 167:1574–1582.

West, L.T., L.P. Wilding, J.K. Landeck, and F.G. Calhoun. 1984. Soil survey of the ICRISAT Sahelian Center, Niger, West Africa. Soil and Crop Sciences Department/Trop. Soils, Texas A&M Univ., College Station, TX.

Wiecek, C.S., and A.S. Messenger. 1972. Calcite contributions by earthworms to forest soils in northern Illinois. Soil Sci. Soc. Am. Proc. 36:478–480.

Yerima, B.P.K., L.P. Wilding, C.T. Hallmark, and F.G. Calhoun. 1989. Statistical relationships among selected properties of northern Cameroon Vertisols and associated Alfisols. Soil Sci. Soc. Am. J. 53:1758–1763.

6.12 Ultisols

Larry T. West
University of Georgia

Friedrich H. Beinroth
University of Puerto Rico

6.12.1 Introduction

Ultisols are a group of soils with an argillic or kandic horizon and low base saturation in lower subsoil horizons. These soils may be found on a variety of parent materials and under a range of climatic conditions, though most have developed under forest vegetation in humid climates. Ultisols are generally considered to be less productive than soils in many of the other orders, but in many regions of the world, they are the most productive soils available. With proper management of organic residues, fallow periods, and/or chemical inputs, the productivity of this resource can be enhanced and maintained. This section presents a brief overview of Ultisol genesis and properties. For a more comprehensive review, the reader is referred to Miller (1983) and West et al. (1997).

6.12.2 Classification: Historical and Current

As soil classification has evolved in the United States soils currently considered to be Ultisols have been included with the Red soils, Yellow soils, or Lateritic soils (Marbut, 1928), the Yellow Podzolic or Red Podzolic great soil groups (Baldwin et al., 1938), and the Red-Yellow Podzolic great soil group (Kellogg, 1949). The order Ultisols was introduced in the 7th approximation (Soil Survey Staff, 1960) to include soils with an argillic horizon and base saturation < 35% in lower subsoil horizons. Since publication of *Soil Taxonomy* (Soil Survey Staff, 1975), the classification of Ultisols has been revised to include kandic horizons as criteria for placement in the order and to include soils in any temperature regime. For a comprehensive definition of Ultisols, the reader is referred to the *Keys to Soil Taxonomy* (Soil Survey Staff, 1998). Table 6.63 lists current orders, suborders and great groups in Ultisols (Soil Survey Staff, 1998).

6.12.3 Geographic Distribution

Worldwide, Ultisols are currently estimated to cover about 11,054,000 km² which is 8.4% of the ice-free global landmass (Table 6.1; Fig. 6.67). About 80% of Ultisols are in tropical regions, and about 18% of the tropics is covered by Ultisols (Eswaran, 1993). About 50% of Ultisols are Udults, 35% Ustults, and 12% Aquults with Humults and Xerults comprising the remaining 3% (Table 6.1). In the United States, Ultisols cover about 860,000 km² (Table 6.64), and occur in 30 states and Puerto Rico (Fig. 6.68). The largest concentration of Ultisols in the United States is in the east central, southeast, and south central parts of the country. Ultisols are also extensive on the older islands of Hawaii, along the west coast, and in unglaciated regions of the northeast and north central parts of the country.

6.12.4 Genesis of Ultisols

Soil as an independent natural body whose properties are a function of five soil-forming factors: local climate, parent materials, organisms, relief, and age (Dokuchaev, 1883; Jenny, 1941), is a unifying philosophy in pedology. As such, Ultisol genesis will be briefly discussed in terms of these factors.

6.12.4.1 Climate

The definition of Ultisols implies two rainfall/evapotranspiration conditions. First, there must be some time during the year when evapotranspiration exceeds precipitation, as this appears to be a prerequisite for the formation of an argillic horizon (van Wambeke, 1991). Second, precipitation must exceed the capacity of the soil to retain water during some time of the year so that water percolates through the solum to remove basic cations and maintain low base saturation (< 35% by sum of cations [pH 8.2] which corresponds roughly to 50% base saturation measured at pH 7.0) in the subsoil (Miller, 1983). Most Ultisols occur in regions with mean annual air temperatures above 6 °C. Thus, the climate conducive to the formation of Ultisols is typically humid tropical or humid warm temperate. As many Ultisols have evolved over long time periods, effects of paleoclimate on Ultisol development cannot be ignored.

6.12.4.2 Parent Material

The definition of Ultisols requires the presence of either an argillic or a kandic horizon, both of which contain silicate clays. Therefore, the parent material of Ultisols must be one that contains either phyllosilicates or primary minerals that can weather to produce them. As the vast majority of rocks meet one or both of these criteria, a wide variety of geologic formations are capable of producing parent materials suitable for Ultisol formation. Residual parent materials that have weathered *in situ*

Table 6.63 Listing of suborders, great groups and subgroups in the Ultisols order [Soil Survey Staff, 1998]

Suborder	Great Group	Subgroup
Aquults	Plinthaquults	Kandic, Typic
	Fragiaquults	Aeric Plinthic, Umbric, Typic
	Albaquults	Vertic, Kandic, Aeric, Typic
	Kandiaquults	Acraquoxic, Arenic Plinthic, Arenic Umbric, Arenic, Grossarenic, Plinthic, Aeric, Umbric, Typic
	Kanhaplaquults	Aquandic, Plinthic, Aeric Umbric, Aeric, Umbric, Typic
	Paleaquults	Vertic, Arenic Plinthic, Arenic Umbric. Arenic, Grossarenic, Plinthic, Aeric, Umbric, Typic
	Umbraquults	Plinthic, Typic
	Epiaquults	Vertic, Aeric Fragic, Fragic, Arenic, Grossarenic, Aeric, Typic
	Endoaquults	Arenic, Grossarenic, Aeric, Typic
Humults	Sombrihumults	Typic
	Plinthohumults	Typic
	Kandihumults	Andic Ombroaquic, Ustandic, Andic, Aquic, Ombroaquic, Plinthic, Ustic, Xeric, Anthropic, Typic
	Kanhaplohumults	Lithic, Ustandic, Andic, Aquic, Ombroaqic, Ustic, Xeric, Anthropic, Typic
	Palehumults	Aquandic, Andic, Aquic, Plinthic, Oxyaquic, Ustic, Xeric, Typic
	Haplohumults	Lithic, Aquandic, Aquic, Andic, Plinthic, Oxyaquic, Ustic, Xeric, Typic
Udults	Plinthudults	Typic
	Fragiudults	Arenic, Plinthaquic, Glossaquic, Aquic, Plinthic, Glossic, Humic, Typic
	Kandiudults	Arenic Plinthaquic, Aquic Arenic, Arenic Plinthic, Arenic Rhodic, Arenic, Grossarenic Plinthic, Grossarenic, Acrudoxic Plinthic, Acrudoxic, Plinthaquic, Aquandic, Andic, Aquic, Plinthic, Ombroaquic, Oxyaquic, Sombric, Rhodic, Typic
	Kanhapludults	Lithic, Fragiaquic, Arenic Plinthic, Arenic, Acrudoxic, Plinthaquic, Andic, Aquic, Ombroaquic, Oxyaquic, Fragic, Plinthic, Rhodic, Typic
	Paleudults	Vertic, Spodic, Arenic Plinthaquic, Aquic Arenic, Plinthaquic, Fragiaquic, Aquic, Anthraquic, Oxyaquic, Lamellic, Arenic, Plinthic, Psammentic, Grossarenic Plinthic, Plinthic, Arenic Rhodic, Arenic, Grossarenic, Fragic, Rhodic, Typic
	Rhoduldults	Lithic, Psammentic, Typic
	Hapludults	Lithic Ruptic-Entic, Lithic, Vertic, Fragiaquic, Aquic Arenic, Aquic, Fragic, Oxyaquic, Lamellic, Psammentic, Arenic, Grossarenic, Inceptic, Humic, Typic
Ustults	Plinthustults	Haplic, Typic
	Kandiustults	Acrustoxic, Aquic, Arenic Plinthic, Arenic, Udandic, Andic, Plinthic, Aridic, Udic, Rhodic, Typic
	Kanhaplustults	Lithic, Acrustoxic, Aquic, Arenic, Udandic, Andic, Plinthic, Ombroaquic, Aridic, Udic, Rhodic, Typic
	Paleustults	Typic
	Rhodustults	Lithic, Psammentic, Typic
	Haplustults	Lithic, Petroferric, Aquic, Arenic, Ombroaquic, Plinthic, Kanhaplic, Typic
Xerults	Palexerults	Aquandic, Aquic, Andic, Typic
	Haploxerults	Lithic Ruptic-Inceptic, Lithic, Aquic, Andic, Lamellic, Psammentic, Arenic, Grossarenic, Typic

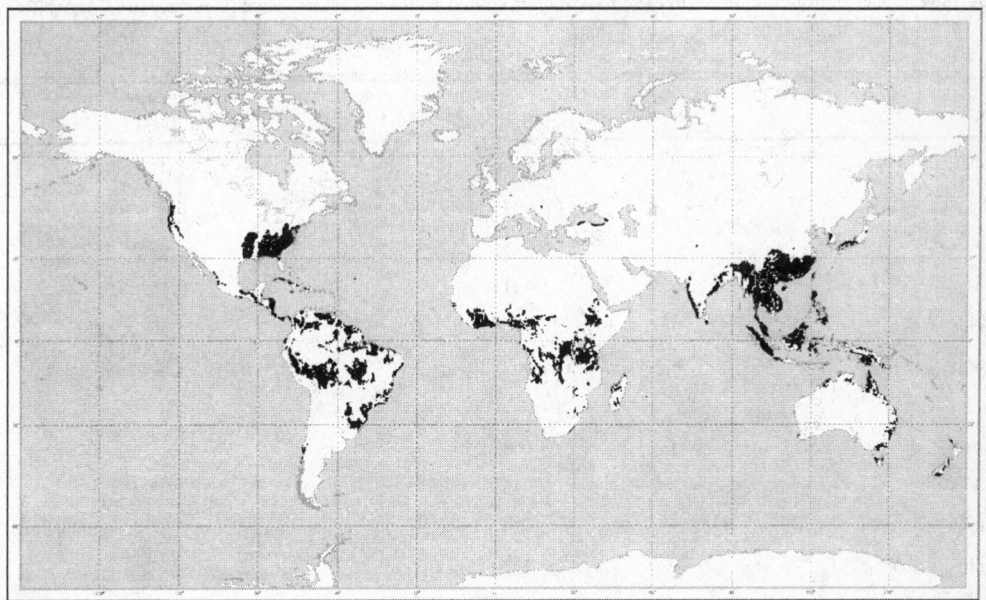

Fig. 6.67 Global distribution of Ultisols [courtesy of UDSA-NRCS, Soil Survey Division, World Soil Resources, 1998]

are usually saprolite. In contrast, transported sediments that form Ultisol parent materials have frequently been preweathered and may have gone through more than one weathering cycle (Fiskell and Perkins, 1970; Allen and Fanning, 1983).

6.12.4.3 Biota
Although Ultisols occur under many tropical and subtropical ecosystems, forests are the dominant vegetation on most natural Ultisol landscapes. The dynamics in forest soils appear to accelerate the formation of argillic horizons through the desiccating effect of tree roots that absorb water, but not colloids in suspension. In Africa and South America, vast areas of Ultisols are under savanna vegetation. It cannot be ascertained, however, whether the soils formed under the present vegetation or under forests in one or more humid paleoclimates.

6.12.4.4 Relief
Ultisols occur on almost all landforms. The one characteristic that Ultisol landforms have in common, however, is geomorphic stability over long periods of time. Geomorphic studies of tropical landscapes invariably show that Ultisols occupy geomorphic positions that are younger and less stable than the surfaces where Oxisols occur, but older and more stable than those of soils in other orders with which they are geographically associated (Beinroth et al., 1974; Lepsch and Buol, 1974; Beinroth, 1981). In temperate regions where Oxisols are absent, Ultisols have developed on the oldest, most stable and highly weathered landscapes such as the central Missouri Ozarks and the coastal plain of the southeastern United States (Scrivner et al., 1966; Daniels et al., 1971).

Table 6.64 Distribution of Ultisols in the conterminous United States and Puerto Rico [Courtesy of USDA-NRCS, Soil Survey Division, National Soil Survey Center]

State	Area (km²)	Percentage of Total
Alabama	97,965	11.4
Arkansas	68,644	8.0
California	8,418	1.0
Delaware	3,421	0.4
Florida	28,106	3.3
Georgia	114,802	13.3
Idaho	17	0.0
Indiana	3,158	0.4
Illinois	44	0.0
Kansas	79	0.0
Kentucky	27,402	3.2
Louisiana	23,395	2.7
Massachusetts	22	0.0
Maryland	16,004	1.9
Missouri	28,534	3.3
Mississippi	49,826	5.8
North Carolina	85,057	9.9
New Jersey	7,617	0.9
New York	579	0.1
Oregon	11,195	1.3
Ohio	7,137	0.8
Oklahoma	16,784	1.9
Pennsylvania	39,375	4.6
Puerto Rico	1,784	0.2
South Carolina	54,720	6.3
Tennessee	44,317	5.1
Texas	24,967	2.9
Virginia	69,910	8.1
Washington	3,574	0.4
West Virginia	25,435	2.9
Total	862,288	100.0

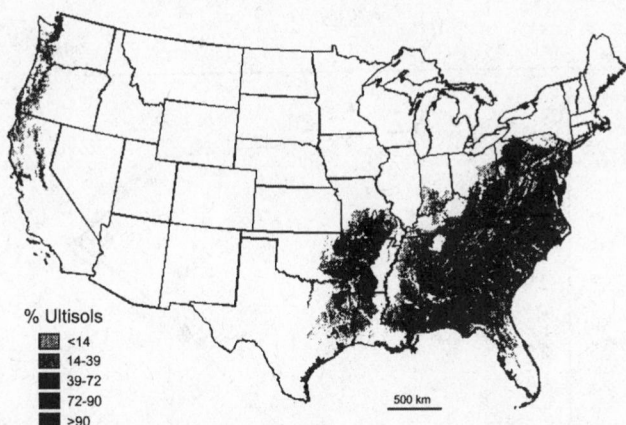

% Ultisols

- <14
- 14-39
- 39-72
- 72-90
- >90

500 km

Fig. 6.68 Distribution of Ultisols in the United States [courtesy of USDA-NRCS, Soil Survey Division, National Soil Survey Center]

6.12.4.5 Time

The time required for the formation of Ultisols depends on the other pedogenic factors, notably weatherability of the parent rock, composition of transported parent materials, and climate and its fluctuations over time. There is much room for variability in these conditions, and consequently, for the age of Ultisols. The Soil Survey Staff (1975) concluded "that formation of the argillic horizon ordinarily requires a few thousands of years." This relatively short time may suffice to produce a minimal expression of an argillic horizon. The development of the thick argillic horizons that typify the Pale great groups of Ultisols, however, would certainly take longer. As such, age of Ultisols has been reported to range from < 12,000 yr for soils on preweathered sediments to > 15 million yr on marine sediments in the Coastal Plain of the southeast United States. (Daniels et al. 1971; Gamble et al., 1970; Miller, 1983).

6.12.5 Morphological Properties

6.12.5.1 Argillic and Kandic Horizons

Other than low base saturation, the one property that distinguishes Ultisols from all but one other soil order (Alfisols, Section E, Chapter 6.11) is the requirement of a clay increase between surface and subsoil horizons in the form of either an argillic or kandic horizon. This clay increase essentially determines the morphology of Ultisols and influences many chemical and physical properties. The depth, thickness, and clay content of the argillic or kandic horizon, however, can vary widely, and this variation has considerable impact on interpretation of these soils for management. Fig. 6.69 illustrates the range in depth and clay content of argillic or kandic horizons that can occur in Ultisols.

In addition to a clay increase, the presence of oriented illuvial clay as clay coatings on pore walls (Fig. 6.70) or bridges between grains is required for an argillic horizon and clay coatings are commonly present in kandic horizons. Illuvial clay is pedogenically significant in terms of land surface age and stability, water movement through the soil, and periods of moisture deficit. In addition, clay coatings may retard water movement between pores and the matrix, slow diffusion of ions into and out of the bulk soil, and have different exchange properties than the bulk soil.

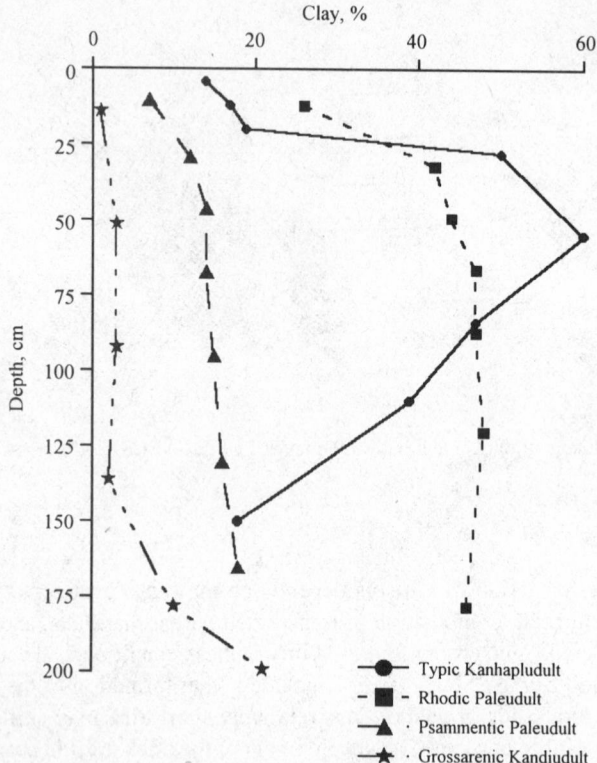

Fig. 6.69 Clay distribution with depth for four subgroups of Ultisols

6.12.5.2 Soil Color

Color of surface horizons in Ultisols is commonly brown to yellowish brown to dark brown as would be expected for soils with relatively low amounts of organic C. However, in cool climates or poorly drained conditions with slow organic matter decomposition, thick, dark, acidic surface horizons

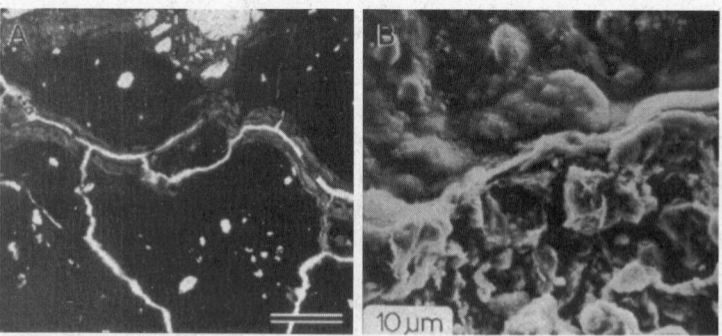

Fig. 6.70 (A) Thin section of micrograph of clay coatings in Btv2 horizon of a Plinthic Kandiudult. Plane-polarized light. Bar length = 0.5 mm; (B) Scanning electron micrograph of clay coating in 2Btx1 horizon of a Typic Fragiudult [Original micrograph from Michael Thompson, Iowa State University]

(umbric epipedons) are common. Many well-drained Ultisols have reddish Bt horizons because of the presence of hematite as a pigmenting agent. Yellowish-brown subsoil horizons, also common in Ultisols, lack hematite because either the parent material lacked mineral precursors for hematite, environmental conditions precluded hematite formation (Schwertmann and Taylor, 1989), or hematite was preferentially reduced and lost under short periods of saturation leaving behind more stable goethite (Macedo and Bryant, 1989; Dobos et al., 1990). Under extended periods of saturation and reduction, Ultisols will develop redox concentrations and depletions or gray matrices as Fe is reduced and mobilized.

6.12.5.3 Plinthite

Plinthite is defined in *Soil Taxonomy* as an Fe-rich, humus-poor mixture of clay with quartz and other diluents (Soil Survey Staff, 1998). This material most often occurs in lower subsoil horizons, is weakly indurated, and is normally low in basic cations and primary silicate minerals other than quartz. Kaolinite is commonly the only clay mineral present with abundance. The reader is referred to van Wambeke (1991) for a complete discussion of the genesis and kinds of plinthite and petroplinthite.

From a pedological perspective, plinthite is an interesting soil feature, but it also has significance in terms of movement of water through soils. Daniels et al. (1978) suggested that 10% platy plinthite would perch water. A similar amount of nodular plinthite was not considered to be water restrictive, but horizons subjacent to the plinthic horizon did restrict water movement. Movement of water and solutes through horizons containing plinthite is primarily restricted to areas of redox depletions (Carlan et al., 1985; Blume et al., 1987; Shaw et al., 1997). Because of water perching by plinthic or subjacent horizons, plinthite in soils suggests that water and solutes are moving from high to low landscape positions as shallow subsurface flow (Hubbard and Sheridan, 1983; Shirmohammadi et al., 1984).

6.12.6 Mineralogical Properties

6.12.6.1 Clay Minerals

The clay fraction of subsoil horizons of most Ultisols is dominated by kaolinite because of the stability of kaolinite as compared to other phyllosilicates (Jackson et al., 1948; Allen and Fanning, 1983). In addition, gibbsite, K mica, K feldspar, and amorphous silica may transform directly to kaolinite without passing through other phyllosilicate intermediates (Garrels and Christ, 1965). As such, kaolinite in soils has been shown to be an early weathering product of biotite and K feldspar in acid gneiss and schist parent materials (Robertus and Buol, 1985; Robertus et al., 1986; Norfleet and Smith, 1989).

In addition to kaolinite, hydroxy Al-interlayered vermiculite (HIV) is often abundant in A and E horizons of Ultisols and may be a major component of upper Bt horizons of these soils (Bryant and Dixon, 1964; Fiskell and Perkins, 1970; Carlisle and Zelazny, 1973; Harris et al., 1980; Karathanasis et al., 1983). Appreciable amounts of mica, smectite, and other 2:1 minerals have also been reported to occur in Ultisols (Nash, 1979; Harris et al., 1984; Karathanasis et al., 1986; Nash et al., 1988). The 2:1 minerals in these Ultisols were inherited from the parent materials instead of forming *in situ* from primary minerals or other phyllosilicates.

A common but minor mineral component of many Ultisols is gibbsite (Carlisle and Zelazny, 1973; Hajek and Zelazny, 1985). Because this mineral is an end product of silicate mineral weathering (Jackson et al., 1948), minor amounts should be expected to be found in Ultisols. Clay fractions of Ultisols developed from acid gneiss and schist in the southeastern United States, however, have been reported to contain as much as 70% gibbsite formed as a product of K feldspar weathering at the rock-saprolite interface (Losche et al., 1970; McCracken et al., 1971; Norfleet and Smith, 1989).

Iron oxides and oxyhydroxides are also common, but minor components in the clay fraction of Ultisols. Goethite is the most common Fe oxide mineral in most Ultisols, but many well drained Ultisols have appreciable amounts of hematite. Hematite formation is favored by warm temperatures and distinct dry seasons (Schwertmann and Taylor, 1989), which are environmental conditions commonly associated with Ultisols.

6.12.6.2 Sand and Silt Minerals

The sand and silt fraction of Ultisols commonly is dominated by quartz, and many Ultisols have >90% quartz and other resistant minerals in their sand and silt separates. However, weakly to moderately developed Ultisols may have appreciable quantities of mica, feldspars, and other weatherable minerals in their coarse fraction (Norfleet and Smith, 1989; Robertus et al., 1986; Bockheim et al., 1996).

6.12.7 Physical Properties

6.12.7.1 Bulk Density, Coefficient of Linear Extensibility (COLE), Water Retention, and Hydraulic Conductivity (K)

Bulk density, COLE, water retention, and hydraulic conductivity of Ultisols vary considerably depending on soil and horizon properties including particle size distribution, organic matter content, mineralogy, macroporosity, parent material, presence of plinthite, and, for surface horizons, management. A horizons of uncultivated Ultisols often have lower bulk density than subsoil horizons,

Table 6.65 Ranges in selected physical properties of Ultisols

Property	Value	References
Bulk Density (Mg m^{-3})		
A horizon	0.54-1.85	Bruce et al. (1983)
Bt horizon	1.05-1.70	Hubbard et al. (1985)
		Quisenberry et al. (1987)
		Southard and Buol (1988a)
		Norfleet and Smith (1989)
COLE (Bt horizons)	0.006-0.157	Karathanasis and Hajek (1985)
		Griffin and Buol (1988)
		Southard and Buol (1988a)
Water retention (cm^3 cm^{-3})		
A horizons		Bruce et al. (1983)
−10 kPa	0.13-0.21	Hubbard et al. (1985)
−30 kPa	0.09-0.18	Quisenberry et al. (1987)
−1,500 kPa	0.05-0.09	
Bt horizons		
−10 kPa	0.20-0.42	
−30 kPa	0.17-0.40	
−1,500 kPa	0.09-0.33	
Saturated Hydraulic Conductivity (μm s^{-1})		
Bt horizons	0.01-13.1	Hubbard et al. (1985)
Btv horizons	0.3-1.3	Southard and Buol (1988b)

but with tillage and the accompanying loss of organic matter, compaction, and aggregate breakdown, the A horizon may become as dense or more dense than underlying Bt horizons (Table 6.65) (Hubbard et al., 1985). Shrink/swell potential for kaolinitic Ultisols is low (Table 6.63) (Hubbard et al., 1985; Southard and Buol, 1988a), but COLE values up to 0.157 have been reported for Bt horizons from Ultisols with smectitic mineralogy (Karathanasis and Hajek, 1985; Griffin and Buol, 1988). However, smectitic Ultisols with high shrink/swell potential often lack wide cracks common to Vertisols and Vertic subgroups of other orders (Karathanasis and Hajek, 1985).

Ultisols that are placed in Typic subgroups and fine-loamy or fine particle size families would generally be considered capable of retaining and supplying sufficient moisture for plant growth in humid climates (Table 6.65). However, water often becomes a limiting factor for crop growth and yield for Ultisols in sandy or coarse-loamy particle size families or that have thick sandy epipedons (Arenic and Grossarenic subgroups).

Most studies of hydraulic properties of Ultisols have found macroporosity to be the best predictor of saturated hydraulic conductivity (Ks) (O'Brien and Buol, 1984; Southard and Buol, 1988b). Preferential flow paths in Ultisols in the Piedmont and Coastal Plain of Georgia have textures similar to the matrix of the horizon, but the preferential flow zones have a significantly greater number and area of pores > 0.05 mm diameter (Table 6.66) (Franklin, 1994; Shaw et al., 1997). Formation of coarse pores in Ultisols is most often attributed to microfabric alteration and channel formation by roots and burrowing organisms (O'Brien and Buol, 1984; Southard and Buol, 1988b; Franklin, 1994; Shaw et al., 1997).

6.12.7.2   Infiltration and Surface Crusting

In many Ultisols, one factor affecting infiltration of water is formation of a surface crust or seal. For 25 Ultisols from the southeastern United States tested under 50–90 mm h^{-1} simulated rainfall, infiltration rate decreased rapidly during the first 10 to 20 mm of the rainfall event, and the infiltration rate at the end of rainfall was less than 10 mm h^{-1} for all of the soils (Miller and Radcliffe, 1992). In general, Ultisols in this region with sandy loam texture and high amounts of water dispersible clay are most prone to crusting. Soils higher in clay tend to take longer to form a surface crust, and sands and loamy sands form only weakly expressed crusts (Miller and Bahruddin, 1986; Radcliffe et al., 1991; Miller and Radcliffe, 1992; Chiang et al., 1993).

Table 6.66   Mean pore area for dyed and undyed areas after break through of methylene-blue for Ultisols from the Piedmont and Coastal Plain of Georgia. [Franklin (1994); Shaw et al. (1997)]

Horizon	Mean Pore Area[†]	
	Dyed	Undyed
	------------%------------	
Typic Kanhapludult (Piedmont)		
A	19.0	3.0
BA	17.7	5.0
Bt	15.8	2.4
Bt	14.4	2.8
Plinthic Kandiudult (Coastal Plain)		
Bt	5.0	1.1
Btv	5.9	0.8
BC	NA[‡]	0.9

† Area of pores > 0.05 mm equivalent circular diameter determined from impregnated polished blocks with image analysis
‡ The BC horizon had insufficient dyed area for evaluation of pore area.

Table 6.67 Comparison of organic C, aggregate stability, infiltration rate, and percentage moist days under conventional and no-till cropping systems [West et al., 1991; Bruce et al., 1992]

Cropping System[H]	Organic C	Aggregate Stability	Infiltration Rate[I]		Moist Days[†]
			residue	residue removed	
	---------g kg^{-1}---------		------------mm h^{-1}------------		% of growing season
CTG	10.4	580	36	22	29
NTG	23.3	890	50	46	49

[H] CTG–Conventional-tilled grain sorghum (*Sorghum bicolor* (L.) Moench); NTG–No-till grain sorghum with clover winter cover.
[I] Infiltration rate at the end of a 1 h of rainfall simulation at approximately 64 mm h^{-1}.
[†] Moist days are defined as the days with soil moisture tension > 0.1 Mpa.

Because energy inputs from rainfall are needed to form surface crusts, crusting is minimal if the soils have residue or vegetative cover and organic matter is maintained in the upper part of the soil. For Ultisols in the Piedmont of Georgia, infiltration rates after 1 h of simulated rainfall were about 40% higher for soils that had been under no-tillage for 5 yr than for soils that had been conventionally tilled (Table 6.67) (Bruce et al., 1992).

6.12.8 Chemical Properties

The requirement that Ultisols have base saturation < 35% in the lower part of the subsoil carries with it certain associated properties including low pH, potentially high Al saturation, appreciable weathering and associated kaolinitic mineralogy, and in many cases, relatively high contents of Fe and Al oxides and oxyhydroxides. This suite of properties has significant impact on the chemical properties of Ultisols and how these properties affect use and management of this soil order.

6.12.8.1 Cation and Anion Exchange Capacity

The CEC of Ultisol horizons depends on the amount and type of clay, Fe, Al, and Mn oxide content, organic matter content, soil solution electrolyte concentration, and pH of the horizon. Reported CEC values for Ultisol horizons range from < 3 to > 20 cmol$_c$ kg^{-1} (Sanchez, 1976; Carlisle et al., 1985; Karathanasis et al., 1986). Generally, Ultisols with kaolinitic mineralogy have mostly pH-dependent charge and low CEC, while those with mixed or smectitic mineralogy have a greater proportion of fixed charge and higher CEC values.

Even when the net charge is negative, many Ultisols have an appreciable positive charge and anion exchange capacity (AEC). Gillman and Sumner (1987), in a study of four Ultisols from the Piedmont of Georgia, reported AEC values at the pH of the soil that ranged from 0.1 to 1.2 cmol$_c$ kg^{-1}, which was attributed to Fe and Al oxides in these soils. For Ultisols from Georgia and South Africa, Grove et al. (1982) reported AEC values ranging from 0.03 to 1.91 cmol$_c$ kg^{-1}. Measurable AEC has also been reported for Ultisols from tropical Australia and Peru (Gillman and Sumpter, 1986; Gillman and Sinclair, 1987).

6.12.8.2 Acidity and Exchangeable Aluminum

Below a pH of about 5.5, Al released from weathering of primary and secondary minerals is present as part of the exchange complex. An Al saturation of greater than 60% (measured as a percentage of the unbuffered CEC or ECEC) has been reported as the threshold that results in soil solution Al concentrations > 1 mg L^{-1} which may reduce yield in Al sensitive crops (Nye et al., 1961; Kamprath,

1970; Sanchez, 1976). Farina and Channon (1991), however, observed yield reductions in maize for Al saturations > 25% of the ECEC.

6.12.8.3 Phosphorus

Phosphate is specifically adsorbed by Fe, Al, and Mn oxides and amorphous or poorly crystalline aluminosilicates. Because of eluviation and weathering, surface horizons of most Ultisols have low contents of these components. Thus, P sorption is generally lower in Ultisols than in Oxisols and Andisols (Sanchez, 1976; van Wambeke, 1991). However, differences in crystallinity of Fe oxides among soils may confound this relationship within and among orders (Pratt et al., 1969; Fox et al., 1971).

6.12.9 Biological Properties

6.12.9.1 Organic Matter

A common misconception is that Ultisols have low contents of organic matter compared to other soil orders. While Ultisols generally have lower amounts of organic C than Mollisols, Oxisols, and Andisols, levels of organic C in Ultisols are similar to those in Alfisols both within and across geographic areas (Sanchez, 1976). There is also little difference in organic C content between Ultisols in temperate and tropical climates (Buol, 1973; Sanchez, 1976).

Organic matter contents in Ultisols decrease rapidly when these soils are converted from forest or pasture to intensive cultivation. For Ultisols in Georgia, Giddens (1957) found that organic matter contents decreased from 20.5 to 16.6 g kg^{-1} in 24 months after conversion of forest to intensive cultivation. Use of no-tillage, cover crops, and residue management, however, can help to restore and maintain organic C levels in Ultisols at precultivation levels. Bruce et al. (1992) reported that use of no-tillage and winter cover crops increased organic C levels in the upper 15 mm of degraded Ultisols from 10.4 to 23.3 g kg^{-1} over a 5 yr period. Associated with the increase in organic C in these soils was a significant increase in water stable aggregates and significant decreases in runoff and interrill soil loss (Table 6.68) (West et al., 1991; Bruce et al., 1992; West et al., 1992).

6.12.9.2 Biologic Populations and Processes

Populations of microorganisms in Ultisols are not appreciably different from soils in other orders. Surface horizons of most Ultisols have acid pH values. Thus, fungi may comprise a greater proportion of the microbial biomass in these soils because of the greater ability of fungi to survive under acid conditions than other microorganisms (Alexander, 1977). Low pH has been reported to reduce rates of nitrification and denitrification in pure cultures (Alexander, 1977). Natural adaptation by

Table 6.68 Interrill soil loss and runoff from three Ultisols with varying tillage and surface cover under 50 mm h^{-1} rainfall intensity for 1 hour on 5–9% slopes [West et al., 1991]

Tillage	Surface condition	----------Interrill soil loss----------			------------Total runoff------------		
		Soil 1	Soil 2	Soil 3	Soil 1	Soil 2	Soil 3
		--------------g m^{-2}-------------			------------------mm-----------------		
Conventional	Crusted	0.46	0.56	0.33	16	23	17
	Tilled	0.72	0.79	0.48	18	23	16
No-till	Bare	0.14	0.19	0.07	3	3	6
	Residue	0.07	0.03	0.02	1	1	1

microorganisms to local field conditions, however, probably alters the effects of pH and other environmental conditions observed in the laboratory.

6.12.10 Conclusions

Worldwide, Ultisols are a widespread soil resource, which because of their extensive weathering, low base saturation, and associated properties, are often considered to be unproductive. It is true that Ultisols, in general, have lower native fertility than Mollisols and Alfisols, and in areas where these orders are abundant, Ultisols are the less desirable soils. In many regions of the world, however, Ultisols are dominant or are the most productive soils available. Only with a thorough understanding of the genesis, properties, and response to management of these soils can the productivity of this valuable resource be maintained and enhanced.

6.12.11 References

Alexander, M.A. 1977. Introduction to soil microbiology. 2nd Ed. John Wiley and Sons, New York, NY.

Allen, B. L., and D.S. Fanning. 1983. Composition and soil genesis. p. 141–192. *In* L.P. Wilding, N.E. Smeck, and G.F. Hall (eds.) Pedogenesis and Soil Taxonomy I. Concepts and interactions. Elsevier, Amsterdam, Netherlands.

Baldwin, M., C.E. Kellogg, and J. Thorp. 1938. Soil classification. p. 979–1001. *In* Soils and Men. USDA Yearbook of Agriculture, US Government Printing Office, Washington, DC.

Beinroth, F. H. 1981. Some highly weathered soils of Puerto Rico, 1. Morphology, formation and classification. Geoderma 27:1–73.

Beinroth, F. H., G. Uehara, and H. Ikawa. 1974. Geomorphic relationships of Oxisols and Ultisols on Kauai, Hawaii. Soil Sci. Soc. Am. Proc. 38:128–131.

Blume, L.J., H.F. Perkins, and R.K. Hubbard. 1987. Subsurface water movement in an upland Coastal Plain soil as influenced by plinthite. Soil Sci. Soc. Am. J. 51:774–779.

Bockheim, J.G., J.G. Marshall, and H.M. Kelsey. 1996. Soil-forming processes and rates on uplifted marine terraces in southwestern Oregon, USA. Geoderma 73:39–62.

Bruce, R.R., J.J. Dane, V.L. Quisenberry, N.L. Powell, and A.W. Thomas. 1983. Physical characteristics of soils in the Southern Region: Cecil. Southern Coop. Series Bull. 267.

Bruce, R.R., G.W. Langdale, L.T. West, and W.P. Miller. 1992. Soil surface modification by biomass inputs affecting rainfall infiltration. Soil Sci. Soc. Am. J. 56:1614–1620.

Bryant, J.P., and J.B. Dixon. 1964. Clay mineralogy and weathering of red-yellow podzolic soil from quartz mica schist in the Alabama Piedmont. Clays Clay Miner. 12:509–521.

Buol, S.W. 1973. Soil genesis, morphology, and classification. p. 1–38. *In* P.A. Sanchez (ed.) A review of soils research in tropical Latin America. NC Agric. Exp. Sta. Tech. Bull. 219.

Carlan, W.L., H.F. Perkins, and R.A. Leonard. 1985. Movement of water in a Plinthic Paleudult using a bromide tracer. Soil Sci. 139:62–66.

Carlisle, V.W., and L.W. Zelazny. 1973. Mineralogy of selected Florida Paleudults. Soil Sci. Soc. Fla. Proc. 33:136–139.

Carlisle, V.W., M.E. Collins, F. Sodek, III, and L.C. Hammond. 1985. Characterization data for selected Florida soils. Soil Sci. Res. Rep. 85-1. Univ. FL, Gainesville, FL.

Chiang, S.C., D.E. Radcliffe, and W.P. Miller. 1993. Hydraulic properties of surface seals in Georgia soils. Soil Sci. Soc. Am. J. 57:1418–1426.

Daniels, R. B., E.E. Gamble, and J.G. Cady. 1971. The relation between geomorphology and soil morphology and genesis. Adv. Agron. 23:51–88.

Daniels, R.B., H.F. Perkins, and E.E. Gamble 1978. Morphology of discontinuous phase plinthite and criteria for its field identification in the southeastern United States. Soil Sci. Soc. Am. J. 42:944–949.

Dobos, R.R., E.J. Ciolkosz, and W.J. Waltman. 1990. The effect of organic carbon, temperature, time and redox conditions on soil color. Soil Sci. 150:506–512.

Dokuchaev, V. V. 1883. Russian Chernozem (Russkii chernozem). [Translated from Russian by N. Kaner.] Israel Program for Scientific Translations, 1967. Available from US Department Commerce, Springfield, VA.

Eswaran, H. 1993. Assessment of global resources: Current status and future needs. Pedologie 43:19–39.

Farina, M.P.W., and P. Channon. 1991. A field comparison of lime requirement indices for maize. Plant Soil 134:127–135.

Fiskell, J.G.A., and H.F. Perkins. 1970. Selected coastal plain soil properties. Univ. FL Southern Coop. Ser. Bull. 148.

Fox, R.L., S.M. Hasan, and R.C. Jones. 1971. Phosphate and sulfate sorption by Latosols. Proc. Int. Symp. Soil Fert. Eval. 1:857–864.

Franklin, D.H. 1994. Morphological evaluation and quantification of flow paths in a Georgia Piedmont soil. M.S. Thesis, University of Georgia, Athens, GA.

Gamble, E. E., R.B. Daniels, and W.D. Nettleton. 1970. Geomorphic surfaces and soils in the Black Creek Valley, Johnston County, North Carolina. Soil Sci. Soc. Am. Proc. 34:276–281.

Garrels, R.M., and C.L. Christ. 1965. Solutions, minerals and equilibria. Harper and Row, New York, NY.

Giddens, J.E. 1957. Rate of loss of carbon from Georgia soils. Soil Sci. Soc. Am. Proc. 21:513–515.

Gillman, G.P., and D.F. Sinclair. 1987. The grouping of soils with similar charge properties as a basis for agrotechnology transfer. Aust. J. Soil Res. 25:275–285.

Gillman, G.P., and M. E. Sumner. 1987. Surface charge characterization and soil solution composition of four soils from the Southern Piedmont in Georgia. Soil Sci. Soc. Am. J. 51:589–594.

Gillman, G.P., and A.S. Sumpter. 1986. Surface charge characteristics and lime requirements of soils derived from basaltic, granitic, and metamorphic rocks in high rainfall tropical Queensland. Aust. J. Soil Res. 24:173–192.

Griffin, R.W., and S.W. Buol. 1988. Soil and saprolite characteristics of Vertic and Aquic Hapludults derived from Triassic Basin sandstones. Soil Sci. Soc. Am. J. 52:1094–1099.

Grove, J.H., C.S. Fowler, and M.E. Sumner. 1982. Determination of the charge character of selected acid soils. Soil Sci. Soc. Am. J. 46:32–38.

Hajek, B.F., and L.W. Zelazny. 1985. Problems associated with Soil Taxonomy mineralogy placement in the nonglaciated humid region. p. 87–93. *In* J.A. Kittrick (ed.) Mineral Classification of soils. Soil Sci. Soc. Am. Spec. Publ. 16. Soil Science Society of America, Madison, WI.

Harris, W.G., S.S. Iyengar, L.W. Zelazny, J.C. Parker, D.A. Lietzke, and W.J. Edmonds. 1980. Mineralogy of a chronosequence formed in New River alluvium. Soil Sci. Soc. Am. J. 44:862–868.

Harris, W.G., L.W. Zelazny, and J.C. Baker. 1984. Depth and particle size distribution of talc in a Virginia Piedmont Ultisol. Clays Clay Miner. 32:227–230.

Hubbard, R.K., and J.M. Sheridan. 1983. Water and nitrate-nitrogen losses from a small upland, Coastal Plain watershed. J. Environ. Qual. 12:291–295.

Hubbard, R.K., C.R. Berdanier, H.F. Perkins, and R.A. Leonard. 1985. Characteristics of selected upland soils of the Georgia Coastal Plain. USDA-ARS, ARS-37. US Government Printing Office, Washington, DC.

Jackson, M.L., S.A. Tyler, A.L. Willis, B.A. Bourbeau, and R.P. Pennington. 1948. Weathering sequence of clay-size minerals. I. Fundamental generalizations. J. Phys. Coll. Chem. 52:1237–1260.

Jenny, H. 1941. Factors of soil formation. McGraw-Hill Book Co., New York, NY.

Kamprath, E.J. 1970. Exchangeable aluminum as a criterion for liming leached mineral soils. Soil Sci. Soc. Am. Proc. 34:252–254.

Karathanasis, A.D., and B.F. Hajek. 1985. Shrink-swell potential of montmorillonitic soils in udic moisture regimes. Soil Sci. Soc. Am. J. 49:159–166.

Karathanasis, A.D., F. Adams, and B.F. Hajek. 1983. Stability relationships in kaolinite, gibbsite, and Al-hydroxy interlayered vermiculite soil systems. Soil Sci. Soc. Am. J. 47:1247–1251.

Karathanasis, A.D., G.W. Hurt, and B.F. Hajek. 1986. Properties and classification of montmorilonite-rich Hapludults in the Alabama Coastal Plains. Soil Sci. 14: 76–82.

Kellogg, C.E. 1949. Preliminary suggestions for the classification and nomenclature of great soil groups in tropical and equatorial regions. Comm. Bur. Soil Sci. Tech. Commun. 46:76–85.

Lepsch, I. F. and S.W. Buol. 1974. Investigation in an Oxisol-Ultisol toposequence in S. Paulo State, Brazil. Soil Sci. Am. Proc. 38:491–496.

Losche, C.K., R.J. McCracken, and C.B. Davey. 1970. Soils of steeply sloping landscapes in the southern Appalachian Mountains. Soil Sci. Soc. Am. Proc. 34:473–478.

Macedo, J., and R.B. Bryant. 1989. Preferential microbial reduction of hematite over goethite in a Brazilian Oxisol. Soil Sci. Soc. Am. J. 51:690–698.

Marbut, C.F. 1928. A scheme for soil classification. Proc. First Int. Cong. Soil Sci. 4:1–31.

McCracken, R.J., E.J. Penderson, L.E. Aull, C.I. Rich, and T.C. Peele. 1971. Soils of the Hayesville, Cecil and Pacolet series in the southern Appalachian and Piedmont regions. NC State Univ. South. Coop. Ser. Bull. 157.

Miller, B. J. 1983. Ultisols. p. 283–323 *In* L. P. Wilding, N. E. Smeck, and G. F. Hall (ed.) Pedogenesis and Soil Taxonomy II. The soil orders. Elsevier, Amsterdam, Netherlands.

Miller, W.P., and M.K. Bahruddin. 1986. Relationship of soil dispersibility to infiltration and erosion of Southeastern soils. Soil Sci. 142:235–240.

Miller W.P., and D.E. Radcliffe. 1992. Soil crusting in the southeastern United States. p. 233–266. *In* M.E. Sumner and B.A. Stewart (ed.) Soil crusting: Chemical and physical processes. Lewis Publishers, Boca Raton, FL.

Nash, V.E. 1979. Mineralogy of soils developed on Pliocene-Pleistocene terraces of the Tombigbee River in Mississippi. Soil Sci. Soc. Am. J. 43:616–623.

Nash, V.E., D.E. Pettry, and M.N. Sudin. 1988. Mineralogy and chemical properties of two Ultisols formed in glauconitic sediments. Soil Sci. 145:270–277.

Norfleet, M.L., and B.R. Smith. 1989. Weathering and mineralogical classification of selected soils in the Blue Ridge Mountains of South Carolina. Soil Sci. Soc. Am. J. 53:1771–1778.

Nye, P., D. Craig, N.T. Coleman, and J.L. Ragland. 1961. Ion exchange equilibrium involving aluminum. Soil Sci. Soc. Am. Proc. 25:14–17.

O'Brien, E.L., and S.W. Buol. 1984. Physical transformations in a vertical soil-saprolite sequence. Soil Sci. Soc. Am. J. 48:354–357.

Pratt, P.F., F.F. Peterson, and C.S. Holzley. 1969. Qualitative mineralogy and chemical properties of a few soils from Sao Paulo, Brazil. Turrialba 19:491–496.

Quisenberry, V.H., D.K. Cassel, J.H. Dane, and J.C. Parker. 1987. Physical characteristics of soils of the Southern Region: Norfolk, Dothan, Wagram, and Goldsboro. Clemson Univ. Southern Coop. Series Bull. 263.

Radcliffe, D.E., L.T. West, R.K. Hubbard, and L.E. Asmussen. 1991. Surface sealing in Coastal Plains loamy sands. Soil Sci. Soc. Am. J. 55:223–227.

Robertus, R.A., and S.W. Buol. 1985. Intermittency of illuviation in Dystrochrepts and Hapludults from the Blue Ridge and Piedmont provinces of North Carolina. Geoderma 36:277–291.

Robertus, R.A., S.B. Weed, and S.W. Buol. 1986. Transformations of biotite to kaolinite during saprolite-soil weathering. Soil Sci. Soc. Am. J. 50:810–819.

Sanchez, P.A. 1976. Properties and management of tropical soils. John Wiley and Sons, New York, NY.

Schwertmann, U., and R.M. Taylor. 1989. Iron oxides. p. 379–438. *In* J.B. Dixon and S.B. Weed (ed.) Minerals in soil environments, 2nd edition. Soil Science Society of America, Madison, WI.

Scrivner, C.L., J.C. Baker, and B.J. Miller. 1966. Soils of Missouri: A guide to their identification and interpretation. C823, Extension Div., Univ. of Missouri, Colombia, MO.

Shaw, J.N., L.T. West, and C.C. Truman. 1998. Hydraulic properties of soils with water restrictive horizons in the Georgia Coastal Plain. Soil Sci.162:875–885.

Shirmohammadi, A., W.G. Knisel, and J.M. Sheridan. 1984. An approximate method for partitioning daily streamflow data. J. Hydro. 74:335–354.

Soil Survey Staff. 1960. Soil classification, A comprehensive system, 7th approximation. USDA, US Government Printing Office, Washington, DC.

Soil Survey Staff. 1975. Soil Taxonomy. A basic system of soil classification for making and interpreting soil surveys. USDA-SCS. US Government Printing Office, Washington, DC.

Soil Survey Staff. 1998. Keys to soil taxonomy. 8th ed. USDA-NRCS, US Government Printing Office, Washington, DC.

Southard, R.J., and S.W. Buol. 1988a. Subsoil blocky structure formation in some North Carolina Paleudults and Paleaquults. Soil Sci. Soc. Am. J. 52:1069–1076.

Southard, R.J., and S.W. Buol. 1988b. Subsoil saturated hydraulic conductivity in relation to soil properties in the North Carolina Coastal Plain. Soil Sci. Soc. Am. J. 52:1091–1094.

van Wambeke, A. 1991. Soils of the tropics: Properties and appraisal. McGraw-Hill, New York, NY.

West, L.T., F.H. Beinroth, M.E. Sumner, and B.T. Kang. 1997. Ultisols: Characteristics and impacts on society. Adv. Agron. 63:179–236.

West, L.T., W.P. Miller, G.W. Langdale, R.R. Bruce, J.M. Laflen, and A.W. Thomas. 1991. Cropping system effects on interrill soil loss in the Georgia Piedmont. Soil Sci. Soc. Am. J. 55:460–466.

West, L.T., W.P. Miller, R.R. Bruce, G.W. Langdale, J.M Laflen, and A.W. Thomas. 1992. Cropping system and consolidation effects on rill erosion in the Georgia Piedmont. Soil Sci. Soc. Am. J. 56:1238–1243.

6.13   Oxisols

Friedrich H. Beinroth
University of Puerto Rico

Hari Eswaran
USDA-NRCS

Goro Uehara
University of Hawaii

Paul F. Reich
USDA-NRCS

6.13.1   Introduction

The deep, red and highly weathered soils of the tropics have long fascinated pedologists, particularly those from the temperate region. The uniqueness of these soils lies not only in their properties, but also in their geographic distribution being confined almost exclusively to the tropics. As this account is intentionally concise, ample references are provided for those seeking additional or more detailed information.

6.13.2   Historical Background

In the early literature, the highly weathered soils of the tropics were identified as red soils, red loams or red earths. In an historical report on travels in South India, Buchanan (1807) described soil material used for construction and called it laterite (Buchanan, 1807). The term was adopted by the pedologic community and is still used today with a wide variety of meanings (Alexander and Cady, 1962; Maignen, 1966). In 1949, a group of scientists proposed the term Latosols (Cline, 1975) that soon became popular and, by the midfifties, the concept of Latosols as highly weathered soils with a low negative charge was firmly established. Several soil classification systems developed during the fifties and sixties recognized such soils in a separate class with the nomenclature shown in Table 6.69.

The term Oxisol originated around 1954 (Smith, 1963; 1965) during the development of *Soil Taxonomy* (Soil Survey Staff, 1975). In 1978, an International Committee on the Classification of Oxisols (ICOMOX) was formed (Buol and Eswaran, 1988) to initiate discussions that would lead to an improved classification. Through circular letters and four international soil classification workshops held in Brazil (1976), Malaysia and Thailand (1978), Rwanda (1981), and Brazil (1986), the current state of knowledge of these soils was assembled and documented (Camargo and Beinroth, 1978; Beinroth and Paramanathan, 1978; Beinroth and Panichapong, 1978; Beinroth et al., 1983; Beinroth et al., 1986).

Table 6.69   Synonyms for Oxisols in other classification systems

Synonym	Classification System	Source
Sols Ferrallitiques	French	Aubert, 1958
Latosols	Brazilian	Bennema et al., 1959
Solos Ferraliticos	Portuguese	Botelho da Costa, 1954, 1959
Oxysols	Ghana	Charter, 1958
Ferralsols		FAO/UNESCO, 1971-1976
Rotlehm	German	Harrassowitz, 1930
Laterites		Maignen, 1966
Red Earths	British	Robinson, 1951
Kaolisols	Belgian	Tavernier and Sys, 1965

6.13.3 Concept and Classification of Oxisols

6.13.3.1 Perceptions

The rationale of the taxon of Oxisols is based largely on the concept of Latosols which was developed to designate "all zonal soils having their dominant characteristics associated with low silica:sesquioxide ratios, low base exchange capacity, low activity of clays, and low content of weatherable minerals" (Kellogg, 1949; Cline, 1975). These characteristics were selected because they were thought to manifest the effects of advanced pedogenesis under tropical conditions.

6.13.3.2 Definition

Simplified from the *Keys to Soil Taxonomy* (Soil Survey Staff, 1998), Oxisols may be defined as those soils that fail to meet the criteria definitive for Gelisols, Histosols, Spodosols, and Andisols but have either (1) an oxic horizon within 150 cm of the mineral soil surface and no kandic horizon within this depth, or (2) 40% or more clay in the surface horizon and a kandic horizon within 100 cm of the mineral soil surface that meets the weatherable mineral properties of the oxic horizon.

The definition of Oxisols has undergone significant changes since it was first published in *Soil Taxonomy* (Soil Survey Staff, 1975). Consequently, more soils now qualify for Oxisols; this should be kept in mind when consulting the earlier literature. It should also be mentioned that not all intensively weathered soils are Oxisols. Highly weathered soils with a marked clay increase below a light textured surface horizon are by definition excluded from the Oxisols and are classified as kandic great groups of Ultisols and Alfisols (van Wambeke, 1992; Soil Survey Staff, 1998).

6.13.3.3 Diagnostic Criteria

Key to the identification of Oxisols is the presence of either an oxic horizon or a kandic horizon underlying a surface horizon with $\geq 40\%$ clay. The key properties of the oxic horizon are its charge characteristics and negligible amounts of weatherable minerals. The charge characteristics are defined by the magnitude of the variable charge (CEC by $NH_4OAc < 16$ cmol$_c$ kg^{-1} clay), and the permanent charge estimated by the Effective Cation Exchange Capacity (ECEC < 12 cmol$_c$ kg^{-1} clay). The low weatherable mineral requirement ensures that there are few weatherable minerals that could release plant nutrients. The particle size class of the oxic horizon is sandy loam or finer and its upper boundary is diffuse. To be diagnostic, the oxic horizon must have a minimum thickness of 30 cm and occur within 150 cm of the mineral soil surface.

The kandic horizon (Moormann, 1985) shares some of the properties of the oxic and argillic horizons. Like the argillic horizon, clay increases with depth but the charge characteristics are those of the oxic horizon. To be diagnostic for Oxisols, the kandic horizon must meet the weatherable mineral requirements of the oxic horizon, occur beneath a surface horizon that has $\geq 40\%$ clay, and have its upper boundary within 100 cm of the soil surface.

Suborders of Oxisols are differentiated on the basis of soil moisture regimes, aquic conditions, the presence of a histic epipedon, and soil color. Defining criteria for great groups are the kandic and sombric horizons, plinthite, acric properties, and base saturation. The 14 differentiae used to establish subgroups are: andic-like properties, Al+Fe percentage, aquic conditions, CEC, pH (pH$_{KCl}$ −pH$_{H_2O}$), histic epipedon, kandic horizon, lithic contact, organic C content, petroferric contact, plinthite, soil color, soil depth, and the sombric horizon.

At the family level, criteria include particle size class, mineralogy class, and soil temperature regimes. Soil mineralogy is an important factor and eight mineralogical classes are recognized (Table 6.70). When the soil has $> 18\%$ Fe$_2$O$_3$ or gibbsite in the fine earth fraction (< 2 mm), the mineralogical class is determined by the amounts of each of these. If there is < 18% Fe_2O_3 (12.6% Fe) or gibbsite, the composition of the clay fraction is considered.

6.13.3.4 Taxa
The class of Oxisols has five suborders, 22 great groups, 213 subgroups, an estimated 400 families, and perhaps as many as 1,000 soil series. The great groups and possible subgroups of the Oxisols are presented in Table 6.71.

6.13.4 Geography of Oxisols

6.13.4.1 Global Extent and Geographic Distribution
The Oxisols of the world comprise about 9.8 million ha (Table 6.1), approximately equivalent to 7.5% of the global and 25% of the tropical land area. The distributions of Oxisol suborders in South America, Africa and globally are illustrated in Figs. 6.72, 6.73 and 6.74. Over 95% of the Oxisols occur in the tropics (Table 6.2) where Udox occupy 13.2% and Ustox 7.9% of the land area. South America has the largest extent of Oxisols and is home to 57% of the world's Oxisols, most of them occurring in Brazil. In Africa, Zaire probably has the largest areas of Oxisols. In Southeast Asia, Oxisols only occur in small isolated areas with the largest presumed to be in Borneo (Kalimantan). In Oceania, there are small areas of Oxisols in Australia and on the islands of the Pacific and Caribbean Basins.

6.13.4.2 Extent and Geographic Distribution in the United States
As expected, Oxisols have not been encountered in the conterminous United States and only occur on the islands of Hawaii and Puerto Rico where they account for about 4.3 and 6.7% of the land area, respectively (Table 6.72).

6.13.5 Formation and Landscape Relationships of Oxisols

6.13.5.1 Factors of Oxisol Formation
Soil is the cumulative result of the interaction of pedogenetic processes which are controlled by the factors of soil formation. Although the factorial approach to understanding soil formation has serious conceptual and operational limitations (Smeck et al., 1983; Wilding, 1994), it remains a unifying philosophy in pedology. The formation of Oxisols is, therefore, discussed here in the context of this paradigm.

Table 6.70 Mineralogy classes of Oxisols

Mineralogy Class	Fe_2O_3	Gibbsite	Kaolinite	Halloysite
Ferritic	> 40%			
Gibbsitic	0-40%	> 40%		
Sesquic	18-40%	18-40%		
Ferruginous	18-40%	0-18%		
Allitic	0-18%	18-40%		
Kaolinitic	0-18%	0-18%	> 50%	
Halloysitic	0-18%	0-18%	< 50%	> 50%
Mixed	0-18%	0-18%	< 50%	< 50%

Table 6.71 Listing of suborders, great groups, and subgroups in the Oxisols order [Soil Survey Staff, 1998]

Suborder	Great Group	Subgroups
Aquox	Acraquox	Plinthic, Aeric, Typic
	Plinthaquox	Aeric, Typic
	Eutraquox	Histic, Plinthic, Aeric, Humic, Typic
	Haplaquox	Histic, Plinthic, Aeric, Humic, Typic
Torrox	Acrotorrox	Petroferric, Lithic, Typic
	Eutrotorrox	Petroferric, Lithic, Typic
	Haplotorrox	Petroferric, Lithic, Typic
Ustox	Sombriustox	Petroferric, Lithic, Humic, Typic
	Acrustox	Aquic Petroferric, Petroferric, Aquic Lithic, Lithic, Anionic Aquic, Anionic, Plinthic, Aquic, Eutric, Humic Rhodic, Humic Xanthic, Humic, Rhodic, Xanthic, Typic
	Eutrustox	Aquic Petroferric, Petroferric, Aquic Lithic, Lithic, Plinthaquic, Plinthic, Aquic, Kandiustalfic, Humic Inceptic, Inceptic, Humic Rhodic, Humic Xanthic, Humic, Rhodic, Xanthic, Typic
	Kandiustox	Aquic Petroferric, Petroferric, Aquic Lithic, Lithic, Plinthaquic, Plinthic, Aquic, Humic Rhodic, Humic Xanthic, Humic, Rhodic, Xanthic, Typic
	Haplustox	Aquic Petroferric, Petroferric, Aquic Lithic, Lithic, Plinthaquic, Plinthic, Aqueptic, Aquic, Oxyaquic, Inceptic, Humic Rhodic, Humic Xanthic, Humic, Rhodic, Xanthic, Typic
Perox	Sombriperox	Petroferric, Lithic, Humic, Typic
	Acroperox	Aquic Petroferric, Petroferric, Aquic Lithic, Lithic, Anionic, Plinthic, Aquic, Humic Rhodic, Humic Xanthic, Humic, Rhodic, Xanthic, Typic
	Eutroperox	Aquic Petroferric, Petroferric, Aquic Lithic, Lithic, Plinthaquic, Plinthic, Aquic, Kandiudalfic, Humic Inceptic, Inceptic, Humic Rhodic, Humic Xanthic, Humic, Rhodic, Xanthic, Typic
	Kandiperox	Aquic Petroferric, Petroferric, Aquic Lithic, Lithic, Plinthaquic, Plinthic, Aquic, Andic, Humic Rhodic, Humic Xanthic, Humic, Rhodic, Xanthic, Typic
	Haploperox	Aquic Petroferric, Petroferric, Aquic Lithic, Lithic, Plinthaquic, Plinthic, Aquic, Andic, Humic Rhodic, Humic Xanthic, Humic, Rhodic, Xanthic, Typic
Udox	Sombriudox	Petroferric, Lithic, Humic, Typic
	Acrudox	Aquic Petroferric, Petroferric, Aquic Lithic, Lithic, Anion Aquic, Anionic, Plinthic, Aquic, Eutric, Humic Rhodic, Humic Xanthic, Humic, Rhodic, Xanthic, Typic
	Eutrudox	Aquic Petroferric, Petroferric, Aquic Lithic, Lithic, Plinthaquic, Plinthic, Aquic, Kandiudalfic, Humic Inceptic, Inceptic, Humic Rhodic, Humic Xanthic, Humic, Rhodic, Xanthic, Typic
	Kandiudox	Aquic Petroferric, Petroferric, Aquic Lithic, Lithic, Plinthaquic, Plinthic, Aquic, Andic, Humic Rhodic, Humic Xanthic, Humic, Rhodic, Xanthic, Typic
	Hapludox	Aquic Petroferric, Petroferric, Aquic Lithic, Lithic, Plinthaquic, Plinthic, Aquic, Inceptic, Andic, Humic Rhodic, Humic Xanthic, Humic, Rhodic, Xanthic, Typic

Climate

With respect to rainfall, the definition of Oxisols implies two conditions. First, in the case of Oxisols with kandic horizons, there must be some time during the year when evapotranspiration exceeds precipitation as this appears to be a prerequisite for the formation of the kandic horizon (van Wambeke, 1992). Second, precipitation must exceed the capacity of the soil to retain water during some time of the year so that water percolates through the solum (Miller, 1983). This causes the leaching of soluble weathering products and favors the residual concentration of kaolinite and sesquioxides (van Wambeke et al., 1983). The wide variety of tropical climates that meet these conditions corresponds to the udic, perudic and ustic soil moisture regimes of *Soil Taxonomy* (Soil

Fig. 6.71   Distribution of Oxisols in South America

Survey Staff, 1998). The few Torrox that now have an aridic soil moisture regime are considered relics of a more humid climate in the past.

Regarding temperature, most Oxisols areas have mean annual air temperatures > 15 °C with minimal annual fluctuations and, therefore, have isothermic or isohyperthermic soil temperature regimes. A few areas in southern Brazil and South Africa may have a thermic temperature regime. The climate conducive to the formation of Oxisols is thus typically humid tropical. As most Oxisols have evolved over geologic time periods, consideration of the paleoclimate is essential to understanding and analyzing their development.

Parent Material
A distinction must be made between transported or allochthonous and residual or autochthonous parent materials. Autochthonous parent materials have weathered *in situ*, whereas allochthonuos materials have been transported by fluviocolluvial processes. The sediments that form the parent material of Oxisols have frequently been preweathered and may have gone through more than one weathering and transport cycle. Where Oxisol development occurs *in situ*, the parent material is usually saprolite,which is a highly weathered parent rock that, unless it is collapsed, still preserves most of the original rock fabrics. In terms of area, the allochthonous Oxisols are more extensive than

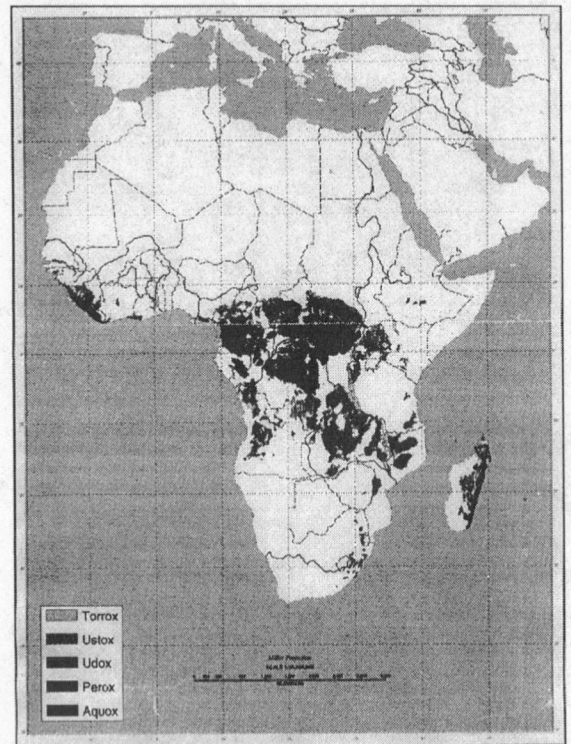

Fig. 6.72   Distribution of Oxisols in Africa

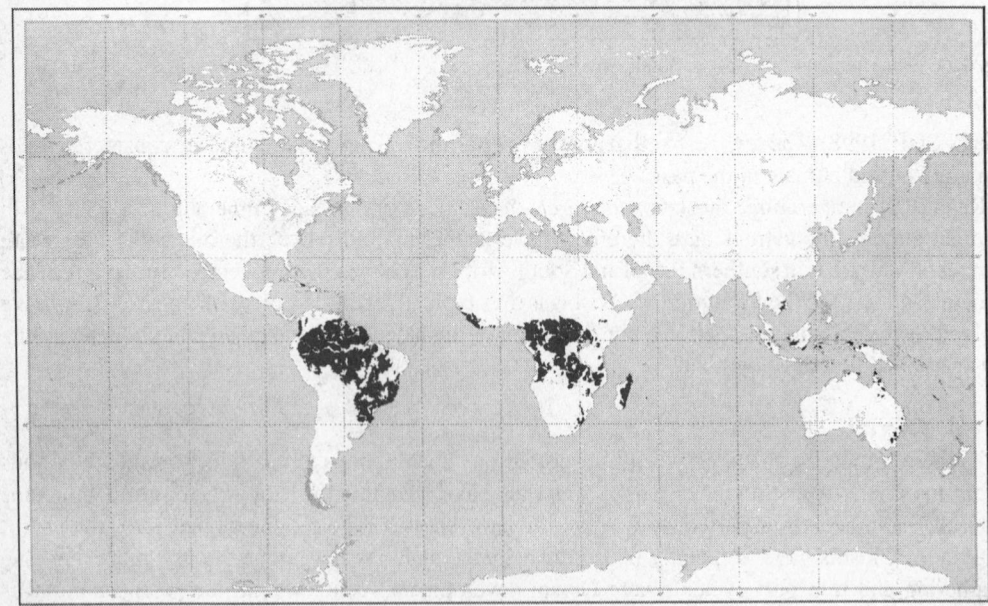

Fig. 6.73   Global distribution of Oxisols [Courtesy of USDA-NRCS, Soil Survey Division, World Soil Resources, 1998]

the autochthonous (Eswaran and Tavernier, 1980). The allochthonous Oxisols on the Brazilian and Guayanian Shields (Lepsch and Buol, 1974; Lepsch et al., 1977) and the Central African Plateau (Ruhe, 1956) are formed in sediments on mid to upper Tertiary surfaces and are generally as deep as the sediments.

Biota

There appear to be no clear cause and effect relationships between vegetation and the geography of Oxisols as they occur under both rainforest and savannas (van Wambeke et al., 1983). Termites and

Table 6.72 Oxisol series of Hawaii and Puerto Rico

| Soil Series | | Classification | | Hawaii | | Puerto Rico | |
Hawaii	Puerto Rico	Subgroup	Code	Area (ha)	%	Area (ha)	%
	Delicias	Rhodic Haplustox	CCEN			280	0.03
	Moteado	Humic Haplaquox	DADD			0	0.00
Molokai		Typic Eutrotorrox	DBBC	14,143	0.84		
Molokai variant		Typic Eutrotorrox	DBBC	578	0.03		
	Matanzas	Lithic Eutrustox	DCCD			1,786	0.20
Wahiawa		Kandiustalfic Eutrustox	DCCH	8,566	0.51		
Helemano		Rhodic Eutrustox	DCCN	11,250	0.67		
Lihue		Rhodic Eutrustox	DCCN	6,300	0.38		
Niu		Rhodic Eutrustox	DCCN	1,368	0.08		
Mahana		Inceptic Haplustox	DCEJ	4,918	0.29		
Makapili		Humic Haplustox	DCEM	963	0.06		
Halii		Anionic Acroperox	DDBE	1,968	0.12		
Kapaa		Anionic Acrudox	DEBF	9,198	0.55		
Pooku		Anionic Acrudox	DEBF	3,154	0.19		
	Nipe	Anionic Acrudox	DEBF			896	0.10
Kunuweia		Typic Acrudox	DEBO	325	0.02		
	Cotito	Lithic Eutrudox	DECD			280	0.03
Hanamaulu		Humic Kandiudox	DEDK	3,150	0.19		
Puhi		Humic Kandiudox	DEDK	5,205	0.31		
Lawai		Typic Kandiudox	DEDN	664	0.04		
	Daguey	Typic Kandiudox	DEDN			6,393	0.72
	Zarzal	Typic Kandiudox	DEDN			0	0.00
	Rosario	Lithic Hapludox	DEED			1,158	0.13
	Los Guineos	Inceptic Hapludox	DEEH			32,553	3.67
	Limones	Humic Hapludox	DEEL			993	0.11
	Catalina	Rhodic Hapludox	DEEM			108	0.01
	Bayamon	Typic Hapludox	DEEO			9,336	1.05
	Coto	Typic Hapludox	DEEO			5,262	0.59
Total Oxisols				71,750	4.28	59,045	6.66
Total Land Area				1,677,308	100.00	886,216	100.00

ants, however, may play an important role in the formation of Oxisols. Lee and Wood (1971) studied the termite activity in soils and its effect on modifications in the solum. van Wambeke (1992) cites a French study which reports that the amount of soil that termites displace varies between 300 and 1,000 kg ha^{-1} yr^{-1} and that their mounds typically comprise 250 m^3. Leaf cutter ants not only transport soil particles upward but carry large amounts of plant tissue below the surface. A comprehensive account of the role of termites and the mesofauna in tropical pedogenesis has been compiled by van Wambeke (1992).

Relief

Oxisols occur on many landforms, including uplands, backslopes, pediments, interfluves, and river and marine terraces. The one characteristic that the loci of autochthonous Oxisols have in common is geomorphic stability over usually long periods of time. Commonly, this implies slope gradients of 60% or less. Small gradients, however, are not necessarily an indication of stability or geomorphic age as a level surface may be a recent floodplain or a Tertiary peneplain.

Time

The time required for the formation of Oxisols obviously depends on the other pedogenetic factors, notably the weatherability of the parent rock, the composition of transported parent materials, and the climate and its fluctuations over time. There is much scope for variability in these conditions, and consequently, in the age of Oxisols.

For Oxisols developed in transported sediments, their geologic age determines the actual maximum time available for soil formation. In Puerto Rico, for example, Oxisols have developed in preweathered materials on marine terraces of Quaternary age, but the autochthonous Oxisols of the interior commonly occur on Pliocene or Miocene surfaces that have been exposed to subaerial weathering for as long as 15 million years (Beinroth, 1981). Oxisol landscapes of comparable age have been reported in Africa, South America and Australia (Miller, 1983).

6.13.5.2   Processes of Oxisol Formation

Various processes combine to produce, either concurrently or sequentially, the unique features of Oxisols. Prominent among these processes are those that (1) lead to the intensive weathering of primary minerals and the removal of soluble weathering products, the formation of 1:1 lattice clays, and the residual accumulation of sesquioxides, and (2) cause an increase of silicate clay in subsurface horizons. The first process, collectively known as laterization, is of paramount importance in the formation of most Oxisols. It involves desilication, ferrallization, ferritization and allitization and causes the chemical migration of silica out of the solum and the relative concentration of sesquioxides in the soil. Formation of Fe coatings or aggregation at the expense of quartz grains is a common process in the tropical environment (Padmanabhon and Mermut, 1996).

Clay increase with depth that is diagnostic for the kandic horizon may result from: (1) the process known as lessivage that causes clay illuviation and results in a clay maximum or bulge in the subsoil; (2) vertical downward translocation of clay without accumulation in an illuvial horizon. French pedologists refer to this process that leads to lighter textured surface horizons as *appauvrissement* or impoverishment (van Wambeke, 1992); (3) clay depletion in the soil surface may be caused by the selective removal of fine particles from the surface soil by erosion or mesofauna; (4) in Oxisols that are seasonally flooded, ferrolysis can cause the destruction of clay in the topsoil (Brinkman, 1970); (5) as postulated by Simonson (1949), *in situ* formation of clay in the B horizon may occur; and (6) lithological discontinuities may account for textural changes in the solum. In the past, the process of podzolization has been associated with Oxisols, but the evidence for it is intangible.

The formation of plinthite is often considered an extreme manifestation of laterization. The Soil Survey Staff (1998) characterizes plinthite as an Fe-rich, humus-poor mixture of sesquioxides, clay, quartz and other diluents that commonly appear as dark red mottles in platy, polygonal, or reticulate patterns and generally forms in a horizon that is saturated with water for some time during the year. In a moist soil, plinthite is soft enough to be cut with a spade, but it changes irreversibly to ironstone or petroplinthite when exposed to repeated wetting and drying. Plinthite (Eswaran and Raghumohan, 1973) is a diagnostic feature of many soils which are hydromorphic or have gone through a hydromorphic phase during their evolution. Hardening of plinthite takes place slowly when the ground watertable is lowered and the soil surface ground cover is removed. If the surface soil horizons are eroded, the underlying plinthite is exposed and hardens rapidly to form petroplinthite (Sys, 1968; Eswaran and Raghumohan, 1973). A related feature is the development of a petroferric contact, which is an abrupt boundary between soil material and an underlying layer of cemented petroplinthite gravel that is hard and impermeable to both roots and water. The definition, genesis, and kinds of plinthite and petroplinthite have been discussed in detail by van Wambeke (1992).

Gleization which is a process of importance in the formation of some Oxisols, refers to the reduction of Fe and Mn under seasonally anaerobic soil conditions, and produces bluish to greenish gray matrix colors with or without yellowish brown, brown or black mottles and ferric and manganiferous concretions. These redoximorphic features are striking characteristics of the aquic subgroups.

The transformation of raw organic material into soil organic matter (SOM) known as humification occurs in all Oxisols, but is of particular importance in the humic and histic subgroups. A related process is the illuviation of humus that results in the formation of the sombric horizon which is a dark colored subsurface horizon found in soils on old geomorphic surfaces of Central Africa and parts of South America (Eswaran and Tavernier, 1980). Although its origin and genesis are still being debated, it is used as a diagnostic horizon in *Soil Taxonomy* because it is a distinctive feature in an otherwise nondescript soil (Eswaran et al., 1986). Oxisols with sombric horizons are restricted to the cool high plateaus of the tropics at altitudes between 1,400 and 3,000 m above sea level that have isothermic or colder temperature regimes and a udic soil moisture regime (van Wambeke, 1992).

In summary, a broad range of environmental determinants, and pedogenetic processes and mechanisms may be involved in the formation of Oxisols. Yet, there is no single set of formative factors, processes and mechanisms that could account for the formation of all Oxisols. The fact that the causative conditions may not have been the same or may have operated at different intensities over time, and may have occurred simultaneously or sequentially, adds complexity to Oxisol genesis.

6.13.5.3 Landscape Relationships

The geomorphic evolution of the landscape is an important factor that is more important in the formation of Oxisols than in many other kinds of soil. As pointed out by Daniels et al. (1971), the occurrence of Oxisols, as that of other soils, is controlled by the interaction of geomorphic and other formative factors and the resulting rates and degrees of expression of pedogenic processes. Beinroth et al. (1974), Lepsch and Buol (1974), and Lepsch et al. (1977) provide illustrative examples of landscape relationships of Oxisols in Hawaii and Brazil that invariably show that Oxisols occupy geomorphic positions that are older and more stable than the surfaces where Ultisols and Inceptisols occur with which they are geographically associated. However, the recent introduction of the kandic horizon, which may be diagnostic for both Ultisols and Oxisols, has blurred the geomorphic boundary between the two orders.

6.13.6   Properties

6.13.6.1   Macromorphology

Compared with the often strikingly horizonated soils of other orders, the field morphology of most Oxisols is visually rather uniform. They nevertheless have some distinguishing attributes. Color is a prominent feature of Oxisols. The surface horizon of most lowland Oxisols is a thin, light colored ochric epipedon. Oxisols at high elevations (> 1,000 m) frequently have a dark-colored, humus-rich surface horizon (Ruhe, 1956), which may qualify for a mollic or umbric epipedon. Many Oxisols of Central Africa also have a dark colored layer, the sombric horizon, in the subsoil. As Table 6.73 indicates, subsurface colors range from light grey in the Aquox to various hues of red in the upland Oxiols and are generally a function of the Fe content of the original material or rock (Eswaran and Sys, 1970). Color is also related to the kind of Fe minerals, with goethite producing yellow colors and hematite red colors. Presence of colloidal organic matter darkens the soil. If there is a fluctuating groundwater table, mottles or plinthite may form in the oscillation zone. If the soil remains saturated with water for long periods, reduction and removal of the Fe results in a whitish horizon.

The texture of Oxisols may vary from sandy loam to clay. A characteristic feature is that the structural elements are very weak; when the soil is gently pressed between the thumb and forefinger, the material collapses or fails abruptly. This is probably a good field indicator for an oxic horizon.

Many Oxisols have stone lines with the stones being quartz or petroplinthite gravel (Ruhe, 1956). The stone line is a mark of a lithologic discontinuity indicating that the material above the line was deposited or formed at a different period than the material below. Stone lines frequently suggest that the soils are formed on transported deposits and point to the allochthonous nature of the material (van Wambeke, 1992). Some Oxisols have multiple stone lines.

6.13.6.2   Mineralogy and Micromorphology

The unique physical and chemical properties of Oxisols result mainly from the mineralogical composition of the colloidal fraction (Uehara and Keng, 1975; Herbillon, 1980). To have a CEC < 16 $cmol_c$ kg^{-1} clay, the clay fraction must be dominated by low activity clays such as kaolinite. Iron oxyhydroxide minerals, such as goethite, hematite, and ferrihydrite are usually associated with kaolinite, but in some Oxisols, the Fe minerals predominate (Jones et al., 1982), particularly in soils belonging to ferruginous or ferritic families (Table 6.68). The Fe minerals have a high positive charge (Jones and Uehara, 1973), which accounts for the special physical and chemical properties discussed later.

Gibbsite is present as a secondary mineral in many Oxisols (Eswaran et al., 1977). Weathering of primary minerals releases Si and Al; Si is lost in the soil solution while the Al crystallizes as gibbsite, mostly as nodules. The gibbsite crystals in the nodules are well crystallized as shown in the SEM micrograph taken at a magnification of x 2,500 (Fig. 6.74). The crystals are euhedral and twinning is common. Typically, gibbsite crystals have the size of fine silt. Oxisols usually have more gibbsite in the silt than in the clay fraction. It is for this reason that gibbsitic families are defined on the basis of the amount of gibbsite in the fine earth (< 2 mm) fraction.

The other frequent mineral in Oxisols is goethite (Eswaran et al., 1978). Plinthite, laterite or petroplinthite frequently exhibit characteristic forms of goethite aggregates in thin sections as illustrated by the SEM micrograph (x 10,000) in Fig. 6.75. The goethite crystals have a typical lenticular shape and appear welded together, which gives the petroplinthic material its strength. Goethite and hematite have different habits and show different crystal forms.

In most Oxisols, the fabric is homogenous without too many specific entities like those illustrated previously (Buol and Eswaran, 1978). In some oxic horizons, a thin lining of ferriargillans (yellow

Table 6.73 Physical properties of selected pedons representing Oxisol suborders

Classification	Depth (cm)	Horizon	Bulk Density	Water Retention 1/3 bar	15 bar	WRD	Particle Size Sand	Silt	Clay	Soil Color
			g cm⁻³	%			%			
Typic Acraquox (Brazil)	0-10	A1	1.3	32.5	26.9	0.1	33.1	10.5	56.4	10YR 6/1
	10-30	Ag	1.4	21.8	17.2	0.1	34.5	11.6	53.9	10YR 7/1
	30-48	Bog1	1.3	28.7	21.1	0.1	25.1	8.8	66.1	10YR 7/1
	48-77	Bog2	1.3	29.4	23.1	0.1	44.5	13.5	42.0	10YR 8/2
	77-90	Bov	1.4	27.3	21.9	0.1	62.3	11.8	25.9	10YR 5/8
Typic Eutrotorrox (Hawaii)	0-23	Ap			21.5		17.1	40.2	42.7	2.5YR 2/4
	23-50	Bo1	1.3	29.6	21.9	0.1	10.4	41.7	47.9	2.5YR 3/4
	50-87	Bo2	1.4	28.2	22.1	0.1	24.1	30.1	45.8	2.5YR 3/4
	87-123	Bo3	1.4	28.9	21.6	0.1	18.8	34.7	46.5	2.5YR 3/4
	123-150	Bo4	1.3	30.9	20.4	0.1	11.2	39.9	48.9	5YR 3/3
Humic Rhodic Eutrustox (Brazil)	0-25	Ap			23.8		18.0	41.2	40.8	2.5YR 3/2
	25-40	AB	1.2	32.4	23.3	0.1	16.8	37.9	45.3	2.5YR 3/2
	40-64	Bo1	1.2	30.8	24.0	0.1	10.9	25.1	64.0	2.5YR 3/2
	64-110	Bo2	1.1	32.2	24.6	0.1	13.3	28.4	58.3	2.5YR 3/2
	110-210	Bo3	1.1	31.4	24.6	0.1	18.5	38.1	43.4	2.5YR 3/2
Typic Kandiperox (Indonesia)	0-10	Ap1	0.9	42.0	26.5	0.1	13.2	28.9	57.9	5YR 3/3
	10-21	Ap2	0.9	39.5	26.6	0.1	11.4	28.7	59.9	5YR 4/3
	21-51	Bo1	1.0	48.0	31.8	0.2	7.0	20.7	72.3	5YR 3/4
	51-81	Bo2	0.9	50.9	32.6	0.2	6.1	18.5	75.4	5YR 3/4
Anionic Acrudox (Puerto Rico)	0-28	A1	1.1	35.4	26.5	0.3	9.2	36.3	53.8	2.5YR 2/4
	28-46	B1	1.2	26.7	22.8	0.2	7.4	34.9	54.5	2.5YR 2/4
	46-71	Bo1	1.1	34.4	24.8	0.4	9.8	30.6	57.7	2.5YR 2/4
	71-97	Bo2	1.3	35.7	25.9	0.4	23.3	21.0	59.6	7.5YR 3/8
	97-120	Bo3	1.4	31.6	26.4	0.2	17.0	23.3	55.7	7.5R 3/4
	120-155	Bo4	1.3	29.8	24.5	0.1	19.2	27.2	59.7	7.5R 3/4
Humic Sombriudox (Rwanda)	0-15	A	1.0	26.5	11.3	0.2	58.2	8.6	33.2	7.5YR 3/3
	15-40	B1	1.4	15.4	11.1	0.1	56.3	9.3	34.4	7.5YR 2/4
	40-66	Bo1	1.3	17.9	12.4	0.1	52.6	8.9	38.5	7.5YR 2/4
	66-91	Bh1	1.3	21.6	15.2	0.1	42.8	8.4	48.8	7.5YR 2/4
	91-121	Bh2	1.3	22.7	16.7	0.1	43.2	7.1	49.7	7.5YR 2/4
	121-150	Bo2	1.4	19.9	15.4	0.1	42.7	7.3	50.0	7.5YR 2/4

Source: USDA/Natural Resources Conservation Service, National Soil Survey Laboratory

coatings on the void walls) may be present. Ultrathin sections under TEM suggest that the combination of random orientation of clay particles, organic matter, and aggregates of Fe-bearing minerals accounts for the isotropic nature of the aggregates under a petrographic microscope (Santos et al., 1989). The presence of the clay skins is evidence of the transitional nature of the soil and that clay

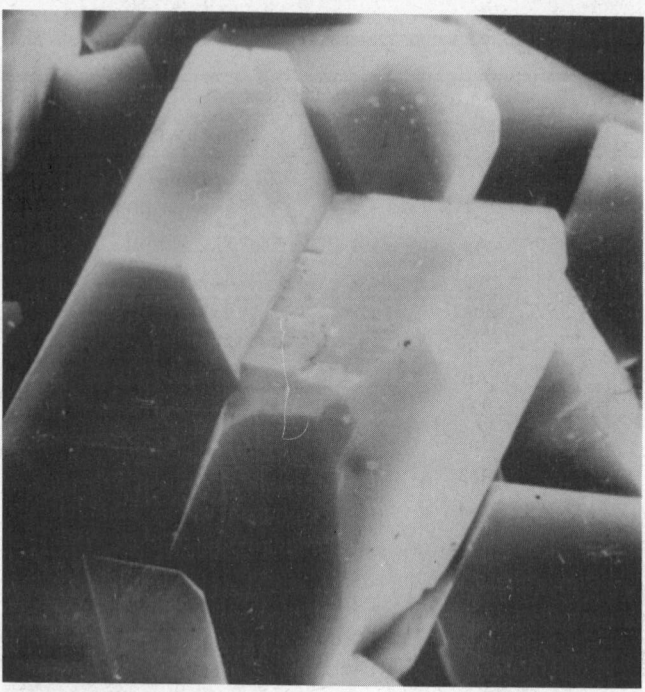

Fig. 6.74 Scanning electron micrograph of a gibbsite nodule in a Gibbsiudox from Malaysia; the gibbsite crystals are typically euhedral and fine and medium silt size; magnification x 10,000.

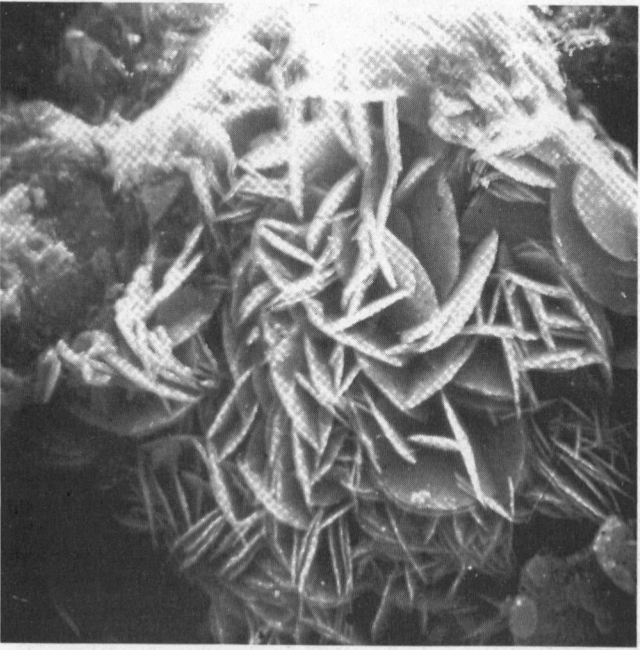

Fig. 6.75 Scanning electron micrograph of goethite in a laterite fragment. The crystals have a lenticular habit with split edges; laterite sample from Angadipuram, Kerela, India, which is the type locality of the Buchanan laterite; magnification x 25,000

illuviation and accumulation was an important process. The transitional nature is indicated by the kandi prefix in the soil name.

Unless eroded, the surface horizons of Oxisols have a relatively high organic matter content with high biological activity including the presence of fungal hyphae and fruiting bodies, which are generally indicators of good soil quality. Typical for the drier tropics is the presence of large termite nests, which can reach 5 m in height. Some species of termites are subsoil dwellers and their galleries may extend several meters into the soil. Bioturbation of the soil is, therefore, an important soil-forming process in tropical soils.

Oxic horizons have a friable consistency. When a large soil clod is gradually crushed in the hands, the material disaggregates and small rounded bodies become evident. These features, which are only observed in Oxisols, have been referred to as pedovites or soil eggs by Belgian pedologists. The excellent and stable structure of Oxisols, and their high macroporosity resulting in rapid infiltration rates, make them resistant to erosion.

6.13.6.3 Chemistry and Physics

The chemistry and physics of Oxisols are inextricably linked to the surface charge characteristics of minerals in the clay fraction. Unlike in most other soils, the surface charge of minerals in Oxisols varies in magnitude and sign. Oxisols, however, are not the only soils dominated by variable charge minerals. Andisols, Ultisols, and Histosols are even better examples of soils with variable surface charge so that the attribute that makes Oxisols unique is the low permanent negative charge of the silicate clay minerals in the clay fraction. Important physical and chemical properties of selected pedons representing Oxisol suborders are summarized in Table 6.74.

The sign of the charge is readily established by the sign of the difference in pH (ΔpH) in the following equation:

$$\Delta pH = pH_{KCl} - pH_{H_2O}$$

Oxisols with a positive ΔpH are rare, but not difficult to find if one knows where to look for them. They are rare because positive ΔpH values almost always occur in the subsoil. They rarely or almost never occur in surface horizons because negatively charged SOM masks the positive charge in the mineral-organic mixture. Organic matter, like the mineral fraction in the oxic horizon, has variable charge characteristics. The difference lies in their respective points of zero charge (p.z.c.) being < pH 3 for organic matter and > 7 for oxic materials. Since soil pH values rarely fall below 3, SOM is net negatively charged in most soils.

The p.z.c. for material in the oxic horizon is highly variable ranging from pH 3 to 6. As a rule, it increases as organic C decreases and the silica/sesquioxide ratio of the clay fraction decreases. Hematite (Fe_2O_3), for example, has a p.z.c. of 8.5 (Parks and de Bruyn, 1962). In this sense, Oxisols are products of desilication, and the end products of desilication are the oxides and hydrous oxides of Fe and Al.

Over 60 yr ago, Mattson (1928, 1932) showed that p.z.c. increased as the silica/sesquioxide ratio decreased. He also showed that the pH of the gel shifted toward the p.z.c. on leaching with distilled water, which he called isoelectric weathering. This concept is useful in explaining the chemical and physical behavior of Oxisols and related soils dominated by low activity clays. This concept is best illustrated with an example, and one of the best examples is the Nipe soil of Puerto Rico and Cuba (Anionic Acrudox in Table 6.72). The isoelectric properties of the Nipe soil are mainly determined by the p.z.c., which is measured as pH_o (pH at which positive and negative charges on variable charge surfaces are equal).

Table 6.75 Chemical properties of selected pedons representing Oxisol suborders

Classification	Horizon	pHo	H₂O	KCl	Δ pH	OC‡	Free Fe	ECEC	CEC 7	CEC 8.2	CEC 7	CEC 8.2	Al Sat. ECEC
						%	%	------cmol$_c$ kg^{-1}------			------%------		%
Typic Acraquox (Brazil)	A1	4.3	4.8	4.5	−0.3	2.4	0.3	1.9	6.8	13.6	3.0	1.0	89.0
	Ag	4.3	4.9	4.6	−0.3	1.6	0.2	1.2	4.8	10.1	4.0	2.0	83.0
	Bog1	5.3	5.5	5.4	−0.1	0.9	0.4	0.1	2.0	6.3	5.0	2.0	0.0
	Bog2	6.8	6.0	6.4	0.4	0.6	1.3	0.2	1.4	5.3	14.0	4.0	0.0
	Bov	6.8	6.0	6.4	0.4	0.4	1.4	0.1	1.0	4.9	2.0	1.0	0.0
Typic Eutrotorrox (Hawaii)	Ap	5.9	6.7	6.3	−0.4	1.0	12.2	8.9	9.4	16.1	95.0	55.0	0.0
	Bo1	6.1	6.5	6.3	−0.2	0.7	11.9	8.1	8.2	14.8	99.0	55.0	0.0
	Bo2	6.3	7.5	6.9	−0.6	0.2	9.9	8.9	9.4	14.9	95.0	60.0	0.0
	Bo3	6.2	6.8	6.5	−0.3	0.2	8.6	9.3	10.2	14.9	91.0	62.0	0.0
	Bo4	6.0	7.0	6.5	−0.5	0.3	8.5	7.9	8.3	13.2	95.0	60.0	0.0
Humic Rhodic Eutrustox (Brazil)	Ap	5.4	6.6	6.0	−0.6	2.8	14.3	15.8	17.6	26.7	90.0	59.0	0.0
	AB	5.9	6.5	6.2	−0.3	2.2	14.7	13.7	15.4	23.6	89.0	58.0	0.0
	Bo1	5.4	6.8	6.1	−0.7	1.2	14.3	8.0	9.0	15.9	89.0	50.0	0.0
	Bo2	5.7	6.9	6.3	−0.6	0.9	14.9	6.1	6.6	13.5	92.0	45.0	0.0
	Bo3	5.9	7.1	6.5	−0.6	0.5	14.8	4.1	4.3	11.1	95.0	35.0	0.0
Typic Kandiperox (Indonesia)	Ap1	4.1	4.9	4.5	−0.4	2.0	5.6	5.9	16.9	25.5	30.0	20.0	15.0
	Ap2	3.8	4.8	4.3	−0.5	1.5	5.7	4.4	15.4	22.9	15.0	10.0	48.0
	Bo1	4.4	5.2	4.8	−0.4	1.9	5.6	5.4	15.1	23.4	34.0	22.0	6.0
	Bo2	4.9	5.3	5.1	−0.2	0.5	5.8	6.0	14.3	21.3	42.0	28.0	0.0
Anionic Acrudox (Puerto Rico)	A1	3.5	5.1	4.3	−0.8	6.0	13.0	7.9	25.4	34.8	11.0	8.0	17.7
	B1	3.8	5.0	4.4	−0.6	2.0	12.9	1.7	12.1	21.5	1.0	0.0	52.9
	Bo1	4.4	5.0	4.7	−0.3	1.3	16.5	0.0	8.2	15.7	0.0	0.0	0.0
	Bo2	6.2	5.2	5.7	0.5	0.9	19.2	0.0	6.4	12.8	0.0	0.0	0.0
	Bo3	6.7	5.5	6.1	0.6	0.7	23.1	0.2	5.3	12.1	2.0	0.0	0.0
	Bo4	7.1	5.7	6.4	0.7	0.6	25.7	0.0	3.8	12.8	0.0	0.0	0.0
Humic Sombriudox (Rwanda)	A	3.3	4.5	3.9	−0.6	1.7	1.2	3.8	10.0	13.7	6.0	4.0	84.0
	B1	3.5	4.5	4.0	−0.5	1.3	1.4	3.0	7.7	12.3	4.0	2.0	90.0
	Bo1	3.4	4.6	4.0	−0.6	1.2	1.8	3.8	9.6	14.6	5.0	3.0	87.0
	Bh1	3.4	4.6	4.0	−0.6	1.4	2.2	5.3	12.3	19.9	7.0	5.0	83.0
	Bh2	3.2	4.6	3.9	−0.7	1.6	2.2	5.9	15.5	25.4	5.0	3.0	88.0
	Bo2	3.2	4.6	3.9	−0.7	1.1	1.8	4.7	11.6	17.3	5.0	3.0	87.0

Source: USDA/Natural Resources Conservation Service, National Soil Survey Laboratory
† ΔpH = pH in KCl minus pH in H₂O, pHo = pH of soil at the zero point of charge where positive and negative charges are equal
‡ OC = Organic carbon content

When pH = 0, pH values measured in M KCl and water are identical indicating that when the material is net negatively charged, pH_o is always lower than pH_{KCl}, and *vice versa*. This does not apply to materials with significant amounts of permanent charge minerals.

The Nipe soil, like all acric Oxisols, is a cation exchanger in the surface horizon and an anion exchanger in the subsoil. Charge characteristics of the Nipe soil and three other Oxisols are illustrated in Fig. 6.76. The anion exchange capacity (AEC) of the subsoil can lead to unexpected consequences. In Hawaii, for example, 3–11 Mg NO_3-N ha^{-1} have been measured in the subsoil and deep saprolite underlying Oxisols and Ultisols (Deenik, 1997), which explains the low NO_3^- levels in the groundwater underlying these soils even after nearly a century of intensive farming. Although the NO_3 remains trapped above the water table, pesticides banned decades ago continue to enter the groundwater. Had the soil minerals been of the permanent charge type, the NO_3-N would have reached the groundwater many years ago.

In cation impoverished Oxisols such as the Nipe soil, lime is often added as a Ca fertilizer rather than an amendment. But because lime raises pH and increases negative charge, Ca^{++} ions remain in the limed layer and do not move to the impoverished subsoil. To circumvent this problem, gypsum and magnesium sulfate are favored over calcite or dolomite whenever the aim is to raise subsoil Ca and Mg. The factors determining movement of surface-applied amendments in variable charge soils are discussed in detail by Sumner (1995).

The acidic organic matter near the surface and the basic oxides in the subsoil determine, in essence, the chemical properties of the Nipe soil. In most Oxisols, the desilication process has not progressed as far as in the Nipe series. In such instances, the subsoil pH will be lower than that of the Nipe soil because the PZC will be lowered by the higher Si content.

An opposing process that counteracts desilication is humification of the desilicated weathering products. Humic acids, like silicic acid, have a PZC below three. At pH levels normally encountered in soils, humus and silica are net negatively charged so that they have strong affinities for positively charged sesquioxides. In one sense, organic matter adsorption on oxide surface has nearly the same effect as resilication of the oxides (Uehara, 1995). The surface of quartz is chemically similar to silicic acid, but its low specific surface renders it virtually inert. Desilicated Oxisols can be rejuvenated by additions of soluble silicates. Large crop responses to additions of calcium silicate above those obtained from similar lime applications have been reported (Plucknett, 1971).

The moisture characteristics curves of well-aggregated Oxisols show two major desorption zones. Water in the interaggregate pores drains rapidly between 0 and -0.01MPa. Another desorption zone occurs when the intraaggregate pores begin to drain at about -15MPa (Sharma and Uehara, 1968a, 1968b), which results in a bimodal pore size distribution (Tsuji et al., 1975; Bui et al., 1989). Some have referred to Oxisols as behaving like aggregated sands because of their stable aggregates and high macroporosity. This description tells only half the story. The aggregates that remain nearly water saturated beyond the wilting point impart additional properties to Oxisols. The water-saturated aggregates increase volumetric heat capacity and lower thermal diffusivity. Crops such as pineapple respond to practices that raise subsoil temperature. In Hawaii, plastic sheets used to increase the effectiveness of soil fumigants also aid the crop by raising subsoil temperature (Ekern, 1967). The intraaggregate water retained at high negative pressures also affects tillage operations. This intraaggregate water is freed under the shearing action of tillage implements and causes the soil to adhere to the implement. Some farmers solve this problem by bolting a teflon sheet onto the implement's shearing surface.

In summary, the physics and chemistry of Oxisols are strongly influenced by the extent to which desilication has occurred. Basic and ultrabasic parent materials readily desilicate and produce soils

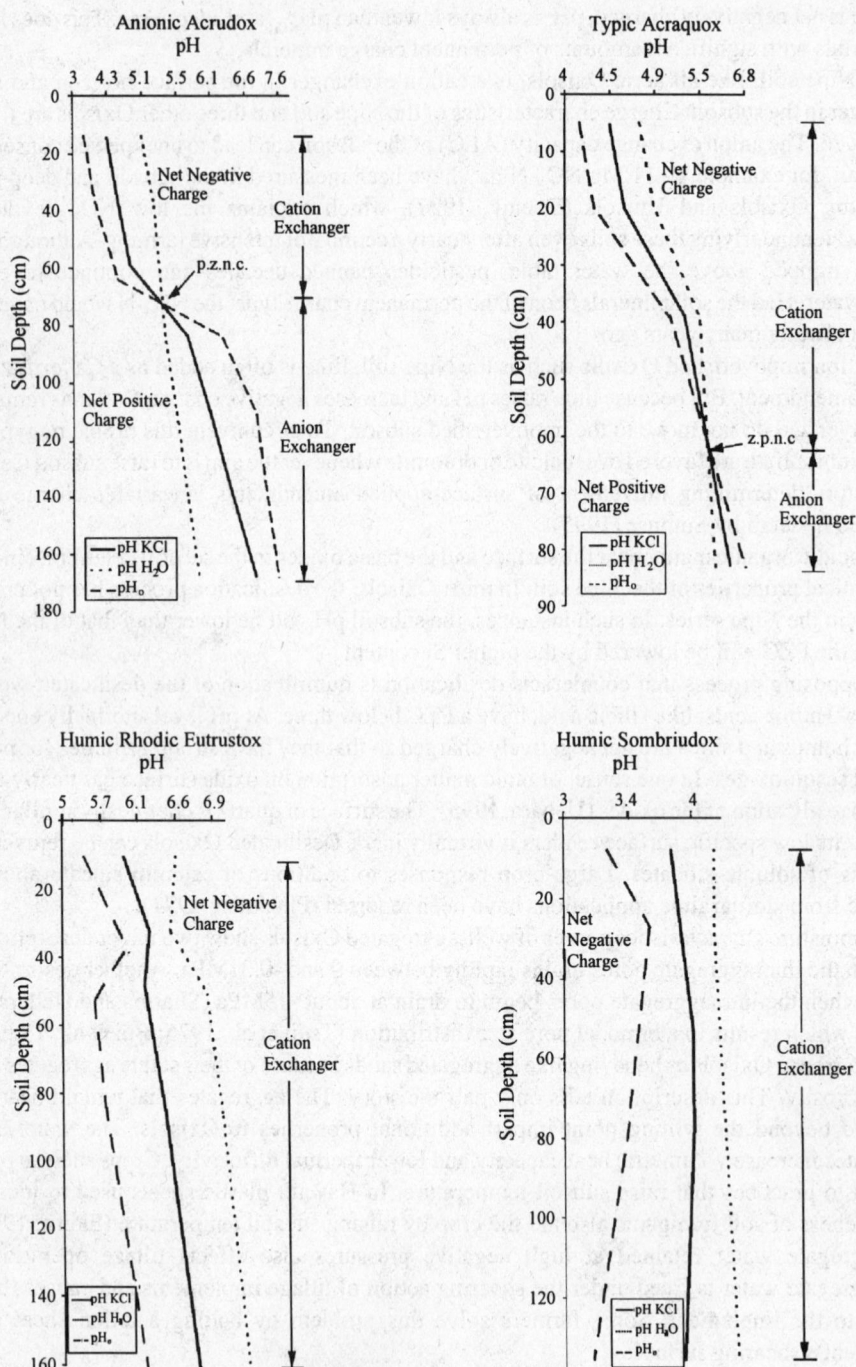

Fig. 6.76 Surface charge characteristics of Oxisols of different suborders

that approach the central concept of Oxisols. Desilication is also aided by warm and humid conditions so that when Oxisols occur outside the tropics, they are almost always associated with ultrabasic rocks.

6.13.7 Conclusions

Oxisols occupy about 25% of the land area of the tropics where they are the single most extensive soil type. Yet, historically they have been perceived as agriculturally unproductive and problematic for management; under low input agriculture, yields are in fact low, risk is high, and the potential for land degradation is also high (Sanchez and Salinas, 1981). The negative notion of the poor agronomic performance derives substance from the inherent chemical constraints of Oxisols, which include, to varying degrees, a low nutrient retention capacity, anion adsorption, Ca deficiency, and Mn and Al toxicity (van Wambeke, 1974; Sanchez, 1976). It is also true that the inputs required to correct these constraints may be economically prohibitive for many farmers, and in places where better endowed soils are available, the Oxisols are, therefore, at a distinct comparative disadvantage (van Wambeke, 1992). Nevertheless, with science-based management that employs the tools and techniques of modern agriculture, Oxisols can be managed to be both economically and sustainably productive. Consequently, no less an authority than Charles E. Kellogg stated that "some day the most productive agriculture of the world will be mostly in the tropics, especially in the humid parts" (Kellogg, 1967). Presumably, this assessment is based on the realization that the favorable physical attributes of Oxisols outweigh their chemical limitations. While the latter can be amended with purchased inputs, good soil structure cannot be bought.

Oxisols are the dominant soils in the humid tropical forest ecosystem, a pristine environment that constitutes an enormous reservoir of sequestered C and unique ecological niches of great biological diversity, in addition to being a resource for food, timber, medicine, and other products for people. Yet, the resource-poor farmers invading these areas practice shifting cultivation, and slash and burn agriculture has become the most extensive form of agriculture in the tropics. As a result, over 15 million ha of forests are being burned annually and some plants and animals are lost permanently when their habitats are destroyed. Moreover, the resilience of these ecosystems is so low that complete regeneration may not be achieved.

Oxisols constitute a major land resource and one of the few remaining frontiers for agricultural development, particularly in Africa and South America, and also support the largest areas of tropical forests. Their conversion to agricultural land will be at the expense of the forest with the concomitant loss of biodiversity and negative impact on global climate. It is imperative, therefore, to develop viable alternatives to traditional agricultural systems. If this challenge can be met successfully, the rewards are not only to provide a means for millions of people to extricate themselves from poverty, but also to ensure the survival of the tropical forests. Landuse policies guided by an understanding of the nature, properties, and ecological functions of Oxisols are critical to sustain the integrity and productivity of these land resources.

6.13.8 References

Alexander, L.T., and J.G. Cady. 1962. Genesis and hardening of laterite in soils. USDA Tech. Bull. 1282.

Aubert, G. 1958. Classification des sols. Compte Rendu, Reunion Sous-comité, Brazzaville, Congo.

Beinroth, F. H., G. Uehara, and H. Ikawa. 1974. Geomorphic relationships of Oxisols and Ultisols on Kauai, Hawaii. Soil Sci. Soc. Am. Proc. 38:128–131.

Beinroth, F. H. 1981. Some highly weathered soils of Puerto Rico. 1: Morphology, formation and classification. Geoderma 27:1–73.

Beinroth, F.H., and S. Paramanathan (ed.). 1978. Proceedings of the Second International Soil Classification Workshop. Part I, Malaysia. Department of Land Development, Bangkok, Thailand.

Beinroth, F.H., and S. Panichapong (ed.). 1978. Proceedings of the Second International Soil Classification Workshop. Part II, Thailand. Dept. of Land Development, Bangkok, Thailand.

Beinroth, F.H., H. Neel, and H. Eswaran (ed.). 1983. Proceedings of the Fourth International Soil Classification Workshop, Rwanda. Part 1: Papers. Part 2: Field trip background and soil data. Agric. Editions 4, ABOS-AGDC, Brussels, Belgium.

Beinroth, F.H., M.N. Camargo, and H. Eswaran (ed.). 1986. Proceedings of the Eighth International Soil Classification Workshop: Characterization, Classification, and Utilization of Oxisols. EMBRAPA-SNLCS, Rio de Janeiro, Brazil.

Bennema, J., R.C. Lemos, and L. Vetturs. 1959. Latosols in Brazil. III Inter-African Soils Conference, Dalaba. I:273–281.

Botelho da Costa, J.V. 1954. Sure quelques questions de nomenclature des sols des regions tropicales. Compte Rendu Conf. Int. Sols Africains, Leopoldville, Congo. 2:1099–1103.

Botelho da Costa, J.V. 1959. Ferralitic, tropical fersiallitic and tropical semi-arid soils: Definitions adopted in the classification of the soils of Angola. III Inter-African Soils Conference, Dalaba. I:317–319.

Brinkman, R. 1970. Ferrolysis, a hydromorphic soil forming process. Geoderma 3:199–206.

Buchanan, F. 1807. A journey from Madras through the countries of Mysore, Kanara and Malabar. East India Co., London, 2:436–461.

Bui, E.N., A.R. Mermut, and M.C.D. Santos. 1989. Microscopic and ultramicroscopic porosity of an Oxisol as determined by image analysis and water retention. Soil Sci. Soc. Am. J. 53:661–665.

Buol, S.W., and H. Eswaran. 1978. Micromorphology of Oxisols. p.325–328. In M. Delgado (ed.), Proc. Vth. Int. Work. Meet. Soil Micromorph., Granada, Spain.

Buol, S.W., and H. Eswaran. 1988. International Committee on Oxisols: Final Report. USDA-SCS Soil Management Support Services, Tech. Monogr. 17. Washington, DC.

Camargo, M.N., and F.H. Beinroth (ed.). 1978. Proc. First International Soil Classification Workshop. EMBRAPA-SNLCS, Rio de Janeiro, Brazil.

Charter, C.F. 1958. Report on the environmental conditions prevailing in Block A, Southern Province, Taganyika Territory, with special reference to the large-scale mechanized production of ground-nuts. Ghana Department of Soil and Land Use Survey, Occasional Paper 1.

Cline, M.G. 1975. Origin of the term Latosol. Soil Sci. Soc. Amer. Proc. 39:162.

Daniels, R. B., E.E. Gamble, and J.G. Cady. 1971. The relation between geomorphology and soil morphology and genesis. Adv. Agron. 23:51–88.

Deenik, J. 1997. Liming effects on nitrate adsorption in soils with variable charge clays, and implications for groundwater contamination. M.S. Thesis, University of Hawaii, Honolulu, HI.

Ekern, P.C. 1967. Soil moisture and soil temperature changes with the use of black vapor barrier mulch and their influence on pineapple (Ananas comosus, (l) Merr.) Soil Sci. Soc. Amer. Proc. 31:270–275.

Eswaran, H., and N.G. Raghumohan. 1973. The micro-fabric of petroplinthite. Soil Sci. Soc. Amer. Proc. 37:79–81.

Eswaran, H., G. Stoops, and C. Sys. 1977. The micromorphology of gibbsite forms in soils. J. Soil Sci. 28:136–143.

Eswaran, H., and C. Sys. 1970. An evaluation of the free iron in tropical basaltic soils. Pedologie 20:62–85.

Eswaran, H., C.H. Lim, V. Sooryanarayanan, and N. Daud. 1978. Scanning electron microscopy of secondary minerals in Fe-Mn glaebules. p. 851–866. In M. Delgado (ed). Proc. Vth. Inter. Work. Meet. Soil Micromorph., Granada, Spain.

Eswaran, H., and R. Tavernier. 1980. Classification and genesis of Oxisols. p. 427–442. In B.K.G. Theng (ed.) Soils with variable charge. New Zealand Soil Science Society, Lower Hutt, New Zealand.

Eswaran, H., H. Ikawa, and J.M. Kimble. 1986. Oxisols of the world. p. 90–123. In Proc. Int. Symp. Red Soils. Publ. Science Press, Beijing, China.

FAO-UNESCO, 1971–1976. Soil Map of the World. FAO, Rome, Italy.

Harrassowitz, H. 1930. Böden der Tropischen Region. Laterit und allitischer Rotlehm. E. Blanck (ed.) Handbuch der Bodenlehre. 3:387–536.

Herbillon, A. 1980. Mineralogy of Oxisols and oxic materials. p. 109–126. In B.K.G. Theng (ed.) Soils with variable charge. New Zealand Society of Soil Science, Lower Hutt, New Zealand.

Jones, R.C., W.H. Hudnall, and W.S. Sakai. 1982. Some highly weathered soils of Puerto Rico, Part 2: Mineralogy. Geoderma 27:75 –137.

Jones, R.C., and G. Uehara. 1973. Amorphous coatings on mineral surfaces. Soil Sci. Soc. Amer. Proc. 37:792–798.

Kellogg, C.E. 1949. Preliminary suggestions for the classification and nomenclature of great soil groups in tropical and equatorial regions. CAB Soil Sci. Tech. Commun. 46:76–85.

Kellogg, C.E., 1967. Comment. p. 232–233. *In* H.M. Southworth and B.F. Johnston (ed.) Agricultural development and economic growth. Cornell University Press, Ithaca, NY.

Lee, K. E., and T.G. Wood. 1971. Termites and soils. Academic Press, London, UK.

Lepsch, I.F., and S.W. Buol. 1974. Investigations in an Oxisol-Ultisol toposequence in Sao Paulo State, Brazil. Soil Sci. Soc. Amer. Proc. 38:491–496.

Lepsch, I.F., S.W. Buol, and R.B. Daniels. 1977. Soils-landscape relationships in the Occidental Plateau of Sao Paulo State, Brazil. Soil Sci. Soc. Amer. J. 41:109–115.

Maignen, R. 1966. Review of research on laterite. UNESCO Paris, France.

Mattson, S. 1928. The electrokinetics and chemical behavior of the aluminosilicates. Soil Sci. 25:289–311.

Mattson, S. 1932. The laws of soil colloidal behavior: IX. Amphoteric reactions and isoelectric weathering. Soil Sci. 34:209–240.

Miller, B.J. 1983. Ultisols. p.283–323. *In* L.P. Wilding, N.E. Smeck, and G.F. Hall (ed.) Pedogenesis and soil taxonomy, II. The soil orders. Elsevier Scientific Publishing Co., Amsterdam, Netherlands.

Moormann, F.R. 1985. Excerpts from the circular letters of the International Committee on Low Activity Clay Soils (ICOMLAC). USDA-SCS Soil Management Support Services, Tech. Monogr. 8.

Padmanabhon, E. and A.R. Mermut. 1996. Submicroscopic structure of Fe coatings on quartz grains in tropical environments. Clays Clay Min. 44:801–910.

Parks, G. A., and P.L. de Bruyn. 1962. The zero point of charge of oxides. J. Phys. Chem. 66:967–973.

Plucknett, D.L. 1971. The use of soluble silicates in Hawaiian agriculture. Univ. of Queensland Paper 1:203–223.

Robinson, G.W. 1951. Soils: Their origin, constitution, and classification. Thomas Murby and Sons, London, UK.

Ruhe, R.V. 1956. Landscape evolution in the High Ituri, Belgian Congo. INEAC Ser. Sci. 66.

Sanchez, P., and J.G. Salinas. 1981. Low-input technology for managing Oxisols and Ultisols in tropical America. Adv. Agron. 34:279–406.

Sanchez, P. 1976. Properties and management of soils in the tropics. John Wiley and Sons, New York, NY.

Santos, M.C.D., A.R. Mermut, and M.R. Ribeiro. 1989. Submicroscopy of clay microaggregates in an Oxisol from Pernambuco, Brazil. Soil Sci. Soc. Am. J. 53:1895–1901.

Sharma, M.L., and G. Uehara. 1968a. Influence of soil structure on water relations in Low Humic Latosols: I. Water retention. Soil Sci. Soc. Amer. Proc. 32:766–770.

Sharma, M.L., and G. Uehara. 1968b. Influence of soil structure on water relations in Low Humic Latosols: II. Water Movement. Soil Sci. Soc. Amer. Proc. 32:770–774.

Simonson, R. W. 1949. Genesis and classification of Red-Yellow Podzolic soils. Soil Sci. Soc. Am. Proc. 14:316–319.

Smeck, N. E., E.C. A. Runge and E.E. Mackintosh. 1983. Dynamics and genetic modelling of soil systems. p. 51–81. *In* L. P. Wilding, N. E. Smeck, and G. F. Hall (ed.) Pedogenesis and Soil Taxonomy, I. Concepts and interactions. Elsevier Scientific Publishing Co., Amsterdam, Netherlands.

Smith, G.D. 1963. Objectives and basic assumptions of the new classification system. Soil Sci. 96:6–16.

Smith, G.D. 1965. Lectures on soil classification. Pedological Society Spec. Bull. 4. Ghent, Belgium.

Soil Survey Staff. 1975. Soil taxonomy: A basic system of soil classification for making and interpreting soil surveys. USDA Agric. Handb. 436, US Government Printing Office, Washington, DC.

Soil Survey Staff. 1998. Keys to soil taxonomy, 8th Ed. 1996. USAD-NRCS. US Government Printing Office, Washington, DC.

Sumner, M.E. 1995. Amelioration of subsoil acidity with minimum disturbance. p. 147–186. *In* N.S. Jayawardane and B.A. Stewart (ed.) Subsoil management techniques. Lewis Publishers, Boca Raton, FL.

Sys, C. 1968. Suggestions for the classification of tropical soils with lateritic materials in the American classification. Pedologie 18:189–198.

Tavernier, R., and C. Sys. 1965. Classification of the soils of the Republic of Congo. Pedologie VOL:91–136.

Tavernier, R., and H. Eswaran. 1972. Basic concepts of weathering and soil genesis in the humid tropics. 2nd. ASEAN Soils Conf.1:383–392. Jakarta, Indonesia.

Tsuji, G.Y., R.T. Watanabe, and W.S. Saki. 1975. Influence of soil microstructure on water characteristics of selected Hawaiian soils. Soil Sci. Soc. Amer. Proc. 39:28–33.

Uehara, G., and J. Keng. 1975. Management implications of soil mineralogy in Latin America. p. 351–363. *In* E. Bornemizsa and A. Alvarado (ed.) Soil management in tropical America. NC State University, Raleigh, NC.

Uehara, G. 1995. Management of isoelectric soils of the humid tropics. p. 271–278. *In* R. Lal, J. Kimble, E. Levine and B.A. Stewart (ed.) Soil management and greenhouse effect. CRC Press, Inc., Boca Raton, FL.

van Wambeke, A. 1974. Management properties of Ferralsols. FAO Soils Bull. 23.

van Wambeke, A., H. Eswaran, A.J. Herbillon, and J. Comerma. 1983. Oxisols. p. 325–354. *In* L.P. Wilding, N.E. Smeck, and G.F. Hall (ed.) Pedogenesis and Soil Taxonomy, II. The Soil Orders. Elsevier Scientific Publishing Co., Amsterdam, Netherlands.

van Wambeke, A. 1992. Soils of the tropics: Properties and appraisal. McGraw-Hill, New York, NY

Wilding, L.P. 1994. Factors of soil formation: Contributions to pedology. p. 15–30. *In* R.J. Luxmoore (ed.). Factors of soil formation: A fiftieth anniversary perspective. Soil Sci. Soc. Am. Spec. Publ. 33. Soil Science Society of America, Madison, WI.

7

Land Evaluation for Landscape Units

J. Bouma
Wageningen Agricultural University

7.1 Introduction

Considering the increasing world population, recent international studies have explored the potential world food supply and associated environmental quality issues, among them land degradation (Penning de Vries et al., 1995). At another scale, new farming systems are being developed using modern information technology, summarized as Precision Agriculture (Robert et al., 1996; Ciba Foundation, 1997). Regional planners consider alternative land use scenarios combining agricultural production with nature conservation and multifunctional land use. In all cases, use of land forms the core of the problem to be studied, and even though socioeconomic considerations play a crucial role, properties of the land are still of prime interest. Consideration of actual and potential land use as a function of land properties fits under the broad umbrella of Land Evaluation as advocated by FAO (1976, 1983). Land evaluation can be realized with descriptive, qualitative methods, but increasingly quantitative simulation models for crop production and solute fluxes are used and they need to be fed with representative soil data. This often occurs mechanistically with no attention to natural soil dynamics or landscape relationships. Soil survey has much to offer, but this expertise has to be packaged and presented more effectively than at present. Aside from the variation in space, there is also the need to consider variation in time, be it days, growing seasons or decades. Increasingly, land evaluation is realized in close interaction with the stakeholders, ranging from farmers to planners and politicians. Innovative developments in soil survey and its evaluation, emphasizing interdisciplinary approaches at different spatial and temporal scales will be discussed.

7.2 Developments in Land Evaluation

7.2.1 Definitions

Land evaluation is defined as follows (FAO, 1976): the process of assessment of land performance when used for specified purposes, involving the execution and interpretation of surveys and studies of land use, vegetation, landforms, soils, climate and other aspects of land in order to identify and make a comparison of promising kinds of land use in terms applicable to the objectives of the evaluation.

Clearly, evaluating the performance of land has been the topic of many studies in the past in different disciplines. There is, for instance, a large literature on land use planning from a regulatory and social perspective, but here the focus will be on agroecology. Adopting the international FAO approach has the advantage of using a widely known procedure in which definitions are so broad that they offer many opportunities for expansion and modification. Two elements stand out when considering the definition of land evaluation: (1) Performance can only be assessed when a specific land use has been defined. In other words, judgement cannot be made as to performance in general, but only for specific types of land use. Land may, for instance, function quite well as a campground but poorly when growing a wheat crop. (2) Attention is not only paid to current land use, but also to potential forms of land use, which may be more or less promising depending on the objectives of the evaluation.

Land is defined as (FAO, 1976): an area of the Earth's surface, the characteristics of which embrace all reasonably stable, or predictably cyclic attributes of the biosphere, vertically above and below this area including those of the atmosphere, the soil and underlying geology, the hydrology, the plant and animal populations and the results of past and present human activity to the extent that these attributes exert a significant influence on present and future uses of the land by man.

A land unit (FAO, 1983) is an area of land possessing specified land characteristics and/or land qualities which can be demarcated on a map. Land is often represented as a georeferenced land unit on soil maps, which present additional data on climate in a soil survey report.

The broad term, Land Use, is specified in terms of Land Utilization Types (LUTs) which define a particular type of land use in varying degrees of detail, but usually including listings of inputs and outputs. In the context of rainfed agriculture, a LUT refers to a crop, crop combination or cropping system with a specified technical and socioeconomic setting (FAO, 1983). When combining the Land Unit with the LUT, one obtains the so-called Land Use System (LUS) defined as: a specified LUT practiced on a given land unit and associated with inputs, outputs and possible land improvements (FAO, 1983). Recent work in Costa Rica has defined such LUSs in terms of the type of technology (T) being used in each particular production system. They refer, therefore, not to a LUS but to a LUST, an example of which is provided in Table 7.1.

A key element in land evaluation is the assessment of land performance. This is done, in principle, by comparing the requirements of a particular type of land use with what the land has to offer. When the two match, the land is suitable for a particular LUST. When they do not to varying degrees, suitability is less. Land suitability is correspondingly defined as: the fitness of a given type of land for a specified kind of land use (FAO, 1976). This matching process, which is central in land evaluation, is handled by defining land qualities and land characteristics. Land qualities are complex attributes of

Table 7.1 A listing of data documenting a Land Utilization Type with Specified Technology (LUST) for maize, sown 15 January on a fertile, well-drained soil, typical for the Neguev area of the Atlantic Zone of Costa Rica. [Adapted from Jansen and Schipper, 1995]

Operation	Date	Labor	Equipment	Materials
Land preparation	1/31/91	20 hours	machete	
Herbicide application	1/2/92	10 hours	knapsack sprayer	2 L Gramoxome
Sowing	1/15/92	10 hours	planting stick	20 kg local variety maize seed
Fertilizer application	1/30/92	10 hours		50 kg ammonium nitrate
Harvest	5/15/92	50 hours		100 bags dry cobs

land which act in a manner distinct from the actions of other land qualities in its influence on the suitability of land for a specified kind of use. The matching process is realized by expressing both land use requirements and what the land has to offer in terms of land qualities, and by comparing the two expressions.

Although this is not mentioned by FAO (1976), land qualities usually cannot be measured directly. Examples are the moisture supply capacity, the workability and the trafficability of land. They vary over the years and are determined by land behavior over extended periods of growing seasons. Modern methods will be discussed later to determine land qualities. However, in the older land evaluation work, attempts were made to find proxies for land qualities (Land Characteristics) which are attributes of land that can be measured or estimated. One may think of texture, organic matter, carbonate content, etc. in this context.

The basic elements of classical land evaluation have now been introduced. One has land which is being used for a particular purpose in a particular way. Not only current land use should be looked at, but also other possible forms of land use which are of interest. One wants to assess land performance for these different alternative forms of land use, and does so by comparing land requirements for each alternative form of land use with what the land has to offer. This matching process is made possible by defining important land qualities, often defined in terms of land characteristics in different classes. The overall analysis results in statements as to relative suitabilities of a given piece of land for a series of land use systems. Much practical experience with this system is reflected in the work of Sys et al. (1991) who provide excellent case studies. Some examples to further illustrate the procedure and to discuss some underlying concepts will be examined.

7.2.2 An Example

An example given by FAO (1983) addresses land suitability for sorghum (Table 7.2) and may serve to illustrate some basic decision steps to be taken which have general validity. Suitability of the land unit being considered is expressed in terms of highly, moderately, marginally and not suitable. Three land qualities are distinguished: rooting conditions, oxygen and nutrient availability. This selection reflects the expert judgement of the land evaluator. Quite possibly, other evaluators would have selected other land qualities. Because land qualities, as mentioned here, cannot be measured directly, land characteristics are used as proxies, called Diagnostic Factors and are drainage class (soil survey reports), effective soil depth (to be estimated from structure descriptions in soil survey reports) and soil reaction (pH). Clearly, some selections reflect lack of good data. Soil drainage classes are very broadly defined and only remotely related to O_2 diffusion. The effective soil depth is more direct but does not consider any particular demands by sorghum; it could apply to any crop. The pH, finally, is

Table 7.2 Land qualities and land characteristics (diagnostic factors) for a sorghum Land Utilization Type (LUT), expressed in terms of crop requirements [From FAO, 1983]

Crop Requirement			Factor Rating			
Land Quality	Diagnostic Factor	Unit	Highly suitable s1	Moderately suitable s2	Marginally suitable s3	Not suitable n
Oxygen availability	Soil drainage class	Class	Well drained/ Excessive	Moderately well drained	Imperfectly drained	Poorly/Very poorly drained
Rooting conditions	Effective soil depth	cm	> 120	50-120	30-50	< 30
Nutrient availability	Soil reaction	pH	5.5-7.5	4.8-5.5 and 7.5-8.0	4.5-5.5 and 8.0-8.5	< 4.5 and > 8.5

a very general indicator for nutrient availability and its selection reflects an apparent desperate lack of information on the soil fertility status. Finally, the evaluator has to couple classes of the diagnostic factors somehow to demands by the sorghum crop (crop requirements) to arrive at the suitability classes. In summary, there are three moments when important decisions have to be made based on expert knowledge and data from the literature: (1) selection of land qualities, (2) selection of land characteristics to describe the land qualities, and (3) selection of gradations of these characteristics to form suitability classes based on estimated crop requirements.

7.3 Beyond Classical Land Evaluation

The classical land evaluation scheme has been widely applied and, in many cases, successfully. Its application has been facilitated by automated computer-driven decision support systems (Rossiter, 1990). Four problems have, however, become clear over the years: (1) Even though the need to define objectives for any land evaluation has been stressed from the start, the development of a mechanistic approach has been observed (Sys, 1991) in which land suitability is defined for a large number of LUSTs; it is not clear who is asking the questions or, worse, whether the answers being provided address the questions that are really being raised. (2) Defining land qualities in terms of land characteristics has become a rather rigid qualitative procedure, even, and particularly, in automated computer-driven decision support systems, allowing little input from modern process-driven land research. (3) The procedure is almost exclusively driven by the properties of the land, and even though the importance of socioeconomic conditions is acknowledged, little is done to take these conditions into account. Land use and its possible changes are usually more a reflection of socioeconomic developments in society than of differences in soil suitabilities for different forms of land use. Moreover, land units, as distinguished in earth science, hardly ever correspond with legal units on which decisions are made. For example, a farmer farms a field with different land units most often, not a single land unit. A district or county where land use decisions may be made may cut through different land units, etc. (4) The procedure was implicitly defined as being scale independent. Most applications of classical land evaluation have been at the regional level, but many land use questions are raised at farm or field level or at the continental or world level. Not only are the questions then quite different [see (1) above] but procedures to be followed should be different as well.

What is now needed is a better evaluation of questions being asked at different spatial scales. Next, proper procedures need to be defined to deal with these questions, realizing a wide variety of stakeholders are involved. And, finally, the proper phase of the land in determining land use decisions needs to be defined. Certainly, decisions are not made for land units, but for georeferenced surfaces on the earth that may contain many land units.

7.4 What is the Question?

Current land use patterns in Europe are more a reflection of the common agricultural policy of the European Union than of relative suitabilities of different land units for different types of land use. Questions vary a great deal. Let us analyze some of them. (1) A farmer wants to know how he can obtain a high yield of profitable crops at minimal cost. Cutting costs to be achieved by, for example, precision application of fertilizers and biocides or minimum tillage is increasingly important. He certainly is not interested to hear that his land is moderately suitable for wheat growing. He will already know that. He wants quantitative information in terms of what to do, and when, where, and how to do it. (2) An environmentalist will ask how use of agrochemicals can be reduced or even abolished, thus avoiding leaching into ground or surface water. To answer such questions, detailed process-based simulation models may have to be applied to estimate the adsorption of agrochemicals as a function of management. Of course, the ideal is to combine the desires under (1) with those under (2), which is the basic concept of precision agriculture. (3) A regional planner may want to formulate alternative land use options within a region. Here, the existing land evaluation procedure may be useful but data are not specific enough to allow quantitative tradeoffs among the various options. (4) A policy maker may see options for land use in a region or a country, formulated by procedures under (3). His question, however, is how attractive options can indeed be realized? How can stakeholders be influenced to do the right thing? Special taxes, bonuses, subsidies? Clearly, land evaluation does not primarily focus on such issues, but they are increasingly important for land use and should, therefore, be considered.

Considering the range of questions that may be encountered, one may distinguish three broad approaches that have been used to answer them. The approaches can be applied, in principle, at any hierarchal scale: (1) The prediction of future land use based on extrapolation of existing trends. The type of question to be answered is: What will be the likely land use changes if trends in land use are extrapolated to the near future? The past is used as a measure for the future; optimization of future land use, considering tradeoffs between contrasting objectives, is not possible and land use changes may be predicted that are not feasible from a biophysical point of view, particularly when applied at regional or higher level (Veldkamp and Fresco, 1996). (2) The exploratory approach which defines a number of realistic land use options for the area to be considered. The stakeholder makes a choice. The type of question to be answered is: What are the options for land use? How can one optimize land use for certain objectives and what are the tradeoffs between these objectives? Whether or not such options are realized depends on the stakeholders. The exploratory approach does not predict, but explores what has been called a window of opportunity. The approach is sometimes criticized because what is agroecologically possible may never be realized in agricultural production systems where socioecenomic factors play a major role, particularly when applied at regional and higher level (FAO, 1976; Latesteijn, 1995). (3) Identification of policy instruments to realize particular land use options. The type of question to be answered: What are effective policy instruments to induce changes in land use applied at all scales? An example for the farm level was reported by Kruseman et al. (1995).

Each of the three approaches mentioned above can be implemented in different ways, but one way is to develop decision support systems which take the stakeholder by the hand and follow the necessary procedures step by step. This will help to implement the procedures, but it is not enough to publish a study without actively involving the stakeholders, be they farmers, planners or politicians. Decision support systems can be applied in the three cases, but most applications are at farm level (Bouma, 1997a). Bouma (1993) has discussed five case studies on land use at different scales, showing that each was associated with different questions, while the procedures to be used in answering the questions varied very much. The different procedures that one can use to answer the stakeholders will now be discussed. Note that these procedures define selections of methods which fit into the extrapolative, exploratory and predictive modes mentioned above.

7.5 What is the Proper Procedure?

A diagram (Fig. 7.1) has been helpful to illustrate various research procedures (Hoosbeek and Bryant, 1992; Bouma and Hoosbeek, 1996) (Section E, Chapter 3). They considered two perpendicular axes, one ranging from qualitative to quantitative and the other from empirical to mechanistic. Different research approaches occur within the plane thus obtained: K1 represents user knowledge; K2 represents expert knowledge; K3 represents knowledge to be obtained through semiquantitative models, in which real soil processes are not known; K4 represents knowledge through quantitative

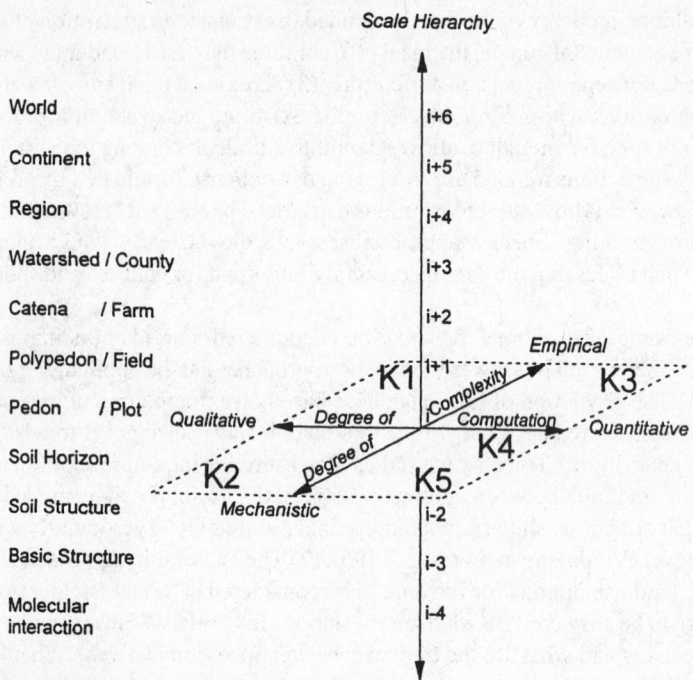

Fig. 7.1 Scale diagram showing a series of hierarchial scales (i levels) and modeling approaches expressed in terms of four characteristics, which are summarized in terms of knowledge levels K1-K5 [Reprinted from Bouma and Hoosbeek, 1996. Soil Sci. Soc. Am. Spec. Pub. 45:1-15 with permission of the Soil Science Society of America].

models where processes are characterized in general terms; and K5 represents the same, but processes are described in great detail which can imply that the entire soil/crop system cannot be characterized anymore and attention is focused on one aspect only. The vertical axis represents the scale hierarchy, where the pedon level (the individual soil) occupies the central position (i level). Higher levels are indicated as i+, while lower levels are i–. The scale in Fig. 7.1 ranges from molecular interaction (i–4) to the world level (i+6).

One can now place the classic land evaluation in the scheme at scale hierarchy of watershed or region while the knowledge level is K2. For other questions raised above, one needs different hierarchies and knowledge levels. For example, the farmer and environmentalist would require the i+1 field scale and a K4 knowledge level to get the necessary quantitative answers for their questions. The regional planner would operate at level i+4 and would need a K3 knowledge level, because the K2 level would be too descriptive not allowing a quantitative tradeoff analysis. A planner would be smart, though, to combine K2 with K3, by restricting the more detailed analyses of K3 to areas where a simpler K2 analysis could not provide answers. For example, Van Lanen et al. (1992) made a land evaluation for Europe in which potential for crop growth was established. They first screened out strongly sloping land and land with shallow bedrock using a K2 approach. Then, in the remaining 40% of the land area, they ran a K3 simulation model to predict crop growth. The K2 approach for these soils (land is moderately suitable for wheat) would not have been satisfactory. This scheme introduces the possibility of combining approaches. A last example will be provided to further illustrate this attractive application of the scale diagram.

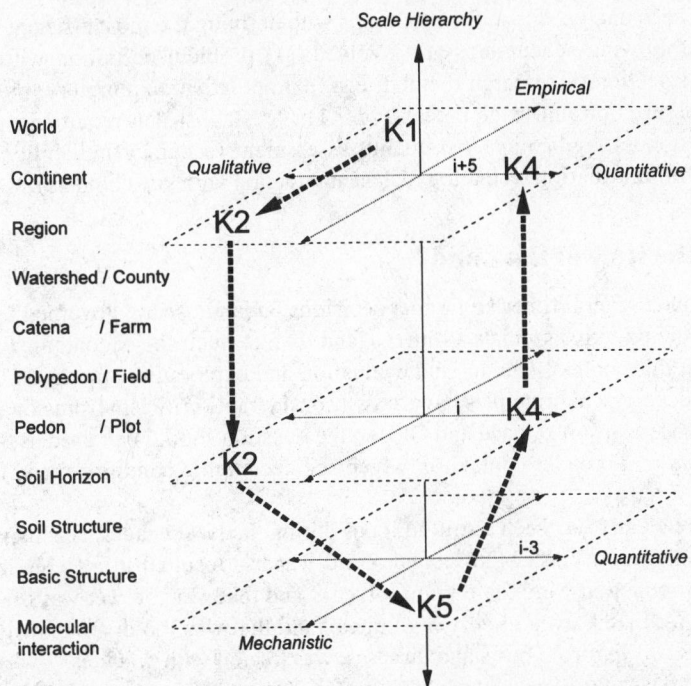

Fig. 7.2 Scale diagram for the study of De Vries et al. (1992) on acidification of European forests. Three hierarchial scales are distinguished. At each scale, different methods corresponding to different knowledge levels are applied, forming a knowledge chain [Reprinted from Bouma and Hoosbeek, 1996. Soil Sci. Soc. Am. Spec. Pub. 45:1-15 with permission of the Soil Science Society of America].

De Vries et al. (1992) did a study to determine the possible impact of acid rain on soil acidification. They dealt with nonagricultural areas without fertilization. They divided Europe (scale i+5) (Fig. 7.2) into grids and for each grid they determined the dominant soil type, using the soil map of Europe (K2 knowledge). Then, they selected a limited number of soil units (level i) that were considered to be representative for European soils (using K2 knowledge). In these soils, they made some detailed measurements of weathering rates (scale i–4; knowledge level K5). Next, this knowledge was scaled back up, resulting in what is assumed to be effective K4 knowledge at the European level. It would have been impossible to make the detailed K5 measurements of weathering rates in all European soils. By using expert knowledge at different scales, measurements could be made more efficient. Ideally, measurements should always include a measure for reliability and accuracy. An overall K5 approach for all soils would have the highest reliability, but would be too expensive. One must know how much is lost in terms of reliability when one goes from K5 to K4, K3 and K2. Decisions as to what to do can only be based on this type of information. The lines in Fig. 7.2 represent a so-called research chain (Section E, Chapter 3) demonstrating how a given problem can be analyzed by combining knowledge at different scales.

Before embarking on any land evaluation activity, one must first analyze the question being raised very carefully in close interaction with the stakeholder. Does one want to see what it means when trends from the past are extended? Does one want to explore alternative options? Does one want to define policy measures focused on the realization of one of these options? Or does one want to define a specific decision support system for the land user, guiding him to the right decisions? Once the question has thus been analyzed, one then proceeds with defining the most efficient research chain. Specifically, the following seven steps are involved: (1) problem definition with interaction of stakeholders; (2) selection of research method, e.g., extrapolative, exploratory, predictive, policy oriented; (3) model selection and development (Fig. 7.1); (4) establish data requirements (existing data or new data); (5) model application; (6) quality assessment (accuracy, reliability, risk); and (7) presentation (role of information technology, close interaction with stakeholders).

7.6   What is the Role of the Land?

It has been mentioned several times so far that decisions on land use are governed by many factors beyond those that are directly associated with the land. In fact, such socioeconomic factors are often most important. In this context, classic land evaluation and exploratory studies on land use which were based on agroecological principles have drawn considerable criticism from nonagriculturalists. This chapter is focused on soil science and GIS, so the question should be raised as to the particular significance of land in a broader context, in which socioeconomic conditions tend to set trends in society at large.

First of all, there clearly has been a shift in focus during the last decades. Exclusive emphasis on food production and security after the second world war has resulted in a technology explosion: problems in food production were there to be solved. Land that was too wet was drained, land that was too poor was fertilized, even at very high rates; land that was too dry was irrigated and land, where crops were suffering from pests and diseases, was treated with biocides.

Initially, these measures were taken with only food production in mind and this has led to considerable pollution of land and water locally. Later, concern for the environment played an increasingly important role, and as this process was unfolding, and as a balance had to be struck between agricultural production on the one hand and environmental quality of soil and water on the other, the importance of agroecological features of the land increased dramatically. There used to be the technology-driven spirit: anything can be done anywhere. Now one realizes again that a sand will never be a clay and that the natural dynamics of soils in different agroecological zones are the basis for developing sustainable production systems that are in harmony with nature and the environment.

However, one needs to take a new look at the way in which one presents our soils to international interdisciplinary research groups that work on sustainability and global change (Bouma, 1994; Bouma and Hoosbeek, 1996). The classical land evaluation approach was land centered and provided few contact points with other disciplines. Besides, it was based on a descriptive K2 approach (which was quite innovative at the time, and still quite relevant for many applications). Now, one is in a position to use modern K4 and K5 simulation methods to provide a quantitative analysis of agricultural production systems, including important tradeoffs between production and environmental quality. A strong emphasis should be placed on providing various options for any given land unit from which the stakeholder may choose.

There is one more point: soils do not occur in random patterns in a landscape. They are formed by geomorphological processes and soil-forming factors that differ significantly in different agroecological zones. Much work in soil science has been done on soil classification in grouping of soils that are comparable in their basic soil properties. Such groupings should be to describe soil behavior in terms of a window of opportunity for any given soil series (which is the lowest hierarchial unit of classification). One expects every soil series to present a characteristic window (a quantitative and scientific expression of the conviction that a sand will never act like a clay no matter what a farmer does!). Even though different forms of management will lead to different soil conditions even within the same soil series, the range of conditions (the window) will be characteristically different for each series. As this issue will not be further explored here, refer to van Lanen et al. (1987, 1992) and Droogers and Bouma (1997) for further details. Also, the fact that mapping units in the field have quite some internal variability that should be considered when making interpretations (Mausbach and Wilding, 1991; Bouma et al., 1996b; Young et al., 1997) must be emphasized.

7.7   Interaction with Stakeholders in the Information Age

Interaction with stakeholders has been emphasized many times in the above sections during problem definition and the research process including final reporting. In the past much research has been top down and supply oriented. The researcher had an impression as to what the problem was that needed to be investigated, and he or she pressed forward using a favorite model, expert system or data gathering technique. In the end, the results of the study were presented to the stakeholders. To be sure, there are many examples of fine and effective research being executed in this way that have led to successful implementation. However, there are also too many examples of research efforts that were less successful, ending up in a desk, covered with dust. The challenge, now, is to involve stakeholders to the extent that research is being executed jointly and with constant interaction, having the effect that the end result of the work is also experienced as a joint product.

A good example of poorly focused attention on stakeholders is the exclusive focus of land evaluation on land mapping units, while areas of land for which decisions have to be made usually do not correspond with the boundaries of land units. A farmer is faced with this fields that often are composed of different types of soil. What is being done in terms of management is largely determined by the proportion of these different land units, wet parts of the field cause a delay in tillage and sowing, while the drier parts would have allowed this much earlier. Precision agriculture attempts to base management on such differences occurring within fields. Again, at higher scale hierarchies land units do not necessarily correspond with units of management. Recently, therefore, resource management domains have been defined, which are relatively homogeneous in terms of agroecological properties and socioeconomic conditions (Dumanski and Craswell, 1997). Modern information technology has an important role to play in stimulating interaction with stakeholders. Visualization of alternative land use patterns associated with different options is a very powerful tool to involve stakeholders. A picture says more than a thousand words. Interactive computer technology allows, for instance, joint generation of alternative land use scenarios with all associated input data by researchers and stakeholders. Also, the abovementioned improvement of results when moving from K2 to K3, K4 and K5 approaches can be visualized as well by showing the accuracy of the land use maps obtained (Bouma, 1997b). The stakeholder can decide whether or not the costly improvement of the product is worth the cost. As Bouma (1993) has pointed out, several problems can be solved well with a relatively cheap K2 approach. This may be scientifically less challenging, but it is quite important from a practical point of view.

7.8   Case Studies on Field and Farm Scale

Four case studies will be discussed, two for The Netherlands, one for Costa Rica, and one for Niger. The seven steps for project planning, as introduced above, will be the framework for the discussion and the selected research chain will be analyzed. Project details will be reviewed broadly and reference will be made to more detailed source publications.

7.8.1 Precision Agriculture: How to Obtain Appropriate Soil Data

7.8.1.1 Problem Definition

Precision agriculture is receiving increasing attention internationally because fine tuning of agricultural management to the needs of plants within spatially variable fields can be attractive from both an economic and ecological point of view (Robert et al., 1996; Ciba Foundation, 1997). Many farmers' fields in The Netherlands are quite heterogeneous. Prime agricultural lands in alluvial areas are stratified and sandy and more clayey spots alternate at small distances. We wanted to explore how modern precision fertilization can help to fine tune management with the objective of maximizing production while minimizing adverse environmental side effects, particularly nitrate pollution of groundwater.

7.8.1.2 Selection of Research Method

Research results from the past could not be used, as they did not consider spatial variability patterns within fields but, rather, implicitly assumed fields to be homogeneous. There was no room for extrapolation. We decided to focus on an exploratory approach because of the complex nature of plant growth and N dynamics which are difficult to simulate. Moreover, any suggestion by research should be based on calculated results for many years, reflecting variable weather conditions. Prediction under such conditions is difficult and it is more in line with a farmer's practice to present a number of options from which he can choose. Results may ultimately have policy implications, but only after more primary research at farm level.

7.8.1.3 Model Selection

The need to define N dynamics clearly in soil in relation to crop development requires the use of a K4 model with a mechanistic and quantitative character. The WAVE model was used (Vanclooster et al., 1994).

7.8.1.4 Data Requirements

Even though a complete 1:50,000 soil map is available for The Netherlands, we lack the necessary 1:5,000 maps for individual fields. Besides, the concept of representative soil profiles is used for mapping units and individual borings taken during mapping are not georeferenced. We, therefore, decided to do a new soil survey using a minimal amount of soil borings. The procedure, as discussed by Finke (1993) and Verhagen et al. (1995), consists of: (1) making an exploratory soil survey; (2) using geostatistics to estimate optimal boring distances; and (3) making the soil survey (50 m by 50 m grid). In addition, the soil water regime has to be simulated, requiring hydraulic conductivity and moisture retention data. Using morphometric criteria, we defined a limited number of functional horizons with identical hydraulic properties that were measured with modern techniques at K5 level. (Finke and Bosma, 1993; Verhagen et al., 1995).

7.8.1.5 Model Application

The model was run for weather conditions in several years, different crops and a wide variety of N fertilization scenarios (Verhagen, 1997). Runs were made for point data and patterns were obtained by using interpolation techniques.

7.8.1.6 Quality Assessment

This study used measured hydraulic characteristics. We have not investigated the error which would have occurred when using estimated data, nor do we know the error associated with calculations using measured hydraulic data and other data associated with crop growth and N transformations, using methodology of Heuvelink (1998) and Bouma et al. (1996a,b).

7.8.1.7 Presentation

By using a spatial interpolation technique (indicator kriging), patterns are obtained within the field defining crop growth and leaching of NO_3^- in terms of probabilities of occurrence. The maps in Fig. 7.3 show the maximum allowable quantity of total N in the rootzone which can be present at the end of the growing season (Sept. 15 in The Netherlands) so as to stay under the threshold value for NO_3^- leaching (35 kg N ha^{-1} yr^{-1} for Dutch conditions) (Verhagen and Bouma, 1998). The calculations were based on 20 yr of weather data for the period Sept. 15–April 15 when precipitation statistically exceeds evapotranspiration. Calculations reflect a probability of exceeding 20%. In other words, values indicated apply to 80% of conditions encountered over the years, which are considered representative. Maps, such as presented in Fig. 7.3, appealed to our farmers. They offer a final target for application of N fertilizer in the growing season when leaching is normally very limited due to crop uptake. They do not set hard values, but reflect uncertainties associated with varying weather conditions over the years.

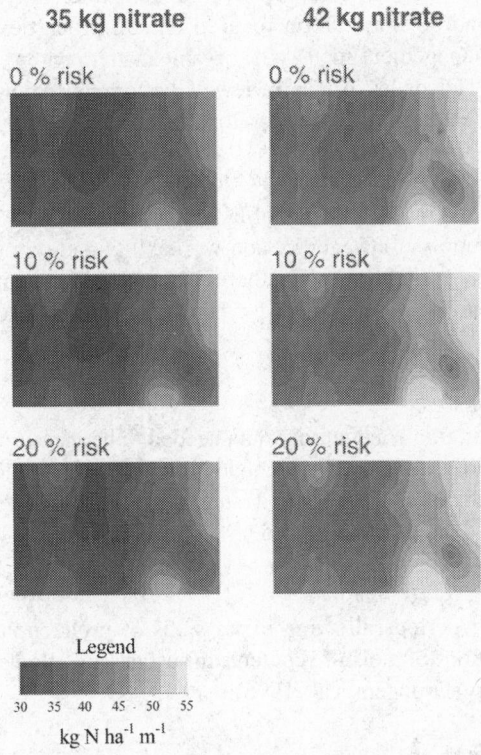

Fig. 7.3 Target maps of the N profile in a Dutch farmer's field (kg ha^{-1} m^{-1}) at September 15 at the start of the wet season with a precipitation surplus. Maps were obtained by geostatistical interpolation of simulated leaching values for 65 points for a period of 30 years. Two threshold leaching values of 35 kg N ha^{-1} and 42 kg N ha^{-1} are considered. The first is an average value for the country, the second relates to the northern part of The Netherlands. Use of indicator kriging for interpolation allows expression in terms of risk levels (only three shown) of exceeding these thresholds [Reprinted from Verhagen and Bouma, 1998. Defining threshold values for residual N-levels. Geoderma 85:199-213 with kind permission of Elsevier Science - NL, Amsterdam, The Netherlands].

7.8.2 Organic Farming: Changing Properties of Soil Series

7.8.2.1 Problem Definition

The need to create agricultural production systems that are economically productive as well as ecologically sustainable has led to much research on the effects of alternative soil management systems on soil conditions. The conventional agronomic approach can be used in which plots are tested, each with different types of treatments. Conclusions about differences among treatments are reached through conventional statistical analyses. Another possibility is to study farms that have applied certain types of management for considerable periods of time. Following this approach we compared soil conditions on an organic and a conventional farm with the objective of establishing the effects of the different management systems on soil properties, so that the resulting effects on sustainability could be assessed (Droogers et al., 1996; Droogers and Bouma, 1997).

7.8.2.2 Selection Research Method

Common soil properties such as bulk density, porosity, moisture retention and hydraulic conductivity reflect the effects of short-term management. Tillage or soil traffic under wet conditions in a given year may lead to compaction, puddling and structure degradation which may not occur in the same soil where soil traffic is avoided. We, therefore, focused on soil properties that are not significantly influenced by short-term management such as the organic matter content, which is affected by long-term management in terms of decades. Again in view of the complex nature of the processes involved (see above case study), we focused on an exploratory approach.

7.8.2.3 Model Selection

Soil conditions can best be expressed in terms of land qualities (FAO, 1976) which are the moisture supply capacity, nutrient supply, trafficability and workability and represent a K2 approach. This, however, is descriptive and not diagnostic. We, therefore, decided to use a deterministic, quantitative K4 model, as in the above study, which allows good characterization of soil moisture regimes and N transformations.

7.8.2.4 Data Requirements

Moisture retention and hydraulic conductivity data needed to be measured because estimates could not be made for the specific treatments. To adequately represent the effects of different contents of organic matter on N mineralization, we measured rate constants to be used in the N module of WAVE. Thus, a relatively high data demand materialized.

7.8.2.5 Model Application

Runs were made for 30-yr periods allowing expressions of production and leaching of NO_3^- in probabilistic terms. Predictions of moisture contents in surface soil allowed estimates to be made for trafficability and workability (Droogers et al., 1996).

7.8.2.6 Quality Assessment

Runs for 1995 were compared with measured data, showing that the model performed satisfactorily. No thorough sensitivity or error propagation analysis was performed.

7.8.2.7 Presentation

Results were presented in graphical form. The increased organic matter content resulted in a 20% increase of potential productivity (Droogers and Bouma, 1996). However, trafficability and workability decreased because the higher moisture contents during the year resulted in shorter periods with adequate trafficability and workability (Droogers et al., 1996). Indeed, strong compaction was observed in fields of the organic farm where conventional plowing to 30 cm depth was used. Based on the results obtained, we advocated a new soil tillage system incorporating shallow plowing or minimum tillage. This appealed to our farmers. We also addressed the question as to how production and NO_3^- leaching could be balanced, using a probabilistic graphical expression (Fig. 7.4). This graph was also used to define soil quality in terms of production and leaching of NO_3^-, illustrating that the user must choose the level of risk he or she is willing to take. This risk relates to yields (probabilities that a certain yield is exceeded) and to NO_3^- leaching (probabilities among the years that the threshold value for NO_3^- leaching is exceeded).

7.8.3 Land Evaluation on Field Level in a Data and Resource Poor Environment

7.8.3.1 Problem Definition
Farmers in the Sahelian region are faced with a high heterogeneity in their fields and low yields, partly because there are no funds available for fertilizers (Brouwer and Bouma, 1997). The problem is how to define soil conditions in such a way that they are helpful to farmers in devising management systems that give them at least some economic output. Few data on soils are available and resources to obtain data are very limited.

7.8.3.2 Selection of Research Method
Many studies in the area have reported results of farming systems research where farmers were asked to give their opinion on the production system (K1 level). Contradictory statements at the K1 level indicate the need for expert advise (K2 level) to be based on scientific principles. More advanced methodologies (K3, K4, K5) may be needed, but so far lack of data and funds for research make such approaches less realistic. Again, an exploratory approach is most suitable while an extrapolative approach is useful to describe conditions in the future without changes in management.

7.8.3.3 Model Selection
A combination of K1 and K2 research approaches is most realistic, particularly because of lack of data and the complex, highly heterogeneous agroecological environment.

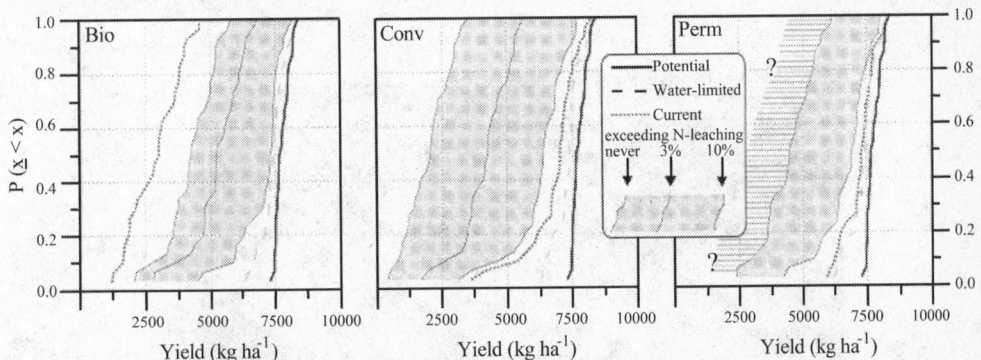

Fig. 7.4 Cumulative probability function of simulated yields for the defined land use scenarios Bio (biological farming), Conv (conventional farming) and Perm (permanent meadow), all for the same soil series. Probabilities were obtained by using 30 yrs of climatic data, and are expressed for three probabilities that Dutch thresholds for nitrate leaching are exceeded [Reprinted from Droogers and Bouma, 1997. Soil Sci. Soc. Am. J. 61:1704-1710 with permission of the Soil Science Society of America].

7.8.3.4 Data Requirements

An important mistake can be made when research concepts are blindly applied to Sahelian conditions. In the past, we have seen application of classical soil fertility experiments with a single dose effect approach. Of course, at some point in the future farmers may be able to buy chemical fertilizers but not at this time. It is, therefore, much better to focus on direct observations of crop growth at field level and to study local management practices.

7.8.3.5 Model Application

Spatial variability in crop growth patterns is very high (Fig. 7.5). Local knowledge and expert knowledge are activated to explain these differences and to see how local knowledge can be used to increase the efficiency of the production system. This approach requires a new paradigm in research, in which the researcher has to ignore many of the standard approaches of the resource-rich research environment. In this work, we observed a heterogeneous infiltration of precipitation due to the microrelief. This, in turn, results in stronger leaching in the micro lows and lower crop growth in normal years. Also, termites are locally active, improving growing conditions. (Mando, 1997). Soils are subject to strong surface crusting and use of crop residues may reduce this phenomenon while also reducing wind erosion (Sterk et al., 1996). Organic manure can be placed selectively on micro highs and micro lows, thereby improving crop uptake and resource efficiency (Bouma et al., 1997).

7.8.3.6 Quality Assessment

Use of mainly K2 knowledge implies that it is difficult to express risk and accuracy in quantitative terms. This has, therefore, not been attempted.

7.8.3.7 Presentation

The key aspect here is the integrated character of expert knowledge. Heterogenity of crop growth can be explained by soil expertise, with a focus on physical, chemical, and biological soil conditions (Brouwer and Bouma, 1997). Placement of organic manure on the micro highs and restricting applications to manure produced overnight rather than accumulated for three nights, could

Fig. 7.5 Spatial variability of peanut growth in the Sahelian region, which is due to fertility remnants of tree growth, termite activity and differences in soil acidity due to preferential infiltration of water [Brouwer and Bouma, 1997]

significantly improve nutrient efficiency (Bouma et al., 1997). Maintenance of some soil cover by crop residues reduces crusting and promotes termite activity. This integrated package can be presented to the farmers and extension agents, preferably on a real farm.

7.8.4 Alternative Land Use Scenarios in Guacimo, Costa Rica

7.8.4.1 Problem Definition
Large areas of tropical forest have been cut in Costa Rica in the past few decades and various types of agricultural land use are found now in the treeless areas. The greater part is occupied by extensive grassland or by cropland while other areas are used for banana or flower plantations. Even though land use planning is not very advanced in a pioneer country such as Costa Rica, the question has been raised by national politicians and agricultural scientists as to how rational land use scenarios can be developed. This case study covers a 58,000 ha region study in the Guacimo area.

7.8.4.2 Selection of Research Method
Clearly, quite contrasting demands have to be considered when defining the most suitable type of land use in a region. The carrying capacity of the land has to be considered, particularly in this area where fertile, young Andisols occur next to old, leached and infertile Ultisols and Oxisols. Which crops can be grown where? Because of high precipitation (4,000 mm yr^{-1}), the natural drainage capacity of the soil is, therefore, a very important property, but other factors are very important as well. Prices and production costs of various crops determine the profitability of any agricultural activity. Is there a market for products and can the market be reached? Predictive classical land evaluation following a K2 approach is not satisfactory because relative suitabilities of the various land units for different land utilization types only cover part of the problem. We, therefore, selected a more quantitative, exploratory approach using expert knowledge to describe different land utilization types (K2) and simple simulation models in a linear programming procedure (K3) which allowed tradeoffs between contrasting demands (Schipper et al., 1995). Thus, alternative land use scenarios were developed which could be visualized on maps using GIS equipment.

7.8.4.3 Model Selection
Linear programming was used to develop different land use scenarios including tradeoffs between contrasting requirements, using the USTED methodology (Stoorvogel et al., 1995).

7.8.4.4 Data Requirements
The analysis combines expertise from soil science, agronomy and agricultural economics. Clearly, each discipline has many data to offer, leading to quantities that cannot be handled. Each discipline was, therefore, asked to cut down its data to the bare essentials, an interesting exercise. For soils, only three main soil types were distinguished although the 1:150,000 soil map showed at least 60 different types. Alternative land use patterns are defined for the different land units in terms of LUSTs, including a specification of the amount of labor and technology being applied. A detailed example for a maize LUST was shown in Table 7.1. Actual and potential LUSTs are defined. This study compared 122 LUSTs and 9 farm types on the basis of size and soil type.

7.8.4.5 Model Applications
Analysis with USTED included a base run for a reference year and additional runs for changing conditions. Policy interventions are translated into possible changes in the socioeconomic or biophysical environment and in coefficients for the linear programming model.

7.8.4.6   Presentation

Coupling the USTED model to GIS (Stoorvogel, 1997) is helpful in communicating results of the calculations in terms of georeferenced patterns. An example is given in Fig. 7.6 which shows the base scenario and a scenario that was calculated after assuming that the price of palm heart would decrease by 25%. Calculations show that palm heart will be replaced by cassava, pasture and tree plantations, resulting in a net decrease in farm income of 12%. The selected cassava, pasture and tree plantation technologies use significantly less biocide than the palm heart system, resulting in a 64% decrease in the defined biocide index. Other scenarios explore the effects of the introduction of an environmental tax and of biocide regulation. Experience in Costa Rica has shown that coupling the USTED methodology with GIS is highly effective when communicating results of alternative land use scenarios to stakeholders.

7.8.5   The World Food Crisis: Assessment of Food Security

7.8.5.1   Problem Definition

The world population is expected to double by the year 2030. Enough food is produced now, even though its distribution is poor, leaving 800 million people with inadequate food supplies. Many studies have been made on the potential of the Earth to feed its people. We will refer here to the recent work by Penning de Vries et al. (1995) because they used a modern land evaluation procedure and made some interesting decisions on procedures that correspond to the global level of detail. This study tried to answer questions about global food security by considering three population growth scenarios and two types of agriculture, one with high and the other with low external inputs. The reader is referred to the above publication for the interesting results of the study.

7.8.5.2   Selection of Research Method

Research at this scale can only be exploratory in a broad sense. Emphasis was placed on exploring agricultural production potentials. Whether or not these are reached is realistically considered to be beyond the scope of the work as it will depend on socioeconomic and political decisions. What is offered are characteristic windows of opportunity.

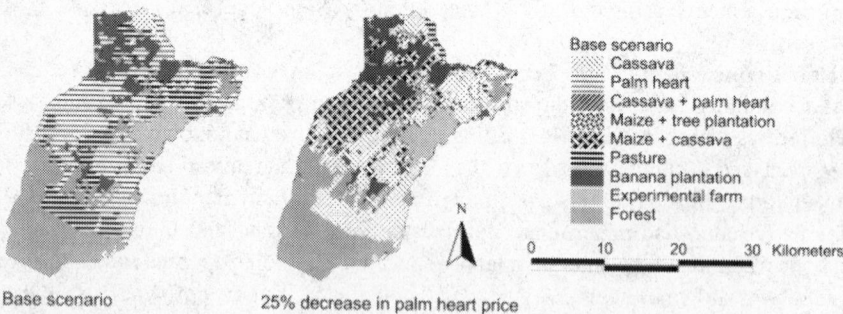

Fig. 7.6   Two alternative land use scenarios for the Guacimo area in Costa Rica. A base scenario is provided as well as predicted land use patterns following a 25% decrease in palm heart price (one arbitrarily selected example of many scenario runs). A linear programming technique has been used to express tradeoffs between different land use scenarios in quantitative terms [Stoorvogel, 1997].

7.8.5.3 Model Selection

A K2 approach might appear to be realistic at first sight. However, no expert can oversee the immensely diverse production conditions prevailing in different areas of the world. The authors chose, therefore, a K3 approach using a very simple model to calculate crop yields based on available radiation, water and nutrients. Thus, a universal approach was introduced allowing comparisons among all calculated values. They assigned all possible crops to a cereal (wheat in moderate climates and rice in the tropics) or grass equivalent to obtain a so-called grain equivalent (GE), which makes comparisons relatively easy.

7.8.5.4 Data Requirements

The NASA database was used for 1° by 1° grid cells, which occupy a 110 km by 110 km area at the equator. Confining attention here to soils for each grid cell, the authors selected values for slope, soil phase and soil texture. Selections were based on the 1:5 million FAO World Soil Map and other available data. Clearly, choices were highly arbitrary. Cells often contain different land units, each representing major soil associations with many soil types. Soil texture of the dominant association was applied to the entire cell. Dominant soils in each cell were considered to be well drained, homogeneous without layers or cracks, 60 cm deep and without runoff. Degradation of soils was ignored, and so were current land use patterns. For the world, 15,500 land units had to be considered along with 700 climatic zones.

7.8.5.5 Model Application

The model was run for all cells, divided into 15 major regions which were distinguished for the UN population study, and for different scenarios representing population increase and management types. Coupling with GIS allowed production of maps.

7.8.5.6 Error Estimation

Estimates for grid cells could have been compared with estimates obtained using more reliable basic data derived from large-scale soil maps. This, however, has not been done. The accuracy of the resulting data can, therefore, not be determined, which illustrates the broad exploratory character of the exercise. Still, the alternative is to have nothing at all and we prefer such a rough estimate, which can be improved upon, over a lack of any estimate.

7.8.5.7 Presentation

GIS was used to present maps for the different scenarios, which was effective in communicating the major results of the study; a problem was perceived for southeast Asia, where food shortages are likely in the future even at low rates of population increase and with high input agriculture.

7.8.6 New Thrusts for Soil Survey and Land Evaluation

7.8.6.1 Technological Developments

Many new technological developments will change soil studies in the years to come. Low cost global positioning systems (GPS) allow accurate positioning of observers and equipment anywhere in the world. New satellites will strongly increase the opportunity to observe the surface of the Earth in more detail than ever before using an increased number of diagnostic features. Remote sensing from satellites or airplanes will allow better evaluation of crop conditions allowing improved yield predictions. In addition, on-the-go yield monitoring is becoming well established and will revolutionize the assessment of the production capacity of the land by providing a continuous record for many years. In turn, measured differences in yield will give rise to specific research on the underlying causes, which may be many.

Technological developments not only occur above, but also below the soil surface. Many new sensors are being developed to be used for the continuous registration of soil water and solute contents. Already, time domain reflectometry is widely used to measure soil water contents, replacing neutron probes. Transducer tensiometers allow accurate, instant registration of pressure heads in the soil within the range of interest for plant growth. Sensors for N contents in soil are being developed in the context of precision agriculture. Information technology not only allows rapid registration, but also transmission to central computers. Indeed, some study sites resemble patients in intensive care.

Opportunities to interpret soil data have dramatically increased as well. A wide range of models is available to predict soil water contents and solute fluxes, even in heterogeneous soils that swell and shrink and may show hydrophobic behavior. Also, expert systems have been perfected allowing a more efficient application of the vast body of knowledge residing with users of the land. Geographical information systems, finally, allow integration of different types of data, coupling of databases with models and construction of digital terrain models with unique opportunities to visualize landscape processes. How will this new technology be used to improve our land evaluation practices? The above case studies already indicated some approaches, but much more is to follow.

7.8.6.2 Future Land Evaluation Based on Soil Survey

In many, so-called developed countries standard soil surveys of agricultural lands at scale 1:20000 to 1:50000 have been completed. Revisions of old maps are being made, but funds are often not available to do this. Rather, surveys for specific purposes are increasingly commissioned. Then, databases rather than classical soil maps are requested to be used for a wide variety of applications, using modeling techniques or other means of interpretation. Such databases may contain georeferenced multitemporal remote sensing data and results of monitoring procedures.

The main question increasingly centers on the focus and the objectives of any particular soil study. With so much data available, there is a considerable risk that the user will be completely overwhelmed. International conferences during the last decade have, however, put some important issues on the international research agenda. An example is the need for sustainable land management which combines objectives for economically and socially acceptable production with environmental requirements (FAO, 1993). Also, definitions of soil quality (Karlen et al., 1997) and land quality (Pieri et al., 1995) will be the focus of much soil research, while requiring interdisciplinay approaches involving high data demands and much data manipulation. The above case studies on precision agriculture in a high and low tech environment and the effects of organic farming on soil behavior are meant to illustrate these types of modern studies. There is, however, one approach using existing soil surveys that can be particularly useful in this context. Here, delineated areas of a particular soil series are studied in terms of soil properties as influenced by soil management (Bouma, 1994; Droogers and Bouma, 1997).

7.8.6.3 Phenoforms of Soil Series: Effects of Management Observed in the Field

Droogers and Bouma (1997) studied a prime agricultural soil in The Netherlands (a Typic Fluvaquent). They suggested the terms genoform for the genetic soil name and phenoform for the management variant, of which three were studied, one resulting from 60 yr of organic farming, (BIO) one from modern intensive arable farming (CONV) and one from continuous grassland (PERM).

They used the validated WAVE model, illustrated above, to calculate crop yields for wheat as a function of a wide range of N fertilization scenarios. By using yield calculations for a 30-yr period with a wide range of weather conditions, they could express both yields and NO_3^- leaching to the groundwater in probabilistic terms (Fig. 7.4). This, in turn, allowed the identification of a soil quality indicator for yield-expressing effects on the environment and risk. This indicator was defined by dividing actual calculated production (x 100) by potential production where the latter is characteristic for each agroecological zone (Bouma and Droogers, 1998).

We believe that distinguishing well-defined phenoforms of existing soil series can be quite effective in extending the use of existing soil surveys. These, of course, define soil suitabilities for a wide variety of land uses without specifying the effects of management. Such effects are of major interest to many users. Finally, this approach can be used at different scales once a database is filled with a number of phenotypes for any given soil series. Larger landscape units on small scale maps are still being defined in terms of their internal composition, for example, 30% soil series x, 50% series y and 20% series z. Then the effects of different types of management on yield and environmental side effects can still be expressed in a probabilistic manner, be it non-georeferenced.

7.9 References

Bouma, J. 1993. Soil behavior under field conditions: Differences in perception and their effects on research. Geoderma 60:1–15.

Bouma, J. 1994. Sustainable land use as a future focus of pedology. Soil Sci. Soc. Amer. J. 58:645–646.

Bouma, J. 1997a. Precision agriculture: Introduction to the spatial and temporal variability of environmental quality. p. 5–13. *In* Precision agriculture: Spatial and temporal variability of environmental quality. John Wiley and Sons, Chichester, UK.

Bouma, J. 1997b. Role of quantitative approaches in soil science when interacting with stakeholders. Geoderma 78:1–12.

Bouma, J., and P. Droogers. 1998. A soil quality indicator for production, considering environment and risk. Geoderma 85:103–110.

Bouma, J., and M.R. Hoosbeek. 1996. The contribution and importance of soil scientists in interdisciplinary studies dealing with land. p. 1–15. *In* R.J. Wagenet and J. Bouma (ed.) The role of soil science in interdisciplinary research. Soil Sci. Soc. Amer. Spec. Publ. 45. Soil Science Society of America, Madison, WI.

Bouma, J., J. Verhagen, J. Brouwer, and J.M. Powell. 1997. Using systems approaches for targeting site specific management on field level p. 25–37. *In* M.J. Kropff, P.S. Teng, P.K. Aggerwal, J. Bouma, B.A.M. Bouman, J.W. Jones and H.H. van Laar (ed.) Applications of systems approaches at the field level. Kluwer Academic Publishers, Dordrecht, Netherlands.

Bouma, J., H.W.G. Booltink, and P.A. Finke. 1996a. Use of soil survey data for modeling solute transport in the vadose zone. J. Environ. Qual. 25:519–526.

Bouma, J., H.W.G. Booltink, P.A. Finke, and A. Stein. 1996b. Reliability of soil data and risk assessment of data applications. p. 63–81. *In* W.D. Nettleton, A.G. Hornsby, R.B. Brown and T.L. Coleman (ed.) Data reliability and risk assessment in soil interpretations. Soil Sci. Soc. Amer. Special Pub. 47. Soil Science Society of America, Madison, WI.

Brouwer, J., and J. Bouma. 1997. Soil and crop growth variability in the Sahel. ICRISAT Sahelian Cent. Infor. Bull. 49.

CIBA Foundation. 1997. Precision agriculture: Spatial and temporal variability of environmental quality. John Wiley and Sons, Chichester, UK.

De Vries, W., M. Posch, G.J. Reinds, and J. Kamari. 1992. Critical loads and their exceedance on forest soils in Europe. Winand Staring Cent. Rep. 58. Wageningen, Netherlands.

Droogers, P., and J. Bouma. 1996. Effects of ecological and conventional farming on soil structure as expressed by water-limited potato yield in a loamy soil in The Netherlands. Soil Sci. Soc. Amer. J. 60:1552–1558.

Droogers, P., and J. Bouma. 1997. Soil survey input in exploratory modeling of sustainable soil management practices. Soil Sci. Soc. Amer. J. 61:1704–1710.

Droogers, P., A. Fermont, and J. Bouma. 1996. Effects of ecological soil management on workability and trafficability of a loamy soil in the Netherlands. Geoderma 73:131–145.

Dumanski, J., and E.T. Craswell. 1997. Needs and potential uses of Resource Management Domains. *In* J.K. Syers, and J. Bouma (ed.). International Workshop on Resource Management Domains. IBSRAM-IRRI, Bangkok, Thailand.

FAO. 1976. A framework for land evaluation. FAO Soils Bull. 32.

FAO. 1983. Guidelines: Land evaluation for rainfed agriculture. FAO Soils Bull. 52.

FAO 1993. FESLM: An international framework for evaluating sustainable land management. FAO World Res. Rep. 73.

Finke, P.A. 1993. Field scale variability of soil structure and its impact on crop growth and nitrate leaching in the analysis of fertilizing scenarios. Geoderma 60:89–109.

Finke, P.A., and W.J.P. Bosma. 1993. Obtaining basic simulation data for a heterogeneous field with stratified marine soils. Hydrol. Proc. 7:63–75.

Heuvelink, G.B.M. 1998. Uncertainty analysis in environmental modelling under a change of spatial scale. Nutr. Cycl. Agroecosys. 50:255–264.

Hoosbeek, M.R., and R. Bryant. 1992. Towards the quantitative modeling of pedogenesis: A review. Geoderma 55:183–210.

Jansen, D.M., and R.A. Schipper. 1995. A static, descriptive approach to quantify land use systems. Neth. J. Agric. Sci. 43:31–47.

Karlen, D.L., M.J. Mausbach, J.W. Doran, R.G. Cline, R.F. Harris and G.E. Schuman. 1997. Soil Quality: A concept, definition and framework for evaluation. Soil Sci. Soc. Amer. J. 61:4–10.

Kruseman, G., R. Ruben, H. Hengsdijk, and M.K. van Ittersum. 1995. Farm household modeling for estimating the effectiveness of price instruments in land use policy. Neth. J. Agric. Science 43:111–124.

Latesteijn, H.C. van. 1995. Scenarios for land use in Europe: Agro-ecological options within socio-economic boundaries. p. 43–65. *In* J. Bouma, A. Kuyvenhoven, B.A.M. Bouman, J.C. Luyten and H.G. Zandstra (ed.) Eco-regional approaches for sustainable land use and food production. Kluwer Academic Publishers, Dordrecht, Netherlands.

Mando, R.A. 1997. The impact of termites and mulch on the water balance of crusted Sahelian soil. Soil Tech. 11:121–139.

Mausbach, M.J., and L.P. Wilding. 1991. Spatial variability of soils and landforms Soil Sci. Soc. Am. Spec. Pub. 28. Soil Science Society of America, Madison, WI.

Penning de Vries, F.W.T., H. van Keulen, and J.C. Luyten. 1995. The role of soil science in estimating global food security in 2040. p. 17–37. *In* R.J. Wagenet and J. Bouma (ed.) The role of soil science in interdisciplinary research. Soil Sci. Soc. Am. Spec. Publ. 45. Soil Science Society of America, Madison, WI.

Pieri, C., J. Dumanski, A. Hamblin, and A.Young. 1995. Land quality indicators. World Bank Disc. Pap. 315.

Robert, P.C., R.H. Rust, and W.E. Larson. 1996. Precision agriculture. American Society of Agronomy, Madison, WI.

Rossiter, D.G. 1990. ALES: A framework for land evaluation using a microcomputer. Soil Use Manag. 6:7–20.

Schipper, R.A., D.M. Jansen, and J.J. Stoorvogel. 1995. Sub-regional linear programming models in land use analysis: A case study of the Neguev settlement. Costa Rica. Neth. J. Agric. Sci. 43:83–111.

Sterk, G., L. Hermann, and A. Bationo. 1996. Wind-blown nutrient transport and soil productivity changes in southeast Niger. Land Degrad. Devel. 7:325–335.

Stoorvogel, J.J., 1997. Using GIS and models for decision support in Costa Rica farming. p.119–129. *In* A. Stein (ed.) Data in action. Proc. Sem. Ser. Grad. Sch. Prod. Ecol., Wageningen Agricultural University, Wageningen, Netherlands.

Stoorvogel, J.J., R.A. Schipper, and D.M. Jansen. 1995. USTED: A methodology for a quantitative analysis of land use scenarios. Neth. J. Agric. Sc. 43:5–19.

Sys, C., E. van Ranst, and J. Debaveye. 1991. Land evaluation. Part I: Principles in land evaluation and crop production calculations. Part II. Methods in land evaluation. International Training Center for Post Graduate Soil Scientists, University of Gent, Gent, Belgium.

Van Lanen, H.A.J., M.J.D. Hack ten Broeke, J. Bouma, and W.J.M. de Groot. 1992. A mixed qualitative/quantitative physical land evaluation methodology. Geoderma 55:37–54.

Vanclooster, M., P. Viane, J. Diels, and K. Christiaens. 1994. WAVE. A mathematical model for simulating water and agrochemicals in the soil and vadose environment: Reference and user manual (release 2.0). Institute for Land and Water Management, University of Leuven, Leuven, Belgium.

Veldkamp, A., and L.O. Fresco. 1996. CLUE-CR: An integrated multi-scale model to simulate land use change scenarios in Costa Rica. Ecol. Model. 91:231–248.

Verhagen, J. 1997. Site-specific fertilizer application for potato production and effects on N-leaching using dynamic simulation modeling. Agric. Ecosyst. Environ. 66:165–175.

Verhagen, J., H.W.G. Booltink, and J. Bouma. 1995. Site specific management: Balancing production and environmental requirements at farm level. Agric. Syst. 49:369–384.

Verhagen, J., and J. Bouma. 1998. Defining threshold values for residual N-levels. Geoderma. 85:199–213.

Young, F.J., R.D. Hammer, and F. Williams. 1997. Estimation of map unit composition from transect data. Soil Sci. Soc. Am. J. 61:854–861.

Perhaps our most precious and vital resource, both physical and spiritual, is that most common matter underfoot which we scarsely even notice and sometimes call "dirt," but which is, in fact, the mother-lode of all terrestrial life and the purifying medium wherein wastes are decomposed and recycled, and productivity is generated
-Daniel Hillel

Soil Mineralogy

Joseph W. Stucki
University of Illinois

Fundamental to the existence of soil is its inorganic mineral fraction. Indeed, of all soil constituents only this one is required, or the soil ceases to be soil. Organic matter, bacteria, and chemicals often enhance various properties of the soil, but without the inorganic minerals it is reduced to a soilless medium incapable of the diverse range of functions that are often taken for granted. A correct and complete understanding of the soil must, therefore, of necessity require an understanding of soil mineralogy.

A common misconception is that soil minerals are rather inert, unchanging, and nonlabile. After all, this is the dead fraction, sometimes referred to as dirt; whereas the organic matter derives from living things. Such a view is borne of ignorance. The chapters in this section, however, quickly dispel this view by painting a very different picture. They will show that the soil minerals are diverse in their origins, structures, and chemistry. They provide active chemical surfaces where many reactions are catalyzed and basic behaviors of the soil are born. They react with organic matter, invoking mutual alterations in the behavior of both themselves and the organic matter. Life-sustaining liquid water is held in their interstitial pores, and even retained in that state at very low temperatures to preserve plant life during hard winter months. Soil minerals come in a wide range of particles sizes, which add texture and body to the soil, and vary in their degree of crystallinity. Their colors are also diverse, ranging from intense reddish brown to very light gray, or even green, yellow, or blue. They are dynamic in their properties, constantly changing with climate, time, and other environmental conditions. Some changes are rapid; others, slow. These attributes bring the soil to life, as it were, and create the framework within which the plant and animal kingdoms spring forth and are sustained through every season and in every clime.

Minerals, the heart of the soil, are a dynamic and essential resource, classified according to their properties. Presented in the following chapters are descriptions of how minerals are formed and transformed. Specifics are given regarding the properties and behavior of a few of them, including phyllosilicates, the plate-shaped minerals; iron and aluminum oxides, which represent the more highly weathered mineral constituents; and amorphous minerals, which display active chemical properties but lack high order in their crystals. These are among the most important of the soil minerals. As such, knowledge of their characteristics will provide the basis from which a more complete understanding of soil physical and chemical properties may be acquired.

1

The Alteration and Formation of Soil Minerals by Weathering

G.J. Churchman
CSIRO, Glen Osmond, Australia

1.1 Introduction

A mineral is an element or chemical compound that is normally crystalline and that has been formed as a result of geological processes (Gaines et al., 1997). Soil results from weathering, or the combined actions of the hydrosphere, biosphere and atmosphere on rock materials. The minerals found in soils are a combination of minerals that have originated in rocks, minerals that have been altered from their original state in the rocks by weathering processes and also new minerals that have been formed during the weathering processes. It follows that soil minerals are either inherited or transformed from the minerals in rocks or else they may be newly formed, i.e., neoformed, or authigenic. However, a mineral in a soil above or surrounding a particular rock formation may not have formed from the local rock; instead, it may have been transported from another parent rock source, cm, m, or km distant or even from another continent or island. This type of mineral is said to be detrital in origin.

This chapter aims to summarize the state of knowledge of processes involved in the alteration and formation, and thus, the nature of minerals formed, during the development of soils. Among all publications on this subject, that of Dixon and Weed (1989) gives probably the most comprehensive and up-to-date compilation of information about minerals in soils. However, the products of weathering are found in sediments and sedimentary rocks as well as in soils, as covered in Weaver (1989) and Chamley (1989). The focus of both is on clay minerals as major products of rock weathering, as is the earlier volume by Millot (1970), and a later book by Velde (1992). Other important background is provided by books by Loughnan (1969) and Carroll (1970), who view the process mainly from the viewpoint of the breakdown of the parent rocks. Drever (1985) examines the processes of weathering, as does the more recent volume edited by White and Brantley (1995). Relevant, if somewhat broader, information is also summarized in general texts on soils (McLaren and Cameron, 1996; Rowell, 1994; Sposito, 1989) as well as Birkeland (1974), who deals more specifically with weathering, and Garrels and Mackenzie (1971) on sedimentary rocks. Together, these and other sources have enabled an assessment of the extent to which consensus has been reached on the processes leading to soil minerals. Where they have indicated the existence of important controversies, papers representing opposing viewpoints are cited more specifically. Papers pioneering important discoveries and understandings are also cited. In addition, recent literature is cited to indicate continuing and developing controversies and new insights into soil mineral origins.

1.2   Weathering

In weathering, chemical, physical and biological forces act upon rocks and their component minerals in an open, aqueous system. Physical processes, including those of grinding, thermal expansion, glacial plucking, salt crystallization, freezing, wetting and drying and erosion by wind and water, can all cause the breakdown of rocks. However, these processes cause little alteration of minerals. It is difficult to distinguish between purely chemical and purely physical processes of weathering; some forms of physical weathering, e.g., frost heaving and erosion by wind may require some chemical preconditioning before they can occur (Garrels and Mackenzie, 1971). By the same token, parent materials of soils often comprise eroded, transported and comminuted debris, or deposits, rather than solid bedrock (McLaren and Cameron, 1996). Nonetheless, chemical effects, including those with a biological origin, dominate mineral alteration processes. Chemical effects include acidification, chelation, oxidation and reduction, and hydration and dehydration. Weathering is most often a physicochemical process. The main agent of weathering is water and its main effect is dissolution of the minerals.

It is tempting to describe the weathering process as involving an alteration of primary minerals in rocks into secondary minerals in soils. However, the established usage for these terms precludes such a simple description. The primary minerals are generally held to be those of magmatic, hydrothermal and metamorphic origin and the secondary minerals as those formed as a result of weathering (Loughnan, 1969). Secondary minerals are commonly clay minerals, reflecting their usual occurrence as fine, clay sized minerals, although secondary minerals can also occur within coarse fractions (Ugolini et al., 1996). The aluminosilicate clay minerals are often accompanied as secondary products of weathering by a variety of oxides, hydroxides and oxyhydroxides. Nonetheless, weathering does not always involve the alteration of primary minerals into secondary minerals. Although igneous rocks make up most of the earth's crust, about 80% of the surfaces of continents are covered with sedimentary rocks (Garrels and Mackenzie, 1971). These latter comprise products of earlier surface processes, principally weathering. They, therefore, include secondary minerals, both in separate particles and also as chemical crusts and cements as well as biogenic materials, e.g., carbonates and silica (Doner and Lynn, 1989; Drees et al., 1989). Hence, secondary minerals often constitute the reagents in weathering reactions. Following initial alteration of primary minerals, secondary minerals are continually recycled by weathering, although metamorphism or melting to form magma, and hence, new igneous rocks, may eventually occur.

Remembering that weathering often causes the alteration of the secondary mineral products from previous weathering cycles as well as of fresh primary minerals, it is helpful nonetheless to consider changes which occur when primary minerals *per se* are altered by this process. Table 1.1 shows the most common primary minerals in soils. With rare exceptions such as rutile (TiO_2), they contain Si bonded to O. In weathering, minerals react with water.

All minerals dissolve in water, but the pathways and rates of dissolution vary widely between different types of minerals. They may dissolve congruently to yield their components in solution in the same proportion in which they occur in the mineral. Salts such as halite (NaCl) dissolve in this manner. Most often, they dissolve incongruently, so that the solid residue has a different composition from the starting mineral. They undergo hydrolysis. In a completely open system, where soluble products are continually removed from the reactants, dissolution can continue until the mineral has been completely dissolved, or else transformed into a less soluble product. Most commonly in soils, weathering leads to a gradation of products in both space and time, resulting in soil profiles in which there are variations with depth.

Table 1.1 Common primary minerals in soils

Quartz	SiO_2	
Microcline	$KAlSi_3O_8$	(Feldspars)
Orthoclase	$KAlSi_3O_8$	(Feldspars)
Sodic plagioclase (albite)	$NaAlSi_3O_8$	(Feldspars)
Calcic plagioclase (anorthite)	$CaAl_2Si_2O_8$	(Feldspars)
Muscovite	$KAl_2(AlSi_3O_{10})(OH,F)_2$	(Micas)
Biotite	$K(Mg,Fe^{II})_3(Al,Si_3)O_{10}(OH,F)_2$	(Micas)
Chlorite	$(Mg,Fe^{II})_{10}Al_2(Si,Al)_8O_{20}(OH,F)_{16}$	
Hornblende	$NaCa_2(Mg,Fe^{II})_4(Al,Fe^{III})(Si,Al)_8O_{22}(OH,F)_2$	(Amphibole)
Augite	$(Ca,Mg,Fe,Al)_2(Al,Si)_2O_6$	(Pyroxene)
Serpentine	$Mg_6Si_4O_{10}(OH)_8$	
Tourmaline	$Na(Mg,Fe^{II},Al,Mn,Li)_3Al_6(BO_3)_3(Si_6O_{18})(OH,F)_4$	
Garnets	$(Fe^{II},Mg,Mn, \text{ or } Ca)_3(Fe^{III},Al, \text{ or } Cr)_2Si_3O_{12}$	
Epidote	$(Ca_2Fe^{III}Al_2O)(Si_2O_7)SiO_4(OH)_4$	
Olivines	$(Mg,Fe^{II})_2SiO_4$	
Sphene	$CaTiSiO_5$	
Zircon	$ZrSiO_4$	
Apatite	$Ca_5(F,Cl)(PO_4)_3$	
Ilmenite	$FeTiO_3$	
Rutile	TiO_2	

In most cases, hydrolysis removes the more soluble ions from minerals. Because of the input of various acids, including dissolved CO_2 from the atmosphere as well as from plant and microbial sources, and also organic acids from such biological weathering agents as fungi, lichens and moss, hydrolysis almost invariably occurs in an acid solution. Hence, the small, mobile, highly polarizing (hydrous) ion H_3O^+ often plays an important part in weathering reactions. However, some ions are particularly insoluble at the pH of most soil-forming environments. Among these is Al, which plays such a central part in weathering reactions that the bonding of Al has been claimed to be "a unifying principle in soil science" (Jackson, 1963). The solubility of Al in relation to pH is plotted in Fig. 1.1. Many primary minerals contain Al (Table 1.1), and therefore, their hydrolysis leads to an incongruent dissolution of other elements besides Al, including the ubiquitous Si, to leave a solid product which is more aluminous than the primary mineral.

A considerable number of primary minerals also contain Fe, usually in the Fe(II) form. Except when under water (e.g., hydromorphic), soils generally form in an aerated, oxidizing environment, at least for a large part of the time. Consequently, Fe(II) is oxidized to Fe(III), where it forms compounds with O that are particularly insoluble. As a result of the relative solubilities of different ions in most earth surface environments, weathering tends to lead to a concentration of Al and Fe(III) in solid products. Compared with Al and Fe(III), Si is relatively soluble, generally as silica SiO_2, or its hydrated form H_4SiO_4 in most soil-forming environments. Hence, the general tendency in weathering is toward desilication. Al-rich aluminosilicate clay minerals such as kaolinite, $Al(OH)_3$ (gibbsite), and also Fe oxides and oxyhydroxides, are the major end products of these processes.

However, weathering takes place in an exceptional diversity of both macro- and microenvironments. Al and even Fe may be soluble in extremely acid environments. Around the roots of living plants, the pH can be as low as 2 (Loughnan, 1969; Garrels and Mackenzie, 1971). More often, it is organic complexing agents which facilitate the mobility of Al and Fe.

In addition, drainage is often impeded. In this case, the environment can be reductive rather than oxidative, so that Fe(II) is the favored form of Fe and its mobility is enhanced. Furthermore, while the environment remains open to weathering agents and dissolution of primary minerals occurs, the

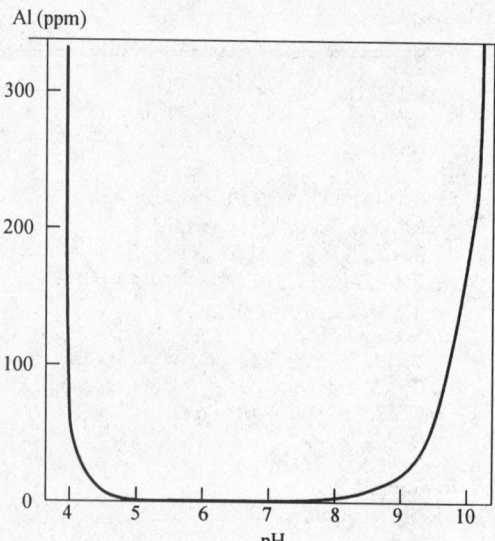

Fig. 1.1 The solubility of aluminum in relation to pH [After McLean, 1976. Chemistry of soil aluminum. Commun. Soil Sci. Plant Anal. 7:619-636. Copyright with permission from Marcel Dekker]

soluble products accumulate. Solution pH often rises. These conditions are conducive to neogenesis, especially of complex aluminosilicate clay minerals such as smectites and palygorskite.

The common primary aluminosilicates (feldspars, olivines, pyroxenes, amphiboles and micas) (Table 1.1) can be considered as salts of H_4SiO_4 and of bases comprised of the appropriate cations (Carroll, 1970; Chamley, 1989). Hydrolysis caused by weathering is a chemical reaction between the aluminosilicate salt and water to form an acid (H_4SiO_4) and a base (secondary aluminosilicate). The hydrolysis of K-feldspar to kaolinite is described in Equation [1.1].

$$2KAlSi_3O_8 + 11H_2O \rightarrow Al_2Si_2O_5(OH)_4 + 4H_4SiO_4 + 2K^+ + 2OH^-$$

K – feldspar   Kaolinite   Silicic acid   [1.1]

Usually, water involved in weathering reactions contains H^+ ions and organic ligands, as well as other reagents which aid the breakdown of primary minerals. When H^+ ions are present, the hydrolysis reaction of K-feldspar becomes:

$$3KAlSi_3O_8 + 6H_2O + 7H_2O \rightarrow Al_2Si_2O_5(OH)_4 + 7H_4SiO_4 + 3K^+$$ [1.2]

1.3 Driving Forces in Mineral Alteration

Minerals are altered by hydrolysis in water, usually containing H^+ ions, organic chelating agents and other soluble reagents, because they are out of equilibrium with the environment at the earth's surface. On the basis of observations of a number of weathering situations, Goldich (1938) drew up a stability series for the common primary minerals which is given in Fig. 1.2. Recently, Franke and Teschner-Steinhardt (1994) have largely confirmed Goldich's (1938) series experimentally, by comparing the rates of dissolution of a number of minerals under similar conditions over 8 years. The stability of the

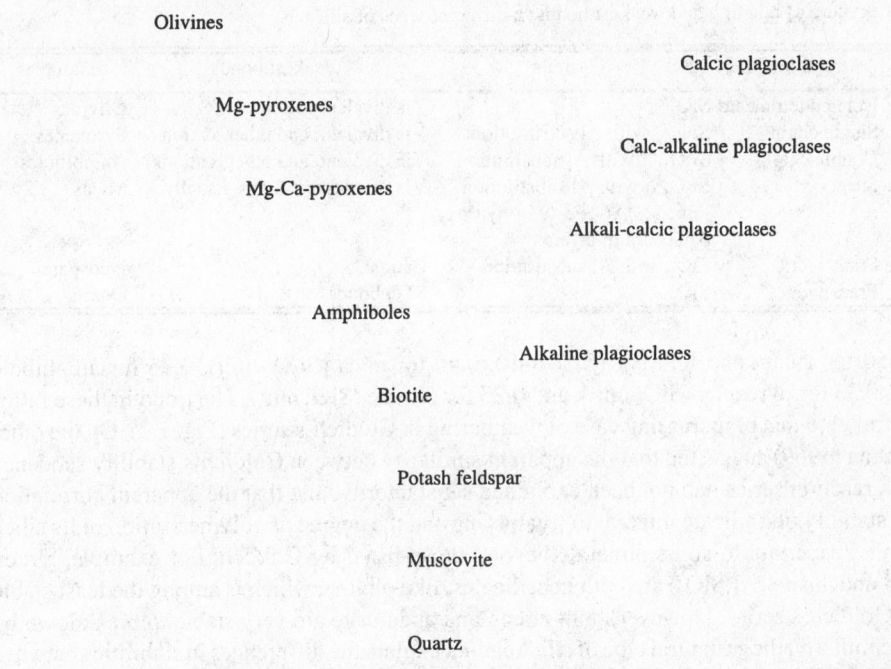

Olivines

Calcic plagioclases

Mg-pyroxenes

Calc-alkaline plagioclases

Mg-Ca-pyroxenes

Alkali-calcic plagioclases

Amphiboles

Alkaline plagioclases

Biotite

Potash feldspar

Muscovite

Quartz

Fig. 1.2 Stability series for the common primary minerals [After Goldich, 1938. A study in rock weathering. J. Geol. 46:17-58. Copyright with permission of University of Chicago Press]

major rock-forming minerals to mild acid hydrolysis decreased in the order: muscovite > alkali-feldspar > biotite > amphibole/pyroxene > plagioclase > nepheline > olivine > calcite (Franke and Teschner-Steinhardt, 1994). Goldich (1938) observed that the stability series shown in Fig. 1.2 had the minerals in an identical order to Bowen's reaction series, which describes the order in which the minerals crystallized out of a magma as temperature fell (Bowen, 1922). According to Goldich, the correspondence between the two series was explainable by reasoning that each of the minerals was out of equilibrium with the earth's surface environment to a greater extent relative to the other minerals when it crystallized at a higher temperature. The higher the formation temperature of a silicate mineral, the more susceptible it would be to breakdown by weathering at the earth's surface.

The alteration of a particular mineral begins with the disruption of its weakest bond. The bonding in all of the common classes of silicates is based on silica tetrahedra, but varies according to: (1) the manner in which the tetrahedra are joined, (2) the extent of substitution of Si in the tetrahedra by lower-charged Al, and (3) the extent of incorporation and location of ions, which inevitably are cations, in the structures in order to balance the (negative) charge consequent upon substitution of Si by Al in tetrahedra.

The identity of the charge-balancing cation can play an important role in governing the susceptibility of the minerals to alteration. Table 1.2 describes the nature of bonding in the common classes of silicates and also silica, and identifies the weakest bond in the structure. The relative susceptibility of the different primary silicates to alteration may be explained by the nature of the different kinds of bonds in the minerals (Sposito, 1989). Because the Si-O bond is more highly covalent than typical metal-O bonds, the relative resistance of the primary minerals to decomposition by weathering is related to the Si:O molar ratio of its fundamental silicate structural unit. These molar

Table 1.2 The nature of bonding and weakest bonds in different types of silicates

Name	Structural type	Formula	Weakest bonds	Examples
Neosilicates	Isolated tetrahedra	SiO_4^{2-}	Via divalent cations	Olivines
Inosilicates	Single chains	SiO_3^{2-}, with Al substitution	Via divalent, and other, cations	Pyroxenes
Inosilicates	Double chains	$Si_4O_{11}^{6-}$, with Al substitution	Via divalent, and other, cations	Amphiboles
Phyllosilicates	Sheet	$Si_2O_5^{2-}$, with Al substitution, joined to Al-, Fe-, Mg-hydroxy octahedra in layers	Via interlayer cations, usually K^+	Micas
Tectosilicates	Framework	SiO_2, with Al substitution	Cations	Feldspars
Tectosilicates	Framework	SiO_2	Si-O bonds	Quartz

ratios are 0.50 for quartz and feldspar (SiO_2 unit), 0.40 for mica (Si_2O_5 unit), 0.36 for amphibole (Si_4O_{11} unit), 0.33 for pyroxene (SiO_3 unit), and 0.25 for olivine (SiO_4 unit). The trend in these ratios is broadly parallel to that of increasing ease of weathering in Goldich's series (Fig. 1.2). On the other hand, Loughnan (1969) suggested that the apparent similarity between Goldich's stability sequence and Bowen's reaction series had not been explained satisfactorily, and that the apparent correlation between the stability of a silicate mineral to weathering and the degree of polymerization of its silica tetrahedra did not pertain to some minerals beyond those listed by Goldich. For example, zircon ($ZrSiO_4$) and andalusite (Al_2SiO_5) are both neosilicates, like olivine, which is among the least stable minerals in Goldich's series. However, both zircon and andalusite are very stable to breakdown by weathering. Similarly, the structural type of silica cannot explain the differences in stabilities between the different feldspars, albite and anorthite, which are both tectosilicates, nor those between the different micas, muscovite and biotite, which are both phyllosilicates.

Within primary minerals, ions are susceptible to exchange by other ions in the leaching solutions. The breakdown of micas, through their transformation to expansible 2:1 minerals, especially vermiculite and some smectites, typically occurs in the first instance through exchange of the interlayer cation, most commonly K, by a hydrated cation, e.g., Mg^{2+} or Ca^{2+}. Under acid conditions, which are normal in weathering, the H^+ ion could effect some simple transformation of micas to vermiculite and/or smectite, but may also bring about concomitant breakdown of the aluminosilicate structure, especially at pH values low enough for the solubilization of Al (Fanning et al., 1989).

Weathering usually occurs in an acidic environment, and commonly, also in the presence of a range of organic compounds. Hence, the breakdown of silicates such as feldspars, olivines, pyroxenes and amphiboles with charge-balancing ions within their structural framework, as well as that of micas, is often enhanced by the displacement of these ions by protons and/or their complexation by organic ligands. The strong effects of complexing organic acids are shown by their capacity to alter the order of mineral stabilities from those in Goldich's (1938) series, which has otherwise been found to be generally applicable (Huang, 1989).

Iron in minerals often plays a particularly important role in weathering. According to Millot (1970), the most important reactions in weathering involve Fe. Reference to Table 1.1 shows that Fe is quite common in primary minerals, especially those with a high susceptibility to weathering (Fig. 1.2). Along with the less abundant Mn, Fe generally occurs in its reduced form in primary minerals. Upon exposure to the atmosphere, Fe(II) and also Mn(II) become oxidized. The change to Fe(III) results in an imbalance in the overall electrostatic charge of the crystal structure. This imbalance is corrected by the loss of other ions from the structure and the mineral becomes particularly susceptible to further breakdown by hydrolysis. Among other effects, that of the oxidation of Fe leads to the almost ubiquitous occurrence of Fe(III) oxides and oxyhydroxides in soils. Brinkman (1970) characterized as ferrolysis, a cyclical process of alternating reduction and oxidation, leading to clay decomposition.

Exchangeable Fe^{2+} ions are produced from Fe(III) oxides during the reduction phase, which occurs when soils are wet, while oxidation on drying leads to Fe(III) oxides, together with exchangeable H^+, which attacks the clay mineral by hydrolysis.

Regardless of the composition or structure of the mineral, it has been found universally that the degree of alteration of primary minerals depends upon the intensity of leaching of the mineral by water. The difference between the composition of the aqueous solution and the solid mineral provides the chemical potential energy for the alteration of minerals. The rate of leaching, by dictating the rate of removal of dissolved solutes, thereby influences not only the rate, but also the course, of mineral alteration. Given sufficient leaching, almost all rocks will leave a residue which is largely composed of the relatively insoluble oxides of Al, Fe and Ti (Garrels and Mackenzie, 1971). As will be shown in the next and subsequent sections, the nature of the weathering products also depends very much upon the extent of leaching.

1.4 Processes and Products of Mineral Weathering

Until the publication of the pioneering paper (Hendricks and Fry, 1930) on the examination of fine fractions of soils (soil colloids) by X-ray diffraction, most workers had regarded these materials, which were indistinct or not visible in optical microscopes, to be amorphous. The application of this one highly relevant technique was to revolutionize the understanding of the products of mineral weathering (Cady and Flach, 1997). The revelation that clay sized components of soils were crystalline, confirming the predictions of some farsighted early workers (Ross, 1928) led to much work on their identification, mainly with X-ray diffraction. This work was facilitated with the advent of goniometers, which gave quicker results than cameras and soon largely supplanted them for the collection of X-ray diffraction data. Although X-ray diffraction remains the central tool in the identification and characterization of clay sized minerals, both aluminosilicate clay minerals *sensu stricto* and also associated oxides, hydroxides and oxyhydroxides, and advances in understanding of these materials have been closely associated with the application of new instrumental techniques to their study.

A weathering sequence for the clay size minerals in soils and sediments that was put forward in 1948 (Jackson et al., 1948) (Table 1.3) has largely stood the test of time as a description of the relative stabilities of these minerals, although not all types of clay minerals had even been identified at that time (Kittrick, 1986). It should be noted that Jackson et al. (1948) (Table 1.3) included primary minerals, as well as secondary products of weathering in their proposed weathering sequence. These authors wished to show the complexity of the weathering system so that, while they conceded the fundamental principle that a specific parent mineral was the source of each colloidal secondary clay mineral, the whole system was seen as a summation of such binary transformations which occurred simultaneously. As observed in Section 1.2, their pioneering approach was a recognition of the reality of the continual cycling of mineral alteration and formation by weathering. In subsequent studies, however, many researchers have explored the nature of the binary transformations. They have sought to link particular secondary products to their primary source minerals and also to the alteration and formation processes involved. Their various observations will now be discussed from the perspectives of the source minerals, their products, and also the processes.

1.4.1 From Olivines, Pyroxenes, and Amphiboles

Olivines, which are among the most unstable minerals in terms of Goldich's (1938) reaction series for primary minerals, release Mg^{2+}, Fe^{2+} (or Mn^{2+} if manganous) easily on weathering to give, first, serpentine, then a smectite (Loughnan, 1969, Nahon et al., 1982). Using optical microscopy, Craig and

Table 1.3 Weathering sequence of clay-size minerals in soils and sedimentary deposits [After Jackson et al., 1948. Weathering sequence of clay-size minerals in soils and sediments. J. Phys. Coll. Chem. 52:1237-1260 with permission of the American Chemical Society]

1.	Gypsum (also halite, etc.)
2.	Calcite (also dolomite, aragonite, etc.)
3.	Olivine-hornblende (also diopside, etc.)
4.	Biotite (also glauconite, chlorite, antigorite, etc.)
5.	Albite (also anorthite, microcline, stilbite, etc.)
6.	Quartz (also cristobalite, etc.)
7.	Illite (also muscovite, sericite, etc.)
8.	Hydrous mica-intermediates
9.	Montmorillonite (also beidellite, etc.)
10.	Kaolinite (also halloysite, etc.)
11.	Gibbsite (also boehmite, etc.)
12.	Hematite (also goethite, limonite, etc.)
13.	Anatase (also rutile, ilmenite, corundum, etc.)

Loughnan (1964) identified a brownish rim on surfaces of weathering olivine from the field as iddingsite. With X-ray diffraction, they found that it contained a trioctahedral smectite which they supposed to be saponite, and also goethite. Later, using electron microscopy, Eggleton (1984) confirmed the identification of iddingsite as saponite with goethite. However, the observation that an aluminous dioctahedral smectite formed from the Mg-olivine, forsterite Mg_2SiO_4, containing very little Al (Nahon et al., 1982) suggests that secondary minerals may receive inputs from more than one primary mineral; in this particular case, a pyroxene (enstatite) was the likely source of Al (Nahon et al., 1982). When olivines are subjected to stronger leaching, a variety of products results, including poorly ordered smectite, kaolinite, halloysite, and various oxides, hydroxides and oxyhydroxides of Fe and, if present in the olivine, also of Mn (Huang, 1989).

Pyroxenes, exemplified by augite in Goldich's (1938) series (Table 1.2), are slightly more stable than olivines. They tend to weather through loss of Mg, Ca and Fe(II) to chlorite and/or smectite, although formation of calcite may occur when the rate of removal of Ca exceeds that of the complete breakdown of the pyroxene (Loughnan, 1969). Transmission electron microscopy enabled Eggleton and Boland (1982) to detect the stepwise alteration of orthopyroxene first to talc and then to smectite. However, the products of the weathering of augite have been found to be related to intensity of weathering, which dictates the nature of the soil environment (Loughnan, 1969; Eswaran, 1979). With strong leaching there is a breakdown of these initial complex products to the simpler compounds, kaolinite, Fe(III) oxides and anatase (Loughnan, 1969). Amphiboles, represented by hornblende in the same series (Table 1.2), appear to weather by much the same path as the slightly less stable pyroxenes (Huang, 1989).

There have been many laboratory studies of the initial stages of alteration of olivines and also pyroxenes, particularly using X-ray photoelectron spectroscopy XPS (Schott and Berner, 1985; Casey et al., 1993; Seyama et al., 1996). Together these workers, in the papers cited and also earlier publications, have found that divalent cations tend to be lost relative to silica at low pH, that oxidation can lead to the formation of a surface film of hydrated Fe(III) oxide, and that at high pH, silica tends to be lost preferentially while the films of Fe(III) oxyhydroxides grow on surfaces. They, and also earlier workers posed the question whether or not a secondary surface layer may also form on these minerals, in the same way that a layer had been found to form on laboratory-altered feldspars (Section 1.4.2). The general consensus is that any layer formed this way is only thin (probably only a few atoms thick) and does not act as a protective layer controlling diffusion from the weathering olivines, pyroxenes and also amphiboles (Huang, 1989). According to Huang (1989), there was also no evidence for the formation of a protective layer in natural weathering. It is likely that biological and biochemical processes in the soil prevent the formation of such a layer (Huang, 1989).

SEM studies summarized by Huang (1989) showed that weathering of pyroxenes and amphiboles largely occurred by the formation, enlargement and coalescence of etch pits developed on dislocation outcrops. Seyama et al. (1996), also using SEM, found that leached olivine grains showed an etched structure. Dissolution of these minerals proceeded heterogeneously. High resolution transmission electron microscopy (HRTEM) has revealed more details about the alteration of olivines, pyroxenes and amphiboles, as discussed in relation to the weathering of the feldspars (Section 1.4.2).

1.4.2   From Feldspars

Many minerals have been reported as the alteration products of feldspars (Allen and Hajek, 1989). These include micas, which form from K feldspars (Carroll, 1970), and also from plagioclases (Millot, 1970), by a process that is sometimes known as sericitization. The micaceous products are often described as sericites (Millot, 1970, Carroll, 1970), although the name sericite has also been given to a wide variety of micaceous minerals (Whitten and Brooks, 1972). With electron microscopy, Eggleton and Buseck (1980) detected micaceous phases, either illite or an interstratified illite-smectite, forming in vacuoles within microcline. Smectites have also been reported to form from feldspars in a number of studies (see Allen and Hajek, 1989). Allen and Hajek (1989) have pointed out that the lack of Mg and Fe in feldspars would mean that smectites formed from them would need to be beidellitic. Nettleton et al. (1970) identified a smectite formed directly from a feldspar as a beidellite-montmorillonite. Arocena et al. (1993), using X-ray microdiffraction (with a microcamera), light and electron microscopy and also measurements of layer charge by alkylammonium saturation, identified a number of 2:1 phyllosilicates with vermiculitic and/or smectitic character forming from sericites derived from feldspars by weathering, although the particular identifications of the weathering products should be regarded as tentative (Laird, 1994; Arocena et al., 1994).

Most often it has been considered that feldspars weather to give kaolinite or halloysite, while gibbsite (Allen and Hajek, 1989) and quartz (Estoule-Choux et al., 1995) have also been identified as forming from feldspars. In several weathering profiles (Eswaran and Wong, 1978; Calvert et al., 1980; Anand et al., 1985), the weathering of feldspars has been found to give rise to all of halloysite, kaolinite and gibbsite. The use of scanning electron microscopy (SEM) in all three studies cited enabled the detection of close associations of one or more of each of these secondary minerals with the primary feldspars. Using SEM, Eswaran and Wong (1978) identified kaolinite crystals that were pseudomorphic after feldspar, rather than hexagonal, like typical kaolinites. They further identified gibbsite crystals that were also pseudomorphic after feldspar. Furthermore, the different secondary minerals, halloysite, kaolinite and/or gibbsite sometimes occur together in close proximity (i.e., within a few μm) to each other (Calvert et al., 1980; Anand et al., 1985). Eswaran and Wong (1978), Calvert et al. (1980) and Anand et al. (1985) all concluded that the properties of micro-environments determined the particular minerals formed there. Anand et al. (1985) suggested that the ease of drainage within grains of feldspar dictated the secondary product formed. If drainage was very good, gibbsite could form. This explanation received support from considerations of the groundwater composition in relation to thermodynamic mineral stability fields (Section 1.7). Using SEM, Estoule-Choux et al. (1995) observed that newly formed quartz was associated with altered microcline. These authors were also able to find a thermodynamic explanation for the association, with congruent dissolution of microcline leading to quartz alone (without kaolinite) over a wide range of pH.

The artificial weathering of feldspars has been studied intensively (Huang, 1989; Kawano and Tomita, 1994; Blum and Stillings, 1995; Oelkers and Schott, 1995; Muir and Nesbitt, 1997). There has been much attention paid to the occurrence, nature and role of leached layers on feldspar surfaces. Huang (1989) concluded that leached layers, which may form on altering feldspars, are not thick enough to inhibit transport and that no continuous layer of secondary precipitates forms on weathered feldspar surfaces. As in the case of olivines, pyroxenes and amphiboles, dissolution of feldspars was mainly seen to occur at sites of excess energy like dislocations, rather than by uniform attack on the whole surface. Huang (1989) considered that heterogeneous attack explained the apparent disagreement between mass balance calculations (Wollast and Chou, 1985) which had indicated the occurrence of a surface layer, on the one hand, and XPS measurements (Berner and Holdren, 1979), which had failed to prove the existence of such a layer, on the other. However, later studies, using secondary ion mass spectrometry (SIMS) (Muir and Nesbitt, 1997), elastic recoil detection (ERD) and Rutherford backscattering (RBS) (Casey et al., 1988, 1989) and also XPS (Muir et al., 1989, 1990; Hellmann et al., 1990) indicated formation of a dealkalized leached layer which was as deep as 1000 Å and was especially thick at low pH. Kawano and Tomita (1994) treated albite in water at 150-225 °C and found alteration at the surface of the mineral to an amorphous phase, which showed some of the characteristics of allophane, and which later detached. Subsequently, smectite, and also small amounts of K-mica, were formed. A kinetic study of dissolution rates of anorthite showed that this particular feldspar broke down in a different way from alkali feldspars (Oelkers and Schott, 1995). Because anorthite has an Si:Al ratio of 1, the removal of Al from anorthite leaves completely detached Si tetrahedra. By contrast, the removal of Al from alkali feldspars, where the Si:Al ratio is 3, leaves Si tetrahedra which are still partially linked. Using SIMS depth profiles, Muir and Nesbitt (1997) have developed a mechanism for the leaching of plagioclase feldspars whereby Ca and Na are preferentially depleted first from within a layer which extends several hundred Å below the mineral surface, to be followed closely by Al, and eventually Si. The extent and nature of the depletion depends on the nature of the contacting solution. Depletion is aided by the availability of ligands to bind cations and is retarded by the presence of cations (Muir and Nesbitt, 1997). Nonetheless, studies of the alteration of feldspars at pHs between 5 and 8, comparable to pHs encountered in most soils, have generally shown only thin leached layers forming (Blum and Stillings, 1995). In general, if the rate of dissolution of feldspars is less than that of silica, no Si-rich layer will form (Blum and Stillings, 1995).

Nevertheless, as pointed out by Hochella and Banfield (1995), there are many aspects to the design of laboratory experiments for weathering feldspars and other minerals, which means that results are not easily applied to field situations. On the other hand, field studies on a watershed scale, such as those of Velbel (1985, 1992) produce results which may not have much relevance to field situations outside those for which they were obtained. According to Hochella and Banfield (1995), the use of HRTEM provides the key to understanding how minerals break down and how and where secondary minerals form. Among other findings from the use of this approach, these authors point to the common recognition that feldspars are often turbid as a result of having minute vesicles filled with fluid even before they enter the weathering regime. This effect reflects the high microporosity exhibited by unweathered feldspars (Worden et al., 1990). HRTEM studies have shown, further, that alteration often occurs preferentially at crystal defects in feldspars (Wilson and McHardy, 1980; Holdren and Speyer, 1987) and also that both feldspar dissolution and the formation of secondary phases occur throughout the primary mineral, not just at grain boundaries (Banfield and Eggleton, 1990; Banfield et al., 1991). HRTEM has enabled the same generalizations to be made for the alteration of olivines, pyroxenes and amphiboles (Hochella and Banfield, 1995). Using SEM and associated energy dispersive X-ray analysis (EDXRA), Inskeep et al. (1993) showed the development of etch pits and secondary coatings as weathering progressed. A kaolin layer silicate was identified in a secondary coating by EDXRA. These authors suggest that secondary coatings are an integral part of primary mineral weathering under natural conditions. Generally, as discussed previously, they have not been observed in laboratory weathering experiments. Effects such as wetting and drying, high solid/solution ratios and long water residence times are typical of the field situation but have not been reproduced in the many laboratory studies (Inskeep et al., 1993). However, it is these features of the field situation, among others, which lead to the results that are characteristic of the natural weathering of feldspars and also other minerals.

1.4.3 From Micas

The formation of phyllosilicate clay minerals, with flat sheets, from the neosilicate olivines, with independent silica tetrahedra, from the inosilicate pyroxenes and amphiboles with chains of tetrahedra, or from the tectosilicate feldspars with continuous frameworks of silica and alumina tetrahedra, clearly involves substantial structural rearrangements. As a result, they are very unlikely to undergo transformation within the solid phase to give phyllosilicates. Phyllosilicates are most likely to form by neogenesis from solutions produced by the dissolution of silicates of the other structural types, as will be discussed later. By contrast, the alteration of the phyllosilicate primary micas to clay minerals with the same structural arrangement is much more straightforward. In the initial stages, at least, the weathering of the micas involves replacement of the cations in the interlayer region between the 2:1 aluminosilicate layers and charge reduction within the layers. As a result, layers are held together less tightly than in the micas and expansion occurs. Potassium occupies the interlayers in both of the common micas, biotite and muscovite. Upon its complete replacement by a hydrated divalent cation such as Mg or Ca, the basal spacing changes from 1.0 nm for the micas to 1.4 nm for the hydrated product, vermiculite. Oxidation of Fe(II) to Fe(III) occurs along with the loss and replacement of interlayer K, especially from biotite, and results in loss of (negative) charge from the layers. Muscovite contains little Fe, but also suffers some loss of charge upon its alteration to vermiculite, so charge reduction occurs by other means in addition to the oxidation of Fe(II) (Weaver, 1989). In the case of biotite, the amount of Fe oxidized during vermiculitization is much greater than the decrease in layer charge (Newman and Brown, 1966). Evidently, other changes besides the oxidation of Fe(II) occur within layers of the mineral during its alteration to vermiculite. However, the mechanism of layer charge reduction of micas, although well studied, is not well understood (Fanning et al., 1989). Using infrared spectroscopy to examine biotites that had been weathered either naturally or artificially, Farmer et al. (1971) found evidence for losses on weathering of both protons from hydroxy groups and also Fe from octahedral sites. Both hypotheses have been confirmed by others (Fanning et al., 1989; Douglas, 1989). Other possible structural changes that have been canvassed include loss of hydroxyl ions (Stucki and Roth, 1977), incorporation of protons into the layer structure (Raman and Jackson, 1966; Leonard and Weed, 1967) and exchange of Si for Al in the tetrahedral sheet (Jackson, 1964; Sridhar and Jackson, 1974; Vicente et al., 1977).

Complete transformation of micas to vermiculite may occur rarely, especially in soils. Instead, only a portion of the K in the interlayer region of micas is usually removed. Many minerals characterized as illites may originate in this way from micas which have first been ground or otherwise been reduced in particle size. There has been much debate about the definition of illites. Illite was originally proposed by Grim et al. (1937) as "a general term for the mica-like clay minerals occurring in argillaceous sediments" which also "showed substantially no expanding-lattice characteristics". In practice, however, minerals described as illites often contain less K and more water than muscovite or biotite and often also contain some expanding layers (Grim, 1968; Wentworth, 1970; Norrish and Pickering, 1983; Weaver, 1989). Fine illites which occurred in two-layer elementary particles comprising alternate swelling and non-swelling layers were isolated from soils by Laird and Nater (1993). Illites which are depleted in K relative to their parent micas occur commonly. Together these observations probably reflect the conclusions reached in many reports that (1) K is lost more readily from fine grained micas than from coarser micas, but also (2) small mica particles often retain a small but significant portion of their K that is highly recalcitrant against replacement, both in nature and in laboratory treatments (Fanning et al., 1989).

Hence, basal spacings often show expansion beyond 1.0 nm, but they do not reach the vermiculite spacing of 1.4 nm. This is because expanded, K-depleted vermiculite layers have become interstratified with the original mica layers. Although sometimes characterized as illites, as has been noted, the products are often recognized instead as interstratified mica (or illite)-vermiculites. These may be random, in which case there is no ordering in the stacking of the layers. Sometimes, however, the ordering of mica and vermiculite layers in the interstratified phase is regular. More often than not in regular interstratifications, there is an alternation of the two types of layers in equal amounts in each crystal. As a result, the basal spacing of the phase, denoting the repeat unit, is a sum of the spacings of the component mica (1.0 nm) and vermiculite (1.4 nm) units, i.e., 2.4 nm.

The XRD peaks for the basal planes of soil clay fractions shown in Fig. 1.3 are all from regular interstratifications in a New Zealand podzolized soil (Churchman, 1978). Only that for the lowest (uA_3) horizon shown represents a mica-vermiculite. Expansion of the samples from the other horizons to 2.7-2.8 nm upon glycerol solvation indicates a regular interstratification of mica and a smectite, which is formed in an analogous manner to those of mica and vermiculite (see later).

Regularity in interstratifications has been rationalized by supposing that the replacement of the K ions in one interlayer by a hydrated divalent ion with a longer bond to the silicate sheet than the K ion has would strengthen the bonding of K within the adjacent interlayer (Bassett, 1959). Hence the next K ions, which are held most weakly, are contained in the interlayers beyond those adjacent to the depleted interlayer and these are replaced by further hydrated divalent ions. Norrish (1973) has provided an explanation for the effect by proposing that structural hydroxyl groups shift toward the opened interlayer region as K is removed from it. They, and particularly their protons, therefore, become more distant from the K in adjacent closed interlayers, hence increasing the stability of these particular K ions against removal. Norrish's (1973) explanation is generally accepted, although untested (Fanning et al., 1989). Regular interstratifications of micas with vermiculite, and also with other transformation products of micas, tend to form in soils in colder climates such as in upland Yugoslavia (Gjems, 1970), Scotland (Wilson, 1970), Scandinavia (Kapoor, 1973) and South Island, New Zealand (Churchman, 1980), where only a slight chemical potential provides the driving force for a slower, gentler transformation than takes place in warmer climates. Irregular interstratifications result from partial transformations of micas in the warmer climates.

The genesis of regularly interstratified phases of micas with their transformation products such as vermiculite necessarily involve the substantially complete removal of K from (generally alternate) interlayers, by the so-called layer weathering model of Jackson et al. (1952). According to the alternative-edge weathering model of Mortland (1958), transformation of micas takes place simultaneously in different interlayers from their edges and from the edges of fractures. This model has gained credence from a number of studies of artificially weathered micas which have revealed fraying and expansion of the edges of mica flakes. These flakes are often large and it is thought that, while edge weathering may be common in particles which are larger than clay size, layer weathering is most common with clay sized micas (Fanning et al., 1989). In some cases, both mechanisms may occur together (Fanning et al., 1989).

Trioctahedral micas, including biotite and phlogopite, weather more readily than dioctahedral micas, including muscovite and also most illites, in agreement with Goldich's series (Fig. 1.2). It has been known for some time (Mortland et al., 1956) that plant growth, by extracting K from the mineral, can effect the transformation of biotite to vermiculite. This process can occur rapidly, within a few weeks (Hinsinger et al., 1992). However, dioctahedral vermiculites are more common than their trioctahedral counterparts (Jackson, 1959). The greater abundance of the dioctahedral vermiculites than the trioctahedral varieties reflects the low stabilities of the latter. Furthermore, there is a tendency toward an increase in the dioctahedral character of vermiculites relative to their parent biotites as a result of loss of ions from the octahedral sites and migration of some Al from tetrahedral to octahedral sites during weathering (Douglas, 1989). In addition, conditions of active leaching, pHs between 4.6 and 5.8, low organic matter contents, and frequent wetting and drying all favor the deposition of Al-hydroxy species in the interlayers of vermiculites (and also smectites) (Rich, 1968). These conditions occur frequently in soils. The hydroxy-Al interlayered vermiculite which results is sometimes known as pedogenic chlorite, as well as by many other names (Barnhisel and Bertsch, 1989). The interlayers so formed greatly stabilize vermiculites against further breakdown (Douglas, 1989). Nevertheless, hydroxy-Al interlayers may be destroyed, at least partially, by certain laboratory treatments, including

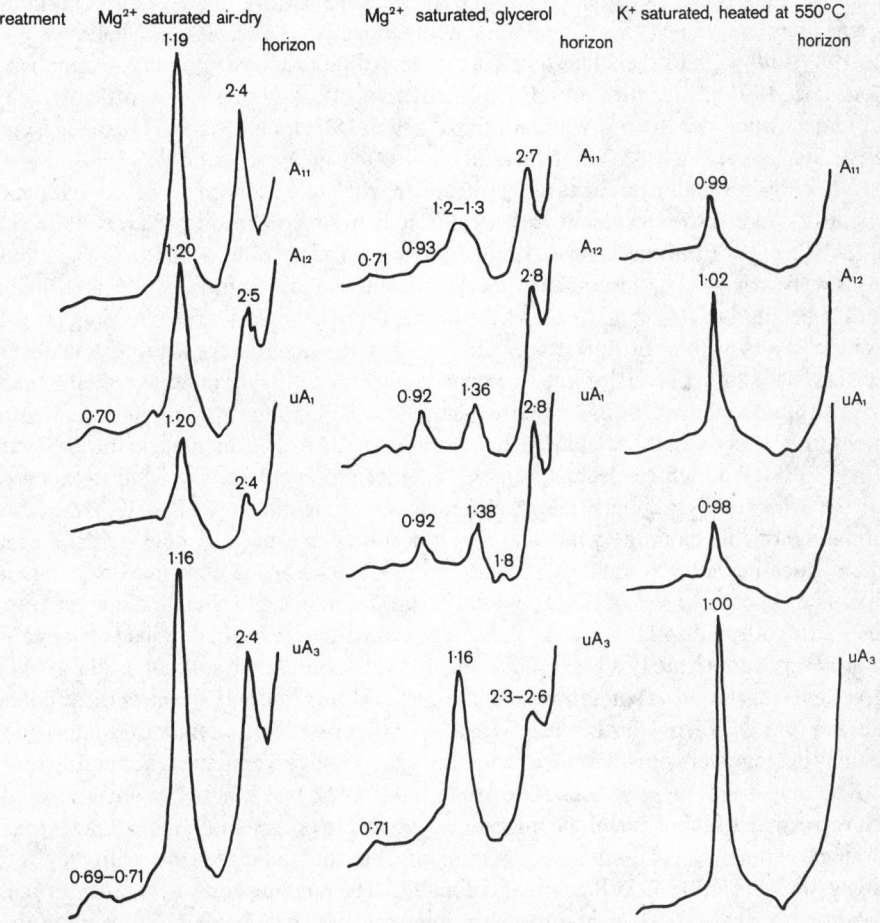

Fig. 1.3 Peaks in X-ray diffraction patterns for the basal planes of regularly interstratified phases in the clay fractions of different upper (A) horizons of a podzolized soil following diagnostic treatments. Peak spacings in nm [Reprinted from Churchman, 1978. Studies on a climosequence of soils in tussock grasslands. NZ J. Sci. 21:467-480 with permission of the DSIR, New Zealand]

the once commonly used citrate-dithionite-bicarbonate CDB pretreatment for soil clays to remove free oxides, and hence, increase the intensities of peaks in X-ray diffraction (Mehra and Jackson, 1960). Citrate complexes with Al and extracts it from the interlayers. When analyses of the clay fractions of the various horizons of a set of soils formed on loess deposits containing micas (and chlorites) in New Zealand were carried out both before and after CDB treatments, it was found that pedogenic chlorite was often abundant in untreated soils (Churchman and Bruce, 1988). Although its content was very often decreased as a result of CDB treatment, some of the hydroxy-Al interlayers, and hence pedogenic chlorite, sometimes remained. Soils could be distinguished from one another on the basis of both the relative abundances of pedogenic chlorite and its relative stability to CDB treatment. Both reflect the courses of weathering in the different soils and provide information on weathering processes that would have been unavailable if analyses had been carried out only after CDB treatments or even without any pretreatments.

Micas may be further transformed beyond vermiculite to give a smectite. Micas may also transform directly to smectites (Aragoneses and García-González, 1991). Formation of smectites by transformation of micas occurs in the eluviated horizons of some podzols under acidic conditions. The resulting smectites are likely to be beidellitic (Ross and Mortland, 1966; Churchman, 1980; Borchardt, 1989), although further alteration to a montmorillonite appears possible (Aragoneses and García-González, 1991). Smectites are also formed from micas within soil profiles which have developed under strong weathering without podzolization (Stoch and Sikora, 1976; Egashira and Tsuda, 1983; Senkayi et al., 1983; Singh and Gilkes, 1991; van Wesemael et al., 1995). Smectites, along with other transformation products of (biotite) mica, viz., illite, "chlorite", mixed-layer minerals and vermiculite, were also considered to have formed in Antarctic soils (Boyer, 1975). When analyzed, the smectites from these various situations were found to be beidellitic. Their mode of formation, as well as those of the smectites formed as a result of podzolization, contrasts strongly with those of most smectites, which tend to form where drainage is poor and pH is relatively high (see later).

Vermiculite layers within an interstratification with mica layers may also be transformed to smectite layers, while other layers of mica within the same mixed-layer mineral remain apparently unaltered. In a climosequence of soils studied by Churchman (1978), the main products of weathering of a mica-chlorite schist were those of the transformation of mica. The minerals in the clay fractions of the surface soils fell into three clearcut groups: i.e., mica present as 75-90% of the total clay sized minerals in the 3 driest soils; regularly interstratified mica-vermiculite present as 70-90% of this total in soils in the intermediate rainfall zone, and also in a soil at an especially cold site; and regularly interstratified mica-beidellite present as virtually all (95-100%) of the clay sized minerals in two wetter soils. The upper three sets of XRD patterns in Fig. 1.3 indicate the mica-beidellites formed in one of these latter soils. Subsoil B horizons of the two soils with mica-beidellites had clay fractions in which the minerals were virtually all (85-100%) mica-vermiculite. These soils were all formed under tussock grassland vegetation. A soil formed under forest in close proximity to one of those containing mica-beidellite had weathered further to give an A horizon with clay minerals comprising 50% discrete beidellite, together with 45% mica-beidellite and 5% mica-vermiculite (Churchman, 1980). When the K_2O contents of these various A horizons and also the two B horizons are compared (Fig. 1.4), it can be seen that K contents of the mica-rich group were consistently higher than those of the soils with mica-vermiculite as dominant clay minerals, but these and the soils with clay fractions dominated by mica-beidellite had very similar K contents. The K content only decreased further when discrete beidellite appeared in place of some of the interstratified mica-beidellite. The transformation from vermiculite layers to smectite layers within the interstratified minerals would have involved a decrease in the charge of the appropriate layers, but it effected no further discernible loss of K from the interlayers of the associated mica.

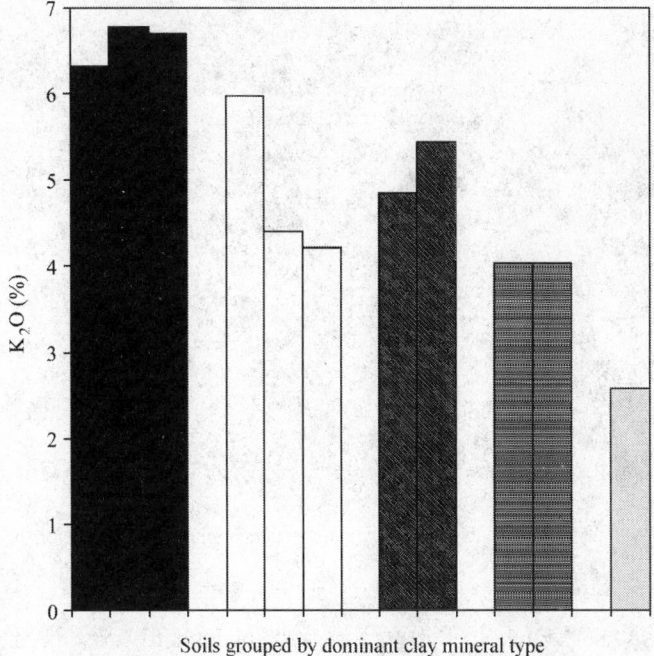

Fig. 1.4 K$_2$O contents of clay fractions of soils from a climosequence and associated vegesequence in relation to their dominant clay mineralogy. Groups of soil clays comprise, from l. to r.: 1. Predominantly mica, from A horizons (3 soils); 2. Predominantly mica-vermiculite, from A horizons (3 soils); 3. Predominantly mica-beidellite, from A horizons (2 soils); 4. Predominantly mica-beidellite, from B horizons (2 soils); and 5. Beidellite (50%), together with mica-beidellite (45%), from an A horizon. [Data from Churchman, 1978; 1980]

In general, the distinction between vermiculites and smectites may not always be clearcut, especially for these types of minerals in soils (Douglas, 1989). Vermiculite has been defined as a phyllosilicate with a charge of between 0.6 and 0.9 per formula unit (Bailey, 1980). In practice, however, vermiculites are usually distinguished from smectites by their inability to expand upon solvation with glycerol or ethylene glycol. Problems arise because, on one hand, some phyllosilicates which expand on solvation with these liquids have the charge of vermiculites (Douglas, 1989), while on another, ethylene glycol can expand layers with a higher charge than is possible with glycerol (Srodon and Eberl, 1980). The problem of nomenclature and definition is quite common for clay minerals, as will be seen in the discussion on halloysite and kaolinite, for example.

Under strong weathering, micas are broken down to form other mineral phases in addition to their vermiculitic and smectitic transformation products. Kaolinite is often formed from micas under strong weathering, such as has occurred in tropical climates in Nigeria (Ojanuga, 1973), in Tertiary weathering in Europe (Stoch and Sikora, 1976), in the formation of lateritic soils in Western Australia (Gilkes and Suddhiprakarn, 1979) and also in deeply weathered soils in the continental United States (Rebertus et al., 1986). The appearance in scanning electron micrographs of pseudomorphs of kaolin after the micas points to their genesis from the micas. The occurrence of tubular crystals of halloysite, as well as kaolinite crystals on weathering mica flakes (Eswaran and Hong, 1976) shows that halloysite may also form from micas. In Fig. 1.5, a scanning electron micrograph shows the development of kaolinite on the edges and also between opened lamellae of large flakes of mica. The studies cited have

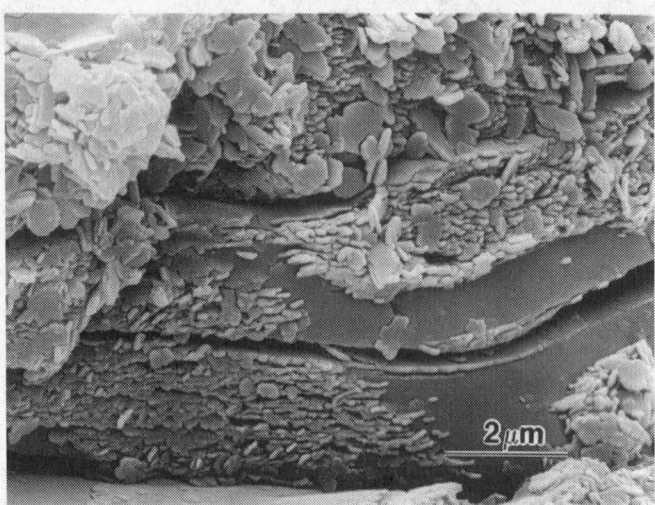

Fig. 1.5   Scanning electron micrograph showing the development of kaolinite crystals from flakes of mica at Birdwood in South Australia

all been concerned with the alteration of biotite to a kaolin mineral. Robertson and Eggleton (1991), using optical microscopy, SEM and TEM, have demonstrated the close association of kaolinite with muscovite. Under TEM, the layers of the two minerals are seen to be in intimate contact, consistent with topotactic alteration of the muscovite to kaolinite.

Iron oxhydroxides have also been identified as products of the weathering of micas. Mostly this has been done by microscopic examination of isolated weathering flakes. Using an optical microscope, Sousa and Eswaran (1975) identified goethite on weathered biotite flakes which was pseudomorphic after the biotite. When they examined the flakes more closely using SEM, these workers observed the simultaneous pseudomorphic alteration of biotite to both kaolinite and geothite. Within one soil profile, Eswaran and Hong (1976) found flakes deep in the profile that had been altered to halloysite and others nearby that had been altered to kaolinite, suggesting these kaolin minerals as the initial products of biotite weathering. They found biotite flakes which were covered with aggregates of goethite higher in the profile. Higher still, where weathering would have been most advanced, they identified vermiculite. The occurrence of this product of the early weathering of mica was attributed to its preservation by coatings of goethite. Gilkes and Suddhiprakarn (1979) found that exfoliated grains of weathered biotite included vermiculite, interstratified mica-vermiculite (or hydrobiotite), kaolinite, goethite and also hematite, as well as gibbsite. Banfield and Eggleton (1988) found that dissolution of biotite occurred even at the earliest stages of weathering. Dissolution helped to open up the structure (Banfield and Eggleton, 1988). Hence kaolinite and goethite were formed, albeit in minor amounts, at the same time as vermiculitization was proceeding. Banfield and Eggleton (1988) observed that goethite formed from a poorly crystalline Fe oxyhydroxide precursor. At the later stages, formation of kaolinite and goethite was the dominant process. Using analytical electron microscopy, Aodjit et al. (1996) observed gels which were rich in Fe and Al on the surfaces of weathering micas. XPS analyses of biotite dissolving in acid solutions showed increasing depletion of Al, K and Fe as leaching proceeded (Seyama et al., 1997). Using Mössbauer spectroscopy, Rice and Williams (1969) showed that, as weathering proceeded, the Fe(II) content of flakes of biotite decreased while that of Fe(III) stayed constant. Aldridge and Churchman (1991) also used Mössbauer spectroscopy, combined with extractable Fe data, to study changes in the nature and location of Fe within the climosequence of soils that was characterized mineralogically by Churchman (1978). In this sequence, where the feldspar remains fresh and weathering products of only mica and, in minor proportions, also chlorite, appear, the changes in Fe can be largely attributed to Fe within mica and its weathering products. The proportion of Fe(II) within aluminosilicate layers decreased with weathering. The clay fractions of the two soils with dominant mica-beidellite showed no Fe(II) within aluminosilicate layers. There was no trend in the proportion of Fe(III) within the layers but the proportion of Fe in oxyhydroxides increased. The results of these two studies with Mössbauer spectroscopy support the indications from laboratory studies of weathering (Farmer et al., 1971) that weathering of micas leads to the release, as Fe (III) oxyhydroxides, of Fe which had been present in micaceous layers as Fe(II). Titanium dioxide may also form as a result of the release of structural Ti during the decomposition of biotite, although information on this particular transformation appears to be sparse (Milnes and Fitzpatrick, 1989).

1.4.4 From Chlorites

Chlorites occur most often as trioctahedral minerals in low grade metamorphic rocks and as early weathering products of ferromagnesian minerals like augite, hornblende and biotite. They are not common in soils. This is mainly due to their low stabilities. They do not appear in Goldich's stability series (Fig. 1.2), but are included in Jackson et al.'s (1948) weathering sequence at the same stage as biotite. As phyllosilicates, they tend to alter initially through loss of their interlayer species, which are hydroxides of Mg, most commonly, but also of Fe and other cations. The initial stages of their alteration are analogous to those of micas. Products of the initial stages of alteration of chlorites have been found to include a randomly interstratified hydrous phase (Churchman, 1980; Weaver, 1989, Murakami et al., 1996), regularly interstratified chlorite-vermiculite (Johnson, 1964; Herbillon and Makumbi, 1975; Churchman, 1980, Murakami et al., 1996), vermiculite (Loveland and Bullock, 1975; Murakami et al., 1996), nontronite (Herbillon and Makumbi, 1975) and kaolinite (Herbillon and Makumbi, 1975; Cho and Mermut, 1992; Murakami et al., 1996). Using both SEM and TEM, Cho and Mermut (1992) found that halloysite was more often formed than kaolinite from a ferruginous chlorite. Churchman (1980) also identified a regular interstratification of chlorite and swelling chlorite among the weathering products of chlorites. Swelling chlorite is a form of chlorite that has been depleted of some, but not all, of its interlayer hydroxides (Stephen and MacEwan, 1951; Bain and Russell, 1981). Its regular interstratification with chlorite was characterized by Stephen and MacEwan (1951).

Separation of the weathering products of chlorites from those of associated rock minerals has often been difficult, but Righi et al. (1993) applied high gradient magnetic separation for this purpose and then identified a mineral which was partly chloritized by hydroxy-Mg interlayers, a randomly interstratified chlorite-smectite and also an interstratified illite-smectite as products of the weathering of chlorite. However, deduction of the weathering pathways is complicated by the close association of chlorite with mica in the parent material. Chlorite is commonly associated with other minerals, particularly micas, in parent rocks. Even though it may appear that only chlorite has weathered from these associations, loss of some K from micas, for example, could have occurred without the clear formation of any secondary phases from the associated minerals. The released ions such as K^+ may well participate in subsequent secondary phases, e.g., illite-smectites which are mainly derived from chlorite but which also receive essential components from minerals other than chlorite. Carnicelli et al. (1997) observed the formation of illite-vermiculite during the weathering of chlorite in a Spodosol and also inferred the involvement of K^+ from micas. These workers found that a low-charge dioctahedral smectite also formed from chlorite in this soil. IR spectroscopy led them to characterize chlorite weathering by the oxidation, then the expulsion of interlayer Fe(II), and the expulsion of Mg from both interlayers and the octahedral sheets of the aluminosilicate layers. Iron very likely became oxidized, then expelled, from the octahedral sheets, and Al was incorporated in these sheets. The hypothesized process mirrors that of the weathering of micas in acid leaching, particularly in podzols (Douglas, 1989) and also confirms indications from the laboratory alteration of chlorite (Ross and Kodama, 1974). Iron oxyhydroxides have been observed in association with weathered chlorites or their alteration products (Herbillon and Makumbi, 1975; Anand and Gilkes, 1984; Cho and Mermut, 1992; Murakami et al., 1996; Carnicelli et al., 1997). In several studies, chlorite was considered to decompose, generally after some interstratification, to give usually unspecified nonsilicates (McKeague and Brydon, 1970; Bain, 1977; Churchman, 1980). Because trioctahedral chlorite is not necessarily a primary mineral, it may be formed simultaneously with some of its alteration products, so that they occur together. This appears to be the case for a chlorite, a regularly interstratified chlorite-vermiculite and a (nontronitic) smectite in soils formed on serpentinite (Ducloux et al., 1976). There is general consensus that trioctahedral chlorites and even their mixed layer derivatives are found only in soils where there has been only slight weathering.

1.4.5   From Volcanic Parent Materials

However, by no means all secondary minerals in soils can be attributed to the alteration of just one type of parent material. In the formation of soils from volcanic parent materials, which are very important in many parts of the world, one of the most productive parent minerals, volcanic glass, is non-crystalline. It dissolves rapidly. Together with other components of the volcanic ejecta, which can include ferromagnesian minerals and feldspars, its dissolution leads to solutions with different compositions, but especially containing dissolved silica and their cations. It is from these solutions that the secondary products form by neogenesis and crystallize out as new solid phases.

It is the occurrence of clay minerals with short-range order which especially characterizes almost all tephra-derived soils (Brindley, 1977; Lowe, 1986). Allophane is the most common of these minerals. Its origin, structure, and stability have been the subject of much investigation and debate (Fieldes, 1955; Parfitt et al., 1983; Lowe, 1986; Wada, 1989, 1995). The lack of long-range crystalline order in allophane has meant that its identification and characterization have had to be performed by other techniques besides X-ray diffraction. Infrared spectroscopy, differential thermal analyses (DTA) and (transmission) electron microscopy were at the forefront early in studies of this mineral (Fieldes, 1955). They have been supplemented by such modern techniques as electron diffraction (Wada and Yoshinaga, 1969) small angle neutron scattering (Hall et al., 1985), ^{29}Si and ^{27}Al nuclear magnetic resonance (NMR) (Goodman et al., 1985) and X-ray photoelectron spectroscopy (XPS) (He et al., 1995). Yet quantification of allophane and the determination of its all important Al:Si elemental ratios are routinely carried out virtually universally by extractable element determinations which are based on methods developed as long ago as 1922 by Tamm (1922) (Parfitt and Henmi, 1980; Wada, 1989).

Since it was first identified (Yoshinaga and Aomine, 1962), imogolite has also been found in some volcanically derived materials. It has better long-range order than allophane, as indicated by several peaks in XRD (Yoshinaga and Aomine, 1962). It is generally accepted that most forms of allophane in soils have a structure which is based on that of imogolite (Cradwick et al. 1972), with a defect gibbsitic Al-octahedral sheet framework (Parfitt and Henmi, 1980; Lowe, 1986; Wada, 1989). Parfitt and Henmi (1980) and Parfitt et al. (1980) suggested that allophanes in soils in New Zealand may be of either of two types: (1) Al rich, with Al:Si ratios near 2:1; these are imogolite like, or proto-imogolite allophanes, and (2) Si rich with Al:Si ratios near 1:1; these are halloysite-like allophanes (Parfitt, 1990). Imogolite *per se* is more aluminous than allophane. It may form by desilication of allophane or directly from weathering solutions, often in close association with allophane (Wada, 1989).

Allophane forms as a result of the weathering of rhyolitic, andestic and basaltic parent materials (Fieldes, 1968; Wada, 1980,1989; Churchman and Tate, 1987). Halloysite can form from these same materials (Fieldes and Swindale, 1954; Fieldes, 1968; Churchman and Theng, 1984; Chamley, 1989). Reasonably recently, there has been a shift in understanding about the genetic relationship between allophane and halloysite in weathered products of volcanic materials. Fieldes (1955) suggested that allophane altered to halloysite with time. The corollary of this, that allophanes tend to have a shorter residence time than halloysites in soils, was generally accepted for the next 20 years, at least. However, reports of long-lived allophane (Ward, 1967; Tonkin, 1970) and of allophane in well-weathered soils (Wada et al., 1986), as well as of halloysites in young soils (Hay, 1960; Bleeker and Parfitt, 1974), suggested that allophanes may not always transform to halloysite with time and also that halloysites may form directly from volcanic materials in some instances. It was also observed that halloysite formed where there was a thick overburden to provide silica, which was presumed to resilicate allophane (Mejia et al., 1968; Aomine and Mizota, 1973). Halloysite formed instead of allophane from volcanic materials where there was poor drainage (Aomine and Wada, 1962; Dudas and Harward, 1975).

In Parfitt et al. (1983) and subsequent papers, Parfitt and co-workers explained these apparently conflicting reports by proposing that the concentration of Si in soil solutions [Si] governs the nature of the secondary mineral that is formed. If [Si] is relatively high, the more siliceous phase (halloysite) may be formed, although allophane with the same Al:Si ratio as halloysite could also result. If [Si] is relatively low, the more aluminous allophane (Al:Si ≈ 2:1) is the favored authigenic phase. When [Si] is very low, gibbsite may form. Lowe (1986, 1995) has summarized the critical levels proposed for [Si] by several workers as follows:

[Si] $\geq$ 10 g m^{-3}, halloysite, or Si-rich allophane favored

[Si] $\leq$ 10 g m^{-3}, Al-rich allophane favored

[Si] $\leq$ 1 g m^{-3}, gibbsite favored.

The environmental conditions rather than the composition of primary minerals are of paramount importance in controlling the course of clay mineral genesis in tephras (Lowe, 1986). The more important conditions include degree of leaching, which is mainly governed, in turn, by rainfall and drainage, the thickness of tephra and depth of burial, and organic cycle effects (Lowe, 1986). Parfitt et al. (1984) estimated that at least 250 mm yr^{-1} of drainage water were needed to enable the formation of allophane. Different forms of allophane can form within the same soil profile. Jongmans et al. (1994) found Al-rich allophane and imogolite in a B horizon, but Si-rich allophane in the C horizon of a soil (Hapludand) formed on andesitic pyroclastics in humid tropical Guadeloupe. In soils formed from basalt in a xeric moisture regime in South Australia, Lowe et al. (1996) also generally found Al-rich allophane in upper horizons but Si-rich allophane in lower horizons. A limestone layer provided an impedance to drainage underneath these lower horizons. Lowe et al. (1996) reasoned that drainage through the upper horizons was adequate to enable the aluminous form of the mineral to form there, but drainage through the lower horizons was insufficient for it to form at depth. However, neither electron microscopy nor the formamide expansion pretreatment for distinguishing dehydrated halloysite from kaolinite in XRD (Churchman et al., 1984) showed that there was any halloysite in these soils (Lowe et al., 1996). On the other hand, smectite, which is strongly siliceous (Al:Si < 1:2) compared to both forms of allophane and also kaolin minerals, was identified at depth in one of the soils, suggesting that drainage was particularly poor in this case. Takahashi et al. (1993) also studied the clay minerals formed in soils from volcanic materials in a xeric moisture regime in northern

California. Gibbsite, imogolite and halloysite coexisted in these soils. Usually, imogolite (or allophane), on the one hand, and halloysite, on the other, are found in soils from volcanic materials in concentrations which are in inverse relationship to each other, as in the various studies by Parfitt and co-workers in New Zealand. Solubility considerations suggested that imogolite or Al-rich allophane were the favored minerals at the Si concentrations in these soils. It was suggested that wide seasonal fluctuations in the moisture contents, leading to a wide range of concentrations of Si could have led to conditions which favored the formation of each of gibbsite, imogolite (or 2:1 allophane) and halloysite at different times in each year (Takahashi et al., 1993). This reflected the changing position of soil solutions within stability fields during the year (Section 1.7).

One of the possibilities canvassed by Takahashi et al. (1993) is the resilication of allophane to halloysite in the solid phase as had been suggested by earlier workers (Fieldes, 1955; Mejia et al., 1968; Aomine and Mizota, 1973). However, the accepted structure of Al-rich allophane, which is based on that proposed for imogolite by Cradwick et al. (1972), cannot alter directly to that of halloysite. This is because the isolated tetrahedra in imogolite occur on the interior surface of hollow spherules with their apices pointing away from outer gibbsitic units, the reverse of their situation in the kaolinite structure (Lowe, 1986). Nonetheless, other mineral phases may form subsequent to the formation of allophane, but in the same weathering system. The use of XRD, TEM with energy dispersive X-ray analysis EDX, and also XPS, showed that the experimental alteration of rhyolitic glass (obsidian) at slightly raised temperatures (up to 200 °C) led to the formation of allophane within hours, followed by that of a smectite within a few days (Kawano et al., 1993)

The organic C cycle plays a vital role in governing the availability of Al, the other main elemental component of allophane and halloysite other than Si (Wada, 1989). When there is very active accumulation of humus, an Al-humus complex may be formed at the expense of allophane. Al-humus typically forms in A_1 horizons, in warm and humid, well-drained conditions under a grass vegetation (Lowe, 1986). It has been found in many Japanese soils. Opaline silica tends to occur alongside Al-humus (Wada, 1989).

The organic C cycle can have quite a rapid effect on the form of Al in soils. After only 30 years under bracken fern following forest clearance, there was a significant change in the secondary minerals in an Andisol from dominant allophane to dominant Al-humus complexes, consistent with a decrease in pH from 5.2 to 4.6 along with an increase in the content of organic C (Johnson-Maynard et al., 1997). When tephra from the 1980 eruption of Mt. St. Helens was weathered for 10 yr, neither allophane, imogolite nor opaline silica was identified in the weathered material (Dahlgren et al., 1997a). The incorporation of Al into Al-humus complexes and also into the interlayers of 2:1 layer silicates appears to have inhibited the formation of other minerals (Dahlgren et al., 1997a). Jackson (1963) coined the phrase "antigibbsite effect" to describe the inhibition of gibbsite formation through the formation of hydroxy-Al interlayers in vermiculites and smectites. The inhibition of allophane formation by competition for Al for the formation of these types of interlayers and/or of Al-humus has an analogous "antiallophane" effect.

Solution pH is important in allophane formation. Allophane forms at pH ≥ 4.8 (Lowe, 1995). While environmental conditions usually prevail over those of composition in dictating the mineral product formed on weathering volcanic parent materials, composition of parent minerals may also affect that of products (Lowe, 1986, 1995). For example, rhyolitic ash may supply more Si than andesitic ash, due to its more siliceous composition. However, rhyolitic glass weathers more slowly than andesitic glass (Lowe, 1995), so that the rate of formation of allophane is affected. Lowe (1995) has expressed some of the major factors affecting the formation of Al-rich and Si-rich allophane and halloysite from andesitic and rhyolitic ashes in cartoon form, as reproduced in Fig. 1.6.

Fig. 1.6 Cartoon summarizing the main parameters involved in the formation of Al-rich allophane versus halloysite or Si-rich allophane. Al-humus complexes tend to form where pH is < 4.8 and organic matter is abundant [Reprinted from Lowe, 1995. G.J. Churchman et al. Clays controlling the environment. Copyright with permission of CSIRO Publications, Melbourne, Australia]

1.4.6 By Podzolization

While desilication commonly occurs in mineral alterations by weathering, with Al being largely preserved in the solid phase, the process of podzolization involves the mobilization of Al, and often also Fe, relative to Si. A fully developed podzol consists of an upper eluvial E horizon which is pale colored due to the dominance of residual quartz grains and an accumulation of sesquioxides of Fe and Al in a lower B_s horizon (McLaren and Cameron, 1996). Often, organic matter also accumulates in a lower horizon. The mechanism by which material is transported from upper to lower horizons has been the topic of much debate and some controversy. It was widely accepted for many years that Fe and Al were transported within podzolizing soils as organic complexes (Russell, 1973). In an alternative mechanism put forward by Farmer et al. (1980) and Anderson et al. (1982), a soluble Al silicate is the predominant mobile form by which Al is transported to B_s and lower horizons in podzols, where it is deposited as imogolite and/or allophane. In support of their theory, Farmer et al. (1980) argued that imogolite and allophane had been identified in the lower horizons of podzols in Scotland (Tait et al., 1978), and that a soluble aluminosilicate (proto-imogolite) could be produced in the laboratory (Farmer et al., 1977; Farmer et al., 1979; Wada et al., 1979). Ross and Kodama (1979) and McKeague and Kodama (1981) also identified imogolite in podzols in Canada. Young et al. (1980) identified allophane in a New Zealand podzol from non-volcanic parent materials and there have been many more reports of allophane and/or imogolite in the lower horizons of podzols since then (Wada, 1989). Nonetheless, Ugolini et al. (1977) collected the solid material migrating through a podzol and identified it as organic matter together with some metals, including Fe and Al. Even afer Farmer et al. (1980), other authors (Buurman and van Reeuwijk, 1984) have continued to maintain the validity of the model based on organic complex formation. Childs et al. (1983) found that allophane occurred in lower B horizons in a number of podzolized soils in New Zealand, and that there was insufficient organic matter in the soils to account for the movement of Al as organic complexes. Their data are consistent with Farmer et al.'s (1990) inorganic mechanism for the transport of Al, but Childs et al.

(1983) point out that transport by an organic complex may also occur. Ugolini and Dahlgren (1987, 1991) propose that Al is transported down the profile of podzols as an organic complex and then imogolite and/or allophane are formed *in situ* from soil solutions in lower B horizons. As part of the continuing debate, Freeland and Evans (1993) suggested that their finding that allophane forms in cutans and on coatings early in the podzolization process suggests that allophane forms first, followed by organic complexes, in contradiction to Ugolini and Dahlgren's (1987, 1991) proposed mechanism. These authors called for more study of the mechanism, including a consideration of the effects of different parent materials and strengths of leaching.

Podzolization can also have a particular effect on minerals in eluvial horizons, apart from the extreme, where minerals break down to quartz alone. Leaching and the associated acidification often lead to the transformation of micas to vermiculite, beidellitic smectite and the interstratifications of these phases with micas, as discussed earlier. McDaniel et al. (1995) reported the occurrence of beidellite in E horizons of podzolized soils (Spodosols) formed in volcanic ash. Although weathering of detrital micas is not ruled out, there was no vermiculite. Beidellite may have formed authigenically in the E horizons of these soils (McDaniel et al., 1995). The dominant mineral in the eluvial horizon of a humus-iron podzol in Scotland contained a dioctahedral mica-vermiculite which incorporated an amorphous organo-Al-Fe complex in its interlayer (Bain and Fraser, 1994). It appears to incorporate an illuvial feature, viz., the metal-organic complex within a mineral formed in an eluvial horizon. Møberg and Borggaard (1991) found that layer lattice silicates, gibbsite and also Fe oxyhydroxides in eluvial horizons of podzols were eventually dissolved under the aggressive conditions established there by acid organic complexing agents, among other factors.

1.4.7   Kaolinite

Kaolinite formation is particularly favored by warm humid climates, but it can also form in cool temperate climates provided rainfall is adequate (Dixon, 1989; Weaver, 1989). Kaolinites form instead of halloysites on volcanic parent materials under heavy rainfall and strong drainage (Kantor and Schwertmann, 1974; Chamley, 1989). Kaolinites vary greatly in their degrees of structural order and their chemical composition. Soil kaolinites are often very disordered (Dixon, 1989). However, the measurement of their degree of ordering (sometimes referred to as crystallinity) may present problems. Several methods have been suggested, involving techniques such as XRD, DTA and IR (Churchman and Theng, 1984; Garcia-Talegon et al., 1994). Because of the poor resolution of some of their peaks in XRD patterns, soil kaolinites (or kaolins, in general) are often difficult to assess by the Hinckley index based on XRD peaks, which is one of the standard measures of order for kaolinites (Hughes and Brown, 1979). It has been considered by many that the better crystalline kaolins are found at depth in well-drained soils formed over granitic or sedimentary rocks (Hughes and Brown, 1979). Hughes and Brown (1979) developed an empirical "crystallinity" index for soil kaolins in order to help explain the influence of genetic factors on their structural order. For a set of kaolins from Nigeria, it appeared that the presence of weatherable minerals in the soil hindered the crystallization of well-ordered kaolins (Hughes and Brown, 1979). This may result from an inhibition of the crystallization of well-ordered kaolins in the presence of alkaline earth cations (Millot, 1970). Iron commonly occurs as an impurity in the structures of kaolinite, probably in quite small proportions (7-14 g kg^{-1}) (Dixon, 1989). There appears to be a relationship between impurity Fe content and structural order (Dixon, 1989). Kaolinites are very common in rocks and deposits, where they may form hydrothermally or are residua from weathering under a previous climate and it may be difficult to distinguish kaolinite formed authigenically by weathering under present conditions from inherited and detrital kaolinites. Garcia-Talegon et al. (1994) have studied the kaolin minerals in deeply weathered profiles in terms of their ordering as measured by XRD, DTA and IR techniques. More ordered kaolinite appeared to be inherited while disordered kaolinite and also dehydrated halloysite probably formed under prevailing conditions.

1.4.8   Halloysite

While a common constituent of soils formed on volcanic materials, halloysite has also been identified as forming from other types of parent materials. However, in studies of halloysite, it is critical that its distinction from kaolinite be established. Unfortunately, this is not always easy (Churchman and Carr, 1975; Bailey, 1990). Before the application of X-ray diffraction, distinctions between halloysite and kaolinite were made on the basis that halloysite had a higher water content (MacEwan, 1947; Churchman and Carr, 1975). Otherwise, the two minerals have an identical chemical composition, although Bailey (1990) wondered whether very accurate analyses might reveal systematic differences. X-ray diffraction showed differences between the two minerals if halloysite was hydrated; these became less clear when it was dehydrated, a process which occurs easily and irreversibly. However, electron microscopy apparently showed differences that were very clear from its first application to the two minerals (Alexander et al., 1943). Kaolinite showed hexagonal plates, while halloysite showed a fibrous, usually tubular morphology. Samples examined later showed other shapes for halloysites, particularly spheroids (Dixon, 1989), but also particles that were more or less flat (Kunze and Bradley, 1964; Carson and Kunze, 1970; Wilke et al, 1978; Wada and Mizota, 1982; Tazaki, 1982; Noro, 1986). Tubular particles, which otherwise had the characteristics of kaolinites, were also identified (Souza Santos et al., 1965). Distinctions based on shape became unclear. For this reason, and because other techniques such as DTA and IR also failed to provide a clear demarcation between at least some kaolinites and some halloysites, Churchman and Carr (1975) proposed a return to the original distinction made on the basis of the presence, or else prior occurrence, of interlayer water in halloysite. However, occurrence of non-platy shapes in electron micrographs has usually provided the identifying feature of halloysites in weathering studies. Even so, there has been some trend toward the use of intercalation tests for detecting halloysite that has become dehydrated, whether in the field or during preparation. These various tests (Wada, 1961; Range et al., 1969; Churchman et al., 1984) rely upon the easier expansion with a polar molecule of the interlayers of halloysite compared to those of kaolinite; they, therefore, provide a test for the prior intercalation of water by halloysite. Furthermore, while the shapes of halloysite particles may not always enable their clearcut distinction from kaolinites, the origin of the different shapes may have genetic significance (Bailey, 1990; Soma et al., 1992).

With the proviso in mind that different workers have defined halloysites by different methods, one notes that halloysite has been identified as forming from the following types of non-volcanic rocks, among others: granites (Parham, 1969; Torrent and Benayas, 1977; Eswaran and Wong, 1978; Wilke et al., 1978; Anand et al., 1985; Nagasawa and Noro, 1987; Churchman, 1990; Robertson and Eggleton, 1991; Cho and Mermut, 1992; Garcia-Talegon et al., 1994), gneissic granite, granitic gneiss (Calvert et al., 1980), gneiss (Eswaran and Hong, 1976), dolerite (Anand and Gilkes, 1984), schist (Hewitt and Churchman, 1982), greywacke (Churchman et al., 1984, Churchman, 1990), greenstone (Norrish, 1995) and a wide variety of rock types, including granites, granodiorites, gabbros, shales and amphibolites (Romero et al., 1992). Generally, halloysite appears as tubes in the weathering products from non volcanic rocks, as it often also does when formed from volcanic materials (Fieldes, 1968; Parham, 1969; Saigusa et al., 1978; Churchman and Theng, 1984). However, halloysite often appears also in spheroidal particles when it is formed from volcanic parent materials in Japan (Sudo and Takahashi, 1956 and Saigusa et al., 1978), Guatemala (Askenasy et al., 1973), Mexico (Dixon and

McKee, 1974), Italy (Quantin et al., 1988), New Zealand (Kirkman, 1975), and Australia (Ward and Roberts, 1990). Other different morphologies for halloysites formed by weathering include crumpled lamellar shapes from weathered pumice in Japan (Wada and Mizota, 1982), crinkly film in weathered volcanic ash in Japan (Tazaki, 1982); thin platy flakes showing some hexagonal morphology and also some tendency to roll into tubes in weathered granite in Germany (Wilke et al., 1978), and somewhat similar very thin, small flakes described as newformed in weathered andesite from Guadeloupe (van Oort et al., 1990). Explanations of the different shapes have largely centered around the effects of structural Fe impurities, on the one hand, and the particular origin of the crystals, on the other.

The rolling of layers in halloysites has generally been ascribed to the intrinsic misfit in b dimensions between the larger tetrahedral sheet and the smaller octahedral sheet (Bates et al., 1950). In halloysite, the tetrahedral sheet is considered to roll around the octahedral sheet in order to relieve the resultant strain. In kaolinite, by contrast, alternate silica tetrahedra rotate in opposite directions to correct the misfit (Radoslovich, 1963). In the case of halloysite, rolling occurs rather than tetrahedral rotation either because the water in the interlayers blocks tetrahedral rotation (Bailey, 1990), or because rolling encounters less resistance from Si-Si repulsion than does tetrahedral rotation in order to correct the same amount of misfit (Singh, 1996). Substitution of the larger Fe(III) for Al in the octahedral sheet would alleviate the misfit between layers. An increase in structural Fe accompanies changes in the morphologies of halloysite particles, first from long tubular, then to short tubular, and ultimately to tabular forms (Churchman and Theng, 1984; Noro, 1986; Bailey, 1990) consistent with an increase in the b dimension of the octahedral layer due to the substitution of Al by Fe. Fig. 1.7 shows transmission electron micrographs of halloysites with different structural Fe contents which illustrate the suggested relationship between structural Fe and morphology. Thus, there appears to be a progression for halloysite particles from long tubular to tabular via shorter tubular shapes. Spheroidal shapes do not

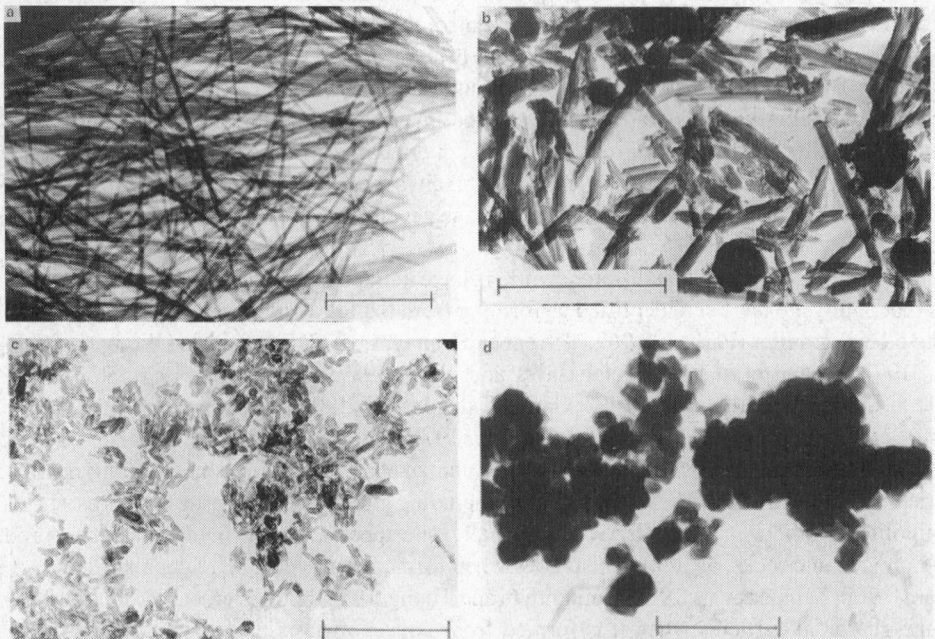

Fig. 1.7 Transmission electron micrographs of halloysites with different structural Fe_2O_3 contents: (a) Patch Clay (0.11%), (b) Te Akatea (0.85%), (c) Hamilton (1.52%), (d) Te Puke (3.21%). The length of the bar represents 1 μm.

appear to fit the same series. They exhibit a wide range of contents of structural Fe, from virtually zero (Johnson et al., 1990) through moderate values similar to those for short tube halloysites (Noro, 1986) to quite high values (Churchman and Theng, 1984; Nagasawa and Noro, 1987). In any case, using XPS, Soma et al. (1992) found that the content of Fe in different layers was different within spheroidal as well as tubular halloysites. This means that conclusions could not necessarily be drawn about the effects of Fe upon morphology on the basis of bulk analyses alone. Nagasawa and Miyazaki (1976) had observed that halloysite from volcanic glass appeared in "rounded, scroll-shaped and/or irregular shapes sometimes associated with short tubes". By contrast, halloysite from feldspar appeared to be tubular (Visconti et al., 1956; Bates, 1962; Parham, 1969; Nagasawa and Miyazaki, 1976). Bailey (1990) has suggested that the prevalence of spheroidal forms of halloysite in alteration products of volcanic glass results from specific conditions of their growth and the nature of the starting conditions rather than for structural reasons. The high dissolution rate of glass may lead to supersaturation of solutions which results in spheroids on precipitation. However, the common association of the different shapes together (Saigusa et al., 1978; Churchman and Theng, 1984) and the apparent transformation of spheroidal forms to tubular forms *in situ* (Sudo and Yotsumoto, 1977) suggest that there is no clearcut distinction between the conditions which favor these two forms of halloysite.

There are several unsolved aspects of halloysite which mean that the genesis of the many different manifestations of this mineral are also not fully understood. Bailey (1990) has suggested, among others, that the reasons why there is interlayer water in halloysite but not kaolinite may reflect a higher layer charge which has not yet been detected consistently in analyses. He suggested that some of the higher charge could originate from the presence of tetrahedral Al in halloysite but not in kaolinite. However, Newman et al. (1994) have since found, using ^{27}Al-NMR, that kaolinites and halloysites both have similar, though vanishingly small, contents of tetrahedral Al. Bailey (1990) also suggested that prismatic forms of halloysite constitute a separate type from each of spheroidal, tubular and platy types, while Chukhrov and Zvyagin (1966) considered the prismatic forms to be the primary forms of halloysite. These elongated crystals do not have any voids; hollow tubular halloysites form from them by spontaneous dehydration or as a result of sample preparation, according to this view. Kirkman (1981) also observed polygonal forms in an halloysite but attributed this distinctive shape to the effect of the dehydration process. Using isotherms for adsorption and desorption by N_2 gas, Churchman et al. (1995) found that halloysite samples with mainly large particles (~ 0.1 μm in width) had few, if any mesopores (diameters between 2 and 15 nm), regardless of whether the particles were tubular, spheroidal or blocky in shape. Halloysites with mainly smaller particles, by contrast, had abundant micropores. These were cylindrical and TEM showed that they originated from the central holes in tubular particles. These findings are consistent with Chukhrov and Zvyagin's (1966) thesis that the larger particles lack voids. However, this observation may be explainable by other means, including the likelihood that greater curvature on layers in smaller particles leads to the central voids observed therein. The reasons why there are variations in structural Fe contents are also intriguing. Soma et al. (1992) suggest that variations between adjacent layers within crystals in their Fe contents reflect variations in the compositions of solutions from which the layers crystallized. The compositions of layers could provide information on the chemistry of micro-environments in which mineral formation took place. One of the most important conclusions of Bailey's (1990) review was that the structural differences between halloysite and kaolinite are substantial enough to make it impossible to transform from one to the other without complete recrystallization.

There have been several studies suggesting a genetic relationship between halloysite and kaolinite. Some (Eswaran and Wong, 1978; Calvert et al., 1980) have indicated that there may be a pattern in deep weathering profiles in granitic rocks whereby halloysite forms early and kaolinite later in the same profile. Using both electron microscopy and also two intercalation tests (with XRD), Churchman and Gilkes (1989) studied the pattern of appearance and disappearance of different forms of halloysite and also of kaolinite at different depths in weathering profiles in Western Australia. Hydrated tubular halloysite dominated the clay fraction of the saprolite close to unweathered dolerite. With further weathering corresponding to distance from the rock up the profile, the halloysite progressively became more dehydrated, then, when fully dehydrated, progressively more difficult to intercalate. Halloysite tubes tended to split and unroll as weathering continued. Kaolinite as hexagonal platy particles appeared later than halloysite in the course of weathering. Halloysite tubes and kaolinite plates appeared to form by separate pathways. There was a similar pattern in profiles on granite. Although no transformation between tubular and platy particles was observed, Churchman and Gilkes (1989) suggested that fully dehydrated tubular particles may constitute intermediates between the two mineral types. Under strong weathering, such as occurred in the old, well leached landscapes of much of Australia, dehydrated tubular kaolins which are unresponsive to intercalation by polar molecules appear to be quite common. They often occur in association with platy kaolinite particles. Janik and Keeling (1993) have successfully used Fourier transform infrared spectroscopy (FTIR) in association with partial least squares analysis (PLS) for the quantitative analysis of the tubular kaolins, which they call halloysite. Comparisons with SEM analyses of the same set of samples showed that this approach offers a rapid tool for the quantitative distinction of the kaolin minerals on the grounds of their morphology. In other studies of Australian weathering profiles, both Robertson and Eggleton (1991) and Singh and Gilkes (1992) show clear evidence from electron microscopy and electron diffraction for the transformation of kaolinite plates to halloysite tubes. Such a transformation appears to contradict both the trends seen in weathering profiles (Eswaran and Wong, 1978; Calvert et al., 1980; Churchman and Gilkes, 1989) and also thermodynamic considerations (Section 1.7). Different parent materials are involved in each case (plagioclase feldspars in Robertson and Eggleton's study and micas in Singh and Gilkes') so the nature of the micro-environment are more likely to be the common feature. In both of these studies and those of the deeply weathered profiles, it is agreed that the essential requirement for the formation of halloysite is the incorporation of water into the interlayers. Hence halloysites are often most common in the lower, wetter parts of weathering profiles (Churchman, 1990).

1.4.9   Neogenetic Vermiculite and Illite

While vermiculite and illite are generally regarded as the products of transformation of primary micas, as already discussed, there have been some reports of the formation of both by neogenesis. Vermiculite was seen to replace feldspars (Smith, 1962), while Barshad and Kishk (1969) considered that dioctahedral vermiculite, which occurred in soil derived from rocks containing no mica, was formed by precipitation from gels.

Whereas increasing intensity of weathering toward the surface in a soil profile generally leads to loss of mica/illite at the expense of its breakdown products, particularly vermiculite, many soils in Australia show an enhancement of this mineral in surface horizons (Norrish and Pickering, 1983). Similar observations have been made for soils in Iran (Mahjoory, 1975) and for some semi-arid soils in North America (Nettleton et al., 1973). Neogenesis of illite may have occurred within these soils. Norrish and Pickering (1983) observed that an illite with a high Fe content which is green in color occurs in a deposit near Lake Eyre in arid Australia. It appears to have formed authigenically in a lacustrine environment. Its neogenetic origin is suggested by the very uniform shape and size of its particles, which are almost all $\sim 0.07~\mu m$ in size (Fig. 1.8a). Norrish and Pickering (1983) pointed out, furthermore, that illites within a number of Australian soils show similar shapes and size, as seen in Fig. 1.8b for a soil located ~300 km from Lake Eyre. This type of soil illite, which also has a high Fe

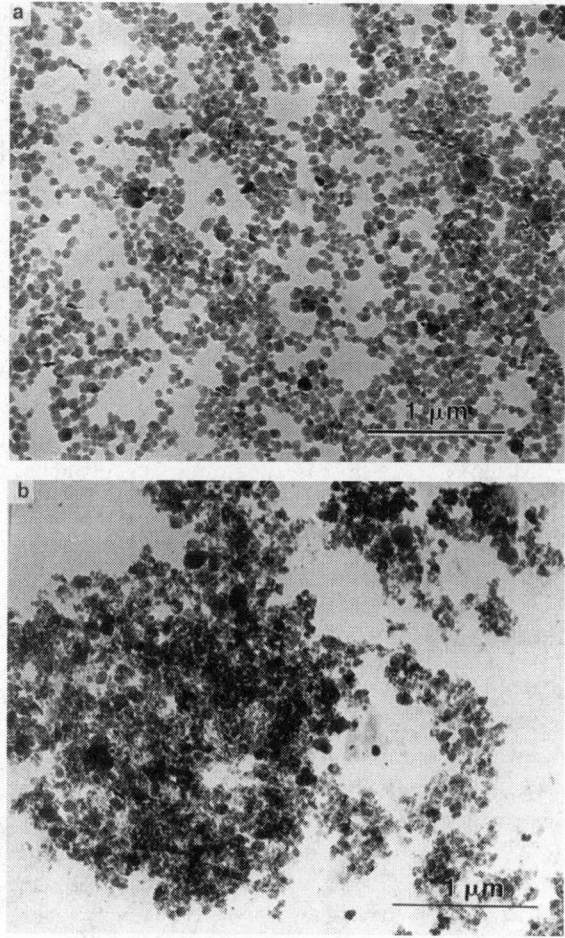

Fig. 1.8 Transmission electron micrographs of (a) Muloorina illite from a deposit of lacustrine origin, and (b) Wilallooka illite from a soil. The lenght of the bar represents 1μm.

content (Norrish and Pickering, 1983), may have formed within the soil by a dissolution-precipitation, i.e., neogenetic process similar to that in the lakes or may be detritus from illite of lacustrine origin. A mineral described as palagonite by Singer (1974) which occurred alongside halloysite in soils from basic pyroclastic layers on the Golan Heights was further examined by Berkgaut et al. (1994) and found to consist of very thin crystallites of a randomly interstratified illite-smectite with ~ 70% illite layers which was likely to have formed authigenically. It was considered that this particular mineral formed in the dry seasons in the Mediterranean climate while halloysite and also kaolinite formed during the wet seasons.

Some earlier suggestions of a neogenetic origin for micas in Hawaii (Swindale and Uehara, 1966) are thought to be unlikely in view of later research which indicated that eolian transport of fine mica was more likely. An eolian route was suggested by the discovery that oxygen isotope ratios of quartz in Hawaiian soils were consistent with their transportation in tropospheric dust from continental sources (Rex et al., 1969). Other fine materials such as micas would have accompanied the quartz

(Syers et al., 1969), especially since its source was likely to have been in widespread sedimentary rocks like shales which were of mixed igneous and low temperature origin (Churchman et al., 1976).

1.4.10   Smectites

Although some smectites, notably beidellites, can form by transformation of other 2:1 phyllosilicates (micas and vermiculite) in leaching, acidic environments, it is poorly drained, alkaline conditions that characterize the typical environments for the formation of smectites in soils. Especially in North America, Europe, North and South Asia, and Australia, there are vast expanses of smectitic soils that have formed under these kinds of conditions (Borchardt, 1989). Smectites formed by weathering are dioctahedral as a general rule (Borchardt, 1989). Even a Mg-rich serpentinite yielded a dioctahedral, Fe-rich smectite rather than a saponite (Wildman et al., 1968). Saponite was identified in the saprolite of a weathered ultrabasic rock but it became unstable in fissures and was replaced by nontronite (Fontanaud and Meunier, 1983). Although it is sometimes present (Sawhney and Jackson, 1958), nontronite is less common than montmorillonite and beidellite in soils, probably because Al is virtually ubiquitous (Borchardt, 1989). In any case, nontronite is likely to be susceptible to attack by the complexing acids produced by roots and microorganisms in the biologically active zones of soils (Farmer, 1997). Soil smectites mainly fall within the montmorillonite and beidellite ranges of octahedral layer compositions and tetrahedral charge (Borchardt, 1989; Weaver, 1989). They generally have more tetrahedral Al and more octahedral Fe than bentonitic montmorillonite (Wilson, 1987; Weaver, 1989), suggesting a tendency toward a ferri-beidellitic composition and structure. However, individual smectites may be heterogeneous in composition. Thus, the layers of Fe-rich smectites occurring in a lateritic profile in Brazil could be described as true solid solutions between nontronite and beidellite end members (Petit et al., 1992).

For smectites to form by neogenesis, solutions of the products of hydrolysis and dissolution of primary, and other secondary, minerals must contain relatively high Si and some Mg and/or Fe (Weaver, 1989). The main conditions for their formation are the opposite of those for the formation of kaolinite (Section 1.4.7). Some of these conditions have been confirmed and closely defined by laboratory syntheses (Harder, 1972, 1977; Siffert, 1978). For montmorillonite to form, pH needed to be above 7.5. Its formation occurred in the laboratory when negatively charged silica combined with a positively charged hydroxide. At pH > 7.5, a positively charged hydroxide was obtained by coprecipitating Mg with Al. Pure aluminum hydroxides are positively charged only at pH< 7.7 and form kaolins if combined with silica at lower pH (Siffert, 1978). Hence, some Mg was necessary for montmorillonite to form. In nature, montmorillonites form by neogenesis from the dissolved products of volcanic ash. This is the origin of deposits of bentonite. Farmer (1997) formed ferruginous beidellites with similar compositions to the smectites in Vertisols by means of the cyclic reduction and oxidation of heated solutions containing Al, Fe(II) and H_4SiO_4, adjusted to pH 8.5 with $Ca(OH)_2$, and with $CaCO_3$ as a buffer. There was no Mg. Soil beidellites usually have substantial Mg in octahedral layers (Wilson, 1987), although some with very little Mg have been reported (Ben-Dor and Singer, 1987). Smectites often form in association with calcareous rocks and formations which can provide a high pH (Fieldes and Swindale, 1954) and also often are an impediment to drainage (Lowe et al., 1996). Basic rocks provide a ready source of the cations necessary for the formation of smectites by neogenesis, but smectites can form from many types of rocks and minerals, provided both that they can provide the necessary constituents for smectites and also that the conditions suit the accumulation of the necessary cations and silica.

Since smectites can be either inherited from parent materials, transformed from other 2:1 aluminosilicates or crystallized from solutions by neogenesis, assessment of their origin can present problems (Borchardt, 1989). Fiore (1993) suggested that a smectite found in a pyroclastic flow deposit in Italy, while not strictly inherited from the parent material, had instead formed by hydrothermal alteration during and/or soon after the emplacement of the flow as a result of the presence of trapped hot fluids. A smectite occurring alongside trioctahedral vermiculite, as well as its parent biotite, and also a hydroxy-interlayered 2:1 aluminosilicate in an arid soil provoked further curiosity because it occurred in both silt and sand fractions as well as the clay fractions (Boettinger and Southard, 1995). As especially fine grained minerals in the main, smectites are quite uncommon in coarser fractions of soils, but Boettinger and Southard (1995) consider that an Al-rich dioctahedral smectite formed in this case from solutions of weathering products and that some was cemented by opaline silica into coarse aggregates. Reid et al. (1996) found that, while all soils along a geomorphic and salinity gradient transect in an arid area contained smectites, these included both beidellitic and montmorillonitic components. There appeared to be multiple origins for the smectites in the soils. Beidellites probably weathered from shales, and while some montmorillonite was inherited, the chemistry of the groundwater was consistent with some having been neoformed. Although inherited, smectites may themselves be altered by weathering. A study of smectites formed in different granitic saprolites by analytical HRTEM led Aodjit et al. (1995) to conclude that the particular process by which smectites were formed, and also the type of smectite formed, was dependent on the nature of the microsystem defined by each primary mineral and its situation with respect to forces promoting weathering. Thus, where leaching and hydrolysis occurred, biotite evolved into mixed layer minerals and beidellite by transformation (Aodjit et al., 1995). However, in a contrasting confined system, micas gave way to montmorillonite, probably by neogenesis following their dissolution (Aodjit et al., 1995). Righi et al. (1995) found that a low charge beidellite in the parent material was converted to high charge beidellite by weathering during the formation of a Vertisol.

1.4.11   Interstratified Minerals

Smectite layers often occur within interstratifications with other phyllosilicates. Weaver (1989) states that "it is quite possible that the great majority of (phyllosilicates) are composed of interstratified layers of different composition." Weaver (1989) also notes that mixed layer illite-smectite is by far the most abundant of these. The abundance of this type of interstratified phase, which encompasses a wide range of possible combinations of the pure components, largely derives from the common formation of illite-smectites from smectites during burial diagenesis and in hydrothermal systems. However, this type of mineral can also form in weathering by the breakdown of micas, the opposite process to that occurring in burial diagenesis. It has also come to be recognized relatively recently that interstratifications involving smectites and kaolinites occur in soils worldwide, including Scotland (Wilson and Cradwick, 1972), Canada (Kodama et al., 1976), various parts of Africa (Kovda et al., 1977; Herbillon et al., 1981; Delvaux et al., 1988; Bühmann and Grubb, 1991), Australia (Norrish and Pickering, 1983; Churchman et al., 1994), El Salvador (Yerima et al., 1985), India (Tomar et al., 1985), Japan (Miura et al., 1988), China (Xie, 1987), and Ohio (Jaynes et al., 1989). However, many of these types of minerals may have been overlooked, as interstratification generally leads only to a broadening and weakening of peaks in XRD patterns. These peaks may be virtually invisible in the presence of peaks for other clay minerals, especially other random mixed layer phases (Herbillon and Makumbi, 1975; Norrish and Pickering, 1983). Most of those described in these and other papers have been more kaolinitic than smectitic in composition (see summary in Churchman et al., 1994), but Bühmann and Grubb (1991) and Churchman et al. (1994) also found examples with up to 60% and 70% of smectite layers, respectively. A higher CEC than expected for kaolinites can indicate the presence of these minerals. Several minerals which had been identified as halloysites with platy morphologies (Carson and Kunze, 1970; Wada et al., 1987) have later been reinterpreted by other workers as kaolin-smectites (Sakharov and Drits, 1973; Parfitt and Churchman, 1988). By the same token, although many characteristics of some clay minerals, especially their chemical composition, DTA/TGA traces and FT-IR spectra apparently typified them as smectites, close scrutiny, especially of their XRD patterns and ^{27}Al MAS NMR spectra showed them to be kaolin-smectites with high smectite contents (Cuadros et al., 1994). Halloysite-smectite interstratifications may also occur (Delvaux et al., 1988, 1989, 1990, 1992; Delvaux and Herbillon, 1995). These have been characterized by a range of techniques, including chemical composition, electron microscopy, selectivity for K^+, FTIR and electron spin resonance (ESR), as well as XRD. ESR assisted in the identification of the smectite layers as belonging to the beidellitic-nontronitic series (Delvaux et al., 1990). In XRD, glycol solvation of samples leads to an expansion of peaks for 1.0 nm spacings, while heating results in low angle asymmetry on 0.7 nm peaks, as in glycol-solvated kaolinite-smectites. Delvaux et al. (1990) suggested that the minerals described as embryonic halloysites by Wada and Kakuto (1985) and Wada et al. (1987) had the essential characteristics of halloysite-smectites.

Kaolin-smectites commonly occur in the intermediate zones between the black and red end members of red-black toposequences of soils, which contain smectite and kaolinite, respectively. Such toposequences can result in landscapes where there are poorly drained hollows (with black, smectitic soils) and well-leached humps or slopes (with red, kaolinitic soils). They often occur on basaltic parent materials (Bühmann and Grubb, 1991). The typical pattern of occurrences of smectites and kaolinites within this landscape pattern has been observed in several localities, including in Australia (Ferguson, 1954), Africa (Kantor and Schwertmann, 1974), Arizona (Johnson et al., 1962) and India (Chaterjee and Rathore, 1976). The appearance of kaolinite-smectite interstratified phases within these landscapes (Herbillon et al., 1981; Bühmann and Grubb, 1991; Delvaux et al., 1995) implies that they are intermediate in the alteration of smectite to kaolinite or, conceivably, *vice versa*. The changes within weathering profiles indicating an effect of time has further elucidated their genetic pathways, even if the effect of time is often complicated by abrupt changes of parent materials and/or detrital material and conditions. Minerals identified as 1:1-2:1 interstratified phases were found in the most strongly weathered section of a polygenetic soil profile, which occurs at the boundary between loess overlying a paleosol (Jaynes et al., 1989). Vermiculite-smectite layers, with various amounts of Al-hydroxy interlayering, occurred lower in the profile, suggesting they were the precursors of the 1:1-2:1 interstratified phases (Jaynes et al., 1989). Churchman et al. (1994) found a pattern in a number of Australian soils in which kaolinite layers always became more abundant in relation to those of smectite in the (randomly) interstratified phase toward the surface of the profiles. In the deep profile of a soil formed on basalt, Churchman et al. (1994) found by modeling the XRD patterns with the NEWMOD computer program of R.C. Reynolds (Fig. 1.9) that the very fine clay fraction < 0.09 μm of the deepest sample at 1.8-1.9 m was composed of 60% of a kaolin-smectite in which 40% of the layers were kaolinite and also 40% of discrete kaolinite. There was a change with distance up the profile so that the very fine clay fraction of the sample at 0.1-0.2 m was composed instead of 40% of a kaolin-smectite which had 60% of kaolinite layers. There was 60% of discrete kaolinite as well in this upper sample. It appears that there had been a desilication of the interstratified phase with increased extent of weathering, with some of the product appearing as additional kaolinite layers within the interstratified phase and some as kaolinite *per se*. The idea that desilication occurs in the alteration of smectite to kaolinite is supported by the observation by Watanebe et al. (1992) that opal C-T appeared during the early stages of the formation of kaolinite from smectite in an acid clay deposit.

Although the direction of weathering from smectite to kaolinite appears to be well established, the mechanism of the conversion is still open to question (Barnhisel and Bertsch, 1989). Earlier geological observations (Altschuler et al., 1963) and syntheses (Poncelot and Brindley, 1967) suggested the

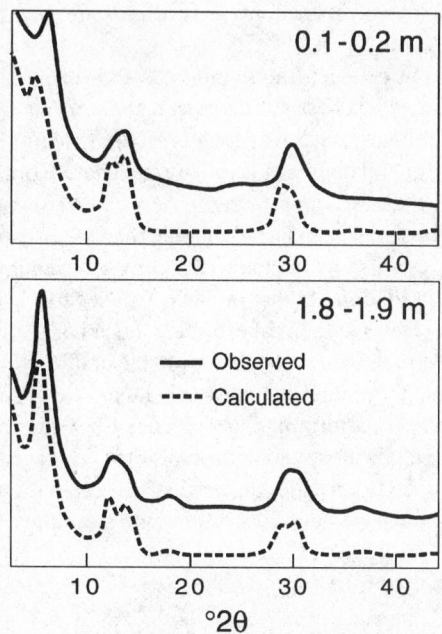

Fig. 1.9   X-ray diffraction pattern of the basal planes of kaolin-smectites in very fine clay fractions (< 0.09μm) from near the surface and also at depth of a soil in comparison with peak profiles modelled by the NEWMOD computer program which gave the best fit in each case. The best fit for the sample from 0.1-0.2m was modelled with 60% kaolinite + 40% kaolinite-smectite where kaolinite (k): smectite (s) was 60:40 in random interstratification. For the sample at 1.8-1.9m, the best fit was modelled with 40% kaolinite + 60% kaolinite-smectite where k:s was 40:60 [Reprinted from Churchman et al., 1994. Aust. J. Soil Res. 32:805-822 with permission of CSIRO Publications, Melbourne, Australia].

possibility of transformations occurring from smectite to the kaolinite via an hydroxy-Al interlayered species in the former. A transformation from hydroxy-Al interlayers in smectites to kaolinite would involve the inversion of silica tetrahedra after they had been stripped from 2:1 layers (Karathanasis and Hajek, 1983). However, Wada and Kakoto (1983,1989) observed an interlayering of vermiculite by a 1:1 kaolin-like phase rather than by a hydroxy-Al species The 1:1 aluminosilicate phase in the interlayers of this mineral was sodium citrate soluble (Wada et al., 1991). Synthetic work had led Lou and Huang (1988) to contend that interlayers in 2:1 expandable aluminosilicates in soils could contain Si as well as Al. The adsorption of precursors of kaolin minerals as interlayer phases in smectites as well as in vermiculites may provide a more credible route for the alteration of smectites first to interstratified kaolinite-smectites and eventually to kaolinites than their transformation within the solid state via hydroxy-Al interlayers. To enable the formation of kaolinite from hydroxy-Al interlayered minerals, dissolution of 2:1 layers, or at least part of these layers, would probably occur to provide Si in solution which could combine with hydroxy-Al. Dissolution-precipitation is almost certainly involved in the process.

Halloysite-smectites form under different conditions than those for the formation of kaolinite-smectites (Delvaux et al., 1995). Delvaux et al. (1995) described their method of formation as syngenesis or coneogenesis. According to these authors, they have been found invariably in volcanic ash soils. Halloysite-smectites form under similar conditions to smectites but where some more leaching occurs. With even more leaching, halloysite would form (Glassmann and Simonson, 1985;

Smith et al., 1987). The products formed in these cases are highly dependent on the micro-environmental conditions (Delvaux et al., 1995).

Interstratifications may occur between kaolin minerals and vermiculite layers, as well as smectite layers (Jaynes et al., 1989). Indeed, interstratifications in general can involve more than two types of layers (Sawhney, 1989). In this context, it is worthwhile recalling Millot's (1970) contention about the various possibilities for clay mineral phases or layers, viz., "These words are without doubt applied to a reality with more nuances than the rigorous definition of these words would have us believe." Furthermore, interstratifications may be closer to the rule than the exception in soils. The widespread illites, for instance, may be regarded as interstratifications of mica layers with vermiculitic layers which have formed by the degradation of other mica layers (Section 1.4.3). Very recent work by Ma (Ma, 1996; Ma and Eggleton, 1996) with high resolution TEM has shown the common presence of a small number only of 2:1 layers at the terminations of stacks of kaolinite layers from a number of sources. It is demonstrated that terminating impurity layers can be of two types, i.e., smectite-like or pyrophyllite-like. The presence of terminating smectite-like (but not pyrophyllite-like) layers corresponded to higher CEC values for some of the kaolinites. The occurrence of these layers, which were usually not detectable in XRD, provides an alternative explanation to isomorphous substitution for a permanent charge, and hence elevated CECs for some kaolinites.

1.4.12 Palygorskite and Sepiolite

As soil minerals, palygorskite and sepiolite tend to appear in dry regions (Allen and Hajek, 1989). Sepiolite is much less common than palygorskite in modern soils (Singer, 1989). Palygorskite almost invariably forms in calcareous soils, which are mildly alkaline and saline (Singer, 1989). Since smectite formation is also favored by this particular set of conditions, it is not surprising that it commonly occurs in association with smectites. Paquet and Millot (1972) found that palygorskite altered to give smectite when the mean annual precipitation exceeded 300 mm. While understanding of the required conditions for the formation of smectites has been aided by successful syntheses of these minerals in the laboratory (Section 1.4.10), no reproducible syntheses of palygorskite have been reported (Singer, 1989). On the other hand, sepiolite has been obtained reproducibly by synthesis (Siffert and Wey, 1962; Priesinger, 1963). It required high activities of Mg and Si and a high pH. Palygorskite would also require some Al and Fe (Singer, 1989). A review of the literature by Singer (1989) indicates three possible routes for the formation of palygorskites in soils as follows: (1) from the influence of fluctuating saline or alkaline groundwater; (2) at a textural transition, e.g., in a cutan, when not present in the body of the soil (Singer and Norrish, 1974); or (3) in or below calcretes, calcareous crusts and caliches. Verrecchia and Le Coustumer (1996) provide a very comprehensive listing of reports of palygorskite occurrences in calcretes.

They may form by alteration of smectites (Singer, 1989). Although the evidence for this transition in soils is scant (Singer, 1989), electron microscopy shows it occurs by diagenesis in ocean sediments (Tazaki et al., 1986a, 1987a). Precipitation from solution, often onto another mineral for nucleation, probably is a common source for palygorskite. Its tendency to form at boundary horizons has been explained by retention of water in the transition zones so that solutions in these zones remain saturated longer than the bulk of the soil, thereby enhancing palygorskite formation from solution (Yaalon and Wieder, 1976). Magnesium, which is required for palygorskite formation, may originate from dolomite, from the conversion of high Mg calcite to low Mg calcite, or as a result of the precipitation of Ca as gypsum (Singer, 1989). Observations that silica with a secondary origin can occur adjacent to palygorskite in some environments (Singer, 1989) suggest that soluble Si had been available for the formation of palygorskite as well as silica. In several of these instances, calcite partially replaces quartz, thereby releasing soluble silica (Singer, 1989). Verrecchia and Le Coustumer (1996) considered that palygorskite is the last step in the cycle of mobile silica. It has also been proposed that palygorskite formed in an alkaline environment in a basin where the products of the breakdown of illite and chlorite had accumulated (Galán et al., 1975; Galán and Castillo, 1984; Dias et al., 1997).

1.4.13 Neogenetic Silica

Since desilication is common in mineral alteration as a consequence of leaching, there may be an opportunity during weathering for the mobile silica phase to re-precipitate as an authigenic form of silica. Authigenic quartz is relatively common in soils as chemically precipitated overgrowths, commonly associated with carbonates (Drees et al., 1989). Quartz appeared to form by crystallization from solutions from weathered microcline (Estoule-Choux et al., 1995) (Sections 1.4.2 and 1.7). The observation of a systematic increase in O isotopic ratios of quartz samples from soils and also shales (Sridhar et al., 1975; Churchman et al., 1976; Clayton et al., 1978) indicates that the smallest particles are authigenic. Silica cements are common in soils in semi-arid environments, where they give rise to hardpans or duric horizons (Chadwick et al., 1987). Electron-optical methods showed that secondary silica was closely associated with pseudomorphs of kaolinite after mica (Singh and Gilkes, 1993). A decrease in volume consequent upon the weathering of micas and feldspars to kaolinite created voids which may be at least partly filled by silica (Chadwick et al., 1987; Singh and Gilkes, 1993). In the extreme, massive silcretes can form at the expense of clay minerals in an acidic environment by a mechanism which is not yet well understood (Thiry and Simon-Coinçon, 1996). Secondary silica, often in association with Fe oxyhydroxides, may also occur as a bonding agent in the less stable fragipans (Harlan et al., 1977; Marsan and Torrent, 1989). It appears that organic matter enhances the dissolution of silica, while silicate coatings and the presence of soluble Al retard its dissolution (Drees et al., 1989). Experimental work has demonstrated the synthesis of quartz at earth surface conditions (Mackenzie and Gees, 1971; Harder, 1971). Various forms of opal also occur in soils, usually as minor components. These include opal-CT, which is usually inherited from rocks (Munk and Southard, 1993), and biogenic opal (opal-A), which is nearly ubiquitous in the form of plant-derived opal phytoliths or, in marine environments, as sponge spicules, diatoms and radiolaria (Drees et al., 1989). Plant opal occurs in greater amounts where there are silica-accumulating plants growing on soils with labile silica rich parent materials. Andisols often provide suitable soils for the development and accumulation of plant opals (Drees et al., 1989). Silica in plant phytoliths can undergo a rapid turnover, with a high proportion of biogenic silica becoming recycled through plants (Alexandre et al., 1997). Opals may be transformed to quartz in soils (Drees et al., 1989).

1.4.14 Iron Oxides and Oxyhydroxides

Iron oxides and oxyhydroxides occur in nearly all soils (Allen and Hajek, 1989). Whereas Fe occurs predominantly as Fe(II) in primary minerals, it exists largely as Fe(III) in soils. It may be oxidized upon release from minerals by hydrolysis or it becomes oxidized within the structure of the primary minerals, so that the minerals becomes more susceptible to breakdown, whereupon the Fe(III) hydrolyzes (Schwertmann and Taylor, 1989). The resulting oxides and oxyhydroxides (commonly referred to collectively as iron oxides) have extremely low solubility products (K_{sp} of $[Fe][OH]^3$ ranges from 10^{-39} to 10^{-44}) (Schwertmann and Taylor, 1989). pH is seldom low enough in soils for simple redissolution to occur. Remobilization occurs instead mainly by means of reduction.

The nature of the Fe oxides which are formed by weathering is generally dependent more on the environmental conditions than on the particular structure of the primary mineral from which the Fe was released (Cornell and Schwertmann, 1996). Goethite (α-FeOOH) is the most widespread Fe oxide in soils. It is often the sole pedogenic Fe oxide in soils formed in cool and temperate climates (Schwertmann and Taylor, 1989). Hematite (α-Fe_2O_3) occurs in soils in warmer climates, but almost always in association with goethite (Schwertmann and Taylor, 1989). These two Fe oxides require contrasting conditions for their formation. Goethite formation is favored by low temperatures, high soil moisture and relatively high organic matter contents. Hematite tends to form under higher temperatures, in drier situations and with less organic matter than is associated with goethite formation. It imparts its characteristic red color to many soils in the tropics and also to the terra rossa soils, which are common in areas with Mediterranean climates (Boero et al., 1992). As a result, terra rossa soils have been considered to be relicts formed under warmer paleoclimates (Boero et al., 1992). However, hematite may also form under cooler climates, provided there is a marked dry period during the year (Bresson, 1974; Schwertmann et al., 1982). Boero et al. (1992) maintain that terra rossa soils in various locations in Italy and their constituent hematite have formed, along with associated Al interlayered vermiculite, under their present Mediterranean climate. Although cooler and/or moister soils would tend to contain considerable organic matter, organic matter may have a direct effect which favors goethite, hence the common occurrence of yellow, i.e., goethitic, topsoils over red, i.e., hematitic subsoils (Schwertmann and Taylor, 1989).

The conditions under which the different Fe oxides have been synthesized in the laboratory owe much to field observations but also aid in understanding mechanisms of formation in the field (Schwertmann and Cornell, 1991; Schwertmann et al., 1995). Thus, for instance, it has been found that the conversion of the poorly ordered Fe oxide (ferrihydrite) gave a maximum of goethite at a pH near 4 and a maximum of hematite at pH 7-8 (Schwertmann and Taylor, 1989). Field observations have shown a similar trend in relative occurrences of the two mineral types with pH (Kämpf and Schwertmann, 1982). Occurrence of hematite in a soil on basalt but not in an adjacent shale suggests that a high rate of release of Fe from parent materials appears to favor hematite formation (Schwertmann and Taylor, 1989). These and other field and laboratory observations can be satisfied by a mechanism for goethite formation which involves nucleation and crystal growth from a solution of Fe^{3+}, $Fe(OH)_2^+$ and $Fe(OH)_4^-$ ions (Schwertmann and Taylor, 1989). The ions may be supplied by any source of Fe. By contrast, hematite forms from Fe oxides with short-range order, particularly ferrihydrite, by transformations in the solid state such as aggregation, followed by coalescence, dehydration and internal structural rearrangement (Schwertmann and Taylor, 1989).

Model calculations have also led to predictions that hematite would be favored over goethite by low water activity, as occurs in small pores (Cornell and Schwertmann, 1996). As discussed in Section 1.7, a small crystal size should also favor hematite rather than goethite. Aluminum substitution, while common in all Fe oxides, can occur to a greater extent in goethite than in hematite, according to synthetic studies (Cornell and Schwertmann, 1996). Where goethite and hematite occur closely together, goethite always has more Al in the structure than hematite, indicating that Al goes preferentially into the goethite (Fontes and Weed, 1991; Cornell and Schwertmann, 1996). Therefore, it is likely that a higher concentration of Al in the weathering system should enhance the formation of goethite over hematite. However, these various calculations and predictions still require validation in both the laboratory and the field (Cornell and Schwertmann, 1996).

Ferrihydrite ($Fe_5HO_8.4H_2O$) is widespread in soils, especially Andisols (Parfitt et al., 1988), the B horizons of podzols (McBride et al, 1983) and in placic horizons (Campbell and Schwertmann, 1984). Its further occurrence in such situations as ochreous precipitates formed upon the emergence of waters containing Fe(II), in weathering crusts resulting from lichen growth on rocks and in bog and lake Fe ores (Schwertmann and Taylor, 1989), indicates that its formation is favored by rapid oxidation of Fe. It also tends to form instead of FeOOH phases (goethite and lepidocrocite) if silicates or organic matter are present in high concentrations. Silicates, organic matter, and phosphates in laboratory syntheses inhibit the formation of the oxyhydroxides, but they also inhibit the further transformation of ferrihydrite to more ordered oxides. Many soil ferrihydrites contain significant amounts of Si (Childs, 1992). However, a recent study using electron-optical techniques as well as XRD and chemical analyses of the products of weathering induced by a lichen on volcanic rock (Adomo et al., 1997) has shown that while ferrihydrites are formed, both hematite and goethite also occur. The particular products of weathering were influenced by the different conditions of pH, humidity and redox potential at different microsites in the rock-lichen interfaces (Adomo et al., 1997). While hematite appears to form from ferrihydrite, a general lack of ferrihydrite alongside hematite in tropical and subtropical soils suggests that its transformation to hematite is rapid by comparison with its initial formation (Schwertmann and Taylor, 1989). Ramanaidou et al. (1996) also found no ferrihydrite in association with hematite (and goethite) in duricrusts from lateritic weathering in Brazil but suggested that hematite may have developed without any precursor. If the organic C content is particularly high, all Fe is complexed by the organic matter and no oxides form (Cornell and Schwertmann, 1996).

Lepidocrocite (γ-FeOOH) occurs in reductomorphic soils alongside goethite. It appears that Fe(II) is necessary for lepidocrocite to form. It occurs as a result of the oxidation of Fe(II); hence, it commonly forms where there is a seasonal alternation of reducing and oxidizing conditions (Cornell and Schwertmann, 1996). Goethite, by contrast, can form from either Fe(II) or Fe(III) in solution. While lepidocrocite is often found in association with goethite, it is metastable relative to goethite and can be transformed, via solution, into the latter, both in the laboratory and also, apparently, in soils (Cornell and Schwertmann, 1996). However, synthesis studies have shown that carbonate ions favor the formation of goethite, so that goethite, but not lepidocrocite, forms in calcareous soils (Cornell and Schwertmann, 1996).

Maghemite (γ-Fe$_2$O$_3$) occurs quite widely, especially in the tropics and subtropics. It is most often found near the surface. It is likely to have formed when other Fe oxides in soils are heated between 300 and 425 $^\circ$C in the presence of organic matter (Schwertmann and Taylor, 1989). Laboratory work suggests that dehydration of lepidocrocite is a possible source, but field relationships suggest that this transformation is unlikely in nature (Schwertmann and Taylor, 1989). Instead, forest fires would have provided suitable conditions for the formation of maghemite from Fe oxides in general (Schwertmann and Fechter, 1984; Anand and Gilkes, 1987a; Schwertmann and Taylor, 1989). Maghemite may also form through a topotactic reaction by the natural oxidation of primary magnetite (Fontes and Weed, 1991).

Akaganeite (ß-FeOOH) forms readily in the laboratory under typical soil conditions (acidic solutions, < 30 $^\circ$C, with Al present) (Singh and Kodama, 1994) but it does not seem to have been identified unequivocally in soils. However, although chloride ions appeared to be essential for the formation of akageneite, lepidocrocite and goethite formed when nitrate was present rather than chloride (Singh and Kodama, 1994). Nevertheless, these authors also showed that goethite formation was suppressed if Al was in a polymeric rather than a monomeric form. The simplifications required for syntheses in the laboratory may overlook the complex factors responsible for the selection of phases which are favored under natural conditions. Yet another iron oxyhydroxide, feroxyhite (δ'-FeOOH) can be formed in the laboratory (Schwertmann and Taylor, 1989), but reports of its appearance in soils are rare (Cornell and Schwertmann, 1996).

Most iron oxides contrast with most aluminosilicates in that they can form rapidly and also can dissolve quite quickly when conditions suit. The mobilization of Fe usually takes place through reduction, although the formation of complexes with organic compounds can also redissolve the oxides. Iron becomes immobilized as oxides when conditions change from reductive to oxidative. At the submicroscopic level, Fe oxides can be seen as laminar coatings on other minerals including quartz (Padmanabhan and Mermut, 1996). Using TEM, with associated EDX analyses, Padmanabhan and

Mermut (1996) observed that Fe oxides had nucleated on the surfaces of dissolving quartz grains where they formed barriers to the interiors of the porous weathered quartz grains, trapping other minerals such as kaolinite within, and also provided nuclei for the further development of Fe oxides. The mobilization of Fe by reduction commonly results from microbial activity (Robert and Berthelin, 1986; Schwertmann and Taylor, 1989; Robert and Chenu, 1992). This occurs only in anaerobic environments, such as water-saturated soils via microbial metabolic activity which involves the concomitant oxidation of decomposing biomass (Cornell and Schwertmann, 1996). The combined oxidation and reduction reaction consumes protons and is favored by a low pH. Robert and Berthelin (1986) consider bacteria to be the main agent for ferrihydrite formation. A group of microorganisms (iron bacteria) are able to oxidize Fe(II) in aqueous solutions. By another mechanism, some microorganisms can bring about the metabolic oxidation of organic ligands that complex Fe(III) in solution (Schwertmann and Taylor, 1989).

Many Fe oxides are only poorly ordered. Their identification and the understanding of their formation processes have been aided by the application of a variety of instrumental techniques, particularly Mössbauer spectroscopy (Kodama et al., 1977; Childs et al., 1979) and differential X-ray diffraction DXRD (Schulze, 1981; Wang et al., 1993). With color being an important characteristic of Fe oxides, their objective analysis by diffuse reflectance spectroscopy (Malengreau et al., 1996) also offers a promising approach. A likely consequence of the application of novel techniques is the detection of previously unidentified phases which can play a role in the genesis of Fe oxides in soils. Thus, for instance, Trolard et al. (1997) have recently reported on the identification of a green rust mineral in a reductomorphic soil by using Mössbauer and Raman spectroscopies. Green rusts, which are Fe(II)-Fe(III) hydroxides, had previously been synthesized in the laboratory, but not identified in soils (Schwertmann and Taylor, 1989, Hansen et al., 1995). Using magnetic susceptibility measurements, Fine et al. (1993) have found evidence for the pedogenesis of maghemite in columns of loess in China.

1.4.15   Aluminum Hydroxides and Oxyhydroxides

Among the aluminum hydroxides and oxyhydroxides, gibbsite [Al(OH)$_3$] is the most common in soils (Hsu, 1989). Gibbsite forms under strong weathering. It was considered to be the end product of weathering in Jackson et al.'s (1948) sequence (Table 1.3). It occurs in old and young soils. In the latter case, it occurs when leaching is particularly strong. In soils on volcanic ash, gibbsite was formed in preference to allophane or halloysite when rainfall was very high and drainage effective. In either case, gibbsite forms because of the low availability of Si. Either extreme desilication has occurred as in some old, highly weathered soils, or Si concentration in solution has been kept low by rapid throughflow of water. In some mountain soils in South Carolina, gibbsite formed where there was extensive leaching (Norfleet et al., 1993). Otherwise kaolinite formed. Consideration of the thermodynamics of mineral-solution relationships and also of reaction kinetics led Norfleet et al. (1993) to conclude that the residence time of Si especially was the determining factor for secondary mineral formation in these soils. Gibbsite may also be favored if parent materials are Al rich (Lowe, 1986). In the humid tropics, gibbsite tends to develop in the highlands and kaolinite in the lowlands (Herbillon et al., 1981; Weaver, 1989). Preformed gibbsite may be dissolved, possibly by organic matter, when inherited in a soil (Wang et al., 1981). Other oxides and oxyhydroxides of aluminum, such as boehmite (AlOOH) and corundum (Al$_2$O$_3$) (Anand and Gilkes, 1987b) and also less well-ordered phases, e.g., χ-alumina (Singh and Gilkes, 1995) and tohdite (5Al$_2$O$_3$.H$_2$O) (Tilley and Eggleton, 1994) have been reported at the surfaces of lateritic profiles. These may have formed either from the dehydration of gibbsite, or in its stead under conditions which have inhibited the crystallization of gibbsite. For example, there appeared to be a close correlation between the contents of corundum and that of maghemite in lateritic duricrusts in Western Australia (Anand and Gilkes, 1987b). These authors considered that maghemite probably formed by dehydration of Fe oxyhydroxides in bush fires (Section 1.4.14) and an analogous origin for corundum from the dehydration of gibbsite and/or boehmite seems likely.

1.4.16 Titanium and Zirconium Minerals

Anatase has formed authigenically in several locations (Milnes and Fitzpatrick, 1989). It commonly forms from the weathering of silicates containing Ti. Ilmenite oxidizes, then hydrolyzes to give pseudorutile. Pseudorutile, which may contain both $Fe(II)$ and $Fe(III)$, and Mn in addition to Ti in oxide form (Grey and Reid, 1975; Anand and Gilkes, 1984; Babu et al., 1994) can then dissolve to give either rutile (Grey and Reid, 1975) or anatase (Anand and Gilkes, 1984). Using magnetic fractionation to separate products of different stages of alteration of ilmenite, and Mössbauer spectroscopy for characterization, Babu et al. (1994) showed that an ilmenite had weathered to give only pseudorutile. However, in the weathered beach sands studied by Grey and Reid (1975), hematite formed alongside rutile, while in the pallid zone of the lateritic weathering profile studied by Anand and Gilkes (1984), Fe oxides were lost, but halloysite, kaolinite and gibbsite crystallized out in pores alongside the anatase. According to Anand and Gilkes (1984), the lower density of anatase than rutile may have enabled it to crystallize in the presence of other ions in soil solution within the weathering profile while the low concentrations of other ions in solutions from beach sands would allow the denser rutile to form there. Titanomaghemite is formed in soils by the topotactic oxidation of primary titanomagnetite (Fitzpatrick and le Roux, 1975). There is no firm evidence for the authigenic formation of Zr minerals or for the weathering of zircon (Milnes and Fitzpatrick, 1989).

1.4.17 Highly Soluble Minerals

The more soluble minerals, including halides and sulfates, occur in soils in arid regions of the world, including Antarctica (McCraw, 1962; Gibson, 1962). They often form by dissolution, then precipitation upon evaporation (Doner and Lynn, 1989). Certain sulfate minerals, particularly jarosite and related phases and gypsum can also form by the oxidation of sulfides (Section 1.4.20). Carbonates, especially calcite, can form in soils in a similar way, although they are most often inherited from parent materials (Doner and Lynn, 1989). However, they are often also present as coatings on or cements between other particles in soils (Doner and Lynn, 1989). Distinction between neogenetic origins on the one hand, and either inheritance or detrital origins on the other is difficult, although pedogenic calcite is nearly always in the form of silt sized crystals (Doner and Lynn, 1989). ^{14}C dating showed that carbonates were formed authigenically in soils on basalt in Israel, but after soil-forming processes were completed (Graef et al., 1997). It has not usually been considered that dolomite can be authigenic in soils (Doner and Lynn, 1989), although Sherman et al. (1962) reported its pedogenic formation. More recently, using O and C isotopic analyses and radiocarbon dating, Kohut et al. (1995) concluded that dolomite had formed by direct precipitation in the form of very small particles in a sulfatic saline soil in Canada.

1.4.18 Manganese Oxides

Manganese oxides are widespread in soils, usually in minor amounts, and they are often poorly ordered (McKenzie, 1989). Among these oxides in soils, birnessite and lithiophorite have been identified most often. The manganese oxides occur as dispersed small particles, as coatings on other soil particles and on ped surfaces, and in concretions, pans and nodules, commonly with Fe oxides

(McKenzie, 1989; Allen and Hajek, 1989). They are most likely to have formed by both biological and nonbiological oxidation of Mn(II), most often to Mn(IV), as in birnessite (McKenzie, 1989).

1.4.19 Phosphate Minerals

Apatite is the major source of P in rocks (Whitten and Brooks, 1972; Lindsay et al., 1989). Apatite $[Ca_5(PO_4)_3(OH)]$ with various degrees of substitution of OH by F and/or CO_3 weathers to give various phosphate minerals, including gorceixite (Ba-Al), florencite (Ce-Al), crandallite (Ca-Al), plumbogummite (Pb, Al and other cations), vivianite (Fe), wavellite (Al), wardite (Na-Al), rhabdophane (rare earths), fluellite (Al, together with a high F content), as well as a myriad of other possibilities (Norrish, 1968; Adams et al., 1973; Tazaki et al., 1986b, 1987b; Banfield and Eggleton, 1989; Lindsay et al., 1989). Electron microscopy showed that apatite in weathered granite was replaced by newly crystallized phosphate phases with morphologies ranging from euhedral crystals to donut shapes. These latter shapes bore a close resemblance to apatite which had been synthesized using microorganisms, and Banfield and Eggleton (1989) raised the possibility that they could have had a microbiological origin.

1.4.20 Sulfide Minerals

Sulfide minerals form by the bacterial reduction of sulfate originating in sea water in the presence of organic matter (Doner and Lynn, 1989). Usually, amorphous FeS is formed first and is then converted to pyrite (FeS_2), and mackinawite (Fe_9S_8) and greigite (Fe_3S_4. HCO_3^-) are released in the process.

These sulfide minerals oxidize readily on aeration to give sulfuric acid, a low pH, and the sulfate minerals jarosite $[KFe_3(OH)_6(SO_4)_2]$ and/or natrojarosite $[NaFe_3(OH)_6(SO_4)_2]$. Using electron-optical methods, Fitzpatrick et al. (1992) identified the poorly ordered oxyhydroxysulfate mineral, schwertmannite in saline sulfidic soils in South Australia. Gypsum may also form if carbonates are present. Jarosite is metastable in soils and hydrolyzes to goethite (van Breeman, 1982). Acid sulfate soils formed in this way are most common in coastal margins (Dost, 1973; Kittrick, 1982), but may also form on inland sulfidic sediments (Carson and Dixon, 1983; Fitzpatrick et al., 1996).

1.4.21 Pyrophyllite, Talc and Zeolites

Pyrophyllite and talc are generally rare in soils (Zelazny and White, 1989). It appears that pyrophyllite in soils is most often inherited from rocks. Velde (1968) suggested that the decomposition of organic matter in sediments caused the reduction of Fe(III) to Fe(II), with the resulting formation of illite and chlorite rather than pyrophyllite. Talc and related minerals may form as a weathering product of pyroxenes or amphiboles (Zelazny and White, 1989). It weathers rapidly to nontronite or Fe oxides, but some may be preserved by occlusion.

Zeolites occur in soils most commonly where the parent rocks are zeolitic, suggesting inheritance (Ming and Mumpton, 1989). However, it also appears likely that they have been neoformed from strongly alkaline solutions in salt-affected soils. Analcime has been reported to form this way in nonvolcanic environments (Baldar and Whittig, 1968; Frankart and Herbillon, 1970), while other Na rich zeolites such as phillipsite, natrolite and chabazite have formed in addition to analcime in soils with a volcanic origin (Hay, 1978; Gibson et al., 1983).

1.5   Occurrence of Clay Minerals in Relation to Soil Types

In terms of their genesis, soils have generally been considered to be zonal, reflecting annual temperature and precipitation, or azonal, where their formation has been controlled by local geological, geochemical or biological conditions (Loughnan, 1969; Carroll, 1970; Chamley, 1989; Weaver, 1989). Zonal soils reflect climatic zones while azonal soils are immature. Some soils have also been considered to be intrazonal. These have mature profiles but reflect a dominant local condition such as the nature of the parent material or relief (Loughnan, 1969; Weaver, 1989). Garrels and Mackenzie (1971) following Marbut (1928) consider that there are three genetic soil types. These are

(1) Pedalfers, which are named by the dominance of Al and Fe in their weathering products, hence "pedAlFer". These form from rocks with a cover of vegetation under a plentiful supply of rainfall. They tend to form in temperate, humid climates under a forest cover. Bacterial decomposition of plant and animal remains produces much CO_2, which attacks primary minerals, releasing alkali and alkali earth cations and also silica into solution. Aluminosilicate and Fe(III) oxides result, the latter often being concentrated in B horizons. Often, this process is described as podzolization.

(2) Pedocals, which are named by analogy after their tendency to accumulate Ca (and also Mg). They form from rocks with a scrub or grass cover under relatively low rainfall. They are calcareous, with a dominantly montmorillonitic clay mineralogy.

(3) Laterites, which form under sparse vegetation and high rainfall where the temperature is high. In this case, bacterial decomposition is rapid and CO_2 escapes into the atmosphere. Primary minerals are leached by high volumes of nearly pure water. Even quartz may be dissolved and the weathered residue is rich in Fe(III) and Al oxides, oxyhydroxides and hydroxides.

According to a simpler generalization into only two groups by Carroll (1970), pedalfer soils are those from which soluble elements are leached and alumina and stable elements remain. Podzolization, lateritization or, with impeded drainage, peat formation and gleization can each take place in the formation of different soils within this group. Calcification is the major process occurring in the formation of pedocal soils, though there may also be salinization in some situations.

For zonal soils, a number of authors (Carroll, 1970; Millot, 1970; Chamley, 1989) following Millar et al. (1958) have identified the processes and summarized the products of weathering by both soil type and the nature of secondary minerals on the basis of the climate zones in which they have arisen, with the proviso that zones are not necessarily well defined and intermediate conditions may exist. A synthesis of these is given in Table 1.4.

Common azonal types include halomorphic or saline soils containing salts, but usually little weathering of primary minerals, and also hydromorphic or gleyed soils, which often also show little

Table 1.4 Mineral weathering in zonal soils in relation to climate zones

Climate zone	Dominant processes	Typical soil types (common names)	Typical secondary minerals
Perpetually cold	Mainly physical weathering	Tundra	Illite, chlorite, mixed-layer minerals, soluble minerals
Arid	Mainly physical weathering	Desert soils	Illite, chlorite, mixed-layer minerals, soluble minerals, palygorskite
Cool-wet	Acidic hydrolysis	Podzolised	Illite, vermiculite, smectite, mixed-layer minerals (upper); poorly ordered minerals, oxides (lower)
Moist temperate	Transformation	Brown earths	Illite, vermiculite, smectite, mixed-layer minerals
Continental: both hot & cold, semiarid	Neoformation	Chernozems	Smectites
Warm, semiarid	Dissolution and precipitation	Calcretes, (silcretes)	Carbonates, palygorskite, smectites, (silica)
Hot, both wet & dry	Leaching, oxidation	Tropical red earths	Kaolinite, iron oxides
Hot and wet	Strong leaching, oxidation	Laterites	Gibbsite, iron oxides, kaolinite

alteration of primary minerals (Chamley, 1989). Soils formed in volcanic environments are also azonal (Chamley, 1989). Their secondary minerals principally include minerals with short-range order, especially allophane, imogolite, and ferrihydrite, or halloysite and smectites, depending on rainfall and drainage and the specific nature of parent materials (e.g., whether rhyolitic or basaltic), or kaolinite and gibbsite, where there has been strong leaching.

While older genetic classifications of soils reflect the pattern described in Table 1.4, the now commonly used Soil Taxonomy leads to only broad correspondences between soil orders and constituent clay minerals, except in some cases (Weaver, 1989). Entisols, Aridisols and Mollisols typically show little alteration to secondary minerals. Inceptisols include a range of clay mineral types, though these may often occur through inheritance. Alfisols can include a wide range of secondary minerals, from kaolinite to smectites. Spodosols arise from podzolization, and often include the typical combination of transformed micaceous minerals in upper, eluvial horizons and aluminosilicates with short-range order as well as oxides and oxyhydroxides in lower illuvial horizons. Andisols, as soils from volcanic ash, often contain allophane. Vertisols most often contain smectites. Ultisols reflect moderately strong leaching, so contain well-developed, relatively stable, secondary minerals, including kaolinite and 2:1 minerals with hydroxy-Al interlayers. Oxisols, as products of strong leaching and oxidation, have clay mineralogies typified by sesquioxides, gibbsite and kaolinite. Most minerals in Oxisols and Ultisols are authigenic (Allen and Hajek, 1989). In Soil Taxonomy, however, mineralogy is considered specifically at the soil family level, rather than by order. In soils in the United States, Puerto Rico and the Virgin Islands, the mineralogy of the most common family was mixed and then by a considerable margin, smectitic, followed by carbonatic and siliceous (Allen and Hajek, 1989).

1.6   Effect of Environmental Factors on the Formation of Clay Minerals

Following early work by Dokuchaev in Russia and Hilgard in the United States, Jenny (1941) formulated an equation to relate properties of soils to the major soil-forming factors, isolated as apparently individual variables (Birkeland, 1974). It is

$$s = f(cl, o, r, p, t, \dots\dots\dots\dots\dots,)$$ [1.3]

where s denotes the soil property of interest; cl, the climatic factor; o, the biotic factor; r, the topographic factor; p, the parent material; and t, the time factor. The equation allows, through the dots, for other, locally important, factors to be included. The equation was recently described (Hudson, 1992) as a powerful paradigm in soil science. The relationship becomes too complex to allow it to be solved unless one factor is allowed to vary while others are kept constant. This is done by means of establishing monosequences of soils, over which other factors in addition to the chosen variable factor either vary very little over the sequence or their variation has negligible effect on the soil property of interest (Birkeland, 1974). While the assumptions involved in the establishment of monosequences are often questionable in natural, highly variable systems (e.g., Crocker, 1952), they can be valuable for understanding the relative roles played by the different factors in waethering (jenny, 1958; Walker, 1965). Appempts may be made to solve Equation [1.3] quantitatively, as in the case of the time function, when the relationship derived from a chronosequence is termed a chronofunction and rates of soil processes can be deduced (Schaetzl et al., 1994). For the case of the formation of secondary minerals by weathering, when s represents degree of alteration of primary minerals or development of secondary minerals, quantitative expression of even the property s may be problematic. This is because the most common methods that have been used for the quantitative analysis of minerals in soils have been subject to substantial uncertainties (Hughes et al., 1984; Whitton and Churchman, 1987). On the other hand, current needs including the use of full pattern methods, especially the Rietveld method, point to increasing accuracy in quantitative analyses of minerals in soils (Bish, 1994).

The validity of quantitative analyses notwithstanding, monosequences have aided our understanding of the relative importance of soil factors and of components of these factors. Indeed, conclusions about conditions leading to the alteration and formation of minerals by weathering, which have already been discussed, have use an applied monosequence approach whenever two situations have been compared and it has been considered that one particular factor dominates all other possible inputs, and thereby, controls the changes identified.

The often close identification of the products of weathering with climate zones in zonal soils, as discussed in the previous section, means that climate is very often a major factor affecting mineral changes. Birkeland (1974) reported that there have been many studies relating clay mineral formation to climate. Generally, these have shown that the amount of precipitation was the key climatic factor, provided drainage was adequate. In hawaii, smectite predominated below ~1,000 mm precipitation, kaolinite between 1,000 and 2,000 mm, and Fe and Al compounds under even higher precipitation (Sherman, 1952). Also in Hawaii, Hay and Jones (1972) found a particular ash layer had weathered to smectite at 250-650 mm precipitation and to gibbsite at 3,700 mm. Across California, Barshad (1966) found a pattern of smectite occurring where precipitation was ≤ 1,000, and gibbsite where precipitation was less than this value. Kaolinite and hallosite occurred over a wide range of precipitation, but were predominant above ~ 500 mm. Illite and vermiculite also occurred in some soils (Barshad, 1966). In these studies, generally, higher rainfall led to a greater loss of silica (Birkeland, 1974). By contrast, Churchman (1978, 1980) studied mineral changes at an earlier stage of weathering within a climosequence in South Island, New Zealand that had been established mainly for soil biochemical studies (Molloy and Blakemore, 1974). This particular sequence of soils, which formed on an extensive formation of schist under native tussok vegetation since the last glaciation but encompass a wide range in mean annual precipitation (from 350 to 5,000 mm), shows only minor mineral alterations consistent with generally cool temperatures. In particular, feldspars remained fresh across the sequence, so that the mineral changes are almost exclusively those of micas and the less abundant chlorites. Across the sequence in surface soils, the micas became transformed to give regular interstratifications of the parent micas with first, vermiculite and then beidellite-smectite. The chlorites also became transformed by interstratifiaction but then decomposed to non-phyllosilicates. It was shown that the amount of mean annual precipitation largely controlled mineral development, although an especially cold temperature regime apparently slowed its rate in a soil at one site. The earlier stages of mineral alteration occurred mainly as a result of displacement of K from mica interlayers while the more advanced stages of weathering involved podzolization occurring by acid hydrolysis in two of the soils. Climate may not always affect weathering directly, by increasing throughput of leaching water, for example. It may also act through its effect on the amount and nature of organic matter and on microbial processes. Thus, the high organic C contents of the two most highly developed members of the sequence (Molloy and Blakemore, 1974) are likely to have led to the acidifiaction of these soils and their consequent podzolization, while the location of one of them at high altitude, and hence its low mean annual temperature (3.8 °C) has probably contributed to a slowing of the rate of microbial activity (Theng et al., 1986). A low rate of microbial decomposition in this particular soil has apparently led to the incorporation of a great deal of organic matter between the smectitic layers in the mica-beidellite which dominates the clay fraction of this soil. As an alternative approach to that of monosequences, Folkoff and Meentenmeyer (1985) carried out a statistical cluster analysis of the secondary minerals in the A horizons of 99 United States soils in relation to a suite of climatic variables for the 99 sites. Climatic variables showed some, but limited ability to predict clay mineral assemblages. The precipitation of the wettest month and Arkley's leaching index, which is based on potential evapotranspiration, accounted for most of the predictive power. However, other factors, particularly parent material, influenced clay mineral compositions to a greater extent than climate.

Studies of the effect of biotic factors on soil mineral development in the context of Jenny's (1941) equation have mostly concentrated on the comparative effects of different types of vegetation. Campbell (1974) found that the clay minerals in soils formed from the weathering of mice, chlorite and feldspar in South Island, New Zealand were differenc according to their distance from native, southern beech (*Nothofagus fusca*) trees. Furthest from trees, where pH was > 4.5 and organic matter levels relatively low, 2:1 aluminosilicates with hydroxy- Al interlayers formed together with kaolinite, allophane and gibbsite. Close to both living trees and also decomposing stumps, pH dropped below 4.5 and hydroxy-Al interlayers, gibbsite and allophane all dissolved, giving dominant vermiculite or smectite in the soils. Harrison et al. (1990) studying another sequence in South Island, New Zealand, have suggested that an historical change from forest to grassland has led to the development of Al-rich secondary phases. In Denmark, Madsen and Nørnberg (1995) found that the degree of weathering of sandy soils, indicated mainly by the extent of alteration of primary minerals, was similar under each of heather, oak and spruce, but greater under each of these species than under grass. In Tuscany, smectite was identified in an E horizon that has developed under a 50-yr-old plantation of Corsican pine (*Pinus nigra* Arn. ssp. larico) (Certini et al., 1998). The formation of smectites in E horizons indicates pdozolization (Section 1.4.6). By contrast, Certini et al. (1998) found little or no indication of secondary clay minerals under a neighboring 70-yr-old plantation of silver fir (*Abies alba* Mill.). The soil under fir trees also showed no development of an E horizon. The pH of stemflow in the pines was substantially lower than that in the firs (Certini et al., 1998).

Earlier discussions on red-black toposequences and the contrast in topographic situations conducive to the formation of kaolinite, on the one hand, and smectite, on the other, show the importance of the topographic factor in mineral formation. Although high elevation generally brings a lower temperature implying slower reactions, its effects on weathering can be surprising. In a toposequence on calcareous micaschist, it was found that podzolization, with alteration of chlorite to smectite occurred at high altitudes (above 2,000 m) irrespective of vegetation, but not at lower altitudes except under Ericacae vegetation (Buurman et al., 1976). There was less alteration of minerals at lower altitudes. The authors suggested that high precipitation and low temperature at high altitudes increased carbonate solubility. As in the red-black soil sequences, toposequences may provide an opportunity to study the effect of different drainage conditions. In a catena over only 100 m on serpentinite, little alteration of the serpentine to low charge vermiculite had occurred at the summit and backslope positions (Bonifacio et al., 1997). In footslope positions, however, alteration to smectite had occurred. Dahlgren et al. (1997b) studied changes in an elevational transect in California. Although there were changes such as the formation of gibbsite and hydroxy-Al interlayered minerals that were consistent with increasing desilication with altitude, these authors attributed them mainly to the effects of climate changes accompanying those in altitude. Climate changes must inevitably accompany altitudinal changes except over very short catenas, such as those studied by Bonifacio et al. (1997) where changes in topography are likely to indicate drainage effects most strongly.

The effect of parent materials is most apparent when there is a close relationship between the primary and secondary minerals, as seen earlier. Thus, Barshad (1966) found that illite occurred only when soils in California were formed on felsic, rather than mafic igneous rocks, due presumably to the requirement of illite for K (Birkeland, 1974). In the same set of soils, smectite persisted under higher precipitation on mafic than on felsic parent materials reflecting the higher content of cations in the mafic rocks (Birkeland, 1974). Furthermore, an effect of parent material may arise from grain size as well as composition, with coarse grained igneous rocks weathering more rapidly than finer grained rocks (Birkeland, 1974). The larger intergranular surface area of fine grained rocks may mean that more energy is required to bring about their disintegration (Birkeland, 1974).

Studies of soil chronosequences have been popular. Reviewing the literature to that time, Birkeland (1974) considered that "clay minerals that form in soils from parent materials low in clay content probably form mineral assemblages stable in that environment, and therefore, variation in the minerals with age generally is not found and not expected." On the other hand, Birkeland also considered that soils formed from parent materials which already contained clay minerals were likely to show changes with time as the clay minerals adjusted to the new environmental conditions. Clays formed by transformation from primary minerals like micas, which, although not clay minerals *sensu stricto*, are also phyllosilicates, could also change with time as they adjust to the weathering environment. Usually, however, chronosequences are subject to the two limitations described by Rabenhorst (1997): (1) small data sets due to the limitation of what is available in nature, and (2) variability within the data sets due to geographical variations in other soil-forming factors. Nevertheless, Bain et al. (1993) observed the destruction of chlorite and the formation of vermiculite at the expense of mica within a chronosequence between 80 and 13,000 years BP. Read et al. (1996) showed a trend in the alteration of primary minerals with time by observing feldspars from a set of terraces of the River Thames in England. Burt and Alexander (1996) showed increased development with time on glacier moraines in Alaska by observing trends in the ratio of oxalate extractable to dithionite extractable Fe. These indicated the development of spodic character, even before the formation of new crystalline clay minerals occurred. Bockheim et al. (1996) found progressive development of weathering in marine terraces in Oregon with time. The changes include the transformation, neoformation and translocation of clays. Bakker et al. (1996) found changes occurring with time in both the alteration of primary minerals and formation of secondary minerals in an evolutionary sequence of between 1850 and ~ 120,000 years in age on rhyolitic tephra deposits in New Zealand. The younger soils had few secondary minerals while halloysite, gibbsite, kaolinite, vermiculite and crystalline Fe oxides occurred in older soils. However, some of the products had the effect of altering drastically such features of the soils as drainage. The soils became polygenetic. It was concluded that the results confirmed the view that the older the soil, the more polygenetic it is likely to be.

In a series of studies, Churchman (1978, 1980, 1993) determined the clay mineralogies of soils from two climosequences, two vegesequences, a parent material sequence, and two chronosequences to help explain the factors governing the transformation of mica over a large part of South Island, New Zealand. Fig. 1.10 shows a plot of the proportion of mica layers which have been expanded to vermiculite or smectite layers in the clay fractions of upper A horizons of these various soils. It was concluded from the integrated results that (1) rainfall, as measured by Mean Annual Precipitation (MAP) was the main climatic factor affecting mineral transformations; (2) a MAP of 100-800 mm was required to obtain vermiculitic layers, within regularly interstratified mica-vermiculite, on schist under tussock grassland; (3) vermiculitic layers were formed from mica in < 1000 yr under heavy rainfall (MAP > 10 m); (4) the transformations of mica occurred more rapidly in granite than in greywacke; and (5) full expansion of layers (to beidellite, without interstratified mica) occurred only under forest.

The expectation that soil clay mineralogy should reflect environmental factors can also be used to test whether the clay mineral suite in soils could have derived from present day environmental conditions. If the clay minerals are unlikely to have formed in present day conditions, they may be inherited from the products of weathering in that location under earlier conditions or they may have been transported from another location. Thus, for instance, the widespread occurrence of kaolinite in

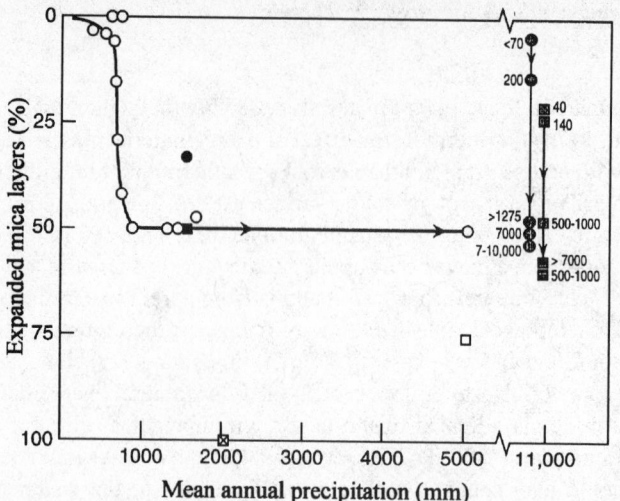

Fig. 1.10 Relationship between proportion of expanded mica layers to mean annual precipitation in a number of sequences in South Island, New Zealand. Circles are for soils under *Chionochloa* tussock grassland. Squares are for soils under native *Nothofagus* beech forest. Open symbols are for climosequences. Grey symbols are for chronosequences. Filled symbols are for soils on greywacke. A crossed square is for a soil on granite under *Nothofagus* beech forest. The ages of soils (years) in the chronosequences are indicated alongside symbols. [Data from Churchman, 1978, 1980 and unpublished]

soils from several different parent materials over a broad range of climates which included dry areas in China, generally indicated inheritance of this mineral from weathering that had occurred under much moister paleoclimatic conditions (Singer, 1993). Furthermore, saprolites and lower horizons of some of these soils contained gibbsite, while vermiculite or Al (and Fe) interlayered vermiculite appeared in all soils, but not in saprolites. This apparent inversion of the sequence of weathering was explained by a polygenetic origin for the soils, with saprolites having already been weathered under moister climates than prevail at present (Singer, 1993).

In conclusion, it is noted that, although monosequences have had some success in explaining the important factors in mineral alteration and formation on weathering, they can only be applied easily to surface horizons. Differentiation within profiles, which is a distinguishing characteristic of all except immature soils, means that each profile can be regarded as constituting an approximate chronosequence, although other factors also vary within profiles. These include biotic and (micro) climate factors, and in polygenetic soils, parent materials.

1.7 Explanations and Predictions of Mineral Development from Bulk Solution Compositions

As a pioneer in the application of thermodynamic concepts to the stabilities and formation of minerals in soils, Kittrick (1967) wrote: "Fundamentally a mineral is a package for its elements. It will persist in nature only as long as it is the most stable package for those elements in its environment. However one might choose to quantitatively characterize the structural stability of a mineral, it would seem impossible to understand weathering without some measure of the environment as well. For mineral weathering, the environment is essentially the activities of common ions." In practice, as Kittrick (1973) stated: "In theory at least, each clay mineral forms under solution conditions where it is the least soluble of the clay minerals competing for that group of elements. Thus, because of the stability of other minerals, the stability field (i.e., set of conditions under which it is stable) of a particular mineral represents a much more restricted chemical environment than does a solution saturated with respect to that mineral alone." The thermodynamic or thermochemical approach to the prediction of the stable assemblage of minerals in a soil system involves constructing stability lines for all possible minerals in the soil and combining them in a stability diagram. Stability lines for minerals are constructed from equations for their dissolution or precipitation reactions in aqueous solutions, e.g., for kaolinite $[Al_2Si_2O_5(OH)_4]$:

$$Al_2Si_2O_5(OH)_4 + 6H^+ \rightarrow 2Al^{3+} + 2H_4SiO_4 + H_2O \qquad [1.4]$$

The equilibrium constants (K) for mineral dissolution reactions like these represented by Equation [1.4] are given by the ratio of the product of the activities of reaction products to that of the product of the activities of the reactants. For the reaction in Equation [1.4]:

$$K_{kaol} = \frac{\left[Al^{3+}\right]^2 \left[H_4SiO_4\right]^2 \left[H_2O\right]}{\left[Kaolinite\right]\left[H^+\right]} \qquad [1.5]$$

where [ ] represent activities. In practice, these parentheses are not normally used. In logarithmic form, Equation [1.5] becomes:

$$\log K_{kaol} = 2\log Al^{3+} + 2\log H_4SiO_4 + 6pH \qquad [1.6]$$

At standard temperature and pressure, K becomes K^o. Log K^o values are generally calculated from thermodynamic data. This is because the standard free energy change for the reaction ΔG^o_r is related to log K^o as follows:

$$\Delta G^o_r = \sum \Delta G^o_f(products) - \Delta G^o_f(reactants) =$$
$$-RT \ln K^o = -5.707 \log K^o \left(kJ\ mol^{-1}\right) \qquad [1.7]$$

Stability diagrams for mineral assemblages consist of a set of lines relating expressions for functions of the activities of appropriate solution species to one another for each mineral species likely to be stable in the assemblage. In soil systems, they are very often expressions for (log Al^{3+} + 3 pH) against log H_4SiO_4. For kaolinite, the expression is

$$\log Al^{3+} + 3pH = -\log H_4SiO_4 + 0.5\log K^o_{kaol} \qquad [1.8]$$

In Fig. 1.11 (Percival, 1985), a stability diagram for kaolinite, halloysite, imogolite and gibbsite has been constructed assuming formulas for these four minerals which include no substitutions for Al or Si. Log K_o values were taken from Lindsay (1979) for kaolinite, halloysite and gibbsite and the value was calculated for imogolite by Percival (1985) from the experimental stability data of Farmer and Fraser (1982). Fig. 1.11 indicates that kaolinite is always the most stable mineral relative to halloysite and imogolite. It is also more stable than gibbsite except at very low silica activities. Thermodynamics, therefore, indicates that where there is sufficient silica, halloysite, imogolite and gibbsite will ultimately dissolve in the soil solution and reprecipitate as kaolinite. It also indicates that imogolite is more stable than halloysite, except at high activities of silica.

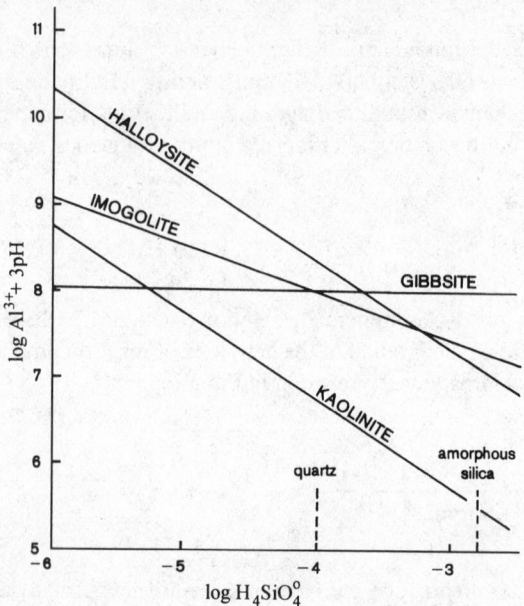

Fig. 1.11 Stability of kaolinite, halloysite and imogolite compared with that of gibbsite [Reprinted from Percival, 1985 with permission od Landcare Resaerch NZ Ltd]

An alternative approach to the construction of stability diagrams takes into account the low concentration of Al in most soil solutions. In this case, equilibrium Al ion activities are excluded but relationships are shown among other ions which are common to a group of minerals. In constructing this type of diagram, the aim is to establish the conditions or boundary lines under which different pairs of minerals are in equilibrium with each other. When microcline (Mi) and muscovite (Mu), for example, are in equilibrium with each other, they are also in equilibrium with the solution they both contact. Hence, their individual dissolution reactions apply simultaneously. If these reactions are combined as simultaneous equations so that Al^{3+} is eliminated, a reaction is obtained as follows:

$$3KAlSi_3O_8 + 2H^+ + 12H_2O \leftrightarrow KAl_2(Si_3O_{10})(OH)_2 + 2K^+ + 6H_4SiO_4$$

microline   muscovite   [1.9]

The expression for $\log K^o$ for this equilibrium becomes:

$$\log K^o_{Mi-Mu} = 2\log K^+ + 6\log H_4SiO_4 + 2pH \qquad [1.10]$$

or,

$$\log K^+ + pH = -3\log H_4SiO_4 + 0.5\log K^o_{Mi-Mu} \qquad [1.11]$$

where $\log K^o_{Mi-Mu} = 3 \log K^o_{Mi} - \log K^o_{Mu}$ (from Equation [1.9]).

Equation [1.11] defines the shared boundary of the stability fields of the two minerals when it is plotted on a $(\log K^+ + pH)$ versus $\log H_4SiO_4$ diagram, as shown along with the boundary lines between the stability field of other commonly associated minerals in Fig. 1.12. Each type of mineral is

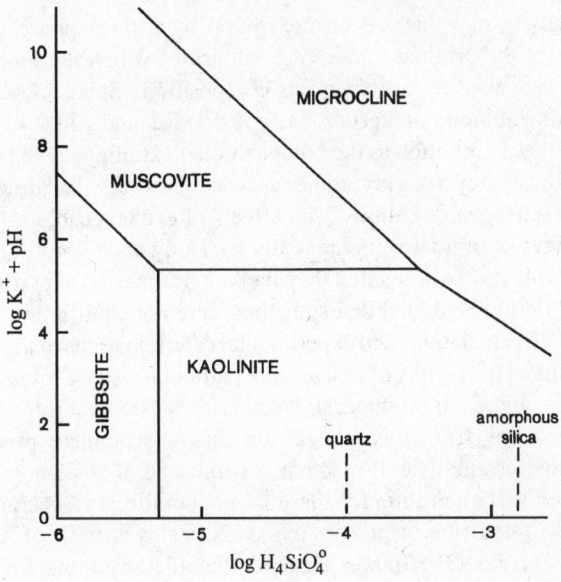

Fig. 1.12   Various stability fields in the (log K+ + pH) vs log H_4SiO_4 stability diagram [Reprinted from Percival, 1985 with permission of Landcare Research NZ Ltd]

considered to be stable for solution compositions within its own field, but unstable for compositions falling within the field of another mineral.

This approach relies upon a number of assumptions. One of the major assumptions is that the thermodynamic data used for the construction of the stability diagrams are reliable. In particular, the validity of these diagrams is based on the accuracy of ΔG°_f values for minerals and soluble species in their adjacent solutions. These values are obtained by either calorimetry or solubility methods (Kittrick, 1986). Solubility methods have been employed often, especially for the more complex phyllosilicates. In these methods, the products of the dissolution of a mineral into its soluble components, i.e., its congruent dissolution at standard temperature and pressure are analyzed and a solubility product K° thereby determined. ΔG°_f for the mineral can then be calculated using Equation [1.7].

Rai and Kittrick (1989) considered that the thermochemical data for simple soil minerals such as quartz, hematite, gibbsite and several primary minerals were good. However, the variability in reported values shows that those for the phyllosilicates may be less reliable. Among the reasons for this are the difficulty of reaching and establishing equilibrium, a lack of congruency in the dissolution, the presence of impurities, and the inherent variability of the composition of minerals like smectites and chlorites.

The attainment of equilibrium has been established by approaches from both under- and oversaturation of solutions in the soluble components of the minerals (Kittrick, 1971). Kaolinite has been shown to dissolve congruently (Kittrick, 1966; May et al., 1986). However, the more complex smectite minerals may dissolve incongruently to give two secondary phases, one more aluminous and another more siliceous, than the original smectites (Churchman and Jackson, 1976; May et al., 1986). According to these authors, the activity of the solutes may be controlled by one of these secondary phases rather than by the smectite. Rosenberg et al. (1984) showed that the composition of a solution in contact with muscovite may have been controlled by a mica phase which was of different composition than that of the original muscovite. Impurities which are more soluble than the clay minerals themselves can also control solution composition. Some clay minerals, at least, are heterogeneous, non-stoichiometric materials (Fritz, 1985; Rai and Kittrick, 1989; Petit et al., 1992; Essene and Peacor, 1995). In addition to the compositional variations both between and within their constituent aluminosilicate layers, clay minerals can also vary because of differences in the composition of their exchangeable cations. The effects of exchangeable cations on smectites upon calculations of ΔG°_{f} have been treated theoretically by Tardy and Garrels (1974) and Mattigod and Sposito (1978). May et al. (1986) noted that they had not been tested experimentally up to that time. However, May et al. (1986) concluded that smectites were not equilibrium phases, thereby casting doubt on the value of both calculations and experimental determinations of the effects of exchangeable cations on thermochemical properties of at least this particular type of mineral.

Application of this approach to natural weathering systems brings further difficulties. A fundamental problem arises from the recognition that geochemical processes are irreversible (Helgeson, 1968). However, the overall reaction is composed of steps involving partial equilibria. Provided there has been sufficient time for these partial equilibria to become established, stability diagrams can be used to describe or predict the stable states at each of the stages of an overall weathering reaction. Sposito (1985) has claimed that the reactions investigated by chemical thermodynamics "must either be at some stable point of development long before data are taken or be so unfavorable kinetically that the reactants can be assumed to be perfectly stable species to a high degree of approximation." According to Sposito (1985), these requirements are met reasonably well for natural soils. On the other hand, changes often occur rapidly in the composition of soil solutions. Furthermore, increasing recognition of the role of biological agents in causing the alteration of minerals (Section 1.8) means that minerals can also undergo rapid changes in composition, especially at their interfaces with soil solutions. It would appear that reactions in soils can be described as being in a stable point of development only if they are viewed macroscopically and without consideration of biological and consequent organic influences.

A further problem in the application of the approach arises from the limited number of axes allowed in the presentation of stability diagrams. Only three dimensions, and hence, three independent expressions of the variables can be represented this way. As a consequence, assumptions need to be made about the concentrations (activities) of elements which are not represented by axes. Other problems include the variations of ΔG°_{f} values with such factors as particle size and degree of crystallinity and hydration (Huang, 1989). For the Fe minerals, calculations by Langmuir and Whittemore (1971) predicted that goethite is more stable than hematite for particles with sizes > 76 nm, but hematite is more stable than goethite for smaller sized particles. Sposito (1985) suggested the possibility of a spectrum of values for kaolinites which reflects the continuum of their structural disorder. For montmorillonites, Kittrick (1971) suggested a spectrum of solubility lines, each corresponding to different ΔG°_{f} values which reflected different compositions for the mineral. Furthermore, when Al is included in stability diagrams, it is usually as Al^{3+}. However, it more commonly occurs in soil solutions as either monomeric or polymeric hydroxy ions and also in organic complexes. Consideration of the speciation of Al can lead to quite different conclusions regarding the stable minerals present in a system from those reached when only simple Al ions or hydroxy-Al ions are assumed to be present (Calvo and Alvarez, 1992). In addition, stability diagrams are constructed to represent only those selected phases which are assumed to be the stable phases controlling solution compositions. Thus, for instance, a stability diagram presented by Keller (1970) plots the compositions of solutions in contact with a number of illites within the kaolinite area in a diagram which also contains areas for K-feldspar, K-mica and gibbsite. However, Keller (1970) contends that the K-mica field needs to be extended into the kaolinite area to include illites. Similarly, pore waters in Great Lakes sediments plotted nominally across the kaolinite-montmorillonite boundary line, but some of these could instead be placed within the stability area for an amorphous ferroaluminum silicate, when this phase was considered as a possible component (Nriagu, 1978). These considerations aside, stability diagrams are also generally limited insofar as they represent only inorganic species and transformations. The effects of uncertainties in ΔG^o_f values and in mineral composition upon stability diagrams can now be estimated with a computer through the use of an interactive spreadsheet (Biddle et al., 1995).

Natural waters that have been in contact with rocks can be sampled simply using springs (Feth et al., 1964) and stream waters (Bricker et al., 1968). Sampling soil solutions is more problematic. The soil solution cannot be analyzed *in situ* at present. Hence, it must be extracted from soils for analysis with minimal alteration to its composition. A number of methods have been devised for this purpose, while an alternative approach involves the use of solution approximations, for which water is added to soil to provide a solution for analysis, albeit a poor representation of the soil solution, thanks to ion exchange, precipitation and dilution reactions induced by dilution (Percival, 1985). Table 1.5 from Percival (1985) gives some of the common methods for obtaining soil solutions. In addition to the difficulties in obtaining a true representation of the solution in contact with soil minerals from solutions extracted from soils by methods including those shown in Table 1.5, changes in both the composition and temperature of soil solutions can occur hourly, daily and also seasonally. There were seasonal changes in the placement of soil solution compositions within different stability fields according to Henderson et al. (1976), Zabowski and Ugolini (1992), Takahashi et al. (1993) (Section 1.4.5), and Ugolini (1995). The latter author also pointed out the significant shifts in stability lines for minerals which can occur with seasons where large temperature changes occur, as in alpine environments.

In general, thermodynamics indicates whether reactions are possible. However, it does not indicate whether these reactions will occur. As pointed out by Rai and Kittrick (1989), diaspore plots below the line for gibbsite in a stability diagram. Hence, it appears to be more stable than gibbsite. However, gibbsite is common in soils while diaspore is rare. Either the thermodynamic data are unreliable or diaspore forms much more slowly than gibbsite (Rai and Kittrick, 1989). Thermodynamics also indicates those minerals which can coexist in equilibrium, e.g., gibbsite and kaolinite and also kaolinite and montmorillonite, and those minerals which cannot, e.g., gibbsite and K-feldspar and also gibbsite and montmorillonite (Birkeland, 1974; Weaver, 1989). It can be used to help indicate the origin of soil minerals (Weaver, 1989). Provided the assumptions involved in constructing the diagrams are reasonable and providing solutions sampled are good representations of those in steady-state contact with soil minerals, the placement of solution compositions within stability areas for

Table 1.5 Methods for obtaining soil solutions [From Percival, 1985 with permission from Landcare Research NZ Ltd]

| Soil Sample in Laboratory | | Soil Sample in Field |
Actual Solution	Solution Approximation	
(1) Suction	(1) Saturation paste	(1) Lysimeters
(2) Displacement by liquid	(2) 1:1 soil:water extract	(2) Suction samplers
(3) Centrifugation	(3) 1:5 (and other ratios)soil:water extract	
(4) Compaction		
(5) Equilibrium extract (Adsorbent containing soil)		

particular minerals indicates the minerals which will form from those solutions at equilibrium. If the soil is dominated by minerals of this latter type, it is probable that the mineral was neoformed from that particular solution. On the other hand, if the dominant minerals are of a different type, there is a strong possibility that they were inherited. The future course of change may see the solutions alter to come to equilibrium with the minerals, or, alternatively, if the composition of the solution is controlled by other inputs, the minerals may alter toward type(s) in equilibrium with the solution (Weaver, 1989).

It has been found that many natural waters had compositions which fell within the stability field for kaolinite (Feth et al., 1964; Bricker et al., 1968; Williams et al., 1986). This suggests that kaolinite is the stable end product of weathering of the minerals involved that encompassed a wide range. Kaolinite was not always present in the weathered materials in contact with these waters. However, some of the results could be in doubt if thermodynamic data led to inaccurate representation of stability fields. For soils in a catena, solute activities fell into areas on stability diagrams for either kaolinite or montmorillonite when both of these minerals were present suggesting a tendency for the dissolution of montmorillonite and the formation of kaolinite in some cases (moderately well to well-drained, acid horizons) and the reverse process, with montmorillonite forming at the expense of kaolinite in other cases (poorly drained or calcareous horizons) (Weaver et al., 1971). These authors found that the $SiO_2:Al_2O_3$ molar ratios of the so-called reactive fractions of the fine clay ($< 0.2 \mu m$) fractions of these soils correlated well with the mineral stability areas into which their matrix solutions fell. The soils with matrix solutions plotting in the montmorillonite stability area had $SiO_2:Al_2O_3$ molar ratios of 3.1-3.9, while those plotting in the kaolinite area ranged from 2.1-2.9. Veen (1973) used stability diagrams to deduce that montmorillonite was forming at the expense of kaolinite in some Australian soils containing both minerals. The volume increase consequent upon the change from a kaolinitic to a montmorillonitic mineralogy explained the occurrence of gilgai features in the soils.

The thermodynamic approach can be used to assist in understanding mineral alteration and formation. In a 3-dimensional stability diagram based on functions of Al, Mg and H_4SiO_4 activities together with pH, Kittrick (1973) was unable to locate a triple point where montmorillonite, vermiculite and amorphous silica would coexist. Instead, solutions went to the triple point for montmorillonite, magnesite and amorphous silica. Therefore, Kittrick (1973) hypothesized that mica derived vermiculites are fast forming unstable intermediates. Furthermore, in Section 1.4.2, location of the composition of groundwaters within mineral stability fields helped Anand et al. (1985) to explain the particular secondary products formed from feldspars in different micro-environments. Estoule-Choux et al. (1995) also made use of mineral stability fields to explain how congruent dissolution of microcline could lead to quartz alone over a wide range of pH, and hence, to explain the observation of apparently newly formed quartz crystals in association with the altered K-feldspar (Section 1.4.2).

In the case of the possible genetic relationship between halloysite and kaolinite, thermodynamic considerations appear, at first sight, to indicate that since ΔG^o_f for halloysites are usually less negative than those for kaolinites (Lindsay, 1979), halloysites are unstable relative to kaolinites, as shown in Fig. 1.11. Many weathering sequences, particularly those within profiles (Eswaran and Wong, 1978; Calvert et al., 1980; Churchman and Gilkes, 1989) are consistent with this indication from thermodynamics. Nevertheless, the visual evidence (Robertson and Eggleton, 1991; Singh and Gilkes, 1992) suggests that halloysite sometimes forms at the expense of kaolinite denoting a contradiction of the thermodynamics. However, if the apparent transformation instead involves dissolution and then precipitation of neoformed halloysite on the surfaces of the remaining kaolinite, its occurrence may represent a change in solution conditions away from those favoring kaolinite toward those favoring halloysite. On the other hand, the ΔG^o_f values for kaolinites and halloysites may not be very different from each other, allowing either to form under very similar conditions. Karathanasis and Hajek (1983) suggested that kaolinites which precipitated from supersaturated solutions appeared to have ΔG^{o}_{f} values approaching those of halloysite.

Thermodynamic considerations of soil solution ion activities enabled Dahlgren and Ugolini (1989) to argue that imogolite could not have formed in upper B horizons of Spodosols in volcanic ash as solutions there were undersaturated in imogolite. However, soil solutions in the B_s and C horizons appeared to be in equilibrium with imogolite, and also gibbsite. As a result, Dahlgren and Ugolini (1989) contended that podzolization in these soils does not occur as a result of transport of Al by imogolite, as advocated by Farmer et al. (1980) and other workers (Section 1.4.6).

There has long been debate about the thermodynamic status of clay minerals (Zen, 1962; Lippmann, 1982; Bohn and Kittrick, 1984; Fritz, 1985; Aja and Rosenberg, 1992, 1996; Essene and Peacor, 1995, 1997). Aja and Rosenberg (1992) summarized the main points for discussion as to whether or not: (1) complex aluminosilicates of variable composition are true thermodynamic phases; (2) illites and smectites are stable or metastable phases; and (3) equilibria among the minerals can be determined by the solubility method. Leaving aside the latter point, which has already been discussed, there is contention over the extent to which both non-stoichiometric minerals such as illites, smectites and also chlorites, which commonly exhibit heterogeneity both within and between crystals (Essene and Peacor, 1995), as well as those minerals which are identified as mixed layer types, can be regarded as single phases. Lippmann (1982) believes that they cannot be so regarded and contends that they are metastable. They are less stable than mechanical mixtures of minerals like pyrophyllite, muscovite, quartz and biotite, according to Lippmann (1982). Lippmann (1982) also contends that other clay minerals such as kaolinite and gibbsite are metastable, so that their so-called stability fields are rather fields of metastable existence. Fritz (1985) views smectites, illites and chlorites as solid solution phases. Bohn and Kittrick (1984) point out that, as components of open systems, "the dissolution of solid mixtures such as soils, rocks and sediments probably cannot attain complete equilibrium." Essene and Peacor (1995, 1997) maintain that such heterogeneous systems as illites and smectites cannot be regarded as thermodynamically stable phases. Nonetheless, Aja and Rosenberg (1992, 1996) consider that they may be treated experimentally as if each of them were distinct phases with unique stability fields.

It may be concluded generally from the examples given in this section that the treatment of each of the clay minerals as stable phases with unique stability fields has led to some successes in understanding processes of mineral alteration and formation. However, it must also be remembered at the same time that there are very many assumptions inherent in this approach. Any predictions of the future course of mineral changes on weathering have to be qualified by the uncertainties surrounding the validity of these assumptions. However, a recognition of the overall irreversibility of weathering reactions has also led to a reaction path approach, whereby the products of successive partial equilibria are identified (Helgeson, 1969; Fritz, 1985; Suarez and Goldberg, 1994; Steinmann et al., 1994). Computers have been used for the calculations involved.

In general, kinetics must be considered alongside thermodynamics. If any system is left for long enough and if drainage is not seriously impeded, virtually all parent materials, including quartz, can be dissolved (Loughnan, 1969), and it is likely that the kaolinite-gibbsite-iron oxide suite of minerals will dominate the products (Weaver, 1989).

1.8 Processes of Mineral Alteration and Formation by Weathering: Synthesis and New Insights

The above discussion reveals that consensus has largely been reached on a number of aspects of the alteration and formation of minerals by weathering which are summarized below.

Transformations within the solid phase provide the mechanism for many common alterations of 2:1 phyllosilicates like those of micas to vermiculite or to interstratifications between these two mineral types. Loss of K from the interlayer regions typifies these transformations. However, they also involve the oxidation of structural Fe as well as a loss of at least part of the Fe. When smectites are formed by acid leaching of vermiculite or the original micas, this also occurs by solid-state transformation. There is an analogous transformation of chlorites to their expanded products, which may be vermiculites, smectites or swelling chlorites, or interstratifications of these types of layers with the chlorites.

Pseudomorphs of a mineral after its antecedent mineral often occur. It is likely that this takes place by the crystallization of the neoformed secondary phase out of a solution formed by the hydrolysis of the original mineral onto the surfaces of parent primary or precursor secondary minerals, which act as templates for nucleation. There may be a topotactic or epitaxial association of the product and parent phases. When the two phases are of different silicate structural types, it is unlikely that the products are formed by solid-state transformation of the parent minerals.

When 2:1 aluminosilicates alter to 1:1 aluminosilicates, this may occur through an interstratification of the two end-member types. The mechanism for the alteration remains unknown but it appears possible that it occurs via a neoformed phase which develops from a compound that is adsorbed from solution within the interlayers of the 2:1 mineral. Hydroxy-metal cations, particularly of Al, are strongly attracted to the interlayers of expanded 2:1 minerals. The precursor adsorbed aluminosilicate species in this case may be similar to the soluble aluminosilicate species which can participate as one of the possible mechanisms in the transport of Al from upper to lower horizons of soils in podzolization.

In soils from volcanic parent materials, Al-organic complexes and allophane may constitute long-term metastable phases as well as other mineral phases such as halloysite and smectites. Alteration of allophane to the more Si-rich halloysite, if it occurs, is likely to take place by dissolution and reprecipitation. It is unlikely that allophane would transform to halloysite in the solid phase. It is also likely that changes between halloysite and kaolinite or *vice versa* occur via dissolution and reprecipitation, even though these two minerals have essentially identical aluminosilicate layer compositions and structures.

The flow of water through mineral assemblages has a very strong influence on the nature of the products which result from their hydrolysis. Compared with Al, Si is usually the mobile element in mineral alteration. Under strong leaching, usually involving high rainfall but also good drainage, all Si is removed and gibbsite forms in the residue. Al-rich allophane and/or imogolite may also form, particularly from volcanic glass, as a result of strong leaching.

When leaching is less efficient as occurs with lower rainfall and/or poorer drainage, halloysite may result as the fast forming phase. Kaolinite also forms but apparently at a slower rate. It can form alongside halloysite or in its place. When even less leaching occurs, especially where drainage is impeded, smectites form. Under these conditions, pH usually rises. The formation of smectite is favored particularly where ferromagnesian minerals are altered and/or calcareous minerals occur. The formation of palygorskite is favored by high pH and high Mg concentrations. Secondary silica can precipitate from solution, often on to the surfaces of other minerals, particularly in acidic environments and where the formation of other minerals has left an excess of Si.

During the podzolization process, when Al becomes the mobile phase relative to Si due to a low pH and/or strong chelation by organic compounds, Al-rich allophane and/or imogolite also form. They occur either in soluble, mobile form (proto-imogolite) or by precipitation from the dissolved Al-rich components, but lower in the profile from the location of the altering minerals.

Iron and Mn oxides and oxyhydroxides result from the oxidation of the reduced form of these elements in primary minerals. Oxidation either occurs within the primary minerals, in which case the resultant oxidized products help cause the breakdown of the mineral structure, or it takes place upon the release of the reduced elements from the primary minerals.

While precipitation as the most important aspect of climate and drainage as the most important aspect of topography are usually the main agents controlling mineral alteration and the nature of the products formed, vegetation as one aspect of the biotic factor is most important in dictating the pH and providing complexing organic agents. However, the nature of the parent material is ultimately vital in governing the nature of the elemental components available for the formation of new minerals.

The minerals formed by weathering are usually in only metastable equilibrium with their environments, however, and a long enough time can apparently lead to similar end products, comprising oxides, oxyhydroxides and hydroxides of Fe and Al from all types of parent materials. Higher temperatures can enhance this process of desilication.

Broadly speaking, studies of the alteration of primary minerals and the formation of secondary minerals have mainly focused on characteristics of bulk solids. Included among more recent insights is the recognition that the processes of dissolution, adsorption and precipitation which are important in the alteration and formation of minerals take place at surfaces (Hochella and Banfield, 1995). Their understanding requires a study of "chemistry in two dimensions" (Hochella and Banfield, 1995) rather than of the bulk chemistry of the solids. To further quote Hochella and Banfield (1995), it is "chemistry in a massive defect array." Since surface defects have been found to be so important as centers of both dissolution and nucleation, surface analysis techniques such as atomic force microscopy (AFM), X-ray photoelectron spectroscopy (XPS) and low energy electron diffraction (LEED) are likely to play an important role in research into these processes (Nagy, 1995). Generally, there is a role in these studies for the high energy X-ray techniques, including X-ray absorption spectroscopy (XAS), Auger and extended X-ray absorption fine structure (EXAFS), as well as XPS (Nagy, 1995). For Hochella and Banfield (1995), however, the key technique is HRTEM.

One of the general results from such studies is that mineral surfaces are heterogeneous at the microscale (Banfield and Hamers, 1997). They have microtopography, which is likely to be similarly important for governing their reactivity as atomic structure and chemical composition (Hochella and Banfield, 1995; Banfield and Hamers, 1997). The surface heterogeneity which has become evident at the micro-level may help to explain observations that different secondary phases can form alongside one another in some mineral assemblages, as has been noted earlier.

Surface area is one of the important measures with a bearing on the rates, particularly, of mineral dissolution. White et al. (1996) showed that BET surface areas of bulk soils and also their aluminosilicate mineral fractions increased with the age of the soils over a series of soils formed on an age succession of terraces. The changes in surface areas indicated increasing surface roughness. When rates of mineral weathering were estimated taking the increases in surface roughness into account, they were between 10^3-10^4 times slower than those estimated assuming low surface roughness. In addition, Hochella and Banfield (1995) note that traditional measures of surface area, such as by adsorption of N_2 by the BET method, are not necessarily relevant for describing the surface areas of weathering minerals. Internal surface areas may be overlooked, yet important changes can occur inside quite small pores in altering minerals. Furthermore, since not all parts of the surface undergo changes on weathering, reactive surface areas may have more relevance than total surface areas.

While emphasizing the vital role of mineral surfaces in comparison with the bulk minerals, Hochella and Banfield (1995) also point out that reactions occur at interfaces and that the properties of the solution side of the interface also differ from those of the bulk solutions. Water has very different physical properties at its interface with solids than in its bulk (Hochella and Banfield, 1995).

Nevertheless, the improved understanding of the nature of surfaces and of their interfaces with solutions has largely been applied to the process of mineral dissolution, rather than those of adsorption, precipitation, and the growth of new phases (Nagy, 1995). However, the two aspects of mineral weathering sequences, i.e., dissolution of parent minerals, on the one hand, and adsorption, precipitation, and growth of new phases, on the other, may not be independent of each other. Products of weathering can form in close association with weathering surfaces (Eggleton, 1975; Banfield et al., 1991) so that they affect the rate and course of further weathering (Eggleton, 1975; Casey, 1995). Nevertheless, much remains to be known about neoformation processes (Nagy, 1995). A few studies (e.g., Tazaki and Fyfe, 1987; Romero et al., 1992) have identified clay precursors, often as gels, but these require more characterization. Mechanisms of nucleation, including the roles of topotactic and epitaxial associations of secondary and primary mineral surfaces (Hochella and Banfield, 1995; Nagy, 1995) also need further study. Another intriguing possibility that has been raised by Ma (1996) and Ma and Eggleton (1996) is that interstratified layers may be very common, if only in the form of very few foreign layers at the ends of stacks of layers of an otherwise pure mineral. This requires wider studies with HRTEM.

Studies of the alteration and formation of soil minerals have also largely concentrated on processes involving inorganic agents. However, new insights have arisen from a recent upsurge in studies of the role of biological and biochemical agents in the alteration and formation of minerals by weathering. There has long been recognition that biota and their metabolic products participate in weathering reactions. Biota are one of the factors of soil formation in Equation [1.3]. Organic compounds from the activities and decomposition of biota were traditionally thought responsible for the transport processes in podzolization and have long been known to complex metals to form Al-humus complexes, for example. In addition, the role of microbes in reducing Fe and Mn in soils is well accepted. A series of papers in a publication edited by Huang and Schnitzer (1986) emphasized both the complexity and the weathering abilities of biological and biochemical agents, and particularly organic acids. Robert and Chenu (1992) reviewed the literature from the next few years on this topic. Many recent studies have attributed biological agents and their products with a more ubiquitous role in weathering than was previously ascribed to them. Recent microbiological literature summarized by Casey (1995) points to the involvement of bacteria in weathering, either directly or indirectly, particularly by the production of reactive ligands. According to Barker et al. (1997), microorganisms (1) disaggregate rocks physically; (2) produce acids and chelating ligands, accelerating weathering rates; (3) stabilize soils, increasing the exposure of minerals to the corrosive actions of acids and metabolites; (4) produce extracellular polymers which moderate water potential, maintain diffusion channels, act as ligands or chelators and serve as nucleation sites for the neoformation of minerals; and (5) absorb nutrients, hence they lower solution saturation and thereby enhance weathering.

In the course of developing a wide ranging model of pedogenesis whereby proton donors in soil solutions are seen as having a central role in determining the products of weathering, Ugolini and Sletten (1991) considered organic acids to be among the most important of these proton donors. The organic acids originate from metabolic products of plants and microbes and as residues of their decomposition. Some of the mineral acids in soils also have biological origins. Nitric acid can originate from nitrification reactions by microbes while H_2CO_3 is formed through the dissolution of CO_2 produced by the respiration of plants and animals and H_2SO_4 can also result from the microbial oxidation of various forms of S. For these reasons, Ugolini and Sletten (1991) see a central role for the biota in weathering.

The evident importance of organic agents including acids and chelating ligands in weathering apparently imposes a further limitation on the relevance of mineral solubility studies such as those discussed in Section 1.7. These studies and the predictions and explanations arising from them are based on inorganic components alone. On the other hand, Nesbitt (1997) argues that the major effect of microbes is to enhance weathering through the supply of acids and strong complexing ligands which nonetheless act through inorganic pathways. Even so, the biota undoubtedly influence the kinetics of mineral alteration and formation reactions very strongly. The kinetics, i.e., the rates of the processes, govern the status of mineral phases whether they are metastable or unstable, and hence the biota have a profound effect on weathering.

Nevertheless, the general implications for this particular insight of Ugolini and Sletten (1991) extend beyond that of the relative importance of biota as one of the soil-forming factors. Ugolini and Sletten (1991) point out that placing emphasis on proton donors in soil solutions constitutes a process-oriented approach to weathering rather than the deterministic approach which is based on soil-forming factors. The latter approach has the fundamental flaw of a lack of independence of the various factors from one another. In comparing Ugolini and Sletten's (1991) approach with that promulgated by Jenny (1941), one may leave aside for a moment the issue of the relative importance of studies of parent and product solid phases, on the one hand, and those of the solutions which drive reactions and within which they largely take place, on the other. If one does, a question of the relevance of comparisons of individual pedons of soils still remains. Recent studies in different parts of the world (Fritsch et al., 1992; Fritsch and Fitzpatrick, 1994) have instead used catenas as the basis for developing soil-water-landscape models. Very recently, Sommer and Schlichting (1997) have synthesized these and other catena-based approaches to hypothesize that there are three archetypes of catenas based on their mobilization and hydrological regimes. These are (1) transformation catenas, in which there are no gains or losses of the soil components under study; (2) leaching catenas, in which there are losses from at least part of the catena but these are not matched by gains in other parts; and (3) accumulation catenas, with gains in at least part of the catena but no compensating losses elsewhere in the catena. Without entering a debate about the validity of these classifications, it can be seen that studies of catenas may enable a process-oriented approach to be taken to mineral alteration and formation. Classifications by catena type could provide associations of mineralogical compositions of soils within the different parts of each catena, which would be related to one another by the processes occurring within the catena. These relationships are likely to be more logical than any drawn between the mineralogical compositions of individual soil pedons and their taxonomic classifications.

Returning to the particular influence of biota at the smaller scale of changes which are detected with modern surface analytical instruments at the surfaces of minerals, it can be seen that microorganisms both initiate and enhance changes to minerals on weathering. This is not surprising, given that microorganisms interact strongly with mineral surfaces and that surfaces and their interactions with solutions play a key role in changes to minerals on weathering. The importance of microorganisms in this regard derives partly from the fact that bacteria are ubiquitous on the Earth's surface (Barker et al., 1997). Furthermore, they and products of both microbial and plant activity and decomposition such as polysaccharides are often associated closely with minerals in soils, as seen in Fig. 1.13. In addition, fungi commonly coexist with bacteria and algae often occur on rock surfaces (Barker et al., 1997).

Dicarboxylic acids that have been characterized in soils (Hue et al., 1986) bring about strong enhancement of rates of dissolution (Amrhein and Suarez, 1988) by forming ring complexes with metals on mineral surfaces (Casey, 1995). In addition, Urrutia and Beveridge (1994) showed that aluminosilicate minerals can be formed by nucleation on bacteria. This occurs even in the presence of organic acids and metals (Urrutia and Beveridge, 1995). Tazaki (1997), using electron microscopy and XRD, showed that bacteria in freshwater could bring about the crystallization of layer silicates giving rise to peaks for 14, 10 and 7 Å spacings in XRD, typical of aluminosilicates, and also to oxides of Fe and Mn. In both laboratory experiments and natural environments, a wide variety of minerals

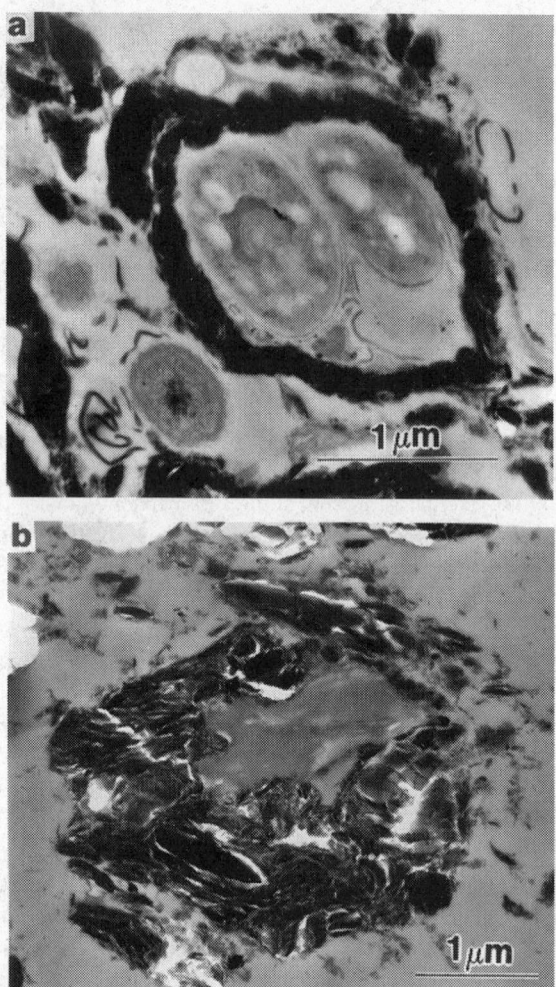

Fig. 1.13 Transmission electron micrographs of ultrathin sections from an Alfisol from South Australia showing darker colored clay particles (a) tangentially surrounding bacteria, and (b) within a microaggregate enclosing extracellular polysaccharide material, which has been highlighted by staining. [From Churchman and Foster, 1994. Trans. 15th World Congr. Soil Sci. 8a:17-34 with permission of the International Soil Science Society]

have been found to form by precipitation on bacterial cells, including some which have since been preserved in fossils (Ferris, 1997). It can be seen that some of these new insights are the outcome of the application of new and more powerful instrumental techniques to weathering systems. The development of unforeseen analytical and investigative instruments is inevitable. Hence one can expect new but unpredictable insights and understanding concerning the alteration and formation of minerals by weathering to continue to arise in the future.

Acknowledgments
I wish to thank Dr. Will Gates of CSIRO Land and Water and Mr John Keeling of Primary Industries and Resources, South Australia, for their comments on the manuscript. I am grateful to Dr Harry Percival of Landcare New Zealand Ltd and Dr. David Lowe of the University of Waikato for

permission to reuse published figures. I also thank Dr. Keith Norrish of CSIRO Land and Water, Dr. Ralph Foster formerly of CSIRO Division of Soils, and Mr. John Keeling for providing unpublished electron micrographs. I am grateful to Mr. Bob Schuster of CSIRO Land and Water for drawing a number of the figures. I am especially grateful to my family for their forbearance while I devoted much of my spare time over a necessarily short time period to the preparation of this chapter.

1.9 References

Adams, J.A., D.T. Howarth, and A.S. Campbell. 1973. Plumbogummite minerals in a strongly weathered New Zealand soil. J. Soil Sci. 24:224–232.

Adomo, P., C. Colombo, and P. Violante. 1997. Iron oxides and hydroxides in the weathering interface between *Stereocaulon vesuvianum* and volcanic rock. Clay Min. 32:453–461.

Aja, S.U., and P.E. Rosenberg. 1992. The thermodynamic status of compositionally-variable clay minerals: A discussion. Clays Clay Min. 40:292–299.

Aja, S.U., and P.E. Rosenberg. 1996. The thermodynamic status of compositionally-complex clay minerals: Discussion of clay mineral thermometry - a critical perspective. Clays Clay Min. 44:560–568.

Aldridge, L.P., and G.J. Churchman. 1991. The role of iron in the weathering of a climosequence of soils derived from schist. Aust. J. Soil Res. 29:387–398.

Alexander, L.T., G.T. Faust, S.B. Hendricks, H. Insley, and H.F. McMurdie. 1943. Relationship of the clay minerals halloysite and endellite. Amer. Miner. 28:1–18.

Alexandre, A., J.-D. Meunier, F. Colin, and J.-M. Koud. 1997. Plant impact on the biogeochemical cycle of silicon and related weathering processes. Geochim. Cosmochim. Acta 61:677–682.

Allen, B.L., and B.F. Hajek. 1989. Mineral occurrence in soil environments. p. 199–278. *In* J.B. Dixon and S.B. Weed (ed.). Minerals in soil environments. 2nd Ed. Soil Science Society of America, Madison, WI.

Altschuler, Z.S., E.J. Dwornik, and H. Kramer. 1963. Transformation of montmorillonite to kaolinite during weathering. Science 141:148–152.

Amrhein, C., and D.L. Suarez. 1988. The use of a surface complexation model to describe the kinetics of ligand-promotion dissolution of anorthite. Geochim. Cosmochim. Acta 52:2785–2793.

Anand, R.R., and R.J. Gilkes. 1984. Weathering of ilmenite in a lateritic pallid zone. Clays Clay Miner. 32:363–374.

Anand, R.R., and R.J. Gilkes. 1987a. Iron oxides in lateritic soils from western Australia. J. Soil Sci. 38:607–622.

Anand, R.R., and R.J. Gilkes. 1987b. The association of maghemite and corundum in Darling Range laterites, Western Australia. Aust. J. Soil Res. 25:303–311.

Anand, R.R., R.J. Gilkes, T.M. Armitage, and J.W. Hillyer. 1985. Feldspar weathering in lateritic saprolite. Clays Clay Miner. 33:31–43.

Anderson, H.A., M.L. Berrow, V.C. Farmer, A. Hepburn, J.D. Russell, and A.D. Walker. 1982. A re-assessment of podzol formation processes. J. Soil Sci. 33:125–136.

Aomine, S., and C. Mizota. 1973. Distribution and genesis of imogolite in volcanic ash soils of northern Kanto, Japan. p. 207–213. *In* J.M. Serratosa (ed.). Proc. Int. Clay Conf. Madrid, Spain.

Aomine, S., and K. Wada. 1962. Differential weathering of volcanic ash and pumice resulting in formation of hydrated halloysite. Amer. Miner. 47:1024–1048.

Aoudjit, H., M. Robert, and F. Elsass. 1995. Genesis, organisation and properties of clays formed in saprolites and soils on granites. p. 367–372. *In* G.J. Churchman, R.W Fitzpatrick, and R.A. Eggleton (ed.) Clays controlling the environment. Proc. 10th Int. Clay Conf. CSIRO Publishing, Melbourne, Australia.

Aoudjit, H., F. Elsass, D. Righi, and M. Robert. 1996. Mica weathering in acidic soils by analytical electron microscopy. Clay Miner. 31:319–332.

Aragoneses, F.J., and M.T. García-González. 1991. High-charge smectite in Spanish "Raña" soils. Clays Clay Miner. 39:211–218.

Arocena, J.M., S. Pawluk, and M.J. Dudas. 1993. Sericites in feldspars as source of 2:1 phyllosilicates in selected sandy soils. Soil Sci. Soc. Am. J. 57:1634–1640.

Arocena, J.M., S. Pawluk, and M.J. Dudas. 1994. Reply to "Comments on Sericites in feldspars as source of 2:1 phyllosilicates in selected sandy soils." Soil Sci. Soc. Am. J. 58:1846–1847.

Askenasy, P.E., J.B. Dixon, and T.R. McKee. 1973. Spheroidal halloysite in a Guatemalan soil. Soil Sci. Soc. Am. Proc. 37:799–803.

Babu, D.S.S., K.A. Thomas, P.N. Mohan Das, and A.D. Damodaran. 1994. Alteration of ilmenite in the Manavalakurichi deposit, India. Clays Clay Miner. 42:567–571.

Bailey, S.W. 1980. Summary of recommendations of AIPEA nomenclature committee. Clays Clay Miner. 28:73–78.

Bailey, S.W. 1990. Halloysite-a critical assessment. Sci. Géolog. 86:89–98.

Bain, D.C. 1977. The weathering of ferruginous chlorite in a podzol from Argyllshire, Scotland. Geoderma 17:193–208.

Bain, D.C., and A.R. Fraser. 1994. An unusually interlayered clay mineral from the eluvial horizon of a humus-iron podzol. Clay Miner. 29:69–76.

Bain, D.C., and J.D. Russell. 1981. Swelling minerals in a basalt and its weathering products from Morven, Scotland: II. Swelling chlorite. Clay Miner. 16:203–212.

Bain, D.C., A. Mellor, M.S.E. Robertson-Rintoul, and S.T. Buckland. 1993. Variations in weathering processes and rates with time in a chronosequence of soils from Glen Feshie, Scotland. Geoderma 57:275–293.

Bakker, L., D.J. Lowe, and A.G. Longmans. 1996. A micromorphological study of pedogenic processes in an evolutionary soil sequence formed on Late Quaternary rhyolitic tephra deposits, North Island, New Zealand. Quaternary Intern. 34-36:249–261.

Baldar, N.A., and L.D. Whittig. 1968. Occurrence and synthesis of soil zeolites. Soil Sci. Soc. Am. Proc. 32:235–238.

Banfield, J.F., and R.A. Eggleton. 1988. A transmission electron microscope study of biotite weathering. Clays Clay Miner. 36:46–70.

Banfield, J.F., and R.A. Eggleton. 1989. Apatite replacement and rare earth mobilization, fractionation and fixation during weathering. Clays Clay Miner. 37:113–127.

Banfield, J.F., and R.A. Eggleton. 1990. Analytical transmission electron microscope studies of plagioclase, muscovite and K-feldspar weathering. Clays Clay Miner. 38:77–89.

Banfield, J.F., and R.J. Hamers. 1997. Processes at minerals and surfaces with relevance to microorganisms and prebiotic synthesis. p. 81–122. In J.F. Banfield and K.H. Nealson (ed.). Geomicrobiology: Interactions between microbes and minerals. Reviews in Mineralogy Vol. 35. Mineralogical Society of America, Washington, DC.

Banfield, J.F., B.J. Jones, and D.R. Veblen. 1991. An AEM-TEM study of weathering and diagenesis, Albert Lake, Oregon. Parts I and II. Geochim. Cosmochim. Acta 55:2781–2810.

Barker, W.W., S.A. Welch, and J.F. Banfield. 1997. Biogeochemical weathering of silicate minerals. p. 391–428. In J.F. Banfield, and K.H. Nealson (ed.) Geomicrobiology: Interactions between microbes and minerals. Reviews in Mineralogy Vol. 35. Mineralogical Society of America, Washington, DC.

Barnhisel, R.I., and P.M. Bertsch. 1989. Chlorites and hydroxy-interlayered vermiculite and smectite. p. 729–788. In J.B. Dixon and S.B. Weed (ed.) Minerals in soil environments. 2nd Ed. Soil Science Society of America, Madison, WI.

Barshad, I. 1966. The effect of a variation in precipitation on the nature of clay mineral formation in soils from acid and basic igneous rocks. p. 167–173. In L. Heller and A. Weiss (ed.) Proc. Int. Clay Conf. Jerusalem Vol. 1. Israel Program for Scientific Translations, Jerusalem, Israel.

Barshad, I., and F.M. Kishk. 1969. Chemical composition of soil vermiculite clays as related to their genesis. Contrib. Miner. Petrol. 24:136–155.

Bassett, W.A. 1959. The origin of the vermiculite deposit at Libby, Montana. Am. Miner. 44:282–299.

Bates, T.E., F.A. Hildebrand, and A. Swineford. 1950. Morphology and structure of endellite and halloysite. Am. Miner. 6:237–248.

Bates, T. E. 1962. Halloysite and gibbsite formation in Hawaii. Clays Clay Miner. 9:315–328.

Ben-Dor, E., and A. Singer. 1987. Optical density of Vertisol clay suspensions. Clays Clay Miner. 35:311–317.

Berkgaut, V., A. Singer, and K. Stahr. 1994. Palagonite reconsidered: Paracrystalline illite-smectites from regoliths on basic pyroclastics. Clays Clay Miner. 42:582–592.

Berner, R.A., and G.R. Holdren, Jr. 1979. Mechanism of feldspar weathering. II. Observations of feldspars from soils. Geochim. Cosmochim. Acta 43:1173–1186.

Biddle, D.L., H.J. Percival, and D.J. Chittleborough. 1995. An interactive spreadsheet for graphing mineral stability diagrams. Comput. Geosci. 21:175–185.

Birkeland, P.W. 1974. Pedology, weathering, and geomorphological research. Oxford University Press, New York, NY.

Bish, D.L. 1994. Quantitative X-ray diffraction analysis of soils. p. 267–295. In J.E. Amonette and L.W. Zelazny (ed.). Quantitative methods in soil mineralogy. Soil Science Society of America, Madison, WI.

Bleeker, P., and R.L. Parfitt. 1974. Volcanic ash and its clay mineralogy at Cape Hoskins, New Britain, Papua New Guinea. Geoderma 11:123–135.

Blum, A.E., and L.L. Stillings. 1995. Feldspar dissolution kinetics. p. 291–351. *In* A.F. White and S.L. Brantley (ed.). Chemical weathering rates of silicate minerals. Reviews in Mineralogy Vol. 31. Mineralogical Society of America, Washington, DC.

Bockheim, J.G., J.G. Marshall, and H.M. Kelsey. 1996. Soil-forming processes and rates on uplifted marine terraces in southwestern Oregon, USA. Geoderma 73:39–62.

Boero, V., A. Promoli, P. Melis, E. Barberis, and E. Ardiuno. 1992. Influence of climate on the iron oxide mineralogy of terra rossa. Clays Clay Miner. 40:8–13.

Boettinger, J.L., and R.J. Southard. 1995. Phyllosilicate distribution and origin in Aridosols on a granitic pediment, Western Mojave Desert. Soil Sci. Soc. Am. J. 59:1189–1198.

Bohn, H.L., and J.A. Kittrick. 1984. Stability diagram of Ca-Na-K plagioclase with authigenic minerals. Chem. Geol. 43:181–186.

Bonifacio, E., E. Zanini, V. Boero, and M. Franchini-Angela. 1997. Pedogenesis in a soil catena on serpentinite in northwestern Italy. Geoderma 75:33–51.

Borchardt, G. 1989. Smectites. p. 675–727. *In* J.B. Dixon and S.B. Weed (ed.) Minerals in soil environments. 2nd Ed. Soil Science Society of America, Madison, WI.

Bowen, N.L. 1922. The reaction principle in petrogenesis. J. Geol. 30:177–198.

Boyer, S.J. 1975. Chemical weathering of rocks on the Lassiter Coast, Antarctic Peninsula, Antarctica. NZ J. Geol. Geophys. 18:623–628.

Bresson, L.M. 1974. A study of integrated microscopy: Rubefaction under wet temperate climate in comparison with mediterranean rubefaction: p. 526–541. *In* G.K. Rutherford (ed.) Soil microscopy. Limestone Press, Kingston, Canada.

Bricker, O.P., A.E. Godfrey, and E.T. Cleaves. 1968. Mineral-weathering interaction during the chemical weathering of silicates. p. 128–142. *In* Trace inorganics in water. Advances in Chemistry Series 73. American Chemical Society, Washington, DC.

Brindley, G.W., 1977. Aspects of order-disorder in clay minerals: A review. Clay Sci. 5:103–112.

Brinkman, R. 1970. Ferrolysis, a hydromorphic soil forming process. Geoderma 3:199–206.

Bühmann, C., and P.L.C. Grubb. 1991. A kaolin-smectite interstratification sequence from a red and black complex. Clay Min. 26:343–358.

Burt, R., and E.B. Alexander. 1996. Soil development on moraines of Mendenhall Glacier, southeast Alaska. 2. Chemical transformations and soil micromorphology. Geoderma 72:19–36.

Buurman, P., and L.P. Van Reeuwijk. 1984. Proto-imogolite: the process of podzol formation. J. Soil Sci. 35:447–452.

Buurman, P., L. van der Plas, and S. Slager. 1976. A toposequence of alpine soils on calcareous mica schists, northern Adula region, Switzerland. J. Soil Sci. 27:395–410.

Cady, J.G., and K.W. Flach. 1997. History of soil mineralogy in the United States Department of Agriculture. Adv. GeoEcol. 29:211–240.

Calvert, C.S., S.W. Buol, and S.B. Weed. 1980. Mineralogical transformations of a vertical rock-saprolite-soil sequence in the North Carolina Piedmont. Soil Sci. Soc. Am. J. 44:1096–1112.

Calvo, R., and E. Alvarez. 1992. Aluminium activity in soil solution and mineral stability in soils from Galicia (NW Spain). Clay Min. 27:325–330.

Campbell, A.S. 1974. The influence of red beech (*Nothofagus Fusca*) on clay mineral genesis. Trans. 10th Int. Congr. Soil Sci. VI (I):60–67.

Campbell, A.S., and U. Schwertmann. 1984. Iron oxide mineralogy of placic horizons. J. Soil Sci. 35:569–582.

Carnicelli, S., A. Mirabella, G. Cecchini, and G. Sanesi. 1997. Weathering of chlorite to a low-charge expandable mineral in a Spodosol on the Apennine Mountains, Italy. Clays Clay Miner. 45:28–41.

Carroll, D. 1970. Rock weathering. Monographs in geoscience. Plenum Press, New York, NY.

Carson, C.D., and J.B. Dixon. 1983. Mineralogy and acidity of an inland acid sulfate soil of Texas. Soil Sci. Soc. Am. J. 47:828–833.

Carson, C.D., and G.W. Kunze. 1970. New occurrence of tabular halloysite. Soil Sci. Soc. Am. Proc. 34:538–540.

Casey, W.H. 1995. Surface chemistry during the dissolution of oxides and silicate minerals. p. 185–217. *In* D.J. Vaughan and R.A.D. Pattrick (ed.) Mineral surfaces. Chapman and Hall, London, UK.

Casey, W.H., M.F. Hochella, and H.R. Westrich. 1993. The surface chemistry of manganiferous silicate minerals as inferred from experiments on tephroite (Mn_2SiO_4). Geochim. Cosmochim. Acta 57:785–793.

Casey, W.H., H.R. Westrich, and G. W. Arnold. 1988. Surface chemisty of labradorite feldspar reacted with aqueous solutions at pH = 2, 3 and 12. Geochim. Cosmochim. Acta 52:821–832.

Casey, W.H., H.R. Westrich, G. W. Arnold, and J.F. Banfield. 1989. The surface chemistry of dissolving labradorite feldspar. Geochim. Cosmochim. Acta 53:2795–2807.

Certini, G., F.C. Ugolini, G. Corti, and A. Agnelli. 1998. Early stages of podzolization under Corsican pine (*Pinus nigra* Arn. *spp. laricio*). Geoderma 83:103–125.

Chadwick, O.A., D.M. Hendricks, and W.D. Nettleton. 1987. Silica in Duric soils: I. A depositional model. Soil Sci. Soc. Am. J. 51:975–982.

Chamley, H. 1989. Clay Sedimentology. Springer-Verlag, Berlin, Germany.

Chaterjee, R.K., and G.S. Rathore. 1976. Clay mineral composition, genesis and classification of some soils developed from basalts in Madhya Pradesh. J. Ind. Soc. Soil Sci. 24:144–157.

Childs, C.W. 1992. Ferrihydrite: A review of structure, properties and occurrence in relation to soils. Zeitschrift für Pflanzenernährung und Bodenkunde 155:441–448.

Childs, C.W., B.A. Goodman, and G.J. Churchman. 1979. Application of Mössbauer spectroscopy to the study of iron oxides in some red and yellow/brown soil samples from New Zealand. p. 555–565. *In* M.M. Mortland and V.C. Farmer (ed.) International Clay Conference 1978. Developments in Sedimentology 27, Elsevier, Amsterdam, Netherlands.

Childs, C.W., R.L. Parfitt, and R. Lee. 1983. Movement of aluminium as an inorganic complex in some podzolised soils, New Zealand. Geoderma 29:139–155.

Cho, Hi Doo, and A.R. Mermut. 1992. Evidence for halloysite formation from weathering of ferruginous chlorite. Clays Clay Miner. 40:608–619.

Chukhrov, F.V., and B.B. Zvyagin. 1966. Halloysite, a crystallochemically and mineralogically distinct species. p. 11–25. *In* L. Heller and A. Weiss (ed.) Proc. Int. Clay Conf., Jerusalem Vol. 1. Israel Program for Scientific Translations.

Churchman, G.J. 1978. Studies on a climosequence of soils in tussock grasslands. 21. Mineralogy. NZ J. Sci. 21:467–480.

Churchman, G.J. 1980. Clay minerals formed from micas and chlorites in some New Zealand soils. Clay Min. 15:59–76.

Churchman, G.J. 1990. Relevance of different intercalation tests for distinguishing halloysite from kaolinite in soils. Clays Clay Miner. 38:591–599

Churchman, G.J. 1993. Use of the soil factor equation for determining the major causes of weathering of micas. O-33. *In* Abstracts, 10th International Clay Conference, Adelaide, Australia.

Churchman, G.J., and J.G. Bruce. 1988. Relationships between loess deposition and mineral weathering in some soils in Southland, New Zealand. p. 11–31. *In* D.N. Eden and R.J. Furkert (ed.). Loess: Its distribution, geology and soils. A.A. Balkema, Rotterdam, Netherlands.

Churchman, G.J., and R.M. Carr. 1975. The definition and nomenclature of halloysites. Clays Clay Miner. 23:382–388.

Churchman, G.J., and R.C. Foster. 1994. The role of clay minerals in the maintenance of soil structure. Tran. 15th World Congr. Soil Sci. 8a:17–34.

Churchman, G.J., and R.J. Gilkes. 1989. Recognition of intermediates in the possible transformation of halloysite to kaolinite. Clay Min. 24:579–590.

Churchman, G.J., and M.L. Jackson. 1976. Reaction of montmorillonite with acid aqueous solutions: Solute activity control by a secondary phase. Geochim. Cosmochim. Acta 40:1521–1529.

Churchman, G.J., and K.R. Tate. 1987. Stability of aggregates of different size grades in allophanic soils from volcanic ash in New Zealand. J. Soil Sci. 38:19–27.

Churchman, G.J., and B.K.G. Theng. 1984. Interactions of halloysites with amides: Mineralogical factors affecting complex formation. Clay Min. 19:161–175.

Churchman, G.J., R.N. Clayton, K. Sridhar, and M.L. Jackson. 1976. Oxygen isotopic composition of aerosol-sized quartz in shales. J. Geophys. Res. 81:381–386.

Churchman, G.J., T.J. Davy, L.A.G. Aylmore, R.J. Gilkes, and P.G. Self. 1995. Characteristics of fine pores in some halloysites. Clay Min. 30:89–98.

Churchman, G.J., P.G. Slade, P.G. Self, and L.J. Janik. 1994. Nature of interstratified kaolin-smectites in some Australian soils. Aust. J. Soil Res. 32:805–822.

Churchman, G.J., J.S Whitton, G.G.C. Claridge, and B.K.G. Theng. 1984. Intercalation method using formamide for differentiating halloysite from kaolinite. Clays Clay Miner. 32:241–248.

Cornell, R.M., and U. Schwertmann. 1996. The iron oxides. VCH, Weinheim. Germany.

Clayton, R. N., M.L. Jackson, and K. Sridhar. 1978. Resistance of quartz silt to isotopic exchange under burial and intense weathering conditions. Geochim. Cosmochim. Acta 42:1517–1522.

Cradwick, P.D.G., V.C. Farmer, J.D. Russell, C.R. Masson, K. Wada, and N. Yoshinaga. 1972. Imogolite, a hydrated aluminium silicate of tubular structure. Nat. Phys. Sci. 240:187–189.

Craig, D.C., and F.C. Loughnan. 1964. Chemical and mineralogical transformations accompanying the weathering of basic volcanic rocks from New South Wales. Aust. J. Soil Res. 2:218–234.

Crocker, R.L. 1952. Soil genesis and the pedogenic factors. Quart. Revi. Biol. 27:139–168.

Cuadros, J., A. Delgado, A. Cardenete, E. Reyes, and J. Linares. 1994. Kaolinite/montmorillonite resembles beidellite. Clays Clay Miner. 42:643–651.

Dahlgren, R.A., and F.C. Ugolini. 1989. Formation and stability of imogolite in a tephritic Spodosol, Cascade Range, Washington, U.S.A. Geochim. Cosmochim. Acta 53:1897–1904.

Dahlgren, R.A., J.P. Dragoo, and F.C. Ugolini. 1997a. Weathering of Mt. St. Helens tephra under a cryic-udic climatic regime. Soil Sci. Soc. Am. J. 61:1519–1525.

Dahlgren, R.A., J.L. Boettinger, G.L. Huntington, and R.G. Amundson. 1997b. Soil development along an elevational transect in the western Sierra Nevada, California. Geoderma 78:207–236.

Delvaux, B., and A.J. Herbillon. 1995. Pathways of mixed-layer kaolin-smectite formation in soils. p. 457–461. *In* G.J. Churchman, R.W Fitzpatrick, and R.A. Eggleton (ed.) Clays controlling the environment. Proc. 10th Int. Clay Conf. CSIRO Publishing, Melbourne, Australia.

Delvaux, B., A.J. Herbillon, J.E. Dufey, G. Burtin, and L. Vielvoye. 1988. Adsorption sélective du potassium sur certaines halloysites (10 Å) de sols tropicaux développés sur roches volcaniques. Signification minéralogique. Compt. Rend. Acad. Sci., Paris. T. 307, série II:311–317.

Delvaux, B., A.J. Herbillon, and L. Vielvoye. 1989. Characterization of a weathering sequence of soils derived from volcanic ash in Cameroon. Taxonomic, mineralogical and agronomic implications. Geoderma 45:375–388.

Delvaux, B., A.J. Herbillon, J.E. Dufey, L.Vielvoye, and M.M. Mestagh. 1990. Surface properties and clay mineralogy of hydrated halloysitic soil clays. Parts I and II. Clay Min. 25:129–160.

Delvaux, B., D. Tessier, A.J. Herbillon, G. Burtin, A.-M. Jaunet, and L. Vielvoye. 1992. Morphology, texture, and microstructure of halloysitic soil clays as related to weathering and exchangeable cation. Clays Clay Miner. 40:446–456.

Dias, I., I. Gonzalez, S. Prates, and E. Galán. 1997. Palygorskite occurrences in the Portuguese sector of the Tagus basin: a preliminary report. Clay Min. 32:323–328.

Dixon, J.B. 1989. Kaolin and serpentine group minerals. p. 467–525. *In* J.B. Dixon and S.B. Weed (ed.). Minerals in soil environments. 2nd Ed. Soil Science Society of America, Madison, WI.

Dixon, J.B., and T.R. McKee. 1974. Internal and external morphology of tubular and spheroidal halloysite particles. Clays Clay Miner. 22:127–137.

Dixon, J.B., and S.B. Weed . 1989. Minerals in soil environments. 2nd Ed. Soil Science Society of America, Madison, WI.

Doner, H.E., and W.C. Lynn. 1989. Carbonate, halide, sulfate, and sulfide minerals. p. 279–330. *In* J.B. Dixon and S.B. Weed (ed.) Minerals in soil environments. 2nd Ed. Soil Science Society of America, Madison, WI.

Dost, H. (ed.). 1973. Acid sulphate soils. ILRI Pub. 18, Vol II. International Institute for Land Reclamation and Improvement. Wageningen, Netherlands.

Douglas, L.A. 1989. Vermiculites. p. 635–674. *In* J.B. Dixon and S.B. Weed (ed.) Minerals in soil environments. 2nd Ed. Soil Science Society of America, Madison, WI.

Drees, L.R., L.P. Wilding, N.E. Smeck, and A.L. Senkayi. 1989. Silica in soils: Quartz and disordered silica polymorphs. p. 913–974. *In* J.B. Dixon and S.B. Weed (ed.). Minerals in soil environments. 2nd Ed. Soil Science Society of America, Madison, WI.

Drever, J.I. (ed.). 1985. The chemistry of weathering. NATO ASI Series 149. D. Reidel, Dordrecht, Netherlands.

Ducloux, J., A. Meunier, and B. Velde. 1976. Smectite, chlorite and a regular interlayered chlorite-vermiculite in soils developed on a small serpentinite body, Massif Central, France. Clay Min. 11:121–135.

Dudas, M.J., and M.E. Harward. 1975. Weathering and authigenic halloysite in soil developed in Mazama ash. Soil Sci. Soc. Am. Proc. 39:561–566.

Egashira, K., and S. Tsuda, 1983. High-charge smectite found in weathered granitic rocks of Kyushu. Clay Sci. 6:67–81.

Eggleton, R.A. 1975. Nontronite topotaxial after hedenburgite. Am. Miner. 60:1063–1068.

Eggleton, R.A. 1984. Formation of iddingsite rims on olivine: A transmission electron microscopy study. Clays Clay Min. 32:1–11.

Eggleton, R.A., and J.N. Boland. 1982. Weathering of enstatite to talc through a sequence of transitional phases. Clays Clay Miner. 30:11–20.

Eggleton, R.A., and P.R. Buseck. 1980. High resolution electron microscopy of feldspar weathering. Clays Clay Miner. 28:173–178.

Essene, E.J., and D.R. Peacor. 1995. Clay mineral thermometry: A critical perspective. Clays Clay Miner. 43:540–553.

Essene, E.J., and D.R. Peacor. 1997. Illite and smectite: metastable, stable or unstable? Further discussion and a correction. Clays Clay Min. 45:116–122.

Estoule-Choux, J., J. Estoule, and D. Hallalouche. 1995. Congruent dissolution of microcline and epitaxial growth of 'skeletal' quartz during weathering of a granite from the Central Hoggar (Algeria). p. 373–377. In G.J. Churchman, R.W Fitzpatrick, and R.A. Eggleton (ed.) Clays controlling the environment. Proc. 10th Int. Clay Conf. CSIRO Publishing, Melbourne, Australia.

Eswaran, H. 1979. The alteration of plagioclases and augites under different environmental conditions. J. Soil Sci. 30:547–555.

Eswaran, H., and C.B. Wong. 1978. A study of a deep weathering profile on granite in Peninsular Malaysia. Parts I, II, and III. Soil Sci. Soc. Am. J. 42:144–158.

Eswaran, H., and Y.Y. Hong. 1976. The weathering of biotite in a profile on gneiss in Malaysia. Geoderma 16:9–20.

Fanning, D.S., V.Z. Keramidas, and M.A. El-Desoky. 1989. Micas. p. 551–634. In J.B. Dixon and S.B. Weed (ed.) Minerals in soil environments. 2nd Ed. Soil Science Society of America, Madison, WI.

Farmer, V.C. 1997. Conversion of ferruginous allophanes to ferruginous beidellites at 95°C under alkaline conditions with alternating oxidation and reduction. Clays Clay Miner. 45:591-597.

Farmer, V.C., and A.R. Fraser. 1982. Synthetic imogolite, a tubular hydroxyaluminium silicate. p. 547-553. In M.M. Mortland and V.C. Farmer (ed.). Intern. Clay Confer. 1978. Elsevier, Amsterdam, Netherlands.

Farmer, V.C., A.R. Fraser, and J.M. Tait. 1977. Synthesis of imogolite. J. Chem. Soc. Chem. Comm. pp. 462–463.

Farmer, V.C., A.R. Fraser, and J.M. Tait. 1979. Characterization of the chemical structures of natural and synthetic aluminosilicate gels and sols by infrared spectroscopy. Geochim. Cosmochim. Acta 43:1417–1420.

Farmer, V.C., J.D. Russell, and M.L. Berrow. 1980. Imogolite and proto-imogolite allophane in spodic horizons: Evidence for a mobile aluminium silicate complex in podzol formation. J. Soil Sci. 31:673–684.

Farmer, V.C., J.D. Russell, W.J. McHardy, A.C.D. Newman, J.L. Ahlrichs, and J.Y.H. Rimsaite. 1971. Evidence for loss of protons and octahedral iron from oxidised biotites and vermiculites. Miner. Mag. 38:121–137.

Ferguson, J.A. 1954. Transformations of clay minerals in black earths and red loams of basaltic origin. Aust. J. Ag. Res. 5:98–108.

Ferris, F.G. 1997. Formation of authigenic minerals by bacteria. p. 187–208. In J.M. McIntosh and L.A.Groat (ed.). Biological-mineralogical interactions. Mineralogical Association of Canada Short Course Volume 25. Mineralogical Association of Canada, Ottawa, Canada.

Feth, J.H., C.E. Roberson, and W.L. Polzer. 1964. Sources of mineral constituents in water from granitic rocks in Sierra Nevada, California and Nevada. Geological Survey Water-Supply Paper 1535-I. US Government Printing Office, Washington DC.

Fieldes, M. 1955. Clay mineralogy of New Zealand soils. Part II: Allophane and related mineral colloids. NZ J. Sci. Tech. 37:336–350.

Fieldes, M. 1968. Clay Mineralogy. p. 22–39. In New Zealand Soil Bureau 1968: Soils of New Zealand. Part 2. NZ Soil Bur. Bull. 26 (2).

Fieldes, M. and L.D. Swindale. 1954. Chemical weathering of silicates in soil formation. NZ J. Sci. Tech. B36:140–154.

Fine, P., M.J. Singer, K.L. Verosub, and J. TenPas. 1993. New evidence for the origin of ferrimagnetic minerals in loess from China. Soil Sci. Soc. Am. J. 57:1537–1542.

Fiore, S. 1993. The occurrences of smectite and illite in a pyroclastic deposit prior to weathering: implications on the genesis of 2:1 clay minerals in volcanic soils. App. Clay Sci. 8:249–259.

Fitzpatrick, R.W., and J. Le Roux. 1975. Pedogenic and soil solution studies on iron-titanium oxides. p. 585–599. In S.W. Bailey (ed.) Proc. Int. Clay Conf. Mexico City. Applied Publishing Ltd., Wilmette, IL.

Fitzpatrick, R.W., E. Fritsch, and P.G. Self. 1996. Interpretation of soil features produced by ancient and modern processes in degraded landscapes. V. Development of saline sulfidic features in non-tidal seepage areas. Geoderma 69:1–29.

Fitzpatrick, R.W., R. Naidu, and P.G. Self. 1992. Iron deposits and microorganisms in saline sulfidic soils with altered soil water regimes in South Australia. p. 263–286. In H.C.W. Skinner and R.W. Fitzpatrick (ed.). Biomineralization processes of iron and manganese. Catena Suppl. 21.

Folkoff, M.E., and V. Meentemeyer. 1985. Climate control of the assemblages of secondary clay minerals in the A-horizon of United States soils. Earth Surf. Process. Landf. 10:621–633.

Fontanaud, A., and A. Meunier. 1983. Mineralised facies of a weathered serpentinized lherzolite from the Pyrenees, France. Clay Min. 18:77-88.

Fontes, M.P.F., and S.B. Weed. 1991. Iron oxides in selected Brazilian Oxisols: I. Mineralogy. Soil Sci. Soc. Am. J. 55:1143-1149.

Frankart, R.P., and A.J. Herbillon. 1970. Présence et genèse d'analcime dans les sols sodique de la Basse Ruzizi (Burundi). Bull. Groupe Franc. Argiles 22:79-89.

Franke, W.A., and R. Teschner-Steinhardt. 1994. An experimental approach to the sequence of the stability of rock-forming minerals towards chemical weathering. Catena 21:279-290.

Freeland, J.A.. and C.V. Evans. 1993. Genesis and profile development of Success soils, northern New Hampshire. Soil Sci. Soc. Am. J. 57:183–191.

Fritsch, E., and R.W. Fitzpatrick. 1994. Interpretation of soil features produced by ancient and modern processes in degraded landscapes. I. A new method for constructing conceptual soil-water-landscape models. Aust. J. Soil Res. 32:889–907.

Fritsch, E., E. Peterschmitt, and A.J. Herbillon. 1992. A structural approach to the regolith: Identification of structures, analysis of structural relationships and interpretations. Sciences Géologiques Bulletin 45:77–97.

Fritz, B. 1985. Multicomponent solid solutions for clay minerals and computer modeling of weathering processes. p. 19–34. *In* J.I. Drever (ed.) The chemistry of weathering. Reidel, New York, NY.

Gaines, R.V., H.W. Skinner, E.F. Foord, B. Mason, and A. Rosenzweig. 1997. Dana's new mineralogy. John H. Wiley and Sons, New York, NY.

Galán, E., and A. Castillo. 1984. Sepiolite-palygorskite in Spanish Tertiary basins: Genetical patterns in continental environments. p. 87–124. *In* A. Singer and E Galán (ed.) Palygorskite-sepiolite, Occurrence, genesis and uses. Development in Sedimentology 37. Elsevier, Amsterdam, Netherlands.

Galán, E., J.M. Brell, A. La Iglesia, and H.S. Robertson. 1975. The Cáceres palygorskite deposit, Spain. p. 81–94. *In* S.W. Bailey (ed.). Proc. Int. Clay Conf., MexicoCity. Applied Publishing, Wilmette, IL.

Garcia-Talegon, J., E. Molina, and M.A. Vicente. 1994. Nature and characteristics of 1:1 phyllosilicates from weathered granite, central Spain. Clay Min. 29:727–734.

Garrels, R.M., and F.T. Mackenzie. 1971. Evolution of sedimentary rocks. W.W. Norton and Co., New York, NY.

Gibson, E.K., S.J. Wentworth, and D.S. McKay. 1983. Chemical weathering and diagenesis of a cold desert soil from Wright Valley, Antarctica: An analog of Martian weathering provinces. J. Geophys. Res. 88:A912–A928.

Gibson, G.W. 1962. Geological investigations in southern Victoria Land, Antarctica. 8. Evaporite salts in the Victoria Valley region. NZ J. Geol. Geophys. 5:361–374.

Gilkes, R.J., and A. Suddhiprakarn. 1979. Biotite alteration in deeply weathered granite. I. Morphological, mineralogical, and chemical properties. Clays Clay Miner. 27:349–360.

Gjems, O. 1970. Mineralogical composition and pedogenic weathering of the clay fraction in podzol weathering profiles in Zalesine, Yugoslavia. Soil Sci. 110:237–243.

Glassmann, J.R., and G.M. Simonson. 1985. Alteration of basalt in soils of Western Oregon. Ssoil Sci. Soc. Am. J. 49:262–273.

Goldich, S.S. 1938. A study in rock weathering. J. Geol. 46:17–58.

Goodman, B.A., J.D. Russell, B. Montez, E. Oldfield, and R.J. Kirkpatrick. 1985. Structural studies of imogolite and allophanes by aluminum-27 and silicon-29 nuclear magnetic resonance spectroscopy. Phys. Chem. Min. 12:342–346.

Graef, F., A. Singer, K. Stahr, and R. Jahn. 1997. Genesis and diagenesis of Paleosols from Pliocene volcanics on the Golan Heights. Catena 30:149–167.

Grey, I.E., and A.F. Reid. 1975. The structure of pseudorutile and its role in the natural alteration of ilmenite. Am. Mineralogist 60:898–906.

Grim, R.E. 1968. Clay mineralogy. McGraw-Hill, New York, NY.

Grim, R.E., R.H. Bray, and W.F. Bradley. 1937. The mica in argillaceous sediments. Am. Mineralogist 22:813–829.

Hall, P.L., G.J. Churchman, and B.K.G. Theng. 1985. Size distribution of allophane unit particles in aqueous suspensions. Clay Clay Min. 33:345–349.

Hansen, H.C.B., O.K. Borggaard, and J. Sørensen. 1995. Evaluation of the free energy of formation of Fe(II)-Fe(III) hydroxide-sulphate (green rust) and its reduction of nitrite. Geochim. Cosmochim. Acta 58:2599–2608.

Harder, H. 1971. Quartz and clay mineral formation at surface temperatures. Mineralog. Soc. Jap. Spec. Pap. 1:106–108.

Harder, H. 1972. The role of magnesium in the formation of smectite minerals. Chem. Geol. 10:31–39.

Harder, H. 1977. Clay mineral formation under lateritic weathering conditions. Clay Min. 12:281–288.

Harlan, P.W., D.P. Franzmeier, and C.B. Roth. 1977. Soil formation on loess in southwestern Indiana. II. Distribution of clay and free iron oxides and fragipan formation. Soil Sci. Soc. Am. J. 42:99–103.

Harrison, R. R.S. Swift, A.S. Campbell, and P.J. Tonkin. 1990. A study of two soil development sequences located in a montane area of Canterbury, New Zealand, I. Clay mineralogy and cation exchange properties. Geoderma 47:261–282.

Hay, R.L. 1960. Rate of clay formation and mineral alteration in a 4000-year-old volcanic ash soil on St. Vincent, B.W.I. Am. J. Sci. 258:354–368.

Hay, R.L. 1978. Geological occurrence of zeolites. p. 135–143. In L.B. Sand and F.A. Mumpton (ed.). Natural zeolites: Occurrence, properties, use. Pergamon Press Inc., Elmsford, NY.

Hay, R.L., and B.F. Jones. 1972. Weathering of basaltic tephra on the island of Hawaii. Geol. Soc. Am. Bull. 83:317–322.

He, H., T.L. Barr, and J. Klinowski. 1995. ESCA and solid-state NMR studies of allophane. Clay Min. 30:201–209.

Helgeson, H.C. 1968. Evaluation of irreversible reactions in geochemical processes involving minerals and aqueous solutions - I. Thermodynamic relations. Geochim. Cosmochim. Acta 32:853–877.

Helgeson, H.C. 1969. Thermodynamics of hydrothermal systems at elevated temperatures and pressures. Am. J. Sci. 267:729–804.

Hellmann, R., C.H. Egglestone, H.F. Hochelle, Jr., and D.A Crerar. 1990. The formation of leached layers on albite surfaces during dissolution under hydrothermal conditions. Geochim. Cosmochim. Acta 54:1267–1282.

Henderson, J.H., H.E. Doner, R.M. Weaver, J.K. Syers, and M.L. Jackson. 1976. Cation and silica relationships of mica weathering in calcareous Harps soil. Clays Clay Miner. 24:93–100.

Hendricks, S.B., and W.H. Fry. 1930. The results of X-ray and microscopical examinations of soil colloids. Soil Sci. 29:457–479.

Herbillon, A.J., and M.N. Makumbi. 1975. Weathering of chlorite in a soil derived from a chlorito-schist under humid tropical conditions. Geoderma 13:89–104.

Herbillon, A.J., R. Frankart, and L. Vielvoye. 1981. An occurrence of interstratified kaolinite-smectite minerals in a red-black soil toposequence. Clay Min. 16:195–201.

Hewitt, A.E., and G.J. Churchman. 1982. Formation, chemistry, and mineralogy of soils from weathered schist, Eastern Otago, New Zealand. NZ J. Sci. 25:253–269.

Hinsinger, P., B. Jaillard, and J.E Dufey. 1992. Rapid weathering of a trioctahedral mica by the roots of ryegrass. Soil Sci. Soc. Am. J. 56:977–982.

Hochella, M.L., Jr., and J.F. Banfield. 1995. Chemical weathering of silicates in nature: a microscopic perspective with theoretical considerations. p. 353–406. In A.F. White and S.L. Brantley (ed.) Chemical weathering rates of silicate minerals. Reviews in Mineralogy Vol. 31. Mineralogical Society of America, Washington, DC.

Holdren, G.R. Jr., and P.M. Speyer. 1987. Reaction rate-surface area relationships during the early stages of weathering: II. Data on eight additional feldspars. Geochim. Cosmochim. Acta 51:2311–2318.

Hsu, P.H. 1989. Aluminum hydroxides and oxyhydroxides. p. 331–378. In J.B. Dixon and S.B. Weed (ed.) Minerals in soil environments. 2nd Ed. Soil Science Society of America, Madison, WI.

Huang, P.M. 1989. Feldspars, olivines, pyroxenes, and amphiboles. p. 975–050. In J.B. Dixon and S.B. Weed (ed.) Minerals in soil environments. 2nd Ed. Soil Science Society of America, Madison, WI.

Huang, P.M., and M. Schnitzer. (ed.). 1986. Interactions of soil minerals with natural organics and microbes. Soil Sci. Soc. Am. Sp. Publ. 17, Soil Science Society of America, Madison, WI.

Hudson, B.D. 1992. The soil survey as a paradigm-based science. Soil Sci. Soc. Am. J. 56:836–841.

Hue, N.V., G.R. Craddock, and F. Adams. 1986. Effect of organic acids on aluminum toxicity in subsoils. Soil Sci. Soc. Am. J. 50:28–34.

Hughes, J.C., and G. Brown. 1979. A crystallinity index for soil kaolins and its relation to parent rock, climate and soil maturity. J. Soil Sci. 30:557–563.

Hughes, R.E., D.M. Moore, and H.D. Glass. 1984. p. 330–359. In J.E. Amonette and L.W. Zelazny (ed). Quantitative methods in soil mineralogy. Soil Science Society of America, Madison, WI.

Inskeep, W.P., J.L. Clayton, and D.W. Mogk. 1993. Naturally weathered plagioclase grains from the Idaho Batholith: Observations using scanning electron microscopy. Soil Sci. Soc. Am. J. 57:851–860.

Jackson, M.L. 1959. Frequency distribution of clay minerals in major great soil groups as related to the factors of soil formation. Clays Clay Miner. 6:133–143.

Jackson, M.L. 1963. Aluminum bonding in soils: A unifying principle in soil science. Soil Sci. Soc. Am. Proc. 27:1–10.

Jackson, M.L. 1964. Chemical composition of soils. p. 71–141. *In* F.E. Bear (ed.). Chemistry of the soil. Reinhold Publishing Corp., New York, NY.

Jackson, M.L., Y. Hseung, R.B. Corey, E.J. Evans, and R.C. Van den Heuval, 1952. Weathering of clay-size minerals in soils and sediments II. Chemical weathering of layer silicates. Soil Sci. Soc. Am. Proc. 16:3–6.

Jackson, M.L., S.A. Tyler, A.L. Willis, G.A. Bourbeau, and R.P. Pennington. 1948. Weathering sequence of clay-size minerals in soils and sediments. I: Fundamental generalizations. J. Phys. Coll. Chem. 52:1237–1260.

Janik, L.J., and J.L.Keeling. 1993. FT-IR partial least-squares analysis of tubular halloysite in kaolin samples from the Mount Hope kaolin deposit. Clay Min. 28:265–378.

Jaynes, W.F., J.M. Bigham, N.E. Smeck, and M.J. Shipitalo. 1989. Interstratified 1:1-2:1 mineral formation in a polygenetic soil from southern Ohio. Soil Sci. Soc. Am. J. 53:1888–1894.

Jenny, H. 1941. Factors of soil formation. McGraw-Hill, New York, NY.

Jenny, H. 1958. Role of the plant factor in the pedogenic functions. Ecol. 39:5–16.

Johnson, L.J. 1964. Occurrence of regularly interstratified chlorite-vermiculite as a weathering product of chlorite in a soil. Am. Miner. 49:556–572.

Johnson, S.L., S. Guggenheim, and A.F. Koster van Groos. 1990. Thermal stability of halloysite by high-pressure differential thermal analysis. Clays Clay Miner. 38:477–484.

Johnson, W.M., J.G. Cady, and M.S. James. 1962. Characteristics of some brown Grumosols of Arizona. Soil Sci. Soc. Am. Proc. 26:389–393.

Johnson-Maynard, J.L., P.A. McDaniel, D.E. Ferguson, and A.L. Falen. 1997. Chemical and mineralogical conversion of Andisols following invasion by bracken fern. Soil Sci. Soc. Am. J. 61:549–555.

Jongmans, A.G., F. van Oort, P. Buurman, A.-M. Jaunet, and J.D.J. van Doesburg. 1994. Morphology, chemistry, and mineralogy of isotropic aluminosilicate coatings in a Guadeloupe Andisol. Soil Sci. Soc. Am. J. 58:501-507.

Kämpf, N., and U. Schwertmann. 1982. Goethite and hematite in a climosequence in southern Brazil and their application in classification of kaolinitic soils. Geoderma 29:27–39.

Kantor, W., and U. Schwertmann. 1974. Mineralogy and genesis of clays in red-black soil sequences on basic igneous rocks in Kenya. J. Soil Sci. 25:67–78.

Kapoor, B.S. 1973. The formation of 2:1-2:2 intergrade clays in some Norwegian podzols. Clay Min. 10:79–86.

Karathanasis, A.D., and B.F. Hajek. 1983. Transformation of smectite to kaolinite in naturally acid soil systems: structural and thermodynamic considerations. Soil Sci. Soc. Am. J . 47:158–163.

Kawano, M., and K. Tomita. 1994. Growth of smectite from leached layer during experimental alteration of albite. Clays Clay Miner. 42:7–17.

Kawano, M., K. Tomita, and Y. Kamino. 1993. Formation of clay minerals during low temperature experimental alteration of obsidian. Clays Clay Miner. 41:431–441.

Keller, W.D. 1970. Environmental aspects of clay minerals. J. Sed. Petrol. 40:788–854.

Kirkman, J.H. 1975. Clay mineralogy of some tephra beds of Rotorua area, North Island, New Zealand. Clay Min. 10:437–449.

Kirkman, J.H. 1981. Morphology and structure of halloysite in New Zealand tephras. Clays Clay Miner. 29:1–9.

Kittrick, J.A. 1966. Free energy of formation of kaolinite from solubility measurements. Am. Miner. 51:1457–1466.

Kittrick, J.A. 1967. Gibbsite-kaolinite equilibria. Soil Sci. Soc. Am. Proc. 31:314–316.

Kittrick, J.A. 1971. Montmorillonite equilibria and the weathering environment. Soil Sci. Soc. Am. Proc. 35:815–820.

Kittrick, J.A. 1973. Mica-derived vermiculites as unstable intermediates. Clays Clay Miner. 21:479–488.

Kittrick, J.A. (ed.) 1982. Acid sulfate weathering. Special Publication 10. Soil Science Society of America, Madison, WI.

Kittrick, J.A. (ed.) 1986. Soil mineral weathering. Van Nostrand Reinhold, New York, NY.

Kodama, H., J.A. McKeague, R.J. Tremblay, J.R. Gosselin, and M.G. Townsend. 1977. Characterization of iron oxide compounds in soils by Mössbauer and other methods. Can. J. Earth Sci. 14:1–15.

Kodama, H., N.M. Miles, S. Shimoda, and J.E. Brydon. 1976. Mixed-layer kaolinite-montmorillonite from soils near Dawson, Yukon Territory. Can. Miner. 14:159–163.

Kohut, C., K. Muehlenbachs, and M.J. Dudas. 1995. Authigenic dolomite in a saline soil in Alberta, Canada. Soil Sci. Soc. Am. J. 59:1499-1504.

Kovda, V.A., B.P. Gradusov, and T.L. Bystritskaya. 1977. p. 148–164. *In* Ossobennosti glinistykh mineralov pochv vostochno-afrikanskovo rifta. Aridnye pochvy, ikh geneziz, geokhimya, ispol'zovanie. (Akademiya Nauk SSSR).

Kunze, G.W., and W.F. Bradley. 1964. Occurrence of a tabular halloysite in a Texas soil. Clays Clay Min. 12:523–527.

Laird, D.A. 1994. Comments on "Sericites in feldspars as source of 2:1 phyllosilicates in selected sandy soils." Soil Sci. Soc. Am. J. 58:1846.

Laird, D.A., and E.A. Nater. 1993. Nature of the illitic phase associated with randomly interstratified smectite/illite in soils. Clays Clay Min. 41:280–287.

Langmuir, D., and D.O. Whittemore. 1971. Variations in the stability of precipitated ferric oxyhydroxides. p. 209–234. In R.F. Gould (ed.) Nonequilibrium systems in natural water chemistry. Advances in Chemistry Series 106.

Leonard, R.A., and S.B. Weed. 1967. Influence of exchange ions on the b-dimension of dioctahedral vermiculite. Clays Clay Min. 15:149–161.

Lindsay, W.L. 1979. Chemical equilibria in soils. Wiley-Interscience, New York, NY.

Lindsay, W.L., P.L.G. Vlek, and S.H. Chien. 1989. Phosphate minerals. p. 1089–1130. In J.B. Dixon and S.B. Weed (ed.). Minerals in soil environments. 2nd Ed. Soil Science Society of America, Madison, WI.

Lippmann, F. 1982. The thermodynamic status of clay minerals. p. 475–485. In H. van Olphen and F. Veniale (ed.) Proceedings of the International Clay Conference 1981. Elsevier, Amsterdam, Netherlands.

Lou, G., and P.M. Huang. 1988. Hydroxy-aluminosilicate interlayers in montmorillonite: Implications for acidic environments. Nature 335:625–627.

Loughnan, F.C.1969. Chemical weathering of the silicate minerals. Elsevier, New York, NY.

Loveland, P.J., and P. Bullock. 1975. Crystalline and amorphous components of the clay fractions in brown podzolic soils. Clay Min. 10:451–469.

Lowe, D.J. 1986. Controls on the rates of weathering and clay mineral genesis in airfall tephras: A review and New Zealand case study. p. 265–330. In S.M. Colman and D.P. Dethier (ed.) Rates of chemical weathering of rocks and minerals. Academic Press, Orlando, FL

Lowe, D.J. 1995. Teaching clays: from ashes to allophane. p. 19–23. In G.J. Churchman, R.W Fitzpatrick and R.A. Eggleton (ed.) Clays controlling the environment. Proc. 10th Int.Clay Conf. CSIRO Publishing, Melbourne, Australia.

Lowe, D.J., G.J. Churchman, R.H. Merry, R.W. Fitzpatrick, M.J. Sheard, and W.H Hudnall. 1996. Holocene basaltic volcanogenic soils of the Mt Gambier area, South Australia, are unusual globally: what do they tell us? Australian and New Zealand National Soils Conference 1996. Vol. 2, Oral Papers, p.153–154.

Ma C. 1996. The ultra-structure of kaolin. Unpublished PhD Thesis, The Australian National University, Canberra, Australia.

Ma, C., and R.A. Eggleton. 1996. Surface layer type of kaolinite. p. 9. In Abstracts, 15th Australian Clay Minerals Society Conference.

MacEwan, D.M.C. 1947. The nomenclature of the halloysite minerals. Miner. Mag. 28:36–44.

Mackenzie, F.T., and R. Gees. 1971. Quartz: synthesis at earth-surface conditions. Science 173:533–535.

Madsen, H.B., and P. Nørnberg. 1995. Mineralogy of four sandy soils developed under heather, oak, spruce and grass in the same fluvioglacial deposit in Denmark. Geoderma 64:233–256.

Mahjoory, R.A. 1975. Clay mineralogy, physical, and chemical properties of some soils in arid regions of Iran. Soil Sci. Soc. Am. Proc. 39:1157–1164.

Malengreau, N., A. Bedidi, J.-P. Muller, and A.J. Herbillon. 1996. Spectroscopic control of iron oxide dissolution in two ferralitic soils. Eur. J. Soil Sci. 47:13–20.

Marbut, C.F. 1928. A scheme for classification. Proc.1st Int. Congr. Soil Sci. 4:1–31.

Marsan, F.A., and J. Torrent. 1989. Fragipan bonding by silica and iron oxides in a soil from northwestern Italy. Soil Sci. Soc. Am. J. 53:1140–1145.

Mattigod, S.V., and G. Sposito. 1978. Improved method for estimating the standard free energies for formation ($\Delta G^{\circ}_{f, 298.15}$) of smectites. Geochim. Cosmochim. Acta 42:1753–1762.

May, H.M., D.G. Kinniburgh, P.A. Helmke, and M.L. Jackson. 1986. Aqueous disssolution, solubilities and thermodynamic stabilities of common aluminosilicate clay minerals: Kaolinite and smectites. Geochim. Cosmochim. Acta 50:1667–1677.

McBride, M.B., B.A. Goodman, J.D. Russell, A.R. Fraser, V.C. Farmer, and D.P.E. Dickson. 1983. Characterization of iron in alkaline EDTA and NH_4OH extracts of podzols. J. Soil Sci. 34:824–840.

McCraw, J.D. 1962. Volcanic detritus in Taylor Valley, Victoria land, Antarctica. NZ J. Geol. Geophys. 5:740–745.

McDaniel, P.A., A.L. Falen, K.R. Tice, R.C. Graham, and S.E. Fendorf. 1995. Beidellite in E horizons of northern Idaho Spodosols formed in volcanic ash. Clays Clay Min. 43:525–532.

McKeague, J.A., and J.E. Brydon. 1970. Mineralogical properties of ten reddish Brown soils from the Atlantic provinces in relation to parent materials and pedogenesis. Can. J. Soil Sci. 50:47–55.

McKeague, J.A., and H. Kodama. 1981. Imogolite in cemented horizons of some British Columbia soils. Geoderma 25:189–197.

McKenzie, R.M. 1989. Manganese oxides and hydroxides. p. 439–465. *In* J.B. Dixon and S.B. Weed (ed.). Minerals in soil environments. 2nd Ed. Soil Science Society of America, Madison, WI.

McLaren, R.G., and K.C. Cameron. 1996. Soil science. 2nd Ed. Oxford University Press, Auckland, New Zealand.

McLean, E.O. 1976. Chemistry of soil aluminum. Comm. Soil Sci. Plant Anal. 7:619–636.

Mehra, O.P., and M.L. Jackson. 1960. Iron oxide removal from soils and clays by a dithionite-citrate system buffered with sodium bicarbonate. Clays Clay Min. 7:317–327.

Mejia, G., H. Kohnke, and J.L. White. 1968. Clay mineralogy of certain soils of Columbia. Soil Sci. Soc. Am. Proc. 32:665–670.

Millar, C.E., L.M. Turk, and H.D. Foth 1958. Fundamentals of soil science. 3rd Ed. John H. Wiley & Sons, New York, NY.

Millot, G. 1970. Geology of clays. Springer-Verlag, New York, NY.

Milnes, A.R., and R.W. Fitzpatrick. 1989. Titanium and zirconium minerals. p. 1131–1205. *In* J.B. Dixon and S.B. Weed (ed.). Minerals in soil environments. 2nd Ed. Soil Science Society of America, Madison, WI.

Ming, D.W., and F.A. Mumpton. 1989. Zeolites in Soils. p. 873–911. *In* J.B. Dixon and S.B. Weed. (ed.) Minerals in soil environments. 2nd Ed. Soil Science Society of America, Madison, WI.

Miura, K., S. Araki, and K. Kyuma. 1988. Genesis of soils derived from various types of parent rock in southwestern Japan. II. Mineralogical characteristics. Soil Sci. Plant Nut. 34:17–29.

Møberg, J.P., and O.K. Borggaard. 1991. Disintegration of the clay fraction in the eluvial horizons of some Danish Podzolised soils. Comm. Soil Sci. Plant Anal. 22:1079–1092.

Molloy, L.F., and L.C. Blakemore. 1974. Studies on a climosequence of soils in tussock grasslands. I. Introduction, sites and soils. NZ J. Sci. 17:233–255.

Mortland, M.M., 1958. Kinetics of potassium release from biotite. Soil Sci. Soc. Am. Proc. 22:503–508.

Mortland, M.M., K. Lawton, and G. Uehara. 1956. Alteration of biotite to vermiculite by plant growth. Soil Sci. 82:477–481.

Muir, I.J., and H.W. Nesbitt. 1997. Reactions of aqueous anions and cations at the labradorite-water interface: Coupled effects of surface processes and diffusion. Geochim.Cosmochim. Acta 61:265–274.

Muir, I.J., G.M. Bancroft, and H.W. Nesbitt. 1989. Characteristics of altered labradorite surfaces by SIMS and XPS. Geochem. Cosmochim. Acta 53:1235–1241.

Muir, I.J., G.M. Bancroft, W. Shotyk, and H.W. Nesbitt. 1990. A SIMS and XPS study of dissolving plagioclase. Geochim. Cosmochim. Acta 54:2247–2256.

Munk, L.P., and R.J. Southard. 1993. Pedogenic implications of opaline pendants in some California Late-Pleistocene Palexeralfs. Soil Sci. Soc. Am. J. 57:149–154.

Murakami.,T., H. Isobe, T. Sato, and T. Ohnuki. 1996. Weathering of chlorite in a quartz-chlorite schist: I. Mineralogical and chemical changes. Clays Clay Min. 44:244–256.

Nagasawa, K., and S. Miyazaki. 1976. Mineralogical properties of halloysite as related to its genesis. p. 257–265. *In* S.W. Bailey (ed.). Proc. Int. Clay Conf. 1975, Mexico City. Applied Publishing, Wilmette, IL.

Nagasawa, K.,and H. Noro. 1987. Mineralogical properties of halloysites of weathering origin. Chem. Geol. 60:145–149.

Nagy, K.L. 1995. Dissolution and precipitation kinetics of sheet silicates. p. 173–233. *In* A.F. White and S.L. Brantley (ed.). Chemical weathering rates of silicate minerals. Rev. Miner. Vol. 31. Mineralogical Society of America, Washington, DC.

Nahon, D., F. Colin,and Y. Tardy. 1982. Formation and distribution of Mg, Fe, Mn-smectites in the first stages of the lateritic weathering of forsterite and tephroite. Clay Min. 17:339–348.

Nesbitt, H.W. 1997. Bacterial and inorganic weathering processes and weathering of crystalline rocks. p. 113–142. *In* J.M. McIntosh and L.A.Groat (ed.). Biological-mineralogical interactions. Mineralogical Association of Canada Short Course Volume 25. Mineralogical Association of Canada.

Nesbitt, H.W., and I.J. Muir. 1988. SIMS depth profiles of weathered plagioclase and processes affecting dissolved Al and Si in some acidic soil solutions. Nature 334:336–338.

Nettleton, W.D., K.W. Flach, and R.E. Nelson. 1970. Pedogenic weathering of tonalite in southern California. Geoderma 4:387–402.

Nettleton, W.D., R.E. Nelson, and K.W. Flach. 1973. Formation of mica in surface horizons of dryland soils. Soil Sci. Soc. Am. Proc. 37:473–478.

Newman, A.C.D.. and G. Brown, 1966. Chemical changes during the alteration of micas. Clay Min. 6:297-310.

Newman, R.H., C.W. Childs, and G.J. Churchman. 1994. Aluminium coordination and structural disorder in halloysite and kaolinite by ^{27}Al NMR spectroscopy. Clay Min. 29:305–312.

Norfleet, M.L., A.D. Karathanasis, and B.R. Smith. 1993. Soil solution composition relative to mineral distribution in Blue Ridge Mountain soils. Soil Sci. Soc. Am. J. 57:1375–1380.

Noro, H. 1986. Hexagonal platy halloysite in an altered tuff bed, Komaki City, Aichi prefecture, Central Japan. Clay Min. 21:401–415.

Norrish, K. 1968. Some phosphate minerals of soils. Trans. 9th Int. Congr. Soil Science, Adelaide, Australia, Vol. II:713–723.

Norrish, K. 1973. Factors in the weathering of mica to vermiculite. p. 417–432. In J.M. Serratosa (ed.) Proceedings of the 1972 International Clay Conference. Div. de Ciencas, Madrid, Spain.

Norrish, K. 1995. An unusual fibrous halloysite. p. 275–284. In G.J. Churchman, R.W Fitzpatrick and R.A. Eggleton. (ed.). Clays controlling the environment. Proc. 10th Int. Clay Conf. CSIRO Publishing, Melbourne, Australia.

Norrish, K., and J.G. Pickering. 1983. Clay minerals. p. 281–308. In Soils: An Australian viewpoint. CSIRO Australia; Melbourne/Academic Press, London, UK.

Nriagu, J.O. 1978. Dissolved silica in pore waters of Lakes Ontario, Erie, and Superior sediments. Limnology and Oceanography 23:53–67.

Oelkers, E.H., and J. Schott. 1995. Experimental study of anorthite dissolution and the relative mechanism of feldspar hydrolysis. Geochimica et Cosmochimica Acta 59:5039–5053.

Ojanuga, A.G. 1973. Weathering of biotite in soils of a humid tropical climate. Soil Sci. Soc. Am. Proc. 37:644–646.

Padmanabhan, E., and A.R. Mermut. 1996. Submicroscopic structure of Fe-coatings on quartz grains in tropical environments. Clays Clay Min. 44:801–810.

Paquet, H., and G. Millot. 1972. Geochemical evolution of clay minerals in the weathered products in soils of Mediterranean climate. p. 199–206. In J.M. Serratosa (ed.) Proceedings of the International Clay Conference, Madrid. Spain.

Parfitt, R.L. 1990. Allophane in New Zealand - a review. Aust. J. Soil Res. 28:343–360.

Parfitt, R.L., and G.J. Churchman. 1988. Clay minerals and humus complexes in five Kenyan soils derived from volcanic ash-a discussion. Geoderma 42:365–367.

Parfitt, R.L., and T. Henmi. 1980. Structure of some allophanes from New Zealand. Clays Clay Min. 28:285–294.

Parfitt, R.L., C.W. Childs, and D.N. Eden. 1988. Ferrihydrite and allophane in four andepts from Hawaii and implications for their classifications. Geoderma 41:223–241.

Parfitt, R.L., R.J. Furkert, and T. Henmi. 1980. Identification and structure of two types of allophane from volcanic ash soils and tephra. Clays Clay Min. 28:328–334.

Parfitt, R.L., M. Russell, and G.E. Orbell. 1983. Weathering sequence of soils from volcanic ash involving allophane and halloysite, New Zealand. Geoderma 29:41–57.

Parfitt, R.L., M. Saigusa, and J.D. Cowie. 1984. Allophane and halloysite formation in a volcanic ash bed under different moisture conditions. Soil Sci. 138:360–364.

Parham, W.E. 1969. Formation of halloysite from feldspar: Low temperature, artificial weathering versus natural weathering. Clays Clay Min. 17:13–22.

Percival, H.J. 1985. Soil solutions, minerals and equilibria. NZ Soil Bur. Sci. Rep. 69.

Petit, S., T. Prot, A. Decarreau, C. Mosser, and M.C. Toledo-Groke. 1992. Crystallochemical study of a population of particles in smectites from a lateritic weathering profile. Clays Clay Min. 40:436–445.

Poncelot, G.M., and G.W. Brindley. 1967. Experimental formation of kaolinite from montmorillonite at low temperatures. Am. Mineralog. 52:1161–1173.

Priesinger, A. 1963. Sepiolite and related compounds: Its stability and applications. Clays Clay Miner. 10:365–371.

Quantin, P., J. Gautheyrou, and P. Lorenzoni. 1988. Halloysite formation through in situ weathering of volcanic glass from trachytic pumices, Vico's Volcano, Italy. Clay Min. 23:423–437.

Rabenhorst, M.C. 1997. The chrono-continuum: an approach to modeling pedogenesis in marsh soils along transgressive coastlines. Soil Sci. 162:2–9.

Radoslovich, E.W. 1963. The cell dimensions and symmetry of layer-lattice silicates. VI. Serpentine and kaolin morphology. Am. Miner. 48:368–378.

Rai, D., and J.A. Kittrick. 1989. Mineral equilibria and the soil system. p. 161–198. In J.B. Dixon and S.B. Weed (ed.) Minerals in soil environments. 2nd Ed. Soil Science Society of America, Madison, WI.

Raman, K.V., and M.L. Jackson. 1966. Layer charge reduction in clay minerals of micaceous soils and sediments. Clays Clay Min. 14:53–68.

Ramanaidou, E., D. Nahon, A. Decarreau, and A.J. Melfi. 1996. Hematite and goethite from duricrusts developed by lateritic chemical weathering of Precambrian banded iron formations, Minas Gerais, Brazil. Clays Clay Min. 44:22–31.

Range, K.J., A. Range, and A. Weiss. 1969. Fire-clay type kaolinite or fire-clay mineral? Experimental classification of kaolinite-halloysite minerals. p. 3–13. *In* L. Heller (ed.) Proceedings of the International Clay Conference, Tokyo. Vol 1. Israel University Press, Jerusalem, Israel.

Read, G., R.A. Kemp, and J. Rose. 1996. Development of a feldspar weathering index and its application to a buried soil chronosequence in southeastern England. Geoderma 74:267–280.

Rebertus, R.A., S.B. Weed, and S.W. Buol. 1986. Transformations of biotite to kaolinite during saprolite-soil weathering. Soil Sci. Soc. Am. J. 50:810–819.

Reid, D.A., R.C. Graham, L.A. Douglas, and C. Amrhein. 1996. Smectite mineralogy and charge characteristics along an arid geomorphic transect. Soil Sci. Soc. Am. J. 60:1602–1611.

Rex, R.W., J.K. Syers, M.L. Jackson, and R.N. Clayton. 1969. Eolian origin of quartz in soils of Hawaiian Islands and in Pacific pelagic sediments. Science 163:277–279.

Rice, C.M., and J.M. Williams. 1969. A Mössbauer study of biotite weathering. Miner. Mag. 37:210–215.

Rich, C.I. 1968. Hydroxy interlayers in expansible layer silicates. Clays Clay Min. 16:15–30.

Righi, D., S. Petit, and A. Bouchet. 1993. Characterization of hydroxy-interlayered vermiculite and illite/smectite interstratified minerals from the weathering of a chlorite in a Cryorthod. Clays Clay Min. 41:484–495.

Righi, D., F. Terribile, and S. Petit. 1995. Low-charge to high-charge beidellite conversion in a Vertisol from south Italy. Clays Clay Min. 43:495–502.

Robert, M., and J. Berthelin. 1986. Role of biological and biochemical factors in soil mineral weathering. p. 453–495. *In* P.M. Huang, and M. Schnitzer. (ed.). Interactions of soil minerals with natural organics and microbes. Soil Sci. Soc. Am. Spec. Pub. 17.

Robert, M., and C. Chenu. 1992. Interactions between soil minerals and microorganisms. p. 307-404. *In* G. Stotsky and J.-M. Bollag (ed.) Soil biochemistry. Marcel Dekker, New York, NY.

Robertson, I.D.M., and R.A. Eggleton. 1991. Weathering of granitic muscovite to kaolinite and halloysite and of plagioclase-derived kaolinite to halloysite. Clays Clay Min. 39:113–126.

Romero, R., M. Robert, F. Elsass, and C. Garcia. 1992. Evidence by transmission electron microscopy of weathering microsystems in soils developed from crystalline rocks. Clay Min. 27:21–33.

Rosenberg, P.E., J.A. Kittrick, and J.R. Alldredge. 1984. Composition of the controlling phase in muscovite equilibrium solubility. Clays Clay Min. 32:480–482.

Ross, C.S., 1928. The mineralogy of clays. p. 555-561. *In* Proc.1st Int. Congr. Soil Sci. Comm. V and VI, Misc. Pap.

Ross, G.J., and H. Kodama. 1974. Experimental transformation of a chlorite into a vermiculite. Clays Clay Miner. 22:205–211.

Ross, G.J., and H. Kodama. 1979. Evidence for imogolite in Canadian soils. Clays Clay Min. 27:297–300.

Ross, G.J., and M.M. Mortland. 1966. A soil beidellite. Soil Sci. Soc. Am. Proc. 39:337–343.

Rowell, D.L. 1994. Soil science: Methods and applications. Longman, Harlow, UK.

Russell, E. W. 1973. Soil conditions and plant growth. 10th Ed. Longman, London, UK.

Saigusa, M., S. Shoji, and T. Kato. 1978. Origin and nature of halloysite in Ando soils from Towada tephra, Japan. Geoderma 20:115–129.

Sakharov, B.A., and V. Drits. 1973. Mixed-layer kaolinite-montmorillonite: a comparison of observed and calculated diffraction patterns. Clays Clay Miner. 21:15–17.

Sawhney, B.L. 1989. Interstratification in layer silicates. p. 789–828. *In* J.B. Dixon and S.B. Weed (ed.) Minerals in soil environments. 2nd Ed. Soil Science Society of America, Madison, WI.

Sawhney, B.L., and M.L. Jackson. 1958. Soil montmorillonite formulas. Soil Sci. Soc. Am. Proc. 22:115–118.

Schaetzl, R.J., L.R. Barrett, and J.A. Winkler. 1994. Choosing models for soil chronofunctions and fitting them to data. Europ. J. Soil Sci. 45:219–232.

Schott, J., and R.A. Berner. 1985. Dissolution mechanisms of pyroxenes and olivines during weathering. p. 35–53. *In* J.I. Drever (ed.) The chemistry of weathering. Reidel, New York, NY.

Schulze, D.G. 1981. Identification of soil iron oxides minerals by differential X-ray diffraction. Soil Sci. Soc. Am. J. 45:437–440.

Schwertmann, U., and R.M. Cornell. 1991. Iron oxides in the laboratory. VCH Verlag, Weinheim, Germany.

Schwertmann, U., and H. Fechter. 1984. The influence of aluminium on iron oxides. XI. Aluminium substituted maghemite in soils and its formation. Soil Sci. Soc. Am. J. 48:1462–1463.

Schwertmann, U., and R.M. Taylor. 1989. Iron oxides. p. 379–438. *In* J.B. Dixon and S.B. Weed (ed.) Minerals in soil environments. 2nd Ed. Soil Science Society of America, Madison, WI.

Schwertmann, U., E. Murad, and D.G. Schulze. 1982. Is there a Holocene reddening (hematite formation) in soils of axeric temperate areas? Geoderma 27:209–223.

Schwertmann, U., H. Fechter, R.M. Taylor, and H. Stanjek. 1995. A lecture and demonstration for students on iron oxide formation. p. 11–14. *In* G.J. Churchman, R.W Fitzpatrick and R.A. Eggleton. (ed.) Clays Controlling the Environment. Proc. 10th Int. Clay Conf. CSIRO Publishing, Melbourne, Australia.

Senkayi, A.L., J.B. Dixon, L.R. Hossner, and B.E. Viani. 1983. Mineralogical transformations during weathering of lignite overburden in east Texas. Clays Clay Miner. 31:49–56.

Seyama, H., M. Soma, and A. Tanaka. 1996. Surface characterization of acid-leached olivines by X-ray photoelectron spectroscopy. Chem. Geol. 129:209–216.

Seyama, H., J. Sato, A. Tanaka, M. Soma, and M. Tsurumi. 1997. Surface alteration of biotite dissolving in acid solutions. p. 1–4 *In* Proc. 7th Europ. Conf. Appl. Surf. Interf. Analy.

Sherman, G.D. 1952. The genesis and morphology of the alumina-rich laterite clays. p. 154-161. *In* Problems of clay and laterite genesis. American Institute of Mining and Metallurgical Engineering, New York, NY.

Sherman, G.D., L.G. Schultz, and L.J. Alway. 1962. Dolomite in soils of the Red River Valley, Minnesota. Soil Sci. 94:304–313.

Siffert, B. 1978. Genesis and synthesis of clays and clay minerals: Recent developments and future prospects. p. 337–347. *In* M.M. Mortland and V.C. farmer (ed.) International Clay Conference 1978. Developments in Sedimentology 27. Elsevier, Amsterdam, Netherlands.

Siffert, B., and R. Wey. 1962. Synthèse d'une sépiolite à la température ordinaire. Compt. Rend. Acad.s Sci. Franc. 254:1460–1463.

Singer, A. 1974. Mineralogy of a palagonite material from the Golan Heights, Israel. Clays Clay Miner. 22:231–240.

Singer, A. 1989. Palygorskite and sepiolite group minerals. p. 829–872. *In* J.B. Dixon and S.B. Weed (ed.) Minerals in soil environments. 2nd Ed. Soil Science Society of America, Madison, WI.

Singer, A. 1993. Weathering patterns in representataive soils of Guanxi Province, south-east China, as indicated by detailed clay mineralogy. J. Soil Sci. 44:173–188.

Singer, A., and K. Norrish. 1974. Pedogenic palygorskite occurrences in Australia. Am. Mineralog. 59:508–517.

Singh, Balbir. 1996. Why does halloysite roll? - A new model. Clays Clay Miner. 44:191–196.

Singh, Balbir, and R.J. Gilkes. 1992. The electron optical investigation of the alteration of kaolinite to halloysite. Clays Clay Miner. 40:212–229.

Singh, Balbir, and R.J. Gilkes. 1993. The recognition of amorphous silica in indurated soil profiles. Clay Min. 28:461–474.

Singh, Balbir, and R.J. Gilkes. 1995. The natural occurrence of χ-alumina in lateritic pisolites. Clay Min. 30:39–44.

Singh, Balwant, and R.J. Gilkes 1991. A potassium-rich beidellite from a lateritic pallid zone in Western Australia. Clay Min. 26:233–244.

Singh, S.S., and H. Kodama. 1994. Effect of the presence of aluminum ions in iron solutions on the formation of iron oxyhydroxides (FeOOH) at room temperature under acidic environment. Clays Clay Miner. 42:606–613.

Smith, K.L., A.R. Milnes, and R.A. Eggleton. 1987. Weathering of basalt: Formation of iddingsite. Clays Clay Miner. 35:418–428.

Smith, W.W. 1962. Weathering of some Scottish basic igneous rocks with reference to soil formation. J. Soil Sci. 13:202–215.

Soma, M., G.J. Churchman, and B.K.G. Theng. 1992. X-ray photoelectron spectroscopic analysis of halloysites with different composition and particle morphology. Clay Min. 27:413–421.

Sommer, M., and E. Schlichting. 1997. Archetypes of catenas in respect to matter-a concept for structuring and grouping catenas. Geoderma 76:1–33.

Sousa, E.C., and H. Eswaran. 1975. Alteration of micas in the saprolite of a profile from Angola. A morphological study. Pedologie 25:71–79.

Souza Santos, P. de, H. de Souza Santos, and G.W. Brindley. 1965. Mineralogical studies of kaolinite-halloysite clays-III. A fibrous kaolin mineral from Piedade, Sao Paulo, Brazil. Amer. Mineralog. 50:619–628.

Sposito, G. 1985. Chemical models of weathering in soils. p. 1–18. *In* J.I. Drever (ed.) The chemistry of weathering. NATO ASI Series 149. D. Reidel, Dordrecht, Netherlands.

Sposito, G. 1989. The chemistry of soils. Oxford University Press, New York, NY.

Sridhar, K., and M.L. Jackson. 1974. Layer charge decrease by tetrahedral cation removal and silicon incorporation during natural weathering of phlogopite to saponite. Soil Sci. Soc. Am. Proc. 38:847–851.

Sridhar, K., M.L. Jackson, and R.N. Clayton. 1975. Quartz oxygen isotopic stability in relation to isolation from sediments and diversity of source. Soil Sci. Soc. Am. Proc. 42:1209–1213.

Srodon, J., and D.D. Eberl. 1980. The presentation of x-ray data for clay minerals. Clay Min. 15:317–320.

Steinmann, P., P.C. Lichtner, and W. Schotyk. 1994. Reaction path approach to mineral weathering reactions. Clays Clay Miner. 42:197–206.

Stephen, I., and D.M.C. MacEwan. 1951. Some chlorite minerals of unusual type. Clay Minerals Bulletin 1:157–161.

Stoch, L., and W. Sikora. 1976. Transformation of micas in the process of kaolinitization of granites and gneisses. Clays Clay Miner. 24:156–162.

Stucki, J.W., and C.B. Roth. 1977. Oxidation reduction mechanism for structural iron in nontronite. Soil Sci. Soc. Am. J. 41:808–814.

Suarez, D.L., and S. Goldberg. 1994. Modeling soil solution, mineral formation and weathering. *In* R.B. Bryant and R.W. Arnold (ed.) Quantitative modeling of soil forming processes. Soil Sci. Soc. Am. Spec. Pub. 39:37–60.

Sudo, T., and H. Takahashi. 1956. Shapes of halloysite particles in Japanese clays. Clays Clay Miner. 4:67–79.

Sudo, T., and H. Yotsumoto. 1977. The formation of halloysite tubes from spherulitic halloysite. Clays Clay Miner. 25:155–159.

Swindale, L.D., and G. Uehara. 1966. Ionic relationships in the pedogenesis of Hawaiian soils. Soil Sci. Soc. Am. Proc. 30:726–730.

Syers, J.K., V.E. Berkheiser, M.L. Jackson, R.N. Clayton, and R.W. Rex. 1969. Eolian sediment influence on pedogenesis during the Quaternary. Soil Sci. 107:421–427.

Tait, J.M., N. Yoshinaga, and B.D. Mitchell. 1978. The occurrence of imogolite in some Scottish soils. Soil Sci. Plant Nut. 24:141–151.

Takahashi, T., R. Dahlgren, and P. van Susteren. 1993. Clay mineralogy and chemistry of soils formed in volcanic materials in the xeric moisture regime of northern California. Geoderma 59:131–150.

Tamm, O. 1922. Eine methode Zur Bestimmung de anorganischen Komponente des Gelcomplexes im Boden. Meddelanden fran Statens Skogsforsokanstalt Stockholm 19:387–404.

Tardy, Y., and R.M. Garrels. 1974. A method of estimating the Gibbs energies of formation of layer silicates. Geochim. Cosmochim. Acta 38:1101–1106.

Tazaki, K. 1982. Analytical electron microscopic studies of halloysite formation processes-morphology and composition of halloysite. p. 573–584. *In* H. van Olphen and F. Veniale (ed.). Proc. Int. Clay Conf. Bologna, Pavia. Elsevier, Amsterdam, Netherlands.

Tazaki, K. 1997. Biomineralization of layer silicates and hydrated Fe/Mn oxides in microbial mats: An electron microscopical study. Clays Clay Min. 45:203–212.

Tazaki, K., and W.S. Fyfe. 1987. Formation of primitive clay precursors on K-feldspar under extreme leaching conditions. p. 53–58. *In* L.G. Schultz, H. van Olphen and F.A. Mumpton (ed.) Proc. Int. Clay Conf., Denver, CO 1985. The Clay Minerals Society, Bloomington, IN.

Tazaki, K., W.S. Fyfe, and G.R. Heath. 1986a. Palygorskite formed on montmorillonite in North Pacific deep-sea sediments. Clay Sci. 6:197–216.

Tazaki, K., W.S. Fyfe, and C.B. Dissanayake. 1986b. Weathering of phosphatic marble to exploitable apatite deposit, Sri Lanka. App. Geochem. 1:287–300.

Tazaki, K., W.S. Fyfe, M. Tsuji, and K. Katayama. 1987a. TEM observation of the smectite-to-palygorskite transition in deep Pacific sediments. App. Clay Sci. 2:233–240.

Tazaki, K., W.S. Fyfe, and C.B. Dissanayake. 1987b. Weathering of apatite under extreme conditions of leaching. Chem. Geol. 60:151–162.

Theng, B.K.G., G.J. Churchman, and R.H. Newman. 1986. The occurrence of interlayer clay-organic complexes in teo New Zealand soils. Soil Sci. 142:262–266.

Thiry, M., and R. Simon-Coinçon. 1996. Tertiary paleoweatherings and silcretes in the southern Paris Basin. Catena 26:1–26.

Tilley, D.B., and R.A. Eggleton. 1994. Tohdite ($5Al_2O_3.H_2O$) in bauxites from Northern Australia. Clays Clay Min. 42:485–488.

Tomar, K.P., J. Podder, and S. Kumar. 1985. Kaolinite-smectite-vermiculite interstratification in Indo-Gangetic alluvial soils of Meerut. Clay Res. 4:14–23.

Tonkin, P.J. 1970. Contorted stratification with clay lobes in volcanic ash beds, Raglan-Hamilton region, New Zealand. Earth Sci. J. 4:129–140.

Torrent, J., and J. Benayas. 1977. Origin of gibbsite in a weathering profile from granite in West-Central Spain. Geoderma 19:37–49.

Trolard, F., J.-M. Génin, M. Abdelmoula, G. Bourrié, B. Humbert, and A. Herbillon. 1997. Identification of a green rust mineral in a reductomorphic soil by Mössbauer and Raman spectroscopies. Geochim. Cosmochim. Acta 61:1107–1111.

Ugolini, F.C. 1995. Formation and stability of smectite in Findley Lake Spodosol. p. 61-73. *In* S. Fiore (ed.) Incontri Scientifici, Vol. 1. Instituto di Ricerca sulle Argille.

Ugolini, F.C., and R. Dahlgren. 1987. The mechanism of podzolization as revealed by soil solution studies. p. 195-203. *In* D. Righi and A. Chauvel (ed.). Podzols and podzolisation. AFES et INRA, Paris, France.

Ugolini, F.C., and R.A. Dahlgren. 1991. Weathering environments and occurrence of imogolite/allophane in selected Andisols and Spodosols. Ssoil Sci. Soc. Am. J. 55:1166–1171.

Ugolini, F.C., and R.S. Sletten. 1991. The role of proton donors in pedogenesis as revealed by soil solution studies. Soil Sci. 151:59–75.

Ugolini, F.C., H. Dawson, and J. Zachara. 1977. Direct evidence of particle migration in the soil solution of a podzol Science 198:603–605.

Ugolini, F.C., G. Corti, A. Agnelli, and F. Piccardi. 1996. Mineralogical, physical and chemical properties of rock fragments in soil. Soil Sci. 161:521–540.

Urrutia, M.M., and T.J. Beveridge. 1994. Formation of fine-grained silicate minerals and metal precipitates by a bacterial surface (*Bacillus subtilis*). Chemical Geol. 116:261–280.

Urrutia, M.M., and T.J. Beveridge. 1995. Formation of short-range ordered aluminosilicates in the presence of a bacterial surface (*Bacillus subtilis*) and organic ligands. Geoderma 65:149–165.

van Breeman, N. 1982. Genesis, morphology and classification of acid sulfate soils in coastal plains. p. 95–108. *In* J.A. Kittrick (ed.) 1982. Acid sulfate weathering. Soil Sci. Soc. Am Spec. Pub. 10.

van Oort, F., A.G. Jongmans, A.-M. Jaunet, J.D.J. van Doesburg, F. Elsass, and T.C.J. Feijtel. Characterization of clay formation in thin sections. A case study on halloysite neoformation in weathered pyroclastic parent rock in Guadeloupe (F.W.I.). Trans. 14th Int. Congr. Soil Sci. VII:100–105.

van Wesemael, B., J.M. Verstraten, and J. Sevink. 1995. Pedogenesis by clay dissolution on acid, low-grade metamorphic rocks under mediterranean forests in southern Tuscany (Italy). Catena 24:105–125.

Veen, A.W.L. 1973. Evaluation of clay mineral equilibria in some clay soils (Usterts) of the brigalow lands. Aust. J. Soil Res. 11:167–184.

Velbel, M.A. 1985. Geochemical mass balances and weathering rates in forested watersheds of the southern Blue Ridge. Am. J. Sci. 285:904–930.

Velbel, M.A. 1992. Geochemical mass balances and weathering rates in forested watersheds of the southern Blue Ridge. III. Cation budgets and the weathering rate of amphibole. Am. J. Sci. 292:58–78.

Velde, B. 1968. The effect of chemical reduction on the stability of pyrophyllite and kaolinite in pelitic rocks. J. Sed. Petrol. 38:13–16.

Velde, B. 1992. Introduction to clay minerals. Chapman and Hall, London, UK.

Verrecchia, E.P., and M-N. Le Coustumer. 1996. Occurrence and genesis of palygorskite and associated clay minerals in a Pleistocene calcrete complex, Sde Boqer, Negev Desert, Israel. Clay Min. 31:183–202.

Vicente, M.A., M. Razzaghe, and M. Robert. 1977. Formation of aluminium hydroxy vermiculite (intergrade) and smectite from mica under acidic conditions. Clay Min. 12:101–112.

Visconti, Y.S., B.N.F. Nicot, and E. Goulart de Andrade. 1956. Tubular morphology of some Brazilian kaolins. Am. Mineralog. 41:67–76.

Wada, K. 1961. Lattice expansion of kaolin minerals by treatment with potassium acetate. Am. Mineralog. 46:78–91.

Wada, K. 1980. Mineralogical characteristics of Andisols. p. 87–107. *In* B.K.G. Theng (ed.) Soils with variable charge. New Zealand Society of Soil Science, Lower Hutt, New Zealand.

Wada, K. 1989. Allophane and Imogolite. p. 1051–1087. *In* J.B. Dixon and S.B. Weed. (ed.) Minerals in soil environments. 2nd Ed. Soil Science Society of America, Madison, WI.

Wada, K. 1995. Structure and formation of non- and para-crystalline aluminosilicate clay minerals: A review. p. 443–448. *In* G.J. Churchman, R.W Fitzpatrick and R.A. Eggleton. (ed.). Clays controlling the environment. Proc. 10th Int. Clay Conf. CSIRO Publishing, Melbourne, Australia.

Wada, K., and Y. Kakuto 1983. Intergradient vermiculite-kaolin mineral in a Korean Ultisol. Clays Clay Miner. 31:183–190.

Wada, K., and Y. Kakuto. 1985. Embryonic halloysites in Ecuador soils derived from volcanic ash. Soil Sci. Soc. Am. J. 49:276–278.

Wada, K., and Y. Kakuto. 1989. "Chloritized" vermiculite in a Korean Ultisol studied by ultramicrotomy and transmission electron microscopy. Clays Clay Miner. 37:263–268.

Wada, K., and N. Yoshinaga. 1969. The structure of imogolite. Am. Miner. 54:50–71.

Wada, K., Y. Kakuto, and H. Ikawa. 1986. Clay minerals, humus complexes, and classification of four "Andepts", Maui, Hawaii. Soil Sci. Soc. Am. J. 50:1007–1013.

Wada, K., Y. Kakuto, and F.N. Muchena. 1987. Clay minerals and humus complexes in five Kenyan soils derived from volcanic ash. Geoderma 39:307–321.

Wada, K., Y. Kakuto, M.A. Wilson, and J.V. Hanna. 1991. The chemical composition and structure of a 14 Å intergradient mineral in a Korean Ultisol. Clay Min. 26:449–461.

Wada, S.-I., and C. Mizota. 1982. Iron-rich halloysite (10Å) with crumpled lamellar morphology from Hokkaido, Japan. Clays Clay Miner. 30:315–317.

Wada, S.-I., A. Eto, and K. Wada. 1979. Synthetic allophane and halloysite. J. Soil Sci. 30:347–355.

Walker, T.W. 1965. The significance of phophorus in pedogenesis. p. 295–316. *In* E.G. Hallsworth and D.V. Crawford (ed.) Experimental pedology. Butterworths, London, UK.

Wang, C., G.J. Ross, and H.W. Rees. 1981. Characteristics of residual and colluvial soils developed on granite and of the associated pre-Wisconsin landforms in north-central New Brunswick. Can. J. Earth Sci. 18:487–494.

Wang, H.D., G.N. White, F. T. Turner, and J.B. Dixon. 1993. Ferrihydrite, lepidocrocite and goethite in coatings from East Texas vertic soils. Soil Sci. Soc. Am. J. 57:1381–1386.

Ward, C.R., and F.I. Roberts. 1990. Occurrence of spherical halloysite in bituminous coals of the Sydney Basin, Australia. Clays Clay Miner. 38:50–506.

Ward, W.T. 1967. Volcanic ash beds of the lower Waikato Basin, North Island, New Zealand. NZ J. Geol. Geophy. 10:1109–1135.

Watanebe, T., Y. Sawada, J.D. Russell, W.J. McHardy, and M.J. Wilson. 1992. The conversion of montmorillonite to interstratified halloysite-smectite by weathering in the Omi acid clay deposit. Clay Min. 27:159–173.

Weaver, C.E. 1989. Clays, muds and shales. Developments in sedimentology 44. Elsevier, Amsterdam, Netherlands.

Weaver, R.M., M.L. Jackson, and J.K. Syers. 1971. Magnesium and silicon activities in matrix solutions of montmorillonite-containing soils in relation to clay mineral stability. Soil Sci. Soc. Am. Proc. 35:823–830.

Wentworth, S.A. 1970. Illite. Clay Sci. 3:140–155.

White, A.F., A.E. Blum, M.S. Schulz, T.D. Bullen, J.W. Harden, and M.L. Peterson. 1996. Chemical weathering rates of a soil chronosequence on granitic alluvium: Quantification of mineralogical and surface area changes and calculation of primary silicate reaction rates. Geochim. Cosmochim.Acta 60:2533–2550.

White, A.F., and S.L. Brantley (ed.).1995. Chemical weathering rates of silicate minerals. Reviews in Mineralogy Vol. 31. Mineralogical Society of America, Washington, DC.

Whitten, D.G.A., and J.R.V. Brooks. 1972. A dictionary of geology. Penguin Books, Harmondsworth, UK.

Whitton, J.S., and G.J. Churchman. 1987. Standard methods for mineral analysis of soil survey samples for characterisation and classification in NZ Soil Bureau. NZ Soil Bureau Scientific Report 79.

Wildman, W.E., M.L. Jackson, and L.D. Whittig. 1968. Serpentinite rock dissolution as a function of carbon dioxide pressure in aqueous solution. Am. Miner. 53:1252–1263.

Wilke, B.-M., U. Schwertmann, and E. Murad. 1978. An occurrence of polymorphic halloysite in granite saprolite of the Bayerischer Wald, Germany. Clay Min. 13:67–80.

Williams, A.G., L. Ternan, and M. Kent. 1986. Some observations on the chemical weathering of the Dartmoor granite. Earth Surf. Proc. Landf. 11:557–574.

Wilson, M.J. 1970. A study of weathering in a soil derived from a biotite-hornblende rock. I. Weathering of biotite. Clay Min. 8:291–303.

Wilson, M.J. 1987. Soil smectites and related interstratified minerals:recent developments. p. 167–173. *In* L.G. Schultz, H. van Olphen and F.A. Mumpton (ed.) Proc. Int. Clay Conf., Denver, CO 1995. The Clay Minerals Society, Bloomington, IN.

Wilson, M.J., and P.D. Cradwick. 1972. Occurrence of interstratified kaolinite-montmorillonite in some Scottish soils. Clay Min. 9:435–437.

Wilson, M.J., and W. J. McHardy. 1980. Experimental etching of a microcline perthite and implications regarding natural weathering. J. Microscopy 120:291–302.

Wollast, R., and L. Chou. 1985. Surface reactions during the early stages of weathering of albite. Geochim. Cosmochim. Acta 56:3113–3122.

Worden, R.H., F.D.L. Walker, I. Parsons, and W.L. Brown. 1990. Development of microporosity, diffusion channels and deuteric coarsening in perthitic alkali feldspars. Contrib. Miner. Petrol. 104:507–515.

Xie, Z. 1987. Clay mineralogy of dark brown forest soils in the Xiao Higgan Mountains. Acta Pedol. Sinica B24:18–26.

Yaalon, D.H., and M. Wieder. 1976. Pedogenic palygorskite in some arid brown (Calciorthod) soils of Israel. Clay Min. 11:73–80.

Yerima, B.P.K., F.G. Calhoun, A.L. Senkayi, and J.B. Dixon. 1985. Occurrence of interstratified kaolinite-smectite in El Salvador Vertisols. Soil Sci. Soc. Am. J. 49:462–466.

Yoshinaga, N., and S. Aomine. 1962. Imogolite in some Ando soils. Soil Sci. Plant Nut. 8:22–29.

Young, A.W., A.S. Campbell, and T.W. Walker. 1980. Allophane isolated from a podzol developed in a non-vitric parent material. Nature 284:46–48.

Zabowski, D., and F.C. Ugolini. 1992. Seasonality in the mineral stability of a subalpine Spodosol. Soil Sci. 154:497–507.

Zelazny, L.W., and G.N. White. 1989. The pyrophyllite-talc group. p. 527–550. *In* J.B. Dixon and S.B. Weed (ed.) Minerals in soil environments. 2nd Ed. Soil Science Society of America, Madison, WI.

Zen, E-an. 1962. Problem of the thermodynamic status of the mixed-layer minerals. Geochim. Cosmochim. Acta 26:1055–1067.

2

Phyllosilicates

C.G. Olson
USDA-NRCS, Lincoln, NE

M.L. Thompson
Iowa State University

M.A. Wilson
USDA-NRCS, Lincoln, NE

2.1 Introduction

Phyllosilicates are one of many subclasses of the silicate mineral class (Fig. 2.1). Common minerals associated with these subclasses are listed in Table 2.1. Common elements found in silicate minerals, their ionic charge, and radii are given in Table 2.2. Together Tables 2.1 and 2.2 illustrate that by volume, an aluminosilicate mineral is predominantly O. The other ions fit into smaller voids among these O ions.

The basic structural feature of all minerals in these subclasses is the SiO_4 tetrahedron linked by the sharing of three of four O anions to form sheets with a pseudohexagonal network in the *ab* crystallographic plane. The sheet has the composition $(Si_2O_5)^{2-}$. The three shared O^{2-} ions are termed basal O^{2-} ions while the fourth apical O^{2-} is not shared with another SiO_4 tetrahedron and may bond to other structures. The tetrahedra are interconnected in sheet-like structures. These sheets are commonly referred to as silica or *tetrahedral sheets* (Fig. 2.2). Cations in each tetrahedral sheet are in fourfold coordination, i.e., each Si cation is bonded to four O^{2-} ions arranged in each tetrahedron. Aluminum may replace as many as half of the Si, producing a formula composition such as $(AlSi_3O_{10})^{5-}$ or $(Al_2Si_2O_{10})^{6-}$. Minerals in which tetrahedral sheets of silica dominate are often referred to as *layer silicates*.

In most phyllosilicates, the tetrahedral sheet is combined with another sheet-like grouping of cations such as Al^{3+}, Mg^{2+}, or Fe^{3+} in sixfold-coordination with O^{2-} and OH^- anions. Sixfold coordination means that the negatively charged anions are arranged around the positively charged cations in an octahedral pattern; one anion at each corner of an octahedron surrounding a central cation. Adjacent octahedra share anions to form a planar network referred to as an *octahedral sheet* (Fig. 2.3a,b).

Phyllosilicates may be divided into two groups of composite tetrahedral-octahedral arrangements: (1) 1:1 structures, or (2) 2:1 structures. Different combinations of tetrahedral and octahedral sheets produce different mineral structures. In each tetrahedral sheet, the free apical O^{2-} ion of each SiO_4 tetrahedron is located above and at the center of a pyramid whose triangular base is formed by the other

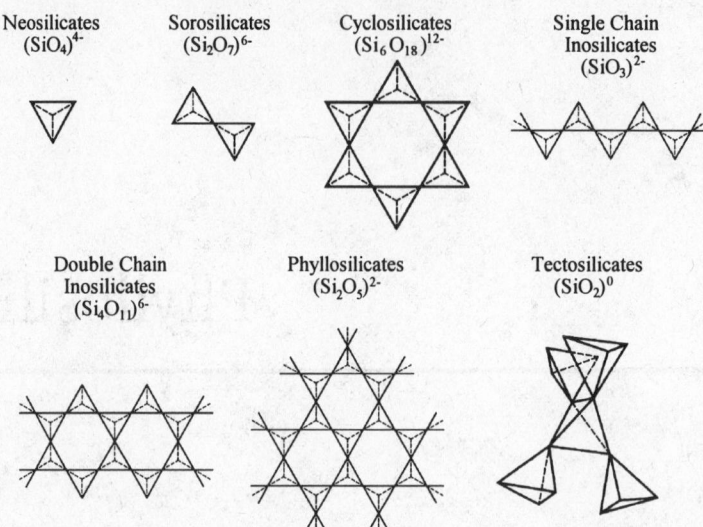

Fig. 2.1 Classification of silicate minerals [Adapted from Schulze, 1989 with permission of the Soil Science Society of America]

three oxygen atoms. The apical oxygen ions form hexagonal rings in the a-b direction with the same spacing as the silicon ions. This hexagonal pattern corresponds roughly to that of the hydroxyl ions on the surface of an octahedral sheet. This makes it possible for the octahedral and tetrahedral sheets to be linked by a common sharing of the oxygen and hydroxyl ions from their individual sheets. When only one surface of an octahedral sheet is shared with a tetrahedral sheet, a 1:1 layer silicate results. If both surfaces are shared, a 2:1 mineral occurs. Composite octahedral-tetrahedral sheets are always stacked in the *c* axis direction. Phyllosilicates have perfect basal cleavage that occurs between the composite layers.

Phyllosilicates may be structurally classified into dioctahedral and trioctahedral types (Fig. 2.3a,b). This classification is based on the cation valence and configuration of the octahedral site in two- or three-layer structures. There are two ways to fill an octahedral site, depending on the cation valence. A divalent cation (Ca^{2+}, Mg^{2+}) when placed into the octahedral site produces a trioctahedral arrangement; three of every three octahedral sites are occupied by a cation (Fig. 2.3a). The sheet is electrically neutral and has a unit formula of $X(OH)_2$ where X represents the divalent cation. If a trivalent cation (e.g., Al^{3+}, Fe^{3+}) is placed into an octahedral site, a dioctahedral arrangement results; two of every three octahedral sites are occupied by a cation (Fig. 2.3b). The sheet is still electrically balanced.

Dioctahedral and trioctahedral clay minerals may be compared with one another in terms of the location of charge that arises from substituting one cation for another within the oxygen network. Several common schematics representing models of the tetrahedral and octahedral layers are illustrated in Fig. 2.4. Packed sphere and ball-and-stick models will be referred to in the following discussion of phyllosilicates.

Phyllosilicates strongly influence chemical and physical properties of soils because of their small particle size, high surface area, and cation exchange properties. Among the phyllosilicate minerals important to soil development, the most common are discussed here: mica, vermiculite, chlorite, smectite, interstratified minerals, kaolinite, and talc and pyrophyllite. With the exception of talc and

Table 2.1 Classification of silicate minerals [From Schulze, 1989 with permission of the Soil Science Society of America]

Silicate Subclass	Shared Oxygen per Silicon	Mineral	Ideal Formula
Nesosilicates	0	Olivine	$(Mg,Fe)_2SiO4$
		Zircon	$ZrSiO4$
		Sphene	$CaTiO(SiO_4)$
		Andalusite	Al_2SiO_5
		Sillimanite	
		Kyanite	
		Staurolite	$Fe_2Al_9O_6(SiO_4)_4(O,OH)_2$
Sorosilicate	1	Epidote	$Ca_2(Al,Fe)Al_2O(SiO_4)(Si_2O_7)(OH)$
Cyclosilicates	2	Beryl	$Be_3Al_2(Si_6O_{18})$
		Tourmaline	$(Na,Ca)(Li,Mg,Al)(Al,Fe,Mn)_6(BO_3)_3(Si_6O_{18})(OH)_4$
Inosilicates (Single Chains)	2	Pyroxenes	
		Augite	$(Ca,Na)(Mg,Fe,Al)(Si,Al)_2O_6$
		Enstatite	$MgSiO_3$
		Hypersthene	$(Mg,Fe)SiO_3$
		Diopside	$CaMgSi_2O_6$
		Hedenbergite	$CaFeSi_2O_6$
Inosilicates (Double Chains)	2.5	Amphiboles	
		Hornblende	$(Ca,Na)_{2-3}(Mg,Fe,Al)_5Si_6(Si,Al)_2O_{22}(OH)_2$
		Tremolite	$Ca_2Mg_5Si_8O_{22}(OH)_2$
		Actinolite	$Ca_2(Mg,Fe)_5Si_8O_{22}(OH)_2$
		Cummingtonite	$(Mg,Fe)_7Si_8O_{22}(OH)_2$
		Grunerite	$Fe_7Si_8O_{22}(OH)_2$
Phyllosilicates	3	Micas	
		Muscovite	$KAl_2(AlSi_3O_{10})(OH)_2$
		Biotite	$K(Mg,Fe)_3(AlSi_3O_{10})(OH)_2$
		Phlogopite	$KMg_3(AlSi_3O_{10})(OH)_2$
		Chlorites	$(Mg,Fe)_3(Si,Al)_4O_{10}(OH)_2(Mg,Fe)_3(OH)_6$
		Other Clay Minerals	
		Talc	$Mg_3Si_4O_{10}(OH)_2$
		Pyrophyllite	$Al_2Si_4O_{10}(OH)_2$
		Kaolinite	$Al_2Si_4O_{10}(OH)_2$
		Smectite	variable
		Vermiculite	variable
		Serpentines	
		Antigorite	$Mg_3Si_2O_5(OH)_4$
		Chrysotile	$Mg_3Si_2O_5(OH)_4$
Tectosilicates	4	Feldspars	
		Orthoclase	$KAlSi_3O_8$
		Albite	$NaAlSi_3O_8$
		Anorthite	$CaAl_2Si_2O_8$
		SiO_2 Group	
		Quartz	SiO_2
		Tridymite	
		Cristobalite	
		Zeolites	
		Analcime	$NaAlSi_2O_6 \cdot H_2O$
		Clinoptilolite	$(Na_3K_3)(Al_6Si_{30}O_{72})_{24} \cdot H_2O$

Table 2.2   Elements in layer silicates [From Schulze, 1989. J.B. Dixon and S.B. Weed. Minerals in soil environments, with permission of the Soil Science Society of America]

Element	Ionic Charge	Radius (nm) Nonhydrated	Coordination Number
Li	+1	0.068	4, 6
Na	+1	0.098	8
K	+1	0.133	8 - 12
Mg	+2	0.074	6
Ca	+2	0.104	8
Al	+3	0.057	4, 6
Si	+4	0.039	4
Fe (+2)	+2	0.080	6
Fe (+3)	+3	0.067	6
O	-2	0.136	-
Rb	+1	0.147	8 - 12

pyrophyllite, these minerals are commonly formed at near surface-temperatures as products of weathering and sedimentary processes (Section F, Chapter 1). Some of these minerals also form during hydrothermal activity but their occurrence is less common in the soil profile. Weaver and Pollard (1973) and Nemecz (1981) provide chemical compositions for numerous phyllosilicates.

Collectively, these minerals are known as clay minerals. The term clay is defined for soils as a naturally occurring material composed primarily of fine-grained minerals (< 2 μm), which is generally plastic at appropriate water contents and will harden when fired (Glossary of Soil Science Terms, 1997). Clay minerals are specific fine-grained minerals, most of which occur in the clay size-fraction. The classification of these minerals is illustrated in Table 2.3, adapted from Bailey (1980a). Grim (1968) discusses another classification based solely on the morphology of clay minerals. Table 2.4 provides chemical composition and properties. Although presented here mineral by mineral, clay minerals do not usually occur independently in soils. Rather, they occur in nature in various assemblages. The identification and modeling of clay minerals in soils using single-phase and monomineralic relationships derived from reference and ideal species both confound and simplify the understanding of their interactions within the soil environment. Without these single-phase comparisons, interpretation would be nearly impossible. However, in employing these procedures, one must exercise a cautious awareness of the limitations of these comparisons to a multiphase, multimineralic soil environment.

2.2   Characteristics of Phyllosilicate Clays

2.2.1   The 1:1 Clay Minerals

2.2.1.1   Kaolinite
The kaolinite structure consists of one octahedral and one tetrahedral sheet with a combined thickness of 0.7 nm (Fig. 2.5). Kaolinite is dioctahedral and the layer is electrically neutral. Other kaolin minerals, dickite and nacrite, have the same basic 0.7-nm unit layer thickness but stacking sequences of layers along the c axis differ. A few structural properties are listed in Table 2.5. More detailed structural properties are discussed in Dixon and Weed (1989) and Zvyagin and Drits (1996).

Kaolinite in soils is both authigenic (formed in place) and allogenic (formed elsewhere). It can form from the weathering of primary and secondary minerals, subsequent to the release of Si^{4+} and Al^{3+}. In most soils, however, kaolinite is inherited from the weathering of older sediments.

Tetrahedral Sheet

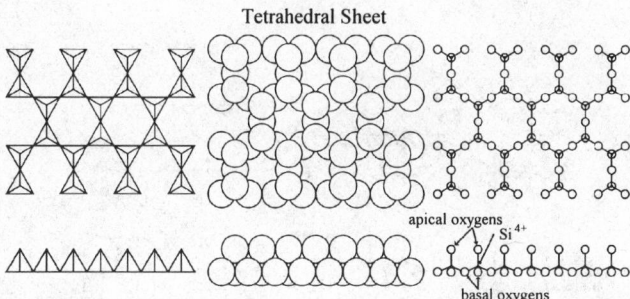

Fig. 2.2 The tetrahedral sheet of a silicate mineral [Adapted from Schulze, 1989 with permission of the Soil Science Society of America]

Geochemical weathering of parent materials in the southeastern United States is shown by an increasing kaolinite content with depth and formation of kaolinite from primary minerals in weathering rock. Kaolinite is commonly found in weathered surface soils and volcanic soils where environments of formation have been somewhat acidic. In Soil Taxonomy, kaolinite-rich soils classify as Ultisols and Oxisols, and, occasionally Andisols (Soil Survey Staff, 1998). Kaolinite is also a common clay mineral of buried soils.

Because the 1:1 layer has little permanent charge and its surface area is small, cation exchange capacities (Table 2.6) of soils containing kaolinite are usually low, perhaps as little as 5–10 cmol$_c$ kg^{-1} although commonly less. Fertility of these soils may also be low. There is little, if any, isomorphous substitution to provide permanent, structural charge. It is difficult to prove why this is so. There is some evidence for occasional substitution of Fe in octahedral positions. It is also possible that some negative charge at the Al-OH surface could develop, depending on pH. But it is very difficult to separate measured CEC in kaolinite from the effects of impurities in the sample.

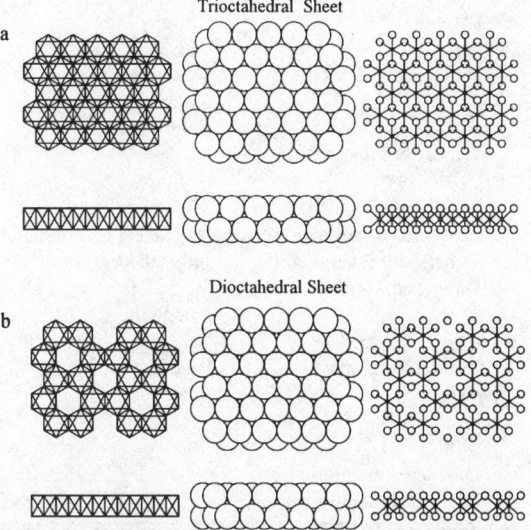

Fig. 2.3 The octahedral sheet of a silicate mineral: (a) the trioctahedral, (b) the dioctahedral structure [Adapted from Schulze, 1989 with permission of the Soil Science Society of America]

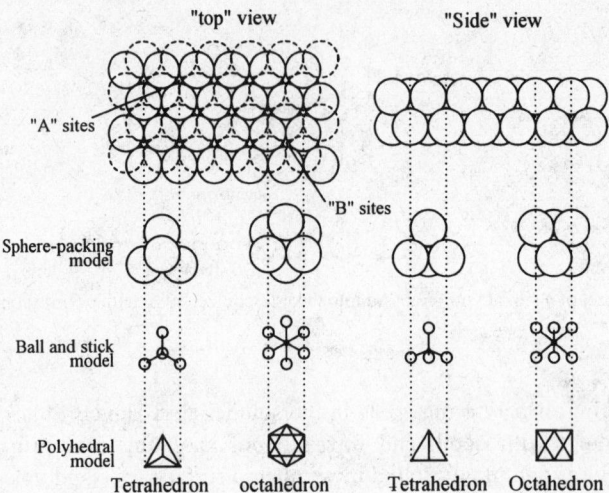

Fig. 2.4 Schematic representations of models for tetrahedral and octahedral layers [Adapted from Schulze, 1989 with permission of the Soil Science Society of America]

Table 2.3 Classification of phyllosilicates in soils

Group	Layer Type	Subgroup	Mineral Species‡	Charge/unit formula†
Kaolin	1:1	Kaolin	Kaolinite	~ 0
			Halloysite	
Pyrophyllite	2:1	Pyrophyllite	Pyrophyllite	~ 0
Talc	2:1	Talc	Talc	~ 0
Smectite	2:1	Dioctahedral smectites	Montmorillonite	0.2-0.6
			Beidellite	
			Nontronite	
		Trioctahedral smectites	Saponite	
			Hectorite	
			Sauconite	
Vermiculite	2:1	Dioctahedral vermiculite	Dioctahedral vermiculite	0.6 -0.9
		Trioctahedral vermiculite	Trioctahedral vermiculite	
Mica	2:1	Dioctahedral micas	Muscovite	~ 1.0
			Paragonite	
			Celadonite	
		Trioctahedral micas	Biotite	
			Phlogopite	
			Lepidolite	
Illite		Illite	Illite	~ 0.6 - 0.9
Chlorite	2:1	Dioctahedral chlorites		variable
		Trioctahedral chlorites		

† Brindley and Brown (1980)
‡ Example species only

Table 2.4 Chemical and structural properties of clay minerals [From Brindley and Brown, 1980. Crystal structure of clay minerals and their X-ray identification. Copyright Mineralogical Society, London, UK]

Mineral Name	Formula	Layers per Unit Cell	b parameter dimension (nm)	Layer Thickness (nm)
Kaolinite	$Al_2Si_2O_5(OH)_4$	1	0.893	0.714
Dickite	$Al_2Si_2O_5(OH)_4$	2	0.894	0.714
Nacrite	$Al_2Si_2O_5(OH)_4$	6	0.894	0.714
Halloysite	$Al_2Si_2O_5(OH)_4 \cdot 2H_2O$	O	0.890	0.96 - 1.00
Pyrophyllite	$Al_2(Si_4)O_{10}(OH)_2$	2	0.89	0.914
Muscovite	$K(Al_2)(Si_3Al)O_{10}(OH)_2$	2	0.902	0.998
Illite	$K_{0.5}(Al_{1.2}Fe_{0.2}Mg_{0.3}Fe_{0.1})(Si_{3.3}Al_{0.7})O_9(OH)_3$	---	0.905	0.993
Montmorillonite	$(Al_{1.77}Fe_{0.03}Mg_{0.2})(Si_{3.74}Al_{0.26})O_{10}(OH)_2 + M^+_{0.46}$	---	0.898	variable
Beidellite	$(Al_{1.96}Fe_{0.04})(Si_{3.46}Al_{0.54})O_{10}(OH)_2 + M^+_{0.54}$		0.893	variable
Nontronite	$(Al_{0.08}Fe_{1.84}Mg_{0.08})(Si_{3.72}Al_{0.28})O_{10}(OH)_2 + M^+_{0.36}$		0.911	variable
Talc	$Mg_3(Si_4)O_{10}(OH)_2$	2	0.910	0.926
Biotite	$(K_{0.8}Na_{0.2})(MgFe^{+2}_{1.4}Al_{1.5}Fe^{+3}_{0.2})(Si_{2.8}Al_{1.2})O_{10.4}(OH)_{1.6}$		0.924	1.005
Hectorite	$(Mg_{2.67}Li_{0.33})(Si_{3.89}Al_{0.05})O_{10}(OH)_2 + M^+_{0.5}$		0.916	variable
Vermiculite	$(Mg_{2.42}Fe_{0.25})(Si_{2.75}Al_{1.25})O_9(OH)_3$		0.920	about 1.4
Chlorite	$(Mg_{1.9}Fe_{0.4}Al_{0.7})(Si_{2.55}Al_{1.45})O_{10.5}(OH)_{7.5}$		0.921	1.406

The anion exchange capacity (AEC) of kaolinite is small but attributable to positive charges that develop at sites on the edges of crystals where O accepts a proton under acid conditions (i.e., pH < point of zero charge [PZC]). Sorption of anions such as SO_4^{2-} and HPO_4^{2-} at these sites has been studied extensively. Depending on the pH, some of the AEC may come from Al^{3+} sorbed to the kaolinite surface.

Kaolinite particles are often > 1μm in size and have a platy or book-like appearance when viewed with an electron microscope. Although not considered an expanding clay mineral, kaolinite will swell with the intercalation of some small polar molecules such as formamide, hydrazine, or urea. Recently, Singh and Mackinnon (1996) proposed a method to transform kaolinite to tubular halloysite.

Commercial Use of Kaolin Minerals

Numerous kaolin industry mines have operated in the Piedmont region of the United States for decades. Kaolin minerals are used in the paper industry for coatings and fillers, in pottery and porcelain and as fillers and binders in pharmaceuticals.

2.2.1.2 Halloysite

Halloysite is similar in structure to kaolinite except for a layer of water molecules that is intercalated within the 1:1 layer (Fig. 2.5). It differs in morphology from kaolinite in that it is often tubular (Giese, 1988; Bailey, 1989; Singh, 1996). Bates (1958) also described lath-shaped and polyhedral forms of halloysite, and Bailey (1989) described a platy halloysite. Spherical halloysite has been identified in many countries including Japan (Sudo et al., 1981), Mexico (Dixon and McKee, 1974), and the United States (McBride et al., 1976).

Halloysite is found primarily in soils of volcanic origin and typically weathers to kaolinite. If not properly handled during sampling and analysis, it is difficult to distinguish halloysite from kaolinite; once dehydrated, the layer spacing decreases to close to that of kaolinite. Although halloysite is reported to persist more consistently under moist conditions (Dixon, 1989), Chadwick et al. (1994) reported the presence of poorly crystalline halloysite in soils along a Hawaiian climatic transect in all but the wettest of sites.

Fig. 2.5 Sphere packing and ball-and-stick models for structures of gibbsite, kaolinite, halloysite·2H₂O, pyrophyllite, mica, vermiculite and smectite, chlorite, and hydroxy-interlayered vermiculite and hydroxy-interlayered smectite [Adapted from Schulze, 1989 with permission of the Soil Science Society of America]

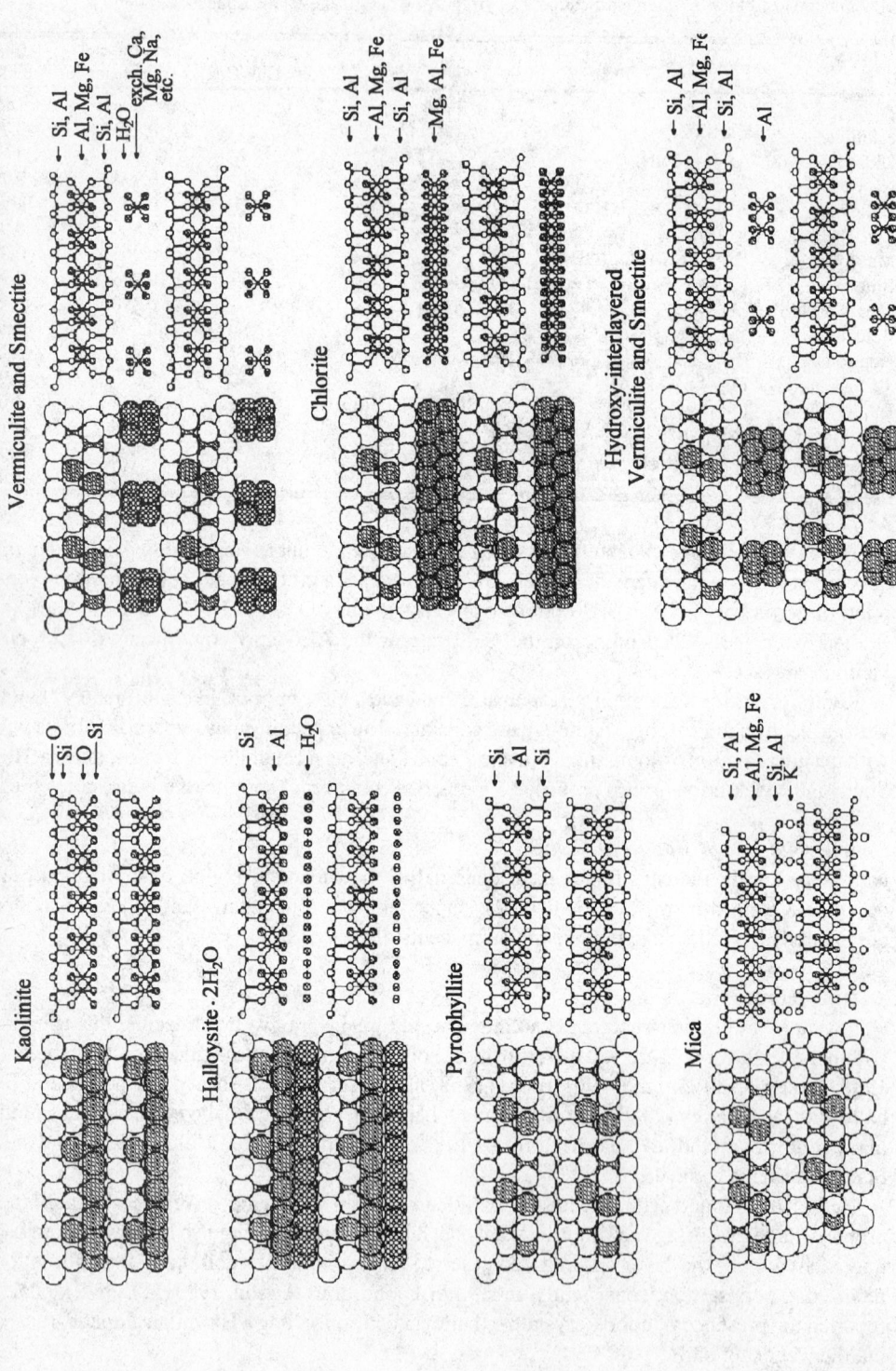

Table 2.5 Structural location of cations in soil clay minerals [Velde, 1995; Sparks, 1995]

Layer Type	Tetrahedral Sheet		Octahedral Sheet†		Dominant Interlayer
	Dominant	Other	Dominant	Other	
0.7 nm					
Kaolins					
Kaolinite	Si		Al		---
Halloysite	Si		Al		$2H_2O$
Dickite	Si		Al		---
1.0 nm					
Pyrophyllite	Si		Mg		---
Talc	Si		Al		---
Smectites					
Dioctahedral					
Montmorillonite	Si	Al	Al	Mg, Fe†	Ca, Na
Beidellite	Si	Al	Al	Fe	Na, Ca
Nontronite	Si		Fe^{3+}, Mg	Al	Ca, K
Trioctahedral					
Saponite	Si	Al	Mg	Al	Ca, Na
Hectorite	Si	Al	Mg	Li	
Illite	Si	Al	Al, Fe^{3+}	Fe^{2+}, Mg	K
Vermicultes					
Dioctahedral	Si	Al	Al	Mg, Fe^{3+}	Hydroxy-ion
Trioctahedral	Si	Al	Mg		
Micas					
Dioctahedral					
Muscovite	Si	Al	Al		K
Paragonite	Si	Al	Al		Na
Celadonite	Si	Al	Al	Fe, Mg	K,Ca
Trioctahedral					
Biotite	Si	Al	Mg, Fe, Al		K
Phlogopite	Si	Al	Mg		K
Lepidolite	Si	Al	Al	Fe,Ti,Mg, Mn, Li	K,Ca,Na

† Fe with or without charge designation indicates that either Fe^{2+} or Fe^{3+} may be present.

2.2.2 The 2:1 Clay Minerals

2.2.2.1 Pyrophyllite and Talc

Pyrophyllite and talc occur in soils rarely and only where weathering is minimal (Zelazny and White, 1989). However, their ideal structure is of great importance in the discussion of clay minerals. Of all the 2:1 phyllosilicates, these minerals are closest to the theoretical composition for both octahedral and tetrahedral layers. There is little substitution of one cation for another, and therefore they have minimal layer charge.

The pyrophyllite unit structure $[Al_2Si_4O_{10}(OH)_2]$ consists of two Si tetrahedral sheets and one Al octahedral sheet with a layer thickness of approximately 0.96 nm (Fig. 2.5). Cation composition is almost exclusively Al and Si. Talc $[(R^{2+})_3Si_4O_{10}(OH)_2]$ usually has three divalent ions (e.g., $R = Mg^{2+}$) in the octahedral site with an occasional trivalent ion in the octahedral and tetrahedral sites. Like pyrophyllite, talc has one octahedral sheet and two tetrahedral sheets for a layer thickness of 0.96 nm. Iron(II) often substitutes for Mg^{2+}.

Pyrophyllite and talc have no charge-balancing cations in the interlayer position and they have the weakest of interlayer bonds (van der Waals forces). Prominent basal cleavage results and they can be

Table 2.6   Cation exchange capacities for layer silicates

Clay Mineral	Exchange Capacity
	$(\text{cmol}_c \text{ kg}^{-1})$*
Kaolinite	2 - 8†
Halloysite (2H$_2$O)	5 - 10
Halloyite (4H$_2$O)	40 - 50
Smectite	60 - 100
Illite	10 - 40
Vermiculite	80 - 150
Chlorite	10 - 40

*Prior to 1982 CEC was reported in milliequivalents per 100 grams (meq/100g). 1 meq/100g = 1 cmol$_c$ kg^{-1}
†Data are from Grim (1953) and Miller and Donahue (1995)

pried apart easily. The weak bonding also gives the minerals a slippery feel because the 2:1 layers slide easily over one another. They are frequently used as lubricants. These minerals are also among the softest known minerals, with a Moh's scale hardness of 1 to 2. The 2:1 type structure does not swell, even in the presence of polar organic molecules.

Often the pyrophyllite and talc group is not considered clay minerals because they form in geologic environments quite different from those of most soil clay minerals. These minerals are generally found as secondary alteration products of preexisting silicate minerals, usually occurring in low temperature metamorphic or hydrothermal environments. Talc has been reported as a weathering product of ultramafic rocks, usually forming from pyroxenes (Colin et al., 1981; Nahon and Colin, 1982) or hornblende (Proust, 1982). Talc usually weathers to nontronite (Besnus et al., 1976) or, with the removal of Si and Al, to Fe oxide (Colin et al., 1981; Nahon and Colin, 1982).

2.2.2.2   Smectite

The term smectite is a generic name for 2:1 clay minerals that swell or collapse easily depending on their pressure potential, osmotic potential, and saturating cations. The term smectite does not imply a specific chemistry or structure but rather a range. In older literature, montmorillonite was a synonym for this mineral group, but today montmorillonite is considered a specific smectite mineral. A schematic structure for smectite is illustrated in Fig. 2.5. Structural properties are in Table 2.5.

Nomenclature for smectites has several combinations of structural variables. The structure was first determined by Hoffman et al. (1933). Smectites have an octahedral sheet that shares O ions between two tetrahedral sheets. Cation substitution occurs in both the octahedral and tetrahedral sheets (Table 2.5). Differences in properties and composition that occur as a consequence of these substitutions are used to classify smectites (Borchardt, 1989).

Of all the smectites, montmorillonite is probably the most common in soils. Montmorillonite, a dioctahedral mineral, has the general formula $(M)_x(Al_{2-x}R^{(2+)}{}_x)(Si_{4-x}Al_x)O_{10}(OH)_2 + H_2O$, where M is the interlayer cation and R^{2+} (e.g., Mg^{2+}) is the divalent cation that substitutes for Al^{3+}. Layer charge is usually small, about 0.2 to 0.4 mol$_c$ formula-unit^{-1}, although smectites with layer charges $\leq$ 0.6 mol$_c$ unit^{-1} are also recognized (Clay Mineral Society, Nomenclature Committee, 1984).

Beidellites, also dioctahedral smectites, have the general formula $(M)_xAl_2(Si_{4-x}Al_x)O_{10}(OH)_2 + H_2O$, where M is as above. As with montmorillonite, most of the excess negative charge of beidellites originates in the tetrahedral rather than the octahedral sheet. Although there are several varieties in this group, nontronite, in which Fe^{3+} replaces the Al in the octahedral sheet, is most similar to beidellite. Nontronite, however, is rarely found in soils except as an early weathering product of basic rocks (Sherman et al., 1962).

The dioctahedral smectites have trioctahedral analogs. The trioctahedral analog of montmorillonite has the general formula $M_x(Mg_{3-x}R^{(+)}_x)Si_4O_{10}(OH)_2 + H_2O$. The R^+ cation is Li for the trioctahedral smectite hectorite. The general formula for the trioctahedral analog of beidellite is $M_xR^{(2+)}_3(Si_{4-x}Al_x)O_{10}(OH)_2 + H_2O$. This group is called the saponites, and R^{2+} can be many different divalent cations resulting in a large number of species. In addition, Fe^{3+} substitutes for Al^{3+}. As their chemistry suggests, saponites are formed by the weathering of basic rocks such as peridotites, basalt, and diabase.

Smectites and one other group, the vermiculites, are substantially responsible for the high CEC of soils in which they are present. Foster (1955) compared CEC to swelling capacity of smectites from major smectite localities (Fig. 2.6). Although there seems to be little direct correlation with CEC, Foster (1955) showed a relation between swelling capacity and cation substitution in the octahedral sheet (Fig. 2.7).

Crystalline swelling of minerals in this group is controlled by numerous forces including forces of attraction due to Coulombic and van der Waals interactions between clay layers and forces of repulsion from hydration of interlayer cations and negative surface charge sites (Norrish, 1954; van Olphen, 1965; Kittrick, 1969; and Parker, 1986). These authors have also developed models for crystalline swelling. Laird (1996) recently described a macroscopic energy balance model to explain swelling.

Aylmore and Quirk (1971) and Tessier (1984) suggested that smectite does not exist as discrete particles in the usual sense because the number of smectite layers that associate with one another is a function of the saturating cation as well as the osmotic and pressure potentials of the clay-water system. Instead, smectites that are saturated with divalent cations, such as Ca, are organized as a network of assemblages called quasicrystals (Fig. 2.8). Quasicrystals are composed of substacks which are composed of layers. Water occurs between layers, between substacks, and between quasicrystals. In smectite systems, where monovalent cations such as Na dominate, associations of smectite layers are termed *tactoids*.

Swelling and shrinking of soils with Ca-saturated smectitic clay occur when water enters or leaves the pore spaces between quasicrystals, as distinct from crystalline swelling where the movement of

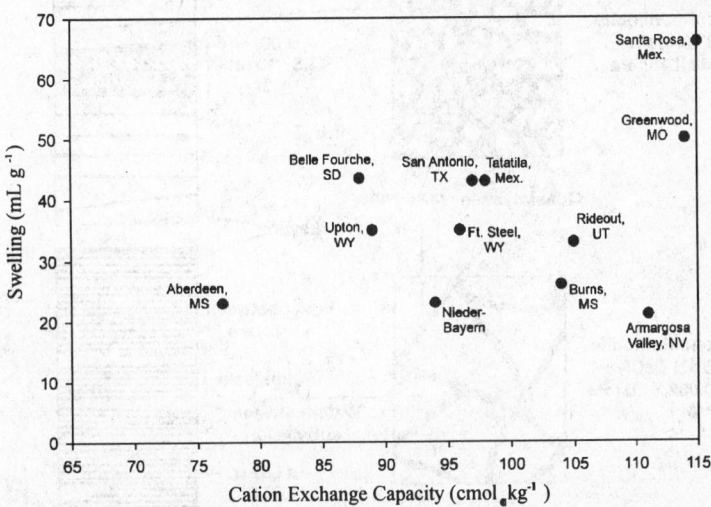

Fig. 2.6 A comparison of cation exchange capacity and swelling for montmorillonitic clay minerals [Adapted from Foster, 1955. Clays Clay Min 3:205-220 with permission]

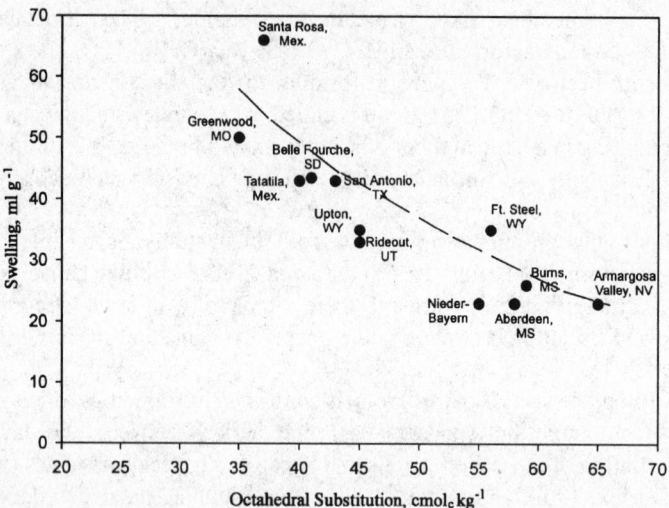

Fig. 2.7 The relation between octahedral substitution in montmorillonites and swelling capacity [Adapted from Foster, 1955. Clays Clay Min 3:205-220 with permission]

water is in and out of interlayer spaces. Significant changes in soil volume are not related to absorption or loss of water at the interlayer scale but to movement of water that occurs at the scale of quasicrystals. One can move water out of the interlayer space of a smectite by oven drying, but that is much more drastic a treatment than generally would occur naturally in soil.

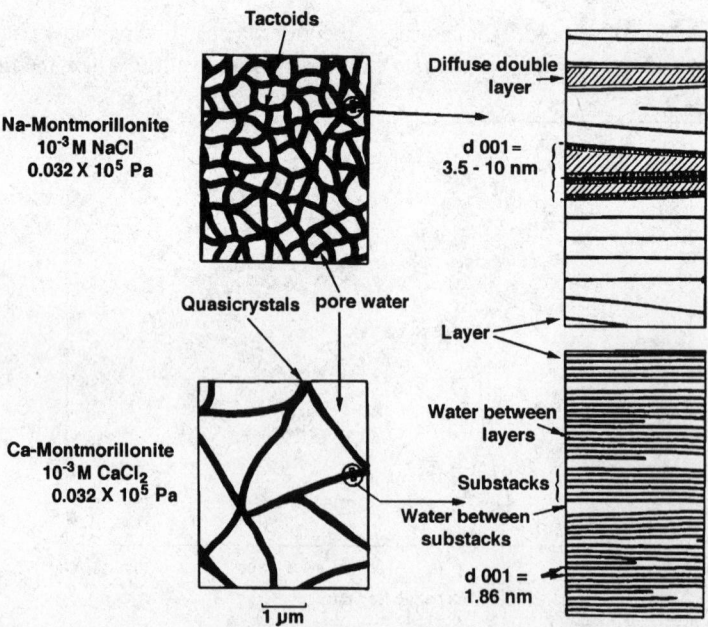

Fig. 2.8 Schematic representation of the microstructures of Ca^{2+} (quasicrystals) and Na^+ (tactoids) smectites prepared with dilute salt solutions [From Tessier, 1984]

The main requirement for soils to shrink and swell significantly is that they have many fine pores in the 1–2 μm diameter range. Pores of this size can accept or release water readily in response to the changes in the pressure potential of soil water. This means that expandable clay minerals such as smectite are not a prerequisite for Vertisol formation. Still, the expansive nature of smectites at the quasicrystal scale causes them to be highly reactive in soil. The tendency of these minerals to expand requires special consideration in geotechnical and foundation engineering.

Smectites occur in soils throughout the United States. They are the most abundant clay mineral in soils derived from material originating west of the Mississippi River (Beavers, 1957). This reflects source areas in the northern Great Plains and Rocky Mountains such as thick sequences of fine-grained clastics and significant volcanic ash deposits. The most common clay minerals in Vertisols are smectites. Nettleton and Brasher (1983) report smectite to be the dominant soil clay mineral or to dominate soil properties in over half of the soils of the desert and semi-arid portions of the United States.

Beidellites are most often reported in acidic soil environments such as Spodosols or Andisols. Ross and Mortland (1966) identified an Al beidellite in northern Michigan and Douglas (1982) reported beidellite in highly acidic soils. Chichester et al. (1969), Dudas and Harward (1975), and McDaniel et al. (1995) reported beidellites in volcanic ash soils. Nettleton et al. (1970) reported beidellites in base-rich sediments under restricted drainage. Cuadros et al. (1994) suggest that some clay minerals identified as beidellites may be mixed-layer kaolinite-montmorillonite as both of these species are essentially chemically identical.

During formation, smectites require an environment high in Si and alkaline earth cations (Borchardt, 1989). Smectite becomes unstable as soil leaching increases. In a soil-leaching environment with low organic matter and moderate acidity (pH ~ 5), smectites weather to 'pedogenic chlorite' (Borchardt, 1989), or, with sufficient Al or Fe, form hydroxy interlayers (Barnhisel and Bertsch, 1989). Refer to the appropriate sections in this chapter for further discussion of chlorites and hydroxy-interlayered clay minerals.

Smectites may alter to kaolinite during long or intense weathering, producing interstratified kaolinite-smectite (Altschuler et al., 1963; Wilson and Cradwick, 1972; Morgan et al., 1979). Sudo and Hayashi (1956) first identified a mixed-layer kaolinite-montmorillonite. More recently, Hughes and Glass (1984) and Hughes et al. (1994) have identified a mixed layer kaolinite-expandable mineral and used it to develop a soil weathering intensity index. They have identified a mixed layer kaolinite-expandable mineral in several buried soils of pre-Wisconsin age.

Commercial Use of Smectites
Smectites have wide use in industry as binding agents for molding sands in the metallurgical castings of foundries, drilling fluids, lubricants and waterproofing agents. They are also used to retard the movement of water and chemical waste and as backfill in radioactive waste repositories. Smectites are adsorbents for removal of oil and grease and for pet litter. In foods, smectites are used to decolor vegetable oil and clarify wine. They are used for water filtration and purification and as catalysts in hydrogenation reactions. Smectites are a major component in the manufacture of bricks, tiles, and other similar building products. Organophilic forms are used in inks, cosmetics and polish. Smectite is a suspending and thickening agent in paint. It is also used to control the gelling of lubricating grease and to increase the thixotropy of fiberglass resins.

2.2.2.3 Vermiculites
Smectites and vermiculites display a continuum of chemical properties. In fact, soil vermiculites have been identified with some characteristics of both smectite and vermiculite. These have been termed

high-charge smectite or low-charge vermiculite. The calculated CEC for trioctahedral vermiculite may be as great as 159 $cmol_c$ kg^{-1} (Table 2.6) and has been estimated to be as great as 200–250 $cmol_c$ kg^{-1} for dioctahedral vermiculite (Jackson, 1975).

The general formula for vermiculite is $M_{x-y}(R^{(2+)}_{3-y}, R^{(3+)}_y)(Si_{4-x}, Al_x)O_{10}(OH)_2 + H_2O$, where x represents the tetrahedral substitution and y, the octahedral substitution (Table 2.5) and M is most often Mg^{2+}. In ambient conditions, 6 water molecules, arranged in two sheets, form an octahedron around each Mg^{2+}. Vermiculites saturated with Ca^{2+} or Mg^{2+} and equilibrated at 54% relative humidity expand to a lesser degree than smectites, i.e., X-ray diffraction (XRD) d-spacings are about 1.4 nm rather than 1.7 nm. The structure for vermiculite is illustrated in Fig. 2.5.

Reference trioctahedral vermiculites show some selectivity for Mg^{2+} over other divalent cations, depending on how much Mg is already present. More generally, monovalent cations with weak hydration energies (K^+, NH_4^+, Rb^+, Cs^+) are preferred by vermiculite surfaces, and they will shut down the structure, possibly trapping some divalent cations. The selective sorption of K^+, NH_4^+, Rb^+, and Cs^+ has been called fixation, although these cations cannot be considered permanently fixed by vermiculite. Still, their release to the soil solution is certainly slowed if they are trapped in the deep interlayer region of a vermiculite particle. Fixation has been interpreted to occur near the edges of particles, and its extent depends on the pH and other ions present.

Vermiculite was one of the later clay minerals to be identified in soils. Vermiculites have been reported in all soil orders of Soil Taxonomy (Soil Survey Staff, 1998) but are most often found in soils of temperate and subtropical climates. Ross et al. (1982) showed that conditions for vermiculite formation are optimal under moderate weathering conditions. Douglas (1989) tabulated occurrences of soil vermiculites. Although generally considered an alteration product of muscovite, biotite or chlorite, Smith (1962) showed that vermiculite replaced feldspar and Barshad and Kishk (1969) reported soil dioctahedral vermiculites as synthesis products from gel precipitates. Alteration of chlorite to vermiculite to smectite has been observed in Spodosols.

Trioctahedral vermiculite forms in soils where chlorite or biotite mica is present. The structure of trioctahedral vermiculite is well understood from the study of macroscopic crystals (Slade et al., 1985; Shirozu and Bailey, 1966); however, that of dioctahedral vermiculite is not (Douglas, 1989). Dioctahedral vermiculite is more common than trioctahedral vermiculite under severe weathering conditions. However, trioctahedral vermiculite may persist under acid conditions when it contains a hydroxy-Al interlayer (Section 2.2.3.2) (Douglas, 1989). Kirkland and Hajek (1972) described a dioctahedral vermiculite found in soils of the southeastern United States. Rich and Obenshain (1955) reported dioctahedral vermiculite in the clay fraction of muscovite-rich soils and assumed that the dioctahedral vermiculite formed from K^+ release during weathering of muscovite. Little is known of its detailed structure.

2.2.2.4 Micas
Micas occur in almost any geologic environment and are abundant in many rocks including shales, slates, phyllites, schists, gneiss, granites, and in sediments derived from these rocks. As clay minerals, they can be derived from preexisting micas by mechanical fragmentation but may also form *in situ*. Micas are 2:1 phyllosilicates having a charge imbalance that is satisfied by a tightly held, nonhydrated, interlayer cation (Fig. 2.5). Bailey (1984) presents a comprehensive review of micas.

Micas have an octahedral sheet sandwiched between two tetrahedral sheets. One of the 4 tetrahedral sites typically contains Al^{3+} rather than Si^{4+}. This results in excess negative charge per formula unit. The negative charge is balanced by monovalent cations, usually K^+, occupying interlayer sites between two tetrahedral-octahedral-tetrahedral layers. The general formula is $M_x(R^{(2+)}_{3-y}, R^{(3+)}_y)(Si_{4-x}, Al_x)O_{10}(OH)_2$ where M is usually K^+ or Na^+. Micas can be either dioctahedral or trioctahedral. The mica group consists of many species because Fe^{2+} and Fe^{3+} substitute for Mg^{2+} and Al^{3+} in octahedral sheets and Na^+ and Ca^{2+} may substitute for K^+. Occasionally, Ba^{2+}, Rb^{2+}, Cs, Sr^+, or NH_4^+ also substitute for K^+ (Nettleton et al., 1970; Speer, 1984), particularly in biotite mica.

Biotite is the most common trioctahedral mica. All cation sites in the octahedral sheet are filled with divalent cations. The total positive charge of K cations in the interlayer space balances the negative charge of the layers. The abundance of Fe^{2+} atoms in the trioctahedral sheet of biotite reduces the stability of biotite in oxidizing environments.

Muscovite is the most abundant dioctahedral mica. Two-thirds of the octahedral cation sites are filled with Al ions. Muscovite usually has only small amounts of structural Fe; ideal muscovite has none. This makes muscovite less sensitive than biotite to oxidation-reduction reactions.

Micas in soil are primarily inherited from parent material. The most common mica groups in rocks and soils are muscovite, biotite, and phlogopite. As previously mentioned, dioctahedral micas such as muscovite are more resistant to weathering than trioctahedral micas such as biotite or phlogopite. Therefore, muscovite is more likely to occur in weathered soils than the trioctahedral micas. The word mica comes from the Latin, *micare,* which means to shine. Soils with abundant mica sparkle when light reflects off the flat, basal cleavage surfaces of mica flakes. Biotite mica is also often present as a component that has partially transformed to other 2:1 minerals. This results in interstratified minerals with poor crystallinity. It is also present in expandable minerals with mica cores (Jackson, 1964).

Mica transforms to expanding 2:1 clay minerals such as vermiculite and smectite when hydrated exchangeable cations replace the primary interlayer cation, e.g., K^+. Upon weathering, the released K^+ is important to plant growth. Exchangeable K^+ not used by plants is retained at cation exchange sites on humus or other clay minerals or becomes entrapped between layers of weathered mica. In the clay fraction of soils, micas with poor crystallinity, lower K^+ content, and higher water contents than macroscopic muscovite are usually called illite.

2.2.2.5 Illite

Illite was a term originally proposed in the 1930s as a group name for clay-size, micaceous minerals (Grim et al., 1937). Today, illite is accepted as a species name or end member of a mineral compositional series (Moore and Reynolds, 1997). In addition, since X-ray methods do not detect interstratification in amounts < 5%, illite is defined as a species containing less than 5% interstratified component (Brindley and Brown, 1980).

Although illite is usually dioctahedral, it differs from muscovite in that the octahedral sheet of illite contains more Fe^{2+} and Mg^{2+} in place of Al^{3+} than muscovite (Table 2.5). The overall layer-charge density of illite is less than that of either biotite or muscovite (Table 2.3). In the tetrahedral sheet of illite, Al^{3+} substitutes less for Si^{4+}. The K^+ content of illite is also lower due to the lower negative layer charge that must be countered. Some illites have NH_4^+ rather than K^+ in the interlayer space (Moore and Reynolds, 1997). The NH_4^+ found in rocks with high organic matter contents (e.g., shales), the assumed source of the N.

Illite is found in the clay fraction of soils and sediments. It is the most abundant clay mineral in sedimentary rock. Illite forms and is stable at the Earth's surface. Shales have both diagenetic and detritally recycled illitic components. Illite also forms during surficial weathering, as well as in hydrothermal and in metamorphic environments.

Layer charge densities reported for illite are $0.6 - 0.8$ mol$_c$ formula-unit^{-1}. In the absence of smectite, Hower and Mowatt (1966) defined illite as having a fixed K^+ content of 0.75 mol$_c$ formula-unit^{-1}. More recently, researchers have determined that this is more likely an average value, with a range from 0.5 to 1.0 (Środoń et al., 1986). This has led a number of researchers to believe that there may be two types of illite (Moore and Reynolds, 1997).

The small particle size of illite gives X-ray powder diffraction patterns a less crystalline appearance when compared to muscovite and biotite. Broadening of XRD peaks representing crystalline disorder is also influenced by irregular stacking, other impurities and imperfect orientation.

2.2.3   The 2:1:1 Clay Minerals

2.2.3.1   Chlorites

Chlorite, like mica, has a 2:1 layer structure with an excess negative charge (Fig. 2.5). Unlike mica, the excess charge is balanced by a positively charged, interlayer hydroxide sheet. Thus, chlorites are described as 2:1:1 minerals where the last 1 refers to the OH sheet which can be dioctahedral or trioctahedral. The general formula is $(R^{2+}, R^{3+})_3{}^{VI}(Si_{4-x}, Al_x)^{IV}O_{10}(OH)_2 \cdot [(R^{2+}, R^{3+})_3{}^{VI}(OH)_6]$ where R is a cation and IV and VI indicate four- and six-fold coordination, respectively. Most chlorites have a trioctahedral 2:1 structure with the most common octahedral cations being Mg^{2+}, Fe^{2+}, and Fe^{3+}. The 2:1 portion is 1.0 nm thick.

The hydroxide sheet is usually composed of $Al(OH)_3$ or $Mg(OH)_2$. Other cations, including Fe^{2+}, Fe^{3+}, Mn, Cr, Cu, V, Li, and Ni, have also been reported in this sheet (Barnhisel and Bertsch, 1989). The hydroxide sheet does not have a plane of atoms shared with an adjacent tetrahedral sheet. The total thickness of the 2:1 layer plus the interlayer hydroxide sheet is ~ 1.4 nm. Minerals of the chlorite group and their nomenclature have been reviewed by Bailey (1975, 1980b, 1988).

The term chlorite has been used for two distinct mineral groups that share a common structure: true chlorite and pedogenic chlorite. True chlorites have a complete interlayer hydroxide sheet and can occur as macroscopic crystals. True chlorite minerals in soils are usually inherited from rocks of low- and medium-grade metamorphic (≤ 400 °C and a few hundred MPa pressure) or igneous origin (Deer et al., 1992). These chlorites also occur as hydrothermal alteration products of pyroxenes, amphiboles or biotite. Chlorites of detrital and authigenic origin are found in sedimentary rocks as well (Deer et al., 1992).

Moore and Reynolds (1997) list five environments for the occurrence of chlorite in low-temperature environments: (1) formation in soil; (2) development in shale under high grade diagenesis; (3) crystallization on the surface of sand grains in porous sandstones; (4) replacement of carbonate grains and matrix in carbonate rocks; and (5) alteration products of odinite in warm shallow, marine waters. Chlorite is unstable relative to other clay minerals in soils, particularly in acidic environments. As a result, quantities of chlorite are small in soils.

Chlorite weathering produces a number of phases. Chlorite has been reported to weather to vermiculite and smectite. Bailey (1988) reports that chlorite may be part of a diagenetic sequence preceding a smectite to illite sequence. Weathering involving the removal of portions of the hydroxide sheet has been reported to result in hydroxy-interlayered minerals or interstratified mineral sequences (Barnhisel and Bertsch, 1989).

Pedogenic chlorite forms by weathering in the soil environment. The interlayer hydroxide sheet is generally not as complete as true chlorite and the mineral itself generally occurs as clay-sized particles. Barnhisel and Bertsch (1989) referred to pedogenic chlorites or Al-chlorites as minerals that maintain their 1.4 nm integrity following heating to 500 – 600 °C. Recent work in the Olympic National Forest of Washington State (Wilson et al., 1996) suggests that pedogenic chlorite can form in the soil profile under fairly extreme leaching conditions.

2.2.3.2   Hydroxy-Interlayered Clay Minerals

Hydroxy-interlayered 2:1 clay minerals contain partially complete interlayered-hydroxide sheets (Fig. 2.5). Rich and Obenshain (1955) first demonstrated that the composition of the interlayer was hydroxy-Al polymers. These 2:1 clay minerals are among the most unusual of the commonly occurring soil minerals because of the great range in composition between end members. Hydroxy-interlayered 2:1 clay minerals such as hydroxy-interlayered vermiculite and hydroxy-interlayered smectites form solid-solution series with pure end members smectite or vermiculite at one end and pedogenic or aluminous chlorite at the other (Barnhisel and Bertsch, 1989). They are also easily transformed. Their composition is entirely dependent on the basic 2:1 clay mineral plus the type and amount of interlayer material. Since there is great variability in the composition of hydroxy-interlayered clay minerals, a general chemical formula has little meaning. However, compositions of minerals from specific geographic locations may be calculated by averaging the analyses from several soil clay samples, as demonstrated by Karathanasis et al. (1983).

Because of the wide range of mineral phases and the possible difficulties in identifying hydroxy-interlayered minerals, a great number of terms have been used for these minerals. Barnhisel and Bertsch (1989) list as many as 15 different names from the literature that describe these minerals.

Hydroxy-interlayered vermiculite is a vermiculite with Al-hydroxy interlayers. The interlayer hydroxide sheet, described in the preceding section on pedogenic chlorite, is incomplete in these minerals as well. Moore and Reynolds (1997) consider this mineral to be a chlorite because its 1.4 nm thickness is unaffected by polar solvents and mild heat treatments. They do cite a difference from the chlorites in that the Al-hydroxy interlayer of hydroxy-interlayered vermiculite can be removed with sodium citrate extraction. The remaining material behaves like vermiculite in response to heat and solvation treatments. Hydroxy-interlayered smectite and hydroxy-interlayered vermiculite are believed to occur in soils as weathering products derived from the deposition of hydroxy-Al polymers within the interlayer spaces of expandable layer silicates and to a lesser extent from chlorite weathering. They have a wide geographic distribution and can be found in several soil orders. Hydroxy-interlayered vermiculites and hydroxy-interlayered smectite are most abundant in Ultisols, Alfisols, and Spodosols (Soil Survey Staff, 1998).

Rich (1968) provided a list of hydroxy-interlayered smectite and hydroxy-interlayered vermiculite occurrences in soils and sediments around the world. He also presented the primary criteria for their environment of formation: viz., moderately acid conditions, low organic matter content, oxidizing conditions, and frequent wetting and drying cycles.

Under intense weathering environments such as acid soil conditions, vermiculites almost always contain a hydroxy-Al interlayer (Rich, 1958). Hydroxy-interlayered vermiculite and hydroxy-interlayered smectite contents are most abundant in the upper solum of acid soils. In these cases, the vermiculites may be trioctahedral (Kato, 1965) or dioctahedral (Rich and Obenshain, 1955). In soils developed from noncarbonatic, mica-bearing parent material, hydroxy-Al interlayer dioctahedral vermiculite is often the most abundant clay mineral in the A and B horizons.

Acidic soil conditions provide a readily available source for Al^{+3} from which to build Al interlayers (Olson, 1988). In acid-rain studies, the role of vermiculite has been explored as a sink for Al ions (Olson, 1988), since hydroxy-interlayered vermiculite is a common constituent of Spodosols in the eastern and northeastern United States.

A mechanism for the transformation of 2:1 to 1:1 clay minerals has long been a question for debate among clay mineralogists. Barnhisel and Bertsch (1989) report several cases which suggest that hydroxy-interlayered vermiculite and hydroxy-interlayered smectite are important precursors to the structural reorganization that must occur in the transition to kaolinite. For example, according to Karathanasis and Hajek (1983), a portion of the Si tetrahedral sheet is released under acidic conditions and associates with a hydroxyl-Al interlayer to form kaolinite-like layers. It is not clear how widespread this mechanism for kaolinite formation is and whether this process occurs in different soils.

Carlisle and Zelazny (1973) and Zelazny et al. (1975) suggest that both kaolinite and vermiculite with hydroxy-Al interlayers form in the profile, but the amount of vermiculite containing hydroxy-Al interlayers is greater than the amount of kaolinite in near-surface horizons, where wetting and drying occur, because it is more resistant to further weathering than kaolinite. It may also be that conditions for kaolinite formation are more favorable in lower parts of the profile (Rich and Obenshain, 1955).

The hydroxy-Al components affect the physico-chemical properties of soil minerals. The degree of interlayer filling and the relative stability of the interlayer components also affect the characteristics and range of physico-chemical properties of hydroxy-interlayered minerals. Hydroxy-Al polymers are effective in stabilizing montmorillonite and preventing or significantly reducing dispersion by Na^+. This, in turn, stabilizes soil structure, a well-known phenomenon that is also attributed to the presence of Al and Fe oxyhydroxides in soils. Tensile strength, liquid limit, shear stress, and shrink/swell properties are all affected by hydroxy-interlayering (Barnhisel and Bertsch, 1989). Zelazny et al. (1980) reported septic tank failures related to transformation of hydroxy-interlayered vermiculite and hydroxy-interlayered smectite to vermiculite/smectite in the drainfields. Soil permeability in the drainfield decreased with the release of hydroxy-Al interlayers.

Hydroxy-interlayered vermiculite and hydroxy-interlayered smectite have a wide range of CEC values. Positively charged hydroxy-Al polymers satisfy charge imbalance but the polymers are not exchangeable; thus CEC is lowered in proportion to how completely the interlayer space is filled (Barnhisel and Bertsch, 1989).

Cation exchange is significantly reduced by the presence of hydroxy-Al interlayers. They can inhibit fixation by preventing layer collapse around an ion such as K^+. Ion selectivity increases as ion fixation decreases (Barnhisel and Bertsch, 1989). Decreasing pH results in increasing effective CEC when some Al-interlayer components dissolve releasing Al^{3+} (Bertsch and Thomas, 1983). Increasing pH can also increase effective CEC due to removal of some portion of the interlayer material as secondary precipitation reactions take place (Barnhisel and Bertsch, 1989).

Hydroxy-interlayers in 2:1 clay minerals enhance the adsorption of anions. Soil scientists have long known that soils with large quantities of Al^{3+} and Fe^{3+} oxides, such as Oxisols, retain large quantities of anions such as phosphate (Barnhisel and Bertsch, 1989). Other factors may be as important to the fixation of phosphate in such soils, but hydroxy interlayers can contribute significantly to anion adsorption.

2.2.4 Interstratified Clay Minerals

Clay minerals are defined as mixed-layer, or interstratified when more than one kind of silicate layer is stacked in the z axis direction perpendicular to the $d(001)$ XRD reflection (i.e., normal to the basal cleavage). As discussed earlier, surfaces of the tetrahedral and octahedral silicate layers are hexagonal networks of either O ions or OH ions. Since the structural geometries are similar for all clay minerals, a simple aggregation of layers of similar structure but differing chemical composition is easily achieved. This, then, accounts for the fact that interstratified minerals are a very common occurrence in soils and sediments. Weaver (1956) reported that 70% of over 6,000 sedimentary rocks that he examined contained mixed-layer clay minerals.

The separation distance between layers depends on the thickness of the sheet, the chemical composition during growth, and in expandable clay minerals such as smectite or vermiculite, the exchangeable cations and the activity of water. Carroll (1970) suggests that mixed-layering occurs in two ways: (1) degradation due to weathering of rocks, sediments, and soils, or (2) diagenesis. Similarly, Sawhney (1989) states that naturally occurring mixed-layer minerals form by (1) hydrothermal alteration, (2) weathering involving partial removal of interlayer K^+ from mica or removal of a hydroxide-interlayer from chlorite, or (3) uptake of K^+, and (4) formation of a brucite-like or gibbsite-like interlayer in expanding layer silicates. MacEwan et al. (1961) explained mixed-layering as a special form of intergrowths.

In the past, even the term chlorite has been used for minerals that are presently classified as interstratified mineral sequences. Some of these minerals with crystalline swelling capacity have been called swelling chlorites (Barnhisel and Bertsch, 1989). For example, Sawhney (1989) describes swelling chlorite as a naturally occurring mineral with a 1.7 nm spacing after glycerol treatment and a 1.4 nm spacing after heating to 500 or 600 °C. Reynolds (1980) and Table 2.7 list examples of naturally occurring interstratified clay minerals. Questions about their influence on nutrient (e.g., K^+ and NH_4^+) fixation and release as well as their origin (transformation or neoformation) are of considerable interest in soil clay mineralogy.

2.2.4.1 Stacking Arrangements for Interstratified Clay Minerals

Regular or Nonrandom Interstratification
For this condition, the occurrence of a given layer depends on the nearest neighbor with a discernible pattern or sequence of layer types also known as *ordering*. The most common variety of ordering with a fixed ratio involves a regular alternation of 2-layer types to produce a repeating sequence ABABAB.... Some regular interstratifications have specific mineral names that have been accepted by the AIPEA Nomenclature Committee (1972) and are listed in Table 2.7. Hydrobiotite, which forms

Table 2.7   Interstratified clay minerals [Reynolds, 1980; Moore and Reynolds, 1997]

Mineral	Layer Types
Regular or regular alternation of layer types - two components	
Aliettite	Talc-trioctahedral smectite
Corrensite	Low-charge trioctahedral smectite-trioctahedral chlorite
	High-charge trioctahedral vermiculite-trioctahedral chlorite
	Trioctahedral chlorite-dioctaheral smectite
Hydrobiotite	Biotite - Vermiculite
Kulkeite	Talc-chlorite
Rectorite	Dioctahedral mica - dioctahedral smectite
Tosudite	Dioctahedral chlorite-smectite
--	Illite-smectite
--	Glauconite-smectite
--	Serpentine-chlorite
--	Kaolinite-smectite
Random or nearly random interstratification - two components	
	Illite-smectite
	Glauconite-smectite
	Mica-vermiculite
	Mica-dioctahedral chlorite
	Smectite-chlorite
	Chlorite-vermiculite
	Kaolinite-smectite
	Mica-chlorite
Three-component systems	
	Illite-chlorite-smectite
	Illite-smectite-vermiculite

from the weathering of biotite, is a regularly interstratified mineral with alternating layers of biotite and vermiculite. Rectorite is composed of illite and smectite in equal proportions and in alternating layers. A combination of illite and smectite layers (nonequal proportions) is called interstratified illite-smectite (I/S). Interstratified minerals with alternating layers of low-charge trioctahedral smectite and chlorite or high-charge trioctahedral vermiculite and chlorite are called corrensite (Table 2.7).

Random Interstratification

In this arrangement, the probability of a given layer occurring at a particular location in the crystal does not depend on the type or pattern of types that precede it in the sequence. Randomly interstratified species are labeled according to the types of layers present (Table 2.7). The most abundant layer type is listed first. For example, an interstratified mix of chlorite and smectite with dominant chlorite layers is referred to as a chlorite-smectite interstratified clay mineral. For partially ordered structures, there is no commonly accepted protocol for nomenclature. Since layers in randomly interstratified mixtures do not have a periodicity, identification becomes a challenge.

Reynolds (1980) and Sawhney (1989) discuss various statistical and theoretical techniques for the identification of these interstratified minerals. Illite/smectite and chlorite/smectite are the first and second most abundant mixed-layer clay minerals in soils and sedimentary rocks (Moore and Reynolds, 1997). Mixed-layer trioctahedral mica/vermiculite is uncommon in sedimentary rocks but is often present in hydrothermal or soil clays (Moore and Reynolds, 1997). Random interstratification is far more common in soils than regular interstratification. Illite-smectite interstratification has been examined in detail because of its importance to the petroleum industry. In the United States, Gulf Coast sediments have abundant sedimentary sequences containing these minerals. Moore and Reynolds (1997) describe illite-smectite diagenesis in some detail.

Recently, kaolinite-smectite has gained interest as a potential index for soil weathering intensity. This interstratified mineral is characterized by a shoulder or asymmetry on the high angle side of the 1.0 nm X-ray peak of the heated sample and a diagnostic peak near 0.8 nm after saturation with K^+ and heating to 300 °C (Schultz et al., 1971). Hughes and Glass (1990,1984), Hughes and Warren (1989) and Hughes et al. (1994) discuss the identification and XRD intensities in paleosols from different climatic regimes. Hughes and Glass (1984) suggest that the weathering of 2:1 clay minerals to kaolinite-smectite can occur in two ways: (1) by nutrient removal by plants, or (2) leaching by percolating acidic water.

Nadeau et al. (1984) and Nadeau (1985) proposed that random or irregularly stratified materials could be mixtures of two or more separate minerals rather than one. Many clay minerals identified by XRD as randomly interstratified minerals might be considered physical mixtures of individual particles of mineral layers. When first dispersed in suspension and then allowed to settle, aggregates of individual particles form sequences that exhibit interparticle diffraction and are perceived by XRD to be interstratified layers. This implies that the appearance of interstratification of some clay minerals on XRD patterns may be an artifact of sample preparation (Nadeau et al., 1984; Nadeau and Bain, 1986). Total chemical analysis interpretations (Laird et al., 1991) also support this premise of separate mineral phases. However, since weathering, clay content, and the percentage of interstratified clay minerals increase with proximity to the soil surface, more study is needed to support a hypothesis that certain interstratified minerals do not form in the weathering environment.

2.3   Instrumental Techniques for Characterization of Phyllosilicates in Soil

There are many techniques available for mineral analysis and identification. Three of the most commonly used methods of analysis are X-ray diffraction (XRD), thermal analysis, and infrared spectroscopy. Interpretation of the results of each technique is illustrated for a subsurface horizon of a Seaton soil, a fine silty, mixed, superactive, mesic, Typic Hapludalf developed in the Late Wisconsinan loess of northeastern Iowa. A review of common procedures to prepare phyllosilicates in soil materials for analyses is presented before the discussion of analytical techniques.

2.3.1 Sample Preparation for Study of Phyllosilicates in Soils

There are many ways to concentrate phyllosilicates in soils in order to facilitate their characterization. In this section, the rationales for four of the most commonly employed pretreatment techniques: carbonate removal, organic matter removal, Fe oxide removal, and size fractionation will be discussed.

Perhaps the single most important step to simplify the characterization of minerals in a soil sample is to separate the sample into discrete size fractions. The removal of organic matter and accessory minerals such as $CaCO_3$ and Fe oxides from the soil both concentrates the silicate minerals and improves the efficiency of subsequent dispersion and fractionation steps. Readers are referred to Jackson (1975), Kunze and Dixon (1986), and Whittig and Allardice (1986) for specific details of the treatment protocols.

2.3.1.1 Carbonate Removal

Many soils contain calcite and other soluble salts which may cement particles together and prevent their complete dispersion. Treatment of an alkaline soil sample with an acid will normally dissolve the calcite and soluble salts, but to minimize unnecessary decomposition of silicate minerals in the process, weak acids are preferred. A solution of Na acetate is most commonly used. The Na cations help to disperse the layer silicates by exchanging for divalent cations that would otherwise bridge two negatively charged layers. The acetate anion buffers the suspension at about pH 5, a pH sufficiently low to dissolve calcite, but not low enough to significantly dissolve silicates.

2.3.1.2 Organic Matter Removal

Samples drawn from soil horizons with abundant organic matter (e.g., > 1% organic C) should be treated to remove the organic compounds. Significant amounts of organic matter may also hold soil particles together and prevent dispersion. In addition, organic matter absorbs X-rays and prevents good orientation of fractionated clay on the sample support used for XRD analysis. Thus, XRD peaks of organic matter-laden samples are not as sharp or as intense as they would otherwise be. Soil organic matter also confounds the interpretation of thermal analyses of soil materials by releasing water over a large temperature range. Finally, organic matter contains functional groups that absorb infrared radiation and may obscure absorbance bands of the minerals of interest.

The most common method to remove organic matter is by oxidation with H_2O_2. The reaction may be represented by the oxidation of a generic organic compound, CH_2O:

$$CH_2O + H_2O_2 \Longleftrightarrow CO_2 + H_2O + 2H^+ \qquad [2.1]$$

This is a strong oxidation reaction that must be conducted in an acidic medium. It is best conducted in an acetate buffer because the pH of the suspension does not drop much below 5 and accidental dissolution of other minerals is minimized. Even after treating the unfractionated soil with H_2O_2, additional treatments of separated clay fractions may be warranted. Organic matter removal with H_2O_2 dissolves some Fe oxides and all Mn oxides, too. For example,

$$MnO_2 + H_2O_2 + 2HOAc \Longleftrightarrow Mn(OAc)_2 + 2H_2O + O_2 \qquad [2.2]$$

Three types of problems may occur during H_2O_2 treatment of soil materials. First, in Ca-rich soils, dawsonite (calcium oxalate) may form as a byproduct of the reaction. Therefore, even if carbonates are not present in the sample, the Na acetate pretreatment is helpful to remove Ca that might precipitate during peroxidation. Second, Fe oxides that are dissolved during the treatment have been shown to reprecipitate and cement together particles in the sample (Sequi and Aringhieri, 1977). Third, the high redox potential and strongly acid conditions may artificially weather some layer silicates, oxidizing structural Fe(II) and lowering the layering charge (Douglas and Feissinger, 1971).

Alternative oxidants include hypobromite or 5% hypochlorite adjusted to pH 9.5. For some soil materials, these oxidants may compare favorably with H_2O_2 for removal of organic matter (Gee and Bauder, 1986), but they may not be as effective as H_2O_2 in soils where clay-organic matter complexes are stabilized by Ca. Gentle heating of the suspension may improve the oxidation reaction with these reagents. Successful organic matter removal also depends on the nature of the organic matter itself. Well-humified organic matter is more resistant to oxidation than are undecomposed plant materials. Typically, chemical oxidation procedures are applied only to A horizon materials (or those with large amounts of organic matter) in order to minimize chemical pretreatments that are not necessary for other horizons.

2.3.1.3   Removal of Free Iron Oxides

In most soils, "free" Fe oxides such as goethite, lepidocrocite, ferrihydrite, or hematite occur in the clay fraction or as coatings on silt or sand grains. If they are present in significant quantities, Fe oxides may bind particles and prevent complete dispersion of silicate particles. Iron oxides may be removed from soil samples with the citrate-bicarbonate-dithionite (CBD) extraction of Mehra and Jackson (1960). This procedure may be performed before or after fractionation of clay.

The principles of the CBD extraction are as follows: ferric ions in the Fe oxide minerals are reduced to Fe^{2+} ions at the low redox potential induced by dithionite (also known as Na hydrosulfite). In neutral or basic systems, the redox potential is very low:

$$4OH^- + S_2O_4^{2-} \Longrightarrow 2SO_3^{2-} + 2H_2O + 2e^- \qquad E(basic) = -1.12V \qquad [2.3]$$

The presence of Fe(II) in the Fe(III) oxide crystal destabilizes its structure, releasing Fe atoms to the extraction solution, where they are complexed by citrate anions to prevent reprecipitation or precipitation as FeS. Sodium bicarbonate is used to buffer the pH of the extract to prevent concomitant acid attack of silicate minerals. Flooding the system with Na^+ ions also assists in maintaining the soil in a dispersed state. Additionally, the temperature must be kept $< 80\,^{\circ}C$ to prevent precipitation of FeS.

2.3.1.4   Size Fractionation

Separating soil materials into discrete size fractions tends to group minerals that are similar to one another. Although phyllosilicates often do occur in the silt and sand fractions of a soil, they are most abundant in soil clay fractions ($< 2\,\mu m$ equivalent spherical diameter); thus clay fractions are the main focus of the present discussion. Soil particles > 2 mm in diameter are removed from soil samples by dry sieving before further fractionation.

Dispersion techniques start with shaking and/or sonification of the sample. Shaking overnight in a chemical dispersing agent with reciprocating shaker is the slow, tried and true method. Sodium carbonate (at pH 9.5) or dilute NaCl is used as the dispersant for soil materials not dominated by sesquioxides. Sonification may be employed with or without a chemical dispersant, although the completeness of dispersion will depend on the amount of organic matter, Fe oxides, or calcite present in the sample.

After the sample is dispersed, particles in the size range of 2 mm to 50 μm diameter are usually fractionated from the bulk dispersed sample by wet sieving. It should be noted that 50 μm is the lower size limit for sand-sized particles in soils classification. The Wentworth classification, often used by geologists, sets the lower size limit at 62 μm. Dispersion of soil materials with large amounts of Fe or Al oxides (soil material that is usually acidic) is most effectively accomplished by increasing the pH, e.g., with slow additions of LiOH (Jones and Malik, 1994).

Stoke's Law provides the theoretical basis for size fractionation of silt and clay particles (Jackson, 1975). It assumes completely dispersed particles, spherical particle shape, and constant specific gravity for particles, temperature, electrolyte level, and fluid viscosity. Some of these assumptions are understood to be incorrect for size fractionation of minerals, e.g., the assumption of spherical shapes and the assumption of constant specific gravity. Particles are allowed to fall through liquid under the acceleration of gravity or by centrifugation.

Sodium is usually the ion of choice for maintaining dispersed clay particles. The large monovalent cation sorbs to clay surfaces loosely and prevents close approach of other particles, thus limiting flocculation. Of course, to maintain a dispersed state the Na ion concentration cannot be too great, for at very high ionic strengths, even Na will coagulate clay particles.

For subsequent analyses and for storage of separated clay fractions, the Na ions at cation exchange sites of dispersed clay are normally replaced by washing the sample with Mg or Ca salts, followed by additional water or acetone washes to remove excess salt. Magnesium or K saturation is preferred for samples that are to be analyzed by XRD.

On the other hand, Ca saturation is chosen for samples to be analyzed in a total chemical assay [for example, by HF dissolution (Bernas, 1968) or by suspension nebulization in inductively coupled plasma (ICP) spectroscopy (Laird et al., 1991)]. Calcium saturation is preferred because it may be used as a direct estimate of CEC. Magnesium might occur in the structure of clays, so total Mg would not estimate CEC as effectively as Ca.

For samples to be used in total chemical analyses, it is especially important that all excess salt ions be removed from the sample. Multiple washes with water, 80% methanol, and finally acetone will remove excess dissolved salts. The completeness of salt removal may be assessed with the $AgNO_3$ test for Cl^-, although determination of the electrical conductivity of the supernatant is probably more effective (Mekonnen et al., 1993).

2.3.1.5 Freeze Drying

Freeze-dried samples are convenient to store, transfer, and weigh for subsequent analyses. In a lyophilizer, the pressure and the temperature of a clay suspension are lowered at the same time so that H_2O sublimes. Before the suspension is placed in the lyophilizer, it is quickly frozen by immersing its container in a dry-ice-in-acetone or -ethanol bath. Quick freezing keeps the size of ice crystals to a minimum. Because ice can recrystallize even when its temperature is < 0 °C, ice crystals might grow and damage some mineral structures. Still, the advantages of freeze-drying samples often outweigh the disadvantages. Other methods of water removal include oven-drying (which produces a hard sample cake) and drying from an acetone slurry (which also consolidates the sample).

2.3.2   X-Ray Diffraction Analyses of Phyllosilicates

2.3.2.1   Principles

X-ray diffraction is a technique to investigate the organization of solids, usually at the atomic scale. Although many newer analytical techniques have been applied to the study of layer silicates in recent years, XRD remains the single most useful approach for identification of minerals. Soil minerals consist of three-dimensional arrays of atoms that are arranged in regularly spaced planes. X-rays (short-wavelength electromagnetic radiation) are well suited to probe these planes. X-ray diffraction in minerals occurs in accordance with Bragg's Law:

$$n\lambda = 2d \sin\theta \qquad\qquad\qquad [2.4]$$

where λ is the wavelength of monochromatic X-rays, d is the interplanar spacing, θ is the critical angle for constructive interference of scattered rays, and n is an integer.

2.3.2.2   Instrumentation

Basic XRD instrumentation consists of a radiation source that includes an X-ray tube and high voltage generator, a goniometer, and a detector with appropriate counting and recording electronics. In practice, a collimated beam of X-ray photons is directed at a mineral or mixture of minerals that is held by the goniometer in a controlled position with respect to the detector (Fig. 2.9). The goniometer moves the sample and the detector through a specified range of angles, during which the intensity of diffracted photons is recorded on a diffractogram. Coherent diffraction (constructive interference) occurs at each angle where Bragg's Law is satisfied (Fig. 2.10). Because λ and θ are known, d is readily calculated.

2.3.2.3   Identification of Layer Silicate Minerals

If the instrumentation is available, XRD is the most efficient and effective method to identify clay minerals that are present in a soil. In general, no two minerals have exactly the same distances between

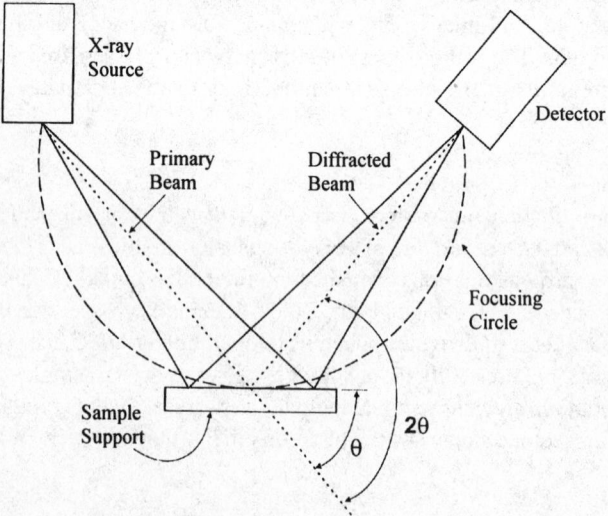

Fig. 2.9   Focusing geometry for X-ray powder diffraction

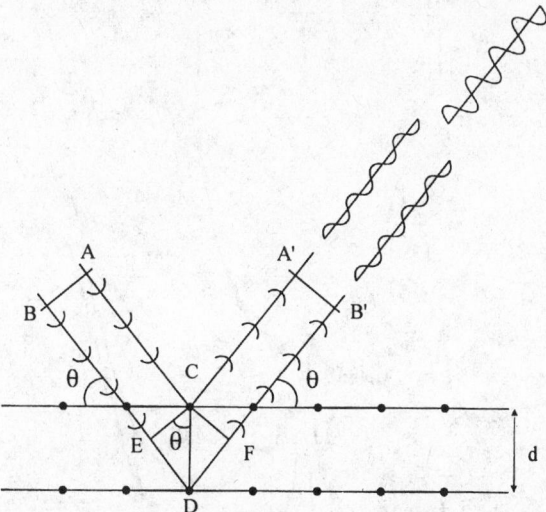

Fig. 2.10   X-ray diffraction and constructive interference from two planes of atoms

planes of atoms; therefore, diffraction angles and patterns of angles are distinctive for individual minerals. On this basis, XRD is a powerful tool for identification of soil minerals in all size fractions. Because the interplanar spacings of some layer silicate minerals are variable and depend on the saturating cation and the presence of water or other polar molecules, thoroughly tested protocols have been developed to assist in the XRD characterization of clay minerals (Whittig and Allardice, 1986). On the other hand, identification of minerals by XRD is not always straightforward because (1) of the low abundance or fine particle size of one or more minerals in a mixture, (2) key diffraction lines of a mineral may be obscured by those of another, or (3) sample pretreatments have altered the interplanar spacings of clay minerals.

From the standpoint of XRD analyses, layer silicates differ from one another primarily in the *z*-dimension. So, unlike most other minerals, layer silicates are often identified by X-ray analyses of samples that are oriented to emphasize those differences. Well-oriented samples can be produced by sedimenting a suspension of clay onto a porous ceramic tile or onto a glass microscope slide. A number of other techniques have been developed for the same purpose (Moore and Reynolds, 1997). In addition, *d*-spacings that are perpendicular to the basal plane of atoms change in response to saturation with different cations, intercalation with organic compounds, and heating. By using consistent and well-planned pretreatments of the sample, the investigator can readily identify layer silicate minerals from the sample XRD pattern. In the discussion that follows, emphasis is placed on the differentiation of layer silicates by analyzing *d*(001) spacings of oriented samples.

Kaolinite
Kaolinite produces a *d*(001) XRD reflection at 0.72 nm that is destroyed after dehydroxylation by heating the mineral at 550 °C for 2 hr (Fig. 2.11 and Fig. 2.12).

Smectite
Smectites such as montmorillonite or beidellite are typically identified by XRD by their characteristic swelling (at the crystalline scale) in polar organic compounds. Magnesium-saturated and air-dried smectites display XRD peaks at about 1.4 nm (Table 2.8). When they are solvated with glycerol or

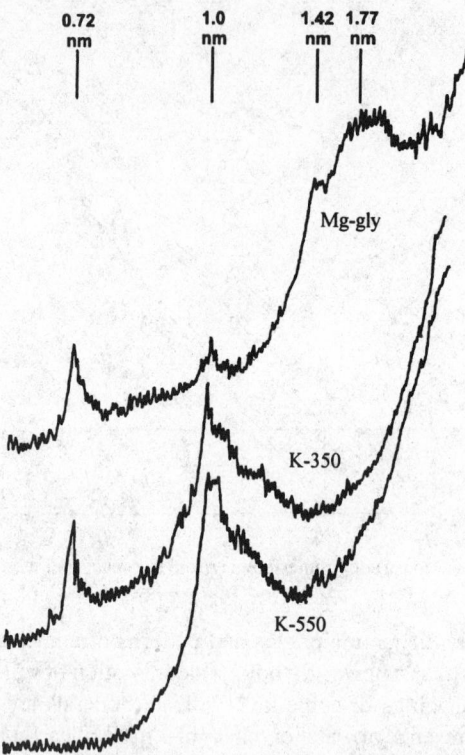

0.72 nm   1.0 nm   1.42 nm   1.77 nm

Mg-gly

K-350

K-550

Fig. 2.11 Characteristic swelling with magnesium-glycerol saturation (Mg-gly) and collapse with potassium saturation and heating to 350 °C or 550 °C (K-350; K-550) for smectites. Kaolinite, illite and vermiculite also occur in this
clay fraction.

ethylene glycol, on the other hand, the organic molecules are sorbed and form two sheets in the
interlayer region. This treatment expands the $d(001)$ spacing to 1.6–1.8 nm, depending on the layer
charge of the smectite layer (Fig. 2.11, 2.12, 2.13); layers with lesser charge tend to expand more.
Potassium saturation of smectite layers collapses them (reversibly) to approximately 1.0-nm $d(001)$
spacings, after excess water molecules have been driven out by heating the sample at 110 °C for 2 hr.

Vermiculite
Like smectites, vermiculite displays $d(001)$ repeat distances of about 1.40–1.45 nm upon Mg saturation.
Because the layer charge of vermiculite is greater than that of smectite, the layers do not expand in the
presence of polar organic solvents (Figs. 2.11, 2.12). The repeat distance of ~1.4 nm represents one sheet
of glycerol in the interlayer position, not two as in smectite. On the other hand, discrete vermiculite crystals
act like smectite layers upon K saturation, collapsing (irreversibly) to $d(001)$ spacings of 1.0 nm.

Illite / Mica
Magnesium-saturated and glycerol-solvated samples of illite display $d(001)$ XRD reflections at about
1.0 nm. This d-spacing does not change upon saturation by different cations, by heat treatments, or by
solvation with polar solvents. Mica responds in a similar manner.

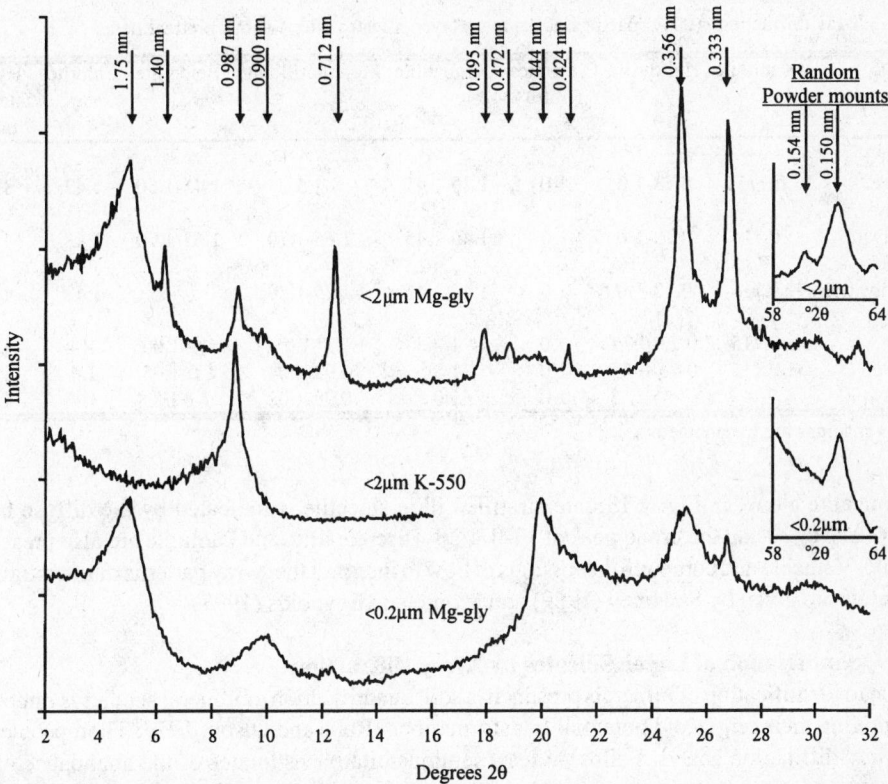

Fig. 2.12 X-ray diffraction patterns of total clay and fine clay from the 70 to 106 cm zone of Seaton silt loam, Dubuque County, Iowa; samples were treated to remove organic matter and Fe oxides, size-fractionated, Mg- or K-saturated, and plated on ceramic tile.

Chlorite/Hydroxy-Interlayered Vermiculite and Smectite

In Mg-saturated samples, chlorite is identified by an XRD peak at 1.4 nm that does not change in response to glycerol or ethylene glycol solvation nor in response to heat treatments. Vermiculite or smectite crystallites that contain hydroxy-Al compounds in their interlayer spaces may produce XRD reflections that range from 1.4 to 1.7 nm, depending on whether glycerol is present or not. Upon heating, water is gradually driven out of hydroxy-interlayered minerals. In weakly interlayered minerals, the $d(001)$ spacing of a K-saturated sample collapses to about 1.0 nm after heating to 550 °C for ≥ 2 hr. On the other hand, when Al interlayers are extensive, complete collapse is not achieved on heating. In fact, the degree to which the layers are propped open by a hydroxy-Al precipitate and resist collapse on increasingly severe heat treatments is a diagnostic indicator of the abundance of hydroxy-interlayered minerals. Fig. 2.11 illustrates XRD patterns of clay fractions with abundant Al-hydroxy interlayering.

Interstratified/Mixed Layer Minerals

The identification of interstratified illite/smectite in soil clay samples is not straightforward because the d-spacings are functions of the proportion of illite and smectite present as well as their crystallinity, which can be quite variable. Rarely are discrete peaks observed; more commonly the analyst must interpret a complex of diffuse bands and broad peaks that are displaced from positions expected for

Table 2.8 X-ray diffraction $d(001)$ spacings of common layer silicates after various pretreatments

	Kaolinite	Halloysite	Clay mica (Illite)	Vermiculite	Montmorillonite	Beidellite	Chlorite	Hydroxy-interlayered minerals
Mg-saturated								
54% relative humidity	0.715‡	0.73-1.0	1.0	1.40-1.45	1.5	1.45-1.50	1.42	1.4-1.5
Ethylene glycol solvation	0.715	0.73-1.0	1.0	1.40-1.45	1.65-1.70	1.65-1.70	1.42	1.4-1.7
Glycerol solvation	0.715	0.73-1.0	1.0	1.40-1.45	1.75-1.80	1.45	1.42	1.4-1.7
K-saturation								
105 °C	0.715	0.73-0.75	1.0	1.0	1.0-1.05	1.0-1.05	1.4	1.1-1.4
350 °C	0.715	0.73-0.75	1.0	1.0	0.98-1.04	1.0-1.05	1.4	1.1-1.4
550 °C	---	---	1.0	1.0	0.98-1.02	1.0	1.4	1.0-1.3

‡All $d(001)$ spacings are in nanometers

illite or smectite alone. In Fig. 2.13, interstratified illite-smectite is indicated by the diffuse hump between 6° and 8° 2θ and the broad peak at 17–18° 2θ. Discrete illite and kaolinite are also present in this sample. Valuable and complete discussions of how to interpret the X-ray patterns of interstratified clay minerals are given by Sawhney (1989) and Moore and Reynolds (1997).

2.3.2.4   Quantification of Layer Silicates by X-ray Diffraction

In addition to identification of minerals present in a soil, quantification of mineral species is important for determining their origin and potential transformations (Ruhe and Olson, 1979; Thompson et al., 1981). X-ray diffraction analyses allow at least semiquantitative estimates of the abundance of the various mineral species present in soils. Although peak area on an X-ray diffractogram is influenced by many factors including particle size, particle orientation, crystal structure, and poorly crystalline impurities, peak area may be related to the quantity of a mineral if possible interferences are accounted

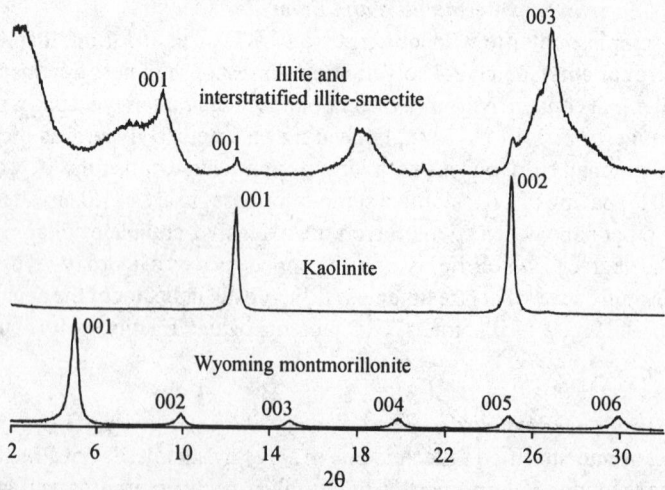

Fig. 2.13 Mg-saturated, glycerol-treated X-ray diffraction patterns of Wyoming montmorillonite, kaolinite, and interstratified illite-smectite; the interstratified illite-smectite sample contains some kaolinite, quartz, and discrete illite.

for by standard addition techniques, internal standards, or comparisons with suitable reference minerals. Detailed discussions of techniques to optimize the quantification of layer silicates by XRD analyses have been presented by Bish (1994) and by Hughes et al. (1994).

2.3.2.5 Structural Organization of Layer Silicate Crystallites: X-ray Techniques

Because of atomic substitutions in both tetrahedrally and octahedrally coordinated positions, most clay minerals in soils deviate from model crystalline structures. The deviations modify both the properties and behavior of the clays, so the deviations are important both to understand and to document. Deviations from ideal structures may be classified as (1) those that occur within octahedral or tetrahedral layers themselves, such as location of octahedral vacancies or rotation of tetrahedra; (2) those that occur in interlamellar positions, such as variations in the distribution of cations, water molecules, or sorbed organic molecules; and (3) those that concern the rotational or translational positions of substacks of layers (Tchoubar, 1984).

X-ray studies are employed to analyze such variations in two ways. Some variations are sufficiently understood that they can be deduced directly from X-ray diffractograms of clay samples. Others require that the effects of the deviations on X-ray patterns be predicted from theory. Then, theoretically calculated X-ray patterns may be compared to actual patterns to test the theoretical hypotheses (Reynolds, 1980).

Crystal defects inside layers may be assessed by XRD after stacking faults have been minimized by strongly ordering the layers with an appropriate choice of saturating cation or with wetting and drying regimes. Then the atomic scattering of X-rays may be measured to determine, for example, the location of vacant sites in octahedral sheets (Amonette, 1988).

2.3.3 Thermal Analysis of Layer Silicate Clay Minerals

Thermal analysis comprises a group of techniques in which the change in mass, temperature, or enthalpy of a material is measured as temperature is varied in a controlled way. Some thermal techniques also measure the dimensions or a mechanical property of a material. People have used controlled temperatures to modify minerals for millennia (e.g., glazing ceramic pots), and some thermal analytical methods have been developed for well over 100 years. Yet thermal analysis is largely empirical. One varies the temperature of a sample, measures changes in the property of interest, and then explains the macroscopic observations by postulating changes on a microscopic scale. Many physical and chemical properties may contribute to a mineral's response to temperature change. Therefore, thermal analyses must be carried out under carefully controlled and standardized conditions.

In studies of layer silicates, thermal analysis is used for the identification and quantification of minerals, as well as for determining thermodynamic and kinetic parameters of reactions that involve minerals. The types of thermal analysis that are commonly applied to studies of layer silicates are thermogravimetric analysis (TGA or TG), differential thermal analysis (DTA), and differential scanning calorimetry (DSC). Other techniques, such as thermodilatometry, evolved gas analysis, and flash pyrolysis, are not routinely used for identification or quantification of minerals and will not be further discussed.

2.3.3.1 Thermal Reactions in Clay Minerals: Overview

As a mineral is heated, the first reaction that occurs is evaporation of water. Water may be lost from many sources in a layer silicate. In some instances, water is also produced by decomposition of structural components of the minerals. Jackson (1975) presented this list of sources of water in minerals: (1) water in micropores, (2) water associated with exterior surfaces of clay mineral quasicrystals or tactoids, (3) water in the interlayer space of expanding clay minerals, (4) water molecules that hydrate interlayer cations, (5) exchangeable H^+ (H_3O^+), (6) water molecules or hydroxyls associated with Al at broken edges of octahedral sheets, (7) hydroxyls associated with Si at broken edges of layers, (8) inner hydroxyls of octahedral sheets, (9) inner surface hydroxyls at layer boundaries of kaolinite or chlorite.

As Table 2.9 suggests, a number of thermal reactions occur in layer silicates other than simple evaporation of water molecules. For example, in the low temperature range (< 400 °C), Mg, Li, or Ni may be relocated from interlayer positions into vacant octahedral sites of montmorillonite. Because such movement changes the layer charge of the montmorillonite, migration of Li into octahedral positions by thermal treatment has been used to distinguish between montmorillonite and beidellite. In addition to a change in layer charge and swelling properties of the mineral, there is an accompanying rearrangement of layer stacking. After Li migration, the stacking of montmorillonite changes to a more ordered pattern so that ditrigonal holes in the basal plane of oxygen atoms directly overlie one another.

In the medium temperature range, i.e., 300–950 °C, dehydroxylation reactions dominate. Dehydroxylation reactions require a proton transfer between two OH groups, and this is thought to be a two-stage reaction. It is believed that the first reaction to occur is dissociation of a free OH group to a proton and an O^{2-} ion. It seems likely that the second step should occur at sites near particle edges where water can be lost more efficiently than from the structural interior of a crystal. So the proton must migrate to that location before the second reaction (the combination of a proton with a second OH to form water) can occur. The overall reaction can be summarized by:

$$OH^- + OH^- \Rightarrow O^{2-} + H_2O \qquad\qquad [2.5]$$

Also in the medium temperature range, exothermic oxidation reactions [e.g., Fe (II) $\Rightarrow$ Fe (III)] are likely to occur. In the high temperature range (> 950 °C), thermal changes are reactions in which the mineral structure decomposes; for example,

$$\text{Dehydroxylated kaolinite} \Rightarrow \text{mullite}, 3Al_2O_3 \cdot 2SiO_2 \qquad\qquad [2.6]$$

These reactions may be either exothermic or endothermic (Table 2.10).

Thermal Gravimetric Analysis

In thermal gravimetric analysis (TGA or TG), one measures the loss of mass upon heating a material. The sample is continuously weighed as it is heated or cooled at a constant rate, and the mass loss is plotted against temperature or time. Table 2.9 presents the origins of mass loss during heating of layer silicate minerals. These rules are applicable primarily to pure minerals, i.e., no organic matter and no

Table 2.9   Sources of mass loss from layer silicates as temperature is increased

Temperature	Dehydration or Dehydroxylation Reactions
25-105 °C	loss of sorbed water and some hydroxyls of poorly crystalline aluminosilicates
105-300 °C	loss of water associated with smectites, vermiculite, halloysite, and poorly crystalline aluminosilicates
300-540 °C	decomposition of hydroxyls of kaolinite, halloysite, nontronite, Fe-beidellite, Fe-vermiculite; some hydroxyls of chlorite; most hydroxyls of poorly crystalline aluminosilicates
540-950 °C	decomposition of hydroxyls of Al, Mg-montmorillonite, hectorite, Mg-vermiculite, muscovite, talc, and the remainder of hydroxyls of chlorite, and poorly crystalline aluminosilicates

Table 2.10 Transformations in layer silicates observed by differential thermal analysis

1. Evaporation of adsorbed species, e.g., water: *endothermic reaction*
2. Loss of structural components, e.g., dehydroxylation: *endothermic reaction*
3. Oxidation, e.g., Fe(II) to Fe(III): *exothermic reaction*
4. Phase changes, e.g., kaolinite ==> mullite: *endothermic or exothermic reaction*

carbonates can be present to confound the patterns by overlap with weight loss from the minerals listed.

Differential Thermal Analysis

In differential thermal analysis (DTA), the temperature difference between a sample and a thermally inert reference material is measured as they are heated together at a controlled rate. Therefore, the measurement is one of energy changes as heat is either taken up by the sample or released by the sample (Table 2.10). Endothermic reactions in the sample are indicated when the sample temperature is less than that of the reference material. If the difference in temperature between the sample and the reference is positive, an exothermic reaction has occurred.

Differential Scanning Calorimetry

The principles of differential scanning calorimetry (DSC) are similar to those of DTA, but in this case, the thermocouples of the instrument are not in direct contact with the sample or the reference material. In addition, the temperature of the sample and the reference material are kept constant. The amount of energy needed to keep the sample and the reference at the same temperature is measured. This technique is more sensitive to minor energy changes in the sample than is DTA. Commonly, the maximum temperature for DSC analysis is about 600 °C because the temperature is limited by the use of Al metal heating blocks.

Sample Preparation for Thermal Analysis

The particle size of samples to be analyzed by thermal techniques is important for two reasons. First, size fractionation of an unknown sample makes it more likely that the sample will have only a small number of minerals, simplifying interpretation of thermal patterns. Second, particle size of the sample will significantly influence the thermal conductivity of the sample when it is packed into the sample cup of the instrument.

Interpretation of patterns derived from samples with vermiculite and smectites is also simplified by first equilibrating saturated samples for 24–48 hr at a standard relative humidity, e.g., 54–56% [e.g., oversaturated $Mg(NO_3)_2$]. This effort standardizes the amount of surface-sorbed water, making water loss more comparable among similar samples. Similarly, the shape of the low-temperature DTA endotherm depends on the hydrated cation.

Both organic matter and Fe oxides can complicate the interpretation of thermal patterns because their oxidation in the range of 100–500 °C can obscure or distort thermal patterns of layer silicates. One alternative is to remove Fe oxides by extracting the sample with a reducing agent such as sodium dithionite (Kunze and Dixon, 1986). Organic matter is usually removed by treatment with H_2O_2. To characterize the phase that was removed, the thermal properties of a sample can be determined before and after removal of Fe oxides and organic matter. Of course, these pretreatments have the potential to oxidize and solubilize Fe atoms in the crystal structure of the sample, as well. Heating samples in an inert atmosphere (N_2) helps to inhibit oxidation of organic compounds during analysis. Several excellent publications (Mackenzie, 1970; Jackson, 1975; Tan et al., 1986; Paterson and Swaffield,

1987; Karathanasis and Harris, 1994) are available that present complete details about sample preparation and instrumental settings for thermal analyses as well as interpretation of thermal analytical data of soil minerals.

2.3.3.2 Thermal Characteristics

To illustrate the interpretation of thermal analyses, differential thermal patterns of 3 reference clay minerals are presented in Fig. 2.14.

Kaolinite Group

The dehydroxylation endotherm of kaolinite and related minerals (dickite and nacrite) occurs at 500–600 °C. Because a weight loss of about 14% occurs with pure kaolinite samples and because other minerals do not contribute to the endothermic peak in this temperature range, the size of the DTA peak (or, preferably, the actual weight loss) can be related to the quantity of kaolinite present in the sample. Kaolin minerals show an additional exotherm at about 900–1,000 °C as the mineral recrystallizes to mullite.

Smectite and Vermiculite Groups

Both smectites and vermiculites show a strong dehydration endotherm upon heating from 25 to 250 °C. At this temperature, water that is outside quasicrystals as well as water that is between substacks of quasicrystals is evaporating. Water that occurs in interlayer positions, either free or directly associated with interlayer cations, is released as the temperature climbs higher toward 250 °C. A dehydroxylation endotherm at about 700 °C and an endotherm or exotherm at about 900 °C are present in differential thermal analyses of Al-smectite and at about 910 °C for Mg-smectite.

Karanthanasis and Harris (1994) have described a technique to quantify the amounts of smectite and vermiculite by thermal analysis, taking advantage of differences in how readily water will

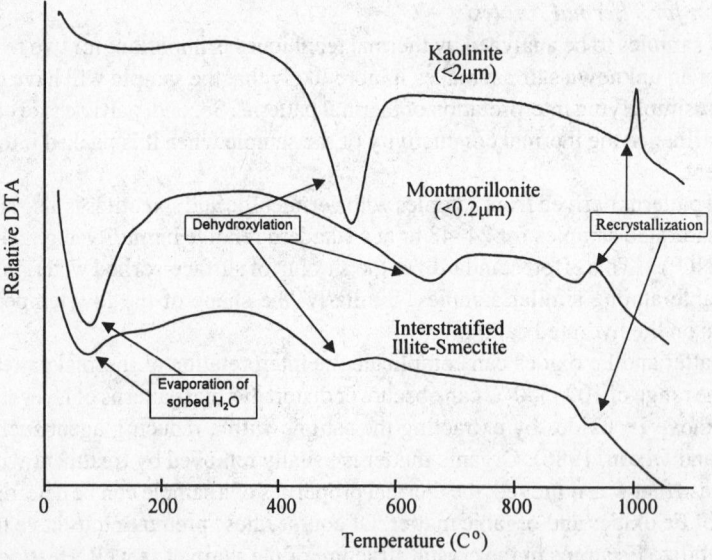

Fig. 2.14 Differential thermal patterns of kaolinite, Wyoming montmorillonite, and interstratified illite-smectite; the interstratified illite-smectite includes some kaolinite; samples were size-fractionated, Mg-saturated, freeze-dried, and equilibrated at 54% relative humidity before analysis.

evaporate from these expandable minerals. When the saturating cation and initial humidity conditions of a sample are well controlled, the loss of water mass over the temperature range of 25 to 250 °C may be used to quantify smectite plus vermiculite. When both vermiculite and smectite are present, a characteristic mass loss of about 2.5% for pure vermiculite from 220 to 250 °C is used to quantify the vermiculite portion of the total 25 – 250 °C loss of water. To be reproducible, the measurement must be conducted on Mg-saturated clay that has been equilibrated at 54–55% relative humidity. The presence of organic matter or Fe oxides can complicate this relationship because both components can also release water in these low-temperature regions.

Clay Mica Group
Although the basic reactions that occur in other layer silicates also occur in micas (i.e., evaporation of water, structural dehydroxylation, and oxidation of Fe), they are not as predictable, and thermal techniques are not readily applied to identify or quantify micaceous minerals.

Identification of Minerals in a Soil Clay Mixture
Fig. 2.15 illustrates the thermal analysis of a soil clay sample. In this sample of the clay fraction from a subsurface horizon in a soil developed in Late Wisconsinan loess, the large DTA endotherm and mass loss on the differential thermal gravimetric (DTG) pattern from 25 to about 225 °C are attributed to evaporation of water from smectite. The mass loss from 225 to 250 °C is attributed to loss of water from vermiculite, and the shallow endotherm at about 450 °C is due to dehydroxylation of kaolinite. Compare this figure with the XRD (Fig. 2.12) and IR (Fig. 2.17) patterns of the same sample.

2.3.4 Infrared Techniques in Layer Silicate Mineralogy
Infrared (IR) radiation is a part of the electromagnetic spectrum that includes wavelengths extending from 1 to 100 μm. Wavelength (λ) is the distance (μm) between the crests of two adjacent waves. Frequency (f) is the number of waves passing a fixed point in a unit of time (sec^{-1}), and wavenumber (n) is the number of waves per unit of length. Therefore,

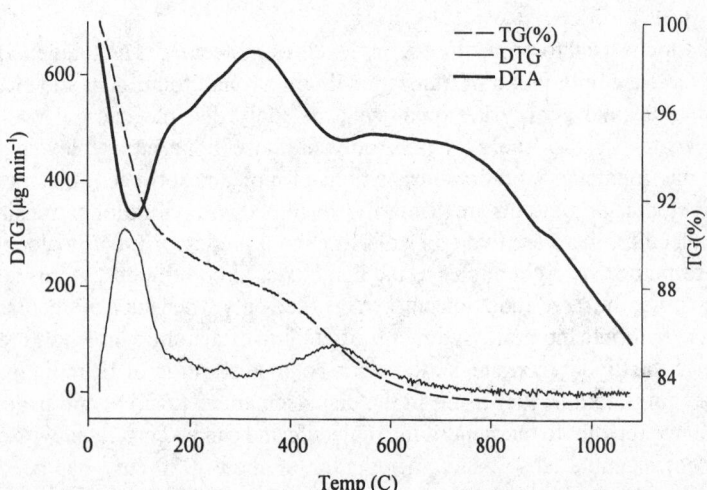

Fig. 2.15 Thermal analyses of the < 2μm soil clay (Seaton silt loam, Dubuque County, Iowa; 70-106 cm); sample was treated to remove organic matter and Fe oxides, size-fractionated, Mg-saturated, freeze-dried, and equilibrated at 54% relative humidity before analysis.

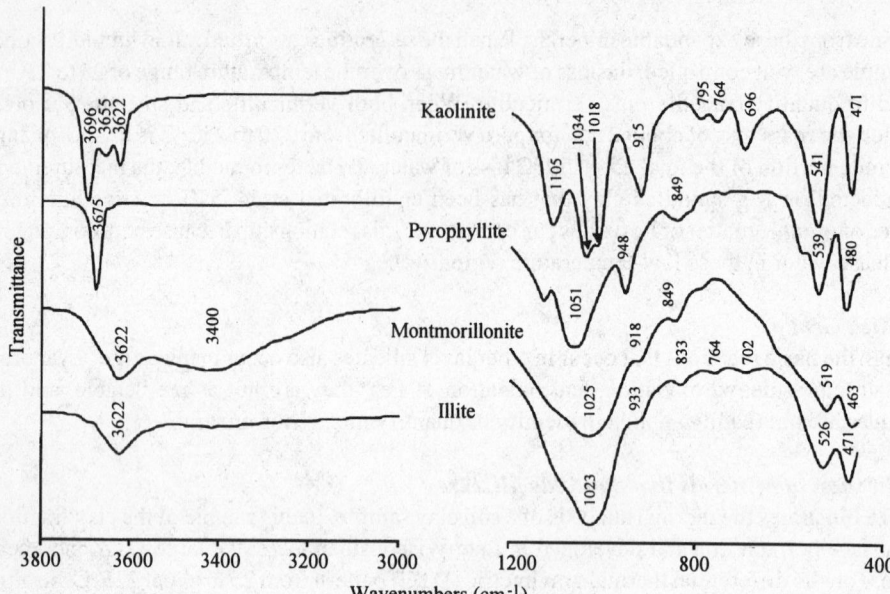

Fig. 2.16 Transmission infrared spectra of layer silicate minerals [Salisbury et al., 1987]

$$c = f\lambda \tag{2.7}$$

and

$$n = 1/\lambda \tag{2.8}$$

where c = the velocity of light, 3×10^{10} cm sec^{-1}. For convenience, the units usually used to discuss the IR spectrum are wavenumbers, but much of the older literature presents IR patterns that are measured in wavelength (μm).

Infrared radiation is used to probe the energy levels of molecules. The total energy of a molecule can be thought of as a combination of translational, vibrational, rotational, and electronic energies. These types of molecular energy are quantized, and only discrete energy levels are permitted. Therefore, molecules absorb energy inputs only at those nonarbitrary levels. Infrared energy corresponds to rotational and vibrational transitions of molecules. Since molecules in solids rarely rotate freely, IR studies of minerals are primarily concerned with vibrational transitions.

In simple molecules, there are two types of vibrational modes: *stretching* along a bond axis and *bending* (or deformation) at right angles to the bond axis. In both modes, atoms are displaced with respect to one another. Infrared radiation can be absorbed only when such molecular scale vibrations result in a change in dipole moment. Symmetric stretching of diatomic molecules (e.g., N_2) does not cause a change in the dipole moment, so there can be no absorption of IR radiation. But molecular stretching or bending motions that result in the displacement of positive and negative charges in a molecule do change the dipole moment of the molecule and can be correlated with absorbance of IR energy levels. For example, absorbance of radiation at about 920 cm^{-1} has been assigned to the bending of OH bonds in hydroxyls that are associated with Al cations in the octahedral sheet of montmorillonite. The specificity of such assignments makes infrared IR spectroscopy a powerful tool in the characterization of minerals in soils.

Recommended texts that offer details of sample preparation techniques for IR analysis as well as interpretations of the infrared patterns of soil minerals include those of van der Marel and Beutelspacher (1976), Russell (1987), and White and Roth (1986).

2.3.4.1 Pleochroism in Infrared Analyses of Layer Silicates

Absorption of IR radiation is at a maximum when the axis of the vibrating bond is parallel to the electric vector of the IR beam (i.e., perpendicular to the beam direction). Conversely, minimum absorption occurs when the electric vector and the bond axis are perpendicular to one another. This dependence of absorption intensity on bond axis is called *pleochroism* and has been exploited to distinguish trioctahedral from dioctahedral layer silicates. Bond axes of the inner OHs in a trioctahedral mineral are perpendicular to the basal oxygen planes of the mineral. Thus, a well-oriented trioctahedral mineral will show variation in absorption of IR radiation by those OHs as the orientation of the sample is changed with respect to the IR beam. Dioctahedral minerals, on the other hand, have inner OHs that are not perpendicular to the basal oxygen plane, so the OH bonds cannot be oriented uniformly with respect to an incident IR beam. Therefore, the angle at which the sample is presented to the IR beam does not affect the intensity of absorbed radiation.

2.3.4.2 Infrared Characteristics of Representative Layer Silicate Minerals

Transmission IR spectra of four representative layer silicate minerals are given in Fig. 2.16. These spectra are presented to demonstrate the assignment of absorption bands to particular types of vibrational modes in minerals.

Kaolinite shows a characteristic IR absorption sequence in the 3,600–3,700 cm^{-1} region. These absorbances correspond to the stretching vibrations of inner surface OHs (3,696 cm^{-1}) and inner OHs (3,620 cm^{-1}). Absorbances at 1,105 and 1,034 cm^{-1} are attributed to antisymmetric Si-O-Si stretching, whereas those at about 1,018 cm^{-1} are attributed to stretching of an O atom bound to both Si and tetrahedral Al. Absorption at 915 cm^{-1} has been assigned to stretching of inner surface OHs by Russell (1987). Absorbances at 696 and 541 cm^{-1} are due to stretching of Si-O-Al bonds.

Pyrophyllite, a dioctahedral mineral, shows a broad OH deformation band at 3,675 cm^{-1} as well as a slight shoulder at about 3,645 cm^{-1} which has been attributed to AlFe^{3+}OH groups (Russell, 1987). Other absorption bands in pyrophyllite can be assigned to similar origins as those that occur at comparable wave numbers in the kaolinite spectrum.

The most distinguishing feature of the *montmorillonite* spectrum is the broad absorption band that ranges from 3,300 to 3,500 cm^{-1} (Fig. 2.16). This band, typically centered about 3,400 cm^{-1}, is due to H-O-H stretching of water molecules present in the interlayer region of montmorillonite. Other prominent bands in the montmorillonite spectrum include a very broad AlOH stretching band at 3,622 cm^{-1}, and OH deformation bands at 918 cm^{-1} (AlAlOH) and 849 cm^{-1} (probably AlMgOH).

Infrared features of *illite* are rarely diagnostic because illite is so variable in chemical composition. The AlOH band at 3,620 cm^{-1} is often coupled with weak absorbance bands at 825–835 and 750–770 cm^{-1}, assigned to AlMgOH deformation and AlO-Si vibrations, respectively.

2.3.4.3 Identification of Minerals in a Soil Clay Mixture

Fig. 2.17 illustrates the characterization of the clay fraction of a soil by IR spectroscopy. Although derived from diffuse reflectance IR measurements, the peaks of this spectrum can be compared directly with IR absorbances in Fig. 2.16. There is considerable overlap in the peaks of different minerals found in the < 2 μm fraction of this soil horizon. The most likely suite of minerals includes kaolinite, quartz, smectite, vermiculite, and illite. These identifications can be compared with those made by XRD (Fig. 2.12) and thermal analysis (Fig. 2.15) for the same sample.

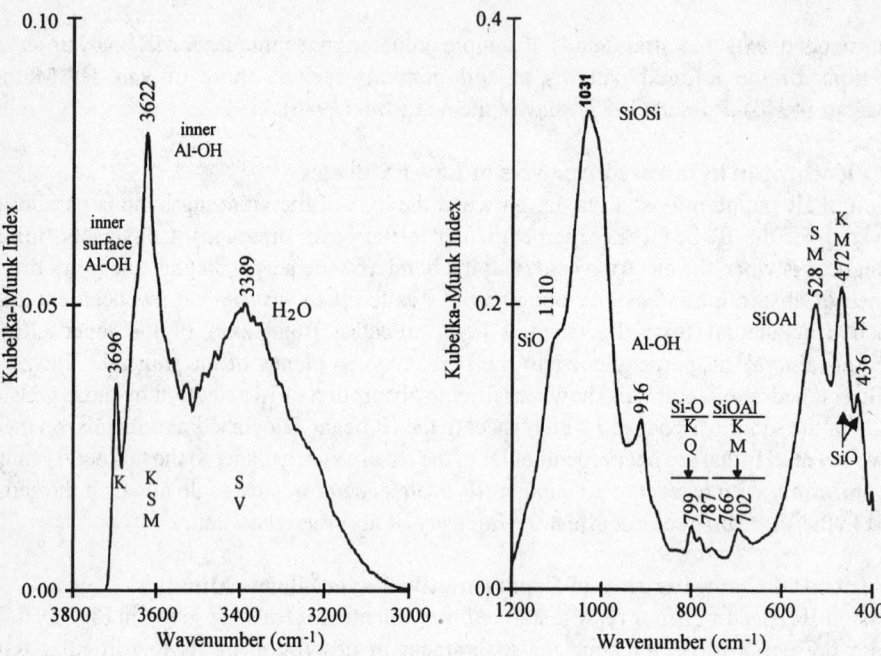

Fig. 2.17 Fourier-transform diffuse reflectance pattern of <2μm soil clay (Seaton soil, Dubuque County, Iowa; 70-106 cm) after removal of organic matter and Fe oxides. Freeze-dried sample was analyzed in a KBr powder mixture. K = kaolinite, S = smectite, M = clay mica, V = vermiculite, Q = quartz

2.4   Clay Minerals and Properties of Soils

Data from eight soil sampling locations (Tables 2.13–2.20) are included to illustrate typical clay mineral suites found in soils. Each site was carefully selected so that a different mineral species would predominate in the clay-size fraction of the soil. Associated soil properties typically reflect the influence from this individual mineral, but clay percentage, other crystalline and noncrystalline clay minerals present, as well as organic matter, influence the resulting measured value of specific properties. Theoretical values, generalized formulas, or derived values often cited in the literature and as discussed earlier in this chapter (e.g., Table 2.6), are for pure mineral systems and do not always reflect what one routinely encounters in analyses of whole soils or even the clay-size fraction of a soil.

2.4.1   Soil Property Indices

2.4.1.1   The CEC/Clay Content Ratio

The CEC/clay content ratio has been used as an auxiliary parameter to assess clay mineralogy and is used for clay activity classes in Soil Taxonomy (Soil Survey Staff, 1996). A simplified categorization that may be useful, if organic matter and amorphous material are accounted for, is presented in Table

Table 2.11   The relation between CEC, clay content and clay activity classes

CEC/Clay Content Ratio	Clay Activity Classes
> 0.7	Montmorillonitic or vermiculitic
0.3 - 0.7	Mixed or illitic or chloritic
< 0.3	Kaolinitic or halloysitic

Table 2.12 Engineering properties: Atterberg Limits for clay minerals [From Mitchell, 1993. Fundamentals of soil behavior. Copyright John Wiley and Sons, New York with permission]

Mineral	Liquid Limit %	Plastic Limit %	Shrinkage Limit %
Montmorillonite†	100 - 900	50 - 100	8.5 - 15
Nontronite†	37 - 72	19 - 27	
Illite	60 - 120	35 - 60	15 - 17
Kaolinite	30 - 110	25 - 40	25 - 29
Hydrated Halloysite†	50 - 70	47 - 60	
Dehydrated Halloysite‡	35 - 55	30 - 45	
Attapulgite	160 - 230	100 - 120	7.6¥
Chlorite§	44 - 47	36 - 40	
Allophane (Undried)	200 - 250	130 - 140	

† Highest values are for monovalent forms, lowest for the di- and trivalent forms.
‡ Highest values for di- and trivalent forms, lowest for monovalent forms.
§ Some chlorites are nonplastic
¥ Data from Lamb and Whitman (1969)

2.11. A more detailed explanation and table is available in Burt (1995). This ratio is useful as an internal check of data and as an estimator of mineralogy when mineralogy data may not be available. Clay activity classes are also currently used in Keys to Soil Taxonomy (Soil Survey Staff, 1998) to provide additional detail concerning clay mineralogical composition in siliceous or mixed family mineralogy classes.

2.4.1.2 The 1500 kPa-Water/Clay Content Ratio

Water retention at 1500 kPa reflects the volume of water filled pores that have an approximate diameter $\leq 0.2\ \mu m$. The 1500 kPa-water/clay content ratio is also a good tool for data assessment. For soils dominated by silicates, this ratio is approximately equal to 0.4. Many soil-related factors cause deviation from this reference point. Low-activity clays lower the ratio and high-activity clays increase the ratio. Organic matter and poorly crystalline materials are known to increase 1500 kPa water retention and thus increase the ratio. Values ≤ 0.6 are generally accepted indices for adequate dispersion (Burt, 1995). Nettleton and Brasher (1983) state that values between 0.3 and 0.6 indicate that clay disperses well during the particle size analyses and that the clay content would be measured accurately by this technique. When the ratio is greater than 0.6 and soil-related factors do not adequately explain the situation, incomplete dispersion is usually a factor.

Table 2.13 Selected soil properties: Cecil soil (S88-GA-219-001)‡

Depth cm	Horizon	Clay %	CEC/Clay content	1500kPa-w/ clay content	CEC[†] cmol$_c$ kg^{-1}	LL[e] %	PI[e] %	COLE[e] cmcm^{-1}	XRD	TGA %	pH CaCl$_2$ H$_2$O	K$_2$O[§]
0-15	Ap	8.6	0.35	0.42	3.0	18	3	0.004	K3,G2,V2,M1¥	K26 G5	6.0 6.3	1.0
56-97	Bt2	47.4	0.13	0.39	6.2	53	29	0.010	K5,G3,V2,M1	K45 G8	4.6 5.0	0.7
137-171	BC	32.2	0.16	0.49	5.2	47	16	0.010	K5,G1,V2	K53 G2	4.2 4.6	0.6

†CEC is reported for the < 2mm fraction
‡Alpha-numeric code is the Soil Survey Laboratory Number, National Soil Survey Center, Lincoln, NE
¥K = Kaolinite, G = Gibbsite, V = Vermiculite, M = Illite, H = Halloysite, Mt = Smectite, T = Talc, P = Pyrophyllite, C = Chlorite; numbers refer to peak height for XRD only: 5 = Very Large, 4 = Large, 3 = Medium, 2 = Small, 1 = Very Small (see Burt, 1995)
eLL=Liquid limit; PI = Plasticity index; COLE=Coefficient of linear extensibility
§K$_2$O is calculated as a % of the <2μm fraction

Table 2.14 Selected soil properties: Retep soil (S91-WA-077-006)[‡]

Depth cm	Horizon	Clay %	O.C.[†] %	CEC/Clay content	1500kPa-w/ clay content	CEC cmol$_c$ kg^{-1}	LL %	PI %	COLE cm cm^{-1}	XRD	K$_2$O %	pH CaCl$_2$	H$_2$O
8-20	AB	12.4	0.97	1.36	0.79	16.9			0.005	H3,Mt1	0.5	5.7	6.3
20-41	2Bt1	13.3	0.70	1.42	0.84	18.9	30	5	0.009	H3,Mt2,M1	0.4	5.7	6.3
41-76	2Bt2	15.3	0.61	1.39	0.78	21.2			0.020	H3,Mt2	0.3	5.7	6.4
76-91	2Cr	13.3	0.13	1.57	1.20	20.9	43	15	0.031	H3,Mt3,M1	0.3	5.6	6.3

[†] O.C. = Organic Carbon
[‡] See Tables 2.13 and Soil Survey Laboratory Staff (1992) for further explanation of abbreviations

When the soil meets the criteria for dispersion, clay content can be used to estimate other properties such as Atterberg Limits (Burt, 1995) which, in turn, may provide some indication of the clay minerals present (Table 2.12). These and other indices and relationships have the greatest value when they can be used as proxies to estimate characteristics of soils where specific analyses are lacking.

2.4.1.3 Example Soils

Cecil, a common soil of the southern Piedmont region in the United States, is classified as a fine, kaolinitic, thermic Typic Kanhapludult (Soil Survey Staff, 1998) in Oconee County, GA (Table 2.13) and is representative of a soil high in kaolinite. The 1500 kPa-water/clay ratio for the Cecil soil suggests that the samples were dispersed well and that the clay content is representative of this soil. From the TGA and XRD analyses, approximately 25% of the total clay content is kaolinite in the Ap horizon and 45–50% is kaolinite below a depth of 50 cm. This distribution also fits well with the determination of Atterberg limits for liquid limit (LL) (Table 2.12) and plasticity index (PI), low CEC, low pH and coefficient of linear extensibility (COLE), all of which are in the range of a soil dominated by kaolinite. The CEC/clay content ratio is less than 0.2, with the exception of the surface horizon. This places the Cecil soil well within the kaolinite clay mineralogy class for soil classification purposes.

Other minerals of lesser amount found in this profile are gibbsite, vermiculite, mica, and below a depth of 50 cm but not shown here, goethite and hematite. Mica flakes large enough to be visible to the eye were described in the field in the lower portion of the profile and minor amounts were detected by XRD in the surface horizons. The increasingly greater amounts of kaolinite with depth dominate the overall CEC and water retention of the clay mineral suite in this soil and tend to lower the values measured for the total assemblage to that which more closely resembles a clay mineral suite composed entirely of kaolinite.

Table 2.15 Selected soil properties: Medary soil (S92-IA-061-002)[†]

Depth cm	Horizon	Clay %	CEC/Clay content	1500kPa/ clay content	CEC cmol$_c$ kg^{-1}	LL %	PI %	COLE cm cm^{-1}	XRD	K$_2$O %	pH CaCl$_2$	H$_2$O
0 - 10	A	39.2	1.06	0.63	41.4	76	33	0.060	Mt3,K3,M2	1.8	6.5	6.6
10 - 20	EB	39.2	0.85	0.60	33.3			0.071	Mt3,K3,M2	1.8	6.4	6.5
20 - 40	Bt1	58.3	0.70	0.38	40.6			0.126	Mt5,K3,M2	1.8	4.2	4.8
40 - 53	Bt2	59.6	0.70	0.41	41.5			0.113	Mt4,K3,M2	1.7	4.1	4.9
53 - 68	Bt3	46.8	0.74	0.46	34.8			0.102	Mt5,K3,M1	1.6	4.2	5.0
68-103	Bt4	29.9	0.79	0.50	23.6			0.032	Mt5,K3,M2	1.4	4.2	4.8
103-138	Bt/C1	31.0	0.77	0.51	23.9			0.032	Mt4,K2,M2	1.7	4.9	5.4
138-148	Bt/C2	34.5	0.79	0.51	27.1	74	46	0.057	Mt4,K3,M2	1.7	7.4	7.8
148-160	Bt/C3	19.1	0.91	0.49	17.3			0.010	Mt4,K2,M2	1.4	7.9	8.2

[†] See Tables 2.13 and 2.14 and Soil Survey Laboratory Staff (1992) for further explanation of abbreviations and methods

Table 2.16 Selected soil properties: Ellijay soil (S85-NC-099-004)[†]

Depth cm	Horizon	Clay %	CEC/Clay content	1500kPa/ clay content	CEC cmol$_c$ kg^{-1}	LL %	PI %	COLE cm cm^{-1}	XRD	pH CaCl$_2$	H$_2$O	K$_2$O %	DTA %
0-10	A	36.9	0.43	0.48	15.8			0.032	K2,T2,V2	5.0	5.5	0.3	K11,Gtr[§]
10-38	Bt1	53.1	0.18	0.37	9.5			0.035	K3,T3	5.2	5.4	0.3	K19,G1
38-86	Bt2	52.0	0.13	0.39	7.0	47	14	0.029	K3,T3,V2	5.6	5.6	0.3	K23,G2
86-132	BC	24.5	0.16	0.96	4.0			0.015	K2,T2,V1	5.9	5.8	0.1	K13,G3
132-150	C1	28.5	0.88	1.06	25.0	82	37	0.023	K1,T2,Mt3	6.1	6.5		Gtr
150-175	C2	26.3	0.88	1.06	23.1				K1,T2,Mt3	5.9	6.0		K01,Gtr

[§]tr = trace
[†]See Tables 2.13 and 2.14 and Soil Survey Laboratory Staff (1992) for further explanation of abbreviations and methods

The Retep soil, a fine-loamy, mixed, superactive, frigid, Vitrandic Argixeroll (Soil Survey Staff, 1998), from Yakima County, WA, is a soil of volcanic ash origin having a significant amount of halloysite (Table 2.14). Optical mineralogy reports 15% or more glass shards, glass aggregates, and glass-coated quartz and opaques in the fine sand fraction. X-ray diffraction indicates that halloysite dominates the clay fraction and smectite is of secondary importance. Both CEC and Atterberg limits are influenced by the amount as well as the type of clay present in the sample. The values of these properties are too low to indicate the presence of much smectite. Most of the samples were air-dried prior to analyses. Dehydrated halloysite would have been the only form detected. Clay content is low, and comparison of clay content to the 1500 kPa-water/clay ratio, suggests that there may have been a problem with complete dispersion of the sample. This ratio is usually higher with increasing organic matter or increasing amounts of poorly crystalline materials in the soil. Both organic matter content and poorly crystalline materials are suspected causes of poor dispersion in this volcanic ash soil. Therefore, using the ratio techniques discussed above to estimate general clay mineralogy class activities is not effective here. When the 1500 kPa-water/clay ratio is > 0.6, a derived clay value may be calculated to determine the CEC/clay ratio under these conditions. In this case, an adjustment is made for organic carbon and the clay content is calculated as:

$$\%\text{Clay} = 2.5(W_{15} - \%OC) \qquad [2.9]$$

where W_{15} is the weight percentage of water retained at 1500 kPa suction on a < 2 mm basis, clay is the weight percentage of clay on a < 2 mm basis, and OC is organic C content. See Burt (1995) for further discussion of these estimation techniques.

The Medary soil, a fine, smectitic, mesic Vertic Hapludalf (Soil Survey Staff, 1998) from a stream terrace in Dubuque County, Iowa, is representative of a soil with significant quantities of smectite

Table 2.17 Selected soil properties: Redondo soil (S78-NM-043-004)[†]

Depth cm	Horizon	Clay %	CEC/Clay content	1500kPa/ clay content	CEC cmol$_c$ kg^{-1}	LL %	PI %	COLE cm cm^{-1}	XRD	K$_2$O %	DTA %
6 - 20	A2	10.1	4.44	1.13	44.8				P3,M1,K1	3.5	
57 - 70	B1	16.1	3.07	0.76	49.5	34	2	0.011	P3,M2,K1	3.1	
70 - 112	B2t	12.8	4.39	1.09	56.2			0.006	P3,M2,K1	3.0	K13
112 - 150	2Bt2	12.0	3.84	1.33	46.1	34	3		P3,M2,K1	2.7	
210 - 270	3Bt5	17.9	3.84	1.01	68.8	40	8	0.009	P4,M2,K1	2.7	

[†] See Tables 2.13 and 2.14 and Soil Survey Laboratory Staff (1992) for further explanation of abbreviations and methods

Table 2.18 Selected soil properties: Trappist soil (S91-KY-1350-001)[†]

Depth cm	Horizon	Clay %	CEC/Clay content	1500kPa/ clay content	CEC cmol$_c$ kg^{-1}	LL %	PI %	COLE cm cm^{-1}	XRD	K$_2$O %
0 - 4	OA				74.7				M2	1.7
13 - 37	Bt1	57.9	0.25	0.37	14.6	62.1[§]	42.9[§]	0.011	M4,K1	4.5
37 - 49	Bt2	65.2	0.25	0.36	16.6	68.7	50.2		M5,K1	5.1
49 - 63	B/C	64.4	0.24	0.37	15.2	67.9	49.4		M5,K1	5.1

[§]Values for liquid limit and plasticity index are calculated values for this profile. Equations are from Burt (1995).
[†]See Tables 2.13 and 2.14 and Soil Survey Laboratory Staff (1992) for further explanation of abbreviations and methods

(Table 2.15). The 1500 kPa-water/clay content ratio is generally < 0.6 indicating that the soil dispersed well during particle size analysis. Measured Atterberg limits are low for smectite and were not completed on horizons where the greatest amount of smectite was detected by XRD. Kaolinite was strongly represented in the horizons analyzed, and the Atterberg limits are more indicative of this mineral. CEC is low for a smectitic soil, but there are significant amounts of kaolinite and some mica in the clay fraction. Total analysis shows K$_2$O amounts to be nearly triple those found in the kaolinitic Cecil and the halloysitic Retep soils, ≤ 1.8 versus ≤ 0.5%, respectively. The K$_2$O percentages suggest illite is present in the soil clay. This would also influence the lower Atterberg limits, CEC, and other soil properties measured on a < 2 mm basis. The analytical procedures used in standard soil characterization techniques (Burt, 1995; Soil Survey Laboratory, 1992) do not allow identification of interstratified clay minerals although illite/smectite interstratification may well be present in this soil.

The Ellijay soil, a fine, ferruginous, mesic Rhodic Kanhapludalf from Jackson County, NC (Soil Survey Staff, 1998), is a soil formed in weathered igneous, ultrabasic talc-rich rock (Table 2.16). With the exception of the C horizon, this soil disperses well. Talc and kaolinite are the primary clay minerals which account for the low CEC and relatively low plasticity index. Liquid limit is higher in the residual C horizon. This soil is developed in an acid environment. Vermiculite was detected by XRD in the surface horizon. K$_2$O percentages are very low and similar to those of the Cecil soil.

The Redondo soil, an ashy, pumiceous, glassy Vitrandic Eutrocryept (Soil Survey Staff, 1998), has pyrophyllite, mica, and kaolinite as a clay mineral suite (Table 2.17). The 1500 kPa-water/clay content ratio is high. Because the parent materials are derived from pumice and tuff, this ratio is likely high due to poorly crystalline materials rather than being solely an indicator of poor dispersion. Pumice and tuff are rigid materials with clay size pores. Similarly, CEC/clay content is also very high and not illustrative of the clay mineral suites present. Atterberg limits are low and representative of kaolinite or pyrophyllite with essentially no layer charge. K$_2$O percentages are high as are extractable bases, particularly K (not presented here), reflective of the mica constituent in a slightly acid environment.

Table 2.19 Selected soil properties: Fickle soil (S93-NM-031-005)[†]

Depth cm	Horizon	Clay %	CEC/Clay content	1500kPa-W/ clay content	CEC cmol$_c$kg^{-1}	LL %	PI %	COLE cm cm^{-1}	XRD
5 - 15	Bt	45.2	0.83	0.36	37.6				V4,K2,M1
15- 41	Btss	32.2	1.01	0.35	32.5			0.041	V4,K2,Mt3
41- 53	Btk1	25.0	1.07	0.34	26.8				V4,K2,Mt4
53-147	Btk2	34.1	1.16	0.38	39.7				V5,K2

[†] See Tables 2.13 and 2.14 and Soil Survey Laboratory Staff (1992) for further explanation of abbreviations and methods

Table 2.20 Selected soil properties: Gudgery soil (S80-CA-045-001)[†]

Depth cm	Horizon	Clay %	CEC/Clay content	1500kPa-W/ clay content	CEC $cmol_c kg^{-1}$	LL %	PI %	COLE $cm\ cm^{-1}$	XRD	K_2O %
30 - 58	A2	30.1	0.62	0.39	18.6	40	13	0.021	C3,M2	3.6
89 - 119	Bt2	32.9	0.49	0.29	16.1	42	19	0.028	C5,M3	
7-178	Bt4	36.0	0.62	0.33	22.4	45	20	0.022	C5,M2	0.9

[†]See Tables 2.13 and 2.14 and Soil Survey Laboratory Staff (1992) for further explanation of abbreviations and methods

The Trappist soil, a clayey-skeletal, illitic, mesic Typic Haplohumult (Soil Survey Staff, 1998) from Lewis County, KY, is representative of an *in situ* soil developed in an illitic shale bedrock (Table 2.18). The 1500 kPa-water/clay content ratio indicates a well-dispersed soil. In this case then, clay contents can be used to estimate Atterberg limits from equations in Burt (1995). These fall within a range of illite and kaolinite. This soil has largely a two-component clay mineral suite of illite and kaolinite, a common occurrence in black shales. CEC/clay content ratio is low for an illite-rich soil but illustrative of an illitic-kaolinitic clay mineral suite. K_2O values indicate quantities of K^+ common in illites. CEC is well within the range attributed to illite (Table 2.6) when compared per unit clay.

The Fickle soil, a fine, vermiculitic, mesic Vertic Argiustoll (Soil Survey Staff, 1998) from McKinley County, NM, is developed in sandstone and shale alluvium having vermiculite and smectite as well as kaolinite and some illite in the clay mineral assemblage (Table 2.19). This soil disperses well but has a high CEC/clay content ratio. This agrees well with the XRD determination that suggests the presence of a significant vermiculite component. CEC values are low for a uniquely smectite assemblage but reasonable for the clay mineral assemblage described. COLE values are low for smectite but within the range of vermiculite.

The Gudgrey soil, a fine-loamy, mixed, superactive, mesic, Humic Haploxerept (Soil Survey Staff, 1998) from Mendocino County, CA, is developed in schist and chlorite-rich phyllite (Table 2.20). This soil is representative of a chlorite-illite clay mineral assemblage. K_2O values indicate the presence of illite in the solum. This soil dispersed well and has CEC/clay content ratios in the smectite and mixed classes for soil mineralogy. Both Atterberg limits and CEC are in the range for a chlorite-illite clay mineral suite.

Having presented this information, it is clear that interpretation of clay mineral behavior from analyses of the clay size fraction of a natural soil certainly is anything but straightforward. A few generalizations might be drawn from these and other comparable soil analyses. First, one cannot neglect the total amount of clay in soils. Soils generally contain much less than 100% clay. The clay mineral component upon which one focuses may comprise only a small portion of the sample. Soil properties need to be expressed in accordance with these proportions. Secondly, pure mineral species seldom, if ever, are present. Even if they were, organic material, carbonates, Fe oxides and oxyhydroxides, non-crystalline material and other factors contribute both to response of the clay mineral in its natural environment as well as to its behavior during laboratory analysis. These factors all influence the values of properties measured.

Minerals such as smectite seem to have a dominating influence on properties. Even a small amount greatly influences CEC in soils otherwise dominated by low-charge minerals such as kaolinite. Low-activity minerals such as kaolinite may be present in many soil clay fractions without altering the overall influence of the dominant mineral species. In many temperate region soils, clay minerals are often closely associated with one another, such as kaolinite with goethite and halloysite with smectite. In the latter case, since halloysite is often present in silica-rich volcanic soils, smectite is also a common constituent. Finally, values of soil properties reflect artifacts of the process of measurement as well. For example, the influence of poor dispersion, and subsequently, the underestimation of total clay content directly affects the CEC/clay content ratio. Therefore, values used to classify materials are highly method-dependent.

2.5   Engineering Properties

Mineralogy is fundamental to understanding the engineering behavior of soils. Since this volume was designed to be used primarily by specialists outside of soil science, it is assumed that most engineering properties are familiar to the reader. Therefore, only a brief discussion of common engineering properties relevant to clay minerals is presented here.

In geotechnical engineering, individuals minerals and their properties are not generally measured, rather characteristics of the whole soil that are greatly influenced by the amount and type of clay minerals in soil are measured. These characteristics include plasticity, swelling compression, strength, and fluid interaction and conductivity. Atterberg limits and grain-size analyses are commonly measured to assess these characteristics. Atterberg limits (Table 2.12) reflect both the amount and type of clay minerals. The shrinkage limit is the lower limit of volume change below which further loss of water by evaporation does not reduce the volume.

To identify the type of clay mineral present, geotechnical engineers use a value called activity,

$$A = \frac{PI}{\text{clay content}}$$ [2.10]

where A is activity, PI is plasticity index and percent clay content of the < 2 mm fraction (Table 2.21). This relationship can be represented schematically in Fig. 2.18. The similarity to the clay activity classes in Keys to Soil Taxonomy (Soil Survey Staff, 1998) should be noted. The higher the activity, the more the clay fraction influences the engineering properties of the soil and the more susceptible is the soil to changes in exchangeable cations and pore fluid composition. For example, Mg^{2+} montmorillonite has an activity of 1.24. When Na^+ is the exchangeable ion in montmorillonite, the activity is 7.09. On the other hand, kaolinite activity only varied from 0.30 to 0.41 for six different cations (Mitchell, 1993). These relations provide a useful proxy for estimating the influence that clay mineral assemblages may have on the whole-soil approach to geotechnical engineering.

Table 2.21   Engineering properties: Activities of clay minerals [From Mitchell, 1993. Fundamentals of soil science. Copyright John Wiley and Sons, New York with permission]

Mineral	Activity[†]
Smectite	1 - 7
Illite	0.5 - 1
Kaolinite	0.5
Halloysite (2 H_2O)	0.5
Halloysite (4 H_2O)	0.1

[†] Activity = $\dfrac{\text{Plasticity Index}}{\text{(Percentage} < 2\ \mu m)}$

Plasticity Index is the difference in water content between the liquid limit and the plastic limit.

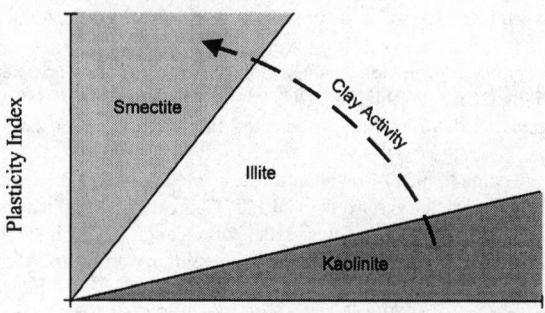

Clay Content, %

Fig. 2.18 The generalized engineering relationship between clay content, plasticity index and clay minerals. The arrow points in the direction of increasing activity as defined in the text.

2.6 References

AIPEA Nomenclature Committee. 1972. Report of Committee AIPEA. Int. Clay Min. Soc. Newsl. 7:8–13.

Altschuler, Z.S., E.J. Dwornik, and H. Kramer. 1963. Transformation of montmorillonite to kaolinite during weathering. Science 141:148–152.

Amonette, J. E. 1988. The role of structural iron in the weathering of trioctahedral micas by aqueous solutions. PhD Dissertation Iowa State University, Ames, IA (Diss. Abstr. AAC8825372).

Aylmore, L.A.G., and, J.P. Quirk. 1971. Domains and quasicrystalline region in clay systems. Soil Sci. Soc. Am. Proc. 35:652–654.

Bailey, S.W. 1975. Chlorites. p. 191–262. *In* J.E. Gieseking (ed.) Soil components. V. 2. Springer-Verlag, New York, NY.

Bailey, S.W. 1980a. Structures of layer silicates. p. 1–123. *In* G.W. Brindley and G. Brown (ed.) Crystal structures of clay minerals and their X-ray identification. Mineralogical Society, London, UK.

Bailey, S.W. 1980b. Summary and recommendations of AIPEA nomenclature committee. Clays Clay Min. 28:73–78.

Bailey, S.W. (ed.). 1984. Micas. Reviews in mineralogy. V. 13. Mineralogical Society of America, Washington, DC.

Bailey, S.W. 1988. Chlorites: Structures and crystal chemistry. *In* S.W. Bailey (ed.) Hydrous phyllosilicates (exclusive of micas). V.19. Mineralogical Society of America, Washington, DC.

Bailey, S.W. 1989. Halloysite - A critical assessment. Sci. Geol. Mem. 86:89–98.

Barnhisel, R.I., and P.M. Bertsch. 1989. Chlorites and hydroxy-interlayered vermiculite and smectite. p. 729–788. *In* J.B. Dixon and S.B. Weed (ed.) Minerals in soil environments. Soil Science Society of America, Madison, WI.

Barshad, I., and F. M. Kishk. 1969. Chemical composition of soil vermiculite clays as related to their genesis. Contrib. Mineral. Petrol. 24:136–155.

Bates, T.F. 1958. Selected electron micrographs of clays and other fine-grained minerals. College of Mineral Industries Circ. 51. The Pennsylvania State University, University Park, PA.

Bates, T.F., F.A. Hildebrand, and A. Swineford. 1950. Morphology and structure of endellite and halloysite. Am. Min. 6:237–248.

Beavers, A. H. 1957. Source and deposition of clay minerals in Peorian loess. Science 126:1285.

Bernas, B. 1968. A new method for decomposition and comprehensive analyses of silicates by atomic absorption spectrometry. Anal. Chem. 40:1682–1686.

Bertsch, P.M., and G.W. Thomas. 1983. The influence of Al exchange equilibria in soil on effective CEC measurements. Agron. Abst. p.151.

Besnus, Y., G. Fusil, C. Janot, M. Pinta, and G. Sieffermann. 1976. Characteristics of some weathering products of chromitic ultrabasic chlorites and chromiferous talc. p. 27–34. *In* S.W. Bailey (ed.) Proc. Int. Clay Conf. 1975. Applied Publ. Ltd., Wilmette, IL.

Bish, D.L. 1994. Quantitative x-ray diffraction analysis of soils. p. 267–295. *In* J.E. Amonette and L.W. Zelazny (ed.) Quantitative methods in soil mineralogy. Soil Science Society of America, Madison, WI.

Borchardt, G. 1989. Smectites. p. 675–728. *In* J.B. Dixon and S.B. Weed (ed.) Minerals in soil environments. Soil Science Society of America, Madison, WI.

Brindley, G.W., and G. Brown (ed.). 1980. Crystal structure of clay minerals and their X-ray identification. Mineralogical Society, London, UK.

Burt, R. 1995. Soil Survey Laboratory information manual. Soil Survey Investigations Rept. N. 45, Version 1.0, USDA, US Government Printing Office, Washington, DC.

Carlisle, V.W., and L.W. Zelazny. 1973. Mineralogy of selected Florida Paleudults. Soil Sci. Soc. Amer. Proc. 36:480–485.

Carroll, D. 1970. Clay minerals: A guide to their X-ray identification. Geol. Soc. Amer. Spec. Pap. 126.

Chadwick, O.A., C.G. Olson, D.M.Hendricks, E.F. Kelly, and R.T. Gavenda. 1994. Quantifying climatic effects on mineral weathering and neoformation in Hawaii. Trans. 15th World Congr. Soil Sci. 8a:94–105.

Chichester, F.W., C.T. Youngberg, and M.E. Harward. 1969. Clay mineralogy of soils formed on Mazama pumice. Soil Sci. Soc. Amer. Proc. 33:115–120.

Clay Mineral Society (CMS) Nomenclature Committee. 1984. Report for 1982 and 1983. Clays Clay Min. 32:239–240.

Colin, F., C. Parron, G. Boequier, and D. Nahon, 1981. Nickel and chromium concentrations by chemical weathering of pyroxenes and olivines. UNESCO Int. Symp. Metallog. Mafic Ultramafic Complexes: East. Mediterr. West. Asia Area. Proc. 1980. 2:56–66.

Cuadros, J., A. Delgado, A. Cardenete, E. Reyes, and J. Linares. 1994. Kaolinite/Montmorillonite resembles beidellite. Clays Clay Min. 42:5:643–651.

Deer, W.A., R.A. Howie, and J. Zussman. 1992. An introduction to the rock-forming minerals. 2nd Ed. Longman Scientific and Technical, Hong Kong.

Dixon, J.B., and T.R. McKee. 1974. Spherical halloysite formation in a volcanic soil in Mexico. Trans. 10th Int. Congr. Soil Sci. VII:115–124.

Dixon, J.B. 1989. Kaolin and serpentine group minerals. p. 467–528. In J.B. Dixon and S.B. Weed (ed.) Minerals in soil environments. Soil Science Society of America, Madison, WI.

Dixon, J.B., and S.B. Weed. 1989. Minerals in soil environments. Soil Science Society of America, Madison, WI.

Douglas, L.A. 1982. Smectites in acidic soils. p. 635–644. In H. van Olphen and F. Veniale (ed.) Proc. Int. Clay Conf. 1981. Elsevier North New York: Holland, Inc.

Douglas, L.A. 1989. Vermiculites. p. 635–674. In J.B. Dixon and S.B. Weed (eds.) Minerals in soil environments. Soil Science Society of America, Madison, WI.

Douglas, L.A., and F. Feissinger. 1971. Degradation of clay minerals by H_2O_2 treatments to oxidize organic matter. Clays Clay Min. 19:67–68.

Dudas, M.J., and M.E. Harward. 1975. Weathering and authigenic halloysite in soil developed in Mazama ash. Soil Sci. Soc. Amer. Proc. 39:561–566.

Foster, M.D. 1955. The relation between composition and swelling in clays. Clays and Clay Min. 3:205–220.

Gee, G.W., and J.W. Bauder. 1986. Particle-size analysis. p. 383–411. In A.Klute. Methods of soil analysis. Part 1. Physical and mineralogical methods. 2nd Ed. Soil Science Society of America, Madison, WI.

Giese, R.F. Jr. 1988. Kaolin minerals: Structures and stabilities. In S.W. Bailey (ed). Hydrous phyllosilicates (exclusive of micas). Rev. Mineral. 19:29–66.

Glossary of Soil Science Terms. 1997. Soil Science Society of America, Madison, WI.

Grim, R.E. 1953. Clay mineralogy. McGraw-Hill Book Co., Inc., New York, NY

Grim, R.E. 1968. Clay mineralogy. McGraw-Hill Book Co., Inc., New York, NY

Grim, R.E., R.H. Bray, and W.F. Bradley. 1937. The mica in argillaceous sediments. Am. Min. 22:813–829.

Hoffman, U.K., K. Endell, and D. Wilm. 1933. Kristallstruktur und quellung von montmorillonit. Z. Krystallogr. 86:340–348.

Hower, J., and T.C. Mowatt. 1966. The mineralogy of illites and mixed illite-montmorillonite. Am. Min. 51:825–854.

Hughes, R.E., and H.D. Glass. 1984. Mixed-layer kaolinite-smectite and heterogeneous swelling smectite as indexes of soil weathering intensity. p. 63. In Program with Abstracts, Clay Mineral Soc. Ann. Mtg. 21. Baton Rouge, LA.

Hughes, R.E., and H.D. Glass. 1990. The nature, detection, occurrences and origin of kaolinite-smectite. p. 65. In Programs with Abstracts, Clay Mineral Soc. Ann. Mtg. 27. Columbia, MO.

Hughes, R.E., and R. Warren. 1989. Evaluation of the economic usefulness of earth materials by X-ray diffraction. p. 47–57. In R.E. Hughes and J.C. Bradbury (ed.) 23rd Proc. Forum Geology Industrial Minerals, Illinois State Geol. Survey, Min. Notes 102.

Hughes, R.E., D.M. Moore, and H.D. Glass. 1994. Qualitative and quantitative analysis of clay minerals in soils. p. 330–359. In J.E. Amonette and L.W. Zelazny (eds.) Quantitative methods in soil mineralogy. Soil Science Society America, Madison, WI.

Jackson, M.L. 1964. Chemical composition of soils. p. 71–141. *In* F.E.Bear (ed.) Chemistry of the soil. Rheinhold Publishing Corp., New York, NY.

Jackson, M. L. 1975. Soil chemical analysis - Advanced course. 2nd Ed. Published by the author. Madison. WI.

Jones, R.C., and H.U. Malik. 1994. Analysis of mineral in oxide-rich soils by x-ray diffraction. p. 296–329. *In* J.E. Amonette and L.W. Zelazny (eds.) Quantitative methods in soil mineralogy. Soil Science Society of America, Madison, WI.

Karathanasis, A.D., and B.F. Hajek. 1983. Transformation of smectite to kaolinite in naturally acid soil systems: Structural and thermodynamic considerations. Soil Sci. Soc. Amer. J. 47:158–163.

Karathanasis, A.D., F. Adams, and B.F. Hajek. 1983. Stability relationships in kaolinite, gibbsite and Al-hydroxy interlayer vermiculite soil systems. Soil Sci. Soc. Amer. J. 47:1247–1251.

Karathanasis, A.D., and W.G. Harris. 1994. Quantitative thermal analysis of soil materials. p. 360–411. *In* J.E. Amonette and L.W. Zelazny (ed.) Quantitative Methods in Soil Mineralogy. Soil Science Society of America, Madison, WI.

Kato, Y. 1965. Mineralogical products of granodiorite at Shinshiro city: (IV) Mineralogical compositions of silt and clay fractions. Soil Sci. Plant Nutr. Tokyo 11:16–27.

Kirkland, D.L., and B.F. Hajek. 1972. Formula derivation of Al-interlayered vermiculite in selected soil clays. Soil Sci. 114:317–322.

Kittrick, J.A. 1969. Interlayer forces in montmorillonite and vermiculite. Soil Sci. Soc. Amer. Proc. 33:217–222.

Klein, C., and C.S. Hurlbut, Jr. 1985. Manual of mineralogy. 20th Ed. John Wiley and Sons, New York, NY.

Kunze, G.W., and J.B. Dixon. 1986. Pretreatment for mineralogical analysis. p. 91–100. *In* A. Klute (ed.), Methods of soil analysis, Part 1. American Society of Agronomy, Madison, WI.

Lagaly, G. and A. Weiss. 1969. Determination of layer charge in mica-type layer silicates. p. 61–80. *In* L. Heller (ed.) Proc. Int. Clay Conf. 1969. Tokyo. Israel Prog. Sci. Transl., Jerusalem, Israel.

Laird, D.A. 1996. Model for crystalline swelling of 2:1 phyllosilicates. Clays Clay Min. 44:553–559.

Laird, D.A., and R.H. Dowdy. 1994. Preconcentration techniques in soil mineralogical analyses. p. 236–266. *In* J.E. Amonette and L.W. Zelazny (ed.) Quantitative methods in soil mineralogy. Soil Science Society of America, Madison, WI.

Laird, D.A., R.H. Dowdy, and R.C. Munter. 1991. Suspension nebulization analysis of clays by inductively coupled plasma-atomic emission spectrometry. Soil Sci. Soc. Am. 55:274–278.

Lamb, T.W., and R.V. Whitman. 1969. Soil Mechanics. John Wiley and Sons, New York, NY.

MacEwan, D.M.C., A.R. Amil, and G. Brown. 1961. Interstratified clay minerals. p. 393–445. *In* G. Brown (ed.) The X-ray identification and crystal structure of clay minerals. 2nd Ed. Mineralogical Society, London, UK.

Mackenzie, R.C. 1970. Differential thermal analysis. Vol. 1. Academic Press, London, UK.

McBride, K.C., J.B. Dixon, and T.R. McKee. 1976. Spheroidal and tubular halloysite in a volcanic deposit of Washoe county, Nevada. 25th Ann. Clay Miner. Conf. Abstr. Corvallis, OR.

McDaniel, P.A., A.L. Falen, K.R. Tice, R.C. Graham, and S.E. Fendorf. 1995. Beidellite in E horizons of northern Idaho Spodosols formed in volcanic ash. Clays Clay Min. 43:5:525–532.

Mehra, O.P., and M.L. Jackson. 1960. Iron oxide removal from soils and clays by a dithionite-citrate system buffered with sodium bicarbonate. Clays Clay Min. 7:317–327.

Mekonnen, E.J., E.A. Nater, and D.A. Laird. 1993. Endpoint determination during excess salt removal by alcohol washing. Soil Sci. Soc. Am. J. 57:874–877.

Miller, R.W., and R.L. Donahue. 1995. Soils in our environment. 7th Ed. Prentice-Hall, NJ.

Mitchell, J.A.1993. Fundamentals of soil behavior. 2nd Ed. John Wiley and Sons, New York, NY.

Moore, D.M., and R.C. Reynolds, Jr. 1997. X-ray diffraction and the identification and analysis of clay minerals. 2nd Ed. Oxford University Press, New York, NY.

Morgan, D.J., D.E. Highley, and D.J. Bland. 1979. A montmorillonite, kaolinite association in the Lower Cretaceous of south-east England. p. 301–310. *In* M.M. Mortland and V.C. Farmer (ed.) Proc. Int. Clay Conf. 1978. Pergamon Press Ltd., Oxford, UK.

Nadeau, P.H. 1985. The physical dimensions of fundamental clay particles. Clay Min. 20:499–514.

Nadeau, P.H., and D.C. Bain. 1986. Composition of some smectites and diagenetic illitic clays and implications for their origin. Clays Clay Min. 34:455–464.

Nadeau, P.H., and J.M. Tait. 1987. Transmission electron microscopy. p. 209–247. *In* M.J. Wilson (ed.) A handbook of determinative methods in clay mineralogy. Blackie, London, UK.

Nadeau, P.H., M.J. Wilson, W.J. McHardy, and J.M. Tait. 1984. Interparticle diffraction: A new concept for interstratified clays. Clay Min. 19:757–769.

Nahon, D.B., and F. Colin. 1982. Chemical weathering of orthopyroxenes under lateritic conditions. Am. J. Sci. 282:1232–1243.

Nemecz, E. 1981. Clay minerals. Akad. Jmiai Kiad., Budapest, Hungary.

Nettleton, W.D., and B.R. Brasher. 1983. Correlation of clay minerals and properties of soils in the western United States. Soil Sci. Soc. Am. J. 47:1032–1036.

Nettleton, W.D., K.W. Flach, and R.E. Nelson. 1970. Pedogenic weathering of tonalite in southern California. Geoderma 4:387–402.

Norrish, K. 1954. The swelling of montmorillonite. Disc. Farad. Soc. 18:120–134.

Olson, C.G. 1988. Clay mineral contribution to the weathering mechanisms in two contrasting watersheds. J. Soil Sci. 39:457–467.

Parker, J.C. 1986. Hydrostatics of water in porous media. p. 209–296. In D.L. Sparks (ed). Soil physical chemistry. CRC Press, Boca Raton, FL.

Paterson, E., and R. Swaffield. 1987. Thermal analysis. p. 99–132. In M.J. Wilson (ed.) A handbook of determinative methods in clay mineralogy. Blackie, London, UK.

Proust, D. 1982. Supergene alteration of hornblende in an amphibolite from Massif Central, France. p. 357–364. In H. van Olphen and F. Veniale (ed.) Proc. Int. Clay Conf. 1982. Elsevier Science Publishing Co., Amsterdam, Netherlands.

Reynolds, R.C. Jr. 1980. Interstratified clay minerals. In B.W. Brindley and G. Brown (ed.) Crystal structures of clay minerals and their X-ray identification. Mineralogical Society, London, UK.

Reynolds, R.C. Jr. 1986. Interstratified clays. In A.G. Cairns-Smith and H. Hartman (ed.) Clay minerals and the origin of life. Cambridge University Press, New York, NY.

Rich, C.I., and S.S. Obenshain. 1955. Chemical and clay mineral properties of a Red-Yellow Podzolic soil derived from muscovite schist. Soil Sci. Soc. Am. Proc. 19:334–339.

Rich, C.I. 1968. Hydroxy interlayers in expansible layer silicates. Clays Clay Min. 16:15–30.

Rich, C.I., 1958. Muscovite weathering in a soil developed in the Virginia Piedmont. Clays Clay Min. 5:203–212.

Ross, G.J., and M.M. Mortland. 1966. A soil beidellite. Soil Sci. Soc. Amer. Proc. 30:337–343.

Ross, G.J., C. Wang., A.I. Ozkan, and H.W. Rees. 1982. Weathering of chlorite and mica in a New Brunswick Podzol developed on till derived from chlorite-mica schist. Geoderma 27:255–267.

Ruhe, R.V., and C.G. Olson. 1979. Estimate of clay mineral content: Additions of proportions of soil clay to constant standard. Application of method to study of loesses in Indiana. Clays Clay Min. 27:322–326.

Russell, J.D. 1987. Infrared methods. p. 133–173. In M.J. Wilson (ed.) A handbook of determinative methods in clay mineralogy. Blackie, London, UK.

Salisbury, J.W., L.S. Walter, and N. Vergo. 1987. Mid-Infrared (2.1 – 25 m) spectra of minerals. 1st Ed. Open-File Report 87-263. US Geological Survey, Reston, VA.

Sawhney, B.L. 1989. Interstratificaiton in layer silicates. p. 789–828. In J.B. Dixon and S.B. Weed (ed.) Minerals in soil environments. Book Series 1. Soil Science Society of America, Madision, WI.

Schulze, D.G. 1989. An introduction to soil mineralogy. p. 1–34. In J.B. Dixon and S.B. Weed (ed.) Minerals in soil environments. Book Series 1. Soil Science Society of America, Madison, WI.

Schultz, L.G., A.O. Shepard, P.D. Blackmon, and H.C. Starkey. 1971. Mixed-layer kaolinite-montmorillonite from the Yucatan Peninsula Mexico. Clays Clay Min. 19:137–150.

Sequi, P., and R. Aringhieri. 1977. Destruction of organic matter by hydrogen peroxide in the presence of pyrophosphate and its effect on specific surface area. Soil Sci. Soc. Am. J. 41:340–342.

Sherman, G.D., H. Ikawa, G. Uehara, and E. Okazaki. 1962. Types of occurrences of nontronite and nontronite-like minerals in soils. Pac. Sci. 16:57–62.

Shirozu, H., and S.W. Bailey. 1966. Crystal structure of a two-layer Mg-vermiculite. Am. Min. 51:1124–1143.

Singh, B. 1996. Why does halloysite roll? - A new model. Clays Clay Min. 44:2:191–196.

Singh, B. and I.D.R. Mackinnon. 1996. Experimental transformation of kaolinite to halloysite. Clays Clay Min. 44:6:825–834.

Slade, P.G., P.A. Stone, and E.W. Radoslovich. 1985. Interlayer structures of two-layer hydrates of Na- and Ca-vermiculites. Clays Clay Min. 33:51–61.

Smith, W.W. 1962. Weathering of some Scottish basic igneous rocks with reference to soil formation. J. Soil. Sci. 13:202–215.

Soil Survey Laboratory Staff. 1992. Soil survey laboratory methods manual. USDA-SCS Soil Sur. Inves. Rep. 42. US Government Printing Office, Washington, DC.

Soil Survey Staff. 1998. Keys to soil taxonomy. 8th Ed. USDA-NRCS. US Government Printing Office, Washington, DC.

Sparks, D.L. 1995. Environmental soil chemistry. Academic Press, New York, NY.

Speer, J.A. 1984. Micas in igneous rocks. Rev. Mineral.13:299–356.

Środoń, J., D.J. Morgan, E.V. Eslinger, D.D. Eberl, and M.R. Karlinger. 1986. Chemistry of illite/smectite and end-member illite. Clays Clay Min. 34:368–378.

Sudo, T., and H. Hayashi. 1956. A randomly interstratified kaolin-montmorillonite in acid clay deposits in Japan. Nature 178:1115–1116.

Sudo, T., S. Shimoda, H. Yotsumoto, and S. Aita. 1981. Electron micrographs of clay minerals. Elsevier North Holland, Inc., New York, NY.

Tan, K.H., B.F. Hajek, and I. Barshad. 1986. Thermal analysis techniques. p. 151–183. *In* A. Klute (ed). Methods of soil analysis Part 1. 2nd Ed. Soil Science Society of America, Madison, WI.

Tchoubar, C. 1984. X-ray studies of defects in clays. Phil. Trans. R. Soc. Lond. A 311:259–269.

Tessier, D. 1984. Etude experimentale de l'organization des materiaux argileux: Hydratation, gonflement et structuration au cours de la desiccation et de la réhumectation. Institut National de la Recherche Agronomique, Versailles, France.

Thompson M.L., N.E. Smeck, and J.M. Bigham. 1981. Parent materials and paleosols in the Teays River Valley, Ohio. Soil Sci. Soc. Am. J. 45:918–925.

van der Marel, H.W., and H. Beutelspacher. 1976. Atlas of infrared spectroscopy of clay minerals and their admixtures. Elsevier, Amsterdam, Netherlands.

van Olphen, H. 1965. Thermodynamics of interlayer adsorption of water in clays. J. Colloid Sci. 20:822–837.

Velde, B. (ed.) 1995. Origin and mineralogy of clays: Clays and the environment. Springer-Verlag, New York, NY.

Weaver, C.E. 1956. The distribution and identification of mixed-layer clays in sedimentary rocks. Amer. Min. 41:3–4:202–221.

Weaver, C.E., and L.D. Pollard. 1973. The chemistry of clay minerals. Elsevier Sci. Publishers, , Netherlands.

White, J.L., and C. B. Roth. 1986. Infrared spectroscopy. p. 291–339. *In* A. Klute (ed.), Methods of Soil Analysis, Part 1. American Society of Agronomy, Madison, WI.

Whittig, L.D., and W.R. Allardice. 1986. X-ray diffraction techniques. p. 331–362. *In* A. Klute (ed.), Methods of soil analysis, Part 1. American Society of Agronomy, Madison, WI.

Wilson, M.A., R. Burt, T.M. Sobecki, R.J. Engel, and K. Hipple. 1996. Soil properties and genesis of pans in till-derived Andisols, Olympic Peninsula, Washington. Soil Sci. Soc. Amer. J. 60:206–218.

Wilson, M.J., and P.W. Cradwick. 1972. Occurrence of interstratified kaolinite-montmorillonite in some Scottish soils. Clay Min. 9:435–437.

Zelazny, L.W., V.W. Carlisle, and F.G. Calhoun. 1975. Clay mineralogy of selected Paleudults from the Lower coastal Plain. Agron. Abst. p. 180.

Zelazny, L.W., D.A. Lietzke, and H.L. Barwood. 1980. Septic tank drainfield failures resulting from mineralogical changes. VA Water Resour. Cent. Bull. 184.

Zelazny, L.W., and G.N. White. 1989. The pyrophyllite-talc group. p. 527–550. *In* J.B. Dixon and S.B. Weed (ed.) Minerals in soil environments.Book Series 1. Soil Science Society of America, Madison. WI.

Zvyagin, B.B., and V.A. Drits. 1996. Interrelated features of structure and stacking of kaolin mineral layers. Clays Clay Min. 44:3:297–303.

3

Oxide Minerals

Nestor Kämpf
Federal University of Rio Grande do Sul, Brazil

Andreas C. Scheinost
University of Delaware

Darrell G. Schulze
Purdue University

3.1 Introduction

Oxide minerals or oxides comprise the oxides, hydroxides, oxyhydroxides, and hydrated oxides of Fe, Mn, Al, Si, and Ti and commonly occur in soils, particularly those in advanced stages of weathering where they can account for as much as 50% of the total soil mass. In the discussion to follow, a polyhedral approach will be used to introduce the mineral species and highlight their major differences and similarities. For additional structural details, the reader is referred to the literature cited. Thereafter, their occurrence and formation in soils, and their influence on soil properties will be discussed, concluding with the major techniques for identifying these minerals in soils.

3.2 Iron Oxides

Iron oxide minerals are ubiquitous in soils and sediments. All Fe oxide minerals are strongly colored and, when finely dispersed throughout the matrix, even small amounts impart bright colors to soils. Iron oxides often cement other soil minerals into stable aggregates, and when they reach massive proportions, such cementations are called laterite or ferricrete. Because iron oxide surfaces have a strong affinity for the oxyanions of P and some transition metals, they play a significant role in the environmental cycling of these elements. A list of the Fe oxides is given in Table 3.1. Recent comprehensive reviews include Schwertmann and Taylor (1989), and Cornell and Schwertmann (1996).

3.2.1 Mineral Phases

The structures of most Fe oxides can be described in terms of close packed arrays of O and OH ions, with Fe occupying interstitial octahedral sites. Structures based on hexagonal close packing (hcp) of the anions are called α-, and those with cubic close packing (ccp), γ-phases (note that this convention is not necessarily followed for oxides of other elements). Both hcp and ccp structures also contain tetrahedral interstices. In two minerals, magnetite and maghemite, Fe is present in some of these tetrahedral sites as well. The basic structural unit for most of the Fe oxides, however, is an octahedron

Table 3.1   Iron oxide minerals: crystallographic properties and d-values for the 6 most intense diffraction lines

Mineral	Chemical formula	Crystal system, Space group	Unit cell dimensions (Å)	Six most intense diffraction lines d-value (Å), relative intensity						Reference[†]
Goethite	α-FeOOH	Orthorhombic Pbnm	a=4.608 b=9.956 c=3.021	4.183 100	2.450 50	2.693 35	1.719 20	2.190 18	2.253 14	29-0713
Lepidocrocite	γ-FeOOH	Ortho-rhombic Cmcm	a=3.88 b=12.53 c=3.07	6.260 100	3.290 90	2.470 80	1.937 70	1.732 40	1.524 40	Cornell and Schwertmann (1996); 08-0098
Akaganéite	β-FeOOH	Monoclinic I2/m	a=10.56 b=3.031 c=10.483 β=90°63'	3.333 100	2.550 55	7.467 40	2.295 35	1.643 35	5.276 30	Post and Buchwald (1991); 34-1266
Schwert-mannite	Fe_8O_8 $(OH)_6 SO_4$	Tetragonal P4/m	a=10.66 c=6.04	2.55 100	3.39 46	4.86 37	1.51 24	2.28 23	1.66 21	Bigham et al. (1994)
Feroxyhyte	δ'-FeOOH	Hexagonal P3ml	a=2.93 c=4.56	2.545 100	2.255 100	1.685 100	1.471 100	4.610 20	1.271 20	Cornell and Schwertmann (1996); 13-0087
Ferrihydrite	$Fe_5HO_8 \cdot$ $4H_2O$	Hexagonal	a=5.08 c=9.4	2.50 100	2.21 80	1.96 80	1.48 80	1.51 70	1.72 50	Towe and Bradley (1967); 29-0712
Hematite	α-Fe_2O_3	Hexagonal R3c	a=5.034 c=13.752	2.700 100	2.519 70	1.694 45	1.840 40	3.684 30	1.486 30	33-0664
Magnetite	Fe_3O_4	Cubic Fd3m	a=8.3967	2.532 40	1.484 30	2.967 30	1.616 20	2.099 12	1.093 100	19-0629
Maghemite	γ-Fe_2O_3	Cubic Fd3m	a=8.35	2.518 100	2.953 40	1.476 30	1.607 20	2.088 15	1.704 10	Goss (1988); 39-1346
Bernalite	$Fe(OH)_3$	Orthorhombic Immm	a=7.544 b=7.560 c=7.558	3.784 100	1.692 17	2.393 16	2.676 15	1.892 10	1.545 9	Birch et al. (1993)
$Fe(OH)_2$	$Fe(OH)_2$	Hexagonal P3ml	a=3.262 c=4.596	4.597 vs[‡]	2.403 vs	2.817 s	1.782 s	1.629 s	1.535 w	Miyamoto (1976)
Green rust I	$Fe(OH)_2$ $Fe(OH)_3Cl$ (variable)	Rhombohedral	a=3.198 c=24.21	8.02 vs	4.01 s	2.701 m	2.408 m	2.037 w	1.487 w	Bernal et al. (1959)
Green rust II	$Fe(OH)_2$ $Fe(OH)_3SO_4$ (variable)	Hexagonal R 3/m	a=3.174 b=10.94	10.92 vs	5.48 s	3.65 s	2.747 m	2.660 ms	2.459 ms	Bernal et al. (1959)

[†] numbers of the format XX-XXXX indicate ICDD file number (ICDD, 1994).

[‡] vs = very stong, s = strong, ms = moderately strong, m = moderate, w = weak

in which each Fe atom is surrounded either by six O or by both O and OH ions. The various Fe oxides differ mainly in the arrangement of these octahedra and in how they are linked.

3.2.1.1   Oxyhydroxides

Goethite (α-FeOOH) consists of double chains of edge-shared octahedra that are joined to other double chains by sharing corners and by hydrogen bonds (Fig. 3.1) (Cornell and Schwertmann, 1996). *Lepidocrocite* (γ-FeOOH) also contains double chains of octahedra, but they are joined by shared edges, resulting in corrugated sheets of octahedra. These corrugated sheets are stacked one on top of the other and are held together by hydrogen bonds (Fig. 3.1) (Ewing, 1935a; Fasiska, 1967). *Akaganéite* (β-FeOOH) consists of double chains that share corners with adjacent chains to give a three-dimensional structure containing tunnels 0.5 nm in cross-section (Fig. 3.1). These tunnels contain Cl^- ions that stabilize the structure (Post and Buchwald, 1991). *Schwertmannite* [$Fe_8O_8(OH)_6(SO_4)$]

Mn: Pyrolusite, MnO_2
Manganite, $\gamma\text{-}MnOOH$
Ti: Rutile, TiO_2

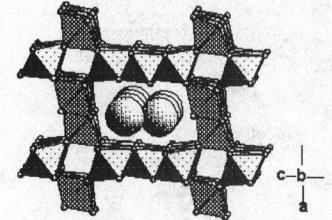

Mn: Romanechite, $Ba_{0.66}(Mn^{4+}Mn^{3+})_5O_{10} \cdot$ 1.34 H_2O

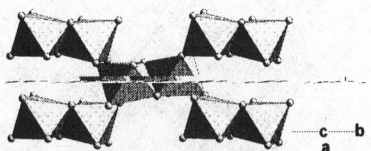

Fe: Goethite, $\alpha\text{-}FeOOH$
Mn: Ramsdellite, MnO_2
Groutite, $MnOOH$
Al: Diaspore, $\alpha\text{-}AlOOH$

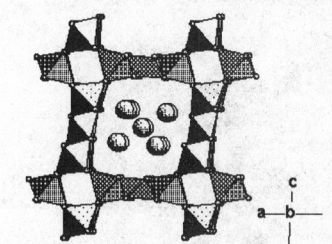

Mn: Todorokite,
$(Na, Ca, K)_{0.3-0.5}(Mn^{4+}Mn^{3+})_6O_{12} \cdot$ 3.5 H_2O

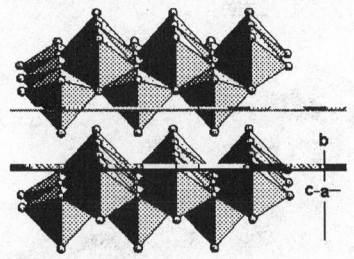

Fe: Lepidocrocite, $\gamma\text{-}FeOOH$
Al: Boehmite, $\gamma\text{-}AlOOH$

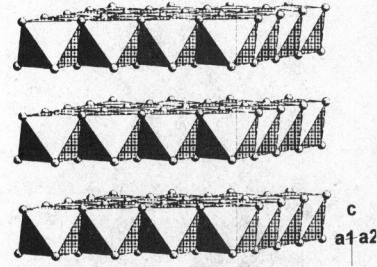

Fe: Feroxyhyte, $\delta'\text{-}FeOOH$ (A=B=Fe)
Mn: Lithiophorite, $(Al,Li) MnO_2 (OH)_2$
(A=Mn, B=Al, Li)
Absolane, $(Co,Ni) MnO_2 (OH)_2$
(A=Mn, B=Co, Ni)
Al: Gibbsite, Bayerite, Nordstrandite,
$Al(OH)_3$, (A=B=Al)

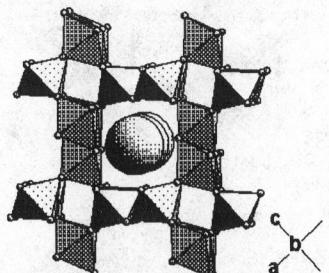

Fe: Akaganéite, $\beta\text{-}FeOOH$
Schwertmannite, $Fe_8O_8(OH)_6(SO_4)$
Mn: Hollandite, $Ba_x(Mn^{4+}Mn^{3+})_8O_{16}$, x=1
Cryptomelane, $K_x(Mn^{4+}Mn^{3+})_8O_{16}$, x=1.3–1.5
Coronadite, $Pb_x(Mn^{4+}Mn^{3+})_8O_{16}$, x=1–1.4

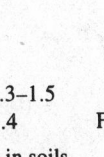

Fe: Green rust, $Fe^{3+}_2Fe^{2+}_4(OH)_{12}(Cl, SO_4, CO_3)$

Fig. 3.1 Structural scheme for oxide minerals in soils

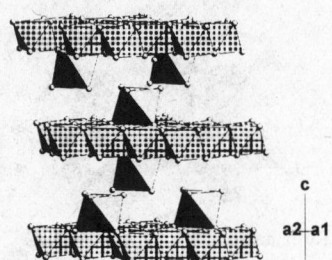

Fe: Chalcophanite, ZnMn$_3$O$_7$·3H$_2$O

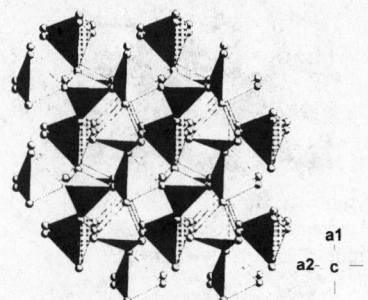

Si: α-Cristobalite, SiO$_2$

Fe: Hematite, α-Fe$_2$O$_3$
Al: Corundum, α-Al$_2$O$_3$
Ti, Fe: Ilmenite, FeTiO$_3$

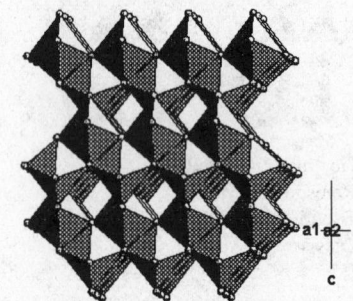

Ti: Anatase, TiO$_2$

Ti: Brookite, TiO$_2$

Fe: Magnetite, Fe$_3$O$_4$
        Maghemite, γ-Fe$_2$O$_3$
Mn: Hausmannite, Mn$_3$O$_4$
Ti, Fe: Titanomagnetite, Fe$_{2-x}$Ti$_x$O$_3$

Fe, Ti: Pseudobrookite, Fe$^{3+}_2$TiO$_2$

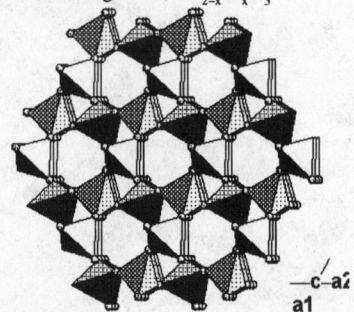

Ti, Fe: Pseudorutile, Fe$_2$Ti$_3$O$_8$(OH)$_2$

Si: α-Quartz, SiO$_2$

Fig. 3.1   (Cont.)

is isostructural with akaganéite, but instead of Cl^-, SO_4^{2-} occupies the tunnels (Fig. 3.1) (Bigham et al., 1990, 1994). *Feroxyhite* (δ'-FeOOH), consists of sheets of edge-sharing octahedra (Fig. 3.1), with the presence of face-sharing octahedra (Manceau and Combes, 1988; Drits et al., 1993a; Manceau and Drits, 1993). In naturally occurring feroxyhite, Fe^{3+} ions are randomly distributed over the octahedral sites, while in synthetic δ-FeOOH the Fe^{3+} ions are distributed in an orderly manner over half of the octahedral sites (Waychunas, 1991). *Ferrihydrite* ($Fe_5HO_8 \cdot 4H_2O$) is a poorly ordered Fe oxide, with a variable degree of ordering. The structure is still being investigated (Drits et al., 1993b; Manceau and Gates, 1997), but can be visualized as a defective hematite structure containing both edge- and face-sharing octahedra (Towe and Bradley, 1967). The Fe^{3+} ions are randomly distributed over the octahedral interstices, with many sites vacant, and more OH^- and H_2O and less Fe^{3+} than in hematite (Cornell and Schwertmann, 1996).

3.2.1.2 Hydroxides
Green rusts are a group of Fe^{2+}-Fe^{3+} hydroxy salts with a structure consisting of sheets of edge-shared $Fe^{2+}(OH)_6$ octahedra in which some of the Fe^{2+} is replaced by Fe^{3+}, creating a positive layer charge. The charge is balanced by anions such as Cl^-, SO_4^{2-}, and CO_3^{2-} located between the octahedral sheets (Fig. 3.1) (Brindley and Bish, 1976; Taylor and McKenzie, 1980). *Bernalite* ($Fe(OH)_3 nH_2O$) (Birch et al., 1993), and a $Fe(OH)_2$ compound are exceedingly rare Fe hydroxides, and are unlikely to occur in soils.

3.2.1.3 Oxides
Hematite (α-Fe_2O_3) consists of sheets of edge-shared octahedra with two-thirds of the available octahedral sites filled with Fe^{3+} ions. The unoccupied sites are regularly arranged to form sixfold rings of occupied octahedra analogous to dioctahedral phyllosilicate sheets (Section F, Chapter 2), and are stacked along the c-axis. Each plane of O is shared by two adjacent dioctahedral sheets. Each octahedron shares three edges with three neighboring octahedra in the same sheet, and a face and six corners with 9 octahedra in adjacent sheets (Fig. 3.1) (Blake et al., 1966; Maslen et al., 1994). *Magnetite* (Fe_3O_4), which differs from most other Fe oxides in that it contains both Fe^{2+} and Fe^{3+} ions, has an inverse spinel structure consisting of octahedral and mixed tetrahedral/octahedral layers stacked along the (111) plane, with Fe^{3+} occupying tetrahedral sites and both Fe^{2+} and Fe^{3+} in octahedral sites (Fig. 3.1). *Titanomagnetites* ($Fe_{3-x}Ti_xO_4$) are solid solutions of magnetite with ulvöspinel (Fe_2TiO_4), in which Ti^{4+} occupies only octahedral sites (Wechsler et al., 1984). *Maghemite* (γ-Fe_2O_3) has the chemical composition of hematite, but has a structure analogous to magnetite (Fig. 3.1). Most or all of the Fe occurs as Fe^{3+} and cation vacancies preserve charge balance caused by the oxidation of Fe^{2+} to Fe^{3+}. The cations are randomly distributed over the tetrahedral and octahedral sites. The vacancies are also randomly distributed but are confined to the octahedral sites. *Titanomaghemites* are oxidation products of titanomagnetites, but their structure needs clarification (Waychunas, 1991). Maghemite-magnetite solid solutions are formed by varying degrees of oxidation of magnetite.

Other metallic cations with an ionic diameter similar to Fe^{3+} (Al, Ni, Ti, Mn, Co, Cr, Cu, Zn, V) may replace the Fe ion in the structure of various Fe oxides. Isomorphous substitution of Al^{3+} for Fe^{3+} occurs most frequently and to the greatest extent (Schwertmann and Carlson, 1994). In addition to M^{3+} cations, M^{2+} and M^{4+} cations may also enter Fe^{3+} oxide structures, but the uptake is usually less than 0.1 mol mol^{-1} (Cornell and Schwertmann, 1996).

The typical shapes of well-crystalline Fe oxide samples are listed in Table 3.2. These shapes, usually observed in samples from synthetic preparations, rocks, ferricretes and bauxites, are frequently less well expressed by soil Fe oxides. Soil hematites usually have a granular texture, and goethites show almost no acicularity (Fig. 3.2), particularly in highly weathered soils (Schwertmann

Table 3.2   Diagnostic criteria for iron oxide minerals

Mineral	Median color Range of hue	Typical crystal shape	DTA events °C	IR bands cm^{-1}	Magnetic hyperfine field (kOe)		
					295 K	77 K	4 K
Goethite	strong yellowish-brown 7.5YR - 2.5Y	Needles, laths	Endotherm 280 - 400	890, 797	382	503	506
Lepidocrocite	moderate orange 5YR - 7.5YR	Laths	Endotherm 300 - 350 Exotherm 370 - 500	1026, 1161, 753			458
Akaganeite	strong brown 2.5YR - 7.5YR	Spindle-shaped rods					489/478/ 473
Schwertmannite	dark orange yellow 5YR - 10YR	Fibers	Endotherm 100 - 300, 650 - 710 Exotherm 540 - 580	3300, 1634, 1186, 1124, 1038, 976, 608			456
Ferrihydrite	brownish orange 2.5YR - 9YR	Spherical	Endotherm 150				470/500
Feroxyhite	strong brown 1.5YR - 5YR	Fibers, needles	Endotherm 250	1110, 920, 790, 670	420	530	535
Hematite	moderate reddish brown 3.5R - 4YR	Hexagonal plates	Nil	345, 470, 540	518	542/535[‡]	542/535[‡]
Maghemite	dark yellowish brown 6 YR - 9.5YR	Cubes	Exotherm 600 - 800[†]	400, 450, 570, 590, 630	500		526
Magnetite	black	Cubes		400, 590	491/460[¶]		

[†] Magnetite converts via maghemite or directly to hematite, depending on particle size.

[‡] With and without Morin transition, respectively.

[¶] For tetrahedral and octahedral Fe, respectively.

and Kämpf, 1985; de Brito Galvão and Schulze, 1996). Thus, morphology alone is unreliable for distinguishing these minerals by transmission electron microscopy (Anand and Gilkes, 1987a; Singh and Gilkes, 1992), although soil lepidocrocites often appear as thin laths very similar to synthetic lepidocrocites (Cornell and Schwertmann, 1996). Soil Fe oxides may range from a few to several hundred nm in length, but data are scarce because of their tendency to occur as microaggregates (Schwertmann, 1988).

3.2.2   Occurrence and Formation

The concentration of Fe oxides in soils ranges between < 1 and > 500 g kg^{-1}, and is related to parent material, degree of weathering, and pedogenic accumulation or depletion processes. Iron oxides may occur evenly distributed in the matrix or concentrated as ferricretes, layers, bands, horizons, nodules, mottles, plinthite, etc.

Iron, which is present as Fe^{2+} in primary minerals (mainly silicates) of most rocks, is released during weathering through protolysis and oxidation processes, hydrolyzes in contact with water and forms Fe^{3+} oxides. In most aerobic environments, the Fe oxides are evenly distributed resulting in homogeneous soil color. Under anaerobic conditions, however, Fe oxides may be reduced and dissolved, thus leading to a heterogeneous distribution of Fe oxides and color (redoximorphic features) (Schwertmann, 1993). The formation of each of the Fe oxide minerals requires specific conditions which have been established for soils by laboratory and field observations (Schwertmann and Taylor, 1989).

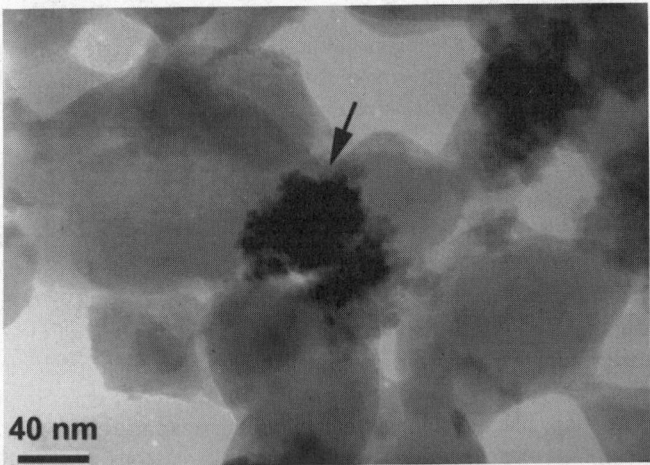

Fig. 3.2 Transmission electron micrograph of the < 0.2 μm fraction of a Rhodic Hapludalf showing a goethite or hematite particle (arrow) surrounded by clean kaolinite particles [Reprinted from de Brito Galvão and Schulze, 1996. Soil Sci. Soc. Am. J. 60:1969-1978 with permission of the Soil Science Society of America]

3.2.2.1 Aerobic Environments

Goethite and hematite are the most common Fe oxides due to their high thermodynamic stability. Their formation pathway starts with Fe^{2+} released by weathering and its immediate oxidation to Fe^{3+}, or by any other source of Fe^{3+} (e.g., dissolution of Fe oxides) which precipitates as ferrihydrite or goethite, depending on which of the solubility products is exceeded first (ferrihydrite $K_{sp} \sim 10^{-39}$ or goethite $10^{-44} < K_{sp} < 10^{-41}$). Hematite forms from its necessary precursor, ferrihydrite, through a solid-state reaction in which individual hematite crystals nucleate and grow within individual ferrihydrite aggregates by a dehydration and rearrangement process. All of the iron in each individual hematite crystal is derived from a single ferrihydrite aggregate (Cornell and Schwertmann, 1996). The transformation of ferrihydrite to goethite, however, proceeds via a dissolution-precipitation process. The presence of face-sharing $Fe(O,OH)_6$ octahedra in ferrihydrite, as in hematite, prevents its direct solid state transformation into goethite, which consists only of edge- and corner-sharing octahedra. Thus, the transformation of ferrihydrite to goethite requires the breakdown of the oxo-bridges of face-sharing octahedra (Combes et al., 1989). Goethite forms from Fe^{3+} ions in solution through a nucleation-crystal growth process. Hence any Fe source (minerals, biological exudates, organic compounds) able to keep a low Fe^{3+} activity in solution will favor goethite. Thus, the formation pathways of hematite and goethite are different, but competitive.

Environmental conditions that favor the formation of ferrihydrite and its subsequent transformation to hematite are a high Fe content in the parent rock (resulting in higher Fe release rate), near neutral pH (pH of minimum solubility of ferrihydrite), higher temperature or lower water activity (favoring the dehydration step) (Torrent et al., 1982), and rapid turnover of biomass (low Fe complexation) (Schwertmann, 1988). Such environmental conditions are usually related to climate (Kämpf and Schwertmann, 1983), landscape (Curi and Franzmeier, 1984; Schwertmann and Latham, 1986), landscape associated with drainage (Macedo and Bryant, 1987; Motta and Kämpf, 1992; Peterschmitt et al., 1996), and soil depth (Bigham et al., 1978; Kämpf and Schwertmann, 1983).

In soils, the transformation rate of ferrihydrite to hematite is probably very rapid, whereas that of ferrihydrite to goethite, which proceeds through dissolution-recrystallization, is slower. This explains the widespread association of ferrihydrite-goethite, while that of ferrihydrite-hematite has been rarely found (Parfitt et al., 1988). There is no pedogenic indication for a solid-state transformation of goethite to hematite by simple dehydration, nor the inverse by simple hydration. However, high temperatures caused by forest or bush fires may transform goethite into hematite, and lepidocrocite into maghemite (Stanjek, 1987) (see maghemite below).

Both hematite and goethite have a similar low solubility (Lindsay, 1979). Under moderate reducing conditions, however, a transformation of red into yellow soils (yellowing or xanthization) may take place due to a preferential dissolution of hematite over goethite (Macedo and Bryant, 1989; Jeanroy et al., 1991; Peterschmitt et al., 1996). The increased resistance of goethite to dissolution is explained by a higher Al for Fe substitution (Torrent et al., 1987; Trolard and Tardy, 1987). The solubility and dissolution of Fe oxides is reviewed by Schwertmann (1991). When Fe oxides are reduced in higher landscape positions, the soluble Fe^{2+} may be transported to lower landscape positions and reprecipitate as Fe oxides (Schwarz, 1994). In aquic environments with prolonged waterlogging, the complete removal of Fe oxides may take place, resulting in soil bleaching (chroma < 2) (Motta and Kämpf, 1991; Peterschmitt et al., 1996). The bleached color is due to the matrix minerals (phyllosilicates, quartz, etc.) in the absence of Fe oxides.

Many species of microorganisms, mainly anaerobic bacteria, are capable of reducing Fe oxides (Lovley, 1995). As heterotrophic organisms, they depend on available biomass for metabolic oxidation so that reduction in soils is most intense in the upper horizons. The process usually involves enzymatic transfer of electrons by microorganisms from the decomposing biomass to Fe^{3+} (Ghiorse and Ehrlich, 1992; Lovley, 1992, 1995).

Some Fe and Mn oxides may be formed by microorganisms in a biologically induced form outside the cell, or matrix mediated by internal cellular precipitation (Ghiorse and Ehrlich, 1992). The most common occurrence of biotic formation of Fe oxides is Fe ochre in field drains (Houot and Berthelin, 1992), and the Fe precipitate, yellow boy, in acid mine drainages (Bigham et al., 1992). Aspects of biomineralization of Fe and Mn in soils are reviewed in Skinner and Fitzpatrick (1992).

3.2.2.2   Reduced Soil Environments
The typical Fe oxides are goethite, lepidocrocite and ferrihydrite, formed by abiotic or biotic processes, while hematite is restricted to mottles and nodules. Goethite and lepidocrocite are widely associated in redoximorphic soils (Schwertmann and Kämpf, 1983; Fitzpatrick et al., 1985; Wang et al., 1993; Anjos et al., 1995). While goethite can form from either Fe^{2+} or Fe^{3+} ions in solution, lepidocrocite in soils seems to require the presence of Fe^{2+} ions (Schwertmann and Taylor, 1989). Thus, lepidocrocite, recognized by its bright orange color, indicates prevailing redoximorphic conditions in a soil profile leading to the formation of Fe^{2+}. Factors, however, that favor the formation of goethite over lepidocrocite are a higher partial pressure of CO_2, normally found closer to roots, the presence of HCO_3^{1-} or CO_3^{2-}, an increasing rate of oxidation, and the presence of Al in the system (Schwertmann and Taylor, 1989; Carlson and Schwertmann, 1990).

Ferrihydrite has been reported in ochreous precipitates from the oxidation of emerging Fe^{2+} containing waters (Schwertmann and Fischer, 1973; Carlson and Schwertmann, 1981; Schwertmann and Kämpf, 1983), in bog and lake Fe ores (Schwertmann et al., 1982), in podzol B horizons (Adams and Kassim, 1984), in placic horizons (Campbell and Schwertmann, 1984), in Andepts (Parfitt et al., 1988), in soil ironpans (Childs et al., 1990), and in constructed wetlands for acid drainage treatment (Karathanasis and Thompson, 1995). These occurrences reflect an environment where Fe^{2+} is rapidly oxidized (abiotically or biotically) in the presence of high concentrations of organic matter and/or Si. These compounds and possibly others hinder the immediate formation of FeOOH phases and the subsequent transformation of ferrihydrite to more stable Fe oxides (Schwertmann, 1966; Cornell,

1987). In contrast, in tropical soils the concentration of ferrihydrite is very low, because of low interference of Si and organics. The presence of ferrihydrite together with FeOOH phases may indicate that environmental conditions are not favorable for crystal development, or that the formations are relatively young, as in paddy soils (Wang et al., 1993) and in constructed wetlands (Karathanasis and Thompson, 1995).

Redoximorphic soils often display greenish blue colors that change rapidly to yellowish brown on exposure to the air. These colors are indicative of the presence of green rusts (Trolard et al., 1997) which occur under reducing and weakly acid to weakly alkaline conditions as intermediate phases in the abiotic formation of goethite, lepidocrocite, and magnetite (Schwertmann and Fechter, 1994).

Schwertmannite has been found in strongly acid sulfate waters associated with mining activities (Bigham et al., 1990, 1992, 1994, 1996a; Fanning et al., 1993), and in an acid alpine stream draining a pyritic schist (Schwertmann et al., 1995a). Most of these occurrences reflect acid environments where bacteria catalyze the oxidation of FeS_2, releasing Fe^{3+} and SO_4^{2-} that, in the pH range 2.8 to 4.0, precipitate as schwertmannite (Bigham et al., 1992, 1996a). At lower pH, jarosite is favored, while at higher pH values goethite and ferrihydrite form (Bigham et al., 1996b). Schwertmannite is metastable, and converts to goethite over time. Bigham et al. (1992) proposed a biogeochemical model for the precipitation of jarosite, schwertmannite, and ferrihydrite, and their conversion to goethite. The model considers the oxidation of Fe^{2+} by *Thiobacillus ferrooxidans* or O_2, the concentration of SO_4^{2-}, the pH, and the presence of other cations (K, Na) in the system.

Magnetite in soils is usually inherited from the parent rock (lithogenic), but both biologically (Fassbinder et al., 1990) and abiotically formed magnetite (Maher and Taylor, 1988) has been reported. The alteration of magnetite to hematite by solid-state transformation, with no evidence for maghemite development is described by Gilkes and Suddhiprakarn (1979) and Anand and Gilkes (1984a). In laboratory studies, particle size determines whether hematite or maghemite forms when magnetite is oxidized below 220 °C (Egger and Feitknecht,1962; Gallagher et al.,1968). Particles < 300 nm in diameter transform to maghemite, while larger particles oxidize to hematite. This may explain why soil maghemites typically occur in the clay fraction.

Maghemite is common in many different soils, especially in the tropics and subtropics, occurring dispersed or concentrated in concretions (Taylor and Schwertmann, 1974; Curi and Franzmeier, 1984; Anand and Gilkes, 1987a; Fontes and Weed, 1991). The two major possible pathways for maghemite formation in soils are the aerial oxidation of lithogenic magnetite (Fontes and Weed, 1991), or the transformation of other pedogenic Fe oxides by heating (between 300 and 425 °C) in the presence of organic compounds (Schwertmann and Fechter, 1984; Anand and Gilkes, 1987b; Stanjek, 1987).

Akaganéite and feroxyhite are both rare minerals. Akaganéite has been found in environments with high chloride concentrations (e.g., 0.1 M), low pH (e.g., 3–4), and high temperature (e.g., 60 °C), as found in some hot springs and volcanic deposits (Schwertmann and Fitzpatrick, 1992). Laboratory synthesis showed that chloride is essential for the crystallization of akaganéite (Shah Singh and Kodama, 1994). The mechanism of feroxyhite formation in nature is unknown. It was observed in rusty precipitates, formed in the interstices of sand grains from rapidly flowing Fe^{2+}-containing water which was quickly oxidized (Carlson and Schwertmann, 1980). A possible association with dominant ferrihydrite in some Typic Hydrandepts of Hawaii is reported by Parfitt et al. (1988). The similarity of the X-ray diffraction (XRD) patterns of feroxyhite and ferrihydrite makes it extremely difficult, however, to distinguish these minerals in natural samples (Carlson and Schwertmann, 1981).

The different types of Fe oxides may show a partial replacement of Fe by other cations conditioned by their availability which, in turn, makes it representative of specific environments like a mineralogical-pedochemical signature. So far, such relationships have been established for Al substitution in goethites, and recently, for V^{3+} in goethite and hematite (Schwertmann and Pfab, 1994, 1996). Medium to high Al substitution (0.15 to 0.33 mole fraction) is usually observed in goethites from environments with low Si and high Al activity, found in highly weathered tropical and subtropical soils, bauxites, and saprolites (Fitzpatrick and Schwertmann, 1982; Schwertmann and Kämpf, 1985; Anand and Gilkes, 1987a; Curi and Franzmeier, 1984; Fontes et al., 1991; Singh and Gilkes, 1992), whereas goethite with low substitution (0–0.15 mole fraction) prevails in slightly acid, eutrophic, soils and redoximorphic soils (Fitzpatrick and Schwertmann, 1982). The occurrence of goethites with highly contrasting Al substitution indicates changes in the weathering rate or soil redox conditions (Motta and Kämpf, 1991). The presence of V^{3+} in goethite and hematite may be used as an indicator for former anoxic environments (Schwertmann and Pfab, 1996).

3.2.3 Influence on Soil Properties

Crystals of Fe oxides in soils are often extremely small, sometimes as small as 5 nm in diameter for poorly crystalline minerals like ferrihydrite, or as small as 150 nm in diameter for better crystalline minerals like goethite and hematite. Structural disorder is common. Fe oxides, therefore, exhibit a large specific surface area (70–250 m^2 g^{-1}). They usually occur in higher amounts in comparison to Mn and Al oxides, they have a high p.z.c. (pH 7–9), and a variable surface charge. All of these factors explain the substantial influence of Fe oxides on the physical and chemical properties of soils.

Significant correlations have been found between the sorption of heavy metals (Cu, Pb, Zn, Cd, Co, Ni, Mn), anions (PO_4, SiO_3, MoO_4, AsO_4, Se_4O, S_4O, and organic anions) and the content and mineralogy of Fe oxides (Section B, Chapter 8).

The Munsell color ranges for the various Fe oxides are listed in Table 3.2. Detailed aspects of that subject are reviewed by Schwertmann (1993) and Cornell and Schwertmann (1996). Basically, reddish soil color (5YR and redder) is due to the presence of hematite (and maghemite in some tropical soils), masking the presence of goethite. Thus, the hematite content determines the redness of a soil (Torrent et al., 1983), while the yellow color due to goethite (between 7.5YR and 2.5Y) is expressed only in the absence of hematite. The presence of lepidocrocite is indicated by an orange color, normally restricted to mottles or localized spots in aquic environments. In surface soils, the colors due to Fe oxides may be masked by black organic compounds. Information about aeration and soil drainage can be inferred from the distribution or absence of different Fe oxide minerals in soils, according to their specific formation conditions described above. Landscape sequences of red soils (hematite and goethite) on well-drained hilltops, through yellow soils (goethite) on moderately drained midslopes, to mottled and gray soils in poorly drained valleys are examples of Fe oxides acting as generic indicators of aerobic and anaerobic environments (Peterschmitt et al., 1996). Localized accumulations of Fe oxides (mottling, plinthite) and bleached matrix colors (chroma < 2), indicative of seasonal to permanent waterlogged soil environments, are used as diagnostic criteria for redoximorphic features and aquic regimes (Section E, Chapter 2).

The aggregating effect of Fe oxides in soils is indicated by significant correlations between the fraction of water-stable aggregates or related structural properties, and the content of Fe oxides (Schwertmann and Taylor, 1989), by EM observations of Fe oxide deposits on kaolinite platelets (Fordham and Norrish, 1979), and by the dispersion of aggregated soils after removal of their Fe oxides with a reducing agent (McNeal et al., 1968). Although an aggregating effect of the Fe oxides in soils is generally accepted, the exact mechanism remains obscure. Thin sections and SEM observations of nodules, concretions, and ferricretes, suggest that cementation develops through the growth of Fe oxide crystals in place between matrix particles, leading to a very stable, nondispersible association (Shafdan et al., 1985). Aggregates, on the other hand, seem to form through an attraction between positively charged Fe oxide particles and negatively charged matrix particles, mainly clay sized phyllosilicates. A typical example of aggregation is the highly stable microaggregates (ground coffee structure) found in Oxisols. Because these aggregates are not dispersible in water, they have a significant influence on the water-holding capacity and the hydraulic conductivity of these soils (Wambeke, 1992). On the other hand, these aggregates can be dispersed by organic ligands (oxalate, citrate) without solubilizing much Fe (Pinheiro-Dick and Schwertmann, 1996).

3.2.4 Identification

3.2.4.1 Field Techniques

Fe oxides show striking colors, ranging between yellow and red, typical of the various forms and helpful in their identification in the field (Table 3.2). Correlations between Munsell hue, and a redness or yellowness index based on hematite/goethite ratios and hematite content, have been established (Torrent et al., 1983). Color alone is, however, not sufficient to unequivocally identify specific Fe oxide minerals. The presence of the magnetic Fe oxides, magnetite and maghemite can be easily detected in soils with a hand magnet (Schulze, 1988).

3.2.4.2 Chemical Dissolution Techniques

Chemical dissolution procedures for Fe oxides, initially proposed to eliminate their interference in the examination of phyllosilicates, were later developed to dissolve, and thus to quantify specific phases (Borggaard, 1988; Parfitt and Childs, 1988). However, the identity of the dissolved phases must be determined by noninvasive techniques, such as X-ray diffraction and Mössbauer spectroscopy. Potassium or Na pyrophosphate solutions have been used for estimating Fe (and Al) organic complexes (McKeague, 1967; Bascomb, 1968). The soil is usually shaken for 16 hr with 0.1 M pyrophosphate at pH 10, but the extraction and clarification procedures are highly technique dependent (Schuppli et al., 1983; Loveland and Digby, 1984). For the extraction of short-range order Fe oxides (predominantly ferrihydrite), a 2 hr (Schwertmann, 1964) or 4 hr (McKeague and Day, 1966) extraction with 0.2 M ammonium oxalate at pH 3 in the dark is widely used. This technique also extracts Fe from organic complexes and poorly crystalline lepidocrocite (Ohta et al., 1993), whereas some magnetites are partly dissolved. Feroxyhite is more resistant to the 2-hr oxalate treatment than ferrihydrite (Carlson and Schwertmann, 1980), and for schwertmannite a 15-min treatment is used (Murad et al., 1994; Bigham et al., 1996a). According to Wang et al. (1993), a shorter oxalate treatment (~10–30 min) is more appropriate to measure ferrihydrite in < 2 μm redoximorphic soil fractions than the much longer treatments commonly used for soils. For the complete dissolution of the pedogenic Fe oxide minerals, without dissolution of other minerals, a treatment with sodium dithionite-citrate-bicarbonate (DCB) is used. The extraction is carried out for 15 min at 75 °C (Mehra and Jackson, 1960), or by shaking for 16 hr with a dithionite-citrate solution at room temperature (Holmgren, 1967). However, as dissolution is affected by particle size, crystallinity (particularly for lithogenic magnetite and hematite) and high Al substitution (goethite and hematite), several DCB treatments may be required.

The ratio of oxalate extractable Fe (Fe_o) to DCB extractable Fe (Fe_d) quantifies the proportion of the more and less active fractions. A high Fe_o/Fe_d ratio and a loss in redness after oxalate treatment gives a first indication of ferrihydrite in a sample. Additionally, it is a useful parameter for characterizing soil properties such as P sorption and pedogenic processes (McKeague and Day, 1966; Blume and Schwertmann, 1969). The ratio of Fe_d to Fe_t (total Fe content by HCl or HF digestion) quantifies the proportion of Fe in primary silicate minerals which has been released by weathering and precipitated as Fe^{3+} oxides, hence estimating Fe sources for potential Fe oxide formation and/or soil weathering stage. Advanced soil weathering stages, indicated by Fe_d/Fe_t ratios > 0.8, are usually shown by Oxisols and Ultisols.

A field test for the presence of Fe^{2+} ions and ferric organic complexes in redoximorphic soils can be performed with a solution of α, α '-dipyridil (Childs, 1981; Kennedy et al., 1982), or with a solution of 1:10-phenanthroline (Richardson and Hole, 1979). Lovley and Phillips (1987) proposed the use of hydroxylamine hydrochloride to determine whether or not Fe^{3+} is available for microbial reduction.

3.2.4.3 X-ray Diffraction and Other Spectroscopic Techniques

The identification of the various Fe oxides requires physical methods, such as XRD, differential thermal analysis (DTA), infrared (IR) (Cambier, 1986), Mössbauer spectroscopy (Murad, 1988, 1996), electron microscopy (EM) (Eggleton, 1988; Boudeulle and Muller, 1988), and magnetic measurements (Coey, 1988). A review of characterization methods for Fe oxides is given by Cornell and Schwertmann (1996) and Schwertmann and Taylor (1989). The most important diagnostic criteria are summarized in Tables 3.1 and 3.2. XRD patterns for the major soil Fe oxides are shown in Fig. 3.3.

For powder XRD, unless one uses a monochromator, $CoK\alpha$ or $FeK\alpha$ radiation should be employed to avoid the high fluorescence background which results when Fe-rich samples are irradiated with $CuK\alpha$ radiation. The sensitivity of XRD is improved by using a differential XRD method (Schulze, 1981), and selective dissolution treatments (Wang et al., 1993). Due to the low concentration of Fe oxides in soils, concentration techniques may be necessary (Schulze, 1988), like particle size separation, magnetic separation (Schulze and Dixon, 1979), and for goethite and hematite in kaolinitic soils, a 5 M NaOH treatment (Norrish and Taylor, 1961; Kämpf and Schwertmann, 1982a). A quantitative determination of Fe oxide minerals in soils is generally possible using the intensities of selected X-ray lines, DTA peaks, or Mössbauer hyperfine sextets. Suitable standards with the same characteristics

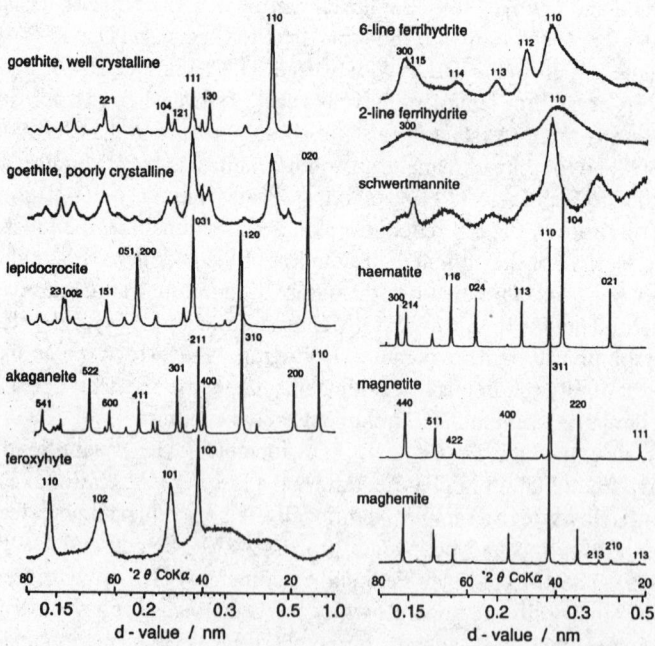

Fig. 3.3 X-ray diffraction patterns for the major soil iron oxide minerals (Co K , peak positions in nanometers) [Reprinted from Cornell and Schwertmann, 1996. The iorn oxides. Copyright Wiley-VCH Verlag, Weinheim, Germany with permission]

may be isolated from soils or can be synthesized. The accuracy of the quantification can be controlled by comparison with the chemically determined Fe oxides (Fe_d) (Kämpf and Schwertmann, 1982b).

The Al substitution in goethite can be calculated from the unit cell c-dimension by Al (mole fraction) $= 17.30-57.20 \cdot c$ (nm) (Schulze, 1984). The c is obtained from the XRD positions $d(110)$ and $d(111)$ by $c = [1/d(111)^2 - 1/d(110)^2]^{-1/2}$. In hematite, Al substitution is given by Al(mole fraction) $= 31.09-61.71 \cdot a$ (nm), where $a = d(110) \cdot 2$, or, preferably, $a = d(300) \cdot 2\sqrt{3}$ (Schwertmann et al., 1979).

3.3 Manganese Oxides

Manganese oxides are usually minor components in soils, but with a significant influence on the soil chemical properties. While Mn is a micronutrient essential for plants and animals, it is the only plant essential element that also frequently occurs in toxic concentration in acid soils. The mineralogy of Mn oxides is complicated by the large number of minerals, and the lack of precise knowledge of some of their structures leading to uncertainties as to whether certain forms should be regarded as distinct mineral species, or merely as variants or mixtures of other mineral forms (Post, 1992; McKenzie, 1989). The poor crystallinity and low concentration of these minerals in most soils is a further challenge for their characterization. A list of Mn oxides is given in Table 3.3. More information can be found in McKenzie (1989) and Graham et al. (1988).

3.3.1 Mineral Phases

Manganese oxides consist mainly of Mn in octahedral coordination. The various minerals differ in the arrangement and linkage of the octahedra. The Mn oxides may be placed into three groups: (1) tectomanganates or tunnel structures; (2) phyllomanganates or layer structures, both groups with mainly tetravalent Mn; and (3) the trivalent oxides or lower oxides.

3.3.1.1 Tectomanganates (tunnel structures)

Tectomanganates are formed from single, double, or wider chains of MnO_6 octahedra, that are linked into a framework, thus forming tunnels (or channels) through the structures. These tunnels are occupied by large foreign cations and water molecules. Pyrolusite, ramsdellite and nsutite are also referred as chain structures because they do not contain tunnels, but are included in the group of tectomanganates because of their structural similarity to the channel or tunnel structure members.

Pyrolusite (MnO_2) consists of single chains of edge-sharing MnO_6 octahedra linked by sharing corners to form 1x1 pseudotunnels (Fig. 3.1) (Baur, 1976). The single rows of unoccupied octahedral sites are not wide enough to accommodate foreign ions or water molecules. High resolution transmission electron microscopy of pyrolusite crystals shows lamellae with the structure of ramsdellite (Yamada et al., 1986). *Ramsdellite* (MnO_2) consists of double chains of MnO_6 octahedra forming 1x2 tunnels (Fig. 3.1) (Post, 1992) that possibly contain water molecules (Potter and Rossman, 1979). *Nsutite* has a structure formed by the intergrowth of pyrolusite-like single chains and ramsdellite-like double chains of octahedra in random fashion (Zwicker et al., 1962; Potter and Rossman, 1979; Turner and Buseck, 1983). Inclusions of todorokite have also been found in nsutite (Turner and Buseck, 1983). *Hollandite, cryptomelane, and coronadite*, sometimes grouped as α-MnO_2, consist of double chains of edge-sharing MnO_6 octahedra linked to form 2x2 tunnels (Fig. 3.1) (Post et al., 1982; Post and Burnham, 1986; Post and Bish, 1989). These minerals have the general formula $A_{0-2}(Mn^{4+}, Mn^{3+})_8(O,OH)_{16}$. The tunnels contain water molecules along with the A cation, which is primarily Ba^{2+} in hollandite, K^+ in cryptomelane, and Pb^{2+} in coronadite. Natural samples usually contain a variety of cations in the tunnels. These large cations are located at specific tunnel sites, and their presence is necessary to prevent the structure from collapsing (Giovanoli, 1985a). The charges of

Table 3.3 Manganese oxide minerals: crystallographic properties and *d*-values for the 6 most intense diffraction lines

Mineral (synthetic equivalent)[†]	Chemical formula	Crystal system, Space group	Unit cell dimen. (Å)	Six most intense diffraction lines *d*-value (Å), relative intensity						Reference[‡]
Pyrolusite (β-MnO$_2$)	MnO$_2$	Tetragonal P4$_2$/mnm	a=4.398 c=2.873	3.11 100	2.407 55	1.623 55	1.306 20	1.304 20	2.110 16	Baur (1976); 24-0735
Ramsdellite	MnO$_2$	Ortho-rhombic Pnma	a=9.27 b=2.866 c=4.533	4.06 100	1.647 57	2.438 49	2.344 46	1.660 34	1.906 31	39-0375
Nsutite (γ-MnO$_2$)	Mn^{4+}Mn^{3+} (O,OH)$_2$	Hexagonal P	a=9.65 c=4.43	1.635 100	4.00 95	2.33 70	2.42 65	2.13 45	1.60 45	Zwicker et al. (1962); 17-0510
Hollandite (α-MnO$_2$)	Ba$_x$ (Mn^{4+}Mn^{3+})$_8$ O$_{16}$ (x=1)	Monoclinic I2/m	a=10.026 b=2.8782 c=9.729 β=91°03'	3.10 100	3.14 88	3.172 40	2.412 37	3.069 32	3.459 23	Post et al. (1982); 38-0476
Cryptomelane	K$_x$ (Mn^{4+}Mn^{3+})$_8$ O$_{16}$ (x=1.3-1.5)	Monoclinic I2/m	a=9.956 b=2.8705 c=9.706 β=90°95'	2.40 100	3.122 51	3.09 45	2.41 40	4.852 32	7.01 30	Post et al. (1982); 44-1386
Coronadite	Pb$_x$ (Mn^{4+}Mn^{3+})$_8$ O$_{16}$ (x=1-1.4)	Monoclinic I2/m	a=9.938 b=2.8678 c=9.834 β=90°39'	3.124 100	3.491 30	2.209 18	2.409 17	6.98 7	1.548 6	Post and Bish (1989); 41-0596
Romanechite	Ba$_{0.66}$ (Mn^{4+}Mn^{3+})$_5$ O$_{10}$·1.34H$_2$O	Monoclinic C2/m	a=13.919 b=2.8459 c=9.678 β=92°39'	2.408 2.882	2.188 100	3.481 85	6.96 60	2.366 55	50	Turner and Post (1988); 14-0627
Todorokite	(Na,Ca)$_{0.3-0.5}$ (Mn^{4+}Mn^{3+})$_6$ O$_{12}$·3.5H$_2$O	Monoclinic P2/m	a=9.764 b=2.8416 c=9.551 β=94°06'	9.55 100	2.399 36	2.388 25	2.355 24	4.77 24	2.345 25	Post and Bish (1988); 38-0475
Chalcophanite	ZnMn$_3$O$_7$ ·3H$_2$O	Trigonal R3	a=7.533 c=20.794	6.93 100	2.228 40	4.07 29	1.590 23	3.507 19	2.550 17	Post and Appleman (1988); 45-1320
Birnessite	(Na,Ca,Mn^{2+}) Mn$_7$O$_4$·2.8 H$_2$O	Monoclinic C2/m	Na: a=5.175 b=2.850 c=7.337 β=103°18'	7.14 100	3.57 27	2.519 14	2.429 13	2.154 7	2.222 5	Post and Veblen (1990); 43-1456
Vernadite (δ-MnO$_2$)	MnO$_2$·nH$_2$O	(Pseudo) tetra-gonal ? I4/m	a=9.866 c=2.844	2.39 100	3.11 60	2.15 60	1.827 40	1.537 40	1.422 40	Chukhrov et al. (1980); 15-0604
Rancieite	(Ca,Mn) Mn$_4$O$_9$·nH$_2$O	Hexagonal P	a=8.68 c=9.00	7.49 100	3.74 14	2.463 10	2.342 6	1.425 4	2.064 2	JCPDS; 22-0718
Buserite	Na$_4$Mn$_{14}$O$_{27}$·21 H$_2$O	Orthorombic	a=17.5 b=30.7 c=10.2	10.1 100	5.01 70	3.34 50	1.46 50	2.56 30	2.47 30	32-1128
Lithiophorite	LiAl$_2$Mn$_2^{4+}$Mn^3 $^+$O$_6$(OH)$_6$	Trigonal R3m	a=2.924 c=28.169	4.71 100	9.43 68	2.371 24	1.880 14	3.143 7	1.453 4	Post and Appleman (1994); 41-1378
Groutite (α-MnOOH)	MnOOH	Orthorhombic Pbnm	a=4.560 b=10.70 c=2.870	4.20 100	2.81 70	2.67 70	2.30 60	1.695 50	2.38 40	Glasser and Ingram (1968); 12-0733
Manganite (γ-MnOOH)	MnOOH	Monoclinic, pseudoortho-rhombic B2$_1$m	a=8.88 b=5.25 c=5.71 β=-90°	3.40 100	2.64 24	1.782 21	2.417 17	1.672 17	2.414 16	Dachs (1963) in Bricker (1965); 41-1379
Feitknechtite (β-MnOOH)	MnOOH	Tetragonal P	a=8.6 c=9.30	4.62 100	2.635 50	2.36 20	1.96 10	1.55 1	1.50 1	Bricker (1965); 18-0804
Hausmannite	Mn$_3$O$_4$	Tetragonal I4$_1$/amd	a=5.7621 c=9.4696	2.847 100	2.768 85	1.544 50	3.089 40	4.924 30	1.799 25	Bricker (1965); 24-0734
Manganosite	Mn^{2+}O	Cubic Fm3m	a=4.4448	2.223 100	2.568 60	1.571 20	1.34 18	0.994 16	0.907 16	Sasaki et al. (1980); 7-0230
Bixbyite (α-Mn$_2$O$_3$)	(Fe,Mn)$_2$O$_3$	Cubic Ia3	a=9.4091	2.716 100	1.663 28	3.842 16	2.352 14	1.418 13	1.845 9	Geller (1971); 41-1442

[†] Designation commonly used in the literature for a synthetic compound with the same crystal structure as the natural mineral are given in parentheses. [‡] Numbers of the format XX-XXXX indicate ICDD file number (ICDD, 1994).

the tunnel cations are balanced by the substitution of Mn^{4+} by Mn^{3+} ions. *Romanèchite* consists of double and triple chains of edge-sharing octahedra that share corners to form a framework containing 2x3 tunnels. The tunnels contain Ba^{2+} (and a variety of other large cations, such as Na, K, Sr) and water molecules (Fig. 3.1) (Turner and Post, 1988). The charges of the tunnel cations are balanced by the substitution of Mn^{4+} by Mn^{3+} ions. High resolution transmission electron microscopy (HRTEM) images show that intergrowths of romanèchite and hollandite are common, and are produced by the sharing of the double chains common to both minerals with double or triple chains occurring at random (Turner and Buseck, 1979). While romanèchite has also been called psilomelane, McKenzie (1989) used it for a structural type, and Waychunas (1991) for mixtures of Mn oxide minerals. *Todorokite* which had its structure and existence as a single mineral questioned until recently (Giovanoli, 1985b; Burns et al., 1985), consists of triple chains of edge-sharing MnO_6 octahedra linked to form large 3x3 tunnels (Post and Bish, 1988) containing Na, Ca, K, Ba, Sr, and water molecules (Fig. 3.1). The octahedra at the edges of the triple chains are larger than those in the middle and therefore probably accommodate the larger, lower valence cations (Mg^{2+}, Mn^{3+}, Cu^{2+}, Ni^{2+}, etc.) found in todorokite samples (Post and Bish, 1988). The occurrence of variable tunnel widths (3x2, 3x3, 3x4, and 3x5) observed in HRTEM images suggests that todorokite represents a family rather than a single mineral (Turner and Buseck, 1981).

3.3.1.2 Phyllomanganates (layer structures)

Chalcophanite has a layer structure composed of sheets of edge-sharing MnO_6 octahedra alternating with planes of Zn cations (but also Mn^{2+}, Ba, Ca, Mg, K, Pb, Cu, etc.) and water molecules (Fig.3.1i) (Post and Appleman, 1988; Ostwald, 1985). One of every seven octahedral sites in the Mn-O sheet is vacant, and the Zn cations are situated above and below the vacancies. The other phyllomanganates are structurally analogous to chalcophanite, with Na^+, Ca^{2+}, K^+, and Mn^{2+} as interlayer cations. *Birnessite* consists of a layer structure analogous to chalcophanite, but with fewer vacancies in the octahedral sheets and with Na, K, or Mg replacing the Zn cations (Fig. 3.1) (Post and Veblen, 1990). The interlayer region contains water molecules as in chalcophanite. *Rancieite* has a layer structure similar to birnessite, with Ca^{2+} as the main interlayer cation, and interlayer water molecules (Bardossy and Brindley, 1978; Potter and Rossman, 1979; Chukhrov et al., 1980). The structure and existence in nature of *buserite*, also know as 10 Å manganite, is still unresolved. Buserite appears to have a layer structure similar to birnessite, with the larger, 10 Å layer spacing probably due to interlayer water (Waychunas, 1991). *Lithiophorite* has a layer structure comprised of sheets of edge- and corner-sharing MnO_6 octahedra alternating with sheets of $(Al, Li)(OH)_6$ octahedra (Fig. 3.1) (Post and Appleman, 1994). The cation sites in the Mn-O octahedral sheet are fully occupied, 2/3 with Mn^{4+} and 1/3 with Mn^{3+}. In the (Al, Li)-OH octahedral sheet, 2/3 of the sites are occupied by Al and 1/3 by Li cations. Lithiophorites with only traces of Li, and others with wide variations in Ni, Co, Cu and Zn concentrations inversely related to the Al content have been found (Ostwald, 1984a). According to Manceau et al. (1987; 1990), Co can occur within the octahedral Mn sheets, while Ni and Cu are located in the (Al, Li)-OH octahedral sheet, probably replacing Li, whereas in asbolane Ni builds partial $Ni(OH)_2$ sheets. Manceau et al. (1987) also found evidence for a mixed layering between lithiophorite and asbolane, as observed by Ostwald (1984a). *Asbolane* has a layer structure with alternating sheets of Mn^{4+}-O octahedra and Co-Ni-OH octahedra (Fig. 3.1) (Chukhrov and Gorshkov, 1981). The Co-Ni sheet may be discontinuous (island-like). The positive charge of the Mn^{4+} sheets is balanced by the negative charge of the Co-Ni sheets. There is hydrogen bonding between the O atoms of the Mn sheets and the OH groups of the Co-Ni sheets. Most asbolanes are fine grained, poorly crystalline minerals. *Vernadite* has been considered a disordered birnessite, lacking regular stacking in the c-axis. However, according to Chukhrov et al. (1980) and Chukhrov and Gorshkov (1981), it is a distinct mineral species with a disordered structure that has some similarity to that of birnessite, represented by layers of hexagonal close-packed O and water molecules in which less than half of the octahedra are occupied by Mn ions. Diffraction patterns of vernadite have only two d-spacings at ~2.4 and ~1.4 Å, whereas birnessite yields additional reflections at 7.0-7.2 Å, 3.5-3.6 Å, and other spacings (Chukhrov et al., 1980).

3.3.1.3 Trivalent Manganese Oxides and Oxyhydroxides

Groutite (α-MnOOH) is isostructural with goethite and ramsdellite (Fig. 3.1) (Glasser and Ingram, 1968). *Feitknechtite* (β-MnOOH) has a structure similar to lepidocrocite (Fig. 3.1), but its structure has not been refined (Waychunas, 1991). *Manganite* (γ-MnOOH) is similar to pyrolusite and rutile (Fig. 3.1) (Huebner, 1976). *Hausmannite* (Mn_3O_4) has a disordered spinel structure analogous to magnetite (Fig. 3.1) (Huebner, 1976). *Manganosite* has the NaCl structure and is of limited occurrence in nature (Huebner, 1976). *Bixbyite* (α-$(Fe,Mn)_2O_3$) has an anion deficient fluorite structure (Geller, 1971), while pure (synthetic) α-Mn_2O_3 is orthorhombic (Huebner, 1976; Waychunas, 1991).

3.3.2 Occurrence and Formation

Many studies have been made on the synthesis of Mn oxides (McKenzie, 1971; Hem and Lind, 1983; Giovanoli, 1985a), and the transformation of one form to another (Rask and Buseck, 1986; Glasser and Smith, 1968; Bricker, 1965; Faulring et al., 1960). Some of these studies, even if more applicable to diagenetic and hydrothermal conditions, and to the formation of ore deposits, may be extrapolated to soil environments. Of the variety of Mn oxides found in terrestrial environments, only a few have been positively identified in soils. For example, nsutite and pyrolusite are common in Mn ore deposits (Zwicker et al., 1962; Varentsov, 1982), but have not been found in soils. Synthesis studies show that large amounts of foreign ions prevent the formation of nsutite, while even small amounts of foreign ions prevent the formation of pyrolusite. In soils, the foreign ions released by mineral weathering probably limit the formation of these Mn oxides, whereas in ore deposits the abundance of Mn would aid in the removal of foreign ions, thus allowing the formation of nsutite and pyrolusite (McKenzie, 1989). According to Ostwald (1984b) the mineral associations birnessite-montmorillonite, chalcophanite-kaolinite and lithiophorite-gibbsite observed in Australian Mn ore deposits may represent special cases of metasomatic replacements of clay minerals by phyllomanganate minerals, whereas the occurrence of vernadite in a wide range of substrates is explained by its formation as a product of rapid microbial oxidation of Mn^{2+} (Chukhrov and Gorshkov, 1981).

Laboratory experiments show that Mn_3O_4 (hausmannite) is the initial product of the chemical oxidation of aqueous Mn^{2+} at 25 $^{\circ}$C in moderately alkaline solutions (pH 8.5–9.5). It converts to β-MnOOH (feitknechtite) and then to γ-MnOOH (manganite) by aging in solution (Hem and Lind, 1983; Murray et al., 1985). At 0 $^{\circ}$C, the initial product is feitknechtite which converts to MnO_2, identified as ramsdellite or birnessite (Hem and Lind, 1983). The formation and transformation of Mn oxides through successively increasing oxidation states are compatible with the common distribution of Mn oxides observed in supergene Mn ores (Bricker, 1965), and in lateritic weathering sequences (Parc et al. (1989). Minor inversions of that Mn oxidation sequence occur, however, due to the presence of foreign ions, thus promoting the formation of cryptomelane and lithiophorite (Fig. 3.4) instead of pyrolusite (Parc et al., 1989). The influence of microorganisms and organic compounds additionally increases the complexity of Mn oxide formation in soil environments.

Manganese is one of the first elements released during weathering of primary minerals, which explains its common accumulation in saprolites. Mobile in solution as Mn^{2+}, its oxidation to Mn^{3+} and Mn^{4+}, and its subsequent precipitation are often accelerated by microorganisms (Ghiorse, 1988). Most of the Mn oxides found in surface environments are compounds of Mn^{4+} with some Mn^{3+}. The typical fine grain size and poor crystallinity of the Mn oxides in soils are probably related to seasonal moisture changes, and the presence of interferring organic and inorganic components in the soil solution.

Manganese oxides are commonly of authigenic origin in soils, formed by direct chemical or bio-chemical precipitation from solution, and by crystallization of poorly organized colloidal sols. However, little is known about the processes involved in their pedogenic formation. In most soils Mn oxides occur as finely dispersed particles, but may be also found as discontinuous black-brown coatings of ped surfaces (mangans), as pore fillings, and as concretions and nodules. Their presence and concentration is more likely in soil environments with alternate reducing and oxidizing conditions, which affect the mobility and the precipitation of Mn (White and Dixon, 1996). Thus, higher accumulations of Mn oxides are more frequent in soils with restricted drainage such as the aquic suborders (Aqualfs, Aquolls, Aquults, etc.), and intergrades (Aquic Hapludalfs, Aquic Dystrochrepts, etc.), and are also commonly found in saprolites. However, Mn accumulations in a soil are not necessarily an indication of a present redox environment, but may reflect former wetter conditions (Allen and Hajek, 1989).

Manganese and Fe oxides are usually associated in nodules with dominant silicates (quartz, clay minerals) that probably act as primary templates for the Mn and Fe oxidation and precipitation (Schwertmann and Fanning, 1976). Golden et al. (1986) demonstrated experimentally that Mn^{4+} oxide minerals can also act as oxidizing agents of Fe^{2+} ions in solution, causing the precipitation of Fe^{3+} oxides in association with Mn oxides. The crystal growth and progressive cementation of the nodules can be explained by an autocatalytic oxidation of Mn^{2+} adsorbed onto the Mn (and Fe) oxides formed (McBride, 1994). Thus, abiotic oxidation accelerates in response to the increased surface available for selective adsorption of Mn^{2+}. The overall process, of Mn oxidation and nodule formation, may also be promoted by Mn oxidizing bacteria (Ehrlich, 1996; Robbins et al., 1992; Leinemann et al., 1997). Furthermore, microorganisms may indirectly control local environments, through elevation of pH and Eh, which favor oxidation of Fe and Mn (Ghiorse and Ehrlich, 1992). Models of biotic oxidation of Mn (and Fe) are described by Ghiorse and Ehrlich (1992) and Ehrlich (1990).

In soil samples from Australia, the abundance of Mn oxides decreased from birnessite and lithiophorite, over hollandite, to only one occurrence of each pyrolusite and todorokite (Taylor et al., 1964). A high frequency of birnessite was also found in soils of other regions (Taylor, 1968; Ross et al., 1976). Todorokite formation from Mn released by the weathering of Mn-substituted siderite in lignite overburden was reported by Senkayi et al. (1986). Birnessite, lithiophorite, and lattice fringes attributed to todorokite, were observed in nodules of a Vertic Argiustoll, while lithiophorite only was found in a Rhodic Paleudult (Uzochukwu and Dixon, 1986). Romanèchite was identified in nodules of a Typic Ochraquult (Robbins et al., 1992), and vernadite associated with feroxyhite was identified in a Scottish Gleyic Cambisol (Birnie and Paterson, 1991). Chukhrov and Gorshkov (1981) as cited by McKenzie (1989) reported that the most common Mn oxide in soils was vernadite, followed by birnessite, cryptomelane, todorokite, hausmannite. Chukhrov and Gorshkov (1981) reported the presence of hausmannite in unspecified soils with pH ~8.2.

3.3.3 Influence on Soil Properties

The wide range of PCZ values reported for synthetic Mn oxides (Healy et al., 1966; Crowther and Dillard, 1983; Oscarson et al., 1983), from 1.5–3.5 for birnessite to 2.8–4.6 for the hollandite group, and 6.4–7.3 for pyrolusite, is probably related to synthesis conditions. In general, most Mn oxides have p.z.c values < 4.0, a high negative charge, a large range of surface areas (5–360 m^2 g^{-1}), and show strong specific adsorption of cations.

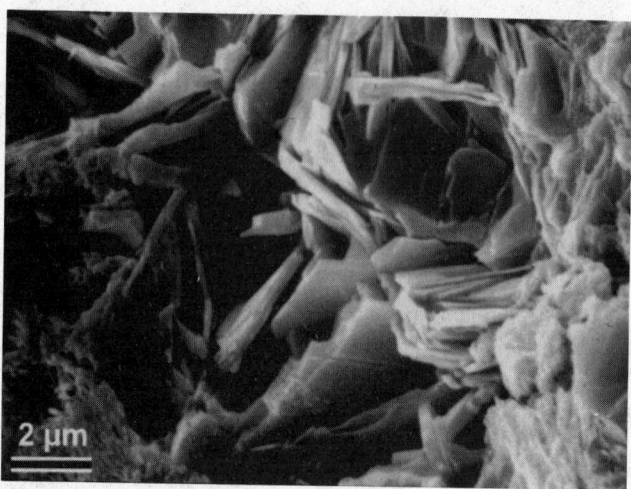

Fig. 3.4 Scanning electron micrograph of the fractured surface of a nodule from a Tropeptic Eutrustox showing crystals of the Mn oxide mineral, lithiophorite [Reprinted from Golden et al., 1993. Aust. J. Soil Res. 31:51-66 with permission]

A high sorption capacity of Mn oxides for metal ions has been found by many authors (see McKenzie, 1989; Murray, 1975). In general, metal ions are sorbed in the increasing order Mg < Ca < Sr < Ba < Ni < Zn < Co < Mn < Cu < Pb (Murray, 1975), leading to the accumulation of relatively high concentrations of heavy metals (Childs, 1975; Sidhu et al., 1977), and actinides in the Mn oxides (Means et al., 1978; Cerling and Turner, 1982).

Manganese oxides are strong inorganic oxidants, which affects the availability or potential hazard of particular metals. In the case of Co adsorption onto birnessite at $pH < 7$, Co^{2+} is oxidized to Co^{3+} by Mn^{4+}, with formation of Mn^{2+} in the process (Crowther and Dillard, 1983). The strong adsorption mechanism has a significant influence on the availability of Co to plants (Adams et al., 1969; McKenzie, 1978).

The Mn oxides have a major influence on the toxicity and bioavailability of As and Cr in terrestrial and aquatic environments. The reduced species As^{3+}, which is more toxic, more soluble and more mobile than the oxidized As^{5+} species, is effectively adsorbed and oxidized by Mn^{4+} oxides at $pH < 6$ (Oscarson et al., 1983; Thanabalasingam and Pickering, 1986; Scott and Morgan, 1995). However, the ability to deplete the concentration of As^{3+} in solution varies between different types of Mn oxides, and is related to the crystallinity, specific surface areas, and p.z.c. of the oxides. In contrast to other transition elements, Cr toxicity and mobility increase with its oxidation state. Thus, the presence of Mn oxides, as oxidizing agents for Cr^{3+} in soil systems, should be considered in disposal studies (Fendorf et al., 1992).

Manganese and Fe oxides also act as final electron acceptors in oxidizing organic compounds, and are consequently dissolved in the process. Organic compounds that form inner sphere complexes with the oxide surface (e.g., catechol) dissolve the Mn oxide more quickly than those compounds that form outer sphere complexes (e.g., hydroquinone) (Stone and Morgan, 1984a, b; McBride, 1987). Lovley and Phillips (1988) found evidence that microorganisms can obtain energy for growth by coupling the oxidation of organic matter to the dissimilatory reduction of Fe^{3+} and Mn^{4+}. They proposed a microbial food-chain model, in which fermentative organisms initially metabolize complex organic matter, and in the next stage a separate group of bacteria oxidizes the fermentation products to CO_2 while reducing Fe^{3+} and Mn^{4+}. The overall process may be an important degradative pathway for organic compounds and the formation of humic material, while the reduction and dissolution of Mn oxides increases the Mn mobility and bioavailability to organisms.

3.3.4   Identification

Accumulations of Mn oxides in soils are readily identified by their characteristic black-brown coloration. The effervescence observed with the addition of H_2O_2 is a usual field criterion that helps to confirm the presence of these minerals. The small amount of Mn oxides in soils, the diffuse nature of the X-ray patterns of some of the minerals, and the coincidence of diagnostic lines with those of associated matrix minerals can make their identification by XRD difficult. Thus, XRD, preferably with Fe Kα radiation, can be used only where natural segregations of the minerals occur, such as nodules, veins or coatings. Even then, a concentration pretreatment may be necessary. Powder XRD patterns for some of the more important Mn oxide minerals are shown by Post (1992) and the most intense XRD lines of the major minerals are listed in Table 3.3. Because of their lower crystallinity, soil Mn oxides may, however, show some deviation from these patterns. A selective dissolution of Mn oxides with acidified hydroxylamine hydrochloride was proposed by Chao (1972), whereas Tokashiki et al. (1986) characterized Mn oxides by using successive sodium hydroxide, hydroxylamine hydrochloride, and dithionite-citrate-bicarbonate (DCB) treatments. Concentration procedures for Mn oxides, selective chemical treatments, identification by XRD and thermal methods are also reported by Uzochukwu and Dixon (1986). A comprehensive compilation of IR powder-absorption spectra of synthetic and naturally occurring Mn oxides is given by Potter and Rossman (1979).

Crystal structure and crystal chemistry refinements have been performed with high resolution transmission electron microscopy (HRTEM), electron microprobe analysis, electron diffraction, powder XRD with Rietveld analysis (Post and Appleman, 1988; Post and Bish, 1988; Post and Veblen, 1990), and X-ray absorption spectroscopy (Manceau et al., 1987, 1992a,b; Manceau and Combes, 1988). HRTEM has led to the discovery of complex intergrowths of variable lattice periodicities in some Mn oxides (Turner and Buseck, 1979, 1981, 1983; Yamada et al., 1986).

3.4   Aluminum Oxides

Of six different Al oxide minerals, gibbsite and, less commonly, boehmite form under soil conditions. Nordstrandite and bayerite have only been identified in restricted geologic environments. Diaspore and corundum are occasionally found in bauxite deposits, but are rare in soils. The relative rarity of Al-oxide minerals in soils, which is in contrast to their structural Fe analogues, may be explained by the competitive formation of alumosilicates and by the difficulty of identifying small amounts of these minerals by X-ray diffraction (Taylor, 1987). More detailed information on Al oxides is given by Hsu (1989).

3.4.1   Mineral Phases

Crystallographic properties for the Al-oxide minerals likely to occur in soils are listed in Table 3.4. Aluminum occurs exclusively in octahedral coordination in the Al oxides, and the structures can be described in terms of the arrangement and linkage of Al-containing octahedra.

3.4.1.1   Hydroxides

Gibbsite (γ-Al(OH)$_3$) consists of sheets of edge-shared octahedra with two-thirds of the available octahedral sites filled with Al^{3+} ions. The occupied octahedral sites form sixfold rings analogous to the phyllosilicate dioctahedral sheet which is stacked along the c-axis, and the OH groups in one sheet reside directly on the top of those in the next, and not in the position of closest packing. The interlayer attractions are weak and are dominated by hydrogen bonding (Fig. 3.1) (Hsu, 1989). *Bayerite*, a second polymorphic form of $Al(OH)_3$, is analogous to gibbsite except that the OH groups of one sheet reside in the depressions between the O atoms of the next sheet and are in close packing (Fig. 3.1) (Hsu, 1989). *Nordstrandite*, a third polymorphic form of $Al(OH)_3$, has a structure in which the stacking of the octahedral sheets alternates between the gibbsite and bayerite arrangements (Hsu, 1989). In addition to these crystalline phases, amorphous $Al(OH)_3$ may occur in soils (Süsser and Schwertmann, 1991)

3.4.1.2   Oxyhydrides

Diaspore (α-AlOOH) is isostructural with goethite (Fig. 3.1) (Ewing, 1935b; Busing and Levy, 1958). *Boehmite* (γ-AlOOH) has the same structure as lepidocrocite (Fig. 3.1). Poorly crystalline boehmite, formerly also called pseudoboehmite, has a restricted number of unit cells along the b-axis (Tettenhorst and Hofmann, 1980).

3.4.1.3   Oxides

Corundum (α-Al_2O_3) is isostructural with hematite (Fig. 3.1) (Maslen et al., 1993).

3.4.2   Occurrence and Formation

Aluminum is a common element in primary silicate minerals, from which it may be released by weathering. Depending on environmental conditions, free Al hydrolyzes, precipitates as Al hydroxide or oxyhydroxide, and crystallizes. The detailed chemistry of Al-oxide formation in soils remains obscure, mainly because of the uncertainties in the formation and the type of hydroxy polymers involved.

Synthesis studies have, so far, given an approximate outline of the probable processes of Al-hydroxide formation. The development of Al-hydroxide polymorphs appears to be related to the rate of precipitation, pH of the system, clay surface, and the nature and concentration of inorganic and organic anions (Huang, 1988; Hsu, 1989). Only synthesis experiments related to pedogenic environments are summarized here. At room temperature, gibbsite forms in acid solutions (pH < 6) under slow hydrolysis, nordstrandite in neutral to alkaline solutions (pH > 7), and bayerite in alkaline solutions under fast hydrolysis (Hsu, 1966; Schoen and Roberson, 1970). These conditions agree with the natural occurrence of gibbsite in highly weathered acidic soils, whereas nordstrandite and bayerite have been found in association with limestone materials. Thermodynamically, gibbsite is the most stable of the three $Al(OH)_3$ polymorphs (Hemingway et al., 1991).

Anions that have strong affinity for Al^{3+}, such as SO_4, CO_3, PO_4, and SiO_3, may interfere with the crystallization of $Al(OH)_3$ (Huang, 1988; Hsu, 1989). Organic ligands like citric, malic, tannic, aspartic, and fulvic acids block the coordination sites of polynuclear Al ions and restrain the hydrolysis, thus inhibiting the crystallization of $Al(OH)_3$ and influencing the nature of the precipitated compound (Kodama and Schnitzer, 1980; Violante and Violante, 1980; Violante and Huang, 1985; Singer and Huang, 1990). This explains why there are no or only small amounts of gibbsite in the A horizon of acidic soils high in organic matter and exchangeable Al, while it is found in greater quantities in deeper horizons.

The pathway for gibbsite formation by direct desilication of primary Al silicates or through clay mineral intermediates (Jackson, 1964) is conditioned by the intensity of leaching, which, in turn, is affected by rainfall, temperature, parent rock, topography, groundwater table, vegetation, and time. Environments with warm temperatures, high rainfall and free drainage favor desilication and leaching of ions, as well as mineralization of organic matter. Oxisols and laterites found in these environments may have significant amounts of gibbsite associated with kaolinite and Fe oxides (Curi and Franzmeier,

1984; Macedo and Bryant, 1987). Gibbsite was found in Dystrochrepts formed from gneiss in North Carolina (Rebertus et al., 1986), and it is frequently a minor component in Ultisols, which are widespread in humid tropical, subtropical, and temperate regions (Allen and Hajek, 1989).

Nordstrandite has been found in nature as crystals radiating into solution cavities in limestone from Guam (Hathaway and Schlanger, 1965), and as small pellets in a limestone soil from Borneo (Wall et al., 1962). Bayerite found in calcareous material in Israel probably formed under hydrothermal conditions (Gross and Heller, 1963). The apparent rarity of bayerite and nordstrandite in soils may reflect identification difficulties, due to their low concentration and/or masking by the presence of gibbsite.

The same factors may be responsible for the rare reports on boehmite in soils (Taylor, 1987). Boehmite has been identified in lateritic materials (Gilkes et al., 1973), and together with diaspore in bauxites (Vgenopoulos, 1984). Boehmite often occurs in bauxite deposits in association with gibbsite from which it is considered to form through partial dehydration by diagenesis or hydrothermal alteration. Hughes et al. (1994) identified boehmite in Native American artifacts from an Ordovician age paleosol in northwestern Illinois. The formation of poorly crystalline boehmite has been attributed to the presence of foreign ions during crystallization (Hsu, 1967). Since poorly crystalline boehmite alters to crystalline $Al(OH)_3$ polymorphs on removal of the foreign ions from the environment, the occurrence of crystalline boehmite in soils other than those derived from bauxite may be less common than believed (Taylor, 1987).

Diaspore has been identified as a surface weathering product formed by the desilication of a kaolinitic clay in Missouri, in association with minor amounts of goethite (Keller, 1978). The possible influence of Fe in the diaspore formation is supported by synthesis experiments, and natural rock weathering and bauxite occurrences (Taylor, 1987). The formation of Fe substituted diaspore is, however, unlikely because the large Fe^{3+} does not fit into the diaspore lattice, in contrast to the substitution of the smaller Al^{3+} in the larger goethite lattice (Davies and Navrotsky, 1983). As confirmation, Fe in a diaspore crystal has been identified as a hematite-like cluster (Hazemann et al., 1992).

Corundum occurs in pegmatites and other rocks associated with nepheline syenites, in high-grade aluminous metamorphic rocks, in aluminous xenoliths in igneous rocks, in metamorphosed bauxitic deposits, and as a detrital mineral in sediments (Deer et al., 1992). Reports of corundum in soils are scarce; because of its high resistance to weathering, it can be found as a residual mineral of igneous or thermally metamorphosed rocks, as in the lateritic bauxites of Australia (Taylor et al., 1983), and in Brazilian Oxisols (Macedo and Bryant, 1987). In some Australian soils derived from corundum-free rocks, the presence of corundum is attributed to the heating of the soil by bush fires (Wells et al., 1989). The corundum may have formed in association with hematite by the dehydroxylation of Al-goethite. The associated crystallization of isostructural hematite may have acted as a template for the nucleation of corundum, thereby reducing the activation energy and enabling its formation at relatively low temperature by the epitaxial growth on hematite.

3.4.3 Influence on Soil Properties

Because of high surface area (100–220 $m^2\ g^{-1}$), high p.z.c. (pH 5–9), and variable surface charge, Al oxides favorably chemisorb heavy metals (Cu, Pb, Zn, Ni, Co, Cd) and anions (PO_4, SiO_3, MoO_4, SO_4, catechols) (Hingston et al., 1974; McBride and Wesselink, 1988; McBride, 1989). The adsorption is related to the reactive (singly coordinated OH^- groups at crystallite edges) rather than total surface area. The high P retention in Oxisols and Ultisols is usually related to the content of gibbsite, in addition to that of Fe oxides (Juo, 1981; Wambeke, 1992). Synthesis studies have shown the effectiveness of $Al(OH)_3$ precipitates in improving the physical properties of soils and clay minerals (see Hsu,

1989). Aluminum oxides, mainly gibbsite, in addition to Fe oxides, are strongly associated with the aggregation of Oxisols and Ultisols, but the mechanism is unclear.

3.4.4 Identification

Crystalline Al oxides can be identified and quantified by XRD, thermal analysis, IR, and EM with a combination of several methods being necessary in soils. Currently, there are no selective dissolution methods for the specific dissolution of crystalline and noncrystalline Al forms in soil samples, but acid NH_4 oxalate treatment (Schwertmann, 1964) is frequently used to estimate the content of undefined, poorly crystalline Al forms in soils. The problems inherent in the identification of poorly crystalline Al oxides in the presence of clay minerals is illustrated by Violante and Huang (1994). By XRD analysis, the detection limit of poorly crystalline boehmite in samples containing kaolinite or montmorillonite was approximately 30 to 40%. When both kaolinite and montmorillonite were present, however, the identification by DTA, IR or TEM often failed even for concentrations of boehmite as high as 50 to 80%.

Under the electron microscope, well-crystallized gibbsite occurs as hexagonal plates (Hsu, 1989), or as elongated hexagonal rods in synthetic crystals (Tait et al., 1983). A rectangular gibbsite particle from a soil is shown in Fig. 3.5. Bayerite frequently occurs as a triangular pyramid, with its long direction perpendicular to the basal plane (Hsu, 1989; Schoen and Roberson, 1970; Violante and Jackson, 1981). Synthetic nordstrandite shows rectangular plates, elongated parallelograms, ill-defined ovoidal particles or clusters of acicular crystals (Schoen and Roberson, 1970; Violante and Jackson, 1979, 1981; Violante et al., 1982; Tait et al., 1983). With the exception of gibbsite, many of these crystal shapes seem to occur only in synthetic Al oxides.

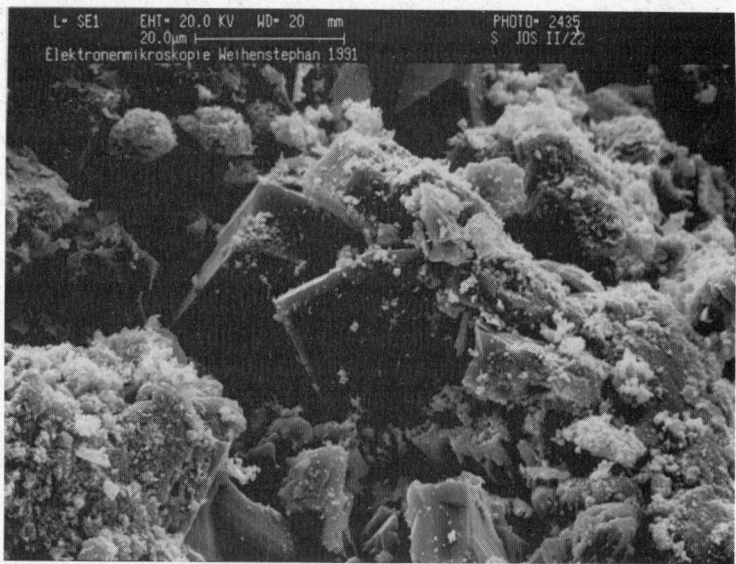

Fig. 3.5 Scanning electron micrograph from a bauxitic zone in a deep lateritic profile from the Jos plateau, Nigeria, showing hematite crystal aggregates on large euhedral gibbsite crystals [Reprinted from Zeese et al., 1994. Mineralogy and stratigraphy of three deep lateritic profiles of the Jos Plateau (Central Nigeria). Catena 21:195-214 with kind permission of Elsevier Science, Amsterdam, Netherlands]

3.4.4.1 X-ray Diffraction

The main XRD peaks of the Al oxides are listed in Table 3.4. For specimens with poor crystallinity, these diffraction peaks may not be completely resolved. Gibbsite in soils can be quantified by XRD, but it will not be detected if its content is below about 50 g kg^{-1} (Jackson, 1969). The strongest peak for boehmite is the d(020) at 6.11 Å, while very fine boehmite particles show a broadening and an apparent line shift of the d(020) peak to larger d-spacings due to a diffraction effect (Tettenhorst and Hoffman, 1980).

3.4.4.2 Thermal Analysis

Differential thermal analysis (DTA) is frequently used for the identification of Al hydroxides and oxyhydroxides (Mackenzie, 1970, 1972; Karathanasis and Harris, 1994). The presence of 10 g kg^{-1} gibbsite can be detected by DTA, and its amount can be estimated by comparing the 300 to 330 °C endothermic peak with those of standards. The peak temperature for bayerite is only slightly lower than that for gibbsite, so that their differentiation in natural samples is impossible (Karathanasis and Harris, 1994). In soil samples, the 300 to 330 °C peak may overlap with that of goethite, requiring an additional DTA curve after the removal of goethite with sodium dithionite (Jackson, 1969). Diaspore yields an endothermic peak at 540 °C, whereas boehmite has a peak which occurs in the range of 450 to 580 °C depending on crystallinity and particle size. The overlap with the kaolinite peak at 550 °C limits the identification of these oxyhydroxides.

3.4.4.3 Infrared Absorption Analysis

Characteristic OH stretching absorption spectra have been used to identify $Al(OH)_3$ and AlOOH polymorphs in monomineralic samples. Gibbsite shows three absorption bands and another doublet,

Table 3.4 Aluminum oxide minerals: crystallographic properties and d-values for the 6 most intense diffraction lines

Mineral	Chemical formula	Crystal system, Space group	Unit cell dimensions (Å)	Six most intense diffraction lines d-value (Å), relative intensity						Reference[†]
Gibbsite	γ-$Al(OH)_3$	Monoclinic P2$_1$/n	a=8.6552 b=5.0722 c=9.7161 β=94°607'	4.85 100	4.371 70	2.385 55	4.32 50	2.45 40	2.05 40	33-0018
Nordstrandite	$Al(OH)_3$	Triclinic P1	a=5.082 b=5.127 c=4.980 α=93°40' β=118°55' γ=70°16'	4.79 100	2.27 30	4.32 25	2.393 25	2.016 25	1.902 20	24-0006
Bayerite	$Al(OH)_3$	Monoclinic P21/a	a=5.062 b=8.671 c=4.713 β=90°27'	2.222 100	4.71 90	4.35 70	1.723 40	3.20 30	1.333 18	20-0011
Diaspore	α-AlOOH	Orthorhombic Pbnm	a=4.396 b=9.426 c=2.844	3.99 100	2.317 56	2.131 52	2.077 49	1.633 43	2.558 30	05-0355
Boehmite	γ-AlOOH	Orthorhombic Amam	a=3.70 b=12.227 c=2.868	6.11 100	3.164 65	2.346 55	1.86 30	1.85 25	1.453 16	21-1307
Corundum	α-Al_2O_3	Rhombohedral R3c	a=4.7592 c=12.992	2.086 100	2.551 98	1.60 96	3.48 72	1.374 57	1.74 48	43-1484

[†] numbers of the format XX-XXXX indicate ICDD file number (ICDD, 1994)

with reported bands for well-crystallized natural gibbsites at 3622, 3627, 3460, 3396, and 3384 cm^{-1}. Each of the absorption bands may shift slightly to lower wavenumbers with decreasing crystallinity (Elderfield and Hem, 1973). The 3622 cm^{-1} absorption band is coincident with that of kaolinite and micaceous clays (Farmer and Russel, 1967), thus precluding its use for the identification of gibbsite. Boehmite and diaspore each show two OH stretching bands, boehmite at 3087 and 3283 cm^{-1}, and diaspore at 2922 and 2990 cm^{-1} (Ryskin, 1974). Raman microprobe spectroscopy offers an alternative procedure in characterizing $Al(OH)_3$ polymorphs in soils (Rodgers, 1993).

3.5 Silicon Oxides

The Si oxides occurring in soils are given in Table 3.5. In nature, SiO_2 polymorphs (quartz, tridymite and cristobalite) occur as both higher temperature or β-phases, and lower temperature or α-phases. In soils, however, only the α-phases are usually found. Opal is a hydrated form of Si of both biogenic and inorganic origin. Microcrystalline quartz, or microquartz, occurs in the form of chalcedony and chert (Heaney, 1994). Microcrystalline fibrous α-quartz, in the form of chalcedony, usually occurs as a secondary infilling of seams and cavities within rocks, sometimes creating concentrically banded agates or geodes. An intergrowth of authigenic microcrystalline α-quartz grains, that range in size from < 1 to 50 μm, occurs in highly silicified sedimentary rocks in the form of chert (Knauth, 1994). More information about Si oxides can be found in Drees et al. (1989), and Heaney et al. (1994).

3.5.1 Mineral Phases

The Si oxides are tectosilicates. The repeating unit is a SiO_4 tetrahedron in which each O is linked to Si atoms of adjacent tetrahedra, forming a three-dimensional framework structure. This contrasts with the Fe, Al, Mn and Ti oxides in which the basic unit is a cation in octahedral coordination. The pattern of the tetrahedral linkage is different for each Si oxide polymorph, and this difference is reflected in their structural, physical, and chemical properties. The structure of the α-phases is closely related to that of their high temperature β-phase equivalents.

Table 3.5 Silica oxide minerals: Crystallographic properties and *d*-values for the 6 most intense diffraction lines

Mineral	Chemical formula	Crystal system, Space group	Unit cell dimensions (Å)	Six most intense diffraction lines[‡] *d*-value (Å), relative intensity						Reference[†]
α-Quartz	SiO_2	Trigonal P3$_1$21	a=4.912	3.342	4.257	1.818	1.542	2.457	2.282	Heaney (1994)
			c=5.403	100	22	14	9	8	8	33-1161
α-Tridymite	SiO_2	Monoclinic C222$_1$	a=10.04	4.08	4.28	3.80	3.24	2.48	2.382	Heaney (1994)
			b=17.28	100	93	68	48	35	21	42-1401
			c=8.20							
			β=91°50'							
α-Cristobal-ite	SiO_2	Tetragonal P4$_1$2$_1$2	a=4.969	4.04	2.487	2.841	3.136	2.467	1.929	Heaney (1994)
			c=6.925	100	13	9	8	4	4	39-1425
Opal (natural)				4.08	2.51	2.86	3.14	1.937	1.878	38-0448
				100	30	10	9	5	5	
Opal-C				~ α-Cristobalite + 4.3						Drees et al. (1989)
Opal-CT				4.10	4.29	2.50	3.34	3.18	2.85	Drees et al. (1989)
				vs	s	s	w	w	w	
Opal-A				4.1	2.0	1.5	1.2			Drees et al. (1989)
				sb	wd	wd	wd			

[†] numbers of the format XX-XXXX indicate ICDD file number (ICDD, 1994)
[‡] vs = very strong, s = strong, w = weak, sb = strong broad, wd = weak diffuse

Quartz consists of paired helical chains of corner-sharing SiO_4 tetrahedra that spiral along the z-axis (Heaney, 1994). The intertwined chains produce open channels parallel to c (z) that appear hexagonal in projection (Fig. 3.1). In *tridymite* and *cristobalite*, the idealized fundamental module consists of a sheet containing six member rings of SiO_4 tetrahedra, with the tetrahedra alternately pointing above and below the plane defined by the basal oxygen atoms (Heaney and Banfield, 1993). Tridymite and cristobalite are differentiated by a different stacking arrangement of this tetrahedral sheet. In tridymite, the sheets are stacked such that the hexagonal rings (ditrigonal and oval rings in α-tridymite structures) lie directly over one another, creating continuous tunnels normal to the sheets (Fig. 3.1). These tunnels account for the lower density of tridymite (2.26 g cm^{-3}) as compared to quartz (2.65 g cm^{-3}). In cristobalite, the stacking involves three sheets that are translated relative to one another such that the hexagonal rings (oval rings in α-cristobalite) do not superimpose (Fig. 3.1). Thus, cristobalite lacks the continuous tunnels normal to the layers as in tridymite, but tunnel structures are formed parallel to the layers. Thus, the density of cristobalite is only slightly higher than that of tridymite (2.32 g cm^{-3}).

Opal is classified into three structural groups based on XRD powder patterns: opal-C (well-ordered α-cristobalite), opal-CT (disordered α-cristobalite, α-tridymite) and opal-A (highly disordered, nearly amorphous) (Jones and Segnit, 1971; Deer et al., 1992), in a sequence of decreasing structural order. In opal-C the stacking is predominantly cristobalitic, which is revealed by weak superstructure XRD reflections characteristic of α-cristobalite. Opal-CT consists of random stackings of α-cristobalite- and α-tridymite-like arrangements, yielding a disordered crystal structure. The microstructure of opal-CT is composed of small spheres < 5 μm in diameter, called lepispheres, which consist of an interpenetrative growth of tiny cristobalite and tridymite blades (Hesse, 1988; Graetsch, 1994). Opal-A is a noncrystalline hydrous Si polymorph (SiO_2 nH_2O) differing from the crystalline polymorphs in that it lacks long-range atomic order. The microstructure of opal-A shows the closest packing of spheres of Si with diameters ranging from 10 to 50 nm (Greer, 1969) with water filling the interstices (Graetsch, 1994). The specific gravity of biogenic opal from soils and plants varies continuously over the range 1.5 to 2.3, with modal values from 2.10 to 2.15 g cm^{-3} (Drees et al., 1989). The broad range is a function of the submicron opal structure, H_2O content, occluded organic matter, and microscopic voids.

Quartz is one of the purest minerals known, and due to its more closed structure is purer than the other Si oxide polymorphs (Drees et al., 1989). Nevertheless, common trace elements in quartz, either interstitial or as substitutions, occur (Al, Ti, Fe, Na, Li, K, Mg, Ca, OH). Both cristobalite and trydimite, which are also the major structural constituents of opal-CT, may accommodate more impurities than quartz because of their more open framework structure. The predominant impurity is Al, with lesser amounts of Fe, Ti, K, Na, Ca, and Mg. Opal-A contains 850 to 950 g kg^{-1} of SiO_2, 40 to 90 g kg^{-1} of bound H_2O, and significant amounts of occluded, chemisorbed, or solid-solution impurities of Al, Fe, Ti, Mn, P, Cu, N, C, alkalies, and alkaline-earths (Drees et al., 1989).

3.5.2 Occurrence and Formation

Quartz is a common constituent in many igneous, sedimentary and metamorphic rocks, and also occurs as a secondary mineral, acting as a cement in sediments. Tridymite is a typical mineral of acid volcanic rocks together with cristobalite that also occurs in basaltic rocks. Trydimite is also a common constituent of highly metamorphosed impure limestones and arkoses adjacent to basic igneous intrusions, while cristobalite occurs in metamorphosed sandstones and sandstone xenoliths in basic rocks. Opal is found in sedimentary, volcanic, and marine environments (Deer et al., 1992). Thus, quartz is the most abundant Si oxide in soil environments, cristobalite occurs in soils developed from volcanic materials, opal may be a significant component of soils, but tridymite occurs only rarely.

Quartz in soils is mainly a primary mineral, inherited from the parent material. The higher stability of quartz compared to other silicate minerals is explained by its crystallization from magma closer to Earth's present surface conditions. The higher stability of quartz relative to the other Si oxide polymorphs is due to a denser packing of the crystal structure and a higher energy required to break the Si-O-Si bond. Quartz in chert forms through a diagenetic transformation from opal-A through opal-CT and opal-C (Hesse, 1988). As demonstrated experimentally, quartz may also precipitate directly (Mackenzie and Gees, 1971). Authigenic quartz in form of chemical precipitations and grain overgrowths frequently occurs in sediments (carbonates, sandstones), and is believed to form in soils, too (Drees et al., 1989).

Quartz is generally concentrated in the sand and silt fractions of soils, with a lower frequency in the coarse clay fraction (0.2-2 μm). The depth distribution of quartz is a function of parent material and degree of weathering (Drees et al. 1989). In relatively undifferentiated soils (Entisols and Inceptisols), distributions of quartz reflect mainly variations in parent material. Quartz may comprise > 90% of the inorganic fraction of Quartzipsamments. In Aridisols, quartz is generally abundant due to restricted weathering while in moderately weathered soils (Spodosols, Alfisols, Mollisols, and Ultisols), the eluvial horizons are enriched in quartz relative to the parent material due to the weathering and removal of less resistant minerals. Conversely, illuvial horizons are lower in quartz than eluvial horizons or parent material due to the dilution by silicate clays, carbonates, or oxides. Highly weathered and leached Oxisols have only very low amounts of quartz left in the clay fraction.

Opal-CT, opal-C, and α-cristobalite are usually limited to specific geographic regions and rock stratigraphic units. Opal-C and α-cristobalite have been reported in Andisols and other soils derived from volcanic material in South America, New Zealand, Central America, and Japan. Opal-CT has been identified in bentonites, siliceous shales, and indurated silicates. As a consequence, soils developed in these materials commonly contain opal-CT. It is also a primary constituent of many cherts, porcelanites, fossil wood, silcretes, and some duripans (Fig. 3.6) (Drees et al., 1989). According to synthesis experiments and evidence from diagenetic environments, opal-A transforms to opal-CT and quartz by a series of dissolution-precipitation reactions (Kastner et al., 1977; Williams and Crerar, 1985; Williams et al., 1985; Cady et al., 1996). The conversion is dependent on temperature, pressure, pH, pore solution chemistry, and specific surface. As the transformation is mediated by clays, oxides and carbonates, it is assumed to occur in soils (Drees et al., 1989). Opal-A may form by both organic and inorganic processes in pedogenic environments. Biogenic opal-A originates from Si accumulated by plants and aquatic organisms, and thus occurs under a wide range of environmental conditions (Jones and Beavers, 1964; Wilding and Drees, 1971).

Biogenic opal is a minor but nearly ubiquitous constituent of soils, with opal phytoliths representing the major form in nonaquatic environments, while sponge spicules, diatoms, and radiolaria are the main opal forms in aquatic environments and limestone residuum (Jones and Beavers, 1964; Wilding and Drees, 1971). Amounts of opal phytoliths in soils commonly range from < 1–30 g kg^{-1}, usually occurring in the 5–20 μm and 20–50 μm size fractions, and decreasing with soil depth (Drees et al., 1989).

Opal-A, which forms when the conditions favor supersaturation of Si in soil solution (Jones and Handreck, 1967), occurs as nodules and as a primary cement for indurated soil horizons (duripans) and silcretes. Such soils usually have pH and moisture regimes favoring solubilization and translocation of Si to lower positions in the soil profile. Geographically, duripans are restricted to materials that provide soluble Si, such as pyroclastics and intermediate and basic igneous rocks (Flach et al., 1969; Soil Survey Staff, 1975). According to Flach et al. (1969), pedogenic transformation of opal to quartz is

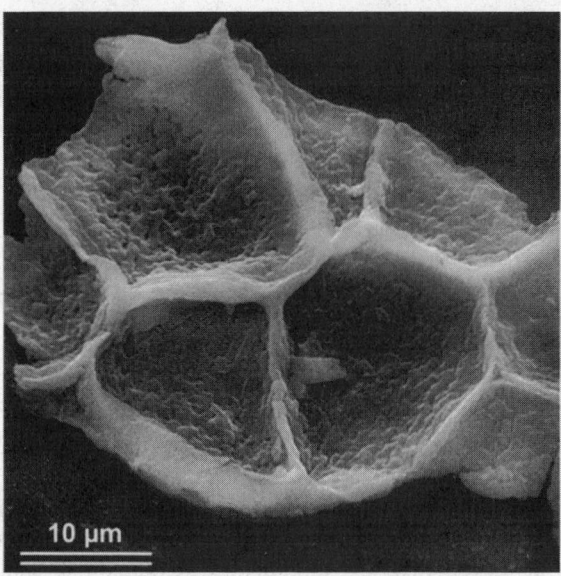

Fig. 3.6 Scanning electron micrograph of an opal phytolith isolated from a forest soil [Reprinted from Drees et al., 1989. Silica in soils. J.B. Dixon and S.B. Weed. Minerals in soil environments, with permission of the Soil Science Society of America]]

believed to occur in duripans and Si-cemented soils. Silica released by weathering of silicate minerals may precipitate as amorphous SiO_2 on soil particles, forming bridges between mineral grains, leading to the hardness and brittleness of fragipans (Franzmeier et al., 1989; Karathanasis, 1989).

At ambient temperature and neutral pH, the solubility of amorphous Si in soils is approximately 50–60 mg Si L^{-1}, and that of quartz is commonly 3–7 mg Si L^{-1}. The solubility of Si oxides is a function of temperature, pH, particle size, chemical composition, and the presence of a disrupted surface layer. For both crystalline and amorphous Si polymorphs, it is essentially constant between the pH limits of 2 and 8.5, increasing rapidly above pH 9 due to the ionization of monosilicic acid.

The presence of organic molecules greatly enhances the dissolution rates of Si oxides. The dissolution of biogenic opal and quartz increases as particle size decreases, due to the increase in surface area (Drees et al., 1989). Chemisorption of metallic ions such as Al, Fe, Mg, Ca, Cu, Pb, and Hg to Si surfaces reduces the dissolution rates of Si due to the formation of relatively insoluble Si coatings (Drees et al., 1989), thus increasing the stability of Si oxides. Oxides of Fe and Al, by acting as a soluble Si sink, greatly increase the dissolution rates of amorphous Si oxides and reactive uncoated quartz surfaces. Dissolution (weathering) of Si oxides is induced by the reduction of Si levels in soil solution by leaching and plant uptake. Dissolution of quartz initiates below approximately 3 mg Si L^{-1} (Kittrick, 1969).

3.5.3 Influence on Soil Properties

Noncemented soils comprised dominantly of quartz are nonplastic due to the weak cohesion (van der Waals forces) that develops between Si particles, and show low water-retention capacity and high hydraulic conductivity. Duripans, cemented by even small amounts of Si, are hard to extremely hard when dry, have high unconfined compressive strength, and resist dispersion in Na hexametaphosphate (Flach et al., 1974).

Due to their essentially uncharged, poorly hydrated surfaces, crystalline Si oxides have relatively small effects on the physicochemical activity affecting the soil-plant relationship, and act as a diluent to the much more reactive clay minerals and sesquioxides. Silica is not essential for the growth of most plants, but orthosilicic acid, H_4SiO_4, has a beneficial effect on the growth of some plants (Jones and Handreck, 1967). Rice and sugarcane, for example, often respond to fertilization with very soluble Si sources such as Si slag.

Due to its ubiquity, abundance, resistance to weathering, and immobility quartz has been often used as an index mineral for weathering and soil formation (Barshad, 1964; Sudom and St. Arnaud, 1971). Extensive leaching may, however, drastically increase the solubility of primary quartz (Little et al., 1978; Asumadu et al., 1988; Pye and Mazzullo, 1994), and authigenic (Breese, 1960) as well as biogenic quartz may form in soils (Wilding and Drees, 1974), thus restricting the use of quartz as an index mineral.

The isotopic composition of quartz (^{18}O), which depends on the temperature of formation, has been used to identify eolian additions to soils (Clayton et al., 1972; Mizota and Matsuhisa, 1995), and the formation of authigenic opaline Si in volcanic ash soils (Mizota et al., 1991).

Opal phytoliths have been used to trace the origin of colluvial sediments (Lutwick and Johnston, 1969), and to identify surface horizons of paleosols (Dormaar and Lutwick, 1969). Depth distributions of sponge spicules may be useful to identify lithological discontinuities, and opal of aquatic organisms (diatoms, sponge spicules, and siliceous shells) provides direct evidence that the parent material of the soil is of marine or lacustrine origin (Jones and Beavers, 1963, 1964; Wilding and Drees, 1968, 1971). Biogenic opal has been extensively used to reconstruct the vegetative history of soils (Drees et al., 1989; Fisher et al., 1995). The ^{13}C/^{12}C isotope ratio of C occluded in opal phytoliths was used to establish the succession of C3 and C4 grasses, providing a quantitative method for monitoring climatic changes (Kelly et al., 1991).

3.5.4   Identification

Silicon oxides can be identified by XRD, scanning electron microscope (SEM), microprobe analysis, IR, and light microscopy. Opal-CT and opal-A may need a concentration by particle size fractionation, specific gravity, and/or differential dissolution. Counting particle size separates under the light microscope is the simplest method of quantifying Si oxides, but is limited in resolution (5 to 10 µm) (Drees et al., 1989).

3.5.4.1   Crystal Form, Habit, and Color

In soils, quartz generally occurs as anhedral grains, rarely exhibiting the characteristic prismatic habit of macrocrystals. Angular grains are generally the result of mechanical fracturing, while rounding is a result of both physical attrition during transport and solution. Overgrowths on quartz grains may suggest long-term stable environments. Particles < 100 µm are commonly shaped like flat plates due to cleavage (Drees et al., 1989). Thus, quartz grain surface morphology may be used to elucidate mineral origin and past and/or present chemical and physical environment.

Opal-CT is commonly observed in sediments as small (< 10 µm) lepispheres. The morphology of opal-A of biogenic origin is closely related to the biological cell or structure in which it originates: opal of forest origin consisting of cellular incrustations with numerous thin sheet structures, opal of grass consisting of solid polyhedral structures resulting from the silicification of the entire cell, and opaline microfossils of sponge spicules, diatoms, and shells (Drees et al., 1989).

Pure Si oxides are colorless, but chemical impurities may impart various colors. Quartz is usually colorless and transparent, or white, with a vitreous luster. Cristobalite is white to milky white and ranges from translucent to opaque. The color and degree of translucency of opal-CT seem to depend on the aggregation of lepispheres. For opal-A, color is not a diagnostic criterion, because it is strongly affected by occluded chemical impurities, and light interference and scattering. In transmitted light, biogenic opal isolated from soils ranges from colorless or light tan to various shades of brown or black (Drees et al., 1989).

3.5.4.2 X-ray Diffraction

The diagnostic parameters for Si oxides are presented in Fig. 3.7 and Table 3.5. Quartz is easily identified by its most intense reflections at 4.26 Å and 3.34 Å, even in mixed mineral assemblages. The accuracy of the quantitative determination of quartz depends on sample preparation and the diffraction peak used (Rowse and Jepson, 1972). For well-ordered or α-cristobalite, the 4.04 Å peak is the main clue in mixed samples provided that there are no feldspars present. Opal-C gives a pattern resembling that of α-cristobalite, except for slight line broadening and minor evidence of trydimite stacking. Opal-CT usually exhibits broad cristobalite peaks at about 4.1 and 2.5 Å, and a peak at 4.3 Å which is attributed to tridymite. The wide variations in the intensity and line profiles make its quanti-

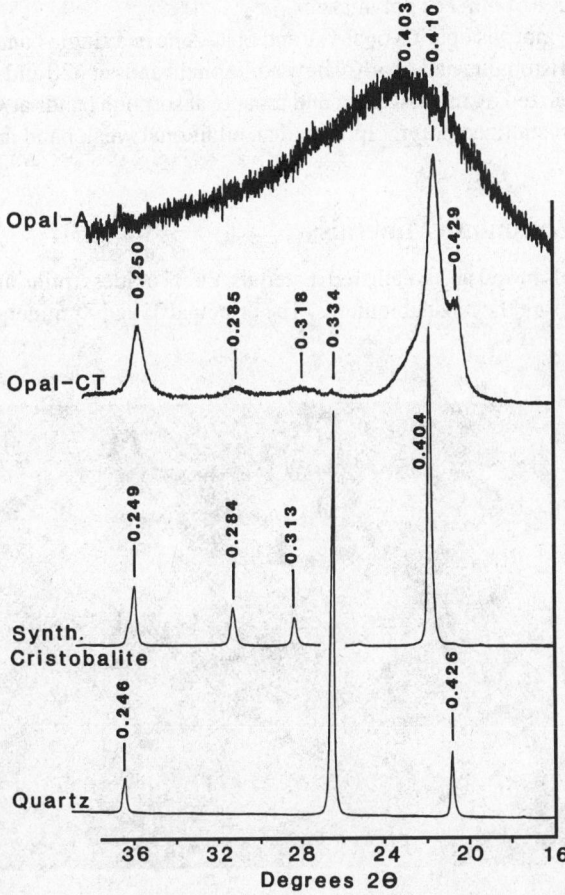

Fig. 3.7 X-ray diffraction patterns for quartz, synthetic cristobalite, opal-CT, and opal-A (Cu Kα radiation, peak position in nanometers) [Reprinted from Drees et al., 1989. J.B. Dixon and S.B. Weed. Minerals in soil environments with permission of the Soil Science Society of America]

fication by XRD difficult. The short-range order in opal-A produces a diffuse, broad X-ray hump centered at about 4.0 Å.

3.5.4.3 Thermal Properties

Quartz exhibits a sharp endothermic DTA peak at about 570 °C (Drees et al., 1989), which may not be evident for microcrystalline quartz (< 0.05 µm). Well-ordered synthetic cristobalite has a characteristic endothermic peak at about 260 °C, whereas opal-CT and opal-A do not yield a characteristic endotherm below 600 °C. According to Rowse and Jepson (1972), DTA is better than either XRD or chemical techniques for detecting small quantities of quartz in clay materials.

3.5.4.4 Infrared Spectroscopy

All tectosilicates are characterized by Si-O-Si stretching vibrations between 950 and 1200 cm^{-1} and by O-Si-O bending vibrations between 400 and 550 cm^{-1}. Quartz is distinguished from the other Si polymorphs by a distinctive absorption band at 692 cm^{-1} and two strong doublets, one at 395 and 370 cm^{-1} and the other at 800 and 780 cm^{-1} (Drees et al., 1989). The latter doublet has been used also for the quantification of quartz (Chester and Green, 1968). This method is, however, restricted to samples < 1µm in particle size and free of other Si polymorphs.

The disordered Si polymorphs opal-C, opal-CT and opal-A have a single band around 800 cm^{-1} in common. Well-ordered cristobalite and opal-C show additional bands at 620 and 380 cm^{-1}. Opal-A of biogenic origin is characterized by much weaker and broader absorption bands at 460 and 1100 cm^{-1} as compared to the better crystallized polymorphs, and an additional weak band at 965 cm^{-1} (Fig. 3.8) (Drees et al., 1989).

3.6   Titanium and Zirconium Minerals

Of all Ti and Zr compounds in soil and weathered materials, the Ti oxides (rutile, anatase and ilmenite), and the Zr silicate (zircon) are the most ubiquitous. The principal Ti and Zr minerals are listed in Table

Fig. 3.8 Transmission electron micrograph of the 2-0.2 µm fraction of a Rhodic Hapludalf showing an anatase or rutile particle (arrow) surrounded by kaolinite and iron oxide particles [Reprinted from de Brito Galvão and Schulze, 1996. Soil Sci. Soc. Am. J. 60:1969-1978 with permission of the Soil Science Society of America]

Table 3.6 Titanium and Zr oxide minerals: crystallographic properties and *d*-values for the 6 most intense diffraction lines

Mineral	Chemical formula	Crystal system, Space group	Unit cell dimensions (Å)	Six most intense diffraction lines *d*-value (Å), relative intensity						Reference[†]
Rutile	TiO_2	Tetragonal P4$_2$/mnm	a=4.5933 c=2.9592	3.247 100	1.687 60	2.487 50	2.188 25	1.624 20	1.36 20	21-1276
Anatase	TiO_2	Tetragonal I4$_1$/amd	a=3.7853 c=9.5139	3.52 100	1.892 35	2.378 20	1.70 20	1.666 20	1.481 14	21-1272
Brookite	TiO_2	Orthorhombic Pbca	a=5.4558 b=9.1819 c=5.1429	3.152 100	2.90 90	3.465 80	1.893 30	1.66 30	2.476 25	29-1360
Ilmenite	$Fe^{2+}TiO_3$	Hexagonal R3	a=5.0884 c=14.093	2.754 100	2.544 70	1.726 55	1.868 40	1.468 35	3.737 30	29-0733
Pseudo-brookite	$Fe^{3+}_2TiO_5$	Orthorhombic Bbmm	a=9.7965 b=9.9805 c=3.7301	3.486 100	2.752 77	4.90 42	2.458 23	1.865 23	2.407 22	41-1432
Pseudorutile	$Fe_2O_3 \cdot nTiO_2 \cdot mH_2O$ $Fe_2Ti_3O_9$	Hexagonal P6$_3$22	a=14.486 c=4.467	3.50 100	2.66 90	2.51 80	1.687 70	3.67 40	3.23 30	29-1494
Ulvöspinel	Fe_2TiO_4	Cubic Fd3m	a=8.393– 8.536	2.573 100	1.509 39	3.018 33	1.642 33	2.134 19	1.742 10	34-0177
Zircon	$ZrSiO_4$	Tetragonal I4$_1$/amd	a=6.604 c=5.979	3.30 100	4.434 45	2.518 45	1.712 40	2.066 20	1.908 14	6-266

[†] numbers of the format XX-XXXX indicate ICDD file number (ICDD, 1994).

3.6. Additional information about Ti and Zr minerals can be found in Milnes and Fitzpatrick (1989), and Lindsley (1976, 1991). Titanomagnetite and titanomaghemite are described in Section 3.2.

3.6.1 Mineral Phases

Titanium occurs primarily in octahedral coordination, and like the Fe and Mn oxides, the structures of the various Ti oxide minerals can be described in terms of the different arrangement of the Ti-containing octahedra (Lindsley, 1976; Waychunas, 1991; Heaney and Banfield, 1993).

Rutile (TiO_2) is isostructural with pyrolusite and manganite, consisting of single chains of edge-sharing TiO_6 octahedra (Fig. 3.1) (Heaney and Banfield, 1993). *Anatase* consists of TiO_6 octahedra that share four O-O edges, two at the top and two at right angles at the bottom. The octahedra outline a three-dimensional framework, rather than distinct chains (Fig.3.1n) (Waychunas, 1991). Schwertmann et al. (1995b) reported the substitution of Ti^{4+} by Fe^{3+} up to 0.1mole fraction in pedogenic and synthetic anatase. Charge compensation is achieved by the incorporation of structural OH. *Brookite* has a more complex structure than rutile or anatase, consisting of deformed TiO_6 octahedra sharing three O-O edges to form staggered, cross-linked chains that are oriented along the c-axis (Fig. 3.1) (Lindsley, 1976). *Ilmenite* ($FeTiO_3$) is almost isostructural with hematite, with one-half of the Fe atoms replaced by Ti, so that the Fe^{3+}-O_3-Fe^{3+} units in hematite become Fe^{2+}-O_3-Ti^{4+} units in ilmenite (Fig. 3.1) (Lindsley, 1976). *Pseudobrookite* (ideally Fe_2TiO_5) has strongly distorted octahedra, but can be described in terms of an ideal ccp anion arrangement (Fig. 3.1) (Waychunas, 1991). There are two types of octahedral sites, M1 and M2 in the ratio 1:2, ideally occupied by Fe^{3+} and Ti^{4+}, respectively. *Pseudorutile* ($Fe_2O_3 \cdot nTiO_2 \cdot mH_2O$; 3 < n < 5 and 1 < m < 2), is a structurally disordered and poorly characterized mineral formed by the alteration of ilmenite. Its structure is based on a hexagonal closest packing of O anions and has been described as an intergrowth of a rutile type structure with a goethite type structure (Fig. 3.1) (Grey et al., 1983). *Zircon* ($ZrSiO_4$) is an orthosilicate with a structure consisting of chains of alternating edge-sharing SiO_4 tetrahedra and [ZrO_8] triangular dodecahedra. The chains are joined laterally by dodecahedra-sharing edges, in an arrangement that produces unoccupied octahedral voids (Speer, 1982).

3.6.2 Occurrence and Formation

Rutile, anatase, ilmenite, and less frequently brookite, are found as accessory minerals in many igneous and metamorphic rocks, and as detrital minerals in sediments, where anatase is often of authigenic origin (Deer et al., 1992). The Ti and Zr minerals in soils may be residual minerals inherited from the parent material, formed through weathering of Ti- and Zr-bearing minerals, or authigenic (Milnes and Fitzpatrick,1989). Rutile, anatase, ilmenite and zircon are commonly residual minerals occurring in the sand and silt fractions of a variety of soils. Ilmenite may weather to pseudorutile and mixtures of rutile, anatase, and Fe oxides. There is evidence of anatase and ilmenite weathering by organic acids in A horizons of Scottish podzols (Berrow et al., 1978), and weathering features were observed on the surface of zircon and rutile grains of Australian Spodic Quartzipsamments (Tejan-Kella et al., 1991).

There are many examples of secondary Ti oxides formed through weathering of primary minerals, in saprolites and in soils. The data on the secondary or authigenic formation of Zr minerals are, however, controversial (Milnes and Fitzpatrick, 1989). The alteration of ilmenite under oxidizing conditions leads to the formation of pseudorutile (Grey and Reid, 1975; Anand and Gilkes, 1984b), but, at this stage, there is no clear evidence that it also takes place under pedogenic conditions (Berrow et al., 1978). Authigenic formation of poorly crystalline anatase as an alteration product of sphene was observed in peaty podzols from Scotland (Berrow et al., 1978). In conclusion, basic aspects of the Ti and Zr minerals, such as whether the minerals are relict or pedogenic, and the conditions responsible for their weathering and formation in soils, are still unresolved (Taylor, 1987; Milnes and Fitzpatrick, 1989).

3.6.3 Influence on Soil Properties

Because of their generally low concentrations in soils, there is little evidence of the effect of Ti and Zr minerals on soil reactivity. Only in tropical soils where their concentrations are much higher may some effects be expected. The broken edge bonds of anatase are hydroxylated and can exhibit variable charge characteristics (Fitzpatrick et al., 1978). The rutile and anatase surface has hydroxyl groups with different reactivities (Tanaka and White, 1982) that can adsorb and retain phosphate and arsenate (Cabrera et al., 1977; Fordham and Norrish, 1983).

Because ilmenite and zircon are generally very resistant in soils, they may be used as reference minerals in weathering and soil genesis studies (Bleeker, 1972; Mitchell, 1975; Claridge and Weatherhead, 1978; Tejan-Kella et al., 1991). The immobility of these minerals must be, however, previously assured (Colin et al., 1993).

3.6.4 Identification

As Ti and Zr minerals can occur in different particle size fractions and exhibit variable crystallinity, a variety of techniques may be necessary to identify and characterize them. Optical microscopy is a useful technique for examining and identifying these minerals, both in undisturbed form in thin sections and as components separated by physical or chemical techniques (Mitchell, 1975). A combination of XRD, IR absorption, DTA, and electron microprobe techniques (Fig. 3.5) is useful for identifying the Ti and Zr minerals in the sand and silt fractions of soils, but may be less satisfactory for the clay fractions due to the interference of layer silicates (Milnes and Fitzpatrick, 1989). Titanium oxides may be concentrated in kaolinitic soils by dissolution of clay minerals in boiling $5M$ NaOH alone (Norrish and Taylor, 1961) or by a combination of boiling $5 M$ KOH and dithionite-citrate-bicarbonate treatments (Zeese et al., 1994). Amorphous Ti oxides can be separated from more crystalline forms by extraction in acid ammonium oxalate (Fitzpatrick et al., 1978). Minor amounts of anatase (> 0.02 %) can be identified by Raman spectroscopy (Murad, 1997). Tejan-Kella et al. (1991) describe a SEM method to characterize microtextural features of zircon and rutile grains. In XRD of the clay fraction, the strong, sharp anatase peak at 3.52 Å (Table 3.6) is often clearly resolved after the interfering 3.59 Å kaolinite line has been eliminated by heating at 550 °C.

Acknowledgments
NK acknowledges CAPES-Brazil for supporting a sabbatical leave at Purdue University. ACS acknowledges the support by the Deutsche Forschungsgemeinschaft. This is journal article no. 15645 of the Purdue Agricultural Research Programs.

3.7 References

Adams, S.N., J.L. Honeysett, K.G. Tiller, and K. Norrish. 1969. Factors controlling the increase of cobalt in plants following the addition of a cobalt fertilizer. Aust. J. Soil Res. 7:29–42.

Adams, W.A., and J.K. Kassim. 1984. Iron oxyhydroxides in soils developed from lower Palaeozoic sedimentary rocks in mid Wales and implications for some pedogenetic processes. J. Soil Sci. 35:117–126.

Allen, B.L., and B.F. Hajek. 1989. Mineral occurrence in soil environments. p. 199–278. *In* J.B. Dixon and S.B. Weed (ed.) Minerals in soil environments. Soil Science Society of America, Madison, WI.

Anand, R.R., and R.J. Gilkes. 1984a. Mineralogical and chemical properties of weathered magnetite grains from lateritic saprolite. J. Soil Sci. 35:559–567.

Anand, R.R., and R.J. Gilkes. 1984b. Weathering of ilmenite in a lateritic pallid zone. Clays Clay Miner. 32:363–374.

Anand, R.R., and R.J. Gilkes. 1987a. Variations in the properties of iron oxides within individual specimens of lateritic duricrust. Aust. J. Soil Res. 25:287–302.

Anand, R.R., and R.J. Gilkes. 1987b. The association of maghemite and corundum in Darling Range laterites, Western Australia. Aust. J. Soil Res. 25:303–311.

Anjos, L.H.C. dos, D.P. Franzmeier, and D.G. Schulze. 1995. Formation of soils with plinthite on a toposequence in Maranhão State, Brazil. Geoderma 64:257–279.

Asumadu, K., R.J. Gilkes, T.M. Armitage, and H.M. Churchward. 1988. The effects of chemical weathering on the morphology and strength of quartz grains-An example from S.W. Australia. J. Soil Sci. 39:375–383.

Bardossy, G., and G.W. Brindley. 1978. Rancieite associated with a karstic bauxite deposit. Amer. Miner. 63:762–767.

Barshad, I. 1964. Chemistry of soil development. p. 1–70. *In* F.E. Bear (ed.) Chemistry of soil. Reinhold Publishing Corporation, New York, NY.

Bascomb, C.L. 1968. Distribution of pyrophosphate-extractable iron and organic carbon in soils. J. Soil Sci. 19:251–268.

Baur, W.H. 1976. Rutile-type compounds. V. Refinement of MnO_2 and MgF_2. Acta Cryst. B32:2200–2204.

Bernal, J.D., D.R. Dasgupta, and A.L. Mackay. 1959. The oxides and hydroxides of iron and their structural interrelationships. Clay Min. Bull 4:15–30.

Berrow, M.L., M.J. Wilson, and G.A. Reaves. 1978. Origin of extractable titanium and vanadium in the A horizon of Scottish podzols. Geoderma 21:89–103.

Bigham, J.M., D.C. Golden, L.H. Bowen, S.W. Buol, and S.B. Weed. 1978. Mössbauer and X-ray evidence for the pedogenic transformation of hematite to goethite. Soil Sci. Soc. Amer. J. 42:979–981.

Bigham, J.M., L. Carlson, and E. Murad. 1994. Schwertmannite, a new iron oxyhydroxy-sulfate from Pyhäsalmi, Finland and other localities. Miner. Mag. 58:641–648.

Bigham, J.M., U. Schwertmann, and G. Pfab. 1996b. Influence of pH on mineral speciation in a bioreactor simulating acid mine drainage. Appl. Geochem. 11:845–849.

Bigham, J.M., U. Schwertmann, and L. Carlson. 1992. Mineralogy of precipitates formed by biogeochemical oxidation of Fe(II) in mine drainage. Catena Suppl. 1:219–232.

Bigham, J.M., U. Schwertmann, L. Carlson, and E. Murad. 1990. A poorly crystallized oxyhydroxide of iron formed by bacterial oxidation of Fe(II) in acid mine waters. Geochim. Cosmochim. Acta 54:2743–2758.

Bigham, J.M., U. Schwertmann, S.J. Traina, R.L. Winland, and M. Wolf. 1996a. Schwertmannite and goethite solubilities and the chemical modelling of iron in acid sulfate waters. Geochim. Cosmochim. Acta 60:2111–2121.

Birch, W.D., A. Pring, A. Reller, and H. Schmalle. 1993. Bernalite, $Fe(OH)_3$, a new mineral from Broken Hill, New South Wales: Description and structure. Amer. Miner. 78:827–834.

Birnie, A.C., and E. Paterson. 1991. The mineralogy and morphology of iron and manganese oxides in an imperfectly-drained Scottish soil. Geoderma 50:219–237.

Blake, R.L., R.E. Hessevick, T. Zoltai, and L.W. Finger. 1966. Refinement of the hematite structure. Amer. Miner. 51:123–129.

Bleeker, P. 1972. The mineralogy of eight latosolic and related soils form Papua-New Guinea. Geoderma 8:191–205.

Blume, H.P., and U. Schwertmann. 1969. Genetic evaluation of the profile distribution of aluminum, iron and manganese oxides. Soil Sci. Soc. Amer. Proc. 33:438–444.

Borggaard, O.K. 1988. Phase identification by selective dissolution techniques. p. 83–98. In J.W. Stucki, B.A. Goodman, and U. Schwertmann (ed.) Iron in soils and clay minerals. D. Reidel Publishing Co., Dordrecht, Holland.

Boudeulle, M., and J.-P. Muller. 1988. Structural characteristics of hematite and goethite and their relationship with kaolinite in a laterite from Cameroon. A TEM study. Bull. Minéral. 111:149–166.

Breese, G.F. 1960. Quartz overgrowths as evidence of silica deposition in soils. Aust. J. Sci. 23:18–20.

Bricker, O. 1965. Some stability relations in the system $Mn-O_2-H_2O$ at 25 C and one atmosphere total pressure. Amer. Miner. 50:1296–1354.

Brindley, G.W., and D.L. Bish. 1976. Green rust: A pyroaurite type structure. Nature 263:353.

Burns, R.G., V.M. Burns, and H.W. Stockman. 1985. The todorokite-buserite problem: Further considerations. Amer. Miner. 70:205–208.

Busing, W.R., and H.A. Levy. 1958. A single crystal neutron diffraction study of diaspore, AlO(OH). Acta Cryst. 11:798–803.

Cabrera, F., L. Madrid, and P. de-Armbarri. 1977. Adsorption of phosphate by various oxides: Theoretical treatment of the adsorption envelope. J. Soil Sci. 28:306–313.

Cady, S.L., H.-R. Wenk, and K.H. Downing. 1996. HRTEM of microcrystalline opal in chert and porcelanite from the Monterey Formation, California. Amer. Miner. 81:1380–1395.

Cambier, P. 1986. Infrared studies of goethites of varying crystallinity and particle size: I. Interpretation of OH and lattice vibration frequencies. Clay Miner. 21:191–200.

Campbell, A.S., and U. Schwertmann. 1984. Iron oxide mineralogy of placic horizons. J. Soil Sci. 35:569–582.

Carlson, L., and U. Schwertmann. 1980. Natural occurrence of feroxyhite (δ'-FeOOH). Clays Clay Miner. 28:272–280.

Carlson, L., and U. Schwertmann. 1981. Natural ferrihydrites in surface deposits from Finland and their association with silica. Geochim. Cosmochim. Acta 45:421–429.

Carlson, L., and U. Schwertmann. 1990. The effect of CO_2 and oxidation rate on the formation of goethite versus lepidocrocite from an Fe(II) system at pH 6 and 7. Clay Miner. 25:65–71.

Cerling, T.E., and R.R. Turner. 1982. Formation of freshwater Fe-Mn coatings on gravel and the behaviour of ^{60}Co, ^{90}Sr, and ^{137}Cs in a small watershed. Geochim. Cosmochim. Acta 46:1333–1343.

Chao, T.T. 1972. Selective dissolution of manganese oxides from soils and sediments with acidified hydroxylamine hydrochloride. Soil Sci. Soc. Amer. J. 36:764–768.

Chester, R., and R.N. Green. 1968. The infra-red determination of quartz in sediments and sedimentary rocks. Chem. Geol. 3:199–212.

Childs, C.W. 1975. Composition of iron-manganese concretions from some New Zealand soils. Geoderma 13:141–152.

Childs, C.W. 1981. Field tests for ferrous iron and ferric-organic complexes (on exchange sites or in water-soluble forms) in soils. Aust. J. Soil Res. 19:175–180.

Childs, C.W., R.W.P. Palmer, and C.W. Ross. 1990. Thick iron oxide pans in soils of Taranaki, New Zealand. Aust. J. Soil Res. 28:245–257.

Chukhrov, F.V., A.I. Gorshkov, E.S. Rudnitskaya, V.V. Beresovskaya, and A.V. Sivtsov. 1980. Manganese minerals in clays: A review. Clays Clay Miner. 28:346–354.

Chukhrov, F.V., and A.I. Gorshkov. 1981. On the nature of some hypergene manganese minerals. Chem. Erde 40:207–216.

Claridge, G.G.C., and A.V. Weatherhead. 1978. Mineralogy of silt fractions of New Zealand soils. N.Z.J. Sci. 21:413–423.

Clayton, R.N., R.W. Rex, J.K. Syers, and M.L. Jackson. 1972. Oxygen isotope abundance in quartz from Pacific pelagic sediments. J. Geophys. Res. 77:3907–3915.

Coey, J.M.D. 1988. Magnetic properties of iron in soil iron oxides and clay minerals. p. 397–466. *In* J.W. Stucki, B.A. Goodman, and U. Schwertmann (ed.) Iron in soils and clay minerals. D. Reidel Publishing Co, Dordrecht, Netherlands.

Colin, F., C. Alarçon, and P. Vieillard. 1993. Zircon: An immobile index in soils? Chem. Geol. 107:273–276.

Combes, J.M., A. Manceau, G. Calas, and J.Y. Bottero. 1989. Formation of ferric oxides from aqueous solutions: A polyhedral approach by X-ray absorption spectroscopy: I. Hydrolysis and formation of ferric gels. Geochim. Cosmochim. Acta 53:583–594.

Cornell, R.M. 1987. Comparison and classification of the effects of simple ions and molecules upon the transformation of ferrihydrite into more crystalline products. Z. Pflanzenernähr. Bodenk. 150:304–307.

Cornell, R.M., and U. Schwertmann. 1996. The iron oxides. Wiley-VCH Verlag, Weinheim, Germany.

Crowther, D.L, and J.G. Dillard. 1983. The mechanism of Co(II) oxidation on synthetic birnessite. Geochim. Cosmochim. Acta 47:1399–1403.

Curi, N., and D.P. Franzmeier. 1984. Toposequence of Oxisols from the Central Plateau of Brazil. Soil Sci. Soc. Amer. J. 48:341–346.

Dachs, H. 1963. Neutronen- und Röntgenuntersuchungen an Manganit, MnOOH. Z. Krist. 118:303–326.

Davies, P.K., and A. Navrotsky. 1983. Quantitative correlations of deviations from ideality in binary and pseudobinary solid solutions. J. Solid State Chem. 46:1–22.

de Brito Galvão, T.C., and D.G. Schulze. 1996. Mineralogical properties of a collapsible lateritic soil from Minas Gerais, Brazil. Soil Sci. Soc. Amer. J. 60:1969–1978.

Deer, W.A., R.A. Howie, and J. Zussman. 1992. An introduction to the rock-forming minerals. 2nd ed. Longman Scientific and Technical, Essex, UK.

Dormaar, J.F., and L.E. Lutwick. 1969. Infrared spectra of humic acids and opal phytoliths as indicators of paleosols. Can. J. Soil Sci. 49:29–37.

Drees, L.R., L.P. Wilding, N.E. Smeck, and A.L. Senkayi. 1989. Silica in soils: Quartz and disordered silica polymorphs. p. 913–974. *In* J.B. Dixon and S.B. Weed (ed.) Minerals in soil environments. 2nd Ed. Soil Science Society of America, Madison, WI.

Drits, V.A., B.A. Sakharov, and A. Manceau. 1993a. Structure of feroxyhite as determined by simulation of X-ray curves. Clay Miner. 28:209–221.

Drits, V.A., B.A. Sakharov, A.L. Salyn, and A. Manceau. 1993b. Structural model for ferrihydrite. Clay Miner. 28:185–207.

Egger, K., and W. Feitknecht. 1962. Über die Oxidation von Fe_3O_4 zu γ-und α-Fe_2O_3. Die differenzthermoanalytische (DTA) und thermogravimetrische (TG) Verfolgung des Reaktionsablaufes an künstlichen Formen von Fe_3O_4. Helv. Chim. Acta 45:2042–2057.

Eggleton, R.A. 1988. The application of micro-beam methods to iron minerals in soils. *In* J.W. Stucki, B.A. Goodman, and U. Schwertmann (ed.) Iron in soils and clay minerals. D. Reidel Publishing Co., Dordrecht, Netherlands.

Ehrlich, H.L. 1990. Geomicrobiology. 2nd Ed. Marcel Dekker, New York, NY.

Ehrlich, H.L. 1996. How microbes influence mineral growth and dissolution. Chemical Geol. 132:5–9.

Elderfield, H., and H.D. Hem. 1973. The development of crystalline structure in aluminum hydroxide polymorphs on aging. Mineral. Mag. 39:89–96.

Ewing, F.J. 1935a. The crystal structure of lepidocrocite. J. Chem. Phys. 3:420–424.

Ewing, F.J. 1935b. The crystal structure of diaspore. J. Chem. Phys. 3:203–207.

Fanning, D.S., M.C. Rabenhorst, and J.M. Bigham. 1993. Colors of acid sulfate soils. p. 91–108. *In* J.M. Bigham, and E.J. Ciolkosz (ed.) Soil color. Soil Sci. Soc. Amer. Spec. Publ. 31. Soil Science Society of America, Madison, WI.

Farmer, V.A., and J.D. Russel. 1967. Infrared absorption spectrometry in clay studies. Clays Clay Miner. 15:121–141.

Fasiska, E.J. 1967. Structural aspects of the oxides and oxidehydrates of iron. Corrosion Sci. 7:833–839.

Fassbinder, J.W.E., H. Stanjek, and H. Vali. 1990. Occurrence of magnetic bacteria in soil. Nature 343:161–163.

Faulring, F.M., W.K. Zwicker, and W.D. Forgeng. 1960. Thermal transformations and properties of cryptomelane. Amer. Miner. 45:946–959.

Fendorf, S.E., M. Fendorf, D.L. Sparks, and R. Gronsky. 1992. Inhibitory mechanisms of Cr(III) oxidation by γ-MnO_2. J. Coll. Interf. Sci. 153:37–54.

Fisher, R.F., C.N. Bourn, and W.F. Fisher. 1995. Opal phytoliths as an indicator of the floristics of prehistoric grasslands. Geoderma 68:243–255.

Fitzpatrick, R.W., and U. Schwertmann. 1982. Al-substituted goethite: An indicator of pedogenic and other weathering environments in South Africa. Geoderma 27:335–347.

Fitzpatrick, R.W., J. le Roux, and U. Schwertmann. 1978. Amorphous and crystalline iron-titanium oxides in synthetic preparation, at near ambient conditions, and in soil clays. Clays Clay Miner. 26:189–201.

Fitzpatrick, R.W., R.M. Taylor, U. Schwertmann, and C.W. Childs. 1985. Occurrence and properties of lepidocrocite in some soils of New Zealand, South Africa and Australia. Aust. J. Soil Res. 23:543–567.

Flach, K.W., W.D. Nettleton, and R.E. Nelson. 1974. The micromorphology of silica-cemented soil horizons in western North America. p. 714–729. In G.K. Rutherford (ed.) Soil microscopy. The Limestone Press, Kingston, Canada.

Flach, K.W., W.D. Nettleton, L.H. Gile, and J.C. Cady. 1969. Pedocementation: Induration by silica, carbonates, and sesquioxides in the Quaternary. Soil Sci. 107:442–453.

Fontes, M.P.F., and S.B. Weed. 1991. Iron oxides in selected Brazilian oxisols: I. Mineralogy. Soil Sci. Soc. Amer. J. 55:1143–1149.

Fontes, M.P.F., L.H. Bowen, and S.B. Weed. 1991. Iron oxides in selected Brazilian Oxisols: II. Mössbauer studies. Soil Sci. Soc. Amer. J. 55:1150–1155.

Fordham, A.W., and K. Norrish. 1979. Electron microprobe and electron microscope studies of soil clay particles. Aust. J. Soil Res. 17:283–306.

Fordham, A.W., and K. Norrish. 1983. The nature of soil particles particularly those reacting with arsenate in a series of chemically treated samples. Aust. J. Soil Res. 21:455–477.

Franzmeier, D.P., L.D. Norton, and G.C. Steinhardt. 1989. Fragipan formation in loess of the midwestern United States. p. 69–97. In N.E. Smeck and E.J. Ciolkosz (ed.) Fragipans: Their occurrence, classification, and genesis. Soil Sci. Soc. Am. Spec. Pub. 24. Soil Science Society of America, Madison, WI.

Gallagher, K.J., W. Feitknecht, and U. Mannweiler. 1968. Mechanism of oxidation of magnetite to γ-Fe$_2$O$_3$. Nature 217:1118–1121.

Geller, S. 1971. Structures of -Mn$_2$O$_3$, (Mn$_{0.983}$Fe$_{0.017}$)$_2$O$_3$ and (Mn$_{0.37}$Fe$_{0.63}$)$_2$O$_3$ and relation to magnetic ordering. Acta Cryst. B27:821–828.

Ghiorse, W.C. 1988. The biology of manganese transforming microorganisms in soil. p. 75–85. In R.D. Graham, R.J. Hannam, and N.C. Uren (ed.) Manganese in soils and plants. Kluwer Academic Publishers, Boston, MA.

Ghiorse, W.C., and H.L. Ehrlich. 1992. Microbial biomineralization of iron and manganese. Catena Suppl. 21:75–99.

Gilkes, R.J., and A. Suddhiprakarn. 1979. Magnetite alteration in deeply weathered adamellite. J. Soil Sci. 30:357–361.

Gilkes, R.J., G. Scholz, and G.M. Dimmock. 1973. Lateritic deep weathering of granite. J. Soil Sci. 24:523–536.

Giovanoli, R. 1985a. Layer structures and tunnel structures in manganates. Chem. Erde 44:227–244.

Giovanoli, R. 1985b. A review of the todorokite-buserite problem: Implications to the mineralogy of marine manganese nodules: Discussion. Amer. Miner. 70:202–204.

Glasser, L.S.D., and I. Smith. 1968. Oriented transformations in the system MnO-O-H$_2$O. Miner. Mag. 36:976–987.

Glasser, L.S.D., and L. Ingram. 1968. Refinement of the crystal structure of groutite, α-MnOOH. Acta Cryst. B24:1233–1236.

Golden, D.C., C.C. Chen, J.B. Dixon, and Y. Tokashiky. 1986. Pseudomorphic replacement of manganese oxides by iron oxide minerals. Geoderma 42:199–211.

Golden, D. C., J. B. Dixon, and Y. Kanehiro. 1993. The manganese oxide mineral, lithiophorite, in an Oxisol from Hawaii. Aust. J. Soil Res. 31: 51–66.

Goss, C.J. 1988. Saturation magnetization, coercivity and lattice parameter changes in the system Fe$_3$O$_4$- -Fe$_2$O$_3$, and their relationship to structure. Phys. Chem. Min. 16:164–171.

Graetsch, H. 1994. Structural characteristics of opaline and microcrystalline silica minerals. Rev. Mineral. 29:209–232.

Graham, R.D., R.J. Hannam, and N.C. Uren. 1988. Manganese in soils and plants. Kluwer Academic Publishers, Dordrecht, Netherlands.

Greer, R.T. 1969. Submicron structure of amorphous opal. Nature 224:1199–1200.

Grey, I. E., C. Li, and J. A. Watts. 1983. Hydrothermal synthesis of goethite-rutile intergrowth structures and their relationship to pseudorutile. Amer. Miner. 68:981–988.

Grey, I.E., and A.F. Reid. 1975. The structure of pseudorutile and its role in the natural alteration of ilmenite. Amer. Mineral. 60:898–906.

Gross, S., and L. Heller. 1963. A natural occurrence of bayerite. Miner. Mag. 33:723–724.

Hathaway, J.C., and S.O. Schlanger. 1965. Nordstrandite (Al$_2$O$_3$·3H$_2$O) from Guam. Amer. Miner. 50:1029–1037.

Hazemann, J.L., A. Manceau, P. Sainctavit, and C. Malgrange. 1992. Structure of the α-Fe$_x$Al$_{1-x}$OOH solid solution. I. Evidence by polarized EXAFS for an epitaxial growth of hematite-like clusters in diaspore. Phys. Chem. Miner. 19:25–38.

Healy, T.W., A.P. Herring, and D.W. Fuerstenau. 1966. The effect of crystal structure on the surface properties of a series of manganese dioxides. J. Coll. Interf. Sci. 21:435–444.

Heaney, P.J. 1994. Structure and chemistry of the low-pressure silica polymorphs. Rev. Mineral. 29:1–40.

Heaney, P.J., and J.A. Banfield. 1993. Structure and chemistry of silica, metal oxides, and phosphates. Rev. Mineral. 28:185–233.

Heaney, P.J., C.T. Prewit, and G.V. Gibbs (eds.). 1994. Silica. Physical behavior, geochemistry and materials applications. Rev. Mineral. 29:375.

Hem, J.D., and C.J. Lind. 1983. Nonequilibrium models for predicting forms of precipitated manganese oxides. Geochim. Cosmochim. Acta 47:2037–2046.

Hemingway, B.S., R.A. Robie, and J.A. Apps. 1991. Revised values for the thermodynamic properties of boehmite, AlO(OH), and related species and phases in the system Al-H-O. Amer. Miner. 76:445–451.

Hesse, R. 1988. Diagenesis # 13. Origin of chert: Diagenesis of biogenic siliceous sediments. Geosci. Canada 15:171–192.

Hingston, F.J., A.M. Posner, and J.P. Quirk. 1974. Anion adsorption by goethite and gibbsite. I. The role of the proton in determining adsorption envelopes. J. Soil Sci. 23:177–192.

Holmgren, G.G.S. 1967. A rapid citrate-dithionite extractable iron procedure. Soil Sci. Soc. Amer. Proc. 31:210–211.

Houot, S., and J. Berthelin. 1992. Submicroscopic studies of iron deposits occurring in field drains: Formation and evolution. Geoderma 52:209–222.

Hsu, P.H. 1966. Formation of gibbsite from aging hydroxy-aluminum solutions. Soil Sci. Soc. Amer. Proc. 30:173–176.

Hsu, P.H. 1967. Effect of salts on the formation of bayerite versus pseudoboehmite. Soil Sci. 103:101–110.

Hsu, P.H. 1989. Aluminum hydroxides and oxyhydroxides. p. 331–378. *In* J.B. Dixon and S.B. Weed (ed.) Minerals in soil environments. Soil Science Society of America, Madison, WI.

Huang, P.M. 1988. Ionic factors affecting aluminum transformations and the impact on soil and environmental sciences. Adv. Agron. 8:1–78.

Huebner, J.S. 1976. The manganese oxides-a bibliographic commentary. Rev. Mineral. 3:SH1–SH17.

Hughes, R.E., D.M. Moore, and H.D. Glass. 1994. Qualitative and quantitative analysis of clay minerals in soils. p. 330–359. *In* J.E. Amonette, and L.W. Zelazny (ed.) Quantitative methods in soil mineralogy. Soil Science Society of America, Madison, WI.

ICDD. 1994. Powder diffraction file 1995, PDF-2 Database Sets 1-45, Minerals. International Centre for Diffraction Data, Newtown Square, PA.

Jackson, M.L. 1964. Chemical composition of soils. p.71–141. *In* F.E. Bear (ed.) Chemistry of the soil. Van Nostrand-Reinhold, New York, NY.

Jackson, M.L. 1969. Soil chemical analysis. Advanced Course. Published by the author. Madison, WI.

Jeanroy, E., J.L. Rajot, P. Pillon, and A.J. Herbillon. 1991. Differential dissolution of hematite and goethite in dithionite and its implication on soil yellowing. Geoderma 50:79–94.

Jones, J.B., and E.R. Segnit. 1971. The nature of opal. I. Nomenclature and constituent phases. J. Geol. Soc. Aust. 18:57–68.

Jones, L.H.P., and K.A. Handreck. 1967. Silica in soils, plants and animals. Adv. Agron. 19:107–149.

Jones, R.L., and A.H. Beavers. 1963. Sponge spicules in Illinois soils. Soil Sci. Soc. Am. Proc. 27:438–440.

Jones, R.L., and A.H. Beavers. 1964. Variation of opal phytolith content among some great soil groups of Illinois. Soil Sci. Soc. Am. Proc. 28:711–712.

Juo, A. S. R. 1981. Chemical characteristics. p. 51–79. *In* D. J. Greenland (ed.) Characterization of soils. Oxford University Press, New York, NY.

Kämpf, N., and U. Schwertmann. 1982a. The 5M NaOH concentration treatment for iron oxides in soils. Clays Clay Miner. 30:401–408.

Kämpf, N., and U. Schwertmann. 1982b. Quantitative determination of goethite and hematite in kaolinitic soils by X-ray diffraction. Clay Miner. 17:359–363.

Kämpf, N., and U. Schwertmann. 1983. Goethite and hematite in a climosequence in southern Brazil and their application in classification of kaolinitic soils. Geoderma 29:27–39.

Karathanasis, A.D. 1989. Solution chemistry of fragipans: Thermodynamic approach to understanding fragipan formation. p. 113–139. *In* N.E. Smeck and E.J. Ciolkosz (ed.) Fragipans: Their occurrence, classification, and genesis. Soil Sci. Soc. Am. Spec. Pub. 24. Soil Science Society of America, Madison, WI.

Karathanasis, A.D., and Y.L. Thompson. 1995. Mineralogy of iron precipitates in a constructed acid mine drainage wetland. Soil Sci. Soc. Amer. J. 59:1773–1781.

Karathanasis, A.D., and W.G. Harris. 1994. Quantitative thermal analysis of soil materials. p. 360–411. *In* J.E. Amonette, and L.W. Zelazny (ed.) Quantitative methods in soil mineralogy. Soil Science Society of America, Madison, WI.

Kastner, M., J.B. Keene, and J.M. Gieskes. 1977. Diagenesis of siliceous oozes-1. Chemical controls on the rate of opal-A to opal-CT transformation-an experimental study. Geochim. Cosmochim. Acta 41:1041–1059.

Keller, W.D. 1978. Diaspore recrystallization at low temperature. Amer. Mineral. 63:326–329.

Kelly, E.F., R.G. Amundson, B.D. Marino, and M.J. Deniro. 1991. Stable isotope ratios of carbon in phytoliths as a quantitative method of monitoring vegetation and climate change. Quater. Res. 35:222–233.

Kennedy, J.A., H.K.J. Powell, and J.M. White. 1982. A modification of Child's field test for ferrous iron and ferric-organic complexes in soils. Aust. J. Soil Res. 20:261–263.

Kittrick, J.A. 1969. Soil minerals in the Al_2O_3-SiO_2-H_2O system and a theory of their formation. Clays Clay Miner. 17:157–167.

Knauth, L.P. 1994. Petrogenesis of chert. Rev. Mineral. 29:233–258.

Kodama, H., and M. Schnitzer. 1980. Effect of fulvic acid on the crystallization of aluminum hydroxide. Geoderma 24:195–205.

Leinemann, C.-P., M. Taillefert, D. Perret, and J.-F. Gaillard. 1997. Association of cobalt and manganese in aquatic systems: Chemical and microscopic evidence. Geochim. Cosmochim. Acta 61:1437–1466.

Lindsay, W.L. 1979. Chemical equilibria in soils. Wiley Interscience, New York, NY.

Lindsley, D.H. 1991. Oxide minerals: Petrologic and magnetic significance. Mineralogical Society of America, Washington, DC.

Lindsley, D.H. 1976. The crystal chemistry and structure of oxide minerals as exemplified by the Fe-Ti oxides. Rev. Mineral. 3:L1–L60.

Little, I.P., T.M. Armitage, and R.J. Gilkes. 1978. Weathering of quartz in dune sands under subtropical conditions in Eastern Australia. Geoderma 20:225–237.

Loveland, P.J., and P. Digby. 1984. The extraction of Fe and Al by 0.1 M pyrophosphate solutions: A comparison of some techniques. J. Soil Sci. 35:243–250.

Lovley, D.R. 1992. Microbial oxidation of organic matter coupled to the reduction of Fe(III) and Mn(IV) oxides. Catena Suppl. 21:101–114.

Lovley, D.R. 1995. Microbial reduction of iron, manganese, and other metals. Adv. Agron. 54:175–231.

Lovley, D.R., and E.J.P. Phillips. 1987. Rapid assay for microbially reducible ferric iron in aquatic sediments. Appl. Environ. Microb. 53:1536–1540.

Lovley, D.R., and E.J.P. Phillips. 1988. Novel mode of microbial energy metabolism: Organic carbon oxidation coupled to dissimilatory reduction of iron or manganese. Appl. Environ. Microb. 54:1472–1480.

Lutwick, L.E., and A. Johnston. 1969. Cumulic soils of the rough fescue prairie-poplar transition region. Can. J. Soil Sci. 49:199–203.

Macedo, J., and R.B. Bryant. 1987. Morphology, mineralogy, and genesis of a hydrosequence of Oxisols in Brazil. Soil Sci. Soc. Amer. J. 51:690–698.

Macedo, J., and R.B. Bryant. 1989. Preferential microbial reduction of hematite over goethite in a Brazilian Oxisol. Soil Sci. Soc. Amer. J. 53:1114–1118.

Mackenzie, F.T., and R. Gees. 1971. Quartz: Synthesis at earth-surface conditions. Science 173:533–535.

Mackenzie, R.C. 1970. Oxides and hydroxides of higher valency elements. p. 271–302. *In* R.C. Mackenzie (ed.) Differential thermal analysis. Vol. 1. Academic Press Ltd., New York, NY.

Mackenzie, R.C. 1972. Soils. p. 267–297. *In* R.C. Mackenzie (ed.) Differential thermal analysis. Vol. 2. Academic Press Ltd., New York, NY.

Maher, B.A., and R.M. Taylor. 1988. Formation of ultrafine-grained magnetite in soils. Nature 336:368–370.

Manceau, A., A.I. Gorshkov, and V.A. Drits. 1992a. Structural chemistry of Mn, Fe, Co, and Ni in manganese hydrous oxides: I. Information from XANES spectroscopy. Amer. Miner. 77:1133–1143.

Manceau, A., A.I. Gorshkov, and V.A. Drits. 1992b. Structural chemistry of Mn, Fe, Co, and Ni in manganese hydrous oxides: II. Information from EXAFS spectroscopy and electron and X-ray diffraction. Amer. Miner. 77:1144–1157.

Manceau, A., and J.M. Combes. 1988. Structure of Mn and Fe oxides and oxyhydroxides: A topological approach by EXAFS. Phys. Chem. Miner. 15:283–295.

Manceau, A., and V.A. Drits. 1993. Local structure of ferrihydrite and feroxyhyte by EXAFS spectroscopy. Clay Miner. 28:165–184.

Manceau, A., and W.P. gates. 1997. Surface structural model for ferrihydrite. Clays Clay Min. 45:448–460.

Manceau, A., P.R. Buseck, D. Miser, J. Rask, and D. Nahon. 1990. Characterization of Cu in lithiophorite from a banded Mn ore. Amer. Miner. 75:490–494.

Manceau, A., S. Llorca, and G. Calas. 1987. Crystal chemistry of cobalt and nickel in lithiophorite and asbolane from New Caledonia. Geochim. Cosmochim. Acta 51:105–113.

Maslen, E.N., V.A. Streltsov, and N.R. Streltsova. 1993. Synchrotron X-ray study of the electron density in α-Al_2O_3. Acta Cryst. B49:973–980.

Maslen, E.N., V.A. Streltsov, and N.R. Streltsova. 1994. Synchrotron X-ray study of the electron density in α-Fe_2O_3. Acta Cryst. B50:435–441.

McBride, M. B. 1989. Reactions controlling heavy metal solubility in soils. Adv. Agron. 10:1–56.

McBride, M.B. 1987. Adsorption and oxidation of phenolic compounds by iron and manganese oxides. Soil Sci. Soc. Amer. J. 51:1466–1472.

McBride, M.B. 1994. Environmental chemistry of soils. Oxford University Press, New York, NY.

McBride, M.B., and L.G. Wesselink. 1988. Chemisorption of catechol on gibbsite, boehmite, and noncrystalline alumina surfaces. Environ. Sci. Technol. 22:703–708.

McKeague, J.A., and J.H. Day. 1966. Dithionite- and oxalate extractable Fe and Al as aids in differentiating various classes of soils. Can. J. Soil Sci. 46:13–22.

McKeague, J.A.. 1967. An evaluation of 0.1M pyrophosphate and pyrophosphate-dithionite in comparison with oxalate as extractants of the accumulation products in Podzols and some other soils. Can. J. Soil Sci. 47:95–99.

McKenzie, R.M. 1971. The synthesis of cryptomelane and some other oxides and hydroxides of manganese. Mineral. Mag. 38:493–502.

McKenzie, R.M. 1978. The effect of two manganese dioxides on the uptake of Pb, Co, Ni, Cu and Zn by subterranean clover. Aust. J. Soil Res. 16:209–214.

McKenzie, R.M. 1989. Manganese oxides and hydroxides. p. 439–465. *In* J.B. Dixon, and S.B. Weed (ed.) Minerals in soil environments. Soil Science Society of America, Madison, WI.

McNeal, B.L., D.A. Layfield, W.A. Norvell, and J.D. Rhoades. 1968. Factors influencing hydraulic conductivity of soils in the presence of mixed salt solution. Soil Sci. Soc. Am. Proc. 32:187–190.

Means, J.L., D.A. Crerar, M.P. Borcsik, and J.O. Duguid. 1978. Adsorption of cobalt and selected actinides by Mn and Fe oxides in soils and sediments. Geochim. Cosmochim. Acta 42:1763–1773.

Mehra, O.P., and M.L. Jackson. 1960. Iron oxide removal from soils and clays by a dithionite-citrate system buffered with sodium bicarbonate. p. 317–342. *In* A. Swineford (ed.) Proc. 7th Clays Clay Miner.Conf. Pergamon Press, Elmsdorf, NY.

Milnes, A.R., and R.W. Fitzpatrick. 1989. Titanium and zirconium minerals. p. 1131–1205. *In* J.B. Dixon and S.B. Weed (ed.) Minerals in soil environments. 2nd Ed. Soil Science Society of America, Madison, WI.

Mitchell, W.A. 1975. Heavy minerals. p. 450–480. *In* J.E. Gieseking (ed.) Soil components. Vol. 2. Inorganic components. Springer-Verlag, Berlin, Germany.

Miyamoto, H. 1976. The magnetic properties of $Fe(OH)_2$. Mat. Res. Bull. 11:329–336.

Mizota, C., and Y. Matsuhisa. 1995. Isotopic evidence for the eolian origin of quartz and mica in soils developed on volcanic materials in the Canary Archipelago. Geoderma 66:313–320.

Mizota, C., M. Itoh, M. Kusakabe, and M. Noto. 1991. Oxygen isotope ratios of opaline silica and plant opal in three recent volcanic ash soils. Geoderma 50:211–217.

Motta, P.E.F. da, and N. Kämpf. 1991. Iron oxide properties as support to soil morphological features for prediction of moisture regimes in Oxisols of Central Brazil. Z. Pflanzenernähr. Bodenk. 155:385–390.

Murad, E. 1988. Properties and behavior of iron oxides as determined by Mössbauer spectroscopy. p. 309–350. *In* J.W. Stucki, B.A. Goodman, and U. Schwertmann (eds.) Iron in soils and clay minerals. D. Reidel Publishing Co., Dordrecht, Netherlands.

Murad, E. 1996. Magnetic properties of microcrystalline iron (III) oxides and related materials as reflected in their Mössbauer spectra. Phys. Chem. Miner. 23:248–262.

Murad, E. 1997. Identification of minor amounts of anatase in kaolins by Raman spectroscopy. Amer. Miner. 82:203–206.

Murad, E., U. Schwertmann, J.M. Bigham, and L. Carlson. 1994. Mineralogical characteristics of poorly crystallized precipitates formed by oxidation of Fe^{2+} in acid sulfate waters. ACS Symp. Ser. 550:190–200.

Murray, J.W. 1975. The interactions of metal ions at the manganese dioxide-solution interface. Geochim. Cosmochim. Acta 39:505–519.

Murray, J.W., J.G. Dillard, R. Giovanoli, H. Moers, and W. Stumm. 1985. Oxidation of Mn(II): Initial mineralogy, oxidation state and aging. Geochim. Cosmochim. Acta 49:463–470.

Norrish, K., and R.M. Taylor. 1961. The isomorphous replacement of iron by aluminium in soil goethites. J. Soil Sci. 12:294–306.

Ohta, S., S. Effendi, N. Tanaka, and S. Miura. 1993. Ultisols of lowland Dipterocarp forest in East Kalimantan, Indonesia. Soil Sci. Plant Nutr. 39:1–12.

Oscarson, D.W., P.M. Huang, W.K. Liaw, and U.T. Hammer. 1983. Kinetics of oxidation of arsenite by various manganese oxides. Soil Sci. Soc. Amer. J. 47:644–648.

Ostwald, J. 1984a. Two varieties of lithiophorite in some Australian deposits. Miner. Mag. 48:383–388.

Ostwald, J. 1984b. The influence of clay mineralogy on the crystallization of the tetravalent manganese layer-lattice minerals. N. Jb. Miner. Mh. 1984:9–16.

Ostwald, J. 1985. Some observations on the chemical composition of chalcophanite. Miner. Mag. 49:752–755.

Parc, S., D. Nahon, Y. Tardy, and P. Vieillard. 1989. Estimated solubility products and fields of stability for cryptomelane, nsutite, birnessite, and lithiophorite based on natural lateritic weathering sequences. Amer. Miner. 74:466–475.

Parfitt, R.L., and C.W. Childs. 1988. Estimation of forms of Fe and Al: A review, and analysis of contrasting soils by dissolution and Mössbauer methods. Aust. J. Soil Res. 26:121–144.

Parfitt, R.L., C.W. Childs, and D.N. Eden. 1988. Ferrihydrite and allophane in four Andepts from Hawaii and implications for their classifications. Geoderma 41:223–241.

Peterschmitt, E., E. Fritsch, J.L. Rajot, and A.J. Herbillon. 1996. Yellowing, bleaching and ferritisation processes in soil mantle of the Western Ghâts, South India. Geoderma 74:235–253.

Pinheiro-Dick, D. and U. Schwertmann. 1996. Microaggregates from oxisols and inceptisols: dispersion through selective dissolutions and physicochemical treatments. Geoderma 74:49–63.

Post, J.E. 1992. Crystal structures of manganese oxide minerals. Catena Suppl. 21:51–73.

Post, J.E., and C.W. Burnham. 1986. Modelling tunnel-cation displacements in hollandites using structure-energy calculations. Amer. Miner. 71:1178–1185.

Post, J.E., and D.E. Appleman. 1988. Chalcophanite, $ZnMn_3O_7 \cdot 3H_2O$: New crystal-structure determinations. Amer. Miner. 73:1401–1404.

Post, J.E., and D.E. Appleman. 1994. Crystal structure refinement of lithiophorite. Amer. Miner. 79:370–374.

Post, J.E., and D.L. Bish. 1988. Rietveld refinement of the todorokite structure. Amer. Miner. 73:861–869.

Post, J.E., and D.L. Bish. 1989. Rietveld refinement of the coronadite structure. Amer. Miner. 74:913–917.

Post, J.E., and D.R. Veblen. 1990. Crystal structure determinations of synthetic sodium, magnesium, and potassium birnessite using TEM and the Rietveld method. Amer. Miner. 75:477–489.

Post, J.E., and V.F. Buchwald. 1991. Crystal structure refinement of akaganéite. Amer. Miner. 76:272–277.

Post, J.E., R.B. von Dreele, and P.R. Buseck. 1982. Symmetry and cation displacements in Hollandites: Structure refinements of hollandite, cryptomelane and priderite. Acta Cryst. B38:1056–1065.

Potter, R.M., and G.R. Rossman. 1979. The tetravalent manganese oxides: Identification, hydration, and structural relationships by infrared spectroscopy. Amer. Miner. 64:1199–1218.

Pye, K., and J. Mazzullo. 1994. Effects of tropical weathering of quartz grain shape. An example from Northeastern Australia. J. Sed. Res. A64:500–507.

Rask, J.H., and P.R. Buseck. 1986. Topotactic relations among pyrolusite, manganite, and Mn_5O_8: A high-resolution transmission electron microscopy investigation. Amer. Miner. 71:805–814.

Rebertus, R.A., S.B. Weed, and S.W. Buol. 1986. Transformation of biotite to kaolinite during saprolite-soil weathering. Soil Sci. Soc. Amer. J. 50:810–819.

Richardson, J.L., and F.D. Hole. 1979. Mottling and iron distribution in a Glossoboralf-Haplaquoll hydrosequence on a glacial moraine in Northwestern Wisconsin. Soil Sci. Soc. Amer. Proc. 43:552–558.

Robbins, E.I., J.P. D'Agostino, J. Ostwald, D.S. Fanning, V. Carter, and R.L. Van Hoven. 1992. Manganese nodules and microbial oxidation of manganese in the Huntley Meadows Wetland, Virginia, USA. Catena Suppl. 21:179–202.

Rodgers, K.A. 1993. Routine identification of aluminum hydroxide polymorphs with the laser Raman microprobe. Clay Miner. 28:85–99.

Ross, S.J. Jr., D.P. Franzmeier, and C.B. Roth. 1976. Mineralogy and chemistry of manganese oxides in some Indiana soils. Soil Sci. Soc. Amer. J. 40:137–143.

Rowse, J.B., and W.B. Jepson. 1972. The determination of quartz in clay minerals: A critical comparison of methods. J. Therm. Anal. 4:169–175.

Ryskin, Y.I. 1974. The vibration of protons in minerals: Hydroxyl, water and ammonium. p. 137–181. In V.C. Farmer (ed.) Infrared spectra of minerals. Mineralogical Society, London, UK.

Sasaki, S., K. Fukino, Y. Takéuchi, and R. Sadanaga. 1980. On the estimation of atomic charges by the X-ray method for some oxides and silicates. Acta Cryst. A36:904–915.

Schoen, R., and C.E. Roberson. 1970. Structures of aluminum hydroxide and geochemical implications. Amer. Mineral. 55:43–77.

Schulze, D.G. 1981. Identification of soil iron oxide minerals by differential x-ray diffraction. Soil Sci. Soc. Amer. J. 45:437–440.

Schulze, D.G. 1984. The influence of aluminum on iron oxides. VIII. Unit cell dimensions of Al substituted goethites and estimation of Al from them. Clays Clay Miner. 32:36–44.

Schulze, D.G. 1988. Separation and concentration of iron-containing phases. p. 63–81. *In* J.W. Stucki, B.A. Goodman, and U. Schwertmann (ed.) Iron in soils and clay minerals. D. Reidel Publishing Co., Dordrecht, Netherlands.

Schulze, D.G., and J.B. Dixon. 1979. High gradient magnetic separation of iron oxides and magnetic minerals from soil clays. Soil Sci. Soc. Amer. J. 43:793–799.

Schuppli, P.A., G.J. Ross, and J. McKeague. 1983. The effective removal of suspended materials form pyrophosphate extracts of soils from tropical and temperate regions. Soil Sci. Soc. Amer. J. 47:1026–1032.

Schwarz, T. 1994. Ferricrete formation and relief inversion: An example from Central Sudan. Catena 21:257–268.

Schwertmann, U. 1964. Differenzierung der Eisenoxide des Bodens durch photochemische Extraktion mit saurer Ammoniumoxalat-Lösung. Z. Pflanzenernähr. Bodenkd. 105:194–202.

Schwertmann, U. 1966. Inhibitory effect of soil organic matter on the crystallization of amorphous ferric hydroxide. Nature 212:645–646.

Schwertmann, U. 1988. Occurrence and formation of iron oxides in various pedoenvironments. p. 267–308. *In* J.W. Stucki, B.A. Goodman, and U. Schwertmann (ed.) Iron in soils and clay minerals. D. Reidel Publishing Co., Dordrecht, Netherlands.

Schwertmann, U. 1991. Solubility and dissolution of iron oxides. Plant Soil 130:1–25.

Schwertmann, U. 1993. Relation between iron oxides, soil color, and soil formation. p. 51–69. *In* J.M. Bigham, and E.J. Ciolkosz (ed.) Soil color. Soil Science Society of America, Madison, WI.

Schwertmann, U., and L. Carlson. 1994. Aluminum influence on iron oxides. XVII. Unit-cell parameters and aluminum substitution of natural goethites. Soil Sci. Soc. Amer. J. 58:256–261.

Schwertmann, U. and D.S. Fanning. 1976. Iron-manganese concretions in hydrosequences of soils in loess in Bavaria. Soil Sci. Soc. Amer. J. 40:731–738.

Schwertmann, U., and H. Fechter. 1984. The influence of aluminium on iron oxides. XI. Aluminium substituted maghemite in soils and its formation. Soil Sci. Soc. Amer. J. 48:1462–1463.

Schwertmann, U., and H. Fechter. 1994. The formation of green rust and its transformation to lepidocrocite. Clay Miner. 29:87–92.

Schwertmann, U., and W.R. Fischer. 1973. Natural amorphous ferric hydroxide. Geoderma 10:237–247.

Schwertmann, U. and R.W. Fitzpatrick. 1992. Iron minerals in surface environments. Catena Suppl. 21:7–30.

Schwertmann, U., and M. Latham. 1986. Properties of iron oxides in some New Caledonian Oxisols. Geoderma 39:105–123.

Schwertmann, U., and N. Kämpf. 1983. Oxidos de ferro jovens em ambientes pedogeneticos brasileiros. Rev. Bras. Cienc. Solo 7:251–255.

Schwertmann, U., and N. Kämpf. 1985. Properties of goethite and hematite in kaolinitic soils of Southern and Central Brazil. Soil Sci. 139:344–350.

Schwertmann, U., and G. Pfab. 1994. Structural vanadium in synthetic goethite. Geochim. Cosmochim. Acta 58:4349–4352.

Schwertmann, U., and G. Pfab. 1996. Structural vanadium and chromium in lateritic iron oxides: Genetic implications. Geochim. Cosmochim. Acta 60:4279–4283.

Schwertmann, U., and R.M. Taylor. 1989. Iron oxides. p.379–438. *In* J.B. Dixon and S.B. Weed (ed.) Minerals in soil environments. Soil Science Society of America, Madison, WI.

Schwertmann, U., D.G. Schulze, and E. Murad. 1982. Identification of ferrihydrite in soils by dissolution kinetics, differential X-ray diffraction and Mössbauer spectroscopy. Soil Sci. Soc. Amer. J. 46:869–875.

Schwertmann, U., J. Friedl, G. Pfab, and A.U. Gehring. 1995b. Iron substitution in soil and synthetic anatase. Clays Clay Miner. 43:599–606.

Schwertmann, U., J. M. Bigham, and E. Murad. 1995a. The first occurrence of schwertmannite in a natural stream environment. Europ. J. Miner. 7:547–552.

Schwertmann, U., R.W. Fitzpatrick, R.M. Taylor, and D.G. Lewis. 1979. The influence of aluminium on iron oxides. II. Preparation and properties of Al substituted hematites. Clays Clay Miner. 27:105–112.

Scott, M.J., and J.J. Morgan. 1995. Reactions at oxide surfaces. 1. Oxidation of As(III) by synthetic birnessite. Environ. Sci. Technol. 29:1898–1905.

Senkayi, A.L., J.B. Dixon, and L.R. Hossner. 1986. Todorokite, goethite, and hematite: Alteration products of siderite in East Texas lignite overburden. Soil Sci. 142:36–42.

Shafdan, H., J.B. Dixon, and F.G. Calhoun. 1985. Iron oxide properties versus strength of ferruginous crust and iron glaebules in soils. Soil Sci. 140:317–325.

Shah Singh, S., and H. Kodama. 1994. Effect of the presence of aluminum ions in iron solutions on the formation of iron oxyhydroxides (FeOOH) at room temperature under acidic environment. Clays Clay Miner. 42:606–613.

Sidhu, P.S., J.L. Sehgal, M.K. Sinha, and N.S. Randhawa. 1977. Composition and mineralogy of iron-manganese concretions from some soils of the Indo-Gangetic plain in northwest India. Geoderma 18:241–249.

Singer, A. and P. M. Huang. 1990. Effects of humic acids on the crystallization of aluminum hydroxides. Clays Clay Miner. 38:47–52.

Singh, B., and R.J. Gilkes. 1992. Properties and distribution of iron oxides and their association with minor elements in the soils of south-western Australia. J. Soil Sci. 43:77–98.

Skinner, H.C.W., and R.W. Fitzpatrick. 1992. Biomineralization. Processes of iron and manganese. Catena Verlag, Cremlingen, Germany.

Soil Survey Staff. 1975. Soil Taxonomy: A basic system of soil classification for making and interpreting soil surveys. Agric. Handb. 436, US Government Printing Office, Washington, DC.

Speer, J.A. 1982. Zircon. Rev. Mineral. 5:67–112.

Stanjek., H. 1987. The formation of maghemite and hematite from lepidocrocite and goethite in a Cambisol form Corsica, France. Z. Pflanzenernähr. Bodenk. 150:314–318.

Stone, A.T., and J.J. Morgan. 1984a. Reduction and dissolution of manganese(III) and manganese(IV) oxides by organics. 1. Reaction with hydroquinone. Environ. Sci. Technol. 18:450–456.

Stone, A.T., and J.J. Morgan. 1984b. Reduction and dissolution of manganese(III) and manganese(IV) oxides by organics. 2. Survey of the reactivity of organics. Environ. Sci. Technol. 18:617–624.

Sudom, M.D., and R.J. St. Arnaud. 1971. Use of quartz, zirconium and titanium as indices in pedological studies. Can. J. Soil Sci. 51:385–396.

Süsser, P., and U. Schwertmann. 1991. Proton buffering in mineral horizons of some acid forest soils. Geoderma 49:63–76.

Tait, J.M., A. Violante, and P. Violante. 1983. Coprecipitation of gibbsite and bayerite with nordstrandite. Clay Miner. 18:95–99.

Tanaka, K., and J. White. 1982. Characterization of species adsorbed on oxidized and reduced anatase. J. Phys. Chem. 86:4708–4714.

Taylor, R.M. 1968. The association of manganese and cobalt in soils: Further observations. J. Soil Sci. 19:77–80.

Taylor, R.M. 1987. Non-silicates oxides and hydroxides. p. 129–201. In A.C.D. Newman (ed.) Chemistry of clays and clay minerals. John Wiley and Sons, New York, NY.

Taylor, R.M., and R.M. McKenzie. 1980. The influence of aluminium on iron oxides. VI. The formation of Fe(II)-Al(III) hydroxy-chlorides, -sulphates, and -carbonates as new members of the pyroaurite group and their significance in soils. Clays Clay Min. 28:179–187.

Taylor, R.M., and U. Schwertmann. 1974. Maghemite in soils and its origin. I. Properties and observations on soil maghemites. Clay Miner. 10:289–298.

Taylor, R.M., R.M. McKenzie, A.W. Fordham, and G.P. Gillman. 1983. Oxide minerals. p. 309–334. In Soils an Australian viewpoint. Academic Press, London, UK.

Taylor, R.M., R.M. McKenzie, and K. Norrish. 1964. The mineralogy and chemistry of manganese in some Australian soils. Aust. J. Soil Res. 2:235–248.

Tejan Kella, M.S., R.W. Fitzpatrick, and D.J. Chittleborough. 1991. Scanning electron microscope study of zircons and rutiles from a podzol chronosequence at Cooloola, Queensland, Australia. Catena 18:11–30.

Tettenhorst, R., and D.A. Hofmann. 1980. Crystal chemistry of boehmite. Clays Clay Miner. 28:373–380.

Thanabalasingam, P., and W.F. Pickering. 1986. Effect of pH on the interaction between As(III) or As(V) and manganese(IV) oxide. Water Air Soil Pollut. 29:205–216.

Tokashiki, Y., J.B. Dixon, and D.C. Golden. 1986. Manganese oxide analysis in soils by a combined x-ray diffraction and selective dissolution methods. Soil Sci. Soc. Amer. J. 50:1079–1084.

Torrent, J., R. Guzman, and M.A. Parra. 1982. Influence of relative humidity on the crystallization of Fe(III) oxides from ferrihydrite. Clays Clay Miner. 30:337–340.

Torrent, J., U. Schwertmann, and V. Barrón. 1987. The reductive dissolution of synthetic goethite and hematite in dithionite. Clay Miner. 22:329–337.

Torrent, J., U. Schwertmann, H. Fechter, and F. Alferez. 1983. Quantitative relationships between soil colour and hematite content. Soil Sci. 136:354–358.

Towe, K.M., and W.F. Bradley. 1967. Mineralogical constitution of colloidal hydrous ferric oxides. J. Coll. Interf. Sci. 24:384–392.

Trolard, F., and Y. Tardy. 1987. The stabilities of gibbsite, boehmite, aluminous goethites and aluminous hematites in bauxites, ferricretes, and laterites as a function of water activity, temperature, and particle size. Geochim. Cosmochim. Acta 51:945–957.

Trolard, F., J.-M. R. Génin, M. Abdelmoula, G. Bourrié, B. Humbert, and A. Herbillon. 1997. Identification of a green rust mineral in a reductomorphic soil by Mössbauer and Raman spectroscopies. Geochim. Cosmochim. Acta 61:1107–1111.

Turner, S., and J.E. Post. 1988. Refinement of the substructure and superstructure of romanèchite. Amer. Miner. 73:1155–1161.

Turner, S., and P.R. Buseck. 1979. Manganese oxide tunnel structures and their intergrowths. Science 203:456–458.

Turner, S., and P.R. Buseck. 1981. Todorokites: A new family of naturally occurring manganese oxides. Science 212:1024–1027.

Turner, S., and P.R. Buseck. 1983. Defects in nsutite (γ-MnO_2) and dry-cell battery efficiency. Nature 304:143–146.

Uzochukwu, G.A., and J.B. Dixon. 1986. Manganese oxide minerals in nodules of two soils of Texas and Alabama. Soil Sci. Soc. Amer. J. 50:1358–1363.

van Wambeke, A. 1992. Soils of the tropics. McGraw-Hill, New York, NY.

Varentsov, I.M. 1982. Groote Eylandt manganese oxide deposits, Australia. Chem. Erde 41:157–173.

Vgenopoulos, A.G. 1984. Genesis of boehmite resp. diaspore in bauxite in dependence of redox equilibrium. Chem. Erde 43:149–159.

Violante, A., and M.L. Jackson. 1979. Crystallization of nordstrandite in citrate systems and in the presence of montmorillonite. p. 517–525. *In* M.M. Mortland and V.C. Farmer (ed.) Proc. Intl. Clay Conf., Elsevier Scientific Publishing Co., Amsterdam, Netherlands.

Violante, A., and M.L. Jackson. 1981. Clay influence on the crystallization of aluminum hydroxide polymorphs in the presence of citrate, sulfate or chloride. Geoderma 25:199–214.

Violante, A., and P. Violante. 1980. Influence of pH, concentration and chelating power of organic anions on the synthesis of aluminum hydroxides and oxyhydroxides. Clays Clay Miner. 28:425–434.

Violante, A., and P.M. Huang. 1985. Influence of inorganic and organic ligands on the formation of aluminum hydroxides and oxyhydroxides. Clays Clay Miner. 33:181–192.

Violante, A., and P.M. Huang. 1994. Identification of pseudoboehmite in mixtures with phyllosilicates. Clay Miner. 29:351–359.

Violante, P., A. Violante, and J.M. Tait. 1982. Morphology of nordstrandite. Clays Clay Miner. 30:431–437.

Wall, J.R.D., E.B. Wolfenden, E.H. Beard, and T. Deans. 1962. Nordstrandite in soil from West Sarawak, Borneo. Nature 196:264–265.

Wang, H.D., G.N. White, F.T. Turner, and J.B. Dixon. 1993. Ferrihydrite, lepidocrocite, and goethite in coatings from East Texas vertic soils. Soil Sci. Soc. Amer. J. 57:1381–1386.

Waychunas, G.A. 1991. Crystal chemistry of oxides and oxyhydroxides. Rev. Mineral. 25:11–68.

Wechsler, B.A., D.H. Lindsley, and C.T. Prewit. 1984. Crystal structure and cation distribution in titanomagnetite ($Fe_{3-x}Ti_xO_4$). Amer. Miner. 69:754–770.

Wells, M.A., R.J. Gilkes, and R.R. Anand. 1989. The formation of corundum and aluminous hematite by thermal dehydroxilation of aluminous goethite. Clay Miner. 24:513–530.

White, G.N., and J.B. Dixon. 1996. Iron and manganese distribution in nodules from a young Texas Vertisol. Soil Sci. Soc. Amer. J. 60:1254–1262.

Wilding, L.P., and L.R. Drees. 1968. Distribution and implications of sponge spicules in surficial deposits in Ohio. Ohio J. Sci. 68:92–99.

Wilding, L.P., and L.R. Drees. 1971. Biogenic opal in Ohio soils. Soil Sci. Soc. Am. Proc. 35:1004–1010.

Wilding, L.P., and L.R. Drees. 1974. Contributions of forest opal and associated crystalline phases to fine silt and clay fractions of soils. Clays Clay Miner. 22:295–306.

Williams, L.A., and D.A. Crerar. 1985. Silica diagenesis, II. General mechanisms. J. Sediment. Petrol. 55:312–321.

Williams, L.A., G.A. Parks, and D.A. Crerar. 1985. Silica diagenesis, I. Solubility controls. J. Sediment. Petrol. 55:301–311.

Yamada, N., M. Ohmasa, and S. Horiuchi. 1986. Textures in natural pyrolusite, β-MnO_2, examined by 1 MV HRTEM. Acta Cryst. B42:58–61.

Zeese, R., U. Schwertmann, G. F. Tietz, and U. Jux. 1994. Mineralogy and stratigraphy of three deep lateritic profiles of the Jos plateau (Central Nigeria). Catena 21:195–214.

Zwicker, W.K., W.O.J.G. Meijer, and H.W. Jaffe. 1962. Nsutite-a widespread manganese oxide mineral. Amer. Miner. 47:246–266.

Poorly Crystalline Aluminosilicate Clays

James Harsh
Washington State University

This chapter covers the non- or poorly crystalline aluminosilicate minerals, commonly known as allophane and imogolite. Little attention will be given to poorly crystalline oxides such as ferrihydrite (Section F, Chapter 3). The poorly crystalline aluminosilicate clays impart special properties to soils that contain them as a result of their solubility, very high specific surface area, variable charge, and unique physical behavior. Although often associated with soils formed from volcanic material (Andisols), allophane and imogolite are found within a wide range of soil orders and derived from a variety of parent materials.

4.1 Poorly Crystalline Materials

Crystalline materials exhibit long-range order over large distances. In other words, the atoms in the crystalline structure show repeating structures over scales of at least μm in length or diameter. Such minerals display narrow X-ray diffraction peaks and their crystal habit can be directly observed at the scale of an optical microscope (nm) or by the naked eye in the case of large crystals. Amorphous materials, on the other hand, exhibit no order, even in the local environment of the atoms. The atoms are arranged in a variety of states and structures where bond lengths, coordination, and geometry vary from site to site. Poorly crystalline materials fall between these two extremes, showing short-range, medium- or limited long-range order. Short-range order is observable by techniques that probe the local environment of each atom, such as nuclear magnetic spectroscopy (NMR) or X-ray photoelectron spectroscopy (XPS). X-ray diffraction (XRD) peaks are indicative of, at least, medium-range order, whereas long-range order is detectable by electron microscopy. Noncrystalline materials can be defined as those with no repeat of structural units in any of the three spatial dimensions; i.e., no well-defined electron or XRD pattern is discernible.

The aluminosilicate clays discussed in this chapter nearly cover the range from amorphous to crystalline materials (Fyfe et al., 1987). Imogolite is a mineral with relatively constant chemical composition ($SiAl_2O_5 \cdot 2H_2O$), only 6-fold coordination for Al, as many as seven XRD peaks, and a distinctive tubular morphology observable under the electron microscope. Thus, imogolite has the major features of a crystalline mineral, but some consider it paracrystalline because each tube displays long-range order only along the length of the tube (Greenland, 1982). Diffraction patterns and electron micrographs, however, show that bundles of imogolite are aligned in an array with a repeat distance of around 2 nm, consistent with the diameter of the tubes (Cradwick et al., 1972). Thus, imogolite displays an added dimension of repeating units in the same way as layered aluminosilicates with regular stacking.

Allophane, on the other hand, refers to a group of noncrystalline aluminosilicate clays with no long-range order, only two diffuse diffraction peaks and variable composition. The end member of this group would be completely disordered allophane, which probably does not exist in nature (Fyfe et al., 1987). In the interest of brevity, allophane, imogolite, and poorly crystalline iron hydrous oxides will be referred to collectively as noncrystalline materials (Parfitt and Dixon, 1991).

Noncrystalline materials are generally metastable with respect to their crystalline counterparts. Imogolite and allophane are more soluble than kaolinite, for example, yet they are common in certain soil environments and may appear long before kaolinite forms. This arises because the interfacial energy between mineral surfaces and an aqueous solution slows nucleation. Poorly crystalline surfaces have lower surface tensions and nucleate more easily, i.e., at a lower saturation index than a crystalline mineral of similar composition. As a result, allophane and imogolite are often found as precursors to kaolinite in weathering soil profiles and may serve as templates for its heterogeneous nucleation (Steefel et al., 1990).

In spite of their wide occurrence as intermediates in soil formation, allophane and imogolite were, until the second half of this century, missed or ignored because of their virtual absence in XRD patterns. When they were found to be important constituents of soils derived from volcanic ash (Taylor, 1933; Yoshinaga and Aomine, 1962), researchers turned to a variety of methods more conducive to their identification, characterization and quantification. These include selective dissolution by oxalic acid to quantify Si and Al, thermoanalytic methods to identify and quantify specific minerals, structural characterization by vibrational spectroscopy (IR and FTIR), and electron microscopy (Wilson, 1987). More recently, short-range structural information has been obtained with solid-state nuclear magnetic resonance (NMR) and X-ray absorption spectroscopy (ESCA). Even the diffuse peaks in XRD patterns are useful to identify and quantify noncrystalline materials (Ruiz Cruz and Moreno Real, 1991). The information obtained from these methods and others will be discussed in the next sections as the nature, occurrence, and properties of allophane and imogolite are considered.

4.2 Allophane and Imogolite: Nature and Occurrence in Soils

4.2.1 Imogolite Characterization and Synthesis

Imogolite is a mineral of fixed composition with a structural formula $(OH)_3Al_2O_3SiOH$. The Al occurs only in octahedral coordination in a gibbsite-like sheet and the Si is coordinated to three O atoms and one OH group. The O atoms are shared with the Al octahedra and the apical OHs point toward the inside of the imogolite tube (Fig. 4.1) (Cradwick et al., 1972). The tubes are typically around 2.3 nm in diameter (2.7–3.2 nm for synthetic imogolite) and the repeat distance along the gibbsite sheet (c-axis) is 0.84 nm. The characteristic XRD lines are shown in Table 4.1.

Information regarding the structure of imogolite has come largely from spectroscopic studies. Infrared spectra give absorption bands indicative of both the gibbsite-like sheet and the orthosilicate anion (Fig.4.2). The Si-O stretching vibration at 960 cm^{-1} is consistent with unpolymerized orthosilicate groups and the OH stretching bands are consistent with Al-OH and Si-OH structures described by Cradwick et al. (1972). In addition, imogolite contains a band at 348 cm^{-1}, which is uniquely indicative of the imogolite structure, specifically, a gibbsite-like sheet with one hydroxyl on

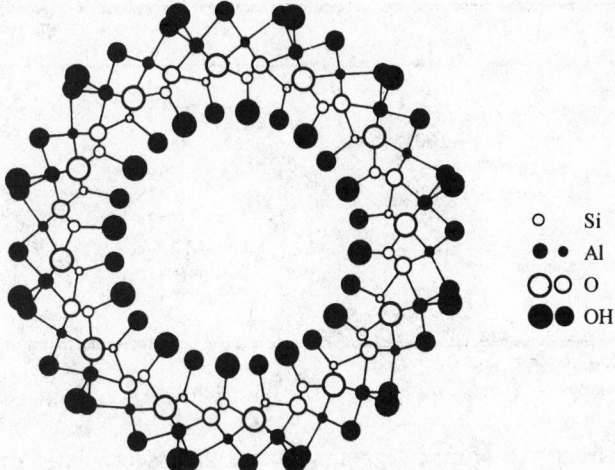

Si
Al
O
OH

Fig. 4.1 Structural model of the imogolite tube in cross-section [Reprinted from Goodman et al., 1985. Phys. Chem. Miner. 12:342–346 with permission of Springer-Verlag GmbH & Co. KG]

each Al replaced by an orthosilicate anion. The tubular morphology is not required for this band as it is also seen in proto-imogolite allophane, which is spherical in shape.

Solid-state NMR spectra also support the above model for imogolite (Goodman et al., 1985; He et al., 1995; Wilson et al., 1986). The ^{27}Al-NMR spectra for various samples of synthetic and natural imogolite show little or no evidence of tetrahedrally coordinated Al (Fig. 4.3). The ^{29}Si resonance line at –78 ppm of imogolite is consistent with Si bound to three Al-O and one OH group and its presence has been used to identify imogolite or proto-imogolite allophane in whole soils (Wilson, 1987).

Electron micrographs of imogolite are the most reliable indicator of its presence in soils or successful synthesis in the laboratory. Both synthetic imogolite and that isolated from a pumice bed show long, bundled tubes, whereas imogolite isolated from a forest soil is somewhat more fragmented (Fig. 4.4). Unlike phyllosilicate clay minerals, imogolite is ultrasonically dispersed from soils using low pH solutions, e.g., pH 4 HCl. Pretreatment usually includes a dithionite-citrate-bicarbonate extraction to remove Fe oxides and H_2O_2 to remove organic matter.

Refluxing an acidic solution of aluminum perchlorate and tetraethyl orthosilicate for five days at 95 °C produces imogolite nearly identical to the natural material isolated from soils or pumice (Farmer et al., 1983). The outside diameter of natural imogolite is close to 2.0 nm, whereas that of the synthetic mineral is near 2.8 nm. The larger diameter may be due, in part, to the high temperature synthesis, because imogolite prepared at room temperature has a diameter of about 2.3 nm (Wada, 1987).

4.2.2   Allophane Characterization and Synthesis

In its most general use, allophane is a group name for noncrystalline clay minerals consisting of silica, alumina, and water in chemical combination (Farmer and Russell, 1990). The further structural and chemical characterization of allophane found in natural environments has led to the identification of three major types of allophane. These have been tentatively named proto-imogolite allophane, halloysite-like (or defect kaolin) allophane and hydrous feldspathoid allophane.

Proto-imogolite allophane is characterized by an Al/Si ratio close to 2, a gibbsite-like sheet of octahedrally coordinated Al with orthosilicate groups sharing three Os with Al. Thus, proto-imogolite allophane has the same short-range order as imogolite, but does not exhibit the tubular morphology.

Table 4.1 Powder X-ray diffraction peaks for proto-imogolite allophane and imogolite

Proto-imogolite allphane		Imogolite	
d (nm)	I	d (nm)	I
1.2vb[1]	70	1.6vb	100
0.43vb	10	0.79	70
0.34vb[#]	100	0.56	35
0.22vb[#]	50	0.44	10
0.19vb[1]	10	0.41	10
0.17	10	0.37	20
0.14[#]	20	0.33vb	65
		0.31	5
		0.26	5
		0.225vb	25

[1]Also present in halloysite-like allophane
[#]Present in all three allophanes

Instead, it first forms fragments of the imogolite structure, then, depending on solution conditions, forms imogolite or spherical allophane particles about 3.5 nm in diameter (Farmer and Russell, 1990). Unlike imogolite, proto-imogolite allophane forms easily at room temperature. Its formation is also favored by below neutral pH. The NMR and FTIR spectra are nearly identical to those for imogolite (Figs. 4.2 and 4.3). The XRD pattern is characterized by very broad lines and lacks several lines indicative of imogolite (Table 4.1). Proto-imogolite allophane is sometimes referred to as Al-rich allophane.

Allophane with an Al/Si ratio closer to one probably has a structure that is closer to kaolinite or halloysite with defects in the tetrahedral sheet. Its IR spectra indicates the presence of polymerized silica tetrahedra and octahedrally coordinated Al (Parfitt et al., 1980). Both IR and NMR spectra suggest that these silica-rich allophanes often contain both the halloysite and imogolite structural units. The halloysite-like feature in the NMR spectrum is characterized by a broad resonance around -86 ppm. Varying amounts of the two types of allophane could account for a large variation in the Al/ Si ratios (from 2.5 to 1) of this allophane when found in weathered pumice (Parfitt, 1990; Farmer and Russell, 1990). Recent studies of the silica-rich allophane by solid-state NMR confirm both the presence of imogolite-like and defect kaolin-like structures. MacKenzie et al. (1991) proposed a model of this allophane in which the isolated orthosilicate groups are associated with the gibbsite sheet of a 1:1 kaolinite layer. The orthosilicate groups penetrate the inner sheet of silica tetrahedra through the holes formed by defects. This model accounts for the range in Al/Si, spherical morphology with diameter around 3.5 nm, and the NMR features. Some Al may substitute for Si in the tetrahedral sites, but these allophanes are dominated by Al in octahedral coordination. Further support for the defect kaolin structure of Si-rich allophane is provided by ESCA (He et al., 1995).

The hydrous feldspathoid, allophane (also Si rich), contains no imogolite units and a significant, if not dominant, amount of tetrahedrally coordinated Al. The basic structure appears to be that of a framework silicate with 1:3 Al for Si substitution. Octahedral Al neutralizes some of the negative charge generated by the substitution and may be associated with the inner surfaces of spherical particles. Particles isolated from a stream deposit in New Zealand are less than 3 nm in diameter and form in CO_2-charged water which increases in pH as it degasses (Wells et al., 1977; Childs et al., 1990). Similar particles can be synthesized in neutral to alkaline solutions and IR and NMR spectra are consistent with a hydrous feldspathoid structure (Farmer et al., 1979; Wada and Wada, 1981). There is no -78 ppm resonance in the ^{29}Si NMR spectra and the changes in NMR spectra with heating

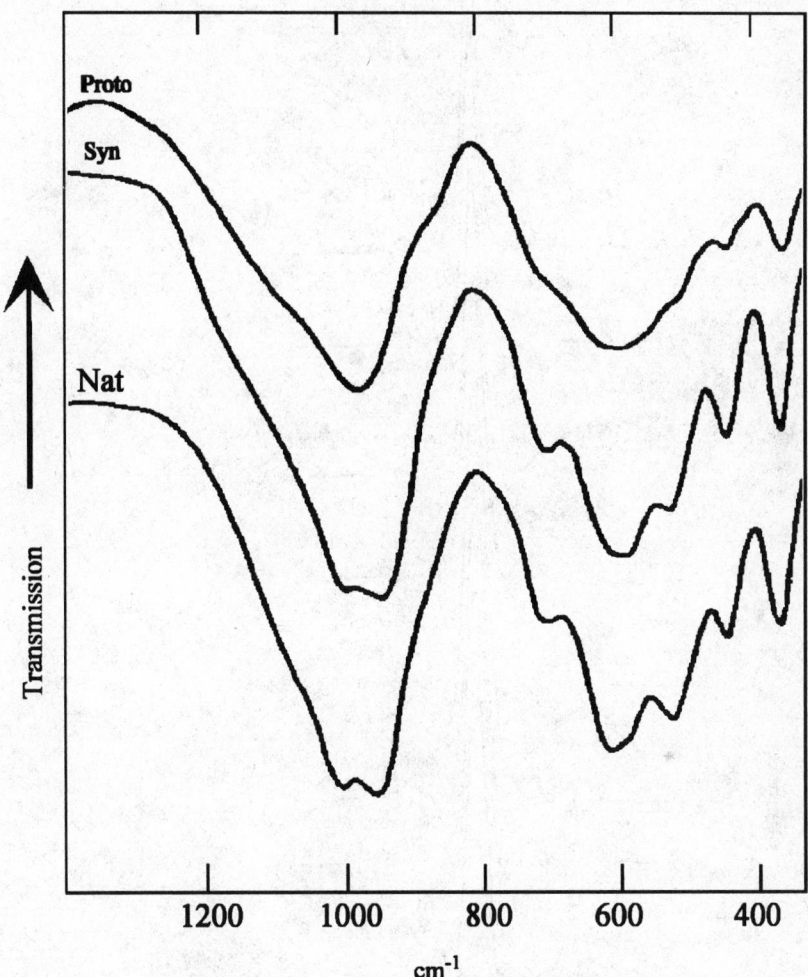

Fig. 4.2 FTIR spectra of natural and synthetic imogolite and allophane [Reprinted from Farmer et al., 1979. Characterization of the chemical structure of natural and synthetic aluminosilicate gels and sols by infrared spectroscopy. Geochim. Cosmochim. Acta. 43:1417-1420 with permission from Elsevier Science]

confirm that the hydrous feldspathoid allophane differs in structure from both proto-imogolite and kaolin defect allophane (MacKenzie et al., 1991).

Allophane can also be identified in soils by electron microscopy, but it is far less distinctive than imogolite (Fig. 4.4). It generally occurs as an amorphous mass of material coating other particles. Often the spheroidal morphology is evident in the Al-rich allophane. The spheroids of Al-rich allophane form aggregates that do not disperse easily. Damage from the electron beam is rapid and care must be taken to obtain good micrographs.

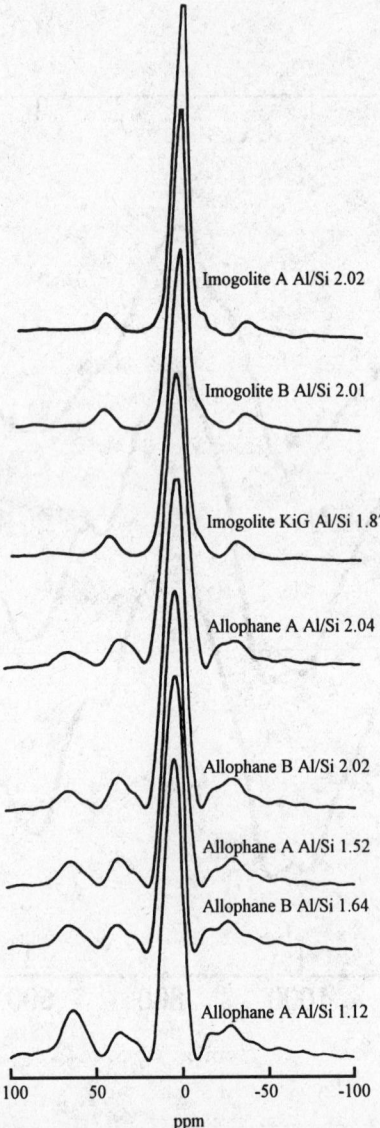

Fig. 4.3 Solid state ^{27}Al NMR spectra of natural and synthetic imogolite and synthetic allophane [Reprinted from Su et al., 1992. Clays Clay Miner. 40:280–286 with permission]

Synthesis of allophane in the laboratory does not require a heating step. The Al/Si ratio and pH of the matrix solution determine the nature of the products. Aluminum-rich allophane, like imogolite, is formed from a solution with Al/Si ~2 and at low pH. The acidic environment keeps Al in octahedral coordination. Decreasing the Al/Si ratio while keeping the pH below neutral, probably favors the defect kaolinite structures where Al is still in octahedral coordination. In nature, such as allophane formation in pumice, there is evidence that proto-imogolite allophane coexists with the defect kaolinite type. Between pH 6 and 7, the hydrous feldspathoid structure predominates with the proportion of tetrahedral Al increasing as the Al/Si ratio decreases. This structure has been found in

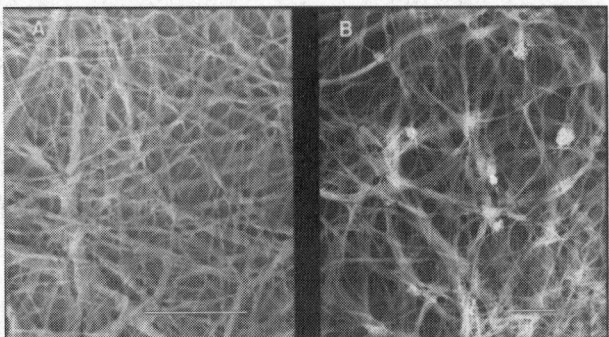

Fig. 4.4   Electron micrographs of (A) synthetic imogolite and natural imogolite from (B) pumice and (C) a forest soil

natural environments, most notable being the Silica Springs deposit in New Zealand (Wells et al., 1977). Increasing pH leads to zeolite-like structures with little or no Al in octahedral coordination. Farmer and Russell (1990) reviewed the synthesis and occurrence of imogolite and the three types of allophane described here.

4.2.3   Occurrence of Imogolite and Allophane in Soils

Allophane and imogolite are often associated with soils derived from volcanic debris, because the rapid release of Al and Si from materials such as volcanic glass results in the precipitation of noncrystalline aluminosilicates. Imogolite was first identified in such soils (Yoshinaga and Aomine, 1962) and allophane was found to give them many of their unique properties. Since then, noncrystalline aluminosilicates have been identified in soils derived from many parent materials, including sandstone, gneiss, granite, and basalt. In basalt, ferrihydrite, a poorly crystalline Fe oxide, often occurs in addition to allophane and imogolite.

The key factor in the formation of allophane and imogolite in soil is sufficient Al and Si in solution. Thus, conditions that lead to rapid weathering of primary minerals, such as high rainfall and low pH, favor their formation. Both imogolite and Al-rich allophane tend to dominate at low pH where Al/Si is high, whereas halloysite and/or Si-rich allophanes form when soluble Si is high ($> 10^{-3.45}$). As a result, low rainfall or xeric moisture regimes, concentrating soluble silica, have been observed to favor halloysite formation at the expense of allophane and imogolite (Parfitt and Wilson, 1985; Takahashi et al., 1993). Noncrystalline minerals can form in Andisols, Inceptisols, Entisols, Spodosols, and Ultisols. Generally, precipitation occurs in horizons where soil organic matter (SOM) is low so that Al exists in inorganic complexes. Also, occurrence is more likely in forested soils than under grasslands. It may be necessary for Al to form hydroxy-complexes before imogolite will form.

The methods described above to characterize the structural properties of noncrystalline aluminosilicates will also serve to reveal their presence in soil. After oxidation of SOM and removal of Fe oxides, allophane and imogolite both disperse at pH 4, allowing separation from clay minerals which tend to flocculate in acid solutions. They can also be concentrated with respect to kaolinite, illite, and many other clays by collecting the < 0.2 m fraction by centrifugation. Allophane may be partially separated from imogolite in an alkaline suspension where imogolite flocculates and the allophane suspension remains stable. After separation has been achieved, FTIR and XRD may be used to characterize the fractions.

Chemical extraction of noncrystalline aluminosilicates and subsequent analysis of Al and Si in the solutions is another way to characterize their presence in soil (Dahlgren, 1994). A common selective dissolution scheme extracts the soil first with Na pyrophosphate, which effectively separates Al and Fe from SOM, but dissolves little from the inorganic fraction. Then, an acid ammonium oxalate solution selectively attacks allophane, imogolite, and poorly crystalline Fe oxides such as ferrihydrite. The Al and Fe in the pyrophosphate extract are subtracted from the oxalate values to estimate Al and Fe associated with the noncrystalline materials. Dividing this corrected Al value by the oxalate extractable Si then gives the average Al/Si ratio of the allophane and imogolite. A ratio near 2 suggests a soil dominated by imogolite and proto-imogolite allophane whereas a smaller value indicates Si-rich allophane is present. Ratios > 2 may occur when other labile sources of Al are present, such as hydroxy-Al interlayers in clay minerals.

4.3   Surface Charge Characteristics of Noncrystalline Aluminosilicates and Variable Charge Soils

4.3.1   Surface Charge Determination

Operational definitions. We will consider two major sources of surface charge on noncrystalline aluminosilicates: permanent structural charge (σ_o) arising from isomorphic substitution of Al in silica tetrahedra and variable surface charge arising from ion associations at surface hydroxyl groups (Sposito, 1984). Surface charge from ion associations with surface hydroxyls arises from complexed protons (σ_H), inner sphere complexes with ions other than H$^+$ (σ_{IS}), and outer sphere complexes (σ_{OS}). A charge balance equation can be written as:

$$\sigma_p = \sigma_o + \sigma_H + \sigma_{IS} + \sigma_{OS} = -\sigma_D$$

where σ_p represents the particle charge or total charge from all sources and σ_D is the diffuse layer charge which consists of the net charge from the swarm of uncomplexed ions around the particle (Sposito, 1984).

With all the Al in imogolite and proto-imogolite allophane in octahedral coordination, there is no permanent structural charge in these minerals. The hydrous feldspathoid allophane, on the other hand, has significant Al for Si substitution. In synthetic allophane of this type, the quantity of tetrahedral Al is not matched by exchangeable cations (Su et al., 1992). In this case, Al present as hydroxy complexes or polymers balances much of the negative σ_o.

The proton charge on variable charge surfaces can be determined by potentiometric titration with H$^+$ and OH$-$. This charge originates from the adsorption or desorption of protons from silanol ($\equiv$Si-OH) or aluminol ($\equiv$Al-OH) groups at the allophane or imogolite surface. Relative amounts of adsorption or desorption are easily determined from the difference between added and remaining protons in solution, but the absolute proton charge on the surface is more difficult to obtain. The point of zero net proton charge (p.z.n.p.c.) can be estimated by the point of zero salt effect (p.z.s.e.) where the proton charge does not change with the ionic strength of the solution. The latter is obtained by finding the crossover point of three or more titrations at different background salt concentrations. If the background salt does not form inner sphere complexes, only the proton charge is determined in the titration. A more elegant method for determining the proton charge without using this approximation is given by Chorover and Sposito (1995).

Outer sphere and diffuse layer cations and anions balance the total negative and positive charges on the surface. Thus, the point of zero net charge (p.z.n.c.) can be determined by ion adsorption across a range of pH in an electrolyte, such as NaCl, that does not form inner sphere complexes with the

Table 4.2 Points of zero charge of synthetic imogolite and allophanes [Reprinted from Su et al., 1992. Clays Clay Miner. 40:280–286 with permission]

| Material | Al/Si | p.z.n.c. in NaCl | | p.z.s.e. | p.z.c. |
		0.10 M	0.01 M		
Imogolite	2.0	8.4	8.4	6.5	> 10
Allophane	2.0	7.9	7.9	8.3	10
Allophane	1.6	5.8	6.7	7.7	9
Allophane	1.2	4.1	5.4	5.9	7.6

aluminol or silanol groups. Finally, electrophoretic mobility provides information about the tendency for ions to form outer sphere complexes with allophane and imogolite. Their mobility when suspended in an aqueous solution subjected to an electric field depends on the ionic strength of the solution and the nature of the surface complexes. The pH at which the particles are stationary in the field is the point of zero charge (p.z.c.).

The charge characteristics of imogolite and some synthetic allophanes as determined by these methods are presented in Table 4.2. Because the p.z.n.c. and p.z.s.e. values of imogolite and Al-rich allophane are all > 6, these materials impart positive charge in acid soils. The Si-rich allophanes, on the other hand, may be either positively or negatively charged depending on soil pH. The Si-rich allophanes also have p.z.s.e. > p.z.n.c., an indicator that permanent negative charge exists in these materials as a result of Al for Si substitution. Imogolite has a p.z.s.e. greater than its p.z.n.c., a fact which could result from permanent positive charge, specific adsorption of the anion, or an experimental artifact (Harsh et al., 1992).

4.3.2 Interaction with Anions and Cations

Inner sphere complexation of anions on allophane and imogolite is common. Phosphate, fluoride, citrate, borate and selenite are known to form inner sphere complexes with noncrystalline aluminosilicates. This reaction contributes to σ_{IS} and can shift the points of zero charge of the material. As a result, more negative charge is likely in the presence of these anions, particularly in acid soils. Some caution should be used in interpreting any strong sorption as surface complexation, however. Veith and Sposito (1977) and Su and Harsh (1993) showed that phosphate and fluoride, respectively, may react with noncrystalline aluminosilicates to form new solid phases. Boron may substitute into tetrahedral sites in coprecipitation with allophane (Su and Suarez, 1997).

In soils containing allophane and imogolite, strong interaction with phosphate can lead to deficiency of this macronutrient in crops. Calcium silicate can be added to such soils to compete for sorption sites and enhance phosphate availability. The ability of noncrystalline materials and allophanic soils to sorb anions by ligand or anion exchange has led to suggestions that they be used to remove contaminants such as phosphate, selenium, technetium, and iodine from waste waters (Wells and Parfitt, 1987; Gu and Schulze, 1991).

Many trace metals form surface complexes with aluminol and silanol groups as evidenced by extensive studies on silica and Al oxides; however, few studies have been performed on allophane and imogolite, *per se* (Clark and McBride, 1984). One study of metal adsorption to a soil dominated by allophane and imogolite in the clay fraction showed the following order of decreasing affinity: Pb>Cu>Zn>Co>Cd>Mg which is quite similar to the selectivity of aluminol groups for divalent metals.

Alkaline earth and alkali metals, halide anions, NO_3^-, SO_4^{2-}, and ClO_4^- will generally exist as outer sphere complexes or in the diffuse layer of allophane and imogolite. Adsorption of exchangeable cations on allophanic soils appears to be similar to that on soils dominated by smectites (Nakahara and Wada, 1994). There have been reports of highly selective K exchange on Andisols, but this could result from trace amounts of illitic minerals or alunite formation in addition to reactions with allophane or imogolite (Espino-Mesa and Hernández-Moreno, 1994). There is circumstantial evidence for specific adsorption of Cl^- on imogolite, as discussed above, but this issue has not been resolved and is not consistent with electrophoretic measurements (Fig. 4.5).

4.4   Interactions of Allophane and Imogolite with Other Soil Constituents

4.4.1   Organic Matter and Iron Oxides

The term Andisol comes from the Japanese words *ando* and *sol*, meaning dark-colored soil. The dark color comes from the fact that soils formed from volcanic debris are often high in organic matter, especially in surface horizons. Noncrystalline materials may play a role in organic matter retention through one or more mechanisms. First, humic substances sorb strongly to imogolite and allophane (Parfitt et al., 1977) and it is possible that sorption occurs through ligand exchange of carboxyl groups to aluminol surface groups. Also, reactive Al released from noncrystalline materials complexes with humic and fulvic acids and leads to precipitation and/or inhibition of degradation (Boudot, 1992; Wada, 1995).

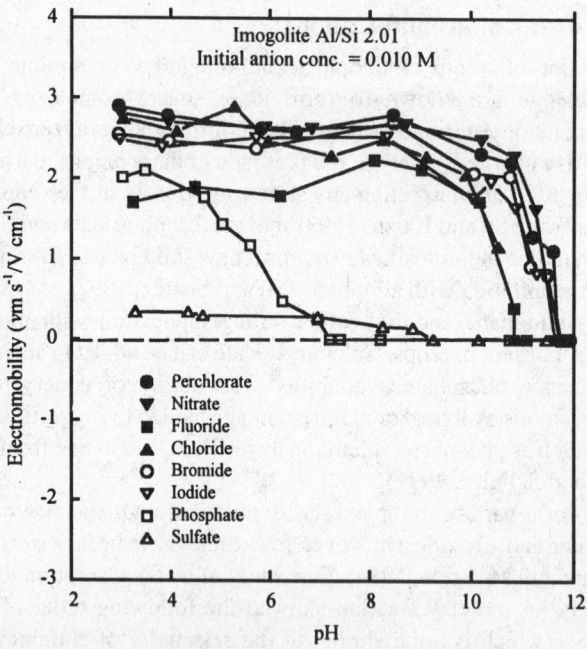

Fig. 4.5   The electrophoretic mobility of imogolite in the presence of various anions [Reprinted from Su and Harsh, 1993. Clays Clay Miner. 41:461–471 with permission]

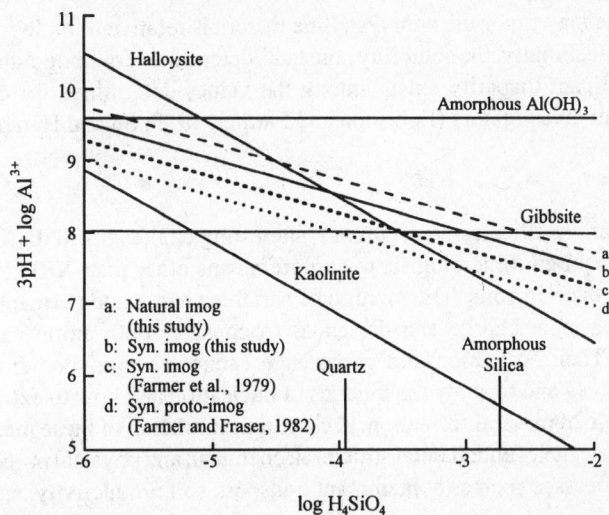

Fig. 4.6 Stability relations for imogolite, proto-imogolite, allophane, crystalline aluminosilicates, and gibbsite [Reprinted from Su and Harsh, 1994. Gibbs free energies of formation at 298 K for imogolite and gibbsite from solubility measurements. Geochim. Cosmochim. Acta 58:1667–1677 with permission from Elsevier Science]

The interaction of organic matter and allophane and allophanic soils also affects the surface charge and sorption properties of the materials. Organic matter can be expected to increase the negative charge on minerals dominated by aluminol surface groups because of the carboxylate groups of humic materials. Treatment of allophanic materials and soils with H_2O_2 to remove SOM increases the p.z.c. (Escudey et al., 1986) and decreases phosphate sorption (Mora and Canales, 1995), respectively. Conversely, adding Fe oxides increases the p.z.c. of allophane rendering the surface more positively charged (Escudey and Galindo, 1983). It is possible the Fe oxides react selectively with silanol groups which have a p.z.n.p.c. ~2, while the Fe and Al oxides, on the other hand, are > 7.

4.4.2 Water

Allophane is characterized by both a high specific surface area and a high value of the relative microporosity. The latter probably results largely from micropores formed from aggregates of particles because the porosity decreases with the shrinkage observed on drying. As a result, water retention in allophanic soils is very high at a given water potential compared to soils dominated by crystalline clays. This, in part, accounts for the high productivity often observed in forested soils high in noncrystalline materials (Martini and Luzuriaga, 1989; Nizeyimana, 1997). Macroporosity is also high in most allophanic soils resulting in good infiltration and hydraulic conductivity. The latter parameter has been found to decrease significantly in soils at pH values where allophane is known to disperse easily (pH 3 and 11) (Nakagawa and Ishiguro, 1994).

The aqueous solubility of proto-imogolite allophane shows them to be metastable with respect to kaolinite at soluble silica activities expected in natural soils (Fig. 4.6). Decreasing soluble silica increases their stability with respect to halloysite, consistent with the finding of halloysite formation at log $(H_4SiO_4) > -3.45$. Fig. 4.5 shows that halloysite could form at even lower Si activity from a thermodynamic standpoint. This incongruity could arise from several difficulties. It may indicate that soil solution analyses performed at a later date may not be relevant to the time of mineral formation.

Kinetic considerations may favor the noncrystalline materials relative to halloysite; this is certainly true relative to kaolinite. Finally, the solubility constants determined for imogolite and allophane may be incorrect. A significant disparity exists among the values determined for different sources of imogolite and different investigators (Lumsdon and Farmer, 1995; Su and Harsh, 1994).

4.5 Conclusion

The noncrystalline aluminosilicates present many interesting challenges to the soil scientists. Their noncrystallinity has required mineralogists to explore means other than XRD to characterize their structure and identify them in soils. The mixture of variable charge and permanent charge sites has renewed interest in variable charge and driven us to consider both anion and cation exchange phenomena in soils. Their formation and persistence cannot be predicted from thermodynamic relations alone, but forces one to study the kinetics of nucleation and how to extrapolate current soil solution data to the time of mineral formation. A careful examination of these materials *in situ* is still needed to determine how their fundamental structural, chemical, and physical properties influence soil behavior critical to processes such as contaminant transport, soil productivity, and soil stability.

4.6 References

Boudot, J.P. 1992. Relative efficiency of complexed aluminum, noncrystalline Al hydroxide, allophane and imogolite in retarding the biodegradation of citric acid. Geoderma 52:29–39.

Childs, C. W., R. L. Parfitt, and R. H. Newman. 1990. Structural studies of Silica Springs allophane. Clay Min. 25:329–341.

Cradwick, P.D.G., V.C. Farmer, J.D. Russell, C.R. Masson, K. Wada and N. Yoshinaga. 1972. Imogolite, a hydrated aluminum silicate of tubular structure. Nature Physical Sci. 240:187–189.

Chorover, J., and G. Sposito. 1995. Surface charge characteristics of kaolinitic tropical soils. Geochim. Cosmochim. Acta 59:875–884.

Clark, C.J., and M.B. McBride. 1984. Chemisorption of Cu(II) and Co(II) on allophane and imogolite. Clays Clay Miner. 32:300–310.

Cradwick, P.D.G., V.C. Farmer, J.D. Russell, C.R. Masson, K. Wada, and N. Yoshinga. 1972. Imogolite, a hydrated aluminum silicate of tubular structure. Nature Phys. Sci. 240:187–189.

Dahlgren, R. A. 1994. Quantification of allophane and imogolite. p. 430–451. J. Amonette and L.W. Zelazny (ed.) *In* Quantitative methods in soil mineralogy Soil Sci. Soc. Am. Misc. Publ., Soil Science Society of America, Madison, WI.

Escudey, M., and G. Galindo. 1983. Effect of iron oxide coatings on electrophoretic mobility and dispersion of allophane. J. Colloid Interface Sci. 93:78–83.

Escudey, M., G. Galindo, and J. Ervin. 1986. Effect of iron oxide dissolution treatment on the isoelectric point of allophanic soils. Clays Clay Miner. 34:108–110.

Espino-Mesa, M., and J.M. Hernández-Moreno. 1994. Potassium selectivity in Andic soils in relation to induced acidity, sulphate status and layer silicates. 16:191–201.

Farmer, V.C., M.J. Adams, A.R. Fraser, and F. Palmieri. 1983. Synthetic imogolite: Properties, synthesis, and possible applications. Clay Minerals 18:459–472.

Farmer, V. C., J. D. Russell, and M. L. Berrow. 1979. Characterization of the chemical structure of natural and synthetic aluminosilicate gels and sols by infrared spectroscopy. Geochim. Cosmochim. Acta 43:1417–1420.

Farmer, V.C., and J.D. Russell. 1990. Structures and genesis of allophanes and imogolite and their distribution in non-volcanic soils. p. 165–178. *In* M.F. De Boodt et al. (ed.) Soil colloids and their associations in aggregates. Plenum Press, New York, NY.

Fyfe, C., L. Evans, W. Chesworth, J. Graham, and M. McBride. 1987. Operation definition of imogolite/allophane phases by multitechnique analysis. p. 829–836. *In* C.R. Rodriguez and Y. Tardy (ed.). Geochemistry and mineral formation in the Earth surface. Cons. Super. Invest. Cient, Spain. CNRS, France.

Goodman, B.A., J.D. Russell, B. Montez, E. Oldfield, and R.J. Kirkpatrick. 1985. Structural studies of imogolite and allophanes by aluminum-27 and silicon-29 nuclear magnetic resonance spectroscopy. Phys. Chem. Minerals 12:342–346.

Greenland, D.H. 1982. Chemistry and the soil environment-surfaces and sorption processes. p. 99–110. *In* K.J. Laidler (ed.) IUPAC frontiers in chemistry. IRRI, Manila, Philippines.

Gu, B., and R.K. Schulze. 1991. Anion retention in soil: Possible application to reduce migration of buried technetium and iodine. NUREG/CR-5464, U.S. Nuclear Regulatory Commission.

Harsh, J.B., S.J. Traina, J. Boyle, and Y. Yang. 1992. Adsorption of cations on imogolite and their effect on surface charge characteristics. Clays Clay Miner, 40:700–706.

He, H., T.L. Barr, and J. Klinowski. 1995. ESCA and solid-state NMR studies of allophane. Clay Minerals. 30:201–209.

Lumsdon, D. G., and V. C. Farmer. 1995. Solubility characteristics: how silicic acid can de-toxify aluminum solutions. Eur. J. Soil Sci. 46:179–186.

MacKenzie, K. J. D., M. E. Bowden, and R. H. Meinhold. 1991. The structure and thermal transformations of allophanes studied by ^{29}Si and ^{27}Al high-resolution solid-state NMR. Clays Clay Miner. 39:337–346.

Martini, J. A., and C. Luzuriaga. 1989. Classification and productivity of six Costa Rica Andepts. Soil Sci. 147:326–338.

Mora, M.L., and J. Canales. 1995. Humin-clay interactions on surface reactivity in Chilean Andisols. Comm. Soil Sci. Plant Anal. 26:2819–2828.

Nakagawa, T., and M. Ishiguro. 1994. Hydraulic conductivity of an allophanic andisol as affected by solution pH. J. Environ. Qual. 23:208–210.

Nakahara, O., and S.-I. Wada. 1994. Ca^{2+} and Mg^{2+} adsorption by an allophanic and a humic Andisol. Geoderma 61:203–212.

Nizeyimana, E. 1997. A toposequence of soils derived from volcanic materials in Rwanda: Morphological, chemical, and physical properties. Soil Science 162: 350–360.

Parfitt, R.L. 1990. Allophane in New Zealand: A review. Aust. *J. Soil Sci.* 28:343 –60.

Parfitt, R., and J. Dixon. 1991. Definitions of allophane: a discussion. Clay Mineral Society News, June, 1991.

Parfitt, R., A.R. Fraser, and V.C. Farmer. 1977. Adsorption on hydrous oxides. III. Fulvic acid and humic acid on goethite, gibbsite, and imogolite. J. Soil Sci. 28:289–296.

Parfitt, R.L., R.J. Furkert, and T. Henmi. 1980. Identification and structure of two types of allophane from volcanic ash soils and tephra. Clays Clay Min. 28:328–334.

Parfitt, R.L. and A.D. Wilson. 1985. Estimation of allophane and halloysite in three sequences of volcanic soils, New Zealand. p. 1–8. *In* E.F. Caldas and D.H. Yaalon (ed.) Volcanic soils. Weathering and landscape relationships of soils on tephra and basalt. Catena Suppl. 7, Catena-Verlag, Braunschweig. Cremling-Destedt, West Germany.

Ruiz Cruz, M.D., and L. Moreno Real. 1991. Practical determination of allophane and synthetic alumina and iron oxide gels by X-ray diffraction. Clay Miner. 26:377–387.

Sposito, G. 1984. The surface chemistry of soils. Clarendon Press, New York, NY.

Steefel, C.I., P. van Capellen, K.L. Nagy, and A.C. Lasaga. 1990. Modeling water-rock interaction in the surficial environment: the role of precursors, nucleation, and Ostwald ripening. p. 322–325. *In* Geochem. Earth Surf. Miner. Formation, 2nd Int. Symp., Aix en Provence, France.

Su, C., and J.B. Harsh. 1993. The electrophoretic mobility of imogolite and allophane in the presence of inorganic anions and citrate. Clays Clay Miner. 41:461–471.

Su, C., and J.B. Harsh. 1994. Gibbs free energies of formation at 298 K for imogolite and gibbsite from solubility measurements. Geochim. Cosmochim. Acta 58:1667–1677.

Su, C., and J.B. Harsh. 1996. Alteration of imogolite, allophane, and acidic soil clays by chemical extractants. Clays Clay Miner. 41:461–471.

Su, C., J.B. Harsh, and P.M. Bertsch. 1992. Sodium and chloride sorption by imogolite and allophanes. Clays Clay Mineral. 40:280–286.

Su, C., and D.L. Suarez. 1997. Boron sorption and release by allophane. Soil Sci. Soc. Am. J. 61:69–77.

Takahashi, T., R. Dahlgren, and P. van Susteren. 1993. Clay mineralogy and chemistry of soils formed in volcanic materials in the xeric moisture regime of northern California. Geoderma 59:131–150.

Taylor, N.H. 1933. Soil processes in volcanic ash-beds. N.Z. J. Sci. Tech. 14:338–352.

Veith, J.A., and G. Sposito. 1977. Reactions of aluminosilicates, aluminum hydrous oxides, and aluminum oxide with o-phosphate: The formation of X-ray amorphous analogs of variscite and montebrasite. Soil Sci. Soc. Am. J. 41:870–876.

Wada, K. 1995. Role of aluminum and iron in the accumulation of organic matter in soils with variable charge. p. 47–58. *In* P.M. Huang, J. Berthelin, J.-M. Bollag, W.B. McGill, and A.L. Page (ed.) Environmental impact of soil component interactions. CRC Press, Boca Raton, FL.

Wada, S.-I., and K. Wada. 1981. Reactions between aluminate ions and orthosilicic acid in dilute, alkaline to neutral solutions. Soil Sci. 132:267–273.

Wells, N., C.W. Childs, and C.J. Downes. 1977. Silica Springs, Tongariro National Park, New Zealand: Analyses of the spring water and characterization of the alumino-silicate deposit. Geochim. Cosmochim. Acta. 41:1497–1506.

Wells, N., and R.L. Parfitt. 1987. Occurences of short-range order clays and their use in pollution control. Proc. Pacific Rim Congr. 87:469–473.

Wilson, M.A., S.A. McCarthy, and P.M. Fredericks. 1986. Structure of poorly ordered aluminosilicates. Clay Minerals 21:879–897.

Wilson, M.J. 1987. A handbook of determinative methods in clay mineralogy. Blackie & Son, Ltd. Glasgow.

Yoshinaga, N., and S. Aomine. 1962. Imogolite in some Ando soils. Soil Sci. Plant Nutr. 8:22–29.

A nation that destroys its soils destroys itself
-Franklin D. Roosevelt

G

Interdisciplinary Aspects of Soil Science

Isaac Shainberg
Volcani Center, Israel

Science has long recognized the complexity of natural systems. This complexity is manifested in the discipline of soil science, from the dynamic of soil water processes to the effects of cultivation on soil properties. In spite of the fact that all processes in natural systems are interdependent and interrelated, scientists soon discovered that the study of such complexity must be simplified, focusing on a reduced number of processes and properties if one is to comprehend essential elements of the system. Given this philosophy, soil scientists have pursued their investigation from a reductionist viewpoint rather than from a more realistic interdisciplinary perspective. While this has led to some understanding of singular processes, there are many interactions that now must be considered. The next scientific steps must therefore treat the uses that arise from interactions between physical, chemical and pedological properties.

The organization of this handbook is an example of the opposing tendencies in approaching the science of soils. Whereas in previous Sections the subdisciplines of soil physics, soil chemistry and soil pedology were presented, the interdisciplinary aspects of soil science will now be discussed. This separation has produced some substantial rewards regarding the basic nature of soils, but the fact remains that relatively few soil scientists cross the subdisciplinary boundaries discussed above to reap the benefits of increased understanding in another area and make an applied contribution in the real world. For example, few soils physicists understand the consequences of soil solution composition on soil hydraulic properties as discussed in Chapters 1 and 2.

Soils are fundamental to the well-being and productivity of both agricultural and natural ecosystems. Because soil is in large but finite supply, the condition of soils in agriculture and the environment is receiving increased attention as an issue of global concern. Degradation of soils by salinity (Chapter 1), sodicity (Chapter 2), unstable structure (Chapter 3), waterlogging (Chapter 4), acid sulfate (Chapter 5), organic and inorganic contaminants (Chapter 6), water and wind erosion (Chapters 7 and 8), application of wastes (Chapter 9), and by tillage (Chapter 10) are discussed in this section which concludes with a critical review of the soil quality concept (Chapter 11) which may enable soil scientists to quantify changes in soil quality. Continuing worldwide soil degradation by erosion, chemicals, acidification and physical abuse requires that attention be paid to soil quality measurements.

1

Salinity

R. Keren
Volcani Center, Israel

1.1 Origin and Distribution of Saline Soils

The history of agriculture has shown that irrigated agriculture cannot survive in perpetuity without adequate salt balance and drainage. The length of time that irrigated agriculture can survive without adequate drainage depends on hydrogeology and water management. Of the world's cultivated lands $(1.5 \times 10^7 \, km^2)$ (Massoud, 1981), about 23% are saline while saline and sodic soils cover about 10% of the total potentially arable land and exist in over 100 countries (Szabolcs, 1989).

Inland saline waters usually contain 500 to 30,000 mg L^{-1} of dissolved solids (EC 0.7–42 dS m^{-1}), while ocean water has an average dissolved solids concentration of 33,000 mg L^{-1} and the Dead Sea, Israel, 270,000 mg L^{-1} (hypersaline). Common sources of saline water for agricultural use are ground or surface waters. The main processes by which soluble salts enter the soil and groundwater include weathering of primary and secondary minerals and application of waters containing salts. The importance of each source depends on the type of soil, climate conditions and agricultural management.

1.1.1 Mineral Weathering

Salt formation takes place during the process of minerals weathering. The accumulation of these soluble salts in groundwater affects water quality which depends on the natural salinity of the soil and the geologic materials with which the water has been in contact. Many soils from arid and semiarid regions and primary minerals such as olivine, hornblende, oligoclase and others contribute substantial amounts of salts from weathering (Rhoades et al., 1968). In all cases, the total salt content of the displaced soil solutions was higher than that of the irrigation waters being applied.

Water within sedimentary strata become increasingly saline with increasing depth (Craig, 1970) with the sequence being SO_4^{2-} rich water near the surface, HCO_3^- water at an intermediate levels and more concentrated Cl^- solutions at greater depths. This water becomes saline through weathering of primary and secondary minerals.

Predictions of soil mineral weathering and soil solution compositions can be made using stability diagrams (Kittrick, 1977; Lindsay, 1979). Mineral solubility is calculated from appropriate mineral equilibrium constants. Limitations of this approach are (1) the availability of adequate data on the solubilities or free energies of formation of various minerals as well as dissolved species present in soil solution, and (2) the paucity of information on the kinetics of minerals dissolution. Thermodynamic data generally are more accurate for the primary minerals involved in increasing the salinity of the soil solution (Rhoades et al., 1968) than for the secondary minerals (Kittrick, 1977). The major ions in the dissolved mineral salts are Na, Ca, Mg, K, Cl, SO_4, HCO_3, CO_3 and NO_3.

1.1.2   Sea Water Intrusion

In coastal areas, salt water intrusion and submergence of the low-lying lands by sea water cause salinization of groundwater and soils, while surface waters could become saline through tidal fluctuations. As the high tide moves into a coastal area, sea water moves into streams and drainage canals and travels inland. This upstream migration of sea water alters the quality of water in streams and drainage canals significantly.

1.1.3   Deposition of Salts by Rainfall

The composition of atmospheric salt deposition varies with distance from the source. The salt is predominantly NaCl at the coast, and becomes dominated by Ca^{2+} and SO_4^{2-} inland. Atmospheric contribution to the salt load of arid lands, which is from 10 to 25% of the total yearly contribution from weathering (Bresler et al., 1982), is often overlooked, but is a factor that must be considered in highly weathered landscapes that have poor drainage.

1.1.4   Deposit and Secondary Salinization

Throughout geologic time, sea water has inundated large areas of the continents which have, subsequently, been uplifted and salts deposited when the water evaporated. Salt bodies, which lie beneath the soil surface, formed when inland lakes evaporated (e.g., Searles Lake, CA). These salts were deposited in Pleistocene and Holocene times during major dry episodes. The salts contain varying mixtures of saline minerals such as halite, trona, and nahcolite (Smith, 1979). The term, fossil salt, has been used to describe the salinity of these deposits which are substantial sources of salinity.

Irrigation with poor quality water, inadequate leaching, seepage from canals, high watertables and high evaporation rates, all contribute to secondary salinization of irrigated soils. These salts will reduce crop yield if they accumulate in the root zone.

The development of waterlogged soils is mainly associated with low-lying lands having poor physical condition and internal drainage. These soils are mainly found in the flood and deltaic plains of rivers and valleys. Salt accumulation depends on the salinities of the applied water and target soil, and the rate at which salts are leached out of the root zone. If a restricting layer exists close to the soil surface, waterlogging occurs and the accumulation of salt associated with waterlogged soils develops within a comparatively short time. In the absence of such a layer and with large capacity vadose zones, irrigation may be practiced centuries before problems with surface drainage arise, if ever. Another source of saline water is irrigation drainage effluent which is often used for irrigation. Although salinity levels may vary, the electrolyte concentrations are usually relatively high.

Under arid or semiarid conditions and in regions of poor natural drainage, salt accumulation is a real hazard. However, salt-affected soils are not confined to semiarid and arid regions. In some other regions, the climate and salt mobility produce saline waters and soil seasonally.

Irrigated agriculture has faced the challenge of sustaining its productivity for generations. Because of natural hydrological and geochemical factors, as well as irrigation-induced activities, soil and water salinity and associated drainage problems continue to plague agriculture.

1.2 Water Quality Criteria for Irrigation

Irrigation water quality is one of the primary considerations ensuring proper water and soil management for crop production. The suitability of a water for irrigation is determined by its potential to cause problems to soils and crops and is related to the management practices needed. The quality of a water for irrigation is determined mainly by the concentration and composition of solutes present. Most soluble salts in water are composed of the cations Na^+, Ca^{2+} and Mg^{2+} and anions Cl^-, SO_4^{2-} and HCO_3^-. Usually, smaller quantities of K^+ and NO_3^- also occur, as do many other ions and molecules (e.g., B). The concentration of HCO_3^- and CO_3^{2-} are pH and CO_2 partial pressure (P_{CO_2}) dependent.

Bicarbonate is an important parameter as it has a tendency to precipitate Ca and Mg as carbonates (Amrhein and Suarez, 1987; Letey et al., 1985). Calcium reacts with the HCO_3^- readily and precipitates as $CaCO_3$. Although Mg is not as easily precipitated as carbonate, it can accelerate the precipitation of $CaCO_3$ by replacing Ca on the exchange complex. As Ca and Mg are precipitated, the relative concentration of Na on the exchange phase increases, resulting in soil dispersion, if the total electrolyte concentration is below the critical flocculation concentration (CFC) of the soil clays. An empirical approach which has been widely used to predict the additional Na hazard associated with $CaCO_3$ precipitation involves calculation of the Residual Sodium Carbonate (RSC) (Eaton, 1950):

$$RSC = \left(CO_3^{2-} + HCO_3^-\right) - \left(Ca^{2+} + Mg^{2+}\right) \qquad [1.1]$$

where all concentrations are in $mmol_c\ L^{-1}$. Although the RSC is still commonly reported in irrigation water analyses, its use cannot be justified (Bower et al., 1968) because it does not quantify $CaCO_3$ precipitation in a manner which can be used for subsequent management decisions (Suarez, 1977; Busenberg and Plummer, 1989; Miyamoto and Pingitore, 1992; Suarez et al., 1992).

The relative abundance of various ions such as Ca^{2+}, SO_4^{2-} and HCO_3^- in irrigation water may affect the concentration of soluble salts in the soil solution due to precipitation or dissolution (common ion effect or changing P_{CO_2}). Expressions proposed to account for these effects include the effective salinity, which represents the total salt concentration of irrigation water minus the fraction expected to precipitate as $CaCO_3$, $MgCO_3$ or $CaSO_4 \cdot 2H_2O$.

In irrigation water, cations must balance anions to maintain electroneutrality. The relative abundance of various anions (such SO_4^{2-}, HCO_3^-) in irrigation water affects the eventual soluble salt concentration and the subsequent cation concentration ratio in the soil solution.

Many problems associated with irrigated agriculture arise from dissolved solute concentration and composition in the water applied. Since all waters contain varying concentrations and different species of solutes, considerable effort has been expended to classify the quality of water in terms of its dissolved solutes composition. The pH of irrigation water is usually not an accepted criterion of water quality because of the disparity in buffer capacity between water and soil.

Salinity is usually expressed as a lumped parameter, e.g., specific electrical conductance (EC) in dS m^{-1}, or total dissolved solids (TDS) in mg L^{-1}. No exact relationship exists between these two measures, but TDS my be approximated by multiplying EC (dS m^{-1}) by a factor varying from 640 in less saline to 800 in hypersaline water.

Generally, the most important water quality parameter for irrigation is the total salt concentration, most commonly measured as EC. The relation between EC (S m^{-1}) and electrolyte concentration (C) of a given salt ($mol_c\ m^{-3}$) is given by:

$$\wedge = EC / C \qquad [1.2]$$

where $\wedge$ is the equivalent conductance (S mol_c^{-1} m^2).

The equivalent conductance can be calculated as:

$$\wedge = \wedge_0 - bC^{1/2} \qquad [1.3]$$

where $\wedge_0$ is the equivalent conductance at infinite dilution for a given electrolyte and b is a constant whose value depends on the salt. For strong electrolytes, the assumption that b is independent of concentration is valid in the low concentration range. By rearranging Equation [1.2] and introducing Equation [1.3], EC is given by:

$$EC = C\wedge_0 - bC^{3/2} \qquad [1.4]$$

The relationships between EC (dS m^{-1}) and electrolyte concentration ($mmol_c$ L^{-1}) for various salt solutions are presented in Fig. 1.1. The EC of an aqueous electrolyte solution increases at a rate of approximately 2% C^{-1}. Although the factor for converting EC (dS m^{-1}) to C ($mmol_c$ L^{-1}) depends on the type of electrolyte (e.g., 7.9 for NaCl), a useful approximation is

$$EC\left(dSm^{-1}\right) \times 10 = C\left(mmol_c \ L^{-1}\right) \qquad [1.5]$$

when $bC^{3/2} << C$.

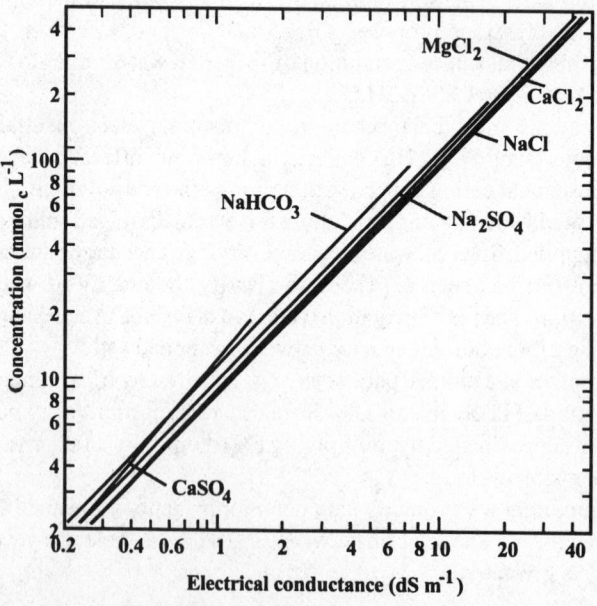

Fig. 1.1 Relationship between electrolyte concentration and specific electrical conductance of various electrolyte solutions at 25 °C [From US Salinity Laboratory Staff, 1954]

Usually, classification schemes to define salinity and sodicity hazards are based on broad generalizations regarding crops to be grown, soil properties, irrigation management, and climate. Initial schemes were developed using various expressions for salinity and sodicity criteria (US Salinity Laboratory Staff, 1954) but problems have been encountered, particularly as far as sodicity is concerned (Sumner and Naidu, 1998; Section G, Chapter 2). A classification based on specific use factors that takes salt tolerance of plants into account was proposed by Bernstein (1967) and Carter (1969). The general water classes in relation to their salt concentrations are presented in Table 1.1. This classification was based only on irrigation water quality and plant sensitivity to salinity.

The salinity of irrigation water can be assessed by relating irrigation water salinity (EC_{iw}), leaching fraction (LF), average root zone salinity and crop salt tolerance (Rhoades, 1982). The relationships between the average salinity in the root zone (EC_e) and EC_{iw} for each LF are presented in Fig. 1.2 in which the horizontal lines indicate crops that can be grown successfully without decreases in yield from salinity. For example, if EC_{iw} is 5 dS m^{-1} and LF is 0.2, only tolerant crops can be grown without decreasing yield. However, if LF is 0.4, moderately tolerant plants can be grown successfully. Thus, assessing the effects of salinity as a water quality parameter depends on the soil, the crop, available water, irrigation management, and the decrease in yield that can be tolerated.

In addition to the total salinity hazard of an irrigation water, the ionic composition must be also considered. A useful index for predicting excessive exchangeable Na is the sodium adsorption ratio (SAR), defined as:

$$SAR = \frac{(Na^+)}{(Ca^{2+})^{1/2}}; \quad 0 \leq SAR \leq \infty \tag{1.6}$$

The activity of Na and Ca ions is given in mmol L^{-1}.

Table 1.1 Water classes in relation to their salt concentration [After Carter, 1969]

Class of Water	Electrical Conductivity dS m^{-1} (25 °C)	Comments
Low salinity	0-0.4	These waters can be used for irrigating most crops grown on most soils with a low probability that salt problems will develop. Some leaching is required, but this generally occurs with normal irrigation practices
Moderate salinity	0.4-1.2	These waters can be used if a moderate amount of leaching occurs. Plants with moderate salt tolerance can be grown in most instances without special practices for salinity control. Producing field beans and potatoes with these waters is hazardous and requires special management practices.
High salinity	1.2-2.25	These waters should not be used on soils with restricted drainage. Plants tolerant to salinity should be grown. Excess water must be applied for leaching.
Very high salinity	2.25-5.0	These waters are suitable for irrigation under special circumstances. Adequate drainage is essential. Only very salt tolerant crops should be grown. Considerable excess water must be applied for leaching.

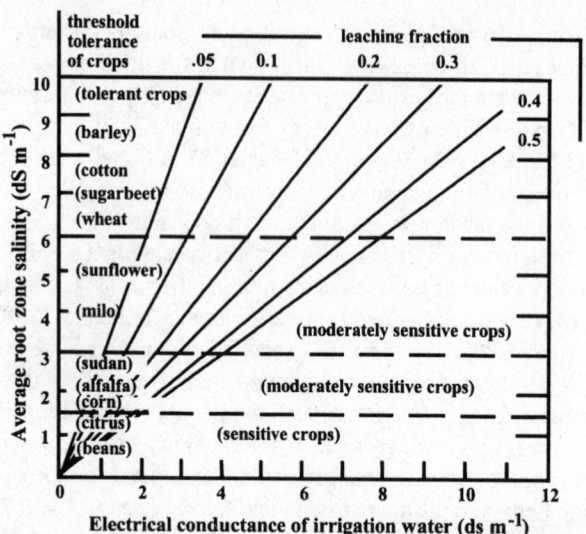

Fig. 1.2   Relationship between average root zone salinity (saturation extract basis), specific electrical conductance of irrigation water, and leaching fraction to use for conditions of conventional irrigation management [From Rhoades, 1982]

The US Salinity Laboratory (1954) assumed that Mg behaves similarly to Ca in the adsorbed phase and defined SAR = $(Na^+)/(Ca^{2+} + Mg^{2+})$

Sposito and Mattigod (1977) distinguished between the true SAR (SAR_t) which includes free ion activities and practical SAR (SAR_p), uncorrected for ion pairs, complexes, or activity coefficients. The relationship between the two parameters is given in Fig. 1.3. The SAR_p is generally less than the SAR_t because of the greater effects of ion pair, complex and activity coefficient corrections on Ca^{2+} and Mg^{2+} than on Na^+. The regression equation between the two parameters is

$$(SAR)_t = 0.08 + 1.115 \, (SAR)_p \qquad\qquad [1.7]$$

The ionic strength (I) of a solution is needed in order to calculate the ion activity coefficient. Calculation of I is complicated by the presence of ion pairs and complexes in soil solutions. However, values of I (mol L^{-1}) can be calculated with sufficient accuracy for most applications from the relation which includes corrections for ion pairs and complexes and is valid for most natural waters (Griffin and Jurinak, 1973):

$$I = 0.013 \, EC \qquad\qquad [1.8]$$

where EC is the specific electrical conductance (dS m^{-1}).

The relationship between ESP and SAR is given by the mass action equation:

$$ESP = \frac{(100K \, SAR)}{(1 + K \, SAR)} \qquad\qquad [1.9]$$

where K is the reaction coefficient (mmol L^{-1})$^{-1/2}$.

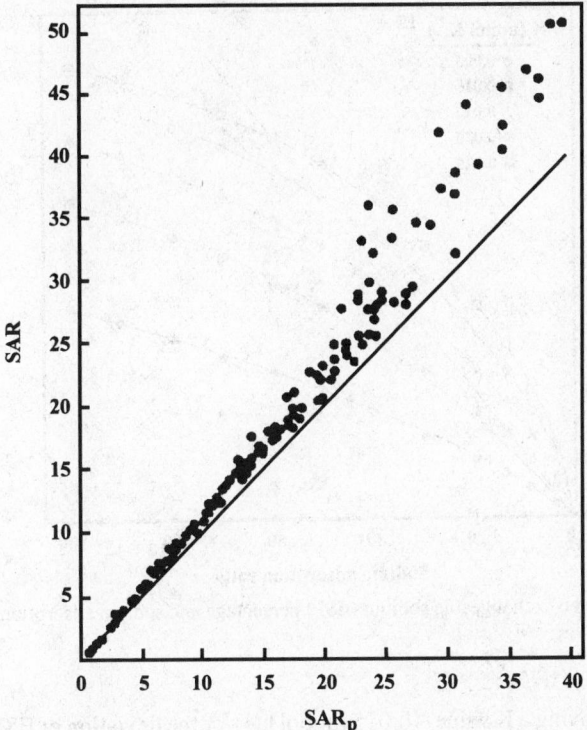

Fig. 1.3 The relationship between the sodium adsorption ratio and the practical sodium adsorption ratio for 161 soil solutions and water extracts.The straight line represents a 1:1 functional relationship between SAR and (SAR)p [Reprinted from Sposito and Mattigod, 1977. Soil Sci. Soc. Am. J. 41:310-315 with permission of the Soil Science Society of America]

When K SAR >>1, 1 can be neglected and

$$ESP \rightarrow 100 \qquad\qquad [1.10]$$

but when K SAR <<1, then K SAR can be neglected in the denominator and

$$ESP \rightarrow 100 \; K \; SAR \qquad\qquad [1.11]$$

Therefore, the Na^+ concentration of an irrigation water alone is not sufficient to estimate the potential sodicity hazard. From Equation [1.9], the effective diagnostic criteria for estimating the sodicity status of a given soil are the SAR of the irrigation water and the reaction coefficient (K) for Na-Ca exchange of the soil. For example, the relationship between ESP and SAR of a soil at equilibrium for various reaction coefficients is given in Fig. 1.4. These calculations indicate that the reaction coefficient (K) for Na-Ca exchange is an important criterion in evaluating water quality for irrigation for a given soil. It is important to note that for soils having $K > 0.01 \; (mmol \; L^{-1})^{-1/2}$, there is only one SAR value that is equal to ESP for each K (the linear line). Positive and negative deviations are observed when SAR is below or above this value, respectively. However, for many

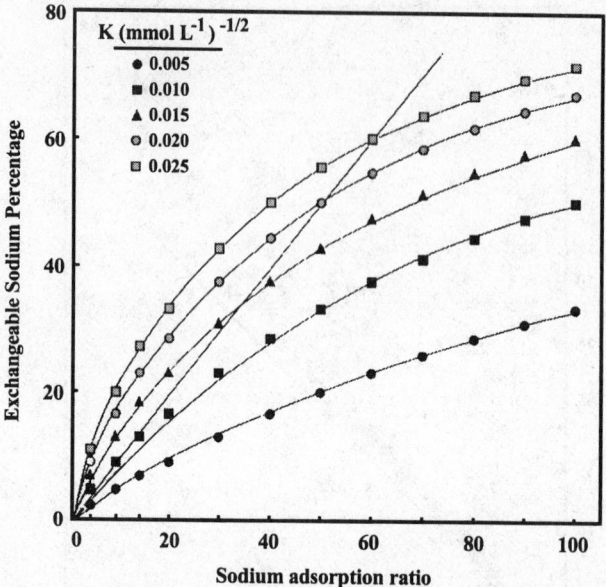

Fig. 1.4   Relationship between exchangeable sodium (ESP) percentage and sodium adsorption ratio (SAR) for various adsorption coefficients

montmorillonitic soils, having a K value ~ 0.015 (mmol L^{-1})$^{-1/2}$, the deviation of ESP from SAR of the soil solution is small up to SAR 30.

Since the SAR is usually determined on an irrigation water or saturation extract of a soil (by adding distilled water), there is an inherent problem of relating these values to the condition that exists in the field. By neglecting the effect of salt precipitation or dissolution, mineral weathering and the uptake of salts by plants, salt concentration in soil solution increases due to root water uptake and water evaporation from bare soil without changing the relative concentrations among the ionic species in soil solution. The SAR value, however, increases proportionally to the square root of the increase in the total concentration. Thus, if the total electrolyte concentration in solution increases by a factor of d, the SAR will increase by the factor of (d)$^{1/2}$. Since the soil ESP increases with the increase in SAR, it is obvious that the resultant ESP would be greater than the ESP expected from the SAR of the irrigation water. This trend has been observed in lysimeters and field experiments (Rhoades, 1968).

It is obvious, therefore, that in attempting to assess the sodicity hazard in soils, texture, clay mineralogy, SAR and the total electrolyte concentration of the percolating solution should be taken into consideration. Other important parameters needed for water quality assessment for irrigation are soil physical properties, pH, precipitation and dissolution processes of primary and secondary minerals and salts in soil, specific toxic ion concentrations in relation to the affinity of soil constituents to adsorb-desorb, as well as soil and irrigation management (e.g., presence of soil amendments such as gypsum, LF, irrigation methods). A comprehensive irrigation water classification scheme incorporating all these factors has not yet been developed.

1.3   Effect of Salinity on Soil Physical Properties

Considerable progress has been made in managing and controlling salinity and sodicity in irrigated lands. However, because of natural hydrological and geochemical factors, as well as irrigation-induced activities, soil and water salinity and associated drainage problems continue to plague agriculture.

The ionic composition and concentration of the soil solution affects soil physical properties. The accumulation of dispersive cations, such as Na (and sometimes Mg and K), promotes clay swelling and/or dispersion altering the geometry of soil pores which, in turn, affects intrinsic soil permeability, water retention and crop productivity. These effects are discussed in detail in Section G, Chapter 2. Suffice it to say that the adverse physical condition imparted to soils by high exchangeable Na percentages (ESP) on the exchange complex can be partially overcome by a sufficiently high concentration of salts.

The hydraulic conductivity (HC) of a soil can be maintained even at high ESP levels provided that the EC of the percolating solution is above a threshold value. Conversely, when irrigation water with very low EC is used, even an ESP = 5 caused a twofold decrease in HC (McIntyre, 1979). Thus, in arid and semiarid regions where irrigation with saline water is practiced and soils are saline and sodic, the deleterious effect of exchangeable Na is more evident during the irrigation with high quality water (or rain) than that during irrigation with saline water.

Infiltration rate (IR) also depends on EC and ESP of the soil (Agassi et al., 1981) but is more sensitive to increasing ESP than the HC, because the energy of impacting raindrops promotes dispersion allowing a seal to form at the surface (Oster and Schroer, 1979). The permeability of this seal decreases with decreasing EC at the soil surface. Thus, when low EC water is applied (rain or snow), salt concentration at the soil surface will be low even in soils containing weatherable minerals such as $CaCO_3$, and clay dispersion takes place. Consequently, sodicity damage to soil will not take place during the irrigation season when EC in the applied water is high enough to counter the dispersive effect of the exchangeable Na, but will occur during the rainy season (or irrigation with high quality water) when EC in the soil decreases to values which allow clay to disperse. Thus, the salinity of the applied water is an important parameter in assessing sodicity hazard (Section F, Chapter 2).

Successful irrigated agriculture, however, requires permanent control of salinity, sodicity and B levels in soil and irrigation water. Reclamation of affected soils by means of improved drainage, chemical treatments and leaching of salts (Keren and Miyamoto, 1990) can reduce natural hazards creating stable conditions for intensive cropping.

1.4 Effects of Salinity on Plants

1.4.1 Crop Tolerance to Salinity

When soluble salt accumulates in a soil to a level harmful to the growth of particular plants, the soil is termed saline. The composition and concentration of salts in the soil solution adversely influence plant growth through (1) osmotic effects, which limit the ability of plants to absorb water from the soil solution; (2) specific ion effects; and (3) changes in soil physical and chemical properties which can have long-term detrimental effects on crop production.

Threshold concentrations for classifying crop tolerance to salinity are (Maas, 1990) (1) sensitive (< 1.5 dS m^{-1}) (bean, clover, carrot, lettuce); (2) moderately sensitive (1.5–3.0 dS m^{-1}) (corn, alfalfa, broccoli, potato); (3) moderately tolerant (3–6 dS m^{-1}) (soybeans, wheat, squash); and (4) tolerant (6–10 dS m^{-1}) (barley, sorghum, sugarbeet) (1 dS m^{-1} ~ 350 mg L^{-1} Cl^-). Salinity refers to concentrations of soluble salts that are so high as to negatively impact plant growth. In the absence of specific ion effects, crop growth reduction due to salinity is related to the osmotic potential, y_o (bar) which is a function of EC_e (dS m^{-1}) at 25 °C as follows:

$$\psi_o = -0.39 \, EC_e \qquad\qquad\qquad\qquad\qquad\qquad\qquad [1.12]$$

This relationship is valid for EC_e in the range 3–30 dS m^{-1}. Because decreasing y_o (increasing absolute value) has the net effect of reducing water availability, plants growing on saline soils often appear to be suffering from drought.

Relative crop yield as a function of EC_e for various crops is given in Fig. 1.5 (Maas and Hoffman, 1977). Relative yield (Y) was calculated from the relation:

$$Y = 100 - B(EC_e - A) \qquad\qquad\qquad\qquad\qquad\qquad [1.13]$$

where EC_e is the electrical conductance of the soil saturation extract solution (dS m^{-1}), A is the salinity threshold (dS m^{-1}), and B is the percent yield decrease per unit salinity increase. Carter (1981) presented salt tolerance data for crops within the four classifications made by Maas and Hoffman (1977). The threshold salinity value and productivity decrease as a percentage of normal yield for each unit increase in EC_e have been summarized by Maas (1990). Such information is useful in selecting crops for growth under anticipated salinity conditions. However, the sensitivity of crops to soil salinity often changes from one stage of growth to the next.

Any assessment of water suitability for irrigation must be made in relation to crop tolerance to salinity, irrigation and soil management (LF, irrigation methods, soil properties, etc.) and yield decreases that can be tolerated. This can be shown (Fig. 1.2) by relating the EC_e in the root zone to EC_{iw} at a LF appropriate for conventional irrigation management (such as furrow or flood) which allows for considerable drying between applications (Rhoades, 1982). Because most plants are relatively salt tolerant during germination, but more sensitive during emergence and early growth, it is imperative to keep salinity in the seedbed low after germination.

Varietal differences in salt tolerance often occur. Rootstocks affect the salt tolerances of tree and vine crops because they regulate the uptake and translocation of potentially toxic ions, such as Na and Cl, to the shoots.

Salt tolerance depends on the method of irrigation and its frequency. As water becomes limiting, plants experience stresses from low matric and osmotic potentials. Available salt tolerance data are

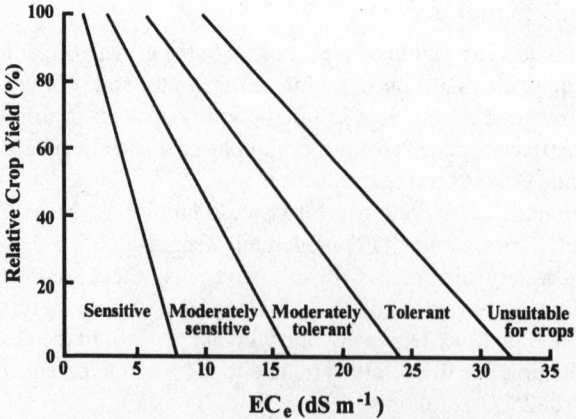

Fig. 1.5 Divisions for classifying crop tolerance to salinity [Reprinted from Maas and Hoffman, 1977. J. Irrig. Drain. Div. ASAE 103:115-134 with permission of the American Society of Civil Engineers]

most appropriate for crops irrigated by furrow or flood irrigation under conventional management. Because salt concentrations in irrigated profiles change constantly (up to severalfold), the plant is most responsive to salinity in that part of the root zone where maximum water uptake occurs.

Sprinkler irrigated crops are potentially subject to additional damage by foliar salt uptake and burn from spray contact on the foliage. Susceptibility to foliar salt injury depends on leaf characteristics and rate of absorption and does not correlate with general salt tolerance. Although injury caused by spray on foliage depends on weather and water stress, relatively little information is available for predicting yield losses. Increased frequency of sprinkling, temperature and evaporation lead to increased salt concentration on the leaves, and damage (Maas, 1986). Climate is a major factor affecting salt tolerance. Most crops can tolerate greater salt stress under cool humid than hot dry conditions. Yield is reduced more by salinity when humidity is low.

Substantial progress has been made in developing empirical models to relate crop yields and irrigation management under saline conditions. Childs and Hanks (1975) extended the production function model of Bresler and Hanks (1969) and Nimah and Hanks (1973b) to consider the effect of salinity and to compute yield. They reported a good correlation between computed and measured transpiration and predictions for crops with different rooting depths, different initial soil salinities and multiyear cropping. Computed and measured data correlated well when levels of salinity were low but yields were overestimated when salinity was high (Wolf, 1977).

An empirical relationship between yield and water potential (matric and osmotic) after converting salt concentration to osmotic potential is (van Genuchten and Hoffman, 1984):

$$Y = \frac{Y_m}{1 + (\Pi / \Pi_{50})^p} \qquad [1.14]$$

where Y is yield, Y_m is maximum yield under nonsaline conditions, Π and Π_{50} are the osmotic potential of the solution and the osmotic potential at which the yield is reduced by 50%, respectively, and p is an empirical constant. This model descibed salt tolerance data as well or better than the model of Maas and Hoffman (1977). Solomon (1985) and Letey et al. (1985) presented a seasonal water-salinity production function based on the response of crops to water and salts. Measured and computed yields for several crops correlated well (Letey et al., 1985; Letey and Dinar, 1986).

The dynamic models described above assume a unique relationship between yield and evapotranspiration (ET) for a given crop and climate that is independent, regardless of whether the water stress leading to the reduced ET is caused by deficit water supply or excess salinity. Letey and Knapp (1991) discussed the usefulness and limitations of these models. Crop-water production functions can be used to evaluate the losses from increased salinity in soil and water, the potential for reuse of saline drainage waters, the demand for water in irrigated agriculture and changes in irrigation and drainage policy.

1.4.2 Specific Molecule and Ion Effects on Plants

In addition to the osmotic and soil structure effects, some dissolved molecules and ions (B, Cl, Na, Al) can cause specific detrimental effects on crops. Toxicity occurs within the crop as the result of uptake and accumulation of these elements within plant tissue. Thus, an excess of these ions in irrigation water may be toxic to various plant physiological processes and may also cause nutritional disorders. In terms of saline soils, B is probably the most important. The effect of salinity on plant physiology and biochemistry is reviewed by Lauchli and Epstein (1990).

1.4.2.1   Boron-Soil Interaction

The main B species likely to occur in soils are $B(OH)_3$ and $B(OH)_4^-$. Boron can be specifically adsorbed by different clay minerals, oxyhydroxides of Al, Fe and Mg, and organic matter, which vary in their adsorption capacities (Keren and Bingham, 1985; Goldberg and Glaubig, 1985). The adsorption depends on the equilibrium B concentration in the soil solution and pH (Bingham et al. 1971; Mezuman and Keren, 1981; Goldberg and Glaubig, 1985). Increasing pH enhances B adsorption by soils, clays and Al and Fe oxides showing a maximum in the alkaline pH range. Boron adsorption isotherms for a given soil at various pH values are presented in Fig. 1.6 (Mezuman and Keren, 1981) as an example. Keren and Bingham (1985) have reviewed the mechanisms and factors that affect the adsorption and desorption of B by soil constituents.

When irrigation water containing B enters soil, B can be adsorbed or desorbed into soil solution, depending on the B concentration in the irrigation water and the level of native adsorbed B. The time required to reach a steady-state B concentration in the soil depends on the B concentration in relation to that before irrigation, the LF, and adsorption capacity of the soil. The time required to reach a steady state ranges from 3 to 150 years (Jame et al., 1982). Three years is adequate for a sandy soil exposed to 10 mg B L^{-1}, and 150 years is required for a clay loam soil at a B concentration of 0.1 mg L^{-1}. Even if irrigation water does not contain toxic concentrations of B, its retention by the soil can lead to accumulation approaching harmful levels in the long term if LF (including rainfall) is insufficient.

1.4.2.2   Boron Soil-Plant Relations

Boron has a marked effect on plants, with the optimum range in soil between deficiency and toxicity being very narrow. Boron deficiency is found primarily in humid regions or in sandy soils, while toxicity occurs most frequently in arid and semiarid regions due to high soil B or B additions in irrigation water.

The threshold B concentration for irrigation water (maximum permissible concentration for a given crop that does not reduce yield or lead to injury symptoms) ranged from as low as 0.3 for sensitive to 2.0 mg L^{-1} for tolerant crops (US Salinity Laboratory Staff, 1954). Yield decreases due to B toxicity can be evaluated from (Bingham et al., 1985):

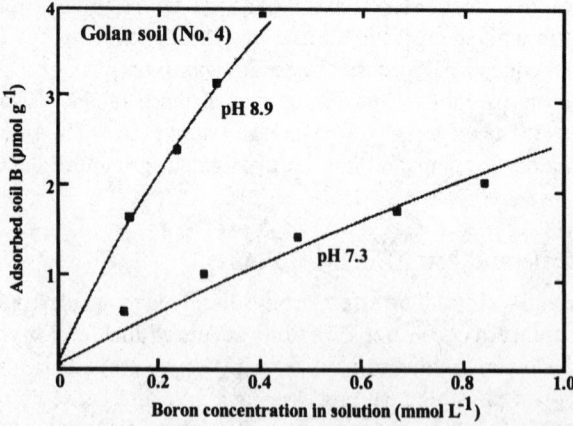

Fig. 1.6   Adsorbed amount of boron on clay soil as a function of the total boron solution concentration in 0.01N BaCl$_2$ solution and of pH [Reprinted from Mezuman and Keren, 1981. Soil Sci. Soc. Am. J. 45:722-726 with permission of the Soil Science Society of America]

$$Y = 100 - m(X - A) \qquad [1.15]$$

where Y is the relative yield (for X ³ A), m is the decrease in yield per unit increase in B concentration, X is the B concentration in the soil solution, and A is the maximum B concentration that does not reduce yield (threshold value). Boron threshold concentrations and the slopes of yield reduction as a function of B concentration for a limited number of crops are given in Table 1.2. A more extensive listing is given by Keren and Bingham (1985).

Boron uptake by plants increases with decreasing clay content for a given amount of B added to soil in irrigation or fertilization (Keren et al. 1985a, 1985b) The reason for this is that B absorption increases with increasing clay content as illustrated in Fig. 1.7 (Mezuman and Keren, 1981). However, when plant B content is plotted against B activity in the soil solution, a linear relationship is obtained (Fig. 1.8) indicating that plants obtain their B solely from that present in the soil solution. Therefore, soil physicochemical characteristics must be taken into consideration when assessing water quality for irrigation in terms of B. Because adsorption sites may act as a pool from which B is supplied to solution, adsorbed B may buffer fluctuations in solution B concentration, which may change insignificantly by changing the soil water content.

Hanks et al. (1983, 1984) carried out long-term research on the use of saline water ($EC_{iw} = 4$ dS m^{-1}, B = 9.4 mg L^{-1}) for irrigation of crops. Over 8 yr, no noticeable decrease in yield of forage crops was

Table 1.2 Boron tolerance limits for agricultural crops

Boron tolerance	Crop species common name	Threshold concentration (mg L^{-1})	Slope[*]
Very sensitive	Lemon	< 0.5	
Sensitive	Avocado	0.5-0.75	
	Grapefruit	0.5-0.75	
	Orange	0.5-0.75	
	Grape	0.5-0.75	
	Onion	0.5-0.75	
	Wheat	0.75-1.0	3.3
	Sunflower	0.75-1.0	
	Bean, Snap	1.0	12
Moderately sensitive	Broccoli	1.0	1.8
	Radish	1.0	1.4
	Potato	1.0-2.0	
	Lettuce	1.3	1.7
Moderately tolerant	Barley	3.4	4.4
	Corn	2.0-4.0	
	Cauliflower	4.0	1.9
Tolerant	Alfalfa	4.0-6.0	
	Sugar beet	4.9	4.1
	Tomato	5.7	3.4
Very tolerant	Cotton	6.0-10.0	
	Celery	9.8	3.2
	Asparagus	10.0-15.0	

[*] % yield reduction resulting from an increase of 1 mg B L^{-1} in the soil solution.

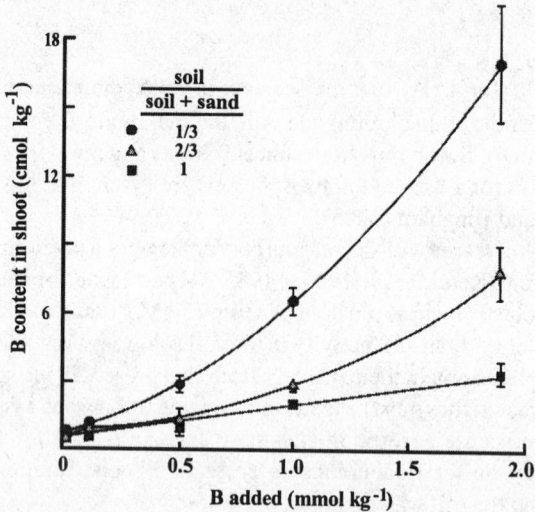

Fig. 1.7   Relationship between boron content in wheat shoots and the amount of boron added to soil, for three soil-sand mixtures [Reprinted from Keren et al. 1985. Soil Sci. Soc. Am. J. 49:1466-1470 with permission of the Soil Science Society of America]

observed but potato, maize, barley and wheat yields decreased after the third year mainly due to B. During the experiment, most of the B was retained in the soil with levels of $7-10 \, mg \, B \, L^{-1}$ in saturation extracts. These data demonstrate the need to consider also the composition of saline water when judging its suitability for irrigation. Boron concentrations in irrigation water should not exceed the values in Table 1.2 for the crops indicated. However, salinity likely reduces the effects of B on plant production. Interaction may occur which could increase the individual tolerance coefficient for B and salinity when a crop is exposed to both sources of stress at the same time (Ferreyra et al., 1997).

1.5   Reclamation of Saline and Boron-Affected Soils

1.5.1   Leaching for Salinity Control

Reclamation of saline soils is essentially a process where soil solution of high salt concentration is displaced by one less concentrated. Consequently, appropriate natural or installed drainage and disposal systems are essential. For flood and sprinkler irrigation the flow of water is downward, and leaching is defined with depth. Lateral flow and salt removal from the main root zone typical of drip irrigation can also be considered as temporary leaching. However, rainfall, flood or sprinkler irrigation must be used to occasionally leach out the salt from the entire rooting profile.

Leaching efficiency is the amount of salt removed from the root zone in drainage water at a given fraction of the irrigation water. This efficiency depends on salt content and distribution in the soil, solute composition, soil structure, and irrigation method and management.

Soil solution concentration and salt transport mechanisms are the main factors controlling leaching efficiency of noninteracting highly soluble salts. For interacting solutes and less soluble salts, chemical and exchange reactions are also important. For leaching of saline soils, Hoffman (1980) proposed the following empirical equation:

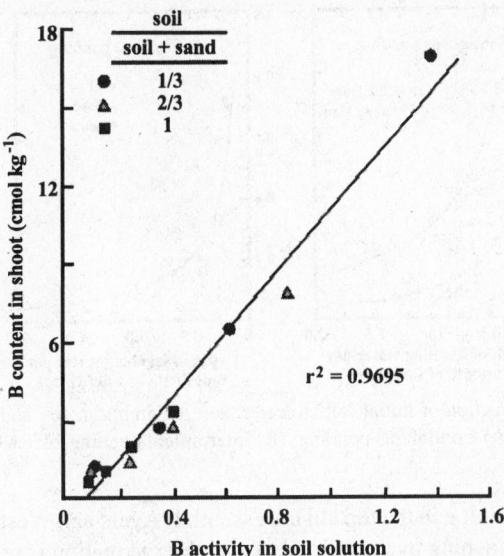

Fig. 1.8 Relationship between boron content in shoot and the boron activity in soil solution, for three ratios of soil-sand mixtures at field capacity water content [Reprinted from Keren et al. 1985. Soil Sci. Soc. Am. J. 49:1466-1470 with permission of the Soil Science Society of America]

$$\left(\frac{C_i}{C_0}\right)\left(\frac{D_1}{D_s}\right) = K \tag{1.16}$$

where C_i and C_0 are final and initial soil salinities, respectively; D_1 and D_s are depth of leaching water applied and depth of soil to be leached, respectively; and K is an empirical efficiency parameter that ranges from 0.1 (sandy loam) to 0.3 (clay). This equation is valid for $D_1/D_s > K$. Relationships between the fraction of initial salt concentration remaining in soil (C/C_0) and depth of leaching water applied per unit depth of soil (D_1/D_s) for continuous and intermittent ponding is presented in Fig. 1.9 (Hoffman, 1980). Under continuous ponding, more water is required for leaching the clay loam than sandy loam soil. However, under intermittent ponding the leaching efficiency of both soils was the same (K = 0.1). Intermittent ponding requires less water than continuous ponding to achieve the same degree of leaching (Miller et al., 1965) and sprinkler irrigation is more efficient than other methods at removing salt from small pores in the soil (Nielsen et al., 1966).

Effective irrigation and leaching require uniformity in water application, infiltration and soil water-holding capacity. Proper management that integrates irrigation methods and scheduling, soil treatments and special cultivation practices is aimed at improving infiltration uniformity. With flood or furrow irrigation, land leveling, increased watering rates, and soil treatments that improve infiltration are recommended practices. Improvement of infiltration uniformity is obtained by reducing application rates below the soil infiltration capacity, addition of ameliorating chemicals that control crust formation, and use of cultivation practices that increase surface storage and reduce local and long distance runoff, such as check dams in the furrow or rough surface disking.

Furrow or drip irrigation leaches salt below and away from the infiltrating areas but salt accumulates at lateral wetting fronts, most commonly in bed ridges, and at the soil surface between emitters. Special cultivation practices that carry salts away from the planting position or adjust

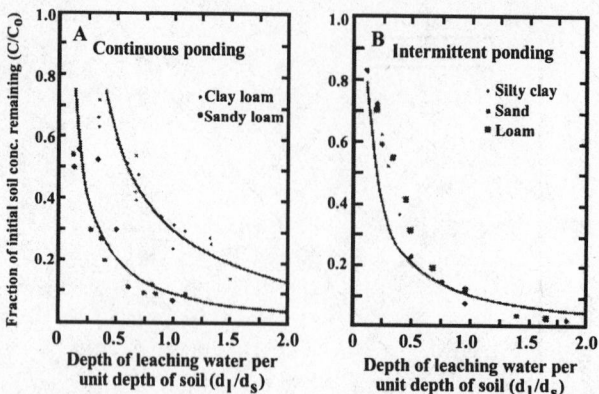

Fig. 1.9 Relationship between fraction of initial salt concentration remaining in soil and depth of leaching water applied for a given depth of soil. (A) Continuous ponding; (B) Intermittent ponding [After Hoffman, 1980]

planting position to salt distribution in the topsoil are essential (Ayers and Westcot, 1985). With both furrow and drip, occasional leaching by rain, flood or sprinkler irrigation is required to remove salt from the entire root zone.

When fields contain several soil types, leaching is more efficient in sandy than clay-rich portions. If leaching is continued until the clay-rich portions are sufficiently leached, an excessive amount of water would have percolated through the sandy areas. This problem is pronounced under ponded leaching where sandy areas are preferentially leached. In fact, the distribution of salinity in basin, border, or furrow irrigated fields coincides closely with soil type (Miyamoto and Cruz, 1986; 1987). To increase efficiency, leaching should be based on the distribution of soil types (Miyamoto and Cruz 1986) or the use of stochastic analysis of solute transport in spatially variable fields to divide a field into small sections that need different amounts of water (Russo, 1984). Dividing highly variable fields into areas of uniform soil type, with irrigation varying according to moisture-holding capacity and infiltration of each area, can significantly increase leaching efficiency.

When salts accumulate in the root zone during the irrigation season, they can be leached by applying water above the amount needed for ET. Over time, salt removal by leaching must equal or exceed salt additions from the applied water. To calculate the leaching requirement (LR), both the irrigation water quality and the crop tolerance to soil salinity must be known (Ayers and Westcot, 1985).

Large changes in soil salinity under nonsteady-state conditions arise from rainfall during and out of the irrigation season, changes in irrigation water quality for significant periods, and heavy intermittent leaching rather than a constant LF at every irrigation. If management can synchronize changes in soil salinity with crop tolerance at different growth stages, the use of brackish water may be increased without detriment. When rainfall can accomplish the required leaching without stressing the crop, additional leaching is not required. This may be the case with tolerant crops in humid climates, or during wet years in semiarid zones. With sensitive crops in humid climates and tolerant crops during dry years in semiarid and arid zones, additional leaching by irrigation may be required.

Leaching into tile or open ditch drains rapidly removes salt directly above and near the drains but removes little midway between the drains. Leaching efficiency decreases as soon as the rapid flow reaches the drains. This uneven leaching is more evident in water-saturated (Luthin et al., 1969) than unsaturated fields, where leaching above the watertable is a one-dimensional process (Talsma, 1967). To leach more efficiently, it is necessary to avoid ponding water near the drains (Miyamoto and

Warrick, 1974) and apply water intermittently to keep the watertable as low as possible (Talsma, 1967).

The high salt-water dilution technique (Reeve and Bower, 1960) which involves successive dilutions of a very saline irrigation water containing divalent cations, can be used when (1) the soil physical conditions have deteriorated and HC of the soil is so low that the time required for reclamation or the amount of amendment required is excessive, or (2) if a sodic soil is to be leached with a water so low in salinity that water infiltration decreases adversely. A nomograph is available for predicting reclamation at various stages of leaching (Reeve and Doering, 1966a). In the early phase, the high salinity of the water prevents clay dispersion and promotes flocculation, and the Ca content provides a source of Ca to replace exchangeable Na so that sodicity is decreased. On dilution with high quality water, the SAR of the irrigation water is reduced by the square root of the dilution factor. To ensure reclamation to the desired depth, D_i/D_s should be about 9 (Reeve and Doering, 1966a). This method is particularly effective for soils with swelling clays that have extremely low hydraulic conductivity (HC). A theoretical analysis of the use of high salt water in reclaiming sodic soils by the addition of a constant quantity of Ca^{2+} in every step in successive dilutions is presented by Misopolinos (1985).

In the reclamation of a highly sodic clay loam soil, Reeve and Doering (1966b) found that infiltration rate (IR) for a saturation gypsum solution was only 2×10^{-8} m s^{-1} and the time for reclamation to 0.9 m was 7 yr of continuous leaching. However, the profile was reclaimed in 3 days with a series of $CaCl_2$ solutions (300–6 mol m^{-3}). The average IR for the $CaCl_2$ solutions was about 150 times that for the gypsum solution. This method has also been shown to reclaim a slowly permeable sodic soil in a humid environment, where HC and IR were increased from 30 to over 100% (Rahman et al., 1974).

The successful application of this method requires a balance between maintaining high EC to reduce reclamation time and low EC to reduce the amount of amendment required. A practical technique is to apply only two-thirds of the solution depth required for exchange equilibria with three dilutions of $CaCl_2$ followed by a gypsum application to satisfy the remaining exchange requirement, and completed by leaching with 0.3 m water m^{-1} soil to be reclaimed. This final step is essential to leach the saline solutions below the root zone.

Leaching reclamation without amendments (Jury et al., 1979; 1987) can be successful when soil drainage is good, adequate leaching water is available and an internal source of Ca exists (e.g., $CaCO_3$). The rate at which $CaCO_3$ dissolves in water depends on the surface area-solution volume ratio, ionic composition of the solution and solid phases, clay affinity for cations, temperature and P_{CO_2}. Amrhein et al. (1985) concluded that transfer of atmospheric CO_2 to solution is an important rate-limiting step in the dissolution kinetics of $CaCO_3$ in soils. The ion activity product (IAP) values expected for calcite (Suarez and Rhoades, 1982) are obtained for $CaCO_3$ particles, isolated from the soils, and in water. On the contrary, a supersaturation with respect to calcite which was observed in solution extracts from calcareous soils (Levy, 1981) appears to be due to the presence of silicates more soluble than calcite, and is not the result of unstable $CaCO_3$ phases (Suarez and Rhoades, 1982). Plummer et al. (1979) concluded that there are several uncertainties in modeling the kinetics of carbonate chemistry; at low pH, the rate depends significantly on the thermodynamic transport constant for H^+ which is not well defined, while reaction site density and controls on pH at the interface between the crystal and the bulk solution are also not well understood.

Soil $CaCO_3$ may dissolve slowly to contribute Ca, especially in the reclamation of saline sodic soils in which its solubility is enhanced (Oster, 1982). Sodic soils commonly contain $CaCO_3$ which dissolves and maintains the soil solution concentration levels above the flocculation value of the soil clays (Alperovitch et al., 1981). This, in turn, reduces ESP and prevents HC from decreasing when exposed to rain. On the other hand, addition of $CaCO_3$ to nonsaline sodic soils is of doubtful value because its dissolution rate is too slow to provide sufficient Ca for exchange, unless an acid or acid former is applied concurrently. Calcareous soils with moderate ESP levels maintain reasonable physical properties through most of the profile but remain susceptible to dispersion near the surface because EC near the surface may be insufficient to maintain physical structure during raindrop impact (Keren, 1991). Under such conditions, an application of a soil amendment (gypsum) on soil surface is necessary to keep IR sufficiently high (Keren and Shainberg, 1981).

1.5.2 Reclamation of Boron-Affected Soils

The presence of excess soluble B in many arid soils is usually attributed to the application of B-containing irrigation waters or the weathering of B-containing materials. Adsorption/desorption of B controls its removal from soils. Griffin and Burau (1974) showed that desorption (Fig. 1.10) was largely from relatively fast reactions on hydroxy Fe, Mg, and Al materials in the clay fraction which are independent of soil texture and initial soil B content. The slowest reaction rate was due to B diffusion from the interior of clay minerals to the solution phase. They did not consider the role of clay minerals and primary B minerals in its desorption and dissolution processes.

If excessive amounts of B accumulate in soils, reclamation can be accomplished by extensive leaching. Leaching experiments (Bingham et al., 1972; Reeve et al., 1955; Rhoades et al., 1970b) show that a large fraction of soil B can be removed by percolating waters but that the remainder persists even after large amounts of water have been applied. The volume of low B water needed to reduce it from toxic to nontoxic levels is two- to threefold greater than is needed for a comparable reduction in Cl (Reeve et al., 1955; Bingham et al., 1972).

The relative decrease in soluble B in field soils during reclamation is given by (Hoffman, 1980):

$$\left(\frac{C}{C_o} \right)\left(\frac{D}{D_s} \right) = 0.6 \tag{1.17}$$

where C and C_o are the final and initial soluble B concentrations and D/D_s is the depth of leaching water per unit depth of soil. Equation [1.17] is independent of the method of water application (sprinkler or ponding).

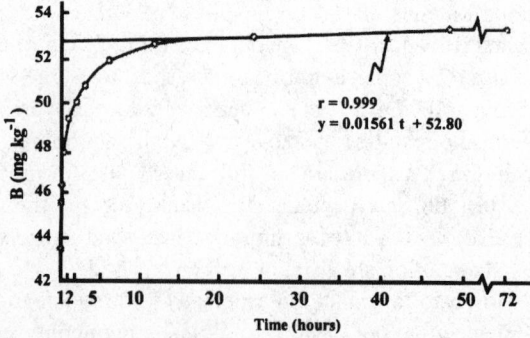

Fig. 1.10 Solution boron concentration in leachate from soil, naturally high in boron, as a function of time [Reprinted from Griffin and Burau, 1974. Soil Sci. Soc. Am. J. 38:892-897 with permission of the Soil Science Society of America]

Since extensive leaching is required for reclaiming soils with excessive B, the reclamation should be pursued while tolerant crops are cultivated; and then when B levels have been reduced, more sensitive crops can be cultivated. Alternatively, high B soils can be reclaimed by adding amendments such as H_2SO_4. Because B adsorption is pH dependent (Keren and Bingham, 1985), acidification effectively increases the B concentration in soil solution thereby making leaching more effective (Prather, 1977). Furthermore, as Si concentration increases with decreasing pH to levels higher than those of B, Si can compete for adsorption sites (Bingham and Page, 1971; McPhail et al., 1972). During leaching, the presence of Si may decrease readsorption of B or even cause B desorption. Applying ~ 3 Mg H_2SO_4 ha^{-1} reduced the amount of water needed to leach the same amount of B fivefold (Prather, 1977).

Lime-induced B deficiency has frequently been observed in acid soils, and is probably due to increased B adsorption resulting from the increased pH and not a direct effect of Ca *per se* (Hatcher and Bower, 1967). In acid soils, increasing rates of B lowered yield and caused increasing typical B toxicity symptoms (Bartlett and Picarelli, 1973). However, in the presence of lime, B toxicity was prevented as soil pH increased above 6.

Native soil B is more difficult to leach than B accumulated from previous irrigations. The reduced B concentrations that follow the leaching of native high B soils may be temporary. Increasing storage times of water in the soil profile results in larger increases in effluent B concentrations (Rhoades et al. 1970b; Bingham et al. 1972). Peryea et al. (1985) showed that this phenomenon, termed B regeneration, is inversely related to the amount of water used for the initial leaching. Boron regeneration is, therefore, of primary concern during the early stages of soil reclamation, when appreciable residual sources of regenerable B are present. The potential for reestablishing phytotoxic concentrations in a soil is a function of the relative completeness of the reclamation processes and the rate of B dissolution after reclamation.

1.6 References

Agassi, M., I. Shainberg, and J. Morin. 1981. Effect of electrolyte concentration and soil sodicity on infiltration rate and crust formation. Soil Sci. Soc. Am. J. 45:848–851.

Alperovitch, N., I. Shainberg, and R. Keren. 1981. Specific effect of magnesium on the hydraulic conductivity of sodic soils. J. Soil Sci. 32:543–554.

Amrhein, C., J.J. Jurinak, and W.M. Moore. 1985. Kinetics of calcite dissolution as affected by carbon dioxide partial pressure, Soil Sci. Soc. Am. J. 49:1393–1398.

Amrhein, C., and D.L. Suarez. 1987. Calcite supersaturation in soils as a result of organic matter mineralization. Soil Sci. Am. J. 51:932–937.

Ayers, R.S., and D.W. Westcot. 1985. Water quality for agriculture. FAO Irrig. Drain. Pap. 29.

Bartlett, R.J., and C.J. Picarelli. 1973. Availability of boron and phosphorus as affected by liming an acid potato soil. Soil Sci. 116:77–83.

Bernstein, L. 1967. Quantitative assessment of irrigation water quality. p. 51-64. *In* Water quality criteria. American Society for Testing Materials Spec. Pub. 416.

Bingham, F.T., and A.L. Page. 1971. Specific character of boron adsorption by an amorphous soil. Soil Sci. Soc. Am. Proc. 35:892–893.

Bingham, F.T., A.L. Page, N.T. Coleman, and K. Flach. 1971. Boron adsorption characteristics of selected amorphous soils from Mexico and Hawaii. Soil Sci. Soc Am. J. 35:546–550.

Bingham, F.T., A.W. Marsh, R. Branson, R. Mahler, and G. Ferry. 1972. Reclamation of salt affected high boron soils in western Kern County. Hilgardia 41:195–211.

Bingham, F.T., J.E. Strong, J.D Rhoades, and R. Keren. 1985. An application of the Maas-Hoffman salinity response model for boron toxicity. Soil Sci. Soc. Am. J. 49:672–674.

Boeseken, J. 1949. The use of boric acid for the determination of the configuration of carbohydrates. Adv. Carbohyd. Chem. 4:189–210.

Bower, C.A., G. Ogata, and J.M. Tucker. 1968. Sodium hazard of irrigation water as influenced by leaching fraction and by precipitation or solution of calcium carbonate. Soil Sci. 106:24–34.

Bresler, E., and R.J. Hanks.1969. Numerical method of estimating simultaneous flow of water and salt in unsaturated soils. Soil Sci. Soc. Am. J. 33:827–832.

Bresler, E., B.L. McNeal, and D.L. Carter. 1982. Saline and sodic soils. Springer-Verlag, New York, NY.

Busenberg, E., and L.N. Plummer. 1989. Thermodynamics of magnesian calcite solid solution at 25 °C and 1 atm total pressure. Geochim. Cosmochim. Acta. 53:1189–1208.

Byrne, R.J., Jr., and D.R. Kester. 1974. Inorganic speciation of boron in seawater. J. Mar. Res. 32:119–127.

Carter, D.L. 1969. Managing moderately saline (salty) irrigation waters. Univ. ID Curr. Inf. Ser. 107.

Childs, E.C., and R.J. Hanks. 1975. Model for soil salinity effects on crop growth. Soil Sci. Soc. Am. Proc. 39:617–622.

Craig, J.R. 1970. Saline water: Genesis and relationship to sediments and host rocks. In R.B. Mattox (ed.) Saline Water. 46th Annual Meeting, American Association for the Advancment of Science, Las Vegas, NV.

Dyrssen, D., and I. Hansson. 1973. Ionic medium effects in seawater. Comparison of acidity constants of carbonic acid in sodium chloride and synthetic seawater. Mar. Chem. 1:137–149.

Eaton, F.M. 1950. Significance of carbonates in irrigation water. Soil Sci. 69:123–133.

Edwards, J.O., G.C. Morrison, V.F. Ross, and J.W. Schultz. 1955. The structure of the aqueous borate ion. J. Am. Chem. Soc. 77:266–268.

Ferreyra, R.E., A.U. Aljaro, R.S. Ruiz, L.P. Rojas, and J.D. Oster. 1997. Behavior of 42 crop species grown in saline soils with high boron concentration. Agri. Water Manag. 34:111–124.

Forsyth, W.G.C. 1950. Studies on the more soluble complexes of soil organic matter. 2. The composition of the soluble polysaccharide fraction. Biochem. J. 46:141–146.

Goldberg, S., and R.A. Glaubig. 1985. Boron adsorption on aluminum and iron oxide minerals. Soil Sci. Soc. Am. J. 49:1374–1379.

Good, C.C.D., and D.M. Ritter. 1962. Alkenylboranes: II. Improved preparative methods and new observations on methylvinylboranes. J. Am. Chem. Soc. 84:1162–1166.

Griffin, R.A., and J.J. Jurinak. 1973. Estimation of activity coefficients from the electrical conductivity of natural aquatic systems and soil extracts. Soil Sci. 116:26–30.

Griffin, R.A., and R.G. Burau. 1974. Kinetic and equilibrium studies of boron desorption from soil. Soil Sci. Soc. Am. J. 38:892–897.

Gu, B., and L.E. Lowe. 1990. Studies on the adsorption of boron on humic acids. Can. J. Soil Sci. 70:305–311.

Hanks, R.J., R.F. Neilson, R.L. Cartee, L.S. Willardson, R.B. Sorenson, A.R. Mitchell, and U. Shani. 1983. Use of saline waste water from electric power plants for irrigation. UT State Univ. Res. Rep. 103.

Hanks, R.J., R.F. Neilson, R.L. Cartee, and L.S. Willardson. 1984. Use of saline waste water from electric power plants for irrigation. p. 473–492. In R.H. French (ed.) Salinity in watercourses and reservoirs. Butterworth Publishers, Boston, MA

Hatcher, J.T., and C.A. Bower. 1967. Adsorption of boron by soils as influenced by hydroxy aluminum and surface area. Soil Sci. 104:422–426.

Hingston, F.J., A.M. Posner, and J.P. Quirk. 1974. Anion adsorption by goethite and gibbsite. II. Desorption of anions from hydrous oxide surfaces. J. Soil Sci. 25:16–26.

Hoffman, G.J. 1980. Guideline for the reclamation of salt-affected soils. p. 49–64. In G.A. O'Connor (ed.) 2nd . Inter-Am. Conf. Salin. Water Manag. Tech. NM State University, Las Cruces, NM.

Huettl, P.J.V. 1976. The pH dependent sorption of boron by soil organic matter. M.S. Thesis, University of Wisconsin, Madison, WI.

Ingri, N. 1963. Equilibrium studies of the polyanions containing B^{III}, Si^{IV}, Ge^{IV} and V^V. Svensk. Kem. Tidskr. 75:199–230.

Ingri, N.G., G. Lagerstorm, M. Frydman, and L.G. Sillen. 1957. Equilibrium studies of polyanions. II. Polyborates in $NaClO_4$ medium. Acta Chem. Scand. 11:1034–1058.

Jame, Y.W., W. Hicholaichuk, A.J. Leyshon, and C.A. Campbell. 1982. Boron concentration in the soil solution under irrigation: A theoretical analysis. Can. J. Soil Sci. 62:461–471.

Jury, W.A.,W.M. Jarrell, and D. Devitt. 1979. Reclamation of saline-sodic soils by leaching. Soil Sci. Soc. Am. J. 43:1100–1106.

Jury, W.A., W.M.Jarrell, and D. Devitt. 1987. Reclamation of saline-sodic soils by leaching, Soil Sci. Soc. Am. J. 51:1092.

Keren, R. 1991. Specific effect of magnesium on soil erosion and water infiltrations. Soil Sci. Soc. Am. J. 55:783–787.

Keren, R., R.G. Gast, and B. Bar-Yosef. 1981. pH-dependent boron adsorption by Na-montmorillonite. Soil Sci. Soc. Am. J. 45:45–48.

Keren, R., and F.T. Bingham. 1985. Boron in water, soils and plants. Adv. Soil Sci. 1:229–276.

Keren, R., and I. Shainberg. 1981. Effect of dissolution rate on the efficiency of industrial and mined gypsum in improving infiltration of a sodic soil. Soil Sci. Soc. Am. J. 45:103–107.

Keren, R., F.T. Bingham, and J.D. Rhoades. 1985a. Plant uptake of boron as affected by boron distribution between liquid and solid phases in soil. Soil Sci. Soc. Am. J. 49:297–302.

Keren, R., F.T. Bingham, and J.D. Rhodes 1985b. Effect of clay content in soil on boron uptake and yield of wheat. Soil Sci. Soc. Am. J. 49:1466–1470.

Keren, R., and R.G. Gast. 1983. pH-dependent boron adsorption by montmorillonite hydroxy-aluminum complexes. Soil Sci. Soc. Am. J. 47:1116–1121.

Keren, R., and U. Mezuman. 1981. Boron adsorption by clay minerals using a phenomenological equation. Clays Clay Min. 29:198–204.

Keren, R., and S. Miyamoto. 1990. Reclamation of saline, sodic and boron-affected soils. p. 410–431. *In* K.K. Tanji (ed.) Agricultral salinity assessment and management. American Society of Civil Engineers, New York, NY.

Keren, R., and G.A. O'Connor. 1982. Effect of excangeable ions and ionic strength on boron adsorption by montmorillonite and illite. Clays Clay Min. 30:341–346.

Kittrick, J.A. 1977. Mineral equilibria and the soil system. p. 1–25. *In* J.B. Dixon and S.B. Weed (ed.) Minerals in soil environments. Soil Science Society of America, Madison, WI.

Konopik, N., and O. Leberl. 1949. Colorimetric determination of pH in the range of 10 to 15. Monatsh. 80:420–429.

Lappert, M.F. 1956. Organic compounds of boron. Chem. Rev. 56:959–1064.

Lauchli, A., and E. Epstein. 1990. Plant responses to saline and sodic conditions. p. 113–137. *In* K.K. Tanji (ed.) Agricultural salinity assessment and management. American Society of Civil Engineers, New York, NY.

Letey, J., and A. Dinar. 1986. Simulated crop-water production functions for several crops when irrigated with saline waters. Hilgardia. 54:1–32.

Letey, J., A. Dinar, and K. Knapp. 1985. Crop-water production function model for saline irrigated waters. Soil Sci. Soc. Am. J. 49:1005–1009.

Letey, J., and K. Knapp. 1991. Crop-water production functions under saline conditions. *In* K.K. Tanji (ed) Agricultural salinity assessment and management. American Society of Civil Engineers, New York, NY.

Levy, D.B., C. Amrhein, M.A. Anderson, and A.M. Daoud. 1995. Coprecipitation of sodium, magnesium and silicon with calcite. Soil Sci. Soc. Am. J. 59:1258–1267.

Levy, R. 1981. Effect of dissolution of alumosilicates and carbonates on ionic activity products of calcium carbonate in soil extracts, Soil Sci. Soc. Am. J. 45:250–255.

Lindsay, W.L. 1979. Chemical equilibria in soils. John Wiley and Sons, New York, NY.

Loomis, W.D., and R.W. Durst. 1992. Chemistry and biology of boron. BioFactors 3:229–239.

Luthin, J.N., P. Fernandez, J. Woerner, and F. Robinson. 1969. Displacement front under ponding leaching. J. Irrig. Drain. Div. ASCE 95:117–125.

Maas, E.V. 1986. Salt tolerance of plants. App. Agric. Res. 1:12–26.

Maas, E.V., 1990. Crop salt tolerance. p. 262–304. *In* K.K. Tanji (ed.) Agricultural salinity assessment and management. American Society of Civil Engineers, New York, NY.

Maas, E.V., and G.J. Hoffman. 1977. Crop salt tolerance-Current assessment. J. Irrig. Drain. Div. ASCE 103:115–134.

Massoud, F.I. 1981. Salt affected soils at global scale and concepts for control. FAO, Rome, Italy.

McIntyre, D.S. 1979. Exchangeable sodium, subplasticity and hydraulic conductivity of some Australian soils. Aust. J. Soil Res. 17:115–120.

McPhail, M., A.L. Page, and F.T. Bingham. 1972. Adsorption interactions of monosilicic and boric acid on hydrous oxides of iron and aluminum. Soil Sci. Soc. Am. Proc. 36:510–514.

Mesmer, R.E., C.F. Baes, Jr., and F.H. Sweeton. 1972. Acidity measurments at elevated temperature. VI. Boric acid equilibria. Inorg. Chem. 11:537–543.

Mezuman, U., and R. Keren. 1981. Boron adsorption by soils using a phenomenological adsorption equation. Soil Sci. Soc. Am. J. 45:722–726.

Miller, R.J., J.W. Biggar, and D.R. Nielsen. 1965. Chloride displacement in Panoche clay loam in relation to water movement and distribution, Water Resour. Res., 1:63–67.

Misopolinos, N.D. 1985. A new concept for reclaiming sodic soils with high-saltwater, Soil Sci. 140:69–74.

Miyamoto, S., and I. Cruz. 1986. Spatial variability and soil sampling for salinity and sodicity apprasial in surface-irrigated orchards. Soil Sci. Soc. Am. J. 50:1020–1025.

Miyamoto, S., and I. Cruz. 1987. Spatial variability of soil salinity in furrow-irrigated torrifluvents. Soil Sci. Soc. Am. J. 51:1019–1025.

Miyamoto, S., and N.E. Pingitore. 1992. Predicting Ca and Mg precipitation in saline solutions following evaporation. Soil Sci. Soc. Am. J. 56:1767–1775.

Miyamoto, S., and A.W. Warrick. 1974. Salt displacement into drain tiles under ponded leaching. Water Resour. Res. 10:275–278.

Nielsen, D.R., J.W. Biggar, and J.N. Luthin. 1966. Desalinization of soils under controlled unsaturated flow conditions. 6th Congr. Int. Comm. Irrig. Drain. 19.15–19.24.

Nimah, M.N., and R.J. Hanks. 1973a. model for estimating soil, water, plant and atmospheric interrelations: I. Description and sensitivity. Soil Sci. Soc. Am. J. 37:522–527.

Nimah, M.N., and R.J. Hanks. 1973b. Model for estimating soil, water, plant and atmospheric interrilations: II. Field test of model. Soil Sci. Soc. Am. J. 37:528–532.

Onak, T.P., H. Landesman, R.E. Williams, and I. Shapiro. 1959. The B11 nuclear magnetic resonance chemical shifts and spin coupling values for various compounds. J. Phys. Chem. 63:1533–1535.

Oster, J.D. 1982. Gypsum usage in irrigated agriculture: A review. Fert. Res. 3:73–89.

Oster, J.D., L.S. Willardson, and G.J. Hoffman. 1972. Sprinkling and ponding techniques for reclaiming saline soils. Trans. ASAE 15:1115–1117.

Oster, J.D., and F.W. Schroer. 1979. Infiltration as influenced by irrigation water quality. Soil Sci. Soc. Am. J. 43:444–447.

Owen, B.B. 1934. The dissociation constant of boric acid from 10 to 50 °C. J. Am. Chem. Soc. 56:1695–1697.

Parks, G.A. 1965. The isoelectric points of solid oxides, solid hydroxydes, and aqueous hydroxo complex systems. Chem. Rev. 65:177–198.

Peryea, F.T., F.T. Bingham, and J.D. Rhoades. 1985. Regeneration of soluble boron by reclaiming high boron soils. Soil Sci. Soc. Am. J. 42:782–786.

Plummer, L.N., D.L. Parkhurst, and T.M.L. Wigley. 1979. Critical review of the kinetics of calcite dissolution and precipitation. Am. Chem. Soc. Symp. Ser. 93:537–573.

Prather, R.J. 1977. Sulfuric acid as an amendment for reclaiming soils high in boron. Soil Sci. Soc. Am. J. 41:1098–1101.

Rahman, M.A., E.A. Hiler, and J.R. Runkles. 1974. High electrolyte water for reclaiming slowly permeable soils. Trans. ASAE 17:129–133.

Reardon, E.J. 1976. Dissociation constants for alkali earth and sodium borate ion pairs from 10 to 50 °C. Chem. Geol. 18:309–325.

Reeve, R.C., and C.A. Bower. 1960. Use of high-salt waters as a flocculant and source of divalent cations for reclaiming sodic soils. Soil Sci. 90:139–144.

Reeve, R.C., and E.J. Doering. 1960. The high salt-water dilution method for reclaiming sodic soils. Soil Sci. Soc. Am. Proc. 39:498–504.

Reeve, R.C., and E.J. Doering. 1966. Field comparison of the high salt-water dilution method and conventional methods for reclaiming sodic soils, 6th Int. Comm. Irrig. Drain. 19.1–19.14.

Reeve, R.C., A.F. Pillsbury, and L.V. Wilcox. 1955. Reclamation of a saline and high boron soil in the Coachella Valley of California. Hilgardia 24:69–91.

Rhoades, J.D. 1968. Leaching requirement for exchangeable sodium control. Soil Sci. Soc. Am. Proc. 32:652–656.

Rhoades, J.D. 1982. Reclamation and management of salt-affected soils after drainage. p. 123–197. In Proc. 1st Ann. West. Prov. Conf. Rational. Water Soil Res. Manag. Lethbridge, Canada.

Rhoades, J.D., R.D. Ingvalson, and J.T. Hatcher. 1970b. Laboratory determination of leachable soil boron. Soil Sci. Soc. Am. Proc. 34:871–875.

Rhoades, J.D., D.B. Krueger, and M.J. Reed. 1968. The effect of soil-mineral weathering on the sodium hazard of irrigation waters. Soils Sci. Soc. Am. J. 32:643–647.

Russo, D. 1984. Satial variability considerations in salinity management. p. 198–219. In I. Shainberg and Y. Shalhevet (ed.) Soil salinity under irrigation. Springer Verlag, Berlin, Germany.

Servoss, R.R., and H.M. Clark. 1957. Vibrational spectra of normal and isotopically labeled boric acid. J. Chem. Phys. 26:1175–1178.

Sims, J.R., and F.T. Bingham. 1968a. Retention of boron by layer silicates, sesquioxides and soil materials: II. Sesquioxides. Soil Sci. Soc. Am. Proc. 32:364–369.

Sims, J.R., and F.T. Bingham. 1968b. Retention of boron by layer silicates sesquioxides and soil materials: III. Iron- and aluminum coated layer silicates and soil materials. Soil. Sci. Soc. Am. Proc. 32:369–373.

Smith, G.J. 1979. Subsurface stratigraphy and geochemistry of late Quaternary evaporites, Searles lake, California. USGS Prof. Pap. 1043.

Solomon, K.H. 1985. Water-salinity-production functions. Trans. ASAE. 28:1975–1980.

Sposito, G. and S.W. Mattigod. 1977. On the chemical formation of the sodium adsorption ratio. Soil Sci. Soc. Am. J. 41:323–329.

Suarez, D.L. 1977. Ion activity products of calcium carbonate in waters below the root zone. Soil Sci. Soc. Am. J. 41:310–315.

Suarez, D.L., and J.D. Rhoades. 1982. The apparent solubility of calcium carbonate in soils, Soil Sci.Soc. Am. J. 46:716–722.

Suarez, D.L., J.D. Wood and I. Abrahim. 1992. Reevaluation of calcite supersaturation in soils. Soil Sci. Soc. Am. J. 56:1776–1784.

Sumner, M.E., and R. Naidu. 1998. Sodic soils: Distribution, properties, management and environmental consequences. Oxford University Press, New York, NY.

Szabolcs, I. 1989. Salt affected soils. CRC Press, Boca Raton, FL.

Talsma, T. 1967. Leaching of tile-drained saline soil. Aust. J.Soil Res. 5:37–46

US Salinity Laboratory Staff, 1954. Diagnosis and improvement of saline and alkali soils. USDA Handb. 60, US Government printing Office, Washington, DC.

van Genuchten, M.Th., and G.J. Hoffman. 1984. Analysis of crop salt tolerance data. p. 258–271. *In* I. Shainberg and J. Shalhevet (eds) Salinity under irrigation. Springer-Verlag. New York, NY.

Wolf, J.K. 1977. The evaluation of a computer model to predict the effects of salinity on crop growth. Ph.D. Thesis UT State University. Logan, UT.

Yermiyahu, U., R. Keren, and Y. Chen. 1988. Boron sorption on composted organic matter. Soil Sci. Soc. Am. J. 52:1309–1313.

Yermiyahu, U., R. Keren, and Y. Chen. 1995. Boron sorption by soil in the presence of composted organic matter. Soil Sci. Soc. Am. J. 59:405–409.

2

Sodicity

G.J. Levy
Volcani Center, ARO, Israel

2.1 Introduction

Soils in numerous areas of the world are adversely affected by the presence of excess Na in the soil solution and as an exchangeable cation. For example, it is estimated that ~50% of the arable land in Australia suffers from sodicity-related problems (Naidu, 1993). Sodium-affected soils exhibit poor soil water and air relations which adversely affect water movement in the soil, root growth and plant production, and make soil difficult to farm when either dry or wet. Consequently, in many cases, such soils have had to be abandoned.

Sodium-affected soils have been referred to in the past as alkali soils. This may have originated from the traditional separation of salt-affected soils into two groups (Szabolcs, 1979): (1) saline soils (soils affected by neutral Na salts, mainly NaCl and Na_2SO_4); and (2) alkali soils (soils affected by Na salts capable of alkaline hydrolysis, including $NaHCO_3$, Na_2CO_3 and Na_2SiO_3). However, at present, Na-affected soils are referred to as sodic soils, with the realization that the level of Na in the soil solution and on the exchange complex, in conjunction with the total electrolyte concentration, determines whether or not a given soil is sodic, and not the type of the Na salt.

Problems associated with sodic soils are expected to increase in the future. The ever increasing need to provide food to an expanding worldwide population, coupled with the increasing demand for good quality water from urban and industrial sectors, results in poorer quality water and soils being used for food production. Wastewater, whether drainage or recycled municipal water, which is significantly more saline and sodic than fresh water, is rapidly becoming a common source for agriculture in many areas in the world. Consequently, understanding sodic soil behavior and its effects on agriculture and the environment is essential in order to properly manage sodic soils for crop production and sustainable agriculture in many parts of the world.

This chapter provides a general review of various aspects related specifically to sodic conditions in soils. For more comprehensive reading on sodic soils, readers are referred to a special issue of the Australian Journal of Soil Research (31[6], 1993) and Sumner and Naidu (1998).

2.2   Sodic Soils: Definition and Distribution

2.2.1   Definition of Sodic Soils

Prior to defining sodic soils it is necessary to define the parameters by which sodicity is evaluated. Two important parameters are generally used. The first is the exchangeable Na percentage (ESP) which describes the level of adsorbed Na in the soil, and is defined as:

$$ESP = 100(Exchangeable\ Na/CEC) \tag{2.1}$$

Where the CEC is normally determined at a reference pH (7.0 or 8.2). Instead of CEC, the sum of exchangeable cations ($Ca + Mg + Na + K + Al = ECEC$) can be substituted in most sodic soil regions. The second parameter reflects the sodicity level of the irrigation water or soil solution and is termed the Na adsorption ratio (SAR) defined as:

$$SAR = [Na]/([Ca+Mg]/2)^{0.5} \tag{2.2}$$

where brackets [ ] reflect cation concentrations in $mmol_c\ L^{-1}$. Thus, SAR has units of $(mmol_c\ L^{-1})^{0.5}$. It should be emphasized that the sodicity hazard of solutions is related to the ratio of Na to the divalent cations present in the water and not to Na concentration alone. In addition, the true SAR (SAR_t) of the solution may differ from that measured (SAR_p) in cases where no corrections for ion pairs or complexes have been made in the determination of ion concentrations (Sposito and Mattigod, 1977). These researchers offered the following empirical relation between the SAR_t and SAR_p:

$$SAR_t = 0.08 + 1.15\ SAR_p \tag{2.3}$$

Furthermore, in arid and semiarid soils irrigation may lead to dissolution/precipitation of $CaCO_3$, depending on the pH of the system. Thus, the SAR of the irrigation water (SAR_{iw}) may differ from the SAR of the drainage water (SAR_{dw}), with the latter being considered as a better indicator for sodicity hazard. Bower et al. (1968) proposed the use of SAR_{iw}, pH and leaching fraction for calculating SAR_{dw} (also known as SAR_{adj}). Rhoades (1968) added to this relationship the contribution of mineral dissolution. Later, Suarez (1981) suggested that SAR_{dw} be calculated based on the concentration of Na and Mg in the irrigation water, leaching fraction and P_{co_2}. The equation proposed by Suarez (1981) is considered more accurate and easier to use than other equations for calculating SAR in the soil and drainage water (Frenkel, 1984).

Exchange reactions take place between the soil solution and the exchange phase. Thus, soil ESP can be estimated from the SAR of saturated paste extracts using the following empirical relationship (USSL Staff, 1954):

$$ESP = 100(-0.0126 + 0.01475\ SAR)/(1 + [-0.0126 + 0.01475\ SAR]) \tag{2.4}$$

or from the nomograph presented in Fig. 2.1. When more dilute extracts are used, such as 1:5 soil:water ratio, then a different relationship holds (Rengasamy et al.,1984):

$$ESP = 1.95\ SAR + 1.8 \tag{2.5}$$

To date there is no widely accepted definition of a sodic soil. The USSL Staff (1954) defined a sodic soil as one whose physical properties are adversely affected by the presence of Na where the ESP is

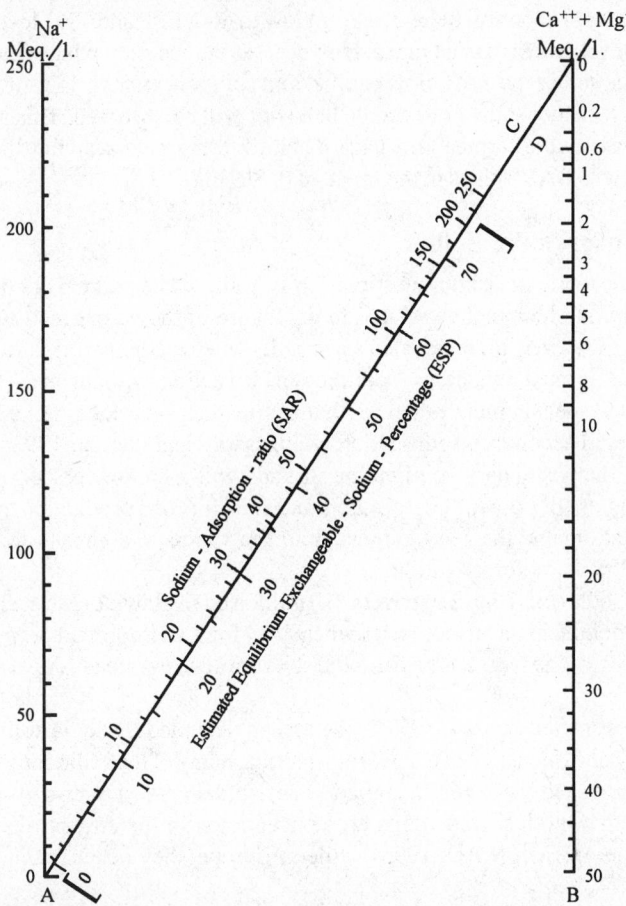

Fig. 2.1 Nomograph for determining the SAR of a saturation extract and estimating the corresponding ESP value of soil at equilibrium with the extract [Courtesy of USSL Staff, 1954]

>15, adding the reservation that this limit must be regarded as somewhat arbitrary and tentative. The study of McNeal and Coleman (1966), who used 7 soils from the western United States, supported the conclusion that ESP 15 can separate sodic and nonsodic behavior of the soil; however, it was added that the electrolyte concentration in the percolating solution must exceed 3 $mmol_c L^{-1}$ in order for this to hold true. Greene et al. (1978) suggested that the threshold value separating sodic from nonsodic soils should depend on soil texture; they proposed ESPs of 10 and 20 for fine and coarse textured soils, respectively. In Australia, McIntyre (1979), who used water with total electrolyte concentration (TEC) of 0.7 $mmol_c L^{-1}$, proposed that soils with ESP > 5 should be considered as Na-affected (sodic) soils. Such differences in the proposed critical ESP levels separating sodic and nonsodic soils arise in part from the fact that in California (USSL Staff, 1954) water with a TEC of 3-10 $mmol_c L^{-1}$ was being considered, whereas in Australia (McIntyre, 1979), the TEC of the water under consideration was only 0.7 $mmol_c L^{-1}$. Consequently, a much higher ESP was required before degradation of the soil in the California studies was observed. The above critical ESP levels were determined based on hydraulic conductivity measurements but studies on soil susceptibility to seal formation have revealed that soils can exhibit sodic behavior even at ESPs of 2-3 (Agassi et al., 1981; Kazman et al., 1983). In view

of the continuous effect of Na on soil behavior, from low to high ESP and TEC levels, no satisfactory decision or a critical level of ESP can be made. However, as will be shown later, the interrelationship between the ESP of the soil or the SAR of its equilibrium solution, and the TEC of the soil solution is important in dictating whether sodic or nonsodic behavior will be observed. This approach was used by Rhoades (1982) to assess the permeability hazard of infiltrating waters having different TEC values when applied to soils with SAR values in the range of 0–30 (Fig. 2.2).

2.2.2 Distribution of Sodic Soils

Prior to considering the world distribution of sodic soils, it should be pointed out that, for most sodic soils, sodicity is a natural phenomenon related to the nature of the parent material and subsequent pedogenic processes. However, there are also sodic soils where sodicity arises from anthropogenic processes, and is thus termed secondary sodification. Irrigation without proper drainage, forest clearing, and other land management practices that can lead to waterlogging, are the main human activities that yield rapid secondary sodification (Rengasamy and Olssen, 1991; Fitzpatrick et al., 1994). It is expected that secondary sodification of soils will increase considerably in the future, because of the growing use of poorer quality waters and soils for food production to meet the demand of the growing population, and the need to reuse drainage waters to compensate for the increasing shortage of good quality water for irrigation.

The distribution of sodic soils (Fig. 2.3) covers 210 million ha worldwide (Bui et al., 1998). The areas of sodic soils on each continent and in countries where they form an important proportion of total area are presented in Table 2.1. The largest areas of sodic soils are concentrated in Australia and the former USSR.

In Australia, it is estimated that 25–30% of the area is occupied by sodic soils with widespread occurrence (Northcote and Skene, 1972). In North America, many of the sodic soils occur on the Great Plains of western Canada and the northern United States, under cold, semiarid to subhumid climates (FAO, 1991). In South America, sodic soils are concentrated in the Argentinian and Paraguayan pampas and northeastern Brazil (FAO, 1991) while in Europe, they occur in the Carpathian basin,

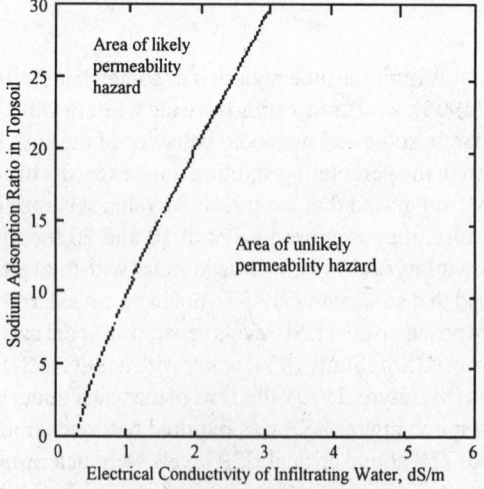

Fig. 2.2   Threshold values of SAR of topsoil and electrical conductivity of infiltrating water associated with the likelihood of substantial losses in permeability [From Rhoades, 1982]

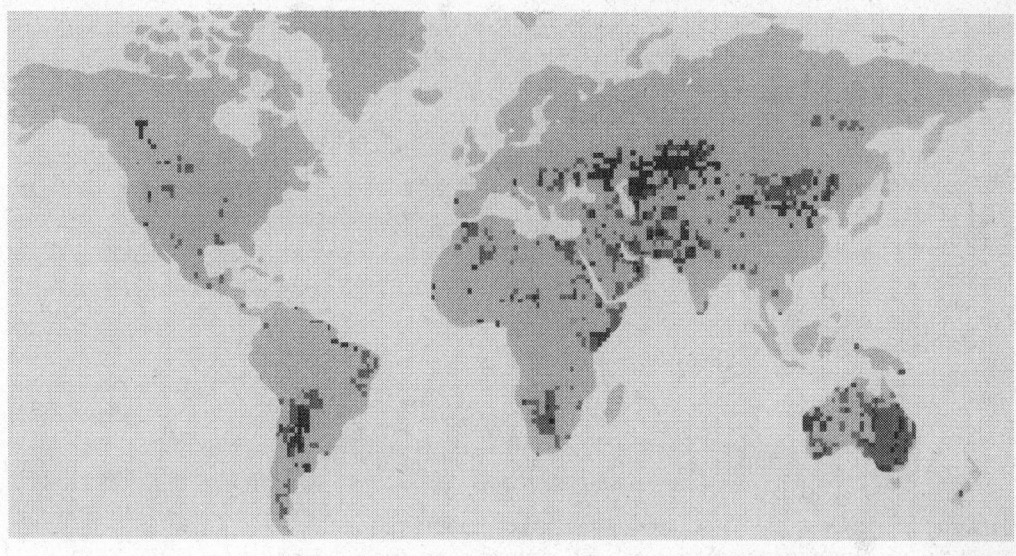

Area Affected

Saline and Sodic Soils		<=5%
(EC. >4 dS/m or ESP >6)		5-50%
Within 1m of the Surface		>50%

Fig. 2.3 World distribution of sodic soils [Reprinted from Bui et al., 1998. M.E. Sumner and R. Naidu. Sodic soils. Copyright Oxford University Press, New York with permission]

Table 2.1 World distribution of sodic soils [From Massoud, 1977]

Continent	Country	Area of sodic soils (thousand ha)
North America	Canada	6,974
	United States	2,590
South America	Argentina	53,139
	Bolivia	716
	Brazil	362
	Chili	3,642
Africa	Botswana	670
	Cameron	671
	Chad	5,950
	Ethiopia	425
	Kenya	448
	Madagascar	1,287
	Namibia	1,751
	Niger	1,389
	Nigeria	5,837
	Somalia	4,033
	Sudan	2,736
	Tanzania	583
	Zambia	863
South Asia	Bangladesh	538
	India	574
	Iran	686
North & Central Asia	China	437
	USSR	119,628
Australia	Australia	339,971

Hungary (Trans-Tisza region and the Danube-Tisza interfulve), Romania, Serbia, Slovakia, Ukraine (near the Black Sea), and the Transcaucasian plain of Georgia and Azerbaijan (Bui et al., 1998). In northern and central Asia, sodic soils are concentrated mainly (1) west of the Caspian Sea in Kazakhstan, Turkmenia, Uzbekistan, Tadzhikistan and Kyrghizia (FAO, 1978); (2) south of the Tien Shan Mountains in the basin of the Xinjiang-Gansu-Qinghai provinces, on the Huang and Huari river plains and in northeastern Manchuria, China; (3) in the Russian Republic of Yakutia; and (4) in the forest-meadow-steppe region of western Siberia (Bui et al., 1998). In southern Asia and the Middle East, sodic soils are widespread and can be found in India, Syria, Iran, Pakistan, and Bangladesh (Bui et al., 1998). In Africa, sodic soils occupy only 0.9% of the total land area and are concentrated in the Kalahari Basin, coastal Tunisia, Chad, Nigeria, Somalia, Sudan, Tanzania and southern Africa (Bui et al., 1998).

The spatial distribution of sodic soils demonstrates that they occur under a range of climates. Among the environmental conditions that promote the formation of sodic soils are the presence of shallow saline groundwater, the occurrence of perched watertables within 1 m of the soil surface, impeded drainage, low slope gradients, and textural discontinuities during deposition of sediments such as eolian, glacial or alluvial materials (Bui et al., 1998).

The outlined distribution of sodic soils and the conditions promoting their formation are based on a pedological or morphological approach. By contrast, a more useful definition of sodic soils is based on land management considerations (Section 2.2.1). Many agricultural soils that exhibit sodic behavior may not fit into the classical sodic soil group, because typical morphological features including columnar subsurface structure are not present. Consequently, the large areas that these soils cover remain largely undetermined.

2.3   Processes Characterizing the Behavior of Sodic Soils

2.3.1   Clay Charge and the Diffuse Double Layer

Interaction between solid and solution phases plays a dominant role in determining physical behavior of soil. Of the various soil constituents, colloidal clay determines much of the physical behavior of soils because of its large specific surface area and charge, which makes it very reactive in physicochemical processes such as swelling and dispersion. These two processes determine, to a large extent, soil microstructure and thus many of its physical properties. A complete discussion of the processes involved in clay dispersion and swelling is presented in Section B, Chapter 6.

2.3.2   Swelling and Dispersion

2.3.2.1   Reference Clay Systems

Upon wetting, swelling takes place in smectites represented most commonly in soils by the mineral montmorillonite. Repulsion forces (swelling pressure) that can be predicted from the diffuse double layer (DDL) theory are responsible (van Olphen, 1977; Bresler et al., 1982). When saturated with Na ions, a thick DDL forms creating high swelling pressures between the clay platelets, resulting ultimately in dispersion of single platelets in dilute solutions (Banin and Lahav, 1968; Shainberg et al., 1971). Conversely, low swelling pressures are observed between Ca-saturated platelets because strong electrical attraction forces between Ca and the clay surfaces (not considered by the DDL theory) prevent complete swelling of Ca montmorillonite, even in distilled water. Instead, Ca platelets aggregate into tactoids or quasicrystals consisting of 4–9 platelets (Aylmore and Quirk, 1959; Blackmore and Miller, 1961) with DDL formation only on the outer and not the internal surfaces of the tactoids (Blackmore and Miller, 1961), which results in a much smaller effective surface area than actually present.

In mixed Na-Ca systems, the ions are not randomly distributed throughout the system as was previously thought (Bresler, 1970), with Ca and Na being located preferentially on inner and outer surfaces, respectively (McAtee, 1961; Shainberg and Otoh, 1968; Bar On et al., 1970; Shainberg et al., 1971). This phenomenon is called demixing (Fig. 2.4). Shainberg et al. (1971) demonstrated that clay swelling is limited at ESP < 15 because of demixing; as ESP increases further, a sharp increase in macroscopic swelling occurs as Na enters the tactoids, and at ESP > 50, swelling in mixed Na-Ca and pure Na systems is similar.

In a stable clay suspension, particles frequently collide because of Brownian motion, but separate again because of DDL repulsive forces. When TEC or valency of the counter ions is increased, DDL forces are reduced, allowing particles to remain together after collision to form flocs that subsequently settle. The minimum TEC at which this occurs is called the critical flocculation concentration (CFC), or simply the flocculation value (FV). For Na and Ca montmorillonite, CFCs are 7–20 mmol$_c$ L^{-1} NaCl and 0.25–1.09 mmol$_c$ L^{-1} CaCl$_2$, respectively (Goldberg and Glaubig, 1987). Increasing pH from 4 to 10 has only a small effect on the CFC of Na montmorillonite, but addition of humic substances causes a marked increase over the pH range 4–8 (Tarchitzky et al., 1993).

For Na and Ca illite, CFC values are 40–50 mmol$_c$ L^{-1} NaCl and 0.25 mmol$_c$ L^{-1} CaCl$_2$, respectively (Arora and Coleman, 1979). At low ESP levels, the CFC of illite is higher than that of montmorillonite because of the smaller attractive forces in illite; illite surfaces are terraced, which leads to mismatching

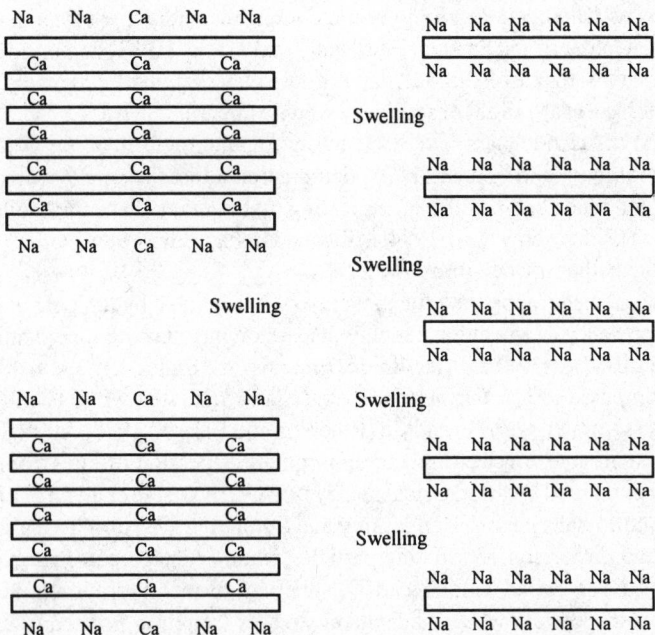

Fig. 2.4 Comparison of particle arrangements in a homoionic Na montmorillonite (right) with that in a Na-Ca system (left) illustrating the formation of tactoids or quasicrystals with demixing of Na and Ca [Reprinted from Sumner, 1993. Sodic soils: New perspectives. Aust. J. Soil Res. 31:683-750 with permission of CSIRO Publications, Melbourne, Australia]

and subsequent poor contact so that the particles cannot approach each other as closely (Oster et al., 1980).

Kaolinite saturated with Na remains flocculated in a salt-free solution at pH < 7 because the positively charged edges interact with the negative basal planes (Schofield and Samson, 1954). Elimination of edge positive charge by increasing the pH, addition of small amounts of various salts such as sodium oxalate, sodium pyrophosphate, sodium polymetaphosphate, sodium alginate, (Schofield and Samson, 1954; Durgin and Chaney, 1984; Frenkel et al., 1992) or introducing small impurities of montmorillonite or illite to the kaolinite suspension (Frenkel et al., 1978; Chiang et al., 1987) brings about dispersion (Goldberg and Forster, 1990). Consequently, kaolinite systems cannot be represented by a unique CFC.

A distinction should be made between swelling and dispersion even though both result from the balance between DDL repulsive forces and van der Waals attractive forces. Swelling is a continuous and reversible process, whose magnitude depends on the the TEC of the ambient solution and the degree of sodicity. Dispersion is not continuous and may occur even at low ESP levels as long as the TEC < CFC. Dispersion is an irreversible process because flocculation by increasing TEC does not result in the original particle associations and orientations.

2.3.2.2 Soil Clay Systems
At low ESP levels, dispersion rather than swelling is the main mechanism for physical degradation of soils. Only when ESP > 15, is the role of swelling dominant in determining soil physical properties (Shainberg and Letey, 1984).

Flocculation and dispersion behavior of soil clays differs significantly from that of pure clay systems, because they usually occur as mixtures and complexes with other minerals, oxides and soil organic matter (SOM). Goldberg and Forster (1990) and Frenkel et al. (1992) demonstrated that the CFC values of soil clays are two- to tenfold higher than for pure clay systems. Conversely, in the presence of hydrous oxides (McNeal et al., 1968) or sparingly soluble minerals such as $CaCO_3$ (Shainberg et al., 1981a,b), clay dispersivity is much less severe. Hence, extrapolation from pure clay to soil systems is extremely difficult. Goldberg and Forster (1990) demonstrated that the smallest differences in CFC were found between reference and soil illites, indicating that in many dispersive soils which contain illite or weathered mica (Rengasamy et al., 1984; Miller and Baharuddin, 1986; Miller et al., 1990), illite plays an important role in their dispersibility.

Clay dispersion is affected not only by the presence of Na but also by the type of complementary divalent cation. The presence of Mg enhances clay dispersion in soils with mixed mineralogy (Ali et al., 1985; Yousaf et al., 1987), as well as in kaolinitic (Emerson and Smith, 1970) and illitic (Rengasamy et al., 1986) soils, compared to Ca. Furthermore, aggregates saturated with Na and Mg disperse at lower ESPs than those saturated with Na and Ca (Emerson and Bakker, 1973; Ali et al., 1985).

Disintegration of aggregates on wetting and subsequent dispersion of clay from the aggregates cannot be adequately explained by the classical theory of DDL repulsive and van der Waals attractive forces developed for colloidal systems. Rengasamy and Sumner (1998) proposed a schematic model to explain swelling and dispersion of soil clay particles when dry aggregates are wetted (Fig. 2.5). Initially, hydration of the adsorbed cations leads to swelling, further hydration of highly sodic soils results in spontaneous dispersion, while mechanical stresses (raindrop impact, tillage, etc.) lead to dispersion even of soils containing predominantly divalent cations. Dehydration of the aggregates, and the buildup of an osmotic pressure by increasing the TEC leads to flocculation of the dispersed particles once more (Rengasamy and Sumner, 1998).

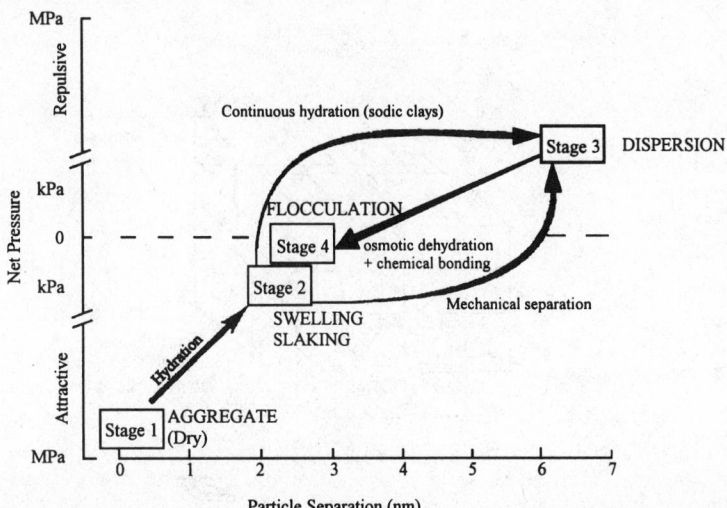

Fig. 2.5 Schematic illustration of processes that take place and intensity of attractive and repulsive forces involved when a dry aggregate of a soil is wetted [Reprinted from Rengasamy and Sumner, 1998. M.E. Sumner and R. Naidu. Sodic soils. Copyright Oxford University Press, New York with permission]

2.4 Aggregate Stability and Organic Matter in Sodic Soils

2.4.1 Aggregation and its Importance

Aggregation in soils is the arrangement of primary particles (sand, silt and clay) into secondary structural units, typically on a mm scale. Soil aggregates thus consist of clay particles, in quasicrystals or domains (Quirk and Aylmore, 1971), interspersed with larger quartzitic primary particles and held together by physicochemical forces that are a product of the chemical and microbiological environment of that particular soil.

Quirk (1978) recalling Bradfield's early statement that "granulation is flocculation plus", emphasized that a necessary condition for the formation of water-stable granules is clay flocculation, followed by stabilization by organic and/or inorganic cementing agents. The principles of flocculation and dispersion are central to the formation of clay quasicrystals and floccules (< 20 μm) that become the building blocks of aggregates. Such floccules are further aggregated by inorganic (clay particles or metal oxides) or organic (humified or residual microbial products) materials into microaggregates, ranging from 20 to 250 μm in size. Macroaggregates, typically > 250 μm, are agglomerates containing microaggregates plus primary sand and silt particles, held together largely by root hairs, fungal hyphae, and large biomolecules (Tisdall and Oades, 1982). The picture in Fig. 2.6, modified from a diagram of Greenland (1979), illustrates these constituents and associations. A further discussion of the role of SOM in structure development is given in Section 2.4.3.

Aggregate resistance to slaking suggests the ability of the long-range binding by SOM to hold smaller particle units together. Slaking is the breakdown of soil aggregates into particles > 20 μm in size as opposed to dispersion, which is the further breakdown of fine soil particles and release of < 2 μm clay. Dispersion of individual clay platelets may follow slaking if the hydration/osmotic forces within microaggregates are sufficient to overcome the attractive forces operating between particles.

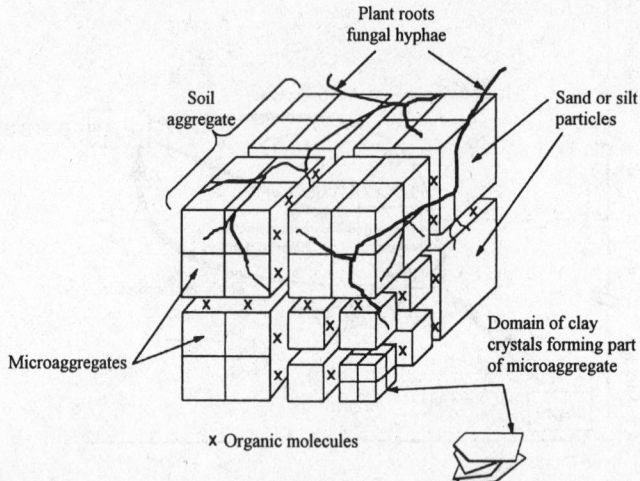

Fig. 2.6 Heirarchial structure of aggregates [Adapted from Greenland, 1979. R. Lal and D.J. Greenland. Soil physical properties and crop production in the tropics. Copyright John Wiley and Sons, New York with permission]

2.4.2   Sodicity Effects on Aggregation

The effect of exchangeable Na in inhibiting macroaggregation may occur by weakening the covalent associations between organic materials and soil minerals, and/or by increasing the osmotic/hydration forces that can cause particle repulsion during wetting. Partial Na saturation affects slaking of both macro- or microaggregates. Using < 2 mm aggregates from three California soils, Abu-Sharar et al. (1987a) showed that aggregate slaking occurred mainly at ESP < 10, and that it was independent of clay dispersion. In an additional study, Abu Sharar et al. (1987b) showed that in soils with a high proportion of macropores, sodic conditions (SAR 10) enhance aggregate slaking rather than clay dispersion. Other studies, however, have suggested that macroaggregate slaking was affected by fractional Na coverage only in the range of ESP > 5 (Aly and Letey, 1990; Shainberg et al., 1992), whereas Coughlan and Loch (1984) and Goldberg et al. (1988) reported no correlation between ESP and aggregate slaking in the low ESP range. Shainberg et al. (1992) demonstrated that slaking of aggregates into microaggregates was rapid and independent of TEC. However, further breakdown of the microaggregates into clay floccules (< 20 μm) was controlled by soil ESP and TEC of the soil solution.

Levy et al. (1993) reported that the stability of 105 μm microaggregates was affected by increasing ESP from 1 to 25 and that increasing sodicity had a similar effect on both aggregate breakdown and clay dispersion. They concluded that dispersion is the driving force controlling breakdown of microaggregates at high ESP.

Rengasamy et al. (1984) made a distinction between spontaneous dispersion where aggregates in suspension are not physically disturbed, and mechanical dispersion where aggregates are shaken in water. High ESP (> 10) and low TEC values are needed for spontaneous dispersion. Under mechanical stress (e.g., impact of raindrops, shear force of flowing water, mechanical energy of cultivation machinery), aggregate breakdown and clay dispersion are observed even at low ESP (< 10) levels (Shainberg et al., 1992). However, in the presence of Al and Fe oxides, high ESP and low TEC have only a small effect in promoting aggregate breakdown and clay dispersion (McNeal et al., 1968; El-Swaify, 1973).

2.4.3 Organic Matter Effects in Sodic Soils

Soil organic matter comprises about 1–10% of a typical soil's mass of which 1–5% is microbial biomass (Nelson and Oades, 1998). Organic matter acts both as a bonding and dispersing agent in soils. The balance between these opposing effects depends upon the nature of the organic materials and their types of interaction with inorganic colloids. With respect to its effects on soil structure and aggregate stability, SOM can be divided into transient and persistent bonding agents.

Transient components of SOM include plant roots and fungal hyphae. They enmesh macroaggregates (Fig. 2.6) and inhibit aggregate breakdown. Transient components are so named because they are present only when plants are growing and fresh organic matter is added, and because they are readily disrupted by mechanical disturbance (Nelson and Oades (1998). Soil stability, and especially that of macroaggregates, has often been related to the presence of transient binding agents in soil (Emerson, 1954; Loveland et al., 1987). High levels of SOM seem to promote resistance to sodic conditions. An inverse relationship has been noted between soil stability (inferred from modulus of rupture) and SOM content at any given ESP, especially for soils containing 10–20% clay (Aylmore and Sills, 1982). Addition of pea straw to soils increased macroaggregate stability at ESP values of 3 to 36% (Barzegar et al., 1996). The favorable effects of transient organic agents on aggregate stabilization can be explained by nonionic bonding mechanisms by which roots, hyphae and natural polysaccharides stabilize aggregates, and which are not always directly affected by interactions between water, clay and exchangeable cations. Nelson and Oades (1998) concluded that the beneficial effects of these transient agents on aggregate stability are most noticeable when ESP is not too high; with adequate TEC to offset its effects, clay content is moderate, Mg content is low, mechanical disturbance is limited, and fresh organic materials are added regularly.

Additional sources of transient SOM are mucilages and sugars produced by plant roots and microorganisms. These materials are transient because they decompose readily (Nelson and Oades, 1998) and act as bonding agents in stabilizing smaller structural units (microaggregates) (Tisdall and Oades, 1982).

The persistent components of SOM include polyanionic colloidal materials (plant remains and microbial products), often referred to as humic substances. Their persistence results from their recalcitrance and association with inorganic materials (Stevenson, 1992). These components play an important role in determining the stability of smaller scale aggregates (microaggregates) through a variety of mechanisms, but especially through their dispersing power. Organic anions are known to increase clay dispersivity, in particular when variable charge minerals are present, mainly by increasing the negative charge density of the soil colloid fraction. There is a large body of literature which documents the influence of organic anions on increasing the CFC of various clay minerals, thus giving rise to their susceptibility to dispersion (Nelson and Oades, 1998). High ESP, the presence of exchangeable Mg or high pH each increase clay dispersivity in the presence of organic anions (Emerson and Smith, 1970). Because humic substances form chelates with Ca (Sposito et al., 1977), the effective SAR of the soil solution increases, and clay dispersion is promoted.

2.5 Hydraulic Properties of Sodic Soils

Sodicity-related degradation of soil structure leads to deterioration in soil water transmission properties, increased susceptibility to crusting, runoff, erosion, and poor aggregate stability. Water transmission properties of soils are commonly characterized by measuring hydraulic conductivity (HC) and/or infiltration rate (IR) which are distinctly different; HC is measured under conditions where the soil surface in undisturbed, while considerable surface disturbance arising from external forces, such as raindrop impact or overhead irrigation, generally occurs when IR is determined. This disturbance often leads to surface seal formation, causing the rate of water movement in the sealed layer to be different from that in the underlying soil.

2.5.1   Hydraulic Conductivity

2.5.1.1   Introduction

The HC of a soil depends on its properties as well as those of the percolating liquid. A detailed discussion of HC is presented in Section A, Chapter 4. A number of models have been proposed to describe the relationships between structure of a porous material as reflected by its pore size distribution and permeability (Kozeny, 1927; Childs and Collis-George, 1950; Marshall, 1958). These types of models have not succeeded in adequately predicting soil HC, in part, because they do not take into account the effect of fluid properties on the soil matrix geometry. Clay swelling and dispersion in response to changes in electrolyte composition and concentration of the percolating solution change the size of conducting pores, and hence HC.

2.5.1.2   Sodicity and Electrolyte Concentration

Fireman and Bodman (1939) and Bodman and Fireman (1950) advanced the concept that the higher the SAR and the lower the TEC of the percolating solution, the greater the reduction in soil HC. Subsequently, Quirk and Schofield (1955) developed the concept of threshold concentration, which they defined as the concentration required to prevent a decrease > 25% in soil permeability for a given soil ESP or SAR of the percolating solution (Fig. 2.7). According to Quirk and Schofield (1955), even a Ca-saturated soil (ESP = 0) may show a reduction in HC, provided that the TEC is below 0.6 $mmol_c$ L^{-1} which has been verified in other studies (Emerson and Chi, 1977). When rain or snow water (< 1 $mmol_c$ L^{-1}) is applied to a soil, even soils with very low ESPs may disperse causing permeability to decrease. The basic approach of Quirk and Schofield (1955) has been extended to a large number of additional soils (McNeal and Coleman, 1966; McNeal et al., 1966, 1968; Yaron and Thomas, 1968; Rhoades and Ingvalson, 1969; Cass and Sumner 1982a,b,c).

The importance of clay swelling and dispersion in controlling the HC has been highlighted by Quirk and Schofield (1955), McNeal et al. (1966), Rowell et al. (1969), Cass and Sumner (1974), Shainberg et al. (1981a), and Radcliffe et al. (1987). Soils begin to swell as the TEC of the soil solution is reduced, but the effect of swelling on HC becomes apparent only at SAR > 10 in medium and heavy textured soils. In light textured soils, the effect of swelling on reducing the size of conducting pores, and hence HC, is hardly noticeable. Because the swelling process is reversible, an increase in HC is observed when TEC is increased again. Clay dispersion, on the other hand, can take place even at low ESPs provided TEC < CFC. When clay dispersion occurs, the dispersed clay particles move through the soil profile and may even cause a complete blockage of conducting pores, and hence irreversible changes in the HC. Consequently, the effects of reducing TEC on the HC are likely to be more severe when clay dispersion is a predominant cause (Shainberg and Letey, 1984).

When Na-affected soils are exposed to rainfall or irrigation water of low TEC, their HC is reduced because the salt concentration in the soil solution is not sufficient to prevent swelling and clay dispersion. For instance, in Australia under rainfed conditions, soils (especially fine textured) often waterlog during the winter rainy season because what little salt is present is leached in the first rains, causing the clay to disperse and seal the soil (Fitzpatrick et al., 1994). In Israel, fields irrigated with water having SAR values as high as 26 with an EC of 4.6 dS m^{-1} showed no permeability problems, because the TEC in the irrigation water was sufficient to prevent the dispersive effect of Na (Frenkel and Shainberg, 1975). However, upon applying distilled water (DW) to simulate winter rainfall, soil HC

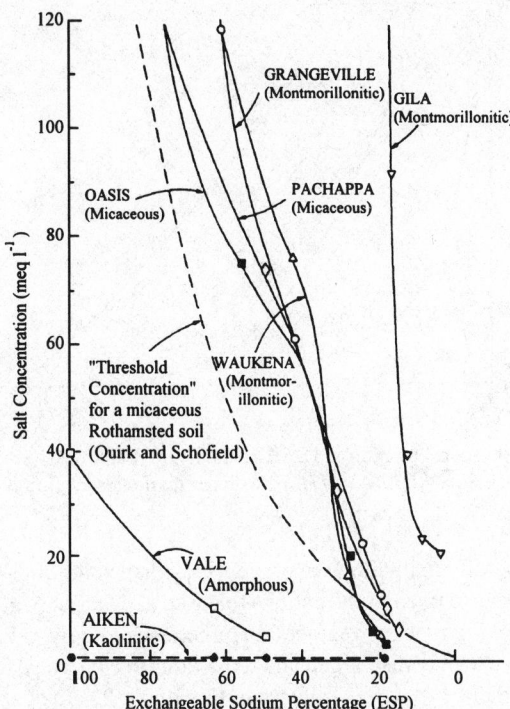

Fig. 2.7 Combinations of salt concentration and ESP required to produce a 25% reduction in hydraulic conductivity for selected soils [Reprinted from McNeal and Coleman, 1966. Soil Sci. Soc. Am. Proc. 20:308-312 with permission of the Soil Science Society of America]

dropped to a small fraction of its initial value of 9–12 mm h^{-1} (Frenkel and Shainberg, 1975). Similar permeability problems arising from a decrease in TEC of the percolating water have been reported for some areas in the Central Valley of California (Mohammed et al., 1979), and for numerous soils in South Africa (Hensley, 1969; Johnston, 1975). Effects of the interaction between SAR, low levels of TEC and soil HC are demonstrated in Fig. 2.8. Maintaining a TEC of 3 mmol$_c$ L^{-1} was sufficient in this case to prevent a decrease in HC for a solution having SAR < 12.

A number of attempts have been made to relate clay swelling to changes in soil HC. Some were more general and included empirical components (McNeal et al., 1968; Lagerwerff et al., 1969) but only that of Russo and Bresler (1977) showed reasonable agreement between calculated and observed data.

2.5.1.3   Soil Properties Affecting Sodic Behavior

Effect of Mineral Weathering

A major factor causing differences among various sodic soils in their susceptibility to hydraulic failure when leached with solutions having low TEC water is their rate of salt release during mineral dissolution (Rhoades et al., 1968; Felhendler et al., 1974; Shainberg et al., 1981a,b; Alperovitch et al., 1985). Mineral dissolution often determines the TEC of percolating solutions and hence soil HC. Sodic soils containing minerals (CaCO$_3$ and a few primary minerals) that readily release soluble electrolytes will not readily disperse when leached with DW at moderate ESP values, because a sufficiently high

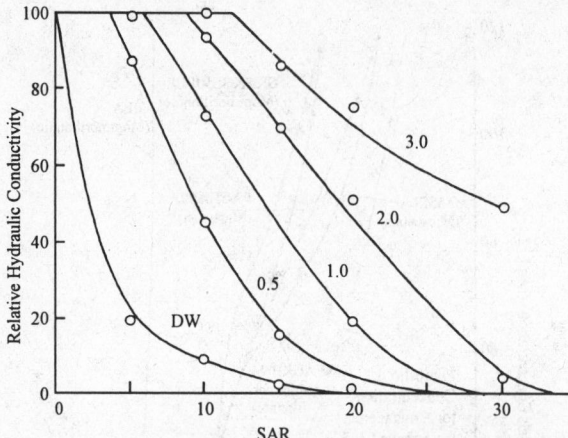

Fig. 2.8   Effects of salt solution concentration and SAR on the HC of Fallbrook soil-sand mixtures [Reprinted from Shainberg et al., 1981. Soil Sci. Soc. Am. J. 45:273-277 with permission of the Soil Science Society of America]

TEC (3 $mmol_c$ L^{-1}) is generally maintained to prevent clay dispersion. In addition, the ESP and SAR values will be reduced because most of the cations released are Ca and Mg. Conversely, soil solution concentrations in soils lacking readily weatherable minerals are likely to be below the CFC, making such soils more susceptible to clay dispersion and reductions in HC. The data presented in Fig. 2.9 for two Israeli soils (Felhendler et al., 1974) illustrate this point. The calcareous soil from Nahal Oz maintained a higher HC when leached with DW than the chemically more stable Netanya sandy loam, which showed high susceptibility to sodic conditions when leached with DW.

Effect of Lime
Sodic soils containing $CaCO_3$ are common in many semiarid and arid regions of the world. Fine $CaCO_3$ particles can improve the physical condition and permeability of sodic soils by (1) acting as a cementing agent in stabilizing soil aggregates (USSL Staff, 1954; Rimmer and Greenland, 1976), and (2) dissolving to maintain the TEC of the soil solution above the CFC to prevent dispersion and HC reduction (Shainberg et al., 1981a; Shainberg and Gal, 1982; Naidu et al., 1993b). The relative importance of each of these mechanisms can be estimated from their effect on IR and crust formation under rainfed conditions (Agassi et al., 1981; Ben-Hur et al., 1985) (Section 2.5.2.2).

Effect of Clay Mineralogy
The type of clay mineral also influences the response of soils to sodic conditions. Soils with high contents of expansible 2:1 layer silicates (e.g., montmorillonite) are the most labile while those high in kaolinite and sesquioxides are the least labile (McNeal and Coleman, 1966; Yaron and Thomas, 1968). Acidic kaolinitic soils have been considered insensitive to changes in soil ESP; however, upon addition of smectitic imputities to these soils, their susceptibility to sodic conditions increased markedly (Frenkel et al., 1978). The HC of nonacidic arid land kaolinitic soils has been reported to decrease significantly upon exposure to sodic conditions (Frenkel et al., 1978; Abu Sharar et al., 1987b). Among the 2:1 layer silicates, montmorillonitic soils have been found to have greater sensitivity to ESP than their vermiculitic counterparts (Rhoades and Ingvalson, 1969).

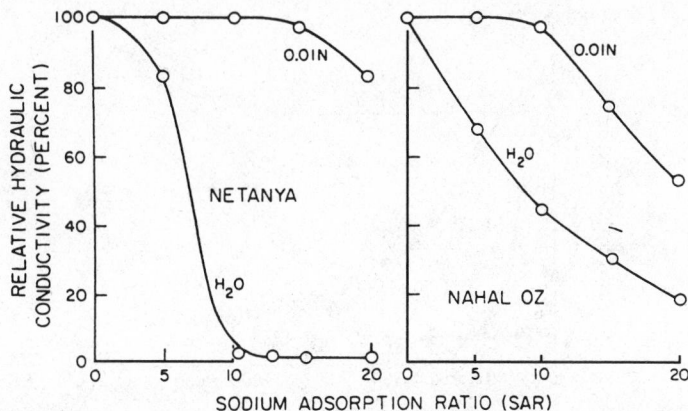

Fig. 2.9 Hydraulic conductivity of a sandy loam (Netanya) soil and a silty loam (Nahal-Oz) soil as a function of the SAR and the concentration of the leaching solutions [Reprinted from Felhendler et al., 1974. Trans. 10th Int. Congr. Soil Sci. I:103-112 with permission of the International Society of Soil Science]

Effect of Soil Texture

Mobile clay in the leachate is usually only observed in sandy soils. When clay content is high, the small size of the conducting pores usually ensures that dispersed clay moves only short distances before it clogs the pores, thus leading to low HC. Therefore, in loams and clays, the dispersion mechanism still operates, but no macroscopic movement of the clay particles is observed. Conversely, in sandy soils, the dispersion mechanism and macroscopic clay movement become evident particularly at higher ESP levels, as was shown by Pupisky and Shainberg (1979). The decline and subsequent recovery in HC for a soil of ESP 20 (Fig. 2.10) can be explained by a change in flow pattern from that of a solution flowing through a matrix of sand covered with clay which starts to disperse and reduce HC, to flow of a clay suspension through a pure sandy matrix having large pores resulting in an increase in HC once more (Pupisky and Shainberg, 1979).

Effect of Exchangeable Magnesium

Although the exchangeable Ca/Mg molar ratio is high in many soils, the reverse is true for some soils. The effect of adsorbed Mg on soil hydraulic properties is a controversial issue. The USSL Staff (1954) grouped Ca and Mg together with respect to their similarities in promoting and maintaining soil structure, while others have found that a Na-Mg saturated soil is structurally less stable (van der Merwe and Burger, 1969) and/or has a lower HC (McNeal et al., 1968) than a Na-Ca saturated soil. Alperovitch et al. (1981, 1985) related the effects of Mg to soil type. In calcareous soils, the presence of Mg enhances the dissolution of $CaCO_3$, thereby producing electrolytes which prevent clay dispersion and HC decay. In noncalcareous soils, Mg causes a decrease in HC of the soils beyond that of a corresponding Na-Ca system (Alperovitch et al., 1981, 1985). Shainberg et al. (1988) concluded that the lower HC values of Na-Mg relative to Na-Ca montmorillonite are related to the effect of Mg on the hydrolysis of montmorillonitic clay. The presence of a high concentrations of Mg at the clay surface slows down the release of octahedral Mg from the mineral lattice (Kreit et al., 1982), thereby lowering the EC of the Na-Mg clay (Shainberg et al., 1988). In kaolinitic soils, this adverse effect of Mg has not been noted (Levy et al., 1989).

A distinction has been made between the direct effect of exchangeable Mg in causing decreases in HC, which has been termed a specific effect, and the inability of Mg in irrigation waters to counter the accumulation of exchangeable Na in soils (McNeal et al., 1968; Chi et al., 1977; Emerson and Chi, 1977;

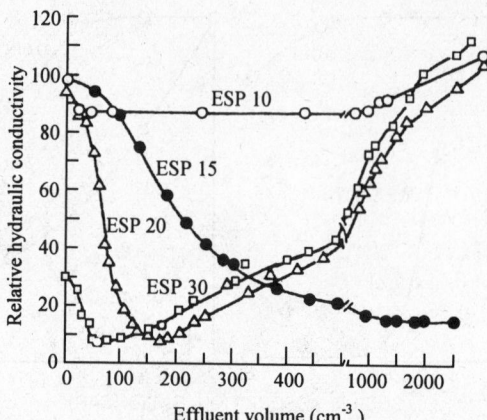

Fig. 2.10   Relative hydraulic conductivity of soils with various ESPs leached with distilled water [Reprinted from Pupisky and Shainberg, 1979. Soil Sci. Soc. Am. J. 43:429-433 with permission of the Soil Science Society of America]

Rahman and Rowell, 1979). Curtin et al. (1994) concluded that the specific effect of Mg was dominant, because the observed adverse effects of Mg on clay dispersion and HC were greater than could be explained simply by a slightly higher exchangeable Na level in Mg systems.

2.5.2   Infiltration Rate

2.5.2.1   Introduction

Soil infiltration rate is defined as the volume flux of water flowing into the profile per unit surface area under any given set of circumstances. Mechanisms controlling changes in the IR over time depend on the mode of water application to the soil. Under conditions where water is supplied to the soil without appreciable energy input, IR depends on the HC of the soil matrix (Section 2.5.1). When water is supplied with appreciable energy impact, the IR decreases from its initial high rate due to the formation of a thin layer (< 2 mm) at the soil surface, termed the surface seal. This seal is characterized by greater density, higher shear strength, finer pores and lower saturated HC than the underlying soil (McIntyre, 1958; Bradford et al., 1987). A structural seal, usually caused by the impact energy of waterdrops, should be distinguished from a depositional seal formed by translocation and deposition of fine soil particles at a discernible distance from their original location (Arshad and Mermut, 1988; Chen et al., 1980). The discussion to follow will focus on the former.

Structural seal formation is due to two complementary mechanisms (McIntyre, 1958; Agassi et al., 1981): (1) physical disintegration and compaction of soil aggregates caused by the impact of the raindrops, and (2) chemical dispersion and movement of clay particles into a region as shallow as 0.1– 0.5 mm depth, where they lodge and clog conducting pores. Such a seal forms during rainfall (Duley, 1939; Epstein and Grant, 1973; Morin and Benyamini, 1977) or sprinkler irrigation (Aarstad and Miller, 1973), and is responsible for the resultant decrease in IR.

2.5.2.2   Effect of Sodicity on Infiltration Rate

Kazman et al. (1983) studied under laboratory conditions the sensitivity of seal formation and IR of smectitic soils to low levels of ESP when exposed to DW (simulating rainwater) (Fig. 2.11). Even at the lowest sodicity (ESP = 1.0), a seal was formed and the IR dropped from an initial value > 100 to a final infiltration rate (FIR) of 7.0 mm h^{-1}. An ESP value of 2.2 was sufficient to cause a further drop in FIR of the sandy loam to 2.4 mm h^{-1}. The amount of rain required to approach the FIR was also affected by ESP (Fig. 2.11). Because raindrop impact energy was the same in all experiments, the differences between IR curves for the various treatments reflect the result of chemical dispersion of soil clay caused by sodicity. The high sensitivity of the soil surface to low ESP values is explained by three factors (Oster and Schroer, 1979; Kazman et al., 1983): (1) the mechanical impact of the raindrops, which enhances chemical dispersion; (2) the absence of a surrounding soil matrix (sand particles), which when present slows clay dispersion and movement; and (3) the almost total absence of electrolytes in the applied DW. With respect to the first factor, it is difficult to separate the mechanical and chemical mechanisms which are complementary. Agassi et al. (1985a) noted that in the absence of the former, the chemical mechanism does not come into effect at low ESP levels (< 5). Evidently, in the lower range of ESP, the chemical mechanism needs some activation energy before it can begin operating at the soil surface. In the case of rainfall, this energy is provided by the impact of the raindrops.

Clay dispersion is highly sensitive to the TEC and SAR of the applied water. This is particularly true in cases where the soil surface is exposed to the mechanical action of falling water droplets, which enhance clay susceptibility to chemical dispersion. Agassi et al. (1981) noted that, the lower the TEC, the faster the rate at which IR decreases (Fig. 2.12). Also for the same TEC, increasing soil ESP results in a sharper decrease in IR and a lower FIR. Similarly, Oster and Schroer (1979) found that TEC greatly affected IR even at low SAR values. They observed an increase in final IR from 2 to 28 mm h^{-1} as TEC in the applied water (SAR 2 to 4.6) increased from 5 to 28 mmol$_c$ L^{-1}. Such results suggest that IR is far more sensitive than HC to the TEC of the applied water (Shainberg and Letey, 1984). Furthermore, the IR values of calcareous and noncalcareous soils are equally sensitive to low levels of ESP (Agassi et al., 1981; Kazman et al., 1983), whereas the HC of calcareous soils is less sensitive to sodic conditions than that of noncalcareous soils (Section 2.5.1.3). Such findings indicate that (1) the rate at which lime dissolves and increases TEC in the soil is insufficient when the soil is exposed to rainfall, and (2) the

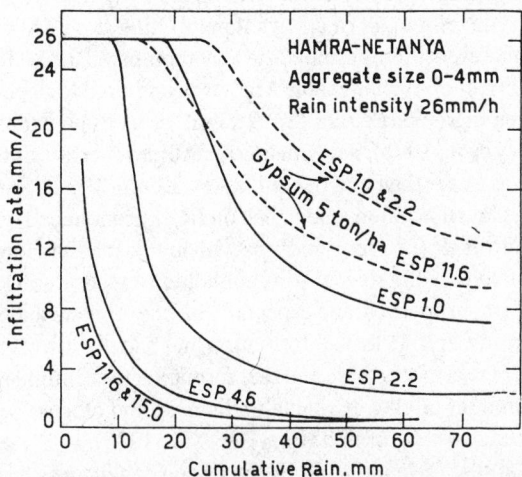

Fig. 2.11 The effects of the soil ESP and phosphogypsum application on the infiltration rate of the Netanya soil as a function of the cumulative rain [Reprinted from Kazman et al., 1983. Effect of low levels of exchangeable Na and applied phosphogypsum on the infiltration rate of various soils. Soil Sci. 35:184-192. Copyright Williams & Wilkins, Baltimore, MD]

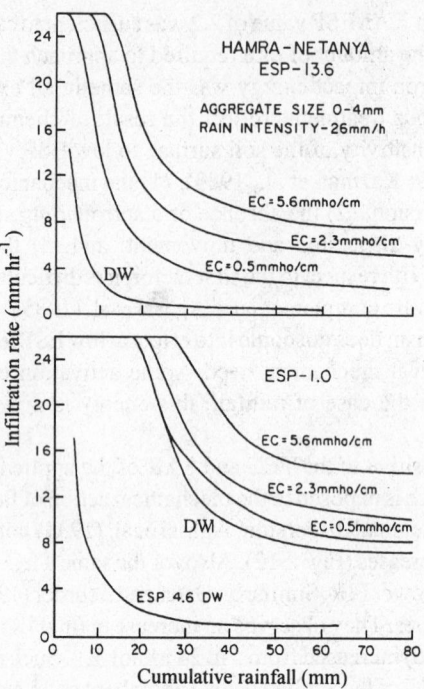

Fig. 2.12 Effect of electrolyte concentration in rain simulation experiments on the infiltration rate of a loess soil [Reprinted from Agassi et al., 1981. Soil Sci. Soc. Am. J. 45:848-851 with permission of the Soil Science Society of America]

role of $CaCO_3$ as a cementing material is negligible, because otherwise both types of soil would have had different IR values (Shainberg and Letey, 1984).

Although Ca and Mg are both divalent cations, their effects as the complementary cation to Na on soil stability are often very different. Elevated exchangeable Mg levels can cause deterioration in soil structure resulting in the development of a Mg solonetz (Ellis and Caldwell, 1935). In addition, Mg enhances dispersion of montmorillonitic and illitic clays compared to Ca (Bakker and Emerson, 1973). With respect to IR, the effect of exchangeable Mg has been found to depend on the kinetic energy of the water droplets. When high energy rain (22 .9 kJ m^{-3}) was used, exchangeable Mg had an effect similar to that of Ca (Levy et al., 1988), but with low to medium energy rainfall (8.0–12.5 kJ m^{-3}), the IR of Mg-Na treated soil was lower than that of Ca-Na soil (Keren, 1990). Such findings suggest that the adverse effect of Mg on clay dispersion, and hence on IR, is pronounced only under conditions where chemical dispersion controls IR (i.e., for raindrops with low to medium kinetic energy).

The susceptibility of sodic soils to seal formation and reduction in IR when exposed to rainfall depends on a number of soil properties, and especially on clay content and mineralogy. Soils with 10–30% clay were the most susceptible to seal formation and had the lowest IR values. With increasing clay content, soil structure was more stable and seal formation was diminished. In soils with lower clay contents (< 10%), the amount of clay available to disperse and clog soil pores was limited and, as a result, a poorly developed seal was formed (Ben-Hur et al., 1985).

Most of the studies on seal formation and IR have been conducted on soils in which the dominant clay minerals are smectites, which are known to be dispersive during both HC and IR determinations.

The effect of exchangeable Na on the HC of kaolinitic soils is quite small (Frenkel et al., 1978; Chiang et al., 1987); however, a number of kaolinitic and illitic soils from the southeastern United States (Miller, 1987; Miller and Scifres, 1988) and South Africa (Levy and van der Watt, 1988; Stern et al., 1991a) have been found to be susceptible to seal formation. Evidently, upon mechanical agitation by the beating action of raindrops, aggregates from kaolinitic soils may disperse (according to the level of exchangeable Na), and consequently, form a seal with resultant low IR. Additionally, Stern et al. (1991a) observed that the susceptibility of kaolinitic soils to seal formation was positively correlated with their contents of smectitic impurities; kaolinitic soils with small amounts of smectite were dispersive and susceptible to seal formation, whereas kaolinitic soils which contained no smectite impurities were less susceptible (Fig. 2.13).

Seal formation and reduction in IR depend strongly on clay dispersion, which is enhanced by increasing soil ESP and/or decreasing TEC. In addition, a strong relation between dispersible clay, IR and soil loss has been reported in numerous studies (Miller and Baharuddin, 1986; So and Cook, 1993). Sumner (1993) compiled a table based on published data showing that, for soils from different regions of the world, IR decreases as SAR/ESP increases, making more water available for runoff which, in turn, tends to increase erosion (Table 2.2). A different view of the relation among soil ESP, IR and erosion was given by Levy et al. (1994), who did not observe a direct relationship between such factors. They argued that with an increase in ESP a lower IR is obtained, but the resultant erosion depends on the balance between the larger volume of runoff and a more dense and structurally more resistant seal; thus the amount of erosion cannot be predicted.

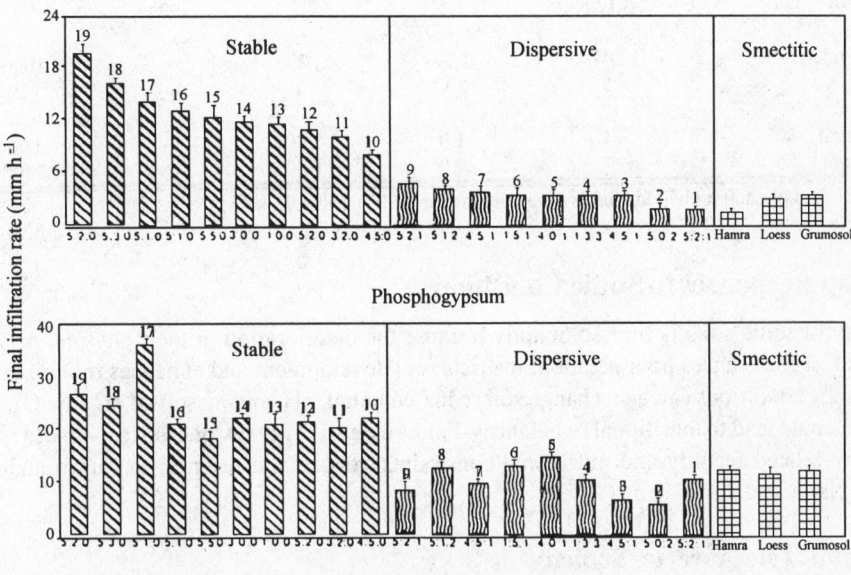

Fig. 2.13 Final IRs of the control and PG-treated stable, dispersive and smectitic soils. The numbers above the columns represent the soil number. The numbers below the columns represent the mineral quantity (5-high, 1-trace) for kaolinite, illite and smectitte. Bars represent standard deviation [Reprinted from Stern et al., 1991. Clay mineralogy effect on rain infiltration, seal formation and soil losses. Soil Sci. 152:455-462. Copyright Williams & Wilkins, Baltimore, MD].

Table 2.2 Effect of water quality and ESP on final infiltration rate (IR), runoff and erosion for soils from a variety of locations [Reprinted from Sumner, 1993. Sodic soils: New perspectives. Aust. J. Soil Res. 31:683-750 with permission of CSIRO Publications, Melbourne, Australia][1]

Soil classification and location	Water quality ($mmol_c$ L^{-1})	ESP/ SAR	Final IR ($mm\ h^{-1}$)	Runoff (%)	Erosion ($kg\ ha^{-1}$)
Typic Rhodoxeralf (a), Israel	DW^2	1.0	7.5	40	
		2.2	2.2	70	
		4.6	0.4	85	
Typic Rhodoxeralf (b), Israel	DW	2.2	5.2		4740
	9.5	2.2	7.3		1070
Calcic Haploxeralf, Israel	DW	1.8	3.0	60	
		6.4	1.0	80	
Typic Chromoxerert, Israel	DW	2.5	2.0	52	
		5.5	1.0	80	
Typic Kandhapludult, Georgia, USA	DW	1.2	8.5	83	157
	5	1.2	23.5	53	130
Typic Halpudult, Georgia, USA	DW	1.2	3.6	92	601
	10	1.2	10.7	77	309
Typic Ochraquult Georgia, USA	DW	1.6	1.6	96	885
	12	1.6	3.1	92	456
Rhodic Paleudult, Georgia, USA	DW	1.37	2		1000
	5	1.37	11		410
Oxic Paleustalf, South Africa	4	1.0	33.0		800
		7.0	10.0		2400

[1] Reference for data of each individual soil is given in Sumner (1993) Table 7.
[2] DW = Distilled water

2.6   Crop Responses to Sodic Conditions

Productivity of sodic soils is limited, mainly because the deterioration in their physical properties (Cairns et al., 1962) causes poor aeration, restricts root development and enhances root diseases. In addition, sodic conditions can also change soil redox potential, pH, and dissolved organic C content, all of which could lead to nutritional imbalances. For an extensive review of crop response to sodicity and sodicity-related fertility and nutritional constraints, readers are referred to Curtin and Naidu (1998) and Naidu and Rengasamy (1993).

2.6.1   Crop Tolerance to Sodicity

Research on crop tolerance to sodicity has been conducted under two different types of conditions, in the presence or in the absence of sodicity-related adverse soil physical conditions. In the United States, research has focused mainly on tolerance to sodicity based solely on nutritional factors (Chang and Dregne, 1955; Pearson and Bernstein, 1958), because in the field, sodicity-induced nutritional imbalances may be obscured by unfavorable physical conditions (Bernstein and Pearson, 1956). Data presented in Table 2.3 indicate that even at relatively low ESP levels (2–10), some crops show signs of Na toxicity. However, most field crops are classified as moderately tolerant or tolerant to sodicity (Table 2.3). Work in India, on the other hand, has been carried out under conditions where soil structure was allowed to be affected by sodicity concurrently (Abrol and Bhumbla, 1979; Singh et al., 1979). Under such conditions, grasses were tolerant, wheat and barley were moderately tolerant, and pea and beans were sensitive to sodicity (Gupta and Abrol, 1990). However, the use of this concept may cause results to be site specific, and hinder their applicability to other locations and soil types. Furthermore, ranking of crops for sodicity tolerance depends on the conditions under which the response was determined. For instance, rice has been found to be highly tolerant under sodic field conditions (Abrol and Bhumbla, 1979), but only moderately tolerant under favorable soil conditions (Table 2.3).

2.6.2 Plant Nutrition in Sodic Soils

Field fertility studies on sodic soils often yield inconclusive results, mainly due to high spatial variability (Toogood, 1978). Consequently, most dependable data on sodicity-related nutritional aspects have been obtained from greenhouse and laboratory studies (Curtin and Naidu, 1998).

Of the three major cations found in the soil in addition to Na (Ca, Mg, K), availability of Mg (Williams and Raupach, 1983) and K (Chhabra, 1985) is not a major concern in sodic soils. Conversely, impaired uptake of Ca is one of the main adverse nutritional effects of sodic soils. Calcium plays a major nutritional and physiological role in plant metabolism. The ratio of Ca to total cation concentration (TCC) is considered to be a better indicator of Ca availability to plants than the actual Ca concentration (Adams, 1974). Carter and Webster (1990) found that Ca/TCC values < 0.1 are often

Table 2.3 Tolerance of various crops to exchangeable sodium percentage (ESP) [From Pearson, 1960]

Tolerance to ESP	Crop	Growth response under field conditions
Extremely sensitive (ESP 2-10)	Deciduous fruits Nuts Citrus Avocado	Sodium toxicity symptoms even at low ESP levels
Sensitive (ESP 10-20)	Beans	Stunted growth at low ESP values even though the physical conditions of the soil may be good
Moderately tolerant (ESP 20-40)	Clover Oats Tall fescue Rice Dallis grass	Stunted growth due to both nutritional factors and adverse soil conditions
Tolerant (ESP 40-60)	Wheat Cotton Alfalfa Barley Tomatoes Beets	Stunted growth usually due to adverse physical conditions of soil
Most tolerant (ESP > 60)	Crested and fairway wheat grass Tall wheat grass Rhodes grass	Stunted growth usually due to adverse physical conditions of soil

found in saturated paste extracts of sodic soils, which for many crops, would result in Ca deficiency (Grieve and Maas, 1988). In addition, high Mg/Ca ratios, which are often found in sodic soils (Williams and Colwell, 1977), can also cause Ca deficiency (Grattan and Grieve, 1992).

Phosphorus does not pose a nutritional problem in sodic soils; some studies even reported that P is more readily available to plants in sodic than in nonsodic soils (Chhabra, 1985; Gupta et al., 1990). Nitrogen, on the other hand, can limit growth in many sodic soils (Cairns et al., 1967; Malhi et al., 1992) because of low rates of N mineralization, denitrification under restricted aeration conditions, volatilization of NH_3 under alkaline conditions, and inhibition of NO_3 uptake by Cl or SO_4 ions which usually are abundant (Curtin and Naidu, 1998).

Poor productivity of sodic soils can also be ascribed, at least in part, to micronutrient deficiencies (Naidu and Rengasamy, 1993). Availability of micronutrients in sodic soils is controlled largely by adsorption and precipitation reactions. Neutral pH values generally enhance adsorption of micronutrients. The composition and concentration of the soil solution also determine plant availability of micronutrients. The extent to which these solution parameters affect the availability has not yet been well established (Curtin and Naidu, 1998). Furthermore, the presence of carbonate minerals which provide active sorption sites for the precipitation of many micronutrients should further reduce their availability (Hodgson et al., 1966).

2.6.3 Ion Toxicity Under Sodic Conditions

Apart from the osmotic effect on plants due to high TEC generally found in sodic soils, high concentrations of Na have been found to be toxic in some woody species (Maas and Hoffman, 1977) and in avocado, citrus and stonefruit trees (Lauchli and Epstein, 1990).

High Na/Ca ratios result in high Na uptake and low uptake of Ca by plants, and thus cause, in addition to Ca deficiency, some other nutritional imbalances and adverse physiological effects (e.g., membane permeability and plant stability). High Na uptake can cause deficiences of elements such as K, Zn, Cu, and Mn (Levitt, 1980). Low uptake of Ca enhances uptake and possible toxicity of elements such as Ni, Pb, Se, Al, and B (Levitt, 1980). Ameliorating sodic soils with gypsum or lime may (1) alleviate heavy metal toxicity, and (2) offset the negative effects created by excessive Na uptake on physiological processes in plants (Rengasamy, 1987).

Boron (B) is fairly soluble in sodic soils with neutral pH; thus B is likely to be toxic rather than deficient in these soils (Naidu and Rengasamy, 1993). Significant boron toxicity in cereals was noted in regions of low rainfall of southern Australia; toxic concentrations of B in excess of $10\,mg\,kg^{-1}$ were recorded in subsurface layers of soils from these regions where sodic conditions prevail (Cartwright et al., 1986). Reclamation of sodic soils having pH > 7 with gypsum and leaching reduces B concentrations in the soil solution to levels within the safe limits, and B is no longer toxic to plants (Gupta and Abrol, 1990).

2.7 Reclamation of Sodic Soils

Reclamation of sodic soils involving improvement in soil structure is essential if productivity is to be maintained. There are two main avenues for reclamation of sodic soils: (1) addition of chemicals that provide a source of Ca to replace exchangeable Na coupled with excess water to leach the Na from the root zone, and (2) addition of organic ameliorants to maintain high levels of SOM to stabilize structure.

2.7.1 Chemical Reclamation

There are two common practices for reclaiming sodic soils by addition of divalent cations: (1) using chemical amendments that release Ca ions to the soil, and (2) application of successive dilutions of a high salt water containing divalent cations. Detailed reviews on the reclamation of sodic soils have been published by Keren (1995) and Oster and Jayawardane (1998).

2.7.1.1 Chemical Amendments

Typical amendments are those containing a source of soluble or potentially soluble Ca when reacted with the soil. Common amendments include gypsum, lime, $CaCl_2$, H_2SO_4, and S.

Gypsum

Gypsum ($CaSO_4 \cdot 2H_2O$) both from natural and anthropogenic sources (byproducts) is the most commonly used amendment for sodic soil reclamation, primarily because of its low cost, reasonable solubility and availability. Gypsum added to a sodic soil can increase permeability by means of both EC and cation exchange effects (Loveday, 1976). The latter is, however, of greater significance if permanent improvements in soil physical properties are desired.

The relative importance of these effects depends on the purpose for which gypsum is used. If gypsum is used in order to maintain high permeability at the soil surface during application of good quality water (i.e., rainwater or other source of water with low TEC), then the electrolyte effect is important. In this case, surface application of gypsum may be worthwhile, and the dissolution rate of gypsum dictates whether the rate of electrolyte supply to the water is sufficient to prevent clay dispersion and swelling. Thus, because of its higher dissolution rate, gypsum from industrial sources is better than from mined (Keren and Shainberg, 1981). In this case, the amount of gypsum required depends on the amount of high quality water applied and the rate of gypsum dissolution, and is relatively independent of the amount of exchangeable Na in the soil profile. Field experiments in Israel in a region of annual precipitation of 350–400 mm indicated that surface application of 5 Mg ha^{-1} of phosphogypsum (gypsum from an industrial source) was effective in maintaining significantly higher infiltration rates and lower runoff levels compared with a nontreated soil (Agassi et al., 1985b; Keren et al., 1983).

Conversely, in soils where the high ESP of the soil profile needs to be reduced, then the cation exchange effect is more important. The amount of gypsum required depends on the amount of exchangeable Na in a selected depth of soil. The amount of exchangeable Na to be replaced per unit land area (Q_{Na}, mol$_c$ ha^{-1}) during reclamation depends on the initial and desired final exchangeable Na levels, CEC, bulk density and depth of soil to be reclaimed. Thus, the amount of gypsum needed to reclaim the soil (gypsum requirement) (USSL Staff, 1954) can be calculated as follows:

$$GR = (8.61 \times 10^{-5}) Q_{Na} \qquad [2.6]$$

Efficiency and rate of exchange, namely, the proportion of Ca that exchanges for adsorbed Na, varies with total salt concentration and ESP, being much greater at initially high ESP values (Chaudhry and Warkentin, 1968; Rhoades, 1982). Removal of Na at ESP < 10 is slow, because part of the applied Ca displaces exchangeable Mg, so efficiency declines to as low as 30% (Loveday, 1976). Efficiency may also be low for fine textured soils because of the slow exchange of Na inside the structural units due to macropore flow (Manin et al., 1982).

The cation exchange process, major ion species, complexation reactions and mineral dissolution have formed the basis of several gypsum requirement models (Dutt and Terkeltoub, 1972; Tanji et al., 1972) which have been field tested and yielded a satisfactory match between calculated and measured data. These models provide a powerful tool for quantitative predictions of water and gypsum required to reclaim a soil to a predetermined level of salinity and sodicity. Recently, Simunek and Suarez (1997) have proposed a one-dimensional multicomponent transport model, UNSTACHEM, that takes into consideration, in addition to the parameters used by the aforementioned models, soil hydaulic properties and water flow velocities to predict the amendment concentration necessary for significant improvement of soil permeability.

Acids and Sulfur

In calcareous soils, addition of acid amendments will react with $CaCO_3$ to produce a soluble source of Ca ($CaSO_4 \cdot 2H_2O$ or $CaCl_2$). In the case of elemental S additions, the process is somewhat more complicated as it requires an initial phase of microbiological oxidation to produce H_2SO_4 (Keren, 1995). The increasing availability of acid waste products, coupled with the need to dispose of them in a safe manner, increases their potential for use as soil amendments (Miyamoto et al., 1975).

Calcium Chloride

Calcium chloride (byproduct from industry) can at times be an economical amendment for soil reclamation. Under certain conditions, such as soils with high ESP, $CaCl_2$ can be an even more efficient amendment than gypsum because of its higher solubility (Alperovitch and Shainberg, 1973) but has only a short-term effect. Prather et al. (1978) suggested that for soils with high ESP and low permeability, a combined application of $CaCl_2$ and gypsum might give a more rapid and more effective reclamation than that of either amendment alone.

Calcium Carbonate

Adding $CaCO_3$ to reclaim sodic soils is only of limited value (Gupta and Abrol, 1990), mainly because of its low dissolution rate which fails to provide an adequate amount of Ca for exchange. However, under certain conditions, naturally occurring carbonate minerals [$CaCO_3$, $MgCO_3$, $CaMg(CO_3)_2$] can play a significant role in reclaiming sodic soils (Inskeep and Bloom, 1986; Nadler et al., 1996). Dissolution of $CaCO_3$ can be enhanced by adding an acid or acid former to the soil (Keren, 1995) or by increasing CO_2 levels in the soil (Simunek and Suarez, 1997). The latter can be achieved by enhancing root activity (Bower and Goertzen, 1958; Rao and Ghai, 1985).

Normally, the beneficial effect of naturally occurring $CaCO_3$ on soil reclamation manifests itself mainly by maintaining the TEC of the soil solution at a level high enough to prevent clay dispersion and subsequent HC reductions (Felhendler et al., 1974; Alperovitch et al., 1981). However, $CaCO_3$ is not effective in maintaining high HC at the soil surface during rainfall because it fails to supply electrolyte fast enough at the soil surface, and thus, IR is impeded (Shainberg and Letey, 1984).

2.7.1.2 Successive Dilutions of High Salt Water

In areas where water is not a limiting factor, reclamation of sodic soils can be achieved by leaching the soil with successive dilutions of high salt water containing divalent cations (Reeve and Bower, 1960). In the early stages, the high TEC prevents clay dispersion and induces flocculation of the soil colloids. Simultaneously, the Ca ions in the water decrease sodicity by replacing exchangeable Na. Upon dilution of the saline water with higher quality water, the SAR of the water is reduced. Reeve and Doering (1966) developed a nomograph that predicts reclamation at any stage of leaching. However, to ensure successful soil reclamation, the depth of water added should be ninefold the depth of soil to be reclaimed (Reeve and Doering, 1966).

Among the various methods of water application during reclamation, sprinkler irrigation was found to be the best in removing salts from small pores (Nielsen et al., 1966) followed by intermittent ponding which requires less water than continuous ponding to obtain a similar degree of leaching (Miller et al., 1965), thereby allowing time for salts to diffuse from soil aggregates to previously well-leached pores prior to the next leaching event.

2.7.2 Biological Reclamation

Addition of organic ameliorants (farm manure, plant residues) is a widely recognized practice for reclaiming sodic soils. Application of organic matter is most effective in ameliorating alkali (soils with high pH) sodic soils. However, its ameliorative effects are generally restricted to the surface horizons where organic matter inputs tend to be concentrated (Nelson and Oades, 1998). Because successful reclamation with organic materials must involve leaching, irrigation must be frequent and in small amounts (Dahiya and Anlauf, 1990). Reclamation under rainfall can be expedited by minimizing runoff and evaporation, and increasing infiltration.

In order to biologically ameliorate sodic soils, they should be cropped continuously. Rice is commonly used, especially in the initial stages of reclamation, because of its tolerance to sodicity as well as the waterlogged conditions which favor reclamation (Abrol and Bhumbla, 1979). When cropping is not possible, pastures, green manure crops and agroforestry can also contribute to soil reclamation (Dahiya and Anlauf, 1990; Singh et al., 1994). A detailed review covering the pros and cons of reclamation while cropping to various pasture grasses and trees is given by Nelson and Oades (1998) who suggest that, for maximum benefit, organic matter and crops must be used in conjunction with chemical amendments and irrigation.

2.8   Management of Sodic Soils

Due to their poor physical and chemical properties, sodic soils suffer from problems of low water permeability, waterlogging, runoff and erosion, slow internal drainage, inadequate aeration, trafficability, and compaction. Thus, in order to maintain sustainable production on sodic soils, proper management practices must be used. Suitable management of Na-affected soils requires special soil management practices, as well as selection of suitable crops and nutritional management (Section 2.6). Examples of successful management practices for sodic soils in 6 different places in the world are presented by Oster and Jayawardane (1998).

2.8.1   Soil   Amendments

2.8.1.1   Gypsum
Prevention of chemical clay dispersion under natural rainfall conditions which can be achieved by applying a readily available slow release source of electrolyte to maintain TEC > CFC in the soil solution is essential when managing sodic soils. Phospho- (PG) and flue gas desulfurization gypsum (FDG) which are byproducts of the phosphate and power generation industries, respectively, are readily available, and meet the above requirements. Spreading PG at rates of 5–10 Mg ha^{-1} on the surface of smectitic (Kazman et al., 1983) and kaolinitic soils (Miller, 1987; Miller and Scifres, 1988) has proved effective in reducing the rate of IR decline and maintaining a higher FIR compared to the control, even on soils of ESP = 1.8 (Fig. 2.14). This, once again, demonstrates that some chemical dispersion takes place even at very low ESP values. Agassi et al. (1986) suggested that the favorable effect of PG on IR should be attributed not only to its effect on EC of the percolating water, but also to (1) physical interference with the continuity of the seal, and (2) partial mulching of the surface protecting it from the beating action of raindrops.

2.8.1.2   Organic Polymers
The use of organic polymers, especially polysaccharides (PSD) and polyacrylamides (PAM) for improving aggregate stability, maintaining high IR and reducing seal formation, has recently been studied extensively (Shainberg and Levy, 1994). Addition of small amounts of polymers (10–20 kg

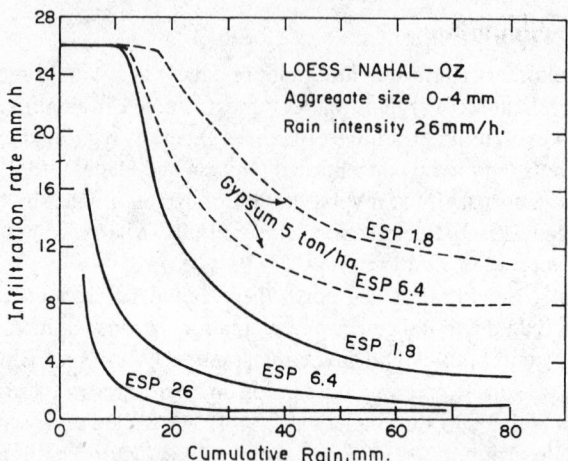

Fig. 2.14   The effects of the soil ESP and phosphogypsum application on the infiltration rate of the Nahal-Oz soil as a function of the cumulative rain [Reprinted from Kazman et al., 1983. Effect of low levels of exchangeable Na and applied phosphogypsum on the infiltration rate of various soils. Soil Sci. 35:184-192. Copyright Williams & Wilkins, Baltimore, MD]

ha^{-1}), either sprayed directly onto the soil surface or added to the applied water, was effective in stabilizing and cementing aggregates together at the soil surface, and hence maintaining high IR values in soils with ESP < 5. This is true under both laboratory (Helalia and Letey, 1988; Levy et al., 1992; Shainberg et al., 1990) and field conditions (Levy et al., 1991; Stern et al., 1992b). The efficacy of anionic polymers in preventing seal formation is enhanced when the soil clay is maintained in a flocculated state (Shainberg et al., 1990), and the resulting final IR could remain as much as tenfold higher than for the control. Of the polymers currently available and under study, anionic PAM which has the longest residual effect, has been found to be the most effective in controlling seal formation and maintaining high IR on soils with ESP < 10 (Levy et al., 1992), but was inconsistent at ESPs of 10–15 depending on water quality and amount of PAM added, and ineffective at ESP > 20 (Levy et al., 1995).

2.8.2   Tillage

Tilling sodic soils, which often results in improvements in physical and hydraulic properties, reduces bulk density and increases total porosity (Blackwell et al., 1991), especially macroporosity (Klute, 1982), improves internal drainage and saturated and unsaturated HC. Consequently, aeration is improved and plant growth is enhanced (Oster and Jayawardane, 1998). However, the beneficial effects of tillage on sodic soils are often not long lasting (Mead and Chan, 1988). Because sodic soils tend to reconsolidate under their own weight when wetted, hydraulic properties deteriorate once more.

Deep tillage (1–2 m) or ripping breaks up hardpans and cemented layers, mixes soil layers and causes, in general, a permanent improvement in soil structure and physical properties. In order to obtain optimal results, plowing should be carried out at the correct soil moisture content, 10% for clay, 5% for loam, and 2.5% (w/w) for sand (Wildman, 1981). However, deep tillage can sometime lead to waterlogging and infiltration problems by bringing sodic subsoil to the surface (McKenzie et al., 1992). Shallow tillage (disking or chiseling) is effective in reducing soil permeability problems associated with surface crusts, hardsetting and compaction. Shallow tillage requires repetition from time to time (Rawitz et al., 1986).

A combined treatment of tillage and gypsum application, termed gypsum slotting, has recently been developed in Australia, in which gypsum at rates sufficient to reclaim the treated soil is added in tilled slots created by tillage (~ 150 mm wide, 0.4 m deep and spaced about 1–2 m apart) (Jayawardane and Blackwell, 1986). The advantages of this technique include efficient removal of excess water from the surface layer and its storage in the subsoil for use during dry periods, reclamation of the soil in the slots, and maintenance of adequate macroporosity. In addition, compaction of the soil in the slots is prevented by the use of bridging implements (Blackwell et al., 1989). Results from field studies indicate that gypsum slotting improves aeration and infiltration and results in higher yields (Jayawardane et al., 1987). Gypsum slotting will not be effective where a shallow watertable exists, or where subsoil permeability is higher than that of the topsoil, or for crops that require prolonged ponding (Jayawardane and Chan, 1994).

An additional method of combining tillage with gypsum application consists of spreading gypsum on the soil surface at a rate calculated to reclaim the sodic soil to a predetermined depth (Equation [2.6]) and tilling to mix the gypsum into the soil, followed by an additional surface application of gypsum to maintain high IR and reduce exchangeable Na at the soil surface (Oster and Jayawardane, 1998).

2.8.3 Cropping

Instead of deep plowing, which at times could aggravate rather than improve soil conditions (Section 2.8.2), vigorous rooted rotation crops can also alleviate subsoil compaction. Hodgson and Chan (1984) showed that using safflower in rotation with cotton improved macroporosity and cotton root growth in a clay soil.

The presence of crops in a sodic soil can improve root zone conditions, especially in area where irrigation supplements rainfall. Jayawardane et al. (1994) showed that growing sunflower in a gypsum slotted soil led to efficient and rapid extraction of water causing the soil to crack; in subsequent irrigation/rainfall events, infiltration increased.

Another alternative to subsoil amelioration is the use of ridge instead of bed farming techniques. With ridges, the plants are located ~ 0.25 m above the neighboring furrows, and thus, aeration is improved, especially in heavy textured soils suffering from waterlogging in the subsoil (Hunter et al., 1980; Hearn and Constable, 1984).

2.9 Environmental Aspects of Sodic Soils

Assessment of Na-related soil degradation has usually been carried out in the context of agriculture and soil productivity. However, sodic behavior poses numerous environmental hazards in agricultural, engineering and urban settings. Detailed reviews of environmental consequences of soil sodicity have been published by Fitzpatrick et al. (1994) and Sumner et al. (1998).

2.9.1 Erosion and Suspended Sediments

Sodic soils are more susceptible to clay dispersion and surface sealing, which generate runoff flow and cause erosion, a serious problem from the crop production point of view. However, despite the fact that ~ 60% or more of the sediments settle out from the runoff water before it reaches a stream or other water body, there is a consensus that the sediments leaving cultivated fields have substantial negative impacts. Crosson (1985) estimated that the cost of offsite damage from cropped land in the

United States exceeded $1 billion annually. Sediments clog drainage ditches and irrigation canals, and decrease the capacity of water reservoirs, thus increasing maintenance costs. In addition, sediments damage or entirely destroy fish spawning areas. Sedimentation of rivers and streams reduces their capacity to carry water and increases the danger of flooding and consequent damage to property and to human health.

The presence of suspended sediments in water increases its turbidity. Under certain conditions common in sodic soils (high pH and ESP), SOM can dissolve in runoff water and enhance clay dispersion and mobilization (Fig. 2.15) (Naidu et al., 1993a), and hence, increase water turbidity. Turbid waters are less attractive for recreational activities and allow less sunlight to pass through, thus negatively affecting aquatic productivity. In addition, turbid water, which is denser than clean water, is more costly to pump. High turbidity in water bodies used for domestic consumption leads to extra costs in purification. Finally, turbid water increases the wear and tear and maintenance costs of equipment through which it passes.

Enhanced clay dispersivity associated with sodic condition in the soil can lead to surface and subsurface erosion processes such as piping and tunneling. The creation of these subsurface channels may result in the formation of deep gullies that further enhances erosion. Records show that tunnel erosion which occurs in many parts of the world, but mainly in kaolinitic and illitic soils (Sumner et al., 1998), requires readily dispersible clay for its development.

2.9.2 Salinity and Pollutants

Sodic soils commonly contain high levels of salts that are transferred to nearby water bodies at a relative fast rate when the soil is cultivated (Fitzpatrick et al., 1995). In Australia and America, high saline-sodic watertables contribute significantly to the salt load of rivers (Sharma and Williamson,

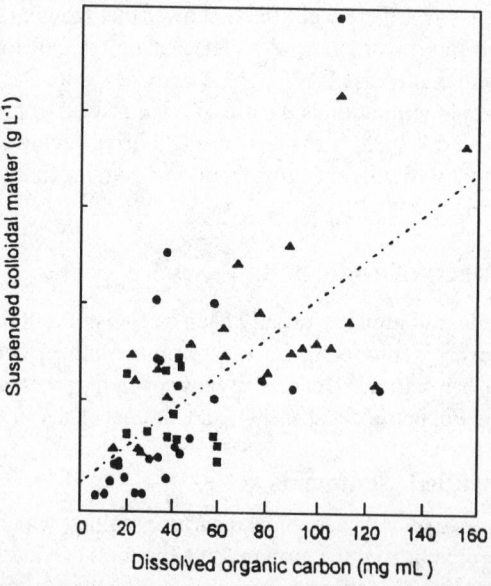

Fig. 2.15 Relationship between dissolved organic carbon (DOC) and sediment loading in subsurface throughflow water under pine (▲), native woodland (■) and pasture (●) [Reprinted from Naidu et al., 1993. Effect of land use on the composition of through flow water immediately above clayey B horizons in the Warren Catchment, South Australia. Aust. J. Exp. Agric. 33:239-244 with permission of CSIRO Publications, Melbourne, Australia]

1984). In addition, sodic soils require proper drainage systems to remove water, which is much more saline than the irrigation water. When this drainage water enters surface water, quality deteriorates, rendering it at times unsuitable for further use in irrigation, livestock watering or cities (Sumner et al., 1998).

In addition to the excessive salt load in sodic soils originating from pedogenic processes and irrigation with saline-sodic waters requiring disposal, there is a salt accumulation resulting from soil reclamation. Most reclamation activities require the addition of Ca salts (e.g., gypsum). In the soil, following exchange reactions, the salts become largely Na salts. In general, Na salts dissolve readily and are easily displaced downward into soil strata and groundwater beneath the irrigated field. Extensive reclamation can also cause shallow saline watertables which often require drainage. The saline drainage water ends up at surface water bodies, increases their salinity, and reduces their value. Hence, reclamation of sodic soils may have some negative environmental consequences that should be weighed carefully when planning and execution of reclamation activities are considered.

Transport of various contaminants by soil colloidal materials whether to ground or surface water bodies in runoff is a serious environmental hazard. Metals that usually are fairly immobile in soil and have strong affinity for soil colloids move together with the mobile colloids (Kaplan et al., 1993). Similarly, insoluble pesticides, herbicides, and some nutrients (P) have a high affinity for soil colloids and can thus be transported with suspended solids to adjacent water bodies and groundwater. The increased concentration of suspended colloids in runoff water due to sodic conditions greatly increases the danger of higher than permitted loadings of these pollutants in ground and surface water bodies.

2.10   Concluding Comments

Many soils world wide contain sodium at levels that may negatively affect their behavior. The area occupied by sodium-affected soils is expected to grow due to increased use of poor quality soils and water for agricultural production. Swelling and dispersion of the soil clay are the most important processes that occur in sodic soils that adversely affect soil structure and soil physical and hydraulic properties. The adverse effects of swelling and dispersion are predominant when the soil is exposed to good (i.e., low TEC) water. Sodicity harms crop production via deterioration in soil physical properties as well as by causing deficiencies of some important nutrients. Sodicity enhances soil erosion which causes both on site (cultivated fields) and off site (environmental) damages. However, sodium-related soil degradation and loss of productivity are, to a certain degree, reversible as sodic soils can be reclaimed. In addition, realization of sodicity hazards and consequent adoption of proper management allow safe and sustainable use of sodium-affected soils, without the danger of them being abandoned.

2.11   References

Aarstad, J.S., and D.E. Miller. 1973. Soil management practices for reducing runoff under center-pivot sprinkler system. J. Soil Water Conserv. 28:171–173.

Abrol, I.P., and D.R. Bhumbla. 1979. Crop responses to differential gypsum applications in a highly sodic soil and the tolerance of several crops to exchangeable sodium under field conditions. Soil Sci. 127:79–85.

Abu-Sharar, T.M., F.T. Bingham, and J.D. Rhoades. 1987a. Stability of soil aggregates as affected by electrolyte concentration and composition. Soil Sci. Soc. Am. J. 51:309–314.

Abu-Sharar, T.M., F.T. Bingham, and J.D. Rhoades. 1987b. Reduction in hydraulic conductivity in relation to clay dispersion and disaggregation. Soil Sci. Soc. Am. J. 51:342–346.

Adams, F. 1974. The soil solution. p. 441–480. *In* E.W. Carson. (ed.) The plant root and its environment. University of Virginia Press, Charlottesville, VA.

Agassi, M., J. Morin, and I. Shainberg. 1985a. Effect of raindrop impact energy and water salinity on infiltration rates of sodic soils. Geoderma 36:263–276.

Agassi, M., J. Morin, and I. Shainberg. 1985b. Infiltration and runoff in wheat fields in the semi-arid region of Israel. Soil Sci. Soc. Am. J. 49:186–190.

Agassi, M., I. Shainberg, and J. Morin. 1981. Effect of electrolyte concentration and soil sodicity on infiltration rate and crust formation. Soil Sci. Soc. Am. J. 45:848–851.

Agassi, M., I. Shainberg, and J. Morin. 1986. Effect of powdered phosphogypsum on the infiltration rate of sodic soils. Irrig. Sci. 7:53–61.

Ali, O.M., M. Yousaf, and J.D. Rhoades. 1987. Effect of exchangeable cation and electrolyte concentration on mineralogy of clay dispersed from aggregates. Soil Sci. Soc. Am. J. 51:896–900.

Alperovitch, N., and I. Shainberg. 1973. Reclamation of alkali soils with $CaCl_2$ solutions. p. 431–440. In A. Hadas et al. (ed.) Physical aspects of soil, water and salts in ecosystems. Springer-Verlag, Berlin, Germany.

Alperovitch, N., I. Shainberg, and R. Keren. 1981. Specific effect of magnesium on the hydraulic conductivity of sodic soils. J. Soil Sci. 32:543–554.

Alperovitch, N., I. Shainberg, and J.D. Rhoades. 1985. Effect of mineral weathering on the response of sodic soils to exchangeble magnesium. Soil Sci. Soc. Am. J. 50:901–904.

Aly, S., and J. Letey. 1990. Physical properties of sodium-treated soil as affected by two polymers. Soil Sci. Soc. Am. J. 54:501–504.

Arora, H. S., and N.T. Coleman. 1979. The influence of electrolyte concentration on flocculation of clay suspensions. Soil Sci. 127:134–139.

Arshad, M.A., and A.R. Mermut. 1988. Micromorphological and physicochemical characteristics of soil crust types in Northwestern Alberta, Canada. Soil Sci. Soc. Am. J. 52:724–729.

Aylmore, L.A.G., and J.P. Quirk. 1959. Swelling of clay-water systems. Nature 183:1752–1753.

Aylmore, L.A.G., and I.D. Sills. 1982. Characterization of soil structure and stability using modulus of rupture-exchangeable sodium percentage relationships. Aust. J. Soil Res. 20:213–224.

Bakker, A.C., and W.W. Emerson. 1973. The comparative effect of exchangeable calcium, magnesium and sodium on some physical properties of red-brown earth subsoils. III. The permeability of Shepperton soil and comparison methods. Aust. J. Soil Res. 11:159–165.

Banin, A., and N. Lahav. 1968. Particle size and optical properties of montmorillonite in suspension. Isr. J. Chem. 6:235–250.

Bar On, P., I. Shainberg, and I. Mochaeli. 1970. The electrophoretic mobility of Na/Ca montmorillonite particles. J. Coll. Interf. Sci. 33:471–472.

Barzegar, A.R., P.N. Nelson, J.M. Oades, and P. Rengasamy. 1996. Organic matter, sodicity and clay type: Influence on aggregation. Soil Sci. Soc. Am. J. 60:583–589.

Ben-Hur, M., I. Shainberg, D. Bakker, and R. Keren. 1985. Effect of soil texture and $CaCO_3$ content on water infiltration in crusted soils as related to water salinity. Irrig. Sci. 6:281–194.

Bernstein, L., and G.A. Pearson. 1956. Influence of exchangeable sodium on the yield and chemical composition of plants: 1. Green beans, garden beets, clover, and alfalfa. Soil Sci. 82:247–285.

Blackmore, A.V., and R.D. Miller. 1961. Tactoid size and osmotic swelling in Ca montmorillonite. Soil Sci. Soc. Am. Proc. 25:169–173.

Blackwell, P.S., N.S. Jayawardane, J. Blackwell, R. White, and R. Horn. 1989. Evaluation of soil recompaction by transverse wheeling of tillage slots. Soil Sci. Soc. Am. J. 53:11–15.

Blackwell, P.S., N.S. Jayawardane, T.W. Greene, J.T. Wood, J. Blackwell, and H.J. Beatty. 1991. Subsoil macropore space of a transitional Red-Brown earth after either deep tillage, gypsum or both. I. Physical effects and short term changes. Aust. J. Soil Res. 29:123–140.

Bodman, G.B., and M. Fireman. 1950. Changes in soil permeability and exchangeable cation status during flow of different irrigation waters. Trans. 4th Int. Congr. Soil Sci.1:397–400.

Bower, C.A., and J.O. Goertzen. 1958. Replacement of adsorbed sodium in soils by hydrolysis of calcium carbonate. Soil Sci. Soc. Am. Proc. 22:33–35.

Bower, C.A., G. Ogata, and J.M. Tucker. 1968. Sodium hazard of irrigation waters as influenced by leaching fraction and by precipitation or solution of calcium carbonate. Soil Sci. 106:29–34.

Bradford, J.M., J.E. Ferris, and P.A. Remley. 1987. Interrill soil erosion processes: I. Effect of surface sealing on infiltration, runoff, and soil splash detachment. Soil Sci. Soc. Am. J. 51:1566–1571.

Bresler, E. 1970. Numerical solution of the equation for interacting diffuse layers in mixed ionic systems with nonsymmetrical electrolyes. J. Coll. Interf. Sci. 33:278–283.

Bresler, E., B.L. McNeal, and D.L. Carter. 1982. Saline and sodic soils: Principles, dynamics, modeling. Springer-Verlag, Berlin, Germany.

Bui, E.N., L. Krogh, R.S. Lavado, F.O. Nachtergaele, T. Toth, and R.W. Fitzpatrick. 1998. p. 35–50. *In* M.E. Sumner and R. Naidu (ed.) Sodic soils. Oxford University Press, New York, NY.

Cairns, R.R., R.A. Milne, and E.W. Bowser. 1962. A nutritional disorder in barley seedlings grown on alkali Solonetz soil. Can. J. Soil Sci. 42:1–6.

Cairns, R.R., W.E. Bowser, R.A. Milne, and P.C. Chang. 1967. Effect of nitrogen fertilization of bromegrass on Solonetzic soils. Can. J. Soil Sci. 47:1–6.

Carter, M.R., and G.R. Webster. 1990. Use of calcium-to-total cation ratio as an index of plant-available calcium. Soil Sci. 149:212–217.

Cartwright, B., B.A. Zarcinas, and L.R. Spouncer. 1986. Boron toxicity South Australian barley crops. Aust. J. Agric. Res. 37:351–359.

Cass, A., and M.E. Sumner. 1974. Use of gypsum and High salt water in reclamation of sodic soil. Trans. 10th Int. Congr. Soil Sci. 10:118–127.

Cass, A., and M.E. Sumner. 1982a. Soil pore structural stability and irrigation water quality: I. Empirical sodiu stability model. Soil Sci. Soc. Am. J. 46:503–506.

Cass, A., and M.E. Sumner. 1982b. Soil pore structural stability and irrigation water quality: II. Sodium stability data. Soil Sci. Soc. Am. J. 46:507–512.

Cass, A., and M.E. Sumner. 1982c. Soil pore structural stability and irrigation water quality: III. Evaluation of soil stability and crop yield in relation to salinity and sodicity. Soil Sci. Soc. Am. J. 46:513–517.

Chang, C.W., and H.E. Dregne. 1955. Effect of exchangeable sodium on soil properties and on growth and cation content on alfalfa and cotton. Soil Sci Soc. Am. Proc. 19:29–35.

Chaudhry, G.H., and B.P. Warkentin. 1968. Studies on exchange of sodium from soils by leaching with calcium sulfate. Soil Sci. 105:190–197.

Chen, J., J. Tarchitzky, J. Morin, and A. Banin. 1980. Scanning electron microscope observations on soil crust and their formation. Soil Sci. 130:49–55.

Chhabra, R. 1985. Crop responses to phosphorus and potassium fertilization of a sodic soil. Soil Sci. Soc. Am. J. 77:699–702.

Chi, C.L., W.W. Emerson, and D.G. Lewis. 1977. Exchangeable calcium, magnesium and sodium and the dispersion of illites in water. I. Characterization of illites and exchange reactions. Aust. J. Soil Res. 15:243–253.

Chiang, S.D., D.E. Radcliffe, W.P. Miller, and K.D. Newman. 1987. Hydraulic conductivity of three southeastern soils as affected by sodium, electrolyte concentration and pH. Soil Sci. Soc. Am. J. 51:1293–1299.

Childs, E.C., and N. Collis-George. 1950. The permeability of porous materials. Proc. Roy. Soc. Ser. A. 201:392–405.

Coughlan, K.J., and R.J. Loch. 1984. The relationship between aggregation and other soil properties in cracking clay soils. Aust. J. Soil Res. 22:59–88.

Crosson, P. 1985. Impact of erosion on land productivity and water quality in the United States. p. 217–236. *In* S.A. El-Swaify, W.C. Moldenhauer, and A. Lo (ed.). Soil erosion and conservation. Soil Conservation Society of America, Ankeny, IA.

Curtin, D., and R. Naidu. 1998. Fertility constraints to plant production. p. 125–142. *In* M.E. Sumner and R. Naidu (ed.) Sodic soils. Oxford University Press, New York, NY.

Curtin, D., H. Steppuhn, and F. Selles. 1994. Effects of magnesium on cation selectivity and structural stability of sodic soils. Soil Sci. Soc. Am. J. 58:730–737.

Dahiya, I.S., and R. Anlauf. 1990. Sodic soils in India: Their reclamation and management. Z. Kulturtech. Landentwick. 31:26–34.

Duley, F.L. 1939. Surface factors affecting the rate of intake of water by soils. Soil Sci. Soc. Am. Proc. 4:60–64.

Durgin, P.B., and J.G. Chaney. 1984. Dispersion of kaolinite by dissolved organic matter from Douglas-fir roots. Can J. Soil Sci. 64:445–455.

Dutt, C. R., and R.W. Terkeltoub. 1972. Prediction of gypsum and leaching requirements for sodium-affected soil. Soil Sci. 114:93–103.

Ellis, J.H., and O.G. Caldwell. 1935. Magnesium clay solonetz. Trans. 3rd Int. Congr. Soil Sci. I:348–350.

El-Swaify, S.A. 1973. Structural changes in tropical soils due to anions in irrigation water. Soil Sci. 24:137–144.

Emerson, W.W. 1954. The determination of the stability of soil crumbs. J. Soil Sci. 5:235–250.

Emerson, W.W., and A.C. Bakker. 1973. The comparative effects of exchangeable calcium, magnesium, and sodium on some physical properties of red-brown earth subsoils. II. The spontaneous dispersion of aggregates in water. Aust. J. Soil Res. 11:151–157.

Emerson, W.W., and C.L. Chi. 1977. Exchangeable Ca, Mg and Na and the dispersion of illites. II. Dispersion of illites in water. Aust. J. Soil Res. 15:255–263.

Emerson, W.W., and B.H. Smith. 1970. Magnesium, organic matter and soil structure. Nature 228:453–454.

Epstein, E., and W.J. Grant. 1973. Soil crust formation as affected by raindrop impact. p. 195–201. *In* A. Hadas, D. Swartzendruber, P.E. Rijtema, M. Fuchs and B. Yaron. (ed.) Physical aspects of soil water and salts in ecosystems. Springer Verlag, New York, NY.

FAO. 1978. Soil map of the world. VIII. North and Central Asia. UNESCO, Paris, France.

FAO. 1991. World soil resources. An explanatory note on the FAO world soil resources map at 1:25000000 scale. World Soil Resources Rep. 66. Food and Agriculture Organization, Rome, Italy.

Felhendler, R., I. Shainberg, and H. Frenkel. 1974. Dispersion and hydraulic conductivity of soils in mixed solution. Trans. 10th Int. Congr. Soil Sci. I:103–112.

Fireman, M., and G.B. Bodman. 1939. The effect of saline irrigation water upon the permeability and base status of soils. Soil Sci. Soc. Am. Proc. 4:71–77.

Fitzpatrick, R.W., S.C. Boucher, R. Naidu, and E. Fritsch. 1994. Environmental consequences of soil sodicity. Aust. J. Soil Res. 32:1069–93.

Fitzpatrick, R.W., S.C. Boucher, R. Naidu, and E. Fritsch. 1995. Environmental consequences of soil sodicity. p. 163–176. *In* R. Naidu, M.E. Sumner and P. Rengasamy (ed.). Australian sodic soils: Distribution, properties and management. CSIRO Publications, Melbourne, Australia.

Frenkel, H. 1984. Reassessment of water quality criteria for irrigation. p. 143–172. *In* I. Shainberg and J. Shalhevet (ed.) Soil salinity under irrigation. Springer-Verlag, Berlin, Germany.

Frenkel, H., and I. Shainberg. 1975. Chemical and hydraulic changes in soils irrigated with brackish water. p.175–198. *In* Irrigation with brackish water. Intern. Symp. Beersheva, Israel.

Frenkel, H., J.O. Goertzen, and J.D. Rhoades. 1978. Effects of clay type and content, exchangeable sodium percentage, and electrolyte concentration on clay dispersion and soil hydraulic conductivity. Soil Sci. Soc. Am. J. 48:32–39.

Frenkel, H., M.V. Fey, and G.J. Levy. 1992. Organic and inorganic anion effects on reference and soil clay critical flocculation concentration. Soil Sci. Soc. Am. J. 56:1762–1766.

Goldberg, S., and R.A. Glaubig. 1987. Effect of saturating cation, pH, and aluminum and iron oxides on the flocculation of kaolinite and montmorillonite. Clay Clay Min. 35:220–227.

Goldberg, S., and H.S. Forster. 1990. Flocculation of reference clays and arid-zone soil clays. Soil Sci. Soc. Am. J. 54:714–718.

Goldberg, S., D.L. Suarez, ad R.A. Glaubic. 1988. Factors affecting clay dispersion and aggregate stability of arid zone soils. Soil Sci. 146:317–325.

Grattan, S.R., and C.M. Grieve. 1992. Mineral element acquisition and growth response of plants in saline environments. Agric. Ecosys. Environ. 38:275–300.

Greene, R.S.B., A.M. Posner, and J.P. Quirk. 1978. A study of the coagulation of montmorillonite and illite suspensions by $CaCl_2$ using the electron microscope. p. 35–40. *In* W.W. Emerson, R.D. Bond and A.R. Dexter (ed.). Modification of soil structure. John Wiley and Sons, New York, NY.

Greenland, D.J. 1979. Structural organization of soils and crop production. p. 47–56. *In* R. Lal, and D.J. Greenland (ed.) Soil physical properties and crop production in the tropics. John Wiley and Sons, New York, NY.

Grieve, C.M., and E.V. Maas. 1988. Differential effects of sodium/calcium ratio on sorghum genotypes. Crop Sci. 28:659–665.

Gupta, R.K., and I.P. Abrol. 1990. Salt-affected soils: Their reclamation and management for crop production. Adv. Soil Sci. 11:223–288.

Gupta, R.K., R.R. Singh, and K.K. Tanji. 1990. Phosphorus release in sodium ion dominated soils. Soil Sci. Soc. Am. J. 54:1245–1260.

Hearn, A.B., and G.A. Constable. 1984. Irrigation for crops in a sub-humid environment. VII. Evaluation of irrigation strategies for cotton. Irrig. Sci. 5:75–94.

Helalia, A.M., and J. Letey. 1988. Cationic polymer effects on infiltration rates with a rainfall simulator. Soil Sci. Soc. Am. J. 52:247–250.

Hensley, M. 1969. Selected properties affecting the irrigable value of some Makatini soils, M.Sc. Agric. Thesis, University of Natal, Pietermaritzburg, South Africa.

Hillel, D. 1980. Applications of soil physics. Academic Press, New York, NY.

Hodgson, A.S., and K.Y. Chan. 1984. Deep moisture extraction and crack formation by wheat and safflower in a Vertisol following irrigated cotton rotations. p. 299–304. *In* J.W. McGarity, E.H. Hoult, and H.B. So. (ed.) The properties and utilization of cracking clay soils. Unversity of New England, Armidale, Australia.

Hodgson, J.F., W.L. Lindsay, and J.F. Trierweiler. 1966. Micronutrient cation complexing in soil solution: II. Complexing of zinc and copper in displaced solution from calcareous soils. Soil Sci. Soc. Am. Proc. 30:723–726.

Hunter, M.N., P.L.M. de Jarbun, and D.E. Byth. 1980. Response of nine soybean lines to soil moisture conditions close to saturation. Aust. J. Exp. Agric. Anim. Husb. 20:339–345.

Inskeep, W.P., and P.R. Bloom. 1986. Calcium carbonate supersaturation in soil solution of calciaquolls. Soil Sci. Soc. Am. J. 50:1431–1437.

Jayawardane, N.S., and J. Blackwell. 1986. Effects of gypsum-enriched slotting on infiltration rates and moisture storage in swelling clay soil. Soil Use Manag. 2:114–118.

Jayawardane, N.S., and K.Y. Chan. 1994. The management of soil physical properties limiting crop production in Australian sodic soils: A review. Aust. J. Soil Res. 32:13–44.

Jayawardane, N.S., J. Blackwell, G. Kirchof, and W.A. Muirhead. 1994. Slotting: A deep tillage technique for ameliorating sodic, acid and other degraded soils and for land treatment of waste. p. 109–146. *In* N.S. Jayawardane and B.A. Stewart (ed.) Subsoil management techniques. Lewis Publishers, Boca Raton, FL.

Jayawardane, N.S., J. Blackwell, and M. Stapper. 1987. Effects of changes in moisture profiles of a transitional Red Brown earth due to surface and slotted gypsum applications. Aust. J. Agric. Res. 38:239–251.

Johnston, M.A. 1975. The effects of different levels of exchangeable sodium on soil hydraulic conductivity. Proc. S. Afr. Sugar Tech. Assoc. 49:142–147.

Kaplan, D.I., P.M. Bertsch, D.C. Adriano, and W.P. Miller. 1993. Soil borne mobile colloids as influenced by water flow and organic carbon. Environ. Sci. Tech. 27:1193–1200.

Kazman, Z., I. Shainberg, and M. Gal. 1983. Effect of low levels of exchangeable Na and applied phosphogypsum on the infiltration rate of various soils. Soil Sci. 35:184–192.

Keren, R. 1990. Water-drop kinetic energy effect on infiltration in sodium-calcium-magnesium soils. Soil Sci. Soc. Am. J. 54:983–987.

Keren, R. 1995. Reclamation of sodic-affected soils. p. 353–374. *In* M. Agassi (ed.) Soil erosion conservation and rehabilitation. Marcel Dekker Inc., New York, NY.

Keren, R., and I. Shainberg. 1981. Effect of dissolution rate on the efficiency of industrial and mined gypsum on surface runoff from loess soil. Soil Sci. Soc. Am. J. 45:103–107.

Keren, R., I. Shainberg, H. Fenkel, and Y. Kalo. 1983. The effect of exchangeable sodium and sypsum on surface runoff from loess soil. Soil Sci. Soc. Am. J. 47:1001–1004.

Keren, R., and M.J. Singer. 1988. Effect of low electrolyte concentration on hydraulic conductivity of sodium/calcium montmorillonite system. Soil Sci. Soc. Am. J. 52:368–373.

Klute, A. 1982. Tillage effects on hydraulic properties of a soil: A review. p. 29–43. *In* P.W. Unger, D.M. van Doren, F.D. Whisle, and E.L. Skidmore (ed.) Predicting tillage effects on soil physical properties and processes. American Society of Agronomy, Madison, WI.

Kozeny, J. 1927. Uber kapillare leitung des wesses in boden. Zitzungsber Akad. Wiss. Wein. 136:271–306

Kreit, J.E., I. Shainberg, and A.J. Herbillon. 1982. Hydrolysis and decomposition of hectorites in dilute solutions. Clays Clay Miner. 30:223–231.

Lagerwerff, J.V., F.S. Nakayama, and M.H. Free. 1969. Hydraulic conductivity related to porosity and swelling of soil. Soil Sci. Soc. Am. Proc. 33:3–11.

Lauchli, A., and E. Epstein. 1990. Plant responses to saline and sodic conditions. p. 113–137. *In* Agricultural salinity assessment and management. Am. Soc. Civil Engineer., New York., NY.

Levitt, J. 1980. Responses of plants to environmental stress. Vol. II. Academic Press, New York, NY.

Levy, G.J., and H.v.H. van der Watt. 1988. Effects of clay mineralogy and soil sodicity on the infiltration rates of soils. S. Afr. J. Plant Soil 5:92–96.

Levy, G.J., M. Agassi, H.J.C. Smith, and R. Stern. 1993. Microaggregate stability of kaolinitic and illitic soils determined by ultrasonic energy. Soil Sci. Soc. Am. J. 57:803–808.

Levy, G.J., N. Alperovitch, A.J. van der Merwe, and I. Shainberg. 1989. The hydrolysis of kaolinitic soils as affected by the type of the exchangeable cation. J. Soil Sci. 40:613–620.

Levy, G.J., M. Ben-Hur, and M. Agassi. 1991. The effect of polyacrylamide on runoff, erosion and cotton yield from fields irrigated with moving sprinkler systems. Irrig. Sci. 12:55–60.

Levy, G.J., J. Levin, M. Gal, M. Ben-Hur, and I. Shainberg. 1992. Polymer effects on infiltration and soil erosion during consecutive simulated sprinkler irrigations. Soil Sci. Soc. Am. J. 56:902–907.

Levy, G.J., J. Levin, and I. Shainberg. 1994. Seal formation and interrill soil erosion. Soil Sci. Soc. Am. J. 58:203–209.

Levy, G.J., J. Levin, and I. Shainberg. 1995. Polymer effects on runoff and soil erosion from sodic soils. Irrig. Sci. 16:9–14.

Levy, G.J., H.v.H. van der Watt, and H.M. du Plessis. 1988. Effect of sodium-magnesium and sodium-calcium systems on soil hydraulic conductivity and infiltration. Soil Sci. 146:303–310.

Loveday J. 1976. Relative significance of electrolyte and cation exchange effects when gypsum is applied to a sodic clay soil. Aust. J. Soil Res. 14:361–371.

Loveland, P.J., J. Hazelden, and R.G. Sturdy. 1987. Chemical properties of salt-affected soils in north Kent and their relationship to soil instability. J. Agric. Sci. 109:1–6.

Maas, E.V., and G.J. Hoffman. 1977. Crop salt tolerance–current assessment. J. Irrig. Drain. Div. ASCE 103:115–134.

Malhi, S.S., D.W. McAndrew, and M.R. Carter. 1992. Effect of tillage and N fertilization of a Solonetzic soil on barley production and some soil properties. Soil Till. Res. 2:95–107.

Manin, M., A. Pissarra, and J.W. van Hoorn. 1982. Drainage and desalinization of heavy clay soil in Portugal. Agric. Water Manag. 5:227–240.

Marshall, T.J. 1958. A relation between permeability and size distribution of pores. J. Soil Sci. 9:1–8.

Massoud, F.I. 1977. Basic principals for prognosis and monitoring of salinity and sodicity. p. 432–454. *In* Proc. Int. Conf. on Management of Saline Water for Irrigation. Texas Technical Univ., Lubbock, TX.

McAtee, J.L. 1961. Heterogeneity in montmorillonites. Clays Clay Miner. 5:279–288.

McIntyre, D.S. 1958. Permeability measurement of soil crusts formed by raindrop impact. Soil Sci. 85:185–189.

McIntyre, D.S. 1979. Exchangeable sodium, subplasticity and hydraulic conductivity of some Australian soils. Aust. J. Soil Res. 17:115–120.

McKenzie, D.C., D.J.M. Hall, T.S. Abbott, A.M. Kay, and J.D. Sykes. 1992. Soil management for irrigated cotton. NSW Agric. Agfact 6:53.

McNeal, B.L., and N.T. Coleman. 1966. Effect of solution composition on soil hydraulic conductivity. Soil Sci. Soc. Am. Proc. 20:308–312.

McNeal, B.L., D.A. Layfield, W.A. Norvell, and J.D. Rhoades. 1968. Factors influencing hydraulic conductivity of soils in the presence of mixed-salt solutions. Soil Sci. Soc. Am. Proc. 32:187–190.

McNeal, B.L., W.A. Norvell, and N.T. Coleman. 1966. Effect of solution composition on soil hydraulic conductivity and on the swelling of extracted soil clays. Soil Sci. Soc. Am. Proc. 30:308–315.

Mead, J.A., and K.Y. Chan. 1988. Effect of deep tillage and seedbed preparation on the growth and yield of wheat on a hardsetting soil. Aust. J. Exp. Agr. 28:491–498.

Miller, R. J., J.W Biggar, and D.R. Nielsen. 1965. Chloride displacement in Panoche clay loam in relation to water movement and distribution. Water Resour. Res. 1:63–73.

Miller, W.P. 1987. Infiltration and soil loss of three gypsum-amended Ultisols under simulated rainfall. Soil Sci. Am. J. 51:1314–1320.

Miller, W.P., and M.K. Baharuddin. 1986. Relationship of soil dispersibility to infiltration and erosion of Southeastern soils. Soil Sci. 142:235–240.

Miller, W.P., and J. Scifres. 1988. Effect of sodium nitrate and gypsum on infiltration and erosion of a highly weathered soil. Soil Sci. 148:304–309.

Miller, W.P., H. Frenkel, and K.D. Newman. 1990. Flocculation concentration and sodium/calcium exchange of kaolinitic soil clays. Soil Sci. Soc. Am J. 54:346–351.

Miyamoto, S. R. L. Prather, and J. L. Stroelhein 1975. Potentially beneficial uses of sulfuric acid in southwestern agriculture. J. Environ. Qual. 4:431–437.

Mohammed, E.T.Y., J. Letey, and R. Branson. 1979. Sulphur compounds in water treatment. Sulph. Agric. 3:7–11.

Morin, J., and Y. Benyamini. 1977. Rainfall infiltration into bare soils. Water Res. Res. 13: 813–817.

Nadler, A., G.J. Levy, R. Keren, and H. Eisenberg. 1996. Sodic calcareous soil reclamation as affected by water chemical composition and flow rate. Soil Sci. Soc. Am. J. 60:252–257.

Naidu, R. 1993. Distribution, properties and management of sodic soils: An introduction. Aust. J. Soil Res. 31:681–682.

Naidu, R., and P. Rengasamy. 1993. Ion interactions and constraints to plant nutrition in Australian sodic soils. Aust. J. Soil Res. 31:801–819.

Naidu, R., R.W. Fitzpatrick, I.O. Hollingsworth, and D.R. Williamson. 1993a. Effect of land use on the composition of through flow water immediately above clayey B horizons in the Warren Catchment, South Australia. Aust. J. Exp. Agric. 33:239–244.

Naidu, R., R.H. Merry, G.J. Churchman, M.J. Wright, R.I. Murray, R.W. Fitzpatrick, and B.A. Zarcinas. 1993b. Sodicity in South Australia: A review. Aust. J. Soil Res. 31:911–929.

Nelson, P.N., and J.M. Oades. 1998. Organic matter, sodicity and soil structure. p. 67–91. *In* M.E. Sumner and R. Naidu (ed.) Sodic soils. Oxford University Press, New York, NY.

Nielsen, D. R., J.W. Biggar, and J.N. Luthin. 1966. Desalinization of soils under controlled unsaturated flow conditions. p. 19.15–19.24. *In* Comm. 6th Int. Congr. Irrig. and Drainage, New Delhi, India.

Northcote, K.H., and J.K.M. Skene. 1972. Australian soils with saline and sodic properties. CSIRO Soil Publ. 27.

Oster, J.D., and N.S. Jayawardane. 1998. Agricultural management of sodic soils. p. 143–165. *In* M.E. Sumner and R. Naidu (ed.) Sodic soils. Oxford University Press, New York, NY.

Oster, J.D., and F.W. Schroer. 1979. Infiltration as influenced by irrigation water quality. Soil Sci. Soc. Am. J. 43:444–447.

Oster, J.D., I. Shainberg, and J.D. Wood. 1980. Flocculation value and gel structure of Na/Ca montmorillonite and illite suspensions. Soil Sci. soc. Am. J. 44:955–959.

Pearson, G.A. 1960. Tolerance of crops to exchangeable sodium. USDA Inf. Bull. 216.

Pearson, G.A., and L. Bernstein. 1958. Influence of exchangéable sodium on yield and chemical composition of plants: II. Wheat, barley, oats, rice, tall fescue, and wheatgrass. Soil Sci. 86:254–261.

Prather, R. J., J.O. Goertzen, O. Rhoades, and H. Frenkel. 1978. Efficient amendment use in sodic soil reclamation. Soil Sci. Soc. Am. J. 42:782–786.

Pupisky, H., and I. Shainberg. 1979. Salt effects on the hydraulic conductivity of a sandy soil. Soil Sci. Soc. Am. J. 43:429–433.

Quirk, J.P. 1978. Some physico-chemical aspects of soil structural stability: A review. p. 3–16. *In* W.W. Emerson, R.D. Bond, and A.R. Dexter (ed.). Modification of soil structure. John Wiley and Sons, New York, NY.

Quirk, J.P., and L.A.G. Aylmore. 1971. Domains and quasi-crystalline regions in clay systems. Soil Sci. Soc. Am. Proc. 35:652–654.

Quirk, J.P., and R.K. Schofield. 1955. The effect of electrolyte concentration on soil permeability. J. Soil Sci. 6:163–178.

Radcliffe, D.E., W.P. Miller, and S.C Chiang. 1987. Effect of soil dispersion on surface run-off in Southern Piedmont soils. p. 1–28. Environmental Resource Center, Georgia Institute of Technology, Atlanta, GA.

Rahman, A.W., and D.L. Rowell. 1979. The influence of Mg in saline and sodic soils: A specific effect or a problem of cation exchange. J. Soil Sci. 30:535–546.

Rao, D.L.N., and S.K. Ghai. 1985. Urease and dehydrogenase activity of alkali and reclaimed soils. Aust. J. Soil Res. 23:661–665.

Rawitz, E., W.B. Hoogmoed, and J. Morin. 1986. The effects of tillage practices on crust properties, infiltration and crop response under semi-arid conditions. p. 278–284. *In* F. Callebaut, D. Gabriels and M. de Boodt (ed.) Assessment of soil surface sealing and crusting. Flanders Research Centre for Soil Erosion and Soil Conservation, Ghent, Belgium.

Reeve, R.C., and C.A. Bower 1960. Use of high-salt waters as a flocculant and source of divalent cations for reclaiming sodic soils. Soil Sci. 90:139–144.

Reeve, R.C., and E.J. Doering. 1966. The high-salt-water dilution method for reclaiming sodic soils. Soil Sci. Soc. Am. Proc. 30:498-504.

Rengasamy, P. 1987. Importance of calcium in irrigation with saline sodic water–a view point. Agric. Water Manage. 12:207–219.

Rengasamy, P. and K.A. Olssen. 1991. Sodicity and soil structure. Aust. J. Soil Res. 29:935–952.

Rengasamy, P., and M.E. Sumner. 1998. Processes involved in sodic behavior. p. 51–67. *In* M.E. Sumner and R. Naidu (ed.) Sodic soils. Oxford University Press, New York, NY.

Rengasamy, P., R.S.B. Greene, and G.W. Ford. 1986. Influence of magnesium on aggregate stability in sodic red-brown earths. Aust. J. Soil Res. 24:229–237.

Rengasamy, P., R.S.B. Greene, G.W. Ford, and A.H. Mehammi. 1984. Identification of dispersive behaviour and the management of red-brown earths. Aust. J. Soil Res. 22:413–431.

Rhoades, J.D. 1968. Mineral weathering correction for estimating the sodium hazard of irrigation waters. Soil Sci. Soc. Am. Proc. 32:648–651.

Rhoades, J.D. 1982. Reclamation and management of salt affected soils after drainage. p. 123–197. *In* Proc. 1st Ann. Western Prov. Conf. Ration. Water Soil Res. Manag. Lethbridge, Alberta, Canada.

Rhoades, J.D., and R.D. Ingvalson. 1969. Macroscopic swelling and hydraulic conductivity properties of four vermiculite soils. Soil Sci Soc. Am. Proc. 33:364–369.

Rhoades, J.D., D.B. Kruger, and M.J. Reed. 1968. The effect of soil mineral weathering on the sodium hazard of irrigaiton waters. Soil Sci. Soc. Am. Proc. 32:643–647.

Rimmer, D.L., and D.J Greenland. 1976. Effect of $CaCO_3$ on the swelling of a soil clay. J. Soil Sci. 27:129–139.

Rowell, D.L., D. Payne, and N. Ahmad. 1969. The effect of the concentration and movment of solutions on the swelling, dispersion and movement of clay in saline and alkali soils. J. Soil Sci. 20:176–188.

Russo, D., and E. Bresler. 1977. Analysis of the saturated and unsaturated hydraulic conductivity in mixed sodium and calcium soil systems. Soil Sci. Soc. Am. J. 41:706–712.

Schofield, R.K., and H.R. Samson. 1954. Flocculation of kaolinite due to the attraction of oppositely charged crystal faces. Disc. Farad. Soc. 18:138–145.

Shainberg, I., and M. Gal. 1982. The effect of lime on the response of soils to sodic conditions. J. Soil Sci. 33:489–498.

Shainberg, I., and J. Letey. 1984. Response of soils to sodic and saline conditions. Hilgardia 52:1–57.

Shainberg, I., and G.J. Levy. 1994. Organic polymers and soil sealing in cultivated soils. Soil Sci. 158:267–273.

Shainberg, I., and H. Otoh. 1968. Size and shape of montmorillonite particles saturated with Na/Ca ions. Isr. J. Chem. 6:251–259.

Shainberg, I., N. Alperovitch, and R. Keren. 1988. Effect of magnesium on the hydraulic conductivity of sodic smectite-sand mixtures. Clays Clay Miner. 36:432–438.

Shainberg, I., E. Bresler, and Y. Klausner. 1971. Studies on Na/Ca montmorillonite systems. I. The swelling pressure. Soil Sci. 111:214–219.

Shainberg, I., G.J. Levy, P. Rengasamy, and H. Frenkel. 1992. Aggregate stability and seal formation as affected by drops' impact energy and soil amendments. Soil Sci. 154:113–119.

Shainberg, I., J.D. Rhoades, and R.J. Prather. 1981a. Effect of low electrolyte concentration on clay dispersion and hydraulic conductivity of a sodic soil. Soil Sci. Soc. Am. J. 45:273–277.

Shainberg, I., J.D. Rhoades, D.L. Suarez, and R.J. Prather. 1981b. Effect of mineral weathering on clay dispersion and hydraulic conductivity of sodic soils. Soil Sci. Soc. Am. J. 45:287–291.

Shainberg, I., D. Warrington, and P. Rengasamy. 1990. Effect of PAM and gypsum application on rain infiltration and runoff. Soil Sci. 149:301–307.

Sharma, M.L., and D.R. Williamson. 1984. Secondary salinization of water resources in Southern Australia. p. 149–163. In R.H. French (ed.) Salinity in water resources and reservoirs. Butterworth Publishers, Boston, MA.

Simunek, J., and D.L. Suarez. 1997. Sodic soil reclamation using multicomponent transport modeling. J. Irrig. Drainage Engin. 123:367–375.

Singh, G., N.T. Singh, and I.P. Abrol. 1994. Agroforestry techniques for the rehabilitation of degraded salt-affected lands in India. Land Degrad. Rehabil. 5:223–242.

Singh, S.B., R. Chhabra, and I.P. Abrol. 1979. Effect of exchangeable sodium on the yield and chemical composition of raya (Brassica juncea L.) Agron. J. 71:767–770.

So, H.B., and G.D. Cook. 1993. The effect of dispersion on the hydraulic conductivity of surface seals in clay soils. Soil Tech. 6:325–330.

Sposito, G., and S.V. Mattigod. 1977. On the chemical foundation of the sodium adsorption ratio. Soil Sci. Soc. Am. J. 41:323–329.

Sposito, G., K.M. Haltzclaw, and C.S. LeVesque-Madore. 1977. Calcium ion complexation by fulvic acid extracts from sewage sludge-soil mixtures. Soil Sci. Soc. Am. J. 42:600–606.

Stern, R., M. Ben-Hur, and I. Shainberg. 1991a. Clay mineralogy effect on rain infiltration, seal formation and soil losses. Soil Sci. 152:455–462.

Stern, R., M.C. Laker, and A.J. van der Merwe. 1991b. Field studies on the effect of soil conditioners and mulch on runoff from kaolinitic and illitic soils. Aust. J. Soil Res. 29:249–261.

Stevenson, F.J. 1992. Humus chemistry: Genesis, composition, reactions. 2nd Ed. John Wiley and Sons, New York, NY.

Suarez, D.L. 1981. Relationship between pH_c and sodium adsorption ratio (SAR) and an alternative method of estimating SAR of soil or drainage water. Soil Sci. Soc. Am. J. 45:469–475.

Sumner M.E. 1993. Sodic soils: New perspectives. Aust. J. Soil Res. 31:683–750.

Sumner, M.E., and R. Naidu. 1998. Sodic soils. Oxford University Press, New York, NY..

Sumner, M.E., W.P. Miller, R.S. Kookana, and P. Hazelton. 1998. Sodicity disperison and environmental quality. p. 167–231. In M.E. Sumner and R. Naidu (ed.) Sodic soils. Oxford University Press, New York, NY.

Szabolcs, I. 1979. Review of research on salt affected soils. UNESCO. Paris France.

Tanji, K. K., L.D. Doneen, G.V. Ferry, and R.S. Ayers. 1972. Computer simulation analysis on reclamation of salt-affected soil in San Joaquin Valley, California. Soil Sci Soc. Am. J. 36:127–133.

Tarchitzky, J., Y. Chen, and A. Banin. 1993. Humic substances and pH effects on sodium-and calcium-montmorillonite flocculation and dispersion. Soil Sci. Soc. Am. J. 57:367–372.

Tisdall, J.M. and J.M. Oades. 1982. Organic matter and water-stable aggregates. J. Soil Sci. 33:141–163.

Toogood, J.A. 1978. Fertility status of Solonetzic soils. p. 32–50. *In* J.A Toogood, and R.R. Cairns (ed.). Solonetzic soils technology and management in Alberta. University of Alberta Bull., Edmonton, Alberta, Canada.

USSL Staff. 1954. Diagnosis and Improvement of saline and alkali soils. Agric. Handb. 60. USDA, Washington, DC.

van der Merwe, A.J., and R. Burger. 1969. The influence of exchangeable cations on certain physical properties of a saline-alkali soil. Agrochemophys. 1:63–66.

van Olphen, H. 1977. An introduction to clay colloid chemistry. 2nd Ed. John Wiley and Sons, New York, NY.

Wildman, W.E. 1981. Managing and modifying problem soils. Univ. CA Coop. Ext. Leaf. 2791.

Williams, C.H., and J.D. Colwell. 1977. Inorganic chemical properties. p. 105–126. *In* J.S. Russell, and E.L. Greacen (ed.) Soil factors in crop production in a semi-arid environment. University of Queensland Press, St. Lucia, Australia.

Williams, C.H., and M. Raupach. 1983. Plant nutrients in Australian soils. *In* Soils: An Australian viewpoint. p. 777–793. Academic Press, London, UK.

Yaron, B., and G.W. Thomas. 1968. Soil hydraulic conductivity as affected by sodic water. Water Res. Res. 4:545–552.

Yousaf, M., O.M. Ali, and J.D. Rhoades. 1987. Dispersion of clay from salt-affected arid land soil aggregates. Soil Sci. Soc. Am. J. 51:920–924.

3

Hardsetting Soils

Chris E. Mullins
University of Aberdeen

3.1 Identification and Definition of Hardsetting Soils

3.1.1 Introduction

Hardsetting soils undergo structural breakdown during wetting and then set to a hard structureless mass during drying. Although many soils behave in this way, only those soils that set hard enough to become difficult or impossible to cultivate are classified as hardsetting. Thus the definition of hardsetting is a practical one that has been framed with cultivation and cropping in mind. Where hardsetting soils exist, their behavior commonly exerts a dominant influence on soil management and crop productivity and may result in serious environmental problems where it results in runoff and erosion.

3.1.2 Origin and History of Hardsetting

The term hardsetting was first used by Northcote (1960, 1979) for use in his classification system of Australian soils, where the term is now widely used. This, and similar terms have also been used elsewhere, for example, in Sweden (Heinonen, 1982) and in India, where Abrol (personal communication) has referred to this type of behavior as self-compaction. However, in the absence of a widely disseminated description of hardsetting behavior and of the processes involved, no generally accepted descriptor has been available until recently, and this has delayed recognition of the widespread nature of hardsetting and of the problems associated with it. A further reason for this delay is that because hardsetting behavior is something that can be triggered or exacerbated by poor soil management and can be ameliorated by good management practices, it cannot easily be related to the soil classification systems described in Section E, Chapters 4–6. Despite this limitation of soil classification, it is quite clear that there are particular soil types that are likely to be or can easily become hardsetting whereas others are not (Section 3.2.1).

   Following an outline and then a more detailed review of hardsetting by Mullins et al. (1987, 1990), in which the distinctive nature of hardsetting behavior was explained, the situation has now changed and a large number of publications on this topic have appeared in the past ten years. These include a special issue on the mechanism and management of hardsetting soils (Blackwell, 1992) and an international conference on the productivity and conservation of sealing, crusting and hardsetting soils (So et al., 1995).

3.1.3   Definition and Identification of a Hardsetting Horizon

The hardsetting horizon of a soil is almost always the A horizon and since the management of such soils is frequently dominated by constraints associated with hardsetting behavior, it is conventional to talk of such soils as hardsetting soils. However, since it is possible to have hardsetting behavior in a B horizon (Chartres et al., 1990), it is important to define hardsetting as a horizon property while bearing in mind that where hardsetting soils are referred to, the implicit assumption is that of a soil with a hardsetting A horizon. The following definition of a hardsetting horizon was submitted to ISSS as a product of a working group at the international symposium on Sealing, Crusting and Hardsetting Soils in Brisbane in 1994 and represents the most specific and generally agreed definition:

> A hardsetting horizon is one that sets to an almost homogeneous mass on drying. It may have occasional cracks, typically at a spacing of $\geq$ 0.1 m. Air dry hardset soil is hard and brittle, and it is not possible to push a forefinger into the profile face. Typically, it has a tensile strength of $\geq$ 90 kN m^{-2}. Soils that crust are not necessarily hardsetting since a hardsetting horizon is thicker than a crust. (In cultivated soils, the thickness of the hardsetting horizon is frequently equal to or greater than that of the cultivated layer.) Hardsetting soil is not permanently cemented and is soft when wet. The clods in a hardsetting horizon that has been cultivated will partially or totally disintegrate upon wetting. If the soil has been sufficiently wetted, it will revert to its hardset state on drying. This can happen after flood irrigation or a single intense rainfall event.

There are a number of points in the above definition worthy of comment because they help to distinguish between hardsetting and other horizons with similar properties which are not hardsetting. Where the soil profile is dry, this definition also permits field identification of a hardsetting horizon from its strength and comparative lack of structure. However, there are also hard horizons that do not soften when wetted, due to irreversible chemical cementation (e.g., plinthite) (Section E, Chapter 2) or other causes, and these need to be excluded during field identification. Hence, it is necessary to test that hard pieces of soil will soften and partially or totally disintegrate when dropped into water in order to confirm that their behavior can be described as hardsetting.

The boundaries between hardsetting and other types of soil behavior are not as distinct, as indicated in Fig. 3.1. There is a continuum of soil behavior, for example, between crusting and hardsetting. Gusli et al. (1995) give examples of a soil that can either develop a crust or hardset, depending on the type of wetting regime to which it has been subjected. In contrast, a soil that has slightly greater wet stability will still be susceptible to crusting but not to hardsetting.

Since there are no sharp boundaries between hardsetting and many other forms of soil behavior, the values for crack spacing and strength in the definition are somewhat arbitrary and are based on current experience. For example, soils with an air dry tensile strength < 90 kN m^{-2} but which comply with the definition in all other respects have been excluded because it is anticipated that their physical limitations (to root growth in drying soil and to cultivation of dry soil) are not sufficiently serious. It may be necessary to review these limits when more field data are available.

Because the behavior of soils that have been identified as having a hardsetting horizon may be altered by soil management, it is always necessary to interpret maps of the distribution of hardsetting soils with caution. This is no different than the problem of interpreting current drainage status from soil maps based on gley morphology. In both cases, the impact of hardsetting or poor drainage status on natural vegetation and the potential limitations to soil management that result may be considerable. However, improved soil management may have greatly reduced the soil's sensitivity to hardsetting, or drainage may have bypassed the causes of poor drainage. Thus, inspection of the particular site in question to examine soil profiles may be necessary to confirm the current state of the soil.

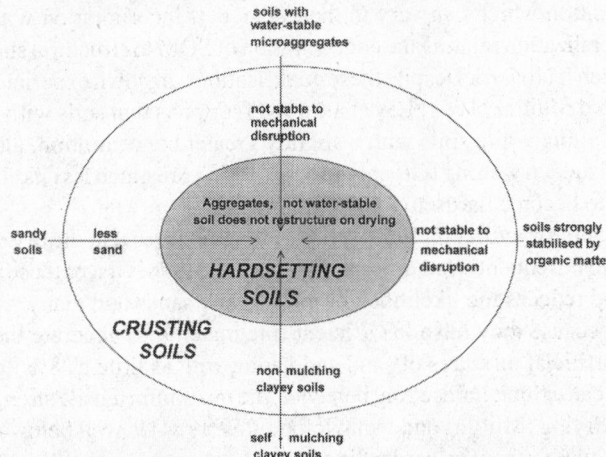

Fig. 3.1 Hardsetting and crusting in relation to soil stability and texture [Reprinted with permission from Mullins, 1997. R. Lal et al. Methods of assessment of soil degradation. Copyright CRC Press, Boca Raton, FL]

3.2 Distribution of Hardsetting Soils

As Isbell (1995) has explained: "it is not surprising that ...soil classifications try to minimize emphasis on A horizon conditions" because their properties are so dependent on agricultural practices. Thus, while no direct relationship between hardsetting soils and soil classification is to be expected, it should be possible to use classification as a general guide to the likelihood of a soil becoming or being susceptible to hardsetting. The extent to which this is the case depends on the system of soil classification and whether or not hardsetting behavior has been clearly identified as a horizon property that can be specifically identified. In practice, although the mechanisms underlying hardsetting behavior are now well understood, none of the major systems of soil classification in current use have a clear and unambiguous horizon definition with properties which correspond precisely and exclusively to hardsetting behavior although the definition used by McDonald et al. (1984) comes close. However, many systems do specifically identify various types of hard horizons and it is to be hoped that future improvements in soil classification will involve incorporation of a description of hardsetting horizons along the lines of that given in Section 3.1.3. In the meantime, this deficiency represents an important limitation to the practical use of soil maps as an indicator of likely soil management and environmental problems and hence of soil quality and the potential sustainability of any particular type of soil management (Section G, Chapter 10).

Despite these problems, it is not difficult to identify those areas and soil types where hardsetting soils are likely to occur, based on the soil properties that predispose a soil to this type of behavior. Therefore, the types of soils that are potentially hardsetting will be identified before considering their worldwide distribution.

3.2.1 Soil Properties Associated with Hardsetting

The relationship between soil texture, organic matter (SOM) content and hardsetting is shown in Fig. 3.1. Because hardsetting behavior results from structural breakdown on wetting, only soils that are sufficiently structurally unstable are likely to be hardsetting. Furthermore, as SOM increases resistance to this kind of breakdown, only soils with less than some critical concentration of SOM are likely to breakdown on wetting. However, SOM consists of a diverse range of materials at different stages of decomposition which can vary in their degree of incorporation within the soil. Thus, the validity of any generalization relating the concentration of SOM to structural stability will be restricted by the datset on which it is based. Despite these qualifications, my own experience in the UK, Australia and Africa (Chan and Mullins, 1994; Ley et al., 1989) suggests that soils with less than 2% SOM are vulnerable to hardsetting while soils with a slightly greater concentration, although not very stable under conditions of sudden wetting (Greenland et al., 1975), are much less likely to undergo sufficient structural collapse to become hardsetting.

In addition to severe structural instability, hardsetting requires that the dry soil has a high strength and only gives rise to infrequent structural cracks, if any at all. The first of these requirements excludes almost all sands and reduces the likelihood of most loamy sands and many silty textured soils from being hardsetting because they have insufficient fine material to generate the required strength on drying. Although artificial mixtures of sand and kaolin with as little as 8% kaolin (which would be classified as sands) can exhibit hardsetting behavior, the maximum tensile strength of 44 kN m^{-2} which they developed on drying (Mullins and Panayiotopoulos, 1984) is well below the critical value of 90 kN m^{-2} required to qualify them as hardsetting.

The second requirement excludes most clays and many clayey soils if they are structured or restructure sufficiently on drying (Young et al., 1991). Furthermore, since a smaller proportion of clay with a large, rather than a small, shrink/swell potential may be needed to cause cracking, it is not suprising that the clay mineralogy of most hardsetting soils is dominated by those minerals (hydrous micas and/or kaolinites) with a smaller shrink/swell potential (Norrish and Pickering, 1983). As a consequence of both requirements, the soils most commonly found to be hardsetting are sandy loams and loams although a range of textures from loamy sand to the borderline between sandy clay and clay can be hardsetting (Mullins et al., 1990). In the Australian red brown earths, hardsetting is associated with a large concentration of silt and fine sand (Cockroft and Martin, 1981; French, 1981), which is not surprising given the tendency for such soils to be particularly structurally unstable.

The above generalizations are subject to two qualifications, both referring to factors other than SOM that can affect structural stability. First, because sodicity is a cause of structural instability, sodic soils are frequently hardsetting. Furthermore, since clayey sodic soils do not tend to form such a close crack spacing on drying as their nonsodic counterparts (Section G, Chapter 2], it is likely that hardsetting may be more common even in sodic soils with a high clay content. Second, some soils may not be hardsetting because they contain sufficient water-stable microaggregates (due to the cementing action of materials such as the oxides and/or oxyhydroxides of Fe and/or Al) to prevent much structural breakdown (Trapnell and Webster, 1986). The hand texture and particle size distribution of such soils do not accurately reflect their physical behavior. Indeed, they can exhibit subplasticity, in which the textural feel of the soil becomes increasingly clayey the longer the soil is remolded due to the progressive destruction of microaggregates. This behavior is common in many oxisols and ultisols which tend not to be hardsetting.

3.2.2 Hardsetting, Soil Classification and Worldwide Distribution

With the exception of the system of Northcote (1960, 1979), hardsetting soils do not appear specifically in systems of soil classification. However, the part of the definition of Alfisols that specifies: "an epipedon that is both massive and hard when dry ..." means that many Alfisols are hardsetting although the definition also encompasses other types of soil behavior. Nevertheless, a soil map of the tropics, where Alfisols are predominantly found in the semiarid parts of Africa and India, probably gives a good indication of their major areas of occurrence. Sodic (natric) (Section G, Chapter 2) soils are also likely to be hardsetting.

In the FAO system of classification (FAO UNESCO, 1974), it is much less easy to determine which soils are likely to be hardsetting. A detailed discussion of the relationship between hardsetting and soil classification is given by Mullins et al. (1990) indicating that luvisols, planosols and solonetz soils are most likely to be hardsetting.

In terms of their worldwide distribution, the low concentration of SOM expected in hardsetting soils means that they are most likely to occur in arid and semi-arid tropical and mediterranean areas. However, in more humid and/or temperate zones, hardsetting behavior is still encountered, particularly in circumstances where the SOM concentration under natural vegetation has been greatly reduced, for example, as a result of an intensive system of horticultural management in which the soil is frequently tilled, a large proportion of the soil surface is kept bare, and roots or above ground parts of the plant are harvested and removed (Young et al., 1991).

Soil types that are hardsetting or potentially hardsetting have been identified in Australia (Isbell, 1995), India (Abrol and Katyal, 1995), the Sahel (Valentin, 1995) and Sweden (Stenberg et al., 1995). Photographs of hardsetting soils are presented in So et al. (1995). Isbell (1995) estimates that hardsetting soils may cover a total of 22–23% of the land area of Australia as either massive sesquioxidic (10%) or duplex soils (12–13%), the latter forming a major part of the Australian wheat belt. Young (1992) has also discussed the occurrence of hardsetting soils in the UK where they are comparatively rare and only tend to occur in cultivated areas. Earlier references to hardsetting soils in Botswana, Brazil, Gambia, India, Sudan, Tanzania, the UK, and the United States are given in Mullins et al. (1990).

3.3 Processes Involved in Hardsetting

An understanding of the processes involved in hardsetting is necessary in order to consider how it affects soil management, productivity, and environmental quality, and how it can be assessed and alleviated. This topic has been covered in many previous reviews (Mullins et al., 1987, 1990; Mullins and Ley, 1995; Mullins, 1997) and the following account is taken from the most recent of these. Hardsetting consists of two distinct processes: structural breakdown of aggregated soil on wetting, and hardening without restructuring on drying.

3.3.1 Structural Breakdown and Collapse

On wetting, the extent of aggregate breakdown can vary from almost complete disintegration to a minor softening at the points of contact. Wetting of a bed of aggregates, for example, can result in softening and coalescence at the regions of contact between aggregates which can change the strength of the dry soil without any visible change in structure (Rice et al., 1997). In flood-irrigated soils, even when there is considerable structural breakdown, collapse (slumping) of the surface can be delayed until after drainage due to the effects of buoyancy (partly due to trapped air). Once the surface has drained, further vertical shrinkage is likely to occur as the soil hardens (Mullins et al., 1992a,b). In order to clearly distinguish between slumping and shrinkage, it is convenient to define slumping as a bulk volume reduction caused by particle or aggregate breakdown and/or rearrangement. However, in practice it may not be possible to distinguish clearly between these two processes.

In addition to the effects of wetting, soil structure can also be destroyed when soils are remoulded or otherwise mechanically disturbed when in a wet state. Such disturbed parts of the profile will also hardset if they do not restructure on drying, as indicated by the considerable literature on puddling of paddy soils (Kirchoff and So, 1995). There is also evidence that some Vertisols can be degraded by cultivation to produce a hardsetting horizon (Seyni-Boukar et al., 1992). Since these soils display this phenomenon in a lower horizon, they should not be described as hardsetting soils but as soils with a hardsetting horizon.

Structural breakdown may also be caused by the combination of raindrop impact and sudden wetting of a bare soil surface. This can cause a surface seal to form which will then harden into a crust (Valentin and Bresson, 1997) if the surface does not restructure on drying (the latter process being referred to as self-mulching when in an aggregated surface tilth). Excluding very sandy, self-mulching clay, and organic matter rich soils, most bare soils are likely to undergo a degree of crusting if suddenly subject to intense rainfall from an initially air dry state. Thus, there is a continuum of soil behavior, ranging from soils that are almost completely water stable and can withstand the pressure of raindrop impact at one extreme, through moderately and severely crusting soils, to hardsetting soils at the other extreme (Fig. 3.1). Indeed, in terms of the processes involved, crusting can be viewed as hardsetting of the soil surface aided by the pressure of raindrop impact.

There is also an interaction between crusting, hardsetting, aggregate size distribution and the type of wetting (Heinonen, 1982; Mullins et al., 1990; Bresson and Moran, 1995; Gusli et al., 1995; Rice et al., 1997). A bed of aggregates subject to intense rainfall may crust while aggregates in the soil beneath retain their structural integrity because they are only slowly wetted. In contrast, flooding can lead to hardsetting of a much thicker layer of soil because water is initially able to infiltrate more quickly, and deeper into the soil (Fig. 3.2). Similarly, the larger the aggregates in a cultivated soil, the greater the depth of soil that can be rapidly wet (and subsequently hardset) by rainfall. In contrast, seedbeds with a greater proportion of small aggregates (< 0.5 mm) are more likely to undergo crusting with a smaller depth of hardset soil underneath. Heinonen (1982, summarized in English in Stenberg et al., 1995) has described three types of hardsetting that have been observed in Swedish fields. These involve complete hardsetting of the cultivated layer, hardsetting in the bottom of the cultivated layer when sustained gentle rainfall has resulted in saturation of this zone, and translocation of slaked fragments to the base of a coarse tilth where they cause hardsetting.

3.3.2 Dispersion

Dispersion of soils is the process in which individual clay particles or clay tactoids spontaneously separate from the soil during wetting. Dispersion is associated with soil sodicity and is sensitive to the exchangeable Na percentage (ESP) of the soil and to the threshold electrolyte concentration (TEC) of the soil solution. A critical ESP of 15 was taken by the US Salinity Laboratory Staff (1954) to define the boundary between sodic and nonsodic soils but a value of 5 has been proposed as more relevant for Australian conditions (McIntyre, 1979). These are only guidelines, because as TEC decreases, ESP at which a clay will just disperse is reduced. Assessment of sodicity and its effect on dispersion are summarized by Rengasamy (1997).

Because sodicity leads to structural instability, many sodic soils are hardsetting (Mullins et al., 1990). Application of gypsum to these soils increases water stable-aggregation and usually results in a marked improvement in other physical properties (macroporosity, hydraulic conductivity, friability, ease of working, reduced soil strength). There are two effects involved, replacement of exchangeable Na with Ca, and increase in the electrolyte concentration. The beneficial effects of gypsum do not persist once Ca is leached from the surface layers. Several factors can reduce the response of soils to gypsum including high concentrations of SOM, covered soils, direct drilling, neutral pH, low concentrations of organic molecules adsorbed on the clay particles, severe drying, and $CaCO_3$. Furthermore, there are also many hardsetting soils that undergo little if any dispersion and are not sodic, but do slake on wetting (Rengasamy, 1997; Young and Mullins, 1991a). These soils do not respond to gypsum.

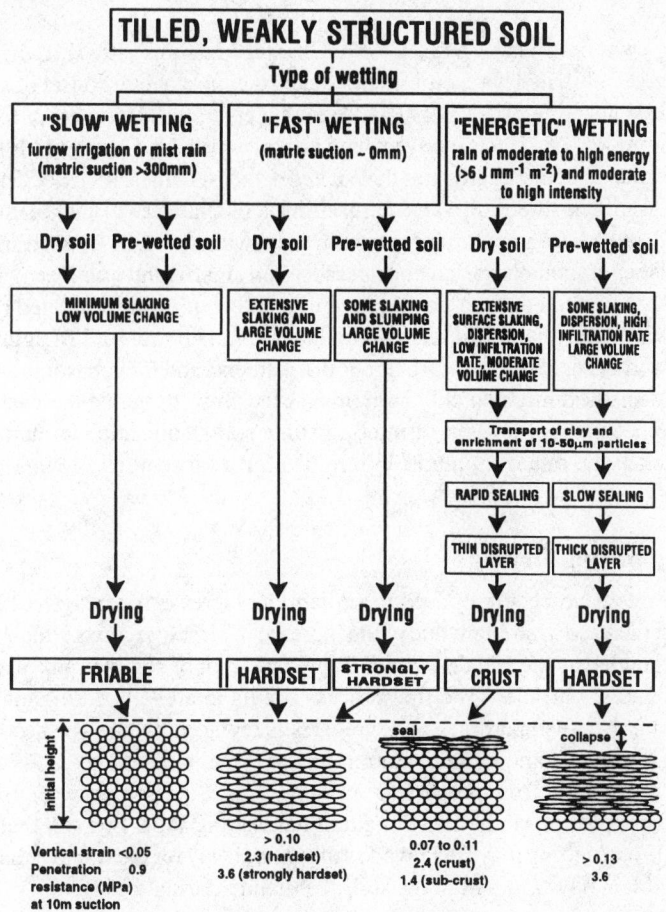

Fig 3.2 Summary of the effects of antecedent water content and type of wetting on the behavior of beds of 5 mm aggregates from a soil that can display both crusting and hardsetting behavior in the field [Reprinted from Gusli et al., 1995. H.B. So et al. Sealing, crusting and hardsetting soils with permission of the Australian Society of Soil Science]

Dispersion has often been used in a wider and less informative sense to include mechanically assisted dispersion (where mechanical disturbance is also involved in providing the energy for particle separation) or even slaking. Because methods for ameliorating dispersible soils often have a limited effect on soils where hardsetting is due to slaking, there is, therefore, a risk of confusion when the term dispersion is used to include any form of structural breakdown including slaking.

3.3.3 Slaking

Slaking, which is the process of fragmentation that occurs when aggregates are suddenly immersed in, or placed in contact with water, occurs because aggregates are not strong enough to withstand the stresses resulting from rapid water uptake. Slaking is affected by antecedent matric potential (ψ_{ma}) rate of wetting, the concentration of SOM, and clay mineralogy.

Dry soil undergoes more slaking than moist soil (Le Bissonnais et al., 1989). Panabokke and Quirk (1957) studied the aggregate stability of two loamy and three clayey soils and found that the aggregates were stable when ψ_{ma} was between -1 and -10 kPa. Chan and Mullins (1994) studied slaking of ten soils. Of these, the three cultivated hardsetting soils slaked the most and the onset of slaking occurred at the highest (i.e., least negative) ψ_{ma} (-10 kPa). Taken together, these two sets of results imply that there may be a maximum ψ_{ma} of -10 kPa above which no soil will slake. Chan and Mullins (1994) also found that the proportion of air dry aggregates that slaked decreased linearly ($r = 0.82$) as their organic carbon content increased. The three cultivated hardsetting soils shared two other features in common. Their proportion of slaked fragments increased linearly with the rate of wetting, and the size distribution of the slaked fragments varied considerably and significantly with ψ_{ma}.

Levy et al. (1997) have studied the effect of prewetting rate, both with simulated rainfall and with tension wetting, on the subsequent infiltration rate under simulated rainfall of aggregate beds of a Typic Chromoxeret and a Calcic Haploxeralf. For both treatments and for both soils, slaking increased and infiltration rate decreased more quickly with time as the prewetting rate increased. Aging of the prewetted aggregate beds for 18 h often resulted in some soil stabilisation so that the subsequent infiltration rate matched the rate of rainfall (33 mm h^{-1}) for a substantially longer period before it started to decline.

3.3.4 Hardening during Drying

Two different processes have been proposed to explain the increase in strength of hardsetting soil during drying: (1) an increase in strength due to the increase in effective stress which results from the decrease in matric potential as the soil dries, and (2) precipitation of soluble salts at zones of contact between aggregates and/or particles. The first process occurs in all soils during the early stages of drying before the water between particles or aggregates is replaced by air. The second process will only occur in soils that release some soluble salts on wetting. Thus hardsetting results from process (1) with an additional contribution from process (2) in some soils. Since process (1) occurs in all soils there is no mystery about the origin of the strength of hardsetting soils. In a soil that has undergone structural breakdown, hardsetting may be viewed as the natural way for the soil to behave. A challenge for the future is to explain how and why some soils restructure during drying.

3.3.5 Strength of Hardsetting Soils

In order to characterize soil degradation, it is necessary to quantify those aspects of soil behavior associated with the deleterious effects of degradation. In the case of hardsetting, these include not only the high strength of the dry soil, but also the rate at which soil strength increases during drying (Ley et al., 1989), since this can result in roots and shoots being unable to penetrate the soil when it is still considerably wetter than the wilting point (Weaich et al., 1992a; Ley et al., 1995). A theory that describes the increase in strength of a moist soil as it dries was first presented by Mullins and Panayiotopoulos (1984), and subsequently, modified to account for the concentration of stress that occurs around soil pores (Mullins et al., 1992b). This has since been widely used to explain the observed strength characteristics of hardsetting, structurally stable, and microaggregated soils (Young and Mullins, 1991b; Mullins et al., 1992a,b; Mtakwa, 1993).

3.3.5.1 Effective Stress and Soil Strength

In an unsaturated soil, assuming that the air in soil pores is at atmospheric pressure, the combination of any externally applied stress and the effect of the matric potential (ψ_m) means that the soil experiences an effective stress (σ') given by:

$$\sigma' = \sigma - \chi\psi_m \qquad\qquad [3.1]$$

where σ is the externally applied normal stress (on any potential plane of failure), and χ is a factor that accounts for the proportion of any surface of failure occupied by water films. Because ψ_m is always negative, the second term in Equation [3.1] has the effect of increasing the effective stress. Near to the soil surface and in undisturbed samples in the laboratory, σ will be small relative to $\chi\psi_m$ once the soil has dried much beyond field capacity (–10 kPa), and can be ignored. Allowing for a contribution c (cohesion) to account for chemical bonds between particles and/or aggregates, Mullins and Panayiotopoulos (1984) proposed that soil tensile strength (Y) could be given as

$$Y = c - \chi\psi_m \tag{3.2}$$

However, this equation needs to be modified to account for the concentration of stress that occurs around pores in the soil, giving

$$Y = c - \left\{\chi\psi_m / f(S)\right\} \tag{3.3}$$

where f(S) is a function that depends on pore shape having a value of 2 for spherical pores and becomes progressively greater for more elongated pores (Snyder and Miller, 1985), and S is the degree of saturation.

Equation [3.3] should apply to all soils and has been successfully used to predict the behavior of both hardsetting and aggregated soils (Mullins et al., 1992b) but requires a knowledge of how c, χ and f(S) vary with S. In a hardsetting soil that has just dispersed or slaked, c is clearly zero, since the soil would not otherwise have broken down. The question is then: What happens to c as the soil dries? The shrinkage due to the action of effective stress is required to pull aggregates and particles close enough to one another for the short-range forces of chemical bonding to become effective. Thus, it seems likely that the dominant contribution to the tensile strength while the soil is still in a fairly wet state (i.e., $\chi > 0.5$) will be due to effective stress. This hypothesis is supported by the observation that the tensile strength of hardsetting soils is initially proportional to effective stress (Fig. 3.3). Given the negligible strength of very wet soils, this shows that it is reasonable to assume that $c \approx 0$ for hardsetting soils when $-\psi_m$ is not too large.

For fairly wet soil (S > 0.5), χ is usually taken as $\approx$ S although χ is likely to become less than S as the soil dries (Mullins and Panayiotopoulos, 1984). In a wet soil, f(S) will take a constant value characteristic of the distribution of pore shapes within the soil. However, as the soil dries and shrinks, new pores, especially elongated cracks, are likely to develop thus increasing the value of f (S). The $-\chi\psi_m$ term represents the contribution of matric potential to effective stress. It can assume very large values in clay soils and is responsible for their shrinkage (Childs, 1969).

Matric potential plays a dual role in determining how soil strength changes during drying. Initially, decreases in ψ_m increase the effective stress and hence the soil strength as already discussed. However, as S decreases, soil water becomes confined to the zones of contact between adjacent aggregates and/ or particles which then experience a normal stress, pulling them together. Consequently, the soil becomes more rigid and stabilized against further volume change. Any further shrinkage can only be accomplished by the development and enlargement of cracks between peds or by internal microcrack development within the soil fabric. The difference between these two types of shrinkage is crucial because it represents the difference between the behavior of clays and other soils that develop structure (or enlarge preexisting structural cracks and flaws) on drying, and hardsetting soils. Mullins et al. (1992b) have suggested that this difference may result from the greater proportion of larger (> 2 μm)

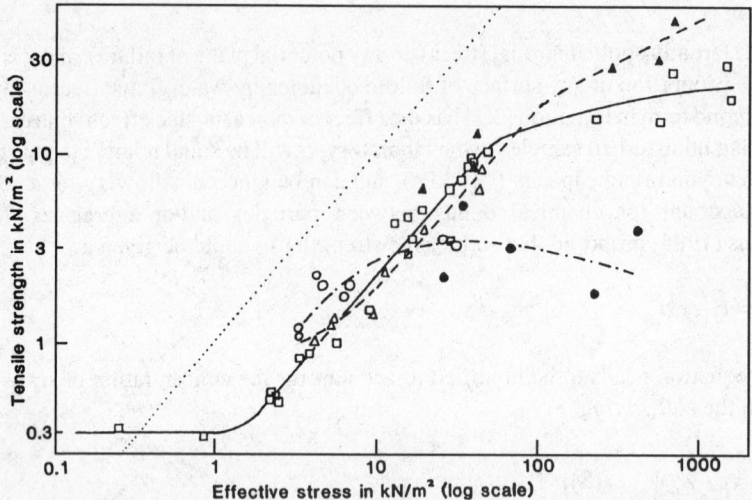

Fig. 3.3   Tensile strength versus effective stress for hardsetting soils from Tatura (triangles), Trangie (squares), and for a water-stable aggregated soil (circles). Lines parallel to the dotted line satisfy equation (3) for c = 0, if f (S) remains constant. The break in the slope of the curves for both hardsetting soils approximately coincides with the point where macroscopic shrinkage ceased [Reprinted from Mullins et al., 1992. Strength development during drying of cultivated, flood-irrigated hardsetting soil. II. Soil Till. Res. 25:129-147. Copyright with permission of Elsevier Science].

particles in hardsetting soils than in clays. These particles may act to limit the extension of cracks as explained by Gordon (1986).

Equation [3.3] seems to work well in predicting the behavior of hardsetting soils down to a ψ_m of at least −80 to −100 kPa (Mullins et al., 1990 and 1992b; Young and Mullins, 1991b; Weaich et al., 1992b). Weaich et al. (1992b) have also shown that the strength characteristic is independent of the rate of drying. At lower potentials, tensile strength continues to increase, but not at the same rate (Ley et al., 1993) (Fig. 3.3) possibly because the soil develops microcracks (Mullins et al., 1992b). The theory of fracture mechanics (Hallett et al., 1995) is probably needed to explain further changes in strength, from this stage down to air dryness.

3.3.5.2   Hardsetting due to Precipitation of Soluble Salts
The effect of precipitation of soluble salts can be expressed in terms of an additional contribution to cohesion that increases as the soil dries. Although a variety of materials have been identified that can cause cementation and increase soil strength (Gilkes et al., 1995), only those that soften on wetting can give rise to hardsetting behavior. Daniel et al. (1988) and Chartres et al. (1990) have shown that soluble Si can contribute to the hardsetting of some subsoils. This may be particularly important in increasing the strength of air-dry soil, particularly in more sandy hardsetting soils that would otherwise not have hardset or have been only weakly hardsetting. While it is probable that this is rather a special case of hardsetting, further work is needed to discover the extent of such behavior.

3.3.5.3   Strength of Air-Dry Hardsetting Soils
The tensile strength of air-dry hardsetting soil can take values approaching 0.2 MN m^{-2} (Ley et al., 1989; Gusli et al., 1994). In comparison to many soils, this is very strong but in comparison to brick or cement (about 5 MN m^{-2}) which are themselves materials with low tensile strength (Gordon, 1986),

it is very weak. In contrast, a material such as high tensile glass fiber can have a tensile strength up to 14 000 MN m^{-2}, approaching the strengths of atomic bonds. Thus, only a tiny portion of the failure plane of a hardsetting soil requires to be secured by chemical bonds to explain their observed strength.

Mtakwa (1993) studied the relationship between the tensile strength of air-dry soil and the ability to cultivate in the dry season. The hardsetting soils studied had a tensile strength of ≥ 90 kN m^{-2} and could not be readily cultivated by hand in the dry season, but other soils that were only slightly weaker could, and frequently were. These tests involved small (40 mm long) undisturbed cylinders of soil and smaller values of tensile strength might well have been obtained on larger samples due to the spatial variability of strength. Furthermore, it is not clear that tensile strength can be uniquely related to ease of cultivation although the value of 90 kN m^{-2} is clearly useful as a rough indication of when cultivation by hand is possible.

3.3.6 Hardsetting and Soil Structure

There are two possible reasons why clays and other structured soils do not display hardsetting behavior. The first is that the soil contains water-stable structural units that do not disintegrate sufficiently on wetting. This is a common scenario, and the main question to ask is: How much structural instability is required for a soil to become hardsetting? The relevant factor here, is whether sufficient slaked or dispersed fragments are produced during wetting to bridge preexisting gaps between aggregates (Section 3.4). In many hardsetting soils, there are more than enough fragments and the whole soil disintegrates into an almost homogeneous mass on wetting, as can be seen in thin section (Mullins et al., 1992b; Bresson and Moran, 1995).

3.3.6.1 Restructuring on Drying

The second possible reason why structured soils do not display hardsetting is that, even if the soil does structurally disintegrate on wetting, it may restructure on drying. For example, Young et al. (1991) found a sharp contrast between the depth and density of cabbage roots growing in a hardsetting soil (to a maximum depth of 300 mm) in comparison to a neighboring soil of identical particle size distribution that also underwent structural collapse on wetting. The latter soil developed small structural cracks that allowed roots to penetrate to > 500 mm. More research is needed to understand the factors and circumstances leading to this type of soil restructuring.

3.3.6.2 Coalescence

Where soil aggregates undergo only partial softening during wetting so that the macropores are retained but the aggregates become joined by partially slaked material, it is possible for the soil to become harder after a cycle of wetting and drying without undergoing complete hardsetting. This process, which is likely to be widespread, has been called coalescence (Cass and Cockroft, 1994, personal communication; Bresson and Moran, 1995). Since it represents a behavior that may border on hardsetting, it is important not only because it may provide an indication of a soil that is potentially hardsetting if degraded or managed differently, but also because such a soil is likely to be sensitive to the type of wetting regime (Rice et al., 1997).

3.4 Methodologies for Assessing Hardsetting Soils

Table 3.1 shows some of the tests that can be used for assessing hardsetting. A distinction is made between methods that characterize the severity of hardsetting behavior, assess the sensitivity of a soil to hardsetting (i.e., its likelihood of hardsetting), or diagnose the likely cause of hardsetting. The results of tests that characterize *in situ* soil behavior depend not only on soil sensitivity to hardsetting

Table 3.1 Tests that can be used to characterize (c), assess (a), or diagnose (d) different aspects of hardsetting [Adapted with permission from Mullins, 1997. R. Lal et al. Methods of assessment of soil degradation. Copyright CRC Press, Boca Raton, FL]

Property	Test	Comment	References
Air-dry strength[c]	Tensile strength of small undisturbed cores	Simple test to characterize strength of dry soil	Ley et al. (1989)
Penetration resistance of < 2 mm wet and dried soil samples[a]	Wet and dry (40 °C) sieved (< 2 mm) soil and test with hand-held penetrometer	Simple test to assess soil sensitivity to hardsetting	Cochrane (1989); Wilson and Proffitt (1995)
Surface strength distribution of air-dry soil[c]	Distribution of PR values measured *in situ* with a small dia. (0.6 mm) flat-tipped probe	Suggested as a test to characterize potential erodibility of hardset and other surfaces by saltating particles	Rice et al. (1997)
Strength characteristic[c] (tensile strength)	Tensile strength as a function of water content or matric suction	Indicates how quickly strength can change during drying	Ley et al. (1989)
Strength characteristic[c] (Penetration resistance, PR)	PR as a function of water content or matric suction	Can be use for modelling effects of hardsetting on emergence and on root growth	Weaich et al. (1992b); Tezera Tsegaye et al. (1995)
Infiltration rate[c]	Measure infiltration rate under simulated field conditions		Loch and Foley (1994)
Structural instability[a & c]	Wet sieving or rainfall simulation and determination of % of material < 125 μm	Used to predict effect of stability on infiltration rate. May also be useful to indicate likely strength behavior	Loch (1994, 1995); Loch and Foley (1994),
Structural instability[a]	Critical antecedent matric potential (ψ_{ma}) for slaking or dispersion	May provide a single number to characterise sensitivity to hardsetting	Chan and Mullins (1994)
Structural instability[d]	Field test of slaking and dispersion of aggregates and of EC	Can be used to indicate whether hardsetting is due to sodicity	Rengasamy (1997)
Friability[a]	Variation of strength of dry aggregates with their size	Relatively insensitive to changes in behavior	Utomo and Dexter (1981); Dexter and Kroesbergen (1985)
Composition and concentration of soil solution[a,d]	EC and SAR or simplified field test	Can be used to indicate likely responsiveness to gypsum	Dellar and Proffitt (1995); Rengasamy (1997)

but also on the preceding set of cultivations, and on wetting and drying events. Consequently, they may be unreliable as indicators of any small changes in soil sensitivity to hardsetting. By the same argument, tests used to assess soil sensitivity to hardsetting cannot be used to indicate the likely severity of hardsetting under field conditions unless those conditions (initial aggregate size distribution, antecedent ψ_m, and type of wetting) are standardized.

3.4.1 Characterization

Methods used to characterize hardsetting can be used to model its effects on infiltration, runoff and erosion (Bridge and Silburn, 1995), ease of cultivation, crop growth (Connolly and Freebairn, 1995), and crop establishment (Mullins et al., 1996).

Measurement of the tensile strength of small undisturbed cores of air-dry soil is simple and requires little equipment, although the soil may have to be wetted to permit sampling and the sample then air dried. This test may provide a rough guide to ease of cultivation although field determination of the energy required for cultivation and the type of tilth produced is preferable. The strength characteristic indicates the rate of increase of soil strength with decreasing water content. However, since factors such as unsaturated hydraulic conductivity of the soil control the rate of soil drying, its main use is to identify the water content or matric potential below which the soil becomes too hard to cultivate or allow root or shoot penetration.

Infiltration rate can be used to determine the rainfall intensity at which runoff is likely to occur; simple ponded infiltration tests will be misleading except for characterizing flood- or furrow-irrigated soils because they will not simulate the type of structural breakdown occurring during rainfall. At present, the more sophisticated techniques using rainfall simulators are probably only useful as a research tool and in modeling studies.

A wide range of tests of soil structural instability have been developed over the past century but are increasingly being considered as unsatisfactory, mainly because the ranking of soils is largely determined by the prewetting and wetting procedures, the antecedent soil water content and the size of aggregates (Valentin and Bresson, 1997). Sealing susceptibility is better predicted when the size distribution of particles and/or fragments released by aggregate breakdown is considered (Le Bissonais, 1990; Loch and Foley, 1994, Roth and Eggert, 1994). Loch (1995) suggests that efficient surface raindrop entry requires pores > 15 μm and that consequently it is particles and slaked fragments < 125 μm that are important in clogging pores and forming a surface resistant to infiltration. In this way, he has been able to relate infiltration rate under rainfall to values of the proportion of material < 125 μm obtained by wet sieving.

3.4.2 Assessment of Sensitivity to Hardsetting

As explained above, the standard wet sieving test for water-stable aggregates is unlikely to give more than a crude indication of soil sensitivity to hardsetting because it involves prewetting a sample from underneath. In this procedure, the rate of wetting will be affected by sample packing, trapped air, and unsaturated hydraulic conductivity. Loch (1995) has suggested that simulated low intensity rainfall is likely to be a more representative and reliable way of prewetting. Thus a modified wet sieving procedure using presized, air-dry field aggregates, wetted by simulated rainfall represents a very promising test for assessing soil sensitivity to hardsetting and deserves further development.

Another sensitivity test that deserves further consideration is the determination of the critical antecedent matric potential (ψ_{ma}) at which aggregates will just start to slake or disperse when dropped into water (Chan and Mullins, 1994). Given a knowledge of the matric potential in different parts of the field profile, this information could be used to indicated which parts of the profile were likely to undergo structural breakdown in different management or irrigation systems.

The friability test is useful in characterizing the ease with which a suitable tilth can be produced by cultivation but is only sensitive to comparatively large changes in soil behavior and is, therefore, of limited use in its present form as a method of assessment.

3.4.3 Diagnostic Testing

Diagnostic tests are useful to indicate whether or not a hardsetting soil can be ameliorated by gypsum treatment. Since sodic soils are responsive to gypsum, a direct test in which the effect of dropping dry aggregates into water and saturated $CaSO_4$ is compared (MacKenzie, 1991, personal communication) may prove to be adequate. However, this may fail to adequately diagnose saline-sodic soils and a more detailed field test using a portable conductivity meter has been described by Rengasamy (1997). Further information on the potential value of gypsum as an ameliorant can be obtained from determination of the sodium adsorption ratio (SAR) and electrical conductivity (EC) of the soil solution (Rengasamy et al., 1984; Dellar and Proffitt, 1995; Rengasamy, 1997).

3.5 Effects of Hardsetting on Plant Growth

Hardsetting has serious implications for root and shoot growth because the penetration resistance (PR) in hardsetting soils is likely to exceed a value of 3 MPa (sufficient to severely impede or halt root growth and prevent emergence of shoots) before the soil has reached wilting point (-1.5 MPa matric potential). For example, Ley et al. (1995) found that the PR of a range of hardsetting soils from Nigeria was near to, or greater than 2 MPa when they had only dried to a matric potential of -100 kPa. Similar results have been found for hardsetting soils from the UK (Young et al., 1991), Australia (Mullins et al., 1992b) and Tanzania (Mtakwa, 1993). Masle and Passioura (1987) have demonstrated the existence of root-shoot signaling that causes a reduction in the extension rate of shoots and is directly triggered by the effect of high mechanical impedance on roots even in the absence of water stress. This indicates that hardsetting may affect shoot growth directly in addition to the knock-on consequences of restricting root growth.

Roots that are growing through and drying a hardsetting soil will progressively increase the mechanical impedance of the soil, and consequently, reduce the rate of root growth. However, whenever the profile is rewetted and the mechanical impedance of the soil is reduced, roots will be able to resume faster growth. Consequently, the overall effect on profile root distribution will depend on the seasonal pattern of rainfall and soil profile wetting.

The effect of root growth restrictions caused by hardsetting on crop growth has been modeled by Bradley and Crout (1994, personal communication) using penetration resistance versus matric potential characteristics for a hardsetting soil from Tanzania (Mtakwa, 1993) and their PARCH (Bradley and Crout, 1993) sorghum crop growth model. The current model does not include any direct feedback effects of impedance on shoot growth, but includes the effects due to restricted root water uptake. The model was run with rainfall records from Botswana and showed that, in comparison to a soil with little if any limitations due to mechanical impedance, there was often little difference in overall grain yield. In some years, the hardsetting soil actually gave substantially greater yields because there was a smaller root and shoot system earlier in the growing season which used less profile water with the result that the later crop development was less severely water stressed. This indicates the complexity of plant-soil interactions in rainfed semi-arid conditions.

Weaich et al. (1992a) have shown how hardsetting is able to prevent seedling emergence and how the effect can be theoretically predicted. Field experiments that simulated farmers' practices of dry and wet planting (i.e., sowing into dry soil before rainfall or sowing into soil that has been recently wetted) on two hardsetting soils in Tanzania (Townend et al., 1996) have also demonstrated that hardsetting can prevent shoot emergence. Hardsetting frequently also imposes major constraints on crop yields due to the constraints on soil management and other problems that it can cause (Section 3.6).

3.6   Management of Hardsetting Soils

It is difficult to overstate the value of soil structure, especially in hardsetting soils. The cost of land degradation in Australia, for example, has variously been estimated at $2 billion (House of Representatives Standing Committee, 1989), and at $685 yr^{-1} (Sinden et al., 1990), the most costly feature of which was soil structural decline, most often in relation to hardsetting.

The potential problems for soil management and crop growth that are associated with hardsetting are shown in Fig. 3.4. Hardsetting presents a problem for soil management not only because it means that the soil cannot be cultivated when dry, but also because, after rainfall or irrigation, hardsetting soils have only a small window of opportunity in which the soil can be cultivated to provide a tilth that is suitable for sowing and good root development. When cultivated in too dry a state, they tend to produce large clods that are not friable and disintegrate into a very fine seedbed with further cultivation (Mullins et al., 1990).

In dryland agriculture, there are major limitations to agricultural productivity caused by hardsetting due to its effect on timing and ease of cultivations, and on crop emergence. Runoff, and hence reduced infiltration can also reduce the profile available water and productivity. Under irrigation, it is possible to overcome some of these limitations, but structural collapse can lead to restricted root aeration and can also restrict water entry. Although not directly due to the properties of the hardsetting A horizon, many hardsetting soils occur in duplex profiles in which the lower horizon has a low saturated conductivity that can result in temporary waterlogging and reduced yields. This behavior is common in Western Australia (Gregory, 1992, personal communication).

Methods for managing hardsetting soils and for their amelioration have been reviewed by Mullins et al. (1990) and described in a number of papers in So et al. (1995). Where sodicity is the cause of hardsetting, gypsum application is a widely used option for soil amelioration (Shainberg et al.,1989)

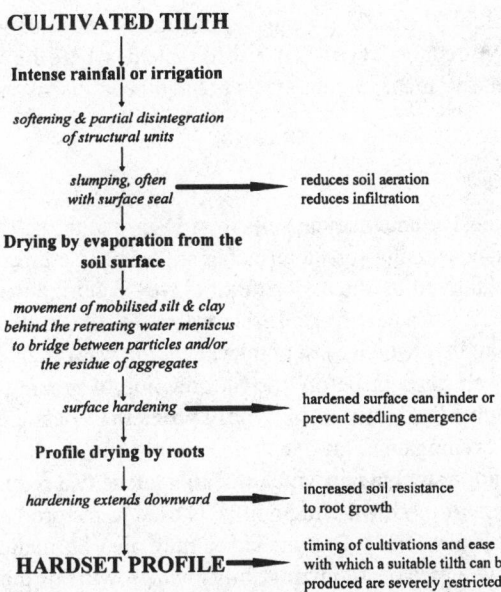

Fig. 3.4   The sequence of possible problems associated with a hardsetting soil [Reprinted from Mullins et al., 1987. Hardsetting soils. Soil Use Manage. 3:70-83 with permission of the British Society of Soil Science]

Table 3.2 Summary of management problems and possible solutions for hardsetting soils

Management system	Problem	Possible solutions
All systems with sodic soils	All hardsetting problems plus poor subsoil drainage	Gypsum application, incorporation. Subsoil gypsum slotting
Rangelands	Scalds, surface structural degradation, poor infiltration, overgrazing	Conservation stocking, and more watering points. Introduced pastures with faster growth.
Crop-pasture rotations	Surface structural damage due to grazing.	Avoid grazing after heavy rainfall.
Dryland agriculture in S & W Australia	Poor infiltration	Use of reduced and/or noninversion tillage. Incorporation of pasture or other soil conditioning crop in the rotation
Dryland agriculture in the semi-arid tropics	Poor infiltration and erosion, poor seedbed conditions, late sowing, poor root growth	Where possible, tillage at the end of previous rainy season or in the dry season to improve infiltration and permit earlier sowing. Zonal tillage, ridge tillage or tied ridges. Incorporation of crop or weed residues. Use of conditioning crops in the rotation.
Irrigated agriculture	Poor infiltration and lateral distribution. Root waterlogging. Crusting after sowing.	Use low intensity irrigation systems or PAM in furrow irrigation water. Crust breaker or seedline mulch.
Irrigated orchards	Poor infiltration and soil aeration. Poor root growth.	Loosen and stabilise subsoil where sodic with gypsum. Use low intensity irrigation systems. Limit herbicide applications to maintain soil organic matter.

(Table 3.2). However, the requirement for repeated applications once the beneficial effects of high electrolyte concentration have declined restricts its use to systems where the crop value can justify the cost. In the following summary, management systems have been discussed in order of increasing intensity of land use.

3.6.1 Rangelands

These represent the least intensive land use; on soils in semi-arid to arid climates, sparse vegetation and low SOM content predispose many soils to crusting and/or hardsetting. Large scalds, devoid of native vegetation and characterized by the exposure of clayey B horizons were remarked upon by Charles Sturt, an early explorer, when he traversed eastern semi-arid Australia in the 1840s. These may have followed periods of drought in which most of the vegetation was removed by native animals and suggest that such features were present before the introduction of grazing (Eldridge et al., 1995). However, overgrazing undoubtedly increases this type of problem by damaging fragile surface layers such as biological crusts of cryptogam (Eldridge and Greene, 1994), reducing SOM, and removing surface vegetation. Conservation stocking, in which stocking intensity is reduced and stocking is more evenly dispersed by provision of more watering points, is used to restore vegetation and ameliorate soil degradation (Eldridge et al., 1995). Pasture cover may also be maintained with the help of introduced species which have a greater colonizing ability and growth rate than native species (Bridge et al., 1983).

3.6.2 Crop/Pasture Rotations

Proffitt et al. (1993) have shown how trampling and grazing by sheep damage the physical properties of hardsetting soils and reduce pasture growth. By removing sheep from the pasture for 2–4 days following heavy rainfall ($\geq$ 10 mm), Proffitt and Bendotti (1995) demonstrated that the surface structural condition could be maintained in a similar state to ungrazed plots.

3.6.3 Dryland Agriculture

Managing hardsetting soils for dryland agriculture depends on finding the most effective balance between the conflicting requirements for cultivations (for weed control, occasional deep loosening, and for temporary increases in infiltration) and the need to maintain stubble coverage and a sufficiently high SOM content to reduce the amount of structural breakdown and maintain infiltration. While it is clear that, as a general rule, cultivations should be kept to a minimum, successful systems that involve zero tillage require better chemical weed control than is usually available.

In climates where crops are grown on stored winter rainfall, early and vigorous crop establishment is important for effective use of soil water. While it is clear that direct drilling can significantly reduce soil susceptibility to hardsetting compared to disk plowing and that stubble retention reduces soil evaporation and increases water harvesting (Wilson and Proffitt, 1995), the loss of weed control and seedling vigor is currently discouraging the adoption of such measures on hardsetting soils in Western Australia. In New South Wales, the problems of poor seedling vigor also continue to negate environmental advantages associated with reduced or zero tillage (Mead and Chan, 1988, 1995; Mason and Fisher, 1986). Since fumigation can produce a 70% increase in grain yield compared to unfumigated direct drilling (Chan et al., 1989), it is clear that soil disturbance due to plowing has important biological effects, including the reduction of soil borne diseases (Roget et al., 1987). Incorporation of a pasture phase (Chan and Khu, 1995) or of crops such as oilseed rape (Packer et al., 1995) into the cropping system in combination with improved tillage systems that involve less mixing than the conventional disk plow, provides alternative means of improving surface aggregate stability and infiltration.

The development of a tillage implement consisting of leading wide sweep shares at shallow depth (to maximize soil disturbance without redistributing SOM), a trailing noninversion ripper (to fracture the soil to greater depth), and a coil or culti packer to firm the soil surface (Mead and Chan, 1995) is one example of the new strategy behind soil management that is targeted on specific soil physical and biological requirements (Tisdall and Adem, 1988). Current requirements for soil disturbance will be further reduced if improved agents and methods of biological control can be successfully applied.

In the semi-arid tropics, hardsetting imposes even greater constraints on farming, especially for farmers with limited or no access to animal or machine power. Management options have been reviewed by Mullins et al. (1990), Abrol and Katyal (1995) for India, Kayombo and Lal (1993) and Hoogmoed (1995) for Africa, and Unger and Cassel (1991) in relation to soil and water conservation. The first constraint is that, if a seedbed and sowing cannot be performed before or immediately after the first substantial rains, then valuable and scarce sowing opportunities are lost. Cultivation at the end of the previous rainy season, or even during the dry season if tractor power is available and the soil is only weakly hardsetting, enable early sowing (Willcocks, 1981). Where seeds are sown with hand implements such as a jembe into untilled land, there is a high risk of runoff because of the high rainfall intensity. Even in tilled soil, high rainfall intensity can quickly cause slumping and surface sealing so that ridging, tied ridging or subsequent cutivations through the crop may be needed to maintain adequate infiltration.

As the availability and use of tractors increases, maintenance of SOM in order to maintain friability and reduce the effects of structural collapse on root growth is likely to assume an increasing importance in the semi-arid tropics. Incorporation of stubble and of local weeds has already been used for this purpose. In more humid and temperate regions, hardsetting soils are less common and the greater availability and persistence of organic residues make it easier to maintain a greater concentration of SOM. Nevertheless, serious crop establishment problems are encountered with hardsetting seedbeds on silty soils in Sweden which may require increased surface SOM concentration from reduced tillage to ameliorate (Stenberg et al., 1995).

3.6.4   Irrigated Systems

Surface collapse, poor infiltration and reduced lateral water movement are major problems for flood- and furrow-irrigated systems in many parts of the world, and can also result in serious erosion (Carter, 1990). Structural collapse and subsequent waterlogging of the cultivated layer can also lead to poor aeration and root growth. Options to reduce these problems include prewetting with water at a low level in the furrow to allow tension wetting of the furrow wall before increasing the flow, use of polyacrylamide in the irrigation water (Lentz et al., 1992; Sojka and Lentz, 1995), and changing to microspray, buried pipe emitters or other low intensity but more expensive irrigation options. Levy et al. (1997) have summarized some of the recent work on effects of prewetting rate and the following ageing period on the subsequent infiltration rate of aggregated soil (Section 3.3.3). This should lead to a more comprehensive and mechanistic understanding of such topics in relation to soil type and aggregate size, and hence underpin development of integrated systems of irrigation management that minimize structural collapse and sealing, thus increasing water-use efficiency and crop yield, and reducing the potential for erosion.

Integrated systems designed to retain a mulch (Tatura system) (Tisdall and Adem, 1988; Mullins et al., 1990) which involve permanent beds with slow wetting from shallow water in the furrows and cultivations at the plastic limit, demonstrate the improvement in mangement flexibility, yield and soil physical properties that can be achieved when hardsetting properties are fully ameliorated. In northwest India and Pakistan, crusting and surface hardening due to rainfall after sowing of flood-irrigated cotton is a problem which can reduce emergence but can be prevented using a seedline mulch of farmyard manure (Agrawal, 1995) or a crust breaker (Montemayor, 1997, personal communication).

3.6.5   Irrigated Orchards

Problems due to hardsetting in irrigated orchards and vineyards are no differerent than in other irrigated situations except that the rotation time of the crop may be tens of years and this places a much greater premium on preventing structural collapse and maintaining a good rooting environment over the long term. A striking example of these limitations and the benefits of reducing them is given by the irrigated orchards in southeastern Australia which were traditionally flood irrigated and suffered from poor root growth due to hardsetting and poor root aeration. With a combination of subsoil incorporation of gypsum, drainage, low intensity sprinkler irrigation and careful management of surface vegetation to preserve structural stability, it is possible to advance the date when production starts and yields of fruit can be more than double that of the flood-irrigated orchards (Cockroft, 1992, personal communication).

Acknowledgments

I would like to acknowledge the many collaborators in Britain, Africa, Australia, and elsewhere, whose ideas and work I have been able to draw on, and the Overseas Development Administration (now DfID) for their continued support of my research on hardsetting.

3.7 References

Abrol, I.P., and J.C. Katyal. 1995. Sealing, crusting and hardsetting soils in Indian agriculture. p. 31–51. *In* H.B. So, G.D. Smith, S.R. Raine, B.M. Schafer and R.J. Loch. (ed.) Sealing, crusting and hardsetting soils: Productivity and conservation. Proc. 2nd. Int. Symp., University of Queensland, Brisbane, Australia.

Agrawal, R.P. 1995. Inproving cotton seedling emergence under surface crusting. p. 363–368. *In* H.B. So, G.D. Smith, S.R. Raine, B.M. Schafer and R.J. Loch. (ed.) Sealing, crusting and hardsetting soils: Productivity and conservation. Proc. 2nd. Int. Symp., University of Queensland, Brisbane, Australia.

Blackwell, P.S. (ed.) 1992. Slaking and hardsetting soils: some research and management aspects. Soil Till. Res. 25:211–261.

Bradley, R., and N. Crout. 1993. The PARCH model for predicting arable resource capture in hostile environments - User guide. University of Nottingham, Sutton Bonington, UK.

Bresson, L.M., and C.J. Moran. 1995. Structural change induced by wetting and drying in seedbeds of a hardsetting soil with contrasting aggregate size distribution. Europ. J. Soil Sci. 46:205–214.

Bridge, B.J., J.J. Mott, R.J. Hartigan, and W.H. Winter. 1983. Improvement of soil structure by the use of sown pasture in the dry savanna woodlands of northern Australia. Aust. J. Soil Res. 21:83–90.

Bridge, B.J., and D.M. Silburn. 1995. Methods for obtaining surface seal hydraulic properties and how they are incorporated into infiltration models. p. 205–221. *In* H.B. So, G.D. Smith, S.R. Raine, B.M. Schafer and R.J. Loch. (ed.) Sealing, crusting and hardsetting soils: Productivity and conservation. Proc. 2nd. Int. Symp., University of Queensland, Brisbane, Australia.

Carter, D.L. 1990. Soil erosion on irrigated lands. p. 1143–1171. *In* B.A. Stewart and D.R. Nielsen. (ed.) Irrigation of agricultural crops. American Society of Agronomy, Madison WI..

Chan, K.Y., and J.F. Khu. 1995. Infiltration problems of a hardsetting soil and possible ameliorative measures. p. 325–330. *In* H.B. So, G.D. Smith, S.R. Raine, B.M. Schafer and R.J. Loch. (ed.) Sealing, crusting and hardsetting soils: Productivity and conservation. Proc. 2nd. Int. Symp., University of Queensland, Brisbane, Australia.

Chan, K.Y., J.A. Mead, W.P. Roberts, and P.T.W. Wong. 1989. The effect of soil compaction and fumigation on poor early growth of wheat under direct drilling. Aust. J. Agric. Res. 42:221–228.

Chan, K.Y., and C.E. Mullins. 1994. Slaking characteristics of some Australian and British soils. Europ. J. Soil Sci. 45:273–283.

Chartres, C.J., J.M. Kirby, and M. Raupach. 1990. Poorly ordered silica and aluminosilicates as temporary cementing agents in hard-setting soils. Soil Sci. Soc. Am. J. 54:1060–1067.

Childs, E.C. 1969. An introduction to the physical basis of soil water phenomena. John Wiley and Sons, London, UK.

Cochrane, H.R. 1989. Simple techniques for analysing the structural stability of hardsetting soils on the farm. Unpublished Manual. University of Western Australia, Perth, WA.

Cockroft, B., and F.M. Martin. 1981. Irrigation. p. 113–149. *In* J.M. Oades, D.G. Lewis and K. Norrish. (ed.) Red-brown earths of Australia. CSIRO Division of Soils, Adelaide, Australia.

Connolly, R.D., and D.M. Freebairn. 1995. Modelling the impact of soil structural degradation on infiltration, water storage, crop growth, and economics of cropping systems. p. 237–246. *In* H.B. So, G.D. Smith, S.R. Raine, B.M. Schafer and R.J. Loch. (ed.) Sealing, crusting and hardsetting soils: Productivity and conservation. Proc. 2nd. Int. Symp., University of Queensland, Brisbane, Australia.

Daniel, H., R.J. Jarvis, and L.A.G. Aylmore. 1988. Hardpan development in loamy sand and its effect upon soil conditions and crop growth. Proc. 11th Int. Soil Till. Res. Org. Conf. 1:233–238.

Dellar, G.A., and A.B.P. Proffitt. 1995. The use of soil SAR and EC to describe the dispersive behavior of hardsetting Western Australian soils. p. 299–304. *In* H.B. So, G.D. Smith, S.R. Raine, B.M. Schafer and R.J. Loch. (ed.) Sealing, crusting and hardsetting soils: Productivity and conservation. Proc. 2nd. Int. Symp., University of Queensland, Brisbane, Australia.

Dexter, A.R., and B. Kroesbergen. 1985. Methodology for determination of tensile strength of soil aggregates. J. Agric. Eng. Res. 31:139–147.

Eldridge, D.J., and R.S.B. Greene. 1994. Microbiotic soil crusts: a review of their roles in soil and ecological processes in the rangelands of Australia. Aust. J. Soil Res. 32:389–415.

Eldridge, D.J., C.J. Chartres, R.S.B. Greene, and J.J. Mott. 1995. Distribution, characteristics and management of sealing, crusting and hardsetting soils under rangeland conditions in Australia. p. 381–399. *In* H.B. So, G.D. Smith, S.R. Raine, B.M. Schafer and R.J. Loch. (ed.) Sealing, crusting and hardsetting soils: Productivity and conservation. Proc. 2nd. Int. Symp., University of Queensland, Brisbane, Australia.

FAO-UNESCO. 1974. Soil Map of the World 1: 5 000 000. I-X. UNESCO, Paris, France.

French, R.J. 1981. Management under low rainfall. p. 97–116. *In* J.M. Oades, D.G. Lewis and K. Norrish. (ed.) Red-brown earths of Australia. Waite Agr. Res. Inst. and CSIRO Division of Soils, Adelaide, Australia.

Gilkes, R.J., A. Patterson, P.J. Gregory, and R.J. Harper. 1995. Microscopic investigation of some chemical mechanisms of soil hardsetting. p. 445–450. *In* H.B. So, G.D. Smith, S.R. Raine, B.M. Schafer and R.J. Loch. (ed.) Sealing, crusting and hardsetting soils: Productivity and conservation. Proc. 2nd. Int. Symp., University of Queensland, Brisbane, Australia.

Gordon, J.E. 1986. The new science of strong materials or why you don't fall through the floor. 2nd Ed. Penguin books, London, UK.

Greenland, D.J., D. Rimmer, and D. Payne. 1975. Determination of the structural stability class of English and Welsh soils, using a water coherence test. J. Soil Sci. 26:294–303.

Gusli, S., A. Cass, D.A. MacLeod, and P.S. Blackwell. 1994. Structural collapse and strength of some Australian soilsin relation to hardsetting. I. Structural collapse on wetting and draining. Europ. J. Soil Sci. 45:15–21.

Gusli, S., A. Cass, D.A. MacLeod, and C.T. Hignett. 1995. Processes that distinguish hardsetting from rain-induced crusting. p. 457–461. *In* H.B. So, G.D. Smith, S.R. Raine, B.M. Schafer and R.J. Loch. (ed.) Sealing, crusting and hardsetting soils: Productivity and conservation. Proc. 2nd. Int. Symp., University of Queensland, Brisbane, Australia.

Hallett, P.D., A.R. Dexter, and J.P.K. Seville. 1995. The application of fracture mechanics to crack propagation in dry soil. Europ. J. Soil Sci. 46:591–599.

Heinonen, R. 1982. Jordens igenslamning och förhårdnande. (Slaking and hardsetting of soils). Special Issue 12. Swedish University of Agricultural Sciences, Uppsala, Sweden.

Hoogmoed, W.B. 1995. The role of tillage in the management of problems in crusting and hardsetting soils. p. 259–277. *In* H.B. So, G.D. Smith, S.R. Raine, B.M. Schafer and R.J. Loch. (ed.) Sealing, crusting and hardsetting soils: Productivity and conservation. Proc. 2nd. Int. Symp., University of Queensland, Brisbane, Australia.

House of Representatives Standing Committee on Environment, Recreation and the Arts. 1989. The effectiveness of land degradation policies and programs. The Parliament of the Commonwealth of Australia, Austalian Government Printing Service, Canberra, Australia.

Isbell, R. F. 1995. Sealing, crusting and hardsetting conditions in Australian soils. p. 15–30. *In* H.B. So, G.D. Smith, S.R. Raine, B.M. Schafer and R.J. Loch. (ed.) Sealing, crusting and hardsetting soils: Productivity and conservation. Proc. 2nd. Int. Symp., University of Queensland, Brisbane, Australia.

Kayombo, B., and R. Lal. 1993. Tillage systems and soil compaction in Africa. Soil Till. Res. 27:35–72.

Kirchoff, G., and H.B. So. 1995. Soil puddling for rice production and its effect on soil structure and water percolation. p. 413–419. *In* H.B. So, G.D. Smith, S.R. Raine, B.M. Schafer and R.J. Loch. (ed.) Sealing, crusting and hardsetting soils: Productivity and conservation. Proc. 2nd. Int. Symp., University of Queensland, Brisbane, Australia.

Le Bissonais, Y. 1990. Experimental study and modelling of soil surface crusting processes. p. 13–28. *In* R.R. Bryan. Soil erosion: Experiments and models. Catena Suppl., Cremlingen-Destedt, Germany.

Le Bissonnais, Y., A. Bruand, and M. Jamagne. 1989. Laboratory experimental study of soil crusting: relation between aggregate breakdown mechanisms and crust structure. Catena 16:377–392.

Lentz, R.D., I. Shainberg, R.E. Sojka, and D.L. Carter. 1992. Preventing irrigation furrow erosion with small applications of polymers. Soil Sci. Soc. Am. J. 57:224–229.

Levy, G.J., J. Levin, and I. Shainberg. 1997. Prewetting rate and ageing effects on seal formation and interrill soil erosion. Soil Sci. 162:131–139.

Ley, G.J., C.E. Mullins, and R. Lal. 1989. Hard-setting behaviour of some structurally weak tropical soils. Soil Till. Res. 13:365–381.

Ley, G.J., C.E. Mullins, and R. Lal. 1993. Effects of soil properties on the strength of weakly structured tropical soils. Soil Till. Res. 28:1–13.

Ley, G.J., C.E. Mullins, and R. Lal. 1995. The potential restriction to root growth in structurally weak Tropical soils. Soil Till. Res. 33:133–142.

Loch, R.J. 1994. A method for measuring aggregate water stability of dryland soils with direct relevance to surface seal development under rainfall. Aust. J. Soil Res. 32:687–700.

Loch, R.J. 1995. Structure breakdown on wetting. p. 113–131. *In* H.B. So, G.D. Smith, S.R. Raine, B.M. Schafer and R.J. Loch. (ed.) Sealing, crusting and hardsetting soils: Productivity and conservation. Proc. 2nd. Int. Symp., University of Queensland, Brisbane, Australia.

Loch, R.J., and J.L. Foley. 1994. Measurement of aggregate breakdown under rain: comparison with tests of water stability and relationships with field measurements of infiltration. Aust. J Soil Res. 32:701–720.

Masle, J., and J.B. Passioura. 1987. The effect of soil strength on the growth of young wheat plants. Aust. J. Plant Physiol. 14:643–656.

Mason, I.B., and R.A. Fischer. 1986. Tillage practices and the growth and yield of wheat in Southern New South Wales: Lockhart, in a 450 mm rainfall region. Aust. J. Exp. Agric. 26:457–468.

McDonald, R.C., R.F. Isbell, J.G. Speight, J. Walker, and M.S. Hopkins. 1984. Australian soil and field survey land handbook. Inkata Press, Melbourne, Australia.

McIntyre, D.S. 1979. Exchangeable sodium, subplasticity and hydraulic conductivity of some Australian soils. Aust. J. Soil Res. 17:115–120.

Mead, J.A., and K.Y. Chan. 1988. Effect of deep tillage and seedbed preparation on the growth and yield of wheat on a hardsetting soil. Aust. J. Exp. Agric. 28:491–498.

Mead, J.A., and K.Y. Chan. 1995. Managing a hardsetting soil by changing tilllage practices. p. 343–348. *In* H.B. So, G.D. Smith, S.R. Raine, B.M. Schafer and R.J. Loch. (ed.) Sealing, crusting and hardsetting soils: Productivity and conservation. Proc. 2nd. Int. Symp., University of Queensland, Brisbane, Australia.

Mtakwa, P.W. 1993. The role of microaggregation in physical edaphology. Ph.D. Thesis (unpublished), Aberdeen University, Aberdeen, UK.

Mullins, C.E. 1997. Hardsetting. p. 109–128. *In* R. Lal, W.H. Blum, C. Valentin and B.A. Stewart. (ed.) Methods of assessment of soil degradation. CRC Press, Boca Raton, FL.

Mullins, C.E., P.S. Blackwell, and J.M. Tisdall. 1992a. Strength development during drying of cultivated, flood–irrigated hardsetting soil. I. Comparison with a structurally stable soil. Soil Till. Res. 25:113–128.

Mullins, C.E., A. Cass, D.A. MacLeod, D.J.M. Hall, and P.S. Blackwell. 1992b. Strength development during drying of cultivated, flood-irrigated hardsetting soil. II. Trangie soil, and comparison with theoretical predictions. Soil Till. Res 25:129–147.

Mullins, C.E., and G.J. Ley. 1995. Mechanisms and characterisation of hardsetting in soils. p. 157–170. *In* H.B. So, G.D. Smith, S.R. Raine, B.M. Schafer and R.J. Loch. (ed.) Sealing, crusting and hardsetting soils: Productivity and conservation. Proc. 2nd. Int. Symp., University of Queensland, Brisbane, Australia.

Mullins, C.E., D.A. MacLeod, K.H. Northcote, J.M. Tisdall, and I.M. Young. 1990. Hardsetting soils: Behavior, occurrence and management. Adv Soil Sci. 11:37–108.

Mullins, C.E., and K.P. Panayiotopoulos. 1984. The strength of unsaturated mixtures of sand and kaolin and the concept of effective stress. J. Soil Sci. 35:459–468.

Mullins, C.E., J. Townend, P.W. Mtakwa, C.A. Payne, G. Cowan, L.P. Simmonds, C.C. Daamen, T. Dunbabin, and R.E.L. Naylor. 1996. EMERGE - Users guide. A model to predict crop emergence in the semi-arid tropics. Plant and Soil Science Department, University of Aberdeen, Aberdeen, UK.

Mullins, C.E., I.M. Young, A.G. Bengough, and G.J. Ley. 1987. Hard-setting soils. Soil Use Manag. 3:79–83.

Norrish, K., and J.G. Pickering. 1983. Clay minerals. p. 281–308. *In* Soils an Australian viewpoint. CSIRO Publications, Melbourne, Australia.

Northcote, K.H. 1960. A factual key for the recognition of Australian soils. Div. Rep. 4/60. CSIRO Division of Soils, Glen Osmond, Australia.

Northcote, K.H. 1979. A factual key for the recognition of Australian soils. 4th Ed. Rellim Technical Publications, Glenside, Australia.

Packer, I.J., J.A. Mead, and T.B. Koen. 1995. The effect of canola on infiltration and surface soil stability. p. 337–342. *In* H.B. So, G.D. Smith, S.R. Raine, B.M. Schafer and R.J. Loch. (ed.) Sealing, crusting and hardsetting soils: Productivity and conservation. Proc. 2nd. Int. Symp., University of Queensland, Brisbane, Australia.

Panabokke, C.R., and J.P. Quirk. 1957. Incipient failure of soil aggregates. J. Soil Sci. 13:60–70.

Proffitt, A.P.B., S. Bendotti, M.R. Howell, and J. Eastham. 1993. The effect of sheep trampling and grazing on soil physical properties and pasture growth for a red-brown earth. Aust. J. Agric. Res. 44:317–331.

Proffitt, A.P.B., and S. Bendotti. 1995. Sheep trampling exacerbates hardsetting on a red-brown earth. p. 401–406. *In* H.B. So, G.D. Smith, S.R. Raine, B.M. Schafer and R.J. Loch. (ed.) Sealing, crusting and hardsetting soils: Productivity and conservation. Proc. 2nd. Int. Symp., University of Queensland, Brisbane, Australia.

Rengasamy, P. 1997. Sodic soils. p. 265–277. *In* R. Lal, W.H. Blum, C. Valentin and B.A. Stewart. (ed.) Methods of assessment of soil degradation. CRC Press, Boca Raton, FL.

Rengasamy, P., R.S.B. Greene, G.W. Ford, and A.H. Mehanni. 1984. Identification of dispersive behavior and the management of red-brown earths. Aust. J. Soil Res. 22:413–431.

Rice, M.A., C.E. Mullins, and I.K. McEwan. 1997. An analysis of soil crust strength in relation to potential abrasion by saltating particles. Earth Surf. Proc. Landf. 22:869–883.

Roget, D.K., N.R. Venn, and A.D. Rovira. 1987. Reduction of Rhizoctonia root rot of direct-drilled wheat by short-term chemical fallow. Aust. J. Exp. Agric. 27:425–430.

Roth, C.H., and T. Eggert. 1994. Mechanisms of aggregate breakdown involved in surface sealing, runoff generation and sediment concentration on loess soils. Soil Till. Res. 32:253–268.

Seyni-Boukar, L., C. Floret, and R. Pontanier. 1992. Degradation of savanna soils and reduction of water available for the vegetation: the case of northern Cameroon Vertisols. Can. J. Soil Sci. 72:481–488.

Shainberg, I., M.E. Sumner, W.P. Miller, M.P.W. Farina, M.A. Pavan, and M.V. Fey. 1989. Use of gypsum on soils: A review. Adv. Soil Sci. 9:1–111.

Sinden, J.A., A.R. Sutas, and T. Yapp. 1990. Damage costs of land degradation: an Australian Perspective. p. 74–87. *In* J.A. Dixon, D.E. James and P.B. Sherman. (ed.) Dryland management: Economic case studies. Earthscan Publications, London, UK.

Snyder, V.A., and R.D. Miller. 1985. Tensile strength of unsaturated soils. Soil Sci. Soc. Am. J. 49:58–65.

So, H.B., G.D. Smith, S.R. Raine, B.M. Schafer, and R.J. Loch (ed.). 1995. Sealing, crusting and hardsetting soils: Productivity and conservation. Proc. 2nd. Int. Symp., University of Queensland, Brisbane, Australia.

Sojka, R.E., and R.D. Lentz. 1995. Net infiltration and soil erosion effects of a few ppm polyacrylamide in furrow irrigation water. p. 349–354. *In* H.B. So, G.D. Smith, S.R. Raine, B.M. Schafer and R.J. Loch. (ed.) Sealing, crusting and hardsetting soils: Productivity and conservation. Proc. 2nd. Int. Symp., University of Queensland, Brisbane, Australia.

Stenberg, M., I. Håkansson, J. von Polgár, and R. Heinonen. 1995. Sealing, crusting and hardsetting soils in Sweden: Occurrence, problems and research. p. 287–292. *In* H.B. So, G.D. Smith, S.R. Raine, B.M. Schafer and R.J. Loch. (ed.) Sealing, crusting and hardsetting soils: Productivity and conservation. Proc. 2nd. Int. Symp., University of Queensland, Brisbane, Australia.

Tezera Tsegaye, C.E. Mullins, and A. Diggle. 1995. Modelling the effect of mechanical impedance on pea (*Pisum sativum* L.) root growth. A comparison between observations and model predictions in a drying soil. New Phytol. 131:179–189.

Tisdall, J.M., and H.H. Adem. 1988. An example of custom prescribed tillage in south eastern Australia. J. Agric. Eng. Res. 40:23–32.

Townend, J., P.W. Mtakwa, C.E. Mullins, and L.P. Simmonds. 1996. Factors limiting establishment of sorghum and cowpea in two contrasting soil types in the semi-arid tropics. Soil Till. Res. 40:89–106.

Trapnell, C.G., and R. Webster. 1986. Microaggregates in red earths and related soils on East and Central Africa, their classification and occurrence. J. Soil Sci. 37:109–123.

Unger, P.W., and D.K. Cassel. 1991. Tillage implement disturbance effects on soil properties related to soil and water conservation: a literature review. Soil Till. Res. 19:363–382.

Utomo, W.H., and A.R. Dexter. 1981. Soil friability. J. Soil Sci. 32:203–213.

US Salinity Laboratory Staff. 1954. Diagnosis and improvement of saline and alkaline soils. USDA Handb. 60, US Government Printing Office, Washington, DC.

Valentin, C. 1995. Sealing, crusting and hardsetting soils. p. 53–76. *In* H.B. So, G.D. Smith, S.R. Raine, B.M. Schafer and R.J. Loch. (ed.) Sealing, crusting and hardsetting soils: Productivity and conservation. Proc. 2nd. Int. Symp., University of Queensland, Brisbane, Australia.

Valentin, C., and L.M. Bresson. 1997. Soil crusting. p. 89–107. *In* R. Lal, W.H. Blum, C. Valentin and B.A. Stewart (ed.) Methods of assessment of soil degradation. CRC Press, Boca Raton, FL.

Weaich, K., K.L. Bristow, and A. Cass. 1992a. Pre-emergent shoot growth of maize under different drying conditions. Soil Sci. Soc. Am. J. 56:1272–1278.

Weaich, K., A. Cass, and K.L. Bristow. 1992b. Use of a penetration resistance characteristic to predict soil strength development during drying. Soil Tilla.Res. 25:149–166.

Willcocks, T.J. 1981. Tillage of clod-forming sandy loam soils in the semi-arid climate of Botswana. Soil Till. Res. 1:323–350.

Wilson, J.M., and A.B.P. Proffitt. 1995. Management options for hardsetting soils in the NE wheatbelt of Western Australia: The farmer's dilemma. p. 293–298. *In* H.B. So, G.D. Smith, S.R. Raine, B.M. Schafer and R.J. Loch (ed.) Sealing, crusting and hardsetting soils: Productivity and conservation. Proc. 2nd. Int. Symp., University of Queensland, Brisbane, Australia.

Young, I.M. 1992. Hardsetting soils in the UK. Soil Till. Res. 25:187–194.

Young, I.M., and C.E. Mullins. 1991a. Water-suspensible solids and structural stability. Soil Till. Res. 19:89–94.

Young, I.M., and C.E. Mullins. 1991b. Factors affecting the strength of undisturbed cores from soils with low structural stability. J. Soil Sci. 42:205–217.

Young, I.M., C.E. Mullins, P.A.Costigan, and A.G. Bengough. 1991. Hardsetting and structural regeneration in two unstable British sandy loams and their influence on crop growth. Soil Till. Res. 19:383–394.

4

Biogeochemistry of Wetlands

K.R. Reddy, E.M. D'Angelo, and W.G. Harris
University of Florida

4.1 Introduction

Wetlands comprise swamps, marshes, fens, peat lands, bogs, potholes, bays, riparian zones, floodplains, and other shallow flooded areas. These areas are located between terrestrial and aquatic systems and often possess the characteristics of both systems. Wetlands have been estimated to occupy about $2.8 \times 10^6 \, km^2$ or 2.2% of the earth's surface (Post et al., 1982); however, the areal extent may be as high as 6% (Mitsch, 1994). One reason for this discrepancy probably lies in the difficulty of accurately defining what is a wetland. Within the conterminous United States, wetlands occupy about 5% of the land surface, and many of these wetlands are among the most productive natural systems. Wetlands contain all critical components of the biosphere because they provide essential ecological functions including habitat for wildlife, groundwater recharge, shoreline stabilization, flood control, and water quality improvement through biogeochemical transformations. Wetland soils specifically function as sinks and as transformers of nutrients, toxic metals, and organics. A major food crop, lowland rice, is grown under waterlogged soil conditions that are similar to wetland soils in physical, chemical, and biological properties. Recently, the impact of wetlands on the production/consumption of greenhouse gases (CO_2, CH_4, N_2O, and methyl sulfides) has been realized.

4.2 Definitions

Although the definition of soil primarily focuses on upland soils, it includes those that undergo periodic or continuous flooding. Depending on scientific disciplines and ecosystems, soils saturated with water are often called flooded, wetland, waterlogged, marsh, paddy or hydric soils, but no universally accepted definition for wetland soils exists among various groups.

The name 'wetland' implies that the lands are located in wet areas. However, defining a 'wetland' is often complex and difficult, because of the diversity in types of wetlands, presence of moisture gradients from clearly defined upland areas to aquatic systems, as well as the observation that many wetlands are not permanently inundated. These difficulties have resulted in the propagation of several legal and nonlegal definitions of wetlands. However, most definitions include reference to three major components of wetlands: (1) hydrology (water), (2) organisms (plants and animals), and (3) hydric soil.

The following definitions are probably the most often used to describe and delineate wetlands from other ecosystems:

Webster's Collegiate Dictionary Definition: "Land or areas (tidal flats or swamps) containing much soil moisture."

USEPA Definition: "Those areas that are inundated or saturated by surface or ground water at a frequency and duration sufficient to support, and that under normal circumstances do support, prevalence of vegetation typically adapted for life in saturated soil conditions. Wetlands generally include swamps, marshes, bogs and similar areas." This is a regulatory definition of wetlands used by EPA and and US Corps of Engineers for administering Section 404 of the Clean Water Act (EPA, 1988).

US Fish and Wildlife Service Definition: Wetlands are lands transitional between terrestrial and aquatic systems where the water table is usually at or near the surface or the land is covered by shallow water. For purposes of this classification, wetlands must have one or more of the following three attributes: (1) at least periodically, the land supports predominately hydrophytes; (2) the substrate is predominately undrained hydric soil; and (3) the substrate is nonsoil and is saturated with water or covered by shallow water at some time during the growing season of each year (Cowardin et al., 1979).

State of Florida Definition: Wetlands are defined as those areas where there is seasonal or intermittent inundation and saturation with periods of dry exposed soil and which is predominately vegetated.

National Research Council Definition: "A wetland is an ecosystem that depends on constant or recurrent, shallow inundation or saturation at or near the surface of the substrate. The minimum essential characteristics of a wetland are recurrent, sustained inundation or saturation at or near the surface and the presence of physical, chemical, and biological features reflective of recurrent, sustained inundation or saturation. Common diagnostic features of wetlands are hydric soils and hydrophytic vegetation. These features will be present except where specific physicochemical, biotic, or anthropogenic factors have removed them or prevented their development" (Lewis, 1995).

Reddy and Patrick (1999) Definition: A much broader definition of a wetland is a biologically active soil or sediment in which the content of water in or the overlapping floodwater is great enough to inhibit oxygen diffusion into the soil/sediment and stimulate anaerobic (oxygen free) biogeochemical processes. This definition does not necessarily contain any reference to plants, but biologically active soil or sediment is necessary, which means the presence of an energy source (usually organic matter). We define wetland soils with or without aquatic vegetation.

Wetlands, as defined by various groups, typically limit wetlands to areas in which there is hydrophytic vegetation. At present this definition is most commonly accepted, although there is considerable disagreement on the boundaries of developing wetlands from upland areas.

4.3   Wetland Soils

Wetland soils form in a variety of climates and parent materials. They represent a broad spectrum of morphological properties and taxonomic classes, and can be dominated by inorganic or organic materials. However, wetland soils have in common the condition of prolonged saturation. The term saturation refers to the condition of zero or positive hydraulic head, during which water would flow into unlined auger holes and a high proportion of pore space would be filled with water. This condition favors certain physical, chemical, biological, and morphological tendencies that help to distinguish wetland soils from their upland counterparts.

4.3.1 Physical Properties

Saturation causes a number of physical changes in soils, including 1) Softening of the soil material as a result of the weakening effect of water on the bonds holding soil particles together as stable aggregates. This physical effect can have several consequences: (i) root penetration by wetland plants is made easier, the soil is much easier to manipulate when wet, an advantage that rice farmers have used for centuries, and (ii) trafficability of land is much poorer when flooded. 2) Flooding alters soil temperature by darkening soil color, thus increasing heat absorption. Increased water content also increases heat conductivity and stabilizes soil temperate to a more constant value compared to upland soils. Saturated soils are often cooler at the surface as a result of evaporation. In wetlands located in cooler climates, the presence of water may prevent soil temperatures from going below $0°C$. 3) Soil dry bulk density (weight of dry soil per unit volume) is usually decreased as a result of flooding. Typical bulk densities of upland and wetland soils range between 0.3–1.5, and 0.1–1.0 cm^{-3}, respectively, with low values typically observed in organic soils. This is due to high water absorption capacity of organic matter and destruction of soil aggregates.

4.3.2 Chemical and Biological Properties

Wetland soil chemistry is strongly influenced by the chemical reduction normally associated with saturation. Upland soils can be transformed into wetland soils as a result of excessive rainfall, rising water table, and high oxygen demand in the soil. Under these conditions, oxidized chemical species are reduced as a result of abiotic and biotic processes. Similarly, improving the drainage of wetland soils can result in the oxidation of many of the reduced compounds in the soils either by chemical or biochemical reactions (Fig. 4.1). The relative abundance of oxidized and reduced chemical species is, therefore, an indicator of the degree of wetness or anaerobic conditions in soils. The elements Fe and Mn are soil components that are particularly influenced by oxidation and reduction. Examples of Fe-bearing minerals likely to be most stable under the reduced conditions of wetlands include pyrite (FeS_2), siderite ($FeCO_3$), vivianite ($Fe_3(PO_4)_2 \cdot 8H_2O$), and jarosite ($K\,Fe_3(SO_4)_2(OH)_6$). The Fe in these minerals is in reduced (Fe(II), "ferrous") form. Common Fe minerals in relatively oxidized soil environments include goethite, lepidocricite, (both FeOOH), and hematite (Fe_2O_3).

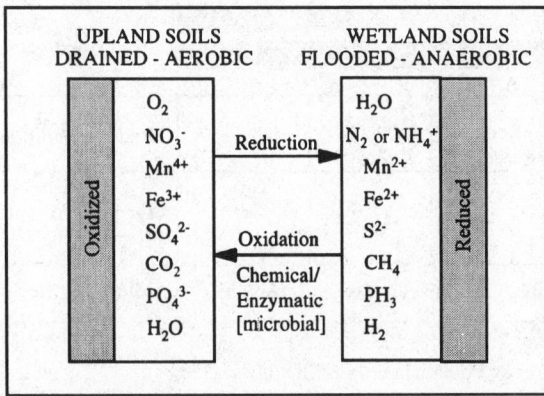

Fig. 4.1 Oxidized and reduced compounds in wetland upland soils

Redox potential is used to measure degree of soil wetness or intensity of soil anaerobic conditions. Analogous to pH (which measures H^+ activity), redox potential (Eh) measures electron (e^-) activity in the soil. Redox potential is defined as the tendency of a pair of chemical species to undergo a transfer of electrons, with one species accepting electrons (reduction) and the other donating electrons (oxidation). Redox reactions involve oxidants, reductants, protons, and electrons, as shown below:

$$a \text{ (oxidants)} + b\,H^+ + n\,e^- = c \text{ (Reductants)} + d\,H_2O \qquad [4.1]$$

The following Nernst equation shows the thermodynamic relationship between redox potential (Eh) and the oxidation-reduction reactions.

$$Eh = E^o - \frac{RT}{nF}\ln\left(\frac{[\text{Reductants}]^c}{[\text{Oxidants}]^a[H^+]^b}\right) \qquad [4.2]$$

where Eh is redox potential, E^o is the standard electrode potential, R is the gas constant, T is the absolute temperature, F is the Faradays constant, n = number of electrons transferred, and [ ] is activity of chemical species in mol L^{-1}.

Redox reactions are critical because they regulate the fate processes of many chemical constituents in wetlands. Relationships among soil hydrologic conditions, redox potential and microbial metabolic activities are depicted in Fig. 4.2. The Eh values of wetland soils range from +700 mv to -300 mv. Negative values represent high electron activity and intense anaerobic conditions typical of permanently waterlogged soils. Under these conditions, there is a low potential for transfer of electrons between oxidized and reduced species, due to lack of oxidized species, such as oxygen. Positive values represent low electron activity and aerobic conditions or moderately anaerobic conditions typical of wetlands in transition zone. Under these conditions, there is a greater potential for electron transfer, due to presence of oxidized species such as oxygen, nitrate, and oxidized forms of Fe and Mn.

An important redox-related process in wetland soils of marshes involves the formation and potential transformation of the mineral pyrite (FeS_2). Saltwater marsh soils tend to have neutral pH and to

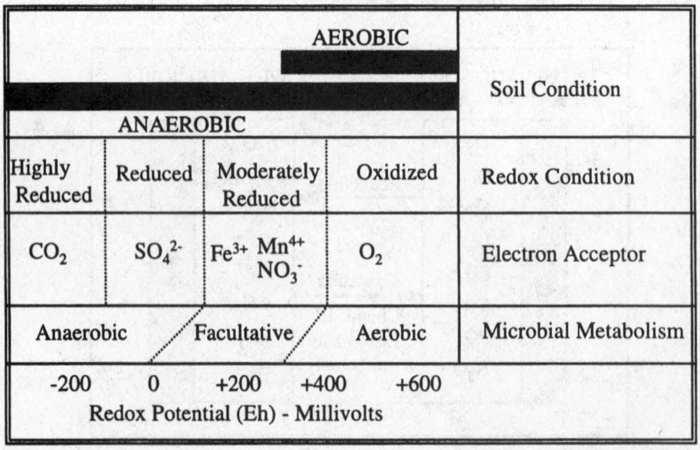

AEROBIC				Soil Condition
ANAEROBIC				
Highly Reduced	Reduced	Moderately Reduced	Oxidized	Redox Condition
CO_2	SO_4^{2-}	Fe^{3+} Mn^{4+} NO_3^-	O_2	Electron Acceptor
Anaerobic		Facultative	Aerobic	Microbial Metabolism
-200 0 +200 +400 +600 Redox Potential (Eh) - Millivolts				

Fig. 4.2 Schematic showing the relationship between hydrologic conditions, redox potential, and metabolic activities of microorganisms in wetlands

support salt tolerant plants (Ponnamperuma, 1972). Pyrite formation occurs as a result of reduction of SO_4^{2-} contained in seawater, high concentrations of Fe(II) in the sediments and rapid accumulation of organic matter that promotes reduction reactions.

$$Fe(OH)_3 + e^- + H^+ \rightleftharpoons Fe(OH)_2 + H_2O \qquad [4.3]$$

$$SO_4^{2-} + 6e^- + 8H^+ \rightleftharpoons S + 4H_2O \qquad [4.4]$$

$$S + 2e^- + 2H^+ \rightleftharpoons H_2S \qquad [4.5]$$

$$Fe(OH)_2 + H_2S \rightleftharpoons FeS + 2H_2O \qquad [4.6]$$

$$FeS + S \rightleftharpoons FeS_2 \text{ (pyrite)} \qquad [4.7]$$

The drainage of pyritic marsh soils poses an environmental problem, because oxidation of pyrite to ferric hydroxide and sulfuric acid results in severe acidity (to pH less than 2). Bacteria involved in oxidizing FeS_2 are *Thiobacillus ferroxidans* and *T. thioxidans*.

The chemical and biological processes of wetland soils are strongly mediated by influences and adaptations of the living communities, as exemplified for the redox processes discussed above. The low oxygen environment typical of wetlands inhibits aerobic microbial activity while stimulating activity of facultative and obligate anaerobes, and favors growth of hydrophytic vegetation adapted to living under anaerobic conditions. The presence and types of wetland and aquatic vegetation suggest the degree of soil wetness and intensity of anaerobic conditions.

4.3.3 Organic Soils

Soils vary greatly in their natural organic matter content. However, organic matter tends to accumulate in wetland soils to a greater extent than upland soils because of a high rate of production relative to rate of decomposition (Mausbach and Richardson, 1994). Thick, dark, organic rich surface layers consisting of slightly to highly decomposed plant remains are therefore common for wetland soils. If these organic rich layers are thick enough such that they essentially comprise the greatest portion of the soil zone, they are referred to as organic soils (Fig. 4.3). The distinction between organic and mineral soils is an arbitrary one, in that depth and organic C content must be specified by a soil taxonomic system. The USDA soil taxonomy specifies an Order of organic soils, Histosols, which is defined as meeting specific depth requirements (in most cases, 40 cm or more) of organic soils material. Organic soil material, in turn, must contain at least 12% organic C if no clay is present, and up to 18% if 60% clay is present (i.e., proportional increase in organic C requirement with increasing clay content). If a layer consisting of organic soil material is thick, but not thick enough to qualify as a Histosol, then it may meet the criteria for a Histic Epipedons ("epipedon" refers to a diagnostic surface or near surface horizon as defined within the USDA soil taxonomic system). Both Histosols and Histic Epipedons are almost exclusively restricted to wetlands. Most organic soils are formed from the accumulation of detritus from hydrophytic vegetation, and transformations of this material into stable humic substances and peat. The organic matter in organic soils is generally most highly decomposed near the soil surface.

The three most extensively occurring suborders of Histosols are distinguished based on the amount of identifiable plant material:

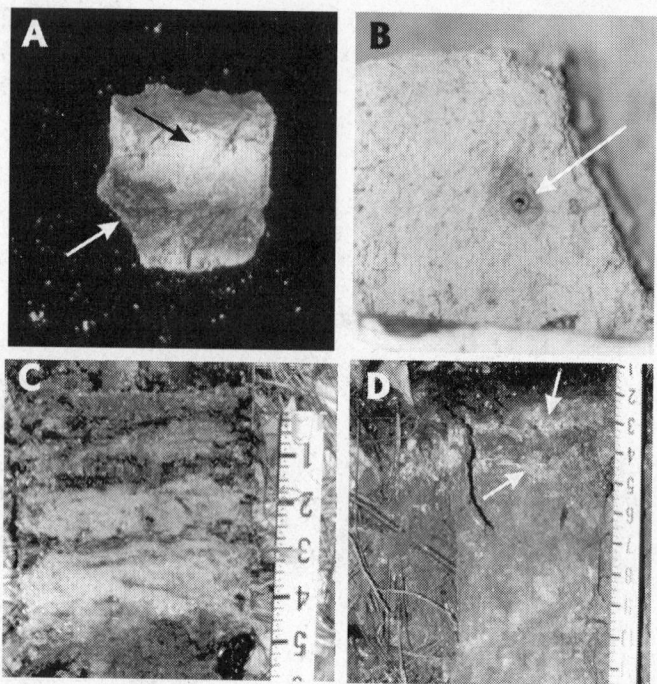

Fig. 4.3 Some features of wetland soils. A. Broken soil aggregate (ped) with redox concentrations in the interior (Dark in photograph but actually red in color; white arrow) and redox depletions near the exterior (gray in color; black arrow); B. Redox concentration (pore lining) illustrating characteristic diffuse boundary with surrounding matrix (white arrow) C. Stratified layers of very dark colored (low value and chroma) sediment alternating with light colored sediment (high value and low chroma) near the soil surface, used as a hydric soil indicator in some regions; D. Light zones near the soil surface (white arrows) produced by redox related stripping of organic matter [Thanks to Mr. Wade Hurt, for photographs used in B, C, and D].

Saprists–about two-thirds of the material well decomposed (muck) and < one-third of the plant material identifiable (peat);

Hemists–about one-half of the material is well decomposed and the other half contains identifiable plant material;

Fibrists–about one-third of the material is well decomposed and > two-thirds of the plant material is identifiable.

A fourth suborder, Folists, is quite restricted in occurrence. It is formed in thick deposits of decomposing leaves and is the only suborder of Histosols not consistently limited to wetland conditions.

Genetically, organic soils are distinguished from mineral soils by processes and conditions affecting the rate and duration of plant detrital accumulations. Long-term dominance by plants that produce abundant below surface biomass under quiescent, shallow water conditions favors the formation of organic soils. Under this scenario, soils rapidly attain intense anaerobic conditions, with redox potential dropping to < -200 mV within a few days after flooding. Organic matter production rates far exceed the oxidation rates, resulting in organic matter accumulation. For example, the sawgrass peat deposits of south Florida have reached depths of several meters in some areas over about a 5000-year period.

Organic soils are far less extensive than mineral soils, but where they do occur they are commonly of considerable economic importance. Many have been artificially drained for agricultural purposes.

Under drained conditions, aerobic biological oxidation of organic matter results in rapid rates of soil subsidence, with rates of about 3 cm yr^{-1} in the Everglades (Stephens, 1969). At current rates of subsidence, land in the Everglades agricultural area will have peat depths less than 100 cm by the year 2000, making them less attractive for productive agriculture. The rate of soil subsidence can be reduced by implementing short-term (1–2 month) or long-term anaerobic conditions, such as through flooding or high water table management strategies.

4.3.4 Mineral Soils

Wetland soils that do not meet the taxonomic requirements for organic soils (e.g., Histosols) are mineral soils. The chemical reduction that accompanies saturated conditions generally results in morphological features that are far more evident in mineral than organic soils, due to the masking effect of organic matter for the latter. Reduction promotes dissolution of redox sensitive Fe and Mn oxides in soils, enabling their mobilization in accordance with principles of mass flow and diffusion unless they encounter an oxidized zone where they reprecipitate as oxides. This redox-induced redistribution provides visual evidence of saturation and of redox gradients within the soil matrix (Fig. 4.3), since both Fe and Mn are strong coloring agents (shades of yellow, brown, red, and black) in their oxide forms. Gray colored zones, in many cases, are attributable to reduction and commonly depletion of these metals.

The process of gray color formation in mineral soils is sometimes referred to as "gleying" or "gleization". Gray coloration corresponds to "low chroma" (e.g., < 2) designations using the conventional Munsell notation for color characterization. A soil horizon dominated by redox-related gray colors is given the subordinate designation "g" (e.g., Btg). High chroma colors of various hues (e.g., brown or red) in zones intermixed with gray colored zones are usually attributable to the oxidation, and subsequent concentration, of these metals. Over time, these zones of concentration may harden to form nodules and concretions. However, such hardened features, unless they have gradual or diffuse boundaries with the surrounding matrix, are indicative of relict rather than contemporary saturation. Features that reflect current wet-dry cycles tend to remain soft.

Historically, the patterns of gray in zones of periodic saturation were called "gray mottles". Gray coloration was used as a diagnostic criterion for saturation in both formal (e.g., in soil taxonomy) and informal (e.g., soil interpretations) contexts in soil science. However, soil scientists recognized that gray coloration is not always the result of chemical reduction. For example, gray colors are sometimes "inherited" from light colored parent material. Also, carbonates and clean sand grains tend to impart gray colors to soils. Hence, a new terminology was sought which would more explicitly and precisely convey the interpretations of saturation and reduction. Such a terminology was developed by the Committee on Soils with Aquic Moisture Regimes (Bouma, 1991). Interpretive morphological terms recommended by the committee were formally adopted for use in a revision of USDA soil taxonomy (Soil Survey Staff, 1996). Collectively, they are referred to as redoximorphic features (Fig. 4.3). Redoximorphic features are now identified in standard soil descriptions, and include the following (Vepraskas, 1992):

1. Redox concentrations: Bodies interpreted as redox-related concentrations of Fe and Mn.
   a. Nodules and concretions: Partially hardened bodies. Concretions have concentric layers in cross-section, while nodules have a uniform internal fabric.
   b. Masses: Soft bodies with reddish or brownish colors.
   c. Pore linings: Zones of Fe and/or Mn accumulation along pore surfaces, as inferred from coloration.

2. Redox depletions: Bodies of low chroma (gray colors) corresponding to zones where (i) Fe and Mn oxides have been depleted through reduction, and in some cases, (ii) clay has been depleted by mobilization due to the loss of the oxide cements.
3. Reduced matrices: Soil matrices where low chroma is the result of chemical reduction of Fe, but not total depletion of Fe. A color change resulting from Fe oxidation occurs within 30 minutes.

Mineral soils, like organic soils, can have thick, dark surface horizons attributable to organic matter accumulation. These horizons may consist entirely of mineral soil material (i.e., below the organic C requirement for organic soil material). For example, many soils of semi arid prairie regions (e.g., Mollisols, USDA taxonomic system) have such surface horizons (called Mollic Epipedons) which are attributable to the climax grass vegetation rather than wetness per se. Alternatively, mineral soils can have organic soil material thick enough to qualify as Histic Epipedons. The distinction between the Mollic Epipedon and a Histic Epipedon, both dark and thick, can be important in wetland delineation (see Section 4.3.5) because the former is not necessarily a wetland indicator while the latter is a very reliable one. However, dark mineral surface horizons can be indicators of wetness in conjunction with immediately underlying redoximorphic features. In humid climates, for instance, surface horizons tend to be thicker and darker in wetlands.

4.3.5   Soils and Wetland Delineation: Hydric Soils Concept

Increased awareness of the important ecological role of wetlands has stimulated policies promoting wetland preservation. It has hence become necessary to delineate wetlands from uplands in a consistent and scientifically sound fashion for jurisdictional purposes. Soils are one of the three key components conventionally used to establish wetland boundaries, the others being vegetation and hydrology (Cowardin et al., 1979). Each component is assessed independently. The soil assessment is particularly critical because soils are the most stable of the three components; e.g., vegetation can be quickly altered and hydrology is sensitive to seasonal climatic fluctuations. The influences of prolonged saturation on soil properties, as previously discussed, are the crux of wetland soil delineation. Recent efforts have been made to document and catalogue soil properties that are most consistently associated with wetland vegetation and hydrology.

"Wetland soils" could simply be defined as soils that occur in wetlands. However, this would be a circular definition if we intend to use soil characteristics as independent criteria in jurisdictional wetland delineation. Delineating wetland soil boundaries requires the specification of soil criteria that document the types of saturated conditions associated with wetland ecosystems. The term "hydric soils" is commonly used in jurisdictional language to apply to soils that meet such criteria. The USDA-NRCS, in conjunction with the National Technical Committee for Hydric Soils, has defined hydric soils as "... soils formed under conditions of saturation, flooding, or ponding long enough during the growing season to develop anaerobic conditions in the upper part" (Federal Register, July 13, 1994). The growing season is the period of the year when soil temperature and moisture are most favorable for microbial activity, and hence anaerobiosis if the soil is saturated. Hydric soils, as defined, would tend to be wet even during the season when evapotranspiration loss is at a maximum.

The USDA hydric soils definition serves as a guideline for local technical limits used in jurisdictional wetlands delineation. However, the practical delineation of hydric soils generally requires some indirect assessment and professional judgement based on readily observable indicators associated with near surface saturation. Field indicators of hydric soils have evolved through collective observations and consensus, and have been formally catalogued by the USDA-NRCS (USDA-NRCS, 1996).

The USDA hydric soil indicators are specified as applying to (i) all soils, (ii) sandy soils, and (iii) loamy and clayey soils. Some are applicable to all USDA Land Resource Regions (USDA, 1981), while others are restricted to certain regions. Most of the indicators reflect pronounced organic matter accumulation at the surface, immediately underlain by redoximorphic features or other local subsurface features consistently indicative of saturation and anaerobiosis.

Delineation of hydric soils requires intensive soil investigation at wetland boundaries, since the interior areas are commonly wet enough that the hydric status is indisputable (i.e., the soils in the wettest areas have almost certainly developed "... anaerobic conditions in the upper part"). The uncertainty of hydric soil identification generally increases with proximity to the upland boundary, increasing the challenge of consistent delineation. The indicators that discriminate most effectively at the boundary are, therefore, the most useful. Soils examination generally requires only a shallow excavation (e.g., about 50 cm) using a tiling spade, since evidence for near surface saturation is being sought.

Detailed interpretive "hydric soils criteria" have been developed by USDA-NRCS (Federal Register, February 24, 1995) for the purpose of identifying soils series that could potentially be hydric soils. Series meeting the criteria are placed on a national Hydric Soils List (USDA, 1991). These hydric soils criteria encompass taxonomic and interpretive criteria linked with individual soil series within the USDA-NRCS soils data base. The criteria are not used in field delineation of hydric soils. The occurrence of a soil series on the USDA Hydric Soils List means only that a soil fitting the series in the field is potentially hydric. In effect, hydric and nonhydric soils could both fit the same series, since hydric soil status per se is not a series criterion.

4.4 Redox Gradients

In wetlands, O_2 is introduced into soils by diffusion through the floodwater, from photosynthetic O_2 production sources, by diffusion and mass flow from the atmosphere through plants into root zone, and by fluctuations in water table depth. Oxygen is crucial in regulating plant and microbial respiration rates, and speciation, mobility, and bioavailability of chemicals in soils. However, when soils are flooded, O_2 exchange between the air-water-soil phases is severely curtailed. For example, O_2 diffusion through water is about 10^4 fold slower than in air. Dissolved O_2 present in the soil porewater is rapidly consumed during aerobic respiration and approaches zero within a few hours depending on soil O_2 demand. Although not related to saturated soil conditions, application of O_2-demanding wastes (animal wastes, composts, sewage sludge) or ammoniacal fertilizer can also result in O_2 depletion in the soil.

When O_2 is limiting, there is a shift in a number of properties that differentiate wetland and upland soils. Typically, there is an accumulation of organic matter, which may reach several meters of peat in some Histosols. Aerobic microbial activity is replaced by predominantly anaerobic activity.

Aerobic organisms use O_2 as their primary electron acceptor during their catabolic activities. In the absence of O_2, aerobic respiration is curtailed and the microbial catabolic activity switches to anaerobic respiration. During this process bacteria obtain energy by oxidizing organic and inorganic compounds through several intermediate steps. The electrons released from these compounds pass through an electron transport chain containing several electron carriers that transfer electrons from energy substrates to terminate electron acceptors.

In wetlands, O_2 cannot always move into the soil rapidly enough to take care of the biological O_2 demand of organisms because of pore blockage by water. Although O_2 supply to the soil is restricted as a result of flooding, its demand remains high. These conditions result in the development of two distinctly different soil layers: (1) an oxidized or aerobic surface soil layer containing O_2, and (2) an underlying reduced anaerobic soil layer where no free O_2 is present (Figs. 4.4 and 4.5). These

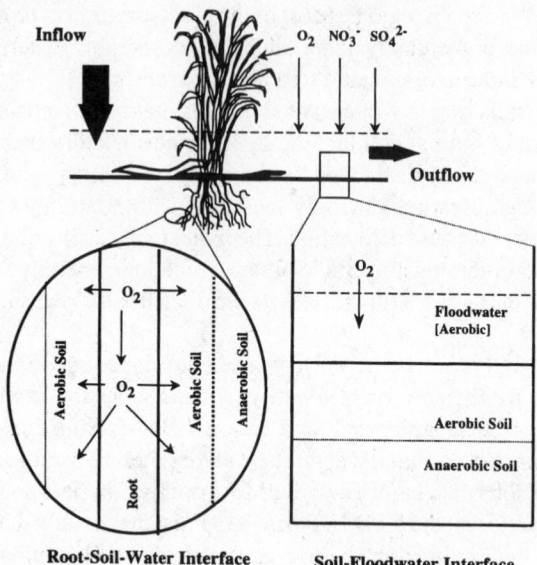

Fig. 4.4 Schematic showing the aerobic-anaerobic interfaces in wetlands

conditions result in unique redox gradients in which aerobic or oxidized conditions occur in two distinct zones, namely, at soil/ floodwater interface, and in the root zone (Fig. 4.4).

Diffusion of O_2 through floodwater maintains aerobic conditions at the soil/floodwater interface which can vary in thickness from few mm to about 2 cm. In wetlands with dense layers of periphyton, production of O_2 during photosynthesis may result in a thick aerobic zone during the day, while consumption of O_2 during respiration may convert this layer to an anaerobic zone during the night. Thus, redox gradients at soil/floodwater interface can have regular daily fluctuations. The juxtaposition of aerobic and anaerobic soil layers has major implications in regulating nitrification/ denitrification reactions, iron oxide regulated precipitation of P, and oxidation of CH_4, sulfides and other reduced species.

The depth of the oxidized and anaerobic zones where a given electron acceptor is stable depends on the inflow of electron acceptors, the availability of organic and inorganic substrates, and hydrologic conditions. The sequential reduction of electron acceptors as a function depth follows the order of O_2 reduction at the soil/floodwater interface, followed by other electron acceptors. The soil depths at which these reductions occur are depicted in Fig. 4.6. It should be noted that different electron acceptors are utilized in the soil simultaneously at different soil depths. Thus, wetlands can be characterized by measuring the concentration of reduced species [Mn(II), Fe(II), S^{2-}] or oxidized species such as (O_2, NO_3^- and SO_4^{2-}).

Wetland plants have unique characteristics of adaptation to anaerobic soil conditions, such as development of internal air spaces (aerenchyma) for transporting O_2 into the root zone. Depending on plant species, these air spaces can occupy up to 60% of the total tissue (Brix, 1996). Oxygen transport in wetland plants occurs through molecular diffusion as a result of partial pressure gradients established within the system. In addition, temperature- and humidity-induced pressurization can move gases through mass flow (Armstrong and Armstrong, 1990; Brix et al, 1992). Low temperatures in colder climates can decrease internal pressurization, thus decreasing overall O_2 transport into the root zone. Reports conflict on the significance of O_2 transport by wetland plants in wastewater

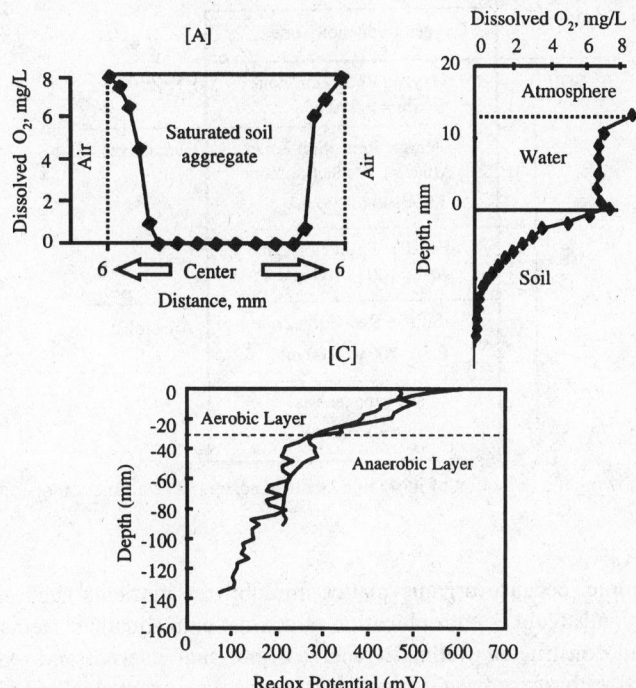

Fig. 4.5 Aerobic-anaerobic interfaces in wetland soils. Dissolved oxygen gradients [A] at the soil-floodwater interface and [B] within a saturated soil aggregate [C] Redox potential of stream sediments

treatment. For example, O_2 transport by *Phragmites* was reported to be in the range of 0.02–12 g m^{-2} day^{-1} (Brix, 1996). This wide range was attributed to seasonality and use of different experimental techniques. Other studies have shown that O_2 release into the root zone supports heterotrophic respiration, BOD reduction, and nitrification (Reddy et al., 1989 a,b).

4.5 Carbon

Compared to upland systems, most wetland soils show an accumulation of organic matter, and therefore, wetlands function as global sinks for C. Accumulation of organic C in wetlands is primarily a result of the balance of two processes: C fixation through photosynthesis and losses through decomposition. Rates of photosynthesis in wetlands are higher than other ecosystems, and rates of decomposition are typically lower due to anaerobic conditions; hence organic matter tends to accumulate. For example, estimated mean primary productivity of a wetland ecosystem was approximately 1,300 g C m^{-2} yr^{-1} (Houghton and Skole, 1990), which is higher than many terrestrial tropical and subtropical ecosystems. Net accumulation of C in some peatland ecosystems was found to be in the range 11 to 105 g C m^{-2} yr^{-1}(Schlesinger, 1997). Whiting and Chanton (1993) estimated that approximately 3% of the net ecosystem production in wetlands escapes to the atmosphere as CH_4. However, these rates are highly variable between wetlands, depending on types of soil and vegetation, hydrology, nutrient inputs, extent of disturbance, presence of toxic substances, solar radiation, length of growing season and other environmental factors. In addition to maintaining proper functioning of wetlands, organic matter storage also plays an important role in protecting other ecosystems and the

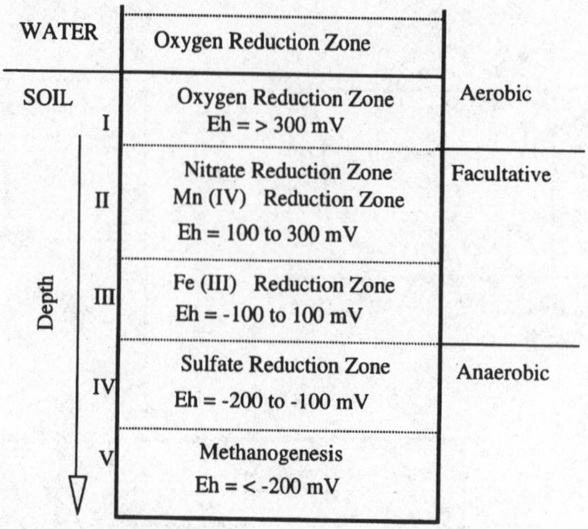

Fig. 4.6 Schematic showing the reduction of inorganic electron acceptors as a function of soil depth (Eh values are based on pH values of 7.)

biosphere. For example, because organic matter immobilizes nutrients such as N, P, and S, its accumulation in wetlands reduces eutrophication of downstream aquatic systems.

Some natural and constructed wetlands receive significant internal and external loadings of organic C, associated with surface water runoff, carbonaceous wastewater, and diurnal tidal events. Anthropogenic inputs of organic C, including oil spills in coastal wetlands, and toxic organics in agricultural and industrial runoff, have not been well quantified in wetlands. Except for indirect effects on primary production, these inputs probably play a minor role in the overall C budget of most wetlands.

Organic C undergoes complex cycling in wetland systems, and the fate of C depends on the specific type of molecule and environmental conditions (Fig. 4.7). Easily degradable (labile) fractions are decomposed to inorganic constituents, while recalcitrant pools are accreted as new peat layers. Organic matter associated with above- and belowground plant biomass, algal and microbial biomass in wetlands consists of complex mixtures of nonhumic substances, including particulate pools (cellulose, hemicellulose, tannins and lignins, proteins), water soluble components (amino acids, sugars, and nucleotide bases), and ether extractable components (lipids, waxes, oils). During frost or natural seasonal die off, nonhumic substances are deposited in the water column and surface soil. Root exudates containing ethanol, carbohydrates, and amino acids may also contribute significant amounts of C to subsurface soils (Mendelssohn et al., 1981). Labile fractions undergo multistep conversions to inorganic constituents including (1) abiotic leaching of water soluble components and fragmentation of tissues into small pieces (< 1 mm), (2) extracellular enzyme hydrolysis of biopolymers (nucleic acids, proteins and cellulose) into monomers, and (3) aerobic and anaerobic catabolism of monomers by heterotrophic microorganisms (Fig. 4.8). The litter layer is a major support structure being a source of energy and nutrients and containing high numbers (10^9–10^{10} g^{-1} dry matter) of bacteria and fungi (Kjoller et al., 1985). Microbial biomass accounts for a significant amount of C in wetland substrates (DeBusk, 1996), and nutrients released from organic matter decomposition are available for plant uptake and growth.

Abiotic leaching of algal and plant biomass is largely complete within days to months after deposition into water, but depends on amount of particulate and structural material. For example, Gaur et al. (1992) measured leaching losses of up to 33% of water hyacinth mass after 4 days immersion in

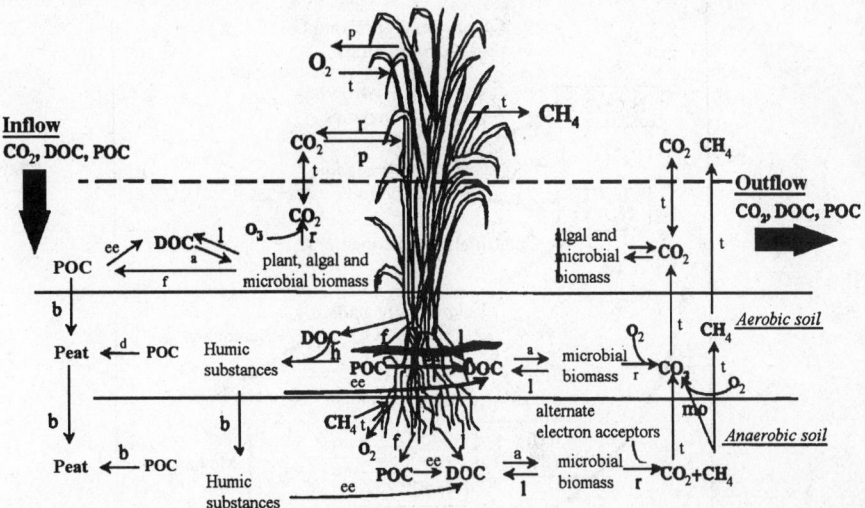

Fig. 4.7 Carbon transformations in the soil-water-plant environment of wetlands. POC = particulate organic carbon; DOC = dissolved organic carbon; a = assimilation; b = burial; ee = extracellular enzyme hydrolysis; f = fragmentation; h = humification; l = leaching; mo = methane oxidation; p = photosynthesis; r = respiration; t = transport

water column. However, Benner et al. (1985) measured only 10–20% losses of lignocellulosic components of plants in 1 month. Particulate materials remaining after initial leaching are typically fragmented by meiofauna before undergoing further decomposition. This step is critical because surface area is increased, allowing microorganisms and enzymes to penetrate tissues.

Extracellular enzymes such as cellulases and proteases which are excreted by plants, fungi, and bacteria, hydrolyze large molecular weight biopolymers into oligomers and monomers that can then be taken up by microorganisms (Fig. 4.8). Enzyme hydrolysis is generally considered the rate-limiting step in organic matter decomposition (Sinsabaugh, 1994). Gritzali et al. (1987) showed about a 15% increase in methanogenesis from anaerobic decomposition of water hyacinth in the presence of cellulase enzymes compared to treatments without enzymes. Because oxidoreductase enzymes such as peroxidases and phenoloxidases involved in lignin oxidation require O_2, lignin degradation is most prevalent in aerobic zones such as soil-water and root-soil interfaces. Major fungal genera involved in decomposition in wetland systems include *Alternaria, Cylindrocarpon, Cladosporium, Penicillium, Fusarium, Trichoderma, Alatospora, Tetacladium, Helicodendronm* and *Helicoon* (Kausik and Hynes, 1971; Given and Dickenson, 1975; Kjoller and Struwe, 1980). Bacteria are dominant decomposers under anaerobic conditions (Alexander, 1977). Important bacterial genera involved in extracellular enzyme hydrolysis of organic matter in wetlands include *Cytophaga, Vibrio, Achromobacter, Bacillus, Micrococcus, Chromobacterium, Streptomyces, Arthrobacter, Actinomyces, Clostridia, Pseudomonas, Flavobacterium, Bacteriodes, Eubacterium, Peptostreptococcus* (Given and Dickenson, 1975; Wheatly et al., 1976; Molongoski and Klug, 1976).

Production and activity of extracellular enzymes are affected by a number of environmental factors including nutrients, pH, O_2 supply, humic and mineral substances, and inorganic cations. For example, Sinsabaugh et al. (1993) found that production of proteases and phosphatases was enhanced in N- and P-limited wetland systems, while lignocellulase activity and organic C mineralization were promoted under nonlimiting nutrient conditions. Fog (1988) concluded from a study of > 60 sites that

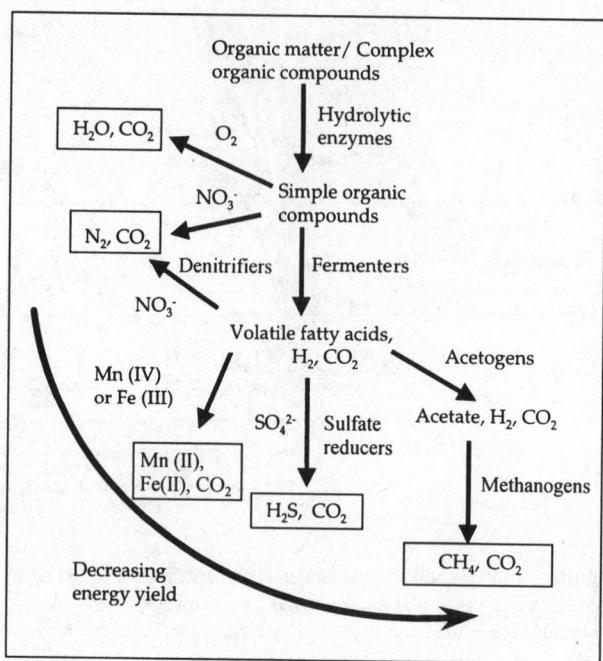

Fig. 4.8 Schematic showing catabolic pathways regulating organic matter decomposition in wetland soils

cellulose degradation was enhanced and lignin degradation was inhibited by inorganic and organic N amendments, probably resulting from N repression of phenoloxidase synthesis. Gordon and Millero (1985) demonstrated that biodegradation rates of low molecular weight organic acids and sugars were decreased due to sorption onto hydroxyapatite mineral surfaces. Goodwin and Zeikus (1987) found that anaerobic decomposition processes in acidic (pH 4.9) bog sediments are decreased, but that C and electron flow pathways are similar to those in neutral sediments. Kim and Wetzel (1993) demonstrated that humic substances complexed and curtailed the activity of extracellular enzymes, which was relieved by the presence of divalent cations such as Ca^{+2} and Mg^{+2}. They speculated that this phenomenon may explain reduced decomposition in soft compared to hard water aquatic systems.

After enzyme hydrolysis, small molecular weight compounds are taken up and utilized as C and energy sources by heterotrophic microorganisms. A multitude of different microorganisms may be involved in this terminal step of decomposition, which depends largely on availability of electron acceptors (Fig. 4.8). Diverse types of microorganisms couple oxidation of organic C substrates (electron donors) and reduction of electron acceptors to energy (ATP) production required for growth. The most common electron acceptors include O_2, NO_3^-, Mn(IV), Fe(III), SO_4^{2-}, and CO_2, which enter wetlands through both internal and external inputs. For example, internal inputs include oxidation of chemical species (NH_4^+, H_2S, Fe(II), and Mn(II)) that diffuse from anaerobic to aerobic soil zones. External inputs of electron acceptors include atmospheric O_2, seawater (SO_4^{2-}), surface water runoff (NO_3^-), and precipitation. Most wetlands contain several electron acceptors, so competition exists for electron donors between microbial groups. Organisms that derive the most energy and have the fastest degradation kinetics outcompete other groups. For these reasons, aerobic bacteria outcompete anaerobes. For example, DeBusk (1996) and D'Angelo and Reddy (1999) determined that aerobic C mineralization was about 3 times faster than under anaerobic conditions. Because recently deposited plant and algal detritus occurs typically in the water column and at the soil surface, most (70–80%) decomposition may be by aerobic fungi and bacteria (Kjoller and Struwe, 1980). Small zones at the root surface may also promote aerobic microbial activity (Schipper and Reddy, 1996).

Root material and aboveground plant and algal biomass that are buried undergo anaerobic decomposition where microorganisms must utilize alternate electron acceptors to oxidize organic matter. Due to energy differences, electron acceptors are utilized in the order $O_2 > NO_3^- > Mn(IV) > Fe(III) > SO_4^{2-} > CO_2$. This phenomenon explains the stratification in microbial activity and chemicals often observed in wetland soil profiles. Slow decomposition rates under anaerobic conditions partially explain why wetlands accumulate organic matter more than upland systems. When wetland soils are drained, rapid rates of aerobic decomposition result in soil subsidence (2–3 cm yr^{-1} in the Florida Everglades) (Snyder, 1987). Under drained conditions, wetlands act as a source of C to the rest of the biosphere.

Carbon compounds that are recalcitrant to aerobic and anaerobic decomposition tend to accumulate in wetlands either as undecomposed plant tissues (peat) or humic substances. Formation of humic substances catalyzed by phenoloxidase enzymes in soils is thought to involve condensation reactions between reactive phenolic groups of tannins and lignins with water soluble nonhumic substances (Francois, 1990). This mechanism accounts for the large molecular weight and heterogenous humic substances that contain significant amounts of N, P, S in their structures. For example, P contained in humic substances may make up between 40–90% of the total P in Histosols (DeBusk et al., 1994). In the absence of O_2, humic substances are resistant to decomposition, and so represent a significant C and nutrient storage in wetlands. Under drained conditions, humic substances are rapidly degraded, which releases nutrients to the bioavailable pool, thereby affecting downstream water quality.

4.6 Nitrogen

Nitrogen is a major nutrient in wetlands and often controls primary productivity. Although for many ecosystems P may limit productivity, under conditions of excessive P supply, ecosystem productivity can be limited by N supply. Nitrogen entering wetlands is present in organic and inorganic forms. The relative proportions of each depend on source and type of water entering these systems. Organic forms are present in dissolved and particulate forms, while inorganic N (NH_4-N and NO_3-N) is present in dissolved forms. Particulate forms are removed through settling and burial, while the removal of dissolved forms is regulated by various biogeochemical reactions functioning in soil and water column (Fig. 4.9). Nitrogen is found at various valence states ranging from -3 (NH_3) to +5 (NO_3^-), and this wide range in oxidation numbers provides the possibility for several oxidation-reduction reactions. Relative rates of these reactions are affected by physicochemical and biological characteristics of soils, organic substrates, and the water column. Detailed reviews of N processes functioning in the soil and the overlying water column of wetland are presented by Reddy and Patrick (1984), Bowden (1987), Johnston (1991), Howard-Williams and Downes (1993), and Reddy and D'Angelo (1994).

Several N transformations effectively process inorganic N through nitrification and denitrification, ammonia volatilization and plant and microbial uptake, and maintain relatively low levels of inorganic N in the water column. The extent of these processes usually increases with N loading. Breakdown of detrital tissue results in release of dissolved organic N to the water column, most of which is resistant to decomposition. Under these conditions, water leaving wetlands may contain elevated levels of N in organic forms. Wetlands usually function as effective sinks for inorganic N, and as sources for organic N.

Net release of NH_4-N by wetland soils is determined by the balance between ammonification and immobilization, which is controlled by the N requirements of microorganisms involved, nature of

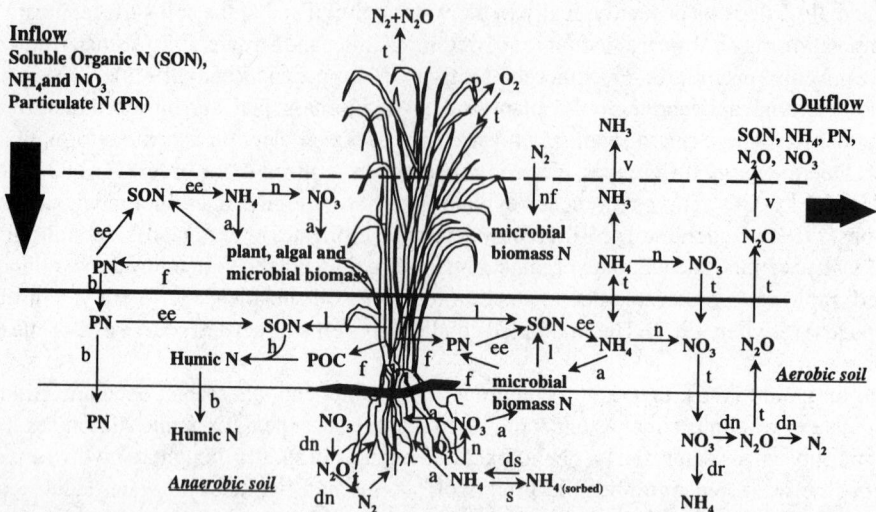

Fig. 4.9 Nitrogen transformations in the soil-water-plant environment of wetlands. POC = particulate organic C; a = assimilation; b = burial; dn = denitrification; dr = dissimilatory nitrate reduction to NH_4^+; ds = desorption; ee = extracellular enzyme hydrolysis; f = fragmentation; h = humification; l = leaching; n = nitrification; nf = nitrogen fixation; p = photosynthesis; s = sorption; t = transport; v = volatilization

organic N, and other soil and environmental factors. Because of low N requirements of anaerobic microorganisms, wetland soils usually accumulate NH_4-N, which supports most of the N requirements of wetland plants. Organic N mineralization can be described as a function of the C/N ratio, extracellular enzyme (such as protease), microbial biomass and soil redox conditions. For example, organic N mineralization was highly correlated with the biomass N of wetland soil maintained under different redox conditions (McLatchey and Reddy, 1998). In another study by Stanford et al. (1973), organic N mineralization of plant detritus was higher in summer than cooler winter months (showing Q_{10} values of about 2 between 15–35 ºC), and temperature effects were greatest for *Sagittaria* compared to *Typha* tissues (Fig. 4.10). These results suggest that measurement of these parameters can indicate the extent of organic N mineralization in many systems.

Ammonia volatilization is an abiotic reaction influenced by the physicochemical characteristics of a wetland water column. Ammonia volatilization is regulated by NH_3 concentration, temperature, vegetation density, air movement above the water surface, mixing in the water column, algal activity and associated pH fluctuations.

$$NH_4^+ + OH^- \rightleftharpoons NH_3\,(aq) + H_2O \qquad\qquad [4.8]$$

$$NH_3\,(aq) \rightleftharpoons NH_3\,(g) \qquad\qquad [4.9]$$

In wetlands, this process can play a significant role if the influent water contains high levels of NH_4-N, and algal photosynthetic activity drives pH above 8.5.

Ammonium oxidation (nitrification) is a two-step process mediated by aerobic chemoautotrophic bacteria, *Nitrosomonas* (NH_4^+ to NO_2^-) and *Nitrobacter* (NO_2^- to NO_3^-). These bacteria derive energy for biosynthesis from oxidation of NH_4^+ and NO_2^-, while using HCO_3^- as a C source.

$$NH_4^+ + 2O_2 \rightarrow NO_3^- + H_2O + 2H^+ \qquad\qquad [4.10]$$

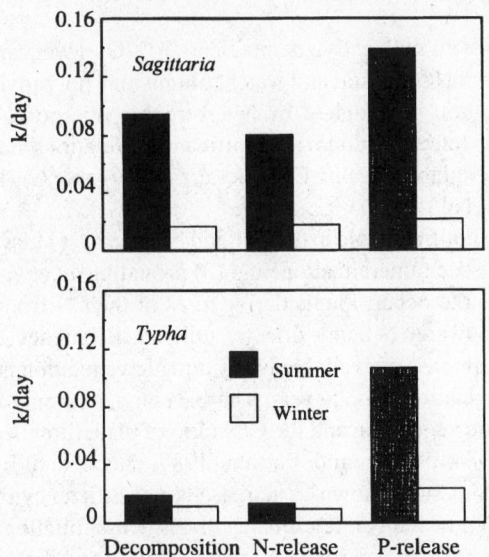

Fig. 4.10 Decomposition of detrital plant tissue and nitrogen and phosphorus release in a constructed wetland. Litter bag technique and loss in dry weight and nutrient composition of detrital tissue were used to determine decomposition and nutrient release. Experiments were conducted for 4 months, starting in July 1992 for summer, and January 1993 for winter [K.R. Reddy, unpublished results]

Since nitrification is an aerobic process, it occurs in the water column, surface aerobic soil layer, and root zone. The relative importance of these zones in overall nitrification depends on O_2 availability and NH_4-N concentration (Reddy and Patrick, 1984). Nitrification rates are reported to be in the range of $0.01–0.161$ g N m^{-2} day^{-1} (mean $= 0.048 \pm 0.044$) (Martin and Reddy, 1997). These values are lower than those reported for mineralization, suggesting that O_2 and NH_4-N availability limit nitrification. Nitrifiers compete with other heterotrophic and lithotrophic organisms for O_2. In wetland environments, O_2 is probably the primary regulator of nitrification.

Nitrate added to or produced in wetlands can be used during assimilatory or dissimilatory NO_3^- reduction. During assimilatory NO_3^- reduction, microorganisms and plants assimilate NO_3^- into their biomass after reduction to NH_4^+ and incorporation into monomers (amino acids) and biopolymers (proteins). Dissimilatory NO_3^- reduction is divided into two main pathways: denitrification (Reaction [4.11]) and dissimilatory NO_3^- reduction to ammonium (Reaction [4.12]). In both cases, NO_3^- is used as an electron acceptor for energy generation by anaerobic bacteria.

$$5C_6H_{12}O_6 + 24\ NO_3^- + 24\ H^+ \rightarrow 30\ CO_2 + 12\ N_2 + 42\ H_2O \qquad [4.11]$$

$$C_6H_{12}O_6 + 3\ NO_3^- + 6H^+ \rightarrow 6CO_2 + 3\ NH_4^+ + 3\ H_2O \qquad [4.12]$$

Denitrification occurs at higher Eh levels (200–300 mV) than dissimilatory reduction (< 0 mV). Denitrification is typically the dominant pathway of NO_3^- removal in wetlands. Reported NO_3^- removal rates by wetlands are in the range of $0.003–1.02$ g N m^{-2} day^{-1} (mean $= 0.097 \pm 0.139$. In most of the wetlands, denitrification rates are usually limited by NO_3^- concentrations and diffusion rates of NO_3^- from aerobic to anaerobic sites (Martin and Reddy, 1997). Denitrification rates are usually higher in soils receiving steady loadings of NO_3^- than soils receiving low or negligible levels (Cooper, 1990;

Gale et al., 1993). In a system with active denitrification, NO_3^- levels are usually maintained at low levels, thus measurement of NO_3^- in soil and water column may not provide reliable indication of this process. Since denitrification is mediated by heterotrophic microorganisms, its rate may also be regulated by available C (electron donor). Significant correlations have been observed between denitrification rates and available organic C (mineralizable organic C) (Burford and Bremner, 1975; Reddy et al., 1982; Gale et al., 1993).

Vegetation can play a significant role in the wetland N cycle by: (1) assimilating N into plant tissue, (2) releasing N through tissue mineralization, and (3) providing an environment in the root zone for nitrification-denitrification to occur. Plants derive most of their N from soil porewater with only a small amount of the floodwater N being directly utilized. Efficiency of N utilization (defined as increase in plant N per unit mass of available N) by aquatic vegetation is highly variable, depending on the type of wetland vegetation (woody versus herbaceous), nutrient loading and temperature. The N-use efficiency of aquatic vegetation and the C/N ratio of plant litter decrease with nutrient loading (Shaver and Melillo, 1984; Reddy and Portier, 1987; Koch and Reddy, 1992). Elevation of temperature from 10 to 25 °C was shown to increase N-use efficiency by *Typha* spp from 5 to 38% (Reddy and Portier, 1987). In temperate climates, most N assimilation occurs during the growing season. During winter months, aboveground biomass is killed and accumulates as detrital tissue. At the same time, a significant portion of N is translocated to below ground biomass. Nitrogen release during decomposition of detrital tissue *Typha* spp was more rapid during summer months than winter. Nitrogen assimilation by herbaceous vegetation is usually short term and usually cycled rapidly within the systems. Unlike herbaceous macrophytes, forested wetlands provide long-term storage in the form of woody biomass. However, a significant portion of nutrients stored in the leaves are returned to the forest floor, and eventually incorporated into the soil (Reddy and DeBusk, 1987). Because forest litter decomposition is slow, turnover periods are much longer. Measurement of plant tissue N and biomass can provide an indication of N removal efficiency by wetlands.

Several studies have used mass balance methods to estimate N retention within a wetland (DeLaune et al., 1989; White and Howes, 1994). These studies showed that N demand of aquatic vegetation, microbial, and periphyton communities is not met by external inputs alone, pointing to the role of internal cycling and turnover. A few studies have quantified the role of remineralization of organic N during plant litter decomposition and belowground biomass turnover. Values of 54 to 95% of plant uptake demand being met through this process have been reported (Hopkinson and Schubauer, 1984; DeLaune et al, 1989; White and Howes, 1994). In a greenhouse study, fertilizer inorganic N was the major source during the early part of the cattail growing season, while soil organic N mineralization supplied plant N requirements during the latter part of the growing season (Reddy and Portier, 1987).

Nitrogen availability to vegetation is controlled by the rate of organic N mineralization, and magnitude of N loss mechanisms such as denitrification and NH_3 volatilization. The limiting processes regulating N loss from wetlands by these mechanisms are NH_4^+ diffusion from anaerobic to aerobic zones, and nitrification in aerobic zones. Denitrification is limited by NO_3^- availability, since most wetlands have abundant supplies of energy. Nitrification-denitrification pathways are key processes that return N to the atmosphere, while biological N_2 fixation completes the N cycle.

4.7   Phosphorus

Unlike C and N, P tends to accumulate in wetlands because there is no significant gaseous loss mechanism. Phosphorus retention by wetlands is regulated by physical (sedimentation and entrainment) and biological mechanisms (uptake and release by vegetation, periphyton and microorganisms) (Fig. 4.11). Several reviews have discussed the mechanisms of P retention in stream and lake sediments, uplands and wetlands (Khalid et al., 1977; Bostrom et al., 1982; Berkheiser et al., 1980; Logan, 1982; Sonzogni et al., 1982; Howard-Williams, 1985; Froelich, 1988; and Reddy and D'Angelo, 1994; Reddy et al., 1999). Phosphorus in influent water is present in soluble and particulate forms, with both forms containing a certain proportion of inorganic and organic pools. Relative proportions of these pools depend on the source and the type of water entering the system. For example, municipal wastewater may contain a large proportion (> 75%) as inorganic P in soluble forms, as compared to effluents from agricultural watersheds. Constructed and riparian (buffer) wetlands can function as effective sediment traps, and P associated with suspended sediments can be effectively removed by wetlands.

Organic P present in wetland soils can be grouped into: (1) orthophosphate ions sorbed onto the surface of P-retaining components (nonoccluded P), (2) P absorbed within the matrices of P-retaining components (occluded P), and (3) phosphate minerals such as apatite (Williams et al., 1971). Several chemical fractionation schemes have been used to identify discrete pools of inorganic P, including exchangeable, Fe/Al bound, and Ca/Mg bound P (Hieltjes and Lijklema, 1980; Olila et al., 1994; Reddy et al., 1995).

The organic P fraction in soils and sediments accounts for a high proportion of total P: 20–60% in mineral soils (Tiessen et al., 1994), 10–70% in lake sediments, and 40–90% in Histosols (DeBusk et al., 1994). Most of the organic P in soils is derived from plant detritus and in part synthesized by soil organisms from inorganic sources (Sanyal and DeDatta, 1991). Organic P in biological tissues and cells can be classified into three groups: (1) inositol phosphates, (2) nucleic acids, and (3) phospholipids (Anderson, 1980), with inositol phosphates comprising up to 60% of the soil organic P (Tate, 1984).

Organic P, both dissolved (DOP) and particulate (POP), can account for a substantial portion of the total P in wetlands. Phosphorus is bound in organic compounds as an ester (C-O-P) (Cosgrove, 1977; Golterman, 1984), a form not readily bioavailable until it undergoes enzyme hydrolysis (Harrison, 1983; Hoppe et al., 1988; Chrost, 1991). Extracellular enzyme activity (such as alkaline and acid phosphatases) increases under inorganic P-limiting conditions which can hydrolyze P esters

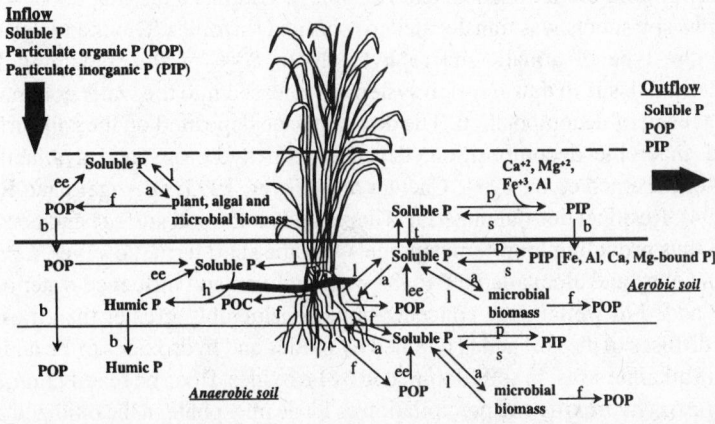

Fig. 4.11 Phosphorus tranformations in the soil-water-plant environment of wetlands. POC = particulate organic C; a = assimilation; b = burial; ee = extracellular enzyme hydrolysis; f = fragmentation; h = humification; l = leaching; p = precipitation; s = solubilization; t = transport

independent of C turnover (McGill and Cole, 1981). Alkaline phosphatase activity was shown to be inversely related to sediment redox potential (Newman and Reddy, 1993). Organic P associated with humic and fulvic acid represents > 40% of total soil P (Brannon and Sommers, 1985). The fulvic acid P constitutes a large fraction of the organic P in most soils and it is likely that this pool is derived from plant litter and recently deposited organic matter (Stewart and Tiessen, 1987). Phosphorus availability from organic P depends on mineralization rate and substrate biodegradability, with both increasing with nutrient loading.

Water column P can be readily removed by periphyton and algal uptake, followed by deposition of dead biomass on the soil surface. Periphyton communities are effective sinks in low P systems. Although in treatment wetlands where P loading is usually high, their overall effect on P removal through direct assimilation may be minimal. Similarly, vegetative uptake can provide either a short- or long-term sink for P, depending on type of vegetation present in the wetland. Phosphorus tied up in detrital plant/algal tissue is rapidly released into the water column during decomposition. However, over the long term, significant portions of the organic P may remain as a part of peat buildup in wetlands. Under anaerobic conditions, these forms of P are relatively resistant to microbial breakdown, and can be considered an important P sink.

Phosphorus uptake by vegetation is maximal during the peak growing season, followed by a decrease or cessation in the fall and winter. During winter periods, a significant portion of the P is translocated to belowground biomass (roots and rhizomes) (Davis and van der Valk, 1983). Phosphorus storage in the aboveground biomass of macrophytes is usually short term, with large amounts of P being released during decomposition of detrital tissue. Phosphorus release rates during summer are 4–6 fold faster than in winter months. Emergent macrophytes which have an extensive network of roots and rhizomes responsible for uptake and storage of P assimilate little P directly from floodwater (Richardson and Marshall, 1986). Davis (1982) measured only 2 to 4% of added floodwater ^{32}P in the living tissue of *Cladium* and *Typha*. Although floodwater P is not assimilated directly by these plants, active uptake of soil porewater P can potentially increase the flux of P from the water column into the soil, thus improving overall P retention. Phosphorus is cycled in soil-water-plant components as follows. Herbaceous vegetation rooted in soils obtains most of its P requirements from the soil porewater and translocates it to aboveground vegetation. Upon maturity and senescence, a substantial portion of the P present in the aboveground vegetation is translocated and stored into belowground biomass (roots and rhizomes). Estimates are that about 45% of the P lost from living *Typha* spp shoots was translocated to roots and rhizomes (Davis and van der Valk, 1983). Depending on the type of aquatic macrophyte, up to 80% of the remaining P stored in the aboveground detrital tissue in nutrient-rich systems is released into the water column either by initial leaching or as a result of decomposition. The detrital tissue deposited on the soil surface is subjected to aerobic and anaerobic decomposition, depending on hydrology and availability of alternate electron acceptors (Caraco et al., 1991; Gachter and Meyer, 1993; D'Angelo and Reddy, 1994 a,b; Koch et al., 1994). Residual detrital material is deposited on the soil surface and becomes an integral part of the soil, thus providing long-term storage. Over the short term, however, rapid turnover rates and cycling can contribute bioavailable P to the water column and influence water quality.

Inorganic P added to wetlands at concentrations considerably greater than those present in the soil porewater diffuses to the soil and is retained by oxides and hydroxides of Fe and Al in acid soils, and by $CaCO_3$ in alkaline soils. In soils dominated by Fe oxides, P can be readily immobilized through sorption by ferric oxyhydroxide and precipitation as ferric phosphate in the oxidized zones at the soil-water interface. The rate of adsorption is controlled by soil pH and Eh, adsorptive surface area (active Fe and Al or $CaCO_3$), and temperature. In calcareous systems, P can be precipitated as Ca mineral bound P, especially when the pH of the floodwater is altered diurnally by algal photosynthetic activity. On the other hand, when soluble P in the soil porewater is higher than in floodwater, diffusion to the overlying water column occurs where the P is either assimilated by planktonic organisms or precipitated as Ca phosphate. In calcareous systems, coprecipitation of P with $CaCO_3$ is a dominant mechanism in immobilization of soluble P. In the water column where pH fluctuates diurnally, with high values during daytime (photosynthetic activity of algae) and low values during night (respiration by algae and bacteria), P can be precipitated and resolubilized (Diaz et al., 1994).

The following empirical relationship was derived for a number of wetland mineral soils and stream sediments in Florida (Reddy et al., 1998):

$$S_{max} = 0.24 \text{ [oxalate Fe + Al]} \qquad [4.13]$$

where S_{max} is maximum P retention capacity, [oxalate Fe + Al] is ammonium oxalate extractable Fe and Al. The above equation is applicable for [Fe +Al] concentration in the range of 0–100 mmoles kg^{-1}. To retain one mole of P, about 4 moles of Fe + Al are needed. Other researchers have reported a significant relationship between extractable Fe and Al, and the P-retention capacity of wetland soils (Richardson, 1985; Walbridge and Struthers, 1993; Gale et al., 1994; Lookman et al., 1995). These relationships suggest that extractable Fe and Al can be used as a predictor of P retention by wetland mineral soils. In sulfate-dominated wetlands, production of sulfides and removal of Fe(II) through precipitation as ferrous sulfide may preclude P retention by Fe(II) compounds (Caraco et al., 1991).

4.8 Metals

Alterations in Eh and pH in wetland soils affect the reactivity and mobility metals. Many oxidized stable metallic compounds are reduced into more soluble and bioavailable forms under wetland conditions. Although solubility of several metals are affected in the wetland environment, two metallic cations (Fe and Mn) play a major role in most wetland mineral soils. Ferric iron and Mn(IV) reduction coupled with organic matter turnover is considered as one of the significant mechanisms in anaerobic habitats (Lovley, 1991; Nealson and Myers, 1992). In wetland soil environments, Fe(III) and Mn(IV) reduction is mediated by enzymatic activity, which couples the use of these compounds as electron acceptors during microbial respiration. However, nonenzymatic reduction of Fe(III) and Mn(IV) has been reported in anaerobic environments (Lovley et al., 1991). Both Fe and Mn are present as solid phases in oxidized form (such as Fe_2O_3, FeOOH, $FePO_4$, and MnO_2) which are reduced to more dissolved forms under anaerobic conditions. Typical Fe(III) reduction under wetland soils conditions and its relationship to Eh and pe is shown below:

$$Fe\,(OH)_3 + 3H^+ + e^- \rightleftharpoons Fe^{2+} + 3H_2O \qquad [4.14]$$

$$Eh = 1.06 - 0.059 \log Fe^{2+} - 0.177 \text{ pH} \qquad [4.15]$$

$$pe = 17.87 + pFe^{2+} - 3\,pH \qquad [4.16]$$

Reduction of metallic compounds is inhibited by the presence of O_2 and to some extent NO_3^-. The reduction of Fe(III) and Mn(IV) has direct impacts on soil chemistry, such as (1) increased water soluble Fe(II) and Mn(II), (2) increased pH, (3) displacement of other cations from the soil exchange complex into porewater, (4) breakdown of organic matter and nutrient release, (5) increased solubility of P and Si, and (6) formation of new minerals (Ponnamperuma, 1972; Gambrell, 1994).

Reduction of Fe(III) and Mn(IV) depends on the supply of electron donors (primarily availability of labile organic matter), and the concentrations of bioavailable oxidized Fe and Mn. Microbial reduction also depends on the ability of bacteria to solubilize solid phases, attach directly to the substrate and transfer electrons, and/or transport the substrate directly into cells as a solid. Reduction capacity of the oxidized Fe compounds decreases with increasing crystallinity: $FePO_4 > Fe(OH)_3 > FeOOH > Fe_2O_3$ (Lovley, 1987).

Iron forms more stable organic metal bonds than Mn which protects it from precipitation reactions. The extent of complexation is greater under wetland soil conditions, because of high concentrations of dissolved organic compounds in porewater. Significant amounts of Fe(II) can be present under aerobic conditions for several days due to complexation with organics.

Similar to Fe and Mn, general chemical forms of trace metals in wetland soil are (1) water soluble forms, such as free ions, inorganic and organic complexes, (2) exchangeable metals, (3) metals precipitated as inorganic compounds, (4) metals complexed with large molecular weight humic materials, (5) metals adsorbed or occluded in precipitated hydrous oxides, (6) metals precipitated as insoluble sulfides, and (7) metals bound within the crystalline lattice structure of primary minerals (Gambrell, 1994).

Wetland soils are effective sinks for many trace metals through precipitation as insoluble metal sulfides (especially in coastal wetlands), and strong association of metals with insoluble humic substances (Gambrell, 1994). However, water table fluctuations or wetland drainage can reverse many of these processes especially by lowering pH and solubilizing metals. For example, in wetland mineral soils, accumulation of Fe(II) displaces many basic cations from the exchange complex, which increases the concentration of basic cations in porewater. Draining these soils removes basic cations from porewater, oxidizes Fe(II) to Fe(III), which results in the release of H^+. This in turn, releases Al and other associated metals. Similarly, metals bound to sulfides can be released under drained conditions when sulfides are oxidized to sulfates, resulting in the release of H^+. Many other metals such as Ca, Mg, Cu, Zn, Pb, and Ni are not directly involved in oxidation-reduction reactions, but their solubilities are affected by changes in pH as a result of flooded and drained conditions.

4.9   Toxic Organics

Toxic organics are naturally occurring or synthetic chemicals, which at low concentrations have adverse effects on growth and activity of plants, animals, and/or microorganisms. Natural and constructed wetlands may receive toxic organics from a number of sources, including (1) direct applications to control aquatic weeds, algae and mosquitoes; (2) accidental spillage, surface runoff, and drainage water from agriculture, urban, and industrial sources; (3) domestic and industrial discharges into natural and constructed wetlands for wastewater improvement; (4) residual toxics remaining in agricultural soils that are converted to wetlands; and (5) *in situ* production by natural processes, mediated by plants and microorganisms.

These sources are likely to contribute vast amounts and multitudes of different types of toxic organics; however, widespread monitoring and toxicological studies have not been conducted to estimate the extent and consequences of contamination (Catallo, 1993). One estimate indicated that about 5000 wetlands and aquatic systems are impacted by agricultural pesticides in the US (Taraban et al., 1993; Cooper, 1993). Many soil processes including microbial respiration (Dusek and Tesarova, 1996), nitrification (Sayler et al., 1982), and cellulose and lignin degradation (Katayama and Kuwatsuka, 1991; McKinley et al., 1982) are negatively impacted by toxic organics, and so pose a serious threat to proper functioning of wetland systems.

Several processes regulate the fate of toxic organics once they enter wetlands, and most information about these processes has been gained through laboratory experiments that simulate wetland systems. In general, toxic organics undergo similar transformation processes as natural organic matter, including aerobic and anaerobic microbial breakdown, vegetative uptake, volatilization, photolysis, abiotic oxidation and hydrolysis, bioaccumulation, sorption, and burial in the soil (Fig. 4.12). Through these processes, many wetlands are capable of attenuating this class of pollutants. However, the extent of these processes depends on the type of compound (water solubility, oxidation state, functional groups, molecular weight), as well as biological and chemical conditions in the soil-water column including pH, Eh, temperature, light intensity, nutrient and electron acceptor availability, type and amount of vegetation, organic matter content, previous history of contamination. For a review of process models refer to Lyman (1982).

Microbial degradation is generally considered the main pathway for attenuation of toxic organics. This task is usually accomplished by sequential enzymatic reactions mediated by many populations working together, rather than by a single species. In wetlands with characteristically diverse populations of microorganisms ranging from aerobes to extreme obligate anaerobes, there is great potential for degradation of many types of chemicals. In many cases, these transformations are beneficial to microorganisms by reducing toxicity or by acting as C sources, nutrients or electron acceptors. Most enzymatic transformations can be grouped into four main categories: hydrolysis, oxidation, reduction and synthesis.

The type of enzymatic transformation involved depends mainly on the type of chemical and its functional groups, as well as environmental conditions that promote the activity of different enzyme systems. In wetland systems, a combination of reactions are typically required to attain complete mineralization of a chemical.

Extracellular enzymes produced by bacteria and fungi hydrolyze many types of toxic organics including organophosphates, phenoxyacetates, phenylureas, acid anilides, and phenylcarbamates (Burns and Edwards, 1980; Nannipieri and Bollag, 1991). Small molecular weight C, N, and P compounds released from these reactions provide nutrients for degrading populations.

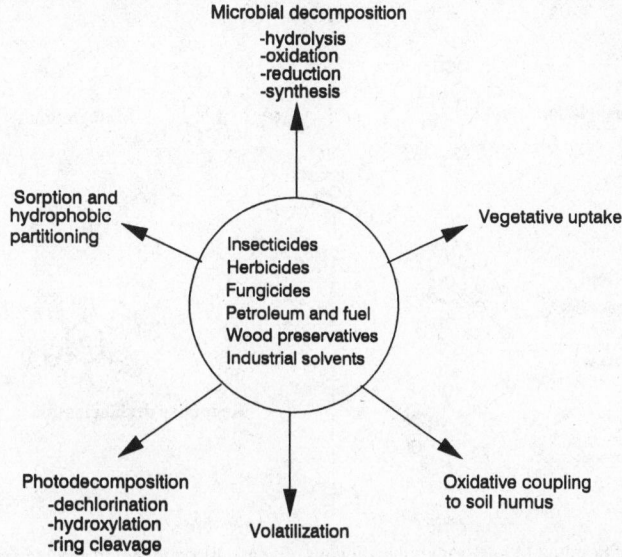

Fig. 4.12 Conceptual diagram of biogeochemical pathways of toxic organics in the environment

Monooxygenase or diooxygenase enzymes incorporate one or two atoms of O_2 into chemicals. Due to the requirement for O_2, this reaction may be limited to aerobic zones such as soil-water or root-soil interfaces. For alkanes and aromatics without functional groups (benzene and toluene), oxygenase activity is typically required before significant rates of degradation occur, which likely explains the recalcitrance of these chemicals in mostly anaerobic zones.

Through oxidative reactions, many toxic organics are eventually utilized as C and energy sources by degrading populations (Fig. 4.13). Monooxygenases and dioxygenases are less effective with highly oxidized chemicals, including highly chlorinated pentachlorophenol, 2,4,5 trichlorophen-oxyacetic acid, hexachlorobenzene, polychlorinated biphenyls, carbon tetrachloride and tetra-chloroethylene which explains the recalcitrance of these chemicals under aerobic conditions. For these chemicals, reductive processes may be more effective microbial transformation processes.

Under extremely anaerobic conditions, reduction reactions are promoted in which electrons from organic matter and intermediate breakdown products (H_2, fatty acids) are transferred to toxic organics (Doong and Wu, 1995; Gibson and Sewell, 1992). These chemicals, therefore, function as electron acceptors, which is coupled to energy generation for degrading populations, much in the same way as aerobes and denitrifiers gain energy from electron transfer to O_2 and NO_3^-. Mohn and Tiedje (1991) found evidence that reductive dechlorination was beneficial to *Desulfomonile tiedjei*, which used 3-chlorobenzoate as an electron acceptor and was coupled to energy (ATP) production. The presence of alternate electron acceptors O_2, NO_3^-, SO_4^{2-} tends to inhibit reduction of toxic organics, since various microbial groups compete with each other for electron equivalents.

One drawback of reduction reactions is that they often result in production of intermediates that are also toxic. For example, Mikesell and Boyd (1988) showed sequential reductive dechlorination of pentachlorophenol to 2,3,4,5-tetrachlorophenol to 3,4,5-trichlorophenol to 3,5-dichlorophenol to 3-chlorophenol and ultimately, CH_4 and CO_2 (Fig. 4.13). As toxic organics become reduced, they may more readily undergo oxidation reactions. Therefore, it has been suggested that alternate reduction and oxidation processes, implemented by flooding or draining soils, or by planting soils with O_2 transporting plants, may promote mineralization compared to one process by itself. Enhanced

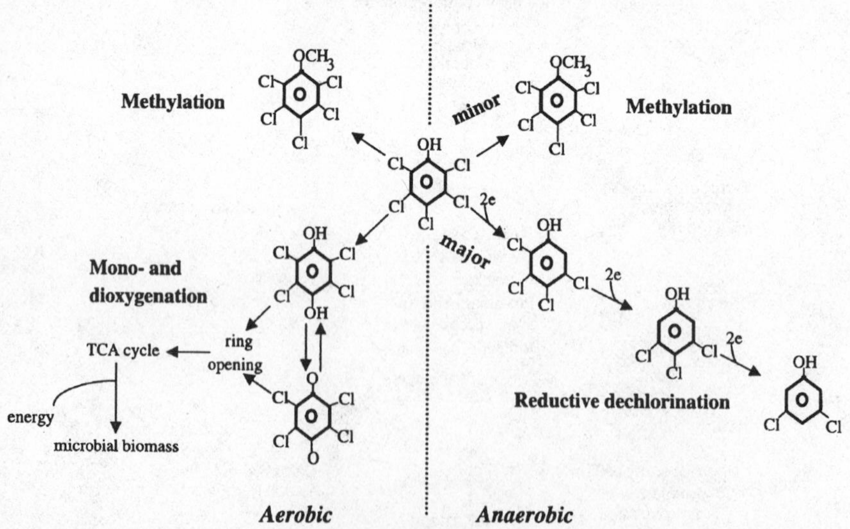

Fig. 4.13 Comparison of dominant transformation pathways of pentachlorophenol in aerobic and anaerobic environments [Modified from Boyd et al., 1989]

degradation by alternate anaerobic-aerobic soil processes has been shown for methoxychlor (Fogel et al., 1982).

In contrast to the above transformations that simplify chemical structures, synthesis reactions result in the addition of functional groups. Methylation of chlorophenols results in the production of volatile chloroanisoles, which likely benefits microorganisms by being less toxic (Neilson et al., 1983) (Fig. 4.13). Chloroanilines and phenolic compounds undergo oxidative coupling, or polymerization with soil humic substances through the enzymatic action of peroxidases, laccases and tyrosinases (Burns and Edwards, 1980; Lassen et al., 1994; Sarkar et al., 1988; Bollag and Liu, 1985) (Fig. 4.14). Shindo and Huang (1984) demonstrated that polymerization of phenols could be abiotically catalyzed by Mn(IV), Fe(III), Al and Si oxides, which also resulted in the formation of humic acids. These processes are thought to be analogous to natural humic substance formation from lignin (polyphenols) and protein molecules, and may benefit microorganisms by removing toxic organics from the bioavailable pool (Allard et al., 1994; Pal et al., 1994).

Synthesis reactions may also result in production of toxic organics in aquatic environments. For example, Asplund et al. (1993) characterized chloroperoxidases in soils that mediated production of chlorinated humic substances. Natural production of organically bound chlorine, mediated by white-rot fungi, may partially explain the widespread genetic capacity of microorganisms to degrade man-made chemicals (Oberg et al., 1997)

Sorption to soils and sediments is an important hurdle to transformation processes, since it reduces the amount in the bioavailable (soluble) pool (Ogram et al., 1985; Bouwer et al., 1994; Mihelcic et al., 1993). Except in the case of extracellular enzymes, which have greater access to soil microsites, desorption and transport of toxic organics to microbial cells are prerequisties for transformation. For strongly sorbed species, such as polychlorinated biphenyls and polyaromatic hydrocarbons with high octanol water partitioning coefficients, desorption may greatly curtail degradation kinetics (Manilal and Alexander, 1991). This process is important in many wetland soils because of accumulation of organic matter, and associated high cation exchange capacity, which promote hydrophobic partitioning and adsorption. Partial removal of toxic organics from the bioavailable pool through

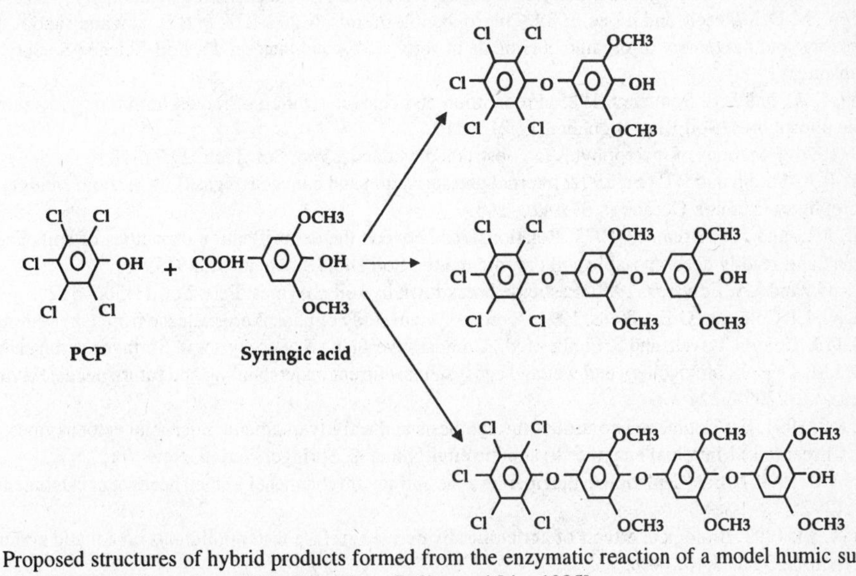

Fig. 4.14 Proposed structures of hybrid products formed from the enzymatic reaction of a model humic substance (syringic acid) with pentachlorophenol [Modified from Bollag and Liu, 1985]

sorption may allow degradation to proceed at otherwise toxic levels (Apajalahti and Salkinoja-Salonen, 1984). On the other hand, anaerobic and neutral pH conditions tend to increase solubility of humic substances of wetland soils, which in turn enhances the solubility of toxic organics (Pardue et al., 1993).

4.10   References

Alexander, M. 1977. Introduction to soil microbiology. John Wiley and Sons., New York, NY.

Allard, A., P. Hynning, M. Remberger, and A. Neilson. 1994. Bioavailability of chlorocatechols in naturally contaminated sediment and of chloroguaiacols covalently bound to C2-guaiacyl residues. Appl. Environ. Microbiol. 60:777–784.

Anderson, G. 1980. Assessing organic phosphorus in soils. p. 411–413. In F.E. Khasawneh, E.C. Sample, and E.J. Kamprath (ed.) The role of phosphorus in agriculture. Soil Science Society of America, Madison, WI.

Apajalahti, J.H.A., and M.S. Salkinoja-Salonen. 1984. Absorption of pentachlorophenol (PCP) by bark chips and its role in microbial PCP degradation. Microb. Ecol. 10:359–367.

Armstrong, J., and W. Armstrong. 1990. Light enhanced convective through flow increases oxygenation in rhizomes and rhizosphere of Phragmites australis (Cav.) Trin. ex. stend. New Phytol. 114:123–130.

Asplund, G., J.V. Christiansen, and A. Grimvall. 1993. A chloroperoxidase-like catalyst in soil: Detection and characterization of some properties. Soil. Biol. Biochem. 25:41–46.

Benner, R.M., M.A. Moran, and R.E. Hodson. 1985. Effects of pH and plant source on lignocellulose biodegradation rate in two wetland ecosystems, the Okeefenokee Swamp and a Georgia salt marsh. Limnol. Oceanogr. 30:489–499.

Berkheiser, V. E., J. J. Street, P. S. C. Rao, and T. L. Yuan. 1980. Partitioning of inorganic orthophosphate in soil-water systems. CRC Crit. Rev. Environ. Cont. 10:179–224.

Bollag, J., and S. Liu. 1985. Copolymerization of halogenated phenols and syringic acid. Pest. Biochem. Physiol. 23:261–272.

Bostrom, B., M. Jansson, and C. Forsberg. 1982. Phosphorus release from lake sediments. Arch. Hydrobiol. Beih. 18:5–59.

Bouma, J. 1991. ICOMAC circular 11, March 15, 1991. Final report of the International Committee for the Classification and Management of Wet Soils. Agricultural University, Wageningen, Netherlands.

Bouwer, E., N. Durant, L. Wilson, W. Zhang, and A. Cunningham. 1994. Degradation at xenobiotics compounds in sutu: capabilities and limits. FEMS Microb. Rev. 15:307–317.

Bowden, W.B. 1987. The biogeochemistry of nitrogen in freshwater wetlands. Biogeochemistry. 4:313–348.

Boyd, S.A., M.D. Mikesell, and J. Lee. 1989. Chlorophenols in soils. p. 209–228. In B.L. Sawhney and K. Brown (ed.). Reactions and movement of organic chemicals in soils. SSSA publication 22. Soil Science Society of America, Madison, WI.

Brannon, C.A., and L.E. Sommers. 1985. Preparation and characterization of model humic polymers containing organic phosphorus. Soil Biol. Biochem. 17:213–219.

Brix, H. 1996. Functions of macrophytes in constructed wetlands. Wat. Sci. Tech. 29:71–78.

Brix, H., B.K. Sorell, and P.T. Orr. 1992. Internal pressurization and convective gas flow in some emergent freshwater macrophytes. Limnol. Oceanogr. 37:1420–1433.

Burford, J.R., and J.M. Bremner, 1975. Relationships between the denitrification capacities of soils and total water soluble and readily decomposable soil organic matter. Soil Biol. Biochem. 7:389–394.

Burns, R.G., and J.A. Edwards. 1980. Pesticide breakdown by soil enzymes. Pest. Sci. 11:506–512.

Caraco, N., J.J. Cole, and G.E. Likens. 1991. A cross-system study of phosphorus release from lake sediments. p. 241–258. In J. Cole, G. Lovett, and S. Findley (ed.) Comparative analysis of ecosystems. Springer Verlag, New York, NY.

Catallo, J.M. 1993. Ecotoxicology and wetland ecosystems: current understanding and future needs. Environ. Toxicol. Chem. 12:2209–2224.

Chrost, R.J. 1991. Environmental control of the synthesis and activity of aquatic microbial ectoenzymes. p. 29–59. In R.J. Chrost (ed.) Microbial enzymes in aquatic environments. Springer-Verlag, New York, NY.

Cooper, A.B. 1990. Nitrate depletion in the riparian zone and stream channel of a small headwater catchment. Hydrobiol. 202:13–26.

Cooper., C.M. 1993. Biological effects of agriculturally derived surface water pollutants on aquatic systems-a review. J. Environ. Qual. 3:402–408.

Cosgrove, D.J. 1977. Microbial transformations in the phosphorus cycle. Adv. Microb. Ecol. 1:95–134.

Cowardin, L.M., V. Cater, F.C. Golet, and E.T. LaRoe. 1979. Classification of wetlands and deepwater habitats of the United States. FWS/OBS-79/31. U.S. Fish and Wildlife Service, Washington, DC.

D'Angelo, E.M., and K.R. Reddy. 1994a. Diagenesis of organic matter in a wetland receiving hypereutrophic lake water. I. Distribution of dissolved nutrients in the soil and water column. J. Environ. Qual. 23:928–936.

D'Angelo, E.M., and K.R. Reddy. 1994b. Diagenesis of organic matter in a wetland receiving hypereutrophic lake water. II. Role of inorganic electron acceptors in nutrient release. J. Environ. Qual. 23:937–943.

D'Angelo, E.M., and K.R. Reddy. 1999. Regulators of heterotrophic microbial potentials in wetland soils. Soil Biol. Biochem. (In press).

Davis, S.M. 1982. Patterns of radiophosphorus accumulation in the Everglades after its introduction into surface water. Tech. Pub. 82–2. South Florida Water Management District. West Palm Beach, FL.

Davis, C.B., and A.G. van der Valk. 1983. Uptake and release of nutrients by lining and decomposing *Typha glauca* Godr., tissues at Eagle Lake, Iowa. Aquat. Bot. 16:75–89.

DeBusk, W.F. 1996. Organic matter turnover along a nutrient gradient in the Everglades. Ph.D. dissertation. University of Florida, Gainesville, FL.

DeBusk, W.F., K.R. Reddy, M.S. Koch, and Y. Wang. 1994. Spatial distribution of soil nutrients in a northern Everglades marsh: Water Conservation Area 2A. Soil Sci. Soc. Am. J. 58:543–552.

DeLaune, R.D., T.C. Feijtel, and W.H. Patrick, Jr., 1989. Nitrogen flows in Louisiana Gulf Coast salt marsh: Spatial considerations. Biogeochem. 8:25–37.

Diaz, O.A., K.R. Reddy, and P.A. Moore. 1994. Solubility of inorganic P in stream water as influenced by pH and Ca concentration. Water Res. 28:1755–1763.

Doong, R., and S. Wu. 1995. Substrate effects on the enhanced biotransformation of polychlorinated hydrocarbon under anaerobic condition. Chemosphere 30:1499–1511.

Dusek, L., and M. Tesarova. 1996. Influence of polychlorinated biphynels on microbial biomass and its activity in grassland soil. Biol. Fertil. Soils 22:243–247.

EPA. 1988. EPA Wetland Identification and Delineation Manual (two volumes). *In* W.S. Sipple (ed.) Office of Wetlands Protection. U.S. Environmental Protection Agency, Washington, DC.

Federal Register, July 13, 1994. Changes in hydric soils of the United States. Washington, D.C.

Federal Register, February 24, 1995. Hydric soils of the United States. Washington, D.C.

Fog, K. 1988. The effect of added nitrogen on the value of decomposition of organic matter. Biol. Rev. 63:433–462.

Fogel, S., R.L. Lancione, and A.E. Sewall. 1982. Enhanced biodegradation of methoxychlor in soil under sequential environmental conditions. Appl. Environ. Microbiol. 44:113–120.

Francois, R. 1990. Marine sediment humic substances:structure, genesis, and properties. Aquatic Sci. 3:41–80.Froelich, P. N. 1988. Kinetic control of dissolved phosphate in natural rivers and estuaries: A primer on the phosphate buffer mechanism. Limnol. Oceanogr. 33:649–668.

Froelich, P.N. 1988. Kinetic control of dissolved phosphate in natural rivers and estuaries: A primer on the phosphate buffer mechanism. Limnol. Oceanogr. 33:649-668

Gale, P.M., I. Devai, K.R. Reddy, and D.A. Graetz. 1993. Denitrification potential of soils from constructed and natural wetlands. Ecol. Eng. 2:119–130.

Gale, P.M., K.R. Reddy, and D.A. Graetz. 1994. Phosphorus retention by wetland soils used for treated wastewater disposal. J. Environ. Qual. 23:370–377.

Gambrell, R.P. 1994. Trace and toxic metals in wetlands: A review. J. Environ. Qual. 23:883–891.

Gatcher, R., and J.S Meyer. 1993. The role of microorganisms in mobilization and fixation of phosphorus in sediments. Hydrobiol. 253:103–121.

Gaur, S., R.K. Singhal, and S.K. Hasija. 1992. Relative contributions of bacteria and fungi to water hyacinth decomposition. Aquatic Botany 43:1–15.

Gibson, S.A., and G.W. Sewell. 1992. Stimulation of reductive dechlorination of tetrachloroethylene in anaerobic aquifer microcosms by additions of short chain organic acids or alcohols. Appl. Environ. Microbiol. 58:1392–1993.

Given, P.H., and C.H. Dickenson. 1975. Biochemistry and microbiology of peats. p. 123–211. *In* E.A. Paul and A.D. McClaren (ed.) Soil biochemistry. Vol. 3. Marcel Dekker, New York, NY.

Glossary of Soil Science Terms. 1997. Soil Science Society of America, Madison, WI. pp. 138.

Golterman, H.L. 1984. Sediments, modifying and equilibrating factors in the chemistry of freshwater. Verh. Int. Ver. Theor. Angew. Limnol. 22:23–59.

Goodwin, S., and J.G. Zeikus. 1987. Ecophysiological adaptations of anaerobic bacteria to low pH: Analysis of anaerobic digestion in aeidic bog sediments. Appl. Environ. Microbiol. 53:57–64.

Gordon, A.S., and F.J. Millero. 1985. Adsorption mediated decrease in the biodegradation rate of organic compounds. Microb. Ecol. 11:289–298.

Gritzali, M., A. Shiralipour, and R.D. Brown. 1987. Cellulase enzymes for reenhancement of methane production from biomass. p. 367–383. In W. Smith and J.R. Frank (ed.) Methane from biomass. Elsevier Applied Science, New York, NY.

Harrison, A.F. 1983. Relationship between intensity of phosphatase activity and physicochemical properties of woodland soils. Soil Biol Biochem. 15:93–99.

Hieltjes, A.H.M., and L. Lijklema. 1980. Fractionation of inorganic phosphates in calcareous sediments. J. Envir. Qual. 9:405–407.

Hopkinson, C.S., and J.P. Schubauer. 1984. Static and dynamic aspects of nitrogen cycling in the salt marsh graminoid Spartina alterniflora. Ecol. 65:961–969.

Hoppe, H.G., S.J. Kim, and K. Gocke. 1988. Microbial decomposition in aquatic environments: Combined processes of extracellular activity and substrate uptake. Appl. Env. Microbiol. 54:784–790.

Houghton, R.A., and D.L. Skole. 1990. Carbon. p. 393–408. In B.L. Turner, W.C. Clark, R.W. Kates, J.F. Richards, J.T. Mathews, and W. B. Meyer (ed.) The earth as transformed by human action. Cambridge University Press, Cambridge, England.

Howard-Williams, C. 1985. Cycling and retention of nitrogen and phosphorus in wetlands: a theoretical and applied perspective. Freshwater Biol. 15:391–431

Howard-Williams, C., and M.T. Downes. 1993. Nitrogen cycling in wetlands. p. 141–167. In T.P. Burt, A.L. Heathwaite, and S.T. Trudgill (ed.) Nitrate: Processes, patterns and management. John Wiley and Sons, New York, NY.

Johnston, C. A. 1991. Sediment and nutrient retention by freshwater wetlands: Effects on surface water quality. CRC Crit. Rev. Environ. Control. 21:491–565.

Kadlec, R. H., and R. L. Knight. 1995. Treatment wetlands. Lewis Publishers, Chelsa, MI.

Katayama, A., and S. Kuwatsuka. 1991. Effect of pesticides on cellulose degradation in soil under upland and flooded conditions. Soil Sci. Plant Nutri. 37:1–6.

Kausik, N.K., and H.B.N. Hynes. 1971. The fate of dead leaves that fall into streams. Arch. Hydrobiol. 68:465–515.

Khalid, R.A., W.H. Patrick, Jr., and R.D. DeLaune. 1977. Phosphorus sorption characteristics of flooded soils. Soil Sci. Soc. Am. J. 41:305–310.

Kim, B., and R.G. Wetzel. 1993. The effect of dissolved humic substances on the alkaline phosphatase and growth of microalgae. Verh. Internat. Verein. Limnol. 25:129–132.

Kjoller, A., and S. Struwe. 1980. Microfungi of decomposing red alder leaves and their substrate utilization. Soil Biol. Biochem. 12:425–431.

Kjoller, A., S. Struwe, and K. Vestberg. 1985. Bacterial dynamics during decomposition of alder litter. Soil Biol. Biochem. 17:463–468.

Koch, M., and K.R. Reddy. 1992. Distribution of soil and plant nutrients along a trophic gradient in the Florida Everglades. Soil Sci. Soc. Am. J. 56:1492–1499.

Koch, M.S. K.R. Reddy, and J. P. Chanton. 1994. Factors controlling seasonal nutrient profiles in a subtropical peatland: Florida Everglades. Soil Sci. Soc. Am. J. 56:1492–1499.

Lakshman, G. 1979. An ecosystem approach to the treatment of wastewaters. J. Environ. Qual. 8:353–361.

Lassen, P., A. Randall, O. Jorgensen, P. Warwick, L. Carlson. 1994. Enzymatically mediated incorporation of 2-chlorophenol and 4-chlorophenol into humic acids. Chemosphere 28:703–710.

Lewis, W.M. 1995. Wetlands: Characteristics and boundaries. National Research Council, National Academy Press, Washington, DC.

Logan, T.J. 1982. Mechanisms for release of sediment-bound phosphate to water and the effects of agricultural land management on fluvial transport of particulate and dissolved phosphate. Hydrobiol. 92:519–530.

Lookman, R., D. Freese, R. Merck, K. Vlassak, and W.H. van Riemsdijk. 1995. Long-term kinetics of phosphate release from soil. Environ. Sci. Technol. 29:1569–1575.

Lovley, D.R. 1987. Organic matter mineralization with the reduction of ferric iron: A review. Geomicrobiol. J. 5:375–399.

Lovley, D.R. 1991. Dissimilatory Fe(III) and Mn(IV) reduction. Microbiol. Rev. 55:259–287.

Lovley, D.R., E.J.P. Phillips, and D.J. Lonergan. 1991. Enzymatic versus monenzymatic mechanisms for Fe(III) reduction in aquatic sediments. Environ. Sci. Technol. 25:1062–1067.

Lyman, W.J. 1982. Handbook of chemical property estimation methods. McGraw-Hill, New York, NY.

Manilal, B., and M. Alexander. 1991. Factors affecting the microbial degradation of phenanthrene in soil. Appl. Microbial. Biotechnol. 35:401–405.

Martin, J.F., and K.R. Reddy. 1997. Interaction and spatial distribution of wetland nitrogen processes. Ecol. Modeling 105:1–21.

Mausbach, M.J., and J.L. Richardson. 1994. Biogeochemical processes in hydric soil formation. p. 68-127. *In* Current topics in wetland biogeochemistry, Vol. 1, Louisiana State University, Wetland Biogeochemistry Institute.

McGill, W.B., and C.V. Cole. 1981. Comparative aspects of organic C, N, S and P cycling through soil organic matter during pedogenesis. Geoderma. 26:267–286.

McKinley, V.L., T.W. Federle, and J.R. Vestal. 1982. Effects of petroleum hydrocarbons on plant litter microbiota in an arctic lake. Appl. Environ. Microbiol. 43:129–135.

McLatchey, G.P., and K.R. Reddy. 1998. Regulation of organic matter decomposition and nutrient release in a wetland soil. J. Environ. Qual. 27:1268–1274.

Mendelssohn, I.A., K.L. McKee, and W.H. Patrick. 1981. Oxygen deficiency in Spartina alterniflora roots: metabolic adaptation to anoxia. Science. 214:439–441.

Metcalf and Eddy, Inc. 1979. Wastewater engineering: Treatment disposal reuse. McGraw-Hill, New York, NY.

Mihelcic, J.R., D.R. Leuking, R.J. Mitzell, and J.M. Stapleton. 1993. Bioavailability of sorbed- and separate phase chemicals. Biodegrad. 4:141–153.

Mikesell, M.D., and S.A. Boyd. 1988. Enhancement of pentachlorophenol degradation in soil through induced anaerobiosis and bioaugmentation with anaerobic sewage sludge. Environ. Sci. Technol. 22:1411–1414.

Mitsch, W.J. 1994. Global wetlands: Old world and new. Elsevier, New York, NY.

Mohn, W.H., and J.M. Tiedje. 1991. Evidence for chemiosmotic coupling of reductive dechlorination and ATP synthesis in *Desulfomonile tiedjei*. Arch. Microbiol. 157:1–6.

Molongoski, J.J., and M.J. Klug. 1976. Characterization of anaerobic heterotrophic bacteria isolated from freshwater lake sediments. Appl. Environ. Microbiol. 31:83–90.

Nannipieri, P., and J.M. Bollag. 1991. Use of enzymes to detoxify pesticide-contaminated soils and waters. J. Environ. Qual. 20:510–517.

Nealson, K.H., and C.R. Myers. 1992. Microbial reduction of manganese and iron: New approaches to carbon cycling. Appl. Environ. Microbiol. 58:439–443.

Neilson, A.H., A. Allard, P. Hynnig, M. Remberger, and L. Lander. 1983. Bacterial methylation of chlorinated phenols and guaiacols: Formation of veratroles from guaiacols and high-molecular-weight chlorinated lignin. Appl. Environ. Microbiol. 45:774–783.

Newman, S., and K.R. Reddy. 1993. Alkaline phosphatase activity in the sediment-water column of a hypereutrophic lake. J. Environ. Qual. 22:832–838.

Oberg, G., Brunberg, H., and O. Hjelm. 1997. Production of organically bound chlorine during descendation of birch wood by common white-rot fungi. Soil Biol. Biochem. 29:191–197.

Ogram, A.V., R.E. Jessup, L.T. Ou, and P.S.C. Rao. 1985. Effects of sorption on biological degradation rates of (2.4-dichloraphenexyl) bacteria acids in soils. Appl. Environ. Microbiol. 49:582–587.

Olila, O.G., K.R. Reddy, and W.G. Harris, Jr. 1994. Forms and distribution of inorganic phosphorus in sediments of two shallow eutrophic lakes in Florida. Hydrobiologia 129:45–65.

Pal, S., J.M. Bollag, and P.M. Huang. 1994. Role of abiotic and biotic catalysts in the transformation of phenolic compounds through oxidation coupling. Soil Biol Biochem. 25:813–820.

Pardue, J.H., P.H. Masscheleyn, R.D. Delaune, and W.H. Patrick., Jr. 1993. Assimilation of hydrophobic chlorinated organics in freshwater wetlands: Sorption and sediment-water exchange. Environ. Sci. Technol. 27:875–882.

Platzer, C., and R. Netter. 1994. Factors affecting nitrogen removal in horizontal flow reed beds. Water Sci. Tech. 29:319–324.

Post, W.M., W.R. Emanuel, P.J. Zinke, and A.G. Stangenbegar. 1982. Soil carbon pools and world life zones. Nature 298:156–159.

Ponnamperuma, F.N. 1972. The chemistry of submerged soils. Adv. Agron. 24:29–96.

Reddy, K.R., and W.H. Patrick, Jr. 1999. Biogeochemistry of wetlands. CRC/Lewis Publishers, Boca Raton, FL (In preparation).

Reddy, K.R., P.S.C. Rao, and R.E. Jessup. 1982. The effect of carbon mineralization and denitrification kinetics in mineral and organic soils. Soil Sci. Soc. Am. J. 46:62–68.

Reddy, K.R., and K.M. Portier. 1987. Nutrient storage capabilities of aquatic and wetland plants. pp. 337–357. *In* K.R. Reddy and W.H. Smith (ed.) Aquatic plants for water treatment and resource recovery. Magnolia Publishers, Inc., Orlando, FL.

Reddy, K.R., and W.H. Patrick, Jr. 1984. Nitrogen transformations and loss in flooded soils and sediments. CRC Crit. Rev. Environ. Cont. 13:273–309.

Reddy, K. R., and E. M. D'Angelo. 1994. Soil processes regulating water quality in wetlands. p 309–324. In W. Mitsch (ed.) Global wetlands-Old world and new. Elsevier Pubishers, New York, NY.

Reddy, K.R., R.H. Kadlec, E. Flaig, and P.M. Gale. 1999. Phosphorus assimilation in streams and wetlands: Crit. Rev. Environ. Sci. Tech. 29:1–64.

Reddy, K. R., and W. F. DeBusk. 1987. Plant nutrient storage capabilities. p. 337–357. In K.R. Reddy and W. H. Smith (ed.) Aquatic plants for water treatment and resource recovery. Magnolia Publishing Inc., Orlando, FL.

Reddy, K.R., W.H. Patrick, Jr., and C.W. Lindau. 1989a. Nitrification-denitrification at the plant-root-sediment interface in wetlands. Limnol. Oceanogr. 34:1004–1013.

Reddy, K.R., E.M. D'Angelo, and T.A. DeBusk. 1989b. Oxygen transport through aquatic macrophytes: The role in wastewater treatment. J. Environ. Qual. 19:261–267.

Reddy, K. R., O. A. Diaz, L. J. Scinto, and M. Agami. 1995. Phosphorus dynamics in selected wetlands and streams of the Lake Okeechobee Basin. Ecol. Eng. 5:183–207.

Reddy, K.R., G.A. O'Connor, and P.M. Gale. 1999. Phosphorus retention capacities of wetland soils and stream sediments impacted by dairy effluent. J. Environ. Qual. 29:1–64.

Richardson, C.J. 1985. Mechanisms controlling phosphorus retention capacity in freshwater wetlands. Science 228:1424–1426.

Richardson, C.J., and P.E. Marshall. 1986. Processes controlling movement, storage and export of phosphorus in a fen peatland. Ecol. Monogr. 56:279–302.

Sanyal, S.K., and S.K. DeDatta. 1991. Chemistry of phosphorus transformations in soils. Adv. Agron. 16:1–32.

Sarkar, J.M., R.L. Malcolm, and J.M. Bollag. 1988. Enzymatic coupling of 2,4-dichlorophenol to stream fulvic acid in the presence of oxidoreductases. Soil Sci. Soc. Am. J. 52:688–694.

Sayler, G.S., M.P. Shiaris, W. Beck, and S. Held. 1982. Effects of polychlorinated biphenyls and environmental biotransformation products on aquatic nitrification. Appl. Environ. Microbiol. 43:949–952.

Schipper, L.A., and K.R. Reddy. 1996. Methane oxidation in the rhizosphere of Sagittaria lancifolia. Soil Sci. Soc. Am. J. 60:611–616.

Schlesinger, W.H. 1997. Biogeochemistry-An analysis of global change. 2nd Ed. Academic Press, Orlando, FL.

Shaver, G.R. and J.M. Melillo. 1984. Nutrient budgets of marsh plants: Efficiency concepts and relation to availability. Ecol. 65:1491–1510.

Shindo, H., and P.M. Huang. 1984. Catalytic effects of manganese (IV), iron (III), aluminum, and silicon oxides on the formation of phenolic polymers. Soil Sci. Soc. Am. J. 48:927–934.

Sinsabaugh, R.L., R.K. Antibus, A.E. Linkens, C.A. McClaugherty, L. Rayburn, D. Repert, and T. Weiland. 1993. Wood decomposition: Nitrogen and phophorus dynamics in relation to extracellular enzyme activity. Ecol. 74:1586–1593.

Sinsabaugh, R.L. 1994. Enzymatic analysis of microbial patterns and processes. Biol. Fertil. Soil 17:69–74.

Snyder, G.H. 1987. Agricultural flooding of organic soils. Univ. FL. IFAS Bull. 870.

Soil Survey Staff, 1996. Keys to soil taxonomy. 7th ed. USDA-NRCS. U.S. Govt. Printing Office, Washington, D.C.

Sonzogni, W. C., S. C. Chapra, D. E. Armstrong, and T. J. Logan. 1982. Bioavailability of phosphorus inputs to lakes. J. Environ. Qual. 11:555–563.

Stanford, G., M.H. Frere, and D.H. Schwaniger. 1973. Temperature coefficient of soil nitrogen mineralization. Soil Sci. 115:321–323.

Stephens, J.C. 1969. Peat and muck drainage problems. J. Irrig. Drainage Div. Proc. Am. Soc. Civil Eng. 95:285–305.

Stewart, J.W.B., and H. Tiessen. 1987. Dynamics of soil organic phosphorus. Biogeochem. 4:41–60.

Suberkropp, K., G.L. Godshalk, and M.J. Klug. 1976. Changes in the chemical composition of leaves during processing in a woodland stream. Ecol. 57:720–727.

Taraban, R.H., O.F. Berry, D.A. Berry, and H.L. Walker, Jr. 1993. Degradation of dicamba by an anaerobic consortium enriched from wetland soil. Appl. Environ. Microbiol. 7:2332–2334.

Tate, K.R. 1984. The biological transformation of P in soil. Plant Soil 76:245–256.

Tiessen, H., J.W.B. Stewart, and A. Oberson. 1994. Innovative soil phosphorus availability indices: Assessing organic phosphorus. p. 143–162. In Soil testing: Prospects for improving nutrient recommendations. Soil Sci. Soc. Am. Spec. Publ. 40.

U.S. Army Corps of Engineers. 1987. USACE wetland delineation manual. U.S. Army Eng. Waterway Exp. Stn. Tech. Rep. Y-87-1.

U.S. Department of Agriculture (USDA). 1981. Land resource regions and major land resource areas of the United States. USDA-SCS Agric. Handbook 286. U.S. Govt. Printing Office, Washington, D.C.

U.S. Department of Agriculture. 1991. Hydric soils of the United States, 3rd. Ed. SCS Misc. Publ.1491.

U.S. Department of Agriculture, Natural Resource Conservation Service (USDA-NRCS). 1996. Field indicators of hydric soils in the United States. Hurt, G.W., P.M. Whited, and R.F. Pringle. 1996. Fort Worth, TX.

Vepraskas, M.J. 1992. Redoximorphic features for identifying aquic conditions. Tech. Bull. 301. North Carolina Agricultural Research Service, North Carolina State University, Raleigh.

Walbridge, M.R., and J.P. Struthers. 1993. Phosphorus retention in non-tidal palustrine forested wetlands of the mid-atlantic region. Wetlands 13:84–94.

Wheatly, R.E., M.P. Greaves, and R.H.E. Inkson. 1976. The aerobic bacterial flora of a raised bog. Soil Siol. Biochem. 8:453–460.

White, D.S., and B.L. Howes. 1994. Translocation, remineralization and turnover of nitrogen in the roots and rhizomes of *Spartina alterniflora*. Am. J. Bot. 81:1225–1234.

Whiting, G.J., and J.P. Chanton. 1993. Primary production control of methane emission from wetlands. Nature. 364: 794–795.

Williams, J.D.H., J.K. Syers, R.F. Harris, and D.E. Armstrong. 1971. Fractionation of inorganic phosphate in calcareous lake sediments. Soil Sci. Soc. Am. Proc. 35:250-255.

5

Acid Sulfate Soils

C. J. Ritsema
Winand Staring Centre, Wageningen, The Netherlands

M. E. F. van Mensvoort
Wageningen Agricultural University

D. L. Dent
University of East Anglia

Y. Tan
CSIRO Land and Water, Canberra, Australia

H. van den Bosch
Winand Staring Centre, Wageningen, The Netherlands

A.L.M. van Wijk
Winand Staring Centre, Wageningen, The Netherlands

5.1 Introduction

Acid sulfate soils generate H_2SO_4 that brings their pH below 4, sometimes as low as 2. The acid leaks into drainage and floodwaters, corrodes steel and concrete, and attacks clay, liberating soluble Al. Drainage waters may also be enriched in heavy metals and As, a toxic cocktail endangering aquatic life and public health. The weathering of sulfidic mine spoil and overburden presents the same problems, so acid sulfate soils are not the concern of soil science alone; agronomy, fisheries and environmental management, land and property development, civil, and mining and water engineering, and public health all have an acute interest.

Knowledge of acid sulfate soils and techniques to manage them have been drawn from geochemistry, mineralogy and microbiology (processes of sulfide accumulation and oxidation), geomorphology, ecology, Quaternary studies, remote sensing and geostatistics (distribution from soil survey), pedology, chemistry and geotechnics (characterization), and water engineering, agronomy and indigenous farmers' knowledge (management). Unfortunately, all these strands do not come together all the time. Fundamental advances have often resulted from almost chance encounters and interdisciplinary collaboration, followed by periods of refinement by conventional disciplinary and institutional groupings working, again, in virtual isolation. Indeed, interdisciplinary work is not only intellectually challenging, it also requires keeping abreast of activity in many disparate fields. The most useful sources include Dost (1973), Dost and van Breemen (1982), Kittrick et al. (1982), and Dent and van Mensvoort (1993).

5.2 Characteristics, Occurrence and Environmental Problems

5.2.1 Characteristics

It is useful to distinguish between sulfidic soils that will oxidize and generate H_2SO_4 if they are drained (potential acid sulfate soils) and acid sulfate soils that have been drained and are generating H_2SO_4 (raw acid sulfate soils) or have passed through the acid generating phase but remain severely acid (ripe acid sulfate soils). Dent (1986) provides an extended description of each.

Sulfidic materials include marine and estuarine sands and clays, gyttja in brackish lakes and lagoons, and peats that originally formed in freshwater but which have been inundated subsequently by brackish water. The common factors are (1) a supply of organic matter; (2) severely reducing conditions brought about by continuous waterlogging; (3) a supply of SO_4^{2-}, usually from tidewater, that is reduced to sulfides by bacteria decomposing the organic matter; and (4) a supply of Fe from the sediment for the accumulation of iron sulfides which make up the bulk of reduced S compounds. These conditions are most abundantly fulfilled in tidal swamps and saltmarshes where, over the last 10,000 yr of continually rising sea levels, thick deposits of sulfidic clay have accumulated (Pons and van Breemen, 1982; Dent and Pons, 1995). Profile 1 (Table 5.1) is an example of an unripe sulfidic clay or Sulfaquent (Soil Survey Staff, 1998) being dark grey, rich in organic remains, with a buttery consistency and stinking of H_2S. It is significant for the development and management of acid sulfate soils that the sulfides are by no means uniformally distributed. Both sands and clays contain grains of pyrite (FeS_2) transported from elsewhere and deposited within the sediment (up to 10 kg m^{-3}). Subsequently, especially in slowly deposited clays, masses of framboids or individual crystals of FeS_2 accumulate *in situ* within rotting roots that provide both organic matter for bacterial decomposition and channels for the diffusion and tidal pumping of SO_4^{2-} and dissolved Fe (Figs. 5.1 and 5.3). Although the matrix may contain only a few embedded grains of pyrite, this network of FeS_2 packed sausages may bring the overall concentration up to 100 kg m^{-3}. Sulfides accumulate only under severely reducing conditions. As sedimentation raises the soil surface above mean sea level, topsoil accumulates under better drainage as a nearly ripe, mottled layer that contains little or no sulfide. Where freshwater conditions succeed brackish water, sulfidic clays may be buried by peat or nonsulfidic alluvium (Diemont et al., 1993).

Drainage of sulfidic soils brings about dramatic changes. If there is insufficient carbonate in the sediment to neutralize the H_2SO_4 generated by oxidation of sulfides, extreme acidity develops within weeks or months. Raw acid sulfate soils (Sulfaquents) (Profile 2) are characterized by a pH < 3.5 and straw yellow mottles and coatings of jarosite around pores and on ped faces. Acid sulfate peats do not have jarosite but often exhibit an inky black subsoil as some of the SO_4^{2-} generated by drainage is reduced to FeS deeper in the profile.

The drainage effluent of raw acid sulfate soils is usually the first outward sign of trouble. It is usually red, from precipitation of ochre, and carries an irridescent film of Fe-oxidizing bacteria; occasionally, it is blue green or crystal clear because dissolved Al precipitates any suspended clay; and it is lethal to fish and invertebrates.

Eventually, when the pyrite in the drained layers is exhausted, jarosite is hydrolyzed to iron oxide, and ramification of roots brings about physical ripening. Ripe acid sulfate soils (Sulfic Fluvaquents) are still very acid and Al saturated. At some depth, there may be a jarosite-mottled horizon overlying the still reduced deep subsoil (Profile 3, Table 5.1). In clays, evolution may take centuries but the

Table 5.1   Representative profiles of acid sulfate soils. Horizon nomenclature follows Diemont et al. (1993).

Profile description 1: Unripe sulfidic clay		
Location		The Gambia floodplain, Sankwia Tenda. 10 m from main river channel, 0.5 m above m.s.l.
Hydrology		Daily tidal flooding, brackish in dry season, fresh for three months in wet season.
Vegetation		*Rhizophora racemosa* forest, 30 m high
Profile described in September, wet season		
G	0-1 cm	Dark greyish brown fluid mud; pH 5.7; EC_e 4 mS cm^{-1}
Grs1	1-14 cm	Black clay; practically unripe; abundant fibrous and aerial roots; pH 7; EC_e 4 mS cm^{-1}; 2.5% S by mass
Grs2	14-76 cm	Dark greenish grey, blackening on exposure to the air; clay; practically unripe; abundant fine fibrous roots occupying 20% soil volume; stinks of H_2S; pH 6.9; EC_e 8 mS cm^{-1}; 5.6% S
Grs3	76-120+ cm	Dark greenish grey, blackening on exposure to the air; clay; practically unripe; many fine to coarse dendritic pores with blackened mangrove remains; stinks of H_2S; pH 6.9; EC_e 8 mS cm^{-1}; 2.6% S
Profile description 2: Raw acid sulfate clay		
Location		By the Ho Phong canal, 2 km SW of Quan Lo, Minhai Province, Vietnam
Hydrology		Freshwater tidal flooding to 0.6 m during wet season; managed brackish water tidal flooding in the dry season
Land use		Rice in wet season, shrimps in dry season
Profile described under 0.3 m floodwater in December (end of wet season)		
Ap	0-15 cm	Very dark grey silty clay; puddled; pH 4.8, EC_e 3.3 mS cm^{-1}
Gj	15-40 cm	Very dark greyish brown clay; nearly ripe, medium blocky structure; few yellow mottles around old root channels and patchy yellow coatings on ped faces; pH 3.7; EC_e 3.0 mS cm^{-1}
Grs	40-130+ cm	Dark grey silty clay; half ripe, structureless; much partly decomposed organic matter; stinks of H_sS; pH 6.1; EC_e 3.0 mS cm^{-1}
Profile description 3: Ripe acid sulfate clay		
Location		Estuarine terrace near Tite, Quinara District, Guinea Bissau
Hydrology		1 m above mean sea level. Flooded with fresh water in the rainy season; groundwater > 3 m below surface in the dry season
Land use		Grassland
Profile described in late June, end of long dry season		
Ah	0-15 cm	Very dark reddish brown clay powdered with ash; ripe, massive; abundant dead fine roots; pH 3.9; EC_e 0.1 mS cm^{-1}
Bg	5-45 cm	Grey with many dark brown to yellowish red mottles; clay; ripe, very hard; coarse prismatic breaking to angular blocky structure; pH 3.7; EC_e 0.1 mS cm^{-1}
Bj	45-70 cm	Reddish grey coarsely mottled brown, finely mottled pale yellow; clay; very coarse to medium prismatic structure; ripe, very hard; pH 3.1; EC_e 0.2 mS cm^{-1}
Gj1	70-160 cm	Brownish grey clay; very coarse prismatic structure with brown, reddish brown and yellow coatings on ped faces; nearly ripe to half ripe; pH 3.3; EC_e 0.1 mS cm^{-1}
Gj2	160-190 cm	Grey clay; practically unripe, structureless; vertical pipes of jarosite
Grs	190-240+ cm	Grey clay; practically unripe; pH 5; EC_e 0.1 mS cm^{-1}

phase of H_2SO_4 generation is short lived in sands on account of their permeability and, generally, lower initial sulfide content.

5.2.2   Occurrence

Acid sulfate soils occupy an area of some 24 million ha worldwide where the topsoil is severely acid or will become so if drained. In addition, there may be as much again thinly covered by peat and

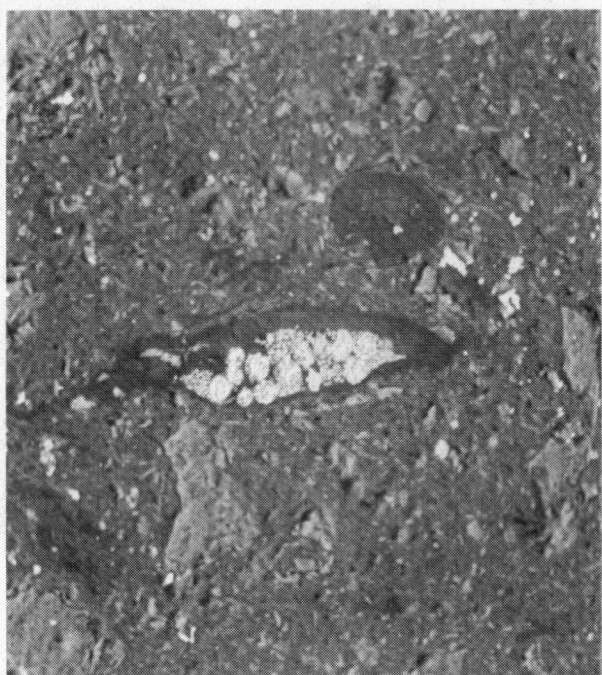

Fig. 5.1   Pyrite framboids packed in a root channel within pyrite-free matrix, Hong Kong harbor [Scanning electron micrograph by N.K. Tovey]

nonsulfidic alluvium (van Mensvoort and Dent, 1997). However, estimates of the extent and distribution of acid sulfate soils suffer more than most from scanty field surveys, even fewer reliable laboratory data and also, variable definition. More significant than their actual area is their location. They are concentrated in otherwise densely settled coastal and floodplains, mostly in the tropics, where development pressures are intense and little suitable alternative land for expansion of farming or urban and industrial development exists. Two-thirds of the known extent is in Vietnam, Thailand, Indonesia, Malaysia and northern Australia.

Survey procedures for acid sulfate soils follow the same sequence of questions as for other natural resources surveys: What is the decision to be made? > What are the criteria? > What are the relevant land attributes and what are their relationships with economic, social and political factors? > What are the relevant soil-specific attributes and what is the required accuracy and precision of these soil data? Table 5.2 lists some of the attributes of acid sulfate soils that most frequently pose limitations to development and conservation options. Unfortunately, acid sulfate soils exhibit enormous spatial variations in all of these characteristics, tied to the dynamic estuarine, deltaic and flood plain environments of which they are a part. Acid sulfate soils also exhibit very significant temporal variability, not least in their defining characteristics of acidity and related toxicities. Almost uniquely, acid sulfate soils export their problems in drainage and floodwaters; consequently, both reliable static soil survey and dynamic chemical/hydrological modeling are required to provide useful information for soil and environmental management.

There is significant spatial variability at every scale between pyrite-plugged pores and the soil matrix, and in-field, local and regional soil patterns. Dent (1986) gives examples of soil patterns at

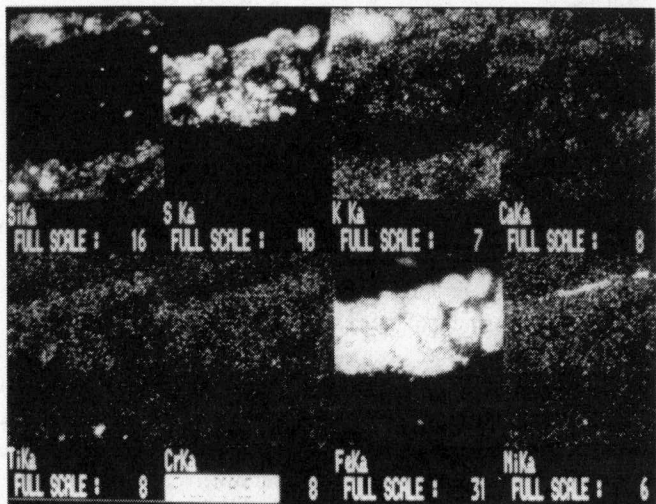

Fig. 5.2 X-ray maps of various elements in the pyrite shown in Fig. 1. Note the association of heavy metals with pyrite. Individual framboids are 20 μm diameter.

different scales that have been unearthed by soil surveys. Because in the tropical, tidal environments surveys are arduous and enlivened by mosquitos and crocodiles, remote sensing is widely used.

The soil surveyor's art can often pick out patterns related to the sedimentary history, such as the common distinction between highly sulfidic, unripe clays or peat in backswamps and riper, coarser textured, maybe even calcareous soils on levees and creek fillings. These patterns may be reflected by the natural vegetation and so can be mapped by air photo interpretation once the local pattern has been established by field transects. Once the land is reclaimed, soil patterns are starkly obvious with acid sulfate soils bare of vegetation. However, the soil pattern is not necessarily visible at the surface because present landforms and vegetation may be only partly and indirectly related to the environment in which sulfides originally accumulated; the area may have been buried by peat or nonsulfidic alluvium, fresh waters may have succeeded brackish and new plant communities succeeded the old.

Where quantitative data are needed on the distribution of total potential acidity or depth of an acid/ potentially acid layer, it is possible to combine mapping based on surface evidence with spatial statistics based on a network of sampling points as used by Ahmed and Dent (1997) in a regional survey. Unless the sampling network is very intensive (and therefore costly), present interpolation techniques like kriging (Section A, Chapter 9) produce spurious patterns unless the landscape is first stratified into natural units; nor do they cope with periodicities in the landscape such as those related to creek patterns, although these periodicities may be revealed by the variograms used to calculate the spatial structure of each attribute. Published studies of spatial variability of acid sulphate soils suggest that > 20% of variability in attributes related to acidity cannot be resolved at any practicable intensity of sampling.

Quantitative methods of analysis are required to support a soil survey program and to provide essential data for modeling the response of the land to various management options. The required analyses must be performed before gross chemical changes take place, or the samples must be dried quickly. Otherwise, the pH may fall from 7 to < 3 within days or weeks. Once dried samples are obtained, standard methods of laboratory analysis may be used. If sophisticated laboratory facilities are available, the crucial data for potential acidity may be estimated from determinations of total S using XRF or an induction furnace, assuming that all the S is pyrite (which it is not) oxidizing according to the equation:

$$FeS_2 + 15/4O_2 + 1/2H_2O \rightarrow Fe(OH)_3 + 2SO_4^{2-} + 4H^+$$ [5.1]

When this equation goes to completion, 4 moles H^+ are released per mole FeS_2 oxidized. The neutralizing capacity of the soil should be determined separately and an assumption has to be made about the extent to which the determined neutralizing capacity comes into effect.

For a field laboratory, a compromise must be reached between the precision beloved of soil chemists and the need for rapid assessment of many soil samples. It is very convenient and, also, useful for subsequent modeling and design of ameliorative treatment, to sample by volume and determine attributes on a volume basis. Dent and Bowman (1996) modified the quick, quantitative method of Konsten et al. (1988) to estimate in sequence, total actual acidity, total potential acidity, Al acidity and residual neutralizing capacity on a volume basis by a combination of titrations with standard alkali and oxidation of the sample with H_2O_2. The method has the advantage that the inherent immediate neutralizing capacity of the soil is taken into account. Alternatively or in combination with direct determination of acidity, samples may be oxidized and sulfate determined by one of many convenient techniques and potential acidity calculated by assuming that the sulfate came from oxidation of FeS_2, as above.

Because of the potentially severe environmental impact of acid sulfate soils, the value of development in prized coastal locations, and the legal/financial costs of error, much effort has recently gone into negotiating protocols for sampling and analysis. Not surprisingly, these negotiations have been political as much as scientific with soil scientists, environmentalists and other factions defending their positions rather than seeking the most cost-effective and reliable solutions. More surprisingly, there appears to have been no statistical testing of sampling intensity of soils, dredged sulfidic sediments or mine spoil against reliability of prediction.

5.2.3 Environmental Problems

Acid sulfate soils cause on- and off-site problems with the latter attracting more attention. On-site problems which have been extensively reviewed (Dent, 1986), include Al toxicity in drained soils, Fe and H_2S toxicity in flooded soils, salinity and nutrient deficiencies. Most agronomic data are from pot experiments and, more recently, growth chamber experiments (Moore and Patrick, 1993) which may be far removed from the uncontrolled situation in the field. Engineering problems include (1) corrosion of steel and concrete; (2) uneven subsidence, low bearing strength and fissuring leading to excessive permeability of unripe soils; (3) blockage of drains and filters by ochre; and (4) the difficulties of establishing vegetation cover on earthworks and restored land.

Off-site problems stem from drainage effluents, earthworks, excavations, and mines. The acid drainage water carries Al released by acid weathering of soil minerals, and heavy metals released by

Table 5.2 Critical characteristics of acid sulfate soils

Actual acidity	Potential acidity
Potential neutralizing capacity	Aluminium and heavy metal toxicity
Hydrogen sulfide toxicity	Salinity
Texture and composition	Ripeness
Transmissivity	Watertable regime
Depth to critical layers	Thickness of critical layers

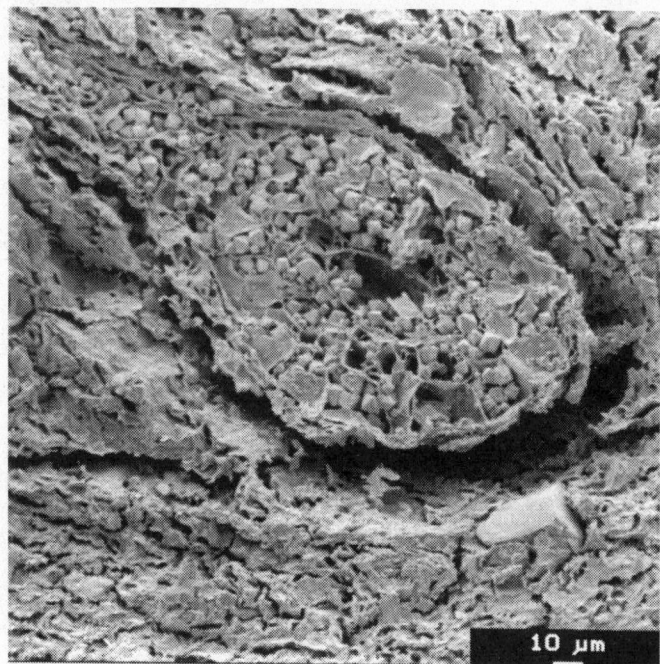

Fig. 5.3   Unripe saline sulfidic clay, Bintang Bolan, The Gambia. Remains of mangrove root in clay matrix

oxidation of sulfide minerals (Fig. 5.2). Toxic drainage waters may be released only episodically, for example, at the onset of the wet season after a period of low watertable during which oxidation has taken place. This makes it difficult sometimes to pinpoint the source of the problem; acid drainage and floodwaters may travel for many km before they are neutralized. Acid sulfate drainage can have disastrous effects on freshwater and estuarine fisheries, especially on invertebrates that are unable to escape. Massive fish kills and ulcerative disease have often been reported in estuarine waters but these have only recently been linked to acid sulfate soils (Callinan et al., 1993; Sammut et al., 1996). The effects on aquatic vegetation are more variable since many species rooting in the reduced mud are little affected; *Nyphar* and *Phragmites* often become dominant in freshwater subject to acid flushes. Even very low concentrations of Al can be hazardous. Concentrations of 1-2 mg Al L^{-1} are toxic to most crops while fish are more susceptible; the few data available indicate fish kills occur at > 0.5 mg L^{-1} with increased mortality from exposure to concentrations of only 0.1 mg L^{-1}. Standards for potable water mostly range between 0.05 and 1.45 mg L^{-1} (Sittig 1994) but are being tightened; for example, Canada, Australia and New Zealand specify 0.005 mg Al L^{-1} (2x10^{-4} mol Al m^{-3}) for water of pH < 6.5. In acid sulfate soils, concentrations of dissolved Al reach 4-80 mol m^{-3} in soil water and 3-10 mol m^{-3} in drainage waters!

The problem of acid effluent is very difficult to deal with. Capping of mines and spoil has proved universally ineffective since this does not prevent continued oxidation of the sulfide. Flooding reduces or prevents oxidation but does not deal with existing acidity unless severely reducing conditions are also reinstated. Because of the dynamic nature of the hazard, depending on management, weather and hydrology as well as soil attributes, modeling is invaluable in designing site-specific management strategies (Section 5.4).

Fig. 5.4 Close-up of Fig. 5.3 showing individual pyrite crystals growing within the skin of the rotting root. Individual crystals are about several μm in diameter [Reprinted from Tovey and Dent, 1998 with permission from the International Society of Soil Science]

5.3   Role of Pyrite

5.3.1   Forms of Reduced Sulfur

Sulfidic soils and sediments contain reduced S as organic sulfur, FeS and a range of metastable polysulfides (collectively known as acid-volatile sulfides) such as greigite (Fe_3S_4) and FeS_2. Their relative proportions depend upon the environment of accumulation but FeS_2 is dominant in marine and brackish water environments where the actual concentration of sulfides is highest. Pyrite itself may occur as loose assembleges of individual crystals (Fig. 5.4) or as dense, spherical clusters (framboids) commonly 10-20 μm diameter (Fig. 5.1).

5.3.2   Formation of Sedimentary Pyrite

The formation of sedimentary sulfides, particularly FeS_2, is well documented (Berner, 1970; Howarth, 1979; Raiswell and Berner, 1985; Marnette et al., 1993; Wilkin and Barnes, 1997). Pyrite formation requires decomposable organic matter and SO_4^{2-} to produce H_2S, Fe to produce metastable Fe sulfides, and an oxidant such as molecular O_2 to transform H_2S to elemental S that can then react with the metastable sulfides to form FeS_2. Quantitative estimates of the rate of FeS_2 accumulation range between 7×10^{-8} to 5×10^{-1} mol S dm^{-3} yr^{-1} (Goldhaber and Kaplan, 1982). In marine sediments, the amount of decomposable organic matter is limiting, in freshwater systems FeS_2 formation is limited by the supply of SO_4^{2-}, but optimum conditions occur in tidal mangrove swamps (Section 5.1).

5.3.3   Pyrite Oxidation

Several possible pathways have been proposed for the oxidation of FeS_2: a sequence of chemical, electrochemical and bacteria-catalyzed reactions (Howie, 1979; Lowson, 1982). Mainly bacteria catalyzed oxidation of FeS_2 will be discussed since chemical reactions are very slow and the presence of bacteria enhances the oxidation processes by orders of magnitude.

Microorganisms play a very important role in the oxidation of FeS_2 (Colmer and Hinkle, 1947). The microorganisms involved are Fe- or S-oxidizing bacteria, chiefly *Thiobacillus ferrooxidans*. Bacteria present in the soil derive energy for growth from that released during the oxidation of FeS_2. In doing so, they catalyze a series of chemical reactions and, under certain conditions, speed up the oxidation process considerably. The role of microorganisms has been classified as either direct or indirect (Lundgren and Silver, 1980). The direct role involves the attachment of microorganisms to the surface of FeS_2, which results in pitting of the mineral surfaces. This direct attack causes surface corrosion of insoluble minerals allowing metals otherwise locked inside mineral particles to dissolve. The indirect role involves the oxidation of pyritic minerals by the products of microbial metabolism. The indirect role of bacteria is believed to enhance the oxidation process by orders of magnitude (Lundgren and Silver, 1980). In the presence of Fe-oxidizing bacteria, the Fe^{2+} can be oxidized to Fe^{3+}, thereby establishing a Fe^{2+}/Fe^{3+} reaction cycle. The metabolism causes the oxidation of a substrate and at the same time, consumes other nutrients to support growth. The energy source for bacteria (*T. ferrooxidans*) is Fe^{2+}, which is oxidized with the release of electrons (Colmer and Hinkle, 1947; Brock and Darland, 1972):

$$2Fe^{2+} \rightarrow 2Fe^{3+} + 2e^-$$
[5.2]

These electrons are available for the reduction of molecular O_2.

$$1/2O_2 + 2e^- + 2H^+ \rightarrow H_2O$$
[5.3]

The electrons transferred during this process produce sufficient energy for the formation of ATP. The C needs of microorganisms can be met by CO_2 fixation (Gale and Beck, 1967) which is coupled to the production of energy from Fe^{2+} oxidation. Few studies have reported whether or not the Fe-grown bacteria can adapt to the use of organic substrates. Other nutrients required for bacterial growth include N, P, S and trace metals.

Although the role of bacteria in catalyzing pyritic oxidation has been established, the population density of bacteria has been treated as irrelevant when modeling FeS_2 oxidation. Okereke and Stevens (1991) found that the Fe^{2+} oxidation rate increased with bacterial (*T. ferrooxidans*) cell concentrations. Because the growth rate of bacteria depends on the concentration of limiting substrates, it may be an important factor affecting the oxidation of pyritic materials. Although numerous mathematical models have been proposed to describe the relationship between the specific growth rate and substrate concentration, the Monod equation is mainly used in soil and groundwater literature despite the recent progress in modeling bacterial growth. This is partly due to the low substrate concentrations and slow growth rates in the soil-water environment. Under these conditions, the Monod equation performs satisfactorily, not warranting the added complexity of any improved models.

5.3.4   Pyrite Oxidation Rates

The overall primary oxidation of FeS_2 to $FeSO_4$ is as follows:

$$FeS_2 + H_2O + 7/2O_2 \rightarrow FeSO_4 + H_2SO_4 + 1440kJ$$
[5.4]

This equation is most often used to determine the conversion ratio of O_2 and S (Pantelis and Ritchie, 1991; Bronswijk et al., 1993). The oxidation of FeS_2 as shown in Reaction [5.4] depends on the supply of O_2, the availability of water, and the physical properties of FeS_2 for the reaction to proceed and generates acid and releases heat; consequently, the acidity and temperature of the surrounding solution would affect the overall reactions. The oxidation of FeS_2 is ultimately determined by the supply of O_2. Models for the description of FeS_2 oxidation are often based on the assumption that all other constituents required for the oxidation process are freely available except for O_2, which is supplied through the porous material from the atmosphere (Dent and Raiswell, 1982; Davis and Ritchie, 1986; Pantelis and Ritchie, 1991; Bronswijk et al., 1993). The rate of pyritic oxidation is often assumed to be a linear function of the dissolved O_2 concentration (Bartlett, 1973; Braun et al., 1974) but recently, the Michaelis-Menton equation has been adopted (Liu et al., 1987; Tan, 1996).

Temperature, which influences both chemical and microbial oxidation, is an important factor in determining the oxidation rate of pyritic materials. Biological oxidation only occurs between 0 to 55 °C (optimum 25-45 °C) (Lundgren and Silver, 1980) but chemical oxidation can take place above this temperature. Jaynes et al. (1984) modeling acid generation in coarse mine spoil, took account of rates of diffusion of both O_2 and Fe^{3+} and, also, the activity of the bacteria generating Fe^{3+} which was estimated from available energy and deviations from ideal temperature, solution pH and O_2 concentration. Pantelis and Ritchie (1992) introduced a ceiling temperature (100 °C) above which microorganisms cease to be effective as catalysts in FeS_2 oxidation. The influence of temperature on oxidation rate follows the empirical Arrhenius equation (Ahonen and Tuovinen, 1991). Because the pyritic oxidation reaction is exothermic, temperature rises depending on the rate of reaction and thermal properties of the bulk soil. In acid sulfate soils, temperature profiles might be used to determine pyritic oxidation rates but because pyritic layers are typically shallow (1-2 m below the surface), distinguishing contributions from pyritic oxidation and solar radiation may be difficult.

5.4   Modeling Soil Physical and Chemical Processes in Acid Sulfate Soils

5.4.1   Introduction
Soil and water management are the key factors for sustainable use of acid sulfate soils for agriculture and natural habitat. Numerous complex physical and chemical processes contribute to the genesis of acid sulfate soils and determine the magnitude and rate of acidification or deacidification and the production/leaching of (toxic) compounds. Because of this complexity, prediction of the effects of soil and water management and their practical consequences are only feasible by using simulation models such as the one for acid sulfate soils (SMASS).

5.4.2   Model Principles

5.4.2.1   General
The SMASS model consists of a number of mutually linked submodels in which the various physical and chemical processes occurring in acid sulfate soils are described mathematically (Fig. 5.5). To solve these equations, the soil profile is divided into compartments which may be of variable size. The initial physical and chemical conditions in each compartment and values for the boundary conditions must be given as model input for the complete simulation period (Fig. 5.5). The physical and chemical conditions in each compartment, together with the water and solute fluxes at the boundary of the soil system are computed at selected time intervals in the sequence illustrated in Fig. 5.5: (1) The water transport submodel computes vertical water transport giving the water content profile which

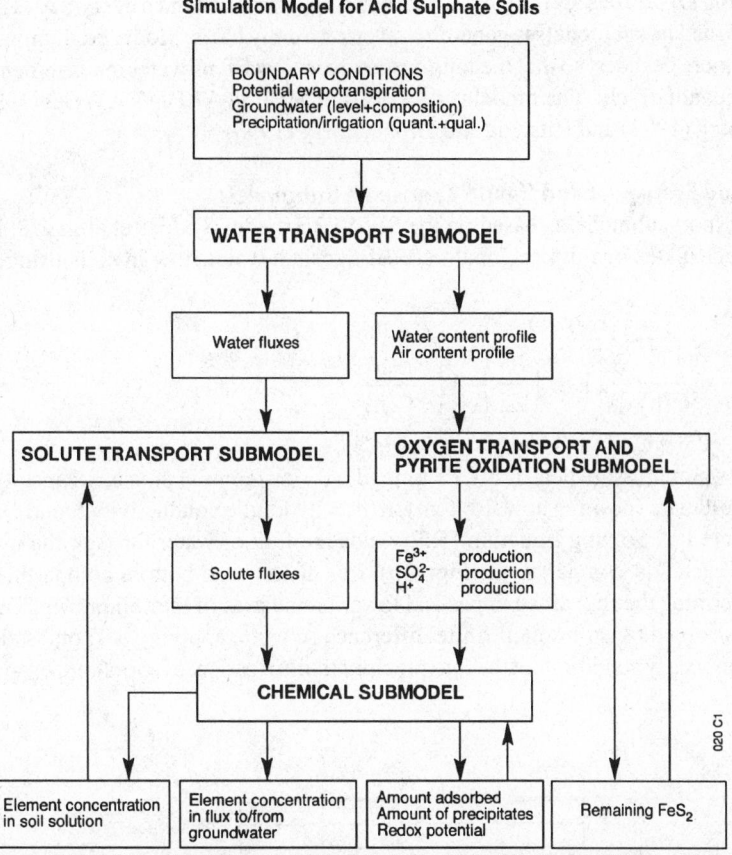

Simulation Model for Acid Sulphate Soils

Fig. 5.5 Structure of Simulation Model for Acid Sulfate Soils (SMASS) [Reprinted from Bronswijk et al., 1995. Evaluation of water management strategies for acid sulfate soils using a simulation model: A case study in Indonesia. Agric. Water. Man. 27:125–142 with kind permission of Elsevier Science - NL, Amsterdam, Netherlands]

complements that of air. (2) In the O_2 transport and FeS_2 oxidation submodel, air contents are used to compute O_2 diffusion coefficients in the air-filled macropores. Oxygen consumption values in the soil are calculated from FeS_2 and organic matter contents, and subsequently, the O_2 content profile in the macropores is computed. (3) Depending on the O_2 concentration at a given depth, the rate of FeS_2 oxidation at that depth is calculated in the O_2 transport and FeS_2 oxidation submodel. For each depth, the oxidized amount of FeS_2 is converted into amounts of H^+, Fe^{3+}, SO_4^{2-} produced, with the remaining FeS_2 being used for calculations in the next time step. (4) The solute transport submodel computes solute fluxes between soil compartments in accordance with the calculated water fluxes (from step 1). (5) In the chemical submodel, first the production/consumption terms for the nonequilibrium processes (such as Fe reduction) are calculated. Then the total concentrations of each chemical component are calculated in the soil compartments by summing the production/consumption terms, the inflow/outflow (from step 4), and the total amounts at the end of the previous time step for each component. From these total concentrations, the equilibrium concentrations in the soil solution, the composition of the exchange complex, and the amount of minerals and precipitates are computed for each compartment.

Time steps for computations of the water and solute transport submodels are in the order of hours. Pyrite oxidation, O_2 profiles and chemical equilibria are computed once every day. The output of the SMASS model and its submodels is generally given on a daily basis. Model predictions can be carried out for one or more decades, so that the long-term effects of various water management strategies can be evaluated quantitatively. The model is described in detail in AARD/LAWOO (1993), Bronswijk and Groenenberg (1993) and Ritsema and Groenenberg (1993).

5.4.2.2 Water Transport and Solute Transport Submodels

The water transport submodel is based on the SWATRE model (Feddes et al., 1978; Belmans et al., 1983) which calculates one-dimensional vertical transient water flow in soils using the basic flow equation:

$$\frac{\partial h}{\partial t} = \frac{1}{C(h)} \frac{\partial}{\partial z} \left[K(h) \left(\frac{\partial h}{\partial z} \right) + 1 \right] - \frac{S(h)}{C(h)} \qquad [5.5]$$

where h = soil water pressure head (cm), t = time (d), C = differential moisture capacity $d\theta/dh$ (cm^{-1}), z = vertical coordinate (positive upward) (cm), K(h) = hydraulic conductivity (cm d^{-1}), and S = water uptake by roots (d^{-1}). Solving Equation [5.5] yields the flux of water through the upper and lower boundaries of each soil compartment (Fig. 5.6). For the top and bottom compartments, boundary conditions determine the flux at the upper and lower boundaries of the soil profile. The complete set of equations is solved by an implicit finite difference scheme, applying a Thomas algorithm. With respect to the boundary conditions at the top (precipitation/irrigation, evaporation, evapotranspiration)

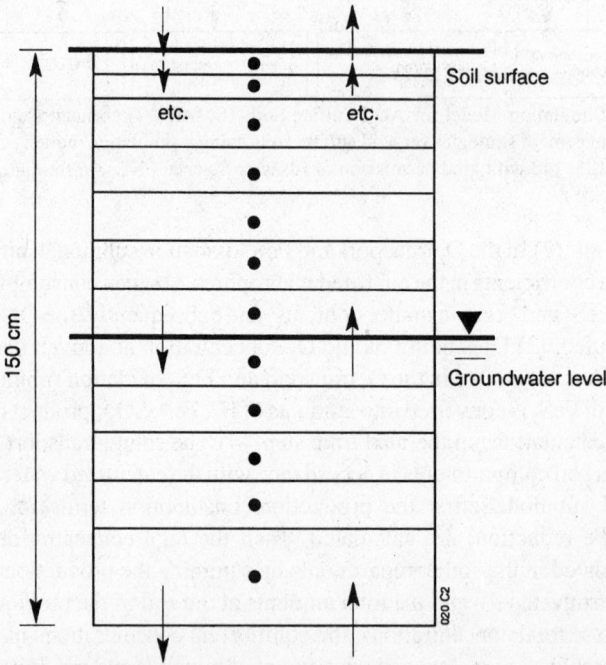

Fig. 5.6 Schematization of a soil profile, as applied in SMASS. Arrows indicate water and solute fluxes

and bottom of the soil system (groundwater level, pressure heads, free drainage, fluxes) various options are possible, which makes the model flexible and generally applicable.

The solute transport submodel is based on the TRANSOL model (Kroes, 1991). Within the soil compartments (Fig. 5.6), complete mixing of the solutes has been assumed. For each compartment, a mass conservation equation is formulated according to:

$$V(n,t)\frac{dC(n,t)}{dt} + C(n,t)\frac{dV(n,t)}{dt} + q_{out} \cdot C_{out} = q_\epsilon \cdot C_\epsilon(n,t) \qquad [5.6]$$

where $C(n,t)$ = concentration of a solute in layer n at time t (mol cm^{-1}), $V(n,t)$ = volume of water in layer n at time t (cm), q_{in} = incoming flux in layer n (cm d^{-1}), q_{out} = outgoing flux from layer n (cm d^{-1}), C_{in} = concentration of incoming flux (mol cm^{-1}), C_{out} = concentration of outgoing flux (mol cm^{-1}).

5.4.2.3   Oxygen Transport and Pyrite Oxidation Submodel

In the model, it is assumed that in natural soils with pH $\geq$ 3, FeS$_2$ is mainly oxidized by O$_2$ as follows (AARD/LAWOO, 1993; Bronswijk et al., 1993):

$$FeS_2 + 3\ 3/4O_2 + 1/2H_2O \rightarrow Fe^{3+} + 2SO_4^{2-} + H^+ \qquad [5.7]$$

The fate of the Fe^{3+}, SO$_4^{2-}$ and H$^+$ produced is considered in the chemical submodel. Neglecting oxidation by Fe^{3+} is justifiable because Fe^{3+} has a low solubility at pH >3. Furthermore, oxidation of FeS$_2$ by Fe^{3+} in natural soils will be inhibited by organic complexation of Fe^{3+} as was found for sewage sludge (Pichtel et al., 1989), and Fe^{3+} concentration must be sustained by the oxidation of Fe^{2+} by O$_2$; consequently, O$_2$ is still the driving force for FeS$_2$ oxidation. In the model, two main O$_2$ diffusion processes are distinguished: (1) vertical through air-filled macropores, and (2) lateral from macropores into the saturated soil matrix (Fig. 5.7). The two processes interact at the walls of the macropores where gaseous O$_2$ dissolves in the matrix soil solution. The equilibrium between gaseous and dissolved O$_2$ at the walls of the macropores is described by Henry's law:

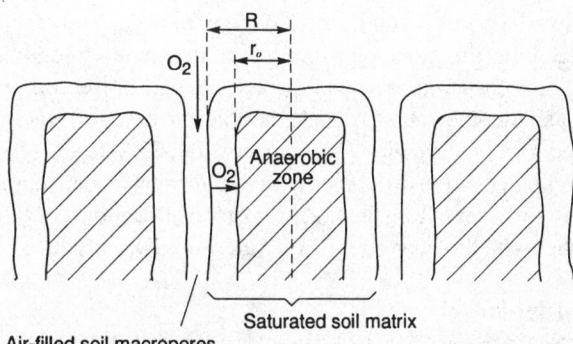

Fig. 5.7 Model approach for the distribution of oxygen in a structured acid sulfate soil, as applied in SMASS. Pyrite is still present in the anaerobic zones. *R* is the radius of the soil aggregates (m), r$_o$ is the radius of the anaerobic zone (m) [Reprinted from Bronswijk et al., 1993. J. Environ. Qual. 22:544–554 with permission of the American Society of Agronomy]

$$[O_2]_{air} = K_H * [O_2]_{water} \qquad\qquad [5.8]$$

Where K_H is Henry's constant dependent on temperature ($K_H = 29.7$ at 20 °C; $K_H = 52$ at 30 °C).
  In SMASS, the gaseous O_2 concentration profile in the air-filled pores is calculated by:

$$\frac{\partial}{\partial x}\left(D_s(\varepsilon_g)\frac{\partial C_a(x)}{\partial x}\right) = \alpha_v \qquad\qquad [5.9]$$

where $C_a(x)$ = O_2 concentration in air-filled macropores (m³ O_2 m⁻³ air), $D_s(\varepsilon_g)$ = O_2 diffusion coefficient in air-filled pores (m² d⁻¹), x = distance (m), α_v = O_2 consumption rate in the soil (m³ O_2 m⁻³ soil d⁻¹), and ε_g = air-filled porosity. The relation between diffusion coefficient (D_s), and air content (ε_g) is described in the model by:

$$D_s(\varepsilon_g) = F\left(1-\left(1-\varepsilon_g\right)^{2/3}\right)D_o \qquad\qquad [5.10]$$

where D_s = O_2 diffusion coefficient in the atmosphere (m² d⁻), F = an empirical tortuosity factor (dimensionless) and D_o = O_2 diffusion coefficient in air (m² d⁻¹). To solve Equation [5.9], the O_2 consumption term α must be quantified. In the model, O_2 is consumed by decomposition of organic matter and oxidation of FeS_2. According to Bronswijk and Groenenberg (1993) and Bronswijk et al. (1993), the steady-state diffusion equation for the aerobic part of the soil matrix can be written as:

$$D_w \frac{d^2C_w(x)}{dx^2} = \frac{0.311262X_{FeS2}}{\rho d}\sqrt{C_w(x)} + Q \qquad\qquad [5.11]$$

where D_w = O_2 diffusion coefficient in the soil matrix (m² d⁻¹), C_w = dissolved O_2 concentration (kg m⁻³), X_{FeS2} is the FeS_2 content (kg m⁻³), ρ = density of FeS_2 (kg m⁻³), d = average diameter of the FeS_2 crystals (m), and Q = O_2 consumption rate by organic matter (kg m⁻³ d⁻¹). According to Christensen et al. (1986), Q is assumed to be constant. The first term on the right hand side of this equation expresses the disappearance of FeS_2 crystals by oxidation and is derived by combining the equal diameter reduction model (Swartzendruber and Barber, 1965) with the rate law for FeS_2 oxidation (McKibben and Barnes, 1986). Solving this equation with appropriate boundary conditions, yields expressions for the steady-state O_2 concentration profile in the aerobic part of the soil matrix, for the radius of the (an)aerobic zone, and for the total O_2 consumption per unit soil volume. Subsequently, this O_2 consumption is used to solve Equation [5.9] numerically. The steady-state O_2 consumption by FeS_2 oxidation is then computed as the difference between the total O_2 consumption and that consumed by organic matter (constant). Next, from the stoichiometry of Reaction [5.7], the amount of oxidized FeS_2 is computed, together with the quantities of Fe^{3+}, H^+, and SO_4^{2-} produced.

5.4.2.4 Chemical Submodel
The important chemical processes occurring in acid sulfate soils are listed in Table 5.3. Pyrite oxidation is computed in the O_2 transport and FeS_2 oxidation submodel while all other chemical processes are modeled in the chemical submodel where changes in solution concentration, adsorbed amounts, and amounts of minerals and precipitates as a result of cation exchange and precipitation/dissolution

Table 5.3 Chemical processes and their effects in acid sulfate soils. Processes included in SMASS are indicated

Process	Effects in SMASS	Included
Rate limited		
Pyrite oxidation	acidification, produces Fe^{3+} and SO_4^{2-}	+
Iron oxidation	acidification, lowers Fe^{2+}-concentration	-
Iron reduction	de-acidification, raises concentration of Fe^{2+}	+
Sulfate reduction	de-acidification, raises sulfide concentration	-
Weathering of primary minerals	produces base cations, consumes protons	-
Weathering of secondary minerals/precipitates	consumes protons, regulates Fe and Al^{3+} concentrations	0
Instantaneous		
Cation exchange	buffers pH and determines concentrations of Ca^{2+} and Mg^{2+}	0
Ion association	raises equilibrium concentrations, especially of Al^{3+}	0

reactions are calculated. Because cation exchange and precipitation/dissolution reactions are related to ion activities, ion association is considered as well. For each time step and compartment, total amounts of each chemical component in solution and on the exchange complex are calculated from the amounts at the end of the previous time step, released into the soil solution due to FeS_2 oxidation, and transported to or from that compartment. From these total amounts, new concentrations, new amounts adsorbed and new amounts of minerals and precipitates are calculated.

Ion association and cation exchange are modeled as equilibrium processes, while precipitation and dissolution of minerals and precipitates are modeled as kinetic processes. The equations describing these different processes are solved simultaneously. In various applications of the SMASS model precipitation/dissolution has been modeled for $Al(OH)SO_4$ (jurbanite), $Al(OH)_3$ (gibbsite), FeOOH (goethite), $CaCO_3$ (calcite), $CaMg(CO_3)_2$ (dolomite), and $CaSO_4$ (gypsum). If necessary for a specific situation, other minerals can easily be included as well. It has been assumed that the dissolution/precipitation rate of these minerals and precipitates is proportional to the difference between the actual activity product (Q_i) and its particular solubility product (K_i) (Kachanoski et al., 1992). The rate expression is formulated as follows:

$$\frac{dB_i}{dt} = K(Q_i - K_i) - \frac{\theta}{\rho}$$ [5.12]

where B_i = amount of a precipitate (mol kg^{-1}), K = a rate constant, ρ = soil dry bulk density (kg m^{-3}) and θ = volumetric water content. This differential equation is approximated with finite differences.

Because the concentrations of NO_3^- in acid sulfate soils are often small and amounts of Mn(III/IV)-oxides are generally negligible, the most likely electron acceptor in redox reactions is Fe^{3+}. Reduction of ferric oxide is described by:

$$Fe(OH)_3 + 1/4CH_2O + 2H^+ \rightarrow Fe^2 + 1/4CO_2 + 11/4H_2O$$ [5.13]

in which $Fe(OH)_3$ represents any reducible ferric oxide and CH_2O represents organic matter. In SMASS, Fe reduction starts when a soil layer is saturated with water. The model distinguishes between reducible and non-reducible ferric oxide. Reducible ferric oxide, which is formed when Fe^{3+} precipitates, is transferred into non-reducible ferric oxide according to:

$$\Delta\left[Fe(OH)_3^R\right] = -k_1(pH)\left[Fe(OH)_3^R\right]\delta t$$ [5.14]

in which $\left[Fe(OH)_3^R\right]$ = amount of reducible ferric oxide (mol kg^{-1}) and k_1(pH) = rate constant (h^{-1}). In the model the pH dependency of the rate constant k_1(pH) is described by an exponential function which fits the data of Schwertmann and Murad (1983). If reducible ferric oxide is present in a soil layer which is saturated with water, the reduction of reducible ferric oxide is described by:

$$\Delta\left[Fe(OH)_3^R\right] = -k_2\Delta t$$ [5.15]

in which k_2 = rate constant (mol kg^{-1} h^{-1}). New amounts of reducible ferric oxide are calculated by combining Equations [5.14] and [5.15] and adding the amount of precipitated $Fe(OH)_3$ calculated in the precipitation/dissolution subroutine. From the amount of ferric oxide reduced, the amounts of Fe^{2+}, OH^- and HCO_3^- produced are calculated using the stoichiometry of Reaction [5.13]. Cation exchange is modeled according to the Gaines-Thomas expression (Bolt, 1967).

5.4.3   Model Validation

5.4.3.1   Comparison of Model with Column Experiments

Ritsema et al. (1992) and Ritsema and Groenenberg (1993) conducted column experiments to study basic physical and chemical processes in acid sulfate soils for model calibration and validation. Several undisturbed soil columns (1 m length, 25 cm diam) containing potential and actual acid sulfate soils were subjected to different hydrological conditions. Every fortnight, the complete water balance, O_2 concentration, redox potential, chemical composition of soil moisture extracted from porous cups at five depths and element concentrations in drainage and ponding water were measured. In addition, the initial and final soil properties such as texture, organic matter, FeS_2, $CaCO_3$, and hydraulic characteristics were determined.

The time span during which the model simulations were carried out was similar to that of the column experiments, while applying the same hydrological regimes, i.e., drainage, submergence and leaching/irrigation. The daily measured actual evaporation (1-1.7 mm d^{-1}) of the columns was given as the top boundary condition in the model validation. Furthermore, daily groundwater levels and quality inside the columns were given as bottom boundary conditions. Ionic concentrations, pH and redox potentials measured at the start of the monitoring period in the columns were used as initial conditions for different model compartments. Measured values of CEC at different depths were used as model input. Other required input parameters and their values used have been given in AARD/LAWOO (1993) and Ritsema and Groenenberg (1993).

Fig. 5.8 summarizes the results of the model validation. All data available on pH, Al^{3+}, Mg^{2+} and SO_4^{2-} from four columns have been plotted against values computed for corresponding days using SMASS. Calculated and measured pH agree reasonably well within the range pH 2.5 to 7. The scattering of pH values is obviously increased by the unreliable pH paper measurements (which can be recognized as vertical columns of symbols at pH 3, 3.5, 4.5 and 5). Measured and calculated Al^{3+} and Mg^{2+} concentrations exhibit more variability. The main reason may be that the soil in the model was uniformly leached, not accounting for macropore flow, leading to an overestimation of leaching efficiency. Measured and predicted SO_4^{2-} concentrations are closely related. In conclusion, in spite of

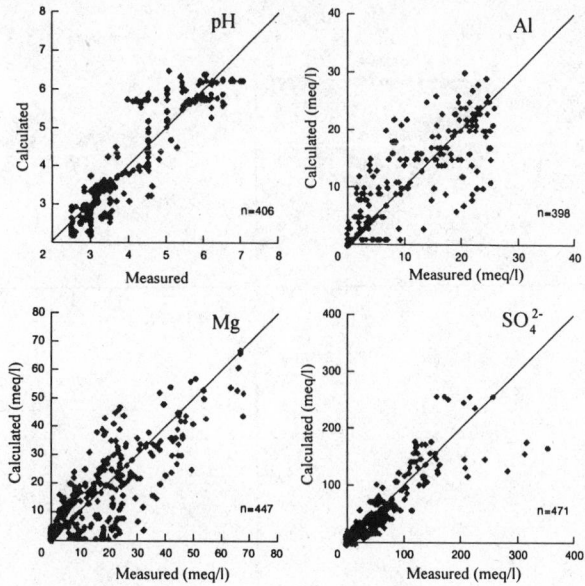

Fig. 5.8 Comparison of model simulations of pH, Al^{3+}, Mg^{2+} and SO_4^{2-} concentrations with measurements at depths of 5, 45 and 85 cm in four columns with two undisturbed acid sulfate soils subjected to drainage, submergence and leaching with fresh and brackish water [Reprinted from van Wijk et al., 1994 with permission of the International Society of Soil Science]

some noise, the relationship between the measured and computed major elements can be described by a 1:1 line over a very wide ranges of concentrations which indicates that the model is widely applicable.

5.4.3.2 Comparison of Model with Field Monitoring

On Pulau Petak, South Kalimantan, Indonesia, monitoring plots were installed (AARD/LAWOO, 1993) to study the main physical and chemical processes in acid sulfate soils under varying seasonal conditions and different water management alternatives. The sites were subjected to either tidal influence (daily flooding with brackish water) or little/no tidal influence (dropping watertable in dry season, rising watertable/flooded with both good and poor quality water in wet season). The same physical and chemical data as in the column experiments were collected every fortnight.

At the Tabunganen site, groundwater levels were high throughout the year, and therefore, all chemical processes occurred in the upper 30 cm of the soil. Consequently, only the field observations and model computations for 5 and 25 cm depth are presented (Fig. 5.9). SMASS computed the oxidation of FeS_2 upon aeration of the soil during the dry period in the first year of field measurements as well. The computed decrease in pH and the corresponding increase in Al^{3+} and SO_4^{2-} concentrations around day 250 agreed with the field measurements. The leaching process in the successive wet period (starting around day 350) was shown by both computed and observed increased pH and decreased SO_4^{2-} concentration. The model predicted aeration and FeS_2 oxidation in the dry period of the second year, starting around day 580 with a decrease in pH at 5 cm depth, and an increase in SO_4^{2-} concentration. In the second dry period, the model predicted an increase in SO_4^{2-} concentrations at 25 cm depth due to leaching of compounds from the topsoil. In reality, however, concentrations at 25 cm depth were much more stable. Part of the SO_4^{2-} produced in the topsoil was, possibly, leached horizontally into the field ditches. As a result, the subsoil received less material from the topsoil than

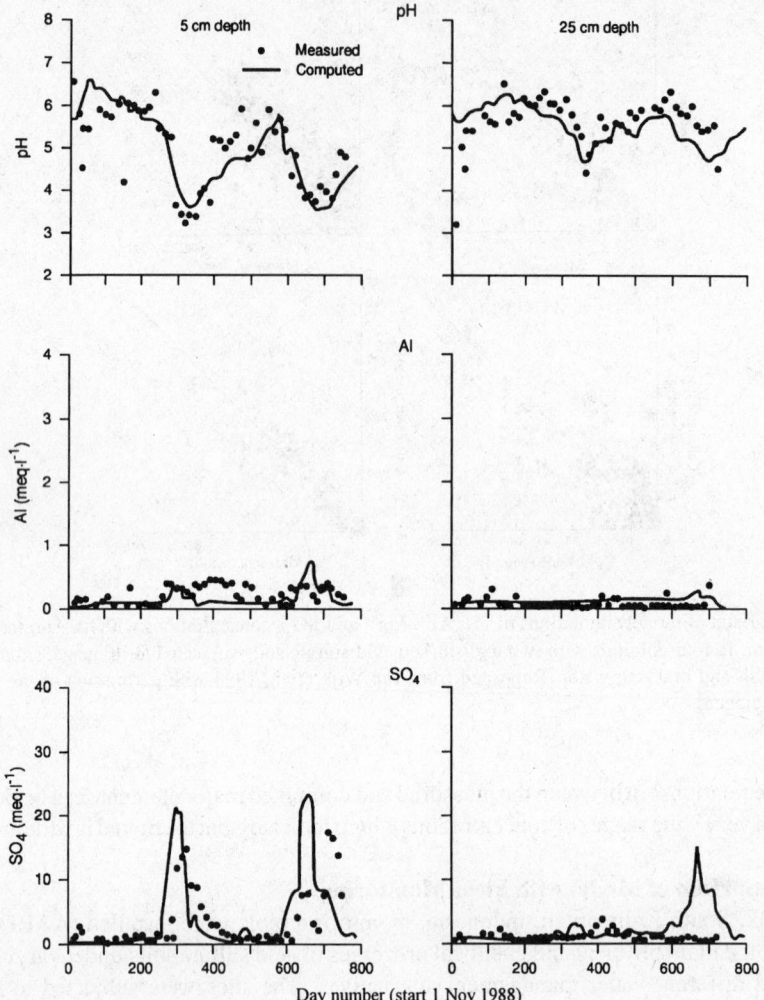

Fig. 5.9   Comparison of model simulations of pH, Al^{3+} and SO_4^{2-} concentrations with measurements at depths of 5 (left-hand side) and 25 cm (right-hand side) from a monitoring field plot on a potential acid sulfate soil (Tabunganen, Indonesia) subjected to daily flooding with brackish water [Reprinted from van Wijk et al., 1994 with permission of the International Society of Soil Science]

was computed with the one-dimensional model. In general, however, there was a good agreement between modeled and actual conditions.

5.4.4   Model Application

5.4.4.1   General

The SMASS model had been applied by Bronswijk et al. (1995) to evaluate the environmental consequences of various water management strategies for the Barambai area, South Kalimantan, Indonesia, which is a a poorly drained backswamp, used for rice cultivation. It lies above springtide level and is not subjected to tidal flooding. Tidal movement only slightly affects the groundwater level. The area is submerged for about 7 months of the year. The soil is a potential acid sulfate soil (Typic

Sulfaquent) with a peaty topsoil (15 cm) and FeS_2 starting from 15 cm downward. Every year, *in situ* acidification due to FeS_2 oxidation occurs in the dry season because the groundwater level drops into the pyritic layer. Rice yields are very low (a few hundred kg ha^{-1}) because of acidification and release of toxic elements such as Fe and Al in the dry season (Kselik et al., 1993). Because FeS_2 is present close to the surface, and water for water conservation is not available during the dry season, new water management strategies should generally aim at improved drainage and deeper groundwater levels to enhance removal of FeS_2 from the soil profile by oxidation. The following water management strategies were selected for further evaluation by Bronswijk et al. (1995): (1) continuation of the present water management; (2) a drainage strategy with a groundwater level of 40 cm below the surface throughout the year; (3) a drainage strategy with a groundwater level of 60 cm below the surface throughout the year; (4) a moderate drainage strategy with a groundwater level of 40 cm below the surface during the dry season (145 days) and above the surface in the wet season (220 days); (5) a moderate drainage strategy with a groundwater level of 60 cm below the surface during the dry season (145 days) and above the surface in the wet season (220 days); (6) a combined drainage/ irrigation strategy with a groundwater level of 40 cm below the surface throughout the year (equal to strategy 2), combined with irrigation with good quality water at the end of the wet season (90 days, pH 5, 30 mm d^{-1}) which is available in the tertiary canals at the end of the wet season, but has not yet been used for irrigation. The strategies 1 to 5 were evaluated for the future development of the FeS_2 profile in the soil. The strategies 1, 2 and 6 were evaluated for the course of pH and Al concentration at a depth of 5, 25 and 65 cm (Fig. 5.10).

For each of the six selected water management strategies, a period of 10 years was simulated, starting with the initial FeS_2, organic matter, pH$^-$ and Al^{3+} profiles, which were measured at the beginning of the field monitoring at Barambai (11/88). The top boundary condition for the simulations consisted of daily precipitation and potential evapotranspiration measured in the field between 1988 and 1990. To arrive at a period of 10 years, these 2-year measurements were repeated 5 times, because of lack of reliable long-term weather records. In the case of strategy 6, an irrigation of 30 mm d^{-1} with good quality water was added to the top of the model, in addition to the daily precipitation and potential evapotranspiration, during 90 days at the end of the wet season. The bottom boundary condition applied for each of the 6 strategies was according to the drainage strategies. For strategy 1, groundwater level fluctuations measured in the field during 2 years were repeated 5 times.

5.4.4.2 Pyrite Oxidation as a Function of Water Management

Fig. 5.10 shows the computed future FeS_2 profiles for the water management strategies investigated. The intensive drainage strategies 2 and 3 (groundwater levels 40 and 60 cm below surface, respectively) will result in a rapid decrease in FeS_2 content of the topsoil. Within 5 years, nearly all FeS_2 above the groundwater table will be oxidized. The moderate drainage strategies 4 and 5 (groundwater levels 40 and 60 cm below surface for only part of the year, respectively) result in a slower oxidation of FeS_2 in the topsoil. After 5 years, considerable amounts of FeS_2 are still present in the topsoil, but after 10 years, nearly all FeS_2 will be oxidized. However, the present water management strategy (Strategy 1) results in the slowest oxidation, which even after 10 years will leave considerable amounts of FeS_2 in the topsoil.

In Fig. 5.11, the change in the depth to FeS_2 over the next 10 years shows that drainage plays a major role in the future development of the FeS_2 profiles. The most extreme drainage strategy (Strategy 3), will lead to a FeS_2-free topsoil of 40 and 60 cm in 3.3 and 4.5 years, respectively, while Strategy 2 will produce 40 cm of FeS_2-free topsoil in 2.8 years. Obviously, the strategies in which the water table is below the soil surface for only part of the year (Strategies 1,4,5) require a longer time for the topsoil to become FeS_2 free. Under the present water management (Strategy 1), FeS_2 content

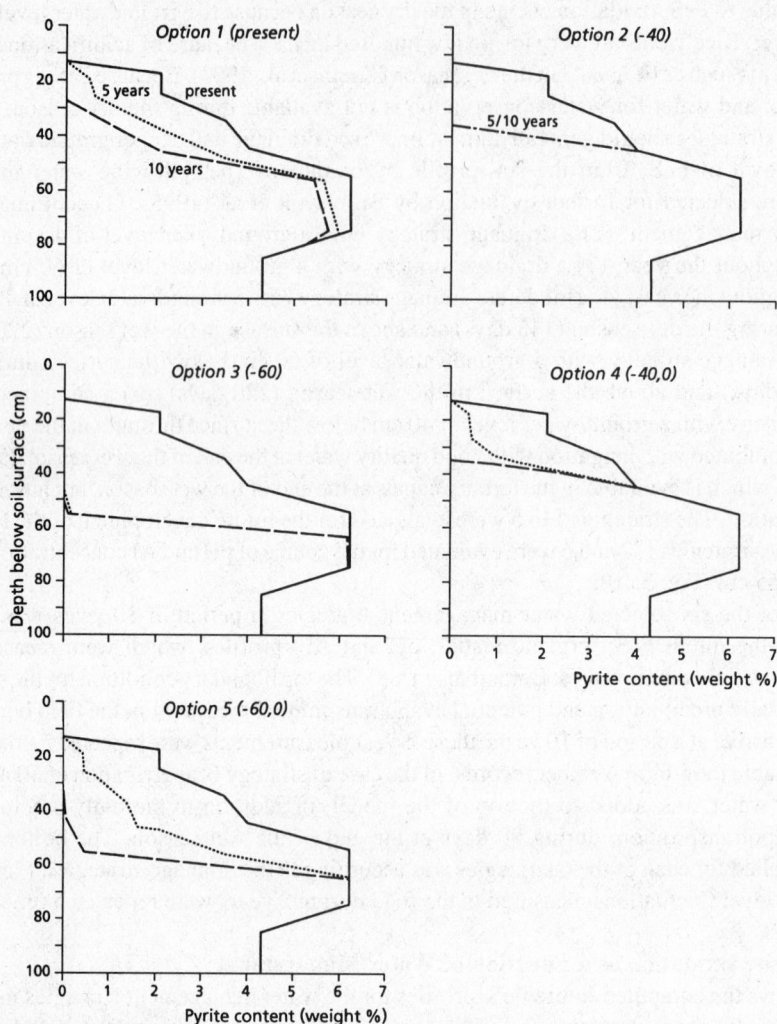

Fig. 5.10 Computed pyrite profiles in the Barambai soil at present, and after 5 and 10 years of various water management strategies (strategy 1 = present water management; 2 = groundwater level -40 cm whole year; 3 = groundwater level -60 cm whole year; 4 = groundwater level -40 cm dry period, 0 cm wet period; 5 = groundwater level -60 cm dry period, 0 cm wet period; 6 = groundwater level -40 cm + irrigation) [Reprinted from Bronswijk et al., 1995. Evaluation of water management strategies for acid sulfate soils using a simulation model: A case study in Indonesia. Agric. Water Man. 27:125-142 with kind permission of Elsevier Science - NL, Amsterdam, Netherlands]

decreases slowly (30 cm in 10 years). From Fig. 5.11, it can be concluded that deeper groundwater levels do not automatically result in a more rapid FeS_2 oxidation in the topsoil.

The various water management strategies should also be evaluated from an environmental point of view. Different strategies will result in different FeS_2 oxidation rates (Fig. 5.12). The right-hand axis gives the H^+ produced (mol m^{-2}) soil, assuming, as a first approximation, that oxidation of 1 mol FeS_2 releases 4 mol H^+ (Reaction 5.1) (Dent, 1986). However, only part of the acidity will be transported

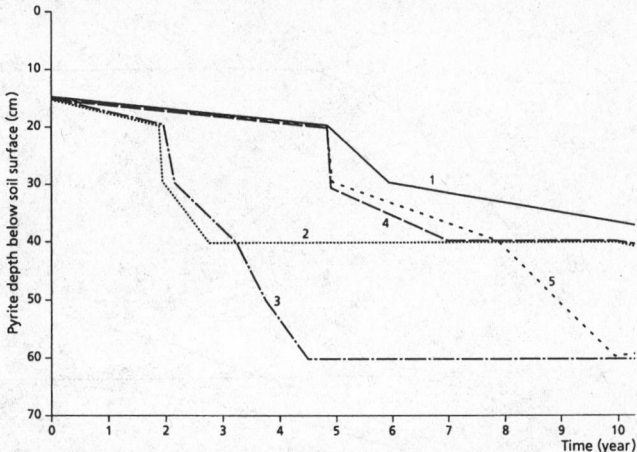

Fig. 5.11 Computed course of the pyrite depth in the Barambai soil over the next 10 years for various water management strategies (1 = present water management; 2 = groundwater level -40 cm whole year; 3 = groundwater level -60 cm whole year; 4 = groundwater level -40 cm dry period, 0 cm wet period; 5 = groundwater level -60 cm dry period, 0 cm wet period) [Reprinted from Bronswijk et al., 1995. Evaluation of water management strategies for acid sulfate soils using a simulation model: A case study in Indonesia. Agric. Water Man. 27:125-142 with kind permission of Elsevier Science - NL, Amsterdam, Netherlands]

to ground and surface waters, while the remainder will be buffered and retained in the soil profile. Nevertheless, the amounts involved are enormous and must have a negative effect on the environment. This has to be modeled by taking account of the buffer capacity and transport mechanisms of the receiving aquatic system (Dent and Bowman, 1996). Acidity produced in the intensive drainage Strategy 3 is highest, and lowest in the moderate drainage strategies. Once the FeS_2 is depleted from temporarily aerated layers, acid production stops. The more intensely drained situations show shorter periods of acid production. In Strategies 2, 3, and 6, further acid production would stop after about 5 years but under the present water management, acid generation will still continue even after 10 years.

5.4.4.3 Soil and Water Quality

If the present water management is continued for another 10 years (Strategy 1), the pH will rapidly decrease every dry season, due to *in situ* FeS_2 oxidation, while the reverse will occur in the wet season as a result of leaching and reduction (Fig. 5.13a). Concurrently, Al^{3+} concentrations will also change (Fig. 5.13b) with maximum concentrations in the dry season being 10 $mmol_c$ Al L^{-1} while toxic levels for rice are ~ 2 $mmol_c$ Al L^{-1} (Smilde, 1989). Consequently, after 10 years, soil conditions will remain as poor as they are at present. Water management Strategy 2 (groundwater level at 40 cm) will result in pH values as low as 2.4 in the whole profile in the dry seasons of the first 2 years but after about 3 years, the FeS_2 in the top 40 cm will be completely oxidized (Fig. 5.11) and the pH in the topsoil will slowly rise to ~ 4, mainly because of leaching by rainfall. The predicted Al^{3+} concentrations show corresponding trends. After 10 years, a less unfavorable soil will be achieved if Strategy 2 rather than the present water management is implemented. The main disadvantage of Strategy 2, however, is the strong acidification during the initial years. Water management Strategy 6 (identical to Strategy 2 plus irrigation) shows a moderate improvement (Fig. 5.13) with pH and Al^{3+} concentrations, especially in the topsoil, slightly better than with Strategy 2 in the first 3 years. After complete oxidation of FeS_2 (4-5 years), pH increases and Al^{3+} decreases more rapidly to acceptable levels with irrigation.

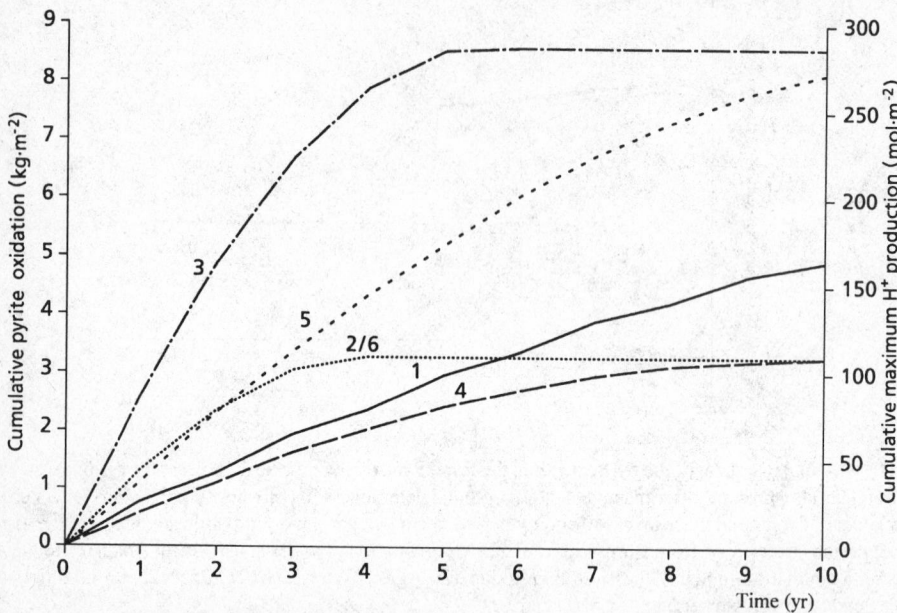

Fig. 5.12   Computed pyrite oxidation and acid production in the Barambai soil over the next 10 years for various water management strategies (1 = present water management; 2 = groundwater level -40 cm whole year; 3 = groundwater level -60 cm whole year; 4 = groundwater level -40 cm dry period, 0 cm wet period; 5 = groundwater level -60 cm dry period, 0 cm wet period) [Reprinted from Bronswijk et al., 1995. Evaluation of water management strategies for acid sulfate soils using a simulation model: A case study in Indonesia. Agric. Water Man. 27:125-142 with kind permission of Elsevier Science - NL, Amsterdam, Netherlands]

5.4.4.4   Concluding Remarks

The examples presented show how the SMASS model can be used to predict long-term changes over time and depth in element concentrations and FeS_2 in soil and drainage water for different water management strategies. The model allows estimation of the progress and direction of acid sulfate soil genesis as a function of management and its agricultural and environmental consequences. For instance, periods during which Fe and Al concentrations exceed toxic levels to crops and elements in drainage water contaminate surface waters can be predicted. The consequences of various water management strategies have been predicted without discussion of their implementation in practice. The success of promising strategies depends on technical, economic and social factors as well. The feasibility of their implementation must be subsequently assessed by technical, economic and social evaluations.

Compared to simple and empirical models, process-based models such as SMASS are often more accurate and much more widely applicable but require more data. In SMASS the system parameters, such as hydraulic functions, are soil-specific and have to be measured or obtained from pedotransfer functions. Chemical system parameters, such as cation exchange coefficients, rate constants for precipitation/dissolution and redox reactions are not soil specific and can be obtained from the literature and model calibration. Initial conditions such as water, FeS_2 and organic matter contents, pH and element concentrations in the soil solution can be obtained from field measurements and chemical analysis. Boundary or variable conditions, such as groundwater levels, daily weather records and concentrations in irrigation, precipitation and groundwater are site specific and are sometimes available or have to be measured.

Another disadvantage of simulation models is that they may be too complicated for use by nonspecialists. To overcome this problem, simulation models can be used to fit complex processes into simple relationships or to derive rules of thumb for specific situations. Fig. 5.14 presents an example of such an approach of moving from complex models to simple tools. In the Barambai area, a reasonable relationship exists between average annual groundwater level in the pyritic layer and the computed annual FeS_2 oxidation rate. Measurements from various laboratory and field experiments in Indonesia (AARD/LAWOO, 1993), have also been included in Fig. 5.14, further illustrating the validity of the relationship. A quick estimation of annual FeS_2 oxidation can now be made quite simply

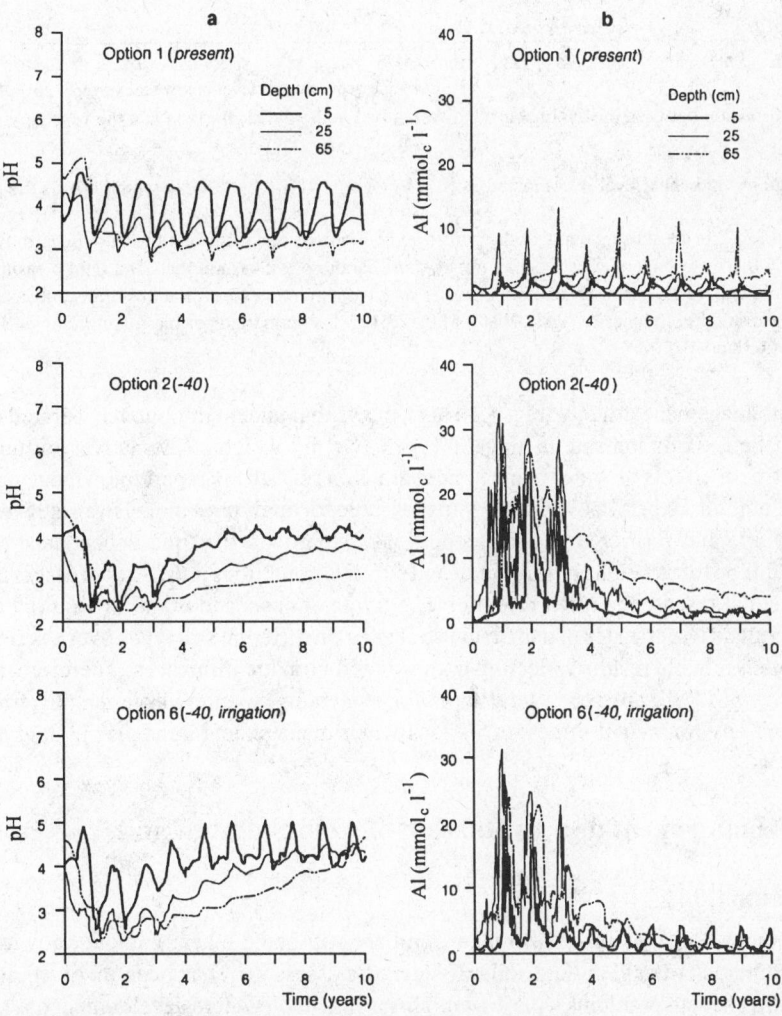

Fig. 5.13 Computed course of pH (a) and aluminum concentration (b) at three depths in the Barambai soil over the next 10 years for various water management strategies (1 = present water management; 2 = groundwater level -40 cm whole year; 6 = groundwater level -40 cm + irrigation) [Reprinted from Bronswijk et al., 1995. Evaluation of water management strategies for acid sulfate soils using a simulation model: A case study in Indonesia. Agric. Water Man. 27:125-142 with kind permission of Elsevier Science - NL, Amsterdam, Netherlands]

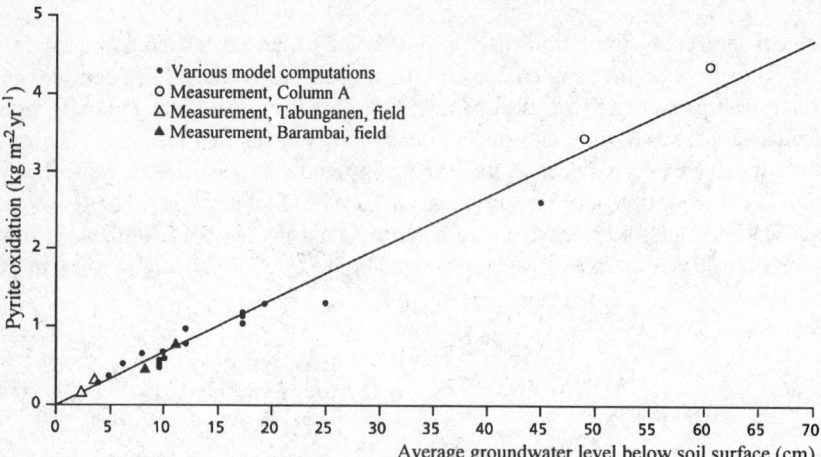

Fig. 5.14 Relationship between the average groundwater level in the pyritic horizon and the yearly pyrite oxidation

rate. The average groundwater level is computed as $\left(\sum_{n=1}^{n=365} \left(h(n) - Z_{FeS_2} \right) \right) / 365$, in which h(n) is the groundwater

level at day n, and Z_{FeS2} is the starting depth of the pyritic layer. For days that the groundwater table is above the pyritic layer h(n)-Z_{FeS2} = 0. The graph contains model computations, laboratory measurements and field measurements [Reprinted from Bronswijk et al., 1995. Evaluation of water management strategies for acid sulfate soils using a simulation model: A case study in Indonesia. Agric. Water Man. 27:125-142 with kind permission of Elsevier Science - NL, Amsterdam, Netherlands]

for any water management strategy. Fig. 5.14 also shows that under continuously aerated conditions the maximum FeS$_2$ oxidation rate is about 1% FeS$_2$ yr^{-1} by weight. Alternatives to modeling for prediction of future effects of water management are either based on expert knowledge or extensive field experimentation. Expert knowledge is indispensable for applying a model in an area, but remains mainly qualitative and restricted to local conditions. Moreover, the time dependence of reaction processes in relation to variations in soil, water and climatic conditions is difficult to assess in this way. Field experimental research is often prohibitive in terms of costs and duration required to conduct experiments with different water management strategies on different soil types over a series of years, in order to obtain reliable results reflecting both soil and climatic influences. Therefore, simulation models, such as SMASS, are very useful tools for evaluating in an environmentally friendly way agricultural and environmental effects of various water management strategies in acid sulfate soil areas.

5.5   Use, Management and Evaluation of Acid Sulfate Land

5.5.1   Introduction

Most management decisions have been taken, and are still being taken, without knowledge of the unique characteristics of acid sulfate soils. If scientific assessment has been involved at all, it has usually been after severe problems have arisen. This is inevitable where development has taken place without previous experience of the problems and where problems are encountered downstream of the acid sulfate land. The development of assessment methods has focused on improvements in speeds and precision of specific determinations rather than dealing with soil heterogeneity and proper data interpretation.

Although methods of reclamation and land use/management have the single purpose of enabling and improving agricultural production, they may have adverse environmental effects which have only recently received any attention. In Australia, drained sugarcane fields release acid into surface waters after heavy rain, killing or at least causing ulcerative disease outbreaks in fish due to immune system failure (Callinan et al., 1993). On the Plain of Reeds, Mekong delta, Minh et al. (1997) found that more Al^{3+} was leached into the surface water early in the rainy season from pineapples and yams (*Dioscorea alata*) in raised beds than rice fields (Fig. 5.15). During a 4-month period, rice, yam and pineapple fields released 2.2, 41 and 44 kmol Al^{3+} ha^{-1}, respectively.

Existing methods (van Mensvoort, 1996; Xuan, 1993; Dent, 1992) allow problems to be reliably diagnosed and on-site reclamation and precautionary watertable management undertaken to avoid, or at least minimize, acid generation. Despite their obnoxious nature and associated toxicities, a wide range of land uses is found on acid sulfate soils. For generations farmers in tropical coastal wetlands such as the Mekong delta of Viet Nam (Xuan et al., 1982; Xuan, 1993; Tri, 1996) have, with great perseverance and skill and often at great cost, developed methods to coexist with the toxicities of acid sulfate soils. Soil and water management is the key to solving the problems of acid sulfate soils (van Breemen, 1976; Tuong, 1993). Surface and watertable management, leaching, narrow drain spacing, maintenance of downward water flow, good leveling, timely plowing and puddling are the most important measures to consider.

5.5.2   Land Use on Potential Acid Sulfate Soils

5.5.2.1   Mangrove Forest Exploitation

Dramatic disappearance of mangrove (*Rhizophora* spp.) forests for charcoal production on potential acid sulfate soils and replacement by shrimp farming has been reported from the Mekong delta (Durang,

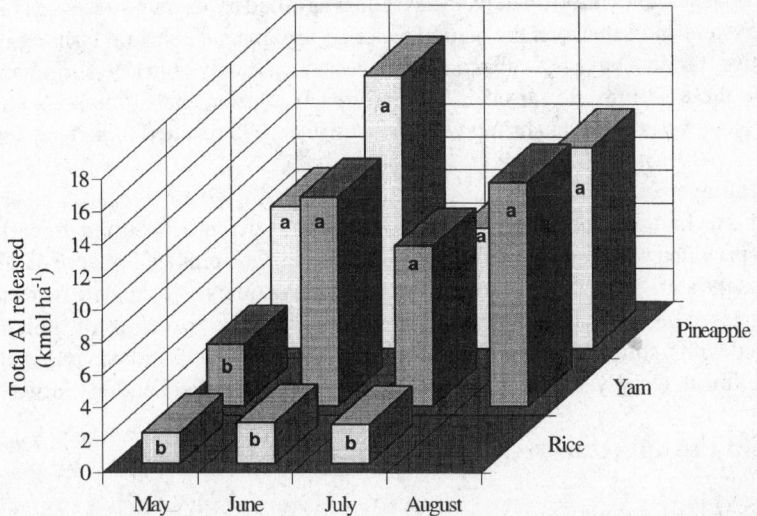

Fig. 5.15 Monthly mean amount of aluminium released to the surrounding canal network from pineapple and yam raised beds and rice fields. In the same months, means (represented by column height) with the same letters are not significantly different at the 5% level by DMRT. [Reprinted from Minh et al., 1997. Contamination of surface water as affected by land use in acid sulfate soils in the Mekong river delta, Vietnam. Agric. Ecosyst. Environ. 61:19-27 with kind permission of Elsevier Science - NL, Amsterdam, Netherlands]

1995), but is likely to have happened elsewhere as well. Besides cutting, mangrove forests are commonly overfished, resulting in the disappearance of a large number of aquatic species. Assessment of the severity of the acid sulfate hazard on other land uses may provide a powerful argument for retaining and protecting the natural ecosystem.

5.5.2.2   Fish or Shrimp Ponds

Brinkman and Singh (1982) developed a method of reclamation for fishponds on potential acid sulfate soils in the Philippines. Using the salt water tides, they drained the ponds for forced oxidation of FeS_2, plowed the pond bottom to promote oxidation, allowed salt water to reenter to dissolve the acid formed, redrained the acid water at low tide, and repeated the process. After 3-5 treatments, the pond bottom was free of acidity. Dikes were also washed with salt water. Rice and fish yields increased by 2.1 Mg ha^{-1} and 400 kg ha^{-1} yr^{-1}, respectively, on reclaimed land (Ligue and Singh, 1988) giving farmers financial independence.

5.5.2.3   The Rice and Shrimp System

This system (Xuan et al., 1986) is applied on both potential (where it was developed) and actual acid sulfate soils in the Mekong delta. The monsoon climate with its contrasting dry and wet seasons allows farmers in coastal lands to exploit the fresh and salt water periods for rice and shrimp production. Land located next to a creek or a canal with large tidal fluctuations is surrounded with a deep ditch and dike of excavated material. At the onset of the rainy season, the water is quickly drained to reduce the salinity so that the rice seedbed is ready early. Rice is sown or transplanted with reported yields of 4 Mg ha^{-1} of paddy. By that time the rains have stopped and surface water becomes saline, the fields are flooded with salt water and kept submerged or almost submerged throughout the dry season. In the original version of the system, shrimp were caught from natural fry, resulting in yields up to 690 kg ha^{-1} yr^{-1}. Fresh sediment was deposited on the field annually from tidal water, raising the land and filling the ditch. This material was used to raise the surrounding dike, which thus is composed of good alluvial material without acidity. Unfortunately, the system was killed by its own success as too many people applied the system and exhausted the natural waters of fry. Because shrimp farming is economically more attractive, farmers have now abandoned the system and only cultivate shrimp using purchased fry. Because these shrimp are raised in the early rainy season, the farmers also abandoned rice cultivation. However, declining shrimp yields since 1994/1995 resulted in poverty for many.

5.5.2.4   Drainage

The aim of this technique (Xuan et al., 1982) is to accelerate salt removal during the early rainy season. Each field is provided with a surrounding shallow ditch and dike, plus a number of parallel field drains and a gate. At the start of the rainy season, farmers drain salt quickly from their fields at low tide, and close their gates at high tide. The system enables farmers to have two crops of short duration rice or rice combined with a short-duration upland crop such as watermelons or beans. Banjarese farmers in Southern Kalimantan apply a similar shallow drainage system called handils (Sarwani et al., 1993)

5.5.3   Land Use on Actual Acid Sulfate Soils

5.5.3.1   Extensive Exploitation

This activity consists of harvesting reeds for matting, and exploitation of indiginous and planted *Melaleuca leucodendron* for timber, firewood and oil extraction (Brinkman and Xuan, 1991) but few options are available on raw acid sulfate soils with severe acidity at shallow depth.

5.5.3.2 Rice

Rice is by far the most important crop cultivated on acid sulfate soils. In the Mekong delta, soils with a sulfuric horizon deeper than 50 cm have been used for many years for traditional or deep water rice cultivation (yielding about 1 Mg ha^{-1}). Until recently (~10 years), inland freshwater Sulfaquepts were scarcely used for cultivation. After the economic reforms of 1986, settlers moved into the uncultivated lands. Coolegem (1996) estimated that from 1987 to 1995, 53% of the land in a district dominated by raw acid sulfate soils had changed from wasteland or *Melaleuca leucodendron* forest to irrigated rice during the dry season (28% of the entire district). Husson and Phung (1994) reported that the relative altitude (a few cm) of fields in relation to the water levels in the growing season was responsible for significant differences in yields (Fig. 5.16). Higher areas suffer from difficulties in maintaining adequate water depth to reduce oxidation and thus acidification, while the lower areas suffer from problems associated with reduction at depth (Fe^{2+} and H$_2$S toxicity). Many elevation differences may occur within one field, causing great heterogeneity in crop growth. Hanhart et al. (1997) and Hanhart and Ni (1993) developed a set of water management measures for such rice land, recommending good leveling, narrow drain spacing (30-60 m) to create a downward flow of water through the surface soil and brief drying of the surface soil at flowering to avoid reduction at depth. Yields obtained were 3.6-3.8 Mg ha^{-1} with the economic break-even point being 2.15 Mg ha^{-1}. Drainage from the rice fields carries an estimated 5 kmol H$^+$ ha^{-1} crop^{-1} into the surface water, so irrigation and drainage waters should be kept separate. In areas with saline surface water in the dry season, traditional rice is cultivated. Moderate fertilizer levels of 100 kg N ha^{-1} and 60 kg P ha^{-1} are recommended, together with a small lime application to improve Ca^{2+} availability (Ren et al., 1993). Poor relationships between uptake and activity of Ca^{2+}, Mg^{2+}, Fe^{2+} in the soil solution have been found for rice on acid sulfate soils, so Moore and Patrick (1993) recommend instead using the charge fraction of these ions in the soil solution, which shows much better relationships with plant uptake.

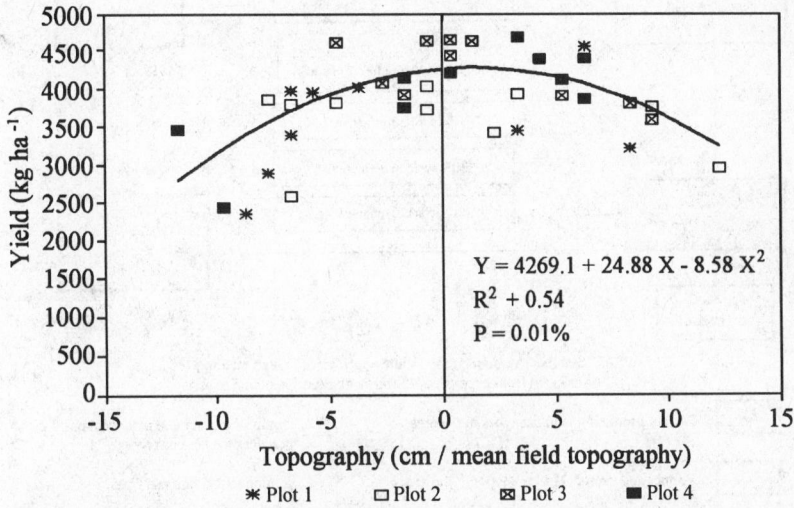

Fig. 5.16 Rice yield as a function of microtopography in Thanh Tan district, Mekong delta, Vietnam. N = 84 [Reprinted Husson and Phung, 1994 with permission of Fonds voor Ontwikkelingssamenwerking, Brussels, Belgium]

5.5.3.3 Balanta System

In West Africa (Senegal, Guinea Bissau) farmers use the so-called *Balanta* system for rice cultivation (van Gent and Ukkerman, 1993) in which land is ridged annually to a height in accordance with the expected freshwater flood of the rainy season. The ridging and annual turning speed up removal of soluble toxins from the surface soil, but the system is in decline because of the heavy labor demand and low financial returns.

5.5.3.4 Annual Upland Crops

Yam, sweet potato, cassava, and sugarcane are the most important. These crops precede rice cultivation in places where no proper water supply for rice in the dry season is available. Farmers make raised beds by piling up the surface soil before the start of the rainy season. Beds are leached strongly during the wet season. Crops are planted immediately after the recession of the flood. Moisture is supplied by residual water from the soil, water in the ditches between the beds and from capillary rise.

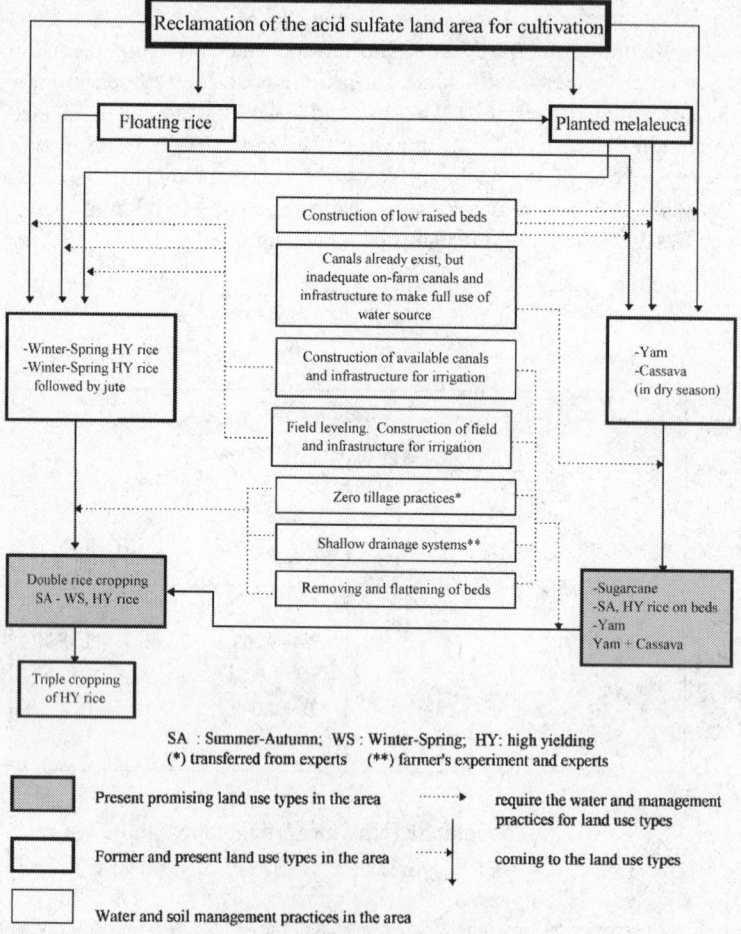

Fig. 5.17   History of land use in Thanh Tan district, Long An province, Mekong delta, Vietnam [Reprinted from Tri, 1996 with permission]

For yam, vegetative growth is luxuriant, but yields are modest (~ 6 Mg ha^{-1}, which is 15 Mg ha^{-1} on an area without beds). The disappointing yields are attributed to excess moisture near harvest when the rainy season has started again (Durang, 1995). After some 3-4 years, the land is leveled and, provided a source of irrigation water is available, rice is cultivated in a double crop system (Tri, 1996).

5.5.3.5 Perennial Upland Crops

Sugarcane, pineapple, and cashew nut, which all require well drained soil in flood season, are grown on beds raised well above the level of the flood. Sugarcane on acid sulfate soils suffers from poor tillering, short and thin canes, with total yields averaging ~ 50 Mg ha^{-1} (Derevier, 1991) while yields on nonacid soils of the Mekong delta can reach 120 Mg ha^{-1}. Although Al^{3+} toxicity was suspected to be the cause, no correlation between soil Al^{3+} and plant nutrient content could be established (Nga et al., 1993). For pineapple, fertilization, and mulching with grass and a sufficient height above the flood are recommended (Tri, 1996; Sen et al., 1987) leading to yields of 40-50 Mg ha^{-1}, even in areas where irrigation is limited.

5.5.4 Integrated Soil, Water and Crop Management

Tri (1996) presented an historical account of land use on the Mekong delta (Fig. 5.17) showing the soil water management measures (raised bed construction, plowing, puddling, leaching, construction of an irrigation and drainage system, field water management during the rice crop) that were taken as changes in land use were being made. This illustrates clearly how soil and water management are of primary importance for proper management of acid sulfate soils. The assessments were mainly made by the farmers themselves, but the integration of different soil and water management practices can only be made by the community of land users. Land use evolved through their learning process. Significantly, the problems perceived first were those of drainage and water supply, and only later watertable management in recognition of acidity problems.

5.5.5 Land Evaluation

Tuong et al. (1991) used hydrological factors to determine land quality in the Mekong delta for rice cultivation (double or triple cultivation, modern or traditional rice). The hydrological land qualities selected were (1) fresh water flood depth and flood duration during the wet season, (2) rainfall amount and distribution emphasizing the occurrence of dry spells in the early rainy season, (3) tidal influence, (4) periods during which saline surface water is present in the dry season, and (5) the adequacy of canals for irrigation. For instance, flood depths of < 30, < 60 and > 60 cm allow modern, short-duration, traditional and deep water rice cultivation, respectively. Tidal influence is judged for its potential in irrigation without pumping; a long period under saline surface water may leave insufficient time for double-cropped rice too.

From a land evaluation study of the acid sulfate soil part of the Mekong delta which, in addition to hydrology, also included soil conditions and used a wider range of land use types, van Mensvoort et al. (1993) conclude that making fresh water available improves the suitability for rice on all acid sulfate land, and that well-constructed raised beds make moderately acid soils (sulfuric horizon > 50 cm) highly suited for pineapple and sugarcane.

Using the same but more detailed methodology, Tri et al. (1993) evaluated two Mekong delta districts, one only partly hampered by acid sulfate soil problems, and the other dominated by raw acid sulfate soils. On moderately acid land, land utilization types (LUTs) such as annual sugarcane in the dry season followed by rice in the wet season, cashew intercropped with pineapple, double rice cultivation and *Melaleuca* forestry could be recommended. On raw acid sulfate soils with much more difficult hydrological conditions, only single or double rice, or yam could be recommended.

The success or failure of the LUTs depends, however, on farmers' skills. Present land use and farmer skills form, in fact, the basis for both land evaluation studies. In the studies of Tri (1996) and van Mensvoort (1996), acid sulfate soil knowledge at various levels (farmers, experts and specialists) is brought together. All knowledge levels are valued equally, but farmer-level knowledge seems to have been underused in the past. Farmer-level knowledge is built on generations of experience in using the land and not only generated interesting land use systems (rice/shrimp, yam, sweet potatoes, pineapples etc) but, also, explained how they were historically developed, under what conditions or in what sequence they can be used, and what management measures need to be taken (e.g., leaching, mulching, no-till, raised bed construction). The expert-level knowledge such as fertilizer recommendations based on field experiments (Ren et al., 1993), refinement of water management measures (Tuong, 1993; Hanhart et al., 1997; Hanhart and Ni, 1993), and variety selection (Xuan, 1993) can be regarded as complementary, reinforcing that of farmers. Finally, true specialist knowledge of a more fundamental nature can also contribute to improved land use, e.g., by better survey methods to improve land use planning (Bregt et al., 1993, Burrough et al., 1988), by getting a better grasp of soil-plant relationships through improved methods of chemical characterization (Moore and Patrick, 1993), or by modeling the processes in acid sulfate soils (Bronswijk and Groenenberg, 1993). To date, the community of acid sulfate soil specialists has taken rather little account of the interests of the largest and most vulnerable group of land users, the farmers, and is only now assembling detailed knowledge into coherent and accessible decision support systems (Dent and Dawson, 1998).

5.6   References

AARD/LAWOO. 1993. Acid sulfate soils in the humid tropics: Simulation model of physical and chemical processes to evaluate water management strategies. Acid sulfate soils in the humid tropics. Vol. 3. International Institute for Land Reclamation and Improvement, Wageningen, Netherlands.

Ahmed F.B., and D.L. Dent. 1997. Resurrection of soil surveys. Case study of acid sulphate soils in The Gambia II. Added value from spatial statistics. Soil Use Manag. 13:37–61.

Ahonen, L., and O. Tuovinen. 1991. Temperature effects on bacterial leaching of sulfide minerals in shake flask experiments. Appl. Environ. Microbiol. 57:138–145.

ASSMAC. 1998. Acid sulfate soil manual. NSW Acid Sulfate Management Advisory Committee, Wollongbar, NSW, Australia.

Bartlett, R.W. 1973. A combined pore diffusion and chalcopyrite dissolution kinetics model for *in situ* leaching of a fragmented copper porphyry. p. 331–374. *In* D.J.I. Evans and R.S. Shoemaker (ed.) International Symposium on Hydrometallurgy. Chicago, Illinois.

Belmans, C., J.G. Wesseling, and R.A. Feddes. 1983. Simulation of the water balance of a cropped soil: SWATRE. J. Hydrol. 63:271–286.

Berner, R.A. 1970. Sedimentary pyrite formation. Am. J. Sci. 268:1–23.

Bolt, G.H. 1967. Cation exchange equations in soil science. A review. Neth. J. Agric. Sci. 15:81–103.

Braun, R.L., A.E. Lewis, and M.E. Wadsworth. 1974. In-place leaching of primary sulfide ores: Laboratory leaching data and kinetics model. Metall. Trans. 5:1717–1726.

Bregt, A.K., J.A.M. Janssen, and P. Alkasuma. 1993. Survey strategies for acid sulfate soils. p. 43–50. *In* D.L. Dent and M.E.F. van Mensvoort (ed.) Selected Papers of the Ho Chi Minh City Symposium on Acid Sulfate Soils. ILRI Publication 53, ILRI, Wageningen, Netherlands.

Brinkman, R., and V.P. Singh. 1982. Rapid reclamation of fishponds on acid sulfate soils. p. 318–330. *In* H. Dost and N. van Breemen (ed.) Proceedings of the Bangkok Symposium on Acid Sulfate Soils. ILRI Pub. 31, International Institute for Land Reclamation and Improvement, Wageningen, Netherlands.

Brinkman, W.J., and V. Xuan. 1991. *Melaleuca leucodendron* s.l., a useful and versatile tree for acid sulfate soils and some other poor environments. Int. Tree Crop J. 6 :61–274

Brock, T.D., and G.K. Darland. 1972. Limits of microbial existence: Temperature and pH. Science 169:1316–1318.

Bronswijk, J.J.B., J.E. Groenenberg, C.J. Ritsema, A.L.M. van Wijk, and K. Nugroho. 1995. Evaluation of water management strategies for acid sulfate soils using a simulation model: A case study in Indonesia. Agric. Water Man. 27:125–142.

Bronswijk, J.J.B., and J.E. Groenenberg. 1993. A simulation model for acid sulfate soils I: Basic principles. p. 341–357. *In* D.L. Dent and M.E.F. van Mensvoort (ed.) Selected Papers of the Ho Chi Minh City Symposium on Acid Sulfate Soils. Pub. 53. International Institute for Land Reclamation and Improvement, Wageningen, Netherlands.

Bronswijk, J.J.B., K. Nugroho, I.B. Aribawa, J.E. Groenenberg, and C.J. Ritsema. 1993. Modeling of oxygen transport and pyrite oxidation in acid sulfate soils, J. Environ. Qual. 22:544–554.

Burrough, P.A., M.E.F. van Mensvoort, and J. Bos. 1988. Spatial analysis as a reconnaissance survey technique: An example from acid sulfate regions in the Mekong delta, Vietnam. p. 68–91. *In* H. Dost (ed.) Selected Papers of the Dakar Symposium on Acid Sulfate Soils. ILRI Pub. 44. International Institute for Land Reclamation and Improvement, Wageningen, Netherlands.

Callinan R.B., G.C. Frazer, and M. Melville. 1993. Seasonally recurrent fish mortalities and ulcerative outbreaks associated with acid sulfate soils in Australian estuaries. p. 403–410. *In* D.L. Dent and M.E.F. van Mensvoort (ed.) Selected Papers of the Ho Chi Minh City Symposium on Acid Sulfate Soils. ILRI Pub. 53. International Institute for Land Reclamation and Improvement, Wageningen, Netherlands.

Cathles, L.M., and J.A. Apps. 1975. A model of the dump leaching process that incorporates oxygen balance, heat balance, and air convection. Metall. Trans. B. 6B:617–624.

Christensen, T.H., B.L. Parker, and J.C. Refsgaard. 1986. A model for the unsaturated zone; oxygen transport and consumption model. Report Danish Hydraulic Institute, Horsholm, Denmark.

Colmer, A.R., and M.E. Hinkle. 1947. The role of microorganisms in acid mine drainage: A preliminary report. Science 106:253–256.

Coolegem, L. 1996. Land use changes in the Plain of Reeds, Vietnam (1987-1995). MSc Thesis, Department of Soil Science and Geology, Wageningen Agricultural University, Wageningen, Netherlands.

Davis, G.B., and A.I.M. Ritchie. 1986. A model of oxidation in pyritic mine wastes: Part 1. Equations and approximate solution. Appl. Math. Model. 10:314–322.

Dent, D.L., 1986. Acid sulfate soils: A baseline for research and development. ILRI Publ. 39. International Institute for Land Reclamation and Improvement, Wageningen, Netherlands.

Dent, D.L. 1992. Reclamation of acid sulfate soils. Adv. Soil Sci. 17:79–121.

Dent D.L., and G.M. Bowman. 1996. Quick, quantitative assessment of the acid sulphate hazard. CSIRO Soils Div. Rep. 128.

Dent, D.L., and B. Dawson. 1998. The acid test. Expert system for the identification, assessment and management of acid sulphate soils. CD:ROM Users Manual. EWMAN Ltd, Norwich, UK.

Dent D.L., and M.E.F. van Mensvoort. 1993. Selected Papers of the Ho Chi Minh City Symposium on Acid Sulphate Soils. ILRI Pub. 53. International Institute for Land Reclamation and Improvement, Wageningen, Netherlands.

Dent D.L., and L.J. Pons. 1995. Acid sulphate soils: A world view. Geoderma 67:263–276.

Dent D.L., and R.W. Raiswell. 1982. Quantitative models to predict the rate and severity of acid sulphate development. p. 73–95. *In* H. Dost, and N. van Breemen. 1982. Proceedings of the Bangkok Symposium on Acid Sulfate Soils. ILRI Pub. 31. International Institute for Land Reclamation and Improvement, Wageningen, Netherlands.

Derevier, A. 1991. Mission d'evaluation des possibilites de culture de la canne a sucre au Vietnam. Institut de Recherches Agronomiques Tropicales er des Cultures Vivrieres, Montpellier, France.

Diemont W.H., L.J. Pons, and D.L. Dent. 1993. Standard profiles of acid sulfate soils. p. 51–60. *In* D.L. Dent and M.E.F. van Mensvoort (ed.) Selected Papers of the Ho Chi Minh City Symposium on Acid Sulphate Soils. International Institute for Land Reclamation and Improvement, Wageningen, Netherlands.

Dost H. 1973. Acid sulphate soils. ILRI Pub. 18. Institute for Land Reclamation and Improvement, Wageningen, Netherlands.

Dost, H., and N. van Breemen. 1982. Proceedings of the Bangkok Symposium on Acid Sulfate Soils. ILRI Pub. 31. International Institute for Land Reclamation and Improvement, Wageningen, Netherlands.

Durang, T. 1995. Yam cultivation on acid sulfate soils. MSc thesis. Department of Soil Science and Geology, Wageningen Agricultural University, Wageningen, Netherlands.

Feddes, R.A., P.J. Kowalik, and H. Zaradny. 1978. Simulation of field water use and crop yield. PUDOC, Wageningen, Netherlands.

Gale, N.L., and J.V. Beck, 1967. Evidence for the Calvin cycle and hexose monophosphate pathway in *Thiobacillus ferrooxidans*. J. Bacteriol. 94:1052–1060.

Goldhaber M.B., and I.R. Kaplan. 1982. Controls and consequences of sulfate reduction rates in recent marine sediments. p. 19–36. *In* J.A. Kittrick, D.S. Fanning, and L.R. Hossner. 1982. Acid sulfate weathering. Soil Sci. Soc. Am. Spec. Pub. 10, Soil Science Society of America, Madison, WI.

Hanhart, K., and D.V. Ni. 1993. Water management of rice fields at Hoa An, Mekong delta, Vietnam. p.161–176. *In* D.L. Dent and M.E.F. van Mensvoort (ed.) Selected Papers of the Ho Chi Minh City Symposium on Acid Sulfate Soils. ILRI Pub. 53, International Institute for Land Reclamation and Improvement, Wageningen, Netherlands.

Hanhart, K., D. V. Ni, N. Bakker, F. Bil, I. Postma, and M.E.F. van Mensvoort. 1997. Surface water management under varying drainage conditions for rice on an acid sulfate soil in the Mekong delta, Vietnam. Agric. Water Manag. 33:99–116.

Howarth, R.W. 1979. Pyrite: its rapid formation in a salt marsh and its importance in ecosystem metabolism. Science. 203:49–51.

Howie, F.M.P. 1979. Curation of palaeontological collections. Spec. Pap. Palaeon. 22:103.

Husson, O., and M.T. Phung. 1994. Synthetic report, Research methodology and main technical results. Framing systems research project, Plain of Reeds. IAS/FOS, Fund for Development Cooperation, Brussels, Belgium.

Jaynes D.B., A.S. Rogowski, and H.B. Pionke. 1984. Acid drainage from reclaimed coal strip mines. I. Model description. Water Resour. Res. 20:233–242.

Kachanoski, R.G., K.K. Tanji, L.T. Rollins, L.D. Whittig, and R. Fujii. 1992. Dissolution kinetics of $CaCO_3$: CARKIN-1, a computer model. Soil Sci. 153:13–24.

Kittrick J.A., D.S. Fanning, and L.R. Hossner. 1982. Acid sulfate weathering. Soil Sci. Soc. Am. Spec. Pub. 10. Soil Science Society of America, Madison, WI.

Konsten C.J.M., W. Andriesse, and R. Brinkman. 1988. A field laboratory method to determine total potential and actual acidity in acid sulphate soils. p. 106–134. *In* H. Dost (ed.) Selected Papers on the Dakar Symposium on Acid Sulphate Soils. ILRI Pub. 44. International Institute for Land Reclamation and Improvement, Wageningen, Netherlands.

Kroes, J.G. 1991. Transol 2.1, A dynamic model for transport and transformation of solutes in soils. User's guide. Internal report 110. DLO Winand Staring Centre, Wageningen, Netherlands.

Kselik, R.A.L., K.W. Smilde, H.P. Ritzema, K. Subagyono, S. Saragih, Mauliana Damanik, and H. Suwardjo. 1993. Integrated research on water management, soil fertility and cropping systems on acid sulfate soils in South Kalimantan, Indonesia. p. 177–194. *In* D.L. Dent and M.E.F. van Mensvoort (ed). Selected Papers of the Ho Chi Minh City Symposium on Acid Sulfate Soils. ILRI Publ. 53. International Institute for Land Reclamation and Improvement, Wageningen, Netherlands.

Ligue, D.L., and V.P. Singh. 1988. Social and economic status of farmers in acid sulfate soil areas in the Philippines. p. 238–249. *In* H. Dost (ed.) Selected Papers of the Dakar Symposium on Acid Sulfate Soils. ILRI Pub. 44. International Institute for Land Reclamation and Improvement, Wageningen, Netherlands.

Liu, M.S., R.M.R. Branion, and D.W. Duncan. 1987. Oxygen transfer to *Thiobacillus* cultures. p. 375–384. *In* P.R. Norris and D.P. Kelly (ed.) Biohydrometallurgy: Proceedings of the International Symposium, Warwick, UK.

Lowson, R.T. 1982. Aqueous oxidation of pyrite by molecular oxygen. Chem. Rev. 82:461–497.

Lundgren, D.G., and M. Silver. 1980. Ore leaching by bacteria. Ann. Rev. Microbiol. 34:263–283.

Marnette, E.C., N. van Breemen, K.A. Hordijk, and T.C. Cappenberg. 1993. Pyrite formation in two freshwater systems in the Netherlands. Geochim. Cosmochim. Acta 57:4165–4177.

McKibben, M.A. and H.L. Barnes. 1986. Oxidation of pyrite in low temperature acidic solutions: rate laws and surface textures. Geochim. Cosmochim. Acta 50: 1509–1520.

Minh, L.Q., T.P. Tuong, M.E.F. van Mensvoort, and J. Bouma. 1997. Contamination of surface water as affected by land use in acid sulfate soils in the Mekong river delta, Vietnam. Agric. Ecosyst. Eviron. 61:19–27.

Moore P.A., and W.H. Patrick, Jr. 1993. Metal availability and uptake by rice in acid sulfate soils. p. 205–224. *In* D.L. Dent and M.E.F. van Mensvoort (ed.) Selected Papers of the Ho Chi Minh City Symposium on Acid Sulfate Soils. Pub. 53. International Institute for Land Reclamation and Improvement, Wageningen, Netherlands.

Nga, T.T., D.V. Ni, and V.T. Xuan. 1993. Cultivation of sugar cane on acid sulfate soils in the Mekong delta. p. 123–128. *In* D.L. Dent and M.E.F. van Mensvoort (ed.) Selected Papers of the Ho Chi Minh City Symposium on Acid Sulfate Soils. ILRI Pub. 53. International Institute for Land Reclamation and Improvement, Wageningen, Netherlands.

Nordstrom, D.K. 1982. Aqueous pyrite oxidation and subsequent formation of secondary iron minerals. p. 37–56. *In* L.R. Hossner, J.A. Kittrick, and D.F. Fanning (ed.) Acid sulfate weathering. Soil Science Society of America. Madison, WI.

Okereke, A., and S.E. Stevens, Jr. 1991. Kinetics of iron oxidation by *Thiobacillus ferrooxidans*. Appl. Environ. Microbiol. 57:1052–1056.

Pantelis, G., and A.I.M. Ritchie. 1991. Macroscopic transport mechanisms as a rate-limiting factor in dump leaching of pyrite ores. Appl. Math. Model. 15:136–143.

Pantelis, G., and A.I.M. Ritchie. 1992. Rate limiting factors in dumps leaching of pyritic ores. Appl. Math. Model. 16: 553–560.

Pichtel, J.R., W.A. Dick, and E.L. McCoy. 1989. Binding of iron from pyritic mine spoil by water soluble organic materials extracted from sewage sludge. Soil Sci. 148:140–148.

Pons, L.J., and N. van Breemen 1982. Factors influencing the formation of potential acidity in tidal swamps. p. 37–51. *In* H. Dost and N. van Breemen. 1982. Proceedings of the Bangkok Symposium on Acid Sulfate Soils. ILRI Pub. 31. International Institute for Land Reclamation and Improvement, Wageningen, Netherlands.

Raiswell, R.W., and R.A. Berner. 1985. Pyrite formation in euxenic and semi euxenic sediments. Am. J. Sci. 285:710–724.

Ren, D.T.T., V.T. Guong, N.M. Hoa, V.Q. Minh, and T.T. Lap. 1993. Fertilisation of nitrogen, phosphorus, potassium and lime for rice on acid sulfate soils. p. 147–154. *In* D.L. Dent and M.E.F. van Mensvoort (ed.) Selected Papers of the Ho Chi Minh City Symposium on Acid Sulfate Soils. ILRI Pub. 53. International Institute for Land Reclamation and Improvement, Wageningen, Netherlands.

Ritsema, C.J., and J.E. Groenenberg. 1993. Pyrite oxidation, carbonate weathering, and gypsum formation in a drained potential acid sulfate soil. Soil Sci. Soc. Am. J. 57:968–976.

Ritsema, C.J., J.E. Groenenberg, and E.B.A. Bisdom. 1992. The transformation of potential into actual acid sulfate soils studied in column experiments. Geoderma 55:259–271.

Sammut, J., R.B. Callinan, and G.C. Fraser. 1996. An overview of the ecological impacts of acid sulfate soils in Australia. p. 140–143. *In* R.J. Smith (ed.) Proc. 2nd Nat. Conf. Acid Sulfate Soils. R.J. Smith & Associates and ASSMAC, Alstonville, Australia.

Sarwani, M., M. Lande, and W. Andriesse. 1993. Farmers' experiences in using acid sulfate soils: Some examples from tidal swampland of southern Kalimantan, Indonesia. p. 113–122. *In* D.L. Dent and M.E.F. van Mensvoort (ed.) Selected Papers of the Ho Chi Minh City Symposium on Acid Sulfate Soils. ILRI Pub. 53. International Institute for Land Reclamation and Improvement, Wageningen, Netherlands.

Schwertmann, U., and E. Murad. 1983. Effect of pH on the formation of goethite and hematite from ferrihydrite. Clays Clay Min. 31:277–284.

Sen, L.N., D.A. Duong, V.T. Guong, N.N. Hung, and T.K. Tinh 1987. Influence of different water management and agronomic practices on pineapple yield in an acid sulfate soil at Binh Son 3 State Farm, Kien Giang. Fin. Rep. Proj. STD 302-NL. Department of Soil Science and Geology, Wageningen Agricultural University, Wageningen, Netherlands.

Sittig, M. 1994. World-wide limits for toxic and hazardous chemicals in air, water and soil. Noyes Publications, Peak Ridge, NJ.

Smilde, K. 1989. Report on a general consultancy mission (soil fertility aspects). Mission Report 20. Research on Acid Sulfate Soils in the Humid Tropics. International Institute for Land Reclamation and Improvement, Wageningen, Netherlands.

Soil Survey Staff. 1998. Keys to soil taxonomy. 8th Ed. USDA-NRCS, Lincoln, NE.

Swartzendruber, D., and S.A. Barber. 1965. Dissolution of limestone particles in soil. Soil Sci. 100:287–291.

Tan, Y. 1996. Comments on Modeling of oxygen transport and pyrite oxidation in acid sulfate soils by Bronswijk et al. J. Environ. Qual. 25:928–930.

Tovey, N.K., and D.L. Dent. 1998. Microstructure and microcosm chemistry of tidal soils. 16th World Congr. Soil Sci. Symp. 3-D. Micromorphology, Montpellier, France.

Tri, L.Q. , N.V. Nhan, H.G.J. Huizing, and M.E.F. van Mensvoort. 1993. Present land use as basis for land evaluation in two Mekong delta districts. p. 299–320. *In* D.L. Dent and M.E.F. van Mensvoort (ed.) Selected Papers of the Ho Chi Minh City Symposium on Acid Sulfate Soils. ILRI Pub. 53, International Institute for Land Reclamation and Improvement, Wageningen, Netherlands.

Tri, L. Q. 1996. Developing management packages for acid sulfate soils based on farmer and expert knowledge. Field study in the Mekong delta, Vietnam. PhD Thesis, Department of Soil Science and Geology, Wageningen Agricultural University, Wageningen, Netherlands.

Tuong, T.P. 1993. An overview of water management on acid sulfate soils. p. 265–281. *In* D.L. Dent and M.E.F. van Mensvoort (ed.) Selected Papers of the Ho Chi Minh City Symposium on Acid Sulfate Soils. ILRI Pub. 53. International Institute for Land Reclamation and Improvement, Wageningen, Netherlands.

Tuong, T.P., C.T. Hoanh, and N.T. Khiem. 1991. Agro-hydrological factors as land qualities in land evaluation for rice cropping patterns in the Mekong Delta, Vietnam. *In* P. Deturck and F.N. Ponnamperuma (ed.) Rice production on acid soils of the tropics. Institute for Fundamental Studies, Kandy, Sri Lanka.

van Breemen, N. 1976. Genesis and solution chemistry of acid sulphate soils in Thailand. Agric. Res. Rep. 848. PUDOC, Wageningen, Netherlands.

van Gent, P. A.M., and R. Ukkerman. 1993. The Balanta rice farming system in Guinea Bissau. p. 103–112. *In* D.L. Dent and M.E.F. van Mensvoort (ed.) Selected Papers of the Ho Chi Minh City Symposium on Acid Sulfate Soils. ILRI Pub. 53. International Institute for Land Reclamation and Improvement, Wageningen, Netherlands.

van Mensvoort, M.E.F. 1996. Soil knowledge for farmers, farmer knowledge for soil scientists: The case of the acid sulfate soils in the Mekong delta, Vietnam. PhD Thesis, Department of Soil Science and Geology, Wageningen Agricultural University, Wageningen, Netherlands.

van Mensvoort, M.E.F., and D.L. Dent. 1997. Acid sulfate soils. p. 301–333. *In* R. Lal et al. (ed.) Methods for assessment of soil degradation. CRC Press, Boca Raton, FL.

van Mensvoort, M.E.F., N.V. Nhan, T.K. Tinh, and L.Q. Tri. 1993. Coarse land evaluation of the acid sulfate soil areas in the Mekong delta based on farmers' experience. p. 321–330. *In* D.L. Dent and M.E.F. van Mensvoort (ed.) Selected Papers of the Ho Chi Minh City Symposium on Acid Sulfate Soils. ILRI Pub. 53. International Institute for Land Reclamation and Improvement, Wageningen, Netherlands.

van Wijk, A.L.M., J.J.B. Bronswijk, J.E. Groenenberg, and C.J. Ritsema. 1994. Model of physical and chemical processes in acid sulfate soils: Principles, validation and application. Trans. 15th World Congr. Soil Sci. 6a:144–163.

Wilkin, R.T., and H.L. Barnes. 1997. Pyrite formation in an anoxic estuarine basin. Am. J. Sci. 297:620–650.

Xuan ,V.T. , N.K. Quang, and L.Q. Tri. 1982. Rice cultivation on acid sulfate soils in the Vietnamese Mekong delta. p. 251–259. *In* H. Dost and N. van Breemen (ed.) Proceedings of the Bangkok Symposium on Acid Sulfate Soils. ILRI Pub. 31. International Institute for Land Reclamation and Improvement, Wageningen, Netherlands.

Xuan, V.T. 1993. Recent advances in integrated land uses on acid sulfate soils. p. 129–136. *In* D.L. Dent and M.E.F. van Mensvoort (ed.) Selected Papers of the Ho Chi Minh City Symposium on Acid Sulfate Soils. ILRI Pub. 53. International Institute for Land Reclamation and Improvement, Wageningen, Netherlands.

Xuan, V.T., N.V. Sanh, D.V. Ni, N.T. Ut, and H.M. Hoang. 1986. The rice-shrimp cropping system on potential and actual acid sulfate soils in the Mekong delta. Paper presented at the Dakar conference on acid sulfate soils. Farming Systems Research and Development Center, Can Tho University, Can Tho, Vietnam.

6

Soils and Environmental Quality

Terry J. Logan
Ohio State University

6.1 Introduction

Soils play a fundamental role in the regulation of pollutants in ecosystems. Soil is the interface for most human activity and is greatly impacted by humans. Soil is an important source of pollution in and of itself (as water and wind eroded sediments), and as a carrier of pollutants. Water runoff and erosion transfer pollutants from terrestrial to fresh water and marine aquatic ecosystems, and particulate aerosols are important short- and long-range carriers of pollutants. Soil is an important sink for pollutants through precipitation, sorption and immobilization reactions (Section B, Chapter 8). The soil sink reduces potentially toxic dissolved concentrations of pollutants that might otherwise contaminate surface and ground water supplies and be taken up by soil and aquatic biota. The soil sink can also reduce the bioavailability of pollutants in soil directly ingested by soil biota, grazing animals and humans (see the section below for a definition of soil pollutant bioavailability).

This chapter encompasses the important classes of soil borne pollutants (nutrients, pesticides and xenobiotics, trace elements, and other chemicals like salts, alkalis, acids and radionuclides), discusses techniques to assess fate of these pollutants in soil, and summarizes approaches for reclamation of chemically degraded soil. The scope of the chapter is that of surface soil, the top 1 m or so in which most human and biological activity occurs. The assumption with this designation is that subsurface and offsite processes involving soil borne pollutants, although important, are not as critical as the processes that occur at the soil surface itself. It is in this zone that the greatest opportunities to manage soil to protect environmental quality occur. This chapter does not deal with the contemporary issue of soil quality, an intrinsic but ill-defined attribute of soil, and one that includes environmental aspects discussed here, yet the source/sink characteristics of soil are clearly one of the characteristics that define soil quality. The topic is discussed more explicitly in Section G, Chapter 11.

6.2 Nutrients

6.2.1 Nitrogen

6.2.1.1 Environmentally Important Nitrogen Forms

Nitrogen, like P, is a macronutrient as well as a pollutant. Nitrogen fluxes in soil are large, with additions to soil as high as 200 kg N ha^{-1} as fertilizer, symbiotic N fixation or acid deposition

(Stevenson, 1986). Additions of organic materials like manures, biosolids and green wastes may also contribute to N soil additions. Losses from soil by volatilization, denitrification, leaching and crop removal are equally large, and there is little accumulation of N in soil beyond seasonal trends, i.e., N is a dynamic element in soil and steady-state conditions are established within the soil system such that soil N levels do not vary much. Nitrogen exists in many physical and chemical forms and in oxidation states that range from -3 to +5 (Table 6.1). Because of the wide range of oxidation states, most N transformations in soil occur as a result of microbial processes.

6.2.1.2   The Environmental Nitrogen Cycle
The N cycle in soil (Fig. 6.1) is dynamic, multiphasic and multicompartmental. The reactions are almost all microbially mediated. From an environmental quality perspective, what is of significance is the effect of large perturbations on components of the cycle leading to increased fluxes to water, air and biota. Important perturbations include flooding and drainage with effects on denitrification, mineralization, and emissions of NO_3^- and NO_x, excessive land application of N fertilizers or organic wastes (e.g., manures, sludges) with increases in NO_3^- leaching; and mineralization of soil organic N associated with land clearing and burning with subsequent leaching losses of NO_3^-.

6.2.1.3   Major Sources of Nitrogen in Soils
The major natural source of N in soil is symbiotic dinitrogen fixation, representing most of the N addition to soils on a global basis (Stevenson, 1986). The importance of this source has grown in the last 20 years or so with expansion of soybean acreage in the United States, Brazil, and China. Major anthropogenic sources of N to soil are synthetic fertilizer applications, acid deposition from fossil fuel combustion, and land application of organic wastes. While N in some previously incinerated organic wastes (primarily municipal refuse and biosolids) are eventually returned to the soil as acid deposition, a significant part of the organic materials being land applied were once landfilled and isolated from the terrestrial ecosystem for decades or more.

6.2.1.4   Major Environmental Impacts of Soil Nitrogen
The diverse chemical forms of N have varied environmental effects (Table 6.1). High NH_3 levels in surface water contribute to fish kills in many areas, and high NO_3^- in drinking water can result in methemoglobinemia in infants. Methemoglobinemia is in fact rare, even though the maximum

Table 6.1 Environmentally important N forms

Phase	Oxidation State	Species	Environmental Significance	Effect
Solid	-3	$R-NH_2$	Source of bound N in sediments	Eutrophication
Aqueous	+5	NO_3^-	Food and water contaminant	Methemoglobinemia Acid rain
Aqueous	-3	NH_3	Surface water contaminant	Fish kills
Aqueous	Mixed	$R-N-N=O$	Food and water contaminant	Nitrosamine; carcinogen
Aqueous	+3	NO_2^-	Food and water contaminant	Methemoglocinemia Reacts with amines to form nitrosamine
Gaseous	-3	NH_3	Volatilization to atmosphere	Acid rain
Gaseous	Mixed	NO_x	Atmospheric contaminant	Greenhouse gas Ozone depletion Acid rain
Gaseous	-3	$(CH_3)_n-NH_2$	Organic matter decomposition byproduct	Odor

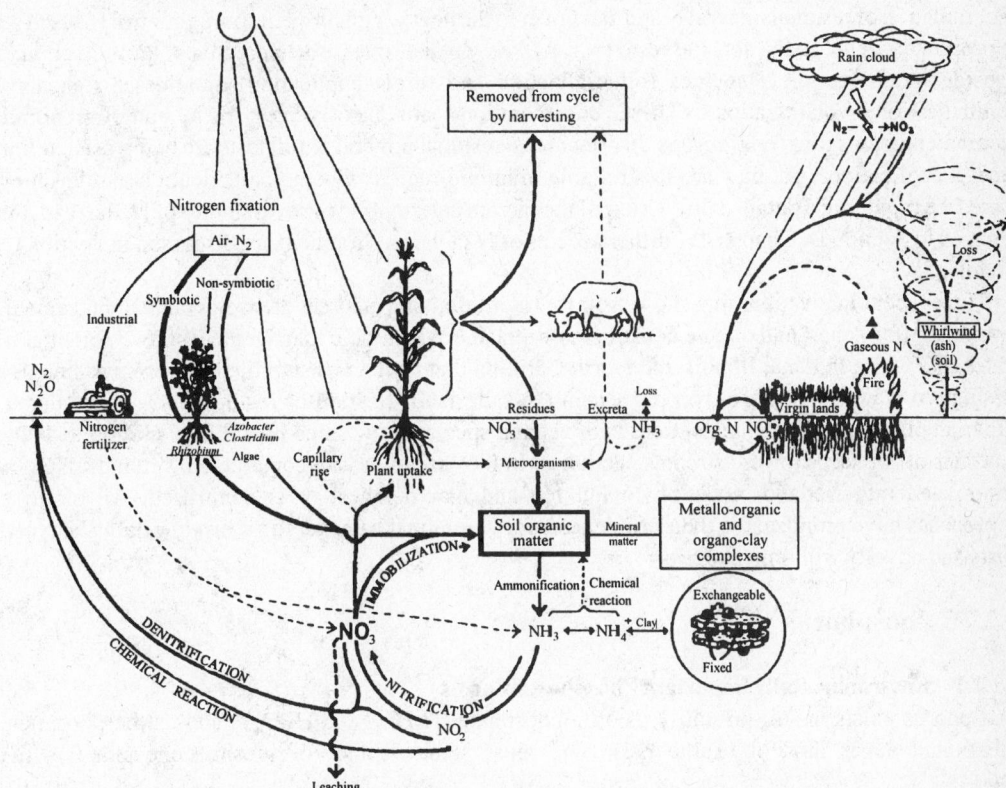

Fig. 6.1 The N cycle in soil [Reprinted with permission of John Wiley & Sons, Inc. from Stevenson, 1986. Cycles of soil: Carbon, nitrogen, phosphorus, sulfur, micronutrients. Copyright John Wiley & Sons, Inc.]

concentration level (MCL) for NO_3–N of 10 mg L^{-1} is routinely exceeded in many parts of the United States, and much of the excess is from N fertilizer plants, feedlots and loafing pens, and distribution facilities not from land applied fertilizer. Nitrous oxide (N_2O) is a source of acid deposition, an ozone depleter, and a greenhouse gas. Nitrous oxide is produced in soil as a product of microbial nitrification and denitrification, but this contribution may be small compared to the burning of fossil fuels; N fertilizer-derived N_2O emissions are estimated to account for 2.5% of global N_2O emissions (Pierzynski et al., 1994).

6.2.1.5 Management of Nitrogen for Environmental Quality Improvement

Managing the N cycle is difficult, as any farmer knows. Because N transformations are diverse, microbially driven, and greatly affected by seasonal climatic differences, it is difficult to manage N in soil in the short term (e.g., a growing season). Because N losses from soil are primarily in gaseous and aqueous forms, restrictions on N losses must be managed by restricting N accumulations and by manipulating N transformations, to the extent possible. Because N in not a conservative nutrient like P (Section 6.3), there are no concerns about long term buildup of N in the soil. Long-term N accumulations can only be achieved with buildup of soil organic matter, a very slow process. Accumulations of mineral soil N in excess of crop removal will carry over to the subsequent year, but not thereafter. This carryover will result in increased leaching losses or losses in tile drainage in humid climates, but may remain in the root zone in drier environments. Source control can be used to limit accumulation of residual mineral N and carryover to the next year. Nitrogen source control is based on appropriate yield goals for harvested crops, or uptake rates for perennials and trees, and considerations for all N sources (mineralization and direct application) and losses (leaching, denitrification, volatilization). These considerations are incorporated into state agronomic recommendations for specific crops. In drier areas, residual mineral N soil tests are being used to fine tune N applications, but they are less reliable in humid regions where denitrification and leaching losses are rapid. For a detailed discussion of the factors determining the availability of N, the reader is referred to Section D, Chapter 1. Further aspects of N in the environment are discussed in Section C, Chapter 5.

Three recent innovations that show promise for minimizing N losses are water table management, riparian buffer zones, and on-site constructed wetlands. Water table management shows potential in states like Ohio, Indiana, Illinois and North Carolina that have extensive tile drainage systems. By raising the water table after crop harvest in the fall, denitrification of residual NO_3 is promoted. Riparian plantings of perennials, trees and wetland species show some potential to assimilate NO_3^- that runs off or seeps toward streams and lakes. In the case of on-site constructed wetlands, NO_3^- is assimilated into wetland biota (phytoplankton and macrophytes) or is denitrified. All of these approaches have promise, but their efficiencies for N removal have yet to be firmly established, and costs and novelty will limit adoption.

6.2.2   Phosphorus

6.2.2.1   Environmentally Important Phosphorus Forms
Phosphorus exists in soil primarily as orthophosphate ($H_2PO_4^-$ and HPO_4^{2-}) and as the phosphate substituted sugar inositol (Table 6.2). The most stable isomer of inositol phosphate is the

Table 6.2   Environmentally important P forms

Phase	Form	Species	Environmental Significance
Solid	Apatite	Fluorapatite	Very stable forms in soil; may be inherited from the
		Chlorapatite	parent material or formed *in situ*
Solid	Other Ca solids	$CaHPO_4.2H_2O$	Relatively high solubility; unstable in soil; secondary
		$Ca(H_2PO_4)_2$	products of high P loading in slightly acidic to neutral
		$CaH(PO_4)_3.2.5H_2O$	soils.
Solid	Al, Fe solids	$FePO_4.2H_2O$	Strengite; amorphous; not identified in soil
		$AlPO_4.2H_2O$	Variscite; amorphous forms are common
		$Fe_3(PO4)_2.8H_2O$	Vivianite; formed under reducing conditions; more
			soluble than variscite
			Al, Fe phosphates are dominant at pH < 6
Solid	Organic P	Inositol phosphates	The hexaphosphate, phytic acid, is most stable;
			represents the majority of organic P in soils
		Phospholipids	Less than 5 % of organic P in soils
Solid	Sorbed	Surface complexes with Al,	Can be displaced by strong ligands like F^- or organic
		Fe oxides, other minerals,	acids; a significant part of the phytoavailable pool
		and organic matter	(higher plants and algae).
Solution	Orthophosphate	HPO_4^{2-} & $H_2PO_4^-$	The most phytoavailable P form; in dynamic equilibrium
			with sorbed P
Solution	Organic P	Mono- and di-esters	May represent 25% or more of dissolved P
Solution	Pyrophosphate	$P_2O_7^{4-}$	Pyrophosphate is a microbial by-product; degrades to
			orthophosphate.
Solution	Colloidal	Sorbed to Fe oxides, clays	Nonfilterable; may over estimate orthophosphate with
			acid molybdate method.

hexaphosphate, phytic acid. Other minor P forms are pyrophosphate ($P_2O_7^{4-}$), a microbial byproduct or organic P mineralization, and phospholipids like lecithin. Orthophosphate forms highly coordinated bonds with the macro metals in soil (Al, Fe and Ca); these associations occur through inner sphere surface complex formation on Al and Fe oxides and on calcite, and by precipitation of Al, Fe and Ca compounds depending on soil pH (Fig. 6.2).

6.2.2.2   The Environmental Phosphorus Cycle
The environmental P cycle in soils (Fig. 6.2) must include as an important component, the transport of dissolved and particulate P to surface waters because of the role of P in eutrophication.

6.2.2.3   Major Sources of Phosphorus
The major sources of P in soil are native P, P added in fertilizer, and that added in wastes like manures and biosolids. Native soils contain about 300 to 1000 mg kg⁻¹ total P. Soil P content increases with finer texture and SOM content, and decreases with age. Chemical fertilizer applications over the last 50 yr or so have probably increased the total P content of soils by 100–200 mg kg⁻¹ reflecting the excess of P applications over crop removal. Because municipal biosolids and manures are normally applied at N utilization rates on grain or pasture crops, P application will exceed crop removal and contribute to increased soil P concentrations. The amount of land applied biosolids in the United States is low, probably less than 3×10^6 dry Mg yr⁻¹, containing 60,000 dry Mg of P. In contrast, USDA (1997) estimates that about 2.3×10^6 Mg P is applied to land in the United States in livestock manure. Locally important sources of P are phosphogypsum mine tailings in Florida and North Carolina.

6.2.2.4   Major Environmental Impacts of Soil Phosphorus
The primary environmental impact of soil P is eutrophication of lakes and streams from runoff and erosion. Phytoplankton in surface waters respond to increased phytoavailable P levels because P is often the growth limiting nutrient. Phytoplankton respond rapidly to dissolved orthophosphate, but

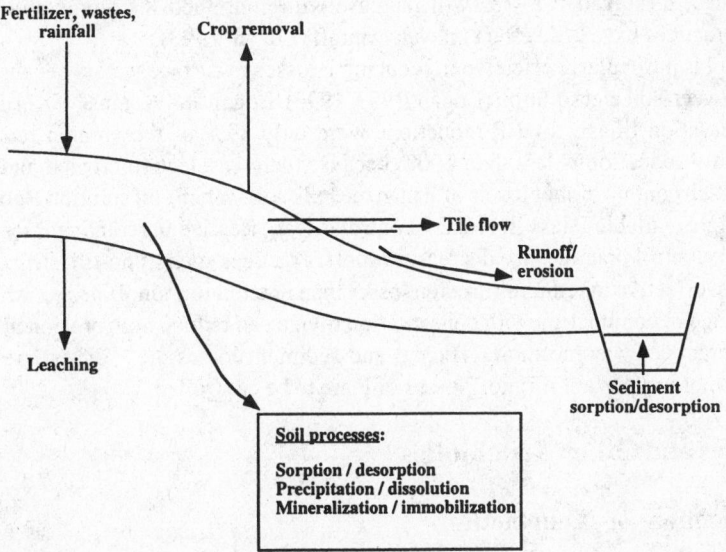

Fig. 6.2 The environmental P cycle

are also able to utilize sediment bound P by desorption and dissolution processes. An average algal bioavailability of sediment bound P of 20–40% has been found by various researchers, using direct bioassay or extraction techniques (Sonsogni et al., 1982).

6.2.2.5 Management of Phosphorus for Environmental Quality Improvement

Because soil P is conservative (not readily lost from soil as volatile emissions or transformed) and because P is strongly retained by the solid phase, the primary management strategies for controlling P discharges to surface waters are source control and erosion prevention (Logan, 1990). Phosphorus is applied to soil as fertilizer for crop production, and the plant available P fraction is readily estimated by soil test. Source control involves managing P applications so that soil test levels are in the range recommended locally for specific crops but are not in excess of those ranges. The National Research Council (1993) reported that in 1990, 42, 18 and 4 states had 30, 50 and > 70% of soil samples testing high to very high in available P, respectively. Some have argued that it is good management to be in the high category to ensure that P will never limit crop yields, but it should be noted that most soil fertility recommendations that define levels of available nutrients as adequate are conservative; in other words, responses to added P are often not obtained even when soil test levels are less than adequate. Therefore, there can be no justification for maintaining soil test levels in the high to very high category. It should also be noted that very high soil test P levels are very slow to decline with continuous cropping with no P added (McCollum, 1991; National Research Council, 1993). Further details on the bioavailability of P are presented in Section D, Chapter 2.

Source control is virtually impossible when soils are used to recycle organic wastes like manures and biosolids, and application rates are normally based on N utilization. At these application rates, P additions exceed crop removal by severalfold and the excess accumulates in the soil and contributes to increases in total and bioavailable P. If sorption is the primary mechanism by which P is retained by soil, then accumulations of P will progressively exceed soil P sorption capacity. The result is that plant available P will increase with P additions faster (nonlinear) than total P concentrations (linear increase with P additions). Since plant available P is more highly correlated to dissolved P and to algal available P than is total P, excess P applications have an exponential effect on eutrophication. Even if erosion control is practiced, dissolved P losses will increase with continued P accumulation since most erosion control practices have little effect on water runoff (Logan, 1993).

Erosion control is particularly effective in reducing P losses to surface waters because of the high P partitioning between soil and solution (Logan, 1993, 1995). Logan and Adams (1981) demonstrated that under conservation tillage total P reductions were only 89% of the erosion reductions. The efficiency of total P reduction is less than 100% because there is selective erosion and transport of clay and SOM which contain higher levels of P than the bulk soil. When soil solution P concentrations are high due to high available P levels, erosion control is less effective in reducing P losses in runoff unless the erosion control practice also decreases runoff. Practices such as no-till, strip cropping and terracing are more effective in reducing erosion losses than in reducing runoff *per se*, while measures such as tile drainage in conjunction with conservation tillage can reduce both erosion and runoff. As erosion control practices become more efficient and sediment losses are < 1 Mg ha^{-1} yr^{-1}, source control becomes more important if runoff losses of P are to be controlled.

6.3 Pesticides and Other Xenobiotics

6.3.1 Classification of Xenobiotics

Examples of xenobiotic compounds of environmental importance in soil are presented in Table 6.3. The pesticides are probably the most studied of the xenobiotics in soil, particularly after revisions of

Table 6.3 Classification of xenobiotics in soils

Class	Types	Examples
Pesticides	Insecticides	DDT, dieldrin, carbofuran
	Herbicides	2,4-D, atrazine, alachlor
	Nematicides	Ethylene bromide, aldicarb, DCP
	Fungicides	Chlorothalonil
Polyaromatic hydrocarbons (PAHs)		Benzo(a)pyrene
		Xylene, phenanthrene
Polychlorinated aromatics	Polychlorinated biphenyls (PCBs)	Various Arachlors
	Phenols	Pentachlorophenol (PCP)
	Polychlorinated dioxins and furans	2,3,7,8-TCDD
Solvents	BTEX	Benzene, toluene,
Hydrocarbons	Alkanes	Gasoline, kerosene, diesel oil, fuel oil
Others	Surfactants	Linear alkyl sulfonate (LAS)
	Silicones	Polydimethyl silicone (PDMS)
	Synthetic fats	Olestra
	Plastics	Bis-2-ethyl phthalate

the Federal Insecticide, Fungicide and Rodenticide Act required pesticide manufacturers to present data on soil pesticide behavior when applying for product registration (Severn and Ballard, 1990). Although many of the most studied compounds are no longer manufactured, principles learned from these materials on the behavior of xenobiotics in soil have been applied to other classes of xenobiotics.

6.3.2 Major Sources of Xenobiotics

The major sources of xenobiotics in soil are applied agricultural pesticides and those spilled at production and distribution facilities, petroleum from leaking tanks, spills in manufacturing, distribution and transportation, and land disposal of xenobiotic containing wastes. Atmospheric sources are important in relatively pristine areas, and account for widespread dispersal of xenobiotics like PCBs, but unit area loadings from atmospheric deposition are low.

6.3.3 Behavior of Xenobiotics

Xenobiotics can occur in soils in solid, dissolved and gaseous phases and all undergo transformations (microbial and abiotic) (Fig. 6.3). The fate of xenobiotics in soil can be predicted with some accuracy by knowing selected properties of the compound and the soil. Important xenobiotic properties are water vapor partitioning measured by Henry's Law constant (K_H), water solubility, soil organic carbon-water partition coefficient (K_{OC}), biodegradation half-life ($t_{1/2}$) (normally assumes that biodegradation follows first order kinetics) and field dissipation half-life (an integration of all processes leading to loss of the compound, and normally determined in field experiments). Important soil properties are SOM content, pH and soil physical properties.

6.3.4 Major Environmental Impacts of Xenobiotics

Xenobiotic levels in soil are generally low (< 100 mg kg^{-1}), particularly those of the pesticides, unless more concentrated by spills, at production facilities, near leaking petroleum tanks, or in areas of concentrated waste disposal. Xenobiotics can be toxic to soil biota, inhibit seed germination, and suppress plant growth. They may also contaminate crops and livestock, and children through direct

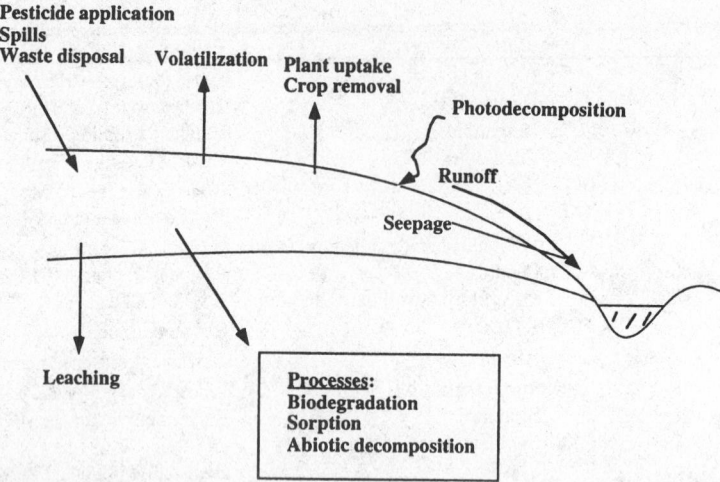

Pesticide application
Spills
Waste disposal   Volatilization   Plant uptake
                                        Crop removal
                                                        Photodecomposition

                                                        Runoff

                                        Seepage

Leaching

Processes:
Biodegradation
Sorption
Abiotic decomposition

Fig. 6.3   Xenobiotic soil cycle

ingestion of contaminated soil. Crop damage is unusual except for some persistent herbicides; pesticide use labels identify carryover phytotoxicity potential. Crop contamination (by uptake) is uncommon, particularly for those compounds with log $K_{OC} > 3$. Livestock contamination can occur at contaminated sites when animals graze pastures by ingestion of soil either directly or indirectly on the vegetation. Likewise, young children will eat soil and dirt that may be contaminated. Increasingly, impacts of xenobiotics and other toxic substances (Section 6.5) on soil microorganism activity are being studied, and a range of effects have been documented. Very mobile long-lived xenobiotics (high water solubility, low K_{OC}, long $t_{1/2}$) also have the potential to migrate from soils to surface and ground water.

6.3.5   Management of Xenobiotics for Environmental Quality Improvement

The primary approach to managing xenobiotics in soil is to utilize the ability of soil microorganisms to degrade these compounds. Degradation half-life is used in screening for persistent compounds in pesticide registration and regulation under FIFRA (Severn and Ballard, 1990), and enhanced biodegradation is used to remove petroleum and other xenobiotic residues from soil. Enhancement may involve the addition of nutrients to stimulate microbial activity, modification of soil conditions (adjusting pH, irrigation), and addition of selected or modified organisms with enhanced degradation capacity.

6.4   Trace Elements

6.4.1   Identification of the Trace Elements in Soils

The trace elements in soils are defined as those with average total concentrations < 100 mg kg^{-1} (Table 6.4). They include plant and animal nutrients, trace metals and trace oxyanions. It should be noted that the popular term heavy metal is incorrect for many of the trace elements because they may not be heavy (generally defined as having densities > 5 g cm^{-3}) and they may not be metallic (e.g., F).

Table 6.4 Classification of trace elements in soils

Symbol	Element	Dominant Soil Species	Average Elemental Soil Content ($mg\ kg^{-1}$)	Effects on Plant and Animal Health
Ag	Silver	Ag^+	0.05	Animal toxicity
As	Arsenic	AsO_4^{3-}	6	Toxic to plants and animals
B	Boron	$B(OH)_3$	20	Plant nutrient; phytotoxic
Be	Beryllium	Be^{2+}	0.3	Toxic to plants and animals
Bi	Bismuth	Bi^{3+}	0.2	Toxic to plants and animals
Cd	Cadmium	Cd^{2+}	0.35	Toxic to animals
Cl	Chlorine	Cl^-	100	Plant nutrient; phytotoxic
Co	Cobalt	Co^{2+}	8	Plant and animal nutrient
Cr	Chromium	Cr^{3+}, CrO_4^{2-}	70	Animal nutrient; Cr^{VI} is carcinogenic
Cs	Cesium	Cs^+	4	No reported toxicity
Cu	Copper	Cu^{2+}	30	Animal and plant nutrient; toxic to plants and animals
Hg	Mercury	$Hg^{2+}, (CH_3)_2Hg$	0.06	Toxic to animals
Mo	Molybdenum	MoO_4^{2-}	1.2	Animal and plant nutrient; toxic to plants and animals
Ni	Nickel	Ni^{2+}	50	Animal and plant nutrient; toxic to plants and animals
Pb	Lead	Pb^{2+}	19	Toxic to plants and animals
Rb	Rubidium	Rb^+	67	Phytotoxic
Sb	Antimony	Sb^{3+}	1	Toxic to animals
Se	Selenium	SeO_4^{2-}	0.4	Animal nutrient; toxic to plants and animals
Sn	Tin	Sn^{4+}	4	Animal nutrient; toxic to animals
Tl	Thallium	Tl^{3+}	0.2	Toxic to animals
U	Uranium	U^{4+}, UO_2^{2+}	2.7	Toxic to animals
V	Vanadium	$(H_nV_{10}O_{28})^{(6-n)-}$	90	Animal nutrient; toxic to plants and animals
W	Tungsten	WO_4^{2-}	1.5	Unknown
Zn	Zinc	Zn^{2+}	70	Animal and plant nutrient; toxic to plants and animals

6.4.2 Major Sources of Trace Elements

The trace elements are naturally occurring, and their abundance in soil reflects the trace element composition of the parent rocks. In general, the enrichment ratio (ER) of trace elements in soil relative to parent rocks is near 1 (Sposito, 1989), suggesting that these elements are not enriched in soil by weathering and SOM accumulation, as is the case for C, N, and S, nor are they preferentially lost from soil by weathering and leaching as are elements such as Ca, Sr and Rb. The trace element content of surface soil can vary naturally by several orders of magnitude, and levels in some soils exceed regulatory limits for contaminated soils. As important as the total contents of the trace elements in soil are, their chemical forms determine their fate and bioavailability. The trace nutrients may be extensively deficient in soils of the world, and trace element fertilizers are often used in modern high yield production agriculture, but the major environmental concerns with these elements have been their potential toxicity to soil biota and the food chain. Some elements like Se are seriously deficient in forages and have to be supplemented in animal feed to avoid white muscle disease, yet Se may also be toxic to livestock in areas grazing high Se (seleniferous) soils. Anthropogenic trace element sources have caused most of the extensive high-level soil contaminations. These sources include mining, ore smelting, waste disposal, and indiscriminate use of trace element pesticides and fertilizers.

6.4.3 Behavior of Trace Elements in Soil

The behavior of trace elements in soil is different for the metals and oxyanions. The metals have higher solubility and lower retention by mineral and organic binding sites as soil pH decreases, while the opposite is true for the oxyanions which are anions at normal soil pH values. The exception to this is B which exists in soil solution as the neutral species $B(OH)_3^o$ at pH < 9.7. The trace metals can also form strong solution complexes with inorganic ligands like Cl^- and SO_4^{2-} and with dissolved humic and fulvic substances which increases their apparent solubility.

Trace elements form coprecipitates with macroelement solids, which may control their solubility in many cases. Until the advent of molecular-level solid-phase analytical techniques such as X-ray absorption fine structure (XAFS) spectroscopy, it was virtually impossible to determine the nature of trace element coprecipitation in mineral solids. XAFS permits nearest neighbor coordination chemistry to be determined, thereby allowing the extent and nature of trace element substitution in mineral solids to be determined.

Trace element solubility in soil is generally very low with soil solution activities of 10^{-6} M. The oxyanions tend to have higher solubilities than the metals and B, in particular, is quite soluble and readily leached from soil. Although trace element solubility is low, plant uptake of trace elements is high enough under contaminated soil conditions and favorable pH to cause excessive uptake and phytotoxicity or food-chain contamination. Further details of trace element partitioning reactions between solid and liquid phases are discussed in Section B, Chapter 8 and Section D, Chapter 2.

6.4.4   Major Environmental Impacts of Trace Elements

Trace elements can be deficient in plants, deficient in the animals grazing forage, phytotoxic, toxic in forage grazed by livestock, toxic to grazing livestock from direct soil ingestion, a contaminant of the food chain from excessive plant uptake, accumulated by small children from direct soil ingestion, and toxic to soil biota, particularly bacteria. Mobile trace elements such as As, Se and B, can contaminate surface and ground waters.

6.4.5   Management of Trace Elements for Environmental Quality Improvement

This discussion is limited to those cases where trace element concentrations in soil have the potential to cause toxicity or food-chain contamination. It is assumed here that good agronomic practice will assure that adequate levels of plant and animal trace nutrients are provided through fertilization or mineral supplements. There are two general approaches to dealing with excessive levels of contaminants in soil: (1) lowering the concentrations to acceptable levels, and (2) immobilization of the contaminant *in situ* so that the mobility and bioavailability of the contaminant are reduced. A useful way to decide which approach is likely to be the most fruitful is to consider whether the contaminant is predominantly mobile or immobile. More mobile contaminants should be remediated using approach 1 and more immobile contaminants by approach 2 (Section 6.10). Most of the trace elements, and particularly the trace metals, are predominantly immobile, while most of the oxyanions are relatively mobile, as is the metal Cd.

Very high levels of trace elements in soil in localized areas may be addressed by removal and reburial in a landfill, but this is often an impractical option. The trace metals can be immobilized by raising pH above 6–7 to enhance precipitation, sorption and SOM complexation. Plant uptake of Cd can be reduced by addition of Zn because Zn effectively competes with Cd when the Zn:Cd ratio is > 100. Oxyanions, on the other hand are least mobile at lower pH where there are reactive oxides of Fe and Al to form inner or outer sphere complexes with the trace element ligands. Trace metals, primarily Pb, have been effectively immobilized by the addition of phosphate as soluble salts or as solid phase apatite. In the case of apatite addition, a Pb analog of apatite, pyromorphite is formed.

6.5   Other Chemicals

6.5.1   Salts, Acids, Alkalis, Radionuclides

Miscellaneous chemical compounds may be present at levels that impact soil processes, plant growth and food-chain contamination. These include (1) soluble salts from saline water sources (irrigation, seeps, flooding), from waste disposal (e.g., food processing brines), from mining (brine associated with oil and gas drilling), and from road salt for deicing; (2) acids (both organic and mineral) from spills, chemical production facilities, and from waste disposal; (3) alkalis from spills, chemical production facilities, and waste disposal (e.g., alkaline fly ash, cement kiln dust, lime kiln dust, improper disposal of alkaline stabilized biosolids); and (4) radionuclides from mine tailings, nuclear production facilities, atmospheric nuclear bomb testing, and accidental atmospheric emissions from nuclear power plants (e.g., Chernobyl).

6.6   Chemical Speciation, Bioavailability, and Risk Assessment

6.6.1   Use of Chemical Speciation in Environmental Soil Assessment

Chemicals in soil can exist both in inorganic and organic forms in solid, liquid and gaseous phases. The solid phase includes primary and secondary minerals, undecomposed and stable SOM and biomass, and sorbed species. The liquid phase (predominantly water but also organic fluids like solvents and petroleum) includes dissolved ions, neutral inorganic species, dissolved organic matter (humic and fulvic substances and discrete organic compounds like organic acids), dissolved xenobiotics (e.g., pesticides), small polymers (e.g., hydroxy Al, detergents), and microparticulates (e.g., colloidal Fe oxide). The soil atmosphere contains normal soil gases (N_2, O_2, CO_2, NO_x), NH_3, volatile organics (pesticides, volatile amines, gasoline), and under reducing conditions, H_2S, H_2 and CH_4. Components of the solid, liquid and gaseous phases interact dynamically by chemical, physical and biological processes.

Current models cannot fully predict chemical speciation because of the enormous complexity of these dynamic processes, but a combination of chemical speciation modeling and advanced analytical techniques has given soil chemists the tools to predict soil chemical speciation more accurately than ever before.

6.6.1.1   Aqueous Phase Speciation

A general axiom in soil chemistry, particularly with respect to the trace elements and other inorganic elements like Al, is that the free (uncomplexed) form of the element is most bioavailable. While this primarily refers to biological uptake from soil solution, knowledge of chemical activity is also important in predicting specific chemical reactions such as precipitation and complexation. Estimation of chemical activity can be made for some species with potentiometric methods (ion selective electrodes), anodic stripping voltammetry, fluorescence spectrometry, colorimetry, electrophoresis, or membrane dialysis. Another approach is to use total solution composition data with thermodynamic equilibrium association models like GEOCHEM (Sposito and Mattigod, 1980) or MINTEQ (Allison et al., 1990) to calculate the distribution of chemical species. Combined with empirical equations to calculate activity coefficients (e.g., Davies equation), free ion activities can be estimated.

An important consideration in the speciation of soil solutions is to have a complete chemical analysis. Not only should all major metals and ligands be analyzed but also important minor constituents. Concentration of dissolved organic C (DOC) is rarely measured, but DOC is often the major ligand complexing polyvalent metals in soil solutions. Preanalytical simulations of soil solution speciation using GEOCHEM or MINTEQ can show if a minor metal or ligand is likely to affect the speciation, but highly reactive metals and ligands (e.g., Cu^{2+} or F^-) should always be considered. Preanalytical chemical speciation can be effectively used to determine which analytes should be measured.

6.6.1.2   Solid-Phase Speciation

Solid-phase soil chemical constituents exist as sorbed species, as discrete mineral phases, as humic substances, as coprecipitated mineral phases, or as metal-organic complexes. These constituents do not exist as separate entities in soil, but rather as complex mixtures. Attempts to treat soil as simple mixtures of mineral and organic components have been unsuccessful because they fail to account for these interactions. Likewise, it is virtually impossible to physically or chemically separate soil solid-phase components for further characterization. Cleaning up the soil by selective chemical treatment (e.g., removal of organic matter with H_2O_2) results in artifacts that make extrapolation of analytical results to real systems highly uncertain. What can be done is to use physical concentration techniques to produce fractions that are enriched with the chemical of interest. These include particle size fractionation, density gradient separation, and magnetic separation. These methods are less invasive and destructive than chemical treatment and are often necessary for the study of trace constituents in soil. For example, it is very difficult to identify crystalline solids in soil with X-ray diffraction (XRD) at concentrations < 1% by weight, yet toxic trace element concentrations in soil can be < 0.1% by weight.

Solid-phase speciation can be divided into three general components with very different analytical requirements and challenges: (1) sorbed species, (2) inorganic solid phases, and (3) organic solid phases. Molecular information on sorption of ions, complexes and organics onto soil solid phases is limited. Tools have included infrared (IR), electron spin resonance (ESR) and fluorescence spectroscopy. Modern fourier-transform IR (FTIR) and lifetime fluorescence spectrophotometry methods have increased the sensitivity and precision of surface species characterization, but defining the reacting surfaces is still difficult for real soil systems. Electron microscopy and atomic force microscopy (AFM) provide physical information on the nature of surface sorbed species from which molecular knowledge can be deduced. As molecular-level information becomes available for sorbed species, they can be modeled with chemical affinity computational methods like the surface complexation models (Sposito, 1984) which treat the solid-phase sorbent as a chemically reactive phase that interacts with specific solutes. This approach provides molecular specificity that is not available with physically based models like the diffuse double layer model.

Identification of inorganic solid phases has traditionally required XRD for crystalline solids and thermal methods such as thermogravimetric analysis (TGA) and differential thermal analysis (DTA) for amorphous materials. Improvement in XRD detection and peak analysis in the last decade or so has permitted the identification of poorly crystalline solids like ferrihydrite and schwertmannite, and advanced analytical electron microscopes permit XRD analysis of single grains in complex mixtures. Scanning transmission electron microscopy and electron microprobe analysis have been used to identify weathering products of trace element inorganic solids in soils and mine wastes. Weathering products are often observed as coatings or rinds on the weathering particles.

An exciting new analytical technique for characterization of inorganic solids is X-ray absorption spectroscopy (XAS) (Brown, 1990). In this method, coherent X-ray beams from synchrotron sources are used to probe soil samples. The resulting spectrum can be refined to provide information on next neighbor atomic coordination in solids. This is one of the only tools that can provide molecular information on trace element coprecipitation in solids. The power of the technique is limited in that it requires relatively homogeneous samples. Soils by their very nature are intimate physicochemical associations of solids, and soils have to be fractionated or otherwise cleaned up to produce reasonably homogeneous solid phases prior to analysis. The problems with chemical cleanup procedures have been discussed above, and only physical fractionation and enrichment techniques should be employed.

Solid-phase characterization of SOM remains an intractable problem. Some advances have been made with solid-state NMR, but molecular information on inorganic and organic contaminant interactions with solid-phase SOM still eludes us. Unresolved questions include understanding the nature of bound pesticide residues (sorbed, polymerized, physically protected), and the long-term stability of organically bound trace elements.

6.7   Concept and Use of Bioavailability in Environmental Soil Assessment

What has long been known in soil fertility is becoming appreciated in environmental soil chemistry, namely, that the total content of an element or compound in soil is not a reliable indicator of its bioavailability. Bioavailability, like mobility, is dependent on the chemical speciation of the element or compound. Bioavailability can be inferred from chemical speciation analysis, but the most accurate assessment of bioavailability is direct or indirect bioassay. Bioassays for soil contaminants include plant growth and uptake (pot and field), assays of soil biota function and contaminant concentration, and feeding trials in which the soil is fed to laboratory animals or target species. A novel substitute to direct animal feeding is *in vitro* extractability of soil contaminants by simulated stomach fluids (Davis et al., 1992). An extensive treatment of bioavailability assessment is presented in Section D, Chapter 4. Measurement of changes in contaminant bioavailability are a necessary test of efficacy for reclamation methods that immobilize rather than remove soil contaminants.

6.8   Soil Exposure Assessment in Environmental Risk Assessment

Soil is a major pathway for human, animal and ecological exposure to environmental contaminants. Contaminants may move from soil to target organisms through a number of pathways. For human exposure, for example, important pathways include

Soil → Crop → Human
Soil → Crop → Livestock → Human
Soil → Livestock → Human
Soil → Surface water → Human
Soil → Surface water → Fish → Human
Soil → Ground water → Human
Soil → Air → Human
Soil → Human

For each soil contaminant, assumptions have to be made as to the extent of exposure for each step in each pathway (e.g., amount of different crops eaten, amount of water consumed, amount of soil ingested), and transfer coefficients have to be developed for each step. These transfer coefficients (e.g., crop tissue concentration per unit soil contaminant concentration) are measures of mobility and bioavailability of the contaminant in each matrix of the pathway (e.g., soil, air, water, biological tissue) and must be determined from soil and contaminant specific measurements.

Table 6.5 Approaches for remediation of chemically degraded soil

Contaminant Type	Mobility	Volatility	Degradation/ Alteration Potential	Contaminant Level	Remediation Approach
Nitrate	High	Low	High	High	Pump and treat, phytoremediation
				Moderate	Phytoremediation
Phosphate	Low	Low	Low	High	Removal, immobilization
				Moderate	Phytoremediation or immobilization
Xenobiotics	Low	Low	High	High	Removal, microbial stimulation
	Low	Low	Low	Moderate	Microbial stimulation, washing
	Low	High	Low	High	Removal
	High	Low	Low	Moderate	Microbial stimulation, dilution, washing
				High	Removal, venting, heating
				Moderate	Venting, mixing
				High	Washing, pump and treat
				Moderate	Washing
Trace elements	Low	Low	Low	High	Removal, immobilization
	Moderate	Low	Low	Moderate	Immobilization
				High	Washing, phytoremediation, immobilization
				Moderate	Phytoremediation, immobilization

6.9   Remediation of Chemically Degraded Soil

Mandated soil remediation must attain action levels of the soil contaminant that were determined by risk assessment, by policy (no net degradation), or best available technology. The action level may be given in terms of a total or mobile soil concentration (e.g., TCLP extractable), or by a biologically available fraction as determined by bioassay. Depending on the nature and chemical form of the contaminant, and the action level, a remediation strategy can be proposed and then tested at the bench or field scale. Strategies for remediation of chemically degraded soil are summarized in Table 6.5. Another important consideration in selection of a remediation approach is the extent of the contamination which will greatly affect feasibility and cost. As this increases, more *in situ* methods will have to be used.

6.10   References

Allison, J.D., D.S. Brown, and K.J. Novo-Gradac. 1990. MINTEQA2/PRODEFA2: A geochemical assessment model for environmental systems. Version 3.00. EPA-600/3-91-921. US Environmental Protection Agency, Athens, GA.

Brown, G.E. 1990. Spectroscopic studies of chemisorption reaction mechanisms at oxide-water interfaces. p. 309–364. *In* M.F. Hochella and A.F. White (ed.) Mineral-water interface chemistry. Mineralogical Society of America, Washington, DC.

Davis, A., M.V. Ruby, and P.D. Bergstrom. 1992. Bioavailability of arsenic and lead in soils from the Butte, Montana, mining district. Environ. Sci. Tech. 26:461–468.

Logan, T. J. 1990. Agricultural best management practices: Implications for groundwater protection. J. Soil Water Conserv. 45:201–206.

Logan, T.J. 1993. Agricultural best management practices for water pollution control: current issues. Agric. Ecosys. Environ. J. 46:223–231.

Logan, T.J. 1995. Water quality. p. 311–336. *In* A. Ward and W. Elliott (ed). Principles of hydrology. Lewis Publishers, Boca Raton, FL.

Logan, T.J., and J.R. Adams. 1981. The effects of conservation tillage on phosphate transport from agricultural land. Lake Erie Management Study Tech. Rep. Ser., US Corps of Engineers, Buffalo, NY.

McCollum, R.E. 1991. Buildup and decline in soil phosphorus: 30-year trends on a Typic Umbraquult. Agron. J. 83:77–85.

National Research Council. 1993. Soil and Water Quality. An agenda for agriculture. National Academy Press, Washington, DC.

Pierzynski, G.M., J.T. Sims, and G.F. Vance. 1994. Soils and environmental quality. Lewis Publishers, Boca Raton, FL.

Severn, D.J., and G. Ballard. 1990. Risk/benefit and regulations. p. 467–491. *In* H.H. Cheng (ed.) Pesticides in the soil environment: Processes, impacts, and modeling. Soil Science Society of America, Madison, WI.

Sonsogni, W., S. Chapra, D.E. Armstrong, and T.J. Logan. 1982. Bioavailability of phosphorus inputs to lakes. J. Environ. Qual. 11:555–563.

Sposito, G. 1984. The surface chemistry of soils. Oxford University Press. New York, NY.

Sposito, G. 1989. The chemistry of soils. Oxford University Press, New York, NY.

Sposito, G., and S.V. Mattigod. 1980. GEOCHEM: A computer program for the calculation of chemical equilibria in soil solutions and other natural water systems. Kearney Foundation of Soil Science, University of California, Riverside, CA.

Stevenson, F.J. 1986. Cycles of soil: Carbon, nitrogen, phosphorus, sulfur, micronutrients. John Wiley and Sons, New York, NY.

USDA. 1997. Agricultural utilization of municipal, animal and industrial wastes. USDA-ARS, Beltsville, MD.

7

Water Erosion

Dino Torri and Lorenzo Borselli
Institute for Soil Genesis and Ecology, Florence, Italy

7.1 Introduction

Soil erosion refers to a series of processes leading to soil depletion *in situ* and the export of sediment toward downstream areas. It will be described by the myriad of processes through which it acts and evolves. Whenever possible mathematical relationships describing the processes will be given in order to allow readers to make their own evaluations. Factors affecting erosion are introduced and described via the same relationships. The chapter will try to identify the main processes and their interrelationships with soil and soil properties and does not intend to be a soil conservation manual. Some extremely important factors in the definition of soil behavior but somewhat external to soil such as vegetation, land use and wildfire will be briefly discussed under appropriate headings which summarize their effects. Only a few basic references will be made to soil erosion models because they are numerous and there is no clear evidence of any of them performing better. Moreover, when making a model, it is the comprehension of processes rather than the sum of their parts that truly matters from the standpoint of soil science. A paragraph will discuss synthetic soil indices which can be used for assessing soil erosion susceptibility. Finally, some effects of soil erosion on certain soil characteristics will be analyzed.

Soil erosion by water takes place through three main processes: (1) detachment of soil particles, aggregates, clods and large soil volumes from the soil mass; (2) movement of detached material (e.g., by gravity or by overland flow); and (3) deposition. Processes are also subdivided based on other criteria into (1) diffuse surface erosion (e.g., splash and interrill erosion), (2) linear erosion (rill and gully), (3) subsurface erosion (piping), and (4) shallow mass movements. In the following discussion, primary soil particles and aggregates (usually < 5 mm) will be often grouped together under the term grain.

7.2 Wetting of the Soil

The first reactions of soil to rain drops take place as soon as water begins to wet the soil. Dispersion, which can vary in intensity, is always active. Dispersion takes place at the soil surface mainly because infiltration and runoff water are usually low in ions. Ignoring dispersion has led to incorrect conclusions usually in the direction of lower runoff intensities, lower erodibility, longer time to ponding and time to runoff intervals (Shainberg and Levy, 1996). Consequently, data collected by means of rainfall experiments in the past, when tap water was commonly used, are to be considered with caution but unfortunately, they are often the only results available. The slaking of aggregates also takes place on dry soils during the first 15-30 minutes of rain. Slaking is due to air which is trapped in the aggregates being compressed by the water as it is driven into the soil by matrix forces. The velocity of air diffusion through water (particularly in micropores) is lower than that at which water penetrates into the aggregates. This causes local increased pressure that can break bonds. The process is extremely rapid; a submerged dry clod can disintegrate in several minutes. Obviously, it ends once the soil is sufficiently wet. The presence of slaking explains why erosion is often more intense when the soil is dry before rainfall. The processes involved in slaking have been discussed in detail by Rengasamy and Sumner (1997).

Because some of the forces that keep the soil particles together depend on water content, wetting has an effect. The force (T) per unit surface with which a grain is linked to the surrounding grains, decreases as the soil is wetted and may be dramatic if the soil is dry because of the sudden drop in soil water potential as moisture increases (Section A, Chapter 3). As dispersion, slaking and other processes continue to detach grains, T decreases. In other words, T decreases as detachment is replaced by redetachment (Rose et al., 1990). T may even increase because of armoring, grain packing, etc.

7.3 Drop Impact

Drop impact is responsible for most of the detached material, changes in surface morphology and sealing. Falling drops accelerate until a maximum velocity value (V_F) is reached (Laws, 1941). Final fall velocity (cm s^{-1}) varies with drop diameter (Φ, mm) approximately as follows:

$$V_F = 938\tanh(-0.44\Phi) \tag{7.1}$$

When the drop hits the soil surface, it compresses the soil and removes particles and aggregates entrained by a crown of lateral droplets that are ejected laterally and upward. The angle of ejection changes with the depth of the water film that covers the soil surface. The relationship (Poesen and Torri, 1988) between angle of ejection (Θ, degree) and water depth (h, cm) is as follows:

$$\tan\Theta = 11.4h^{0.84} + 0.6 \tag{7.2}$$

In the absence of any water film, θ seems to vary inversely to soil resistance (T, kPa, measured with a Swedish falling cone) as empirically found by Al Durrah and Bradford (1982):

$$\Theta = 40.6T^{-0.42} \tag{7.3}$$

The velocity of the lateral jets that will eventually evolve into ejected droplets may reach 8 times the velocity of the impinging drop (Engels, 1955). The potential of a drop to detach material can be estimated in various ways. The kinetic energy at the moment of impact on the soil surface is most widely used:

$$D_S = k_s[E - E_0] \tag{7.4}$$

where D_s is soil detachment (kg m^{-2}) and E is the kinetic energy (kJ m^{-2}) at impact, E_0 is defined as the threshold energy to initiate the detachment process (Sharma et al., 1991), and k_s is soil detachability (kg kJ^{-1}). The general trends in k_s obtained from the data of Poesen and Savat (1981) and Torri et al. (1987) are shown in Fig. 7.1. Two peaks of maximum soil detachability occur, one at very high clay content, and the other where median grain size diameter (D_{50}) values coincide with silt size, at low clay content. Detachability decreases with increasing D_{50}.

Other relationships have been developed but require soil databases richer in measured parameters. Examples proposed by Sharma et al. (1991) and Torri et al. (1998), respectively, are

$$k_s = \exp(3.877 - 0.349\ln(T) - 0.037C)$$ [7.5]

and

$$E_0 = 0.013T + 0.005C$$

$$k_s = 0.0036\frac{\delta_s D_{50}}{1.5T_R}\exp(1.31\tan\alpha + 0.067C)$$ [7.6]

and

$$E_0 = 0$$

where T is the shear strength measured using a Swedish falling cone (kPa), δ_s is the grain dry bulk density (kg m^{-3}), C is clay content (%), T_R is the shear strength measured using a pocket torvane (kPa), D_{50} in the median grain diameter (mm) and α is the local slope angle of the soil surface. After detachment the soil particles and aggregates are ejected and follow a parabolic path. Generally, the amount of splashed material that is deposited at a certain distance from the impacted area decreases

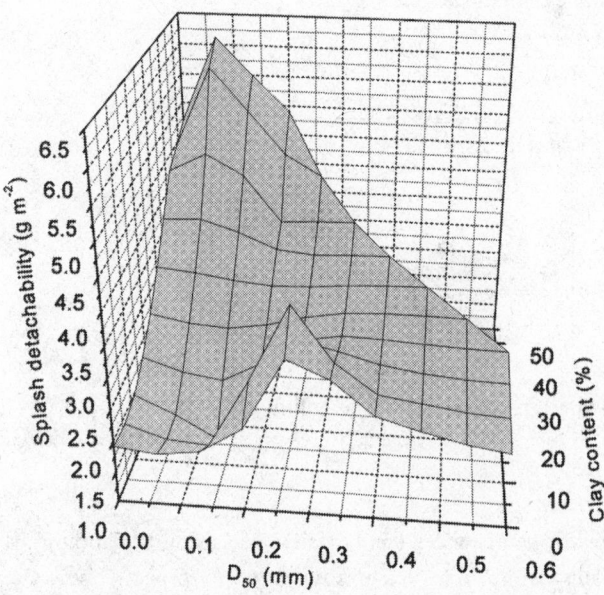

Fig. 7.1 Trends between splash detachability, median grain diameter and clay content

exponentially with distance (Poesen and Savat, 1981; Torri et al. 1987). The distance of ejection (X, m) is related to the median diameter (D_{50}, mm) by (Poesen and Savat, 1981):

$$X = 0.03D_{50}^{-0.218}$$ [7.7]

Equation [7.7] is only indicative of a trend but does not account for the effects of slope gradient and the presence of water at the soil surface. Using several assumptions, Poesen and Savat (1981) proposed:

$$X = 0.0192D_{50}^{-0.218} \pm 0.301\sin\alpha$$ [7.8]

where the sign is positive for downslope movement and negative for upslope movement.

Because the soil surface is usually rough, the drop hits a surface whose gradient is usually different from that of the field or hillslope. The local gradient (over the scale of a few mm) causes the drop to vary its detaching power and most of the soil grains to be ejected in the local downslope direction. The increase in detachment is due to the fact that gravity acts as a force resisting detachment only when the grains are ejected upslope. Falling vertically on to a horizontal surface, a drop splits uniformly in all directions. When the surface is inclined, the ejection of droplets is mainly downslope. Under such conditions, detachment rate increases with slope gradient as described by Torri and Poesen (1992):

$$k_s = \frac{1}{\pi}\int_0^{2\pi}\frac{1 - \dfrac{(\mu - \sin\mu)}{2\pi}}{\cos\alpha\tan\varphi - \sin\alpha\cos\varpi'}\,d\varpi$$ [7.9]

where φ is the angle of friction between soil grains, ϖ, ϖ' and μ are auxiliary variables (the integral is to be solved by numerical methods) and

$$\varpi' = \arccos\left[\frac{\cos\varpi}{\sqrt{(\cos^2\alpha\sin^2\varpi + \cos^2\varpi)}}\right]$$

while

$$\mu = 2\arccos\left[\frac{\sin\alpha\cos\varpi}{\sqrt{0.5(\cos^2\alpha\sin^2\varpi + \cos^2\varpi)}}\right]$$

The net flux of splashed grains (F) is defined as follows:

$$F = k_dX_d - k_uX_u$$ [7.10]

where the up- and downslope distances can be calculated using Equation [7.8] and k_d and k_u from Equation [7.9], integrating from $^{3\pi}/_2$ to $^\pi/_2$ and from $^\pi/_2$ to $^{3\pi}/_2$, respectively.

Empirical relationships for F have been proposed by several authors, an example being that of Poesen and Savat (1981):

$$F = k_s E\left[0.301\sin\alpha + 0.0192 D_{50}^{-0.218}(1 - e^{-2.42\sin\alpha})\right]$$ [7.11]

Poesen (1985) adjusted this equation to take the effects of oblique rains on F into consideration. The energy of impact (E) is not equal to raindrop energy unless the soil is bare and no water is interposed between the soil surface and the drop because water standing (or flowing) over the soil surface reduces drop impact energy (Palmer, 1963) described by Torri et al. (1987) as follows:

$$E = E_d e^{-\beta_H h}$$ [7.12]

where E_d is the energy of the drops just before impacting the water surface and β_H (cm^{-1}) is an empirical coefficient depending on soil surface characteristics (linked to grain roughness) for which only few data exist (Table 7.1). It can be estimated from (Torri et al., 1998):

$$\beta_H = 0.36\ln\left(6\frac{D_{95}}{D_{50}}\right)$$ [7.13]

where D_{95} is the grain diameter below which 95% of soil particles and aggregates occur (by weight).

7.4 Overland Flow Erosion

Water flow detaches and transports soil particles and aggregates. The forces resisting detachment are the grain weight, its angle of friction with surrounding grains, and the cohesive forces with which each grain sticks to the soil mass. Detachment operates by drag and lift forces which are roughly proportional to the flow velocity squared. Flow, in turn, is characterized by its unit discharge rate (Q, m^2 s^{-1}), velocity (V, m s^{-1}) and depth (h, m). The Manning equation provides links between these variables:

$$Q = \frac{1}{n} h^{\frac{5}{3}} \tan^{0.5}\alpha$$ [7.14]

$$V = \frac{1}{n} h^{\frac{2}{3}} \tan^{0.5}\alpha$$ [7.15]

Table 7.1 Values of β_h [after Torri et al., 1987]

Text. class	β_H (cm^{-1})
Sandy loam	0.9-1.0
Clay	1.5-2.8
Silty clay	1.9-3.2

where n is the hydraulic resistance which depends on the soil surface conditions such as grain size, tortuosity, vegetation, clod size, etc. The most commonly used equation to calculate n is that of Strickler (1923):

$$n = 0.0594 D_{50}^{\frac{1}{6}} \qquad [7.16]$$

When aggregates and clods dominate the surface, Equation [7.16] cannot be used. Gilley and Finkner (1991) and Mwendera and Feyen (1992) found the following relationships to be suitable although they give markedly different results:

$$n = \frac{0.172 RR^{0.74}}{Re^{0.282}} \qquad [7.17a]$$

$$n = 0.0056 \, e^{0.1361 \, RR} \qquad [7.17b]$$

where Re is the Reynolds number (Re = Q/γ), γ is the kinematic viscosity of water [$m^2 \, s^{-1}$]) and RR (mm) which is a measure of soil surface random roughness (i.e., surface irregularity such as clods, which cannot be perceived as oriented in any preferential direction) RR is obtained by calculating the standard deviation of relative soil elevations at the surface measured with sticks or laser equipped profilometers (Zobeck and Onstad, 1987). A new promising and alternative approach to model friction resistance is based on the ratio between a microrelief height and the flow depth (inundation ratio) (Lawrence, 1997).

Equations [7.17ab] account for the effect of rock fragments on flow velocity (incorporated in the RR value). When RR is not available, rock fragments can be accounted for with other empirical relationships such as that reported by Poesen et al. (1994) where if n' is Manning's hydraulic resistance (Equation [7.16]) using grain size data for the fine fraction, the presence of a rock fragment cover (S_R,%) modifies the n value as follows:

$$n = n' \, \exp(0.017 S_R) \qquad [7.18]$$

Flow velocity varies with depth following a logarithmic law when the flow is fully turbulent:

$$V_{(z)} = \frac{2.3}{c_k} U_s \log\left(\frac{30.2z}{K_N}\right) \qquad [7.19]$$

where c_K is von Karman's universal constant, z is the vertical coordinate (ranging up to h), U_S is the flow shear velocity in the vicinity of the bed, and K_N is Nikuradze's sand roughness which can be calculated from

$$K_N = (26.6n)^6 \qquad [7.20]$$

This is valid if n is entirely due to grain roughness. As the flow is usually turbulent, the local flow velocity can be either larger or smaller than the mean velocity at that depth (z). This gives rise to a distribution of velocities at each z value. The standard deviations of the horizontal and the vertical instantaneous velocity distributions can be estimated as follows (Naden, 1987):

$$\sigma_x(z) = 0.16V_{(z)}\left[\frac{K_N}{z}\right]^{0.65} \qquad\qquad [7.21]$$

and

$$\sigma_z = 0.17\sigma_x \qquad\qquad [7.22]$$

As a consequence of these distributions of instantaneous velocity, detaching forces, being proportional to the instantaneous velocity squared, are similarly distributed in time. Consequently, particle detachment occurs even if the average value of the detaching forces is smaller than that of the bonding forces (Naden, 1987; Nearing, 1991). This is further stressed by the fact that the forces keeping a particle at rest are spatially distributed.

A different and more practical approach is based on describing flow erosive power by flow shear stress (τ, Pa) and flow stream power (U_p, W m^{-2}) which are calculated as follows:

$$\tau = \rho_w gh \tan \alpha = \rho_w U_S^2 \qquad\qquad [7.23]$$

$$U_p = \rho_w gQ \tan \alpha \qquad\qquad [7.24]$$

where ρ_w is the density of the fluid (kg m^{-3}) and g is the acceleration of gravity (m s^{-2}). Equation [7.23] derives from considerations based on statics, while Equation [7.24] represents the loss of potential energy per unit of time and surface when the fluid is discharging at constant rate (Q) in time and space.

Studying flow detachment in isolation from other processes, Nearing et al. (1991) found the following empirical relationship:

$$D_f = 2.8\exp\left(356\tan\alpha + 264h + 0.734D_M\right)10^{-6} \qquad\qquad [7.25]$$

where D_f is the sediment detachment in kg m^{-2} s^{-1}, h is in m, and D_M is the mean weight diameter of the sediment (mm). Torri et al. (1998) found a different relationship which was calculated over larger ranges of slopes and soil types:

$$D_f = 0.38\frac{\rho_g D_{50}}{T_R} U_p \exp\left[0.0064\frac{C}{D_{50}} + 7.1\tan\alpha - 1.3\left(\frac{\rho_g - \rho_w}{\rho_w}\right)\right] \qquad\qquad [7.26]$$

where ρ_g is the mean dry bulk density of the soil grains (aggregates and loose single particles, kg m^{-3}), T_R is in kPa, D_{50} in mm and D_f is expressed in kg m^{-2} s^{-1}. These two equations were developed using only slightly overlapping sets of data. When one equation is extrapolated toward the other's range of validity, they miss each other's trend. Differences mostly depend on differences in methodologies, one of which (Torri et al., 1998) is in agreement with Shield's criterion for grain entrainment.

Once a grain is entrained by the flow, the processes leading to sedimentation begin. The trajectory followed by the grains can be quite complex; some grains are transported in contact with the soil surface, by rolling or sliding, others in suspension. Consequently, sedimentation laws, such as

Stokes', are extremely poor approximations of what really happens. The concept of flow transport capacity is a means for overcoming the problem. Transport capacity is defined as the amount of sediment that a given flow can transport over an infinite length when detachment and deposition rates are constant. Many (> 10) equations have been proposed mainly for rivers and channels. Govers (1990) expressly developed for rill conditions a set of equations which can be interpolated as follows:

$$T_c = 0.0176 Q \rho_g \left(1 + 7.4 e^{120 D_{50}}\right) \left(100 V \sin\alpha - 0.4\right)^{0.28 \ln(130 D_{50})} \qquad [7.27]$$

where T_c is transport capacity in kg s^{-1} m^{-1}.

Once transport capacity is known, net erosion rate due to overland flow scouring can be assessed as a transport deficit. In this case, the amount of soil that will be detached in excess of what is deposited (D_{fN}) is given by:

$$D_{fN} = \beta_T Q \left(T_c - T_a\right) \qquad [7.28]$$

where T_a is the present sediment load and β_T is a coefficient of proportionality strictly linked to the various detachability coefficients such as D_f from Equation [7.26]. When Equation [7.28] gives negative values, then sedimentation is dominant. The same concept expressed in Equation [7.28] has been modelled in other ways (Foster et al., 1989; Woolhiser et al., 1990):

$$D_{fN} = D_f \left(1 - \frac{T_a}{T_c}\right) \qquad [7.29]$$

The general picture is further complicated by the fact the raindrop-detached soil is easily entrained by the flow because the particles are already detached. As a result, splashed particles compete with overland flow detachment because they add to T_a.

As already said, part of the drops' kinetic energy is spent on impact with the overland flow layer which causes an increase in turbulence. Kinnel (1990) developed a set of equations to account for the raindrop effect summarized as:

$$D_f = \beta_{if} I V \lambda(h, r) \qquad [7.30]$$

where β_{if} is the soil susceptibility to be eroded by a shallow flow impacted by raindrop, I is rain intensity, V mean flow velocity and λ is a complex function of the flow depth and the distribution of raindrop sizes (r).

7.4.1 Rill and Interrill Erosion

Overland flow tends to rapidly coalesce into rivulets whose paths are unstable and can quickly change due to local deposition or erosion episodes. They wind around small protuberances until they reach local micro impluvia where they merge with other rivulets forming well-defined flow patterns that can eventually develop into rills. The interrill area has its own slope gradient (that is not parallel to rills or main local micro impluvia), length and hydraulic roughness. Hence, different equations have been proposed for dealing with the two situations. For example, interrill erosion rate (A_i) was expressed as follows by Liebenow et al. (1990):

$$A_i = K_i I^2 \left(1.05 - 0.85 e^{-\sin\alpha}\right) \qquad\qquad [7.31]$$

where K_i is the interrill erodibility and I is the rainfall intensity.

Rill erosion is often treated by using Equation [7.29], and balancing erosion and deposition within the rill bed with the erosion inflow from interrill areas.

Rills are cut into the soil when the flow is sufficiently erosive. Smerdon and Beasley (1961) found that critical flow shear stress (Equation [7.23]), at which incision begins, increases linearly with clay content. Others found that critical shear velocity for rill excavation increases with soil shear strength measured with a torvane (Rauws and Govers, 1988). Similar relationships can be found in the literature relative to cohesive channels. Obviously other soil characteristics such as high contents of exchangeable Na or smectite can reverse the situation and a clay soil would behave as a silt. Such effects are only partially reflected in the measurements of soil shear strength because of the crude way in which it is measured at the soil surface.

7.4.1.1 Rock Fragment Effects on Interrill and Rill Erosion

When nonerodible (or scarcely erodible) grains, such as rock fragments, are present, the soil surface is carved by erosion into pedestals surmounted by grains. Erosion can continue until a pavement of rock fragments protects the soil from any further loss. The first direct effect of rock fragments is that of intercepting raindrops. Consequently, drop detachment decreases proportionally to $(1-S_R)$ where S_R is the fraction of soil surface covered by rock fragments. A further effect is that of concentrating runoff between stones, locally increasing the flow capacity for detachment and transport. More specifically, water flowing around isolated rock fragments excavates a horseshoe-shaped incision on the upstream side of the stone due to whirls (Poesen et al., 1994). The rock fragment can be considered isolated from other fragments until they cover < 20-25% of the surface. After such a threshold, eddies interfere and tend to disperse their energy onto nearby nonerodible rock fragments rather than erodible grains. As a consequence, erosion initially increases and then decreases while the rock fragment cover increases.

If rills are included, then the processes become more complicated. Rill erosion is decreased by the rock fragments because flow velocity decreases (Equation [7.18]), while part of the flow shear stress is spent on scarcely erodible rock fragments. Moreover, an erosion pavement can be easily generated, protecting the soil surface. In conclusion, at a rill-interrill scale (i.e., on areas between 10 and 10000 m^2), rock fragment cover causes erosion to decrease.

7.5 Pipe Erosion

Pipes develop in particular situations. Some are linked to karst processes, involving soil chemical dissolution, and removal by subsurface flow while others are linked to excess subsurface water that evacuates through cracks or macropores in which the water can flow fast enough to transport sediment. Subsoil sodicity further promotes clay dispersion and transport. These processes result in a pipe or karst-like depression. The formation of pipes is facilitated by the presence of cracks or of impervious layers. If a soil horizon overlaying a less permeable layer has adequate horizontal hydraulic conductivity, water flows rapidly and becomes erosive. In other words, everything that promotes rapid accumulation of water beneath the soil surface (e.g., tectonic joints, microfaults, cracks, biological macropores, etc.) or rapid water extrusion (e.g., artificial and natural trenches such as gullies, landslides, abrupt increase of hillslope gradient, etc.) contributes to promote tunnel scouring and seepage erosion. Tunnel scouring takes place when subsurface water flows sufficiently fast to entrain soil particles. Seepage erosion takes place when the water head can be dissipated at a scarp or at any other discontinuity, detaching particles and causing the discontinuity to retreat. An overview of pipe erosion is presented by Jones and Bryan (1997) illustrating the various situations in which pipes can develop and their significance.

Assuming that Darcy's law is valid at a macroscopic scale, the work done by water during seepage exerts a seepage force on the soil particles. The seepage force exerted per unit area normal to the flux is computed as

$$J' = \gamma_w H_s \qquad [7.32]$$

where J' is the seepage pressure, γ_w ($= \rho_w g$) is the weight of water per unit volume, H_s is the local hydraulic head evaluated from equipotential water lines. The local hydraulic gradient may by easily computed as local hydraulic head loss (piezometric level) for unit length (L):

$$i = \frac{H_s}{L} \qquad [7.33]$$

The seepage force per unit volume is given by (Whitlow, 1990):

$$J = i\gamma_w \qquad [7.34]$$

The flow of water in soil and the relative seepage pressure per unit of volume (J') reduce the intergranular contact pressure locally (vertical effective stress) and soil shear strength. In some circumstances, the seepage pressure reduces the vertical effective stress and soil shear strength to zero (quick condition) when the critical gradient (i_c) is reached:

$$i_c = \frac{\gamma_s - \gamma_w}{\gamma_w} \qquad [7.35]$$

where γ_s is the weight of the saturated soil mass per unit volume.

When $i_c \approx 1.0$, Equation [7.35] indicates that a floating condition arises which has been reported for loose and medium sandy soils with local saturated flow. Vukovic and Pusic (1992) present experimental and empirical criteria to define the critical gradient for quick condition and for the process of suffusion (washing away of fine particles from a porous medium by seepage flow) which produces an internal selective erosion. They present an experimentally derived chart in which quick condition and suffusion critical gradients are functions of the coefficient of uniformity $\eta = D_{60}/D_{10}$ (D_{60} and D_{10} are 60th and 10th percentiles of the grain size distribution) that may be approximated by the following function:

$$i_c = 0.22 + 1.03e^{-\frac{\eta}{8.5}} \qquad [7.36]$$

A further criterion to evaluate the risk of suffusion in incoherent soils is the critical flow velocity which arises when the following condition is fulfilled:

$$V_c = \frac{K^{0.365}}{200} \qquad [7.37]$$

where V_c is the critical flow velocity for suffusion (m s^{-1}), and K is the hydraulic conductivity (m s^{-1}). For Darcian flow, the effective pore flow velocity (V) may be assessed from equipotential water lines and the following relationship:

$$V = \frac{Ki}{n}$$ [7.38]

where n is the soil porosity. This equation must be modified to take slope into account. If soil cohesion is not considered, the critical gradient is computed as (van Rhee and Bezujen, 1992):

$$i_c = -(1-n)\rho_{rel}\frac{\sin(\phi - \alpha)}{\sin \phi}$$ [7.39]

where the negative sign in the second term indicates an outflow condition,

$$\rho_{rel} = \frac{(\rho_s - \rho_w)}{\rho_w}$$

is the relative grain density, ϕ is the angle of soil internal friction, and α is the slope angle.

The most relevant process in cohesive soils is erosion due to bypass flow (i.e., flow through biopores, shrinkage cracks, tension cracks, joints). During intense rainstorms, runoff may be intercepted by the macropores and circulate in the net of fissures. Erosion will occur following modalities similar to those described in Section 7.4 for overland flow.

Pipe erosion becomes prevalent when dispersion affects soil behavior. The susceptibility of soil to pipe erosion is assessed by the increase in size of a standardized hole made in remolded samples at high flow rates (pinhole test) (Sherard et al., 1976). A basic classification for evaluating the risk of pipe erosion on the basis of the dispersibility potential of a soil is given by Fitzpatrick et al. (1995) for soils in Southeastern Australia. A soil is at risk of pipe erosion when the following condition is fulfilled:

$$EC < 0.062ESP + 0.33$$ [7.40]

where EC is the electrical conductivity at 25 °C in a 1:5 soil:water suspension and ESP is the exchangeable Na percentage.

7.6 Shallow Mass Movement

Shallow mass movements are important in water erosion because (1) they are responsible for the evolution of rill and gully slopes during and between rainstorms; (2) the large majority of mass movements in cultivated fields are due to excess water in the soil profile in the vicinity of more impervious layers; and (3) mass movements may trigger gullies and pipes and *vice versa*.

Local mass movements of soil result from local severe degradation processes and are usually associated with other forms of erosion such as rills, pipes and gullies. In agriculture, mass movements inhibit farm operations and land use by loss of accessibility, exposure of infertile subsoil and unprotected soil surface layers to splash and rill erosion, and net downslope movement of the soil mass. Gravity is the principal force producing slides, slips, slumps, flows and landslides. At a particular water content soil becomes unstable. Landslides and mass movements are usually classified by type and velocity of movement (Varnes, 1978). Rapid movements of soil mass over a distinct sliding surface are termed landslides. Slower long-term deformations having a series of sliding surfaces and exhibiting viscous movement are termed creep.

Generally mass movement occurs when the weight (shear stress) of the surface material on the slope exceeds the retaining ability (shear strength) of the material. Risk of mass movement is increased by erosion or excavation undermining the foot of a slope, loss of stabilizing roots through removal of vegetation and increase in pore water pressure within the soil profile. Increased pore water pressure or greater water absorption may weaken intergranular bonds, reducing internal friction, therefore lessening the cohesive strength of the soil and ultimately the slope stability.

Usually slope stability grade is determined quantitatively by the ratio between available shear strength of soil mass and imposed shear stress computed along the assumed sliding surface:

$$F_s = \frac{T}{S} \qquad [7.41]$$

where F_s is the stability factor and depends on the shape of the sliding surface and on the type of movement, T is the shear strength (the sum of resisting forces), S is the shear stress (the sum of driving forces).

Values of $F_s < 1$, $F_s > 1$ and $F_s = 1$ correspond to unstable, stable and metastable slopes. Considering the spatial and temporal variability of soil properties, values of $F_s > 1.2$ are considered satisfactory for stable conditions in a conservative analysis. Usually, sliding surfaces are supposed to be circular, planar or generic. Two-dimensional analysis is commonly used to evaluate overall slope stability (Huang, 1983). Basic inputs are slope transect, interface profile of the discontinuities present in the soil mass, mechanical properties of each stratum on the slope portion being considered, and the profile of the final phreatic surface separating saturated from unsaturated soil.

Mass movements may be classified using the following relationship (Janbu, 1973):

$$d_L = \frac{d}{L} \qquad [7.42]$$

where d is the maximum depth of the sliding surface and L the length of the sliding mass (Fig. 7.2). Landslides characterized by $d_L < 0.05$ are considered planar movements with sliding surfaces subparallel to the slope. Higher values are typical of circular or pseudocircular sliding surfaces.

For shallow mass movements, the infinite slope approach assuming planar movements (Fig. 7.3) is commonly used. The stability factor may be evaluated from

$$F_s = \frac{c' + \left(h\gamma_s - h_w\gamma_w\right)\cos^2\beta \tan\phi'}{h\gamma_s \sin\beta \cos\beta} \qquad [7.43]$$

where h is the depth of the soil layer or the slip surface in the topsoil used to evaluate the stability, h_w is the height of the water table above the slip surface ß is the local slope, ø' is the angle of soil internal friction, c' is the internal cohesion of the soil, γ_s and γ_w are the weights of soil and water per unit volume, respectively.

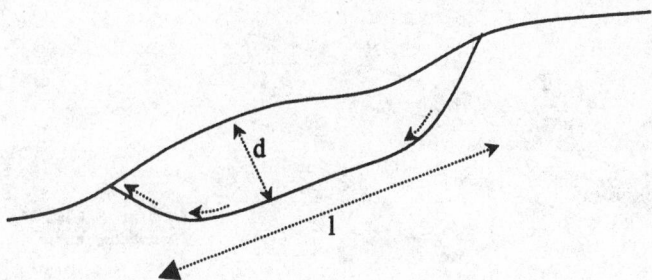

Fig. 7.2 Shallow mass movement: definition of the d and l parameters

Equation [7.43] applies to planar sliding of the soil mass where sliding surface and seepage flow are approximately parallel to the land surface. Such conditions only occur in some instances, whereas the infinite slope criterion is used for evaluating the stability over large areas or watersheds. Some computer models use the infinite slope method under probabilistic and deterministic approaches (LISA) (Hammond et al., 1992).

The $\emptyset'$ and c' are the soil strength parameters commonly measured in the laboratory on undisturbed soil cores, based on the Mohr-Coulomb soil strength criterion (Whitlow, 1990) or estimated indirectly from empirical correlations with penetrometer or vane test values measured in the field (Huang, 1983, Whitlow, 1990).

Special situations such as concave-convex slopes, gully walls, tension cracks, thawing slopes, and vegetation effects require modifications to Equation [7.43]. Alternatively, limit equilibrium methods should be employed for nonplanar slopes with tension cracks (Janbu, 1973). Several features such as evidence of previous mass movement, poor vegetation cover and linear features across slopes related to cracks or minor terracing in the landscape can indicate whether a certain area deserves study in order to assess the risk of mass movement. Old or dormant landslides characterized by long, uneven hummocky slopes and cracks and fissures and bent tree trunks on steep slopes may also indicate a history of mass movement.

7.7 Gully Erosion

When a linear incision becomes too deep or too large to be easily removed by normal tillage operations, the channel is upgraded from a rill to a gully. If the gully is removed shortly after its formation (which is common in cultivated fields), then it is called ephemeral. For a sharper definition of gully formation, independent of tillage equipment, any channel with ephemeral runoff and maximum cross-sectional area equal or above $0.1\ m^2$ is a gully. This definition includes both the incision disappearing gradually upslope and those ending abruptly with a step. A classification of channel heads was proposed by Dietrich and Dunne (1993) on the basis of the morphology of the head: (1) gradual through small steps (< 0.1 m); (2) large steps (0.1-1 m); (3) small headcuts (1-10 m); and (4) large headcuts (> 10 m). Here gully head is used to indicate where incision is evident even in the absence of flow. Gullies have been classified in several ways, following genetic or morphologic criteria. They can be described as digitated if they repeatedly bifurcate while retreating, until no dominant channel, is clearly visible; axial when there is a main channel; or frontal gullies, if the width of the retreating head is evidently larger than the length of the gully (Morgan, 1995).

Gullies are generated by concentrated overland flow sufficiently erosive to excavate deep or large channels which generally happens where several concentrated flow lines merge or where a concentrated flow of relatively clear water (e.g., an outflow from a forest, pasture or paved road)

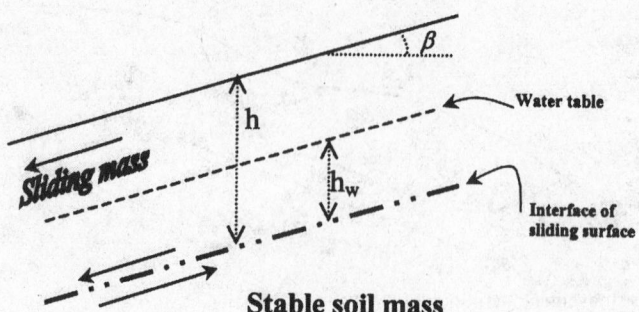

Fig. 7.3   Shallow mass movement: parameters relevant for assessing slope stability

intersects a cultivated field. In addition, gullies form in highly dispersible soils or when pipes and tunnels collapse after reaching a critical size. This frequently occurs in sodic soils (Sumner et al., 1997). Furthermore, gullies may form because of local base-level lowering due to human activities or bed-channel scouring. A retreating gully head can intercept unincised flow paths and cause tributary gullies to form. More generally, gullies may develop where the flow path intersects an earth bank (such as road or terrace bank, lynchets, etc.) (Poesen, 1993).

Gully erosion is characterized by movement of the gully head upslope. Gully sidewalls are usually modeled as rills, slumps and mud flows. The large quantities of sediment produced by head, wall and bed erosion tend to rapidly fill the transport capacity of the gully, reducing further bed scouring downstream. Hence gullies are often characterized by an upstream segment where erosion is active and a downstream portion where sedimentation prevails. Where gully walls are made of easily weathered materials, early pedogenetic processes allow the development of Entisols. The contact between the weathered and relatively unweathered layers corresponds to a surface of discontinuity where permeability decreases abruptly, favoring soil saturation by water. As soon as soil formation reaches a certain depth that depends on local slope and soil shear strength values, new mass movements rejuvenate the gully walls. The deep incision of permanent gullies can trigger seepage erosion anywhere along its length. This, in turn, can cause the formation of new tributary gullies. If the tributaries of different orders are numerous, they intercept one another. In this case, residual surfaces of the original slope are confined between interconnected gullies (Jones and Bryan, 1997).

Topics of major interest for gully erosion are whether and where gullies form in a given landscape, how fast they will erode and how far upslope they can retreat. Their location in the landscape is determined by local precipitation (drier areas require larger catchments than wetter ones), mean valley gradient and local gradient at gully head (Dietrich and Dunne, 1993). They found a trend between channel head drainage area and local valley gradient at the channel head for several rangeland areas. Their observations show that a threshold exists for channel head location which depends on the area (A) of the drainage basin and the valley gradient (s) at gully head. Several sets of observations of different areas are discussed by Poesen et al. (1998). The general relationship expressing the threshold is as follows:

$$A \geq as^{-b} \qquad\qquad [7.44]$$

where the exponent b varies between 0.3 and 0.4 and the constant a depends on precipitation, land use, etc.

From this relationship, the minimum distance between gully head and divide can be deduced. Land use and soil erodibility play an important role in defining the threshold area. Retreat rates of gully heads depend on the amount of water that the gully head drains, that is to say, that gully heads retreat at a rate proportional to the catchment area (A). Seginer (1966) found the following relationship for gully retreat rate (G_R) in south Israel:

$$G_R = cA^b \tag{7.45}$$

The empirical constant c depends on local characteristics (soil, land use, topography, precipitation) while the exponent b is generally close to 0.5. Stocking (1981) reported relationships for southern Africa. A second threshold can be expected to exist in relation to where gullies end, due to the massive sedimentation that counteracts gully development (Poesen et al., 1998).

Usually local equilibrium is established between the erosion processes that make sidewalls retreat and those that further scour the gully bed. Hence, local relationships between width, depth and cross-section of gullies can be established. For example, in the United States, gullies are characterized by a width to depth ratio close to 3 on cohesive material and 1.75 in noncohesive sediments (Reid and Dunne, 1996). Poesen et al. (1998) showed that the depth of an ephemeral gully increases rapidly beyond a critical slope (>3-4%) unless an erosion-resistant horizon (e.g., Bt, plough pan, fragipan, etc.) is present at shallow depth in the profile. In this case the gully width increases with the steepening of slopes.

7.8 Vegetation, Land Leveling and Wildfires

This paragraph will summarize the main processes through which vegetation affects erosion. Then some lines will be dedicated to a particular aspect of land use (i.e., land leveling) because part of its implications for soil erosion are usually neglected. Then a further paragraph will be dedicated to wildfires. The items selected have been chosen for the strong influences they have on the soil, soil properties and soil distributions. Obviously, the presence of temporary bare soil surfaces and artificially generated soil roughness surfaces usually increases the amount of erosion and modifies the general physical, chemical and microbiological properties of the soil, thus affecting the dynamic response of the soil to water erosion processes. These type of effects are described in Morgan (1995) and Haan et al. (1994); hence they will be omitted here.

7.8.1 Vegetation

Vegetation has direct and indirect influence on the erosion process. The primary action is the shield effect of crop canopy. It usually reduces the total kinetic energy of the rainfall that is dissipated on the soil surface. The leaves and stems intercept part of the raindrops so that only a fraction reaches the ground surface directly. The intercepted part is either stored on the plant (usually less than 0.2 mm) or transported to the ground as stem flow (which can be erosive) or redrop from leaves and branches. The redrop drops are generally from 3.5 to 4 mm in size (Brandt, 1989) and their impacting energy depends on the distance from the leaves and branches to the ground. Consequently, the most effective action is due to dead leaves and branches laying on the soil surface. Depressing impacting energy has the effect of maintaining infiltration at high values, thus reducing overland flow and erosion. Moreover, the presence of vegetation at the soil surface increases hydraulic resistance, further reducing flow transport and erosion potential. This is generally coupled to some trap efficiency; grassed strips along stream banks are one of the main resources for combating pollution.

Vegetation has the effect of reducing shallow mass movements; roots add resistance to the soil if they pass through the sliding surface, fixing the potentially sliding portion to the underneath layers. Sliding potential may increase if trees are rooted only above the sliding surface as previously discussed.

A long-term effect of vegetation is the enrichment of the organic matter component of the soil, which, in a healthy soil, becomes a ligand-increasing soil aggregate stability. Frequent tillage operations or other sources of disturbances (such as overgrazing) heavily interfere with the positive effect of vegetation and make the soil particularly erodible.

7.8.2   Land Leveling

Since the beginning of agriculture, farmers have acted to adapt plants and slope morphology to their needs. Today, reshaping the slope morphology has become a sort of *conditio sine qua non* for growing new crops in large areas of the world. The earth movements change the slope gradients, the concentration lines of surface and subsurface water fluxes, the watershed shapes and sizes. The depth affected by earth movement often reaches the C and R soil horizons. Generally, upslope soils and soils on convexities are cut and the material moved to concavities or down the slope. Hence soil depth and soil characteristics are modified within the slope. This modifies the spatial variability of runoff generation and often promotes larger runoff concentrations. Rill and ephemeral gully formation is thus generally enhanced. The concavities that have been filled with additional material increase their instability; subsurface water fluxes concentrate water in the same concavity, but in this case, the weight of the soil column above the potential sliding surface is larger, enhancing the risk of shallow mass movements. Increased frequencies of ephemeral gully generation and of shallow mass movement trigger further land leveling, establishing positive feedback mechanisms. Row-crops can intensify the mechanism because each row acts as a preferential flow path further concentrating runoff where rows end.

7.8.3   Wildfire and Soil Erosion

A wildfire is a dramatic event that modifies soil physicochemical parameters and the soil behavior. The intensity of the modification is always related to temperatures developed at the soil surface. As temperature rises, heating increases, causing decreases in clay and silt contents (Giovannini and Lucchesi, 1997). The heat developed by the fire promotes the aggregation of clay and silt particles into sand sized grains resistant to the chemical and physical actions of water. Particle organization may change and soil structure passes from fine to coarse granular up to blocky or massive (Giovannini and Lucchesi, 1997). In addition, if temperature reaches sufficiently high levels, clays are denatured and lose their charge rendering them less dispersible (Rengasamy and Sumner, 1997). On the other hand, high temperatures can cause the distillation of organic matter. The nonvolatile compounds, in particular, move downward into the soil profile until they condense in a cooler portion of the soil. Under certain conditions of combustion and soil moisture content, a subsurface hydrophobic layer develops (De Bano et al., 1976; Giovannini and Lucchesi, 1983). The net effect of these mechanisms is a decrease in organic matter content in the soil surface layer (Giovannini and Lucchesi, 1997), often associated with an increase in water permeability. However, if a hydrophobic layer forms, water permeability decreases dramatically and overland flow production is enhanced.

If the colloidal fraction of the soil decreases, its erodibility will probably increase as the relationship between erodibility, clay and organic matter content indicates (Fig. 7.4). If the increased erodibility is coupled with an increased runoff production potential, then more erosion is to be expected. Branson et al. (1981) report values from the literature which indicate that the predominant effect is toward infiltration reduction and erosion increase. In particular, soil loss (on an annual base) after a moderate to intense fire may increase by something between 2 and 10 Mg ha^{-1}, with exceptional values of more than 100 up to 500 Mg ha^{-1} in a fine sandy loam over pumice.

7.9 Soil Erosion Models

The first model able to estimate soil loss over a wide range of situations was the Universal Soil Loss Equation (USLE) (Wischmeier and Smith, 1978) applicable to the United States, east of the Rocky Mountains. The model which was recently revised (RUSLE) is as follows (Haan et al., 1994):

$$A = KRLSCP \tag{7.46}$$

where A is the average annual soil loss over a 20–22 year period, K is soil erodibility, R is rainfall erosivity, L is the field length factor, S is the gradient factor, C is the crop and management factor and P is the effect of the soil protective practices. Soil erodibility is determined experimentally as the soil loss per unit R measured on a 22.1 m long plot with a 9% uniform slope gradient which must be kept under continuous fallow with frequent up and down cultivations to remove weeds and erosion features. Erosivity is defined as

$$R = 0.01 \sum EI_{30} \tag{7.47}$$

where E is raindrop kinetic energy over the storm (MJ ha^{-1}) before hitting either the soil surface or vegetation, and I_{30} (mm h^{-1}) is the maximum rain intensity in 30 minutes during the storm. In other parts of the world, maximum intensities over shorter periods were found better suited. Rainfall kinetic energy can be estimated from rainfall characteristics using a plethora of equations which are different because rainfall characteristics, such as drop size distribution, depend on regional and local conditions (e.g., convective/advective rains, summer or winter storms, prevalence of intense winds or delicate breezes, air temperature, etc.). Moreover, the accuracy with which rain is or was recorded is not the same (this affects energy calculations because of the fractal nature of rain intensity). Examples of these equations are

$$E = 29 \sum I_j \Delta t_j \left[1 - 0.72 e^{-0.05 I_j} \right] \tag{7.48a}$$

$$E = 36 \sum I_j \Delta t_j \left[1 - 0.56 e^{-0.34 I_j} \right] \tag{7.48b}$$

$$E = \sum I_j \Delta t_j \left[1.9 + 6.11 \log_{10} T + 11.0 \log_{10} I_j \right] \tag{7.48c}$$

where E is in J m^{-2}, I_j is rain intensity during the period Δt_j, and T is the air temperature close to soil surface during the rainstorm. Equation [7.48a] comes from the RUSLE while [7.48b] (Kinnel, 1981) estimates larger energies than [7.48a] and [7.48c] (Zanchi and Torri, 1980). Equation [7.48c] modifies kinetic energy estimates using mean air temperature during the storm in an attempt to account for the different characteristics of winter and spring rains in a Mediterranean climate.

The slope and length factors (Equation [7.46]) are necessary because overland flow is not explicitly involved in the model. Increasing slope gradient and field length causes overland flow to be more aggressive in both transport and detachment capacities. The following equations are to calculate L and S:

$$L = \left(\frac{x}{22.13} \right)^m \tag{7.49}$$

and

$$S = 3.0\sin^{0.8}\alpha + 0.56 \qquad \text{if } x < 4.57m$$
$$S = 10.8\sin\alpha + 0.03 \qquad \text{if } x < 4.57, \sin\alpha < 0.09 \qquad\qquad [7.50]$$
$$S = 16.8\sin\alpha - 0.50 \qquad \text{if } x < 4.57, \sin\alpha \geq 0.09$$

where x is the field length and m an exponent depending on α and the rill to interrill erosion ratio (ß):

$$m = B\frac{\beta}{1+\beta} \qquad\qquad [7.51]$$

where $B = 0.5$, 1 or 2 for soils with low, moderate or high susceptibility to erosion, respectively. Other equations and algorithms are available for erosion due to thawing, for concave, convex, and segmented slopes as well as for the use of RUSLE in small basins. The other two factors (C and P) were originally tabulated but are now subdivided into a series of subfactors.

The USLE approach to soil erosion description is empirical and based on statistical relationships undermining its universal applicability. Other models have been developed since then improving the erosion and the catchment description. Generally, these models contain equations for interrill and rill detachment and transport and for sedimentation. The description of the basin using a set of cascading planes or a grid of square cells is necessary to guarantee the conservation of matter from one element to the other; this is achieved using a balance equation generally referred to as continuity equation.

The first generation of models mainly applies USLE to small catchments on a daily basis; in other terms, USLE was applied to more complex morphologies (from regular fields to slopes and catchments) and to shorter time intervals (from year totals to daily totals). Examples of models of this type are CREAMS, GLEAMS and EPIC (see papers in Boardman and Favis-Mortlock, 1998 for extended bibliography). A different approach brought researchers to model soil erosion on the basis of a detailed process description. The basic equations used in such models are of the types discussed in the previous sections. Among these type of models are WEPP and EUROSEM (see Boardman and Favis-Mortlock, 1998). While the former gives daily totals, the latter describes only what may happen within single rainstorms and produces event sedigraph and average rill evolution. Changing the representation of the basin from cascading planes to raster produces differences in the models' capabilities to reproduce real situations. One example is LISEM (Boardman and Favis-Mortlock, 1998), where the cascading planes of EUROSEM are substituted by a raster representation.

The hydrological model implemented imposes limitations to the erosion processes described in a model. Usually, water fluxes in the soil are only approximated (Richards' equation in rarely used) so that runoff due to soil saturation or water exfiltration, pipe erosion and part of gully erosion cannot be described.

7.10   Soil Characteristics and Erosion

In the previous sections, many equations contain soil characteristics as either bulk parameters, or clay content, median particle or aggregate diameter and density, rock fragment cover, soil cohesion, etc. In addition to these parameters, others are included that describe soil surface conditions such as roughness or local gradient as experienced by the impacting drop. All these parameters can be considered to act at the process scale but this is not the scale at which soil erosion is usually perceived. In order to appreciate general trends, field scale is more valid. The USLE specifically offers the chance of examining the effect of soil on erosion using one parameter, namely, soil erodibility. This soil composite characteristic summarizes the medium term effect of the soil profile on soil erosion. Erodibility was observed to vary with basic soil characteristics (Figs. 7.4 and 7.5) such as clay, organic matter and geometric mean diameter (D_G) (Torri et al., 1997). In Fig. 7.4, an upper envelope surface is drawn under which the observed soil erodibility values (K) lie. It is clear that there are ranges of clay fraction (0.10-0.15) and organic matter content (1-2%) at which soil erodibility peaks. These peaks are logical because in a coarse textured soil, particles are too heavy to entrain and transport and infiltration is relatively large, all of which decrease erodibility. On the other hand, cohesive forces increase with clay content which reduces erodibility while increasing organic matter content makes aggregates more resistant to sealing and consequently decreases runoff and erosion. The effect of grain size on soil erodibility is shown in Fig. 7.5, where organic matter content has been replaced by the logarithm of the geometric mean diameter (D_G) of the grain size distribution:

$$D_g = \frac{1}{2}\sum_j F_j Ln\left(d_j d_{j-1}\right)$$

[7.52]

where d_j is the upper and d_{j-1} the lower limit of each size class (lower size for clay being $d_0 = 0.005$ μm, by definition) and F_j is the frequency (by weight). Once D_G, organic matter and clay content are known, a maximum K-value can be selected as the minimum value between the two deducible from Figs. 7.4 and 7.5.

When erodibility is examined using selected datasets, other relationships are possible such as those used in the RUSLE. Those relationships indicate that soils with good granular structure (high Fe oxide and organic matter contents) are less erodible. The positive effect of aggregate stability in reducing erodibility has been confirmed by several authors (Rousseva, 1989) who found direct relationships between K and aggregate stability. Consequently, soil aggregate stability indices (such as the number of drops needed to break and aggregate) can be used as a surrogate for erodibility.

Another important soil characteristic is rock fragment content (S_R). Poesen et al. (1994) examined its effect on erodibility on 78 rock fragment-rich soils ($S_R > 10\%$). The ratio (R_R) between the measured erodibility and erodibility calculated using the algorithms proposed by Torri et al. (1997) was observed to decrease as follows (Poesen et al., 1994):

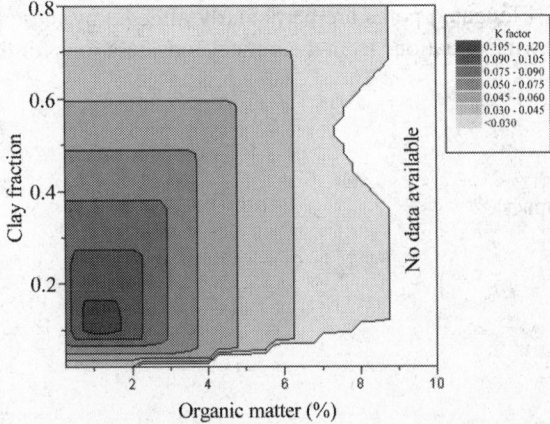

Fig. 7.4 Maximum soil erodibility in relation to the content of organic matter and clay

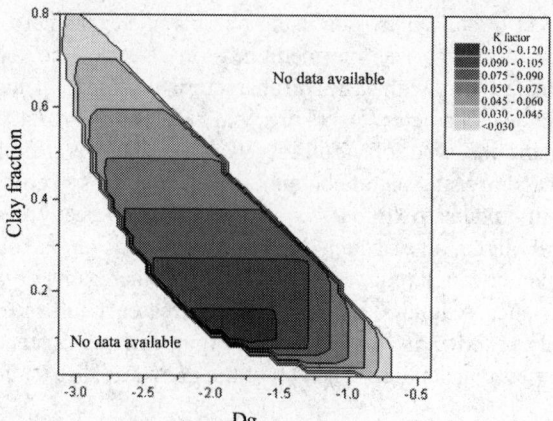

Fig. 7.5 Maximum soil erodibility in relation to the geometrical mean diameter and clay

$$R_R = e^{-0.04(S_R - 10)}$$ [7.53]

Since the data are dispersed, estimations of maximum and minimum values for each rock fragment cover are needed. Of the data, 82% are wrapped within the two following curves (Poesen et al., 1994):

$$R_{R,MAX} = 1.11 - 0.011S_R$$ [7.54a]

$$R_{R,MIN} = e^{-0.15(S_R - 10)}$$ [7.54b]

The equations in this section indicate that surface erosion can be intense if overland flow has enough energy to carry detached soil grains. Hence, the greater the runoff production and the higher the runoff peaks, the higher the soil loss. Consequently, the processes that increase runoff production increase erosion. Since raindrop impact, deposition of detached particles and infilling of pores with transported sediment cause a sealed layer to develop, the processes linked to splash and overland flow are of paramount importance. Elongated pores are particularly affected by these processes. Panini et al. (1997) derived the following equations from theoretical considerations for the processes involved: depositional seal,

$$\Delta P = bD$$ [7.55]

structural seal (drop impact),

$$\Delta P_E = P_{ES0}\left(1 - e^{-C_E E}\right)$$ [7.56]

pore infilling,

$$\Delta P_E = P_{E0}\left(1 - e^{-C_D D}\right)$$ [7.57]

where Δ denotes a decrease; P_E is elongated porosity (%); P_{E0} and P_{ES0} are initial values of elongated porosity and elongated porosity where structural seals develop; b, C_E, and C'_D constants determined by best fitting techniques; E is the cumulative energy released by drop impacts and D is the cumulative detachment by drop impact (which is used in Equation [7.55] and [7.56] as surrogates for cumulative deposition and cumulative sediment flow towards pores, repectively). The reduction in porosity causes a reduction in hydraulic conductivity as illustrated in the relationships found by Brakensiek and Rawls (1983):

$$K_c(t) = K_{fc} + \left(K_{ic} + K_{fc}\right)e^{\left[-CB\left(\frac{RR}{4}\right)E(t)\right]}$$ [7.58]

where $K_c(t)$ is the saturated conductivity of the crust, K_{fc} is the final crust saturated conductivity, K_{ic} is the initial crust saturated conductivity, B is the fraction of the soil surface exposed to rainfall, RR is the random roughness, and C is a constant defined by

$$C = \frac{\ln\left[\dfrac{K_{fc}}{K_{ic} - K_{fc}}\right]}{E_o}$$ [7.59]

where E_o is the cumulative rainfall kinetic energy until $K_{ic} = 2\,K_{fc}$. E(t) is a function that may be derived by Equations [7.48a,b,c]. It calculates the cumulative kinetic energy of the rainfall in relation to time and allows the description of variations in crust conductivity. The opposite effect, i.e., an increase in the mean infiltration rate over a field, can result from erosion, particularly rills (Poesen, 1984). Soil export is the opposite of deposition and consequently, crusts are thinner and less developed. Rills cut through the crust and allow water to infiltrate into a new soil surface, protected from raindrop impact by the presence of the concentrated overland flow. The spatial distribution of different types of seals with different thicknesses causes an apparent increase in infiltration rate with increasing rainfall intensity or, more precisely, overland flow depth (Fox et al., 1998). The most impervious surfaces will produce runoff even during low intensity storms while the most permeable ones will contribute overland flow only several times per year.

7.11 Conclusions

Water erosion is characterized by many processes due to raindrops, overland flow, pipe flow, mass movements. The main equations describing those processes highlight the main soil characteristics involved directly: grain size, colloidal content, chemical composition of soil solution, structural stability, infiltration capacity, rock fragment content, soil surface roughness, soil profile characteristics, etc. Then factors such as slope morphology, catchment size, flow concentration lines, vegetation, land use and land use techniques are the co-factors that can control or speed up erosion. In order to use soils and land sustainably and keep them for future generations, soil erosion processes must be understood. Current erosion models are tools for helping experts to devise possible conservation measures even if they must be used with caution, bearing in mind they are just approximations.

7.12 References

Al Durrah, M.M., and J.M. Bradford. 1982. Parameters for describing soil detachment due to single waterdrop impact. Soil Sci. Soc. Am. J. 46:836–840.

Boardman, J., and D.T. Favis-Mortlock. 1998. Modeling soil erosion by water: Some conclusions. p. 515–517. *In* NATO ASI Series I. Global environmental change. Springer-Verlag, Berlin, Germany.

Brandt, C.J. 1989. The size distribution of throughfall drops under vegetation canopies. Catena 16:507–524.

Branson, F. A., G.F. Gifford, K.G. Renard, and R.F. Hadley. 1981. Rangeland hydrology. Kendal Hunt Publishing Company, Dubuque, IA.

Brakensiek. D.L., and W.J. Rawls. 1983. Agricultural management effects on soil water processes. Part II: Green and Ampt parameters for crusting soils. Trans. ASAE 26:1753–1757.

De Bano, L.F., S.M. Savage, and D.A. Hamilton. 1976. The transfer of heat and hydrophobic substances during burning. Soil Sci. Soc. Am. J. 40:779–782.

Dietrich, W.E., and T. Dunne. 1993. The channel head. p. 177–219. *In* K. Beven and M.J. Kirkby (ed.) Channel network hydrology. John Wiley and Sons, Cichester, UK.

Engels, O.G. 1955. Waterdrop collisions with solid surfaces. J. Res. Nat. Bur. Stand. 54: 281–298.

Fitzpatrick, R.W., S.C. Boucher, R. Naidu, and E. Fritsch. 1995. Environnmental consequence of soil sodicity. p. 163–176. *In* R. Naidu, M.E. Sumner and P. Rengasamy (ed.) Australian sodic soils: Distribution, properties and management. CSIRO Publications, Melbourne, Australia.

Foster, G.R., L.J. Lane, M.A. Nearing, S.C. Finkner, and D.C. Flanagan. 1989. Erosion component. p. 276. *In* L.J. Lane and M.A. Nearing (ed.) USDA - Water erosion prediction project: Hillslope profile model documentation. NERSL Rep. 2. USDA-ARS National Soil Erosion Research Laboratory, West Lafayette, IN.

Fox, D.M., Y. Le Bissonais, and P. Quetin. 1998. The implications of spatial variability in surface hydraulic resistance for infiltration in a mound and depression microtopography. Catena 32:101–114.

Gilley, J.E., and S.C. Finkner. 1991. Hydraulic roughness coefficient as affected by random roughness. Trans. ASAE 34:897–903.

Giovannini, G., and S. Lucchesi. 1983. Effect of fire on hydrophobic and cementing substances of soil aggregates. Soil Sci. 136:231–236.

Giovannini, G., and S. Lucchesi. 1997. Modifications induced in soil physico-chemical parameters by experimental fires at different intensities. Soil Sci. 162:479–486.

Govers, G. 1990. Empirical relationships on the transporting capacity of overland flow: a laboratory study. IAHS Spec. Pub. 189:45–63.

Haan, C.T., B.J. Barfield, and J.C. Hayes. 1994. Design hydrology and sedimentology for small catchments. Academic Press, San Diego, CA.

Hammond, C., D. Hall, S. Miller, and P. Swetik. 1992. Level I Stability Analysis (LISA): documentation for vers. 2.0. USDA For. Serv. Gen. Tech. Rep. INT-285.

Huang, Y.H., 1983. Stability analysis of earth slopes. Van Nonstrand Reinhold Company. New York, NY.

Janbu, N. 1973. Slope stability computation. p. 47–86. *In* H. Hirshfield and H.G. Poulos (ed.) Embankment dam engineering: Casagrande memorial volume. John Wiley and Sons, New York, NY.

Jones, J.A.A., and R.B Bryan (ed.). 1997. Piping erosion. Geomorph. 20.

Kinnel, P.I.A., 1981. Rainfall intensity-kinetic energy relationships for soil loss prediction. Soil Sci. Soc. Am. J. 45:153–155.

Kinnel, P.I.A., 1990. The mechanics of raindrop induced flow transport. Aust. J. Soil Res. 28:497–516

Laws, J.O. 1941. Measurements of fall velocity of water-drops and rain-drops. Am. Geoph. Un. Trans. 22:709–721.

Liebenow, A.M., W.J. Elliot, J.M. Laflen, and K.D. Kohl. 1990. Interrill erodibility: collection and analysis of data from cropland soil. Trans. ASAE 33:1882–1888.

Morgan, R.P.C., 1995. Soil erosion and conservation. 2nd Ed. Longman, London, UK.

Mwendera, E.J., and J. Feyen. 1992. Estimation of depression storage and Manning's resistance coefficient from random roughness measurements. Geoderma 52:235–250.

Naden, P. 1987. An erosion criterion fof gravel bed rivers. Earth Surf. Proc. Landf. 12:83–93.

Nearing, M.A. 1991. A probabilistic model of soil detachment by shallow turbulent flow. Trans. ASAE 34:81–85.

Nearing, M.A., J.M. Bradford, and S.C. Parker. 1991. Soil detachment by shallow flow at low slopes. Soil Sci. Soc. Am. J. 55:339–344.

Palmer, R. 1963. The influence of a thin water layer on waterdrop impact forces. IAHS Pub. 65:141–148.

Panini, T., D. Torri, S. Pellegrini, M. Pagliai, and M.P. Salvador Sanchis. 1997. A theoretical approach to soil porosity and sealing developement using simulated rainstorms. Catena 31:199–218.

Poesen, J. 1984. The influence of slope angle on infiltration rate and Hortonian overland flow volume. Z. Geom. Suppl. Bd. 49:117–131.

Poesen, J. 1985. An improved splash transport model. Z. Geomorph. 29:193–211.

Poesen, J. 1993. Gully topology and gully control measures in the European loess belt. p. 221–239. *In* S. Wicherek (ed.) Farm land erosion in temperate plains environment and hills. Elsevier Science Publishers B.V., Amsterdam, Netherlands.

Poesen, J., and J. Savat. 1981. Detachment and transportation of loose sediments by raindrop splash: part II: Detachability and transportability measuraments. Catena 8:19–41.

Poesen, J., and D. Torri. 1988. The effect of cup size on splash detachment and transport measurements - Part I: Field measurements. Catena Suppl. 12:113–126.

Poesen, J.W., D. Torri, and K. Bunte. 1994. Effects of rock fragments on soil erosion by water at different spatial scales: a review. Catena, 23:141–166.

Poesen, J., K. Vandaele, and B. Van Wesemarl. 1998. Gully erosion: Importance and model implication. *In* J. Boardman and D. Favis-Mortlock (ed.). Modeling soil erosion by water. NATTO ASI Series Vol. 55. Springer Verlag, Berlin, Germany.

Rauws, G., and G. Govers. 1988. Hydraulics and soil mechanical aspects of rill generation in agricultural soils. J. Soil Sci. 39:111–124.

Reid, L.M., and T. Dunne. 1996. Rapid evaluation of sediment budgets. Catena Verlag, Reiskirchen, Germany.

Rengasamy, P., and M.E. Sumner. 1997. Processes involved in sodic behavior. *In* M.E. Sumner and R. Naidu (ed.) Sodic soils. Oxford University Press, New York, NY.

Rose, C.W., P.B. Hairsine, A.P.B. Proffitt, and R.K. Misra. 1990. Interpreting the role of soil strength in erosion processes. Catena Suppl. 17:153–165.

Rousseva, S. 1989. A laboratory index for soil erodibility assessment. Soil Tech. 2:287–299.

Seginer, I. 1996. Gully development and sediment yield. J. Hydrol. 4:236–252.

Shainberg, I., and G.J. Levy. 1996. Infiltration and seal formation processes. p. 1–24. *In* M. Agassi (ed.) Soil erosion, conservation and rehabilitation. Marcel Dekker, New York, NY.

Sharma, P.P., S.C. Gupta, and W.J. Rawls. 1991. Soil detachment by single raindrops of varying kinetic energy. Soil Sci. Soc. Am. J. 55:301–307.

Sherard, J.L., L.P. Dannigan, R.S. Decker, and E.F. Steele. 1976. Pinhole test for identifying dispersive soil. J. Geotec. Eng. Div. ASCE 102:69–85.

Smerdon, E.T., and R.P. Beasley. 1961. Critical tractive forces in cohesive soils. Agric. Eng. 26–29.

Stocking, M.A., 1981. Causes and prediction of the advance of gullies. p. 30–41. *In* Proc. South-East Asian Reg. Symp. Probl. Soil Eros. Sedimen. Bangkok, Thailand.

Sumner, M.E., W.P. Miller, R.S. Kookana, and P. Hazelton. 1997. Sodicity, dispersion, and environmental quality. p. 167–190. *In* M.E. Sumner and R. Naidu (ed.) Sodic soils. Oxford University Press, New York, NY.

Torri, D., and L. Borselli. 1991. Overland flow and soil erosion: some processes and their interactions. Catena Suppl. 19:129–137.

Torri, D., and J. Poesen. 1992. The effect of soil surface slope on raindrop detachment. Catena 19:561–578

Torri, D., J. Poesen, and L. Borselli. 1997. Predictability and uncertainty of the soil erodibility factor using a global dataset. Catena 31:1–22.

Torri, D., R. Ciampalini, and P. Accolti Gil. 1998. The role of soil aggregates in soil erosion processes. p. 247–258. *In* J. Boardmann and D.T. Favis-Mortlock (ed.) Modeling soil erosion by water. Springer-Verlag, Berlin, Germany.

Torri, D., M. Sfalanga, and M. Del Sette. 1987. Splash detachment: runoff depth and soil cohesion. Catena 14:149–155.

van Rhee, C., and A. Bezujen. 1992. Influence of seepage on stability of sandy slopes. J. Geotech. Eng. 118:1263–1240.

Varnes, D.J. 1978. Slope movement: Type and processes. p. 11–33. *In* R.L. Shuster and R.J. Kriziek (ed.) Landslides: Analysis and control. Trasportation Research Board Spec. Rep. 176.

Vukovic, M., and M. Pusic. 1992. Soil stability and deformation due to seepage. Water Resource Publications, Littleton, CO.

Whitlow, R. 1990. Basic soil mechanics. Longman. Hong Kong.

Wischmeier, W.H., and D.D. Smith. 1978. Predicting rainfall erosion losses. USDA Agric. Handb. 537.

Woolhiser, D.A., R.E. Smith, and D.C. Goodrich. 1990. KINEROS: A kinematic runoff and erosion model: Documentation and User Manual. USDA-ARS Pub. ARS-77.

Zanchi, C., and D. Torri. 1980. Evaluation of rainfall energy in central Italy. p. 133–142. *In* M. de Boodt and D. Gabriels (ed.) Assessment of erosion. Wiley, London, UK.

Zobeck, T.M., and C.A. Onstad. 1987. Tillage and rainfall effects on random roughness. A review. Soil Till. Res. 9:1–20.

8

Wind Erosion

D.W. Fryrear
Custom Products (Retired from USDA-ARS, Big Spring, TX)

8.1 Introduction

Wind erosion is a basic geomorphological process that erodes the land surface and shapes yardangs, ventifacts, pedestal rocks, blowouts, ergs, dunes, lag deposits, desert pavement and desert armor (Thornbury, 1957). On every continent, eolian features document the persistence and magnitude of wind forces. While the transport capacity of the wind is much less than that of water, wind erosion can remove the entire nutrient-rich soil surface regardless of field size or location. In the wind erosion process, fine particles are removed leaving coarse particles on the eroded surface. As the surface continues to erode, the remaining soil becomes less fertile. The loss of fertile soil material from agricultural lands degrades the soils in the source area and intensifies wind erosion. The resulting dust clouds contaminate the atmosphere in the transport area and impact citizenry throughout the depositional area. It is not uncommon for the source and depositional areas to be thousands of km apart.

In 1790 Deane used hedges and locust trees and in 1824 Drown used the addition of 50 to 80 mm of clay to blowy sand soil surfaces in efforts to reduce wind erosion in the United States (McDonald, 1971). The basic principle of deep plowing to increase the clay content of the surface soil is still used to control wind erosion on thousands of ha of sandy soil in the Southern Great Plains (Chepil et al., 1963a). The benefits of deep plowing persist for several years, depending on management practices. Straw barriers in a checkerboard pattern (Xu et al., 1982), raised fences (Liu et al., 1983), or forest belts, also called shelterbelts (Dong et al., 1983), have been used in China to protect railroads and highways from blowing sand. Shelterbelts are used in India to reduce wind speeds, erosion, and evaporation losses (Gupta et al., 1981). The success of shelterbelts depends on dominant erosive wind directions, sufficient annual rainfall to sustain the trees, and other potential uses of the trees (lumber, pulpwood, firewood, etc.). Any system that reduces the speed of the wind at the soil surface can significantly reduce soil erosion (Fig. 8.1).

Wind erosion concerns can be grouped into three categories: environmental, agricultural, and transportation (Greeley and Iverson, 1985). Environmental concerns include effects of dust on human health, visibility, and climate. Agricultural concerns include the impact of soil erosion by wind on the ability of agricultural production to be sustained in the longterm. Transportation concerns include reduced visibility from blowing soil on highways, railroads, and airports.

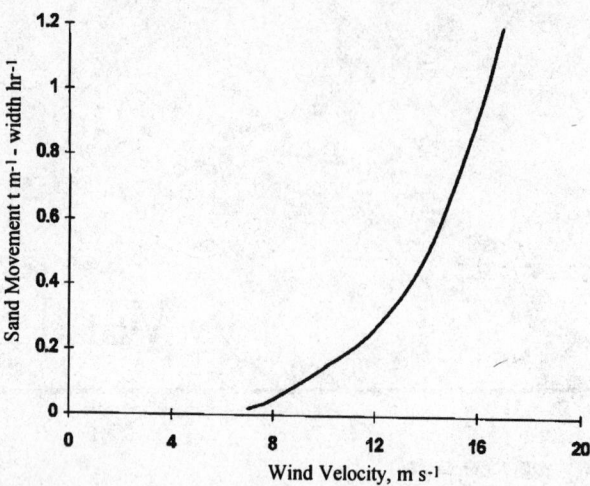

Fig. 8.1   Transport flux of sand as a function of wind velocity [Reprinted from R.A. Bagnold, 1941. The physics of blown sand and desert dunes (Fig. 22) with kind permission of Kluwer Academic Publishers]

8.2   Mechanics of Wind Erosion Processes

Wind is defined as "the motion of the atmosphere relative to the surface" (Greeley and Iverson, 1985). As wind flows over the land surface, the surface exerts a drag such that an atmospheric boundary layer approximately one meter thick is formed. This atmospheric boundary layer impacts the detachment (also called deflation), movement, and deposition of soil particles. When soils erode, the wind within this layer is defined as turbulent. This turbulent characteristic facilitates the transfer of momentum from the wind to the soil surface and exerts a drag or shear effect on the surface. This drag can be computed from wind speeds as follows.

$$\tau = \rho(U_*)^2 \qquad\qquad [8.1]$$

where

τ = wind drag, dynes m^{-2}
ρ = air density, gm m^{-3}
U_* = drag speed, m s^{-1}

Drag speed can be computed by

$$U_* = \frac{U_z}{5.75 \, \log \dfrac{z}{k}} \qquad\qquad [8.2]$$

where

U_* = drag speed, m s^{-1}

U_z = wind speed at height z, m s^{-1}
z = height where U_z is measured, m
k = aerodynamic surface roughness, m

Drag speed can also be determined graphically using the relationships in Fig. 8.2. The wind speed (U_z) is plotted as a function of the log of the height (z). From this plot the aerodynamic surface roughness (k) and drag speed (U_*) can be determined as the intercept and slope, respectively.

The relationship between speed and height (Equation [8.2] and Fig. 8.2) is valid for clear air. As the concentration of transported material increases, the relationships are no longer valid (Anderson et al., 1991).

Weather systems that erode soils vary in their time frames, space scale, and predictability (Table 8.1).

The dust devil is a very common sight in semiarid regions but the time and spatial scales are very short and very small, respectively. The haboob is a very dramatic cloud of dust that appears to roll over the landscape. While haboobs can "blot out the sun" for short periods of time they usually do not cover near the area or last as long as dust storms produced by downslope, frontal, or cyclogenic weather systems. Thirty years of dust storm data at Big Spring, Texas were analyzed and the average number of storm days was 26 per year with an average of 59.8 hours with dust per year (Fryrear, 1981). Severe storms with visibilities of less than 300 meters produced an estimated 54.9% of the total dust load with 11% of the total dust hours. Most of the dust storms included in these analyses resulted from frontal weather systems.

8.2.1 Wind Speed

Wind erosion is possible when the velocity of the wind at the soil surface exceeds the threshold velocity required to move the least stable soil particle. The detached particle may move a few millimeters before finding a more protected site on the landscape. The wind speed required to move this least stable particle is called the static threshold (Bagnold, 1941). If the wind speed increases, soil

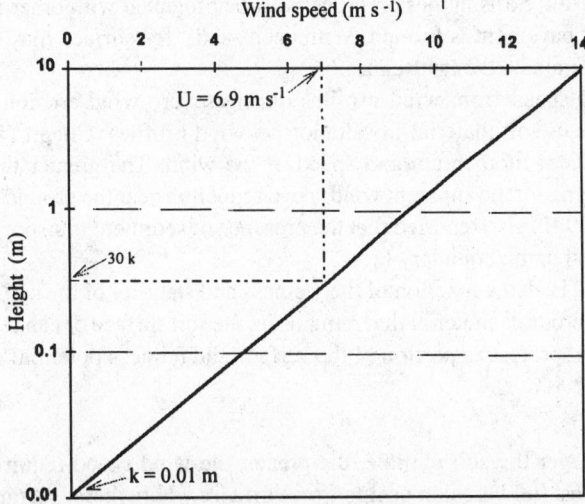

Fig. 8.2 Determining aerodynamic roughness (k) and drag speed (U_*) graphically from plot of wind speed and height of wind speed measurement ($U_* = 6.9/8.493 = 0.8124$ m s^{-1})

Table 8.1 Hierarchy of weather-duststorm systems

Duststorm	Weather System	Duration Time (hr)	Aerial Scale (km)	Predictability
1. Dust Devil	Microtemperature differences	0.1-0.5	0.01-0.50	Observe only in real time
2. Haboob	Thunderstorm downdraft into dry air, i.e., gravity flow	0.05-6	25-75	1-12 hr
3. Severe mountain downslope windstorm	Complex terrain enhancement of downward transport of midtropospheric momentum	0.5-18	25-250	12-36 hr
4. Frontal	Gravity flow, pressure gradient with dynamic assist	1-8	500-1000	24-48 hr
5. Cyclogenic				
a. Low level jet	Boundary layer thermal differences, momentum transfer within shallow adiabatic plantary boundary layer.	6-12	500-1000	24-48 hr
b. Upper level jet	Deep adiabatic heating of troposphere through dynamic subsidence	8-24	500-1000	24-72 hr
c. Surface storm circulation	Deeper-gradient winds	8-18	50-150	12-36 hr
d. Severe mountain downslope windstorm	(Mechanism No. 3 above)*			

movement begins. When soil movement is sustained, the dynamic threshold speed has been attained and the quantity of soil that can be transported by the wind varies as the cube of the velocity (Bagnold, 1941; Chepil and Woodruff, 1963). Many transport equations have been reported that include a term for wind speed (Greeley and Iverson, 1985; Sorensen, 1991).

Soil roughness, erodibility, wetness, and crop residue quantity and orientation are parameters that impact the vulnerability of soils to wind erosion. If the erodible soil surface is covered with vegetation or residues from a previous crop, the force of the wind is transferred to the nonerodible cover and erosion is controlled. Unfortunately, rainfall in arid and semiarid regions may be insufficient to vegetatively protect the soil. Soils in these regions must be protected with other methods or the soil may erode until a desert pavement is formed. With deep sands, the surface may never stabilize and active sand dunes will dominate the landscape.

Computation of drag speed from wind profiles during severe wind erosion events is difficult because of the influence of soil material in saltation on wind profiles. Chepil (1941) reported that saltation movement reduces momentum and speed of the wind. The greater the concentration of moving sand grains, the greater the apparent wind speed reduction near the ground (Chepil and Milne, 1941). Anderson et al. (1991) also reported that the presence of sediment transport in the windstream influences the wind speed in the boundary layer.

Threshold wind speed is also a function of the wetness and stability of the soil surface (Saleh and Fryrear, 1995). Any nonerodible material that remains on the soil surface or can resist the force of the wind during a wind event protects a portion of the surface and reduces potential erosion.

8.2.2 Soil

For mineral soils, the larger the soil particle, the greater the wind speed required to dislodge and transport the particle. Most field erosion occurs from surfaces where there is a range in particle size. For the same wind, the transport rate may be twice as large for particles of mixed size as for particles of uniform size. The speed and turbulence of the wind, density ratio of particles to air and surface roughness dictate the largest soil particle that can be eroded (Chepil and Woodruff, 1963; Batt and Peabody, 1994; Greeley and Iverson, 1985). The erosion of soil aggregates by impacting sand particles increases with the square of the particle velocity. Increasing the size of the abrading particles from 125 to 715 μm gave a small increase in erosion of the soil aggregates (Hagen, 1984). Electromagnetic and static forces influence detachment and transport of small particles. When coupled with cohesive forces, μm size particles may coalesce into larger particles less susceptible to transport by wind.

Wind-eroded particles are normally transported in one of three modes: creep, saltation, or suspension (Chepil and Woodruff, 1963). The specific mode of transport depends on the speed and turbulence of the wind and soil surface conditions. Creep particles 1000 to 2000 μm in diameter roll along the surface. The creep component of the total soil flux can be a significant portion or almost zero. The texture and condition of the soil surface determine if large soil aggregates can roll over the soil or become trapped in microdepressions.

Saltating particles 100 to 1000 μm in diameter move in a series of short hops depending on surface roughness, particle size, and wind speed. For speeds slightly above threshold, a "saltation size" particle may roll on the surface in a creep mode. For high wind speeds, "saltation size" particles may be transported at a considerable height in a mode normally called suspension. The saltating particles are largely responsible for the abrasion of soil aggregates and accelerate the erosion process until the wind has attained its maximum transport capacity.

Suspended particles range from 1 to 100 μm in diameter, move into the upper atmosphere, and are subject to long-range transport. Data suggest that relatively large particles may be transported considerable distances (Betzer et al., 1988). For most wind-eroded surfaces, the majority of the mass being transported is close to the soil surface. That portion in suspension may represent a small percentage of the total mass being transported but may have a major impact on air quality (Saxton, 1995, 1996; Stetler and Saxton, 1995).

8.2.2.1 Soil Erodible Fraction

For most agricultural fields, the soil is composed of sand, silt, and clay particles. Organic matter and $CaCO_3$ are components important in relating soil texture to susceptibility to wind erosion (Fryrear et al., 1994). Using the standard compact rotary sieve (Chepil, 1962) and a sample of the surface 20 mm of dry topsoil, the erodible fraction (EF) of the soil can be determined. Several thousand samples were analyzed to develop the following equation:

$$EF = \frac{29.09 + 0.31\ Sa + 0.17\ Si + 0.33\ Sa/Cl - 2.59\ OM - 0.95 CaCO_3}{100} \qquad [8.3]$$

where

EF	=	erodible fraction (0.077 to 0.822)
Sa	=	sand % (5.5 to 93.6)
Si	=	silt % (0.5 to 69.5)
Cl	=	clay % (1.2 to 53.0)
OM	=	organic matter % (0.18 to 4.79)
$CaCO_3$	=	calcium carbonate % (0.0 to 25.2)

The range of values in the dataset is in parentheses. The percentage of sand, silt, and clay are based on a dispersed mechanical analysis procedure where clay particles are smaller than 2 μm, silt particles are 2 to 60 μm, and sand particles are 60 to 600 μm (Taylor, 1960).

With this equation, the potential erodibility of a mineral soil can be estimated from the physical and chemical properties. Equation [8.3] allows scientists to evaluate wind erosion control practices on small plots. If the soil surface is covered with erodible material, theoretically, erosion will continue as long as the wind speed is above the threshold. Large sand dunes can result, but even in desert dune regions, the interdunal area is normally composed of more resistant clay or rock material.

8.2.2.2   Soil Roughness

In most agricultural systems, the farmer tills the soil to prepare a seedbed, control weeds, or improve rainwater infiltration. Tilling normally increases soil surface roughness. Soil roughness may consist of uniformly spaced ridges or furrows and randomly distributed soil aggregates. The protection from soil ridges persists longer than random roughness, but soil ridge roughness is sensitive to wind direction. When the wind is perpendicular to ridges 0.06 to 0.25 m high, erosion may be controlled (Fig. 8.3). When the wind is parallel to the ridge, the ridge is not effective in controlling wind erosion (Saleh, 1994). Aerodynamic roughness with ridges or soil aggregates controls or, at least, reduces soil erosion until rainfall smooths the surface (Fryrear, 1984; Saleh et al., 1997). Roughness also reduces dust emissions from agricultural fields compared to dust from a flat, dry, loose soil (Batt and Peabody, 1994).

8.2.3 Plants

8.2.3.1 Seedlings

Emerging crop seedlings are sensitive to wind damage. In fact, each year thousands of ha of crops are replanted because of windblown sand damaging or destroying the young crop (Woodruff, 1956b; Skidmore, 1966; Fryrear and Armbrust, 1969; Armbrust et al., 1969; Fryrear, 1971; Armbrust, 1972;

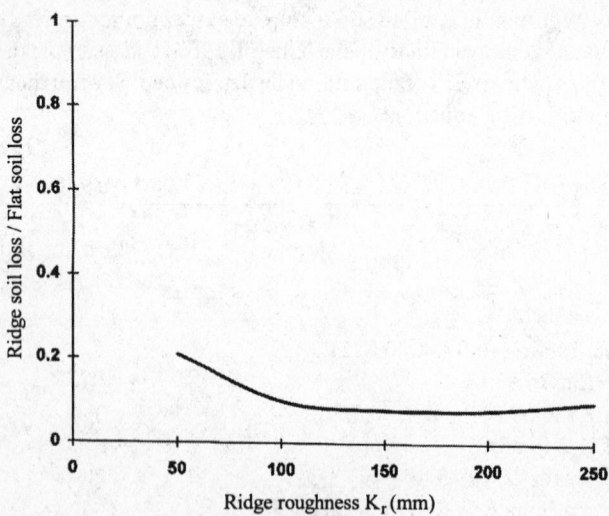

Fig. 8.3   Soil loss ratio for various soil ridge roughness (k_r) values when wind is perpendicular to soil ridge [Fryrear, 1984]

Armbrust, 1984). The first four weeks after planting are critical to many crops. Some crops are so sensitive that a twenty-minute exposure to windblown sand may destroy 75% of the seedlings (Table 8.2). As the plants mature, they are less sensitive to wind damage and the developing plant canopy affords some protection against wind erosion (Fryrear and Downes, 1975; Armbrust, 1977). Windblown sand injury delays maturity, increases susceptibility to diseases, and impacts the physiology of plants (Claflin et al., 1973; Armbrust et al., 1974; Armbrust, 1982).

8.2.3.2 Growing Plants

As crops mature, they can tolerate more sand injury and can reduce the wind speed at the soil surface (Fryrear et al., 1975; Bilbro, 1989, 1991, 1992). However, some crops such as cabbage or tobacco may be of inferior quality if sand is contained within the plant leaves (Fryrear and Downes, 1975; Armbrust, 1979).

8.2.3.3 Flat Residues

When crops are harvested, some plant material remains in the field. This residue or stubble has value as a source of nutrients for the soil, as animal feed, as building material or fuel, and as protection against wind erosion. Studies were conducted to determine the relationship between the proportion of soil surface covered with plant material and soil erosion (Fryrear, 1985; Bilbro et al., 1994). The available published data were combined into a single value that expresses soil loss ratio (SLR_f) as a function of flat soil cover. The SLR_f is the soil erosion with residue cover divided by the soil erosion without residue cover. Fig. 8.4 illustrates that covering a small portion of the soil surface can significantly reduce wind erosion. These studies assume that the flat residues are uniformly distributed over the soil surface.

8.2.3.4 Standing Residues

When small grains, corn, sunflowers, soybeans, or sorghum are harvested, a portion of the crop residue remains standing in the field. Standing residues are at least six times more effective than the same mass of material flat on the soil surface (Englehorn et al., 1952; van de Ven et al., 1989; Bilbro and Fryrear, 1994) (Figs. 8.5 and 8.6). These results emphasize that management practices that maintain residues in an erect position are more effective in controlling wind erosion than practices that flatten the residues or incorporate residues into the soil. The silhouette of a standing crop residue is the vertical area of the plants in one square meter of ground area. The effect of silhouette area on soil erosion is expressed as a soil loss ratio of standing residues (SLR_s). SLR_s is the soil loss with standing residues divided by the soil loss when the surface is bare.

Table 8.2 Exposure time to windblown sand for 25, 50, or 75% survival [Fryrear and Downes, 1975]

Crop	Survival (%)		
	25	50	75
	Exposure time in minutes		
Carrots	11.0	7.5	3.9
Onions	31.0	21.4	11.0
Peppers	12.0	7.9	3.8
Cabbage	14.8	10.1	5.4
Cucumbers	21.2	14.1	7.1
Southern Peas	49.3	32.7	16.0
Cotton	18.3	12.4	6.6

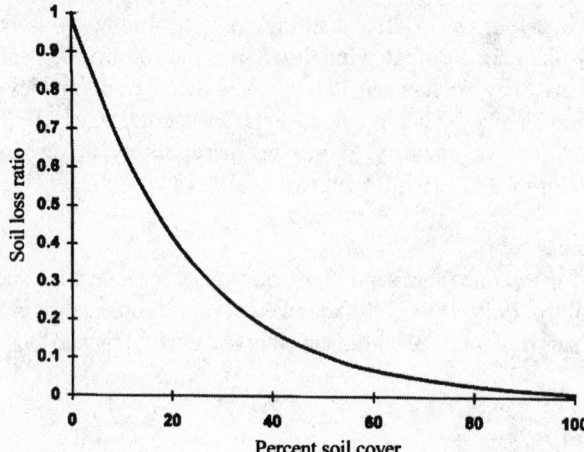

Fig. 8.4   Soil loss ratio (soil loss with partial cover divided by soil loss for same soil without any cover) as a function of percent soil cover with nonerodible elements such as flat residues or rocks

8.2.4   Man

Although wind erosion is a normal process, it can be enhanced by the irresponsible activities of man (Chepil and Woodruff, 1955). When fragile soil or plant resources in arid regions are abused by man's activities, restoration requires many years or may be impossible. The abuse may be the result of surface disturbance from overgrazing, over cultivation, vegetative burning, woodcutting, salinization, or the use of animal dung as fuel instead of fertilizer (Middleton, 1987). When effective management practices are used, the soil, crop, and climatic resources can be utilized indefinitely. The most effective management practice must be flexible enough to respond to variable climatic conditions.

Strzelecki (1845) noted that the clearing of forest, woodlands, or grasslands reduced soil fauna and organic matter and impacted soil temperature and soil permeability. In addition, man-induced changes in surface and atmospheric conditions may impact local and regional climatic conditions (Saxton, 1995, 1996; Stetler and Saxton, 1995; Zobeck et al., 1996; Williams and Balling, 1996). The impact of man may be overshadowed by natural fluctuations in the climatic system.

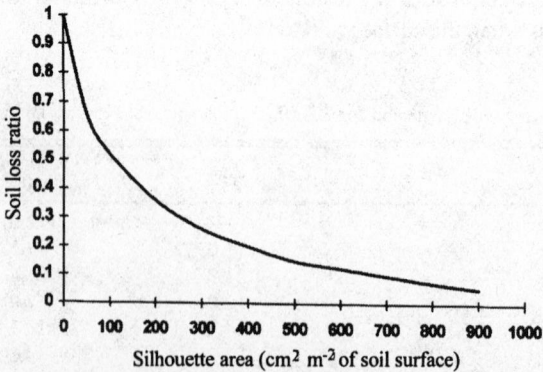

Fig. 8.5   Soil loss ratio (soil loss with partial cover divided by soil loss for same soil without any cover) as a function of silhouette area [Bilbro and Fryrear, 1994]; silhouette area is the profile area of all standing residues in a one meter square area.

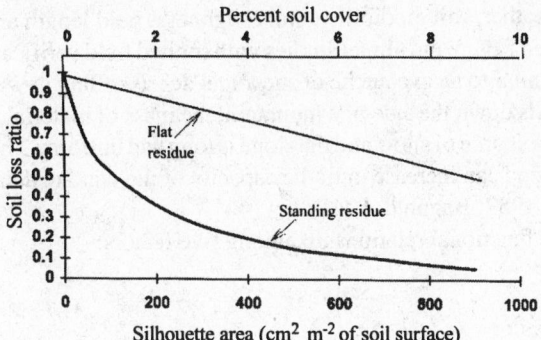

Fig. 8.6 Comparison of soil loss ratio with the same quantity of residue standing as silhouette and flat on the soil surface [Bilbro and Fryrear, 1994]

8.3 Modeling the Wind Erosion Process

The wind erosion process is a complex interaction of weather-soil-plants-man. To evaluate the effect of any single factor, the interactions between factors must be understood. One method of expanding our understanding is to model the process. It is doubtful that the physical process will be completely modeled, but the major factors contributing to wind erosion can be modeled.

Wind erosion models can be physically (also called process), empirically, or theoretically based. The application of empirically based models is limited to the range of conditions under which the model was developed. Also, empirically based models require carefully controlled experiments to develop the essential coefficients. Theoretically and physically based models may be widely adapted but require that the theory and assumptions imbedded within the model are correct. Because of the complex interactions involved in wind erosion, the physical processes may not be well understood. For example, an analytical model developed by Sorensen (1991) utilizes processes for trajectory of the particles, wind modification, and grain bed collisions. Formulae were derived for transport rate and flux of the sand grains from the surface into the air. In developing a conceptual transport model, McEwan and Willetts (1993) concluded that inaccurate wind speeds and sand trap inefficiencies may have been responsible for scatter in wind tunnel data. They also cautioned wind erosion scientists that extrapolation of sand transport rate formulae from the uniform laboratory wind tunnel conditions to field conditions may not be justified.

An empirically derived differential equation of dust concentrations in a thin layer near the ground to predict dust concentrations at the top of the surface layer was developed by Berkofsky and McEwan (1994). Unfortunately, it neglects horizontal transport which is a major component of the wind erosion process; however, they recognized that the detachment plus a source term was inversely proportional to surface roughness.

While the above studies have been valuable sources of new approaches and have expanded our understanding of the wind erosion process, they have not been combined with other input parameters to develop a complete wind erosion model. Essentially all the models developed to date are a combination of all three approaches. The user must determine which model produces the desired output with the available input data.

8.3.1 Wind Erosion Equation (WEQ)

WEQ (Woodruff and Siddoway, 1965) has been widely used to plan conservation systems and estimate annual soil erosion. WEQ was one of the first models to estimate soil erosion in the field using annual input values for weather, soil erodibility, soil roughness, field length and crop residues. The WEQ, which was developed from wind tunnel studies with limited field verification, assumes that the wind erosion process is similar to an avalanche of snow and debris rolling down a mountain (Chepil, 1957). As an avalanche rolls down the side of a mountain, the mass of material in the avalanche may increase as long as there is a source of snow and the slope is long and uniform. The quantity of material being transported by the wind can increase until the capacity of the wind to detach and transport soil has been reached (Chepil, 1957; Bagnold, 1941).

The WEQ consists of a functional relationship among five factors:

$$E = f(IKCLV)$$
[8.4]

where

E = annual soil erosion, Mg ha^{-1}
f = functional relationship
I = soil erodibility, Mg ha^{-1} yr^{-1}
K = soil roughness factor
C = climatic factor
L = unprotected field length, m
V = vegetation factor expressed as small grain equivalent.

WEQ estimates soil loss (note units of E and I) for an unprotected field length (L). If soil loss is divided by field length, the units are mass per unit width or an expression of transport mass.

Concerns that the climatic factor did not describe wind erosion in high or low rainfall regions were addressed and a new climatic factor was developed (Woodruff and Armbrust, 1968; Skidmore, 1986). The WEQ was modified for shorter time periods, but as stated by Bondy et al. (1980, p. 176), "No experimental data base exists for using the wind erosion equation for periods of less than one year."

WEQ was coupled with the Erosion Productivity Impact Calculator (EPIC) (Williams et al., 1984) to estimate daily wind erosion soil losses (Cole et al., 1983). The C, I, K and V factors in WEQ are annual values, but in the EPIC daily wind erosion model they are considered constant for a single day. Comparisons of estimated wind erosion with EPIC were verified with WEQ estimates.

8.3.2 Pasak

A soil loss model reported by Pasak in 1973 was intended as a single event model (Holy, 1980):

$$E = 22.02 - 0.72P - 1.69V + 2.64Rr$$
[8.5]

where

E = erodibility, kg ha^{-1}
P = nonerodible particles, %
V = relative soil moisture
Rr = wind speed, km hr^{-1}

The lack of inputs for crop residues and soil roughness limits the use of this model. Also wind speed and soil moisture conditions are not constant during an erosion event.

8.3.3   Texas Erosion Analysis Model (TEAM)

TEAM developed by Gregory et al. (1988) is a computer program developed to simulate wind profile development and soil movement over multiple field lengths. The basic equation is

$$X = C\left(SU_*^2 - U_{*_t}^2\right)U_*\left(1 - e^{-0.000169\,A_a\,IL_f}\right) \qquad [8.6]$$

where

X	=	rate of soil movement at length L_f, $M\,L^{-1}\,T^{-1}$
$C(SU_*^2 - U_{*_t}^2)U_*$	=	maximum rate of soil movement which occurs when the surface is covered with fine noncohesive material
C	=	a constant which depends on width sampled and units used for U_*
L_f	=	length of unprotected field in the wind direction
A_a	=	abrasion adjustment factor
I	=	soil erodibilitiy factor involving soil shear strength and angle
S	=	surface cover factor which expresses the amount of detachment energy at the top of the cover which is transferred to the soil
U_*	=	shear velocity
U_{*_t}	=	threshold shear velocity

The abrasion adjustment factor is computed using the equation:

$$A_a = \left(1 - A_l\right)\left(1 - e^{-0.0072 e^{0.00079\,IL_f}}\right) \qquad [8.7]$$

where

A_l = lower limit of abrasion effect (assumed 0.23).

8.3.4   Wind Erosion Assessment Model (WEAM)

This process-based model developed in Australia is a combination of established and theoretical results (Shao et al., 1996). It synthesized research results on sand drift and dust entrainment, approximated quantitative assessment of wind erosion, and evaluated the current limits to our knowledge of the wind erosion processes. The WEAM model does not predict the evolution of soil surface properties due to wind erosion, natural weathering processes, or management interventions. WEAM treats erosion of a dry, bare soil in a given wind. To model daily or annual erosion, it is essential that changes in soil and crop conditions be described by the model.

When coupled with large computers, WEAM does have the capability of estimating the movement of large dust clouds across complex landscapes. Input requirements are such that GIS or other information databases are essential to utilize this model effectively. The lack of management intervention inputs limits WEAM's usefulness for field or short period erosion estimates. It does express the regional or continental impacts of wind erosion.

8.3.5   The Wind Erosion Prediction System (WEPS)

The WEQ did not describe wind erosion in high rainfall areas or in extremely dry regions. The climatic factor (Woodruff and Armbrust, 1968; Skidmore, 1986) and management periods (Bondy et al., 1980) for WEQ were modified but never validated with field erosion data. As erosion measuring equipment (Fryrear, 1986) and procedures (Fryrear et al., 1991) were developed, it became apparent that there were questions concerning the use of WEQ. To improve erosion estimates, available technology must be assembled into a new model called the Wind Erosion Prediction System (WEPS). WEPS will be a process-based model with a continuous, daily time step that simulates weather, field conditions, and wind erosion. WEPS is modular in design so the model can be upgraded with improved technology (Hagen, 1991). The WEPS model simulation must begin when there is no growing crop and must cover at least a single year. Heterogeneous regions are simulated by dividing into homogenous subregions where soil type, biomass, and management are similar (Hagen et al., 1995). WEPS will permit subhourly estimates of soil movement and will describe changes in surface conditions during erosion events. Development is continuing.

8.3.6   Revised Wind Erosion Equation (RWEQ)

To expedite the availability of new technology, the RWEQ effort was suggested. The RWEQ model requires simple inputs to estimate long-term soil loss from agricultural lands. RWEQ has been field tested against measured transport data for a broad range of soil, field, and climatic conditions. A physically based mass transport equation was derived by Stout (1990) for RWEQ. This equation is based on the concept that a self-balancing mechanism controls wind erosion and is the major relationship in RWEQ. RWEQ utilizes the principle that wind erosion is the result of the erosive force of the wind exceeding resisting forces. The resisting forces include the soil inherent wind erodibility, the surface roughness, and the protection from flat or standing plant residues.

With the self-balancing concept, momentum is transferred from the wind to the soil surface (Owens, 1964). The equation derived by Stout (1990) is

$$\frac{Q_x}{Q_{max}} = 1 - e^{-\left(\frac{x}{b}\right)} \qquad [8.8]$$

where

Q_x = mass being transported by wind at field length x, kg m^{-1}
Q_{max} = maximum transport capacity of wind over that field surface, kg m^{-1}
x = field length, m
b = field length where wind attains 63.2% of Q_{max}, m

Equation [8.8] assumes if the wind speed is above threshold, the erosion process detaches and picks up soil particles at an increasing rate across a field. Equation [8.8] is a negative exponential decay curve that gives increasing transport mass as field length increases. The highest soil loss per unit area occurs immediately downwind of the nonerodible upwind boundary, assuming $Q_x = 0$ at x = 0 (Owens, 1964). While developed for transport layers close to the soil surface, Stout's (1990) basic equation was used to describe the total mass being transported between the soil surface and a height of 2 meters. Equation [8.8] was modified to the form:

$$Q_x = Q_{max}\left(1 - e^{-\left(\frac{x}{s}\right)^2}\right)$$ [8.9]

where

s = inflection point where the slope changes from positive to negative.

Equation [8.9] gives a sigmoid curve where the mass being transported is zero at the upwind boundary, then increases until the wind attains 63.2% of its capacity at field length "s" (Fig. 8.7)(Fryrear and Saleh,1996). Beyond field length "s" the rate of increase in transport mass is limited by the capacity of the wind to detach and transport additional soil. Therefore, the rate of increase in mass being transported diminishes until the maximum transport capacity has been achieved. The basic relationship between field length and transport mass in RWEQ reflects that as the quantity of material being transported approaches the maximum transport capacity of the wind, any additional material picked up must result in the deposition of a portion of the original load. The field length required for the wind to become saturated varies with the soil surface conditions and the velocity of the wind. Rarely does the wind become saturated before surface conditions or wind velocity changes. With measurements of transport mass and field lengths, the coefficients Q_{max} and "s" can be determined using statistical least square analysis. When estimating transport mass for noninstrumented fields, the field conditions and weather are the only inputs available.

The input data required to operate RWEQ include factors for weather (WF), soil (EF, SCF), crop (COG), and tillage (K). Measured values for the input factors were correlated with computed values for Q_{max} and "s" from single erosion events. These events were for fields of the same size and shape. The soil surface was flat and dry (SW = 1) with no snow cover (SD = 1).

From this analysis, the relationship between the input parameters and Q_{max} is described as:

$$Q_{max} = 109.8(WF \times EF \times SCF \times K \times COG)$$ [8.10]

where

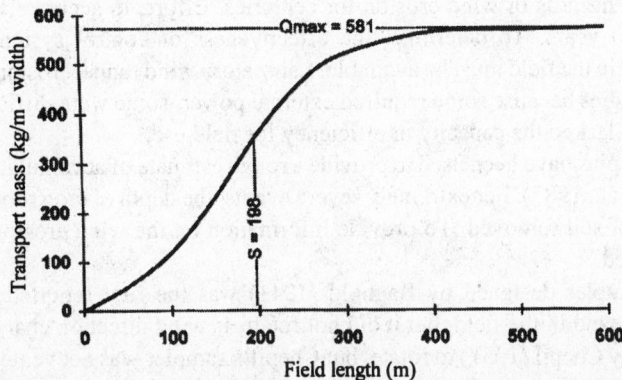

Fig. 8.7 Measured mass flux (20 minute periods, sampler 12.5 mm wide x 50.8 mm high) from western Canada [Chepil and Milne, 1941]

Q_{max} = maximum transport capacity (kg m–width^{-1})
WF = weather factor in (kg m-width^{-1}) (contains SW and SD values)
EF = soil erodible fraction
SCF = soil crust factor
K = soil roughness
COG = vegetation factor including flat residues, standing residues, and canopy.
SW = soil wetness
SD = snow depth

The slope "s" of the empirical relationship in [8.9] is also computed from the input factor using the equation:

$$s = 150.71(WF \times EF \times SCF \times K \times COG)^{-0.3711}$$ [8.11]

where

s = critical field length where Q_x is 63.2% of Q_{max}, m

When input data are available, the estimated Q_{max} and "s" values for single storms are close to the measured values (Table 8.3). With the coefficients from equations [8.10] and [8.11] as input for Equation [8.9], soil erosion can be computed for circular fields up to 60 ha and for square or rectangular fields up to 145 ha. The WF is time based, so erosion can be computed for any time interval when the remaining parameters are constant. Within the RWEQ computer program, the WF is adjusted for rainfall and snow cover, crop residues are decomposed by water and temperatures (Steiner et al.,1994), and soil roughness is decayed by rainfall amount and intensity (Saleh et al., 1997).

The same coefficients used to estimate soil erosion for a single event are used to estimate erosion from the entire erosion season (Table 8.4). With the RWEQ computer program, erosion is computed for a field length along the dominant wind direction. The transport mass from Equation [8.9] is important in evaluating potential plant injury from blowing sand. The critical field length from Equation [8.9] can be used to plan field barrier systems or field widths that minimize soil erosion.

8.4 Wind Erosion Measuring Systems: Field Instrumentation

Man has lived with the hazards of wind erosion for centuries. Efforts to accurately measure wind erosion span almost 60 years. To determine the effectiveness of control systems, methods of measuring wind erosion in the field must be available. Laboratory wind tunnel erosion samplers were not suitable for field studies because some required external power, some were difficult to construct and maintain, and some lacked the capacity or efficiency for field use.

Permanent elevation pins have been used to provide a rough estimate of accumulated erosion over many years (Gibbons et al., 1983). For extremely severe events, the depth of crop roots exposed is an indicator of the depth of soil removed. To provide information on the wind erosion process, field measurements are needed.

The vertical slot sampler designed by Bagnold (1941) was the first reported instrument for collecting wind-eroded sand in the field, but it did not rotate as wind direction changed. Bagnold's sampler was modified by Chepil (1957) to rotate, but Chepil's sampler was not vented nor suited for highly erodible soils. Later samplers were vented and rotated into the wind (Fryberger et al., 1979;

Table 8.3   Wind factor (WF), soil erodible fraction (EF), soil crust factor (SCF), soil roughness (K), flat and standing residues (COG), maximum transport capacity (Q_{max}), and critical field length (s) for selected erosion events. Sites are coded Big Spring, Texas (BS); Mabton, Washington (MW); Elkhart, Kansas (EK); Kennett, Missouri (KM); and Eads, Colorado (EC)

Site	Date	WF	EF	SCF	K	COG	Q_{max}	s
							kg m^{-1}	m
BS	27-01-90*	2.3	.64	.77	.95	.90	112	123
BS	29-01-90	2.8	.64	.77	.95	.90	133	88
BS	08-02-90	.6	.64	.77	.95	.90	96	289
BS	06-03-90	2.8	.64	.77	.95	.90	226	149
BS	29-03-93**	3.6	.77	.77	1.00	.96	402	84
MW	02-04-91	8.4	.79	.91	.82	.43	168	43
EK	09-03-92	41.9	.70	.65	.91	.65	1430	98
KM	13-03-93	15.3	.85	.90	.85	1.00	751	109
EC	12-03-91	179.9	.26	.21	.80	.48	648	179

* Includes January 27th and 28th, 1990 wind data.
** Includes March 28th and 29th, 1993 wind data.

Merva and Peterson, 1983), but had a metal edge that slid over the eroding soil surface. This metal lip presented a problem in collecting surface creep.

Fryrear (1986) developed the BSNE (Big Spring Number Eight) sampler that collects airborne material above the soil surface. The BSNE does not contact the soil surface and is the first field wind-aspirated sampler to accurately sample eroded material. Extensive field wind erosion data have been collected from these samplers (Fryrear et al., 1991). Wind-eroded material can be collected from a height of 0.05 meters to any accessible height. The BSNE's sampling efficiency averaged over 89% for washed sand or sieved soil (sieved to remove all aggregates larger than 250 mm) and wind velocities of 10.4 to 15.7 m s^{-1} (Fryrear, 1986). For particles less than 10 μm, the efficiency was 40% (Shao et al., 1993). Samples from field erosion equipment can be used to validate wind erosion models and to support assumptions on the wind erosion process (Fryrear et al., 1991; Fryrear and Saleh, 1993).

Stout and Fryrear (1989) developed a surface creep and saltation sampler for collecting eroded material from the soil surface to a height of 0.2 meters. Stout's sampler has an efficiency of 98% in wind tunnel tests with fine sand and the rotating portion does not contact the soil surface. The sampler is suited for field use and has operated successfully in extremely erodible soil; however, the sampler must be lowered as the soil surface erodes. The combination of the BSNE and Stout's sampler permits the measurement of eroding material from the soil surface to any accessible height. For most studies, the BSNE samplers are located from 0.05 to 1.0 meters above the soil surface.

With the development of good field erosion measuring equipment (Fryrear, 1986; Fryrear et al., 1991), significant improvements have been made in describing wind erosion mechanics. Procedures have been developed to analyze field erosion data (Fryrear and Saleh, 1993) and interpret the results (Fryrear et al., 1991; Vories and Fryrear, 1991). Wind erosion has been measured in numerous fields (ranging from 2.5 to 145 ha) to test a working wind erosion model on a variety of weather, crop, soil, and farming systems. With the measurements of field erosion, it is possible to establish baseline conditions for future comparisons. The field measurements with the BSNE verified that soil mass transport varies as the cube of the wind velocity (Table 8.4). Mass transport appears to increase with field length until the wind stream becomes saturated. Mass transport as high as 1231 kg m-width^{-1} has been measured from a single erosion event (Table 8.3–Elkhart, Kansas 09/03/1992). This level of transport is sufficient to destroy most crop seedlings within 15 minutes. Measured soil losses from an entire erosion season varied from 0.00 to 28.9 kg m^{-2} (Table 8.4).

8.4.1   Threshold Velocities

Visual observations can be used to identify when soil erosion begins (i.e., when threshold velocity has been exceeded). Since it is not always possible to be in the field when erosion begins, Gillette and Stockton (1986) developed a piezo-electric quartz crystal sensor to detect the exact minute soil movement begins. With this device, the number of particles and the momentum of the particles being transported by the wind can be measured. This sensor, called SENSIT, also permits the researcher to accurately determine the duration of wind erosion events and the intensity of blowing sand during the erosion event. Coupled with wind direction sensors, the direction of erosive winds can be determined.

8.5   Major Control Systems

The resources available influence the type of practices used to control wind erosion. The major practices include surface residues, soil roughening, wind barrier/shelterbelts, cover crops, and soil amendments. Because the climate is so variable, no single practice is equally efficient each year nor suited for all regions of the country. The recommendations are to understand the principles of erosion control practices and use combinations of practices. For example, surface residues can be very effective, except in the more arid climates where normal residue levels may not be sufficient to provide complete protection. Also, in some regions termites may consume all dead surface residues. Soil roughening may be effective until rainfall or irrigation has destroyed surface aggregates. Wind barriers or shelterbelts can provide excellent protection, but like soil ridges they are not effective when the wind is parallel to the barrier. Trees require many years before they are tall or dense enough to provide complete protection from perpendicular winds.

8.5.1   Crop Residues

The effectiveness of crop residues in controlling soil erosion can be expressed as a percentage of the soil surface that is covered. After small grains, corn, or sorghum are harvested, there is usually sufficient residue in the field to protect the soil. When these same residues have value as fodder for animals, fuel (Williams and Balling, 1996), or building materials, there are economic pressures to remove more and more of the residue. The relationship between soil erosion and soil covered with residues flat on the soil surface illustrate that covering even 30% of the surface reduces soil losses by 70% (Fig. 8.4). The challenge is to leave sufficient residue in the field to maintain the soil resource.

The impact of standing stubble has been extensively researched (Englehorn et al., 1952; Siddoway et al., 1965; Skidmore et al., 1966). van de Ven et al. (1989) developed a silhouette index to describe the relationship between standing plant material (stubble) and wind erosion (Fig. 8.5). This way of describing the effect of standing residue is currently used in the RWEQ wind erosion model. Standing residues present a silhouette to the wind that reduces the wind stress at the soil surface. Standing residues are about six times more effective in reducing wind erosion than the same mass lying flat on the soil surface (Fig. 8.6). The combined effect of crop residues and soil roughness can be expressed as the product of the coefficients (Horning et al., 1996).

Plant seedlings may be damaged or destroyed by blowing sand (Fryrear and Downes, 1975). As seedlings develop and mature, the taller plants provide some protection to the soil surface. This protection has been related to the portion of the soil surface covered with plant canopy (Bilbro and Fryrear, 1994; Armbrust and Bilbro, 1997). As relationships that include rigidity, leafiness, and canopy height are improved, these will be incorporated into the new models.

Table 8.4 Comparison of total soil loss measured on instrumented sites and soil loss estimated with RWEQ for various time periods

Site	Time Period		Soil Loss	
	Beginning	End	Measured	Estimated
			-------------------kg m^2-------------------	
Akron, CO	10/27/88	05/26/89	0.83	6.70
Akron, CO	10/20/89	04/25/90	1.10	0.38
Eads, CO	10/30/90	05/07/91	2.43	1.57
Eads, CO	09/15/91	04/15/92	0.67	0.81
Crown Point, IN	01/10/90	12/31/90	31.21	23.42
Crown Point, IN	01/01/91	12/31/91	23.95	11.23
Crown Point, IN	01/01/92	06/04/92	0.42	0.67
Elkhart, KS	02/27/90	12/30/90	0.29	0.63
Elkhart, KS	01/01/91	12/30/91	1.32	3.41
Elkhart, KS	01/01/92	10/15/92	15.50	20.73
Elkhart, KS	01/01/93	05/25/93	2.06	13.72
Crookston, MN	11/07/88	11/01/89	0.21	0.00
Crookston, MN	11/02/89	05/06/90	0.32	0.00
Swan Lake, MN	04/11/91	12/31/91	1.14	2.51
Swan Lake, MN	01/01/92	12/31/92	0.13	0.04
Swan Lake, MN	01/01/93	12/31/93	0.00	0.00
Kennett, MO	12/02/92	06/17/93	13.73	5.42
Kennett, MO	11/08/93	05/05/94	0.64	1.75
Havre, MT	10/28/92	05/05/93	0.01	4.44
Havre, MT	10/19/93	03/30/94	0.01	0.52
Lindsey, MT	10/18/90	05/21/91	0.03	1.97
Lindsey, MT	10/08/91	04/08/92	0.09	2.02
Scobey, MT	10/03/88	05/10/89	4.68	7.37
Scobey MT Fallow	10/04/89	04/21/90	1.34	0.07
Scobey, MT Stubble	10/04/89	04/21/90	0.39	0.00
Sidney, NE	10/25/88	05/24/89	0.52	0.52
Sidney, NE	10/24/89	04/24/90	0.38	0.04
Sidney, NE	10/31/90	05/07/91	2.29	4.44
Portales, NM	11/24/94	04/06/95	0.01	0.09
Fargo, ND	12/06/94	05/07/95	0.00	0.00
Fargo, ND	10/24/95	05/07/96	0.00	0.00
Fargo, ND	10/24/96	05/15/97	0.00	0.00
Big Spring, TX	01/12/89	05/03/89	21.54	25.26
Big Spring, TX	01/05/90	05/04/90	20.96	17.06
Big Spring, TX	01/28/91	05/15/91	0.10	0.11
Big Spring, TX	03/16/93	06/01/93	28.78	29.85
Big Spring, TX	01/06/94	05/18/94	17.16	11.23
Big Spring, TX	01/11/95	05/15/95	26.29	39.15
Big Spring, TX	01/12/96	05/16/96	3.99	10.51
Big Spring, TX	01/23/97	05/23/97	13.63	11.99
Martin-C, TX #2	01/24/95	06/06/95	0.30	0.20
Martin-C, TX #3	01/11/95	05/23/95	0.80	0.52
Martin-C, TX #4	02/11/95	05/19/95	0.30	0.43
Plains E, TX	11/15/94	06/03/95	2.20	1.82
Plains B, TX	12/13/94	05/24/95	1.60	0.36
Plains, TX E	12/12/95	06/04/96	3.83	0.61
Plains, TX B (800m)	12/13/95	05/29/96	2.02	0.22
Plains, TX B (1600m)	12/13/95	05/29/96	1.55	0.54
Mabton, WA	12/13/90	04/28/91	3.68	2.98
Prosser, WA #1	12/03/91	03/25/92	0.17	0.13
Prosser, WA #2	06/10/92	06/15/93	0.32	1.05

8.5.2   Soil Roughness

Managing crop residues is an important method of controlling wind erosion but if crops other than small grains, corn, or sorghum are grown, residue production may not be sufficient to protect fragile soils. During drought periods or when residue levels are not adequate to protect erodible soils, other erosion control methods must be employed.

Roughening the soil with tillage implements can be an effective alternative (Fryrear, 1984) (Fig. 8.3). Soil roughening can be effective if performed before the soils begin to erode (Woodruff et al., 1957). Timing is critical, and may be more important than the implement used to roughen the soil (Chepil et al., 1952; Woodruff and Chepil, 1956).

The impact of effective tillage operations was described by Chepil and Woodruff (1954). They related that when potential soil loss was 25 Mg ha^{-1}, a single lister operation reduced potential soil loss to 0.25 Mg ha^{-1}. The same magnitude was reported by Fryrear (1969).

Since tillage can be very effective in controlling wind erosion, methods of describing soil roughness (Zobeck and Onstad, 1987; Potter et al., 1990), tillage effects (Zobeck and Popham, 1992; Saleh et al., 1997) and methods of measuring surface roughness (Zobeck and Onstad, 1987; Saleh, 1993) have been developed.

8.5.3   Wind Barriers/Shelterbelts

Hedges (McDonald, 1971), trees (Woodruff and Zingg, 1955; Woodruff, 1956a), and annual crops (Fryrear, 1963) have been used as barriers to reduce wind speeds to the lee of the barrier. The expression that the barrier protects the soil surface ten times the height of the barrier is correct for only one wind speed, barrier density, and surface condition. Field research is underway to develop a wind barrier model that relates leeward wind reduction and the shape and extent of the protected zone to wind velocity and barrier density (J.D. Bilbro, Personal communication).

8.5.4   Soil Amendments

Controlling wind erosion requires the use of the available soil and plant resources. In many regions wind erosion is closely associated with drought conditions. During prolonged droughts plant residue levels decline and soil roughness techniques are ineffective. If the drought continues over multiple years, annual crop barriers may not grow and trees may die. The options available for reducing wind erosion become very limited. To provide one additional option, research has been and is being conducted on soil amendments (Chepil, 1954, 1955; Chepil et al., 1963b; Armbrust and Dickerson, 1971; Lyles et al., 1969). The problems include cost of the amendment, excessive quantities needed, and limited effectiveness.

8.6   Conclusions

Wind erosion is one of the basic geological processes that shape and mold the landscape. Deep deposits of rich loess soil are evidence that wind erosion and deposition have played a major role in modifying the earth's surface. These loess soils are vital agricultural regions in modern agriculture.

Research on wind erosion began in the 1700s and continues today. This research is aimed at understanding the mechanics of wind erosion, measuring the products of erosion, modeling the processes, and understanding interactions. Field measuring methods have been developed to describe plant residues and soil roughness conditions.

Modeling wind erosion begins with WEQ. New models (WEAM, TEAM, RWEQ, and WEPS) are much more flexible and reflect new science not available in WEQ. These models represent the current knowledge of wind erosion mechanics and provide the cornerstone for evaluating wind erosion control systems. These systems take advantage of the available soil, plant, management, and weather resources.

Field measuring techniques have progressed to the point where it is possible to detect the moment soil erosion begins and mass is being transported. With these field samples the vertical and horizontal distribution of eroded material within an eroding field can be described. These field data permit the verification of estimated erosion.

The new models permit designing systems that effectively utilize the available resources and efficiently control wind erosion.

8.7 References

Anderson, R.S., M. Sorensen, and B.B. Willetts. 1991. A review of recent progress in our understanding of the aeolian sediment transport. Acta Mechanica [Suppl] 1: 1–19.

Armbrust, D.V. 1972. Recovery and nutrient content of sandblasted soybean seedlings. Agron. J. 64:707–708.

Armbrust, D.V. 1977. Chemical tests to evaluate plant sandblast damage. Agron. J. 69:523–525.

Armbrust, D.V. 1979. Wind- and sandblast-damage to tobacco plants at various growth stages. Tobacco Sci. XXIII:117–119 and Tobacco Int. 181:63–65.

Armbrust, D.V. 1982. Physiological responses to wind and sandblast damage by grain sorghum plants. Agron. J. 74:133–135.

Armbrust, D.V. 1984. Wind and sandblast injury to field crops: Effect of plant age. Agron. J. 76:991–993.

Armbrust, D.V., and J.D. Bilbro. 1997. Relating plant canopy characteristics to soil transport capacity by wind. Agron. J. 89:157–162.

Armbrust, D.V., and J.D. Dickerson. 1971. Temporary wind erosion control: Cost and effectiveness of 34 commercial materials. J. Soil and Water Cons. 26:154–157.

Armbrust, D.V., J.D. Dickerson, and J.K. Greig. 1969. Effect of soil moisture on the recovery of sandblasted tomato seedlings. J. Am. Soc. Hort. Sci. 94:214–217.

Armbrust, D.V., G.M. Pawlsen, and R. Ellis. 1974. Physiological responses to wind- and sand blast-damaged winter wheat plants. Agron. J. 66:421–423.

Bagnold, R.A. 1941. The physics of blown sand and desert dunes. Methuen, London, England.

Batt, R.G., and S.A. Peabody II. 1994. Dust sweep-up experiments. Defense Nuclear Agency, Alexandria, VA, Tech. Rep. 94.

Berkofsky, L. and I. McEwan. 1994. The prediction of dust erosion by wind: An interactive model. Boundary-layer Meteorol. 67:385–406.

Betzer, P.R., K.L. Carder, R.A. Duce, J.T. Merrill, N.W. Tindale, M. Uematsu, D.K. Costello, R.W. Young, R.A. Feely, J.A. Breland, R.E. Bernstein, and A.M. Greco. 1988. Long-range transport of giant mineral aerosol particles. Nature 336(6199):568–571.

Bilbro, J.D. 1989. Evaluation of sixteen fall-seeded cultivars for controlling wind erosion. J. Soil and Water Cons. 44:228–231.

Bilbro, J.D. 1991. Relationships of cotton dry matter production and plant structural characteristics for wind erosion modeling. J. Soil and Water Cons. 46:381–384.

Bilbro, J.D. 1992. Sunflower dry matter production and plant structural relationships for wind erosion modeling. J. Soil and Water Cons. 47:194–197.

Bilbro, J.D., and D.W. Fryrear. 1994. Wind erosion losses as related to plant silhouette and soil cover. Agron. J. 86:550–553.

Bilbro, J.D., E.L. Harris, and O.R. Jones. 1994. Erosion control with sparse residue. USDA-ARS, Cons. Res. Rep. 37. p. 30–32.

Bondy, E., L. Lyles, and W.A. Hayes. 1980. Computing soil erosion by periods using wind-energy distribution. J. Soil and Water Cons. 35:173–176.

Chepil, W.S. 1941. Relation of wind erosion to the dry aggregate structure of a soil. Sci. Agr. 21:488–507.

Chepil, W.S. 1954. The effect of synthetic conditioners on some phases of soil structure and erodibility by wind. Soil Sci. Soc. Am. Proc. 18:386–391.

Chepil, W.S. 1955. Effects of asphalt on some phases of soil structure and erodibility by wind. Soil Sci. Soc. Am. Proc. 19:125–128.

Chepil, W.S. 1957. Width of field strips to control wind erosion. Kan. Ag. Exp. Sta. Tech. Bull. No. 92, Dec.

Chepil, W.S. 1962. A compact rotary sieve and the importance of dry sieving in physical soil analysis. Soil Sci. Soc. Am. Proc. 26:4–6.

Chepil, W.S., E. Burnett, and F.L. Duley. 1963a. Management of sandy soils in the Central United States. USDA Farmers' Bul. No. 2195, Nov. 1963.

Chepil, W.S., C.L. Englehorn, and A.W. Zingg. 1952. The effects of cultivation on erodibility of soils by wind. Soil Sci. Soc. Am. Proc. 16:19–21.

Chepil, W.S., and R.A. Milne. 1941. Wind erosion of soil in relation to roughness of surface. Soil Sci. 52:417–433.

Chepil, W.S., and N.P. Woodruff. 1954. Estimations of wind erodibility of field surfaces. J. Soil and Water Cons. 9:257–265, 285.

Chepil, W.S., and N.P. Woodruff. 1955. How to reduce dust storms. Kans. Agric. Expt. Stn. Cir. No. 318, p. 11.

Chepil, W.S., and N.P. Woodruff. 1963. The physics of wind erosion and its control. Adv. Agron. 15:211–301.

Chepil, W.S., N.P. Woodruff, F.H. Siddoway, D.W. Fryrear, and D.V. Armbrust. 1963b. Vegetative and non-vegetative materials to control wind and water erosion. Soil Sci. Soc. Am. Proc. 27:86–89.

Claflin, L.E., D.L. Stuteville, and D.V. Armbrust. 1973. Windblown soil in the epidemiology of bacterial leaf spot of alfalfa and common light of beans. Phytopath. 63:1417–1419.

Cole, G.W., L. Lyles, and L.J. Hagen. 1983. A simulation model of daily wind erosion soil loss. Tran. ASAE 26:1758–1765.

Dong, G., G. Zou, C. Li, and F. Cheng. 1983. Preliminary observation of the efficiency of the wind-preventing and sand-resisting forest belt in the western part of the Great Bend of the Huanghe River. J. Desert Res. 3:9–19.

Englehorn, C.L., A.W. Zingg, and N.P. Woodruff. 1952. The effects of plant residue cover and clod structure on soil loss by wind. Soil Sci. Soc. Am. Proc. 16:19–21.

Fryberger, S.G., T.S. Ahlbrandt, and S. Andrews. 1979. Origin, sedimentary features, and significance of low-angle eolian "sand sheet" deposits, Great Sand Dunes National Monument and Vicinity, Colorado. J. Sed. Petrol. 49:733–746.

Fryrear, D.W. 1963. Annual crops as wind barriers. Trans. ASAE 6:340–342, 352.

Fryrear, D.W. 1969. Reducing wind erosion in the Southern Great Plains. TX Ag. Exp. Sta. Misc, Pub. 929.

Fryrear, D.W. 1971. Survival and growth of cotton plants damaged by windblown sand. Agron. J. 63:638–642.

Fryrear, D.W. 1981. Dust storms in the Southern Great Plains. Trans. ASAE 24:991–994.

Fryrear, D.W. 1984. Soil ridges-clods and wind erosion. Trans. ASAE 28:781–784.

Fryrear, D.W. 1985. Soil cover and wind erosion. Trans. ASAE 28:781–784.

Fryrear, D.W. 1986. A field dust sampler. J. Soil and Water Cons. 41:117–120.

Fryrear, D.W., and D.V. Armbrust. 1969. Cotton gin trash for wind erosion control. Tex. Ag. Exp. Sta. Misc, Pub. 928.

Fryrear, D.W., D.V. Armbrust, and J.D. Downes. 1975. Plant response to wind erosion damage. Proc. 30th Annual Meeting SCSA, Aug. 10–13, 1974, San Antonio, Texas.

Fryrear, D.W., and J.D. Downes. 1975. Consider the plant in planning wind erosion control systems. Trans. ASAE 18:1070–1072.

Fryrear, D.W., Krammes, C.A., Williamson, D.L., and Zobeck, T.M. 1994. Computing the wind erodible fraction of soils. J. Soil Water Cons. 49:183–188.

Fryrear, D.W., and A. Saleh. 1993. Field wind erosion: Vertical distribution. Soil Sci. 155(4):294–300.

Fryrear, D.W., and A. Saleh. 1996. Wind erosion: Field length. Soil Sci. 161(6):398–404.

Fryrear, D.W., J.E. Stout, L.J. Hagen, and E.D. Vories. 1991. Wind erosion: Field measurement and analysis. Trans. ASAE 34:155–160.

Gibbons, R.P., J.M Tromble, J.T. Hennessy, and M. Gardenas. 1983. Soil movement in mesquite dunelands and former grasslands of Southern New Mexico from 1939 to 1980. J. Range Manag. 36:145–148.

Gillette, D.A., and P.H. Stockton. 1986. Mass, momentum, and kinetic energy fluxes of saltating particles. p. 88–103. In W. G. Nickling (ed.) Aeolian Geomorphology, Binghamton Symp. Allen and Unwin, Boston, MA.

Greeley, R., and J.D. Iverson. 1985. Wind as a geological process on Earth, Mars, Venus, and Titan. Cambridge Univ. Press, New York, NY.

Gregory, J.M., J. Borrelli, and C.B. Fedler. 1988. TEAM: Texas erosion analysis model. p. 88–103. In Proc. of 1988 Wind Erosion Conference, Texas Tech University, Lubbock, TX.

Gupta, J.P., R.K. Aggarwal, and N.P. Raikhy. 1981. Soil erosion by wind from bare sandy plains in western Rajasthan, India. Arid Envir. 4:15–20.

Hagen, L.J. 1984. Soil aggregate abrasion by impacting sand and soil particles. Trans. ASAE 27:805–808, 816.

Hagen, L.J. 1991. A wind erosion prediction system to meet user needs. J. Soil Water Cons. 46:106–111.

Hagen, L.J., L.E. Wagner, and J. Tatarko. 1995. Wind Erosion Prediction System (WEPS). BETA Release 95–08 USDA-ARS, Manhattan, KS.

Holy, M. 1980. Environmental sciences and applications. Vol. 9: Erosion and environment. Pergamon Press, New York, NY.

Horning, L.B., L.D. Stetler, and K.E. Saxton. 1996. Surface residue and roughness for wind erosion protection. ASAE Paper 96-2116. American Society of Agricultural Engineering, Phoenix, AZ.

Liu, X., Y. Lin, D. He, and F. Chen. 1983. A study on the experiment of the under-leading fence in wind tunnel. J. Desert Res. 3:25–34.

Lyles, L., D.V. Armbrust, J.D. Dickerson, and N.P. Woodruff. 1969. Spray-on adhesives for temporary wind erosion control. J. Soil Water Cons. 24:190–193.

McDonald, A. 1971. Early American soil conservationists. USDA Misc. Pub. 449.

McEwan, I.K., and B.B. Willetts. 1993. Sand transport by wind: A review of the current conceptual model. *In* K. Pye (ed.). The dynamics and environmental context of aeolian sedimentary systems. Spec. Publ. Geol. Soc. 72:7–16.

Merva, K.R., and G. Peterson. 1983. Wind erosion sampling in the North Central Region. ASAE Paper 83-2133, American Society of Agricultural Engineering, Phoenix, AZ.

Middleton, N.J. 1987. Desertification and wind erosion in the western Sahel: The example of Mauritania. School of Geography, Univ. of Oxford Res. Pap. 40

Owens, P.R. 1964. Saltation of uniform sand grains in air. J. Fluid Mech. 20:225–242.

Potter, K.N., T.M. Zobeck, and L.J. Hagen. 1990. A microrelief index to estimate soil erodibility by wind. Trans. ASAE 33:151–155.

Saleh, A. 1993. Soil roughness measurement: Chain method. J. Soil Water Cons. 48:527–529.

Saleh, A. 1994. Measuring and predicting ridge-orientation effect on soil surface roughness. Soil Sci. Soc. Am. J. 58:1228–1230.

Saleh, A., and D.W. Fryrear. 1995. Threshold wind velocities of wet soils as affected by windblown sand. Soil Sci. 160:304–309.

Saleh, A., D.W. Fryrear, and J.D. Bilbro. 1997. Aerodynamic roughness prediction from soil surface roughness measurements. Soil Sci. 162:205–210.

Saxton, K.E. 1995. Wind erosion and its impact on offsite air quality in the Columbia Plateau: an integrated research plan. Trans. ASAE 38:1031–1038.

Saxton, K.E. 1996. Agricultural wind erosion and air quality impacts: a comprehensive research program. J. Altern. Agric. 11:64–70.

Shao, Y., G.H. McTainsh, J.F. Leys, and M.R. Raupach. 1993. Efficiencies of sediment samplers for wind erosion measurement. Aust. J. Soil Res. 31:519–532.

Shao, Y., M.R. Raupach, and J.F. Leys. 1996. A model for predicting aeolian sand drift and dust entrainment on scales from paddock to regions. Aust. J. Soil Res. 34:309–342.

Siddoway, F.H., W.S. Chepil, and D.V. Armbrust. 1965. Effect of kind, amount, and placement of residues on wind erosion control. Trans. ASAE 8:327–331.

Skidmore, E.L. 1966. Wind and sandblast injury to seedling green beans. Agron. J. 58:311–315.

Skidmore, E.L. 1986. Wind erosion climatic erosivity. Climate Change 9:195–208.

Skidmore, E.L., L.L. Nossaman, and N.P. Woodruff. 1966. Wind erosion as influenced by row spacing, row direction, and grain sorghum population. Soil Sci. Soc. Am. Proc. 30:505–509.

Sorensen, M. 1991. An analytic model of wind-blown sand transport. Acta Mechanica [Suppl] 1:67–81.

Steiner, J.L., H.M. Schomberg, C.L. Douglas, and A.L. Black. 1994. Standing stem persistence in no-tillage small-grain fields. Agron. J. 86:76–81.

Stetler, L.D., and K.E. Saxton. 1995. Fugitive dust (PM_{10}) emissions and eroded soil relations from agricultural fields on the Columbia Plateau. Particulate matter: Health and regulatory issues. In Proc. AWMA International Conference on Particulate Matter, Pittsburgh, PA April 46, 1995. Air Waste Management Association, Pittsburgh, PA.

Stout, J.E. 1990. Wind erosion within a simple field. Trans. ASAE 33:1597–1600.

Stout, J.E., and D.W. Fryrear. 1989. Performance of a windblown particle sampler. Trans. ASAE 32: 2041–2045.

Strzelecki, P.E. De. 1845. Physical description of New South Wales and Van Diemen's Land. Longiman, Brown, Green, and Longmans, London, England.

Taylor, D.W. 1960. Fundamentals of soil mechanics. 11th Ed. John Wiley and Sons, New York, NY.

Thornbury, W.D. 1957. Principles of geomorphology. John Wiley and Sons, New York, NY.

van de Ven, T.A.M., D.W. Fryrear, and W.P. Spaan. 1989. Vegetation characteristics and soil loss by wind. J. Soil Water Cons. 44:347–349.

Vories, E.D., and D.W. Fryrear. 1991. Vertical distribution of wind-eroded soil over a smooth, bare field. Trans. ASAE 34:1763–1768.

Williams, J.R., C.A. Jones, and P.T. Dyke. 1984. A modeling approach to determining the relationship between erosion and soil productivity. Trans. ASAE 27:129–144.

Williams, M.A.J., and R.C. Balling. 1996. Interactions of desertification and climate. Arnold Publishers, London, UK.

Woodruff, N.P. 1956a. The spacing interval for supplemental shelterbelts. J. Forestry 54:115–122.

Woodruff, N.P. 1956b. Wind-blown soil abrasive injuries to winter wheat plants. Agron. J. 48:499–504.

Woodruff, N.P., and D.V. Armbrust. 1968. A monthly climatic factor for the wind erosion equation. J. Soil Water Cons. 23:103–104.

Woodruff, N.P., and W.S. Chepil. 1956. Implements for wind erosion control. Agric. Eng. 37:751–754, 758.

Woodruff, N.P., W.S. Chepil, and R.D. Lynch. 1957. Emergency chiseling to control wind erosion. KS. Agric. Exp. Stn. Tech. Bull. 90.

Woodruff, N.P., and F.H. Siddoway. 1965. A wind erosion equation. Soil Sci. Soc. Am. Proc. 29:602–608.

Woodruff, N.P., and A.W. Zingg. 1955. A comparative analysis of wind tunnel and atmospheric air flow patterns about single and successive barriers. Trans. Am. Geophys. Union 36:203–208.

Xu, J., Z. Pei, and R. Wang. 1982. A research on the width of the protection belt of half-hidden straw checkerboard barriers. J. Desert Res. 2:16–22.

Zobeck, T.M, J.E. Stout, T.E. Gill, and D.W. Fryrear. 1996. Airborne dust and sediment measurements in agricultural fields. International Conference on Air Pollution from Agricultural Operations in Kansas City, MO, February 7–9, 1996. Midwest Plan Service, Iowa State University, Ames, IA.

Zobeck, T.M., and C.A. Onstad. 1987. Tillage and rainfall effects on random roughness: A review. Soil Tillage Res. 9:1–20.

Zobeck, T.M., and T.W. Popham. 1992. Influence of microrelief, aggregate size and precipitation on soil crust properties. Trans. ASAE 35:487–492.

9

Land Application of Wastes

D.M. Miller
University of Arkansas

W.P. Miller
University of Georgia

9.1 A Brief History and Overview of Land Application

Prior to the introduction and widespread use of commercial fertilizers in the early 20th century, application of manures and other byproducts to soils was an important way of maintaining soil fertility and productivity. Prior to Liebig's discovery of the role of soil-derived nutrients in plant growth, such applications were made based on the empirical observation that plant growth was stimulated by such applications. The importance of these materials in agricultural production has declined during this century in response to the lower cost of commercial (inorganic) fertilizer per mass of nutrient applied, and thus the term "waste" has been applied to manures and other materials, since these materials may have a negative (i.e., disposal) value in lieu of their use as soil amendments. Current environmental regulations regarding release of nutrients into the environment, and the construction of landfill facilities have revived interest in land application of waste materials to crop and forest lands as a way of avoiding disposal costs, as well as recycling nutrients to soils as an alternative to commercial fertilizers. The organic materials present in many of these byproducts also has the potential to build humus levels in soils, and thereby improve long-term productivity.

A wide range of materials derived from both agricultural and industrial activities has recently been proposed for land application. From the point of view of the producer of these materials, they are wastes if they have no or negative value. If, however, they can be shown to have positive growth effects on cultivated plants that justify their land application, and do not have adverse environmental impacts with such a use, they might rightly be termed byproducts rather than wastes, which has been proposed in a recent publication on the subject (Powers, 1998). The focus of current research is to provide justification for the environmentally safe use of such materials in agriculture as substitutes for commercial fertilizers, and to avoid the cost of landfilling of a potential resource.

9.1.1 Historical Perspective

The oldest stable agricultural systems known, from both the West and East, employed additions of organic matter to soil to maintain productivity, those materials being primarily animal manures and nightsoil (human excrement). Asian agriculture dating from 2,500 BCE employed these materials extensively; about the same time, Greek and later Roman agriculture used manures as well, but did not find human waste as suitable, perhaps due to its tendency to cause disease (Tisdale et al., 1993).

During the Middle Ages, agriculture appeared to follow the Roman model, and continued use of organic materials (manure, food waste, and perhaps human excrement) over time created distinctive topsoil materials high in C and extractable nutrients (plaggens). Similar soil horizons have been found in Amazonia at about the same time associated with village sites of long habitation, presumably enriched with food processing wastes added intentionally (or not) to garden plots near habitations.

Industrialization in the West during the 18th and 19th centuries brought many changes to agriculture, which had been operating in Europe and America with only modest change since Rome. The emphasis on tillage in order to allow plant uptake of humus continued to encourage manure additions, but Liebig's clarification of the matter in the 1860s opened the way to widespread use of fertilizers for their nutrient, rather than their humus content.

Prior to the development of industrial processes to manufacture inorganic fertilizers in the later 19th century, imported guano (sea bird manure) from Chile and the Pacific islands was a major fertilizers used in row crop production in Europe, and particularly in the southern United States for cotton production as an N and P source (Wines, 1985). Once the Haber process was developed to manufacture ammoniacal salts on an industrial scale, and phosphatic mineral deposits were discovered and mined along the eastern seaboard and Florida, the rapidly depleting guano reserves were abandoned.

During the period 1920–1940, inexpensive supplies of inorganic nutrients coupled with other changes greatly increased yields of crops and encouraged specialization in farming, which often meant separation of row crop and animal production. In confined production facilities, manures became wastes, and an increasing disposal problem arose. In the 1970s it became clear that stockpiled or overapplied manures caused water (surface and groundwater) contamination with N and P.

The environmental movement of the 1970s initiated the modern emphasis on environmental quality and recycling which, coupled with economic constraints on producers of a variety of byproduct materials, has led to a new focus on utilization of a wide range of these materials on agricultural lands.

9.1.2   Types of Byproducts Considered

Animal manures, largely from cattle, poultry, and swine production, still make up the bulk of byproduct materials applied to soils. The increasing centralization and industrialization of animal production has led to problems with distribution of such manures back to cropland, primarily due to high transport cost relative to cost of equivalent inorganic nutrients. Despite the long history of use, rates of application are currently being studied to determine water quality impacts and optimal application rates for crop production.

Domestic waste, the nightsoil of pre-industrial days, is now handled in wastewater treatment plants and emerges as municipal sewage solids (sewage sludge), also referred to as biosolids. This material is increasingly land applied, either in dedicated land-treatment facilities or on farmers' fields. Governmental regulation is encouraging application to land through publication of guidelines for acceptable nutrient and heavy metal loadings, thereby avoiding landfilling of this material. Other domestic wastes include yard trimmings (grass clippings, chipped limbs, etc.), now commonly composted or mixed with other byproducts instead of consuming landfill space.

Industrial byproducts are increasingly being viewed as soil amendments due to their negative value to industries and the potential capacity to substitute for costly inorganic fertilizers. Some industrial byproducts such as blast furnace slags, wood ash and kiln dusts have been used locally for some time as lime substitutes due to their alkaline nature; metal refining baghouse dusts are commonly used to formulate micronutrient fertilizers. Other materials such as coal fly ash, byproduct gypsums, food processing and pulp and paper manufacturing byproducts are being considered for land application based on either alkalinity, nutrient content, and/or organic matter content. With these new materials, there is a greater uncertainty concerning possible environmental hazards from trace metal and organic contaminants, and this has prompted ongoing research.

The Resource Conservation and Recovery Act (RCRA) and the Clean Water Act (CWA) and subsequent USEPA rulings contain regulations governing the definitions of waste types, discharge or loading limits for pollutants, and management practices for land application of biosolids (Walker et al., 1997). In particular, hazardous waste is defined as materials that are corrosive, ignitable or explosive, or toxic, the latter based on the leaching potential of trace metals. While under certain circumstances materials defined as hazardous may be land applied, in general, these materials cannot be used as soil amendments.

9.1.3 Rationale for Land Application

Land application in the sense used here refers to addition of byproducts to soil for the purposes of improving crop growth, either through addition of nutrients or improvement of soil physical or chemical properties, either in the short or longer term. Materials used for land application should be studied and documented to be safe and effective for use in agriculture, with no overriding constraint to their use. This is distinctly different from land treatment, which suggests that materials are being applied to land primarily in order to dispose of (treat) them. There is a long history of land treatment of waste materials such as biosolids, secondary municipal waste water, and industrial solids and waste waters on dedicated treatment facilities, which are managed and monitored for such uses. The design and operation of such facilities are typically under permit from governmental regulatory agencies, and are not discussed in this review.

The key to land application is the idea of safe, beneficial reuse, which is the major criteria used by the USEPA and many state environmental agencies to approve materials for land application. Animal manures are typically not closely scrutinized due to their long history of use and accepted value in agriculture, although increasing use of hormones and high metal levels as feed additives has raised some questions. Other materials, however, require testing to document the absence of significant levels of contaminants, and to demonstrate some positive effect on crop yield or potential soil productivity. This latter criterion may be addressed by demonstrating improvement in long-term productive potential by increasing soil water-holding capacity, improving soil structure or tilth, or by overall increase in nutrient levels that may not affect yield immediately.

The environmental issue has been partially addressed by the adoption of the 40 CFR 503 ruling, which specifies acceptable concentrations and annual and cumulative loadings of nine contaminant metals applied to crop land in municipal biosolids (USEPA, 1993). While this ruling has been applied to other types of byproducts, it is not clear that the specifications apply to inorganic types of wastes. Some further study is necessary to demonstrate the safety of these loadings for other types of wastes. The European Union (EU) has adopted much more stringent standards than those of USEPA, and a considerable debate is in progress on this topic.

A final note of caution; there is considerable economic incentive for industries generating byproducts of questionable safety and/or agronomic value to market these materials, often at subsidized low pricing schedules, for agricultural use. Documentation accompanying such materials may contain incomplete and/or misleading data. Most states do not require materials marketed as fertilizer or lime to be analyzed for contaminants, and instances of hazardous wastes being land applied as fertilizer have been noted in the popular press (Wilson, 1997). In the absence of federal or state regulation or guidelines relating to land application, those involved in producing, handling, and marketing byproducts must be aware of the hazards, and take necessary precautions, particularly obtaining complete analyses for all contaminants, to assure public safety is not compromised.

9.2   Properties of Common Land-Applied Byproducts

9.2.1   Agricultural Materials

Strictly speaking, the single most important type of agricultural byproduct applied to soils is crop residues. However, crop residues are not often classified as a waste material because they are naturally returned to the soil during the course of tillage operations, and because it is universally acknowledged that returning residues to the soil enhances soil productivity. Therefore, crop residues will not be discussed further in this chapter.

By virtually any standard the most important agricultural byproducts applied to soils are animal manures. Because of the huge amounts applied and their relatively high nutrient content, it could even be argued that agricultural animal manure is the single most important land-applied byproduct of agricultural or nonagricultural origin. Huge quantities of agricultural animal manures are produced each year in the United Sates and throughout the developed countries of the world. In 1978, a USDA task force estimated that just over 0.5 billion Mg (fresh wt.) of manure, or 39% of all the agricultural animal manure generated in the United Sates, was produced in confined areas and was thus available for land application. Brady and Weil (1996) estimate that over 1 billion Mg (fresh wt.) of animal manure is now produced annually in confined livestock and poultry operations in the United Sates.

The geographic distribution of animal manure production in the United Sates, as well as total quantities produced, has changed markedly since the middle of this century. Prior to the 1940s, animals and crops were for the most part raised together on millions of small farms scattered throughout the world. In this precommercial fertilizer era, manure was a valuable resource. It was faithfully collected, stored and applied to crop land at rates that seldom supplied nutrients in excess of crop needs, primarily because there was an appropriate balance between the number of animals produced and the amount of land available. With the advent of chemical fertilizers, manure was no longer an economical source of plant nutrients, so small farmers divested themselves of their livestock. The massive crops of corn, wheat and sorghum that could be produced inexpensively using the new chemical fertilizers became an economical source of feed grains for a new, more efficient form of animal agriculture that was characterized by economy of scale, high density, confined production and centralized processing and distributing facilities. Higher land prices, increased cost of labor and other factors have prompted animal production facilities to locate in areas not well suited to row crop agriculture. The end result is a concentration of animals in specific regions of the country where limited areas of crop land cannot accommodate the quantities of manure produced. The situation is exemplified by the poultry industry in the southeastern United States (AK, AL, GA, NC, DE, MD) where intensive poultry production is generating more manure than the region's sparse crop lands can accommodate.

The chemical and physical properties of the manures produced by today's animal agriculture vary widely. A property of particular importance as far as storage and handling of the manure is concerned is water content (or, conversely, solids content) which is sometimes referred to as consistency. Manures containing less than 5% solids are considered liquids, while those containing between 5 and 12% solids are referred to as slurries; those containing between 12 and 20% solids are semisolid; and those containing more than 20% solids are termed solid. When they are first excreted, many agricultural animal manures are classed as semisolids. Liquid manures and manure slurries typically are produced when water is used to wash semisolid manures out of the animal rearing areas into pits or lagoons for storage. Left undisturbed in a lagoon, solids will settle out, creating a situation in which a manure slurry is overlain by liquid manure. Solid manures are produced either by drying of slurries or semisolids, or by mixing of manures with a carbonaceous bedding material such as straw or wood chips. An example is chicken broiler litter, which is a physical mixture of poultry (broiler) manure and either rice hulls, wood chips or sawdust.

The chemical composition of agricultural animal manure is highly variable and depends on many factors including (1) type of animal, (2) nature and composition of feed, (3) nature and composition of added bedding material, (4) manure/bedding ratio, and (5) the manner in which the manure was collected, stored and handled prior to land application. Representative chemical composition data for a number of different types of animal manures are presented in Table 9.1. Poultry and swine manures generally have higher N and P contents than either beef or dairy manure. In addition, poultry litters tend to have lower N and P contents than poultry manures because of the diluting effect of the carbonaceous bedding materials present in litters. These data also illustrate the fact that when animal manures are applied to soil, significant quantities of nutrients other than N and P are added to the soil. For example, application of 4 Mg ha^{-1} (dry weight) of swine manure having a composition similar to that shown in Table 9.1 would provide, in addition (kg ha^{-1}) to 252 N and 80 P, 100 K, 180 Ca, 40 Mg, 4.4 Fe, 1.56 Zn, and 0.6 Cu.

It should be emphasized that the data in Table 9.1, while representative of actual analyses, are for comparative purposes and should not be used to calculate actual manure application rates. The variability in the chemical makeup of manures is so high that the use of average data to calculate manure application rates can result in significant over or under application of nutrients. In actual practice, a sample of the manure to be land applied should be analyzed to determine its exact chemical composition, which should be used in application rate calculations. Sims (1995) has discussed both the need for and problems associated with obtaining accurate chemical analyses of agricultural animal manures.

9.2.2 Forestry Products

Processing of forest products generates large amounts of byproducts, most of which are currently landfilled, incinerated or stockpiled near production facilities. The production of pulp for papermaking and as an industrial feedstock by the Kraft and related processes is the major regional source of byproducts, with industries principally located in the Southeast (AL, GA, SC, TN), Northwest (OR, WA), New England (MA, NY, PA), and Pacific Northwest (WA, OR). Total byproduct production in the United States was approximately 13 million Mg (dry wt.) in 1989, with

Table 9.1 Representative chemical composition of animal manures

Manure	Tot. N	P	K	Ca	Mg	Fe	Zn	Cu
	--------------------------g kg^{-1}--------------------------					----------------mg kg^{-1}----------------		
Poultry[a]	52	22	20	81	10	890	1032	155
Poultry[b]	40	16	23	23	5	1000	315	473
Swine[c]	63	20	25	45	10	1100	390	150
Beef[d]	19	7	20	13	7	5000	8	2
Dairy[e]	24	7	21	14	8	1800	165	30

[a]Layer manure (no bedding); after Edwards and Daniel (1992) and Sims and Wolf (1994)
[b]Broiler litter; after Edwards and Daniel (1992) and Brady and Weil (1996)
[c]After Choudhary et al. (1996) and Brady and Weil (1996)
[d]After Eghball and Power (1994). [e]After Brady and Weil (1996)

about 50% of this being organic-based sludges and the remainder inorganic materials such as byproduct lime and boiler ash (Unwin, 1997; NCASI, 1997).

During the pulping process, chemical dissolution of lignin is employed to separate the cellulose for further processing. The solid byproducts produced consist of inorganic residues from the wood and processing chemicals (grits, dregs), reject cellulose and wood knots, spent alkali used in regenerating the pulping chemicals (lime mud) and ash from boilers fired with wood byproduct and/or fossil fuel. The reject cellulose taken directly from the pulping digestor is termed primary sludge; often this material is sluiced to a lagoon and mixed with other byproduct streams (ash and alkali byproduct), and may undergo some forced aerobic decomposition. This latter material is referred to as secondary or activated sludge. Pulp and paper mills use steam generated in boilers which typically burn wood byproduct, bark, reject pulp and accessory fossil fuels, generating ash that is variable depending on the fuel source. A variety of proprietary processes are used in pulp and paper making, and individual plants differ substantially in their handling of byproduct streams; some mills combine all byproduct streams prior to disposal. Therefore, byproducts are extremely heterogenous and typically vary in composition within the same mill over time as operating conditions change.

Some general properties of the major classes of pulp and paper byproducts are given in Table 9.2. Ranges are shown rather than average values, given the extreme variation in products between different mills. Both primary and secondary pulp sludges are largely composed of organic materials (reject wood and pulp), while the other classes of byproducts are primarily inorganic. Nutrient contents are in general low, particularly for N, P and K. Some ashes and the lime muds are quite alkaline ($pH_{H2O} > 12$), and have good potential as agricultural lime substitutes. Calcium carbonate equivalences (CCE) for lime muds are often > 90%, and in the range 10–65% for primary sludges which often contain free $CaCO_3$ (Morris et al., 1997).

Trace element contents, including regulated metals such as Hg, Cd, and As, are quite variable but typically quite low in most of these materials. Ash from boilers using nonwood fuel (coal, shredded tires, oil) are not reported here, but are often higher in trace elements such as As and Zn. Dioxin content

Table 9.2   Chemical composition of pulp and paper mill waste products

Element	Primary Sludges	Secon./Mixed Sludges	Wood Ash
		g kg^{-1} by wt.	
N	0.7-20 (3)	6-90 (23)	0.3-3
P	0.1-4 (0.2)	0.4-17 (4)	0.6-16
K	0.1-10 (2)		23-162
Ca	0.3-210 (14)		211-466
Mg	0.2-19 (1.6)		25-117
		mg kg^{-1} dry wt.	
Cu	3.9-1590		3-210 (75)
Zn	13-3780 (188)		63-2200 (443)
Co	<0.1-9.7		0.5-20 (8.7)
Mo	<2-14		<0.1-114 (15)
As	<0.05-8.3 (1.2)		3-63 (23)
Pb	<0.05-880 (28)		22-220 (66)
Hg	0.001-3.5 (0.35)		<0.1-2.8 (0.4)
Cr	3-2250 (42)		3-130 (39)
Ni	1.3-133		<1-97 (24)
Cd	0.1-56 (1.2)		<0.01-21 (5)

Mean values in parenthesis where available. Sludge data from NCASI (1984); Morris et al. (1997); Camberato et al. (1997) Ash data from Morris et al. (1997); Someshwar (1996)

of ash and pulp byproduct from mills where Cl_2 process bleaching is used has been a subject of concern. Studies have shown significant levels of dioxin or furan compounds in some bleached kraft mills where Cl_2 bleaching is used, but this method is rapidly being replaced by other technologies (Camberato et al., 1997; Miner, 1997). Older stockpiled sludges in lagoons may locally contain higher levels of dioxins and furans. Except where salt laden (i.e., salt water-saturated wood) was used as a feedstock, boiler ash from pulp and paper mills is quite low in these compounds (Someshwar, 1996).

9.2.3 Industrial Byproducts

There is obviously a large range of byproduct materials produced by industries, some of which may have potential for use as soil amendments. Many have no appreciable nutrient, organic or alkali content that might benefit crop production, or have levels of toxic contaminants that rule out land application. Several classes of byproducts described in this section have been used, or have been suggested for use on soils. Several types have been tested for crop response or agronomic efficacy or for potential environmental impacts, but certainly much work remains to be done in this area. The use of any of these materials is not recommended in the absence of careful chemical characterization and consultation with agricultural and environmental experts or state officials.

9.2.3.1 Slags

Iron and steel making slags are perhaps the best known industrial byproducts historically used in agriculture. These materials are produced by addition of lime to furnaces to remove Si impurities, and are composed primarily of $CaSiO_3$. Blast furnace (Fe) slags have a high CCE value (80–100%), and have been used for many years locally as lime substitutes in the northeastern United States and Europe. Basic (steel) slags are produced from high P Fe ores, and have a high P content (4–8%) as well as considerable CCE (40–60%; Tisdale et al., 1993). Electric furnace slags have intermediate CCE and P values.

In 1995, 21 million Mg were produced in the United States, with about 60% being Fe slags and 40% steel or electric furnace slag (USGS, 1997). While in the past, much of the basic slag was land applied as a P source (10 million Mg in 1974), current use is low, with only 0.25 million Mg land applied worldwide in 1995. In the United States. production is centralized in the industrial Midwestern states (IL, IN, OH, PA), and competing uses (cement production, lightweight aggregate) have increased prices to \$30–\$100 Mg^{-1} (USGS, 1997), largely pricing agricultural uses out of the available market.

No environmental studies dealing with contaminant (i.e., heavy metal) contents of slags or their fate in soils appear in the literature; it is likely that certain metals alloyed in steel (Ni, Cr, Mo) may be present in the slag and should be carefully monitored in any land application.

9.2.3.2 Fly Ash

Coal-burning electric-generating plants produce fly ash in all industrial countries. In the United States in 1993, utilities produced nearly 55 million Mg of fly ash, as well as 20 million Mg of bottom ash and boiler slag materials (Am. Coal Ash Assoc., 1993). Fly ash is a fine (typically < 0.05 mm diam) largely silicious material trapped in electrostatic precipitators prior to discharge of flue gases from smoke stacks. Nationally, about 25% of this material is reused as a cement additive or used in structural fills, roadbases, and other engineering applications; the remainder is stored on site in lagoons or landfills. While fly ash has traditionally been exempt from regulation, new landfill construction is likely to require liners and monitoring of leachates to prevent groundwater contamination with metals, which has been observed at a number of existing facilities. These higher disposal costs have spurred an interest in use of fly ash as a soil amendment.

The composition of coal fly ash differs significantly from that of wood ash (Table 9.2) due to the nature of the fuel; coal has a greater ash content due to inclusions of minerals in the coal as it was deposited and then modified over geologic time. Clay minerals deposited within the coal bearing strata give rise to high Si and Al contents, and Fe precipitated as pyrite (FeS) within the coal adds Fe and a range of associated trace metals. Coals mined in KY, IN, AL, WV, IL, and PA tend to be higher in pyrite (high S, bituminous coals), with a resulting ash that is acidic to neutral in pH, high in Fe, and relatively low in Ca and Mg, classified as Class F in the ASTM system (Mattigod et al., 1990). Ash from sub-bituminous and lignite western United States coals (low S) has lower Fe and associated metals, and higher Ca, Mg and pH values (Class C), and is more pozzolanic (hardens irreversibly, similar to portland cement), and therefore is more desirable as á cement additive. Representative chemical analyses are given in Table 9.3.

The Class F ashes, constituting about 50% of the annual production in the United States, have limited nutrient value; major nutrient concentrations are low, and studies of availability indicate that K in particular is not readily water soluble, limiting its effect as a fertilizer. These ashes typically have low CCE (1–10%) and, therefore, have limited potential as lime substitutes. A wide range of microelements are present that have a potential benefit for agricultural use. However, some Class F ashes have As levels that are a cause for concern environmentally. Using USEPA metal limits for biosolids (Section 9.5.1) as a comparison, As concentrations above the 75 mg kg^{-1} ceiling occurred in about 50% of ashes in this group. In comparison, Class C ashes are much higher in Ca, Mg, and K, and have a significant liming potential due to the higher CCE values (20–60%). Trace metals are overall lower, with only 1 sample in 9 having >75 mg kg^{-1} As. Boron is elevated in all ashes relative to levels in soils, particularly in Class C ashes. Boron phytotoxicity may injure plants temporarily, although this

Table 9.3   Chemical analysis (total digestion) of 9 Class C and 49 Class F fly ashes

Major Elements	Class C ashes (9)			Class F ashes (49)		
	Mean	Min	Max	Mean	Min	Max
	----------------------g kg^{-1}--------------------			--------------------g kg^{-1}--------------------		
Al	68	51	104	110	63	144
Fe	51	29	95	80	17	174
Si	137	93	177	231	198	282
P	3.6	1.0	6.3	1.4	0.8	2.6
K	6.3	2.7	11.6	15.4	3.1	25.3
Ca	172	101	225	32	2	115
Mg	21.9	4.4	41.8	8.9	3	27
Trace Elements						
	----------------------mg kg^{-1}--------------------			--------------------mg kg^{-1}--------------------		
As	39.9	8.0	96.0	93	8.0	391
B	nd	nd	nd	425	55.0	1108
Cd	nd	nd	nd	1.0	0.1	3.2
Co	nd	nd	nd	47	14	120
Cr	66.9	40.0	123.0	171	25	651
Cu	116.4	45.0	223.0	133	48	242
Pb	45.6	19.0	92.0	63	13.0	273
Mo	14.6	5.3	32.0	29.8	3.7	139
Ni	54.0	34.0	78.0	121	14.0	309
Se	11.4	6.0	14.2	11.4	1.0	47
V	nd	nd	nd	259	58	470
Zn	266.0	25.0	658.0	270	27	2050

Data combined from Ainsworth and Rai (1987) and Miller and Sumner (1997)
nd = no data available

effect often disappears as B is leached out of the root zone by rainfall. There are no human health effects of B in food crops or water supplies.

9.2.3.3   Gypsum ($CaSO_4 \cdot 2H_2O$)

This is produced as a byproduct by a number of industries, most notably the phosphate fertilizer industry and electric generating utilities as well as by pigment processing and electroplating industries. Nearly 1 billion Mg of phosphogypsum (PG) from P fertilizer processing are stockpiled in Florida and the Mississippi Valley, with an annual production of 40 million Mg, mostly in central Florida (Miller and Sumner, 1997). In addition, electric utilities are producing increasing amounts of flue gas desulfurization gypsum (FDG) in scrubbers designed to remove SO_2 from stack gases in response to amendments to the Clean Air Act. About 1 million Mg were produced in 1996 from forced air scrubbers that fully oxidize SO_2 to gypsum, with projections for increases in the future. An additional 15 million Mg of mixed desulfurization byproducts containing calcium sulfite ($CaSO_3$), $CaSO_4 \cdot 2H_2O$, CaO and fly ash are produced by utilities using a variety of scrubber technologies (Korcak, 1995). These materials have a wide range of physical and chemical properties, based on the scrubber design and chemistry used.

Some typical analyses for various gypsiferous byproducts are given in Table 9.4. Fluidized bed combustion (FBC) material is produced from a dry scrubber that injects lime directly into the furnace, one common technology that produces a mixed byproduct composed of calcium sulfite/sulfate material containing unreacted lime. It has a variable composition depending on the boiler design often containing considerable residual CaO or $Ca(OH)_2$ and some fly ash (evidenced by the Si content). No CCE data were located, but it would likely be a reasonable lime substitute. Phosphogypsum and FDG are relatively pure materials, approaching the theoretical 23% Ca and 19% S of $CaSO_4 \cdot 2H_2O$. Impurities in PG are largely Si sand and a minor amount of fly ash in FDG. Trace elements are quite low in all materials, with B being the exception in the case of FBC and FDG materials.

Table 9.4   Chemical composition of various gypsum containing byproducts

	Fluidized bed combustion byproduct	Phosphogypsum	Flue gas gypsum
Macroelements	----------------------------------- g kg^{-1} dry wt.-----------------------------------		
P	0.1-0.5	8-16	<1-2
K	0.5-8	0.5-2	1
Ca	240-460	180-235	210-235
Mg	1-16	0.4-2	0.2-2
S	72-140	128-181	166-186
Si	nd	15-50	0.7
Al	3-20	1-20	0.3-2
Microelements	----------------------------- mg kg^{-1} dry wt.-----------------------------		
B	95-170	< 3	75
Mo	0.1-0.3	2-11	1.3
Cu	12-20	9	8.0
Zn	12-20	nd	36
Ni	13-20	10	9.7
Pb	1.5-7.5	15	<0.1
Cd	0.5	4	0.01
Cr	9-23	20	10
Se	0.2-0.6	1-2	2.9
Hg	nd	0.3-0.4	0.01-0.4
As	nd	1-4	3.1

Summarized from Korcak (1995), Miller (1995) and Alcordo and Rechcigl (1995)

Both PG and FDG have a history of use in agriculture; in the southeastern United States, byproduct $CaSO_4 \cdot 2H_2O$ materials are applied to peanuts as a Ca supplement, and various types of $CaSO_4 \cdot 2H_2O$ have been used in the western United States, Australia and South Africa on high Na soils to improve physical properties. Higher application rates (up to 10 Mg ha^{-1}) have been shown to improve root penetration on acid subsoils with no enhanced metal uptake on a range of crops (Shainberg et al., 1989; Sumner, 1993; Ritchey et al., 1995). Field studies using FDG have demonstrated limited uptake of metals over three years on a range of crops from 10 Mg ha^{-1} application rates on three soils in Georgia (Sumner et al., 1996; Miller and Sumner, 1997). The USEPA, however, has restricted the use of PG containing > 10.0 pCi ^{226}Ra g^{-1} due to trace contamination with radionculides of the ^{238}U decay series, principally ^{226}Ra and ^{222}Rn gas, which carry certain health risks (Alcordo and Rechcigl, 1995; Miller and Sumner, 1997). The risk assessments on which this decision were based involve inhalation hazards to persons living on highly amended soils, and are being challenged by fertilizer and phosphate industry groups. Currently, PG containing >10 pCi g^{-1} of total radiation, which is most of the stockpile and production in Florida, is restricted.

9.2.3.4 Flue and Kiln Dusts

These are fine airborne particulates collected from manufacturing facilities, particularly metal processing/refining and cement manufacturing. Flue dusts often refer to fines from metal processing (especially electric furnaces), and are primarily ferrous with admixtures of heavy metals; something in the range of 5 million Mg yr^{-1} are produced in the United States, most of which is either recycled within the facility or landfilled. If such dusts do not contain contaminant metals (Cd, Pb, etc.) they are a valuable source of trace nutrients, and significant quantities are used by the micronutrient industry as a source of Cu and Zn fertilizer (Wyatt, 1997). However, many flue dusts and similar products contain contaminant metals (up to 20% heavy metals such as Pb, Cd and Zn) and are classified as either hazardous wastes or are derived from hazardous wastes, thus restricting their use. It should be noted that several instances of ill-informed use of flue dusts as liming materials and/or micronutrient sources have resulted in Zn and Pb poisoning of crop land to the point of near sterilization (Davis et al., 1995b), and the use of such materials should be carefully monitored for metal loading rates applied to soils.

Cement kiln dust is a very fine mixed calcium silicate/oxide/carbonate material produced during kilning of portland cement. Roughly 6 million Mg yr^{-1} are collected in baghouses at cement kilns, with 80% being disposed of on the mill site in unlined landfills. About 1 million Mg are shipped off-site as a byproduct additive/stabilizer (for hazardous waste or biosolid stabilization) or as a lime substitute for use in agriculture. No specific data on the latter use appears to be available, nor any agronomic studies on use of the material. In 1992 USEPA ordered a study of kiln dusts prompted by the increasing use of byproduct or waste fuels, including some hazardous petroleum and solvent byproducts, as an accessory fuel in heating the kilns, which must reach very high temperatures in order to calcine the cement. In a report to Congress under RCRA, data indicated elevated levels of dioxins, furans, metals, and/or radionuclides in some kiln dust samples. A preliminary ruling in 1994 (59 Fed. Reg. 709, 1994, and 40 CFR Section 266) stated that kiln operators burning hazardous wastes must provide analysis of kiln dusts in order to determine whether they constitute hazardous waste. Until further analytical data on contaminant levels and their fates in soils are available, caution is clearly indicated in use of kiln dusts as lime substitutes.

9.2.4 Municipal Solid Waste and Biosolids

No other byproduct material has been as extensively studied for its agricultural value and environmental impact as has biosolids, the solid residual material resulting from sewage digestion at municipal wastewater treatment facilities. Biosolids were used in Europe in the 1950s and '60s on sewage farms to grow a variety of crops, and excessive uptake of heavy metals was eventually observed. Many studies conducted during the 1970s showed that high rates of sludge from industrial cities containing high metal levels did result in enhanced plant uptake of Cd and other metals. Regulations followed that limited industrial metal inputs into water treatment plants, which considerably reduced metal levels in sludges (Chang et al., 1987). USEPA has encouraged land application of biosolids in order to recycle nutrients and avoid landfill costs, and has provided regulations to define clean sludges for unrestricted use on crop land (Section 9.5.1). About 7.5 million Mg of biosolids (dry wt.) are generated annually in the United States, and currently 25% is land applied on dedicated disposal sites or farmers' fields; the remainder is landfilled or incinerated.

Nearly 210 million Mg of municipal solid waste (MSW), including household garbage and some municipal and inert industrial wastes, are produced annually in the United States, and finding ways to reduce landfilling costs of this material is a growing industry. Technological improvements in separation of plastic, glass, and metal from the MSW stream have suggested that the remaining 40–60% of the waste, which is largely paper and food waste, might be land applied. There are a number of firms with operational separation and composting units in the United States, subsidized by municipalities hoping to avoid capital expenses of building new subtitle D (lined and monitored) landfills. Current production levels of compost (derived from MSW and/or manures) are estimated at 8 million Mg yr^{-1}, but are likely to increase dramatically in the future (Barker, 1997, Stratton et al., 1995).

Biosolids are a good source of N and P for agronomic crops (Table 9.5), and many examples of their use on agronomic crops, forest trees, horticultural crops, and plants grown on reclaimed lands have been reported in the literature. Typically, use is based on N content as is true of animal manures, although P may be oversupplied in some cases. Some inorganic N is present, but because most must be mineralized by soil organisms, release occurs over several years. Potassium is low in sludges, and must be added as supplemental K fertilizer. Sludges have a variable effect on soil pH, depending on how they are processed (some are lime stabilized, and thus have considerable alkalinity); most sludges not containing free lime will acidify soil over time due to acid generated by nitrification, and pH must be monitored and adjusted. No similar data on MSW compost is available, but N and P are likely to be only 50% of the levels in biosolids, and mineralization rates are appreciably slower (Barker, 1997).

Trace elements in biosolids have been extensively documented in the United States and Europe, given the historical interest in land application of this material (Table 9.5). Mercury and Cd are the most toxic of the metals typically monitored in biosolids, and are elevated in some sludges compared to background levels in soils (< 0.05–1 mg kg^{-1}). Other metals listed in Table 9.5 can be hazardous when sludges of higher concentration are added at high rates to soils over a period of years, and any sludge being considered for land application should be analyzed. Potential hazards are discussed in Section 9.5.2. Regulatory limits have been established in the United States and the EU, as discussed in Section 9.5.1.

9.2.5 Mixing and Composting of Byproducts

Composting refers to the process of stockpiling byproduct materials under conditions which favor decomposition of organic materials to form a more stable product for use on soils. Carbon is lost as CO_2 during composting, resulting is a loss of bulk and concentration of nutrients (especially N) compared to the original material (Barker, 1997). The heat release accompanying composting kills many pathogens, and is an approved process for producing Class A biosolid material for land

Table 9.5 Major and trace elements in sewage sludge and MSW compost

	United Kingdom	Biosolids United States	Australia	MSW
		-----------------------g kg^{-1}-----------------------------		
N		1-200 (37)	21-82 (39)	
P		1-150 (23)	11-89 (30)	
K		0.2-65 (4.5)	0.9-4 (2)	
Ca		1-250 (45)	10-43 (19)	
Mg		0.3-25 (6)	2-7 (4)	
		-----------------------mg kg^{-1}-----------------------------		
As	< 2-123 (6)	0.3-316 (6)	3-21 (9)	1-5 (2.6)
Cd	< 2-152 (9)	0.7-8,220 (7)	3-21 (9)	1-13 (2.9)
Co	< 2-620 (10)			
Cr	4-23,000 (197)	2-3,750 (40)	81-1,620 (805)	8-130 (35)
Cu	69-6,100 (590)	7-3,120 (460)	571-1,427 (915)	31-620 (154)
Hg	< 2-140 (4)	0.2-47 (4)	1-5 (3)	0.5-3.7 (1.3)
Mo	< 2-154 (5)	2-68 (11)		
Ni	9-930 (61)	2-980 (29)	25-179 (94)	7-101 (25)
Pb	43-2,600 (398)	9-1,670 (106)	140-579 (320)	22-913 (215)
Se	< 2-15 (3)	0.5-70 (5)	0.5-11 (3)	
Zn	280-27,600 (1,140)	38-68,000 (725)	834-1,826 (1,300)	152-1,360 (503)

UK data from Smith (1996); US data from Hue (1995); Australian data from McLaren and Smith (1996); MSW data from Epstein et al. (1992)

application. Composting also eliminates odor from some materials, and improves handling characteristics for land spreading. Composting is common for processing of MSW and biosolids prior to land application.

A wide range of literature is available on formulating and processing compost, as well as describing the microbiology involved; these specifics are beyond the scope of this review. It should be noted, however, that composting can nearly always benefit materials destined for land application, due to the changes noted above. For some materials with active pathogens (particularly biosolids and MSW), wide C:N ratios, and/or objectionable odors, the advantages are significant, and if the material is to be used by home gardeners or near urban areas, are crucial to the successful marketing of the material (Barker, 1997; Stratton et al., 1995).

The concept of mixing byproduct materials to obtain more optimum physical, chemical or nutritional characteristics, in combination or apart from composting, has been used in formulating biosolid mixes with ash and cement kiln dust (Logan et al., 1997). Blending of organic (sludge) and inorganic (ash) materials has advantages in balancing nutrient content, adjusting pH, optimizing water content, and immobilizing contaminant metals. Currently, the additional costs of handling, blending, and storage of mixtures and composts are likely greater than the market value; costs of a manure/pulp mill sludge composting operation were estimated at \$50–120 Mg^{-1} (NCASI, 1984). Subsidies by contributing industries, as well as increases in demand and cost of landfilling, may make such operations feasible in the future. Hyatt (1995) has summarized information on potential marketing of composts in agriculture, and concludes that improving compost quality (particularly N content and availability) should increase farmer demand and establish pricing at a reasonable level for compost producers.

9.3 Soil Fertility Considerations in Land Application

The primary beneficial effect of land application of many byproducts is the supply of essential plant nutrients to crops, particularly macronutrients (N, P, K). There are instances where the supply of other elements (Ca, S, or micronutrients) may be considered as the primary effect (such as in the case of gypsum), or the liming value may be appreciable. Additionally, some applications (particularly of organic materials) may increase soil organic matter (SOM), improve soil structure, or increase water-holding capacity. Assessing the type and extent of beneficial effect to be achieved with a given type and rate of application is largely at the discretion of state officials in departments of agriculture or environmental regulation. It should be reiterated that a beneficial effect must be demonstrated or inferred for a given type of application, or such application may be construed as simply disposal.

9.3.1 Nutrient Content and Bioavailability

As mentioned in the previous section, most byproducts contain significant quantities of several macro- and micronutrients, but it is the N content that is used most often as the basis for calculating application rates. This has historically been true because of the high N requirements of most crops coupled with the low N-supplying power of most soils. In order to understand the reasoning behind N-based application rates, it is necessary to know something about the chemistry of N in manures.

Most organic byproducts contain two forms of N: organic and inorganic N. Organic N is chemically combined with C, H and O while inorganic N occurs as NO_3^- and NH_4^+. The process by which bacteria oxidize NH_4^+ to NO_3^- is called nitrification. Both NO_3^- and NH_4^+ can be taken up and used by plants, but organic N cannot. However, both NH_4^+ and NO_3^- can be lost in large quantities from the soil, the former by denitrification (conversion to gaseous N_2O and N_2) and the latter by NH_3 volatilization. Denitrification is a microbially mediated process which occurs under reducing (wet) conditions, while NH_3 volatilization is an abiotic process which occurs at high pH. The organic N fraction, on the other hand, is not readily lost from the soil or byproduct through these processes, but many microorganisms convert it into inorganic N by mineralization. Finally, inorganic N can be converted into organic N by microorganisms that take up NH_4^+ and NO_3^- ions and convert these into N-containing organic compounds. This process, called immobilization, temporarily reduces the amount of plant-available N in the soil. The losses resulting from NH_3 volatilization and denitrification, on the other hand, are permanent. A detailed discussion of the N cycle in soils and byproducts is presented by Brady and Weil (1996), Paul and Clark (1989), Sims and Wolf (1994), and in Section C, Chapter 5.

Most of the N in organic byproducts is organic N. According to Cabrera and Gordillo (1995), the relative proportion of N in the inorganic form in beef, swine and poultry manure samples is approximately 8.1%, 5.8% and 23%, respectively. In contrast to this, they also report that when manures are mixed with water and stored as a liquid or slurry, mineralization occurs to such an extent that in aged slurries and liquids there is more inorganic N than organic N. In virtually all manures the inorganic N pool is dominated by NH_4^+, although instances of high NO_3^- levels have also been reported.

Sims (1986) has proposed the following model for estimating plant-available N (PAN) from soil applied manure:

$$PAN = A_iN_i + P_mN_o \qquad [9.1]$$

where N_i and N_o are the quantities of inorganic and organic N in the manure, A_i is the fraction of inorganic N available for plant uptake, and P_m is the fraction of organic N that is mineralizable. The ability of this model to accurately predict the amount of plant-available N that will result from application of an organic byproduct material to soil is highly dependent on the accuracy of the values assigned to the coefficients A_i and P_m. Naturally, the values assigned to N_i and N_o will also have a major impact on the calculated PAN, but it is possible, at least in theory, to accurately measure these quantities. Much of the N in manures may be lost prior to land application during storage of the manure, which will decrease the values of both N_i and N_o. The USDA Agricultural Waste Management Field Handbook (USDA, 1992) contains estimates of N, P and K losses for several different types of animal manure stored under a variety of conditions. Obviously, organic byproducts should be sampled and analyzed immediately prior to land application in order to account for storage losses not only of N, but of other nutrients as well.

There is a great deal of uncertainty involved in assigning values to the coefficients A_i and P_m. The coefficient A_i would have a value of 1 if there were no losses of inorganic N following application of the byproduct, but this is seldom the case. Depending on a host of environmental and management variables such as soil type and moisture, temperature, and method of byproduct application (e.g., incorporated versus unicorporated), losses of NO_3^- by leaching (Davis et al., 1995a) and denitrification (Johnson and Wolf, 1995) and losses of NH_4^+ by NH_3 volatilization (Thompson et al., 1990) and immobilization can cause signifcant losses (> 50%) of the manure's inorganic N pool. The proportion of the organic N (P_m) that is mineralizable will likewise depend on a variety of factors, including temperature and moisture conditions, the C:N ratio of the byproduct, and the amount of N in the urea or uric acid forms, forms particularly susceptible to mineralization. In practice, P_m is not used *per se* in the calculation of application rates. Rather, the mineralization rate of organic N is represented by so-called decay series, which give the percentage of the organic N that is mineralized as a function of time after initial application of the byproduct. The data in Table 9.5 are representative of typical decay series for organic N in several byproduct materials. These data illustrate the following general concepts: (1) in fresh manures, a large percentage (> 75%) of the organic N is mineralized during the first year after application; (2) rates of organic-N mineralization are generally higher for poultry manures than for swine or cattle manures; and (3) storage of manures causes decreased rates of mineralization following application because the bulk of the easily mineralizable fractions of the organic-N pool have already been mineralized during storage. While data such as those in Table 9.5 can provide general guidance in nutrient management planning, local Cooperative Extension Service or Natural Resource Conservation Service offices should be contacted for additional information that is tailored to specific, local conditions.

Table 9.6 Cumulative mineralization of organic N in a variety of byproducts at several times after initial application

Material	Cumulative N mineralized after		
	Year 1	Year 2	Year 3
		%	
Poultry manure (layer)	80	82	83
Swine manure (fresh)	75	79	81
Swine manure (aged)	60	66	68
Lagoon effluent	40	46	49
Paper mill sludge (second.)	30	45	50
Biosolids (anerobic)	20	30	35
Biosolids (aerobic)	15	25	30
Compost	10	15	20

USDA (1992), NCASI (1984), King (1986).

For biosolids, N mineralization rates vary considerably with treatment process; nitrogen in anaerobic sludges tends to mineralize more rapidly than in aerobic sludges (King, 1986; Table 9.6), but both release N less rapidly than manures from organic compounds. This is related to the partial decomposition that takes place during wastewater treatment, during which N in incorporated into more recalcitrant compounds. Composting of manures and organic compounds leads to a similar result that N availability declines to lower levels (Cabrera and Gordillo, 1995).

In computing application rates based on N content, mineralization rates in Table 9.6 should be used to adjust rates for N availability, also accounting for any inorganic N present and N mineralized from previous years' applications to the same field. Losses of N from volatilization of stored manures can be appreciable, so reanalysis for N should be done if the material is stored for appreciable time periods.

Byproducts such as pulp and paper sludges, waste paper, and composts of various materials that contain low amounts of N (< 1%) with C:N ratios of > 50 probably have little potential to mineralize reasonable amounts of N for crop production. These materials will, in fact, immobilize N from the soil and may result in serious N deficiencies in crops planted soon after application. Additional (inorganic fertilizer) N should be applied to these soils to prevent N deficiency, or a period of several months should be allowed before planting to allow for decomposition to recycle N back into available forms.

Byproduct application rates are usually based on the N requirement of the crop that is being fertilized. However, many byproducts are also good sources of other macro- and micronutrients, particularly P, Zn and Cu. Perhaps because they contain so many different plant essential elements in highly available forms, manures have been found to be much more effective than synthetic fertilizers at restoring agronomic productivity to eroded and otherwise disturbed soils (Whitney et al., 1950; Aina and Egolum, 1980; Frye et al., 1985; Dormaar et al., 1988; Larney and Janzen, 1996; Larney and Janzen, 1996). While some of the proven restorative ability of manures and other byproducts may be due to improvements in the physical and biological properties of degraded soils, a majority of the aforementioned authors cite improved availability of N and P as the likely mechanism. Freeze et al. (1993) have pointed out that the enhanced value imparted by its restorative capabilities should make it economically feasible to transport manure farther than would be the case if it were to be applied strictly as an NPK source to nondegraded soils.

The primary use of gypsum on soil is the addition of Ca, which can be a limiting nutrient on sandy and/or very acidic soils. Shainberg et al. (1989), Sumner (1993) and Ritchey et al. (1995) have reviewed the literature on the use of gypsum on acidic soils to increase available Ca in the subsoil in order to promote deeper rooting and water extraction by a range of crops. Yield increases are routinely obtained on a range of soils and crops with single applications of up to 10 Mg ha^{-1}, and the increases are maintained over > 10 yr. Such applications give a good economic return over time, and show no adverse environmental effects.

9.4 Effects of Land Application on Soil Properties

Improvements in soil properties such as structure (aggregation), permeability, water-holding capacity, and CEC often do not have an immediate or consistent effect on crop yields, although soil scientists probably would agree that they are desirable. Byproduct additions, particularly organic materials, have the potential to affect soil properties, but the effects may not be apparent over a short time period.

9.4.1 Physical Properties

Land application of byproducts generally has been shown to improve soil physical properties (Wallingford et al., 1985), although there are several reports describing studies in which animal manures had either no effect or deleterious effects on some soil physical properties. Hileman (1967), for example, found no changes in either soil organic C content or bulk density as a result of 3 yr of poultry litter applications to a silt loam soil at rates as high as 18 Mg ha^{-1} annually. Similarly, Sharpley et al. (1992) found no consistent effect of long-term (up to 15 yr) poultry or swine manure applications on soil physical properties. Weil and Kroontje (1979), on the other hand, reported that after 5 yr of incorporating poultry litter at an annual rate of 110 Mg ha^{-1} the bulk density of the plow layer of a clay loam soil decreased from 1.1 to 0.8 g cm^{-1}. However, these authors also reported that a layer of partially decomposed manure had accumulated at the bottom of the plow layer which greatly reduced the water percolation rate. Cattle manure has been shown to improve soil aggregation (Hafez, 1974), to reduce soil bulk density (Tiarks et al., 1974), and to improve water-holding capacity of soils (Unger and Stewart, 1974).

A comprehensive study of the effects of long-term manure applications on soil physical properties was conducted by Schjønning et al. (1994) in Denmark. Farm yard manure (FYM) or inorganic fertilizer was applied for 90 yr to plots on a sandy loam soil. The FYM treatments had significantly higher organic C contents, lower bulk densities and decreased compactabilities compared to the inorganic fertilizer treatments. The results of this study provide particularly compelling evidence for the beneficial effects which long-term manure applications have on soil physical properties. It is expected that other types of organic byproducts might have similar effects.

In a review of literature on effects of biosolid application on physical properties, increases in soil organic C of 0.5–2.5% were noted over 2–5 yr at higher rates (20–100 Mg ha^{-1}) of sludge additions (Hill and James, 1995). Increases in C are typically correlated with lower bulk density, higher aggregation and aggregate stability, and improved water retention. Preliminary results on a sandy soil in Georgia showed much improved aggregate stability and water-holding capacity with primary and secondary pulp mill sludge added at rate of 40 Mg ha^{-1} (Cochran et al., 1997).

Infiltration and hydraulic conductivity of sludge-amended soils is more variable, as time and rate of application can have a large effect, and variability is often high. Freshly applied sludges may lower saturated hydraulic conductivity at the surface due to pore clogging, but incorporated sludge additions over time tend to improve infiltration due to better aggregation and macroporosity, leading to less surface crusting under rainfall (Hill and James, 1995).

Shiralipour et al. (1992) reviewed the literature on the impacts of MSW compost on soil physical properties. Typically, high (> 100 Mg ha^{-1}) rates of application have been added to degraded or very sandy soils. In most cases, SOM levels and water retention increased and bulk density decreased with compost additions.

Several studies have suggested that fly ash applications may be used to increase water-holding capacities of very sandy soils at high rates of application (100–300 Mg ha^{-1}) due to addition of fine particles (Ghodrati et al., 1995). Gypsum is widely used in semi arid regions on sodic soils to improve aggregation and permeability, and byproduct gypsum should perform equally well as the mined type mostly in use (Shainberg et al., 1989). Surface-applied gypsum also improves infiltration rates of weathered acidic soils over several rainfall events due to its flocculating effect on soil clays (Miller, 1987). Radcliffe et al. (1986) demonstrated that given sufficient time to move into the subsoil, surface applications of gypsum reduced the penetration resistance of subsoil hardpan layers.

9.4.2   Biological Properties

There do not appear to be a large number of published reports on the effects of manure applications on soil biological properties. Because of the great diversity of soil microrganisms and the complexity of the interactions within the soil microbial community, studies tend to focus on how manure applications affect either a single group of soil organisms or the soil microbial biomass as a whole. For example, Christie and Beattie (1989) found that soil microbial biomass C and N decreased with increasing rates of applied swine manure slurry, but increased with increasing rates of applied cattle manure slurry. Eight years of pig slurry or solid cattle manure applications were reported to increase numbers of bacteria, actinomycetes and fungi in soils (N'Dayegamiye and Côté, 1989), while 19 yr of pig and cattle slurry additions to grassland significantly decreased vesicular arbuscular (VA) mycorrhiza infection of plant roots (Christie and Kilpatrick, 1992). The reduction in VA mycorrhiza infection apparently caused by application of manure may not be a serious cause for concern, as the impact of mycorrhizal infection on plant nutrition is greatest in low fertility soils, and declining mycorrhizal populations in nutrient-rich, manured soils may have little or no impact on plant growth. Martensson and Witter (1990) concluded that the changes which they observed in populations of N-fixing soil organisms brought about by applications of manure were, in fact, caused by manure-induced changes in soil pH.

A potential concern in byproduct application is the possibility of chronic toxicity to soil microbial populations caused by low levels of added contaminants. Such concerns have been voiced particularly in the European scientific community. Although no systematic body of information is currently available to evaluate this possibility, data from Europe suggests there are levels of some trace metals that do inhibit overall microbial activity and specific groups of organisms (Smith, 1996). Typically, these levels are greater than regulatory limits on metal loadings of soils (Section 9.5.1), but in certain cases (Zn) (Smith, 1996), lower metal levels may inhibit some soil microbes.

A recurring question is the safety of land-applied biosolids in terms of disease vectors. Treatment methods approved by USEPA to eliminate pathogenic organisms from biosolids to create a Class A or unregulated material include heat sterilization, alkali treatment to high pH, ozonation, and composting. The available data support the contention that these are effective treatments when executed properly, although the study of microbial and virus agents in sludge is perhaps the most difficult aspect of assessing risk to the public, due to the difficulty of quantifying these organisms (Angle, 1994). The public is likely to fear sludge application as a disease vector, and attribute disease to locally applied sludge; it will be difficult to sort fact from fiction in these cases, but further research in this area can address these fears. Good on-farm sludge management, including proper site selection, controlling dust or mist during application, immediate incorporation into the soil, and prevention of runoff from treated fields, will minimize any such dangers.

9.4.3 Chemical Properties

The impact of byproduct amendments on soil chemical properties depends strongly on the types and rates of byproducts applied. Materials might be separated into those that are largely organic, those that are inorganic and alkaline, and gypsiferous byproducts.

Organic materials (manures, biosolids, pulp mill sludge, MSW compost) typically increase soil organic C, and therefore increase CEC due to the high negative charge on organic matter. The amount of charge varies with the level of C increase, as well as the pH, since the CEC is of the variable charge type. Increases in CEC have been measured in amended soils in proportion to the C increases, but have also been noted to decline as decomposition processes reduce the organic C levels in years following a single application (Hill and James, 1995). The capacity of soils to specifically (strongly) adsorb metal cations is similarly a function of organic matter, and partially explains the lower plant availability of metals added in biosolids compared to inorganic salts (Smith, 1996).

Organic byproducts also cause short-term shifts in soil pH and Eh (redox potential) due to the highly reducing nature of many organic byproducts and microbial activity (and hence O_2 depletion) resulting from their addition. Anaerobically digested biosolids and uncomposted animal manures added at high rates (> 20 Mg ha^{-1}) may result in temporary pH increases of 1–2 units, and decreases in Eh to 0 v or below within a week of application. In a study of swine manure added at 100 mt ha^{-1}, considerable solubilization of Fe and Mn oxides was observed during this period, although trace metal solubility did not increase appreciably due to adsorption onto organic matter (Miller et al., 1985). Within 2–3 weeks, Eh and pH returned to near initial levels, and management practices to avoid problems with such changes are obvious.

Inorganic, alkaline byproducts such as fly and wood ash, kiln dusts, and lime obviously impact soil pH, and rates should be adjusted based on the CCE of the material to reach a desired pH level. Some of these materials (such as alkaline fly ashes) also contain considerable soluble salts, which initially increase electrical conductivities (EC) of soil pore waters when land applied, occasionally injuring seedlings. These salts are rapidly leached out of the root zone within a few rainfall events, and typically have no residual effect on plant growth.

Gypsum is soluble in water (2.5 g L^{-1}) and this solubility has been used to great advantage on soils with acidic subsoils to encourage deeper rooting of a range of crop plants (Shainberg et al., 1989). High application rates (10–20 Mg ha^{-1}) of byproduct gypsum result in appreciable increases in exchangeable Ca, with a decrease in exchangeable Al in subsoil horizons of Ultisols and Oxisols over a several-year period. Overall, CEC also increased possibly by the ligand exchange of SO$_4^{2-}$ for OH$^-$ (Souza and Ritchey, 1986); pH in the subsoil is not appreciably changed. The increase in base saturation and Ca availability increased root distribution down to 1 m, resulting in greater water extraction and yield improvements for alfalfa of up to 100% compared to unamended control plots (Shainberg et al., 1989; Sumner, 1993). Both PG and FDG are similarly soluble, and have identical effects on subsoil chemistry and crop yields.

9.5 Environmental Aspects of Byproduct Applications

For many in the public and regulatory sectors, the primary question regarding land application of byproducts centers on public safety and human health. Without a clear demonstration that levels of contaminants added in a given byproduct application will have acceptably minimal effects on human and ecological health, such application should not be made.

Unfortunately, even within the scientific community, no universal consensus exists as to what materials and rates are safe for use in agriculture. Fundamental questions regarding plant uptake of elements, the impacts of contaminants on microbial communities in soils, and the solubility and rate of movement of contaminants in soils are still under active investigation. A significant body of research was recently summarized by the USEPA in formulating rules for biosolid use, and was used to set limits for safe use of that material. However, European countries have set (in many cases) more stringent rules (Smith, 1996). For many types of potentially land-applied materials there are no available environmental guidelines.

In the United States, state environmental agencies are largely responsible for implementation of federal regulations regarding land application. Where no such regulations exist, state agencies develop their own guidelines or consider materials on a case by case basis. State departments of agriculture are involved in registering byproducts as fertilizers or lime substitutes, and sometimes do not effectively interact with state environmental agencies to ensure safe use of these materials. There are examples of quite hazardous materials being land applied where state agencies have failed to coordinate (Wilson, 1997).

Given the fact that some industries may view land application of marginally useful or environmentally questionable materials as a low cost alternative to landfilling, it behoves land owners to ask byproduct producers and state agencies about materials being applied, including information on potential benefits, analysis of contaminants, and rates of application. Colleges of agriculture at state land grant universities are often further sources of information about land application.

9.5.1 Governmental Regulations Concerning Land Application

The USEPA regulates the handling, disposal and land application of wastes under Congressional authority, and through rule making associated with legislative acts. Nearly all such regulation deals with industrial and municipal byproducts, with agricultural byproducts being expressly exempt (this exemption may change as manure applications are becoming increasingly linked with water quality degradation). State departments of agriculture and/or environmental affairs implement federal regulations, and may modify them in some cases. The European Union (EU) has also adopted resolutions, implemented by member countries, regulating handling and use of wastes.

The primary legislation affecting industrial wastes in the United States is the Resource Conservation and Recovery Act (RCRA) of 1976 which defines waste and, in particular, the concept of hazardous waste, which must be specially handled and typically cannot be land applied (40 CFR 264-5). Hazardous waste is liquid or solid material derived from a manufacturing process which has one of the following characteristics: (1) corrosivity, typically due to pH > 12 or pH < 2 of liquid or liquid derived from a solid; (2) ignitability or explosivity largely associated with organic solvents and petroleum wastes; or (3) toxicity, as determined by trace contaminant levels of leachates as measured in an acetate buffer at pH 5.2 (Toxicity Characteristic Leaching Procedure, TCLP). Certain wastes are categorically listed as hazardous, while others are excluded, and exemptions may be granted by petition to USEPA. The presence of certain toxic organics and radionuclides automatically classifies a waste as hazardous. Hazardous wastes fall under Subtitle C of RCRA, and must be transported and disposed of in a proscribed manner (e.g., hazardous waste landfills); nonhazardous waste falls under Subtitle D, and can be disposed of in less secure (but still lined and monitored) landfills, and may be potentially land applied.

Municipal solid wastes, FDG, most kiln dusts (not containing restricted trace contaminants), most pulp and paper byproducts (except highly alkaline residues or high dioxin sludges), and many other more or less inert byproducts are classified nonhazardous. Biosolids are exempt, being covered under special rules of 40 CFR 503. Fly ash is also exempt, but may be restricted in use if it fails the TCLP. Flue dusts typically fail the TCLP, but may be used as micronutrient fertilizers if Pb and Cd are low. Phosphogypsum has been regulated by USEPA due to its radionuclide content in a ruling that is still under litigation. Animal manures and most agricultural byproducts are currently exempt and unrestricted.

Biosolid use on land is regulated separately under authority of the Clean Water Act in 40 CFR 503, revised in 1993 in a careful review of available data on hazards associated with land application of this material (USEPA, 1993). Processing guidelines were established to reduce disease vectors in sludge by specifying treatments such as alkali treatment, sterilization, or composting. Sludges treated in this manner contain minimum numbers of pathogenic microbes, and are considered Class A. Class A sludges can be land applied without permit or restriction, although records must be kept on amounts applied to individual fields. Class B sludges, not treated to decrease disease vectors, must be applied by permit and more carefully managed and monitored. Management practices such as injection or incorporation of sludge into soil are also specified.

Maximum safe metal levels were a point of much debate in codifying 40 CFR 503. The approach used was to assess risk to human populations based on pathways such as direct soil injestion (children), contamination of food (vegetable and animal) by plant uptake, and groundwater pollution by leaching (Ryan, 1994). The pathway with the greatest risk (in terms of cancer incidence or toxic reaction risk

Table 9.7 USEPA and EU limits for metals in biosolids and loading rates for land application of biosolids (annual and lifetime)

	USEPA			EU		
Element	Metal[1] concentrations	Annual loading	Cumulative loading	Metal[4] concentrations	Annual loading	Cumulative loading
	------mg kg^{-1}------	----------------kgha^{-1}--------------		------mg kg^{-1}------	----------------kg ha^{-1}--------------	
As	41-75	2	41	-----	(0.7)[5]	(100)[5]
Cd	39-85	1.9	39 {21}[6]	20-40	0.15	2-6
Cu	1500-4300	75	1500	1000-1750	12	100-280
Cr	[3000][2]	150	3000	-----	(15)	(800)
Pb	300-840	15	300	750-1200	15	100-600
Hg	17-57	0.85	17	16-25	0.1	2-3
Mo	[18-75][3]	[0.9][3]	[18][3]	-----	(0.2)	(8)
Ni	[420][2]	21	420	300-400	3	60-150
Se	[100][2]	5	100	-----	(0.15)	(6)
Zn	2800-7500	140	2800	2500-4000	30	300-600

From USEPA 40 CFR 503; Smith (1996).
1-lower value is for high quality unrestricted use; higher value is ceiling for application.
2-limits for these metals are tentative
3- limits on Mo are being removed by USEPA
4-lower value for soils at pH < 6, higher value for pH > 7
5-values in parenthesis are UK limits; no EU values specified for these metals
6- USEPA is considering reducing Cd loading to 21 kg ha^{-1} at USDA's request

of an assumed highly exposed individual) for a given metal was chosen to limit loadings of that metal during land application. The results were maximum metal concentrations in sludges for high quality sludges that can be used without restriction, and ceiling levels for other sludges above which land application is not permitted; annual and cumulative (lifetime) loading rates (kg ha^{-1}) are also specified for a given site (Table 9.7). USEPA regulations are formulated based on a maximum 50 Mg ha^{-1} application rate for high quality sludge, and a 1000 Mg ha^{-1} lifetime application.

The 40 CFR 503 limits are considerably higher than previous regulations, and are, in general, greater by a factor of 5–15 for most metals than EU soil-loading rates, which were arrived at in a more conservative estimation of risk (EU Directive 86/278/EEC; Smith, 1996; Table 9.7). Interestingly, metal concentrations in sludges do not differ significantly between the two, but maximum rates of application in Mg ha^{-1} allowed (computed as [loading in kg ha^{-1}]/[concentration in mg kg^{-1}]×1000) differ significantly. Using the lower metal concentration limits for EU sludges and corresponding annual metal loading rates, only about 5–10 Mg ha^{-1} yr^{-1} of sludge can be applied, compared to 50 Mg ha^{-1} yr^{-1} for the USEPA limits. Depending on N and P needs of the crop on a given site and the available nutrients in the sludge, more than 10 Mg ha^{-1} may be needed to meet annual nutrient needs, therefore requiring supplemental inorganic fertilizer be added (Section 9.3.1). Rates based on N or P required would seldom exceed the 50 Mg ha^{-1} limit for metals based on USEPA values.

Metal limits in 40 CFR 503 are often applied to the utilization of other byproducts for land application. This is probably justified for other organic byproducts such as MSW compost or pulp and paper sludges, which have a similar matrix and impact on soil chemistry. However, for inorganic materials such as gypsum, fly ash, and byproduct dusts with a largely inorganic matrix, this could lead to increases in leachable or plant-available metals above safe levels, due to the fact that metals in an organic matrix are generally less soluble (Section 9.5.2).

9.5.2 Soil and Food-Chain Contamination

Contamination of soil and subsequent uptake by plants and transfer to animals are major risk pathways for metals and organic contaminants in byproducts applied to land. A large body of information is available concerning such risks for biosolid applications, summarized recently in a number of books and review articles (Chang et al., 1987; Hue, 1995; Chaney, 1994; Smith, 1996). For other types of byproducts, there is no systematic evaluation of contaminant effects, but there are data that allow some interpretation of potential risks, particularly using the experience with biosolids as a guiding example.

9.5.2.1 Cadmium

This is perhaps the most important potential contaminant of food supplies on amended lands. Being relatively soluble in soils (particularly at low soil pH) and chemically similar to Zn, it is readily taken up by crop plants and is quite toxic to humans. Poisonings through eating contaminated rice have been documented in Japan, and ingestion limits have been established in the United States and Europe; cereal grains in the EU must contain less than 0.1 mg kg^{-1} Cd to be marketed (Smith, 1996). The 40 CFR 503 limit of 39 mg kg^{-1} Cd in biosolids has been criticized as overly liberal, and was, in fact, based on a risk pathway involving direct ingestion of sludge or soil sludge mixtures by children. Chaney (1994) cites studies that indicate that Cd uptake by plants on sludge-amended soil is seldom excessive due to the high Zn levels usually present, but also notes that some United States grain exceeds the EU Cd limit, and that lower Cd loadings might be desirable. He indicates that 21 mg kg^{-1} Cd in sludge (and a 21 kg ha^{-1} loading limit) may be necessary to limit plant uptake to acceptable levels. These data are being reviewed by USEPA.

9.5.2.2 Zinc, Copper, and Nickel

These elements are toxic to plants before they accumulate sufficient tissue concentrations to affect animals or humans. As a result, over application tends to kill or stunt plants, preventing poisoning of animals or humans consuming them. Zinc is readily translocated to plant shoots, while Cu and Ni are toxic to plant roots without much movement to leaves. None of the elements moves appreciably into plant seeds. All are more soluble, and therefore toxic, at pH < 6, and therefore, careful monitoring of pH can limit plant availability on amended soils. The 40 CFR 503 limits for these metals are quite high, although many byproducts contain appreciable concentrations. In organic materials, solubility is limited by complexation by organic matter; in inorganic materials such as flue dusts or fly ash, solubilities may be significantly higher (Hue, 1995; Ainsworth and Rai, 1987). Solubility decreases with time after application due to reactions with soil mineral surfaces.

Because swine and poultry manures may contain elevated Cu and Zn due to feed supplementation, repeated applications can result in elevated levels of total and extractable Cu (Mullins et al., 1982; Anderson et al., 1991; del Castilho et al., 1993; Kingery et al., 1994) and Zn (Christie and Beattie, 1989; Kingery et al., 1994) in soils, but it is still not clear whether this is a significant problem in terms of plant uptake and subsequent transfer through the food chain.

9.5.2.3 Lead, Chromium and Mercury

These are not taken up by plants to any degree, and thus are not limited by risk pathways that involve plants. These elements are strongly adsorbed by soil and are quite insoluble. Rather, direct soil ingestion by children living on amended soils is the limiting pathway for these metals. This scenario assumes 1–6 yr old children ingest 200 mg day^{-1} of contaminated soil, increasing risks of Pb-related neurological damage (Smith, 1996). Similarly, animals in grazing pastures may ingest considerable amounts of byproduct directly as soil and material coating leaves, and thus are exposed to these metals even without direct plant uptake. For these metals, 40 CFR 503 limits should be fairly readily transferred to other byproducts, given that no soil or plant chemistry is involved in the risk assessment.

9.5.2.4 Arsenic, Selenium, and Molybdenum

These are seldom present in high concentrations in biosolids, but may be present in wood and fly ash and in various byproduct dusts. Fly ash from high S coals in particular has elevated As contents, as well as appreciable Se and Mo (Table 9.3). These elements are anionic in soils, which means they are more soluble at higher pH (opposite of the metallic trace elements) due to strong adsorption on Fe oxides at low pH. Both Se and Mo are required trace elements for animals, and Se has been cited as deficient in human diets in the eastern United States (Adriano, 1986). Both are toxic at higher concentrations, as is As. Levels of As or Se in food or feed crops above 0.25–1 mg kg^{-1} are considered a potential hazard. Arsenic is widely applied as a pesticide, and orchard and cotton lands often contain 10–30 mg kg^{-1} As, although normal background levels in soils are < 1 mg kg^{-1}. Arsenic is similar in chemistry to P and thus taken up by plants, although a wide range of factors influence uptake. Molybdenum is only a concern for grazing ruminants, which suffer Mo-induced Cu deficiency if the Cu/Mo ratio of forage falls below 2 (Chang et al., 1987).

9.5.2.5 Trace Organic Contaminants

These are present in many types of byproducts, and in sufficient quantities may classify an individual material as hazardous under RCRA. Safe levels in byproducts for land application have not been determined, although future revisions of the 40 CFR 503 rule for biosolids may address limits for specific compounds. Rule making on these contaminants has been limited by difficulty and cost of analysis, and limited studies in the literature on occurrence, reactions, and transformations in the environment (Hue, 1995).

Smith (1996) has summarized types of toxic organics likely to be found in biosolids. The most important classes, each of which contains an array of individual compounds, are polychlorinated dibenzo-*p*-dioxins and furans (PCDD/Fs), polynuclear aromatic hydrocarbons (PAHs), and polychlorinated biphenyls (PCBs). All are moderately high molecular weight, nonvolatile and insoluble compounds that adsorb strongly to soil, show little tendency to be taken up by plants, and degrade slowly in the soil environment. Biosolids typically contain < 1 mg kg^{-1} of PCBs and PAHs, and considerably less PCDD/Fs.

Pulp and paper mill sludges, as well as wood and fly ash, have been suspected of containing dioxins and furans (PCDD/Fs) originating from Cl_2 treatment during pulp bleaching and from combustion, particularly in the presence of Cl salts. A 1988 survey of 104 kraft mill sludges derived from Cl_2 bleaching processes gave TCDD (dioxin) values averaging 70 ng kg^{-1} and 600 ng kg^{-1} TCDF (furans) (Camberato et al., 1997; Miner, 1997). Provisional standards for materials to be land applied have been set at 10–20 ng kg^{-1} based on toxic equivalent (TEQ) of 2,3,7,8-TCCD, the most toxic of the dioxin compounds. Non-Cl_2 bleaching methods are being widely adopted by the paper industry, and result in TEQs uniformly < 10 ng kg^{-1}. Interestingly, six processed municipal biosolids contained from 13 to 75 ng kg^{-1} TEQs, significantly higher than the clean technology pulp sludges (Miner, 1997).

Fly ash from boilers burning wood and fossil fuels at pulp mills may contain significant dioxins depending on boiler configuration and fuel type (Someshwar et al., 1996). Computed TCDD TEQs mean values ranged from 20–400 ng kg^{-1}, with one data set averaging 1,800 ng kg^{-1} when salt-saturated wood was used as fuel.

Overall, considerably more study on organic contaminants is required, and USEPA is likely to consider organics in revisions of the 40 CFR 503 rule, despite difficulties in obtaining data and assessing risks for these compounds (Smith, 1996).

9.5.2.6 Nitrate and Ammonium

These are included here as they can cause toxicities to plant and animals on byproduct-amended soils. Manures (especially poultry) or biosolids may contain appreciable inorganic NH_4, or rapidly mineralize N as NH_4 under optimal conditions. This NH_4 can be converted to NH_3 gas at high pH > 7. Ammonia is quite toxic to plants, and may cause appreciable damage to crops, especially seedlings (USDA, 1992). In addition, high levels of NH_3 can cause NO_2^-, which is very toxic to plants, to accumulate in soil (Schumann and Sumner, 1998).

Nitrate may accumulate in plant tissues during certain times of the year in response to high N levels in sludge- or manure-amended soils. Forage crops on soils heavily amended with biosolids have been found to contain 400 mg kg^{-1} N as NO_3. Ruminants are quite susceptible to NO_3 poisoning, and significant mortality of cows fed with hay at such NO_3 levels has been observed.

9.5.2.7 Antibiotics, Pesticides and Hormones

These are commonly fed to animals and find their way into the soil when manure is land applied. In their review of this subject, Sims and Wolf (1994) list 13 antibiotics, 5 coccidiostats and 7 larvacides that have been added to poultry diets. One of the only studies to specifically implicate antibiotics from manures in causing poor plant growth is that of Tietjen (1975). The environmental effects of adding manure containing antibiotics, pesticides and hormones to the soil are yet to be fully evaluated, and more research is needed in this area.

9.5.3 Water Quality Considerations

The adverse impacts of improper land application of byproducts on both surface and groundwater quality can be severe. However, by far the greatest incidence of such impacts involves N and P in animal manures, rather than trace contaminants in industrial byproduct, and such problems are commonly traced to poor management or over-application of manures. On the other hand, many instances of water contamination can be traced to disposal of wastes in landfills or stockpiling of byproducts at the point of production. The reasonable reuse of these materials on crop land, where appropriate, would therefore reduce the likelihood of water contamination.

Applications of animal manure can cause leaching of high levels of NO_3^- to groundwater and enrichment of surface waters with P contained in runoff. Instances of groundwater pollution with NO_3^- in England are now so common that legislation is being considered to limit N applications to crop land to < 200 kg ha^{-1} (Smith, 1996). Both United States and UK limits are 10 mg NO_3-N L^{-1} in groundwater, and soils receiving high rates of manures or high N sludges often release NO_3^- rapidly enough to pollute shallow aquifers, especially on sandy soils.

In recent years, as point sources of water pollution have been identified and eliminated, the importance of nonpoint source water pollution has become more apparent. Agricultural activities are believed to cause a considerable amount of the nonpoint source pollution of both surface and groundwater in the United States and other developed countries. Furthermore, land application of animal manures is now suspected of being a major P source, which is frequently the nutrient limiting biological productivity in freshwater lakes (Sharpley and Rekolainen, 1997). The recent outbreak of *Pfiesteria*-like organisms in Chesapeake Bay, whose toxins are believed to have caused memory loss in humans and the death of thousands of fish (Martin and Cooke, 1994), may be linked to elevated P levels in the bay, which is surrounded by an area of intensive animal production (MacIlwain, 1997).

The nutrient that is of primary concern with regard to water pollution resulting from land application of animal manures is P (Sharpley et al., 1994; Daniel et al., 1994). The relative amounts of N and P contained in many animal manures are such that when application rates of manure are calculated based on the N requirement of the crop, the amount of P added often exceeds the P requirement of the crop. This results in accumulation of P in the soil. While most soils have the capacity to fix, or immobilize, large amounts of P, this capacity is finite and can be exceeded if N-based manure applications are made repeatedly over a period of years. As the soil's ability to fix P declines, a greater proportion of the P in the soil exists in a water-soluble form (as the orthophophate anion), which is highly susceptible to edge of field loss in runoff. Except in very sandy soils which have low P-fixation capacities, leaching of P to groundwater is generally not considered to be a major loss mechanism of P from manure-amended soils.

Loss of N from fields receiving animal manures is also a concern. While most of the N in animal manures is in the form of organic N compounds or NH_4^+, mineralization and nitrification can result in the formation of large amounts of NO_3^- following manure application if the soil is warm and well-aerated (Johnson and Wolf, 1995). Unlike the $H_2PO_4^-$ anion, NO_3^- is not fixed in appreciable quantities by most soils and is, therefore, more likely to be leached through the soil to groundwater (Unwin, 1986; Adams et al., 1994).

In addition to N and P, it has been shown that runoff from manure-amended soils can also contain pesticides and hormones. Pote et al. (1994) found concentrations of cyromazine, a feed-through larvacide, as high as 101 g L^{-1} in runoff from plots that had been treated with 17.7 Mg ha^{-1} of caged layer manure. For up to 1 yr after the manure was applied cyromazine was not detected in water samples collected from lysimeters placed at a depth of 60 cm in the soil.

Nichols et al. (1997) found elevated levels of the hormone 17ß-estradiol in runoff from poultry litter-amended pastures. The effects which compounds such as these may have on aquatic ecosystems are not currently known.

The potential for pollution of surface waters resulting from land application of animal manures can be minimized if certain management practices are followed (Daniel et al., 1994). The primary factors affecting quality of runoff from animal manure-amended fields are application rate, method of application and timing of application (Sharpley, 1997). In the case of P, to this list should be added the soil P status (or the degree to which the P-fixation capacity is already saturated) because it is well documented that the amount of P in the surface soil directly influences the amount of P in runoff (Daniel et al., 1997). Research has shown that concentrations of N and P in runoff increase as the manure application rate increases (Edwards and Daniel, 1993) and that runoff N and P concentrations decrease as the interval of time between manure application and the rainfall event increases (Westerman and Overcash, 1980; Edwards and Daniel, 1993). Incorporation of manure into the soil generally decreases concentrations of N and P in runoff (Mueller et al., 1984), although Nichols et al. (1994) reported that shallow (2–3 cm) incorporation of poultry litter applied to established fescue pasture had no effect on runoff N and P concentrations.

With regard to trace metals and organic contaminants in land-applied products, leaching of contaminants to groundwater or transport to surface waters is unlikely with good management of amended soils (Smith, 1996). This assumes application rates have been chosen based on limited metal loadings (less than 40 CFR 503 limits), that trace organics are limited to mg kg^{-1} levels or less, and that soils are properly limed and runoff and erosion controlled. The solubility of contaminants under these conditions is so low that movement by leaching is very slow, and with control of surface runoff, little export to surface water is expected.

9.6 Recommendations

1. Before considering land-applying a byproduct material, check with state departments of agriculture and/or environmental divisions to see if the product has been registered or approved for use on land, or is likely to be unsuitable (classified as hazardous waste based on testing or known levels of contaminants).

2. Byproducts have a wide range of properties that affect potential agronomic value and environmental limitations, even within one class of materials. A current, accurate analysis of important chemical properties of the material (nutrient content, liming value, contaminant levels) is the most important piece of information needed to make decisions on use and application rates. Ask the producer to provide this information, or be ready to assume the cost of such an analysis.

3. Byproducts are applied based on one of the following rationales: (1) as a nutrient source, typically either for N or P; (2) as a liming agent, determined by the CCE; or (3) as a soil conditioner, in order to increase soil organic matter, water-holding capacity, etc. Application rates should be calculated for (1) and (2) based on (available) nutrients supplied or lime required to reach a target pH. For (3), specific rates are difficult to determine agronomically, but should be environmentally conservative (see 4. below).

4. Compute annual loading rates of metals or other contaminants resulting from the application; for biosolids, use the 40 CFR 503 regulations to insure loadings are safe. For other byproducts, use 40 CFR 503 limits as guidelines. Keep accurate records of loadings of metals and nutrients for individual fields over time, and do not exceed 40 CFR 503 lifetime cumulative limits.

5. Use good management practices on fields to prevent chemical (metals, organics) or microbial (pathogen) losses through runoff or erosion. Incorporate applications immediately, and follow soil and water conservation practices.

6. Soil test for pH and nutrients annually at cooperative extension service or private laboratories. Maintain pH > 6 with lime additions, especially with organic byproducts, to minimize metal solubility. Monitor soil P and metal levels, and reduce applications if soil test levels exceed high ratings over time. Check soluble (NO_3^- and NH_4^+) N levels in soils periodically to ensure that excess soluble N is not accumulating in soil.

9.7 References

Adams, P.L., T.C. Daniel, D.R. Edwards, D.J. Nichols, D.H. Pote and H.D. Scott. 1994. Poultry litter and manure contributions to nitrate leaching through the vadose zone. Soil Sci. Soc. Amer. J. 58:1206–1211.

Adriano, D. 1986. Trace elements in the terrestrial environment. Springer-Verlag, New York, NY.

Aina, P.O., and E. Egolum. 1980. The effect of cattle feedlot manure and inorganic fertilizer on the improvement of subsoil productivity of two soils. Soil Sci. 129:212–217.

Ainsworth, C.C., and D. Rai. 1987. Chemical characterization of fossil fuel combustion wastes. EPRI Rpt. EA-5321. Electric Power Research Institute, Palo Alto, CA.

Alcordo, I.S., and J.E. Rechcigl. 1995. Phosphogypsum and other byproduct gypsums. p. 365–426. In J. E. Rechcigl (ed.) Soil amendments and environmental quality. Lewis Publishers, Boca Raton, FL.

American Coal Ash Association. 1993. Coal combustion byproduct (CCB) production and use: 1966–1993. American Coal Ash Assoc., Alexandria VA.

Anderson, M.A., J.R. McKenna, D.C. Martens, S.J. Donohue, E.T. Kornegay, and M.D. Lindemann. 1991. Long-term effects of copper rich swine manure application on continuous corn production. Commun. Soil Sci. Plant Anal. 22:993–999.

Angle, J.S. 1994. Sewage sludge: Pathogenic considerations. p. 35–39. In C.E. Clapp, W.E. Larson, and R.H. Dowdy (ed.) Sewage sludge: Land utilization and the environment. Soil Science Society of America, Madison, WI.

Barker, A.V. 1997. Composition and uses of compost. p. 140–163. In J.E. Rechcigl and H.C. MacKinnon (ed.) Agricultural uses of byproducts and wastes. American Chemical Soc., Washington, DC.

Brady, N.C., and R. Weil. 1996. The nature and properties of soils (11th Ed.). Prentice Hall, Upper Saddle River, NJ.

Cabrera, M.L., and R.M. Gordillo. 1995. Nitrogen release from land-applied manures. pp. 393–403. *In* K. Steele (ed.) Animal waste and the land-water interface. CRC Press, Boca Raton, FL.

Camberato, J.J., E.D. Vance, and A.V. Someshwar. 1997. Composition and land application of paper manufacturing residuals. p. 185–203. *In* J.E. Rechcigl and H. MacKinnon (ed.) Agricultural uses of byproducts and wastes. American Chemical Society, Washington, DC.

Chaney, R.L. 1994. Trace metal movement: Soil-plant systems and bioavailability of biosolids-applied metals. p. 27–32. *In* Clapp, E.E., W.E. Larson, and R.H. Dowdy (ed.) Sewage sludge: Land utilization and the environment. Soil Science Society of America, Madison, WI.

Chang, A.C., T.D. Hinesly, T.E. Bates, H.E. Doner, R.H. Dowdy, and J.A. Ryan. 1987. Effects long-term sludge application on accumulation of trace elements by crops. p. 53–66. *In* Land Application of Sludge. Lewis Publishers, Chelsea, MI.

Chang, C., T.G. Sommerfeldt, and T. Entz. 1990. Rates of soil chemical changes with eleven annual applications of cattle feedlot manure. Can. J. Soil Sci. 70:673–680.

Choudhary, M., L.D. Bailey, and C.A. Grant. 1996. Review of the use of swine manure in crop production: Effects on yield and composition and on soil and water quality. Waste Manag. Res. 14:581–595.

Christensen, B.T. 1986. Ammonia volatilization loss from surface applied animal manure. *In* A.D. Kofoed, J.H. Williams and P. L'Hermite (eds.). Efficient land use of sludge and manure. Elsevier Scientific Publishers, London, UK.

Christie, P., and J.A.M. Beattie. 1989. Grassland soil microbial biomass and accumulation of potentially toxic metals from long-term slurry application. J. Appl. Ecol. 26:597–607.

Christie, P., and D.J. Kilpatrick. 1992. Vesicular-arbuscular mycorrhiza infection in cut grassland following long-term slurry application. Soil Biol. Biochem. 24:325–331.

Cochran, R., W. Miller, and L. Morris. 1997. Decomposition of pulp mill biosolids and effects on soil properties. Agron. Abstr., p. 275.

Daniel, T.C. O.T. Carton, and W.L. Magette. 1997. Nutrient management planning. p. 297–309. *In* H. Tunney et al. (eds.) Phosphorus loss from soil to water. CAB International, Wallingford, UK.

Daniel, T.C., A.N. Sharpley, D.R. Edwards, R. Wedepohl, and J.L. Lemunyon. 1994. Minimizing surface water eutrophication from agriculture by phosphorus management. J. Soil Water Conserv. 49:30–38.

Davis, J.G., G. Vellidis, R.K. Hubbard, J.C. Johnson, G.L. Newton, and R.R. Lowrance. 1995a. Nitrogen uptake and leaching in a no-till forage rotation irrigated with liquid dairy manure. p. 405–410. *In* K. Steele (ed.) Animal waste and the land-water interface. CRC Press, Boca Raton, FL.

Davis, J.G., G. Weeks, and M.B. Parker. 1995b. Use of deep tillage and liming to reduce zinc toxicity in peanuts grown on flue dust contaminated land. Soil Technol. 8:85–95.

del Castilho, P., W.J. Chardon, and W. Salomons. 1993. Influence of cattle-manure slurry application on the solubility of cadmium, copper, and zinc in a manured, acidic, loamy sand soil. J. Environ. Qual. 22:689–696.

Dormaar, J.F., C.W. Lindwall, and G.C. Kozub. 1988. Effectiveness of manure and commercial fertilizer in restoring productivity of an artificially eroded Dark Brown Chernozemic soil under dryland conditions. Can. J. Soil Sci. 68:669–679.

Edwards, D.R., and T.C. Daniel. 1992. Environmental impacts of on-farm poultry waste disposal: A review. Biores. Technol. 41:9–33.

Edwards, D.R., and T.C. Daniel. 1993. Drying interval effects on runoff from fescue plots receiving swine manure. Trans. ASAE 36:1673–1678.

Edwards, D.R., and T.C. Daniel. 1993. Effects of poultry litter application rate and rainfall intensity on quality of runoff from fescuegrass plots. J. Environ. Qual. 22:361–365.

Eghball, B., and J.F. Power. 1994. Beef cattle feedlot manure management. J. Soil Water Conserv. 49:113–122.

Epstein, R., R.L. Chaney, C. Henry, and T.J. Logan. 1992. Trace elements in municipal solid waste compost. Biomass Bioenergy 3:227–238.

Freeze, B.S., C. Webber, C.W. Lindwall, and J.F. Dormaar. 1993. Risk simulation of the economics of manure application to restore eroded wheat cropland. Can. J. Soil Sci. 73:267–274.

Frye, W.W., O.L. Bennett, and G.J. Buntley. 1985. Restoration of crop productivity on eroded or degraded soils. p. 335–356. *In* R.F. Follett and B.A. Stewart (ed.) Soil erosion and crop productivity. American Society of Agronomy, Madison, WI.

Ghodrati, M., J.T. Sims, and B.L. Vasilas. 1995. Evaluation of fly ash as a soil amendment for the Atlantic Coastal Plain: 1. Soil hydraulic properties and elemental leaching. Water Air Soil Pollut. 81:349–361.

Hafez, A.A.R. 1974. Comparative changes in soil physical properties induced by admixtures of manures from various domestic animals. Soil Sci. 118:53–59.

Hilemann, L. 1967. The fertilizer value of broiler litter. AK Agric. Exp. Sta. Rep. Ser. 158.

Hill, R.L., and B. R. James. 1995. Influence of waste amendments on soil properties. p. 311–326. *In* J. E. Rechcigl (ed.) Soil amendments and environmental quality. Lewis Publishers, Boca Raton, FL.

Hue, N.V. 1995. Sewage sludge. p. 199–247. *In* J. E. Rechcigl (ed.) Soil amendments and environmental quality. Lewis Publishers, Boca Raton, FL.

Hyatt, G.W. 1995. Economic, scientific, and infrastructure basis for using municipal composts in agriculture. p. 19–72. *In* D.L. Karlen, R.J. Wright, and W.O. Kemper (ed.) Agricultural utilization of urban and industrial byproducts. American Society of Agronomy, Madison, WI.

Johnson, W.F., Jr., and D.C. Wolf. 1995. Nitrogen transformations in soil amended with poultry litter under aerobic conditions followed by anaerobic periods. p. 27–34. *In* K. Steele (ed.) Animal waste and the land-water interface. CRC Press, Boca Raton, FL.

King, L.D. (Ed.). 1986. Agricultural use of municipal and industrial sludges in the southern U.S. NC State Univ. Southern Coop. Ser. Bull. 314.

Kingery, W.L., C.W. Wood, D.P. Delaney, J.C. Williams, and G.L. Mullins. 1994. Impact of long-term land application of broiler litter on environmentally related soil properties. J. Environ. Qual. 23:139–148.

Korcak, R. F. 1995. Utilization of coal combustion byproducts in agriculture and horticulture. p. 107–130. *In* D.L. Karlen, R.J. Wright, and W.O. Kemper (ed.) Agricultural utilization of urban and industrial byproducts. American Society of Agronomy, Madison, WI.

Kriesel, W., C.S. McIntosh, and W.P. Miller. 1994. The potential for beneficial reuse of sewage-sludge and coal combustion byproducts. J. Environ. Manag. 42: 299–315.

Larney, F.J., and H.H. Janzen. 1996. Restoration of productivity to a desurfaced soil with livestock manure, crop residue, and fertilizer amendments. Agron. J. 88:921–927.

Logan, T.J., B.J. Lindsay, and S. Titko. 1997. Characteristics and standards for processed biosolids in the manufacture and marketing of horticultural fertilizers and soil blends. p. 63–71. *In* J.E. Rechcigl and H. MacKinnon (ed.) Agricultural uses of byproducts and wastes. American Chemical Society, Washington, DC.

MacIlwain, C. 1997. Scientists close in on "cell from hell" lurking in Chesapeake Bay. Nature 389:317–318.

McLaren, R.G., and C.J. Smith. 1996. Issues in the disposal of industrial and urban wastes. p. 183–212. *In* R. Naidu et al. (ed.) Contaminants and the soil environment in the Australasia-Pacific region. Kluwer Academic Publishers, Dordrecht, Netherlands.

Mårtensson, A.M., and E. Witter. 1990. Influence of various soil amendments on nitrogen-fixing soil microorganisms in a long-term field experiment, with special reference to sewage sludge. Soil Biol. Biochem. 22:977–985.

Martin, A., and G.D. Cooke. 1994. Health risks in eutrophic water supplies. Lake Line 14:24–26.

Martin, J.H., Jr. 1997. The Clean Water Act and animal agriculture. J. Environ. Qual. 26:1198–1203.

Mattigod, S.V., D. Rai, L.E. Eary, and C.C. Ainsworth. 1990. Geochemical factors controlling the mobilization of inorganic constituents from fossil fuel combustion residues: I. Review of major elements. J. Environ. Qual. 19:188–201.

Miller, W. P. 1987. Infiltration and soil loss of three gypsum-amended Ultisols under simulated rainfall. Soil Sci. Soc. Am. J. 51:1314–1320.

Miller, W.P. 1995. Environmental considerations in land application of byproduct gypsums. p. 183–208. *In* D.L. Karlen, R.J. Wright, and W.O. Kemper (ed.) Agricultural utilization of urban and industrial byproducts. American Society of Agronomy, Madison, WI.

Miller, W. P., D. C. Martens, and L. W. Zelazny. 1985. Effects of manure amendment on soil chemical properties and hydrous oxides. Soil Sci. Soc. Am. J. 49:856–861.

Miller, W.P., and M.E. Sumner. 1997. Agricultural and industrial uses of byproduct gypsums. p. 226–239. *In* J.E. Rechcigl and H.C. MacKinnon (ed.) Agricultural uses of byproducts and wastes. American Chemical Society, Washington, DC.

Miner, R.A. 1997. National solid and hazardous waste regulatory initiatives of interest of the forest products industry. Proc. NCASI 1997 Southern Regional Mtg., Asheville, NC, June 1997. NCASI, Research Triangle Park, NC.

Moore, P.A., Jr., and D.M. Miller. 1994. Decreasing phosphorus solubility in poultry litter with aluminum, calcium and iron amendments. J. Environ. Qual. 23:325–330.

Morris, L.A., W.P. Miller, M.E. Sumner, and W.L. Nutter. 1997. Characterization and utilization of pulp mill wastes for forest land application. Proc. NCASI 1997 Southern Regional Mtg., Asheville, NC, June 1997. NCASI, Research Triangle Park, NC.

Morrison, J.L. 1969. Distribution of arsenic from poultry litter in broiler chickens, soil, and crops. J. Agric. Food Chem. 17:1288–1290.

Mueller, D.H., R.C. Wendt, and T.C. Daniel. 1984. Phosphorus losses as affected by tillage and manure application. Soil Sci. Soc. Amer. J. 48:901–905.

Mullins, G.L., D.C. Martens, W.P. Miller, E.T. Kornegay, and D.L. Hallock. 1982. Copper availability, form, and mobility in soils from three annual copper-enriched hog manure applications. J. Environ. Qual. 11:316–323.

NCASI. 1984. Land application and related utilization of pulp and paper mill sludges. NCASI Bulletin 439. NCASI, Research Triangle Park, NC.

NCASI. 1997. Solid waste management and disposal practices in the U.S. paper industry. NCASI Tech. Bulletin 641. NCASI, Research Triangle Park, NC.

N'Dayegamiye, A., and D. Côté. 1989. Effect of long-term pig slurry and solid cattle manure application on soil chemical and biological properties. Can. J. Soil Sci. 69:39–45.

Nichols, D.J., T.C. Daniel, P.A. Moore, Jr., D.R. Edwards, and D. H. Pote. 1997. Runoff of estrogen hormone 17ß-estradiol from poultry litter applied to pasture. J. Environ. Qual. 26:1002–1006.

Nichols, D.J., T.C. Daniel, and D.R. Edwards. 1994. Nutrient runoff from fescue pasture after incorporation of poultry litter and inorganic fertilizer. Soil Sci. Soc. Amer. J. 58:1224–1228.

Paul, E.A., and F.E. Clark. 1989. Soil microbiology and biochemistry. Academic Press, San Diego, CA.

Pote, D.H., T.C. Daniel, D.R. Edwards, J.D. Mattice, and D.B. Wickliff. 1994. Effect of drying and rainfall intensity on cyromazine loss from surface-applied caged-layer manure. J. Environ. Qual. 23:101–104.

Powers, J.E. 1998. Land application of byproducts. American Society of Agronomy, Madison, WI.

Radcliffe, D.E., R.L. Clark, and M.E. Sumner. 1986. Effect of gypsum and deep-rooting perennials on subsoil mechanical impedance. Soil Sci. Soc. Am. J. 50:1566–1570.

Ritchey, K.D., D. de Sousa, C.M. Feldhake, and R.B. Clarke. 1995. Improved water and nutrient uptake from subsurface layers of gypsum-amended soils. p. 157–182. In D.L. Karlen, R.J. Wright, and W.O. Kemper (ed.) Agricultural utilization of urban and industrial byproducts. American Society of Agronomy, Madison, WI.

Ryan, J.A. 1994. Utilization of risk assessment in development of limits for land application of municipal sewage sludge. p. 55–65. In C.E. Clapp, W.E. Larson, and R.H. Dowdy (ed.) Sewage sludge: Land utilization and the environment. Soil Science Society of America, Madison, WI.

Schjønning, P., B.T. Christensen, and B. Cartensen. 1994. Physical and chemical properties of a sandy loam receiving animal manure, mineral fertilizer or no fertilizer for 90 years. Europ. J. Soil Sci. 45:257–266.

Schumann, A., and M.E. Sumner. 1998. Potential ammonia and nitrate toxicity from flyash-sewage-poultry manure mixtures. Proc. World Congress of Soil Science No. 1289. Montpellier, France.

Shainberg, I., M. E. Sumner, W. P. Miller, M. P. Farina, M. A. Pavan, and M. V. Fey. 1989. Use of gypsum on soils: A review. Adv. Soil Sci. 9:1–111.

Sharpley, A.N., B.J. Carter, B.J. Wagner, S.J. Smith, E.L. Cole and G.A. Sample. 1992. Impact of long-term swine and poultry manure application on soil and water resources in eastern Oklahoma. OK Agric. Exp. Sta. Tech.1 Bull. T-169.

Sharpley, A.N. 1997. Rainfall frequency and nitrogen and phosphorus runoff from soil amended with poultry litter. J. Environ. Qual. 26:1127–1132.

Sharpley, A.N., S.C. Chapra, R. Wedepohl, J.T. Sims, T.C. Daniel, and K.R. Reddy. 1994. Managing agricultural phosphorus for protection of surface waters: Issues and options. J. Environ. Qual. 23:437–451.

Sharpley, A.N., and S. Rekolainen. 1997. Phosphorus in agriculture and its environmental implications. p. 1–53. In H. Tunney et al. (ed.) Phosphorus loss from soil to water. CAB International, Wallingford, UK.

Shiralipour, A., D.B. McConnell, and W.H. Smith. 1992. Physical and chemical properties of soils as affected by municipal solid waste compost application. Biomass Bioenergy 3:261–266.

Shreve, B.R., P.A. Moore, Jr., T.C. Daniel, D.R. Edwards, and D.M. Miller. 1995. Reduction of phosphorus in runoff from field-applied poultry litter using chemical amendments. J. Environ. Qual. 24:106–111.

Sims, J.T. 1986. Nitrogen transformations in a poultry manure amended soil: temperature and moisture effects. J. Environ. Qual. 15:59–67.

Sims, J.T. 1995. Characteristics of animal wastes and waste-amended soils: An overview of the agricultural and environmental issues. pp. 1–13. In K. Steele (ed.) Animal waste and the land-water interface. CRC Press, Boca Raton, FL.

Sims, J.T., and D.C. Wolf. 1994. Poultry waste management: agricultural and environmental issues. Adv. Agron. 52:1–83.

Smith, S.R. 1996. Agricultural recycling of sewage sludge and the environment. CAB International, Wallingford, UK.

Someshwar, A.V. 1996. Wood and combination wood-fired boiler ash characterization. J. Environ. Qual. 25:962–972.

Sommerfeldt, T.G., and C. Chang. 1987. Soil-water properties as affected by twelve annual applications of cattle feedlot manure. Soil Sci. Soc. Amer. J. 51:7–15.

Souza, D.M.G., and K.D. Ritchey. 1986. Uso do gesso no solo de Cerrado. Ann. Sem. Uso Fosfogesso Agricultura p. 119–144.

Stratton, M.L., A.V. Barker, and J.E. Rechcigl. 1995. Compost. p. 249–310. *In* J.E. Rechcigl (ed). Soil amendments and environmental quality. Lewis Publishers, Boca Raton, FL.

Sumner, M.E. 1993. Gypsum and acid soils: The world scene. Adv. Agron. 51:1–32.

Sumner, M.E., W.P. Miller, D. E. Radcliffe, C. S. Hoveland, U. Kukier, C. Fauzia, and R. Arnold. 1996. Use of byproduct flue gas desulfurization gypsum as an ameliorant for Southeastern soils. Final Rpt., DOE Clean Coal Technol. Proj. Dept. Of Energy, Washington, DC.

Tiarks, A.E., A.P Mazurak, and L. Chesnin. 1974. Physical and chemical properties of soil associated with heavy applications of manure from cattle feedlots. Soil Sci. Soc. Amer. Proc. 38:826–830.

Tietjen, C. 1975. Influence of antibiotics and growth promoting feed additives on the manuring effect of animal excrements in pot experiments with oats. p. 328–330. *In* Managing livestock wastes: Proc. Third International Symposium, Champaign-Urbana, IL.

Tisdale, S.L., W.L. Nelson, J.D. Beaton, and J.L. Havlin. 1993. Soil fertility and fertilizers. 5th Ed. Macmillan Publishers, New York, NY.

Thompson, R.B., B.F. Pain, and D.R. Lockyer. 1990. Ammonia volatilization from cattle slurry following surface aoolication to grassland. Plant Soil 125:109–115.

Unger, P.W. and B.A. Stewart. 1974. Feedlot waste effects on soil conditions and water evaporation. Soil Sci. Soc. Amer. Proc. 38:954–957.

Unwin, J.P:. 1997. Results of the NCASI 1995 wastewater and solid waste survey. Proc. NCASI 1997 Southern Regional Mtg., Asheville, NC, June 1997. NCASI, Research Triangle Park, NC.

Unwin, R.J. 1986. Leaching of nitrate after application of organic manures. Lysimeter studies. p. 158–167. *In* A.D. Kofoed et al. (ed.) Efficient land use of sludge and manure. Elsevier Scientific Publishers, London, UK.

USDA. 1978. Improving soils with organic wastes. USDA Special Task Force Report. USDA, Washington, DC.

USDA. 1992. Agricultural waste management field handbook. National Engineering Manual, Part 651. United States Department of Agriculture, Washington, DC.

USEPA. 1993. Part 503 Standards for the use or disposal of sewage sludge. Fed. Reg. 58:9387–9404.

USGS. 1997. Commodity minerals information: Iron and steel slags. Http://minerals.er. usgs.gov/minerals/0pubs/commodity/iron_&steel_slag.

Walker, J.M., R.M. Southworth, and A.B. Rubin. 1997. U.S. Environmental Protection Agency regulations and other stakeholder activities affecting agricultural use of byproducts and wastes. p. 28–49. *In* J.E. Rechcigl and H.C. MacKinnon (ed.) Agricultural uses of byproducts and wastes. American Chemical Society, Washington, DC.

Wallingford, G.W., W.L. Powers and L.S. Murphy. 1985. Present knowledge on the effects of land application of animal waste. p. 580–586. *In* Managing livestock wastes. Proc. Third International Symposium, Champaign-Urbana, IL.

Weil, R.R., and W. Kroontje. 1979. Physical condition of a Davidson clay loam after five years of heavy poultry manure applications. J. Environ. Qual. 8:387–392.

Westerman, P.W., and M.R. Overcash. 1980. Short-term attenuation of runoff pollution potential for land-applied swine and poultry manure. p. 289–292. *In* Livestock waste - A renewable resource. American Society of Agricultural Engineering, St. Joseph, MI.

Whitney, R.S., R. Gardener, and D.W. Robertson. 1950. The effectiveness of manure and commercial fertilizer in restoring productivity of subsoils exposed to leveling. Agron. J. 42:239–245.

Wilson, D. 1997. Fear in the fields, Part 1 and 2. Seattle Times, July 3, 1997.

Wines, R.A. 1985. Fertilizer in America. Temple University Press, Philadelphia, PA.

Wyatt, J.M. 1997. Byproduct usage in fertilizer micronutrients. p. 255–267. *In* J.E. Rechcigl and H.C. MacKinnon (ed.) Agricultural uses of byproducts and wastes. American Chemical Society, Washington, DC.

10

Conservation Tillage

J.M. Bradford
USDA-ARS, Weslaco, TX

G.A. Peterson
Colorado State University

10.1 Introduction

Written on a stone wall in the Iowa State University library is the statement "Where tillage begins other arts follow"–Daniel Webster. Soil tillage as developed over the centuries permitted farmers to grow more and better crops by loosening and mixing soil and controlling weeds. With the introduction of herbicides in the 1940s and 1950s and the continuous development of new herbicides, the need to mechanically control weeds has slowly diminished. And with the development of new no-till (NT) planters, the need to prepare seedbeds through plowing and disking has decreased. Originally the moldboard plow was a necessity to produce high crop yields, but with plowing (clean tillage) came accelerated water and wind erosion, extreme organic matter oxidation, and in general accelerated land degradation. With excessive erosion, government conservation compliance regulations became necessary to reduce erosion on highly erodible lands. To meet these regulations and in an attempt to increase profit, farmers have slowly over the last 30 yr switched to conservation tillage systems.

Conservation tillage is a broad term used to define any tillage system which has as its primary objective reduction of soil and water losses. However, the primary reason that farmers in many regions of the United States change from conventional tillage to some form of conservation tillage is to reduce farm input costs and increase profits (Smart and Bradford, 1996). Other secondary reasons include increased water storage and insulation of the soil surface from temperature extremes. Conservation tillage reduces traffic operations over the field, which decreases not only soil compaction but also input costs for fuel, tractors, labor, and other equipment. In some cases input costs are actually increased because herbicides cost more than the tillage they replace. In these situations, the farmer adopts the new technology because yields increase at a faster rate than the input costs. Timeliness of operations such as planting or cultivation is improved by using conservation tillage in some situations. With the introduction of more effective herbicides and improvements in sprayer design, such as the hooded sprayer, weed control without tillage can be accomplished, even in the subtropical areas where temperatures seldom reach freezing (Smart and Bradford, 1997). Possibly the most important benefit of conservation is the soil restoration processes associated with conservation tillage (Langdale et al., 1992) or in the words of Crovetto (1996): "The great economic advantage of no-till is that low value soils can be brought into higher production without erosion risks".

Even though in this chapter we discuss the soil science principles of conservation tillage, we believe that the ultimate goal of conservation tillage should be to adapt NT practices to all local soil and climatic combinations. However, if weeds cannot be controlled with herbicides, then a cultivator must be used. If soil compaction becomes a problem, then deep chiseling or zone tillage is needed. If cold soil temperature and excess water at spring planting time are problems, then a portion of the residue from the seeding row must be moved in the fall (while banding fertilizer). With time, as the soil quality improves, or as the farmer becomes more familiar with the total system, soil problems are reduced. Many of the failures reported for NT studies have been associated with short-term experiments of 2 to 5 yr, use of improper planting equipment, and lack of timeliness of operations. Results from the first three years of a NT study often have limited meaning because both the soils and operators are having to adjust to a grossly different environment. We believe that the major benefits of NT agriculture can be assessed only after it has been in place for a decade or more. After all, if it has taken over 100 to 150 yr of maximum tillage to degrade and change our soils, why should one expect that we would see the complete picture regarding NT after only a few years. Truly, adoption of no-till is an investment in the soils of our future.

10.2 Definitions

In this chapter the authors make no distinction between conservation tillage, minimum tillage, or reduced tillage. Types of conservation tillage include no-tillage, ridge tillage, mulch tillage, and zone tillage. The definitions set forth by the Conservation Technology Information Center are as follows (Schertz and Becherer, 1994; Hill, 1996).

No-Till. The soil is left undisturbed from harvest to planting except for nutrient injection. Planting or drilling is accomplished in a narrow seedbed or slot made by coulters, row cleaners, disk openers, in-row chisels, or rototillers. Weed control is accomplished primarily with herbicides. Cultivation may be used for emergency weed control.

Strip-till is considered a modification of no-till. It is used to place nutrients in bands 12–20 cm deep in the fall, to reduce soil compaction or consolidation, and/or to create a warmer and drier soil environment in the spring, especially in northern latitudes.

Ridge Tillage. The soil is left undisturbed from harvest to planting except for nutrient injection. Planting is completed in a seedbed prepared on ridges with sweeps, disk openers, coulters, or row cleaners. Residue is left on the surface between ridges. Weed control is accomplished with herbicides and/or cultivation. Ridges are rebuilt during cultivation.

Mulch Tillage. The soil is disturbed prior to planting. Tillage tools such as chisels, field cultivators, disks, sweeps, or blades are used. Weeds are controlled with herbicides and/or cultivation.

Crop Residue Management. Any tillage and planting system that uses no-till, ridge tillage, mulch tillage, or another system designed to retain all or a portion of the previous crop's residue on the soil surface. The portion required depends on other conservation practices that may be included in the farmer's total conservation plan.

10.3 Carbon Cycle and No-till

In contrast to tilled agriculture, NT-soil management greatly alters the C cycle because placement of the plant residue C additions into the soil is greatly changed. The traditional moldboard plow system is the antithesis of NT. Moldboard plowing inverts and incorporates all of the previous crop residue into the soil in one operation. In this case, the microbial populations responsible for residue decomposition are stimulated by the instantaneous addition of large amounts of organic C. No matter what system is used, however, decomposition rates are controlled by soil water conditions, soil temperature, and inorganic N supply. If the soil water content is above the wilting point and the soil temperature is above 4 °C, inorganic N supply is the limitation. In tilled systems, if appropriate amounts of N fertilizer are added, the soil organisms can complete decomposition of the aboveground crop residue as well as the root mass in the surface soil layer within a few months after plowing. Often the surface soil is disked before plowing and/or the crop residues are shredded to enhance farm implement function. This operation creates more residue surface area for the soil organisms and hastens the decomposition rate. Large quantities of CO_2 are evolved from soils shortly after plowing. Soils usually are tilled several more times after plowing to prepare a fine, firm seedbed for planting. These operations stir the soil and residue mixture, and further stimulate decomposition of the organic C if the soil contains adequate water. Tillage also hastens soil drying which retards decomposition if no rainfall or irrigation occurs.

Continued use of the moldboard plow over long periods of time decreases soil aggregate size, which permits organism access to C that was once physically inaccessible within aggregates. Thus organic C that took literally centuries to store is released as CO_2 within a few years. For example, when the temperate zones of North America were first cultivated, as much as 10^9 Mg of C were released as CO_2 (Paustian et al., 1997).

When soils that have been managed with maximum tillage, such as a moldboard plow, are converted to NT, a dramatic change occurs in the way organic C is introduced into the soil system, both in terms of timing and relative position. Furthermore, the soil under NT management remains wetter for longer periods. Picture the new system (NT) as having very little residue mixing with the soil. Crop residues are on the surface and decomposition occurs much more slowly than under tillage because much less residue surface area is exposed to the soil organisms, even if the producer shreds the residue. The root mass in the surface soil also remains basically intact, again reducing surface area exposed to decomposition. Soil surfaces are generally moist enough, especially in the spring and early summer, that water availability does not limit decomposition. However, the residue cover, especially in northern latitudes, keeps the soil cooler for longer than a bare surface soil, thus slowing decomposition. Under NT, nitrogen supply is more limiting in decomposition because the organisms only have access to the inorganic N in a much smaller soil volume than in a tilled system. Obviously the root mass decomposers do have access to more N than do the surface residue decomposers, but overall residue decomposition is delayed relative to a tilled soil.

After a period of 5 to 10 yr, the amount of surface residues in a NT system appears to reach an equilibrium for a given soil and climate combination (Sherrod et al., 1996). Resident residue levels increase in cooler relative to warmer climates, due to slower decomposition rates in the former. Within climatic zones, soil type also affects residue levels. Excessively drained soils have lower residue levels relative to well-drained soils primarily because of the production of less biomass. At the other extreme, poorly drained soils have maximum residue levels because biomass production is high and because the cooler soil environment in the wet soil delays decomposition, which in turn is accentuated by the heavy residue cover.

What then is the net effect of no-till on the C cycle? Generally speaking, the total C sequestered in the surface residue plus the organic C in the uppermost soil layer (0 to 5 cm) is higher compared to the same soil under tilled conditions. Furthermore, substantial amounts of C still reside as crop residue in the upper 2.5 cm of soil. Ortega (1995) found residue C amounts within the upper 2.5 cm of soil equal to that present on the soil surface.

Organic C also tends to increase under NT relative to tilled systems. However, the rate of increase is slow and is usually confined to the upper soil layers with little effect on deeper layers (Dick, 1983). Changes in soil C after 18 yr were 0% to −11% for NT, −12% to −23% for minimum till, and −14% to −23% for CT. Organic P also was higher in surface soil layers with NT than with other tillages.

Wood and Edwards (1992) found that NT, relative to conventional tillage (CT), had C and N concentrations that were 67% and 66% higher, respectively, than plow tillage in the 0–10 cm depth after 10 yr. Macroaggregate (> 250 mm) numbers under NT were greater and more stable than those under CT, probably due to greater biological activity in NT. Microbiological activity was more episodic in the tilled system, which had a drier surface soil than NT. Conventional tillage also caused more physical disruption coupled with less production of aggregate stabilizing materials. In a dryer climate (eastern Colorado), Wood et al. (1991) concluded that higher equilibrium levels of soil C and N could be maintained by increasing cropping intensity, which increased C input to the system. These changes were measurable even after only four years of NT. Less inorganic NO_3 also was present in the intensive system indicating that the N was subject to less loss from leaching or denitrification.

The most recent data from Colorado show that in some NT cropping systems, organic matter contents have increased in the upper 5 cm of soil, but have declined in the 5–10 cm layer. The net change for the 0 to 10 cm soil depth generally has been negative after seven years of NT, especially in systems that include fallow. Under climate and soil situations that permit continuous cropping, adoption of NT is likely to increase soil organic matter much more. This means that the more humid the climate (high precipitation coupled with low potential evapotranspiration), the greater the probability that NT will allow soil organic matter to increase. Conversely, under climates with low precipitation and high evapotranspiration which usually require fallow periods in the cropping systems, there would be less chance of increasing soil organic matter; and possibly a net loss of C, even under NT management, may occur.

No-till adoption will promote restoration of some of the soil C lost when soils were broken from their native condition. The amount of sequestration, however, is small and slow relative to the losses from the original soil organic matter level. It will take years of NT practices to increase soil C because the amounts of available N and P required to provide the correct C:N and C:P ratio for C sequestration are limited. For example, producers usually apply adequate N for the crop, but purposely plan to leave no excess in the soil system. Therefore, little N is available to balance the high C input from crop residue. The same situation can occur with P, especially in soils with low P availability. Paustian et al. (1997) provide an excellent modern summary of the effects of tillage on soil organic matter. They conclude that reduction in tillage, and especially NT, provides an environment that is conducive to storage of C in soils. However, if reduction in tillage adversely affects crop yield, the effects on soil organic matter can be the reverse. In the next section, we will address why practices such as NT can reduce plant yield in some environments.

10.4   Residue Management and Decomposition

Residue management in conservation tillage systems is much different than in conventional systems. As mentioned in the C cycle section, the incorporation of residues into the soil is minimized in conservation tillage systems. Farm implements are specifically designed to leave the maximum amount of residue on the soil surface in contrast to plow systems which intentionally incorporate all materials. The amount of residue remaining on the soil surface after one pass of various implements is presented in Table 10.1. Note that machines which invert and/or vigorously stir the soil, like the moldboard plow and the tandem disk, leave less than 5% and 60%, respectively. Repeated passes with a tandem disk result in essentially no residue on the soil surface. In contrast, slicing action implements, such as the chisel plow and the sweep plow, leave 50% to 90% and 85% to 90%, respectively. Managers thus have some control over residue cover by the implement choices they make. No-till operations rely on herbicidal weed control and no primary tillage is used. Therefore, NT systems have the maximum possible surface residue cover.

Table 10.1   Effect of various implements on surface residue cover remaining after one pass of the machine [Fenster et al., 1965; Hill et al., 1994]

Tillage and Planting Implements	Percent Remaining after One Pass of the machine
Moldboard Plow	0-10
Machines which Fracture Soil	
Paraplow	80-90
V-Ripper/ 30-35 cm deep w/50 cm spacings	70-90
Chisel Plows	
Straight points	60-80
Twisted points	50-70
Sweeps & Field Cultivators	
Sweeps - 60 cm or wider	75-90
Sweeps - 30-50 cm	60-80
Duckfoot points	60-70
Disk Machines	
One-way disk	55-80
Tandem or Offset disk	
25 cm or greater blade spacing	25-50
18 - 25 cm blade spacing	30-70
Harrows & Packers	
Springtooth	60-80
Spike tooth	70-90
Flex-tine tooth	75-90
Roller harrow (cultipacker)	60-80
Packer roller	90-95
Row Cultivators	
Sweeps	75-90
Finger wheel cultivator	65-75
Rolling disk cultivator	45-55
Ridge-till cultivator	20-40
Drills & Planters	
Conventional drill with double disk openers	85-95
No-till drill with no coulters	85-95
No-till drill with coulters	75-85
Conventional row crop planter - double disk openers	85-95
No-till row crop planter - no coulters	85-90
No-till row crop planter - with coulters	65-90
Strip-till planter	60-80
Ridge-till planter	60-80
Natural weathering	
Over-winter following summer harvest of small grain	70-90
Over-winter following fall harvest of summer crop	80-95

Decomposition rates are directly affected by amount of residue incorporation. Schomberg et al. (1994) reported that both surface and buried residue immobilized N, but surface residues immobilized N for 3 times as long as did buried residue, which is a direct indication of residue decomposition status. Maximum decomposition rates occurred when the residues were completely incorporated into the soil, but rates were at a minimum in NT systems. In their study, sorghum and wheat residues decomposed at about the same rate, and decomposition rates were highest under the wettest regimes. Decomposition rates of residues in NT and other conservation systems are highly dependent on surface soil water content and soil temperature. Highest decomposition rates occur when soil temperatures are between 20 to 30 °C and the soil water content is near field capacity just under the residue cover. In climates where atmospheric water demand is low, relative humidity is high, warm temperatures prevail, and a long growing season is present, decomposition of surface residues is maximized. Peterson et al. (1995), working across an potential evapotranspiration (ET) gradient in Colorado under NT conditions, showed that loss of surface residue from corn planting to corn harvest varied from 32% in the lowest to 62% in the highest ET zone. The higher temperatures in the high ET zone and the longer warm season promoted faster decomposition despite the drier conditions. Obviously, temperature controlled decomposition more than relative humidity of the air at the soil surface.

The objective of conservation tillage farming is to maintain surface residue cover throughout the growing season, and particularly during the preplant and seedling stages of a crop cycle because at these times the soil is most vulnerable to erosion by either wind or water. Once the crop canopy provides cover, the residue cover is secondary until after crop harvest. The importance of a residue cover after harvest depends on the particular climate. If erosion potential is great after harvest, residue cover is obviously needed. Some crops such as soybean leave less cover (6 Mg ha^{-1}) in comparison to crops like corn (10 Mg ha^{-1}). Furthermore, the fragility of the soybean residue results in 20% to 30% degradation over winter in contrast to 5% to 20% for the less fragile corn stalk residue (Hill et al., 1994).

Planting equipment often dictates how much tillage is needed for seedbed preparation. Prior to the advent of conservation tillage, seedbeds had to be free of residue so that planters could operate properly. Today, NT planters for both small grain and row crops are commercially available. These planters do minimal disturbance and leave most of the residue on the soil surface (Table 10.1).

10.5   Biological Activity

Biological activity in NT soils is generally higher than in CT soil for several reasons. First, there is a continuous, more uniform supply of C as an organism energy source because the residue C is not introduced all at once in NT compared to CT. Second, the surface soil water content remains higher for longer periods of time in reduced tillage systems, particularly in NT, in comparison to CT which allows organisms, both flora and fauna, to function for longer periods of time. Blevins et al. (1984) summarized the NT influence on soil biology to which the reader is referred to obtain a historic perspective.

High residue levels favor macroorganisms such as the earthworm. Teotia et al. (1950) clearly demonstrated that residue mulch at the soil surface was critical to maintaining earthworm populations in cultivated systems. The number of worms under mulch tillage was three times that of a conventionally plowed soil with the same crop rotation. Although mulch rates of 14 Mg ha^{-1} produced the greatest amount of worm activity, rates < 3.5 Mg ha^{-1} still improved soil structural stability. No-till is superior to cultivated systems for promoting earthworm activity (Mackay and Kladivko, 1985; de St. Remy and Daynard, 1982).

Strong interrelationships between earthworm activity and soil physical properties result in increases in water infiltration. Edwards et al. (1988) observed runoff values of < 2 and 700 mm yr^{-1} for watersheds that had been in continuous NT and CT corn production, respectively, for 20 yr in an area receiving over 1,000 mm yr^{-1} of precipitation. This difference was attributed to the presence of earthworm burrows in the NT system, which increased earthworm populations and allowed their burrows to remain intact from season to season. Their intact burrows provide continuous pathways for water infiltration into soil. Furthermore, the earthworm burrows have diameters much larger than the usual soil pore size, which promotes rapid intake of large volumes of water. This water is immediately transmitted past the surface soil and is stored in a position where evaporation is greatly reduced.

Edwards et al. (1989) reported that the rainstorm event that promoted the most flow in the burrows was a high intensity short duration event (3.6 mm in 0.3 hr). This indicates that burrows would enhance seasonal water infiltration substantially in climates where high intensity thunderstorms are prevalent. Trojan and Linden (1994) showed that earthworm activity could more than double water penetration to NT versus CT soil at a high rainfall intensity.

Obviously, NT promotes increased biological activity in terms of earthworm populations which, in turn, greatly alters water infiltration soil properties. Reduced runoff, in turn, results in decreased erosion potential; the rapid transfer of water to depth in the profile conserves water. A negative aspect of this rapid transfer is the movement of soluble fertilizers and pesticides to positions where they are less effective for the purposes for which they were applied. Furthermore, shallow groundwater supplies could be readily contaminated in the process.

Biological activity at the microscale also is greatly affected by adoption of reduced and NT systems, and these effects can be either positive or negative. On the positive side, Dick (1984) observed increased activity of many enzymes in the top 7.5 cm of soil after 18 yr of NT practice in Ohio, which was related to organic C contents. Furthermore, the herbicides and pesticides used in NT management did not appear to adversely affect enzyme activity. Other evidence for increased biological activity under NT is provided by Staley et al. (1988), who reported that soil biomass-C reached a maximum in surface soil after just 1 yr under NT. Biomass-C approached levels found in pasture environments, and equilibrated in 10 yr at a level that was 30% greater than CT.

Processes involving N also are excellent indicators of microbiological activity and several authors have made interesting observations regarding contrasts among NT and CT systems. Lamb et al. (1987) showed that N_2 fixation potentials in wetter NT soils were twofold greater than in drier plowed soils. Even though the quantities of N_2 fixed were small (0.33 kg ha^{-1} yr^{-1}), their data do show the distinct difference in microbial environments which persist in NT compared to CT. Bremer and van Kessel (1992) found that the increased microbial biomass present under NT reduced potential N losses during periods of low crop demand because N was incorporated into organic compounds that were less subject to loss through leaching and denitrification. Wood et al. (1990) reported that adopting NT after long-term use of CT quickly increased the amount of surface residue and the amount of potentially mineralizable N. When they used more intense cropping systems in conjunction with NT, they obtained the highest potential N mineralization.

Possible negative effects of microbial activity on soil-plant systems include greater loss of N from urea fertilizers, greater denitrification, and production of phytotoxic compounds. Dick et al. (1991) showed that soils with a 25-yr history of NT had increased urease activity at the surface, making urea-based fertilizers more susceptible to N volatilization losses. Rice and Smith (1982) showed that denitrification potential increased under NT because of the more conducive environment provided by the large amounts of surface C coupled with the high biological activity at the soil surface and the wetter soil conditions under residue cover.

Phytotoxic compounds synthesized by microorganisms during residue decomposition were identified many years ago. No-till environments have regenerated interest in these compounds and in the processes involved in their production. The slower residue decomposition, the prolonged production period, the colder soil temperatures all occurring in or on the soil surface and very near the emerging plant seedlings create enhanced phytoxicity problems in NT environments. Buchanan and Kin (1993) clearly demonstrated the prolonged nature of the decomposition period under NT in contrast to tillage systems. Working in a warm North Carolina environment, they found that after 50 weeks 45% and 20% of the C remained under NT and CT systems, respectively. Similar results were obtained for wheat and soybean stalk residues. Cochran et al. (1977) demonstrated that surface residues of lentil, pea, wheat, barley, bluegrass all produced phytotoxic compounds at some point during the spring season. Toxicities were worst when wet weather and low temperatures prevailed. Lentil and pea materials were more toxic to wheat seedlings than were the other compounds, but fortunately, toxin production was short lived. Martin et al. (1990) found that microbially synthesized phytotoxic materials from corn residues were more damaging to corn seedlings than were the byproducts of soybean and oat residue decomposition. Thus corn planted back into corn residues would be more prone to phytotoxin damage than corn planted into soybean or oat residues. These findings illustrate that crop rotation may be more necessary with NT than with CT systems just to avoid potential seedling damage.

10.6 Soil Fertility and Nutrition

10.6.1 Nitrogen

All soil biological processes are highly governed by the organic and inorganic portions of the N cycle. The N cycle is inextricably linked to the C cycle. Whether or not mineral N is present in the soil for plant uptake is dependent on what is happening within the C cycle. Since N is the most plant-growth-limiting nutrient, and since it has been thoroughly demonstrated that conversion to no-till and other reduced till systems alters the C cycle, it is important to understand the specific effects of tillage systems on N availability to plants.

10.6.1.1 Immobilization of Fertilizer N

All conservation tillage systems tend to immobilize N as compared to conventional tillage, but conservation tillage systems do not immobilize more N than conventional systems; they only keep the N immobilized longer. Net N mineralization is eventually equal in all systems, but the timing of N release is delayed when residues are either slowly incorporated into the soil or totally left on the soil surface. Rice and Smith (1984) as far back as 1984 recognized that no-till immobilized fertilizer N. They reported that within the first 35 d after application there was 19% and 11% immobilization, respectively, for no-till and conventional till. They surmised that using subsurface N-fertilization techniques would minimize immobilization in NT. According to Schomberg et al. (1994) surface and buried residue both immobilized N, but surface residue immobilized the N three times longer than did buried residue. Net immobilization was longer than 1 yr for surface residues and about 0.33 yr for buried. Residue type, either sorghum or wheat, made little difference in terms of decomposition rate. Smith and Sharpley (1990) and Smith and Sharpley (1993) studied effects of residue type and placement on N immobilization and mineralization. They concluded that surface residues caused temporary N immobilization and that the C:N ratio of the materials governed the length of time of immobilization. Alfalfa had the maximum mineralization followed by peanut, soybean, oat, sorghum, wheat, and corn in decreasing order of mineralization. Although surface placed residues had less N mineralized than buried, the difference between placements was minor compared to differences among types of residue. Under dryland conditions in New Mexico with very low initial soil organic matter content, Christensen et al. (1994) found that after using no-till for 5 yr, soils had less inorganic N and more organic C and N than a soil under stubble mulch tillage. Their unfertilized grain sorghum was very deficient in no-till systems compared to the stubble mulch treatment. Soil organic C increased with time under both no-till and stubble mulch systems. They did not reach a point in the 5-yr study where no-till soil mineralized as much N as did the stubble-mulched soil.

Although immobilization of N delays N availability to plants, the impact on N fertilizer management does not appear to be as significant as once thought. Bundy et al. (1992) did find that corn yields in no-till systems where no N was applied were lower than for other tillage systems they tested, suggesting that N mineralization was lower in that system and that immobilization was greater. However, their data showed that one can use a variety of N-application methods across tillage systems with little difference in outcome, and that placement was a minor issue. Maskina et al. (1993) observed that increased residue levels increased grain yields in a Nebraska tillage experiment, and additional residue increased stover production and N uptake of subsequent corn crops. Nitrogen immobilization did not seem to be a factor. Tillage per se, no-till vs. disked, had no affect on grain yield in any year. Vigil and Kissel (1991) developed equations to predict net N mineralization for a season based on N content and/or N + lignin/N ratio of the residue and concluded that net N mineralization would not occur if the C:N ratio of the residue was above 40.

10.6.1.2 Soil N and C Changes with Time in No-Till

Shifting to reduced and no-tillage systems also affects formation, size distribution, and strength of soil aggregates. Changes in aggregate properties alter the rates at which both physical and biological processes proceed. When soils are not tilled, macroaggregates that form from natural processes tend to remain longer and have an opportunity to become strengthened. Beare et al. (1994a,b) found that the macroaggregates actually protected soil organic matter from decomposition. In their study 19% of the total mineralizable C and N in no-till treatments came from these larger aggregates, while only 10% and 5% of the C and N, respectively, came from them under conventional tillage. The fact that macroaggregates protect more of the total soil organic matter in no-till systems compared to conventional till may explain why conversion to no-till results in a need for higher N-fertilizer rates at the outset. Apparently a new steady state is reached and the immobilization capacity due to macroaggregate formation is sated. In fact, Follett and Schimel (1989) inferred from data collected from a 15-yr no-till experiment that less tillage helped conserve mineral N. In other words, the higher, but temporary, immobilization of N in no-till systems reduced the opportunity for both leaching and denitrification losses of mineral N.

How long does the net immobilization period of no-till last before mineralization processes catch up? Rice et al. (1984) observed that after 10 yr in no-till, N mineralization was equal to that of conventional till. These authors suggest that the net immobilization phase when no-till is adopted is transitory. Both their field data and lab incubations verified the change with time. Working within a wheat-fallow system in western Nebraska, Lamb et al. (1985) reported that for the first 5 to 7 yr after adoption of reduced till and/or no-till, less nitrate accumulated during the fallow period than in plowed soils. Approximately 7 yr after adoption, both no-till and reduced till (stubble mulch) soils had fallow nitrate accumulations equal to plowed soils and continued to have equal contents in subsequent years.

10.6.1.3 Fertilizer N Conservation and Efficiency

Perhaps of more concern than immobilization of the N fertilizer is the potential loss of N from urea fertilizers that are applied directly to a no-till, residue-covered soil. Stecker et al. (1993a, 1993b) working in Missouri reported substantial N losses from urea-based N fertilizers in no-till systems when the fertilizer was not injected below the soil surface. When N was injected into the soil below the residue, time and source of application had no effect on yield of corn. Kolberg et al. (1996) working in a more arid climate, eastern Colorado, found that placing N below the residues was important for fall-planted wheat, but that placement had no effect for corn. Climatic conditions, hot and dry at fall wheat planting time, apparently stimulated N losses from urea compounds that were placed on the residue. At corn planting time the weather was cooler and wetter, and apparently the loss mechanisms were not active enough to be a problem.

10.6.2 Phosphorus

Reduced and no-till systems tend to keep soils cooler than clean-tilled systems (see other sections of this chapter for an explanation). In the case of spring-planted crops in northern latitudes, this results in slower plant root growth in the reduced till systems. Since P is a nonmobile nutrient and uptake depends greatly on root interception, cold soils can create temporary P deficiencies. Starter fertilizers placed near the seed at planting are required in these cases. Failure to correct the temporary P deficiency stunts plant growth and delays development so that the crop may fail to mature before fall frost occurs. Simple application of a P-containing starter fertilizer usually solves the problem. Stecker et al. (1988), working in a wheat-fallow system, demonstrated that placement of P fertilizer with the seed was more beneficial under no-till than for plowed and/or reduced tillage (stubble mulch)systems. Apparently the cooler soil temp under high residue conditions in the spring was responsible for this difference.

10.7   Water and Wind Erosion

Soil erosion by water is a process of particle detachment, transport, and deposition. Sediment must first be detached from the soil mass before it can be transported. For interrill areas, erosion rates are controlled by the soil detaching capacity of impacting raindrops and the soil transporting capacity of surface flow. For rills, erosion rates are determined by the detachment and transport capacity of flowing water. Water erosion is reduced, therefore, by limiting either the detachment or transport processes; however, total erosion control is best achieved by eliminating runoff or reducing the flow rate so that detached particles cannot be transported.

Infiltration, surface water storage, and erosion have all been shown to be directly affected by tillage and surface residue cover. Erosion is reduced under conservation tillage systems due to (1) a greater crop residue cover that protects the soil surface from direct impact of raindrops, (2) increased consolidation and greater surface layer soil strength, (3) increased tortuosity and hydraulic resistance of the runoff flow paths, (4) decreased capacity of runoff to detach and transport soil particles, (5) less runoff due to greater infiltration, and (6) increased deposition of sediment in small ponded areas created by surface residue (Gilley et al., 1986; Brown et al., 1989; McGregor et al., 1990). Norton and Brown (1992) reported that both interrill and rill erosion rates were significantly lower for older consolidated ridges in a ridge tillage system compared to freshly formed ridges. The mean interrill erosion rates were 40% and 59% lower in the older ridges compared to the freshly formed ridges for Hoytville silty clay (fine, illitic, mesic Mollic Ochraqualf) and Rossmoyne silt loam (fine-silty, mixed, mesic Aquic Fragiudalf), respectively. For both soils, the rill erosion rate for the older ridges was reduced 72% compared to the freshly formed ridges. West et al. (1992) determined that no-tillage with residue removed resulted in a 60% to 70% decrease in rill erodibility compared to conventionally tilled systems. They attributed the reduction to an increase in organic C from 9.3 to 12.9 g kg^{-1} and water-stable aggregates from 50% to 76% and an increase in soil strength due to consolidation. Greater aggregate stability and size for NT compared to CT soils were reported by others (Bruce et al., 1990; Cambardella and Elliott, 1993; Drees et al., 1994; Beare et al., 1994; Lal et al., 1994). Increased aggregation is due to increased organic matter, greater surface residue mulch, and increased biological activity (Doran, 1980).

Interrill soil loss on a moderately well-drained, Saybrook silt loam (fine-silty, mixed, mesic Typic Argiudoll) under NT for 15 yr was reduced to 0.01 kg m^{-2} h^{-1} compared to 2.45 kg m^{-2} h^{-1} for CT (Bradford and Huang, 1994). Runoff was 12 mm h^{-1} for the NT condition and 68 mm h^{-1} for the tilled soil condition (Table 10.2). The Saybrook soil contained 15% clay, 76% silt, and 9% sand. NT consistently decreased soil loss by 80% to 90% relative to CT (moldboard plow) across six sampling periods on a Griswold silt loam (fine-loamy, mixed, mesic Typic Argiudoll) near Arlington, WI (Andraski et al., 1985). In contrast, NT increased runoff by 56% compared to CT in a semiarid environment of northwest Texas (Jones et al., 1994b). The smooth consolidated surface of the NT that encouraged runoff resisted erosion, and sediment loss was 54% less than for the tilled treatment.

Soil detachment rate by surface flow in rills is calculated as a linear function of hydraulic shear stress by the equation:

$$D_c = K_r(\tau - \tau_c) \qquad [10.1]$$

where D_c is the detachment capacity (kg m^{-2} s^{-1}), K_r is the rill erodibility (s m^{-1}), τ_c is the critical shear stress (Pa), and τ is the shear stress of flowing water (Pa), given by:

$$\tau = \delta r_h S \qquad [10.2]$$

where δ is the specific weight of water (N m^{-3}), S is the rill slope, and r_h is hydraulic radius (defined as the ratio of flow cross-section area to the wetted perimeter of flow). Foster et al. (1980) estimated that the critical shear stress required to initiate erosion in rills in a NT soil is about 10 to 15 times greater than that in a freshly tilled soil. Rill erodibility (K_r) of a freshly tilled NT ($K_r = 5.41$ mg s^{-1} N^{-1}) was significantly greater than that for an undisturbed NT ($K_r = 1.70$ mg s^{-1} N^{-1}) for a Saybrook silt loam (fine silty, mixed, mesic Typic Argiudoll) in north central Illinois (Table 10.3) (King et al., 1995). The soil and site was the same as that reported by Bradford and Huang (1994) for the interrill study. The critical shear stress (τ_c) of the freshly tilled NT ($\tau = 1.24$ Pa) was much less than that for the undisturbed NT ($\tau_c = 7.14$ Pa). Intermediate values for K_r and τ_c were found for the undisturbed NT with residue removed, thus indicating the importance of residue in slowing runoff, dissipating forces of surface flow, and reducing raindrop impact forces due to greater ponding and greater flow depth. If a critical discharge, slope steepness, or slope length is exceeded, massive rilling will occur in a NT soil, but these critical values are much greater in a NT than in a CT soil.

Surface runoff is necessary to transport detached particles under both interrill and rill erosion processes. By controlling runoff, erosion can be controlled. Runoff from a 0.5-ha watershed of Rayne silt loam (fine-loamy, mixed, mesic, Typic Hapludaults) at Coshocton, Ohio, that was farmed for 20 yr in continuous NT corn, averaged < 2 mm yr^{-1}. Average rainfall was > 1,000 mm yr^{-1} (Edwards et al., 1988). Storms with 1-min rainfall intensities > 76 mm h^{-1} were required before any surface runoff could be generated (Dick et al., 1989). The high infiltration rates were attributed to the large macropores that were created by earthworm activity and old root channels. With CT practices, these large macropores are destroyed to the tillage depth each year.

Table 10.2 Runoff and interrill soil loss rates on Saybrook silt loam for a rainfall intensity of 100 mm h^{-1} [Bradford and Huang, 1994]

Tillage/residue cover treatment	Runoff	Soil loss
	mm h^{-1}	kg m^{-2} h^{-1}
No-till, residue	12.3	0.01
No-till, residue removed	73.8	0.94
No-till, residue removed (dried, crusted)	85.0	1.96
Till, residue	10.7	0.02
Till, residue removed	67.8	2.45
Till, residue removed (dried, crusted)	78.6	3.77

Table 10.3 Comparison of rill erodibility and critical hydraulic shear stress for no-tillage and conventional tillage sites [King et al., 1995]

Treatments	Rill erodibility	Critical shear
	mg s^{-1} m^{-1}	Pa
Aged tilled, conventional	12.2	3.08
Freshly tilled, conventional	8.79	1.55
Freshly tilled, no-till	5.41	1.24
Undisturbed, residue removed	3.62	1.97
Undisturbed, no-till	1.73	7.14

The effectiveness of any tillage method for controlling erosion ultimately depends upon the amount of crop residue left on the soil surface, especially the amount remaining after planting. Residues left on the surface are much more effective for erosion control than incorporated residues (Meyer and Mannering, 1963; Wischmeier and Smith, 1978); however, in some climates such as the semiarid southwestern United States, and with certain crops such as cotton, it is difficult to produce adequate residue cover to protect the soil from water erosion.

Surface residue cover is also one of the cheapest and most effective measures to protect soil against erosive winds. In wind tunnel experiments, soil loss due to wind erosion was reduced 57% with a 20% covering (wood dowels) and 95% with a 50% cover (Fryrear, 1985). Covering a sandy Sahelian soil (sandy, siliceous, isohyperthermis Psammentic Paleustalk) surface with 1,500 kg ha^{-1} of flat millet stalks reduced sediment transport by wind by 63.6%, and a cover of 1,000 kg ha^{-1} reduced sediment transport by 42.2% (Sterk and Spaan, 1997). Effectiveness of the residue cover depends on the quantity, kind, and orientation in relation to the soil surface (Siddoway et al., 1965; Lyles and Allison, 1976; 1981). Standing stubble is more effective in reducing erosion than flat plant cover. Lyles and Allison (1981) reported that among the seven crops and orientations tested in a wind tunnel, standing winter wheat residue was the most effective for wind erosion protection and flat random sunflowers the least effective. To provide the same wind erosion protection, 5.5 and 8.7 times more standing grain sorghum and corn stubble (weight basis), respectively, than standing wheat stubble was needed (Lyles and Allison, 1976). In regions where CT systems have not been successful in controlling wind erosion, the primary reason is the low amount of plant residues produced. Continued use of NT for several years, however, allows residue to accumulate because of positive feedbacks in the system. As more water is saved in a semiarid environment and soil organic C is increased, for example, crop biomass production is increased, which in turn leaves more residue cover.

The protective action of residue cover is most important during crop establishment and early crop growth. In sandy soils of the southern Great Plains and the Rio Grande Valley of Texas, sandblasting and burying of young seedling crops requires that cotton and grain sorghum be reseeded about 2 in 5 yr. Data on comparisons of susceptibility of soils to wind erosion under specific residue coverings are not available. In the Lower Rio Grande Valley, however, we observe much less airborne soil particles under NT compared to mulch tillage. In northern Mexico along the Rio Grande River, wind erosion was much greater under CT compared to NT systems (de Quattro, 1997). In some areas, soil degradation due to wind erosion had progressed to the point that the damage was irreversible. By managing residue cover through conservation tillage, soil erosion by wind is greatly reduced.

10.8 Surface Sealing, Crusting, and Seedling Emergence

Seal formation on a freshly cultivated soil exposed to the beating action of falling raindrops is due to two mechanisms: (1) the breakdown of soil aggregates by the impact force of the raindrops, and (2) a physiochemical dispersion of the soil clays. The initial moisture condition of superficial soil aggregates at the beginning of the rainstorm affects the resistance of aggregates to breakdown or dispersion (le Bissonnais, 1990). For initially dry superficial aggregates, aggregate breakdown under rainfall is mainly due to slaking. The dry surface slakes rapidly under wetting and forms a seal of lower permeability. If the soil surface has a high water content before rainfall, the degree of aggregate breakdown and surface sealing is low and results primarily from the mechanical impact of raindrops (Bradford and Huang, 1994). Wetting and drying of newly formed crusts (dry seals) normally act to weaken the crust and it may largely disappear as the soil surface granulates. Even though drying forms new aggregates, their size and stability are less than those of the original surface, and the final infiltration rates at the end of subsequent storms are lower than the values at the end of the first storm (Ben-Hur et al., 1985).

Soils crusts reduce seedling emergence, but its final effect on yield is crop dependent. For example, in the coarser textured soils in the Lower Rio Grande Valley of Texas, recently seeded cotton under CT systems often is replanted due to crusting following an intense rainstorm. The same is true for many of the soils in the Southeast (Miller and Radcliffe, 1992). Lower seedbed water content has a detrimental effect on plant emergence and stand establishment (Karlen, 1989).

The extent of crusting and its effect on seedling emergence are highly dependent upon the tillage system and the amount of residue cover. Maximum crusting occurs in soil tilled with a moldboard plow followed by several diskings. Crust formation is minimized with a conservation tillage system that maintains high crop residue cover percentages.

In addition to the positive effects of plant residue cover on increasing seedling emergence, high surface residue levels can create the negative effect of reduced and uneven seedling emergence, particularly in the cooler, wetter climates. In regions where spring planting is delayed by snow melt and warming of the soil, seedling emergence is less under conservation tillage compared to conventional tillage unless crop residue is partially removed from the row. Uneven crop emergence and reduced stand density was reported in the northern Corn Belt in Wisconsin (Carter and Barnett, 1987), in Iowa (Mock and Erbach, 1977), in Minnesota (Ford and Hicks, 1992), and in Illinois (Nafziger et al., 1991). By partially removing the crop residue and forming small ridges during fall application of fertilizer, Kinsella (personal communcation, 1997) has found soil temperatures at planting time to be warmer, seeding zone to be somewhat drier, and seedling emergence earlier and greater over no-tillage systems where residue is not removed. Row-zone tillage in which narrow bands are tilled to satisfy the seedbed requirements allows large portions of the field surface area between rows to be managed to satisfy the infiltration, runoff, and erosion needs (Burwell et al., 1968).

10.9 Compaction

Soil compaction is defined as the volume change produced by momentary load application caused by rolling, tamping, or vibration. It is the process by which soil grains are rearranged into closer contact to decrease void space. Soil compaction involves an expulsion of air without significant change in the amount of water in the soil mass. Compaction can result from external forces such as tractor tires and tillage implements or from internal forces (sometimes termed consolidation) resulting from increased drying, overburden, or changes in groundwater table. The usual effects of compaction are increased bulk density, soil strength, or probe penetration and decreased compressibility and permeability. Drying or dessication of soils results in an increased bulk density but it also results in cracking and promotion of macropores if expanding lattice clays are present.

One of the farmers' primary concern in switching to NT systems is the problem of soil compaction. Bulk density and soil strength are generally less for plowed than for NT soil (Mielke et al., 1986; Bruce et al., 1990; Hill, 1990; Cassel et al., 1995). Bulk densities were higher in four midwestern United States soils in the NT surface 75 mm soil depth, no differences in two, and less bulk densities in only one NT soil (Mielke et al., 1986). At the 75–150 mm depth, bulk density was greater in no-till for only two soils, and at the 75–150 mm depth, there were no differences in bulk density among tillage treatments. Clay contents were similar, ranging from 23% to 36%, and years in NT ranged from 6 to 13 yr. After 28 yr of NT, bulk density of a Wooster silt loam (fine-loamy, mixed, mesic Typic Fragiudalf) at Wooster, Ohio was lower for continuous corn than all other tillage (moldboard plow and chisel plow) and crop rotation treatments (Lal et al., 1994). In poorly structured soils with low organic matter, long-term NT can lead to higher soil strengths that may limit root growth and crop yields. In most soils, however, even though strength and bulk density under no-tillage conditions are often greater than under conventional tillage conditions, crop yields are seldom reduced (Radcliffe et al., 1988). Detrimental effects of greater compaction are often offset by the more positive effects of increased infiltration, reduced soil temperature, reduced soil water evaporation, and greater water storage. Also roots in NT systems grow down existing root channels, worm holes, or fractures between soil structural units, and root growth may actually be greater (Merrill et al., 1996). Degree of soil compaction is gradually reduced with time with NT. Macroporosity and earthworm channels are generally greater in NT soils (Drees et al., 1994). In the subtropical, semiarid regions of Texas, abundance of earthworms are found in NT soils, whereas earthworm activity is absent in CT soils (Smart, personal communication, 1977). Even in colder semiarid regions such as North Dakota, changing to a no-tillage system greatly enhances earthworm populations (Deibert and Utter, 1994). The opportunity for plow pans or other mechanically compacted layers to reform is sharply reduced due to reduced tractor size, less tillage, and fewer trips across the field. Furthermore, compaction of soil surface layers by raindrop forces or surface sealing is less under conservation tillage due to protection of surface soil layers by greater residue cover. Consolidation of surface layers due to desiccation is reduced with greater residue cover under conservation tillage because of less soil water evaporation.

The degree of reconsolidation and compaction that occurs when plowing is eliminated depends upon the crop rotation, amount of surface residue, control of wheel-traffic, climatic factors such as rainfall and temperature, and soil variables. Soil variables include soil texture and structure, clay mineralogy, organic matter content, and others. Compaction due to tractor tires and consolidation from desiccation are more of a problem in coarser textured soils with massive structure and low organic matter levels, especially with monocultures of low residue crops such as cotton and soybeans. For example, in the highly weathered soils of the southeastern United States, reduced yields due to soil compaction have been reported (Touchton and Johnson, 1992), whereas in the deep loessial soil regions of Illinois and Indiana, compaction is less of a problem (Kinsella, personal communication, 1997).

Soil compaction problems must be corrected by mechanical means before starting a NT system (Radcliffe et al., 1988), or crop yields will be reduced for several years until biological forces loosen the soil. During the first few years of conversion to NT, yields in some soils are suppressed due to compaction. To reduce soil reconsolidation or if traffic compaction occurs, a paratill or similar tool which loosens the soil from beneath leaving the soil surface relatively smooth and the previous crop residue at the surface, is used (Unger, 1993). Soil compaction in the upper horizons can also be reduced with strip tillage in the row (Radcliffe et al., 1989; Raper et al., 1994) or while applying fertilizer with fertilizer knifes. By controlling traffic patterns, compaction is generally confined to the wheel track areas (Larney and Kladivko, 1989; Liebig et al., 1993). About 70% to 90% of traffic compaction occurs during the first pass of a tractor and machinery. By not controlling traffic patterns, therefore, total compaction in the field is increased. With time, soil compaction is gradually alleviated by biological means from the roots and surface residue. After three years of NT in a Willacy fine sandy loam in south Texas, soil structure was improved, yields were equal to conventional tillage, and the compaction problem was gradually reduced.

Extensive cultivation has led to soil degradation and a deterioration of soil physical conditions. With increasing organic matter and improved soil structure resulting from conservation tillage, the soil physical and biological environment improves and compaction problems decrease (Dao, 1996).

10.10   Cover Crops

Cover crops may play an important role within a conservation tillage system by protecting the soil against wind and water erosion, increasing soil organic matter, improving soil structure and other soil physical properties, altering soil temperature and soil moisture, and increasing fertility by recycling nutrients and providing biologically fixed nitrogen (in the case of leguminous cover crops). Cover crops also can improve weed control through competition and allelopathic effects and may improve environmental quality through the protection of surface and groundwater. Efficient use of cover crops in management systems, however, depends upon the purpose of the cover crop and on climatic factors, such as temperature, length of growing season, and rainfall.

Cover crops are more adaptable to farming systems in the southeastern United States because of greater rainfall and need for erosion control. In northern climates, cover crops are often not used because they generally leave the soils cooler in the spring, compete for water, and tend to shorten the growing season In the semiarid regions of the United States, management of cover crops to minimize water stress and possible yield reduction on subsequent crops is a major constraint to farmer acceptance. Farmers are also reluctant to use cover crops because of the additional cost for establishment and termination. However, cover crops are an effective means of controlling wind erosion in semiarid regions.

Intense management is required to achieve success with cover crops. Timing to kill the cover crop in the spring is critical to yield of the crop due to factors such as soil drying and warming, insect and weed pressure, and planter performance. Late April or early May kill dates consistently resulted in greater corn yield than earlier kill dates in Maryland (Clark et al., 1997). Summer soil water conservation by cover crop residues was more important than spring moisture depletion by growing cover crops in determining final corn yield.

10.11   Water Conservation

Residue management and conservation tillage effects on soil water storage and usage are highly dependent upon climate and soil type. Soil water content is almost always greater at planting in NT than in CT. This is due to increased infiltration and water storage during the noncrop period in soils with adequate residue cover, greater soil water loss caused by cultivation, and reduced soil water evaporation from better insulation by surface residues with NT. The relative benefits of residue management in conserving water increase from the wetter, more humid regions of the United States into the drier, semiarid regions.

In humid or subhumid regions, additional soil water resulting from conservation tillage has increased crops yields over CT, particularly when soil water becomes limited during the growing season. Conservation tillage normally decreases the frequency and intensity of short midseason droughts. In midsummer, surface soil temperatures are often several degrees cooler under plant residue than those under bare surfaces, resulting in less evaporation in the cooler, residue covered surface. Wagger and Denton (1989) reported yield increases for NT corn and soybeans of 32% and 43%, respectively, compared to CT practices on a Pacolet sandy clay loam (clayey, kaolinitic, thermic

Typic Hapludult) in North Carolina. Yield increases were attributed primarily to greater soil moisture availability as a result of reduced runoff. Total water use efficiency (units of crop produced per unit of water used) for corn grain production increased from 177 kg ha^{-1} cm^{-1} for CT systems to 215 kg ha^{-1} cm^{-1} for NT in the 4th year of a tillage study on a Hiwassee clay loam (clayey, kaolinitic, thermic Rhodic Kanhapludult) in North Carolina (Wagger and Cassel, 1993). For corn silage, water use efficiency was 286 kg ha^{-1} cm^{-1} for CT and 344 kg ha^{-1} cm^{-1} for NT. Both corn grain and silage yields under NT were about 20% greater than yields under CT.

In some cases, cooler, wetter soils can lead to slower crop development and lower yields. In Ohio on a very poorly drained Hoytville silty clay loam (fine, illitic, Mollic Ochraqualf), the NT treatment averaged 13% lower yield than the average of plowed treatments for 10 yr of continuous corn (van Doren et al., 1976).

In semiarid regions, where moisture is a greater limiting factor for plant growth, increased moisture conservation from NT normally produces greater crop yields. Water content in a silty clay loam (Udic Boroll) in north central Alberta, Canada was greater on NT compared to CT plots. During May of the third barley crop, water contents in the 0–15 and 15–30 cm depths were 40.2% and 38.3%, respectively, for NT and 31.8% and 32.1% for CT with straw removed (Nyborg and Malhi, 1989). Dalal (1989) reported increased and deeper water movement under NT compared with that under CT in a Hermitage clay (very fine, montmorillonitic, Udic Pellustert) in Queensland, Australia. Clay content of the soil was 65% and mean annual rainfall was 685 mm.

Maintaining surface residues via stubble mulching in Great Plains wheat-fallow systems more than doubled precipitation use efficiency from 92 to 140 kg ha^{-1} cm^{-1} (Greb, 1979). However, conversion to NT has not led to further increases in productivity in wheat-fallow systems. NT systems store water more rapidly than tilled systems during the early part of the fallow period, and usually no-tilled fallow systems have as much stored water by early spring as they do if one fallows for an additional 3 months (Peterson et al., 1996). It has been clearly demonstrated from North Dakota to Texas that addition of spring crops to the system is the only way to efficiently use the extra precipitation stored with NT. Conversion from wheat-fallow to systems such as wheat-corn or sorghum-fallow has increased water use efficiency from 60 kg ha^{-1} cm^{-1} with wheat-fallow to 86 kg ha^{-1} cm^{-1} for wheat-corn or sorghum-fallow (Peterson et al., 1996). Not only do the more intensive cropping systems result in a 43% increase in water use efficiency, they greatly reduce soil erosion by both water and wind.

Amount of rainfall is not the only issue involved in water conservation with NT. Precipitation distribution is equally important as one compares effectiveness of NT in the semiarid winter wheat production areas of the Pacific Northwest (average annual precipitation is 290 mm) to the Great Plains environment. In the Mediterranean climate of the Pacific Northwest, farmers often experience increased water loss during the fallow period with NT compared to conventional tillage. Soil water storage was lower for NT compared to stubble mulch tillage and bare soil (moldboard plow) tillage (Schillinger and Bolton, 1993). The soil was a Walla Walla (coarse silty, mixed, mesic, Typic Haploxeroll) silt loam. They conclude that the NT practice lost soil water at a faster rate than either stubble mulch or bare soil tillage during the summer months because of soil capillary continuity. The continuity of the capillary channels from the subsoil to the soil surface was effectively broken by tillage in the stubble mulch and bare soil systems. The major reason for the difference in outcomes for NT fallow in the Great Plains versus those in the Pacific Northwest is the long dry period (no summer rainfall) in the Northwest compared to mostly summer rainfall in the Plains. Hammel (1995) reported reduced wheat yields under NT in the Palouse region of northern Idaho in the Pacific Northwest. Average annual rainfall was 660 mm, and soil was a Palouse (fine-silty, mixed, mesic Pachic Ultic Haploxeroll) silt loam. Depth and amount of water extracted by winter wheat grown under NT were considerably less compared with the CT (moldboard plow) and minimum (chisel plow) systems. He attributed the decrease in profile water extraction under NT to decreased root function, possibly due to soilborne diseases, higher surface layer impedance, or soil temperature. Dao (1993) in Oklahoma also reported increased water storage in a Bethany silt loam (fine, mixed, thermic Pachic Paleustoll) for continuous NT wheat from 1983 through 1991. Mean annual precipitation at El Reno is 822 mm.

Tanaka (1989) also evaluated NT practices for wheat production on a Williams loam (fine-loamy, mixed, Typic Argiboroll) near Sidney, Montana. He concluded that the response of spring wheat yields to tillage treatments varies yearly depending upon soil water storage after fallow and early spring temperatures. He states "that during years with minimal soil water storage after fallow and above average early spring temperatures, chemical fallow has the potential to produce more spring wheat than stubble-mulch fallow, but during years with abundant soil water storage after fallow and below average spring temperatures, stubble-mulch fallow has the potential to produce more spring wheat than chemical fallow." Based on a 4-yr average, water use efficiency is greater for stubble mulch and reduced tillage plots compared to chemical fallow. Because of warmer soil temperatures under chemical fallow later in the growing season, greater water stress causes lower yields.

Greater surface residues reduce soil desiccation resulting from dry, strong winds that occur during the winter months, from south Texas to western Canada. In south Texas and northern Mexico, greater water contents at planting time in NT are often the difference between adequate seedling emergence and no germination. Conservation tillage practices that increase water storage have been developed for cotton production in the central Texas High Plains (Harman et al., 1989) and for dryland sorghum production in Bushland, Texas (Unger, 1991b; Jones et al., 1994a). In the semiarid regions of the United States during prolonged dry periods, cultivated soils with little residue cover and NT soils with 30% or more residue cover will both dry out with time to the same water level. In some instances, water contents in the upper layers in NT are less than in CT because of rapid capillary drying in NT. Surface cultivation under CT will decrease this drying in some soils. The best solution to this problem is to avoid fallow and plant summer crops, such as sorghum in the Texas environment.

Reviews of tillage systems and soil and water conservation were written by Smika and Unger (1986), Lal (1989), Unger (1990), Logan et al. (1991), and Unger (1991a).

10.12 Soil by Climate Interactions

Surface residues have major benefits to agriculture; the foremost of which are (1) improved water conservation, and (2) decreased soil erosion potential. The degree of benefit is controlled by the particular soil and climate scenario in question. Surface residues also may have effects on agroecosystems that are detrimental to plant growth and require management adjustments to ameliorate. For example, soil temperature is reduced by residue cover and this can be a negative factor for spring planted crops in northern latitudes.

The interactive effects of surface residue accumulation with soil and climate factors determine whether the residue effects are positive, neutral, or even negative. Primary issues are (1) problems with cold soil temperatures, excessively wet soil conditions, high runoff situations; (2) interactions affecting crop rotation choices and weed control; and (3) using surface residue management in conjunction with other conservation measures.

10.12.1 Problems Related to Cold Soils and Excessive Soil Water

In cold and humid climates, lower crop yields obtained with NT are often attributed to lower soil temperatures and excess soil moisture during the early growth of spring crops, especially when crop residues are not removed aside from the row (Fausey and Lal, 1989). McCalla and Duley (1946) showed that corn residue mulches decreased soil temperatures in midsummer (July–August) at Lincoln, NE by 3 to 7 °C as mulch rate varied from 5 to 20 Mg ha^{-1}. They were attempting to lower summer temperatures to decrease evaporation of water. This exemplifies an ameliorating effect of residue on a negative process, summer evaporation. In contrast, Al-Darby and Lowery (1987), showed that soil cover of 55–87% in NT conditions lowered soil temperatures by as much as 4 °C compared to CT under spring conditions in Wisconsin. Corn seedling emergence (100%) was delayed by 2 to 8 days with NT compared to CT systems. Although emergence was delayed and plant growth slower during the early parts of the season, final corn grain yields were not significantly different for any system tested in their environment. The authors concluded that the additional water savings in the NT system compensated for the early season delays in plant growth. Also in Wisconsin, Carter and Barnett (1987) reported that for continuous corn until NT, soil temperatures were cooler, emergence percentage and stand establishment were lower, and phenological development was delayed compared with corn produced using a moldboard plow system. Soil temperature at the 2-cm depth for barley plots in the semiarid region of southern Alberta, Canada, averaged 13.9 °C in NT and 14.7 °C under CT in spring (May) but these differences disappeared later in the growing season (Carefoot et al., 1990). These differences did not affect grain yield. Residue cover does delay plant development. For example, Fortin and Pierce (1990) reported that corn grown in residue-covered plots required 3.5 more days to reach the VT stage in 1987 and 8 more days in 1988. They concluded that evaluation of treatment effects on plants should be done by growth stages, and not by calendar day. Producers must be prepared to deal with delayed growth of spring crops in reduced and NT systems because of low soil temperatures. For example, delayed emergence may reduce crop competition and allow weed growth to flourish; these weeds use water and nutrients, like N, and could ultimately decrease crop yield. Therefore, proper attention to weed control, planting dates, cultivar selection and starter fertilizer are vital when one knows cold stress will be present. In general, the need for intense management increases whenever additional stresses are imposed.

Soil properties that affect water infiltration, permeability, and drainage must always be properly assessed when making residue management decisions. Factors that cause soils to remain wetter longer generally cause them to be colder. Again, this is only a problem with spring-planted crops, but at northern latitudes that affects spring cereals, soybean, corn, and grain sorghum. The latter two species, corn and grain sorghum, are particularly sensitive compared to the small grains. Research in the Corn Belt has shown that NT management on poorly drained soils has resulted in lower yields compared to CT. Long-term research has shown, however, that continued use of NT and choice of disease-resistant cultivars have overcome the negative response. Grain yields after 18 years of NT are now equal or greater than for CT fields. Grain yield responses to NT on well-drained soil have increased with time under the new system (Dick et al., 1991). Producers with poorly drained soils should expect lower productivity in the initial years after switching to high residue systems. It may be best to avoid complete NT in these situations. Better choices may be ridging and bedding which reduce adverse effects of high residue systems on poorly drained soils.

Soil texture, slope, and drainage classification are the key factors in making decisions about tillage system choice (Peterson, 1994). Poorly drained soils are not easily adapted to high residue systems and may be better managed with limited tilled systems than with strict NT. Coarse textured soils, even when poorly drained, are less of a problem than fine textured soils in the same drainage category. Well-drained soils are good candidates for NT and reduced till if they are silt loam or coarser in texture. Silty clay loam and silty clay soils may present problems for planting equipment if they are wet, and therefore, would be best managed with a more highly disturbed system; perhaps ridge till. All steeply sloping soils regardless of textural class are best managed with reduced or NT systems because their runoff and erosion potentials are high.

10.12.2   Weed Control Problems Related To Soil, Climate, and Tillage Choice

Weed control problems usually increase, at least in initial years, when farmers convert to high residue management systems. With less burial of weed seeds, greater weed emergence seems to be the norm. Grassy weeds like downy brome (*Bromus tectorum*) are particularly favored when tillage is reduced. Furthermore, substitution of sweep tillage for mixing and inverting operations (disks and plows) decrease ease of grassy weed control. Weeds that have growth cycles that are in synchrony with the crop cycle are the major offenders in these cases. Again using downy brome as an example, it germinates at about the same time as winter wheat planting. This type of weed is especially difficult to handle in monoculture systems like continuous wheat or wheat-fallow.

Crop rotation is one of the best ways to combat weed problems in any system and this especially true in reduced and no-till situations. By changing crop type one can control weeds at a different point in their growth cycle. For example, following wheat, a cool season crop, with crops like sunflower, millet and corn (warm season crops) gives an opportunity to control weeds that are a problem for wheat.

Herbicides substitute for part or all of the tillage operations in high residue systems. Close adherence to the labels of these products is essential for maximum weed control benefits to be achieved. Required rates of material can vary with soil texture and organic matter content. In general, the finer textured soils and those of higher organic matter content require higher rates of soil-applied herbicides in order to get weed control. Soil clays and organic matter inactivate these herbicides and therefore rates must be adjusted as described by the manufacturer. Caution also must be exercised in conjunction with herbicide carryover to the next crop. Coarse textured, low organic matter and high pH soils are particularly susceptible to carryover problems. Again, applying them at levels no higher than label specifications will minimize these potential problems.

Conversion to high residue systems will dictate that growers acquire herbicide management skills if they have not already done so. Correct sprayer operation from calibration to nozzle type selection is a key ingredient to success in residue management systems. New technology, where herbicides are applied selectively over a landscape, is on the horizon and will greatly aid in judicious use of herbicide materials.

10.12.3   Surface Residues and Other Conservation Measures

Maximizing snow catch is a vital conservation measure in dryer areas, since snow constitutes 20–50% of the annual precipitation. It represents a valuable resource for agriculture. Stubble height is a management tool used to maximize snow catch. Taller stubble retains more snow, but also can increase spring melt runoff. Soils which freeze deeply prior to snowfall and remain frozen and nonconductive to water while under snow cover provide the worst scenario for meltwater runoff. In warmer climates like the central and southern Plains snowmelt is more easily captured because soils do not freeze as deeply and often thaw under the snow, and substantial snowfall occurs after spring thawing begins. Greb (1979) reports that the efficiency of storing meltwater is often double that of water received as rain.

Contour farming in conjunction with residue maintenance further maximizes water capture from either rain or snowmelt. Terraces also are effective devices for water capture, but are less necessary on land with good residue cover and contour farming. Obviously, slopes >7–8% would still be best managed with terraces in addition to residue maintenance.

Maintaining residues on the soil surface is an invaluable management tool. It enhances water capture and retention, but the cost of storing the additional water with improved residue management can decrease profit if producers are not prepared to use the water.

Finally, producers should be aware that residues may have negative effects in terms of colder soil temperatures during spring planting. Appropriate system choices including planting method, crop choice, starter fertilizer decisions, etc. can help ameliorate the potentially negative cold temperature effects.

10.13　Conclusions

No-tillage systems represent the most dramatic change in soil management in modern history of agriculture. Soil disturbance to plant a crop is no longer necessary. Historically agriculturalists have had to prepare their whole field as a seedbed, even though seeds are only placed in a small fraction of the soil. Now most of the field can be left undisturbed and only a small portion prepared as a seedbed. Quite obviously these changes decrease labor and power requirements for the farmer. By not disturbing the majority of the soil during the prior to crop planting, soils can be kept in a more "natural setting". When soil perturbations are minimized, soil aggregate integrity is maintained and the physical, chemical and biological components of the soil respond positively. Erosion is minimized, nutrients are retained, and the soil biological processes can proceed with fewer interruptions. The greatest benefits of NT agriculture still lie ahead in a time when farmers, technology transfer persons, and scientists have learned how to fully exploit this ecologically beneficial system.

10.14　References

Al-Darby, A.M., and B. Lowery. 1987. Seed zone soil temperature and early growth with three conservation tillage systems. Soil Sci. Soc. Amer. J. 51:768–774.

Andraski, B.J., D.H. Mueller, and T.C. Daniel. 1985. Effects of tillage and rainfall simulation date on water and soil losses. Soil Sci. Soc. Am. J. 49:1512–1517.

Beare, M.H., M.L. Cabrera, P.F. Hendrix, and D.C. Coleman. 1994a. Aggregate-protected and unprotected organic matter pools in conventional- and no-tillage soils. Soil Sci. Soc. Am. J. 58:787–795.

Beare, M.H., P.F. Hendrix, and D.C. Coleman. 1994b. Water-stable aggregates and organic matter fractions in conventional- and no-tillage soils. Soil Sci. Soc. Am. J. 58:777–786.

Ben-Hur, M., I. Shainberg, D. Bakker, and R. Keren. 1985. Effect of soil texture and $CaCO_3$ content on water infiltration in crusted soil as related to water salinity. Irrig. Sci. 6:281–294.

Blevins, R.L., M.S. Smith, and G.W. Thomas. 1984. Changes in soil properties under no-tillage. p. 190–230. In R.E. Phillips and S.H. Phillips (ed.) No-tillage agriculture: Principles and practices. Van Nostrand Reinhold Co. New York, NY.

Bradford, J.M., and C. Huang. 1992. mechanisms of crust formation: Physical components. p. 55–72. In M.E. Sumner and B.A. Stewart (ed.) Soil crusting: Chemical and physical processes. Lewis Publishers, Boca raton, FL.

Bradford, J. M., and C. Huang. 1994. Interrill soil erosion as affected by tillage and residue cover. Soil Till. Res. 31:353–361.

Bremer, E., and C. van Kessel. 1992. Seasonal microbial biomass dynamics after addition of lentil and wheat residues. Soil Sci. Soc. Am. J. 56:1141–1146.

Brown, L.C., G.R. Foster, and D.B. Beasley. 1989. Rill erosion as affected by incorporated crop residue and season consolidation. Trans. ASAE 32:1967–1978.

Bruce, R.R., G.W Langdale, and A.L. Dillard. 1990. Tillage and crop rotation effect on characteristics of a sandy surface soil. Soil Sci. Soc. Am. J. 54:1744–1747.

Buchanan, M., and L.D. Kin. 1993. Carbon and phosphorus losses from decomposing crop residues in no-till and conventional till agroecosystems. Agron. J. 85:631–638.

Bundy, L.G., T.W. Andraski, and T.C. Daniel. 1992. Placement and timing of nitrogen fertilizers for conventional and conservation tillage corn production. J. Prod. Agric. 5:214–221.

Burwell, R.E., L.L. Sloneker, and W.W. Nelson. 1968. Tillage influences water intake. J. Soil Water Cons. 23:185–187.

Cambardella, C.A., and E.T. Elliott. 1993. Carbon and nitrogen distribution in aggregates from cultivated and native grassland soils. Soil Sci. Soc. Am. J. 57:1071–1076.

Carefoot, J.M., M. Nyborg, and C.W. Lindwall. 1990. Tillage-induced soil changes and related grain yield in a semi-arid region. Can. J. Soil Sci. 7:203–214.

Carter, P.R., and K.H. Barnett. 1987. Corn-hybrid performance under conventional and no-tillage systems after thinning. Agron. J. 79:919–926.

Cassel, D.K., C.W. Raczkowski, and H.P. Denton. 1995. Tillage effects on corn production and soil physical conditions. Soil Sci. Soc. Am. J. 59:1436–1443.

Christensen, N.B., W.C. Lindemann, E. Salazar-Sosa, and L.R. Gill. 1994. Nitrogen and carbon dynamics in no-till and stubble mulch tillage systems. Agron. J. 86:298–303.

Clark, A.J., A.M. Decker, J.J. Meisinger, and M.S. McIntosh. 1997. Kill date of vetch, rye, and a vetch-rye mixture: I. Cover crop and corn nitrogen. Agron. J. 89:427–434.

Cochran, V.L., L.F. Elliott, and R.I. Papendick. 1977. The production of phytotoxins from surface crop residues. Soil Sci. Soc. Am. J. 41:903–908.

Crovetto, C. 1996. Stubble over the soil. Am. Soc. Agron., Madison, WI.

Dalal, R.C. 1989. Long term effects of no-tillage, crop residue, and nitrogen application on properties of a Vertisol. Soil Sci. Soc. Am. J. 53:1511–1515.

Dao, T.H. 1993. Tillage and winter wheat residue management effects on water infiltration and storage. Soil Sci. Soc. Am. J. 57:1586–1595.

Dao, T.H. 1996. Tillage system and crop residue effects on surface compaction of a Paleustoll. Agron. J. 88:141–148.

Deibert, E.J., and R.A. Utter. 1994. Earthworm populations related to soil and fertilizer management practices. Better Crops 78:9–11.

de Quattro, J. 1997. Sustaining agriculture in drought years. Agric. Res. 45:4–9.

de St. Remy, E.A., and T.B. Daynard. 1982. Effects of tillage methods on earthworm populations in monoculture corn. Can. J. Soil Sci. 62:699–703.

Dick, W.A. 1983. Organic carbon, nitrogen, and phosphorus concentrations and pH in soil profiles as affected by tillage intensity. Soil Sci. Soc. Am. J. 47:102–107.

Dick, W.A. 1984. Influence of long-term tillage and crop rotation combinations on soil enzyme activities. Soil Sci. Soc. Am. J. 48:569–574.

Dick, W.A., E.L. McCoy, W.M. Edwards, and R. Lal. 1991. Continuous application of no-tillage to Ohio soils. Agron. J. 83:65–73.

Dick, W.A., R.J. Roseberg, E.L. McCoy, W.M. Edwards, and F. Haghiri. 1989. Surface hydrologic response of soils to no-tillage. Soil Sci. Soc. Am. J. 53:1520–1526.

Doran, J.W. 1980. Soil microbial and biochemical changes associated with reduced tillage. Soil Sci. Soc. Am. J. 44:765–771.

Drees, L.R., A.D. Karathanasis, L.P. Wilding, and R.L. Blevins. 1994. Micromorphological characteristics of long-term no-till and conventionally tilled soils. Soil Sci. Soc. Am. J. 58:508–517.

Edwards, W.M., M.J. Shipitalo, L.B. Owens, and L.D. Norton. 1989. Water and nitrate movement in earthworm burrows within long-term no-till cornfields. J. Soil Water Conserv. 44:240–242.

Edwards, W.M., L.D. Norton, and C.E. Redmond. 1988. Characterizing macropores that affect infiltration into nontilled soil. Soil Sci. Soc. Am. J. 52:483–487.

Fausey, N.R., and R. Lal. 1989. Drainage-tillage effects on Crosby-Kokomo soil association in Ohio. Soil Tech. 2:371–383.

Fenster, C.R., N.P. Woodruff, W.S. Chepil, and F.H. Siddoway. 1965. Performance of tillage implements in a stubble mulch system: III. Effects of tillage sequences on residues, soil cloddiness, weed control, and wheat yield. Agron. J. 57:52–55.

Follett, R.F., and D.S. Schimel. 1989. Effect of tillage practices on microbial biomass dynamics. Soil Sci. Soc. Am. J. 53:1091–1096.

Ford, J.H., and D.R. Hicks. 1992. Corn growth and yield in uneven emerging stands. J. Prod. Agric. 5:185–188.

Fortin, M.C., and F.J. Pierce. 1990. Developmental and growth effects of crop residues on corn. Agron. J. 82:710–715.

Foster, G. R., C. B. Johnson, and J. D. Nowlin. 1980. A model to estimate sediment yield from field sized areas: Application to planning and management for control of nonpoint source pollution. p. 193–281. *In* Creams: A field scale model for chemicals, runoff, and erosion from agricultural management systems, USDA Conservation Res. Rep. 26.

Fryrear, D.W. 1985. Soil cover and wind erosion. Trans. ASAE 28:781–784.

Gilley, J.E., S.C. Finkner, R.G. Spomer, and L.N. Mielke. 1986. Runoff and erosion as affected by corn residue: Part I. Total losses. Trans. ASAE 29:157–160.

Greb, B.W. 1979. Reducing drought effects on croplands in the west-central Great Plains. U.S.D.A. Info. Bul. 420.

Hammel, J.E. 1995. Long-term tillage and crop rotation effects on winter wheat production in northern Idaho. Agron. J. 87:16–22.

Harman, W.L., G.J. Michels, and A.F. Wiese. 1989. A conservation tillage system for profitable cotton production in the central Texas High Plains. Agron. J. 81:615–618.

Hill, R.L. 1990. Long-term conventional and no-tillage effects on selected soil physical properties. Soil Sci. Soc. Am. J. 54:161–166.

Hill, P.R. 1996. Conservation tillage: A checklist for U.S. farmers. Conservation Technology Information Center, West Lafayette, IN.

Hill, P.R., K.J. Eck, and J.R. Wilcox. 1994. Managing crop residue with farm equipment. Purdue University Agronomy Guide AY-280.

Jones, O.R., V.L. Hauser, and T.W. Popham. 1994a. No-tillage effects on infiltration, runoff, and water conservation on dryland. Trans. ASAE 37:473–479.

Jones, O.R., S.J. Smith, and L.M. Southwick. 1994b. Tillage system effects on water conservation and runoff water quality-Southern High Plains Drylands. p. 67–76. *In* Proc. of the Great Plains Residue Mgmt. Conf., Great Plains Agricultural Council.

Karlen, D.L. 1989. Tillage and planting system effects on corn emergence from Norfolk loamy sand. Appl. Agric. Res. 4:190–195.

King, K. W., D. C. Flanagan, L. D. Norton, and J. M. Laflen. 1995. Rill erodibility parameters influenced by long-term management practices. Trans. ASAE 38:159–164.

Kolberg, R.L., N.R Kitchen, D.G. Westfall, and G.A. Peterson. 1996. Cropping intensity and nitrogen management impact of dryland no-till rotations in the semiarid western Great Plains. J. Prod. Agric. 9:517–522.

Lal, R. 1989. Conservation tillage for sustainable agricultural: Tropical vs temperature environments. Adv. Agron. 42:85–196.

Lal, R., A.A. Mahboubi, and N.R. Fausey. 1994. Long-term tillage and rotation effects on properties of a central Ohio soil. Soil Sci. Soc. Am. J. 58:517–522.

Lamb, J.A., J.W. Doran, and G.A. Peterson. 1987. Nonsymbiotic dinitrogen fixation in no-till and conventional wheat-fallow systems. Soil Sci. Soc. Am. J. 51:356–361.

Lamb, J.A., G.A. Peterson, and C.R. Fenster. 1985. Fallow nitrate accumulation in a wheat-fallow rotation as affected by tillage system. Soil Sci. Soc. Am. J. 49:1441–1446.

Langdale, G.W., L.T. West, R.R. Bruce, W.P. Miller, and A.W. Thomas. 1992. Restoration of eroded soil with conservation tillage. Soil Tech. 5:81–90.

Larney, F.J., and E.J. Kladivko. 1989. Soil strength properties under four tillage systems at three long-term study sites in Indiana. Soil Sci. Soc. A. J. 53:1539–1545.

le Bissonnais, Y. 1990. Experimental study and modeling of soil surface crusting processes. Catena Supp.17:13–28.

Liebig, M.A., A.J. Jones, L.N. Mielke, and J.W. Doran. 1993. Controlled wheel traffic effects on soil properties in ridge tillage. Soil Sci. Soc. Am. J. 57:1061–1066.

Logan, T.J., R. Lal, and W.A. Dick. 1991. Tillage systems and soil properties in North America. Soil Till. Res. 20:241–270.

Lyles, L., and B.E Allison. 1976. Wind erosion: The protective role of simulated standing stubble. Trans. ASAE 19:61–64.

Lyles, L, and B.E. Allison. 1981. Equivalent wind-erosion protection from selected crop residues. Trans. ASAE 24:405–408.

Mackay, A.D., and E.J. Kladivko. 1985. Earthworms and rate of breakdown of soybean and maize residues in soil. Soil Biol. Biochem. 17:851–857.

Martin, V.L., E.L. McCoy, and W.A. Dick. 1990. Allelopathy of crop residues influences corn seed germination and early growth. Agron. J. 82:555–560.

Maskina, M.S., J.F. Power, J.W. Doran, and W.W. Wilhelm. 1993. Residual effects of no-till crop residues on corn yield and nitrogen uptake. Soil Sci. Soc. Am. J. 57:1555–1560.

McCalla, T.M., and F.L. Duley. 1946. Effect of crop residues on soil temperature. Agron. J. 38:75–89.

McGregor, K.C., C.K. Mutchler, and M.J.M. Romkens. 1990. Effects of tillage with different crop residues on runoff and soil loss. Trans. ASAE 33:1551–1556.

Merrill, S.A., A.L. Black, and A. Bauer. 1996. Conservation tillage affects root growth of dryland spring wheat under drought. Soil Sci. Soc. Am. J. 60:575–83.

Meyer, L.D., and J.V. Mannering. 1963. Crop residues as surface mulches for controling erosion on sloping land under intensive cropping. Trans. ASAE 6:322–323, 327.

Mielke, L.N., J.W. Doran, and K.A. Richard. 1986. Physical environment near the surface of plowed and no-tilled soils. Soil Till. Res. 7:355–366.

Miller, W. P., and D. E. Radcliffe. 1992. Soil crusting in the southeastern United States. p. 233–266. *In* M. E. Sumner and B. A. Stewart (eds.) Soil crusting: Chemical and physical processes. Lewis Publishers, Boca Raton, FL.

Mock, J.J., and D.C. Erbach. 1977. Influence of conservation tillage environments on growth and productivity of corn. Agron. J. 69:337–340.

Nafziger, D., P.R. Carter, and E.E. Grahan. 1991. Response of corn to uneven emergence. Crop Sci. 31:811–815.

Norton, L.D., and L.C. Brown. 1992. Time-effect of water erosion for ridge tillage. Trans. ASAE 35:473–478.

Nyborg, M., and S.S. Malhi. 1989. Effect of zero and conventional tillage on barley yield and nitrate nitrogen content, moisture and temperature of soil in north-central Alberta. Soil Till. Res. 15:1–9.

Ortega, R.A. 1995. Residue distribution and potential C and N mineralization in no-till dryland agroecosystems. M.S. Thesis. Colorado State University, Fort Collins, CO.

Paustian, K., H.P. Collins, and E.A. Paul. 1997. Management controls on soil carbon. p. 15–49. *In* E.A. Paul, K. Paustian, E.T. Elliott, and C.V. Cole (eds.) Soil organic matter in temperate agroecosystems. Lewis Publishers. Boca Raton, FL.

Peterson, G.A. 1994. Interactions of surface residues with soil and climate. p. 9–12. *In* W. C. Moldenhauer, and A.L. Black (ed.) Crop residue management to reduce erosion and improve soil quality. Northern Great Plains. USDA Conserv. Res. Rep. 38.

Peterson, G.A., A.J. Schlegel, D.L. Tanaka, and O.R. Jones. 1996. Precipitation use efficiency as affected by cropping and tillage systems. J. Prod. Agr. 9:180–186.

Peterson, G.A., D.G. Westfall, L. Sherrod, R. Kolberg, and D. Poss. 1995. Sustainable dryland agroecosystem management. CO Agric. Expt. Stn. Tech. Bul. TB95-1.

Radcliffe, D.E., G. Manor, R.L. Clark, L.T. West, G.W. Langdale, and R.R. Bruce. 1989. Effect of traffic and in-row chiseling on mechanical impedance. Soil Sci. Soc. Am. J. 53:1197–1201.

Radcliffe, D.E., E.W. Tollner, W.L. Hargrove, R.L. Clark, and M.H. Golabi. 1988. Effect of tillage practices on infiltration and soil strength of a Typic Hapludult soil after ten years. Soil Sci. Soc. Am. J. 52:798–804.

Raper, R.L., D.W. Reeves, E.C. Burt, and H.A. Torbert. 1994. Conservation tillage and traffic effects on soil condition. Trans. ASAE 37:763–768.

Rice, C.W., and M.S. Smith. 1982. Denitrification in no-till and plowed soils. Soil Sci. Soc. Am. J. 46:1168–1173.

Rice, C.W., and M.S. Smith. 1984. Short-term immobilization of fertilizer nitrogen at the surface of no-till and plowed soils. Soil Sci. Soc. Am. J. 48:295–297.

Rice, C.W., M.S. Smith, and R.L. Blevins. 1984. Soil nitrogen availability after long-term continuous no-tillage and conventional tillage corn production. Soil Sci. Soc. Am. J. 50:1206–1210.

Schertz, D.L., and J. Becherer. 1994. Terminology. p.3. *In* B.A. Stewart and W.C. Moldenhauer (eds.) Crop residue management to reduce erosion and improve soil quality. Northern Great Plains. USDA Conser. Res. Rep. 37.

Schillinger, W.F., and F.E. Bolton. 1993. Fallow water storage in tilled vs. untilled soils in the Pacific Northwest. J. Prod. Agric. 6:267–269.

Schomberg, H.H., J.L. Steiner, and P.W. Unger. 1994. Decomposition and nitrogen dynamics of crop residues: Residue quality and water effects. Soil Sci. Soc. Am. J. 58:372–381.

Sherrod, L., G.A. Peterson, and D.G. Westfall. 1996. No-till rotational residue dynamics across an ET gradient. Agron. Abstracts 88:282.

Siddoway, F.H., W.S. Chepil, and D.V. Armbrust. 1965. Effect of kind, amount, and placement of residue on wind erosion control. Trans. ASAE 8:327–331.

Smart, J.R., and J.M. Bradford. 1996. No-tillage and reduced tillage cotton production in south Texas. Proc. Beltwide Cotton Conf. 2:1397–1400.

Smart, J.R., and J.M. Bradford. 1997. Cotton weed management in no-till maize and sorghum stubble. Proc. Beltwide Cotton Conf. 1:801–802.

Smika, D.E., and P.W. Unger. 1986. Effect of surface residues on soil water storage. Adv. Soil Sci. 5:111–138.

Smith, S.J., and A.N. Sharpley. 1990. Soil nitrogen mineralization in the presence of surface and incorporated crop residues. Agron. J. 82:112–116.

Smith, S.J., and A.N. Sharpley. 1993. Nitrogen availability from surface-applied and soil-incorporated crop residues. Agron. J. 85:776–778.

Staley, T.E., W.M. Edwards, C.L. Scott, and L.B. Owens. 1988. Soil microbial biomass and organic component alterations in a no-tillage chronosequence. Soil Sci. Soc. Am. J. 52:998–1005.

Stecker, J.A., D.D. Buchholz, R.G. Hanson, N.C. Wollenhaupt, and K.A. McVay. 1993a. Application placement and timing of nitrogen solution for no-till corn. Agron. J. 85:645–650.

Stecker, J.A., D.D. Buchholz, R.G. Hanson, N.C. Wollenhaupt, and K.A. McVay. 1993b. Broadcast nitrogen sources for no-till continuous corn and corn following soybean. Agron. J. 85:893–897.

Stecker, J.A., D.H. Sander, F.N. Anderson, and G.A. Peterson. 1988. Phosphorus fertilizer placement and tillage in a wheat-fallow cropping sequence. Soil Sci. Soc. Am. J. 52:1063–1068.

Sterk, G., and W.P. Spaan. 1997. Wind erosion control with crop residues in the Sahel. Soil Sci. Soc. Am. J. 61:911–917.

Tanaka, D.L. 1989. Spring wheat plant parameters as affected by fallow methods in the northern Great Plains. Soil Sci. Soc. Am. J. 53:1506–1511.

Teotia, S.P., F.L. Duley, and T.M. McCalla. 1950. Effect of stubble mulching on number and activity of earthworms. NE Agric. Expt. Stn. Res. Bul. 165.

Touchton, J.T., and J.W. Johnson. 1992. Soybean tillage and planting method effects on yield of double-cropped wheat and soybeans. Agron. J. J. 74:57–59.

Trojan, M.D., and D.R. Linden. 1994. Tillage, residue, and rainfall effects on movement of an organic tracer in earthworm-affected soils. Soil Sci. Soc. Am. J. 58:1489–1494.

Unger, P.W. 1990. Conservation tillage systems. Adv. Soil Sci. 13:27–68.

Unger, P.W. 1991a. Reduced tillage systems. p. 387–422. In J. Skujins (ed.) Semiarid lands and deserts: Soil resource and reclamation. Marcel Dekker, Inc., New York, NY.

Unger, P.W. 1991b. Ontogeny and water use of no-tillage sorghum cultivars on dryland. Agron J. 83:961–968.

Unger, P.W. 1993. Reconsolidation of a Torrertic Paleustoll after tillage with a Paratill. Soil Sci. Soc. Am. J. 57:195–199.

van Doren, D.M., G.B. Triplet, and J.E. Henry. 1976. Influence of long term tillage, crop rotation, and soil type combinations on corn yield. Soil Sci. Soc. Am. J. 40:100–105.

Vigil, M.F., and D.E. Kissel. 1991. Equations for estimating the amount of nitrogen mineralized from crop residues. Soil Sci. Soc. of Am. J. 55:757–767.

Wagger, M.G., and D.K. Cassel. 1993. Corn yield and water-use efficiency as affected by tillage and irrigation. Soil Sci. Soc. Am. J. 57:229–234.

Wagger, M.G., and H.P. Denton. 1989. Tillage effects on grain yields in a wheat, double-crop soybean, and corn rotation. Agron. J. 81:493–498.

West, L.T., W.P. Miller, R.R. Bruce, G.W. Langdale, J.M. Laflen, and A.W. Thomas. 1992. Cropping system and consolidation effects on rill erosion in the Georgia Piedmont. Soil Sci. Soc. Am. J. 56:1238–1243.

Wischmeier, W.H., and D.D. Smith. 1978. Predicting rainfall erosion losses-A guide to conservation planning. USDA Agric. Handbook 537. Washington, D.C.

Wood, C.W. and J.H. Edwards. 1992. Agroecosystem management effects on soil carbon and nitrogen. Agric. Ecosyst. Environ. 39:123–138.

Wood, C.W., D.G. Westfall, G.A. Peterson, and I. C. Burke. 1990. Impacts of cropping intensity on carbon and nitrogen mineralization under no-till dryland agroecosystems. Agron. J. 82:1115–1120.

Wood, C.W., D.G. Westfall, and G.A. Peterson. 1991. Soil carbon and nitrogen changes on initiation of no-till cropping systems. Soil Sci. Soc. Am. J. 55:470–476.

11

Soil Quality

Michael J. Singer and Stephanie Ewing
University of California, Davis

11.1 Introduction

11.1.1 Why Soil Quality?

Soils are fundamental to the well-being and productivity of agricultural and natural ecosystems. Soil quality is a concept being developed to characterize the usefulness and health of soils. In the United States, soil quality includes soil fertility, potential productivity, resource sustainability and environmental quality. In Canada and Europe, contaminant levels and their effects are central to soil quality (Moen, 1988; Denneman and Robberse, 1990; Cairns, 1991; Sheppard et al., 1992). A general definition of soil quality is the degree of fitness of a soil for a specific use (Gregorich et al., 1994). Other definitions of soil quality have been published, and are reviewed later in this chapter. The existence of multiple definitions suggests that the soil quality concept continues to evolve.

Recent attention has focused on the sustainability of human uses of soil, based on concerns that soil quality may be declining (Fournier, 1989; Parr et al., 1990; Doran et al., 1996; Boehn and Anderson, 1997). We use sustainable to mean that a use or management of soil will sustain human well-being over time. Lal (1995) described the land resources of the world (of which soil is one component) as "finite, fragile, and nonrenewable," and reported that only about 22% (3.26 billion ha) of the total land area on the globe is suitable for cultivation and at present, only about 3% (450 million ha) has a high agricultural production capacity. Because soil is in large but finite supply, and some soil components cannot be renewed within a human time frame, the condition of soils in agriculture and the environment is an issue of global concern (Howard, 1993; FAO, 1997). Concerns include soil losses from erosion, maintaining agricultural productivity and system sustainability, protecting natural areas, and adverse effects of soil contamination on human health (Crosson and Stout, 1983; Siegrist, 1989; Denneman and Robberse, 1990; Haberern, 1992; Howard, 1993; Sims et al., 1997). Parr et al. (1992) state, "...soil degradation is the single most destructive force diminishing the world's soil resource base." Soil quality guidelines are intended to protect the ability of ecosystems to function properly (Cook and Hendershot, 1996). The US National Academy of Sciences (NAS, 1993) suggests that the United States should adopt a national policy "...that seeks to conserve and enhance soil quality...".

Useful evaluation of soil quality requires agreement about why soil quality is important, how it is defined, how it should be measured, and how to respond to measurements with management, restoration, or conservation practices. Because determining soil quality requires one or more value judgments and because we have much to learn about soil, these issues are not easily addressed.

Definitions of soil quality have been based both on human uses of soil and on the functions of soil within natural and agricultural ecosystems. For purposes of this chapter, we are discussing soil quality within the context of managed agricultural ecosystems. To many in agriculture and agricultural research, productivity is analogous to soil quality. Maintaining soil quality is also a human health concern because air, groundwater and surface water consumed by humans can be adversely affected by mismanaged and contaminated soils, and because humans may be exposed to contaminated soils in residential areas. Contamination may include heavy metals, toxic elements, excess nutrients, volatile and nonvolatile organics, explosives, radioactive isotopes and inhalable fibers (Sheppard et al., 1992).

Soil quality is not determined by any single conserving or degrading process or property, and soil has both dynamic and relatively static properties that also vary spatially (Bouma and Bregt, 1989; Bouma, 1997; Carter et al., 1997). Gregorich et al. (1994) state that "soil quality is a composite measure of both a soil's ability to function and how well it functions, relative to a specific use." Increasingly, contemporary discussion of soil quality includes the environmental cost of production and the potential for reclamation of degraded soils.

Reasons for assessing soil quality in an agricultural or managed system may be somewhat different than reasons for assessing soil quality in a natural ecosystem. In an agricultural context, soil quality may be *managed*, to maximize production without adverse environmental effect, while in a natural ecosystem, soil quality may be *observed*, as a baseline value or set of values against which future changes in the system may be compared.

11.1.2   Land Quality

Land is a broad concept, including the atmosphere, the earth's surface and subsurface, hydrology, plants, animals, people and their interactions (FAO 1976, 1995, 1997). Soil is a part of land (McRae and Burnham, 1981; Davidson, 1986; Brinkman, 1997) and soil quality a subset of land quality. Land quality has been defined as "a complex attribute of land which acts in a distinct manner in its influence on the function of land for a specific kind of use." (FAO 1976, 1995; Rossiter, 1996; Sombroek, 1997). "Soil quality" is more restrictive than "land quality" but frequently incorporates the same emphasis on use (Carter et al., 1997).

Several cautions have been expressed by Sombroek (1997) and others (FAO, 1997) about land quality evaluations that are equally applicable to soil quality evaluations. One is that "land qualities are not absolute values, but have to be assessed in relation to the functions of the land use that one has in mind" (Sombroek, 1997). A second is that "...different indicators are needed to track changes in each of the land's main components (and their subdivisions) and that the data and information needs are so diverse, ranging from farmers to politicians, that a single, core set of indicators is probably not possible to develop over the short term." Finally, "Single indicators [of soil or land quality] may be meaningless until the related factors in the system as a whole are clearly understood" (Brinkman, 1997). This chapter reviews (1) concepts and definitions of soil quality for agriculture, (2) characteristics of soil that may be used in assessing soil quality, and (3) methods of integrating these characteristics into soil quality indices. Some guidelines are given for interpreting soil quality measurements and conclusions are drawn from the material presented.

11.1.3 Soil Quality is Not a New Idea

Soil quality has historically been equated with agricultural productivity, and thus is not a new idea. Soil conservation practices to maintain soil productivity are as old as agriculture itself, with documentation dating to the Roman Empire (Jenny, 1961). The Storie Index (Storie, 1932, 1964) and USDA Land Capability Classification (Klingebiel and Montgomery, 1973) were developed to separate soils into different quality classes. Soil quality is implied in many decisions farmers make about land purchases and management, and in the economic value rural assessors place on agricultural land for purposes of taxation.

Beginning in the 1930s, soil productivity ratings were developed in the United States and elsewhere to help farmers select crops and management practices that would maximize production and minimize erosion or other adverse environmental effects (Huddleston, 1984). These rating systems are important predecessors of recent attempts to quantitatively assess soil quality. In the 1970s, attempts were made to identify and protect soils of the highest productive capacity by defining "prime agricultural lands" (Miller, 1979; Reganold and Singer, 1979; Singer et al., 1979).

An idea related to soil quality is "carrying capacity". Carrying capacity is the number of individuals that can be supported in a given area (Budd, 1992). Soils with high productivity have high carrying capacity, and are considered to be high quality. Sustainability implies that a system does not exceed its carrying capacity over time. Recent attempts to define soil quality and develop indices to measure it have many of the properties of the earlier soil productivity ratings (Doran and Parkin, 1994; Doran and Jones, 1996; Snakin et al., 1996; Seybold et al., 1997).

11.1.4 Soil Quality Versus Air and Water Quality

Cox (1995) calls for national goals for soil quality that "... recognize the inherent links between soil, water and air quality." Haberern (1992) suggests that the decade of the 1990s is the time to study the soil as we have recognized and studied air quality and water quality in the preceding two decades. Air and water quality standards are generally based on maximum allowable concentrations of materials hazardous to human health. They are specified and enforced by regulators according to public uses of these resources. The result is that changes in air and water quality are now monitored to protect human health. With few exceptions, soil quality standards have not been set, nor have regulations been created regarding maintenance of soil quality.

To the extent that soil has been the disposal site of hazardous wastes, as well as a pathway by which contamination or other applied chemicals may present a human health risk, sporadic regulation of soil quality (in terms of contamination) does exist in the United States. These regulations are in the form of laws regulating hazardous waste (e.g., Comprehensive Environmental Response, Compensation, and Liabilities Act, CERCLA; and Resource Conservation and Recovery Act, RCRA), toxic substances (Toxic Substances Control Act, TSCA), and pesticides (Federal Insecticide, Fungicide, and Rodenticide Act, FIFRA). However, these standards are often contradictory, inconsistent with each other and with current methods of assessing risk.

For example, in the United States, federal regulations supporting CERCLA (40 CFR) is a list of "hazardous substances" and the levels in various media (e.g., soil, water) to which the Environmental Protection Agency (EPA) must respond with a cleanup effort. However, EPA has fielded considerable controversy about contaminant levels and chemical forms that legitimately constitute a human health risk. Target cleanup levels have also been subject to debate and legislation.

Soil quality assessment requires definition of a "clean" soil (Sims et al., 1997). From this point of view, good quality soil has been defined as posing "...no harm to any normal use by humans, plants or animals; not adversely affecting natural cycles or functions; and not contaminating other components of the environment" (Moen, 1988). The parallel to air and water quality is easy to draw on a conceptual level, but designation of soil quality standards is significantly complicated by soil variability and heterogeneity (Smith et al., 1993).

11.1.5 Soil Quality and Soil Health

Soil health and soil quality are often used synonymously (Warkentin, 1995; Singer and Warkentin, 1996). Concepts of soil health are briefly reviewed here before proceeding to further define soil quality, because some farmers prefer the term soil health to soil quality. When farmers in Wisconsin were asked how they recognized a healthy soil, they listed a total of 97 different soil properties (Romig et al., 1995). Most frequently mentioned were biological and physical properties of surface horizons (Table 11.1). In addition to soil properties, farmers used plant, animal and human health, and water properties to judge the health of their soils.

Soil health has been described as integral to the concept of sustainable agriculture (Reganold et al., 1990), although soil is not inherently "unhealthy" if it is not a good agricultural soil. Soil health may be considered as the state of the soil at a particular time, equivalent to the "dynamic" soil properties that change in the short term (Carter et al., 1997). In contrast, soil quality may be considered as soil usefulness for a particular purpose over a longer time scale, equivalent to "intrinsic" or "static" soil quality (Carter et al., 1997). Examples of dynamic soil properties are organic matter content, the number or diversity of organisms, and microbial constituents or products.

Soil health may be evaluated by comparing the present condition of a soil with set reference points or baseline values that reflect the soil's overall quality (Granatstein and Bezdicek, 1992; Snakin et al., 1996; Boehn and Anderson, 1997; Seybold et al., 1997). Granatstein and Bezdicek (1992) consider whether the standard reference state of a soil in agricultural systems should be "native" soil conditions or conditions given maximum agronomic, environmental and economic performance. Seybold et al. (1997) consider three potential baseline values (reference points) for soil quality: the native, the intensively managed, and the altered (reclaimed) states. However, they do not specify what reference levels are "healthy". Given that native conditions may be unknown or difficult to determine, more appropriate standard conditions for agricultural systems could be based on high-yielding, low environmental impact systems. For natural systems, observed baseline values, considering previous human influences, are appropriate.

Historical analyses of land use and soil property trends might be useful as a means of assessing soil degradation and appropriate baseline values (Lindert et al., 1996a,b). Using the present condition of intensively managed and altered (reclaimed) sites clearly would be much easier than establishing reference or baseline values. However, it is the most highly managed soils, experiencing the greatest losses of topsoil and decreases in organic matter, that raise the most concern that modern intensive agriculture is not sustainable. Furthermore, accepting the state of intensively managed or altered soils as a baseline may confuse the issue of soil health versus soil quality. Intensively managed and/or altered soils may be "healthy" in terms of high productivity but may not be of higher "intrinsic" quality than other soils managed at a lower intensity. "Dynamic" properties that are easily degraded by poor soil management may be the most sensitive indicators of the degree to which soil health has been changed, provided an initial state is defined. Conversely, dynamic properties that are subject to management may be poor measures of intrinsic quality.

Long-term research plots in Canada (Beke et al., 1994; Bremer et al., 1994; Biederbeck et al., 1996), Denmark (Schjønning et al., 1994), England (Jenkinson et al., 1994; Johnston et al., 1986), Germany (Merker, 1956), New Zealand (Murata et al., 1995), Sweden (Gerzabeck et al., 1997), and the United States (Odell et al. 1984) have been used to establish baseline values of soil properties against which changes due to management were measured. Not surprisingly, these studies have shown

Table 11.1 Soil health properties derived from farmer interviews [From Romig et al., 1995. J. Soil Water Conserv. 50:229-236 with permission of the Soil and Water Conservation Society of America]

Rank	Soil health property	Description Healthy	Unhealthy
1	organic matter	as high as possible, at soil's potential, manure, compost > 3%, 2%, 7-8%, putting more back	rough, lack of organic matter, less, low
2	crop appearance	green, healthy, uniform, lush, dense stand, tall, larger, sturdy, stout, proper color, darker, good crop	yellow, stunted corn, small, poor color, poorer, lack of green, light green, streaks in field
3	erosion	would not erode, water, & wind not taking soil, prevented, stays in place, less, slowed down, delayed	blows sooner, washes, topsoil's lost, erodes more, clouds of dust, ravines, runs bad, any, easier
4	earthworms	fishing & red worms present, see after rain, a lot, angle worms, see holes & castings, see during plowing	not there, don't work, can't find, no holes, lack of, killed by insecticides or anhydrous, void
5	drainage	water goes away, fast, better, no ponding, moves through, takes a lot of rain, drains properly, dries out	tight, waterlogged, drains too fast, ponding, no outlet for water, won't drain, slop, poor, saturated
6	tillage ease	one pass & ready, breaks up, mellow, easier, smooth, crumbles, flows, plow a gear faster, minimum	never works down, needs more disking, lumps, slabs, shiney, pulls hard, worked wet, overworked
7	soil structure	won't roll out of hand, crumbly, loose, holds together, granular	hard, doesn't hold together, lumpy falls apart, massive, cloddy, lumpy, clumpy, tight, compacted, powder
8	pH	7.0, 6.7-6.8, 6.2-6.7, balanced, neutralize	< 6.0, high, nothing works, wrong, too low, high acidity
9	soil test	up to recommendations, high, elevated, complete, where it belongs, every year or two, stay up with soil test	law of minimum at work
10	yield	9.5-11.5 Mg ha^{-1} corn, 4 Mg ha^{-1} soybeans, 30-40% higher, + 0.7 Mg ha^{-1}, better 5 year average, significantly higher	7 Mg ha^{-1} corn, 9.5 Mg ha^{-1} corn, 2.5 Mg ha^{-1} soybeans, 20-50% less, don't get much off, down, reduced, low
11	compaction	doesn't pack down, not compacted, stays loose, not out there when wet	compacted, plow layer, packs down, hardpan, plowsoil, tight, can't get into it, packed
12	infiltration	water doesn't stand, absorbs, water moves into soil, soaks, rapid, no ponding, fast, spongy	water runs off soil, sits on top, water stands, doesn't absorb, puddles, nonporous
13	soil color	dark, black, dark brown, gray, holds dark color	orange, brown, light, white, red, blue-gray, subsoil color, bleached, sandy colored, light brown, pale, anemic, gray
14	nitrogen	put on less, manure, as required, compost, slurry, more available, organic N, organic matter	too much N, chemical N, commercial fertilizers burn ground, anhydrous, sludge
15	water retention	holds moisture, get by with less, retains more, moisture travels, gives and takes water freely, conserving	too much water, doesn't hold water, drys out, too wet or dry, droughty, stays wet, runs out of moisture, poor
16	phosphorus	as required	-------
17	nutrient deficiency	has what it needs, no shortage of elements, no spots on leaves	yellow, purple, discoloration in leaves, lodging, crop falls off, stripping, brown streaks, firings on bottom, blight
18	decomposition	breaks down, decays, rots in 4-5 months, manure part of soil in 1 year, disappears, not too fast, 2/3 gone in year	see stalks from last year, doesn't break down, manure plows up next year
19	potassium	as required	-------
20	roots	larger, spread out, grown down, white, deep, numerous, good penetration, full, lots of feeders, branched out	don't penetrate, undeveloped, balled up, grow crossways, discolored, diseased, at hard angles, shallow, short

that cropping practices change soil physical, chemical and biological properties. For example, Gerzabeck et al. (1997) found that 40 years of manure addition to a clay loam soil near Uppsala, Sweden increased the soil organic matter content significantly over similar soils without manure but with mineral N applications.

11.2   Definitions of Soil Quality

Among the authors and organizations defining soil quality are Larson and Pierce (1991), Arshad and Coen (1992), Karlen et al. (1992), Doran and Parkin (1994) and Karlen et al. (1997). The next section reviews some of the definitions and soil characteristics used to define soil quality. The reader should understand that the definition of soil quality and selection of soil characteristics needed to quantify soil quality are continuing to evolve. For example, Bouma (1989a) recognized that an essential problem with definitions that produce carefully limited suitability classes is that empirical decisions must be made to separate the classes along what is essentially a continuum. That is, if soil organic matter is part of a soil quality definition, where on the continuum of soil organic matter content does one draw the line between a high quality and low quality soil? Does high organic matter content always indicate high soil quality? These are non-trivial questions under discussion by the soil science community.

11.2.1   Functional Approach

Carter et al. (1997) suggest a framework for evaluating soil quality that includes (1) describing each soil function on which quality is to be based, (2) selecting soil characteristics or properties that influence the capacity of the soil to provide each function, (3) choosing indicators of characteristics that can be measured, and (4) using methods that provide accurate measurement of those indicators.

The following soil functions appear frequently in the soil science literature: (1) soil maintains biological activity/productivity (Doran and Parkin, 1994; Snakin et al., 1996; Karlen et al., 1997), serves as medium for plant/crop growth (Larson and Pierce, 1991; Arshad and Coen, 1992; Karlen et al., 1992), supports plant productivity/yield (Arshad and Coen, 1992; Karlen et al., 1997), supports human/animal health (Doran and Parkin, 1994; Karlen et al., 1997); (2) partitions and regulates water/ solute flow through environment (Larson and Pierce, 1991; Arshad and Coen, 1992; Karlen et al., 1997); (3) serves as an environmental buffer/filter (Larson and Pierce, 1991), maintains environmental quality (Arshad and Coen, 1992; Doran and Parkin, 1994; Karlen et al., 1997); and (4) cycles nutrients, water, energy and other elements through the biosphere (Anderson and Gregorich, 1984; Karlen et al., 1997). Clearly, these functions are interrelated. Later in this chapter, discussion focuses on the first and third functions (productivity and environmental buffering) as encompassing those aspects of soil quality most debated in the literature.

11.2.2   Examples

Larson and Pierce (1991) defined soil quality as "the capacity of a soil to function within the ecosystem boundaries and interact positively with the environment external to that ecosystem." Three soil functions are considered essential: provide a medium for plant growth, regulate and partition waterflow through the environment, and serve as an effective environmental filter.

Arshad and Coen (1992) define soil quality as "the sustaining capability of a soil to accept, store and recycle water, minerals and energy for production of crops at optimum levels while preserving a healthy environment." They discuss terrain, climate and hydrology as site factors that contribute to soil quality and suggest that socioeconomic factors such as land use, operator and management should be included in a soil quality analysis. This approach is consistent with the FAO approach to land quality analysis (FAO, 1997).

Karlen et al. (1992) define soil quality as "the ability of the soil to serve as a natural medium for the growth of plants that sustain human and animal life." Their definition is based on the role of soil quality in the long-term productivity of soil and maintenance of environmental quality.

Doran and Parkin (1994) defined soil quality as "the capacity of a soil to function within ecosystem boundaries to sustain biological productivity, maintain environmental quality, and promote plant and animal health." Gregorich et al. (1994) define soil quality as "a composite measure of both a soil's ability to function and how well it functions relative to a specific use" or "the degree of fitness of a soil for a specific use."

The Soil Science Society of America Ad Hoc Committee on Soil Health proposed that soil quality is "the capacity of a specific kind of soil to function, within natural or managed ecosystem boundaries, to sustain plant and animal productivity, maintain or enhance water and air quality, and support human health and habitation" (Karlen et al., 1997). This definition requires that five functions must be evaluated to describe soil quality: (1) sustaining biological activity, diversity, and productivity; (2) regulating and partitioning water and solute flow; (3) filtering, buffering, degrading, immobilizing and detoxifying organic and inorganic materials, including industrial and municipal byproducts and atmospheric deposition; (4) storing and cycling nutrients and other elements within the earth's biosphere; and (5) providing support of socioeconomic structures and protection for archeological treasures associated with human habitation.

No soil is likely to successfully provide all of these functions, some of which occur in natural ecosystems and some of which are the result of human modification. We can summarize by saying that soil quality depends on the extent to which soil functions to benefit humans. Thus, for food production or mediation of contamination, soil quality means the extent to which a soil fulfills the role we have defined for it. Within agriculture, high quality equates to maintenance of high productivity without significant soil or environmental degradation.

11.3   Measuring Soil Quality

11.3.1   Considerations in Selecting Soil Characteristics

The Glossary of Soil Science terms produced by the Soil Science Society of America (1996) states that soil quality is an inherent attribute of a soil that is inferred from soil characteristics or indirect observations. To proceed from a dictionary definition to a measure of soil quality, a minimum dataset (MDS) of soil characteristics that represents soil quality must be selected and quantified (Larson and Pierce, 1991, 1994; Gregorich et al, 1994; Papendick et al., 1995). The MDS may include biological, chemical or physical soil characteristics (Table 11.2). For agriculture, the measurement of properties should lead to a relatively simple and accurate way to rank soils based on potential plant production without soil degradation. Unfortunately, commonly identified soil quality parameters may not correlate well with yield (Reganold, 1988).

In the next section, we consider these four points concerning the selection and quantification of soil characteristics: (1) soil characteristics may be desirable or undesirable, (2) soil renewability involves judgment of the extent to which soil characteristics can be controlled or managed, (3) rates of change in soil characteristics vary, and (4) there may be significant temporal or spatial variation in soil characteristics.

11.3.1.1   Components

Components of soil quality definitions may include desirable and undesirable characteristics. Desirable soil characteristics may either be the presence of a property that benefits soil productivity

Table 11.2   Minimum data set for soil quality analyses [From Kennedy and Papendick, 1995. J. Soil Water Conserv. 50:243-248 with permission of the Soil and Water Conservation Society of America]

Organic matter
Aggregation
Bulk density
Depth to hardpan
Electrical conductivity
Fertility
Respiration
pH
Soil test
Yield
Infiltration
Mineralizable nitrogen potential
Water holding capacity

and/or other important soil functions, or the absence of a property that is detrimental to these functions. A soil characteristic may include a range of values that contributes positively to quality and a range that contributes negatively. Soil pH, for example, may be a positive or negative characteristic depending on its value. Larson and Pierce (1991) suggest that ranges of property values can be defined as optimal, suboptimal or superoptimal. A pH range of 6 to 7.5 is optimal for production of most crops. Outside of this range, pH is suboptimal and soil quality is lower than at the optimal pH range. The complexity of the soil quality concept is illustrated by the fact that the choice of optimal pH range is crop or use dependent.

Letey (1985) suggested that identification of a range of water content that is nonlimiting to plant productivity might be a good way of assessing the collective effect of soil physical characteristics that contribute to crop productivity. For soils of decreasing quality, the width of the nonlimiting water range decreases.

Undesirable soil characteristics may be either the presence of contaminants or a range of values of soil characteristics that contribute negatively to soil quality. The presence of chemicals that inhibit plant root growth or the absence of nutrients that result in low yields or poor crop quality are examples of undesirable soil characteristics that lower soil quality.

11.3.1.2 Renewable Resource
The extent to which soil is viewed as a renewable resource shapes our approach to soil quality. "Soil" in this context is the natural, three-dimensional, horizonated individual, not something created by earth moving machinery. For the purpose of assessing human impact on sustainability of soil quality, it may be appropriate to use only those soil properties that are slowly or nonrenewable. Shorter term assessments may be based on those properties that change rapidly and are subject to easy management.

Willis and Evans (1977) argued that soil is not renewable over the short term based on studies that suggest that 30 to more than 1,000 years are required to develop 25 mm of surface soil from parent material by natural processes. Jenny (1980) also argued that soil is not renewable over the time scale to which humans relate. Howard (1993) suggests defining soil quality based on undisturbed natural soils and to set quality standards based on changes in soils which cannot be reversed naturally or by ecological approaches.

The renewability of soil depends on the soil property considered. For example, once soil depth is reduced by wind or water erosion so that it is too shallow to support crops, it is not renewable within a human or management time frame.

Some important soil characteristics are slowly renewable. Organic matter, most nutrients and some physical properties may be renewed through careful long-term management. Certain chemical properties (pH, salinity, N, P, K content) may be altered to a more satisfactory range for agriculture within a growing season or two, while removal of unwanted chemicals may take much longer.

11.3.1.3 Temporal Variation

No soil property is permanent, but rates and frequency of change vary widely among properties. Soil properties also vary with ecosystem, arguably depending most on climate. In rangelands, for example, temporal variability is high and relatively unpredictable due to the strong dependence of soil properties on soil wetness (Herrick and Whitford, 1995). Variability in soil wetness is not restricted to rangelands and may be an especially important determinant of microbial community structure and function in both irrigated and rainfed agricultural systems.

Arnold et al. (1990) suggest that changes in soil properties can be nonsystematic, periodic, or trend. Nonsystematic changes are short term and unpredictable. Periodic are predictable and trend changes tend to be in one direction over time. Carter et al. (1997) distinguish between dynamic soil properties that are most subject to change through human use and are strongly influenced by agronomic practices, and intrinsic or static properties that are not subject to rapid change or management.

Examples of dynamic soil characteristics are the size, membership, distribution, and activity of a soil's microbiological community; the soil solution composition, pH, and nutrient ion concentrations, and the exchangeable cation population. Soils respond quickly to changes in conditions such as water content. As a result, the optimal frequency and distribution of soil measurements vary with the property being measured.

Soil mineralogy, particle size distribution and soil depth are static soil quality indicators. Although changes occur continuously, they are slow under natural conditions. Organic matter content may be a dynamic variable, but the chemical properties of organic matter may change only over periods on the order of 100 to 1,500 years depending on texture.

Soil properties that change quickly present a problem because many measurements are needed to know the average value and to determine if changes in the average indicate improvement or degradation of soil quality. Conversely, properties that change very slowly are insensitive measures of short-term changes in soil quality. Papendick et al. (1995) argue that the MDS required for soil quality analysis includes a mix of "dynamic" and relatively "static" properties.

11.3.1.4 Spatial Variation

A soil quality assessment must specify area. One could use the pedon (the three-dimensional soil individual) as the unit of measure, or a soil map unit, a landscape, a field or an entire watershed. The choice will depend to some degree on what property is of interest and the spatial variability of the property. Karlen et al. (1997) propose that soil quality can be evaluated at scales ranging from points to regional, national and international (Figure 11.1). They suggest that the more detailed scales provide an opportunity to "understand" soil quality while larger scale approaches provide interdisciplinary monitoring of soil quality and changes in soil quality. Pennock et al. (1994) discuss scaling up data from discrete sampling points to landscape and regional scales.

11.3.2 Examples of Soil Quality Characteristics

Here, we describe some of the physical, chemical, and biological characteristics for assessing soil quality within the context of agricultural production (Table 11.3). Methods for indexing soil quality using these measurements are discussed in Section 11.4. At present, no consensus exists as to

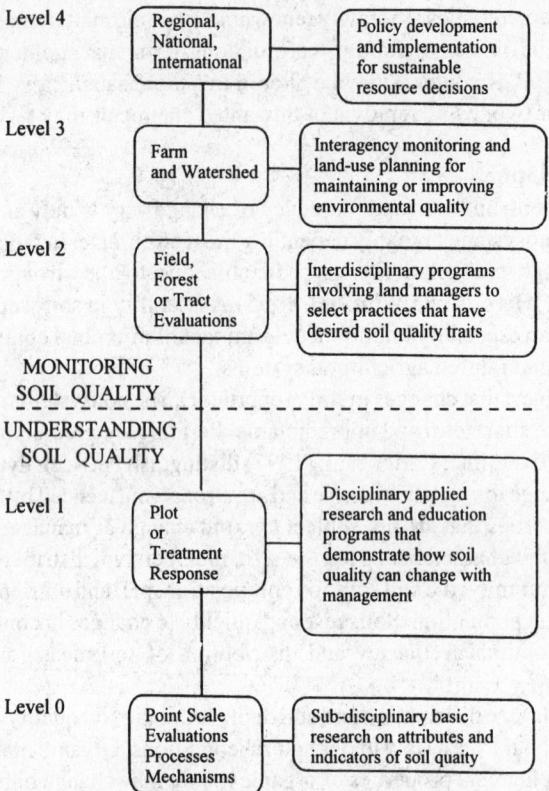

Fig. 11.1   Scales of soil quality assessment [Reprinted from Karlen et al., 1997. Soil Sci. Soc. Am. J. 61:4-10 with permission of the Soil Science Society of America]

thresholds or specific values of each property or the most appropriate combination of dynamic or static properties that describe different levels of soil quality.

11.3.2.1   Physical Characteristics

Soil physical characteristics (Table 11.4) are a necessary part of soil quality assessment because they often cannot be easily improved (Karlen and Cambardella, 1996; Wagenet and Hutson, 1997). Larson and Pierce (1991) summarize the physical indicators of soil quality as those properties that influence crop production by determining: (1) whether a soil can accommodate unobstructed root growth and provide pore space of sufficient size and continuity for root penetration and expansion, (2) the extent to which the soil matrix will resist deformation, and (3) the capacity of soil for water supply and aeration. Factors such as effective rooting depth, porosity or pore size distribution, bulk density, hydraulic conductivity, soil strength and particle size distribution capture these soil functions (Table 11.4).

Reganold and Palmer (1995) use texture, color, dry and moist consistence, structure type, a structure index, bulk density of the 0-5 cm zone, penetration resistance of 0 to 20 and 20 to 40 cm zones and topsoil thickness as physical determinants of soil quality. Letey (1985) suggests that structure, texture, bulk density, and profile characteristics affect management practices in agriculture but are not directly related to plant productivity. He proposes that water potential, oxygen diffusion

Table 11.3 Soil characteristics and soil quality [From Larson and Pierce, 1991 with permission from International Board for Soil Research and Management]

Soil attributes	Elements of soil quality - ability to:				
	accept, hold and release nutrients	accept, hold and release water	promote growth	provide a suitable soil biotic habitat	resist degradation
Surface horizons					
Organic matter					
Total	X	X	X	X	X
Labile	X		X	X	X
nutrient supply	X		X	X	X
Surface soil texture	X	X	X	X	
Surface soil depth	X	X	X		
Surface soil structure		X	X		X
Surface soil pH	X	X	X	X	X
Surface soil electrical conductivity	X		X	X	X
Limiting horizon					
Subsoil texture	X	X	X	X	X
Subsoil depth	X	X	X		X
Subsoil structure		X	X		X
Subsoil pH	X		X		X
Subsoil electrical conductivity	X		X	X	X

rate, temperature, and mechanical resistance directly affect plant growth, and thus are the best indicators of the physical quality of a soil for production.

Soil tilth, a poorly defined term that describes the physical condition of soil, also may be an indicator of a soil's ability to support crops. Farmers may assess soil tilth by kicking a soil clod. More formal measurements to describe soil tilth include bulk density, porosity, structure, roughness and aggregate characteristics (Karlen et al., 1992). Many of the processes that contribute to soil structure, aggregate stability, bulk density and porosity are not well understood, making soil tilth a difficult parameter to quantify.

Soil depth is an easily measured and independent property that provides direct information about a soil's ability to support plants. Effective soil depth is the depth available for roots to explore for water and nutrients. Layers that restrict root growth or water movement include hard rock, naturally dense soil layers such as fragipans, petrocalcic and, petroferric horizons, duripans, and human-induced layers of high bulk density such as plow pans and traffic pans. Effective soil depth is a problem for agricultural use of over 50% of soils in Africa (Eswaran et al., 1997). Soil depth requirements vary with crop or species. Many vegetable crops, for example, are notably shallow rooted while grain crops and some legumes like alfalfa are deep rooted. Variation will be even greater in unmanaged, natural systems. Wheat yield in Colorado was shown to decrease from 2,700 to 1,150 kg ha^{-1} over a 60-yr period of cultivation primarily due to decrease in soil depth (Bowman et al., 1990).

11.3.2.2 Chemical Characteristics

Assessment of soil quality based on soil chemistry, whether the property is a contaminant or part of a healthy system, requires a sampling protocol, a method of chemical analysis, an understanding of how its chemistry affects biological systems and interacts with mineral forms, methods for location of possible contamination, and standards for soil characterization. Some soil chemical properties suggested as soil quality indicators are listed in Table 11.5.

Table 11.4   Physical indicators of soil quality

Aeration
Aggregate stability
Bulk density
Clay mineralogy
Color
Consistence (dry, moist, wet)
Depth to root limiting layer
Hydraulic conductivity
Oxygen diffusion rate
Particle size distribution
Penetration resistence
Pore connectivity
Pore size distribution
Soil strength
Soil tilth
Structure type
Temperature
Total porosity
Water-holding capacity

Nutrient availability depends on soil physical and chemical processes, such as weathering and buffering, and properties such as organic matter content, CEC and pH. At low and high pH, for example, some nutrients become unavailable to plants and some toxic elements become more available. Larson and Pierce (1991) chose those chemical properties that either inhibit root growth or that affect nutrient supply due to the quantity present or the availability. Reganold and Palmer (1995) used chemical parameters related to nutrient availability as measures of soil quality, including CEC, total N and P, pH and extractable P, S, Ca, Mg and K. Karlen et al. (1992) suggest that total and available plant nutrients, and nutrient cycling rates, should be included in soil quality assessments.

Soil properties may be severely compromised by intended or unintended human additions of chemical compounds and soil productivity reduced if unwanted chemicals exceed safe thresholds. Data are required to determine whether or not a site is significantly polluted and if it requires clean-up (Beck et al., 1995; Sims et al., 1997). International standard methods have been created to maintain the quality of measurements (Hortensius and Welling, 1996). A difficult determination is the level of each chemical that is considered an ecological risk. Beck et al. (1995) provide a list of levels for organic chemicals adopted by The Netherlands and Canada. EPA uses similar lists for compounds considered hazardous (e.g., 40 CFR). Sims et al. (1997) argue that clean and unclean are two extremes of a continuum and that it is more appropriate to define the physical, chemical and biological state of the soil as acceptable or unacceptable.

In The Netherlands, soil quality reference values have been created for heavy metals and organic chemicals based on a linear relationship with soil clay and organic matter content. The Dutch Ministry of Housing, Physical Planning and Environment has used the maximum of a range of reference values for a given substance as a provisional reference value for good soil quality (Howard, 1993; de Bruijn and de Walle, 1988).

11.3.2.3   Biological Characteristics

The focus of many soil quality definitions is soil biology (Table 11.6). Soil supports a diverse population of organisms, ranging in size from viruses to large mammals, that usually interacts positively with plants and other system components (Berry, 1994; Paul and Clark, 1996). However,

Table 11.5   Chemical indicators of soil quality

Base saturation percentage
Cation exchange capacity
Contaminant availability
Contaminant concentration
Contaminant mobility
Contaminant presence
Electrical conductivity
ESP
Nutrient cycling rates
pH
Plant nutrient availability
Plant nutrient content
SAR

some soil organisms such as nematodes, bacterial and fungal pathogens reduce plant productivity. Many proposed soil quality definitions focus on the presence of beneficial rather than absence of detrimental organisms, although both are critically important. Various measures of microbial community viability have been suggested as measures or indices of soil quality. Community level studies consider species diversity and frequency of occurrence of species.

Visser and Parkinson (1992) found that diverse soil microbiological criteria may be used to indicate deteriorating or improving soil quality. They suggested testing the biological criteria for soil quality at three levels: population, community and ecosystem. Microorganisms and microbial communities are dynamic and diverse, making them sensitive to changes in soil conditions (Kennedy and Papendick, 1995). Their populations include fungi, bacteria including actinomycetes, protozoa, and algae. Soil microorganisms form crucial symbiotic relationships with plants, including mycorrhizal infection for P and N acquisition and bacterial infection for fixation of atmospheric N.

Authors emphasizing use of biological factors as indicators of soil quality often equate soil quality with relatively dynamic properties such as microbial biomass, microbial respiration, organic matter mineralization and denitrification, and organic matter content (Karlen et al., 1992; Visser and Parkinson, 1992; Reganold et al., 1993; Franzluebbers et al., 1995; Yakovchenko et al., 1996; Boehn and Anderson, 1997; Franzluebbers and Arshad, 1997), or soil microbial C, phospholipid analyses and soil enzymes (Franzluebbers et al., 1995; Jordan et al., 1995; Gregorich et al., 1997), or total organic C and N (Karlen et al., 1992; Franzluebbers and Arshad, 1995; Franco-Vizcaino, 1997; Gregorich et al., 1997). Visser and Parkinson (1992) question the suitability of enzyme assays for microbial activity and soil quality assessments. Waksman (1927), who studied measurements of soil microorganisms that could indicate soil fertility, found that physical and chemical factors as well as soil biology were needed to predict soil fertility.

Meso- and macrofauna populations have also been considered as part of soil quality definitions (Berry, 1994). One could choose to use presence or absence of a particular species or population of a particular species as a measure of soil quality. Stork and Eggleton (1992) discuss species richness as a powerful indicator of invertebrate community and soil quality, although determining the number of species is a problem. They suggest that keystone species, taxonomic diversity at the group level, and species richness of several dominant groups of invertebrates can be used as part of a soil quality definition.

Measuring soil fauna populations involves decisions about which organisms to measure and how to measure them. An example is the earthworm population, the size of which is frequently mentioned as an important measure of soil quality. Measurement choices include numbers of organisms per

Table 11.6 Microbial indicators of soil quality [From Kennedy and Papendick, 1995. J. Soil Water Conserv. 50:243-248 with permission of the Soil and Water Conservation Society of America]

Organic carbon
Microbial biomass
C and N
Total biomass
Bacterial
Fungal
Potentially mineralizable N
Soil respiration
Enzymes
Dehydrogenase
Phosphatase
Arlysulfatase
Biomass C/total organic carbon
Respiration/biomass
Microbial community fingerprinting
Substrate utilization
Fatty acid analysis
Nucleic acid analysis

volume or weight of soil, number of species, or a combination of numbers of organisms and species. Reganold and Palmer (1995) use total earthworms per square meter, total earthworm weight (g m^{-2}) and average individual earthworm weight as biological indicators of soil quality.

Measurement of one or more components of the N cycle including ammonification, nitrification and nitrogen fixation, may be used to assess soil fertility and soil quality (Visser and Parkinson, 1992). Presumably, high rates of N turnover may infer a dynamic and healthy soil biological community. In contrast, low soil quality or poor soil health may be inferred from lack of N turnover. The interpretation of N turnover rates is highly dependent on the kinds of substrates added to soils and climate variables such as soil temperature and moisture. One needs to be careful when comparing N turnover rates within soils and among different soils to be sure that the cause of differences is a soil quality parameter and not natural variability. Presence of pesticide residues, for example, may reduce N turnover rate. In such an instance, both the presence of the pesticide and the N turnover rate would be needed to determine that the soil quality had been impaired.

11.3.3   Relating Soil Quality to Soil Function

Of the interrelated and broadly defined soil functions listed in Section 11.2.1, agricultural production and environmental buffering are considered here. Production incorporates use of and need for functioning soil resources in agriculture, and environmental buffering incorporates the direct and indirect effects of human use on ecosystem function and human health.

11.3.3.1   Agricultural Production

Worldwide agriculture is the most extensive human land use, and soil characteristics are a critical determinant of agricultural productivity. Agriculture includes irrigated and rainfed cultivated cropland, permanent crops such as orchards and vineyards, irrigated pasture, range, and forestry. Each cropping system has distinct soil and soil management conditions for optimal production. It has been suggested that soil productivity is the net resultant of soil degradation processes and soil conservation practices (Parr et al., 1990). An appropriate definition of soil quality and the criteria necessary to evaluate and monitor soil quality is a step toward "the development of systematic criteria of sustainability" (NRC 1993). Issues to be considered when discussing soil quality for agriculture include (1) How are productivity and sustainability related? (2) Is the cropping system in question cultivated or non-cultivated? (3) Is the cropping system in question an irrigated or dryland system? Sustainability of agricultural systems is critical to human welfare and is an a subject of research and debate (Letey, 1994). High productivity and sustainability must be converging goals if the growing human population is to be fed without destroying the resources necessary to produce food. Sustainability implies that a system is at a desirable steady state. Thermodynamically, soil is an open system through which matter and energy flow and a steady state is characterized by a minimum production of entropy (Andiscott, 1995). Ellert et al. (1997) review related literature on ways of assessing soil function on an ecosystem scale, commenting that the complexity and organization of living systems, which seem to defy the second law of thermodynamics (increasing disorder/entropy), may provide a means to broadly assess ecosystem function.

The purpose of agriculture is to provide products for human sustenance and by definition is not sustainable unless the nutrients removed in the products are returned to the soil. Many of the arguments about the sustainability of agricultural systems relate to the form in which nutrients are most sustainably returned. No agricultural system will be sustainable in the long run without management that considers nutrient cycling and energy budgets. The more intense the agricultural system, the more energy and resources must be expended to maintain the system.

The relative quality of a soil for agriculture can depend on the resources available to farmers. In the United States, resources may be readily available for management of dynamic soil properties such as nutrient or water status. In other countries, farmers may be resource poor, and agricultural systems are generally low input, meaning that large-scale irrigation is absent, use of fertilizers, pesticides, and herbicides is minimal, and high energy, mechanized equipment is not available (Eswaran et al., 1997). This means, for example, that soil quality for agriculture will be more dependent on climate than if the same soils were part of a highly managed, irrigated system (Eswaran et al., 1997). Similarly, sustainability is more dependent on maintenance of dynamic soil properties because resources may not exist to remedy losses.

It is difficult to overstate the importance of irrigation to food production. One-third of the total global harvest of food comes from the 17% (250 million ha) of the world's cropland that is irrigated (Hoffman et al., 1990; Gleick, 1993); three-quarters of which are in developing countries (Tribe, 1994). India, China, the former Soviet Union, the United States and Pakistan have the greatest area of irrigated land. Should soil quality criteria be the same for irrigated and dryland agriculture? Sojka (1996) suggests that the arid and semi-arid soils that support most irrigated agriculture have thin erodible surfaces, characteristics that would classify such soils as having poor quality. Yet under irrigation, they feed much of the world. Without irrigation, for example, in many African soils, moisture stress becomes a significant factor limiting production, and the water-holding capacity of a soil becomes crucial (Eswaran et al., 1997). This suggests that a standard set of criteria based on potential productivity is not a sufficient definition of soil quality.

Soils that are not cultivated are a much larger component of agriculture, broadly defined, than those that are cultivated. About 65% of the land in the United States is forest (284 million ha) or range land (312 million ha), with only about 284 million ha cultivated (NRC, 1994). Herrick and Whitford (1995) suggest that range land soils, which often serve multiple uses, present unique challenges and opportunities for assessing soil quality because spatial and temporal variability are higher than in cropped systems.

On range lands and forest lands, food, fiber, timber production, biomass for fuel, wildlife, biodiversity, recreation, and water supply are all potential uses that may have diverse criteria for quality soils. Herrick and Whitford (1995) give the example of a thick O horizon that may be an indicator of good timber production but has no predictive value of soil quality for the rancher.

The National Research Council (NRC, 1994) recommends that range land health be determined using three criteria: degree of soil stability and watershed function, integrity of nutrient cycles and energy flows, and presence of functioning recovery mechanisms. Soil erosion by wind and water and infiltration or capture of precipitation were selected as processes that could be used as indicators of soil stability and watershed function. Specific indicators or properties need to be related to these two broad processes. The amount of nutrients available, the speed with which nutrients cycle, and measures of the integrity of energy flow through the system were considered fundamental components of range land health. Finally, the capacity of range land ecosystems to react to change depends on recovery mechanisms that result in capture and cycling of nutrients, capture of energy, conservation of nutrients, energy and water, and resilience to change. Specific indicators include status of vegetation, age class and distribution.

The evaluation of land quality for forestry is a well-known practice. Indices range from quantitative through semi-quantitative to qualitative. Quantitative evaluations, such as site index, use regression equations to predict tree height at a predetermined tree age based on soil and climate data. Qualitative evaluations assign land to classes based on soil and climate properties.

11.3.3.2 Environmental Buffering

In soil science, the term "buffer" refers collectively to processes that constrain shifts in the dissolved concentration of any ion when it is added to or removed from the soil system (Singer and Munns, 1996). Soils "buffer" nutrients as well as contaminants and other solutes, via sorption to or incorporation into clay and organic materials. The extent to which a soil immobilizes or chemically alters substances that are toxic, thus effectively detoxifying them, reflects "quality" in the sense that humans or other biological components of the system are protected from harm. This is the basis for the European concept of soil quality (Moen, 1988; Siegrist, 1989; Denneman and Robberse, 1990). Lack of soil function in this category is reflected as direct toxicity or as contamination of air or water. Identifying substances that qualify as "contaminants" can be challenging because some, such as nitrates and phosphates, are important plant nutrients as well as potential water pollutants. An example is agricultural runoff containing NO_3 or soluble P (Yli-Halla et al., 1995). This chapter does not attempt a comprehensive review of research in this area, which is covered in an earlier chapter, but instead presents a few sample articles pertinent to this aspect of soil quality.

Holden and Firestone (1997) define soil quality in this context as "the degree to which the physical, chemical, and biological characteristics of the soil serve to attenuate environmental pollution." Howard (1993) defines the ecological risk of a chemical in the environment as "the probability that a random species in a large community is exposed to a concentration of the chemical greater than its no-effect level." The extent to which a soil is capable of reducing the probability of exposure is a measure of its quality.

A well-studied example of a common soil contaminant is Pb (McBride et al., 1997). Although legislated limits may be on a concentration basis in soil (e.g., 500 µg kg^{-1}), risk assessment techniques have attempted to account for the chemical form of Pb present, as well as the observed relative relationship between the amount of Pb present in soil and blood levels in local residents (Bowers and Gauthier, 1994). Critics have questioned analytical techniques used to determine bioavailable levels of Pb in soil, as well as the degree to which toxicity data account for its chemical fate and ecologically damaging properties (Cook and Hendershot, 1996). Natural variability of soils and variation within a soil series make average values or average background values inadequate for soil quality assessments.

In addition, bioaccumulation and toxicity need to be considered when establishing levels of toxicants that may not be exceeded in a "high quality" soil for a given use (Traas et al. 1996; Jongbloed et al., 1996).

Another example is the effect of heavy metals such as Cr(VI) on soil biological properties. Based on a study of three New Zealand soils of contrasting texture, organic matter content, and CEC, Speir et al. (1995) propose an "ecological dose value" that represents the inhibitory effects of a heavy metal (in this case, Cr(VI)) on the kinetics of soil biological properties, and serves as a generic index for determination of permissible concentration levels for heavy metals in soils.

11.4 Indices of Soil Quality

A single soil characteristic is of limited use in evaluating differences in soil quality (Reganold and Palmer, 1995). Using more than one quantitative variable requires some system for combining the measurements into a useful index (Halvorson et al., 1996). The region, crop, or general soil use for which an index was created will likely limit its effectiveness outside the scope of its intended application. Even an index designed only to rate productivity is not likely to be useful for all crops and soils, leading Gersmehl and Brown (1990) to advocate regionally targeted systems.

Rice is a good example of a crop requiring significantly different soil properties than other crops. It is a food staple for a large proportion of the world population. Approximately 146 million ha were in rice production in 1989 (FAO, 1989) mainly (90%) in Asia. High quality soils for paddy rice may be poor quality for most other irrigated and dryland crops because they may be saline or sodic, and high in clay with slow infiltration and permeability. These physical and chemical properties often constrain production of other crops. Although they are not reviewed here, various land suitability classifications specifically for rice have been developed since the turn of the century (Dent, 1978).

Examples of several soil quality indexing systems are presented in the following sections. To some extent, recent attempts to enumerate the factors of soil quality resemble Jenny's (1941) introduction of the interrelated factors of soil formation.

11.4.1 Nonquantitative Systems

An index is categorized here as nonquantitative if it does not combine evaluated parameters into a numerical index that rates soils along a continuous scale. Examples are the USDA Land Capability Classification and the US Bureau of Reclamation (USBR) Irrigation Suitability.

11.4.1.1 USDA Land Capability Class

The purpose of the Land Capability Classification (LCC) was to place arable soils into groups based on their ability to sustain common cultivated crops that do not require specialized site conditioning or treatment (Klingebiel and Montgomery, 1973). Nonarable soils, unsuitable for long-term, sustained cultivation, are grouped according to their ability to support permanent vegetation, and according to the risk of soil damage if mismanaged.

The LCC combines three rating values at different levels of abstraction: capability class, subclass, and unit. At the most general level, soils are placed in eight classes according to whether they (a) are capable of producing adapted plants under good management (classes I to IV), (b) are capable of producing specialized crops under highly intensive management involving "elaborate practices for soil and water conservation" (classes V to VII), or (c) do not return on-site benefits as a result of management inputs for crops, grasses or trees without major reclamation (Klingebiel and Montgomery, 1973). The four possible limitations/hazards under the subclass rating are erosion hazard, wetness, rooting zone limitations and climate. The capability unit groups soils that have about the same responses to systems of management and have longtime estimated yields that do not vary by more than 25% under comparable management.

An example of the criteria used in California for placing soils into the eight capability classes, as well as the values of the variables that define the capability classes is shown in Table 11.7. The issue of critical limits is a difficult one in soils because of the range of potential uses and the interactions among variables (Arshad and Coen, 1992).

Several studies have shown that lands of higher LCC have higher productivity than lands of lower LCC (Patterson and Mackintosh, 1976; van Vliet et al., 1979; Reganold and Singer, 1984). In a study of 744 alfalfa, corn, cotton, sugar beet and wheat growing fields in the San Joaquin Valley of California, those with LCC ratings between 1 and 3 had significantly lower input/output ratios than fields with ratings between 3.01 and 6 (Reganold and Singer, 1984). This suggests that use of the LCC system provides an economically meaningful assessment of soil quality for agriculture.

11.4.1.2 USBR Irrigation Suitability

This was a frequently used system of land evaluation for irrigation in the Western US during the period of rapid expansion of water delivery systems (USBR, 1953; Maletic and Hutchings, 1967; FAO, 1979; McRae and Burnham, 1981). It combines social and economic evaluations of the land with soil and other ecological variables to determine if the land has the productive capacity, once irrigated, to repay the investment necessary to bring water to an area. It recognizes the unique importance of irrigation to agriculture and the special qualities of soils that make them irrigable.

11.4.2 Quantitative Systems

Quantitative systems result in a numerical index, typically with the highest number being assigned to the best quality soils. Systems may be additive, multiplicative or more complex functions. They have two important advantages over nonquantitative systems: (1) they are easier to use with GIS and other automated data retrieval and display systems, and (2) they typically provide a continuous scale of assessment. No single national system is presently in use but several state or regional systems exist.

11.4.2.1 Storie Index Rating (SIR)

Although he considered the productivity of the land to be dependent on 32 soil, climate and vegetative properties (Table 11.8), only nine properties were used in the SIR, because incorporating a greater number of factors made the system unwieldy. The nine factors are soil morphology (A), surface texture (B), slope (C), and six variables (X_i) that rate drainage class, sodicity, acidity, erosion, microrelief and fertility; rated from 1% to 100%. These are converted to their decimal value and multiplied together (Storie, 1932; Wier and Storie, 1936; Storie, 1964) (Equation [11.1]):

$$SIR = \left(A \times B \times C \times \prod_{i=1}^{6} Xi \right) \times 100 \qquad [11.1]$$

Values for each factor were derived from Storie's experience mapping and evaluating soils in California, and in soil productivity studies in cooperation with the California Agricultural Experiment Station cost-efficiency projects relating to orchard crops, grapes and cotton. In describing the SIR, Storie (1932, 1964) explicitly mentioned "soil quality". Soils that were deep, had no restricting subsoil horizons, and held water well had the greatest potential for the widest range of crops.

Table 11.7 Land capability classes in California [USDA-NRCS]

Capability class	Effective soil depth (inches)[1]	Climate Thornthwaite 1948 indices (inches) Irr. ET, 32 F	Climate Dry 4ETa	Surface layer texture Irrigated	Surface layer texture Dryland	Permeability[2]	Drainage Class[3]	Available water capacity[4]	Slope A[5,6,7]	Slope B[5,6,7]	Erosion hazard	Flooding hazard	Salinity EC x 10 °C[8]	Sodicity ESP[8]	Toxic substances[9]	Frost free season
I	≥ 40	≥ 20	≥ 20	Sandy loam thru clay loam	Sandy loam thru clay loam	Mod. Rapid thru mod. Slow	Well or mod. Well and > 60"	≥ 7.5 in. or Ave. AWC ≥ 0.13 In/In.	≤ 2%	≤ 2%	None or slight	None or rare	< 4 mmhos (none)	None	None	≥ 140 days
II	≥ 40	≥ 14	≥ 16	Loamy sand thru clay, may be gravelly	Sandy loams thru clays, may be gravelly	Rapid thru slow	Somewhat poorly thru somewhat excessively and > 36"	≥ 5.0 in. or Ave. AWC ≥ 0.08 In/In.	≤ 5%	≤ 8%	None thru mod.	None thru occasional	< 8 mmhos	< 25	None or slight	≥ 100 days
III	≥ 20	≥ 10	≥ 12	Any, may be gravelly or cobbly	Sandy loams thru clays, may be gravelly or cobbly	Rapid thru slow	Poorly thru excessively and > 20"	≥ 3.5 in. or Ave. AWC ≥ 0.06 In/In.	≤ 8%		None thru 15% high	None thru occasional	< 16 mmhos	< 50	None thru mod	≥ 80 days
IV	≥ 10	≥ 6	≥ 8	Any, may be very gravelly, very cobbly or stony[10]	Loamy sand thru clay, very gravelly, very cobbly or stony[10]	Any	Poorly thru excessively and > 20"	≥ 2.5 in. or Ave. AWC ≥ 0.04 In/In.	< 15%	25%	Any	None thru frequent[11]	< 16 mmhos	< 50	None thru mod.	≥ 50 days
V	≥ 20	≥ 6	≥ 8	Any, may be ext. gravelly, ext. cobbly or very stony	Any, may be ext. gravelly, ext. cobbly or very stony	Any	Any	≥ 3.0 in.	< 2%	< 2%	None or slight	Any	< 8 mmhos	< 25	None or slight	Any
VI[12]	≥ 10	≥ 4	≥ 6	Any, may be ext. gravelly, ext. cobbly or very stony	Any, may be ext. gravelly, ext. cobbly or very stony	Any	Any	≥ 2.0 in.	< 25%	50%	Any	Any	Dry land < 16 mmhos Irr. any	Dry land < 25 Irr. < 50	Dry land slight Irr. Slight thru mod.	Any
VII[13]	Any	---	≥ 2	Any	Any	Any	Any	≥ 1.0 in.	< 50%	< 75%	Any	Any	Any	Any	Any	Any
VIII[14]	Any	Any	Any	Any	Any	Any	Any	Any	Any	Any	Any	Any	Any	Any	Any	Any

[1] Clay pans with permeabilities less than 0.06 in/hr will be treated as limiting the effective depth
[2] Permeability of the least permeable subsurface horizon
[3] Depth to water table during growing season
[4] Available moisture between field capacity and wilting point
[5] Use erosion hazard to help determine upper slope percent
[6] In existing mapping units 9% and 30% can be substituted for 8% and 25%
[7] Column A is used for soils with K factors of 0.37 or greater, soils subject to rill and gully erosion, such as soils formed from granitic parent material or with clay pans; other soils are in group B

[8] For salts and sodicity to be a major limitation, there should be other soil limitations such as slow permeabilities or high water tables
[9] Such as boron and magnesium that leach with difficulty
[10] Coarse fragments interfere with tillage, but do not prevent cropping
[11] Frequent flooding that does not prevent normal cropping
[12] Range and woodland mechanical practices can be aplied to class VI land
[13] Range and woodland mechanical practices are impractical on class VII land
[14] Class VIII lands have limitations that preclude their use for commercial plant production and restrict their use to recreation, water supply or aesthetic purposes

Table 11.8   Soil and land properties for land evaluation [Storie, 1964]

Property Group	Individual Property
Surface conditions	Physiographic position
	Slope
	Microrelief
	Erosion deposition
	External drainage, runoff
Soil physical conditions	Soil color
	Soil depth
	Soil density and porosity
	Soil permeability
	Soil texture
	Stoniness
	Soil structure
	Soil workability-consistence
	Internal drainage
	Water-holding capacity
	Plant-available water
Soil chemical conditions	Organic matter
	Nitrogen
	Reaction
	Calcium carbonate, bases
	Exchange capacity
	Salts: Cl, SO_4, Na
	Toxicities, e.g., B
	Available P
	Available K
	Minor elements, e.g., Zn, Fe
	Fertility
Mineralogical conditions	Mineralogy
Climate	Precipitation
	Temperature
	Growing season
	Winds
Vegetative cover	Natural vegetation

The usefulness of the SIR as a soil quality index would be greatest if there was a statistically significant relationship between SIR values and an economic indicator of land value. Reganold and Singer (1984) found that area-weighted average SIR values between 60 and 100 for 744 fields in the San Joaquin Valley of California had lower but statistically insignificant input/output ratios than fields with indices < 60. The lack of statistical significance does not mean that better quality lands could not be farmed at economically lower cost or at higher cost and higher output than the lower quality lands.

11.4.2.2   Productivity Index (PI)

The productivity index model (PI) was developed to evaluate soil productivity in the top 100 cm, especially with reference to potential productivity loss due to soil erosion (Neill, 1979; Kiniry et al., 1983; Pierce et al., 1983). The PI model rates soils on the sufficiency for root growth based on potential available water storage capacity, bulk density, aeration, pH, and electrical conductivity. A value from zero to one is assigned to each property describing the importance of that parameter for root development. The product of these five index values is used to describe the fractional sufficiency of any soil layer for root development.

Pierce et al. (1983) modified the PI to include the assumption that nutrients were not limiting and that climate, management and plant differences are constant. A number of authors found that it is useful to various degrees (Pierce et al., 1984; Gantzer and McCarty, 1987; Lindstrom et al., 1992).

11.4.2.3 Soil Quality Index (SQI)
Parr et al. (1992) suggest that a SQI could take the form of Equation [11.2]:

$$SQI = f(SP, P, E, H, ER, BD, FQ, MI) \qquad [11.2]$$

where SQI is a function of soil properties (SP), potential productivity (P), environmental factors (E), human and animal health (H), erodibility (ER), biological diversity (BD), food quality and safety (FQ) and management inputs (MI). Determination of the specific measurable indicators of each variable and the interactions among these diverse variables is a daunting task. Moreover, the mathematical method of combining these factors, as well as the resulting value that would indicate a high quality soil, is not specified. The inclusion of variables BD, FQ and MI make this a land quality index as suggested by FAO (1997).

11.4.2.4 Soil Quality (Q)
Larson and Pierce (1991) defined soil quality (Q) as the state of existence of soil relative to a standard or in terms of a degree of excellence. They argue that defining Q in terms of productivity is too limiting and does not serve us well. Rather, Q is defined as the sum of individual soil qualities q_i and expressed as Equation [11.3]:

$$Q = f(q_i \cdots q_n) \qquad [11.3]$$

These authors do not identify the best subset of properties or their functional and quantitative relationship, but do suggest that a MDS should be selected from those soil characteristics in which changes are measurable and relatively rapid (i.e., "dynamic" properties), arguing that it is more important to know about changes in soil quality (dQ) than the magnitude of Q (Larson and Pierce, 1991). Changes in soil quality are a function of changes in soil characteristics (q_i) over time (t):

$$dQ = f\left[\left(q_{i,t} - q_{i,t_0}\right) \cdots \left(q_{n,t} - q_{n,t_0}\right)\right] \qquad [11.4]$$

If dQ/dt is ≥ 0, the soil or ecosystem is improving relative to the standard at time t_0. If dQ/dt <0, soil degradation is occurring. Time zero can be selected to meet management needs or goals. If there is a drastic change in management, time zero can be defined as prior to the change. If a longer time period of comparison is considered more appropriate, properties of an uncultivated or pristine soil could be used.

The MDS recommended by Larson and Pierce (1991) includes N mineralization potential or P buffering capacity, total organic C, labile organic C, texture, plant-available water capacity, structure (bulk density is recommended as a surrogate variable), strength, maximum rooting depth, pH and EC. In instances when data are unavailable, pedotransfer functions (Bouma, 1989a,b) can be used to estimate values of soil characteristics. These estimates can then be used as part of the minimum dataset to estimate soil quality or changes in soil quality brought about by management.

Although this is a quantitative system, some qualitative judgments are needed to make decisions about changes in soil quality. In particular, interpretation of the meaning of magnitude of changes in a characteristic or the number of characteristics to change from time zero to the time of the measurement is qualitative. The authors do not address how large a change in pH, soil depth, bulk density or organic C represents serious soil degradation, or the values that define soil as high or low quality.

11.4.2.5 Soil Quality Index (QI)

Karlen et al. (1994) developed QI based on a 10-year crop residue management study. QI is based on four soil functions: (1) accommodating water entry, (2) retaining and supplying water to plants, (3) resisting degradation, and (4) supporting plant growth. Numerous properties were measured and values normalized based on standard scoring functions. One function is based on the concept that more of a property is better, one that less is better and the third that an optimum is better. Lower threshold values receive a score of zero, upper threshold values receive a score of one, and baseline values receive a score of one-half.

Priorities are then assigned to each value. For example, aggregate stability was given the highest weight among factors important in water entry. After normalizing, each value is then multiplied by its weighting factor (wt) and products are summed (Equation [11.5]):

$$QI = q_{we}(wt) + q_{wt}(wt) + q_{rd}(wt) + q_{spg}(wt)$$ [11.5]

Subscripts refer to the four main functions described earlier. It should also be noted that resisting degradation (rd) and sustaining plant growth (spg) are assigned secondary and tertiary levels of properties that themselves are normalized and weighted before a final value is calculated and incorporated into Equation [11.5].

The resulting index resulted in values between zero and one. Of the three systems in the study, the one with the highest rate of organic matter return to the soil had the highest index value, and the soil with the lowest had the lowest value. The authors suggest that this demonstrates the usefulness of the index for monitoring the status and change in status of a soil as a function of management. They also suggest that the index and the soil characteristics that go into the index may change as the index is refined (Karlen et al. 1994).

11.4.2.6 Other Indices

Snakin et al. (1996) developed an index of soil degradation that assigns three separate values from one to five reflecting the degree to which a soil's physical, chemical, and biological properties are degraded, as well as the rate of degradation.

The Canadian soil capability classification system is similar to the older US systems and is quantitative. In a study in southwestern Ontario, Patterson and Mackintosh (1976) found that high gross returns per ha were three times as likely if the productivity index of land, based on the soil capability classification, was between 90 and 100 than if it fell between 80 and 89.

Smith et al. (1993) and Halvorson et al. (1996) propose a multiple-variable indicator transform procedure to combine values or ranges of values that represent the best estimate of soil quality. Their system converts measured data values into a single value according to specified criteria. They do not attempt to define soil quality or specify what soil characteristics are to be used. They combine this procedure with kriging to develop maps that indicate the probabilities of meeting a soil quality criterion on a landscape level. Critical threshold values must be known, assumed, or determined in order to separate different soil qualities.

Numerous additive productivity rating systems have been developed for specific states, as reviewed by Huddleston (1984). In these systems, soil properties are assigned numerical values according to their expected impact on plant growth. The index is usually calculated as the sum of the values assigned to each property with 100 the maximum value. Huddleston (1984) notes advantages and disadvantages to such a system which are similar to those for many of the soil quality indices previously discussed. Additive systems become complex as the number of factors, cropping systems, and soil and climatic conditions increases. A unique problem of subtractive systems (one in which 100 is the starting point and values are deducted for problem conditions) is that negative values result when multiple factors are less than satisfactory.

11.5 Conclusions

Soil quality is a concept being developed to characterize the usefulness and health of soils, because soils are fundamental to the well-being and productivity of agricultural and natural ecosystems. It is a compound characteristic that cannot be directly measured. Many definitions of soil quality can be found in the literature and no set of soil characteristics has been universally adopted to quantify definitions. Soil quality is often equated with agricultural productivity and sustainability. An approach toward developing soil quality definitions is one that assesses soil quality in the context of a soil's potential to perform given functions in a system; e.g., maintains productivity, partitions and regulates water and solute flow through an ecosystem, serves as an environmental buffer, and cycles nutrients, water, and energy through the biosphere. Air and water quality standards are usually based on maximum allowable concentrations of materials hazardous to human health. A definition of soil quality based on this concept would encompass only a fraction of the important roles soils play in agriculture and the environment but could be essential to soil remediation.

To proceed from a definition to a measure of soil quality, a minimum dataset of soil characteristics that represent soil quality must be selected and quantified. Many soil physical, chemical and biological properties have been suggested to separate soils of different quality. These include desirable and undesirable properties. Desirable soil characteristics may either be the presence of a property that benefits crop productivity and environmental buffering and/or other important soil functions, or the absence of a property that is detrimental to these functions. In particular, absence of contaminants is an important soil quality characteristic. In selecting characteristics, it is necessary to recognize that some soil properties are static, in the sense that they change slowly over time and others are dynamic. In addition, spatial and temporal variability of soil properties must be considered when selecting the properties used to assess soil quality.

A single soil property is of limited use in evaluating soil quality. Qualitative and quantitative soil quality indices have been suggested that combine quantitative values of soil properties. Quantitative systems may be additive, multiplicative or more complex functions. Regardless of the definition or suite of soil variables chosen to define and quantify soil quality, it is critical to human welfare that soils be managed to provide for human health and well-being while minimizing soil and environmental degradation.

11.6 References

Anderson, D.W., and E.G. Gregorich. 1984. Effect of soil erosion on soil quality and productivity. p. 105-113. *In* Soil erosion and degradation. Proc. 2nd Ann. Western Prov. Conf. Rational. Water Soil Res. Manag. Sask., Saskatoon, Canada.

Andiscott, T.M. 1995. Entropy and sustainability. Europ. J. Soil Sci. 46:161-168.

Arnold, R.W., I. Zaboles, and V.C. Targulian (ed.). 1990. Global soil change. Report of an IIASA-ISSS-UNEP task force on the role of soil in global change. International Institute for Applied Systems Analysis, Laxanberg, Austria.

Arshad, M.A., and G.M. Coen. 1992. Characterization of soil quality: Physical and chemical criteria. Am. J. Altern. Agr. 7:25-31.

Beck, A.J., S.C. Wilson, R.E. Alcock, and K.C. Jones. 1995. Kinetic constraints on the loss of organic chemicals from contaminated soils: Implications for soil-quality limits. Critical Rev. Environ. Sci. Technol. 25:1-43.

Beke, G.J., H.H. Janzen, and T. Entz. 1994. Salinity and nutrient distribution in soil profiles of long-term crop rotations. Can. J. Soil Sci. 74:229-234.

Berry, E.C. 1994. Earthworms and other fauna in the soil. p. 61-90. *In* J.L. Hatfield and B .A. Stewart (ed.) Soil biology: effects on soil quality. Lewis Publishers, Boca Raton, FL.

Biederbeck, V.O., C.A. Campbell, H.U. Krainetz, D. Curtain, and O.T. Bouman. 1996. Soil microbial and biochemical properties after ten years of fertilization with urea and anhydrous ammonia. Can. J. Soil Sci. 76:7-14.

Boehn, M.M., and D.W. Anderson. 1997. A landscape-scale study of soil quality in three prairie farming systems. Soil Sci. Soc. Am. J. 61:1147-1159.

Bouma, J. 1989a. Using soil survey data for quantitative land evaluation. Adv. Soil Sci. 9:177-213.

Bouma, J. 1989b. Land qualities in space and time. p. 3-13. *In* J. Bouma and A.K. Bregt (ed.) Land qualities in space and time. Pudoc, Wageningen, Netherlands.

Bouma, J. 1997. Soil environmental quality: A European perspective. J. Environ. Qual. 26:26-31.

Bouma, J., and A.K. Bregt (ed.). 1989. Land qualities in space and time. Pudoc, Wageningen, Netherlands.

Bowers, T.S., and T.D. Gauthier. 1994. Use of the output of a lead risk assessment model to establish soil lead cleanup levels. Environ. Geochem. Health 16:191-196.

Bowman, R.A., J.D. Reeder, and G.E. Schuman. 1990. Evaluation of selected soil physical, chemical and biological parameters as indicators of soil productivity. Proc. Int. Conf. on Soil Quality in Semi-arid Ag. 2:64-70. Univ. of Saskatchewan, Saskatoon, Canada.

Bremer, E., H.H. Janzen, and A.M. Johnston. 1994. Sensitivity of total, light fraction and mineralizable organic matter to management practices in a Lethbridge soil. Can. J. Soil Sci. 74:131-138.

Brinkman, R. 1997. Land quality indicators: aspects of land use, land, soil and plant nutrients. p. 95-104. *In* Land quality indicators and their use in sustainable agriculture and rural development. FAO Land and Water Bull. 5. Rome, IT.

Budd, W.W. 1992. What capacity the land? J. Soil Water Conserv. 47:28-31.

Cairns, J. Jr. 1991. Restoration ecology; a major opportunity for ecotoxecologists. Environ. Toxicol. Chem. 10:429-432.

Carter, M.R., E.G. Gregorich, D.W. Anderson, J.W. Doran, H.H. Janzen, and F.J. Pierce. 1997. Concepts of soil quality and their significance. *In* E.G. Gregorich and M. Carter (ed.) Soil quality for crop production and ecosystem health. Elsevier Science Publishers, Amsterdam, Netherlands.

Cook, N., and W.H. Hendershot. 1996. The problem of establishing ecologically based soil quality criteria: The case of lead. Can J. Soil Sci. 76:335-342.

Cox, C. 1995. Soil quality: Goals for national policy. J. Soil Water Conserv. 50:223.

Crosson, P.R., and A.J. Stout. 1983. Productivity effects of cropland erosion in the United States. Resources for the Future. Washington, DC.

Davidson, D. (ed.). 1986. Land Evaluation. Van Nostrand Reinhold Co., New York, NY.

de Bruijn, P.J., and F.B. de Walle. 1988. Soil standards for soil protection and remedial action in the Netherlands. *In* K. Wolf, W.J. van den Brink, and F.J. Colon (ed.) Contaminated soil '88, Vol. I. Kluwer, Dordrecht, Netherlands.

Denneman, C.A.J., and J.G. Robberse. 1990. Ecotoxicological risk assessment as a base for development of soil quality criteria. p. 157-164. *In* F. Arendt, M. Hinsenveld and W.J. van den Brink (ed.) Contaminated soil '90. Proc. Intl. KfK/TNO Conf. on Contaminated Soil, Karlsruhe, Germany, Kluwer Academic Publishers, Dordrecht, Netherlands.

Dent, F.J. 1978. Land suitability classification. p. 273-293. *In* Soils and rice. International Rice Research Institute, Manila.

Doran, J.W., and A.J. Jones (ed.). 1996. Methods for assessing soil quality. Soil Sci. Soc. Am. Spec. Pub. 49.

Doran, J.W., and T.B. Parkin. 1994. Defining soil quality for a sustainable environment. *In* J.W. Doran, D.C. Coleman, D.F. Bezdicek, and B.A. Stewart (ed.) Soil Sci. Soc. Am. Spec. Pub. 35, Soil Science Society of America, Madison, WI.

Doran, J.W., M. Sarrantonio, and M.A. Liebig. 1996. Soil health and sustainability. Adv. Agron. 56:1-54.

Ellert, B.H., M.J. Clapperton, and D.W. Anderson. 1997. An ecosystem perspective of soil quality. *In* E.G. Gregorich, and M. Carter (ed.) Soil quality for crop production and ecosystem health. Elsevier Science Publishers, Amsterdam, The Netherlands.

Eswaran, H., R. Almaraz, E. van den Berg, and P. Reich. 1997. An assessment of the soil resources of Africa in relation to productivity. Geoderma 77:1-18.

Food and Agricultural Organization. 1976. A framework for land evaluation. FAO Soils Bulletin 32. FAO, Rome, Italy.

Food and Agricultural Organization. 1979. Land evaluation criteria for irrigation. Report of an expert consultation. World Soil Resour. Rep. 50. FAO, Rome, Italy.

Food and Agricultural Organization. 1989. FAO yearbook. FAO, Rome, Italy.

Food and Agricultural Organization. 1995. Planning for sustainable use of land resources: towards a new approach. W.G. Sombroek and D. Sims (ed.) Land and Water Bull. 2. FAO, Rome, Italy.

Food and Agricultural Organization. 1997. Land quality indicators and their use in sustainable agriculture and rural development. Land and Water Bull. 5. FAO, Rome, Italy.

Fournier, F. 1989. The effect of human activity on soil quality. p. 25-32. *In* J. Bouma and A.K. Bregt (ed.). Land qualities in space and time. Pudoc, Wageningen, Netherlands.

Franco-Vizcaino, E. 1997. Comparative soil quality in maize rotations with high or low residue diversity. Biol. Fertil. Soils 24:32-38.

Franzluebbers, A.J., and M.A. Arshad. 1997. Soil microbial biomass and mineralizable carbon of water-stable aggregates. Soil Sci. Soc. Am. J. 61:1090-1097.

Franzluebbers, A.J., D.A. Zuberer, and F.M. Hons. 1995. Comparison of microbiological methods for evaluating quality and fertility of soil. Biol. Fertil. Soils. 19:135-140.

Gantzer, C.J., and T.R. McCarty. 1987. Predicting corn yields on a claypan soil using a soil productivity index. Trans. ASAE 30:1347-1352.

Gersmehl, P.J., and D.A. Brown. 1990. Geographic differences in the validity of a linear scale of innate soil productivity. J. Soil Water Conserv. 45:379-382.

Gerzabeck, M.H., F. Pichlmayer, H. Kirchman, and G. Haberhauer. 1997. The response of soil organic matter to manure amendments in a long-term experiment at Ultuna, Sweden. Europ. J. Soil Sci. 48:273-282.

Gleick, P.H. 1993. An introduction to global fresh water issues. *In* P.H. Gleick, (ed.) Water in Crisis. Oxford University Press. New York, NY.

Granatstein, D., and D.F. Bezdicek. 1992. The need for a soil quality index: Local and regional perspectives. Am. J. Alt. Agric. 7:12-16.

Gregorich, E.G., M.R. Carter, D.A. Angers, C.M. Monreal, and B.H. Ellert. 1994. Towards a minimum data set to assess soil organic matter quality in agricultural soils. Can. J. Soil Sci. 74:367-386.

Gregorich, E.G., M.R. Carter, J.W. Doran, C.E. Pankhurst, and L.M. Dwyer. 1997. Biological attributes of soil quality. *In* E.G. Gregorich and M. Carter (ed.) Soil quality for crop production and ecosystem health. Elsevier Science Publishers, Amsterdam, Netherlands.

Haberern, J. 1992. A soil health index. J. Soil Water Conserv. 47:6.

Halvorson, J.J., J.L. Smith, and R.I. Papendick. 1996. Integration of multiple soil parameters to evaluate soil quality: a field example. Biol. Fertil. Soils 21:207-214.

Herrick, J.E., and W.G. Whitford. 1995. Assessing the quality of rangeland soils: Challenges and opportunities. J. Soil Water Conserv. 50:237-242.

Hoffman, G.J., T.A. Howell, and K.H. Solomon. 1990. Introduction. *In* G.J. Hoffman, T.A. Howell, and K.H. Solomon (ed.) Management of farm irrigation systems. ASAE, St. Joseph, MI.

Holden, P.A., and M.K. Firestone. 1997. Soil microorganisms in soil cleanup: How can we improve our understanding? J. Environ. Qual. 26:32-40.

Hortensius, D., and R. Welling. 1996. International standardization of soil quality measurements. Commun. Soil Sci. Plant Anal. 27:387-402.

Howard, P.J.A. 1993. Soil protection and soil quality assessment in the EC. Sci. Total Env. 129:219-239.

Huddleston, J.H. 1984. Development and use of soil productivity ratings in the United States. Geoderma 32:297-317.

Jenkinson, D.S., N.J. Bradbury, and K. Coleman. 1994. How the Rothamsted classical experiments have been used to develop and test models for the turnover of carbon and nitrogen in soil. p. 117-138. *In* R.A. Leigh and A.E. Johnston (ed.) Long-term experiments in agricultural and ecological sciences. CAB International, Wallingford, UK.

Jenny, H. 1941. Factors of soil formation. A system of quantitative pedology. McGraw-Hill, New York, NY.

Jenny, H. 1961. E.W. Hilgard and the birth of modern soil science. Industrie Grafiche V. Lischi & Figli, Pisa, Italy.

Jenny, H. 1980. The soil resource. Springer-Verlag, New York, NY.

Johnston, A.E., K.W.T. Goulding, and P.R. Poulton. 1986. Soil acidification during more than 100 years under permanent grassland and woodland at Rothamsted. Soil Use Manage. 2:3-10.

Jongbloed, R.H., T.P. Traas, and R. Luttik. 1996. A probabilistic model for deriving soil quality criteria based on secondary poisoning of top predators. II. Calculations for dichlorodiphenyltrichloroethane (DDT) and cadmium. Ecotox. Env. Safety 34:279-306.

Jordan, D., R.J. Kremer, W.A. Bergfield, K.Y. Kim, and V.N. Cacino. 1995. Evaluation of microbial methods as potential indicators of soil quality in historical agricultural fields. Biol. Fertil. Soils 19:297-302.

Karlen, D.L., and C.A. Cambardella. 1996. Conservation strategies for improving soil quality and organic matter storage. Adv. Soil Sci. CRC Press, Inc., Boca Raton, FL.

Karlen, D.L., N.S. Eash, and P.W. Unger. 1992. Soil and crop management effects on soil quality indicators. Amer. J. Altern. Agric. 7:48-55.

Karlen, D.L., M.J. Mausbach, J.W. Doran, R.G. Cline, R.F. Harris, and G.E. Schuman. 1997. Soil quality: A concept, definition and framework for evaluation. Soil Sci. Soc. Am. J. 61:4-10.

Karlen, D.L., N.C. Wollenhaupt, D.C. Erbach, E.C. Berry, J.B. Swan, N.S. Eash, and J.L. Jordahl. 1994. Crop residue effects on soil quality following 10-years of no-till corn. Soil Tillage Res. 31:149-167.

Kennedy, A.C., and R.I. Papendick. 1995. Microbial characteristics of soil quality. J. Soil Water Conserv. 50:243-248.

Kiniry, L.N., C.L. Scrivner, and M.E. Keener. 1983. A soil productivity index based upon predicted water depletion and root growth. Univ. Missouri, CO. Agr. Ag. Expt. Sta. Bull. 1051.

Klingebiel, A.A., and P.H. Montgomery. 1973. Land-capability classification. Agriculture Handbook No. 210. Soil Conservation Service USDA, Washington, DC.

Lal, R. 1995. Trends in world agricultural use: Potential and constraints. p. 521-536. In R. Lal and B.A. Stewart (ed.) Soil management, experimental basis for sustainability and environmental quality. CRC Press, Boca Raton, FL.

Larson, W.E., and F.J. Pierce. 1991. Conservation and enhancement of soil quality. p. 175-203. In Evaluation for sustainable land management in the developing world. Int. Board Soil Res. and Management, Bangkok, Thailand.

Larson, W.E., and F.J. Pierce. 1994. The dynamics of soil quality as a measure of sustainable management. p. 37-51. In J.W. Doran, D.C. Coleman, D.F. Bezdicek, and B.A. Stewart (ed.) Defining soil quality for a sustainable environment. Soil Sci. Soc. Am. Spec. Pub. 35.

Letey, J. 1985. Relationship between soil physical properties and crop production. Adv. Soil Sci. 1:227-294.

Letey, J. 1994. Is irrigated agriculture sustainable? p. 23-37. In Soil and water science: Key to understanding our global environment. Soil Sci. Soc. Am. Spec. Pub. 41.

Lindert, P.H., J. Lu, and W. Wu. 1996a. Trends in the soil chemistry of North China since the 1930s. J. Environ. Qual. 25:1168-1178.

Lindert, P.H., J. Lu, and W. Wu. 1996b. Trends in the soil chemistry of South China since the 1930s. Soil Sci. 161:329-342.

Lindstrom, M.J., T.E. Schumacher, A.J. Jones, and C. Gantzer. 1992. Productivity index model comparison for selected soils in North Central United States. J. Soil Water Conserv. 47:491-494.

Maletic, J.T., and T.B. Hutchings. 1967. Selection and classification of irrigable lands. p. 125-173. In R.M. Hagan et al. (ed.) Irrigation of agricultural lands. Soil Science Society of America, Madison, WI.

McBride, M., S. Sauve, and W. Hendershot. 1997. Solubility control of Cu, Zn, Cd, and Pb in contaminated soils. Eur. J. Soil Sci. 48:337-346.

McRae, S.G., and C.P. Burnham. 1981. Land Evaluation. Clarendon Press, Oxford, UK.

Merker, J. 1956. Untersuchungen an den Ernten und Boden des versuches Ewiger Roggebau in Halle (Saale). Kühn-Archiv 70:154-215.

Miller, F.P. 1979. Defining, delineating and designating uses for prime and unique agricultural lands. p. 291-318. In M.T. Beatty, G.W. Petersen, and L.D. Swindale (ed.) Planning the uses and management of land. Soil Science Society of America, Madison, WI.

Moen, J.E.T. 1988. Soil protection in The Netherlands. p.1495-1503. In K. Wolf, W. J. van den Brink, and F.J. Colon (ed.) Contaminated soil '88. Kluwer Academic Publishers, Dordrecht, Netherlands.

Murata, T., M.L. Nguyen, and K.M. Goh. 1995. The effects of long-term superphosphate application on soil organic matter content and composition from an intensively managed New Zealand pasture. Europ. J. Soil Sci. 46:257-264.

National Academy of Sciences. 1993. Soil and water quality: An agenda for agriculture. National Academy Press, Washington, DC.

National Research Council. 1993. Soil and Water Quality: An agenda for agriculture. Committee on Long-range Soil and Water Conservation, Board on Agriculture, National Academy Press, Washington, DC.

National Research Council. 1994. Rangeland health. New methods to classify, inventory, and monitor rangelands. National Academy Press, Washington, DC.

Neill, L.L. 1979. An evaluation of soil productivity based on root growth and water depletion. M.S. Thesis. University of Missouri, Columbia, MO.

Odell, R.T., S.W. Melsted, and W.M. Walker. 1984. Changes in organic carbon and nitrogen of Morrow plot soils under different treatments, 1904-1973. Soil Sci. 137:160-171.

Papendick, R.I., J.F. Parr, and J. van Schilfgaarde. 1995. Soil quality: New perspective for a sustainable agriculture. *In* Proc. International Soil Conservation Organization. New Delhi, India. 1994.

Parr, J.F., B.A. Stewart, S.B. Hornick, and R.P. Singh. 1990. Improving the sustainability of dryland farming systems: A global perspective. Adv. Soil Sci. 13:1-8.

Parr, J.F., R.I. Papendick, S.B. Hornick, and R.E. Meyer. 1992. Soil quality: Attributes and relationship to alternative and sustainable agriculture. Am. J. Altern. Agric. 7:5-11.

Patterson, G.T., and E.E. Mackintosh. 1976. Relationship between soil capability class and economic returns from grain corn production in Southwestern Ontario. Can. J. Soil Sci. 56:167-174.

Paul, E. A., and F. E. Clark. 1996. Soil microbiology and biochemistry. 2nd Ed. Academic Press, Inc., New York, NY.

Pennock, D.J., D.W. Anderson, and E. De Jong. 1994. Landscape-scale changes in indicators of soil quality due to cultivation in Saskatchewan, Canada. Geoderma 64:1-19.

Pierce, F.J., W.E. Larson, R.H. Dowdy, and W.A.P. Graham. 1983. Productivity of soils: Assessing long-term changes due to erosion. J. Soil Water Conserv. 38:39-44.

Pierce, F.J., W.E. Larson, R.H. Dowdy, and W.A.P. Graham. 1984. Soil productivity in the corn belt: An assessment of erosion's long term effects. J. Soil Water Conserv. 39:131-136.

Reganold, J.P. 1988. Comparison of soil properties as influenced by organic and conventional farming systems. Am. J. Altern. Agric. 3:144-155.

Reganold, J.P., and A.S. Palmer. 1995. Significance of gravimetric versus volumetric measurements of soil quality under biodynamic, conventional, and continuous grass management. J. Soil Water Conserv. 50:298-305.

Reganold, J.P., A.S. Palmer, J.C. Lockhart, and A. N. Macgregor. 1993. Soil quality and financial performance of biodynamic and conventional farms in New Zealand. Sci. 260:344-349.

Reganold, J.P. R.I. Papendick, and J. F. Parr. 1990. Sustainable agriculture. Sci. Am. 262:72-79.

Reganold, J.P., and M.J. Singer. 1979. Defining prime farmland by three land classification systems. J. Soil Water Conserv. 34:172-176.

Reganold, J.P., and M.J. Singer. 1984. Comparison of farm production input/output ratios of two land classification systems. J. Soil Water Conserv. 39:47-53.

Romig, D.E., M.J. Garlynd, R.F. Harris, and K. McSweeney. 1995. How farmers assess soil health and quality. J. Soil Water Conserv. 50:229-236.

Rossiter, D.G. 1996. A theoretical framework for land evaluation. Geoderma 72:165-190.

Schjønning, P., B.T. Christensen, and B. Carstensen. 1994. Physical and chemical properties of a sandy loam receiving animal manure, mineral fertilizer or no fertilizer for 90 years. Europ. J. Soil Sci. 45:257-268.

Seybold, C.A., M.J. Mausbach, D.L. Karlen, and H.H. Rogers. 1998. Quantification of soil quality. p. 387–404. *In* R. Lal, J.M. Kimble, R.F. Follett and B.A. Stewart (ed.) Soil processes and the carbon cycle. CRC Press, Boca Raton, FL.

Sheppard, S.C., C. Gaudet, M.I. Sheppard, P.M. Cureton, and M.P. Wong. 1992. The development of assessment and remediation guidelines for contaminated soils, a review of the science. Can. J. Soil Sci. 72:359-394.

Siegrist, R.L. 1989. International review of approaches for establishing cleanup goals for hazardous waste contaminated land. The Agricultural Research Council of Norway. Institute of Georesources and Pollution Research, Oslo, Norway N-1432 Aas-NHL.

Sims, J.T., S.D. Cunningham, and M.E. Sumner. 1997. Assessing soil quality for environmental purposes: Roles and challenges for soil scientists. J. Environ. Qual. 26:20-25.

Singer, M. J., and D. N. Munns. 1996. Soils: An introduction. 3rd Ed. Prentice-Hall, Inc., NJ.

Singer, M.J., K.K. Tanji and J.H. Snyder. 1979. Planning uses of cultivated cropland and pastureland. p. 225-271. *In* M.T. Beatty, G.W. Petersen, and L.D. Swindale (ed.) Planning the uses and management of land. Soil Science Society of America, Madison, WI.

Singer, M.J., and B.P. Warkentin. 1996. Soils in an environmental context: an American perspective. Catena 27:179-189.

Smith, J.L., J.J. Halvorson, and R.I. Papendick. 1993. Using multiple-variable indicator kriging for evaluating soil quality. Soil Sci. Soc. Am. J. 57:743-749.

Snakin, V.V., P.P. Krchetov, T.A. Kuzovnikova, I.O. Alyabina, A.F. Gurov, and A.V. Stepichev. 1996. The system of assessment of soil degradation. Soil Tech. 8: 331-343.

Soil Science Society of America. 1996. Glossary of Soil Science Terms. Soil Science Society of America, Madison, WI.

Sojka, R.E. 1996. A PAM Primer: A brief history of PAM and PAM-issues related to irrigation. p. 11-20. *In* R.E. Sojka and R.D. Lentz (ed.) Managing irrigation-induced erosion and infiltration with polyacrylamide. Univ. ID Misc. Pub. 101-96.

Sombroek, W.G. 1997. Land resource evaluation and the role of land-related indicators. p. 9-17. *In* Land quality indicators and their use in sustainable agriculture and rural development. FAO Land and Water Bull. 5. FAO, Rome, Italy.

Speir, T.W., H.A. Kettles, A. Parshotam, P.L. Searle, and L.N.C. Vlaar. 1995. A simple kinetic approach to derive the ecological dose ED (50) value for the assessment of Cr(VI) toxicity to soil biological properties. Soil Biol. Biochem. 27:801-810.

Storie, R. E. 1932. An index for rating the agricultural value of soils. CA Agr. Exp. Sta. Bull. 556.

Storie, R.E. 1964. Handbook of soil evaluation. Associated Students Store, U. Cal. Berkeley, CA.

Stork, N.E., and P. Eggleton. 1992. Invertebrates as determinants and indicators of soil quality. Am. J. Altern. Agric. 7:38-47.

Traas, T.P., R. Luttik, and R.H. Jongbloed. 1996. A probabilistic model for deriving soil quality criteria based on secondary poisoning of top predators I. Model description and uncertainty analysis. Ecotoxicol. Env. Safety 34:264-278.

Tribe, D. 1994. Feeding and greening the world, the role of agricultural research. CAB International, Wallingford, UK.

USBR. 1953. Land classification handbook. USDI, Bur. Recl. Pub. V, Part 2.

van Vliet, L.J.P., E.E. Mackintosh, and D.W. Hoffman. 1979. Effects of land capability on apple production in Southern Ontario. Can. J. Soil Sci. 59:163-175.

Visser, S., and D. Parkinson. 1992. Soil biological criteria as indicators of soil quality: Soil microorganisms. Am. J. Altern. Agric. 7:33-37.

Wagenet, R.J., and J.L. Hutson. 1997. Soil quality and its dependence on dynamic physical processes. J. Environ. Qual. 26:41-48.

Waksman, S.A. 1927. Principles of soil microbiology. The Williams and Wilkins Co., Baltimore, MD.

Warkentin, B.P. 1995. The changing concept of soil quality. J. Soil Water Conserv. 50:226-228.

Wier, W.W., and R.E. Storie. 1936. A rating of California soils. Bull. 599. CA Agr. Exp. Sta.

Willis, W.O., and C.E. Evans. 1977. Our soil is valuable. J. Soil Water Conserv. 32:258-259.

Yakovchenko, V., L.J. Sikora, and D.D. Kaufman. 1996. A biologically based indicator of soil quality. Biol. Fertil. Soils 21:245-251.

Yli-Halla, M., H. Hartikainen, P. Ekholm, E. Turtola, M. Puustinen, and K. Kallio. 1995. Assessment of soluble phosphorus load in surface runoff by soil analyses. Agric. Ecosys. Environ. 56:53-62.

The soil must be man's most treasured posses-
sion: so he who tends the soil wisely and with
care is assuredly the foremost among men.
                                  -Sir George Stapleton

H

Soil Databases

Marion F. Baumgardner
Purdue University

Soil Scientists–Generators of Data

Soil scientists, like all other scientists and generators of large quantities of data, have been challenged and aided enormously during the past few decades by the simultaneous evolution/revolution in methods and instrumentation for data acquisition, storage, retrieval, analysis, interpretation, manipulation, modeling, simulation, accessibility, and distribution. No previous generation has had so complete and remarkable an array of tools to extract useful, timely information for the monitoring and assessment of the Earth system, its components and processes.

Scope

In developing the outline and scope of this section on Soil Databases, it was determined early on that the scope must be limited. It is recognized that almost every soil scientist collects, stores, analyzes and interprets data. Examples are almost endless, from the soil physicist conducting fundamental research on the movement of fluids through porous media to the terrestrial ecosystem specialist who is studying the processes of land degradation on a continental or global scale.

The decision was made to focus attention in Section H on databases which describe soils in their natural environment. These databases may be descriptive and have quantitative data with spatial dimensions (soil mapping units), or they may be point data which describe and quantify specific properties of a soil profile at a specific location/address and time (date) on the Earth's surface.

With the rapid development and availability of powerful data acquisition systems, computational facilities and the rapidly emerging technology and use of geographic information systems (GIS) and global positioning systems (GPS), soil scientists are challenged to integrate these technologies into their efforts to provide more objective decision-support systems in the management and monitoring of our valuable soil resources.

The Need for Global Compatibility

Many reasons have been given for the need for a universal system of soil classification and survey and a wide range of spatial scales, from detailed (local) to very generalized (continental to global). Some of these reasons are discussed in detail in different Chapters in this Section. Two major reasons are suggested here. The first relates to issues related to global change. It has been only during the past two decades that environmental concerns have "reached the top" of both the scientific and political agendas. The idea is finally being accepted that the Earth is a system, and that any natural or human-induced change in one portion of the system may have an impact on the entire system. Although many soil scientists have espoused this concept for many years, the global soil science community until now has been unable to provide a universally compatible soil map of the world which will meet the spatial scales and details required by the global modeling community. Globally compatible databases are essential for establishing the current status of Earth's components and processes and for monitoring rates of change in the status of these resources, including soils.

A second important need for globally compatible soil databases is for guidance in the transfer of known technology and research observations from one area of the Earth to any other area with similar soils, climate and other related environmental similarities.

Organization

This section consists of six chapters. Chapter 1 by F.O. Nachtergaele, FAO, Rome, presents the concepts, history, development and future thrust of the World Soil Map (scale 1:5M) produced collaboratively by two agencies of the United Nations: the Food and Agriculture Organization (FAO) and the U.N. Educational, Scientific, and Cultural Organization (UNESCO).

Chapter 2, prepared by V.W.P. van Engelen, ISRIC, Wageningen, The Netherlands, describes the SOTER Project. This is a collaborative project initiated by the International Society of Soil Science to develop and use a universal legend in generating a World Soils and Terrain Digital Database at a spatial scale of 1:1M. Other collaborators in this project are the International Soil Reference and Information Centre, FAO, and the United Nations Environment Programme (UNEP).

Another small scale soil database, WISE, a 1° by 1° Resolution Global Soil Database, is described in Chapter 3. This database was developed by N. H. Batjes at ISRIC, Wageningen, The Netherlands.

Canadian soil scientists, R. Coote and B. MacDonald, provided Chapter 4 in which they describe and illustrate the soil information systems of Canada. Chapter 5 is a description of soil survey databases of the United States submitted by D. J. Lytle, Natural Resources Conservation Service, US Department of Agriculture.

Although each of the first five chapters gives some illustrations of the use of the soil databases which are described, Chapter 6 is a special chapter on the applications of soil databases prepared by G. Petersen and his colleagues at the Pennsylvania State University.

Many other countries, regions and organizations have activities related to the development of soil databases. It was beyond the scope and time limitations of this Handbook to prepare an exhaustive survey and description of all soil databases which include cartographic and descriptive data of soil maps at all scales. An attempt has been made to present representative concepts and methods which are being used in several countries in the development of relational databases at a range of spatial scales.

Many national and international organizations and agencies have made soils databases accessible on the internet. Additions and changes in electronic accessibility of soils data are occurring continuously, but the following worldwide web sites may be useful in locating soil databases of interest. Many of these sites have linkages with other global, national and regional databases.

Soils Data, Agency	Website (http:)
GLASOD, SOTER, other soil databases	
International Soil Reference and Information Centre	
Wageningen, The Netherlands	www.isric.nl
Canadian Soil Information System	
Agriculture and Agri-Food Canada	
Ottawa, Canada	www.res.agr.ca/CANSIS
Digital Soil Map of the World	
Food and Agriculture Organization	
Rome, Italy	www.fao.org/catalog/New/products
World Soil Resources	
Natural Resources Conservation Service	
United States Department of Agriculture	
Washington, DC	www.nhq.nrcs.usda.gov/WSR/

1

From the Soil Map of the World to the Digital Global Soils and Terrain Database: 1960-2002

F.O. Nachtergaele
FAO, Rome, Italy

1.1 Introduction

Institutes and organizations involved in applied research at a global scale, for example, climate change and the greenhouse effect, or studies such as "Agriculture Towards 2010" (Alexandratos, 1995), have a definite need for soil information because these data are a crucial input in models that simulate crop growth and calculate anticipated yields and water balance, or assess the environmental impact of different land-use practices.

An overview of a number of model examples and their soil data requirements are summarized in Table 1.1.

1.2 The Soil Map(s) of the World

At the global level, only two relatively large-scale soil maps exist: a 1:10 million scale map prepared by Kovda and coworkers, and the 1:5 million scale FAO-Unesco Soil Map of the World (FAO, 1971-1981). In addition, a number of simplifications and transformations of the latter map exist, such as those produced by USDA (using the USDA Soil Taxonomy as a classification system), the simplified 1:15 million scale world map of landscapes (Milanova and Kuslin, 1993) and the FAO 1:25 million scale World Soil Resources Map (FAO, 1993).

It is generally accepted that the 1:5 million scale FAO-Unesco map is the most appropriate source of soil information for studies at a continental, regional or global nature. A brief history of this map follows.

The International Society of Soil Science (ISSS) at its Seventh Congress, Madison, WI, USA, in 1960 recommended that soil maps of continents and large regions be published. As a followup, FAO and UNESCO decided in 1961 to prepare a Soil Map of the World at 1:5,000,000 scale.

Table 1.1 Uses of soil information in applied research [After Scholes et al., 1995]

Model examples	Key soil parameters used
Biogeochemical models	C, N, P, water retention, depth, acidity, clay, sand and stone content
Plant response models	
Agricultural models	
Sediment yield models	Texture, water retention and transmission, depth, C, erodibility
Water balance models	Water retention and transmission
Trace gas models	C, N, texture, pH, redox potential
Landform history	Soil type, isotopes
CO_2, CH_4 and N_2O inventories	C, bulk density, depth, soil moisture regime
Climate models	Water and heat capacity, surface reflectance
Environmental impact models	Soil fertility, soil erodibility
Agro-ecological zoning	Soil type, texture, slope, soil phase

The project started in 1961 and was completed over a span of 20 yr. It is the fruit of worldwide collaboration between innumerable soil scientists. Successive drafts of the soil map and of the legend were prepared from a compilation of existing material, combined with systematic field identification and correlation. The first draft of the Soil Map of the World was presented to the Ninth Congress of the ISSS in Adelaide, Australia, in 1968. In accordance with the recommendation of the Congress that the Soil Map of the World should be published at the earliest possible date, the first sheets (South America) were issued in 1971. The results of field correlation in different parts of the world and the various drafts of the legend were published as issues of FAO World Soil Resources Reports (FAO, 1971-1981).

With the rapidly advancing computer technology and the expansion of Geographical Information Systems during the eighties, several attempts were made to digitize the Paper Soil Map of the World. The first effort was carried out by ESRI (1984) in vector format and contained a number of different layers of land resource-related information (vegetation, geology), often incomplete and not fully elaborated. This version was distributed until 1991 by UNEP-GRID, when it was replaced by the official FAO Digitized Soil Map of the World, which focused on soil information only and allowed an analysis by individual country. The ten diskettes cost US$ 350 and a special reduction was given for developing countries. It required an Arc-Info GIS system to be fully exploitable.

A first rasterized version of the soil map was prepared by Zöbler using the ESRI (1984) map as a base and using $1° \times 1°$ grid cells. Only the dominant (FAO) soil unit in each cell was indicated. This digital product gained some popularity because of its simplicity, particularly in the United States.

In 1993, FAO and ISRIC combined efforts to produce a raster map with a 30' x 30' cell size in the interest of the WISE (World Inventory of Soil Emissions) project (Batjes et al., 1995). This database contains the distribution of up to 10 different soil units and their percentages in each cell. In 1996, FAO produced its own raster version which had the finest resolution with a 5' x 5' cell size (9 km x 9 km at the equator) and which had a full database completely corresponding with the paper map in terms of soil units, topsoil texture, slope class and soil phase. This version is available on CD-ROM (US$350) and contains in addition to the Vector Map discussed above a large number of databases and digital maps of derived soil properties (pH, OC, C/N, soil moisture storage capacity, soil depth, etc.)(Appendix 1.1).

An overview of the publication stages of the paper Soil Map and its digitized version is given in Table 1.2.

Table 1.2   Important dates in the development of the Soil Map of the World

1960	ISSS recommends the preparation of the soil maps of Continents
1961	FAO and Unesco start the Soil Map of the World project.
1971	Publication of the first sheet of the paper map (South America)
1981	Publication of the last sheet of the paper map (Europe)
1984	ESRI digitizes the map and other information in Vector format
1989	Zöbler produces a 1° x 1° raster version
1991	FAO produces an Arc/Info Vector map including country boundaries
	ISRIC produces a 30' x 30' raster version
1993	FAO produces a 5' x 5' raster version with derived soil properties
1996	ISRIC produces a 30' x 30' raster version with derived soil properties

1.3   Information Contained in the Soil Map of the World

1.3.1   Direct information contained in the Soil Map

The Soil Map of the World contains direct information on the composition of each mapping unit in terms of the soil type that is dominant, associated or included, the topsoil texture of the dominant soil type (three classes: coarse, medium and fine), the dominant slope class of the unit (three classes: 0-8%, 8–30% and > 30%) and the eventual soil phase present (saline, sodic, petrocalcic phases, etc.). This is illustrated in Fig. 1.1a.

This direct soil information can be expanded on the basis of rules worked out for the interpretation of the Soil Map of the World. These were elaborated in the Agro-ecological Zones studies that FAO carried out in several countries, such as Mozambique (FAO, 1982), China (FAO, 1995), Bangladesh (Brammer and Antoine, 1988) and Kenya (FAO, 1991). Using these rules, Fig. 1.1a can be transformed into Fig. 1.1b, which contains assumed quantified soil information.

1.3.2   Indirect and Derived Information: Pedo- and Taxotransfer Functions

A pedotransfer function is a mathematical relationship between two or more soil parameters which shows a reasonable high level of statistical confidence. This relationship is used to facilitate the estimation of a non-measured soil parameter from one or more measured ones.

The use of pedotransfer functions and some correlations are well-known, e.g., between pH and base saturation, between CEC of the soil and the clay and organic matter contents, or between salinity, pH and ESP (Nachtergaele, 1976). Pedotransfer functions are exclusively used when soil profile data are available and as such are of less interest for making interpretations on the basis of the Soil Map of the World. They will become very important when a soil profile-linked SOTER (Soil and Terrain Database) approach is followed (Section 1.5).

A taxotransfer function is the estimation of soil parameters based on modal soil characteristics of soil units, as derived from a combination of their classification name or taxon (which by definition often implies a certain range for a number of properties), expert knowledge and empirical rules, and a statistical analysis of a large number of soil profiles belonging to the same taxon.

This approach was first used and discussed (but not published) by FAO in 1992. Recently a large number of soil parameters have been derived in this way, particularly by FAO on the CD-ROM version of the Soil Map of the World (FAO, 1996) and by Batjes (1995 and 1996) for the WISE data set.

The use of a taxotransfer function for pH, for instance, would lead, in the examples of Figs. 1.1a and 1.1b, to the following:

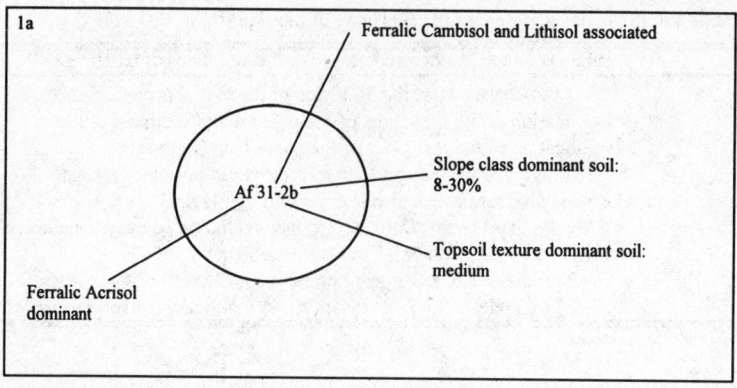

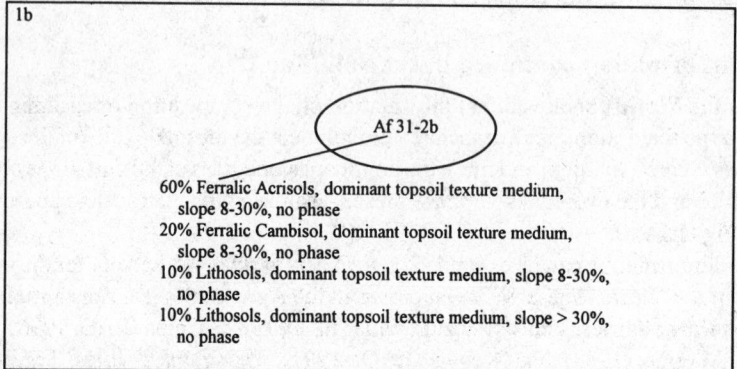

Fig. 1.1   (a) Direct soil information contained in the Soil Map of the World; (b) derived soil information and composition of the mapping unit

Af31 - 2b:
    60% Af - medium textured: pH class 4.5 - 5.5
    20% Bf - medium textured: pH class 4.5 - 5.5
    20% I - medium textured: pH class 5.5 - 7.2
Conclusion: 80% pH class 4.5 - 5.5, 20% pH class 5.5 - 7.2.

1.4   Advantages and Disadvantages of the Soil Map of the World

It should be realized that the Soil Map of the World was made based on soil survey results and technology available in the sixties. Since then, a wealth of new soil information has become available in many (but not all) countries (see Table 1.3). Therefore, it is to be expected that some of the information contained in the map is of uneven quality, often outdated and sometimes completely wrong (Bouwman, 1990).

The technology of survey aids such as remote sensing, Geographical Positioning Systems (GPS) and the computerized storage and retrieval of soil data in soil profile databases and Geographical Information Systems (GIS) has also made tremendous progress.

Two National Soil Maps at 1:1 million scale, the one for Kenya (Kenya Soil Survey, 1980, and KENSOTER, Kenya Soil Survey, 1995) and the one for Botswana (De Wit and Nachtergaele, 1991), were recently compared with the soil information contained in the 1:5 million scale Soil Map of the

Table 1.3 Status of national soil maps worldwide*

Status	Countries		Area covered	
	Number	%	'000 km²	%
Complete coverage at 1:1 M scale or larger	109	68	42,261	31
Not complete	49	32	94,783	69
Total	158	100	137,044	100

* More detailed information by individual country for some regions is given in Appendix 2.

World (Van Waveren, 1995a and 1995b). The findings confirmed with sometimes surprising accuracy, the dominant soil units in the Botswana Hardveld (Table 1.4). However, a rather poor correlation was found between the two scales for textural distribution (Table 1.5) in Botswana, and the distribution of Ferralsols and Solonetzes in Kenya.

The greatest disadvantage of soil maps in general, as perceived by modelers and geostatisticians, is that boundaries drawn on a map are based on expert opinion and are highly dependent on the soil classification applied. Ideally, nonsoil scientists would prefer remotely sensed and randomly sampled data that could be stored in a geographical information system which, by some geostatistical krigging technique, would generate an infinite number of thematic maps.

Although some of this criticism is valid, the alternative as described above is an unworkable proposition on a worldwide scale. This purely mechanistic approach is also unsound, because it considers soils in isolation from their major pedogenetic factors, and requires a large (and uneconomical) sampling density to cater to soil variability.

Another weak point of this profile-based approach is that structural and morphological soil characteristics are often ignored and become highly dependent on laboratory analysis. Comparative benchmark testing of analytical techniques has shown that most laboratory measurements (with the exception of pH and electrical conductivity) have a variability of 20% or more between different laboratories (Nachtergaele, 1977). This should serve as a warning that highly accurate results derived from soil analytical information are also highly suspect.

This does not mean that soil profile information should not be collected, but it emphasizes that the information on its own and out of its landscape context is of little value. This is precisely why the SOTER approach, which marries the best of classical soil survey methodology with sophisticated modern technology approaches, is promoted by ISRIC, UNEP and FAO.

1.5 The World Soils and Terrain Database

SOTER is another initiative of the ISSS and the approach was adopted at the 13th World Congress of Soil Science in 1986. Under a UNEP project, the SOTER methodology was developed in close

Table 1.4 Distribution of major soil groups in the Hardveld (Botswana)

Major Soil Group	1:5 M	1:1 M
Luvisols	44%	38%
Arenosols	21%	20%
Cambisols	12%	9%
Vertisols	5%	1%
Gleysols	5%	0%
Lithosols/Rock	11%	3%
Regosols	1%	18%
Nitosols	1%	1%
Other		10%

Table 1.5 Slope class and textural class distribution in Botswana derived from two maps at different scales

Slope Class	1:1 M	1:5 M	Textural Class	1:1 M	1:5 M
a 0-8%	96%	78%	1 Coarse	79%	49%
b 8-30%	4%	21%	2 Medium	18%	41%
c > 30%	0%	1%	3 Fine	2%	10%
			N/A	1%	

cooperation with the Land Resources Research Center of Canada, FAO and ISSS. After initial testing in three areas, involving five countries (Argentina, Brazil, Uruguay, USA, Canada), the methodology was endorsed by the ISSS Working Group on World Soils and Terrain Digital Database (DM). It was further refined and, in 1993, the Procedures Manual for Global and National Soils and Terrain Digital Databases was jointly published by UNEP, ISSS, FAO and ISRIC, thus obtaining international recognition. The Procedures Manual is available in English, French and Spanish (UNEP/ISRIC/FAO/ISSS, 1995a,b,c).

The Soil and Terrain Digital Database (SOTER) program provides an orderly arrangement of natural resource data in such a way that these data can be readily accessed, combined and analyzed from the point of view of potential use and production, in relation to food requirements, environmental impact and conservation. Basic in the SOTER approach is the mapping of areas with a distinctive, often repetitive pattern of landform, morphology, slope, parent material and soils at 1:1 million scale (SOTER units). Each SOTER unit is linked through a Geographic Information System with a computerized database containing all available attributes on topography, landform and terrain, soils, climate, vegetation and land use. In this way, each type of information or each combination of attributes can be displayed spatially as a separate layer or overlay or in tabular form.

The SOTER concept was primarily developed for application at country (national) scale and national SOTER maps have been prepared, with ISRIC's assistance, for Uruguay (1:1 M), Kenya (1:1 M) and Hungary (1:500,000), whereas in Jordan and Syria this exercise is still ongoing. Other countries in which the methodology has been applied include Bolivia, Ethiopia, Argentina, the Gambia and Myanmar; moreover, there are a number of proposals for an expansion of this application to many more countries.

The original idea of SOTER was to develop this system worldwide at an equivalent scale of 1:1 M in order to replace the paper Soil Map of the World (Sombroek, 1984). However, it soon became obvious that the resources were lacking to tackle and complete this huge task in a reasonable time frame. However, this still remains the long term objective pursued on a country by country basis, mainly by ISRIC and UNEP.

In the early 1990s, FAO recognized that a rapid update of the Soil Map of the World would be a feasible option if the original map scale of 1:5 M were retained, and started, together with UNEP, to fund national updates at 1:5 M scale of soil maps in Latin America and Northern Asia. At the same time, FAO tested the physiographic SOTER approach in Asia (van Lynden, 1994), Africa (Eschweiler, 1993), Latin America (Wen, 1993), and the former Soviet Union and Mongolia (Stolbovoy, 1996).

These parallel programs of ISRIC, UNEP and FAO merged together in mid 1995, when at a meeting in Rome the three major partners agreed to join all resources and work toward a common world SOTER-shell approach covering the globe at 1:5 M scale by the 17th ISSS Congress of 2002. The ongoing and planned activities are summarized in Table 1.6. At 1:5 M scale, the world SOTER database contains several layers of information, as illustrated in Appendix 3, while these data are stored in a relational database, as illustrated in Fig. 1.2.

Table 1.6 Operational plan for the World SOTER-shell: 1995–2002

Region	Status	Main Agencies Involved	Publication
Latin America and the Caribbean	Ongoing	ISRIC, UNEP, FAO, CIATT national soil institutes	Dec. 1996
East and South Africa	Ongoing	FAO	June 1998
North and Central Asia	Ongoing	IIASA, Dokuchaev Institute, Academia Sinica, FAO, ISRIC	Dec. 1997
Southeast Asia	Project Proposal submitted	Awaits funding (FAO, ISRIC)	?
Near East	Ongoing	FAO (Saudi Arabia, Yemen, Lebanon, Morocco, Mauritania)	Dec. 1998
West Africa	Project Proposal submitted	Awaits ADB funding (ISRIC, UNEP)	Dec. 1999
Europe	Proposal discussed	EU Soils Bureau/Eastern Europe (SOVEUR Project)	?
USA and Canada	---	---	?
Australia	---	---	?

1.6 Soil Profile Databases

Several soil profile databases exist which contain georeferenced soil profile morphological and/or analytical information. These databases are of major importance for the development of pedotransfer functions (Section 1.3.2). If used unstratified and outside their natural context, they may cause extremely misleading statements and conclusions.

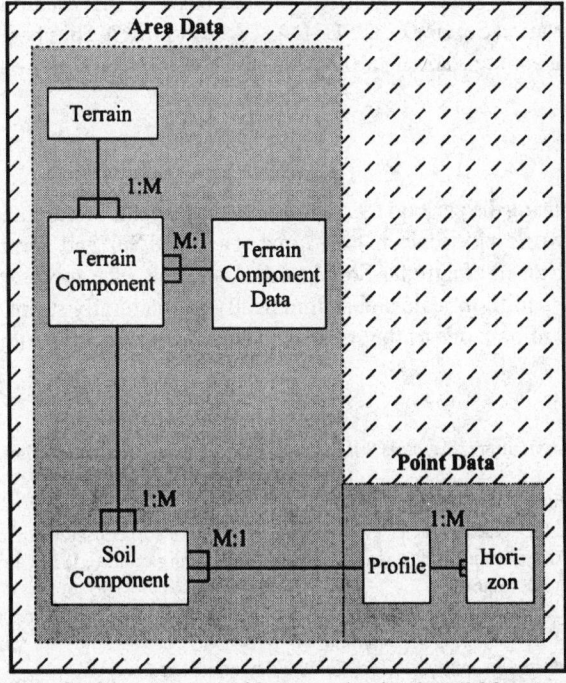

Fig. 1.2 SOTER attribute database structure with area and point data (1:M = one to many, M:1 = many to one relations)

Table 1.7 The global pedon database as compiled by ISRIC [Oldeman, 1996, personal communication]

Broad Geographic Region	In WISE	FAO[1]	In Homogenized Database (Int. set) NRCS[2]	ISIS[3]	Total
Africa	1,799	93	204	18	315
S., W. and Northern Asia	522	24	44	0	68
China, India, Indonesia, The Philippines	553	45	129	106	280
Australia and Pacific Islands	122	28	27	0	55
Europe	492	5	2	0	7
North America	266	14	144	0	158
Latin America and the Caribbean	599	41	114	86	241
Total	4,353	250	664	210	1,124

[1]FAO: Food and Agriculture Organization of the United Nations.
[2]NRCS: Natural Resources Conservation Service of the US Department of Agriculture.
[3]ISIS: International Soil Information System (ISRIC, Wageningen).

Many of these databases are national datasets. The most important set in number of profiles stored is the one held by the Natural Resources Conservation Service of the United States with more than 20,000 profiles. Other countries also have important soil datasets, such as Botswana (more than 2,500 profiles), Yemen (300 profiles), Mali (600 profiles), to name only a few.

Two other soil datasets are worth mentioning in this respect, as they store mainly analytical data and not soil morphological data. One is that built up by the European Union (Working Group on Soil Information System Development). The other, developed by Zinke et al. (1984), concentrates on georeferenced data of carbon and nitrogen and is rather thin on other parameters, which makes its use difficult for more general applications.

The Global Pedon Database, built up by ISRIC on the basis of information received from the Natural Resources Conservation Service of the USDA, FAO and ISRIC, has been extensively tested on internal consistency and contains about 1,100 profiles. Table 1.7 gives the composition by region of this dataset.

1.7 Conclusions

Good progress was made during the period 1993-1996 on the systematic storage and correlation of soil resource information on a worldwide scale, in the format of a global Soil and Terrain Database as well as in the establishment of an international Soil Profile Database. The long-term aim of a global coverage equivalent to a 1:1 million scale and a climatically and naturally stratified soil profile data information system should remain one of the soil science community's major objectives.

1.8 References

Alexandratos, N. 1995. World agriculture: Towards 2010. An FAO Study. FAO and John Wiley and Sons, New York, NY.

Batjes, N.H. 1995. A global data set of soil pH properties. ISRIC Tech. Pap. 27.

Batjes, N.H., E.M. Bridges, and F.O. Nachtergaele. 1995. World inventory of soil emission potentials: Development of a global soil database of process controlling factors. p. 110-115. S. Peng et al. (ed) In Climatic change and rice. Springer Verlag, Heidelberg, Germany.

Batjes, N.H. 1996. A global data set of derived soil properties by FAO-Unesco soil unit (Version 1.0). Draft. ISRIC, Wageningen, Netherlands.

Bouwman, A.F. 1990. Soils and the Greenhouse effect. John Wiley and Sons Ltd., UK

Brammer, H. and J. Antoine. 1988. Resources appraisal of Bangladesh for agricultural development. BGD/81/035. FAO Tech. Rep. 1.

De Wit, P.V., and F.O. Nachtergaele. 1991. An explanatory note to the 1:1 M scale soil map of Botswana. Field Document No. 30. UNDP/FAO BOT/85/011. Ministry of Agriculture, Gaborone, Botswana.

Eschweiler, H. 1993. Draft physiographic map of Latin America. AGLS Working Paper. FAO, Rome, Italy.

ESRI. 1984. Final Report UNEP/FAO World and Africa GIS Database. Environmental Systems Research Institute, Redlands, CA.

FAO. 1982. Land resource inventory of Mozambique. FAO/AGOA MOZ/75/011. Field Document No. 35. Maputo, Mozambique.

FAO. 1991. Agro-ecological land resources assessment for agricultural development planning. A Case study of Kenya. Main Report and 9 Annexes. World Soil Resources Report 71, 71/1 - 71/9. FAO, Rome, Italy.

FAO. 1991. The digitized soil map of the world. World Soil Resources Report 67 (10 diskettes). FAO, Rome, Italy.

FAO. 1993. World soil resources. An explanatory note on the FAO World Soil Resources Map at 1:25,000,000 scale. World Soil Resources Report No. 66, Rev. 1. FAO, Rome, Italy.

FAO. 1994. China land resources, use and productivity assessment. Main Report. Beijing, China.

FAO. 1996. The digitized soil map of the world including derived soil properties. CD-ROM. FAO, Rome, Italy.

FAO. 1971-1981. The FAO-Unesco Soil Map of the World. Legend and 9 volumes. Unesco, Paris, France.

Kenya Soil Survey.1980. Exploratory soil map and agro-climatic map of Kenya. Scale: 1:1,000,000. Kenya Soil Survey. Ministry of Agriculture, Nairobi, Kenya.

Kenya Soil Survey. 1995. 1:1 M scale soil and terrain database of Kenya (KENSOTER). Ministry of Agriculture, Nairobi, Kenya.

Kovda, V.A. 1977. The Soil Map of the World Scale 1:10,000,000, USSR Academy of Science Publishing House, Moscow, Russia.

Milanova, E.V., and A.V. Kuslin. 1993. The world map of present-day landscapes. Department of World Physical Geography and Geoecology. Moscow State University, Moscow, Russia.

Nachtergaele, F.O. 1976. Studies of some saline and sodic soils in Sudan. FAO/UNDP Project Tech. Rep. 24.

Nachtergaele, F.O. 1977. Preliminary results of the benchmark soil sample analysis in Manila and Cebu soil characterization sections. Lab. Info. Pap. 10. AGO PHI/74/003. Bureau of Soils, Manila, Philippines.

Scholes, R.J., D. Skole, and J.S. Ingram. 1995. A global database of soil properties: Proposal for implementation. IGBP-DIS Working Paper No. 10, IGBP, Paris, France.

Sombroek, W.G. 1984. Towards a global soil resources inventory at scale 1:1 million. Discussion Paper. International Soil Reference and Information Center. Wageningen, Netherlands.

Stolbovoy, V. 1996. Draft physiographic map of the former Soviet Union and Mongolia at scale 1:5 million. AGLS Working Paper. FAO, Rome, Italy.

UNEP/ISSS/ISRIC/FAO. 1995a. Global and national soils and terrain digital databases (SOTER) procedures manual. World Soil Resources Report 74. FAO, Rome, Italy.

UNEP/ISSS/ISRIC/FAO. 1995b. Bases digitales de datos de suelos y terreno a nivel mundial y nacional (SOTER). Manual de Procedimientos. Edici n revisada. Informe sobre Recursos Mundiales de Suelos 74. FAO, Rome, Italy.

UNEP/ISSS/ISRIC/FAO. 1995c. Base de données numériques sur les sols et le terrain au niveau mondial et national (SOTER). Rapport sur les ressources mondiales en sols no. 74. FAO, Rome, Italy.

van Lynden, G.W.J. 1994. Draft physiographic map of Asia. FAO/ISRIC Working Paper. FAO, Rome, Italy.

van Waveren, E. 1995a. A comparison of soil inventories of Kenya at Scales 1:5 M and 1:1 M. AGLS Draft Working Paper. FAO, Rome, Italy.

van Waveren, E. A. 1995b. Comparison of soil inventories of Botswana on the FAO Soil Map of the World and the soil map of Botswana at Scale 1:1 M. AGLS Draft Working Paper. FAO, Rome, Italy.

Wen, Ting-tiang. 1993. Draft physiographic map of Latin America. AGLS Working Paper. FAO, Rome, Italy.

Zinke, P.J., A.G. Stangenberger, W.M. Post, W.R. Emanuel, and J.S. Olson. 1984. Worldwide organic soil carbon and nitrogen data. ORNL TM-8857. Oak Ridge National Laboratory, Oak Ridge, TN.

Zöbler, L. 1989. A world soil hydrology file for global climate modeling. *In* Proceedings, International Geographic Information Systems (IGIS) Symposium: The Research Agenda, November 15-18, 1987, Arlington, Virginia, Association of American Geographers, Washington, DC.

Appendix 1 Some options contained in FAO's Soil Map of the World CD-ROM

Inventory, derived and interpreted data	
Soil unit - phase - topsoil texture - topography	1
Effective soil depth	2
Soil drainage classes	3
Soil moisture storage capacity	4
Permafrost	5
Soil suitability	6
Fertility attributes (FCC)	16
Soil characteristics (Topsoil/subsoil)	
pH	7
Organic carbon	8
Nitrogen (%)	9
C/N ratio	10
CEC soil	11
CEC clay	12
Base saturation	13
Texture class	14
Organic carbon pool (kg.m-$_\eta$)	15
Fertility attributes (Fertility capability classification)	
Topsoil textural class (S, L, C, O)	1
Gravel presence	2
Subsoil textural class (only if different from topsoil)	3
Excess wetness (g)	4
More than 3 months dry (d)	5
Low nutrient retention (e)	6
High Al saturation of exchange complex (a)	7
Moderate soil acidity (h)	8
High P-fixation due to iron (i)	9
High P-fixation due to amorphous material (x)	10
Vertic properties and Vertisols (v)	11
Low ability to supply K (k)	12
Calcareous soils (b)	13
Presence of soluble salts (s)	14
High levels of sodium (n)	15
Potential acid sulphate soils (c)	16

Appendix 2 Country soil maps

Country	1:1 M	1:500 000	1:250 000	> 1:250,000	Partial information	Year
Africa (south of the Sahara)						
Angola					*	
Benin				CPCS		1978
Botswana	FAO					1990
Burkina Faso		CPCS				1976
Burundi			Local			1980
Cameroon		S. Tax				1991
Cape Verde					*	
Cent. African Rep.				CPCS		1978
Chad	CPCS					
Comoros					*	
Congo	CPCS					1976
Côte d'Ivoire				CPCS		
Equat. Guinea					*	
Ethiopia	FAO					1988
Gabon				CPCS		1977
Gambia				Local		1976
Ghana					*	
Guinea					*	
Guinea Bissau					*	
Kenya	SOTER					1988
Lesotho			Local			1983
Liberia		FAO				1990
Madagascar	CPCS					1968
Malawi			FAO			1991
Mali		S. Tax				1983
Mauritius				CPCS		1984
Mozambique	FAO					1995
Namibia					*	
Niger					*	
Nigeria		Local				1981
Rwanda				S. Tax		1990
Sao Tome & Principe					0	
Senegal				CPCS		1980
Seychelles					*	
Sierra Leone					*	
South Africa			Local			
Swaziland			Local			1968
Tanzania					*	
Togo				CPCS		1979
Uganda	FAO					1988
Zaire					*	
Zambia	FAO					1991
Zimbabwe	Local					1979
Asia and the Pacific						
Afghanistan					*	
American Samoa					*	
Australia					*	
Bangladesh			S. Tax			1985
Bhutan					*	
Brunei					*	
Cook Islands				Local		1980
China People's Rep.	Local					1992

Appendix 2 Country soil maps (cont.)

Country	1:1 M	1:500 000	1:250 000	> 1:250,000	Partial information	Year
Asia and the Pacific (cont.)						
D.P.R. Korea					*	
Fiji		Local				1981
Hong Kong					*	
India		S. Tax				1992
Indonesia					*	
Iran					*	
Japan	Local					1990
Cambodia	FAO					1977
Kiribati					*	
Laos					*	
Malaysia		Local				1980
Maldives					*	
Mongolia					*	
Myanmar					*	
Nauru					*	
New Caledonia	CPCS					1975
Nepal	Local					1978
Nive Island					*	
New Zealand	Local					1975
Philippines				S. Tax		1980
Papua New Guinea	Local					1988
Korea, Rep. of	S. Tax					1983
Samoa				Local		1963
Singapore				Local		1977
Solomon Isl.			Local			1974
Sri Lanka		S. Tax				1971
Thailand	S. Tax					1979
Tokelau Isl.					*	
Tonga				Local		1972
Trust Terr. of Pacif. Isl.					0	
Tuvalu				Local		1987
Vanuatu				CPCS		1978
Vietnam	Local					1960
Northern Africa and the Near East Region						
Algeria					*	
Barhain			Local			1975
Egypt	FAO					1976
Iraq	S. Tax					198?
Israel		Local				1975
Jordan	FAO					198?
Lebanon	S. Tax					1985
Libya					*	
Mauritania	S. Tax					
Morocco					*	
Oman			S. Tax			1990
Qatar				Local		1976
Saudi Arabia				S. Tax		1988
Somalia	FAO					1996
Sudan	FAO					1996
Syria		SOTER				1996
United Arab Emirates					0	
Yemen					*	

S. Tax = Soil Taxonomy; CPCS = ORSTOM; FAO = FAO-UNESCO; Local = Local classification

Appendix 3 Major elements contained in the world soil and terrain database

1.	**MAJOR LANDFORMS**	
	Level land (< 8%, < 100 m/km local relief):	
	LA	Valley floor, elongated landform with gentle length gradient
	LD	Depression, surrounded on at least three sides by higher and steeper land
	LF	Low gradient footslope, bordering steeper land
	LP	Plain
	LT	Plateau, surrounded on at least three sides by lower and steeper land
	Sloping land	
	SG	Uniform slopes: 8 - 30%, > 50 m/slope unit
	SH	Hills: 8 - 30%, > 50 m/slope unit
	SL	Mountainous highland: 8 - 30%, > 600 m/2 km
	Steep land	
	TM	Mountains: slope > 30%, > 600 m/2 km local relief
	Complex landforms	
	CV	Valleys, too narrow to map lateral slopes and valley bottom separately
	Other	
	WA	Water bodies
2.	**TERRAIN COMPONENTS**	
2.1	**SLOPE CLASS**	
	w	0 - 2% Flat, wet
	a	0 - 2% Flat
	b	2 - 8% Undulating
	c	8 - 15% Rolling
	d	15 - 30% Moderately steep
	e	30 - 60% Steep
	f	> 60% Very steep
2.2	**SUFFIX1, SUFFIX2** (additional landform information)	
	i	Intermontane basin
	r	Ridges
	k	Karst
	d	Dunes
	l	Lava fields
2.3	**HYPSOMETRY**	
	1	Level land, < 300 m above sea level
	2	Level land, 300-600 m above sea level
	3	Level land, 600-1500 m above sea level
	4	Level land, 1500-3000 m above sea level
	5	Level land, > 3000 m above sea level
	10	Sloping land, < 200 m above local base level
	11	Sloping land, 200-400 m above local base level
	12	Sloping land, 400-800 m above local base level
	12*	Sloping land, > 800 m above local base level
	13	Steep land, 600-1500 m above local base level
	14	Steep land, 1500-3000 m above local base level
	15	Steep land, 3000-5000 m above local base level
	16	Steep land, > 5000 m above local base level

2

SOTER: The World Soils and Terrain Database

V.W.P. van Engelen
International Soil Reference and Information Centre
Wageningen, Netherlands

2.1 Introduction

The increasing pressure on land resources, leading to degradation and pollution of these resources, and a reduced productive capacity calls for a system that can store detailed information on land resources in such a way that these data can be accessed, combined and analyzed from various points of view be it potential use, food productivity, environmental impact or conservation (Oldeman and van Engelen, 1993; Arnold, 1995). On a global basis such a system is still lacking. The only global database on soils is the FAO-Unesco Soil Map of the World at scale 1:5 million (1974-1981). However, a global coverage that includes quantified attributes on soils and terrain is still absent. As a consequence, modeling exercises and extrapolation at a global level still rely on databases with only a limited set of profile data and a coarse spatial resolution (Zobler, 1986).

Improvements at the global level are the release of the digital version of the Soil Map of the World (FAO, 1995) and the WISE database by the International Soil Reference and Information Centre (Batjes, 1995; Batjes et al., 1995; Section H, Chapter 3). Although these databases allow for quantification of soil parameters, the spatial distribution of the soils is still based on survey data used for the compilation of the Soil Map of the World in the 1960s and 1970s. At a national level, the development of soil databases has expanded enormously (Section H, Chapter 5).

This chapter will elaborate on the development of the World Soils and Terrain Database (SOTER). After a brief description of the historical background of SOTER, the rationale and objectives of such an activity, an overview of the concept and general approach of the methodology are given, followed by an explanation of the contents and the structure of the database. Applications that have been developed so far will also be discussed.

2.2 Background and Objectives

2.2.1 Historical Background

During the mid eighties, the emergence of computers and GIS opened the ways to start a system that could (1) handle an extensive number of quantified soil and terrain attributes, and (2) cover the globe in a higher resolution than any existing coverage.

The discussion paper "Towards a Global Soil Resources Inventory at Scale 1:1 M" (Sombroek, 1984), was an initiator for the International Society of Soil Science (ISSS) to convene a workshop of international experts on soils and related disciplines in January 1986 in Wageningen, The Netherlands, where the "Structure of a Digital International Soil Resources Map Annex Data Base" (Baumgardner and Oldeman, 1986) was being discussed as an answer to the need to update and expand the soil database used for the 1:5 million FAO/Unesco Soil Map of the World. Based on the findings and recommendations of this workshop, a project proposal was written for SOTER, a World Soils and Terrain Digital Database at a scale of 1:1 million (ISSS, 1986b).

The idea was further elaborated at a workshop in 1987 where the United Nations Environment Programme (UNEP) granted funding for development of the methodology. As this was the first effort to compile a global dataset making use of new techniques, the initiative borrowed heavily from existing activities at national level. The Land Resources Research Centre of Agriculture Canada, already having its Canadian Soil Information System (CanSIS) (Agriculture Canada, 1983) at scale 1:1 million operational at that time, was asked to build on their experience and to develop a global methodology to be applied at the same scale. A first version of a universal legend annex procedures manual was published in 1988 with financial support from UNEP (Shields and Coote, 1988).

The methodology was applied first in a pilot area, covering parts of Argentina, Brazil and Uruguay. With the experience gained in this demonstration area and also in a test in North America on the border between the US and Canada (Shields, 1992), the methodology was further refined during a number of workshops (ISSS, 1988 and 1989) and a revised version was issued (Shields and Coote, 1989). Results related to activities for the development of SOTER with testing in two pilot areas in South and North America were discussed during the International Soil Congress in Kyoto, Japan (Baumgardner, 1990; Scoppa et al., 1990; Shields and Coote, 1990). Several new versions of the Procedures Manual were published in the following years (ISRIC, 1990a and 1991) and finally, a fifth was issued together with FAO in 1993 (van Engelen and Wen, 1993; FAO, 1993). This became the basis for further work during the ensuing years with only minor modifications in the 1995 edition. The last version was also published in French and Spanish.

2.2.2   Objectives

A wealth of soils and terrain information has been collected in recent decades. However, when it comes to using this information, two situations are common: (1) data exist but the decision makers ignore their existence or do not have access to them; and (2) data exist, are accessible, but are incomprehensible to the decision makers; because data are of variable quality without indications of reliability, only highly specialized experts can make an interpretation.

To overcome these problems, SOTER intends to make data on soil and terrain resources more accessible, answer queries quickly and transform basic data into more understandable information for a wide array of users: resource managers, policy/ decision makers and the scientific community at large. It will do this by utilizing information technology to establish a World Soils and Terrain Database, containing digitized map units and their attribute data (ISSS, 1986b). Maps with reliability indications can also be produced. Nevertheless, experts in natural resources are still needed for the operation of such a system. Where no soils and terrain information is available, SOTER can form the framework for storage and interpretation of such data.

2.3   The SOTER Approach

2.3.1   General Concept

Basic in the SOTER approach is the mapping of areas of land (SOTER units) with a distinctive, often repetitive, pattern of landform, lithology, surface form, slope, parent material, and soil (van Engelen and Wen, 1995). Each SOTER unit represents a unique combination of terrain and soil characteristics. Fig. 2.1 shows the representation of a SOTER unit in the database and gives an example of a SOTER map, with polygons that have been mapped at various levels of differentiation.

The mapping of land characteristics has evolved from the idea that land (in which terrain and soil occur) incorporates processes and systems of interrelationships between physical, biological (and social) phenomena evolving through time. This idea was initially developed in Russia and Germany. Similar integrated concepts were used later in the land systems approach in Australia (Christian and Stewart, 1953) and its successors (McDonald et al., 1990; Gunn et al., 1990). SOTER follows this line of thought.

The SOTER mapping approach resembles physiographic soil mapping. The main difference lies in the stronger emphasis SOTER places on the terrain-soil relationship as compared to what is commonly done in traditional soil mapping. This will be true particularly at smaller mapping scales. At the same time, SOTER adheres to rigorous data entry formats necessary for the construction of a universal terrain and soil database. As a result of this approach, the data accepted by the database will be standardized and will have the highest achievable degree of reliability based on available historic data.

Although SOTER has initially been designed for use at a scale of 1:1 million, the methodology is also applicable at larger scales connected with the development of national soil and terrain databases. A first testing of such a detailed database was carried out in the Sao Paulo State of Brazil at a scale of 1:100,000 (de Oliviera and van den Berg, 1992). Later, it was used at the same scale in Argentina and Uruguay. It is the intention to use the approach to construct a natural resources information system for the study of land degradation in areas in Egypt, Lebanon, Syria and Tunisia at scale 1:100,000.

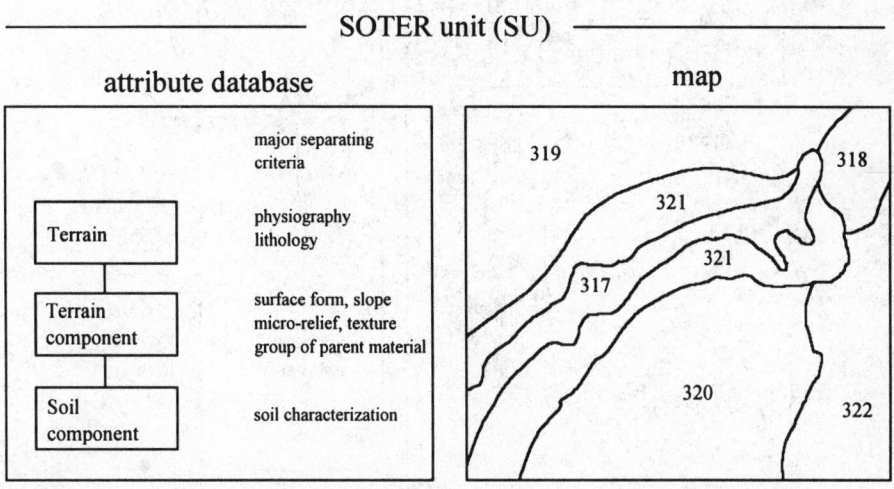

Fig. 2.1   Relations between a SOTER unit and its component parts and major separating criteria

2.3.2   Database Structure

When mapping spatial phenomena, two types of data can be distinguished: (1) geometric data, i.e., the location and extent of an object represented by a point, line or surface, and topology (shapes, neighbors and hierarchy of delineations); and (2) attribute data, i.e., characteristics of the object.

In the SOTER database, soils and terrain information consists of a geometric component, which indicates the location and topology of SOTER units, and an attribute part that describes the non-spatial SOTER unit characteristics. The geometry is stored in that part of the database that is handled by Geographic Information System (GIS) software, while the attribute data are stored in a separate set of attribute files, manipulated by a Relational Database Management System (RDBMS) of which a detailed database structure definition is given by Tempel (1994). A unique label attached to both the geometric and attribute databases connect these two types of information for each SOTER unit (Fig. 2.2). The geometric database contains information on the delineations of the SOTER unit as well as topographic map data.

An effective and flexible tool for storing and managing non-spatial attributes of SOTER is a relational database (Pulles, 1988). The data are stored in tables, whose records are related to each other through the specific identification fields (primary keys) which form the links between the various subsections of the database. A schematic representation of the structure of the attribute database is presented in Fig. 2.3.

The attribute database consists of sets of files for use in a RDBMS. The attributes of the terrain unit and terrain component are either directly available or can be derived from other parameters during the compilation of the database. Only for horizon data, two types of attributes can be distinguished: mandatory and non-mandatory, depending on their importance for most interpretations. All non-

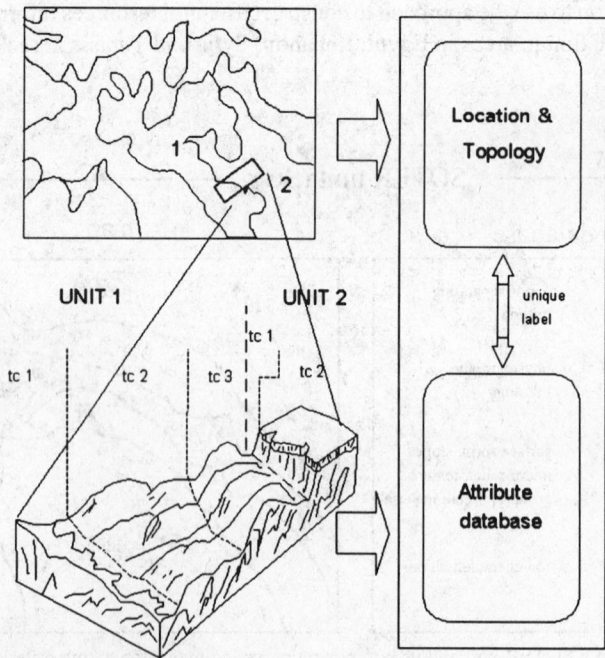

Fig. 2.2 SOTER units, their terrain components (tc), attributes and location

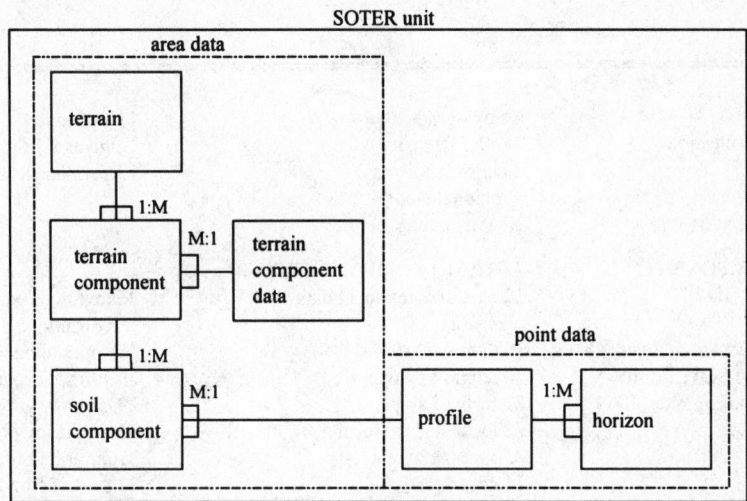

Fig. 2.3 SOTER attribute database structure with area and point data (1:M = one to many, M:1 = many to one relation[s])

spatial attributes of the SOTER units are listed in Table 2.1. The listing for the soil component attributes is compatible with the data set that is stored in the FAO-ISRIC Soil Database (FAO, 1989).

2.4   Status

2.4.1   Implementation

The SOTER methodology has now been applied and tested at various scales and in different regions of the world. Examples are shown below and in Fig. 2.4.

2.4.1.1   Continental Level

FAO, UNEP and ISRIC have started an initiative to replace the FAO-Unesco Soil Map of the World (SMW) by a 1:5 million SOTER database. The oldest sheets had the highest priority in updating. Therefore, Latin America, Central America and the Caribbean were selected to be covered by a 1:5 million SOTER database. Contributions came from national soil survey institutes and were technically coordinated by ISRIC. The first version is now available (FAO et al., 1998).

For other continents, work is under way to replace the SMW by a 1:5 million SOTER database (Section H, Chapter 1). As funding for such a major activity is still not secure, a step-by-step approach has been chosen. Some of the themes that are required in such a database have been compiled separately and can form the building blocks of a comprehensive SOTER database. This is the case in the former USSR, and SE Asia where the terrain and soil information or only terrain data, respectively, are already available in the SOTER format.

2.4.1.2   National Level

National SOTER databases at a scale of 1:1 million have been completed by the national soil institutes in Uruguay and Kenya, while a substantial part of northern Argentina is also covered. Hungary, Jordan and Syria have operational SOTER databases at a scale of 1:500,000.

Table 2.1   Nonspatial attributes of SOTER unit

TERRAIN

1 SOTER unit_ID	6 slope gradient	11 dissection
2 year of data collection	7 relief intensity	12 general lithology
3 map_ID	8 major landform	13 permanent water surface
4 minimum elevation	9 regional slope	
5 maximum elevation	10 hypsometry	

TERRAIN COMPONENT | TERRAIN COMPONENT DATA
14 SOTER unit_ID | 18 terrain component data-ID

		26 texture group non-
		consolidated parent material
15 terrain component number	19 dominant slope	27 depth to bedrock
16 proportion of SOTER unit	20 length of slope	28 surface drainage
17 terrain component data_ID	21 form of slope	29 depth to groundwater
	22 local surface form	30 frequency of flooding
	23 average height	31 duration of flooding
	24 coverage	32 start of flooding
	25 surface lithology	

SOIL COMPONENT | HORIZON (* = mandatory)
33 SOTER unit ID | 63 profile_ID*

33 SOTER unit ID	63 profile_ID*	96 soluble K^+
34 terrain component number	64 horizon number*	97 soluble Cl^-
35 soil component number	65 diagnostic horizon*	98 soluble SO_4^{2-}
36 proportion of SOTER unit	66 diagnostic property*	99 soluble HCO_3^-
37 profile_ID	67 horizon designation	100 soluble CO_3^{2-}
38 number of reference profiles	68 lower depth*	101 exchangeable Ca^{2+}
39 position in terrain	69 distinctness of transition	102 exchangeable Mg^{2+}
component	70 moist colour*	103 exchangeable Na^+
40 surface rockiness	71 dry colour	104 exchangeable K^+
41 surface stoniness	72 grade of structure	105 exchangeable Al^{3+}
42 types of erosion/deposition	73 size of structure elements	106 exchangeable acidity
43 area affected	74 type of structure*	107 CEC soil*
44 degree of erosion	75 abundance of coarse fragments*	108 total carbonate equiv.
45 sensitivity to capping	76 size of coarse fragments	109 gypsum
46 rootable depth	77 very coarse sand	110 total carbon*
47 relation with other soil	78 coarse sand	111 total nitrogen
components	79 medium sand	112 P_2O_5
	80 fine sand	113 phosphate retention
PROFILE	81 very fine sand	114 Fe dithionite
48 profile_ID	82 total sand*	115 Al dithionite
49 profile database_ID	83 silt*	116 Fe pyrophosphate
50 latitude	84 clay*	117 Al pyrophosphate
51 longitude	85 particle size class	118 clay mineralogy
52 elevation	86 bulk density*	
53 sampling date	87 moisture content at various tensions	
54 lab_ID	88 hydraulic conductivity	
55 drainage	89 infiltration rate	
56 infiltration rate	90 pH H_2O*	
57 surface organic matter	91 pH KCl	
58 classification FAO	92 electrical conductivity	
59 classification version	93 soluble Na^+	
60 national classification	94 soluble Ca^{2+}	
61 Soil Taxonomy	95 soluble Mg^{2+}	
62 phase		

2.4.1.3 Subnational Level
In various small regions, SOTER databases have been compiled at a scale of 1:250,000 to 1:100,000, first as a test of the applicability of the methodology (de Oliveira and van den Berg, 1992), and later as an operational activity. There are too many to mention all regional and national initiatives that emerged in the recent past.

2.4.2 Utilization

2.4.2.1 Water Erosion Risk
The importance of water erosion in the process of land degradation has led to the development of a SOTER application that assesses the erosion risk. For that purpose, a computer programme has been designed that can compute the relative erosion losses by SOTER units. The programme, called SWEAP (van den Berg and Tempel, 1995), uses existing erosion models, with slight modifications to adapt them to the specified data definitions of SOTER, such as the Universal Soil Loss Equation (Wishmeier and Smith, 1978) and the Soil Loss Estimation Model for Southern Africa (Ewell and Stocking, 1982). The program uses terrain and soil data from the SOTER database at scales of 1:1 million or greater. Climate and land use data are to be taken from the same database. The model has been applied amongst others in Kenya (van Engelen et al., 1997).

For use at smaller scales, a qualitative assessment has been developed (Batjes, 1996). Soil erodibility has been determined by an index derived from topsoil texture, structure and soil depth. Slope gradients were taken directly from the SOTER database while rainfall erosivity was derived from Agro-Ecological Zones and tabular data.

2.4.2.2 Land Evaluation
An automated procedure for qualitative land evaluation was developed and implemented in the Automated Land Evaluation System (ALES) designed by Rossiter and van Wambeke (1993) on the basis of the Framework of Land Evaluation (FAO, 1976). The objective was to produce an application that allows for a quick separation of potentially suitable from nonsuitable SOTER units for the intended land use, indicating constraints to different kinds of land use.

Several land utilization types (LUT) have been implemented in the model, ranging from rainfed cultivated maize and sorghum under low technology and input levels in Kenya (Mantel, 1995), toward rainfed cultivated wheat under medium technology and input levels (Mantel and van Engelen, 1997). These LUTs are characterized by 11 land use requirements and evaluated by matching land use requirements with the corresponding land qualities. For the assessment of the land quality water availability during the growing season, a simple water balance model (WATSAT) has been designed based on the work of Doorenbos and Kassam (1979) and Driessen and Konijn (1992).

2.4.2.3 Salinity Status
The salinity status of the soils in the SOTER database can be assessed and their estimated yield for a wide range of crops can be calculated with a program that compares salt data of the soil by either direct measurement of soluble salts or by deduction in case of their absence (Rotmans and Riezebos, 1997).

2.4.2.4 Land Degradation and Food Productivity
The impact of erosion on productivity of a land use system, given the variability in natural conditions (e.g., soils, landform and climate), has been studied in three countries with a SOTER database, situated in two regions; South America (Uruguay and part of Argentina) and East Africa (Kenya) (Mantel and

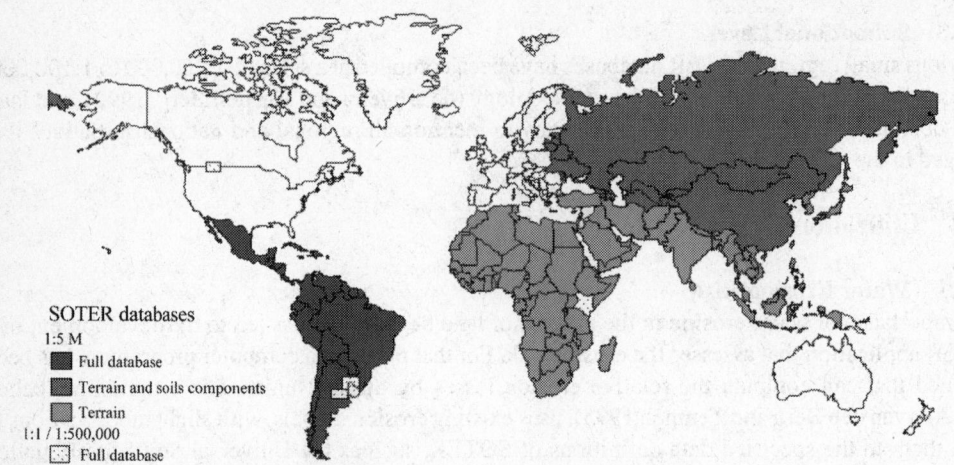

Fig. 2.4 Status of SOTER databases (1997)

van Engelen, 1997). This application made a link between land evaluation and water erosion models on one side, and a crop growth simulation model on the other side.

The study areas are characterized by different types of land use and occur in highly varying agro-ecological conditions. A chain of models has been used to study the impact of erosion on crop production now and in the future. Soils and terrain attributes have been analyzed in various models and were linked to a Geographical Information System (GIS), permitting spatial analysis. For each mapping unit suitable for the land use, the potential yield before and after an erosion scenario of 20 years was calculated. The impact of change in soil properties influencing crop performance, induced by removal of topsoil through sheet erosion, has been analyzed in this study. In the two countries in Latin America, soil erosion affected mostly the physical properties of the soils, resulting in 25-50% decline in potential yields. In Kenya, the largest yield reduction was mainly due to loss in soil fertility.

2.5   Discussion and Conclusions

There is need for a natural resource information system that can satisfy the needs of a wide array of users. As new soil survey reports become more complex, the task of extracting the required information gets more arduous and often impossible for those users who are not technically trained in soils. So far, traditional soil maps and their digitized versions irrespective of scale do not offer the possibility of delivering the required information to the users, as most of the data collected during the surveys cannot be displayed on the map or are lacking in the legend. In SOTER, almost all original field data can be stored to create a geographically referenced natural resources information that has the capability of extracting a range of single-value maps or data and interpretative products in response to the specific information and data needs of a wide diversity of decision and policy makers.

At a continental level, it is expected that a 1:5 million SOTER database will replace the current global soil map of FAO-Unesco (1974-1981) by the year 2002. At a national level, a global coverage will be more difficult to reach. It will depend on the success of existing SOTER activities: is SOTER really delivering the goods the users are requesting?

2.6 References

Agriculture Canada. 1983. The Canada soil information system (CanSIS). Manual for describing soils in the field. Land Resources Research Centre, Ottawa, Canada.

Arnold R.W. 1995. Role of soil survey in obtaining a global carbon budget. p. 257–263. *In* R. Lal, J. Kimble, E. Levine and B.A Stewart (ed.) Soils and global change. Lewis Publishers, Boca Raton, FL.

Batjes, N.H. 1995. World Inventory of Soil Emission Potentials (WISE 2.1) - Profile database user manual and coding protocols. Technical Paper 26. ISRIC, Wageningen, Netherlands.

Batjes, N.H. 1996. A qualitative assessment of water erosion risk using the 1:5M SOTER database for northern Argentina , south-east Brazil and Uruguay. Working Paper and Preprint 96/04. ISRIC, Wageningen.

Batjes, N.H., E.M. Bridges and F.O. Nachtergaele. 1995. World inventory of soil emission potentials: Development of a global soil data base of process-controlling factors. p. 102–115. *In* S. Peng et al. (ed.) Climate change and rice. Springer Verlag, Heidelberg, Germany.

Baumgardner, M.F. 1990. A 1:1 M world soils and terrain digital database: implementing a concept. Trans. 14th Int. Congr. Soil Sci. V:113–119.

Baumgardner, M.F., and L.R. Oldeman. 1986. Proceedings of an International Workshop on the Structure of a Digital International Resources Map Annex Data Base. 20-24 January 1986, Wageningen, ISSS, Wageningen, Netherlands.

de Oliviera, J.B., and M. van den Berg. 1992. Applications of the SOTER methodology to a semi-detailed survey (1:1000,000) in Piricaba region, Sao Paulo State, Brazil. SOTER Rep. 6. ISSS-ISRIC, Wageningen, Netherlands.

Doorenbos J., and A.H. Kassam. 1979. Yield response to water. FAO Irrig. Drain. Pap. 33.

Driessen, P.M., and N.T. Konijn. 1992. Land use system analysis. Pudoc, Wageningen, Netherlands.

Ewell, H.A., and M.A. Stocking. 1982. Developing a simple yet practical method of soil loss estimation. Trop. Agric. 59:43–48.

FAO. 1976. A framework for land evaluation. FAO Soils Bull. 32.

FAO. 1989. FAO-ISRIC soil data base. FAO World Soil Resour. Rep. 64.

FAO. 1993. Global and national soils and terrain digital databases (SOTER). Procedures manual. FAO World Soil Resour. Rep. 74.

FAO. 1995. Digital Soil Map of the World and derived soil properties. Vers. 3.5. FAO, Rome, Italy.

FAO, ISRIC, UNEP, and CIP. 1998. Soil and terrain database for Latin America and the Caribbean, 1:5 million scale. Land and Water Digital Media Series 5. FAO, Rome, Italy.

FAO-Unesco. 1974-1981. Soil Map of the World 1:5,000,000. Vol. I-X. Unesco, Paris, France.

Gunn, R.H., J.A. Beattie, R.E. Reid, and R.H.M. van de Graaff. 1990. Australian soil and land survey handbook. Inkata Press, Melbourne, Australia.

ISRIC. 1990a. SOTER Procedures manual for small-scale map and database compilation. 3rd vers. V.W.P. van Engelen and J.H.M. Pulles (ed) Working paper and preprint 90/2. ISRIC, Wageningen, Netherlands.

ISRIC. 1990b. Proceedings of the international workshop on procedures manual revisions for the Global Soils and Terrain Digital Database, Wageningen 24-26 April 1990. N.H. Batjes (ed.) Working paper and preprint 90/5. ISRIC, Wageningen, Netherlands.

ISRIC. 1991. The SOTER manual. Procedures for small scale digital map and database compilation of soil and terrain conditions. V.W.P. van Engelen and J.H.M. Pulles (ed.) Working paper and preprint 91/3. ISRIC, Wageningen, Netherlands.

ISSS. 1986a. Proceedings of an international workshop on the structure of a digital international soil resources map annex database. M.F. Baumgardner and L.R. Oldeman (ed.) SOTER Report 1, ISSS, Wageningen, Netherlands.

ISSS. 1986b. Project proposal "World soils & terrain digital database at a scale 1:1 M (SOTER)". M.F. Baumgardner (ed.). ISSS, Wageningen, Netherlands.

ISSS. 1988. Proceedings of the First Regional Workshop on a Global Soils and Terrain Digital Database and Global Assessment of Soil Degradation. SOTER Report 3. ISSS, Wageningen, Netherlands.

ISSS. 1989. Proceedings of the Second Regional Workshop on a Global Soils and Terrain Digital Database (12-16 December 1988, Porto Alegre). W.L. Peters (ed.). SOTER Report 4. ISSS, Wageningen, Netherlands.

Mantel, S. 1995. The automated land evaluation system applied to SOTER. With an example from West Kenya. Working paper and Preprint 95/03. ISRIC, Wageningen, Netherlands.

Mantel, S., and V.W.P. van Engelen. 1997. The impact of land degradation on food productivity. Case studies of Argentina, Kenya and Uruguay. Report 97/01. ISRIC, Wageningen, Netherlands.

McDonald, R.C., R.F. Isbell, J.G. Speight, J. Walker, and M.S. Hopkins. 1990. Australian soil and land survey. Field handbook. 2nd Ed. Inkata Press, Melbourne, Australia.

Oldeman, L.R., and V.W.P. van Engelen. 1993. A world soils and terrain digital database (SOTER) - An improved assessment of land resources. Geoderma 60:309–325.

Pulles, J.H.M. 1988. A model for a soils and terrain digital database. Working Paper & Preprint 88/8. ISRIC, Wageningen, Netherlands.

Rossiter, D.G., and A.R. van Wambeke. 1993. Automated land evaluation system (ALES). Ver. 4 user's manual. SCAS Teaching Series T93-2. Cornell University, Ithaca, NY.

Rotmans, A., and O. Riezebos. 1997. SOSA, SOTER salt status assessment programme. Beta version, ISRIC, Wageningen, Netherlands.

Scoppa, C., J. Salazar, R. DiGiacomo, R. Irutia, P. Fasolo, J. Olmos, J. Potter, J. Molfino, and C. Alvarez. 1990. Compiling, generating and correlating a common soils and terrain database for Argentina, Brazil and Uruguay. Trans. 14th Int. Congr. Soil Sci. V:126–131.

Shields, J.A. 1992. Preliminairy report for North American Pilot Area of the SOTER project (NASOTER). Agriculture Canada, USDA and ISSS, Ottawa, Canada.

Shields, J.A., and D.R. Coote. 1988. SOTER procedures manual for small scale map and data base compilation. Working Paper and Preprint 88/2. ISRIC, Wageningen, Netherlands.

Shields, J.A., and D.R. Coote. 1989. SOTER procedures manual for small scale map and data base compilation and procedures for interpretation of soil degradation status and risk (Revised version). Working Paper and Preprint 89/3, ISRIC, Wageningen, Netherlands.

Shields, J.A., and D.R. Coote. 1990. Development, documentation and testing of the Soil and Terrain (SOTER) database and its use in the Global Assessment of Soil Degradation (GLASOD). Trans. 14th Int. Congr. Soil Sci. V:120–125.

Sombroek, W.G. 1984. Towards a global soil resource inventory at scale 1:1 million. Working Paper and Preprint 84/4. ISRIC, Wageningen, Netherlands.

Tempel, P. 1994. Global and national soils and terrain digital databases (SOTER). Database structure. Working Paper and Preprint 94/05. ISRIC, Wageningen, Netherlands.

van den Berg, M., and P. Tempel. 1995. SOTER water erosion assessment programme. Documentation vers. 1.5. SOTER Report 7. ISSS, UNEP, ISRIC, Wageningen, Netherlands.

van Engelen, VW.P., and T.T. Wen. 1993. Global and national soils and terrain digital databases (SOTER). Procedures manual. ISSS-UNEP-FAO-ISRIC, Wageningen, Netherlands.

van Engelen, V.W.P., and T.T. Wen. 1995. Global and national soils and terrain digital databases (SOTER). Procedures manual (revised ed.). ISSS-UNEP-FAO-ISRIC, Wageningen, Netherlands.

van Engelen, V.W.P., J.W. Resink, and P.T. Gicheru. 1997. Kenya: A study using the SOTER methodology. p. 114–119. In N. Middleton and D. Thomas (ed.) World atlas of desertification. 2nd Ed. Arnold, London, UK.

Wischmeier, W.H., and D.D. Smith. 1978. Predicting rainfall erosion losses. USDA Agric. Handb. 537.

Zobler, L. 1986. A world soils file for global climate. NASA Tech. Memo. 87-202. Goddard Institute for Space Studies, New York, NY.

3

Development of a 0.5° by 0.5° Resolution Global Soil Database

N.H. Batjes
International Soil Reference and Information Centre
Wageningen, Netherlands

3.1 Introduction

Soils are an essential part of natural ecosystems and necessary for the growth of human food, animal fodder, fiber and timber crops. Formed at the Earth's surface, the soil is the layer in which naturally radiatively active gases, such as CH_4 and N_2O, are exchanged with the atmosphere and in which contaminants tend to accumulate. The close environmental links between soils, crops and water mean that if soils are degraded or polluted, the foodchain and drinking water supplies are likely to be affected also (Arnold et al., 1990; Houghton et al., 1990; Stigliani, 1991).

The availability of geographic databases of parameters such as vegetation cover, climate and soil properties and of the driving socioeconomic variables have created opportunities for extrapolation and modeling at the global level (Prentice et al., 1992; Hagen et al., 1993; Matthews, 1993; Zuidema et al., 1995). These databases are needed at scales and resolutions that reflect the effects on regional and global processes which are relevant to modeling goals (Wessman, 1992). However, many of the soil datasets currently available for global environmental research were based of necessity on limited soil profile data and coarse resolution spatial data (Zobler, 1986). A number of less than satisfactory results from General Circulation Models (GCM) have been attributed to inadequate parameterization of values used; for example, soil water-holding capacity has been estimated from the topsoil texture only, the contribution from the subsoil being ignored (Wessmann, 1992). Consequently, there is a need for up-to-date and uniformly described soil data for impact and scenario studies at the regional and global levels (Oldeman and van Engelen, 1993; Arnold, 1995; FAO, 1995; Madsen and Jones, 1995), notably in relation to data reliability and risk assessment in soil interpretations (Mausbach and Wilding, 1991; Burrough, 1992; Nettleton et al., 1996).

During the last decade, the availability of national and regional soil databases has increased greatly (Section H, Chapters 1, 2, 4, 5). At the global level, the situation has also improved with the release of the digital Soil Map of the World by the Food and Agriculture Organization (FAO), and the WISE (World Inventory of Soil Emission Potentials) database by the International Soil Reference and Information Centre (ISRIC). In the future, the challenge for soil scientists will be to update these databases using the best available data, either from historical sources or new surveys, in order to maintain uniform sets of primary and derived soil data that are readily applicable by user groups in the research and resource planning areas.

This chapter first outlines the rationale for developing the 0.5° by 0.5° WISE soil database at ISRIC, a short-duration project entitled World Inventory of Soil Emission Potentials (Batjes et al., 1995). Thereafter, the structure and contents of the database are presented, specifying the source materials used for data compilation. Finally, the scope of the WISE database in global research is discussed, while simultaneously identifying some of the remaining uncertainties.

3.2   Background and Rationale

Soil database development activities at ISRIC started in the mid-eighties when the information held in its World Soil Reference Collection was computerized (Batjes et al., 1994). A second thrust has been to develop a methodology for updating both the area and attribute data on world soils in SOTER, the Soils and Terrain Digital Database project (Oldeman and Van Engelen, 1993). SOTER was initiated in 1986 at the request of the International Society of Soil Science, during the 13th International Congress at Hamburg. First, this has been done in recognition of the fact that parts of the 1:5,000,000 scale Soil Map of the World (FAO-Unesco, 1974-1981) are now out of date (Sombroek, 1990). Second, SOTER was launched in response to the increased demand for quantitative data for land use planning and environmental research (Bouma and Bregt, 1989; Matson and Ojima, 1990; Ingram and Gregory, 1996).

From a technical and scientific standpoint, world soil coverage in a 1:5,000,000 SOTER should require about 5 to 10 years (Section H, Chapter 2). In practice, however, the speed of database implementation will be largely dictated by donor funding. Meanwhile, modellers will require actualized, georeferenced and quantified data considered representative for the major soils of the world. This demand has been met through WISE. In spite of its somewhat ambiguous name, the WISE database was conceived especially for a geographical quantification of main soil factors that control processes of global change, including fluxes of greenhouse gases and effects of land degradation on global food production.

3.3   Database Development

3.3.1   Structure

The integrated WISE database is comprised of two main components (Fig. 3.1): (1) the area data in a digital file which specifies the type and relative extent of the component soil units (FAO-Unesco, 1974) of each terrestrial 0.5° by 0.5° degree grid-cell, and (2) the attribute data which comprise a selection of soil profile data for the 106 soil units shown on the digital grid map, to which are attached subfiles listing the analytical methods and source of primary data used (Fig. 3.2).

3.3.2   Area Data

Small scale maps encompass a marked degree of data integration, the aim being to simplify the geographical distribution of soils to a regionally representative pattern of spatially dominant soils. In the context of the WISE project, the predictive use of the 1:5,000,000 scale Soil Map of the World was of significant importance. Although parts of this map, compiled in the 1960s and 1970s (FAO-Unesco, 1974-1981), are now out-of-date, the digital version (FAO, 1995) will remain the best available source on the global distribution of soils pending completion of a 1:5,000,000 scale SOTER database (Section H, Chapter 2).

The areas of soil units depicted on the 5' by 5' version of the digital Soil Map of the World were used as a cartographic basis for the preparation of a 0.5° by 0.5° resolution database for WISE. First, the type

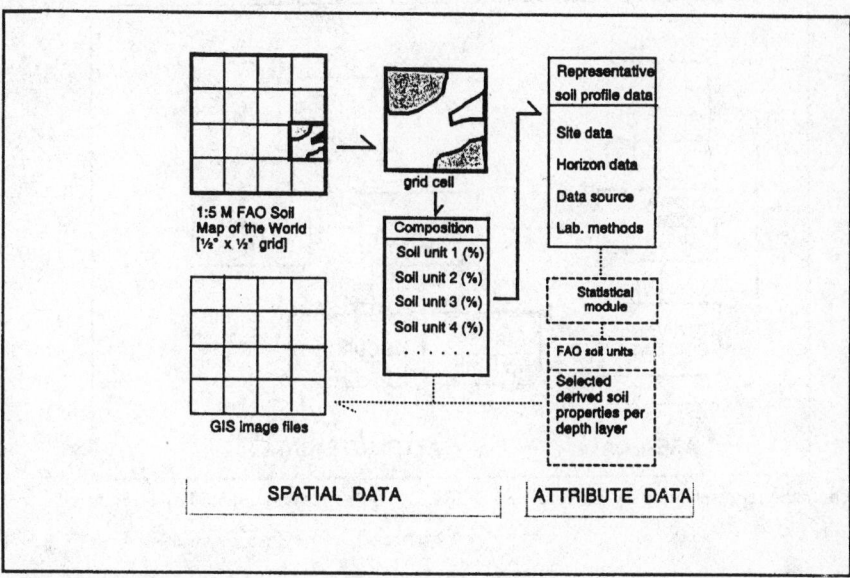

Fig. 3.1 Schematic representation of the WISE database [Reprinted from Batjes, 1997. Soil Use Manag. 13:9–16 with permission]

and relative area of the component soil units occurring at the centre of each 5' by 5' grid cell have been determined, using FAO's (1995) composition rules for soil mapping units. Subsequently, the information for the 36 component 5' by 5' cells of a particular 0.5° by 0.5° grid have been analyzed in order to identify the dominant soil units (up to 10), and their relative extent, in each half degree grid. The algorithms for this part of the work have been prepared by FAO Land and Water Development Division staff in close cooperation with the WISE project.

3.3.3 Point Data

Prior to developing the WISE database structure, the major soil factors needed for a wide range of global environmental studies were reviewed. The selected attributes may be considered in three groups: general information, physical data and chemical soil data (Table 3.1). These attributes, which are common to both the European Soil Database (Madsen and Jones, 1995) and proposals for an International Geosphere-Biosphere Programme Data Information System (IGBP-DIS) Global Soil Database (Scholes et al., 1995), appear to have gained wide support among the scientific community.

In the process of collating materials for compilation of the WISE profile database the quality and validity of the original data had to be evaluated carefully, while at the same time recognizing that these are the only materials available. Four general criteria have been defined for accepting profiles: (1) completeness and apparent reliability of data, (2) traceability of source of data, (3) classifiable in the FAO-Unesco (1974) legend, and (4) georeferenced within defined limits. In the systematic compilation of the soil profile data, special attention has been given to documenting the laboratory methods by which the various analytical results were obtained. In conjunction with this work, Vogel (1994) carried out a review of the comparability of different soil analytical methods in order to permit a screening by analytical procedures during the statistical analyses of the data.

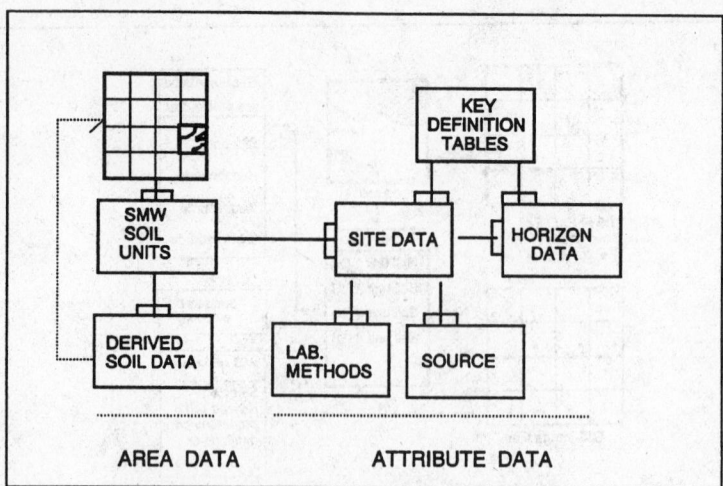

Fig. 3.2   Schematic representation of the SITE and HORIZON data files and related key-attribute files

The soil profile data in WISE were compiled from five sources: (1) ISRIC's Soil Information System (ISIS), (2) FAO's Soil Database (SDB), (3) the soil analytical data set of the Natural Resources Conservation Service (NRCS) of the United States of America, (4) profile descriptions chosen by national soil survey organizations to be representative for the units of the Soil Map of the World present in their countries, and (5) data gathered from survey monographs held in ISRIC's library. Holdings of the ISIS, SDB and NRCS databases were offloaded into the uniform WISE format using an automated transfer facility (Tempel, 1994). All digitally transferred data have been subjected to routine quality checks similar to those applied to the profile data entered manually (Batjes, 1995). In addition, the inferred quality of the original profile descriptions has been estimated in WISE.

3.3.4   Geographic and Taxonomic Coverage

The current version of WISE holds 4353 profiles which originate from Africa (1799), South America and the Caribbean (599), South East Asia (553), South West and North Asia (522), Europe (492), North America (266), and Australia and the Pacific Islands (122). Fig. 3.3 shows that the distribution of profiles throughout the world is not uniform; important geographic gaps are China, the former Soviet Union, and the northern Territories of Canada. Conversely, it is not surprising that few soil profile descriptions are available from the arid, central parts of the Sahara.

Table 3.2 lists the classification (FAO-Unesco, 1974) of the profiles currently held in WISE. As most profile descriptions have been obtained from routine type of soil surveys, complete horizon datasets are not always available for all the considered attributes. Consequently, the number of observations for each of these attributes will vary between soil units, and with the depth range considered.

If it is assumed that at least 10 profile descriptions will be needed for a satisfactory characterization of a soil unit represented on the 1:5,000,000 scale Soil Map of the World and considering analytical results obtained with different methods, in a range of laboratories, are not necessarily comparable (Pleijsier, 1989; Vogel, 1994), it follows that additional profile descriptions are needed for: all Greyzems and Podzoluvisols, gleyic Chernozems, plinthic members of Ferralsols and Gleysols, gelic members of Gleysols, Histosols and Regosols, shallow Lithosols and Rankers, luvic Kastanozems,

Table 3.1 Soil attributes considered in the WISE profile database [Reprinted from Batjes, 1994. J. Geophys. Res. 99:16,479–16,489 with permission of the American Geophysical Union]

Site Data	Horizon Data
WISE_ID (Unique identifier of profile)	WISE_ID + horizon_NO (unique reference number
Soil classification and source	for horizon within a profile)
FAO-Unesco classification (1974 legend)	
phase	General attributes
topsoil texture class	horizon designation
FAO-Unesco classification (1990 rev. legend)	depth, top
phase	depth bottom
USDA subgroup level classification	matrix color (dry and moist)
edition (year) of Soil Taxonomy	mottling
local classification	presence of roots
source of data	
name of laboratory where analyses were made	Chemical attributes*
soil profile description status	organic carbon
date of description	total N
	available P
Location	pH-H_2O
country	pH-KCl
location of soil profile, descriptive	pH-$CaCl_2$
latitude (deg/min/s)	electrical conductivity (EC)
longitude (deg/min/s)	free $CaCO_3$
altitude	exchangeable Ca^{2+}
	exchangeable Mg^{2+}
General site data	exchangeable Na^+
major landform	exchangeable K^+
landscape position	exchangeable Al^{3+} + H^+ (exchangeable acidity)
aspect	exchangeable Al^{3+} (exchangeable aluminum)
slope	cation exchange capacity (CEC)
drainage class	effective CEC (at field pH)
groundwater depth	base saturation (as percent of CEC)
effective soil depth	
parent material	Physical attributes*
Köppen climate classification	structure type
land use	particle size distribution:
natural vegetation	weight % sand
	weight % silt
	weight % clay
	stone and gravel content
	bulk density
	volume percent water held at specified suctions
	hydraulic conductivity at specified suctions

WISE, World Inventory of Soil Emission Potentials
*Analytical methods are specified in a separate key-attribute file

ferric Podzols, gelic Regosols, mollic Solonetz, mollic and humic Planosols, gypsic Xerosols, takyric and haplic Yermosols, takyric and mollic Solonchaks. These under-represented legend units account for about 28% of the world land area; shallow and rocky Lithosols account for about 40% of this total.

In the absence of adequate profile data, or their limited accessibility due to strict rules that need to be followed by the data users (Jones and Buckley, 1997), the geographical and taxonomic spread of profiles in WISE is the best that could be achieved in the project period available for data compilation. The short-term duration of projects such as WISE, however, largely obscures the ongoing nature of

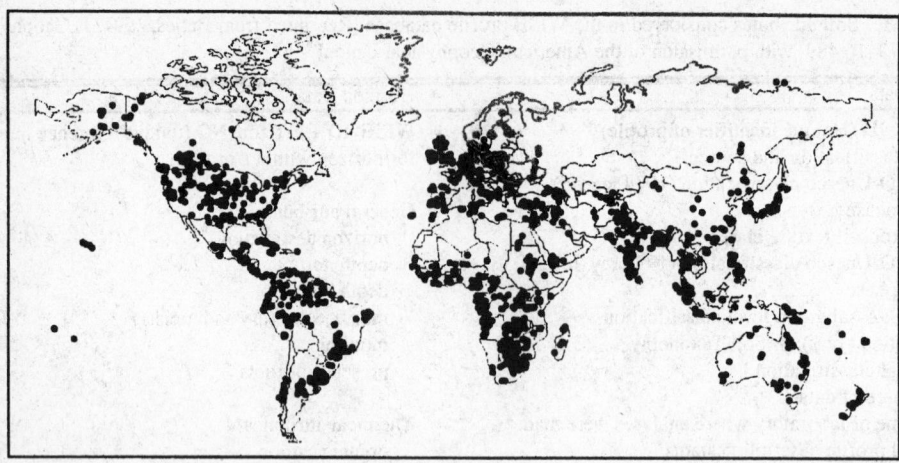

Fig. 3.3   Geographic distribution of soil profiles held in WISE

soil data collection activities. With larger numbers of representative profiles, the effectiveness of the derived datasets will be improved, gradually allowing for a better estimation of median values and confidence intervals for the major soil attributes.

3.4   Applications

The integrated WISE database, in combination with auxiliary sources of the main biophysical and socioeconomic variables (Kineman and Ohrenschall, 1993), can be used for a range of environmental studies at resolutions coarser than 0.5° by 0.5°. Such studies include assessments of crop production potentials, soil gaseous emission potentials, land degradation by water (Batjes, 1996c) and pollution, and studies of global carbon pools (Batjes, 1996b). Depending on the topics under investigation (Table 3.3) and the spatial and temporal resolution required, the various primary soil data may have to be aggregated and analyzed in different ways (Batjes, 1997; Batjes et al., 1997). The resulting secondary files of derived soil data will provide a uniform basis for production of thematic maps, using GIS, and input for auxiliary modeling activities.

Physical soil attributes are seldom collected on a routine basis, particularly water retention versus tension relationships, unsaturated hydraulic conductivity and bulk density, because they are difficult and expensive to measure. Thus they are underrepresented in the WISE soil profile database and most other digital soil databases. The solution then is to estimate these parameters from the available measured soil data (Table 3.4) using pedotransfer (Bouma and van Lanen, 1987; Tietje and Tapkenhinrichs, 1993; Batjes, 1996a). Scholes et al. (1995) presented a tabular overview of primary and secondary soil data needed in relation to various core activities of the IGBP. These applications include biogeochemical, plant response, water balance, trace gas, and environmental impact models.

An important development has been the adoption by the Global Soil Data Taskgroup of the IGBP-DIS for a subset of the WISE profile database as a foundation for some of their modeling work (Scholes et al., 1995). The full profile complement of WISE has been used in a collaborative activity of FAO, the International Institute for Applied Systems Analysis (IIASA) and ISRIC to generate uniform files of derived soil properties for use in regional and global Agro-Ecological Zoning studies. The derived attributes can be linked to both the vector and 5' by 5' versions of the digital Soil Map of the World (FAO, 1995), using the code for soil unit and topsoil textural class as the logical link. The

Table 3.2 FAO-Unesco (1974) soil units represented in WISE

FAO soil unit*
A: Acrisols
Af=124 Ag=21 Ah=71 Ao=68 Ap=36
B: Cambisols
Bc=30 Bd=91 Be=140 Bf=47 Bg=49 Bh=49 Bk=115 Bv=45 Bx=17
C: Chernozems
Cg=0 Ch=24 Ck=32 Cl=14
D: Podzoluvisols
Dd=5 De=5 Dg=1
E: Rendzinas
C=35
F: Ferralsols
Fa=24 Fh=50 Fo=85 Fp=8 Fr=44 Fx=50
G: Gleysols
Gc=15 Gd=63 Ge=90 Gh=33 Gm=47 Gp=4 Gx=7
H: Phaeozems
Hc=25Hg=18 Hh=73 Hl=92
I: Lithosols
I=8
J: Fluvisols
Fc=141 Jd=32 Je=167 Jt=26
K: Kastanozems
Kh=13 Kk=14 Kl=1
L: Luvisols
La=28 Lc=109 Lf=114 Lg=101 Lk=145 Lo=148 Lp=12 Lv=17
M: Greyzems
Mg=1 Mo=7
N: Nitosols
Nd=25 Ne=43 Nh=13
O: Histosols
Od=35 Oe=11 Ox=4
P: Podzols
Pf=2 Pg=15 Ph=20 Pl=11 Po=29 Pp=12
Q:Arenosols
Qa=12 Qc=184 Qf=89 Ql=36
R: Regosols
Rc=28 Rd=35 Re=54 Rx=2
S: Solonetz
Sg=17 Sm=5 So=42
T: Andosols
Th=90 Tm=28 To=16 Tv=31
U: Rankers
U=8
V: Vertisols
Vc=152 Vp=148
W: Planosols
Wd=10 We=22 Wh=1 Wm=8 Ws=21 Wx=0
X: Xerosols
Xh=20 Xk=19 Xl=88 Xy=8
Y: Yermosols
Yh=9 Yk=13 Yl=17 Yt=1 Yy=15
Z: Solonchaks
Zg=21 Zm=3 Zo=47 Zt=2

*For definitions of soil unit codes see FAO-Unesco (1974); total = 4353 profiles

Table 3.3 Examples of soil attributes needed for selected environmental studies* [Reprinted with permission from Batjes et al., 1995. World inventory of soil emission potentials. S. Peng (ed.) Climate change and rice. Copyright Springer-Verlag GmbH & Co. KG]

Topic	Soil Processes	Relevant Soil Factors
Crop production	water release, weathering, cation exchange	fertility status, salinity, soil moisture characteristics and rootable depth
Pollution by heavy metals	adsorption, solubility	Organic matter content, pH, $CaCO_3$ content, water balance and salinity
Acidification	weathering, base exchange	$CaCO_3$, exchangeable bases, cation exchange capacity (CEC) and mineralogy
CH_4 emission	methanogenesis and methane oxidation	Organic matter content, C/N ratio, pH, sulphate content, texture, N content, nutrient status, drainage class (redox potential) and soil structure/porosity
N_2O emission	denitrification and nitrification	Organic matter content, texture, pH, soil moisture characteristics, soil drainage nutrient status and soil structure/porosity

*Also needed from auxiliary sources are data on climate, land use/vegetation, management practices (e.g., fertilizer application; organic matter amendments; irrigation/drainage practices; loading with pollutants) and information on the main controlling socio-economic factors

preliminary subset (Batjes et al., 1997), complemented with information derived directly from the Soil Map of the World (e.g., drainage, depth), is being circulated amongst a group of experts to fill the remaining gaps. Subsequent to this data completion and peer-review process, the validated digital file is intended to provide a uniform set of derived soil properties for modeling at a regional and global scale. In a preceding study, the profile database has been used to create a set of median, derived soil properties for use with the 0.5° by 0.5° grid version of Soil Map of the World adopted for WISE (Batjes, 1997). Contrary to what has been the case for the 5' by 5' resolution version, the composition rules for topsoil textural class, slope and phases of FAO (1995) could not be maintained for WISE; this was largely in view of the added uncertainty of these rules when used at a spatial resolution of 0.5° by 0.5° (i.e., 36 times lower than for 5' by 5' Soil Map of the World).

3.5   Discussion and Conclusions

Global soil databases, such as the digital Soil Map of the World and WISE, are needed to study broad questions arising at the supranational level, for instance soil gaseous emission potentials, soil vulnerability to processes of environmental degradation, and possible impacts of land degradation on strategic food production. A key question that will arise in such studies is: What variation in soil data is acceptable for applications at a particular scale and resolution? This is an aspect worthy of further research. Regional needs for soil information will be geared to resolving problems that can now be well defined, necessitating more detailed spatial and attribute data than are available in WISE. Examples of such databases are the USDA Soil Database (Section H, Chapter 5) and the European Soil Database (Le Bas and Jamagne, 1996). Data sharing should be encouraged by good metadata and network information resource tools, which provide browsing, searching, and management capabilities for information distributed through computer networks (Medyckyj-Scott et al., 1996).

Uncertainty in databases, particularly global ones, can arise from different sources. These include errors in the original materials used in database compilation, arbitrary assumptions for generating new

Table 3.4 Examples of secondary information that can be derived from measured soil parameters held in WISE [After Ingram, 1993]

Primary Soil Parameters	Derived Data
Organic C, total N, bulk density and gravel content by horizon	organic C and N pools
$\theta(h)$, bulk density and rooting depth	water-holding capacity
Particle size distribution, moisture content and heat capacity of soil constituents	heat conductance
CEC, OC content, clay content	clay mineralogy
CEC and exchangeable bases	soil nutrient status
pH	soil acidity; inferred toxicities (e.g., Al^{3+})

spatial data, and inconsistencies in translation of the original classification systems to record data into a uniform format (Goodchild, 1994; Kiiveri, 1997). Additional uncertainties are associated with the scaling up of site data to a global level (Rosswall et al., 1988; Walker, 1994). The various types of uncertainties are difficult to evaluate and will vary among datasets and models used (Leemans and van der Born, 1995; Pan et al., 1995).

Although attempts have been made in the past to use the printed Soil Map of the World as a basis for determining the nature of the soil cover for modeling purposes (Zobler, 1986), the resolution used was so coarse that the potential of the map as a source of information has only been partially exploited (Webb et al., 1991; Bachelet and Neue, 1993; Bouwman et al., 1993). Being based on center point sampling of the original Soil Map of the World on a 1° by 1° grid, the Zobler map largely ignores the natural spatial heterogeneity of soils within a grid cell and introduces a degree of uncertainty that is probably unacceptable (Arnold, 1996). In this respect, the griding procedure developed by FAO staff as part of the WISE project has greatly increased the amount of information about the world soil pattern which can be derived from the 1:5,000,000 scale (0.5° by 0.5° resolution) Soil Map of the World. Regional and continental studies at 1:5,000,000 scale that require a spatial resolution finer than 0.5° by 0.5°, however, should be based on the vector version of the digital Soil Map of the World, and ultimately on its successor SOTER.

So far the WISE database has assembled, in a uniform format, more than 4350 representative soil profiles with their morphological, chemical and physical attributes. A subset of 1125 of these profiles, all analyzed according to compatible methods, has been made available in the public domain for global modeling work by IGBP-DIS and other user groups. Recognizing that most potential users are not interested in the raw data per se but in more complex derived parameters relevant to their specific applications, ISRIC has made available in the public domain several sets of derived soil data on a 0.5° by 0.5° grid. Nonetheless, many of the opportunities provided by the existence of the WISE database still remain to be exploited within collaborative projects.

While there is no physical limit for the maximum number of profiles that can be accommodated in WISE, continued database maintenance and update are currently limited by a shortage of funds. In case of an eventual expansion of the WISE profile database, the long-term goal for data acquisition should be to exclusively collect soil profiles which have been described and sampled according to uniform standards (FAO-ISRIC, 1990; Soil Survey Staff, 1993) and analyzed in recognized laboratories which use comparable analytical methods. With larger numbers of profiles by soil unit in WISE, the effectiveness of the derived datasets will be improved, gradually allowing better estimation of medians and confidence intervals for the major soil attributes. In this activity, special attention should be paid to the under-represented soil units and regions of the world.

Operational, technical and scientific problems associated with the accuracy and regional representativeness of the various spatial and attribute data in global databases are well-known, yet difficult to remedy in studies based on available historic data. At this moment, there is an unfortunate tendency to overlook ground surveys of soil and terrain conditions, and the supporting laboratory analyses of soils, at representative sites. Compensation and recognition for data collectors and data custodians should become standard practice to encourage sustained, detailed data collection and harmonization of storage and analysis methods.

An important scientific concern with soil databases compiled from historic data is that they may not provide timely enough information for some applications of environmental change. Global monitoring systems are needed that register changes in soil quality with time in conjunction with changes in the driving biophysical and socioeconomic forces, for example, the Global Terrestrial Observing System (GTOS) conceived by FAO, UNEP, UNESCO, WMO and ICSU/IGBP. The global representativeness of the experimental and transect sites in terms of the soils and agro-ecological units represented can be derived from available soil databases, providing the geographic basis for extrapolation of site data and model output to a regional and global level.

Finally, free and unfettered access to and use of digital datasets will become increasingly difficult for a scientific research community that is being pressured into using accrued intellectual capital and data holdings as a measure of funding future activities. Data accessibility issues are complex because of technical, organizational and institutional aspects of data distribution. Of these, ownership, access rights and payment for use of data will probably be the most complex to resolve to the satisfaction of all parties concerned.

Acknowledgments

The WISE database was developed at ISRIC with initial sponsorship from The Netherlands National Research Programme on Global Air Pollution and Climate Change (NRP project 851039). Special thanks are due to Dr. E.M. Bridges (ISRIC) for contributions in developing the profile database, Dr. Nachtergaele (FAO) for developing the spatial griding algorithms, and the many organizations or individuals who provided representative soil profile descriptions for inclusion in WISE.

3.6   References

Arnold, R.W. 1995. Role of soil survey in obtaining a global carbon budget. p. 257-263. *In* R. Lai, J. Kimble, E. Levine and B.A. Stewart (ed.) Soils and global change. Lewis Publishers, Boca Raton, FL.

Arnold, R.W. 1996. Soil survey reliability: Minimizing the consumer's risk. p. 13–20. *In* W. D. Nettleton, A.G. Hornsby, R.B. Brown, and T.L. Coleman (ed.) Data reliability and risk assessment in soil interpretations. Soil Sci. Soc. Am. Spec. Pub. 47, Soil Science Society of America, Madison, WI.

Arnold, R.W., I. Szabolcs, and V.O. Targulian. 1990. Global soil change. International Institute for Applied Systems Analysis, Laxenburg, Austria.

Bachelet, D., and H.U. Neue. 1993. Methane emissions from wetland rice areas of Asia. Chemosph. 26:219-237.

Batjes, N.H. 1995. World inventory of soil emission potentials (WISE 2.1): Profile database user manual and coding protocols. Tech. Pap. 26. International Soil Reference and Information Centre, Wageningen, Netherlands.

Batjes, N.H. 1996a. Development of a world data set of soil water retention properties using pedotransfer rules. Geoderma 71:31-52.

Batjes, N.H. 1996b. Total carbon and nitrogen in the soils of the world. Europ. J. Soil Sci. 47:151-163.

Batjes, N.H. 1996c. Global assessment of land vulnerability to water erosion on a ½° by ½° grid. Land Degrad. Develop. 7:353-365.

Batjes, N.H. 1997. A world data set of derived soil properties by FAO-Unesco soil unit for global modeling. Soil Use Manag. 13:9-16.

Batjes, N.H., E.M. Bridges, and F.O. Nachtergaele, 1995. World inventory of soil emission potentials: Development of a global soil database of process-controlling factors. p. 102-115. *In* S. Peng, K.T. Ingram, H.U. Neue and L.H. Ziska (ed.) Climate change and rice. Springer-Verlag, Heidelberg, Germany.

Batjes, N.H., V.W.P. van Engelen, J.H. Kauffman, and L.R. Oldeman. 1994. Development of soil databases for global modeling. Trans. 15th World Congr. Soil Sci. 6a:40-57.

Batjes, N.H., G. Fischer, F.O. Nachtergaele, V.S. Stolbovoy, and H.T. van Velthuizen. 1997. Soil data derived from WISE for use in global and regional AEZ-studies (Ver. 1-0). Report IR-97-025. FAO/IIASA/ISRIC, International Institute for Applied Systems Analysis, Laxenburg, Austria.

Bouma, J., and H.A.J. van Lanen. 1987. Transfer functions and threshold values: From soil characteristics to land qualities. p. 106-110. *In* K.J. Beek, P.A. Burrough, and D.E. McCormack (ed.) Quantified land evaluation procedures. Publication 6, International Institute for Aerospace Survey and Earth Sciences (ITC), Enschede, Netherlands.

Bouma, J., and A.K. Bregt. 1989. Land qualities in space and time. PUDOC, Wageningen, Netherlands.

Bouwman, A.F., I. Fung, E. Matthews, and J. John, 1993. Global analysis of the potential for N_2O production in natural soils. Global Biogeochem. Cycl. 7:557-597.

Burrough, P.A. 1992. Soil variability: A late 20th century view. Soils Fert. 56:529-562.

FAO-ISRIC, 1990. Guidelines for soil description, 3rd Ed. Food and Agriculture Organization, Rome, Italy.

FAO-Unesco. 1974. Soil map of the world. Vol. 1. Legend. United Nations Educational, Scientific, and Cultural Organization, Paris, France.

FAO-Unesco. 1990. Soil map of the world: Revised legend (with corrections). Tech. Pap. 20. International Soil Reference and Information Centre, Wageningen, Netherlands.

FAO. 1995. Digital soil map of the world and derived soil properties (Ver. 3.5). Food and Agriculture Organization, Rome, Italy.,

FAO-Unesco, 1974-1981. Soil map of the world, 1:5,000,000. Vol. 1 to 10. United Nations Educational, Scientific, and Cultural Organization, Paris, France.

Goodchild, M.F. 1994. Sharing imperfect data. p. 102-110. *In* A. Singh (ed.) Proc. UNEP and IUFPRO International Workshop in cooperation with FAO on developing large environmental databases for sustainable development. GRID Information Series No. 22, UNEP/GRID, Sioux Falls, IA.

Hagen, N., H.B. Kleeberg, and O. Niekamp. 1993. Parameter estimation for ecosystem models with special regard to target functions. Model. Geo-Biosphe. Proc. 2-293-325.

Houghton, T.J., G.J. Jenkins, and J.J. Ephraums. 1990. Climate change: The IPCC's scientific assessment. Intergovernmental Panel on Climate Change, Cambridge University Press, Cambridge, UK.

Ingram, J. 1993. IGBP-DIS / GCTE global soils database workshop (October 8-9, 1992, Silsoe). IGBP-DIS Working Paper 7. International Geosphere-Biosphere Program Data and Information System, Paris, France.

Ingram, J., and P. Gregory. 1996. Global change and terrestrial ecosystems GCTE activity 3.3: Effects of global change on soils. Report 12. GCTE, Wallingford, UK.

Jones, R.J.A., and B. Buckley. 1997. European soil database: Information access and data distribution procedures. Directorate General Joint Research Centre, Space Applications Institute, European Soil Bureau, Ispra, Italy.

Kliveri, H.T. 1997. Assessing, representing and transmifting positional uncertainty in maps. Int. J. Geograph. Info. Syst. 11:33-52.

Kineman, J.J, and M.A. Ohrenschall. 1993. Global ecosystems database (Ver. 1.0). United States Department of Commerce, National Oceanic and Atmospheric Administration, National Geophysical Data Centre and World Data Centre-A for Solid Earth Geophysics, Boulder, CO.

Le Bas, C., and N. Jamagne. 1996. Soil databases to support sustainable development. Report EU16371. Joint Research Centre of the European Commission (EC/JRC) and Institut National de la Recherche Agronomique (INRA), Orléans, France.

Leemans, R., and G.J. van der Born. 1995. Determining the potential distribution of vegetation, crops and agricu;tural productivity. Water Air Soil Poll. 76:133-161.

Madsen, H.B., and R.J.A. Jones. 1995. The establishment of a soil profile analytical database for the European Union. p. 55-63. *In* D. King, R.J.A. Jones and A.J. Thomasson (ed.) European land information systems for agro-environmental monitoring. Official Publications of the European Communities, Luxembourg.

Matson, P.A., and D.S. Ojima. 1990. Terrestrial blosphere exchange with global atmospheric chemistry. Global Change Report 13. International Geosphere-Biosphere Programme (IGBP), Stockholm, Sweden.

Matthews, E. 1993. Global geographical databases for modeling trace gas fluxes. Int. J. Geo. Info. Syst. 7:124-142.

Mausbach, M.J., and L.P. Wilding. 1991. Spatial variability of soils and landforms. Soil Sci. Soc. Am. Spec. Pub. 28. Soil Science Society of America, Madison, WI.

Medyckyj-Scott, D., M. Cuthbertson, and I. Newman, 1996. Discovering environmental data: Metadatabases, network information resource tools and the GENIE system. Int. J. Geo. Info. Syst. 10:65-84.

Nettleton, W.D., A.G. Hornsby, R.B. Brown, and T.L. Coleman. 1996. Data reliability and risk assessment in soil interpretations. Soil Sci. Soc. Am. Spec. Pub. 47, Soil Science Society of America, Madison, WI.

Oldeman L.R., and V.W.P. van Engelen. 1993. A world soils and terrain digital database (SOTER): An improved assessment of land resources. Geoderma 60:309-325.

Pan, Y., A.D. Mcguire, D.W. Kicklighter, and J.M. Melillo. 1995. The importance of climate and soils for estimates of net primary production: A sensitivity analysis with the terrestrial ecosystem model. Global Change Biol. 96:5-23.

Pleijsier, K. 1989. Variability in soil data. p. 89-99. In J. Bouma and A. K. Bregt (ed.) Land qualities in space and time. PUDOC, Wageningen, Netherlands.

Prentice, I.C., W. Cramer, S.P. Harrison, R. Lleemans, R.A. Monserud, and A.M. Solomon. 1992. A global biome model based on plant physiology and dominance, soil properties and climate. J. Biogeogr. 19:117-134.

Rosswall, T., R.G. Woodmansee, and P. G. Risser. 1988. Scales and global change: Spatial and temporal variability in biospheric and geospheric processes. John Wiley, New York, NY.

Scholes, R.J., D. Skole, and J.S. Ingram. 1995. A global database of soil properties: Proposal for implementation. IGBP-DIS Working Paper 10. International Geosphere-Biosphere Program Data and Information System, Paris, France.

Soil Survey Staff. 1993. Soil survey manual. USDA Handbk. 18, USDA, Washington, DC.

Sombroek, W.G. 1990. Geographic quantification of soils and changes in their properties. p. 225-224. In A.F. Bouwman (ed.) Soils and the greenhouse effect. John Wiley, New York, NY.

Stigliani, W.M. 1991. Chemical time bombs: Definition, concepts, and examples. Executive Report 16. International Institute for Applied Systems Analysis, Laxenburg, Austria.

Tempel, P. 1994. Data transfer between disparate soil databases. Working Paper 94/08. ISRIC, Wageningen, Netherlands.

Tietje, O., and M. Tapkenhinrichs. 1993. Evaluation of pedo-transfer functions. Soil Sci. Soc. Am. J. 57:1088-1095.

Vogel, A.W. 1994. Comparability of soil analytical data: Determinations of CEC, organic carbon, soil reaction, bulk density and volume percent water at selected pF values by different methods. Working Paper 94/07. Wageningen, Netherlands.

Walker, B.H. 1994. Landscape to regional-scale: Responses of terrestrial ecosystems to global change. Ambio 24:67-73.

Webb, R.S., C.E. Rosenzweig, and E.R. Levine. 1991. A global data set of soil particle size properties. NASA Tech. Mem. 4286. Goddard Institute for Space Studies, New York, NY.

Wessmann, C.A. 1992. Spatial scales and global change: Bridging the gap from plots to GCM grid cells. Ann. Rev. Ecol. Syst. 23:175-200.

Zobler, L. 1986. A world soils file for global climate. NASA Tech. Memo. 87-802. Goddard Institute for Space Studies, New York. NY.

Zuidema, G., G.J. van der Born, J. Alcamo, and G.J.J. Kreileman. 1995. Simulation of global land cover changes as affected by economic factors and climate. Water Air Soil Poll. 76:163-198.

4

The Canadian Soil Database

D.R. Coote
Agricultural Watersheds Associates, Stittsville, Canada

K.B. MacDonald
Agriculture and Agri-Food Canada, Ottawa

4.1 Canadian Soil Database

Since 1972, the Canadian Soil Information System (CanSIS) has supported research activities of Agriculture and Agri-Food Canada (AAFC) by building the National Soil DataBase (NSDB) and acting as a source of GIS products and expertise through its personnel, GIS systems and procedures. Within AAFC, soil data are the responsibility of the Research Branch, through CanSIS, working via a network of federal staff located in each of the provinces (Canadian Land Resource Network - CLRN). CanSIS is part of the Eastern Cereals and Oilseeds Research Centre (ECORC) and is located on the Central Experimental Farm in Ottawa. CanSIS handles soil, geographic, climate, land use and crop yield data. The soil data are part of the National Soil Database (NSDB). The system and GIS procedures are implemented using Arc/Info software (Environment System Research Institute, Redland, CA). The following sections deal with federally held GIS-based soil and related data, as well as with the various complimentary provincial datasets in those provinces having the capability to use GIS. Much of the following information is updated at http://res.agr.ca/CANSIS/NSDB/.

4.2 The National Soil Database (NSDB)

The NSDB serves as the national archive for land resource information that was collected by federal and joint federal-provincial field surveys, or created by land data analysis projects. In most cases, each province or territory maintains a duplicate archive of data pertinent to its area, frequently supplemented by data of local interest. These archives are the responsibility of various departments, usually Agriculture, Environment or Natural Resources, of the respective province or territory.

The NSDB is the set of computer readable files which contain soil, landscape, and climatic data for all regions of Canada. The NSDB includes GIS coverages at a variety of scales. The majority of effort has been devoted to the compilation and management of soils information. Information about the characteristics and distribution of Canadian soils is compiled at three levels of detail ranging from very broad national coverage to quite detailed localized soil inventory projects. The broad categories of soils information in the NSDB are as follows:

Soil Map of Canada: compilation scale 1:5 million; coverage includes the entire land mass of Canada; originally produced in the early 1980s; much of the data will be replaced by data generalized from the Soil Landscapes of Canada.

Soil Landscapes of Canada (SLC): compilation scale 1:1 million; coverage includes the entire land mass of Canada; data include major soil and landscape attributes.

Detailed Soil Surveys: compilation scale varies (1:20,000 to 1:250,000); coverage includes much of the significant agricultural areas of Canada; data content varies; printed provincial indexes are available.

Datasets created by land analysis projects include:

Land Potential Database (LPDB): compilation scale 1:5 million using Soils of Canada polygons; coverage includes the entire land mass of Canada; encompasses extensive databases of modelled and actual crop yield, climate, etc.

National Ecological Framework: compilation scale 1:2 million; coverage includes the entire land mass of Canada; polygons are nested groupings of Soil Landscapes of Canada polygons; data include all of the Soil Landscapes of Canada database, as well as data that are provided by other federal and provincial agencies.

4.3   Soil Map of Canada/Land Potential Database (LPDB)

The Land Potential Database for Canada (LPDB) contains data about soil, climate, land degradation and modeled and actual crop yields for the major economically important crops in Canada (Kirkwood et al., 1983). This information is summarized for each of the 755 soil map units (polygons) designated by the Soils of Canada map (Clayton et al., 1977) at a scale of 1:5 million. This data base was developed as part of a global assessment of agroecological crop production potentials (FAO, 1978). The soil map unit is the common variable (relate variable) for each of the datasets within the Land Potential Database.

4.3.1   Land Potential Database Files

CLIMATE5180   Monthly mean values of climate variables, based on the 1951-1980 climate normals.

GROWSEASCLIM   Mean daily values of climate variables during the growing season.

SOILINVENTORY   Summary of soil information from Soils of Canada, Volume 2, Inventory (Clayton et al., 1977).

ACRI   Agroclimatic Resource Index

SOILCLIM   Summary of soil climate information from Soils of Canada, Volume 2, Inventory (Clayton et al., 1977).

ACTUALYIELDS   Actual yields for selected crops, area-weighted from crop reporting district data.

DEGRADATION   Variables pertaining to soil degradation.

CONSTRAINTFREE   Modeled crop yields (fresh weight grain) for selected crops.

CLI   Canada Land Inventory (CLI) capability for agriculture.

PROVPROPORTION   Proportion of mapped polygon by province.

SOILSUITABILITY   Percent of polygon that is well suited, suited and not suited for growing selected crops.

Associated file:

LANDFLEX   Physical land variable that describes the opportunity for land use diversification.

4.4   Soil Landscapes of Canada (SLC)

4.4.1   General Overview

Starting in 1982, the federal Soil Inventory program undertook a project to compile a computerized database to record the attributes of the soil and land for the whole country and to prepare maps from this information at a scale of 1:1 million. This SLC project, which used existing more detailed maps where available and LANDSAT imagery and some field work for other regions, represents an updated and more detailed inventory of Canadian land resources for national and broad regional planning activities and a replacement for the older and more general Soils of Canada. The objective of the SLC project was to assemble consistent and comparable information about the soil and land nationwide based on permanent natural attributes (Shields et al., 1991).

The SLC map and database are intended to serve the following purposes: (1) to assess the productivity of the land nationally or over large areas; (2) to find areas that have actual or potential problems affecting land use, such as salinity or susceptibility to erosion and to assess the severity; (3) to locate general areas that may be suitable for particular types of land use, which can be selected for more detailed investigations; (4) to apply general research findings and agrotechnology procedures that are successful in one part of the country to other areas that have similar attributes; (5) to link soil and land information with other databases, such as information on climate, economics, or census, for assessing land use on a regional, national, or even an international scale; and (6) to educate students of soil geography at colleges and universities. In addition to these purposes, the data have served as a framework to support other databases, such as Canada's Ecological Land Classification System.

4.4.2   SLC Versions

The creation of the SLC database has taken a number of years. Updates and corrections to boundaries, attributes and file structures have taken place over time. The version number reflects these updates. The number before the decimal place identifies the version of the polygon boundaries, while the number after the decimal place refers to the version of the attribute data for those polygons. New versions are released as major structural or attribute changes are implemented. The following versions have been released: Version 1.0 (1991), 2.0 (1994), 2.1 (1995), 2.2 (1996).

In version 1.0 of the SLC database, attributes are described in two extended legend files. The extended legend files have generalized soil attributes for dominant and subdominant soil landscapes in each SLC polygon. The SLC extended legend properties were defined by local soil experts in each region, in terms of a few broad classes for each attribute.

The full array of attributes that describe a distinct type of soil and its associated landscape, such as surface form, slope, water table depth, permafrost and lakes, is called a soil landscape. SLC polygons may contain one or more distinct soil landscape components and may also contain small but highly contrasting inclusion components. The location of these components within the polygon is not defined.

Subsequently, the SLC database has been revised and improved. From version 2.0 on, the general structure of data describing SLC polygons is organized such that each polygon is described by a Polygon Attribute Table (PAT) and a series of components (CMP) which occupy a specified proportion of the polygon area, as well as by a Landscape Attribute Table (LAT). Each soil component is further characterized by two additional files, the Soil Names File (SNF) and Soil Layer File (SLF), which contain modal values of physical and chemical properties for each soil. This data structure is similar in concept to the structure used for detailed soil maps (scales 1:20,000 to 1:125,000) where a Soil Map Unit File (SMUF) is used to identify up to three component soil series and their extent in each detailed map polygon. These are linked, via common soil codes, to the same provincial SNF and SLF files used for SLC level assessments.

4.4.3   SLC Component Table (Ver. 2.2)

Filename: {covername}.CMP
Description: The Component table describes soil landscapes which are components of a polygon.
Contents:

Name	Description
SL	Soil landscape (relate key)
CMP	Component (relate key)
PERCENT	Percent of polygon occupied by component
KINDMAT	Kind of surface material
VEGET	Vegetation cover and/or land use
PMDEP	Parent material mode of deposition
CFRAG	Coarse fragment content of the control section
ROOTDP	Rooting depth (unrestricted)
DRAIN	Drainage class
DEVEL	Soil development (classification)
CALC	Parent material calcareous class
LOCSF	Local surface form
SLOPE	Slope gradient
SNF	Soil names file (relate key necessary to identify province)
SOILCODE	Soil name code (relate key)
MODIFIER	Soil modifier (relate key)

Relationships:
(1)   A record in this file is unique with respect to SL & CMP.
(2)   SL & CMP are required to relate (1:many-N) to the Carbon Layer File.
(3)   SNF & SOILCODE & MODIFIER are required to relate (N:1) to the Soil Names File (SNF) and the Soil Layer File (SLF).
(4)   The Landscape Attribute Table (LAT) contains information about water bodies (information which previously was found in this file).

While the data model does not restrict the number of components or the proportion of the polygon that they occupy, restrictions have been recommended from a subject matter and organization standpoint. Note that the dominant component for some polygons will be a non-soil (e.g., ice or rock), and analyses involving dominant soil components should take this fact into account.

4.4.4   SLC Landscape Attribute Table (Ver. 2.2)

Filename: {covername}.LAT
Description: The Landscape Attribute Table contains attributes of the polygon, and attributes that apply to the entire landscape represented by that polygon. All areas are in ha.

Contents:

Name	Description
SL	Soil landscape (relate key)
LAND_AREA	Area applicable to the component table
POLYGON_AREA	Area of the entire polygon
WATER_TOTAL	Area of water (including oceans)
WATER_FRESH	Area of total fresh water
HYDRO_OCEAN	Area of ocean water shown in the HYDRO* coverage
HYDRO_FRESH	Area of fresh water shown on the HYDRO coverage
SMALL_WATER	Area of small water bodies not shown in the HYDRO coverage
SMALL_LAND	Area of small islands not shown in the HYDRO coverage

* The HYDRO coverage is a non-soil database that describes water bodies throughout Canada.

Relationships:
A record in this file is unique with respect to SL

4.5 Detailed Soil Surveys

Detailed soil surveys from all provinces and territories of Canada are held by the National Soil Database (NSDB) of CanSIS if they have been digitized. Not all soil maps published in Canada are digitized and available in a GIS format. Progress is being made in digitizing all soil surveys, but it is unlikely this will be completed before the end of the century. In addition, many of the maps and databases are held by the applicable provincial governments, which are also taking steps to complete digitization in collaboration with the NSDB.

The basic elements of the detailed soil survey digital files are the graphic files and the Soil Map Unit File (SMUF) which defines the contents of each map polygon in terms of the component soil series and associated slope and drainage characteristics. As with the revised version of the SLC map, the components of the detailed soil polygons are further described through links to the Soil Names File (SNF) and the Soil Layer File (SLF).

4.5.1 Soil Map Unit File (SMUF)

Filename: {covername}.SMUF
Description: The Soil Map Unit File identifies the soil codes and slopes which are components of a polygon.
Contents:

Name	Description
PROVINCE	Province code
MAPUNITNOM	Map unit number
SOIL_CODE1	Code for soil with largest percentage area
MODIFIER1	Modifier code for this soil
EXTENT1	Percent of map unit occupied by this soil
SOIL_CODE2	Code for soil with second largest percentage area
MODIFIER2	Modifier code for this soil
EXTENT2	Percent of map unit occupied by this soil
SOIL_CODE3	Code for soil with third largest percentage area

MODIFIER3	Modifier code for this soil
EXTENT3	Percent of map unit occupied by this soil
SLOPEP1	Slope of soil 1
SLOPEP2	Slope of soil 2
SLOPEP3	Slope of soil 3
STONE1	Stoniness of soil 1
STONE2	Stoniness of soil 2
STONE3	Stoniness of soil 3
DATE	Date information was compiled

Relationships:

(1)   A record in this file is unique with respect to PROVINCE & MAPUNITNOM.

(2)   MAPUNITNOM is required for the linkage between geographic location (polygons) and soil landscape attributes in the associated attribute tables.

(3)   PROVINCE + SOIL_CODE + MODIFIER are required to relate (1:N) to the SNF and SLF.

Limitations:

The proportion of the polygon occupied by each separate component is identified. However, their location within the polygon is not defined. While the data model does not restrict the number of components or the proportion of the polygon that they occupy, restrictions have been recommended from a subject matter/organization standpoint. The model shown restricts the number of components to 3, but it is intended to remove this restriction in the near future. Note that the dominant component for some polygons will be a non-soil (i.e., water, ice or rock). Analyses involving dominant soil components should take this fact into account.

4.5.2   The Soil Attribute Files

Each province maintains a computerized listing of all the soil series defined within it and lists modal characteristics for the soil (in the SNF record) and for the soil horizons (in the SLF). Copies of these files are maintained in the NSDB and, through linkages and foreign keys, they form an integral part of various of the detailed and SLC maps and some other thematic coverages in the NSDB. In most cases the data in the SNF/SLF are calculated or interpreted from field samples and laboratory analysis; however, in some cases, the data have been estimated from soils of similar derivation and location.

The attributes associated with these files are standardized across Canada and were selected as the most appropriate for computerized models and interpretations of land use, environmental sustainability and crop production. The contents of these files are described in the following sections.

4.5.2.1   Soil Names File (SNF)

Description: The Soil Names File contains information that applies to the entire soil.

The list of attributes for SNF is

Name	Description
PROVINCE	Province code (relate key)
SOIL_CODE	Soil code (relate key)
MODIFIER	Soil code modifier (relate key)
LU	Land use
SOILNAME	Soil name
KIND	Kind of soil
WATERTBL	Watertable characteristics
ROOTRESTRI	Soil layer that restricts root growth

RESTR_TYPE	Type of root restricting layer
DRAINAGE	Soil drainage class
MDEP1	Mode of deposition - first
MDEP2	Mode of deposition - second
MDEP3	Mode of deposition - third
ORDER	Soil order
S_GROUP	Soil subgroup
G_GROUP	Soil great group
PROFILE	Header for Detail II file
DATE	Revision date
SLFNA	Availability of soil layer data
SOIL_KEY	Linkage key (relate key)
LAYER_KEY	Linkage key (relate key)

Relationships:

(1)   A record in this file is unique with respect to PROVINCE + SOIL_CODE + MODIFIER + LU.

(2)   There is one SNF per province.

(3)   SOIL_CODE + MODIFIER + LU is not necessarily equivalent between provinces.

(4)   This file is linked to the SLF by PROVINCE + SOIL_CODE + MODIFIER + LU and to SMUFs by PROVINCE + SOIL_CODE + MODIFIER.

4.5.2.2   Soil Layer File (SLF)

Description: The Soil Layer File is designed to handle attributes which vary in a vertical direction. The list of attributes for SLF is

Name	Description
PROVINCE	Province code (relate key)
SOIL_CODE	Soil code (relate key)
MODIFIER	Modifier code for this soil (relate key)
LU	Land use code
LAYER_NO	Horizon number
HZN_LIT	Horizon lithological discontinuity
HZN_MAS	Master horizon
HZN_SUF	Horizon suffix
HZN_MOD	Horizon modifier
UPDEPTH	Upper depth
LDEPTH	Lower depth
COFRAG	Coarse fragments
COFRAG#	Number of observations for coarse fragments

(Note: For all of the remaining mean values, the number of observerations is recorded.)

DOMSAND	Dominant sand fraction
FSAND	Very fine sand
TSAND	Total sand
TSILT	Total silt
TCLAY	Total clay
ORGCARB	Organic carbon
PHCA	pH in calcium chloride
PH2	pH as per project report

BASES	Base saturation
CEC	Cation exchange capacity
KSAT	Saturated hydraulic conductivity
KP0	Water retention @ 0 kP
KP10	Water retention @ 10 kP
KP33	Water retention @ 33 kP
KP1500	Water retention @ 1500 kP
BD	Bulk density
EC	Electrical conductivity
CACO3	Calcium carbonate equivalent
VONPOST	Von Post estimate of decomposition
WOOD	Woody material
DATE	Date of revision

Note: Methods are specified in the SOIL.DOC file of the workspace.

Relationships:

(1) A record in this file is unique with respect to PROVINCE + SOIL_CODE + MODIFIER + LAYER_NO + LU.

(2) This file is linked to SNF by PROVINCE + SOIL_CODE + MODIFIER + LU and to SMUF by PROVINCE + SOIL_CODE + MODIFIER.

4.6   National Ecological Framework

The National Ecological Framework was developed as a collaborative project between Agriculture and Agri-Food Canada and Environment Canada, and involved all of the provincial departments of Agriculture, Environment and/or Natural Resources. Its purpose is to provide a hierarchial framework for a wide range of planning and analytical activities that need to reflect ecological (including soil) considerations. These include State of the Environment Reporting, wildlife habitat monitoring, climate change studies, acid rain assessments and forest inventories, to name a few. The Framework has been well accepted within Canada, and is now linked directly to the North American Ecological map (Ecological Stratification Working Group, 1995).

All of the data available in the SLC database are accessible to the National Ecological Framework, since the SLC polygons are nested within the framework. Ecodistricts are groupings of approximately 5 to 20 SLC polygons. The SLC data are accessed at this level. In addition, Ecodistricts are defined in terms of area, land cover (from AVHRR imagery), and presence of permafrost (from Natural Resources Canada).

At the broadest level, the land mass of Canada has been subdivided into 15 terrestrial Ecozones which are further subdivided into Ecoregions and Ecodistricts. At the Ecoregion level (there are 217 Ecoregions in Canada), data are available on regional physiography, land cover, population, temperature and precipitation. There are an average of approximately 5 Ecodistricts in each Ecoregion.

4.7   Provincial Soil Databases

4.7.1   Prince Edward Island (PEI)

The PEI Department of Agriculture and Forestry has available on their GIS (CARIS) soils maps at a scale of 1:10000 (survey intensity was at a scale of 1:20000 but base information is at a scale of

1:10000) for the entire province. In addition, they have the forest inventory data for 1935, 1980 and 1990; contour lines 2 m interval (1977 DTM); and watershed boundaries for the province.

4.7.2 New Brunswick

The Land Resources Branch of the New Brunswick Department of Agriculture and Rural Development maintains the CanSIS soil maps in CARIS format. In addition, they have prepared maps for the southeast of the Province and the upper Saint John River valley showing the soil and landscapes evaluated for suitability for potato production. They have also been overlaid with maps of cleared and forested land cover, water and man-made features. Other maps and data from special projects have also been prepared.

4.7.3 Ontario

Soil map digital data for the Province of Ontario is maintained by the GIS unit of the Ontario Ministry of Natural Resources. In general, these datasets are copies of the county level soil map information archived in the NSDB. The province also retains digital map records of agricultural land use as compiled from a 1983 survey. The responsibility for official plans for land management has recently been transferred to Ontario municipalities. As a consequence, several Ontario municipalities are developing GIS databases and are actively contributing to the update of soils information for inclusion. It is likely that this trend will continue and that GIS soil data in Ontario will be stored and used locally.

4.7.4 Manitoba

The only GIS data base that may not be available through CanSIS is the digital Rural Municipality (RM) database that has recently been compiled at a scale of 1:125,000. This was part of a coordinated project for the three prairie provinces (Manitoba, Saskatchewan and Alberta), supported by the Prairie Farm Rehabilitation Administration (PFRA), and agency of Agriculture and Agri-Food Canada. The databases are available in dxf format from the PAMAP GIS, supported by the Manitoba government.

4.8 Non-GIS Soil Databases

While there are many non-GIS soil databases in Canada that have been developed over the years, either for soil survey data prior to the GIS era or for special purposes, one recent database is especially important. This is the Soil Quality Benchmark Site database. It has been developed to contain all the data collected under a ten-year project initiated in 1989 to quantitatively measure changes in soil quality under current agricultural practices in all the important agricultural regions of Canada (Wang et al., 1995).

The database contains the baseline data measured at either grid points or along transects of the benchmark fields at the initiation of the project. It also contains the annual and once-per-rotation data monitored during the ten-year project. Twenty-three sites were selected across Canada between 1989 and 1992, and each was initially characterized by two detailed soil profiles, as well as by measurements taken at grid points (usually every 20 m) and transect lines running from the crest to the foot of representative slopes. Subsequent measurements have been made on an annual basis (e.g., crop yields), and at the end of each rotation cycle (e.g., bulk density).

4.8.1 The Soil Quality Benchmark Site Database

A relational database was designed for soil quality evaluation. With a host of data types on a variety of measured entities, the main goal was to attain efficient data storage that would support reasonably simple manipulation and retrieval. The Benchmark Site database achieved this goal by using small files developed in dBASE IV (version 1.5). Each file contains similar types and landscape entities. Currently the files contain baseline, reference and on-going data. Results of repeat measurements are entered into files like those containing the baseline data so that temporal comparisons can be made. New data are checked and refined before being appended to the database; occasionally old data are updated if corrections or calculated values need to be added.

The dBASE files that comprise the database system are listed and briefly described below. File name extensions, always .DBF (but sometimes including .DBT and others) are omitted. File names that begin with BS indicate baseline data. Most files contain data from all the benchmark sites, if appropriate and available. Extracting data by site (and other filters) can be done quite easily.

Name	Description
SITEINFO	Reference file. General information about each benchmark site including identification, location, agroecological region, major soils and landform, potential degradation problems, type of management, site manager, and farm cooperator.
BSPTINFO	Baseline and reference data. Landscape information about the field sampling points, e.g., slope position and shape, soil series/variant, and map unit.
BSTOPO	Baseline and reference data. Spatial data for creating contour maps and locating field sampling points; relative or real world coordinates, including elevation, in m.
BSDESCR	Baseline and reference data. Descriptions of the soil horizons that were sampled (colour, texture, structure, etc.).
BSSLCHEM	Baseline data. Routine chemical data (pH, total C, total N, $CaCO_3$ equiv., CEC and exchangeable cations, available P and K) on all samples.
BSPTSIZ	Baseline data. Particle size and surface area on selected samples.
BSEALFE	Baseline data. Extractable Al and Fe, analysed by various methods on selected samples (mainly done on humid region soils, esp. podzols).
BSECSEL	Baseline data. Electrical conductivity, soluble cations and SAR on selected samples.
BSTTLELM	Baseline data. Total analysis for at least 14 elements (Al, Ca, Co, Cr, Cu, Fe, K, Li, Mg, Mn, Na, Ni, Pb and Zn) on selected samples.
BSNO3_N	Baseline data. Nitrate-N data on selected deep samples from selected sites.
BSCS137	Baseline data. [137]Cesium counts (per unit weight and unit area) for selected samples and sites. Includes bulk density by the Kubiena box method.
BSSLMINE	Baseline or reference data. Mineralogical analysis (semi-quantitative results) on clays from selected samples.
BSEM38	Baseline data. Electrical conductivity (0-60 and 0-120 cm) derived from electromagnetic inductance at selected points and selected sites.
BSMSTRN	Baseline or reference data. Moisture retention at 0, 10, 30, 60, 100, 330 and 1500 cm water column equivalent determined on cores, and at 4 and 15 bars determined on ground samples. Includes bulk density by the core method.
BSAGREG	Baseline data. Dry aggregate analysis (rotary sieve) from selected sites.
BSKSAT	Baseline data. Saturated hydraulic conductivity, measured by Guelph Permeameter for 2 or 3 depths at selected field points.
BSPTRMST	Baseline data. Penetrometer resistance and moisture content (dated) for 3 or 4 depths at selected field points. Spring and fall results are included at some sites to compare moist and dry seasons.
BSTHWRM	Baseline data. Earthworm counts and weights for selected horizons at selected sites (mainly humid region sites).

BSBIOPRT   Baseline data. Biopore and root counts for selected depths and selected sites (mainly humid region sites).

YLDINFO   On-going reference data. Annual information on crop type, harvest notes, and the file that contains the yield data for each site.

GRAINYLD   On-going data. Grain and residue yield (kg ha^{-1}), harvest index (%), and residue:grain ratio for grain crops (i.e., seed-bearing crops such as cereals, oilseeds, etc. where the seed is separated from the rest of the aboveground dry matter) by site, sampled field point, year and crop type.

FORAGYLD   On-going data. Dry matter yield (kg ha^{-1}) of forage crops by site, sampled field point, year and type of crop.

Note 1: Yields of other types of crop (e.g., sugar beets, sweet corn, potatoes) are reported in different yield files from the grain crops because harvesting methods and yield parameters differ.

Note 2: Climate data from Campbell Scientific monitoring stations (installed at some sites) will be added to the database in the near future. Hourly, daily and monthly summary files are planned.

4.9 References

Clayton, J.S., W.A. Ehrlich, D.B. Cann, J.H. Day, and I.B. Marshall. 1977. Soils of Canada. Vols. 1 and 2. Canada Department of Agriculture, Ottawa, Canada.

Ecological Stratification Working Group. 1995. A national ecological framework for Canada. Agriculture and Agri-Food Canada, Centre for Land and Biological Resources Research and Environment Canada, State of the Environment Directorate, Ecozone Analysis Branch, Ottawa-Hull, Canada.

FAO. 1978. Report on the agro-ecological zones project. Vol. 1. Methodology and results for Africa. World Resource Report 48. FAO, Rome, Italy.

Kirkwood, V., J. Dumanski, A. Bootsma, R.B. Stewart, and R. Muma. 1983. The Land Potential Database for Canada, User's Handbook. Tech. Bull. 1983-4E. Research Branch, Agriculture Canada.

Shields, J.A., C. Tarnocai, K.W.G. Valentine, and K.B. MacDonald. 1991. Soil landscapes of Canada procedures manual and user's handbook. Agriculture Canada Pub. 1868/E.

Wang, C., L.J. Gregorich, H.W. Rees, B.D. Walker, D.A. Holmstrom, E.A. Kenney, D.J. King, L.M. Kozak, W. Michalyna, M.C. Nolin, K.T. Webb, and E.F. Woodrow. 1995. Benchmark sites for monitoring agricultural soil quality. p. 31–40. *In* D.F. Acton and L.J. Gregorich (ed.) The health of our soils. Centre for Land and Biological Resources Research, Research Branch, Agriculture and Agri-Food Canada, Ottawa, Canada.

5

United States Soil Survey Database

D.J. Lytle
USDA-NRCS

5.1 Introduction

Soil surveys have been conducted in the United States since about 1886. In 1953, the Secretary of Agriculture established the National Cooperative Soil Survey (NCSS), a group of federal and state agencies and universities loosely knit through cooperative agreements, in part to accelerate the completion of soil surveys. The USDA Natural Resources Conservation Service (NRCS), formerly the Soil Conservation Service (SCS), provides federal leadership to the NCSS. Soil surveys and supporting pedological studies conducted through the NCSS are complete for 76% of the conterminous United States and about 11% of Alaska. About 94% of the private land in the United States has a detailed, generally 1:12,000 to 1:24,000 scale, soil survey. This information is available in approximately 2,200 county level soil survey reports and databases. A digital soil survey at 1:250,000 scale for the conterminous United States and 1:1 million scale for Alaska is also available. Each year between 50 and 100 more new soil surveys with digital spatial and attribute data, and corresponding soil survey reports are developed and published. For many of the existing county level soil surveys, the maps have not been digitized. However, an intensive effort has begun to digitize all of these surveys.

Beginning in the late 1960s and early 1970s, NCSS soil scientists recognized the need and potential utility for automating the large quantity of soil survey information that was being generated. In the more than 30 years since, the NCSS has been an innovative leader in the adaptation and use of information technology, and, as a result, the demand for soil survey information is greater than it has ever been. A National Soil Information System (NASIS) has recently been introduced to improve the acquisition, management and distribution of these data.

5.2 Description of Soil Survey Databases

The NCSS develops and maintains six defined sets of data. These are the Soil Characterization Record (SCR), Map Unit Record (MUR), Taxonomic Unit Record (TUR), Soil Survey Geographic Database (SSURGO), State Soil Geographic Database (STATSGO), and the National Soil Geographic Database (NATSGO). Each set of data was developed for a specific need or purpose. A simplified schema of how these data are related is presented in Fig. 5.1. Information contained within these databases has been reviewed and certified in accordance with NCSS data quality standards (Soil Survey Staff, 1995a, 1996a; Hudson, 1992).

5.2.1   Soil Characterization Record (SCR)

This database stores the laboratory characterization of about 20,000 pedons of United States soils and about 1,100 pedons from other countries. Morphological descriptions are available for about 15,000 of these pedons. These data were primarily produced by the Soil Survey Laboratory at the National Soil Survey Center (NSSC) and at previous SCS laboratories located in Beltsville, MD and Riverside, CA. Sample preparation and analytical procedures are carried out according to standard methods (Soil Survey Staff, 1996b). The NSSC has also produced a guide for interpreting the information (Soil Survey Staff, 1995a) which is linked to the Map Unit Record (MUR) by the soil series component, map unit and taxonomic classification. These pedons are sampled by NRCS scientists and NCSS cooperators at universities, state and other federal agencies. Sites are usually selected to represent a central concept of a soil series, and over time, as they are sampled and characterized, these pedons bracket a range of soil properties for a soil series or landscape. These data are used to populate the MUR and for research purposes, such as environmental model development.

Programming for the development of this database began at the NRCS soil survey laboratory as early as 1966 by automating the calculation and reporting of bulk density, particle size, chemical, and mineralogical analysis. These early efforts evolved into the development of a Laboratory Information Management System that collects, controls quality, calculates and reports laboratory analysis.

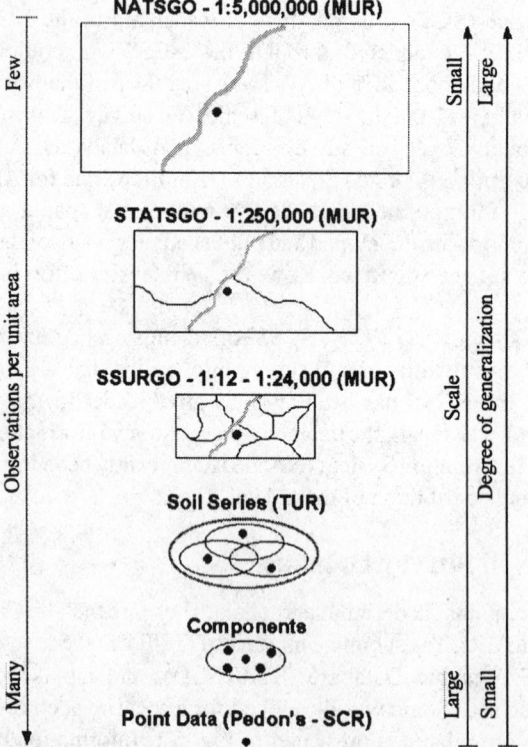

Fig. 5.1   Spatial data relationships [Modified from Waltman, 1997, personal communication]

As part of the NCSS many universities have developed soil survey laboratory characterization data for the soil surveys in their states. Institutions with significant databases include the University of California Davis, Clemson University, Colorado State University, University of Florida, Kansas State University, University of Idaho, University of Illinois, Louisiana State University, University of Maine, Michigan Technological University, Mississippi State University, University of Missouri, Montana State University, Ohio State University, Oregon State University, Pennsylvania State University, University of Tennessee, Texas A&M University, Virginia Polytechnic Institute & State University, and West Virginia University. These data are being integrated into NASIS and made available over the World Wide Web (WWW).

5.2.2   Map Unit Record (MUR)

Information in the MUR is developed from observations made during field mapping, including reconnaissance observations, transects, grid samples, field measurements of soil properties and, as previously mentioned, laboratory analysis of soils. The MUR includes physical, chemical and morphological properties, and component percentages, interpretation ratings, and performance data stored for a potentially unlimited number of soil map unit components (portions of soil series). This information on components is stored as a component object which can have an unlimited number of horizons. Component objects are grouped together in the mapping process to form soil map units, called data map units, which are linked to map units in a soil survey legend through a correlation table that allows different map units to be linked to the same data map unit. This allows perfect joins between soil survey areas and multiple legends for the same geographic area. This capability is useful in correlation across political boundaries and for retention of old symbols on maps. The legend contains the map unit symbol, map unit name, type (consociation, complex, etc.), area and correlation notes. This structure provides the capability of having a set of data available for each delineation (polygon), point or line feature on the soil survey map. The relationship between horizon properties, horizons, components, and map unit polygons or area features and point and line features is presented in Fig. 5.2. Component objects and other attributes are assigned to polygons, points or lines. Table 5.1 provides

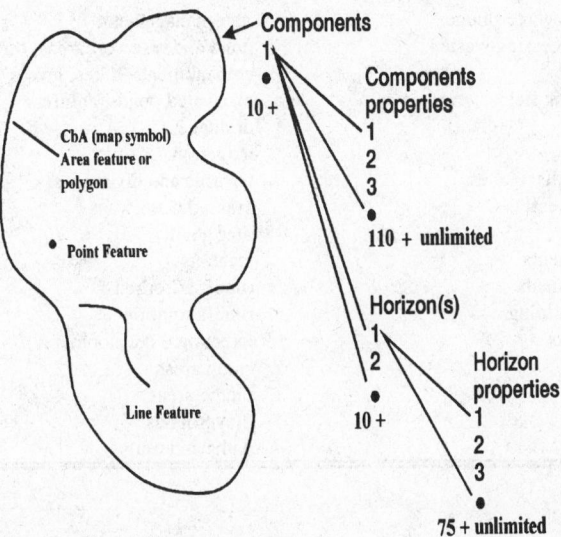

Fig. 5.2   Data relationships

Table 5.1   NASIS soil data elements

Component data

Component: name, percent, kind	Runoff
Taxonomic classification	Soil moisture status
Geomorphic description	Soil loss tolerance (T factor)
Parent material	Erosion class
Climatic data (AAT, APT, FFS)	Wind erodibility, index, group
Land use: earth cover kind	Restrictive layers: depth, hardness, kind
Slope: low, high, length, aspect	Subsidence depth: total and initial
Elevation	Surface rock fragments
Flooding: frequency, duration, months of year	Potential frost action
Ponding: depth, duration, frequency	Soil slippage
Drainage class	Albedo dry
Hydric condition	Soil temperature: monthly, depth
Hydrologic group	

Horizon data

Horizon designation and depths	Soil reaction (pH), water, $CaCl_2$
USDA texture	Extractable Al and acidity
Erosion factors: K, Kf, critical shear stress	Free iron oxides
Rock frag: percent, size, kind, shape, hardness	Cation exchange capacity: CEC, ECEC
Percent of soil passing #4, 10, 40, 200 sieves	Electrical conductivity
Percent sand, silt, and clay and fractions	Sodium absorption ratio
Liquid limit and plasticity index	Gypsum
UNIFIED and AASHTO class	Calcium carbonate equivalent
Linear extensibility	Permeability
Particle density	Water content, 1/3, 1/10, 15 bar
Bulk density	Available water capacity
Excavation difficulty class	Saturated hydraulic conductivity
Organic matter percent	Soil morphology data elements (future additions)

Interpretive and performance data

Waste management	Construction material:
-waste water disposal by irrigation	-roadfill
-waste water treatment (slow)	-sand
-waste water treatment (overland flow)	-gravel
-waste water treatment (rapid infiltration)	-topsoil
-land application of sewage sludge	Water management:
-manure and food processing waste	-pond and reservoir areas
Sanitary facilities:	-embankments, dikes, levees
-septic tank absorption, fields	-excavated ponds, aquifers
-sewage lagoon areas	-drainage
-sanitary landfills	-irrigation
-daily cover for landfills	-terraces and diversions
Building site development:	-grassed waterways
-shallow excavations	Water quality:
-dwellings w/o basements	-leaching
-dwellings with basements	-runoff adsorbed
-small commercial buildings	-runoff solution
-local roads and streets	Recreational development:
-lawns	-camp areas
-landscaping	-picnic areas
-golf fairways	-playgrounds
	-paths and trails

Table 5.1 Cont.

Land capability	Wildlife habitat suitability
Highly erodible land	-burrowing mammals and reptiles
CRP rental rate	-crawfish aquiculture
Prime farmland	-desertic herbaceous plants
Soil rating for plant growth (SRPG)	-domestic grasses and legumes
Crop yields, common crops	-freshwater wetland plants
Forestry limitations & ratings:	-grain and seed crops
-haul road and landing construction	-herbaceous tundra
-hand planting suitability	-irrigated domestic grasses and legumes
-harvest equipment operability	-irrigated freshwater wetland plants
-log landing suitability	-irrigated grain and seed crops
-mechanical planting suitability	-irrigated saline water wetland plants
-mechanical site preparation (deep)	-riparian herbaceous plants
-mechanical site preparation (surface)	-riparian shrubs, vines and trees
-potential erosion hazard (off road/trail)	-saline water plants
-potential erosion hazard (roads/trails)	-sedge-grass tundra
-road suitability - natural surface	-tussock tundra
-soil rutting hazard	-upland conifer trees
Tree species: existing/potential	-upland decidious trees
Ordination symbol	-upland desertic herbaceous
Site index	-upland mixed conifer/decidious trees
Range & woodland vegetation	-upland shrub tundra
-plant species, cover percent	-upland shrubs and vines
-percentage composition	-upland wild herbaceous plants
-potential production	Windbreak suitability

a listing of the physical, chemical, and morphological properties and other data stored either for horizons, components or map units. To date, there are over 300 data elements.

The NASIS design accommodates the easy addition of new data elements and allows these data to be stored according to the land use. This ability to store use-dependent data is an important new capability in NASIS. Most of the chemical and physical properties in the MUR vary according to land use. When GIS users have land use data to use in combination with the digitized soil map, then soil data with a corresponding range in properties for a specific land use can be selected, and better soil interpretations can be made. The MUR is developed in conjunction with soil survey maps generally at scales of 1:12,000 to 1:31,680. The MUR is the database from which most soil survey publications are produced and is provided to NRCS field office computers for use in natural resource planning. MUR data are also used by landowners, private and public land managers, federal, state, and local land use planners, environmental modelers, regulators and others.

5.2.3 Taxonomic Unit Record (TUR)

The taxonomic unit record is composed of two datasets, the Soil Series Classification Database (SC) and the Official Series Description. The SC contains the taxonomic classification of each soil series identified in the United States together with other information about the soil series, such as office of responsibility, series status, dates of origin and establishment, and geographic areas of usage. The SC is maintained by the soils staff of the NRCS Major Land Resources Area (MLRA) Offices. Additions and changes are continually being made, resulting from on-going soil survey work and refinement of the soil classification system. As the database is updated, the changes are immediately available to the user, so the data retrieved are always the most current. The SC is available through the Internet World Wide Web (WWW) which provides the capability of viewing the contents of individual series records, querying the database on any data element and producing a report with the selected soils or national reports with all soils in the database. The standard reports available allow the user to display the soils by series name or by taxonomic classification.

OSD is a term applied to the soil description that defines a specific soil series. The OSDs are in a text format and describe general and detailed information about each recognized soil in the United States, including location, author's initials, introductory paragraph, taxonomic classification, detailed soil profile description, location of the typical soil profile, range in characteristics, competing series, geographic setting, geographically associated soils, drainage and permeability, use and vegetation, distribution and extent, where and when the series was established, remarks, and additional data. While doing survey work, field soil scientists have the existing official soil series descriptions that are applicable to their soil survey areas. Other official soil series descriptions that include soils in adjacent or similar survey areas are also commonly used. Thus, OSDs serve as a national standard for identifying and classifying soils and as the basis for the placement of a soil series in the Soil Taxonomy. Soil series names are also commonly used to name soil map units. Scientists in other disciplines, such as agronomists, horticulturists, engineers, planners, and extension specialists also use the descriptions to learn about the properties of soils in a particular area. OSDs are available for the more than 18,000 soil series recognized in the United States and are maintained and made available over the WWW.

5.2.4   Soil Survey Geographic Database (SSURGO)

SSURGO is the digitized version of detailed county level soil surveys. The SSURGO spatial data are compiled, digitized, and archived in 7.5-minute topographic quadrangle units at scales of 1:12,000 to 1:31,680. The minimum-sized area delineated on a map ranges from about 0.61 to 4.05 ha at these scales. The mapping bases used meet national map accuracy standards and are either orthophotoquads, quarter-quads, or US Geological Survey 7.5-minute quadrangles. SSURGO attribute information is obtained from the MUR (Soil Survey Staff, 1995b). SSURGO is used exclusively in GIS systems, for the same purpose and by the same users as the MUR. An example of a SSURGO interpretative map is presented in Fig. 5.3. SSURGO and other vector digitized information are available for a portion of the United States. Starting in 1997, the NRCS began an accelerated effort to digitize all detailed soil surveys and a presidential initiative provided funding to complete the work.

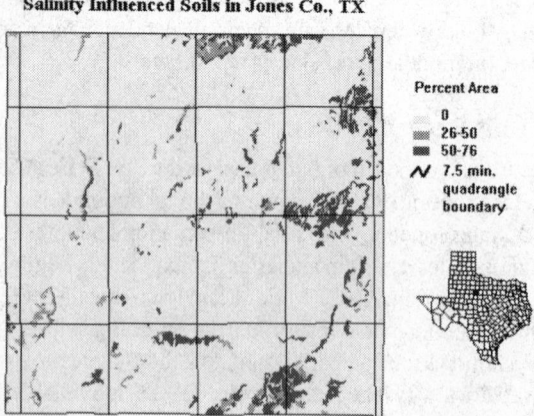

Fig. 5.3   SSURGO interpretative map

Efforts to digitize soil surveys in the United States started in 1971 with the grid cell Map Information Assembly and Display System (MIADS). Soil surveys were digitized in 4.05 and 10.12 ha grids, often along with land use information for the same grid cell. Beginning in 1976, Computervision® software was used to vector digitize soil surveys by either scanning or manual methods. Currently, soil surveys are vector digitized by scanning, using a variety of commercial and public domain software.

5.2.5 State Soil Geographic Database (STATSGO)

STATSGO was designed to be used for broad planning and management exclusively in GIS systems (Bliss and Reybold, 1989). An example of a STATSGO interpretative map is presented in Fig. 5.4. Typically, it is used for state or region-wide analysis to identify natural resource and environmental problem areas, prime farmland, potential productivity, global change modeling (carbon maps, etc.) and other uses (Lytle et al., 1996). STATSGO has also been used to develop resource (staffing and funding) allocation models for NRCS. Several federal agencies such as the US Environmental Protection Agency, US Geological Survey, and Agriculture Research Service, and many university researchers use STATSGO extensively for small scale modeling and assessment projects.

STATSGO was developed as the result of NCSS cooperator efforts to create regional general soil maps. The idea for STATSGO was generated at an NCSS work planning conference in 1978. Work on STATSGO began in the mid 1980s and was completed in the mid 1990s. STATSGO maps were made by generalizing detailed county soil survey maps, which were then compiled and digitized from US Geological Survey 1:250,000 base maps. STATSGO attribute data are obtained from the MUR. The STATSGO map units consist of up to 21 soil series components. NRCS staff assigned a percent composition to each of these components from information in the detailed county level soil surveys. The minimum-size area delineated on a STATSGO map is about 607 ha (Soil Survey Staff, 1994).

5.2.6 National Soil Geographic Database (NATSGO)

NATSGO is information derived from the Natural Resources Inventory (NRI) (Nusser and Goebel, 1997). This NRI soil information is linked to two digital spatial datasets; the first is the Major Land

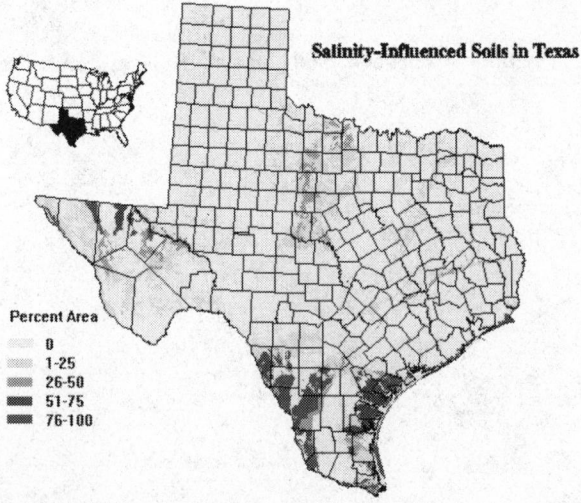

Fig. 5.4 STATSGO interpretative map

Resource Area (MLRA) map of the United States (USDA, 1981), digitized at a scale of 1:7.5 million, and the second is the digitized National Resources Inventory (NRI) points. The attribute dataset is derived from MUR and linked to NRI points. Each point has an assigned expansion factor so that it can statistically represent the area of each MLRA or be used to present information for the point itself. An example of a NATSGO interpretative map using this latter procedure is presented in Fig. 5.5. NATSGO which is used exclusively in GIS systems, is designed to be used for national, regional, and multi-state resource appraisal, planning and monitoring.

5.2.7   Data Access

Each of the databases described can be accessed over the WWW at (http://www.statlab.iastate.edu/soils/nasdaf/). This national site contains data, metadata, standards, and linkages to soil information and information on the NCSS and the NRCS Soil Survey Program.

5.3   The Evolution of Soil Survey Map Unit Databases

5.3.1   The First Generation

The NRCS first established a national soil database in the early 1970s through a cooperative agreement with the Statistical Laboratory at Iowa State University (ISU). ISU was chosen because of its long history of cooperative work with NRCS dating back to the 1940s. Programming work for a soil database began in 1972 with automation of the soil interpretations records or Form SOILS-5 which was primarily used as an input form to generate tables on engineering uses of soils for soil survey reports. Computer programs were developed to store, check, and print the data. The record for the Cecil soil series (NC0018) was the first one stored on the ISU mainframe in 1973. In 1974, the generation of manuscript tables of soil properties for inclusion in NCSS Soil Survey Reports was introduced. Initially, all data processing was done locally at ISU with hard copy through the mail. The SOILS-5 forms and the SOILS-6 forms, used to enter map unit information for the surveys, were mailed from NRCS offices to ISU for processing. Printed copies of revised records and generated tables were mailed back to NRCS. This automated table generation system replaced a very tedious, time-consuming manual process of creating tables for the published reports.

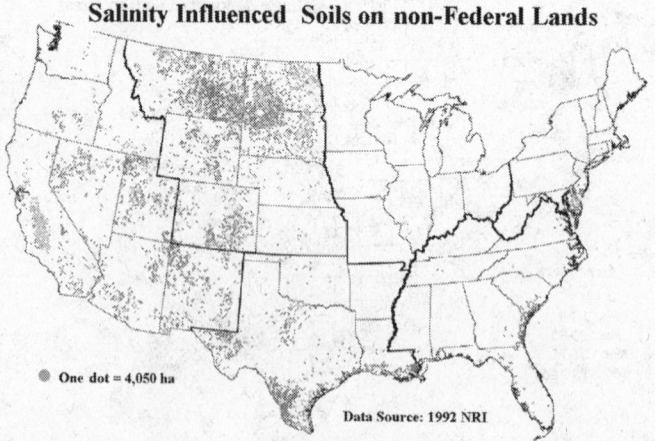

Salinity Influenced Soils on non-Federal Lands

One dot = 4,050 ha

Data Source: 1992 NRI

Fig. 5.5   NATSGO interpretative map

With the availability of this useful product came a much greater interest in getting data into the system. The capability to automatically generate soil interpretations for selected uses from the soil property data stored in the database using programmed criteria was added in 1977. After 1977, other enhancements were developed, including the addition of the Official Series Description (OSD) and Soil Series Classification (SC) databases.

Computerization in NRCS offices for processing soil survey data began in 1977 with Linolex word processing equipment in NRCS regional offices to process manuscript tables received on magnetic tape from ISU. The equipment was used to edit the tables for final publication. Remote access to ISU from NRCS began in the early 1980s with Harris Remote Job Entry equipment in both state and regional offices. Communication was through 4800 baud dial-up commercial ports. This was a time of significant change as batch software had to be redesigned for remote usage and data entry. Processing and printing of manuscripts shifted from ISU to NRCS offices.

5.3.2 The Second Generation

Work on the second generation was begun in 1978 with an NRCS effort to develop a computer program to rate soils for prime and important farmland to create maps for the Colorado Important Farmlands project. This project required rating about 4500 soil map units in Colorado, using national criteria for prime farmland and state criteria for farmlands of state importance and unique lands. The most difficult aspect was making consistent ratings across soil survey areas. The program evaluated 10 soil characteristics and was fairly accurate in its ratings placement. However, it required a large database to make the ratings, and the effort required to develop the database made the project unfeasible. The need for this large database which also had to be readily accessible and easy to manipulate resulted in the development of concepts for the second generation of soil information management.

These concepts were first documented in 1980 in the first technical report for the Colorado Soil Resource Information System (SRIS). SRIS demonstrated the feasibility of integrating several natural resource databases into a common, easy to use data environment. SRIS included (1) a soil map unit component database, (2) a Soil Interpretation database, (3) a Pedon Characterization database, (4) a Climatology database, (5) a Plant database, and (6) a soil management component and schema for the data and a description of the system. SRIS was the first effort to manage soil data using a new technology called Database Management Systems (DBMS). This new information system allowed questions to be answered that related to more than one natural resource. It facilitated easy access to soil information and allowed the data to be managed independently of the application software that accessed it, in contrast to the SOIL-5 database which required a computer program to be written for each unique request. SRIS used the System 2000 DBMS which provided a query language that could be understood and used by anyone with a basic understanding of soil information. In 1982, the soil database was implemented using MicroRim®, later renamed Rbase®, relational database management (RDBMS) software.

As an outgrowth of the SRIS effort, NRCS established the software development staff at Fort Collins, Colorado in 1985. The mission of this staff was to develop computer software to assist the NRCS field office. In 1987, this effort produced the Computer Assisted Management and Planning System (CAMPS) field office software and the State Soil Survey Database (SSSD). SSSD, which used Prelude® RDBMS software, was the culmination of the SRIS effort.

With the release of SSSD, NRCS equipped their state offices with UNIX computers. The SSSD software allowed NRCS state offices to manage their portion of the soil survey databases, which were downloaded from ISU via telecommunications. The primary function of the first release of SSSD was to clean up soil data and provide a download of the Map Unit Interpretation Record (MUIR) database to CAMPS. The first release of SSSD provided the ability to develop reports through standard database queries and manage non-technical soil descriptions. With this software release also came the recognition that a soil scientist (soil dataset manager) position was needed at each NRCS state office to manage the soil information system.

With SSSD, the NRCS state offices had the capability of editing the soil map unit property and interpretation (MUIR) data at ISU to more accurately represent local conditions and to then return a copy of the edited data to ISU. This capability provided for a national collection of MUIR data in 1993. SSSD releases in 1988 through 1994 added additional software. In 1988, the Pedon Description Program version 1.0, and the Official Soil Series management and soil reports modules were released. In 1989, the Soil Survey Geographic (SSURGO) database: Geographic Resource Analysis Support System (GRASS) geographic information system interface was released. In 1989, a UNIX mail system called SoilNet and an automated version of the SOILS-6 form, which was used to record map unit data and facilitate the downloading and managing of MUIR data from ISU, were released. In 1991, the Soil Survey Schedule module was released which provided management, scheduling and record-keeping software for NRCS state and national offices to use in production soil survey efforts. In 1993, the Hydric Soils, Range Site, and MUIR incremental update modules were released.

Although table generation remained the primary purpose, the MUIR database soon began to be used for more than developing soil interpretation tables for reports. NRCS began to use the database to answer questions on a wide range of soil-related issues across the United States, for example, extent of salt-affected soils, soil loss tolerance and erosion potential for determination of highly erodible land, and most recently, hydric soils to aid in identification of wetlands. The uses of the soil database continued to expand and change until it became apparent in 1988 that SSSD and MUIR could not meet these changing needs. New information systems technology was also available that could advance the use of soil survey information.

However, the MUIR soil database system was remarkable in that it was able to evolve in many ways over time, but still kept its basic system design for about 25 years, until it was retired in 1996. At that time, MUIR contained data from about 2,900 survey areas and included approximately 250,000 soil map units. Implementation of the replacement system, NASIS, was begun in 1994. The SSSD system was retired in 1997 after the soil information in the MUIR database was converted to the MUR database in NASIS.

5.3.3 The Third Generation

Development of NASIS began with the analysis and documentation of the business of soil survey from beginning to end (Ernstrom and Lytle, 1993). Teams from various levels in the NCSS are established to complete the requirements analysis. Using structured systems analysis, these teams documented requirements that are passed on to contract software programmers. This analysis documented the important shift of the NCSS from producing static, printed soil survey reports to providing a dynamic database of soil information that could service a wide range of needs.

To meet these needs, a field data collection system was needed that would ensure the integrity and completeness of the data collected, including its geographic coordinates. The system needed to provide accurate and complete soil survey information to users based on what was observed during the soil survey process. Implicit in this idea was the ability to describe accurately the variability of soils and their properties as they occur in the landscape. This new system had to provide for a continuous update of the database as new information is gathered, so that one version of these data was available to users at the field, state and national levels.

NASIS had to provide a means for developing interpretations criteria and generating soil interpretations by a variety of scientists based on local to national requirements. For example, at the local level, there might be a need for an interpretation of suitability for animal waste disposal, and at the national level there may be a need for a soil productivity index. To ensure consistency, these interpretations must be applied to only one nationally consistent version of the data. The system had to provide for effective and efficient data delivery which included easy access by both internal and external (non-NCSS) users. This information needed to be delivered with common data structure, data dictionary definitions and appropriate metadata so that users understood the information and would use it appropriately.

5.3.3.1 NASIS System Objectives

Many hours of analysis (discussion) early on and numerous follow-up meetings identified the following specific system objectives (Soil Survey Staff, 1991):

- To put automated tools in the hands of front-line field office staff.
- One-time data entry, so that data could be retrieved by multiple software modules in various computer programs.
- Simplistic data entry in same format as collected.
- Validations to ensure proper entry of data and algorithms to provide default values.
- Automated procedures for correlation and quality assurance.
- Flexibility of the system to adapt to changes in procedure, standards, new data needs and policy.
- Capability to aggregate large-scale digital soil maps to smaller scales based on user-defined criteria.
- Data manipulation and retrieval option for all databases and software modules to include modeling capability.
- Provide single property values or representative values in addition to ranges to be used in models.
- Capability to provide confidence limits and reliability of map unit data.
- Continuous update of national, state and field office soil survey databases.
- Field access to state and national databases at the discretion of the users.
- Permanent storage of all soil survey documentation.
- Capability to transfer data files between various kinds of equipment.
- Two-way linkages to other natural resource databases.
- Software modules that are interactive, menu driven, and user friendly.
- Training on how to use the new system.

5.3.3.2 Software Design

NASIS was designed using an isolated design philosophy that separates the user interface, the NASIS engine, the INFORMIX engine, and the data dictionary-driven concepts so that changes can be easily made without reprogramming the entire system. This approach has allowed the easy addition of new capabilities and permitted the programmer to reuse previously developed codes.

NASIS was developed in the C+ programming language using the X Window system, a UNIX-based graphics window system. Similar to the Microsoft Windows® application that comes on many personal computers, the X Window system is a graphical user interface (GUI). This GUI allows users to manipulate the application by means of graphical objects, such as buttons, scroll bars, icons, and so forth. A pointing device, typically a mouse, is used. Because NASIS operates in this windows environment, several applications or processes can be run at the same time, each appearing in a separate pane of the window on the screen. NASIS users see a cascading or overlapping that mirrors the hierarchical structure of the data in the soil database.

5.3.3.3 Security

INFORMIX was selected as the NASIS database management software (DBMS) largely because of its security features. This proprietary design enabled the construction of a system that prevents most accidental or intentional corruption of data. NASIS allows data to be owned by individual users or groups of users so that only qualified scientists can edit or create data. The owner of an object has the authority to change data as needed. Individual or group ownership can be established based on needs.

5.3.3.4 Software, Hardware and Network Requirements

NASIS software requires the same configurations regardless of the user. At the 17 Major Land Resource Area offices where soil data are intensively managed, NASIS software use requires the full version of INFORMIX software, a high speed micro- or minicomputer and INTERNET access. At other NRCS field or state offices, university, state agency or federal agency sites with less intensive use, it requires only a high speed personal computer and X terminal emulation software for INTERNET access to NASIS. NASIS is operational in about 60 field, state, MLRA and national sites, with plans to connect 200 NRCS soil survey project offices and other NCSS cooperators at federal, state and university offices.

5.3.3.5 Software Development and Implementation Status

As with the SSSD software, NASIS software has been released in successive yearly versions. Version 1.0 was released in 1994. It provided for the validation and conversion of MUIR to MUR, security system and controls, an operational data dictionary, editors for areas, legends, data map units and on-line help. Version 2.0 of NASIS released in 1995 provided Cut, Copy, and Paste functions for data objects, a query editor and manager, global assign functions, report generation and an enhanced on-line help system. Version 3.0 released in 1996 provided the capability to generate representative values for a range of soil properties, calculation and validation routines for data, the ability to create interpretations criteria and generate interpretations, and the ability to produce a wide range of reports including map unit descriptions. The version released in 1997 provided for the replacement of the national MUIR with MUR, centralized storage of MUR data via the INTERNET at ISU and data sharing via INTERNET. It provided downloads to the NRCS field office computing system (FOCS), Geospatial databases, and public access to the MUR though the WWW.

These releases of NASIS software have dealt primarily with the development, management, etc. of MUR data; however, the NASIS version released in 1998 incorporates capabilities for input and access to soil site information.

5.4 Standards for Soil Survey Data

Under the Office of Management and Budget Circular A-16, revised 1990, lead agency responsibility to coordinate soil data-related activities is assigned to the USDA NRCS. This departmental, government-wide leadership for soil data coordination is carried out under the policy guidance and oversight of the Federal Geographic Data Committee (FGDC).

NRCS is responsible for developing and implementing a plan to coordinate soil data-related activities among federal and non-federal agencies. A Soil Data Subcommittee was established to assist the NRCS to coordinate federal and non-federal interests in soil data. The Subcommittee consists of representatives designated by the Federal agencies that collect, or finance the collection of, soil data as part of their mission or have direct application of these data through legislated mandate. The Subcommittee is chaired by the member from the NRCS, Soil Survey Division. The subcommittee coordinates: (1) exchange of information and transfer of data; (2) the establishment and implementation of standards for quality, content, and transferability; and (3) the coordination of the identification of requirements and the collection of spatial data to minimize duplication of effort where practical and economical. Specific responsibilities of the Subcommittee include the following:

* Facilitate the coordination of agencies' activities related to soil data and the exchange of soil data, by formal and informal means.
* Facilitate collection and compilation of information for soil data activities. Participate in the development and evaluation of data definitions and standards used by the United Nations and other international organizations and make recommendations for their inclusion in soil data as appropriate.
* Identify ways in which soil data from any sources may be included in the National Spatial Data Infrastructure.
* Recommend any changes in the definition of soil data to more accurately describe its scope.
* Assist the NRCS in establishing and publishing standards and specifications for the data (for example, incorporating conversion to metric units) and assist in establishing priorities for soil data production.
* Assist in the development and adoption of common standards of content, format, and accuracy for soil data for use by all federal agencies and to encourage use by non-federal organizations, to increase its interchangeability and enhance its potential for multiple use.
* Promote government-wide use of defined and published spatial data transfer standards for soil data.
* Facilitate the economic and efficient application of soil data through the sharing of experiences involving applications.
* Support higher-order or cross-cutting activities established or recognized by the FGDC.
* Encourage agencies to schedule the disposition of soil data in coordination with the National Archives and Records Administration in order to provide for the permanent preservation of historically valuable data.

The Subcommittee may create work groups or task groups, or further subdivisions as appropriate to carry out its activities and meet its responsibilities. A work group of this subcommittee has been assigned the responsibility of developing and maintaining the Definitions, Terminology, and Content portion of the Soil Geographic Data Standard. The Subcommittee chairperson coordinates activities with other FGDC subcommittees, working groups, and other components by participating in the FGDC's Coordination Group. The FGDC subcommittees and working groups are listed in Table 5.2. To date, standards have been created for the soils data dictionary, soil survey digitizing and metadata. The Keys to Soil Taxonomy (Soil Survey Staff, 1998), the National Soil Survey Handbook (Soil Survey Staff, 1996a) and the Soil Survey Manual (Soil Survey Staff, 1993) have also been designated as standards. These standards are accessible at (http://www.statlab.iastate.edu/soils/nssc).

Table 5.2 FGDC subcommittee and Working Groups

Federal Geographic Data Committee	
Subcommittees	Working Groups
Base Cartographic Data	Data Clearinghouse
Bathymetric Data	Earth Cover
Cadastral Data	Facilities
Cultural and Demographic Data	Historical Data
Geodetic Control Data	Standards
Geologic Data	Sample Inventory and Monitoring of Natural Resources and the Environment
Ground Transportation	
International Boundaries and Sovereignty	
Soils Data	
Vegetation Data	
Water Data	
Wetlands Data	

5.5   Research and Development Opportunities

Research, development and testing are being conducted in several areas that hold a great deal of promise for further use of information system technology in soil survey. The first area is in field data collection. Pilot projects are underway in which digitized soil survey information is used in conjunction with a grid-point sampling scheme and global positioning systems in order to provide information on the variability of soils and soil properties in the landscape. Testing is beginning with hand-held field devices that will provide on-screen mapping (digitizing) and collection of soil data. GIS is being used in the field for pre-mapping and quality assurance of mapping. GIS tools and procedures are being used to generalize large-scale mapping to smaller scales. When SSURGO is complete for the United States, these tools can be used to create the replacements for STATSGO and NATSGO or a user view of these data at any other scale.

The largest challenge for the future will be keeping up with the rapid changes in technology and making the necessary cultural changes in the NRCS and NCSS. These technologies are still not well understood by most NRCS and NCSS staff. For instance, GIS is often thought of as a mapping tool rather than an analytical tool for modeling or problem solving. However, before these concepts are well learned, GIS will likely merge with RDBMS and the analytical capabilities of GIS will be better realized. It is likely that several more years will pass before maps (geospatial) and soil property (attribute) data are thought of as one dataset.

Many have been slow to recognize how the methods of developing, managing and delivering soil survey are changing as a result of information technology. Many have also been slow to recognize the rapid adoption of GIS technology and the resulting increased demand for digital soil surveys. Many have been slow to recognize the value of information as a commodity, especially the value of soil survey information. The United States has a $5 billion replacement cost investment in the soil survey. Information technology will be increasingly used to manage and capitalize on this investment. Finally, information technology and soil survey information will increasingly be in demand to help provide solutions as the United States population and the demand on natural resources increases.

Acknowledgments

The author gratefully acknowledges the contribution of the following individuals to this chapter and to the development of the U.S. Soil Information System: Harvey P. Terpstra, Statistical Laboratory,

Ames, IA, and David L. Anderson, Gale W. TeSelle, Russell J. Kelsea, James R. Fortner, Ricky J. Bigler, Sharon W. Waltman, Adrian Smith, Jon D. Vrana, Fred E. Minzenmayer, Ken E. Harward, Ellis C. Benham and Dennis Darling, NRCS.

5.6 References

Bliss, N.B., and W.U. Reybold. 1989. Small-scale digital soil maps for interpreting natural resources. J. Soil Water Cons. 44:30–34.

Bliss, N.B. 1989. A natural resource database: Techniques for linking the Major Land Resource Area Map, the 1982 National Resources Inventory and the Soil Interpretation Record Databases in a Geographic Information System. EROS Data Center, Sioux Falls, SD.

Ernstrom, D.J., and D.J. Lytle. 1993. Enhanced soils information systems from advances in computer technology. Geoderma 60:327–341.

Hudson, B.D. 1992. The soil survey as paradigm-based science. Soil Sci. Soc. Am. J. 56:836–841.

Lytle, D.J., N.B. Bliss, and S.W. Waltman. 1996. Interpreting the state soil geographic database. p. 49–52. *In* M.F. Goodchild, et al. (ed.) GIS and environmental modeling: Progress and research issues. GIS World Books, Fort Collins, CO.

Nusser, S.M., and J.J. Goebel. 1997. The National Resources Inventory: A long-term multi-resource monitoring program. Environ. Ecol. Stats. 4:181–204.

Soil Survey Staff. 1991. National soil information system - soil interpretation and information dissemination subsystem. Draft requirements statement. USDA-NRCS, National Soil Survey Center, Lincoln, NB.

Soil Survey Staff. 1993. Soil survey manual. USDA-NRCS Agric. Handb. 18. US Government Printing Office, Washington, DC.

Soil Survey Staff. 1994. State soil geographic (STATSGO) database data use information. USDA Misc. Pub. 1492.

Soil Survey Staff. 1995a. Soil survey laboratory information manual. Soil Survey Investigations Report 45, Ver. 1.0, USDA-NRCS, US Government Printing Office, Washington, DC.

Soil Survey Staff. 1995b. Soil survey geographic (SSURGO) database data use information. USDA Misc. Pub. 1527.

Soil Survey Staff. 1996a. National soil survey handbook. USDA-NRCS, US Government Printing Office, Washington, DC.

Soil Survey Staff. 1996b. Soil survey laboratory methods manual. Soil Survey Investigations Report 42, Ver. 3.0, USDA-NRCS, US Government Printing Office, Washington, DC.

Soil Survey Staff. 1998. Keys to soil taxonomy. USDA-NRCS, US Government Printing Office, Washington, DC.

United States Department of Agriculture. 1981. Land resource regions and major land resource areas of the United States. USDA-SCS Agric. Handb. 296. USDA-NRCS, Washington, DC.

Waltman, S.W. 1997. Personal communication.

6

The Use of Soil Databases in Resource Assessments and Land Use Planning

G.W. Petersen, Egide Nizeyimana, D.A. Miller and B.M. Evans
Pennsylvania State University

6.1 Introduction

Information about soil properties and quality is important in understanding the functionality of major Earth ecosystems. Furthermore, soils play a major role in the distribution of plants, water and nutrients in terrestrial environments. Scientists in every aspect of environmental sciences need to understand soil processes to some extent for accurate assessment and prediction of resources and land use planning. A knowledge of soil properties and related factors assists in technology transfer, decision making and in the evaluation of resource programs and policies (Arnold, 1984). As a result, soil scientists are being called upon more often than before to solve problems associated with environmental degradation and land use planning.

The availability of remote sensing data and digital elevation models (DEM) and the increased capability of computers to handle large volumes of data during the last few years have prompted soil scientists to arrange soil survey information into soil databases. Today, soil databases contain soil characterization, quality, and interpretative data necessary for applications such as land evaluations and hydrologic and ecosystem modeling. The development and use of automated computer systems such as geographic information systems (GIS), spatial decision support systems (SDSS) and expert systems (ES) in recent years require that soil data and other environmental attributes be incorporated for more efficient and effective landscape analyses.

An historic overview of soil databases has been provided in Section H, Chapter 5. Soil databases evolved in the soil science community in order to organize soil attributes in a coherent manner for efficient storage and handling by the user. Early, non-digital soil databases consisted of tabular information on soil properties held in a generalized and classified form (Burrough, 1991) and were commonly referred to as tabular soil databases (Nielsen et al., 1996). The use of this type of database, however, is limited due to its poor flexibility when querying data for different, and particularly intense, spatial analyses. Furthermore, soil boundaries have to be redrawn manually every time new data analyses are performed for different interpretation maps. Geographic (e.g., digital) soil databases evolved in order to include both spatial and nonspatial descriptions and are designed to work exclusively within a GIS. They consist of spatial soils data, which may be represented by points (e.g., pedon locations), lines or polygons (e.g., soil mapping unit delineations), and attribute data that describe the characteristics of the spatial features (Burrough, 1989). The attribute data are stored in relational databases, whereas soil mapping units are stored in a topological arc-node or grid cell format, depending on the GIS software application (Burrough, 1991). Geographic soil databases are presently popular due to the wide availability of GIS software packages and environmental data sources in digital form. In this case, a GIS is used to store and retrieve existing database information, to analyze and integrate soil with other environmental data from various sources, and finally to generate interpretative maps from attribute data. GIS has become an effective and efficient technology for scientists, managers and decision makers to address multidisciplinary and complex environmental monitoring, assessment and management programs.

A complete account of the uses of soil databases in all areas of environmental science is not possible given the scope of this chapter. Rather, the intent is to present various ways these models and utility systems, such as GIS and GIS-based software, are parameterized using soil databases. The discussion will emphasize their uses in agricultural planning and management and related activities such as hydrology and water quality assessments.

6.2   Land Use Planning

Land use planning consists of making decisions about the use of land and resources (FAO, 1993). It is primarily conducted to achieve the best use of land, and its implementation is often driven by future people's needs in terms of productivity and environmental sustainability. Land use planning may not be as important in sparsely populated countries and communities. In this case, the analysis is straightforward and is usually aimed at finding the best locations for each of the potential land uses at hand. However, it is important in highly populated communities due to conflicts that arise between competing uses and interests of users in these areas (Brinkman, 1994). Planning activities are tailored to making the optimal uses of the limited land resources. Land use planning is conducted at different levels of decision-making: national, district, local government or farm (FAO, 1993). The national level is primarily concerned with national priorities as driven by land use policy and legislation. District level land use planning usually deals with development projects while local planning may be carried out at the community or watershed level. Farm level land use planning usually involves determining best management practices for different fields in the farm. The land use planning process differs from project to project depending on the goals and availability of data. Most projects involve sequentially organized concepts, evaluation and comparison of alternative land uses in terms of their suitability and impacts on the environment, and finally design and implementation of the plan. Soil databases are used in the land evaluation step of the land use planning process. A schematic representation of the land use planning process is shown in Fig. 6.1.

6.2.1   Land Evaluation

Land evaluation has been defined as the process of assessing or predicting the performance of land for specific purposes (FAO, 1976). The land is viewed here as an attribute of place and can range in size from a small parcel to a region. Soil attributes needed in this type of analysis may be obtained from soil databases and are typically used alone or in conjunction with other land characteristics to derive the distribution of land suitability, limitations or potential ratings for various land use types. These databases and associated attribute data may be part of a GIS, a GIS/model linkage, SDSS and/or ES. The analysis results may be presented in tabular or graphical form and are intended to provide key information necessary for land users or managers in making meaningful decisions about management,

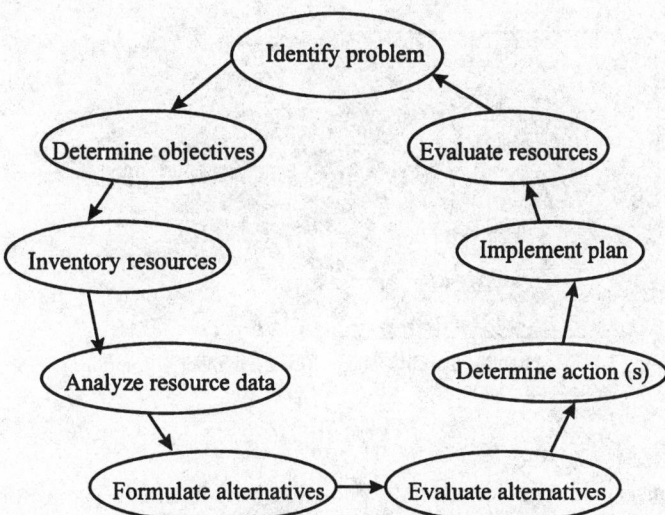

Fig. 6.1 THE USDA-NRCS farm planning process [Reprinted from Drungil et al., 1995. ASAE Pub. 5-95:177-185 with permission of the American Society of Agricultural Engineers]

conservation, and/or land use planning. As a result, the land evaluation is the most important part of the land use planning process (Whitley and Xiang, 1993).

The land evaluation is performed using soil interpretative groupings, productivity ratings or crop growth models. Soil interpretations involve determining land suitabilities or limitations for specific uses. They are generated using ranges of environmental land characteristics including soil properties, landscape-derived data, etc. and are often part of national soil surveys and many spatial soil databases. Crop growth models, on the other hand, consist of computer-assisted assessments and/or modeling of land qualities. In addition to the biophysical aspects of land evaluations, potential land uses are also evaluated in terms of their ability to provide a commodity, based on economical and social factors characterizing the land.

6.2.1.1 Soil Interpretations

Soil interpretations are usually predicted using numerical and descriptive soil information provided in soil surveys. These interpretations are arranged into classes of relative soil suitability or limitations for a specific use (Soil Survey Staff, 1993). Several soil interpretative generalizations have been developed for agricultural applications and are often part of soil survey reports and soil databases. The most well-known examples are the Land Capability Classification (LCC) of the USDA-NRCS (Klingebiel and Montgomery, 1961) and the soil Fertility Capability Classification (FCC) of the Food and Agriculture Organization (FAO) (Sanchez et al., 1982).

The LCC system was developed to assist farmers in planning farm layouts, crop rotation and other soil management practices (van Diepen et al., 1991). It consists of eight soil capability classes varying from most suitable (I) to least suitable (VIII) soils for agricultural production (Klingebiel and Montgomery, 1961). Soils occurring in the same class share the same limitations and risks when used for crop growth, and therefore, should respond similarly to soil conservation and management needs. Each class, which may have different units, contains subclasses that indicate the main limitation.

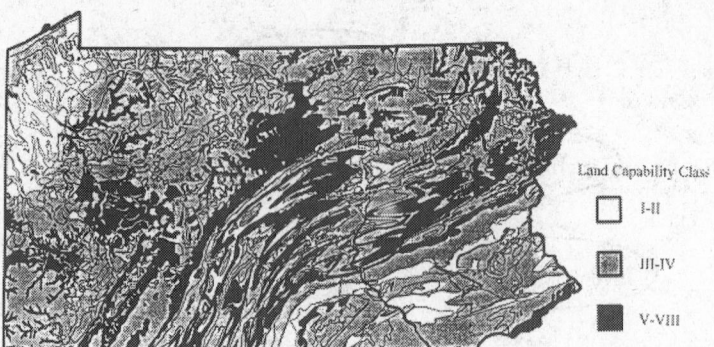

Fig. 6.2 Pennsylvania STATSGO Land Capability Classes (LCC); number of limitions to crop production increases from I to VIII.

Soils in the same capability unit are similar enough to be suited to the same crops, require similar management and have similar productivity levels. For example, soils in the land capability subclass IIIe have severe limitations that reduce the choice of plants, require special conservation practices or both. The susceptibility to erosion (e) is the dominant limitation. A land capability class is provided for each soil series in the Map Unit Interpretations Records (MUIR), Soil Survey Geographic (SSURGO) and the State Soil Geographic (STATSGO) database attribute tables. A GIS display of land capability classes derived from STATSGO for the state of Pennsylvania is shown in Fig. 6.2.

The FCC system, on the other hand, was designed to group soils that respond similarly to the same soil management practices (McQuaid et al., 1995). It is a soil fertility-based ranking system that assigns condition modifiers (e.g., low CEC, acidity, salinity) to each soil mapping unit. Groupings are defined according to the number of chemical and physical limitations that the soils have for proper plant growth. The FCC is part of the FAO/UNESCO digital soil databases and has been used in conjunction with country-wide soil survey systems.

While most of the soil interpretations were developed for plant production, soil ratings for non-agricultural applications are also available in soil databases. In the United States, county soil survey reports, MUIR, SSURGO and STATSGO contain soil interpretation tables of soil suitability and/or limitation ratings for each soil mapping unit for various specific uses such as septic tank absorption fields, sanitary landfills, etc. MUIR and SSURGO databases, where available, are currently used by local governments and private consultants for resource assessments, land use planning and management.

6.2.1.2   Agricultural Productivity Ratings and Indices

Soil interpretation schemes described above have been used in some instances as measures of soil productivity (Liu and Craul, 1991). However, farm productivity assessments have mostly been carried out using soil productivity ratings and indices. Soils are compared on a relative scale (0-100) based on predicted crop yields of a specific plant as a percentage of standard yields which equal 100% (Soil Survey Staff, 1993). This approach has some advantage since ratings for several crops and different soils can be combined to produce a general rating for a large area. Most county soil survey reports in the United States contain soil productivity ratings for major crops and common trees in that county. These data may be entered into a GIS and can be retrieved and used in land evaluations just like any other attribute data.

The level of agricultural productivity for different tracts of a farm can also be determined using ranking/indexing classifications of soil mapping units based on soil properties known to affect crop growth. In this type of classification, soils are ranked in terms of their relative productivity based on a combination of soil properties and qualities relating to productivity potential. As a result, productivity indices are not affected by changes in technology as compared to productivity ratings which are based on plant yields. A soil productivity index model can be multiplicative, additive or a combination of both. In any of these models, a soil parameter is given ranges of values with corresponding ratings that show its suitability for crop production. These are multiplied or added for each mapping unit to create a single index value.

The best known example of a soil productivity ranking system is the Storie Index (Storie, 1937). This index was recently programmed in C by personnel of the Iowa State University Statistical Laboratory under a cooperative agreement with the USDA-NRCS National Soil Survey Center (NSSC) as the Soil Ratings for Plant Growth (SRPG) model (Sinclair, 1996). The SRPG is a multiplicative index aimed at classifying soils in terms of their ability to produce fiber, vegetative growth and grains for commodity crops. Soil properties used in the model are derived from soil databases and include organic C, bulk density, clay content, available water capacity, pH, calcium carbonate, CEC, texture, rock fragments, etc. Soil climatic data (moisture and temperature regimes) and landscape features were also incorporated into the final stages of the SRPG calculations. The resulting classification is a relative index of land quality for each soil series. Results can be graphically displayed on a farm, county, state or national basis using GIS.

Another example of an often used soil productivity ranking/indexing model is the Productivity Index (Pierce et al., 1983). The index ranks soils in terms of their suitability for root growth and development. It assumes that crop yields are a function of root growth, including rooting depth, which affects the soil environment. The model is a multiplicative function that predicts soil productivity changes due to long-term soil erosion based on soil parameters (e.g., pH, available water capacity, bulk density) derived from the Soil Interpretation Record (SIR) SOILS-5 data. This dataset is part of the SSURGO and STATSGO databases. The model was developed for the north central United States region but can be applied to other areas after proper calibration of its variables.

6.2.1.3   Regression and Crop Growth Models

Biophysical land evaluations are commonly conducted using regression and crop growth simulation models. Statistical methods are used to develop empirical relationships between crop yields and land and climate characteristics of the area. Model coefficients are computed for a specific land location by regressing a sample of yields to corresponding climate (e.g., temperature, rainfall, radiation) and soil characteristics known to affect crop yields. An example of a statistical model relevant to land evaluation is that of Olson and Olson (1986). The authors evaluated various parameters and found that the best fit regression model of yield estimates for areas in the state of New York was a function of rainfall, number of growing degree days, basic cation status and organic C content of soils. Regression models are site specific and should be calibrated for land areas in different environments. In addition to simple and multiple regressions, crop yields may be estimated using more complex statistical approaches such as multivariate analysis and principal components.

During the past few years, scientists have used simulation models to estimate biophysical land suitabilities. The modeling approach to land evaluations in land use planning efforts offers some advantages over conventional and statistical methods because the effect and cost associated with temporal changes in land and/or climatic attributes can also be determined. Potential land uses can be compared on the basis of crop yields predicted under future climate changes, long-term soil degradation, etc. Similarly, those factors known to affect the socio-economic aspects of land evaluations, such as future population growth and technology changes, can be used to predict the capacity of the land to sustain future food demands.

A number of process-oriented crop simulation models that have potential for use in land evaluation have been published in the literature. Examples are models of de Witt and van Keulen (1987) and Weir et al. (1984), which are aimed at simulating plant growth and subsequently crop yields from agricultural lands. The use of models in land evaluations has been limited due to the time involved in compiling the large volume of input data and in parameterizing the model, particularly for large areas. The integration of these models with GIS and similar systems, however, may reduce significantly the processing time and resources.

Sharifi and van Keulen (1994) have developed a SDSS for land use planning at the farm level. In this system, planning, crop simulation and linear programming models were integrated with a GIS. A Decision Support System for Agrotechnology Transfer (DSSAT) has also been developed by the International Benchmark Sites Network for Agrotechnology Transfer (IBSNAT). IBSNAT includes many United States universities (e.g., University of Hawaii, University of Florida) and the University of Puerto Rico (IBSNAT, 1989). The system integrates data from crops, soils, and weather as one standardized input with several crop models to simulate multi-year yields under different crop management strategies at farm and regional levels. Two GIS-based interfaces, the Agricultural and Environmental Geographic Information System (AEGIS) and AEGIS/WIN were also developed by the University of Florida and linked to the DSSAT (Engel et al., 1996; Luyten and Jones, 1997). AEGIS uses ARC/INFO and AEGIS/WIN uses ARCVIEW software. The complete system consists of databases, crop models, expert systems and GIS (Fig. 6.3).

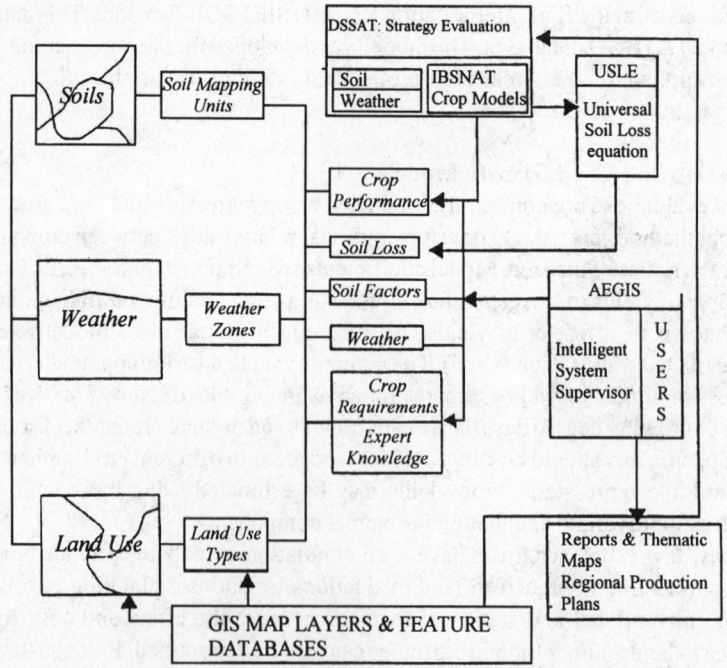

Fig. 6.3   AEGIS decision support system [Adapted from IBSNAT, 1992 with permission]

6.2.2 Standard Land Evaluation Systems

Standard land evaluation systems contain two procedures: biophysical and economic. The biophysical suitability analyses use any of the land capability classification schemes, agricultural productivity ratings or modeling approaches described above. Biophysical land suitability classes may also be derived using land qualities (e.g., erosion hazard, soil water storage) or locally measured/observed land characteristics from sources such as soil databases, remote sensing, geologic maps and DEMs (Wagenet et al., 1991). In this case, numerical values are assigned to suitability ranges of each relevant parameter and land quality. They are combined to yield an overall rating. Suitability classes for ranges of various soil, landscape and hydrologic properties can be found in manuals such as the National Soils Handbook (USDA, 1993) or derived by the user, based on the knowledge of local soil characteristics and other relevant parameters. An economic suitability analysis, on the other hand, includes an evaluation of a given site's location in relation to major highways and urban land, and social and economic restrictions of the municipality, county or state.

A GIS-based land use evaluation system, the Comprehensive Resource Inventory and Evaluation System (CRIES), has been developed (Schultink, 1987). CRIES uses a grid-cell approach to delineate agro-ecological production zones within a GIS and estimates crop production potential for existing and alternative land uses. It has two interlinked components: a geographic information system (CRIES-GIS) and the agro-economic information system (CRIES-AIS). The CRIES-GIS component has modules for model parameterization, result display, and other routine GIS-based analyses. The CRIES-AIS has modules for water balance modeling, yield predictions, and linear programming, among others. The system was developed primarily for land use planning applications within the farming systems framework in developing countries.

Other systems that are commonly used in land evaluations are the Automated Land Evaluation System (ALES) (Rossiter, 1990) and the Land Evaluation Computer System (LECS) (Wood and Dent, 1983). Both systems are based on the FAO framework of land evaluation. ALES is a PC-based computer program designed to allow land evaluators to build their own expert systems based on the local knowledge of the land. LECS has been implemented in the FAO's Agricultural Planning Toolkit (APT) (Rossiter, 1996) and the Integrated Land and Watershed Management Information System (ILWIS) (Elbersen et al., 1988) of the International Institute for Aerospace Survey and Earth Sciences (ITC), Enschede, The Netherlands. Land evaluation systems may also be part of an SDSS or ES designed particularly to aid in the decision-making process.

6.2.3 Site Suitability Assessments

A site suitability analysis typically involves the assessment of the level of affinity that a specific site has for a particular land use. An example would be the determination of the most appropriate location for a water treatment plant in a subdivision. Interpretations based only on soil surveys are rarely adequate for such non-agricultural uses (Soil Survey Division Staff, 1993). The analysis often integrates local information on soils, landforms, current land uses, geology and location in relation to such factors as accessibility, sources of water for irrigation and associated social and political restrictions of the subdivision and state. Furthermore, the proposed use may have its own limitations that should be taken into account. An analysis for suitable sites for land application of sewage sludge should, for example, consider the physical, chemical and biological properties of the waste.

GIS has proved to be an efficient tool in site selection studies. Its database structure, spatial overlay modeling and data analysis capabilities are powerful bases for locational and class assignment operations using data from different sources. A good example of the use of GIS for locating appropriate sites for forest land application of sewage waste is provided by Hendrix and Buckley

(1992). The authors derived physical site suitability ratings for an area in Vermont based on the EPA guidelines (USEPA, 1981) and merged them with social and political restrictions of the state and counties to derive a land applicability classification. Similar approaches have also been used to locate sites for solid waste disposal (Weber et al., 1990; Karthikeyan et al., 1993).

Many systems have been developed over the years in the agricultural community and in many parts of the world to assist the site evaluation process. An example is the USDA-NRCS Land Evaluation and Site Assessment (LESA) system (Wright et al., 1983). LESA was developed to aid planners and public officials at the city, township, county or state level in determining the agricultural quality and economic viability of a farm (Dunford et al., 1983; Stamm et al., 1987). The land evaluation portion of LESA determines soil productivity levels, farm size and agricultural sales volume; the site assessment portion deals with factors such as location, amount of non-agricultural land, zoning restrictions, etc. (van Horn et al., 1989; Daniels, 1990). GIS-based LESA systems have also been developed and implemented at the county and state levels (Williams, 1985; Ferguson et al., 1991). LESA and similar systems have been incorporated into many county, state and national programs in the United States, such as the Purchase of Agriculture Conservation Easement program (PACE) (Christensen et al., 1988; Clayville, 1994). PACE has been used in farm prioritization programs such as farmland preservation, taxation and real estate valuation.

6.3   Conservation Planning and Nutrient Management

Land degradation is becoming a major issue in many parts of the world, particularly in areas of high agricultural pressure. As a result, a variety of models and decision support systems (DSS) have been developed in many countries to help farmers and land use planners assess the extent of land degradation and subsequently develop appropriate management plans. Most of these models and systems use soil databases as sources of input parameters. GIS has been used to generate thematic maps of soil degradation. These interpretative maps are used as a preliminary step in land evaluation for field nutrient management and conservation planning activities. For example, a spatial representation of slope, soil pH, organic C, Al content, water-holding capacity and/or texture can help identify areas of major deficiencies or limitations in the farm. However, the distribution of values of a single parameter may not be enough to explain differences in land degradation. More often, an interpretative GIS layer showing the distribution of soil degradation hazard classes is created by map overlay analyses of soil variables (e.g., texture, structure, organic C content), vegetation (land use/cover), and slope.

Several field-scale hydrologic and water quality simulation models have been developed primarily to serve as soil conservation and nutrient management planning tools. Their primary role is to evaluate existing and alternative management practices to control water erosion and nonpoint source pollution in surface and subsurface waters. An example is the Chemical, Runoff, and Erosion from Agricultural Management Systems (CREAMS) model (Knisel, 1980). The model essentially provides relative comparisons of sediment and pollutant loads from alternative management practices in a particular field or farm. Furthermore, many of these models have been linked to GIS. In this case, the model parameterization and display functions are handled through a graphical user interface (GUI). Model input data are derived from geographic soil databases either developed as part of the integration or from existing ones such as SSURGO. SSURGO has been used, for example, to provide input data via a GIS linkage to field-scale water quality models such as Chemical Movement through Layered Soils (CMLS), Leaching and Chemistry Estimation (LEACHM) (Wilson et al., 1996; Inskeep et al., 1996) and the Nitrate Leaching Economic Analysis Package (NLEAP) (Shaffer et al., 1996).

The parameterization of field-scale water quality models has also been accomplished through the use of DSSs. In addition to sediment and nutrient assessments, these systems evaluate different scenarios necessary for making management recommendations. For example, the USDA-NRCS has developed a DSS, the Field Office Computer System (FOCS), for field officers to help farmers address erosion and water quality problems using sound management plans (Kuenstler et al., 1995). FOCS contains a series of models and decision support tools in a UNIX environment; it uses the NRCS soil databases as the main source of model input data. FOCS results may be utilized by farmers to adjust management practices for distinct fields in a given farm or may be used by NRCS field officers to help farmers set priorities while providing technical assistance for nutrient management planning (Kuenstler et al., 1995). The development of a link between the National Soil Information System (NASIS) described earlier (Section H, Chapter 5) and FOCS is anticipated in the near future to allow export of soil attributes from any soil database in NASIS into the FOCS structure. Three of the tools included in the FOCS, the Revised Universal Soil Loss Equation (RUSLE), the Field Office Water Quality (FOWQ) and the Hydrologic Unit Water Quality (HUWQ), are described below.

RUSLE is an improved and computerized version of the Universal Soil Loss Equation (USLE) (Wishmeier and Smith, 1978). The USLE was originally designed to compute long-term average soil losses caused by field-scale sheet and rill water erosion. The equation,

$$A = R \times K \times LS \times C \times P \qquad [6.1]$$

is used to predict the long-time average annual soil loss (A) as a function of rainfall and runoff (R), soil erodibility (K), slope length and steepness (LS), cover and management (C), and conservation practice (P). The soil erodibility factor (K) quantifies the susceptibility of soil particles to detachment and movement by water and can be derived from MUIR and SSURGO. A comparison of RUSLE results with soil loss tolerance values can help one choose appropriate control and conservation measures needed to reduce soil erosion.

The FOWQ tool presently allows nutrient screening, nutrient management planning, pesticide screening and farmstead assessment. As an example, the nutrient screening tool serves to identify those fields in a farm that have high potential for N and P losses from applied N-fertilizers based on site characteristics and current management practices. The values of soil characteristics (e.g., K factor, organic C, permeability rate, pH) needed in the sensitivity evaluation of mapping units in a farm are currently extracted from MUIR or SSURGO through a user interface.

The HUWQ tool includes, among others, two field-scale water quality models, the Erosion Productivity Impact Calculator (EPIC) and Groundwater Loading Effects of Agricultural Management System (GLEAMS). Soil parameters are derived from soil databases such as MUIR and SSURGO and input to models through the user interface (Drungil et al., 1995). Simulation results are often used to rank different fields in a farm in terms of their potential for N and P losses.

6.4 Environmental Resource Assessments and Modeling

Most resource assessment and modeling studies are performed at the watershed level. Mathematical models that are commonly used to estimate water quantity and quality in watersheds are based on the distribution of physical parameters such as soils, land use/cover type, and geometric characteristics of the drainage area. Gathering information on these parameters, particularly for large drainage basins, has always been a problem for hydrologists because of the work involved in collecting aerial photographs and topographic maps necessary for complete coverage, and the extensive field work required to determine the distribution of soils and geometric properties of the basin. At present, though, databases are used extensively to provide model input data to estimate water amounts and quality. Results are often used to support watershed management and planning activities.

6.4.1   Groundwater and Stream Recharge

The degree to which precipitation falling in a given watershed ends up as surface or subsurface flow is a function of soil parameters that affect water infiltration rates, such as texture, structure, organic C, and bulk density (Gray et al., 1970). Soil moisture and groundwater recharges are often simulated using soil infiltration models (Khan and Liang, 1989). A number of physically based infiltration equations have been developed over the years to estimate the amount of rainfall/irrigation water that percolates into the soil. Most of these equations have been incorporated into hydrology and water quality models. Examples of these approximations are the Green-Ampt (Green and Ampt, 1911), Holtan (Holtan, 1961), Horton (Horton, 1940) and Phillip two-term (Phillip, 1957) equations. The soil infiltration descriptors in these equations are, however, based on water flow properties.

Driven by the need to make hydrologic modeling more operational, scientists later developed procedures for estimating these parameters using soil properties readily available in county soil survey reports and soil databases (Rawls et al., 1983). For example, parameters that govern infiltrated water flow in soils to a drainage tile or groundwater in the Area Nonpoint Source Watershed Environmental Response Simulation (ANSWERS) model (Beasley and Huggins, 1991) are estimated using the Holtan infiltration equation. This equation, which requires variables for total porosity, field capacity, steady-state infiltration rate, etc. for each soil series, is solved using empirical equations based on soil characteristics (e.g., permeability rate, bulk and particle densities, and texture) readily available in soil databases such as MUIR and SSURGO. Similarly, the parameter for the amount of water that percolates through the vadose zone to sewer systems in the Storm Water Management Model (SWMM) (Huber and Dikinson, 1988) is determined using the Green-Ampt equation. To facilitate the use of this equation in hydrologic modeling, the various Green-Ampt infiltration parameters (total porosity, effective porosity, wetting front capillary pressure, and hydraulic conductivity) have been estimated for USDA soil textural classes (Rawls et al., 1983; Vereecken et al., 1989), an attribute present in most soil databases.

The amount of runoff water contributed to stream flow has been commonly estimated using mass balance procedures and rainfall-runoff models. In the first case, surface water runoff is computed as the difference between precipitation and losses (evapotranspiration, infiltration, etc.). In the second case, various soil, landscape and land use characteristics are simplified to represent the whole watershed or large subunits of it as a uniform homogeneous system (Novotny, 1981). The infiltration rate and other losses are expressed in terms of soil and land cover/use characteristics. One example of this approach is the Curve Number (CN) procedure. CN values represent the amount of rainfall that moves through the watershed as runoff after subtracting the combined effect of soil, land cover, management practice, antecedent moisture condition, and hydrologic conditions (Stuebe and Johnson, 1990). The CN is related to infiltration amounts by the relationship:

$$S = 1000/CN - 100 \qquad\qquad\qquad\qquad [6.2]$$

where S is the potential loss of water by the soil over the drainage area.

Most soil characteristics related to surface runoff (e.g., texture, structure, permeability, and hydraulic conductivity) are reflected in the Hydrologic Soil Group (HSG), a rating scheme previously developed by the USDA (1986). The HSG code reflects the natural drainage properties of a given soil and has values ranging from A (well-drained soils with high infiltration rates) to D (poorly drained soils with very slow infiltration rates). Soils with an HSG code of A have a low potential for surface runoff, but may pose a severe pollution hazard to groundwater. The reverse is true for soils with an HSG code of D (Young et al., 1982). A CN value is assigned to each category depending on the land cover type according to USDA-NRCS tables (USDA, 1986). In the past, CNs have been estimated manually using HSG values in county soil survey reports. This process is difficult and time consuming, especially for large drainage basins. Today, CN values are more often determined using SSURGO-derived HSGs and land cover/use classes from remotely sensed imagery (Nizeyimana, 1995). The GIS overlay and subsequent analysis of CN determinations can be performed using vector coverages or on a per pixel basis using gridded layers of HSGs and land use types. In this case, an HSG boundary layer is derived from a soil database such as SSURGO and combined with a land use class coverage using GIS analytical functions (Stuebe and Johnson, 1990). A CN is then assigned to each HSG/land use complex contained in the resulting attribute table. The watershed area-weighted CN is determined based on the areal percentage distribution of each combination of HSG/land use.

6.4.2 Water Quality and Nutrient Management

Surface hydrologic processes in watersheds are the driving forces behind erosion and sedimentation and the transport of nutrients and other pollutants to various surface and groundwater bodies. The infiltration equations and/or CN methods described above form the basis for estimating runoff volumes in many water quality models such as ANSWERS, Simulation for Water Resources in Rural Basins-Water Quality (SWRRBWQ) (Arnold et al., 1990) and AGNPS (Young et al., 1994). On the other hand, the USLE originally developed for field assessments of soil erosion has also been extensively applied to watersheds. The simplicity of this equation and the spatial variability exhibited by its parameters makes it well-suited for estimating sediment production and transport in watersheds. The USLE is also used in many watershed hydrology/water quality models, including AGNPS and SWRRBWQ.

Soil databases have been used extensively in watershed-based assessments of water quality. They are used to provide values for soil parameters (e.g., HSG, K factor, infiltration parameters) needed for estimating the magnitude of sediment, nutrients, pesticides and other pollutants in surface and subsurface media. The simulation of pollutants in these models is accomplished by taking into account pollutant dissolution characteristics in water and/or its adsorption properties on sediments. Water quality model parameterization using soil databases is usually performed through an interface that links either the model and the soil database or the model and a GIS. In the latter case, the soil database is an integral part of the GIS.

The SWRRBWQ model has an accompanying soil database. The soil characterization data required for input to SWRRBWQ are derived from a subset of the NRCS SIR SOILS-5 (Arnold and Sammons, 1988). This soil database, which is provided on computer diskettes, is accessed through an interactive data entry interface written in FORTRAN 77. The user enters the soil series name and the system searches the database automatically until a match is found. It then displays the information for that series and automatically formats the data into the model input data file.

Another example of a model/soil database link is the Database Analyzer and Parameter Estimator (DBAPE) interface that was developed by the USEPA (Imhoff et al., 1990). The interface is an interactive computer program specifically designed to provide soil information for unsaturated zone models such as the Pesticide Root Zone Model (PRZM) and the Risk of Unsaturated/Saturated Transport of Chemicals (RUSTIC) model (Carsel and Jones, 1990; Carsel et al., 1991). The soil database consists of a subset of the SIR SOILS-5. The soil database, in this case, is an integral part of the interface.

Most often, however, watershed model parameters are derived from a soil database consisting of a digitized soil map and associated attributes from soil survey reports. The soil geographic database thus produced is similar to SSURGO. The GIS may be used to prepare model input data using its spatial analysis and database management utilities, but the model and the GIS are run separately. Such GIS-based model parameterization has been accomplished for the USDA-Water Erosion Prediction Project (WEPP) (Savabi et al., 1995) and the Hydrological Simulation Program FORTRAN (HPSF) (Al-Abed and Whiteley, 1995) models.

The most recent and popular approach, however, consists of developing GIS techniques that partially or fully parameterize existing models through a GUI where the user interacts with various modules to select data input and module utilities and to display graphical and/or tabular representations (Petersen et al., 1997). In this manner, the GIS has been integrated with many widely used water quality models including ANSWERS (de Roo et al., 1989; João and Walsh, 1992), AGNPS (Tim and Jolly, 1994; Yoon, 1996) and SWRRBWQ (Evans et al., 1992; Drungil et al., 1995).

The GIS/model linkage may also be part of an SDSS or ES. A SDSS is a customized GIS application for users, particularly decision makers, who wish to take advantage of the problem-solving capabilities of GIS packages and modeling without having to become proficient in using all of their operations and functions (Petersen et al., 1997). The system differs from traditional GIS analysis because it is intended to provide selected tools via a customized GUI to design and model alternatives, select and evaluate these alternatives, and subsequently display results in maps, tables, and/or diagrams. In most straightforward applications, the interface and utility modules are built using the programming facilities provided by various GIS packages. However, SDSSs often combine GIS technology with modeling and programming facilities offered by computer languages such as C, FORTRAN, and Pascal (Petersen et al., 1995). Unlike SDSSs, ESs give potential solutions to the user. They provide not only data analysis capabilities and decision-making information, but also expert knowledge and reasoning rules to manipulate and evaluate information for specific uses (van der Vlugt, 1989).

6.5   Resource Assessments and Planning at Regional and Global Scales

Several GIS-based systems have been developed in different countries and regions for land evaluations. In addition to databases, GIS analysis and associated programming tools, some land resource information systems (LRIS) have modeling capabilities and decision-making tools useful in land resources assessment and evaluations. The FAO Agro-Ecological Zones (AEZ) is an example of such a system in which various GIS layers are developed from database attributes and combined on a per-grid cell basis to derive model input data for yield simulations, land suitability and productivity analyses, and land degradation (Koohafkan and Antoine, 1997). The land optimization modeling is performed via decision support tools. Different steps of the AEZ methodology for land use planning are presented in Fig. 6.4.

Land use planning at regional and global scales has typically been geared toward assessing land qualities for use in environmental assessments and management. Many national and international planning agencies have concentrated their efforts on developing soil information systems (SIS) and GIS databases to aid in the analysis, display, and/or dissemination of spatial data to users. In this section, SIS and GIS databases are briefly described and their use in various aspects of land use assessments and modeling is discussed.

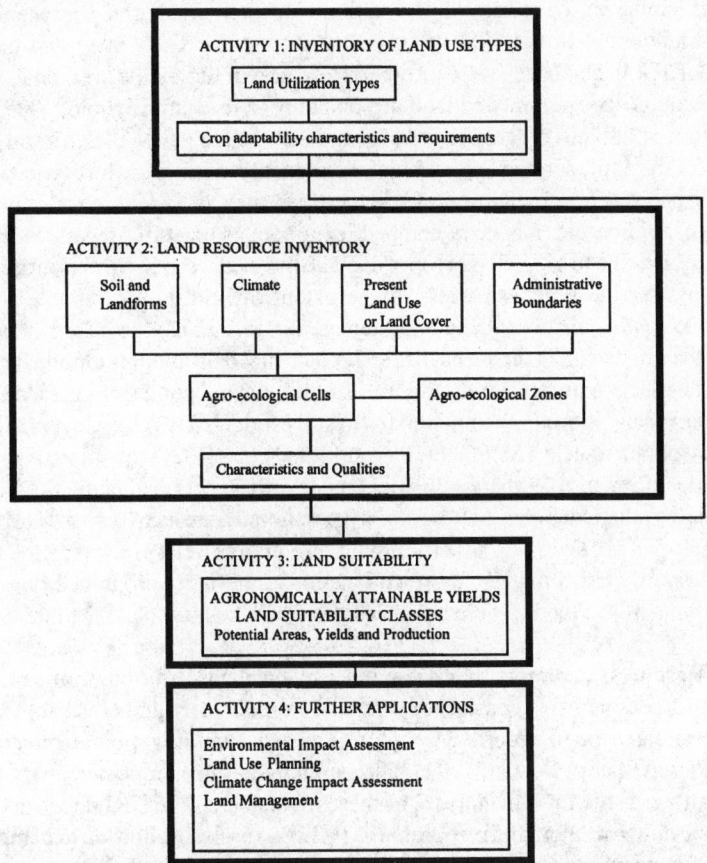

Fig. 6.4 FAO Agro-Ecological Zoning (AEZ) methodology [Adapted from Koohafkan and Antoine, 1997 with permission of the Food and Agriculture Organization, Rome, Italy]

6.5.1 Soil Quality

Traditional methods for assessing and mapping soil quality (i.e., soil erosion, soil salinity, organic matter depletion, acidification) are based on observations and laboratory data collected at field and/or watershed scales. Most soil quality assessment tools, such as the USLE equation, were originally designed to operate at field scales. Today, however, characteristics used in soil quality assessments can be mapped for large areas using existing small-scale soil and terrain databases, remotely sensed data and other data sources (Nizeyimana and Petersen, 1997). Regional and global databases are typically developed by extrapolating field- and watershed-based data to larger mapping units or by providing links between the spatial information and tables containing characterization data and interpretation records (Petersen et al., 1997). In general, land quality information resulting from environmental assessments using these databases helps governments and international agencies understand the regional and global scope of environmental problems such as soil degradation and water pollution (Bliss and Waltman, 1994). Examples of land quality assessments based on soil databases at these scales are provided below.

A number of countries have developed nationwide soil databases for use in land quality assessments and modeling. In the U.S., two soil databases, the STATSGO and the National Soil Geographic (NATSGO) database, were designed for use in statewide, regional and nationwide environmental resource assessment and land use planning (Bliss and Reybold, 1989). STATSGO is distributed to users as GIS coverages for individual states, while the NATSGO soil map exists as a single coverage for the entire United States. More detailed information on the two datasets is provided in Section H, Chapter 5. The distribution of soil qualities such as available water holding capacity, organic C storage, acidity, etc. has been mapped using ranges of attribute values from STATSGO (Miller and White, 1998) and NATSGO (Kern, 1995; Bliss et al., 1995) for the entire country. These GIS-derived maps provide a general view of the extent of limiting or suitable soil attributes to agricultural production in the country and are often used in further analyses for land use assessments and planning. In the same vein, Petersen et al. (1998) determined the magnitude and distribution of soil productivity loss under urban/industrial land use for the selected conterminous United States using GIS analysis and overlay of urban thematic maps to soil productivity rating layers. Soil productivity distributions were obtained using SRPG model results and STATSGO mapping units (Sinclair et al., 1996). Nationwide GIS maps of urban/industrial land use were obtained using the United States Air Force Defense Meteorological Satellite Program's Operational Linescan System (DMSP/OLS) nighttime imagery (Imhoff et al., 1996). Soils in the highest productivity class in the United States represent 5.2% of the total land. Unfortunately, these soils are also the most urbanized ones; the degree of urbanization within soil productivity classes increases from the least (1.4%) to the most productive soils (4.5%).

The National Resources Inventory (NRI) is another spatial dataset that has tremendous potential for use in soil quality assessments. It consists of a database of land use/land cover information and soil characteristics that have been recorded at georeferenced sampling points on nonfederal lands throughout the United States (Goebel, 1992). The sampling has been repeated every five years since 1982, and thus, allows for temporal analyses of these parameters. The NRI data can be mapped on a county, hydrologic unit, or major land resource area (MLRA) basis. In addition to being a major source of data for soil and water management, the NRI was linked to the SIR SOILS-5 database. For example, Burkart et al. (1994) derived a GIS map showing the distribution of net soil loss in midwestern states using two NRI data variables, the soil loss tolerance (T factor) and USLE-derived average annual soil loss estimates. Kellogg et al. (1994) derived nitrate and pesticide leaching indices for the whole nation based on SIR associated with the NRI data. Kern (1995) used NATSGO MLRAs as a geographic base and the 1992 NRI-SIR data to estimate geographic patterns of soil water-holding capacity for the conterminous United States. Empirical regression models were used in conjunction with soil physical properties including soil classification, rock fragment, sand, clay and organic C contents, depth to indurated layers, and depth-to-bedrock to obtain spatially distributed estimates of soil water retention. Soil databases have been developed for similar applications in Canada (Kirkwood et al., 1996), The Netherlands (DeVries and Denneboom, 1993) and Australia (Bui et al., 1996). In the latter case, the authors used a geographic soil database to determine regional salinization risk resulting from forest clearing in the North Queensland area.

At the continental scale, a soils map of Europe (1:1,000,000 scale) previously compiled and prepared by Tavernier (1985) has been digitized under the Coordinated Information on the European Environment (CORINE) program (Platou et al., 1989). Its legend is similar to that of the FAO world soil map. Each mapping unit is linked to a soil topological unit that identifies the main soil whose attribute data are in a relational database. Soil and terrain parameters from this database were used, for example, to generate GIS maps of soil erosion risk for the southern region of the European Community (Bonfils, 1989).

Finally, at the global scale, a soil database has been developed for the entire world (1:5,000,000 scale) by digitizing the FAO soil map of the world (FAO, 1995). This GIS coverage has been used to derive land qualities (available water storage, soil productivity potentials, soil degradation potential, etc.) for the United States (Kern, 1995) and Africa (Eswaran et al., 1997). In a similar vein, the development of a more detailed digital database, the World Soils and Terrain (SOTER), has begun under the auspices of the United Nations/FAO, the International Soil Science Society (ISSS), and the International Soil Reference and Information Center (ISRIC, 1993). SOTER was designed at the 1:1,000,000 scale and accommodates most soil classification systems. Map units in this database are described by one to three terrain components and one to three soils whose characteristics are presented in separate attributes files (Baumgardner, 1994).

Due to the present need for assessing human-induced soil degradation, a Global Assessment of Soil Degradation (GLASOD) map that uses SOTER soil and terrain data was produced at a scale of 1:10,000,000 through a collaborative project between the United Nations Environment Program (UNEP) and ISRIC (Oldeman et al., 1990). GLASOD indicates the spatial distribution of water erosion, wind erosion and physical deterioration such as soil compaction. The GLASOD digital map is being developed by digitizing map units and recording attribute data in a GIS. It was, however, generated using incomplete data since SOTER is not expected to be completed for another 15 to 20 years. It does not, therefore, provide sufficient detail for many applications. Nevertheless, the information provided is crucial at the present time to policy and decision makers for establishing priority programs.

A collaborative effort between FAO, ISRIC, the International Geosphere-Biosphere Program (IGBP) and NRCS is also underway to develop a global soils dataset for use in global change research (Global Soils Data Task, 1995). The objective of this project is to develop and link an international pedon database to a spatial representation and to derive spatial and statistical parameters needed in global change modeling.

6.5.2 Nonpoint Source Pollution

Since the early 1990s, there has been a considerable interest in statewide and regional nonpoint source pollution assessments and modeling. The emergence of new technologies (e.g., GIS, remote sensing, GPS, digital databases) supports the integration of model input data on field and watershed processes and properties for analysis over larger land areas. This type of analysis normally involves the development of watershed vulnerability indices and ranking schemes using data on land use, soil, watershed characteristics and pollutant sorption properties. State and regional planning agencies often need this kind of information to identify the most vulnerable or critical areas to which mitigation projects can be targeted in order to control or reduce nonpoint pollution.

Considerable information exists on the use of GIS in the regional assessment of agricultural nonpoint pollution. Examples of GIS-based statewide analyses include the evaluation and ranking of surface and groundwater watersheds for nutrient and pesticide pollution potential in Pennsylvania (Hamlett et al., 1992; Hamlett et al., 1995) and the identification of watersheds most affected by nonpoint source pollution in Virginia (Flagg et al., 1990). In these studies, STATSGO-derived properties were used in combination with other spatial data sources to compute plant nutrient and pesticide loads in surface and groundwater from agricultural fields. An example of a GIS-based indexing/screening approach to statewide assessment of nonpoint pollution is shown in Fig. 6.5. In

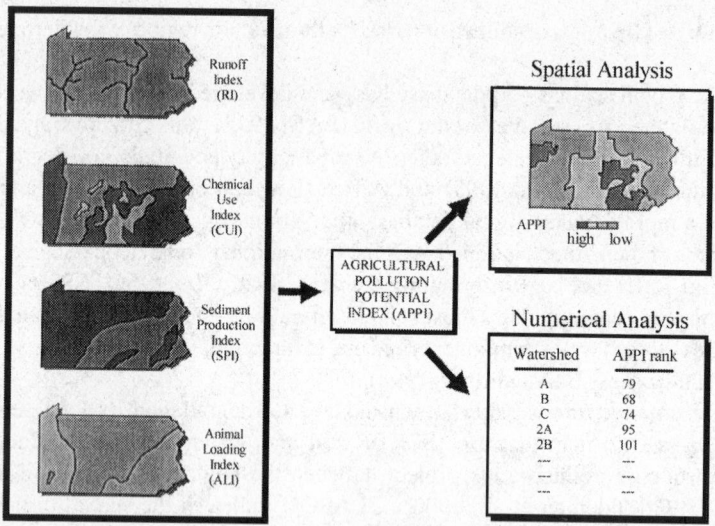

Fig. 6.5   Model components for agricultural pollution potential predictions

other instances, however, STATSGO and similar environmental databases were linked to hydrology and water quality models to evaluate the leaching of agrochemicals in Montana (Wilson et al., 1996), New York (Wagenet et al., 1993), the northeastern United States (Bleecker et al., 1995) and the whole United States (Srinivasan et al., 1995). An approach to integrating a regional environmental database with a water quality model is shown in Fig. 6.6. STATSGO-based soil interpretation parameters have also been combined with population and septic system densities from the 1990 United States Census to estimate N loadings delivered from septic systems to surface water in Pennsylvania and to rank watersheds in terms of levels of vulnerability (Nizeyimana et al., 1996). Similar GIS-based projects, in which soil data were derived from generalized environmental databases, have been completed in Montana (Nielsen et al., 1990), Rhode Island (Baker and Panciera, 1990) and Hawaii (Khan and Liang, 1989).

Over the last 5 to 10 years, there has been an increasing number of SDSSs developed specifically for environmental resource assessment and management at a regional scale. These systems are often designed and implemented for the analysis and management of river basins and regional environmental planning and management. For example, a UNIX-based SDSS was developed to parameterize a water quality model that simulates water flow, sediment and nutrient transport, and to evaluate different scenarios of management practices within a basin (Engel et al., 1991). The system was developed by integrating the AGNPS model with a GIS through a set of unstructured modules written in C language. Although the main component of this SDSS is the GIS/model interface, visualization and interpretative tools have also been incorporated to guide the decision-making process.

Another example of an SDSS is the Lake Okeechobee Agricultural Decision Support System (LOADS). It was developed primarily to assist in regional environmental planning of the Lake Okeechobee watershed in Florida (Negahban et al., 1996). System users are allowed to alter land uses and management practices and view environmental and economic effects resulting from the changes. This SDSS consists of a regional GIS-based model and a hydrologic/nutrient transport model, both linked to a GUI.

6.5.3 Climate and Hydrology Modeling

Over the past several decades, the climate and hydrology modeling communities have been developing increasingly sophisticated approaches to quantify the exchange of energy, water, and C between the land surface and the atmosphere in parameterizations commonly referred to as Soil-Vegetation-Atmosphere Transfer Schemes (SVATS). Sellers et al. (1997) trace the development of these models and show that each generation has successively added increased detail concerning the physical, biological, and chemical processes responsible for the fluxes of energy, water and C in the earth system. As these models have developed, a growing awareness of the importance of soil properties and processes has led to the increasing inclusion of these factors into the models. The specification of soil hydrologic processes, and ultimately the distribution of soil moisture, is a critical element in most of these models.

Most of the modeling applications in the area of climate and hydrology require an understanding of the surface and subsurface nature of the soil environment. Soil moisture is the primary state variable controlling the partitioning of atmospheric heat and moisture at the land surface. Knowledge of the amount and location of water in the soil is required for a variety of space and time scales in order to determine water and energy balances. Unfortunately, soil moisture is the most poorly measured and documented component related to water and energy balances at continental to global scales.

Lack of soil moisture measurements and uncertainty in the distribution of soil moisture over large areas has prompted the development of soil hydrology model components in SVATS that use soil physical property information in conjunction with pedotransfer functions. Such functions are used to determine soil hydraulic properties and process parameters for water movement in the soil profile to allow modeling of the soil moisture component. A number of soil databases have been developed or enhanced to address the need for soil physical and hydraulic characteristics in soil hydrology models. These efforts have built on existing soils information at the national and international levels. For example, Webb et al. (1993) produced a global dataset of soil profile physical properties by combining the FAO/UNESCO soil map of the world with the World Soil Data File (Zobler, 1986). These data specify the top and bottom depths and percentages of sand, silt, and clay of individual soil horizons for 106 soil mapping units. Similarly, Miller and White (1998) developed a multi-layer soil characteristics dataset for the conterminous United States (CONUS-SOIL) for use in regional climate and hydrology modeling. CONUS-SOIL is based on the STATSGO data for the conterminous United States and provides soil physical and hydraulic properties including soil textural and rock fragment classes, depth-to-bedrock, bulk density, porosity, rock fragment volume, particle size fractions (sand, silt, and clay), available water storage capacity, and HSGs. Data coverages in CONUS-SOIL are available as vector or raster (1 km grid resolution) map coverages in several formats tailored for regional-scale hydrologic modeling. Batjes (1997) developed a world dataset of derived properties from a statistical analysis of the more than 4500 soil profiles held in the World Inventory of Soil Emission (WISE), a database previously developed by ISRIC. A range of soil physical and chemical properties is included and can be linked to the FAO digital soil map of the world through a legend code (FAO, 1995).

6.5.4 Ecosystem Assessment and Modeling

The assessment of ecosystems at the global scale has primarily involved the determination of primary production for various biomes at the land surface (Field et al., 1995). Estimates of the Net Primary Productivity (NPP) are important to land use planning at regional and global scales because they provide a better understanding of (1) how ecosystems function and their effects on the environment (Cramer et al., 1997), and (2) the impact and consequences of global C cycle perturbations, land use and climate changes on ecosystem structure and food security (Olson and Prince, 1996). Terrestrial

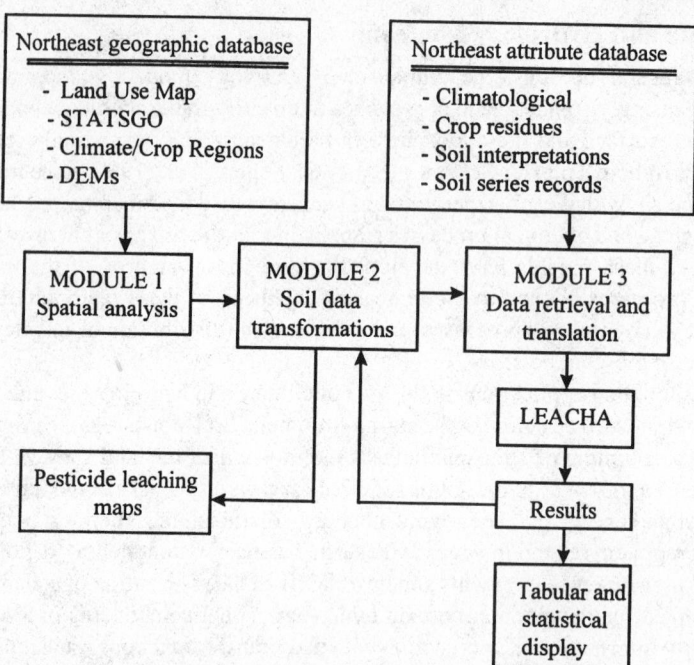

Fig. 6.6 Approach to modeling nonpoint source pollution within a GIS at a regional scale [After Bleecker et al., 1995. J. Soil Water Conserv. 50:388-394 with permission of the Soil Conservation Society of America]

NPP is defined as the net atmospheric C uptake by vegetation through the process of net photosynthesis minus dark respiration. Plants remove C from the atmospheric pool. Half of this C is returned to the atmosphere during plant respiration and the remaining amount, known as NPP, is incorporated in plant matter. Humans depend on this fraction for their needs because it provides food, fuel and fiber.

NPP is estimated by extrapolating field measurements to the region and biosphere using vegetation maps or by modeling the plant productivity. Where ecosystem processes are a primary focus, information on the soil C reservoir is also desired for model input. An example is the Carnegie-Stanford-Ames (CASA) biosphere model (Potter et al., 1993). The model runs on a monthly time step and integrates remotely sensed data, spatial databases of soils and climate data to simulate global net ecosystem production. In this case, NPP is estimated via a submodel as a product of intercepted photosynthetically active radiation (IPAR) and light utilization efficiency modified by temperature and soil moisture. A schematic representation of the integration of submodels and datasets in the CASA model is indicated in Fig. 6.7.

Soil databases have also been compared for their ability to provide accurate input parameters to ecosystem models. Pan et al. (1996) investigated how alternative input parameters of temperature, precipitation, solar radiation and soil texture to the Terrestrial Ecology Model (TEM) affect estimates of NPP for the conterminous United States. Their work compared databases developed by Zobler (1986) and FAO/CSRC (1974), and a dataset derived from the work of Kern (1995). They found the Zobler and Kern datasets to contain finer soil textures when compared to the FAO/CSRC data set. Use of the Kern-derived dataset yielded a 3% higher NPP than that of the FAO/CSRC due to higher rates of N cycling caused by enhanced volumetric soil moisture contents. Use of the Zobler (1986) attributes

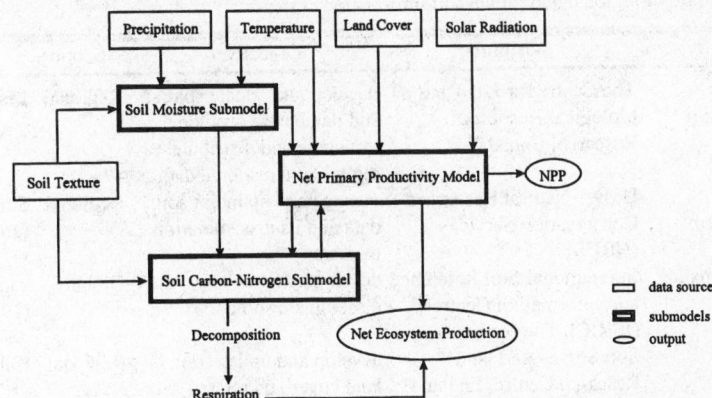

Fig. 6.7   Model framework for NPP determinations in the CASA model [Adapted from Potter et al., 1993]

had little effect when compared to the other two datasets. Potter et al. (1994) evaluated the use of the Zobler (1986) data set in a terrestrial C cycle model. They tested the assumption that global soil database quality was critical in the prediction of ecosystem controls on C cycling at large spatial scales. Their model sensitivity analysis suggested that predicted soil C storage is highly sensitive to texture. A spatially uniform, fine texture setting yielded the highest soil C pool size. This was consistent over the entire global gradient of climate conditions.

Lathrop et al. (1995) reported the use of STATSGO-derived available water-holding capacity in a forest ecosystem model for the northeast United States. They reported significant challenges regarding the within-unit heterogeneity of STATSGO data and the need to quantify this variability for use in ecosystem modeling. Similarly, Zheng et al. (1996) compared available water capacity estimated from topography to STATSGO data in Montana.

6.5.5   Availability and Distribution of Soil Databases

A number of Soil Information Systems (SIS) have been developed in recent years to create soil and terrain databases and make spatial data available to users for application in land use planning, environmental impact assessments, land and water quality analyses, etc. SIS are multi-purpose systems that integrate geographic databases and GIS tools to analyze, retrieve, record, report and display relationships between data. The type of data in these databases and the complexity of GIS analyses in SIS depend on the intended use of the system. SIS often contain data from different sources and scales of application but possess the same coordinate system and reference datum for easy integration. They are usually designed so that they can be accessed by many users. Some have been linked to the World Wide Web (WWW) for quick and easy access via various Internet browsers or file transfer protocol (FTP) clients.

During the last few years, scientists have created regional and global SIS. These systems have served primarily to develop and update spatial and non-spatial databases of soil parameters and associated attributes (terrain, land use, etc.) and to distribute them to users. These systems are developed primarily using popular database design software such as INFORMIX and ORACLE aided by programming languages (e.g., FORTRAN, C, Perl). Table 6.1 shows some of the most recent soil (or land) information systems and their primary functions.

Table 6.1 Examples of regional and global soil information systems

Soil Information System	Origin	Objectives	Scope	Source
Canadian Soil Information System (CANSIS)	The Centre for Land and biological Resources Research, Canada	develop and update spatial soil databases; provide data exchange and distribute digital land resource data	Regional	MacDonald and Valentine (1992)
National Soil Information System (NASIS)	USDA-Natural Resources Conservation Service (NRCS)	manage and maintain soil data and its dissemination to users	Regional	Soil Survey Staff (1991)
ISRIC Soil Information System (ISIS)	International Soil Reference and Information Centre (ISRIC), The Netherlands	develop soil databases and disseminate soil data	Global	van der Ven et al. (1995)
Land Information System (LandIS)	Soil Survey and land Research Centre, England	develop and update soil, land cover and related data and its dissemination	Regional	Hallett et al. (1996)

6.6 Future Trends and Considerations

The progress made to date on integrating soil information into land use planning, ecosystem, climate, and hydrology models is commendable. However, much work remains to improve the overall quality and usefulness of soil databases for land use planners and modelers. Areas for further investigation include soil survey data availability and scaling issues, among others. Formal soil survey programs have been active in many parts of the world for the past century or more. However, many areas remain without soil survey and characterization information. A primary objective of soil survey organizations worldwide should be the completion of at least intermediate scale mapping for all parts of the globe. Detailed and intermediate scale map products, where available, form the basis for generalized regional, continental, and eventually global scale soil databases. Successive refinement of existing regional and global soil databases will only be possible by improvement and availability of source products of higher resolution and quality. Consistent methodology and standards between countries are important factors in ensuring an accurate regional or global soil database. If such consistency is applied, the needs of continental and regional scale modelers could also be met within a coordinated global context. The ongoing effort of SOTER, described earlier, should help to address many of these concerns (Oldeman and van Engelen, 1993).

Most soil surveys are still conducted at the detailed (pedon) scale. No generally accepted theory has been developed for consistent upscaling of soil information from detailed, pedon-scale soil surveys to regional and global scales. The development of high quality regional and global scale soil databases will require focused research efforts to address this issue. Failure to improve our understanding of the basic scaling issues involved in moving from the pedon to greater scales will stymie all efforts to produce the best possible small-scale databases.

The demand for GIS and GIS-based analysis systems in environmental assessments and land use planning will increase in the future as high resolution digital datasets and databases become available. Detailed spatial datasets combined with RS, GPS, geostatistics and the capability of modern computers to handle large volumes of data will increase GIS applications in land use planning and management at scales finer than the farm level.

GIS has been highly regarded as a centralized data analysis, management, and planning tools for precision farming. The concept of site-specific farming is to determine meaningful farm management units based on spatial variability of soil properties within fields and to efficiently use agricultural inputs. Its aim is to optimize crop production with minimum impact to the environment.

Acknowledgments
This work was financially supported in part by the National Aeronautics and Space Administration (NASA), Earth Science Enterprise/Earth Observing System (ESE/EOS) under Grant NAG 5-3856.

6.7 References

Al-Abed, N., and H.R. Whiteley. 1995. Modeling water quality and quantity in the quantity in the lower portion of the Grand River watershed, Ontario. p.213-222. *In* C. Heatwole (ed.) Water quality modeling. ASAE Pub. 5-95. ASAE, St. Joseph, MI.

Arnold, J.G., and N.B. Sammons. 1988. Decision support system for selecting inputs to a basin scale model. Water Resour. Bull. 24:749-759.

Arnold, J.G., J.R. Williams, A.D. Dicks, and N.B. Sammons. 1990. SWRBB. A basin scale simulation model for soil and water resources management. Texas A&M University Press, College Station, TX.

Arnold, R.W. 1984. Soil database management. p. 135-146. *In* Proc. Int. Symp. Minimum Data Sets Agrotech. Transf. Pantacheru, India. March 21-26, 1984. ICRISAT Center, India.

Baker, C.P., and E.C. Panciera. 1990. A geographic information system for groundwater protection planning. J. Soil Water Conserv. 45:246-248.

Batjes, N.H. 1997. A world dataset of derived soil properties by FAO-UNESCO soil unit for global modeling. Soil Use Manag. 13:9-16.

Baumgardner, M.F. 1994. A world soils and terrain digital database: Linkages. p. 718-727. *In* Trans. 15th World Congr. Soil Sci. 6a: Comm. V: Symposia.

Beasley, D.B., and L.F. Huggins. 1991. ANSWERS. Area nonpoint source watershed environment response simulation, User's Manual. 2nd. Ed. Agric. Eng. Dept. Pub. 5. University of Georgia, Athens, GA.

Bleecker, M., S.D. DeGloria, J.L. Hutson, R.B. Bryant, and R.J. Wagenet. 1995. Mapping atrazine leaching potential with integrated environmental databases and simulation models. J. Soil Water Conserv. 50:388-394.

Bliss, N.B., and W.U. Reybold. 1989. Small-scale digital soil maps for interpreting natural resources. J. Soil Water Conserv. 44:30-34.

Bliss, N.B. and S.W. Waltman. 1994. Modeling variations of land qualities at regional and global scales using geographic information systems. p. 644-661. *In* Trans. 15th World Congr. Soil Sci. Vol. 6a: Comm. V: Symposia.

Bliss, N.B., S.W. Waltman, and G.W. Petersen. 1995. Preparing a soil carbon inventory for the United States using geographic Information systems. p. 275-295. *In* R. Lal et al. (ed.) Soils and global change. Lewis Publishers, Boca Raton, FL.

Bonfils, P. 1989. Une évaluation du risque d'érosion dans les pays du sud de la communauté européene (Le Programme CORINE). Sci. Sol 27:33-36.

Brinkman, R.. 1994. Recent developments in land use planning, with special reference to FAO. p.13-21. *In* L. O. Fresco et al. (ed.) The future of the land: Mobilizing and integrating knowledge for land use options. John Wiley and Sons. Chichester, UK.

Bui, E.N., K.R.J. Smettem, C. Moran, and J. Williams. 1996. Use of soil survey information to assess regional salinization risk using a geographic information system. J. Environ. Qual. 25: 433-439.

Burkart, M.R., S.L. Oberle, M.J. Hewitt, and J. Pickus. 1994. A framework for regional agro-ecosystems characterization using the National Resources Inventory. J. Environ. Qual. 23:866-874.

Burrough, P.A. 1989. Matching databases and quantitative models in land resource assessment. Soil Use Manag. 5:3-8.

Burrough, P.A. 1991. Soil information systems. *In* D.J. Maguire et al. (ed.) Geographic information systems. Principles and applications. Vol. 2: Applications. Longman Scientific and Technical, Essex, UK.

Carsel, R.F., J.C. Imhoff, J.L. Kittle, Jr., and P.R. Hummel. 1991. Development of a database and model parameter analysis system for agricultural soils. J. Environ. Qual. 20:642-647.

Carsel, R.F., and R.L. Jones. 1990. Using soil and meteorological databases in unsaturated zone modeling of pesticides. Groundwater Monit. Rev. 10:96-101.

Christensen, G.R., W.W. Budd, J.P. Reganold, and F.R. Steiner. 1988. Farmland protection in Washington State: An analysis. J. Soil Water Conserv. 43:82-89.

Clayville, B.L. 1994. Applications of geographic information systems to Pennsylvania's Agricultural Conservation Easement Purchase program. M.S. thesis. PA State University, University Park, PA.

Cramer, W., B. Moore III, and D. Sahagian. 1997. Global net primary productivity: A model intercomparison. pp. 11-12. The International Geosphere-Biosphere Programme (IGBP)- Global Analysis, Interpretation, and modeling task Force (GAIM) Newsletter 1, Winter 1997.

Daniels, T. 1990. Using LESA in a purchase of development right program. J. Soil Water Conserv. 45:617-621.

de Roo, A.P.J., L. Hazelhoff, and P.A. Burrough. 1989. Soil erosion modeling using 'ANSWERS' and geographic information systems. Earth Surf. Proc. Land. 14:517-532.

DeVries, F., and J. Denneboom. 1993. The digital version of the Dutch soil map of The Netherlands. SC-DLO Tech Doc. 1. Winand Staring Center, Wageningen, Netherlands.

de Witt, C.T., and H. van Keulen. 1987. Modeling production of field crops and its requirements. Geoderma 40:253-265.

Drungil, C.E.C., W.F. Geter, and R.G. Shepherd. 1995. Watershed scale water quality modeling in the NRCS HUWQ project. p. 177-185. In C. Heatwole (ed.) Water quality modeling. ASAE Pub. 5-95. ASAE, St Joseph, MI.

Dunford, R.W., R.D. Roe, F.R. Steiner, W.R. Wagner, and L.E. Wright. 1983. Implementing LESA in Whitman Co., Washington. J. Soil Water Conserv. 38:87-89.

Elbersen, G.W., W. Ismangun, D.S. Sutaatmadja, and À.A. Solihin. 1988. Small scale soil survey and automated land evaluation. ITC J. 1:51-59.

Engel, T., J.W. Jones, and G. Hoogenboom. 1996. Spatial analysis and visualization of DSSAT crop simulation results using AEGIS/WIN. p. 297-402. In C. Lokhorst et al. (ed.) Information and communication technology applications in agriculture: State of the art and future perspectives,: Agro-Informaticareenks, Wageningen, Netherlands.

Engel, B.E., R. Srinivasan, and C. Rewerts. 1991. A spatial decision support system for modeling and managing agricultural nonpoint source pollution. In M.F. Goodchild et al. (ed.) Environmental modeling with GIS. Oxford University Press, New York, NY.

Eswaran, H., R. Almarez, E. Van Den Berg, and P. Reich. 1997. An assessment of soil resources of Africa in relation to soil productivity. Http://www.nhq.nrcs.usda.gov/ WSR/papers/africa.html. World Soil Resources, Soil Survey Division, USDA NRCS Washington, DC.

Evans, B.M., J. Grimm, L. Thornton, and P. Saunders. 1992. Linking GIS and Hydrologic modeling. p. 499–504. IN Proc. Water Forum '92, Baltimore, MD.

FAO. 1976. A framework for land evaluation. FAO Soils Bull. 55.

FAO. 1993. Guidelines for land use planning. FAO Devel. Ser. 1.

FAO. 1995. Digital soil map of the world and derived soil properties (Vers. 3.5). CD-ROM. FAO, Rome, Italy.

FAO/CSRC. 1974. Soil map of the world. 1:5,000,000. UNESCO, Paris, France. 1/2 ° digitization by Complex Systems Research Center, University of New Hampshire, Durham, NH.

Ferguson, C.A., R.L. Bowen, and M.A. Kahn. 1991. A statewide LESA system for Hawaii. J. Soil Water Conserv. 46:263-267.

Field, C.B., J.T. Randerson, and C.M. Malmström. 1995. Global net primary production: combining ecology and remote sensing. Remote Sens. Environ. 51: 74-88.

Flagg, J.M., W.C. Hession, and V.O. Shanholtz. 1990. Geographic information systems and water quality models as state level nonpoint source pollution control management tools. p. 74-77. In Proc. application of geographic information systems, simulation models, and knowledge-based systems for land use management. Virginia Polytechnic Institute and State University, Blacksburg, VA.

Global Soils Data Task. 1995. Global change data sets. A collaborative project. FAO, ISRIC, IGBP, NRCS. Lincoln, NE.

Goebel, J.J. 1992. Description of the national resources inventory. p. A1-A8. In R.L. Kellogg et al. (ed.) Agricultural chemical use and ground water quality: Where are the potential problem areas. USDA-SCS, Washington, DC.

Gray, D.M., D.I. Norium, and J.M. Wigham. 1970. Infiltration and physics of flow of water through media. In D. M. Gray (ed.) Handbook of the principles of hydrology. Section V. National Research Council of Canada, Ottawa, Canada.

Green, W.H., and C.A. Ampt. 1911. Studies of soil physics. Flow of water and air through soils. J. Agric. Sci. 4:1-24.

Hallett, S.H., R.J.A. Jones, and C.A. Keay. 1996. Environmental information systems developments for planning sustainable land use. Int. J. Geog. Info. Syst. 10:47-64.

Hamlett, J.M., D.A. Miller, R.L. Day, G.W. Petersen, G.M. Baumer, and J.M. Russo. 1992. Statewide GIS-based ranking of watersheds for agricultural pollution prevention. J. Soil Water Conserv. 47:399-404.

Hamlett, J.M., G.W. Petersen, S. Harrison, and S.R. Messier. 1995. GIS pesticide vulnerability assessment procedure for Pennsylvania groundwater sub-basins. Prelim. Rep. PA Dept. Environ. Resour.. Bureau Soil Water Conservation ERRI, PA State University, University Park, PA.

Hendrix, W.G., and J.A. Buckley. 1992. Use of a geographic information system for selection of sites for land application of sewage waste. J. Soil Water Conserv. 92:271-275.

Holtan, H.N. 1961. A concept of infiltration estimates in watershed engineering. ARS USDA-ARS Pap. 41-51.

Horton, R.E. 1940. An approach toward a physical interpretation of infiltration capacity. Soil Sci. Soc. Am. Proc. 5:399-419.

Huber, W.C., and R.E. Dikinson. 1988. Storm water management model (SWMM). Vers. 4. User's Manual. EPA/600/3-88/001a. USEPA. Athens, GA.

IBSNAT Project. 1989. Decision support system for agrotechnology transfer (DSSAT)- User's Guide. IBSNAT project, University of Hawaii, Honolulu, HI.

IBSNAT Project. 1992. Linking DSSAT to a geographic information system, IBSNAT. Agrotechn. Trans. 15:1-6

Imhoff, J.C., R.F. Carsel, J.L. Kittle, and P.R. Hummel. 1990. Database analyzer and parameter estimator (DBAPE) interactive computer program user's manual. EPA/600/3-89/083. USEPA, Athens, GA.

Imhoff, M.L., W.T. Lawrence, C.D. Elvidge, T. Paul., E. Levine, M.V. Privalsky, and V. Brown. 1996. Using nighttime DMSP/OLS images of city lights for estimating the impact of urban land use on soil resources in the U.S. Remote Sens. Environ. 59: 105-117.

Inskeep, P.P., J.M. Wraith, J.P. Wilson, R.D Snyder, R.E. Macur, and H.M. Gaber. 1996. Input parameter and model resolution effects on predictions of solute transport. J. Environ. Qual. 25:453-462.

International Soil Reference and Information Center (ISRIC). 1993. Global and National soils and terrain database (SOTER). Procedures Manual. World Soil Resources Report 74, FAO, Rome, Italy.

João, E.M., and S.J. Walsh. 1992. GIS implications for hydrologic modeling: Simulation of nonpoint pollution generated as a consequence of watershed development scenarios. Comput. Environ. Urban Syst. 16:43-63.

Karthikeyan, K.G., H.A. Elliott, and R.C. Brandt. 1993. Using a geographic information system for siting water treatment sludge monofills. p. 315-325. In J.K. Mitchell (ed.) Proc. Intern. Symp. Integr. Resour. Manag. Landsc. Modif. Environ. Protect. ASAE, St. Joseph, MI.

Kellogg, R.L., M.S. Maizel, and D.W. Goss. 1994. The potential for leaching of agrochemicals used in crop production: A national perspective. J. Soil Water Conserv. 49: 294-298.

Kern, J.S. 1995. Geographic patterns of soil water-holding capacity in the conterminous United States. Soil Sci. Soc. Am. J. 59:1126-1133.

Khan, M.A., and T. Liang. 1989. Computer-aided regional mapping of pesticide runoff. ASAE Microf. 89-2661. ASAE, St. Joseph, MI.

Kirkwood, V., J. Dumanski, A. Bootsma, R.B. Stewart, and R. Muma. 1996. Land Resource Research Centre Research Branch, Agriculture Canada. Ottawa, Ontario. Techn. Bull. 1983-4E. LRRC Contribution No. 86-29.

Klingebiel, A.A., and P.H. Montgomery. 1961. Land capability classification. USDA-SCS Agric. Handb. 210.

Knisel, W. G. 1980. CREAMS: A field-scale model for chemical, runoff and erosion from agricultural management systems. USDA-SCS Conserv. Res. Rep. 26.

Koohafkan, A.P., and J. Antoine. 1997. Application of Agro-Ecological zoning and GIS tools for land use planning, food security strategy development and land vulnerability assessments. Unpublished Report.

Kuenstler, W.F., D. Ernstrom, and E. Seeley. 1995. Nutrient management-building tools to predict potential nutrient losses. p.459-467. In C. Heatwole (ed.) Water quality modeling. ASAE Publ. 5-95. ASAE, St Joseph, MI.

Lathrop, R.G. Jr., J.D. Abler, and J.A. Bognar. 1995. Spatial variability of digital soil maps and its impact on regional ecosystem modeling. Ecol. Model. 82:1-10.

Liu, R., and P.J. Craul. 1991. A GIS based soil capability classification for agriculture in Yuisu county, China. p. 99-106. In Annual Convention of the ACSM-ASPRS. Vol. 4. Techn. Pap. ACSM-ASPRS, Bethesda, MD.

Luyten, J.C., and J.W. Jones. 1997. AEGIS+: A GIS-based graphical user-interface for defining spatial crop management strategies and visualization of crop simulation results. Agron. Abstr. 89:23.

MacDonald, K.B., and K.W.G. Valentine. 1992. CanSIS/NSDB A general description. CLBRR Contrib. No. 92-35. Agriculture Canada, Ottawa, Canada.

McQuaid, B.F., S.W. Buol, and J.A. Kelley. 1995. Application of the soil fertility capability classification in soil survey reports. Soil Surv. Horiz. 36:117-121.

Miller, D.A., and R.A. White. 1998. A conterminous United States multilayer soil characteristics dataset for regional climate and hydrology modeling. Earth Interactions, 2. [Available on-line at http://EarthInteractions.org].

Negahban, B., C. Fonyo, K.L. Campbell, J.W. Jones, W.G. Boggess, G. Kiker, E. Hamouda, E. Flaig, and H. Lal. 1996. LOADSS: A GIS-based decision support system for regional environmental planning. p. 277-282. In M.F. Goodchild et al. (ed.) GIS and environmental modeling: Progress and research issues. GIS World Inc., Fort Collins, CO.

Nielsen, D. A., R. J. Biggler, T. Sobecki, and D. J. Lytle. 1996. Application of soil survey attribute data to GIS pollution assessment models. p. 175-183. *In* D.L. Corwin and K. Loague (ed.) Applications of GIS to the modeling of non-point source pollutants in the vadose zone. Soil Sci. Soc. Am. Spec. Pub. 48. Soil Science Society of America, Madison, WI.

Nielsen, G.A., J. M. Shapiro, P.A. McDaniel, R.D. Snyder, and C. Montagne. 1990. MAPS: A GIS for land resource management in Montana. J. Soil Water Conserv. 45:450-453.

Nizeyimana, E. 1995. Evaluation of digital elevation models, remotely sensed data and soil databases as sources of input parameters to SCS runoff model using a Geographic Information System. M.S. Thesis. PA State University, University Park, PA.

Nizeyimana, E., and G.W. Petersen. 1997. Remote sensing applications to soil degradation assessments. p. 393-405. *In* R. Lal et al. (ed.) Methods for assessment of soil degradation. CRC Press, Boca Raton, FL.

Nizeyimana, E., G.W. Petersen, M.C. Anderson, B.M. Evans, J.M. Hamlett, and G.M. Baumer. 1996. Statewide GIS/census data assessment of nitrogen loadings from septic systems in Pennsylvania. J. Environ. Qual. 25:346-354.

Novotny, V. 1981. A review of hydrologic and water quality models used for simulation of agricultural pollution. *In* V. Novotny and G. Chesters (ed.) Handbook of nonpoint pollution. Sources and management. Van Nostrand Reinhold Company, New York, NY.

Oldeman, L.R., and V.W.P. van Engelen. 1993. A world soils and terrain digital database (SOTER) - An improved assessment of land resources. Geoderma 60:309-325.

Oldeman, L.R., R.T.A. Hakkeling, and W.G. Sombroek. 1990. World map of the status of human-induced soil degradation: An explanatory note. Global assessment of soil degradation (GLASOD). ISRIC/UNEP, Wageningen, Netherlands.

Olson, K.R., and G.W. Olson. 1986. Use of multiple regression analyses to estimate average corn yields using selected soils and climatic data. Agric. Syst. 20:105-120.

Olson, D., and S.D. Prince. 1996. Global primary production data initiative update. The International Geosphere-Biosphere Programme (IGBP): A study of global change of the International Council of Scientific Unions. Global Change Newsl. 27.

Pan, Y., A.D. McGuire, D.W. Kicklighter, and J.M. Melillo. 1996. The importance of climate and soils for estimates of net primary production: a sensitivity analysis with the terrestrial ecosystem model. Glob. Change Biol. 2:5-23.

Petersen, G.W., J.C. Bell, K. McSweeney, G.A. Nielsen, and P.C. Robert. 1995. Geographic information systems in Agronomy. Adv. Agron. 55:67-111.

Petersen, G.W., E. Nizeyimana, and B.M. Evans. 1997. Application of GIS to land degradation assessments. p. 377-391. *In* R. Lal et al. (ed.) Methods for assessment of soil degradation. CRC Press, Boca Raton, FL.

Petersen, G.W., E. Nizeyimana, M.L. Imhoff, and W.T. Lawrence. 1998. Assessing soil productivity losses due to urbanization using remote sensing and soil productivity models. *In* Proc. 16th World Congr. Soil Sci. Comm. 35: Symposia. International Society of Soil Science, Montpellier, France.

Phillip, J.R. 1957. The theory of infiltration: 4. Sorptivity and algebraic equations. Soil Sci. 84:257-265.

Pierce, F.J., W.E. Larson, R.H. Dowdy, and W.E.P. Graham. 1983. Productivity of soils: Assessing long-term changes due to erosion. J. Soil Water Conserv. 38:131-136.

Platou, S.W., A.H. Norr, and H.B. Madsen. 1989. Digitization of the EC soil map. p. 132-45. *In* R.J.A. Jones and B. Biagi (ed.) Computerization of land use data. EUR 11151 EN., Office for Official Publications of the European Community, Luxembourg.

Potter, C.S., P.A. Matson, and P.M. Vitousek. 1994. Evaluation of soil database attributes in a terrestrial carbon cycle model: Implications for global change research. p. 277-298. *In* W.K. Michener et al. (ed.) Environmental information management and analysis. Taylor and Francis, London, UK.

Potter, C. S., J.T. Randerson, C.B. Field, P.A. Matson, P.M. Visoutek, H.A. Mooney, and S.A. Klouster. 1993. Terrestrial ecosystem production: A process model based on global satellite and surface data. Global Biogeoch. Cycl. 7: 811-841.

Rawls, W.J., D.L. Brakensiek, and N. Miller. 1983. Green-Ampt infiltration parameters from soils data. J. Hydraul. Eng. 109:62-70.

Rossiter, D.G. 1990. ALES: A framework for land evaluation using a microcomputer. Soil Use Manag. 6:7-20.

Rossiter, D.G. 1996. A theoretical framework for land evaluation. Geoderma 72:165-190.

Sanchez, P.A., W. Couto, and S.W. Buol. 1982. The fertility capability soil classification system: Interpretation, applicability and modification. Geoderma 27:283-309.

Savabi, M.R., D.C. Flanagan, and B.A. Engel. 1995. Application of WEPP and GIS-GRASS to a small watershed in Indiana. J. Soil Water Conserv. 50:477-483.

Schultink, G. 1987. The CRIES resource information system: Computer-aided land resource evaluation for development planning and policy analysis. Soil Surv. Land Eval. 7: 47-62.

Sellers, P.J., R.E. Dickinson, D.A. Randall, A.K. Betts, F.G. Hall, J.A. Berry, G.J. Collatz, A.S Denning, H.A. Mooney, C.A. Nobre, N. Sato, C.B. Field, and A. Henderson-Sellers. 1997. Modeling the exchanges of energy, water, and carbon between continents and the atmosphere. Science 275:502-509.

Shaffer, M.J., M.D. Hall, B.K. Wylie, and D.G. Wagner. 1996. NLEAP/GIS approach for identifying and mitigating regional nitrate-nitrogen leaching. p. 283-294. *In* D.L. Corwin and K. Loague (ed.) Applications of GIS to the modeling on non-point source pollutants in the vadose zone. Soil Sci. Soc. Am. Spec. Pub. 48. Soil Science Society of America, Madison, WI.

Sharifi, M.A., and H. van Keulen. 1994. A decision support system for land use planning at farm enterprise level. Agric. Syst. 45: 239-257.

Sinclair, H.R. 1996. Soil ratings for plant growth. Agron. Abstr. 88:279.

Sinclair, H.R., W.J. Waltman, H.P. Terpstra, and D.R. Margentan. 1996. Soil ratings for plant growth. USDA-NRCS National Soil Survey Center, Lincoln, NE.

Soil Survey Staff. 1991. National Soil Information System (NASIS): Soil interpretation and information dissemination subsystem. Draft requirements statement. USDA-NRCS, National Soil Survey Center. Lincoln, NE.

Soil Survey Division Staff. 1993. Soil survey manual. USDA Agric. Handb. 18. US Government Printing Office, Washington, DC.

Srinivasan, R., J.G. Arnold, R.S. Muttiah, and P.T. Dyke. 1995. Plant and hydrologic simulation for the conterminous U.S. using SWAT and GIS. Hydrol. Sci. Technol. 11:160-168.

Stamm, T., R. Gill, and K. Page. 1987. Agricultural land evaluation and site assessment in Latah County, Idaho. Environ. Manag. 11:379-388.

Storie, R.E. 1937. An index for rating agricultural value of soils. Bull. 556. Univ. of California, Berkley, CA.

Stuebe, M.M., and D.M Johnson. 1990. Runoff volume estimation using GIS techniques. Water Resour. Bull. 26:611-620.

Tavernier, R. 1985. Soil map of the European Communities. 1: 1 000000. Off. Publ. of the European Communities. Luxembourg.

Tim, U.S., and R. Jolly. 1994. Evaluating agricultural nonpoint source pollution using integrated geographic information systems and hydrology/water quality models. J. Environ. Qual. 23:25-35.

US Department of Agriculture. 1986. Urban hydrology for small watersheds. USDA-SCS Tech. Rel. 55. US Government Printing Office, Washington, DC.

US Department of Agriculture. 1993. National soils handb. USDA-SCS. US Government Printing Office, Washington, DC.

US Environmental Protection Agency. 1981. Process design manual for land treatment of municipal wastes. PB 299655/1. Cincinnati, OH.

van der Ven, T., P. Tempel, and J. Verhagen. 1995. ISRIC Soil Information System ISIS 4.0. User Manual. Revised ed. Techn. Pap. 15. International Soil Reference and Information Centre. Wageningen, Netherlands.

van der Vlugt, M. 1989. The use of a GIS-based decision support system in physical planning. p. 459-467. *In* Proc. GIS/LIS'89. Vol. 2. American Society for Photogrammetry and Remote Sensing, Bethesda, MD.

van Diepen, C.A., H. Van Keulen, J. Wolf, and J.A.A. Berkhout. 1991. Land evaluation: From intuition to quantification. Adv. Soil Sci. 15:140-204.

van Horn, T.G., G.C. Steinhardt, and J.E. Yahner. 1989. Evaluating the consistency of results for the agricultural land evaluation and site assessment (LESA) system. J. Soil Water Conserv. 44:615-620.

Vereecken, H., J. Maes, J. Feyen, and P. Darius. 1989. Estimating the soil moisture retention characteristics from texture, bulk density and carbon content. Soil Sci. 148:389-403.

Wagenet, R.J., R. B. Bryant, S.D. DeGloria, and R.G. Perrit. 1993. Mapping ground water contamination potential using integrated simulation modeling and GIS. Final rep. Coop. Res. Res. Serv., USDA, Washington, DC.

Wagenet, R.J., J. Bouma, and R.B. Grossman. 1991. Minimum data sets for use of soil survey information in soil interpretative models. p. 161-182. *In* M.J. Mausbach and L.P. Wilding (ed.) Spatial variabilities of soils and landforms. Soil Sci. Soc. Am. Spec. Pub. 28. Soil Science Society of America, Madison, WI.

Webb, R.S., C.E. Rosenzweig, and E.R. Levine. 1993. Specifying land surface characteristics in general circulation models: Soil profile data set and derived water-holding capacities. Glob. Biogeoch. Cycl. 7:97-108.

Weber, R.S., J. Jenkins, and J.J. Leszkiewicz. 1990. Application of geographic information system technology to landfill site selection. p. 421-430. *In* Proc. of the application of geographic information systems, and knowledge-based systems for land use management. VA Polytechnic and State University, Blacksburg, VA.

Weir, A.H., P.L. Bragg, J.R. Porter, and J.H. Rayner. 1984. A winter wheat crop simulation model without water and nutrient limitations. J. Agric. Sci. 102: 311-382.

Whitley, D.L., and W.N. Xiang. 1993. A GIS-based method for integrating expert knowledge into land suitability analysis. p. 24-37. *In* URISA Ann. Proc. Value of LIS/GIS: Real and perceived GIS technology and its relevance to real world problems facing your organization. Vol. II. University of Wisconsin Press, Madison, WI.

Williams, T.H., 1985. Implementing LESA on a geographic information system-a case study. Photogram. Eng. Remote Sens. 5:1923-1932.

Wilson, J.P., W.P. Inskeep, J.M. Wraith, and R.D. Snyder. 1996. GIS-based solute transport modeling applications: Scale effects of soil and climate data input. J. Environ. Qual. 25:433-439.

Wishmeier, W.H., and D.D. Smith. 1978. Predicting Rainfall Erosion Losses-A Guide to Conservation Planning. USDA Agric. Handb. 537.

Wood, S.R., and F.J. Dent. 1983. LECS: A land evaluation computer system. AGOF/INS/78/006 Vols. 5 and 6. Ministry of Agriculture, Government of Indonesia, Bogor, Indonesia.

Wright, L.E., W. Zitzmann, K. Young, and R. Googins. 1983. LESA-agricultural land evaluation and site assessment. J. Soil Water Conserv. 38:82-86.

Yoon, J. 1996. Watershed-scale nonpoint source pollution modeling and decision support system based on a model-GIS-RDBM linkage. p. 99-108. *In* C.A. Hallan et al. (ed.) Proc. Ann. Symp. GIS Water Resour. AWRA, Fort Lauderdale, FL.

Young, R. A., M. A. Otterby, and A. Roos. 1982. An evaluation system to rate feedlot pollution potential. USDA-ARS, Washington, DC.

Young, R.A., D.D. Bosch, and W.P. Anderson. 1994. Agricultural Non-Point Source Pollution Model, Version 4.02. AGNPS User's Guide. USDA. Agric. Res. Serv. Conserv. Res. Rel. 35.

Zheng, D., E.R. Hunt, Jr., and S.W. Running. 1996. Comparison of available soil water capacity estimated from topography and soil series information. Land. Ecol. 11:3-14.

Zobler, A. 1986. A world soil file for global climate modeling. NASA Tech. Mem. 87802. National Aeronautics and Space Administration, Goddard Space Flight Center, Greenbelt, MD.

Index